BFO750 CAS:7440-41-7 HR:3
BERYLLIUM
DOT: UN 1966/UN 1567
mf: Be mw: 9.01

PROP: A silvery-white, relatively soft, lustrous metal ductile at red heat. Unreactive to H_2O and air; dissolves vigorously in dil acids. Be Reacts with aq alkalis or H_2. Mp: 1287–1292°, bp: 2970°, d: 1.85.

SYNS: BERYLLIUM · 9 □ BERYLLIUM COMPOUNDS, n.o.s. (UN 1566) (DOT) □ BERYLLIUM, powder (UN, 1567) (DOT) □ GLUCINIUM □ GLU-CINIUM □ RCRA WASTE NUMBER P015

TOXICITY DATA WITH **REFERENCE**
dnd-esc 30 μmol/L MUREAV 89,95,81
dni-nml-ivn 30 μmol/kg PHMCAA 12,298,70
dnd-hmn:hla 30 μmol/L MUREAV 89,95,81
dnd-mus:ast 30 μmol/L MUREAV 89,95,81
itr-rat TDLo: 13 mg/kg:NEO ENVRAL 21,63,80
ivn-rbt TDLo: 20 mg/kg:ETA LANCAO 1,463,50
ihl-hmn TCLo: 300 mg/m³:PUL AEHLAU 9,473,64
ivn-rat LD50: 496 μg/kg LAINAW 15,176,66

CONSENSUS REPORTS: NPT 7th Annual Report on Carcinogens, IARC Cancer Review: Group 1 IMEMDT 58,41,93; Human Sufficient Evidence IMEMDT58,41,93 Animal Sufficient Evidence IMEMDT 1,17,72; Animal Sufficient Evidence IMEMDT 23,143,80; Animal Sufficient Evidence IMEMDT 58,41,93. Beryllium and its compounds are on the Community Right-To-Know List. Reported in EPA TSCA Inventory.

OSHA PEL: TWA 0.002 mg(Be)/m³; STEL 0.005 mg(BE)/m³/30M; CL 0.025 mg(Be)/m³
ACGIH TLV: TWA 0.002 mg/m³, Suspected Human Carcinogens
DFG TRK: Animal Carcinogen, Suspected Human Carcinogen. Grinding of beryllium metal and alloys: 0.005 mg/m³ calculated as beryllium in that portion of dust that can possibly be inhaled; other beryllium compounds: 0.002 mg/m³ calculated as beryllium in that portion of dust that can possibly be inhaled.
NIOSH REL: CL not to exceed 0.0005 mg(Be)/m³
DOT CLASSIFICATION: 6.1; *Label:* Poison (UN 1566); DOT Class: 6.1; *Label:* Poison, Flammable Solid (UN 1567)

SAFETY PROFILE: Confirmed carcinogen with experimental carcinogenic, neoplastigenic, and tumorgenic data. A deadly poison by inhalation: lung fibrosis, dyspnea, and weight loss. Human mutation data reported. See also BERYLLIUM COMPOUNDS. A moderate fire hazard in the form of dust or powder, or when exposed to flame or by spontaneous chemical reaction. Slight explosion hazard in the form of powder or dust. Incompatible with halocarbons. Reacts incandescently with fluorine or chlorine. Mixtures of the powder with CCl_4 or trichloroethylene will flash or spark on impact. When heated to decomposition in air it emits very toxic fumes of BeO. Reacts with Li and P.

For occupational chemical analysis use NIOSH: Beryllium, 7102; Elements, 7300.

HR: – Hazard Rating indicating relative hazard for toxicity, fire, and reactivity with 3 denoting the worst hazard level. *See introduction: paragraph 3, p. xi.*

CAS: – The American Chemical Society's Chemical Abstracts Service number. A complete CAS number cross-index is located in Section 2. *See introduction: paragraph 4, p. xi.*

Cited References: – A code which represents the cited reference for the toxicity data. Complete bibliographic citations in order are listed in Section 4. *See introduction: paragraph 16, p. xxii.*

Consensus Reports: – Supply additional information to enable the reader to make knowledgeable evaluations of potential chemical hazards. *See introduction: paragraph 17, p. xxiii.*

Safety Profiles: – These are text summaries of the toxicity, flammability, reactivity, incompatibility, and other dangerous properties of the entry. *See introduction: paragraph 19, p. xxv.*

Analytical Methods: – References to OSHA and NIOSH occupational analytical methods. *See introduction: paragraph 19, p. xxv.*

Sax's
Dangerous
Properties of
Industrial Materials

Eleventh Edition

Sax's Dangerous Properties of Industrial Materials

Eleventh Edition

RICHARD J. LEWIS, SR.

A JOHN WILEY & SONS, INC., PUBLICATION

Library of Congress Cataloging-in-Publication Data is available.

Lewis, Richard J., Sr.
    Sax's dangerous properties of industrial materials—11th ed.
    ISBN 0-471-47662-5

Printed in the United States of America.

10 9 8 7 6 5 4 3 2 1

Dedicated to Grace Ross Lewis.
Her effort and advice on every aspect of the
design and preparation made this edition possible.

Welcome to William Joseph Herrmann,
a new generation to carry the torch.

Acknowledgments

I extend thanks to Bob Esposito for his encouragement
and to Shirley Thomas for her attention to all phases
of the production of this work.

Contents

Sax's Dangerous Properties of Industrial Materials, Eleventh Edition is also available as a CD-ROM.
To order:
Call toll-free: 1-800-225-5954
e-mail: consumers@wiley.com

Preface

This Eleventh edition of *Dangerous Properties of Industrial Materials* includes comprehensive hazard information on the substances encountered in the workplace. Theobjective of the work is to promote safety by providing the most up-to-date hazard information available.

Over two-thirds of the entries have been revised for this edition. There are 25,962 entries in this volume; 2,597 are new to this volume. Preference was given in selection of new entries to those listed in the EPA TSCA Inventory. These are reported to be used in commerce in the United States.

Numerous synonyms have been added to assist in locating the many materials that are known under a variety of systematic and common names. The synonym cross-index contains 141,700 entries consisting of the entry name as well as each synonym. This index should be consulted first to locate a material by name. Synonyms are given in English as well as other major languages such as French, German, Dutch, Polish, Japanese, and Italian.

Over 1,100 additional entries have had physical and chemical properties added. Whenever available, physical descriptions, formulas, molecular weights, melting points, boiling points, explosion limits, flash points, densities, autoignition temperatures, and the like have been supplied.

A court order has vacated the OSHA Air Standards set in 1989 and contained in 29CFR 1910.1000. OSHA has decided to enforce only pre-1989 air standards. We have elected to include both the Transitional Limits that went into effect on December 31, 1992, and the Final Rule limits, that went into effect September 1, 1989. These represent the current best judgment as to appropriate workplace air levels. While they may not be enforceable by OSHA, they are better guides than the OSHA Air Standards adopted in 1969. OSHA has stated that it "...continues to believe that many of the old limits which

it will now be enforcing are out of date (they predate 1968) and are not sufficiently protective of employee health based on current scientific information and expert recommendations. In addition, many of the substances for which OSHA has no PELs present serious health hazards to employees".

The following classes of data are new or have been updated for all entries for which they apply:

1. New to this edition are immediately dangerous to life or health concentrations (IDLHs) for 1,035 substances. These values are defined in the Properties section of the Introduction.

2. ACGIH TLVs and BEIs reflect the latest recommendations and now include intended changes.

3. German MAK and BAT reflect the latest recommendations.

4. NTP 10th Annual Report on Carcinogens entries are identified.

5. CAS numbers are provided for additional entries.

Each entry concludes with a Safety Profile, a textual summary of the hazards presented by the entry. The discussion of human exposures includes target organs and specific effects reported. Carcinogenic and reproductive assessments have been completely revised for this edition.

Fire and explosion hazards are briefly summarized in terms of conditions of flammable or reactive hazard. Where feasible, fire-fighting materials and methods are discussed. Materials that are known to be incompatible with an entry are listed here.

Also included in the safety profile are comments on disaster hazards that serve to alert users of materials to the dangers that may be encountered on entering storage premises during a fire or other emergency. Although the presence of water, steam, acid fumes, or powerful vibrations can cause the decomposition of many materials into dangerous compounds, of particular concern are high temperatures (such as those resulting from a fire) because these can cause many otherwise mild chemicals to emit highly toxic gases or vapors such as NO_x, SO_x, acids, and so forth, or evolve vapors of antimony, arsenic, mercury, and the like.

Every effort has been made to include the most current and complete information. The author welcomes comments or corrections to the data presented.

Richard J. Lewis, Sr.

Introduction

The list of potentially hazardous materials includes drugs, food additives, preservatives, ores, pesticides, dyes, detergents, lubricants, soaps, plastics, extracts from plant and animal sources, plants and animals that are toxic by contact or consumption, and industrial intermediates and waste products from production processes. Some of the information refers to materials of undefined composition. The chemicals included are assumed to exhibit the reported toxic effect in their pure state unless otherwise noted. However, even in the case of a supposedly "pure" chemical, there is usually some degree of uncertainty as to its exact composition and the impurities that may be present. This possibility must be considered in attempting to interpret the data presented because the toxic effects observed could in some cases be caused by a contaminant. Some radioactive materials are included but the effect reported is the chemically produced effect rather than the radiation effect.

For each entry the following data are provided when available: the DPIM code, hazard rating, entry name, CAS number, DOT number, molecular formula, molecular weight, line structural formula, a description of the material and physical properties, and synonyms. Following this are listed the toxicity data with references for reports of primary skin and eye irritation, mutation, reproductive, carcinogenic, and acute toxic dose data. The Consensus Reports section contains, where available, NTP 8th Annual Report on Carcinogens notation, IARC reviews, NTP Carcinogenesis Testing Program results, EPA Extremely Hazardous Substances List, the EPA Genetic Toxicology Program, and the Community Right-To-Know List. We also indicate the presence of the material in the update of the EPA TSCA inventory of chemicals in use in the United States. The next grouping consists of the U.S. Occupational Safety and Health Administration's (OSHA) permissible exposure levels, the American Conference of Governmental Industrial Hygienists' (ACGIH) Threshold Limit Values (TLVs), German Research Society's (MAK) values, National Institute for Occupational Safety and Health (NIOSH) recommended exposure levels, and U.S. Department of Transportation (DOT) classifications. Each entry concludes with a Safety Profile that discusses the toxic and other hazards of the entry. The Safety Profile concludes with the OSHA and NIOSH occupational analytical method, referenced by method name or number.

1. *DPIM Entry Code* identifies each entry by a unique code consisting of three letters and three numbers, for example, AAA123. The first letter of the entry code indicates the alphabetical position of the entry. Codes beginning with "A" are assigned to entries indexed with the A's. Each listing in the cross-indexes is referenced to its appropriate entry by the DPIM entry code.

2. *Entry Name* is the name of each material, selected, where possible, to be a commonly used designation.

3. *Hazard Rating* (**HR:**) is assigned to each material in the form of a number (1, 2, or 3) that briefly identifies the level of the toxicity or hazard. The letter "D" is used where the data available are insufficient to indicate a relative rating. In most cases a "D" rating is assigned when only in-vitro mutagenic or experimental reproductive data are available. Ratings are assigned on the basis of low (1), medium (2), or high (3) toxic, fire, explosive, or reactivity hazard.

The number "3" indicates an LD50 below 400 mg/kg or an LC50 below 100 ppm; or that the material is explosive, highly flammable, or highly reactive.

The number "2" indicates an LD50 of 400–4,000 mg/kg or an LC50 of 100–500 ppm; or that the material is flammable or reactive.

The number "1" indicates an LD50 of 4000–40,000 mg/kg or an LC50 of 500–5000 ppm; or that the material is combustible or has some reactivity hazard.

4. *Chemical Abstracts Service Registry Number* (*CAS:*) is a numeric designation assigned by the American Chemical Society's Chemical Abstracts Service and uniquely identifies a specific chemical compound. This entry allows one to conclusively identify a material regardless of the name or naming system used.

5. *DOT:* indicates a four-digit hazard code assigned by the U.S. Department of Transportation. This code is recognized internationally and is in agreement with the United Nations coding system. The code is used on

transport documents, labels, and placards. It is also used to determine the regulations for shipping the material.

6. *Molecular Formula (mf:)* or *atomic formula (af:)* designates the elemental composition of the material and is structured according to the Hill System (see *Journal of the American Chemical Society*, 22(8): 478–494, 1900), in which carbon and hydrogen (if present) are listed first, followed by the other elemental symbols in alphabetical order. The formulas for compounds that do not contain carbon are ordered strictly alphabetically by element symbol. Compounds such as salts or those containing waters of hydration have molecular formulas incorporating the CAS dot-disconnect convention. In this convention, the components are listed individually and separated by a period. The individual components of the formula are given in order of decreasing carbon atom count, and the component ratios given. A lowercase "x" indicates that the ratio is unknown. A lower case "n" indicates a repeating, polymer-like structure. The formula is obtained from one of the cited references or a chemical reference text, or derived from the name of the material.

7. *Molecular Weight (mw:)* or *atomic weight (aw:)* is calculated from the molecular formula, using standard elemental molecular weights (carbon = 12.01).

8. *Structural Formula* is a line formula indicating the structure of a given material.

9. *Properties (PROP:)* are selected to be useful in evaluating the hazard of a material and designing its proper storage and use procedures. A definition of the material is included where necessary. The physical description of the material may refer to the form, color, and odor to aid in positive identification. When available, the boiling point, melting point, density, vapor pressure, vapor density, and refractive index are given. The flash point, autoignition temperature, and lower and upper explosive limits are included to aid in fire protection and control. An indication is given of the solubility or miscibility of the material in water and common solvents. Unless otherwise indicated, temperature is given in Celsius, pressure in millimeters of mercury. Levels identified as "IDLH:" indicate concentrations that meet the definition of "immediately dangerous to life or health concentrations" (IDLHs). These are definded according to the NIOSH Respirator Decision Logic (DHHS [NIOSH] Publication No. 87-108, NTIS Publication No. PB-91-151183). It is a situation "that poses a threat of exposure to airborne contaminants when that exposure is likely to cause death or immediate or delayed permanent adverse health effects or prevent escape from such an environment."

10. *Synonyms* for the entry name are listed alphabetically. Synonyms include other chemical names, common or generic names, foreign names (with the language in parentheses), or codes. Some synonyms consist in whole or in part of registered trademarks. These trademarks are not identified as such. The reader is cautioned that some synonyms, particularly common names, may be ambiguous and refer to more than one material.

11. *Skin and Eye Irritation Data* lines include, in sequence, the tissue tested (skin or eye); the species of animal tested; the total dose and, where applicable, the duration of exposure; for skin tests only, whether open or occlusive; an interpretation of the irritation response severity when noted by the author; and the reference from which the information was extracted. Only positive irritation test results are included.

Materials that are applied topically to the skin or to the mucous membranes can elicit either (a) systemic effects of an acute or chronic nature or (b) local effects, more properly termed "primary irritation." A primary irritant is a material that, if present in sufficient quantity for a sufficient period of time, will produce a nonallergic, inflammatory reaction of the skin or of the mucous membrane at the site of contact. Primary irritants are further limited to those materials that are not corrosive. Hence, concentrated sulfuric acid is not classified as a primary irritant.

a. Primary Skin Irritation. In experimental animals, a primary skin irritant is defined as a chemical that produces an irritant response on first exposure in a majority of the test subjects. However, in some instances compounds act more subtly and require either repeated contact or special environmental conditions (humidity, temperature, occlusion, etc.) to produce a response.

The most standard animal irritation test is the Draize procedure (*Journal of Pharmacology and Experimental Therapeutics*, 82: 377–419, 1944). This procedure has been modified and adopted as a regulatory test by the Consumer Product Safety Commission (CPSC) in 16 CFR 1500.41 (formerly 21 CFR 191.11). In this test a known amount (0.5 mL of a liquid, or 0.5 g of a solid or semisolid) of the test material is introduced under a one-square-inch gauze patch. The patch is applied to the skin (clipped free of hair) of 12 albino rabbits. Six rabbits are tested with intact skin and six with abraded skin. The abrasions are minor incisions made through the stratum corneum but are not sufficiently deep to disturb the dermis or to produce bleeding. The patch is secured in place with adhesive tape, and the entire trunk of the animal is wrapped with an impervious material, such as rubberized cloth, for a 24-hour period. The animal is immobilized during exposure. After 24 hours the patches are removed and the resulting reaction evaluated for erythema, eschar, and edema formation. The reaction is again scored at the end of 72 hours (48 hours after the initial reading), and the two readings are averaged. A material producing any degree of positive reaction is cited as an irritant.

As the modified Draize procedure, described previously has become the standard test specified by the U.S. government, nearly all of the primary skin irritation data either strictly adheres to the test protocol or involves only simple modifications to it. When test procedures other than those described previously are reported in the literature, appropriate codes are included in the data line to indicate those deviations.

The most common modification is the lack of occlusion of the test patch, so that the treated area is left open to the atmosphere. In such cases the notation "open" appears in the irritation data line. Another frequent modification involves immersion of the whole arm or whole body in the test material or, more commonly, in a dilute aqueous solution of the test material. This type of test is often conducted on soap and detergent solutions. Immersion data are identified by the abbreviation "imm" in the data line.

The dose reported is based first on the lowest dose producing an irritant effect and second on the latest study published. The dose is expressed as follows:

(1) Single application by the modified Draize procedure is indicated by only a dose amount. If no exposure time is given, then the data are for the standard 72-hour test. For test times other than 72 hours, the dose data are given in milligrams (or another appropriate unit)/duration of exposure, for example, 10 mg/24H.

Category	Code	Skin Reaction (Draize)
Slight (Mild)	MLD	Well-defined erythema and slight edema (edges of area well defined by definite raising)
Moderate	MOD	Moderate-to-severe erythema and moderate edema (area raised approximately 1 mm)
Severe	SEV	Severe erythema (beet redness) to slight eschar formation (injuries in depth) and severe dema (raised more than 1 mm and extending beyond area of exposure)

(2) Multiple applications involve administration of the dose in divided portions applied periodically. The total dose of test material is expressed in milligrams (or another appropriate unit)/duration of exposure, with the symbol "I" indicating intermittent exposure, for example, 5 mg/6D-I.

The method of testing materials for primary skin irritation given in the Code of Federal Regulations does not include an interpretation of the response. However, some authors do include a subjective rating of the irritation observed. If such a severity rating is given, it is included in the data line as mild ("MLD"), moderate ("MOD"), or severe ("SEV"). The Draize procedure employs a rating scheme that is included here for informational purposes only, because other researchers may not categorize irritation response in this manner.

b. Primary Eye Irritation. In experimental animals, a primary eye irritant is defined as a chemical that produces an irritant response in the test subject on first exposure. Eye irritation study procedures that Draize developed have been modified and adopted as a regulatory test by CPSC in 16 CFR 1500.42. In this procedure, a known amount of the test material (0.1 mL of a liquid, or 100 mg of a solid or paste) is placed in one eye of each of six albino rabbits; the other eye remains untreated, serving as a control. The eyes are not washed after instillation and are examined at 24, 48, and 72 hours for ocular reaction. After the recording of ocular reaction at 24 hours, the eyes may be further examined following the application of fluorescein. The eyes may also be washed with a sodium chloride solution (U.S.P. or equivalent) after the 24-hour reaction has been recorded.

A test is scored positive if any of the following effects are observed: (1) ulceration (besides fine stippling); (2) opacity of the cornea (other than slight dulling of normal luster); (3) inflammation of the iris (other than a slight deepening of the rugae or circumcorneal injection of the blood vessel); (4) swelling of the conjunctiva (excluding the cornea and iris) with eversion of the eyelid; or (5) a diffuse crimson-red color with individual vessels not clearly identifiable. A material is an eye irritant if four of six rabbits score positive. It is considered a nonirritant if none or only one of six animals exhibits irritation. If intermediate results are obtained, the test is performed again. Materials producing any degree of irritation in the eye are identified as irritants. When an author has designated a substance as either a mild, moderate, or severe eye irritant, this designation is also reported.

The dose reported is based first on the lowest dose producing an irritant effect and second on the latest study published. Single and multiple applications are indicated as described previously under "Primary Skin Irritation." Test times other than 72 hours are noted in the dose. All eye irritant test exposures are assumed to be continuous, unless the reference states that the eyes were washed after instillation. In this case, the notation "rns" (rinsed) is included in the data line.

Because Draize procedures for determining both skin and eye irritation specify rabbits as the test species, most of the animal irritation data are for rabbits, although any of the species listed in Table 1 may be used. We have endeavored to include as much human data as possible, since this information is directly applicable to occupational exposure, much of which comes from studies conducted on volunteers (for example, for

TABLE 1. Species
(With assumptions for toxic dose calculation from nonspecific data*)

Species	Abbrev.	Age	Weight	Consumption Food (g/day)	(Approx.) Water (mL/day)	1 ppm in Food Equals, in (mg/kg/day)	Approximate Gestation Period (days)
Bird-type not specified	brd		1 kg				
Bird-wild bird species	bwd		40 g				
Cat, adult	cat		2 kg	100	100	0.05	64 (59-68)
Child	chd	1-13 Y	20 kg				
Chicken, adult	ckn	8 W	800 g	140	200	0.175	
Cattle	ctl		500 kg	10,000		0.02	284 (279-290)
Duck, adult (domestic)	dck	8 W	2.5 kg	250	500	0.1	
Dog, adult	dog	52W	10 kg	250	500	0.025	62 (56-68)
Domestic animals (Goat, Sheep)	dom		60 kg	2,400		0.04	G: 152 (148-156) S: 146 (144-147)
Frog, adult	frg		33 g				
Guinea Pig, adult	gpg		500 g	30	85	0.06	68
Gerbil	grb		100 g	5	5	0.05	25 (24-26)
Hamster	ham	14W	125 g	15	10	0.12	16 (16-17)
Human	hmn	Adult	70 kg				
Horse, Donkey	hor		500 kg	10,000		0.02	H: 339 (333-345) D: 365
Infant	inf	0-1 Y	5 kg				
Mammal (species unspecified in reference)	mam		200 g				
Man	man	Adult	70 kg				
Monkey	mky	2.5 Y	5 kg	250	500	0.05	165
Mouse	mus	8 W	25 g	3	5	0.12	21
Non-mammalian species	nml						
Pigeon	pgn	8 W	500 g				
Pig	pig		60 kg	2,400		0.041	114 (112-115)
Quail (laboratory)	qal		100 g				
Rat, adult female	rat	14W	200 g	10	20	0.05	22
Rat, adult male	rat	14W	250 g	15	25	0.06	
Rat, adult	rat	14W	200 g	15	25		
Rat, weanling	rat	3 W	50 g	15	25	0.3	
Rabbit, adult	rbt	12 W	2 kg	60	330	0.03	31
Squirrel	sql		500 g	44			
Toad	tod		100 g				
Turkey	trk	18 W	5 kg				
Woman	wmn	Adult	50 kg	270			

*Values given in Table 1 are within reasonable limits usually found in the published literature and are selected to facilitate calculations for data from publications in which toxic dose information has not been presented for an individual animal of the study. See, for example, *Association of Food and Drug Officials, Quarterly Bulletin*, volume 18, page 66, 1954; Guyton, *American Journal of Physiology*, volume 150, page 75, 1947; *The Merck Veterinary Manual*, 5th Edition, Merck&Co., Inc., Rahway, NJ, 1979; and The UFAW *Handbook on the Care and Management of Laboratory Animals*, 4th Edition, Churchill Livingston, London, 1972. Data for lifetime exposure are calculated from the assumptions for adult animals for the entire period of exposure. For definitive dose data, the reader must review the referenced publication.

cosmetic or soap ingredients) or from persons accidentally exposed. When accidental exposure, such as a spill, is cited, the line includes the abbreviation "nse" (nonstandard exposure). In these cases it is often very difficult to determine the precise amount of the material to which the individual was exposed. Therefore, for accidental exposures an estimate of the concentration or strength of the material, rather than a total dose amount, is generally provided.

12. *Mutation Data* lines include, in sequence, the mutation test system utilized, the species of the tested organism (and, where applicable, the route of administration or cell type), the exposure concentration or dose, and the reference from which the information was extracted.

A mutation is defined as any heritable change in genetic material. Unlike irritation, reproductive, tumorigenic, and toxic dose data, which report the results of whole-animal studies, mutation data also include studies on lower organisms such as bacteria, molds, yeasts, and insects, as well as in-vitro mammalian cell cultures. Studies of plant mutagenesis are not included. No attempt is made to evaluate the significance of the data or to rate the relative potency of the compound as a mutagenic risk to humans.

Each element of the mutation line is discussed as follows:

a. Mutation Test System. Several test systems are used to detect genetic alterations caused by chemicals. Additional test systems may be added as they are reported in the literature. Each test system is identified by the three-letter code shown in parentheses. For additional information about mutation tests, the reader may wish to consult the *Handbook of Mutagenicity Test Procedures*, edited by B.J. Kilbey, M. Legator, W. Nichols, and C. Ramel (Amsterdam: Elsevier Scientific Publishing Company/North-Holland Biomedical Press, 1977).

(1) The Mutation in Microorganisms (mmo) System utilizes the detection of heritable genetic alterations in microorganisms that have been exposed directly to the chemical.

(2) The Microsomal Mutagenicity Assay (mma) System utilizes an in-vitro technique that allows enzymatic activation of promutagens in the presence of an indicator organism in which induced mutation frequencies are determined.

(3) The Micronucleus Test (mnt) System utilizes the fact that chromosomes or chromosome fragments may not be incorporated into one or the other of the daughter nuclei during cell division.

(4) The Specific Locus Test (slt) System utilizes a method for detecting and measuring rates of mutation at any or all of several recessive loci.

(5) The DNA Damage (dnd) System detects the damage to DNA strands, including strand breaks, crosslinks, and other abnormalities.

(6) The DNA Repair (dnr) System utilizes methods of monitoring DNA repair as a function of induced genetic damage.

(7) The Unscheduled DNA Synthesis (dns) System detects the synthesis of DNA during usually nonsynthetic phases.

(8) The DNA Inhibition (dni) System detects damage that inhibits the synthesis of DNA.

(9) The Gene Conversion and Mitotic Recombination (mrc) System utilizes unequal recovery of genetic markers in the region of the exchange during genetic recombination.

(10) The Cytogenetic Analysis (cyt) System utilizes cultured cells or cell lines to assay for chromosomal aberrations following the administration of the chemical.

(11) The Sister Chromatid Exchange (sce) System detects the interchange of DNA in cytological preparations of metaphase chromosomes between replication products at apparently homologous loci.

(12) The Sex Chromosome Loss and Nondisjunction (sln) System measures the nonseparation of homologous chromosomes at meiosis and mitosis.

(13) The Dominant Lethal Test (dlt). A dominant lethal is a genetic change in a gamete that kills the zygote produced by that gamete. In mammals, the dominant lethal test measures the reduction of litter size by examining the uterus and noting the number of surviving and dead implants.

(14) The Mutation in Mammalian Somatic Cells (msc) System utilizes the induction and isolation of mutants in cultured mammalian cells by identification of the gene change.

(15) The Host-Mediated Assay (hma) System uses two separate species, generally mammalian and bacterial, to detect heritable genetic alteration caused by metabolic conversion of chemical substances administered to host mammalian species in the bacterial indicator species.

(16) The Sperm Morphology (spm) System measures the departure from normal in the appearance of sperm.

(17) The Heritable Translocation Test (trn) measures the transmissibility of induced translocations to subsequent generations. In mammals, the test uses sterility and reduced fertility in the progeny of the treated parent. In addition, cytological analysis of the F1 progeny or subsequent progeny of the treated parent is carried out to prove the existence of the induced translocation. In *Drosophila*, heritable translocations are detected genetically using easily distinguishable phenotypic markers, and these translocations can be verified with cytogenetic techniques.

(18) The Oncogenic Transformation (otr) System utilizes morphological criteria to detect cytological differences between normal and transformed tumorigenic cells.

(19) The Phage Inhibition Capacity (pic) System utilizes a lysogenic virus to detect a change in the genetic characteristics by the transformation of the virus from noninfectious to infectious.

(20) The Body Fluid Assay (bfa) System uses two separate species, usually mammalian and bacterial. The test substance is first administered to the host, from whom body fluid (for example, urine, blood) is subsequently taken. This body fluid is then tested in-vitro, and mutations are measured in the bacterial species.

b. Species. Those test species that are peculiar to mutation data are designated by the three-letter codes as follows:

	Code	Species
Bacteria	bcs	*Bacillus subtilis*
	esc	*Escherichia coli*
	hmi	*Haemophilus influenzae*
	klp	*Klebsiella pneumoniae*
	sat	*Salmonella typhimurium*
	srm	*Serratia marcescens*
Molds	asn	*Aspergillus nidulans*
	nsc	*Neurospora crassa*
Yeasts	smc	*Saccharomyces cerevisiae*
	ssp	*Schizosaccharomyces pombe*
Protozoa	clr	*Chlamydomonas reinhardi*
	Eug	*Euglena gracilis*
	omi	other microorganisms
Insects	dmg	*Drosophila melanogaster*
	dpo	*Drosophila pseudo-obscura*
	grh	grasshopper
	slw	silkworm
	oin	other insects
Fish	sal	salmon
	ofs	other fish

If the test organism is a cell type from a mammalian species, the parent mammalian species is reported, followed by a dash and the cell type designation. For example, human leukocytes are coded "hmn-leu." The various cell types currently cited in this edition are as follows:

In the case of host-mediated and body-fluid assays, both the host organism and the indicator organism are given as follows: host organism/indicator organism, for example, "ham/sat" for a test in which hamsters were exposed to the test chemical and *S. typhimurium* was used as the indicator organism.

For in-vivo mutagenic studies, the route of administration is specified following the species designation, for example, "mus-orl" for oral administration to mice. See Table 2 for a complete list of routes cited. The route of administration is not specified for in-vitro data.

Designation	Cell Type
ast	Ascites tumor
bmr	bone marrow
emb	embryo
fbr	fibroblast
hla	HeLa cell
kdy	kidney
leu	leukocyte
lng	lung
lvr	liver
lym	lymphocyte
mmr	mammary gland
ovr	ovary
spr	sperm
tes	testis
oth	other cell types not listed above

c. Units of Exposure. The lowest dose producing a positive effect is cited. The author's calculations are used to determine the lowest dose at which a positive effect was observed. If the author fails to state the lowest effective dose, two times the control dose will be used. Ideally, the dose should be reported in universally accepted toxicological units such as milligrams of test chemical per kilogram of test animal body weight. Although this is possible in cases where the actual intake of the chemical by an organism of known weight is reported, it is not possible in many systems using insect and bacterial species. In cases where a dose is reported or where the amount can be converted to a dose unit, it is normally listed as milligrams per kilogram (mg/kg). However, micrograms (æg), nanograms (ng), or picograms (pg) per kilogram may also be used for convenience of presentation. Concentrations of gaseous materials in air are listed as parts per hundred (pph), per million (ppm), per billion (ppb), or per trillion (ppt).

Test systems using microbial organisms traditionally report exposure data as an amount of chemical per liter (L) or amount per plate, well, disc, or tube. The amount may be on a weight (g, mg, æg, ng, or pg) or molar (millimole (mmol), micromole (æmol), nanomole (nmol), or picomole (pmol)). These units describe the exposure concentration rather than the dose actually taken up by the test species. Insufficient data currently exist to permit the development of dose amounts from this information. In such cases, therefore, the material concentration units that the author used are reported.

Because the exposure values reported in host-mediated and body-fluid assays are doses delivered to the host organism, no attempt is made to estimate the exposure concentration to the indicator organism. The exposure values cited for host-mediated assay data are in units of milligrams (or other appropriate units of weight) of material administered per kilogram of host body weight, or in parts of vapor or gas per million (ppm) parts of air (or other appropriate concentrations) by volume.

13. *Toxicity Dose Data* lines include, in sequence, the route of exposure; the species of animal studied; the toxicity measure; the amount of material per body weight or concentration per unit of air volume and, where applicable, the duration of exposure; a descriptive notation of the type of effect reported; and the reference from which the information was extracted. Only positive toxicity test results are cited in this section.

All toxic-dose data appearing in the CD-ROM are derived from reports of the toxic effects produced by individual materials. For human data, a toxic effect is defined as any reversible or irreversible noxious effect on the body, any benign or malignant tumor, any teratogenic effect, or any death that has been reported to have resulted from exposure to a material via any route. For humans, a toxic effect is any effect that was reported in the source reference. There is no qualifying limitation on the duration of exposure or for the quantity or concentration of the material, nor is there a qualifying limitation on the circumstances that resulted in the exposure. Regardless of the absurdity of the circumstances that were involved in a toxic exposure, it is assumed that the same circumstances could recur. For animal data, toxic effects are limited to the production of tumors, benign (neoplastigenesis) or malignant

Table 2. Routes of Administration to, or Exposure of, Animal Species to Toxic Substances

Route	Abbreviation	Definition
Eyes	eye	Administration directly onto the surface of the eye. Used exclusively for primary irritation data. See *Ocular*.
Intraaural	ial	Administration into the ear
Intraarterial	iat	Administration into the artery
Intracerebral	ice	Administration into the cerebrum
Intracervical	icv	Administration into the cervix
Intradermal	idr	Administration within the dermis by hypodermic needle
Intraduodenal	idu	Administration into the duodenum
Inhalation	ihl	Inhalation in chamber, by cannulation, or through mask
Implant	imp	Placed surgically within the body location described in reference
Intramuscular	ims	Administration into the muscle by hypodermic needle
Intraplacental	ipc	Administration into the placenta
Intrapleural	ipl	Administration into the pleural cavity by hypodermic needle
Intraperitoneal	ipr	Administration into the peritoneal cavity
Intrarenal	irn	Administration into the kidney
Intraspinal	isp	Administration into the spinal canal
Intratracheal	itr	Administration into the trachea
Intratesticular	itt	Administration into the testes
Intrauterine	iut	Administration into the uterus
Intravaginal	ivg	Administration into the vagina
Intravenous	ivn	Administration directly into the vein by hypodermic needle
Multiple	mul	Administration into a single animal by more than one route
Ocular	ocu	Administration directly onto the surface of the eye or into the conjunctival sac. Used exclusively for systemic toxicity data.
Oral	orl	Per os, intragastric, feeding, or introduction with drinking water
Parenteral	par	Administration into the body through the skin. Reference cited is not specific about the route used. Could be ipr, scu, ivn, ipl, ims, irn, or ice.
Rectal	rec	Administration into the rectum or colon in the form of enema or suppository
Subcutaneous	scu	Administration under the skin
Skin	skn	Application directly onto the skin, either intact or abraded. Used for both systemic toxicity and primary irritant effects.
Unreported	unr	Dose, but not route, is specified in the reference.

(carcinogenesis); the production of changes in the offspring resulting from action on the fetus directly (teratogenesis); and death. There is no limitation on either the duration of exposure or on the quantity or concentration of the dose of the material reported to have caused these effects.

The report of the lowest total dose administered over the shortest time to produce the toxic effect was given preference, although some editorial liberty was taken so that additional references might be cited. No restrictions were placed on the amount of a material producing death in an experimental animal nor on the time period over which the dose was given.

Each element of the toxic dose line is discussed as follows:

a. Route of Exposure or Administration. Although many exposures to materials in the industrial community occur via the respiratory tract or skin, most studies in the published literature report exposures of experimental animals in which the test materials were introduced primarily through the mouth by pills, in food, in drinking water, or by intubation directly into the stomach. The abbreviations and definitions of the various routes of exposure reported are given in Table 2.

b. Species Exposed. Because the effects of exposure of humans are of primary concern, we have indicated, when available, whether the results were observed in man, woman, child, or infant. If no such distinction was made in the reference, the abbreviation "hmn" (human) is used. However, the results of studies on rats or mice are the most frequently reported and hence provide the most useful data for comparative purposes. The species and abbreviations used in reporting toxic dose data are listed alphabetically in Table 1.

c. Description of Exposure. In order to describe the administered dose reported in the literature, six abbreviations are used. These terms indicate whether the dose caused death (LD) or other toxic effects (TD) and whether it was administered as a lethal concentration (LC) or toxic concentration (TC) in the inhaled air. In general, the term "Lo" is used where the number of subjects studied was not a significant number from the population or the calculated percentage of subjects showing an effect was listed as 100. The definition of terms is as follows:

Toxic Dose TDLo-Low--the lowest dose of a material introduced by any route, other than inhalation, over any given period of time and reported to produce any toxic effect in humans or to produce carcinogenic, neoplastigenic, or teratogenic effects in animals or humans.

TCLo--Toxic Concentration Low--the lowest concentration of a material in air to which humans or animals have been exposed for any given period of time that has produced any toxic effect in humans or produced a carcinogenic, neoplastigenic, or teratogenic effect in animals or humans.

LDLo--Lethal Dose Low--the lowest dose (other than LD50) of a material introduced by any route, other than inhalation, over any given period of time in one or more divided portions and reported to have caused death in humans or animals.

LD50--Lethal Dose Fifty--a calculated dose of a material that is expected to cause the death of 50% of an entire defined experimental animal population. It is determined from the exposure to the material, by any route other than inhalation, of a significant number from that population. Other lethal dose percentages, such as LD1, LD10, LD30, and LD99, may be published in the scientific literature for the specific purposes of the author. Such data would be published if these figures, in the absence of a calculated lethal dose (LD50), were the lowest found in the literature.

LCLo--Lethal Concentration Low--the lowest concentration of a material in air, other than LC50, that has been reported to have caused death in humans or animals. The reported concentrations may be entered for periods of exposure that are less than 24 hours (acute) or greater than 24 hours (subacute and chronic).

LC50--Lethal Concentration Fifty--a calculated concentration of a material in air, exposure to which for a specified length of time is expected to cause the death of 50% of an entire defined experimental animal population. It is determined from the exposure to the material of a significant number from that population.

The following table summarizes the previous information.

d. Units of Dose Measurement. As in almost all experimental toxicology, the doses given are expressed in terms of the quantity administered per unit body weight, or quantity per skin surface area, or quantity per unit volume of the respired air. In addition, the duration of time over which the dose was administered is also listed, as needed. Dose amounts are generally expressed as milligrams (thousandths of a gram) per kilogram (mg/kg). In some cases, because of dose size and its practical presentation in the file, grams per kilogram (g/kg), micrograms (millionths of a gram) per kilogram (æg/kg), or nanograms (billionths of a gram) per kilogram (ng/kg) are used. Volume measurements of dose were converted to weight units by appropriate calculations. Densities were obtained from standard reference texts. Where densities were not readily available, all liquids were assumed to have a density of 1 g/mL. Twenty drops of liquid are assumed to be equal in volume to 1 mL.

All body weights have been converted to kilograms (kg) for uniformity. For those references in which the dose was reported to have been administered to an animal of unspecified weight or a given number of animals in a group (for example, feeding studies) without weight data, the weights of the respective animal species were assumed to be those listed in Table 1 and the dose is listed on a per-kilogram body-weight basis. Assumptions for daily food and water intake are found in Table 1 to allow approximation doses for humans and species of experimental animals in cases in which the dose was originally reported as a concentration in food or water. The values presented are selections that are reasonable for the species and convenient for dose calculations.

Category	Exposure Time	Route of Exposure	Toxic Effects Human	Toxic Effects Animal
TDLo	Acute or chronic	All except inhalation	Any non-lethal	CAR, NEO, ETA, TER, REP
TCLo	Acute or chronic	Inhalation	Any non-lethal	CAR, NEO, ETA, TER, REP
LDLo	Acute or chronic	All except Inhalation	Death	Death
LD50	Acute	All except inhalation	Not applicable	Death (statistically determined)
LCLo	Acute or chronic	Inhalation	Death	Death
LC50	Acute	Inhalation	Not applicable	Death (statistically determined)

Concentrations of a gaseous material in air are generally listed as parts of vapor or gas per million parts of air by volume (ppm). However, parts per hundred (pph or percent), parts per billion (ppb), or parts per trillion (ppt) may be used for convenience of presentation. If the material is a solid or a liquid, the concentrations are listed preferably as milligrams per cubic meter (mg/m^3) but may, as applicable, be listed as micrograms per cubic meter ($æg/m^3$), nanograms per cubic meter (ng/m^3), or picograms (trillionths of a gram) per cubic meter (pg/m^3) of air. For those cases in which other measurements of contaminants are used, such as the number of fibers or particles, the measurement is spelled out.

Where the duration of exposure is available, time is presented as minutes (M), hours (H), days (D), weeks (W), or years (Y). Additionally, continuous (C) indicates that the exposure was continuous over the time administered, such as ad-libitum feeding studies or 24-hour, 7-day-per-week inhalation exposures. Intermittent (I) indicates that the dose was administered during discrete periods, such as daily or twice weekly. In all cases, the total duration of exposure appears first after the kilogram body weight and a slash, and is followed by descriptive data; for example, 10 mg/kg/3W-I indicates ten milligrams per kilogram body weight administered over a period of three weeks, intermittently in a number of separate, discrete doses. This description is intended to provide the reader with enough information for an approximation of the experimental conditions, which can be further clarified by studying the reference cited.

e. Frequency of Exposure. Frequency of exposure to the test material depends on the nature of the experiment. Frequency of exposure is given in the case of an inhalation experiment, for human exposures (where applicable), or where CAR, NEO, ETA, REP, or TER is specified as the toxic effect.

f. Duration of Exposure. For assessment of tumorigenic effect, the testing period should be the life span of the animal, or should extend until statistically valid calculations can be obtained regarding tumor incidence. In the toxic dose line, the total dose causing the tumorigenic effect is given. The duration of exposure is included to give an indication of the testing period during which the animal was exposed to this total dose. For multigenerational studies, the time during gestation when the material was administered to the mother is also provided.

g. Notations Descriptive of the Toxicology. The toxic dose line thus far has indicated the route of entry, the species involved, the description of the dose, and the amount of the dose. The next entry found on this line when a toxic exposure (TD or TC) has been listed is the toxic effect. Following a colon will be one of the notations found in Table 3. These notations indicate the organ system affected or special effects that the material produced, for example, TER = teratogenic effect. No attempt was made to be definitive in reporting these effects because such definition requires detailed qualification that is beyond the scope of this CD-ROM. The selection of the dose was based first on the lowest dose producing an effect and second on the latest study published.

14. *Reproductive Effects Data* lines include, in sequence, the reproductive effect reported, the route of exposure, the species of animal tested, the type of dose, the total dose amount administered, the time and duration of administration, and the reference from which the information was extracted. Only positive reproductive effects data for mammalian species are cited. Because of differences in the reproductive systems among species and the systems' varying responses to chemical exposures, no attempt is made to extrapolate animal data or to evaluate the significance of a substance as a reproductive risk to humans.

Each element of the reproductive effects data line is discussed as follows:

a. Reproductive Effect. For human exposure, the effects are included in the safety profile. The effects include those reported to affect the male or female reproductive systems, mating and conception success, fetal effects (including abortion), transplacental carcinogenesis, and post-birth effects on parents and offspring.

b. Route of Exposure or Administration. See Table 2 for a complete list of abbreviations and definitions of the various routes of exposure reported. For reproductive effects data, the specific route is listed either when the substance was administered to only one of the parents or when the substance was administered to both parents by the same route. However, if the substance was administered to each parent by a different route, the route is indicated as "mul" (multiple).

c. Species Exposed. Reproductive effects data are cited for mammalian species only. Species abbreviations are shown in Table 1. Also shown in Table 1 are approximate gestation periods.

d. Type of Exposure. Only two types of exposure, TDLo and TCLo, are used to describe the dose amounts reported for reproductive effects data.

e. Dose Amounts and Units. The total dose amount that was administered to the exposed parent is given. If the substance was administered to both parents, the individual amounts to each parent have been added together and the total amount shown. Where necessary, appropriate conversion of dose units has been made. The dose amounts listed are those for which the reported effects are statistically significant. However, human case reports are cited even when no statistical tests can be performed. The statistical test is that used by the author. If no statistic is reported, a Fisher's Exact Test is applied with significance at the 0.05 level, unless

TABLE 3. Notations Descriptive of the Toxicology

Notation	Effects (not limited to effects listed)
ALR	Allergic systemic reaction such as might be experienced by individuals sensitized to penicillin.
BAH	Behavioral--includes wakefulness, euphoria, hallucinations, coma, etc.
BCM	Blood clotting mechanism effects--any effect that increases or decreases clotting time.
BLD	Blood effects--effect on all blood elements, electrolytes, pH, proteins, oxygen carrying or releasing capacity.
BPR	Blood pressure effects--any effect that increases or decreases any aspect of blood pressure.
CAR	Carcinogenic effects--see paragraph 15 in text.
CNS	Central nervous system effects--includes effects such as headaches, tremor, drowsiness, convulsions, hypnosis, anesthesia.
COR	Corrosive effects--burns, desquamation.
CUM	Cumulative effects--where material is retained by the body in greater quantities than is excreted, or the effect is increased in severity by repeated body insult.
CVS	Cardiovascular effects--such as an increase or decrease in the heart activity through effect on ventricle or auricle; fibrillation; constriction or dilation of the arterial or venous system.
DDP	Drug dependence effects--any indication of addiction or dependence.
ETA	Equivocal tumorigenic agent--see text.
EYE	Eye effects--irritation, diplopmia, cataracts, eye ground, blindness by effects to the eye or the optic nerve.
GIT	Gastrointestinal tract effects--diarrhea, constipation, ulceration.
GLN	Glandular effects--any effect on the endocrine glandular system.
IRR	Irritant effects--any irritant effect on the skin, eye, or mucous membrane.
MLD	Mild irritation effects--used exclusively for primary irritation data.
MMI	Mucous membrane effects--irritation, hyperplasia, changes in ciliary activity.
MOD	Moderate irritation effects--used exclusively for primary irritation data.
MSK	Musculoskeletal effects--such as osteoporosis, muscular degeneration.
NEO	Neoplastic effects--see text.
PNS	Peripheral nervous system effects.
PSY	Psychotropic effects--exerting an effect upon the mind.
PUL	Pulmonary system effects--effects on respiration and respiratory pathology.
RBC	Red blood cell effects--includes the several anemias.
REP	Reproductive effects--see text.
SEV	Severe irritation effects--used exclusively for primary irritation data.
SKN	Skin effects--such as erythema, rash, sensitization of skin, petechial hemorrhage.
SYS	Systemic effects--effects on the metabolic and excretory function of the liver or kidneys.
TER	Teratogenic effects--nontransmissible changes produced in the offspring.
UNS	Unspecified effects--the toxic effects were unspecific in the reference.
WBC	White blood cell effects--effects on any of the cellular units other than erythrocytes, including any change in number or form.

the author makes a strong case for significance at some other level.

Dose units are usually given as an amount administered per unit body weight or as parts of vapor or gas per million parts of air by volume. There is no limitation on either the quantity or concentration of the dose, or the duration of exposure reported to have caused the reproductive effect(s).

f. Time and Duration of Treatment. The time when a substance is administered to either or both parents may significantly affect the results of a reproductive study, because there are differing critical periods during the reproductive cycles of each species. Therefore, to provide some indication of when the substance was administered, which should facilitate selection of specific data for analysis by the user, a series of up to four terms follows the dose amount. These terms indicate to which parent(s) and at what time the substance was administered. The terms take the general form:

(uD male/vD pre/w-xD preg/yD post)

where u = total number of days of administration to male prior to mating

v = total number of days of administration to female prior to mating

w = first day of administration to pregnant female during gestation

x = last day of administration to pregnant female during gestation

y = total number of days of administration to lactating mother after birth of offspring

If administration is to the male only, then only the first of the above four terms is shown following the total dose to the male, for example, 10 mg/kg (5D male). If administration is to the female only, then only the second, third, or fourth term, or any combination thereof, is shown following the total dose to the female, for example:

10 mg/kg (3D pre)

10 mg/kg (3D pre/4-7D preg)

10 mg/kg (3D pre/4-7D preg/5D post)

10 mg/kg (3D pre/5D post)

10 mg/kg (4-7D preg)

10 mg/kg (4-7D preg/5D post)

10 mg/kg (5D post) (NOTE: This example indicates administration was only to the lactating mother, and only after birth of the offspring.)

If administration is to both parents, then the first term and any combination of the last three terms are listed, for example, 10 mg/kg (5D male/3D pre/4-7D preg). If administration is continuous through two or more of the above periods, the above format is abbreviated by replacing the slash (/) with a dash (-). For example, 10 mg/kg (3D pre-5D post) indicates a total of 10 mg/kg

administered to the female for three days prior to mating, on each day during gestation, and for five days following birth. Approximate gestation periods for various species are shown in Table 1.

g. Multigeneration Studies. Some reproductive studies entail administration of a substance to several consecutive generations, with the reproductive effects measured in the final generation. The protocols for such studies vary widely. Therefore, because of the inherent complexity and variability of these studies, they are cited in a simplified format as follows. The specific route of administration is reported if it was the same for all parents of all generations; otherwise the abbreviation "mul" is used. The total dose amount shown is that administered to the F0 generation only; doses to the Fn (where n = 1, 2, 3, etc.) generations are not reported. The time and duration of treatment for multigeneration studies are not included in the data line. Instead, the dose amount is followed by the abbreviation ("MGN"), for example, 10 mg/kg (MGN). This code indicates a multigeneration study, and the reader must consult the cited reference for complete details of the study protocol.

15. *Carcinogenic Study Result.* Tumorigenic citations are classified according to the reported results of the study to aid the reader in selecting appropriate references for in-depth review and evaluation. The classification ETA (equivocal tumorigenic agent) denotes those studies reporting uncertain, but seemingly positive, results. The criteria for the three classifications are listed as follows. These criteria are used to abstract the data in individual reports on a consistent basis and do not represent a comprehensive evaluation of a material's tumorigenic potential to humans.

The following nine technical criteria are used to abstract the toxicological literature and classify studies that report positive tumorigenic responses. No attempts are made either to evaluate the various test procedures or to correlate results from different experiments.

(1) A citation is coded "CAR" (carcinogenic) when review of an article reveals that all the following criteria are satisfied:

(a) There is a statistically significant increase in the incidence of tumors in the test animals. The statistical test is that used by the author. If no statistic is reported, a Fisher's Exact Test is applied with significance at the 0.05 level, unless the author makes a strong case for significance at some other level.

(b) A control group of animals is used and the treated and control animals are maintained under identical conditions.

(c) The sole experimental variable between the groups is the administration or nonadministration of the test material (see (10) that follows).

(d) The tumors consist of autonomous populations of cells of abnormal cytology capable of invading and destroying normal tissues, or the tumors metastasize as confirmed by histopathology.

(2) A citation is coded "NEO" (neoplastic) when review of an article reveals that all the following criteria are satisfied:

(a) There is a statistically significant increase in the incidence of tumors in the test animals. The statistical test is that used by the author. If no statistic is reported, a Fisher's Exact Test is applied with significance at the 0.05 level, unless the author makes a strong case for significance at some other level.

(b) A control group of animals is used and the treated and control animals are maintained under identical conditions.

(c) The sole experimental variable between the groups is the administration or nonadministration of the test material.

(d) The tumors consist of cells that closely resemble the tissue of origin, that are not grossly abnormal cytologically, that may compress surrounding tissues, but that neither invade tissues nor metastasize; or

(e) The tumors produced cannot be classified as either benign or malignant.

(3) A citation is coded "ETA" (equivocal tumorigenic agent) when some evidence of tumorigenic activity is presented, but one or more of the criteria listed in (1) or (2) previously are lacking. Thus, a report with positive pathological findings, but with no mention of control animals, is coded "ETA."

(4) Because an author may make statements or draw conclusions based on a larger context than that of the particular data reported, papers in which the author's conclusions differ substantially from the evidence presented in the paper are subject to review.

(5) All doses except those for transplacental carcinogenesis are reported in one of the following formats.

(a) For all routes of administration other than inhalation: cumulative dose is reported in milligrams (or another appropriate unit)/killogram/duration of administration.

Whenever the dose reported in the reference is not in the units discussed herein, conversion to this format is made. The total cumulative dose is derived from the lowest dose level that produces tumors in the test group.

(b) For inhalation experiments: concentration is reported in parts per million (or milligrams/cubic meter)/total duration of exposure.

The concentration refers to the lowest concentration that produces tumors.

(6) Transplacental carcinogenic doses are reported in one of the following formats:

(a) For all routes of administration other than inhalation, cumulative dose is reported in milligrams/killogram/(time of administration during pregnancy).

The cumulative dose is derived from the lowest single dose that produces tumors in the offspring. The test chemical is administered to the mother.

(b) For inhalation experiments, concentration is reported in parts per million (or milligrams/cubic meter)/(time of exposure during pregnancy).

The concentration refers to the lowest concentration that produces tumors in the offspring. The mother is exposed to the test chemical.

(7) For the purposes of this listing, all test chemicals are reported as pure, unless stated to be otherwise by the author. This does not rule out the possibility that unknown impurities may have been present.

(8) A mixture of compounds whose test results satisfy the criteria previously mentioned in (1), (2), or (3) is included if the composition of the mixture can be clearly defined.

(9) For tests involving promoters or initiators, a study is included if the following conditions are satisfied (in addition to the criteria previously mentioned in (1), (2), or (3)):

(a) The test chemical is applied first, followed by an application of a standard promoter. A positive control group in which the test animals are subjected to the same standard promoter under identical conditions is maintained throughout the duration of the experiment. The data are only used if positive and negative control groups are mentioned in the reference.

(b) A known carcinogen is first applied as an initiator, followed by application of the test chemical as a promoter. A positive control group in which the test animals are subjected to the same initiator under identical conditions is maintained throughout the duration of the experiment. The data are used only if positive and negative control groups are mentioned in the reference.

16. *Cited Reference* is the final entry of the irritation, mutation, reproductive, tumorigenic, and toxic dose data lines. This is the source from which the information was extracted. All references cited are publicly available. No governmental classified documents have been used for source information. All references have been given a unique six-letter CODEN character code (derived from the American Society for Testing and Materials *CODEN for Periodical Titles* and the CAS *Source Index*), which identifies periodicals, serial publications, and individual published works. For those references for which no CODEN was found, the corresponding six-letter code includes asterisks (*) in the last one or two positions following the first four or five

letters of an acronym for the publication title. Following the CODEN designation (for most entries) are: the number of the volume, followed by a comma; the page number of the first page of the article, followed by a comma; and a two-digit number, indicating the year of publication in the twentieth century. When the cited reference is a report, the report number is listed. Where contributors have provided information on their unpublished studies, the CODEN consists of the first three letters of the last name, the initials of the first and middle names, and a number sign (#). The date of the letter supplying the information is listed. All CODEN acronyms are listed in alphabetical order and defined in the CODEN Section.

17. *Consensus Reports* lines supply additional information to enable the reader to make knowledgeable evaluations of potential chemical hazards. Two types of reviews are listed: (a) International Agency for Research on Cancer (IARC) monograph reviews, which are published by the United Nations World Health Organization (WHO); and (b) the National Toxicology Program (NTP).

a. Cancer Reviews. In the U.N. International Agency for Research on Cancer (IARC) monographs, information on suspected environmental carcinogens is examined, and summaries of available data with appropriate references are presented. Included in these reviews are synonyms, physical and chemical properties, uses and occurrence, and biological data relevant to the evaluation of carcinogenic risk to humans. The monographs in the series contain an evaluation of approximately 1200 materials. Single copies of the individual monographs (specify volume number) can be ordered from WHO Publications Centre USA, 49 Sheridan Avenue, Albany, NY 12210, telephone (518) 436-9686.

The format of the IARC data line is as follows. The entry "IARC Cancer Review:" indicates that the carcinogenicity data pertaining to a compound have been reviewed by the IARC committee. The committee's conclusions are summarized in three words. The first word indicates whether the data pertain to humans or to animals. The next two words indicate the degree of carcinogenic risk as defined by IARC.

For experimental animals the evidence of carcinogenicity is assessed by IARC and judged to fall into one of four groups defined as follows:

(1) Sufficient Evidence of carcinogenicity is provided when there is an increased incidence of malignant tumors: (a) in multiple species or strains; (b) in multiple experiments (preferably with different routes of administration or using different dose levels); or (c) to an unusual degree with regard to the incidence, site, or type of tumor, or age at onset. Additional evidence may be provided by data on dose-response effects.

(2) Limited Evidence of carcinogenicity is available when the data suggest a carcinogenic effect but are limited because: (a) the studies involve a single species, strain, or experiment; (b) the experiments are restricted by inadequate dosage levels, inadequate duration of exposure to the agent, inadequate period of follow-up, poor survival, the use of too few animals, or inadequate reporting; or (c) the neoplasms produced often occur spontaneously and, in the past, have been difficult to classify as malignant by histological criteria alone (for example, lung adenomas and adenocarcinomas, and liver tumors in certain strains of mice).

(3) Inadequate Evidence is available when, because of major qualitative or quantitative limitations, the studies cannot be interpreted as showing either the presence or absence of a carcinogenic effect.

(4) No Evidence applies when several adequate studies are available that show that within the limitations of the tests used, the chemical is not carcinogenic.

It should be noted that the categories *Sufficient Evidence* and *Limited Evidence* refer only to the strength of the experimental evidence that these chemicals are carcinogenic and not to the extent of their carcinogenic activity nor to the mechanism involved. The classification of any chemical may change as new information becomes available.

The evidence for carcinogenicity from studies in humans is assessed by the IARC committees and judged to fall into one of four groups defined as follows:

(1) Sufficient Evidence of carcinogenicity indicates that there is a causal relationship between the exposure and human cancer.

(2) Limited Evidence of carcinogenicity indicates that a causal relationship is credible, but that alternative explanations, such as chance, bias, or confounding, could not adequately be excluded.

(3) Inadequate Evidence, which applies to both positive and negative evidence, indicates that one of two conditions prevailed: (a) there are few pertinent data; or (b) the available studies, while showing evidence of association, do not exclude chance, bias, or confounding.

(4) No Evidence applies when several adequate studies are available that do not show evidence of carcinogenicity.

This cancer review reflects only the conclusion of the IARC committee based on the data available for the committee's evaluation. Hence, for some substances there may be a disparity between the IARC determination and the information on the tumorigenic data lines (see paragraph 15). Also, some substances previously reviewed by IARC may be reexamined as additional data become available. These substances will contain multiple IARC review lines, each of which is referenced to the applicable IARC monograph volume.

An IARC entry indicates that some carcinogenicity data pertaining to a compound have been reviewed by the IARC committee. It indicates whether the data pertain to humans or to animals and whether the results of the determination are positive, suspected, indefinite, or negative, or whether there are no data.

This cancer review reflects only the conclusion of the IARC committee, based on the data available at the time of the committee's evaluation. Hence, for some materials there may be disagreement between the IARC determination and the tumorigenicity information in the toxicity data lines.

b. NTP Status. The notation "NTP 8th Annual Report on Carcinogens" indicated that the entry is listed on the seventh report made to the U.S. Congress by the National Toxicology Program (NTP) as required by law. This listing implies that the entry is assumed to be a human carcinogen.

Another NTP notation indicates that the material has been tested by the NTP under its Carcinogenesis Testing Program. These entries are also identified as National Cancer Institute (NCI), which reported the studies before the NCI Carcinogenesis Testing Program was absorbed by NTP. To obtain additional information about NTP, the Carcinogenesis Testing Program, or the status of a particular material under test, contact the Toxicology Information and Scientific Evaluation Group, NTP/TRTP/NIEHS, Mail Drop 18-01, P.O. Box 12233, Research Triangle Park, NC 27709.

c. EPA Extremely Hazardous Substances List. This list was developed by the U.S. Environmental Protection Agency (EPA) as required by the Superfund Amendments and Reauthorization Act of 1986 (SARA). Title III, Section 304 requires notification by facilities of a release of certain extremely hazardous substances. These 402 substances were listed by the EPA in the *Federal Register* of November 17, 1986.

d. Community Right-To-Know List. This list was developed by the EPA as required by the Superfund Amendments and Reauthorization Act of 1986 (SARA). Title III, Sections 311–312 require manufacturing facilities to prepare Material Safety Data Sheets and notify local authorities of the presence of listed chemicals. Both specific chemicals and classes of chemicals are covered by these sections.

e. EPA Genetic Toxicology Program (GENE-TOX). This status line indicates that the material has had genetic effects reported in the literature during the period 1969–1979. The test protocol in the literature is evaluated by an EPA expert panel on mutations, and the positive or negative genetic effect of the substance is reported. To obtain additional information about this program, contact GENE-TOX Program, USEPA, 401 M Street, SW, TS796, Washington, DC 20460, telephone (202) 260-1513.

f. EPA TSCA Status Line. This line indicates that the material appears on the chemical inventory prepared by the Environmental Protection Agency in accordance with provisions of the Toxic Substances Control Act (TSCA). Materials reported in the inventory include those that are produced commercially in or are imported into this country. The reader should note, however, that materials already regulated by the EPA under FIFRA and by the Food and Drug Administration under the Food, Drug, and Cosmetic Act, as amended, are not included in the TSCA inventory. Similarly, alcohol, tobacco, and explosive materials are not regulated under TSCA. TSCA regulations should be consulted for an exact definition of reporting requirements. For additional information about TSCA, contact EPA, Office of Toxic Substances, Washington, DC 20402. Specific questions about the inventory can be directed to the EPA Office of Industry Assistance, telephone (800) 424-9065.

18. *Standards and Recommendations* section contains regulations by agencies of the U.S. government or recommendations by expert groups. "OSHA" refers to standards promulgated under Section 6 of the Occupational Safety and Health Act of 1970. "DOT" refers to materials regulated for shipment by the Department of Transportation. Because of frequent changes to and litigation of federal regulations, it is recommended that the reader contact the applicable agency for information about the current standards for a particular material. Omission of a material or regulatory notation from this edition does not imply any relief from regulatory responsibility.

a. OSHA Air Contaminant Standards. The values given are for the revised standards that were published in January 13, 1989 and were scheduled to take effect from September 1, 1989 through December 31, 1992. These are noted with the entry "OSHA PEL:" followed by "TWA" or "CL," meaning either time-weighted average or ceiling value, respectively, to which workers can be exposed for a normal 8-hour day, 40-hour work week without ill effects. For some materials, TWA, CL, and Pk (peak) values are given in the standard. In those cases, all three are listed. Finally, some entries may be followed by the designation "(skin)." This designation indicates that the compound may be absorbed by the skin and that, even though the air concentration may be below the standard, significant additional exposure through the skin may be possible.

b. ACGIH Threshold Limit Values. The American Conference of Governmental Industrial Hygienists

(ACGIH) Threshold Limit Values are noted with the entry "ACGIH TLV:" followed by "TWA" or "CL," meaning either time-weighted average or ceiling value, respectively, to which workers can be exposed for a normal 8-hour day, 40-hour work week without ill effects. The notation "CL" indicates a ceiling limit that must not be exceeded. The notation "skin" indicates that the material penetrates intact skin, and skin contact should be avoided even though the TLV concentration is not exceeded. STEL indicates a short-term exposure limit, usually a 15-minute time-weighted average, which should not be exceeded. Biological Exposure Indices (*BEI:*) are, according to the ACGIH, set to provide a warning level ". . .of biological response to the chemical, or warning levels of that chemical or its metabolic product(s) in tissues, fluids, or exhaled air of exposed workers. . . ."

The latest annual TLV list is contained in the publication *Threshold Limit Values and Biological Exposure Indices*. This publication should be consulted for future trends in recommendations. The ACGIH TLVs are adopted in whole or in part by many countries and local administrative agencies throughout the world. As a result, these recommendations have a major effect on the control of workplace contaminant concentrations. The ACGIH may be contacted for additional information at Kemper Woods Center, 1330 Kemper Meadow Drive, Cincinnati, OH 45240.

c. DFG MAK. These lines contain the German Research Society's Maximum Allowable Concentration values. Those materials that are classified as to workplace hazard potential by the German Research Society are noted on this line. The MAK values are also revised annually and discussions of materials under consideration for MAK assignment are included in the annual publication together with the current values. *BAT:* indicates Biological Tolerance Value for a Working Material which is defined as, ". . .the maximum permissible quantity of a chemical compound, its metabolites, or any deviation from the norm of biological parameters induced by these substances in exposed humans." *TRK:* values are Technical Guiding Concentrations for workplace control of carcinogens. For additional information, write to Deutsche Forschungsgemeinschaft (German Research Society), Kennedyallee 40, D-5300 Bonn 2, Federal Republic of Germany. The publication *Maximum Concentrations at the Workplace and Biological Tolerance Values for Working Materials Report No. 34* can be obtained from VCH Publishers, Inc., 303 N.W. 12th Ave, Deerfield Beach, FL 33442-1788 or Verlag Chemie GmbH, Buchauslieferung, P.O. Box 1260/1280, D-6940 Weinheim, Federal Republic of Germany.

d. NIOSH REL. This line indicates that a NIOSH criteria document recommending a certain occupational exposure has been published for this compound or for a class of compounds to which this material belongs. These documents contain extensive data, analysis, and references. The more recent publications can be obtained from the National Institute for Occupational Safety and Health, U.S. Department of Health and Human Services, 4676 Columbia Pkwy., Cincinnati, OH 45226.

e. DOT Classification. This is the hazard classification according to the U.S. Department of Transportation (DOT) or the International Maritime Organization (IMO). This classification gives an indication of the hazards expected in transportation, and serves as a guide to the development of proper labels, placards, and shipping instructions. The basic hazard classes include compressed gases, flammables, oxidizers, corrosives, explosives, radioactive materials, and poisons. Although a material may be designated by only one hazard class, additional hazards may be indicated by adding labels or by using other means as directed by DOT. Many materials are regulated under general headings such as "pesticides" or "combustible liquids" as defined in the regulations. These are not noted here, as their specific concentration or properties must be known for proper classification. Special regulations may govern shipment by air. This information should serve *only as a guide*, because the regulation of transported materials is carefully controlled in most countries by federal and local agencies. Because there are frequent changes to regulations, it is recommended that the reader contact the applicable agency for information about the current standards for a particular material. United States transportation regulations are found in 40 CFR, Parts 100 to 189. Contact the U.S. Department of Transportation, Materials Transportation Bureau, Washington, DC 20590.

19. *Safety Profiles* are text summaries of the reported hazards of the entry. The word "experimental" indicates that the reported effects resulted from a controlled exposure of laboratory animals to the substance. Toxic effects reported include carcinogenic, reproductive, acute lethal, and human nonlethal effects, skin and eye irritation, and positive mutation study results.

Human effects are identified either by *human* or more specifically by *man, woman, child*, or *infant*. Specific symptoms or organ systems effects are reported when available.

Carcinogenicity potential is denoted by the words "confirmed," "suspected," or "questionable." The substance entries are grouped into three classes based on experimental evidence and the opinion of expert review groups. The OSHA, IARC, ACGIH, and DFG

MAK decision schedules are not related or synchronized. Thus, an entry may have had a recent review by only one group. The most stringent classification of any regulation or expert group is taken as governing.

Class I--Confirmed Carcinogens

These substances are capable of causing cancer in exposed humans. An entry was assigned to this class if it had one or more of the following data items present:

a. an OSHA regulated carcinogen

b. an ACGIH assignment as a human or animal carcinogen

c. a DFG MAK assignment as a confirmed human or animal carcinogen

d. an IARC assignment of human or animal sufficient evidence of carcinogenicity, or higher

e. NTP 8th Annual Report on Carcinogens

Class II--Suspected Carcinogens

These substances may be capable of causing cancer in exposed humans. The evidence is suggestive, but not sufficient to convince expert review committees. Some entries have not yet had expert review, but contain experimental reports of carcinogenic activity. In particular, an entry is included if it has positive reports of carcinogenic endpoint in two species. As more studies are published, many Class II carcinogens will have their carcinogenicity confirmed. On the other hand, some will be judged noncarcinogenic in the future. An entry was assigned to this class if it had one or more of the following data items present:

a. an ACGIH assignment of suspected carcinogen

b. a DFG MAK assignment of suspected carcinogen

c. an IARC assignment of human or animal limited evidence

d. two animal studies reporting positive carcinogenic endpoint in different species

Class III--Questionable Carcinogens

For these entries there is minimal published evidence of possible carcinogenic activity. The reported endpoint is often neoplastic growth with no spread or invasion characteristic of carcinogenic pathology. An even weaker endpoint is that of equivocal tumorigenic agent (ETA). Reports are assigned this designation when the study was defective. The study may have lacked control animals, may have used a very small sample size, often may lack complete pathology reporting, or may suffer many other study design defects. Many of these studies were designed for other than carcinogenic evaluation, and the reported carcinogenic effect is a by-product of the study, not the goal. The data are presented because some of the substances studied may be carcinogens. There are insufficient data to affirm or deny the possibility. An entry was assigned to this class if it had one or more of the following data items present:

a. an IARC assignment of inadequate or no evidence

b. a single human report of carcinogenicity

c. a single experimental carcinogenic report, or duplicate reports in the same species

d. one or more experimental neoplastic or equivocal tumorigenic agent reports

Fire and explosion hazards are briefly summarized in terms of conditions of flammable or reactive hazard. Materials that are incompatible with the entry are listed here. Fire and explosion hazards are briefly summarized in terms of conditions of flammable or reactive hazard. Fire-fighting materials and methods are discussed where feasible. A material with a flash point of 100°F or less is considered dangerous; if the flash point is from 100 to 200°F, the flammability is considered moderate; if it is above 200°F, the flammability is considered low (the material is considered combustible).

Also included in the safety profile are disaster hazards comments, which serve to alert users of materials, safety professionals, researchers, supervisors, and firefighters to the dangers that may be encountered on entering storage premises during a fire or other emergency. Although the presence of water, steam, acid fumes, or powerful vibrations can cause many materials to decompose into dangerous compounds, we are particularly concerned with high temperatures (such as those resulting from a fire) because these can cause many otherwise inert chemicals to emit highly toxic gases or vapors such as NO_x, SO_x, acids, and so forth, or evolve vapors of antimony, arsenic, mercury, and the like.

The Safety Profile concludes with the OSHA and NIOSH occupational analytical methods, referenced by method name or number. The OSHA Manual of Analytical Methods can be ordered from the ACGIH, Kemper Woods Center, 1330 Kemper Meadow Drive, Cincinnati, OH 45240. The NIOSH Manual of Analytical Methods is available from NIOSH Publications Office, 4676 Columbia Parkway, Cincinnati, OH 45226.

Key to Abbreviations

abs − absolute
ACGIH − American Conference of Governmental Industrial Hygienists
af − atomic formula
alc − alcohol
alk − alkaline
amorph − amorphous
anhyd − anhydrous
approx − approximately
aq − aqueous
atm − atmosphere
autoign − autoignition
aw − atomic weight
BEI − ACGIH Biological Exposure Indexes
bp − boiling point
b range − boiling range
CAS − Chemical Abstracts Service
cc − cubic centimeter
CC − closed cup
CL − ceiling concentration
COC − Cleveland open cup
compd(s) − compound(s)
conc − concentration, concentrated
contg − containing
cryst − crystal(s), crystalline
d − density
D − day(s)
decomp − decomposition
deliq − deliquescent
dil − dilute
DOT − U.S. Department of Transportation
EPA − U.S. Environmental Protection Agency
eth − ether
(F) − Fahrenheit
FCC − Food Chemical Codex
FDA − U.S. Food and Drug Administration
flam − flammable
flash p − flash point
fp − freezing point
g − gram
glac − glacial
gran − granular, granules
H − hour(s)
HR: − hazard rating
htd − heated
htg − heating
hygr − hygroscopic
IARC − International Agency for Research on Cancer
immisc − immiscible
incomp − incompatible
insol − insoluble

IU − International Unit
kg − kilogram (one thousand grams)
L − liter
lel − lower explosive limit
liq − liquid
M − minute(s)
m^3 − cubic meter
mf − molecular formula
mg − milligram
misc − miscible
mL − milliliter
mm − millimeter
mod − moderately
mp − melting point
mppcf − million particles per cubic foot
mw − molecular weight
μ, u − micro
μg − microgram
n − refractive index
ng − nanogram
NIOSH − National Institute for Occupational Safety and Health
nonflam − nonflammable
NTP − National Toxicology Program
OBS − obsolete
OC − open cup
org − organic
ORM − other regulated material (DOT)
OSHA − Occupational Safety and Health Administration
Pa − Pascals
PEL − permissible exposure level
pet − petroleum
pg − picogram (one trillionth of a gram)
Pk − peak concentration
pmole − picomole
powd − powder
ppb − parts per billion (v/v)
pph − parts per hundred (v/v)(percent)
ppm − parts per million (v/v)
ppt − parts per trillion (v/v)
prac − practically
prep − preparation
PROP − properties
refr − refractive
rhomb − rhombic
S, sec − second(s)
sl, slt − slight
sltly − slightly
sol − soluble
soln − solution
solv(s) − solvent(s)
spar − sparingly

spont − spontaneous(ly)
STEL − short-term exposure limit
subl − sublimes
TCC − Tag closed cup
tech − technical
temp − temperature
TLV − Threshold Limit Value
TOC − Tag open cup
TWA − time weighted average
uel − upper explosive limit
unk − unknown, unreported
ULC, ulc − Underwriters Laboratory Classification
USDA − U.S. Department of Agriculture
vac − vacuum
vap − vapor

vap d − vapor density
vap press − vapor pressure
visc − viscosity
vol − volume
W − week(s)
Y − year(s)
% − percent(age)
> − greater than
< − less than
<= − less than or equal to
=> − greater than or equal to

° − degrees of temperature in Celsius (centigrade)
°F − temperature in Fahrenheit

Sax's Dangerous Properties of Industrial Materials

Eleventh Edition

Section 1
DOT Guide Number
Cross-Index

NA 0349 see GJU600
NA 0473 see BAO900
NA 0473 see LDP000
NA 0473 see MMP000
NA 1051 see HHS000
NA 1247 see MLH750
NA 1361 see CDI000
NA 1361 see CDI250
NA 1361 see CDJ000
NA 1463 see CMK000
NA 1499 see SIT750
NA 1549 see AQE000
NA 1549 see AQK000
NA 1556 see DFP200
NA 1557 see ARI000
NA 1557 see ARJ100
NA 1574 see CAM300
NA 1574 see CAM500
NA 1649 see TCF000
NA 1665 see NMS000
NA 1707 see TEL750
NA 1760 see TGH250
NA 1760 see TGH252
NA 1778 see SCO500
NA 1807 see PHS250
NA 1811 see PKU250
NA 1829 see SOR500
NA 1831 see SOI520
NA 1911 see DDI450
NA 1956 see HDF050
NA 1967 see PAK230
NA 1986 see AFJ000
NA 1987 see AFJ000
NA 1993 see DHE800
NA 1993 see DHE900
NA 1993 see FOP000
NA 1999 see ARO500
NA 2212 see ARM250
NA 2212 see ARM260
NA 2212 see ARM268
NA 2212 see ARM275
NA 2212 see ARM280
NA 2215 see MAK900
NA 2672 see ANK250
NA 2762 see AFK250
NA 2783 see DXH325
NA 2783 see MNH000
NA 2783 see PAK000
NA 2783 see TCF250
NA 2783 see TIQ250
NA 2809 see MCW250
NA 2811 see SBT200
NA 2821 see PDN750
NA 2845 see MOC000

NA 2920 see DEV200
NA 2922 see SHR000
NA 2927 see EOP600
NA 2927 see EOR000
NA 2949 see SHR000
NA 3018 see MNH000
NA 3018 see TCF250
NA 9163 see ZTJ000
NA 9202 see CBW750
NA 9206 see MOB399
NA 9260 see AGX000
NA 9269 see TLB750
UN 0004 see ANS500
UN 0027 see ERF500
UN 0027 see PLL750
UN 0028 see ERF500
UN 0028 see PLL750
UN 0072 see CPR800
UN 0074 see DUR800
UN 0075 see DJE400
UN 0076 see DUY600
UN 0079 see HET500
UN 0081 see DYG000
UN 0110 see GJU600
UN 0114 see TEF500
UN 0118 see CPR800
UN 0129 see LCM000
UN 0130 see LEE000
UN 0133 see MAW250
UN 0135 see MDC000
UN 0143 see NGY000
UN 0144 see NGY000
UN 0146 see NMB000
UN 0147 see NMQ500
UN 0151 see PBT050
UN 0153 see PIC800
UN 0154 see PID000
UN 0155 see TML325
UN 0208 see TEG250
UN 0209 see TMN490
UN 0214 see TMK500
UN 0215 see TML000
UN 0219 see SMP500
UN 0220 see UTJ000
UN 0222 see ANN000
UN 0224 see BAI000
UN 0226 see CQH250
UN 0234 see SGP550
UN 0235 see PIC500
UN 0236 see PIC750
UN 0282 see NHA500
UN 0284 see GJU600
UN 0285 see GJU600
UN 0292 see GJU600

UN 0293 see GJU600
UN 0318 see GJU600
UN 0340 see CCU250
UN 0341 see CCU250
UN 0342 see CCU250
UN 0343 see CCU250
UN 0372 see GJU600
UN 0385 see NFJ000
UN 0391 see CPR800
UN 0394 see SMP500
UN 0402 see PCD500
UN 0411 see PBC250
UN 0452 see GJU600
UN 0483 see CPR800
UN 0484 see CQH250
UN 0490 see NMP620
UN 1001 see ACI750
UN 1002 see AFG250
UN 1003 see AFG250
UN 1005 see AMY500
UN 1006 see AQW250
UN 1008 see BMG700
UN 1009 see TJY100
UN 1010 see BOP100
UN 1011 see BOR500
UN 1013 see CBU250
UN 1014 see CBV250
UN 1015 see CBV000
UN 1016 see CBW750
UN 1017 see CDV750
UN 1018 see CFX500
UN 1020 see CJI500
UN 1022 see CLR250
UN 1023 see HHJ500
UN 1026 see COO000
UN 1027 see CQD750
UN 1028 see DFA600
UN 1029 see DFL000
UN 1032 see DOQ800
UN 1033 see MJW500
UN 1035 see EDZ000
UN 1036 see EFU400
UN 1037 see EHH000
UN 1038 see EIO000
UN 1039 see EMT000
UN 1040 see EJN500
UN 1041 see EJO000
UN 1045 see FEZ000
UN 1046 see HAM500
UN 1048 see HHJ000
UN 1049 see HHW500
UN 1050 see HHL000
UN 1052 see HHU500
UN 1053 see HIC500

UN 1055 see IIC000
UN 1060 see MFX600
UN 1061 see MGC250
UN 1063 see MIF765
UN 1064 see MLE650
UN 1065 see NCG500
UN 1066 see NGP500
UN 1067 see NGU500
UN 1069 see NMH000
UN 1070 see NGU000
UN 1072 see OQW000
UN 1073 see OQW000
UN 1075 see LGM000
UN 1076 see PGX000
UN 1077 see PMO500
UN 1079 see SOH500
UN 1080 see SOI000
UN 1081 see TCH500
UN 1082 see CLQ750
UN 1083 see TLD500
UN 1085 see VMP000
UN 1086 see VNP000
UN 1087 see MQL750
UN 1088 see AAG000
UN 1089 see AAG250
UN 1090 see ABC750
UN 1091 see ABC750
UN 1092 see ADR000
UN 1093 see ADX500
UN 1098 see AFV500
UN 1099 see AFY000
UN 1100 see AGB250
UN 1104 see AOD725
UN 1104 see AOD735
UN 1106 see AOJ000
UN 1106 see PBV500
UN 1108 see AOI800
UN 1110 see MGN500
UN 1112 see AOL250
UN 1114 see BBL250
UN 1123 see BPU750
UN 1123 see BPV000
UN 1123 see BPV100
UN 1125 see BPX750
UN 1126 see BMX500
UN 1128 see BRK000
UN 1129 see BSU250
UN 1130 see CBB500
UN 1131 see CBV500
UN 1134 see CEJ125
UN 1135 see EIU800
UN 1136 see CMY900
UN 1143 see COB250
UN 1144 see COC500

UN 1586 see CNN500	UN 1671 see PDN750	UN 1755 see CMK000	UN 1838 see TGH350
UN 1587 see CNL000	UN 1672 see PFJ400	UN 1756 see CMJ560	UN 1839 see TII250
UN 1589 see COO750	UN 1673 see PEY000	UN 1757 see CMJ560	UN 1840 see ZFA000
UN 1591 scc DEP600	UN 1673 scc PEY250	UN 1758 see CML125	UN 1841 see AAG500
UN 1592 see DEP800	UN 1673 see PEY500	UN 1759 see FBI000	UN 1845 see CBU250
UN 1593 see MJP450	UN 1674 see ABU500	UN 1760 see FBI000	UN 1846 see CBY000
UN 1594 see DKB110	UN 1677 see ARD250	UN 1761 see DBU800	UN 1847 see PLT250
UN 1595 see DUD100	UN 1678 see PKV500	UN 1762 see CPE500	UN 1848 see PMU750
UN 1597 see DUQ180	UN 1679 see PLC175	UN 1763 see CPR250	UN 1858 see HDF000
UN 1597 see DUQ200	UN 1680 see PLC500	UN 1764 see DEL000	UN 1859 see SDF650
UN 1597 see DUQ400	UN 1683 see SDM100	UN 1765 see DEN400	UN 1860 see VPA000
UN 1597 see DUQ600	UN 1684 see SDP000	UN 1766 see DGF200	UN 1862 see COB750
UN 1599 see DUY600	UN 1685 see ARD750	UN 1767 see DEY800	UN 1862 see EHO200
UN 1600 see PGN100	UN 1685 see SEY100	UN 1768 see PHF250	UN 1865 see PNQ500
UN 1603 see EGV000	UN 1686 see SEY200	UN 1769 see DFF000	UN 1866 see RDP000
UN 1604 see EEA500	UN 1686 see SEY500	UN 1770 see BNG750	UN 1868 see DAE400
UN 1605 see EIY500	UN 1687 see SFA000	UN 1770 see BNH000	UN 1869 see MAC750
UN 1606 see IGN000	UN 1688 see HKC500	UN 1771 see DYA800	UN 1870 see PKY250
UN 1607 see IGO000	UN 1689 see SGA500	UN 1773 see FAU000	UN 1872 see LCX000
UN 1608 see IGM000	UN 1690 see SHF500	UN 1775 see FDD125	UN 1873 see PCD250
UN 1611 see HCY000	UN 1691 see SME500	UN 1775 see HHS600	UN 1884 see BAO000
UN 1613 see HHS000	UN 1692 see SMN500	UN 1776 see PHJ250	UN 1885 see BBX000
UN 1614 see HHS000	UN 1692 see SMO500	UN 1777 see FLZ000	UN 1886 see BAY300
UN 1616 see LCV000	UN 1695 see CDN200	UN 1779 see FNA000	UN 1887 see CES650
UN 1617 see ARC750	UN 1697 see CEA750	UN 1780 see FOY000	UN 1888 see CHJ500
UN 1617 see LCK000	UN 1698 see PDB000	UN 1781 see HCQ000	UN 1889 see COO500
UN 1617 see LCK100	UN 1699 see CGN000	UN 1782 see HDE000	UN 1891 see EGV400
UN 1618 see LCL000	UN 1701 see XRS000	UN 1783 see HEO000	UN 1892 see DFH200
UN 1620 see LCU000	UN 1702 see TBP750	UN 1784 see HFX500	UN 1894 see PFN100
UN 1621 see LIC000	UN 1704 see SOD100	UN 1786 see HHV000	UN 1895 see MCU750
UN 1622 see ARD000	UN 1705 see TCF260	UN 1787 see HHI500	UN 1897 see PCF275
UN 1623 see MDF350	UN 1708 see TGQ500	UN 1788 see HHJ000	UN 1898 see ACO500
UN 1624 see MCY475	UN 1708 see TGQ750	UN 1789 see HHL000	UN 1902 see DNK800
UN 1625 see MDF000	UN 1708 see TGR000	UN 1790 see HHU500	UN 1905 see SBN500
UN 1626 see PLU500	UN 1709 see TGL750	UN 1791 see SHU500	UN 1907 see SEE000
UN 1627 see MDE750	UN 1710 see TIO750	UN 1792 see IDS000	UN 1908 see SFT500
UN 1629 see MCS750	UN 1712 see ZDS000	UN 1793 see PHE500	UN 1910 see CAU500
UN 1629 see MDE250	UN 1713 see ZGA000	UN 1794 see LDY000	UN 1911 see DDI450
UN 1630 see MCW500	UN 1714 see ZLS000	UN 1798 see HHM000	UN 1913 see NCG500
UN 1631 see MCX500	UN 1715 see AAX500	UN 1799 see NNE000	UN 1914 see BSJ500
UN 1636 see MDA250	UN 1716 see ACD750	UN 1800 see OBI000	UN 1915 see CPC000
UN 1637 see MDC500	UN 1717 see ACF750	UN 1801 see OGE000	UN 1916 see DFJ050
UN 1638 see MDC750	UN 1718 see ADF250	UN 1802 see PCD250	UN 1917 see EFT000
UN 1639 see MCV250	UN 1722 see AGB500	UN 1803 see HJH500	UN 1918 see COE750
UN 1640 see MDF250	UN 1723 see AGI250	UN 1804 see TJA750	UN 1919 see MGA500
UN 1641 see MCT500	UN 1724 see AGU250	UN 1805 see PHB250	UN 1921 see PNL400
UN 1642 see MDA500	UN 1725 see AGX750	UN 1806 see PHR500	UN 1922 see PPS500
UN 1643 see NCP500	UN 1726 see AGY750	UN 1808 see PHT250	UN 1923 see CAN000
UN 1644 see MCU000	UN 1727 see ANJ000	UN 1809 see PHT275	UN 1928 see MLE000
UN 1646 see MCU250	UN 1728 see PBY750	UN 1810 see PHQ800	UN 1931 see ZGJ100
UN 1647 see BNM750	UN 1729 see AOY250	UN 1812 see PLF500	UN 1931 see ZIJ100
UN 1648 see ABE500	UN 1730 see AQD000	UN 1813 see PLJ500	UN 1932 see ZOA000
UN 1650 see NBE500	UN 1731 see AQD000	UN 1814 see PLJ500	UN 1935 see COI500
UN 1651 see AQN635	UN 1732 see AQF250	UN 1816 see PNX250	UN 1938 see BMR750
UN 1653 see NDB500	UN 1733 see AQC500	UN 1817 see PPR500	UN 1939 see PHU000
UN 1654 see NDN000	UN 1736 see BDM500	UN 1818 see SCQ500	UN 1940 see TFJ100
UN 1656 see NDP400	UN 1737 see BEC000	UN 1819 see AHG000	UN 1941 see DKG850
UN 1657 see NDR000	UN 1738 see BEE375	UN 1821 see SEG800	UN 1942 see ANN000
UN 1658 see NDR500	UN 1739 see BEF500	UN 1823 see SHS000	UN 1951 see AQW250
UN 1659 see NDS500	UN 1741 see BMG500	UN 1823 see SHS500	UN 1952 see EJO000
UN 1660 see NEG100	UN 1742 see BMG750	UN 1824 see SHS000	UN 1954 see DFB400
UN 1661 see NEN500	UN 1744 see BMP000	UN 1824 see SHS500	UN 1958 see DGL600
UN 1661 see NEO000	UN 1745 see BMQ000	UN 1825 see SIN500	UN 1959 see VPP000
UN 1661 see NEO500	UN 1746 see BMQ325	UN 1827 see TGC250	UN 1961 see EDZ000
UN 1662 see NEX000	UN 1747 see BSR000	UN 1829 see SOR500	UN 1962 see EIO000
UN 1663 see NIF000	UN 1748 see HOV500	UN 1830 see SOI500	UN 1963 see HAM500
UN 1663 see NIF010	UN 1749 see CDX750	UN 1832 see SOI500	UN 1964 see HHJ500
UN 1664 see NMO500	UN 1750 see CEA000	UN 1833 see SOO500	UN 1965 see HHJ500
UN 1664 see NMO525	UN 1751 see CEA000	UN 1834 see SOT000	UN 1966 see BFO750
UN 1664 see NMO550	UN 1752 see CEC250	UN 1835 see TDK500	UN 1966 see HHW500
UN 1669 see PAW500	UN 1753 see CKM250	UN 1836 see TFL000	UN 1969 see MOR750
UN 1670 see PCF300	UN 1754 see CLG500	UN 1837 see TFO000	UN 1971 see MDQ750

UN 2478 see IKG800
UN 2478 see IKH000
UN 2478 see IKH099
UN 2478 see OBG000
UN 2478 see TGM750
UN 2478 see THY750
UN 2478 see TKJ250
UN 2478 see XSS260
UN 2480 see MKX250
UN 2481 see ELS500
UN 2482 see PNP000
UN 2485 see BRQ500
UN 2487 see PFK250
UN 2488 see CPN500
UN 2489 see MJP400
UN 2490 see BII250
UN 2491 see EEC600
UN 2493 see HDG000
UN 2495 see IDT000
UN 2496 see PMV500
UN 2497 see SJF000
UN 2498 see FNK025
UN 2501 see TND250
UN 2502 see VBA000
UN 2503 see ZPA000
UN 2505 see ANH250
UN 2506 see ANJ500
UN 2507 see CKO750
UN 2508 see MRD500
UN 2509 see PKX750
UN 2511 see CKS750
UN 2512 see ALT000
UN 2512 see ALT250
UN 2514 see PEO500
UN 2515 see BNL000
UN 2516 see CBX750
UN 2517 see CFX250
UN 2521 see KFA000
UN 2522 see DPG600
UN 2524 see ENY500
UN 2525 see DJT200
UN 2526 see FPW000
UN 2527 see IIK000
UN 2528 see IIW000
UN 2529 see IJU000
UN 2530 see IJW000
UN 2531 see MDN250
UN 2535 see MMA250
UN 2538 see NHP990
UN 2541 see TBE000
UN 2542 see THX250
UN 2545 see HAC000
UN 2547 see SJZ100
UN 2548 see CDX250
UN 2552 see HDA000
UN 2553 see NAH600
UN 2554 see CIU750
UN 2555 see CCU250
UN 2556 see CCU250
UN 2557 see CCU250
UN 2558 see BNI000
UN 2564 see TII250
UN 2565 see DGT600
UN 2567 see SJA000
UN 2572 see PFI000
UN 2573 see TEJ100
UN 2574 see TNP500
UN 2576 see PHU000
UN 2579 see PIJ000
UN 2580 see AGX750
UN 2581 see AGY750
UN 2582 see FAU000

UN 2587 see QQS200
UN 2591 see XDS000
UN 2599 see FOO562
UN 2603 see COY000
UN 2606 see MPI750
UN 2607 see ADR500
UN 2608 see NIY000
UN 2610 see THN000
UN 2611 see CKR500
UN 2612 see MOU830
UN 2614 see IMW000
UN 2615 see EPC125
UN 2616 see IOI000
UN 2617 see MIQ745
UN 2618 see VQK650
UN 2619 see DQP800
UN 2621 see ABB500
UN 2622 see GGW000
UN 2626 see CDU000
UN 2628 see PLG000
UN 2629 see SHG500
UN 2630 see SJT500
UN 2642 see FIC000
UN 2643 see MHR250
UN 2644 see MKW200
UN 2646 see HCE500
UN 2647 see MAO250
UN 2648 see NIE600
UN 2649 see BIK250
UN 2650 see DFU000
UN 2651 see MJQ000
UN 2655 see PLH750
UN 2656 see QMJ000
UN 2657 see SBR000
UN 2658 see SBO500
UN 2659 see SFU500
UN 2661 see HCL500
UN 2662 see HIH000
UN 2664 see DDP800
UN 2666 see EHP500
UN 2668 see CDN500
UN 2670 see TJD750
UN 2671 see AMI000
UN 2671 see AMI250
UN 2671 see AMI500
UN 2673 see CEH250
UN 2674 see DXE000
UN 2676 see SLQ000
UN 2677 see RPZ000
UN 2678 see RPZ000
UN 2679 see LHI100
UN 2680 see LHI100
UN 2681 see CDD750
UN 2682 see CDD750
UN 2684 see DIY800
UN 2685 see DJI400
UN 2686 see DHO500
UN 2687 see DGU200
UN 2688 see BNA825
UN 2689 see CDT750
UN 2692 see BMG400
UN 2698 see TDB000
UN 2699 see TKA250
UN 2708 see MHV750
UN 2710 see DWT600
UN 2711 see DDJ900
UN 2713 see ADJ500
UN 2716 see BST500
UN 2717 see CBA750
UN 2719 see BAI750
UN 2720 see CMJ600
UN 2721 see CNJ900

UN 2723 see MAE000
UN 2724 see MAS900
UN 2725 see NDG000
UN 2726 see NDG550
UN 2727 see TEK750
UN 2728 see ZSA000
UN 2729 see HCC500
UN 2733 see AOE200
UN 2733 see BPY000
UN 2733 see BPY250
UN 2733 see DAG600
UN 2733 see HBL600
UN 2733 see HFK000
UN 2733 see OEK010
UN 2733 see PBV505
UN 2734 see AOE200
UN 2734 see BPY000
UN 2734 see BPY250
UN 2734 see DAG600
UN 2734 see HBL600
UN 2734 see HFK000
UN 2734 see PBV505
UN 2738 see BQH850
UN 2739 see BSW550
UN 2740 see PNH000
UN 2746 see CBX109
UN 2750 see DGG400
UN 2751 see DJW600
UN 2752 see EBQ700
UN 2752 see EKM200
UN 2761 see AFK250
UN 2761 see DHB400
UN 2785 see MPV400
UN 2785 see TET900
UN 2789 see AAT250
UN 2790 see AAT250
UN 2798 see DGE400
UN 2799 see PFW200
UN 2799 see PFW210
UN 2802 see CNJ950
UN 2802 see CNK500
UN 2803 see GBG000
UN 2805 see LHH000
UN 2806 see LHM000
UN 2812 see AHG000
UN 2815 see AKB000
UN 2817 see ANJ000
UN 2818 see ANT000
UN 2819 see PBW750
UN 2820 see BSW000
UN 2822 see CKW000
UN 2823 see COB500
UN 2826 see CLJ750
UN 2829 see HEU000
UN 2830 see LHK000
UN 2831 see MIH275
UN 2835 see SEM500
UN 2837 see SEG800
UN 2838 see VNF000
UN 2839 see AAH750
UN 2840 see BSU500
UN 2841 see DCH200
UN 2842 see NFY500
UN 2845 see EOQ000
UN 2849 see CKP600
UN 2851 see BMG800
UN 2853 see MAF600
UN 2854 see COE000
UN 2855 see ZIA000
UN 2858 see ZOA000
UN 2859 see ANY250
UN 2862 see VDU000

UN 2864 see PLK810
UN 2865 see OLS000
UN 2869 see TGG250
UN 2870 see AHG875
UN 2871 see AQB750
UN 2872 see DDL800
UN 2873 see DDU600
UN 2874 see FPU000
UN 2875 see HCL000
UN 2876 see REA000
UN 2877 see ISR000
UN 2879 see SBT500
UN 2906 see IMG000
UN 2907 see CCK125
UN 2931 see VEZ000
UN 2933 see CKT000
UN 2936 see TFK250
UN 2937 see PDE000
UN 2938 see MHA750
UN 2943 see TCS500
UN 2945 see MHV000
UN 2946 see ALS990
UN 2948 see AID500
UN 2950 see MAC750
UN 2951 see OPE000
UN 2952 see ASL750
UN 2956 see TML750
UN 2965 see BMH000
UN 2966 see MCN250
UN 2967 see SNK500
UN 2968 see MAS500
UN 2969 see CCP000
UN 2970 see BBS300
UN 2972 see DVF400
UN 2973 see DRO400
UN 2975 see TFS750
UN 2976 see TFT500
UN 2977 see UOJ000
UN 2978 see UOJ000
UN 2979 see UNS000
UN 2980 see URS000
UN 2981 see URA200
UN 2984 see HIB005
UN 2984 see HIB010
UN 2989 see LCV100
UN 3022 see BOX750
UN 3023 see MKJ250
UN 3023 see OFE030
UN 3054 see CPB625
UN 3055 see AJU250
UN 3056 see HBB500
UN 3057 see TJX500
UN 3064 see NGY000
UN 3071 see CPW300
UN 3071 see FPM000
UN 3071 see HBD500
UN 3071 see HES000
UN 3071 see IMU000
UN 3071 see LBX000
UN 3071 see PBM000
UN 3071 see PML500
UN 3071 see TGO700
UN 3071 see TGP000
UN 3071 see TGP250
UN 3080 see AGJ000
UN 3080 see BMT150
UN 3080 see CHL250
UN 3080 see CKA750
UN 3080 see CKB000
UN 3080 see COI250
UN 3080 see FHC200
UN 3080 see FLR100

UN 3080 see IKG800	UN 3080 see OBG000	UN 3080 see TKJ250	UN 3083 see PCF750
UN 3080 see IKH000	UN 3080 see TGM750	UN 3080 see XIJ000	UN 3136 see CBY750
UN 3080 see IKH099	UN 3080 see THY750	UN 3080 see XSS260	UN 3149 see PCL500

CAS Number
Cross-Index

50-00-0 see FMV000	50-99-7 see GFG000	52-43-7 see AFS500	54-47-7 see PII100
50-01-1 see GKY000	51-02-5 see INT000	52-46-0 see AQO000	54-49-9 see HNB875
50-02-2 see SOW000	51-03-6 see PIX250	52-49-3 see BBV000	54-62-6 see AMG750
50-03-3 see HHQ800	51-05-8 see AIT250	52-51-7 see BNT250	54-64-8 see MDI000
50-04-4 see CNS825	51-06-9 see AJN500	52-52-8 see AJK250	54-71-7 see PIF250
50-06-6 see EOK000	51-12-7 see BET000	52-53-9 see IRV000	54-77-3 see DTO000
50-07-7 see AHK500	51-14-9 see MJS550	52-60-8 see DJR800	54-80-8 see INS000
50-09-9 see ERE000	51-15-0 see FNZ000	52-62-0 see PBT000	54-84-2 see CMP900
50-10-2 see ORQ000	51-17-2 see BCB750	52-66-4 see PAP500	54-85-3 see ILD000
50-11-3 see DJO800	51-18-3 see TND500	52-67-5 see MCR750	54-86-4 see NDW500
50-12-4 see MKB250	51-20-7 see BOL000	52-68-6 see TIQ250	54-87-5 see SEE250
50-13-5 see DAM700	51-21-8 see FMM000	52-76-6 see NNV000	54-88-6 see TLE750
50-14-6 see VSZ100	51-28-5 see DUZ000	52-78-8 see ENX600	54-91-1 see BHJ250
50-18-0 see CQC650	51-30-9 see IMR000	52-85-7 see FAB600	54-92-2 see ILE000
50-19-1 see HNJ000	51-34-3 see SBG000	52-86-8 see CLY500	54-95-5 see PBI500
50-21-5 see LAG000	51-35-4 see HNT525	52-88-0 see MGR500	54-96-6 see DCD000
50-22-6 see CNS625	51-40-1 see NNO699	52-89-1 see CQK250	55-03-8 see LFG050
50-23-7 see CNS750	51-41-2 see NNO500	52-90-4 see CQK000	55-18-5 see NJW500
50-24-8 see PMA000	51-42-3 see AES000	53-02-1 see UVJ460	55-21-0 see BBB000
50-27-1 see EDU500	51-43-4 see VGP000	53-03-2 see PLZ000	55-22-1 see ILC000
50-28-2 see EDO000	51-44-5 see DER600	53-06-5 see CNS800	55-27-6 see NNP050
50-29-3 see DAD200	51-45-6 see HGD000	53-10-1 see VJZ000	55-31-2 see AES500
50-31-7 see TIK500	51-46-7 see TBJ000	53-16-7 see EDV000	55-37-8 see DST200
50-32-8 see BCS750	51-48-9 see TFZ275	53-18-9 see DIG800	55-38-9 see FAQ900
50-33-9 see BRF500	51-50-3 see DCT050	53-19-0 see CDN000	55-43-6 see DCR200
50-34-0 see HKR500	51-52-5 see PNX000	53-21-4 see CNF000	55-48-1 see ARR500
50-35-1 see TEH500	51-55-8 see ARR000	53-36-1 see DAZ117	55-51-6 see BIA750
50-36-2 see CNE750	51-56-9 see HGH150	53-38-3 see MNC125	55-52-7 see PDN000
50-37-3 see DJO000	51-57-0 see MDT600	53-39-4 see AOO125	55-55-0 see MGJ750
50-39-5 see HNY500	51-58-1 see NIJ400	53-41-8 see HJB050	55-56-1 see BIM250
50-41-9 see CMX700	51-60-5 see DQY909	53-43-0 see AOO450	55-57-2 see PEO000
50-44-2 see POK000	51-61-6 see DYC400	53-44-1 see MBW775	55-63-0 see NGY000
50-47-5 see DSI709	51-62-7 see BBK750	53-46-3 see DJM800	55-65-2 see GKQ000
50-48-6 see EAH500	51-63-8 see BBK500	53-46-3 see XCJ000	55-68-5 see MCU750
50-49-7 see DLH600	51-64-9 see AOA500	53-59-8 see CNF400	55-80-1 see DUH600
50-50-0 see EDP000	51-65-0 see FLD100	53-60-1 see PMI500	55-81-2 see MFC500
50-52-2 see MOO250	51-66-1 see AAR250	53-64-5 see BKP300	55-86-7 see BIE500
50-53-3 see CKP250	51-67-2 see TOG250	53-69-0 see DQI600	55-91-4 see IRF000
50-54-4 see QHA000	51-68-3 see DPE000	53-70-3 see DCT400	55-92-5 see MFX560
50-55-5 see RDK000	51-71-8 see PFC500	53-79-2 see AEI000	55-93-6 see DSU000
50-56-6 see ORU500	51-73-0 see DWX600	53-84-9 see CNF390	55-97-0 see HEA000
50-59-9 see TEY000	51-74-1 see HGE000	53-85-0 see DGK100	55-98-1 see BOT250
50-60-2 see PDW400	51-75-2 see BIE250	53-86-1 see IDA000	56-04-2 see MPW500
50-62-4 see HFF500	51-77-4 see GDG200	53-89-4 see BCP650	56-05-3 see DGJ175
50-63-5 see CLD250	51-78-5 see ALU500	53-94-1 see HIU500	56-10-0 see AJY250
50-65-7 see DFV400	51-79-6 see UVA000	53-95-2 see HIP000	56-12-2 see PIM500
50-67-9 see AJX500	51-80-9 see TDR750	53-96-3 see FDR000	56-17-7 see CQJ750
50-69-1 see RJA700	51-82-1 see DRQ650	54-04-6 see MDI500	56-18-8 see AIX250
50-70-4 see SKV200	51-83-2 see CBH250	54-05-7 see CLD000	56-23-5 see CBY000
50-71-5 see AFT750	51-84-3 see CMF250	54-06-8 see AES639	56-24-6 see TMI250
50-76-0 see AEB000	51-85-4 see MCN500	54-11-5 see NDN000	56-25-7 see CBE750
50-78-2 see ADA725	51-93-4 see EQC600	54-12-6 see TNW500	56-29-1 see ERD500
50-79-3 see DER400	51-98-9 see ABU000	54-16-0 see HLJ000	56-33-7 see DWN150
50-81-7 see ARN000	52-01-7 see AFJ500	54-21-7 see SJO000	56-34-8 see TCC250
50-84-0 see DER100	52-21-1 see SOV100	54-25-1 see RJA000	56-35-9 see BLL750
50-89-5 see TFX790	52-24-4 see TFQ750	54-30-8 see NOC000	56-36-0 see TIC000
50-90-8 see CFJ750	52-26-6 see MRO750	54-31-9 see CHJ750	56-37-1 see BFL300
50-91-9 see DAR400	52-28-8 see CNG500	54-35-3 see PAQ200	56-38-2 see PAK000
50-98-6 see EAW500	52-31-3 see TDA500	54-36-4 see MCJ370	56-40-6 see GHA000
50-98-6 see EAX000	52-39-1 see AFJ875	54-42-2 see DAS000	56-41-7 see AFH625

83-86-3 see PIB250
83-88-5 see RIK000
83-89-6 see ARQ250
83-93-2 see TEF775
83-98-7 see MJH900
84-01-5 see CLY750
84-02-6 see PMF250
84-04-8 see CJL500
84-08-2 see PAJ750
84-11-7 see PCX250
84-12-8 see PCY300
84-15-1 see TBC640
84-16-2 see DLB400
84-17-3 see DAL600
84-19-5 see DHB500
84-21-9 see SPB100
84-36-6 see RCA200
84-51-5 see EGL500
84-57-1 see DFQ200
84-58-2 see DEX400
84-60-6 see DMH600
84-61-7 see DGV700
84-65-1 see APK250
84-66-2 see DJX000
84-68-4 see DEQ400
84-69-5 see DNJ400
84-74-2 see DEH200
84-75-3 see DKP600
84-76-4 see DVJ000
84-77-5 see DGX600
84-79-7 see HLY500
84-80-0 see VTA000
84-83-3 see FBV050
84-86-6 see ALI000
84-89-9 see ALI240
84-96-8 see AFL500
84-97-9 see PCK500
85-00-7 see DWX800
85-00-7 see EJC025
85-01-8 see PCW250
85-02-9 see BDB750
85-22-3 see PAT850
85-31-4 see TFJ500
85-32-5 see GLS750
85-34-7 see TIY500
85-36-9 see AAM875
85-40-5 see TDB100
85-41-6 see PHX000
85-43-8 see TDB000
85-44-9 see PHW750
85-52-9 see BDL850
85-60-9 see BRP750
85-68-7 see BEC500
85-70-1 see BQP750
85-71-2 see MOD000
85-73-4 see PHY750
85-79-0 see DDT200
85-82-5 see FAG080
85-83-6 see SBC500
85-84-7 see FAG130
85-86-9 see OHI200
85-91-6 see MGQ250
85-98-3 see DJC400
86-00-0 see NFP500
86-13-5 see BDI000
86-21-5 see TMJ750
86-26-0 see PEG000
86-28-2 see EHG025
86-29-3 see DVX200
86-30-6 see DWI000
86-34-0 see MNZ000
86-35-1 see EOL100

86-40-8 see XAK000
86-50-0 see ASH500
86-51-1 see DNZ200
86-52-2 see CIP750
86-53-3 see NAY090
86-54-4 see HGP495
86-55-5 see NAV490
86-56-6 see DSU400
86-57-7 see NHQ000
86-60-2 see ALI300
86-65-7 see NBE850
86-72-6 see CBN100
86-73-7 see FDI100
86-74-8 see CBN000
86-75-9 see HOE100
86-85-1 see MLH000
86-86-2 see NAK000
86-87-3 see NAK500
86-88-4 see AQN635
86-92-0 see THA300
86-93-1 see PGJ750
86-95-3 see QNA000
86-96-4 see QEJ800
86-97-5 see ALJ500
86-98-6 see DGJ250
87-00-3 see HGK550
87-01-4 see DPJ800
87-02-5 see AKI000
87-08-1 see PDT500
87-09-2 see AGK250
87-10-5 see THW750
87-11-6 see ABI250
87-12-7 see BOD600
87-13-8 see EEV200
87-17-2 see SAH500
87-18-3 see BSH100
87-19-4 see IJN000
87-20-7 see IME000
87-25-2 see EGM000
87-29-6 see API750
87-31-0 see DUR800
87-33-2 see CCK125
87-39-8 see AFU000
87-41-2 see PHW800
87-42-3 see CKV500
87-44-5 see CCN000
87-47-8 see PPQ625
87-48-9 see BNL750
87-51-4 see ICN000
87-52-5 see DYC000
87-56-9 see MRU900
87-59-2 see XMJ000
87-60-5 see CLK200
87-61-6 see TIK100
87-62-7 see XNJ000
87-63-8 see CLK227
87-65-0 see DFY000
87-66-1 see PPQ500
87-68-3 see HCD250
87-69-4 see TAF750
87-76-3 see TLN150
87-78-5 see MAW100
87-79-6 see SKV400
87-82-1 see HCA385
87-84-3 see CPB550
87-85-4 see HEC000
87-86-5 see PAX250
87-87-6 see TBQ500
87-89-8 see IDE300
87-90-1 see TIQ750
87-97-8 see DEE300
87-99-0 see XPJ000

88-02-8 see TGO800
88-04-0 see CLW000
88-05-1 see TLG500
88-06-2 see TIW000
88-09-5 see DHI400
88-10-8 see DIW400
88-12-0 see EEG000
88-14-2 see FQF000
88-15-3 see ABI500
88-18-6 see BSE460
88-19-7 see TGN250
88-21-1 see SNO100
88-24-4 see MJN250
88-26-6 see IFX200
88-27-7 see DEA100
88-27-7 see FAB000
88-29-9 see ACL750
88-30-2 see TKD400
88-32-4 see BQI010
88-35-7 see AOS500
88-38-0 see BIU600
88-41-5 see BQW490
88-44-8 see AKQ000
88-45-9 see PEY800
88-51-7 see AJJ250
88-58-4 see DEC800
88-60-8 see BQV600
88-61-9 see XJJ000
88-62-0 see AMT000
88-63-1 see PFA250
88-64-2 see AHQ300
88-65-3 see BMT800
88-67-5 see IEE000
88-68-6 see AKH800
88-69-7 see IQX100
88-72-2 see NMO525
88-73-3 see CJB750
88-74-4 see NEO000
88-75-5 see NIF010
88-82-4 see TKQ250
88-84-6 see GKO000
88-85-7 see BRE500
88-88-0 see PIE530
88-89-1 see PID000
88-96-0 see BBO500
88-99-3 see PHW250
89-00-9 see QOJ200
89-02-1 see DUR800
89-05-4 see PPQ630
89-19-0 see BQX250
89-25-8 see NNT000
89-28-1 see SAU480
89-32-7 see PPQ635
89-37-2 see DVB850
89-40-7 see NIX100
89-52-1 see AAJ150
89-55-4 see BOE500
89-57-6 see AMM500
89-58-7 see NMS520
89-61-2 see DFT400
89-63-4 see KDA050
89-65-6 see SAA025
89-68-9 see CLJ800
89-69-0 see TIT750
89-72-5 see BSE000
89-73-6 see SAL500
89-78-1 see MCF750
89-80-5 see MCG275
89-81-6 see MCF000
89-82-7 see POI615
89-83-8 see TFX810
89-84-9 see DMG400

89-86-1 see HOE600
89-93-0 see MHM510
89-94-1 see MPO000
89-98-5 see CEI500
90-00-6 see PGR250
90-01-7 see HMK100
90-02-8 see SAG000
90-03-9 see CHW675
90-04-0 see AOV900
90-05-1 see GKI000
90-11-9 see BNS200
90-12-0 see MMB750
90-13-1 see CIZ000
90-15-3 see NAW500
90-16-4 see BDH000
90-17-5 see TIT000
90-20-0 see AKH000
90-22-2 see VBK000
90-26-6 see NMV300
90-27-7 see PEP250
90-30-2 see PFT250
90-33-5 see MKP500
90-34-6 see PMC300
90-39-1 see SKX500
90-41-5 see BGE250
90-42-6 see LBX100
90-43-7 see BGJ250
90-44-8 see APM500
90-45-9 see AHS500
90-46-0 see XBJ000
90-47-1 see XBS000
90-49-3 see PFB350
90-50-6 see CMQ100
90-51-7 see AKH800
90-64-2 see MAP000
90-65-3 see PAP750
90-69-7 see LHY000
90-72-2 see TNH000
90-80-2 see GFA200
90-81-3 see EAW100
90-82-4 see POH000
90-84-6 see DIP400
90-87-9 see HII600
90-89-1 see DIW000
90-90-4 see BMU170
90-94-8 see MQS500
90-98-2 see DES000
91-01-0 see HKF300
91-02-1 see PGE760
91-04-3 see HLX950
91-08-7 see TGM800
91-10-1 see DOJ200
91-15-6 see PHY000
91-16-7 see DOA200
91-17-8 see DAE800
91-19-0 see QRJ000
91-20-3 see NAJ500
91-21-4 see TCU500
91-22-5 see QMJ000
91-23-6 see NER000
91-33-8 see BDE250
91-38-3 see CIB700
91-40-7 see PEG500
91-44-1 see DIL400
91-49-6 see BPU500
91-51-0 see LFT100
91-53-2 see SAV000
91-55-4 see DSI850
91-56-5 see ICR000
91-57-6 see MMC000
91-58-7 see CJA000
91-59-8 see NBE500

224-98-6 see NAZ000	298-46-4 see DCV200	303-98-0 see UAH000	317-52-2 see HEG000
225-51-4 see BAW750	298-51-1 see THK600	304-06-3 see PGH000	317-64-6 see DQK000
226-36-8 see DCS400	298-57-7 see CMR100	304-17-6 see IRG000	317-83-9 see CPK550
226-47-1 see DDC800	298-59-9 see RLK000	304-20-1 see HGP500	317-97-5 see FLG200
229-87-8 see PCX300	298-81-7 see XDJ000	304-28-9 see BGP250	318-03-6 see BDB500
230-27-3 see BDC000	298-83-9 see NFQ020	304-43-8 see BGK250	318-22-9 see TKH000
234-17-3 see MJR775	298-83-9 see NMK100	304-55-2 see DNV800	318-98-9 see ICC000
235-06-3 see THT400	298-93-1 see DUG400	304-59-6 see SJK385	319-84-6 see BBQ000
238-04-0 see NAU530	298-96-4 see TMV500	304-81-4 see PDB750	319-85-7 see BBR000
238-84-6 see NAU600	299-11-6 see MNO500	304-84-7 see DKE200	319-86-8 see BFW500
239-01-0 see BCG250	299-20-7 see EDW200	305-03-3 see CDO500	319-94-8 see PAV780
239-60-1 see DDB200	299-26-3 see AME500	305-15-7 see DGC850	320-67-2 see ARY000
239-64-5 see DCX800	299-27-4 see PLG800	305-33-9 see IGG700	320-72-9 see DGK200
239-67-8 see BDA750	299-28-5 see CAS750	305-53-3 see SIN000	321-14-2 see CLD825
240-39-1 see PPI815	299-29-6 see FBK000	305-72-6 see KFK300	321-25-5 see FHQ000
240-44-8 see BCI250	299-39-8 see SKX750	305-80-6 see BBO250	321-30-2 see AEH500
243-17-4 see BCJ800	299-42-3 see EAW000	305-84-0 see CCK665	321-38-0 see FKK000
244-63-3 see NNR300	299-45-6 see PKT000	305-85-1 see DNG000	321-54-0 see CIK750
257-07-8 see DDE200	299-61-6 see GBU700	305-97-5 see LGU000	321-55-1 see DFH600
258-76-4 see DUP000	299-75-2 see TFU500	306-07-0 see BEX500	321-60-8 see FGI050
260-94-6 see ADJ500	299-84-3 see RMA500	306-12-7 see OOK100	321-64-2 see TCJ075
262-12-4 see DDA800	299-85-4 see DGD800	306-23-0 see HNG700	321-98-2 see EAW200
262-20-4 see PDQ750	299-86-5 see COD850	306-37-6 see DSF800	324-93-6 see AKC500
262-38-4 see TBI775	300-08-3 see AQU000	306-40-1 see CMG250	325-23-5 see BCI750
271-89-6 see BCK250	300-37-8 see DNG400	306-52-5 see TIP300	325-69-9 see FLE000
273-53-0 see BDI500	300-42-5 see DBA800	306-53-6 see MKU750	326-43-2 see PGG355
274-09-9 see MJR000	300-54-9 see MRW250	306-67-2 see GEK000	326-52-3 see FKT050
274-98-6 see THT280	300-57-2 see AFX000	306-83-2 see TJY500	326-61-4 see PIX000
275-02-5 see THT275	300-62-9 see BBK000	306-94-5 see PCG700	327-57-1 see AJC950
275-51-4 see ASP600	300-68-5 see THL575	307-34-6 see OAP100	327-97-9 see CHK175
277-06-5 see DHB820	300-76-5 see NAG400	308-48-5 see PCG755	327-98-0 see EPY000
280-57-9 see DCK400	300-87-8 see DSK950	309-00-2 see AFK250	328-04-1 see CJD650
283-24-9 see ARX300	300-88-9 see CEI250	309-29-5 see ENL100	328-38-1 see LER000
283-56-7 see TJK750	300-92-5 see AHA275	309-36-4 see MDU500	328-74-5 see BLO250
283-60-3 see TMO750	301-04-2 see LCV000	309-43-3 see SBN000	328-84-7 see DGO300
283-66-9 see DCK700	301-11-1 see LBO000	311-28-4 see TBL000	329-01-1 see TKJ250
285-67-6 see EBO100	301-12-2 see DAP000	311-44-4 see CHF600	329-21-5 see BSG000
286-20-4 see CPD000	301-13-3 see TNI300	311-45-5 see NIM500	329-56-6 see NNP000
286-75-9 see CPR835	302-01-2 see HGS000	311-47-7 see CLV375	329-63-5 see AES625
287-23-0 see COW000	302-15-8 see MKN500	311-89-7 see HAS000	329-65-7 see EBB500
287-29-6 see SCF550	302-17-0 see CDO000	312-30-1 see DKH250	329-71-5 see DVA000
287-92-3 see CPV750	302-22-7 see CBF250	312-45-8 see HAQ000	329-89-5 see ALL250
288-13-1 see POM500	302-23-8 see PMG600	312-73-2 see TKB800	329-98-6 see TGO300
288-32-4 see IAL000	302-27-2 see ADH750	312-93-6 see BFW325	329-99-7 see MIT600
288-47-1 see TEU300	302-33-0 see DIG400	313-06-4 see DAZ115	330-54-1 see DXQ500
288-88-0 see THS855	302-40-9 see DHU900	313-67-7 see AQY250	330-55-2 see DGD600
288-94-8 see TEF650	302-41-0 see PJA140	313-74-6 see DQJ400	330-64-3 see DNS200
289-14-5 see EJO500	302-48-7 see BGX775	313-93-9 see EPZ000	330-68-7 see FFH000
289-80-5 see OJW200	302-49-8 see EHV500	313-94-0 see BSR500	330-95-0 see NCW100
289-95-2 see PPO750	302-66-9 see MNM500	313-95-1 see FMF000	331-39-5 see CAK375
290-37-9 see POL490	302-70-5 see CFA750	313-96-2 see MFQ750	331-54-4 see CFX300
290-87-9 see THR525	302-72-7 see AFH600	314-03-4 see MOO750	331-87-3 see FFI000
291-21-4 see TLS500	302-79-4 see VSK950	314-04-5 see DKJ200	331-91-9 see FHQ010
291-64-5 see COX500	302-83-0 see TAL490	314-13-6 see BGT250	332-14-9 see PDI500
294-93-9 see COD475	302-84-1 see SCA350	314-19-2 see AQP500	332-54-7 see FHQ100
297-76-7 see EQJ500	302-95-4 see SGE000	314-35-2 see CNR125	332-69-4 see BMN350
297-78-9 see OAN000	302-96-5 see AOO401	314-40-9 see BMM650	332-97-8 see EKK500
297-88-1 see MDO760	303-04-8 see DFM000	314-42-1 see BNM000	333-18-6 see EIW000
297-90-5 see MKR250	303-21-9 see DVG200	315-18-4 see DOS000	333-20-0 see PLV750
297-97-2 see EPC500	303-25-3 see MAX275	315-22-0 see MRH000	333-25-5 see DEW000
299-97-4 see PGX300	303-26-4 see NNK500	315-30-0 see ZVJ000	333-29-9 see DXN600
298-00-0 see MNH000	303-33-3 see HAL500	315-37-7 see TBF750	333-36-8 see HDC000
298-02-2 see PGS000	303-34-4 see LBG000	315-72-0 see DCV800	333-40-4 see DTV400
298-03-3 see DAO500	303-45-7 see GJM000	315-80-0 see DCW800	333-41-5 see DCM750
298-04-4 see DXH325	303-47-9 see CHP250	316-05-2 see QCS900	333-49-3 see TFH800
298-06-6 see PHG500	303-49-1 see CDU750	316-14-3 see MGW000	333-93-7 see POK325
298-07-7 see BJR750	303-53-7 see PMH600	316-41-6 see BFN625	334-22-5 see BHN750
298-12-4 see GIQ000	303-54-8 see DPW600	316-42-7 see EAN000	334-44-1 see FIY000
298-14-6 see PKX100	303-69-5 see DYB600	316-46-1 see FMN000	334-48-5 see DAH400
298-18-0 see DHB600	303-70-8 see DPX400	316-49-4 see MGV500	334-50-9 see SLG600
298-39-5 see SLZ000	303-75-3 see HKS900	316-81-4 see TFM100	334-56-5 see FHL000
298-45-3 see HLT000	303-81-1 see SMB000	317-34-0 see TEP500	334-62-3 see FHN000

334-64-5 see FHM000
334-71-4 see FIA000
334-88-3 see DCP800
335-57-9 see PCH000
335-67-1 see PCH050
335-76-2 see PCG725
337-28-0 see EFA000
337-47-3 see SOX500
338-66-9 see DKH825
339-43-5 see BSM000
340-56-7 see MDT250
340-57-8 see MIH925
341-69-5 see OJW000
341-70-8 see DHF600
342-69-8 see MPU000
342-95-0 see BHT250
343-75-9 see DUK200
343-89-5 see FFG000
343-94-2 see AJX250
344-07-0 see PBE100
344-62-7 see TKB350
344-72-9 see AMU550
345-78-8 see POH250
346-18-9 see PKL250
348-67-4 see MDT730
349-37-1 see DKH000
350-03-8 see ABI000
350-12-9 see RCK730
350-30-1 see CHI950
350-46-9 see FKL000
350-87-8 see DKH100
351-61-1 see TKA950
351-63-3 see DKG980
351-65-5 see DRL000
352-21-6 see AKF375
352-32-9 see FMC000
352-93-2 see EPH000
353-03-7 see EKI000
353-13-9 see FMO000
353-16-2 see FHC200
353-17-3 see FHD000
353-18-4 see CON500
353-21-9 see FIZ000
353-36-6 see FIB000
353-42-4 see BMH000
353-50-4 see CCA500
353-59-3 see BNA250
354-06-3 see TJY000
354-13-2 see TIJ175
354-15-4 see DKI900
354-21-2 see TIM000
354-23-4 see TJY750
354-25-6 see TCH150
354-32-5 see TJX500
354-33-6 see PBD400
354-58-5 see TJE100
354-93-8 see PCH300
355-02-2 see PCH290
355-43-1 see PCH100
355-66-8 see OBK100
355-80-6 see OBU000
356-12-7 see FDD150
356-18-3 see DFM025
356-27-4 see HAY000
356-69-4 see PCH275
356-69-4 see PCH350
357-07-3 see ORG100
357-08-4 see NAH000
357-09-5 see FJF100
357-56-2 see AFJ400
357-57-3 see BOL750
358-21-4 see PCG760

358-52-1 see POA250
358-74-7 DJJ400
359-06-8 see FFR000
359-11-5 see TKA400
359-29-5 see TIP400
359-40-0 see OLK000
359-46-6 see TJX750
359-48-8 see PCO250
359-83-1 see DOQ400
360-53-2 see HAY500
360-54-3 see MKK750
360-68-9 see DKW000
360-70-3 see NNE550
360-89-4 see OBO000
360-97-4 see AKK250
361-09-1 see SFW000
361-37-5 see MLD250
362-02-7 see MIJ300
362-05-0 see HKH600
362-06-1 see HKI200
362-07-2 see MEL785
362-74-3 see COV625
363-03-1 see PEL750
363-13-3 see BEH000
363-17-7 see FER000
363-20-2 see TIH800
363-24-6 see DVJ200
363-42-8 see TLQ000
363-49-5 see HIK500
364-62-5 see AJH000
364-71-6 see FFT000
364-76-1 see FKK100
364-98-7 see DCQ700
365-26-4 see HKH500
366-18-7 see BGO500
366-29-0 see BJF600
366-70-1 see PME500
366-71-2 see MKN750
366-93-8 see BHN000
367-12-4 see FKT100
367-25-9 see DKF700
367-51-1 see SKH500
368-39-8 see TJL600
368-43-4 see BBT250
368-47-8 see PGO500
368-53-6 see TKB775
368-68-3 see DKG100
368-97-8 see DKI400
369-57-3 see BBO325
370-14-9 see FMS875
370-81-0 see COF675
371-28-8 see CGZ000
371-29-9 see FIN000
371-40-4 see FFY000
371-41-5 see FKV000
371-47-1 see SIE000
371-62-0 see FIE000
371-67-5 see TJY275
371-69-7 see EKK550
371-78-8 see BLQ325
371-86-8 see PHF750
372-09-8 see COJ500
372-18-9 see DKF800
372-48-5 see FLT100
372-64-5 see BLO325
372-70-3 see FJA100
372-91-8 see FHC000
373-02-4 see NCX000
373-14-8 see FJA000
373-44-4 see OCU200
373-88-6 see TKA750
373-91-1 see TKD375

375-01-9 see HAW100
375-22-4 see HAX500
375-73-5 see PCG625
376-18-1 see HCO000
376-50-1 see DJT100
376-53-4 see PCG600
376-89-6 see HDC300
377-38-8 see TCJ000
378-44-9 see BFV750
379-52-2 see TMV850
379-79-3 see EDC500
381-73-7 see DKF200
382-10-5 see HDC450
382-21-8 see OBM000
382-67-2 see DBA875
383-73-3 see BLO000
386-17-4 see IEU100
388-72-7 see FFZ000
389-08-2 see EID000
390-64-7 see PEV750
391-57-1 see FEE000
391-70-8 see TNV625
392-56-3 see HDB000
392-83-6 see BOJ750
392-95-0 see CGM200
393-52-2 see FGP000
393-75-9 see CGM225
394-69-4 see FLV060
395-28-8 see VGF000
395-47-1 see TKF525
396-01-0 see UVJ450
396-32-7 see FLV070
398-32-3 see FKZ000
399-24-6 see FKQ100
400-44-2 see CHK750
400-99-7 see NMQ100
401-78-5 see BOJ500
402-26-6 see TKF530
402-31-3 see BLO270
402-51-7 see TKF535
402-71-1 see THH450
404-42-2 see DKF170
404-72-8 see ISL000
404-82-0 see PDM250
404-86-4 see CBF750
405-22-1 see FPE100
405-86-7 see FHV000
406-20-2 see MKE000
406-23-5 see AGG750
406-90-6 see TKB250
407-25-0 see TJX000
407-83-0 see FHF000
407-98-7 see FFW000
407-99-8 see FLR100
408-35-5 see SIZ025
409-21-2 see SCQ000
420-04-2 see COH500
420-12-2 see EJP500
420-23-5 see DRY289
420-46-2 see TJY900
420-52-0 see TKF775
421-17-0 see TKB300
421-20-5 see MKG250
421-53-4 see TJZ000
422-01-5 see PBF600
422-02-6 see CJI525
422-05-9 see PBE750
422-55-9 see CHK800
422-56-0 see DFW850
422-61-7 see PBF300
422-63-9 see PBE500
422-64-0 see PBF000

423-62-1 see IEU075
424-40-8 see DJK100
425-51-4 see MCB375
425-87-6 see CLR000
427-00-9 see DKX600
427-01-0 see GCG300
427-45-2 see TNG050
427-51-0 see CQJ500
428-59-1 see HDF050
429-30-1 see TJX250
430-66-0 see TJY950
431-03-8 see BOT500
431-46-9 see MFR500
431-63-0 see HDE050
431-97-0 see BLO300
432-04-2 see TNN775
432-60-0 see AGF750
433-27-2 see EFK500
434-03-7 see GEK500
434-05-9 see PMC700
434-07-1 see PAN100
434-13-9 see LHW000
434-16-2 see DAK600
434-22-0 see NNX400
434-64-0 see PCH500
436-30-6 see MKD750
436-40-8 see DCN000
437-38-7 see PDW500
437-74-1 see XCS000
438-41-5 see MDQ250
438-60-8 see DDA600
438-67-5 see EDV600
439-14-5 see DCK759
439-25-8 see MKD500
440-17-5 see TKK250
440-58-4 see AAI750
441-38-3 see BCP500
442-51-3 see HAI500
443-30-1 see DOT600
443-48-1 see MMN250
443-79-8 see IKX010
444-27-9 see TEV000
445-29-4 see FGH000
446-35-5 see DKH900
446-52-6 see FFY100
446-72-0 see GCM350
446-86-6 see ASB250
447-05-2 see PPJ900
447-14-3 see TKH310
447-25-6 see DAR150
447-31-4 see CFJ100
451-40-1 see PEB000
452-06-2 see AMH000
452-35-7 see EEN500
452-71-1 see FJR900
452-86-8 see DNE200
453-13-4 see DKI800
453-18-9 see MKD000
453-20-3 see HOI200
454-29-5 see HGI300
454-41-1 see ALF600
455-14-1 see TKB750
455-16-3 see TGO500
455-80-1 see HHM500
455-90-3 see NMO700
456-27-9 see NEZ200
456-59-7 see DNU100
456-88-2 see FLD000
457-60-3 see NCJ500
457-87-4 see EGI500
458-24-2 see ENJ000
458-37-7 see ICC800

458-88-8 see PNT000	470-90-6 see CDS750	483-55-6 see HMI000	495-73-8 see BDD000
459-02-9 see FKG000	471-03-4 see BIH500	483-57-8 see FBP300	496-06-0 see IGF300
459-22-3 see FLC000	471-25-0 see PMT275	483-63-6 see EHO500	496-11-7 see IBR000
459-44-9 see TGM450	471-29-4 see MKI750	483-84-1 see FBZ100	496-15-1 see ICS300
459-72-3 see EKG500	471-34-1 see CAT775	484-20-8 see MFN275	496-41-3 see CNU875
459-80-3 see GCW000	471-35-2 see TDP250	484-23-1 see OJD300	496-67-3 see BNP750
459-99-4 see FIM000	471-46-5 see OLO000	484-47-9 see TMS750	496-72-0 see TGM250
460-07-1 see ACB250	471-53-4 see GIE000	484-78-6 see HJD000	496-74-2 see TGN000
460-12-8 see BOQ625	471-77-2 see NBU800	485-19-8 see SLX500	497-18-7 see CBS500
460-19-5 see COO000	471-95-4 see BON000	485-31-4 see BGB500	497-19-8 see SFO000
460-35-5 see TJY200	472-54-8 see NNT500	485-35-8 see CQL500	497-25-6 see OMM000
460-40-2 see TKH020	473-41-6 see BSQ250	485-47-2 see DMV200	497-26-7 see MJH775
461-05-2 see CCK655	474-07-7 see LFT800	485-49-4 see BFX530	497-38-1 see NNK000
461-56-3 see FLR000	474-25-9 see CDL325	485-50-7 see CBF550	497-39-2 see DDX000
461-58-5 see COP125	474-74-8 see HND200	485-51-8 see CNR745	497-56-3 see DUT000
461-72-3 see HGO600	474-86-2 see ECW000	485-71-2 see CMP910	497-76-7 see HIH100
461-78-9 see CLY250	475-08-1 see CBK125	485-89-2 see OPK300	498-02-2 see HLQ500
461-89-2 see THR750	475-20-7 see LID100	486-17-9 see BRS000	498-23-7 see CMS320
462-06-6 see FGA000	475-25-2 see HAO600	486-25-9 see FDO000	498-24-8 see MDI250
462-08-8 see AMI250	475-26-3 see FHJ000	486-84-0 see MPA050	498-66-8 see NNH500
462-18-0 see TJJ300	475-80-9 see MJR780	486-89-5 see RHZ100	498-67-9 see DFL100
462-27-1 see FIH100	475-81-0 see TDI475	487-10-5 see ASN500	498-81-7 see MCE100
462-72-6 see FHA000	475-83-2 see NOE500	487-19-4 see NDX300	498-94-2 see ILG100
462-73-7 see FHB000	476-32-4 see CDL000	487-26-3 see FBW150	499-04-7 see AQT625
462-74-8 see FGY100	476-56-2 see FPD050	487-53-6 see DHO600	499-06-9 see MDJ748
462-94-2 see PBK500	476-60-8 see LEX200	487-54-7 see SAN200	499-12-7 see ADH000
462-95-3 see EFT500	476-66-4 see EAI200	487-68-3 see MDJ745	499-30-9 see PDF525
463-04-7 see AOL500	476-70-0 see DNZ100	487-79-6 see KAJ200	499-44-5 see IRR000
463-40-1 see OAX100	477-27-0 see ADE000	487-89-8 see FNO100	499-75-2 see CCM000
463-51-4 see KEU000	477-29-2 see DAN375	487-93-4 see DPG109	500-00-5 see MCE500
463-58-1 see CCC000	477-30-5 see MIW500	488-10-8 see JCA100	500-22-1 see NDM100
463-71-8 see TFN500	477-47-4 see PIE100	488-17-5 see DNE000	500-28-7 see MIJ250
463-82-1 see NCH000	477-73-6 see GJI400	488-23-3 see TDM250	500-34-5 see EQP500
463-88-7 see VQR300	478-08-0 see LIN100	488-41-5 see DDP600	500-38-9 see NBR000
464-07-3 see BRU300	478-15-9 see TDI750	488-81-3 see RIF000	500-55-0 see AQO250
464-10-8 see NMQ000	478-43-3 see RHZ700	489-84-9 see DSJ800	500-64-1 see GJI250
464-41-5 see BMD300	478-84-2 see BNM250	489-98-5 see PIC800	500-72-1 see OLW000
464-45-9 see NCQ820	478-99-9 see LJI000	490-02-8 see ARO000	500-92-5 see CKB250
464-48-2 see CBB000	479-13-0 see COF350	490-31-3 see RLP000	501-22-4 see HNF100
464-49-3 see CBB250	479-18-5 see DNC000	490-78-8 see DMG600	501-30-4 see HLH500
464-72-2 see TEA600	479-20-9 see ARQ600	490-79-9 see GCU000	501-36-0 see TKP200
465-12-3 see EBB800	479-23-2 see CMC000	490-91-5 see IQF000	501-53-1 see BEF500
465-16-7 see OHQ000	479-45-8 see TEG250	490-97-1 see SHG600	501-68-8 see BEG000
465-19-0 see BOM655	479-50-5 see DHU000	491-07-6 see IKY000	502-26-1 see SLL400
465-39-4 see BOM650	479-92-5 see INY000	491-26-9 see NDR100	502-37-4 see HOW100
465-42-9 see CBF760	480-16-0 see MRN500	491-35-0 see LEL000	502-39-6 see MLF250
465-65-6 see NAG550	480-18-2 see DMD000	491-36-1 see QFA000	502-42-1 see SMV000
465-69-0 see FKF100	480-22-8 see APH250	491-59-8 see CML750	502-44-3 see LAP000
465-73-6 see IKO000	480-30-8 see SKS700	491-70-3 see TDD500	502-47-6 see CMT125
465-84-9 see SMM100	480-40-0 see DMS900	491-92-9 see RHZ000	502-49-8 see CPS250
466-06-8 see POB500	480-41-1 see NBP350	492-08-0 see PAB250	502-55-6 see BJU000
466-11-5 see DBC550	480-54-6 see RFP000	492-17-1 see BGF109	502-56-7 see NMZ000
466-24-0 see PIC250	480-63-7 see IKN300	492-18-2 see SIH500	502-72-7 see CPU250
466-40-0 see IKZ000	480-68-2 see NET550	492-39-7 see NNM510	502-85-2 see HJS500
466-49-9 see ARP050	480-78-4 see PJG135	492-41-1 see NNM000	503-01-5 see ILK000
466-81-9 see EDG500	480-79-5 see IDG000	492-80-8 see IBB000	503-09-3 see EBU000
466-99-9 see DLW600	480-81-9 see SBX500	492-88-6 see NOC100	503-17-3 see COC500
467-22-1 see CBQ625	481-06-1 see SAU500	492-94-4 see FPZ000	503-20-8 see FFJ000
467-36-7 see TES500	481-39-0 see WAT000	493-52-7 see CCE500	503-28-6 see ASN400
467-60-7 see DWK400	481-42-5 see PJH610	493-53-8 see ADA750	503-30-0 see OMW000
467-63-0 see TJK000	481-49-2 see CCX550	494-03-1 see BIF250	503-38-8 see TIR920
468-28-0 see LIU000	481-72-1 see DMU600	494-19-9 see DKY800	503-41-3 see DXI500
468-61-1 see DHQ200	481-85-6 see MMC250	494-38-2 see BJF000	503-49-1 see HMC000
469-21-6 see DYE500	482-15-5 see OGI075	494-47-3 see FPS000	503-74-2 see ISU000
469-59-0 see JCS000	482-27-9 see IMO500	494-52-0 see AON875	503-80-0 see BJJ250
469-61-4 see CCR500	482-41-7 see MKD250	494-97-3 see NNR500	503-87-7 see TFJ750
469-62-5 see DAB879	482-44-0 see IHR300	494-98-4 see PQB500	504-03-0 see LIQ550
469-65-8 see HJO500	482-49-5 see DYB000	495-18-1 see BCL500	504-15-4 see MPH500
469-79-4 see KFK000	482-54-2 see CPB120	495-45-4 see MPL000	504-17-6 see MCK500
469-81-8 see MRN675	482-89-3 see BGB275	495-48-7 see ASO750	504-20-1 see PGW250
470-67-7 see IKC100	483-04-5 see AFG750	495-54-5 see DBP999	504-24-5 see AMI500
470-82-6 see CAL000	483-18-1 see EAL500	495-69-2 see HGB300	504-29-0 see AMI000

504-60-9 see PBA100
504-63-2 see PML250
504-75-6 see IAT000
504-88-1 see NIY500
504-90-5 see TFS500
505-14-6 see TFH600
505-22-6 see DVP600
505-29-3 see DXI550
505-44-2 see MPN100
505-57-7 see HFA500
505-60-2 see BIH250
505-66-8 see HGI900
505-71-5 see DUU800
505-71-5 see DUV800
505-75-9 see CMN000
506-12-7 see HAS500
506-30-9 see EAF000
506-32-1 see AQS750
506-59-2 see DOR600
506-61-6 see PLS250
506-63-8 see DQR200
506-64-9 see SDP000
506-65-0 see GIW189
506-68-3 see COO500
506-77-4 see COO750
506-78-5 see COP000
506-82-1 see DQW800
506-85-4 see FOS050
506-87-6 see ANE000
506-93-4 see GLA000
506-96-7 see ACD750
507-02-8 see ACO500
507-09-5 see TFA500
507-16-4 see SNT200
507-19-7 see BQM250
507-20-0 see BQR000
507-25-5 see CBY500
507-28-8 see TEA300
507-40-4 see BRP500
507-42-6 see THU500
507-55-1 see DFW830
507-60-8 see SBF500
507-63-1 see PCH325
507-70-0 see BMD000
508-54-3 see HJX500
508-59-8 see PAM175
508-65-6 see EBL500
508-75-8 see CNH780
508-77-0 see CQH750
509-09-1 see PBF250
509-14-8 see TDY250
509-15-9 see GCK000
509-20-6 see ADG500
509-67-1 see TCY750
509-86-4 see COY500
510-13-4 see MAK500
510-15-6 see DER000
510-90-7 see SFG700
511-09-1 see EDB100
511-12-6 see DLK800
511-13-7 see CMW700
511-46-6 see CKE000
511-55-7 see PEM750
512-13-0 see TLW000
512-16-3 see COW700
512-24-3 see CCZ000
512-48-1 see DJU200
512-56-1 see TMD250
512-64-1 see EAD500
512-69-6 see RBA100
512-85-6 see ARM500
513-10-0 see TLF500

513-12-2 see EPI000
513-31-5 see DDS200
513-36-0 see CIU500
513-37-1 see IKE000
513-38-2 see IIV509
513-42-8 see IMW000
513-44-0 see IIX000
513-48-4 see IEH000
513-49-5 see BPY100
513-77-9 see BAJ250
513-78-0 see CAD800
513-79-1 see CNB475
513-81-5 see DQT150
513-85-9 see BOT000
513-86-0 see ABB500
513-88-2 see DGG500
513-92-8 see TDE250
514-10-3 see AAC500
514-61-4 see MDM350
514-65-8 see BGD500
514-73-8 see DJT800
514-78-3 see CBE800
514-85-2 see VSK975
515-42-4 see SJH050
515-64-0 see SNJ350
515-69-5 see BGO775
515-83-3 see TIO000
515-96-8 see OLM310
516-21-2 see COX400
516-90-5 see LHW100
516-95-0 see EBA100
517-09-9 see ECV000
517-16-8 see EME500
517-25-9 see TMM500
517-28-2 see HAP500
517-66-8 see ERE150
517-85-1 see DCR800
517-92-0 see CML600
518-34-3 see TDX830
518-47-8 see FEW000
518-75-2 see CMS775
518-82-1 see MQF250
519-23-3 see EAI850
519-37-9 see HLC000
519-65-3 see DVU100
519-87-9 see PDX500
519-88-0 see DDT300
520-07-0 see AQN250
520-09-2 see DYC700
520-14-9 see GFC200
520-18-3 see ICE000
520-26-3 see HBU000
520-36-5 see CDH250
520-45-6 see MFW500
520-52-5 see PHU500
520-53-6 see HKE000
520-68-3 see EAC500
520-85-4 see MBZ150
521-10-8 see AOO475
521-11-9 see MJE760
521-18-6 see DME500
521-24-4 see DLK000
521-31-3 see AJO290
521-35-7 see CBD625
521-74-4 see DDS600
521-78-8 see SOX550
522-00-9 see DIR000
522-12-3 see QCJ000
522-16-7 see ARS250
522-23-6 see MDU750
522-25-8 see POL000
522-27-0 see BJW810

522-40-7 see DKA200
522-48-5 see VRZ000
522-60-1 see HHR700
522-70-3 see BLX750
522-75-8 see DNT300
523-44-4 see FAG010
523-50-2 see FQC000
523-80-8 see AGE500
523-86-4 see SMC500
523-87-5 see DYE600
524-38-9 see HNS600
524-40-3 see RJK100
524-42-5 see NBA000
524-83-4 see DWE800
524-89-0 see MEL725
525-02-0 see BEM750
525-05-3 see HNB000
525-26-8 see CMX840
525-64-4 see FDM000
525-66-6 see ICB000
525-79-1 see FPT100
525-82-6 see PER700
526-08-9 see AIF000
526-18-1 see DYE700
526-26-1 see SML500
526-55-6 see ICS000
526-62-5 see BGW750
526-73-8 see TLL500
526-75-0 see XKJ000
526-84-1 see DMW200
526-99-8 see GAR000
527-07-1 see SHK800
527-09-3 see CNM100
527-17-3 see TDM810
527-53-7 see TDM500
527-60-6 see MDJ740
527-62-8 see AJM525
527-72-0 see TFM600
527-73-1 see NHG000
527-84-4 see IRN300
528-21-2 see TKN250
528-29-0 see DUQ400
528-43-8 see BFX520
528-44-9 see TKU700
528-45-0 see DUR550
528-48-3 see FBW000
528-53-0 see DAM400
528-58-5 see COI750
528-74-5 see DFO000
528-76-7 see DUR200
528-92-7 see IQX000
528-94-9 see ANT600
528-97-2 see BQH250
529-05-5 see DRV000
529-19-1 see TGT500
529-33-9 see TDI350
529-34-0 see DLX200
529-44-2 see HDW150
529-65-7 see EFQ500
529-84-0 see MJV800
530-31-4 see ANM250
530-35-8 see EJR300
530-43-8 see CDP700
530-50-7 see DWD100
530-57-4 see SPE700
530-78-9 see TKH750
530-91-6 see TCX750
531-18-0 see HDY000
531-53-3 see DUG700
531-59-9 see MEK300
531-72-6 see TFA350
531-76-0 see BHT750

531-82-8 see AAL750
531-85-1 see BBX750
531-86-2 see BBY000
531-95-3 see ECW700
532-03-6 see GKK000
532-11-6 see AOO490
532-27-4 see CEA750
532-28-5 see MAP250
532-32-1 see SFB000
532-34-3 see BRT000
532-43-4 see TET500
532-49-0 see DDW000
532-54-7 see ILH000
532-62-7 see AIT750
532-76-3 see COU250
532-77-4 see OKK100
532-82-1 see PEK000
532-91-2 see MEC550
532-94-5 see SHN500
533-06-2 see CBK500
533-23-3 see EHY600
533-28-8 see IJZ000
533-31-3 see MJU000
533-45-9 see CHD750
533-51-7 see SDU000
533-58-4 see IEV000
533-73-3 see BBU250
533-74-4 see DSB200
533-75-5 see TNV550
533-87-9 see TKP000
533-96-0 see SJT750
534-07-6 see BIK250
534-13-4 see DSK900
534-15-6 see DOO500
534-16-7 see SDN200
534-17-8 see CDC750
534-22-5 see MKH000
534-33-8 see ACN250
534-52-1 see DUS700
534-76-9 see EQS500
535-55-7 see MCU500
535-65-9 see GEW750
535-75-1 see HDS300
535-77-3 see IRN400
535-80-8 see CEL290
535-83-1 see TKL890
535-87-5 see DBQ190
535-89-7 see CCP500
536-17-4 see DOT800
536-21-0 see AKT000
536-25-4 see AKF000
536-29-8 see DFX400
536-33-4 see EPQ000
536-38-9 see CJJ100
536-43-6 see BPR500
536-46-9 see DTM000
536-50-5 see TGZ000
536-59-4 see PCI550
536-60-7 see CQI250
536-69-6 see BSI000
536-74-3 see PEB750
536-80-1 see IFC000
536-90-3 see AOV890
537-00-8 see CCY500
537-05-3 see PDN500
537-12-2 see DVV500
537-45-1 see CHQ750
537-46-2 see PFP850
537-92-8 see ABI750
537-98-4 see FBP200
538-02-3 see CPV609
538-03-4 see ARL000

538-04-5 see CHO750
538-07-8 see BID250
538-08-9 see DBJ200
538-23-8 see TMO000
538-28-3 see BEU500
538-32-9 see BFN125
538-41-0 see ASK925
538-42-1 see SFX000
538-43-2 see GGA950
538-62-5 see DVY950
538-65-8 see BQV500
538-71-6 see DXX000
538-74-9 see BEY900
538-75-0 see MDR800
538-79-4 see MDM750
538-93-2 see IIN000
539-03-7 see CDZ100
539-17-3 see DPO200
539-21-9 see AHI875
539-32-2 see BSJ550
539-35-5 see CMP885
539-47-9 see FPK050
539-48-0 see PEX250
539-71-9 see ELO000
539-88-8 see EFS600
539-90-2 see BSW500
540-07-8 see PBW450
540-08-9 see OFE020
540-09-0 see TJF250
540-10-3 see HCP700
540-18-1 see AOG000
540-23-8 see TGS750
540-36-3 see DKG000
540-37-4 see IEC000
540-38-5 see IEW000
540-42-1 see PMV250
540-47-6 see CQE750
540-49-8 see DDN950
540-51-2 see BNI500
540-54-5 see CKP750
540-59-0 see DFI210
540-61-4 see GHI000
540-63-6 see EEB000
540-67-0 see EMT000
540-69-2 see ANH500
540-72-7 see SIA500
540-73-8 see DSF600
540-80-7 see BRV760
540-84-1 see TLY500
540-88-5 see BPV100
541-02-6 see DAF350
541-09-3 see UPS000
541-19-5 see BJI000
541-22-0 see DAF600
541-25-3 see CLV000
541-28-6 see IHU200
541-33-3 see BRQ050
541-41-3 see EHK500
541-42-4 see IQQ000
541-47-9 see MHT500
541-53-7 see DXL800
541-58-2 see DUG200
541-59-3 see MAM750
541-64-0 see FPY000
541-66-2 see FMX000
541-69-5 see PEY750
541-73-1 see DEP599
541-85-5 see EGI750
541-91-3 see MIT625
541-95-7 see MOU500
542-18-7 see CPI400
542-46-1 see CMU850

542-54-1 see MNJ000
542-55-2 see IIR000
542-56-3 see IJD000
542-58-5 see CGO600
542-59-6 see EJI000
542-62-1 see BAK750
542-63-2 see DIV000
542-69-8 see BRQ250
542-75-6 see DGG950
542-76-7 see CKT250
542-78-9 see PMK000
542-85-8 see ELX525
542-88-1 see BIK000
542-90-5 see EPP000
542-92-7 see CPU500
543-15-7 see HBB000
543-20-4 see SNG000
543-21-5 see ACJ250
543-38-4 see AKD500
543-39-5 see MLO250
543-49-7 see HBE500
543-53-3 see PPB550
543-59-9 see PBW500
543-63-5 see BRS750
543-67-9 see PNQ750
543-80-6 see BAH500
543-81-7 see BFP000
543-82-8 see ILM000
543-86-2 see IHQ000
543-87-3 see ILW100
543-90-8 see CAD250
543-94-2 see SME000
544-13-8 see TLR500
544-16-1 see BRV500
544-17-2 see CAS250
544-25-2 see COY000
544-40-1 see BSM125
544-47-8 see CEQ800
544-62-7 see GGA915
544-63-8 see MSA250
544-76-3 see HCO600
544-85-4 see DYC900
544-92-3 see CNL000
544-97-8 see DUO200
545-06-2 see TII750
545-55-1 see TND250
545-91-5 see DWH600
545-93-7 see QCS000
546-46-3 see ZFJ250
546-48-5 see PAP110
546-68-9 see IRN200
546-71-4 see ENQ000
546-80-5 see TFW000
546-88-3 see ABB250
546-89-4 see LGO100
546-93-0 see MAC650
547-32-0 see MRM250
547-44-4 see SNQ550
547-57-9 see MRL100
547-58-0 see MND600
547-63-7 see MKX000
547-64-8 see MLC600
547-91-1 see IEP200
547-95-5 see MIG850
548-00-5 see BKA000
548-26-5 see BNK700
548-39-0 see PCJ225
548-42-5 see AEY375
548-43-6 see EAJ000
548-57-2 see SBE500
548-61-8 see THP000
548-62-9 see AOR500

548-68-5 see DHY400
548-73-2 see DYF200
548-77-6 see TKP050
548-83-4 see GAZ000
548-93-6 see AKE750
549-18-8 see EAI000
549-49-5 see QJJ100
550-24-3 see EAJ600
550-28-7 see AKL625
550-33-4 see RJF000
550-34-5 see NAH800
550-44-7 see MOC300
550-49-2 see AER300
550-53-8 see IPY500
550-70-9 see TMX775
550-74-3 see PIE000
550-82-3 see HNG500
550-90-3 see LIQ000
550-99-2 see NCW000
551-06-4 see ISN000
551-08-6 see BRQ100
551-11-1 see POC500
551-16-6 see PCU500
551-36-0 see AMM750
551-58-6 see SOW500
551-74-6 see MAW750
551-92-8 see DSV800
552-16-9 see NFG500
552-30-7 see TKV000
552-41-0 see PAC250
552-46-5 see NBF000
552-80-7 see DUC300
552-86-3 see FQI000
552-89-6 see NEU500
552-94-3 see SAN000
553-24-2 see AJQ250
553-26-4 see BGO600
553-27-5 see AOQ875
553-30-0 see DBN400
553-53-7 see NDU500
553-54-8 see LGW000
553-68-4 see IAC000
553-69-5 see PGG350
553-79-7 see NIY525
553-84-4 see PCI750
553-97-9 see MHI250
554-00-7 see DEO290
554-12-1 see MOT000
554-13-2 see LGZ000
554-14-3 see MPV000
554-18-7 see AOO800
554-35-8 see GFC100
554-68-7 see TJO050
554-70-1 see TJT775
554-76-7 see SLP600
554-84-7 see NIE600
554-92-7 see TKW750
554-99-4 see MJV000
555-06-6 see SEO500
555-15-7 see NGC000
555-16-8 see NEV500
555-21-5 see NIJ000
555-30-6 see DNA800
555-31-7 see AHC600
555-37-3 see BRA250
555-43-1 see GGU400
555-57-7 see MOS250
555-60-2 see CKA550
555-65-7 see BMM600
555-77-1 see TNF250
555-84-0 see NDY000
555-89-5 see NCM700

555-96-4 see BEQ000
556-12-7 see FPF000
556-18-3 see AIC825
556-22-9 see GII000
556-24-1 see ITC000
556-52-5 see GGW500
556-56-9 see AGI250
556-61-6 see ISE000
556-64-9 see MPT000
556-65-0 see TFF500
556-67-2 see OCE100
556-72-9 see PBJ750
556-82-1 see MHU110
556-88-7 see NHA500
556-89-8 see NMQ500
556-90-1 see IAP000
556-97-8 see CLV500
557-04-0 see MAJ030
557-05-1 see ZMS000
557-07-3 see ZJS000
557-09-5 see ZEJ000
557-11-9 see AGV000
557-17-5 see MOU830
557-18-6 see DJO100
557-19-7 see NDB500
557-20-0 see DKE600
557-21-1 see ZGA000
557-30-2 see EEA000
557-34-6 see ZBS000
557-36-8 see OFA200
557-40-4 see DBK000
557-48-2 see NMV760
557-66-4 see EFW000
557-91-5 see DDN800
557-93-7 see BOA250
557-98-2 see CKS000
557-99-3 see ACM000
558-13-4 see CBX750
558-17-8 see TLU000
558-25-8 see MDR750
558-30-5 see IIQ600
559-11-5 see DLY850
560-53-2 see SMN002
561-27-3 see HBT500
561-43-3 see IBP200
561-78-4 see NEB000
562-09-4 see CIS000
562-10-7 see PGE775
562-74-3 see TBD825
562-94-7 see EQL600
562-95-8 see TJS500
563-12-2 see EEH600
563-25-7 see DDY800
563-41-7 see SBW500
563-43-9 see EFU050
563-45-1 see MHT250
563-46-2 see MHT000
563-47-3 see CIU750
563-52-0 see CEV250
563-54-2 see DGG800
563-58-6 see DGG750
563-63-3 see SDI800
563-68-8 see TEI250
563-71-3 see FBH100
563-80-4 see MLA750
564-00-1 see DHB800
564-25-0 see DYE425
564-36-3 see EDA600
564-94-3 see FNK150
565-33-3 see AKQ250
565-59-3 see DTI200
565-74-2 see BNM100

565-80-0 see DTI600
566-09-6 see SKR500
566-28-9 see ONO000
566-48-3 see HJB200
567-47-5 see HMX000
568-69-4 see TNJ750
568-70-7 see HMF500
568-75-2 see HMF000
568-81-0 see DQK200
569-34-6 see MEF800
569-57-3 see CLO750
569-58-4 see AGW750
569-59-5 see PDD000
569-61-9 see RMK020
569-64-2 see AFG500
569-65-3 see HGC500
569-77-7 see TDD500
570-22-9 see IAM000
571-20-0 see AOO405
571-22-2 see HJB100
571-60-8 see NAN500
572-48-5 see DXO000
573-20-6 see VTA100
573-56-8 see DVA200
573-58-0 see SGQ500
573-83-1 see PLQ775
574-25-4 see TFJ825
574-66-3 see BCS400
575-38-2 see NAN508
575-44-0 see NAN507
576-08-9 see LAR100
576-24-9 see DFX500
576-26-1 see XLA000
576-42-1 see PLJ350
576-55-6 see TBJ500
576-68-1 see MAW500
577-11-7 see DJL000
577-19-5 see NFQ080
577-55-9 see DNN800
577-59-3 see NEL450
577-66-2 see EEO000
577-71-9 see DVA400
577-85-5 see HLC600
577-91-3 see PDM750
578-32-5 see DRK800
578-46-1 see NMP550
578-54-1 see EGK500
578-57-4 see BMT400
578-66-5 see AML250
578-94-9 see PDB000
578-95-0 see ADI825
579-07-7 see PGA500
579-10-2 see MFW000
579-56-6 see VGA300
579-66-8 see DIS650
580-17-6 see AMK725
580-22-3 see AMK700
580-48-3 see CDQ325
580-74-5 see DVU000
581-08-8 see ABG350
581-12-4 see NCQ900
581-28-2 see AHS000
581-29-3 see ADJ375
581-64-6 see AKK750
581-88-4 see IKB000
581-89-5 see NHQ500
582-08-1 see ASN750
582-17-2 see NAO500
582-22-9 see PGB760
582-25-2 see PKW760
582-52-5 see DVO100
582-60-5 see DQM100

582-61-6 see BDL750
583-03-9 see BQJ500
583-15-3 see MCX500
583-39-1 see BCC500
583-52-8 see PLN300
583-57-3 see DRF800
583-58-4 see LJB000
583-59-5 see MIR000
583-60-8 see MIR500
583-63-1 see BDC250
583-68-6 see BOG500
583-69-7 see BNL260
583-75-5 see MHR500
583-78-8 see DFX850
583-80-2 see MQI000
583-91-5 see HMR600
584-02-1 see IHP010
584-03-2 see BOS250
584-08-7 see PLA000
584-09-8 see RPB200
584-26-9 see ADC750
584-42-9 see SIT850
584-48-5 see DVB820
584-79-2 see AFR250
584-84-9 see TGM750
584-93-0 see BOL250
584-94-1 see DSE509
585-08-0 see PCX000
585-54-6 see AQZ900
585-79-5 see NFQ090
585-88-6 see MAO285
586-06-1 see DMV800
586-11-8 see DVA600
586-38-9 see AOU500
586-62-9 see TBE000
586-76-5 see BMU100
586-77-6 see BNF250
586-78-7 see NFQ100
586-92-5 see POQ500
586-95-8 see POR810
586-98-1 see POR800
587-02-0 see EOH100
587-15-5 see DGV200
587-63-3 see DLR000
587-65-5 see PER300
587-84-8 see DVY000
587-85-9 see DWD800
587-98-4 see MDM775
588-07-8 see CDZ050
588-16-9 see AAQ750
588-22-7 see DFY500
588-42-1 see TJL250
588-59-0 see SLR000
589-16-2 see EGL000
589-18-4 see MHB250
589-32-2 see TCD300
589-38-8 see HEV500
589-41-3 see HKQ025
589-43-5 see DSE600
589-59-3 see ITA000
589-79-7 see EKJ500
589-82-2 see HBF000
589-90-2 see DRG200
589-92-4 see MIR625
589-93-5 see LJA000
589-98-0 see OCY100
590-00-1 see PLS750
590-01-2 see BSJ500
590-12-5 see DDN820
590-14-7 see BOA000
590-17-0 see BMS100
590-19-2 see BOP250

590-21-6 see PMR750
590-28-3 see PLC250
590-29-4 see PLG750
590-46-5 see CCH850
590-63-6 see HOA500
590-69-2 see DDU700
590-86-3 see MHX500
590-88-5 see BOR750
590-92-1 see BOB250
590-96-5 see HMG000
591-08-2 see ADD250
591-09-3 see ACS750
591-10-6 see BJJ500
591-11-7 see MKH500
591-12-8 see AOO750
591-12-8 see MKH250
591-17-3 see BOG300
591-21-9 see DRG000
591-23-1 see MIQ750
591-24-2 see MIR600
591-27-5 see ALS990
591-31-1 see AOT300
591-33-3 see ABG250
591-35-5 see DFY450
591-47-9 see MIR750
591-50-4 see IEC500
591-51-5 see PFL500
591-60-6 see BPV250
591-62-8 see EHA100
591-76-4 see MKL250
591-78-6 see HEV000
591-80-0 see PBQ750
591-87-7 see AFU750
591-89-9 see PLU500
591-97-9 see CEU825
592-01-8 see CAQ500
592-04-1 see MDA250
592-05-2 see LCU000
592-31-4 see BSS250
592-35-8 see BQP250
592-41-6 see HFB000
592-42-7 see HCR500
592-45-0 see HCR000
592-56-3 see AAG100
592-57-4 see CPA500
592-62-1 see MGS750
592-65-4 see IJO000
592-76-7 see HBJ000
592-79-0 see FFV000
592-82-5 see ISD100
592-84-7 see BRK000
592-85-8 see MCU250
592-87-0 see LEC500
592-88-1 see AGS250
593-12-4 see FKS000
593-14-6 see FKT000
593-29-3 see PLS775
593-53-3 see FJK000
593-54-4 see MOB000
593-56-6 see MNG500
593-57-7 see DQG600
593-59-9 see EGM150
593-60-2 see VMP000
593-63-5 see CEC500
593-68-0 see EON000
593-70-4 see CHI900
593-74-8 see DSM450
593-75-9 see MKX500
593-77-1 see MKQ875
593-78-2 see MKT000
593-79-3 see DUB200
593-81-7 see TLF750

593-82-8 see DSG000
593-84-0 see TFF250
593-85-1 see GKW100
593-88-4 see TLH150
593-89-5 see DFP200
593-90-8 see TLM775
593-91-9 see TLM750
593-92-0 see VPF200
594-09-2 see TMD275
594-10-5 see TLH100
594-19-4 see BRR750
594-20-7 see DGF900
594-27-4 see TDV750
594-31-0 see DGO800
594-34-3 see DDQ150
594-36-5 see CII250
594-42-3 see PCF300
594-44-5 see EEC000
594-65-0 see TII000
594-70-7 see NFQ500
594-71-8 see CJE250
594-72-9 see DFU000
594-84-3 see DFE100
595-21-1 see SMN275
595-33-5 see VTF000
595-48-2 see MPN250
595-57-3 see CCP750
596-03-2 see DDO200
596-11-2 see EDH000
596-27-0 see CNX400
596-43-0 see TNP600
596-51-0 see GIC000
597-25-1 see DST800
597-49-9 see TJP550
597-63-7 see TCE750
597-64-8 see TCF750
597-71-7 see NNR400
597-88-6 see DJT000
597-96-6 see MKL350
598-02-7 see DJW500
598-14-1 see DFH200
598-16-3 see THV100
598-26-5 see DSL289
598-31-2 see BNZ000
598-32-3 see MQL250
598-38-9 see DFG000
598-49-2 see CDY825
598-50-5 see MQJ000
598-52-7 see MPW600
598-55-0 see MHZ000
598-57-2 see NHN500
598-58-3 see MMF500
598-63-0 see LCP000
598-64-1 see DRQ600
598-72-1 see BOB000
598-73-2 see BOJ000
598-74-3 see AOE200
598-77-6 see TJB250
598-78-7 see CKS750
598-79-8 see CEE500
598-92-5 see CJC800
598-94-7 see DUM150
598-99-2 see MQC150
599-52-0 see HCE000
599-61-1 see SOA000
599-64-4 see COF400
599-69-9 see DUH000
599-71-3 see BDW100
599-79-1 see PPN750
600-05-5 see DDS500
600-14-6 see PBL350
600-25-9 see CJE000

624-72-6 see DKH300
624-74-8 see DNE500
624-76-0 see IEL000
624-79-3 see ELT000
624-83-9 see MKX250
624-84-0 see FNN000
624-85-1 see ELM500
624-90-8 see MGR750
624-91-9 see MMF750
624-92-0 see DRQ400
625-17-2 see DEC200
625-22-9 see DEC000
625-27-4 see MNK100
625-28-5 see ITD000
625-33-2 see PBR500
625-45-6 see MDW275
625-48-9 see NFY550
625-49-0 see NHY250
625-50-3 see EFM000
625-52-5 see EQD875
625-53-6 see EPR600
625-55-8 see IPC000
625-58-1 see ENM500
625-69-4 see PBL000
625-76-3 see DUW507
625-84-3 see DTV300
625-86-5 see DSC000
625-95-6 see IFG100
626-01-7 see IEB000
626-02-8 see IEV010
626-16-4 see DGP200
626-17-5 see PHX550
626-19-7 see IMI000
626-35-7 see ENN100
626-38-0 see AOD735
626-48-2 see MQI750
626-55-1 see BOC510
626-56-2 see MOH000
626-58-4 see MOH250
626-60-8 see CKW250
626-64-2 see HOB100
626-67-5 see MOG500
626-82-4 see BRK900
626-86-8 see ELC600
626-93-7 see HEU600
627-03-2 see EEK500
627-06-5 see PNX550
627-11-2 see CGU199
627-12-3 see PNG250
627-13-4 see PNQ500
627-18-9 see BNY750
627-19-0 see PCA250
627-21-4 see PCA300
627-22-5 see CET250
627-26-9 see COQ750
627-30-5 see CKP600
627-31-6 see TLR050
627-41-8 see MFN250
627-44-1 see DJO400
627-45-2 see EKK600
627-49-6 see DJW000
627-53-2 see DJY700
627-54-3 see DKB150
627-63-4 see FOY000
627-72-5 see DGP000
627-75-8 see AQV500
627-92-9 see IHT000
627-93-0 see DOQ300
628-02-4 see HEM500
628-13-7 see POR750
628-16-0 see HCV850
628-17-1 see IET500

628-20-6 see CEU300
628-21-7 see TDQ400
628-28-4 see BRU780
628-32-0 see EPC125
628-33-1 see EFJ000
628-36-4 see DKJ600
628-37-5 see DJU400
628-39-7 see EIN550
628-41-1 see CPA750
628-46-6 see IKN400
628-52-4 see CMJ000
628-63-7 see AOD725
628-68-2 see DJD750
628-73-9 see HER500
628-76-2 see DFX200
628-77-3 see PBH125
628-81-9 see EHA500
628-83-1 see BSN500
628-85-3 see DWU000
628-86-4 see MDC000
628-87-5 see IBB100
628-89-7 see DKN000
628-91-1 see DWW500
628-92-2 see COY250
628-94-4 see AEN000
628-96-6 see EJG000
629-03-8 see DDO800
629-04-9 see HBN100
629-05-0 see OGI025
629-09-4 see HEG200
629-11-8 see HEP500
629-13-0 see DCL600
629-14-1 see EJE500
629-15-2 see EJF000
629-17-4 see EJC035
629-20-9 see CPS500
629-25-4 see LBN000
629-27-6 see OFA100
629-31-2 see EAO200
629-33-4 see HFQ100
629-35-6 see DEE000
629-38-9 see MIE750
629-40-3 see OCW050
629-50-5 see TJH500
629-59-4 see TBX750
629-60-7 see TJH750
629-62-9 see PAY750
629-64-1 see HBO000
629-70-9 see HCP100
629-78-7 see HAS100
629-82-3 see OEY000
630-08-0 see CBW750
630-10-4 see SBV000
630-16-0 see TBJ560
630-18-2 see PJA750
630-20-6 see TBQ000
630-56-8 see HNT500
630-60-4 see OKS000
630-64-8 see HAN800
630-72-8 see TMK250
630-76-2 see TEA750
630-93-3 see DNU000
631-06-1 see DVP400
631-07-2 see EOL000
631-27-6 see GHR609
631-40-3 see TED250
631-41-4 see TCB500
631-60-7 see MDE250
631-61-8 see ANA000
631-64-1 see DDJ100
631-67-4 see DUG450
632-14-4 see TMJ250

632-21-3 see TBN300
632-22-4 see TDX250
632-58-6 see TBT100
632-69-9 see RMP175
632-79-1 see TBJ700
632-99-5 see MAC250
633-03-4 see BAY750
633-59-0 see CLX250
633-65-8 see BFN600
633-96-5 see CMM220
634-19-5 see TCR300
634-60-6 see AML600
634-66-2 see TBN740
634-90-2 see TBN745
634-93-5 see TIJ750
634-95-7 see DKD650
635-22-3 see CJA185
635-39-2 see GLK000
635-41-6 see TKX250
635-65-4 see HAO900
635-85-8 see DOI400
635-93-8 see CLD800
636-04-4 see PGL250
636-09-9 see DKB160
636-21-5 see TGS500
636-23-7 see DCE000
636-26-0 see TFQ650
636-47-5 see SLI300
636-53-3 see IMK100
636-97-5 see NFH000
637-01-4 see TDT250
637-03-6 see PEG750
637-07-0 see ARQ750
637-12-7 see AHH825
637-23-0 see TNS200
637-32-1 see CKB500
637-39-8 see NEI700
637-53-6 see TFA250
637-56-9 see PDL750
637-60-5 see MNU500
637-61-6 see CHQ500
637-92-3 see EHA550
638-03-9 see TGS250
638-07-3 see EHH100
638-10-8 see MIP800
638-16-4 see THS250
638-21-1 see PFV250
638-23-3 see CBR675
638-29-9 see VBA000
638-38-0 see MAQ000
638-49-3 see AOJ500
638-53-9 see TJI250
638-65-3 see SLL500
639-14-5 see GMG000
639-58-7 see CLU000
640-15-3 see PHI500
640-19-7 see FFF000
640-68-6 see VBU000
641-13-4 see DCY900
641-16-7 see TDY600
641-38-3 see AGW476
641-81-6 see AQO500
642-15-9 see DUO350
642-44-4 see AFW500
642-65-9 see DBF200
642-72-8 see BCD750
642-78-4 see SLJ050
643-22-1 see EDJ500
643-28-7 see INW100
643-79-8 see PHV500
643-84-5 see MAO600
644-06-4 see AEX850

644-08-6 see MJH905
644-26-8 see AOM000
644-31-5 see ACC250
644-35-9 see PNS250
644-62-2 see DGM875
644-64-4 see DQZ000
644-97-3 see DGE400
645-05-6 see HEJ500
645-08-9 see HJC000
645-12-5 see NGH500
645-15-8 see BLA600
645-35-2 see HGE750
645-43-2 see GKU000
645-48-7 see PGM750
645-55-6 see NEO510
645-56-7 see PNS500
645-59-0 see HHP100
645-62-5 see EKR000
645-88-5 see ALQ650
646-04-8 see PBQ250
646-06-0 see DVR800
646-07-1 see IKT000
646-13-9 see IJN100
646-20-8 see HBD000
646-25-3 see DAG650
650-42-0 see PMD350
650-51-1 see TII500
650-52-2 see BLO280
651-48-9 see SNV100
652-04-0 see MHF250
652-67-5 see HID350
654-42-2 see DSG700
655-35-6 see CBR500
655-86-7 see PDB600
657-24-9 see DQR600
657-27-2 see LJO000
657-84-1 see SKK000
659-70-1 see ITB000
659-86-9 see IEZ800
660-27-5 see DNM400
660-68-4 see DIS500
661-11-0 see FIX000
661-18-7 see FFX000
661-54-1 see TKH030
661-69-8 see HEE500
661-95-0 see DDO450
662-01-1 see DFM110
662-50-0 see HAX000
663-25-2 see HAY200
663-97-8 see UWJ200
664-95-9 see CPR000
665-66-7 see AED250
667-29-8 see TJX780
667-49-2 see TJX825
668-37-1 see DHY200
669-49-8 see XSS900
670-40-6 see POH500
670-54-2 see EEE500
671-04-5 see CGI500
671-16-9 see PME250
671-35-2 see FMO100
671-51-2 see ASH750
672-06-0 see MIA500
672-66-2 see DTN896
672-76-4 see TFV750
673-04-1 see BJP250
673-06-3 see PEC500
673-19-8 see BSG250
673-22-3 see HLR600
673-31-4 see PGA750
674-81-7 see NKH000
674-82-8 see KFA000

814-29-9 see TNE750	838-85-7 see PHE300	873-58-5 see SIN650	922-68-9 see MKI550
814-49-3 see DIY000	838-88-0 see MJO250	873-74-5 see COK125	922-89-4 see DBG900
814-67-5 see PML300	838-95-9 see DUC000	873-83-6 see AMW250	923-01-3 see COJ800
814-68-6 see ADZ000	841-06-5 see INQ000	874-21-5 see SJS500	923-26-2 see HNV500
814-78-8 see MKY500	841-18-9 see SMR500	874-60-2 see TGU020	923-34-2 see TJR750
814-80-2 see CAT600	841-67-8 see TEH520	874-66-8 see FPX050	924-16-3 see BRY500
814-82-4 see MDD500	841-73-6 see BQW825	875-22-9 see RFK000	924-42-5 see HLU500
814-91-5 see CNN755	842-00-2 see EPI300	876-83-5 see MKV500	924-43-6 see BRV750
814-94-8 see TGE250	842-07-9 see PEJ500	877-22-5 see HJB500	924-46-9 see MNA000
815-58-7 see TIJ500	843-23-2 see SMT500	877-24-7 see HIB600	925-15-5 see DWV800
815-84-9 see LEA000	843-34-5 see HIK000	877-31-6 see NJV000	925-83-7 see SBK000
816-43-3 see LHE475	846-35-5 see BCG000	877-66-7 see TGO550	926-06-7 see IPY000
816-57-9 see NLO500	846-49-1 see CFC250	879-39-0 see TBS000	926-53-4 see PNH650
816-80-8 see TMD650	846-50-4 see CFY750	879-72-1 see DPU400	926-56-7 see MNI000
817-09-4 see TNF500	848-21-5 see NNR125	880-52-4 see AEE100	926-57-8 see DEU650
817-87-8 see EGN000	848-53-3 see HGI100	881-03-8 see MMN800	926-64-7 see DOS200
817-99-2 see CBK000	848-75-9 see MLD100	881-99-2 see BLL825	926-65-8 see IRS000
818-08-6 see DEF400	849-55-8 see DNU200	882-09-7 see CMX000	926-82-9 see DSD400
818-23-5 see HBO790	849-99-0 see DGT500	882-33-7 see PEW250	926-93-2 see MLJ500
818-61-1 see ADV250	850-52-2 see AGW675	882-35-9 see EIT100	927-07-1 see BSD250
818-72-4 see OGI050	850-57-7 see TEN100	883-40-9 see DVZ100	927-49-1 see ULA000
819-17-0 see CAT700	851-68-3 see MCI500	886-06-6 see PIV500	927-62-8 see BRB300
819-35-2 see NKX300	853-23-6 see HJB250	886-50-0 see BQC750	927-68-4 see BNI600
820-54-2 see MJL000	853-34-9 see KGK000	886-74-8 see CJQ250	927-73-1 see CEV300
820-75-7 see DCO800	853-35-0 see DLJ700	886-86-2 see EFX500	927-80-0 see EEL000
821-06-7 see DDL600	855-19-6 see CLG900	891-33-8 see TOE150	928-04-1 see ACJ500
821-08-9 see HCU500	855-22-1 see DME525	892-17-1 see MJE500	928-45-0 see BRV325
821-09-0 see PBR000	857-95-4 see BHY500	892-20-6 see TMV775	928-51-8 see CEU500
821-10-3 see DEV400	859-18-7 see LGE000	892-21-7 see NGA700	928-55-2 see EOY500
821-14-7 see ASN250	860-22-0 see FAE100	894-09-7 see IFO000	928-65-4 see HFX500
821-48-7 see BHO250	860-39-9 see BLA800	894-52-0 see DLI400	928-70-1 see PLK580
821-55-6 see NMY500	860-79-7 see CQH325	894-56-4 see CDR500	928-72-3 see DXE200
822-06-0 see DNJ800	863-61-6 see VTA650	897-15-4 see DPY200	928-95-0 see HFD500
822-12-8 see SIN900	865-04-3 see MEK700	897-55-2 see DQD000	928-96-1 see HFE000
822-16-2 see SJV500	865-21-4 see VKZ000	900-95-8 see ABX250	929-06-6 see AJU250
822-36-6 see MKU000	865-28-1 see TOF825	902-83-0 see DHW200	929-37-3 see DJG400
822-38-8 see EJQ100	865-33-8 see PLK850	903-46-8 see DLJ730	929-73-7 see DXW060
822-87-7 see CFG250	865-37-2 see DOQ750	905-30-6 see PPU780	930-22-3 see EBJ500
823-40-5 see TGM100	865-47-4 see PKY750	908-35-0 see MJJ000	930-37-0 see GGW600
823-87-0 see SIW550	866-55-7 see DEZ000	909-39-7 see IDF000	930-55-2 see NLP500
824-11-3 see TNI750	866-82-0 see CNK625	910-06-5 see TMV000	930-68-7 see CPD250
824-72-6 see PFW100	866-84-2 see PLB750	911-45-5 see CMX500	930-73-4 see MOZ000
824-78-2 see SIV600	866-87-5 see TEF250	912-60-7 see NOA500	931-97-5 see HKA000
824-79-3 see SKJ500	866-97-7 see TEA250	915-67-3 see FAG020	932-52-5 see AMW100
825-51-4 see DAF000	867-13-0 see EIC000	917-54-4 see MLD000	932-64-9 see NMP620
826-10-8 see MDQ500	867-27-6 see DAO800	917-57-7 see VPU000	932-83-2 see NKI000
826-36-8 see TDT770	867-36-7 see DFF400	917-58-8 see PLE575	933-48-2 see TLO510
826-39-1 see MQR500	867-55-0 see LHL000	917-61-3 see COI250	933-52-8 see TDO250
826-62-0 see BFY000	868-14-4 see PKU600	917-64-6 see MLE250	933-75-5 see TIV500
826-81-3 see HMQ600	868-18-8 see BLC000	917-69-1 see CNA300	934-00-9 see PPQ550
827-52-1 see PER750	868-59-7 see EHU600	917-70-4 see LAW000	934-32-7 see AIG000
827-61-2 see AAE250	868-77-9 see EJH000	918-00-3 see TJB775	934-73-6 see CKH000
827-94-1 see DDQ450	868-85-9 see DSG600	918-04-7 see AAH500	935-02-4 see PGD100
828-00-2 see ABC250	869-01-2 see BSA250	918-37-6 see HET675	935-92-2 see POG400
828-26-2 see HEL000	869-24-9 see CLQ250	918-52-5 see DVD200	935-95-5 see TBS750
829-65-2 see TGV250	869-29-4 see ADR250	918-54-7 see TMM000	936-02-7 see HJL100
830-03-5 see ABS750	869-50-1 see CIL900	918-85-4 see MNL250	936-05-0 see MMN500
830-81-9 see NAU600	870-08-6 see DVL400	918-99-0 see BLR625	936-49-2 see PFJ300
830-96-6 see ICS200	870-24-6 see CGP250	919-16-4 see TKU500	936-52-7 see CPY800
831-52-7 see PIC500	870-62-2 see HEB000	919-28-8 see ABX150	937-14-4 see CJI750
831-61-8 see EKM100	870-72-4 see SHI500	919-30-2 see TJN000	937-25-7 see FKF800
831-71-0 see NGC400	871-22-7 see DDT400	919-31-3 see CON250	937-40-6 see MHP250
832-06-4 see BPI300	871-25-0 see AJY500	919-54-0 see DIX000	938-71-6 see CJC625
832-69-9 see MNN400	871-31-8 see EGM500	919-76-6 see AHO750	938-73-8 see EEM000
832-92-8 see MDI750	871-58-9 see PKY850	919-86-8 see DAP400	938-81-8 see NJI700
834-12-8 see MPT500	871-67-0 see HEF400	920-37-6 see CEE750	939-23-1 see PGE750
834-24-2 see SLQ900	872-35-5 see IAO000	920-46-7 see MDN899	939-27-5 see ENL862
834-28-6 see PDF250	872-36-6 see VOK000	920-66-1 see HDC500	939-48-0 see IOD000
835-31-4 see NAH500	872-50-4 see MPF200	921-03-9 see TII550	940-14-7 see VPZ333
835-64-3 see HNI000	872-85-5 see IKZ100	921-09-5 see TBO760	940-31-8 see PDV725
836-30-6 see NFY000	873-32-5 see CEM000	922-63-4 see EFS700	940-41-0 see PFE500
838-57-3 see ENO100	873-51-8 see DGB875	922-67-8 see MOS875	940-69-2 see DXN709

1116-01-4 see PHQ250
1116-54-7 see NKM000
1116-76-3 see DVL000
1117-37-9 see EIF100
1117-41-5 see TMJ100
1117-55-1 see HFS600
1117-71-1 see MHR790
1117-94-8 see DGQ600
1117-99-3 see TIH825
1118-12-3 see BSS310
1118-14-5 see TMI000
1118-27-0 see LGC000
1118-39-4 see AAW500
1118-42-9 see DVV200
1118-46-3 see BSR250
1118-58-7 see MNH750
1118-92-9 see DTE200
1119-16-0 see MQJ500
1119-22-8 see DXR500
1119-34-2 see AQW000
1119-68-2 see PBX250
1119-69-3 see DGS200
1119-85-3 see HFC000
1119-94-4 see DYA810
1119-97-7 see TCB200
1120-01-0 see HCP900
1120-04-3 see OBG100
1120-10-1 see CJH500
1120-21-4 see UJS000
1120-46-3 see LDQ000
1120-48-5 see DVJ600
1120-71-4 see PML400
1121-30-8 see HOB500
1121-31-9 see MCQ700
1121-47-7 see FPR000
1121-53-5 see BFJ850
1121-70-6 see SIM100
1121-92-2 see SMV500
1122-10-7 see DDP400
1122-17-4 see DFN700
1122-54-9 see ADA365
1122-56-1 see CPB050
1122-58-3 see DQB600
1122-60-7 see NFV500
1122-62-9 see PPH200
1122-82-3 see ISJ000
1122-90-3 see ARJ755
1122-91-4 see BMT700
1123-54-2 see AIB340
1123-61-1 see DFP800
1123-85-9 see HGR600
1123-91-7 see MHI300
1124-11-4 see TDU800
1124-31-8 see PLM700
1124-33-0 see NJA500
1125-27-5 see DFJ800
1125-78-6 see TCY000
1125-88-8 see DOG700
1126-34-7 see AIF250
1126-48-3 see HLN800
1126-78-9 see BQH850
1126-79-0 see BSF750
1127-76-0 see ENL860
1128-05-8 see MPQ900
1128-08-1 see DLQ600
1128-67-2 see MHJ250
1129-41-5 see MIB750
1129-50-6 see BSV500
1129-88-0 see REA200
1130-51-4 see DJN430
1130-69-4 see NLQ500

1131-18-6 see PFQ350
1132-20-3 see CEC100
1132-39-4 see PGH250
1133-64-8 see NJK150
1134-23-2 see EHT500
1134-35-6 see DQS100
1134-47-0 see BAC275
1135-99-5 see DWO400
1136-45-4 see MNV000
1136-84-1 see TCY275
1137-41-3 see AIR250
1137-42-4 see HJN100
1137-79-7 see BGF899
1138-80-3 see CBR125
1139-30-6 see CCN100
1141-37-3 see BHP125
1141-88-4 see DXJ800
1142-70-7 see BOR000
1143-38-0 see DMG900
1145-73-9 see DUB800
1145-90-0 see CJN800
1146-95-8 see DWF200
1147-55-3 see HLE000
1149-99-1 see LIO600
1154-59-2 see TBV000
1155-03-9 see ZUA000
1155-38-0 see MGZ000
1155-49-3 see PNB250
1156-19-0 see TGJ500
1160-36-7 see TKP850
1161-13-3 see CBR220
1162-06-7 see TMT000
1162-65-8 see AEU250
1163-19-5 see PAU500
1163-36-6 see CMV400
1164-33-6 see HEL500
1164-38-1 see ELF500
1165-39-5 see AEV000
1165-48-6 see DNV000
1166-52-5 see DXX200
1169-26-2 see HEI000
1170-02-1 see EIV100
1172-18-5 see DAB800
1172-82-3 see HMQ000
1173-26-8 see CNS650
1173-88-2 see MNV250
1174-83-0 see PHN250
1176-08-5 see DTO600
1176-75-6 see PEU545
1177-87-3 see DBC400
1178-29-6 see MQR200
1178-99-0 see MFF625
1179-69-7 see TEZ000
1181-54-0 see MND500
1182-87-2 see EAQ050
1184-53-8 see MIP250
1184-57-2 see MLG000
1184-58-3 see DOQ700
1184-64-1 see CBW200
1184-78-7 see TLE100
1184-85-6 see MLH760
1184-88-9 see SGM600
1185-55-3 see MQF500
1185-57-5 see FAS700
1186-09-0 see TJU800
1187-00-4 see BKM500
1187-33-3 see DEH300
1187-46-8 see EGV450
1187-58-2 see MOS900
1187-59-3 see MGA300
1187-93-5 see TKG800
1189-71-5 see CLG250

1189-85-1 see BQV000
1190-16-5 see COR325
1190-53-0 see BOM750
1190-53-0 see BQL000
1190-76-7 see BOX600
1190-93-8 see ACQ250
1191-15-7 see DNI600
1191-16-8 see DOQ350
1191-50-0 see SIO000
1191-79-3 see BAI800
1191-80-6 see MDF250
1191-96-4 see EHU500
1192-22-9 see ECC600
1192-27-4 see DCP200
1192-28-5 see CPW750
1192-30-9 see TCS550
1192-62-7 see ACM200
1192-75-2 see BJP450
1192-89-8 see PFM200
1193-02-8 see AIF750
1193-24-4 see PPP100
1193-54-0 see DFN800
1193-81-3 see CPL125
1194-02-1 see FGH100
1194-65-6 see DER800
1195-08-0 see FOJ060
1195-16-0 see TCZ000
1195-67-1 see ASK500
1195-69-3 see DOV600
1195-79-5 see TLW250
1196-57-2 see QSJ100
1197-16-6 see CCI550
1197-18-8 see AJV500
1197-40-6 see FPX028
1197-55-3 see AID700
1198-27-2 see ALK000
1198-55-6 see TBU500
1198-77-2 see SKO575
1198-97-6 see PGF800
1199-03-7 see QSJ000
1199-18-4 see HKF875
1199-85-5 see CIF250
1200-00-6 see ILB000
1200-89-1 see TCU650
1204-06-4 see ICO000
1204-59-7 see AIW250
1204-78-0 see MGF750
1204-79-1 see AIW000
1208-52-2 see MJP750
1210-05-5 see DVV800
1210-35-1 see DDA100
1210-56-6 see MLJ100
1211-28-5 see PNS000
1211-40-1 see ALM000
1212-29-9 see DGV800
1214-39-7 see BDX090
1215-16-3 see BHO500
1215-83-4 see CMW500
1218-34-4 see ADE075
1218-35-5 see OKO500
1219-20-1 see OOE100
1220-83-3 see SNL800
1220-94-6 see AKP250
1221-56-3 see SKM000
1222-05-5 see GBU000
1222-57-7 see ZUA200
1222-98-6 see NFS505
1223-31-0 see BLA750
1223-36-5 see CJN750
1224-64-2 see NIM000
1224-78-8 see AFQ700
1225-20-3 see IGC100

1225-55-4 see POF250
1225-60-1 see AEG625
1225-65-6 see DYB800
1227-61-8 see DIB600
1227-61-8 see MCH250
1229-29-4 see AEG750
1229-35-2 see MDT500
1229-55-6 see CMS238
1230-33-7 see DON700
1231-93-2 see EQJ100
1233-89-2 see BHV500
1235-13-8 see NNX300
1235-82-1 see BGD750
1238-54-6 see HIE525
1239-04-9 see PMD325
1239-29-8 see AOO300
1239-31-2 see HJB225
1239-45-8 see DBV400
1241-94-7 see DWB800
1244-76-4 see VSF000
1246-87-3 see DND400
1248-18-6 see NAP100
1249-84-9 see ARX800
1250-95-9 see EBM000
1252-18-2 see BFC400
1253-28-7 see GEK510
1257-78-9 see PME700
1258-84-0 see PMD525
1263-89-4 see APP500
1263-94-1 see RDK050
1264-51-3 see ACH750
1264-72-8 see PKD300
1270-21-9 see PLK860
1271-19-8 see DGW200
1271-28-9 see NDA500
1271-29-0 see TGH500
1271-54-1 see BGY700
1271-55-2 see ABA750
1277-43-6 see BIR529
1291-32-3 see ZTK400
1300-14-7 see AOQ250
1300-21-6 see DFF800
1300-32-9 see FQJ050
1300-71-6 see XKA000
1300-72-7 see XJJ010
1300-73-8 see XMA000
1301-14-0 see DEE600
1302-52-9 see BFO500
1302-59-6 see THH000
1302-74-5 see EAL100
1302-76-7 see AHF500
1302-78-9 see BAV750
1303-00-0 see GBK000
1303-11-3 see ICG100
1303-13-5 see NDE030
1303-18-0 see ARJ750
1303-28-2 see ARH500
1303-33-9 see ARI000
1303-36-2 see ARH800
1303-39-5 see ZDJ000
1303-86-2 see BMG000
1303-96-4 see SFE500
1303-96-4 see SFF000
1304-28-5 see BAO000
1304-29-6 see BAO250
1304-56-9 see BFT250
1304-76-3 see BKW600
1304-82-1 see BKY000
1304-85-4 see BKW100
1305-62-0 see CAT225
1305-78-8 see CAU500
1305-79-9 see CAV500

1305-99-3 see CAW250
1306-05-4 see FDH000
1306-06-5 see HJE100
1306-19-0 see CAH500
1306-23-6 see CAJ750
1306-25-8 see CAJ800
1306-38-3 see CCY000
1307-82-0 see SJJ500
1307-86-4 see CNC233
1307-96-6 see CND125
1308-04-9 see CND825
1308-06-1 see CND020
1308-14-1 see CMH260
1308-31-2 see CMI500
1308-38-9 see CMJ900
1309-32-6 see COE000
1309-37-1 see IHC450
1309-37-1 see IHF000
1309-38-2 see IHC550
1309-42-8 see MAG750
1309-48-4 see MAH500
1309-60-0 see LCX000
1309-64-4 see AQF000
1310-53-8 see GDS000
1310-53-8 see GEC000
1310-58-3 see PLJ500
1310-65-2 see LHI100
1310-73-2 see SHS000
1310-73-2 see SHS500
1310-82-3 see RPZ000
1312-03-4 see MDG000
1312-43-2 see ICI100
1312-73-8 see PLT250
1312-81-8 see LBA100
1313-13-9 see MAS000
1313-27-5 see MRE000
1313-30-0 see SIM200
1313-60-6 see SJC500
1313-82-2 see SJY500
1313-85-5 see SJT000
1313-96-8 see NEA050
1313-97-9 see NCC000
1313-99-1 see NDF500
1314-04-1 see MQU510
1314-05-2 see NDF400
1314-06-3 see NDH500
1314-12-1 see TEL040
1314-13-2 see ZKA000
1314-18-7 see SMK500
1314-20-1 see TFT750
1314-22-3 see ZLJ000
1314-24-5 see PHT500
1314-32-5 see TEL050
1314-34-7 see VEA000
1314-35-8 see TOC750
1314-36-9 see YGA000
1314-41-6 see LDS000
1314-56-3 see PHS250
1314-60-9 see AQF750
1314-61-0 see TAF500
1314-62-1 see VDU000
1314-62-1 see VDZ000
1314-80-3 see PHS000
1314-84-7 see ZLS000
1314-85-8 see PHS500
1314-86-9 see TEC000
1314-87-0 see LDZ000
1314-96-1 see SMM000
1314-98-3 see ZNJ100
1315-04-4 see AQF500
1317-25-5 see AHA135
1317-33-5 see MRD750

1317-34-6 see MAT500
1317-35-7 see MAV550
1317-36-8 see LDN000
1317-37-9 see IHN000
1317-38-0 see CNO250
1317-39-1 see CNO000
1317-40-4 see CNQ000
1317-42-6 see CNE200
1317-43-7 see NBT000
1317-60-8 see HAO875
1317-63-1 see LFW000
1317-65-3 see CAO000
1317-70-0 see OBU100
1317-80-2 see RSP100
1317-86-8 see SLQ100
1317-95-9 see TMX500
1317-98-2 see VAD100
1318-02-1 see ZAT300
1318-16-7 see BAR900
1319-43-3 see BFP755
1319-69-3 see PLG810
1319-77-3 see CNW500
1319-80-8 see OAH100
1320-37-2 see DGL600
1320-67-8 see MJW750
1320-94-1 see TCQ250
1321-12-6 see NMO510
1321-16-0 see TCJ100
1321-31-9 see EEL100
1321-38-6 see DNK200
1321-64-8 see PAW750
1321-65-9 see TIT500
1321-67-1 see NAW000
1321-74-0 see DXQ740
1321-94-4 see MMB500
1322-38-9 see THV800
1322-48-1 see BEF750
1322-78-7 see PGP750
1322-98-1 see DAJ000
1323-39-3 see SLL000
1324-28-3 see DLO900
1324-54-5 see VGP100
1324-55-6 see DFN450
1326-22-3 see DCE800
1326-82-5 see CMS250
1327-41-9 see AGZ998
1327-53-3 see ARI750
1327-85-1 see CMS257
1328-53-6 see PJQ100
1329-86-8 see THV000
1330-20-7 see XGS000
1330-38-7 see COF420
1330-43-4 see DXG035
1330-61-6 see IKL000
1330-78-5 see TNP500
1331-07-3 see CKQ250
1331-09-5 see MJH750
1331-11-9 see EFH000
1331-12-0 see PNL100
1331-17-5 see PNK000
1331-22-2 see MIR250
1331-24-4 see IQM000
1331-28-8 see CLE600
1331-29-9 see EHY500
1331-31-3 see EHH500
1331-41-5 see HKU500
1331-43-7 see DIY200
1331-45-9 see EHQ000
1331-50-6 see TLV250
1332-21-4 see ARM250
1332-29-2 see TGE300
1332-37-2 see IHG100

1332-40-7 see CNK559
1332-58-7 see KBB600
1332-94-1 see LAS000
1333-22-8 see CNM600
1333-39-7 see HJH500
1333-53-5 see IRL000
1333-74-0 see HHW500
1333-82-0 see CMK000
1333-83-1 see SHQ500
1333-86-4 see CBT750
1334-78-7 see TGK250
1335-09-7 see MKJ750
1335-31-5 see MDA500
1335-32-6 see LCH000
1335-66-6 see IKJ000
1335-87-1 see HCK500
1335-88-2 see TBR000
1336-21-6 see ANK250
1336-36-3 see PJL750
1336-80-7 see FBC100
1336-93-2 see MAS820
1337-59-3 see EAN750
1337-81-1 see VQK550
1338-02-9 see NAS000
1338-16-5 see IHL000
1338-23-4 see MKA500
1338-24-5 see NAR000
1338-32-5 see PJQ000
1338-39-2 see SKV000
1338-41-6 see SKV150
1338-43-8 see SKV100
1338-58-5 see AEA250
1339-93-1 see DKL875
1341-24-8 see CDN505
1341-49-7 see ANJ000
1343-78-8 see CNF050
1343-90-4 see MAJ000
1344-00-9 see SEM000
1344-09-8 see SCN700
1344-28-1 see AHE250
1344-32-7 see TIL250
1344-34-9 see NCL000
1344-40-7 see LCV100
1344-43-0 see MAT250
1344-48-5 see MDG750
1344-57-6 see UPJ000
1344-67-8 see CNJ950
1344-81-6 see CAX800
1344-95-2 see CAW850
1345-04-6 see AQL500
1345-05-7 see LHX000
1345-07-9 see DDI200
1345-25-1 see IHC500
1390-65-4 see CCK590
1391-36-2 see LEX100
1391-41-9 see EAQ500
1391-62-4 see FMR500
1391-75-9 see GES000
1391-82-8 see GJU800
1393-38-0 see SMW500
1393-48-2 see TFQ275
1393-62-0 see AAD000
1393-63-1 see APE050
1393-72-2 see CDK500
1393-89-1 see GJY000
1394-04-3 see XFS600
1395-21-7 see BAB750
1397-84-8 see AFI625
1397-89-3 see AOC500
1397-94-0 see AQM250
1398-61-4 see CDM750
1399-70-8 see HAN500

1399-80-0 see DYA600
1400-17-5 see KHU136
1400-61-9 see NOH500
1401-55-4 see CDM250
1401-55-4 see MQV250
1401-55-4 see QBJ000
1401-55-4 see SOY500
1401-55-4 see TAD750
1401-63-4 see TFC500
1401-69-0 see TOE600
1402-38-6 see AEA500
1402-61-5 see AEC200
1402-68-2 see AET750
1402-82-0 see AOB875
1402-83-1 see AOO925
1402-88-6 see ARN250
1403-17-4 see LFF000
1403-22-1 see CBC750
1403-27-6 see CCN250
1403-28-7 see CCN500
1403-29-8 see CCN750
1403-47-0 see DYF400
1403-51-6 see EQT000
1403-66-3 see GCO000
1403-71-0 see HAH800
1403-74-3 see HGB500
1403-89-0 see LFL500
1403-92-5 see LIU500
1404-01-9 see MRY000
1404-04-2 see NCE000
1404-15-5 see NMV500
1404-24-6 see PKC250
1404-26-8 see PKC500
1404-64-4 see SKX000
1404-88-2 see TOG500
1404-90-6 see VFA000
1404-93-9 see VFA050
1404-96-2 see VQU450
1405-10-3 see NCG000
1405-20-5 see PKC700
1405-35-2 see AAC000
1405-36-3 see CBF675
1405-39-6 see CMQ725
1405-41-0 see GCS000
1405-69-2 see GIA100
1405-86-3 see GIG000
1405-87-4 see BAC250
1405-90-9 see CBC500
1405-97-6 see GJO000
1406-05-9 see PAQ000
1406-11-7 see PKC000
1406-16-2 see VSZ095
1406-65-1 see CKN000
1407-03-0 see AMY700
1407-47-2 see AOO900
1414-45-5 see NEB050
1415-73-2 see BAF825
1420-03-7 see PMS800
1420-04-8 see DFV600
1420-07-1 see DRV200
1420-40-2 see DAF800
1420-53-7 see CNG750
1421-14-3 see PMM000
1421-23-4 see BEA850
1421-28-9 see DNU310
1421-49-4 see DEC900
1421-63-2 see TKO250
1421-68-7 see AHL500
1421-85-8 see EBP500
1421-89-2 see DOZ100
1422-07-7 see CNF750
1423-11-6 see TKK050

1423-97-8 see EDU000
1424-27-7 see AAS750
1424-48-2 see THD875
1425-61-2 see DME200
1425-67-8 see CCG000
1426-40-0 see FGW100
1429-30-7 see PCU000
1432-14-0 see MIH900
1432-75-3 see NEC000
1435-55-8 see HIG500
1436-34-6 see HFC100
1438-30-8 see NCP875
1438-94-4 see FPX100
1442-74-6 see BKF900
1444-64-0 see PES000
1445-75-6 see DNQ875
1445-79-0 see TLT000
1446-17-9 see AQT250
1448-16-4 see BIU750
1448-23-3 see GEW700
1449-05-4 see GIEO5O
1450-14-2 see HED425
1450-63-1 see TEA400
1453-58-3 see MOX010
1453-82-3 see ILC100
1455-77-2 see DCF200
1456-28-6 see DTA000
1457-46-1 see PGG500
1459-93-4 see IML000
1461-22-9 see CLP500
1461-23-0 see TIC250
1461-25-2 see TBM250
1463-08-7 see BNF750
1464-33-1 see MFN600
1464-42-2 see SBU725
1464-43-3 see DWY800
1464-43-3 see SBP600
1464-44-4 see PDO300
1464-53-5 see BGA750
1464-69-3 see VQA150
1465-25-4 see NBH500
1465-26-5 see BHV000
1467-79-4 see DRF600
1468-26-4 see THT350
1468-37-7 see DUN600
1468-95-7 see APG600
1470-35-5 see UVJ400
1470-37-7 see BNW750
1470-94-6 see IBU000
1474-80-2 see DJA300
1476-23-9 see AGJ000
1476-53-5 see NOB000
1477-19-6 see BBJ500
1477-20-9 see PFA600
1477-40-3 see ACQ258
1477-42-5 see AKQ500
1477-55-0 see XHS800
1477-57-2 see BIX250
1477-79-8 see DPM200
1478-61-1 see HCZ100
1480-19-9 see HAF400
1481-70-5 see GJQ100
1482-50-4 see PMA500
1484-12-4 see MIE300
1484-13-5 see VNK100
1485-00-3 see MJR800
1487-49-6 see MKP250
1488-25-1 see OKS250
1491-09-4 see BCF750
1491-10-7 see BCF500
1491-41-4 see HMV000
1491-59-4 see ORA100

1492-02-0 see GFM200
1492-54-2 see MHF000
1492-55-3 see MHE750
1492-93-9 see BHP750
1493-13-6 see TKB310
1493-23-8 see FLH100
1495-50-7 see COO825
1496-02-2 see FGH050
1496-94-2 see TNA800
1497-16-1 see ASA750
1498-40-4 see EOQ000
1498-51-7 see EOR000
1498-64-2 see MRI000
1498-88-0 see TLU500
1499-54-3 see BBB250
1500-91-0 see EIL200
1500-94-3 see ACI550
1501-84-4 see AJU625
1502-47-2 see HAQ700
1502-95-0 see DBH400
1504-55-8 see MIO750
1504-74-1 see MEJ750
1505-26-6 see DSI000
1505-95-9 see IOU000
1506-02-1 see FBW110
1508-65-2 see OPK000
1510-31-2 see DKJ225
1511-62-2 see BND800
1514-42-7 see FFS100
1515-76-0 see ABM250
1516-17-2 see AAU750
1516-27-4 see HGI600
1516-32-1 see BSO500
1518-15-6 see BIY600
1518-16-7 see TBW750
1518-58-7 see DJD000
1519-47-7 see XSS000
1520-71-4 see TNA750
1520-78-1 see TLU175
1527-12-4 see PGL270
1528-74-1 see DUS000
1529-30-2 see TJV250
1530-48-9 see AGT750
1532-19-0 see CMV475
1537-62-8 see CGY825
1538-09-6 see BFC750
1541-60-2 see MHJ750
1544-46-3 see FFE000
1544-68-9 see ISM000
1551-44-6 see CPI300
1552-12-1 see CPR825
1553-34-0 see THL500
1553-36-2 see BGX500
1553-60-2 see IJG000
1555-58-4 see MHQ750
1557-57-9 see DGS300
1558-25-4 see CIY325
1559-35-9 see EKX500
1561-49-5 see DGV650
1563-01-5 see DIJ300
1563-66-2 see CBS275
1563-67-3 see DLS800
1564-53-0 see APH600
1565-94-2 see BLD810
1568-80-5 see SLD570
1569-01-3 see PNB500
1569-02-4 see EFF500
1569-69-3 see CPB625
1570-45-2 see ELU000
1570-64-5 see CFE000
1571-30-8 see HOE000
1571-33-1 see PFV500

1575-72-0 see AGQ100
1576-35-8 see THE250
1576-87-0 see PBP800
1579-40-4 see THC600
1582-09-8 see DUV600
1585-74-6 see CGC200
1586-92-1 see EER000
1587-41-3 see DFE550
1589-47-5 see MFK800
1589-49-7 see MFL000
1591-30-6 see BGF250
1592-23-0 see CAX350
1594-56-5 see DVF800
1596-09-4 see CPP050
1596-52-7 see DVE000
1596-84-5 see DQD400
1597-82-6 see PAL600
1599-49-1 see MNM450
1600-27-7 see MCS750
1600-31-3 see PIE525
1600-37-9 see PAY200
1600-44-8 see TDQ500
1603-40-3 see ALC000
1603-41-4 see AMA010
1604-01-9 see CNH730
1605-18-1 see IMX100
1605-51-2 see EIL000
1605-58-9 see DJQ200
1606-67-3 see PON000
1606-83-3 see BST900
1607-17-6 see PCR150
1607-30-3 see DQW600
1607-57-4 see BOK500
1608-26-0 see HEK100
1609-06-9 see COM085
1609-47-8 see DIZ100
1610-17-9 see EGD000
1610-18-0 see MFL250
1615-75-4 see FOO553
1615-80-1 see DJL400
1616-88-2 see MEO000
1617-90-9 see VLF000
1618-08-2 see DCP775
1619-34-7 see QUJ990
1622-32-8 see CGO125
1622-61-3 see CMW000
1622-62-4 see FDD100
1622-79-3 see FKI000
1623-14-9 see ECI300
1623-19-4 see THN750
1623-24-1 see PHE500
1623-99-0 see PGI100
1624-01-7 see OCM100
1624-02-8 see BLS700
1627-73-2 see BAZ000
1628-58-6 see DQC400
1629-58-9 see PBR250
1631-29-4 see CKC500
1631-82-9 see CIS625
1632-16-2 see EKR500
1633-83-6 see BOU250
1634-02-2 see TBM750
1634-04-4 see MHV859
1634-73-7 see AMV375
1634-78-2 see OPK250
1636-70-0 see TMK200
1637-24-7 see HFQ000
1638-22-8 see BSE450
1639-09-4 see HBD500
1639-60-7 see PNA500
1639-66-3 see SOD050
1639-79-8 see BLG280

1640-89-7 see EHU000
1641-74-3 see EQQ100
1642-54-2 see DIW200
1643-19-2 see TBK500
1643-20-5 see DRS200
1645-07-4 see FHP100
1645-75-6 see HDD000
1646-26-0 see ACC100
1646-75-9 see MLX800
1646-87-3 see MLW750
1646-88-4 see AFK000
1649-08-7 see DFA000
1649-18-9 see FLU000
1653-19-6 see DEU400
1653-64-1 see MJT000
1656-16-2 see BLI250
1656-48-0 see OQQ000
1660-94-2 see MJT100
1661-03-6 see MAI250
1663-35-0 see VPZ000
1665-48-1 see XVS000
1665-59-4 see DJV300
1666-13-3 see BLF500
1667-11-4 see CIH825
1668-19-5 see DYE409
1668-54-8 see MGH800
1669-83-6 see BKJ650
1670-14-0 see BBM750
1671-82-5 see MML750
1672-46-4 see DKN300
1674-62-0 see MHP400
1674-96-0 see ARR875
1675-54-3 see BLD750
1677-46-9 see HMR500
1678-25-7 see BBR750
1679-07-8 see CPW300
1679-09-0 see MHS550
1680-21-3 see TJQ100
1682-39-9 see AKC250
1684-42-0 see ADI750
1688-71-7 see AJC750
1689-82-3 see HJF000
1689-83-4 see HKB500
1689-84-5 see DDP000
1689-89-0 see HLJ500
1689-99-2 see DDM200
1693-71-6 see THN250
1694-09-3 see FAG120
1694-20-8 see NMC050
1696-17-9 see BCM250
1696-20-4 see ACR750
1698-53-9 see DGE800
1698-60-8 see PEE750
1701-69-5 see PMX300
1701-71-9 see PNV755
1701-73-1 see VBA100
1702-17-6 see DGJ100
1703-58-8 see BOU500
1704-04-7 see AJY000
1704-62-7 see DPA600
1705-85-7 see MIN750
1706-01-0 see MKC750
1707-14-8 see MNV750
1707-95-5 see ONY000
1708-35-6 see HBC000
1708-39-0 see BBA000
1709-50-8 see DIU400
1709-70-2 see TMJ000
1711-06-4 see TGU010
1711-42-8 see DVV550
1712-64-7 see IQP000
1713-07-1 see AAJ125

2014-27-9 see FAC100
2014-29-1 see MMA525
2014-30-4 see DHI875
2014-31-5 see FAC185
2014-32-6 see FAC170
2014-33-7 see FAC175
2015-28-3 see PCO500
2016-36-6 see CMG000
2016-56-0 see DXW050
2016-57-1 see DAG600
2019-14-9 see DSH000
2019-69-4 see DJL700
2021-19-4 see BSM400
2021-21-8 see MIS250
2022-85-7 see FHI000
2023-60-1 see DRY800
2023-61-2 see DRZ000
2027-17-0 see IQN100
2028-52-6 see BNK350
2028-63-9 see EQM600
2028-76-4 see NEZ100
2031-67-6 see MQD750
2032-04-4 see DCQ550
2032-59-9 see DOR400
2032-65-7 see DST000
2033-24-1 see DRP200
2033-94-5 see DMX800
2034-22-2 see THV450
2035-89-4 see FMV300
2035-99-6 see IHP500
2038-03-1 see AKA750
2039-06-7 see DWO950
2039-87-4 see CLE750
2041-74-9 see BKP325
2041-76-1 see EEV100
2043-24-5 see MLY500
2044-64-6 see DOP000
2044-73-7 see POR790
2045-18-3 see BHD300
2045-19-4 see BHG100
2045-41-2 see BIH000
2045-52-5 see PEN000
2047-14-5 see AAT000
2047-30-5 see CEH500
2049-95-8 see AOF000
2049-96-9 see PBV800
2050-08-0 see SAK000
2050-24-0 see DJP000
2050-43-3 see ABO758
2050-46-6 see CCP900
2050-47-7 see BHJ000
2050-67-1 see DET825
2050-68-2 see DET850
2050-92-2 see DCH200
2051-60-7 see CGM750
2051-61-8 see CGM770
2051-62-9 see CGM780
2051-78-7 see AFY250
2051-79-8 see AKR750
2051-85-6 see CMP600
2051-89-0 see BBY250
2052-14-4 see BSL250
2052-15-5 see BRR700
2052-46-2 see DTM400
2052-49-5 see TBK750
2053-25-0 see EQN750
2055-46-1 see TLS000
2058-46-0 see HOI000
2058-52-8 see CIS750
2058-62-0 see MPY000
2058-66-4 see ENB000
2058-67-5 see EOH500

2061-56-5 see TKH050
2062-77-3 see TKK750
2062-78-4 see PIH000
2062-84-2 see FLK100
2064-23-5 see DPK500
2067-58-5 see BIG250
2068-78-2 see LEZ000
2072-68-6 see GJW000
2074-50-2 see PAJ250
2077-46-5 see TJD600
2078-54-8 see DNR800
2079-00-7 see BLX500
2079-89-2 see AMB750
2082-84-0 see DAJ500
2083-91-2 see DQE900
2085-83-8 see FAC150
2085-85-0 see FAC166
2086-83-1 see BFN500
2088-72-4 see DRB600
2089-46-5 see BHR750
2092-16-2 see CAY250
2092-55-9 see HLI000
2092-56-0 see CMS150
2094-99-7 see IKG800
2095-06-9 see DKM120
2095-58-1 see AFY750
2096-42-6 see ARP000
2097-18-9 see VQK600
2097-19-0 see PGH750
2100-17-6 see PBP750
2104-09-8 see ALP000
2104-64-5 see EBD700
2104-96-3 see BNL250
2109-64-0 see DED200
2111-75-3 see DKX100
2113-47-5 see PDY000
2113-54-4 see PDY250
2113-57-7 see BMW290
2114-00-3 see BOB550
2114-11-6 see AGA750
2114-18-3 see CGU000
2114-33-2 see ABU800
2116-65-6 see BFG750
2117-36-4 see CLP250
2117-78-4 see EFL600
2120-70-9 see PDR000
2121-12-2 see LJB800
2122-19-2 see MJZ000
2122-70-5 see ENL900
2122-77-2 see TIX000
2122-86-3 see NGK000
2126-70-7 see SIK000
2126-84-3 see MBV750
2127-10-8 see DXL200
2130-56-5 see BFX250
2131-55-7 see ISH000
2131-64-8 see ISJ100
2133-34-8 see ASC500
2134-15-8 see SKS400
2135-17-3 see FDD075
2138-43-4 see IOK000
2139-00-6 see AJI000
2139-25-5 see PCJ325
2139-93-7 see DBH100
2142-68-9 see CEB000
2142-94-1 see FNC000
2144-08-3 see TKN800
2144-45-8 see DDH200
2149-70-4 see OIU900
2150-48-3 see DIS200
2150-58-5 see DPS600
2150-60-9 see APG700

2152-34-3 see IBM000
2152-44-5 see VCA000
2153-26-6 see TBE500
2153-98-2 see NNO000
2153-98-2 see NNW500
2154-02-1 see MDV000
2154-68-9 see TDV300
2154-71-4 see BLP300
2155-70-6 see THZ000
2155-71-7 see DEB200
2156-27-6 see MOA600
2156-56-1 see SGG000
2156-71-0 see CKN675
2156-72-1 see BQY275
2156-96-9 see DAI500
2157-19-9 see HCC550
2158-03-4 see PIL525
2158-04-5 see PDK750
2158-76-1 see BKB100
2162-74-5 see BJE550
2163-69-1 see CPT000
2163-77-1 see HJE500
2163-80-6 see MRL750
2164-08-1 see CPR600
2164-09-2 see DFO800
2164-17-2 see DUK800
2167-23-9 see DEF600
2169-38-2 see LHT400
2169-44-0 see TKX700
2169-64-4 see THM750
2169-75-7 see EAG100
2173-44-6 see FAC060
2173-47-9 see FAC179
2173-57-1 see NBJ000
2175-91-9 see DSB400
2176-62-7 see PAY250
2177-18-6 see EEE800
2179-16-0 see BMM125
2179-57-9 see AGF300
2179-59-1 see AGR500
2179-92-2 see TIB250
2180-92-9 see BSI250
2181-04-6 see PKZ000
2181-42-2 see TMG500
2185-78-6 see TDW000
2185-86-6 see VKA600
2185-92-4 see BGE325
2185-98-0 see ARQ000
2186-24-5 see GGY175
2188-67-2 see PBV750
2190-04-7 see OAP300
2191-10-8 see CAD750
2192-20-3 see HOR470
2192-21-4 see DHS200
2196-13-6 see TFK000
2197-57-1 see DSU310
2198-53-0 see ABP760
2198-54-1 see ABP770
2198-58-5 see AJO250
2198-61-0 see IHU100
2200-44-4 see DSL200
2203-57-8 see HAY650
2205-73-4 see TFK300
2205-78-9 see DBG300
2206-89-5 see ADT000
2207-69-4 see EDB000
2207-75-2 see AFQ750
2207-76-3 see HID000
2207-85-4 see IBP309
2209-86-1 see BIK500
2210-25-5 see INH000
2210-28-8 see PNP750

2210-63-1 see MQY400
2211-64-5 see HKA109
2212-05-7 see CKA200
2212-10-4 see CIO000
2212-67-1 see EKO500
2212-99-9 see MAW875
2213-00-5 see MLE600
2213-63-0 see DGJ950
2213-84-5 see DYD400
2215-33-0 see DWX100
2215-77-2 see PDR605
2216-12-8 see NIT500
2216-33-3 see MNC500
2216-34-4 see MNC750
2216-45-7 see XQJ700
2216-51-5 see MCG250
2216-68-4 see MME250
2217-15-4 see TAF760
2217-44-9 see TDE750
2217-82-5 see SFC600
2218-96-4 see DUR425
2219-30-9 see PAP550
2219-82-1 see BRU790
2223-82-7 see DUL200
2223-93-0 see OAT000
2224-15-9 see EEA600
2224-44-4 see NFR600
2226-96-2 see HOI300
2227-13-6 see CKL750
2227-17-0 see DAE425
2227-79-4 see BBM250
2231-57-4 see TFE250
2233-00-3 see TJC050
2234-13-1 see OAP000
2235-12-3 see HEZ000
2235-25-8 see EME100
2235-43-0 see PBP200
2235-46-3 see DHI600
2235-54-3 see SOM500
2235-59-8 see IQB000
2237-30-1 see AIR125
2237-92-5 see BPA500
2238-07-5 see DKM200
2239-92-1 see IOK100
2240-14-4 see EOM000
2243-27-8 see OES300
2243-42-7 see PDR603
2243-62-1 see NAM000
2243-76-7 see NEY000
2243-94-9 see TMM600
2244-11-3 see MDL500
2244-16-8 see CCM100
2244-21-5 see PLD000
2252-95-1 see TCH100
2255-17-6 see PHD750
2257-09-2 see ISP000
2260-00-6 see MPV790
2260-08-4 see ADD875
2265-64-7 see DBC510
2266-22-0 see FLV000
2269-06-9 see TDI500
2270-40-8 see AOP250
2271-93-4 see HEF500
2273-43-0 see BSL500
2273-45-2 see DTH400
2273-46-3 see DVV000
2274-11-5 see EIP000
2274-74-0 see CDS500
2275-14-1 see PDC750
2275-18-5 see IOT000
2275-23-2 see MJG500
2275-61-8 see MGE100

2275-81-2 see BGY500
2276-52-0 see PBP300
2276-90-6 see IGD000
2277-19-2 see NNA325
2277-92-1 see DMZ000
2278-22-0 see PCL750
2278-50-4 see CMU475
2279-64-3 see PFP500
2279-76-7 see CLU250
2281-78-9 see CDY100
2282-34-0 see AON250
2282-89-5 see FAC157
2283-08-1 see HMX500
2284-30-2 see BFI400
2288-13-3 see MPI650
2288-32-6 see MND575
2294-47-5 see DCL125
2295-31-0 see TEV500
2295-58-1 see TKP100
2297-30-5 see AOO410
2298-13-7 see MPY750
2301-52-2 see PHY400
2302-84-3 see FOJ000
2302-93-4 see ABH250
2303-16-4 see DBI200
2303-17-5 see DNS600
2303-47-1 see DNW800
2304-30-5 see TBM000
2304-85-0 see PON350
2305-05-7 see OES100
2306-33-4 see MRI275
2307-49-5 see TIK000
2307-55-3 see DAB815
2307-68-8 see SKQ400
2307-97-3 see NNU500
2310-17-0 see BDJ250
2312-35-8 see SOP000
2312-73-4 see BEA100
2312-76-7 see DUU600
2314-17-2 see BSN325
2314-97-8 see IFM100
2315-02-8 see AEX000
2315-20-0 see DKK100
2315-36-8 see DIX400
2318-18-5 see DMX200
2319-96-2 see MGV750
2321-07-5 see FEV000
2321-53-1 see MDQ770
2338-05-8 see FAW100
2338-12-7 see NFJ000
2338-18-3 see AKL000
2338-29-6 see TBW000
2338-76-3 see TKF000
2343-36-4 see FFK200
2345-24-6 see NCO200
2345-26-8 see GDI000
2346-00-1 see DLV900
2347-47-9 see DFE235
2348-79-0 see DPN300
2348-82-5 see MEY800
2349-07-7 see HFQ550
2349-67-9 see AKM000
2350-24-5 see EKU100
2353-33-5 see ARY125
2353-45-9 see FAG000
2354-61-2 see FLL000
2363-58-8 see EBD500
2364-87-6 see THH500
2365-26-6 see IRV300
2365-30-2 see CHE750
2365-48-2 see MLE750
2367-25-1 see MKE250

2373-98-0 see DMI400
2374-05-2 see BOL303
2374-14-3 see FLV100
2375-03-3 see USJ100
2378-02-1 see HDF075
2379-55-7 see DTY700
2379-81-9 see DUP100
2379-90-0 see AKO350
2380-63-4 see POM600
2381-15-9 see MGW740
2381-16-0 see MGW500
2381-18-2 see BBB750
2381-19-3 see MHL500
2381-21-7 see MOX875
2381-31-9 see MGW250
2381-34-2 see MHF500
2381-39-7 see MHM000
2381-40-0 see DQI800
2381-85-3 see AJN250
2382-43-6 see MIM300
2382-96-9 see MCK900
2385-81-1 see FPQ100
2385-85-5 see MQW500
2385-87-7 see LJM000
2386-25-6 see ACI500
2386-52-9 see SDR500
2386-87-0 see EBO050
2386-90-5 see BJN250
2387-59-9 see CCH125
2390-56-9 see AJP300
2390-59-2 see EQG550
2391-03-9 see DXG100
2391-30-2 see CMM750
2392-39-4 see DAE525
2393-53-5 see EDV500
2396-43-2 see TNC100
2396-63-6 see HAV500
2396-68-1 see BSO100
2398-95-0 see PGW600
2398-96-1 see TGB475
2399-48-6 see TCS100
2399-85-1 see TMX750
2400-59-1 see DKO400
2401-85-6 see DUS600
2402-78-0 see DGI800
2402-79-1 see TBU000
2402-95-1 see CKW325
2404-52-6 see TFO750
2404-78-6 see TJT900
2406-25-9 see DEF090
2407-43-4 see EKL500
2409-35-0 see FAC163
2409-55-4 see BQV750
2413-38-9 see FMO150
2417-77-8 see BNO500
2417-90-5 see BOB500
2417-93-8 see PNP275
2418-14-6 see DNV610
2422-79-9 see MGX500
2422-88-0 see BQY250
2423-66-7 see QSA000
2424-09-1 see SBN525
2424-75-1 see DPE200
2425-06-1 see CBF800
2425-10-7 see XTJ000
2425-25-4 see DIW600
2425-41-4 see MRF200
2425-66-3 see CJD750
2425-74-3 see BRJ750
2425-79-8 see BOS100
2425-85-6 see MMP100
2426-07-5 see DHD800

2426-08-6 see BRK750
2426-54-2 see DHT125
2428-04-8 see TNG275
2429-73-4 see CMN800
2429-74-5 see CMO500
2429-80-3 see ADG000
2429-81-4 see CMO820
2429-82-5 see CMO800
2429-84-7 see CMO870
2430-22-0 see ILJ000
2430-27-5 see PNX600
2431-50-7 see TIL360
2431-96-1 see PEU100
2432-90-8 see PHW550
2432-99-7 see AMW000
2435-16-7 see HBP425
2435-64-5 see SNW800
2435-76-9 see DCQ600
2436-66-0 see MPF750
2436-90-0 see CMT050
2437-25-4 see DXT400
2437-29-8 see MAK600
2438-32-6 see PJJ325
2438-49-5 see DTK800
2438-51-9 see PCZ000
2438-72-4 see BPP750
2438-88-2 see TBR250
2439-01-2 see ORU000
2439-10-3 see DXX400
2439-35-2 see DPB300
2439-77-2 see AOT750
2439-99-8 see BLG250
2440-22-4 see HML500
2440-29-1 see BSV750
2440-45-1 see BJT250
2441-88-5 see FAR200
2441-97-6 see CFG300
2442-10-6 see ODW028
2443-39-2 see ECD500
2444-90-8 see BLD800
2445-07-0 see USJ075
2445-76-3 see HFV100
2445-83-2 see MIP775
2446-84-6 see DQH509
2447-57-6 see AIE500
2448-39-7 see AJJ500
2448-68-2 see PII200
2449-49-2 see DSO800
2450-71-7 see POA000
2451-62-9 see GGY150
2454-11-7 see FNK040
2454-37-7 see AKR000
2457-76-3 see CEG750
2460-49-3 see DFL720
2460-77-7 see DDV550
2461-15-6 see GGY100
2463-45-8 see CBW400
2463-53-8 see NNA300
2463-84-5 see NFT000
2465-27-2 see IBA000
2465-29-4 see ADK000
2466-76-4 see ACO250
2467-10-9 see TBS100
2467-12-1 see TMO550
2467-13-2 see TJR500
2467-15-4 see TJK250
2468-21-5 see CCP925
2469-34-3 see SBY000
2469-55-8 see OPC100
2470-73-7 see MKQ000
2472-17-5 see ARR250
2475-31-2 see ICU135

2475-33-4 see CMU770
2475-44-7 see BKP500
2475-45-8 see TBG700
2475-46-9 see MGG250
2479-46-1 see REF070
2481-94-9 see OHJ875
2482-80-6 see EQI600
2485-10-1 see FEN000
2487-01-6 see BRU750
2487-40-3 see TFL500
2487-90-3 see TLB750
2488-01-9 see PEW725
2489-52-3 see FLY200
2489-77-2 see TMH750
2490-89-3 see CBI500
2491-06-7 see MPI100
2491-52-3 see NET500
2491-76-1 see CGD250
2492-26-4 see SIG500
2493-04-1 see NGD600
2494-55-5 see PMW750
2494-56-6 see BSY300
2495-54-7 see HIQ500
2496-92-6 see ISD000
2497-07-6 see OQS000
2497-18-9 see HFE100
2497-21-4 see HFE200
2497-34-9 see ELD500
2498-27-3 see HNN500
2498-75-1 see MGV250
2498-76-2 see MGV000
2498-77-3 see MGU750
2499-58-3 see HBL100
2499-95-8 see ADV000
2504-18-9 see MHW750
2504-55-4 see AJK800
2507-91-7 see KFA100
2508-18-1 see FIT200
2508-19-2 see TMK800
2508-20-5 see NKF500
2508-23-8 see NBD000
2508-86-3 see AML500
2510-95-4 see DVX600
2511-00-4 see EHT600
2511-10-6 see DJR700
2514-52-5 see HCE600
2514-53-6 see EJD600
2517-98-8 see ABR250
2519-30-4 see BMA000
2521-01-9 see EGT000
2522-81-8 see PCV350
2523-37-7 see MKC800
2524-02-9 see TFO600
2524-03-0 see DTQ600
2524-04-1 see DJW600
2525-21-5 see MPG000
2528-16-7 see MRK100
2528-36-1 see DEG600
2529-46-6 see MDQ075
2530-10-1 see DUG425
2530-46-3 see DXY300
2530-83-8 see ECH000
2530-85-0 see TLC250
2530-87-2 see TLC300
2530-99-6 see AGS500
2531-84-2 see MNN520
2532-17-4 see IEE050
2532-49-2 see PNQ250
2532-50-5 see BLM750
2533-82-6 see MGQ750
2536-18-7 see DUW503
2536-31-4 see CDT000

2536-91-6 see MGD500
2537-36-2 see TCD000
2538-85-4 see HLI100
2539-17-5 see TBQ290
2539-53-9 see IKO100
2540-82-1 see DRR200
2541-68-6 see FJP000
2541-69-7 see MGW750
2545-59-7 see TAH900
2545-60-0 see PLQ760
2548-87-0 see ODQ400
2549-67-9 see EGN500
2549-90-8 see EKY000
2549-93-1 see BGT750
2550-26-7 see PDF800
2550-40-5 see CPK700
2550-75-6 see BFY100
2551-62-4 see SOI000
2552-89-8 see AFH275
2554-06-5 see TCB600
2556-10-7 see EES380
2556-73-2 see MHY750
2562-38-1 see NFV530
2564-35-4 see GLS100
2564-65-0 see BBF500
2564-83-2 see TDT800
2567-14-8 see TJB800
2567-83-1 see TCD002
2568-25-4 see MNT600
2569-58-6 see PEK100
2570-26-5 see PBA000
2571-22-4 see TOE175
2571-86-0 see CNR725
2578-28-1 see SBU800
2578-75-8 see FQJ000
2579-20-6 see BGT500
2580-78-1 see BMM500
2581-34-2 see NFV000
2581-69-3 see NIK000
2583-80-4 see DXG025
2586-57-4 see CMO600
2586-58-5 see CMO810
2586-60-9 see CMP000
2587-75-9 see ABX175
2587-76-0 see TMO585
2587-81-7 see TJU150
2587-82-8 see THY850
2587-84-0 see DED400
2587-90-8 see MIW250
2588-03-6 see TEZ200
2588-04-7 see TEZ100
2589-01-7 see TKL250
2589-02-8 see BIY000
2589-15-3 see SCD500
2589-47-1 see DNB000
2591-57-3 see ENI175
2591-86-8 see FOH000
2592-28-1 see MFF550
2592-62-3 see HCK000
2592-85-0 see DNU325
2592-95-2 see HJN650
2593-15-9 see EFK000
2594-20-9 see IPA000
2595-54-2 see DJI000
2597-03-7 see DRR400
2597-09-3 see AGR125
2597-54-8 see ACL000
2597-93-5 see EME050
2597-95-7 see MLG750
2598-25-6 see CJK000
2598-31-4 see HOE200
2598-70-1 see MNU100

2598-71-2 see MFG200
2598-72-3 see MNR100
2598-73-4 see CJP500
2598-74-5 see MNP500
2598-75-6 see CJK100
2598-76-7 see MNP400
2600-55-7 see DVC200
2602-46-2 see CMO000
2605-44-9 see CAG775
2606-85-1 see MHL250
2606-87-3 see FJO000
2610-05-1 see CMN750
2610-11-9 see CMO885
2610-86-8 see WAT209
2611-61-2 see UVA150
2611-82-7 see FMU080
2612-33-1 see CKU625
2614-06-4 see TEH510
2614-76-8 see BJY825
2616-10-1 see BGV800
2619-97-8 see GHE000
2621-62-7 see DGB480
2622-08-4 see TJF750
2622-21-1 see VNZ990
2622-26-6 see PIW000
2623-22-5 see ECT700
2623-23-6 see MCG750
2624-03-5 see EDB200
2624-17-1 see SGB550
2624-31-9 see PLO000
2624-43-3 see FBP100
2624-44-4 see DIS600
2626-34-8 see DQO650
2627-95-4 see TDQ075
2628-16-2 see ABW550
2628-17-3 see VQA200
2629-78-9 see BLM000
2631-37-0 see CQI500
2631-40-5 see MIA250
2631-68-7 see TJE200
2633-54-7 see TIR250
2634-33-5 see BCE475
2636-26-2 see COQ399
2637-34-5 see TFP300
2637-37-8 see QOJ100
2638-94-0 see ASL500
2639-63-6 see HFM700
2641-56-7 see DIV600
2642-37-7 see ICD000
2642-50-4 see MFR000
2642-71-9 see EKN000
2642-98-0 see CML800
2644-70-4 see HGV000
2646-17-5 see TGW000
2646-38-0 see CDL375
2648-61-5 see DEN200
2650-18-2 see FMU059
2651-85-6 see DJH500
2652-77-9 see DWA500
2654-47-9 see EME600
2655-14-3 see DTN200
2655-19-8 see DEG400
2656-72-6 see HBO600
2664-63-3 see TFJ000
2665-12-5 see TJF500
2665-30-7 see MOB699
2666-14-0 see TNL750
2666-17-3 see CMM070
2668-24-8 see TIP630
2668-92-0 see NAQ500
2669-32-1 see DRU400
2669-35-4 see TID100

2674-91-1 see DSK600
2675-35-6 see BKI300
2675-77-6 see CJA100
2679-01-8 see MJU250
2680-03-7 see DOP800
2682-20-4 see MLB700
2686-99-9 see TMD000
2687-25-4 see TGY800
2687-91-4 see EPC700
2687-94-7 see OFU100
2688-84-8 see PDR490
2691-41-0 see CQH250
2696-84-6 see PNE000
2696-92-6 see NMH000
2697-60-1 see AJS950
2697-65-6 see AJS900
2698-38-6 see EMR000
2698-41-1 see CEQ600
2699-79-8 see SOU500
2700-89-2 see DFS200
2702-72-9 see SGH500
2703-13-1 see MOB599
2705-87-5 see AGC500
2706-28-7 see CMM758
2706-47-0 see TNC725
2706-47-0 see TNC750
2706-50-5 see AOA750
2706-66-3 see EDB125
2709-56-0 see FMO129
2713-09-9 see FFS000
2719-13-3 see NIQ550
2719-23-5 see TEW100
2719-32-6 see ISI000
2720-73-2 see PKV100
2724-69-8 see MOA500
2726-03-6 see CMW250
2728-04-3 see DKD000
2731-16-0 see DBA200
2732-09-4 see FNR000
2735-04-8 see DNY500
2736-21-2 see MFS500
2736-23-4 see DGK900
2736-73-4 see DGH600
2736-80-3 see DVD000
2738-06-9 see EIL050
2738-18-3 see DSD800
2740-04-7 see DNV200
2743-38-6 see DDE300
2746-81-8 see PMI250
2747-31-1 see DTK600
2748-88-1 see MSB500
2749-79-3 see HCV000
2750-76-7 see RKA000
2752-65-0 see CBA125
2757-18-8 see TEM399
2757-28-0 see TKS500
2757-90-6 see GFO080
2758-06-7 see BNK325
2758-42-1 see DGA100
2759-71-9 see CQJ250
2761-24-2 see PBZ000
2763-96-4 see AKT750
2764-72-9 see DWW730
2767-41-1 see DFC200
2767-47-7 see DUG800
2767-54-6 see BOI750
2767-55-7 see DJB000
2767-61-5 see BOK750
2767-80-8 see TNI600
2768-02-7 see TLD000
2771-13-3 see MFC600
2773-92-4 see DNX400

2778-04-3 see EAS000
2778-41-8 see TDX400
2778-42-9 see TDX300
2781-10-4 see BJQ250
2781-11-5 see DIV300
2782-57-2 see DGN200
2782-70-9 see BES250
2782-91-4 see TDX000
2783-12-2 see NFQ550
2783-17-7 see DXW800
2783-94-0 see FAG150
2783-96-2 see SGQ000
2784-73-8 see ACR600
2784-86-3 see HIJ400
2784-89-6 see NEM350
2784-94-3 see BKF250
2785-87-7 see MFM750
2787-93-1 see HEW100
2788-26-3 see PHA565
2797-51-5 see AJI250
2799-07-7 see TNR475
2801-68-5 see DOK600
2802-70-2 see FJI510
2804-00-4 see FLJ000
2807-30-9 see PNG750
2808-86-8 see TBT750
2809-21-4 see HKS780
2812-73-9 see CLJ750
2813-95-8 see ACE500
2814-20-2 see IQL000
2814-77-9 see CJD500
2818-88-4 see MHI500
2818-89-5 see DQO600
2820-51-1 see NDP400
2823-90-7 see FFO000
2823-91-8 see FFN000
2823-93-0 see FFM000
2823-94-1 see FFP000
2823-95-2 see FFQ000
2824-10-4 see FFL000
2825-00-5 see DTU200
2825-15-2 see NHE100
2825-60-7 see FDB000
2825-82-3 see TLR675
2827-46-5 see TNJ825
2828-42-4 see PEQ500
2832-19-1 see CHO775
2832-40-8 see AAQ250
2834-84-6 see AKD250
2835-39-4 see ISV000
2835-95-2 see AKJ750
2835-99-6 see AKZ000
2836-32-0 see SHT000
2840-26-8 see AIA250
2842-37-7 see PFB500
2842-38-8 see CPG125
2845-82-1 see PPN500
2847-65-6 see TMV800
2850-61-5 see DDV250
2852-07-5 see TIL350
2854-16-2 see ALB250
2855-19-8 see DXU400
2856-75-9 see MOM750
2857-03-6 see TIB750
2858-66-4 see PAO500
2862-16-0 see DAE200
2865-19-2 see DEA000
2865-70-5 see CJM750
2867-47-2 see DPG600
2869-09-2 see MMD000
2869-10-5 see MMD250
2869-12-7 see MJA750

3251-08-9 see AJC375
3251-23-8 see CNM750
3251-56-7 see NHA000
3252-43-5 see DDJ400
3253-62-1 see CNH785
3254-63-5 see PHD250
3254-66-8 see DNR309
3254-89-5 see CCR875
3257-28-1 see CMP620
3258-02-4 see HKA270
3261-53-8 see GES100
3263-31-8 see CMU815
3264-31-1 see BFZ160
3264-82-2 see PBL500
3266-23-7 see EBJ100
3266-39-5 see TJC500
3267-78-5 see TNB000
3268-19-7 see BER500
3268-27-7 see PFL200
3268-49-3 see MPV400
3268-49-3 see TET900
3268-87-9 see OAJ000
3270-02-8 see AFJ450
3270-78-8 see AQN625
3270-86-8 see DJS800
3271-05-4 see MEW760
3271-76-9 see CMU810
3272-96-6 see CDZ000
3274-28-0 see OCU100
3275-64-7 see MGN600
3276-41-3 see NJX000
3277-26-7 see TDP775
3278-35-1 see EQH000
3278-43-1 see THY200
3279-46-7 see MIB500
3279-95-6 see HKS775
3282-30-2 see DTS400
3282-32-4 see PET800
3282-85-7 see DKD400
3286-46-2 see BKJ325
3286-98-4 see DEG150
3287-99-8 see BDY000
3289-23-4 see BOU300
3289-28-9 see CPB075
3290-92-4 see TLX250
3290-99-1 see AOV500
3292-42-0 see AIF800
3295-92-9 see MOS400
3296-90-9 see DDQ400
3299-38-5 see DHM100
3300-67-2 see CLF150
3301-75-5 see BAX250
3308-51-8 see CQN125
3308-64-3 see HMA500
3309-77-1 see CEQ500
3309-87-3 see FOR000
3314-35-0 see ALX750
3315-16-0 see SDO525
3316-24-3 see NMR100
3316-46-9 see THM300
3319-31-1 see TJR600
3321-80-0 see MON250
3322-93-8 see DDM300
3323-53-3 see HEG120
3327-22-8 see CHQ300
3328-25-4 see CMD250
3328-33-4 see PGZ915
3328-84-5 see EID150
3329-56-4 see BPG500
3333-52-6 see TDW250
3333-67-3 see NCY500
3333-85-5 see PGF100

3337-71-1 see SNQ500
3338-24-7 see PHG750
3340-93-0 see MGU500
3340-94-1 see MGT500
3342-64-1 see EDV555
3342-98-1 see HMA000
3342-99-2 see MIK750
3343-07-5 see MIM250
3343-08-6 see MIL250
3343-10-0 see DAL200
3343-12-2 see DML800
3343-15-5 see LJI100
3343-28-0 see PHX100
3343-29-1 see PIA100
3344-12-5 see IKX200
3344-14-7 see MKC250
3344-70-5 see DDN300
3347-22-6 see DLK200
3351-28-8 see MIM500
3351-30-2 see MIN250
3351-31-3 see MIN000
3351-32-4 see MIM750
3352-87-2 see DJD500
3353-08-0 see MIU750
3365-94-4 see BBY500
3366-61-8 see ACC000
3366-65-2 see PDA250
3367-28-0 see CHW250
3367-29-1 see CHV750
3367-30-4 see CHW500
3367-31-5 see CGM450
3367-32-6 see CHV500
3368-13-6 see PGJ000
3369-52-6 see HDK200
3369-56-0 see ANG135
3374-04-7 see BHN500
3374-05-8 see NAG450
3375-22-2 see DFN330
3375-31-3 see PAD300
3380-34-5 see TIQ000
3382-99-8 see DBK800
3383-96-8 see TAL250
3385-03-3 see FDD085
3385-21-5 see DBQ800
3385-78-2 see TLT775
3388-04-3 see EBO000
3391-83-1 see OKW100
3391-86-4 see ODW000
3393-34-8 see HLF000
3398-09-2 see AIC000
3398-46-7 see EDC585
3398-48-9 see SMA500
3398-69-4 see NKD500
3401-80-7 see DGK250
3403-42-7 see MFK750
3404-63-5 see EGV600
3405-32-1 see TBO762
3405-45-6 see DEE400
3407-94-1 see DBN000
3412-76-8 see DDG400
3413-58-9 see ELC500
3413-72-7 see MQB750
3414-47-9 see TKJ750
3414-62-8 see HIT100
3414-63-9 see NNE525
3415-90-5 see CIQ600
3416-21-5 see MIL500
3420-97-1 see EEY550
3424-82-6 see DEV900
3425-46-5 see PLS000
3425-61-4 see PBX325
3426-01-5 see DWL600

3426-43-5 see BGW100
3426-62-8 see DOR800
3426-89-9 see PHE850
3428-24-8 see DGJ200
3432-10-8 see AMO250
3437-84-1 see DNJ000
3440-75-3 see TED750
3441-14-3 see CMO870
3441-64-3 see SPA000
3443-15-0 see AJH129
3444-13-1 see MDF500
3445-11-2 see HLB380
3454-11-3 see MRK250
3457-98-5 see PEI250
3458-22-8 see YCJ000
3459-20-9 see GHK200
3463-21-6 see EFJ600
3465-73-4 see DEB400
3465-74-5 see DEA600
3465-75-6 see BIS500
3468-11-9 see DNE400
3468-63-1 see DVB800
3470-17-5 see BBU125
3471-05-4 see TCJ825
3471-31-6 see MES850
3473-12-9 see PDK500
3475-63-6 see TLU750
3476-50-4 see TLO000
3477-28-9 see CHV250
3477-94-9 see CKQ500
3478-44-2 see ERE200
3478-94-2 see MOL250
3482-37-9 see TLN750
3483-12-3 see DXO800
3484-68-2 see RAG400
3485-14-1 see AJJ875
3485-82-3 see SKH000
3486-30-4 see ADE675
3486-30-4 see ERG100
3486-35-9 see ZEJ050
3486-66-6 see CNR100
3486-86-0 see BOR250
3495-42-9 see CMA500
3495-54-3 see AHA125
3497-00-5 see PFW200
3497-00-5 see PFW210
3504-13-0 see MRK000
3505-38-2 see CGD500
3506-09-0 see PNB800
3506-23-8 see POK250
3511-19-1 see DGL800
3513-92-6 see DKB170
3518-05-6 see DQH600
3518-65-8 see CHY000
3520-72-7 see CMS145
3521-84-4 see BGB315
3522-94-9 see TLT200
3524-62-7 see MHE000
3524-68-3 see PBC750
3527-05-7 see MGS925
3529-82-6 see ISO000
3531-19-9 see CGL325
3531-43-9 see TDE775
3535-75-9 see AMJ250
3536-49-0 see CMM062
3542-36-7 see DVN300
3543-75-7 see CQM750
3544-23-8 see MFF500
3544-24-9 see AID625
3544-35-2 see CJN250
3544-94-3 see CDP725
3545-67-3 see CLD100

3545-88-8 see DBY600
3546-10-9 see CME250
3546-11-0 see DRI400
3546-29-0 see SEH600
3546-41-6 see PQC500
3547-04-4 see BIM775
3547-33-9 see OGA000
3553-80-8 see EIB000
3555-11-1 see PAU600
3558-60-9 see PFD325
3560-78-9 see TKI750
3562-15-0 see IIM100
3562-84-3 see DDP200
3563-36-8 see SCB500
3563-57-3 see AGR100
3563-76-6 see PJA130
3563-92-6 see ZVA000
3564-09-8 see FAG018
3564-14-5 see CMP880
3564-27-0 see CMP882
3564-66-7 see TMK100
3565-15-9 see SHX500
3565-26-2 see NLR000
3566-00-5 see MIB250
3566-10-7 see ANZ000
3566-25-4 see HGI525
3567-08-6 see IIY100
3567-38-2 see PGE000
3567-65-5 see CMM320
3567-66-6 see CMS228
3567-69-9 see HJF500
3567-72-4 see TIT100
3567-76-8 see DKR200
3567-79-1 see TIT050
3568-23-8 see IDA500
3568-56-7 see ENI500
3569-10-6 see VAG400
3569-57-1 see CKU750
3569-58-2 see HOJ150
3570-10-3 see MNC150
3570-54-5 see TCJ500
3570-58-9 see CHC750
3570-61-4 see SKL000
3570-75-0 see NDY500
3570-80-7 see FEV100
3570-93-2 see CPB650
3571-74-2 see AGT250
3572-06-3 see AAR500
3572-35-8 see BJA825
3572-47-2 see DVM000
3572-74-5 see DSQ600
3572-80-3 see COV500
3574-96-7 see CJR550
3575-31-3 see OEQ000
3577-01-3 see CCR890
3581-11-1 see DTQ800
3582-17-0 see DKH200
3583-47-9 see DFF600
3585-32-8 see DUI709
3586-14-9 see MNV770
3586-69-4 see NES000
3589-21-7 see TAL000
3589-22-8 see DEL400
3590-07-6 see EGU000
3590-84-9 see TDY790
3590-84-9 see TDY795
3590-93-0 see DAN388
3593-24-6 see BSR930
3594-15-8 see DVN909
3597-20-4 see BIB750
3597-21-5 see EAK100
3597-91-9 see HLY400

3598-37-6 see AAF750
3599-28-8 see SIF450
3599-32-4 see CCK000
3604-87-3 see EAB600
3605-01-4 see TNR485
3607-48-5 see IDR000
3608-75-1 see PIB925
3610-02-4 see HJN550
3610-27-3 see AAV500
3613-30-7 see DSM800
3613-73-8 see TCQ260
3613-82-9 see TLQ750
3613-89-6 see HEF300
3613-97-6 see RCZ200
3614-57-1 see IGH700
3614-69-5 see FMU409
3615-21-2 see DGO400
3615-24-5 see INM000
3622-76-2 see VOF000
3622-84-2 see BQJ650
3624-87-1 see AJA500
3624-96-2 see BFX125
3625-18-1 see DQU200
3626-28-6 see CMO830
3627-49-4 see PDO900
3632-91-5 see MAG000
3634-83-1 see XIJ000
3635-74-3 see DOZ000
3639-66-5 see DOQ600
3644-11-9 see MEX300
3644-32-4 see NII200
3644-37-9 see BGK000
3644-61-9 see MRW125
3646-61-5 see DFE700
3647-17-4 see MFF250
3647-19-6 see CJR250
3647-69-6 see CHE000
3648-18-8 see DVJ800
3648-20-2 see DXQ400
3648-21-3 see HBP400
3648-36-0 see CMM850
3653-48-3 see SIL500
3655-88-7 see MPP750
3658-48-8 see BJQ709
3658-77-3 see HKC575
3663-42-1 see OKU300
3665-51-8 see HMW500
3666-67-9 see DDH600
3669-32-7 see CMQ475
3670-09-5 see BFH100
3671-00-9 see MPD250
3671-71-4 see FDZ000
3671-78-1 see FDX000
3674-13-3 see EHY050
3675-13-6 see BOW800
3676-91-3 see TDP525
3680-02-2 see MQM750
3681-71-8 see HFE150
3681-93-4 see GFC050
3683-12-3 see PFP600
3684-97-7 see MMT750
3685-84-5 see AAE500
3686-43-9 see DLY700
3686-69-9 see PNA200
3687-13-6 see DGD400
3687-31-8 see LCK100
3687-61-4 see CPN750
3687-67-0 see CMU320
3688-08-2 see ISK000
3688-11-7 see PFO750
3688-35-5 see AJE000
3688-53-7 see FQN000

3688-66-2 see CNG250
3688-79-7 see MEB750
3688-85-5 see CIP500
3689-24-5 see SOD100
3689-50-7 see AFL750
3689-76-7 see CDY325
3689-77-8 see BHQ300
3690-04-8 see AMF375
3690-12-8 see AKO750
3690-18-4 see MPB500
3690-28-6 see PDW800
3690-50-4 see MOW500
3690-53-7 see DHU700
3691-16-5 see VMA000
3691-35-8 see CJJ000
3691-71-2 see EPF800
3691-74-5 see GBB500
3691-78-9 see BBU625
3692-90-8 see PMN250
3693-22-9 see DDB600
3693-53-6 see MJT750
3694-45-9 see CEQ750
3694-52-8 see NIM525
3695-77-0 see TMT500
3696-23-9 see CKL050
3696-23-9 see UTA450
3696-28-4 see DXN300
3697-24-3 see MIN500
3697-25-4 see AAE000
3697-27-6 see DRE800
3697-30-1 see EGO500
3698-54-2 see TDY000
3700-89-8 see CBS800
3701-40-4 see NAO600
3703-76-2 see CMX800
3703-79-5 see BQF250
3704-09-4 see MQS225
3704-42-5 see NIW400
3706-77-2 see DXN850
3710-30-3 see OBK000
3710-84-7 see DJN000
3714-62-3 see TBR500
3715-67-1 see EJA100
3715-90-0 see THJ825
3715-92-2 see NKK500
3717-88-2 see FCB100
3719-37-7 see BBD980
3721-28-6 see PET750
3724-65-0 see COB500
3724-89-8 see AKB500
3727-31-9 see MIT700
3730-54-9 see EHY700
3731-39-3 see DUH800
3731-51-9 see ALB750
3732-82-9 see HEK050
3732-90-9 see MJF000
3733-45-7 see ACD250
3733-63-9 see BBV750
3734-17-6 see DTO200
3734-33-6 see DAP812
3734-48-3 see HCN000
3734-60-9 see APV750
3734-67-6 see CMM300
3734-95-0 see PHK250
3734-97-2 see AMX825
3735-23-7 see MNO750
3735-33-9 see PHL750
3735-81-7 see TJE885
3735-90-8 see PDC850
3736-26-3 see HIF575
3736-86-5 see THL750
3736-90-1 see IFA100

3736-92-3 see DWJ400
3737-00-6 see BNA880
3737-09-5 see DNN600
3737-22-2 see IKC070
3737-33-5 see ELQ600
3737-35-7 see DHW400
3737-41-5 see TBV300
3737-66-4 see UAG075
3737-72-2 see UAG000
3740-52-1 see NII500
3741-12-6 see CJF000
3741-14-8 see NGA500
3741-32-0 see EIU900
3741-38-6 see COV750
3748-13-8 see DNL700
3750-18-3 see HDY100
3750-26-3 see PMC750
3751-44-8 see BGV750
3757-31-1 see BSX600
3758-54-1 see CCP675
3759-07-7 see DRM000
3759-61-3 see NBH200
3759-92-0 see FPI100
3761-41-9 see DSS400
3761-42-0 see DSS800
3761-53-3 see FMU070
3762-27-4 see BFD760
3763-39-1 see VPF150
3763-72-2 see BGQ100
3765-65-9 see TBI600
3766-55-0 see AGT000
3766-60-7 see CKF750
3766-81-2 see MOV000
3768-60-3 see BJJ125
3771-19-5 see MCB500
3772-23-4 see BRQ000
3772-26-7 see IRG100
3772-51-8 see PHG600
3772-76-7 see MDU300
3773-37-3 see DSS600
3775-55-1 see ALM250
3775-85-7 see EJD000
3775-90-4 see BQD250
3777-12-6 see PMW800
3777-69-3 see AOJ600
3778-73-2 see IMH000
3778-76-5 see TGJ150
3779-45-1 see FOJ050
3781-28-0 see FLN000
3785-34-0 see BHD250
3786-76-3 see DLB650
3787-28-8 see TIJ250
3789-77-3 see DRC800
3792-59-4 see SCC000
3794-64-7 see HAW000
3794-83-0 see TEE250
3804-89-5 see SIJ000
3805-65-0 see BNV760
3808-42-2 see RAG300
3810-74-0 see SLY500
3810-80-8 see LIB000
3810-81-9 see BKS810
3811-04-9 see PLA250
3811-06-1 see CBQ125
3811-10-7 see BCP685
3811-49-2 see MEC250
3811-73-2 see MCQ750
3813-05-6 see BAV000
3817-11-6 see HJQ350
3818-69-7 see HEF200
3818-70-0 see MFM100
3818-88-0 see EAI875

3818-90-4 see CKI000
3819-00-9 see PII500
3820-53-9 see DTH800
3820-67-5 see GGQ050
3823-94-7 see MQF200
3825-26-1 see ANP625
3836-23-5 see NNQ000
3837-54-5 see DRU000
3837-55-6 see DSW600
3840-18-4 see THB100
3844-45-9 see FAE000
3844-60-8 see DRN400
3844-63-1 see NKL000
3848-24-6 see HEQ200
3851-22-7 see ZGW100
3852-09-3 see MFL400
3852-72-0 see OPO100
3853-74-5 see MEV600
3854-14-6 see DOV850
3855-45-6 see TKG525
3858-83-1 see AID650
3861-47-0 see DNG200
3861-73-2 see ADE750
3861-99-2 see CGM500
3862-21-3 see DUE600
3862-73-5 see TJX900
3863-59-0 see HHQ875
3867-15-0 see PIN225
3868-31-3 see ONW150
3868-61-9 see EAQ815
3871-82-7 see MNN250
3877-86-9 see EAH100
3878-19-1 see FQK000
3878-45-3 see TMU100
3883-43-0 see DFE600
3886-80-4 see LBR100
3886-90-6 see DTC200
3886-91-7 see DTH700
3891-30-3 see SLM000
3898-08-6 see DWO000
3898-45-1 see PDK300
3900-31-0 see FDB100
3902-71-4 see HOM270
3913-02-8 see BSA500
3913-71-1 see DAI350
3915-83-1 see NCO500
3917-15-5 see AGV250
3921-94-6 see DOV825
3921-96-8 see PPU850
3921-98-0 see EEN700
3921-99-1 see DWR300
3922-00-7 see PPU800
3922-90-5 see OHM900
3926-62-3 see SFU500
3930-19-6 see SMA000
3931-89-3 see TFN750
3938-45-2 see DQX300
3941-06-8 see PDV700
3942-54-9 see CKF000
3943-74-6 see VFF100
3947-65-7 see NCE500
3949-14-2 see MFO250
3952-78-1 see AFM400
3952-98-5 see AGH125
3953-10-4 see EGZ000
3958-19-8 see TMQ600
3958-60-9 see NFM700
3963-79-9 see AIC500
3963-80-2 see HAA400
3963-95-9 see MDO250
3965-55-7 see BKN000
3967-55-3 see DFI800

3978-86-7 see DLV800
3979-76-8 see MHO000
3982-20-5 see PIA000
3982-91-0 see TFO000
3983-39-9 see HNO500
3983-40-2 see HNP000
3991-73-9 see OFQ050
3999-01-7 see LGF900
3999-10-8 see PGJ500
4000-16-2 see THN800
4002-76-0 see DCU400
4003-94-5 see NMC000
4005-51-0 see AMR250
4008-48-4 see NHF500
4013-92-7 see IGG300
4016-11-9 see EBQ700
4016-11-9 see EKM200
4016-14-2 see IPD000
4019-40-3 see TIP750
4024-34-4 see TGJ475
4026-97-5 see CMS126
4027-14-9 see TIF500
4027-17-2 see COI000
4028-00-6 see OBI100
4028-15-3 see DUU700
4028-32-4 see BHB950
4032-26-2 see DWY000
4033-46-9 see CCF250
4035-89-6 see TNJ300
4039-32-1 see LGX000
4043-88-3 see DAL375
4044-65-9 see PFA500
4048-33-3 see AKD800
4051-27-8 see DHE000
4052-92-0 see CLG100
4053-08-1 see CJA050
4053-40-1 see LEM000
4055-39-4 see MQX500
4055-40-7 see MQX750
4062-60-6 see DEC100
4063-41-6 see DQF700
4064-09-9 see HAN550
4065-45-6 see HLR700
4067-16-7 see PBD000
4070-75-1 see DDE100
4074-88-8 see ADT250
4074-90-2 see HEO150
4075-79-0 see PDY500
4075-81-4 see CAW400
4075-96-1 see TMA600
4076-02-2 see DNU860
4076-40-8 see MHL750
4079-68-9 see DJX800
4080-31-3 see CEG550
4081-23-6 see SAN300
4083-64-1 see IKG925
4084-27-9 see DAR700
4091-14-9 see VLU220
4091-39-8 see CEU800
4091-50-3 see CNH375
4093-35-0 see VCK100
4095-45-8 see PDB250
4096-33-7 see HIE000
4097-22-7 see DHA325
4097-47-6 see IOX000
4097-89-6 see NEI800
4098-71-9 see IMG000
4100-38-3 see DRM600
4104-14-7 see BIM000
4104-85-2 see THO250
4106-66-5 see DDB800
4109-96-0 see DGK300

4110-66-1 see BGY050
4113-57-9 see CLH625
4114-31-2 see EHG000
4116-10-3 see CEA100
4118-51-8 see IPT000
4119-81-7 see BGY135
4119-82-8 see BGY130
4120-77-8 see AAM250
4120-78-9 see PCY500
4124-30-5 see DEM825
4128-71-6 see PEH750
4128-83-0 see EIP100
4135-11-9 see PKC550
4147-05-1 see HKH550
4147-10-8 see NNP100
4147-51-7 see EPN500
4148-16-7 see LJD500
4149-06-8 see DKR600
4149-60-4 see HAS050
4150-34-9 see TMW000
4154-69-2 see BOV625
4163-15-9 see CQG750
4164-06-1 see ARJ800
4164-07-2 see ARJ760
4164-28-7 see DSV200
4166-00-1 see MRN000
4166-09-0 see PFS600
4168-79-0 see TDU000
4169-04-4 see PNL300
4170-30-3 see COB250
4171-11-3 see DWI300
4171-13-5 see ENJ500
4175-38-6 see BBN100
4176-16-3 see VAG300
4176-53-8 see ALS000
4180-23-8 see PMR250
4185-47-1 see NFW000
4186-64-5 see AGU300
4189-47-3 see BMS250
4193-55-9 see BGW000
4194-69-8 see ICH000
4196-86-5 see PBC000
4197-25-5 see PCJ200
4203-77-4 see CMU825
4205-90-7 see DGB500
4205-91-8 see CMX760
4206-61-5 see DJE200
4208-80-4 see CMM890
4212-43-5 see PCO100
4212-94-6 see ING400
4213-30-3 see BHP250
4213-32-5 see BHU750
4213-34-7 see BHU500
4213-40-5 see BIA000
4213-41-6 see BIC600
4213-44-9 see CLD500
4213-45-0 see QDS000
4219-24-3 see HFD000
4222-21-3 see CIK825
4222-26-8 see CIB725
4222-27-9 see CIB625
4223-11-4 see VMU000
4224-69-5 see MHS300
4224-87-7 see MIF762
4230-21-1 see PDS750
4230-97-1 see AGM500
4231-35-0 see DJX600
4232-84-2 see THO550
4234-79-1 see MDT625
4235-09-0 see PCY750
4238-84-0 see LJH000
4239-06-9 see TDG500

4241-73-0 see GFG070
4242-33-5 see FIP999
4245-76-5 see MML250
4245-77-6 see ENU000
4246-51-9 see DJD800
4247-19-2 see DHC600
4247-30-7 see ECA000
4248-21-9 see EKT600
4248-66-2 see COS909
4248-77-5 see MDR275
4251-89-2 see EHJ000
4253-22-9 see DEI200
4253-34-3 see MQB500
4254-22-2 see ACN875
4255-24-7 see AGQ250
4261-14-7 see BDX100
4261-68-1 see CGV600
4268-36-4 see MOV500
4269-15-2 see AKB900
4269-88-9 see MJC750
4274-06-0 see MIH500
4275-28-9 see TFI100
4279-76-9 see PDR250
4281-67-8 see CPP800
4288-84-0 see CHS250
4297-61-4 see SBF525
4298-10-6 see AQR250
4298-16-2 see DAR100
4301-50-2 see FDB200
4302-87-8 see AJR000
4309-66-4 see AMO000
4310-69-4 see EHI500
4312-87-2 see FMZ000
4312-97-4 see SFV250
4314-63-0 see LJJ000
4317-14-0 see AMY000
4318-42-7 see IRG025
4321-69-1 see CMM400
4323-43-7 see HFG600
4328-17-0 see TMM775
4331-22-0 see VJP800
4331-29-7 see BCC050
4331-98-0 see DGT000
4332-51-8 see BFL100
4333-62-4 see MKU100
4335-09-5 see CMO840
4335-90-4 see DBH900
4336-19-0 see DUF400
4337-66-0 see MRH300
4337-86-4 see ONI300
4339-69-9 see FQB000
4342-03-4 see DAB600
4342-30-7 see SAM000
4342-36-3 see BDR750
4343-68-4 see NMH275
4345-03-3 see TGJ060
4345-16-8 see NNN000
4346-51-4 see ALL800
4350-09-8 see HOA600
4350-09-8 see HOO000
4351-10-4 see HFR500
4351-70-6 see CGO525
4353-28-0 see PBO250
4354-45-4 see MOQ500
4354-73-8 see CPM800
4354-76-1 see PJJ225
4356-33-6 see SKS150
4360-12-7 see AFH250
4360-47-8 see CMQ500
4361-80-2 see AQP750
4362-40-7 see CIL800
4363-03-5 see AIV750

4364-06-1 see PGA775
4366-50-1 see ENM000
4368-28-9 see FOQ000
4368-56-3 see CMM080
4372-46-7 see HMQ575
4373-34-6 see AFH000
4375-07-9 see EBB600
4375-11-5 see IBC000
4376-20-9 see MRI100
4377-73-5 see BDD200
4378-73-8 see OMC000
4384-81-0 see SGR600
4386-79-2 see MJH250
4387-13-7 see BJW800
4388-03-8 see BFX325
4388-07-2 see BLV200
4388-82-3 see CPO500
4390-16-3 see MRK750
4394-00-7 see NDX500
4394-77-8 see HMA600
4394-85-8 see FOA100
4394-93-8 see BJD250
4395-92-0 see IRA000
4403-90-1 see APL500
4408-81-5 see PNJ100
4409-11-4 see CEQ900
4410-48-4 see AFH280
4412-09-3 see MRV000
4413-13-2 see BPK750
4413-31-4 see MIF763
4414-88-4 see BCC000
4415-51-4 see CKI825
4418-26-2 see SGD000
4418-61-5 see AMR000
4418-66-0 see DMN000
4420-65-9 see MRE600
4420-67-1 see VCA100
4420-74-0 see TLC000
4420-79-5 see BHB750
4424-06-0 see CMU820
4425-78-9 see CBS250
4426-51-1 see MQG500
4427-56-9 see IQJ000
4430-31-3 see OBV200
4431-24-7 see EIQ200
4431-47-4 see ITD020
4432-31-9 see MRT150
4434-05-3 see CNV500
4435-53-4 see MHV750
4435-60-3 see PQB800
4437-18-7 see BNQ500
4437-22-3 see FPX025
4437-85-8 see BOT200
4439-24-1 see IIP000
4439-84-3 see AKG500
4441-17-2 see TMX250
4442-79-9 see CPL250
4442-85-7 see CPB500
4449-51-8 see CPT750
4455-26-9 see MND125
4457-71-0 see MNI750
4457-90-3 see PBE000
4460-32-6 see DNE800
4460-46-2 see PER500
4461-41-0 see CEV000
4461-48-7 see MNK750
4463-22-3 see ACD000
4465-92-3 see BHW500
4465-94-5 see CQN000
4466-76-6 see BCS000
4466-77-7 see TDS750
4467-21-4 see CNV050

4468-02-4 see ZIA750	4582-21-2 see NNI000	4722-99-0 see RJU000	4922-98-9 see ALY675
4468-39-7 see MEE150	4584-46-7 see DRC000	4724-58-7 see DPH000	4928-02-3 see AJO280
4468-48-8 see PEA500	4587-15-9 see BHX300	4727-50-8 see KHU050	4928-41-0 see EIB500
4468-93-3 see EGX000	4589-97-3 see HLI400	4727-62-2 see PIK375	4929-77-5 see DHZ100
4470-72-8 see BHC500	4591-46-2 see TDY075	4728-82-9 see AGC250	4933-19-1 see POL750
4471-27-6 see EIP050	4593-81-1 see DDY000	4731-77-5 see BLB250	4940-11-8 see EMA600
4472-06-4 see ASF750	4593-82-2 see TDF500	4732-59-6 see PQC200	4948-28-1 see PIH050
4474-17-3 see CON825	4599-60-4 see MCH525	4736-60-1 see EQD150	4949-20-6 see PBA300
4474-24-2 see CMM092	4602-84-0 see FAB800	4743-57-1 see CCH150	4951-97-7 see EII700
4474-91-3 see IKX030	4611-02-3 see CLZ000	4744-10-9 see DOM200	4955-90-2 see GCU050
4479-75-8 see SBW965	4611-05-6 see CNF109	4746-36-5 see ARY625	4956-14-3 see MNW780
4482-55-7 see FAR050	4617-17-8 see TFF200	4746-61-6 see GHK500	4956-15-4 see DYB250
4484-61-1 see MET000	4618-24-0 see BQG600	4747-82-4 see TJG500	4956-31-4 see CIT000
4484-72-4 see DYA800	4619-66-3 see MIW750	4747-90-4 see DNK100	4964-14-1 see TLH170
4485-12-5 see LHQ100	4620-70-6 see BRH300	4749-28-4 see NIY200	4964-27-6 see DIS850
4485-25-0 see BGU500	4621-04-9 see IOO300	4750-57-6 see MKG800	4965-17-7 see TBI200
4486-44-6 see IRF500	4629-58-7 see NIM575	4752-02-7 see NIL500	4968-09-6 see PMH550
4488-22-6 see BGB750	4630-95-9 see PAB750	4755-59-3 see CMW600	4972-29-6 see BBR390
4492-96-0 see DYE200	4634-47-3 see NNI100	4756-45-0 see HLS500	4979-32-2 see DGU800
4494-26-2 see FNK033	4636-83-3 see BJK750	4756-75-6 see TMQ550	4981-24-2 see DEA400
4498-32-2 see DCW600	4637-56-3 see HIY500	4757-55-5 see DNW700	4981-48-0 see PJB000
4499-01-8 see DGN400	4638-03-3 see AGB750	4759-48-2 see VSK955	4984-82-1 see CPV500
4499-40-5 see CMG300	4638-25-9 see TMI750	4760-34-3 see MNU000	4985-15-3 see DPH600
4499-86-9 see TED000	4638-44-2 see ASI300	4762-14-5 see IOF100	4988-64-1 see MCQ500
4500-29-2 see DMT400	4648-33-3 see MFB600	4764-17-4 see MJQ775	4994-16-5 see PES500
4502-10-7 see AKE000	4648-54-8 see TMF100	4771-80-6 see CPC650	4998-76-9 see CPA775
4503-12-2 see CNS200	4653-73-0 see BSF275	4774-73-6 see DCL350	5001-51-4 see CAT650
4504-27-2 see ASD500	4654-26-6 see BJS300	4779-94-6 see AKT500	5003-48-5 see SAN600
4509-11-9 see ECP000	4656-04-6 see BLR140	4781-76-4 see AKS000	5003-71-4 see AJA000
4511-19-7 see TIY800	4657-20-9 see DWC650	4786-20-3 see COC300	5005-71-0 see PIR000
4514-19-6 see MHI000	4657-93-6 see AAF250	4795-29-3 see TCS500	5008-52-6 see LIN050
4515-18-8 see NLH500	4658-28-0 see ASG250	4800-34-4 see NBF600	5014-35-7 see DHP400
4515-30-4 see CJC000	4662-17-3 see DNF450	4800-94-6 see CBO250	5016-18-2 see DAN450
4524-95-2 see MGR600	4664-00-0 see POQ750	4801-58-5 see PIO925	5022-29-7 see EOR500
4525-33-1 see DRJ850	4664-01-1 see POR000	4803-27-4 see API000	5026-74-4 see ONE100
4525-46-6 see BFM750	4665-48-9 see DCU200	4806-61-5 see EHR000	5029-46-9 see TML100
4526-20-9 see MJD100	4671-75-4 see TCB000	4807-55-0 see MPH750	5034-77-5 see IAN000
4528-34-1 see NFY570	4672-49-5 see BKM125	4808-30-4 see HCA700	5035-67-6 see TID250
4528-52-3 see DAZ120	4676-82-8 see TCS750	4812-22-0 see NHE000	5036-03-3 see EAE500
4531-35-5 see TCG000	4678-44-8 see GJU300	4812-23-1 see NFR500	5042-54-6 see CMM800
4533-39-5 see NFW500	4678-45-9 see GJU310	4812-25-3 see NHT000	5047-01-8 see DUY232
4535-87-9 see CGX250	4680-78-8 see FAE950	4812-40-2 see HMJ000	5051-16-1 see NAD000
4535-90-4 see MIG250	4681-36-1 see FIK000	4822-44-0 see MCK000	5052-75-5 see MMJ960
4536-23-6 see HEN500	4682-94-4 see FQL200	4824-78-6 see EGV500	5053-08-7 see DAI200
4540-66-3 see DXL400	4684-94-0 see CKN375	4826-62-4 see DXU280	5054-57-9 see INU000
4542-46-5 see MCO000	4685-14-7 see PAI990	4829-56-5 see DAK500	5055-20-9 see BKB750
4543-32-2 see HIY300	4685-14-7 see PAJ100	4831-62-3 see MMQ250	5064-31-3 see SIP500
4543-33-3 see TDY120	4685-18-1 see FQQ500	4835-11-4 see DEC699	5075-13-8 see EOM100
4544-15-4 see PIM000	4688-40-8 see SFB150	4836-08-2 see ADY750	5076-19-7 see TLY175
4547-69-7 see BBL825	4691-65-0 see DXE500	4836-09-3 see MPT250	5086-74-8 see TDX750
4548-53-2 see FAG050	4692-94-8 see BDF100	4845-05-0 see CPE125	5090-37-9 see MHJ500
4549-40-0 see NKY000	4693-02-1 see NHK600	4845-14-1 see FNX000	5090-54-0 see VAG250
4549-43-3 see MMT500	4695-62-9 see FAM300	4845-49-2 see DWL800	5096-17-3 see CHI825
4549-44-4 see EHC000	4696-54-2 see BAI825	4845-58-3 see MCP250	5096-18-4 see MME750
4551-59-1 see FAJ100	4696-76-8 see BAU270	4848-63-9 see BLF750	5096-19-5 see BCH250
4551-59-1 see FAJ150	4697-14-7 see TGA520	4849-32-5 see DUM800	5096-21-9 see TLD250
4553-62-2 see MQK000	4698-96-8 see PFU600	4854-84-6 see AJJ770	5096-24-2 see DLM600
4553-89-3 see CMP500	4699-26-7 see DQM400	4856-87-5 see HEG050	5096-57-1 see TCJ800
4559-86-8 see TBM850	4700-56-5 see CFY500	4856-95-5 see MRQ250	5100-91-4 see OOA000
4561-43-7 see PFA750	4702-38-9 see IKH669	4861-79-4 see MES980	5102-83-0 see CMS208
4562-27-0 see ORS050	4702-64-1 see DBT000	4872-26-8 see BSQ500	5103-71-9 see CDR675
4562-27-0 see PPP140	4704-77-2 see MRF275	4880-82-4 see GEO600	5103-74-2 see CDR800
4562-36-1 see GEU000	4707-32-8 see LBC500	4881-17-8 see DUO700	5104-49-4 see FLG100
4564-87-8 see CBT250	4711-74-4 see HED000	4885-02-3 see DFQ000	5106-46-7 see PCM750
4567-22-0 see TDV250	4711-95-9 see ECG200	4891-54-7 see SPF000	5107-01-7 see AFS640
4568-81-4 see DSC100	4711-96-0 see CPD300	4897-25-0 see CIQ100	5107-49-3 see FDA875
4568-82-5 see DNF500	4713-59-1 see DDN200	4901-51-3 see TBS500	5110-69-0 see IEM300
4568-83-6 see EID250	4715-23-5 see AAI115	4904-61-4 see COW935	5116-24-5 see HMB550
4569-77-1 see ALS100	4719-04-4 see THR820	4911-55-1 see PBB000	5117-17-9 see DJY800
4569-88-4 see CMM770	4719-75-9 see TEE300	4914-36-7 see ACP300	5118-29-6 see AEG875
4570-11-0 see DYB100	4720-09-6 see AOO375	4916-38-5 see EQC500	5119-48-2 see EDB300
4570-41-6 see AIS600	4722-68-3 see MJH500	4920-79-0 see CJA200	5123-63-7 see DHM000

5743-04-4 see CAD275	5862-77-1 see EFC300	5967-42-0 see MDJ000	6065-18-5 see NIA000
5743-18-0 see CAK750	5863-35-4 see NHP500	5967-73-7 see MDP770	6065-19-6 see NHZ000
5743-27-1 see CAM600	5868-05-3 see NCW300	5968-79-6 see APE625	6066-49-5 see BSH500
5746-40-7 see AKA900	5869-25-0 see FDE200	5970-32-1 see MCU000	6071-27-8 see DAG750
5746-86-1 see NNS500	5870-29-1 see CPZ125	5970-45-6 see ZCA000	6071-81-4 see COP400
5748-26-5 see IRC050	5870-82-6 see TJL700	5974-19-6 see MQN000	6078-07-5 see HJA600
5749-78-0 see PMU100	5870-93-9 see HBN150	5975-73-5 see BGS750	6080-56-4 see LCJ000
5756-69-4 see HMP000	5874-97-5 see MDM800	5976-47-6 see CEF250	6087-56-5 see DCP300
5760-50-9 see ULS400	5878-19-3 see MDW300	5976-61-4 see HKH850	6088-91-1 see IEG000
5760-73-6 see LBV000	5878-19-3 see MFL100	5976-95-4 see TAF775	6091-11-8 see AQQ250
5763-61-1 see VIK050	5882-45-1 see DIS720	5977-35-5 see AFQ575	6091-44-7 see HET000
5766-67-6 see EJB000	5882-48-4 see DPN800	5978-92-7 see HOR500	6098-44-8 see ABL000
5774-35-6 see BGX850	5888-33-5 see IHX700	5980-33-6 see HMB000	6098-46-0 see BDP000
5776-49-8 see PHA750	5888-51-7 see EQF150	5980-86-9 see CFD250	6104-30-9 see IIV000
5779-79-3 see AAF000	5888-61-9 see THU000	5985-35-3 see HMH500	6106-07-6 see SKJ325
5786-21-0 see CMY650	5890-18-6 see LBO100	5985-35-3 see MDV250	6106-46-3 see SBH520
5786-68-5 see QWJ500	5892-15-9 see AIT000	5986-38-9 see DTC990	6109-70-2 see QUS000
5786-77-6 see TLQ250	5892-41-1 see AAD875	5987-82-6 see OPI300	6109-97-3 see AJV250
5787-50-8 see BSE700	5892-48-8 see SEG000	5988-31-8 see AIT500	6111-78-0 see MGX250
5787-73-5 see NCL300	5894-60-0 see HCQ000	5988-51-2 see DPA200	6112-39-6 see MHJ000
5788-49-8 see DTU825	5897-16-5 see CAQ600	5988-91-0 see TCY300	6112-76-1 see MCQ100
5789-17-3 see IAT100	5902-51-2 see BQT750	5989-27-5 see LFU000	6117-91-5 see BOY000
5796-14-5 see DYC200	5902-52-3 see BQU500	5989-54-8 see MCC500	6120-10-1 see DQF000
5796-89-4 see PCO150	5902-76-1 see MLG250	5989-77-5 see DLL000	6130-87-6 see TDW300
5797-06-8 see PCM550	5902-79-4 see MLF500	5991-71-9 see DKV700	6130-92-3 see AMQ750
5798-79-8 see BMW250	5902-85-2 see HKO012	6000-43-7 see GHK000	6130-93-4 see TDS000
5800-19-1 see MQQ000	5902-95-4 see CAM000	6000-44-8 see GHG000	6131-99-3 see HKC550
5803-62-3 see HLF550	5902-97-6 see TJK900	6000-82-4 see TME255	6145-29-5 see SIL575
5805-95-8 see HLN135	5903-13-9 see MME809	6001-93-0 see DGD100	6146-52-7 see NHK500
5806-00-8 see BKS812	5905-52-2 see LAL000	6002-77-3 see PCC475	6147-53-1 see CNA500
5806-84-8 see BKX500	5906-35-4 see HEG400	6004-24-6 see CDF750	6150-94-3 see MAI700
5807-94-3 see DHA500	5907-38-0 see MDM500	6004-98-4 see HFG000	6151-25-3 see QCA175
5809-59-6 see HJQ000	5910-75-8 see DXQ000	6009-67-2 see DIF200	6152-43-8 see DPD400
5810-11-7 see CGH810	5910-77-0 see TNN760	6009-81-0 see MRP100	6152-43-8 see DTO800
5810-88-8 see DJK400	5910-85-0 see HAV450	6011-14-9 see GHI100	6152-95-0 see PAP100
5814-20-0 see TDL250	5910-87-2 see NMV775	6011-62-7 see IGG600	6153-33-9 see MBW100
5826-73-3 see DRB400	5910-89-4 see DTU400	6012-97-1 see TBV750	6155-81-3 see DSL000
5826-91-5 see DJX400	5913-76-8 see CNG675	6014-43-3 see SMN200	6156-78-1 see MAQ250
5827-03-2 see DKB600	5913-82-6 see DOX000	6018-32-2 see ARA000	6158-73-2 see TMU775
5827-05-4 see EPH500	5915-41-3 see BQB000	6018-89-9 see NCX500	6159-44-0 see UQT700
5827-58-7 see DHA600	5917-61-3 see MJO800	6019-02-9 see NDN100	6159-55-3 see VGA000
5830-30-8 see DQX200	5921-54-0 see EMT600	6023-26-3 see HIS000	6163-73-1 see TLA600
5831-08-3 see MEU750	5926-26-1 see CIY750	6027-28-7 see BQA750	6164-98-3 see CJJ250
5831-09-4 see DRH400	5926-90-9 see GGY125	6028-07-5 see HAI300	6165-01-1 see BRS500
5831-10-7 see DRH200	5928-69-8 see LFW300	6029-87-4 see FOT000	6168-86-1 see ODY000
5831-11-8 see TLP000	5929-01-1 see BDV750	6030-03-1 see EMS000	6169-12-6 see AQE500
5831-12-9 see MEV000	5929-09-9 see BBU750	6030-80-4 see ECW500	6175-45-7 see PFH260
5831-16-3 see DLI200	5933-35-7 see NIJ450	6032-29-7 see PBM750	6180-21-8 see LHC000
5831-17-4 see DMF600	5934-20-3 see DPI000	6033-07-4 see HNQ000	6186-91-0 see CHJ250
5834-17-3 see MDZ000	5934-69-0 see DAB807	6034-59-9 see TNV700	6190-33-6 see BFN750
5834-25-3 see DDC000	5936-28-7 see HGQ500	6035-40-1 see NBP275	6191-22-6 see ALR250
5834-81-1 see PFN500	5939-37-7 see HLQ100	6036-95-9 see MCJ250	6192-13-8 see NBX500
5834-84-4 see PNH800	5942-95-0 see CBH500	6040-19-3 see COE180	6192-29-6 see BRS250
5834-96-8 see CJR300	5945-40-4 see MQR765	6044-68-4 see DOM600	6192-36-5 see PDM100
5836-10-2 see PNH750	5949-18-8 see SJH000	6046-93-1 see CNI325	6192-44-5 see HZL250
5836-28-2 see TDF750	5949-44-0 see TBG050	6047-01-4 see CQL750	6197-30-4 see ODY150
5836-29-3 see EAT600	5950-69-6 see HHI000	6047-17-2 see TIX750	6198-57-8 see VGU000
5836-73-7 see DEQ000	5952-41-0 see CDN510	6051-87-2 see NAU525	6199-67-3 see COE190
5836-85-1 see DLR200	5954-50-7 see DSA800	6052-82-0 see DKB100	6202-23-9 see DPX800
5837-17-2 see DLU600	5954-90-5 see MOB750	6055-19-2 see CQC500	6211-32-1 see YCA000
5837-78-5 see TGA800	5956-60-5 see BFN550	6055-52-3 see HEO100	6217-24-9 see DVY100
5840-95-9 see EMC000	5956-63-8 see BFO100	6055-69-2 see ADH875	6219-71-2 see CEG625
5843-82-3 see OEM000	5957-75-5 see TCM000	6062-26-6 see CLO000	6219-89-2 see AMT300
5847-48-3 see GHQ000	5959-42-2 see DWB000	6064-83-1 see PHA575	6225-06-5 see DUN200
5847-52-9 see TIC750	5959-52-4 see ALJ000	6065-01-6 see NHV500	6237-23-6 see NFW535
5847-55-2 see DEJ250	5959-56-8 see ALK500	6065-04-9 see NHU000	6237-24-7 see NFW525
5851-43-4 see IOB500	5959-98-8 see CHD700	6065-09-4 see NHX000	6238-69-3 see NKO000
5853-29-2 see CCX125	5960-88-3 see DFG700	6065-10-7 see NHW500	6240-55-7 see DFU400
5854-93-3 see AFH750	5962-42-5 see PHA560	6065-11-8 see NHW000	6247-46-7 see CMU800
5857-37-4 see CHK000	5964-24-9 see SKH150	6065-13-0 see NHB500	6249-65-6 see HNK500
5857-94-3 see AJH500	5965-13-9 see DKX000	6065-14-1 see NHB000	6251-69-0 see ECA100
5858-81-1 see CMS155	5967-09-9 see BGQ000	6065-17-4 see NHD000	6258-06-6 see DKR000

6262-51-7 see PAV810
6264-93-3 see AJJ800
6268-32-2 see NLG500
6269-50-7 see HKE500
6270-33-3 see FPM300
6272-74-8 see LBD100
6273-75-2 see PDL500
6275-69-0 see DBM400
6279-54-5 see BRC750
6279-87-4 see BIG600
6280-15-5 see MKI250
6280-75-7 see DQQ600
6280-99-5 see DED500
6281-23-8 see NGI000
6283-24-5 see ABQ000
6283-25-6 see CJA180
6283-63-2 see DJV250
6285-05-8 see CKT500
6285-34-3 see DAJ450
6285-57-0 see NFI240
6288-93-3 see PNW800
6292-55-3 see FDS000
6292-91-7 see MJS750
6294-34-4 see BIB800
6294-93-5 see CLS000
6295-12-1 see HJI000
6295-57-4 see CCH750
6296-45-3 see CHF500
6300-37-4 see CMP090
6303-21-5 see PGY250
6304-07-0 see DHK200
6304-33-2 see TMQ250
6305-04-0 see HNE500
6305-18-6 see MFH900
6305-43-7 see DDK600
6307-17-1 see NIM630
6307-82-0 see CJC515
6310-09-4 see CDN525
6317-18-6 see MJT500
6318-57-6 see HJE400
6320-14-5 see CMM840
6325-54-8 see CIG250
6325-93-5 see NFB000
6332-68-9 see BAL275
6334-11-8 see TLH000
6334-30-1 see THO500
6334-96-9 see OPE040
6336-12-5 see TEP750
6341-85-1 see DHD200
6358-06-1 see AJI260
6358-29-8 see CMO875
6358-53-8 see DOK200
6358-64-1 see CGC050
6358-69-6 see TNM000
6358-85-6 see DEU000
6360-54-9 see CMO825
6361-21-3 see CJA500
6362-79-4 see DEK200
6365-83-9 see BPG250
6366-20-7 see MEB500
6366-23-0 see MGY250
6366-24-1 see CIG500
6368-72-5 see EOJ500
6369-59-1 see TGM400
6370-43-0 see OHK000
6371-76-2 see CMS160
6373-20-2 see CMU750
6373-74-6 see SGP500
6375-55-9 see CMM759
6376-26-7 see DHP200
6377-18-0 see CDK250
6378-65-0 see HFQ500

6379-37-9 see OEK500
6379-46-0 see DVI600
6379-69-7 see TJE750
6380-21-8 see PMS250
6381-61-9 see ANT500
6381-77-7 see EDE700
6382-01-0 see MRK525
6385-02-0 see SIF425
6392-46-7 see DBI800
6393-42-6 see DVI100
6397-02-0 see CMM100
6407-29-0 see DDC100
6411-75-2 see AKD600
6413-10-1 see EFR100
6414-38-6 see DLQ000
6416-57-5 see NBG500
6416-68-8 see TGE155
6419-19-8 see NEI100
6420-06-0 see ASM100
6420-47-9 see BRE750
6422-83-9 see TGY770
6422-86-2 see BJS500
6422-99-7 see HEG130
6423-43-4 see PNL000
6424-34-6 see TGU500
6424-75-5 see CMM090
6424-76-6 see CMU500
6428-31-5 see CMN240
6428-38-2 see CMN300
6441-77-6 see CMM000
6443-91-0 see MDX000
6448-90-4 see DNY800
6452-54-6 see AGP250
6452-62-6 see TDI250
6452-71-7 see CNR500
6452-73-9 see AGP000
6452-73-9 see THK750
6453-98-1 see FQO050
6459-94-5 see CMM330
6465-92-5 see PHC750
6470-09-3 see CCA125
6471-49-4 see NAY000
6483-64-3 see DOO400
6483-86-9 see DVR400
6484-52-2 see ANN000
6485-34-3 see CAM750
6485-39-8 see MAS800
6485-40-1 see CCM120
6485-91-2 see DHF800
6493-05-6 see PBU100
6493-69-2 see SLI335
6504-77-4 see AKN750
6505-75-5 see CJF500
6506-37-2 see NHH000
6509-08-6 see EBK500
6512-83-0 see SBU150
6514-85-8 see LEF300
6515-09-9 see TJC900
6515-38-4 see HOL125
6522-40-3 see DGL875
6522-86-7 see DGN800
6533-00-2 see NNQ500
6533-68-2 see SBG600
6533-73-9 see TEJ000
6540-76-7 see IDV000
6542-37-6 see OMM300
6547-08-6 see MPI200
6548-29-4 see CMO880
6556-11-2 see HFG550
6558-78-7 see BRZ000
6569-51-3 see BMB500
6569-69-3 see OLM000

6576-51-8 see DXG600
6577-41-9 see OLW400
6580-41-2 see ASP750
6581-06-2 see QVA000
6592-85-4 see HGR000
6592-90-1 see DQV200
6596-50-5 see SBU650
6600-31-3 see DGT400
6601-20-3 see DBL600
6607-45-0 see DGK800
6607-49-4 see VNA000
6607-53-0 see MQM200
6610-08-8 see NLA500
6611-01-4 see TDQ225
6618-03-7 see TNA500
6619-97-2 see XPJ300
6620-60-6 see BGC625
6622-76-0 see MPW700
6623-41-2 see AMW750
6623-66-1 see NFV000
6624-70-0 see AEQ500
6627-34-5 see DFT100
6628-83-7 see ALF750
6629-04-5 see COJ750
6629-29-4 see NFV540
6630-99-5 see AMS000
6632-39-9 see NIQ600
6632-68-4 see DPN400
6637-88-3 see CMO860
6639-36-7 see NFV700
6639-99-2 see EDT500
6640-24-0 see CKF800
6645-46-1 see CCK660
6649-23-6 see LFA000
6652-04-6 see PFT600
6654-31-5 see AGG250
6659-60-5 see BOU700
6659-62-7 see ECI600
6667-75-0 see BLH315
6671-96-1 see ASA800
6673-35-4 see ECX100
6683-19-8 see DED100
6696-47-5 see OHO000
6696-58-8 see THI000
6700-56-7 see MIE600
6707-60-4 see OKW110
6708-14-1 see BFX545
6708-69-6 see BJP500
6711-48-4 see TDP750
6714-29-0 see IAN100
6724-53-4 see PCH800
6728-21-8 see AGK750
6728-26-3 see PNC500
6732-77-0 see MFG525
6734-80-1 see SIL550
6735-35-9 see PNH775
6736-03-4 see PGY600
6740-88-1 see KEK200
6742-07-0 see AQR500
6746-27-6 see TCW800
6746-59-4 see ENK500
6753-98-6 see HGM550
6754-13-8 see HAK300
6775-25-3 see TGK225
6784-25-4 see BJW750
6785-70-2 see NDY565
6786-32-9 see BDN500
6789-80-6 see HFA515
6789-88-4 see HFL500
6795-16-0 see HKB300
6795-23-9 see AEW000
6795-60-4 see NCI525

6797-13-3 see EGR500
6804-07-5 see FOI000
6805-41-0 see EDK875
6807-96-1 see TKO500
6809-93-4 see BCH800
6809-94-5 see BFZ180
6809-95-6 see BFZ100
6810-26-0 see BGI250
6813-90-7 see BFZ110
6813-91-8 see MFX550
6813-92-9 see HNT550
6813-93-0 see BFZ130
6813-95-2 see BFZ140
6818-18-4 see DCF600
6818-37-7 see BJY800
6832-13-9 see DCN200
6832-16-2 see MIX750
6834-92-0 see SJU000
6834-98-6 see FPC000
6836-11-9 see EAI050
6837-24-7 see CPQ275
6837-93-0 see AKH500
6837-97-4 see DFC600
6842-15-5 see PMP750
6843-30-7 see DOU700
6846-35-1 see IBL000
6846-50-0 see TLZ000
6856-01-5 see EQT100
6856-43-5 see CPM750
6863-58-7 see BRH760
6863-58-7 see OPE030
6864-37-5 see BGT800
6865-35-6 see BAO825
6865-68-5 see TNG150
6866-10-0 see CJG000
6869-17-6 see HAN900
6869-51-8 see RFZ000
6870-67-3 see JAK000
6872-06-6 see MKV800
6874-80-2 see CAZ125
6875-10-1 see VIZ200
6884-59-9 see SAG100
6887-42-9 see DAM410
6889-41-4 see CDS100
6892-68-8 see DXN350
6893-02-3 see LGK050
6893-20-5 see MER000
6893-24-9 see MEQ750
6898-43-7 see BCC250
6900-06-7 see AEQ750
6901-97-9 see IFW000
6902-77-8 see GCM300
6902-91-6 see GDO200
6914-98-3 see PEL650
6915-00-0 see CJR580
6915-15-7 see MAN000
6918-51-0 see DUB000
6921-27-3 see POA500
6921-35-3 see EBQ500
6923-22-4 see MRH209
6923-52-0 see AQJ750
6925-01-5 see BBN850
6926-39-2 see CLH500
6926-40-5 see BOG250
6928-74-1 see DXQ850
6933-90-0 see IMX150
6935-63-3 see SLS500
6937-66-2 see MCT000
6938-06-3 see NDT500
6938-94-9 see DNL800
6939-83-9 see TNH850
6943-65-3 see EJA000

7446-32-4 see AQJ250
7446-34-6 see SBT000
7446-70-0 see AGY750
7447-39-4 see CNK500
7447-40-7 see PLA500
7447-41-8 see LHB000
7447-44-1 see ALB500
7448-86-4 see ONW100
7450-97-7 see OKW000
7452-79-1 see EMP600
7456-24-8 see DUC400
7458-65-3 see HDS200
7460-84-6 see SLK500
7466-54-8 see AOV250
7467-29-0 see MLP770
7473-98-5 see HMQ100
7476-08-6 see COK500
7476-91-7 see PDF500
7481-89-2 see DHA350
7483-25-2 see DDY400
7487-28-7 see BJN500
7487-88-9 see MAJ250
7487-94-7 see MCY475
7488-55-3 see TGF010
7488-56-4 see SBR000
7488-70-2 see TFZ300
7491-74-9 see NNE400
7492-29-7 see IRW000
7492-37-7 see BEN800
7492-44-6 see BQV250
7492-66-2 see CMS323
7492-67-3 see CMT300
7492-70-8 see BQP000
7493-57-4 see PDD400
7493-74-5 see AGQ750
7493-78-9 see AOG750
7495-45-6 see EES300
7495-93-4 see CAD500
7496-02-8 see NFT400
7499-32-3 see DLT000
7505-62-6 see BBC750
7506-77-6 see DJW100
7506-80-1 see HKS550
7511-54-8 see EHM000
7517-76-2 see CPB150
7518-35-6 see MAW800
7519-36-0 see NLL500
7521-80-4 see BSR000
7525-62-4 see EPG000
7526-26-3 see MDQ825
7529-27-3 see EJE000
7530-07-6 see OFI000
7531-39-7 see DJW200
7532-52-7 see ALM750
7532-60-7 see BQH800
7532-85-6 see DXR450
7535-34-4 see MKK250
7538-45-6 see MCO250
7539-12-0 see AGS000
7542-37-2 see NCF500
7546-30-7 see MCW000
7548-44-9 see BAT795
7548-46-1 see BAT800
7549-37-3 see DOE000
7550-35-8 see LGY000
7550-45-0 see TGH350
7553-56-2 see IDM000
7554-65-6 see MOX000
7558-63-6 see MRF000
7558-79-4 see SJH090
7558-80-7 see SJH100
7560-83-0 see MJC750

7562-61-0 see UWJ100
7563-42-0 see CAV000
7568-37-8 see MHY550
7568-93-6 see HNF000
7570-25-4 see VLY300
7570-26-5 see DUV720
7572-29-4 see DEN600
7575-48-6 see TMG750
7576-88-7 see AJR600
7578-36-1 see BMG750
7580-37-2 see TDH250
7580-67-8 see LHH000
7585-39-9 see COW925
7585-41-3 see CMS148
7599-79-3 see CMX750
7601-54-9 see SJH200
7601-55-0 see DUM000
7601-87-8 see DRN600
7601-89-0 see PCE750
7601-90-3 see PCD250
7613-16-3 see OCW000
7616-83-3 see MCY755
7616-83-3 see MDG200
7616-94-6 see PCF750
7621-86-5 see ALV050
7631-86-9 see SCI000
7631-89-2 see ARD750
7631-90-5 see SFE000
7631-95-0 see DXE800
7631-98-3 see DXZ000
7631-99-4 see SIO900
7632-00-0 see SIQ500
7632-04-4 see SJB100
7632-10-2 see DBB000
7632-50-0 see ANF800
7632-51-1 see VEF000
7633-57-0 see NKN000
7635-51-0 see DTN800
7637-07-2 see BMG700
7641-77-2 see HEC000
7644-67-9 see ASO501
7645-25-2 see ARC750
7646-69-7 see SHO500
7646-78-8 see TGC250
7646-79-9 see CNB599
7646-85-7 see ZFA000
7646-93-7 see PKX750
7647-01-0 see HHL000
7647-01-0 see HHX000
7647-01-0 see HHX500
7647-10-1 see PAD500
7647-14-5 see SFT000
7647-15-6 see SFG500
7647-17-8 see CDD000
7647-18-9 see AQD000
7647-19-0 see PHR750
7648-01-3 see EPF500
7650-84-2 see PNI600
7651-40-3 see AGL875
7651-91-4 see DFO900
7652-64-4 see BLG400
7654-03-7 see NCQ100
7659-86-1 see EKW300
7660-25-5 see LFI000
7660-71-1 see MDM000
7664-38-2 see PHB250
7664-39-3 see HHU500
7664-41-7 see AMY500
7664-93-9 see SOI500
7664-93-9 see SOI530
7664-98-4 see DWV000
7665-72-7 see BRK800

7672-94-8 see CLH800
7673-09-8 see TNE775
7680-73-1 see MOY875
7681-11-0 see PLK500
7681-28-9 see GAX000
7681-34-7 see SAQ000
7681-38-1 see SEG800
7681-49-4 see SHF500
7681-52-9 see SHU500
7681-53-0 see SHV000
7681-55-2 see SHV500
7681-57-4 see SII000
7681-65-4 see COF680
7681-76-7 see MMN750
7681-82-5 see SHW000
7681-93-8 see PIF750
7682-90-8 see BAS000
7683-59-2 see DMV600
7689-03-4 see CBB870
7691-02-3 see TDQ050
7693-26-7 see PLJ250
7693-52-9 see NFQ200
7695-91-2 see TGJ055
7697-37-2 see NED500
7697-37-2 see NEE500
7697-46-3 see PGF900
7698-91-1 see MGL600
7698-97-7 see FAO200
7699-31-2 see DJL600
7699-41-4 see SCL000
7699-43-6 see ZSJ000
7700-17-6 see COD000
7704-34-9 see SOD500
7704-71-4 see MAF750
7704-99-6 see ZRA000
7705-07-9 see TGG250
7705-08-0 see FAU000
7705-12-6 see IHB675
7718-54-9 see NDH000
7718-98-1 see VEP000
7719-09-7 see TFL000
7719-12-2 see PHT275
7720-78-7 see FBN100
7721-01-9 see TAF000
7722-06-7 see AMN300
7722-64-7 see PLP000
7722-73-8 see HGN000
7722-84-1 see HIB010
7722-84-1 see HIB050
7722-86-3 see PCN750
7722-88-5 see TEE500
7723-14-0 see PHO500
7723-14-0 see PHP010
7726-95-6 see BMP000
7727-15-3 see AGX750
7727-18-6 see VDP000
7727-21-1 see DWQ000
7727-37-9 see NGP500
7727-43-7 see BAP000
7727-54-0 see ANR000
7732-18-5 see WAT259
7733-02-0 see ZNA000
7738-94-5 see CMH250
7739-33-5 see TKU000
7745-89-3 see CLK230
7751-31-7 see BHQ750
7757-79-1 see PLL500
7757-81-5 see SJV000
7757-82-6 see SJY000
7757-83-7 see SJZ000
7757-93-9 see CAW100
7758-01-2 see PKY300

7758-02-3 see PKY500
7758-05-6 see PLK250
7758-09-0 see PLM500
7758-16-9 see DXF800
7758-19-2 see SFT500
7758-23-8 see CAW110
7758-88-5 see CDA750
7758-89-6 see CNK250
7758-94-3 see FBI000
7758-95-4 see LCQ000
7758-97-6 see LCR000
7758-98-7 see CNP250
7758-99-8 see CNP500
7759-35-5 see HMB650
7761-45-7 see MQR100
7761-88-8 see SDS000
7763-77-1 see CGX000
7764-50-3 see DKV175
7765-88-0 see MGD210
7770-47-0 see MKW000
7771-44-0 see EAF050
7772-76-1 see ANR750
7772-98-7 see SKI000
7772-99-8 see TGC000
7773-01-5 see MAR000
7773-06-0 see ANU650
7773-34-4 see DJB200
7774-29-0 see MDD000
7774-29-0 see MDD250
7774-41-6 see ARC500
7774-65-4 see MCF515
7774-82-5 see TJJ400
7775-09-9 see SFS000
7775-11-3 see DXC200
7775-14-6 see SHR500
7775-19-1 see SII100
7775-27-1 see SJE000
7775-41-9 see SDQ000
7776-33-2 see SBU710
7778-18-9 see CAX500
7778-39-4 see ARB250
7778-43-0 see ARC000
7778-44-1 see ARB750
7778-50-9 see PKX250
7778-53-2 see PLQ410
7778-54-3 see HOV500
7778-66-7 see PLK000
7778-74-7 see PLO500
7778-77-0 see PLQ405
7778-80-5 see PLT000
7778-83-8 see PNH250
7779-27-3 see HDW000
7779-41-1 see AFJ700
7779-80-8 see IIS000
7779-86-4 see ZGJ100
7779-86-4 see ZIJ100
7779-88-6 see ZJJ000
7779-90-0 see ZJS400
7780-06-5 see IOO000
7782-39-0 see DBB800
7782-41-4 see FEZ000
7782-44-7 see OQW000
7782-49-2 see SBO500
7782-49-2 see SBP000
7782-50-5 see CDV750
7782-61-8 see IHC000
7782-63-0 see FBO000
7782-64-1 see MAS750
7782-65-2 see GEI100
7782-68-5 see IDK000
7782-75-4 see MAH775
7782-77-6 see NMR000

7782-78-7 see NMJ000
7782-79-8 see HHG500
7782-82-3 see SBO200
7782-87-8 see PLQ750
7782-89-0 see LGT000
7782-92-5 see SEN000
7782-94-7 see NEG000
7782-99-2 see SOO500
7783-00-8 see SBO000
7783-06-4 see HIC500
7783-07-5 see HIC000
7783-08-6 see SBN500
7783-18-8 see ANK600
7783-20-2 see ANU750
7783-28-0 see ANR500
7783-30-4 see MDC750
7783-33-7 see NCP500
7783-35-9 see MDG500
7783-36-0 see MDG250
7783-40-6 see MAF500
7783-41-7 see ORA000
7783-42-8 see TFL250
7783-46-2 see LDF000
7783-47-3 see TGD100
7783-48-4 see SMI500
7783-49-5 see ZHS000
7783-50-8 see FAX000
7783-53-1 see MAW000
7783-54-2 see NGW000
7783-55-3 see PHQ500
7783-56-4 see AQE000
7783-60-0 see SOR000
7783-61-1 see SDF650
7783-64-4 see ZQS000
7783-66-6 see IDT000
7783-70-2 see AQF250
7783-71-3 see TAF250
7783-79-1 see SBS000
7783-80-4 see TAK250
7783-81-5 see UOJ000
7783-82-6 see TOC550
7783-85-9 see IGL200
7783-91-7 see SDN500
7783-92-8 see SDN399
7783-93-9 see SDV000
7783-95-1 see SDQ500
7784-08-9 see SDM100
7784-13-6 see AGZ000
7784-18-1 see AHB000
7784-19-2 see THQ500
7784-21-6 see AHB500
7784-24-9 see AHF200
7784-27-2 see AHD900
7784-30-7 see PHB500
7784-31-8 see AHG800
7784-33-0 see ARF250
7784-34-1 see ARF500
7784-35-2 see ARI250
7784-37-4 see MDF350
7784-40-9 see LCK000
7784-41-0 see ARD250
7784-42-1 see ARK250
7784-44-3 see DCG800
7784-45-4 see ARG750
7784-46-5 see SEY500
7785-20-8 see NCY060
7785-33-3 see GDO000
7785-84-4 see SKM500
7785-87-7 see MAU250
7786-17-6 see MJO750
7786-29-0 see MNC175
7786-30-3 see MAE250

7786-34-7 see MQR750
7786-61-0 see VPF100
7786-67-6 see MCE750
7786-81-4 see NDK500
7787-32-8 see BAM000
7787-36-2 see PCK000
7787-47-5 see BFQ000
7787-49-7 see BFR500
7787-52-2 see BFR750
7787-55-5 see BFT100
7787-56-6 see BFU500
7787-62-4 see BKW750
7787-68-0 see BKY600
7787-69-1 see CDC500
7787-71-5 see BMQ325
7788-97-8 see CMJ560
7788-98-9 see ANF500
7788-98-9 see NCQ550
7788-99-0 see CMG850
7789-00-6 see PLB250
7789-02-8 see CMJ610
7789-04-0 see CMK300
7789-06-2 see SMH000
7789-09-5 see ANB500
7789-12-0 see SGI500
7789-17-5 see CDE000
7789-18-6 see CDE250
7789-20-0 see HAK000
7789-21-1 see FLZ000
7789-23-3 see PLF500
7789-24-4 see LHF000
7789-25-5 see NMH500
7789-26-6 see NMU000
7789-27-7 see TEK000
7789-29-9 see PKU250
7789-29-9 see PKU500
7789-30-2 see BMQ000
7789-33-5 see IDN200
7789-38-0 see SFG000
7789-40-4 see TEI750
7789-41-5 see CAR375
7789-42-6 see CAD600
7789-43-7 see CNB250
7789-46-0 see IGP000
7789-47-1 see MCY000
7789-51-7 see SBN550
7789-57-3 see THX000
7789-59-5 see PHU000
7789-60-8 see PHT250
7789-61-9 see AQK000
7789-67-5 see TGB750
7789-69-7 see PHR250
7789-75-5 see CAS000
7789-77-7 see CAT210
7789-78-8 see CAT200
7789-79-9 see CAT250
7789-80-2 see CAT500
7789-82-4 see CAT750
7789-89-1 see TJB000
7789-90-4 see TJC750
7789-92-6 see TJM250
7789-99-3 see MGN300
7790-01-4 see MFX000
7790-03-6 see COM125
7790-07-0 see AEP250
7790-12-7 see AKD625
7790-21-8 see PLO750
7790-28-5 see SJB500
7790-29-6 see RQA000
7790-30-9 see TEK500
7790-47-8 see TGD750
7790-58-1 see PLU000

7790-59-2 see PLR750
7790-76-3 see CAW450
7790-78-5 see CAE425
7790-79-6 see CAG250
7790-80-9 see CAG550
7790-84-3 see CAJ250
7790-86-5 see CCY750
7790-91-2 see CDX750
7790-92-3 see HOV000
7790-93-4 see CDU000
7790-94-5 see CLG500
7790-98-9 see PCD500
7790-99-0 see IDS000
7791-03-9 see LHM875
7791-07-3 see SJB450
7791-10-8 see SMF500
7791-11-9 see RPF000
7791-12-0 see TEJ250
7791-13-1 see CNB800
7791-18-6 see MAE500
7791-20-0 see NDA000
7791-21-1 see DEL600
7791-23-3 see SBT500
7791-25-5 see SOT000
7791-26-6 see URA000
7791-27-7 see PPR500
7795-91-7 see EJJ000
7796-16-9 see TIV275
7797-83-3 see BCI800
7803-49-8 see HLM500
7803-51-2 see PGY000
7803-52-3 see SLQ000
7803-55-6 see ANY250
7803-57-8 see HGU501
7803-62-5 see SDH575
7803-63-6 see ANJ500
7803-68-1 see TAI750
8000-03-1 see EDQ000
8000-25-7 see RMU000
8000-26-8 see PIH400
8000-26-8 see PIH500
8000-27-9 see CCR000
8000-28-0 see LCC000
8000-28-0 see LCD000
8000-29-1 see CMT000
8000-34-8 see CMY475
8000-34-8 see CMY510
8000-42-8 see CBG500
8000-46-2 see GDA000
8000-48-4 see EQQ000
8000-66-6 see CCJ625
8000-68-8 see PAL750
8000-78-0 see GBU825
8000-95-1 see CAK800
8001-15-8 see EAM500
8001-23-8 see SAC000
8001-25-0 see OIQ000
8001-26-1 see LGK000
8001-28-3 see COC250
8001-29-4 see CNU000
8001-30-7 see CNS000
8001-31-8 see CNR000
8001-35-2 see CDV100
8001-50-1 see TBC500
8001-54-5 see AFP250
8001-58-9 see CMY825
8001-61-4 see CNH792
8001-64-7 see CNH775
8001-67-0 see CNG800
8001-74-9 see CDK750
8001-78-3 see HHW502
8001-79-4 see CCP250

8001-85-2 see BMA750
8001-88-5 see BGO750
8001-95-4 see AGW300
8002-01-5 see AER750
8002-03-7 see PAO000
8002-05-9 see PCR250
8002-05-9 see PCS250
8002-06-0 see PDO275
8002-09-3 see PIH750
8002-26-4 see TAC000
8002-33-3 see TOD500
8002-46-8 see LGK225
8002-65-1 see NBS300
8002-66-2 see CDH500
8002-68-4 see JEA000
8002-73-1 see OHJ130
8002-74-2 see PAH750
8002-75-3 see PAE500
8002-80-0 see WBL100
8002-89-9 see TEQ250
8003-03-0 see ARP250
8003-05-2 see MDH500
8003-08-5 see POF275
8003-19-8 see DGG000
8003-22-3 see CMS240
8003-34-7 see POO250
8004-13-5 see PFA860
8004-86-2 see HGE925
8004-87-3 see MQN025
8004-92-0 see CMM510
8005-78-5 see CMM780
8006-25-5 see EDC565
8006-28-8 see SEE000
8006-38-0 see ADH500
8006-39-1 see TBD500
8006-44-8 see CBC175
8006-54-0 see LAU520
8006-61-9 see GBY000
8006-64-2 see TOD750
8006-75-5 see DNU400
8006-77-7 see PIG740
8006-78-8 see LBK000
8006-80-2 see OHI000
8006-81-3 see YAT000
8006-82-4 see BLW250
8006-83-5 see HOX000
8006-84-6 see FAP000
8006-87-9 see OGY220
8006-90-4 see PCB250
8006-99-3 see CDL500
8007-00-9 see PCP750
8007-00-9 see PCQ000
8007-01-0 see RNA000
8007-01-0 see RNF000
8007-02-1 see LEH000
8007-04-3 see HGK725
8007-08-7 see GEQ000
8007-11-2 see OJO000
8007-12-3 see OGQ100
8007-20-3 see CCQ500
8007-43-0 see SKV170
8007-45-2 see CMY800
8007-45-2 see CMY805
8007-46-3 see TFX500
8007-46-3 see TFX750
8007-47-4 see FBS200
8007-48-5 see OGQ150
8007-56-5 see HHM000
8007-70-3 see AOU250
8007-75-8 see BFO000
8007-80-5 see CCO750
8007-87-2 see COE175

8008-20-6 see KEK000
8008-26-2 see OGM850
8008-45-5 see NOG500
8008-46-6 see OGU000
8008-51-3 see CBB500
8008-52-4 see CNR735
8008-53-5 see EEH580
8008-56-8 see LEI000
8008-56-8 see LEJ000
8008-57-9 see OGY000
8008-74-0 see SCB000
8008-79-5 see SKY000
8008-80-8 see SLG650
8008-88-6 see BAE100
8008-94-4 see LFN300
8011-47-0 see CBB375
8011-48-1 see PIH775
8011-76-5 see SOV500
8012-34-8 see MCV750
8012-54-2 see ARI500
8012-74-6 see LIC000
8012-89-3 see BAU000
8012-95-1 see MQV750
8012-96-2 see IGF000
8013-01-2 see YAK050
8013-07-8 see FCC100
8013-10-3 see JEJ000
8013-17-0 see IDH200
8013-38-5 see CNG827
8013-75-0 see FQT000
8013-76-1 see BLV500
8013-77-2 see SAX000
8013-86-3 see CQJ000
8013-90-9 see IFV000
8014-09-3 see PAM783
8014-13-9 see COF325
8014-17-3 see OHJ150
8014-19-5 see PAE000
8014-29-7 see OGY200
8014-95-7 see SOI520
8015-01-8 see MBU500
8015-12-1 see EEH520
8015-14-3 see LJE000
8015-18-7 see AMK250
8015-19-8 see DNX500
8015-29-0 see MDL750
8015-30-3 see EAP000
8015-35-8 see DAB000
8015-38-1 see OKS500
8015-43-8 see DVJ500
8015-54-1 see DXP800
8015-55-2 see AFM375
8015-64-3 see AOO760
8015-65-4 see WBJ650
8015-73-4 see BAR250
8015-79-0 see OGK000
8015-86-9 see CCK640
8015-88-1 see CCL750
8015-90-5 see OGL100
8015-91-6 see CMQ510
8015-92-7 see CDH750
8015-97-2 see CMY500
8016-03-3 see DAC400
8016-14-6 see HHW560
8016-20-4 see GJU000
8016-23-7 see GKM000
8016-26-0 see LAC000
8016-31-7 see LII000
8016-37-3 see MSB775
8016-38-4 see NCO000
8016-45-3 see PIG730
8016-63-5 see OGQ200

8016-68-0 see SBA000
8016-69-1 see SED500
8016-85-1 see TAD500
8016-87-3 see TAE500
8016-88-4 see TAF700
8016-96-4 see VJU000
8017-59-2 see SLY200
8018-01-7 see DXI400
8018-15-3 see MCS000
8020-83-5 see DAP900
8021-27-0 see AAC250
8021-28-1 see CBB900
8021-29-2 see FBV000
8021-39-4 see BAT850
8022-00-2 see MIW100
8022-15-9 see LCA000
8022-37-5 see ARL250
8022-56-8 see SAE500
8022-56-8 see SAE550
8022-81-9 see BMA600
8022-91-1 see HGF000
8022-96-6 see OGM200
8023-53-8 see AFP750
8023-75-4 see NBP000
8023-80-1 see GCM000
8023-85-6 see CCQ750
8023-88-9 see CNT350
8023-89-0 see EAI500
8023-91-4 see GBC000
8023-94-7 see HGN500
8023-95-8 see HAK500
8024-12-2 see VIK500
8024-14-4 see DAJ800
8024-32-6 see ARW800
8024-37-1 see COG000
8025-81-8 see SLC000
8028-45-3 see ORS100
8028-73-7 see ARE500
8028-73-7 see ARE750
8028-75-9 see ARE250
8028-89-5 see CBG125
8029-29-6 see BAF250
8029-68-3 see IAD000
8030-30-6 see NAH600
8030-55-5 see GME000
8030-78-2 see TAC300
8031-00-3 see FMS000
8031-03-6 see MQV000
8031-42-3 see DKL200
8032-32-4 see PCT250
8034-17-1 see EAL000
8036-63-3 see FCN515
8037-19-2 see TGH750
8039-98-3 see MRM300
8042-47-5 see MQV875
8047-15-2 see SAV500
8047-24-3 see VFA200
8047-28-7 see EDC575
8047-67-4 see IHG000
8047-99-2 see ENY200
8048-31-5 see TEO750
8048-52-0 see DBX400
8049-17-0 see FBG000
8049-47-6 see PAF600
8049-62-5 see LEK000
8050-07-5 see OIM000
8050-07-5 see OIM025
8050-09-7 see RNU100
8050-89-3 see OJK340
8051-02-3 see VHZ000
8051-03-4 see ZKJ000
8052-16-2 see AEA750

8052-41-3 see SLU500
8052-42-4 see ARO500
8052-42-4 see ARO750
8052-42-4 see PCR500
8053-21-2 see PMM100
8053-39-2 see TNO275
8054-43-1 see CDB760
8055-33-2 see TBF650
8056-51-7 see NNL500
8056-92-6 see EQK100
8057-51-0 see EDK000
8057-62-3 see PAM782
8059-82-3 see RCA275
8059-83-4 see BOO650
8061-51-6 see LFQ500
8061-52-7 see CAT675
8062-14-4 see CMP250
8063-06-7 see COF750
8063-14-7 see CBD750
8063-18-1 see SLV500
8063-77-2 see CBV250
8064-08-2 see CCS675
8064-18-4 see TNX375
8064-35-5 see MLX850
8064-38-8 see TGJ350
8064-42-4 see MAO880
8064-66-2 see MCA500
8064-76-4 see EEH575
8064-77-5 see BAV350
8064-79-7 see IGI000
8064-90-2 see TKX000
8065-36-9 see BTA250
8065-48-3 see DAO600
8065-71-2 see GEW725
8065-83-6 see CAY950
8065-91-6 see CNV750
8066-01-1 see MLC000
8066-21-5 see ZJS300
8066-27-1 see FAQ930
8066-33-9 see PBT500
8067-24-1 see DLL400
8067-82-1 see AFK500
8067-85-4 see AOC875
8067-98-9 see FAO300
8068-28-8 see SFY500
8068-44-8 see CMX847
8069-64-5 see TEQ000
8069-76-9 see DVI800
8070-50-6 see EJO000
8072-20-6 see CKL500
8073-61-8 see AHK300
8075-78-3 see DAP810
8077-30-3 see BAP260
8077-38-1 see TKU680
9000-01-5 see AQQ500
9000-02-6 see AHJ000
9000-07-1 see CCL250
9000-21-9 see FPQ000
9000-28-6 see GLY000
9000-29-7 see GLY100
9000-30-0 see GLU000
9000-36-6 see KBK000
9000-36-6 see SLO500
9000-40-2 see LIA000
9000-55-9 see PJJ000
9000-64-0 see BAF000
9000-65-1 see THJ250
9000-69-5 see PAO150
9000-70-8 see PCU360
9000-94-6 see AQN550
9001-00-7 see BMO000
9001-05-2 see CCP525

9001-13-2 see CMY725
9001-33-6 see FBS000
9001-37-0 see GFG100
9001-47-2 see GFO070
9001-62-1 see GGA800
9001-67-6 see NCQ200
9001-73-4 see PAG500
9001-92-7 see BAC020
9001-98-3 see RCZ100
9002-01-1 see SLW450
9002-04-4 see TFU800
9002-07-7 see TNW000
9002-13-5 see UTU550
9002-18-0 see AEX250
9002-60-2 see AES650
9002-61-3 see CMG675
9002-62-4 see PMH625
9002-64-6 see PAK250
9002-67-9 see LIU300
9002-68-0 see FMT100
9002-69-1 see RCK740
9002-70-4 see SCA750
9002-72-6 see PJA250
9002-83-9 see KDK000
9002-84-0 see TAI250
9002-86-2 see PKQ059
9002-88-4 see PJS750
9002-89-5 see PKP750
9002-89-5 see PKP800
9002-92-0 see DXY000
9002-92-0 see EEU000
9002-92-0 see LBS000
9002-92-0 see LBT000
9002-92-0 see LBU000
9002-92-0 see LBU050
9002-93-1 see PKF500
9002-98-6 see PJX800
9002-98-6 see PJX825
9002-98-6 see PJX835
9002-98-6 see PJX845
9003-00-3 see ADY250
9003-01-4 see ADV900
9003-04-7 see SJK000
9003-05-8 see PJK350
9003-07-0 see PKI250
9003-07-0 see PMP500
9003-08-1 see MCB050
9003-11-6 see PJH630
9003-11-6 see PJK150
9003-11-6 see PJK151
9003-13-8 see BRP250
9003-17-2 see PJL350
9003-20-7 see AAX250
9003-22-9 see AAX175
9003-27-4 see PJY800
9003-29-6 see PJL400
9003-34-3 see PJR500
9003-36-5 see FMW333
9003-39-8 see PKQ250
9003-39-8 see PKQ500
9003-39-8 see PKQ750
9003-39-8 see PKR000
9003-39-8 see PKR250
9003-39-8 see PKR500
9003-39-8 see PKR750
9003-39-8 see PKS000
9003-53-6 see SMQ500
9003-54-7 see ADY500
9003-55-8 see SMR000
9003-56-9 see ADX740
9003-63-8 see PJL500
9003-74-1 see TBC550

9003-98-9 see EDK650
9004-06-2 see EAG875
9004-07-3 see CML880
9004-10-8 see IDF300
9004-17-5 see IDF325
9004-24-4 see AGW700
9004-32-4 see SFO500
9004-34-6 see CCU150
9004-38-0 see CCU050
9004-51-7 see IGT000
9004-53-9 see DBD800
9004-54-0 see DBC800
9004-54-0 see DBD000
9004-54-0 see DBD200
9004-54-0 see DBD400
9004-54-0 see DBD600
9004-54-0 see DBD700
9004-57-3 see EHG100
9004-62-0 see HKQ100
9004-64-2 see HNV000
9004-65-3 see HNX000
9004-66-4 see IGS000
9004-67-5 see MIF760
9004-70-0 see CCU250
9004-74-4 see MFJ750
9004-74-4 see MFK000
9004-74-4 see MFK250
9004-81-3 see PJY000
9004-82-4 see SIB500
9004-83-5 see PKE350
9004-86-8 see PKE750
9004-87-9 see OJD230
9004-95-9 see PJT300
9004-96-0 see PJY100
9004-96-0 see PJY250
9004-98-2 see OIG000
9004-98-2 see OIG040
9004-98-2 see OIK000
9004-98-2 see PJW500
9004-98-2 see PKE500
9004-99-3 see PJV250
9004-99-3 see PJV500
9004-99-3 see PJV750
9004-99-3 see PJW000
9004-99-3 see PJW250
9004-99-3 see PJW750
9005-00-9 see SLM500
9005-00-9 see SLN000
9005-08-7 see PJU500
9005-08-7 see PJU750
9005-25-8 see SLJ500
9005-27-0 see HLB400
9005-32-7 see AFL000
9005-34-9 see ANA300
9005-35-0 see CAM200
9005-36-1 see PKU700
9005-37-2 see PNJ750
9005-38-3 see SEH000
9005-46-3 see SFQ000
9005-49-6 see HAQ500
9005-64-5 see PKG000
9005-64-5 see PKL000
9005-65-6 see PKL100
9005-66-7 see PKG500
9005-67-8 see PKL030
9005-70-3 see TOE250
9005-71-4 see SKV195
9005-79-2 see GHK300
9005-81-6 see CCT250
9006-00-2 see PJH500
9006-04-6 see ROH900
9006-26-2 see EJM950

9006-42-2 see MQQ250
9006-59-1 see AFI780
9007-12-9 see TFZ000
9007-13-0 see CAW500
9007-16-3 see ADV950
9007-28-7 see CMF600
9007-35-6 see SDI750
9007-40-3 see COC875
9007-43-6 see CQM325
9007-49-2 see DAZ000
9007-73-2 see FBB000
9007-81-2 see FOO600
9007-83-4 see IBP250
9007-92-5 see GEW875
9008-54-2 see CMS218
9008-57-5 see GHU000
9008-97-3 see PIB575
9008-99-5 see PJA200
9009-54-5 see PKL500
9009-65-8 see POD750
9009-86-3 see RJK000
9010-06-4 see SEH450
9010-41-7 see TNP550
9010-53-1 see UVJ475
9010-66-6 see ZAT100
9010-85-9 see IIQ500
9010-98-4 see PJQ050
9011-04-5 see HCV500
9011-05-6 see UTU500
9011-06-7 see CGW300
9011-09-0 see VPK333
9011-13-6 see SEA500
9011-14-7 see PKB500
9011-18-1 see DBD750
9011-93-2 see LJQ000
9012-59-3 see CBA000
9012-63-9 see VJZ200
9014-01-1 see BAC000
9014-02-2 see NBV500
9014-89-5 see PJV100
9014-90-8 see SNY100
9014-92-0 see TAX500
9015-68-3 see ARN800
9015-73-0 see DHW600
9016-00-6 see PJR000
9016-01-7 see OJM400
9016-45-9 see NND500
9016-45-9 see PKF000
9016-45-9 see TAW250
9016-45-9 see TAW500
9016-45-9 see TAX000
9016-45-9 see TAX250
9016-87-9 see PKB100
9031-11-2 see GAV100
9032-08-0 see AOM125
9034-34-8 see HHK000
9034-40-6 see LIU370
9036-06-0 see PMJ100
9036-19-5 see GHS000
9036-19-5 see IAH000
9036-19-5 see OFM000
9036-19-5 see OFM900
9036-19-5 see OFO000
9036-19-5 see OFQ000
9036-19-5 see OFU000
9036-66-2 see AQR800
9038-95-3 see GHY000
9038-95-3 see MNE250
9038-95-3 see UBS000
9038-95-3 see UCA000
9038-95-3 see UCJ000
9038-95-3 see UDA000

9038-95-3 see UDJ000
9038-95-3 see UDS000
9038-95-3 see UEA000
9038-95-3 see UEJ000
9038-95-3 see UFA000
9039-53-6 see UVS500
9039-61-6 see RDA350
9040-12-4 see RJK050
9041-08-1 see HAQ550
9041-29-6 see OJD232
9041-93-4 see BLY780
9045-81-2 see PKQ100
9046-56-4 see VGU700
9047-13-6 see AOM150
9048-46-8 see HGL930
9049-05-2 see CAO250
9049-76-7 see HNY000
9050-36-6 see MAO300
9051-57-4 see ANO600
9056-38-6 see NMB000
9060-10-0 see BLY760
9061-82-9 see SFP000
9062-04-8 see CBS600
9067-32-7 see HGN600
9072-62-2 see PKK100
9074-07-1 see SBI860
9076-25-9 see PCC000
9078-78-8 see GIX350
9080-17-5 see ANT000
9080-79-9 see SFO100
9082-00-2 see PJX900
9082-07-9 see SFW300
9084-06-4 see BLX000
9086-60-6 see CCH000
9087-69-8 see HAH900
9087-70-1 see PAF550
10004-44-1 see HLM000
10008-90-9 see CLU500
10010-36-3 see FLY000
10017-37-5 see MGJ400
10021-35-9 see TFC570
10022-31-8 see BAN250
10022-50-1 see NMT500
10022-60-3 see AEP500
10022-68-1 see CAH250
10022-70-5 see SHU525
10023-25-3 see DKS800
10024-58-5 see DAH450
10024-66-5 see MAR260
10024-70-1 see MEF500
10024-74-5 see BKQ500
10024-78-9 see EMY000
10024-89-2 see MRQ600
10024-90-5 see MET100
10024-93-8 see NBY000
10024-97-2 see NGU000
10025-09-9 see PBI300
10025-64-6 see ZKS100
10025-65-7 see PJE000
10025-67-9 see SON510
10025-68-0 see SBS500
10025-69-1 see TGC275
10025-70-4 see SMH525
10025-73-7 see CMJ250
10025-74-8 see DYG600
10025-76-0 see ERA500
10025-77-1 see FAW000
10025-78-2 see TJD500
10025-82-8 see ICK000
10025-85-1 see NGQ500
10025-87-3 see PHQ800
10025-91-9 see AQC500

10025-97-5 see IGJ499
10025-98-6 see PLN750
10025-99-7 see PJD250
10026-03-6 see SBU000
10026-04-7 see SCQ500
10026-06-9 see TGC282
10026-07-0 see TAJ250
10026-08-1 see TFT000
10026-10-5 see UQJ000
10026-11-6 see ZPA000
10026-12-7 see NEA000
10026-13-8 see PHR500
10026-17-2 see CNC100
10026-18-3 see CNE250
10026-22-9 see CND010
10026-24-1 see CNE150
10027-06-2 see NNI150
10028-15-6 see ORW000
10028-18-9 see NDC500
10028-22-5 see FBA000
10028-70-3 see TAN755
10029-04-6 see ELI500
10031-13-7 see LCL000
10031-18-2 see MCX750
10031-26-2 see IGQ000
10031-27-3 see TAK600
10031-30-8 see CAY100
10031-37-5 see CMK425
10031-43-3 see CNN000
10031-53-5 see ERC000
10031-58-0 see TDS500
10031-59-1 see TEL750
10031-82-0 see EEL500
10031-96-6 see EQS100
10032-00-5 see AAY500
10032-02-7 see GDG000
10032-13-0 see HFQ575
10032-15-2 see HFR200
10034-81-8 see PCE000
10034-85-2 see HHI500
10034-93-2 see HGW500
10034-96-5 see MAU300
10035-10-6 see HHJ000
10036-47-2 see TCI000
10038-98-9 see GDY000
10039-33-5 see DVM600
10039-54-0 see OLS000
10040-45-6 see SJJ175
10042-59-8 see PNN250
10042-76-9 see SMK000
10042-84-9 see SEP500
10042-88-3 see TAM000
10043-01-3 see AHG750
10043-09-1 see BJN750
10043-18-2 see BPL750
10043-35-3 see BMC000
10043-52-4 see CAO750
10043-67-1 see AHF100
10043-67-1 see PKU725
10043-84-2 see MAS815
10043-92-2 see RBA000
10045-34-8 see AGM060
10045-86-0 see FAZ500
10045-94-0 see MDF000
10045-95-1 see NCB000
10048-13-2 see SLP000
10048-32-5 see PAJ500
10048-95-0 see ARC250
10049-03-3 see FFD000
10049-04-4 see CDW450
10049-05-5 see CMJ300
10049-06-6 see TGG750

10049-07-7 see RHK000
10049-08-8 see RRZ000
10049-14-6 see UOC300
10051-06-6 see PBD500
10058-07-8 see PPN000
10058-20-5 see SMZ000
10058-44-3 see FAZ525
10059-74-2 see MFF635
10060-08-9 see CAP750
10060-12-5 see CMK450
10061-01-5 see DGH200
10061-02-6 see DGH225
10070-95-8 see BHR500
10072-24-9 see QCS875
10072-25-0 see DFH000
10072-50-1 see PKD050
10075-36-2 see IHO200
10083-53-1 see MGL500
10085-81-1 see BCH750
10086-50-7 see IJA000
10087-89-5 see DWL400
10094-34-5 see BEL850
10097-26-4 see BSA750
10097-28-6 see SDH000
10099-57-7 see IRC000
10099-58-8 see LAX000
10099-59-9 see LBA000
10099-60-2 see LBB000
10099-66-8 see LIW000
10099-67-9 see LIY000
10099-70-4 see BOX250
10099-71-5 see MAL000
10099-72-6 see MAL500
10099-73-7 see MAL750
10099-74-8 see LDO000
10099-76-0 see LDW000
10101-41-4 see CAX750
10101-50-5 see SJC000
10101-53-8 see CMK415
10101-68-5 see MAU750
10101-83-4 see SKC000
10101-88-9 see TNM750
10101-89-0 see SJH300
10101-97-0 see NDL000
10101-98-1 see NDK990
10102-06-4 see URA200
10102-17-7 see SKI500
10102-18-8 see SJT500
10102-20-2 see SKC500
10102-23-5 see SBN505
10102-40-6 see DXE875
10102-43-9 see NEG100
10102-44-0 see NGR500
10102-45-1 see TEK750
10102-49-5 see IGN000
10102-50-8 see IGM000
10102-53-1 see ARB000
10103-50-1 see ARD000
10103-60-3 see ARD600
10103-62-5 see CAM222
10103-89-6 see QTS100
10108-56-2 see BQW750
10108-61-9 see BSH600
10108-64-2 see CAE250
10108-73-3 see CDB000
10112-91-1 see MCY300
10117-38-1 see PLT500
10118-76-0 see CAV250
10118-90-8 see MQW250
10119-31-0 see ZPJ000
10121-94-5 see MNS000
10123-62-3 see COM100

10124-36-4 see CAJ000
10124-37-5 see CAU000
10124-43-3 see CNE125
10124-48-8 see MCW500
10124-50-2 see PKV500
10124-53-5 see MAK000
10124-56-8 see SHM500
10124-65-9 see PLK750
10125-76-5 see NJM400
10125-85-6 see DTO100
10127-36-3 see BMM550
10137-69-6 see CPE500
10137-73-2 see CQB275
10137-74-3 see CAO500
10137-80-1 see EKT000
10137-87-8 see EMO000
10137-90-3 see ENX000
10137-96-9 see EJK000
10137-98-1 see EJL000
10138-01-9 see ERC550
10138-17-7 see DRP000
10138-21-3 see DFK400
10138-34-8 see EBK000
10138-39-3 see AGF500
10138-41-7 see ECX500
10138-44-0 see EEV150
10138-47-3 see EGJ000
10138-52-0 see GAH000
10138-59-7 see HDS250
10138-60-0 see HFF000
10138-62-2 see HGG000
10138-63-3 see PFC100
10138-79-1 see TNF000
10138-87-1 see EGJ500
10138-89-3 see TLA000
10139-98-7 see DAZ140
10140-08-6 see DJA600
10140-75-7 see DBF875
10140-84-8 see DGC000
10140-87-1 see DFG159
10140-89-3 see DGH800
10140-91-7 see CGH675
10140-94-0 see CHH125
10140-97-3 see CJG750
10140-99-5 see CLP800
10141-00-1 see PLB500
10141-05-6 see CNC500
10141-07-8 see COD100
10141-15-8 see COM250
10141-19-2 see VQU000
10141-22-7 see DFR400
10143-20-1 see DSJ200
10143-22-3 see DTG200
10143-23-4 see DTI400
10143-31-4 see DWV600
10143-38-1 see DYH100
10143-53-0 see DJF000
10143-54-1 see DJF600
10143-56-3 see DJG200
10143-60-9 see DJK600
10143-66-5 see DOB200
10143-67-6 see AAG750
10158-43-7 see DTM800
10161-34-9 see HKI100
10161-84-9 see SBU900
10161-85-0 see DJA325
10163-15-2 see DXD600
10163-83-4 see DYE650
10165-13-6 see CEI340
10165-33-0 see AKM250
10168-80-6 see ECY500
10168-81-7 see GAL000

10168-82-8 see HGH100
10169-00-3 see ANL500
10169-02-5 see CMM325
10171-76-3 see BJM750
10171-78-5 see HNX800
10176-39-3 see FND100
10176-39-3 see PBG200
10182-82-8 see HOC500
10187-79-8 see MMK000
10187-86-7 see AKY000
10190-99-5 see SEY050
10192-29-7 see ANE250
10192-30-0 see ANB600
10193-36-9 see SCN600
10193-95-0 see BKM000
10196-18-6 see NEF500
10199-89-0 see NFS550
10210-36-3 see GJG000
10210-68-1 see CNB500
10212-25-6 see COW900
10213-09-9 see DFW800
10213-10-2 see SKO000
10213-15-7 see MAH250
10213-74-8 see EGY000
10213-75-9 see EKZ000
10215-25-5 see TEU000
10215-33-5 see BPS500
10217-52-4 see HGU500
10218-83-4 see DDQ100
10222-01-2 see DDM000
10232-90-3 see DFJ500
10232-91-4 see EGY500
10232-92-5 see DMM600
10232-93-6 see DOF000
10233-88-2 see GJE000
10235-09-3 see PHN500
10238-21-8 see CEH700
10241-05-1 see MRD500
10242-13-4 see MNE275
10256-92-5 see TCJ025
10262-69-8 see LIN800
10264-25-2 see VAG100
10265-92-6 see DTQ400
10290-12-7 see CNN500
10294-33-4 see BMG400
10294-34-5 see BMG500
10294-40-3 see BAK250
10294-41-4 see CDB250
10294-46-9 see CNO350
10294-48-1 see PCF775
10294-54-9 see CDE500
10294-66-3 see PLW000
10294-70-9 see TGD500
10308-44-8 see MAO280
10308-82-4 see AKD375
10308-83-5 see DBW100
10308-84-6 see GLC000
10308-90-4 see DUX600
10309-37-2 see BAD625
10309-79-2 see MHN750
10309-97-4 see HNK900
10310-21-1 see CHK300
10310-32-4 see EPW600
10310-38-0 see EIU000
10311-84-9 see DBI099
10312-83-1 see MDW250
10318-23-7 see DPP400
10318-26-0 see DDJ000
10319-70-7 see AHN000
10322-73-3 see EDU600
10325-39-0 see DEU300
10325-94-7 see CAH000

10326-21-3 see MAE000
10326-24-6 see ZDS000
10326-27-9 see BAK020
10328-51-5 see CPL100
10328-92-4 see MKW300
10329-95-0 see MRU080
10331-57-4 see DFD000
10335-79-2 see NAV000
10339-31-8 see NGY700
10339-55-6 see ELZ000
10347-38-3 see BLK750
10347-81-6 see MAW850
10350-81-9 see PMY000
10356-76-0 see FHO000
10356-92-0 see DAS400
10361-03-2 see SII500
10361-37-2 see BAK000
10361-44-1 see BKW250
10361-76-9 see PLP750
10361-79-2 see PLX750
10361-80-5 see PLY250
10361-82-7 see SAR500
10361-83-8 see SAT200
10361-84-9 see SBC000
10361-91-8 see YDJ000
10361-92-9 see YES000
10361-93-0 see YFJ000
10361-95-2 see ZES000
10369-17-2 see HEJ375
10371-86-5 see MEL775
10377-48-7 see LHR000
10377-60-3 see MAH000
10377-66-9 see MAS900
10377-98-7 see LAO000
10379-14-3 see CFG750
10380-28-6 see BLC250
10380-77-5 see COE125
10387-13-0 see BIJ750
10389-72-7 see DRC600
10390-80-4 see EBD000
10397-75-8 see TDQ230
10402-33-2 see AGL000
10402-52-5 see PGB750
10402-53-6 see ECU550
10402-90-1 see PFB000
10405-02-4 see KEA300
10405-27-3 see DKF400
10415-75-5 see MDE750
10415-87-9 see PFR200
10416-59-8 see TMF000
10419-79-1 see CHG375
10420-90-3 see HCS100
10421-48-4 see FAY200
10421-48-4 see IHB900
10427-00-6 see FPV000
10428-19-0 see TBL500
10429-82-0 see EAK000
10431-47-7 see SBO100
10431-86-4 see BSY000
10433-59-7 see EAK500
10435-35-5 see ALN800
10436-39-2 see TBT500
10443-70-6 see EHL500
10447-38-8 see DWM400
10448-09-6 see HBA600
10453-86-8 see BEP500
10453-89-1 see CML650
10457-58-6 see BQY000
10457-59-7 see IOR000
10457-90-6 see BNU725
10465-10-8 see DJI350
10465-27-7 see SAR000

10466-65-6 see PLQ000
10467-10-4 see EMA000
10473-64-0 see DWL500
10473-70-8 see CKI500
10473-98-0 see DWL525
10474-14-3 see MBV720
10476-81-0 see SMF000
10476-85-4 see SMG500
10476-86-5 see SMJ500
10476-95-6 see AAW250
10477-72-2 see PCU425
10478-42-9 see MMT300
10482-16-3 see MKR000
10484-36-3 see AOK000
10486-00-7 see SJB350
10488-36-5 see TGJ250
10497-05-9 see PHE750
10504-99-1 see DWM600
10510-77-7 see EOE000
10519-11-6 see DAF100
10519-12-7 see DAF150
10526-15-5 see BJS550
10534-86-8 see HBX000
10535-87-2 see PEF500
10537-47-0 see DED000
10540-29-1 see NOA600
10543-95-0 see HDA000
10544-63-5 see EHO200
10544-72-6 see NGU500
10545-99-0 see SOG500
10546-24-4 see MME500
10552-94-0 see NLQ000
10557-85-4 see IEI600
10563-26-5 see TBI700
10563-29-8 see AME100
10563-70-9 see TDL000
10578-16-2 see DKG600
10580-52-6 see VDA000
10580-77-5 see BJN000
10584-98-2 see DDY600
10588-01-9 see SGI000
10589-74-9 see PBX500
10592-13-9 see HGP550
10593-16-5 see DPH430
10595-95-6 see MKB000
10598-82-0 see NFO700
10599-70-9 see ACI400
10599-90-3 see CDO750
10603-03-9 see MQK100
10605-21-7 see MHC750
10606-42-5 see EMI550
11001-74-4 see QQS075
11002-18-9 see CBF500
11002-21-4 see CCO000
11002-90-7 see ART500
11003-24-0 see ARY750
11003-38-6 see CBF635
11004-30-1 see SKQ515
11005-02-0 see ORI300
11005-63-3 see SMN000
11005-70-2 see CCX625
11005-92-8 see CCX175
11005-94-0 see CNW100
11006-22-7 see FCA000
11006-31-8 see TAB300
11006-33-0 see PGQ500
11006-33-0 see PGQ750
11006-64-7 see ITG000
11006-70-5 see OIS000
11006-76-1 see VRF000
11006-90-9 see EAE400
11011-73-7 see BML750

11013-97-1 see MKK600
11014-59-8 see LAU400
11015-37-5 see MRA250
11016-29-8 see PJJ350
11016-71-0 see RQF350
11016-72-1 see RQK000
11018-93-2 see TER250
11024-24-1 see DKL400
11028-39-0 see MCE275
11028-71-0 see CNH625
11029-61-1 see GJO025
11030-13-0 see ITH000
11031-48-4 see SAX500
11032-05-6 see TAK750
11032-12-5 see TOE750
11032-41-0 see DLL100
11041-12-6 see CME400
11042-64-1 see OJY200
11043-90-6 see TGF060
11043-98-4 see MQX000
11043-99-5 see MQX250
11048-13-8 see NBR500
11048-92-3 see GME300
11050-62-7 see IKV000
11051-88-0 see CNF159
11052-01-0 see EDC600
11052-70-3 see MRA275
11052-72-5 see OIM100
11052-79-2 see PAC600
11054-63-0 see TNX650
11054-70-9 see LBF500
11055-06-4 see FPD000
11056-06-7 see SHM000
11056-12-5 see CMS230
11067-81-5 see TED550
11067-82-6 see SKF600
11068-91-0 see NCY135
11069-19-5 see DEV200
11069-34-4 see MGS500
11071-15-1 see AQH000
11072-93-8 see EDL500
11077-03-5 see PAF000
11078-23-2 see CNH800
11080-14-1 see NCQ520
11081-39-3 see PKB775
11082-38-5 see TNM850
11084-85-8 see SJH095
11085-39-5 see FPA000
11094-61-4 see ECX000
11096-42-7 see NND000
11096-82-5 see PJN250
11097-67-9 see AEC175
11097-69-1 see PJN000
11097-82-8 see GCO200
11099-03-9 see CMS236
11100-14-4 see PJN750
11100-45-1 see EAV025
11103-72-3 see RSK100
11103-86-9 see PLW500
11104-28-2 see PJM000
11104-30-6 see DAP822
11105-11-6 see TOD000
11105-19-4 see NCX525
11107-01-0 see TOB750
11111-23-2 see LHX350
11111-49-2 see HDB500
11112-10-0 see AQB250
11113-75-0 see NDL050
11114-18-4 see LFN000
11114-20-8 see CCL350
11114-46-8 see FBD000
11114-92-4 see CNA750

11115-82-5 see EAT800
11116-31-7 see BLY250
11116-32-8 see BLY500
11118-72-2 see AQM000
11119-62-3 see ADG125
11120-29-9 see PJO250
11121-08-7 see KDA025
11121-57-6 see PII150
11121-96-3 see IGL100
11126-42-4 see PJP800
11133-76-9 see IGL120
11133-98-5 see CNI600
11135-81-2 see PLS500
11138-49-1 see AHG000
11138-66-2 see XAK800
11138-87-7 see LAZ000
11139-76-7 see AFM800
11141-16-5 see PJM250
12001-26-2 see MQS250
12001-28-4 see ARM275
12001-29-5 see ARM268
12001-47-7 see BKV250
12001-65-9 see HEI650
12001-79-5 see VSZ500
12001-85-3 see NAT000
12001-89-7 see DGR200
12002-03-8 see COF500
12002-19-6 see MCV250
12002-26-5 see TDJ100
12002-43-6 see GEO000
12002-53-8 see BQT600
12003-96-2 see AHH125
12005-86-6 see SHM000
12007-25-9 see MAD250
12007-33-9 see BMH659
12007-46-4 see SIX550
12007-56-6 see CAN250
12007-97-5 see MRC650
12008-41-2 see DXF200
12008-61-6 see ANG125
12009-21-1 see BAP750
12009-31-3 see BAP500
12010-12-7 see BFQ750
12010-53-6 see PJH775
12010-67-2 see BKY250
12011-67-5 see IGQ750
12011-76-6 see DAC450
12012-50-9 see PLW200
12012-95-2 see AGQ000
12013-55-7 see CAX000
12013-56-8 see CAR750
12013-82-0 see TIH250
12014-28-7 see CAI125
12014-93-6 see DEK600
12016-80-7 see CNC240
12018-18-7 see NDA100
12018-19-8 see ZFA100
12018-40-5 see CMJ910
12018-95-0 see CNM650
12020-65-4 see ERC500
12024-21-4 see GBS050
12025-32-0 see GEI000
12027-67-7 see ANH600
12029-98-0 see IDS300
12030-85-2 see PLL250
12030-88-5 see PLE260
12031-80-0 see LHO000
12032-88-1 see MAV250
12033-59-9 see SBT100
12033-88-4 see TEE000
12034-12-7 see SJZ100
12035-39-1 see NDL500

12035-50-6 see VAC000
12035-51-7 see NDB875
12035-52-8 see NCY100
12035-71-1 see HAK050
12035-72-2 see NDJ500
12036-02-1 see OKI000
12036-10-1 see RSF875
12037-82-0 see PHQ750
12038-67-4 see RGK000
12039-52-0 see TEL500
12040-44-7 see AFI900
12041-76-8 see DET000
12041-87-1 see TDX860
12042-91-0 see AHA000
12044-52-9 see PJE750
12044-65-4 see MBU790
12045-01-1 see DGQ300
12047-11-9 see BAL625
12047-27-7 see BAP502
12047-79-9 see BAN500
12048-51-0 see BKY300
12049-50-2 see CAY300
12053-67-7 see CDF000
12054-48-7 see NDE000
12056-53-0 see PLW150
12057-24-8 see LHM770
12057-74-8 see MAI000
12057-92-0 see MAT750
12058-74-1 see SKG500
12058-85-4 see SJI500
12060-00-3 see LED000
12063-27-3 see IHN050
12064-62-9 see GAP000
12067-99-1 see PHU750
12068-03-0 see SIK460
12068-85-8 see IGV000
12069-68-0 see BAP800
12069-69-1 see CNJ750
12070-08-5 see TGG000
12070-09-6 see UOB100
12070-12-1 see TOB500
12070-14-3 see ZQC200
12070-27-8 see BAH750
12071-29-3 see SME100
12071-33-9 see UOC200
12071-83-9 see ZMA000
12075-68-2 see TJP775
12079-65-1 see CPV000
12079-66-2 see CDD625
12081-88-8 see PLG825
12089-29-1 see BGY720
12107-76-5 see DDU000
12108-13-3 see MAV750
12111-24-9 see CAY500
12116-66-4 see HAE500
12122-15-5 see NCY525
12122-67-7 see EIR000
12124-97-9 see ANC250
12124-99-1 see ANJ750
12125-01-8 see ANH250
12125-02-9 see ANE500
12125-03-0 see PLS760
12125-09-6 see PHA000
12125-56-3 see NDE010
12125-61-0 see BMM075
12125-77-8 see COY100
12126-59-9 see ECU750
12126-59-9 see PMB000
12133-60-7 see CDW500
12134-29-1 see TIH750
12136-15-1 see TKW000
12136-83-3 see SIP000

12136-85-5 see RQF000	12442-63-6 see CMB500	13004-56-3 see QWJ000	13154-66-0 see PNV775
12137-13-2 see NDJ475	12504-13-1 see SML000	13007-92-6 see HCB000	13164-03-9 see XPJ100
12139-70-7 see CNC750	12505-77-0 see DDI500	13009-91-1 see TNK000	13168-78-0 see NLF300
12142-88-0 see NDL425	12510-42-8 see EDC650	13009-99-9 see MAC000	13171-21-6 see FAB400
12161-82-9 see BFO250	12519-36-7 see ZGW200	13010-07-6 see NHS000	13171-22-7 see DTP600
12164-01-1 see TNO250	12529-66-7 see DNW200	13010-08-7 see NLC000	13171-25-0 see YCJ200
12164-12-4 see SJU500	12540-13-5 see CNI500	13010-10-1 see NLC500	13172-31-1 see DXH350
12164-94-2 see ANA750	12542-36-8 see GJM259	13010-20-3 see NMA500	13177-25-8 see FDE100
12165-69-4 see PHT750	12542-85-7 see MGC230	13010-47-4 see CGV250	13177-29-2 see NGA600
12167-74-7 see CAW120	12550-17-3 see AQI000	13013-17-7 see PNB790	13181-17-4 see BNK400
12172-73-5 see ARM262	12550-82-2 see BIS300	13014-18-1 see PAY550	13183-79-4 see MPQ250
12173-10-3 see CMV850	12558-71-3 see YFA100	13018-50-3 see BKS825	13185-22-3 see NGG550
12174-11-7 see PAE750	12558-92-8 see MDB250	13019-22-2 see DAI400	13189-98-5 see HNV050
12179-04-3 see SKC550	12602-23-2 see CNB495	13021-50-6 see EET600	13194-48-4 see EIN000
12192-57-3 see ART250	12604-53-4 see FBE000	13025-29-1 see CBV600	13195-76-1 see TKR750
12194-11-5 see BIR500	12604-58-9 see FBP000	13025-29-1 see IKG400	13219-97-1 see MNY800
12198-93-5 see ORU900	12605-70-8 see NCX515	13029-08-8 see DET800	13224-31-2 see DBC100
12201-85-3 see MAK400	12607-70-4 see NCY600	13029-44-2 see DHB550	13225-10-0 see MKI125
12206-14-3 see SKF000	12607-92-0 see AGX125	13037-20-2 see EOK550	13230-04-1 see DSV600
12208-54-7 see ANO900	12607-93-1 see TAH750	13037-55-3 see CKJ750	13232-74-1 see HET350
12217-79-7 see CMP070	12609-89-1 see MRW800	13037-86-0 see HBP285	13235-16-0 see CAW000
12218-77-8 see TFU000	12622-79-6 see CMK750	13042-02-9 see HFB600	13240-06-7 see ERB000
12219-87-3 see CMM200	12623-78-8 see AEC000	13045-94-8 see SAX200	13242-44-9 see DOY600
12222-78-5 see CMP075	12626-36-7 see CAI600	13046-06-5 see THS500	13244-33-2 see AIA500
12224-02-1 see BHC750	12626-81-2 see LED100	13048-33-4 see HEQ100	13244-35-4 see CIX200
12225-18-2 see CMS214	12627-35-9 see PAR250	13055-82-8 see DNA600	13252-14-7 see HDC435
12225-26-2 see CMS220	12629-02-6 see VSK000	13056-98-9 see PEU500	13253-44-6 see MDT800
12228-13-6 see DVT800	12634-34-3 see MAB750	13058-67-8 see LIN000	13254-34-7 see FON200
12232-67-6 see BFR000	12638-07-2 see CNB825	13061-80-8 see HFF300	13256-06-9 see DCH600
12232-97-2 see BKW500	12640-73-2 see MAB800	13063-43-9 see DFN850	13256-07-0 see AOL000
12232-99-4 see SFD000	12640-89-0 see SBT200	13065-64-0 see DQA710	13256-11-6 see MNU250
12234-64-9 see CMM120	12642-23-8 see PJP750	13067-93-1 see CON300	13256-12-7 see DVE400
12235-21-1 see FAL050	12645-45-3 see IGJ300	13071-27-7 see DUO600	13256-13-8 see NKF000
12236-46-3 see CMB750	12645-49-7 see IHB677	13071-79-9 see BSO000	13256-15-0 see NJQ000
12236-86-1 see CMS224	12645-50-0 see IHB800	13073-35-3 see AKB250	13256-19-4 see NKM500
12244-57-4 see GJC000	12653-52-0 see ASN390	13073-86-4 see TCN250	13256-21-8 see NKQ500
12244-59-6 see PLM575	12656-69-8 see DXI300	13074-00-5 see HOM259	13256-22-9 see NLR500
12245-39-5 see CPR840	12656-85-8 see MRC000	13074-65-2 see HFO600	13256-23-0 see NLH000
12255-10-6 see NCY125	12663-46-6 see COW750	13074-85-6 see SAS000	13256-32-1 see DTB200
12255-11-7 see GEK300	12672-29-6 see PJM750	13074-91-4 see DYG800	13257-44-8 see NNE100
12255-80-0 see NDJ399	12674-11-2 see PJL800	13078-15-4 see CKI850	13261-62-6 see DPJ600
12256-33-6 see NDJ400	12674-40-7 see TFT100	13078-75-6 see DOK000	13265-01-5 see DUM100
12258-22-9 see HBI500	12675-92-2 see CNA600	13080-06-3 see BRJ125	13265-60-6 see DOP200
12261-99-3 see PIX300	12679-83-3 see SPE500	13080-89-2 see SNZ000	13266-07-4 see ABX325
12263-85-3 see MGC225	12681-83-3 see IGL110	13082-24-1 see DQA700	13270-61-6 see HDW200
12264-18-5 see CAR800	12684-33-2 see SCF500	13083-37-9 see MCQ800	13271-93-7 see DUL500
12265-93-9 see SHK500	12685-29-9 see CAD290	13084-45-2 see BJD750	13274-42-5 see TED600
12266-58-9 see BGR500	12687-98-8 see RMU100	13084-46-3 see DDI000	13275-42-8 see BNV765
12267-44-6 see RQF100	12688-25-4 see JDS000	13084-47-4 see BJD500	13275-68-8 see EGI000
12271-71-5 see ZDJ100	12706-94-4 see APF000	13086-63-0 see SDW500	13277-59-3 see NEE000
12273-50-6 see PLJ000	12709-98-7 see LDM000	13087-53-1 see IGC000	13279-22-6 see PEH250
12275-13-7 see EIR500	12710-02-0 see LFP000	13092-75-6 see SDJ000	13279-24-8 see EJN400
12275-58-0 see SGM100	12712-28-6 see BIK750	13092-75-6 see SDR759	13280-07-4 see CEV800
12291-11-1 see BLL250	12718-69-3 see TOC000	13093-88-4 see LEO000	13284-86-1 see SIC500
12298-43-0 see HAF375	12728-25-5 see DBA400	13098-39-0 see HDE500	13286-32-3 see DIU800
12298-68-9 see PLW285	12737-87-0 see PJO750	13101-58-1 see DCQ200	13289-18-4 see HAN600
12328-03-9 see TJU600	12738-76-0 see TAN000	13104-21-7 see MHR150	13292-46-1 see RKP000
12331-76-9 see PLU750	12758-40-6 see CCF125	13104-70-6 see HGL600	13292-87-0 see MPL250
12336-95-7 see CMJ565	12765-82-1 see SNQ700	13106-47-3 see BFP750	13295-76-6 see NGG600
12359-48-7 see AQJ600	12768-78-4 see CMM100	13106-76-8 see ANM750	13296-94-1 see BNS800
12380-95-9 see TIH000	12770-50-2 see BFP250	13107-39-6 see MPK750	13302-06-2 see TIF000
12394-14-8 see CAT235	12770-99-9 see DDE050	13115-28-1 see NHR500	13302-08-4 see TMW500
12397-35-2 see CCB500	12772-56-4 see LHA000	13115-40-7 see FMU039	13311-57-4 see CPI000
12400-16-7 see BFR250	12772-68-8 see LJP500	13116-53-5 see TBT300	13311-84-7 see FMR050
12401-86-4 see SIN500	12774-81-1 see TBN100	13118-10-0 see AEY400	13312-42-0 see BCT250
12407-86-2 see TMC750	12788-93-1 see ADF250	13121-70-5 see CQH650	13312-52-2 see NBV200
12408-07-0 see TJC800	12789-03-6 see CDR760	13121-71-6 see ABW600	13316-70-6 see CDF500
12412-52-1 see SBW600	12789-46-7 see PBW750	13127-50-9 see MHD250	13319-75-0 see BMG800
12427-38-2 see MAS500	12794-92-2 see AGY000	13138-21-1 see DCN875	13323-62-1 see DEJ000
12430-27-2 see CDE125	12798-63-9 see OKK500	13138-45-9 see NDG000	13324-20-4 see CMS222
12436-28-1 see SHN000	13001-46-2 see PLK600	13147-25-6 see ELG500	13327-32-7 see BFS250
12439-96-2 see VEZ100	13002-65-8 see TAC880	13150-00-0 see SIC000	13329-71-0 see IQT000

13331-27-6 see NEY500
13344-50-8 see NLS000
13345-21-6 see BCX250
13345-23-8 see BCW750
13345-25-0 see BCT750
13345-50-1 see MCA025
13345-58-9 see HKU000
13345-60-3 see MEW000
13345-61-4 see FNT000
13345-62-5 see CIN750
13345-64-7 see TLK750
13347-42-7 see CFI750
13354-35-3 see PGL000
13355-00-5 see PLK800
13356-08-6 see BLU000
13360-45-7 see CES750
13360-63-9 see EHA050
13361-31-4 see IIQ100
13361-32-5 see AGC150
13362-04-4 see DTC300
13365-38-3 see DOS800
13366-73-9 see PHV250
13367-92-5 see DQG700
13380-94-4 see OPC000
13382-33-7 see HAH000
13394-86-0 see BLV250
13395-16-9 see BGR000
13396-41-3 see SKE550
13400-13-0 see CDD500
13402-08-9 see ACX500
13402-51-2 see BFK750
13403-01-5 see DVV109
13406-60-5 see DHZ050
13410-01-0 see DXG000
13411-16-0 see NDY400
13412-64-1 see DGE200
13413-18-8 see HEW150
13419-46-0 see DGP125
13422-55-4 see VSZ050
13422-81-6 see TDJ300
13424-46-9 see LCM000
13425-22-4 see ALX250
13425-94-0 see ARM000
13426-91-0 see DBU800
13435-12-6 see TME750
13441-36-6 see PFY200
13442-07-4 see HIV000
13442-08-5 see HIV500
13442-09-6 see HIW000
13442-10-9 see HIW500
13442-11-0 see CHM500
13442-12-1 see CHN000
13442-13-2 see DFM200
13442-14-3 see CCF750
13442-15-4 see HIX500
13442-16-5 see HIY000
13442-17-6 see DVE200
13444-71-8 see PCJ250
13444-85-4 see NGW500
13444-89-8 see NMJ400
13444-90-1 see NMT000
13445-50-6 see DWP229
13446-10-1 see PCJ750
13446-18-9 see NEE100
13446-30-5 see MAK250
13446-34-9 see MAR250
13446-48-5 see ANO250
13446-49-6 see PLL125
13446-73-6 see RPK000
13446-74-7 see RPP000
13446-75-8 see RPU000
13448-22-1 see ODY100

13449-22-4 see BPX500
13450-90-3 see GBM000
13450-92-5 see GDW000
13453-06-0 see ANV750
13453-07-1 see GIW176
13453-30-0 see TEJ100
13453-57-1 see LCQ300
13453-78-6 see LHN000
13454-96-1 see PJE250
13455-36-2 see CND920
13455-50-0 see HGG500
13461-01-3 see AAF800
13463-30-4 see LEC000
13463-39-3 see NCZ000
13463-40-6 see IHG500
13463-41-7 see ZMJ000
13463-67-7 see TGG760
13464-37-4 see SEY200
13464-38-5 see SEY150
13464-42-1 see SEY100
13464-82-9 see ICJ000
13464-97-6 see HGZ000
13465-07-1 see HHZ000
13465-33-3 see MCX700
13465-73-1 see BOE750
13465-77-5 see HCH500
13465-95-7 see PCD750
13466-78-9 see CCK500
13467-82-8 see BSD100
13470-26-3 see VEK000
13472-08-7 see ASM025
13472-33-8 see SJD500
13472-45-2 see SKN000
13473-90-0 see AHD750
13474-03-8 see AEM750
13476-05-6 see ECZ000
13477-00-4 see BAJ500
13477-09-3 see BAM250
13477-17-3 see CAI000
13477-20-8 see CAJ100
13477-21-9 see CAJ500
13477-34-4 see CAU250
13478-00-7 see NDG500
13478-10-9 see FBJ000
13478-18-7 see MRD800
13478-33-6 see CND900
13478-34-7 see CNI750
13479-19-1 see EGA600
13479-21-5 see EGA650
13479-29-3 see HOP000
13482-49-0 see TAM500
13483-18-6 see BIJ250
13484-13-4 see EBR500
13492-01-8 see PET500
13494-80-9 see TAJ000
13494-80-9 see TAJ600
13494-90-1 see GBS000
13494-91-2 see GBS100
13494-98-9 see YFS000
13495-01-7 see HAO000
13497-05-7 see SJN000
13497-91-1 see TBP500
13508-53-7 see DRC400
13510-48-0 see BFR300
13510-49-1 see BFU250
13517-10-7 see BMH500
13517-17-4 see SFW500
13517-26-5 see SJM500
13517-49-2 see SPA650
13520-61-1 see NDJ000
13520-74-6 see SCR100
13520-83-7 see URS000

13520-90-6 see VEK100
13520-92-8 see ZPS000
13520-96-2 see POS500
13523-86-9 see VSA000
13529-51-6 see HHH000
13529-75-4 see LHE525
13530-65-9 see ZFJ100
13536-84-0 see UQA000
13536-85-1 see FFB000
13537-18-3 see TFW500
13537-21-8 see CMI300
13537-32-1 see PHJ250
13537-45-6 see HGU100
13539-59-8 see AQN750
13547-70-1 see MRG070
13548-38-4 see CMJ600
13551-87-6 see NHH500
13551-92-3 see NHI000
13552-44-8 see MJQ100
13553-79-2 see RKU000
13556-50-8 see DCL100
13560-89-9 see DAI460
13561-08-5 see BJO000
13567-11-8 see AHI550
13569-63-6 see RGP000
13569-65-8 see RHP000
13573-18-7 see SKN000
13589-15-6 see MIH250
13590-71-1 see PGZ950
13590-82-4 see CDB400
13590-97-1 see DXX800
13593-03-8 see DJY200
13596-11-7 see AHA130
13597-72-3 see PHB550
13597-95-0 see BFU000
13597-99-4 see BFT000
13598-00-0 see PCV400
13598-15-7 see BFS000
13598-36-2 see PGZ899
13598-37-3 see ZGJ050
13598-47-5 see SHO000
13598-56-6 see UPA000
13600-88-9 see THQ250
13600-98-1 see SFX750
13601-02-0 see DCJ600
13601-08-6 see TBI000
13601-19-9 see SHE350
13603-07-1 see MMY750
13607-48-2 see BGL000
13609-67-1 see HHQ825
13614-98-7 see MQW100
13619-98-2 see PGF112
13629-82-8 see DOV200
13637-63-3 see CDX250
13637-65-5 see CHK250
13637-76-8 see LDS499
13637-84-8 see SOT500
13642-52-9 see SKW000
13647-35-3 see EBY600
13654-09-6 see PCC480
13654-91-6 see CHL250
13655-52-2 see AGW250
13655-95-3 see MJW875
13657-68-6 see GDO100
13669-70-0 see FAL000
13674-84-5 see TNG000
13674-87-8 see FQU875
13675-27-9 see THM910
13676-54-5 see BKL800
13679-74-8 see MPR300
13681-87-3 see CNL300
13682-73-0 see PLC175

13684-56-5 see EEO500
13684-63-4 see MEG250
13693-09-9 see XEJ300
13693-11-3 see TGH252
13696-04-3 see DXG450
13701-67-2 see DDI600
13706-09-7 see TFO500
13706-10-0 see TFO250
13706-86-0 see MKL300
13707-88-5 see AGW000
13709-32-5 see PCM250
13709-61-0 see XFA000
13710-19-5 see CLK325
13713-13-8 see DQB309
13717-04-9 see PMX250
13718-26-8 see SKP000
13721-39-6 see SIY250
13738-63-1 see FKM100
13738-70-0 see PGN100
13743-07-2 see HKW500
13746-66-2 see PLF250
13746-89-9 see ZSA000
13746-98-0 see TEK800
13749-37-6 see BOI000
13749-94-5 see MPS270
13752-51-7 see OPQ100
13754-23-9 see BLF250
13754-56-8 see DVT400
13755-29-8 see SKE000
13755-38-9 see SIW500
13758-99-1 see PPT325
13759-17-6 see HAD500
13759-83-6 see SAT000
13762-14-6 see CNC250
13762-26-0 see ZQJ000
13762-51-1 see PKY250
13762-80-6 see HHA000
13764-35-7 see PDP500
13764-49-3 see EQO500
13765-19-0 see EQP500
13766-26-2 see PJL325
13767-90-3 see PLD250
13768-38-2 see OKG000
13768-67-7 see YDS800
13769-43-2 see PLK810
13770-16-6 see MAT899
13770-61-1 see ICI000
13770-89-3 see NDK000
13770-96-2 see SEM500
13775-53-6 see TNK450
13776-58-4 see XFS000
13779-41-4 see PHF250
13780-03-5 see CAN000
13780-57-9 see SOF000
13782-01-9 see PLI750
13782-22-4 see SKL500
13783-04-5 see TGG625
13791-92-9 see CIR275
13798-24-8 see TAM100
13812-39-0 see HGY000
13814-96-5 see LDE000
13814-98-7 see SMI000
13815-28-6 see ANI000
13815-31-1 see ANI500
13816-33-6 see IOD050
13820-40-1 see ANE750
13820-41-2 see ANV800
13820-45-6 see CLO700
13820-81-0 see PAR750
13820-83-2 see HBX500
13820-91-2 see PJD750
13822-05-4 see CKA500

14698-29-4 see OOG000	14976-57-9 see FOS100	15307-79-6 see DEO600	15611-84-4 see PLN050
14708-14-6 see NDC000	14977-61-8 see CML125	15308-34-6 see NNT100	15617-30-8 see TJE300
14709-62-7 see ZJJ400	14985-19-4 see TFX100	15318-45-3 see MPN000	15622-65-8 see MRB250
14721-18-7 see CMH270	14986-48-2 see CMJ355	15336-18-2 see HCM050	15625-89-5 see TLX175
14722-22-6 see IEO000	14986-60-8 see FFA000	15336-58-0 see SDJ025	15630-89-4 see SJB400
14722-38-4 see MIO770	14986-84-6 see HEY500	15336-81-9 see DKM130	15640-93-4 see TCG500
14722-38-4 see MIO800	15005-90-0 see CMG800	15336-82-0 see ECI200	15652-38-7 see DAE600
14733-73-4 see BMU750	15016-15-6 see SPB800	15337-60-7 see BAI770	15660-29-4 see ANX800
14737-08-7 see CFG500	15017-02-4 see DXP400	15339-36-3 see MAR750	15662-33-6 see RSZ000
14737-91-8 see MEJ775	15020-57-2 see DUC600	15347-57-6 see LCG000	15663-27-1 see PJD000
14742-53-1 see BII000	15029-32-0 see MRR750	15351-05-0 see BTA325	15663-42-0 see PAS859
14745-52-9 see MGR900	15044-98-1 see AGM800	15356-70-4 see MCG000	15669-07-5 see FBG200
14745-61-0 see CHO800	15044-99-2 see AGM802	15364-94-0 see MAU000	15676-16-1 see EPD500
14746-03-3 see BPA250	15056-34-5 see THQ750	15372-34-6 see BIS750	15686-63-2 see EDV700
14751-74-7 see HIR000	15061-57-1 see CNO325	15375-94-7 see HAP000	15686-71-2 see ALV000
14751-76-9 see HIR500	15083-53-1 see DPU800	15383-68-3 see SDN100	15687-18-0 see CKG000
14751-87-2 see FDU875	15086-94-9 see BMO250	15387-10-7 see AQN500	15687-27-1 see IIU000
14751-89-4 see TND750	15089-03-9 see MEC332	15388-46-2 see TNE000	15699-18-0 see NCY050
14751-90-7 see HIP500	15089-07-3 see CEL600	15411-45-7 see DTJ159	15702-63-3 see BKJ260
14753-13-0 see MMR250	15090-10-5 see BEW750	15421-84-8 see DIO200	15702-65-5 see BKJ275
14753-14-1 see CJF250	15090-12-7 see BEW500	15422-00-1 see DIQ200	15707-23-0 see ENF200
14760-99-7 see DRN300	15090-13-8 see BEL500	15424-14-3 see BDM600	15708-41-5 see EJA379
14763-77-0 see CNL250	15090-16-1 see BEN750	15430-91-8 see DHA375	15709-62-3 see BLS500
14765-30-1 see MOU800	15091-30-2 see DDT000	15438-31-0 see FBN000	15710-39-1 see PAU700
14774-78-8 see CKC325	15096-52-3 see SHF000	15442-77-0 see BIX500	15712-13-7 see TBN150
14779-78-3 see AOI250	15104-03-7 see MMZ000	15446-08-9 see CKK500	15718-71-5 see EGT500
14781-32-9 see BIU125	15105-92-7 see SJN650	15448-47-2 see PNL650	15721-02-5 see TBO000
14787-38-3 see RBF200	15112-89-7 see TNH300	15451-93-1 see DIE350	15721-33-2 see DNO200
14788-78-4 see JDA000	15114-92-8 see DNU350	15457-05-3 see NIX000	15727-43-2 see DCW000
14807-75-1 see DXN400	15119-62-7 see NIW450	15457-77-9 see PJI250	15736-98-8 see SHJ500
14807-96-6 see TAB750	15120-17-9 see ARD500	15457-87-1 see TFT250	15742-33-3 see HBW000
14807-96-6 see TAB775	15121-11-6 see CKQ750	15460-48-7 see DGB800	15764-24-6 see EEW100
14808-60-7 see SCJ500	15121-84-3 see NIM560	15465-08-4 see SLI330	15769-72-9 see EES000
14813-29-7 see MRZ200	15130-85-5 see IDD000	15467-20-6 see DXF000	15770-21-5 see PPY300
14816-16-1 see MOC275	15131-55-2 see BIS200	15468-32-3 see SCK000	15773-47-4 see DJN800
14816-18-3 see BAT750	15131-84-7 see CML830	15475-56-6 see MDV750	15790-54-2 see PNV800
14816-67-2 see SKW500	15139-76-1 see OJK325	15481-65-9 see BKJ700	15791-78-3 see CMP060
14817-09-5 see DAJ400	15148-80-8 see BQB250	15481-70-6 see DCE400	15804-19-0 see QRS000
14838-15-4 see NNM500	15151-00-5 see AGR750	15489-16-4 see AQH500	15805-73-9 see VNK000
14838-45-0 see HNT075	15154-19-5 see MLP300	15493-35-3 see SLY000	15825-70-4 see MAW250
14842-81-0 see MAI600	15154-36-6 see AIN000	15500-65-9 see AOO403	15826-16-1 see DYA000
14848-01-2 see CEO000	15154-37-7 see AIM250	15500-66-0 see PAF625	15826-37-6 see CNX825
14857-34-2 see EES200	15158-64-2 see ERB500	15500-66-0 see PAF630	15829-53-5 see MDF750
14860-49-2 see CMW459	15158-67-5 see CDA500	15501-74-3 see SBY200	15842-89-4 see CJT125
14860-53-8 see TEC250	15165-67-0 see DGA880	15503-86-3 see RFU000	15860-21-6 see CPH500
14861-06-4 see VNU000	15172-86-8 see BGP750	15503-87-4 see RFP100	15860-31-8 see FEM100
14873-10-0 see BJY000	15176-29-1 see EHV200	15511-81-6 see NOH100	15861-05-9 see FFU000
14874-86-3 see ANH300	15180-03-7 see DBK400	15520-10-2 see MNI515	15873-56-0 see CDY600
14881-92-6 see ZHJ000	15181-46-1 see HIC600	15521-65-0 see BJK250	15875-13-5 see TNH250
14882-18-9 see SAI600	15188-09-7 see VQU333	15529-90-5 see CLQ500	15876-67-2 see DXG800
14883-62-6 see DGO250	15189-51-2 see GIZ100	15534-95-9 see BKH650	15879-93-3 see GFA000
14884-42-5 see CMK275	15191-85-2 see SCN500	15535-69-0 see DED800	15886-83-6 see OCW025
14885-29-1 see IGH000	15194-98-6 see CAM300	15535-79-2 see DVL200	15886-84-7 see TLR250
14899-36-6 see DBC525	15213-49-7 see PLD600	15536-01-3 see EON600	15893-52-4 see DMW300
14901-07-6 see IFX000	15214-89-8 see ADS300	15538-67-7 see THU275	15904-73-1 see PDP600
14901-08-7 see COU000	15216-10-1 see NJL000	15545-48-9 see CIS250	15905-86-9 see UPA100
14907-98-3 see YAG500	15219-97-3 see OLK200	15546-11-9 see BKO250	15914-23-5 see DRE200
14913-33-8 see DEX000	15227-42-6 see DFF500	15546-12-0 see BJR250	15918-62-4 see MEA700
14915-37-8 see BLG600	15230-48-5 see DFL600	15546-16-4 see BHK250	15922-78-8 see HOC000
14917-59-0 see BMV250	15233-65-5 see DON200	15547-17-8 see EPK000	15923-42-9 see HDQ500
14918-35-5 see DBB400	15237-44-2 see AEM250	15565-25-0 see BEK500	15930-94-6 see CMK500
14919-77-8 see SCA400	15242-96-3 see CMH300	15567-46-1 see MMY000	15932-80-6 see MCF500
14925-39-4 see BMT000	15246-55-6 see TNG100	15568-57-7 see ABN800	15932-89-5 see HME000
14929-11-4 see SDY500	15263-52-2 see BHL750	15571-58-1 see DVM800	15942-48-0 see MIA000
14930-96-2 see CQM125	15267-04-6 see AMN500	15572-56-2 see INL000	15954-91-3 see CAF750
14931-40-9 see HGU025	15267-95-5 see CIY500	15588-95-1 see SLU600	15954-94-6 see LDD000
14932-06-0 see BIT750	15271-41-7 see CFF250	15589-00-1 see DOG600	15954-98-0 see ZGS100
14938-35-3 see AOM250	15275-07-7 see IGX550	15589-31-8 see TAL575	15964-31-5 see IPR000
14938-42-2 see DLJ600	15284-15-8 see MDP240	15597-43-0 see ISF000	15968-05-5 see TBO600
14940-68-2 see ZSS000	15284-39-6 see MQM775	15598-34-2 see PPC100	15972-60-8 see CFX000
14949-00-9 see AMR750	15285-42-4 see BFA250	15605-28-4 see NMI000	15973-99-6 see DVF000
14959-86-5 see DXU830	15301-48-1 see BCE825	15611-43-5 see CKN250	15978-91-3 see DCF760

15978-93-5 see DNF000
15980-15-1 see OLY000
16006-09-0 see IIE000
16009-13-5 see HAQ050
16017-38-2 see CML325
16018-21-6 see MNP450
16028-14-1 see TET600
16032-41-0 see DPT200
16033-21-9 see MOA725
16033-21-9 see PFS500
16034-77-8 see IDJ550
16034-99-4 see BMT300
16037-91-5 see AQH800
16037-91-5 see AQI250
16039-54-6 see CNC242
16039-55-7 see CAG750
16051-77-7 see ISC500
16053-71-7 see DLF000
16054-41-4 see BIX125
16056-11-4 see TMB000
16064-14-5 see CLC750
16065-83-1 see CMJ580
16066-38-9 see DWV400
16069-36-6 see DGV100
16071-86-6 see CMO750
16071-96-8 see BKM530
16078-34-5 see ACO320
16079-88-2 see BNA300
16088-56-5 see DVQ600
16088-62-3 see ECE700
16090-02-1 see CMP200
16091-18-2 see DVK200
16102-24-2 see DGW875
16110-13-7 see BBH250
16111-27-6 see NNL400
16111-62-9 see DJK800
16118-49-3 see CBL500
16120-70-0 see BRK100
16128-42-0 see IAT400
16130-58-8 see COJ625
16136-32-6 see BBW250
16142-27-1 see MRU075
16143-89-8 see MFF750
16154-78-2 see HDS225
16169-16-7 see NIR050
16169-17-8 see HIT600
16176-02-6 see COT000
16179-44-5 see HKQ500
16187-03-4 see PBQ300
16187-15-8 see BOX825
16203-97-7 see APH500
16212-28-5 see TJC100
16215-49-9 see BSC800
16219-75-3 see ELO500
16219-98-0 see NKQ000
16219-99-1 see NKQ100
16222-66-5 see TEM100
16224-33-2 see BQT500
16227-10-4 see BPU000
16230-71-0 see DWI400
16238-56-5 see BNR000
16239-84-2 see ALN500
16240-52-1 see TDO100
16248-90-1 see TGB600
16260-59-6 see DFL800
16268-87-4 see AJS250
16277-49-9 see MEL600
16282-67-0 see DKH830
16291-96-6 see CDI250
16299-13-1 see MEX285
16301-26-1 see ASP000
16310-68-2 see TCO000

16320-04-0 see ENX575
16322-14-8 see CBO750
16322-19-3 see NFD800
16327-90-5 see BCI300
16329-92-3 see BJV635
16329-93-4 see BJV630
16331-85-4 see CAK350
16338-97-9 see NJV500
16338-99-1 see HBP000
16339-01-8 see MMY250
16339-04-1 see ELX500
16339-05-2 see BRY250
16339-07-4 see NKW500
16339-12-1 see DSZ000
16339-14-3 see NLY000
16339-16-5 see CIQ500
16339-18-7 see NKL500
16339-21-2 see MMX750
16354-47-5 see MEU500
16354-48-6 see MJX000
16354-50-0 see EMM500
16354-52-2 see DIT800
16354-53-3 see DOA000
16354-54-4 see MOU250
16354-55-5 see EMN000
16356-11-9 see ULA100
16357-59-8 see EFJ500
16361-01-6 see DKY200
16368-97-1 see BJR625
16371-55-4 see TEO100
16377-01-8 see IDA525
16378-21-5 see PJA190
16378-22-6 see TMK150
16395-80-5 see MNA250
16409-43-1 see RNU000
16409-45-3 see MCG500
16409-46-4 see MCG900
16413-88-0 see TCG250
16417-75-7 see DIM400
16423-68-0 see FAG040
16430-32-3 see APM800
16431-49-5 see PGN045
16432-36-3 see TNN500
16448-54-7 see IHC100
16449-54-0 see AHA875
16454-60-7 see NCB500
16456-56-7 see MRF500
16468-98-7 see ENE000
16470-24-9 see CMP120
16478-59-4 see HAP100
16481-54-2 see PJJ250
16484-77-8 see CIR325
16484-86-9 see DHG100
16485-39-5 see OMS400
16488-48-5 see BQA020
16491-24-0 see HCS700
16495-13-9 see BFC225
16501-01-2 see MEQ000
16503-95-0 see DOF900
16508-97-7 see HKA130
16509-79-8 see ZEA000
16518-17-5 see PLL200
16524-23-5 see TGV000
16532-79-9 see BNV750
16543-55-8 see NLD500
16550-39-3 see MBV775
16561-29-8 see PGV000
16561-55-0 see ZLJ100
16566-62-4 see NAU500
16566-64-6 see NAU000
16568-02-8 see AAH000
16580-04-4 see BOT600

16587-71-6 see AOH750
16590-41-3 see CQF099
16595-80-5 see LFA020
16606-02-3 see TIL275
16607-77-5 see ODQ300
16607-80-0 see CPR500
16610-44-9 see MNR300
16634-82-5 see CEB500
16646-44-9 see TDG250
16648-69-4 see OQU000
16650-10-5 see TBQ275
16669-59-3 see IIE100
16672-39-2 see DJF400
16672-87-0 see CDS125
16675-05-1 see PGU250
16676-29-2 see NAH100
16680-47-0 see ECW520
16690-44-1 see MER250
16694-30-7 see KGK150
16699-07-3 see MMR750
16699-10-8 see MMS750
16711-32-3 see PIU200
16714-23-1 see MGR800
16714-68-4 see PAY000
16719-32-7 see BOQ500
16721-80-5 see SHR000
16725-53-4 see TCA400
16726-46-8 see CLW650
16731-55-8 see PLR250
16747-33-4 see EIK500
16752-77-5 see MDU600
16755-07-0 see RIU000
16757-80-5 see MHH750
16757-81-6 see MHM250
16757-82-7 see MHG500
16757-83-8 see MHG750
16757-84-9 see DQO200
16757-85-0 see DQN000
16757-86-1 see DQN200
16757-87-2 see DQN800
16757-88-3 see DQN400
16757-89-4 see DQO400
16757-90-7 see DQN600
16757-91-8 see DQO000
16757-92-9 see TLM500
16758-26-2 see DVY050
16758-28-4 see EKC990
16760-11-5 see POU250
16760-12-6 see PPB000
16760-13-7 see AMQ250
16760-14-8 see AMQ000
16760-18-2 see AMP500
16760-22-8 see PPA750
16760-23-9 see PPB250
16773-42-5 see OJS000
16781-80-9 see CIN000
16785-81-2 see VEA100
16800-47-8 see THM250
16802-49-6 see AMP750
16802-50-9 see POU000
16808-85-8 see FDV000
16812-54-7 see NDL100
16813-36-8 see NJY000
16816-67-4 see PAG150
16822-88-1 see FMR080
16824-81-0 see LDA500
16825-72-2 see IFS350
16825-74-4 see IFQ775
16830-14-1 see NKO425
16830-15-2 see ARN500
16842-03-8 see CNC230
16845-29-7 see RSA000

16846-24-5 see JDS200
16853-85-3 see LHS000
16870-90-9 see HOP259
16871-71-9 see ZIA000
16871-90-2 see PLH750
16872-09-6 see DEJ500
16872-11-0 see FDD125
16872-11-0 see HHS600
16879-01-9 see DLS275
16881-77-9 see MJE900
16883-45-7 see TDJ500
16887-79-9 see ILF000
16893-05-3 see TBO776
16893-06-4 see TBO777
16893-85-9 see DXE000
16893-93-9 see PLI250
16899-81-3 see DMV500
16919-27-0 see PLI000
16919-58-7 see ANF250
16919-73-6 see PLA750
16921-30-5 see PLR000
16921-96-3 see IDQ000
16923-95-8 see PLG500
16924-00-8 see PLH000
16924-32-6 see DRJ200
16925-39-6 see CAX250
16930-93-1 see HCS600
16930-96-4 see HFX000
16940-66-2 see SFF500
16940-81-1 see HDE000
16941-12-1 see CKO750
16941-92-7 see HIA500
16949-15-8 see LHT000
16949-65-8 see MAF600
16960-39-7 see MPS250
16961-83-4 see SCO500
16962-07-5 see AHG875
16962-40-6 see ANI250
16967-79-6 see ECT600
16971-82-7 see DEK500
16974-11-1 see GJU050
16984-48-8 see FEX875
16986-24-6 see NBV100
16987-02-3 see CKT100
16993-94-5 see BBX250
17003-75-7 see FHW000
17003-79-1 see FMV050
17003-82-6 see FHZ000
17008-69-4 see POG275
17010-21-8 see CAG500
17010-59-2 see CIL710
17010-61-6 see DFD400
17010-62-7 see EPU000
17010-63-8 see EPT500
17010-64-9 see DJB400
17010-65-0 see EOI000
17012-89-4 see MIK250
17012-91-8 see BBG750
17013-01-3 see DXD800
17013-07-9 see AHB750
17013-35-3 see EMO875
17013-37-5 see EJY000
17013-41-1 see DDV400
17014-71-0 see PLP250
17018-24-5 see MPY250
17024-17-8 see NIB600
17024-18-9 see NIC000
17024-19-0 see NIC100
17025-30-8 see DTY200
17026-81-2 see AJT750
17029-22-0 see PLH500
17031-32-2 see CNO500

17040-19-6 see DAP600
17043-56-0 see MNM750
17057-98-6 see DRY100
17058-53-6 see BLF600
17066-89-6 see DHS400
17068-78-9 see ARM266
17070-44-9 see MEJ250
17074-42-9 see NFW100
17074-44-1 see NFW200
17083-63-5 see PLI500
17083-85-1 see TCB750
17088-21-0 see EEF000
17088-72-1 see PAR600
17088-73-2 see DMN200
17090-79-8 see MRE225
17090-93-6 see SEZ355
17091-40-6 see IIM050
17095-24-8 see RCU000
17097-76-6 see CAT125
17099-70-6 see AHB400
17099-80-8 see EJA410
17099-81-9 see HIA000
17100-11-7 see EJA390
17109-49-8 see EIM000
17117-34-9 see NEW700
17124-74-2 see DAZ135
17125-80-3 see BAO750
17126-65-7 see ELX527
17132-74-0 see HGI050
17140-68-0 see DIH600
17140-78-2 see DYB400
17140-81-7 see NGE780
17146-95-1 see PBP400
17156-85-3 see TBJ475
17156-88-6 see DRJ400
17157-48-1 see BMR000
17160-71-3 see FDA880
17167-73-6 see AAH100
17168-82-0 see AQQ100
17168-83-1 see DBY300
17168-85-3 see THP250
17169-60-7 see FBD500
17174-98-0 see COK300
17176-77-1 see DDH400
17177-50-3 see ONE050
17185-68-1 see AQQ125
17194-00-2 see BAM500
17194-82-0 see HNG550
17210-48-9 see DNY400
17224-08-7 see DKF600
17224-09-8 see TCH250
17226-43-6 see GGA050
17230-87-4 see CLS250
17230-88-5 see DAB830
17236-22-5 see IFG000
17242-52-3 see PKV000
17243-39-9 see BCH300
17243-64-0 see EOA500
17243-64-0 see EOB000
17247-77-7 see DLX800
17256-39-2 see CGJ280
17264-01-6 see MIS750
17268-47-2 see DOX400
17274-12-3 see SKL700
17278-93-2 see MJA000
17284-75-2 see BQH500
17297-82-4 see ORI400
17300-62-8 see MAJ775
17306-43-3 see NHG200
17309-87-4 see DSI800
17311-31-8 see DVQ800
17321-77-6 see CDV000

17333-74-3 see TGM550
17333-83-4 see CEJ500
17333-84-5 see CEJ250
17333-86-7 see BBN250
17339-60-5 see BJG100
17351-75-6 see BLU600
17351-87-1 see AKZ300
17359-54-5 see BDI100
17360-35-9 see OOY100
17369-59-4 see PNO500
17372-87-1 see BNH500
17380-19-7 see COP550
17380-21-1 see COP525
17381-88-3 see DBJ400
17397-89-6 see ECE500
17400-65-6 see DPQ400
17400-68-9 see DQE400
17400-69-0 see MJF500
17400-70-3 see MJF750
17401-48-8 see DQH800
17406-45-0 see THG250
17416-17-0 see DPQ800
17416-18-1 see DPQ600
17416-20-5 see MJG000
17416-21-6 see DQE600
17418-58-5 see AKI750
17427-00-8 see ABV250
17431-55-9 see CHU050
17433-31-7 see ADA000
17433-39-5 see ADD000
17449-96-6 see CMW750
17455-13-9 see COD500
17455-23-1 see DGT300
17463-44-4 see AKP500
17466-45-4 see PCU350
17471-59-9 see TDU750
17471-82-8 see GKG300
17476-04-9 see LGS000
17489-40-6 see MAN400
17493-73-1 see CGI200
17496-59-2 see DEL800
17498-10-1 see DJN875
17501-44-9 see PBL750
17505-25-8 see FPI200
17508-17-7 see DVC700
17511-60-3 see TJG600
17513-40-5 see MGY500
17518-47-7 see NBN000
17523-77-2 see PLN500
17524-18-4 see THP750
17526-17-9 see DMT200
17526-24-8 see BBH500
17526-74-8 see GGY000
17548-36-6 see HCY500
17549-30-3 see DJD200
17557-23-2 see NCI300
17560-51-9 see ZAK300
17563-48-3 see BRA550
17573-21-6 see BCY250
17573-23-8 see DKU000
17575-20-1 see LAT000
17575-21-2 see LAT500
17575-22-3 see LAU000
17575-26-7 see TKO300
17576-63-5 see FKO000
17576-88-4 see BNE600
17590-01-1 see AOB300
17595-59-4 see INE025
17596-45-1 see BMB150
17597-95-4 see HFG700
17598-02-6 see DAM830
17598-65-1 see DBH200

17599-02-9 see DPC200
17599-08-5 see AIM000
17599-09-6 see AIU000
17601-12-6 see BLE250
17605-71-9 see MJU750
17605-83-3 see THG300
17606-31-4 see NCN650
17607-20-4 see BGW710
17608-59-2 see NKC000
17615-73-5 see AOT125
17617-13-9 see DID800
17617-23-1 see FMQ000
17617-45-7 see PIE510
17617-46-8 see DMB300
17622-94-5 see EJM000
17639-93-9 see CKT000
17650-86-1 see AHL000
17650-98-5 see CAK285
17654-88-5 see MCP500
17660-58-1 see DJB100
17661-50-6 see TCB100
17667-23-1 see DVO809
17672-21-8 see HJC500
17673-25-5 see PGS250
17676-08-3 see LJM600
17689-16-6 see TEY750
17692-34-1 see HHK050
17692-39-6 see PDU250
17697-53-9 see ASP510
17697-55-1 see ASP500
17700-09-3 see NMQ025
17702-41-9 see DAE400
17702-57-7 see FNE500
17710-62-2 see CJU125
17710-63-3 see HLK900
17710-64-4 see MIA800
17719-22-1 see MPT300
17721-94-7 see BRZ200
17721-95-8 see DTA400
17730-82-4 see BTA000
17737-65-4 see CMX770
17750-93-5 see TCJ775
17751-20-1 see DEX600
17754-90-4 see DIJ230
17756-81-9 see COM830
17760-93-9 see PJP300
17766-26-6 see SKN100
17766-62-0 see BQK830
17766-63-1 see PFX600
17766-66-4 see CKJ100
17766-68-6 see MFH760
17766-70-0 see MFH770
17766-74-4 see THF300
17766-75-5 see THF310
17766-77-7 see TKY300
17766-79-9 see TEX220
17773-41-0 see HMR550
17780-75-5 see CMY000
17784-47-3 see ADJ750
17786-31-1 see CNB510
17794-13-7 see TAO750
17796-82-6 see CPQ700
17804-35-2 see BAV575
17804-49-8 see PMF540
17814-73-2 see MOR250
17822-71-8 see DHP550
17822-72-9 see DHP450
17822-73-0 see DHP500
17822-74-1 see DHQ800
17831-71-9 see ADT050
17861-62-0 see NDG550
17869-27-1 see AMG500

17874-34-9 see DQV000
17878-69-2 see MPJ050
17887-09-1 see AMG200
17893-55-9 see NIV100
17902-23-7 see FLZ050
17918-11-5 see PEH500
17924-92-4 see ZAT000
17927-57-0 see CFK325
17948-33-3 see RRP100
17958-39-3 see CQI525
17958-73-5 see CMP100
17959-11-4 see MDX250
17959-12-5 see MLX000
17960-21-3 see MMJ955
17962-31-1 see PBI700
17969-20-9 see CKK250
17977-68-3 see FOE000
17982-67-1 see EML500
18010-40-7 see BOO000
18024-11-8 see TLN133
18046-21-4 see CKI750
18048-06-1 see DJA330
18051-18-8 see MIQ400
18067-13-5 see DAB750
18109-81-4 see BOR350
18127-01-0 see BMK300
18139-02-1 see FHX000
18139-03-2 see BJU500
18156-74-6 see TMF250
18162-48-6 see BQS300
18172-33-3 see CQD000
18179-67-4 see MFO000
18181-70-9 see IEN000
18181-80-1 see IOS000
18186-71-5 see DXU200
18197-22-3 see FNL000
18204-79-0 see TMF625
18207-29-9 see NLD800
18230-75-6 see TMF125
18237-15-5 see AME750
18237-16-6 see AJB500
18244-91-2 see TJU500
18252-65-8 see DEU115
18262-71-0 see PIV550
18264-75-0 see ALN750
18264-88-5 see DMA400
18268-70-7 see FCD500
18268-70-7 see PJC250
18273-30-8 see BJE250
18278-44-9 see MOB500
18279-20-4 see DVU600
18279-21-5 see DKP200
18283-93-7 see TBJ300
18296-44-1 see VAD200
18307-23-8 see SBY100
18309-32-5 see VIP000
18312-12-4 see IDZ100
18312-66-8 see SBV100
18318-83-7 see HFA620
18323-44-9 see CMV675
18355-50-5 see BGM000
18355-54-9 see EKD000
18356-02-0 see MPJ000
18361-48-3 see LEV025
18365-12-3 see PNH500
18378-89-7 see MQW750
18380-68-2 see DQG400
18406-41-2 see EIT200
18409-46-6 see HCR600
18413-14-4 see ELC000
18417-89-5 see SAU000
18429-70-4 see DQJ600

20246-69-9 see CAD550
20247-50-1 see EEB225
20248-45-7 see TNH500
20265-96-7 see CJR200
20265-97-8 see AOX500
20268-51-3 see NEW600
20268-52-4 see CEJ000
20275-19-8 see BGH000
20281-00-9 see CDE325
20300-26-9 see GJM025
20302-25-4 see POV500
20308-90-1 see MRT200
20311-78-8 see EAO500
20324-33-8 see TNA000
20325-40-0 see DOA800
20333-40-8 see BRF550
20344-15-4 see MCH535
20354-26-1 see BGD250
20369-63-5 see TID150
20373-56-2 see DEW400
20389-01-9 see DNG800
20398-06-5 see EEE000
20404-94-8 see DDO400
20405-19-0 see MIH750
20408-97-3 see GFK000
20424-95-7 see CNL800
20427-56-9 see RSK000
20427-59-2 see CNM500
20436-27-5 see OGI100
20485-57-8 see MKC775
20519-92-0 see TLT100
20537-88-6 see AMD000
20539-85-9 see BJO125
20548-54-3 see CAY000
20562-02-1 see SKS000
20562-03-2 see CDG500
20570-96-1 see BEQ250
20589-63-3 see NIB500
20600-96-8 see TDY100
20611-21-6 see BBT000
20620-82-0 see NAK700
20624-25-3 see SGJ500
20627-28-5 see DQJ800
20627-31-0 see DQK800
20627-32-1 see TLK000
20627-33-2 see TLJ750
20627-34-3 see TLK500
20627-73-0 see NEV510
20633-84-5 see LIU450
20645-04-9 see ZTS000
20652-39-5 see DNP600
20661-60-3 see NJH500
20667-12-3 see SDU500
20675-51-8 see PBW400
20679-58-7 see BHD150
20680-07-3 see MHS375
20684-29-1 see FAC130
20685-78-3 see SPE000
20689-96-7 see ENS500
20691-83-2 see DSN200
20691-84-3 see CCE750
20697-04-5 see CLF000
20706-25-6 see PNA225
20719-22-6 see POX750
20719-23-7 see EKB000
20719-34-0 see TAP000
20719-36-2 see TAP250
20731-44-6 see NIA700
20737-02-4 see SDS500
20738-78-7 see DWX200
20740-05-0 see HFN500
20743-57-1 see BGF000

20762-60-1 see PKW000
20762-98-5 see FQJ025
20777-39-3 see LCA100
20777-49-5 see DKV160
20780-49-8 see DTE800
20794-96-1 see CHT500
20816-12-0 see OKK000
20819-47-0 see DBA100
20819-54-9 see TMD400
20820-44-4 see NHK650
20820-80-8 see EIF500
20829-66-7 see EIV700
20830-75-5 see DKN400
20830-81-3 see DAC000
20839-15-0 see ACZ000
20839-16-1 see PGS500
20846-00-8 see NIY025
20854-03-9 see TGY250
20856-57-9 see CDP750
20859-73-8 see AHE750
20866-13-1 see HGL630
20917-34-4 see DSL400
20917-49-1 see OBY000
20917-50-4 see OCA000
20919-99-7 see HBD650
20921-41-9 see MNX850
20921-50-0 see PGQ000
20929-99-1 see BJW250
20930-00-1 see BJW500
20930-10-3 see DVY900
20941-65-5 see EPJ000
20977-05-3 see EDM000
20977-50-8 see FGV000
20982-36-9 see BHQ000
20982-74-5 see NMV740
20986-33-8 see IMM000
20991-79-1 see TEK300
21000-42-0 see DWX000
21001-46-7 see TCM400
21019-30-7 see MLD140
21019-39-6 see HKB600
21035-25-6 see BCD325
21038-21-1 see TCV460
21041-93-0 see CNC238
21041-95-2 see CAG525
21059-46-1 see CAM675
21062-28-2 see WCJ750
21064-50-6 see MHD000
21070-32-6 see BRW750
21070-33-7 see BRM750
21075-41-2 see MNS250
21082-50-8 see EME000
21083-47-6 see DWN600
21085-56-3 see MNO775
21087-64-9 see MQR275
21107-27-7 see TLP275
21109-95-5 see BAP250
21109-99-9 see AAS300
21124-09-4 see MRI250
21124-13-0 see MQB250
21140-85-2 see MFG600
21145-77-7 see EEE100
21149-87-1 see EMP500
21150-21-0 see AHI523
21150-22-1 see AHI520
21150-23-2 see IBL100
21172-28-1 see ALG500
21187-98-4 see DBL700
21208-99-1 see CQC250
21209-02-9 see MIT250
21224-57-7 see EOK500
21224-77-1 see EFC000

21224-81-7 see THB250
21230-20-6 see BBR380
21232-47-3 see TBN550
21233-74-9 see EHM200
21239-57-6 see TBN200
21243-26-5 see DLO200
21247-98-3 see BCV250
21248-00-0 see BMU500
21248-01-1 see CEK500
21254-73-9 see CPU750
21255-83-4 see BMP500
21256-18-8 see OLW600
21259-20-1 see FQS000
21259-75-6 see BLQ525
21259-76-7 see TFK270
21260-46-8 see BKW000
21267-72-1 see CDS275
21280-29-5 see LAQ100
21282-96-2 see AAY750
21284-11-7 see COB000
21288-85-7 see TBH300
21299-86-5 see INU200
21308-79-2 see MNF250
21322-39-4 see EIY700
21327-74-2 see BGQ325
21340-68-1 see MIO975
21342-26-7 see TKQ300
21351-39-3 see UTU600
21351-79-1 see CDD750
21361-93-3 see NDP500
21362-69-6 see MCH600
21368-68-3 see CBA800
21372-60-1 see TKH025
21380-82-5 see ACI375
21393-59-9 see IGX000
21413-28-5 see DWJ300
21416-67-1 see RCA375
21416-87-5 see PIK250
21436-96-4 see XOJ000
21436-97-5 see TLG750
21447-39-2 see BHR400
21447-86-9 see BHQ760
21447-87-0 see BIA100
21450-81-7 see AAS500
21452-14-2 see AKR500
21457-22-7 see DME700
21466-07-9 see BNV500
21466-08-0 see NNU000
21476-57-3 see BJI125
21482-59-7 see TDP275
21498-08-8 see LIA400
21535-47-7 see BMA625
21548-32-3 see DHH200
21554-20-1 see AJQ500
21556-79-6 see HJS450
21561-99-9 see MMY500
21564-17-0 see BOO635
21567-21-5 see HNB550
21572-61-2 see TBO778
21586-21-0 see MNI525
21587-39-3 see MEC330
21590-92-1 see AHP125
21593-23-7 see CCX500
21595-62-0 see PPC250
21600-42-0 see DTV200
21600-43-1 see DJY000
21600-45-3 see HKV000
21600-51-1 see CBP250
21609-90-5 see LEN000
21615-29-2 see BKB000
21621-73-8 see MLW630
21621-75-0 see MLW600

21621-78-3 see IRX100
21626-24-4 see IGX875
21638-36-8 see MMJ000
21642-82-0 see PBS750
21642-83-1 see PBS500
21644-95-1 see AKK000
21645-51-2 see AHC000
21649-57-0 see CBO000
21650-02-2 see ARX150
21658-26-4 see NIQ500
21662-09-9 see DAI360
21667-01-6 see BHW250
21679-31-2 see TNN250
21689-84-9 see COI050
21704-44-9 see ELY575
21704-46-1 see CBI675
21705-02-2 see AHI510
21708-94-1 see EAF100
21709-44-4 see DXN830
21711-65-9 see NLA000
21711-71-7 see NJM100
21715-46-8 see CGQ500
21722-83-8 see EHS000
21725-46-2 see BLW750
21727-09-3 see PAX775
21736-83-4 see SLI325
21738-42-1 see OLT000
21739-91-3 see CQK600
21757-82-4 see TIL800
21787-81-5 see CHJ700
21794-01-4 see RQP000
21800-49-7 see KBB800
21813-99-0 see FQR100
21816-42-2 see DSY000
21820-82-6 see DMB000
21829-25-4 see AEC750
21842-58-0 see DLQ800
21848-62-4 see NNR200
21884-44-6 see LIV000
21888-96-0 see DBE000
21892-31-9 see PBD300
21894-02-0 see XJJ005
21898-19-1 see VHA350
21905-27-1 see AGQ775
21905-32-8 see HNX600
21905-40-8 see CKA575
21908-53-2 see MCT500
21916-66-5 see POB250
21917-91-9 see MNO250
21923-23-9 see CLB022
21928-82-5 see MMT250
21947-75-1 see CKS099
21947-76-2 see CKS100
21961-08-0 see HFY000
21961-30-8 see AJE333
21962-24-3 see PNJ500
21970-53-6 see DQD200
21985-87-5 see PBM100
21987-62-2 see DAZ110
21988-57-8 see MJE790
21988-58-9 see MJD300
21992-92-7 see MPI225
22029-76-1 see IFT500
22040-99-9 see CCJ385
22041-26-5 see CCH890
22041-28-7 see CCH300
22041-33-4 see CCE820
22041-34-5 see CCE800
22041-39-0 see DTV330
22041-44-7 see DQY925
22042-96-2 see DJG700
22047-25-2 see ADA350

22047-49-0 see OFU300
22052-01-3 see HOP300
22056-53-7 see MPO800
22059-60-5 see RSZ600
22071-15-4 see BDU500
22072-35-1 see TLT800
22086-53-9 see CGN325
22089-22-1 see TNT500
22098-38-0 see NIG500
22103-31-7 see TAG875
22120-39-4 see DST400
22128-43-4 see MKR550
22131-79-9 see AGN000
22131-80-2 see MDI225
22137-01-5 see AJZ000
22143-50-6 see MLD050
22144-77-0 see ZUS000
22150-76-1 see BGD075
22151-75-3 see HOO875
22187-44-6 see TCO200
22188-15-4 see NMG000
22189-32-8 see SKY500
22199-08-2 see SNI425
22199-30-0 see CMX820
22204-24-6 see POK575
22204-53-1 see MFA500
22204-91-7 see MON600
22205-30-7 see DVM400
22205-45-4 see CNP750
22205-57-8 see CDC250
22208-25-9 see ELJ600
22212-55-1 see EGS000
22223-55-8 see MQH100
22224-92-6 see FAK000
22225-32-7 see FDU000
22232-71-9 see MBV250
22235-85-4 see AGM250
22236-53-9 see CJC750
22238-17-1 see THY000
22248-79-9 see RAF100
22251-01-0 see FDT000
22254-24-6 see IGG000
22260-42-0 see COF825
22260-51-1 see BNB325
22261-92-3 see ADF600
22262-18-6 see MPI205
22262-19-7 see MPN275
22287-69-0 see PEK050
22295-11-0 see TDU250
22295-99-4 see HCA000
22298-04-0 see DLO000
22298-29-9 see BFV760
22304-30-9 see ASA000
22316-47-8 see CIR750
22323-45-1 see ZJA000
22326-31-4 see SNU500
22345-47-7 see GJS200
22346-43-6 see HLX000
22349-59-3 see DTJ200
22373-78-0 see MRE230
22374-89-6 see PEE300
22388-72-3 see HBY000
22392-07-0 see DKB175
22393-63-1 see BOL315
22394-38-3 see TEX210
22398-80-7 see ICI300
22413-78-1 see LJB150
22432-68-4 see TBQ255
22433-76-7 see QPS200
22439-58-3 see ACH090
22445-73-4 see TDX835
22451-06-5 see TAI800

22457-23-4 see EMP550
22464-76-2 see HOV503
22471-42-7 see HFA000
22480-64-4 see BNX000
22494-42-4 see DKI600
22504-72-9 see DUD900
22506-53-2 see DUW120
22514-23-4 see FMU225
22537-48-0 see CAG600
22545-60-4 see AEG250
22560-50-5 see SFX730
22571-95-5 see SPB500
22575-95-7 see DOF200
22578-17-2 see SJA500
22578-86-5 see BNG310
22583-29-5 see DSC800
22585-64-4 see BOL310
22590-50-7 see BFX650
22591-21-5 see DGF300
22604-10-0 see TAI050
22608-53-3 see TMD625
22609-73-0 see NDY600
22637-13-4 see BKJ800
22653-19-6 see TDP500
22668-01-5 see HKW455
22670-79-7 see PEJ250
22691-91-4 see DSV289
22692-30-4 see FHZ200
22708-05-0 see BAO400
22710-42-5 see BHB400
22713-35-5 see DQP125
22722-03-8 see ELY550
22722-98-1 see SGK800
22729-75-5 see FLZ025
22750-53-4 see CAD325
22750-56-7 see CDC125
22750-65-8 see BGR250
22750-69-2 see DCM600
22750-85-2 see DPR200
22750-86-3 see DPR400
22750-93-2 see EOD000
22751-18-4 see NFX500
22751-23-1 see NIJ200
22751-24-2 see NFA500
22754-97-8 see RPB100
22755-01-7 see OOS000
22755-07-3 see DCW200
22755-15-3 see AAM650
22755-25-5 see SIW625
22755-34-6 see TNE275
22755-36-8 see TLE500
22756-36-1 see RPB150
22760-18-5 see POB300
22771-17-1 see DGV900
22771-18-2 see MRH212
22781-23-3 see DQM600
22788-18-7 see CCE250
22791-18-0 see SLD550
22791-23-7 see BCJ100
22791-33-9 see BCJ005
22797-20-2 see DLG000
22798-21-6 see DEU160
22817-48-7 see PIT625
22817-49-8 see MRU090
22826-61-5 see TBJ250
22830-45-1 see FBL000
22832-87-7 see MQS560
22839-47-0 see ARN825
22862-76-6 see AOY000
22868-13-9 see SGR500
22885-98-9 see MPQ500
22898-01-7 see SKE100

22900-79-4 see DOO900
22904-40-1 see LDB000
22916-47-8 see MQS550
22933-76-2 see DGL825
22935-72-4 see DJW900
22936-03-4 see MEL525
22936-17-0 see DJP520
22936-20-5 see DIV700
22936-34-1 see EEQ200
22936-44-3 see ACU300
22936-75-0 see ARW775
22936-75-0 see DTS700
22936-86-3 see CQI750
22941-83-9 see DJO420
22941-96-4 see DJW890
22942-02-5 see DJW880
22942-43-4 see DJB420
22942-83-2 see KFK150
22952-87-0 see ASG675
22953-41-9 see BHG000
22953-53-3 see BHV750
22953-54-4 see BHW000
22958-08-3 see ASE875
22960-71-0 see EEX500
22966-79-6 see EDR500
22966-83-2 see MPC250
22967-92-6 see MLF550
22975-76-4 see FQD130
23007-85-4 see TCV500
23031-25-6 see TAN100
23031-32-5 see TAN250
23031-36-9 see THI500
23031-38-1 see POD875
23043-52-9 see ABX810
23067-13-2 see EDI500
23079-28-9 see TNJ285
23093-74-5 see BON400
23103-98-2 see DOX600
23107-11-1 see DAL100
23107-12-2 see DAL400
23107-96-2 see BDS500
23109-05-9 see AHI625
23110-15-8 see FOZ000
23111-70-8 see BEB500
23115-33-5 see BLR500
23125-28-2 see DJB460
23129-50-2 see HEY000
23135-22-0 see DSP600
23139-00-6 see MNA650
23139-02-8 see BNX125
23155-02-4 see PHA550
23165-19-7 see SMP450
23180-57-6 see PAC000
23181-80-8 see AKD550
23182-46-9 see PIX800
23184-66-9 see CFW750
23188-23-0 see DCP650
23188-89-8 see TIA800
23191-75-5 see FOO875
23192-42-9 see HCP070
23209-59-8 see CAX260
23210-56-2 see IAG600
23210-58-4 see IAG625
23214-92-8 see AES750
23217-81-4 see AAP750
23217-86-9 see AMV750
23217-87-0 see AAQ000
23233-88-7 see BOL325
23239-41-0 see SGB500
23239-51-2 see RLK700
23245-64-9 see NHR200
23246-96-0 see RJZ000

23248-53-5 see SNT100
23255-03-0 see NHK850
23255-69-8 see FQR000
23255-93-8 see HGO500
23256-30-6 see NGG000
23256-50-0 see GKO750
23257-56-9 see LFO000
23257-58-1 see LFG100
23261-20-3 see DCI600
23273-02-1 see DOX200
23279-53-0 see AAP500
23279-54-1 see EFP000
23282-20-4 see NMV000
23291-96-5 see DAL350
23292-52-6 see TJX375
23292-85-5 see TMW250
23314-84-3 see MJG100
23315-05-1 see EAG000
23319-66-6 see TNI500
23324-72-3 see MKR500
23327-57-3 see NBS500
23332-31-2 see AAS310
23339-04-0 see DRX600
23344-17-4 see SMB850
23362-09-6 see MPJ100
23383-11-1 see FBJ075
23389-74-4 see HEW050
23389-74-4 see PBX800
23395-20-2 see HMS500
23399-90-8 see TIY250
23412-26-2 see THA600
23414-72-4 see ZLA000
23420-61-3 see XOS500
23422-53-9 see DSO200
23424-47-7 see CBN200
23435-31-6 see DON400
23436-19-3 see IIG000
23444-65-7 see HLJ650
23445-00-3 see SOZ100
23452-05-3 see AGW500
23456-94-2 see DSA600
23471-13-8 see EED600
23471-23-0 see MHS250
23476-83-7 see POC000
23483-74-1 see BNL275
23488-38-2 see TBK250
23489-00-1 see EKN100
23489-01-2 see ENY100
23489-02-3 see ENX100
23489-03-4 see DAJ300
23491-45-4 see MOD500
23491-52-3 see BGZ100
23495-12-7 see EJK500
23505-41-1 see DIN600
23509-16-2 see BAR750
23510-39-6 see CHG400
23518-13-0 see HIS140
23518-25-4 see POC525
23521-13-3 see DUA200
23521-14-4 see DUA400
23522-05-6 see GJM300
23526-02-5 see BAC125
23535-89-9 see DEH800
23537-16-8 see RRA000
23541-50-6 see DAC200
23560-59-0 see HBK700
23564-05-8 see PEX500
23564-06-9 see DJV000
23572-32-9 see DIF300
23581-62-6 see EAV100
23590-99-0 see IDE050
23593-75-1 see MRX500

23595-00-8 see TKF699
23597-98-0 see CKK050
23599-75-9 see RAF400
23604-99-1 see SBF510
23605-05-2 see APN000
23605-74-5 see HEQ600
23606-32-8 see SDL500
23609-20-3 see COM900
23611-64-5 see MPS600
23611-66-7 see BOG255
23611-67-8 see DFC300
23611-68-9 see DDM820
23611-69-0 see NFW210
23647-94-1 see PEW550
23657-27-4 see TNO300
23668-11-3 see PAB500
23668-76-0 see DNR200
23674-86-4 see DKJ300
23679-20-1 see DBL000
23696-28-8 see HKV100
23705-25-1 see PLN100
23710-76-1 see ITD015
23712-05-2 see DGA800
23734-06-7 see CCL109
23734-88-5 see AKC625
23745-86-0 see PLG000
23746-34-1 see PKX500
23753-67-5 see BNU125
23757-42-8 see AMQ500
23777-55-1 see MJB300
23777-63-1 see DCF750
23777-80-2 see HCA275
23779-99-9 see TKG000
23783-98-4 see PGX275
23784-96-5 see CFB825
23795-03-1 see DWW200
23826-71-3 see MED100
23828-92-4 see AHJ500
23834-96-0 see BND325
23840-95-1 see ZTK300
23844-24-8 see BCL250
23885-72-5 see CDP350
23902-87-6 see CKI180
23902-88-7 see BQK800
23902-89-8 see CKI040
23902-91-2 see CKI185
23903-11-9 see DWK900
23904-72-5 see CKI190
23904-74-7 see CKI050
23904-87-2 see CKI090
23904-88-3 see CKI080
23917-55-7 see CKI030
23918-98-1 see POJ300
23920-57-2 see CKI070
23924-78-9 see CCC120
23930-19-0 see AFK875
23947-60-6 see BRI750
23950-58-5 see DTT600
23965-86-8 see NNQ050
23978-09-8 see LFQ000
24017-47-8 see THT750
24026-35-5 see BDT500
24027-84-7 see BCX500
24031-96-7 see DXI450
24040-34-4 see CEC700
24049-18-1 see AGN250
24050-10-0 see PNW300
24050-16-6 see MPR000
24050-58-6 see DWK700
24063-71-6 see IKG100
24066-82-8 see ELX530
24089-00-7 see MND100

24094-93-7 see CMJ850
24095-80-5 see BLP325
24096-53-5 see DGF000
24124-25-2 see LGJ000
24143-08-6 see AAL500
24143-17-7 see OMK300
24151-93-7 see PIX775
24166-13-0 see CMY525
24167-38-2 see DHU910
24167-76-8 see SGM000
24168-96-5 see IKN200
24211-64-1 see CAZ050
24211-81-2 see UVS700
24219-97-4 see MQS220
24220-18-6 see DPS200
24221-86-1 see EAW995
24234-06-8 see ELA600
24237-00-1 see GMG100
24245-27-0 see DWC625
24259-89-0 see DVY300
24264-08-2 see DJN700
24268-87-9 see NFW430
24268-89-1 see NFW460
24270-51-7 see IAK200
24279-91-2 see BGX750
24280-93-1 see MRX000
24282-51-7 see HOI400
24283-57-6 see CFC500
24291-69-8 see DFE570
24291-70-1 see DFE560
24292-52-2 see HBU400
24301-86-8 see PEF750
24301-89-1 see DWB300
24305-27-9 see TNX400
24307-26-4 see MCH540
24308-84-7 see ZDS500
24325-70-0 see MNB100
24342-55-0 see HGI200
24342-56-1 see MIQ725
24345-16-2 see AQN650
24346-78-9 see HBP250
24353-58-0 see IIE200
24356-60-3 see HMK000
24356-66-9 see AEH100
24356-94-3 see DAM300
24358-29-0 see CGJ250
24365-47-7 see LEX400
24365-61-5 see PGF125
24378-32-3 see PAY600
24380-21-0 see EDK000
24382-04-5 see MAN800
24393-94-0 see DLY300
24397-89-5 see AEA109
24407-55-4 see BKP200
24423-68-5 see DNL200
24425-13-6 see BQJ750
24426-36-6 see IPG000
24458-48-8 see NHK800
24463-15-8 see PON400
24471-48-5 see HMQ550
24477-37-0 see DBE885
24496-65-9 see BDD600
24518-45-4 see DEN820
24519-85-5 see DLK700
24526-64-5 see NMV700
24529-88-2 see SED525
24536-75-2 see TLA525
24543-59-7 see BMA650
24544-04-5 see DNN630
24549-06-2 see MJY000
24554-26-5 see NGM500
24560-98-3 see OAT100

24565-13-7 see DXB500
24570-10-3 see LHX499
24570-11-4 see LHX600
24570-12-5 see LHX510
24579-91-7 see CHN750
24589-78-4 see MQG750
24596-38-1 see IOT875
24596-39-2 see BRB450
24596-41-6 see BRB460
24600-36-0 see FMU000
24602-86-6 see DUJ400
24613-89-6 see CMI250
24615-84-7 see HGO700
24621-17-8 see ZQB100
24623-77-6 see AHC250
24632-47-1 see NDY350
24632-48-2 see NDY370
24643-97-8 see ELL550
24656-22-2 see PIC100
24671-21-4 see CKA580
24683-26-9 see ELI550
24684-41-1 see MJD750
24684-42-2 see DLU200
24684-49-9 see MEV250
24684-56-8 see DLP600
24684-58-0 see ABN250
24689-89-2 see CGW275
24690-46-8 see DTT400
24699-40-9 see CNE375
24704-64-1 see AHE875
24729-96-2 see CMV690
24735-35-1 see ZTS100
24748-25-2 see TDH775
24771-52-6 see PBH150
24781-13-3 see SAK500
24800-44-0 see TMZ000
24801-88-5 see IKG900
24813-03-4 see BHP150
24815-24-5 see TLN500
24815-93-8 see SDZ370
24817-51-4 see PDF780
24818-79-9 see AHA150
24829-11-6 see DHD400
24848-81-5 see TMO500
24851-98-7 see HAK100
24853-80-3 see ASC250
24854-67-9 see DHC000
24869-88-3 see DOS300
24884-69-3 see MMP200
24891-23-4 see AIN100
24891-41-6 see TLK600
24902-02-1 see PNE800
24909-09-9 see DLC000
24928-15-2 see PGS750
24928-17-4 see PGT250
24934-91-6 see CDY299
24936-38-7 see PJK400
24936-68-3 see PKE300
24937-72-2 see MAM500
24937-79-9 see DKH600
24938-16-7 see MDN505
24938-91-8 see TJJ250
24939-03-5 see PJY750
24951-05-1 see PIX750
24955-83-7 see DPT300
24961-39-5 see BNO750
24961-49-7 see DMK400
24973-25-9 see DOJ800
24991-55-7 see DOM100
25007-79-8 see DCD050
25013-15-4 see VQK650
25013-16-5 see BQI000

25014-41-9 see PJK360
25035-37-4 see BBO750
25036-33-3 see PKP250
25038-54-4 see PJK750
25038-54-4 see PJY500
25038-59-9 see PKF750
25044-01-3 see IMW500
25046-79-1 see PMG000
25047-48-7 see CMV375
25053-63-8 see EFT100
25056-70-6 see TIX250
25057-89-0 see MJY500
25059-80-7 see EHL600
25061-59-0 see MBV710
25067-59-8 see VNK200
25068-38-6 see EBF500
25068-38-6 see EBG000
25068-38-6 see EBG500
25068-38-6 see ECL000
25068-38-6 see IPM000
25068-38-6 see IPN000
25068-38-6 see IPO000
25068-38-6 see IPP000
25074-67-3 see MRF575
25081-01-0 see CLF500
25086-25-3 see PKQ070
25086-29-7 see VQK595
25086-81-1 see TFX795
25086-89-9 see AAU300
25090-71-5 see DAT000
25090-72-6 see DAT600
25090-73-7 see DAT200
25103-09-7 see ILR000
25103-12-2 see TKT000
25103-58-6 see DXT800
25104-37-4 see OJV600
25122-46-7 see CMW300
25122-57-0 see CMW400
25134-21-8 see NAC000
25136-53-2 see GGA912
25136-55-4 see DRO800
25136-85-0 see PJL600
25144-78-9 see NNS600
25147-05-1 see CBT125
25147-91-5 see TLC600
25148-26-9 see CIN500
25151-00-2 see BSO750
25152-84-5 see DAE450
25152-85-6 see HFE500
25154-52-3 see NNC500
25154-54-5 see DUQ180
25155-18-4 see MHB500
25155-23-1 see XLS100
25155-25-3 see BHL100
25155-30-0 see DXW200
25161-91-5 see FBP400
25167-31-1 see CEI325
25167-67-3 see BOW250
25167-70-8 see TMA250
25167-83-3 see TBS250
25167-93-5 see CJA950
25168-04-1 see NMS000
25168-05-2 see CLK130
25168-15-4 see TGF200
25168-24-5 see BKK250
25168-26-7 see ILO000
25174-65-6 see BFE770
25174-66-7 see CKI020
25182-84-7 see NIY050
25190-06-1 see GHY100
25191-20-2 see PJQ400
25201-35-8 see PAV250

25212-88-8 see MDN600	25389-94-0 see KAV000	25724-34-9 see EEX100	26027-38-3 see NNB300
25214-70-4 see FMW330	25395-31-7 see GGA100	25724-35-0 see AHT875	26037-72-9 see IFL000
25230-72-2 see TNV575	25395-41-9 see HLT300	25724-50-9 see DUJ000	26041-59-8 see AOO400
25232-42-2 see VPP100	25402-50-0 see IDP000	25724-54-3 see HEQ120	26043-11-8 see NDD000
25234-79-1 see DSF100	25405-85-0 see PGT000	25724-58-7 see PHW500	26045-95-4 see PAY610
25238-02-2 see CLS125	25410-64-4 see LHX515	25724-60-1 see BJO075	26049-68-3 see HHD500
25240-93-1 see DUV000	25410-69-9 see LHX498	25726-97-0 see TIS100	26049-69-4 see DSG400
25251-03-0 see PCC750	25413-02-9 see EGV550	25726-99-2 see DDU800	26049-70-7 see HHE000
25252-96-4 see PME100	25413-64-3 see PNF750	25727-08-6 see PGY300	26049-71-8 see HHB000
25255-23-6 see RQP100	25417-20-3 see NBS700	25735-67-5 see AOM750	26062-79-3 see DTS500
25264-93-1 see HFA650	25425-12-1 see CMS500	25736-79-2 see UIA000	26072-78-6 see DBK120
25265-19-4 see PAN250	25429-29-2 see PAV600	25736-79-2 see UIJ000	26076-87-9 see DCC125
25265-19-4 see PAN500	25430-97-1 see DJL200	25736-79-2 see UIS000	26087-47-8 see BKS750
25265-71-8 see DWS500	25439-20-7 see PEB775	25748-74-7 see PKO750	26094-13-3 see OHU100
25265-77-4 see TEG500	25447-65-8 see DLK780	25749-08-0 see NCY051	26095-59-0 see OKO400
25267-15-6 see PJQ250	25448-25-3 see IKM025	25756-29-0 see MIT610	26096-99-1 see CGI125
25267-55-4 see CNQ375	25450-02-6 see LHX500	25764-08-3 see CDB325	26097-80-3 see CBA100
25284-83-7 see MDR775	25451-15-4 see FAG200	25782-30-3 see PAO160	26099-09-2 see PJY850
25287-52-9 see PFF750	25455-73-6 see SDV700	25791-96-2 see NCT000	26129-32-8 see CJP750
25295-51-6 see DXB400	25474-92-4 see DXH400	25796-01-4 see NBO525	26134-62-3 see LHM000
25301-02-4 see TDN750	25480-76-6 see CIQ625	25805-16-7 see PKP000	26140-60-3 see TBD000
25310-48-9 see MLX300	25481-21-4 see EJA500	25808-74-6 see LDG000	26148-68-5 see AJD750
25311-71-1 see IMF300	25483-10-7 see TCG450	25812-30-0 see GCK300	26155-31-7 see MRN260
25312-34-9 see IFT400	25486-91-3 see DUF000	25834-80-4 see BGU100	26156-56-9 see HBV000
25316-40-9 see HKA300	25486-92-4 see DLT400	25843-45-2 see ASP250	26157-96-0 see DCL300
25321-09-9 see DNN709	25487-36-9 see HGL650	25843-64-5 see IAK100	26159-34-2 see NBO550
25321-14-6 see DVG600	25487-66-5 see CCI590	25852-70-4 see BSS000	26160-83-8 see TJG550
25321-41-9 see XJA000	25496-72-4 see GGA925	25853-45-6 see DAR930	26162-66-3 see MRN600
25322-20-7 see TBP750	25496-72-4 see GGR200	25854-41-5 see SGP600	26163-27-9 see SDP550
25322-68-3 see PJT000	25498-49-1 see MEV500	25855-92-9 see BNP850	26166-37-0 see DAP875
25322-68-3 see PJT200	25512-42-9 see DET700	25868-47-7 see DWA400	26171-23-3 see TGJ850
25322-68-3 see PJT225	25513-46-6 see GFM310	25869-00-5 see FAS800	26171-78-8 see MCG910
25322-68-3 see PJT230	25523-79-9 see XES000	25869-93-6 see HAE000	26172-55-4 see CIN100
25322-68-3 see PJT240	25526-93-6 see DAR200	25870-02-4 see SDV500	26183-52-8 see PKE390
25322-68-3 see PJT250	25535-16-4 see PMT000	25875-51-8 see RLK890	26196-45-2 see ALM100
25322-68-3 see PJT500	25546-65-0 see XQJ650	25876-07-7 see BRG700	26219-22-7 see TLA500
25322-68-3 see PJT750	25550-58-7 see DUY600	25876-47-5 see FOV000	26225-59-2 see MBX000
25322-68-3 see PJU000	25551-13-7 see TLL250	25877-27-4 see DPI710	26227-73-6 see MEE100
25322-68-3 see PJV000	25564-22-1 see PBW600	25889-63-8 see DRY000	26241-10-1 see PLD710
25322-69-4 see PKI500	25566-92-1 see DGR400	25898-71-9 see HFZ000	26248-87-3 see TJC870
25322-69-4 see PKI550	25567-67-3 see CGL750	25899-50-7 see PBQ275	26249-01-4 see SFB100
25322-69-4 see PKI750	25586-43-0 see MRG050	25910-37-6 see NFA600	26249-12-7 see DDJ900
25322-69-4 see PKJ500	25596-24-1 see TMG775	25913-34-2 see TGE165	26249-20-7 see UAK100
25322-69-4 see PKK000	25601-84-7 see DOL800	25916-47-6 see ADW100	26259-45-0 see BQC250
25323-30-2 see DFH800	25604-70-0 see BMY825	25917-89-9 see MMG200	26259-90-5 see TIR990
25323-68-6 see TIM100	25604-71-1 see CHR325	25917-90-2 see MFB300	26261-57-4 see DSB300
25323-74-4 see PME650	25606-41-1 see PNI250	25928-94-3 see ECK500	26264-05-1 see BBS275
25324-56-5 see TGE500	25614-03-3 see BNB250	25930-79-4 see TME500	26264-06-2 see CAR790
25329-35-5 see PAV800	25614-78-2 see ACL250	25931-01-5 see PKL750	26266-57-9 see MRJ800
25331-92-4 see AFH550	25627-41-2 see POD800	25932-11-0 see SJH025	26266-68-2 see EKQ600
25332-09-6 see PMS900	25639-42-3 see MIQ745	25938-97-0 see BMU150	26270-58-6 see DEW200
25332-39-2 see CKJ000	25639-45-6 see DKK400	25939-05-3 see TJA100	26270-59-7 see DOD600
25332-39-2 see THK880	25640-78-2 see IOF200	25951-54-6 see PKQ000	26270-60-0 see DOC800
25333-83-9 see PMX500	25641-53-6 see SGP550	25952-35-6 see TCQ275	26270-61-1 see DON000
25339-09-7 see IKC050	25646-71-3 see MDQ850	25952-53-8 see EAE100	26270-62-2 see DHP600
25339-17-7 see IKK000	25646-77-9 see AKZ100	25953-06-4 see DIO300	26283-13-6 see CDD250
25339-56-4 see HBK350	25650-97-9 see ISD033	25954-13-6 see ANG750	26295-56-7 see TJX300
25339-57-5 see BOP100	25655-41-8 see PKE250	25956-17-6 see FAG100	26305-03-3 see PCB750
25340-17-4 see DIU000	25677-40-1 see PBX350	25961-87-9 see HBP275	26308-28-1 see POL475
25340-18-5 see TJO750	25679-28-1 see PMR000	25961-89-1 see HFT550	26309-95-5 see AOD000
25351-18-2 see TLY250	25682-07-9 see AJU875	25967-29-7 see FMR100	26311-45-5 see PBW000
25354-97-6 see HFP500	25683-07-2 see POK400	25987-94-4 see SDX200	26328-00-7 see IRQ000
25355-59-3 see MFG400	25687-87-0 see PEO600	25988-97-0 see DOR500	26328-04-1 see VGK000
25355-61-7 see MMW775	25704-18-1 see SJK375	25991-93-9 see TIS500	26328-53-0 see AOA050
25376-45-8 see TGL500	25704-81-8 see APK500	26000-17-9 see LJB100	26351-01-9 see AHB625
25377-72-4 see AOI800	25708-11-6 see DPH500	26002-80-2 see PDR700	26354-18-7 see OIY000
25377-73-5 see DXV000	25710-89-8 see XEJ000	26006-71-3 see SKC100	26367-75-9 see AAI110
25382-52-9 see MJU350	25711-26-6 see BLL500	26011-40-5 see HLQ050	26375-23-5 see PKM250
25383-99-7 see SJV700	25717-80-0 see MRN275	26011-83-6 see DOK400	26377-04-8 see TIF600
25384-17-2 see AGV890	25721-38-4 see PID100	26016-98-8 see CAW376	26377-29-7 see PHI250
25387-67-1 see CBB875	25723-52-8 see EIW500	26016-99-9 see DXF600	26377-33-3 see BNG300
25389-94-0 see KAM000	25724-33-8 see DXR400	26020-35-9 see THH555	26388-47-6 see ROU500

28399-17-9 see EQR000
28407-37-6 see CMO650
28416-66-2 see PCK639
28434-00-6 see AFR750
28434-01-7 see BEP750
28434-86-8 see BGT000
28438-99-5 see FPQ900
28472-18-6 see TLT250
28519-06-4 see DAJ200
28522-57-8 see MQD250
28538-70-7 see NKU400
28553-12-0 see PHW585
28557-25-7 see KHU100
28558-28-3 see EAE000
28577-62-0 see DEU375
28597-01-5 see PGP600
28610-84-6 see PMD825
28613-21-0 see CBH770
28616-48-0 see DIV200
28622-66-4 see DMA000
28622-84-6 see DLB800
28625-02-7 see TEJ750
28643-80-3 see SIO788
28652-04-2 see DCH800
28657-80-9 see CMS130
28660-63-1 see DDX600
28660-67-5 see BLH309
28675-08-3 see DFE800
28675-83-4 see EDC560
28728-55-4 see IFT300
28743-45-5 see NFW425
28767-61-5 see TDY650
28772-56-7 see BMN000
28777-70-0 see TIA130
28781-86-4 see EQB500
28785-06-0 see PNE500
28801-69-6 see TIF250
28802-49-5 see DSB800
28808-62-0 see FOM200
28832-64-6 see ALR750
28839-49-8 see MPP250
28842-05-9 see DQH000
28846-35-7 see DSG320
28846-36-8 see MKN600
28846-37-9 see ADQ560
28846-38-0 see DSG340
28846-39-1 see DEC797
28846-40-4 see PIN200
28846-41-5 see MRR112
28846-42-6 see MRR115
28846-43-7 see MKA300
28846-44-8 see ADQ550
28853-06-7 see EOB100
28860-95-9 see CBQ500
28894-74-8 see DAD700
28895-91-2 see ACR400
28907-45-1 see CGG500
28907-84-8 see ALI500
28911-01-5 see THS800
28920-43-6 see FEI100
28930-30-5 see BLR750
28950-34-7 see SOO000
28961-43-5 see TLX100
28965-70-0 see DCF725
28968-07-2 see CFH825
28981-97-7 see XAJ000
28985-91-3 see EEA550
28995-89-3 see TDY500
29023-82-3 see CGH800
29023-83-4 see BID800
29023-84-5 see ADQ610
29023-85-6 see ADQ600

29025-14-7 see BPI125
29025-67-0 see CKM750
29026-74-2 see INA200
29050-86-0 see HNH475
29053-27-8 see CIL500
29063-00-1 see DQG100
29067-70-7 see BSV250
29082-74-4 see OAP050
29091-05-2 see CNE500
29091-21-2 see DWS200
29098-15-5 see DGN000
29110-48-3 see GKU300
29110-68-7 see DUS500
29119-58-2 see IPS400
29122-56-3 see BQT000
29122-60-9 see BQT250
29122-68-7 see TAL475
29128-41-4 see DSI489
29135-62-4 see HET700
29144-42-1 see CDF250
29149-32-4 see BRA600
29171-20-8 see LFY333
29173-31-7 see MLJ000
29177-84-2 see EKF600
29181-23-5 see AGO750
29193-35-9 see BDW750
29202-04-8 see DFD200
29214-60-6 see EKS150
29216-28-2 see MCJ400
29222-39-7 see PEL000
29232-93-7 see DIN800
29256-79-9 see MRF600
29285-24-3 see PLM250
29291-35-0 see NLP000
29291-35-8 see NKG450
29316-05-0 see PBV600
29342-61-8 see DCW400
29343-52-0 see HNB600
29362-48-9 see ACM245
29382-82-9 see FBZ000
29385-43-1 see MHK000
29387-86-8 see PML260
29393-20-2 see BAC175
29396-39-2 see CPQ650
29407-84-9 see EBF200
29411-74-3 see SBU700
29446-15-9 see DEX120
29450-63-3 see PAV300
29457-72-5 see LHN100
29462-18-8 see BAV625
29471-80-5 see BJP425
29477-83-6 see LJC000
29507-62-8 see EJA400
29510-76-7 see CFL300
29523-51-1 see BKJ600
29553-26-2 see MPN750
29560-58-5 see EEI050
29560-84-7 see CEF500
29575-02-8 see BHK500
29590-42-9 see ILK100
29597-36-2 see CCR510
29611-03-8 see AEW500
29621-88-3 see SBU303
29632-98-2 see SHQ600
29653-30-3 see DRK300
29671-92-9 see CHJ599
29674-96-2 see MRJ250
29676-95-7 see DLU800
29689-14-3 see CMJ100
29701-07-3 see KBA100
29714-87-2 see DTF200
29727-70-6 see AMP000

29734-68-7 see DLL600
29761-21-5 see IKL100
29767-20-2 see EQP000
29790-52-1 see NDR000
29804-22-6 see ECB200
29806-73-3 see OFG100
29847-17-4 see MPJ800
29868-97-1 see GCE500
29876-14-0 see NDW525
29883-15-6 see AOD500
29886-96-2 see COU525
29903-04-0 see DUP200
29908-03-0 see AEM310
29911-28-2 see DWS600
29914-92-9 see DWV200
29929-77-9 see COG250
29964-84-9 see IKM000
29968-64-7 see FDY000
29968-68-1 see BGN000
29968-75-0 see ABO750
29973-13-5 see EPR000
29975-16-4 see CKL250
29984-33-6 see AQQ905
30007-39-7 see HAJ700
30007-47-7 see BNT000
30015-57-7 see TAL275
30026-92-7 see BRR250
30031-64-2 see BOP750
30034-03-8 see CCR925
30041-69-1 see DPR000
30077-45-3 see CGV275
30096-82-3 see DIM600
30099-72-0 see BLL000
30129-29-4 see VAQ200
30129-30-7 see CCM050
30136-13-1 see PNB750
30168-23-1 see OBW100
30184-88-4 see BLP250
30185-90-1 see DIM200
30192-67-7 see BLQ250
30194-63-9 see AMJ500
30213-35-5 see KGK140
30223-48-4 see FDE000
30233-80-8 see AQU500
30233-81-9 see AQU750
30236-29-4 see SNH125
30256-73-6 see ICZ200
30256-74-7 see ICZ150
30286-75-0 see ONI000
30299-08-2 see CMV700
30310-80-6 see HNB500
30335-72-9 see MFB350
30345-27-8 see HMD000
30345-28-9 see HMD500
30366-55-3 see MDA750
30392-41-7 see BLV125
30406-18-9 see HBF550
30458-69-6 see AKT520
30460-34-5 see GJU315
30460-36-7 see GJU330
30484-77-6 see FDD080
30486-72-7 see AJA650
30507-70-1 see TBX300
30516-87-1 see ASE900
30517-65-8 see DAZ100
30525-89-4 see PAI000
30533-63-2 see PLD700
30544-47-9 see HKK000
30544-72-0 see CLW625
30558-43-1 see OLT100
30560-19-1 see DOP600
30562-34-6 see GCI000

30578-37-1 see MQS100
30586-10-8 see DFX000
30618-84-9 see GGR300
30638-08-5 see CND940
30645-13-7 see POA100
30652-11-0 see DSI250
30653-83-9 see AJC000
30674-80-7 see IKG700
30685-43-9 see MJD300
30719-67-6 see GJM035
30746-58-8 see TBO780
30748-29-9 see PEW000
30777-19-6 see BCJ300
30812-87-4 see EHY000
30833-53-5 see MRI775
30835-61-1 see DLR600
30835-65-5 see DLT900
30846-35-6 see FMW343
30914-89-7 see TKJ500
30947-30-9 see BJK560
30950-27-7 see PCJ000
30956-43-5 see MKM750
30957-47-2 see BJD375
30994-24-2 see PLC800
31012-29-0 see BDQ250
31031-74-0 see PCV750
31044-86-7 see TNC825
31052-46-7 see DEJ400
31055-59-1 see SDP200
31058-64-7 see TBH000
31065-88-0 see COP500
31078-10-1 see ITD018
31080-39-4 see NMX500
31083-55-3 see DVT459
31088-06-9 see PAE260
31112-62-6 see MQR300
31121-93-4 see IAB100
31122-82-4 see BBO600
31126-81-5 see RHK850
31136-61-5 see LJE100
31185-56-5 see MQE000
31185-58-7 see MOA750
31217-72-8 see EMV000
31218-83-4 see MKA000
31225-17-9 see DGD085
31232-27-6 see WAJ000
31237-72-2 see CKO800
31242-93-0 see CDV175
31242-94-1 see TBP250
31248-66-5 see EDF000
31272-21-6 see AJR400
31279-70-6 see ILB150
31282-04-9 see AQB000
31301-20-9 see CKI600
31304-44-6 see SEG650
31305-88-1 see EBH500
31323-50-9 see OKS100
31329-57-4 see NAE000
31330-22-0 see TKB275
31332-72-6 see HFA225
31335-41-8 see DFS600
31340-36-0 see HDD500
31364-55-3 see MMX500
31375-17-4 see MCF525
31385-05-4 see NEH610
31391-27-2 see PAV750
31397-22-5 see PGN030
31416-87-2 see AKS500
31430-15-6 see FDA887
31430-18-9 see OJD100
31431-39-7 see MHL000
31431-43-3 see CQE325

31477-60-8 see CCW725	32362-68-8 see DQP400	33077-70-2 see HJN700	33629-47-9 see BQN600
31499-72-6 see DLP800	32373-17-4 see BBE000	33082-92-7 see MHI600	33631-41-3 see REK330
31501-11-8 see HFE510	32385-11-8 see SDY750	33089-61-1 see MJL250	33631-47-9 see REK350
31508-00-6 see PAV620	32388-21-9 see OMK000	33094-66-5 see NFD700	33634-75-2 see SIZ100
31540-62-2 see FIW000	32426-11-2 see OES400	33094-74-5 see NHN550	33656-20-1 see CKI060
31551-45-8 see DUW300	32446-40-5 see AON500	33098-26-9 see BRB750	33671-46-4 see CKA000
31566-31-1 see OAV000	32449-92-6 see GFM000	33098-27-0 see BRC000	33691-06-4 see IAG300
31571-52-5 see HOJ100	32458-57-4 see EQN700	33100-27-5 see PBO000	33693-04-8 see BQC500
31581-09-6 see XSS923	32479-72-4 see ALC600	33113-10-9 see NCC100	33697-73-3 see TGJ625
31632-68-5 see PFP000	32504-14-6 see MNO780	33125-86-9 see TDG760	33755-46-3 see DBC575
31635-40-2 see AEF500	32522-68-2 see DJJ393	33125-97-2 see HOU100	33765-68-3 see ELF100
31642-65-6 see TGX500	32524-44-0 see NBO500	33132-61-5 see BRE250	33765-80-9 see ELF110
31647-36-6 see MHH200	32527-15-4 see HCV600	33132-71-7 see BQS250	33770-60-4 see DFC800
31671-16-6 see DWO600	32534-81-9 see PAT830	33132-75-1 see AJB000	33786-89-9 see CEJ200
31674-58-5 see MDT900	32536-52-0 see OAF200	33132-85-3 see BQS000	33804-48-7 see DPQ200
31677-93-7 see WBJ500	32568-89-1 see DTH100	33132-87-5 see AGF000	33813-20-6 see EJQ000
31698-14-3 see COW875	32588-76-4 see EIT150	33145-10-7 see IIU100	33854-16-9 see DVJ100
31699-72-6 see ALR500	32598-13-3 see TBO700	33175-34-7 see DGD075	33855-47-9 see BMW000
31706-95-3 see HMB700	32598-14-4 see PAV630	33196-65-5 see TDO500	33857-23-7 see AMX400
31709-32-7 see TNB750	32607-15-1 see TEE750	33204-76-1 see CMS234	33857-26-0 see DAC800
31717-87-0 see COX000	32607-31-1 see PLU600	33207-59-9 see MEC600	33857-28-2 see TIL630
31750-48-8 see DBA710	32613-12-0 see BJK780	33212-68-9 see DSM289	33868-17-6 see MMX000
31751-59-4 see DWN100	32617-22-4 see PLD575	33213-65-9 see EAQ800	33876-97-0 see MRU076
31785-60-1 see DLX300	32665-36-4 see EQY600	33229-34-4 see HKN875	33878-50-1 see KBB700
31793-07-4 see PJA220	32674-23-0 see PGZ900	33232-39-2 see DPE100	33904-55-1 see NJS500
31810-89-6 see DBQ220	32707-10-1 see BKW850	33236-07-6 see TCQ700	33911-28-3 see RDZ885
31817-29-5 see MFJ200	32740-01-5 see DQI400	33237-74-0 see FBP850	33918-12-6 see MEY200
31820-22-1 see NKU570	32750-98-4 see TKD350	33244-00-7 see PEK675	33942-87-9 see MJA500
31823-54-8 see SEO600	32752-13-9 see PMF535	33245-39-5 see FDA900	33942-88-0 see MHE250
31828-50-9 see SBN450	32754-35-1 see MCS100	33265-79-1 see PIN275	33944-90-0 see SBU500
31842-01-0 see IDA400	32755-26-3 see CJI809	33266-07-8 see ACR050	33953-73-0 see BCX750
31868-18-5 see MQR760	32762-51-9 see BNU800	33271-65-7 see CNL310	33956-49-9 see CNG760
31874-15-4 see PFF000	32764-43-5 see AJL125	33286-22-5 see DNU600	33965-80-9 see CHU000
31876-38-7 see MRE250	32766-75-9 see HMU000	33310-62-2 see NEH600	33972-75-7 see IHB680
31895-22-4 see TFH750	32767-68-3 see III000	33330-91-5 see CCC325	33979-03-2 see HCC800
31897-92-4 see BNO000	32774-16-6 see HCD100	33342-05-1 see GEW780	33979-15-6 see CMV950
31897-98-0 see AMC500	32780-64-6 see HMM500	33364-51-1 see BJE000	34014-18-1 see BSN000
31898-11-0 see BNO250	32784-82-0 see IBQ100	33372-39-3 see BKC250	34018-28-5 see LCO000
31906-04-4 see LJE200	32787-44-3 see MLF520	33372-40-6 see DNB600	34031-32-8 see ARS150
31924-91-1 see RQP050	32795-47-4 see NMV725	33384-03-1 see PDI250	34037-79-1 see DBK600
31927-64-7 see MGP250	32795-84-9 see BMT750	33388-24-8 see EHH700	34044-10-5 see TAH675
31932-35-1 see MOY800	32807-28-6 see MIH600	33389-33-2 see DLY200	34099-73-5 see BMC250
31959-87-2 see MKI000	32808-51-8 see CPJ000	33389-36-5 see HKO000	34114-98-2 see DAD600
31981-99-4 see THH380	32808-53-0 see CPJ250	33390-21-5 see LJB700	34123-59-6 see IRA050
31982-00-0 see THH360	32809-16-8 see PMF750	33396-37-1 see MCI750	34131-99-2 see IOO310
31984-14-2 see THH355	32838-28-1 see BPF825	33399-00-7 see BMO310	34140-59-5 see TKU675
32006-07-8 see TGG600	32852-21-4 see FNB000	33400-47-4 see DAB200	34148-01-1 see CMV500
32017-76-8 see BOF250	32852-92-9 see PDR200	33401-94-4 see TCW750	34149-92-3 see PKP500
32059-15-7 see GKO800	32854-75-4 see LBD000	33402-03-8 see HNC000	34177-12-3 see DWO875
32061-49-7 see CNL625	32861-85-1 see DGA850	33406-36-9 see BKO825	34183-22-7 see PMJ525
32064-64-5 see MKN300	32865-01-3 see DNW759	33419-42-0 see EAV500	34200-96-9 see OKS150
32065-38-6 see THH375	32871-90-2 see BEK250	33419-68-0 see SAC875	34202-69-2 see HDA500
32066-79-8 see TGM500	32887-03-9 see MCB550	33421-40-8 see ALY000	34205-21-5 see UTA300
32078-95-8 see PKK775	32889-48-8 see PMF600	33423-92-6 see TBO800	34211-26-2 see PPC000
32093-26-8 see DNU330	32891-29-5 see MBV100	33433-82-8 see CAY675	34214-51-2 see CHJ000
32106-51-7 see SKE500	32921-15-6 see EFC800	33433-82-8 see VCK200	34255-03-3 see CMX920
32116-24-8 see ACT300	32921-23-6 see CPI375	33445-03-3 see CCO675	34256-82-1 see CGO550
32116-25-9 see ACT330	32941-23-4 see EFC700	33447-90-4 see DSU100	34257-95-9 see DLO875
32139-72-3 see TIL600	32941-30-3 see HBF600	33453-19-9 see CQF125	34289-01-5 see MFF575
32144-31-3 see DKM100	32949-37-4 see HMN000	33453-96-2 see OPQ200	34291-02-6 see BSX325
32179-45-6 see MIH300	32954-58-8 see FQL100	33467-73-1 see HFE516	34314-31-3 see TIL255
32210-23-4 see BQW500	32976-87-7 see DQL400	33467-74-2 see HFE650	34320-82-6 see TJA500
32222-06-3 see DMJ400	32976-88-8 see ENT000	33499-84-2 see DUN300	34333-07-8 see TNG250
32226-65-6 see CEL000	32986-56-4 see TGI250	33531-34-9 see CIK250	34346-90-2 see TLA250
32226-69-0 see QUJ300	32988-50-4 see VQZ000	33531-59-8 see MIE250	34346-96-8 see BNQ750
32228-97-0 see PFE900	33005-95-7 see SOX400	33543-31-6 see MKC500	34346-97-9 see BNQ250
32238-28-1 see PKM500	33017-08-2 see TKE550	33550-22-0 see TID000	34346-98-9 see BMX250
32248-37-6 see PIV750	33020-34-7 see IPW000	33564-30-6 see CCS510	34346-99-1 see BNQ000
32266-10-7 see HFG650	33060-69-4 see CBM875	33564-31-7 see DKF125	34375-78-5 see DLJ000
32266-82-3 see CAR875	33069-62-4 see TAH775	33568-99-9 see BKL000	34381-68-5 see AAE125
32283-21-9 see BCJ125	33073-60-8 see CHF000	33605-67-3 see CCK550	34388-29-9 see MNG525
32345-29-2 see DJV700	33077-69-9 see EEN600	33614-49-2 see TNK500	34419-05-1 see MJJ250

37094-61-4 see TKD390	37558-19-3 see PGT800	38304-52-8 see IAO100	38940-51-1 see BDU250
37106-97-1 see CML870	37571-88-3 see DLB850	38304-91-5 see DCB000	38951-85-8 see CLH550
37106-99-3 see PMI000	37574-47-3 see BCV500	38357-93-6 see ITD100	38957-41-4 see EAN700
37125-93-2 see MKW150	37574-48-4 see HJN500	38363-32-5 see PAP230	38964-22-6 see DEX130
37132-72-2 see FOM000	37627-60-4 see BRQ300	38363-40-5 see PAP225	38965-69-4 see XCS680
37139-99-4 see OHK200	37636-51-4 see DAS600	38380-02-8 see PAV610	38965-70-7 see XCS700
37141-32-5 see NIS298	37640-57-6 see THO750	38380-07-3 see HCC700	38998-91-3 see BJM700
37150-27-9 see BDE000	37661-08-8 see BAB250	38380-08-4 see HCC600	39002-10-3 see BEO500
37151-16-9 see NIJ350	37670-00-1 see HEE600	38402-95-8 see BLT775	39013-93-9 see CGY000
37169-10-1 see DFU800	37677-14-8 see IKT100	38411-22-2 see HCC590	39032-87-6 see ICW150
37172-05-7 see EQN300	37680-65-2 see TIL258	38434-77-4 see ENT500	39047-21-7 see BJG000
37187-22-7 see PBL600	37686-84-3 see DLR100	38444-81-4 see TIL262	39067-39-5 see CML620
37203-43-3 see OAD095	37686-85-4 see DLR150	38444-86-9 see TIL260	39070-08-1 see MNB250
37205-85-9 see MND000	37699-43-7 see DSX800	38444-93-8 see TBO520	39071-30-2 see CQD250
37205-87-1 see NCG800	37717-82-1 see BFV900	38457-67-9 see TIQ300	39079-58-8 see DHC400
37205-87-1 see NCG820	37723-78-7 see IGA000	38460-95-6 see UMJ000	39082-00-3 see CEA050
37205-87-1 see NCG830	37724-43-9 see MOF750	38462-04-3 see ARM750	39108-14-0 see AGX300
37205-87-1 see NCG840	37724-45-1 see MLR500	38483-28-2 see MJU150	39133-31-8 see TKU650
37206-20-5 see IJB100	37724-47-3 see MOG000	38514-71-5 see ALM500	39148-24-8 see AHH800
37209-31-7 see DBB500	37750-86-0 see DQM800	38521-66-3 see PNF050	39156-41-7 see DBO400
37221-23-1 see PJJ500	37753-10-9 see SOD000	38524-82-2 see EPC175	39195-82-9 see DSS900
37224-57-0 see CMK400	37762-06-4 see ZAK200	38527-91-2 see EEE200	39196-18-4 see DAB400
37226-23-6 see COE100	37764-25-3 see DBI300	38539-23-0 see ABN500	39197-62-1 see ENP000
37227-61-5 see NCX510	37764-28-6 see EBY500	38562-01-5 see POC750	39202-39-6 see GLQ000
37229-14-4 see LHX360	37793-22-9 see MNB000	38571-73-2 see GGI000	39227-58-2 see TIL620
37230-84-5 see HAC800	37795-69-0 see DLQ400	38589-14-9 see MOJ500	39227-61-7 see PAV820
37231-28-0 see MCB525	37795-71-4 see MJE250	38613-77-3 see PCB400	39227-62-8 see HCF150
37235-82-8 see DDT250	37853-59-1 see EEB250	38615-43-9 see BRK830	39236-46-9 see GDO800
37239-28-4 see ADZ125	37882-31-8 see POB000	38640-62-9 see BKL600	39254-48-3 see PGH500
37244-00-1 see VIZ100	37893-02-0 see FDA890	38641-94-0 see GIQ100	39274-39-0 see XEJ100
37244-85-2 see TAI600	37894-46-5 see CHI200	38668-83-6 see TNK400	39277-41-3 see VRP000
37244-86-3 see VGA100	37902-85-5 see MKB320	38692-98-7 see ELD100	39277-47-9 see AEX750
37246-64-3 see TER100	37924-13-3 see TKF750	38693-06-0 see HIS110	39283-42-6 see RCF000
37247-90-8 see BGB250	37940-57-1 see BGM100	38700-88-8 see CKI612	39293-24-8 see ACP000
37268-68-1 see PJL100	37971-35-0 see PHA300	38704-36-8 see ELM600	39296-38-3 see TAL335
37270-71-6 see EQM500	37971-99-6 see XBA000	38726-90-8 see DCO200	39300-45-3 see AQT500
37280-35-6 see CBF680	37994-82-4 see BCY000	38726-91-9 see DCN600	39300-88-4 see GMA000
37280-56-1 see LEX000	38001-34-2 see DXS000	38748-32-2 see TNC175	39301-00-3 see AFI750
37286-64-9 see MKS250	38005-31-1 see IDU000	38753-50-3 see NIR550	39306-82-6 see PJC500
37297-87-3 see MCW350	38029-10-6 see POM800	38777-13-8 see PMY310	39324-65-7 see HBT100
37299-86-8 see RGZ100	38035-28-8 see DRL600	38780-35-7 see DGU709	39340-46-0 see ERC800
37300-23-5 see ZFJ125	38048-87-2 see DRX800	38780-36-8 see BIS250	39346-76-4 see SFO600
37306-44-8 see THS822	38078-09-0 see DIR875	38780-37-9 see DGT200	39360-54-8 see ECC750
37312-62-2 see SCA625	38082-89-2 see PQC525	38780-39-1 see DGC600	39367-89-0 see DIA800
37317-41-2 see PJP000	38083-17-9 see CJN900	38780-42-6 see DEU200	39387-42-3 see BBA625
37321-09-8 see AQP885	38090-84-5 see DJO410	38787-96-1 see PMP250	39389-47-4 see DXG400
37324-23-5 see PJN500	38090-92-5 see DST100	38802-82-3 see DNT000	39390-62-0 see EBH430
37324-24-6 see PJO000	38092-76-1 see HKM175	38819-28-2 see AFI500	39391-39-4 see APT750
37329-49-0 see TOC250	38094-02-9 see DRS600	38820-59-6 see HEQ610	39403-67-3 see MRU756
37333-40-7 see BIT250	38105-25-8 see ABS250	38821-53-3 see SBN440	39404-28-9 see SAU350
37338-91-3 see TBF350	38105-27-0 see ABO250	38838-26-5 see ACG250	39409-64-8 see TOE200
37339-90-5 see LEK100	38115-19-4 see PHP500	38860-48-9 see MED000	39413-47-3 see BFV250
37341-99-4 see WAT211	38134-58-6 see CGU100	38860-52-5 see NFL500	39425-26-8 see MQN250
37346-96-6 see PLV250	38139-15-0 see PAS879	38870-89-2 see MDW780	39426-77-2 see ACG125
37350-58-6 see MQR144	38178-38-0 see DEX100	38892-09-0 see SGQ100	39430-27-8 see NCY530
37353-63-2 see PJO500	38178-99-3 see XBA100	38906-58-0 see ELL600	39432-81-0 see FOO560
37364-06-0 see CNI610	38194-50-2 see SOU550	38911-59-0 see NNX650	39456-76-3 see LIJ000
37388-50-4 see PHP250	38198-35-5 see PPW000	38914-96-4 see CGR000	39456-78-5 see SAU475
37394-32-4 see PGW000	38222-35-4 see DXS200	38914-97-5 see CGX750	39457-24-4 see BAS500
37394-33-5 see IDB000	38230-32-9 see DAL010	38915-00-3 see BHZ000	39464-66-9 see PKE850
37415-55-7 see PGV750	38232-63-2 see MCX000	38915-14-9 see EHJ500	39465-37-7 see TGH520
37415-56-8 see BSX750	38234-12-7 see HIJ000	38915-22-9 see CGX500	39472-31-6 see KBU000
37415-62-6 see SIN850	38237-74-0 see AMW500	38915-28-5 see IAE000	39482-21-8 see TMF600
37425-13-1 see AGU750	38237-76-2 see AJD250	38915-40-1 see BPI500	39491-47-9 see AEK000
37430-50-5 see DLH900	38241-20-2 see NLF000	38915-49-0 see CGP500	39515-41-8 see DAB825
37514-30-0 see MEU300	38241-21-3 see NLT000	38915-50-3 see CGR250	39515-51-0 see PDR600
37517-28-5 see APS750	38246-95-6 see HNG100	38915-59-2 see CHH000	39543-79-8 see BAU255
37517-30-9 see AAE100	38252-74-3 see BQQ250	38915-61-6 see CGR500	39543-80-1 see ACN320
37519-65-6 see SBA875	38252-75-4 see BRO500	38915-61-6 see CGR750	39543-84-5 see ACN300
37557-89-4 see BEM250	38285-49-3 see MHX000	38925-90-5 see CGS750	39543-94-7 see ACN310
37558-16-0 see PGT100	38293-27-5 see ONU100	38940-46-4 see PFY750	39544-02-0 see ELI600
37558-17-1 see PGT750	38302-26-0 see LEX250	38940-50-0 see BDS750	39552-01-7 see HLK600

39562-16-8 see ENO200
39562-22-6 see MEP600
39562-70-4 see EMR600
39565-00-9 see NML100
39597-90-5 see EBQ550
39603-48-0 see MJN300
39603-53-7 see NLM500
39603-54-8 see ORS000
39617-48-6 see AJM550
39637-16-6 see DGB400
39638-73-8 see HKA250
39660-55-4 see OBS800
39687-95-1 see MKX575
39699-08-6 see NAO525
39702-40-4 see CNQ300
39718-89-3 see MDP850
39718-99-5 see NFH100
39729-81-2 see TBH250
39735-49-4 see NEW550
39753-42-9 see MFR775
39754-64-8 see PDU750
39765-80-5 see NMV755
39794-99-5 see MNX900
39801-14-4 see MRI750
39807-15-3 see BRA300
39831-55-5 see APT000
39834-38-3 see DQL600
39845-47-1 see EJV400
39864-15-8 see NMX550
39884-52-1 see NLE000
39884-53-2 see NKU600
39885-14-8 see MEW250
39895-81-3 see HLC550
39900-38-4 see CCR524
39920-56-4 see AAU500
40036-79-1 see TCA750
40054-69-1 see EQN600
40068-20-0 see PCV775
40088-47-9 see TBJ550
40101-31-3 see BNX300
40112-23-0 see GJM030
40164-67-8 see ABY300
40180-04-9 see TGA600
40193-47-3 see DKA000
40199-26-6 see CKF040
40202-39-9 see CQB250
40225-02-3 see SAB800
40237-34-1 see DUG889
40267-72-9 see GDG100
40283-68-9 see BMD250
40283-91-8 see BFC000
40283-92-9 see BFC500
40284-08-0 see AEG000
40284-10-4 see AEG500
40321-76-4 see PAV850
40334-69-8 see BIQ250
40343-30-4 see HKW350
40343-32-6 see HKW345
40358-04-1 see NMO400
40373-39-5 see BCJ025
40373-42-0 see DQM650
40373-43-1 see MNE300
40373-44-2 see MNF300
40378-58-3 see MON500
40428-75-9 see SNF000
40487-42-1 see DRN200
40502-72-5 see BEH250
40507-23-1 see THH350
40507-78-6 see IBR200
40527-42-2 see PIW500
40548-68-3 see NLT500
40560-76-7 see TGN100

40561-27-1 see COR600
40568-90-9 see MHG250
40576-21-4 see PIR200
40576-23-6 see MOH310
40576-25-8 see PDI550
40580-75-4 see DFE229
40580-89-0 see NJK500
40596-69-8 see KAJ000
40626-35-5 see HBU415
40630-63-5 see OCW200
40661-97-0 see MLG500
40665-92-7 see CMX880
40666-04-4 see IAD100
40709-23-7 see AHP375
40711-41-9 see BRL750
40713-31-3 see CGQ250
40762-15-0 see DKV400
40778-40-3 see PMC600
40816-40-8 see SKS100
40828-46-4 see TEN750
40847-64-1 see DKR400
40853-53-0 see IQH000
40853-56-3 see AAV250
40876-98-0 see SGJ550
40910-49-4 see ELZ050
40911-07-7 see BRN500
40942-73-2 see OOO100
40951-13-1 see DLT800
40953-35-3 see DCP880
40959-16-8 see TCT250
41024-90-2 see CDM625
41029-45-2 see PMS775
41083-11-8 see THT500
41084-90-6 see DGQ650
41093-93-0 see DAB925
41096-46-2 see EQD000
41098-56-0 see TGE152
41109-80-2 see PIK075
41136-03-2 see ALY500
41136-22-5 see CCS560
41155-82-2 see HKW300
41191-04-2 see THR500
41198-08-7 see BNA750
41208-07-5 see MRU760
41217-05-4 see DQG000
41219-25-4 see TFE300
41219-30-1 see CLH820
41219-31-2 see CLH810
41226-18-0 see DHQ000
41226-20-4 see DOC000
41239-48-9 see DKB165
41240-93-1 see DIP800
41262-21-9 see TLH250
41287-56-3 see BOD500
41287-72-3 see BOD000
41288-00-0 see MPA250
41296-95-1 see BPJ000
41363-50-2 see PFR350
41365-24-6 see DJC200
41372-08-1 see MJE780
41394-05-2 see ALA500
41407-89-0 see PLQ300
41409-50-1 see DAE625
41422-43-9 see TDQ000
41427-34-3 see COS750
41448-29-7 see EHN500
41451-75-6 see BOL500
41459-10-3 see CBO800
41481-90-7 see PAR799
41483-43-6 see BRJ000
41506-14-3 see NNC700
41510-23-0 see CCW500

41519-23-7 see HFE520
41542-50-1 see DHZ000
41572-59-2 see DEX800
41575-87-5 see DCF800
41575-94-4 see CCC075
41593-31-1 see DKW200
41595-29-3 see PEX275
41598-07-6 see POC275
41621-49-2 see BAR800
41643-22-5 see CKH100
41643-23-6 see CKF100
41643-24-7 see CKF767
41663-73-4 see AJI650
41663-84-7 see NHN700
41699-09-6 see MHE500
41708-72-9 see TGI699
41708-76-3 see ICD100
41735-28-8 see MMU000
41735-29-9 see ENS000
41735-30-2 see ENN500
41753-43-9 see PAF450
41762-18-9 see THQ100
41772-23-0 see ALK250
41787-75-1 see CHQ400
41814-78-2 see MQC000
41825-28-9 see LDA000
41826-92-0 see CNG980
41886-31-1 see DNK300
41890-92-0 see DLR300
41892-01-7 see DRS400
41920-59-6 see BFC300
41941-50-8 see EAZ600
41948-17-8 see AQS875
41956-77-8 see CCI500
41956-90-5 see HFW500
41992-22-7 see SLD800
41999-58-0 see SFO700
42011-48-3 see NGN500
42013-48-9 see DTN775
42013-69-4 see AJW750
42017-07-2 see LHF625
42028-27-3 see DLM000
42028-33-1 see HJE575
42069-72-7 see CJD630
42135-22-8 see NGB500
42145-91-5 see EGC500
42149-31-5 see DRK400
42200-33-9 see CNR675
42238-29-9 see NFA000
42242-58-0 see BCL100
42242-59-1 see BCK800
42242-72-8 see CBK250
42279-29-8 see PAW250
42281-59-4 see CQF079
42296-74-2 see HCQ600
42310-84-9 see BLD000
42340-32-9 see MFF560
42355-01-1 see DCI900
42373-04-6 see CMM870
42397-64-8 see DVD600
42397-65-9 see PER250
42436-07-7 see HFE625
42461-89-2 see BOA750
42471-28-3 see NDY800
42472-93-5 see MOC750
42472-96-8 see OOM300
42489-15-6 see PLV000
42489-15-6 see SKF500
42498-58-8 see EBW000
42509-80-8 see PHK000
42520-97-8 see DEV300
42540-40-9 see FOD000

42542-07-4 see MJR750
42576-02-3 see MJB600
42579-28-2 see NKJ000
42583-55-1 see CCK575
42604-12-6 see FMV200
42609-52-9 see DQQ700
42615-29-2 see AFO500
42713-66-6 see NJT000
42723-79-5 see PPH174
42754-15-4 see PPH145
42754-16-5 see PPH140
42754-17-6 see PPH168
42754-18-7 see PPH176
42754-19-8 see PPH150
42754-20-1 see CNV600
42754-21-2 see PPH155
42754-22-3 see PPH185
42754-23-4 see CPK100
42790-35-2 see EOL700
42794-63-8 see MQY125
42794-87-6 see TCG750
42824-34-0 see HAA310
42839-36-1 see CAY875
42840-17-5 see MFO500
42864-78-8 see BFW400
42873-80-3 see PCA400
42874-01-1 see NGB600
42874-03-3 see OQU100
42884-33-3 see ALJ250
42920-39-8 see BKO835
42924-53-8 see MFA300
42959-18-2 see SDH670
42971-09-5 see EGM100
42978-42-7 see BDP500
42978-43-8 see ACU500
42978-66-5 see TMZ100
43033-72-3 see ACQ260
43047-59-2 see CFP750
43047-99-0 see DEX200
43071-52-9 see MEC320
43085-16-1 see PEE600
43087-91-8 see ALV500
43121-43-3 see CJO250
43130-12-7 see EGM200
43143-11-9 see OIU850
43192-03-6 see HMI100
43200-80-2 see ZUA450
43210-67-9 see FAL100
43222-48-6 see ARW000
43229-80-7 see FNE100
44992-01-0 see OOM500
45173-31-7 see TKT100
45776-10-1 see HEW000
46061-25-0 see NJO300
46231-41-8 see PDS500
46355-07-1 see IRC100
46817-91-8 see VKA875
46941-74-6 see THJ600
47897-65-4 see AEW750
48134-75-4 see CQI550
48145-04-6 see PER250
48163-10-6 see LBQ000
49538-98-9 see DNT200
49540-32-1 see NJW700
49558-46-5 see NIW300
49561-54-8 see WBA600
49575-13-5 see MPL600
49642-61-7 see EFO000
49720-72-1 see DBA175
49720-83-4 see PDL100
49773-64-0 see BJE600
49780-10-1 see NMV750

49796-88-5 see DVT600
49828-25-3 see CJD400
49830-98-0 see PIN100
49842-07-1 see TGI500
49850-29-5 see AMS625
49852-84-8 see CIH000
49852-85-9 see BNP000
49860-08-4 see AMN275
50264-69-2 see DEL200
50264-86-3 see CEQ625
50264-96-5 see GGR100
50265-78-6 see BQI270
50274-95-8 see CJC610
50277-65-1 see PMT300
50285-70-6 see NLX500
50285-71-7 see DRV600
50285-72-8 see DJP600
50308-82-2 see CIP000
50308-83-3 see CIO750
50308-84-4 see CIP250
50308-86-6 see DSR600
50308-87-7 see DSR800
50308-88-8 see MEW750
50308-89-9 see MEW500
50308-94-6 see QOJ250
50309-02-9 see DBW000
50309-03-0 see DCC200
50309-11-0 see MLT250
50309-16-5 see AJW500
50309-17-6 see MLT750
50309-20-1 see AAL000
50311-48-3 see TKE525
50335-03-0 see CDG750
50335-09-6 see EOP500
50355-74-3 see TJA000
50355-75-4 see DUI800
50357-45-4 see PBJ800
50370-12-2 see DYF700
50416-18-7 see AJE250
50424-93-6 see MMI250
50425-34-8 see DSS000
50425-35-9 see MLU750
50454-68-7 see TGJ875
50465-39-9 see TGJ000
50468-61-6 see DDC050
50471-44-8 see RMA000
50475-76-8 see LGQ000
50485-03-5 see ACP750
50510-11-7 see AJI500
50510-12-8 see AJM750
50512-35-1 see IRU500
50528-97-7 see XGA725
50539-74-7 see DRL100
50563-36-5 see DSM500
50570-59-7 see BHN250
50577-64-5 see SDK000
50585-39-2 see DEX090
50585-41-6 see TBJ510
50585-46-1 see TBO810
50588-13-1 see BGR325
50594-66-6 see CLS075
50594-77-9 see CLR300
50601-74-6 see EKS200
50602-44-3 see THG600
50645-52-8 see IHJ000
50650-74-3 see AQQ000
50657-29-9 see OHQ300
50662-24-3 see LHM850
50673-06-8 see DLH890
50673-08-0 see DLH830
50673-09-1 see DLH840
50673-10-4 see DGF130

50673-11-5 see DGF100
50679-08-8 see TAI450
50684-32-7 see MNS550
50700-72-6 see VGU075
50707-40-9 see EPW000
50722-38-8 see ACH075
50764-78-8 see EPM600
50765-87-2 see PMB500
50767-79-8 see FAS100
50782-69-9 see EIG000
50802-21-6 see TGD125
50802-69-2 see DLH850
50802-70-5 see DLH870
50809-32-0 see TFZ100
50814-62-5 see EAG500
50815-73-1 see FOO562
50816-18-7 see DAI450
50831-29-3 see TDI300
50832-74-1 see ALN250
50838-36-3 see MQA000
50846-45-2 see BAC325
50847-92-2 see DUG600
50864-67-0 see BAO300
50865-01-5 see DXF700
50880-57-4 see HLX600
50892-23-4 see CLW250
50892-99-4 see FPK100
50901-84-3 see ALA000
50901-87-6 see ALA250
50903-99-6 see NDY675
50906-29-1 see PHY275
50908-62-8 see AEJ000
50916-04-6 see IAT200
50916-11-5 see ITD025
50916-12-6 see OMM100
50922-29-7 see ZFJ130
50924-49-7 see BMM000
50933-33-0 see GJK000
50935-04-1 see CCK625
50976-02-8 see DGW400
51003-81-7 see DPY600
51003-83-9 see BRC500
51012-33-0 see TGA375
51013-18-4 see MPB600
51022-69-6 see COW825
51022-70-9 see SAF400
51022-74-3 see IGD100
51023-56-4 see CCW750
51025-94-6 see CQK100
51029-30-2 see FEH000
51034-39-0 see CKP720
51042-51-4 see NHR620
51077-50-0 see DMD100
51085-52-0 see NMP600
51104-87-1 see DEU509
51109-89-8 see NOA700
51110-01-1 see SKS600
51118-04-8 see TDY080
51131-85-2 see HKH000
51138-16-0 see NIL625
51155-15-8 see TET780
51165-36-7 see SBL500
51200-87-4 see DTG750
51207-31-9 see TBO850
51218-45-2 see MQQ450
51218-49-6 see PMB850
51234-28-7 see OJI750
51235-04-2 see HFA300
51249-05-9 see ALZ000
51257-84-2 see LEJ700
51264-14-3 see ADL750
51286-83-0 see PLV275

51308-52-2 see DRV890
51308-53-3 see DRV870
51308-55-5 see PPH125
51308-56-6 see PPH115
51308-57-7 see PPH110
51308-58-8 see DRV885
51308-59-9 see PPH135
51308-60-2 see DRV880
51308-61-3 see DRV875
51308-62-4 see DRV930
51308-64-6 see BRD600
51308-65-7 see PPH170
51308-66-8 see PPH165
51308-67-9 see DRV920
51308-68-0 see PPH160
51308-69-1 see PPH180
51308-70-4 see PPH172
51308-71-5 see DRV900
51308-72-6 see MLO350
51308-73-7 see EMT100
51308-74-8 see HBO800
51308-75-9 see BKS770
51308-76-0 see BJK660
51308-77-1 see CKF540
51308-78-2 see CKF545
51308-79-3 see BNX055
51308-80-6 see BNX050
51312-42-6 see SJJ000
51317-24-9 see LDP000
51321-79-0 see PGY750
51325-35-0 see DBF400
51333-22-3 see BOM520
51338-27-3 see IAH050
51340-26-2 see DAQ002
51366-35-9 see CAM100
51371-34-7 see IHO700
51379-04-5 see DRV910
51379-05-6 see PPH130
51410-44-7 see HKI000
51460-26-5 see AER666
51473-23-5 see LEP000
51481-10-8 see VTF500
51481-61-9 see TAB250
51481-65-3 see MQS200
51487-69-5 see LAU500
51503-61-8 see ANS250
51505-90-9 see AJU700
51542-33-7 see NKR000
51566-62-2 see CMU000
51579-82-9 see AIU500
51580-86-0 see THS050
51593-70-5 see BCS500
51622-02-7 see DKO000
51627-14-6 see APT250
51628-36-5 see BEP600
51628-37-6 see PDR664
51628-56-9 see BEP650
51628-95-6 see PDR675
51628-96-7 see PDS925
51629-13-1 see PDR660
51629-15-3 see PDR665
51629-37-9 see PDR680
51629-48-2 see PDR670
51629-54-0 see PDR668
51629-58-4 see CIL920
51629-59-5 see CIL910
51629-74-4 see MLO301
51629-79-9 see MLO320
51630-04-7 see PDV360
51630-12-7 see PFR150
51630-33-2 see PDV330
51630-58-1 see FAR100

51632-16-7 see PDR610
51635-81-5 see EDX000
51707-55-2 see TEX600
51715-75-4 see PMN600
51753-57-2 see PDB350
51762-05-1 see CCS530
51764-33-1 see IFP800
51775-17-8 see CGX625
51775-36-1 see THH575
51781-06-7 see CCL800
51781-21-6 see MQT550
51786-53-9 see XOS000
51787-42-9 see TLI500
51787-43-0 see TLI750
51787-44-1 see TDL500
51787-75-8 see DUS100
51799-29-2 see NDY360
51821-32-0 see HKT200
51839-24-8 see BAP802
51857-25-1 see TDY085
51877-12-4 see DDN100
51877-74-8 see PCK075
51898-39-6 see AET250
51909-61-6 see GAA100
51922-16-8 see GAA120
51938-12-6 see NLN000
51938-13-7 see BRX750
51938-14-8 see BRO000
51938-15-9 see BRY000
51938-16-0 see MKP000
51938-42-2 see SKR875
51940-44-4 see PIZ000
51954-76-8 see FMY050
51961-45-6 see AMF600
52032-20-9 see PCH320
52049-26-0 see CHA500
52061-60-6 see IQE000
52093-21-7 see MQS579
52096-16-9 see NEN300
52096-22-7 see BCD125
52098-13-2 see HIS100
52098-14-3 see HIS120
52098-15-4 see HIS130
52098-16-5 see BES300
52098-17-6 see MDX300
52098-18-7 see MDX310
52098-56-3 see EPM000
52112-09-1 see BLN750
52112-66-0 see AIX750
52112-67-1 see AJL750
52112-68-2 see AJF750
52125-53-8 see EJV000
52129-71-2 see DUV400
52137-03-8 see PJU250
52152-93-9 see CCS550
52157-57-0 see CIO275
52162-18-2 see AAM000
52171-37-6 see DOD400
52171-41-2 see DOD200
52171-42-3 see DOF800
52171-92-3 see DRE400
52171-93-4 see DLI000
52171-94-5 see DUF200
52175-10-7 see POJ100
52195-07-0 see PIJ600
52205-73-9 see EDT100
52207-48-4 see DXC900
52207-83-7 see AGH500
52207-87-1 see TGH690
52208-11-4 see DVR500
52210-18-1 see IFX300
52212-02-9 see PII250

54531-52-1 see BCJ150
54546-26-8 see BSR600
54567-24-7 see ASE500
54573-23-8 see CPW325
54578-28-8 see TEQ750
54578-28-8 see TEQ800
54579-17-8 see EDI000
54593-27-0 see BJG500
54593-83-8 see FOL050
54597-56-7 see MRJ600
54605-45-7 see IDJ500
54622-43-4 see DYE550
54634-49-0 see NLP600
54643-52-6 see HOB000
54662-30-5 see DAW350
54693-46-8 see HMK050
54708-51-9 see CJW500
54708-68-8 see CKD250
54739-18-3 see FMR550
54746-50-8 see BPD000
54749-90-5 see CLX000
54767-75-8 see SOU600
54774-45-7 see PCK065
54774-46-8 see PCK055
54774-47-9 see PCK085
54779-53-2 see AJS875
54818-88-1 see DDD400
54824-17-8 see MQX775
54824-20-3 see MAB055
54827-17-7 see TDM800
54844-65-4 see HAQ200
54849-39-7 see ILR120
54851-13-7 see TKF723
54856-23-4 see BFV350
54889-82-6 see PNE750
54897-62-0 see ELE500
54897-63-1 see NKE000
54925-45-0 see AEL500
54927-63-8 see SCA000
54965-21-8 see VAD000
54965-24-1 see TAD175
54976-93-1 see SFV300
55011-46-6 see NMK000
55028-70-1 see AQT575
55028-71-2 see ECW550
55042-15-4 see UOA000
55042-51-8 see MFN700
55044-04-7 see ELN600
55049-48-4 see SFX725
55077-30-0 see AAC875
55079-83-9 see REP400
55080-20-1 see ABO500
55090-44-3 see NKU000
55102-43-7 see ROF200
55102-44-8 see RFU600
55112-89-5 see BIB250
55118-19-9 see DPB200
55123-66-5 see AAL300
55124-14-6 see BJE325
55134-13-9 see NBO600
55158-44-6 see CNK700
55165-33-8 see BET750
55179-31-2 see SCF525
55208-38-3 see PAS835
55216-04-1 see BIO500
55217-61-3 see AKL750
55219-65-3 see CJN300
55248-17-4 see SHU515
55254-34-7 see SDX800
55256-53-6 see ARU250
55257-88-0 see DXY750
55268-74-1 see BGB400

55268-75-2 see CCS600
55283-68-6 see ENE500
55285-05-7 see MRU050
55285-14-8 see CCC280
55290-64-7 see OMY825
55294-15-0 see EAE675
55297-95-5 see TET800
55297-96-6 see DAR000
55299-24-6 see CGB000
55308-37-7 see MFF650
55308-57-1 see PEU650
55308-64-0 see EFC259
55309-14-3 see MFJ110
55335-06-3 see TJE890
55345-04-5 see FDN100
55365-87-2 see EPC150
55380-34-2 see DRO000
55391-23-6 see MQC300
55391-24-7 see MQC320
55391-31-6 see DRQ500
55391-34-9 see MID870
55398-24-8 see BDO199
55398-25-9 see BDQ000
55398-26-0 see BDO500
55398-27-1 see EOJ000
55398-86-2 see MRG000
55448-20-9 see MAS250
55467-31-7 see CCX725
55477-20-8 see CEV750
55477-27-5 see CHO250
55489-49-1 see CFL750
55489-49-1 see CFR000
55509-78-9 see SBN510
55511-98-3 see RCA300
55512-33-9 see FAQ600
55514-14-2 see MLX250
55520-67-7 see VOA550
55528-90-0 see VAG200
55541-30-5 see DBC500
55556-85-9 see NLK000
55556-86-0 see DTA800
55556-88-2 see DRN800
55556-91-7 see NLL000
55556-92-8 see NLU500
55556-93-9 see NLK500
55556-94-0 see MJH000
55557-00-1 see DVE600
55557-02-3 see NKH500
55566-30-8 see TDI000
55576-67-5 see NBW600
55589-62-3 see AAF900
55600-34-5 see CMX845
55608-09-8 see PMB300
55609-55-7 see PEU525
55620-97-8 see EAB100
55621-29-9 see BRN250
55634-91-8 see ZTS200
55635-13-7 see FBP350
55636-92-5 see HKS600
55644-07-0 see DGS700
55644-07-0 see DGS800
55651-31-5 see DLP400
55651-36-0 see DLP200
55661-38-6 see ALF500
55673-54-6 see AJD375
55688-38-5 see HLK500
55689-65-1 see OMU000
55700-98-6 see CQD900
55719-85-2 see PFD250
55720-07-5 see HOE150
55720-09-7 see BKJ300
55720-10-0 see NFY100

55720-11-1 see NHY600
55720-13-3 see NFS800
55720-14-4 see PDN800
55720-15-5 see NHY700
55720-16-6 see NHR640
55720-17-7 see NFT100
55720-18-8 see NFQ300
55721-11-4 see HJS900
55726-47-1 see EAU075
55738-54-0 see DPL000
55758-10-6 see NHR600
55764-18-6 see OOK000
55769-64-7 see CNW125
55779-06-1 see FOK000
55792-21-7 see PIO750
55798-64-6 see BCA250
55814-41-0 see INE050
55818-96-7 see TDH500
55837-18-8 see IJH000
55837-27-9 see PDW250
55837-29-1 see TGF175
55838-67-0 see EBH420
55843-86-2 see IAY000
55852-84-1 see BAC260
55861-78-4 see ISR200
55864-04-5 see IRA200
55864-39-6 see ALG375
55870-36-5 see BIT350
55870-64-9 see EBC000
55898-33-4 see LJP000
55902-04-0 see AGL750
55904-02-4 see AMH850
55921-66-9 see ALA750
55936-75-9 see MEG500
55936-76-0 see MEG000
55936-77-1 see ABQ500
55936-78-2 see MEH250
55939-60-1 see SHT500
55941-39-4 see AAJ500
55963-33-2 see SLJ550
55965-84-9 see ISD066
55981-23-2 see CEK875
55984-51-5 see NKV000
56001-43-5 see NCN800
56011-02-0 see IHV050
56047-14-4 see RHK250
56073-07-5 see BGO100
56073-10-0 see TAC800
56090-02-9 see HEK550
56092-91-2 see COQ333
56105-46-5 see IIW100
56124-62-0 see TJX350
56139-33-4 see BQU000
56172-46-4 see DTE000
56173-18-3 see HHE500
56179-80-7 see ECR259
56179-83-0 see DLZ000
56180-94-0 see AAD900
56183-20-1 see MEB000
56187-09-8 see HCN100
56217-89-1 see NDY380
56217-90-4 see NDY390
56219-04-6 see HCP032
56222-04-9 see FAJ200
56222-35-6 see NLP700
56235-95-1 see NKC500
56238-63-2 see SFQ300
56239-24-8 see CHB000
56267-87-9 see TJT500
56275-41-3 see DFB400
56280-76-3 see PHT000
56283-74-0 see DAK000

56287-19-5 see TLM250
56287-41-3 see CFO750
56287-74-2 see AEW625
56299-00-4 see POF800
56302-13-7 see SAY950
56305-04-5 see TNR625
56316-37-1 see PNG500
56320-22-0 see ARJ100
56329-42-1 see ZJA100
56370-81-1 see TBG600
56375-33-8 see NJO000
56386-98-2 see PKN750
56390-16-0 see FGQ000
56391-56-1 see SBD000
56391-57-2 see NCP550
56393-22-7 see HHD000
56395-66-5 see HLF500
56400-60-3 see DSG200
56411-66-6 see DDN700
56411-67-7 see DHC309
56420-45-2 see EBB100
56425-91-3 see IRN500
56433-01-3 see CJC600
56455-90-4 see DEK000
56463-61-7 see AKT850
56463-62-8 see AKO430
56463-65-1 see AKC560
56463-70-8 see AKT830
56463-73-1 see AJI330
56463-76-4 see AKT800
56464-05-2 see DPL950
56464-19-8 see AJR100
56464-20-1 see AKA600
56470-64-5 see AOO150
56480-06-9 see HCA600
56484-47-0 see BMK550
56488-59-6 see BSE750
56501-30-5 see DXX100
56501-31-6 see EPJ600
56501-32-7 see EOB200
56501-33-8 see EKN600
56501-34-9 see ENX875
56501-35-0 see DXY700
56501-36-1 see DXX875
56516-72-4 see NKG500
56530-47-3 see DAU600
56530-48-4 see DXU600
56530-49-5 see DAS800
56533-30-3 see DKF620
56534-02-2 see CDR820
56538-00-2 see CGQ280
56538-01-3 see CGQ400
56538-02-4 see CGX325
56573-85-4 see TIC500
56583-56-3 see KGK400
56602-09-6 see DAT400
56605-16-4 see SLD900
56606-38-3 see PPP000
56611-65-5 see PHV725
56614-97-2 see BBG200
56622-38-9 see BKO800
56631-46-0 see PKO000
56641-03-3 see CDV625
56641-38-4 see CDR860
56643-49-3 see AMD250
56654-52-5 see DEF000
56654-53-6 see BRE000
56668-59-8 see NDD500
56683-54-6 see HCP060
56713-63-4 see MNT750
56717-11-4 see AGE750
56726-04-6 see DAU400

56731-15-8 see VRP780
56741-95-8 see AIY850
56743-33-0 see CCD625
56749-17-8 see WCB100
56764-40-0 see LCZ000
56767-15-8 see DWA700
56773-42-3 see TCD100
56775-88-3 see ZBA500
56776-01-3 see BQE250
56776-25-1 see DFO600
56779-19-2 see PKN500
56795-65-4 see BRL500
56795-66-5 see PNO000
56796-20-4 see CCS350
56797-40-1 see HCP022
56833-74-0 see SMC375
56839-43-1 see EAV700
56856-83-8 see AAW000
56863-02-6 see BKF500
56877-15-7 see DSO500
56892-30-9 see BCX000
56892-31-0 see BCY500
56892-32-1 see BCY750
56892-33-2 see BCZ000
56894-91-8 see BIJ500
56927-39-0 see DSH200
56929-36-3 see DXI200
56937-68-9 see PGV500
56939-74-3 see CNH285
56943-26-1 see TJE050
56961-60-5 see BBC000
56961-62-7 see EGO000
56961-65-0 see MJY250
56970-24-2 see MEA000
56973-16-1 see TCI500
56974-46-0 see PEU000
56974-61-9 see GAD400
56986-35-7 see BPV325
56986-36-8 see BRX500
56986-37-9 see BPV500
57011-48-0 see DCM550
57021-61-1 see XLS300
57057-83-7 see TIP600
57061-73-1 see DOF650
57061-77-5 see DOF630
57063-29-3 see CPJ600
57074-51-8 see HOT000
57082-24-3 see CCN050
57109-90-7 see CDQ250
57117-24-5 see NKX000
57117-31-4 see PAW100
57117-41-6 see PAV860
57117-44-9 see HCH425
57137-10-7 see EEE650
57149-07-2 see MFG515
57164-87-1 see ADN000
57164-89-3 see ADN250
57165-71-6 see TKD000
57170-08-8 see MFJ105
57176-66-6 see PIE800
57176-67-7 see PIE830
57197-43-0 see SNL850
57228-01-0 see NMV460
57229-41-1 see ABY150
57267-78-4 see ANL100
57277-87-9 see NGE777
57281-35-3 see CPG500
57282-49-2 see LJM800
57285-10-6 see IFT100
57294-74-3 see AJK625
57296-63-6 see IBQ400
57312-03-5 see CKL200

57314-55-3 see MRU757
57333-96-7 see SBL600
57377-32-9 see HOT200
57383-74-1 see MCJ300
57404-88-3 see DLC800
57414-02-5 see MMD750
57420-66-3 see EQD600
57421-56-4 see SDR350
57432-60-7 see EDA875
57432-61-8 see MJV750
57444-62-9 see RDP100
57449-30-6 see COC825
57455-37-5 see UJA200
57465-28-8 see PAV700
57472-68-1 see DWS650
57495-14-4 see KGK100
57512-42-2 see DCL800
57520-17-9 see GLQ100
57524-15-9 see TGJ885
57530-25-3 see AMZ125
57541-72-7 see DFW200
57541-73-8 see DDQ800
57543-54-1 see ACQ760
57543-55-2 see ACQ730
57543-56-3 see ACQ700
57543-57-4 see MEC340
57543-75-6 see MEU275
57544-34-0 see CEM300
57554-34-4 see CBG375
57558-46-0 see SBN300
57566-47-9 see FPM150
57576-44-0 see APU500
57583-34-3 see MQH500
57583-54-7 see REA050
57590-20-2 see PBJ875
57598-00-2 see HJG000
57608-59-0 see FCC000
57619-29-1 see PMJ550
57625-97-5 see DVZ050
57629-90-0 see BPM500
57631-93-3 see PMD850
57644-54-9 see BKX800
57644-85-6 see NJJ500
57645-49-5 see DBA250
57645-91-7 see CMV325
57647-13-9 see BJA809
57647-35-5 see BGC500
57647-53-7 see BJH000
57648-21-2 see TGB175
57651-82-8 see HJV000
57653-85-7 see HCF000
57665-49-3 see BJA750
57667-51-3 see AMV752
57667-51-3 see AMV752
57670-85-6 see SOR200
57693-13-7 see BMC800
57707-64-9 see ASE000
57716-89-9 see MPN600
57726-65-5 see DWF700
57729-86-9 see DBA520
57743-92-7 see OOS100
57773-65-6 see LIU353
57775-22-1 see EQO000
57775-27-6 see PIK625
57781-14-3 see HAG325
57801-81-7 see LEJ600
57808-65-8 see CFC100
57808-66-9 see DYB875
57816-08-7 see DKY400
57835-92-4 see NJA100
57837-19-1 see MDM100
57846-03-4 see BNR325

57852-57-0 see DAN100
57859-47-9 see HFE518
57865-37-9 see HLJ600
57872-78-3 see AEK750
57872-80-7 see AEM000
57878-77-0 see IAM040
57891-85-7 see PLJ780
57897-99-1 see SBU600
57910-39-1 see BQI400
57910-79-9 see TAH950
57933-19-4 see MKD800
57936-22-8 see NIG600
57943-81-4 see CMV250
57948-13-7 see DFQ400
57962-60-4 see MNY750
57966-95-7 see COM300
57982-77-1 see LIU420
57982-78-2 see BOM530
57998-68-2 see ASK875
58001-89-1 see MCA775
58011-68-0 see POM275
58013-12-0 see NDW520
58013-13-1 see KGK120
58013-14-2 see KGK130
58030-91-4 see DLC400
58039-73-9 see PDV300
58048-24-1 see AMH500
58048-25-2 see AEJ500
58048-26-3 see AMH750
58050-46-7 see BEB000
58050-49-0 see ENL500
58066-85-6 see HCP750
58066-96-9 see ELG100
58086-32-1 see ADB250
58098-08-1 see SNI000
58100-26-8 see CQK125
58128-20-4 see CMS226
58131-55-8 see NFF000
58138-08-2 see TJK100
58139-32-5 see MMW250
58139-33-6 see DTN400
58139-34-7 see MMW750
58139-35-8 see MNV500
58139-48-3 see MRU000
58144-61-9 see PFL100
58152-03-7 see AKJ000
58164-88-8 see AQE250
58166-91-9 see TNC600
58169-97-4 see MMV250
58194-38-0 see ART000
58200-70-7 see HAR150
58209-98-6 see AEU500
58229-88-2 see DVN600
58240-55-4 see THQ000
58240-57-6 see EKW200
58243-85-9 see FNK200
58253-49-9 see OBA100
58253-49-9 see PKE400
58270-08-9 see DFE469
58302-42-4 see SDP600
58306-30-2 see BKN250
58338-59-3 see DBQ125
58344-21-1 see BCJ090
58344-24-4 see TFM625
58344-42-6 see DSH600
58350-08-6 see NHN600
58391-12-1 see AFN950
58429-99-5 see DQL200
58430-00-5 see DQK600
58430-01-6 see TLI250
58430-94-7 see TLT500
58431-24-6 see ABT500

58451-82-4 see BJM500
58451-85-7 see BJM250
58451-87-9 see BIV000
58484-07-4 see CHA000
58494-43-2 see CHA750
58501-21-6 see FPC200
58513-59-0 see TFQ600
58546-54-6 see SBE450
58551-69-2 see CCC110
58567-11-6 see FMV100
58580-55-5 see PAG050
58581-89-8 see ASC125
58654-67-4 see MLO800
58654-67-4 see MND075
58658-27-8 see AHT250
58682-45-4 see BMS750
58683-84-4 see MFB250
58695-41-3 see CPG000
58696-86-9 see ANM625
58705-49-0 see BHL800
58718-68-6 see PAE875
58763-27-2 see CIU325
58763-31-8 see BPI625
58776-08-2 see PFJ780
58785-63-0 see LIF000
58786-99-5 see BPG325
58802-08-7 see PAV840
58810-48-3 see CGI550
58812-37-6 see CML860
58814-86-1 see AEC625
58817-05-3 see AOI500
58821-98-0 see DAW300
58845-80-0 see LIG000
58886-98-9 see DLC200
58911-02-7 see TCY500
58911-30-1 see CLR825
58917-39-8 see CNS300
58917-67-2 see BCU250
58917-67-2 see DMP900
58917-67-2 see DMQ000
58917-67-2 see DMQ600
58917-91-2 see DMR000
58941-14-3 see DFL709
58955-81-0 see CKH120
58955-82-1 see MNV765
58955-83-2 see MFG520
58955-84-3 see PFT700
58955-85-4 see MNV760
58957-92-9 see DAN000
58958-60-4 see ISC550
58968-53-9 see TMX550
58970-76-6 see BFV300
58989-02-9 see HKY000
59010-86-5 see CKK530
59017-64-0 see IGD200
59034-32-1 see TKB325
59034-34-3 see CFX625
59086-92-9 see PGU750
59122-46-2 see MJE775
59128-97-1 see HAG800
59130-69-7 see HCP550
59133-39-0 see ALA300
59160-29-1 see LFO300
59163-97-2 see MID750
59177-62-7 see CID000
59177-64-9 see BNN250
59177-70-7 see MOW000
59177-76-3 see MOW250
59177-78-5 see BFG500
59177-85-4 see BNN500
59178-29-9 see TGC285
59182-63-7 see CPP000

59183-17-4 see DFE300
59183-18-5 see ACH125
59198-70-8 see DKF130
59209-40-4 see POF550
59229-75-3 see MMG100
59230-57-8 see IOF050
59230-81-8 see BNR250
59261-17-5 see DVQ759
59277-89-3 see AEC700
59291-65-5 see HCC830
59297-18-6 see SLQ625
59327-98-9 see ASG625
59330-98-2 see CPJ525
59333-90-3 see ERE100
59348-49-1 see AJI530
59348-62-8 see DCP700
59355-75-8 see MFX600
59392-53-9 see MAK350
59405-47-9 see POS250
59413-14-8 see HNZ000
59417-86-6 see FGI100
59419-71-5 see BGT150
59443-94-6 see ALQ635
59464-43-6 see BBK800
59467-70-8 see MQT525
59483-61-3 see CJA150
59485-08-4 see NII100
59512-21-9 see BDT750
59536-65-1 see FBU000
59544-89-7 see TKH250
59547-52-3 see LGK100
59557-05-0 see MCG850
59558-23-5 see THB000
59607-71-5 see MFB500
59643-84-4 see PMY750
59652-20-9 see MIY250
59652-21-0 see MIY000
59653-73-5 see TBC450
59665-11-1 see MFX500
59669-26-0 see LBF100
59680-34-1 see AEC250
59690-88-9 see ALY750
59697-12-0 see TCD600
59703-84-3 see SJJ200
59729-32-7 see CMS258
59748-51-5 see NEM000
59748-95-7 see MDY000
59749-17-6 see FLH315
59749-18-7 see CKF850
59749-19-8 see DGE320
59749-20-1 see DGE315
59749-21-2 see FLH200
59749-22-3 see CKF800
59749-23-4 see DGE310
59749-24-5 see DGE300
59749-27-8 see FLH300
59749-28-9 see CKF810
59749-29-0 see PMH930
59749-30-3 see PMH933
59749-31-4 see FLH325
59749-33-6 see PMH915
59749-34-7 see PMH912
59749-35-8 see DGE325
59749-36-9 see PMH920
59749-37-0 see DGE305
59749-38-1 see PMH917
59749-39-2 see PMH923
59749-40-5 see FLH310
59749-42-7 see PMH927
59749-44-9 see PMH910
59749-45-0 see FLH320
59749-46-1 see FGI025

59749-49-4 see CBT600
59749-50-7 see FLH150
59749-51-8 see FLH400
59756-60-4 see FMQ200
59766-02-8 see DDF400
59766-31-3 see FQU200
59803-98-4 see BMM575
59804-37-4 see TAL485
59808-78-5 see TBO770
59820-43-8 see HKK100
59863-59-1 see DFW600
59865-13-3 see CQH100
59897-92-6 see EIL600
59901-90-5 see ABW250
59901-91-6 see HOE500
59917-23-6 see HLM600
59917-39-4 see VGU750
59921-81-2 see FGW000
59928-80-2 see MEI500
59935-47-6 see FEP000
59937-28-9 see MAO275
59943-31-6 see CIT625
59960-30-4 see NJR500
59960-90-6 see DHO000
59965-27-4 see EGS500
59978-65-3 see NDE500
59985-27-2 see AMJ750
59988-01-1 see ADM500
59989-92-3 see DGQ900
60012-89-7 see BLG325
60029-23-4 see FMY100
60040-13-3 see BQA100
60047-17-8 see LFY500
60050-37-5 see BNX250
60062-60-4 see PNX500
60075-64-1 see ELQ050
60075-65-2 see IEF100
60075-67-4 see AOJ950
60075-68-5 see HFQ520
60075-69-6 see HFQ510
60075-70-9 see IEF200
60075-72-1 see EKW100
60075-74-3 see EEQ100
60075-76-5 see ELQ200
60075-83-4 see CBW600
60075-86-7 see CBW100
60084-10-8 see RJF500
60086-22-8 see CMX860
60102-37-6 see PCQ750
60145-64-4 see AKD875
60153-49-3 see MMS200
60160-75-0 see DTM200
60166-93-0 see IFY000
60168-88-9 see FAK100
60172-01-2 see PIG500
60172-03-4 see PIG250
60172-05-6 see PPD500
60172-05-6 see SMW000
60172-15-8 see POT500
60177-39-1 see CIF775
60207-90-1 see PMS930
60207-93-4 see EDW100
60207-93-4 see VFA100
60223-52-1 see TDJ000
60240-47-3 see MAS300
60254-65-1 see HFP650
60254-66-2 see HBN700
60254-67-3 see OEU100
60254-68-4 see NNB600
60254-95-7 see HJN875
60268-85-1 see BCU000
60268-85-1 see DMQ800

60268-86-2 see BMK560
60315-52-8 see MNE400
60318-52-7 see TJF350
60325-46-4 see SOU650
60345-95-1 see MDF100
60357-67-7 see DFN350
60364-26-3 see PMS825
60391-92-6 see NKI500
60397-73-1 see BKN500
60398-22-3 see MID250
60414-81-5 see NJM500
60444-92-0 see BPG750
60448-19-3 see BCW250
60452-14-4 see BTA500
60462-51-3 see HJA000
60476-14-4 see KGK500
60476-17-7 see FOO700
60479-97-2 see BTA125
60494-19-1 see BKO500
60504-57-6 see DAY835
60504-95-2 see DAM000
60510-57-8 see PGN400
60525-15-7 see ZBA525
60539-20-0 see DXR800
60548-62-1 see BOU100
60550-91-6 see COW500
60553-18-6 see FIK875
60560-33-0 see COS500
60561-17-3 see SNH150
60566-40-7 see AHT000
60573-88-8 see AKG250
60580-30-5 see DWU400
60589-06-2 see OKU200
60597-20-8 see TLT750
60599-38-4 see NJN000
60605-72-3 see BJK000
60607-34-3 see OMG000
60616-74-2 see MAG500
60616-85-5 see MQW300
60617-06-3 see CNR550
60617-12-1 see EAQ600
60633-76-3 see COK659
60634-59-5 see SNB500
60662-79-5 see DWC100
60672-60-8 see PBG100
60676-83-7 see AJK500
60676-86-0 see SCK600
60676-90-6 see ZRS000
60687-93-6 see CMP905
60706-43-6 see MOE250
60706-49-2 see MOP500
60706-52-7 see MOQ000
60719-83-7 see CJX000
60719-84-8 see AOD375
60723-51-5 see BOM000
60731-46-6 see CBR300
60735-64-0 see DHI876
60748-45-0 see ARS000
60762-50-7 see GFO200
60763-49-7 see CMS125
60764-83-2 see MCP000
60767-85-3 see TBM100
60784-40-9 see CHE325
60784-41-0 see EIP500
60784-42-1 see PNJ000
60784-43-2 see TDQ250
60784-44-3 see PBG750
60784-46-5 see CHB750
60784-47-6 see CHC250
60784-48-7 see CHA250
60789-89-1 see MNP300
60828-78-6 see TBB775

60837-57-2 see APE300
60842-44-6 see DJY100
60851-34-5 see HCH450
60864-95-1 see DML200
60883-74-1 see BMX000
60961-52-6 see LIU342
60967-88-6 see BBD250
60967-89-7 see BBD500
60967-90-0 see BBF000
60968-01-6 see DKS600
60968-08-3 see DKS400
60996-85-2 see CAL075
61001-01-2 see MFG254
61001-02-3 see MFG256
61001-03-4 see MFG252
61001-04-5 see THA100
61001-05-6 see CKA625
61001-06-7 see BNW600
61001-08-9 see MNU600
61001-09-0 see FLG150
61001-11-4 see CKA600
61001-12-5 see DGD090
61001-13-6 see NIR075
61001-14-7 see EFC625
61001-15-8 see PNA600
61001-16-9 see AGP300
61001-18-1 see MFF620
61001-19-2 see TGY300
61001-20-5 see MFF600
61001-21-6 see IAK300
61001-31-8 see PEU600
61001-36-3 see PGE350
61001-40-9 see MFH800
61001-42-1 see PGE300
61005-12-7 see BSF250
61034-40-0 see NJL850
61036-64-4 see TAI400
61136-62-7 see CDC375
61136-68-3 see MAQ700
61137-63-1 see CHD500
61166-04-9 see MMI700
61166-05-0 see MMI710
61177-45-5 see PLB775
61203-01-8 see MHS400
61227-05-2 see TET250
61242-71-5 see NNG500
61262-53-1 see EEB230
61288-13-9 see BMQ800
61318-91-0 see SNH480
61319-99-1 see DRW100
61336-70-7 see AOA100
61337-67-5 see HDS210
61379-65-5 see RKZ100
61413-38-5 see FKE000
61413-39-6 see MEJ500
61413-61-4 see HMG500
61417-04-7 see MDX350
61417-05-8 see MDY250
61417-08-1 see CED750
61417-10-5 see ADL000
61422-45-5 see CCK630
61424-17-7 see HJS400
61431-86-5 see SLN300
61432-55-1 see MNU300
61443-57-0 see DML000
61445-55-4 see MIF250
61447-07-2 see MMI640
61461-77-6 see FCN531
61462-73-5 see CED500
61471-62-3 see DUJ800
61477-94-9 see PJA170
61481-19-4 see BQK850

61483-80-5 see IDE100
61490-68-4 see TDD750
61499-28-3 see HNX500
61503-59-1 see CNR250
61514-68-9 see SDW100
61532-76-1 see NHN575
61556-82-9 see TDR250
61583-29-7 see DFT033
61583-30-0 see CJY250
61614-71-9 see TDF250
61691-82-5 see ABN725
61695-69-0 see ONE000
61695-70-3 see VMF000
61695-72-5 see ONC000
61695-74-7 see ONG000
61702-43-0 see ALO500
61706-44-3 see ALX500
61711-25-9 see AFJ375
61718-82-9 see FMR575
61724-64-9 see CMM830
61734-86-9 see NLE500
61734-88-1 see ELM000
61734-89-2 see BSB500
61734-90-5 see BSB750
61735-77-1 see FGI000
61735-78-2 see DKG400
61738-03-2 see MEO500
61738-04-3 see MEV750
61738-05-4 see MEP500
61740-00-9 see AET630
61750-69-4 see DTE100
61785-70-4 see NIV000
61785-72-6 see MFH750
61785-73-7 see THD275
61788-32-7 see HHW800
61788-33-8 see PJP250
61788-72-5 see FAB920
61788-85-0 see CCP300
61788-85-0 see CCP305
61788-85-0 see CCP310
61788-90-7 see CNF325
61789-01-3 see FAB900
61789-17-1 see GKG400
61789-30-8 see CNF175
61789-36-4 see NAR200
61789-40-0 see CNF185
61789-51-3 see NAR500
61789-64-8 see NAR100
61789-80-8 see QAT550
61789-91-1 see JDJ300
61789-92-2 see MBU777
61790-14-5 see NAS500
61790-53-2 see DCJ800
61790-81-6 see LAU560
61791-12-6 see CCP330
61791-14-8 see EEJ000
61791-24-0 see EEJ500
61791-24-0 see EEK000
61791-24-0 see EEK025
61791-26-2 see EEK050
61791-36-4 see TAC075
61791-39-7 see TAC050
61791-42-2 see SFY000
61792-05-0 see HLO400
61792-11-8 see LEH100
61825-94-3 see OKY100
61827-42-7 see PKE370
61827-74-5 see AJO000
61840-09-3 see HDF100
61840-39-9 see OKO200
61848-66-6 see DAD075
61848-70-2 see DAD040

61866-12-4 see CHF250
61869-08-7 see PAL650
61907-23-1 see APL250
61912-76-3 see LIN600
61925-70-0 see BQX000
61931-80-4 see HGI585
61941-56-8 see AHK625
61947-30-6 see DNJ000
61949-76-6 see PCK050
61949-77-7 see PCK070
62018-89-7 see HKS500
62018-90-0 see HJS000
62018-91-1 see BRB000
62018-92-2 see BRX250
62037-49-4 see MIG500
62046-37-1 see DET600
62046-63-3 see PAI750
62064-66-8 see BEY800
62078-98-2 see CGY600
62093-52-1 see CKA100
62116-25-0 see BMS000
62126-20-9 see CNQ250
62133-36-2 see TJP780
62169-70-4 see TAL350
62178-60-3 see NLY500
62207-76-5 see EIS000
62209-22-7 see COG750
62214-31-7 see BDM700
62229-50-9 see EBA275
62229-70-3 see PAC500
62232-46-6 see MGE200
62258-26-8 see MIG750
62303-19-9 see BDN125
62314-67-4 see DML400
62316-46-5 see NMG500
62346-96-7 see DQP500
62355-03-7 see COG500
62362-59-8 see SBZ000
62374-53-2 see BKJ250
62375-91-1 see MMH000
62382-21-2 see MLL150
62382-23-4 see MLX820
62399-48-8 see APK750
62421-98-1 see MNB300
62422-00-8 see DAS500
62441-54-7 see FDB300
62450-06-0 see TNX275
62450-07-1 see ALD500
62476-59-9 see SFV650
62501-24-0 see ACF150
62524-93-0 see BLH400
62536-49-6 see SLO000
62559-75-5 see CJQ800
62571-86-2 see MCO750
62573-57-3 see PBX750
62573-59-5 see NJP000
62581-31-1 see MOZ200
62593-23-1 see NJQ500
62601-71-2 see POE300
62602-94-2 see OJG550
62610-77-9 see MDN150
62613-15-4 see DWD300
62637-93-8 see TLE250
62641-66-1 see PNL500
62641-67-2 see NJX500
62641-68-3 see ELH000
62666-20-0 see CKA125
62669-70-9 see RGP600
62669-77-6 see RGW100
62681-13-4 see EFR500
62700-64-5 see MPT600
62765-90-6 see IHK100

62765-93-9 see NCS000
62778-13-6 see BJJ000
62783-48-6 see NKR500
62783-49-7 see NKS000
62783-50-0 see NKS500
62796-23-0 see EDD500
62822-49-5 see HNT200
62851-48-3 see AFM000
62851-59-6 see ERC600
62861-56-7 see CKS325
62861-57-8 see MEH775
62865-26-3 see CMS212
62865-36-5 see DFQ500
62893-20-3 see CCS369
62907-78-2 see SNJ400
62912-45-2 see PBH500
62912-47-4 see PBI000
62912-51-0 see OPQ000
62924-70-3 see FDD078
62928-11-4 see IGG775
62929-91-3 see PME600
62952-06-1 see ARP125
62959-43-7 see MAK275
62967-27-5 see IKI000
62973-76-6 see TJF000
62987-05-7 see BLH325
63007-70-5 see CFJ000
63018-40-6 see BBC500
63018-49-5 see BBI750
63018-50-8 see BBJ000
63018-56-4 see BBJ250
63018-57-5 see BBH750
63018-59-7 see BBH000
63018-62-2 see BBF750
63018-63-3 see BNF315
63018-64-4 see BQI500
63018-67-7 see CIG000
63018-68-8 see COM000
63018-69-9 see BBB500
63018-70-2 see MGY000
63018-94-0 see TLH050
63018-98-4 see ACC750
63018-99-5 see AOE750
63019-08-9 see DKA800
63019-09-0 see TMF750
63019-12-5 see EHD000
63019-14-7 see DSV000
63019-23-8 see DRE600
63019-25-0 see DQL800
63019-29-4 see EEV000
63019-32-9 see HBN000
63019-34-1 see HFL000
63019-42-1 see SNX000
63019-46-5 see CME000
63019-50-1 see BAY250
63019-51-2 see CGW750
63019-52-3 see CGH250
63019-53-4 see EHL000
63019-57-8 see DHI800
63019-59-0 see DQC200
63019-60-3 see DQC000
63019-65-8 see DBF000
63019-67-0 see AKC000
63019-68-1 see FEI500
63019-69-2 see MEB250
63019-70-5 see TDM000
63019-72-7 see MEK750
63019-73-8 see SMS000
63019-76-1 see DUM600
63019-77-2 see NIK500
63019-78-3 see MOY500
63019-81-8 see AKI500

63019-82-9 see POQ000
63019-93-2 see DOV400
63019-97-6 see MGF500
63019-98-7 see MGE000
63020-03-1 see MFA250
63020-21-3 see DDD800
63020-25-7 see MIQ250
63020-27-9 see EEX000
63020-32-6 see PNF000
63020-33-7 see PNI500
63020-37-1 see MPO250
63020-39-3 see TDL750
63020-45-1 see BBG500
63020-47-3 see INZ000
63020-48-4 see IOA000
63020-53-1 see IOF000
63020-60-0 see MEN750
63020-61-1 see MEU250
63020-69-9 see DRH800
63020-76-8 see MIV250
63020-91-7 see CGK500
63021-00-1 see DRU600
63021-11-4 see OIC000
63021-32-9 see BBM500
63021-33-0 see EHW000
63021-35-2 see EHW500
63021-43-2 see MSB000
63021-45-4 see NBC000
63021-46-5 see NMF000
63021-47-6 see NMF500
63021-48-7 see NMD500
63021-49-8 see NME000
63021-50-1 see NME500
63021-51-2 see NNA000
63021-62-5 see SMU000
63021-67-0 see OBW000
63021-86-3 see NIY700
63035-21-2 see NCG700
63039-89-4 see EMM000
63039-90-7 see NOE550
63039-95-2 see HBF500
63040-01-7 see TLH500
63040-02-8 see TLH750
63040-05-1 see TLH350
63040-09-5 see MIM000
63040-20-0 see QQA000
63040-21-1 see TBI750
63040-24-4 see HLR500
63040-25-5 see AKN250
63040-27-7 see CGK750
63040-30-2 see PEB250
63040-32-4 see TMG000
63040-43-7 see PGU500
63040-44-8 see PGU000
63040-49-3 see DOB600
63040-53-9 see BCQ750
63040-54-0 see DCY600
63040-55-1 see FNK000
63040-56-2 see FNU000
63040-57-3 see FNV000
63040-58-4 see FNQ000
63040-63-1 see DPO800
63040-64-2 see DPP000
63040-65-3 see DPP200
63040-98-2 see DKM800
63041-00-9 see BOI250
63041-01-0 see DKM400
63041-05-4 see MJD000
63041-07-6 see MRI500
63041-14-5 see MQD000
63041-15-6 see MQD500
63041-19-0 see ILG000

63918-85-4 see PFI750	63979-26-0 see ALW000	64028-99-5 see TLY200	64048-13-1 see DJB600
63918-89-8 see DFK200	63979-37-3 see DMF400	64029-07-8 see MLL100	64048-70-0 see BCT500
63918-91-2 see HFT600	63979-62-4 see DNL400	64029-08-9 see MLO900	64048-90-4 see DGO600
63918-97-8 see LEE000	63979-65-7 see BSL750	64029-10-3 see OPC025	64048-94-8 see AKF750
63918-98-9 see EFG500	63979-84-0 see SNG500	64036-46-0 see DJV800	64048-98-2 see TNO000
63919-00-6 see DNI200	63979-86-2 see SJW300	64036-72-2 see MGD750	64048-99-3 see DJN400
63919-01-7 see FID000	63979-95-3 see BHD000	64036-79-9 see MID900	64049-00-9 see MID900
63919-09-5 see DRL300	63980-13-2 see TKH500	64036-86-8 see LIP000	64049-02-1 see IET000
63919-14-2 see ZRJ000	63980-18-7 see TGW750	64036-91-5 see BKD750	64049-03-2 see TBG710
63919-17-5 see DCF720	63980-19-8 see TGW500	64037-07-6 see AIX500	64049-04-3 see SGE500
63919-18-6 see SJQ500	63980-20-1 see DKD200	64037-08-7 see AIY250	64049-07-6 see PER600
63919-19-7 see BLN200	63980-27-8 see TGX000	64037-09-8 see AIY000	64049-11-2 see CLV250
63919-21-1 see CND000	63980-44-9 see BIG750	64037-10-1 see AJF250	64049-21-4 see HBN600
63919-22-2 see SIW000	63980-59-6 see DPA000	64037-11-2 see AJG750	64049-22-5 see HFP600
63929-61-3 see LIU325	63980-61-0 see PHO250	64037-12-3 see AJM500	64049-23-6 see IHQ100
63934-40-7 see HJR000	63980-62-1 see IJB000	64037-13-4 see AKL500	64049-27-0 see HLO350
63934-41-8 see HJR500	63980-89-2 see PHM750	64037-14-5 see AKN000	64049-28-1 see MCS250
63937-14-4 see MDC500	63981-09-9 see HHF500	64037-15-6 see AKQ750	64049-29-2 see MJM250
63937-26-8 see SFR000	63981-20-4 see BGP500	64037-20-3 see CGD270	64049-56-5 see OBI200
63937-27-9 see AGL500	63981-21-5 see OOK175	64037-50-9 see EIJ000	64049-71-4 see PDF510
63937-29-1 see PIO900	63981-49-7 see MCY500	64037-51-0 see MJJ500	64049-85-0 see TGH650
63937-30-4 see AOP510	63981-53-3 see HJZ000	64037-53-2 see CES250	64050-01-7 see TGH655
63937-31-5 see AOP800	63981-92-0 see HJY500	64037-54-3 see DEU200	64050-03-9 see HOL000
63937-32-6 see BQM750	63982-03-6 see HAU500	64037-56-5 see BQL500	64050-15-3 see EFI000
63937-47-3 see ODO000	63982-15-0 see FIS000	64037-57-6 see BQL750	64050-20-0 see DSO400
63938-10-3 see CLH000	63982-25-2 see PEQ000	64037-65-6 see AGJ750	64050-23-3 see CGH500
63938-16-9 see CMQ000	63982-32-1 see MIC250	64037-73-6 see ASM000	64050-25-5 see BCS450
63938-20-5 see YFA000	63982-40-1 see MID000	64038-09-1 see ALW500	64050-44-8 see CCJ000
63938-21-6 see DSM600	63982-47-8 see DPL900	64038-10-4 see SHN275	64050-46-0 see BIJ000
63938-24-9 see ACP500	63982-49-0 see MIA775	64038-38-6 see DRB800	64050-54-0 see DOL400
63938-26-1 see LJK000	63982-52-5 see BKR000	64038-39-7 see FHR000	64050-77-7 see HNR000
63938-27-2 see EDE000	63989-69-5 see IGO000	64038-40-0 see TLI000	64050-79-9 see HNR500
63938-28-3 see EDE500	63989-75-3 see BDV250	64038-41-1 see CCG250	64050-81-3 see HNS000
63938-62-5 see CEO125	63989-79-7 see AJJ750	64038-48-8 see HOQ000	64050-83-5 see HNS500
63938-82-9 see PBH075	63989-82-2 see DUT200	64038-49-9 see HOQ000	64050-95-9 see TGH685
63938-92-1 see CBW650	63989-83-3 see DUT900	64038-55-7 see DCP600	64051-06-5 see HNN000
63938-93-2 see CCH250	63989-84-4 see DUU200	64038-56-8 see DUI400	64051-08-7 see HNO000
63941-74-2 see IFS400	63989-85-5 see DUU400	64038-57-9 see MLF000	64051-12-3 see ABV500
63942-42-7 see BKT300	63990-56-7 see DWY400	64039-27-6 see TFJ250	64051-16-7 see BED500
63942-43-8 see BIQ660	63990-57-8 see DCI100	64043-53-4 see CMR250	64051-18-9 see HNP500
63942-44-9 see BKU120	63990-88-5 see BFT750	64043-55-6 see TDE000	64051-20-3 see HNQ500
63943-38-4 see PJQ800	63990-96-5 see AOF500	64046-00-0 see ANT250	64057-51-8 see CHE250
63944-10-5 see NFW400	63991-14-0 see AJV850	64046-01-1 see SJR500	64057-52-9 see DDD600
63949-98-4 see DCC500	63991-23-1 see AKA250	64046-07-7 see TGH680	64057-57-4 see SJB000
63950-89-0 see BHA000	63991-26-4 see MGH250	64046-11-3 see TGH670	64057-58-5 see TIR750
63951-03-1 see ARJ900	63991-31-1 see MKS000	64046-12-4 see TGH665	64057-70-1 see MJR500
63951-06-4 see BGQ300	63991-43-5 see MQC250	64046-20-4 see TGH675	64057-75-6 see MNQ250
63951-08-6 see BJN850	63991-48-0 see AHT900	64046-38-4 see TGH660	64057-79-0 see CBV750
63951-09-7 see BKH000	63991-49-1 see AHT950	64046-46-4 see MES940	64058-13-5 see AFX500
63951-11-1 see CIL700	63991-57-1 see ECG500	64046-47-5 see CJD625	64058-14-6 see AGA000
63951-12-2 see PFB850	63991-70-8 see MIX000	64046-48-6 see PGB770	64058-26-0 see BKE750
63951-29-1 see TMJ300	63992-02-9 see SJR000	64046-56-6 see DWY600	64058-30-6 see COX250
63951-45-1 see VHF000	63992-41-6 see SFV500	64046-59-9 see TGV500	64058-36-2 see LAG030
63951-48-4 see DTH600	64005-62-5 see AOL750	64046-61-3 see AAT500	64058-43-1 see HEJ400
63956-71-8 see BJM650	64005-84-1 see BKO770	64046-62-4 see ALQ750	64058-44-2 see HNK950
63956-95-6 see FPW100	64005-91-0 see BIM100	64046-67-9 see MES930	64058-54-4 see OOO000
63957-11-9 see PNA300	64011-26-3 see TNC250	64046-68-0 see CAX400	64058-57-7 see AIE125
63957-36-8 see PDL000	64011-34-3 see TGR250	64046-79-3 see QDJ000	64058-65-7 see AKF500
63957-37-9 see ALT750	64011-35-4 see TGR500	64046-93-1 see AQC250	64058-72-6 see ABK000
63957-38-0 see ALU000	64011-36-5 see TGR750	64046-96-4 see CAL500	64058-74-8 see BGQ250
63957-39-1 see ALU250	64011-37-6 see MCY250	64046-97-5 see SJG000	64058-92-0 see TBD250
63957-41-5 see ARB270	64011-39-8 see BLQ750	64046-99-7 see CBK750	64059-02-5 see DXJ400
63957-48-2 see DDF600	64011-43-4 see HNE450	64047-26-3 see MCS500	64059-26-3 see LAH000
63957-59-5 see HNK550	64011-44-5 see MEA250	64047-27-4 see EJA150	64059-29-6 see BEA250
63968-64-9 see ARL375	64011-46-7 see ECE550	64047-30-9 see TLY000	64059-33-2 see TEY300
63976-07-8 see ALL000	64011-53-6 see BKH400	64047-79-6 see NIF100	64059-42-3 see CIQ000
63977-41-3 see DKR800	64011-62-7 see CPG625	64047-82-1 see NIH000	64059-47-8 see PEW600
63977-49-1 see DQS600	64011-64-9 see BAL100	64047-83-2 see NIH500	64059-53-6 see PPN250
63978-55-2 see BHP500	64019-93-8 see DWP559	64048-05-1 see SFJ500	64060-01-1 see PJG125
63978-73-4 see TJI500	64024-08-4 see UTA000	64048-06-2 see PLE750	64070-10-6 see AQH250
63978-93-8 see NBG000	64025-05-4 see SJQ000	64048-07-3 see SIM000	64070-11-7 see AQG500
63978-98-3 see BFA930	64025-06-5 see PBL250	64048-08-4 see HLQ000	64070-12-8 see AQG750

64070-13-9 see DHA400
64070-14-0 see AOW750
64070-15-1 see AOX000
64070-83-3 see ARE000
64082-34-4 see SKQ000
64082-35-5 see LHK000
64082-43-5 see MGX750
64083-05-2 see LCE000
64083-08-5 see MAB050
64090-82-0 see AOW500
64091-90-3 see MMS250
64091-91-4 see MMS500
64092-22-4 see NDN050
64092-23-5 see BHB000
64092-48-4 see ZUA300
64093-79-4 see NBW000
64109-72-4 see EJT575
64128-32-1 see NND100
64131-03-9 see SHU510
64140-51-8 see DWR200
64153-17-9 see PFP800
64153-23-7 see PFP730
64187-24-2 see CBR200
64187-25-3 see CBR175
64187-27-5 see CBR215
64187-42-4 see CBR235
64187-43-5 see CBR225
64241-34-5 see CAK275
64245-83-6 see SPA500
64245-99-4 see MJS250
64246-03-3 see BKR500
64246-07-7 see DOV800
64246-13-5 see BKR750
64246-17-9 see MJS500
64249-01-0 see AOT255
64253-30-1 see OPG050
64253-71-0 see BAR830
64253-73-2 see RHA125
64265-57-2 see EML600
64267-45-4 see DXA900
64267-46-5 see NMV600
64267-66-9 see CEO110
64285-06-9 see AOO120
64296-43-1 see HBA259
64309-76-8 see AKL250
64314-28-9 see BAR825
64314-52-9 see DAM800
64318-79-2 see CDB775
64363-96-8 see DAM430
64365-11-3 see CDI000
64365-16-8 see QAT510
64366-96-7 see VRP790
64387-77-5 see MLM700
64387-78-6 see MLM600
64431-68-1 see MAB250
64440-87-5 see CMN125
64441-42-5 see BRK250
64461-82-1 see TGH600
64461-98-9 see HCF100
64466-47-3 see DOA810
64466-48-4 see DOA815
64466-49-5 see DOA820
64466-50-8 see DOA830
64474-06-2 see MRN050
64475-85-0 see MQV900
64485-93-4 see CCR950
64490-92-2 see SKJ350
64491-74-3 see PEW750
64502-82-5 see RHA150
64506-49-6 see IMS300
64508-90-3 see BHY625
64521-13-7 see DLE600

64521-14-8 see MBV500
64521-15-9 see DLF200
64521-16-0 see ECP500
64529-56-2 see ENJ600
64532-94-1 see IRA100
64532-96-3 see HFU700
64544-07-6 see CCS625
64550-80-7 see BAD000
64551-89-9 see DMO500
64552-25-6 see MLC500
64553-70-4 see PAQ900
64598-80-7 see DLE000
64598-81-8 see DLE200
64598-82-9 see DMP200
64598-83-0 see DMO600
64603-91-4 see TCU550
64604-09-7 see TCA250
64604-91-7 see AJU900
64622-45-3 see IAB000
64624-44-8 see GFU200
64653-03-8 see BIW250
64675-27-0 see PMM400
64686-82-4 see AJD500
64693-33-0 see BJP325
64694-79-7 see ACH080
64709-48-4 see DKS909
64709-49-5 see DKT000
64709-50-8 see DKT200
64719-39-7 see QBS000
64741-44-2 see GBW000
64741-46-4 see NAQ560
64741-49-7 see MQV755
64741-50-0 see MQV815
64741-51-1 see MQV785
64741-52-2 see MQV810
64741-53-3 see MQV780
64741-54-4 see NAQ520
64741-55-5 see NAQ540
64741-56-6 see RDK200
64741-57-7 see GBW005
64741-58-8 see GBW025
64741-59-9 see DXG840
64741-61-3 see DXG810
64741-62-4 see CMU890
64741-63-5 see NAQ550
64741-67-9 see RDK075
64741-68-0 see NAQ530
64741-71-5 see VRU300
64741-80-6 see RDK100
64741-83-9 see NAQ580
64741-87-3 see NAQ570
64741-88-4 see MQV850
64741-89-5 see MQV855
64741-96-4 see MQV845
64741-97-5 see MQV852
64742-03-6 see MQV860
64742-04-7 see MQV859
64742-05-8 see MQV862
64742-10-5 see MQV863
64742-11-6 see MQV857
64742-16-1 see AQW500
64742-17-2 see MQV872
64742-18-3 see MOI500
64742-19-4 see MQV770
64742-20-7 see MQV765
64742-21-8 see MQV775
64742-44-5 see PCS260
64742-45-6 see PCS270
64742-46-7 see DXG830
64742-47-8 see KEK100
64742-52-5 see MQV790
64742-53-6 see MQV800

64742-54-7 see MQV795
64742-54-7 see MQV796
64742-55-8 see MQV805
64742-56-9 see MQV840
64742-63-8 see MQV820
64742-64-9 see MQV835
64742-65-0 see MQV825
64742-68-3 see MQV776
64742-69-4 see MQV777
64742-70-7 see MQV778
64742-71-8 see MQV779
64742-80-9 see DXG820
64742-81-0 see KEK110
64742-86-5 see GBW010
64742-88-7 see ILR150
64742-95-6 see SKS350
64755-14-2 see BMK325
64771-59-1 see MCX600
64781-77-7 see DCQ500
64789-67-9 see LIU341
64808-48-6 see LHZ600
64817-78-3 see MSA750
64819-51-8 see BQI300
64838-75-1 see DMP000
64854-98-4 see CBD500
64854-99-5 see CBD250
64862-96-0 see BKB325
64891-12-9 see PBX100
64891-34-5 see LIU345
64902-72-3 see CMA700
64910-63-0 see BSM250
64920-31-6 see DLD200
64925-80-0 see PGR775
64953-12-4 see LBH200
64977-44-2 see FJY000
64977-46-4 see FKA000
64977-47-5 see FKB000
64977-48-6 see FKC000
64977-49-7 see FKD000
64988-06-3 see EMF600
65002-17-7 see MCO775
65036-47-7 see BNG125
65039-20-5 see CIS325
65041-92-1 see ACS500
65043-22-3 see IBW500
65057-90-1 see TAC500
65057-91-2 see TAC750
65072-01-7 see AHR600
65072-04-0 see VIZ000
65089-17-0 see CLW500
65094-73-7 see NFW435
65098-93-3 see TEO800
65141-46-0 see NDL800
65146-47-6 see AIW500
65176-75-2 see HOL200
65184-10-3 see TAL560
65195-55-3 see ARW150
65210-28-8 see CPG250
65210-29-9 see MNP250
65210-30-2 see DPH200
65210-31-3 see DIC800
65210-32-4 see MLP750
65210-33-5 see MOI500
65210-37-9 see BKQ250
65216-94-6 see COM500
65229-18-7 see HMQ500
65232-69-1 see TJL775
65235-63-4 see BND250
65235-79-2 see SDJ500
65268-91-9 see DHU400
65271-80-9 see MQY090
65272-47-1 see MKJ000

65277-42-1 see KFK100
65296-81-3 see CCJ350
65313-33-9 see CIH900
65313-34-0 see DIV400
65313-35-1 see TNR500
65313-36-2 see EIY550
65313-37-3 see TDP300
65400-79-5 see MPR250
65400-81-9 see AAL250
65405-73-4 see GDM100
65405-76-7 see HFE300
65405-77-8 see SAJ000
65405-84-7 see TDO255
65445-59-2 see DTA600
65445-60-5 see CJG375
65445-61-6 see CJG500
65454-27-5 see MCB100
65454-61-7 see TAI200
65496-97-1 see SDX630
65497-24-7 see POF525
65520-53-8 see HOH000
65521-60-0 see PLW275
65546-74-9 see ROA400
65561-73-1 see DNA300
65567-32-0 see EOL050
65573-02-6 see IBQ075
65589-70-0 see ADQ700
65597-24-2 see CDX800
65597-25-3 see TJX650
65606-61-3 see TLT300
65654-08-2 see MEA800
65654-13-9 see TEN725
65664-23-5 see PLU575
65666-07-1 see SDX625
65700-59-6 see DAU200
65700-60-9 see DAU000
65701-06-6 see CBY800
65701-07-7 see BBO625
65734-38-5 see ACN500
65763-32-8 see DLD800
65792-56-5 see NLN500
65793-50-2 see AIS250
65860-38-0 see TCU000
65894-78-2 see AIS700
65899-73-2 see TGF050
65907-30-4 see FPO200
65928-58-7 see SMP400
65954-42-9 see HKY700
65979-81-9 see ZKS000
65980-75-8 see PAY300
65986-79-0 see ABR500
65986-80-3 see ENR500
65996-92-1 see CMY900
65996-93-2 see CMZ100
65997-15-1 see PKS750
66007-89-4 see SFP500
66009-08-3 see BIW500
66017-91-2 see PNR250
66063-05-6 see UTA400
66064-11-7 see AIZ000
66071-96-3 see CNR980
66085-59-4 see NDY700
66104-23-2 see MPU500
66104-24-3 see BFP500
66147-68-0 see ADP750
66147-69-1 see ADP000
66215-27-8 see CQE400
66217-76-3 see MHB000
66230-04-4 see FAR150
66231-56-9 see CNH650
66232-25-5 see BHS250
66232-28-8 see BHT000

70103-81-0 see MEH000
70103-82-1 see MEI250
70114-87-3 see MKG275
70124-77-5 see COQ385
70132-29-5 see BJS600
70134-26-8 see DEU100
70142-16-4 see BMQ500
70145-54-9 see HLH000
70145-55-0 see DWZ100
70145-56-1 see HLU000
70145-60-7 see PEO750
70145-80-1 see AMJ625
70145-82-3 see MFJ000
70145-83-4 see MFD250
70161-10-3 see MBZ120
70193-21-4 see BPT300
70213-45-5 see ART125
70224-81-6 see MDH000
70244-12-1 see DXS300
70247-32-4 see COS825
70247-50-6 see SIL600
70247-51-7 see SDR400
70277-99-5 see BNE325
70278-00-1 see DEO500
70288-86-7 see ITD875
70299-48-8 see ENA500
70301-54-1 see DLV400
70301-64-3 see DLS000
70301-64-3 see MOG250
70301-68-7 see DLV600
70303-46-7 see EPR200
70303-47-8 see BSO200
70319-10-7 see AFI950
70324-20-8 see SDP025
70324-23-1 see DTY600
70324-35-5 see PLD730
70343-15-6 see DUX560
70348-66-2 see PNH550
70356-03-5 see CCR850
70384-29-1 see PCB000
70415-59-7 see NKU520
70441-82-6 see MBZ000
70441-84-8 see FNW100
70441-85-9 see MBY500
70443-38-8 see DNC800
70458-96-7 see BAB625
70476-82-3 see MQY100
70490-99-2 see ABL875
70501-82-5 see NLU480
70536-17-3 see AFI850
70548-53-7 see DBW200
70561-82-9 see DBW600
70601-54-6 see PFP840
70601-60-4 see PFP820
70601-68-2 see PFP780
70608-72-9 see HKG600
70648-26-9 see HCH400
70657-70-4 see MFL800
70664-49-2 see ASF625
70682-61-0 see TNI100
70701-62-1 see IDE400
70711-40-9 see BKB300
70711-41-0 see DBE875
70715-91-2 see ING300
70715-92-3 see ABR125
70729-68-9 see TCE375
70786-64-0 see DSY800
70786-72-0 see HKB650
70816-59-0 see TEF600
70851-61-5 see MIW060
70858-14-9 see VLU210
70907-61-8 see CPP750

70950-00-4 see GIT000
70951-83-6 see TCN300
70983-41-4 see ZQS100
70992-03-9 see TBG750
71002-67-0 see BRW500
71016-15-4 see MHW350
71030-37-0 see HKG650
71033-52-8 see PFP760
71033-59-5 see PFP720
71033-66-4 see PFP700
71033-69-7 see PFP740
71033-80-2 see LIU319
71033-82-4 see LIU315
71033-93-7 see LIU321
71033-95-9 see LIU323
71034-07-6 see LIU343
71048-99-2 see SEO550
71073-91-1 see ADW750
71108-02-6 see BJK670
71108-04-8 see DRV300
71108-05-9 see DSP700
71108-06-0 see DSP710
71108-08-2 see DXN820
71108-14-0 see DXN500
71108-21-9 see DST210
71250-00-5 see CDD325
71251-30-4 see MRE750
71267-22-6 see NJK300
71277-79-7 see DXD875
71292-84-7 see TCH325
71330-43-3 see BAD800
71359-62-1 see HAY059
71359-64-3 see HDC425
71373-14-3 see CCP950
71382-50-8 see BCV125
71392-29-5 see CDM500
71400-33-4 see DUV020
71400-34-5 see DUV030
71422-67-8 see CDS800
71435-43-3 see DKW400
71439-68-4 see BGW325
71441-09-3 see AQZ400
71441-28-6 see AQZ200
71441-30-0 see AQZ300
71447-49-9 see LIU380
71463-34-8 see AOC275
71500-21-5 see AGU400
71507-79-4 see EBY100
71522-58-2 see FOJ100
71561-11-0 see DES500
71598-10-2 see NLD000
71609-22-8 see HKB000
71615-27-5 see EDQ600
71628-96-1 see MCB600
71631-15-7 see CMS135
71653-63-9 see RSZ375
71677-48-0 see MMZ800
71697-59-1 see DFG300
71700-95-3 see VIZ500
71712-04-4 see BJK580
71731-58-3 see TGF075
71751-41-2 see ARW200
71752-66-4 see NJO500
71752-67-5 see NMA000
71752-68-6 see NLX000
71752-69-7 see NKO400
71752-70-0 see NKK000
71767-91-4 see HOS500
71771-90-9 see DAP850
71785-87-0 see NKC300
71799-98-9 see PPH100
71827-03-7 see DKR700

71839-05-9 see SOB550
71840-26-1 see MOB275
71850-03-8 see DAG700
71856-48-9 see MGS700
71859-30-8 see HCH485
71878-19-8 see PKL150
71901-54-7 see DTI709
71939-10-1 see PLR175
71963-77-4 see ARL425
71964-72-2 see DQK900
72007-81-9 see PFR300
72017-28-8 see NAY500
72017-60-8 see AQC000
72028-04-7 see IGO500
72040-09-6 see CGL125
72041-34-0 see DGW300
72044-13-4 see MDA100
72050-78-3 see DBE625
72064-79-0 see AAF625
72066-32-1 see MLF750
72074-66-9 see DMS400
72074-67-0 see DMS200
72074-68-1 see DMS600
72074-68-1 see ECS500
72074-69-2 see ECS000
72076-47-2 see AAI800
72079-76-6 see NIM640
72117-72-7 see COW780
72122-60-2 see MRR775
72150-17-5 see SLI200
72178-02-0 see CLS050
72207-94-4 see EQF100
72209-26-8 see AEJ250
72209-27-9 see AEK500
72214-01-8 see ELB400
72239-53-3 see FQB100
72254-58-1 see ALE750
72275-67-3 see FOL000
72299-02-6 see MIS500
72320-60-6 see EOS100
72378-89-3 see TGF000
72385-44-5 see TIT275
72432-14-5 see AOY270
72443-10-8 see FOS300
72479-13-1 see NKU370
72479-23-3 see NKE100
72485-26-8 see DLD000
72490-01-8 see EOE200
72492-12-7 see SLE880
72497-31-5 see PIA400
72505-63-6 see ENU500
72505-65-8 see IQS000
72505-66-9 see BSA000
72505-67-0 see NKN500
72558-82-8 see CCQ200
72571-82-5 see PII350
72586-67-5 see MFX725
72586-68-6 see DTN875
72589-96-9 see CAE375
72590-77-3 see HHQ850
72595-96-1 see DFG400
72595-97-2 see DFO400
72595-99-4 see DFO200
72596-00-0 see BDU000
72596-01-1 see DGO200
72596-02-2 see DFV000
72667-36-8 see ADQ500
72674-05-6 see AFN500
72676-73-4 see THH490
72676-74-5 see THH480
72676-77-8 see THH460
72676-78-9 see THH470

72704-51-9 see IGD075
72732-50-4 see DAD500
72738-89-7 see ADL250
72738-90-0 see MPL750
72738-98-8 see ADN500
72739-00-5 see ADK750
72812-40-9 see ISR100
72820-33-8 see DIL200
72841-24-8 see FAO050
72850-64-7 see BEG300
72869-73-9 see TAD250
72894-24-7 see EFC650
72917-35-2 see DSH500
72957-64-3 see DDN150
72977-18-5 see AIQ875
72985-54-7 see BJV625
72985-56-9 see CER825
73018-98-1 see ASG700
73032-94-7 see SKS650
73074-20-1 see AMI750
73080-51-0 see IHR200
73085-26-4 see BKM100
73090-70-7 see EBB700
73106-12-4 see AAK500
73118-22-6 see ACQ000
73118-23-7 see BKG500
73118-24-8 see BKG750
73121-56-9 see EAU200
73128-65-1 see MDF050
73160-32-4 see MQM150
73176-66-6 see DUE750
73180-00-4 see HKG680
73239-98-2 see NKO600
73245-91-7 see DAB630
73250-68-7 see BDG100
73263-81-7 see EIB600
73289-27-7 see BJL100
73297-70-8 see GJY100
73309-75-8 see AGW625
73341-53-4 see HIU550
73341-70-5 see GMC000
73341-73-8 see MAB400
73343-67-6 see AOY300
73343-69-8 see HNK800
73343-70-1 see HNK700
73343-71-2 see DHR900
73343-72-3 see DNF550
73343-74-5 see DDP300
73361-47-4 see IHH300
73372-63-1 see MRJ200
73379-85-8 see KHK100
73419-42-8 see CAK250
73452-31-0 see BLQ850
73452-32-1 see BSR825
73454-79-2 see MHW250
73459-03-7 see HLV955
73473-54-8 see CPX625
73487-24-8 see NJO200
73506-32-8 see HGW100
73506-39-5 see SGS000
73507-40-1 see MEL800
73526-98-4 see BJU350
73529-24-5 see PON750
73529-25-6 see EEF500
73561-96-3 see MBW780
73599-90-3 see CEE850
73599-91-4 see BMT130
73599-92-5 see HME100
73599-93-6 see HMK150
73599-94-7 see CEE825
73599-95-8 see BMT100
73622-67-0 see DBH800

73637-11-3 see CIF750	73926-87-1 see CET000	74137-66-9 see EDB150	75236-19-0 see EPC950
73637-16-8 see ACA750	73926-88-2 see CHL000	74137-68-1 see EDC520	75240-09-4 see AKB050
73639-62-0 see CIL775	73926-89-3 see CIC000	74139-77-8 see MRW080	75240-12-9 see AMG100
73651-49-7 see DKX800	73926-90-6 see TJU850	74193-14-9 see DHA200	75240-14-1 see AIV700
73663-86-2 see CLH100	73926-91-7 see DGK400	74195-73-6 see RGP450	75240-16-3 see ALY100
73665-15-3 see MRW500	73926-92-8 see DGK600	74203-42-2 see POV750	75240-20-9 see AKB070
73666-84-9 see TEF725	73926-94-0 see CLG000	74203-45-5 see NDF000	75240-22-1 see AKA650
73671-86-0 see DJP700	73927-60-3 see EHX000	74203-58-0 see PGA250	75318-62-6 see BGO325
73680-58-7 see AHB250	73927-86-3 see DEI400	74203-59-1 see MOD750	75318-64-8 see BGL450
73684-69-2 see MRW750	73927-87-4 see DWW400	74203-61-5 see SJK475	75318-65-9 see BGO200
73688-63-8 see APM250	73927-88-5 see BSP000	74220-04-5 see SBV500	75318-76-2 see CLN325
73688-85-4 see DPO275	73927-89-6 see TMV825	74222-97-2 see SNW550	75321-19-6 see TMN000
73693-97-7 see PIM750	73927-90-9 see QSJ800	74223-64-6 see MQR400	75321-20-9 see DVD400
73696-62-5 see ALG250	73927-91-0 see TID750	74252-25-8 see IDA100	75348-40-2 see DHQ600
73696-64-7 see BKU000	73927-92-1 see TNB500	74273-75-9 see DCM700	75348-49-1 see PPV750
73696-65-8 see ENH000	73927-93-2 see TIE000	74278-22-1 see KHU000	75389-89-8 see DJP100
73698-75-6 see AIK500	73927-94-3 see IEF000	74317-46-7 see MNA300	75410-87-6 see ICZ100
73698-76-7 see AIK750	73927-95-4 see TIE500	74339-98-3 see DKX875	75410-89-8 see BMK634
73698-77-8 see AIL000	73927-96-5 see QTJ000	74340-04-8 see DKX900	75411-83-5 see NKU500
73698-78-9 see AIL250	73927-97-6 see TDO000	74356-00-6 see CCS371	75442-13-6 see DAR900
73713-75-4 see EAN500	73927-98-7 see TIG000	74381-53-6 see LEZ300	75444-63-2 see YBS500
73728-78-6 see TDJ250	73927-99-8 see TNC000	74444-58-9 see DND000	75464-10-7 see PMW760
73728-79-7 see TLF000	73928-00-4 see TKT850	74444-59-0 see ECT000	75464-11-8 see BSY400
73728-82-2 see AKF250	73928-01-5 see CLE500	74465-36-4 see DMR400	75464-12-9 see DMX900
73747-22-5 see BEL550	73928-02-6 see MFP500	74465-38-6 see DMT000	75524-40-2 see CHJ625
73747-29-2 see AJO750	73928-03-7 see MPJ250	74465-39-7 see DMS800	75530-68-6 see NDY650
73747-51-0 see BFZ170	73928-04-8 see MPJ500	74469-00-4 see ARS125	75567-58-7 see NHP050
73747-53-2 see ICW100	73928-11-7 see DCJ000	74512-62-2 see PNR800	75625-24-0 see HCA650
73747-54-3 see MQE100	73928-12-8 see EJW600	74518-94-8 see NIM605	75659-26-6 see ACH300
73758-18-6 see TJJ750	73928-18-4 see TJS750	74518-95-9 see DRV700	75662-22-5 see DGW450
73758-56-2 see DEK400	73928-21-9 see TJT250	74548-80-4 see DVX400	75679-01-5 see AKS100
73771-13-8 see SJJ190	73940-79-1 see SJP500	74578-38-4 see UNJ810	75680-27-2 see NMV200
73771-52-5 see BJA000	73940-85-9 see DKC600	74578-69-1 see CCS588	75738-58-8 see CCS300
73771-72-9 see FJU000	73940-86-0 see TMI100	74634-56-3 see DKX930	75757-64-1 see SCD800
73771-73-0 see FJV000	73940-87-1 see BLS900	74686-30-9 see WBA700	75775-83-6 see ADP500
73771-74-1 see FJW000	73940-88-2 see TIE250	74738-17-3 see FAQ220	75841-84-8 see MBV735
73771-79-6 see DNC600	73940-89-3 see TIF750	74749-73-8 see DUO800	75867-00-4 see FAO220
73771-81-0 see QIS000	73940-90-6 see CHU750	74749-74-9 see GEQ450	75881-16-2 see NKU550
73785-34-9 see DBG200	73941-35-2 see THV500	74758-13-7 see DPI600	75881-17-3 see NJI900
73785-40-7 see NJS300	73944-13-5 see AIW800	74758-19-3 see BJH500	75881-19-5 see NKU580
73790-27-9 see MOK000	73953-53-4 see HHA100	74764-40-2 see IIG600	75881-20-8 see NKW800
73791-29-4 see AIY750	73954-17-3 see BCA375	74764-93-5 see MNU750	75881-22-0 see MMX200
73791-32-9 see TMA750	73963-72-1 see CMP825	74782-23-3 see OKS200	75884-37-6 see FQQ100
73791-39-6 see ALV100	73973-02-1 see PDB300	74789-25-6 see EMC100	75888-03-8 see FPO100
73791-40-9 see BRQ800	73986-52-4 see DFW700	74790-08-2 see SLE875	75889-62-2 see DIU500
73791-41-0 see DGA425	73986-95-5 see DGA400	74806-04-5 see CCK510	75965-74-1 see MFB400
73791-42-1 see CKD800	73987-00-5 see ACH375	74816-28-7 see CFM000	75965-75-2 see MFB410
73791-43-2 see IPS100	73987-16-3 see TLO600	74816-32-3 see CFU250	75993-65-6 see MPI800
73791-44-3 see MNU050	73987-51-6 see TGK500	74847-35-1 see PPQ650	76002-91-0 see ECE800
73791-45-4 see MOT800	73987-52-7 see EEP000	74861-59-9 see ANB700	76014-80-7 see OOO300
73803-48-2 see CHK825	73989-17-0 see ARW050	74886-24-1 see CNH525	76014-81-8 see MMS300
73806-23-2 see OMA000	73990-29-1 see AHI500	74920-78-8 see EKL250	76050-42-5 see PIB300
73806-49-2 see THJ750	74007-80-0 see OPE100	74926-97-9 see BRR500	76050-49-2 see FMR700
73815-11-9 see MEW800	74011-58-8 see EAU100	74926-98-0 see IQD000	76059-11-5 see DVS100
73816-43-0 see BKJ500	74037-18-6 see DEU125	74927-02-9 see BKQ750	76059-13-7 see DVS300
73816-74-7 see CHG250	74037-31-3 see ASA250	74938-11-7 see HKF350	76059-14-8 see DVS400
73816-75-8 see DGO000	74037-60-8 see BHB500	74940-23-1 see HIE600	76069-32-4 see TAB785
73816-77-0 see IBF000	74038-45-2 see DHH600	74940-26-4 see HIE700	76095-16-4 see EAO100
73825-59-9 see DAQ000	74038-78-1 see DSK300	74940-61-7 see DUI900	76123-46-1 see CAT685
73825-85-1 see SNS100	74039-01-3 see AIW750	74955-23-0 see HIE570	76145-76-1 see THG700
73825-87-3 see DMG100	74039-02-4 see AMO750	75016-34-1 see NKT100	76175-45-6 see TEY600
73826-58-1 see DGN600	74039-78-4 see BLT500	75016-36-3 see NKT090	76180-96-6 see AKT600
73834-77-2 see ALQ100	74039-79-5 see BLT750	75034-93-4 see DKQ650	76206-36-5 see NBI000
73840-42-3 see MOA250	74039-80-8 see BLU250	75038-71-0 see PLU550	76206-37-6 see NBM500
73870-33-4 see NLV500	74039-81-9 see BLU500	75084-21-8 see CEB800	76206-38-7 see BRY750
73873-83-3 see OPY000	74050-97-8 see HAG300	75084-22-9 see NFL200	76263-73-5 see POM700
73909-20-3 see TCQ550	74051-80-2 see CDK800	75084-25-2 see DEN300	76298-68-3 see ACY700
73926-79-1 see TGE750	74070-46-5 see CJD300	75157-06-1 see NEH700	76306-39-3 see DOG800
73926-80-4 see DDF000	74093-43-9 see SDM550	75184-71-3 see PNV900	76306-40-6 see DRS700
73926-81-5 see DDG600	74115-01-8 see CLH160	75198-31-1 see NGI800	76319-15-8 see PAM800
73926-83-7 see THU250	74137-64-7 see EDB025	75219-46-4 see BFV325	76379-66-3 see DXU250
73926-85-9 see BIV900	74137-65-8 see EDB020	75225-51-3 see LII050	76379-67-4 see OAX050

76429-97-5 see DWP950	77650-95-4 see DJX300	77966-77-9 see BRI500	78219-63-3 see PDJ250
76429-98-6 see DCM875	77680-87-6 see SED700	77966-79-1 see DDU200	78232-23-2 see HEG010
76487-32-6 see ALU875	77698-19-2 see NKL300	77966-80-4 see DHT000	78232-24-3 see HEG020
76487-65-5 see DVY875	77698-20-5 see NLE400	77966-81-5 see DIC400	78246-54-5 see HLX925
76541-72-5 see CMU875	77732-09-3 see MFC100	77966-82-6 see DHK800	78265-89-1 see CAB500
76547-98-3 see LGM400	77791-20-9 see DHN800	77966-83-7 see DIK800	78265-90-4 see CAB750
76549-16-1 see MRR900	77791-27-6 see DIN400	77966-84-8 see DNM600	78265-91-5 see CAB250
76556-13-3 see ANG625	77791-37-8 see END500	77966-85-9 see DTR800	78265-95-9 see HLR000
76578-14-8 see QMA100	77791-38-9 see EOU500	77966-90-6 see PIN000	78265-97-1 see PFF500
76615-66-2 see EEB200	77791-40-3 see MDK250	77966-93-9 see CGI750	78279-14-8 see AKO300
76631-42-0 see NMV480	77791-41-4 see MRT250	77967-05-6 see BGH500	78279-15-9 see AKN800
76648-01-6 see TAA400	77791-42-5 see MLS250	77967-24-9 see DHL200	78280-29-2 see DII600
76674-14-1 see BJW600	77791-43-6 see MLS750	77967-25-0 see DHL400	78280-31-6 see AMN000
76674-21-0 see FMR200	77791-53-8 see BPR000	77984-94-2 see CFR500	78281-06-8 see HLE650
76706-59-7 see ALL300	77791-55-0 see BQA500	77985-00-3 see BGF050	78302-38-2 see BNF300
76706-97-3 see CEV830	77791-57-2 see CFR250	77985-16-1 see CFN500	78302-39-3 see BNF310
76706-98-4 see CEV850	77791-58-3 see CFL500	77985-17-2 see CKU000	78308-32-4 see AQZ150
76706-99-5 see CEV840	77791-63-0 see DEY000	77985-21-8 see DHM400	78308-37-9 see BQH750
76714-88-0 see DGC100	77791-64-1 see DFF200	77985-23-0 see DHV200	78308-53-9 see MKS750
76738-28-8 see DFQ100	77791-66-3 see DIS000	77985-24-1 see DHV400	78313-59-4 see BKS800
76738-62-0 see TMK125	77791-67-4 see DIC000	77985-25-2 see DHV600	78329-75-6 see SCA550
76749-37-6 see ACS250	77791-69-6 see ONQ000	77985-27-4 see PIO000	78329-87-0 see BQN250
76790-18-6 see NBC500	77824-42-1 see AQE300	77985-28-5 see PIO250	78329-88-1 see BSI750
76790-19-7 see BPQ250	77824-43-2 see AQE320	77985-29-6 see PIO500	78329-97-2 see EIA000
76822-96-3 see VRU000	77824-44-3 see AQE305	77985-30-9 see PPW750	78330-02-6 see ELK000
76824-35-6 see FAB500	77846-96-9 see CKU250	77985-31-0 see PPX000	78336-03-5 see FJF200
76828-34-7 see ECJ100	77848-20-5 see THI300	77985-32-1 see PPX250	78338-31-5 see DTA690
76849-19-9 see PMN550	77855-81-3 see MQT600	78003-71-1 see SKS800	78338-32-6 see DTA700
76858-53-2 see MHX100	77879-90-4 see GEO200	78100-57-9 see VFP200	78343-32-5 see FJI500
77094-11-2 see AJQ600	77893-24-4 see MRJ700	78109-79-2 see BKC500	78350-94-4 see LHM750
77162-70-0 see NFW350	77922-38-4 see TME260	78109-80-5 see BPJ500	78354-52-6 see MBU775
77174-66-4 see CMW550	77944-89-9 see CIK500	78109-81-6 see BPJ750	78371-75-2 see XWS000
77182-82-2 see ANI800	77945-03-0 see DIA000	78109-87-2 see DII400	78371-84-3 see BIP500
77227-69-1 see CHJ400	77945-09-6 see DHX600	78109-88-3 see EEY000	78371-85-4 see BIP750
77234-90-3 see DAP880	77966-20-2 see BPJ250	78109-90-7 see PGE500	78371-90-1 see CLK500
77248-43-2 see HFU550	77966-25-7 see BPY500	78110-10-8 see CLB250	78371-91-2 see CLK750
77248-44-3 see XWS100	77966-26-8 see DHJ800	78110-23-3 see MDK000	78371-92-3 see CLL250
77248-45-4 see MLX830	77966-27-9 see DHK000	78110-37-9 see CEY250	78371-93-4 see CLL000
77251-47-9 see CPD625	77966-28-0 see PPU750	78110-38-0 see ARX875	78371-94-5 see CLL750
77251-49-1 see AEY406	77966-30-4 see AFW250	78111-17-8 see OHK100	78371-95-6 see CLL500
77255-40-4 see CML820	77966-31-5 see BDY250	78128-69-5 see BKD000	78371-96-7 see CLM000
77267-47-1 see DLH880	77966-32-6 see BEE750	78128-80-0 see BRO250	78371-97-8 see CLM500
77267-48-2 see IRC060	77966-34-8 see BEJ250	78128-81-1 see CPM250	78371-98-9 see CLM250
77267-49-3 see INE065	77966-38-2 see CEX500	78128-83-3 see MJF250	78371-99-0 see CLM750
77267-50-6 see MKM800	77966-40-6 see CFH000	78128-84-4 see HND000	78372-00-6 see CLN000
77267-52-8 see NAP525	77966-41-7 see CFI000	78128-85-5 see PGC750	78372-01-7 see CLN500
77267-59-5 see DLH860	77966-42-8 see CFL000	78168-93-1 see AJR200	78372-02-8 see CLO500
77267-60-8 see DLH820	77966-43-9 see CFR750	78173-90-7 see SON520	78372-03-9 see DIB000
77276-08-5 see INE062	77966-44-0 see CFS000	78173-91-8 see ARS135	78372-04-0 see DIJ200
77280-91-2 see HFK500	77966-45-1 see CFS250	78173-92-9 see ARS130	78372-05-1 see DPE800
77280-93-4 see NHE550	77966-46-2 see CFM250	78186-37-5 see MQO250	78372-06-2 see DPG400
77287-90-2 see COW675	77966-47-3 see CFM500	78186-39-7 see IRX050	78393-38-1 see CEP250
77314-23-9 see HIU600	77966-48-4 see CFM750	78186-61-5 see MLJ750	78393-39-2 see CLN250
77327-05-0 see DHA300	77966-49-5 see CFN000	78193-30-3 see SON525	78415-72-2 see MQU600
77337-54-3 see PNM650	77966-51-9 see CFU500	78194-09-9 see CIB650	78431-47-7 see POM710
77372-67-9 see ACE250	77966-52-0 see CFW250	78218-16-3 see CGY500	78432-19-6 see DVD300
77372-68-0 see BPQ000	77966-53-1 see CFW000	78218-37-8 see CGF500	78441-84-6 see GKW050
77402-03-0 see MGA400	77966-54-2 see CGC000	78218-38-9 see CIE500	78455-93-3 see NJO150
77405-29-9 see EIM100	77966-55-3 see CGD000	78218-40-3 see CKO000	78491-02-8 see IAS100
77430-23-0 see MPS300	77966-56-4 see CGE750	78218-42-5 see CIU250	78499-27-1 see DLI650
77439-76-0 see CFL100	77966-58-6 see CGN250	78218-43-6 see BQN500	78538-74-6 see MPA065
77458-01-6 see CKJ200	77966-59-7 see CGP380	78218-49-2 see TMU750	78587-05-0 see CJU275
77469-44-4 see BIF625	77966-61-1 see CHS750	78219-33-7 see EOY100	78600-25-6 see ABX800
77491-30-6 see SAD100	77966-62-2 see CID250	78219-35-9 see MOI250	78649-41-9 see IFS500
77492-93-4 see NMR300	77966-63-3 see CIT750	78219-38-2 see PFD500	78776-28-0 see DEC775
77500-04-0 see AJQ675	77966-67-7 see CKN750	78219-45-1 see PFD750	78822-08-9 see KDA000
77501-90-7 see FIW100	77966-68-8 see CKT750	78219-52-0 see PNT500	78831-88-6 see DWP900
77503-17-4 see FQQ400	77966-70-2 see DDM600	78219-53-1 see PNT750	78859-36-6 see ALF650
77523-56-9 see ACQ790	77966-71-3 see DHS800	78219-57-5 see EOV000	78907-15-0 see BQG850
77536-66-4 see ARM260	77966-72-4 see DIC200	78219-58-6 see EOW000	78907-16-1 see BPY625
77536-67-5 see ARM264	77966-73-5 see DIK600	78219-61-1 see MON000	78919-11-6 see DMS500
77536-68-6 see ARM280	77966-75-7 see BEJ500	78219-62-2 see PDJ000	78937-12-9 see CDE400

87188-51-0 see BPI400	90035-12-4 see CKA030	92662-79-8 see NGC570	96806-35-8 see NKJ100
87209-80-1 see MQY350	90035-14-6 see HOI245	92662-80-1 see NGC550	96811-96-0 see DPI750
87233-62-3 see EAL050	90043-86-0 see AKD775	92665-29-7 see CCS527	96827-85-9 see OKY300
87237-48-7 see HAG850	90045-36-6 see GEQ100	92760-57-1 see TEW000	96860-89-8 see DJX350
87414-49-1 see BSX150	90293-48-4 see DRL425	92768-71-3 see AEY410	97043-02-2 see EEK600
87425-02-3 see MGD100	90293-50-8 see MES550	92768-72-4 see AEY420	97170-07-5 see CML832
87625-62-5 see POI100	90293-52-0 see DOV890	92784-30-0 see BGW400	97170-07-5 see TCU750
87767-48-4 see DXI560	90293-53-1 see DOV860	92823-03-5 see FCN100	97170-08-6 see CML835
87773-08-8 see MJT025	90293-54-2 see DOV870	92880-08-5 see AEY415	97170-41-7 see HOH200
87818-31-3 see CMP955	90293-55-3 see DOV880	92880-10-9 see AEY430	97194-20-2 see DBV200
87820-88-0 see GJU200	90293-56-4 see BKO100	93023-34-8 see EIF450	97196-24-2 see XVJ000
88026-65-7 see SPD100	90320-57-3 see SKW800	93088-18-7 see VIK200	97205-35-1 see AKR100
88069-49-2 see PIF600	90350-40-6 see MOR600	93164-88-6 see AAD250	97629-28-2 see ALM140
88150-42-9 see AMY100	90370-29-9 see FQD140	93165-23-2 see VKA650	97676-32-9 see RHZ600
88193-04-8 see DRI900	90466-79-8 see BJA200	93165-85-6 see BAD300	97702-94-8 see DIB200
88208-15-5 see NLY750	90566-09-9 see BGS825	93195-54-1 see AEY425	97702-95-9 see DIB800
88208-16-6 see NJY500	90583-38-3 see SOP600	93251-89-9 see DGL700	97702-97-1 see DIE400
88254-07-3 see COP765	90584-32-0 see DFJ100	93333-82-5 see BOV810	97702-98-2 see DII800
88255-01-0 see NCP600	90729-15-0 see DWQ850	93334-51-1 see SLO100	97702-99-3 see DIJ000
88266-67-5 see LFJ500	90742-91-9 see DFM875	93348-17-5 see LGK375	97703-02-1 see DIK200
88283-41-4 see DGJ160	90842-58-3 see BPM690	93356-94-6 see IPL500	97703-04-3 see DIK400
88321-09-9 see OMY925	91100-28-6 see DDD900	93405-68-6 see DEY400	97703-08-7 see DIM800
88338-63-0 see CDM575	91216-69-2 see BHJ625	93407-11-5 see DPO100	97703-11-2 see DIN200
88485-37-4 see FMR600	91219-87-3 see DRV250	93431-23-3 see BLD325	97780-06-8 see MJW900
88598-54-3 see DMS550	91254-96-5 see NJA200	93572-42-0 see LGK375	97805-00-0 see VIA875
88671-89-0 see MRW775	91259-16-4 see NFI200	93673-39-3 see BAW150	97845-62-0 see POK100
88678-67-5 see POO800	91269-98-6 see HNK585	93763-70-3 see PCJ400	97864-38-5 see DFE000
88746-71-8 see ARP625	91297-11-9 see DBL300	93780-95-1 see DMS410	97886-45-8 see POQ600
88845-25-4 see AQZ100	91308-70-2 see NJJ950	93780-98-4 see DMS420	97901-22-9 see IAT300
88847-89-6 see HKA770	91308-71-3 see AGM125	93793-83-0 see HNT100	97919-22-7 see CMA600
88969-41-9 see DLX100	91315-15-0 see AFJ850	94110-08-4 see AJE350	97945-32-9 see BFY400
88973-46-0 see ONW200	91315-88-7 see MRN625	94168-98-6 see RJZ100	98046-72-1 see HIT610
89021-88-5 see GFW050	91336-54-8 see MKK500	94213-68-0 see SEY075	98046-73-2 see HIT620
89022-11-7 see CBT175	91423-46-0 see AFS100	94218-75-4 see IDG100	98114-60-4 see ELE720
89022-12-8 see MDY300	91441-48-4 see XDJ025	94238-00-3 see GBU750	98114-61-5 see HNV150
89194-77-4 see YGA700	91465-08-6 see LAS200	94361-06-5 see CQJ150	98114-62-6 see HMB565
89213-87-6 see HGL680	91480-86-3 see AKO400	94362-44-4 see DCQ650	98114-63-7 see ELE700
89365-50-4 see HNI600	91480-88-5 see AKO100	94408-08-9 see DEC795	98114-64-8 see HNV100
89367-14-6 see MMR800	91480-89-6 see AKC550	94421-68-8 see HKR600	98151-91-8 see NFI230
89367-15-7 see MMR810	91480-90-9 see AJO800	94593-91-6 see CMS127	98151-92-9 see BCS550
89367-92-0 see IGE100	91480-92-1 see AJM600	94764-55-3 see NFV710	98225-48-0 see BMM150
89398-07-2 see BFK370	91480-97-6 see HMN100	94855-74-0 see DSP800	98271-51-3 see MOL300
89565-68-4 see TNU100	91480-98-7 see MEX275	94857-19-9 see TDX320	98459-16-6 see CCW800
89591-51-5 see DUO300	91481-02-6 see AJI300	94944-79-3 see DIQ125	98477-03-3 see COP752
89784-60-1 see POK500	91481-03-7 see AKO450	94948-59-1 see HGL920	98565-18-5 see MID850
89837-93-4 see OOO050	91481-04-8 see DTO300	95004-22-1 see UAG025	98644-24-7 see ZAT050
89911-78-4 see NBR100	91503-79-6 see FJT100	95058-81-4 see GCK500	98717-16-9 see RMA600
89911-79-5 see NOC400	91513-55-2 see DEU150	95237-86-8 see OAP400	98730-04-2 see DEN500
89930-55-2 see OAD100	91513-56-3 see CGN400	95266-40-3 see EHU550	98819-68-2 see PFF350
89947-76-2 see HLE750	91524-15-1 see FLV050	95273-83-9 see VKZ100	99020-76-5 see NEM100
89957-52-8 see BFW010	91598-91-3 see HOB050	95282-98-7 see ABL625	99026-65-0 see HOB025
89958-12-3 see SAF300	91599-74-5 see BAV400	95465-99-9 see EHY100	99071-30-4 see HNK575
89970-80-9 see HJS910	91648-24-7 see AGQ050	95524-59-7 see EMY100	99152-10-0 see ADE100
89970-81-0 see HJS930	91674-71-4 see DEM400	95619-40-2 see DTS625	99422-01-2 see EPR100
89970-82-1 see POQ330	91682-96-1 see SLQ650	95719-24-7 see MRG060	99520-55-5 see EQN225
89970-83-2 see POQ300	91724-16-2 see SME500	95719-25-8 see PFR600	99520-58-8 see IBZ100
89970-84-3 see POQ310	91832-40-5 see AMS650	95719-26-9 see CPM300	99520-64-6 see DLK750
89970-85-4 see POQ320	91845-41-9 see PCR000	95770-03-9 see CGE250	99591-73-8 see DVO920
89985-01-3 see DGC050	92065-18-4 see AJD810	95840-06-5 see OKS630	99614-02-5 see OJD150
89997-47-7 see SAF000	92065-77-5 see AKY880	95860-07-4 see HLN125	99686-99-4 see MMP113
90000-16-1 see IBJ050	92065-82-2 see ALQ642	95896-78-9 see TLT768	99687-00-0 see MMP110
90015-77-3 see AAJ600	92065-83-3 see DPN500	95918-50-6 see CIM100	99711-15-6 see CFA250
90028-00-5 see DKL300	92065-85-5 see OOO500	96081-07-1 see DYB300	99759-48-5 see HBU410
90029-72-4 see CQM400	92065-91-3 see ALQ640	96231-64-0 see OJC000	99814-12-7 see SDY675
90029-73-5 see AEH300	92077-78-6 see CMP810	96525-23-4 see MGJ775	99832-61-8 see DKV710
90034-97-2 see TCU120	92078-92-7 see HMB560	96563-06-3 see TLS600	99834-93-2 see MFX650
90034-98-3 see TCU110	92113-31-0 see HID100	96741-20-7 see CML812	99900-94-4 see ASI100
90034-99-4 see HOI265	92129-95-8 see WBS730	96741-21-8 see CML814	99999-42-5 see CIC500
90035-01-1 see TCU100	92145-26-1 see MGU550	96741-21-8 see DLT640	100178-16-3 see THI252
90035-05-5 see TCU130	92180-79-5 see TLT771	96790-39-5 see CML831	100215-34-7 see NCT500
90035-06-6 see BNX045	92202-07-8 see PJH615	96790-40-8 see CML834	100242-24-8 see HNK600
90035-11-3 see BNW300	92456-72-9 see CCO700	96806-34-7 see NKJ050	100325-51-7 see DLB500

107538-05-6 see MBW778	112839-32-4 see FPQ050	119034-07-0 see OFE100	124617-84-1 see ACU130
107555-93-1 see PAT820	112885-41-3 see MRU253	119034-08-1 see NIV200	124617-85-2 see ABU100
107564-21-6 see DBT550	112945-52-5 see SCH002	119034-09-2 see NIV150	124737-31-1 see DOU100
107572-59-8 see BJE700	113124-69-9 see TCU660	119034-14-9 see PGH650	124824-14-2 scc MNW200
107638-88-0 see GFM250	113136-77-9 see COV133	119034-15-0 see PGH600	125276-72-4 see DCS821
107667-60-7 see ZEJ100	113698-18-3 see MEY100	119034-17-2 see CKJ775	125316-60-1 see HOL145
107746-52-1 see MEL100	113698-22-9 see MEV800	119034-18-3 see CKJ765	125317-39-7 see VLF400
107868-30-4 see MJL275	113698-48-9 see NGE130	119034-20-7 see BNX420	125930-50-9 see AGV880
107910-75-8 see GBU200	113698-50-3 see BKU150	119034-21-8 see BNX400	125974-08-5 see CFB700
108171-26-2 see PAH800	113779-16-1 see BCI265	119034-23-0 see SNL920	126268-14-2 see MPF300
108171-27-3 see PAH810	113852-37-2 see HNS550	119034-24-1 see SNL900	126426-74-2 see BKS780
108212-75-5 see CAY700	114088-58-3 see PHA725	119168-77-3 see CGH750	126463-66-9 see PIB650
108278-70-2 see NKO900	114119-92-5 see FEM050	119301-53-0 see DSG300	126833-17-8 see FAO250
108278-73-5 see MEH760	114369-43-6 see CKI790	119422-08-1 see BOS050	126936-19-4 see MFJ030
108278-74-6 see MEH100	114451-41-1 see NCQ530	119446-68-3 see THS900	126956-10-3 see MFJ010
108278-75-7 see MEI300	114654-31-8 see SNV050	119515-38-7 see MOT100	126983-60-6 see MFN550
108283-47-2 see ASE120	114704-98-2 see REF265	119823-35-7 see TNP512	126983-61-7 see MEX400
108319-06-8 see TAK850	114899-77-3 see EAE050	119823-36-8 see TNP515	127019-52-7 see CEO120
108320-78-1 see HMH050	114949-22-3 see AEC300	120042-13-9 see ASG300	127019-55-0 see NFL600
108320-79-2 see HMH100	114977-28-5 see TAH800	120162-55-2 see ASH275	127087-87-0 see NNC650
108354-47-8 see AKT650	115086-54-9 see CKW385	120373-24-2 see IRR050	127087-87-0 see NND200
108419-32-5 see AAT550	115362-13-5 see EBH900	120500-96-1 see EAV050	127277-53-6 see CPB065
108451-26-9 see DIU433	115436-72-1 see RLF400	120720-15-2 see GIA050	127311-83-5 see MKK800
108682-50-4 see TKB294	115566-02-4 see BLK600	120868-66-8 see CKW355	127380-62-5 see HDR800
108682-51-5 see TKB289	115609-71-7 see AMV770	120928-09-8 see FAK200	127502-68-5 see IQQ500
108682-53-7 see TKB287	115659-47-7 see PCG635	121010-10-4 see CPY850	127545-69-1 see ACS100
108682-54-8 see TKB296	115722-23-1 see NAE900	121153-48-8 see OKS620	127697-56-7 see FLZ080
108682-55-9 see TKB290	115722-24-2 see AQW110	121153-49-9 see OKS600	128202-32-4 see TDB300
108682-57-1 see TKB292	115722-25-3 see AQW100	121268-17-5 see SKL600	128202-33-5 see HOI222
108698-12-0 see TKB298	115970-17-7 see DKV720	121514-80-5 see TNJ400	128345-62-0 see RBF450
108778-72-9 see RCA500	116248-39-6 see MMP125	121776-33-8 see FPZ200	128639-02-1 see CCK520
108910-63-0 see AEY402	116255-48-2 see BNA920	122007-85-6 see PJB810	128758-36-1 see CJT800
108910-64-1 see AEY404	116316-70-2 see UVS600	122111-03-9 see GCK100	128758-37-2 see CKC600
108944-67-8 see NBW100	116355-83-0 see MAB500	122129-97-9 see DMX300	128999-90-6 see DEZ100
109293-97-2 see DKI300	116355-84-1 see FPA500	122130-63-6 see NJP300	129117-54-0 see DSM480
109322-04-5 see CHR200	116397-83-2 see MRW100	122185-09-5 see BLQ550	129134-95-8 see CEK400
109460-96-0 see MIQ300	116425-35-5 see AET600	122322-18-3 see CFL225	129134-96-9 see CQB300
109509-25-3 see CHR850	116580-64-4 see NBR900	122322-19-4 see CFC600	129217-90-9 see MRU752
109581-93-3 see TAA900	116788-55-7 see IDB200	122322-20-7 see BOC600	129286-36-8 see BCP530
109651-74-3 see CHR700	116897-98-4 see DWY100	122322-21-8 see BOC650	129286-37-9 see FGG900
109835-10-1 see HLX900	116929-00-1 see MIJ275	122322-22-9 see BNA350	129407-28-9 see TAI220
110011-81-9 see TLC850	116962-64-2 see AIX320	122322-23-0 see CJU150	129722-12-9 see AQW600
110068-86-5 see POH700	116962-65-3 see AIX300	122322-24-1 see CKW330	129927-33-9 see UTU700
110143-05-0 see DHA385	116962-66-4 see DNC300	122322-26-3 see BND775	130005-62-8 see AJQ300
110147-48-3 see MMA600	116978-88-2 see AJO550	122349-91-1 see AMV780	130063-43-3 see DMK633
110335-28-9 see CGP375	117559-89-4 see DOF430	122453-73-0 see BNA600	130175-14-3 see MRW785
110429-62-4 see CMV485	117568-24-8 see ACI640	122587-20-6 see DFT053	130209-82-4 see XAA500
110439-07-1 see DEC785	117629-85-3 see AOT260	122587-22-8 see MJB550	130525-77-8 see HKQ700
110516-60-4 see HND150	117718-60-2 see TEX550	122647-32-9 see IAB200	130536-39-9 see QNJ200
110553-27-0 see MHR025	117823-31-1 see FOJ033	122795-43-1 see GAD500	130607-26-0 see HGO550
110559-84-7 see NKE120	117900-35-3 see DNE900	122931-48-0 see DOM700	130773-02-3 see NCP525
110559-85-8 see NLE440	117906-15-7 see CKW370	123298-15-7 see MLX275	130892-66-9 see DEX270
110690-43-2 see BDN600	117929-12-1 see DLJ100	123298-28-2 see DTN125	130892-67-0 see TIL700
111041-98-6 see NGE100	117929-13-2 see TCQ400	123312-89-0 see DLV850	130933-92-5 see EBO990
111042-01-4 see DLU750	117929-14-3 see NFW220	123318-82-1 see CFJ200	131081-40-8 see SDI100
111070-55-4 see NGE160	117929-15-4 see DVD900	123333-56-2 see MEB900	131147-89-2 see DTM900
111070-57-6 see DWH900	118200-96-7 see ECC702	123497-99-4 see BAF850	131147-93-8 see DTK420
111109-77-4 see DWS900	118359-59-4 see GFO072	123533-90-4 see FPM200	131206-84-3 see CKW360
111479-05-1 see IPI400	118428-37-8 see PIG800	123577-49-1 see HLV960	131229-60-2 see ACU144
111639-04-4 see EDJ100	118428-38-9 see PIG804	123671-92-1 see TDB250	131229-61-3 see ACU148
111686-79-4 see RCK750	119033-84-0 see CKK700	123794-07-0 see PPT600	131229-62-4 see ACU136
111841-85-1 see IOF075	119033-85-1 see CKK600	123794-11-6 see EET200	131229-64-6 see ACU156
111858-68-5 see NLW600	119033-87-3 see MFJ020	123794-12-7 see HKW460	131543-22-1 see NAP545
111955-14-7 see NLC600	119033-91-9 see MFH955	123794-13-8 see MLP400	131748-56-6 see CKW380
111974-72-2 see QCJ300	119033-92-0 see MFH945	123941-02-6 see DTK400	131860-33-8 see ASP525
111988-43-3 see CKW335	119033-98-6 see CKJ770	124088-59-1 see BEM400	131929-60-7 see LEK500
111988-49-9 see CKW410	119033-99-7 see CKJ760	124149-19-5 see EAE700	131983-72-7 see TNP260
112309-61-2 see CIM325	119034-00-3 see MFH960	124428-11-1 see BGM070	132050-00-1 see TDR800
112309-62-3 see DFP550	119034-01-4 see MFH950	124439-07-2 see EAN900	132295-56-8 see CNI700
112410-23-8 see TAI175	119034-03-6 see BPQ330	124496-00-0 see CIR600	132295-57-9 see CNI800
112573-73-6 see SDY625	119034-04-7 see BPQ300	124533-50-2 see FLL200	132298-15-8 see DTY650
112725-15-2 see HMJ600	119034-06-9 see OFE200	124533-68-2 see FLZ090	132299-20-8 see MLS800

301644-22-4 see CJR215
301644-23-5 see CJR230
301644-24-6 see CJR220
301644-25-7 see CJR210
301644-26-8 see BNV754
301644-27-9 see BNV752
302542-40-1 see MNX300
302542-42-3 see IEI700
302542-44-5 see DSQ810
302542-49-0 see MNX310
302542-50-3 see DSQ830
302542-51-4 see IEI730
302542-57-0 see MNX320
302542-60-5 see DSQ840

302542-63-8 see IEI740
302561-65-5 see DTN150
302959-28-0 see FKK045
302959-30-4 see MEY900
302959-32-6 see DSU300
314238-30-7 see TLP300
314238-31-8 see TLT755
314238-32-9 see TLT763
314238-33-0 see TLT760
314238-34-1 see TLT757
314238-35-2 see TLT150
315706-65-1 see TLA650
315706-66-2 see CKU300
315706-67-3 see FLR050

315706-68-4 see BOB570
315706-69-5 see DGI630
315706-70-8 see PNH522
315706-71-9 see MOU820
315706-72-0 see NIY600
315706-73-1 see PNH533
315706-74-2 see CKJ125
315706-75-3 see FLL100
315706-76-4 see BNX330
315706-77-5 see DGE500
315706-78-6 see PGB850
315706-79-7 see MNW790
315706-80-0 see NIV050
315706-81-1 see PNH527

316172-58-4 see TLQ100
316172-59-5 see TLP800
326800-75-3 see DLJ050
326800-76-4 see FKK035
326800-79-7 see DLR700
326800-80-0 see DLI630
339152-94-2 see CFL120
339152-95-3 see CFB730
339152-96-4 see BNQ050
357210-17-4 see AKZ400
730771-71-0 see HKI075

Section 3
Synonym Cross-Index

1080 see SHG500
3A see SMQ500
A 00 see AGX000
11A see NHG000
A-19 see BGY140
A 21 see DMV600
A-36 see DAE600
A 42 see TED500
A65 see PJA140
A 66 see PMA750
A 71 see CHG000
A 95 see AGX000
A 99 see AGX000
9AA see AHS500
A 007 see BKI300
A-101 see CGA500
A-105 see BGY130
A-139 see BDC750
A 143 see TLC300
A-157 see AGM802
A 162 see MQD750
A 171 see TLD000
A 172 see TNJ500
A-20D see GLU000
A 260 see AQZ150
A-310 see ALF250
A 348 see PAH500
A 350 see AJH129
A 350 see DKR200
A 361 see ARQ725
A 363 see DOR400
A 3-80 see SMQ500
A 435 see SNQ550
A 446 see LFW300
A 468 see DLS800
A-502 see SNJ000
A 585 see PIK625
593-A see PIK075
688A see DDG800
A 688 see PDT250
A 820 see BQN600
883A see PHA550
A 884 see DBA800
A-980 see CEW500
A 995 see AGX000
A 999 see AGX000
D-90-A see CGL250
A 1100 see TJN000
A 1141 see GGS000
1212A see GLU000
A-1348 see PEC250
A 1390 see MNW150
A 1530 see AQF000
A 1582 see AQF000
A 1803 see BHJ250
A-1981 see DTO200
1A-4OA see AKE250
A 2275 see DBJ100
A 2297 see TLP750
A-2371 see MQW750

2814-A see NCP875
A 3322 see TEO250
A 3615 see AHK300
A 4077 see FMP000
A-4700 see EDK875
A-4760 see EDM000
A 4766 see DSM500
A-4828 see TNT500
4 4942 see IMH000
A 5089 see DSM500
A 5160 see XOS500
A 6413 see AAC000
72-A34 see BQN600
A-8103 see BHJ250
A-8506 see TFQ275
D-40TA see CKL250
A 10846 see DUD800
A-11025 see RDF000
A 11032 see UVA000
A-12223 see PHK000
A-12253 see TAL350
A 19120 see BEX500
A-27053 see CBR500
A-32686 see POB500
A 3733A see HAL000
A 3823A see MRE225
A 41304 see DBA875
A-45975 see TEF700
A-48257 see BOM600
A 5071A see AHK300
A 5072A see AHK300
A-56268 see MJV775
A 60-20R see PJS750
A 60-70R see PJS750
A 62254 see TAK850
A 71100 see DNC100
A 80198 see DAN388
A-91033 see DQA400
A 53930C see RAG400
734571A see DGB500
A 83543A see LEK500
A7301153 see HLF500
A93-29311 see LBF100
A 1 (sorbent) see AHE250
A 15 (polymer) see AAX175
A 100 (pharmaceutical) see IGS000
AA-9 see DXW200
AA/2 see XPJ300
AA-497 see TCU000
AA 1099 see AGX000
AA1199 see AGX000
AAB see PEI000
AAoC see AAY600
AACAPTAN see CBG000
AACIFEMINE see EDU500
A. AESTIVALIS see PCU375
AAF see FDR000
2-AAF see FDR000
AAFERTIS see FAS000
AA223 LEDERLE see AFI625
AALINDAN see BBQ500

AAMANGAN see MAS500
A. AMOREUXI VENOM see AOO250
A. AMURENSIS see PCU375
AAMX see ABA250
AAN see AAY000
A. ANNUA see PCU375
AAOT YELLOW see CMS210
AAP see AIB300
AAPROTECT see BJK500
AARANE see CNX825
AARARRE see CNX825
AAT see AIC250
AAT see PAK000
o-AAT see AIC250
AATACK see TFS350
AATP see PAK000
AATREX see ARQ725
AATREX 4L see ARQ725
AATREX 80W see ARQ725
AATREX NINE-O see ARQ725
A. AUSTRALIS HECTOR VENOM see AOO265
AAVOLEX see BJK500
AAZIRA see BJK500
2-AB see BPY000
AB-15 see ALV750
AB-42 see COH250
AB 109 see DVU300
AB 206 see MQU525
AB 35616 see CDQ250
1A-4B-20A see AIS700
ABACIL see CDT250
ABACIN see TKX000
ABACTRIM see TKX000
ABADOL see AMS250
ABADOLE see AMS250
ABAMECTIN see ARW200
ABAMN see ABM500
ABAR see LEN000
ABASIN see ACE000
ABATE see TAL250
ABATHION see TAL250
ABAVIT see PFP500
ABBOCILLIN see BDY669
ABBOCORT see HHQ800
ABBOLEXIN see PGG350
ABBOMEEN E-2 see DMT400
ABBOMEEN E-2 AEROSOL see DMT400
ABBOTT-22370 see TKX250
ABBOTT-28440 see DEW400
ABBOTT-30360 see FIW000
ABBOTT 30400 see PAP000
ABBOTT-35616 see CDQ250
ABBOTT 36581 see BOR350
ABBOTT 38414 see VGU700
ABBOTT 40566 see TCM250
ABBOTT-43326 see MQT550
ABBOTT 44090 see PNR750
ABBOTT-44747 see FOL000
ABBOTT-45975 see FPP100
ABBOTT-468 11 see CCS550

ABBOTT ANTIBIOTIC M259 see AAC000
ABBOTT'S A.P. 43 see DIF200
ABC 8/3 see OLM300
ABC 12/3 see TEQ175
ABCID see SNN300
ABC LANATOSIDE COMPLEX see LAU400
ABD see AID650
ABECARNIL see IOF075
A. BELLADONNA see AHI635
ABELMOSCHUS MANIHOT (Linn.) Medik., extract see HGA550
ABENSANIL see HIM000
ABEREL see VSK950
ABESON NAM see DXW200
ABESTA see RDK000
ABG 3034 see BDX090
ABG 6215 see EOE200
AB 50912 HEMIHYDROCHLORIDE see CCS300
ABICEL see CCU150
ABICOL see RDK000
ABIES ALBA OIL see AAC250
ABIES OIL see CBB900
ABIETIC ACID see AAC500
ABIETIC ACID, METHYL ESTER see MFT500
ABIGUANIL see AHO250
ABILIT see EPD500
ABIOL see HJL500
ABIROL see DAL300
"A" BLASTING POWDER see ERF500
ABMINTHIC see DJT800
ABN-Δ⁸-THC see HNE400
ABOL see DOX600
ABOVIS see AAC875
1-N-6-ABP see NET100
3-N-6-ABP see NET120
ABPH see DHL850
ABRACOL S.L.G see OAV000
ABRAMYCIN see TBX000
ABRAREX see AHE250
ABRIAL LAVANDIN OIL see LCA000
ABRICYCLINE see TBX000
ABRIN see AAD000
ABRINS see AAD000
ABRODEN see SHX000
ABRODIL see SHX000
ABROMA AUGUSTA Linn., root extract see UIS300
ABROMEEN E-25 see CPG125
ABROMINE see GHA050
ABROVAL see BNP750
ABRUS PRECATORIUS L., seed kernel extract see AAD100
ABRUS PRECATORIUS LINN., EXTRACT see JCA150
ABRUS PRECATORIUS L. (SEED) see RMK250
ABRUS PRECATORIUS OIL see AAD125
ABS see AFO500
ABS (pyrolysis products) see ADX750
ABSENTOL see TLP750
ABSETIL see TLP750
ABSIN see ACE000
ABSINTHIUM see ARL250
ABSOLUTE ETHANOL see EFU000
ABSOLUTE FRENCH ROSE see RMP000
ABSOLUTE MIMOSA see MQV000
ABSONAL see BDJ600
ABSONAL V see BDJ600
ABSORBABLE GELATIN SPONGE see PCU360
ABSTENSIL see DXH250
ABSTINYL see DXH250
ABTS see ASH375
ABURAMYCIN see OIS000
ABURAMYCIN B see CMK650
5-AC see ARY000
AC 8 see PJS750
AC 12 see AQZ150
AC 220 see MPP000
AC-223 see MHO200

AC 394 see PJS750
AC 680 see PJS750
AC-R-11 see BHJ500
AC-1075 see AQQ750
AC 1198 see PMO250
AC 1220 see PJS750
AC 1370 see CCS525
AC 220J see MPP000
AC 2770 see FND100
AC 2770 see PBG200
AC 3092 see BFA000
AC 3422 see EEH600
AC 5223 see DXX400
AC 5230 see ADA725
AC 12402 see BFY100
AC-12682 see DSP400
AC 18133 see EPC500
AC 18682 see IOT000
AC-18,737 see EAS000
AC 24055 see DUI000
AC 26,691 see CQL250
AC 38023 see FAB600
A-43064 see DXN900
AC 47031 see PGW750
AC 47470 see DHH400
AC 52160 see TAL250
AC 64475 see DHH200
AC 84777 see ARW000
AC 85258 see DFE469
AC 92553 see DRN200
AC 1692-40 see DKK100
AC 217300 see HGP525
AC 222705 see COQ385
AC 252214 see IAH100
AC 801757 see CGH750
AC 921000 see BSO000
ACABEL see CDG250
ACACIA see AQQ500
ACACIA DEALBATA GUM see AQQ500
ACACIA (EXTRACT) see AAD250
ACACIA FARNESIANA (Linn.) Willd., extract excluding roots see AAD500
ACACIA GUM see AQQ500
ACACIA MOLLISSIMA TANNIN see MQV250
ACACIA PALIDA (PUERTO RICO) see LED500
ACACIA SENEGAL see AQQ500
ACACIA SYRUP see AQQ500
ACACIA VILLOSA see AAD750
ACADYL see BJZ050
ACAJOU (HAITI) see MAK300
A. CALIFORNICA see HGL575
ACALMID see CKE750
ACALO see CKE750
ACAMOL see HIM000
ACAMYLOPHENINE see NOC000
ACAMYLOPHENINE DIHYDROCHLORIDE see AAD875
A. CANADENSIS see PAM780
ACANTHIFOLICIN, 9,10-DEEPITHIO-9,10-DIDEHYDRO- see OHK100
ACANTHIFOLICIN, 9,10-DEEPITHIO-9,10-DIDEHYDRO-35-METHYL- see MND300
ACANTHOPHIA ANTARCTICUS VENOM see ARU875
ACAPRIN see PJA120
ACAR see DER000
ACARABEN 4E see DER000
ACARAC see MJL250
ACARACIDE see SOP500
ACARALATE see PNH750
ACARBOSE see AAD900
ACARFLOR see CJU275
ACARICYDOL E 20 see CJT750
ACARIFLOR see CJU275
ACARIN see BIO750
ACARITHION see TNP250
ACAROL see IOS000
ACARON see CJJ250
ACAVYL see BJZ000
ACCEL see BEA100

ACCELERATE see DXD000
ACCELERATOR CDC see CNL300
ACCELERATOR CZ see CPI250
ACCELERATOR EFK see ZHA000
ACCELERATOR L see BJK500
ACCELERATOR OTOS see OPQ100
ACCELERATOR THIURAM see TFS350
ACCELERINE see DSY600
ACCEL R see BKU500
ACCENT see MRL500
ACCICURE HBS see CPI250
ACCLAIM see FAQ200
ACCOBOND 3524 see MCB050
ACCOBOND 3900 see MCB050
ACCOBOND 3903 see MCB050
ACCO FAST RED KB BASE see CLK225
ACCO NAPHTHOL AS see CMM760
ACCONEM see DHH200
ACCOSPERSE CYAN GREEN G see PJQ100
ACCOSPERSE TOLUIDINE RED XL see MMP100
ACCO SULFUR BORDEAUX A-CF see CMS257
ACCO SULFUR RED BROWN B-CF see CMS257
ACCO SULFUR RED BROWN R-CF see CMS257
ACCOTHION see DSQ000
ACCOTHION O-ANALOG see PHD750
ACCUCOL see PPN750
ACCUSAND see SCK600
ACCUZOLE see SNN500
ACD 7029 see MIB000
AC-DI-SOL NF see SFO500
ACDRILE see MBX800
8:9-ACE-1:2-BENZANTHRACENE see BAW000
4,10-ACE-1,2-BENZANTHRACENE see AAE000
ACEBUTOLOL see AAE100
(±)-ACEBUTOLOL see AAE100
dl-ACEBUTOLOL see AAE100
ACEBUTOLOL HYDROCHLORIDE see AAE125
ACECARBROMAL see ACE000
ACECLIDINE see AAE250
ACECLIDIN-HCL see QUS000
ACECOBARB see SBN400
ACECOLINE see ABO000
ACECOLINE see CMF250
ACEDAPSONE see SNY500
ACEDIST see BNV500
ACEDOXIN see DKL800
ACEDRON see BBK500
ACE-E 50 see EHP700
ACE-EE see EHP700
ACEF see CCS250
ACEFEN see AAE500
ACEGLUTAMIDE ALUMINUM see AGX125
ACEITE CHINO (CUBA) see TOA275
ACELAN, combustion products see ADX750
ACELYSIN see ARP125
ACEMETACIN see AAE625
ACENAPHTHALENE see AAE750
ACENAPHTHANTHRACENE see AAF000
5-ACENAPHTHENAMINE see AAF250
ACENAPHTHENE see AAF275
ACENAPHTHENEDIONE see AAF300
ACENAPHTHO(1,2-b)PHENANTHRENE see NAU530
13H-ACENAPHTHO(1,8-ab)PHENANTHRENE see DCR600
ACENAPHTHYLENE see AAF500
ACENAPHTHYLENE, 1,2-DIHYDRO- see AAF275
1,2-ACENAPHTHYLENEDIONE see AAF300
ACENIT see CGO550
ACENOCOUMARIN see ABF750
ACENOCOUMAROL see ABF750
ACENOCUMAROL see ABF750
ACENOKUMARIN see ABF750
ACENTERINE see ADA725

ACEOTHION see DSQ000
ACEPHAT (GERMAN) see DOP600
ACEPHATE see DOP600
ACEPHATE-MET see DTQ400
ACEPHENE see DPE000
ACEPRAMINE see AJD000
ACEPREVAL see AAF625
ACEPROMAZINA see ABH500
ACEPROMAZINE see ABH500
ACEPROMAZINE MALEATE see AAF750
ACEPROMETAZINE see AAF800
ACEPROMETHAZINE see AAF800
ACEPROMIZINA see ABH500
ACEPROSOL see AOO800
ACEPYRENE see CPX500
ACEPYRYLENE see CPX500
ACERDOL see CAV250
ACESAL see ADA725
ACESULFAME K see AAF900
ACESULFAME POTASSIUM see AAF900
ACETAAL (DUTCH) see AAG000
ACETACID RED B see HJF500
ACETACID RED J see FMU070
ACETACID RED 2BR see FAG020
3-ACETAET 4ᵝ-PROPANOATE
LEUCOMYCIN V see LEV025
ACETAGESIC see HIM000
ACETAL see AAG000
ACETAL see ADA725
ACETALDAZINE see AAG100
ACETALDEHYD (GERMAN) see AAG250
ACETALDEHYDAZINE see AAG100
ACETALDEHYDE see AAG250
ACETALDEHYDE, 4-(9-ACRIDINYL)-2-
METHYL-3-THIOSEMICARBAZONE see
MKA300
ACETALDEHYDE, AMINE SALT see
AAG500
ACETALDEHYDE AMMONIA see AAG500
ACETALDEHYDE, AZINE see AAG100
ACETALDEHYDE BIS(2-
METHOXYETHYL)ACETAL see AAG750
ACETALDEHYDE, CHLORO-, HYDRATE
see CDY600
ACETALDEHYDE DIBUTYL ACETAL see
DDT400
ACETALDEHYDE,(1,3-DIHYDRO-1,3,3-
TRIMETHYL-2H-INDOL-2-YLIDENE)- see
FBV050
ACETALDEHYDE DIMETHYL ACETAL see
DOO600
ACETALDEHYDE, ((3,7-DIMETHYL-2,6-
OCTADIENYL)OXY)-, (E)- see GDM100
ACETALDEHYDE, ((3,7-DIMETHYL-6-
OCTENYL)OXY)- see CMT300
ACETALDEHYDE, DIPROPYL ACETAL see
AAG850
ACETALDEHYDE-DI-n-PROPYL ACETAL
see AAG850
ACETALDEHYDE, 2-(2-
ETHOXYETHOXY)ETHYL 3,4-
(METHYLENEDIOXY)PHENYL ACETAL
see MJS550
ACETALDEHYDE ETHYL cis-3-HEXENYL
ACETAL see EKS100
ACETALDEHYDE ETHYL HEXYL ACETAL
see EKS120
ACETALDEHYDE,
ETHYLIDENEHYDRAZONE see AAG100
ACETALDEHYDE ETHYL LINALYL
ACETAL see ELZ050
ACETALDEHYDE ETHYL PHENETHYL
ACETAL see EES380
ACETALDEHYDE ETHYL 2-
PHENYLETHYL ACETAL see EES380
ACETALDEHYDE-N-FORMYL-N-
METHYLHYDRAZONE see AAH000
ACETALDEHYDE, (HEXYLOXY)-,
DIMETHYL ACETAL see HFG700
ACETALDEHYDE, HYDROXY- see GHO100
ACETALDEHYDE, METHOXY-(8CI,9CI) see
MDW250

ACETALDEHYDE-N-METHYL-N-
FORMYLHYDRAZONE see AAH000
ACETALDEHYDE METHYLHYDRAZONE
see AAH100
ACETALDEHYDE, N-
METHYLHYDRAZONE see AAH100
ACETALDEHYDE OXIME see AAH250
ACETALDEHYDE, PHENETHYL PROPYL
ACETAL see PDD400
ACETALDEHYDE SODIUM BISULFITE see
AAH500
ACETALDEHYDE SODIUM SULFITE see
AAH500
ACETALDEHYDE, TETRAMER see TDW500
ACETALDEHYDE, TRICHLORO-(9CI) see
CDN550
ACETALDEHYDE, TRIMER see PAI250
ACETAL DIETHYLIQUE (FRENCH) see
AAG000
ACETALDOL see AAH750
ACETALDOXIME see AAH250
ACETALE (ITALIAN) see AAG000
ACETALGIN see HIM000
ACETAL R see PDD400
ACETAMIDE see AAI000
ACETAMIDE, N-(2-ACETOXYETHYL)-N-
ETHYL- see ABN800
ACETAMIDE, N-(4'-(ACETYLAMINO)(1,1'-
BIPHENYL)-4-YL)-N-HYDROXY-(9CI) see
HKB000
ACETAMIDE, N-
((ACETYLAMINO)METHYL)-2-CHLORO-N-
(2,6-DIETHYLPHENYL)- see ABY300
ACETAMIDE, N-(ACETYLOXY)-N-(1,1'-
BIPHENYL)-4-YL-(9CI) see ABJ750
ACETAMIDE, N-9-ACRIDINYL- see ABX810
ACETAMIDE, N-(1-ADAMANTYL)- see
AEE100
ACETAMIDE, N-(2-(5-AMINO-7-BROMO-4-
CHLORO-2H-BENZOTRIAZOL-2-YL)-5-
(BIS(2-METHOXYETHYL)AMINO)-4-
METHOXYPHENYL)- see ABX833
ACETAMIDE, N-(2-(5-AMINO-7-BROMO-4-
CHLORO-2H-BENZOTRIAZOL-2-YL)-5-((2-
CYANOETHYL)ETHY LAMINO)-4-
METHOXYPHENYL)- see ABX836
ACETAMIDE, 2-AMINO-N-(2-(2,5-
DIMETHOXYPHENYL)-2-
HYDROXYETHYL)-,
MONOHYDROCHLORIDE, (±)- (9CI) see
MQT530
ACETAMIDE, 2-AMINO-N-(β-HYDROXY-
2,5-DIMETHOXYPHENETHYL)-,
MONOHYDROCHLORIDE, (±)- see MQT530
ACETAMIDE, 2-AMINO-N-(1-METHYL-1,2-
DIPHENYLETHYL)-
,MONOHYDROCHLORIDE, (+−)- see
RCK750
ACETAMIDE, N-((4-
AMINOPHENYL)SULFONYL)-,
MONOSODIUM SALT (9CI) see SNQ000
ACETAMIDE, 2-(2-
BENZOTHIAZOLYLOXY)-N-METHYL-N-
PHENYL- see BDG100
ACETAMIDE, N,N-BIS(2-
(ACETYLOXY)ETHYL)- see BGQ050
ACETAMIDE, N,N-BIS(2-CHLOROETHYL)-
see BHN300
ACETAMIDE, N,N-BIS(2-
HYDROXYETHYL)-, DIACETATE see
BGQ050
ACETAMIDE, N-(5-(BIS(2-
METHOXYETHYL)AMINO)-2-((2-BROMO-
4,6-DINITROPHENYL)AZO)-4-
METHOXYPHENYL)- see BNG300
ACETAMIDE, 2-BROMO- see BMR025
ACETAMIDE, N-(2-((2-BROMO-4,6-
DINITROPHENYL)AZO)-5-((2-
CYANOETHYL)ETHYLAMINO)-4-METHO
XYPHENYL)- see BNG310
ACETAMIDE, N-(3-BROMOPHENYL)- see
BMR050

ACETAMIDE, N-(4-BROMOPHENYL)- see
BMR100
ACETAMIDE, 2-CHLORO-N-(2-
CYANOETHYL)- see COM830
ACETAMIDE, 2-CHLORO-N-(2,6-DIETHYL-
4-HYDROXYPHENYL)- see CEC260
ACETAMIDE, 2-CHLORO-N-(2,6-DIETHYL-
4-OXO-2,5-CYCLOHEXADIEN-1-
YLIDENE)- see CEC270
ACETAMIDE, 2-CHLORO-N-(2,6-
DIETHYLPHENYL)-N-(2-
PROPOXYETHYL)- see PMB850
ACETAMIDE, 2-CHLORO-N-(2,6-
DIMETHYL-4-OXO-2,5-CYCLOHEXADIEN-
1-YLIDENE)- see CGH820
ACETAMIDE, 2-CHLORO-N-(2,6-
DIMETHYLPHENYL)-N-(2-
METHOXYETHYL)- see DSM500
ACETAMIDE, 2-CHLORO-N-(2,6-
DIMETHYLPHENYL)-N-((1-
METHYLETHOXY)METHYL)- (9CI) see
CGI200
ACETAMIDE-2-CHLORO-N-(2,6-
DIMETHYLPHENYL)-N-((2-
METHYLPROPOXY)METHYL)-(9CI) see
IIE200
ACETAMIDE, 2-CHLORO-N-(2,6-
DIMETHYLPHENYL)-N-(TETRAHYDRO-2-
OXO-3-FURANYL)- see CGI550
ACETAMIDE, 2-CHLORO-N-
(ETHOXYMETHYL)-N-(2-ETHYL-6-
METHYLPHENYL)- see CGO550
ACETAMIDE, 2-CHLORO-N-(2-ETHYL-6-
METHYLPHENYL)-N-(2-METHOXY-1-
METHYLETHYL)-, mixture with 2-((4-
CHLORO-6-(ETHYLAMINO)-1,3,5-TRIAZIN-
2-YL)AMINO)-2-
METHYLPROPANENITRILE see MQQ475
ACETAMIDE, 2-CHLORO-N-
HYDROXYMETHYL- see CHO775
ACETAMIDE, 2-CHLORO-N-
(ISOPROPOXYMETHYL)-N-(2,6-XYLYL)- see
CGI200
ACETAMIDE, 2-CHLORO-N-(2-METHOXY-
3,6-DIMETHYLPHENYL)-N-((1-
METHYLETHOXY)METHYL)- see CIB650
ACETAMIDE, 2-(p-CHLOROPHENOXY)-N-
(2-(DIMETHYLAMINO)ETHYL)- see CJN800
ACETAMIDE, 2-(4-CHLOROPHENOXY)-N-
(2-(DIMETHYLAMINO)ETHYL)-(9CI) see
CJN800
ACETAMIDE, 2-CHLORO-N-PHENYL- see
PER300
ACETAMIDE, 2-CYANO-N-
((ETHYLAMINO)CARBONYL)-2-
(METHOXYIMINO)- see COM300
ACETAMIDE, N,N'-(1,4-
CYCLOHEXYLENEDIMETHYLENE)BIS(2-
(1-AZIRIDINYL)- see CPL100
ACETAMIDE, DICHLORO- see DEM300
ACETAMIDE, 2,2-DICHLORO-N-(α-
(HYDROXYMETHYL)-p-
NITROPHENACYL)-, (+−)- see AAI110
ACETAMIDE,2,2-DICHLORO-N-(1-
(HYDROXYMETHYL)-2-(4-
NITROPHENYL)-2-OXOETHYL)- see
AAI110
ACETAMIDE,2,2-DICHLORO-N-(1-
(HYDROXYMETHYL)-2-(4-
NITROPHENYL)-2-OXOETHYL)-, (+−)- see
AAI110
ACETAMIDE, 2,2-DICHLORO-(8CI,9CI) see
DEM300
ACETAMIDE, N-(2,5-DICHLOROPHENYL)-
(9CI) see DGB480
ACETAMIDE, 2-(DIETHYLAMINO)-N-(1,3-
DIMETHYL-4-(o-FLUOROBENZOYL)-5-
PYRAZOLYL)-, MONOHYDROCHLORIDE
see AAI118
ACETAMIDE, N,N-DIETHYL-2-PHENOXY-
see RCZ200

ACETAMIDE, N,N-DIETHYL-2-PHENYL-
see PEU100

ACETAMIDE, N,N-DIETHYL-N'-(1,2,3,4-
TETRAHYDRO-1-NAPHTHYL)- see TGI725

ACETAMIDE, N-(4,5-DIHYDRO-4,5-DIOXO-
1-PYRENYL)- see AAM680

ACETAMIDE, N-(10,12-DIHYDRO-10-
OXOISOINDOLO(1,2-B)QUINAZOLIN-2-
YL)- see ACB300

ACETAMIDE, N-(2,7-
DIMETHYLDIPYRIDO(1,2-A:3',2'-
D)IMIDAZOL-4-YL)- see AAJ600

ACETAMIDE, N-(3,5-DIMETHYL-4-
HYDROXYPHENYL)- see DOO900

ACETAMIDE, N-(2,6-DIMETHYLPHENYL)-
2-METHOXY-N-(2-OXO-3-OXAZOLIDI
NYL)- see MFC100

ACETAMIDE, N,N-DIMETHYLTHIO- see
DUG450

ACETAMIDE, N-DODECYL- see LBR100

ACETAMIDE, N-ETHENYL-N-PHENYL- see
VLU220

ACETAMIDE, N-(2-ETHOXYPHENYL)-(9CI)
see ABG350

ACETAMIDE, 2-ETHOXY-N-(5,6,7,9-
TETRAHYDRO-10-(METHYLTHIO)-9-OXO-
1,2,3-TRIMETHOXYBENZO(a)HEPTALEN-
7-YL)-, (S)- see EEK600

ACETAMIDE, N-9H-FLUOREN-2-YL-N-
NITROSO- see FEM050

ACETAMIDE, N-(2-FLUOROBENZOYL)-
1,3-DIMETHYL-1H-PYRAZOL-5-YL)-2-((3-(2-
METHYL-1-
PIPERIDINYL)PROPYL)AMINO)-, (Z)-2-
BUTENEDIOATE (1:2) see AAI125

ACETAMIDE, N-(9-(p-
FLUOROPHENYLIMINO)FLUOREN-2-YL)-
see FLG200

ACETAMIDE, N-(HEXAHYDRO-2,4-
METHANO-4H-FURO(3,2-B)PYRROL-3-YL)-
N-METHYL-, (2R-(2-α,3-α,3A-β,4-α,6A-β))- see
ACP300

ACETAMIDE, N-(4-HYDROXY-3,5-
DIMETHYLPHENYL)- see DOO900

ACETAMIDE, 2,2'-((2-
HYDROXYETHYL)IMINO)BIS(N-(α,α-
DIMETHYLPHENETHYL)-N-METHYL-,
HYDROCHLORIDE see EAN650

ACETAMIDE, 2,2'-((2-
HYDROXYETHYL)IMINO)BIS(N-(1,1-
DIMETHYL-2-PHENYLETHYL)-N-
METHYL-, HCL see EAN650

ACETAMIDE, 2-HYDROXY-N-(2-
HYDROXY-1-(HYDROXYMETHYL)-2-(4-
NITROPHENYL)ETHYL)-, (R-(R*,R*))- see
CDP350

ACETAMIDE, 2-(HYDROXYIMINO)-N-
PHENYL-(9CI) see GIK100

ACETAMIDE, N-(4-((2-HYDROXY-5-
METHYLPHENYL)AZO)PHENYL)- see
AAQ250

ACETAMIDE, 2-(p-HYDROXYPHENYL)- see
HNG550

ACETAMIDE, N-(3-HYDROXY-1-
PYRENYL)- see HOB025

ACETAMIDE, N-(6-HYDROXY-1-
PYRENYL)- see HOB050

ACETAMIDE, N-(3-
ISOTHIOCYANATOPHENYL)-(9CI) see
ISG000

2-ACETAMIDE-4-MERCAPTOBUTYRIC
ACID γ-THIOLACTONE see TCZ000

ACETAMIDE, N-(2-(5-METHOXY-1H-
INDOL-4-YL)ETHYL)- see MES870

ACETAMIDE, N-(2-(2-METHOXY-1-
NAPHTHALENYL)-1-METHYLETHYL)- see
MFA355

ACETAMIDE, N-METHYL-N-(2,3,3A,6A-
TETRAHYDRO-2,4-METHANO-4H-
FURO(3,2-B)PYRROL-3-YL)-, (2R-(2-α,3-α,3A-
β,4-β,6A-β))- see DAK450

ACETAMIDE,N-(1,1A,3,3A,4,5,5,5A,5B,6-
DECACHLOROOCTAHYDRO-2-
HYDROXY-1,3,4-METHENO-1H-
CYCLOBUTA(CD)PENTALEN-2-YL)- see
AAI115

ACETAMIDE, N-(5-NITRO-2-THIAZOLYL)-
see ABY900

ACETAMIDE, N-PHENYL- see AAQ500

ACETAMIDE, N-PHENYL-N-VINYL- see
VLU220

ACETAMIDE, N-1-PYRENYL- see AAM650

ACETAMIDE, N,N'-2,6-PYRIDINEDIYLBIS-
see POR300

ACETAMIDE, N-6-QUINOLINYL- see
QPS200

ACETAMIDE, N-6-QUINOLYL- see QPS200

ACETAMIDE, N-SULFANILYL-, SODIUM
deriv see SNQ000

ACETAMIDE, N-SULFANILYL-, N-SODIUM
deriv see SNQ000

5-ACETAMIDE-1,3,4-THIADIAZOLE-2-
SULFONAMIDE see AAI250

ACETAMIDE, N-(2-THIAZOLYL)- see
TEW100

ACETAMIDE, α-TRICHLORO- see TII000

ACETAMIDE, 2,2,2-TRIFLUORO-N-(2-(2-
METHOXY-5-METHYLPHENYL)ETHYL)-
see MEX265

ACETAMIDE, 2,2,2-TRIFLUORO-N-(2-(2-
METHOXY-1-NAPHTHALENYL)-1-
METHYLETHYL)- see MFA375

ACETAMIDE, N-(2-
(TRIFLUOROMETHYL)PHENYL)-(9CI) see
TKB350

ACETAMIDINE, N,N'-DIPHENYL- see
DVW750

ACETAMIDINE HYDROCHLORIDE see
AAI500

3-ACETAMIDO-5-(ACETAMIDOMETHYL)-
2,4,6-TRIIODOBENZOIC ACID see AAI750

5-ACETAMIDO-1-
ACETOXYNAPHTHALENE-2-SULFONIC
ACID PYRIDINIUM SALT see AAI800

9-ACETAMIDOACRIDINE see ABX810

N-(p-(9-(3-
ACETAMIDOACRIDINYL)AMINO)PHENYL
)METHANESULFONAMIDE see AAJ000

3-ACETAMIDO-5-AMINO-2,4,6-
TRIIODOBENZOIC ACID see AAJ125

4-ACETAMIDOANILINE see AHQ250

p-ACETAMIDOANILINE see AHQ250

p-ACETAMIDOAZOBENZENE see PEH750

p-ACETAMIDOBENZALDEHYDE
THIOSEMICARBAZONE see FNF000

ACETAMIDOBENZENE see AAQ500

p-ACETAMIDOBENZENEARSONIC ACID,
SODIUM SALT, TETRAHYDRATE see
ARA000

p-ACETAMIDOBENZENESTIBONIC ACID
SODIUM SALT see SLP500

4'-ACETAMIDOBENZIDINE see ACC000

2-ACETAMIDOBENZOIC ACID see AAJ150

o-ACETAMIDOBENZOIC ACID see AAJ150

((p-ACETAMIDOBENZOYL)OXY)
TRIBUTYLSTANNANE see TIB750

ACETAMIDOBIPHENYL see PDY000

4-ACETAMIDOBIPHENYL see PDY500

N-(4'-
ACETAMIDOBIPHENYLYL)ACETOHYDRO
XAMIC ACID see HKB000

2-ACETAMIDO-4,5-BIS-
(ACETOXYMERCURI)THIAZOLE see
AAJ250

4-ACETAMIDOBUTYRIC ACID see AAJ350

4-ACETAMIDO-4-CARBOXAMIDO-n-(N-
NITROSO)BUTYLCYANAMIDE see AAJ500

3-ACETAMIDODIBENZFURANE see
DDC000

3-ACETAMIDODIBENZOFURAN see
DDC000

3-ACETAMIDODIBENZTHIOPHENE see
DDD600

3-ACETAMIDODIBENZTHIOPHENE
OXIDE see DDD800

7-ACETAMIDO-6,7-DIHYDRO-10-
HYDROXY-1,2,3-TRIMETHOXY-
BENZO(a)HEPTALEN-9(5H)-ONE see
ADE000

7-ACETAMIDO-6,7-DIHYDRO-1,2,3,10-
TETRAMETHOXY-BENZO(a)HEPTALEN-
9(5H)-ONE see CNG938

4-ACETAMIDO-2,7-
DIMETHYLDIPYRIDO(1,2-A:3',2'-
D)IMIDAZOLE see AAJ600

2-ACETAMIDO-4,5-DIMETHYLOXAZOLE
see AAJ750

2-ACETAMIDO-4,5-DIPHENYLOXAZOLE
see AAK000

ACETAMIDOETHANE see EFM000

2-ACETAMIDOETHANOL see HKM000

4-ACETAMIDO-2-ETHOLXBENZOIC ACID
METHYL ESTER see EEK100

1-ACETAMIDO-4-ETHOXYBENZENE see
ABG750

2,2'-((5-ACETAMIDO-2-
ETHOXYPHENYL)IMINO)DIETHANOL see
BKB000

4-(2-
ACETAMIDOETHYLDITHIO)BUTANESUL
FINATE SODIUM see AAK250

3-ACETAMIDOFLUORANTHENE see
AAK400

2-ACETAMIDOFLUORENE see FDR000

N-(2-ACETAMIDOFLUOREN-1-YL)-N-
FLUOREN-2-YL ACETAMIDE see AAK500

1-(N-ACETAMIDOFLUOROMETHYL)-
NAPHTHALENE see MME809

4-ACETAMIDO-2'-HYDROXY-5'-
METHYLAZOBENZENE see AAQ250

3-ACETAMIDO-4-HYDROXY-
PHENYLARSONIC ACID see ABX500

7-ACETAMIDO-10-HYDROXY-1,2,3-
TRIMETHOXY-6,7-
DIHYDROBENZO(a)HEPTALEN-9(5H)-
ONE see ADE000

ACETAMIDOMALONIC ACID DIETHYL
ESTER see AAK750

2-ACETAMIDO-4-MERCAPTOBUTYRIC
ACID THIOLACTONE see TCZ000

l-α-ACETAMIDO-β-
MERCAPTOPROPIONIC ACID see ACH000

2-(3-ACETAMIDO-5-N-METHYL-
ACETAMIDO-2,4,6-
TRIIODOBENZAMIDO)2-DEOXY-d-
GLUCOSE see MQR300

6-ACETAMIDO-4-METHYL-1,2-
DITHIOLO(4,3-B)PYRROL-5(4H)-ONE see
ABI250

7-ACETAMIDO-1-METHYL-4-(p-(p-((1-
METHYLPYRIDINIUM-4-
YL)AMINO)BENZAMIDO)
ANILINO)QUINOLINIUM DI-p-
TOLUENESULFONATE see AAL000

5-ACETAMIDO-1-METHYL-3-(5-NITRO-2-
FURYL)-s-TRIAZOLE see MMK000

3-ACETAMIDO-5-METHYLPYRROLIN-4-
ONE(4,3-D)-1,2-DITHIOLE see ABI250

2-ACETAMIDO-N-(3-METHYL-2-
THIAZOLIDINYLIDENE)ACETAMIDE see
AAL250

2-(2-ACETAMIDO-4-
METHYLVALERAMIDO)-N-(1-FORMYL-4-
GUANIDINOBUTYL)-4-
METHYLVALERAMIDE see LEX400

(S)-2-(2-ACETAMIDO-4-
METHYLVALERAMIDO)-N-(1-FORMYL-4-
GUANIDINOBUTYL)-4-METHYL-
VALERAMIDE see AAL300

4-ACETAMIDO-3-NITROANISOLE see
NEK100

p-ACETAMIDONITROBENZENE see
NEK000

5-ACETAMIDO-3-(5-NITRO-2-FURYL)-6H-
1,2,4-OXADIAZINE see AAL500

2-ACETAMIDO-4-(5-NITRO-2-FURYL)THIAZOLE see AAL750
2-ACETAMIDO-5-(NITROSOCYANAMIDO)VALERAMIDE see AAM000
4-ACETAMIDO-4-PENTEN-3-ONE-1-CARBOXAMIDE see PMC750
p-ACETAMIDOPHENACYL CHLORIDE see CEC000
3-ACETAMIDOPHENANTHRENE see PCY500
9-ACETAMIDOPHENANTHRENE see PCY750
2-ACETAMIDOPHENATHRENE see AAM250
2-ACETAMIDOPHENOL see HIL000
3-ACETAMIDOPHENOL see HIL500
4-ACETAMIDOPHENOL see HIM000
m-ACETAMIDOPHENOL see HIL500
o-ACETAMIDOPHENOL see HIL000
p-ACETAMIDOPHENOL see HIM000
1-(4-ACETAMIDOPHENOXY)-3-ISOPROPYLAMINO-2-PROPANOL see ECX100
4'-(ACETAMIDO)PHENYL-2-ACETOXYBENZOATE see SAN600
p-ACETAMIDOPHENYL ACETYLSALICYLATE see SAN600
1-(p-ACETAMIDOPHENYL)-3,3-DIMETHYLTRIAZENE see DUI000
3-ACETAMIDOPHENYL ISOTHIOCYANATE see ISG000
2-ACETAMIDO-4-PHENYLOXAZOLE see AAM500
4-ACETAMIDOPHENYLSULFONYL CHLORIDE see AAM600
p-ACETAMIDOPHENYLSULFONYL CHLORIDE see AAM600
1-ACETAMIDOPYRENE see AAM650
1-ACETAMIDOPYRENE-4,5-QUINONE see AAM680
1-ACETAMIDOPYREN-3-OL see HOB025
1-ACETAMIDOPYREN-6-OL see HOB050
4-ACETAMIDOPYRIDINE see AAM750
1-ACETAMIDO-2-PYRROLIDINONE see NNE400
6-ACETAMIDOQUINOLINE see QPS200
4-ACETAMIDOSTILBENE see SMR500
trans-4-ACETAMIDOSTILBENE see SMR500
2-ACETAMIDO-5-SULFONAMIDO-1,3,4-THIADIAZOLE see AAI250
ACETAMIDOTHIADIAZOLESULFONAMIDE see AAI250
2-ACETAMIDOTHIAZOLE see TEW100
α-ACETAMIDO-γ-THIOBUTYROLACTONE see TCZ000
3-ACETAMIDOTOLUENE see ABI750
p-ACETAMIDOTOLUENE see ABJ250
3-ACETAMIDO-2,4,6-TRIIODOBENZOIC ACID see AAM875
3-ACETAMIDO-2,4,6-TRIIODOBENZOIC ACID SODIUM SALT see AAN000
2-(3-ACETAMIDO-2,4,6-TRIIODO-5-(N-METHYLACETAMIDO)BENAZMIDO)-2-DEOXY-d-GLUCOPYRANOSE see MQR300
2-(3-ACETAMIDO-2,4,6-TRIIODO-5-(N-METHYLACETAMIDO)BENAZMIDO)-2-DEOXY-d-GLUCOSE see MQR300
2-((4-(3-ACETAMIDO-2,4,6-TRIIODOPHENOXY)BUTOXY)METHYL)BUTYRIC ACID SODIUM SALT see AAN250
2-(2-(3-ACETAMIDO-2,4,6-TRIIODOPHENOXY)ETHOXY)ACETIC ACID SODIUM SALT see AAN500
2-(2-(3-ACETAMIDO-2,4,6-TRIIODOPHENOXY)ETHOXY)BUTYRIC ACID SODIUM SALT see AAN750
2-((2-(3-ACETAMIDO-2,4,6-TRIIODOPHENOXY)ETHOXY)METHYL)PROPIONIC ACID SODIUM SALT see AAO000

2-(2-(3-ACETAMIDO-2,4,6-TRIIODOPHENOXY)ETHOXY)PROPIONIC ACID SODIUM SALT see AAO250
2-(2-(3-ACETAMIDO-2,4,6-TRIIODOPHENOXY)ETHOXY)-2-(o-TOLYL)ACETIC ACID SODIUM SALT see AAO500
2-(2-(3-ACETAMIDO-2,4,6-TRIIODOPHENOXY)ETHOXY)-2-(p-TOLYL)ACETIC ACID SODIUM SALT see AAO750
2-(2-(3-ACETAMIDO-2,4,6-TRIIODOPHENOXY)ETHOXY)VALERIC ACID SODIUM SALT see AAP000
2-((3-(3-ACETAMIDO-2,4,6-TRIIODOPHENOXY)PROPOXY)METHYL)BUTYRIC ACID SODIUM SALT see AAP250
2-(3-ACETAMIDO-2,4,6-TRIIODOPHENYL)BUTYRIC ACID see AAP500
2-(3-ACETAMIDO-2,4,6-TRIIODOPHENYL)PROPIONIC ACID see AAP750
2-(3-ACETAMIDO-2,4,6-TRIIODOPHENYL)VALERIC ACID see AAQ000
ACETAMIN see SNY500
ACETAMINE DIAZO BLACK RD see DCJ200
ACETAMINE RUBINE B see CMP080
ACETAMINE SCARLET B see ENP100
ACETAMINE YELLOW 2R see DUW500
ACETAMINE YELLOW CG see AAQ250
m-ACETAMINOANILINE see AHQ000
p-ACETAMINOBENZYLIDENETHIOSEMICARBAZONE see FNF000
4-ACETAMINOBEZENESULFON see SNQ600
3-ACETAMINODIBENZOTHIOPHENE see DDD600
ACET-o-AMINOFENOL (CZECH) see HIL000
p-ACETAMINOFENYL-2-HYDROXYETHYLSULFON see HLB500
p-ACETAMINOFENYL-β-HYDROXYETHYLSULFON see HLB500
2-ACETAMINOFLUORENE see FDR000
2-ACETAMINO-4-(5-NITRO-2-FURYL)THIAZOLE see AAL750
4-ACETAMINO-2-NITROPHENETOLE see NEL000
ACETAMINOPHEN see HIM000
2-ACETAMINOPHENANTHRENE see AAM250
3-ACETAMINOPHENANTHRENE see PCY500
9-ACETAMINOPHENANTHRENE see PCY750
2-ACETAMINOPHENOL see HIL000
p-ACETAMINOPHENOL see HIM000
1-(4-ACETAMINOPHENYL)-3,3-DIMETHYLTRIAZENE see DUI000
trans-4-ACETAMINOSTILBENE see SMR500
ACETAMOX see AAI250
2'-ACETANAPHTHONE, 5',6',7',8'-TETRAHYDRO-3,5',5',6',8',8'-HEXAMETHYL- see FBW110
ACETANHYDRIDE see AAX500
ACETANIL see AAQ500
ACETANILID see AAQ500
ACETANILIDE see AAQ500
ACETANILIDE, 2-ACETYL- see AAY000
ACETANILIDE, p-BROMO- see BMR100
ACETANILIDE, 3'-BROMO- see BMR050
ACETANILIDE, 4'-BROMO- see BMR100
ACETANILIDE, 2-CHLORO- see PER300
ACETANILIDE, 2-CHLORO-2',6'-DIMETHYL-N-ISOBUTOXYMETHYL- see IIE200
ACETANILIDE, 2-CHLORO-2',6'-DIMETHYL-N-(2-OXOTETRAHYDRO-3-FURYL)- see CGI550

ACETANILIDE, 2',5'-DICHLORO- see DGB480
ACETANILIDE, 2'-ETHOXY- see ABG350
ACETANILIDE, 4'-(2-HYDROXYETHYLSULFONYL)- see HLB500
ACETANILIDE, 4-PHENYL-, N-ACETATE (ester) see ABJ750
ACETANILIDE-4-SULFONIC ACID, 3-AMINO- see AHQ300
ACETANILIDE-p-SULFONYL CHLORIDE see AAM600
ACETANILIDE, 2-TRIFLUOROMETHYL- see TKB350
ACETANILIDE, N-VINYL-(8CI) see VLU220
m-ACETANISIDIDE see AAQ750
o-ACETANISIDIDE see AAR000
p-ACETANISIDIDE see AAR250
p-ACETANISIDIDE, 2-NITRO- see NEK100
ACETANISOLE (FCC) see MDW750
ACETARSIN see ACN250
ACETARSOL see ABX500
ACETARSONE see ABX500
ACETARSONE DIETHYLAMINE SALT see ACN250
ACETATE d'AMYLE (FRENCH) see AOD725
ACETATE-AS see HHQ800
ACETATE BLUE G see TBG700
ACETATE BRILLIANT BLUE 4B see MGG250
ACETATE de BUTYLE (FRENCH) see BPU750
ACETATE de BUTYLE SECONDAIRE (FRENCH) see BPV000
ACETATE C-7 see HBL000
ACETATE C-8 see OEG000
ACETATE C-11 see UMS000
ACETATE C-12 see DXV400
ACETATE de CELLOSOLVE (FRENCH) see EES400
21-ACETATECORTICOSTERONE see CNS650
ACETATE CORTISONE see CNS825
ACETATE de CUIVRE (FRENCH) see CNI250
16-ACETATE DIGITOXIN see ACH375
(Z)-ACETATE-9-DODECEN-1-OL see GJU050
ACETATE de l'ETHER MONOETHYLIQUE de l'ETHYLENE-GLYCOL (FRENCH) see EES400
ACETATE de l'ETHER MONOMETHYLIQUE de l'ETHYLENE-GLYCOL (FRENCH) see EJJ500
ACETATE d'ETHYLGLYCOL (FRENCH) see EES400
ACETATE FAST ORANGE R see AKP750
ACETATE FAST PINK 3B see DBX000
ACETATE FAST RED 2B see AKE250
ACETATE FAST RUBINE B see CMP080
ACETATE FAST SCARLET B see ENP100
ACETATE FAST YELLOW G see AAQ250
ACETATE FAST YELLOW 5RL see CMP090
ACETATE of 4-(HYDROXYPHENYL)-2-BUTANONE see AAR500
ACETATE d'ISOBUTYLE (FRENCH) see IIJ000
ACETATE d'ISOPROPYLE (FRENCH) see INE100
ACETATE of LIME see CAL750
ACETATE de METHYLE (FRENCH) see MFW100
ACETATE de METHYLE GLYCOL (FRENCH) see EJJ500
ACETATE P.A. see AGQ750
ACETATE PHENYLMERCURIQUE (FRENCH) see ABU500
ACETATE de PLOMB (FRENCH) see LCV000
1-ACETATE-1,2,3-PROPANETRIOL see GGO000
ACETATE de PROPYLE NORMAL (FRENCH) see PNC250
ACETATE RED VIOLET R see DBP000

ACETATE-REPLACING FACTOR see DXN800
ACETATE de TRIMETHYLPLOMB (FRENCH) see ABX125
ACETATE de TRIPHENYL-ETAIN (FRENCH) see ABX250
ACETATE de TRIPROPYLPLOMB (FRENCH) see ABX325
ACETATE TURQUOISE BLUE B see DMM400
ACETATE de VINYLE see VLU250
ACETATE YELLOW 6G see MEB750
ACETATO(2-AMINO-5-NITROPHENYL)MERCURY see ABQ250
(ACETATO)(p-AMINOPHENYL)MERCURY see ABQ000
(ACETATO)BIS(HEPTYLOXY)PHOSPHINYLMERCURY see AAR750
(ACETATO)BIS(HEXYLOXY)PHOSPHINYLMERCURY see AAS000
ACETATO di CELLOSOLVE (ITALIAN) see EES400
(ACETATO)(DIETHOXYPHOSPHINYL)MERCURY see AAS250
ACETATO(p-(DIETHYLAMINO)PHENYL)MERCURY see AAS300
ACETATO(p-(DIMETHYLAMINO)PHENYL)MERCURY see AAS310
ACETATO((5-HYDROXYMERCURI)-2-THIENYL)MERCURY see HLQ000
(ACETATO-o)HYDROXY-mu-2,5-THIOPHENEDIYLDIMERCURY see HLQ000
ACETATO(2-METHOXYETHYL)MERCURY see MEO750
ACETATO di METIL CELLOSOLVE (ITALIAN) see EJJ500
(ACETATO-O)ETHYLMERCURY see EMD000
ACETATOPHENYLMERCURATE(1-) AMMONIUM SALT see PFO550
(ACETATO)PHENYLMERCURY see ABU500
(ACETATO)PHENYL-MERCURY mixed with CHLOROETHYL-MERCURY (19:1) see PFO500
ACETATO di STAGNO TRIFENILE (ITALIAN) see ABX250
(ACETATO)(2,3,5,6-TETRAMETHYLPHENYL)MERCURY see AAS500
(ACETATO)(8-THEOPHYLLINYL)MERCURY see TEP750
(ACETATO)(TRIMETAARSENITO)DICOPPER see COF500
ACETATOTRIPHENYLSTANNANE see ABX250
ACETAZINE see ABH500
ACETAZOLAMID see AAI250
ACETAZOLAMIDE see AAI250
ACETAZOLAMIDE SODIUM see AAS750
ACETAZOLAMIDE SODIUM SALT see AAS750
ACETAZOLEAMIDE see AAI250
ACETCARBROMAL see ACE000
ACETDIMETHYLAMIDE see DOO800
ACETDRON see AOB250
ACETEIN see ACH000
ACETENE see EIO000
ACETETHYLANILIDE see EFQ500
ACETEUGENOL see EQS000
ACETHION see DIX000
ACETHION AMIDE see AAT000
ACETHIONE see DIX000
ACETHROPAN see AES650
ACETHYDRAZIDE see ACM750
ACETHYDROXAMSAEURE (GERMAN) see ABB250
ACETHYLPROMAZIN see ABH500
ACETIC ACID see AAT250
ACETIC ACID (aqueous solution) (DOT) see AAT250

ACETIC ACID-4-((ACETOXYMETHYL)NITROSAMINO)BUTYL ESTER see ABM500
ACETIC ACID-2-((ACETOXYMETHYL)NITROSAMINO)ETHYL ESTER see ABN750
ACETIC ACID-4-((ACETOXYMETHYL)NITROSAMINO)PROPYL ESTER see ABV750
ACETIC ACID, (N-ACETYL-N-(4-BIPHENYL)AMINO) ESTER see ABJ750
ACETIC ACID (N-ACETYL-N-(2-FLUORENYL)AMINO) ESTER see ABL000
ACETIC ACID(N-ACETYL-N-(4-FLUORENYL)AMINO)ESTER see ABO500
ACETIC ACID (N-ACETYL-N-(2-PHENANTHRYL)AMINO)ESTER see ABK250
ACETIC ACID-(N-ACETYL-N-(p-STYRYLPHENYL)AMINO) ESTER see ABW500
ACETIC ACID, (ACETYLTHIO)-(9CI) see ACQ250
ACETIC ACID ALLYL ESTER see AFU750
ACETIC ACID-3-ALLYLOXYALLYL ESTER see AAT500
ACETIC ACID, 4-ALLYLOXY-3-CHLOROPHENYL-, compounded with 2-AMINOETHANOL see MDI225
ACETIC ACID AMIDE see AAI000
ACETIC ACID, AMINO-sec-BUTYL- see IKX000
ACETIC ACID, 2,2'-(((4-(AMINOCARBONYL)PHENYL)ARSINIDENE)BIS(THIO))BIS-(9CI) see TFA350
ACETIC ACID, 2,2'-((4-(AMINOCARBONYL)PHENYL)ARSINIDENE)BIS(THIO)BIS-, DISODIUM SALT see SKH100
ACETIC ACID, AMINOOXO-, HYDRAZIDE see OLM310
ACETIC ACID, (p-AMINOPHENYL)- see AID700
ACETIC ACID, AMMONIUM SALT see ANA000
ACETIC ACID, AMYL ESTER see AOD725
ACETIC ACID, AMYL ESTER see AOD750
ACETIC ACID, ANHYDRIDE (9CI) see AAX500
ACETIC ACID, ANHYDRIDE with NITRIC ACID (1:1) see ACS750
ACETIC ACID, ANHYDRIDE WITH DIETHYL PHOSPHATE see MJD100
ACETIC ACID ANILIDE see AAQ500
ACETIC ACID, 2-(1-AZIRIDINYL)-, METHYL ESTER see MGR900
ACETIC ACID, BARIUM SALT see BAH500
ACETIC ACID, BENZ(a)ANTHRACENE-7,12-DIMETHANOL DIESTER see BBF750
ACETIC ACID, BENZ(a)ANTHRACENE-7-METHANOL ESTER see BBH500
ACETIC ACID, BENZOYL-, ETHYL ESTER see EGR600
ACETIC ACID BENZYL ESTER see BDX000
ACETIC ACID, BIS(4-CHLOROPHENOXY)-, 1-METHYL-4-PIPERIDINYL ESTER see MON600
ACETIC ACID, BIS(p-CHLOROPHENOXY)-, 1-METHYL-4-PIPERIDYL ESTER see MON600
ACETIC ACID, (((3,5-BIS(1,1-DIMETHYLETHYL)-4-HYDROXYPHENYL)METHYL)THIO)-, 2-ETHYLHEXYL ESTER see BJK600
ACETIC ACID, 2,2'-((BIS(PHENYLMETHYL)STANNYLENE)BIS(THIO))BIS-, DIISOOCTYL ESTER (9CI) see EDC560
ACETIC ACID, ((3,5-BIS(TRIFLUOROMETHYL)PHENYL)HYDRAZONO)CYANO-, METHYLESTER see COQ376

ACETIC ACID, compd. with BORON FLUORIDE (BF3) (8CI) see BMG750
ACETIC ACID, BROMO-, 2-BUTENE-1,4-DIYL ESTER see BHD150
ACETIC ACID, 2-BROMO-2-CYANO, ETHYL ESTER see EGV450
ACETIC ACID, BROMO-, 1-METHOXYISOPROPYL ESTER see MES930
ACETIC ACID, (2-BROMOPROPIONAMIDO)-, ETHYL ESTER see EGV550
ACETIC ACID-1,3-BUTADIENYL ESTER see ABM250
ACETIC ACID-2-BUTOXY ESTER see BPV000
ACETIC ACID-2-(tert-BUTYL)-4,6-DINITRO-m-TOLYL ESTER see BRU750
ACETIC ACID n-BUTYL ESTER see BPU750
ACETIC ACID-tert-BUTYL ESTER see BPV100
ACETIC ACID, 3-BUTYL-5-METHYL-TETRAHYDRO-2H-PYRAN-4-YL ESTER see MHX000
ACETIC ACID (sec-BUTYLNITROSAMINOMETHYL) ESTER see BPV500
ACETIC ACID-1-(BUTYLNITROSOAMINO)BUTYL ESTER see BPV325
ACETIC ACID, CADMIUM SALT see CAD250
ACETIC ACID, CADMIUM SALT, DIHYDRATE see CAD275
ACETIC ACID, CALCIUM MAGNESIUM SALT see CAT685
ACETIC ACID, CALCIUM STRONTIUM SALT (4:1:1) see CAX400
ACETIC ACID, C$_{7-9}$-BRANCHED ALKYL ESTERS, C$_8$-rich see AAT550
ACETIC ACID, CEDROL ESTER see CCR250
ACETIC ACID CHLORIDE see ACF750
ACETIC ACID, CHLORO-, tert-BUTYL ESTER see BQR100
ACETIC ACID, CHLORO-, 1,1-DIMETHYLETHYL ESTER see BQR100
ACETIC ACID, CHLORO-, ESTER WITH DIMETHYL(2,2,2-TRICHLORO-1-HYDROXYETHYL)PHOSPHONATE see DUI900
ACETIC ACID, CHLORO-, 2-(2-HYDROXYETHOXY)ETHYL ESTER see HKI600
ACETIC ACID, CHLORO-, 1-METHOXYISOPROPYL ESTER see MES940
ACETIC ACID, (4-CHLORO-2-METHYLPHENOXY)-, POTASSIUM SALT see MIH800
ACETIC ACID, ((4-CHLORO-2-METHYL)PHENYL)THIO- see CLO600
ACETIC ACID, 2-(p-CHLOROPHENYL)-2-ISOPROPYL-, CYANO(p-PHENOXYPHENYL)METHYL ESTER see COQ387
ACETIC ACID, CHLORO-, 3-PHENYLPROPYL ESTER see PGB770
ACETIC ACID, 3-CHLOROPROPYLENE ESTER see CIL900
ACETIC ACID, ((4-CHLORO-o-TOLYL)OXY)-, POTASSIUM SALT see MIH800
ACETIC ACID, CHLORO-, 2,2,2-TRICHLORO-1-(DIMETHOXYPHOSPHINYL)ETHYL ESTER see DUI900
ACETIC ACID, CHROMIUM (2+) SALT (8CI, 9CI) see CMJ000
ACETIC ACID, CINNAMYL ESTER see CMQ730
ACETIC ACID, CITRONELLYL ESTER see AAU000
ACETIC ACID, COBALT(2+) SALT see CNC000

ACETIC ACID, COBALT(3+) SALT see CNA300

ACETIC ACID, COBALT(2+) SALT, TETRAHYDRATE see CNA500

ACETIC ACID, CUPRIC SALT see CNI250

ACETIC ACID, CYANO-, ALLYL ESTER see AGC150

ACETIC ACID, CYANO-, ISOBUTYL ESTER see IIQ100

ACETIC ACID, CYANO-, 2-METHYLPROPYL ESTER see IIQ100

ACETIC ACID, 2-CYANO-2-OXO-, METHYL ESTER,2-(3,5-BIS(TRIFLUOROMETHYL)PHENYL)HYDRAZONE see COQ376

ACETIC ACID, CYANO-, 2-PROPENYL ESTER see AGC150

ACETIC ACID, 1-CYANO-2,2,2-TRIFLUORO-1-METHYLETHYLESTER see ABR700

ACETIC ACID, CYANO-, TRIPHENYLSTANNYL ESTER see TMV825

ACETIC ACID, (1,2-CYCLOHEXYLENEDINITRILO)TETRA- see CPB120

ACETIC ACID, CYCLOHEXYLETHYL ESTER see EHS000

ACETIC ACID, (1,2-CYCLOHEXYLIDENEDINITRILO)TETRA-, CADMIUM COMPLEX, TRANS- see CAD900

ACETIC ACID, 9-DECENYL ESTER see DAI450

ACETIC ACID, DIAMINOPROPANOLTETRA- see DCB100

ACETIC ACID, ((DIBENZYLSTANNYLENE)DITHIO)DI-, DIISOOCTYL ESTER see EDC560

ACETIC ACID, DIBROMO- see DDJ100

ACETIC ACID, 2,2'-((DIBUTYLSTANNYLENE)BIS(THIO))BIS-, DINONYL ESTER see DEH650

ACETIC ACID, ((DIBUTYLSTANNYLENE)DITHIO)DI-, DINONYL ESTER (8CI) see DEH650

ACETIC ACID, DICHLORO-, ANHYDRIDE WITH DIETHYLHYDROGEN PHOSPHATE see DEM400

ACETIC ACID, (3,4-DICHLOROPHENOXY)-see DFY500

ACETIC ACID, 2,4-DICHLOROPHENOXY-, 4-CHLORO-2-BUTENYL ESTER see KHU025

ACETIC ACID, (2,4-DICHLOROPHENOXY)-, ETHYL ESTER see EHY600

ACETIC ACID, (2,4-DICHLOROPHENOXY)-, 2-ETHYLHEXYLESTER see DBB480

ACETIC ACID, (2,4-DICHLOROPHENOXY)-, METHYL ESTER see DAA825

ACETIC ACID, (2,4-DICHLOROPHENOXY)-, MIXT. WITH N-ETHYL-N'-(1-METHYLETHYL)-6-(METHYLTHIO)-1,3,5-TRIAZINE-2,4-DIAMINE see AHK300

ACETIC ACID, 2,4-DICHLOROPHENOXY-, OCTYL ESTER see OEU050

ACETIC ACID, (DIETHOXYPHOSPHINYL)-, ETHYL ESTER (9CI) see EIC000

ACETIC ACID, (DIETHYLAMINO)OXO-, HEPTYL ESTER see HBN700

ACETIC ACID, (DIETHYLAMINO)OXO-, HEXYL ESTER see HFP650

ACETIC ACID, (DIETHYLAMINO)OXO-, NONYL ESTER see NNB600

ACETIC ACID, (DIETHYLAMINO)OXO-, OCTYL ESTER see OEU100

ACETIC ACID, ((2,3-DIHYDRO-6,7-DICHLORO-2-METHYL-1-OXO-2-PHENYL-1H-INDEN-5-YL)OXY)-, (±)- see IBQ400

ACETIC ACID DIHYDROTERPINYL ESTER see DME400

ACETIC ACID, (((3,4-DIHYDROXY-2-ANTHRAQUINONYL)METHYL)IMINO)DI-see AFM400

ACETIC ACID, (3,4-DIHYDROXYPHENYL)-see HGI980

ACETIC ACID, (3,4-DIMETHOXYPHENYL)-see HGK600

ACETIC ACID, DIMETHYL- see IJU000

ACETIC ACID DIMETHYLAMIDE see DOO800

ACETIC ACID, 2-(DIMETHYLAMINO)ETHYL ESTER see DOZ100

ACETIC ACID-α-(DIMETHYLAMINOMETHYL)BENZYL ESTER see ABN700

ACETIC ACID-1,3-DIMETHYLBUTYL ESTER see HFJ000

ACETIC ACID-2,6-DIMETHYL-m-DIOXAN-4-YL ESTER see ABC250

ACETIC ACID-1,1-DIMETHYLETHYL ESTER see BPV100

ACETIC ACID-3,7-DIMETHYL-6-OCTEN-1-YL ESTER see AAU000

ACETIC ACID, (2,4-DINITRO-6-s-BUTYLPHENYL) ESTER see ACE500

ACETIC ACID, (4,6-DINITRO-2-s-BUTYLPHENYL) ESTER see ACE500

ACETIC ACID-4,6-DINITRO-o-CRESYL ESTER see AAU250

ACETIC ACID, ((2,5-DIOXO-3-CYCLOHEXEN-1-YL)THIO)- see DVR250

ACETIC ACID, DIPHENYL-, 2-(BIS(2-HYDROXYETHYL)AMINO)ETHYL ESTER, HYDROCHLORIDE see DVW900

ACETIC ACID, 9-DODECEN-1-YL ESTER, (E)- see DXU822

ACETIC ACID, DODECYL ESTER see DXV400

ACETIC ACID, ESTER with N-4-BIPHENYLYLACETOHYDROXAMIC ACID see ABJ750

ACETIC ACID ESTER with N-(FLUOREN-3-YL)ACETOHYDROXAMIC ACID see ABO250

ACETIC ACID, ESTER with N-(FLUOREN-4-YL)ACETOXYHYDROXAMIC ACID see ABO500

ACETIC ACID, ESTER with N-NITROSO-2,2'-IMINODIETHANOL see NKM500

ACETIC ACID ESTER with N-(2-PHENANTHRYL)ACETOHYDROXAMIC ACID see ABK250

ACETIC ACID-ESTER with N-(p-STYRYLPHENYL)ACETOHYDROXAMIC ACID see ABW500

ACETIC ACID, ESTER with N-(p-STYRYLPHENYL)ACETOHYDROXAMIC ACID see SMT000

ACETIC ACID, ESTER WITH DIMETHYL (2,2,2-TRICHLORO-1-HYDROXYETHYL)PHOSPHONATE see CDN510

ACETIC ACID, ETHENYL ESTER see VLU250

ACETIC ACID ETHENYL ESTER HOMOPOLYMER see AAX250

ACETIC ACID ETHENYL ESTER POLYMER with CHLORETHENE (9CI) see AAX175

ACETIC ACID ETHENYL ESTER, POLYMER with 1-ETHENYL-2-PYRROLIDINONE see AAU300

ACETIC ACID, ETHOXY- see EEK500

ACETIC ACID-2-ETHOXYETHYL ESTER see EES400

ACETIC ACID, (2-(N-ETHYLACETAMIDO)ETHYL) ESTER see ABN800

ACETIC ACID, (ETHYLENEDINITRILO)TETRA-, ALUMINUM(III)COMPLEX see EJA390

ACETIC ACID, (ETHYLENEDINITRILO)TETRA-, ALUMINUMSODIUM SALT see EJA400

ACETIC ACID, (ETHYLENEDINITRILO)TETRA-, CALCIUM (II) COMPLEX see CAR800

ACETIC ACID, (ETHYLENEDINITRILO)TETRA-, CALCIUM DISODIUM SALT see CAR780

ACETIC ACID, (ETHYLENEDINITRILO)TETRA-, CHROMIUM(III)COMPLEX see EJA410

ACETIC ACID, (ETHYLENEDINITRILO)TETRA-, DIPOTASSIUM SALT see EEB100

ACETIC ACID, (ETHYLENEDINITRILO)TETRA-, YTTRIUM complex see YFA100

ACETIC ACID, (ETHYLENEDINITRILO)TETRA-, ZINC(II) COMPLEX see ZGS100

ACETIC ACID, ETHYLENE ETHER see VLU250

ACETIC ACID α-ETHYLEXYL ESTER see OEE000

ACETIC ACID-1-(ETHYLNITROSAMINO)ETHYL ESTER see ABT500

ACETIC ACID, FLUORO-, ALUMINUM SALT see AHB100

ACETIC ACID, FLUORO-, COPPER(II) SALT see CNL800

ACETIC ACID, FLUORO-, THALLIUM(I) SALT see TEK100

ACETIC ACID FURFURYL ESTER see FPM100

ACETIC ACID GERANIOL ESTER see DTD800

ACETIC ACID, GLACIAL see AAT250

ACETIC ACID-3-HEPTANOL ESTER see AAU500

ACETIC ACID-2,4-HEXADIEN-1-OL ESTER see AAU750

ACETIC ACID, HEXYLENE GLYCOL see HFQ000

ACETIC ACID HEXYL ESTER see HFI500

ACETIC ACID, (2-HYDROXYETHOXY)-, γ-LACTONE (6CI,7CI) see DVQ630

ACETIC ACID-3-HYDROXYHEPTYL ESTER see AAU500

ACETIC ACID, (o-((2-HYDROXY-3-HYDROXYMERCURI)PROPYL)CARBAMOYL)PHENOXY- see NCM800

ACETIC ACID, (4-(4-HYDROXY-3-IODOPHENOXY)-3,5-DIIODOPHENYL)-,DIETHANOLAMINE SALT see TKR100

ACETIC ACID, (4-(4-HYDROXY-3-IODOPHENOXY)-3,5-DIIODOPHENYL)-, compounded with 2,2'-IMINODIETHANOL see TKR100

ACETIC ACID, (p-HYDROXYPHENYL)- see HNG600

ACETIC ACID, ((2-HYDROXY-1,3-TRIMETHYLENE)DINITRILO)TETRA- see DCB100

ACETIC ACID, ((2-HYDROXYTRIMETHYLENE)DINITRILO)TETRA-(8CI) see DCB100

ACETIC ACID, (1H-INDOL-3-YL)-, 2-DIMETHYLAMINOETHYLESTER, HYDROCHLORIDE see DPG133

ACETIC ACID, 3-INDOLYL ESTER see IDA600

ACETIC ACID, IODO-, DIESTER WITH ETHYLENE GLYCOL see EIS100

ACETIC ACID, IODO-, ETHYLENE ESTER see EIS100

ACETIC ACID, IODO-, 2-HYDROXYETHYL ESTER see HKS800

ACETIC ACID, IODO-, MONOESTER with ETHYLENE GLYCOL see HKS800

ACETIC ACID, IRON(2+) SALT see FBH000

ACETIC ACID, ISOBUTYL ESTER see IIJ000

ACETIC ACID, ISOCYANATO-, ETHYL ESTER see ELS600

ACETIC ACID, ISOPENTYL ESTER see IHO850

ACETIC ACID, ISOPROPENYL ESTER see MQK750

ACETIC ACID ISOPROPYL ESTER see INE100

ACETIC ACID-2-ISOPROPYL-5-METHYL-2-HEXEN-1-YL ESTER see AAV250

ACETIC ACID, ISOTHIOCYANATO-, ETHYL ESTER see ELX530

ACETIC ACID, LEAD SALT see LCG000

ACETIC ACID LEAD(2+) SALT see LCV000

ACETIC ACID, LEAD(2+) SALT TRIHYDRATE see LCJ000

ACETIC ACID LINALOOL ESTER see LFY600

ACETIC ACID, LITHIUM SALT see LGO100

ACETIC ACID, MAGNESIUM SALT see MAD000

ACETIC ACID MANGANESE(II) SALT (2:1) see MAQ000

ACETIC ACID-p-MENTHAN-8-OL ESTER see DME400

ACETIC ACID, MERCAPTO-, ACETATE (8CI) see ACQ250

ACETIC ACID, MERCAPTO-, p-CARBAMOYLDITHIOBENZENEARSONITE (2:1), DISODIUM SALT see SKH100

ACETIC ACID, MERCAPTO-, DIESTER WITH p-CARBAMOYLDITHIOBENZENEARSONOUS ACID, DISODIUM SALT see SKH100

ACETIC ACID, MERCAPTO-, 1,2-ETHANEDIYL ESTER see MCN000

ACETIC ACID, 2-(MERCAPTOMETHYLTHIO)-, ETHYL ESTER, S-ESTER with O-ETHYL METHYLPHOSPHONOTHIOATE see EMC100

ACETIC ACID, MERCAPTO-, MONOESTER WITH GLYCEROL see GGR300

ACETIC ACID, MERCAPTO-, MONOESTER WITH 1,2,3-PROPANETRIOL see GGR300

ACETIC ACID, MERCAPTOPHENYL-, ETHYL ESTER, DIMETHYL PHOSPHATE (6CI) see PDW800

ACETIC ACID, MERCAPTOPHENYL-, ETHYL ESTER, S-ESTERWITH O,O-DIMETHYL PHOSPHOROTHIOATE see PDW800

ACETIC ACID, MERCURY(2+) SALT see MCS750

ACETIC ACID-3-METHOXYBUTYL ESTER see MHV750

ACETIC ACID, 2-(2-(2-METHOXYETHOXY)ETHOXY)ETHYL ESTER see AAV500

ACETIC ACID, 2-(2-METHOXYETHOXY)ETHYL ESTER see MIE750

ACETIC ACID, 2-METHOXY-1-METHYLETHYL ESTER see PNL265

ACETIC ACID, METHOXY((1-OXO-2-PROPENYL)AMINO)-, METHYL ESTER see MGA400

ACETIC ACID, 2-METHOXYPROPYL ESTER see MFL800

ACETIC ACID, 2-METHYLAMYL ESTER see MGN300

ACETIC ACID METHYL ESTER see MFW100

ACETIC ACID-1-METHYLETHYL ESTER (9CI) see INE100

ACETIC ACID-2-METHYL-6-METHYLENE-7-OCTEN-2-YL ESTER see AAW500

ACETIC ACID METHYLNITROSAMINOMETHYL ESTER see AAW000

ACETIC ACID, 3-(METHYLNITROSAMINO)-2-OXOPROPYL ESTER see HMJ600

ACETIC ACID, (METHYL-ONN-AZOXY)METHYL ESTER see MGS750

ACETIC ACID, 2-METHYL-2,4-PENTANEDIOL DIESTER see HFQ000

ACETIC ACID, 2-METHYLPENTYL ESTER see MGN300

ACETIC ACID, α-METHYL-PHENETHYL ESTER see ABU800

ACETIC ACID-4-METHYLPHENYL ESTER see MNR250

ACETIC ACID-2-METHYL-2-PROPENE-1,1-DIOL DIESTER see AAW250

ACETIC ACID-2-METHYLPROPYL ESTER see IIJ000

ACETIC ACID-1-METHYLPROPYL ESTER (9CI) see BPV000

ACETIC ACID, MONOGLYCERIDE see GGO000

ACETIC ACID MYRCENYL ESTER see AAW500

ACETIC ACID, 1-NAPHTHYL-, 2-DIMETHYLAMINOETHYL ESTER,HYDROCHLORIDE see DPH430

ACETIC ACID 1-NAPHTHYL ESTER see NAU600

ACETIC ACID, (2-NAPHTHYLOXY)- see NBJ700

ACETIC ACID, 1-NAPHTHYLOXY-, 2-DIMETHYLAMINOETHYLESTER, HYDROCHLORIDE see AAW600

ACETIC ACID, 2-NAPHTHYLOXY-, 2-DIMETHYLAMINOETHYLESTER, HYDROCHLORIDE see DPH440

ACETIC ACID, NICKEL(2+) SALT see NCX000

ACETIC ACID, NICKEL(+2) SALT, TETRAHYDRATE see NCX500

ACETIC ACID, NITRILOTRI-, DISODIUM SALT, IRON(II) CHELATE see DXF100

ACETIC ACID, NITRILOTRI-, DISODIUM SALT, MONOHYDRATE see NHK850

ACETIC ACID, NITRILOTRI-, IRON(III) chelate see IHC100

ACETIC ACID, NITRILOTRI-, MERCURY(II) COMPLEX see MDF050

ACETIC ACID, NITRILOTRI-, MONOSODIUM SALT see NEH800

ACETIC ACID, (p-NITROBENZOYL)-, ETHYL ESTER see ENO100

ACETIC ACID, NITRO-, ETHYL ESTER see ENN100

ACETIC ACID, (p-NITROPHENYL)- see NII510

ACETIC ACID, ((4-NITROPHENYL)AMINO)OXO-(9CI) see NHX100

ACETIC ACID, (p-NITROPHENYLSELENYL)- see NIV100

ACETIC ACID, OCTYL ESTER see OEG000

ACETIC ACID, ((OCTYLSTANNYLIDYNE)TRITHIO)TRI-, TRIS(2-ETHYLHEXYL) ESTER see OGI000

ACETIC ACID, OXIME see ABB250

ACETIC ACID, (3-OXO-1(3H)-ISOBENZOFURANYLIDENE)-(9CI) see CCH150

ACETIC ACID, OXO-, METHYL ESTER see MKI550

ACETIC ACID, (4-OXO-2-THIAZOLIDINYLIDENE)-, BUTYL ESTER (9CI) see BPI300

ACETIC ACID, 2,2'-OXYBIS-(9CI) see ONQ100

ACETIC ACID, OXYBIS-, DIMETHYL ESTER see DTH500

ACETIC ACID, 2,2'-OXYBIS-, DI-2-PROPENYL ESTER see DKN100

ACETIC ACID, OXYDI- see ONQ100

ACETIC ACID, OXYDI-, DIALLYL ESTER see DKN100

ACETIC ACID, OXYDI-, DIMETHYL ESTER see DTH500

ACETIC ACID, OXYDIETHYLENE ESTER see DJD750

ACETIC ACID PALLADIUM SALT see PAD300

ACETIC ACID, PALLADIUM(2+) SALT see PAD300

ACETIC ACID, PHENOXY-, 2-CHLOROETHYL ESTER see CHG050

ACETIC ACID 2-PHENOXYETHYL ESTER see EJL600

ACETIC ACID, PHENYL-, BUTYL ESTER see BQJ350

ACETIC ACID, PHENYL-, 2-CHLOROETHYL ESTER see CHG070

ACETIC ACID, PHENYL-, 3,7-DIMETHYL-2,6-OCTADIENYL ESTER, (E)-(8CI) see GDM400

ACETIC ACID, PHENYL-, 3,7-DIMETHYL-6-OCTENYL ESTER see CMU050

ACETIC ACID, PHENYL-, 1,5-DIMETHYL-1-VINYL-4-HEXENYL ESTER (8C1) see LGC050

ACETIC ACID-2-PHENYLETHYL ESTER see PFB250

ACETIC ACID PHENYLHYDRAZONE see ACX750

ACETIC ACID, PHENYLMERCURY DERIV. see ABU500

ACETIC ACID PHENYLMETHYL ESTER see BDX000

ACETIC ACID, PHENYL-, SODIUM SALT see SFA200

ACETIC ACID, PHOSPHONO-, TRIETHYL ESTER (9CI) see EIC000

ACETIC ACID-2-PROPENYL ESTER see AFU750

ACETIC ACID, (PROPYLENEDINITRILO)TETRA- see PNJ100

ACETIC ACID, n-PROPYL ESTER see PNC250

ACETIC ACID-1-(PROPYLNITROSAMINO)PROPYL ESTER see ABT750

ACETIC ACID, SAMARIUM SALT see SAR000

ACETIC ACID, SELENOCYANO-, BUTYL ESTER see BSL350

ACETIC ACID, SELENOCYANO-, POTASSIUM SALT see PLS100

ACETIC ACID, SILVER(1+) SALT see SDI800

ACETIC ACID, SODIUM SALT see SEG500

ACETIC ACID, SULFO-, 1-DODECYL ESTER, SODIUM SALT see SIB700

ACETIC ACID, SULFO-, DODECYL ESTER, S-SODIUM SALT (7CI) see SIB700

ACETIC ACID, 9-TETRADECEN-1-YL ESTER, (Z)- see TCA400

ACETIC ACID, THIOCYANATO-, PROPYL ESTER see TFF150

ACETIC ACID, THIO-, S-(p-NITROPHENYL) ESTER see NIW450

ACETIC ACID, TRIANHYDRIDE with ANTIMONIC ACID see AQJ750

ACETIC ACID, ((TRIBUTYLSTANNYL)THIO)-, ISOOCTYL ESTER see ILR110

ACETIC ACID, TRICHLORO-, compounded with 1,1-DIMETHYL-3-PHENYLUREA (1:1) see FAR050

ACETIC ACID, TRICHLORO-, compounded with N,N-DIMETHYL-N'-PHENYLUREA (1:1) (9CI) see FAR050

ACETIC ACID, TRICHLORO-, ETHYLENE ESTER (2:1) see EJD600

ACETIC ACID, TRICHLORO-, mixed with UREA, 1,1-DIMETHYL-3-PHENYL-(1:1) see FAR050

ACETIC ACID, TRIETHYLENE GLYCOL DIESTER see EJB500

ACETIC ACID, TRIFLUORO-, 4-BROMOPHENYL ESTER see BNX500

ACETIC ACID, TRIFLUORO-, p-BROMOPHENYL ESTER see BNX500

ACETIC ACID, TRIFLUORO-, SODIUM SALT see SKL100
ACETIC ACID (m-TRIMETHYLAMMONIO)PHENYL ESTER METHYLSULFATE see ABV500
ACETIC ACID, ((TRIMETHYLSTANNYL)THIO)-, ISOOCTYL ESTER see ILR120
ACETIC ACID, glacial or acetic acid solution, >80% acid, by weight (UN 2790) (DOT) see AAT250
ACETIC ACID solution, >10% but not >80% acid, by weight (UN 2790) (DOT) see AAT250
ACETIC ACID-VETIVEROL ESTER see AAW750
ACETIC ACID VINYL ESTER see VLU250
ACETIC ACID, VINYL ESTER, CHLOROETHYLENE COPOLYMER see PKP500
ACETIC ACID, VINYL ESTER, POLYMER with CHLOROETHYLENE see AAX175
ACETIC ACID VINYL ESTER POLYMERS see AAX250
ACETIC ACID VINYL ESTER, POLYMER with 1-VINYL-2-PYRROLIDINONE see AAU300
ACETIC ACID, ZINC SALT see ZBS000
ACETIC ACID, ZINC SALT, DIHYDRATE see ZCA000
ACETIC ALDEHYDE see AAG250
ACETIC ANHYDRIDE see AAX500
ACETIC CHLORIDE see ACF750
ACETIC-4-CHLOROANILIDE see CDZ100
ACETIC ETHER see EFR000
ACETIC OXIDE see AAX500
ACETIC PEROXIDE see PCL500
ACETICYL see ADA725
ACETIDIN see EFR000
ACETILARSANO see ACN250
3-((3-ACETIL-4-(3-tert-BUTILAMINO)-2-HIDROXIPROPOXI)FENIL)-1,1-DIETILUREA HCl (SPANISH) see SBN475
ACETILE DIAZO BLACK N see DPO200
ACETILSALICILICO see ADA725
ACETILUM ACIDULATUM see ADA725
ACETIMIDIC ACID see AAI000
ACETIN see GGO000
ACETIN, DI- see GGA100
ACETIN, MONO- see MRE300
ACETIODONE see AAN000
ACETIROMATE see ADD875
ACETISAL see ADA725
ACETISOEUGENOL see AAX750
ACETKARBROMAL see ACE000
ACETMIN see AOB500
ACETOACETAMIDE, N-(2-BENZOTHIAZOLYL)- see BDF100
ACETOACETAMIDE, 2-CHLORO-N-ETHYL- see CEA100
ACETOACETAMIDOBENZENE see AAY000
2-(ACETOACETAMIDO)BENZOTHIAZOLE see BDF100
ACETOACETANILID see AAY000
ACETOACETANILIDE see AAY000
ACETOACETANILIDE, 4-CHLORO- see CEA050
ACETOACETANILIDE, o-CHLORO- see AAY600
ACETOACETANILIDE, p-CHLORO- see AAY250
ACETOACETANILIDE, 2'-CHLORO- see AAY600
ACETOACETANILIDE, 2',4'-DIMETHYL- see OOI100
o-ACETOACETANISIDE see ABA500
o-ACETOACETANISIDIDE, 4'-CHLORO- see AAY300
ACETOACET-o-ANISIDIN (CZECH) see ABA500
ACETOACET-o-CHLORANILIDE see AAY600

ACETOACET-p-CHLORANILIDE see AAY250
ACETOACET-o-CHLOROANILIDE see AAY600
ACETOACET-p-CHLOROANILIDE see AAY250
ACETOACET-4-CHLORO-2-METHYLANILIDE see AAY300
ACETOACET-2,4-DIMETHYLPHENYL see OOI100
ACETOACETIC ACID ANILIDE see AAY000
ACETOACETIC ACID-o-ANISIDIDE see ABA500
ACETOACETIC ACID BUTYL ESTER see BPV500
ACETOACETIC ACID, 4-CHLORO-, ETHYL ESTER see EHH100
ACETOACETIC ACID, 4-CHLORO-, METHYL ESTER see MIH600
ACETOACETIC ACID, CYCLOHEXYL ESTER see CPP100
ACETOACETIC ACID-3,7-DIMETHYL-2,6-OCTADIENYL ESTER see AAY500
ACETOACETIC ACID, ETHYL ESTER see EFS000
ACETOACETIC ACID, 2-HYDROXYETHYL ESTER, ACRYLATE (8CI) see AAY750
ACETOACETIC ACID, 2-PHOSPHONO-, TRIETHYL ESTER see EHY700
ACETOACETIC ACID, TRIESTER WITH 2-ETHYL-2-(HYDROXYMETHYL)-1,3-PROPANEDIOL see ELJ600
ACETOACETIC ACID, 1,1,1-TRIHYDROXY-METHYLPROPANE TRIESTER see ELJ600
ACETOACETIC ANILIDE see AAY000
ACETOACETIC ESTER see EFS000
ACETOACETIC METHYL ESTER see MFX250
o-ACETOACETOCHLORANILIDE see AAY600
ACETOACETONE see ABX750
ACETOACETONITRILE, 2-PHENYL-(8CI) see PEA500
p-ACETOACETOPHENETIDIDE see AAZ000
ACETOACETOPHENONE see BDJ800
2-ACETOACETOXYETHYL ACRYLATE see AAY750
ACETOACETO-m-XYLIDIDE see OOI100
2,4-ACETOACETOXYLIDIDE see OOI100
2',4'-ACETOACETOXYLIDIDE see OOI100
ACETOACET-p-PHENETIDIDE see AAZ000
ACETOACET-o-TOLUIDIDE see ABA000
ACETOACET-m-XYLIDIDE see ABA250
ACETOACET-m-XYLIDIDE see OOI100
2-ACETOACETYLAMINOANISOLE see ABA500
((ACETOACETYL)AMINO)BENZENE see AAY000
2-ACETOACETYLAMINOTOLUENE see ABA000
ACETOACETYLANILINE see AAY000
ACETOACETYL-o-ANISIDE see ABA500
ACETOACETYL-o-ANISIDINE see ABA500
ACETOACETYL-o-ANISINE see ABA500
ACETOACETYL-2-CHLOROANILIDE see AAY600
ACETOACETYL-2-METHYLANILIDE see ABA000
ACETOACETYL-m-XYLIDIDE see OOI100
p-ACETOAMINOANILINE see AHQ250
o-ACETOAMINOBENZOIC ACID see AAJ150
ACETOAMINOFLUORENE see FDR000
ACETO-m-AMINOTOLUENE see ABI750
ACETOANILIDE see AAQ500
ACETO-m-ANISIDIDE see AAQ750
ACETO-p-ANISIDIDE see AAR250
ACETOARSENITE de CUIVRE (FRENCH) see COF500
ACETO AZIB see ASL750

ACETOBUTOLOL HYDROCHLORIDE see AAE125
ACETO-CAUSTIN see TII250
ACETOCHLOR see CGO550
ACETOCID see SNQ710
ACETO-CORT see HHQ800
ACETO DIPP see NBL000
ACETO DNPT 40 see DVF400
ACETO DNPT 80 see DVF400
ACETO DNPT 100 see DVF400
ACETOFENATE see TIL800
ACETOFERROCENE see ABA750
2-ACETOFLUORENE see FEI200
ACETOGUAIACONE see HLQ500
ACETOHEXAMIDE see ABB000
ACETO HMT see HEI500
ACETOHYDRAZIDE see ACM750
ACETOHYDROXAMIC ACID see ABB250
ACETOHYDROXAMIC ACID, N-(4'-ACETAMIDOBIPHENYL-4-YL)- see HKB000
ACETOHYDROXAMIC ACID, N-(4-BIPHENYLYL)-, ACETATE see ABJ750
ACETOHYDROXAMIC ACID, FLUOREN-2-YL-o-GLUCURONIDE see HIQ500
ACETOHYDROXAMIC ACID, N-(7-IODOFLUOREN-2-YL)-O-MYRISTOYL- see MSB100
ACETOHYDROXAMIC ACID, 2-(2-PHENYLACETAMIDO)- see HIY300
ACETOHYDROXIMIC ACID see ABB250
ACETOHYDROXIMIC ACID, THIO-, METHYL ESTER see MPS270
ACETOIN see ABB500
ACETOL see ADA725
ACETOL (1) see ABC000
ACETOMENAPHTHONE see VTA100
ACETOMESIDIDE see TLD250
ACETOMETHOXAN see ABC250
ACETOMETHOXANE see ABC250
ACETOMETHYLANILIDE see MFW000
ACETOMORFINE see HBT500
ACETOMORPHINE see HBT500
ACETONANIL see PJQ750
ACETONANIL see TLP500
ACETONANYL see TLP500
1-ACETONAPHTHALENE see ABC475
β-ACETONAPHTHALENE see ABC500
ACETONAPHTHONE see ABC500
1-ACETONAPHTHONE see ABC475
2-ACETONAPHTHONE see ABC500
1'-ACETONAPHTHONE see ABC475
2'-ACETONAPHTHONE see ABC500
α-ACETONAPHTHONE see ABC475
β-ACETONAPHTHONE see ABC500
2'-ACETONAPHTHONE, 5',6',7',8'-TETRAHYDRO-3',5',5',6',8',8'-HEXAMETHYL-see EEE100
ACETONCIANHIDRINEI (ROUMANIAN) see MLC750
ACETONCIANIDRINA (ITALIAN) see MLC750
ACETONCYAANHYDRINE (DUTCH) see MLC750
ACETONCYANHYDRIN (GERMAN) see MLC750
ACETON (GERMAN, DUTCH, POLISH) see ABC750
ACETONE see ABC750
ACETONE ANIL see TLP500
ACETONE BIS(ETHYL SULFONE) see ABD500
ACETONE-o-CARBANILOYLOXIME see PEQ500
ACETONE CHLOROFORM see ABD000
ACETONECYANHYDRINE (FRENCH) see MLC750
ACETONE CYANOHYDRIN (ACGIH,DOT) see MLC750
ACETONE-1,3-DICHLORO-1,1,3,3-TETRAFLUOROACETONE see DGL400
ACETONE DIETHYL KETAL see ABD250
ACETONE DIETHYLSULFONE see ABD500

ACETONE, HEXACHLORO- see HCL500
ACETONE OILS (DOT) see ABC750
ACETONE-OXIME-N-PHENYLCARBAMATE see PEQ500
ACETONE OXIME PHENYLURETHANE see PEQ500
ACETONE PEROXIDE see ABE000
ACETONE SEMICARBAZONE see ABE250
ACETONIC ACID see LAG000
ACETONITRIL (GERMAN, DUTCH) see ABE500
ACETONITRILE see ABE500
ACETONITRILE, (p-AMINOBENZOYL)-(6CI,7CI) see ALQ635
ACETONITRILE, AMINO-, MONOHYDROCHLORIDE (9CI) see GHI100
ACETONITRILE, BENZOYLPHENYL- see OOK200
ACETONITRILE, BROMO- see BMS100
ACETONITRILE, (p-CHLOROPHENYL)- see CEP300
ACETONITRILE IMIDAZOLE-5,7,7,12,14,14-HEXAMETHYL-1,4,8,11-TETRAAZA-4,11-CYCLOTETRADECA DIENE IRON(II) PERCHLORATE see ABE750
ACETONITRILE, 2,2'-IMINOBIS-(9CI) see IBB100
ACETONITRILE, IMINODI- see IBB100
ACETONITRILE, ((α-METHYLPHENETHYL)AMINO)PHENYL- see AOB300
ACETONITRILE, NITRILOTRI- see NEI600
ACETONITRILE, 2,2',2"-NITRILOTRIS- see NEI600
ACETONITRILE, (m-NITROPHENYL)- see NFN500
ACETONITRILE, (o-NITROPHENYL)- see NFN600
ACETONITRILE, PHENYLACETO see PEA500
ACETONITRILE, (PHENYLTHIO)- see COP755
ACETONITRILETHIOL see MCK300
ACETONKYANHYDRIN (CZECH) see MLC750
ACETONOXIME see ABF000
ACETONYL see ADA725
ACETONYL ACETONE see HEQ500
p-ACETONYLANISOLE see AOV875
3-(α-ACETONYLBENZYL)-4-HYDROXYCOUMARIN see WAT200
3-(α-ACETONYLBENZYL)-4-HYDROXYCOUMARIN SODIUM SALT see WAT220
ACETONYL BROMIDE see BNZ000
ACETONYL CHLORIDE see CDN200
3-(α-ACETONYLFURFURYL)-4-HYDROXYCOUMARIN see ABF500
3-(α-ACETONYL-p-NITROBENZYL)-4-HYDROXY-COUMARIN see ABF750
3-(α-ACETONYL-p-NITROBENZYL)-4-HYDROXY-COUMARIN see ABF750
1-ACETONYL-1-NITROSOUREA see OOO300
p-ACETONYLOXYBENZENEARSONIC ACID SODIUM SALT see SJK475
2-ACETONYLPIPERIDINE see PAO500
ACETO PAN see PFT250
ACETO PBN see PFT500
(S)-ACETOPHAN see SDY625
ACETOPHEN see ADA725
ACETO-p-PHENALIDE see ABG750
ACETOPHENAZINE see ABG000
ACETOPHENAZINE MALEATE see ABG000
p-ACETOPHENETIDE see ABG750
m-ACETOPHENETIDIDE see ABG250
o-ACETOPHENETIDIDE see ABG350
p-ACETOPHENETIDIDE see ABG750
ACETO-p-PHENETIDIDE see ABG750
ACETOPHENETIDIN see ABG750
ACETOPHENETIDINE see ABG750
ACETO-4-PHENETIDINE see ABG750
ACETOPHENETIN see ABG750

ACETOPHENONE see ABH000
ACETOPHENONE, 2-ANILINO-4'-(BENZYLOXY)-2-PHENYL- see AOR635
ACETOPHENONE, 2-ANILINO-4'-(2-(DIETHYLAMINO)ETHOXY)-2-PHENYL-, HYDROCHLORIDE see AOR700
ACETOPHENONE, 2-BROMO-4'-CHLORO- see CJJ100
ACETOPHENONE, 4'-tert-BUTYL-2',6'-DIMETHYL-3',5'-DINITRO- see ACE600
ACETOPHENONE, 2-CHLORO-4'-HYDROXY-(6CI,7CI) see HNE500
ACETOPHENONE, 2-CHLORO-m-NITRO- see CJA140
ACETOPHENONE, 2-CHLORO-2-PHENYL-(8CI) see CFJ100
ACETOPHENONE, α-CHLORO-α-PHENYL- see CFJ100
ACETOPHENONE, 2-DIAZO- see PET800
ACETOPHENONE, 3',4'-DIHYDROXY-2-(ISOPROPYLAMINO)-, HYDROCHLORIDE see DMV500
ACETOPHENONE, 2,2-DIPHENYL-(6CI,7CI,8CI) see TMS100
ACETOPHENONE, 2'-HYDROXY-(8CI) see HIN500
ACETOPHENONE, 4'-HYDROXY-, o-(BUTYLCARBAMOYL)OXIME, o-ESTER WITH o,o-DIETHYL PHOSPHOROTHIOATE see DIV700
ACETOPHENONE, 4'-HYDROXY-, o-(DIMETHYLCARBAMOYL)OXIME, o-ESTER WITH o,o-DIETHYL PHOSPHOROTHIOA see DJB420
ACETOPHENONE, 4'-HYDROXY-, o-(ISOPROPYLCARBAMOYL)OXIME, o-ESTER WITH o,o-DIETHYL PHOSPHOROTHIO see DJP520
ACETOPHENONE, 4'-HYDROXY-, o-(METHYLCARBAMOYL)OXIME, o-ESTER WITH o,o-DIETHYL PHOSPHOROTHIOATE see DJO420
ACETOPHENONE, 2-HYDROXY-2-PHENYL- see BCP250
ACETOPHENONE, 4'-METHOXY-3'-METHYL- see MET100
ACETOPHENONE, 4'-NITRO- see NEL600
ACETOPHENONE, OXIME see ABH150
ACETOPHENONE, 3'-PHENOXY- see PDR200
ACETOPHENONE, 4'-PHENYL- see MHP500
ACETOPHENONE THIOSEMICARBAZONE see ABH250
ACETOPHOS see DIW600
ACETOPROMAZINE see ABH500
ACETOPROPIONIC ACID see LFH000
ACETOPROPYL ALCOHOL see ABH750
ACETOPURPURINE 8B see CMO880
3-ACETOPYRIDINE see ABI000
ACETOPYRROTHINE see ABI250
ACETOQUAT CPB see HCP800
ACETOQUAT CPC see CCX000
ACETOQUAT CTAB see HCQ500
ACETOQUINONE BLUE L see TBG700
ACETOQUINONE BLUE R see TBG700
ACETOQUINONE LIGHT GOOSEBERRY RL see AKE250
ACETOQUINONE LIGHT GREEN BLUE JL see DMM400
ACETOQUINONE LIGHT HELIOTROPE NL see DBP000
ACETOQUINONE LIGHT ORANGE JL see AKP750
ACETOQUINONE LIGHT PINK RLZ see AKO350
ACETOQUINONE LIGHT PURE BLUE R see MGG250
ACETOQUINONE LIGHT RUBINE BLZ see CMP080
ACETOQUINONE LIGHT SCARLET BLZ see ENP100

ACETOQUINONE LIGHT VIOLET N see AKP250
ACETOQUINONE LIGHT YELLOW see AAQ250
ACETOQUINONE LIGHT YELLOW 4JLZ see AAQ250
ACETOQUINONE LIGHT YELLOW 2RZ see DUW500
ACETOSAL see ADA725
ACETOSALIC ACID see ADA725
ACETOSALIN see ADA725
ACETO SDD 40 see SGM500
ACETOSPAN see AQX500
ACETOSULFAMIN see SNQ710
ACETO TETD see TFS350
2-ACETOTHIENONE see ABI500
ACETOTHIOAMIDE see TFA000
2-ACETOTHIOPHENE see ABI500
ACETO TMTM see BJL600
ACETOTOLUIDE see ABI750
4-ACETOTOLUIDE see ABJ250
o-ACETOTOLUIDE see ABJ000
p-ACETOTOLUIDE see ABJ250
m-ACETOTOLUIDIDE see ABI750
o-ACETOTOLUIDIDE see ABJ000
p-ACETOTOLUIDIDE see ABJ250
O-ACETOTOLUIDIDE, 2-CHLORO-N-(ETHOXYMETHYL)-6'-ETHYL-(8CI) see CGO550
o-ACETOTOLUIDIDE, α,α,α-TRIFLUORO- see TKB350
ACETOVANILLONE see HLQ500
ACETOVANILONE see HLQ500
ACETOVANYLLON see HLQ500
ACETOXIME see ABF000
ACETOXON see DIW600
N-ACETOXY-4-ACETAMIDOBIPHENYL see ABJ750
N-ACETOXY-2-ACETAMIDOFLUORENE see ABL000
ACETOXY(2-ACETAMIDO-5-NITROPHENYL)MERCURY see ABK000
N-ACETOXY-2-ACETAMIDOPHENANTHRENE see ABK250
N-ACETOXY-4-ACETAMIDOSTILBENE see ABW500
N-ACETOXY-4-ACETAMIDOSTILBENE see SMT000
2-ACETOXY-4'-ACETAMINO)PHENYLBENZOATE see SAN600
N-ACETOXY-4-ACETYLAMINOBIPHENYL see ABJ750
N-ACETOXY-2-ACETYLAMINOFLUORENE see ABL000
N-ACETOXY-N-ACETYL-2-AMINOFLUORENE see ABL000
N-ACETOXY-2-ACETYLAMINOPHENANTHRENE see ABK250
trans-N-ACETOXY-4-ACETYL-AMINOSTILBENE see ABL250
2-ACETOXYACRYLONITRILE see ABL500
α-ACETOXYACRYLONITRILE see ABL500
19-ACETOXY-Δ^{1,4}-ANDROSTADIENE-3,17-DIONE see ABL625
3-ACETOXY-8-AZAXANTHINE see ABL650
2-ACETOXYBENZOIC ACID see ADA725
o-ACETOXYBENZOIC ACID see ADA725
6-ACETOXY-BENZO(a)PYRENE see ABL750
N-α-ACETOXY-N-BENZYLNITROSAMINE see ABL875
N-ACETOXY-4-BIPHENYLACETAMIDE see ABJ750
3-β-ACETOXY-BIS NOR-Δ^5-CHOLENIC ACID see ABM000
1-ACETOXY-1,3-BUTADIENE see ABM250
N-(4-ACETOXYBUTYL)-N-(ACETOXYMETHYL)NITROSAMINE see ABM500
N-(α-ACETOXY)BUTYL-N-BUTYLNITROSAMINE see BPV325

17-β-ACETOXY-19-NOR-17-α-PREGN-4-EN-20-YN-3-ONE see ABU000

17-β-ACETOXY-19-NOR-17-α-PREGN-4-EN-20-YN-3-ONE see ABU000

N-ACETOXY-N-OCTYLOXYBENZAMIDE see ABU100

2-ACETOXYPENTANE see AOD735

N-ACETOXY-4-PHENANTHRYLACETAMIDE see ABK250

3-ACETOXYPHENOL see RDZ900

4-(p-ACETOXYPHENYL)-2-BUTANONE see AAR500

ACETOXYPHENYLMERCURY see ABU500

4-ACETOXYPHENYL METHYL CARBINOL see ABU600

(p-ACETOXYPHENYL)METHYL CARBINOL see ABU600

2-ACETOXY-1-PHENYLPROPANE see ABU800

3-ACETOXYPHENYLTRIMETHYLAMMONIUM IODIDE see ABV250

3-ACETOXYPHENYLTRIMETHYLAMMONIUM METHYLSULFATE see ABV500

2-ACETOXY-N-(3-(m-(1-PIPERIDINYLMETHYL)PHENOXY)PROPYL)ACETAMIDE HYDROCHLORIDE see HNT100

3-β-ACETOXYPREGN-6-ENE-20-CARBOXYLIC ACID see ABM000

17-α-ACETOXYPREGN-4-ENE-3-β-OL-20-ONE and MESTRANOL (20:1) see ABV600

17-β-ACETOXY-5-α-17-α-PREGN-2-ENE-20-YNE see PMA450

ACETOXYPROGESTERONE see PMG600

17-ACETOXYPROGESTERONE see PMG600

17-α-ACETOXYPROGESTERONE see PMG600

1-ACETOXYPROPANE see PNC250

2-ACETOXYPROPANE see INE100

3-ACETOXYPROPENE see AFU750

N-(3-ACETOXYPROPYL)-N-(ACETOXYMETHYL)NITROSAMINE see ABV750

N-(α-ACETOXY)PROPYL-N-N-PROPYLNITROSAMINE see ABT750

3-ACETOXYQUINUCLIDINE GLAUCOSTAT see AAE250

1'-ACETOXYSAFROLE see ACV000

1'-ACETOXYSAFROLE-2',3'-OXIDE see ABW250

15-ACETOXYSCIRPENDIOL see ECT700

15-ACETOXYSCIRPEN-3,4-DIOL see ECT700

15-ACETOXYSCIRPENOL see ECT700

N-ACETOXY-N-(4-STILBENYL)ACETAMIDE see ABW500

4-ACETOXYSTYRENE see ABW550

p-ACETOXYSTYRENE see ABW550

4-ACETOXYTOLUENE see MNR250

p-ACETOXYTOLUENE see MNR250

α-ACETOXYTOLUENE see BDX000

ACETOXYTRIBUTYLPLUMBANE see THY850

ACETOXYTRIBUTYLSTANNANE see TIC000

ACETOXY-TRICHLORFON see CDN510

ACETOXYTRICYCLOHEXYLSTANNANE see ABW600

ACETOXYTRIETHYLPLUMBANE see TJU150

ACETOXYTRIETHYLSTANNANE see ABW750

ACETOXYTRIETHYLTIN see ABW750

ACETOXYTRIHEXYLSTANNANE see ABX000

ACETOXYTRIHEXYLTIN see ABX000

ACETOXYTRIISOPROPYLSTANNANE see TKT750

ACETOXYTRIMETHYLPLUMBANE see ABX125

ACETOXYTRIMETHYLSTANNANE see TMI000

ACETOXYTRIOCTYLSTANNANE see ABX150

3-ACETOXY-1,2,4-TRIOXOLANE see VLU310

ACETOXYTRIPENTYLSTANNANE see ABX175

ACETOXYTRIPHENYLLEAD see TMT000

ACETOXY-TRIPHENYL-STANNAN (GERMAN) see ABX250

ACETOXYTRIPHENYLSTANNANE see ABX250

ACETOXY-TRIPHENYL-STANNANE see ABX250

ACETOXYTRIPHENYLTIN see ABX250

ACETOXYTRIPROPYLPLUMBANE see ABX325

ACETOXYTRIPROPYLSTANNANE see TNB000

ACETOXYTRIS(β,β-DIMETHYLPHENETHYL)STANNANE see TMK200

3-ACETOXYXANTHINE see HOP300

ACETOZALAMIDE see AAI250

ACETO ZDBD see BIX000

ACETO ZDED see BJK500

ACETO ZDMD see BJK500

ACET-p-PHENALIDE see ABG750

ACETPHENARSINE see ABX500

ACETPHENETIDIN see ABM000

ACET-p-PHENETIDIN see ABG750

p-ACETPHENETIDIN see ABG750

ACETRIZOATE SODIUM see AAN000

ACETRIZOIC ACID see AAM875

ACETRIZOIC ACID SODIUM SALT see AAN000

ACET-THEOCIN see TEP000

ACETYLACETANILIDE see AAY000

α-ACETYLACETANILIDE see AAY000

ACETYLACETONATE-1,5-CYCLOOCTADIENE RHODIUM see CPR840

ACETYL ACETONE see ABX750

ACETYL ACETONE PEROXIDE with >9% by weight active oxygen (DOT) see PBL600

2-ACETYLACETOPHENONE see BDJ800

α-ACETYLACETOPHENONE see BDJ800

N-(ACETYLACETYL)ANILINE see AAY000

O-ACETYL-1-ACETYLBENZOCYCLOBUTENE OXIME see BFZ130

3-ACETYLACONITINE HYDROBROMIDE see ABX800

3-ACETYLACROLEIN see KGA200

ACETYL ADALIN see ACE000

ACETYLADRIAMYCIN see DAC000

5-(ACETYLAMINO)-1-(ACETOXY)-2-NAPHTHALENESULFONIC ACID, COMPOUND WITH PYRIDINE (1:1) see AAI800

3-(ACETYLAMINO)-5-((ACETYLAMINO)METHYL)-2,4,6-TRIIODOBENZOIC ACID see AAI750

3-(ACETYLAMINO)-5-(ACETYLMETHYLAMINO)-2,4,6-TRIIODO-BENZOIC ACID see MQR350

2-((3-(ACETYLAMINO)-5-(ACETYLMETHYLAMINO)-2,4,6-TRIIODOBENZOYL)AMINO)-2-DEOXY-d-GLUCOSE see MQR300

5-(ACETYLAMINO)-1-(ACETYLOXY)-2-NAPHTHALENESULFONIC ACID COMPD. WITH PYRIDINE (1:1) see AAI800

9-ACETYLAMINOACRIDINE see ABX810

3-ACETYLAMINOANILINE see AHQ000

4-(ACETYLAMINO)ANILINE see AHQ250

m-(ACETYLAMINO)ANILINE see AHQ000

p-(ACETYLAMINO)ANILINE see AHQ250

4-ACETYLAMINOAZOBENZENE see PEH750

p-ACETYLAMINOBENZALDEHYDE THIOSEMICARBAZONE see FNF000

ACETYLAMINOBENZENE see AAQ500

N-ACETYL-p-AMINOBENZENEARSONIC ACID, SODIUM SALT, TETRAHYDRATE see ARA000

N-ACETYL-4-AMINOBENZENESULFONAMIDE see SNQ710

4-(ACETYLAMINO)BENZENESULFONYL CHLORIDE see AAM600

2-(ACETYLAMINO)BENZOIC ACID see AAJ150

N-ACETYLAMINOBENZOIC ACID see AAJ150

4-(ACETYLAMINO)BENZOIC ACID with 2-(DIMETHYLAMINO)ETHANOL (1:1) see DOZ000

2-ACETYLAMINOBIPHENYL see PDY000

3-ACETYLAMINOBIPHENYL see PDY250

4-ACETYLAMINOBIPHENYL see PDY500

2-(2-(ACETYLAMINO)-4-(BIS(2-METHOXYETHYL)AMINO)-5-METHOXYPHENYL)-5-AMINO-7-BROMO-4-CHLORO-2H- see ABX833

γ-ACETYLAMINOBUTYRIC ACID see AAJ350

N-((ACETYLAMINO)CARBONYL)-2-BROMO-2-ETHYLBUTANAMIDE see ACE000

N-((ACETYLAMINO)CARBONYL)-α-ETHYLBENZENEACETAMIDE see ACX500

2-(2-(ACETYLAMINO)-4-(N-(2-CYANOETHYL)ETHYLAMINO)-5-METHOXYPHENYL)-5-AMINO-7-BROMO-4-CHLORO- see ABX836

3-ACETYLAMINODIBENZOFURAN see DDC000

2-ACETYLAMINODIBENZOTHIOPHENE see DDD400

3-ACETYLAMINODIBENZOTHIOPHENE see DDD600

3-ACETYLAMINODIBENZOTHIOPHENE-5-OXIDE see DDD800

2-ACETYLAMINO-9,10-DIHYDROPHENANTHRENE see DMA400

4-ACETYLAMINO-p-DIPHENYLBENZENE see TBD250

2-ACETYLAMINOETHANOL see HKM000

S-(2-(ACETYLAMINO)ETHYL)-O,O-DIMETHYL PHOSPHORODITHIOATE see DOP200

3-ACETYLAMINO-FLUORANTHEN see AAK400

3-ACETYLAMINOFLUORANTHENE see AAK400

2-ACETYLAMINO-FLUOREN (GERMAN) see FDR000

4-ACETYLAMINOFLUOREN (GERMAN) see ABY000

4-ACETYLAMINOFLUORENE see ABY000

N-ACETYL-2-AMINOFLUORENE see FDR000

2-ACETYLAMINOFLUORENE (OSHA) see FDR000

2-ACETYLAMINO-9-FLUORENOL see ABY150

2-ACETYLAMINOFLUORENONE see ABY250

2-ACETYLAMINO-9-FLUORENONE see ABY250

2-ACETYLAMINO-9-(p-FLUOROPHENYLIMINO)FLUORENE see FLG200

4-ACETYLAMINO-2-HYDROXYMERCURIBENZOIC ACID, SODIUM SALT see SJP500

3-ACETYLAMINO-4-HYDROXYPHENYLARSONIC ACID see ABX500

(3-(ACETYLAMINO)-4-HYDROXYPHENYL)ARSONINE (9CI) see ABX500

3-ACETYLAMINO-5-N-METHYL-ACETYLAMINO-2,4,6-TRIIODOBENZOYL GLUCOSAMINE see MQR300

N-((ACETYLAMINO)METHYL)-2-CHLORO-N-(2,6-DIETHYLPHENYL)ACETAMIDE see ABY300

6-(ACETYLAMINO)-4-METHYL-1,2-DITHIOLO(4,3-B)PYRROL-5(4H)-ONE see ABI250

5-ACETYLAMINO-N-METHYL-2,4,6-TRIIODOISOPHTHALAMIC ACID see IGD000

5-ACETYLAMINO-N-METHYL-2,4,6-TRIIODOISOPHTHALAMIC ACID SODIUM SALT see IGC100

2-ACETYLAMINO-4-(5-NITRO-2-FURYL)THIAZOLE see AAL750

2-ACETYLAMINO-5-NITROTHIAZOLE see ABY900

5-(ACETYLAMINO)-4-OXO-5-HEXENAMIDE see PMC750

p-(ACETYLAMINO)PHENACYL CHLORIDE see CEC000

2-ACETYLAMINOPHENANTHRENE see AAM250

3-ACETYLAMINOPHENANTHRENE see PCY500

9-ACETYLAMINOPHENANTHRENE see PCY750

N-(o-(ACETYLAMINO)PHENETHYL)-1-HYDROXY-2-NAPHTHALENECARBOXAMIDE see HIJ500

2-(ACETYLAMINO)PHENOL see HIL000

3-(ACETYLAMINO)PHENOL see HIL500

m-(ACETYLAMINO)PHENOL see HIL500

o-(ACETYLAMINO)PHENOL see HIL000

p-ACETYLAMINOPHENOL see HIM000

N-ACETYL-2-AMINOPHENOL see HIL000

N-ACETYL-p-AMINOPHENOL see HIM000

p-N-ACETYLAMINOPHENYLACETYLSALICYLATE see SAN600

p-ACETYLAMINOPHENYL DERIVATIVE of NITROGEN MUSTARD see BHO500

ACETYL-p-AMINOPHENYLSTIBINSAURES NATRIUM (GERMAN) see SLP500

1-ACETYLAMINOPYRENE see AAM650

2-ACETYLAMINOPYRENE see PON500

N-ACETYL-1-AMINOPYRENE see AAM650

4-ACETYLAMINOPYRIDINE see AAM750

6-ACETYLAMINOQUINOLINE see QPS200

4-ACETYLAMINOSTILBENE see SMR500

trans-4-ACETYLAMINOSTILBENE see SMR500

4-ACETYLAMINO-p-TERPHENYL see TBD250

2-ACETYLAMINO-1,3,4-THIADIAZOLE-5-SULFONAMIDE see AAI250

2-ACETYLAMINOTHIAZOLE see TEW100

4-(ACETYLAMINO)TOLUENE see ABJ250

3-(ACETYLAMINO)-2,4,6-TRIIODOBENZOIC ACID see AAM875

5-ACETYLAMINO-2,4,6-TRIIODO ISOPHTHALIC ACID DI-(N-METHYL-2,3-DIHYDROXYPROPYLAMIDE) see ACA125

3-(2-(3-ACETYLAMINO-2,4,6-TRIIODOPHENOXY)ETHOXY)-2-ETHYLPROPIONIC ACID see IGA000

2-((2-(3-(ACETYLAMINO)-2,4,6-TRIIODOPHENOXY)ETHOXY)METHYL)BUTANOIC ACID see IGA000

N-ACETYL-N-(3-AMINO-2,4,6-TRIIODOPHENYL)-β-AMINOISOBUTYRIC ACID see IDJ550

3-(ACETYL-(3-AMINO-2,4,6-TRIIODOPHENYL)AMINO)-2-METHYLPROPANOIC ACID see IDJ550

N-ACETYL-N-(3-AMINO-2,4,6-TRIIODOPHENYL)-2-METHYL-β-ALANINE see IDJ550

ACETYL ANHYDRIDE see AAX500

ACETYLANILINE see AAQ500

3-ACETYLANILINE see AHR500

4-ACETYLANILINE see AHR240

m-ACETYLANILINE see AHR500

N-ACETYLANILINE see AAQ500

ACETYL-p-ANISIDINE see AAR250

4-ACETYLANISOLE see MDW750

p-ACETYLANISOLE see MDW750

9-ACETYL-1,7,8-ANTHRACENETRIOL see ACA750

10-ACETYL-1,8,9-ANTHRACENETRIOL see ACA750

10-ACETYLANTHRALIN see ACA750

ACETYLANTHRANILIC ACID see AAJ150

N-ACETYLANTHRANILIC ACID see AAJ150

ACETYL-l-ARGININE, NITROSATED see AAM000

ACETYLARSAN see ACN250

N-ACETYLARSANILIC ACID, SODIUM SALT, TETRAHYDRATE see ARA000

ACETYLATED MONO- and DIGLYCERIDES see ACA900

ACETYLATED MONOGLYCERIDES see ACA900

ACETYLATED SUDAN BLACK B see PCJ200

ACETYL AZIDE see ACB000

1-ACETYLAZIRIDINE see ACB250

N-ACETYLBATRACYLIN see ACB300

(4-(p-(p-ACETYLBENZAMIDO)ANILINO)-6-AMINO-1-METHYLQUINOLINIUM)-p-AMIDINOHYDRAZONE-p-TOLUENESULFONATE-MONO-p-TOLUENSULFONATE see ACB750

ACETYLBENZENE see ABH000

α-ACETYLBENZENEACETONITRILE see PEA500

1-(p-ACETYLBENZENESULFONYL)-3-CYCLOHEXYLUREA see ABB000

N-ACETYLBENZIDINE see ACC000

1-ACETYLBENZOCYCLOBUTENE OXIME see BFZ160

2-ACETYLBENZOFURAN see ACC100

5-ACETYL BENZO(C)PHENANTHRENE see ACC750

ACETYL BENZOYL ACONINE see ADH750

ACETYLBENZOYLMETHANE see BDJ800

ACETYL BENZOYL PEROXIDE (solid) see ACC250

2-ACETYL-3:4-BENZPHENANTHRENE see ACC750

4-ACETYLBIPHENYL see MHP500

p-ACETYLBIPHENYL see MHP500

N-ACETYL-4-BIPHENYLHYDROXYLAMINE see ACD000

1-ACETYL-3,3-BIS(4-(ACETYLOXY)PHENYL)-1,3-DIHYDRO-2H-INDOL-2-ONE see ACD500

N-(N-ACETYL-3-(p-(BIS(2-CHLOROETHYL)AMINO)PHENYL)ALANYL-3-PHENYLALANINE) ETHYL ESTER see ACD250

1-ACETYL-3,3-BIS(p-HYDROXYPHENYL)OXINDOLE DIACETATE see ACD500

ACETYL BROMIDE see ACD750

ACETYLBROMODIETHYLACETYLCARBAMIDE see ACE000

N-ACETYL-N-BROMODIETHYLACETYLCARBAMIDE see ACE000

N-ACETYL-N-BROMODIETHYLACETYLUREA see ACE000

N-ACETYL-N'-α-BROMO-α-ETHYLBUTYRYLCARBAMIDE see ACE000

1-ACETYL-3-(2-BROMO-2-ETHYLBUTYRYL)UREA see ACE000

1-ACETYL-3-(α-BROMO-α-ETHYLBUTYRYL)UREA see ACE000

o-ACETYL-N-(p-BUTOXYPHENYLACETYL)HYDROXYLAMINE see ACE250

3-(3-ACETYL-4-(3-tert-BUTYLAMINO-2-HYDROXYPROPOXY)PHENYL)-1,1-DIETHYLHARNSTOFF HCl (GERMAN) see SBN475

3-(3-ACETYL-4-(3-tert-BUTYLAMINO-2-HYDROXYPROPOXY)PHENYL)-1,1-DIETHYLUREA HYDROCHLORIDE see SBN475

ACETYL BUTYL CITRATE see THX100

o-ACETYL-2-sec-BUTYL-4,6-DINITROPHENOL see ACE500

O-ACETYL-2-sec-BUTYL-4,6-DINITROPHENOL see ACE500

2-ACETYL-5-tert-BUTYL-4,6-DINITROXYLENE see ACE600

3-ACETYL-5-sec-BUTYL-4-HYDROXY-3-PYROLIN-2-ONE,MONOSODIUM SALT see ADC250

3-ACETYL-5-sec-BUTYL-4-HYDROXY-3-PYROLIN-2-ONE SODIUM SALT see ADC250

3-ACETYL-5-sec-BUTYL-4-HYDROXY-3-PYRROLIN-2-ONE see VTA750

1-(2-ACETYL-4-n-BUTYRAMIDOPHENOXY)-2-HYDROXY-3-ISOPROPYLAMINOPROPANE see AAE100

dl-1-(2-ACETYL-4-BUTYRAMIDOPHENOXY)-2-HYDROXY-3-ISOPROPYLAMINOPROPANE HYDROCHLORIDE see AAE125

ACETYLBUTYRYL see HEQ200

3'-o-ACETYLCALOTROPIN see ACF000

ACETYLCAPROLACTAM see ACF100

N-ACETYLCAPROLACTAM see ACF100

o-ACETYL-N-CARBOXY-N-PHENYL-HYDROXYLAMINE ISOPROPYL ESTER see ING400

ACETYLCARBROMAL see ACE000

ACETYL CARENE see ACF150

ACETYL CEDRENE see ACF250

ACETYL CHLORIDE see ACF750

ACETYL CHLORIDE, (2,4-BIS(1-METHYLBUTYL)PHENOXY)- see DCI100

ACETYL CHLORIDE, 2,4-DI-2-AMYLPHENOXY- see DCI100

ACETYL CHLORIDE, DIPHENYL- see DVX500

N-ACETYL-N-(2-CHLORO-6-METHYLPHENYL)-1-PYRROLIDINEACETAMIDE MONOHYDROCHLORIDE see CLB250

4-ACETYL-4-(3-CHLOROPHENYL)-1-(3-(4-METHYLPIPERAZINO)-PROPYL)PIPERIDINE TRIHYDROCHLORIDE see ACG125

1-(3-(3-ACETYL-4-(p-(CHLOROPHENYL)PIPERIDINO)PROPYL)-4-METHYLPIPERAZINE) TRIHYDROCHLORIDE see ACG125

ACETYLCHOLINE see CMF250

ACETYLCHOLINE BROMHYDRATE see CMF260

ACETYLCHOLINE BROMIDE see CMF260

ACETYLCHOLINE CHLORIDE see ABO000

ACETYLCHOLINE HYDROBROMIDE see CMF260

ACETYLCHOLINE HYDROCHLORIDE see ABO000

ACETYL CHOLINE ION see CMF250

ACETYLCHOLINIUM CHLORIDE see ABO000

ACETYLCINOBUFAGIN see CMS126

ACETYLCITRIC ACID, TRIBUTYL ESTER see THX100

ACETYLCOLAMINE see HKM000

N-ACETYL COLCHINOL see ACG250

N-ACETYL-COLCHINOL (GERMAN) see ACG250

2-ACETYLCOUMARONE see ACC100

ACETYL CYCLOHEXANEPERSULFONATE see ACG300

ACETYL CYCLOHEXANESULFONYL PEROXIDE, >82% wetted with <12% water (DOT) see ACG300
4-ACETYL-N-((CYCLOHEXYLAMINO)CARBONYL)-BENZENESULFONAMIDE see ABB000
ACETYL CYCLOHEXYLSULFONYL PEROXIDE see ACG300
ACETYLCYSTEINE see ACH000
N-ACETYLCYSTEINE see ACH000
N-ACETYL-l-CYSTEINE see ACH000
N-ACETYL-N-CYSTEINE see ACH000
N-ACETYL-l-CYSTEINE (9CI) see ACH000
3-ACETYLDEOXYNIVALENOL see ACH075
ACETYL-10-DEOXYSEPACONITINE see LBD000
N-ACETYL-7H-DIBENZO(c,g)CARBAZOLE see ACH080
2-ACETYLDIBENZOTHIOPHENE see ACH090
ACETYL-1,1-DICHLOROETHYL PEROXIDE see ACH125
1-ACETYL-3-(2,2-DICHLOROETHYL)UREA see ACH250
14-ACETYLDICTYOCARPINE see ACH300
1-ACETYL-9,10-DIDEHYDRO-N,N-DIETHYL-6-METHYLERGOLINE-8-β-CARBOXAMIDE BITARTRATE see ACP500
1-ACETYL-9,10-DIDEHYDRO-N-ETHYL-6-METHYLERGOLINE-8-β-CARBOXAMIDE see ACP750
3B-o-(4-o-ACETYL-2,6-DIDEOXY-3-C-METHYL-α-l-ARABINOHEXOPYRANOSYL)-7-METHYL-OLIVOMYCIN D see CMK650
16-ACETYLDIGITALINUM VERUM see ACH375
ACETYLDIGITOXIN-α see ACH500
α-ACETYLDIGITOXIN see ACH500
ACETYLDIGITOXIN-β see ACH750
ACETYLDIGOXIN-α see ACI000
α-ACETYLDIGOXIN see ACI000
β-ACETYLDIGITOXIN see ACH750
ACETYLDIGOXIN-β see ACI250
β-ACETYLDIGOXIN see ACI250
ε-ACETYLDIGOXIN (GERMAN) see DKN600
3-ACETYL-1,5-DIHYDRO-4-HYDROXY-5-(1-METHYLPROPYL)-2H-PYRROL-2-ONE see VTA750
3-ACETYL-1,5-DIHYDRO-4-HYDROXY-5-(1-METHYLPROPYL)-2H-PYRROL-2-ONE SODIUM SALT see ADC250
3-ACETYL-10-(3-DIMETHYLAMINOPROPYL)PHENOTHIAZINE see ABH500
2-ACETYL-10-(3-(DIMETHYLAMINO)PROPYL)PHENOTHIAZINE, MALEATE see AAF750
ACETYLDIMETHYLARSINE see ACI375
2-ACETYL-3,5-DIMETHYL-4-ETHYL-PYRROLE see EIL200
3-ACETYL-2,5-DIMETHYLFURAN see ACI400
3-ACETYL-2,4-DIMETHYL-PYRROLE see ACI500
3-ACETYL-2,5-DIMETHYL-PYRROLE see ACI550
N-ACETYLDIPHENYLAMINE see PDX500
N-ACETYL-N-(4,5-DIPHENYL-2-OXAZOLYL)ACETAMIDE see ACI629
ACETYL DIPTEREX see CDN510
10-ACETYLDITHRANOL see ACI640
ACETYLEN see ACI750
ACETYLENE see ACI750
ACETYLENE, dissolved (DOT) see ACI750
ACETYLENE BLACK see CBT750
(ACETYLENECARBONYLOXY)TRIPHENYLTIN see TMW600
ACETYLENECARBOXYLIC ACID see PMT275

ACETYLENECARBOXYLIC ACID METHYL ESTER see MOS875
ACETYLENE CHLORIDE see ACJ000
ACETYLENE COMPOUNDS and ALKYNES see ACJ125
ACETYLENEDICARBOXAMIDE see ACJ250
ACETYLENEDICARBOXYLIC ACID DIAMIDE see ACJ250
ACETYLENEDICARBOXYLIC ACID, DIMETHYL ESTER see DOP400
ACETYLENEDICARBOXYLIC ACID MONOPOTASSIUM SALT see ACJ500
ACETYLENE DICHLORIDE see DFI210
trans-ACETYLENE DICHLORIDE see ACK000
ACETYLENE, METHYL- see MFX590
ACETYLENE, PHENYL- see PEB750
ACETYLENE TETRABROMIDE see ACK250
ACETYLENE TETRACHLORIDE see TBQ100
ACETYLENE TRICHLORIDE see TIO750
ACETYL ENHEPTIN see ABY900
ACETYLENOGEN see CAN750
N-ACETYL ETHANOLAMINE see HKM000
ACETYL ETHER see AAX500
N-ACETYL ETHYL CARBAMATE see ACL000
N-ACETYLETHYL-2-cis-CROTONYLCARBAMIDE see ACL250
ACETYL ETHYLENE see BOY500
ACETYLETHYLENEIMINE see ACB250
N'-ACETYL ETHYLNITROSOUREA see ACL500
ACETYL ETHYL TETRAMETHYL TETRALIN see ACL750
ACETYLETHYL TETRAMETHYLTETRALIN see ACL750
ACETYLEUGENOL see EQS000
N-ACETYL-p-FENYLENDIAMIN (CZECH) see AHQ250
N-ACETYL-m-FENYLENEDIAMIN (CZECH) see AHQ000
ACETYLFERROCENE see ABA750
1-ACETYLFERROCENE see ABA750
2-ACETYLFLUORENE see FEI200
N-ACETYL-N-9H-FLUOREN-2-YL-ACETAMIDE see DBF200
ACETYL FLUORIDE see ACM000
ACETYL FLUORIDE, FLUORO- see FFS100
21-ACETYL-6-α-FLUORO-16-α-METHYLPREDNISOLONE see PAL600
ACETYLFORMALDEHYDE see PQC000
ACETYLFORMIC ACID see PQC100
ACETYLFORMYL see PQC000
ACETYLFURAN see ACM200
2-ACETYLFURAN see ACM200
16-ACETYLGITOXIN see ACM250
ACETYLGITOXIN-α see ACM245
N-ACETYL-l-GLUTAMINE ALUMINUM SALT see AGX125
6-o-ACETYL GRAYANOTOXIN 1 see GJU315
6-ACETYL-GRAYANOTOXIN I see GJU315
5-ACETYL-1,1,2,3,3,6-HEXAMETHYLINDAN see HEJ400
7-ACETYL-1,1,3,4,4,6-HEXAMETHYL-1,2,3,4-TETRAHYDRONAPHTHALENE see FBW110
7-ACETYL-1,1,3,4,4,6-HEXAMETHYLTETRALIN see FBW110
6-ACETYL-1,1,2,4,4,7-HEXAMETHYLTETRALINE see EEE100
N-ACETYLHOMOCYSTEINE THIOLACTONE see TCZ000
N-ACETYLHOMOCYSTEINTHIOLAKTON (GERMAN) see TCZ000
ACETYL HYDRAZIDE see ACM750
N-ACETYLHYDRAZINE see ACM750
ACETYL HYDROPEROXIDE see PCL500
ACETYLHYDROQUINONE see DMG600
2-ACETYLHYDROQUINONE see DMG600
ACETYLHYDROXAMIC ACID see ABB250

N-ACETYL-4-HYDROXY-m-ARSANILIC ACID see ABX500
N-ACETYL-4-HYDROXY-m-ARSANILIC ACID, CALCIUM SALT see CAL500
N-ACETYL-4-HYDROXYARSANILIC ACID compounded with DIETHYLAMINE (1:1) see ACN250
N-ACETYL-4-HYDROXY-m-ARSANILIC ACID DIETHYLAMINE SALT see ACN250
N-ACETYL-4-HYDROXY-m-ARSANILIC ACID SODIUM SALT see SEG000
2-ACETYL-7-(2-HYDROXY-3-sec-BUTYLAMINOPROPOXY)BENZOFURAN see ACN310
2-ACETYL-4-(2-HYDROXY-3-tert-BUTYLAMINOPROPOXY)BENZOFURAN see ACN300
2-ACETYL-7-(2-HYDROXY-3-tert-BUTYLAMINOPROPOXY)BENZOFURAN see ACN320
17-β-ACETYL-17-HYDROXYESTR-4-ENE-3-ONE HEXANOATE see GEK510
2-ACETYL-10-(3-(4-(β-HYDROXYETHYL)PIPERAZINYL)PROPYL)PHENOTHIAZINE see ABG000
2-ACETYL-10-(3-(4-(β-HYDROXYETHYL)PIPERIDINO)PROPYL)PHENOTHIAZINE see PII500
2-ACETYL-7-((2-HYDROXY-3-ISOPROPYLAMINO)PROPOXY)BENZOFURAN HYDROCHLORIDE see BAU255
3'-ACETYL-4'-(2-HYDROXY-3-(ISOPROPYLAMINO)PROPOXY)BUTYRANILIDE see AAE100
3'-ACETYL-4'-(2-HYDROXY-3-(ISOPROPYLAMINO)PROPOXY)BUTYRANILIDE HYDROCHLORIDE see AAE125
(±)-N-(3-ACETYL-4-(2-HYDROXY-3-((1-METHYLETHYL)AMINO)PROPOXY)PHENYL)BUTANAMIDE see AAE100
N-ACETYL-N'-(p-HYDROXYMETHYL)PHENYLHYDRAZINE see ACN500
2-ACETYL-5-HYDROXY-3-OXO-4-HEXENOIC ACID Δ-LACTONE see MFW500
N-ACETYL-S-(2-HYDROXYPHENYLETHYL)-l-CYSTEINE see ACN600
N-ACETYL-S-(2-HYDROXY-1-PHENYLETHYL)-l-CYSTEINE see ACN700
5-ACETYL-8-HYDROXYQUINOLINE see HOE200
ACETYL HYPOBROMITE see ACN875
2-ACETYL-7-(2-HYROXY-3-ISOPROPYLAMINOPROPOXY)BENZOFURAN see HLK600
ACETYLIDES see ACO000
1-ACETYLIMIDAZOLE see ACO250
N-ACETYLIMIDAZOLE see ACO250
ACETYLIN see ADA725
ACETYL-3-INDOLE see ICY100
3-ACETYLINDOLE see ICY100
5-ACETYLINDOLE see ACO300
ACETYL-5-INDOLE see ACO300
3-ACETYLINDOLE-5-CARBONITRILE see COP550
5-ACETYLINDOLINE see ACO320
ACETYL IODIDE see ACO500
ACETYLISOEUGENOL see AAX750
ACETYL ISONIAZID see ACO750
N-ACETYLISONIAZID see ACO750
1-ACETYL-2-ISONICOTINOYLHYDRAZINE see ACO750
N-ACETYLISONICOTINYLHYDRAZIDE see ACO750
N-ACETYL-N'-ISONICOTINYL HYDRAZIDE see ADA000
ACETYLISOPENTANOYL see MKL300
ACETYL ISOVALERYL see MKL300
3-o-ACETYLJERVINE see JDA000
ACETYLKAPROLAKTAM see ACF100
ACETYLKIDAMYCIN see ACP000

ACETYL KITASAMYCIN see MRE750
(2-ACETYLLACTOYLOXYETHYL)TRIMETHYLAMMONIUM HEMI-1,5-NAPHTHALENEDISULFONATE see AAC875
(2-ACETYLLACTOYLOXYETHYL)TRIMETHYLAMMONIUM 1,5-NAPHTHALENEDISULFONATE see AAC875
ACETYL-LANATOSID A (GERMAN) see ACP250
ACETYLLANATOSIDE A see ACP250
ACETYLLANDROMEDOL see AOO375
N-ACETYL-l-LEUCYL-N-(4-((AMINOIMINOMETHYL)AMINO)-1-FORMYLBUTYL)-l-LEUCINAMIDE (9CI) see LEX400
N-ACETYL-l-LEUCYL-l-LEUCYL-l-ARGINAL see LEX400
N-ACETYLLOLINE see ACP300
d-1-ACETYL LYSERGIC ACID DIETHYLAMIDE see AFJ450
1-ACETYLLYSERGIC ACID DIETHYLAMIDE BITARTRATE see ACP500
1-ACETYLLYSERGIC ACID ETHYLAMIDE see ACP750
d-1-ACETYL LYSERGIC ACID MONOETHYLAMIDE see ACP750
ACETYLMANDELIC ACID-(2-(DIMETHYLAMINO)-1-PHENYL)ETHYL ESTER see ACQ000
β-ACETYLMANDELOYLOXY-β-PHENYLETHYL DIMETHYLAMINE see ACQ000
ACETYL MERCAPTAN see TFA500
ACETYLMERCAPTOACETIC ACID see ACQ250
N-ACETYL-3-MERCAPTOALANINE see ACH000
α-1-ACETYLMETHADOL see ACQ258
l-α-ACETYLMETHADOL see ACQ258
levo-α-ACETYLMETHADOL see ACQ258
l-α-ACETYLMETHADOL HYDROCHLORIDE see ACQ260
ACETYLMETHIONINE see ACQ275
N-ACETYLMETHIONINE see ACQ275
ACETYL-dl-METHIONINE see ACQ270
N-ACETYL-l-METHIONINE see ACQ275
dl-N-ACETYLMETHIONINE see ACQ270
N-ACETYL-dl-METHIONINE see ACQ270
1-ACETYL-17-METHOXYASPIDO-SPRIDINE see ARP050
3-ACETYL-6-METHOXY-2H-1-BENZOPYRAN see ACQ700
3-ACETYL-7-METHOXY-2H-1-BENZOPYRAN see ACQ730
3-ACETYL-8-METHOXY-2H-1-BENZOPYRAN see ACQ760
2-ACETYL-7-METHOXYNAPHTHO(2,1-b)FURAN see ACQ790
N-ACETYL-5-METHOXYTRYPTAMINE see MCB350
4-(N-ACETYL-N-METHYL)AMINO-4'-(N',N'-DIMETHYLAMINO)AZOBENZENE see DPQ200
N'-ACETYL-N'-METHYL-4'-AMINO-N,N-DIMETHYL-4-AMINOAZOBENZENE see DPQ200
4-(N-ACETYL-N-METHYL)AMINO-4'-N-METHYLAMINOAZOBENZENE see MLK750
N-ACETYL-METHYLANILINE see MFW000
ACETYL METHYL BROMIDE see BNZ000
ACETYLMETHYLCARBAMIC ACID 1,3-BENZODIOXOL-4-YL ESTER see BCJ025
ACETYLMETHYLCARBAMIC ACID 2,2-DIMETHYL-1,3-BENZODIOXOL-4-YL ESTER see BCJ005

N-ACETYL-N-(METHYLCARBAMOYLOXY)-N'-METHYLUREA see CBG075
ACETYL METHYL CARBINOL see ABB500
ACETYLMETHYLCHOLINE see MFX560
ACETYL-β-METHYLCHOLINE see MFX560
o-ACETYL-β-METHYLCHOLINE CHLORIDE see ACR000
2-ACETYL-2-METHYL-1,3-DITHIOLANE see ACR050
α-ACETYL-6-METHYLERGOLINE-8-β-PROPIONAMIDE see ACR100
(2-(3-ACETYL-2-METHYL-1-INDOLIZINYL)-2-METHYLETHYL)DIMETHYLHEXYLAMMONIUMBROMIDE see ACR120
(2-(3-ACETYL-2-METHYL-1-INDOLIZINYL)-2-METHYLETHYL)DIMETHYLPENTYLAMMONIUMBROMIDE see ACR130
(2-(3-ACETYL-2-METHYL-1-INDOLIZINYL)-2-METHYLETHYL)DIMETHYL(1,4-XYLYL)AMMONIUM CHLORIDE see ACR140
N-ACETYL-N-(2-METHYL-4-((2-METHYLPHENYL)AZO)PHENYL)ACETAMIDE see ACR300
ACETYL-METHYL-NITROSO-HARNSTOFF (GERMAN) see ACR400
ACETYLMETHYLNITROSOUREA see ACR400
N'-ACETYL-METHYLNITROSOUREA see ACR400
3-ACETYL-10-(3'-N-METHYL-PIPERAZINO-N'-PROPYL)PHENOTHIAZIN see ACR500
3-ACETYL-6-METHYL-2,4-PYRANDIONE see MFW500
3-ACETYL-6-METHYLPYRANDIONE-2,4 see MFW500
3-ACETYL-6-METHYL-2H-PYRAN-2,4(3H)-DIONE see MFW500
N'-ACETYL-N'-MONOMETHYL-4'-AMINO-N-ACETYL-N-MONOMETHYL-4-AMINOAZOBENZENE see DQH200
N'-ACETYL-N'-MONOMETHYL-4'-AMINO-N-MONOMETHYL-4-AMINOAZOBENZENE see MLK750
6-ACETYLMORPHINE see ACR600
6-O-ACETYLMORPHINE see ACR600
4-ACETYLMORPHOLINE see ACR750
N-ACETYLMORPHOLINE see ACR750
3-ACETYL-10-(3'-MORPHOLINOPROPYL)PHENOTHIAZIN see MMA600
N-ACETYL-N-MYRISTOYLOXY-2-AMINOFLUORENE see ACS000
(N-ACETYL-d-β-NALL-d-PCL-PHE2-d-PHE3-d-ARG6-PHE7-ARG8-d-ALA10)NH2 GNRH see ACS100
1-ACETYLNAPHTHALENE see ABC475
2-ACETYLNAPHTHALENE see ABC500
β-ACETYLNAPHTHALENE see ABC500
o-ACETYL-N-(2-NAPHTHOYL)HYDROXYLAMINE see ACS250
N-ACETYL-1-NAPHTHYLAMINE see NAK000
N-ACETYL-2-NAPHTHYLHYDROXYLAMINE see NBD000
N-ACETYLNEOMYCIN see ACS375
8-ACETYLNEOSOLANIOL see ACS500
ACETYL NITRATE see ACS750
ACETYL NITRITE see ACT000
p-ACETYLNITROBENZENE see NEL600
2-ACETYL-5-NITROFURAN see ACT250
N'-ACETYL-5-NITRO-2-FUROHYDRAZIDE see NGH600
o-ACETYL-N-((5-NITRO-2-FURYL)FORMIMIDOYL)HYDROXYLAMINE see NGC500

O-ACETYL NITROHYDROXYAMINE see NEC500
O-ACETYL-N-NITROHYDROXYLAMINE see NEC500
ACETYLNITROPEROXIDE see PCL750
ACETYL(P-NITROPHENYL)SULFANILAMIDE see SNQ600
2-ACETYL-4-NITROPYRROLE see ACT300
2-ACETYL-5-NITROPYRROLE see ACT330
1-ACETYL-4-NITROSO-2,6-DIMETHYLPIPERAZINE see NJI900
2-ACETYL-5-NITROTHIOPHENE see NML100
ACETYLON FAST BLUE G see TBG700
ACETYLON FAST PINK B see AKE250
ACETYLON FAST RED VIOLET R see DBP000
N-ACETYL-S-(1 OR 2-PHENYL-2-HYDROXYETHYL)CYSTEINE see ACN700
ACETYL OXIDE see AAX500
5-ACETYLOXINE see HOE200
7-ACETYL-5-OXO-5H-(1)BENZOPYRANO(2,3-b)PYRIDINE see ACU125
(3-β,5-α)-3-(ACETYLOXY)ANDROSTAN-17-ONE see HJB225
2-(ACETYLOXY)BENZOIC ACID see ADA725
2-(ACETYLOXY)BENZOIC ACID 4-(ACETYLAMINO)PHENYL ESTER see SAN600
2-(ACETYLOXY)BENZOIC ACID, mixed with 3,7-DIHYDRO-1,3,7-TRIMETHYL-1H-PURINE-2,6-DIONE and N-(4-ETHOXYPHENYL)ACETAMIDE see ARP250
N-(ACETYLOXY)-N-(1,1'-BIPHENYL)-4-YLACETAMIDE see ABJ750
2-(ACETYLOXY)-3-BROMO-N-(4-BROMOPHENYL)-5-CHLORO-BENZENECARBOTHIOAMIDE see BOL325
N-(ACETYLOXY)-N-BUTOXYBENZAMIDE see ACU130
N-(ACETYLOXY)-N-BUTOXY-4-CHLOROBENZAMIDE see ACU136
N-(ACETYLOXY)-N-BUTOXY-4-METHOXYBENZAMIDE see ACU144
N-(ACETYLOXY)-N-BUTOXY-4-METHYLBENZAMIDE see ACU148
N-(ACETYLOXY)-N-BUTOXY-4-NITROBENZAMIDE see ACU156
2-(ACETYLOXY)-N-(4-CHLOROPHENYL)-3,5-DIIODOBENZAMIDE see CGB250
17-(ACETYLOXY)-6-CHLOROPREGNA-4,6-DIENE-3,20-DIONE see CBF250
ACETYLOXYCYCLOHEXIMIDE see ABN000
(11-β,16-α)-21-(ACETYLOXY)-16,17-(CYCLOPENTYLIDENEBIS(OXY))-9-FLUORO-11-HYDROXYPREGNA-1,4-DIENE-3,20-DIONE see COW825
3-(ACETYLOXY-3,7-DIHYDRO)-1H-PURINE-2,6-DIONE see HOP300
10-(ACETYLOXY)-1,8-DIHYDROXY-9(10H)-ANTHRACENONE see ACI640
(11-β)-21-(ACETYLOXY)-11,17-DIHYDROXY-PREGN-4-ENE-3,20-DIONE (9CI) see HHQ800
2-(ACETYLOXY)-4-((3,4-DIMETHYL-5-ISOXAZOLYL)IMINO)-1(4H)-NAPHTHALENONE see ACU200
2-ACETYLOXY-N-(3,4-DIMETHYL-5-ISOXAZOLYL)-1,4-NAPHTHOQUINONE-4-IMINE see ACU200
4-ACETYLOXY-12,13-EPOXY-3,7,15-TRIHYDROXY-(3-α,4-β,7-β)-TRICHOTHEC-9-EN-8-ONE see FQR000
(16-β,17-β)-17-(ACETYLOXY)-16-ETHYL-ESTR-4-EN-3-ONE (9CI) see ELF110
(2-(ACETYLOXY)ETHYL)NITROSOCARBAMIC ACID ETHYL ESTER see EFR500

2-(ACETYLTHIO)-N,N,N-TRIMETHYLETHANAMINIUM IODIDE see ADC300
ACETYL THIOUREA see ADD250
1-ACETYL-2-THIOUREA see ADD250
p-ACETYLTOLUENE see MFW250
N-ACETYL-p-TOLUIDIDE see ABJ250
ACETYL-o-TOLUIDINE see ABJ000
ACETYL-p-TOLUIDINE see ABJ250
N-ACETYL-m-TOLUIDINE see ABI750
ACETYL TRIBUTYL CITRATE see ADD400
ACETYL TRIBUTYL CITRATE see THX100
ACETYL TRIETHYL CITRATE see ADD750
N-ACETYL-N-(2,4,6-TRIIODO-3-AMINOPHENYL)-β-AMINOISOBUTYRIC ACID see IDJ550
ACETYLTRIIODOTHYRONINE FORMIC ACID see ADD875
N-ACETYL TRIMETHYLCOLCHICINIC ACID see ADE000
N-ACETYL TRIMETHYLCOLCHICINIC ACID METHYL ETHER see CNG938
3-ACETYL-2,4,5-TRIMETHYL-PYRROLE see ADE050
ACETYL-l-TRP see ADE075
ACETYLTRYPTOPHAN see ADE075
ACETYL-l-TRYPTOPHAN see ADE075
N-ACETYLTRYPTOPHAN see ADE075
N-ACETYL-l-TRYPTOPHAN see ADE075
(S)-N-ACETYLTRYPTOPHAN see ADE075
ACETYLURETHANE see ACL000
ACETYLUREUM see PEC250
N-(N-ACETYLVALYL)-N-NITROSOGLYCINE see ACE100
ACETYL YELLOW G see CMM758
ACETYLZIRCONIUM, ACETONATE see PBL750
AC GA see PJS750
ACH see CMF250
ACHANIA, flower extract see ADE125
ACH CHLORIDE see ABO000
ACHILLEIC ACID see ADH000
ACHIOTE see APE100
ACHLESS see TKH750
ACHLETIN see HII500
ACHROCIDIN see ABG750
ACHROMYCIN see TBX000
ACHROMYCIN see TBX250
ACHROMYCIN HYDROCHLORIDE see TBX250
ACHROMYCIN (PURINE DERIVATIVE) see AEI000
ACHTL see TCZ000
AC 2197 HYDROCHLORIDE see DBA475
ACIBILIN see TAB250
ACICLOVIR see AEC700
ACID see DJO000
2B ACID see AJJ250
ACIDAL BLACK 10B see FAB830
ACIDAL BRIGHT PONCEAU 3R see FMU080
ACIDAL BRILLIANT RED 2G see CMM300
ACIDAL CARMINE V see CMM062
ACIDAL FAST ORANGE see HGC000
ACIDAL GREEN G see FAE950
ACID ALIZARINE PURE BLUE B see CMM080
ACID ALIZARINE PURE BLUE R see CMM080
ACID ALIZARINE RED B see CMG750
ACID ALIZARINE SAPPHIRE SE see APG700
ACID ALIZARINE SKY BLUE B see CMM090
ACID ALIZARINE VIOLET see HLI000
ACID ALIZARINE VIOLET B see HLI000
ACID ALIZARINE VIOLET N see HLI000
ACID ALIZARIN RED B see CMG750
ACIDAL LIGHT GREEN SF see FAF000
ACIDAL NAVY BLUE 3BR see FAB830
ACIDAL PONCEAU G see FMU070
ACID AMARANTH see FAG020
ACID AMIDE see NCR000

ACID AMMONIUM CARBONATE see ANB250
ACID AMMONIUM FLUORIDE see ANJ000
ACID AMMONIUM SULFATE see ANJ500
ACID ANTHRACENE RED G see CMG750
ACID ANTHRACENE RED G see CMM325
ACID ANTHRACENE RED GA-CF see CMM325
ACID ANTHRACENE YELLOW GR see CMM759
ACID ANTHRAQUINONE BLUE see APG700
ACID ANTHRAQUINONE BRILLIANT BLUE see CMM092
ACID ANTHRAQUINONE PURE BLUE see CMM090
ACID BLACK 1 see FAB830
ACID BLACK 10A see FAB830
ACID BLACK 10B see FAB830
ACID BLACK 12B see FAB830
ACID BLACK 10BA see FAB830
ACID BLACK BASE M see FAB830
ACID BLACK BRX see FAB830
ACID BLACK BX see FAB830
ACID BLACK H see FAB830
ACID BLACK JVS see FAB830
ACID BLACK 4BN see FAB830
ACID BLACK 10BN see FAB830
ACID BLACK 4BNU see FAB830
ACID BLUE 1 see ADE500
ACID BLUE 3 see CMM062
ACID BLUE 7 see ADE675
ACID BLUE 9 see FMU059
ACID BLUE 41 see CMM070
ACID BLUE 43 see APG700
ACID BLUE 62 see CMM080
ACID BLUE 78 see CMM090
ACID BLUE 80 see CMM092
ACID BLUE 87 see COF420
ACID BLUE 92 see ADE750
ACID BLUE 129 see CMM100
ACID BLUE 185 see CMM120
ACID BLUE A see ADE750
ACID BLUE BLACK 10B see FAB830
ACID BLUE BLACK B see FAB830
ACID BLUE BLACK BG see FAB830
ACID BLUE BLACK DOUBLE 600 see FAB830
ACID BLUE O see ERG100
ACID BLUE V see ADE500
ACID BLUE W see FAE100
ACID BRIGHT AZURE Z see ADE500
ACID BRIGHT RED see CMM300
ACID BRILLIANT BLUE ANTHRAQUINONE see CMM092
ACID BRILLIANT BLUE RAWL see CMM092
ACID BRILLIANT BLUE VF see ADE500
ACID BRILLIANT BLUE Z see ADE500
ACID BRILLIANT GREEN BS see ADF000
ACID BRILLIANT GREEN SF see FAF000
ACID BRILLIANT PINK B see FAG070
ACID BRILLIANT RED see CMM300
ACID BRILLIANT RUBINE 2G see HJF500
ACID BRILLIANT SCARLET 3R see FMU080
ACID BRILLIANT SKY BLUE Z see ADE500
ACID BUTYL PHOSPHATE see ADF250
ACID CALCIUM PHOSPHATE see CAW110
ACID CARBOYS, EMPTY see ADF500
ACID CHLORIDE see DCI100
ACID CHROME BLACK ET see EDC625
ACID CHROME BLACK ETN see EDC625
ACID CHROME BLUE BA see HJF500
ACID CHROME ORANGE GR see CMP882
ACID CHROME RED A see CMG750
ACID CHROME RED B see CMG750
ACID CHROME VIOLET K see HLI000
ACID CHROME VIOLET N see HLI000
ACID CHROME YELLOW GG see SIT850
ACID CHROME YELLOW 2GW see SIT850
ACID COPPER ARSENITE see CNN500
ACIDE ACETIQUE (FRENCH) see AAT250

ACIDE ACETYLSALICYLIQUE (FRENCH) see ADA725
ACIDE ACETYL SELENO-2 BENZOIQUE see SBU100
ACIDE ANISIQUE (FRENCH) see MPI000
ACIDE ARSENIEUX see ARI750
ACIDE ARSENIQUE LIQUIDE (FRENCH) see ARB250
ACIDE BENZOIQUE (FRENCH) see BCL750
ACIDE BENZOYL-2-PHENYLACETIQUE (FRENCH) see BDS500
ACIDE BROMACETIQUE (FRENCH) see BMR750
ACIDE BROMHYDRIQUE (FRENCH) see HHJ000
l'ACIDE BUCLOXIQUE (FRENCH) see CPJ000
ACIDE BUCLOXIQUE CALCIUM (FRENCH) see CPJ250
ACIDE CACODYLIQUE (FRENCH) see HKC000
ACIDE CARBOLIQUE (FRENCH) see PDN750
ACIDE CHLORACETIQUE (FRENCH) see CEA000
ACIDE CHLORHYDRIQUE (FRENCH) see HHL000
ACIDE N-(p-CHLOROBENZYL)PYROGLUTAMIQUE see CKF800
ACIDE 2-(4-CHLORO-2-METHYL-PHENOXY)PROPIONIQUE (FRENCH) see CIR500
ACIDE p-CHLOROPHENYL-2-THIAZOLE-ACETIQUE-4 (FRENCH) see CKK250
ACIDE CHROMIQUE (FRENCH) see CMH250
ACIDE CRESYLIQUE (FRENCH) see CNW500
ACIDE CYANACETIQUE (FRENCH) see COJ500
ACIDE CYANHYDRIQUE (FRENCH) see HHS000
l'ACIDE (CYCLOHEXYL-4, CHLORO-3, PHENYL)-4,OXO-4, BUTYRIQUE CALCIUM (FRENCH) see CPJ250
ACIDE DEHYDROCHOLIQUE (FRENCH) see DAL000
ACIDE N-(DICHLORO-2',4' BENZYL)PYROGLUTAMIQUE see DGE300
ACIDE N-(DICHLORO-2',6'-BENZYL)PYROGLUTAMIQUE see DGE305
ACIDE N-(DICHLORO-3',4'-BENZYL)PYROGLUTAMIQUE see DGE310
ACIDE-2,4-DICHLORO PHENOXYACETIQUE (FRENCH) see DAA800
ACIDE-2-(2,4-DICHLORO-PHENOXY) PROPIONIQUE (FRENCH) see DGB000
ACIDE DIMETHYLARSINIQUE (FRENCH) see HKC000
ACIDE DIMETHYL-ETHYL-ALLENOLIQUE ETHER METHYLIQUE (FRENCH) see DRU600
ACIDE DIPHENYLETHOXYACETIQUE (FRENCH) see EES300
ACIDE DIPHENYLHYDROXYACETIQUE see BBY990
ACIDE DISELINO SALICYLIQUE see SBU150
ACIDE ETHYLENEDIAMINETETRACETIQUE (FRENCH) see EIX000
ACIDE ETHYL-8 OXO-5 PIPERAZINYL-2 DIHYDRO-5,8 PYRIDO(2,3-d)PYRIMIDINE-6 CARBOXYLIQUE see PIZ000
ACIDE 1-ETIL-7-METIL-1,8-NAFTIRIDIN-4-ONE-3-CARBOSSILICO (ITALIAN) see EID000
ACIDE FLUORHYDRIQUE (FRENCH) see HHU500

ACPC see AJK250
ACP GRASS KILLER see TII500
AC 8 (POLYMER) see PJS750
ACQUINITE see ADR000
ACQUINITE see CKN500
ACRAFIX FHN see EAZ600
ACRALDEHYDE see ADR000
ACRAMINE RED see DBN000
ACRAMINE YELLOW see AHS750
ACRAMIN SCARLET LDCN see DTV360
ACRAMOLL W see VPK333
ACRANIL see ADI750
ACRANIL DIHYDROCHLORIDE see ADI750
ACRANIL HYDROCHLORIDE see ADI750
ACREX see CBW000
ACRIBEL, combustion products see ADX750
ACRICHINE see ARQ250
ACRICHINE see CFU750
ACRICID see BGB500
ACRIDAN see ADI775
ACRIDANE see ADI775
ACRIDAN, 9-METHOXY-9-((4-METHYL-1-
PIPERAZINYL)METHYL)- see MEX285
ACRIDANONE see ADI825
9-ACRIDANONE see ADI825
2-ACRIDINAMINE see AHS000
9-ACRIDINAMINE see AHS500
3-ACRIDINAMINE (9CI) see ADJ375
9-ACRIDINAMINE, N-(5-(4-((2-
CHLOROETHYL)ETHYLAMINO)PHENOX
Y)PENTYL)- see EHI600
9-ACRIDINAMINE
MONOHYDROCHLORIDE see AHS750
9-ACRIDINAMINE, 1,2,3,4-TETRAHYDRO-
(9CI) see TCJ075
ACRIDINE see ADJ500
ACRIDINE, 9-ACETAMIDO- see ABX810
ACRIDINE, 9-(2,2-BIS(2-
CHLOROETHYL)HYDRAZINO)-,
MONOHYDROCHLORIDE see BID800
ACRIDINE-9-CARBOXAMIDE, N,N-
DIETHYL-1,2,3,4-TETRAHYDRO- see
ADJ550
1-ACRIDINECARBOXAMIDE, N-(2-
(DIMETHYLAMINO)ETHYL)- see DPB220
4-ACRIDINECARBOXAMIDE, 9-((2-
METHOXY-4-
((PROPYLSULFONYL)AMINO)PHENYL)AM
INO)-, HYDROCHLORIDE see MFN000
ACRIDINE-9-CARBOXAMIDE, 1,2,3,4-
TETRAHYDRO-N,N-DIETHYL- see ADJ550
ACRIDINE, 6-CHLORO-9-((4-
(DIETHYLAMINO)-1-
METHYLBUTYL)AMINO)-2-METHOXY-,
DIMETHANESULFONATE see QCS900
ACRIDINE, 2-CHLORO-9-(2,2-
DIMETHYLHYDRAZINO)- see CGH800
ACRIDINE, 9-(2-((2-
CHLOROETHYL)AMINO)ETHYLAMINO)-
6-CHLORO-2-METHOXY-,
DIHYDROCHLORIDE, HYDRATE see
QCS875
ACRIDINE, 9-((2-((2-
CHLOROPROPYL)AMINO)ETHYL)AMINO)
-2-METHOXY-, DIHYDROCHLORIDE,
HEMIHYDRATE see CKU250
2,6-ACRIDINEDIAMINE see DBN000
3,6-ACRIDINEDIAMINE see DBN600
3,9-ACRIDINEDIAMINE (9CI) see ADJ625
3,6-ACRIDINEDIAMINE,
MONOHYDROCHLORIDE (9CI) see
PMH250
3,6-ACRIDINEDIAMINE SULFATE (2:1) see
DBN400
3,6-ACRIDINEDIAMINE SULPHATE see
DBN400
ACRIDINE, 9-(2,2-DIBUTYLHYDRAZINO)-,
MONOHYDROCHLORIDE see DEC797
ACRIDINE, 9,10-DIHYDRO-(9CI) see ADI775
ACRIDINE, 9-((3-
(DIMETHYLAMINO)PROPYL)AMINO)-3-
NITRO- see NFW525

ACRIDINE, 9-((3-
(DIMETHYLAMINO)PROPYL)AMINO)-2-
NITRO-, DIHYDROCHLORIDE see NFW535
ACRIDINE, 9-(2,2-
DIMETHYLHYDRAZINO)- see DSG320
ACRIDINE, 9-(2,2-
DIMETHYLHYDRAZINO)-,
MONOHYDROCHLORIDE see DSG330
ACRIDINE, 9-(2,2-
DIMETHYLHYDRAZINO)-,
MONO(METHYL SULFATE) see DSG340
ACRIDINE, 9-((3-
(HEXYLAMINO)PROPYL)AMINO)-1-
NITRO-, DIHYDROCHLORIDE see HFK500
ACRIDINE HYDROCHLORIDE see ADJ750
ACRIDINE, 9-(1-METHYLHYDRAZINO)-,
MONOHYDROCHLORIDE see MKN600
ACRIDINE MONOHYDROCHLORIDE see
ADJ750
ACRIDINE, 9-(MORPHOLINOAMINO)- see
MRR112
ACRIDINE, 9-(MORPHOLINOAMINO)-,
MONO(METHYL SULFATE) see MRR115
ACRIDINE, 9-(MORPHOLINOCARBONYL)-
1,2,3,4-TETRAHYDRO- see MRR760
ACRIDINE, 9-((2-
MORPHOLINOETHYL)AMINO)- see
MRT200
ACRIDINE MUSTARD see ADJ875
ACRIDINE ORANGE see BAQ250
ACRIDINE ORANGE see BJF000
ACRIDINE ORANGE FREE BASE see BJF000
ACRIDINE ORANGE NO see BAQ250
ACRIDINE ORANGE R see BAQ250
ACRIDINE, 9-(PIPERIDINOAMINO)- see
PIN200
ACRIDINE RED see ADK000
ACRIDINE RED 3B see ADK000
ACRIDINE RED, HYDROCHLORIDE see
ADK000
ACRIDINE, 1,2,3,4-TETRAHYDRO-5-
CHLORO-9-MORPHOLINO-,
HYDROCHLORIDE see CLH100
ACRIDINE, 1,2,3,4-TETRAHYDRO-9-
(MORPHOLINOCARBONYL)- see MRR760
ACRIDINE, 1,2,3,4-TETRAHYDRO-9-
(PIPERIDINOCARBONYL)- see PIU100
ACRIDINE YELLOW see DBT400
ACRIDINE YELLOW BASE see DBT200
ACRIDINE YELLOW G see DBT400
ACRIDINIUM, 3,6-BIS(DIMETHYLAMINO)-
10-(PHENYLMETHYL)- see BDX033
ACRIDINIUM CHLORIDE see ADJ750
ACRIDINIUM, 3,6-DIAMINO-10-METHYL-,
CHLORIDE, HYDROCHLORIDE see
TNV700
ACRIDINO(2,1,9,8-klmna)ACRIDINE see
ADK250
ACRIDINO(2,1,9,8-klmna)ACRIDINE
SULFATE see DCK200
9(10H)-ACRIDINONE (9CI) see ADI825
4'-(9-ACRIDINYLAMINO)-2'-
AMINOMETHANESULFONANILIDE see
ADK750
4'-(9-ACRIDINYLAMINO)-3'-
AMINOMETHANESULFONANILIDE see
ADL000
4'(9-
ACRIDINYLAMINO)HEXANESULFONANI
LIDE see ADL250
4'-(9-
ACRIDINYLAMINO)METHANESULFON-m-
ANISIDE MONOHYDROCHLORIDE see
ADL500
4'-(9-
ACRIDINYLAMINO)METHANESULPHON-
m-ANISIDIDE see ADL750
4'-(9-ACRIDINYLAMINO)-2'-
METHOXYMETHANESULFONANILIDE
see ADM000

4'-(9-ACRIDINYLAMINO)-3'-
METHOXYMETHANESULFONANILIDE
see ADL750
4'-(9-ACRIDINYLAMINO)-3'-
METHOXYMETHANESULFONANILIDE
see ADM250
N-(4-(ACRIDINYL-9-AMINO)-3-
METHOXYPHENYL)ETHANESULFONAMI
DE METHANESULFONATE see ADM500
N-(4-(9-ACRIDINYLAMINO)-3-
METHOXYPHENYL)METHANESULFONA
MIDE see ADM000
N-(4-(9-ACRIDINYLAMINO)-3-
METHOXYPHENYL)METHANESULFONA
MIDE compounded with LACTIC ACID see
AOD425
4'-(9-ACRIDINYLAMINO)-2'-
METHYLMETHANESULFONANILIDE see
ADN000
4'-(9 ACRIDINYLAMINO)-3'-
METHYLMETHANESULFONANILIDE see
ADN250
4'-(9-
ACRIDINYLAMINO)METHYLSULFONYL-
m-ANISIDINE see ADL750
4'-(9-ACRIDINYLAMINO)-2'-
NITROMETHANESULFONANILIDE see
ADN500
N-(p-(ACRIDIN-9-
YLAMINO)PHENYL)BUTANESULFONAMI
DE, HYDROCHLORIDE see ADO250
N-(p-(9-ACRIDINYLAMINO)PHENYL)-1-
ETHANESULFONAMIDE see ADO500
N-(p-(ACRIDIN-9-YLAMINO)PHENYL)-
ETHANESULFONAMIDE,
HYDROCHLORIDE see ADO750
N-(p-(ACRIDIN-9-
YLAMINO)PHENYLHEXANESULFONAMI
DE) HYDROCHLORIDE see ADP000
N-(p-(ACRIDIN-9-
YLAMINO)PHENYL)METHANESULFONA
MIDE HYDROCHLORIDE see ADP500
N-(p-(ACRIDIN-9-
YLAMINO)PHENYL)PENTANESULFONAM
IDE HYDROCHLORIDE see ADP750
N-(p-(9-ACRIDINYLAMINO)PHENYL)-1-
PROPANESULFONAMIDE see ADQ000
N-(p-(ACRIDIN-9-
YLAMINO)PHENYL)PROPANESULFONAM
IDE HYDROCHLORIDE see ADQ250
N'-9-ACRIDINYL-N-(2-CHLOROETHYL)-N-
ETHYL-1,3-PROPANEDIAMINE
DIHYDROCHLORIDE see CGX750
N-(9-ACRIDINYL)-N'-(2-CHLOROETHYL)-
1,3-PROPANEDIAMINE see ADQ500
N'-9-ACRIDINYL-N,N-DIMETHYL-1,4-
BENZENEDIAMINE see DOS800
2-(1-(9-
ACRIDINYL)HYDRAZINO)ETHANOL
MONOHYDROCHLORIDE see ADQ550
4-(9-ACRIDINYL)-2-METHYL-3-
THIOSEMICARBAZIDE see ADQ560
4-(9-ACRIDINYL)-2-METHYL-3-
THIOSEMICARBAZONE ACETONE see
ADQ600
4-(9-ACRIDINYL)-3-THIOSEMICARBAZIDE
see ADQ610
ACRIDONE see ADI825
9-ACRIDONE see ADI825
9(10H)-ACRIDONE see ADI825
ACRIFLAVIN see DBX400
ACRIFLAVINE see ADQ700
ACRIFLAVINE see XAK000
ACRIFLAVINE NEUTRAL see XAK000
ACRIFLAVINE mixture with PROFLAVINE
see DBX400
ACRIFLAVINIUM CHLORIDE see DBX400
ACRIFLAVINIUM CHLORIDUM see DBX400
ACRIFLAVON see DBX400
ACRIFLAVON see XAK000
ACRILAFIL see ADY500
ACRINAMINE see ARQ250

ACRYLONITRILE POLYMER with CHLOROETHYLENE see ADY250
ACRYLONITRILE POLYMER with STYRENE see ADY500
ACRYLONITRILE-STYRENE COPOLYMER see ADY500
ACRYLONITRILE-STYRENE POLYMER see ADY500
ACRYLONITRILE-STYRENE RESIN see ADY500
ACRYLONITRILE, 3-(p-TOLYLSULFONYL)- see THD875
ACRYLONITRILE, 2,3,3-TRICHLORO- see TJC100
ACRYLONITRILE, TRICHLORO-(8CI) see TJC100
ACRYLOPHENONE see PMQ250
ACRYLOPHENONE, 4'-METHOXY- see ONW100
ACRYLOPHENONE, 3-(5-NITRO-2-FURYL)-3',4',5'-TRIMETHOXY- see NGN600
2-ACRYLOXYETHYLDIMETHYLSULFONIUM METHYL SULFATE see ADY750
ACRYLOYL CHLORIDE see ADZ000
ACRYLOYLNOVOCAINE see PME100
8-ACRYLOYLOXY-7-BROMO-5-CHLOROQUINOLINE see BNA900
2-(ACRYLOYLOXY)ETHANOL see ADV250
β-(ACRYLOYLOXY)PROPIONIC ACID see HGO700
ACRYLOYL PROCAINAMIDE MONOMER see PMD850
ACRYLOYL PROCAINE MONOMER see PME100
ACRYLSAEUREAETHYLESTER (GERMAN) see EFT000
ACRYLSAEUREMETHYLESTER (GERMAN) see MGA500
ACRYLYL CHLORIDE see ADZ000
ACRYLYLPROCAINE see PME100
ACRYPET see PKB500
ACRYSOL A 1 see ADV900
ACRYSOL A 3 see ADV900
ACRYSOL A 5 see ADV900
ACRYSOL AC 5 see ADV900
ACRYSOL ASE-75 see ADV900
ACRYSOL WS-24 see ADV900
ACS see ADY500
ACS see AJD000
AC 1370 SODIUM see CCS525
ACT see AEB000
ACTAEA (VARIOUS SPECIES) see BAF325
ACTAMER see TFD250
ACTASAL see CMG000
ACTEDRON see BBK000
ACTELIC see DIN800
ACTELLIC see DIN800
ACTELLIFOG see DIN800
ACTEMIN see AOB500
ACTEROL see NHH000
ACTH see AES650
ACTHAR see AES650
ACTI-AID see CPE750
ACTICARBONE see CBT500
ACTICEL see SCH002
ACTI-CHLORE see CDP000
ACTIDILAT see TMX775
ACTIDIONE see CPE750
ACTIDIONE TGF see CPE750
ACTIDOL see TMX775
ACTIDONE see CPE750
ACTIHAEMYL see ADZ125
ACTILIN see NCF000
ACTINE see EHP000
ACTINIC RADIATION see AEA000
ACTINOBOLIN see AEA109
ACTINOCHRYSIN see AEA750
ACTINOGAN see AEA250
ACTINOLITE ASBESTOS see ARM260
ACTINOMYCIN see AEA500
ACTINOMYCIN 23-21 see AEA625

ACTINOMYCIN 1048A see AEC000
ACTINOMYCIN 2104L see AEB750
ACTINOMYCIN BV see AEC200
ACTINOMYCIN C see AEA750
ACTINOMYCIN D see AEB000
ACTINOMYCINDIOIC D ACID, DILACTONE see AEB000
ACTINOMYCIN DV see AEC200
ACTINOMYCIN D, 3^A)-(4-HYDROXY-l-PROLINE)- see AEC185
ACTINOMYCIN I see AEB000
ACTINOMYCIN J1 see AEC200
ACTINOMYCIN K see AEB500
ACTINOMYCIN L see AEB750
ACTINOMYCIN S see AEC000
ACTINOMYCIN S3 see AEC175
ACTINOMYCIN S3 see AEC200
ACTINOMYCIN-V see AEC200
ACTINOMYCIN X2 see AEC200
ACTINOMYCIN X0(β) see AEC185
ACTINOSPECTACIN DIHYDROCHLORIDE PENTAHYDRATE see SKY500
ACTINOXANTHIN see AEC250
ACTINOXANTHINE see AEC250
ACTIOL see MBX800
ACTISPRAY see CPE750
ACTITHIAZIC ACID see CCI500
ACTITHIAZIC ACID see CMP885
ACTIVATED ALUMINUM OXIDE see AHE250
ACTIVATED ATTAPULGITE see PAE750
ACTIVATED CARBON see CBT500
ACTIVATED CARBON see CDI000
ACTIVE ACETYL ACETATE see EFS000
ACTIVE DICUMYL PEROXIDE see DGR600
ACTIVE METHIONINE see AEM310
ACTIVE VALERIC ACID see MHS600
ACTIVIN see AEC300
ACTIVIN see DYF450
ACTIVOL see ALT250
ACTOL see NDX500
ACTON see AES650
ACTONAR see AES650
ACTOR Q see DVR200
ACTOZINE see BCA000
ACTOZINE see PGA750
ACTRAPID see IDF300
AC-17 TRIHYDRATE see AER666
ACTRIL see HKB500
AC-TRY see ADE075
ACTYBARYTE see BAP000
ACTYLOL see LAJ000
ACULEACIN A see AEC625
ACUPAN see NBS500
ACYCLOGUANOSINE see AEC700
ACYCLOGUANOSINE SODIUM (OBS.) see AEC725
ACYCLOVIR see AEC700
ACYCLOVIR SODIUM SALT see AEC725
ACYLANID see ACH500
ACYLATE see ING400
ACYLATE-1 see ING400
ACYLFULVENE, 6-(HYDROXYMETHYL)- see HLU600
ACYLPYRIN see ADA725
ACYTOL see LAJ000
5-ACZ see ARY000
AD see AEB000
AD 1 see AGX000
AD1M see AGX000
AD 32 see TJX350
AD 41 see TJX300
AD 122 see OEM000
AD-205 see DSH000
AD-810 see BCE750
AD-1590 see DLI650
ADAB see DPO200
ADAKANE 12 see DXT200
ADALAT see AEC750
ADALGUR see GGQ050
ADALIN see BNK000
ADAM AND EVE see ITD050

1-ADAMANTAMINE see TJG250
1-ADAMANTANAMINE see TJG250
ADAMANTANAMINE HYDROCHLORIDE see AED250
1-ADAMANTANAMINE HYDROCHLORIDE see AED250
1-ADAMANTANEACETIC ACID-2-(DIETHYLAMINO)ETHYL ESTER, ETHYL IODIDE see AED750
1-ADAMANTANEACETIC ACID-3-(DIMETHYLAMINO)PROPYL ESTER, ETHYL IODIDE see AEE000
ADAMANTINE HYDROCHLORIDE see AED250
N-(1-ADAMANTYL)ACETAMIDE see AEE100
S-((N-1-ADAMANTYLAMIDINO)METHYL)HYDROGEN THIOSULFATE, HYDRATE (4:1) see AEE250
ADAMANTYLAMINE HYDROCHLORIDE see AED250
1-ADAMANTYLAMINE HYDROCHLORIDE see AED250
1-(1-ADAMANTYLAMINO)-2,2,2-TRIFLUORO-1-(TRIFLUOROMETHYL)ETHANOLSESQUIHYDRATE see AEE500
5-(1-ADAMANTYL)-2,4-DIAMINO-6-ETHYLPYRIMIDINE ETHYLSULFONATE see AEF000
5-(1-ADAMANTYL)-2,4-DIAMINO-6-METHYLPYRIMIDINE ETHYLSULFONATE see AEF250
N-1-ADAMANTYL-N-(2-(DIMETHYLAMINO)ETHOXY)ACETAMIDE HYDROCHLORIDE see AEF500
2,2'-(1,3-ADAMANTYLENE)N,N,N',N'-TETRAMETHYL-ETHYLAMINE DIHYDROCHLORIDE see BJG750
3-(1-ADAMANTYL)-1-(2-FLUOROETHYL)-1-NITROSOUREA see AEF600
N-(2-ADAMANTYL)-2-MERCAPTOACETAMIDINE HYDROCHLORIDE see AEG000
S-(N-(1-ADAMANTYLMETHYLAMIDINO)METHYL)PHOSPHOROTHIOATE MONOSODIUM SALT see AEG129
N-(1-ADAMANTYLMETHYL)-2-MERCAPTOACETAMIDINE HYDROCHLORIDE see AEG250
(2-(2-ADAMANTYLOXY)ETHYL)TRIETHYL-AMMONIUMIODIDE see DHP000
(2-(1-ADAMANTYLOXY)PROPYL)DIMETHYLETHYLAMMONIUM IODIDE see DPU600
3-(1-ADAMANTYL)PROPYLAMINE HYDROCHLORIDE see AMC500
N-(3-(1-ADAMANTYL)PROPYL)-2-MERCAPTOACETAMIDINE HYDROCHLORIDE HYDRATE (10:10:3) see AEG500
ADAME see DPB300
ADAMQUAT 80 MC see OOM500
ADAMSITE see PDB000
ADAMYCIN see HOH500
ADANON see MDO750
ADANON HYDROCHLORIDE see MDP000
ADANON HYDROCHLORIDE see MDP750
ADANTON HYDROCHLORIDE see AEG625
ADAPIN see AEG750
ADAPRIN see VTA100
ADAPTOL see AEG875
ADC AURAMINE O see IBA000
ADC BRILLIANT GREEN CRYSTALS see BAY750
ADCHEM GMO see GGR200
ADC MALACHITE GREEN CRYSTALS see AFG500
ADC PERMANENT RED TONER R see CJD500

ADC RHODAMINE B see FAG070
ADC TOLUIDINE RED B see MMP100
ADDEX-THAM see TEM500
ADDISOMNOL see BNK000
ADDITIN 30 see PFT250
ADDITIVE TI see IKG925
ADDUKT
HEXACHLORCYKLOPENTADIENU S
CYKLOPENTADIENEM (CZECH) see
HCN000
2-ADE see PDR490
A1-DEHYDROMETHYLTESTERONE see
DAL300
ADEKA FC 450 see DIV300
ADEKATOL SO 160 see AFJ160
ADELFAN see RDK000
ADELFA (PUERTO RICO) see OHM875
ADELPHANE see RDK000
ADELPHIN see RDK000
ADELPHIN-ESIDREX-K see RDK000
ADEMETIONINE see AEM310
ADEMINE see UVJ450
ADEMOL see TKG750
ADENINE see AEH000
ADENINE ARABINOSIDE see AEH100
ADENINE ARABINOSIDE see AQQ900
ADENINE ARABINOSIDE
MONOPHOSPHATE see AQQ905
ADENINE ARABINOSIDE 5'-
MONOPHOSPHATE see AQQ905
ADENINE, 9-BENZYL- see BDX100
ADENINE, N-BENZYL- see BDX090
ADENINE, N-BENZYL-9-(TETRAHYDRO-
2H-PYRAN-2-YL)-(8CI) see BEA100
ADENINE DEOXYRIBONUCLEOSIDE see
DAQ200
ADENINE DEOXYRIBOSE see DAQ200
ADENINE-FLAVIN DINUCLEOTIDE see
RIF100
ADENINE-FLAVINE DINUCLEOTIDE see
RIF100
ADENINE, N-FURFURYL- see FPT100
ADENINE-NICOTINAMIDE
DINUCLEOTIDE see CNF390
ADENINE-1-N-OXIDE see AEH250
ADENINE PROPENAL see AEH300
ADENINE-RIBOFLAVIN DINUCLEOTIDE
see RIF100
ADENINE-RIBOFLAVINE
DINUCLEOTIDE see RIF100
ADENINE RIBOSIDE see AEH750
ADENINE SULFATE see AEH500
ADENINIMINE see AEH000
ADENINSULFAT see AEH500
ADENIUM (VARIOUS SPECIES) see DBA450
ADENOCK see ZVJ000
ADENOHYPOPHYSEAL GROWTH
HORMONE see PJA250
ADENOHYPOPHYSEAL LUTEOTROPIN
see PMH625
ADENOSIN (GERMAN) see AEH750
ADENOSINE see AEH750
β-ADENOSINE see AEH750
β-d-ADENOSINE see AEH750
ADENOSINE, 5'-((l-3-AMINO-3-
CARBOXYPROPYL)METHYLSULFONIO)-5'-
DEOXY-, HYDROXIDE, INNER SALT see
AEM310
ADENOSINE, 3'-AMINO-3'-DEOXY- see
AJK800
ADENOSINE-3'-(α-AMINO-p-
METHOXYHYDROCINNAMAMIDO)-3'-
DEOXY-N,N-DIMETHYL see AEI000
ADENOSINE-5'-CARBOXAMIDE see AEI250
ADENOSINE, 2-CHLORO- see CEF100
ADENOSINE CYCLIC MONOPHOSPHATE
see AOA130
ADENOSINE-3',5'-CYCLIC
MONOPHOSPHATE see AOA130
ADENOSINE CYCLIC-3',5'-PHOSPHATE see
AOA130

ADENOSINE-5'-(N-
CYCLOBUTYL)CARBOXAMIDE see AEI500
ADENOSINE-5'-(N-
CYCLOPENTYL)CARBOXAMIDE see
AEI750
ADENOSINE-3',5'-CYCLOPHOSPHATE see
AOA130
ADENOSINE-5'-(N-
CYCLOPROPYL)CARBOXAMIDE see AEJ000
ADENOSINE-5'-(N-
CYCLOPROPYL)CARBOXAMIDE-N'-
OXIDE see AEJ250
ADENOSINE-5'-(N-
CYCLOPROPYLMETHYL)CARBOXAMIDE
see AEJ500
ADENOSINE, 2'-DEOXY-, 5'-
(DIHYDROGEN PHOSPHATE), POLYMERS
see PJQ400
ADENOSINE, 1,2-DIHYDRO-2'-DEOXY-2-
OXO- see HKA750
ADENOSINE 5'-(DIHYDROGEN
PHOSPHATE), SODIUM SALT see AEM750
ADENOSINE, 1,2-DIHYDRO-2-OXO- see
IKS450
ADENOSINE-5'-(N-(2-
(DIMETHYLAMINO)ETHYL))CARBOXAMI
DE see AEJ750
ADENOSINE-5'-(N,N-
DIMETHYL)CARBOXAMIDE HYDRATE see
AEK000
ADENOSINE DIPHOSPHATE see AEK100
ADENOSINE 5'-DIPHOSPHATE see AEK100
ADENOSINE DIPHOSPHORIC ACID see
AEK100
ADENOSINE 5'-DIPHOSPHORIC ACID see
AEK100
ADENOSINE-5'-(N-ETHYL)CARBOXAMIDE
HEMIHYDRATE see AEK250
ADENOSINE-5'-(N-
ETHYL)CARBOXAMIDE-N'-OXIDE see
AEK500
ADENOSINE-5'-(N-
HEXYL)CARBOXAMIDE HEMIHYDRATE
see AEK750
ADENOSINE-5'-(N-(2-
HYDROXYETHYL))CARBOXAMIDE see
AEL000
ADENOSINE-5'-(N-
ISOPROPYL)CARBOXAMIDE see AEL250
ADENOSINE-5'-(N-
METHOXY)CARBOXAMIDE HYDRATE see
AEL500
ADENOSINE-5'-(N-
METHYL)CARBOXAMIDE HEMIHYDRATE
see AEL750
ADENOSINE, N-METHYL-, mixed with
SODIUM NITRITE (1:4) see SIS650
ADENOSINE 3'-MONOPHOSPHATE see
SPB100
ADENOSINE-5'-MONOPHOSPHATE see
AOA125
ADENOSINE-3',5'-MONOPHOSPHATE see
AOA130
ADENOSINE-5'-MONOPHOSPHATE
POTASSIUM SALT see AEM500
ADENOSINE 5'-MONOPHOSPHATE
SODIUM SALT see AEM750
ADENOSINE-5-MONOPHOSPHORIC ACID
see AOA125
ADENOSINE-5'-MONOPHOSPHORIC ACID
see AOA125
ADENOSINE-5'-MONOPHOSPHORIC ACID
POTASSIUM SALT see AEM500
ADENOSINE, 2-(4-NITROPHENYL)- see
NIJ350
(ADENOSINE)PENTAAMMINERUTHENIU
M(3+) TRIBROMIDE see AEL800
ADENOSINE 5'-(PENTAHYDROGEN
TETRAPHOSPHATE), 5'-5'-ESTER WITH
ADENOSINE see AEM300
ADENOSINE PHOSPHATE see AOA125
ADENOSINE 3'-PHOSPHATE see SPB100

ADENOSINE-5'-PHOSPHATE see AOA125
ADENOSINE-3',5'-PHOSPHATE see AOA130
ADENOSINE-5'-PHOSPHATE POTASSIUM
SALT see AEM500
ADENOSINE-5'-PHOSPHORIC ACID see
AOA125
ADENOSINE-5'-PHOSPHORIC ACID
POTASSIUM SALT see AEM500
ADENOSINE-5'-(N-
PROPYL)CARBOXAMIDE see AEM000
ADENOSINE PYROPHOSPHATE see
AEK100
ADENOSINE 5'-PYROPHOSPHATE see
AEK100
ADENOSINE 5'-PYROPHOSPHORIC ACID
see AEK100
ADENOSINE 5'-
(TETRAHYDROGENTRIPHOSPHATE),
DISODIUM SALT see AEM100
ADENOSINE-5'-
(TETRAHYDROGENTRIPHOSPHATE)
SODIUM SALT see AEM250
ADENOSINE 5'-TETRAPHOSPHATE, 5'-
ESTER WITH ADENOSINE see AEM300
ADENOSINE, 5'-(TRIHYDROGEN
DIPHOSPHATE) (9CI) see AEK100
ADENOSINE, 5'-(TRIHYDROGEN
PYROPHOSPHATE) see AEK100
ADENOSINE 5'-(TRIHYDROGEN
PYROPHOSPHATE), 5'-5'-ESTER with
RIBOFLAVINE see RIF100
ADENOSINE TRIPHOSPHATE see ARQ500
ADENOSINE-5'-TRIPHOSPHATE see
ARQ500
ADENOSINE TRIPHOSPHATE DISODIUM
see AEM100
ADENOSINE-5'-TRIPHOSPHORIC ACID see
ARQ500
ADENOSYLMETHIONINE see AEM310
S-ADENOSYLMETHIONINE see AEM310
l-S-ADENOSYLMETHIONINE see AEM310
ADENOVITE see AOA125
S-ADENOXYL-l-METHIONINE see AEM310
ADENYL see AOA125
ADENYLDEOXYRIBOSIDE see DAQ200
ADENYLIC ACID see AOA125
3'-ADENYLIC ACID see SPB100
tert-ADENYLIC ACID see AOA125
5'-ADENYLIC ACID, 2'-DEOXY-,
HOMOPOLYMER see PJQ400
5'-ADENYLIC ACID, HOMOPOLYMER,
COMPLEX with 5'-URIDYLIC ACID
HOMOPOLYMER(1:1) (9CI) see PJK400
5'-ADENYLIC ACID, POLYMERS,
COMPLEX with 5'-URIDYLICACID
POLYMERS (1:1) see PJK400
5'-ADENYLIC ACID POTASSIUM SALT see
AEM500
5'-ADENYLIC ACID, SODIUM SALT see
AEM750
5'-ADENYLPHOSPHORIC ACID see AEK100
ADENYLPYROPHOSPHORIC ACID see
ARQ500
ADEPHOS see ARQ500
ADEPSINE OIL see MQV750
ADERGON see AOR500
ADERMINE see PPK250
ADERMINE HYDROCHLORIDE see PPK500
ADETOL see ARQ500
ADETPHOS see AEM100
ADHERE see MIQ075
ADHESIVE 502 see EHP700
ADIABEN see CKK000
ADIAZINE see PPP500
ADIGOSIDE see AEM800
ADILACTETTEN see AEN250
ADINE 0102 see PCC480
ADINOL T see SIY000
ADIPAMIDE see AEN000
ADIPAN see AOB250
ADIPAN see BBK000

ADIPAN HEXAMETHYLENDIAMINU see HEG120
ADIPARTHROL see AOB250
ADIPEX see MDQ500
ADIPEX see MDT600
ADIPHENIN see DHX800
ADIPHENIN see THK000
ADIPHENINE see DHX800
ADIPIC ACID see AEN250
ADIPIC ACID, AMMONIUM SALT (8CI) see HEO120
ADIPIC ACID, BIS(2-(2-BUTOXYETHOXY)ETHYL) ESTER see DDT500
ADIPIC ACID BIS(3,4-EPOXY-6-METHYLCYCLOHEXYLMETHYL) ESTER see AEN750
ADIPIC ACID, BIS(2-ETHOXYETHYL) ESTER see BJO225
ADIPIC ACID BIS(2-ETHYLHEXYL) ESTER see AEO000
ADIPIC ACID-3-CYCLOHEXENYLMETHANOL DIESTER see AEO250
ADIPIC ACID DIALLYL ESTER see AEO500
ADIPIC ACID DIAMIDE see AEN000
ADIPIC ACID, DIBUTOXYETHYL ESTER see BHJ750
ADIPIC ACID DIBUTYL ESTER see AEO750
ADIPIC ACID DI-(3-CARBOXY-2,4,6-TRIIODOANILIDE) DISODIUM see BGB325
ADIPIC ACID DIDECYL ESTER (mixed isomers) see AEP000
ADIPIC ACID, DIESTER with 6-METHYL-3-CYCLOHEXENE-1-METHANOL see BKR100
ADIPIC ACID-(DI-2-(2-ETHYLBUTOXY)ETHYL) ESTER see AEP250
ADIPIC ACID DI(2-ETHYLBUTYL) ESTER see AEP500
ADIPIC ACID DIETHYL ESTER see AEP750
ADIPIC ACID, DIHEXYL ESTER see HEO200
ADIPIC ACID, DI(2-HEXYLOXYETHYL) ESTER see AEQ000
ADIPIC ACID DIHYDRAZIDE see AEQ250
ADIPIC ACID DIISOPENTYL ESTER see AEQ500
ADIPIC ACID DIISOPROPYL ESTER see DNL800
ADIPIC ACID DINITRILE see AER250
ADIPIC ACID DI-2-PROPYNYL ESTER see AEQ750
ADIPIC ACID, DIVINYL ESTER see HEO150
ADIPIC ACID, compd. with 1,6-HEXANEDIAMINE (1:1) see HEG120
ADIPIC ACID, 1,6-HEXANEDIAMINE SALT see NOH100
ADIPIC ACID, 6-METHYL-3-CYCLOHEXENYL-METHANOL DIESTER see BKR100
ADIPIC ACID, METHYL VINYL ESTER see MQL000
ADIPIC ACID, MONOETHYL ESTER see ELC600
ADIPIC ACID NITRILE see AER250
ADIPIC ACID, POLYMER with 1,4-BUTANEDIOL and METHYLENEDI-p-PHENYLENE ISOCYANATE see PKM250
ADIPIC ACID, POLYMER with 1,4-BUTANEDIOL, METHYLENEDI-p-PHENYLENE ISOCYANATE and 2,2'-(p-PHENYLENEDIOXY)DIETHANOL see PKM500
ADIPIC ACID, POLYMER with ETHYLENE GLYCOL and METHYLENEDI-p-PHENYLENE ISOCYANATE see PKL750
ADIPIC ACID, UREA mixed with CARBOXYMETHYLCELLULOSE ACIDS see AER000
ADIPIC DIAMIDE see AEN000
ADIPIC DIHYDRAZIDE see AEQ250

ADIPIC KETONE see CPW500
ADIPINIC ACID see AEN250
ADIPINSAEURE-DI-(3-CARBOXY-2,4,6-TRIJOD-ANILID) DINATRIUM (GERMAN) see BGB325
ADIPIODONE MEGLUMINE see BGB315
ADIPLON see PPO000
ADIPODINITRILE see AER250
ADIPOL 2EH see AEO000
5,5'-(ADIPOLYDIMINO)BIS(2,4,6-TRIIODO-N-METHYLISOPHTHALAMIC ACID) see IDJ500
ADIPONITRILE see AER250
ADIPOSETTIN see NNW500
ADIPRAZINE see HEP000
ADITYL see ACE000
ADJUDETS see BBK500
ADK (CZECH) see AGL500
ADLUMIDINE see AER300
(−)-ADLUMIDINE see CBF550
(+)-ADLUMIDINE see AER300
d-ADLUMIDINE see AER300
l-ADLUMIDINE see CBF550
ADM see AES750
ADMA see AOP510
ADMA 2 see DRR800
ADMER PB 02 see PMP500
ADMEX 741 see FAB920
ADMEX 746 see FAB920
ADM HYDROCHLORIDE see HKA300
ADMIRE see CKW400
ADMUL see OAV000
ADNEPHRINE see VGP000
ADO see AGX000
2-ADO see DDB600
ADOBACILLIN see AIV500
ADOBIOL see AER500
ADOGEN 364 see AHP765
ADOGEN 412 see LBX075
ADOGEN 442 see QAT550
ADOGEN 444 see HCQ525
ADOGEN 448 see QAT550
ADOGEN 464 see QAT565
ADOGEN 471 see TAC300
ADOGENEN 142 see OBC000
ADOGEN 442-100 P see QAT550
ADOL see CPR600
ADOL see HCP000
ADOL see OAX000
ADOL see OBA000
ADOL 34 see OBA000
ADOL 68 see OAX000
ADOL 80 see OBA000
ADOL 85 see OBA000
ADOL 90 see OBA000
ADOL 320 see OBA000
ADOL 330 see OBA000
ADOL 340 see OBA000
ADOL (PESTICIDE) see CPR600
ADONAL see EOK000
ADONA TRIHYDRATE see AER666
ADONA TRIHYDRATE see AER666
ADONIDIN see AER750
ADONIS (VARIOUS SPECIES) see PCU375
ADONITOL see RIF000
ADOPON see NOC000
ADORM see TDA500
ADP see AEK100
5'-ADP see AEK100
ADPHEN see DKE800
ADP (NUCLEOTIDE) see AEK100
ADR see HKA300
A. DRACONTIUM see JAJ000
ADRAN see IIU000
ADRAXONE see AES639
ADRENAL see VGP000
ADRENAL CORTEX HORMONE see AES650
ADRENALEX see CNS800
1-ADRENALIN see VGP000
ADRENALIN BITARTRATE see AES000
ADRENALIN CHLORIDE see AES500
d-ADRENALINE see AES250

dl-ADRENALINE see EBB500
l-(+)-ADRENALINE see AES250
ADRENALINE ACID TARTRATE see AES000
(−)-ADRENALINE ACID TARTRATE see AES000
ADRENALINE BITARTRATE see AES000
(−)-ADRENALINE BITARTRATE see AES000
1-ADRENALINE BITARTRATE see AES000
1-ADRENALINE-d-BITARTRATE see AES000
1-ADRENALINE CHLORIDE see AES500
(−)-ADRENALINE HYDROCHLORIDE see AES500
1-ADRENALINE HYDROCHLORIDE see AES500
(±)-ADRENALINE HYDROCHLORIDE see AES625
dl-ADRENALINE HYDROCHLORIDE see AES625
ADRENALINE HYDROGEN TARTRATE see AES000
(−)-ADRENALINE HYDROGEN TARTRATE see AES000
1-ADRENALINE HYDROGEN TARTRATE see AES000
ADRENALINE TARTRATE see AES000
(−)-ADRENALINE TARTRATE see AES000
l-ADRENALINE TARTRATE see AES000
ADRENALIN HYDROCHLORIDE see AES500
ADRENALIN-MEDIHALER see VGP000
ADRENALONE see MGC350
ADRENAMINE see VGP000
ADRENAN see VGP000
ADRENAPAX see VGP000
ADRENASOL see VGP000
ADRENATRATE see VGP000
ADRENOCHROME see AES639
ADRENOCHROME SULFONATE AC 17 TRIHYDRATE see AER666
ADRENOCORTICOTROPHIC HORMONE see AES650
ADRENOCORTICOTROPHIN see AES650
ADRENOCORTICOTROPIC HORMONE see AES650
ADRENOCORTICOTROPIN see AES650
ADRENODIS see VGP000
ADRENOHORMA see VGP000
ADRENOMEDULLIN (HUMAN) see HGL670
ADRENOMONE see AES650
ADRENON see MGC350
ADRENONE see MGC350
ADRENOR see NNO500
ADRENOTROPHIN see AES650
ADRENUTOL see VGP000
ADRESON see CNS800
ADRESON see CNS825
ADREVIL see PEU000
ADRIACIN see HKA300
ADRIAMICINA see MDO250
ADRIAMYCIN see AES750
ADRIAMYCIN see HKA300
ADRIAMYCIN-HCl see AES750
ADRIAMYCIN, HYDROCHLORIDE see HKA300
ADRIAMYCIN-14-OCTANOATEHYDROCHLORIDE see AET250
ADRIAMYCIN SEMIQUINONE see AES750
ADRIANOL see SPC500
ADRIBLASTIN see HKA300
ADRIBLASTINA see AES750
ADRIBLASTINE see HKA300
ADRINE see VGP000
ADRIXINE see BBK500
ADROIDIN see PAN100
ADRONAL see CPB750
ADROYD see PAN100
ADRUCIL see FMM000
ADS see FBP350
ADSORBONAC see CAR780
ADULSIN see MIF760

ADUMBRAN see CFZ000
ADUVEX 248 see HND100
ADVANTAGE see CCC280
ADVASTAB 46 see HND100
ADVASTAB 47 see DMW250
ADVASTAB 401 see BFW750
ADVASTAB 405 see MJO500
ADVASTAB 800 see TFD500
ADVASTAB 802 see DXG700
ADVASTAB DBTM see DEJ100
ADVASTAB 17 MO see BKK750
ADVASTAB PS 802 see DXG700
ADVASTAB T290 see DEJ100
ADVASTAB T340 see DEJ100
ADVAWAX 140 see OAV000
ADVENTAN see FQJ100
ADYNOL see ARQ500
AE see AGX000
AE 0047 see WAT225
AE0047 see WAT230
(S)-AE0047 see NDY550
(S)-(+)-AE 0047 see NDY550
AEAMN see ABN750
AED see CQJ750
AEDURID see EHV200
A-ENDOSULFAN-α see EAQ810
AENH (GERMAN) see ENV000
A 38414 (ENZYME) see VGU700
AEORLIN see BQF500
AEP 1 see SFO100
AERBRON see POF500
AERO see MCB000
AERO-CYANAMID see CAQ250
AERO CYANAMID GRANULAR see CAQ250
AERO CYANAMID SPECIAL GRADE see
CAQ250
AERO CYANATE see PLC250
AERO liquid HCN see HHS000
AEROL 1 (pesticide) see TIQ250
AEROLITE 300 see UTU500
AEROLITE A 300 see UTU500
AEROLITE FFD see UTU500
AEROLITE MF 15 see MCB050
AEROMATT see CAT775
AEROMONAS HYDROPHILA A₃
ENDOTOXIN see AET500
AEROSEB-DEX see SOW000
AEROSEB-HC see CNS750
AEROSIL see SCH002
AEROSOL GPG see DJL000
AEROSOL MA80 see DKP800
AEROSOL of THERMOVACUUM
CADMIUM see CAK000
AEROSPORIN see PKC500
AEROTEX 92 see MCB050
AEROTEX 3700 see MCB050
AEROTEX GLYOXAL 40 see GIK000
AEROTEX M 3 see MCB050
AEROTEX MW see MCB050
AEROTEX RESIN MW see MCB050
AEROTEX UM see MCB050
AEROTHENE MM see MJP450
AEROTHENE TT see MIH275
AEROXANTHATE see PKV100
AEROXANTHATE 343 see SIA000
AEROXANTHATE 350 see PKV100
AERUGIDIOL see AET600
AES-2Mg see MAI500
AESCIN (GERMAN) see EDK875
α-AESCIN see EDL000
β-AESCIN see EDL500
AESCIN SODIUM SALT see EDM000
AESCIN TRIETHANOLAMINE SALT see
EDM500
AESCULETIN DIMETHYL ETHER see
DRS800
AESCULUS (VARIOUS SPECIES) see HGL575
AESCUSAN see EDK875
α-AESCUSAN see EDL000
β-AESCUSAN see EDL500
AESCUSAN SODIUM SALT see EDM000
AES-MG see MAI500

AESTOCIN see DPE200
AET see AJY250
AET-2HBR see AJY250
AET BROMIDE see AJY250
AET DICHLORIDE see AJY500
AET DIHYDROBROMIDE see AJY250
AETHALDIAMIN (GERMAN) see EEA500
AETHANETHIOL (GERMAN) see EMB100
AETHANOL (GERMAN) see EFU000
AETHANOLAMIN (GERMAN) see EEC600
AETHER see EJU000
2-AETHINYLBUTANOL see EQL000
1-AETHINYL-CYCLOHEXYL-
ALLOPHANAT-(1) see EQL600
AETHIONIN see EEI000
AETHISTERON see GEK500
AETHOBROMID DES α,α-
DICYCLOPENTYLESSIGSAEURE-β'-
DIAETHYLAMINO AETHYLESTER
(GERMAN) see DGW600
AETHON see ENY500
AETHOPHYLLINUM see HLC000
AETHOPROPRAPAZIN see DIR000
AETHOSUXIMIDE (GERMAN) see ENG500
AETHOXEN see VMA000
2-(2-AETHOXY-AETHOXY)-AETHY-3,6,9-
TRIOXA-UNDECAN see MJS550
2-AETHOXY-AETHYLACETAT (GERMAN)
see EES400
1-AETHOXY-2-AMINO-4-NITROBENZOL
see NIC200
1-AETHOXY-4-(1-CAETO-2-
HYDROXYAETHYL)-NAPHTALAENE
SUCCINATE see SNB500
3-(AETHOXYCARBONYLAMINOPHENYL)-
N-PHENYL-CARBAMAT (GERMAN) see
EEO500
S-AETHOXY-CARBONYLTHIAMIN
HYDROCHLORID (GERMAN) see EEQ500
2-AETHOXY-6,9-
DIAMINOACRIDINLACTAT (GERMAN) see
EDW500
1-p-AETHOXYPHENYL-3-
DIAETHYLAMINO-INDAN CITRAT
(GERMAN) see DHR800
p-AETHOXYPHYLHARNSTOFF (GERMAN)
see EFE000
AETHOXYSILATRAN see EFJ600
5-AETHOXY-3-TRICHLORMETHYL-1,2,4-
THIADIAZOL (GERMAN) see EFK000
AETHUSA CYNAPIUM see FMU200
AETHYLACETAT (GERMAN) see EFR000
AETHYL ACETOXYMETHYLNITROSAMIN
(GERMAN) see ENR500
AETHYLACRYLAT (GERMAN) see EFT000
AETHYL-AETHANOL-NITROSOAMIN
(GERMAN) see ELG500
AETHYLALKOHOL (GERMAN) see EFU000
AETHYLAMINE (GERMAN) see EFU400
2-AETHYLAMINO-4-sek.BUTYLAMINO-6-
CHLOR-1,3,5-TRIAZIN see THR600
4-AETHYLAMINO-2-tert-BUTYLAMINO-6-
METHYLTHIO-s-TRIAZIN (GERMAN) see
BQC750
2-AETHYLAMINO-5-BUTYL-4-YL-
DIMETHYLSULFAMAT (GERMAN) see
BRJ000
2-AETHYLAMINO-4-CHLOR-6-
ISOPROPYLAMINO-1,3,5-TRIAZIN
(GERMAN) see ARQ725
2-AETHYLAMINO-6-CHLOR-4-METHYL-4-
PHENYL-4H-3,1-BENZOXAZIN (GERMAN)
see CGQ500
2-AETHYLAMINO-4-ISOPROPYLAMINO-6-
CHLOR-1,3,5-TRIAZIN (GERMAN) see
ARQ725
2-AETHYLAMINO-3-PHENYL-NOR-
CAMPHAN (GERMAN) see EOM000
AETHYLANILIN (GERMAN) see EGK000
AETHYLBENZOL (GERMAN) see EGP500
AETHYLBUTYLKETON (GERMAN) see
EHA600

AETHYL-N-BUTYL-NITROSOAMIN
(GERMAN) see EHC000
AETHYL-tert-BUTYL-NITROSOAMIN
(GERMAN) see NKD500
AETHYLCARBAMAT (GERMAN) see
UVA000
AETHYLCHLORID (GERMAN) see EHH000
AETHYL-CHLORVYNOL see CHG000
1-AETHYL-CYCLOHEXANOL-(1)
(GERMAN) see EHR500
AETHYL-2-(3',5'-DIJOD-4'-OXYBENZOYL)-3
CUMARON see EID200
O-AETHYL-S-(2-
DIMETHYLAMINOAETHYL)-
METHYLPHOSPHONOTHIOATE
(GERMAN) see EIF500
O-AETHYL-S,S-DIPHENYL-
DITHIOPHOSPHAT (GERMAN) see EIM000
S-AETHYL-N,N-
DIPROPYLTHIOLCARBAMAT (GERMAN)
see EIN500
1,1'-AETHYLEN-2,2'-BIPYRIDINIUM-
DIBROMID see EJC025
AETHYLEN-BIS-THIURAMMONOSULFID
(GERMAN) see ISK000
AETHYLENBROMID (GERMAN) see EIY500
AETHYLENCHLORID (GERMAN) see
EIY600
AETHYLENECHLORHYDRIN (GERMAN)
see EIU800
AETHYLENEDIAMIN (GERMAN) see
EEA500
AETHYLENGLYKOLAETHERACETAT
(GERMAN) see EES400
AETHYLENGLYKOLMETHYLAETHERAC
ETAT (GERMAN) see EJJ500
AETHYLENGLYKOL-
MONOMETHYLAETHER (GERMAN) see
EJH500
AETHYLENIMIN (GERMAN) see EJM900
AETHYLENIMINO-2-OXYBUTEN
(GERMAN) see VMA000
AETHYLENOXID (GERMAN) see EJN500
AETHYLENSULFID (GERMAN) see EJP500
N-AETHYLFORMAMID see EKK600
AETHYLFORMIAT (GERMAN) see EKL000
AETHYLHARNSTOFF und
NATRIUMNITRIT (GERMAN) see EQE000
AETHYLHARNSTOFF und NITRIT
(GERMAN) see EQE000
S-AETHYL-N-HEXAHYDRO-1H-
AZEPINTHIOLCARBAMAT (GERMAN) see
EKO500
1-AETHYLHEXANOL (GERMAN) see
EKQ000
AETHYLIDENCHLORID (GERMAN) see
DFF809
AETHYLIS see EHH000
AETHYLIS CHLORIDUM see EHH000
2-(O-AETHYL-N-
ISOPROPYLAMINDOTHIOPHOSPHORYLO
XY)-BENZOSAEURE-ISOPROPYLESTER
(GERMAN) see IMF300
AETHYL-ISOPROPYL-NITROSOAMIN
(GERMAN) see ELX500
AETHYLMERCAPTAN (GERMAN) see
EMB100
AETHYLMETHYLKETON (GERMAN) see
MKA400
2-AETHYL-6-METHYL-N-(1-METHYL-2-
METHOXYAETHYL)-CHLORACETANILID
(GERMAN) see MQQ450
O-AETHYL-O-(3-METHYL-4-
METHYLTHIOPHENYL)-
ISOPROPYLAMIDO-PHOSPHORSAEURE
ESTER (GERMAN) see FAK000
3-β-AETHYL-1-METHYL-4-PHENYL-4-α-
PIPERIDYLPROPIONAT HYDROCHLORID
(GERMAN) see NOE550
3-β-AETHYL-1-METHYL-4-PHENYL-4-α-
PROPIONYLOXYPIPERIDIN
HYDROCHLORID (GERMAN) see NOE550

N-AETHYL-N'-NITRO-N-NITROSOGUANIDIN (GERMAN) see ENU000
O-AETHYL-O-n(4-NITROPHENYL)-PHENYL-MONOTHIOPHOSPHONAT (GERMAN) see EBD700
AETHYLNITROSO-HARNSTOFF (GERMAN) see ENV000
AETHYLNITROSOURETHAN (GERMAN) see NKE500
3-AETHYL-PENTANOL-(3) see TJP550
5-AETHYL-5-PENTYL-(2')-BARBITURSAEURE (GERMAN) see PBS250
O-AETHYL-S-PHENYL-AETHYL-DITHIOPHOSPHONAT (GERMAN) see FMU045
S-AETHYL-N-PHENYL-DITHIOCARBAMAT see EOK550
5-AETHYL-5-PHENYL-HEXAHYDROPYRIMIDIN-4,6-DION (GERMAN) see DBB200
4-AETHYL-1-PHOSPHA-2,6,7-TRIOXABICYCLO(2.2.2)OCTAN (GERMAN) see TNI750
4-AETHYL-1-PHOSPHA-2,6,7-TRIOXABICYCLO(2.2.2)OCTAN-1-OXID (GERMAN) see ELJ500
AETHYL-4-PICOLYLNITROSAMIN (GERMAN) see NLH000
N-AETHYLPIPERIDIN (GERMAN) see EOS500
N-(1-AETHYLPROPYL)-3,4-DIMETHYL-2,6-DINITROANILIN (GERMAN) see DRN200
N-(1-AETHYLPROPYL)-2,6-DINITRO-3,4-XYLIDIN (GERMAN) see DRN200
AETHYLPROPYLKETON (GERMAN) see HEV500
AETHYLRHODANID (GERMAN) see EPP000
AETHYLSENFOEL (GERMAN) see ISK000
1-N-AETHYLSISOMICIN see SBD000
S-2-AETHYLSULFINYL-1-METHYL AETHYL-O,O DIMETHYL-MONOTHIOPHOSPHAT see DSK600
O-AETHYL-O-(2,4,5-TRICHLORPHENYL)-AETHYLTHIONOPHOSPHONAT (GERMAN) see EPY000
AETHYLTRICHLORPHON (GERMAN) see EPY600
N-AETHYL-N-(2,4,6-TRIJOD-3-AMINOPHENYL)-SUCCINAMIDSAEURE (GERMAN) see AMV375
AETHYLURETHAN (GERMAN) see UVA000
AETHYL-VINYL-NITROSOAMIN (GERMAN) see NKF000
AETHYLZINNTRICHLORID (GERMAN) see EPS000
AETINA see EPQ000
AETIVA see EPQ000
AETM (GERMAN) see ISK000
AETT see ACL750
AF see XAK000
AF2 see FQO000
AF 5 see SNY100
AF 10 see FMW330
AF-45 see EQN320
AF 101 see DXQ500
AF 260 see AHC000
AF 425 see CKI000
AF 594 see CPN750
AF 864 see BBW500
AF 983 see BAV325
AF 1161 see CKJ000
AF 1161 see THK880
AF 1890 see DEL200
AF 2071 see BAV275
AF 2259 see IJJ000
70H-2AAF see HIK500
AF-2 (preservative) see FQN000
AFASTOGEN BLUE 5040 see DNE400
AFATIN see BBK500
AFAXIN see VSK600

AFBI see AEU250
A.F. BLUE No. 1 see FMU059
A.F. BLUE No. 2 see FAE100
AFCF see FAS800
AFCOLAC B 101 see SMR000
AFCOLENE see SMQ500
AFCOLENE 666 see SMQ500
AFCOLENE S 100 see SMQ500
2AAF DIMER see AAK500
AFESIN see CKD500
AFFIRM see ARW200
A.F. GREEN No. 2 see FAF000
A.F. GREEN NO. 1 see FAE950
AF 438 HYDROCHLORIDE see OOE100
AFIBRIN see AJD000
AFICIDA see DOX600
AFICIDE see BBQ500
A-FIL CREAM see TGG760
AFILINE see MPU250
AFI-PHYLLIN see DNC000
AFI-TIAZIN see PDP250
AFKO-HIST see WAK000
AFKO-SAL see SAH000
AFL see AEW500
AFL 1081 see FFF000
AFL 1082 see FFH000
AFLATOXICOL see AEW500
AFLATOXICOL B see AET630
AFLATOXICOL NATURAL EPIMER see AEW500
AFLATOXICOL, UNNATURAL see AET630
AFLATOXIN see AET750
AFLATOXIN B see AEU250
AFLATOXIN B1 see AEU250
AFLATOXIN B2 see AEU750
AFLATOXIN B1 DICHLORIDE see AEU500
AFLATOXIN B1-2,3-DICHLORIDE see AEU500
AFLATOXIN G1 see AEV000
AFLATOXIN G2 see AEV500
AFLATOXIN G1 mixed with AFLATOXIN B1 see AEV250
AFLATOXIN M1 see AEW000
AFLATOXIN Ro see AEW500
AFLATOXIN RO' see AET630
A. FLAVA see HGL575
AFLIX see DRR200
AFLON see TAI250
AFLOQUALONE see AEW625
A. FLOS-AQUAE TOXIN see AON825
AFLOXAN see POF550
AFLUON see MAF500
AFNOR see CJJ000
AFNOR K-C25NW see CNB825
AFNOR ZFENC45-36 see IGL100
A.F. ORANGE No. 1 see FAG010
A.F.ORANGE No. 2 see TGW000
A. FORDII see TOA275
AFOS see DJI000
AFRAZINE see AEX000
A.F. RED No. 1 see FAG018
A.F. RED No. 5 see XRA000
AFRICAN COFFEE TREE see CCP000
AFRICAN LILAC TREE see CDM325
AFRIDOL BLUE see AEW750
AFRIN see AEX000
AFRIN HYDROCHLORIDE see AEX000
AFTATE see TGB475
AF 1312/TS see CEQ625
A 76 (FUEL) see GCC200
AFUNGIN see RCK730
A.F. VIOLET No 1 see FAG120
A.F YELLOW No. 2 see FAG130
A.F. YELLOW No. 3 see FAG135
A.F. YELLOW NO. 4 see FAG140
A.F. YELLOW NO. 5 see FAG150
AG 3 see CBR500
AG 3 see CBT500
AG 5 see CBT500
8 AG see AJO500
AG 307 see AMS800
AG-629 see SLE880

AG. 5895 see VGA100
AG 58107 see TAI600
AG 3 (ADSORBENT) see CBT500
AG 5 (ADSORBENT) see CBT500
AGALITE see TAB750
AGALLO FORTE see MEP250
AGALLOL see MEP250
AGALLOLAT see MEP250
AGALOL see MEP250
AGAR see AEX250
AGAR-AGAR see AEX250
AGAR AGAR FLAKE see AEX250
AGAR-AGAR GUM see AEX250
AGARICUS BISPORUS MUSHROOM LECTIN see SKW800
AGARIN see AKT750
AGARITINE see GFO080
AGASTEN see FOS100
AGATE see SCI500
AGATE see SCJ500
AGC see GFA000
AGEDAL see DPH600
AGE see AGH150
AGEFLEX AMA see AGK500
AGEFLEX BGE see BRK750
AGEFLEX CGE see GGS000
AGEFLEX EGDM see BKM250
AGEFLEX FA-2 see DHT125
AGEFLEX FA-10 see IKL000
AGEFLEX FM-1 see DPG600
AGEFLEX FM-4 see BQD250
AGEFLEX FM 246 see DXY200
AGEFLEX n-HA see ADV000
AGEFLOC WT 20 see DTS500
AGELFLEX FM-10 see IKM000
AGENAP see NAR000
AGENE see NGQ500
AGENT 504 see DAI600
AGENT AT 539 see ODY150
AGENT AT 717 see PKQ250
AGENT BLUE see HKC000
AGENT ORANGE see AEX750
AGENT RD-510 see PKE850
AGERATOCHROMENE see AEX850
AGERITE see AEY000
AGERITE see BLE500
AGERITE 150 see HKF000
AGERITE ALBA see AEY000
AGERITEDPPD see BLE500
AGERITE ISO see HKF000
AGERITE MA see PJQ750
AGERITE POWDER see PFT500
AGERITE RESIN D see TLP500
AGERITE WHITE see NBL000
AGEROPLAS see BKB250
AGGLUTININ see AAD000
AGIDOL see BFW750
AGIDOL 3 see DEA100
AGIDOL 3 see FAB000
AGIDOL 7 see MJN250
AGIDOL 42 see DEE300
AGIDOL AF-2 see EEA550
AGIL see IPI400
AGILENE see DRK600
AGILENE see PJS750
AGIOLAN see VSK600
AGI TALC, BC 1615 see TAB750
AGKISTRODON ACUTUS VENOM see HGM600
AGKISTRODON CONTORTRIX CONTROTRIX VENOM see SKW775
AGKISTRODON CONTORTRIX MOKASEN VENOM see NNX700
AGKISTRODON CONTORTRIX VENOM see AEY125
AGKISTRODON PISCIVORUS PISCIVORUS VENOM see EAB200
AGKISTRODON PISCIVORUS VENOM see AEY130
AGKISTRODON RHODOSTOMA VENOM see AEY135
A. GLABRA see HGL575

AGLICID see BSQ000
AGLUMIN see DIS600
AGM-9 see TJN000
AGOCHROMOL ORANGE AS see CMP882
AGOFOLLIN see EDR000
AGORAL see PDO750
AGOSTILBEN see DKA600
AGOTAN see PGG000
AGOVIRIN see TBG000
AGP see AKC625
AGR 307 see AMS800
AGRAZINE see PDP250
AGREFLAN see DUV600
AGRIA 1050 see DSQ000
AGRIBON see SNN300
AGRIBROM see BNA300
AGRICIDE MAGGOT KILLER (F) see
CDV100
AGRICULTURAL LIMESTONE see CAO000
AGRIDIP see CNU750
AGRIFLAN 24 see DUV600
AGRIMEK see ARW200
AGRI-MYCIN see SLY500
AGRIMYCIN 17 see SLW500
AGRION see SGH500
AGRISIL see EPY000
AGRISOL G-20 see BBQ500
AGRISTREP see SLY500
AGRISYNTH B2D see BOX300
AGRITAN see DAD200
AGRITOL see BAC040
AGRITOX see CIR250
AGRITOX see EPY000
AGRIYA 1050 see DSQ000
AGRIZAN see CNK559
A-GRO see MNH000
AGROCERES see HAR000
AGROCIDE see BBQ500
AGROCLAVINE see AEY375
AGROFORM see UTU500
AGROFOROTOX see TIQ250
AGROMET see DTN125
AGROMICINA see TBX000
AGRONAA see NAK500
AGRONEXIT see BBQ500
AGROSAN see ABU500
AGROSAND see ABU500
AGROSAN GN 5 see ABU500
AGROSIL LR see SCN700
AGROSIL S see SCN700
AGROSOL see MLF750
AGROSOL S see CBG000
AGROTECT see DAA800
AGROTHION see DSQ000
AGROXONE see CIR250
AGROXONE 3 see SIL500
AGROXONE 4 see MIH800
AGROX 2-WAY and 3-WAY see CBG000
AGRYPNAL see EOK000
AGSTONE see CAO000
AGUATHOL see DXD000
AH see DBM800
AH-42 see TEO250
AH 289 see CDR000
AH 501 see PAJ000
AH 1932 see PFY105
AH 2526 see VJZ050
AH 3232 see CDQ250
AH 3365 see BQF500
AH 19065 see RBF400
AH 5158A see HMM500
AHA see ABB250
AHCOCID CARMINE 2G see CMM300
AHCOCID FAST SCARLET R see FMU070
AHCO DIRECT BLACK GX see AQP000
AHCOQUINONE BLUE IR BASE see
HOK000
AHCOQUINONE RED S see SEH475
AHCOQUINONE SKY BLUE B see CMM090
AHCOVAT BLUE BCF see DFN300
AHCOVAT BRILLIANT VIOLET 2R see
DFN450

AHCOVAT BRILLIANT VIOLET 4R see
DFN450
AHCOVAT BROWN AR see CMU780
AHCOVAT BROWN AR-BN see CMU780
AHCOVAT BROWN BR see CMU770
AHCOVAT NAVY BLUE BR see VGP100
AHCOVAT OLIVE ARN see DUP100
AHCOVAT OLIVE R see DUP100
AHCOVAT PRINTING BLUE 2BD see
ICU135
AHCOVAT PRINTING GOLDEN YELLOW
see DCZ000
AHCOVAT PRINTING NAVY BLUE XSA see
VGP100
AHCOVAT PRINTING ORANGE R see
CMU815
AHCOVAT RUBINE R see CMU825
AHCTL see TCZ000
AHE POI (HAWAII) see EAI600
AH-289 HYDROCHLORIDE see CDR250
A. HIPPOCASTANUM see HGL575
AHMT (PERFUME) see EEE100
AHOUAI des ANTILLES see YAK350
AHR 85 see GKK000
AHR 233 see MFD500
AHR 376 see AEY400
AHR-619 see ENL100
AHR-619 see SLU000
AHR-1680 see DMB000
AHR-1767 see ENC600
AHR-3053 see CCH125
AHR 3219 see EQN600
AHR 5850D MONOHYDRATE see AHK625
AHTN see EEE100
AHYDOL (RUSSIAN) see TMJ000
A 66 HYDROCHLORIDE see MNV750
A 446 HYDROCHLORIDE see LFW300
A-HYDROCORT see HHR000
AHYGROSCOPIN-B see VGZ000
AHYPNON see MKA250
AI3-09071 see NAQ000
AI3-14631 see GKW100
AI3-14678 see DDM425
AI3-17256 see POR300
AI 318284 see MRQ750
AI3-18285 see CBF725
AI3-18581 see TNI300
AI3-20871 see MJS550
AI 3-22542 see DKC800
AI 3-22641 see FQQ500
AI3-25449 see AKK250
AI3-25722 see BHL800
AI3-26057 see COM800
AI 3-27040 see AAI115
AI3-28253 see NIR100
AI3-28673 see NHN700
AI 3-29024 see AFR750
AI3-29128 see PHK000
AI3-29158 see AHJ750
AI 3-29183 see DEG150
AI3-29750 see THI500
AI 3-29759 see PCH315
AI3-29832 see HCY600
AI3-32960 see HCS600
AI3-34872 see CNG760
AI3-34886 see ECB200
AI3-35349 see TJF400
AI3-35584 see TCA300
AI3-35770 see HDR700
AI3-35937 see HCP050
AI3-35966 see ISZ000
AI3-36161 see AEY402
AI3-36174 see AEY404
AI3-36206 see MEL700
AI3-36401 see HJG100
AI3-36420 see ODO000
AI3-36537 see CPI350
AI3-36542 see CPI380
AI3-36543 see MLM500
AI3-36558 see MLM600
AI3-36559 see MLM700
AI3-36561 see DSP650

AI3-36563 see TCV400
AI3-36564 see MLL650
AI3-36565 see MLL655
AI3-36566 see MLL660
AI3-36570 see TCV375
AI3-36728 see OAP070
AI3-37135 see AEY406
AI3-37220 see CPD625
AI3-37220 see CPD630
AI3-38274 see AEY410
AI3-38275 see AEY415
AI3-38276 see AEY420
AI3-38282 see AEY425
AI3-38283 see AEY430
AI3-50172 see HEF500
AI 3-51254 see MGE100
AI 3-51408 see DCD050
AI3-52713 see CQE400
AI3-70087 see HKQ500
AI3-70643 see DTE100
AI3-70736 see BJK650
AI3-38192A see AEY435
AI3-38194A see AEY440
AI3-38221A see AFA000
AI3-38352A see AFA100
AI3-38354A see AFA110
AI3-38355A see AFA115
AI3-38357A see AFA120
AI3-38359A see AFA125
AI3-38360A see AFA130
AI3-38361A see AFA135
AI3-38420B see AFA140
AI3-38421B see AFA145
AI3-38422B see AFA150
AI3-38423B see AFA155
AI3-38427A see AFA160
AI3-38599A see AFA165
AI3-38862A see AFA170
AI3-39048A see AFA175
AI3-39049A see AFA180
AI3-39050A see AFA185
AI3-39051A see AFA190
AIA see PNR800
AI3-36329-A see MIS500
AI3-35714-AGA see PBX100
AI3-35716-AGB see VAG100
AIB see MGB000
AIBN see ASL750
AIC see AKK250
AI3-20685GC see AFA200
AICA see AKK250
TRANS-AID see ANW750
AI 93 (FUEL) see GCC200
AI3-36175-Ga see EMY100
AI3-20784-GC see DJL700
AIGLONYL see EPD500
A-250-II see CBF680
A 11725 II see MRW750
AIL du CANADA (CANADA) see WBS850
AIMAX see MLJ500
AIMSAN see DRR400
AIMSAN see MEG250
AIP see AHE750
AIPTASIA PALLIDA VENOM see SBI800
AIPYSURUS LAEVIS VENOM see SBI880
AIPYSURUS LAEVIS VENOM (AUSTRALIA)
see AFG000
AIR, refrigerated liquid see AFG250
AIRBRON see ACH000
AIRDALE BLUE IN see FAE100
AIREDALE BLACK BHD see CMN800
AIREDALE BLACK ED see AQP000
AIREDALE BLACK 2BG see FAB830
AIREDALE BLUE D see CMO500
AIREDALE BLUE 2BD see CMO000
AIREDALE BLUE FFD see CMN750
AIREDALE BLUE RL see ADE750
AIREDALE BLUE RWD see CMO600
AIREDALE BROWN BSD see CMO820
AIREDALE BROWN MD see CMO800
AIREDALE CARMOISINE see HJF500
AIREDALE GREEN BD see CMO840

AIREDALE GREEN BWD see CMO830
AIREDALE ORANGE II see CMM220
AIREDALE RED FD see CMO870
AIREDALE RED KD scc CMO885
AIREDALE RED PGM see CMM320
AIREDALE RED RM see NAO600
AIREDALE SCARLET 3BD see CMO875
AIREDALE SCARLET 4BD see CMO870
AIREDALE SCARLET GM see CMM325
AIREDALE VIOLET ND see CMP000
AIREDALE YELLOW E see SGP500
AIREDALE YELLOW 3GM see CMM759
AIREDALE YELLOW T see FAG140
AIR-FLO GREEN see CNN500
AIRLOCK see CAU500
AIRONE see ZMA000
AIR, compressed (UN 1002) (DOT) see AFG250
AIR, refrigerated liquid (cryogenic liquid) (UN 1003) (DOT) see AFG250
AIR, refrigerated liquid (cryogenic liquid) non-pressurized (UN 1003) (DOT) see AFG250
AISELAZINE see HGP500
AISEMIDE see CHJ750
AISI 332 see IGL100
A. ITALICUM see ITD050
AITC see AGJ250
AI3-26730-X see TJK800
AIZEN ACID PHLOXINE PB see ADG250
AIZEN AMARANTH see FAG020
AIZEN ASTRA PHLOXINE FF see CMM840
AIZEN AURAMINE see IBA000
AIZEN BRILLIANT ACID BLUE AFH see ERG100
AIZEN BRILLIANT ACID PURE BLUE VH see ADE500
AIZEN BRILLIANT BLUE FCF see FMU059
AIZEN BRILLIANT SCARLET 3RH see FMU080
AIZEN CATHILON ORANGE GL see CMM820
AIZEN CATHILON ORANGE GLH see CMM820
AIZEN CATHILON PINK FG see CMM850
AIZEN CATHILON PINK FGH see CMM850
AIZEN CATHILON RED GTLH see BAQ750
AIZEN CATHION YELLOW 3GH see CMM890
AIZEN CATHION YELLOW 3GL see CMM890
AIZEN CATHION YELLOW 3GLH see CMM890
AIZEN CHROME VIOLET BH see HLI000
AIZEN CRYSTAL VIOLET see AOR500
AIZEN CRYSTAL VIOLET EXTRA PURE see AOR500
AIZEN DIAMOND GREEN GH see BAY750
AIZEN DIRECT BLACK BH see CMN800
AIZEN DIRECT BLUE 2BH see CMO000
AIZEN DIRECT BORDEAUX GH see CMO872
AIZEN DIRECT BROWN MH see CMO800
AIZEN DIRECT DARK GREEN BH see CMO830
AIZEN DIRECT DEEP BLACK EH see AQP000
AIZEN DIRECT DEEP BLACK GH see AQP000
AIZEN DIRECT DEEP BLACK RH see AQP000
AIZEN DIRECT FAST RED FH see CMO870
AIZEN DIRECT GREEN BH see CMO840
AIZEN DIRECT SKY BLUE 5BH see CMO500
AIZEN EOSINE GH see BNH500
AIZEN EOSINE GH see BNK700
AIZEN ERYTHROSINE see FAG040
AIZEN FOOD BLUE No. 2 see FAE000
AIZEN FOOD GREEN No. 3 see FAG000
AIZEN FOOD ORANGE No. 1 see FAG010
AIZEN FOOD ORANGE No. 2 see TGW000
AIZEN FOOD RED No. 5 see XRA000
AIZEN FOOD VIOLET No 1 see FAG120
AIZEN FOOD YELLOW NO. 5 see FAG150

AIZEN MAGENTA see MAC250
AIZEN MALACHITE GREEN see AFG500
AIZEN METANIL YELLOW see MDM775
AIZEN METHYLENE BLUE BH see BJI250
AIZEN NAPHTHOL ORANGE I see FAG010
AIZEN ORANGE I see FAG010
AIZEN PONCEAU RH see FMU070
AIZEN PRIMULA BROWN BRLH see CMO750
AIZEN PRIMULA RED 4BH see CMO885
AIZEN PRIMULA SCARLET 4BSH see CMO870
AIZEN PRIMULA TURQUOISE BLUE GL see COF420
AIZEN RHODAMINE BH see FAG070
AIZEN RHODAMINE 6GCP see RGW000
AIZEN SOT BLACK 6 see PCJ200
AIZEN TARTRAZINE see FAG140
AIZEN URANINE see FEW000
AIZEN VICTORIA BLUE BOH see VKA600
AJAN see NBS500
AJAX GMO see GGR200
AJAX, LEMON (scouring powder) see AFG625
AJI CABALLERO (PUERTO RICO) see PCB275
AJI de GALLINA (PUERTO RICO) see PCB275
AJI GUAGUAO (CUBA) see PCB275
AJINOMOTO see MRL500
AJI PICANTE (PUERTO RICO) see PCB275
AJMALICINE see AFG750
AJMALICINE HYDROCHLORIDE see AFH000
AJMALICINE MONOHYDROCHLORIDE see AFH000
AJMALINE see AFH250
AJMALINE BIS(CHLOROACETATE) (ester) HYDROCHLORIDE see AFH275
AJMALINE HYDROCHLORIDE see AFH280
AJO see WBS850
AK-33X see MAV750
AK 214-82 see MDN600
AK, flower extract see CAZ075
'AKA'AKAI (HAWAII) see WBS850
'AKA'AKAI-PILAU (HAWAII) see WBS850
AKADAMA see CAT775
AK (ADSORBENT) see CBT500
AKAR see DER000
AKARITHION see TNP250
AKARITOX see CKM000
AKEBIA SAPONIN PD see HAK075
AKEBOSIDE STC see HAK075
AKEE see ADG400
AKERSTOX see AQV990
AKETDRIN see AOB250
AKF-94 see DNU330
AKHNOT see WAT000
AKI see ADG400
AKINETON see BGD500
AKINETON HYDROCHLORIDE see BGD750
AKINOPHYL see BGD500
AKINOPHYL see BGD750
AKIRIKU RHODAMINE B see FAG070
AKLAVIN see DAY835
AKLOMIDE see AFH400
AKLOMIX see AFH400
AKLOMIX-3 see HMY000
AKLONIN (GERMAN) see PCV750
AKLONINE see PCV750
AKNADI EXTRACT see SLO100
AKOIN HYDROCHLORID (GERMAN) see PDN500
AKOTIN see NCQ900
AK PS see AFH500
AKRA, flower extract see CAZ075
AKRICHIN see ARQ250
AKROCHEM ETU-22 see IAQ000
AKROFOL see SFV250
AKROLEIN (CZECH) see ADR000
AKROLEINA (POLISH) see ADR000
AKRO-MAG see MAH500

AKRO-ZINC BAR 85 see ZKA000
AKRYLAMID (CZECH) see ADS250
AKRYLONITRYL (POLISH) see ADX500
AKRYLOYLOXYETHYLESTER KYSELINY ACETOCTOVE see AAY750
AKSA, combustion products see ADX750
AKTAMIN see NNO500
AKTAMIN HYDROCHLORIDE see NNP000
AKTEDRIN see AOB250
AKTEDRON see AOB500
AKTIKON see ARQ725
AKTIKON PK see ARQ725
AKTINIT A see ARQ725
AKTINIT PK see ARQ725
AKTINIT S see BJP000
AKTIVAN see FNF000
AKTIVEX see CCX000
AKTIVIN see CDP000
AKTON see DIX600
AKULON see PJY500
AKVAZIN see GCC200
AKYL DIMETHYL BENZYL AMMONIUM SACCHARINATE see BBA625
AKYPOROX O 50 see PJY100
AKYPOSAL TLS see SON000
AKZO CHEMIE MANEB see MAS500
AL-50 see RDP300
AL-100 see ZVJ000
AL-1021 see FGV000
AL 1076 see PFJ000
AL-1612 see AFH550
'ALA-AUMOE (HAWAII) see DAC500
ALABASTER see CAX750
ALABASTER NO. 3 see FAG150
ALACHLOR (USDA) see CFX000
ALACIL see AFW500
ALACINE see PFC750
ALAIXOL II see OAT000
AL-ALCHILI (ITALIAN) see DNI600
ALAMANDA MORADA FALSA (PUERTO RICO) see ROU450
ALAMINE 4 see DXW000
ALAMINE 6 see HCO500
ALAMINE 7 see OBC000
ALAMINE 11 see OHM700
ALAMINE 7D see OBC000
ALAMINE 308 see DVL000
ALAMINE 336 see AHP765
ALAMINE 336 see DVL000
ALAMINE 336S see AHP765
ALAMON see HOR470
ALAMOS see CJR300
ALANAP see NBL200
ALANAP 1 see NBL200
ALANAP-3 see SIO500
ALANAP 10G AT see NBL200
ALANAPE see NBL200
ALANE see AHB500
ALANEX see CFX000
d-ALANINAMIDE, N-ACETYL-3-(1-NAPHTHALENYL)-d-ALANYL-4-CHLORO-d-PHENYLALANYL-d-PHENYLALANY l-L-SERYL-l-TYROSYL-d-ARGINYL-l-PHENYLALANYL-l-ARGINYL-l-PROLYL- see ACS100
ALANINE see AFH625
(+-)-ALANINE see AFH600
l-ALANINE see AFH625
(S)-ALANINE see AFH625
dl-ALANINE see AFH600
l-(+)-ALANINE see AFH625
α-ALANINE see AFH625
l-α-ALANINE see AFH625
dl-α-ALANINE see AFH600
β-ALANINE, N-(AMINOCARBONYL)-N-NITROSO-, METHYL ESTER see MEH100
ALANINE, 3-(((3-AMINO-4-HYDROXYPHENYL)PHENYLARSINO)THI O)- see AKI900
ALANINE, 3-AZIDO-, l- see ASE100
l-ALANINE, 3-AZIDO- see ASE100

ALANINE, 3-AZIDO-, tert-BUTYL ESTER, dl-
see ASE120

dl-ALANINE, 3-AZIDO-, 1,1-
DIMETHYLETHYL ESTER see ASE120

dl-ALANINE, 3-AZIDO-2-METHYL- see
ASG300

d-ALANINE, N-BENZOYL-N-(3-CHLORO-4-
FLUOROPHENYL)-, ISOPROPYL ESTER see
FBW135

ALANINE, N-BENZOYL-N-(3,4-
DICHLOROPHENYL)-, ETHYLESTER, l- see
KBB700

ALANINE, N-BENZOYL-N-(3,4-
DICHLOROPHENYL)-, ETHYL ESTER, DL-
see EGS000

ALANINE, N-BENZYLOXYCARBONYL-3-
PHENYL-, VINYL ESTER, l- see CBR235

ALANINE, N-CARBOXY-3-PHENYL-, N-
BENZYL 1-VINYL ESTER, l- see CBR235

ALANINE, 3-(p-CHLOROPHENYL)-, dl- see
FAM100

ALANINE, 3-CYANO- see COJ800

ALANINE, N-(2,4-DICHLORO-
PHENOXYACETYL)-3-PHENYL-, dl- see
DFZ100

β-ALANINE see AFH650

β-ALANINE, N-(((((2,3-DIHYDRO-2,2-
DIMETHYL-7-
BENZOFURANYL)OXY)CARBONYL)METH
YLAMINO)THIO)-N-(1-METHYLETHYL)-,
ETHYL ESTER see AKC570

β-ALANINE, N-(2,4-DIHYDROXY-3,3-
DIMETHYL-1-OXOBUTYL)-, (R)-(9CI) see
PAG300

DL-ALANINE, N-(2,6-DIMETHYLPHENYL)-
N-(METHOXYACETYL)-, METHYL ESTER
(9CI) see MDM100

ALANINE, N-(2,6-DIMETHYLPHENYL)-N-
METHOXY-, METHYL ESTER see DTN125

dl-ALANINE, N-(2,6-DIMETHYLPHENYL)-
N-METHOXY-, METHYL ESTER see
DTN125

ALANINE, 3,3'-DISELENOBIS-(9CI) see
SBP600

ALANINE, 3,3'-DISELENODI-, d- see SBU300

ALANINE, 3,3'-DISELENODI-, l- see SBU303

ALANINE, 3,3'-DISELENODI-, dl- see
SBU200

ALANINE, 3,3'-DISELENODI-, meso- see
DWY900

ALANINE, 3-(p-FLUOROPHENYL)-, dl- see
FLD100

ALANINE, N-l-γ-GLUTAMYL-3-
(METHYLENECYCLOPROPYL)- see
HOW100

ALANINE, 3-(4-(4-HYDROXY-3,5-
DIIODOPHENOXY)-3,5-DIIODOPHENYL)-,
MONOSODIUM SALT, PENTAHYDRATE,
see SKJ325

ALANINE, 3-(4-(4-HYDROXY-3-
IODOPHENOXY)-3,5-DIIODOPHENYL)-, l-
see LGK050

l-ALANINE, γ-
(HYDROXYMETHYLPHOSPHINYL)-l-α-
AMINOBUTYRYL-l-ALANYL-,
MONOSODIUM SALT see SEO550

ALANINE, 3-((3-HYDROXYPROPYL)THIO)-
see HNV050

ALANINE, 3-((3-HYDROXYPROPYL)THIO)-,
l- see HNV050

β-ALANINE, N-METHYL-N-NITROSO- see
MMT300

ALANINE MUSTARD see BHN500

ALANINE,N-(p-((2,4-DIAMINO-6-
PTERIDYLMETHYL)AMINO)BENZOYL)-,
SODIUM SALT, dl- see DCC500

ALANINE NITROGEN MUSTARD see
BHV250

ALANINE NITROGEN MUSTARD see
PED750

ALANINE, PHENYL- see PEC750

ALANINE, 3-PHENYL- see PEC750

l-ALANINE, PHENYL- see PEC750

ALANINE, N-PIPERONYL- see PIX100

ALANINE, 3-SELENYL-, (+−)- see SBV100

dl-ALANINE, 3-SELENYL-(9CI) see SBV100

ALANINE, 3-SELENYL-, DL (8CI) see SBV100

β-ALANINE, N-(2-((3-
(TRIMETHOXYSILYL)PROPYL)AMINO)ET
HYL-, METHYL ESTER see MJE793

β-ALANINOL see PMM250

ALANOSINE see AFH750

l-ALANOSINE see AFH750

4-N-d-ALANYL-2,4-DIAMINO-2,4-
DIDEOXY-l-ARABINOSE see AFI500

β-ALANYL-l-HISTIDINE see CCK665

N-l-ALANYL-3-(5-OXO-7-
OXABICYCLO(4.1.0)HEPT-2-YL)-l-ALANINE
see BAC175

ALANYL-2,6-XYLIDIDE see TGI699

ALA de PICO (MEXICO) see RBZ400

ALAR see DQD400

ALAR-85 see DQD400

ALATHON see PJS750

ALATHON see PKF750

ALATHON 14 see PJS750

ALATHON 15 see PJS750

ALATHON 5B see PJS750

ALATHON 1560 see PJS750

ALATHON 6600 see PJS750

ALATHON 7026 see PJS750

ALATHON 7040 see PJS750

ALATHON 7050 see PJS750

ALATHON 7140 see PJS750

ALATHON 7511 see PJS750

ALATHON 71XHN see PJS750

ALAUN (GERMAN) see AGX000

ALAZIN see PFC750

ALAZINE see PFC750

ALAZOPEPTIN see AFI625

ALBACAR see CAT775

ALBACAR 5970 see CAT775

ALBA-DOME see AEY000

ALBAFIL see CAT775

ALBAGEL PREMIUM USP 4444 see BAV750

ALBAGLOS see CAT775

ALBAGLOS SF see CAT775

ALBALITH see ZNJ100

ALBALON LIQUIFILM see NCW000

ALBAMINE see SNQ710

ALBAMIX see SMB000

ALBAMYCIN see NOB000

ALBAMYCIN see SMB000

ALBAMYCIN SODIUM see NOB000

ALBARICOQUE (SPANISH) see AQP890

ALBEGO see CFY250

ALBEMAP see BBK500

ALBENDAZOLE SULFONE see PNV900

ALBENDAZOLE (USDA) see VAD000

ALBEXAN see SNM500

ALBIGEN A see PKQ250

ALBIOTIC see LGD000

ALBITOCIN see AFI750

ALBOLINE see MQV750

ALBOMYCIN A1 see GJU800

ALBON see SNN300

ALBONE see HIB050

ALBONE 35 see HIB010

ALBONE 50 see HIB010

ALBONE 70 see HIB010

ALBONE 35CG see HIB010

ALBONE 50CG see HIB010

ALBONE 70CG see HIB010

ALBORAL see DCK759

ALBOSAL see SNM500

AL BROMOHYDRATE see AGY000

ALBSAPOGENIN see GMG000

ALBUCID see SNQ710

ALBUCIDE see SNQ000

ALBUCID SOLUBLE see SNQ000

ALBUMAID XP see AHR600

ALBUMIN see AFI850

ALBUMIN, EGG see AFI780

ALBUMIN MACRO AGGREGATES see
AFI850

ALBUTEROL see BQF500

ALCA see AHA135

ALCALASE see BAC000

ALCANFOR see CBB250

ALCHLOQUIN see CHR500

ALCIAN BLUE see AFI900

ALCIDE see CDW450

ALCIDE LD see SFT500

ALCLOFENAC EPOXIDE see AFI950

ALCLOFENAC SODIUM SALT see AGN250

ALCLOMETASONE DIPROPIONATE see
AFI980

ALCLOPHENAC see AGN000

ALCLOXA see AHA135

ALCOA 331 see AHC000

ALCOA F 1 see AHE250

ALCOA SODIUM FLUORIDE see SHF500

ALCOBAM NM see SGM500

ALCOBAM ZM see BJK500

ALCOBON see FHI000

AL-CO-CURE IBA see IHX700

ALCODET 218 see PJV100

ALCODET 260 see PJV100

ALCODET MC 2000 see PKE350

ALCOGUM see ADV900

ALCOGUM see MDN600

ALCOGUM L 21 see MDN600

ALCOHOL see EFU000

ALCOHOL, anhydrous see EFU000

ALCOHOL, dehydrated see EFU000

ALCOHOL C-8 see OEI000

ALCOHOL C-9 see NNB500

ALCOHOL C-10 see DAI600

ALCOHOL C-11 see UMA000

ALCOHOL C-11, see UNA000

ALCOHOL C-12 see DXV600

ALCOHOL C-16 see HCP000

ALCOHOL, DENATURED see AFJ000

ALCOHOLS, C7-9 see LGF875

ALCOHOLS, C6-12 see AFJ100

ALCOHOLS, C9-11 see AFJ150

ALCOHOLS, C8-10, ETHOXYLATED see
AFJ172

ALCOHOLS, C12-13, ETHOXYLATED see
AFJ155

ALCOHOLS, C12-15, ETHOXYLATED see
AFJ160

ALCOHOLS, C12-16, ETHOXYLATED see
AFJ165

ALCOHOLS, C14-15, ETHOXYLATED see
AFJ168

ALCOHOLS, C16-18, ETHOXYLATED see
AFJ170

ALCOHOLS, C8-10, ETHOXYLATED
PROPOXYLATED see AFJ172

ALCOHOLS, C12-15, ETHOXYLATED
PROPOXYLATED see AFJ175

ALCOHOLS, C11-15-SECONDARY,
ETHOXYLATED see TAZ100

ALCOHOLS, C12-14-SECONDARY,
ETHOXYLATED PROPOXYLATED see
TBA800

ALCOHOLS, N.O.S. see AFJ250

ALCOHOL SULFATE see AFJ375

ALCOHOL SULPHATE see AFJ375

ALCOHOLS, n.o.s. (UN 1987) (DOT) see
EFU000

ALCOHOLS, toxic, n.o.s. (UN 1986) (DOT) see
EFU000

ALCOID see AFJ400

ALCOOL ALLILCO (ITALIAN) see AFV500

ALCOOL ALLYLIQUE (FRENCH) see
AFV500

ALCOOL AMILICO (ITALIAN) see IHP000

ALCOOL AMYLIQUE (FRENCH) see
AOE000

ALCOOL BUTYLIQUE (FRENCH) see
BPW500

ALCOOL BUTYLIQUE SECONDAIRE
(FRENCH) see BPW750

ALCOOL BUTYLIQUE TERTIAIRE (FRENCH) see BPX000
ALCOOL ETHYLIQUE (FRENCH) see EFU000
ALCOOL ETILICO (ITALIAN) see EFU000
l'ALCOOL n-HEPTYLIQUE PRIMAIRE (FRENCH) see HBL500
ALCOOL ISOAMYLIQUE (FRENCH) see IHP000
ALCOOL ISOBUTYLIQUE (FRENCH) see IIL000
ALCOOL ISOPROPILICO (ITALIAN) see INJ000
ALCOOL ISOPROPYLIQUE (FRENCH) see INJ000
ALCOOL METHYL AMYLIQUE (FRENCH) see MKW600
ALCOOL METHYLIQUE (FRENCH) see MGB150
ALCOOL METILICO (ITALIAN) see MGB150
ALCOOL PROPILICO (ITALIAN) see PND000
ALCOOL PROPYLIQUE (FRENCH) see PND000
ALCOPAN-250 see PAG200
ALCOPHOBIN see DXH250
ALCOPOL O see DJL000
ALCOTEX 88/05 see PKP750
ALCOTEX 88/10 see PKP750
ALCOWAX 6 see PJS750
ALCUONIUM DICHLORIDE see DBK400
ALCURONIUM CHLORIDE see DBK400
ALD-52 see AFJ450
ALDABAN see TAI450
ALDACOL Q see PKQ250
ALDACTAZIDE see AFJ500
ALDACTIDE see AFJ500
ALDACTONE see AFJ500
ALDACTONE A see AFJ500
ALDADIENE-KALIUM see PKZ000
ALDADIENE-POTASSIUM see PKZ000
ALDANIL see FMW000
ALDECARB see CBM500
ALDECIN see AFJ625
ALDEHYDE-14 see UJJ000
ALDEHYDE ACETIQUE (FRENCH) see AAG250
ALDEHYDE ACRYLIQUE (FRENCH) see ADR000
ALDEHYDE AMMONIA see AAG500
ALDEHYDE B see COU500
ALDEHYDE BUTYRIQUE (FRENCH) see BSU250
ALDEHYDE C-6 see HEM000
ALDEHYDE C-8 see OCO000
ALDEHYDE C-9 see NMW500
ALDEHYDE C-10 see DAG000
ALDEHYDE C-14 see HBN200
ALDEHYDE C-18 see CNF250
ALDEHYDE C-18 see NMV790
ALDEHYDE C-10 DIMETHYLACETAL see AFJ700
ALDEHYDE C-12, MNA see MQI550
ALDEHYDE C-12 MNA DIMETHYL ACETAL see MNB600
ALDEHYDE C-14, MYRISTIC see TBX500
ALDEHYDECOLLIDINE see EOS000
ALDEHYDE C-14 PEACH see HBN200
ALDEHYDE C-14 PURE see UJA800
ALDEHYDE CROTONIQUE (FRENCH) see COB260
ALDEHYDE C-11, UNDECYLENIC see ULJ000
ALDEHYDE-2-ETHYLBUTYRIQUE (FRENCH) see DHI000
ALDEHYDE FORMIQUE (FRENCH) see FMV000
ALDEHYDE ISOVALERIANIQUE see MHX500
ALDEHYDE M.N.A. see MQI550
ALDEHYDE PROPIONIQUE (FRENCH) see PMT750

ALDEHYDES see AFJ800
ALDEHYDINE see EOS000
ALDEHYDODICHLOROMALEIC ACID see MRU900
ALDEIDE ACETICA (ITALIAN) see AAG250
ALDEIDE ACRILICA (ITALIAN) see ADR000
ALDEIDE BUTIRRICA (ITALIAN) see BSU250
ALDEIDE FORMICA (ITALIAN) see FMV000
ALDER BUCKTHORN see MBU825
ALDERLIN see INS000
ALDERLIN HYDROCHLORIDE see INT000
ALDICARBE (FRENCH) see CBM500
ALDICARB OXIME see MLX800
ALDICARB SULFOXIDE see MLW750
ALDICARB (USDA) see CBM500
ALDIFEN see DUZ000
AL-DIISOBUTYL see DNI600
ALDIMORPH see AFJ850
ALDINAMID see POL500
ALDO see CLY500
ALDO-28 see OAV000
ALDO 40 see GGR200
ALDO-72 see OAV000
ALDOCORTEN see AFJ875
ALDOCORTENE see AFJ875
ALDOCORTIN see AFJ875
ALDO HMS see OAV000
ALDOL see AAH750
ALDOMET see DNA800
ALDOMET see MJE780
ALDOMETIL see DNA800
ALDOMETIL see MJE780
ALDOMIN see DNA800
ALDOMIN see MJE780
ALDO MO-FG see GGR200
ALDO MS see OAV000
ALDO MSA see OAV000
ALDO MSLG see OAV000
ALDOMYCIN see NGE500
ALDOSPERSE L 9 see DXY000
ALDOSTERONE see ECA100
(+)-ALDOSTERONE see AFJ875
d-ALDOSTERONE see AFJ875
ALDOSTERONE (8CI) see AFJ875
ALDOXIME see AAH250
2-ALDOXIME PYRIDINIUM-N-METHYL METHANESULPHONATE see PLX250
ALDOXYCARB see AFK000
ALDREX see AFK250
ALDREX 30 see AFK250
ALDRICH see DOJ200
ALDRIN see AFK250
ALDRIN, cast solid (DOT) see AFK250
ALDRINE (FRENCH) see AFK250
ALDRITE see AFK250
ALDROSOL see AFK250
ALDYL A see PJS750
ALECOR see ECU600
ALELAILA (PUERTO RICO) see CDM325
ALENDRONATE see AKF300
ALENDRONATE SODIUM SALT see SKL600
ALENDRONIC ACID see AKF300
ALENTIN see BSM000
ALENTOL see AOB250
ALEPSIN see DNU000
ALERMINE see CLD250
ALERYL see BBV500
ALESTEN see SNQ710
ALETAMINE HYDROCHLORIDE see AGQ250
ALEUDRIN see DMV600
ALEURITES (VARIOUS SPECIES) see TOA275
ALEURITIC ACID, tech see TKP000
ALEVAIRE see TDN750
ALEVIATIN see DKQ000
ALEXAN see AQQ750
ALEXAN see COW900
ALEXANDRIAN LAUREL see MBU780
ALF see AKR500
ALFACALCIDOL see HJV000

ALFACILLIN see PDD350
ALFACOL see TGJ050
ALFACRON see ARY800
ALFACRON see IEN000
ALFADIONE see AFK500
ALFADRYL see PGQ100
ALFALFA MEAL see AFK750
ALFAMAT see GFA000
ALFAMETHRIN see RCZ050
ALFANAFTILAMINA (ITALIAN) see NBE700
ALFA-NAFTYLOAMINA (POLISH) see NBE700
ALFATESINE (FRENCH) see AFK500
ALFATIL see CCR850
ALFA-TOX see DCM750
ALFA-TOX see MEI500
ALFATROFIN see AES650
ALFAXALONE see AFK875
ALFENAMIN see AHA875
ALFENOL 3 see PKF500
ALFENOL 9 see PKF500
ALFENOLU 8 see NNC700
ALFEPROL (RUSSIAN) see AGW250
ALFIBRATE see AHA150
ALFICETYN see CDP250
ALFIDE see SOX875
ALFIMID see DYC800
ALFLORONE see FHH100
ALFOCILLIN see PDD350
ALFOL 8 see OEI000
ALFOL-10 see ODE000
ALFOL 12 see DXV600
ALFOL 810 see DAI800
ALFONAL K see AFK900
ALFONIC 1012-40 see AFJ160
ALFONIC 1216-22 see AFJ165
ALFOXYLATE see RCZ050
ALFUCIN see NGE500
ALFUZOSIN HYDROCHLORIDE see XDJ050
ALGAE, BROWN see AFK920
ALGAE MEAL, DRIED see AFK925
ALGAE, RED see AFK930
ALGAFAN see PNA500
ALGAMON see SAH000
ALGANET see AFK940
ALGAROBA see LIA000
ALGEON 22 see CFX500
ALGERIAN IVY see AFK950
ALGERIL see PMX250
ALGESTONE ACETONIDE see PMH550
ALGESTONE ACETOPHENIDE see DAM300
ALGIAMIDA see SAH000
ALGIDON see MDP750
ALGIL see DAM700
ALGIMYCIN see ABU500
ALGIN see ANA300
ALGIN see CAM200
ALGIN see PKU700
ALGIN see SEH000
ALGIN (polysaccharide) see SEH000
ALGINATE KMF see SEH000
ALGIN GUM see CAO250
ALGINIC ACID see AFL000
ALGINIC ACID HYDROGEN SULFATE SODIUM SALT (9CI) see SEH450
ALGIPON L-1168 see SEH000
ALGISTAT see DFT000
ALGLOFLON see TAI250
ALGOCOR see EID200
ALGODON de SEDA (CUBA, PUERTO RICO) see COD675
ALGOFLON SV see TAI250
ALGOFRENE 22 see CFX500
ALGOFRENE TYPE 1 see TIP500
ALGOFRENE TYPE 2 see DFA600
ALGOFRENE TYPE 5 see DFL000
ALGOFRENE TYPE 6 see CFX500
ALGOFRENE TYPE 67 see ELN500
ALGOL ORANGE RF see CMU815
ALGOLYSIN see MDP750

ALGONEURINA see TES800
ALGOSEDIV see TEH500
ALGOTROPYL see HIM000
ALGRAIN see EFU000
ALGUSI, extract see AHI630
ALGYLEN see TIO750
ALHELI EXTRANJERO (PUERTO RICO) see OHM875
ALIANT see TGX550
ALICIDON see GGQ050
ALICOP see LBX075
ALICYANATE see PLC250
ALIDOCHLOR see CFK000
ALIETTE see AHH800
ALIGUAT 4 see LBX075
ALIMEMAZINE see AFL500
ALIMEMAZINE-S,S-DIOXIDE see AFL750
ALIMEMAZIN HYDROCHLORIDE see TKV200
ALIMET see HMR600
ALIMEZINE see AFL500
ALINAMIN see DXO300
ALINAMIN F see FQJ100
ALINDOR see BRF500
ALIOMYCIN see AFM000
ALIPAL CO 430 see SNY100
ALIPAL CO 433 see SNY100
ALIPAL CO 436 see ANO600
ALIPAL EP see ANO600
ALIPAL EP 110 see ANO600
ALIPAL EP 120 see ANO600
ALIPHATIC and AROMATIC EPOXIDES see AFM250
ALIPHATIC CHLORINATED HYDROCARBONS see CDV250
ALIPORINA see TEY000
ALIPUR see AFM375
ALIPUR-O see CPT000
ALIQUAT 7 see TLW500
ALIQUAT 203 see DGX200
ALIQUAT 206 see DRK200
ALIQUAT 207 see DXG625
ALIQUAT 264 see QAT550
ALIQUAT 336 see MQH000
ALIQUAT 336 see QAT565
ALIQUAT 336N see MQH000
ALIQUAT 336S see QAT565
ALIQUAT 7402 see QAT565
ALIQUAT N 263 see QAT565
ALIQUAT 336-PTC see MQH000
ALISOBUMAL see AGI750
ALITHON 7050 see PJS750
ALITIA S see TES800
ALITON see DXO300
ALIVAL see NMV725
ALIVAL (ANTIDEPRESSANT) see NMV725
ALIZANTHRENE BLUE RC see DFN300
ALIZANTHRENE NAVY BLUE R see VGP100
ALIZANTHRENE NAVY BLUE RT see VGP100
ALIZARIN see DMG800
ALIZARINA see DMG800
ALIZARIN B see DMG800
ALIZARINBORDEAUX see TDD000
ALIZARIN BRILLANT BLUE BS see APG700
ALIZARIN CARMINE (BIOLOGICAL STAIN) see SEH475
ALIZARIN COMPLEXON see AFM400
ALIZARIN COMPLEXONE see AFM400
ALIZARINE see DMG800
ALIZARINE 3B see DMG800
ALIZARINE ACID BLUE B see CMM090
ALIZARINE ACID RED B EXTRA see CMG750
ALIZARINE AZUROL SE see APG700
ALIZARINE B see DMG800
ALIZARINE BLUE A see CMM070
ALIZARINE BLUE AR see CMM070
ALIZARINE BLUE BLACK OCBN see CMP880

ALIZARINE BLUE BLACK OCGN see CMP880
ALIZARINE BLUE BLACK OCGP see CMP880
ALIZARINE BLUE GRL see CMM090
ALIZARINE BLUE OCB see CMP880
ALIZARINE BLUE SE see APG700
ALIZARINE BORDEAUX see TDD000
ALIZARINE BORDEAUX B see TDD000
ALIZARINE BRILLANT BLUE BS see APG700
ALIZARINE BRILLIANT SAPPHIRE R see CMM080
ALIZARINE BRILLIANT SKY BLUE R see CMM080
ALIZARINE BROWN HD see TKN500
ALIZARINE BROWN R see TKN500
ALIZARINE CARMINE INDICATOR see SEH475
ALIZARINE CHROME ORANGE G see NEY000
ALIZARINE CHROME ORANGE GR see CMP882
ALIZARINE CHROME RED BG see CMG750
ALIZARINE CHROME RED G see CMM325
ALIZARINE COMPLEXON see AFM400
ALIZARINE COMPLEXONE see AFM400
ALIZARINE CYANINE GREEN BASE see BLK000
ALIZARINE DIRECT BLUE AR see CMM070
ALIZARINE DIRECT BLUE ARA see CMM070
ALIZARINE DIRECT PURE BLUE R see CMM080
ALIZARINE FAST BLUE 2B see CMM090
ALIZARINE FAST BLUE BE see CMM100
ALIZARINE FAST BLUE R see CMM092
ALIZARINE FAST BLUE RFE see CMM080
ALIZARINE FAST LIGHT BLUE C see CMM090
ALIZARINE FLUORINE BLUE see AFM400
ALIZARINE GR see SIT850
ALIZARINE INDICATOR see DMG800
ALIZARINE LAKE RED 2P see DMG800
ALIZARINE LAKE RED 3P see DMG800
ALIZARINE LAKE RED IPX see DMG800
ALIZARINE LIGHT BLUE AR see CMM090
ALIZARINE LIGHT BLUE SE see APG700
ALIZARINE L PASTE see DMG800
ALIZARINE MILLING BLUE R see CMM092
ALIZARINE MSE see APG700
ALIZARINE NAC see DMG800
ALIZARINE ORANGE see NEY000
ALIZARINE ORANGE 2GN see NEY000
ALIZARINE ORANGE R see NEY000
ALIZARINE ORANGE RD see NEY000
ALIZARINE PASTE 20% BLUISH see DMG800
ALIZARINE PURE BLUE B see CMM090
ALIZARINE R-CF see NEY000
ALIZARINE RED see DMG800
ALIZARINE RED A see SEH475
ALIZARINE RED AS see SEH475
ALIZARINE RED B see DMG800
ALIZARINE RED B2 see DMG800
ALIZARINE RED INDICATOR see SEH475
ALIZARINE RED IP see DMG800
ALIZARINE RED IPP see DMG800
ALIZARINE RED L see DMG800
ALIZARINE RED S (BIOLOGICAL STAIN) see SEH475
ALIZARINE RED S SODIUM SALT see SEH475
ALIZARINE RED SW see SEH475
ALIZARINE RED SZ see SEH475
ALIZARINE RED W see SEH475
ALIZARINE RED WA see SEH475
ALIZARINE RED for WOOL see SEH475
ALIZARINE RED WS see SEH475
ALIZARINE S see SEH475
ALIZARINE SAPPHIRE AR see CMM070
ALIZARINE SAPPHIRE SE see APG700

ALIZARINE SAPPHIROL SE see APG700
ALIZARINE S EXTRA CONC. A EXPORT see SEH475
ALIZARINE S EXTRA PURE A see SEH475
ALIZARINE SKY BLUE B see CMM090
ALIZARINE SKY BLUE BS-CF see CMM090
ALIZARINE SKY BLUE R see CMM080
ALIZARINE SUPRA BLUE R see CMM080
ALIZARINE SUPRA SKY RA see CMM080
ALIZARINE VIOLET 3B BASE see HOK000
ALIZARINE VIOLET N see HLI200
ALIZARINE YELLOW see EAI200
ALIZARINE YELLOW 2G see SIT850
ALIZARINE YELLOW AGP see SIT850
ALIZARINE YELLOW G see SIT850
ALIZARINE YELLOW GG see SIT850
ALIZARINE YELLOW GGW see SIT850
ALIZARINE YELLOW GM see SIT850
ALIZARINE YELLOW P see NEY000
ALIZARINE YELLOW R see NEY000
ALIZARINE YELLOW RW see NEY000
ALIZARIN FLUORINE BLUE see AFM400
ALIZARINKOMPLEXON see AFM400
ALIZARIN LIGHT BLUE SE see APG700
ALIZARINPRIMEVEROSIDE see DMG800
ALIZARIN RED see DMG800
ALIZARIN RED S see SEH475
ALIZARINROT-S see SEH475
ALIZARIN S see SEH475
ALIZARINSULFONATE see SEH475
ALIZARIN YELLOW see SIT850
ALIZARIN YELLOW G see SIT850
ALIZARIN YELLOW GG see SIT850
ALIZARIN YELLOW G SODIUM SALT see SIT850
ALIZAROL ORANGE R see NEY000
ALIZAROL YELLOW GW see SIT850
ALJADEN see CDK800
ALKALIES see AFM500
ALKALI GRASS see DAE100
ALKALINE DEOXYRIBONUCLEASE see EDK650
ALKALINE DNASE see EDK650
ALKALI RESISTANT RED DARK see NAY000
ALKALOID C see AFG750
ALKALOID H 3, from COLCHICUM ANTUMNALE see MIW500
ALKALOID II see AFG750
ALKALOIDS see AFM750
ALKALOID SALTS see AFM750
ALKALOIDS, VERATRUM see VIZ000
ALKALOID V see CPT750
ALKAMID see PJY500
ALKAMON DS see AFM800
ALKANES see AFN250
ALKANNA RED see HLJ650
ALKANNIN see HLJ650
ALKAPOL PEG-200 see PJT000
ALKAPOL PEG-300 see PJT000
ALKAPOL PEG-400 see MFD500
ALKAPOL PEG-600 see PJT000
ALKAPOL PEG-6000 see PJT000
ALKAPOL PEG-8000 see PJT000
ALKAPOL PPG-1200 see PKI500
ALKARAU see RDK000
ALKARSODYL see HKC500
ALKASERP see RDK000
ALKATERIC CAB-A see CNF185
ALKATHENE see PJS750
ALKATHENE 200 see PJS750
ALKATHENE 22 300 see PJS750
ALKATHENE 17/04/00 see PJS750
ALKATHENE ARN 60 see PJS750
ALKATHENE RXDG33 see TAI250
ALKATHENE WJG 11 see PJS750
ALKATHENE WNG 14 see PJS750
ALKATHENE XDG 33 see PJS750
ALKATHENE XJK 25 see PJS750
ALK-AUBS see DXH250
ALKAVERVIR see VIZ000
ALKAZENE 42 see EHY000

α-ALKENESULFONIC ACID see AFN500
ALKENYL DIMETHYLETHYL AMMONIUM BROMIDE see AFN750
ALK-ENZYME see BAC000
ALKERAN see PED750
ALKERAN (RUSSIAN) see BHV000
ALKIRON see MPW500
ALKOFEN B see TIA100
ALKOFEN MBP see MHR050
ALKOHOL (GERMAN) see EFU000
ALKOHOLU ETYLOWEGO (POLISH) see EFU000
ALKOTEX see PKP750
ALKOVERT see PIB250
ALKOXY ALCOHOLS, C8-10, ETHOXYLATED PROPOXYLATED see AFJ172
3-(ALKYLAMINO)PROPIONITRILE see AFO200
C$_{14-30}$ ALKYLAROMATIC DERIVATIVES see BBM100
ALKYL ARYL POLYETHER ALCOHOLS see ARL875
ALKYLARYLPOLYGLYKOLAETHER (GERMAN) see ARL875
ALKYL ARYL SULFONATE see AFO250
ALKYLATE 215 see AFO300
ALKYLATE 315 see LGE200
ALKYLBENZENESULFONATE see AFO500
p-n-ALKYLBENZENESULFONIC ACID DERIVATIVE, SODIUM SALT see AFO750
ALKYLBENZENESULFONIC ACID SODIUM SALT see AFP000
p-N-ALKYLBENZENSULFONAN SODNY (CZECH) see AFO750
ALKYL(C$_{9-11}$) ALCOHOL, ETHOXYLATED see AFN900
ALKYL(C12-C15) ALCOHOL ETHOXYLATED see AFN950
ALKYL(C$_{14-16}$)DIMETHYLBENZYL AMMONIUM CHLORIDES see AFP100
ALKYL(C$_8$H$_{17}$ to C$_{18}$H$_{37}$) DIMETHYL-3,4-DICHLOROBENZYL AMMONIUM CHLORIDE see AFP750
ALKYL(C$_6$H$_{18}$)DIMETHYL-3,4-DICHLOROBENZYLAMMONIUM CHLORIDE see AFP750
ALKYL(C-13) POLYETHOXYLATES(ETHOXY-6) see TJJ250
ALKYL(C-14) POLYETHOXYLATES (ETHOXY-7) see TBY750
ALKYL(C$_{9-15}$)TOLYL METHYLTRIMETHYL AMMONIUM CHLORIDE see DYA600
ALKYL DIMETHYL BENZALKONIUM CHLORIDE see AFP075
ALKYL DIMETHYL BENZALKONIUM SACCHARINATE see BBA625
ALKYL DIMETHYLBENZYL AMMONIUM CHLORIDE see AFP250
ALKYL DIMETHYL 3,4-DICHLOROBENZENE AMMONIUM CHLORIDE see AFP300
ALKYLDIMETHYLETHYLBENZYL AMMONIUM CHLORIDE see BBA500
ALKYLDIMETHYL(PHENYLMETHYL)QUA TERNARY AMMONIUM CHLORIDES see AFP250
ALKYL((ETHYLPHENYL)METHYL)DIMET HYL QUATERNARY AMMONIUM CHLORIDES see BBA500
ALKYLNITRILE see AFQ000
ALKYL PHENOL POLYGLYCOL ETHERS see ARL875
ALKYL PHENOXY POLYETHOXY ETHANOLS see ARL875
ALKYL PHENYL POLYETHYLENE GLYCOL ETHER see AFQ250
ALKYL PYRIDINES R see AFQ500
ALKYROM see AFQ575
ALLACYL see AFW500
ALLAMANDA see AFQ625

ALLAMANDA CATHARTICA see AFQ625
ALL-trans-ANHYDRORETINOL see AFQ700
ALLANTOXANIC ACID, POTASSIUM SALT see AFQ750
ALLBRI ALUMINUM PASTE and POWDER see AGX000
ALLBRI NATURAL COPPER see CNI000
(ALL E)-2-(1,3,5,7,9-DODECAPENTAENYLOXY)ETHANOL see AFQ800
ALLEDRYL see BBV500
(ALL E)-1-(2-HYDROXYPROPOXY)-1,3,5,7,9-DODECAPENTAENE see AFQ820
(+)-ALLELRETHONYL (+)-cis,trans-CHRYSANTHEMATE see AFR250
ALLENE see AFR000
ALLEOSIDE A DIHYDRATE see HAO000
(ALL E)-1-PROPOXY-1,3,5,7,9-DODECAPENTAENE see DXT900
ALLERCLOR see TAI500
ALLERCUR see CMV400
ALLERCURE HYDROCHLORIDE see CMV400
ALLERGAN see CKV625
ALLERGAN 211 see DAS000
ALLERGAN B see BBV500
ALLERGEFON MALEATE see CGD500
ALLERGEN see LJR000
ALLERGEVAL see BBV500
ALLERGICAL see BBV500
ALLERGICAN see CLX300
ALLERGIN see BBV500
ALLERGIN see TAI500
ALLERGINA see BBV500
ALLERGISAN see CLX300
ALLERGISAN see TAI500
ALLERGIVAL see BBV500
ALLERON see PAK000
ALLETHRIN see AFR250
d-ALLETHRIN see AFR250
(+)-cis-ALLETHRIN see AFR500
d-trans ALLETHRIN see AFR250
d-trans ALLETHRIN see BGC750
trans-(+)-ALLETHRIN see AFR750
ALLETHRIN I see AFR250
ALLETHRIN RACEMIC MIXTURE see AFS000
d-ALLETHROLONE CHRYSANTHEMUMATE see AFR750
(+)-ALLETHRONYL (+)-trans-CHRYSANTHEMUMATE see AFR750
ALL-trans-N-ETHYLRETINAMIDE see REK330
ALLETONE see HKC575
ALLEVIATE see AFR250
ALL-trans-FECAPENTAENE 12 see AFS100
ALLICIN see AFS250
ALLIDOCHLOR see CFK000
ALLIE see MQR400
ALLIED GC 9160 see MDT625
ALLIED PE 617 see PJS750
ALLIED WHITING see CAT775
ALLIGATOR LILY see BAR325
ALLIGATOR PEAR OIL see ARW800
ALLILE (CLORURO di) (ITALIAN) see AGB250
ALLIL-GLICIDIL-ETERE (ITALIAN) see AGH150
1-ALLILOSSI-2,3 EPOSSIPROPANO (ITALIAN) see AGH150
ALLILOWY ALKOHOL (POLISH) see AFV500
ALLIONAL see AFT000
ALLISAN see RDP300
ALLIUM SATIVUM Linn., powder see GBU850
ALLIUM (Various Species) see WBS850
ALLOBARBITAL see AFS500
ALLOBARBITONE see AFS500
ALLOCAINE see AIL750
ALLOCAINE see AIT250
ALLOCHRISIN see SEZ400
ALLOCHRISINE see SEZ400

ALLOCHRYSINE see SEZ400
ALLOCLAMIDE see AFS625
ALLOCLAMIDE HYDROCHLORIDE see AFS640
ALLODAN see AFS750
ALLODENE see BBK000
ALLOFENYL see AGQ875
ALLOFERIN see DBK400
ALLO-GLAUCOTOXIGENIN see AFS800
ALLOGLAUCOTOXIGENIN see AFS800
ALLOGLUCOTOXIGENIN see AFS800
ALLO-HELVETICOSIDE see HAN900
ALLOMALEIC ACID see FOU000
ALLOMALEIC ACID DIMETHYL ESTER see DSB600
ALLOMYCIN see AHL000
ALLONAL see AFT000
ALLO-OCIMENOL see LFX000
ALLOPHANIC ACID, 1-ETHYNYLCYCLOHEXYL ESTER see EQL600
1-(p-ALLOPHANOYLBENZYL)-2-METHYLHYDRAZINE HYDROBROMIDE see MKN750
ALLOPHENYLUM see AGQ875
ALLOPREGNAN-3-β-OL-20-ISONICOTINYLHYDRAZONE see AFT125
ALLOPSEUDOCODEINE HYDROCHLORIDE see AFT250
ALLOPURINOL see ZVJ000
ALLOPYDIN see AGN000
ALLORPHINE see AFT500
ALLOTROPAL see EQL000
ALLO-UZARIGENIN see UWJ200
ALLOUZARIGENIN see UWJ200
ALLOXAN see AFT750
ALLOXAN HYDRATE see DMG950
ALLOXAN MONOHYDRATE see DMG950
ALLOXAN MONOHYDRATE see MDL500
ALLOXAN-5-OXIME see AFU000
ALLOXANTIN see AFU250
ALLOXYDIM see ZTS200
ALLOXYDIM-SODIUM see FBP350
2-ALLOXYETHANOL (CZECH) see AGO000
ALLOY 400 see NCX525
ALLOY 800 see IGL100
ALLOY 21-6-9 see IGL110
ALLOZYM see ZVJ000
ALLPHASEM see AGQ875
ALLTEX see CDV100
ALLTOX see CDV100
ALLUMINIO(CLORURO DI) (ITALIAN) see AGY750
ALLUMINIO DIISOBUTIL-MONOCLORURO (ITALIAN) see CGB500
ALLURAL see ZVJ000
ALLURA RED see FAG100
ALLURA RED AC see FAG100
ALLUVAL see BNP750
ALLY see MQR400
ALLY 20DF see MQR400
ALLYL ACETATE see AFU750
ALLYLACETIC ACID see PBQ750
ALLYL ADIPATE see AEO500
ALLYL AL see AFV500
ALLYL ALCOHOL see AFV500
ALLYL ALDEHYDE see ADR000
ALLYLALKOHOL (GERMAN) see AFV500
5-ALLYL-5-(1-(ALLYTHIO)ETHYL)BARBITURIC ACID SODIUM SALT see AFV750
ALLYLAMINE see AFW000
ALLYLAMINE, N,N,2-TRIMETHYL- see TME255
2-(ALLYLAMINO)-6'-CHLORO-o-ACETOTOLUIDIDE HYDROCHLORIDE see AFW250
2-(ALLYLAMINO)-2'-CHLORO-6'-METHYLACETANILIDE HYDROCHLORIDE see AFW250
1-ALLYL-6-AMINO-3-ETHYL-2,4(1H,3H)-PYRIMIDINEDIONE see AFW500

1-ALLYL-6-AMINO-3-ETHYLURACIL see AFW500

p-ALLYLANISOLE see AFW750

ALLYLBARBITAL see AGI750

ALLYLBARBITONE see AGI750

ALLYLBARBITURAL see AFS500

ALLYLBARBITURIC ACID see AGI750

ALLYLBENZENE see AFX000

ALLYL BENZENE SULFONATE see AFX250

5-ALLYL-1,3-BENZODIOXOLE see SAD000

5-ALLYL-5-BENZYL-2-THIOBARBITURIC ACID SODIUM SALT see AFX500

ALLYL-BIS(β-CHLOROETHYL)AMINE HYDROCHLORIDE see AFX750

ALLYL BROMIDE see AFY000

ALLYL BUTANOATE see AFY250

5-ALLYL-5-(2-BUTENYL)-2-THIOBARBITURIC ACID SODIUM SALT see AFY300

5-ALLYL-5-sec-BUTYLBARBITURIC ACID see AFY500

ALLYL-sec-BUTYL THIOBARBITURIC ACID see AFY750

5-ALLYL-5-sec-BUTYL-2-THIOBARBITURIC ACID see AFY750

5-ALLYL-5-sec-BUTYL-2-THIOBARBITURIC ACID SODIUM SALT see AGA000

5-ALLYL-5-(1-BUTYLTHIO)ETHYL)BARBITURIC ACID SODIUM SALT see AGA250

ALLYL BUTYRATE see AFY250

ALLYL CAPROATE see AGA500

ALLYL CAPRYLATE see AGM500

ALLYL CARBAMATE see AGA750

N-ALLYL CARBAMIC ACID-3-DIMETHYLAMINOPHENYL ESTER HYDROCHLORIDE see AGB000

ALLYLCARBAMIC ESTER of m-OXYPHENYLDIMETHYLAMINE HYDROCHLORIDE see AGB000

ALLYLCARBAMIDE see AGV000

ALLYLCATECHOL METHYLENE ETHER see SAD000

ALLYLCHLORID (GERMAN) see AGB250

ALLYL CHLORIDE see AGB250

ALLYL CHLOROCARBONATE see AGB500

ALLYL 2-CHLOROETHYLSULFIDE see AGB600

ALLYL CHLOROFORMATE see AGB500

ALLYL CHLOROFORMATE (DOT) see AGB500

ALLYLCHLOROHYDRIN ETHER see AGB750

ALLYL (3-CHLORO-2-HYDROXYPROPYL) ETHER see AGB750

ALLYL CINERIN see AFR250

ALLYL CINNAMATE see AGC000

ALLYL COMPOUNDS see AGC125

ALLYL CYANIDE see BOX500

ALLYL CYANOACETATE see AGC150

6-ALLYL-α-CYANOERGOLINE-8-PROPIONAMIDE see AGC200

ALLYL CYCLOHEXANEACETATE see AGC250

ALLYL CYCLOHEXANEPROPIONATE see AGC500

3-ALLYL-5-(1-CYCLOHEXEN-1-YL)-1,5-DIMETHYLBARBITURIC ACID see PPP133

5-ALLYL-5-(2-CYCLOHEXEN-1-YL)-2-THIOBARBITURIC ACID see TES500

ALLYL CYCLOHEXYLACETATE see AGC250

2-ALLYL-2-CYCLOHEXYLACETIC ACID-3-(DIETHYLAMINO)-2,2-DIMETHYLPROPYLESTER HYDROCHLORIDE see AGC750

3-ALLYLCYCLOHEXYL PROPIONATE see AGC500

5-ALLYL-5-(2-CYCLOPENTENYL)-2-THIOBARBITURIC ACID see AGD000

N-ALLYL-7,8-DEHYDRO-4,5-EPOXY-3,6-DIHYDROXYMORPHINAN see AFT500

17-α-ALLYL-3-DEOXY-19-NORTESTOSTERONE see AGF750

N-ALLYL-N-DESMETHYLMORPHINE see AFT500

l-ALLYL-(6-DIAZO-5-OXO)-l-NORLEUCYL-(6-DIAZO-5-OXO)-l-NORLEUCINE see AFI625

ALLYL DIGLYCOL CARBONATE see AGD250

6-ALLYL-6,7-DIHYDRO-5H-DIBENZ(c,e)AZEPINE see ARZ000

6-ALLYL-6,7-DIHYDRO-5H-DIBENZ(c,e)AZEPINE PHOSPHATE see AGD500

6-ALLYL-6,7-DIHYDRO-3,9-DICHLORO-5H-DIBENZ(c,e)AZEPINE see AGD750

l-N-ALLYL-7,8-DIHYDRO-14-HYDROXYNORMORPHINONE see NAG550

N-ALLYL-7,8-DIHYDRO-14-HYDROXYNORMORPHINONE, HYDROCHLORIDE see NAH000

6-ALLYL-6,7-DIHYDRO-6-METHYL-5H-DIBENZ(c,e)AZEPINIUM IODIDE see AGE000

N-ALLYL-2,12-DIHYDROXY-1,11-EPOXYMORPHINENE-13 HYDROCHLORIDE see NAG500

1-ALLYL-3,4-DIMETHOXYBENZENE see AGE250

4-ALLYL-1,2-DIMETHOXYBENZENE see AGE250

1-ALLYL-2,5-DIMETHOXY-3,4-METHYLENEDIOXYBENZENE see AGE500

ALLYLDIMETHYLARSINE see AGE625

1-ALLYL-1-(3,7-DIMETHYLOCTYL)PIPERIDINIUM BROMIDE see AGE750

1-ALLYL-1-(3,7-DIMETHYLOCTYL)-PIPERIDIUMBROMID (GERMAN) see AGE750

β-ALLYL-N,N-DIMETHYLPHENETHYLAMINE see AGF000

ALLYLDIOXYBENZENE METHYLENE ETHER see SAD000

5-ALLYL-1,3-DIPHENYLBARBITURIC ACID see AGF250

5-ALLYL-1,3-DIPHENYL-2,4,6(1H,3H,5H)-PYRIMIDINETERIONE see AGF250

ALLYL DISULFIDE see AGF300

ALLYL DISULPHIDE see AGF300

ALLYLE (CHLORURE d') (FRENCH) see AGB250

ALLYL ENANTHATE see AGH250

ALLYLENE see MFX590

17-ALLYL-4,5-α-EPOXY-3,14-DIHYDROXYMORPHINAN-6-ONE see NAG550

17-ALLYL-4,5-α-EPOXY-3,14-DIHYDROXYMORPHINAN-6-ONE HYDROCHLORIDE see NAH000

ALLYL-3,4-EPOXY-6-METHYLCYCLOHEXANECARBOXYLATE see AGF600

ALLYL-2,3-EPOXYPROPYL ETHER see AGH150

ALLYL-9,10-EPOXYSTEARATE see ECO500

ALLYLESTER KYSELINY CHLORMRAVENCI see AGB500

ALLYLESTER KYSELINY METHAKRYLOVE see AGK500

ALLYLESTER KYSELINY MRAVENCI see AGH000

17-α-ALLYL-ESTRATRIENE-4,9,11,17-β-OL-3-ONE see AGW675

ALLYLESTRENOL see AGF750

17-α-ALLYL-4-ESTREN-17-β-OL see AGF750

17-α-ALLYLESTR-4-EN-17-β-OL see AGF750

ALLYLETHER see DBK000

ALLYL ETHER of PROPYLENE GLYCOL see PNK000

1-ALLYL-3-ETHYL-6-AMINOTETRAHYDROPYRIMIDINEDIONE see AFW500

ALLYL ETHYL ETHER see AGG000

2-ALLYL-5-ETHYL-2'-HYDROXY-9-METHYL-6,7-BENZOMORPHAN see AGG250

ALLYL FLUORIDE see AGG500

ALLYL FLUOROACETATE see AGG750

ALLYL FORMATE see AGH000

ALLYL GLUCOSINOLATE see AGH125

α-ALLYL GLYCEROL ETHER see AGP500

ALLYLGLYCIDAETHER (GERMAN) see AGH150

ALLYL GLYCIDYL ETHER see AGH150

4-ALLYLGUAIACOL see EQR500

ALLYL HEPTANOATE see AGH250

ALLYL HEPTOATE see AGH250

ALLYL HEPTYLATE see AGH250

ALLYL HEXAHYDROPHENYLPROPIONATE see AGC500

ALLYL HEXANOATE (FCC) see AGA500

N-ALLYLHEXOBARBITAL see PPP133

8-ALLYL-(±)-1-α-H,5-α-H-NORTHROPAN-3-α-OL see AGM250

ALLYL HOMOLOG of CINERIN I see AFR250

ALLYL HONOLOG of CINERIN I see BGC750

ALLYLHYDRAZINE HYDROCHLORIDE see AGH500

ALLYL HYDROPEROXIDE see AGH750

17-α-ALLYL-17-β-HYDROXY-Δ4-ESTREN see AGF750

17-α-ALLYL-17-β-HYDROXY-4-ESTRENE see AGF750

17-α-ALLYL-17-β-HYDROXYESTR-4-EN-3-ONE see AGM200

4-ALLYL-1-HYDROXY-2-METHOXYBENZENE see EQR500

2-ALLYL-4-HYDROXY-3-METHYL-2-CYCLOPENTEN-1-ONE see AFR750

d,l-2-ALLYL-4-HYDROXY-3-METHYL-2-CYCLOPENTEN-1-ONE-d,l-CHRYSANTHEMUM MONOCARBOXYLATE see AFR250

N-ALLYL-3-HYDROXYMORPHINAN see AGI000

17-α-ALLYLHYDROXY-19-NOR-4-ANTROSTENE see AGF750

l-N-ALLYL-14-HYDROXYNORDIHYDROMORPHINONE see NAG550

1-N-ALLYL-14-HYDROXYNORDIHYDROMORPHINONE HYDROCHLORIDE see NAH000

ALLYLHYDROXYPHENYLARSINE OXIDE see AGQ775

ALLYLIC ALCOHOL see AFV500

ALLYLIDENE DIACETATE see ADR250

ALLYL IODIDE see AGI250

ALLYL α-IONONE see AGI500

ALLYLISOBUTYLBARBITAL see AGI750

ALLYLISOBUTYLBARBITURATE see AGI750

5-ALLYL-5-ISOBUTYLBARBITURIC ACID see AGI750

5-ALLYL-5-ISOBUTYL-2-THIOBARBITURIC ACID SODIUM SALT see SFG700

ALLYL ISOCYANATE see AGJ000

ALLYLISOPROPYLACETYLCARBAMIDE see IQX000

ALLYLISOPROPYLACETYLUREA see IQX000

5-ALLYL-5-ISOPROPYLBARBITURATE see AFT000

ALLYLISOPROPYLBARBITURIC ACID see AFT000

5-ALLYL-5-ISOPROPYLBARBITURIC ACID see AFT000

ALLYLISOPROPYLMALONYLUREA see AFT000

ALLYL ISORHODANIDE see AGJ250

ALLYL ISOSULFOCYANATE see AGJ250

ALLYL ISOTHIOCYANATE see AGJ250

ALLYL ISOTHIOCYANATE, stabilized (DOT) see AGJ250

ALLYL ISOVALERATE see ISV000

ALLYL ISOVALERIANATE see ISV000

3-ALLYL-4-KETO-2-METHYLCYCLOPENTENYL CHRYSANTHEMUMMONOCARBOXYLATE see AFR250

ALLYLLITHIUM see AGJ375

ALLYL MERCAPTAN see AGJ500

α-ALLYLMERCAPTO-α, α-DIETHYLACETAMIDE see AGJ750

2-ALLYLMERCAPTO-2-ETHYLBUTYRAMIDE see AGJ750

2-ALLYLMERCAPTOISOBUTYRAMIDE see AGK000

α-ALLYLMERCAPTOISOBUTYRAMIDE see AGK000

ALLYLMERCAPTOMETHYLPENICILLIN see AGK250

ALLYLMERCAPTOMETHYLPENICILLINIC ACID see AGK250

ALLYL MESYLATE see AGK750

ALLYL METHACRYLATE see AGK500

ALLYL METHANESULFONATE see AGK750

4-ALLYL-1-METHOXYBENZENE see AFW750

5-ALLYL-1-METHOXY-2,3-(METHYLENEDIOXY)BENZENE see MSA500

4-ALLYL-2-METHOXYPHENOL see EQR500

4-ALLYL-2-METHOXYPHENOL ACETATE see EQS000

4-ALLYL-2-METHOXYPHENOL FORMATE see EQS100

4-ALLYL-2-METHOXYPHENYLPHENYLACETATE see AGL000

5-ALLYL-5-(1-METHYLALLYL)-2-THIOBARBITURIC ACID SODIUM SALT see AGL250

5-ALLYL-5-(1-METHYLBUTYL)BARBITURIC ACID see SBM500

5-ALLYL-5(1-METHYLBUTYL)BARBITURIC ACID SODIUM DERIVATIVE see SBN000

5-ALLYL-5-(1-METHYLBUTYL)BARBITURIC ACID SODIUM SALT see SBN000

5-ALLYL-5-(1-METHYLBUTYL)MALONYLUREA see SBM500

5-ALLYL-5-(1-METHYLBUTYL)MALONYLUREA SODIUM SALT see SBN000

5-ALLYL-5-(1-METHYLBUTYL)-2-THIOBARBITURATE SODIUM see SOX500

5-ALLYL-5-(1-METHYLBUTYL)-2-THIOBARBITURIC ACID see AGL375

5-ALLYL-5-(1-METHYLBUTYL)-2-THIO-BARBITURIC ACID SODIUM SALT see SOX500

ALLYL 3-METHYLBUTYRATE see ISV000

2-ALLYL-3-METHYL-2-CYCLOPENTEN-1-ON-4-YL-N,N'-DIMETHYL-KARBAMAT (CZECH) see AGL500

1-ALLYL-3,4-METHYLENEDIOXYBENZENE see SAD000

4-ALLYL-1,2-METHYLENEDIOXYBENZENE see SAD000

N-ALLYLMETHYLHEXABITAL see PPP133

2-ALLYL-3-METHYL-4-HYDROXY-2-CYCLOPENTEN-1-ONE DIMETHYLCARBAMATE see AGL500

5-ALLYL-1-METHYL-5-(1-METHYL-2-PENTYNYL)BARBITURIC ACID SODIUM SALT see MDU500

N-ALLYL-3-METHYL-N-α-METHYLPHENETHYL-6-OXO-1(6H)-PYRIDAZINE ACETAMIDE see AGL750

3-ALLYL-2-METHYL-4-OXO-2-CYCLOPENTEN-1-YL CHRYSANTHEMATE see AFR250

dl-3-ALLYL-2-METHYL-4-OXOCYCLOPENT-2-ENYL dl-cis trans CHRYSANTHEMATE see AFR250

(±)-3-ALLYL-2-METHYL-4-OXO-2-CYCLOPENTENYL 2,2,3,3-TETRAMETHYLCYCLOPROPANE CARBOXYLATE see TAL575

(±)-5-ALLYL-5-(1-METHYL-2-PENTYNYL)-2-THIOBARBITURIC ACID see AGL875

dl-1-ALLYL-1-METHYL-4-PHENYL-4-PIPERIDINOL PROPIONATE HYDROCHLORIDE see AGV890

5-ALLYL-5-(1-METHYLPROPENYL)BARBITURIC ACID see AGM000

5-ALLYL-5-(1-METHYLPROPYL) BARBITURIC ACID see AFY500

5-ALLYL-5-(2'-METHYL-N-PROPYL) BARBITURIC ACID see AGI750

ALLYL MONOSULFIDE see AGS250

ALLYLMORPHINE HYDROCHLORIDE see NAG500

ALLYL MUSTARD OIL see AGJ250

ALLYLNITRILE see BOX500

1-ALLYL-2-NITROIMIDAZOLE see AGM060

4-(ALLYLNITROSAMINO)-1-BUTANOL see HJS400

3-(ALLYLNITROSAMINO)-1,2-PROPANEDIOL see NJY500

1-(ALLYLNITROSAMINO)-2-PROPANONE see AGM125

4-(ALLYLNITROSOAMINO)BUTRIC ACID see CCJ375

N-ALLYL-N-NITROSO-3-BUTENYLAMINE see BPD000

1-ALLYL-1-NITROSOUREA see NJK000

N-ALLYLNORMORPHINE see AFT500

N-ALLYLNORMORPHINE HYDROCHLORIDE see NAG500

17-ALLYL-19-NORTESTOSTERONE see AGM200

17-α-ALLYL-19-NORTESTOSTERONE see AGM200

ALLYL OCTANOATE see AGM500

17-α-ALLYL-4-OESTRENE-17-β-OL see AGF750

ALLYLOESTRENOL see AGF750

2-(1-ALLYLOXYAMINOBUTYLIDENE)-5,5-DIMETHYLMETHOXYCARBONYLCYCLOHEXANE-1,3-DIONE SODIUM SALT see FBP350

2-ALLYLOXYBENZAMIDE see AGM750

o-(ALLYLOXY)BENZAMIDE see AGM750

m-(ALLYLOXY)-N-(BIS(1-AZIRIDINYL)PHOSPHINYL)BENZAMIDE see AGM800

p-(ALLYLOXY)-N-(BIS(1-AZIRIDINYL)PHOSPHINYL)BENZAMIDE see AGM802

ALLYLOXYCARB see DBI800

2-(ALLYLOXY)-4-CHLORO-N-(2-DIETHYLAMINO)ETHYL)BENZAMIDE see AFS625

2-ALLYLOXY-4-CHLORO-N-(2-(DIETHYLAMINO)ETHYL)BENZAMIDE HYDROCHLORIDE see AFS640

(4-ALLYLOXY-3-CHLOROPHENYL)ACETIC ACID see AGN000

(4-(ALLYLOXY)-3-CHLOROPHENYL)ACETIC ACID SODIUM SALT see AGN250

1-ALLYLOXY-3-CHLORO-2-PROPANOL see AGB750

(±)-1-(β-(ALLYLOXY)-2,4-DICHLOROPHENETHYL)IMIDAZOLE see FPB875

2-(ALLYLOXY)-N-(2-(DIETHYLAMINO)ETHYL)-α,α,α-TRIFLUORO-p-TCLUAMIDE see FDA875

o-ALLYLOXY-N,N-DIETHYLBENZAMIDE see AGN500

o-ALLYLOXY-N,N-DIMETHYLBENZAMIDE see AGN750

1-ALLYLOXY-2,3-EPOXY-PROPAAN (DUTCH) see AGH150

1-ALLYLOXY-2,3-EPOXYPROPAN (GERMAN) see AGH150

1-(ALLYLOXY)-2,3-EPOXYPROPANE see AGH150

2-ALLYLOXYETHANOL see AGO000

o-ALLYLOXY-N-(β-HYDROXYETHYL)BENZAMIDE see AGO250

6-ALLYLOXY-2-METHYLAMINO-4-(N-METHYLPIPERAZINO)-5-METHYLTHIOPYRIMIDINE see AGO500

ALLYLOXYPENTABROMOBENZENE see PAU600

2-(o-ALLYLOXYPHENOXY)-2-HYDROXY-N-ISOPROPYL-1-PROPYLAMINE HYDROCHLORIDE see THK750

1-(o-ALLYLOXY)PHENOXY)-3-(ISOPROPYLAMINO)-2-PROPANOL see CNR500

(−)-1-(o-ALLYLOXYPHENOXY)-3-ISOPROPYLAMINO-2-PROPANOLHYDROCHLORIDE see AGP000

1-(o-ALLYLOXYPHENOXY)-3-ISOPROPYLAMINOPROPAN-2-OL HYDROCHLORIDE see THK750

(+)-1-(o-ALLYLOXYPHENOXY)-3-ISOPROPYLAMINO-2-PROPANOLHYDROCHLORIDE see AGO750

3-(ALLYLOXYPHENOXY)-1,2-PROPANEDIOL see AGP250

2-(3-ALLYLOXYPHENYL)IMIDAZO(2,1-A)ISOQUINOLINE see AGP300

2-(m-(ALLYLOXY)PHENYL)IMIDAZO(2,1-A)ISOQUINOLINE see AGP300

5-(o-(ALLYLOXY)PHENYL)-3-(o-TOLYL)-s-TRIAZOLE see AGP400

1-ALLYLOXY-2,3-PROPANEDIOL see AGP500

3-ALLYLOXY-1,2-PROPANEDIOL see AGP500

N,N'-(2-ALLYLOXY-1,3-PROPANEDIOXYSULFINYL)BIS(3-METHYLPHENYLMETHYLCARBAMATE) see AGP600

N,N'-(3-ALLYLOXY-1,2-PROPANEDIOXYSULFINYL)BIS(3-METHYLPHENYLMETHYLCARBAMATE) see AGP610

N,N'-(3-ALLYLOXY-1,2-PROPANEDIOXYSULFINYL)BIS(1-NAPHTHYL METHYLCARBAMATE) see AGP620

3-ALLYLOXYPROPIONITRILE see AGP750

β-ALLYLOXY-PROPIONITRILE see AGP750

ALLYL PALLADIUM CHLORIDE DIMER see AGQ000

ALLYLPENTABROMOPHENYLETHER see PAU600

ALLYL PENTAERYTHRITOL see AGQ050

2-ALLYLPENTANOIC ACID see AGQ100

α-ALLYL PHENETHYLAMINEHYDROCHLORIDE see AGQ250

2-ALLYL PHENOL see AGQ500

o-ALLYL PHENOL see AGQ500

ALLYL PHENOXYACETATE see AGQ750

1-(o-ALLYLPHENOXY)-3-(ISOPROPYLAMINO)-2-PROPANOL see AGW250

1-(o-ALLYLPHENOXY)-3-(ISOPROPYLAMINO)-2-PROPANOL HYDROCHLORIDE see AGW000

ALLYL PHENYLACETATE see PMS500

ALLYL-3-PHENYLACRYLATE see AGC000
ALLYL PHENYL ARSINIC ACID see AGQ775
5-ALLYL-5-PHENYLBARBITURIC ACID see AGQ875
ALLYL PHENYL ETHER see AGR000
1-ALLYL-2-PHENYL-ETHYLAMINE HYDROCHLORIDE see AGQ250
2-ALLYL-2-PHENYL-4-PENTENAMIDE see AGR100
2-ALLYL-2-PHENYL-4-PENTENOIC ACID 2-(DIETHYLAMINO)ETHYL ESTER HYDROCHLORIDE see AGR125
ALLYL PHENYL THIOUREA see AGR250
1-ALLYL-3-PHENYL-2-THIOUREA see AGR250
ALLYL PHOSPHATE see THN750
ALLYL PHOSPHITE see PHN300
ALLYLPRODINE HYDROCHLORIDE see AGV890
ALLYL PROPYL DISULFIDE see AGR500
ALLYLPROPYMAL see AFT000
m-ALLYLPYROCATECHIN METHYLENE ETHER see SAD000
4-ALLYLPYROCATECHOL FORMALDEHYDE ACETAL see SAD000
ALLYLPYROCATECHOL METHYLENE ETHER see SAD000
1-ALLYLQUINALDINUM BROMIDE see AGR750
ALLYLRETHRONYL dl-cis-trans-CHRYSANTHEMATE see AFR250
ALLYLRHODANID (GERMAN) see AGS750
ALLYLSENFOEL (GERMAN) see AGJ250
ALLYL SEVENOLUM see AGJ250
ALLYLSUCCINIC ANHYDRIDE see AGS000
ALLYL SUCROSE see SNH075
ALLYL SULFIDE see AGS250
ALLYL SULFOCYANIDE see AGS750
12-ALLYL-7,7a,8,9-TETRAHYDRO-3,7a-DIHYDROXY-4aH-8,9c-IMINOETHANOPHENANTHRO(4,5-bcd)FURANONE see NAG550
ALLYL-TETRA-HYDROGERANYL-PIPERIDINIUMBROMID (GERMAN) see AGE750
3-(ALLYL-(TETRAHYDRONAPHTHYL)AMINO)-N,N-DIETHYL-PROPIONAMIDE see AGS375
3-(ALLYL-(TETRAHYDRONAPHTHYL)AMINO)-N,N-DIETHYLPROPIONAMIDE HYDROCHLORIDE see AGS375
ALLYLTHEOBROMINE see AGS500
1-ALLYLTHEOBROMINE see AGS500
ALLYLTHIOCARBAMIDE see AGT500
ALLYL THIOCARBONIMIDE see AGJ250
ALLYL THIOCYANATE see AGS750
2-ALLYLTHIO-2-ETHYLBUTYRAMIDE see AGJ750
ALLYLTHIOMETHYLPENICILLIN see AGK250
4-ALLYLTHIOSEMICARBAZIDE see AGT000
2-(ALLYLTHIO)-2-THIAZOLINE see AGT250
1-ALLYLTHIOUREA see AGT500
N-ALLYLTHIOUREA see AGT500
1-ALLYL-2-THIOUREA see AGT500
ALLYL TRENBOLONE see AGW675
ALLYL TRI-N-BUTYLPHOSPHONIUMCHLORIDE see AGT750
ALLYL TRICHLORIDE see TJB600
ALLYL TRICHLOROSILANE see AGU250
ALLYLTRICHLOROSILANE, stabilized (DOT) see AGU250
ALLYLTRIETHYLAMMONIUM IODIDE see AGU300
ALLYLTRIMETHYLAMMONIUM CHLORIDE see HGI600
ALLYL TRIMETHYLHEXANOATE see AGU400

ALLYL 3,5,5-TRIMETHYLHEXANOATE see AGU400
ALLYLTRIPHENYL STANNANE see AGU500
ALLYLTRIPHENYLTIN see AGU500
9-ALLYL-2-(4-(2-TRITYLETHYL)-1-PIPERAZINYL)-9H-PURINEDIMETHANESULFONATE see AGU750
ALLYLUREA see AGV000
1-ALLYLUREA see AGV000
N-ALLYLUREA see AGV000
2-ALLYLVALERIC ACID see AGQ100
4-ALLYLVERATROLE see AGE250
ALLYL VINYL ETHER see AGV250
N-ALLYNORATROPINE see AGM250
ALLYXYCARB see DBI800
(ALL-Z)-5,8,11,14-EICOSATETRAENOIC ACID see AQS750
ALMAZINE see CFC250
ALMECILLIN see AGK250
ALMEDERM see HCL000
ALMEFRIN see SPC500
ALMEFROL see VSZ450
ALMINOPROFEN see MDP850
(±)-ALMINOPROFEN see MGC200
ALMINOPROFENO see MDP850
ALMITE see AHE250
ALMITRINA (SPANISH) see BGS500
ALMOCARPINE see PIF000
ALMOCARPINE see PIF250
ALMOCETAMIDE see SNQ000
ALMOND ARTIFICIAL ESSENTIAL OIL see BAY500
ALMOND OIL BITTER, FFPA (FCC) see BLV500
ALMORA see MAG000
ALNASID see FLD100
ALNERT see MGE200
ALNOVIN see FAO100
ALNOX see AFS500
ALNOXIN see PQC500
ALOBARBITAL see AFS500
ALOCASIA (VARIOUS SPECIES) see EAI600
ALOCHLOR see CFX000
ALODAN see BFY100
ALODAN (GEROT) see DAM700
ALODAN (PESTICIDE) see BFY100
ALOE see AGV875
ALOE-EMODIN see DMU600
ALOGINAN see FOS100
ALON see AHE250
A. LONGIFLORA see BOO700
ALOPERIDIN see CLY500
ALOPERIDOLO see CLY500
ALOSENN see AGV880
ALOSITOL see ZVJ000
ALOTEC see MDM800
ALOXICOLL see AGZ998
ALPEN see AIV500
ALPEN-N see SEQ000
ALPERIDINE HYDROCHLORIDE see AGV890
ALPEROX C see LBR000
ALPHA CHYMAR see CML880
ALPHA-CHYMAR OPHTH see CML880
ALPHACILINA see AOD000
ALPHACILLIN see AOD000
ALPHACROIC ORANGE R see NEY000
ALPHACROIC VIOLET B see HLI000
ALPHACYPERMETHRIN see RCZ050
ALPHADIONE see AFK500
ALPHALIN see VSK600
ALPHA MEDOPA see DNA800
ALPHAMIN see FOS100
ALPHANAPHTHYL THIOUREA see AQN635
ALPHANAPHTYL THIOUREE (FRENCH) see AQN635
ALPHAPRODINE see NEA500
ALPHAPRODINE HYDROCHLORIDE see NEB000
ALPHASOL OT see DJL000

ALPHASONE ACETOPHENIDE see DAM300
ALPHASPRA see NAK500
ALPHASTEROL see VSK600
ALPHAXALONE see AFK875
ALPHAZURINE see FMU059
ALPHAZURINE 2G see ADE500
ALPHAZURINE A. see ADE675
ALPHAZURINE A (6CI) see ERG100
ALPHEBA see AGQ875
ALPHENAL see AGQ875
ALPHENATE see AGQ875
ALPHENOL 8 see NNC700
ALPHEX FIT 221 see PJS750
AL-PHOS see AHE750
ALPINE TALC USP, BC 127 see TAB750
ALPINE TALC USP, BC 141 see TAB750
ALPINE TALC USP, BC 662 see TAB750
ALPINYL see HIM000
ALPRAZOLAM see XAJ000
ALPRENOL HYDROCHLORIDE see AGW000
ALPRENOLOL see AGW250
ALPROSTADIL see POC350
ALPROSTADIL-α-CYCLODEXTRIN CLATHRATE see AGW275
ALQOVERIN see BRF500
ALQUEQUENJE (PUERTO RICO) see JBS100
ALRATO see AQN635
ALRHEUMAT see BDU500
ALRHEUMUM see BDU500
ALROSOL O see OHU200
ALSERIN see RDK000
ALSEROXYLON see AGW300
ALSOL see CHI200
ALSTONINE HYDROCHLORIDE see AGW375
ALTABACTINA see FPI000
ALTADIOL see EDS100
ALTAFUR see FPI000
ALTAN see DMH400
ALTAX see BDE750
ALTAZINE BLACK BH see CMN800
ALTCO SPERSE FAST YELLOW GFN NEW see AAQ250
ALTERNARIOL see AGW476
ALTERNARIOL and ALTERNARIOL MONOMETHYL ETHER (1:1) see AGW550
ALTERNARIOL-9-METHYL ETHER see AGW500
ALTERNARIOL MONOMETHYL ETHER see AGW500
ALTERNARIOL MONOMETHYL ETHER and ALTERNARIOL (1:1) see AGW550
ALTERTON see CKE750
ALTERUNGSSCHUTZMITTEL ZMB see ZIS000
ALTEZOL see AKO500
ALTHESIN see AFK500
ALTHOSE HYDROCHLORIDE see MDP000
ALTICINA see PDD350
ALTO see CQJ150
ALTOCHROME MILLING SCARLET G see CMM320
ALTOCHROME SCARLET G see CMM325
ALTOCYL BRILLIANT BLUE B see MGG250
ALTODEL see POB000
ALTODOR see DIS600
ALTOSID see KAJ000
ALTOSID (GERMAN) see AGW625
ALTOSIDE see AGW625
ALTOSID IGR see KAJ000
ALTOSID SR 10 see KAJ000
ALTOTAL see PQC500
ALTOWHITES see KBB600
ALTOX see AFK250
ALTOZAR see EQD000
ALTRAD see EDO000
ALTRENOGEST see AGW675
ALTRETAMINE see HEJ500
ALUDRINE see DMV600
ALUFENAMINE see AHA875

ALUFIBRATE see AHA150
ALUGAN see BNQ600
ALUGEL 34TN see AHH825
ALULINE see ZVJ000
ALUM see AHF200
ALUM see AHG750
ALUMIGEL see AHC000
ALUMINA see AHE250
β-ALUMINA see AHE250
β-ALUMINA see AHG000
γ-ALUMINA see AHE250
β"-ALUMINA see AHG000
ALUMINA FIBRE see AGX000
ALUMINA HYDRATE see AHC000
ALUMINA HYDRATED see AHC000
α-ALUMINA (OSHA) see AHE250
ALUMINATE(3-), HEXAFLUORO-,
TRISODIUM (8CI) see TNK450
ALUMINATE(3-), HEXAFLUORO-
TRISODIUM, (OC-6-11)- see TNK450
ALUMINATE(1−), TETRAHYDRO-,
SODIUM, (T-4)-(9CI) see SEM500
ALUMINA TRIHYDRATE see AHC000
α-ALUMINA TRIHYDRATE see AHC000
ALUMINIC ACID see AHC000
ALUMINIUM BRONZE see AGX000
ALUMINIUM CARBOXYMETHYL
CELLULOSE see AGW700
ALUMINIUM CELLULOSE GLUCOLATE see
AGW700
ALUMINIUM
CHLOROHYDROXYALLANTOINATE see
AHA135
ALUMINIUM CM-CELLULOSE see AGW700
ALUMINIUM STEARATE see AHH825
ALUMINON see AGW750
ALUMINOPHOSPHORIC ACID see PHB500
ALUMINOSILICATES, ZEOLITES see
ZAT300
ALUMINUM see AGX000
ALUMINUM 27 see AGX000
ALUMINUM A00 see AGX000
ALUMINUM ACEGLUTAMIDE see AGX125
ALUMINUM ACETYLACETONATE see
TNN000
ALUMINUM(III) ACETYLACETONATE see
TNN000
ALUMINUM ACID PHOSPHATE see PHB500
ALUMINUM ALLOY, Al,Be see BFP250
ALUMINUM ALUM see AHG750
ALUMINUM AMMONIUM SULFATE see
AGX250
ALUMINUM AZIDE see AGX300
ALUMINUM BARIUM TITANATE see
AGX350
ALUMINUM BARIUM TITANIUM OXIDE
see AGX350
ALUMINUM BERYLLIUM ALLOY see
BFP250
ALUMINUM BOROHYDRIDE see AGX500
ALUMINUM BOROHYDRIDE (DOT) see
AHG875
ALUMINUM BOROHYDRIDE in devices
(DOT) see AHG875
ALUMINUM BROMHYDROXIDE see
AGY000
ALUMINUM BROMIDE see AGX750
ALUMINUM BROMIDE HYDROXIDE see
AGY000
ALUMINUM BROMIDE, anhydrous (UN 1725)
(DOT) see AGX750
ALUMINUM BROMIDE, solution (UN 2580)
(DOT) see AGX750
ALUMINUM BROMOHYDROL see AGY000
ALUMINUM CALCIUM SILICATE see
AGY100
ALUMINUM CARBIDE see AGY250
ALUMINUM CHLORATE see AGY500
ALUMINUM CHLORHYDRATE see AHA000
ALUMINUM CHLORHYDROL see AHA000
ALUMINUM CHLORHYDROXIDE see
AGZ998

ALUMINUM CHLORHYDROXIDE see
AHA000
ALUMINUMCHLORID (GERMAN) see
AGY750
ALUMINUM CHLORIDE see AGY750
ALUMINUM CHLORIDE (1:3) see AGY750
ALUMINUM CHLORIDE, solution (DOT) see
AGY750
ALUMINUM CHLORIDE, anhydrous (DOT)
see AGY750
ALUMINUM CHLORIDE, BASIC (9CI) see
AGZ998
ALUMINUM CHLORIDE HEXAHYDRATE
see AGZ000
ALUMINUM(III) CHLORIDE,
HEXAHYDRATE see AGZ000
ALUMINUM CHLORIDE HYDROXIDE see
AGZ998
ALUMINUM CHLORIDE HYDROXIDE see
AHA000
ALUMINUM CHLORIDE NITROMETHANE
see AHA125
ALUMINUM CHLORIDE OXIDE see
AGZ998
ALUMINUM CHLORIDE OXIDE see
AHA130
ALUMINUM, CHLORO((2,5-DIOXO-4-
IMIDAZOLIDINYL)UREATO)TETRAHYDR
OXYDI- see AHA135
ALUMINUM CHLOROHYDRATE see
AGZ998
ALUMINUM CHLOROHYDROXIDE see
AHA000
ALUMINUM
CHLOROHYDROXYALLANTOINATE see
AHA135
ALUMINUM,
CHLOROTETRAHYDROXY((2-HYDROXY-
5-OXO-2-IMIDAZOLIN-4-YL)UREATO)DI-
see AHA135
ALUMINUM CLOFIBRATE see AHA150
ALUMINUM COMPOUNDS see AHA175
ALUMINUM DEHYDRATED see AGX000
ALUMINUM DEXTRAN see AHA250
ALUMINUM, DICHLOROETHYL- see
EFU050
ALUMINUM DISTEARATE see AHA275
ALUMINUM DISTEARATE (ACGIH) see
AHA275
ALUMINUM EDTA COMPLEX see EJA390
ALUMINUM ETHYLATE see AHA750
ALUMINUM FLAKE see AGX000
ALUMINUM FLUFENAMATE see AHA875
ALUMINUM FLUORIDE see AHB000
ALUMINUM FLUOROACETATE see AHB100
ALUMINUM FLUOROSILICATE see AHB400
ALUMINUM FLUOROSULFATE, HYDRATE
see AHB250
ALUMINUM FLUORURE (FRENCH) see
AHB000
ALUMINUM FLUOSILICATE (AL2SIF18) see
AHB400
ALUMINUM FORMATE see AHB375
ALUMINUM FOSFIDE (DUTCH) see AHE750
ALUMINUM HEXAFLUOROSILICATE see
AHB400
ALUMINUM HEXAFLUOROSILICATE see
THH000
ALUMINUM HYDRATE see AHC000
ALUMINUM HYDRIDE see AHB500
ALUMINUM HYDRIDE-DIETHYL ETHER
see AHB625
ALUMINUM HYDRIDE-TRIMETHYL
AMINE see AHB750
ALUMINUM HYDROBORATE see AHG875
ALUMINUM HYDROXIDE see AHC000
ALUMINUM(III) HYDROXIDE see AHC000
ALUMINUM HYDROXIDE CHLORIDE see
AHA000
ALUMINUM HYDROXIDE DISTEARATE
see AHA275
ALUMINUM HYDROXIDE GEL see AHC000

ALUMINUM HYDROXIDE OXIDE see
AHC250
ALUMINUM,
HYDROXYBIS(OCTADECANOATO-O)-
(9CI) see AHA275
ALUMINUM, HYDROXYBIS(STEARATO)-
see AHA275
ALUMINUM HYDROXYBROMIDE see
AGY000
ALUMINUM HYDROXYCHLORIDE see
AHA000
ALUMINUM HYDROXYDISTEARATE see
AHA275
ALUMINUM IODIDE see AHC500
ALUMINUM ISOPROPOXIDE see AHC600
ALUMINUM(II) ISOPROPYLATE see
AHC600
ALUMINUM LACTATE see AHC750
ALUMINUM LITHIUM HYDRIDE see
LHS000
ALUMINUM MAGNESIUM PHOSPHIDE see
AHD250
ALUMINUM METAHYDROXIDE see
AHC250
ALUMINUM METAL (OSHA) see AGX000
ALUMINUM METHYL see AHD500
ALUMINUM MONOPHOSPHATE see
PHB500
ALUMINUM MONOPHOSPHIDE see
AHE750
ALUMINUM MONOSTEARATE see AHA250
ALUMINUM, molten (NA 9260) (DOT) see
AGX000
ALUMINUMNATRIUMLACTAT (GERMAN)
see AHF750
ALUMINUM NICOTINATE see AHD650
ALUMINUM NITRATE (DOT) see AHD750
ALUMINUM(III) NITRATE (1:3) see AHD750
ALUMINUM NITRATE NONAHYDRATE
see AHD900
ALUMINUM(III) NITRATE,
NONAHYDRATE (1:3:9) see AHD900
ALUMINUM NITRIDE see AHE000
ALUMINUM OXIDE see AHE250
ALUMINUM OXIDE see EAL100
ALUMINUM OXIDE (2:3) see AHE250
α-ALUMINUM OXIDE see AHE250
β-ALUMINUM OXIDE see AHE250
γ-ALUMINUM OXIDE see AHE250
ALUMINUM OXIDE CHLORIDE see
AHA130
ALUMINUM OXIDE HYDRATE see AHC000
ALUMINUM OXIDE SILICATE see AHF500
ALUMINUM OXIDE TRIHYDRATE see
AHC000
ALUMINUM OXYCHLORIDE see AHA130
ALUMINUM PHOSETHYL see AHH800
ALUMINUM PHOSPHATE see PHB500
ALUMINUM PHOSPHATE (1:1) see PHB500
ALUMINUM PHOSPHATE, solution (DOT)
see PHB500
ALUMINUM PHOSPHIDE see AHE750
ALUMINUM PHOSPHINATE see AHE875
ALUMINUM PICRATE see AHF000
ALUMINUM POTASSIUM ALUM see AHF100
ALUMINUM POTASSIUM ALUM see PKU725
ALUMINUM POTASSIUM DISULFATE see
AHF100
ALUMINUM POTASSIUM DISULFATE see
PKU725
ALUMINUM POTASSIUM SULFATE see
AHF100
ALUMINUM POTASSIUM SULFATE, ALUM
see AHF100
ALUMINUM POTASSIUM SULFATE, ALUM
see PKU725
ALUMINUM POTASSIUM SULFATE,
ANHYDROUS see AHF100
ALUMINUM POTASSIUM SULFATE,
ANHYDROUS see PKU725
ALUMINUM POTASSIUM SULFATE,
DODECAHYDRATE see AHF200

ALUMINUM POWDER see AGX000
ALUMINUM POWDER, coated (UN 1309) (DOT) see AGX000
ALUMINUM POWDER, uncoated (UN 1396) (DOT) see AGX000
ALUMINUM PYRO POWDERS (OSHA) see AGX000
ALUMINUM SESQUIOXIDE see AHE250
ALUMINUM(III) SILICATE (2:1) see AHF500
ALUMINUM SODIUM FLUORIDE see SHF000
ALUMINUM SODIUM FLUORIDE see TNK450
ALUMINUM SODIUM HYDRIDE see SEM500
ALUMINUM SODIUM LACTATE see AHF750
ALUMINUM SODIUM OXIDE see AHG000
ALUMINUM SODIUM SULFATE see AHG500
ALUMINUM STEARATE see AHH825
ALUMINUM STEARATE (ACGIH) see AHA250
ALUMINUM SULFATE see AHG750
ALUMINUM SULFATE (2:3) see AHG750
ALUMINUM SULFATE OCTADECAHYDRATE see AHG800
ALUMINUM SULPHATE see AHG750
ALUMINUM TETRAHYDROBORATE see AGX500
ALUMINUM TETRAHYDROBORATE see AHG875
ALUMINUM THALLIUM SULFATE see AHH000
ALUMINUM-TITANIUM ALLOY (1:1) see AHH125
ALUMINUM TRIACETYLACETONATE see TNN000
ALUMINUM TRIBROMIDE see AGX750
ALUMINUM, TRIBROMOTRIMETHYLDI- see MGC225
ALUMINUM TRICHLORIDE see AGY750
ALUMINUM TRICHLORIDEHEXAHYDRATE see AGZ000
ALUMINUM, TRICHLOROTRIMETHYLDI- see MGC230
ALUMINUM, TRIETHYL- see TJN750
ALUMINUM TRIFLUORIDE see AHB000
ALUMINUM TRIHYDRAT see AHC000
ALUMINUM TRIHYDRIDE see AHB500
α-ALUMINUM TRIHYDRIDE see AHB500
ALUMINUM TRIHYDROXIDE see AHC000
ALUMINUM TRINITRATE see AHD750
ALUMINUM TRINITRATE NONAHYDRATE see AHD900
ALUMINUM TRIPROPYL see AHH750
ALUMINUM, TRIPROPYL- see TMY100
ALUMINUM TRIS(ACETYLACETONATE) see TNN000
ALUMINUM TRIS(O-ETHYL PHOSPHONATE) see AHH800
ALUMINUM, TRIS(2-HYDROXYPROPANOATO-O¹),O²- (9CI) see AHC750
ALUMINUM, TRIS(LACTATO)- see AHC750
ALUMINUM, TRIS(2-METHYLPROPYL)- (9CI) see TKR500
ALUMINUM TRISODIUM HEXAFLUORIDE see TNK450
ALUMINUM TRISTEARATE see AHH825
ALUMINUM (TRIS(2-((3-TRIFLUOROMETHYL)PHENYL)AMINO)BENZOATO-N,o)-o-PYRIN see AHA875
ALUMINUM TRISULFATE see AHG750
ALUMINUM WELDING FUMES (OSHA) see AGX000
ALUMITE see AHE250
ALUM POTASSIUM see AHF100
ALUM POTASSIUM see PKU725
A 21 LUNDBECK see KFK000
ALUNDUM see AHE250

ALUNEX see TAI500
ALUNITINE see AHD650
ALUPENT see MDM800
ALUPHOS see PHB500
ALURAL see BNP750
ALURATE see AFT000
ALURENE see CLH750
ALUSAL see AHC000
ALUTOR M 70 see PKB500
ALUTYL see PGG000
ALUZINE see CHJ750
ALVEDON see HIM000
ALVEOGRAF see HJE100
ALVINOL see CHG000
ALVIT see DHB400
ALVO see OLW600
ALVYL see PKP750
ALYPIN see AHI250
ALYPINE see AHI250
ALYSINE see SJO000
ALZODEF see CAQ250
AM 580 see TDB765
AM-715 see BAB625
AM-2604 A see VRP775
1-A-2-MA (RUSSIAN) see AKM250
AMABEVAN see CBJ000
AMACEL BLUE BNN see MGG250
AMACEL BLUE GG see TBG700
AMACEL BRILLIANT BLUE B see MGG250
AMACEL CERISE B see DBX000
AMACEL DEVELOPED NAVY SD see DCJ200
AMACEL GREEN BLUE B see DMM400
AMACEL GREEN BLUE G see DMM400
AMACEL HELIOTROPE R see DBP000
AMACEL PINK B see AKE250
AMACEL PURE BLUE B see TBG700
AMACEL RUBINE B see CMP080
AMACEL SCARLET GB see ENP100
AMACEL VIOLET 6B see AKP250
AMACEL YELLOW CW see KDA075
AMACEL YELLOW G see AAQ250
AMACEL YELLOW LS see KDA075
AMACEL YELLOW RR see DUW500
AMACID AMARANTH see FAG020
AMACID BLACK 10BR see FAB830
AMACID BLUE FG CONC see FMU059
AMACID BLUE V see ADE500
AMACID BRILLIANT BLUE see FAE100
AMACID CHROME BLUE R see HJF500
AMACID FAST BLUE R see ADE750
AMACID FAST YELLOW RS see CMM759
AMACID FUCHSINE 4B see CMS228
AMACID FUCHSINE 6B see CMM400
AMACID GREEN B see FAE950
AMACID GREEN G see FAF000
AMACID LAKE SCARLET 2R see FMU070
AMACID MILLING RED PGS see CMM320
AMACID MILLING RED PRS see CMM330
AMACID MILLING SCARLET G see CMM325
AMACID ORANGE Y see CMM220
AMACID PHLOXINE see CMM300
AMACID WOOL GREEN S see ADF000
AMACID YELLOW M see MDM775
AMACID YELLOW RG see CMM758
AMACID YELLOW T see FAG140
A. MACULATUM see ITD050
AMADIL see HIM000
AMAFAST TURQUOISE 8GGL see COF420
AMAIZO W 13 see SLJ500
AMAL see AMX750
AMALOX see ZKA000
AMALTI SYRUP see MAO285
AMALTY MR 100 see MAO285
AMANIL BLACK GL see AQP000
AMANIL BLACK WD see AQP000
AMANIL BLUE RW see CMO600
AMANIL BLUE 2BX see CMO000
AMANIL BORDEAUX B see CMO872
AMANIL BROWN MR see CMO800
AMANIL CHLORAMINE RED 8BS see CMO880

AMANIL DEVELOPED BLACK BHSW see CMN800
AMANIL FAST BROWN HP see CMO820
AMANIL FAST RED FS see CMO870
AMANIL FAST RED 8BL see CMO885
AMANIL FAST RED 8BLW see CMO885
AMANIL FAST SCARLET 3B see CMO875
AMANIL FAST SCARLET 4BS see CMO870
AMANIL FAST TURQUOISE see COF420
AMANIL FAST TURQUOISE LB see COF420
AMANIL FAST VIOLET N see CMP000
AMANIL GREEN B see CMO840
AMANIL GREEN LT see CMO830
AMANIL NAPHTHOL AS see CMM760
AMANIL PURPURINE 4B see DXO850
AMANIL RAYON BROWN B see CMO820
AMANIL SKY BLUE see CMO250
AMANIL SKY BLUE see CMO500
AMANIL SKY BLUE 6B see CMN750
AMANIL SUPRA BLUE 9GL see CMO650
AMANIL SUPRA BROWN LBL see CMO750
AMANIL TOLUYLENE ORANGE Y see CMO860
AMANIN see AHI523
AMANINE see AHI523
AMANITA RUBESCENS TOXIN see AHI500
β-AMANITIN see AHI520
ε-AMANITIN see AHI510
γ-AMANITIN see AHI550
α-AMANITIN, 1-l-ASPARTIC ACID-(9CI) see AHI520
α-AMANITIN, 1-l-ASPARTIC ACID-3-(4-HYDROXY-l-ISOLEUCINE)-, (S)- see AHI510
α-AMANITIN, 1-l-ASPARTIC ACID-4-(2-MERCAPTO-l-TRYPTOPHAN)- see AHI523
α-AMANITINE see AHI625
β-AMANITINE see AHI520
γ-AMANITINE see AHI550
α-AMANITIN, 3-(4-HYDROXY-l-ISOLEUCINE)-, (S)- see IBL100
α-AMANITIN (8CI, 9CI) see AHI625
AMANO N-AP see GGA800
AMANTADINE see TJG250
AMANTADINE HYDROCHLORIDE see AED250
AMANTHRENE BLUE BCL see DFN300
AMANTHRENE BRILLIANT VIOLET RR see DFN450
AMANTHRENE BROWN BR see CMU770
AMANTHRENE BROWN R see CMU780
AMANTHRENE GOLDEN YELLOW see DCZ000
AMANTHRENE NAVY BLUE BN see VGP100
AMANTHRENE NAVY BLUE 2B-MF see ICU135
AMANTHRENE NAVY BLUE NEW see ICU135
AMANTHRENE OLIVE R see DUP100
AMANTHRENE ORANGE R see CMU815
AMANTHRENE SUPRA NAVY BLUE BN see VGP100
AMANTHRENE SUPRA NAVY BLUE BNR see VGP100
AMAPALO AMRILLO (PUERTO RICO) see PLW800
AMAPLAST GREEN OZ see BLK000
AMAPLAST RED VIOLET P 2R see DBP000
AMARANT see FAG020
AMARANTH see FAG020
AMARANTHE USP (biological stain) see FAG020
AMARBEL EXTRACT see AHI630
AMAREX see MQQ250
AMARINE see COE190
AMARSAN see ABX500
AMARTHOL AS see CMM760
AMARTHOL FAST ORANGE GC BASE see CEH690
AMARTHOL FAST RED B BASE see NEQ000
AMARTHOL FAST RED TR BASE see CLK220

AMARTHOL FAST RED TR BASE see CLK235
AMARTHOL FAST RED TR SALT see CLK235
AMARTHOL FAST SCARLET G BASE see NMP500
AMARTHOL FAST SCARLET G SALT see NMP500
AMARTHOL FAST SCARLETT GG BASE see DEO295
AMARYLLIS see AHI635
AMASUST see AMX750
AMATIN see HCC500
AMATOL see AHI750
β-AMATOXIN see AHI520
γ-AMATOXIN see AHI550
AMAX see BDG000
AMAZE see IMF300
AMAZOLON see AED250
AMAZON YELLOW X2485 see DEU000
AMB see AOC500
AMBACAMP see BAB250
AMBAM see ANZ000
AMBATHIZON see FNF000
AMBAZON see AHI875
AMBAZONE see AHI875
AMBEN see AIH600
AMBENONIUM CHLORIDE see MSC100
AMBENONIUM DICHLORIDE see MSC100
AMBENYL see BAU750
AMBER see AHJ000
AMBER see THQ550
AMBER ACID see SMY000
AMBERGRIS TINCTURE see AHJ000
AMBEROL ST 140F see AHC000
AMBESIDE see SNM500
AMBESTIGMIN CHLORIDE see MSC100
AMBIBEN see AJM000
AMBILHAR see NML000
AMBINON see CMG675
AMBISTRIN see SLY200
AMBITERIC D 40 see LBU200
AMBITHION see FAO300
AMBLOSIN see AIV500
AMBOCHLORIN see CDO500
AMBOCLORIN see CDO500
AMBOFEN see CDP250
AMBOMYCIN see AFI625
AMBOX see BGB500
AMBRA see AHJ000
AMBRACYN see TBX250
AMBRAMICINA see TBX000
AMBRAMYCIN see TBX000
AMBRATE see EQN300
AMBRETTE SEED LIQUID see AHJ100
AMBRETTE SEED OIL see AHJ100
AMBRETTOLID see OKU100
AMBRETTOLIDE see OKU100
AMBROXOL see AHJ250
AMBROXOL HYDROCHLORIDE see AHJ500
AMBRUNATE see MNQ500
AMBUCETAMID see DDT300
AMBUCETAMIDE see DDT300
AMBUSH see AHJ750
AMBUSH see CBM500
AMBUTYROSIN A see BSX325
AMBUYROSIN A see BSX325
AMBYLAN see AQP885
AMBYTHENE see PJS750
6-AMC see CML800
AMCAP see AOD125
AMCHA see AJV500
trans-AMCHA see AJV500
AMCHEM 70-25 see BQN600
AMCHEM 68-250 see CDS125
AMCHEM A-280 see BQN600
AMCHEM GRASS KILLER see TII250
AMCHEM R 14 see PKL750
AMCHEM 2,4,5-TP see TIX500
AMCHLOR see ANE500
AMCIDE see ANU650

AMCILL see AIV500
AMCILL see AOD125
AMCILL-S see SEQ000
AMCINONIDE see COW825
AMCO see PMP500
AMD see DNA800
AMD see MJE780
AMDEX see BBK500
AMDON GRAZON see PIB900
AMDRAM see DBA800
AMDRO see HGP525
AME see AGW500
AME and AOH (1:1) see AGW550
AMEBACILIN see FOZ000
AMEBAN see CBJ000
AMEBARSONE see CBJ000
AMEBICIDE see EAN000
AMEBIL see CHR500
AMECHOL see ACR000
AMEDEL see BHJ250
AMEDRINE see DBA800
AMEISENATOD see BBQ500
AMEISENMITTEL MERCK see BBQ500
AMEISENSAEURE (GERMAN) see FNA000
AMEPROMAT see MQU750
AMERCIAN CYANAMID 18133 see EPC500
AMERCIDE see CBG000
AMERFIL see PMP500
AMERHOLD DR 25 see EFT100
AMERICAINE see EFX000
AMERICAN ALLSPICE see CCK675
AMERICAN BITTERSWEET see AHJ875
AMERICAN CL-26691 see CQL250
AMERICAN CYANAMID 4,049 see MAK700
AMERICAN CYANAMID 5223 see DXX400
AMERICAN CYANAMID 12,008 see DJN600
AMERICAN CYANAMID 12,503 see POP000
AMERICAN CYANAMID 12880 see DSP400
AMERICAN CYANAMID 18682 see IOT000
AMERICAN CYANAMID 18706 see DNX600
AMERICAN CYANAMID-38023 see FAB600
AMERICAN CYANAMID-43073 see MJG500
AMERICAN CYANAMID 47031 see PGW750
AMERICAN CYANAMID AC 43,064 see DXN600
AMERICAN CYANAMID AC 43,913 see PHN250
AMERICAN CYANAMID AC 52,160 see TAL250
AMERICAN CYANAMID CL-24055 see DUI000
AMERICAN CYANAMID CL-26,691 see CQL250
AMERICAN CYANAMID CL-38,023 see FAB600
AMERICAN CYANAMID CL-43913 see PHN250
AMERICAN CYANAMID CL-47,300 see DSQ000
AMERICAN CYANAMID CL-47470 see DHH400
AMERICAN CYANAMID E.I. 43,913 see PHN250
AMERICAN CYANIMID 24,055 see DUI000
AMERICAN ELDER see EAI100
AMERICAN HELLEBORE see VIZ000
AMERICAN HOLLY see HGF100
AMERICAN LAUREL see MRU359
AMERICAN MANDRAKE see MBU800
AMERICAN MEZEREON see LEF100
AMERICAN NIGHTSHADE see PJJ315
AMERICAN PENICILLIN see BFD250
AMERICAN PENNYROYAL OIL see PAR500
AMERICAN VERATRUM see VIZ000
AMERICAN VERMILION see CJD500
AMERICAN WHITE HELLEBORE see FAB100
AMERICIUM see AHK000
AMERICIUM TRICHLORIDE see AHK250
AMERIZOL see TOA000
AMERLATE P see IPS500
AMERLATE W see IPS500

AMEROL see AMY050
AMEROX OE-20 see PJW500
AMET (GERMAN) see AAI750
AMETANTRONE ACETATE see BKB300
AMETHOCAINE see DQA010
AMETHOCAINE HYDROCHLORIDE see TBN000
AMETHONE HYDROCHLORIDE see DIF200
AMETHOPTERIN see MDV500
AMETHOPTERIN see MDV750
AMETHOPTERIN SODIUM see MDV600
AMETHYST see SCI500
AMETHYST see SCJ500
AMETOTERINA see ABY900
AMETOX see SKI500
AMETRIDIONE see AJR200
AMETRIODINIC ACID see AAI750
AMETRYN-ATRAZINE MIXT. see HBT100
AMETRYNE-ATRAZINE MIXT. see HBT100
AMETRYNE-2,4-D MIXT. see AHK300
AMETYCIN see AHK500
AMEX see BQN600
AMEX 820 see BQN600
AMEZIN see HBT100
AMEZINIUMMETILSULFAT (GERMAN) see MQS100
AMEZINIUM METILSULFATE see MQS100
AMEZIN S 47 see HBT100
AMFENACO (SPANISH) see AIU500
AMFENAC SODIUM MONOHYDRATE see AHK625
AMFETAMINA see AOB250
AMFETAMINE see AOB250
d-AMFETASUL see BBK500
AMFETYLINE HYDROCHLORIDE see CBF825
AMFH see AAH100
AMFIPEN see AIV500
AM-FOL see AMY500
AMFOMYCIN see AOB875
AMGABA see AJC375
AMIANTHUS see ARM250
AMIBIARSON see CBJ000
AMICAL 48 see DNF850
AMICAR see AJD000
AMICARDINE see AHK750
AMICETIN see AHL000
AMICETIN CITRATE see AHL250
AMICHLOPHENE see CJN750
AMICIDE see ANU650
AMICIN see NHG000
AMIDANTEL see DPF200
AMIDATE see HOU100
AMIDAZIN see EPQ000
AMIDAZOPHEN see DOT000
AMIDEFRINE MESYLATE see AHL500
AMIDE of GABA see AJC375
AMIDEPHRINE MESYLATE see AHL500
AMIDEPHRINE MONOMETHANESULFONATE see AHL500
AMIDE PP see NCR000
AMIDES see AHL750
AMIDES, COCO, N,N-BIS(HYDROXYETHYL) see CNF330
AMIDINE BLUE 4B see CMO250
p-AMIDINOBENZOIC ACID BUTYL ESTER see AHL875
p-AMIDINOBENZOIC ACID HEXYL ESTER see AHL880
p-AMIDINOBENZOIC ACID PENTYL ESTER see AHL885
p-AMIDINOBENZOIC ACID PROPYL ESTER see AHL890
N-AMIDINO-2-(2,6-DICHLOROPHENYL)ACETAMIDE HYDROCHLORIDE see GKU300
N-(2-AMIDINOETHYL)-3-AMINOCYCLOPENTANOCARBOXAMIDE see AHN625
N"-(2-AMIDINOETHYL)-4-FORMAMIDO-1,1',1"-TRIMETHYL-(N,4':N',4"-

TERPYRROLE)-2-CARBOXAMIDE
HYDROCHLORIDE see DXG600
N''-(2-AMIDINOETHYL)-4-FORMAMIDO-
1,1',1''-TRIMETHYL-(N,4':N',4''-
TERPYRROLE)-2-CARBOXAMIDE see
SLI300
1-AMIDINOHYDRAZONO-4-
THIOSEMICARBAZONO-2,5-
CYCLOHEXADIENE see AHI875
S-(AMIDINOMETHYL) HYDROGEN
THIOSULFATE see AHN000
AMIDINOMYCIN see AHN625
6-AMIDINO-2-NAPHTHYL 4-
GUANIDINOBENZOATE
DIMETHANESULFONATE see AHN700
3-AMIDINO-1-p-NITROPHENYLUREA
HYDROCHLORIDE see NIJ400
4-AMIDINO-1-(NITROSAMINOAMIDINO)-
1-TETRAZENE see TEF500
4-AMIDINO-1-
(NITROSOAMINOAMIDINO)-1-
TETRAZENE see TEF500
N¹-AMIDINOSULFANILAMIDE see AHO250
AMIDINOUREA see GLS900
1-AMIDINOUREA see GLS900
N-AMIDINOUREA see GLS900
AMIDITHION see AHO750
AMID KWASU FENYLO-DWUALLILO-
OCTOWEGO see AGR100
AMID KYSELINY AKRYLOVE see ADS250
AMID KYSELINY OCTOVE (POLISH) see
AAI000
AMID KYSELINY PROPIONOVE see
PMU250
AMID KYSELINY STAVELOVE (CZECH)
see OLO000
AMID KYSELINY TRICHLOROCTOVE see
TII000
o-AMIDOAZOTOLUOL (GERMAN) see
AIC250
o-AMIDOBENZOIC ACID see API500
AMIDOCHLOR see ABY300
AMIDOCYANOGEN see COH500
l-α-AMIDO-EPSILON-AMIDINO
HEXANOIC ACID
MONOHYDROCHLORIDE
MONOHYDRATE see IDA525
AMIDOFEBRIN see DOT000
AMIDOFOS see COD850
AMIDO-G-ACID see NBE850
AMIDO GREEN V see CMM100
AMIDOHEXYLALKOHOL see AKD800
AMIDOL see AHP000
AMIDOLINE see AHP125
AMIDON see ILD000
AMIDO NAPHTHOL RED 2B see CMM400
AMIDO NAPHTHOL RED 2G see CMM300
AMIDO NAPHTHOL RED 6B see CMM400
AMIDO NAPHTHOL RED G see CMM300
AMIDO NAPHTHOL RED GA see CMM300
AMIDONE see MDO750
AMIDONE HYDROCHLORIDE see MDP000
AMIDON HYDROCHLORIDE see MDP750
AMIDOPEROXYMONOSULFURIC ACID see
HLN100
AMIDOPHEN see DOT000
AMIDOPHENAZONE see DOT000
AMIDOPHOS see COD850
AMIDOPROCAIN see PME000
AMIDOPYRAZOLINE see DOT000
AMIDOPYRIN see DOT000
AMIDO RED 2G see CMM300
AMIDO RED 6B see CMM400
AMIDOSAL see SAH000
AMIDOSULFONIC ACID see SNK500
AMIDOSULFONIC PERACID see HLN100
AMIDOSULFURIC ACID see SNK500
AMIDO SULFURYL AZIDE see AHP250
AMIDOTRIZOATE MEGLUMINE see
AOO875
AMIDOTRIZOIC ACID see DCK000

AMIDOUREA HYDROCHLORIDE see
SBW500
AMIDOX see DAA800
AMIDOXAL see SNN500
AMIDO YELLOW E see SGP500
AMIDO YELLOW EA see SGP500
AMIDO YELLOW EA-CF see SGP500
AMIDRINE see ILM000
AMIDRYL see BBV500
AMID-SAL see SAH000
AMID-THIN see SAH000
AMIFENATSOL HYDROCHLORIDE see
DCA600
AMIFUR see NGE500
AMIKACIN see APS750
AMIKACIN SULFATE see APT000
AMIKAPRON see AJV500
AMIKHELLIN HYDROCHLORIDE see
AHP375
AMIKIN see APT000
AMIKLIN see APT000
AMILAC 3 see MCB050
AMILAN see NOH000
AMILAN CM 1001 see PJY500
AMILAR see PKF750
AMILPHENOL see AON000
AMIMYCIN see OHM900
AMINACRINE see AHS500
AMINACRINE HYDROCHLORIDE see
AHS750
AMINARSON see CBJ000
AMINARSONE see CBJ000
AMINASINE see CKP250
AMINATE BASE see AKC750
AMINAZIN see CKP250
AMINAZINE see CKP250
AMINAZIN MONOHYDROCHLORIDE see
CKP500
AMINE 220 see AHP500
AMINE AB see OBC000
AMINE BB see DXW000
AMINES see AHP750
AMINES, C(18-20)-tert-ALKYL see AHP752
AMINES, C10-18-ALKYL,
HYDROCHLORIDES see AHP755
AMINES, FATTY see AHP760
AMINES, TRI-C8-10-ALKYL- see AHP765
AMINE 2,4,5-T FOR RICE see TAA100
AMINIC ACID see FNA000
AMINICOTIN see NCR000
AMINITROZOL see ABY900
AMINITROZOLE see ABY900
5-AMINOACENAPHTHENE see AAF250
5-((2-AMINOACETAMIDO)METHYL)-1-(4-
CHLORO-2-(o-
CHLOROBENZOYL)PHENYL)-N,N-
DIMETHYL-1H-s-TRIAZOLE-3-
CARBOXAMIDE, HYDROCHLORIDE,
DIHYDRATE see AHP875
2-AMINO-4-ACETAMINIFENETOL
(CZECH) see AJT750
3-AMINOACETANILID (CZECH) see
AHQ000
4'-AMINOACETANILID (CZECH) see
AHQ250
4-AMINOACETANILIDE see AHQ250
m-AMINOACETANILIDE see AHQ000
p-AMINOACETANILIDE see AHQ250
3'-AMINOACETANILIDE see AHQ000
4'-AMINOACETANILIDE see AHQ250
3-AMINOACETANILIDE-4-SULFONIC
ACID see AHQ300
AMINOACETIC ACID see GHA000
AMINOACETONITRILE see GHI000
AMINOACETONITRILE BISULFATE see
AHQ750
AMINOACETONITRILE
HYDROCHLORIDE see GHI100
AMINOACETONITRILE HYDROGEN
SULFATE see AHQ750

AMINOACETONITRILE HYDROSULFATE
see AHQ750
AMINOACETONITRILE SULFATE see
AHR000
2-AMINOACETOPHENONE see AHR250
m-AMINOACETOPHENONE see AHR500
p-AMINOACETOPHENONE see AHR240
3'-AMINOACETOPHENONE see AHR500
4'-AMINOACETOPHENONE see AHR240
β-AMINOACETOPHENONE see AHR500
ω-AMINOACETOPHENONE see AHR250
m-AMINOACETYLBENZENE see AHR500
p-AMINOACETYLBENZENE see AHR240
AMINO ACID SOLUTION, AMIPAREN see
AHR600
2-AMINOACRIDINE see AHS000
3-AMINOACRIDINE see ADJ375
5-AMINOACRIDINE see AHS500
9-AMINOACRIDINE see AHS500
2-AMINOACRIDINE (EUROPEAN) see
ADJ375
3-AMINOACRIDINE (EUROPEAN) see
AHS000
AMINOACRIDINE HYDROCHLORIDE see
AHS750
5-AMINOACRIDINE HYDROCHLORIDE
see AHS750
9-AMINOACRIDINE
MONOHYDROCHLORIDE see AHS750
9-AMINOACRIDINE PENICILLIN see
AHT000
4'-(2-AMINO-9-
ACRIDINYLAMINO)METHANESULFONAN
ILIDE see AHT825
4'-((3-AMINO-9-
ACRIDINYL)AMINO)METHANESULFONA
NILIDE see AHT250
1-AMINOADAMANTANE see TJG250
AMINOADAMANTANE
HYDROCHLORIDE see AED250
1-AMINOADAMANTENE
HYDROCHLORIDE see AED250
1-AMINOADAMATANE see TJG250
2-AMINOADENINE see POJ500
2-AMINOAETHANOL (GERMAN) see
EEC600
2-
AMINOAETHYLISOSELENOURONIUMBR
OMID-HYDROBROMID (GERMAN) see
AJY000
β-AMINOAETHYLISOTHIURONIUM-
CHLORID-HYDROCHLORID (GERMAN)
see AJY500
β-AMINOAETHYL-ISOTHIURONIUM
DIHYDROBROMID(GERMAN) see AJY250
β-AMINOAETHYL-MORPHOLIN
(GERMAN) see AKA750
4-AMINO-6-((2-AMINO-1,6-
DIMETHYLPYRIMIDINIUM-4-YL)AMINO)-
1-METHYL-QUINALDINIUM BIS(METHYL
SULFATE) see AQN625
6-AMINO-2-(3'-
AMINOFENYL)BENZIMIDAZOL
HYDROCHLORID (CZECH) see AHT900
6-AMINO-2-(4'-
AMINOFENYL)BENZIMIDAZOL
HYDROCHLORID (CZECH) see AHT950
4-AMINO-N-
(AMINOIMINOMETHYL)BENZENESULFO
NAMIDE see AHO250
3-AMINO-N-(3-AMINO-3-
IMINOPROPYL)CYCLOPENTANECARBOX
AMIDE see AHN625
3-AMINO-2-(3-AMINO-4-
METHOXYPHENYL)-4H-1-BENZOPYRAN-
4-ONE see DBX100
5-AMINO-4-((4-
AMINOMETHYLPHENYL)IMINO)-2-
METHOXY-2,5-CYCLOHEXADIEN-1-ONE
see AKO300
6-AMINO-4-((3-AMINO-4-(((4-((1-
METHYLPYRIDINIUM-4-

YL)AMINO)PHENYL)AMINO)CARBONYL)P HENYL)AMINO)-1-METHYLQUINOLINIUM),DIIODIDE see AHT850

1-AMINO-3-AMINOMETHYL-3,5,5-TRIMETHYL CYCLOHEXANOL see AHT875

4-AMINO-N-(2'-AMINOPHENYL)BENZAMIDE see DBQ125

6-AMINO-2-(3'-AMINOPHENYL)BENZIMIDAZOLE, DIHYDROCHLORIDE see AHT900

6-AMINO-2-(4'-AMINOPHENYL)BENZIMIDAZOLE DIHYDROCHLORIDE see AHT950

2-AMINOANILINE see PEY250

3-AMINOANILINE see PEY000

4-AMINOANILINE see PEY500

m-AMINOANILINE see PEY000

p-AMINOANILINE see PEY500

3-AMINOANILINE DIHYDROCHLORIDE see PEY750

4-AMINOANILINE DIHYDROCHLORIDE see PEY650

m-AMINOANILINE DIHYDROCHLORIDE see PEY750

p-AMINOANILINE DIHYDROCHLORIDE see PEY650

3-AMINO-p-ANISIC ACID see AIA250

2-AMINOANISOLE see AOV900

3-AMINOANISOLE see AOV890

4-AMINOANISOLE see AOW000

m-AMINOANISOLE see AOV890

o-AMINOANISOLE see AOV900

p-AMINOANISOLE see AOW000

2-AMINOANISOLE HYDROCHLORIDE see AOX250

o-AMINOANISOLE HYDROCHLORIDE see AOX250

4-AMINOANISOLE-3-SULFONIC ACID see AIA500

1-AMINOANTHRACENE see APG050

2-AMINOANTHRACENE see APG100

5-AMINOANTHRACENE see APH700

9-AMINOANTHRACENE see APH700

α-AMINOANTHRACENE see APG050

β-AMINOANTHRACENE see APG100

1-AMINO-9,10-ANTHRACENEDIONE see AIA750

2-AMINO-9,10-ANTHRACENEDIONE see AIB000

1-AMINOANTHRACHINON (CZECH) see AIA750

1-AMINOANTHRAQUINONE see AIA750

2-AMINOANTHRAQUINONE see AIB000

1-AMINO-9,10-ANTHRAQUINONE see AIA750

2-AMINO-9,10-ANTHRAQUINONE see AIB000

α-AMINOANTHRAQUINONE see AIA750

β-AMINOANTHRAQUINONE see AIB000

N-(4-AMINOANTHRAQUINONYL)BENZAMIDE see AIB250

4-AMINOANTIPYRENE see AIB300

AMINOANTIPYRIN see AIB300

AMINOANTIPYRINE see AIB300

4-AMINOANTIPYRINE see AIB300

p-AMINO-N-ANTIPYRINYL-N-METHYLBENZAMIDE see AIN100

4-AMINO-1-ARABINOFURANOSYL-2-OXO-1,2-DIHYDROPYRIMIDIN see AQQ750

4-AMINO-1-ARABINOFURANOSYL-2-OXO-1,2-DIHYDROPYRIMIDINE see AQQ750

4-AMINO-1-β-D-ARABINOFURANOSYL-2(1H)-PYRIMIDINON see AQQ750

4-AMINO-1-β-D-ARABINOFURANOSYL-2(1H)-PYRIMIDINONE see AQQ750

2-AMINO-4-ARSENOSOPHENOL see OOK100

2-AMINO-4-ARSENOSOPHENOL HYDROCHLORIDE see ARL000

AMINOARSON see CBJ000

6-AMINO-8-AZAPURINE see AIB340

AMINOAZOBENZENE see PEI000

4-AMINOAZOBENZENE see PEI000

p-AMINOAZOBENZENE see PEI000

4-AMINO-1,1'-AZOBENZENE see PEI000

4-AMINOAZOBENZENE-3,4'-DISULFONIC ACID see AJS500

4-AMINOAZOBENZENE-3,4'-DISULFONIC ACID DISODIUM SALT see CMM758

4-AMINOAZOBENZENE HYDROCHLORIDE see PEI250

p-AMINOAZOBENZENE HYDROCHLORIDE see PEI250

4'-AMINOAZOBENZENE-4-SULFONIC ACID see AIB350

4-AMINOAZOBENZENE-4'-SULPHONIC ACID see AIB350

4-AMINOAZOBENZOL see PEI000

p-AMINOAZOBENZOL see PEI000

6-AMINO-3,4'-AZODI-BENZENESULFONIC ACID see AJS500

AMINOAZOPHENAZONE see AIB300

2-AMINO-5-AZOTOLUENE see AIC250

p-AMINO-2':3-AZOTOLUENE see AIC000

2'-AMINO-2:5'-AZOTOLUENE see TGW500

4'-AMINO-2,3'-AZOTOLUENE see AIC250

4'-AMINO-2:3'-AZOTOLUENE see AIC250

4'-AMINO-3,2'-AZOTOLUENE see AIC000

4'-AMINO-4,2'-AZOTOLUENE see AIC500

4'-AMINO-4-3'-AZOTOLUENE see TGW750

AMINOAZOTOLUENE (indicator) see AIC250

o-AMINOAZOTOLUENE (MAK) see AIC250

o-AMINOAZOTOLUENO (SPANISH) see AIC250

o-AMINOAZOTOLUOL see AIC250

AMINOBENZ see AKF000

10-AMINOBENZ(a)ACRIDINE see AIC750

4-AMINOBENZALDEHYDE see AIC825

p-AMINOBENZALDEHYDE see AIC825

4-AMINOBENZALDEHYDE THIOSEMICARBAZONE see AID250

p-AMINOBENZALDEHYDETHIOSEMICARB AZONE see AID250

m-AMINOBENZAL FLUORIDE see AID500

2-AMINOBENZAMIDE see AID620

3-AMINOBENZAMIDE see AID625

m-AMINOBENZAMIDE see AID625

o-AMINOBENZAMIDE see AID620

3-AMINO-BENZAMIDE (9CI) see AID625

4-AMINOBENZAMIDINE see AID650

p-AMINOBENZAMIDINE see AID650

1-AMINO-4-BENZAMIDOANTHRAQUINONE see AIB250

5-AMINO-1:2-BENZANTHRACENE see BBC000

10-AMINO-1,2-BENZANTHRACENE see BBB750

AMINOBENZENE see AOQ000

4-AMINOBENZENEACETIC ACID see AID700

(±)-α-AMINO-BENZENEACETIC ACID, DECYL ESTER, HYDROCHLORIDE see PFF500

(±)-α-AMINO-BENZENEACETIC ACID HEPTYL ESTER HYDROCHLORIDE see PFF750

(±)-α-AMINOBENZENEACETIC ACID,3-METHYLBUTYL ESTER HYDROCHLORIDE (9CI) see PCV750

(±)-α-AMINO-BENZENEACETIC ACID NONYL ESTER HYDROCHLORIDE see PFG250

(±)-α-AMINO-BENZENEACETIC ACID OCTYL ESTER HYDROCHLORIDE see PFG500

(±)-α-AMINO-BENZENEACETIC ACID PENTYL ESTER HYDROCHLORIDE see PFG750

4-AMINOBENZENEARSONIC ACID see ARA250

p-AMINOBENZENEARSONIC ACID see ARA250

p-AMINOBENZENEAZODIMETHYLANILIN E see DPO200

p-AMINO BENZENE DIAZONIUMPERCHLORATE see AID750

2-AMINO-p-BENZENEDISULFONIC ACID see AIE000

1-AMINO-2,5-BENZENEDISULFONIC ACID see AIE000

2-AMINO-1,4-BENZENEDISULFONIC ACID see AIE000

2-AMINO-BENZENE-1,4-DISULFONIC ACID see AIE000

(S)-α-AMINOBENZENEPROPANOIC ACID see PEC750

4-AMINOBENZENESTIBONIC ACID see SLP600

p-AMINOBENZENESTIBONIC ACID see SLP600

p-AMINOBENZENESTIBONIC ACID COMPOUND with UREA (3:1) see AIE125

p-AMINOBENZENESULFAMIDE see SNM500

3-(p-AMINOBENZENESULFAMIDO)-6-METHOXYPYRIDAZINE see AKO500

2-(p-AMINOBENZENESULFANAMIDE)-3-METHOXYPYRAZINE see MFN500

p-AMINOBENZENESULFONACETAMIDE see SNQ710

3-AMINOBENZENESULFONAMIDE see MDM760

4-AMINOBENZENESULFONAMIDE see SNM500

m-AMINOBENZENESULFONAMIDE see MDM760

p-AMINOBENZENESULFONAMIDE see SNM500

6-(4-AMINOBENZENESULFONAMIDO)-4,5-DIMETHOXYPYRIMIDINE see AIE500

5-(p-AMINOBENZENESULFONAMIDO)-3,4-DIMETHYLISOOXALE see SNN500

5-(p-AMINOBENZENESULFONAMIDO)-3,4-DIMETHYLISOXAZOLE see SNN500

2-(p-AMINOBENZENESULFONAMIDO)-4,5-DIMETHYLOXAZOLE see AIE750

2-(p-AMINOBENZENESULFONAMIDO)-4,6-DIMETHYLPYRIMIDINE see SNJ000

6-(4-AMINOBENZENESULFONAMIDO)-2,4-DIMETHYLPYRIMIDINE see SNJ350

6-(p-AMINOBENZENESULFONAMIDO)-2,4-DIMETHYLPYRIMIDINE see SNJ350

2-(p-AMINOBENZENESULFONAMIDO)-5-METHOXYPYRIMIDINE SODIUM SALT see MFO000

2-(p-AMINOBENZENESULFONAMIDO)-5-METHYLTHIADIAZOLE see MPQ750

3-(p-AMINOBENZENESULFONAMIDO)-2-PHENYLPYRAZOLE see AIF000

2-p-AMINOBENZENESULFONAMIDOQUINO XALINE see QTS000

2-(p-AMINOBENZENESULFONAMIDO)THIAZ OLE see TEX250

2-AMINOBENZENESULFONIC ACID see SNO100

3-AMINO-BENZENESULFONIC ACID see SNO000

4-AMINOBENZENESULFONIC ACID see SNN600

m-AMINOBENZENESULFONIC ACID see SNO000

o-AMINOBENZENESULFONIC ACID see SNO100

p-AMINOBENZENESULFONIC ACID see SNN600

1-AMINOBENZENE-3-SULFONIC ACID see SNO000

m-AMINOBENZENESULFONIC ACID SODIUM SALT see AIF250

p-AMINOBENZENESULFONYL-2-AMINO-4,5-DIMETHYLOXAZOLE see AIE750

6-(p-AMINOBENZENESULFONYL)AMINO-2,4-DIMETHYLPYRIMIDINE see SNJ350

N-(4-AMINOBENZENESULFONYL)-N'-BUTYLUREA see BSM000

p-AMINOBENZENESULFONYLGUANIDINE see AHO250

p-AMINOBENZENESULFONYLUREA see SNQ550

1-(4-AMINOBENZENESULFONYL)UREA see SNQ550

m-AMINOBENZENESULPHONAMIDE see MDM760

5-(p-AMINOBENZENESULPHONAMIDE)-3,4-DIMETHYLISOXAZOLE see SNN500

2-p-AMINOBENZENESULPHONAMIDO-4,6-DIMETHOXYPYRIMIDINE see SNI500

5-(p-AMINOBENZENESULPHONAMIDO)-3,4-DIMETHYLISOXAZOLE see SNN500

3-p-AMINOBENZENESULPHONAMIDO-7-METHOXYPYRIDAZINE see AKO500

2-p-AMINOBENZENESULPHONAMIDOQUINOXALINE see QTS000

2-(p-AMINOBENZENESULPHONAMIDO)THIAZOLE see TEX250

N-p-AMINOBENZENESULPHONYLGUANIDINE MONOHYDRATE see AHO250

2-AMINOBENZENETHIOL see AIF500

4-AMINOBENZENETHIOL see AIF750

p-AMINOBENZENETHIOL see AIF750

2-AMINOBENZENETHIOL HYDROCHLORIDE see AIF800

2-AMINOBENZIMIDAZOLE see AIG000

4-AMINOBENZIMIDAZOLE see BCC050

2-AMINO-6-BENZIMIDAZOLYL PHENYLKETONE see AIH000

AMINOBENZOIC ACID see AIH600

2-AMINOBENZOIC ACID see API500

3-AMINOBENZOIC ACID see AIH500

4-AMINOBENZOIC ACID see AIH600

m-AMINOBENZOIC ACID see AIH500

o-AMINOBENZOIC ACID see API500

p-AMINOBENZOIC ACID see AIH600

γ-AMINOBENZOIC ACID see AIH600

p-AMINOBENZOIC ACID-2-N-AMYLAMINOETHYL ESTER see PBV750

p-AMINOBENZOIC ACID BUTYL ESTER see BPZ000

p-AMINOBENZOIC ACID-3-(DIBUTYLAMINO)PROPYL ESTER, HYDROCHLORIDE see AIT000

p-AMINOBENZOIC ACID-2-(2-(2-(2-(2-(DIETHYLAMINO)ETHOXY)ETHOXY)ETHOXY)ETHYL ESTER, HYDROCHLORIDE see AIK500

p-AMINOBENZOIC ACID-2-(2-(2-(2-(DIETHYLAMINO)ETHOXY)ETHOXY)ETHOXY)ETHYL ESTER, HYDROCHLORIDE see AIK750

p-AMINOBENZOIC ACID-2-(2-(2-(DIETHYLAMINO)ETHOXY)ETHOXY)ETHYL ESTER, HYDROCHLORIDE see AIL000

p-AMINOBENZOIC ACID-2-(2-(DIETHYLAMINO)ETHOXY)ETHYL ESTER, HYDROCHLORIDE see AIL250

(p-AMINOBENZOIC ACID-3-(β-DIETHYLAMINO)ETHOXY)PROPYL ESTER see AIL500

4-AMINOBENZOIC ACID DIETHYLAMINOETHYL ESTER see AIL750

p-AMINOBENZOIC ACID-2-DIETHYLAMINOETHYL ESTER see AIL750

4-AMINOBENZOIC ACID 2-(DIETHYLAMINO)ETHYL ESTER, HYDROCHLORIDE see AIT250

p-AMINOBENZOIC ACID-2-DIETHYLAMINOETHYL ESTER, HYDROCHLORIDE see AIT250

p-AMINOBENZOIC ACID-N-1-DIETHYLAMINO-1-ISOBUTYLETHANOL METHANESULFONATE see LEU000

p-AMINOBENZOIC ACID-β-DIETHYLAMINOISOHEXYL ESTER METHANESULFONATE see LEU000

p-AMINOBENZOIC ACID-3-(DIETHYLAMINO)PROPYL ESTER HYDROCHLORIDE see AIM000

p-AMINOBENZOIC ACID-N,N-DIETHYLLEUCINOL ESTER METHANESULFONATE see LEU000

p-AMINOBENZOIC ACID-2-(DIISOPROPYLAMINO)ETHYL ESTER, HYDROCHLORIDE see AIT500

p-AMINOBENZOIC ACID-3-(DIISOPROPYLAMINO)PROPYL ESTER HYDROCHLORIDE see AIM250

p-AMINOBENZOIC ACID 3-(DIMETHYLAMINO)-1,2-DIMETHYLPROPYL ESTER, HYDROCHLORIDE see AIT750

p-AMINOBENZOIC ACID-(2-(DIMETHYLAMINO)-1-PHENYL)ETHYL ESTER see AIU250

p-AMINOBENZOIC ACID-2-(DIPROPYLAMINO)ETHYL ESTER HYDROCHLORIDE see AIN000

p-AMINOBENZOIC ACID 3-(DIPROPYLAMINO)PROPYL ESTER, HYDROCHLORIDE see AIU000

4-AMINOBENZOIC ACID ETHYL ESTER see EFX000

o-AMINOBENZOIC ACID, ETHYL ESTER see EGM000

p-AMINOBENZOIC ACID ETHYL ESTER see EFX000

3-AMINOBENZOIC ACID ETHYL ESTER METHANESULFONATE see EFX500

p-AMINO-BENZOIC ACID 1-ETHYL-4-PIPERIDYL ESTER, HYDROCHLORIDE see EOV000

p-AMINOBENZOIC ACID ISOBUTYL ESTER see MOT750

p-AMINOBENZOIC ACID METHYLANTIPYRYL-AMIDE see AIN100

2-AMINOBENZOIC ACID METHYL ESTER see APJ250

o-AMINOBENZOIC ACID METHYL ESTER see APJ250

p-AMINOBENZOIC ACID METHYL ESTER see AIN150

p-AMINOBENZOIC ACID-(2-METHYL-2-(1-METHYL-HEPTYLAMINO))PROPYL ESTER HYDROCHLORIDE see MLO750

p-AMINOBENZOIC ACID-3-(3-METHYLPIPERIDINO) PROPYL ESTER HYDROCHLORIDE see MOK500

p-AMINO-BENZOIC ACID-1-METHYL-4-PIPERIDYL ESTER, HYDROCHLORIDE see MON000

4-AMINOBENZOIC ACID, MODOSODIUM SALT see SEO500

p-AMINOBENZOIC ACID MONOGLYCERYL ESTER see GGQ000

p-AMINO-BENZOIC ACID-1-PHENETHYL-4-PIPERIDYL ESTER HYDROCHLORIDE see PDJ000

2-AMINOBENZOIC ACID-3-PHENYL-2-PROPENYL ESTER see API750

p-AMINOBENZOIC ACID PHOSPHATE see AIQ875

p-AMINOBENZOIC ACID-3-PIPERIDINOPROPYL ESTER HYDROCHLORIDE see PIS000

m-AMINOBENZOIC ACID-2-(2-PIPERIDYL)ETHYL ESTER HYDROCHLORIDE see AIQ880

o-AMINOBENZOIC ACID 2-(2-PIPERIDYL)ETHYL ESTER HYDROCHLORIDE see AIQ885

p-AMINOBENZOIC ACID-2-(2-PIPERIDYL)ETHYL ESTER HYDROCHLORIDE see AIQ890

p-AMINOBENZOIC ACID SODIUM SALT see SEO500

p-AMINOBENZOIC DIETHYLAMINOETHYLAMIDE see AJN500

2-(p-AMINOBENZOLSULFONAMIDO)-4,5-DIMETHYLOXAZOL (GERMAN) see AIE750

6-(4'-AMINOBENZOL-SULFONAMIDO)-2,4-DIMETHYLPYRIMIDIN (GERMAN) see SNJ000

(p-AMINOBENZOLSULFONYL)-2-AMINO-4,6-DIMETHYLPYRIMIDIN (GERMAN) see SNJ000

(p-AMINOBENZOLSULFONYL)-4-AMINO-2,6-DIMETHYLPYRIMIDIN (GERMAN) see SNJ350

(p-AMINOBENZOLSULFONYL)-4-AMINO-2-METHYL-6-METHOXY-PYRIMIDIN (GERMAN) see MDU300

(p-AMINOBENZOLSULFONYL)-2-AMINO-4-METHYLPYRIMIDIN (GERMAN) see ALF250

4-AMINO-BENZOLSULFONYL-METHYLCARBAMAT (GERMAN) see SNQ500

2-AMINOBENZONITRILE see APJ750

3-AMINOBENZONITRILE see AIR125

m-AMINOBENZONITRILE see AIR125

o-AMINOBENZONITRILE see APJ750

4-AMINOBENZONITRILE (9CI) see COK125

p-AMINOBENZONITRILE (8CI) see COK125

p-AMINOBENZOPHENONE see AIR250

3-AMINOBENZO(a)PYRENE see BCS550

3-AMINOBENZO-6,7-QUINAZOLINE-4-ONE see AIS250

3-AMINOBENZO(g)QUINAZOLIN-4(3H)-ONE see AIS250

2-AMINOBENZOTHIAZOLE see AIS500

6-AMINO-2-BENZOTHIAZOLETHIOL see AIS550

3-AMINOBENZOTRIFLUORIDE see AID500

m-AMINOBENZOTRIFLUORIDE see AID500

p-AMINOBENZOTRIFLUORIDE see TKB750

2-AMINOBENZOXAZOLE see AIS600

3-(p-AMINOBENZOXY)-1-DI-n-BUTYLAMINOPROPANE see BOO750

3-(p-AMINOBENZOXY)-1-DI-n-BUTYLAMINOPROPANE SULFATE see BOP000

(p-AMINOBENZOYL)ACETONITRILE see ALQ635

2-(4-AMINOBENZOYL)ACETONITRILE see ALQ635

1-AMINO-4-BENZOYLAMINOANTHRACHINON (CZECH) see AIB250

1-AMINO-4-(BENZOYLAMINO)ANTHRAQUINONE see AIB250

4-AMINO-1-BENZOYLAMINOANTHRAQUINONE see AIB250

p-AMINOBENZOYLAMINOMETHYLHYDROCOTARNINE see AIS625

1-AMINO-4-BENZOYLANTHRAQUINONE-2-SULFONIC ACID see AIS700

2-AMINO-3-BENZOYL BENZENEACETIC ACID see AIU500

2-AMINO-3-BENZOYLBENZENEACETIC ACID SODIUM SALT HYDRATE see AHK625

2-AMINO-5-BENZOYLBENZIMIDAZOLE see AIH000

p-AMINOBENZOYLDIBUTYLAMINOPROPA NOL see BOO750

AMINOBENZOYLDIBUTYLAMINOPROPA NOL HYDROCHLORIDE see AIT000

p-AMINOBENZOYLDIBUTYLAMINOPROPA NOL SULFATE see BOP000

p-AMINOBENZOYLDIETHYLAMINOETHAN OL see AIL750

p-AMINOBENZOYLDIETHYLAMINOETHAN OL HYDROCHLORIDE see AIT250

p-AMINO BENZOYL DIETHYL AMINO PROPANOL HYDROCHLORIDE see AIM000

o-AMINOBENZOYL DI(ISOPROPYLAMINO)ETHANOL HYDROCHLORIDE see IJZ000

p-AMINO BENZOYL DI-ISO-PROPYL AMINO ETHANOL HYDROCHLORIDE see AIT500

p-AMINO BENZOYL DIISOPROPYL AMINO PROPANOL HYDROCHLORIDE see AIM250

p-AMINOBENZOYLDIMETHYLAMINO-1,2-DIMETHYLPROPANOL HYDROCHLORIDE see AIT750

1-AMINO-4-BENZOYL-9,10-DIOXO-9,10-DIHYDROANTHRACENE-2-SULFONIC ACID see AIS700

p-AMINO BENZOYL DI-N-PROPYL AMINO ETHANOL HYDROCHLORIDE see AIN000

p-AMINO BENZOYL DI-N-PROPYL AMINOPROPANOL HYDROCHLORIDE see AIU000

o-AMINOBENZOYLFORMIC ANHYDRIDE see ICR000

β-4-AMINOBENZOYLOXY-β-PHENYLETHYL DIMETHYLAMINE see AIU250

2-AMINO-3-BENZOYLPHENYLACETIC ACID see AIU500

2-AMINOBENZTHIAZOLE see AIS500

4-(4-AMINOBENZYL)ANILINE see MJQ000

α-(α-AMINOBENZYL)BENZYL ALCOHOL HYDROCHLORIDE see DWB000

2-AMINO-6-BENZYLMERCAPTOPURINE see BFL125

2-AMINO-6-BENZYL-MP see BFL125

AMINOBENZYLPENICILLIN see AIV500

d-(−)-α-AMINOBENZYLPENICILLIN see AIV500

d-(−)-α-AMINOBENZYLPENICILLIN SODIUM SALT see SEQ000

AMINOBENZYLPENICILLIN TRIHYDRATE see AOD125

α-AMINOBENZYLPENICILLIN TRIHYDRATE see AOD125

4-AMINO-N-(1-BENZYL-4-PIPERIDYL)-5-CHLORO-o-ANISAMIDE HYDROXYSUCCINATE see AIV625

3-AMINO-1-BENZYL-5H-PYRIDO(4,3-B)INDOLE ACETATE see AIV700

2-AMINO-6-(BENZYLTHIO)PURINE see BFL125

2-AMINOBIPHENYL see BGE250

4-AMINOBIPHENYL see AJS100

o-AMINOBIPHENYL see BGE250

p-AMINOBIPHENYL see AJS100

4'-AMINO-(1,1'-BIPHENYL)-4-CARBONITRILE see AJJ770

4-AMINOBIPHENYL DIHYDROCHLORIDE see BGE300

4-AMINOBIPHENYL ETHER see PDR500

4-AMINO-3-BIPHENYLOL see AIV750

4'-AMINO-4-BIPHENYLOL see AIW000

3-AMINO-4-BIPHENYLOL HYDROCHLORIDE see AIW250

4-AMINO-3-BIPHENYLOL HYDROGEN SULFATE see AKF250

N-(4'-AMINO(1,1'-BIPHENYL)-4-YL)-ACETAMIDE see ACC000

2-AMINO-5-BIPHENYLYLIMIDAZOLE HYDROCHLORIDE see AIW500

2-(4'-AMINO-1,1'-BIPHENYL-4-YL)-2H-NAPHTHO(1,2-d)TRIAZOLE-6,8-DISULFONIC ACID, DIPOTASSIUM SALT see AIW750

5-AMINO(3,4'-BIPYRIDIN)-6-(1H)-ONE see AOD375

2-AMINO-4-(p-(BIS(2-CHLOROETHYL)AMINO)PHENYL)BUTYRI C ACID see AJE000

3-((AMINO(BIS(2-CHLOROETHYL)AMINO)PHOSPHINYL)O XY)PROPANOIC ACID see CCE250

4-AMINO-3,6-BIS((4-((2,4-DIAMINO-5-SULFOPHENYL)AZO)PHENYL)AZO)-5-HYDROXY-2,7-NAPHTHALENEDISULFONIC ACID see AIW800

5-AMINO-1-BIS(DIMETHYLAMIDE)PHOSPHORYL-3-PHENYL-1,2,4-TRIAZOLE see AIX000

5-AMINO-1-BIS(DIMETHYLAMIDO)PHOSPHORYL-3-PHENYL-1,2,4-TRIAZOLE see AIX000

5-AMINO-1-(BIS(DIMETHYLAMINO)PHOSPHINYL)-3-PHENYL-1,2,4-TRIAZOLE see AIX000

AMINOBIS(PROPYLAMINE) see AIX250

AMINOBIS(PROPYLAMINE) see AIX250

4-AMINO-3,6-BIS((4-((3,3',5,5'-TETRAHYDROXY(1,1'-BIPHENYL)-4-YL)AZO)PHENYL)AZO)-5-HYDROXY-2,7-NAPHTHALENEDISULFONIC ACID see AIX300

4-AMINO-3,6-BIS((4-((2,4,6-TRIHYDROXYPHENYL)AZO)PHENYL)AZ O)-5-HYDROXY-2,7-NAPHTHALENEDISULFONIC ACID see AIX320

1-AMINO-4-BROMANTHRACHINON-2-SULFONAN SODNY (CZECH) see DKR000

1-AMINO-2-BROM-4-HYDROXYANTHRACHINON (CZECH) see AIY500

2-AMINO-5-BROMOBENZOXAZOLE see AIX500

2-AMINO-6-BROMOBENZOXAZOLE see AIX750

2-AMINO-5-BROMO-6-CHLOROBENZOXAZOLE see AIY000

2-AMINO-6-BROMO-5-CHLOROBENZOXAZOLE see AIY250

4-AMINO-5-BROMO-N-(2-(DIETHYLAMINO)ETHYL)-o-ANISAMIDE see VCK100

4-AMINO-N-(5-BROMO-4,6-DIMETHYL-2-PYRIMIDINYL)BENZENESILFANILAMIDE see SNH875

1-AMINO-2-BROMO-4-HYDROXYANTHRAQUINONE see AIY500

1-AMINO-2-BROMO-4-(2-(2-HYDROXYETHYL)SULFONYL-4-METHYLPHENYLAMINO)ANTHRAQUINO NE see AIY750

2-AMINO-5-BROMO-6-PHENYL-4(1H)-PYRIMIDINONE see AIY850

N-AMINO-2-(m-BROMOPHENYL)SUCCINIMIDE see AIZ000

1-AMINO-3-BROMOPROPANE HYDROBROMIDE see AJA000

1-AMINO-BUTAAN (DUTCH) see BPX750

1-AMINOBUTAN (GERMAN) see BPX750

1-AMINOBUTANE see BPX750

2-AMINOBUTANE see BPY000

(S)-AMINOBUTANEDIOIC ACID see ARN850

4-AMINOBUTANOIC ACID see PIM500

2-AMINOBUTAN-1-OL see AJA250

2-AMINO-1-BUTANOL see AJA250

4-AMINO-2-(4-BUTANOYLHEXAHYDRO-1H-1,4-DIAZEPIN-1-YL)-6,7-DIMETHOXYQUINAZOLINE HYDROCHLORIDE see AJA375

3-AMINO-2-BUTOXYBENZOIC ACID-2-DIETHYLAMINOETHYL ESTER HYDROCHLORIDE see AJA500

4-AMINO-2-BUTOXY-BENZOIC ACID 2-(DIETHYLAMINO)ETHYL ESTER HYDROCHLORIDE see SPB800

4-AMINO-2-BUTOXY-BENZOIC ACID 2-(DIETHYLAMINO)ETHYL ESTER, MONOHYDROCHLORIDE see SPB800

2-AMINO-n-BUTYL ALCOHOL see AJA250

4-AMINO-N-((BUTYLAMINO)CARBONYL)BENZENESU LFONAMIDE see BSM000

dl-4-AMINO-α-(tert-BUTYLAMINO)-3-CHLORO-5-(TRIFLUOROMETHYL)PHENETHYL ALCOHOL HCl see MAB300

4-AMINO-α-((tert-BUTYLAMINO)METHYL)-3-CHLORO-5-(TRIFLUOROMETHYL)BENZYL ALCOHOL HCl see KGK300

4-AMINO-α-((tert-BUTYLAMINO)METHYL)-3,5-DICHLOROBENZYL ALCOHOL HYDROCHLORIDE see VHA350

4-AMINO-α-((tert-BUTYLAMINO)METHYL)-3,5-DICHLOROBENZYLALKOHOL-HYDROCHLORID (GERMAN) see VHA350

N^1-3-(((4-AMINOBUTYL)AMINO)PROPYL)BLEOMYC INAMIDE see BLY500

p-AMINOBUTYLBENZENE see AJA550

1-AMINO-4-BUTYLBENZENE see AJA550

2-AMINO-5-BUTYLBENZIMIDAZOLE see AJA650

(4-AMINOBUTYL)DIETHOXYMETHYLSILAN E see AJA750

p-AMINO-β-sec-BUTYL-N,N-DIMETHYLPHENETHYLAMINE see AJB000

3-(2-AMINOBUTYL)INDOLE ACETATE see AJB250

3-(4-AMINOBUTYL)INDOLE HYDROCHLORIDE see AJB500

Δ-AMINOBUTYLMETHYLDIETHOXYSILAN E see AJA750

4-AMINO-6-tert-BUTYL-3-METHYLTHIO-as-TRIAZIN-5-ONE see MQR275

4-AMINO-6-tert-BUTYL-3-(METHYLTHIO)-1,2,4-TRIAZIN-5-ONE see MQR275

5-AMINO-N-BUTYL-2-PROPARGYLOXYBENZAMIDE see AJC000

5-AMINO-N-BUTYL-2-(2-PROPYNYLOXY)BENZAMIDE see AJC000

(4-AMINOBUTYL)TRIETHOXYSILANE see AJC250

4-AMINOBUTYRAMIDE see AJC375

4-AMINOBUTYRIC ACID see PIM500

γ-AMINO-N-BUTYRIC ACID see PIM500

γ-AMINOBUTYRIC ACID CETYL ESTER see AJC500

4-AMINOBUTYRIC ACID LACTAM see PPT500

γ-AMINOBUTYRIC ACID LACTAM see PPT500

4-AMINOBUTYRIC ACID METHYL ESTER see AJC625

γ-AMINOBUTYRIC LACTAM see PPT500

γ-AMINOBUTYROLACTAM see PPT500

p-AMINOBUTYROPHENONE see AJC750

4'-AMINOBUTYROPHENONE see AJC750

4-AMINO-2-(4-BUTYRYLHEXAHYDRO-1H-1,4-DIAZEPIN-1-YL)-6,7-DIMETHOXYQUINAZOLINE HYDROCHLORIDE see BON350

AMINO-C ACID see NAO525

AMINOCAINE see AIT250

AMINOCAPROIC ACID see AJD000

α-AMINO-γ-(p-DICHLOROETHYLAMINO)-PHENYLBUTYRIC ACID see AJE000

3-AMINO-1-(3,4-DICHLORO-α-METHYLBENZYL)-2-PYRAZOLIN-5-ONE see EAE675

2-AMINO-4,6-DICHLOROPHENOL see AJM525

1-AMINO-3,3-DI(4-CHLOROPHENYL)CYCLOPENTANE HYDROCHLORIDE see AJM550

5-AMINO-2-(1-(3,4-DICHLOROPHENYL)-ETHYL)-2,4-DIHYDRO-3H-PYRAZOL-3-ONE see EAE675

(4-AMINO-3,5-DICHLOROPHENYL)GLYCOLIC ACID see AJM575

1-(4-AMINO-5-(3,4-DICHLOROPHENYL)-2-METHYL-1H-PYRROL-3-YL)ETHANONE see AJM600

2-AMINO-5-((3,4-DICHLOROPHENYL)THIOMETHYL)-2-OXAZOLINE see AJM750

5-AMINO-9-(DIETHYLAMINO)BENZO(a)PHENOXAZIN-7-IUM SULFATE (2:1) see AJN250

2-AMINO-4-DIETHYLAMINOETHOXYPYRIMIDINE see AJN375

2-AMINO-4-(2-DIETHYLAMINOETHOXY)PYRIMIDINE see AJN375

p-AMINO-N-(2-DIETHYLAMINOETHYL)BENZAMIDE see AJN500

4-AMINO-N-(2-(DIETHYLAMINO)ETHYL)-BENZAMIDE (9CI) see AJN500

p-AMINO-N-(2-(DIETHYLAMINO)ETHYL)BENZAMIDE HYDROCHLORIDE see PME000

4-AMINO-N-(2-(DIETHYLAMINO)ETHYL)BENZAMIDE MONOHYDROCHLORIDE see PME000

p-AMINO-N-(2-DIETHYLAMINOETHYL)BENZAMIDE SULFATE see AJN750

N-(2-AMINO-5-DIETHYLAMINOPHENETHYL)METHANE SULFONAMIDEHYDROCHLORIDE see AJO000

1-AMINO-3-(DIETHYLAMINO)PROPANE see DIY800

2-AMINO-4-γ-DIETHYLAMINOPROPYLAMINO-5,6-DIMETHYLPYRIMIDINE see AJO100

p-AMINODIETHYLANILINE see DJV200

p-AMINO DIETHYLANILINE HYDROCHLORIDE see AJO250

4-AMINODIFENIL (SPANISH) see AJS100

p-AMINODIFENYLAMIN (CZECH) see PFU500

4-AMINODIFENYLETHER see PDR500

N-(4-AMINO-9,10-DIHYDRO-9,10-DIOXO-1-ANTHRACENTY)-BENZAMIDE see AIB250

2-AMINO-1,9-DIHYDRO-9-((2-HYDROXYETHOXY)METHYL)-6H-PURIN-6-ONE see AEC700

2-AMINO-1,9-DIHYDRO-9-((2-HYDROXYETHOXY)METHYL)-6H-PURIN-6-ONE MONOSODIUM SALT see AEC725

6-AMINO-1,2-DIHYDRO-1-HYDROXY-2-IMINO-4-PIPERIDINOPYRIMIDINE see DCB000

(s)-7-AMINO-6,7-DIHYDRO-10-HYDROXY-1,2,3-TRIMETHOXYBENZO(a)HEPTALEN-9(5H)-ONE see TLN750

4-o-(3-AMINO-3,4-DIHYDRO-6-((METHYLAMINO)METHYL)-2H-PYRAN-2-YL)-2-DEOXY-6-o-(3-DEOXY-4-C-METHYL-3-(METHYLAMINO)-β-l-ARABINOPYRANOSYL)-d-STREPTAMINE (2S-cis)-, SULFATE (salt) see GAA120

3-AMINO-1,5-DIHYDRO-5-METHYL-1-β-d-RIBOFURANOSYL-1,4,5,6,8-PENTAAZAACENAPHTHYLENE see TJE870

α-AMINO-2,3-DIHYDRO-3-OXO-5-ISOXAZOLEACETIC ACID see AKG250

7-AMINO-1,3-DIHYDRO-5-PHENYL-2H-1,4-BENZODIAZEPIN-2-ONE see AJO280

5-AMINO-2,3-DIHYDRO-1,4-PHTHALAZINEDIONE see AJO290

(s)-7-AMINO-6,7-DIHYDRO-1,2,3,10-TETRAMETHOXYBENZO(a)HEPTALEN-9(5H)-ONE see TLO000

5-AMINO-1,6-DIHYDRO-7H-v-TRIAZOLO(4,5-d)PYRIMIDIN-7-ONE see AJO500

5-AMINO-1,4-DIHYDRO-7H-1,2,3-TRIAZOLO(4,5-d)PYRIMIDIN-7-ONE (9CI) see AJO500

4-AMINO-3-((4-((4-((2,4-DIHYDROXYPHENYL)AZO)BENZOYL)AMINO)PHENYL) AZO)-5-HYDROXY-6-((2-METHOXYPHENYL)AZO)-1-NAPHTHALENESULFONIC ACID see AJO550

l-2-AMINO-1-(3,4-DIHYDROXYPHENYL)ETHANOL see NNO500

2-AMINO-3-(3,4-DIHYDROXYPHENYL)PROPANOIC ACID see DNA200

2-AMINO-3,5-DIIODOBENZOIC ACID see API800

2-AMINO-4-DI-ISOBUTYLAMINOETHOXYPYRIMIDINE see AJO625

2-AMINO-4-(2-DIISOBUTYLAMINOETHOXY)PYRIMIDINE see AJO625

3-AMINO-2-(2-(DIISOPROPYLAMINO)ETHOXY)BUTYROPHENONEDIHYDROCHLORIDE see AJO750

1-(4-AMINO-5-(3,4-DIMETHOXYPHENYL)-2-METHYL-1H-PYRROL-3-YL)ETHANONE see AJO800

2-AMINO-1-(2,5-DIMETHOXYPHENYL)-1-PROPANOL HYDROCHLORIDE see MDW000

4-AMINO-N-(2,6-DIMETHOXY-4-PYRIMIDINYL)BENZENESULFONAMIDE see SNN300

4-AMINO-N-(4,6-DIMETHOXY-2-PYRIMIDINYL)BENZENESUL-FONAMIDE see SNI500

4-AMINO-N-(5,6-DIMETHOXY-4-PYRIMIDINYL)BENZENESULFONAMIDE see AIE500

1-(4-AMINO-6,7-DIMETHOXY-2-QUINAZOLINYL-4-(2-FURANYLCARBONYL)) PIPERAZINE see AJP000

1-(4-AMINO-6,7-DIMETHOXY-2-QUINAZOLINYL)-4-(2-FURANYLCARBONYL)PIPERAZINE HYDROCHLORIDE see FPP100

1-(4-AMINO-6,7-DIMETHOXY-2-QUINAZOLINYL)-4-(2-FUROYL)PIPERAZINE MONOHYDROCHLORIDE see FPP100

1-(4-AMINO-6,7-DIMETHOXY-2-QUINAZOLINYL)-4-((TETRAHYDRO-2-FURANYL)CARBON YL)PIPERAZINE HCl 2H2O see TEF700

4-AMINO-4'-DIMETHYLAMINOAZOBENZENE see DPO200

4'-AMINO-N,N-DIMETHYL-4-AMINOAZOBENZENE see DPO200

2-AMINO-4-DIMETHYLAMINOETHOXYPYRIMIDINE see AJP125

2-AMINO-4-(2-DIMETHYLAMINOETHOXY)PYRIMIDINE see AJP125

3-AMINO-N-(2-(DIMETHYLAMINO)ETHYL)NAPHTHALIMIDE see NAC600

3-AMINO-7-DIMETHYLAMINO-2-METHYLPHENAZATHIONIUM CHLORIDE see AJP250

3-AMINO-7-DIMETHYLAMINO-2-METHYLPHENAZINE HYDROCHLORIDE see AJQ250

3-AMINO-7-(DIMETHYLAMINO)-2-METHYL-PHENAZINE MONOHYDROCHLORIDE see AJQ250

3-AMINO-7-(DIMETHYLAMINO)-2-METHYL-PHENOSELENAZIN-5-IUM CHLORIDE see SBU950

3-AMINO-7-(DIMETHYLAMINO)PHENOTHIAZIN-5-IUM CHLORIDE see DUG700

3-AMINO-7-(DIMETHYLAMINO)-5-PHENYLPHENAZINIUM CHLORIDE see AJP300

1-AMINO-3-DIMETHYLAMINOPROPANE see AJQ100

AMINODIMETHYLAMINOTOLUAMINOZINE HYDROCHLORIDE see AJQ250

p-AMINODIMETHYLANILINE see DTL800

4-AMINO-2',3-DIMETHYLAZOBENZENE see AIC250

4'-AMINO-2,3'-DIMETHYLAZOBENZENE see AIC250

AMINODIMETHYLBENZENE see XMA000

1-AMINO-2,4-DIMETHYLBENZENE see XMS000

1-AMINO-2,5-DIMETHYLBENZENE see XNA000

3-AMINO-1,4-DIMETHYLBENZENE see XNA000

4-AMINO-1,3-DIMETHYLBENZENE see XMS000

1-AMINO-2,4-DIMETHYLBENZENE HYDROCHLORIDE see XOJ000

1-AMINO-2,5-DIMETHYLBENZENE HYDROCHLORIDE see XOS000

3-AMINO-1,4-DIMETHYLBENZENE HYDROCHLORIDE see XOS000

4-AMINO-1,3-DIMETHYLBENZENE HYDROCHLORIDE see XOJ000

5-AMINO-1,4-DIMETHYLBENZENE HYDROCHLORIDE see XOS000

6-AMINO-5,8-DIMETHYL-9H-CARBAZOL-3-OL see AJQ300

3-AMINO-1,4-DIMETHYL-γ-CARBOLINE see TNX275

2-AMINODIMETHYLETHANOL see IIA000

4-AMINO-6-(1,1-DIMETHYLETHYL)-3-(ETHYLTHIO)-1,2,4-TRIAZIN-5(4H)-ONE see ENJ600

4-AMINO-6-(1,1-DIMETHYLETHYL)-3-(METHYLTHIO)-1,2,4-TRIAZIN-5(4H)-ONE see MQR275

4-AMINO-3',5'-DIMETHYL-4'-HYDROXYAZOBENZENE see AJQ500

3-AMINO-2,8-DIMETHYL-7-(2-HYDROXY-1-NAPHTHYLAZO)-5-PHENYLPHENAZINIUM CHLORIDE see CMM770

2-AMINO-1,5-DIMETHYLIMIDAZO(4,5-B)PYRIDINE see AJQ510

2-AMINO-1,6-DIMETHYLIMIDAZO(4,5-b)PYRIDINE see AJQ520

2-AMINO-3,5-DIMETHYLIMIDAZO(4,5-b)PYRIDINE see AJQ530

2-AMINO-3,4-DIMETHYLIMIDAZO(4,5-f)QUINOLINE see AJQ600

2-AMINO-3,8-DIMETHYLIMIDAZO(4,5-f)QUINOXALINE see AJQ675

2-AMINO-3,8-DIMETHYL-3H-IMIDAZO(4,5-f)QUINOXALINE see AJQ675

4-AMINO-N-(3,4-DIMETHYL-5-ISOXAZOLYL)BENZENESULPHONAMIDE see SNN500

2-AMINO-6-DIMETHYL-4-(p-(p-((p-((1-METHYLPYRIDINIUM-3-YL)CARBAMOYL)PHENYL)CARBABENZA MIDO)ANILINO)PYRIMIDIMIUM), DIIODIDE see AJQ750

4-AMINO-N-(4,5-DIMETHYL-2-OXAZOLYL)BENZENESULFONAMIDE see AIE750

p-AMINO-N,α-DIMETHYLPHENETHYLAMINE see AJR000

2-AMINO-4,5-DIMETHYLPHENOL see AMW750

4-AMINO-1,2-DIMETHYL-5-PHENYLPYRROL-3-YLETHANONE see AJR100

1-(4-AMINO-1,2-DIMETHYL-5-PHENYL-1H-PYRROL-3-YL)ETHANONE see AJR100

1-AMINO-3-(2,2-DIMETHYLPROPYL)-6-(ETHYLTHIO)-1,3,5-TRIAZINE-2,4(1H,3H)-DIONE see AJR200

5-AMINO-1,3-DIMETHYL-4-PYRAZOLYL o-FLUOROPHENYL KETONE see AJR400

(5-AMINO-1,3-DIMETHYL-1H-PYRAZOL-4-YL)(2-FLUOROPHENYL)METHANONE see AJR400

3-AMINO-1,4-DIMETHYL-5H-PYRIDO(4,3-b)INDOLE see TNX275

3-AMINO-1,4-DIMETHYL-5H-PYRIDO(4,3-b)INDOLE ACETATE see AJR500

4-AMINO-N-(2,6-DIMETHYL-4-PYRIMIDINYL)BENZENESULFONAMIDE (9CI) see SNJ350

4-AMINO-N-(4,6-DIMETHYL-2-PYRIMIDINYL)BENZENESULFONAMIDE, MONOSODIUM SALT see SJW500

6-AMINO-2,3-DIMETHYLQUINOXALINE see AJR600

2-AMINO-4,6-DINITROPHENOL see DUP400

2-AMINO-4,6-DINITROTOLUENE see AJR750

4-AMINO-2,6-DINITROTOLUENE see MJG600

4-AMINO-3,5-DINITROTOLUENE see DVI100

1-AMINO-9,10-DIOXO-9,10-DIHYDRO-2-ANTHRACENECARBOXYLIC ACID see AJS000

2-AMINODIPHENYL see BGE250

4-AMINODIPHENYL see AJS100

o-AMINODIPHENYL see BGE250

p-AMINODIPHENYL see AJS100

4-AMINODIPHENYLAMINE see PFU500

p-AMINODIPHENYLAMINE see PFU500

2-AMINODIPHENYLENE OXIDE see DDB600

2-AMINODIPHENYLENOXYD (GERMAN) see DDB800

2-AMINO-1,2-DIPHENYLETHANOL HYDROCHLORIDE see DWB000

2-AMINODIPHENYL ETHER see PDR490

4-AMINODIPHENYL ETHER see PDR500

2-AMINO-6-((1,2-DIPHENYLETHYL)AMINO)-3-PYRIDINECARBAMIC ACID ETHYL ESTER, MONOHYDROCHLORIDE see DAB200

p-AMINODIPHENYLIMIDE see PEI000

2-AMINODIPYRIDO(1,2-a:3',2'-d)IMIDAZOLE see DWW700

2-AMINODIPYRIDO(1,2-a:3',2'-d)IMIDAZOLE HYDROCHLORIDE see AJS225

2-AMINO-4,6-DIPYRROLIDINOTRIAZINE see AJS250

4-AMINO-3,4'-DISULFOAZOBENZENE see AJS500

2-AMINO-1,4-DISULFOBENZENE see AIE000

1-AMINO-3,6-DISULFO-8-NAFTYLESTER KYSELINA p-TOLUENSULFONOVE (CZECH) see AKH500

5-AMINO-1,2,4-DITHIAZOLE-3-THIONE see IBL000

1-AMINODODECANE see DXW000

4-AMINO-1-DODECYLQUINALDINIUM ACETATE see AJS750

AMINODUR see TEP500

9-AMINOELLIPTICINE see AJS875

2-AMINOETANOLO (ITALIAN) see EEC600

AMINOETHANE see EFU400

1-AMINOETHANE see EFU400

2-AMINO-ETHANESELENOL HYDROCHLORIDE see AJS900

2-AMINOETHANESELENOSULFURIC ACID see AJS950

2-AMINOETHANESULFONIC ACID see TAG750

2-AMINOETHANESULFONO-p-PHENETIDINE HYDROCHLORIDE see TAG875

2-AMINOETHANETHIOL see AJT250

2-AMINOETHANETHIOL DIHYDROGEN PHOSPHATE (ESTER) see AKA900

2-AMINO-ETHANETHIOL DIHYDROGEN PHOSPHATE(ester), MONOSODIUM SALT see AKB000

2-AMINOETHANETHIOSULFURIC ACID see AJT500

1-AMINOETHANOL see AAG500

2-AMINOETHANOL compounded with 6-CYCLOHEXYL-1-HYDROXY-4-METHYL-2(1H)-PYRIDINONE (1:1) see BAR800

2-AMINOETHANOL HYDROCHLORIDE see EEC700

β-AMINOETHANOL HYDROCHLORIDE see EEC700

2-AMINOETHANOL (MAK) see EEC600

3-AMINO-4-ETHOXYACETANILIDE see AJT750

4-AMINOETHOXYBENZENE see PDK790

2-AMINO-3-ETHOXYCARBONYL-5-BENZYL-4,5,6,7-TETRAHYDROTHIENO (2,3-c)PYRIDINE HYDROCHLORIDE see AJU000

2-AMINO-3-ETHOXYCARBONYL-6-BENZYL-4,5,6,7-TETRAHYDROTHIENO(2,3-c)PYRIDINE HYDROCHLORIDE see TGE165

2-AMINOETHOXYETHANOL see AJU250

2-(2-AMINOETHOXY)ETHANOL see AJU250

4-AMINO-5-(ETHOXYMETHYL)-2-METHYLPYRIMIDINE see MGG300

5-AMINO-6-ETHOXY-2-NAPHTHALENESULFONIC ACID see AJU500

4-AMINO-N-(6-ETHOXY-3-PYRIDAZINYL)BENZENESULFONAMIDE see SNJ100

1-(1-AMINOETHYL)ADAMANTANE HYDROCHLORIDE see AJU625

α-AMINOETHYL ALCOHOL see AAG500

β-AMINOETHYL ALCOHOL see EEC600

(((2-(2-AMINOETHYL)AMINO)ETHYL)AMINO)ME THYL)PHENOL see AJU700

3-((2-((2-AMINOETHYL)AMINO)ETHYL)AMINO)PR OPIONITRILE see LBV000

(3-(2-AMINOETHYL)AMINOPROPYL)TRIMETH OXYSILANE see TLC500

2-AMINOETHYLAMMONIUM PERCHLORATE see AJU875

o-AMINOETHYLBENZENE see EGK500

1-AMINO-4-ETHYLBENZENE see EGL000

β-AMINOETHYLBENZENE see PDE250

4-(2-AMINOETHYL)-1,2-BENZENEDIOL HYDROCHLORIDE see DYC600

α-(1-AMINOETHYL)BENZENEMETHANOL HYDROCHLORIDE see PMJ500

4-(2-AMINOETHYL)-1,2,3-BENZENETRIOL HYDROCHLORIDE see HKG000

α-(1-AMINOETHYL)-BENZYL ALCOHOL see NNM000

dl-α-(1-AMINOETHYL)BENZYL ALCOHOL see NNM500

α-(1-AMINOETHYL)BENZYL ALCOHOL HYDROCHLORIDE see NNN000

α-(1-AMINOETHYL)BENZYL ALCOHOL HYDROCHLORIDE see PMJ500

3-(2-AMINOETHYL)-1-BENZYL-5-METHOXY-2-METHYLINDOLE HYDROCHLORIDE see BEM750

N-(2-AMINOETHYL)-3,5-BIS(1,1-DIMETHYLETHYL)-4-HYDROXYBENZENEPROPANAMIDE see AJU900

2-AMINOETHYL BROMIDE see BNI650

3-AMINO-9-ETHYLCARBAZOLE see AJV000

3-AMINO-N-ETHYLCARBAZOLE see AJV000

3-AMINO-9-ETHYLCARBAZOLEHYDROCHLORIDE see AJV250

trans-4-AMINOETHYLCYCLOHEXANE-1-CARBOXYLIC ACID see AJV500

4-(2-AMINOETHYL)-6-DIAZO-2,4-CYCLOHEXADIENONE HYDROCHLORIDE see DCQ575

4-(2-AMINOETHYL)DIETHYLENETRIAMINE see NEI800

α-(1-AMINOETHYL)-2,4-DIMETHOXYBENZYL ALCOHOL HYDROCHLORIDE see AJV850

α-(1-AMINOETHYL)-2,5-DIMETHOXYBENZYL ALCOHOL HYDROCHLORIDE see MDW000

2-AMINOETHYL DISULFIDE DIHYDROCHLORIDE see CQJ750

AMINOETHYLENE see EJM900

2-AMINOETHYL ESTER CARBAMIMIDOTHIOIC ACID DIHYDROBROMIDE see AJY250

AMINOETHYLETHANEDIAMINE see DJG600

AMINOETHYL ETHANOLAMINE see AJW000

N-AMINOETHYLETHANOLAMINE see AJW000

N-(2-AMINOETHYL)ETHYLENEDIAMINE see DJG600

6-AMINO-1-ETHYL-4-p-((p-((1-ETHYLPYRIDINIUM-4-YL)AMINO)2-AMINOPHENYL)CARBAMOYL)ANILINO)Q UINOLINIUM DIIODIDE see AJW250

6-AMINO-1-ETHYL-4-(p-(p-((1-ETHYLPYRIDINIUM-4-YL)AMINO)BENZAMIDO)ANILINO)QUIN OLINIUM DIIODIDE see AJW500

6-AMINO-1-ETHYL-4-(p-((p-((1-ETHYLPYRIDINIUM-4-YL)AMINO)PHENYL)CARBAMOYL)ANILIN OQUINOLINIUM) DIBROMIDE see AJW750

β-AMINOETHYLGLYOXALINE see HGD000

1-AMINO-2-ETHYLHEXAN (CZECH) see EKS500

1-α-(1-AMINOETHYL)-m-HYDROXYBENZYL ALCOHOL see HNB875

(−)-α-(1-AMINOETHYL)-m-HYDROXYBENZYL ALCOHOL BITARTRATE see HNC000

1-α-(1-AMINOETHYL)-m-HYDROXYBENZYL ALCOHOL BITARTRATE see HNC000

1-α-(1-AMINOETHYL)-m-HYDROXYBENZYL ALCOHOL HYDROGEN-d-TARTRATE see HNC000

3-(β-AMINOETHYL)-5-HYDROXYINDOLE see AJX500

α-(1-AMINOETHYL)-3-HYDROXY-4-METHYLBENZYL ALCOHOL HYDROCHLORIDE see MKS000
4-(2-AMINOETHYL)IMIDAZOLE see HGD000
β-AMINOETHYLIMIDAZOLE see HGD000
4-(2-AMINOETHYL)IMIDAZOLE BIS(DIHYDROGEN PHOSPHATE) see HGE000
4-(2-AMINOETHYL)IMIDAZOLE DI-ACID PHOSPHATE see HGE000
4-AMINOETHYLIMIDAZOLE HYDROCHLORIDE see HGE500
3-(2-AMINOETHYL)INDOLE see AJX000
(AMINO-2 ETHYL)-3-INDOLE (FRENCH) see AJX000
3-(1-AMINOETHYL)INDOLE HYDROCHLORIDE see LIU100
3-(2-AMINOETHYL)INDOLE HYDROCHLORIDE see AJX250
3-(2-AMINOETHYL)INDOL-5-OL see AJX500
3-(2-AMINOETHYL)INDOL-5-OL CREATININE SULFATE see AJX750
AMINOETHYLISOSELENOURONIUM BROMIDE HYDROCHLORIDE see AJY000
2-β-AMINOETHYLISOTHIOUREA see AJY250
S-β-AMINOETHYLISOTHIOURONIC DIHYDROCHLORIDE see AJY500
2-(β-AMINOETHYL)ISOTHIOURONIUM BROMIDE HYDROBROMIDE see AJY250
2-AMINOETHYLISOTHIOURONIUM DIACETATE see AJY300
2-AMINOETHYLISOTHIOURONIUM DIBROMIDE see AJY250
2-AMINOETHYLISOTHIOURONIUMDICHLORIDE see AJY500
2-AMINOETHYLISOTHIOURONIUM BROMIDE HYDROBROMIDE see AJY250
S-(2-AMINOETHYL)ISOTHIURONIUM BROMIDE HYDROBROMIDE see AJY250
β-AMINOETHYLISOTHIURONIUM BROMIDE HYDROBROMIDE see AJY250
S-(β-AMINOETHYL)ISOTHIURONIUM BROMIDE HYDROBROMIDE see AJY250
2-AMINOETHYLISOTHIOURONIUM DIHYDROBROMIDE see AJY250
2-AMINOETHYL MERCAPTAN see AJT250
4-AMINO-N-ETHYL-m-(β-METHANESULFONAMIDOETHYL)-m-TOLUIDINE see AJY750
3-(2-AMINOETHYL)-5-METHOXYBENZOFURAN HYDROCHLORIDE see AJZ000
6-(2-AMINOETHYL)-5-METHOXYBENZOFURAN HYDROCHLORIDE see AKA000
6-(β-AMINOETHYL)-5-METHOXYBENZOFURANHYDROCHLORIDE see AKA000
α-(1-AMINOETHYL)-4-METHOXYBENZYLALCOHOL HYDROCHLORIDE see AKA250
α-(1-AMINOETHYL)-4-METHOXYBENZYL ALCOHOL HYDROCHLORIDE see AKA250
3-(2-AMINOETHYL)-5-METHOXYINDOLE see MFS400
3-(2-AMINOETHYL)-5-METHOXYINDOLE HYDROCHLORIDE see MFT000
3-(2-AMINOETHYL)-6-METHOXYINDOLE HYDROCHLORIDE see MFS500
3-(2-AMINOETHYL)-5-METHOXY-2-METHYLBENZO-FURAN, HYDROCHLORIDE see MGH750
2-AMINOETHYLMETHYLAMINE see MJW100
2-(AMINOETHYL)-2-METHYL-1,3-BENZODIOXOLE HYDROCHLORIDE see AKA500

2-AMINOETHYL-2-METHYL-1,3-BENZODIOXOLE HYDROCHLORIDE see AKA500
1-(4-AMINO-1-ETHYL-2-METHYL-5-PHENYL-1H-PYRROL-3-YL)ETHANONE see AKA600
3-AMINO-4-ETHYL-1-METHYL-5H-PYRIDO(4,3-B)INDOLE ACETATE see AKA650
N-AMINOETHYLMORPHOLINE see AKA750
2-AMINO-5-ETHYLNONANE see EMY000
4-(1-AMINOETHYL)PHENOL see AKA800
p-(2-AMINOETHYL)PHENOL see TOG250
p-β-AMINOETHYLPHENOL see TOG250
4-(2-AMINOETHYL)PHENOL HYDROCHLORIDE see TOF750
p-(2-AMINOETHYL)PHENOL MONOCHLORIDE see TOF750
S-(2-AMINOETHYL) PHOSPHOROTHIOATE see AKA900
AMINOETHYLPIPERAZINE see AKB000
N-AMINOETHYLPIPERAZINE see AKB000
1-(2-AMINOETHYL)PIPERAZINE see AKB000
N-(2-AMINOETHYL)PIPERAZINE see AKB000
N-(β-AMINOETHYL)PIPERAZINE see AKB000
6-AMINO-3-ETHYL-1-(2-PROPENYL)-2,4(1H,3H-)-PYRIMIDINEDIONE see AFW500
2-AMINO-3-ETHYL-9H-PYRIDO(2,3-B)INDOLE see PPH400
3-AMINO-1-ETHYL-5H-PYRIDO(4,3-B)INDOLE ACETATE see AKB050
3-AMINO-4-ETHYL-5H-PYRIDO(4,3-B)INDOLE ACETATE see AKB070
4-(2-AMINOETHYL)PYROCATECHOL see DYC400
4-(2-AMINOETHYL)PYROCATECHOL HYDROCHLORIDE see DYC600
2-AMINOETHYLSULFONIC ACID see TAG750
5-AMINO-2-ETHYLTETRAZOL see AKB125
2-AMINO-4-(ETHYLTHIO)BUTYRIC ACID see AKB250
2-AMINO-4-(ETHYLTHIO)BUTYRIC ACID see EEI000
l-2-AMINO-4-(ETHYLTHIO)BUTYRIC ACID see AKB250
dl-2-AMINO-4-(ETHYLTHIO)BUTYRIC ACID see EEI000
S-(2-AMINOETHYL)THIOPHOSPHATEMONOSODIUM SALT see AKB500
2-AMINOETHYL-2-THIOPSEUDOUREA DICHLORIDE see AJY500
2-(2-AMINOETHYL)-2-THIOPSEUDOUREA DIHYDROCHLORIDE see AJY500
2-(2-AMINOETHYL)-2-THIOPSEUDOUREA HYDROBROMIDE see AJY250
N-(2-(4-AMINO-N-ETHYL-m-TOLUIDINO)ETHYL)METHANESULFONAMIDE SULFATE (2:3) see MDQ850
3-AMINO-α-ETHYL-2,4,6-TRIIODOHYDROCINNAMIC ACID see IFY100
3'-AMINO-N-ETHYL-2',4',6'-TRIIODOSUCCINANILIC ACID see AMV375
AMINOFENAZONE (ITALIAN) see DOT000
p-AMINOFENETOL see PDK790
5-AMINO-3-FENIL-1-BIS(-DIMETILAMINO)-FOSFORIL-1,2,4-TRIAZOLO (ITALIAN) see AIX000
m-AMINOFENOL (CZECH) see ALS990
p-AMINOFENOL (CZECH) see ALT250
5-AMINO-3-FENYL-1-BIS(DIMETHYL-AMINO)-FOSFORYL-1,2,4-TRIAZOOL (DUTCH) see AIX000
m-AMINOFENYLMOCOVINA HYDROCHLORID (CZECH) see ALY750

AMINOFILINA (SPANISH) see TEP500
1-AMINOFLUORANTHENE see FDE100
8-AMINOFLUORANTHENE see FDE200
AMINOFLUOREN (GERMAN) see FDI000
2-AMINOFLUORENE see FDI000
2-AMINOFLUORENONE see AKB875
4-AMINOFLUORENONE see AKB900
2-AMINO-9-FLUORENONE see AKB875
4-AMINO-9-FLUORENONE see AKB900
2-AMINO-N-FLUOREN-2-YLACETAMIDE see AKC000
2-AMINO-5-FLUOROBENZOXAZOLE see AKC250
4-AMINO-4'-FLUORODIPHENYL see AKC500
α-AMINO-α-(FLUOROMETHYL)IMIDAZOLE-4-PROPIONIC ACID see MRI285
6-AMINO-2-(FLUOROMETHYL)-3-(2-METHYLPHENYL)-4(3H)-QUINAZOLINONE (9CI) see AEW625
6-AMINO-2-FLUOROMETHYL-3-(o-TOLYL)-4(3H)-QUINAZOLINONE see AEW625
2-AMINO-1-(3-FLUOROPHENYL)ETHANOL HYDROBROMIDE see AKS500
(2-AMINO-6-(((4-FLUOROPHENYL)METHYL)AMINO)-3-PYRIDINYL)CARBAMIC ACID ETHYL ESTER MALEATE see FMP100
1-(4-AMINO-5-(3-FLUOROPHENYL)-2-METHYL-1H-PYRROL-3-YL)ETHANONE see AKC550
1-(4-AMINO-5-(4-FLUOROPHENYL)-2-METHYL-1H-PYRROL-3-YL)ETHANONE see AKC560
3-AMINO-2-FLUORO-PROPANOIC ACID HYDROCHLORIDE see FFT100
4-AMINO-5-FLUORO-2(1H)-PYRIMIDINONE see FHI000
AMINOFORM see HEI500
AMINOFORMAMIDINE see GKW000
AMINOFORMAMIDINE HYDROCHLORIDE see GKY000
AMINOFOSTINE see AMD000
AMINOFURACARB see AKC570
2-AMINOGLUTARAMIC ACID see GFO050
l-2-AMINOGLUTARAMIDIC ACID see GFO050
l-2-AMINOGLUTARIC ACID see GFO000
α-AMINOGLUTARIC ACID see GFO000
l-2-AMINOGLUTARIC ACID HYDROCHLORIDE see GFO025
α-AMINOGLUTARIC ACID HYDROCHLORIDE see GFO025
AMINOGLUTETHIMIDE see AKC600
p-AMINOGLUTETHIMIDE see AKC600
AMINOGLUTETHIMIDE PHOSPHATE see AKC625
1-AMINOGLYCEROL see AMA250
AMINOGLYCOL see ALB000
AMINOGUANIDINE see AKC750
AMINOGUANIDINE HYDROCHLORIDE see AKC800
AMINOGUANIDINE SULFATE see AKD250
AMINOGUANIDINE SULPHATE see AKD250
AMINO GUANIDINIUM NITRATE see AKD375
2-AMINO-4-(GUANIDINOOXY)-l-BUTYRIC ACID see AKD500
l,2-AMINO-4-(GUANIDINOOXY)BUTYRIC ACID see AKD500
1-AMINOHEPTANE see HBL600
3-AMINOHEPTANE see HBM500
dl-2-AMINOHEPTANE see HBM490
7-AMINOHEPTANENITRILE see AKD550
2-AMINOHEPTANE SULFATE see AKD600
7-AMINOHEPTANOIC ACID, ISOPROPYL ESTER see AKD625
7-AMINOHEPTANONITRILE see AKD550

3-(7-AMINOHEPTYL)INDOLE ADIPATE see AKD750

1-AMINOHEXAHYDROAZEPINE see HEG400

AMINOHEXAHYDROBENZENE see CPF500

AMINOHEXAHYDROBENZENE HYDROCHLORIDE see CPA775

9-AMINO-2,3,5,6,7,8-HEXAHYDRO-1H-CYCLOPENTA(b)QUINOLINE HYDROCHLORIDE HYDRATE see AKD775

N-AMINOHEXAMETHYLENEIMINE see HEG400

1-AMINOHEXANE see HFK000

2-AMINOHEXANOIC ACID see AJC950

(S)-2-AMINOHEXANOIC ACID see AJC950

ω-AMINOHEXANOIC ACID see AJD000

6-AMINOHEXANOIC ACID CYCLIC LACTAM see CBF700

6-AMINOHEXANOIC ACID HOMOPOLYMER see PJY500

6-AMINOHEXANOL see AKD800

AMINOHEXYL ALCOHOL see AKD800

3-AMINO-4-HOMOISOTWISTANE see AKD875

1-AMINOHOMOPIPERIDINE see HEG400

N-AMINOHOMOPIPERIDINE see HEG400

α-AMINOHYDROCINNAMIC ACID see PEC750

AMINOHYDROQUINONE DIMETHYL ETHER see AKD925

2-AMINO-3-HYDROXYACETOPHENONE see AKE000

1-AMINO-4-HYDROXY-9,10-ANTHRACENEDIONE see AKE250

1-AMINO-4-HYDROXYANTHRAQUINONE see AKE250

4-AMINO-4'-HYDROXYAZOBENZENE see AKE500

2-AMINO-1-HYDROXYBENZENE see ALT000

3-AMINO-1-HYDROXYBENZENE see ALS990

4-AMINO-1-HYDROXYBENZENE see ALT250

3-AMINO-4-HYDROXYBENZENEARSONIC ACID see HJE500

4-AMINO-2-HYDROXYBENZENEARSONIC ACID see HJE400

2-AMINO-3-HYDROXYBENZOIC ACID see AKE750

4-AMINO-2-HYDROXYBENZOIC ACID see AMM250

5-AMINO-2-HYDROXYBENZOIC ACID see AMM500

4-AMINO-2-HYDROXYBENZOIC ACID, 2-(DIETHYLAMINO)ETHYL ESTER, HYDROCHLORIDE (9CI) see AMM750

4-AMINO-2-HYDROXYBENZOIC ACID, 2-(DIMETHYLAMINO)ETHYL ESTER, HYDROCHLORIDE see AMN000

2-AMINO-3-HYDROXYBENZOIC ACID, METHYL ESTER see HJC500

3-AMINO-4-HYDROXYBENZOIC ACID METHYL ESTER see AKF000

α-AMINO-p-HYDROXYBENZYLPENICILLIN TRIHYDRATE see AOA100

4-AMINO-4'-HYDROXYBIPHENOL see AIW000

4-AMINO-3-HYDROXYBIPHENYL see AIV750

4-AMINO-3-HYDROXYBIPHENYL SULFATE see AKF250

4-AMINO-1-HYDROXYBUTANE-1,1-DIPHOSPHONATE see AKF300

4-AMINO-1-HYDROXYBUTANE-1,1-DIPHOSPHONIC ACID see AKF300

4-AMINO-1-HYDROXYBUTANE-1,1-DIYLDIPHOSPHONIC ACID see AKF300

AMINOHYDROXYBUTYLIDENE BIPHOSPHONATE MONOSODIUM SALT TRIHYDRATE see SKL600

(4-AMINO-1-HYDROXYBUTYLIDENE)BISPHOSPHONIC ACID see AKF300

4-AMINO-1-HYDROXYBUTYLIDENE-1,1-BIS(PHOSPHONIC ACID) see AKF300

4-AMINO-3-HYDROXYBUTYRIC ACID see AKF375

l-2-AMINO-3-HYDROXYBUTYRIC ACID see TFU750

γ-AMINO-β-HYDROXYBUTYRIC ACID see AKF375

1-AMINO-4-HYDROXY-5-CHLORANTHRACHINON (CZECH) see AJH250

5-AMINO-2-HYDROXY-2,4,6-CYCLOHEPTATRIEN-1-ONE see AMV800

4-AMINO-2-HYDROXY-2,4,6-CYCLOPHEPTATRIEN-1-ONE see AMV790

2-AMINO-3-HYDROXY-1,7-DIHYDRO-7-METHYL-6H-PURIN-6-ONE see HMD000

2-AMINO-3-HYDROXY-1,7-DIHYDRO-8-METHYL-6H-PURIN-6-ONE see HMD500

3-AMINO-4-HYDROXYDIPHENYL HYDROCHLORIDE see AIW250

3-AMINO-4-(2-HYDROXY)ETHOXYBENZENARSONIC ACID see AKF500

(3-AMINO-4-(2-HYDROXYETHOXY)PHENYL)ARSINE OXIDE see AKF750

(R)-4-(2-AMINO-1-HYDROXYETHYL)-1,2-BENZENEDIOL see NNO500

4-AMINO-N-(2-HYDROXYETHYL)-o-TOLUENESULFONAMIDE see AKG000

2-AMINO-1-HYDROXYHYDROINDENE HYDROCHLORIDE see AKL100

α-AMINO-3-HYDROXY-5-ISOXAZOLEACETIC ACID HYDRATE see AKG250

α-AMINO-3-HYDROXY-5-ISOXAZOLESSIGSAURE HYDRAT (GERMAN) see AKG250

AMINO-(3-HYDROXY-5-ISOXAZOLYL)ACETIC ACID see AKG250

2-AMINO-5-HYDROXYLEVULINIC ACID see AKG500

1-AMINO-4-HYDROXY-2-METHOXY-9,10-ANTHRACENEDIONE see AKO350

1-AMINO-4-HYDROXY-2-METHOXYANTHRAQUINONE see AKO350

4-AMINO-5-HYDROXYMETHYL-2-METHOXYPYRIMIDINE see AKO750

2-AMINO-4-(HYDROXYMETHYLPHOSPHINYL)BUTANOIC ACID MONOAMMONIUM SALT see ANI800

2-AMINO-2-(HYDROXYMETHYL)PROPANE-1,3-DIOL see TEM500

2-AMINO-2-(HYDROXYMETHYL)-1,3-PROPANEDIOL see TEM500

4-AMINO-5-HYDROXY-2,7-NAPHTHALENEDISULFONIC ACID see AKH000

8-AMINO-7-HYDROXY-3,6-NAPHTHALENEDISULFONIC ACID, SODIUM SALT see AKH250

4-AMINO-5-HYDROXY-2,7-NAPHTHALENEDISULFONIC ACID-p-TOLUENESULFONATE (ESTER) see AKH500

4-AMINO-5-HYDROXY-1-NAPHTHALENESULFONIC ACID see AKH750

6-AMINO-4-HYDROXY-2-NAPHTHALENESULFONIC ACID see AKH800

7-AMINO-4-HYDROXY-2-NAPHTHALENESULFONIC ACID see AKI000

3-AMINO-7-((2-HYDROXY-1-NAPHTHYL)AZO)-2,8-DIMETHYL-5-PHENYLPHENAZINIUM CHLORIDE see CMM770

3-AMINO-4-HYDROXYNITROBENZENE see NEM500

3-AMINO-2-HYDROXY-5-NITRO-BENZENESULFONIC ACID see HMY500

3-AMINO-4-HYDROXY-5-NITROBENZENESULFONIC ACID see AKI250

4-AMINO-3-HYDROXY-4'-NITRODIPHENYLHYDROCHLORIDE see AKI500

l-2-AMINO-3-(HYDROXYNITROSAMINO)PROPIONIC ACID see AFH750

2-AMINO-5-HYDROXY-4-OXOPENTANOIC ACID see AKG500

(S)-2-AMINO-5-HYDROXY-4-OXOPENTANOIC ACID see HND250

(S-(4*,s*))-N-(3-AMINO-2-HYDROXY-1-OXO-4-PHENYLBUTYL)-l-LEUCINE see BFV300

3-AMINO-4-HYDROXY-PHENARSINE HYDROCHLORIDE see ARL000

2-AMINO-3-HYDROXYPHENAZINE see ALS100

1-AMINO-4-HYDROXY-2-PHENOXYANTHRAQUINONE see AKI750

(2S-(2-α,5-α,6-β(S*)))-6-((AMINO(4-HYDROXYPHENYL)ACETYL)AMINO)-3,3-DIMETHYL-7-OXO-4-THIA-1-AZABICYCLO(3.2.0)HEPTANE-2-CARBOXYLIC ACID TRIHYDRATE see AOA100

((5-(3-AMINO-4-HYDROXYPHENYL)ARSENO)-2-HYDROXYANILINO)METHANOL SULFOXYLATE SODIUM see NCJ500

(3-AMINO-4-HYDROXYPHENYL)ARSENOUS ACID see OOK100

3-AMINO-4-HYDROXYPHENYLARSINE OXIDE HYDROCHLORIDE see ARL000

3-AMINO-4-HYDROXYPHENYL ARSINOXIDE HYDROCHLORIDE see ARL000

(3-AMINO-4-HYDROXYPHENYL)ARSONOUS DICHLORIDE MONOHYDROCHLORIDE see DFX400

3-(R)-AMINO-2-(s)-HYDROXY-4-PHENYLBUTANOYL-(2)-LEUCINE see BFV300

3-AMINO-4-HYDROXYPHENYL DICHLORARSINE HYDROCHLORIDE see DFX400

(3-AMINO-4-HYDROXYPHENYL)DICHLOROARSINE HYDROCHLORIDE see DFX400

2-AMINO-3-HYDROXYPHENYL METHYL KETONE see AKE000

3-(((3-AMINO-4-HYDROXYPHENYL)PHENYLARSINO)THIO)ALANINE see AKI900

dl-2-AMINO-1-HYDROXY-1-PHENYLPROPANE see NNM500

threo-2-AMINO-1-HYDROXY-1-PHENYLPROPANE see NNM510

1-N-(S-3-AMINO-2-HYDROXYPROPIONYL)BETAMYCIN see AKJ000

1-N-(S-3-AMINO-2-HYDROXYPROPIONYL)-GENTAMICIN B SULFATE see SBE800

1-N-(S-3-AMINO-2-HYDROXYPROPIONYL) GENTAMYCIN B see AKJ000

3-AMINO-4(1-(2-HYDROXY)PROPOXY)BENZENEARSONIC ACID see AKJ250

S-3-AMINO-2-HYDROXYPROPYL SODIUMHYDROGEN PHOSPHOROTHIOATETETRAHYDRATE see AKJ500

l-N-(p-(((-2-AMINO-4-HYDROXY-6-PTERIDINYL)METHYL)AMINO)BENZOYL) GLUTAMIC ACID see FMT000

6-AMINO-N-HYDROXY-3-PYRIDINECARBOXAMIDE see ALL300

4-AMINO-2-HYDROXYPYRIMIDINE see CQM600

N-(4-(((N-(2-AMINO-4-HYDROXY-6-QUINAZOLINYL)METHYL)PROP-2-YNILAMINO)BENZOYL))-l-GLUTAMIC ACID see PMN550

4-AMINO-2-HYDROXYTOLUENE see AKJ750

5-AMINO-7-HYDROXY-1H-v-TRIAZOLO(d)PYRIMIDINE see AJO500

4-AMINO-4'-HYDROXY-2,3',5'-TRIMETHYLAZOBENZENE see AKK000

2-AMINOHYPOXANTHINE see GLI000

5-AMINOIMIDAZOLE-4-CARBOXAMIDE see AKK250

4-AMINO-5-IMIDAZOLECARBOXAMIDE HYDROCHLORIDE see IAL100

5-AMINO-4-IMIDAZOLECARBOXAMIDE UREIDOSUCCINATE see AKK625

5-AMINO-1H-IMIDAZOLE HYDROCHLORIDE see AKL250

α-AMINOIMIDAZOLE-4-PROPIONIC ACID, COBALT(2+) SALT see BJY000

α-AMINO-o-IMINOETHANE HYDROCHLORIDE see AAI500

9-((AMINOIMINOMETHYL)AMINO)-N-(10-((AMINOIMINOMETHYL)AMINO)-1-(3-AMINOPROPYL)-20-HYDROXYDECYL)NONANAMIDE see EQS500

4-((AMINOIMINOMETHYL)AMINO)BENZOIC ACID 2-(METHOXYCARBONYL)PHENYL ESTER see CBT175

4-((AMINOIMINOMETHYL)AMINO)BENZOIC ACID 2-METHOXY-4-(2-PROPENYL)PHENYL ESTER see MDY300

N-(4-((AMINOIMINOMETHYL)AMINO)-1-FORMYLBUTYL)-3-PHENYLALANYL-l-PROLINAMIDE SULFATE, (S)- see PEE100

o-((AMINOIMINOMETHYL)AMINO)-l-HOMOSERINE see AKD500

3-(((2-((AMINOIMINOMETHYL)AMINO)-4-THIAZOLYL)METHYL)THIO)-N-(AMINOSULFONYL)PROPANIMIDAMIDE see FAB500

N-(AMINOIMINOMETHYL)-2,6-DICHLOROBENZENEACETAMIDE HYDROCHLORIDE see GKU300

N-(AMINOIMINOMETHYL)-N'-(4-NITROPHENYL)UREA MONOHYDROCHLORIDE see NIJ400

(2-((AMINOIMINOMETHYL)THIO)ETHYL)TRIETHYLAMMONIUM BROMIDE HYDROBROMIDE see TAI100

2-((AMINOIMINOMETHYL)THIO)-N,N,N-TRIETHYL-ETHANAMINIUM BROMIDE, MONOHYDROBROMIDE (9CI) see TAI100

2-AMINO-4-IMINO-1(4H)-NAPHTHALENONE HYDROCHLORIDE see ALK625

7-AMINO-3-IMINO-3H-PHENOTHIAZINEMONOHYDROCHLORIDE see AKK750

2-(6-AMINO-3-IMINO-3H-XANTHEN-9-YL)BENZOIC ACID METHYL ESTER MONOHYDROCHLORIDE see RGP600

2-AMINOINDANE HYDROCHLORIDE see AKL000

2-AMINOINDAN HYDROCHLORIDE see AKL000

AMINOINDANOL HYDROCHLORIDE see AKL100

5-AMINOINDAZOLE HYDROCHLORIDE see AKL250

α-AMINO-INDOLE-3-PROPRIONIC ACID see TNX000

α'-AMINO-3-INDOLEPROPRIONIC ACID see TNX000

l-α-AMINO-3-INDOLEPROPRIONIC ACID see TNX000

2-AMINO-3-INDOL-3-YL-PROPRIONIC ACID see TNX000

m-AMINOIODOBENZENE see IEB000

2-AMINO-5-IODOBENZOXAZOLE see AKL500

4-AMINO-4'-IODOBIPHENYL see AKL600

2-AMINOISOBUTANE see BPY250

α-AMINOISOBUTANOIC ACID see MGB000

β-AMINOISOBUTANOL see IIA000

2-AMINOISOBUTYRIC ACID see MGB000

α-AMINOISOBUTYRIC ACID see MGB000

α-AMINOISOCAPROIC ACID see LES000

AMINOISOMETRADIN see AKL625

AMINOISOMETRADINE see AKL625

3-AMINOISONAPHTHOIC ACID see ALJ000

α-AMINOISOPROPYL ALCOHOL see AMA500

2-AMINOISOPROPYLBENZENE see INW100

o-AMINOISOPROPYLBENZENE see INW100

d-R-AMINO-3-ISOSSAZOLIDONE (ITALIAN) see CQH000

l-(+)-α-AMINOISOVALERIC ACID see VBP000

d-4-AMINO-3-ISOXAZOLIDINONE see CQH000

d-4-AMINO-3-ISOXAZOLIDONE see CQH000

AMINOKAPRON see AJD000

AMINOMEL see AHR600

dl-2-AMINO-4-MERCAPTOBUTYRIC ACID see HGI300

trans-1-AMINO-2-MERCAPTOMETHYLCYCLOBUTANEHYDROCHLORIDE see AKL750

3-AMINO-4-MERCAPTO-6-METHYLPYRIDAZIN (GERMAN) see ALB625

3-AMINO-4-MERCAPTO-6-METHYLPYRIDAZINE see ALB625

2-AMINO-6-MERCAPTOPURINE see AMH250

2-AMINO-6-MERCAPTOPURINE RIBONUCLEOSIDE see TFJ500

2-AMINO-6-MERCAPTOPURINE RIBOSIDE see TFJ500

2-AMINO-6-MERCAPTO-9(β-d-RIBOFURANOSYL)PURINE see TFJ500

2-AMINO-5-MERCAPTO-1,3,4-THIADIAZOLE see AKM000

5-AMINO-2-MERCAPTO-1,3,4-THIADIAZOLE see AKM000

AMINOMERCURIC CHLORIDE see MCW500

AMINOMESITYLENE see TLG500

2-AMINOMESITYLENE see TLG500

AMINOMESITYLENE HYDROCHLORIDE see TLH000

2-AMINOMESITYLENE HYDROCHLORIDE see TLH000

6-AMINO-1-METALLYL-3-METHYLPYRIMIDINE-2,4-DIONE see AKL625

AMINOMETHANAMIDINE see GKW000

AMINOMETHANAMIDINE HYDROCHLORIDE see GKY000

AMINOMETHANE see MGC250

6-AMINOMETHAQUALONE see AKM125

1-AMINO-2-METHOXYANTHRAQUINONE see AKM250

4-AMINO-3-METHOXYAZOBENZENE see MFB000

4-AMINO-3-METHOXYAZOBENZENE see MFF500

3-AMINO-4-METHOXY BENZANILIDE see AKM500

1-AMINO-2-METHOXYBENZENE see AOV900

1-AMINO-4-METHOXYBENZENE see AOW000

2-AMINO-5-METHOXY BENZENESULFONIC ACID see AIA500

3-AMINO-4-METHOXYBENZOIC ACID see AIA250

2-AMINO-4-METHOXYBENZOTHIAZOLE see AKM750

2-AMINO-5-METHOXYBENZOXAZOLE see AKN000

4-AMINO-4'-METHOXY-3-BIPHENYLOL see HLR500

4-AMINO-4'-METHOXY-3-BIPHENYLOLHYDROCHLORIDE see AKN250

2-AMINO-3-METHOXYDIPHENYLENE OXIDE see AKN500

2-AMINO-3-METHOXYDIPHENYLENOXYD (GERMAN) see MDZ000

4-AMINO-N-(2-METHOXYETHYL)-7-((2-METHOXYETHYL)AMINO-2-PHENYL)-6-PTERIDINECARBOXAMIDE see AKN750

3'-(α-AMINO-p-METHOXYHYDROCINNAMIDO)-3'-DEOXY-N,N-DIMETHYLADENOSINE MONOHYDROCHLORIDE see POK250

3'-(α-AMINO-p-METHOXYHYDROCINNAMIDO)-3'-DEOXY-N,N-DIMETHYLADENOSINE DIHYDROCHLORIDE see POK300

3'-(l-α-AMINO-p-METHOXYHYDROCINNAMIDO)-3'-DEOXY-N,N-DIMETHYLADENOSINE see AEI000

1-AMINO-2-METHOXY-5-METHYLBENZENE see MGO750

2-AMINO-5-METHOXY-2'(or 3')-METHYLINDIAMINE see AKN800

6-AMINO-8-METHOXY-1-METHYL-4-(p-(p-((1-METHYLPYRIDINIUM-4-YL)AMINO)BENZAMIDO)ANILINOQUINOLINIUM) DI-p-TOLUENESULFONATE see AKO000

1-(4-AMINO-5-(4-METHOXY-3-METHYLPHENYL)-2-METHYL-1H-PYRROL-3-YL)ETHANONE see AKO100

7-AMINO-4-(2-METHOXY-p-(p-((1-METHYLPYRIDINIUM-4-YL)AMINO)BENZAMIDO)ANILINO)-1-METHYLQUINOLINIUM)) DIBROMIDE see AKO250

4-AMINO-N-(6-METHOXY-2-METHYL-4-PYRIMIDINYL)BENZENESULFONAMIDE see MDU300

7-AMINO-9-α-METHOXYMITOSANE see AHK500

2-AMINO-1-METHOXYNAPHTHALENE see MFA000

3-AMINO-4-METHOXYNITROBENZENE see NEQ500

2-AMINO-1-METHOXY-4-NITROBENZENE see NEQ500

2-AMINO-5-METHOXY-2'(OR 3')-METHYLINDOANILINE see AKO300

1-AMINO-2-METHOXY-4-OXYANTHRAQUINONE see AKO350

1-(4-AMINO-5-(3-METHOXYPHENYL)-2-METHYL-1H-PYRROL-3-YL)ETHANONE see AKO400

1-(4-AMINO-5-(4-METHOXYPHENYL)-2-METHYL-1H-PYRROL-3-YL)ETHANONE see AKO430

1-(4-AMINO-5-(o-METHOXYPHENYL)-2-METHYL-1H-PYRROL-3-YL)ETHANONE see AKO450

(S)-3'-((2-AMINO-3-(4-METHOXYPHENYL)-1-OXOPROPYL)AMINO)-3'-DEOXY-N,N-DIMETHYLADENOSINE see AEI000

4-AMINO-6-METHOXY-1-PHENYLPYRIDAZINIUM-METHYLSULFAT (GERMAN) see MQS100

4-AMINO-6-METHOXY-1-PHENYLPYRIDAZINIUM METHYL SULFATE see MQS100

4-AMINO-N-(6-METHOXY-3-PYRIDAZINYL)-BENZENESULFONAMIDE see AKO500

4-AMINO-2-METHOXY-5-PYRIMIDINEMETHANOL see AKO750

4-AMINO-2-METHOXY-5-PYRIMIDINEMETHANOL see AKO750

4-AMINO-N-(6-METHOXY-4-PYRIMIDINYL)-BENZENESULFONAMIDE (9CI) see SNL800

4-AMINO-N-(5-METHOXY-2-PYRIMIDINYL)BENZENESULFONAMIDE SODIUM SALT see MFO000

3-AMINO-4-METHOXYTOLUEN (CZECH) see MFQ500

3-AMINO-4-METHOXYTOLUENE see MGO750

3-AMINOMETHYLALIZARIN-N,N-DIACETIC ACID see AFM400

1-AMINO-4-(METHYLAMINO)-9,10-ANTHRACENEDIONE see AKP250

4-AMINO-1-METHYLAMINOANTHRAQUINONE see AKP250

dl-α-AMINO-β-METHYLAMINOPROPIONIC ACID see AKP500

2-AMINO-N-METHYLANILINE see MNU000

4-AMINO-2-METHYLANILINE see TGM000

2-AMINO-4-METHYLANISOLE see MGO750

1-AMINO-2-METHYL-9,10-ANTHRACENEDIONE see AKP750

1-AMINO-2-METHYLANTHRAQUINONE see AKP750

(AMINOMETHYL)BENZENE see BDX750

1-AMINO-2-METHYLBENZENE see TGQ750

2-AMINO-1-METHYLBENZENE see TGQ750

3-AMINO-1-METHYLBENZENE see TGQ500

4-AMINO-1-METHYLBENZENE see TGR000

1-AMINO-2-METHYLBENZENE HYDROCHLORIDE see TGS500

2-AMINO-1-METHYLBENZENE HYDROCHLORIDE see TGS500

β-(AMINOMETHYL)BENZENEPROPANOIC ACID see PEE500

β-(AMINOMETHYL)-BENZENEPROPANOIC ACID HYDROCHLORIDE see GAD000

2-AMINO-5-METHYLBENZENESULFONIC ACID see AKQ000

3-AMINO-4-METHYLBENZENESULFONYLCYCLOHEXYLUREA see AKQ250

2-AMINO-4-METHYLBENZOTHIAZOLE see AKQ500

2-AMINO-6-METHYLBENZOTHIAZOLE see MGD500

2-AMINO-4-METHYL-BENZOTHIAZOLE, HYDROCHLORIDE see MGD750

2-AMINO-5-METHYLBENZOXAZOLE see AKQ750

3-AMINO-α-METHYLBENZYL ALCOHOL see AKR000

m-AMINO-α-METHYLBENZYL ALCOHOL see AKR000

4-AMINOMETHYL-1-BENZYLPYRROLIDIN-2-ONE FUMARATE (2:1) see AKR100

4-((4-AMINO-1-METHYLBUTYL)AMINO)-7-CHLOROQUINOLINE DIHYDROCHLORIDE see CLD100

8-(4-AMINO-1-METHYLBUTYLAMINO)-6-METHOXYQUINOLINE see PMC300

8-((4-AMINO-1-METHYLBUTYL)AMINO)-6-METHOXYQUINOLINE DIPHOSPHATE see AKR250

3-AMINO-3-METHYL-1-BUTYNE see MHX200

2-AMINO-3-METHYL-α-CARBOLINE see ALD750

3-AMINO-1-METHYL-γ-CARBOLINE see ALD500

2-AMINO-4-METHYL-5-CARBOXANILIDOTHIAZOLE see AKR500

β-(AMINOMETHYL)-4-CHLOROBENZENEPROPANOIC ACID see BAC275

β-(AMINOMETHYL)-p-CHLOROHYDROCINNAMIC ACID see BAC275

trans-p-(AMINOMETHYL)CYCLOHEXANECARBOXYLICACID see AJV500

trans-1-AMINOMETHYLCYCLOHEXANE-4-CARBOXYLIC ACID see AJV500

trans-4-AMINOMETHYL-1-CYCLOHEXANECARBOXYLIC ACID see AJV500

trans-4-(((4-(AMINOMETHYL)CYCLOHEXYL)CARBONYL)OXY)BENZENEPROPANOIC ACID see CDF375

trans-4-(((4-(AMINOMETHYL)CYCLOHEXYL)CARBONYL)OXY)-BENZENEPROPANOIC ACID HYDROCHLORIDE see CDF380

4-AMINO-3-METHYL-N,N-DIETHYLANILINEHYDROCHLORIDE see AKR750

2-AMINOMETHYL-2,3-DIHYDRO-4H-PYRAN see AKS000

2-AMINOMETHYL-3,4-DIHYDRO-2H-PYRAN see AKS000

α-(AMINOMETHYL)-3,4-DIHYDROXYBENZYL ALCOHOL see ARL500

α-(AMINOMETHYL)-3,4-DIHYDROXYBENZYL ALCOHOL see ARL750

l-α-(AMINOMETHYL)-3,4-DIHYDROXYBENZYL ALCOHOL see NNO500

2-AMINO-N-(1-METHYL-1,2-DIPHENYLETHYL)ACETAMIDE MONOHYDROCHLORIDE (+−)- see RCK750

2-AMINO-4-METHYLDIPYRIDO(1,2-A:3',2'-D)IMIDAZOLE see AKS100

2-AMINO-6-METHYLDIPYRIDO(1,2-a:3',2'-d)IMIDAZOLE see AKS250

2-AMINO-6-METHYLDIPYRIDO(1,2-a:3',2'-d)IMIDAZOLE HYDROCHLORIDE see AKS275

N-AMINOMETHYLDOPA see CBQ500

α-AMINO-2-METHYLENE-CYCLOPROPANEPROPANOIC ACID (9CI) see MJP500

α-AMINOMETHYLENECYCLOPROPANEPROPIONIC ACID see MJP500

l-α-AMINO-β-METHYLENECYCLOPROPANEPROPIONIC ACID see MJP500

α-AMINO-β-(2-METHYLENECYCLOPROPYL)PROPIONIC ACID see MJP500

2-AMINO-4,5-METHYLENEHEX-5-ENOIC ACID see MJP500

α-AMINOMETHYL-3-FLUOROBENZYLALCOHOL HYDROBROMIDE see AKS500

4-AMINO-10-METHYLFOLIC ACID see MDV500

2-AMINO-6-METHYLHEPTANE see ILM000

6-AMINO-2-METHYLHEPTANE see ILM000

6-AMINO-2-METHYL-2-HEPTANOL HYDROCHLORIDE see HBB000

4-AMINO-2-METHYL-3-HEXANOL see AKS750

β-(AMINOMETHYL)HYDROCINNAMIC ACID see PEE500

β-(AMINOMETHYL)-HYDROCINNAMIC ACID HYDROCHLORIDE see GAD000

α-(AMINOMETHYL)-m-HYDROXYBENZYL ALCOHOL see AKT000

α-(AMINOMETHYL)-p-HYDROXYBENZYL ALCOHOL see AKT250

α-AMINOMETHYL-3-HYDROXYBENZYLALCOHOL HYDROCHLORIDE see AKT500

(±)-α-(AMINOMETHYL)-m-HYDROXYBENZYL ALCOHOL HYDROCHLORIDE see NNT100

5-AMINOMETHYL-3-HYDROXYISOXAZOLE see AKT750

2-AMINO-1-METHYLIMIDAZO(4,5-b)PYRIDINE see AKT510

2-AMINO-3-METHYLIMIDAZO(4,5-b)PYRIDINE see AKT520

2-AMINO-1-METHYLIMIDAZO(4,5-f)QUINOLINE see AKT530

2-AMINO-3-METHYLIMIDAZO(4,5-f)QUINOLINE see AKT600

2-AMINO-3-METHYLIMIDAZO(4,5-f)QUINOLINE DIHYDROCHLORIDE see AKT620

2-AMINO-3-METHYLIMIDAZO(4,5-f)QUINOXALINE see AKT650

5-(AMINOMETHYL)-3-ISOXAZOLOL see AKT750

5-(AMINOMETHYL)-3(2H)-ISOXAZOLONE see AKT750

4-AMINO-N-(5-METHYL-3-ISOXAZOLYL)BENZENESULFONAMIDE see SNK000

5-AMINOMETHYL-3-ISOXYZOLE see AKT750

l-α-AMINO-γ-METHYLMERCAPTOBUTYRIC ACID see MDT750

4-AMINOMETHYL-5-MERCAPTOMETHYL-2-METHYL-3-PYRIDINOL THIO ACETATE HYDROBROMIDE see ADD000

6-AMINO-3-METHYL-1-(2-METHYLALLYL)-2,4(1H,3H)-PYRIMIDINEDIONE see AKL625

6-AMINO-3-METHYL-1-(2-METHYLALLYL)URACIL see AKL625

1-(4-AMINO-2-METHYL-5-(2-METHYLPHENYL)-1H-PYRROL-3-YL)ETHANONE see AKT800

1-(4-AMINO-2-METHYL-5-(3-METHYLPHENYL)-1H-PYRROL-3-YL)ETHANONE see AKT830

1-(4-AMINO-2-METHYL-5-(4-METHYLPHENYL)-1H-PYRROL-3-YL)ETHANONE see AKT850

6-AMINO-3-METHYL-1-(2-METHYL-2-PROPENYL)-2,4(1H,3H)-PYRIMIDINEDIONE see AKL625

4-AMINO-2-METHYL-1-NAPHTHALENOL see AKX500

2-AMINO-3-METHYLNAPHTHO(1,2-D)IMIDAZOLE see AKT900

2-AMINO-3-METHYLNAPHTHO(2,1-D)IMIDAZOLE see AKT900

4-AMINO-2-METHYL-1-NAPHTHOL see AKX500

1-AMINO-2-METHYL-5-NITROBENZENE see NMP500

3-AMINO-4-METHYL-5-(5-NITRO-2-FURYL)-s-TRIAZOLE see AKY000

2-AMINO-4-((E)-2-(1-METHYL-5-NITRO-1H-IMIDAZOL-2-YL)ETHENYL)PYRIMIDINE see TJF000

2-AMINO-6-(1-METHYL-4-NITRO-5-IMIDAZOLYL)MERCAPTOPURINE see AKY250

2-AMINO-6-(1'-METHYL-4'-NITRO-5'-IMIDAZOLYL)MERCAPTOPURINE see AKY250

(E)-2-AMINO-4-(2-(1-METHYL-5-NITROIMIDAZOL-2-YL)VINYL)PYRIMIDINE see TJF000

2-(AMINOMETHYL)NORBORNANE see AKY750

2-AMINO-2-METHYLOL-1,3-PROPANEDIOL see TEM500

2-AMINO-4-METHYLOXAZOLE see AKY875

2-AMINO-1-METHYL-2-OXOETHYL-N-(((METHYLAMINO)CARBONYL)OXY)ETHANIMIDOTHIOATE see AKY880

2-AMINO-3-METHYLPENTANOIC ACID see IKX000

2-AMINO-4-METHYLPENTANOIC ACID see LES000

8-AMINO-7-METHYL-2-PHENAZINOL REACTION PRODUCTS WITH SODIUM SULFIDE see CMS257

2-AMINO-4-METHYLPHENOL see AKY950

4-AMINO-4-METHYLPHENOL see AKZ000

5-AMINO-2-METHYLPHENOL see AKJ750

2-((4-AMINO-3-METHYLPHENYL)ETHYLAMINO)ETHANOL SULFATE see AKZ100

2-AMINO-1-METHYL-6-PHENYLIMIDAZO(4,5-B)PYRIDINE see AKZ200

2-AMINO-3-METHYL-6-PHENYLIMIDAZO(4,5-b)PYRIDINE see AKZ300

4-(AMINOMETHYL)-1-(PHENYLMETHYL)-2-PYRROLIDINONE (E)-2-BUTENEDIOATE (2:1) see AKR100

AMINO-3'-METHYLPHENYLNORHARMAN see AKZ400

9-(4'-AMINO-3'-METHYLPHENYL)-9H-PYRIDO(3,4-B)INDOLE see AKZ400

cis-2-AMINO-5-METHYL-4-PHENYL-1-PYRROLINE see ALA000

trans-2-AMINO-5-METHYL-4-PHENYL-1-PYRROLINE see ALA250

1-(4-AMINO-2-METHYL-5-PHENYL-1H-PYRROL-3-YL)ETHANONE HYDROCHLORIDE see ALA300

8-AMINO-2-METHYL-4-PHENYL-1,2,3,4-TETRAHYDROISOQUINOLINE see NMV700

8-AMINO-2-METHYL-4-PHENYL-1,2,3,4-TETRAHYDROISOQUINOLINE MALEATE see NMV725

4-AMINO-3-METHYL-6-PHENYL-1,2,4-TRIAZIN-5(4H)-ONE see ALA500

2-AMINO-4-METHYL-5-PHOSPHONO-3-PENTENOIC ACID see ALA550

2-AMINO-4-(N-METHYLPIPERAZINO)-5-METHYLTHIO-6-CHLOROPYRIMIDINE see ALA750

1-AMINO-2-METHYLPROPANE see IIM000

2-AMINO-2-METHYLPROPANE see BPY250

2-AMINO-2-METHYL-1,3-PROPANEDIOL see ALB000

2-AMINO-2-METHYLPROPANOIC ACID see MGB000

2-AMINO-2-METHYLPROPANOL see IIA000

1-AMINO-2-METHYL-2-PROPANOL see ALB250

2-AMINO-2-METHYLPROPAN-1-OL see IIA000

2-AMINO-2-METHYL-1-PROPANOL see IIA000

S-2-AMINO-2-METHYLPROPYL DIHYDROGEN PHOSPHOROTHIOATE see ALB500

S-(2-AMINO-2-METHYLPROPYL)PHOSPHOROTHIOATE see ALB500

(−)-α-(AMINOMETHYL)PROTOCATECHUYL ALCOHOL see NNO500

4-AMINO-N[10]-METHYLPTEROYLGLUTAMIC ACID see MDV500

4-AMINO-N[10]-METHYLPTEROYLGLUTAMIC ACID DISODIUM SALT see MDV600

3-AMINO-6-METHYL-4-PYRIDAZINETHIOL see ALB625

2-AMINOMETHYLPYRIDINE see ALB750

2-AMINO-3-METHYLPYRIDINE see ALC000

2-AMINO-4-METHYLPYRIDINE see ALC250

2-AMINO-5-METHYLPYRIDINE see AMA010

2-AMINO-6-METHYLPYRIDINE see ALC500

2-AMINO-1-METHYLPYRIDINIUM p-TOLUENESULFONATE see ALC600

4-((3-AMINO-4-((4-((1-METHYLPYRIDINIUM-4-YL)AMINO)BENZOYL)AMINO)PHENYL)AMINO)-1-METHYLQUINOLINIUM)DIBROMIDE see ALC750

2-AMINO-3-METHYL-9H-PYRIDO(2,3-b)INDOLE see ALD750

3-AMINO-1-METHYL-5H-PYRIDO(4,3-b)INDOLE see ALD500

3-AMINO-7-METHYL-5H-PYRIDO(4,3-B)INDOLE see ALD600

3-AMINO-1-METHYL-5H-PYRIDO(4,3-b)INDOLE ACETATE see ALE750

4-AMINO-N-(4-METHYL-2-PYRIMIDINYL)-BENZENESULFONAMIDE see ALF250

4-AMINO-N-(4-METHYL-2-PYRIMIDINYL)-BENZENESULFONAMIDE MONOSODIUM SALT see SJW475

1-(4-AMINO-2-METHYLPYRIMIDIN-5-YL)METHYL-3-(2-CHLOROETHYL)-3-NITROSOUREA see ALF500

3-((4-AMINO-2-METHYL-5-PYRIMIDINYL)METHYL)-1-(2-CHLOROETHYL)-1-NITROSOUREA HYDROCHLORIDE see ALF500

3-(4-AMINO-2-METHYL-5-PYRIMIDINYL)METHYL-1-(2-CHLOROETHYL)-1-NITROSOUREA see NDY800

N'-((4-AMINO-2-METHYL-5-PYRIMIDINYL)METHYL)-N-(2-CHLOROETHYL)-N-NITROSOUREA HCl see ALF500

3-((4-AMINO-2-METHYL-5-PYRIMIDINYL)METHYL)-5-(2-HYDROXYETHYL)-4-METHYLTHIAZOLIUM CHLORIDE see TES750

3-(4-AMINO-2-METHYLPYRIMIDYL-5-METHYL)-4-METHYL-5,β-HYDROXYETHYLTHIAZOLIUM NITRATE see TET500

4-AMINO-N-METHYL-1,2,5-SELENADIAZOLE-3-CARBOXAMIDE see MGL600

(+−)-2-AMINO-4-(METHYLSELENO)BUTANOIC ACID see SBU800

2-AMINO-4-(METHYLSELENYL)BUTYRIC ACID see SBU725

2-AMINO-4-(METHYLSULFINYL)BUTYRIC ACID see ALF600

2-AMINO-4-(S-METHYLSULFONIMIDOYL)-BUTANOIC ACID (9CI) see MDU100

4-AMINO-6-METHYL-1H-2,5,10,10B-TETRAAZAFLUORANTHENE see ALF650

2-AMINOMETHYLTETRAHYDROPYRAN see ALF750

2-AMINO-N-(3-METHYL-2-THIAZOLIDINYLIDENE)ACETAMIDE see ALG250

2-AMINO-4-(METHYLTHIO)BUTYRIC ACID see MDT750

l(−)-AMINO-γ-METHYLTHIOBUTYRIC ACID see MDT750

5-AMINO-3-METHYLTHIO-1,2,4-OXADIAZOLE see ALG375

4-AMINO-3-METHYLTOLUENE see XMS000

2-AMINO-4-METHYLTOLUENE HYDROCHLORIDE see XOS000

4-AMINO-3-METHYLTOLUENE HYDROCHLORIDE see XOJ000

6-AMINO-2-METHYL-3-(o-TOLYL)-1(3H)-QUINAZOLINONE see AKM125

α-AMINOMETHYL-m-TRIFLUOROMETHYLBENZYL ALCOHOL see ALG500

2-AMINO-4-METHYLVALERIC ACID see LES000

l,2-AMINO-4-METHYLVALERIC ACID see LES000

dl-2-AMINO-3-METHYLVALERIC ACID see IKX010

dl-2-AMINO-4-METHYLVALERIC ACID see LER000

α-AMINO-β-METHYLVALERIC ACID see IKX000

α-AMINO-γ-METHYLVALERIC ACID see LES000

AMINOMETRADINE see AFW500

AMINOMETRAMIDE see AFW500

5-AMINOMITONAFIDE see NAC600

AMINOMIX see AHR600

AMINOMONOPERSULFURIC ACID see HLN100

2-AMINO-6-MP see AMH250

1-AMINONAFTALEN (CZECH) see NBE700

2-AMINONAFTALEN (CZECH) see NBE500

1-AMINONAPHTHALENE see NBE700

2-AMINONAPHTHALENE see NBE500

3-AMINO-2-NAPHTHALENECARBOXYLIC ACID see ALJ000

2-AMINO-1,5-NAPHTHALENEDISULFONIC ACID see ALH000

2-AMINO-4,8-NAPHTHALENEDISULFONIC ACID see ALH250

3-AMINO-1,5-NAPHTHALENEDISULFONIC ACID see ALH250

6-AMINO-NAPHTHALENE-1,3-DISULFONIC ACID see ALH500

7-AMINO-1,3-NAPHTHALENEDISULFONIC ACID see NBE850

7-AMINO-1,5-NAPHTHALENEDISULFONIC ACID see ALH250

1-AMINONAPHTHALENE-4-SULFONIC ACID see ALI000

1-AMINO-6-NAPHTHALENESULFONIC ACID see ALI250

2-AMINO-1-NAPHTHALENESULFONIC ACID see ALH750

4-AMINO-1-NAPHTHALENESULFONIC ACID see ALI000

5-AMINO-1-NAPHTHALENESULFONIC ACID see ALI240

5-AMINO-2-NAPHTHALENESULFONIC ACID see ALI250

7-AMINO-1-NAPHTHALENESULFONIC ACID see ALI300

5-AMINO-2-NAPHTHALENESULFONIC ACID SODIUM SALT see ALI500

7-AMINO-1,3,6-NAPHTHALENETRISULFONIC ACID see ALI750

AMINONAPHTHALENOL see ALJ250
3-AMINO-2-NAPHTHOIC ACID see ALJ000
4-AMINO-1-NAPHTHOIC ACID 2-(DIETHYLAMINO)ETHYL ESTER HYDROCHLORIDE see NAH800
2-AMINO-1H-NAPHTHO(2,3-d)IMIDAZOLE see ALJ100
2-AMINO-3H-NAPHTHO(3,2-d)IMIDAZOLE see ALJ100
2-AMINO-1-NAPHTHOL see ALJ250
5-AMINO-2-NAPHTHOL see ALJ500
8-AMINO-2-NAPHTHOL see ALJ750
1-AMINO-2-NAPHTHOL-3,6-DISULPHONIC ACID SODIUM SALT see AKH250
1-AMINO-2-NAPHTHOL HYDROCHLORIDE see ALK000
1-AMINO-4-NAPHTHOL HYDROCHLORIDE see ALK500
2-AMINO-1-NAPHTHOL HYDROCHLORIDE see ALK250
4-AMINO-1-NAPHTHOL HYDROCHLORIDE see ALK500
2-AMINO-1-NAPHTHOL PHOSPHATE (ESTER) SODIUM SALT see BGU000
AMINONAPHTHOL SULFONIC ACID J see AKI000
AMINONAPHTHOL SULFONIC ACID S see AKH750
2-AMINO-1,4-NAPHTHOQUINONE IMINE HYDROCHLORIDE see ALK625
2-AMINO-1-NAPHTHYL ESTER SULFURIC ACID see ALK750
2-AMINO-1-NAPHTHYLGLUCOSIDURONIC ACID see ALL000
2-AMINO-1-NAPHTHYL HYDROGEN SULFATE see ALK750
2-AMINO-1-NAPHTHYL HYDROGEN SULPHATE see ALK750
2-AMINO-N^6-HYDROXY-2'-DEOXYADENOSINE see AJL200
AMINONICOTINAMIDE see ALL250
6-AMINONICOTINAMIDE see ALL250
6-AMINONICOTINIC ACID AMIDE see ALL250
6-AMINONICOTINOHYDROXAMIC ACID see ALL300
6-AMINO-NICOTINSAEUREAMID (GERMAN) see ALL250
6-AMINONIKOTINSAEUREAMID (GERMAN) see ALL250
7-AMINONITRAZEPAM see AJO280
2-AMINO-4-NITROANILINE see ALL500
4-AMINO-2-NITROANILINE see ALL750
2-AMINO-5-NITROANISOL (CZECH) see NEQ000
2-AMINO-4-NITROANISOLE see NEQ500
2-AMINO-5-NITROANISOLE see NEQ000
3-AMINONITROBENZENE see IEB000
m-AMINONITROBENZENE see NEN500
p-AMINONITROBENZENE see NEO500
1-AMINO-2-NITROBENZENE see NEO000
1-AMINO-3-NITROBENZENE see NEN500
1-AMINO-4-NITROBENZENE see NEO500
2-AMINO-5-NITRO BENZENESULFONIC ACID see NEP500
2-AMINO-5-NITROBENZENESULFONIC ACID AMMONIUM SALT see ALL800
2-AMINO-4-NITRO-BENZOIC ACID see NES500
4-AMINO-4'-NITROBIPHENYL see ALM000
4-AMINO-4'-NITRO-3-BIPHENYLOL HYDROCHLORIDE see AKI500
4-AMINO-3-NITRO-6-CHLOROANILINE see ALM100
4-AMINO-3-NITRO-2,5-DIMETHYLANILINE see ALM120
4-AMINO-3-NITRO-5,6-DIMETHYLANILINE see ALM140
4-AMINO-2-NITRODIPHENYLAMINE see NEM350

4-AMINO-4'-NITRODIPHENYL SULFIDE see AOS750
4-AMINO-3-NITRO-6-FLUOROANILINE see FKL500
2-AMINO-5-(2-(5-NITRO-2-FURYL)-1-(2-FURYL)-VINYL)-1,3,4-OXADIAZOLE see FPI200
2-AMINO-5-(5-NITRO-2-FURYL)-1,3,4-OXADIAZOLE see ALM250
2-AMINO-5-(5-NITRO-2-FURYL)-1,3,4-THIADIAZOLE see NGI500
5-AMINO-2-(5-NITRO-2-FURYL)-1,3,4-THIADIAZOLE see NGI500
2-AMINO-4-(5-NITRO-2-FURYL)THIAZOLE see ALM500
5-AMINO-3-(5-NITRO-2-FURYL)-s-TRIAZOLE see ALM750
3-AMINO-6-(2-(5-NITRO-2-FURYL)VINYL))PYRIDAZINE HYDROCHLORIDE see ALN250
2-AMINO-4-(2-(5-NITRO-2-FURYL)VINYL)THIAZOLE see ALN500
3-AMINO-6-(2-(5-NITRO-2-FURYL)VINYL)-as-TRIAZINE see FPF000
3-AMINO-6-(2-(5-NITRO-2-FURYL)VINYL)-1,2,4-TRIAZINE see FPF000
1-AMINO-3-NITRO GUANIDINE see ALN750
4-AMINO-3-NITRO-5-β-HYDROXYETHYLANILINE see ALN800
4-AMINO-3-NITRO-6-METHOXYANILINE see MFB300
4-AMINO-3-NITRO-5-METHYLANILINE see MMG210
4-AMINO-3-NITRO-6-METHYLANILINE see MMG200
2-AMINO-4-NITROPHENOL see NEM500
2-AMINO-5-NITROPHENOL see ALO000
4-AMINO-2-NITROPHENOL see NEM480
2-AMINO-4-NITROPHENOL SODIUM SALT see ALO500
2-((4-AMINO-2-NITROPHENYL)AMINO)ETHANOL see ALO750
2-AMINO-4-(p-NITROPHENYL)THIAZOLE see ALP000
l-2-AMINO-3-((N-NITROSO)HYDROXYLAMINO)PROPIONIC ACID see AFH750
4-AMINO-4'-NITRO-2,2'-STILBENEDISULFONIC ACID see ALP750
AMINONITROTHIAZOLE see ALQ000
2-AMINO-5-NITROTHIAZOLE see ALQ000
AMINONITROTHIAZOLUM see ALQ000
2-AMINO-4-NITROTOLUENE see NMP500
3-AMINO-4-NITROTOLUENE see NMP550
4-AMINO-2-NITROTOLUENE see NMP000
3-AMINONORHARMAN see ALQ100
AMINONUCLEOSIDE see ALQ625
AMINONUCLEOSIDE PUROMYCIN see ALQ625
1-AMINOOCTADECANE see OBC000
2-AMINOOCTANE see OEK010
l-α-AMINO-4(OR 5)-IMIDAZOLEPROPIONIC ACID see HGE700
l-α-AMINO-4(OR 5)-IMIDAZOLEPROPIONIC ACID MONOHYDROCHLORIDE see HGE800
AMINOOXAMIDE see OLM310
N-AMINOOXAMIDE see OLM310
4-AMINO-β-OXOBENZENEPROPANENITRILE see ALQ635
1-(4-AMINO-4-OXO-3,3-DIPHENYLBUTYL)-1-METHYLPIPERIDINIUM BROMIDE (9CI) see RDA375
(4-((2-AMINO-2-OXOETHYL)AMINO)PHENYL)ARSONIC ACID see CBJ750
N-(2-AMINO-2-OXOETHYL)-2-DIAZOACETAMIDE see CBK000

2-AMINO-2-OXOETHYL-2,2-DIMETHYL-N-(((METHYLAMINO)CARBONYL)OXY)PROP ANIMIDOTHIOATE see ALQ640
2-AMINO-2-OXOETHYL N-(((METHYLAMINO)CARBONYL)OXY)ETH ANIMIDOTHIOATE see ALQ642
α-((2-AMINO-1-OXOPROPYL)AMINO)-5-OXO-7-OXABICYCLO(4.1.0)HEPTANE-2-PROPANOIC ACID see BAC175
N-(p-(((2-AMINO-4-OXO-6-PTERIDINYL)METHYL)-N-NITROSOAMINO)BENZOYL)-l-GLUTAMIC ACID see NLP000
AMINOOXYACETIC ACID see ALQ650
2-AMINOOXYACETIC ACID BUTYL ESTER, HYDROCHLORIDE see ALQ750
1-AMINO-4-OXYANTHRAQUINONE see AKE250
α-(AMINOOXY)-6-BROMO-m-CRESOL see BMM600
AMINOOXYTRIPHENE HYDROCHLORIDE see TNJ750
AMINOPAR see AMM250
6-AMINOPENICILLANIC ACID see PCU500
d-(−)-α-AMINOPENICILLIN see AIV500
AMINOPENTAFLUOROBENZENE see PBD250
2-AMINO-3,4,5,7,8-PENTAMETHYLIMIDAZO(4,5-f)QUINOXALINE see PBI600
AMINOPENTAMIDE see DOY400
dl-AMINOPENTAMIDE HYDROCHLORIDE see ALR250
1-AMINOPENTANE see PBV505
2-AMINOPENTANE see DHJ200
2-AMINOPENTANEDIOIC ACID see GFO000
2-AMINOPENTANEDIOIC ACID HYDROCHLORIDE see GFO025
3-(5-AMINOPENTYL)INDOLE ADIPATE see ALR500
1-AMINOPERHYDROAZEPINE see HEG400
AMINOPERIMIDINE see ALR750
4-AMINO-PGA see AMG750
AMINOPHEN see AOQ000
1-AMINOPHENANTHRENE see ALS000
2-AMINOPHENANTHRENE see PDA500
3-AMINOPHENANTHRENE see PDA500
9-AMINOPHENANTHRENE see PDA750
1-AMINOPHENAZINE see PDB400
3-AMINO-2-PHENAZINOL see ALS100
AMINOPHENAZONE see AIB300
AMINOPHENAZONE see DOT000
4-AMINOPHENAZONE see AIB300
AMINOPHENAZONE mixed with SODIUM NITRITE (1:1) see DOT200
17-(p-AMINOPHENETHYL)-MORPHINAN-3-OL (−)- see ALS250
(−)-17-(m-AMINOPHENETHYL)-MORPHINAN-3-OL, HYDROCHLORIDE see ALS500
(±)-17-(p-AMINOPHENETHYL)MORPHINAN-3-OL, HYDROCHLORIDE see ALS750
1-(p-AMINOPHENETHYL)-4-PHENYLISONIPECOTIC ACID, ETHYL ESTER see ALW750
1-(p-AMINOPHENETHYL)-4-PHENYLPIPERIDINE-4-CARBOXYLIC ACID ETHYL ESTER see ALW750
2-AMINOPHENETOLE see PDK819
4-AMINOPHENETOLE see PDK790
p-AMINOPHENETOLE see PDK790
2-AMINOPHENOL see ALT000
3-AMINOPHENOL see ALS990
4-AMINOPHENOL see ALT250
m-AMINOPHENOL see ALS990
o-AMINOPHENOL see ALT000
m-AMINOPHENOL (DOT) see ALS990
p-AMINOPHENOL (DOT) see ALT250
m-AMINOPHENOL, chlorinated see ALT550

m-AMINOPHENOL ANTIMONYL TARTRATE see ALT750

o-AMINOPHENOL ANTIMONYL TARTRATE see ALU000

p-AMINOPHENOL ANTIMONYL TARTRATE see ALU250

2-AMINOPHENOL-4-ARSONIC ACID DIETHYLAMINE SALT see ACN250

4-AMINOPHENOL HYDROCHLORIDE see ALU500

p-AMINOPHENOL HYDROCHLORIDE see ALU500

o-AMINOPHENOL-OXO(TARTRATO)ANTIMONATE(1-)- see ALU000

p-AMINOPHENOL-OXO(TARTRATO)ANTIMONATE(1-)- see ALU250

4-AMINOPHENOL PHOSPHATE (3:1) (ester) see THO550

p-AMINOPHENOL PHOSPHATE (3:1) (ester) see THO550

p-AMINOPHENOL TARTRATE see ALU750

2-AMINO-3H-PHENOXAZIN-3-ONE see QCJ275

2-AMINOPHENOXAZON see QCJ275

2-AMINOPHENOXAZONE see QCJ275

1-(p-AMINOPHENOXY)-3-(N¹)-(o-METHOXYPHENYL)-N⁴-PIPERAZINYL)PROPANE 2HCl see ALU875

1-(3-(p-AMINOPHENOXY)PROPYL)-4-(o-METHOXYPHENYL)PIPERAZINE DIHYDROCHLORIDE see ALU875

AMINOPHENUROBUTANE see BSM000

N-(p-AMINOPHENYL)ACETAMIDE see AHQ250

7-(d-α-AMINOPHENYL-ACETAMIDO)CEPHALOSPORANIC ACID see CCR890

7-(d-α-AMINOPHENYLACETAMIDO)DESACETOXYCEPHALOSPORANIC ACID see ALV000

7-(d-2-AMINO-2-PHENYLACETAMIDO)-3-METHYL-Δ₃-CEPHEM-4-CARBOXYLIC ACID see ALV000

6-(d(−)-α-AMINOPHENYLACETAMIDO)PENICILLANIC ACID see AIV500

6-(d-α-AMINO PHENYL ACETAMIDO) PENICILLANIC ACID PIVALOYL OXY METHYL ESTER HYDROCHLORIDE see AOD000

4'-(p-AMINOPHENYL)ACETANILIDE see ACC000

4-AMINOPHENYLACETIC ACID see AID700

(p-AMINOPHENYL)ACETIC ACID see AID700

(2S-(2-α,5-α,6-β(S*)))-6-((AMINOPHENYLACETYL)AMINO)-3,3-DIMETHYL-7-OXO-4-THIA-1-AZABICYCLO(3.2.0)HEPTANE-2-CARBOXYLIC ACID mixt. with (2S-(2-α,5-α,6-β))-3,3-DIMETHYL-6-(((5-METHYL-3-PHENYL-4-ISOXAZOLYL)CARBONYL)AMINO)-7-OXO-4-THIA-1-AZABICYCLO(3.2.0)HEPTANE-2-CARBOXYLIC ACID see AOC875

2-(4-AMINOPHENYL)-5-AMINOBENZIMIDAZOLE see ALV050

AMINOPHENYLARSINE ACID see ARA250

p-AMINOPHENYLARSINE ACID see ARA250

p-AMINOPHENYLARSINE OXIDE DIHYDRATE see ALV100

p-AMINOPHENYLARSINIC ACID see ARA250

4-AMINOPHENYLARSONIC ACID see ARA250

(4-AMINOPHENYL)ARSONIC ACID SODIUM SALT see ARA500

p-((p-AMINOPHENYL)AZO)BENZENESULFONIC ACID see AIB350

4-((4-AMINOPHENYL)AZO)-N,N-DIMETHYLBENZENAMINE see DPO200

β-AMINO-α-PHENYLBENZENEETHANOL HDYROCHLORIDE see DWB000

2-(4-AMINOPHENYL)-1H-BENZIMIDAZOL-5-AMINE see ALV050

5-AMINO-2-PHENYLBENZOTHIAZOLE see ALV500

5-AMINO-3-PHENYL-1-BIS (DIMETHYL-AMINO)-PHOSPHORYLE-1,2,4-TRIAZOLE (FRENCH) see AIX000

5-AMINO-3-PHENYL-1-BIS(DIMETHYLAMINO)-PHOSPHORYL-1H-1,2,4-TRIAZOL (GERMAN) see AIX000

4-AMINO-3-PHENYLBUTANOIC ACID see PEE500

1-(4-AMINOPHENYL)-1-BUTANONE see AJC750

4-AMINO-3-PHENYLBUTYRIC ACID see PEE500

1-AMINO-2-PHENYLCYCLOPROPANE SULFATE see PET500

1-m-AMINOPHENYL-2-CYCLOPROPYLAMINOETHANOLDIHYDROCHLORIDE see ALV750

p-AMINOPHENYL DERIVATIVE of NITROGEN MUSTARD see BIG250

1-(4-AMINOPHENYL)-4-(DIETHYLCARBOXAMIDE)-5-METHYL-1,2,3-TRIAZOLE HYDROCHLORIDE see ALW000

4-AMINOPHENYL DISULFIDE see ALW100

1-AMINO-1-PHENYLETHANE see ALW250

1-AMINO-2-PHENYLETHANE see PDE250

2-AMINO-1-PHENYL-1-ETHANOL see HNF000

3-(2-(4-AMINOPHENYL)ETHENYL)BENZENAMINE see DCD100

4-AMINOPHENYL ETHER see OPM000

p-AMINOPHENYL ETHER see OPM000

p-AMINOPHENYL ETHYL ETHER HYDROCHLORIDE see PDL750

2-(p-AMINOPHENYL)-2-ETHYLGLUTARIMIDE see AKC600

2-(p-AMINOPHENYL)-2-ETHYLGLUTARIMIDE PHOSPHATE see AKC625

α-(p-AMINOPHENYL)-α-ETHYLGLUTARIMIDE PHOSPHATE see AKC625

5-(p-AMINOPHENYL)-5-ETHYL-1-METHYLBARBITURIC ACID see ALW500

N-β-(p-AMINOPHENYL)ETHYLNORMEPERIDINE see ALW750

N-(β-(p-AMINOPHENYL)ETHYL)-4-PHENYL-4-CARBETHOXYPIPERIDINE see ALW750

2,2'-((4-AMINOPHENYL)IMINO)BISETHANOL SULFATE HYDRATE see BKF800

4-((4-AMINOPHENYL)(4-IMINO-2,5-CYCLOHEXADIEN-1-YLIDENE)METHYL)-2-METHYLB ENZENAMINE see MAC500

4-((4-AMINOPHENYL)(4-IMINO-2,5-CYCLOHEXADIEN-1-YLIDENE)METHYL)-2-METHYLBENZENAMINE see RMK000

4-((4-AMINOPHENYL)(4-IMINO-2,5-CYCLOHEXADIEN-1-YLIDENE)METHYL)-2-METHYLB ENZENAMINE HCl see MAC250

4-((4-AMINOPHENYL)(4-IMINO-2,5-CYCLOHEXADIEN-1-YLIDENE)METHYL), MONOCHLORIDE see RMK020

4-((p-AMINOPHENYL)(4-IMINO-2,5-CYCLOHEXADIEN-1-YLIDENE)METHYL)-o-TOLUIDINE see RMK000

2-(m-AMINOPHENYL)-3-INDOLECARBOXALDEHYDE, 4-(m-TOLYL)-3-THIOSEMICARBAZONE see ALW900

p-AMINOPHENYL IODIDE see IEC000

p-AMINOPHENYLMERCAPTAN see AIF750

p-AMINOPHENYLMERCURIC ACETATE see ABQ000

3-AMINOPHENYLMETHANE see TGQ500

1-(p-AMINOPHENYL)-2-METHYLAMINOPROPAN (GERMAN) see AJR000

1-(p-AMINOPHENYL)-2-METHYLAMINOPROPANE see AJR000

α-(4-AMINOPHENYL)-β-METHYLAMINO-PROPANE see AJR000

2-(p-AMINOPHENYL)-6-METHYLBENZOTHIAZOLE see MHJ300

2-(p-AMINOPHENYL)-6-METHYLBENZOTHIAZOLYL-7-SULFONIC ACID see ALX000

(AMINOPHENYLMETHYL)-PENICILLIN see AIV500

m-AMINOPHENYL METHYL SULFIDE see ALX100

p-AMINOPHENYL METHYL SULFIDE see AMS675

AMINOPHENYLNORHARMAN see ALX120

2-AMINO-5-PHENYL-OXAZOLINE FORMATE see ALX250

2-AMINOPHENYL PHENYL ETHER see PDR490

4-AMINOPHENYL PHENYL ETHER see PDR500

o-AMINOPHENYL PHENYL ETHER see PDR490

p-AMINOPHENYL PHENYL ETHER see PDR500

2-(p-AMINOPHENYL)-2-PHENYLPROPIONAMIDE see ALX500

1-AMINO-2-PHENYLPROPANE see PGB760

d-2-AMINO-1-PHENYLPROPANE see AOA500

(±)-2-AMINO-1-PHENYLPROPANE SULFATE see AOB250

2-AMINO-1-PHENYL-1-PROPANOL see NNM000

2-AMINO-1-PHENYL-1-PROPANOL HYDROCHLORIDE see NNN000

(±)-2-AMINO-1-PHENYL-1-PROPANOL HYDROCHLORIDE see PMJ500

1-(4-AMINOPHENYL)-1-PROPANONE see AMC000

α-AMINO-β-PHENYLPROPIONIC ACID see PEC750

dl-α-AMINO-β-PHENYLPROPIONIC ACID see PEC500

3-AMINO-1-PHENYL-2-PYRAZOLINE see ALX750

4-AMINO-N-(1-PHENYL-1H-PYRAZOL-5-YL)BENZENESULFONAMIDE see AIF000

1-(4-AMINO-5-PHENYL(1H)-PYRAZOL-3-YL)ETHANONE see ALX879

4-AMINO-5-PHENYL-3-PYRAZOLYL METHYL KETONE see ALX879

2-AMINO-5-PHENYLPYRIDINE see ALY000

1-(3-AMINOPHENYL)-2-(1H)-PYRIDINONE see ALY250

9-(4'-AMINOPHENYL)-9H-PYRIDO(3,4-B)INDOLE see ALX120

3-AMINO-1-PHENYL-5H-PYRIDO(4,3-B)INDOLE ACETATE see ALY100

AMINOPHENYLPYRIDONE see ALY250

1-(3-AMINOPHENYL)-2-PYRIDONE see ALY250

1-m-AMINOPHENYL-2-PYRIDONE see ALY250

1-(m-AMINOPHENYL)-2(1H)-PYRIDONE see ALY250

4-AMINOPHENYLSULFONAMIDE see SNM500

1-(6-AMINO-9H-PURIN-9-YL)-1-DEOXY-N-ETHYLRIBOFURANURONAMIDE HEMIHYDRATE see AEK250

1-(6-AMINO-9H-PURIN-9-YL)-1-DEOXY-N-ETHYLRIBOFURANURONAMIDE-N-OXIDE see AEK500

1-(6-AMINO-9H-PURIN-9-YL)-1-DEOXY-N-HEXYLRIBOFURANURONAMIDE HEMIHYDRATE see AEK750

1-(6-AMINO-9H-PURIN-9-YL)-1-DEOXY-N-(2-HYDROXYETHYL)RIBOFURANURONAMIDE see AEL000

1-(6-AMINO-9H-PURIN-9-YL)-1-DEOXY-N-ISOPROPYLRIBOFURANURONAMIDE see AEL250

1-(6-AMINO-9H-PURIN-9-YL)-1-DEOXY-N-METHOXYRIBOFURANURONAMIDE HYDRATE see AEL500

1-(6-AMINO-9H-PURIN-9-YL)-1-DEOXY-N-METHYLRIBOFURANURONAMIDE HEMIHYDRATE see AEL750

1-(6-AMINO-9H-PURIN-9-YL)-1-DEOXY-N-PROPYLRIBOFURANURONAMIDE see AEM000

β-d-1-(6-AMINO-9H-PURIN-9-YL)-1-DEOXYRIBOFURANURONAMIDE see AEI250

1-(6-AMINO-9H-PURIN-9-YL)-N-(2-(DIMETHYLAMINO)-ETHYL-1-DEOXYRIBOFURANURONAMIDE) see AEJ750

((2-(6-AMINO-9H-PURIN-9-YL)ETHOXY)METHYL)PHOSPHONIC ACID see PHA700

(S)-3-(6-AMINO-9H-PURIN-9-YL)-1,2-PROPANEDIOL see AMH800

(R,S)-3-(6-AMINO-9H-PURIN-9-YL)-1,2-PROPANEDIOL see AMH850

4-AMINOPYRAZOLOPYRIMIDINE see POM600

4-AMINOPYRAZOLO(3,4-d)PYRIMIDINE see POM600

1-AMINOPYRENE see PON000

3-AMINOPYRENE see PON000

AMINOPYRIDINE see POP100

2-AMINOPYRIDINE see AMI000

AMINO-2-PYRIDINE see AMI000

3-AMINOPYRIDINE see AMI250

AMINO-3-PYRIDINE see AMI250

4-AMINOPYRIDINE see AMI500

AMINO-4-PYRIDINE see AMI500

o-AMINOPYRIDINE see AMI000

p-AMINOPYRIDINE see AMI500

m-AMINOPYRIDINE (DOT) see AMI250

α-AMINOPYRIDINE see AMI000

γ-AMINOPYRIDINE see AMI500

3-AMINO-PYRIDINE-2-CARBOXALDEHYDE see AMI600

3-AMINOPYRIDINE-2-CARBOXALDEHYDE THIOSEMICARBAZONE see AMI600

3-AMINOPYRIDINE HYDROCHLORIDE see AMI750

4-AMINOPYRIDINE HYDROCHLORIDE see AMJ000

4-AMINO-PYRIDINEN-OXIDE see AMJ250

4-AMINOPYRIDINE-1-OXIDE see AMJ250

2-((3-AMINO-2-PYRIDINYL)METHYLENE)HYDRAZINECARBOTHIOAMIDE see AMI600

4-AMINO-N-(2-(4-(2-PYRIDINYL)-1-PIPERAZINYL)ETHYL)BENZAMIDE see AMJ500

5-AMINO-5-(4-PYRIDINYL)-2(1H)-PYRIDINONE see AOD375

2-AMINO-9H-PYRIDO(2,3-B)INDOLE see AJD750

3-AMINO-5H-PYRIDO(4,3-b)INDOLE see AMJ600

3-AMINO-9H-PYRIDO(3,4-B)INDOLE see ALQ100

2-AMINO-5-(4-PYRIDYL)-1,3,4-THIADIAZOLEHYDROCHLORIDE see AMJ625

2-AMINOPYRIMIDINE see PPH300

4-AMINO-2(1H)-PYRIMIDINONE see CQM600

4-AMINO-N-2-PYRIMIDINYLBENZENESULFONAMIDE see PPP500

4-AMINO-N-2-PYRIMIDINYL-BENZENESULFONAMIDE MONOSILVER(1+) SALT see SNI425

4-AMINO-N-(2-PYRIMIDINYL)BENZENESULFONAMIDE SILVER SALT see SNI425

2-(2-AMINO-4-PYRIMIDINYLVINYL)QUINOXALINE-N,N'-DIOXIDE see AMJ750

AMINOPYRINE see DOT000

AMINOPYRINE see POP100

AMINOPYRINE-BARBITAL see AMK250

AMINOPYRINE mixed with SODIUM NITRITE (1:1) see DOT200

AMINOPYRINE SODIUM SULFONATE see AMK500

AMINOQUIN see RHZ000

2-AMINOQUINOLINE see AMK700

3-AMINOQUINOLINE see AMK725

8-AMINOQUINOLINE see AML250

4-AMINOQUINOLINE-1-OXIDE see AML500

2-AMINO-4-((2-QUINOXALINYL-N,N'-DIOXIDE)VINYL)PYRIMIDINES see AMJ750

AMINO REDUCTONE see AJL125

2-AMINORESORCINOL HYDROCHLORIDE see AML600

AMINOREXFUMARATE see ALX250

6-AMINO-9-β-d-RIBOFURANOSYL-9H-PURINE see AEH750

2-AMINO-9-β-d-RIBOFURANOSYLPURINE-6,8(1H,9H)-DIONE see ONW150

2-AMINO-9-(β-d-RIBOFURANOSYL)PURINE-6-THIOL see TFJ500

2-AMINO-9-β-d-RIBOFURANOSYL-9H-PURINE-6-THIOL see TFJ500

4-AMINO-1-β-d-RIBOFURANOSYL-2(1H)-PYRIMIDINONE see CQM500

4-AMINO-7-(β-d-RIBOFURANOSYL)-PYRROLO(2,3-D)PYRIMIDINE see TNY500

4-AMINO-7-β-d-RIBOFURANOSYL-7H-PYRROLO(2,3-D)PYRIMIDINE see TNY500

4-AMINO-7-β-d-RIBOFURANOSYL-7H-PYRROLO(2,3-d)PYRIMIDINE-5-CARBONITRILE see VGZ000

4-AMINO-7-β-d-RIBOFURANOSYL-7H-PYRROLO(2,3-d)PYRIMIDINE-5-CARBOXAMIDE see SAU000

2-AMINO-9-β-d-RIBOFURANOSYL-6-SELENO-OH-PURIN-6(1H)-ONE see SBU700

4-AMINO-1-β-d-RIBOFURANOSYL-d-TRIAZIN-2(1H)-ONE see ARY000

4-AMINO-1-β-d-RIBOFURANOSYL-1,3,5-TRIAZIN-2(1H)-ONE see ARY000

5-AMINO-2-β-d-RIBOFURANOSYL-as-TRIAZIN-3(2H)-ONE see AMM000

AMINO-S ACID see AMM125

p-AMINOSALICYLATE SODIUM see SEP000

AMINOSALICYLIC ACID see AMM250

4-AMINOSALICYLIC ACID see AMM250

5-AMINOSALICYLIC ACID see AMM500

m-AMINOSALICYLIC ACID see AMM500

p-AMINOSALICYLIC ACID see AMM250

4-AMINOSALICYLIC ACID-2-(DIETHYLAMINO)ETHYL ESTER HYDROCHLORIDE see AMM750

p-AMINOSALICYLIC ACID, 2-(DIETHYLAMINO)ETHYL ESTER, HYDROCHLORIDE see AMM750

p-AMINOSALICYLIC ACID, 2-(DIMETHYLAMINO)ETHYL ESTER HYDROCHLORIDE see AMN000

p-AMINOSALICYLIC ACID HYDRAZIDE see AMN250

p-AMINOSALICYLIC ACID SODIUM SALT see SEP000

5-AMINOSALICYLIC ACID-o-SULFATE SODIUM SALT see AMN275

p-AMINOSALICYLSAEURE (GERMAN) see AMM500

p-AMINOSALICYLSAEUREDIAETHYLAMINOAETHYLESTER-CHLORHYDRAT (GERMAN) see AMM750

p-AMINOSALICYLSAEURES SALZ (GERMAN) see TMJ750

4-AMINO-1,2,5-SELENADIAZOLE-3-CARBOXAMIDE see AMN300

2-AMINOSELENOAZOLIN (GERMAN) see AMN500

2-AMINOSELENOAZOLINE see AMN500

4-AMINOSEMICARBAZIDE see CBS500

AMINOSIDIN see NCF500

AMINOSIDINE SULFATE see APP500

AMINOSIDINE SULPHATE see APP500

AMINOSIDIN SULFATE see APP500

AMINOSIN see AGT500

AMINOSTERIL KE see AHR600

4-AMINOSTILBENE see SLQ900

p-AMINOSTILBENE see SLQ900

trans-4-AMINOSTILBENE see AMO000

4-AMINO-2-STILBENECARBONITRILE see COS750

2-(p-AMINOSTYRYL)-6-(p-ACETYLAMINOBENZOYLAMINO)QUINOLINE METHOACETATE see AMO250

(AMINOSUBERIC ACID 1,7)-EEL CALCITONIN see CBR300

l-α-AMINOSUCCINAMIC ACID see ARN810

l-AMINOSUCCINIC ACID see ARN850

dl-AMINOSUCCINIC ACID see ARN830

o-AMINOSULFANILIC ACID see PFA250

1-AMINO-2-SULFO-4-(4'-AMINO-3'-SULFOANILINO)ANTHRAQUINONE see SOB600

2,(4'-AMINO-3'-SULFO-1,1'-BIPHENYL-4-YL)-2H-NAPHTHO(1,2,4)TRIAZOLE-6,8-DISULFONIC ACID, TRIPOTASSIUM SALT see AMO750

6-AMINO-5-SULFOMETHYL-2-NAPHTHALENESULFONIC ACID see AMP000

1-AMINO-4-SULFONAPHTHALENE see ALI000

1-AMINO-6-SULFONAPHTHALENE see ALI250

AMINOSULFONIC ACID see SNK500

4-(AMINOSULFONYL)BENZOIC ACID see SNL890

3-(AMINOSULFONYL)-5-(BUTYLAMINO)-4-PHENOXY-3-(AMINOSULFONYL)-5-(BUTYLAMINO)-4-PHENOXYBENZOIC ACID see BON325

3-(AMINOSULFONYL)-4-CHLORO-N-(2,3-DIHYDRO-2-METHYL-1H-INDOL-1-YL)-BENZAMIDE (9CI) see IBV100

5-(AMINOSULFONYL)-4-CHLORO-N-(2,6-DIMETHYLPHENYL)-2-HYDROXY BENZAMIDE (9CI) see CLF325

5-(AMINOSULFONYL)-N-((1-ETHYL-2-PYRROLIDINYL)METHYL)-2-METHOXYBENZAMIDE see EPD500

3-(AMINOSULFONYL)-4-PHENOXY-5-(1-PYRROLIDINYL)-BENZOIC ACID see PDW250

O-(4-(AMINOSULFONYL)PHENYL) O,O-DIMETHYL PHOSPHOROTHIOATE see CQL250

N-(5-(AMINOSULFONYL)-1,3,4-THIADIAZOL-2-YL)ACETAMIDE see AAI250

2-AMINO-5-((4-SULFOPHENYL)AZO)-BENZENESULFONIC ACID see AJS500

4-(4-AMINO-3-SULFOPHENYLAZO)BENZENESULFONIC ACID see AJS500
5-(AMINOSULFURANYL)-4-CHLORO-2-((2-FURNAYLMETHYL)AMINO)BENZOIC ACID see CHJ750
AMINOSYN see AHR600
AMINOSYN RF see AHR600
3,3'-(2-AMINOTEREPHTHALOYBIS(IMINO-p-PHENYLENECARBONYLIMINO))BIS(1-ETHYLPYRIDINIUM), DI-p-TOLUENESULFONATE see AMQ000
3,3'-(2-AMINOTEREPHTHALOYLBIS(IMINO(3-AMINO-p-PHENYLENE)CARBONYLIMINO))BIS(1-ETHYLPYRIDINIUM), DI-p-TOLUENESULFONATE see AMP500
3,3'-(2-AMINOTEREPHTHALOYLBIS(IMINO(3-AMINO-p-PHENYLENE)CARBONYLIMINO))BIS(1-PROPYLPYRIDINIUM), DI-p-TOLUENESULFONATE see AMP750
3,3'-(2-AMINOTEREPHTHALOYLBIS(IMINO-p-PHENYLENECARBONYLIMINO))BIS(1-METHYLPYRIDINIUM), DI-p-TOLUENESULFONATE see AMQ250
o-2-AMINO-2,3,4,6-TETRADEOXY-6-(METHYLAMINO)-α-d-glycero-HEX-4-ENOPYRANOSYL-(1-4)-o-(3-DEOXY-4-C-METHYL-3-(METHYLAMINO)-β-l-ARABINOPYRANOSYL-(1-6))-2-DEOXY-d-STREPTAMINE see GAA100
d-o-2-AMINO-2,3,4,6-TETRADEOXY-6-(METHYLAMINO)-α-d-erythro-HEXOPYRANOSYL-(1-4)-o-(3-DEOXY-4-C-METHYL-3-(METHYLAMINO)-β-l-ARABINOPYRANOSYL-(1-6))-2-DEOXY-STREPTAMINE SULFATE see MQS600
9-AMINO-1,2,3,4-TETRAHYDROACRIDINE see TCJ075
5-AMINO-6,7,8,9-TETRAHYDROACRIDINE (European) see TCJ075
8-AMINO-1,2,3,4-TETRAHYDRO-2-METHYL-4-PHENYLISOQUINOLINE MALEATE see NMV725
2-AMINO-1,2,3,4-TETRAHYDRONAFTALEN (CZECH) see TCY250
5-AMINO-2,2,4,4-TETRAKIS(TRIFLUOROMETHYL)IMIDAZOLIDINE see AMQ500
4-AMINO-2,2,5,5-TETRAKIS(TRIFLUOROMETHYL)-3-IMIDAZOLINE see AMQ500
2-AMINOTETRALIN see TCY250
AMINOTETRALIN (CZECH) see TCY250
2-AMINO-3,4,5,8-TETRAMETHYLIMIDAZO(4,5-F)QUINOXALINE see AMQ600
2-AMINO-3,4,7,8-TETRAMETHYLIMIDAZO(4,5-F)QUINOXALINE see AMQ620
1-AMINO-2,2,6,6-TETRAMETHYLPIPERIDINE see AMQ750
4-AMINO-2,2,6,6-TETRAMETHYLPIPERIDINE see AMQ800
AMINOTETRAZOLE see AMR000
5-AMINOTETRAZOLE see AMR000
5-AMINO-1H-TETRAZOLE see AMR000
AMINOTHIADIAZOLE see AMR250
2-AMINO-1,3,4-THIADIAZOLE see AMR250
2-AMINO-1,3,4-THIADIAZOLEHYDROCHLORIDE see AMR500
2-AMINO-1,3,4-THIADIAZOLE, MONOHYDROCHLORIDE see AMR500
2-AMINO-1,3,4-THIADIAZOLE-5-SULFONAMIDE SODIUM SALT see AMR750

2-AMINO-1,3,4-THIADIAZOLE-5-THIOL see AKM000
5-AMINO-1,3,4-THIADIAZOLE-2-THIOL see AKM000
5-AMINO-1,3,4-THIADIAZOLINE-2-THIONE see AKM000
2-AMINO-Δ²-1,3,4-THIADIAZOLINE-5-THIONE see AKM000
5-AMINO-1,2,3,4-THIATRIAZOLE see AMS000
AMINOTHIAZOLE see AMS250
2-AMINOTHIAZOLE see AMS250
2-AMINO-2-THIAZOLINE see TEV600
4-AMINO-N-2-THIAZOLYLBENZENESULFONAMIDE see TEX250
1-AMINO-2-(4-THIAZOLYL)-5-BENZIMIDAZOLECARBAMIC ACID ISOPROPYL ESTER see AMS625
(6R-(6-α,7-β(Z)))-1-((7-(((2-AMINO-4-THIAZOLYL)((1-CARBOXY-1-METHYLETHOXY)IMINO)ACETYL)AMINO)-2-CARBOXY-8-OXO-5-THIA-1-AZABICYCLO(4.2.0)OCT-2-EN-3-YL)METHYL)-PYRIDINIUM HYDROXIDE, inner salt, PENTAHYDRATE see CCQ200
7-(((2-AMINO-4-THIAZOLYL)(HYDROXYIMINOACETYL)AMINO)-3-ETHENYL-8-OXO-5-THIA-1-AZABICYCLO(4.2.0)OCT-2-ENE-2-CARBOXYLIC ACID, (6R-(6-α,7-β(Z)))- see AMS650
2-(((1-(2-AMINO-4-THIAZOLYL)-2-((2-METHYL-4-OXO-1-SULFO-3-AZETIDINYL)AMINO)-2-OXOETHYLIDENE)AMINO)OXY)-2-METHYLPROPANOIC ACID, (2S-(2-α,3β(Z))))- see ARX875
1-(2-AMINOTHIAZOLYL)-2-(5-NITRO-2-FURYL)ETHYLENE see ALN500
3-AMINOTHIOANISOLE see ALX100
4-AMINOTHIOANISOLE see AMS675
m-AMINOTHIOANISOLE see ALX100
p-AMINOTHIOANISOLE see AMS675
o-AMINOTHIOFENOLAT ZINECNATY (CZECH) see BGV500
α-AMINO-2-THIOPHENEPROPANOIC ACID see TEY250
2-AMINOTHIOPHENOL see AIF500
4-AMINOTHIOPHENOL see AIF750
o-AMINOTHIOPHENOL see AIF500
p-AMINOTHIOPHENOL see AIF750
2-AMINOTHIOPHENOL HYDROCHLORIDE see AIF800
o-AMINOTHIOPHENOL HYDROCHLORIDE see AIF800
6-AMINO-2-THIOURACIL see AMS750
N-AMINOTHIOUREA see TFQ000
N-(4-(((AMINOTHIOXOMETHYL)HYDRAZONO)METHYL)PHENYL)ACETAMIDE see FNF000
2-((2-AMINO-4-TIAZOLYL)METHYL)ISOTHIOURONIUM DICHLORIDE see AMS800
3-AMINOTOLUEN (CZECH) see TGQ500
4-AMINOTOLUEN (CZECH) see TGR000
2-AMINOTOLUENE see TGQ750
3-AMINOTOLUENE see TGQ500
4-AMINOTOLUENE see TGR000
m-AMINOTOLUENE see TGQ500
o-AMINOTOLUENE see TGQ750
p-AMINOTOLUENE see TGR000
α-AMINOTOLUENE see BDX750
ω-AMINOTOLUENE see BDX750
2-AMINOTOLUENE HYDROCHLORIDE see TGS500
4-AMINOTOLUENE HYDROCHLORIDE see TGS750
o-AMINOTOLUENE HYDROCHLORIDE see TGS500

α-AMINO-p-TOLUENESULFONAMIDE, MONOACETATE see MAC000
2-AMINO-p-TOLUENESULFONIC ACID see AMT000
4-AMINOTOLUENE-3-SULFONIC ACID see AKQ000
4-AMINO-o-TOLUENESULFONIC ACID see AMT250
6-AMINO-m-TOLUENESULFONIC ACID see AKQ000
p-AMINO-α-TOLUIC ACID see AID700
3-AMINO-p-TOLUIDINE see TGL750
5-AMINO-o-TOLUIDINE see TGL750
p-(4-AMINO-m-TOLUIDINO)PHENOL see AMT300
p-((4-AMINO-m-TOLYL)AZO)BENZENESULFONAMIDE see SNX000
1-(3-AMINO-p-TOLYLSULFONYL)-3-CYCLOHEXYLUREA see AKQ250
AMINOTRATE PHOSPHATE see TJL250
AMINOTRIACETIC ACID see AMT500
AMINOTRIAZOLE see AMY050
2-AMINOTRIAZOLE see AMY050
3-AMINOTRIAZOLE see AMY050
3-AMINO-s-TRIAZOLE see AMY050
2-AMINO-1,3,4-TRIAZOLE see AMY050
3-AMINO-1,2,4-TRIAZOLE see AMY050
3-AMINO-1H-1,2,4-TRIAZOLE see AMY050
3-AMINO-1,2,4-TRIAZOLE (ACGIH) see AMY050
AMINOTRIAZOLE (PLANT REGULATOR) see AMY050
AMINO TRIAZOLE WEEDKILLER 90 see AMY050
7-AMINO-1H-v-TRIAZOLO(4,5-d)PYRIMIDINE see AIB340
5-AMINO-1H-v-TRIAZOLO(d)PYRIMIDIN-7-OL see AJO500
5-AMINO-v-TRIAZOLO(4,5-d)PYRIMIDIN-7-OL see AJO500
AMINOTRIAZOL-SPRITZPULVER see AMY050
3-AMINO-1-TRICHLORO-2-PENTANOL see AMU000
((3-AMINO-2,4,6-TRICHLOROPHENYL)METHYLENE)HYDRAZIDE BENZENESULFONIC ACID see AMU125
4-AMINO-3,5,6-TRICHLOROPICOLINIC ACID see PIB900
4-AMINO-3,5,6-TRICHLORO-2-PICOLINIC ACID see PIB900
4-AMINO-3,5,6-TRICHLOROPICOLINIC ACID POTASSIUM SALT see PLQ760
3-AMINO-1-TRICHLORO-2-PROPANOL see AMU500
4-AMINO-3,5,6-TRICHLORPICOLINSAEURE (GERMAN) see PIB900
1-AMINOTRICYCLO(3.3.1.1³,⁷)DECANE see TJG250
2-AMINO-4-(TRIFLUOROMETHYL)-5-THIAZOLECARBOXYLIC ACID ETHYL ESTER see AMU550
30-AMINO-3,14,25-TRIHYDROXY-3,9,14,20,25-PENTAAZATRIACONTANE-2,10,13,21,24-PENTAONE see DAK200
3-AMINO-2,4,6-TRIIODO-BENZOIC ACID see AMU625
N-(3-AMINO-2,4,6-TRIIODOBENZOYL)-N-(2-CARBOXYETHYL)ANILINE see AMU750
3-((3-AMINO-2,4,6-TRIIODOBENZOYL)PHENYLAMINO)PROPIONIC ACID see AMU750
2-(3-AMINO-2,4,6-TRIIODOBENZYL)BUTYRIC ACID see IFY100
4-((3-AMINO-2,4,6-TRIIODOPHENYL)ETHYLAMINO)-4-OXO-BUTANOIC ACID see AMV375

3-(3-AMINO-2,4,6-TRIIODOPHENYL)-2-ETHYLPROPANOIC ACID see IFY100
β-(3-AMINO-2,4,6-TRIIODOPHENYL)-α-ETHYLPROPIONIC ACID see IFY100
2-(3-AMINO-2,4,6-TRIIODOPHENYL)VALERIC ACID see AMV750
N-(3-AMINO-2,4,6-TRIJODBENZOYL)-N-PHENYL-β-AMINOPROPIONSAEURE (GERMAN) see AMU750
1-AMINO-2,4,6-TRIMETHYLBENZEN (CZECH) see TLG500
1-AMINO-2,4,5-TRIMETHYLBENZENE see TLG250
2-AMINO-1,3,5-TRIMETHYLBENZENE see TLG500
1-AMINO-2,4,5-TRIMETHYLBENZENE HYDROCHLORIDE see TLG750
2-AMINO-1,3-5-TRIMETHYLBENZENE HYDROCHLORIDE see TLH000
4-AMINO-a,a,4-TRIMETHYLCYCLOHEXANEMETHAMINE see MCD750
AMINOTRI(METHYLENEPHOSPHONIC ACID) see NEI100
AMINOTRI(METHYLENEPHOSPHONIC ACID) PENTASODIUM SALT see PBP200
2-AMINO-3,5,6-TRIMETHYLIMIDAZO(4,5-b)PYRIDINE see AMV752
2-AMINO-3,5,6-TRIMETHYLIMIDAZO(4,5-b)PYRIDINE see AMV752
2-AMINO-3H-3,5,7-TRIMETHYLIMIDAZO(4,5-b)PYRIDINE see AMV754
2-AMINO-3,4,5-TRIMETHYLIMIDAZO(4,5-F)QUINOXALINE see AMV760
2-AMINO-3,4,8-TRIMETHYLIMIDAZO(4,5-F)QUINOXALINE see TLT768
2-AMINO-3,5,7-TRIMETHYLIMIDAZO(4,5-F)QUINOXALINE see AMV770
2-AMINO-3,7,8-TRIMETHYLIMIDAZO(4,5-f)QUINOXALINE see TLT771
AMINOTRIMETHYLOMETHANE see TEM500
AMINOTRI(METHYLPHOSPHONIC ACID) see NEI100
7-AMINO-2,4,6-TRIMETHYLQUINOLINE see AMV780
6-AMINO-2,3,5-TRIMETHYLQUINOXALINE see QQS340
AMINOTRIS(HYDROXYMETHYL)METHANE see TEM500
AMINOTRIS(METHANEPHOSPHONIC ACID) see NEI100
AMINOTRIS(METHYLPHOSPHONIC ACID) see NEI100
4-AMINOTROPOLONE see AMV790
5-AMINOTROPOLONE see AMV800
2-AMINOTROPONE see AJJ800
AMINO-TS-ACID see AMV875
AMINOUNDECANOIC ACID see AMW000
11-AMINOUNDECANOIC ACID see AMW000
11-AMINOUNDECYLIC ACID see AMW000
5-AMINOURACIL see AMW100
6-AMINOURACIL see AMW250
AMINOURACIL MUSTARD see BIA250
AMINOUREA see HGU000
AMINOUREA HYDROCHLORIDE see SBW500
p-AMINO VALEROPHENONE see AMW500
AMINOX see AMM250
2-(4'-AMINOXENYL)NAFTO-α,β-TRIAZOL-6,8-DISULFONAN DRASELNY (CZECH) see AIW750
2-(4'-AMINOXENYL)NAFTO-α,β-TRIAZOL-6,8,3'TRISULFONAN DRASELNY (CZECH) see AMO750
2-AMINO-1,4-XYLENE see XNA000
4-AMINO-1,3-XYLENE see XMS000
2-AMINO-1,4-XYLENE HYDROCHLORIDE see XOS000

4-AMINO-1,3-XYLENE HYDROCHLORIDE see XOJ000
2-AMINO-4,5-XYLENOL see AMW750
3-AMINO-4-(2-(2,6-XYLYLOXY)ETHYL)-4H-1,2,4-TRIAZOLE see AMX000
AMINOZIDE see DQD400
1,2-AMINOZOPHENYLENE see BDH250
AMINUTRIN see LJM700
AMINZOL SOLUBLE see ALQ000
AMIODARONE see AJK750
AMIODOXYL BENZOATE see AMX250
AMIOYL see BCA000
AMIP 15m see TAI250
AMIPAN T see DUO400
AMIPAQUE see MQR300
AMIPAREN see AHR600
AMIPENIX S see AIV500
AMIPHENAZOLE HYDROCHLORIDE see DCA600
AMIPHOS see DOP200
AMI-PILO see PIF250
AMIPOL 6S see LBU200
AMIPOLNE see AAE500
AMIPRESS see HMM500
AMIPROFOS-METHYL see AMX300
AMIPROL see DCK759
AMIPROPHOS see AMX400
AMIPTAN see AHK750
AMIPURIMYCIN HYDRATE see AMX500
AMIPYLO see AMF375
AMIRAL see CJO250
AMISEPAN see SBG500
AMISOL ODE see OHU200
AMISOMETRADIN see AKL625
AMISOMETRADINE see AKL625
AMISTURA P see PIF250
AMISYL see BCA000
AMITAKON see BCA000
AMITAL see AMX750
AMITHIOZONE see FNF000
AMITID see EAI000
AMITIOZON see FNF000
AMITOL see AMY050
AMITON see DJA400
AMITON OXALATE see AMX825
AMITRAZ see MJL250
AMITRAZE see MJL250
AMITRAZ ESTRELLA see MJL250
AMITRENE see BBK250
AMITRENE see BBK500
AMITRIL see AMY050
AMITRIL see EAI000
AMITRIL T.L. see AMY050
AMITRIPTILINE see EAH500
AMITRIPTYLIN (GERMAN) see EAH500
AMITRIPTYLINE see EAH500
AMITRIPTYLINE CHLORIDE see EAI000
AMITRIPTYLINE-N-OXIDE see AMY000
AMITRIPTYLINOXIDE see AMY000
AMITROL see AMY050
AMITROL 90 see AMY050
AMITROLE see AMY050
AMITROL-T see AMY050
AMITRYPTYLINE HYDROCHLORIDE see EAI000
AMI-U-II see AHR600
AMIXICOTYN see NCR000
AMIZIL HYDROCHLORIDE see BCA000
AMIZOL see AMY050
AMIZOL D see AMY050
AMIZOL DP NAU see AMY050
AMIZOL F see AMY050
AMLODIPINE see AMY100
AMLURE see PNE800
AMMAT see ANU650
AMMATE see ANU650
AMMELIDE see MCB000
AMMICARDINE see AHK750
AMMIDIN see IHR300
AMMI-KHELLIN see AHK750
AMMINE PENTAHYDROXO PLATINUM see AMY250

AMMINETRICHLORO-PLATINATE(1⁻) POTASSIUM (SP-4-2) see PJD750
4-AMMINOANTIPIRINA see AIB300
AMMIPURAN see AHK750
AMMISPASMIN see AHK750
AMMIVIN see AHK750
AMMIVISNAGEN see AHK750
AMMN see AAW000
AMMO see RLF350
AMMOFORM see HEI500
AMMOIDIN see XDJ000
AMMONERIC see ANE500
AMMONIA see AMY500
AMMONIA, anhydrous, liquefied (DOT) see AMY500
AMMONIA ANHYDROUS see AMY500
AMMONIA AQUEOUS see ANK250
AMMONIAC (FRENCH) see AMY500
AMMONIACA (ITALIAN) see AMY500
AMMONIA GAS see AMY500
AMMONIAK (GERMAN) see AMY500
AMMONIA SOLUTIONS, relative density <0.880 at 15 degrees C in water, with >50% ammonia (DOT) see AMY500
AMMONIA SOLUTIONS, with >10% but not >35% ammonia (UN 2672) (DOT) see ANK250
AMMONIA SOLUTIONS, with >35% but not >50% ammonia (UN 2073) (DOT) see ANK250
AMMONIATED GLYCYRRHIZIN see AMY700
AMMONIATED GLYCYRRHIZIN see GIE100
AMMONIATED MERCURY see MCW500
AMMONIA WATER 29% see ANK250
AMMONIIUM, BENZYLDIMETHYL(2-(2-(4-(1,1,3,3-TETRAMETHYLBUTYL)TOLYLOXY)ETHOXY)ETHYL)-, CHLORIDE, MONOHYDRATE see MHB500
AMMONIO (DICROMATO DI) (ITALIAN) see ANB500
AMMONIOFORMALDEHYDE see HEI500
2-AMMONIOTHIAZOLE NITRATE see AMZ125
AMMONIUM ACETATE see ANA000
AMMONIUM, (2-(3-ACETYL-2-METHYL-1-INDOLIZINYL)-2-METHYLETHYL)DIMETHYLHEXYL-, BROMIDE see ACR120
AMMONIUM, (2-(3-ACETYL-2-METHYL-1-INDOLIZINYL)-2-METHYLETHYL)DIMETHYLPENTYL-, BROMIDE see ACR130
AMMONIUM, (2-(3-ACETYL-2-METHYL-1-INDOLIZINYL)-2-METHYLETHYL)DIMETHYL(1,4-XYLYL)-, CHLORIDE see ACR140
AMMONIUM ACID ARSENATE see DCG800
AMMONIUM ACID SULFATE see ANJ500
AMMONIUM ADIPATE see HEO120
AMMONIUM-AETHYL-CARBAMOYL-PHOSPHONAT (GERMAN) see ANG750
AMMONIUM ALGINATE see ANA300
AMMONIUM, ALKYL(3,4-DICHLOROBENZYL)DIMETHYL-, CHLORIDE see AFP300
AMMONIUM, ALKYL(C₁₂-C₁₆)DIMETHYLBENZYL-, CHLORIDES see QAT520
AMMONIUM, ALKYLDIMETHYL(3,4-DICHLOROPHENYL)-, CHLORIDE see AFP300
AMMONIUM, ALLYLTRIETHYL-, IODIDE see AGU300
AMMONIUM, ALLYLTRIMETHYL-, CHLORIDE see HGI600
AMMONIUM ALUMINUM FLUORIDE see THQ500
AMMONIUM AMIDOSULFONATE see ANU650
AMMONIUM AMIDOSULPHATE see ANU650

AMMONIUM (3-AMINO-3-CARBOXYPROPYL)METHYLPHOSPHINATE see ANI800

AMMONIUM AMINOFORMATE see AND750

AMMONIUM 2-AMINO-4-(HYDROXYMETHYLPHOSPHINYL)BUTANOATE see ANI800

AMMONIUM (AMINYLENIUM BIS[TRIHYDROBORATE]) see ANA500

AMMONIUM, ANDROSTA-3,5-DIEN-17-β-YLTRIMETHYL-,IODIDE see AOO400

AMMONIUM ARSENATE, solid (DOT) see DCG800

AMMONIUM AURINTRICARBOXYLATE see AGW750

AMMONIUM AZIDE see ANA750

AMMONIUM BENZAMIDOOXYACETATE see ANB000

AMMONIUM-2-(BENZAMIDOOXY)ACETATE see ANB000

AMMONIUM BENZOATE see ANB100

AMMONIUM, BENZYLBIS(2-HYDROXYETHYL)DODECYL-, CHLORIDE see BDJ600

AMMONIUM ((3-N-BENZYLCARBAMOYLOXY)PHENYL)TRIMETHYL METHYLSULFATE see BED500

AMMONIUM, BENZYLDIETHYL((2,6-XYLYLCARBAMOYL)METHYL)-, BENZOATE see DAP812

AMMONIUM, BENZYLDIMETHYLOCTADECYL-, 3-NITROBENZENESULFONATE see BEM400

AMMONIUM, BENZYLHEXADECYLDIMETHYL-, CHLORIDE see BEL900

AMMONIUM, BENZYLTRIETHYL-, CHLORIDE see BFL300

AMMONIUM, BENZYLTRIMETHYL-, METHOXIDE see TLM100

AMMONIUM BICARBONATE (1:1) see ANB250

AMMONIUMBICHROMAAT (DUTCH) see ANB500

AMMONIUM BICHROMATE see ANB500

AMMONIUM BIFLUORIDE see ANJ000

AMMONIUM, (4-BIPHENYLCARBONYLMETHYL)DIMETHYL(2-HYDROXYETHYL)-, BROMIDE see BGF050

AMMONIUM BIS(2,3-DIBROMOPROPYL)PHOSPHATE see MRH214

AMMONIUM BISULFATE see ANJ500

AMMONIUM BISULFIDE see ANJ750

AMMONIUM BISULFITE see ANB600

AMMONIUM BITHIOLICUM see IAD000

AMMONIUM BORANECARBOXYLATE see ANB700

AMMONIUM BOROFLUORIDE see ANH000

AMMONIUM BROMATE see ANC000

AMMONIUM BROMATE (DOT) see ANC000

AMMONIUM BROMIDE see ANC250

AMMONIUM BROMO SELENATE see ANC750

AMMONIUM, BUTYLDIMETHYL(2-(3-ACETYL-2-METHYL-1-INDOLIZINYL)-2-METHYLETHYL)-, IODIDE see BRB200

AMMONIUM, BUTYLTRIPROPYL-, IODIDE see BSR930

AMMONIUM, (2-BUTYRYLOXYETHYL)TRIMETHYL-, IODIDE see BSY300

AMMONIUM CADMIUM CHLORIDE see AND250

AMMONIUM CALCIUM ARSENATE see AND500

AMMONIUM CARBAMATE see AND750

AMMONIUM CARBAZOATE see ANS500

AMMONIUMCARBONAT (GERMAN) see ANE000

AMMONIUM CARBONATE see ANB250

AMMONIUM CARBONATE see ANE000

AMMONIUM, (2-CARBOXYETHYL)DIETHYLMETHYL-, IODIDE,ETHYL ESTER see CCE800

AMMONIUM, (2-CARBOXYETHYL)DIMETHYLETHYL-, IODIDE,ETHYL ESTER see CCE820

AMMONIUM, (2-CARBOXYETHYL)TRIMETHYL-, IODIDE, N-BUTYLESTER see CCF300

AMMONIUM, (2-CARBOXYETHYL)TRIMETHYL-, IODIDE, ETHYLESTER see CCF330

AMMONIUM, (3-CARBOXY-2-HYDROXYPROPYL)TRIMETHYL-, CHLORIDE, (−)- see CCK660

AMMONIUM, (3-CARBOXY-2-HYDROXYPROPYL)TRIMETHYL-, CHLORIDE, (±)- see CCK655

AMMONIUM CARBOXYMETHYL CELLULOSE see CCH000

AMMONIUM, (CARBOXYMETHYL)DODECYLDIMETHYL-, HYDROXIDE, inner salt (7CI,8CI) see LBU200

AMMONIUM, (CARBOXYMETHYL)HEXADECYLDIMETHYL-, HYDROXIDE, inner salt (8CI) see CDF450

AMMONIUM, (CARBOXYMETHYL)TRIMETHYL-, CHLORIDE see CCH850

AMMONIUM, (CARBOXYMETHYL)TRIMETHYL-, CHLORIDE, HYDRAZIDE see GEQ500

AMMONIUM, (CARBOXYMETHYL)TRIMETHYL-, IODIDE, METHYLESTER see CCH890

AMMONIUM, (3-CARBOXYPROPYL)TRIMETHYL-, IODIDE, ETHYLESTER see CCJ385

AMMONIUM, (3-CARBOXYPROPYL)TRIMETHYL-, IODIDE, METHYLESTER see CCJ388

AMMONIUM CHLORATE see ANE250

AMMONIUMCHLORID (GERMAN) see ANE500

AMMONIUM CHLORIDE see ANE500

AMMONIUM CHLOROHEPTENE ARSONATE see CEI250

AMMONIUM, (3-CHLORO-2-HYDROXYPROPYL)TRIMETHYL-, CHLORIDE see CHQ300

AMMONIUM, (2-(p-((2-CHLORO-4-NITROPHENYL)AZO)PHENETHYLAMINO)ETHYL)TRIMETHYL- see BAQ750

AMMONIUM CHLOROPALLADATE(II) see ANE750

AMMONIUM CHLOROPALLADATE(IV) see ANF000

AMMONIUM, (3-(2-CHLOROPHENOTHIAZIN-10-YL)PROPYL)TRIMETHYL-,IODIDE see MIJ300

AMMONIUM CHLOROPLATINATE see ANF250

AMMONIUM CHROMATE see ANF500

AMMONIUM CHROMATE see NCQ550

AMMONIUM CHROMATE(VI) see ANF500

AMMONIUM CHROMATE(VI) see NCQ550

AMMONIUM CHROME ALUMS see ANF625

AMMONIUM CHROMIC SULFATE see ANF750

AMMONIUM CITRATE see ANF800

AMMONIUM CITRATE, DIBASIC (DOT) see ANF800

AMMONIUM CRYOLITE see THQ500

AMMONIUM CYANIDE see ANG000

AMMONIUM, (3-CYANO-2-HYDROXYPROPYL)TRIMETHYL-,CHLORIDE, (+−)- see COL050

AMMONIUM DECAHYDRODECABORATE (2−) see ANG125

AMMONIUM, DECYLDIMETHYLOCTYL, CHLORIDE see OES400

AMMONIUM, DECYLTRIMETHYL-, BROMIDE see DAJ500

AMMONIUM, DIALLYLDIMETHYL-, CHLORIDE, POLYMERS see DTS500

AMMONIUM, (4-((4,6-DIAMINO-m-TOLYL)IMINO)-2,5-CYCLOHEXADIEN-1-YLIDENE)DIMETHYL-, CHLORIDE, MONOHYDRATE see TGU500

AMMONIUM, (3,4-DICHLOROBENZYL)DODECYLDIMETHYL-, CHLORIDE see LBV100

AMMONIUMDICHROMAAT (DUTCH) see ANB500

AMMONIUMDICHROMAT (GERMAN) see ANB500

AMMONIUM DICHROMATE see ANB500

AMMONIUM DICHROMATE(VI) see ANB500

AMMONIUM, DICYCLOHEXYL-, 2-CYCLOHEXYL-4,6-DINITROPHENATE see CPK550

AMMONIUM, DIETHYL(4-((p-(DIETHYLAMINO)PHENYL)(3,6-DISULFO-1-NAPHTHYL)METHYLENE)-2,5-CYCLOHEXADIENYLIDENE)-,HYDROXIDE, INNER SALT, SODIUM SALT see ANG135

AMMONIUM, DIETHYL(m-HYDROXYPHENYL)METHYL-, CHLORIDE see DJN430

AMMONIUM DIFLUORIDE see ANJ000

AMMONIUM DIFLUORIDE mixed with HYDROCHLORIC ACID see ANG250

AMMONIUM DIHEXADECYLDIMETHYL-, CHLORIDE see DRK200

AMMONIUM, (2-(10',11'-DIHYDRO-5'H-AZEPIN-5'-YL)-1-METHYL)ETHYLTRIMETHYL-, IODIDE see DKR800

AMMONIUM, (7-(DIMETHYLAMINO)-3H-PHENOXAZIN-3-YLIDENE)DIMETHYL-, CHLORIDE see DPN900

AMMONIUM, (4-DIMETHYLCARBAMOYLBUTYL)TRIMETHYL-,IODIDE see DQY925

AMMONIUM, DIMETHYLDIOCTADECYL-, CHLORIDE see DXG625

AMMONIUM, DIMETHYLDIOCTYL-, CHLORIDE see DTF820

AMMONIUM DIMETHYL DITHIOCARBAMATE see ANG500

AMMONIUM, N,N-DIMETHYL-N,N',N',N'-TETRAETHYL-N,N'-OXYDIETHYLENEDI-, DITARTRATE see TCD600

AMMONIUM-3,5-DINITRO-1,2,4-TRIAZOLIDE see ANG625

AMMONIUM DISULFATONICKELATE(II) see NCY050

AMMONIUM DNOC see DUT800

AMMONIUM DODECYL SULFATE see SOM500

AMMONIUM-N-DODECYL SULFATE see SOM500

AMMONIUM, DODECYLTRIMETHYL-, BROMIDE see DYA810

AMMONIUM DODECYLTRIMETHYL-, CHLORIDE see LBX075

AMMONIUM ETHYL CARBAMOYLPHOSPHONATE see ANG750

AMMONIUM ETHYL CARBAMOYLPHOSPHONATE solution see ANG750

AMMONIUM, ETHYLTRIMETHYL-, IODIDE see EQC600

AMMONIUM-FERRIC-CYANO-FERRATE(II) see FAS800

AMMONIUM-FERRIC-FERROCYANIDE see FAS800

AMMONIUM FERRIC OXALATE see ANG925

AMMONIUM FERRIOXALATE see ANG925

AMMONIUM FERROUS SULFATE HEXAHYDRATE see IGL200

AMMONIUM FLUOALUMINATE see THQ500

AMMONIUM FLUOBORATE see ANH000

AMMONIUM FLUORIDE see ANH250

AMMONIUM FLUORIDE comp. with HYDROGEN FLUORIDE (1:1) see ANJ000

AMMONIUM FLUOROBERYLLATE see ANH300

AMMONIUM FLUOROBORATE see ANH000

AMMONIUM FLUOROSILICATE (DOT) see COE000

AMMONIUM FLUORURE (FRENCH) see ANH250

AMMONIUM FLUOSILICATE see COE000

AMMONIUM FORMATE see ANH500

AMMONIUMGLUTAMINAT (GERMAN) see MRF000

AMMONIUM GLYCYRRHIZINATE see GIE100

AMMONIUM HEPTAMOLYBDATE see ANH600

AMMONIUM HEXACHLOROPALLADATE see ANF000

AMMONIUM HEXACHLOROPLATINATE(IV) see ANF250

AMMONIUM HEXACHLORORHODATE see HCM050

AMMONIUM HEXACHLORORHODATE(III) see HCM050

AMMONIUM HEXACYANOFERRATE(II) see ANH875

AMMONIUM HEXAFLUOROALUMINATE see THQ500

AMMONIUM HEXAFLUOROFERRATE see ANI000

AMMONIUM HEXAFLUOROSILICATE see COE000

AMMONIUM HEXAFLUOROTITANATE see ANI250

AMMONIUM HEXAFLUOROVANADATE see ANI500

AMMONIUM, HEXAMETHYLENEBIS((CARBOXYMETHYL)DIMETHYL)-, DICHLORIDE, DIDODECYL ESTER see HEF200

AMMONIUM, HEXAMETHYLENEBIS(TRIMETHYL-, DIBENZENESULFONATE see HEG100

AMMONIUM HEXANITRO COBALTATE see ANI750

AMMONIUM (dl-HOMOALANINE-4-YL)METHYLPHOSPHINATE see ANI800

AMMONIUM HYDROFLUORIDE see ANJ000

AMMONIUM HYDROGEN BIFLUORIDE see ANJ000

AMMONIUM HYDROGEN CARBONATE see ANB250

AMMONIUM HYDROGEN DIFLUORIDE see ANJ000

AMMONIUM HYDROGEN FLUORIDE see ANJ000

AMMONIUM HYDROGEN FLUORIDE, solid (UN 1727) (DOT) see ANJ000

AMMONIUM HYDROGEN FLUORIDE, solution (UN 2817) (DOT) see ANJ000

AMMONIUM HYDROGEN SULFATE see ANJ500

AMMONIUM HYDROGEN SULFIDE see ANJ750

AMMONIUM HYDROGEN SULFITE see ANB600

AMMONIUM HYDROSULFIDE see ANJ750

AMMONIUM HYDROSULFIDE, solution (DOT) see ANJ750

AMMONIUM HYDROXIDE see ANK250

AMMONIUM, (m-HYDROXYPHENETHYL)TRIMETHYL-, HYDROXIDE see HNF100

AMMONIUM, (m-HYDROXYPHENYL)TRIMETHYL-, BROMIDE,DECAMETHYLENEBIS(METHYLCARBAMATE) see DAM625

AMMONIUM, (2-HYDROXYPROPYL)TRIMETHYL-, ACETATE(ESTER) see MFX560

AMMONIUM, (2-HYDROXYPROPYL)TRIMETHYL-, CHLORIDE see MIM300

AMMONIUM HYPOPHOSPHITE see ANK500

AMMONIUM HYPOSULFITE see ANK600

AMMONIUM ICHTHOSULFONATE see IAD000

AMMONIUM IMIDOBISSULFATE see ANK650

AMMONIUM IMIDODISULFONATE see ANK650

AMMONIUM IMIDOSULFONATE see ANK650

AMMONIUM IODATE see ANK750

AMMONIUM IODIDE see ANL000

AMMONIUM IRON HEXACYANOFERRATE see FAS800

AMMONIUM IRON(II) SULFATE, HEXAHYDRATE (2:1:2:6) see IGL200

AMMONIUM ISETHIONATE see ANL100

AMMONIUM LANTHANUM NITRATE see ANL500

AMMONIUM LAURYL SULFATE see SOM500

AMMONIUM MAGNESIUM ARSENATE see ANL750

AMMONIUM MAGNESIUM ARSENATE see MAD025

AMMONIUM MAGNESIUM ARSENATE DIHYDRATE see MAD025

AMMONIUM MAGNESIUM CHROMATE see ANM000

AMMONIUM MANDELATE see ANM250

AMMONIUM MERCAPTAN see ANJ750

AMMONIUM MERCAPTOACETATE see ANM500

AMMONIUM, (2-MERCAPTOETHYL)TRIMETHYL-, IODIDE ACETATE see ADC300

AMMONIUM METAVANADATE (DOT) see ANY250

AMMONIUM METHANEARSONATE see MDQ770

AMMONIUM-3-METHYL-2,4,6-TRINITROPHENOXIDE see ANM625

AMMONIUM MOLYBDATE see ANM750

AMMONIUM MOLYBDATE(II) ((NH$_4$)$_6$MO$_7$O$_{24}$) see ANH600

AMMONIUM MONOHYDROGEN SULFATE see ANJ500

AMMONIUM MONOSULFITE see ANB600

AMMONIUM MURIATE see ANE500

AMMONIUM NICKEL SULFATE see NCY050

AMMONIUM NICKEL^{2+} SULFATE HEXAHYDRATE see NCY060

AMMONIUM NITRATE see ANN000

AMMONIUM(I) NITRATE(1:1) see ANN000

AMMONIUM NITRATE, liquid (hot concentrated solution) (UN 2426) (DOT) see ANN000

AMMONIUM NITRATE, with >0.2% combustible substances (UN 0222) (DOT) see ANN000

AMMONIUM NITRATE, with not >0.2% of combustible substances (UN 1942) (DOT) see ANN000

AMMONIUM NITRITE see ANO250

AMMONIUM aci-NITROMETHANE see ANO400

AMMONIUM-N-NITROSOPHENYLHYDROXYLAMINE see ANO500

AMMONIUM NONOXYNOL-4-SULFATE see ANO600

AMMONIUM, OCTADECYLTRIETHYL-, IODIDE see OBI100

AMMONIUM, OCTADECYLTRIMETHYL-,p-FLUOROBENZENESULFONATE see OBI200

AMMONIUM ORTHOPHOSPHITE see ANS250

AMMONIUM OXALATE see ANO750

AMMONIUM OXOFLUOROMOLYBDATE see ANO875

AMMONIUM PARAMOLYBDATE see ANH600

AMMONIUM PARAMOLYBDATE see ANM750

AMMONIUM PARATUNGSTATE HEXAHYDRATE see ANO900

AMMONIUM PENTADECAFLUOROOCTANATE see ANP625

AMMONIUM PENTA PEROXODICHROMATE see ANP000

AMMONIUM PERCHLORATE see ANP250

AMMONIUM PERCHLORATE (DOT) see PCD500

AMMONIUM PERCHLORYL AMIDE see ANP500

AMMONIUM PERFLUOROCAPRILATE see ANP625

AMMONIUM PERFLUOROCAPRYLATE see ANP625

AMMONIUM PERFLUORONONANOATE see HAS050

AMMONIUM PERFLUOROOCTANOATE see ANP625

AMMONIUM-m-PERIODATE see ANP750

AMMONIUM PERMANGANATE see PCJ750

AMMONIUM PEROXO BORATE see ANQ250

AMMONIUM PEROXYCHROMATE see ANQ750

AMMONIUM PEROXYDISULFATE see ANR000

AMMONIUM PERSULFATE see ANR000

AMMONIUM, (4,β-PHENETHYLENEBIS(CARBONYLMETHYL))BIS(DIMETHYL-2-HYDROXYETHYL-, DIBROMIDE see PDF510

AMMONIUM PHENYLDITHIOCARBAMATE see ANR250

AMMONIUM, (p-PHENYLENEBIS(CARBONYLMETHYL))BIS(DIMETHYL-2-HYDROXYETHYL-, DIBROMIDE see PEW600

AMMONIUM, (p-PHENYLENEBIS(OXYTRIMETHYLENE))BIS(ETHYLDIMETHYL-, DIIODIDE see DHA500

AMMONIUM, PHENYLTRIMETHYL-, BROMIDE see TMB000

AMMONIUM PHOSPHATE see ANR500

AMMONIUM PHOSPHATE, DIBASIC see ANR500

AMMONIUM PHOSPHATE, MONOBASIC see ANR750

AMMONIUM PHOSPHIDE see ANS000

AMMONIUM PHOSPHITE see ANS250

AMMONIUM PICRATE see ANS500

AMMONIUM PICRATE, dry or wetted with <10% water, by weight (UN 0004) (DOT) see ANS500

AMMONIUM PICRATE, wetted with not <10% water, by weight (UN 1310) (DOT) see ANS500

AMMONIUM PICRONITRATE see ANS500

AMMONIUM PLATINIC CHLORIDE see ANF250

AMMONIUM POLYSULFIDE (solution) see ANT000

AMMONIUM POLYSULFIDE, solution (DOT) see ANT000

AMMONIUM POTASSIUM HYDROGEN PHOSPHATE see ANT100

AMMONIUM POTASSIUM SELENIDE mixed with AMMONIUM POTASSIUM SULFIDE see ANT250

AMMONIUM POTASSIUM SULFIDE mixed with AMMONIUM POTASSIUM SELENIDE see ANT250

AMMONIUM, 5-α-PREGNAN-3-β,20-α-YLENEBIS(TRIMETHYL-, DIIODIDE see MAO280

AMMONIUM, (2-(10h-PYRIDO(3,2-b)(1,4)BENZOTHIAZIN-10-YL)PROPYL)TRIMETHYL-, BROMIDE see PPH350

AMMONIUM REINECKATE HYDRATE see ANT300

AMMONIUM RHODANATE see ANW750

AMMONIUM RHODANIDE see ANW750

AMMONIUM SACCHARIN see ANT500

AMMONIUM SALICYLATE see ANT600

AMMONIUM SALTPETER see ANN000

AMMONIUM SALTS of PHOSPHATIDIC ACIDS see ANU000

AMMONIUMSALZ der AMIDOSULFONSAEURE (GERMAN) see ANU650

AMMONIUMSALZ des α-METHYL-β-HYDROXY-Δ^{α,β}-BUTYCOLACTAM (GERMAN) see MKS750

AMMONIUM SILICOFLUORIDE see COE000

AMMONIUM SILICON FLUORIDE see COE000

AMMONIUM STEARATE see ANU200

AMMONIUM STEARATE see ANU200

AMMONIUM SULFAMATE see ANU650

AMMONIUM SULFATE (2:1) see ANU750

AMMONIUM SULFATE, and CHROMIC SULFATE, TETRACOSAHYDRATE see ANF625

AMMONIUM SULFHYDRATE see ANJ750

AMMONIUM SULFIDE, solution, red see ANT000

AMMONIUM SULFIDE (POLY-) see ANT000

AMMONIUM SULFOCYANATE see ANW750

AMMONIUM SULFOCYANIDE see ANW750

AMMONIUM SULFOICHTHYOLATE see IAD000

AMMONIUM SULPHAMATE see ANU650

AMMONIUM SULPHATE see ANU750

AMMONIUM-d-TARTRATE see DCH000

AMMONIUM TARTRATE (DOT) see DCH000

AMMONIUM TELLURATE see ANV750

AMMONIUM TETRACHLOROPALLADATE see ANE750

AMMONIUM TETRACHLOROPLATINATE see ANV800

AMMONIUM, TETRADECYLTRIMETHYL-, BROMIDE see TCB200

AMMONIUM, TETRAETHYL-, PERCHLORATE see TCD002

AMMONIUM TETRAFLUOROBERYLLATE see ANH300

AMMONIUM TETRAFLUOROBORATE see ANH000

AMMONIUM TETRAFLUOROBORATE(1-) see ANH000

AMMONIUM, TETRAMETHYL-, HYDROXIDE see TDK500

AMMONIUM TETRANITROPLATINATE(II) see ANW250

AMMONIUM, TETRAPENTYL-, CHLORIDE see TBI200

AMMONIUM TETRAPEROXO CHROMATE see ANW500

AMMONIUM THIOCYANATE see ANW750

AMMONIUM THIOGLYCOLATE see ANM500

AMMONIUM THIOGLYCOLLATE see ANM500

AMMONIUM THIOSULFATE see ANK600

AMMONIUM, TRIBUTYL(2,4-DICHLOROBENZYL)-, CHLORIDE see THY200

AMMONIUM TRICHLOROACETATE see ANX750

AMMONIUM, TRIETHYLHEXADECYL-, BROMIDE (8CI) see CDF500

AMMONIUM TRIFLUOROSTANNITE see ANX800

AMMONIUM, TRIMETHYLTALLOW ALKYL-, CHLORIDES see TAC300

AMMONIUM, TRIMETHYLTETRADECYL-, BROMIDE see TCB200

AMMONIUM, TRIMETHYL(p-UREIDOPHENYL)-, IODIDE see TMJ300

AMMONIUM-2,4,5-TRINITROIMIDAZOLIDE see ANX875

AMMONIUM TRIOXALATOFERRATE(III) see ANG925

AMMONIUM TRISULFIDE see ANT000

AMMONIUM VANADATE see ANY250

AMMONIUM VANADI-ARSENATE see ANY500

AMMONIUM VANADO-ARSENATE see ANY750

AMMONIUMYL, DIBUTYL-, HEXACHLOROSTANNATE(2-) (2:1) see BIV900

AMMONYL BR 1244 see BEO000

AMMONYX see AFP250

AMMONYX 4 see DTC600

AMMONYX 2200 see QAT550

AMMONYX CA SPECIAL see DTC600

AMMONYX CPC see CCX000

AMMONYX DME see EKN500

AMMONYX G see BEL900

AMMONYX KP see OHK200

AMMONYX LO see DRS200

AMMONYX T see BEL900

AMMOPHYLLIN see TEP500

AMN see AOL000

AMNESTROGEN see ECU750

AMNICOTIN see NCR000

AM (NITRIFICATION INHIBITOR) see AJI100

AMNOSED see TDA500

AMNUCOL see SEH000

1A-2MO-4OA see AKO350

AMOBAM see ANZ000

AMOBARBITAL see AMX750

AMOBARBITONE see AMX750

AMOBEN see AJM000

AMOCO 1010 see PMP500

AMOCO 610A4 see PJS750

AMOCO NT-45 PROCESS OIL see DXG830

AMOEBAL see ABX500

AMOENOL see CHR500

AMOGLANDIN see POC500

AMOIL see AON300

AMOKIN see CLD000

AMOLANONE HYDROCHLORIDE see DIF200

AMONAFIDE see NAC600

AMONIAK (POLISH) see AMY500

A. MONTANA see TOA275

AMONYX AO see DRS200

AMOPYROQUIN DIHYDROCHLORIDE see PMY000

A1-MORIN see MRN500

AMORPHAN see NNW500

AMORPHOUS AESCIN see EDK875

AMORPHOUS CROCIDOLITE ASBESTOS see ARM275

AMORPHOUS QUARTZ see SCK600

AMORPHOUS SILICA see DCJ800

AMORPHOUS SILICA see SCK600

AMORPHOUS SILICA DUST see SCH002

AMORPHOUS SILICA FUME see SCH001

AMOSCANATE see AOA050

AMOSENE see MQU750

AMOSITE ASBESTOS see ARM262

AMOSITE (OBS.) see ARM250

AMOSPAN see AMX750

AMOSULALOL HYDROCHLORIDE see AOA075

AMOSYT see DYE600

AMOTRIL see ARQ750

AMOTRIPHENE HYDROCHLORIDE see TNJ750

AMOX see NGS500

AMOXAPINE see AOA095

AMOXEPINE see AOA095

AMOXICILLIN mixed with POTASSIUM CLAVULANATE (2:1) see ARS125

AMOXICILLIN TRIHYDRATE see AOA100

AMOXONE see DAA800

AMP see AOA125

5-AMP see AOA125

cAMP see AOA130

3'-AMP see SPB100

5'-AMP see AOA125

3',5'-AMP see AOA130

AMP (nucleotide) see AOA125

AMPACET E/C see EHG100

AMPAZINE see DQA600

AMPD see ALB000

AMPELOPSIN (FLAVANOL) see HDW125

AMPELOPTIN see HDW125

AMPERIL see AIV500

AMPERIL see AOD125

AMPHAETEX see BBK500

AMPHATE see AOB500

AMPHEDRINE see BBK500

AMPHEDROXY see DBA800

AMPHEDROXYN see DBA800

AMPHENAZOLE HYDROCHLORIDE see DCA600

AMPHENICOL see CDP250

AMPHENIDONE see ALY250

AMPHEPRAMONUM HYDROCHLORIDE see DIP600

AMPHEREX see BBK500

AMPHETAMINE see AOA250

(+)-AMPHETAMINE see AOA500

d-AMPHETAMINE see AOA500

dl-AMPHETAMINE see BBK000

AMPHETAMINE HYDROCHLORIDE see AOA750

AMPHETAMINE PHOSPHATE see AOB500

dl-AMPHETAMINE PHOSPHATE see AOB500

dl-AMPHETAMINE SALT with FINE RESIN see AOB000

(−)-AMPHETAMINE SULFATE see BBK750

(+)-AMPHETAMINE SULFATE see BBK500

d-AMPHETAMINE SULFATE see BBK500

l-AMPHETAMINE SULFATE see BBK750

(±)-AMPHETAMINE SULFATE see AOB250

dl-AMPHETAMINE SULFATE see AOB250

AMPHETAMINIL see AOB300

dl-AMPHETAMINIL see AOB300

AMPHETANE PHOSPHATE see AOB500

AMPHIBOLE see ARM250

AMPHICOL see CDP250

AMPHISOL HYDROCHLORIDE see DCA600

AMPHITOL 24B see LBU200

AMPHITOL 20BS see LBU200

AMPHOIDS S see BBK250

AMPHOJEL see AHC000

AMPHOLAK XCO 30 see IAT250

AMPHOMORONAL see AOC500

AMPHOMYCIN see AOB875

AMPHORDS S see BBK250

AMPHOS see AOB500

AMPHOTERGE K-2 see AOC250

AMPHOTERIC-2 see AOC250

AMPHOTERIC 2 see IAT250

AMPHOTERIC-17 see AOC275

AMPHOTERICIN beta see AOC500

AMPHOTERICIN B see AOC500

AMPHOTERICIN B DEOXYCHOLATE see FPC200

AMPHOTERICIN B, METHYL ESTER HYDROCHLORIDE see AOC750
AMPHOTERICIN B, MIXT. WITH (3-α,5-β,12-α)-3,12-DIHYDROXYCHOLAN-24-OIC ACIDMONOSODIUM SALT see FPC200
AMPHOTERICIN B SODIUM DESOXYCHOLATE see FPC200
AMPHOTERICINE B see AOC500
AMPHOZONE see AOC500
AMPI-BOL see AIV500
AMPICHEL see AOD125
d-AMPICILLIN see AIV500
d-(−)-AMPICILLIN see AIV500
AMPICILLIN A see AIV500
AMPICILLIN ACID see AIV500
AMPICILLIN ANHYDRATE see AIV500
AMPICILLIN, (5-METHYL-2-OXO-1,3-DIOXOLEN-4-YL)METHYL ESTER, HYDROCHLORIDE see LEJ500
AMPICILLIN-OXACILLIN MIXTURE see AOC875
AMPICILLIN PIVALOYLOXYMETHYL ESTER HYDROCHLORIDE see AOD000
AMPICILLIN SODIUM see SEQ000
AMPICILLIN SODIUM SALT see SEQ000
AMPICILLIN TRIHYDRATE see AOD125
AMPICILLIN (USDA) see AIV500
AMPICIN see AIV500
AMPIKEL see AIV500
AMPIKEL see AOD125
AMPIMED see AIV500
AMPINOVA see AOD125
AMPIPENIN see AIV500
AMPLIACTIL see CKP250
AMPLIACTIL MONOHYDROCHLORIDE see CKP500
AMPLICITIL see CKP250
AMPLIGRAM see TEY000
AMPLIN see AOD125
AMPLISOM see AIV500
AMPLIT see IFZ900
AMPLITAL see AIV500
AMPLIVIX see EID200
5'-AMP POTASSIUM SALT see AEM500
AMPROLENE see EJN500
AMPROLIUM see AOD175
AMPROTROPINE PHOSPHATE see AOD250
AMP SODIUM SALT see AEM750
5'-AMP SODIUM SALT see AEM750
AMPS (SULFONIC ACID) see ADS300
AMPY-PENYL see AIV500
AMPYRONE see AIB300
AMPYROX see SBH500
AMRINONE see AOD375
AMRITAMYCIN see MQX250
AMROOD, extract see POH800
AMS see ANU650
AMS 5382 see CNB825
AMS 5656C see IGL110
AMSA see ADL750
m-AMSA see ADL750
m-AMSA see ADM000
AMSACRINE see ADL750
AMSACRINE LACTATE see AOD425
m-AMSA HYDROCHLORIDE see ADL500
m-AMSA LACTATE see AOD425
m-AMSA METHANESULFONATE see ADL750
AMSCO 140 see MQV900
AMSCO H-J see NAH600
AMSCO H-SB see NAH600
AMSECLOR see CDP250
AMSIDINE see ADL750
AMSINCKIA INTERMEDIA see TAG250
AMSINE see ADL750
AMSONIC ACID see FCA100
AMSONIC ACID DISODIUM SALT see FCA200
AMSPEC-KR see AQF000
AMSTAT see AJV500
AMSUBIT see IAD000

AMSULOSIN HYDROCHLORIDE see EFA200
AMSUSTAIN see AOA500
AMSUSTAIN see BBK500
AMTHIO see ANW750
AMUDANE see GKE000
AMUNO see IDA000
AMYBAL see AMX750
AMYDRICAINE see AHI250
AMYGDALIC ACID see MAP000
AMYGDALIN see AOD500
d,l-AMYGDALIN see IHO700
AMYGDALINIC ACID see MAP000
AMYGDALONITRILE see MAP250
n-AMYL ACETATE see AOD725
AMYL ACETATE (DOT) see AOD725
sec-AMYL ACETATE see AOD735
AMYL ACETATE (mixed isomers) see AOD750
AMYL ACETIC ESTER see AOD725
AMYL ACID PHOSPHATE (DOT) see PBW750
AMYL ALCOHOL see AOE000
N-AMYL ALCOHOL see AOE000
sec-AMYL ALCOHOL see PBM750
tert-AMYL ALCOHOL (DOT) see PBV000
AMYL ALCOHOL, NORMAL see AOE000
AMYL ALDEHYDE see VAG000
N-AMYLALKOHOL (CZECH) see AOE000
AMYLAMINE see PBV505
n-AMYLAMINE see PBV505
AMYLAMINE (DOT) see PBV505
iso-AMYLAMINE see AOE200
AMYLAMINE (mixed isomers) see PBV500
AMYLAMINES (DOT) see PBV500
2-N-AMYLAMINOETHYL-p-AMINOBENZOATE see PBV750
AMYLAZETAT (GERMAN) see AOD725
AMYL AZIDE see AOE500
AMYLBARBITONE see AMX750
5-n-AMYL-1:2-BENZANTHRACENE see AOE750
sec-AMYLBENZENE see PBV600
tert-AMYLBENZENE see AOF000
AMYL BENZOATE see IHP100
AMYL BENZOATE see PBV800
4-AMYL-N-BENZOHYDRYLPYRIDINIUM BROMIDE see AOF250
p-AMYLBENZOIC ACID see PBW000
AMYL BIPHENYL see AOF500
AMYL BROMIDE see AOF800
d-AMYL BROMIDE see AOF750
n-AMYL BROMIDE see AOF800
AMYL BUTYRATE see AOG000
n-AMYL BUTYRATE see AOG000
Δ-N-AMYLBUTYROLACTONE see NMV790
γ-N-AMYLBUTYROLACTONE see CNF250
AMYLCAINE see PBV750
AMYL CAPROATE see PBW450
AMYL CAPRONATE see PBW450
AMYLCARBINOL see HFJ500
n-AMYL CHLORIDE see PBW500
AMYL CHLORIDE (DOT) see PBW500
α-AMYL CINNAMALDEHYDE see AOG500
AMYL CINNAMATE see AOG600
AMYL CINNAMIC ACETATE see AOG750
α-AMYLCINNAMIC ALCOHOL see AOH000
α-AMYL CINNAMIC ALDEHYDE see AOG500
α-AMYLCINNAMYL ALCOHOL see AOH000
AMYL CINNAMYLIDENE METHYL ANTHRANILATE see AOH100
6-n-AMYL-m-CRESOL see AOH250
4-tert-AMYLCYCLOHEXANONE see AOH750
AMYLCYCLOHEXYL ACETATE (mixed isomers) see AOI000
AMYLDICHLORARSINE see AOI200
N-AMYLDICHLORARSINE see AOI200
AMYL-p-DIMETHYLAMINOBENZOATE see AOI250
AMYLDIMETHYL-p-AMINO BENZOIC ACID see AOI500

AMYL DIMETHYL PABA see AOI250
AMYLEINE see AOM000
AMYLENE see AOI800
tert-AMYLENE see IHR220
2,4-AMYLENEGLYCOL see PBL000
AMYLENE HYDRATE see PBV000
n-AMYLENE PENTENE see AOI800
AMYLENES, MIXED see AOJ000
AMYLESTER KYSELINY DUSICNE see AOL250
AMYLESTER KYSELINY OCTOVE see AOD725
2-AMYLESTER KYSELINY OCTOVE see AOD735
sek.AMYLESTER KYSELINY OCTOVE see AOD735
AMYL ETHER see PBX000
n-AMYL ETHER see PBX000
AMYLETHYLCARBINOL see OCY100
AMYL ETHYL KETONE see ODI000
AMYLETHYLMETHYLCARBINOL see MND050
AMYL FORMATE see AOJ500
n-AMYL FORMATE see AOJ500
2-AMYLFURAN see AOJ600
AMYL HARMOL HYDROCHLORIDE see AOJ750
o-n-AMYL HARMOL HYDROCHLORIDE see AOJ750
AMYL HEPTANOATE see AOJ900
AMYL HEXANOATE see IHU100
AMYL HEXANOATE see PBW450
n-AMYLHYDRAZINE HYDROCHLORIDE see PBX250
AMYL HYDRIDE (DOT) see PBK250
tert-AMYL HYDROPEROXIDE see PBX325
AMYL HYDROSULFIDE see PBM000
3-AMYL-1-HYDROXY-6,6,9-TRIMETHYL-6H-DIBENZO(b,d)PYRAN see CBD625
AMYL IODIDE see IET500
n-AMYL IODIDE see IET500
N-AMYL-p-IODOBENZYL CARBONATE see AOJ950
AMYLISOEUGENOL see AOK000
AMYL KETONE see ULA000
AMYL LACTATE see AOK250
AMYL LAURATE see AOK500
n-AMYL MERCAPTAN see PBM000
AMYL MERCAPTAN (DOT) see PBM000
tert-AMYLMERCAPTAN see MHS550
6-AMYLMERCAPTOPURINE see PBX900
AMYL METHYL ALCOHOL see AOK750
AMYL METHYL CARBINOL see HBE500
AMYL-METHYL-CETONE (FRENCH) see MGN500
tert-AMYL METHYL ETHER (ACGIH) see PBX400
n-AMYL METHYL KETONE see MGN500
AMYL METHYL KETONE (DOT) see MGN500
n-AMYL-N-METHYLNITROSAMINE see AOL000
AMYL NITRATE see AOL250
n-AMYL NITRITE see AOL500
AMYL NITRITE (DOT) see AOL500
N-(n-AMYL)-N'-NITRO-N-NITROSOGUANIDINE see NLC500
n-AMYLNITROSOUREA see PBX500
1-AMYL-1-NITROSOUREA see PBX500
n-AMYL-N-NITROSOURETHANE see AOL750
AMYLOBARBITAL see AMX750
AMYLOBARBITONE see AMX750
AMYLOCAINE see AOM000
AMYLOFENE see EOK000
AMYLOGLUCOSIDASE see AOM125
AMYLOMAIZE VII see SLJ500
AMYLOPECTINE SULPHATE see AOM150
AMYLOPECTIN, HYDROGEN SULFATE see AOM150
AMYLOPECTIN SULFATE see AOM150

AMYLOPECTIN SULFATE (SN-263) see
AOM150
β-AMYLOSE see CCU150
AMYLOWY ALKOHOL (POLISH) see IHP000
AMYLOXYISOEUGENOL see AOK000
o-AMYLPHENOL see AOM325
4-n-AMYLPHENOL see AOM250
AMYL PHENOL 4T see AON000
2-sec-AMYLPHENOL see AOM500
4-sec-AMYLPHENOL see AOM750
4-tert-AMYLPHENOL see AON000
p-tert-AMYLPHENOL see AON000
α-AMYL-β-PHENYLACROLEIN see AOG500
α-N-AMYL-β-PHENYLACRYL ACETATE see
AOG750
3-sec-AMYLPHENYL-N-
METHYLCARBAMATE see AON250
2-AMYL-3-PHENYL-2-PROPEN-1-OL see
AOH000
AMYL PHTHALATE see AON300
AMYL POTASSIUM XANTHATE see PKV100
AMYL PROPANOATE see AON350
AMYL PROPIONATE see AON350
n-AMYL PROPIONATE see AON350
AMYL SALICYLATE see SAK000
AMYLSINE see PBV750
AMYL SULFHYDRATE see PBM000
AMYL THIOALCOHOL see PBM000
n-AMYL THIOCYANATE see AON500
tert-AMYLTHIOL see MHS550
AMYL TRICHLOROSILANE see PBY750
AMYLTRIETHOXYSILANE see PBZ000
AMYLTRIMETHYLAMMONIUM IODIDE
see TMA500
AMYLUM see SLJ500
AMYL-Δ-VALEROLACTONE see DAF200
AMYLVINYLCARBINOL see ODW000
AMYL VINYL CARBINOL ACETATE see
ODW028
AMYL VINYL CARBINYL ACETATE see
ODW028
AMYL ZIMATE see BJK500
AMYRIS OIL see WBJ650
AMYRIS OIL, ACETYLATED see AON595
AMYRIS OIL, WEST INDIAN TYPE see
AON600
AMYTAL see AMX750
AMYTAL SODIUM see AON750
AN see AAQ500
6-AN see ALL250
AN 23 see DJO800
AN-148 see MDP750
AN-201 see TDC900
AN 1041 see LJR000
AN 1087 see EOY000
AN 1324 see GFM200
ANA see NAK500
6-ANA see ALL250
ANABACTYL see CBO250
ANABAENA FLOS-AQUAE TOXIN see
AON825
ANABASEINE see TCJ825
ANABASIDE HYDROCHLORIDE see PIT650
ANABASIN see AON875
(−)-ANABASIN see AON875
ANABASIN CHLORIDE see PIT650
ANABASINE see AON875
ANABASINE MONOHYDROCHLORIDE see
PIT650
ANABASINE, SULFATE (1:1) see PIV550
ANABASIN HYDROCHLORIDE see PIT650
ANABAZIN see AON875
ANABET see CNR675
ANABOLEEN see DME500
ANABOLEX see DME500
ANABOLIN see DAL300
ANABOLIT see CLG900
ANAC 110 see CNI000
ANACARDONE see DJS200
ANACEL see TBN000
ANACETIN see CDP250
ANACOBIN see VSZ000

ANACORDONE see DJS200
ANADOLOR see AIT250
ANADOMIS GREEN see CMJ900
ANADREX see DBB000
ANADROL see PAN100
ANADROYD see PAN100
ANAESTHETIC ETHER see EJU000
ANAFEBRINA see DOT000
ANAFLEX see UTU500
ANAFLON see HIM000
ANAFRANIL see CDU750
ANAFRANIL see CDV000
ANAGALLIS ARVENSIS LINN., EXTRACT
see JDS100
ANAGESTONE ACETATE mixed with
MESTRANOL (10:1) see AOO000
ANAGIARDIL see MMN250
ANAGYRINE see RHZ100
(−)-ANAGYRINE see RHZ100
ANAHIST see RDU000
ANALEPTIN see HLV500
ANALETIL see GCE600
ANALEXIN see PGG350
ANALEXIN see PGG355
ANALGIN-TEMPIDONE MIXTURE see
TAL335
ANALGIZER see DFA400
(d-ANA⁶))-LHRH ACETATE see LIU307
O-ANALOG of DIMETHOATE see DNX800
ANALUD see PEW000
ANALUX see DPE000
ANAMENTH see TIO750
ANAMID see SAH000
ANANASE see BMO000
ANANDAMIDE see HKR600
ANANDAMIDE (20.4, N-6) see HKR600
ANANSIOL see MNM500
ANAPAC see ABG750
A. NAPELLUS see MRE275
ANAPHRANIL see CDV000
ANAPOLON see PAN100
ANAPRAL see TLN500
ANAPREL see TLN500
ANAPRILIN see ICC000
ANAPRIME see FDD075
ANAPROTIN see DME500
ANARCON see AFT500
ANASCLEROL see VLF000
ANASTERON see PAN100
ANASTERONAL see PAN100
ANASTERONE see PAN100
ANASTRESS see MQU750
ANATASE see OBU100
ANATENSIN see GGS000
ANATENSOL see TJW500
ANATHYLMON see MQU750
ANATOLA see VSK600
ANATOXIN-a see AOO120
ANATOXIN I see AOO120
ANATRAN see ABH500
ANATRAN see HII500
ANATROPIN mixed with MESTRANOL (10:1)
see AOO000
ANAUTINE see DYE600
ANAVAR see AOO125
ANAYODIN see IEP200
ANAYODIN see SHW000
ANAZOLENE, SODIUM see ADE750
ANC 113 see OLW400
ANCAMINE 2049 see BGT800
ANCAMINE TL see MJQ000
ANCEF see CCS250
ANCHOIC ACID see ASB750
ANCHRED STANDARD see IHC450
ANCHUSA ACID see HLJ650
ANCHUSIN see HLJ650
ANCILLIN see AOD125
ANCISTRODON PISCIVORUS VENOM see
AOO135
ANCITABINE see COW875
ANCITABINE HYDROCHLORIDE see
COW900

ANCOBON see FHI000
ANCOLAN see HGC500
ANCOLAN DIHYDROCHLORIDE see
MBX250
ANCOR EN 80/150 see IGK800
ANCORTONE see PLZ000
ANCOTIL see FHI000
ANCOVENIN, 2-l-LYSINE-6-(3-
AMINOALANINE)-10-l-PHENYLALANINE-
12-l-PHENYLALANINE-13-l-V ALINE-15-
(ERYTHRO-3-HYDROXY-l-ASPARTIC
ACID)-, CYCLIC6-19)-IMINE see LEX100
ANCRACK see NBL200
ANCROD see VGU700
ANCYLOL see DNG000
ANCYTABINE see COW875
ANDAKSIN see MQU750
ANDANTOL see AEG625
ANDANTON see OGI075
ANDAXIN see MQU750
ANDERE see BQL000
ANDHIST see RDU000
ANDIAMINE see HFF500
ANDORDRIN DIPROPIONATE see AOO150
ANDRACTIM see DME500
ANDRAMINE see DYE600
ANDRANE see CCR510
ANDREST-5-EN-17-ONE, 3-β-HYDROXY-,
ACETATE HJB250
ANDREZ see SMR000
ANDRIOL see TBG050
ANDROCTONUS AMOREUXI VENOM see
AOO250
ANDROCTONUS AUSTRALIS HECTOR
VENOM see AOO265
ANDRODIOL see AOO475
ANDROFLUORENE see AOO275
ANDROFLUORONE see AOO275
ANDROFURAZANOL see AOO300
ANDROGEN see TBG000
ANDROLIN see TBF500
ANDROLONE see DME500
Δ¹⁰(18))-ANDROMEDENOL see GJU300
ANDROMEDOL see GJU310
ANDROMEDOTOXIN see AOO375
ANDROMETH see MPN500
ANDRONAQ see TBF500
ANDROSAN see MPN500
ANDROSAN see TBG000
ANDROSAN (tablets) see MPN500
ANDROSTA-1,4-DIENE-3,17-DIONE, 6-
METHYLENE- see MJL275
ANDROSTA-2,5-DIENO(2,3-d)ISOXAZOL-
17-OL, 4,4,17-TRIMETHYL-, (17-β)-(9CI) see
HOM259
ANDROSTA-3,5-DIEN-17-ONE, 3-(1-
PYRROLIDINYL)- see PPU780
3,5-ANDROSTADIEN-17-β-
YLTRIMETHYLAMMONIUM IODIDE see
AOO400
ANDROSTA-3,5-DIEN-17-β-
YLTRIMETHYLAMMONIUM IODIDE see
AOO400
ANDROSTALONE see MJE760
ANDROSTANAZOL see AOO401
ANDROSTANAZOLE see AOO401
5-α-ANDROSTAN-3-α,17-β-DIOL, 2-β,16-β-
BIS(1-METHYLPIPERIDINIO)-,
DIBROMIDE see AOO403
5α-ANDROSTAN-3α,17β-DIOL, 2β,16β-
DIPIPECOLINIO-, DIBROMIDE,
DIACETATE see PAF625
ANDROSTANE-2-CARBONITRILE, 4,5-
EPOXY-17-HYDROXY-3-OXO-, (2-α-4-α-5-α-
17-β)- see EBY600
5-α-ANDROSTANE-2-α-CARBONITRILE, 4-
α-5-EPOXY-17-β-HYDROXY-3-OXO- see
EBY600
3-β,17-β-ANDROSTANEDIOL see AOO405
ANDROSTANE-3,17-DIOL, (3-β,5-α,17-β)-
(9CI) see AOO405

5-α-ANDROSTANE-3-β,17-β-DIOL see AOO405
5-α-ANDROSTANE-17-α-METHYL-17-β-OL-3-ONE see MJE760
ANDROSTANOLONE see DME500
5-α-ANDROSTAN-17-β-OL-3-ONE see DME500
5-α-ANDROSTAN-3-α-OL-17-ONE see HJB050
ANDROSTANOLONE PROPIONATE see DME525
ANDROSTAN-17-ONE, 3-(ACETYLOXY)-, (3-β,5-α)- see HJB225
ANDROSTAN-17-ONE, 3-HYDROXY-, (3-α-5-α)- see HJB050
5-α-ANDROSTAN-17-ONE, 3-α-HYDROXY-see HJB050
5-β,14-β-ANDROSTAN-3-ONE, 17-β-HYDROXY-, ACRYLATE see HJB150
ANDROSTAN-3-ONE, 17-HYDROXY-17-METHYL-, (5-α-17-β)-(9CI) see MJE760
5-α-ANDROSTAN-3-ONE, 17-β-HYDROXY-, UNDECANOATE see TBG050
ANDROSTANON-3-α-OL-17-ONE see HJB050
ANDROSTEN see MPN500
Δ⁴-ANDROSTEN-3,17-DIONE see AOO425
ANDROST-2-ENE-2-CARBONITRILE, 3,17-DIHYDROXY-4,17-DIMETHYL-4,5-EPOXY-, (4-α-5-α- 17-β)- see EBH400
ANDROST-5-ENE-3,17-DIOL, DIPROPANOATE, (3-β,17-β)- (9CI) see AOO410
ANDROSTENEDIOL DIPROPIONATE see AOO410
ANDROST-5-ENE-3-β,17-β-DIOL, DIPROPIONATE see AOO410
ANDROSTENEDIONE see AOO425
4-ANDROSTENE-3,17-DIONE see AOO425
Δ-4-ANDROSTENEDIONE see AOO425
Δ⁴-ANDROSTENE-3,17-DIONE see AOO425
ANDROSTENEDIONE ENAMINE see PPU780
4-ANDROSTENE-17-α-METHYL-17-β-OL-3-ONE see MPN500
Δ⁴-ANDROSTENE-17-β-PROPIONATE-3-ONE see TBG000
ANDROST-2-EN-17-OL, 17-METHYL-, (5-α,17-β)- see MGN600
5-α-ANDROST-2-EN-17-β-OL, 17-METHYL-see MGN600
ANDROSTENOLONE see AOO450
ANDROST-4-EN-17β-OL-3-ONE see TBF500
Δ⁴-ANDROSTEN-17(β)-OL-3-ONE see TBF500
ANDROSTENOLONE ACETATE see HJB250
ANDROST-5-EN-17-ONE, 3-(ACETYLOXY)-, (3-β)- see HJB250
ANDROST-4-EN-3-ONE, 17-(ACETYLOXY)-4-CHLORO-, (17-β)- see CLG900
ANDROST-4-EN-3-ONE, 4-CHLORO-17-β-HYDROXY-, ACETATE see CLG900
ANDROST-4-EN-3-ONE, 17-(3-CYCLOPENTYL-1-OXOPROPOXY)-, (17-β)-(9CI) see TBF600
ANDROSTEN-17-ONE, 3-β-HYDROXY-, ACETATE see HJB225
ANDROST-5-EN-17-ONE, 3-β-HYDROXY-, ACETATE see HJB250
ANDROST-4-EN-3-ONE, 17-β-HYDROXY-1-α-7-α-DIMERCAPTO-17-METHYL-, 1,7-DIACETATE see TFK300
ANDROST-5-EN-17-ONE, 3-β-HYDROXY-, HYDROGEN SULFATE see SNV100
ANDROST-4-EN-3-ONE, 17-β-HYDROXY-, (17-β)-, MIXT WITH (17-β)-1,3,5(10)-TRIENE-3,17-DIOL see TBF650
ANDROST-4-EN-3-ONE, 17-((1-OXOUNDECYL)OXY)-, (17-β)- see TBG050
ANDROST-5-EN-17-ONE, 3-(SULFOOXY)-, (3-β)- see SNV100
ANDROSTEROLO see AOO275

ANDROSTERONE see HJB050
cis-ANDROSTERONE see HJB050
ANDROSTESTONE-M see AOO475
ANDROSTESTON-M see AOO475
ANDROTARDYL see TBF750
ANDROTESTON see TBG000
ANDROTEST P see TBG000
ANDROTEX see AOO425
ANDRUSOL see TBF500
ANDRUSOL-P see TBG000
AN-DTPA see TAI200
ANECOTAN see DFA400
ANECTINE see CMG250
ANECTINE see HLC500
ANECTINE CHLORIDE see HLC500
ANELIX see HIM000
ANELMID see DJT800
ANEMONE see PAM780
ANEMONE (VARIOUS SPECIES) see PAM780
ANERGAN see ABH500
ANERTAN see MPN500
ANERTAN see TBG000
ANERTAN (tablets) see MPN500
ANERVAL see BRF500
ANESTACON see DHK400
ANESTACON HYDROCHLORIDE see DHK600
ANESTHENYL see MGA850
ANESTHESIA ETHER see EJU000
ANESTHESIN see EFX000
ANESTHESOL see AIT250
ANESTHETIC COMPOUND No. 347 see EAT900
ANESTHETIC ETHER see EJU000
ANESTHONE see EFX000
ANESTIL see AIT250
ANETAIN see BQA010
ANETHAINE see TBN000
trans-ANETHOL see PMR250
cis-ANETHOLE see PMR000
trans-ANETHOLE see PMR250
ANETHOLE (FCC) see PMQ750
ANETHOLE TRITHIONE see AOO490
ANETHOLTRITHION see AOO490
ANEURAL see MQU750
ANEURIMEC see DXO300
ANEURINE see TES750
ANEURINE DISULFIDE see TES800
ANEUXRAL see MQU750
ANEXOL see CDP000
ANFETAMINA see AOB250
ANFLAGEN see IIU000
ANFOTERICO LB see LBU200
ANFRAM 3PB see TNC500
ANFT see ALM500
ANG 66 see AOP250
ANGECIN see FQC000
ANGEL DUST see AOO500
ANGELICA LACTONE see MKH250
α-ANGELICA LACTONE see AOO750
β-ANGELICA LACTONE see MKH500
Δ¹-ANGELICA LACTONE see MKH500
Δ²-ANGELICA LACTONE see MKH250
β,γ-ANGELICA LACTONE see MKH250
ANGELICA OIL, root see AOO760
ANGELICA ROOT OIL see AOO760
ANGELICA SEED OIL see AOO790
ANGELICIN (coumarin derivative) see FQC000
ANGELICIN (STEROID) see SDZ350
ANGELIKA OEL see AOO760
ANGELI'S SULFONE see AOO800
ANGELI SULFONE see AOO800
ANGEL'S TRUMPET see AOO825
ANGEL TULIP see SLV500
ANGEL WINGS see CAL125
ANGIBID see NGY000
ANGICAP see PBC250
ANGIFLAN see DBX400
ANGIGRAFIN see AOO875
ANGINAL see PCP250

ANGININ see PPH050
ANGININE see NGY000
ANGININE see PPH050
ANGINON see AHI875
ANGINYL see DNU600
ANGIOCICLAN see POD750
ANGIO-CONRAY see IGC100
ANGIO-CONTRIX see IGC100
ANGIOGRAFIN see AOO875
ANGIOKAPSUL see ARQ750
ANGIOKINASE see PJB800
ANGIOLINGUAL see NGY000
ANGIOMIN see XCS000
ANGIOPAC see VLF000
ANGIOTENSIN see AOO900
ANGIOTENSIN II, HUMAN see IKX030
ANGIOTENSIN II, 5-l-ISOLEUCINE- see IKX030
ANGIOTENSIN II (MOUSE) see IKX030
ANGIOTONIN see AOO900
ANGIOXINE see PPH050
ANGITET see PBC250
ANGITRIT see TJL250
ANGLISLITE see LDY000
ANGOLAMYCIN see AOO925
ANGOPRIL see IIG600
ANGORIN see NGY000
ANGORLISIN see ELH600
ANG.-STERANTHREN (GERMAN) see DCR800
ANG-STERANTHRENE see DCR800
ANGUIDIN see AOP250
ANGUIDIN, ACETATE see SBF525
ANGUIDINE see AOP250
ANGUIFUGAN see DJT800
ANHIBA see HIM000
ANHISTABS see WAK000
ANHISTAN see FOS100
ANHISTOL see WAK000
ANHITOL 24B see LBU200
ANHYDRIDE ACETIQUE (FRENCH) see AAX500
ANHYDRIDE ARSENIEUX see ARI750
ANHYDRIDE ARSENIQUE (FRENCH) see ARH500
ANHYDRIDE CARBONIQUE (FRENCH) see CBU250
ANHYDRIDE CARBONIQUE et OXYDE d'ETHYLENE MELANGES (FRENCH) see EJO000
ANHYDRIDE CHROMIQUE (FRENCH) see CMK000
ANHYDRIDE KYSELINY 4-CHLOR-1,2,3,6-TETRAHYDROFTA-LOVE (CZECH) see CFG500
ANHYDRIDE PHTHALIQUE (FRENCH) see PHW750
ANHYDRIDES see AOP500
ANHYDRIDE VANADIQUE (FRENCH) see VDU000
ANHYDRID KYSELINY 3,6-ENDOMETHYLEN-Δ₄-TETRAHYDROFTALOVE (CZECH) see BFY000
ANHYDRID KYSELINY KROTONOVE see COB900
ANHYDRID KYSELINY MASELNE see BSW550
ANHYDRID KYSELINY OCTOVE see AAX500
ANHYDRID KYSELINY TETRAHYDROFTALOVE (CZECH) see TDB000
ANHYDRID KYSELINY TRIFLUOROCTOVE (CZECH) see TJX000
2,2'-ANHYDRO-1-β-d-ARABINOFURANOSYLCYTOSINE HYDROCHLORIDE see COW900
2,2'-ANHYDROARABINOSYLCYTOSINE see COW875
2,2'-ANHYDROARABINOSYLCYTOSINE HYDROCHLORIDE see COW900

ANHYDROARA C see COW875

ANHYDRO-4,4'-BIS(DIETHYLAMINO)TRIPHENYLMETHANOL-2',4"-DISULPHONIC ACID, MONOSODIUM SALT see ADE500

ANHYDRO-4-CARBAMOYL-5-HYDROXY-1-β-d-RIBOFURANOSYL-IMIDAZOLIUMHYDROXIDE see BMM000

2,2'-ANHYDROCYTARABINE HYDROCHLORIDE see COW900

ANHYDROCYTIDINE see COW875

2,2'-ANHYDROCYTIDINE see COW875

2,2'-ANHYDROCYTIDINE HYDROCHLORIDE see COW900

8,9-ANHYDRO-4"-DEOXY-3'-N-DESMETHYL-3'-N-ETHYLERYTHROMYCIN B-6,9-HEMIACETAL see AOP502

ANHYDRO-DIMETHYLAMINO HEXOSE REDUCTONE see AOP510

3,6-ANHYDRO-d-GALACTAN see CCL250

ANHYDROGITALIN see GEU000

ANHYDRO-d-GLUCITOL MONOOCTADECANOATE see SKV150

ANHYDROGLUCOCHLORAL see GFA000

10-(1',5'-ANHYDROGLUCOSYL)ALOE-EMODIN-9-ANTHRONE see BAF825

ANHYDROHEXITOL SESQUIOLEATE see SKV170

ANHYDROHYDROXYPROGESTERONE see GEK500

ANHYDROL see EFU000

ANHYDROMYRIOCIN see AOP750

ANHYDRONE see PCE000

ANHYDRO-PIPERIDINO HEXOSE REDUCTONE see AOP800

ANHYDRORETINOL see AFQ700

ANHYDROSORBITOL STEARATE see SKV150

ANHYDRO-o-SULFAMINEBENZOIC ACID see BCE500

ANHYDROTRIMELLIC ACID see TKV000

ANHYDROUS AMMONIA see AMY500

ANHYDROUS BORAX see DXG035

ANHYDROUS CALCIUM SULFATE see CAX500

ANHYDROUS CHLORAL see CDN550

ANHYDROUS CHLOROBUTANOL see ABD000

ANHYDROUS HYDRIODIC ACID see HHI500

ANHYDROUS HYDROBROMIC ACID see HHJ000

ANHYDROUS HYDROCHLORIC ACID see HHL000

ANHYDROUS IRON OXIDE see IHC450

ANHYDROUS OXIDE of IRON see IHC450

ANHYDROUS SODIUM ARSANILATE see ARA500

ANHYDROVITAMIN A see AFQ700

ANHYDROVITAMIN A1 see AFQ700

trans-ANHYDROVITAMIN A see AFQ700

ANHYDROXYPROGESTERONE see GEK500

ANI see ISN000

ANICON KOMBI see CIR250

ANICON M see CIR250

ANICON P see RBF500

ANIDRIDE ACETICA (ITALIAN) see AAX500

ANIDRIDE CROMICA (ITALIAN) see CMK000

ANIDRIDE CROMIQUE (FRENCH) see CMJ900

ANIDRIDE FTALICA (ITALIAN) see PHW750

ANILANA, combustion products see ADX750

ANILAZIN see DEV800

ANILAZINE see DEV800

l'ANILIDE de l'ACIDE (PYRROLIDINO-N)-3-N-BUTYRIQUE (FRENCH) see MPB500

ANILIDE of (PYRROLIDINO-N)-3-N-BUTYRIC ACID see MPB500

ANILID KYSELINY ACETOCTOVE see AAY000

ANILID KYSELINY SALICYLOVE (POLISH) see SAH500

ANILIN (CZECH) see AOQ000

ANILINA (ITALIAN, POLISH) see AOQ000

ANILINE see AOQ000

ANILINE, p-(ACETOXYMERCURI)-N,N-DIETHYL- see AAS300

ANILINE, p-(ACETOXYMERCURI)-N,N-DIMETHYL- see AAS310

ANILINE, N-ACETYL- see AAQ500

ANILINE ANTIMONYL TARTRATE see AOQ250

ANILINE, ARSENATE see ARB270

ANILINE, p-ARSENOSO- see ARJ755

ANILINE, p-ARSENOSO-N,N-BIS(2-CHLOROETHYL)- see ARJ760

ANILINE, p-ARSENOSO-N,N-BIS(2-HYDROXYETHYL)- see ARJ770

ANILINE, p-ARSENOSO-N,N-DIETHYL- see ARJ800

ANILINE, p-ARSENOSO-, DIHYDRATE see ALV100

p-ANILINEARSONIC ACID see ARA250

ANILINE, 4,4'-AZODI- see ASK925

ANILINE, p-(5-BENZOFURYLAZO)-N,N-DIMETHYL- see BCK800

ANILINE, p-(7-BENZOFURYLAZO)-N,N-DIMETHYL- see BCL100

ANILINE, N-BENZYLIDENE-m-NITRO- see BBA800

ANILINE, 2,6-BIS(ACETOXYMERCURI)-4-NITRO- see BGQ300

ANILINE, N,N-BIS(2-BROMOETHYL)- see BHG100

ANILINE, N,N-BIS(2-CHLOROETHYL)-2,3-DIMETHOXY- see BIC600

ANILINE, N,N-BIS(2-(2,3-EPOXYPROPOXY)ETHOXY)- see BJN850

ANILINE, N,N-BIS(2-(2,3-EPOXYPROPOXY)ETHYL)- see BJN875

ANILINE, N,N-BIS(2-IODOETHYL)- see BKJ600

ANILINE, p-((3-BROMO-4-ETHYLPHENYL)AZO)-N,N-DIMETHYL- see BNK275

ANILINE, p-((4-BROMO-3-ETHYLPHENYL)AZO)-N,N-DIMETHYL- see BNK100

ANILINE, 2-BROMO-4-NITRO- see BNS800

ANILINE, p-(m-BROMOPHENYLAZO)-N,N-DIMETHYL- see BNE600

ANILINE, p-(p-BROMOPHENYLAZO)-N,N-DIMETHYL- see BNV760

ANILINE, p-((3-BROMO-p-TOLYL)AZO)-N,N-DIMETHYL- see BNQ100

ANILINE, p-((4-BROMO-m-TOLYL)AZO)-N,N-DIMETHYL- see BNQ110

ANILINE, 4-BUTYL- see AJA550

ANILINE, N-sec-BUTYL-4-tert-BUTYL-2,6-DINITRO- see BQN600

ANILINE, p-BUTYL-N-(p-METHOXYBENZYLIDENE)- see MEE100

ANILINE, p-((p-BUTYLPHENYL)AZO)-N,N-DIMETHYL- see BRB450

ANILINE, p-((p-(tert-BUTYL)PHENYL)AZO)-N,N-DIMETHYL- see BRB460

ANILINE-3-CARBOXYLIC ACID see AIH500

ANILINE CARMINE POWDER see FAE100

ANILINE CHLORIDE see BBL000

ANILINE, m-CHLORO-N-(2,2-DIFLUOROETHYL)- see CFX300

ANILINE, 4-CHLORO-2,5-DIMETHOXY- see CGC050

ANILINE, 2-CHLORO-N-(2,4-DINITRO-6-(TRIFLUOROMETHYL)PHENYL)-5-(TRIFLUOROMETHYL)- see FDB300

ANILINE, 2-CHLORO-N-(4,6-DINITRO-α-α-α-TRIFLUORO-O-TOLYL)-5-(TRIFLUOROMETHYL)- see FDB300

ANILINE, N-(2-CHLOROETHYL)-N-ETHYL- see EHJ600

ANILINE, m-CHLORO-, HYDROCHLORIDE see CEH690

ANILINE, p-CHLORO-, HYDROCHLORIDE see CJR200

ANILINE, 5-CHLORO-2-MERCAPTO-, HYDROCHLORIDE see CHU100

ANILINE, 2-CHLORO-4-(METHYLSULFONYL)- see CIX200

ANILINE, 2-CHLORO-5-NITRO- see CJA180

ANILINE, 4-CHLORO-2-NITRO- see KDA050

ANILINE, 4-CHLORO-3-NITRO- see CJA185

ANILINE COMPLEX WITH TRINITROBENZENE see BBL100

ANILINE, N-(2-CYANOETHYL)-N-ETHYL- see EHQ500

ANILINE, 2,6-DIBROMO-4-NITRO- see DDQ450

ANILINE, 2,4-DICHLORO- see DEO290

ANILINE, p-DICHLOROARSINO-, HYDROCHLORIDE see AOR640

ANILINE, 2,3-DICHLORO-(7CI,8CI) see DEO210

ANILINE, 2,5-DICHLORO-4-NITRO- see DFT100

ANILINE, 2,6-DIETHYL- see DIS650

ANILINE, N-DIETHYLAMINOETHYL-N-ETHYL-, HYDROCHLORIDE see PFB850

ANILINE, N,N-DIETHYL-, HYDROCHLORIDE see DIS720

ANILINE, N,N-DIETHYL-p-NITROSO- see NJW600

ANILINE, 2,4-DIFLUORO- see DKF700

ANILINE, 2,6-DIISOPROPYL- see DNN630

ANILINE, 2,5-DIMETHOXY- see AKD925

ANILINE, 2,4-DIMETHOXY-, HYDROCHLORIDE see MEA625

ANILINE, N,N-DIMETHYL-p-(4'-CHLORO-3'-METHYLPHENYLAZO)- see CIL710

ANILINE, N,N-DIMETHYL-p-(3,5-DIFLUOROPHENYLAZO)- see DKH100

ANILINE, N,N'-DIMETHYL-4,4'-METHYLENEDI- see MJO000

ANILINE, N,N-DIMETHYL-o-NITRO- see NFW600

ANILINE-2,5-DISULFONIC ACID see AIE000

ANILINE, p,p'-DITHIODI- see ALW100

ANILINE DYES see AOQ500

ANILINE, N-(2,3-EPOXYPROPYL)-N-METHYL- see GGY155

ANILINE, p-ETHOXY- see PDK790

ANILINE, 2-ETHOXY-5-NITRO- see NIC200

ANILINE, m-ETHYL- see EOH100

ANILINE, o-ETHYL-(8CI) see EGK500

ANILINE, p-((o-ETHYLPHENYL)AZO)-N,N-DIMETHYL- see EIF450

ANILINE, 4-FLUORO-3-NITRO- see FKK100

ANILINE-FORMALDEHYDE CONDENSATE see FMW330

ANILINE-FORMALDEHYDE POLYMER see FMW330

ANILINE GREEN see AFG500

ANILINE GREEN see BAY750

ANILINE HYDROCHLORIDE (DOT) see BBL000

ANILINE, 2-(HYDROXYMERCURI)-4-NITRO- see HLO350

ANILINE, p-(ISOBORNYLOXY)- see IHY500

ANILINE, o-ISOPROPYL- see INW100

ANILINE, 2-METHOXY-4-NITRO- see NEQ000

ANILINE, 4,4'-METHYLENEBIS(o-ETHYL)- see MJN100

ANILINE, 2-METHYL-4-NITRO- see MMF780

ANILINE, m-(METHYLTHIO)- see ALX100

ANILINE, p-(METHYLTHIO)- see AMS675

ANILINE MUSTARD see AOQ875

ANILINE, 4-NITRO- see NEO500

ANILINE, N-NITRO- see NEO510
ANILINE OIL see AOQ000
ANILINE OIL DRUMS, EMPTY see AOR000
ANILINE, 2,3,4,5,6-PENTANITRO- see PBM100
ANILINE, 2-PHENOXY- see PDR490
ANILINE, o-PHENOXY-(8CI) see PDR490
ANILINE, p,p'-(m-PHENYLENEDIOXY)DI- see REF070
ANILINE, POLYMER with FORMALDEHYDE (8CI) see FMW330
ANILINE, N,N'-1-PROPEN-1-YL-3-YLIDENEDI-, MONOHYDROCHLORIDE see PFJ500
"ANILINE SALT" see BBL000
p-ANILINESULFONAMIDE see SNM500
ANILINE-4-SULFONIC ACID see SNN600
m-ANILINESULFONIC ACID see SNO000
ANILINE-p-SULFONIC ACID see SNN600
ANILINE-p-SULFONIC AMIDE see SNM500
ANILINE, 4,4'-(SULFONYLBIS(p-PHENYLENEOXY))DI- see SNZ000
ANILINE-p-SULPHONIC ACID see SNN600
ANILINE, 2-(o-TOLYLOXY)- see THB100
ANILINE, 2,3,4-TRIFLUORO- see TJX900
ANILINE, N-(2,2,2-TRIFLUOROETHYL)- see TKA950
ANILINE, 2,4,6-TRINITRO- see PIC800
ANILINE VANADATE, DIHYDRATE see AOR250
ANILINE VIOLET see AOR500
ANILINE VIOLET PYOKTANINE see AOR500
ANILINE YELLOW see PEI000
ANILINIUM CHLORIDE see BBL000
ANILINIUM NITRATE see AOR625
ANILINIUM PERCHLORATE see AOR630
β-ANILINOACROLEIN ANIL HYDROCHLORIDE see PFJ500
p-ANILINOANILINE see PFU500
ANILINOBENZENE see DVX800
2-ANILINOBENZOIC ACID see PEG500
o-ANILINOBENZOIC ACID see PEG500
2-ANILINO-4'-(BENZYLOXY)-2-PHENYLACETOPHENONE see AOR635
4'-(α-ANILINOBENZYLOXY)-2-PHENYLACETOPHENONE see AOR635
α-ANILINO-p-BENZYLOXYPHENYL BENZYL KETONE see AOR635
4-ANILINODICHLOROARSINE, HYDROCHLORIDE see AOR640
2-ANILINO-4'-(2-(DIETHYLAMINO)ETHOXY)-2-PHENYLACETOPHENONE HYDROCHLORIDE see AOR700
α-ANILINO-p-DIETHYLAMINOETHOXYPHENYL BENZYL KETONE HYDROCHLORIDE see AOR700
2-ANILINO-5-(2,4-DINITROANILINO)BENZENESULFONIC ACID SODIUM SALT see SGP500
1-ANILINO-2,5-DISULFONIC ACID see AIE000
ANILINOETHANE see EGK000
2-ANILINOETHANOL see AOR750
(2-ANILINOETHYL)HYDRAZONE DIHYDROCHLORIDE see AOS000
ANILINOMETHANE see MGN750
ANILINONAPHTHALENE see PFT500
1-ANILINONAPHTHALENE see PFT250
2-ANILINONAPHTHALENE see PFT500
2-ANILINO-5-NITROBENZENESULFONIC ACID see AOS500
2-(((2-ANILINO-5-NITROPHENYL)IMINO)METHYL)-4-CHLOROPHENOL see NFT100
2-(((2-ANILINO-5-NITROPHENYL)IMINO)METHYL)-6-METHOXYPHENOL see PDN800

o-(((2-ANILINO-5-NITROPHENYL)IMINO)METHYL)-6-NITROPHENOL see NHR640
4-(((2-ANILINO-5-NITROPHENYL)IMINO)METHYL)PHENOL see NHY600
o-(((2-ANILINO-5-NITROPHENYL)IMINO)METHYL)PHENOL see NHY700
ANILINO (p-NITROPHENYL) SULFIDE see AOS750
4-ANILINOPHENOL see AOT000
p-ANILINOPHENOL see AOT000
3-ANILINOPROPIONITRILE see AOT100
β-ANILINOPROPIONITRILE see AOT100
4-((4-ANILINO-5-SULFO-1-NAPHTHYL)AZO)-5-HYDROXY-2,7-NAPHTHALENEDIFULFONIC ACID TRISODIUM see ADE750
ANILINO-2-SULFONIC ACID see SNO100
ANILINO-o-SULFONIC ACID see SNO100
6-(p-ANILINOSULFONYL)METANILAMIDE see AOT125
ANILINO-o-SULPHONIC ACID see SNO100
ANILITE see AOT250
ANILIX see CKL500
ANILOFOS see AOT255
ANILOPHOS see AOT255
ANIMAG see MAH500
ANIMAL CONIINE see PBK500
ANIMAL GALACTOSE FACTOR see OJV500
ANIMAL OIL see BMA750
ANIMAL STARCH see GHK300
ANIMERT see CKL750
ANIMERT V-10 see CKL750
ANIMERT V-101 see CKL750
ANIMERT V-10K see CKL750
ANIONIC GREG M see AOT260
ANIPRIME see FDD075
ANISALACETONE see MLI400
2-ANISALDEHYDE see AOT525
m-ANISALDEHYDE see AOT300
o-ANISALDEHYDE see AOT525
p-ANISALDEHYDE see AOT530
p-ANISALDEHYDE, 2-HYDROXY- see HLR600
o-ANISAMIDE see AOT750
o-ANISAMIDE, N-((1-ETHYL-2-PYRROLIDINYL)METHYL)-5-(ETHYLSULFONYL)- see EPD100
p-ANISANILIDE, 3-NITRO- see NEP650
(−)-ANISATIN see AOT800
ANISATIN see AOT800
ANISE ALCOHOL see MED500
ANISE CAMPHOR see PMQ750
ANISEED OIL see AOU250
ANISENE see CLO750
ANISE OIL see AOU250
4-ANISIC ACID see AOU600
m-ANISIC ACID see AOU500
o-ANISIC ACID see MPI300
p-ANISIC ACID see AOU600
p-ANISIC ACID, ETHYL ESTER see AOV000
ANISIC ACID HYDRAZIDE see AOV500
o-ANISIC ACID, HYDRAZIDE see AOV250
p-ANISIC ACID, HYDRAZIDE see AOV500
p-ANISIC ACID, METHYL ESTER see AOV750
o-ANISIC ACID, METHYL ESTER (7CI,8CI) see MLH800
p-ANISIC ACID, 2-PHENYLHYDRAZIDE see MEC332
ANISIC ALCOHOL see MED500
ANISIC ALDEHYDE see AOT530
ANISIC HYDRAZIDE see AOV500
ANISIC KETONE see AOV875
2-ANISIDINE see AOV900
4-ANISIDINE see AOW000
m-ANISIDINE see AOV890
o-ANISIDINE see AOV900
p-ANISIDINE see AOW000

m-ANISIDINE ANTIMONYL TARTRATE see AOW500
o-ANISIDINE ANTIMONYL TARTRATE see AOW750
p-ANISIDINE ANTIMONYL TARTRATE see AOX000
o-ANISIDINE HYDROCHLORIDE see AOX250
p-ANISIDINE HYDROCHLORIDE see AOX500
2-ANISIDINE NITRATE see MEA600
o-ANISIDINE NITRATE see NEQ500
m-ANISIDINE, 4-(PHENYLAZO)- see MDY400
ANISKETONE see AOV875
ANIS OEL (GERMAN) see AOU250
p-ANISOL ALCOHOL see MED500
ANISOLE see AOX750
ANISOLE, 2-AMINO-5-NITRO- see NEQ000
ANISOLE, o-BROMO- see BMT400
ANISOLE, p-BROMO- see AOY450
ANISOLE, p-(3-BROMOPROPENYL)-, (E)- see BMT300
ANISOLE, 2,4-DIAMINO-, HYDROGEN SULFATE see DBO400
ANISOLE, 2,4-DIAMINO-, SULFATE see DBO400
ANISOLE, 2-ISOPROPYL-5-METHYL- see MPW650
ANISOLE, o-(2-NITROVINYL)- (7CI,8CI) see NMR100
ANISOLE, 2,4,6-TRINITRO- see TMK300
ANISOMYCIN see AOY000
ANISOPIROL see HAH000
ANISOPYRADAMINE see DBM800
ANISOTROPINE METHOBROMIDE see LJS000
3-p-ANISOYL-3-BROMOACRYLIC ACID, SODIUM SALT see SIK000
(E)-3-p-ANISOYL-3-BROMOACRYLIC ACID SODIUM SALT see CQK600
ANISOYL CHLORIDE see AOY250
N-ANISOYL-GABA see AOY270
ANISOYLHYDRAZINE see AOV500
p-ANISOYLHYDRAZINE see AOV500
3-ANISOYL-2-MESITYLBENZOFURAN see AOY300
β-(p-ANISOYLPHENYL)HYDRAZIDE see MEC332
ANISTADIN see HII500
ANISYL ACETATE see AOY400
2-(p-ANISYL)ACETIC ACID see MFE250
ANISYLACETONE see MFF580
ANISYLACETONITRILE see MFF000
ANISYL ALCOHOL (FCC) see MED500
m-ANISYLAMINE see AOV890
o-ANISYLAMINE see AOV900
p-ANISYLAMINE see AOW000
o-ANISYLAMINE HYDROCHLORIDE see AOX250
ANISYL BROMIDE see AOY450
ANISYL BROMIDE see BMT400
N-(o-ANISYL)-2-(p-BUTOXYPHENOXY)-N-(2(DIETHYLAMINO)ETHYL)ACETAMIDE HYDROCHLORIDE see APA000
ANISYL-N-BUTYRATE see MED750
ANISYL FORMATE see MFE250
ANISYLIDENE ACETONE see MLI400
ANISYL METHYL KETONE see AOV875
ANISYL PHENYLACETATE see APE000
p-ANISYL 3-PYRIDYL KETONE see MED100
p-ANISYL 4-PYRIDYL KETONE see MFH930
2-(p-ANISYLSULFONAMIDO)-5-ISOBUTYL-1,3,4-THIADIAZOLE see IIY100
p-ANISYOL CHLORIDE see AOY250
ANIT see ISN000
ANITSOTROPINE METHYLBROMIDE see LJS000
ANKERBIN see SLJ050
ANKILOSTIN see PCF275
ANN (GERMAN) see AAW000
ANNALINE see CAX750

ANNANOX CK see SCQ000
ANNATTO see APE050
ANNATTO COLORING DYE see APE050
ANNATTO EXTRACT see APE100
ANNONA MURICATA see SKV500
ANNOTTA see APE050
ANNUAL POINSETTIA see EQX000
(6)ANNULENE see BBL250
ANODYNON see EHH000
ANOFEX see DAD200
ANOL see CPB750
ANON BL see LBU200
ANOPROLIN see ZVJ000
A-NORANDROSTANE-2-α-17-α-
DIETHYNYL-2-β,17-β-DIOL see EQN320
ANORDRIN see AOO150
ANOREXIDE see BBK000
ANOVIGAM see EEM000
ANOVLAR 21 see EEH520
ANOX 20 see DED100
β-ANOXIN see DNM400
ANOXOMER see APE300
ANOZOL see DJX000
ANP 2 see AHP755
ANP 235 see DPE000
ANP 246 see CJN750
ANP 3548 see BPM750
ANP 3624 see TGA600
ANP 4364 see DFO600
ANPARTON see ARQ750
AN 1 (PHARMACEUTICAL) see AOB300
235 ANP HYDROCHLORIDE see AAE500
ANPROLENE see EJN500
ANPROLINE see EJN500
ANQI see ALK625
ANQUIL see FLK100
ANQUIL see RDK000
ANSADOL see SAH500
ANSAL see AQN250
ANSAMITOCIN P-4 see APE529
ANSAR see HKC000
ANSAR 157 see MDQ770
ANSAR 160 see HKC500
ANSAR 170 see MRL750
ANSAR 184 see DXE600
ANSAR 560 see HKC500
ANSAR-290 D see TJK900
ANSAR DSMA LIQUID see DXE600
ANSATIN see TKH750
ANSEPRON see PGA750
ANSIACAL see MDQ250
ANSIATAN see MQU750
ANSIBASE ORANGE GC see CEH690
ANSIBASE RED KB see CLK225
ANSIBASES RED RL see MMF780
ANSIL see MQU750
ANSILAN see CGA000
ANSIOLISINA see CFZ000
ANSIOLISINA see DCK759
ANSIOWAS see MQU750
ANSIOXACEPAM see CFZ000
ANSOLYSEN see PBT000
ANSOLYSEN BITARTRATE see PBT000
ANSOLYSEN TARTRATE see PBT000
ANSUL ETHER 181AT see PBO500
ANT-1 see TKH750
ANTABUS see DXH250
ANTABUSE see DXH250
ANTADIX see DXH250
ANTADOL see BRF500
ANTAENYL see DXH250
ANTAETHAN see DXH250
ANTAETHYL see DXH250
ANTAETIL see DXH250
ANTAGE W 400 see MJO500
ANTAGE W 500 see MJN250
ANTAGONATE see TAI500
ANTAGOSAN see PAF550
ANTAGOTHYROID see TFR250
ANTAGOTHYROIL see TFR250
ANTAK see DAI600
ANTAK see ODE000

ANTALCOL see DXH250
ANTALERGAN see WAK000
ANTALLIN see CAR780
ANTALVIC see PNA500
ANTAMINE see WAK000
ANTAN see NAH500
ANTAROX 224 see AFJ172
ANTAROX A-200 see PKF500
ANTARSIN see DNV600
ANTASTEN see PDC000
ANTAZOLINE see PDC000
ANTEBOR see SNQ000
ANTEMOQUA see AJX500
ANTEMOVIS see AJX500
ANTENE see BJK500
ANTEPAR see PIJ500
ANTERGAN see BEM500
ANTERGAN HYDROCHLORIDE see
PEN000
ANTERGYL see SEO500
ANTERIOR PITUITARY GROWTH
HORMONE see PJA250
ANTERIOR PITUITARY LUTEOTROPIN see
PMH625
ANTERON see SCA750
ANTETAN see DXH250
ANTETHYL see DXH250
ANTETIL see DXH250
ANTEX-490 see SCA750
ANTEYL see DXH250
ANTHALLAN HYDROCHLORIDE see
APE625
ANTHANTHREN (GERMAN) see APE750
ANTHANTHRENE see APE750
ANTHANTHRENEQUINONE see DCY900
ANTHANTHRONE see DCY900
ANTHECOLE see PIJ500
ANTHELMYCIN see APF000
ANTHELONE U see UVJ475
ANTHELVET see TDX750
ANTHER see IHV050
ANTHGLUTIN see GFO075
ANTHIO see DRR200
ANTHIOLIMINE see LGU000
ANTHIOMALINE see LGU000
ANTHIOMALINE NONAHYDRATE see
AQE500
ANTHION see DWQ000
ANTHIPHEN see MJM500
ANTHISAN see WAK000
ANTHISAN MALEATE see DBM800
ANTHIUM DIOXIDE see CDW450
ANTHON see TIQ250
ANTHONAPHTHOL AS see CMM760
ANTHOPHYLITE see ARM264
ANTHRA(9,1,2-cde)BENZO(h)CINNOLINE
see APF750
ANTHRA(9,1,2-
cde)BENZO(rst)PENTAPHENE-5,10-
DIONE,15,18-DINITRO- see DUP830
ANTHRA(2,1-d:6,5-d')BISTHIAZOLE-6,12-
DIONE, 2,8-DIPHENYL- see VGP200
6H-ANTHRA(9,1-CD)ISOTHIAZOLE-3-
CARBOXYLIC ACID,6-OXO- see ISD043
ANTHRA(1,9-CD)PYRAZOL-6(2H)-ONE,
7,10-DIHYDRO-2-(2-((2-
HYDROXYETHYL)AMINO)ETHYL)-5-((2-
(METHYLAMINO)ETHYL)AMINO)- see
XDJ025
ANTHRACEN (GERMAN) see APG500
1-ANTHRACENAMINE see APG050
2-ANTHRACENAMINE see APG100
9-ANTHRACENAMINE see APH700
ANTHRACENE see APG500
ANTHRACENE BROWN FD see TKN500
ANTHRACENE BROWN FF see TKN500
ANTHRACENE BROWN G see TKN500
ANTHRACENE BROWN N see TKN500
ANTHRACENE BROWN S see TKN500
ANTHRACENE BROWN WH see TKN500
ANTHRACENE BROWN WL see TKN500
ANTHRACENE-9-CARBINOL see APG600

ANTHRACENE-9-CARBOXYLIC ACID see
APG550
2-ANTHRACENECARBOXYLIC ACID, 9,10-
DIHYDRO-4,5-DIHYDROXY-9,10-DIOXO-
see RHZ700
ANTHRACENE, DIMETHYL- see DQG100
9,10-ANTHRACENEDIONE see APK250
9,10-ANTHRACENEDIONE, 1-AMINO-4-
HYDROXY-(9CI) see AKE250
9,10-ANTHRACENEDIONE, 1-AMINO-4-
HYDROXY-2-METHOXY-(9CI) see AKO350
9,10-ANTHRACENEDIONE, 1-AMINO-4-
(METHYLAMINO)-(9CI) see AKP250
9,10-ANTHRACENEDIONE, 5,8-BIS((2-
PHENYLETHYL)AMINO)-1,4-DIHYDROXY-
see BLE550
9,10-ANTHRACENEDIONE, 1-BROMO-4-
(METHYLAMINO)- see BNN550
9,10-ANTHRACENEDIONE, 2-CHLORO- see
CEI100
9,10-ANTHRACENEDIONE, 2,6-DIAMINO-
see APK850
9,10-ANTHRACENEDIONE, 2,7-DIAMINO-
(9CI) see DBP420
9,10-ANTHRACENEDIONE, 1,5-
DIAMINOBROMO-4,8-DIHYDROXY- see
DBQ220
9,10-ANTHRACENEDIONE, 1,5-
DIAMINOCHLORO-4,8-DIHYDROXY- see
CMP070
9,10-ANTHRACENEDIONE, 1,4-DIAMINO-
2,3-DIHYDRO- see DBR450
9,10-ANTHRACENEDIONE, 1,8-DIAMINO-
4,5-DIHYDROXY- see DBQ250
9,10-ANTHRACENEDIONE, 1,4-DIAMINO-
2-METHOXY-(9CI) see DBX000
9,10-ANTHRACENEDIONE, 1,4-DIAMINO-
5-NITRO-(9CI) see DBY700
9,10-ANTHRACENEDIONE, 1,5-
DICHLORO- see DEO700
9,10-ANTHRACENEDIONE, 1,8-
DICHLORO- see DEO750
9,10-ANTHRACENEDIONE, 1,2-
DIHYDROXY- see DMG800
9,10-ANTHRACENEDIONE, 1,8-
DIHYDROXY-4,5-DINITRO-(9CI) see
DMN400
9,10-ANTHRACENEDIONE, 1,8-
DIHYDROXY-4-((4-(2-
HYDROXYETHYL)PHENYL)AMINO)-5-
NITRO- see CMP060
9,10-ANTHRACENEDIONE, 1,3-
DIHYDROXY-2-(HYDROXYMETHYL)-(9CI)
see LIN100
9,10-ANTHRACENEDIONE, 1,8-
DIHYDROXY-2,4,5,7-TETRANITRO- see
CML600
9,10-ANTHRACENEDIONE, 1,5-
DIPHENOXY- see DVW100
9,10-ANTHRACENEDIONE, 1-METHOXY-
(9CI) see MEA650
9,10-ANTHRACENEDIONE, 1-
(PHENYLTHIO)-(9CI) see PGL000
9,10-ANTHRACENEDIONE, 1,2,5,8-
TETRAHYDROXY-(9CI) see TDD000
9,10-ANTHRACENEDIONE, 1,3,6,8-
TETRAHYDROXY-2-(1-HYDROXYHEXYL)-,
(−)- see HLF550
9,10-ANTHRACENEDIONE, 1,2,3-
TRIHYDROXY-(9CI) see TKN500
9,10-ANTHRACENEDIONE, 1,4,5-
TRIHYDROXY-2-METHYL-(9CI) see FPD050
1,8-ANTHRACENEDISULFONIC ACID, 9,10-
DIHYDRO-9,10-DIOXO-, DISODIUM SALT
see DLJ730
9-ANTHRACENEMETHANOL see APG600
9-ANTHRACENEMETHANOL, ACETATE
see APM800
ANTHRACENE, 9-NITRO- see NES100
ANTHRACENE ORANGE G see CMP882
ANTHRACENE PRINTING BROWN see
TKN500

1-ANTHRACENESULFONAMIDE, 9,10-DIHYDRO-N-(4-(BIS(2-CYANOETHYL)AMINO)PHENYL)-D,10-DIOXO-4-HYDROXY- see BIQ600

2-ANTHRACENESULFONIC ACID, 1-AMINO-4-(4-AMINO-3-SULFOANILINO)-9,10-DIHYDRO-9,10-DIOXO- see SOB600

2-ANTHRACENESULFONIC ACID, 1-AMINO-4-((4-AMINO-3-SULFOPHENYL)AMINO)-9,10-DIHYDRO-9,10-DIOXO- see SOB600

2-ANTHRACENESULFONIC ACID, 1-AMINO-4-(3-((4,6-DICHLORO-s-TRIAZIN-2-YL)AMINO)-4-SULFOANILINO)-9,10-DIHYDRO-9,10-DIOXO- see CMS222

2-ANTHRACENESULFONIC ACID, 1-AMINO-9,10-DIHYDRO-9,10-DIOXO-4-(2,4,6-TRIMETHYLANILINO)-, MONOSODIUM SALT see CMM100

2-ANTHRACENESULFONIC ACID, 1-AMINO-9,10-DIHYDRO-4-(p-(N-METHYLACETAMIDO)ANILINO)-9,10-DIOXO-, MONOSODIUM SALT see CMM070

2-ANTHRACENESULFONIC ACID, 4,8-DIAMINO-9,10-DIHYDRO-1,5-DIHYDROXY-9,10-DIOXO-, MONOSODIUM SALT see APG700

2-ANTHRACENESULFONIC ACID, 9,10-DIHYDRO-1-AMINO-4-BENZOYL-9,10-DIOXO- see AIS700

2-ANTHRACENESULFONIC ACID, 9,10-DIHYDRO-1-AMINO-4-(CYCLOHEXYLAMINO)-9,10-DIOXO-, MONOSODIUM SALT see CMM080

2-ANTHRACENESULFONIC ACID, 9,10-DIHYDRO-3,4-DIHYDROXY-9,10-DIOXO-, MONOSODIUM SALT see SEH475

1,4,9,10-ANTHRACENETETRAOL see LEX200

1,4,9,10-ANTHRACENETETROL (9CI) see LEX200

1,8,9-ANTHRACENETRIOL see APH250

1,8,9-ANTHRACENETRIOL TRIACETATE see APH500

9(10H)-ANTHRACENONE see APM500

9(10H)-ANTHRACENONE, 10-(ACETYLOXY)-1,8-DIHYDROXY- see ACI640

9(10H)-ANTHRACENONE, 1,8-DIHYDROXY- see DMG900

9(10H)-ANTHRACENONE, 1,8-DIHYDROXY-10-(1-OXOBUTYL)- see BSY400

9(10H)-ANTHRACENONE, 1,8-DIHYDROXY-10-(1-OXOETHYL)- see ACI640

9(10H)-ANTHRACENONE, 1,8-DIHYDROXY-10-(1-OXOPROPYL)- see PMW760

9-ANTHRACENYL PHENYL KETONE see APH600

ANTHRACHINON-1,8-DISULFONAN DRASELNY (CZECH) see DLJ600

ANTHRACHINON-1,5-DISULFONAN SODNY (CZECH) see DLJ700

ANTHRACHINON-1-SULFONAN SODNY (CZECH) see DLJ800

ANTHRACHINON-1-SULFONAN SODNY (CZECH) see SER000

ANTHRACIN see APG500

ANTHRACITE PARTICLES see CMY760

1-ANTHRACYLAMINE see APG050

2-ANTHRACYLAMINE see APG100

10-ANTHRACYLAMINE see APH700

ANTHRA-DERM see DMG900

ANTHRADIONE see APK250

ANTHRAFLAVIC ACID see DMH600

ANTHRAFLAVIN see DMH600

ANTHRAGALLIC ACID see TKN500

ANTHRAGALLOL see TKN500

ANTHRALAN BLUE B see CMM070

ANTHRALAN YELLOW RRT see SGP500

ANTHRALIN see APH250

ANTHRALIN see DMG900

1-ANTHRAMINE see APG050

2-ANTHRAMINE see APG100

9-ANTHRAMINE see APH700

ANTHRAMYCIN see API500

ANTHRAMYCIN METHYL ETHER see API125

ANTHRAMYCIN-11-METHYL ETHER see API125

ANTHRA(2,1,9-mna)NAPHTH(2,3-h)ACRIDINE-5,10,15-TRIONE see CMU810

ANTHRANILAMIDE see AID620

ANTHRANILIC ACID see API500

ANTHRANILIC ACID, N-ACETYL- see AAJ150

ANTHRANILIC ACID, N-(2-BENZYLIDENEHEPTYLIDENE)-, METHYL ESTER see AOH100

ANTHRANILIC ACID, N-(3-(p-tert-BUTYLPHENYL)-2-METHYLPROPYLIDENE)-, METHYL ESTER see LFT100

ANTHRANILIC ACID, N-(7-CHLORO-4-QUINOLYL)-, 2,3-DIHYDROXYPROPYL ESTER see GGQ050

ANTHRANILIC ACID, CINNAMYL ESTER see API750

ANTHRANILIC ACID, 3,5-DIIODO- see API800

ANTHRANILIC ACID, LINALYL ESTER see APJ000

ANTHRANILIC ACID, METHYL ESTER see APJ250

ANTHRANILIC ACID, N-METHYL-, HYDRAZIDE see MGQ100

ANTHRANILIC ACID, N-2-NAPHTHYL- see NBF600

ANTHRANILIC ACID, N-o-NITROPHENYL- see NIJ450

ANTHRANILIC ACID, PHENETHYL ESTER see APJ500

ANTHRANILIC ACID, N-(o-TOLYL)- see MNR300

ANTHRANILIMIDIC ACID see AID620

ANTHRANILONITRILE see APJ750

m-ANTHRANILONITRILE see AIR125

ANTHRANILOYLLYCOCTONINE see LJB150

ANTHRANOL CHROME ORANGE GR see CMP882

ANTHRANOL CHROME YELLOW 2G see SIT850

ANTHRANOL CHROME YELLOW R see NEY000

ANTHRANOL CHROME YELLOW 5GS see SIT850

(+)-ANTHRANOYLLYCOCTONINE see LJB150

ANTHRANTHRENE see APE750

ANTHRAPOLE AZ see BQK250

ANTHRA(1,9-cd)PYRAZOL-6(2H)-ONE see APK000

β-ANTHRAQUINOLINE see NAZ000

ANTHRAQUINONE see APK250

9,10-ANTHRAQUINONE see APK250

ANTHRAQUINONE ACID BLUE see APG700

ANTHRAQUINONE, 1-AMINO-4-HYDROXY- see AKE250

ANTHRAQUINONE, 1-AMINO-4-HYDROXY-2-METHOXY- see AKO350

ANTHRAQUINONE, 1,4-BIS((2-HYDROXYETHYL)AMINO)-5,8-DIHYDROXY- see DMM400

ANTHRAQUINONE, 1,4-BIS(METHYLAMINO)- see BKP500

ANTHRAQUINONE BLUE see IBV050

ANTHRAQUINONE BLUE SKY see CMM090

ANTHRAQUINONE BRILLIANT GREEN CONCENTRATE ZH see APK500

ANTHRAQUINONE, 2-BROMO-1,5-DIAMINO-4,8-DIHYDROXY- see BNC800

ANTHRAQUINONE, 1-BROMO-4-(METHYLAMINO)- see BNN550

9,10-ANTHRAQUINONE-2-CARBOXYLIC ACID, 1-NITRO- see NFS502

ANTHRAQUINONE, 2-CHLORO- see CEI100

ANTHRAQUINONE DEEP BLUE see IBV050

ANTHRAQUINONE, 2,6-DIAMINO- see APK850

ANTHRAQUINONE, 2,7-DIAMINO- see DBP420

ANTHRAQUINONE, 1,5-DIAMINOBROMO-4,8-DIHYDROXY- (7CI,8CI) see DBQ220

ANTHRAQUINONE, 1,8-DIAMINO-4,5-DIHYDROXY- see DBQ250

ANTHRAQUINONE, 2,3-DIHYDRO-1,4-DIAMINO- see DBR450

ANTHRAQUINONE, 1,4-DIHYDROXY-5,8-BIS((2-HYDROXYETHYL)AMINO)- see DMM400

ANTHRAQUINONE, 1,8-DIHYDROXY-4-(p-(2-HYDROXYETHYL)ANILINO)-5-NITRO- see CMP060

ANTHRAQUINONE, 1,3-DIHYDROXY-2-HYDROXYMETHYL- see LIN100

ANTHRAQUINONE, 1,8-DIHYDROXY-2,4,5,7-TETRANITRO- see CML600

1,2-ANTHRAQUINONEDIOL see DMG800

1,8-ANTHRAQUINONEDISULFINIC ACID see APK635

1,5-ANTHRAQUINONEDISULFONIC ACID see APK625

ANTHRAQUINONEDISULFONIC ACID, DIPOTASSIUM SALT see DLJ600

ANTHRAQUINONE, 2-(1-HYDROXYHEXYL)-1,3,6,8-TETRAHYDROXY- see HLF550

ANTHRAQUINONE, 1-METHOXY- see MEA650

9,10-ANTHRAQUINONE-2-SODIUM SULFONATE see SER000

2-ANTHRAQUINONESULFONATE SODIUM see SER000

ANTHRAQUINONE-2-SULFONATE SODIUM SALT see SER000

2-ANTHRAQUINONESULFONIC ACID, 1-AMINO-4-BENZOYL- see AIS700

2-ANTHRAQUINONESULFONIC ACID SODIUM SALT see SER000

ANTHRAQUINONE, 1,3,6,8-TETRAHYDROXY-2-(1-HYDROXYHEXYL)- see HLF550

ANTHRAQUINONE, 1-THIOPHENYL see PGL000

ANTHRAQUINONE, 1,4,5-TRIHYDROXY-2-METHYL- see FPD050

α-ANTHRAQUINONYLAMINE see AIA750

β-ANTHRAQUINONYLAMINE see AIB000

ANTHRAQUINONYLAMINOANTHRAQUINONE see IBI000

((N-ANTHRAQUINON-2-YL)AMINOMETHYLENE)DIMETHYLAMMONIUM CHLORIDE see APK750

1,4-ANTHRAQUINONYLDIAMINE see DBP000

1,5-ANTHRAQUINONYLDIAMINE see DBP200

1,5-ANTHRAQUINONYLDIAMINE see DBP400

1,8-ANTHRAQUINONYLDIAMINE see DBP400

2,6-ANTHRAQUINONYLDIAMINE see APK850

2,2'-(1,4-ANTHRAQUINONYLENEDIIMINO)BIS(5-METHYLBENZENESULFONIC ACID) DISODIUM SALT see APL500

1,1'-(ANTHRAQUINON-1,4-YLENEDIIMINO)DIANTHRAQUINONE see APL750

1,1'-(ANTHRAQUINON-1,5-YLENEDIIMINO)DIANTHRAQUINONE see APM000

4,4'-(1,4-ANTHRAQUINONYLENEDIIMINODIPHENYL-1,4-ENEDIOXO)BENZENESULFONIC ACID see APM250

ANTHRA RED G see CMM325

ANTHRARUFIN see DMH200

ANTHRASORB see CBT500

1,8,9-ANTHRATRIOL see APH250

ANTHRAVAT GOLDEN YELLOW see DCZ000

ANTHRAVAT NAVY BLUE BR see VGP100

5,9,14,18-ANTHRAZINETETRONE, 7,16-DICHLORO-6,15-DIHYDRO- see DFN300

5,9,14,18-ANTHRAZINETETRONE, 6,15-DIHYDRO- see IBV050

5,9,14,18-ANTHRAZINETETRONE, 6,15-DIHYDROHYDROXY- see DLO900

ANTHRIMIDE see IBI000

ANTHROGON see FMT100

9-ANTHROIC ACID see APG550

2-ANTHROIC ACID, 9,10-DIHYDRO-4,5-DIHYDROXY-9,10-DIOXO- see RHZ700

2-ANTHROIC ACID, 9,10-DIHYDRO-1-NITRO-9,10-DIOXO- see NFS502

ANTHRONE see APM500

9-ANTHRONOL see APM750

ANTHROPODEOXYCHOLIC ACID see CDL325

ANTHROPODESOXYCHOLIC ACID see CDL325

ANTHROPODODESOXYCHOLIC ACID see CDL325

(6-(3-(9-ANTHRYL)-d-ALANINE))-LHRH ACETATE see LIU307

2-ANTHRYLAMINE see APG100

9-ANTHRYLCARBINOL see APG600

9-ANTHRYLMETHANOL see APG600

9-ANTHRYLMETHYL ACETATE see APM800

ANTHURIUM see APM875

ANTIAETHAN see DXH250

ANTIANGOR see CBR500

α-ANTIARBIN see APN000

ANTIB see FNF000

ANTIBASON see MPW500

ANTIBIOCIN see PDT750

3C ANTIBIOTIC see HLT100

ANTIBIOTIC 289 see LIU500

ANTIBIOTIC 60-6 see CCX725

ANTIBIOTIC 899 see VRA700

ANTIBIOTIC D-45 see SLW475

ANTIBIOTIC 1037 see VGZ000

ANTIBIOTIC 1142 see NCP875

ANTIBIOTIC 1600 see APP500

ANTIBIOTIC 1719 see ASO501

ANTIBIOTIC 66-40 see SDY750

ANTIBIOTIC 833A see PHA550

ANTIBIOTIC 205T3 see NMV500

ANTIBIOTIC 29275 see CBF680

ANTIBIOTIC 33876 see APF000

ANTIBIOTIC 67-694 see RMF000

ANTIBIOTIC 281471 see MRW800

ANTIBIOTIC 6761-31 see TFQ275

ANTIBIOTIC A see API125

ANTIBIOTIC A-246 see FPC000

ANTIBIOTIC A-649 see OIU499

ANTIBIOTIC A 130A see LEJ700

ANTIBIOTIC A-5283 see PIF750

ANTIBIOTIC A 8506 see TFQ275

ANTIBIOTIC A 3733A see HAL000

ANTIBIOTIC A-64922 see OIU499

ANTIBIOTIC A 28695 A see SCA000

ANTIBIOTIC AB 206 see MQU525

ANTIBIOTIC AD 32 see TJX350

ANTIBIOTIC A-250-II see CBF680

ANTIBIOTIC AK PS see AFH500

ANTIBIOTIC AM-2604 A see VRP775

ANTIBIOTIC APM see XFS600

ANTIBIOTIC AT 265 see CEF125

ANTIBIOTIC A-399-Y4 see VGZ000

ANTIBIOTIC AY 22989 see RBK000

ANTIBIOTIC B 41D see MQT600

ANTIBIOTIC B 599 see CMK650

ANTIBIOTIC N-329 B see VBZ000

ANTIBIOTIC B-14437 see SAU000

ANTIBIOTIC B 21085 see GEO000

ANTIBIOTIC B-98891 see MQU000

ANTIBIOTIC BAY-f 1353 see MQS200

ANTIBIOTIC BB-K 8 see APS750

ANTIBIOTIC BB-K8 SULFATE see APT000

ANTIBIOTIC BL-640 see APT250

ANTIBIOTIC BL-S 640 see APT250

ANTIBIOTIC BU 2231A see TAC500

ANTIBIOTIC BU 2231B see TAC750

ANTIBIOTIC C 076B1A see ARW150

ANTIBIOTIC CC 1065 see APT375

ANTIBIOTIC C 076A1A, 5-O-DEMETHYL- see ARW150

ANTIBIOTIC CGP 9000 see CCS530

ANTIBIOTIC DC 11 see TEF725

ANTIBIOTIC DE 3936 see LIF000

ANTIBIOTIC DL 473IT see RKZ100

ANTIBIOTIC E212 see VGZ000

ANTIBIOTIC F-1370A see EDW200

ANTIBIOTIC 1163 F.I. see LIN000

ANTIBIOTIC FN 1636 see PEC750

ANTIBIOTIC FR 1923 see APT750

ANTIBIOTIC G-52 see GAA100

ANTIBIOTIC G-52 SULFATE see APU000

ANTIBIOTIC G-52 SULFATE see GAA120

ANTIBIOTIC HA-9 see TFC500

ANTIBIOTIC K-179 see EDW200

ANTIBIOTIC KA 66061 see SLF500

ANTIBIOTIC KM 208 see BAC175

ANTIBIOTIC KW 1062 see MQS579

ANTIBIOTIC KW-1070 see FOK000

ANTIBIOTIC LA 7017 see MQW750

ANTIBIOTIC M 4365A2 see RMF000

ANTIBIOTIC MA 144A see APU500

ANTIBIOTIC MA 144A1 see APU500

ANTIBIOTIC MA 144A2 see TAH675

ANTIBIOTIC MA 144B2 see TAH650

ANTIBIOTIC MA 144M1 see MAB250

ANTIBIOTIC MA 144S2 see APV000

ANTIBIOTIC MA 144T1 see DAY835

ANTIBIOTIC MA 144U2 see MAX000

ANTIBIOTIC MM 14151 see CMV250

ANTIBIOTIC No. 899 see VRF000

ANTIBIOTIC NSC 70845 see NMV500

ANTIBIOTIC NSC-71936 see CMQ725

ANTIBIOTIC OS 3966A see RMK200

ANTIBIOTIC 20-798RP see DAC300

ANTIBIOTIC PA-93 see SMB000

ANTIBIOTIC PA-105 see OHM900

ANTIBIOTIC PA-106 see AOY000

ANTIBIOTIC PA147 see APV750

ANTIBIOTIC PA 11481 see VRA700

ANTIBIOTIC PA 48009 see LEX100

ANTIBIOTIC PA 114 B1 see VRA700

ANTIBIOTIC Ro 21 6150 see LEJ700

ANTIBIOTIC RI-331 see HND250

ANTIBIOTIC S 15-1A see RAG300

ANTIBIOTIC 6059-S see LBH200

ANTIBIOTIC S 7481F1 see CQH100

ANTIBIOTIC SF 733 see XQJ650

ANTIBIOTIC SF 837 see MBY150

ANTIBIOTIC SF 767B see NCF500

ANTIBIOTIC SF 837 A1 see MBY150

ANTIBIOTIC SF 837 A₁ see MBY150

ANTIBIOTIC SF 2052 SULFATE see DAB630

ANTIBIOTIC 1719 SODIUM SALT see ASO510

ANTIBIOTIC SPA-S 565 see RJZ100

ANTIBIOTIC 66-40 SULFATE see APY500

ANTIBIOTIC T see COB000

ANTIBIOTIC T-1384 see NCP875

ANTIBIOTIC TM 25 see HOH500

ANTIBIOTIC TM 481 see LIF000

ANTIBIOTIC U-12,241 see CMS230

ANTIBIOTIC U 18496 see ARY000

ANTIBIOTIC U 48160 see QQS075

ANTIBIOTICUM PA147 (GERMAN) see APV750

ANTIBIOTIC 44 VI see MRW800

ANTIBIOTIC W-847-A see MCA250

ANTIBIOTIC WR 141 see GJS000

ANTIBIOTIC X 146 see TFQ275

ANTIBIOTIC X 537 see LBF500

ANTIBIOTIC X-465A see CDK250

ANTIBIOTIC XK 41C see MCA250

ANTIBIOTIC XK 62-2 see MQS579

ANTIBIOTIC XS-89 see SMM500

ANTIBIOTIC X465A SODIUM SALT see CDK500

ANTIBIOTIC YL-704 A3 see JDS200

ANTIBIOTIC YL 704 B₁ see MBY150

ANTIBIOTIC YL-704 B3 see LEV025

ANTIBIOTIQUE see NCF000

ANTIBLAGE 78 see BIB800

ANTIBULIT see SHF500

ANTICANITIC VITAMIN see AIH600

ANTICARIE see HCC500

ANTICHLOR see SKI500

ANTI-CHROMOTRICHIA FACTOR see AIH600

anti-CHROMOTRICHIA FACTOR see AIH600

ANTIDEPRIN see DLH600

ANTIDEPRIN HYDROCHLORIDE see DLH630

ANTIDUROL see DAM700

ANTIEGENE MB see BCC500

ANTIETANOL see DXH250

ANTI-ETHYL see DXH250

ANTIETIL see DXH250

ANTIFEBRIN see AAQ500

ANTIFEEDANT 24005 see DUI000

ANTIFEEDING COMPOUND 24,055 see DUI000

ANTIFOAM FD 62 see SCR400

ANTIFOLAN see MDV500

ANTIFORMIN see SHU500

ANTIGENE RDF see PJQ750

ANTI-GERM 77 see BEN000

ANTIGESTIL see DKA600

ANTIHELMYCIN see AQB000

ANTIHEMORRHAGIC VITAMIN see VTA000

ANTIHIST see DBM800

ANTIHISTAL see PDC000

ANTI-INFECTIVE VITAMIN see VSK600

ANTI-INFLAMMATORY HORMONE see CNS750

ANTIKNOCK-33 see MAV750

ANTIKOL see DXH250

ANTIKREIN see PAF550

ANTILEPSIN see DNU000

ANTILIPID see ARQ750

ANTILYSIN see PAF550

ANTILYSINE see PAF550

ANTIMALARINA see ARQ250

ANTIMICINA see ILD000

ANTIMIGRANT C 45 see SEH000

ANTIMILACE see TDW500

ANTIMIT see BIE500

ANTIMOINE FLUORURE (FRENCH) see AQE000

ANTIMOINE (TRICHLORURE d') see AQC500

ANTIMOL see SFB000

ANTIMONIAL SAFFRON see AQF500

ANTIMONIC "ACID" see AQF750

ANTIMONIC ACID, SODIUM SALT see AQB250

ANTIMONIC ACID, TUNGSTEN SALT see TOA800

ANTIMONIC CHLORIDE see AQD000

ANTIMONIC OXIDE see AQF750

ANTIMONIC SULFIDE see AQF500

ANTIMONIO (PENTACLORURO DI) (ITALIAN) see AQD000

ANTIMONIO (TRICLORURO di) see AQC500

ANTIMONIOUS OXIDE see AQF000

ANTIMONITE see SLQ100

ANTIMONITE (mineral) see SLQ100
ANTIMONOUS CHLORIDE see AQC500
ANTIMONOUS CHLORIDE (DOT) see AQC500
ANTIMONOUS FLUORIDE see AQE000
ANTIMONOUS SULFATE see AQJ250
ANTIMONOUS SULFIDE see AQL500
ANTIMONPENTACHLORID (GERMAN) see AQD000
ANTIMONTRICHLORID see AQC500
ANTIMONWASSERSTOFFES (GERMAN) see SLQ000
ANTIMONY see AQB750
ANTIMONY(III) ACETATE see AQJ750
ANTIMONY AMMONIA TRIACETIC ACID see AQC000
ANTIMONY, (o,o-BIS(1-METHYLETHYL)PHOSPHORODITHIOATO-S,S')DIPHENYL-, (T-4)- see BKS780
ANTIMONY, BIS(TRICHLORO) compounded with 1 mole of OCTAMETHYL PYROPHOSPHORAMIDE see AQC250
ANTIMONY BLACK see AQB750
ANTIMONY BUTTER see AQC500
ANTIMONY CHLORIDE see AQC500
ANTIMONY(V) CHLORIDE see AQD000
ANTIMONY CHLORIDE (DOT) see AQC500
ANTIMONY(III) CHLORIDE see AQC500
ANTIMONY COMPOUNDS see AQD500
ANTIMONY DIMERCAPTOSUCCINATE see AQD750
ANTIMONY DIMERCAPTOSUCCINATE(IV) see AQD750
ANTIMONY, (DIPHENYLPHOSPHINODITHIOATO-S,S')DIPHENYL-, (T-4)- see DWJ600
ANTIMONY EMETINE IODIDE see EAM000
ANTIMONY FLUORIDE see AQF250
ANTIMONY(V) FLUORIDE see AQF250
ANTIMONY(III) FLUORIDE (1:3) see AQE000
ANTIMONY GLANCE see AQL500
ANTIMONY GLANCE see SLQ100
ANTIMONY HYDRIDE see SLQ000
ANTIMONY LACTATE see AQE250
ANTIMONY LACTATE, solid (DOT) see AQE250
ANTIMONYL ANILINE TARTRATE see AOQ250
ANTIMONYLBRENZEATECHINDISULFOSAURES NATRIUM (GERMAN) see AQH500
ANTIMONYL-2,4-DIHYDROXY-5-HYDROXYMETHYL PYRIMIDINE see AQE300
ANTIMONYL-2,4-DIHYDROXY PYRIMIDINE see AQE305
ANTIMONYL-7-FORMYL-8-HYDROXYQUINOLINE-5-SULPHONATE see AQE320
ANTIMONY LITHIUM THIOMALATENONAHYDRATE see AQE500
ANTIMONYL POTASSIUM TARTRATE see AQG250
ANTIMONY, compounded with NICKEL (1:1) see NCY100
ANTIMONY NITRIDE see AQE750
ANTIMONY ORANGE see AQL500
ANTIMONY OXIDE see AQF000
ANTIMONY(3+) OXIDE see AQF000
ANTIMONY PENTACHLORIDE see AQD000
ANTIMONY PENTACHLORIDE (DOT) see AQD000
ANTIMONY(V) PENTAFLUORIDE see AQF250
ANTIMONY PENTAFLUORIDE (DOT) see AQF250
ANTIMONY PENTAOXIDE see AQF750
ANTIMONY PENTASULFIDE see AQF500
ANTIMONY PENTOXIDE see AQF750

ANTIMONY PERCHLORIDE see AQD000
ANTIMONY PEROXIDE see AQF000
ANTIMONY POTASSIUM DIMETHYLCYSTEINOTARTRATE see AQG000
ANTIMONY POTASSIUM TARTRATE see AQG250
d-ANTIMONY POTASSIUM TARTRATE see AQG500
l-ANTIMONY POTASSIUM TARTRATE see AQH000
dl-ANTIMONY POTASSIUM TARTRATE see AQG750
meso-ANTIMONY POTASSIUM TARTRATE see AQH250
ANTIMONY POWDER (DOT) see AQB750
ANTIMONY PYROCATECHOL SODIUM DISULFONATE see AQH500
ANTIMONY RED see AQF500
ANTIMONY REGULUS see AQB750
ANTIMONY SESQUIOXIDE see AQF000
ANTIMONY SESQUISULFIDE see AQL500
ANTIMONY SODIUM DIMETHYL CYSTEINO TARTRATE see AQH750
ANTIMONY SODIUM GLUCONATE see AQH800
ANTIMONY(V) SODIUM GLUCONATE see AQI250
ANTIMONY(III) SODIUM GLUCONATE see AQI000
ANTIMONY SODIUM OXIDE-l-(+)-TARTRATE see AQI750
ANTIMONY SODIUM PROPYLENE DIAMINE TETRAACETIC ACID DIHYDRATE see AQI500
ANTIMONY SODIUM TARTRATE see AQI750
ANTIMONY(III) SULFATE (2:3) see AQJ250
ANTIMONY SULFIDE see AQF500
ANTIMONY SULFIDE see AQL500
ANTIMONY SULFIDE (SB$_2$S$_4$) (9CI) see AQJ600
ANTIMONY TARTRATE see AQJ500
ANTIMONY TELLURIDE see AQL750
ANTIMONY THIOANTIMONATE see AQJ600
ANTIMONY TRIACETATE see AQJ750
ANTIMONY TRIBROMIDE see AQK000
ANTIMONY TRIBROMIDE see AQK000
ANTIMONY TRIBROMIDE, solid or solution (DOT) see AQK000
ANTIMONY TRICHLORIDE see AQC500
ANTIMONY TRICHLORIDE, solid (DOT) see AQC500
ANTIMONY TRICHLORIDE, liquid (DOT) see AQC500
ANTIMONY TRICHLORIDE, solution (DOT) see AQC500
ANTIMONY TRICHLORIDE OXIDE see AQK250
ANTIMONY TRIETHYL see AQK500
ANTIMONY TRIFLUORIDE see AQE000
ANTIMONY TRIFLUORIDE, solid or solution (DOT) see AQE000
ANTIMONY TRIHYDRIDE see SLQ000
ANTIMONY TRIIODIDE see AQK750
ANTIMONY TRIMETHYL see AQL000
ANTIMONY TRIOXIDE see AQF000
ANTIMONY TRIPHENYL see AQL250
ANTIMONY TRIPHENYLDICHLORIDE see DGO800
ANTIMONY TRISULFATE see AQJ250
ANTIMONY TRISULFIDE see AQL500
ANTIMONY TRISULFIDE COLLOID see AQL500
ANTIMONY TRITELLURIDE see AQL750
ANTIMONY VERMILION see AQL500
ANTIMONY WHITE see AQF000
ANTIMOONPENTACHLORIDE (DUTCH) see AQD000
ANTIMOONTRICHLORIDE see AQC500
ANTIMOSAN see AQH500

ANTIMUCIN WDR see ABU500
ANTIMYCIN see AQM000
ANTIMYCIN see CMS775
ANTIMYCIN A see AQM250
ANTIMYCIN A1 see DUO350
ANTIMYCIN A3 see BLX750
ANTIMYCIN A4 see AQM260
ANTIMYCOIN see AQM500
ANTINOLO RED B see DNT300
ANTINONIN see DUS700
ANTINOSIN see TDE750
ANTIO see DRR200
ANTIOK S see DXG700
ANTI OX see MJO500
ANTIOXIDANT 1 see MJO500
ANTIOXIDANT 29 see BFW750
ANTIOXIDANT 116 see PFT500
ANTIOXIDANT 330 see TMJ000
ANTIOXIDANT 425 see MJN250
ANTIOXIDANT 736 see TFD000
ANTIOXIDANT 754 ifx200
ANTIOXIDANT AS see TFD500
ANTIOXIDANT D see HLI500
ANTIOXIDANT DBPC see BFW750
ANTIOXIDANT E 702 see MJM700
ANTIOXIDANT HS see PJQ750
ANTIOXIDANT HSL see PJQ750
ANTIOXIDANT LTDP see TFD500
ANTIOXIDANT MB (CZECH) see BCC500
ANTIOXIDANT No. 33 see DEG000
ANTIOXIDANT PBN see PFT500
ANTIOXIDANT TOD (CZECH) see BLB500
ANTIOXIDANT ZMB see ZIS000
ANTIPAR see DHF600
ANTIPAR see DII200
ANTI-PELLAGRA VITAMIN see NCQ900
ANTIPERZ see TII500
ANTIPHEN see MJM500
ANTI-PICA see FDA880
ANTI-PICA see HAF400
ANTIPIRICULLIN see AQM250
ANTIPRESSINE DIHYDROCHLORIDE see DQB800
ANTIPREX 461 see ADV900
ANTIPREX A see ADV900
ANTIPYONIN see SFF000
ANTIPYRINE see AQN000
ANTIPYRINE IODIDE see IEU085
ANTIPYRINE, 4-IODO- see IEU085
ANTIPYRINE SALICYLATE see AQN250
N-ANTIPYRINYL-2-(DIMETHYLAMINO)PROPIONAMIDE see AMF375
N-((ANTIPYRINYLISOPROPYLAMINO)METHYL)NICOTINAMIDE see AQN500
(ANTIPYRINYLMETHYLAMINO)METHANESULFONIC ACID SODIUM SALT see AMK500
(ANTIPYRINYLMETHYLAMINO)METHANESULFONIC ACID SODIUM SALT MONOHYDRATE see MDM500
ANTIRAD see AJY250
ANTIRADON see AJY250
ANTIREN see PIJ000
ANTIRESISTANT see DDW500
ANTIREX see EAE600
ANTIRRHININ see KEA325
ANTI-RUST see SIQ500
ANTISACER see DKQ000
ANTISACER see DNU000
ANTISAL 1a see TGK750
ANTISEPSIN see BMR100
ANTISEPTOL see BEN000
ANTISERUM against the isozyme of LACTATE DEHYDROGENASE see LAO300
ANTISERUM to SPERM SPECIFIC LACTATE DEHYDROGENASE see LAO300
ANTISERUM TO LUTEINIZING HORMONE see LIU302
ANTISOL 1 see PCF275
ANTISTERILITY VITAMIN see VSZ450

ANTISTINE see PDC000
ANTISTOMINUM see BBV500
ANTISTREPT see SNM500
ANTI-STRESS see EQL000
ANTITANIL see DME300
ANTI-TETANY SUBSTANCE 10 see DME300
ANTITHROMBIN III see AQN550
ANTITROMBOSIN see BJZ000
ANTITUBERKULOSUM see ILD000
ANTIULCERA MASTER see CDQ500
ANTI-UV P see HND100
ANTIVERM see PDP250
ANTIVITIUM see DXH250
ANTIXEROPHTHALMIC VITAMIN see VSK600
ANTLERMICIN A see TEF725
ANTOBAN see PIJ500
ANTODYN see GGA950
ANTODYNE see GGA950
ANTOFIN see AFT500
ANTOL see EGV000
ANTOMIN see BBV500
ANTORA see PBC250
ANTORPHINE see AFT500
ANTOSTAB see SCA750
ANTOX see AQF000
ANTOXYLIC ACID see ARA250
ANTOZITE 67 see DQV250
ANTOZITE 67E see DQV250
ANTRACOL see ZMA000
ANTRACROMO BROWN D see TKN500
ANTRAMYCIN see API000
ANTRANCINE 12 see BQI000
ANTRAPUROL see DMH400
ANTRENIL see ORQ000
ANTRENYL see ORQ000
ANTRENYL BROMIDE see ORQ000
ANTRIOPEPTIN (HUMAN α-COMPONENT) see HGL680
ANTROMBIN K see WAT209
ANTRYCIDE see AQN625
ANTRYCIDE METHYL SULFATE see AQN625
ANTRYPOL see BAT000
ANTU see AQN635
ANTUITRIN S see CMG675
ANTURAT see AQN635
ANTURIO see APM875
ANTUSSAN see DBE200
ANTX-a see AOO120
ANTYMON (POLISH) see AQB750
ANTYMONOWODOR (POLISH) see SLQ000
ANTYWYLEGACZ see CMF400
ANU see PBX500
ANURAL see MQU750
ANUSPIRAMIN see BRF500
A. NUTTALLIANA see PAM780
ANVIL see HCN050
ANVITOFF see AJV500
ANXIETIL see MQU750
ANXINE see GGS000
ANXIOLIT see CFZ000
ANZIEF see ZVJ000
ANZON-TMS see AQF000
AO3 see DED100
AO 29 see BFW750
AO-40 see TMJ000
AO 4K see BFW750
AO 90 see AHR600
AO 128 see DNC100
AO 425 see MJN250
AO 754 see IFX200
AO A1 see AGX000
AOAA see ALQ650
A. OBLONGIFOLIA see BOO700
AOH see AGW476
AOH and AME (1:1) see AGW550
AOM see ASP250
AOMB see BCC500
A. OPPOSITIFOLIA see BOO700
AORAL see VSK600
AOS see AFN500

AP see PEK250
4-AP see AMI500
AP-14 see PAM500
17-AP see PMG600
AP 43 see DIF200
AP 50 see AQF000
A5MP see AOA125
9AAP see AHT000
AP-237 see BTA000
AP 407 see AOD250
A1-0109 P see AHE250
A 1588LP see AQF000
6-APA see PCU500
APACHLOR see CDS750
A. PACHYPODA see BAF325
APACIL see AMM250
APADODINE see DXX400
APADON see HIM000
APADRIN see MRH209
APAETP see AMD000
A. PALAESTINUM see ITD050
APAMIDE see HIM000
APAMIDON see FAB400
APAMIN see AQN650
APAMINE see AQN650
APAMINE see DBA800
APAMN see ABV750
APAP see HIM000
APARKAN see BBV000
APARKAZIN see DHF600
APARSIN see BBQ500
APAS see AMM250
APASCIL see MQU750
APATATE DRAPE see TES750
A. PATENS see PAM780
APATITE, HYDROXY see HJE100
APAURIN see DCK759
APAVAP see DGP900
APAVINPHOS see MQR750
APAZONE see AQN750
APAZONE DIHYDRATE see ASA000
APC see ABG750
APC see DBI800
APC (pharmaceutical) see ARP250
APCO 2330 see PEY000
APD see DBB500
'APE (HAWAII) see EAI600
APELAGRIN see NCQ900
APESAN see IPU000
APETAIN see BBK500
APEX 4 see BSL600
APEX 462-5 see TNC500
APEXOL see VSK600
APFO see ANP625
APGA see AMG750
APH see ACX750
APHAMITE see PAK000
APHENYLBARBIT see EOK000
APHIDAN see EPH500
APHILAN-R BASE see VJZ050
APHOLATE see AQO000
APHOS see DVX400
APHOSAL see GFA000
APHOX see DOX600
APHOXIDE see TND250
APHRODINE see YBJ000
APHRODINE HYDROCHLORIDE see YBS000
APHROSOL see YBJ000
APHTIRIA see BBQ500
APIGENIN see CDH250
APIGENIN 8-C-GLUCOSIDE see GFC050
APIGENINE see CDH250
APIGENIN TRIACETATE see THM300
APIGENOL see CDH250
API No. 2 FUEL OIL see DHE800
APIOL see AGE500
APIP see AOP800
APIRACHOL see TGJ150
APIRACOHL see TGJ150
APIROLIO 1431 C see TIM100
APIROLIO 1476 C see PAV600

APK 1 see NCY135
APL see CMG675
APL (hormone) see CMG675
APLAKIL see CFZ000
APLIDAL see BBQ500
APL-LUSTER see TEX000
APLYSIATOXIN see AQO100
APM see XFS600
APN see AQO000
β-APN see AMB750
APNPS see SNQ600
APO see AQO300
APO see TND250
APOATROPIN see AQO250
APOATROPINE see AQO250
APOCAROTENAL see AQO300
β-APO-8'-CAROTENAL see AQO300
APOCHOLIC ACID see AQO500
APOCID MILLING RED G see CMM320
APOCID ORANGE 2G see HGC000
APOCODEINE see AQO750
APOCYNAMARIN see SMM500
APOCYNINE see HLQ500
APOIDINA see CMG675
APOLAN see ARQ750
APOLON B₆ see PII100
APOMINE BLACK GX see AQP000
APOMINE GREEN B see CMO840
APOMORFIN see AQP250
APOMORPHINE see AQP250
APONEURON see AOB300
APONORIN see HII500
APOPEN see PDT500
APOPLON see RDK000
APORMORPHINE see AQP250
APORMORPHINE CHLORIDE see AQP500
6A-β-APORMPHINE-10,11-DIOL HYDROCHLORIDE see AQP500
6A-α-APORPHINE, 9,10-DIMETHOXY-1,2-(METHYLENEDIOXY)- see ERE150
6A-β-APORPHINE-10,11-DIOL see AQP250
6A-α-APORPHIN-1-OL, 2,11-DIMETHOXY-,HYDROCLORIDE see ISD033
APOSULFATRIM see TKX000
APOTHESINE see AQP750
3-α,16-α-APOVINCAMINIC ACID ETHYL ESTER see EGM100
A-POXIDE see MDQ250
APOZEPAM see DCK759
4-APP see POM600
APPA see PHX250
APPALACHIAN TEA see HGF100
APPEX see RAF100
APPLAUD see BOO631
APPLE SEEDS see AQP875
APPLE of SODOM (extract) see AQP800
APPL-SET see NAK500
APPRESINUM see HGP500
APPRESSIN see HGP495
APRALAN see AQP885
APRAMYCIN see AQP885
APRELAZINE see HGP500
APREN S see TES800
APRESAZIDE see HGP500
APRESINE see HGP500
APRESOLIN see HGP495
APRESOLIN see HGP500
APRESOLINE-ESIDRIX see HGP500
APRESOLINE HYDROCHLORIDE see HGP500
APREZOLIN see HGP495
APREZOLIN see HGP500
APRICOT PITS see AQP890
APRIDOL see EQL000
APRIL FOOLS see PAM780
APRINDINE HYDROCHLORIDE see FBP850
APRINOX see BEQ625
APROBARBITAL see AFT000
APROBARBITAL SODIUM see BOQ750
APROBARBITONE see AFT000
APROBARBITONE SODIUM see BOQ750
APROBIT see DQA400

APROCARB see PMY300
APROL 160 see MND100
APROL 161 see MND050
APRON see MDM100
APRON 2E see MDM100
APRONAL see IQX000
APRONALIDE see IQX000
APRON FL see MDM100
APROTININ see PAF550
APROZAL see AFT000
AP-S see ANT000
6-APS see PCU500
APSICAL see RDK000
APSIN VK see PDT750
APTAL see CFD990
APTIN see AGW000
APTROL SULFATE see AQQ000
A. PULSATILLA see PAM780
APURIN see ZVJ000
APURINA see DWW000
APUROL see ZVJ000
APV see CBR000
APYONINE AURAMINE BASE see IBB000
A-200 PYRINATE see AQQ050
AQ 110 see TKX125
AQ 227 see MRU090
AQ 229 see PIT625
AQD see TMJ800
4HAQO see HIY500
AQUA AMMONIA see ANK250
AQUACAL see CAX350
AQUACAT see CNA250
AQUA CERA see HKJ000
AQUACHLORAL see CDO000
AQUACIDE see DWX800
AQUACRINE see EDV000
AQUA-1,2-DIAMINOETHANE DIPEROXO
CHROMIUM(IV) see AQQ100
AQUA-1,2-
DIAMINOPROPANEDIPEROXOCHROMIU
M(IV) DIHYDRATE see AQQ125
AQUAFIL see SCH002
AQUAFINE RED E 9 see CMM760
AQUA FORTIS see NED500
AQUAKAY see MMD500
AQUA-KLEEN see DAA800
AQUALINE see ADR000
AQUALOSE see PKE700
AQUALOSE DL 12 see LAU560
AQUALOSE L 30 see LAU560
AQUA MEPHYTON see VTA000
AQUAMOLLIN see EIV000
AQUAMYCETIN see CDP250
AQUAMYCIN see ACJ250
AQUAPEL (POLYSACCHARIDE) see SLJ500
AQUAPHOR see CLF325
AQUAPLAST see SFO500
AQUA REGIA see HHM000
AQUAREX METHYL see SIB600
AQUARILLS see CFY000
AQUARIUS see CFY000
AQUASOL see VSP000
AQUASYNTH see VSK600
AQUATAG see BDE250
AQUATENSEN see MIV500
AQUATHOL see EAR000
AQUATIN see CLU000
AQUA-VEX see TIX500
AQUAVIRON see TBG000
AQUAZINE see BJP000
AQUEOUS DISPERSION RESIN see EFT100
AQUILIDE A see POI100
AQUINONE see MMD500
AR2 see AGX000
AR 3 see CBT500
AR 10 see DDW500
AR-11 see HNP500
AR-12 see MID000
AR-13 see HNR000
AR-16 see TGH665
AR-17 see HNP000
AR-19 see AGB000

AR-21 see HNQ500
AR-22 see BED250
AR-23 see BED500
AR-25 see HNS500
AR-32 see DQY909
AR-33 see HNQ000
AR-35 see HNS000
AR-41 see DQX800
AR-45 see AQQ250
AR177 see ZNS400
AR 12008 see DIO200
AR 81242 see DEC100
AR 84996 see BRB300
ARA-A see AEH100
ARA-ATP see ARQ500
ARABIC GUM see AQQ500
l-ARABINARIC ACID, 2-DEOXY-2-
METHYL-3,4-DI-C-METHYL-, 1,4-LACTONE
see MRG200
ARABINOCYTIDINE see AQQ750
ARABINOCYTIDINE see AQQ750
9-β-d-ARABINO FURANOSYL ADENINE see
AQQ900
9-(β-d-ARABINOFURANOSYL)ADENINE-5'-
(DIHYDROGEN PHOSPHATE) see AQQ905
9-β-d-
ARABINOFURANOSYLADENINEMONOH
YDRATE see AEH100
9-β-d-ARABINOFURANOSYLADENINE 5'-
TRIPHOSPHATE see ARQ500
1-β-D-ARABINOFURANOSYL-4-AMINO-
2(1H)PYRIMIDINONE see AQQ750
1-β-d-ARABINOFURANOSYL-2,2'-
ANHYDRO-CYTOSINE HYDROCHLORIDE
see COW900
1-ARABINOFURANOSYLCYTOSINE see
AQQ750
1-β-ARABINOFURANOSYLCYTOSINE see
AQQ750
1-(β-D-ARABINOFURANOSYL)CYTOSINE
see AQQ750
1-β-d-ARABINOFURANOSYLCYTOSINE
HYDROCHLORIDE see AQR000
1-β-d-ARABINOFURANOSYLCYTOSINE-5'-
PALMITATE see AQS875
1-β-d-ARABINOFURANOSYLCYTOSINE-5'-
PALMITOYL ESTER see AQS875
N-(1-β-d-ARABINOFURANOSYL-1,2-
DIHYDRO-2-OXO-4-
PYRIMIDINYL)DOCOSANAMIDE see
EAU075
1-β-d-ARABINOFURANOSYL-5-
FLUOROCYTOSINE see AQR250
9-β-d-ARABINOFURANOSYL-9H-PURINE-
6-AMINE MONOHYDRATE see AEH100
1-β-d-ARABINOFURANOSYL-2',3',5'-
TRIACETATE see AQR500
ARABINOGALACTAN see AQR800
(+)-ARABINOGALACTAN see AQR800
ARABINOSYLADENINE see AQQ900
9-ARABINOSYLADENINE see AQQ900
β-d-ARABINOSYLADENINE see AQQ900
ARABINOSYLADENINE
MONOPHOSPHATE see AQQ905
5'-ARABINOSYLADENINE
MONOPHOSPHATE see AQQ905
β-D-ARABINOSYLCYTOSINE see AQQ750
ARABINOSYLCYTOSINE
HYDROCHLORIDE see AQR000
ARABINOSYLCYTOSINE PALMITATE see
AQS875
ARABINOSYL CYTOSINE PALMITATE see
PAE260
ARABITIN see AQQ750
ARABOASCORBIC ACID see SAA025
d-ARABOASCORBIC ACID see EDE600
d-ARABOASCORBIC ACID see SAA025
ARAB RAT DETH see WAT200
ARA-C see AQQ750
ARACET APV see PKP750
ARACHIC ACID see EAF000
ARACHIDIC ACID see EAF000

ARACHIDONIC ACID see AQS750
ARACHIDONOYL ETHANOLAMIDE see
HKR600
N-ARACHIDONOYL-2-
HYDROXYETHYLAMIDE see HKR600
ARACHIDONYLETHANOLAMIDE see
HKR600
ARACHIS OIL see PAO000
ARACID see CJR500
ARACIDE see SOP500
ARA-CP see AQS875
ARA-C PALMITATE see AQS875
ARACTIDINE see AQQ750
ARA-CYTIDINE see AQQ750
ARACYTIDINE-5'-PALMITATE see AQS875
ARACYTIDINE 5'-PALMITATE see PAE260
ARACYTIN see AQQ750
ARAGONITE see CAO000
ARAKONIUM CHLORIDE see AFP300
ARALDIT DY 026 see BOS100
ARALDITE ACCELERATOR 062 see DQP800
ARALDITE ERE 1359 see REF000
ARALDITE 6010 mixed with ERR 4205 (1:1) see
OPI200
ARALDITE HARDENER 972 see MJQ000
ARALDITE HARDENER HY 951 see TJR000
ARALDITE HY 951 see TJR000
ARALEN see CLD000
ARALEN DIPHOSPHATE see CLD250
ARALEN HYDROCHLORIDE see CLD100
ARALEN PHOSPHATE see CLD250
ARALKONIUM CHLORIDE see LBV100
ARALO see PAK000
ARAMINE see HNB875
ARAMINE see HNC000
ARAMITE see SOP500
ARAMITEARARAMITE-15W see SOP500
ARANCIO CROMO (ITALIAN) see LCS000
ARASAN see TFS350
ARATAN see MEP250
ARATEN PHOSPHATE see AQT250
ARATHANE see AQT500
ARATRON see SOP500
ARBAPROSTIL see AQT575
ARBESTAB DSTDP see DXG700
ARBITEX see BBQ500
ARBOCEL see CCU150
ARBOCEL BC 200 see CCU150
ARBOCELL B 600/30 see CCU150
ARBOGAL see DSQ000
ARBOL DEL QUITASOL (CUBA) see
CDM325
ARBOL de PERU (MEXICO) see PCB300
ARBOREN F 11 see FLZ025
ARBORICID see BSQ750
ARBOROL see DUS700
ARBOTECT see TEX000
ARBRE FRICASSE (HAITI) see ADG400
ARBUTIN see HIH100
ARBUZ see PAG500
ARCACIL see PDT750
ARCADINE see BCA000
ARCASIN see PDT750
ARCHIDONATE see AQS750
ARCHIDYN see RKP000
ARCOBAN see MQU750
ARCOMONOL TABLETS see MQY400
ARCOSOLV see DWT200
ARCOTRATE see PBC250
ARCTON see CBY750
ARCTON 0 see CBY250
ARCTON 3 see CLR250
ARCTON 4 see CFX500
ARCTON 6 see DFA600
ARCTON 7 see DFL000
ARCTON 9 see TIP500
ARCTON 22 see CFX500
ARCTON 33 see FOO509
ARCTON 63 see FOO000
ARCTON 114 see FOO509
ARCTON 134A see EEC100
ARCTUVIN see HIH000

ARCUM R-S see RDK000
ARDALL see SJO000
ARDAP see RLF350
ARDEX see BBK500
ARDF 26 see GEW780
ARDF 26 SE see GEW780
ARDUAN see PII250
ARECA CATECHU see BFW000
ARECA CATECHU Linn., nut extract see BFW000
ARECA CATECHU Linn., fruit extract see BFW000
ARECAIDINE see AQT625
ARECAIDINE METHYL ESTER see AQT750
ARECAINE see AQT625
ARECA NUT see AQT650
ARECHIN see CLD250
ARECOLINE see AQT750
ARECOLINE BASE see AQT750
ARECOLINE BROMIDE see AQU000
ARECOLINE HYDROBROMIDE see AQU000
ARECOLINE HYDROCHLORIDE see AQU250
AREDION see CKM000
AREGINAL see EKL000
ARELIZ see PDW250
ARELON see IRA050
ARELON R see IRA050
ARESIN see CKD500
ARESKAP 100 see AQU500
ARESKET 300 see AQU750
ARESKLENE 400 see AQV000
ARESOL see RLU000
ARETAN see MEP250
ARETAN 6 see MEP250
ARETAN-NIEUW see BKS810
ARETIT see ACE500
ARETIT see BRE500
ARETIT (the phenol) see ACE500
AREZIN see CKD500
AREZINE see CKD500
ARFICIN see RKP000
ARFONAD see TKW500
ARFONAD CAMPHORSULFONATE see TKW500
ARFONAD ROCHE see TKW500
ARGAMINE see AQW000
ARGEMONE OIL mixed with MUSTARD OIL see OGS000
ARGENTATE(1-), TRI-MU-IODODIIODOTETRA-, RUBIDIUM see RQF100
ARGENT FLUORURE (FRENCH) see SDQ500
ARGENTIC FLUORIDE see SDQ500
ARGENTIUM CREDE see SDI750
ARGENTOUS OXIDE see SDU500
ARGENTUM see SDI500
ARGEZIN see ARQ725
ARGININE see AQV980
l-ARGININE see AQV980
l-ARGININE, nitrosated see NJU500
5-l-ARGININECYANOGINOSIN LA see AQV990
ARGININE HYDROCHLORIDE see AQW000
d-ARGININE HYDROCHLORIDE see AQV500
l-ARGININE HYDROCHLORIDE see AQW000
ARGININE MONOHYDROCHLORIDE see AQW000
ARGININE, MONOHYDROCHLORIDE, d-see AQV500
l-ARGININE MONOHYDROCHLORIDE see AQW000
d-ARGININE, MONOHYDROCHLORIDE (9CI) see AQV500
8-l-ARGININEOXYTOCIN see AQW125
ARGININE VASOPRESSIN see AQW050
ARGININE-VASOTOCIN see AQW125
8-ARGININE VASOTOCIN see AQW125

l-ARGINYL-l-ASPARAGINYL-l-ARGINYL-l-LEUCYL-l-ISOLEUCYL-l-PROLYL-l-PROLYL-l-PHENYLALANYL-l-TRYPTOPHYL-l-LYSYL-l-THREONYL-l-ARGININAMIDE see AQW100
l-ARGINYL-l-LEUCYL-l-ISOLEUCYL-l-PROLYL-l-PROLYL-l-PHENYLALANYL-l-TRYPTOPHYL-l-LYSINAMIDE see AQW110
ARGIPRESTOCIN see AQW125
ARGIVENE see AQW000
ARGO BRAND CORN STARCH see SLJ500
ARGOBYL see TKP100
ARGOLD see CMP955
ARGON see AQW250
ARGONAL see PFM250
(ARG⁸)OXYTOCIN see AQW125
ARGUAD 12 see LBX075
ARGUAD 12D see LBX075
ARGUAD 12-23 see LBX075
ARGUAD 12-33 see LBX075
ARGUAD 12-50 see LBX075
ARGUAD 12-37W see LBX075
ARGUAD MC 50 see LBX075
ARGUN see AGN000
ARG-VASOTOCIN see AQW125
8-ARG-VASOTOCIN see AQW125
ARHEOL see OGY220
ARIAVIT RED 2G see CMM300
ARIBINE see MPA050
ARICHIN see CFU750
ARIEN see AQW500
ARIGAL C see MCB050
ARILAT see CBM750
ARILATE see BAV575
ARILATE see CBM750
ARINAMINE see DXO300
ARIOTOX see TDW500
ARIPIPRAZOLE see AQW600
ARISAEMA (VARIOUS SPECIES) see JAJ000
ARISAN see CKF750
ARISTAMID see SNJ350
ARISTAMIDE see SNJ350
ARISTOCORT see AQX250
ARISTOCORT ACETONIDE see AQX500
ARISTOCORT DIACETATE see AQX750
ARISTOCORT FORTE PARENTERAL see AQX750
ARISTOCORT SYRUP see AQX750
ARISTODERM see AQX500
ARISTOGEL see AQX500
ARISTOGYN see SNJ350
ARISTOLIC ACID see AQX825
ARISTOLIC ACID METHYL ESTER see MGQ525
ARISTOLICHIA INDICA L., ALCOHOLIC EXTRACT see AQY000
ARISTOLOCHIC ACID see AQY250
ARISTOLOCHIC ACID B see MJR780
ARISTOLOCHIC ACID II see MJR780
ARISTOLOCHIC ACID II SODIUM SALT see SEY075
ARISTOLOCHIC ACID I SODIUM SALT see SEY050
ARISTOLOCHIC ACID IV see MEA700
ARISTOLOCHIC ACID IVA METHYL ETHER see MEA700
ARISTOLOCHIC ACID SODIUM SALT see AQY125
ARISTOLOCHIC ACID, SODIUM SALT see SEY050
ARISTOLOCHINE see AQY250
ARISTOPHYLLIN see DNC000
ARISTOSPAN see AQY375
ARIZOLE see PIH750
ARIZOLE see PMQ750
ARKADE see PFR130
ARKITROPIN see MDL000
ARKLONE P see FOO000
ARKOFIX NG see DTG000
ARKOFIX NM see MCB050
ARKOPAL N-090 see PKF000
ARKOTINE see DAD200

ARKOZAL see BSQ000
ARLACEL 40 see MRJ800
ARLACEL 60 see SKV150
ARLACEL 80 see SKV100
ARLACEL 83 see SKV170
ARLACEL 161 see OAV000
ARLACEL 169 see OAV000
ARLACEL C see SKV170
ARLACIDE G see CDT250
ARLANONE BLUE 2B see ICU135
ARLANTHRENE GOLDEN YELLOW see DCZ000
ARLANTHRENE VIOLET 4R see DFN450
ARLAZOL FAST TURQUOISE BLUE 6GN see COF420
ARLEF see TKH750
ARLIDIN HYDROCHLORIDE see DNU200
ARLOSOL GREEN B see BLK000
ARLOSOL YELLOW S see CMS240
ARMAC 18D see OAP300
ARMAC OD see OAP300
ARMAZAL see COH250
ARMCO 21-6-9 see IGL110
ARMCO IRON see IGK800
ARMEEN 18 see OBC000
ARMEEN L-7 see HBM490
ARMEEN 12D see DXW000
ARMEEN 16D see HCO500
ARMEEN 18D see OBC000
ARMEEN DM-12D see DRR800
ARMEEN DM16D see HCP525
ARMEEN DM 18D see DTC400
ARMEEN O see OHM700
ARMENIAN BOLE see IHC450
ARMINE see ENQ000
ARMODOUR see PKQ059
ARMOFILM see OBC000
ARMOFOS see SKN000
ARMOISE OIL see AQY385
ARMOSTAT 801 see OAV000
ARMOTAN MO see SKV100
ARMOTAN MS see SKV150
ARMOTAN PML-20 see PKG000
ARMOTAN PMO-20 see PKL100
ARMSTRONG'S ACID see AQY400
ARMSTRONG'S S ACID see AQY400
ARMYL see MRV250
ARNATTA see APE050
ARNATTO see APE050
ARNAUDON'S GREEN see CMK300
ARNAUDON'S GREEN (HEMIHEPTAHYDRATE) see CMK300
ARNEEL 8 see OCW100
ARNICA see AQY500
ARNITE A see PKF750
ARNOSULFAN see SNN300
ARNOTTA see APE050
ARO see CBT750
AROALL see SJO000
AROCHLOR 1221 see PJM000
AROCHLOR 1242 see PJM500
AROCHLOR 1254 see PJN000
AROCHLOR 1260 see PJN250
AROCHLOR 5460 see PJP800
AROCLOR see PJL750
AROCLOR 54 see CLD250
AROCLOR 1016 see PJL750
AROCLOR 1016 see PJL800
AROCLOR 1221 see PJL750
AROCLOR 1232 see PJL750
AROCLOR 1232 see PJM250
AROCLOR 1242 see PJL750
AROCLOR 1242 see PJM500
AROCLOR 1248 see PJL750
AROCLOR 1248 see PJM750
AROCLOR 1254 see PJL750
AROCLOR 1254 see PJN000
AROCLOR 1260 see PJL750
AROCLOR 1260 see PJN250
AROCLOR 1262 see PJL750
AROCLOR 1262 see PJN500
AROCLOR 1268 see PJL750

AROCLOR 1268 see PJN750
AROCLOR 2565 see PJL750
AROCLOR 2565 see PJO000
AROCLOR 4465 see PJL750
AROCLOR 4465 scc PJO250
AROCLOR 5442 see PJP750
AROCLOR 5442 see PJP750
AROCLOR 5460 see PJP800
AROFLOW see CBT750
AROFT see AEW625
AROFUTO see AEW625
AROGEN see CBT750
AROLON see ADV900
AROMA BLANCA (CUBA, HAWAII) see LED500
AROMATIC AMINES see AQY750
AROMATIC CASTOR OIL see CCP250
AROMATIC OILS (HYDROCARBONS) see AQZ150
AROMATIC PETROLEUM DERIVATIVE SOLVENT see DXG850
AROMATIC SOLVENT (PETROLEUM) see DXG850
AROMATIC SPIRITS of AMMONIA see AQZ000
AROMATOL see AQZ100
AROMEX see AQZ150
AROMEX see CBT750
AROMOX DMMC-W see DRS200
ARON see ADV900
ARON A 10H see ADV900
ARON ALPHA 402X see EHP700
ARON ALPHA D see EHP700
ARON COMPOUND HW see PKQ059
AROSOL see PER000
AROSURF TA 100 see DXG625
AROTINOIC ACID see AQZ200
AROTINOIC METHANOL see AQZ300
AROTINOID ETHYL ESTER see AQZ400
AROTONE see CBT750
AROVEL see CBT750
AROVIT see VSP000
AROZIN see AOT255
ARPEZINE see PIJ500
ARPHONAD see TKW500
ARQUAD 16 see HCQ525
ARQUAD 18 see TLW500
ARQUAD 16-29 see HCQ525
ARQUAD 16-50 see HCQ525
ARQUAD 18-50 see TLW500
ARQUAD 16-25W see HCQ525
ARQUAD DM14B-90 see TCA500
ARQUAD DM18B-90 see DTC600
ARQUAD DMMCB-75 see AFP250
ARQUAD R 40 see DXG625
ARQUAD T see TAC300
ARQUAD 2HT see QAT550
ARQUAD T-50 see TAC300
ARQUAD 2HT75 see QAT550
ARQUEL see DGM875
ARRESIN see CKD500
ARRESTEN see MQQ050
ARRET see CMG000
ARRET see LIH000
ARRHENAL see DXE600
ARROW see CBT750
ARROWROOT STARCH see SLJ500
ARROW WOOD see MBU825
ARSACETIN see AQZ900
ARSACETIN SODIUM SALT see AQZ900
ARSACETIN SODIUM SALT, TETRAHYDRATE see ARA000
ARSACETIN TETRAHYDRATE see ARA000
9-ARSAFLUORENINIC ACID see ARA100
ARSAFLUORINIC ACID see ARA100
ARSAMBIDE see CBJ000
ARSAMIN see ARA500
ARSAMINOL see SAP500
ARSAN see HKC000
ARSANILIC ACID see ARA250
4-ARSANILIC ACID see ARA250
p-ARSANILIC ACID see ARA250

ARSANILIC ACID, N-ACETYL-, SODIUM SALT see AQZ900
p-ARSANILIC ACID, N,N-BIS(2-CHLOROETHYL)- see BIA300
p-ARSANILIC ACID, N,N-BIS(2-HYDROXYETHYL)- see BKD600
p-ARSANILIC ACID, BISMUTH, SODIUM SALT see BKX500
ARSANILIC ACID, N-(4,6-DIAMINO-s-TRIAZIN-2-YL)-,DISODIUM SALT see SIF450
p-ARSANILIC ACID, N,N-DIETHYL- see DIS775
ARSANILIC ACID, MONOSODIUM SALT see ARA500
ARSANILIC ACID SODIUM SALT see ARA500
ARSAPHENAN see ACN250
ARSECLOR see DFX400
ARSECODILE see HKC500
ARSEN (GERMAN, POLISH) see ARA750
ARSENAMIDE see TFA350
ARSENATE see ARB250
ARSENATE of IRON, FERRIC see IGN000
ARSENATE of IRON, FERROUS see IGM000
ARSENATE of LEAD see LCK000
ARSENENOUS ACID, CALCIUM SALT (2:1) see CAM300
ARSENENOUS ACID, POTASSIUM SALT see PKV500
ARSENIATE de CALCIUM see ARB750
ARSENIATE de MAGNESIUM (FRENCH) see ARD000
ARSENIATE de PLOMB (FRENCH) see ARC750
ARSENIC see ARA750
ARSENIC-75 see ARA750
ARSENIC, metallic (DOT) see ARA750
ARSENIC ACID see ARH500
m-ARSENIC ACID see ARB000
o-ARSENIC ACID see ARB250
ARSENIC ACID, solid (DOT) see ARB250
ARSENIC ACID, solid (DOT) see ARB250
ARSENIC ACID, liquid (DOT) see ARB250
ARSENIC ACID, AMMONIUM MAGNESIUM SALT, HYDRATE (1:1:1:2) see MAD025
ARSENIC ACID ANHYDRIDE see ARH500
ARSENIC ACID, ANILINE SALT see ARB270
ARSENIC ACID, CALCIUM SALT see CAM222
ARSENIC ACID, CALCIUM SALT (2:3) see ARB750
ARSENIC ACID CESIUM SALT see CAK350
ARSENIC ACID, COMPD. WITH BENZENAMINE (1:3) see ARB270
ARSENIC ACID, COPPER(2+) SALT (2:3), TETRAHYDRATE see CNI750
ARSENIC ACID, DISODIUM SALT see ARC000
ARSENIC ACID, DISODIUM SALT, HEPTAHYDRATE see ARC250
ARSENIC ACID (H3-AS-O4), AMMONIUM MAGNESIUM SALT, (1:1:1) see MAD025
ARSENIC ACID, HEMIHYDRATE see ARC500
ARSENIC ACID, LEAD SALT see ARC750
ARSENIC ACID, LEAD(2+) SALT (2:3) see LCK100
ARSENIC ACID, MAGNESIUM SALT see ARD000
ARSENIC ACID, MANGANESE(2+) SALT, HYDRATE (2:3:6) see MAQ700
ARSENIC ACID, METHYLPHENYL-(9CI) see HMK200
ARSENIC ACID (H3AsO4), MONOCESIUM SALT (8CI,9CI) see CAK350
ARSENIC ACID, MONOPOTASSIUM SALT see ARD250
ARSENIC ACID, MONOSODIUM SALT see ARD500
ARSENIC ACID, MONOSODIUM SALT see ARD600

ARSENIC ACID, SODIUM SALT see ARD750
ARSENIC ACID, SODIUM SALT (9CI) see ARD500
ARSENIC ACID, TRICESIUM SALT see CDC375
ARSENIC ACID, TRISODIUM SALT see SEY150
ARSENIC(V) ACID, TRISODIUM SALT, HEPTAHYDRATE (1:3:7) see ARE000
ARSENIC ACID, ZINC SALT see ZDJ000
ARSENICAL DIP see ARE250
ARSENICAL DIP, liquid (DOT) see ARE250
ARSENICAL DUST see ARE500
ARSENICAL FLUE DUST see ARE500
ARSENICAL FLUE DUST see ARE750
ARSENICALS see ARA750
ARSENICALS see ARF750
ARSENIC ANHYDRIDE see ARH500
ARSENIC BISULFIDE see ARF000
ARSENIC BLACK see ARA750
ARSENIC BLANC see ARI750
ARSENIC(III) BROMIDE see ARF250
ARSENIC BUTTER see ARF500
ARSENIC CHLORIDE see ARF500
ARSENIC(III) CHLORIDE see ARF500
ARSENIC COMPOUNDS see ARF750
ARSENIC DICHLOROETHANE see DFH200
ARSENIC DIETHYL see ARG000
ARSENIC DIMETHYL see ARG250
ARSENIC DISULFIDE see ARJ100
ARSENIC FLUORIDE see ARI250
ARSENIC HEMISELENIDE see ARG500
ARSENIC HYDRID see ARK250
ARSENIC HYDRIDE see ARK250
ARSENIC IODIDE see ARG750
ARSENIC OXIDE see ARH500
ARSENIC OXIDE see ARI750
ARSENIC(V) OXIDE see ARH500
ARSENIC(III) OXIDE see ARI750
ARSENIC PENTASULFIDE see ARH250
ARSENIC PENTOXIDE see ARH500
ARSENIC PHOSPHIDE see ARH750
ARSENIC SELENIDE see ARH800
ARSENIC SESQUIOXIDE see ARI750
ARSENIC SESQUISELENIDE see ARH800
ARSENIC SESQUISULFIDE see ARI000
ARSENIC SULFIDE see ARI000
ARSENIC SULFIDE (DOT) see ARJ100
ARSENIC SULFIDE YELLOW see ARI000
ARSENIC SULPHIDE see ARI000
ARSENIC TERSULPHIDE see ARI000
ARSENIC TRIBROMIDE see ARF250
ARSENIC TRIFLUORIDE see ARI250
ARSENIC TRIHYDRIDE see ARK250
ARSENIC TRIIODIDE see ARG750
ARSENIC TRIIODIDE mixed with MERCURIC IODIDE see ARI500
ARSENIC TRIOXIDE see ARI750
ARSENIC TRIOXIDE mixed with SELENIUM DIOXIDE (1:1) see ARJ000
ARSENIC TRISELENIDE see ARH800
ARSENIC TRISULFIDE see ARH500
ARSENIC TRISULFIDE see ARJ100
ARSENIC TRISULFIDE (DOT) see ARI000
ARSENICUM ALBUM see ARI750
ARSENIC YELLOW see ARI000
ARSENIDES see ARJ250
ARSENIGEN SAURE see ARI750
ARSENIOUS ACID see ARI750
ARSENIOUS ACID, CALCIUM SALT see CAM500
ARSENIOUS ACID, SODIUM SALT see ARJ500
ARSENIOUS ACID, SODIUM SALT see SEY500
ARSENIOUS ACID, SODIUM SALT POLYMERS see ARJ500
ARSENIOUS ACID, STRONTIUM SALT see SME500
ARSENIOUS ACID, TRISILVER(1+) SALT see SDM100

ARSENIOUS ACID (H₃AsO₃), TRISODIUM SALT (8CI) see SEY200

ARSENIOUS ACID, ZINC SALT (9CI) see ZDS000

ARSENIOUS CHLORIDE see ARF500

ARSENIOUS and MERCURIC IODIDE, solution (DOT) see ARI500

ARSENIOUS OXIDE see ARI750

ARSENIOUS SULPHIDE see ARI000

ARSENIOUS TRIOXIDE see ARI750

ARSENITE see ARI750

ARSENITE de POTASSIUM (FRENCH) see PKV500

ARSENITE de SODIUM (FRENCH) see SEY500

ARSENIURETTED HYDROGEN see ARK250

ARSENO 39 see ARL000

ARSENO-BISMULAK see BKX500

ARSENOCHOLINE see HLC550

ARSENOLITE see ARI750

ARSENOMARCASITE see ARJ750

1-(p-ARSENOPHENYL)UREA see CBI500

ARSENOPYRITE see ARJ750

ARSENOSAN see ARL000

p-ARSENOSOANILINE see ARJ755

4-ARSENOSOANILINE, DIHYDRATE see ALV100

ARSENOSOBENZENE see PEG750

p-ARSENOSO-N,N-BIS(2-CHLOROETHYL)ANILINE see ARJ760

p-ARSENOSO-N,N-BIS(2-HYDROXYETHYL)ANILINE see ARJ770

p-ARSENOSO-N,N-DIETHYLANILINE see ARJ800

4-ARSENOSO-2-NITROPHENOL see NHE600

p-ARSENOSOPHENOL see HNG800

p-ARSENOSOTOLUENE see TGV100

ARSENOSUGAR see DNB700

ARSENOUS ACID see ARI750

ARSENOUS ACID ANHYDRIDE see ARI750

ARSENOUS ACID, SODIUM SALT (9CI) see SEY500

ARSENOUS ACID, TRISILVER(1+) SALT (9CI) see SDM100

ARSENOUS ACID, TRISODIUM SALT see SEY200

ARSENOUS ANHYDRIDE see ARI750

ARSENOUS BROMIDE see ARF250

ARSENOUS CHLORIDE see ARF500

ARSENOUS FLUORIDE see ARF250

ARSENOUS HYDRIDE see ARK250

ARSENOUS IODIDE see ARG750

ARSENOUS OXIDE see ARI750

ARSENOUS OXIDE ANHYDRIDE see ARI750

ARSENOUS SELENIDE see ARH800

ARSENOUS SULFIDE see ARI000

ARSENOUS TRIBROMIDE see ARF250

ARSENOUS TRICHLORIDE (9CI) see ARF500

ARSENOUS TRIIODIDE (9CI) see ARG750

ARSENOWODOR (POLISH) see ARK250

ARSENOXIDE see ARL000

ARSENOXIDE SODIUM see ARJ900

ARSENPHENOLAMINE HYDROCHLORIDE see SAP500

ARSENTRIOXIDE see ARI750

ARSENWASSERSTOFF (GERMAN) see ARK250

ARSEVAN see NCJ500

ARSINE see ARK250

ARSINE, (p-AMINOPHENYL)DICHLORO-, HYDROCHLORIDE see AOR640

ARSINE, (p-AMINOPHENYL)OXO-, DIHYDRATE see ALV100

ARSINE, AMYLDICHLORO- see AOI200

ARSINE BORON TRIBROMIDE see ARK500

ARSINE, sec-BUTYLDICHLORO- see BQY300

ARSINE, CHLORO(2-CHLOROVINYL)PHENYL- see PER600

ARSINE, (2-CHLOROETHYL)DICHLORO- see CGV275

ARSINE, DICHLOROHEPTYL- see HBN600

ARSINE, DICHLOROHEXYL- see HFP600

ARSINE, DICHLOROISOPENTYL- see IHQ100

ARSINE, DICHLOROPENTYL- see AOI200

ARSINE, DICHLOROPROPYL- see PNH650

ARSINE, DIIODOMETHYL- see MGQ775

ARSINE, DIPHENYLHYDROXY- see DVY100

ARSINE, ETHYL- see EGM150

ARSINE, ETHYLENEBIS(DIPHENYL)- see EIQ200

ARSINE, HYDROXYDIPHENYL- see DVY100

ARSINE OXIDE, ALLYLHYDROXYPHENYL- see AGQ775

ARSINE OXIDE, BUTYLHYDROXYISOPROPYL- see BRQ800

ARSINE OXIDE, (o-CHLOROPHENYL)(3-(2,4-DICHLOROPHENOXY)-2-HYDROXYPROPYL)HYDROXY- see DGA425

ARSINE OXIDE, (m-CHLOROPHENYL)HYDROXY(β-HYDROXYPHENETHYL)- see CKA575

ARSINE OXIDE, (p-CHLOROPHENYL)HYDROXYMETHYL- see CKD800

ARSINE OXIDE, DIBUTYLHYDROXY- see DDV250

ARSINE OXIDE, DIETHYLHYDROXY- see DIS850

ARSINE OXIDE, HYDROXYDIMETHYL-, SODIUM SALT, TRIHYDRATE see HKC550

ARSINE OXIDE, HYDROXY(2-HYDROXYPROPYL)PHENYL- see HNX600

ARSINE OXIDE, HYDROXYISOBUTYLISOPROPYL- see IPS100

ARSINE OXIDE, HYDROXYMETHYLPHENETHYL- see MNU050

ARSINE OXIDE, HYDROXYMETHYLPHENYL- see HMK200

ARSINE OXIDE, HYDROXYMETHYLPROPYL- see MOT800

ARSINE OXIDE, TRIMETHYL- see TLH170

ARSINE, OXO(4-CARBOXY)PHENYL- see CCI550

ARSINE SELENIDE, TRIMETHYL- see TLH250

ARSINE SULFIDE, DIMETHYLDI- see DQG700

ARSINE-TRI-1-PIPERIDINIUM CHLORIDE see ARK750

ARSINETTE see LCK000

ARSINETTE see LCK100

ARSINIC ACID, DIBUTYL-(9CI) see DDV250

ARSINIC ACID, DIMETHYL-(9CI) see HKC000

ARSINIC ACID, DIMETHYL-, SODIUM SALT (9CI) see HKC500

ARSINOSOLVIN see ARA500

ARSINOTRIS PIPERIDINIUM TRICHLORIDE see ARK750

ARSINYL see DXE600

ARSION see CEI250

ARSODENT see ARI750

ARSONAMIDOUS CHLORIDE, N,N-DIPHENYL- see DVY050

ARSONATE liquid see MRL750

ARSONIC ACID see ABX500

ARSONIC ACID, (4-(ACETYLAMINO)PHENYL)-, MONOSODIUM SALT (9CI) see AQZ900

ARSONIC ACID, (4-AMINOPHENYL)-, MONOSODIUM SALT (9CI) see ARA500

ARSONIC ACID, CALCIUM SALT (1:1) see CAM520

ARSONIC ACID, COPPER(2+) SALT (1:1) (9CI) see CNN500

ARSONIC ACID, (4-((4,6-DIAMINO-1,3,5-TRIAZIN-2-YL)AMINO)PHENYL)-, DISODIUM SALT see SIF450

ARSONIC ACID, (4-HYDROXYPHENYL)-, polymer with FORMALDEHYDE see BCJ150

ARSONIC ACID, METHYL-, IRON SALT (9CI) see IHB680

ARSONIC ACID, METHYL-, MONOAMMONIUM SALT see MDQ770

ARSONIC ACID, METHYL-, POTASSIUM SALT see PKV500

ARSONIC ACID, SODIUM SALT (9CI) see ARJ500

ARSONIUM, (2-HYDROXYETHYL)TRIMETHYL- see HLC550

ARSONIUM, (3-HYDROXYPHENYL)DIETHYLMETHYL-, IODIDE, METHYLCARBAMATE see MID900

ARSONIUM, TETRAMETHYL- see TDL100

ARSONIUM, TETRAPHENYL-, CHLORIDE see TEA300

2-(4'-ARSONOANILINO)-4,6-DIAMINO-s-TRIAZINE SESQUISODIUM SALT see SIF450

((p-ARSONOPHENYL)CARBAMOYL)DITHIOCARBAMIC ACID see DXM100

4-ARSONOPHENYLGLYCINAMIDE see CBJ750

p-ARSONOPHENYLUREA see CBJ000

(3-(p-ARSONOPHENYL)UREIDO)DITHIOBENZOIC ACID see ARK800

ARSONOUS DICHLORIDE, ETHYL-(9CI) see DFH200

ARSONOUS DICHLORIDE, METHYL-(9CI) see DFP200

ARSONOUS DICHLORIDE, (1-METHYLPROPYL)-(9CI) see BQY300

ARSONOUS DICHLORIDE, PHENYL-(9CI) see DGB600

ARSONOUS DIIODIDE, METHYL-(9CI) see MGQ775

N-((p-ARSONOPHENYL)CARBAMOYL)DITHIOGLYCINE see DXM100

ARSPHEN see ABX500

ARSPHENAMINE see SAP500

ARSPHENAMINE METHYLENESULFOXYLIC ACID SODIUM SALT see NCJ500

ARSPHENOXIDE see ARL000

ARSYCODILE see HKC550

ARSYNAL see DXE600

ART 2 see CBT500

ARTAM see PGG000

ARTANE see BBV000

ARTANE HYDROCHLORIDE see BBV000

ARTANE TRIHEXYPHENIDYL see BBV000

ARTEANNUIN see ARL375

ARTEGODAN see PAH250

ARTEMETHER see ARL425

ARTEMISIA OIL see ARL250

ARTEMISIA OIL (WORMWOOD) see ARL250

ARTEMISINE see ARL375

ARTEMISININ see ARL375

ARTEMISININELACTOL METHYL ETHER see ARL425

ARTERENOL see NNO500

d-ARTERENOL see ARL500

l-ARTERENOL see NNO500

dl-ARTERENOL see ARL750

(+−)-ARTERENOL BITARTRATE see NNE525

l-ARTERENOL BITARTRATE see NNO699

dl-ARTERENOL HYDROCHLORIDE see NNP050

ARTERIOFLEXIN see ARQ750

ARTERIOVINCA see VLF000

ARTEROCOLINE see ABO000

ARTEROCOLINE see CMF250
ARTEROCYN see PLV750
ARTERODY see BBJ750
ARTEROSOL see ARQ750
ARTES see ARQ750
ARTES see MHO200
ARTEVIL see ARQ750
d-ARTHIN see VSZ100
ARTHODIBROM see NAG400
ARTHO LM see MLH000
ARTHRIPUR see IEU085
ARTHROBID see SOU550
ARTHROCHIN see CLD000
ARTHROCINE see SOU550
ARTHROPAN see CMG000
ARTHRYTIN OXOATE see AMX250
ARTIC see MIF765
ARTIFICIAL ALMOND OIL see BAY500
ARTIFICIAL ANT OIL see FPQ875
ARTIFICIAL BARITE see BAP000
ARTIFICIAL CINNAMON OIL see CCO750
ARTIFICIAL GUM see DBD800
ARTIFICIAL HEAVY SPAR see BAP000
ARTIFICIAL MUSTARD OIL see AGJ250
ARTIFICIAL SILK BLACK G see CMN240
ARTIFICIAL SILK BLACK GN see CMN240
ARTIFICIAL SILK BLACK GR see CMN240
ARTIFICIAL SWEETENING SUBSTANZ
GENDORF 450 see SJN700
ARTISIL BLUE BGL see CMP075
ARTISIL BLUE BSG see MGG250
ARTISIL BLUE GREEN GP see DMM400
ARTISIL BLUE SAP see TBG700
ARTISIL BRILLIANT PINK RFS see AKO350
ARTISIL BRILLIANT ROSE 5BP see DBX000
ARTISIL DIRECT RED 3BP see AKE250
ARTISIL DIRECT YELLOW G see AAQ250
ARTISIL ORANGE 3RP see AKP750
ARTISIL RED 3BP see AKE250
ARTISIL VIOLET 2RP see DBP000
ARTISIL YELLOW FL see KDA075
ARTISIL YELLOW G see AAQ250
ARTISIL YELLOW 2GN see AAQ250
ARTIZIN see BRF500
ARTO-ESPASMOL see DNU100
ARTOLON see MQU750
ARTOMEY see VCK100
ARTOMYCIN see TBX250
ARTONIL see BBW750
ARTOSIN see BSQ000
ARTOZIN see BSQ000
ARTRACIN see IDA000
ARTRIBID see SOU550
ARTRIL 300 see IIU000
ARTRINOVO see IDA000
ARTRIONA see CNS825
ARTRIVIA see IDA000
ARTRIZONE see BRF500
ARTROBIONE see CMG000
ARTROFLOG see HNI500
ARTROPAN see BRF500
A. RUBRA see BAF325
ARUMEL see FMM000
ARUM (Various Species) see ITD050
ARUSAL see IPU000
ARVIN see VGU700
ARVYNOL see CHG000
ARWIN see VGU700
ARWOOD COPPER see CNI000
ARYL ALKYL POLYETHER ALCOHOL see
ARL875
ARYLAM see CBM750
ARYLAN PWS see BBS275
ARZENE see PEG750
AS see AFJ375
AS see CCU250
AS-15 see SLY500
2-ASe see AMN500
A.S. 1.398 see BAC020
AS-4137 see TCD600
AS-17665 see NDY500
AS (surfactant) see AFJ375

ASA see ADA725
A.S.A. see ADA725
ASA 158-5 see PJA130
ASABAINE see DJM800
ASABAINE see XCJ000
ASA COMPOUND see ABG750
A.S.A. EMPIRIN see ADA725
ASAGRAEA OFFICINALIS see VHZ000
ASAGRAN see ADA725
ASAHISOL 1527 see AAX250
ASALIN see ARM000
ASALINE see ARM000
ASAMEDOL see DHS200
ASAMID see ENG500
ASANA see FAR150
ASANA XL see FAR150
ASARON see IHX400
ASARONE see IHX400
(Z)-ASARONE see ARM100
cis-ASARONE see ARM100
ASARONE, trans- see IHX400
trans-ASARONE see IHX400
α-ASARONE see IHX400
β-ASARONE see ARM100
cis-β-ASARONE see ARM100
ASARUM CAMPHOR see IHX400
5-ASA SULFATE SODIUM SALT see AMN275
ASATARD see ADA725
ASAZOL see MRL750
ASB 516 see AAX250
ASBEST (GERMAN) see ARM250
ASBESTINE see TAB750
ASBESTOS see ARM250
7-45 ASBESTOS see ARM268
ASBESTOS (ACGIH) see ARM260
ASBESTOS (ACGIH) see ARM262
ASBESTOS (ACGIH) see ARM264
ASBESTOS (ACGIH) see ARM268
ASBESTOS (ACGIH) see ARM275
ASBESTOS (ACGIH) see ARM280
ASBESTOS, ACTINOLITE see ARM260
ASBESTOS, AMOSITE see ARM262
ASBESTOS, ANTHOPHYLITE see ARM264
ASBESTOS, ANTHOPHYLLITE see ARM266
ASBESTOS, CHRYSOTILE see ARM268
ASBESTOS, CROCIDOLITE see ARM275
ASBESTOS FIBER see ARM250
ASBESTOS, TREMOLITE see ARM280
ASC see AAM600
ASC-4 see THW750
ASC 66825 see CEX800
ASC 67178 see CEX800
ASCABIN see BCM000
ASCABIOL see BCM000
ASCAREX SYRUP see PIJ500
ASCARIDOL see ARM500
ASCARIDOLE see ARM500
ASCARIDOLE (organic peroxide) (DOT) see
ARM500
ASCARISIN see ARM500
ASCARYL see HFV500
ASCENSIL see ALC250
ASCEPTICHROME see MCV000
ASCLEPIN see ACF000
ASCOFURANONE see ARM750
ASCOPHEN see ARP250
ASCORBIC ACID see ARN000
l-ASCORBIC ACID see ARN000
l(+)-ASCORBIC ACID see ARN000
ASCORBIC ACID SODIUM SALT see
ARN125
l-ASCORBIC ACID SODIUM SALT see
ARN125
ASCORBICIN see ARN125
ASCORBIN see ARN125
ASCORBUTINA see ARN000
ASCORBYL PALMITATE see ARN150
ASCORBYL STEARATE see ARN180
ASCORPHYLLINE see HLC000
ASCOSERP see RDK000
ASCOSERPINA see RDK000
ASCOSIN see ARN250

ASCUMAR see ABF750
ASCURIT see IAL200
ASCURON see BJI000
ASEBOTOXIN see AOO375
ASECRYL see GIC000
ASENDIN see AOA095
ASEPSIN see BMR100
ASEPTA HERBAN see HDP500
ASEPTICHROME see MCV000
ASEPTOFORM see HJL500
ASEPTOFORM E see HJL000
ASEPTOFORM P see HNU500
ASEPTOLAN see SAH500
ASEX see SFS000
ASHES (residues) see CMY765
ASHLENE see NOH000
ASIATICOSIDE see ARN500
ASIDON 3 see TEH500
ASILAN see SNQ500
A-1125 (SILANE) see MJE793
A 171 (SILANE DERIVATIVE) see TLD000
ASIPRENOL see DMV600
ASL-603 see BMV750
AS 61CL see ADY500
ASL 8052-001 see MKR125
ASMACORIL see PER700
ASMADION see TEH500
ASMALAR see DMV600
ASMATANE MIST see AES000
ASMATANE MIST see VGP000
ASMAVAL see TEH500
ASM MB see BCC500
ASOZIN see MGQ750
ASP 47 see SOD100
ASP 51 see TED500
ASPALON see ADA725
ASPAMINOL HYDROCHLORIDE see
ARN700
ASPARA see MAI600
ASPARAGIC ACID see ARN850
l-ASPARAGIC ACID see ARN850
ASPARAGINASE see ARN800
l-ASPARAGINASE see ARN800
1-ASPARAGINASE X see ARN800
l-ASPARAGINASI (ITALIAN) see ARN800
ASPARAGINATE CALCIUM see CAM675
l-ASPARAGINE see ARN810
l-ASPARAGINE AMIDOHYDROLASE see
ARN800
ASPARAGINIC ACID see ARN850
l-ASPARAGINIC ACID see ARN850
ASPARA K see PKV600
ASPARTAME see ARN825
ASPARTAT see MAI600
ASPARTIC ACID see ARN850
l-ASPARTIC ACID see ARN850
(l)-ASPARTIC ACID see ARN850
(S)-ASPARTIC ACID see ARN850
dl-ASPARTIC ACID see ARN830
l-(+)-ASPARTIC ACID see ARN850
l-ASPARTIC ACID, N-ACETYL-, DILITHIUM
SALT see DNU330
l-ASPARTIC ACID, COMPD. WITH 18-
DECARBOXY-40-DEMETHYL-3,7-
DIDEOXO-N$^{3'}$-((DIMETHYLAMINO)
ACETYL)-18-(((2-
(DIMETHYLAMINO)ETHYL)AMINO)CARB
ONYL)-3,7-DIHYDROXY-N^{47})-METHYL-5-
OXOCANDICIDIN D, CYCLIC 15,19-
HEMIACETAL (2:1) see DOS400
ASPARTIC ACID, N-(p-((2,4-DIAMINO-6-
PTERIDYLMETHYL)AMINO)BENZOYL)-,
SODIUM SALT see SGE500
ASPARTIC ACID DISODIUM SALT see
SEZ350
l-ASPARTIC ACID, DISODIUM SALT (9CI)
see SEZ350
l-ASPARTIC ACID, POTASSIUM SALT (9CI)
see PKV600
l-ASPARTIC ACID, SODIUM SALT (9CI) see
SEZ355

ASPARTYLPHENYLALANINE METHYL
ESTER see ARN825
N-l-α-ASPARTYL-l-PHENYLALANINE 1-
METHYL ESTER (9CI) see ARN825
ASPATOFORT see AHR600
ASPEGIC see ARP125
ASPERASE see ARN875
ASPERFILLUS ALKALINE PROTEINASE see
SBI860
ASPERGILLIC ACID see ARO000
ASPERGILLIN see ARO250
ASPERGILLOPEPTIDASE B see SBI860
ASPERGUM see ADA725
ASPHALEN SULPHUR BLACK C see CMS250
ASPHALEN SULPHUR BLACK S see CMS250
ASPHALT see ARO500
ASPHALT (CUT BACK) see ARO750
ASPHALT, at or above its Fp (DOT) see
ARO500
ASPHALT FUMES (ACGIH) see ARO500
ASPHALT, PETROLEUM see ARO500
ASPHALT, PETROLEUM see PCR500
ASPHALTUM see ARO500
A. SPICATA see BAF325
ASPICULAMYCIN see ARP000
ASPIDELAPS SCUTATUS NEUROTOXIN
S4C6 see NCQ530
ASPIDO-SPRIDINE, 1-ACETYL-17-
METHOXY-(8CI,9CI) see ARP050
ASPIDO-SPRIN see ARP050
ASPIDO-SPRINE see ARP050
(−)-ASPIDO-SPRINE see ARP050
ASPIRDROPS see ADA725
ASPIRIN see ADA725
ASPIRIN ACETAMINOPHEN ESTER see
SAN600
ASPIRINE see ADA725
ASPIRIN-ISOPROPYLANTIPYRINE see
PNR800
ASPIRIN-dl-LYSINE see ARP125
ASPIRIN LYSINE SALT see ARP125
ASPIRIN-NATRIUM (GERMAN) see ADA750
ASPIRIN, PHENACETIN, and CAFFEINE see
ARP250
ASPIRISINE see ARP125
ASPISOL see ARP125
ASPON see TED500
ASPON-CHLORDANE see CDR750
ASPOR see EIR000
ASPORUM see EIR000
ASPRO see ADA725
ASPRON see HBT500
ASSAM TEA see ARP500
ASSIFLAVINE see DBX400
ASSIMIL see MJE760
ASSIPRENOL see DMV600
ASSUGRIN see SGC000
ASSUGRIN FEINUSS see SGC000
ASSUGRIN VOLLSUSS see SGC000
ASSURE see QMA100
ASTA see CQC650
ASTA 4968 see MRG025
ASTA 5122 see SOD000
ASTA 5160 see KFK125
ASTA B518 see CQC650
ASTA C see CNR750
ASTA CD 072 see CNR825
ASTARIL see SLP600
ASTA Z 4828 see TNT500
ASTA Z 4942 see IMH000
ASTA Z 7557 see ARP625
ASTELTOXIN see ARP640
ASTEMIZOL (GERMAN) see ARP675
ASTEMIZOLE see ARP675
ASTERIC see ADA725
ASTEROMYCIN see ARP000
ASTHENTHILO see DKL800
ASTHMA METER MIST see VGP000
ASTHMA WEED see CCJ825
ASTMAHALIN see VGP000
ASTMAMASIT see DNC000
ASTM B127 see NCX525

ASTM B163-800 see IGL100
ASTM B164-A see NCX525
ASTM B344-60NI, 16CR see NCX515
ASTM XM10 see IGL110
A-STOFF see CDN200
ASTOMIN see MLP250
ASTRA CHRYSOIDINE R see PEK000
ASTRACILLIN see PDD350
ASTRA DIAMOND GREEN GX see BAY750
ASTRAFER see IGT000
ASTRA FUCHSINE B see MAC250
ASTRALON see PKQ059
ASTRANTIAGENIN D see GMG000
ASTRAPHLOXIN see CMM840
ASTRA PHLOXINE see CMM840
ASTRA PHLOXINE G see CMM840
ASTRA PHLOXINE G EXTRA see CMM840
ASTRAPHLOXIN G see CMM840
ASTRATONE see EJQ500
ASTRAZOLO see SNN500
ASTRAZON GOLDEN YELLOW GL see
CMM895
ASTRAZON GOLDEN YELLOW GL-E see
CMM895
ASTRAZON GOLDEN YELLOW GL-FW see
CMM895
ASTRAZON ORANGE see CMM820
ASTRAZON ORANGE G see CMM820
ASTRAZON ORANGE RRL see CMM830
ASTRAZON PINK FG see CMM850
ASTRAZON RED GTL see BAQ750
ASTRAZON RED VIOLET FRR see EFC650
ASTRAZON YELLOW 3G see CMM890
ASTRAZON YELLOW 3GL see CMM890
ASTRESS see CFZ000
ASTRIDINE see CCK125
ASTRINGEN see AHA000
ASTROBAIN see OKS000
ASTROBOT see DGP900
ASTROCAR see DJS200
ASTROLOY see NCY135
ASTROMEL NW 6A see MCB050
ASTROPHYLLIN see DNC000
ASTROTIA STOKESII VENOM see SBI890
ASTRUMAL see PLO500
ASTYN see EHP000
ASUCCIN see SMY000
ASUCROL see CKK000
ASUGRYN see SGC000
ASULAM see SNQ500
ASULAM POTASSIUM SALT see PKV700
ASULFIDINE see PPN750
ASULFOX F see SNQ500
ASULOX see SNQ500
ASULOX 40 see SNQ500
ASUNTHOL see CNU750
A. SUPERBA (AUSTRALIA) VENOM see
ARU750
A. SUPERBA VENOM see ARV625
ASURO see DBD750
ASVERIN-C see ARP875
ASVERIN CITRATE see ARP875
ASVERINE CITRATE see ARP875
AS XVII see KEA300
ASYMMETRICAL TRIMETHYL BENZENE
see TLL750
ASYMMETRIN see FNO000
AT see AGK250
AT see AGW675
AT see AMY050
3,A-T see AMY050
AT 7 see HCL000
o-AT see AIC250
A.T. 10 see DME300
AT-90 see AMY050
AT 101 see HID350
AT 265 see CEF125
AT-290 see PED750
AT 327 see BLV000
AT-581 see ARQ000
AT 717 see PKQ250
AT-2266 see EAU100

ATA see AMY050
ATABRINE see ARQ250
ATABRINE DIHYDROCHLORIDE see
CFU750
ATABRINE HYDROCHLORIDE see CFU750
ATABRON see CDS800
ATACRYL YELLOW 4G see CMM890
ATACRYL YELLOW 4G-FS see CMM890
ATACTIC POLY(ACRYLIC ACID) see
ADV900
ATACTIC POLYPROPYLENE see PMP500
ATACTIC POLYSTYRENE see SMQ500
ATACTIC POLY(VINYL CHLORIDE) see
PKQ059
ATADIOL see CKE750
ATALCO C see HCP000
ATALCO O see OBA000
ATALCO S see OAX000
ATAMASCO LILY see RBA500
ATARA see CJR909
ATARAX see CJR909
ATARAX see HOR470
ATARAX DIHYDROCHLORIDE see HOR470
ATARAX HYDROCHLORIDE see VSF000
ATARAXOID see CJR909
ATARAXOID DIHYDROCHLORIDE see
HOR470
ATARAZOID see CJR909
ATA-Sb see AQC000
ATAZINA see CJR909
ATAZINAX see ARQ725
ATC see TEV000
ATCC No. 20034 see GAV050
ATCC No. 20474 see CBC425
AT 327 CITRATE see ARP875
ATCOTIBINE see ILD000
ATCP see PIB900
ATDA see AMR250
ATDA HYDROCHLORIDE see AMR500
ATEC see ADD750
ATECULON see ARQ750
ATEM see IGG000
ATEMI see CQJ150
ATEMI C see CQJ150
ATEMPOL see EQL000
ATENOLOL see TAL475
ATENSIN see GGS000
ATENSINE see DCK759
ATERAX see CJR909
ATERAX DIHYDROCHLORIDE see HOR470
ATERIAN see AHO250
ATERIOSAN see ARQ750
ATEROCYN see PLV750
ATEROID see SEH450
ATEROSAN see PPH050
ATGARD see DGP900
ATHAPROPAZINE see DIR000
ATHEBRATE see ARQ750
ATHERILINE see ARQ325
ATHEROLINE see ARQ325
ATHEROLIP see AHA150
ATHEROMIDE see ARQ750
ATHEROPRONT see ARQ750
ATHOPROPAZIN see DIR000
ATHRANID-WIRKSTOFF see ARQ750
ATHROMBIN see WAT220
ATHROMBINE-K see WAT200
ATHROMBON see PFJ750
ATHYLEN (GERMAN) see EIO000
ATHYLENGLYKOL (GERMAN) see EJC500
ATHYLENGLYKOL-MONOATHYLATHER
(GERMAN) see EES350
ATHYL-GUSATHION see EKN000
ATHYMIL see BMA625
ATIC VAT BLUE BC see DFN300
ATIC VAT BLUE XRN see IBV050
ATIC VAT BRILLIANT PURPLE 4R see
DFN450
ATIC VAT BROWN R see CMU780
ATIC VAT OLIVE R see DUP100
AT-III see AQN550
ATILEN see DCK759

ATILON see SNY500
ATIPI see ARQ500
ATIRAN see MEP250
ATIRIN see CCS250
ATIVAN see CFC250
ATLACIDE see SFS000
ATLANTIC see CBT750
ATLANTIC ACID FAST BLUE B see CMM090
ATLANTIC ALIZARINE MILLING BLUE RB
see CMM092
ATLANTIC ARTIFICIAL SILK BLACK G see
CMN240
ATLANTIC BLACK BD see AQP000
ATLANTIC BLACK C see AQP000
ATLANTIC BLACK E see AQP000
ATLANTIC BLACK EA see AQP000
ATLANTIC BLACK GAC see AQP000
ATLANTIC BLACK GG see AQP000
ATLANTIC BLACK GXCW see AQP000
ATLANTIC BLACK GXOO see AQP000
ATLANTIC BLACK SD see AQP000
ATLANTIC BLUE 2B see CMO000
ATLANTIC BLUE RW see CMO600
ATLANTIC BORDEAUX B see CMO872
ATLANTIC BRILLIANT YELLOW MN see
CMP050
ATLANTIC BROWN BCW see CMO820
ATLANTIC BROWN BP see CMO820
ATLANTIC BROWN D 3Y see CMO810
ATLANTIC BROWN M see CMO800
ATLANTIC CONGO RED see SGQ500
ATLANTIC DARK GREEN see CMO830
ATLANTIC FAST RED F see CMO870
ATLANTIC FAST TURQUOISE LGA see
COF420
ATLANTIC GREEN 2B see CMO840
ATLANTIC GREEN WT see CMO830
ATLANTICHROME ORANGE G see CMP882
ATLANTICHROME VIOLET B see HLI000
ATLANTICHROME YELLOW 2G see SIT850
ATLANTIC RESIN FAST BLUE see CMN750
ATLANTIC RESIN FAST BROWN BRL see
CMO750
ATLANTIC SCARLET 3B see CMO875
ATLANTIC SKY BLUE A see CMO500
ATLANTIC VIOLET N see CMP000
ATLAS "A" see SEY500
ATLAS G 263 see EKN550
ATLAS G 711 see BBS275
ATLAS G 924 see SLL000
ATLAS G-2133 see DXY000
ATLAS G-2142 see PJY100
ATLAS G-2144 see PJY100
ATLAS G 2146 see HKJ000
ATLAS G 3300 see BBS275
ATLAS G 3802 see PJT300
ATLAS G 3816 see PJT300
ATLASOL SPIRIT RED3 see MMP100
ATLAS WHITE TITANIUM DIOXIDE see
TGG760
ATLATEST see TBF750
AT LIQUID see AMY050
ATLOX 1087 see PKL100
ATLOX 4862 see BLX000
AuTM see GJC000
ATMONIL see AOR500
ATMOS 150 see OAV000
ATMUL 67 see OAV000
ATMUL 84 see OAV000
ATMUL 124 see OAV000
ATNASE see GIA050
ATOCIN see PGG000
ATOMIT see CAO000
ATOMIT see CAT775
ATOMITE see CAT775
ATONIN O see ORU500
ATOPHAN see PGG000
ATOPHAN-NATRIUM (GERMAN) see
SJH000
ATOPHAN SODIUM see SJH000
ATOREL see IDE000
ATORVASTATIN CALCIUM see DMS950

ATOSIL see DQA400
ATOVER see PPH050
ATOXAN see CBM750
ATOXICOCAINE see AIT250
ATOXYL scc ARA500
ATOXYLIC ACID see ARA250
ATP see ARQ500
5'-ATP see ARQ500
ATP (nucleotide) see ARQ500
ATP DISODIUM see AEM100
ATP DISODIUM SALT see AEM100
AT-17 PHOSPHATE see MLP250
ATP Na SALT see AEM250
ATRAL see PJA120
ATRALYMON see ECB200
ATRANEX see ARQ725
ATRANORIN see ARQ600
ATRASINE see ARQ725
ATRATOL see SFS000
ATRATOL 80W see ARQ700
ATRATOL A see ARQ725
ATRATON see EGD000
ATRATONE see EGD000
ATRAVET see AAF750
ATRAVET see ABH500
ATRAXINE see MQU750
ATRAZIN see ARQ725
ATRAZINE see ARQ725
ATRAZINE-AMETRYNE MIXT. see HBT100
ATRED see ARQ725
ATREX see ARQ725
α-hmn ATRIAL NATRIURETIC HORMONE
see HGL680
ATRIOPEPTIN-28 (HUMAN) see HGL680
α-ATRIOPEPTIN (HUMAN) see HGL680
ATRIOPEPTIN-33(RAT), 1-DE-l-LEUCINE-2-
DE-l-ALANINE-3-DEGLYCINE-4-DE-l-
PROLINE-5-DE-l-ARGININE-1 7-l-
METHIONINE- see HGL680
ATRIPHOS see ARQ500
A. TRIPHYLLUM see JAJ000
A. TRISPERMA see TOA275
ATRIVYL see MMN250
ATROCHIN see SBG000
ATROL see DPA000
ATROLEN see ARQ750
ATROMET see HBT100
ATROMET T see HBT100
ATROMID see ARQ750
ATROMIDIN see ARQ750
ATROMID S see ARQ750
ATROPA BELLADONNA see DAD880
ATROPAMIN see AQO250
ATROPAMINE see AQO250
ATROPIN (GERMAN) see ARR000
ATROPINE see ARR000
(−)-ATROPINE see HOU000
ATROPINE METHOBROMIDE see MGR250
ATROPINE METHONITRATE see MGR500
ATROPINE METHYLBROMIDE see MGR250
ATROPINE METHYL NITRATE see MGR500
ATROPINE OCTABROMIDE see OEM000
ATROPINE-N-OCTYL BROMIDE see
OEM000
ATROPINE SULFATE (1:1) see ARR250
ATROPINE SULFATE (2:1) see ARR500
ATROPIN-N-OCTYLBROMID (GERMAN)
see OEM000
ATROPIN SIRAN (CZECH) see ARR500
ATROPINSULFAT (GERMAN) see ARR500
ATROPYLTROPEINE see AQO250
ATROQUIN see SBG000
ATROTON see EGD000
ATROVENT see IGG000
ATROVIS see ARQ750
AT SEFEN see DPE000
ATSETOZIN see ABH500
ATTAC 6 see CDV100
ATTAC 6-3 see CDV100
ATTACLAY see PAE750
ATTACLAY X 250 see PAE750
ATTACOTE see PAE750

ATTAGEL see PAE750
ATTAGEL 40 see PAE750
ATTAGEL 50 see PAE750
ATTAGEL 150 see PAE750
ATTAPULGITE see PAE750
ATTAR ROSE see RNA000
ATTAR of ROSE see RNA000
ATTAR ROSE TURKISH see OHJ000
ATTASORB see PAE750
ATUL ACID BLACK BX see FAB830
ATUL ACID BLACK 10BX see FAB830
ATUL ACID CRYSTAL ORANGE G see
HGC000
ATUL ACID GERANINE 6B see CMM400
ATUL ACID GERANINE G see CMM300
ATUL ACID ORANGE II see CMM220
ATUL ACID SCARLET 3R see FMU080
ATUL CONGO RED see SGQ500
ATUL CRYSTAL RED F see HJF500
ATUL DEVELOPED BLACK BT see CMN800
ATUL DIRECT BLACK E see AQP000
ATUL DIRECT BLUE 2B see CMO000
ATUL DIRECT BROWN CN see CMO810
ATUL DIRECT BROWN MR see CMO800
ATUL DIRECT BROWN MY see CMO800
ATUL DIRECT DARK GREEN P see
CMO830
ATUL DIRECT GREEN B see CMO840
ATUL DIRECT RED 4B see DXO850
ATUL DIRECT SKY BLUE see CMO500
ATUL DIRECT VIOLET N see CMP000
ATUL FAST YELLOW R see DOT300
ATUL INDIGO CARMINE see FAE100
ATUL OIL ORANGE T see TGW000
ATUL OIL RED G see OHI200
ATUL ORANGE R see PEJ500
ATUL PIGMENT RED RS SODIUM SALT see
NAP100
ATUL SULFUR RED BROWN 3B see CMS257
ATUL SULPHUR BLACK GXE see CMS250
ATUL SULPHUR BLACK GXR see CMS250
ATUL SULPHUR BLACK RP see CMS250
ATUL SUNSET YELLOW FCF see FAG150
ATUL TARTRAZINE see FAG140
ATUL VULCAN FAST PIGMENT ORANGE
G see CMS145
ATUMIN see ARR750
ATURBANE HYDROCHLORIDE see ARR875
ATX II see ARS000
ATYSMAL see ENG500
AU 3 see CBT500
AU-95722 see AMD000
l'AUBIER de TILIA SYLVESTRIS (FRENCH)
see SAW300
l'AUBIER de TILLEUL (FRENCH) see
SAW300
AUBYGEL GS see CCL250
AUBYGUM DM see CCL250
AUCUBA JAPONICA see JBS050
AUGMENTIN see ARS125
AUGMENTIN (antibiotic) see ARS125
'AUKO'I (HAWAII) see CNG825
AULES see TFS350
AULIGEN see BJU000
AURAMINE BASE see IBB000
AURAMINE HYDROCHLORIDE see IBA000
AURAMINE (MAK) see IBA000
AURAMINE (MAK) see IBB000
AURAMINE O (BIOLOGICAL STAIN) see
IBA000
AURAMINE YELLOW see IBA000
AURAMYCIN A see ARS130
AURAMYCIN B see ARS135
AURANETIN see ARS250
AURANILE see DKQ000
AURANILE see DNU000
AURANOFIN see ARS150
AURANTHINE see ARS200
AURANTHINE see ARS250
AURANTIA see HET500
AURANTICA see MRN500
AURANTIN see AEA500

AURANTINE see ARS250
AURANTIN (FLAVONE) see ARS250
AURATE(3⁻), BIS(MONOTHIOSULFATO)-,
TRISODIUM, DIHYDRATE see TNK500
AURATE(1-),(2-HYDROXY-3-MERCAPTO-1-
PROPANESULFONATO(2-))-,SODIUM see
SEZ400
AURATE(1-), TETRACHLORO-, SODIUM see
GIZ100
AURATE(1-), TETRACHLORO-, SODIUM,
(SP-4-1)-(9CI) see GIZ100
AUREINE see ARS500
AUREMETINE see ARS750
AUREOCICLINA see CMB000
AUREOCINA see CMA750
AUREOCYCLINE see CMB000
AUREOFUSCIN see ART000
AUREOLIC ACID see MQW750
AUREOLIN see PLI750
AUREOMYCIN see CMA750
AUREOMYCIN A-377 see CMA750
AUREOMYCIN HYDROCHLORIDE see
CMB000
AUREOMYKOIN see CMA750
AUREOTAN see ART250
AUREOTHIN see DTU200
AURIC CHLORIDE see GIW176
AURICIDINE see GJG000
AURINE-TRICARBOXYLATE
d'AMMONIUM (FRENCH) see AGW750
AURINTRICARBOXYLIC ACID
AMMONIUM SALT see AGW750
AURIPIGMENT see ARI000
AURLELIC ACID see MQW750
AUROCIDIN see GJG000
AURO-DETOXIN see GIX350
AUROLIN see GJG000
AUROMOMYCIN see ART125
AUROMYOSE see ART250
AUROPAN see NBU000
AUROPEX see GJG000
AUROPIN see GJG000
AURORA YELLOW see CAJ750
AUROSAN see GJG000
AUROTAN see ART250
1-AUROTHIO-d-GLUCOPYRANOSE see
ART250
AUROTHIOGLUCOSE see ART250
AUROTHION see GJG000
AUROTHIOPROPANOL SODIUM
SULFONATE see SEZ400
AUROVERTIN see ART500
AURUMINE see ART250
AUSOVIT see DXO300
AUSTIOX see TGG760
AUSTIOX R-CR see RSP100
AUSTOCYSTIN D see ARU250
AUSTRACIL see CDP250
AUSTRACOL see CDP250
AUSTRALIAN BROWN SNAKE VENOM see
TEG650
AUSTRALIAN COPPERHEAD SNAKE
VENOM see ARU750
AUSTRALIAN DEATH ADDER SNAKE
VENOM see ARU875
AUSTRALIAN GUM see AQQ500
AUSTRALIAN KING BROWN SNAKE
VENOM see ARV000
AUSTRALIAN KING COBRA SNAKE
VENOM see ARV125
AUSTRALIAN RED-BELLIED BLACK
SNAKE VENOM see ARV250
AUSTRALIAN ROUGH SCALED SNAKE
VENOM see ARV375
AUSTRALIAN TAIPAN SNAKE VENOM see
ARV500
AUSTRALIAN TIGER SNAKE VENOM see
ARV550
AUSTRALIS see ARV000
AUSTRALOL see IQZ000
AUSTRANAL see MCK500
AUSTRAPEN see AIV500

AUSTRAPINE see RDK000
AUSTRASTAPH see SLJ050
AUSTRELAPS SUPERBA (AUSTRALIA)
VENOM see ARU750
AUSTRELAPS SUPERBA VENOM see
ARV625
AUSTRELAPS SUPERBUS VENOM see
ARV625
AUSTRIAN CINNABAR see LCS000
AUSTROMINAL see EOK000
AUSTROVIT PP see NCR000
AUTAN see DKC800
AUTHRON see ART250
AUTOLYZED YEAST EXTRACT see BAD400
AUTOMIN see BBV500
AUTOMOBILE EXHAUST CONDENSATE
see GCE000
AUTOMOTIVE DIESEL OIL see DHE900
AUTOMOTIVE DIESEL OIL see FOP000
AUTOMOTIVE GASOLINE see GCC200
AUTUMN CROCUS see CNX800
AUXEOMYCIN see CMB000
AUXILSON see DBC510
AUXINUTRIL see PBB750
AUXOBIL see DYE700
AV00 see AGX000
AV000 see AGX000
AVACAN see AAD875
AVACAN HYDROCHLORIDE see ARV750
AVADEX see DBI200
AVADEX BW see DNS600
AVADYL see NOC000
AVAGAL see DJM800
AVAGAL see XCJ000
AVAPENA see CKV625
AVAZYME see CML880
AVELLANO (DOMINICAN REPUBLIC) see
TOA275
AVENA SATIVA LINN., EXTRACT see
OAD090
AVENGE see ARW000
AVENTOX see TJL500
AVERANTIN see HLF550
(−)-AVERANTIN see HLF550
AVERMECTIN see ARW050
AVERMECTIN A1A, 4"-(ACETYLAMINO)-
26-(BENZOYLOXY)-5-o-DEMETHYL-4"-
DEOXY-, (4"R)- see ABX830
AVERMECTIN A1A, 4"-(ACETYLAMINO)-5-
o-DEMETHYL-4"-DEOXY-26-((2-
METHOXYETHOXY)METHOXY)-, (4"R)- see
ARW100
AVERMECTIN A1A, 4"-(ACETYLAMINO)-5-
o-DEMETHYL-4"-DEOXY-26-METHOXY-,
(4"R)- see ABX840
AVERMECTIN A1A, 5-o-DEMETHYL-26-
HYDROXY- see HJE525
AVERMECTIN A1A, 5-o-DEMETHYL-26-
HYDROXY-4"-o-((2-
METHOXYETHOXY)METHYL)- see
DAN400
AVERMECTIN A1A, 22,23-DIHYDRO-4"-
(ACETYLAMINO)-5-o-DEMETHYL-4"-
DEOXY-26-((2-
METHOXYETHOXY)METHOXY)-, (4"R)- see
ARW110
AVERMECTIN A1A, 22,23-DIHYDRO-5-o-
DEMETHYL- see DKR700
AVERMECTIN A1A, 22,23-DIHYDRO-5-o-
DEMETHYL-26-((2-
METHOXYETHOXY)METHOXY)- see
DKX200
AVERMECTIN B1A see ARW150
AVERMECTIN B₁ₐ see ARW150
AVERMECTIN B', 4"-DEOXY-4"-
(METHYLAMINO)-, (4"R)-,BENZOATE
(SALT) see EAJ150
AVERMECTIN B(SUB 1) see ARW200
AVERMECTIN B(SUB 1) TECHNICAL
GRADE see ARW200
AVERMIN see AOR500
A. VERNALIS see PCU375

AVERSAN see DXH250
AVERTIN see ARW250
AVERTIN see THV000
AVERZAN see DXH250
AVESYL see GGS000
A/VI see EBY500
AVIAMINE see AHR600
AVIBEST C see ARM268
AVIBON see VSK600
AVICADE see RLF350
AVICEL see CCU150
AVICEL 101 see CCU150
AVICEL 102 see CCU150
AVICEL PH 101 see CCU150
AVICEL PH 105 see CCU150
AVICOL see ARW750
AVICOL see PAX000
AVID EC see ARW200
AVIDIN see GIA100
AVIDINS see GIA100
AVIL-RETARD see TMK000
AVIOMARIN see DYE600
AVIROL 118 CONC see SIB600
AVIROSAN see ARW775
AVIROSAN see DTS700
AVIROSAN see PIX775
AVISUN see PMP500
AVITA see VSK600
AVITEX C see HCP900
AVITEX SF see HCP900
AVITOL see VSK600
AVITROL see AMI500
AVLOCLOR see CLD000
AVLOCLOR see CLD250
AVLON see DBX400
AVLON see XAK000
AVLOSULPHONE see SOA500
AVLOTANE see HCI000
AVM see ARW150
AVOCADO OIL see ARW800
AVOCAN see NOC000
AVOGODRITE see PKY000
AVOKSIN see FMR575
AVOLIN see DTR200
AVOMEC see ARW200
AVOMINE see DQA400
AVON GREEN A-4379 see BAY750
AVOXIN see FMR575
AVOXYL see GGS000
AVT see AQW125
AW-14-2446 see CMW600
AW-14'2333 see HOU059
AW-15'1129 see CKK050
AWD 19-166 see BMB125
AWELYSIN see SLW450
AWPA #1 see CMY825
AW 15 (POLYSACCHARIDE) see HKQ100
AX 363 see CAT775
AXEROPHTHAL see VSK985
AXIOM see DIX600
AXM see ABN000
AXOLE see FPK000
AXURIS see AOR500
AY-5406 see BCA000
AY-6108 see AIV500
AY 8682 see CQH625
AY 9944 see BHN000
AY-11483 see EDU600
AY 21011 see ECX100
AY 22342 see MJG100
AY-22,352 see CIO375
AY 22989 see RBK000
AY 24034 see LIU370
AY-25,329 see NMV750
AY-25,674 see ONQ000
AY 43-004 see DOE100
AY 56012 see OGI075
AY-57,062 see ABH500
AY-61122 see MLJ500
AY 61123 see ARQ750
AY-62014 see BPT750
AY 62021 see CLX250

AY-62022 see MBZ100
AY 64043 see ICB000
AY 64043 see ICC000
AY 136155 see MBZ100
AY 11483-16 see EDU600
AYAA see AAX250
AYAF see AAX250
AYAPANIN see MEK300
AYERMATE see MQU750
AYERST 62013 see TJQ333
AYFIVIN see BAC250
AYUSH-47 see ARX125
8-AZAADENINE see AIB340
9-AZAANTHRACENE see ADJ500
10-AZAANTHRACENE see ADJ500
5-AZA-10-ARSENAANTHRACENE
CHLORIDE see PDB000
12-AZABENZ(a)ANTHRACENE see BAW750
AZABENZENE see POP250
3-AZABENZONITRILE see NDW515
12-AZABENZO(a)PYRENE see BDD600
AZABICYCLANE CITRATE see ARX150
7-AZABICYCLO(2.2.1)HEPTANE, 2-(6-
CHLORO-3-PYRIDINYL)-,
DIHYDROCHLORIDE, exo-(+−)- see EAZ100
2-AZABICYCLO(2.2.1)HEPTANE, 2-
METHYL- see MGR600
3-AZABICYCLO(3.2.2)NONANE see ARX300
3-AZABICYCLO(3.2.2)NONANE, 3-
(CHLOROACETYL)- see CEC100
3-AZABICYCLO(3.2.2)NONANE, 3-
(IODOACETYL)- see IDZ100
1-AZABICYCLO(2.2.2)OCTANE see QUJ400
8-AZABICYCLO(3.2.1)OCTANE-2-
CARBOXYLIC ACID, 3-(4-
CHLOROPHENYL)-8-METHYL-, PHENYL
ESTER, HYDROCHLORIDE, (1R,2S,3S,5S)-
see CKM800
1-AZABICYCLO(2.2.2)OCTAN-3-OL (9CI) see
QUJ990
1-AZABICYCLO(2.2.2)OCTAN-3-OL,
BENZILATE (9CI) see QVA000
1-AZABICYCLO(2.2.2)OCTAN-3-OL,
BENZYLATE (ESTER), HYDROCHLORIDE
see QWJ000
1-AZABICYCLO(3.2.1)OCTAN-6-OL
DIPHENYLACETATE HYDROCHLORIDE
see ARX500
AZABICYCLO(2.2.1)OCTAN-3-OL, HYDROGEN
DIPHENYLACETATE (ester), HYDROGEN
SULFATE (2:1) DIHYDRATE see QVJ000
1-AZABICYCLO(3.2.1)OCTAN-6-OL-9-
FLUORENECARBOXYLATE
HYDROCHLORIDE see ARX750
AZABICYCLOOCTANOL METHYL
BROMIDE DIPHENYLACETATE see
ARX770
1-(3-AZABICYCLO(3.3.0)OCT-3-YL)-3-(p-
TOLYLSULFONYL)UREA see DBL700
AZAC see TAN300
4-AZACHRYSENE see NAU700
AZACITIDINE see ARY000
AZACOSTEROL DIHYDROCHLORIDE see
ARX800
AZACOSTEROL HYDROCHLORIDE see
ARX800
AZACTAM see ARX875
AZACYCLOHEPTANE see HDG000
1-AZACYCLOHEPTANE see HDG000
2-AZACYCLOHEPTANONE see CBF700
AZACYCLOHEXANE see PIL500
AZACYCLONOL HYDROCHLORIDE see
PIY750
β-1-AZACYCLOOCTYLETHYLGUANIDINE
SULFATE see GKS000
1-AZACYCLOOCT-2-YL METHYL
GUANIDINE see GKO800
1-AZA-2,4-CYCLOPENTADIENE see PPS250
AZACYCLOPENTANE see PPS500
AZACYCLOPROPANE see EJM900
AZACYCLOTRIDECAN-2-ONE see COW930
2-AZACYCLOTRIDECANONE see COW930

AZACYTIDINE see ARY000
5-AZACYTIDINE see ARY000
6-AZACYTIDINE see AMM000
5'-AZACYTIDINE see ARY000
5-AZADEOXYCYTIDINE see ARY125
5-AZA-2'-DEOXYCYTIDINE see ARY125
7-AZADIBENZ(a,h)ANTHRACENE see
DCS400
7-AZADIBENZ(a,j)ANTHRACENE see
DCS600
14-AZADIBENZ(a,j)ANTHRACENE see
DCS800
7-AZA-7H-DIBENZO(c,g)FLUORENE see
DCY000
AZADIENO see MJL250
9-AZA-1,17-
DIGUANIDINOHEPTADECANE
TRIACETATE see GLQ100
9-AZAFLUORENE see CBN000
AZAFORM see MJL250
AZAGUANINE see AJO500
AZAGUANINE-8 see AJO500
8-AZAGUANINE see AJO500
2-AZAHYPOXANTHINE see ARY500
1-AZAINDAN see ICS300
1-AZAINDENE see ICM000
3-AZAINDOLE see BCB750
AZAK see TAN300
AZALEA see RHU500
AZALEUCINE see ARY625
4-AZALEUCINE see ARY625
AZALINE see ARM000
AZALOMYCIN F see ARY750
AZALONE see PDC000
AZAMETHIPHOS see ARY800
AZAMETHONE see MKU750
AZAMETHONIUM BROMIDE see MKU750
AZAMETON see MKU750
AZAMIN 4B see DXO850
AZAMINE T 810 see AHP765
AZAN see AJO500
1-AZANAPHTHALENE see QMJ000
2-AZANAPHTHALENE see IRX000
AZANIDAZOLE see TJF000
AZANIL RED SALT TRD see CLK235
AZANIN see ASB250
AZANOL FAST ACID BLACK 10B see
FAB830
3-AZAPENTANE-1,5-DIAMINE see DJG600
AZAPERONE (USDA) see FLU000
AZAPETINE see ARZ000
AZAPETINE PHOSPHATE see AGD500
1-AZAPHENANTHRENE see BDB750
4-AZAPHENANTHRENE see BDC000
5-AZAPHENANTHRENE see PCX300
9-AZAPHENANTHRENE see PCX300
3-AZAPHTHALIMID (GERMAN) see
POQ750
4-AZAPHTHALIMID (GERMAN) see POR000
AZAPICYL see ADA000
AZAPLANT see AMY050
AZAPLANT KOMBI see AMY050
AZAPROPAZON (GERMAN) see AQN750
AZAPROPAZON DIHYDRAT (GERMAN)
see ASA000
AZAPROPAZONE see ASA000
AZAPROPAZONE (anhydrous) see AQN750
AZAPROPAZONE SODIUM see ASA250
AZAPROPAZON NATRIUMSALZ
(GERMAN) see ASA250
8-AZAPURINE, 6-AMINO- see AIB340
8-AZAPURINE, 6-HYDROXY-2-THIO- see
HOJ100
AZAR see TAN300
AZARIBINE see THM750
AZASERIN see ASA500
AZASERINE see ASA500
l-AZASERINE see ASA500
AZASPIRACID see ASA600
6-AZASPIRO(3,4)OCTANE-5,7-DIONE see
ASA750

AZA-6-SPIRO(3,4)OCTANE-DIONE-5,7
(FRENCH) see ASA750
2-AZASPIRO(5.5)UNDEC-7-ENE see ASA800
4-(3-AZASPIRO(5.5)UNDEC-3-YL)-4'-
FLUORO-BUTYROPHENONE
HYDROCHLORIDE see ASA875
AZASTENE see HOM259
AZASTEROL see ARX800
AZATADINE DIMALEATE see DLV800
AZATADINE MELEATE see DLV800
AZATHIOPRINE see ASB250
8-AZATHIOXANTHINE see MLY000
8-AZA-2-THIOXANTHINE see HOJ100
AZATIOPRIN see ASB250
N-(4-AZA-endo-TRICYCLO(5.2.1.5²·⁶)-
DECAN-4-YL)-4-CHLORO-3-
SULFAMOYLBENZAMIDE see CHK825
6-AZAURACIL see THR750
4(6)-AZAURACIL see THR750
6-AZAURACILRIBOSIDE see RJA000
6-AZAURACIL-β-d-RIBOSIDE see RJA000
AZAURIDINE see RJA000
6-AZAURIDINE see RJA000
8-AZAXANTHINE see THT350
AZBOLEN ASBESTOS see ARM264
AZBOLEN ASBESTOS see ARM266
AZDEL see PMP500
AZDID see BRF500
AZELAIC ACID see ASB750
AZELAIC ACID, CYCLIC ETHYLENE
ESTER see EIP050
AZELAIC ACID DI(2-ETHYLHEXYL)ESTER
see BJQ500
AZELAIC ACID DIHEXYL ESTER see
ASC000
AZELASTIN see ASC130
AZELASTINE see ASC125
AZELASTINE HYDROCHLORIDE see
ASC130
AZEPERONE see FLU000
AZEPHEN see ASC250
1H-AZEPIN-1-AMINE, HEXAHYDRO-(9CI)
see HEG400
1H-AZEPINE, 1-
((ETHYLSULFINYL)CARBONYL)HEXAHYD
RO- see MRA800
1H-AZEPINE, HEXAHYDRO-1-AMINO- see
HEG400
1H-AZEPINE, HEXAHYDRO-1-((2-
METHYLCYCLOHEXYL)CARBONYL)- see
HDR700
AZEPINE PHOSPHATE see AGD500
2H-AZEPIN-2-ONE, 1-
ACETYLHEXAHYDRO- see ACF100
AZEPROMAZINE see ABH500
AZETALDEHYDSCHWEFLIGSAUREN
NATRIUMS (GERMAN) see AAH500
3-AZETIDINECARBOXYLIC ACID see
ASC350
l-2-AZETIDINECARBOXYLIC ACID see
ASC500
AZETOCHLOR see CGO550
AZETYLAMINOFLUOREN (GERMAN) see
FDR000
AZG see AJO500
AZIDANIL see TBJ250
AZIDE see SFA000
AZIDES see ASC750
AZIDINE BLUE 3B see CMO250
AZIDITHION see ASD000
AZIDOACETIC ACID see ASD375
AZIDOACETONE see ASD500
AZIDOACETO NITRILE see ASE000
l-AZIDOALANINE see ASE100
dl-AZIDOALANINE tert-BUTYL ESTER see
ASE120
dl-3-AZIDOALANINE tert-BUTYL ESTER see
ASE120
AZIDOBENZENE see PEH000
N-AZIDO CARBONYL AZEPINE see ASE250
AZIDOCARBONYL GUANIDINE
see ASE500

AZIDOCODEINE see ASE875
3'-AZIDO-3'-DEOXYTHYMIDINE see ASE900
AZIDODIMETHYL BORANE see ASF500
2-AZIDO-3,5-DINITROFURAN see ASF625
AZIDODITHIOCARBONIC ACID (DOT) see ASF750
AZIDODITHIOFORMIC ACID see ASF750
(5-α,6-β)-6-AZIDO-4,5-EPOXY-17-METHYL-MORPHINAN-3,14-DIOL see HJE600
2-AZIDOETHANOL NITRATE see ASF800
2-AZIDOETHYL NITRATE see ASF800
AZIDOETHYL NITRATE (DOT) see ASF800
AZIDO FLUORINE see FFA000
AZIDOGLYCEROL see ASG700
2-AZIDO-4-ISOPROPYLAMINO-6-METHYLTHIO-s-TRIAZINE see ASG250
2-AZIDO-4-ISOPROPYLAMINO-6-METHYLTHIO-1,3,5-TRIAZINE see ASG250
3-AZIDO-2-METHYL-dl-ALANINE see ASG300
N-AZIDO METHYL AMINE see ASG500
2-AZIDOMETHYLBENZENEDIAZONIUM TETRAFLUOROBORATE see ASG625
4-AZIDO-N-(1-METHYLETHYL)-6-(METHYLTHIO)-1,3,5-TRIAZIN-2-AMINE see ASG250
4-AZIDO-N-(1-METHYLETHYL)-6-(METHYLTHIO)-1,3,5-TRIAZIN-2-AMINI see ASG250
AZIDOMETHYLPHOSPHONIC ACID, ETHYL METHYL ESTER see EML500
AZIDOMORPHINE see ASG675
3-AZIDO-1,2-PROPANEDIOL see ASG700
5-AZIDOTETRAZOLE see ASH000
AZIDOTHIOCARBONIC ACID see ASF750
AZIDOTHYMIDINE see ASE900
3-AZIDO-1,2,4-TRIAZOLE see ASH250
AZIDOTRIMETHYLSILANE see TMF100
AZIJNZUUR (DUTCH) see AAT250
AZIJNZUURANHYDRIDE (DUTCH) see AAX500
AZIMETHYLENE see DCP800
AZIMIDOBENZENE see BDH250
AZIMINOBENZENE see BDH250
AZIMSULFURON see ASH275
AZINDOLE see BCB750
AZINE see POP250
AZINE BRILLIANT BLUE RW see CMO600
AZINE BROWN M see CMO800
AZINE DARK GREEN BH/C see CMO830
AZINE DEEP BLACK EW see AQP000
AZINE DIAZO BLACK BHK see CMN800
AZINE FAST BLACK D see CMN230
AZINE FAST RED FC see CMO870
AZINE GREEN BX see CMO840
AZINE SKY BLUE 5B see CMO500
AZINFOS-ETHYL (DUTCH) see EKN000
AZINFOS-METHYL (DUTCH) see ASH500
2,2'-AZINOBIS(3-ETHYL-7-BENZOTHIAZOLINESULFONIC ACID), DIAMMONIUM SALT see ASH375
2,2'-AZINO-DI(3-ETHYL-BENZTHIAZOLINE SULPHONIC ACID (6)), AMMONIUM SALT see ASH375
AZINOS see EKN000
AZINOTHRICIN see ASH425
AZINPHOS-AETHYL (GERMAN) see EKN000
AZINPHOS ETHYL see EKN000
AZINPHOS-ETILE (ITALIAN) see EKN000
AZINPHOS METHYL see ASH500
AZINPHOS METHYL, liquid (DOT) see ASH500
AZINPHOS-METILE (ITALIAN) see ASH500
AZIONYL see ARQ750
AZIPROTRYN see ASG250
AZIPROTRYNE see ASG250
AZIRANE see EJM900
AZIRIDIN (GERMAN) see EJM900
AZIRIDINE see EJM900

1-AZIRIDINEACETAMIDE, N,N'-(1,4-CYCLOHEXYLENEDIMETHYLENE)BIS- see CPL100
1-AZIRIDINEACETIC ACID, METHYL ESTER see MGR900
AZIRIDINE, 1-(3-(BIS(2-CHLOROETHYL)AMINO-p-TOLUOYL))- see BHQ760
AZIRIDINE, 1,1'-CARBONYLBIS- see BJP450
1-AZIRIDINECARBOXAMIDE, N,N-DIMETHYL- see EJA100
1-AZIRIDINECARBOXAMIDE, N,N'-HEXAMETHYENEBIS- see HEF500
1-AZIRIDINECARBOXAMIDE, N,N'-1,6-HEXANEDIYLBIS-(9CI) see HEF500
1-AZIRIDINECARBOXAMIDE, N-METHYL- see EJN400
1-AZIRIDINECARBOXANILIDE see PEH250
AZIRIDINE CARBOXYLIC ACID ETHYL ESTER see ASH750
AZIRIDINE, N-CHLORO- see CEI340
1-AZIRIDINE ETHANOL see ASI000
AZIRIDINE, 1,1'-ISOPHTHALOYLBIS(2-METHYL)- see BLG400
AZIRIDINE, 1,1'-(1,3-PHENYLENEDICARBONYL)BIS(2-METHYL)-(9CI) see BLG400
1-AZIRIDINEPROPANENITRILE see ASI250
1-AZIRIDINEPROPANOIC ACID, CINNAMYL ESTER see ASI100
1-AZIRIDINEPROPANOIC ACID, ETHYLENE ESTER see EIP100
1-AZIRIDINEPROPANOIC ACID, 3-PHENYL-2-PROPENYL ESTER see ASI100
1-AZIRIDINEPROPIONIC ACID, CINNAMYL ESTER see ASI100
1-AZIRIDINEPROPIONIC ACID, ETHYLENE ESTER see EIP100
1-AZIRIDINE PROPIONITRILE see ASI250
AZIRIDINE-1,3,5,2,4,6-TRIAZATRIPHOSPHORINE DERIVATIVE see AQO000
AZIRIDINIUM, 1-(2-CHLOROETHYL)-1-METHYL- see MIG800
2-(1-AZIRIDINYL)ACETIC ACID METHYL ESTER see MGR900
1-AZIRIDINYL m-(BIS(2-CHLOROETHYL)AMINO)PHENYL KETONE see ASI300
1-AZIRIDINYL-BIS(DIMETHYLAMINO)PHOSPHINE OXIDE see ASK500
AZIRIDINYL CHLORIDE see CEI340
1-(1-AZIRIDINYL)-N-(m-CHLOROPHENYL)FORMAMIDE see CJR250
6-(1-AZIRIDINYL)-4-CHLORO-2-PHENYLPYRIMIDINE see CGW750
2-(1-AZIRIDINYL)ETHANOL see ASI000
4-AZIRIDINYL-3-HYDROXY-1,2-BUTENE see VMA000
1-(1-AZIRIDINYL)-N-(p-METHOXYPHENYL)FORMAMIDE see MFF250
α-(1-AZIRIDINYLMETHYL)BENZYL ALCOHOL see PEH500
2-(1-AZIRIDINYL)-1-PHENYLETHANOL see PEH500
1-AZIRIDINYL PHOSPHINE OXIDE (TRIS) (DOT) see TND250
1-AZIRIDINYLPHOSPHONITRILE TRIMER see AQO000
AZIRIDINYLQUINONE see ASK875
p-1-AZIRIDINYL-N,N,N',N'-TETRAMETHYLAMINO-PHOSPHINE OXIDE see ASK500
p-1-AZIRIDINYL-N,N,N',N'-TETRAMETHYLPHOSPHINIC DIAMIDE see ASK500
2-(1-AZIRIDINYL)-1-VINYLETHANOL see VMA000
AZIRIDYL BENZOQUINONE see BDC750
AZIRPOTRYNE see ASG250

AZIUM see SFA000
AZIUM see SOW000
AZO-33 see ZKA000
AZOAETHAN (GERMAN) see ASN250
AZOAMINE PINK O see NEQ000
AZOAMINE RED ZH see NEO500
AZOAMINE SCARLET see NEQ500
p-AZOANILINE see ASK925
AZOBASE DCA see DEO295
AZOBASE MNA see NEN500
AZOBENZEEN (DUTCH) see ASL250
AZOBENZEN (CZECH) see COX250
AZOBENZENE see ASL250
AZOBENZENE, 4-ANILINO- see PEI800
AZOBENZENE OXIDE see ASO750
AZOBENZIDE see ASL250
AZOBENZOL see ASL250
4,4'-AZOBISBENZENAMINE see ASK925
AZOBISBENZENE see ASL250
1,1'-AZOBISCARBAMIDE see ASM270
AZOBISCARBONAMIDE see ASM270
AZOBISCARBOXAMIDE see ASM270
2,2'-AZOBIS(2-CYANOPENTANE) see ASM025
4,4'-AZOBIS(4-CYANOPENTANOIC ACID) see ASL500
AZOBIS(CYANOVALERIC ACID) see ASL500
4,4'-AZOBIS(4-CYANOVALERIC ACID) see ASL500
1,1'-AZOBIS(FORMAMIDE) see ASM270
AZOBISISOBUTYLONITRILE see ASL750
α,α'-AZOBISISOBUTYLONITRILE see ASL750
AZOBIS ISOBUTYRAMIDE HYDROCHLORIDE see ASM000
AZOBISISOBUTYRONITRILE see ASL750
2,2'-AZOBIS(ISOBUTYRONITRILE) see ASL750
2,2'-AZOBISISOVALERONITRILE see ASM025
2,2'-AZOBIS(2-METHYLBUTANENITRILE) see ASM025
2,2'-AZOBIS(2-METHYLBUTYRONITRILE) see ASM025
2,2'-AZOBIS(α-METHYLBUTYRONITRILE) see ASM025
2,2'-AZOBIS(2-METHYLPROPANIMIDAMIDE) DIHYDROCHLORIDE see ASM050
2,2'-AZOBIS(2-METHYLPROPIONAMIDINE) DIHYDROCHLORIDE see ASM050
2,2'-AZOBIS(2-METHYLPROPIONITRILE) see ASL750
AZO BLUE see ASM100
AZOCARBONITRILE see DGS300
AZOCARD BLACK EW see AQP000
AZOCARD BLUE 2B see CMO000
AZOCARD BLUE BH see CMN800
AZOCARD DARK GREEN B see CMO830
AZOCARD FAST RED F see CMO870
AZOCARD GREEN B see CMO840
AZOCARD RED 4B see DXO850
AZOCARD RED CONGO see SGQ500
AZOCARD VIOLET N see CMP000
AZOCATALYST M see ASM025
AZOCHLORAMIDE see ASM250
AZOCHROMAL ORANGE R see NEY000
AZOCHROMAL TELESIO RED see CMG750
AZOCHROMOL BLUE BLACK EB see CMP880
AZOCHROMOL YELLOW 5G see SIT850
AZOCINE, OCTAHYDRO-, HYDROCHLORIDE see OBU150
AZOCORD BROWN M see CMO800
AZOCYCLOTIN see THT500
AZO DARK BLUE C 2B see FAB830
AZO DARK BLUE HR see FAB830
AZO DARK BLUE S see FAB830
AZO DARK BLUE SH see FAB830
4,4'-AZODIANILINE see ASK925
AZODIBENZENE see ASL250

AZODIBENZENEAZOFUME see ASL250
AZODICARBAMIDE see ASM270
AZODICARBOAMIDE see ASM270
AZODICARBONAMIDE see ASM270
AZODICARBONAMIDE see ASM300
AZODICARBOXAMIDE see ASM270
AZODICARBOXYLIC ACID DIAMIDE see ASM270
AZODIISOBUTYRONITRILE see ASL750
2,2'-AZODIISOBUTYRONITRILE see ASL750
AZODIISOBUTYRONITRILE (DOT) see ASL750
α,α'-AZODIISOBUTYRONITRILE see ASL750
AZODINE see PDC250
N,N'-(AZODI-4,1-PHENYLENE)BIS(N-METHYLACETAMIDE) see DQH200
AZODIUM see PDC250
AZODOX-55 see ZKA000
AZODRIN-71 see MRH209
"AZODRIN" see ASN000
AZODRIN (OSHA) see MRH209
AZODRIN PESTICIDE see MRH209
AZO DYE-1 see BNG300
AZO DYE-2 see BNG310
AZO DYE No. 6945 see MIH500
AZODYNE see PDC250
AZOENE FAST BLUE BASE see DCJ200
AZOENE FAST GARNET GBCP SALT see MPY750
AZOENE FAST ORANGE GR BASE see NEO000
AZOENE FAST ORANGE RD SALT see CEG800
AZOENE FAST RED B BASE see NEQ000
AZOENE FAST RED KB BASE see CLK225
AZOENE FAST RED 3GL BASE see KDA050
AZOENE FAST RED TR BASE see CLK220
AZOENE FAST RED TR SALT see CLK235
AZOENE FAST SCARLET GC BASE see NMP500
AZOENE FAST SCARLET GC SALT see NMP500
AZOEN FAST SCARLET 2G BASE see DEO295
AZO ETHANE see ASN250
AZOFENE see BDJ250
AZOFENOL 4K see CMP090
AZOFIX BLUE B SALT see DCJ200
AZOFIX SCARLET GG SALT see DEO295
AZOFIX SCARLET G SALT see NMP500
AZOFORMALDOXIME see ASN375
AZOFOS see MNH000
AZO FUCHSINE see CMS228
AZO FUCHSINE 6B see CMM400
AZO-GANTANOL see SNK000
AZO GANTRISIN see PDC250
AZO GANTRISIN see SNN500
AZO GASTANOL see PDC250
AZOGEN BLACK D see CMN230
AZOGEN DEVELOPER A see NAX000
AZOGEN DEVELOPER H see TGL750
AZOGENE ECARLATE R see NEQ500
AZOGENE FAST ORANGE GC BASE see CEH690
AZOGENE FAST ORANGE GCN BASE see CEH690
AZOGENE FAST ORANGE GR see NEO000
AZOGENE FAST RED GL BASE see MMF780
AZOGENE FAST RED NRL SALT see MMF780
AZOGENE FAST RED RL see MMF780
AZOGENE FAST RED TR see CLK220
AZOGENE FAST RED TR see CLK235
AZOGENE FAST SCARLET G see NMP500
AZO GERANINE 2G see CMM300
AZO GERANINE 2GA see CMM300
AZOGNE FAST BLUE B see DCJ200
AZO GRENADINE see CMS228
AZOGROUND AS see CMM760
AZOIC DIAZO COMPONENT 4 see MPY750

AZOIC DIAZO COMPONENT 6 see NEO000
AZOIC DIAZO COMPONENT 9 see KDA050
AZOIC DIAZO COMPONENT 12 see NMP500
AZOIC DIAZO COMPONENT 32 see CLK225
AZOIC DIAZO COMPONENT 37 see NEO500
AZOIC DIAZO COMPONENT 46 see CLK200
AZOIC DIAZO COMPONENT 11 BASE see CLK220
AZOIC DIAZO COMPONENT 11 BASE see CLK235
AZOIC DIAZO COMPONENT 13 BASE see NEQ500
AZOIC RED 36 see MGO750
AZOIMIDE see HHG500
AZOLAN see AMY050
AZOLASTONE see MQY110
AZOLE see AMY050
AZOLE see PPS250
AZOLID see BRF500
AZOLMETAZIN see SNJ000
AZOLYAT A see ASN390
AZO MAGENTA G see CMS228
AZO-MANDELAMINE see PDC250
AZOMETHANE see ASN400
AZO MILLING RED G see CMM325
AZOMINE see PDC250
AZOMINE BLACK BH see CMN800
AZOMINE BLACK EWO see AQP000
AZOMINE BLUE 2B see CMO000
AZOMINE BROWN M see CMO800
AZOMINE GREEN B see CMO840
AZOMYCIN see NHG000
AZOMYCIN RIBOSIDE see NHG200
1,1'-AZONAPHTHALENE see ASN500
2,2'-AZONAPHTHALENE see ASN750
AZONAPHTHOL A see CMM760
AZO NAPHTHOL RED 6B see CMM400
AZONAPHTHOL RED J see CMM300
8-AZONIABICYCLO(3.2.1)OCTANE, 8-(2-FLUOROETHYL)-3-((HYDROXYDIPHENYLACETYL)OXY)-8-METHYL-, BROMIDE, (endo,syn)-, MONOHYDRATE see FMR300
AZONIASPIRO(3-α-BENZILOYLOXY-NORTROPAN-8,1'-PYRROLIDINE)-CHLORIDE see KEA300
AZONIASPIRO COMPOUND XVII see KEA300
AZONUTRIL see AHR600
AZOPHENYLENE see PDB500
AZOPHLOXIN see CMM300
AZOPHLOXINE see CMM300
AZO PHLOXINE GA see CMM300
AZO PHLOXINE GA-CF see CMM300
AZOPHOS see MNH000
AZOPYRIN see PPN750
AZO RED R see FAG020
AZORESORCIN see HNG500
AZO RHODINE 2G see CMM300
AZO RHODINE 6B see CMM400
AZORUBIN see HJF500
AZOSALT R see PFU500
AZOSEMIDE see ASO375
AZOSEPTALE see TEX250
AZOSSIBENZENE (ITALIAN) see ASO750
AZO-STANDARD see PDC250
AZO-STAT see PDC250
AZOSULFIZIN see SNN500
AZOTA CABALLO see GIW200
AZOTE (FRENCH) see NGR500
AZOTHIOPRINE see ASB250
AZOTHOATE see CJR300
AZOTIC ACID see NED500
AZOTO (ITALIAN) see NGR500
AZOTOL A see CMM760
2:3'-AZOTOLUENE see DQH000
AZOTOMYCIN see ASO501
AZOTOMYCIN see ASO510

AZOTOMYCIN SODIUM see ASO510
AZOTOMYCIN SODIUM SALT see ASO510
AZOTOWY KWAS (POLISH) see NED500
AZOTOX see DAD200
AZOTOYPERITE see BIE500
AZOTREX see PDC250
AZOTURE de SODIUM (FRENCH) see SFA000
AZOTU TLENKI (POLISH) see NGT500
AZOXYAETHAN (GERMAN) see ASP000
AZOXYBENZEEN (DUTCH) see ASO750
AZOXYBENZENE see ASO750
AZOXYBENZENE, 4,4'-DICHLORO- see DEP450
AZOXYBENZIDE see ASO750
AZOXYBENZOL (GERMAN) see ASO750
AZOXYDIBENZENE see ASO750
AZOXYETHANE see ASP000
AZOXYISOPROPANE see ASP510
AZOXYMETHANE see ASP250
1-AZOXYPROPANE see ASP500
2-AZOXYPROPANE see ASP510
1,1'-AZOXYPROPANE see ASP500
AZOXYSTROBIN see ASP525
o,o'-AZOXYTOLUENE, 3,3'-DINITRO- see BKS812
p,p'-AZOXYTOLUENE, 3,3'-DINITRO- see BKS814
AZQ see ASK875
AZS see ASA500
AZT see ASE900
AZTEC BPO see BDS000
AZTHREONAM see ARX875
AZTREONAM see ARX875
AZUCENA de MEJICO (MEXICO) see AHI635
AZULENE see ASP600
AZULENE, 1,2,3,4,5,6,7,8-OCTAHYDRO-1,4-DIMETHYL-7-(1-METHYLETHYLIDENE)-, MONOEPOXIDE see EBU100
AZULENO(6,5-B)FURAN-2,5-DIONE,3,3A,4,4A,7A,8,9,9A-OCTAHYDRO-8-METHYL-3-METHYLENE-,(3ar-(3A-α,4A-α,7A-α,8-α,9A-α)) see MQR765
6(1H)-AZULENONE, 2,3,3A,7,8,8A-HEXAHYDRO-1,3A-DIHYDROXY-1,4-DIMETHYL-7-(1-METHYLETHYLIDENE)-, (1S,3AR,8AR)- see AET600
6(1H)-AZULENONE, OCTAHYDRO-1,4-DIHYDROXY-1,4-DIMETHYL-7-(1-METHYLETHYLIDENE)-, (1R,3AR,4S,8AS)- see ZAT050
AZULENO(5,6,7-cd)PHENALENE see ASP750
AZULFIDINE see PPN750
AZULON see DSJ800
AZUPROSTAT see SDZ350
AZUR see RJA000
6-AZUR see RJA000
AZURE A see DUG700
AZUREN see ILD000
AZURENE see BNU725
6-AZURIDINE see RJA000
AZUROL ALIZARINOVY SW (CZECH) see DKX800
AZZURRO DIRETTO 3B see CMO250
B-9 see SNA500
B10 see SFO500
B-12 see VSZ000
23B see DHY400
B-28 see AJO500
B32 see HCL000
B-45 see AOF250
B-71 see PKC000
B 75 see IGS000
B/77 see DNX600
B 193 see MNW200
B 208 see MIH925
B-314 see BEI000
B-322 see BEW750
B-324 see BEW500
B-325 see BEX750
B-329 see BEL500

B-331 see BEN750
B-343 see IHQ100
B 377 see DVC200
B 41D see MQT600
B-436 see PEV750
B-500 see QMJ000
B 518 see CQC650
B 577 see HKK000
B586 see CKE750
B-622 see DEV800
693B see SLP600
B 995 see DQD400
B 1500 see MCA100
B-1,776 see BSH250
B-1843 see BLG500
B 2000 see GHY100
B2734 see NIV000
B2740 see THD275
276-Ba see LIN800
B2772 see MFH750
B2837 see NFF000
B-3015 see SAZ000
B-4130 see AAI750
B 4992 see BKS810
B 5333 see CFY250
B-5477 see EAT810
B-9002 see HMV000
B-10094 see CON300
B 11163 see DTQ800
B-11420 see IGA000
B1-2459 see INE050
B-14437 see SAU000
B-15000 see IFY000
B 17476 see RQF350
B-21085 see GEO200
B-66256 see USJ000
B 66347 see CLW625
B 67347 see CLW625
70-314B see BQN600
B 77488 see BAT750
B 859-35 see NDY550
B 90912 see SAU000
B-98891 see MQU000
Ba 2724 see DGQ400
Ba 2756 see AKK250
Ba 2797 see CAY500
Bi 3411 see CDO000
P-4657-B see NBP500
Ba-16038 see AKC600
Ba 34647 see BAC275
Ba 36278 see SGB500
Ba-39089 see THK750
Ba 49249 see EEH575
B 8509-035 see NDY550
B 10 (polysaccharide) see SFO500
BA see BBC250
BA see BDX090
6-BA see BDX090
BA 59 see MKA270
BA 168 see LIA400
BA 1355 see PDC875
BA 2650 see SNY500
BA 2665 see MGN600
BA 2666 see DBE150
BA 2682 see UVJ460
BA 2696 see IPY500
BA 2726 see DBF800
BA 2759 see TML100
BA 2773 see HGI980
BA 2784 see DRV000
BA5968 see HGP495
BA 5968 see HGP500
Be 724-A see BEQ625
BA 11044 see CDN525
BA 21381 see IAN100
BA 21401 see EPW600
BA 30,803 see BCH750
BA 32644 see NML000
BA 34276 see MAW850
BA 51-090278 see VQA100
BA 51-090462 see DCY400
BA 51-090492 see DAN450

BAAM see MJL250
BABEIRO AMARILLO (PUERTO RICO) see
YAK300
BABIDIUM CHLORIDE see HGI000
BABN see BPV325
BABN see BRX000
B. ABORTUS Bang. LIPOPOLYSACCHARIDE
see LGK350
BA 253 BR see ONI000
Ba-253-BR-L see ONI000
BABROCID see NGE500
Ba 598 BROMIDE HYDRATE see FMR300
BABS see BLX500
BABUL STEM BARK EXTRACT see AAD250
BABURAN see PJA120
BABY PEPPER see ROA300
B(c)AC see BAW750
BACACIL see BAB250
BACAMPICILLIN HYDROCHLORIDE see
BAB250
BACARATE see DKE800
BACCIDAL see BAB625
BACFOR BL see BEO000
BACHEM 9022 see LIU353
BA 32644 CIBA see NML000
B"-ACID see LIU000
BACIFURANE see DGQ500
BACIGUENT see BAC250
BACI-JEL see BAC250
BACILIQUIN see BAC250
BACILLIN see BAC175
BACILLIN see ILD000
BACILLOL see CNW500
BACILLOMYCIN (8CI, 9CI) see BAB750
BACILLOMYCIN R see BAB750
BACILLOPEPTIDASE A see BAC000
BACILLOPEPTIDASE B see BAC000
BACILLUS CEREUS exo-ENTEROTOXIN see
BAB650
BACILLUS Sp. No. 21 POLYSACCHARIDE
see BAB700
BACILLUS SPHAERICUS ALKALINE
PROTEINASE see SCC550
BACILLUS SUBTILIS BPN see BAB750
BACILLUS SUBTILIS CARLSBERG see
BAC000
BACILLUS SUBTILIS NEUTRAL PROTEASE
see BAC020
BACILLUS THURINGENSIS see BAC040
BACILLUS THURINGIENSIS see BAC040
BACILLUS THURINGIENSIS BERLINER see
BAC040
BACILLUS THURINGIENSIS EXOTOXIN
see BAC125
BACILLUS THURINGIENSIS var.
ISRAELENSIS see BAC130
BACILLUS THURINGIENSIS subsp.
ISRAELENSIS POLYPEPTIDE crystal see
BAC135
BACILLUS THURINGIENSIS MORRISONI
EXOTOXIN see BAC140
BACILYSIN see BAC175
BACIMETHRIN see AKO750
BACIMETHRINE see AKO750
BACIMETRIN see AKO750
BACITEK OINTMENT see BAC250
BACITRACIN see BAC250
BACITRACIN METHYLENE
DISALICYLATE see BAC260
BACITRACIN ZINC see BAC265
BACLOFEN see BAC275
BACLON see BAC275
BACL VITAMIN H1 see AIH600
BACMECILLINAM see BAC325
BACO AF 260 see AHC000
BACTERAMID see SNM500
BACTERIODES FRAGILIS ENDOTOXIN see
BAC390
BACTERIODES FRAGILIS ENDOTOXIN,
EXTRACTS see BAC400
BACTESULF see SNN500
BACTICLENS see CDT250

B 2992 ACTIVE PRINCIPLE see MQX250
BACTOCILL see MNV250
BACTOFEN see BDJ600
BACTOL see CHR500
BACTOLATEX see SMQ500
BACTOPEN see SLJ000
BACTOSPEIN see BAC040
BACTRAMIN see TKX000
BACTRAMYL see PAF250
BACTRATYCIN see TOG500
BACTRIM see SNK000
BACTRIM see TKX000
BACTROL see BGJ750
BACTROMIN see TKX000
BACTROVET see SNN300
BACTUCIDE see BAC040
BACTUR see BAC040
BACTYLAN see SEP000
BACULO (PUERTO RICO) see SBC550
BADEN ACID see ALI300
BA-1,2-DIHYDRODIOL see BBD250
BA-3,4-DIHYDRODIOL see BBD500
BA-5,6-DIHYDRODIOL see BBD980
BA-8,9-DIHYDRODIOL see BBE750
BA-10,11-DIHYDRODIOL see BBF000
BA-5,6-cis-DIHYDRODIOL see BBE000
BA-5,6-trans-DIHYDRODIOL see BBD980
BADIL see AOR500
BA-1,2-DIOL-3,4-EPOXIDE-1 see DMP000
BA-3,4-DIOL-1,2-EPOXIDE-1 see DLE000
BA-3,4-DIOL-1,2-EPOXIDE-1 see DLE200
BA-3,4-DIOL-1,2-EPOXIDE-2 see DLE200
BA-10,11-DIOL-8,9-EPOXIDE-1 see BAD000
BA-8,9-DIOL-10,11-EPOXIDE-1 see DMO600
BADISCHE ACID see ALI300
BAF see THR500
BAFHAMERITIN-M see XQS000
BAGALOL see MEP250
BAGAODRYL see BBV500
BAGASSE DUST see BAD250
BAGOLAX see MIF760
BA (GROWTH STIMULANT) see BDX090
BAH4 see GIA050
BAIRDCAT B16 see HCP525
BAISA, EXTRACT see BAD300
BAISHI, EXTRACT see BAD300
BAJATEN see IBV100
BAJKAIN see BMA125
BAKELITE AYAA see AAX250
BAKELITE DFD 330 see PJS750
BAKELITE DHDA 4080 see PJS750
BAKELITE DYNH see PJS750
BAKELITE EHBC see OJV600
BAKELITE EHBM see OJV600
BAKELITE LP 70 see AAX175
BAKELITE LP 90 see AAX250
BAKELITE RMD 4511 see ADY500
BAKELITE SMD 3500 see SMQ500
BAKELITE VLFV see AAX175
BAKELITE VMCC see AAX175
BAKELITE VYNS see AAX175
BAKER'S ANTIFOL see DKC800
BAKER'S ANTIFOLANTE see THR500
BAKER'S ANTIFOL SOLUBLE see THR500
BAKER'S P AND S LIQUID and OINTMENT
see PDN750
BAKERS YEAST EXTRACT see BAD400
BAKERS YEAST GLYCAN see BAD400
BAKING SODA see SFC500
BAKONTAL see BAP000
BAKTAR see TKX000
BAKTHANE see BAC040
BAKTOL see CFD990
BAKTOLAN see CFD990
BAKTONIUM see BEL900
BAKUCHIOL see BAD625
BAL see BAD750
BALAGRIN see BAD800
BALATA see BAE000
BALDRIAN OEL see BAE100
BALDRIAN OIL see BAE100
BALDRISEDON see VAD200

BALFON 7000 see TAI250
BALI VISCOSE BLACK G see CMN240
BALI VISCOSE BLACK N see CMN240
BALMADREN see VGP000
BALS 3A see RNU100
BALSAM APPLE see BAE325
BALSAM APPLE see FPD100
BALSAM CAPTIVI see CNH792
BALSAM FIR OIL see CBB900
BALSAM FIR OIL see FBU850
BALSAM FIR, OREGON see OJK340
BALSAM GURJUN see GME000
BALSAM OF FIR see FBS200
BALSAM, OREGON see OJK340
BALSAM PEAR see FPD100
BALSAM of PERU see BAE750
BALSAM PERU OIL (FCC) see BAE750
BALSAMS, CANADA see FBS200
BALSAMS, COPAIBA see CNH792
BALSAMS, TOLU see BAF000
BALSAM TOLU see BAF000
BAMBERMYCIN see MRA250
BAMBICORT see HHQ800
BAMBUTEROL see BAF100
BAMCH see BGT750
BAMD 400 see MQU750
BAMETAN SULFATE see BOV825
BAMETHANE see BQF250
BAMETHAN SULFATE see BOV825
BAMIFYLLINE HYDROCHLORIDE see
THL750
B-AMIN see TES750
BAMIPHYLLINE HYDROCHLORIDE see
THL750
BAMN see BRX500
BANABIN see CKF500
BANABIN-SINTYAL see CKF500
BANAMITE see TJA100
BANANA OIL see IHO850
BANANOTE see MDW750
BANARAT see WAT211
BANASIL see RDK000
BANCEMINE 115-60 see MCB050
BANCEMINE 125-60 see MCB050
BANCEMINE SM 947 see MCB050
BANCEMINE SM 970 see MCB050
BANCEMINE SM 975 see MCB050
BANCOL see NCN650
BANDANE see BAF250
BANDIS G100 see RNU100
BANDOL see CBQ625
BANDUR see CJD300
BANEBERRY see BAF325
BANEX see MEL500
BANGTON see CBG000
BAN-HOE see CBM000
BANICOL see BEL900
BANISIL see RDK000
BANISTERINE see HAI500
BANLEN see MEL500
BANMINTH see TCW750
BANMINTH II see MRN260
BANNER see PMS930
BANOCIDE see DIW200
BANOL see CGI500
BANOL TUCO SOK see CGI500
BANOL TURF FUNGICIDE see PNI250
BANOMITE see BAF500
BANROT see PEW750
BANTENOL see MHL000
BANTHIN see DJM800
BANTHIN see XCJ000
BANTHINE see XCJ000
BANTHINE BROMIDE see DJM800
BANTHINE BROMIDE see XCJ000
BANTHIONINE see MDT740
BANTROL see HKB500
BANUCALAD (HAWAII) see TOA275
BANVEL see MEL500
BANVEL HERBICIDE see MEL500
BANVEL T see TIK000
BAP see BDX090

BAP see NJM500
6-BAP see BDX090
9-BAP see BDX100
BAPC see BAB250
BAP (GROWTH STIMULANT) see BDX090
2-BAP HYDROCHLORIDE see BEA000
BAPN see AMB500
BAPN FUMARATE see AMB750
BAPTITOXIN see CQL500
BAPTITOXINE see CQL500
BARACOUMIN see BJZ000
BARAMINE see BBV500
BARATOL see TLG000
BARAZAE see SNN500
BARAZAN see BAB625
BARBADOS GOOSEBERRY see JBS100
BARBADOS LILY see AHI635
BARBADOS NUT see CNR135
BARBADOS PRIDE see CAK325
BARBALLYL see AFS500
BARBALOIN see BAF825
BARBAMATE see CEW500
BARBAMIL see AMX750
BARBAMYL see AMX750
BARBAMYL ACID see AMX750
BARBAN see CEW500
BARBANE see CEW500
BARBAPIL see EOK000
BARBASCO see RNZ000
BARBENYL see EOK000
BARBEXACLONE see CPO500
BARBEXACLONUM see CPO500
BARBIDAL see AFS500
BARBIDORM see ERD500
BARBILEHAE (BARBILETTAE) see EOK000
BARBIMON see AMK250
BARBININE see BAF850
BARBIPHENYL see EOK000
BARBITA see EOK000
BARBITAL see BAG000
BARBITAL see BAG500
BARBITAL Na see BAG250
BARBITAL SODIUM see BAG250
BARBITAL SODIUM see BAG500
BARBITAL SOLUBLE see BAG250
BARBITONE see BAG000
BARBITONE see BAG500
BARBITONE SODIUM see BAG250
BARBITURATES see BAG500
BARBITURIC ACID see BAG750
BARBITURIC ACID, 3-ALLYL-5-(1-
CYCLOHEXEN-1-YL)-1,5-DIMETHYL- see
PPP133
BARBITURIC ACID, 5-ALLYL-5-(2-
CYCLOHEXEN-1-YL)-2-THIO-, SODIUM
SALT see SEH600
BARBITURIC ACID, 5-ALLYL-5-ISOBUTYL-
2-THIO-, SODIUM SALT see SFG700
BARBITURIC ACID, 1,3-BIS(2-
PIPERIDINOETHYL)-5-PHENYL-5-
PIPERIDINO- see BLG280
BARBITURIC ACID, 1-(1-(1-CYCLOHEXYL-
N-METHYL-2-PROPANAMINE)-5-ETHYL-5-
PHENYL see CPO500
BARBITURIC ACID, 5-ETHYL-5-(3 OR 6-
OXO-1-CYCLOHEXEN-1-YL)- see OJV600
BARBITURIC ACID, 5-NITRO- see NET550
BARBONAL see EOK000
BARBONIN HYDROCHLORIDE see PAH260
BARBOPHEN see EOK000
BARBOSEC see SBM500
BARBOSEC see SBN000
BARDAC 22 see DGX200
BAR-DEX see AOB500
BARDIOL see EDO000
BARECO POLYWAX 2000 see PJS750
BARECO WAX C 7500 see PJS750
BARENE see DEJ500
m-BARENE see NBV100
o-BARENE see DEJ500
BARHIST see DPJ400
BARIDIUM see PDC250

BARIDOL see BAP000
B. ARIETANS VENOM see BLV075
BARIO (PEROSSIDO di) (ITALIAN) see
BAO250
BARITE see BAP000
BARITOP see BAP000
BARITRATE see PBC250
BARIUM see BAH250
BARIUM ACETATE see BAH500
BARIUM ACETYLIDE see BAH750
BARIUM ALUMINUM TITANATE see
AGX350
BARIUM AZIDE see BAI000
BARIUM AZIDE, dry or wetted with <50%
water, by weight (UN 0224) (DOT) see BAI000
BARIUM AZIDE, wetted with not <50% water,
by weight (UN 1571) (DOT) see BAI000
BARIUM BENZOATE see BAI500
BARIUM BICHROMATE see BAL500
BARIUM BINOXIDE see BAO250
BARIUM BIS(TETRAFLUOROBORATE) see
BAL750
BARIUM BROMATE see BAI750
BARIUM CADMIUM LAURATE see BAI770
BARIUM CADMIUM STEARATE see BAI800
BARIUM CALCIUM TITANATE see BAI810
BARIUM CALCIUM TITANIUM OXIDE see
BAI810
BARIUM CAPRYLATE see BAI825
BARIUM CARBIDE see BAJ000
BARIUM CARBONATE see BAJ250
BARIUM CARBONATE (1:1) see BAJ250
BARIUM CHLORATE see BAJ500
BARIUM CHLORIDE see BAK000
BARIUM CHLORIDE, DIHYDRATE see
BAK020
BARIUM CHLORITE see BAK125
BARIUM CHROMATE (1:1) see BAK250
BARIUM CHROMATE(VI) see BAK250
BARIUM CHROMATE OXIDE see BAK250
BARIUM COMPOUNDS (soluble) see BAK500
BARIUM CYANIDE see BAK750
BARIUM CYANIDE, solid (DOT) see BAK750
BARIUM CYANOPLATINITE see BAL000
BARIUM CYCLOHEXANESULFAMATE see
BAL100
BARIUM DIACETATE see BAH500
BARIUM DIAZIDE see BAL250
BARIUM DIBENZYLPHOSPHATE see
BAL275
BARIUM DICHLORIDE see BAK000
BARIUM DICHLORIDE DIHYDRATE see
BAK020
BARIUM DICHROMATE see BAL500
BARIUM DICYANIDE see BAK750
BARIUM DIHYDROXIDE see BAM500
BARIUM DINITRATE see BAN250
BARIUM DIOXIDE see BAO250
BARIUM DISTEARATE see BAO825
BARIUM FERRATE see BAL625
BARIUM FERRITE see BAL625
BARIUM FLUOBORATE see BAL750
BARIUM FLUORIDE see BAM000
BARIUM FLUOROSILICATE see BAO750
BARIUM FLUOSILICATE see BAO750
BARIUM HEXAFERRITE see BAL625
BARIUM HEXAFLUOROSILICATE see
BAO750
BARIUM HEXAFLUOROSILICATE(2-) see
BAO750
BARIUM HYDRIDE see BAM250
BARIUM HYDROXIDE see BAM500
BARIUM HYPOPHOSPHITE see BAM750
BARIUM IODATE see BAN000
BARIUM METATITANATE see BAP502
BARIUM MONOXIDE see BAO000
BARIUM NITRATE (DOT) see BAN250
BARIUM(II) NITRATE (1:2) see BAN250
BARIUM NITRIDE see BAN500
BARIUM
NITROSYLPENTACYANOFERRATE see
PAY600

BARIUM OCTANOATE see BAI825
BARIUM OCTOATE see BAI825
BARIUM OXIDE see BAO000
BARIUM PERCHLORATE (DOT) see PCD750
BARIUM PERMANGANATE see PCK000
BARIUMPEROXID (GERMAN) see BAO250
BARIUM PEROXIDE see BAO250
BARIUMPEROXYDE (DUTCH) see BAO250
BARIUMPOLYSULFID see BAO300
BARIUM POLYSULFIDE see BAO300
BARIUM PROTOXIDE see BAO000
BARIUM REINECKATE see BAO400
BARIUM RHODANIDE see BAO500
BARIUMSILICOFLUORID see BAO750
BARIUM SILICOFLUORIDE see BAO750
BARIUM SILICON FLUORIDE see BAO750
BARIUM STEARATE see BAO825
BARIUM STYPHNATE see BAO900
BARIUM SULFATE see BAP000
BARIUM SULFIDE see BAO300
BARIUM SULFIDE see BAP250
BARIUM SULFIDE, mixture with SULFUR see
BAP260
BARIUM SUPEROXIDE see BAO250
BARIUM TETRAFLUOROBORATE see
BAL750
BARIUM TETRATITANATE see BAP500
BARIUM THIOCYANATE see BAP300
BARIUM TITANATE(IV) see BAP500
BARIUM TITANATE(IV) see BAP502
BARIUM TITANIUM OXIDE see BAP500
BARIUM TITANIUM OXIDE see BAP502
BARIUM TITANIUM TRIOXIDE see BAP502
BARIUM ZIRCONATE see BAP750
BARIUM ZIRCONIUM OXIDE see BAP750
BARIUM ZIRCONIUM(IV) OXIDE see
BAP750
BARIUM ZIRCONIUM TRIOXIDE see
BAP750
BARIZON see MOV000
BARLENE 125 see DRR800
BARNETIL see EPD100
BARNON PLUS see FBW135
BAROLUB FTO see HOG000
BAROS CAMPHOR see BMD000
BAROSPERSE see BAP000
BAROTRAST see BAP000
BARPENTAL see NBU000
BARQUAT CME-A see EKN550
BARQUAT CT 29 see HCQ525
BARQUAT MB-50 see AFP250
BARQUAT MB 80 see QAT520
BARQUAT SB-25 see DTC600
BARQUINOL see CHR500
BARRAGE see DAA800
BARRA SUPER see BLX000
BARRELIERIN see GJM300
BARRICADE see RLF350
BARSEB HC see CNS750
BAR-TIME see BBK250
BARTOL see EOK000
BARYTA see BAO000
BARYTA WHITE see BAP000
BARYTA YELLOW see BAK250
BARYTES see BAP000
BARYUM FLUORURE (FRENCH) see
BAM000
BAS see BEM750
BAS-083 see MCH540
BAS 263 see LAU500
BAS 0660 see SMP450
BAS 263I see LAU500
BAS-3050 see MNS500
BAS 3220 see DJV000
BAS-3460 see MHC750
BAS 421F see MRQ300
BAS 4239 see CEQ500
BAS 9052 see CDK800
BAS-0660W see SMP450
BAS 2203F see TJJ500
BAS 2900H see CDS275
BAS 2903H see CDS275

BAS 67054 see MHC750
BAS-08300W see MCH540
BAS 08301W see MCH540
BAS 32500F see PEX500
BAS 36801F see MAP300
BAS 42100F see MRQ300
BAS 514 00H see QMS100
BAS85559X see MCH540
BAS 90210H see FBP350
BAS 90210H see ZTS200
BAS 90520H see CDK800
BASACRYL GOLDEN YELLOW X-GFL see
CMM895
BASACRYL RED GL see CMM870
BASAGRAN see MJY500
BASALIN see FDA900
BASAMAIZE see CDS275
BASAMID see DSB200
BASAMID-FLUID see VFU000
BASAMID G see DSB200
BASAMID-GRANULAR see DSB200
BASAMID P see DSB200
BASAMID-PUDER see DSB200
BASANITE see BRE500
BASCHEM 12 see MAG750
BASCOREZ see AAX250
BASCURAT see BOV825
BASE 661 see SMR000
BASECIL see MPW500
BASEDOL see AMS250
BASE LP 12 see DXY000
BASEN see DNC100
BASE OIL see PCR250
BASERGIN see LJL000
BASETHYRIN see MPW500
BASF see UTU500
BAS 352 F see RMA000
BAS 2205-F see DUJ400
BASFAPON see DGI400
BASFAPON B see DGI400
BASFAPON B see DGI600
BASFAPON/BASFAPON N see DGI400
BASF BRILLIANT INDIGO 4B see ICU135
BASF BRILLIANT INDIGO 4BC see ICU135
BASF-GRUNKUPFER see CNK559
BASF III see SMQ500
BASF-MANEB SPRITZPULVER see MAS500
BASFUNGIN see MLX850
BASFUNGINE see MLX850
BASF URSOL 3GA see ALT000
BASF URSOL D see PEY500
BASF URSOL EG see ALS990
BASF URSOL ERN see NAW500
BASF URSOL P BASE see ALT250
BAS-290-H see CDS275
BAS 351-H see MJY500
BAS 392-H see FDA900
BAS 9106 H see FIW100
BASIC ALUMINUM CHLORATE see AHA000
BASIC BERYLLIUM CARBONATE see
BFP755
BASIC BISMUTH NITRATE see BKW100
BASIC BISMUTH SALICYLATE see SAI600
BASIC BLUE 9 see BJI250
BASIC BLUEK see VKA600
BASIC BRIGHT GREEN see BAY750
BASIC BROWN 4 see CMM780
BASIC CHROMIC SULFATE see CMJ565
BASIC CHROMIC SULFATE see NBW000
BASIC CHROMIC SULPHATE see NBW000
BASIC CHROMIUM CARBONATE see
CMJ100
BASIC CHROMIUM SULFATE see CMJ565
BASIC CHROMIUM SULFATE see NBW000
BASIC CHROMIUM SULPHATE see NBW000
BASIC COBALT CARBONATE see BAP800
BASIC COBALT CARBONATE see BAP802
BASIC COPPER CARBONATE see CNJ750
BASIC COPPER CHLORIDE see CNK559
BASIC CUPRIC CARBONATE see CNJ750
BASIC FUCHSIN see MAC250
BASIC FUCHSINE see MAC250

BASIC GREEN 4 see AFG500
BASIC LEAD ACETATE see LCH000
BASIC LEAD CHROMATE see LCS000
BASIC MAGENTA see MAC250
BASIC MAGENTA E-200 see MAC250
BASIC MERCURIC SULFATE see MDG000
BASIC NICKEL CARBONATE see NCY500
BASIC NICKEL(II) CARBONATE see
NCY600
BASIC NICKEL CARBONATE
TETRAHYDRATE see NCY530
BASIC ORANGE 21 see CMM820
BASIC ORANGE 28 CMM830
BASIC ORANGE 3RN see BAQ250
BASIC ORANGE 3RN see BJF000
BASIC PANCREATIC TRYPSIN INHIBITOR
see PAF550
BASIC PARAFUCHSINE see RMK020
BASIC RED 1 see RGW000
BASIC RED 2 see GJI400
BASIC RED 13 see CMM850
BASIC RED 18 see BAQ750
BASIC RED 29 see CMM870
BASIC RHODAMINE YELLOW see RGW000
BASIC RHODAMINIC YELLOW see RGW000
BASIC ROSE 2S see CMM850
BASIC VIOLET 3 see AOR500
BASIC VIOLET 5 see AJP300
BASIC VIOLET 10 see FAG070
BASIC VIOLET 14 see MAC250
BASIC VIOLET BN see AOR500
BASIC VIOLET K see MQN025
BASIC YELLOW 11 see CMM890
BASIC YELLOW 28 see CMM895
BASIC YELLOW K see DBT400
BASIC ZINC CHROMATE see ZFJ100
BASIC ZINC CHROMATE see ZFJ130
BASIC ZIRCONIUM CHLORIDE see ZSJ000
BASIL OIL see BAR250
BASIL OIL, COMOROS TYPE see BAR275
BASIL OIL, EUROPEAN TYPE (FCC) see
BAR250
BASIL OIL EXOTIC see BAR275
BASIL OIL, REUNION TYPE see BAR275
BASIL OIL, SWEET see BAR250
BASINEX see DGI400
BASITAC see INE050
BASKET FLOWER see BAR325
BASLE GREEN see COF500
BASOFORTINA see MJV750
BASOFORTINA see PAM000
BASOLAN see MCO500
BASON see BNK350
BASORA CORRA see BAR500
BASOTECT see MCB050
B. ASPER VENOM see BMI000
BASSA see MOV000
BASSINET (CANADA) see FBS100
BASTA see ANI800
BASTARD ACACIA see GJU475
BASUDIN see DCM750
BASUDIN 10 G see DCM750
BAS 08305 W see MCH540
BAS 08306 W see MCH540
BAS 08307 W see MCH540
BATASAN see ABX250
BATAZINA see BJP000
BATEL see BFW250
BATHYRAN see MCK500
BATILOL see GGA915
BATRACHOTOXIN see BAR750
BATRAFEN see BAR800
BATRIDOL see DMG900
BATRILEX see PAX000
BATROXOBIN see RDA350
B. ATROX VENOM see BMI125
BATTRE AUTOUR (HAITI) see JBS100
BATYL ALCOHOL see GGA915
BAU see CBT500
BAUMYCIN A1 see BAR825
BAUMYCIN A2 see BAR830
BAUXITE see BAR900

BAUXITE RESIDUE see IHC450
BAVISTIN see MHC750
BAX see BAU750
BAX 1400Z see PMO250
BAX 2793Z see THL750
1BA-4-XA (RUSSIAN) see CEL750
BAXACOR see DHS200
BAXARYTMON see PMJ525
BAY 1040 see AEC750
BAY 1470 see DMW000
BAY 1518 see DJB460
BAY 1521 see DPH600
BAY 2353 see DFV400
BAY 3517 see AJV500
BAY 4059 see BOL325
BAY 4503 see PMX250
BAY 4934 see MGQ750
BAY 5097 see MRX500
BAY 5621 see BAT750
BAY 5821 see DJY200
BAY-6228 see RLK750
BAY 7660 see MOC275
BAY 9010 see PMY300
BAY 9015 see DFD000
BAY 9017 see DJR800
BAY 9026 see DST000
BAY 9027 see ASH500
BAY 10756 see DAO600
BAY 11405 see MNH000
BAY 15203 see DAO800
BAY 15203 see MIW100
BAY 15922 see TIQ250
BAY 16225 see EKN000
BAY 18436 see DAP400
BAY 19149 see DGP900
BAY 19639 see DXH325
BAY 21097 see DAP000
BAY 23129 see PHI500
BAY 23323 see OQS000
BAY 23655 see DSK600
BAY 25141 see FAQ800
BAY 25634 see EAT600
BAY 29492 see LIN400
BAY 29493 see FAQ900
BAY 30130 see DGI000
BAY 32394 see TNH750
BAY 32651 see MIB250
BAY 33051 see DRR400
BAY 33819 see BIM000
BAY 34042 see ENI500
BAY 34727 see COQ399
BAY 36743 see EHL670
BAY 37341 see DJR800
BAY 37342 see DST200
BAY 38156 see PHM750
BAY 39731 see MIA250
BAY 41637 see MOV000
BAY 41831 see DSQ000
BAY 42247 see PHD750
BAY 42696 see DQE800
BAY 42903 see EOO000
BAY 44646 see DOR400
BAY 45432 see DNX800
BAY 46131 see ZMA000
BAY 47531 see DFL200
BAY 48130 see DLS800
BAY 50282 see DBI800
BAY 50519 see EOP500
BAY 61597 see MQR275
BAY 62863 see DLS800
BAY 68138 see FAK000
BAY 70143 see CBS275
BAY 70533 see CFC750
BAY 71628 see DTQ400
BAY 75546 see BAS000
BAY 77049 see DJY200
BAY 77488 see BAT750
BAY 79770 see CDP750
BAY-92114 see IMF300
BAY-a 7168 see NDY600
BAY d8815 see DPF200
BAY e 5009 see EMR600

BAY-O-2820 see MRJ200
BAY 105807 see MIA250
BAY A 1040 see AEC750
BAY BUE 1452 see THT500
BAY BUSH (BAHAMAS) see CNH789
BAYCAIN see BMA125
BAYCAINE see BMA125
BAYCALNE see BMA125
BAYCARB see MOV000
BAYCARON see MCA100
BAYCHROM A see CMK415
BAYCHROM F see CMK415
BAYCID see FAQ900
BAYCLEAN see AFP250
BAY COE 3664 see BAS500
BAYCOR see SCF525
BAYCOVIN see DIZ100
BAY DIC 1468 see MQR275
BAY DIC 1559 see ENJ600
BAY-DRW 1139 see ALA500
BAY-E-393 see SOD100
BAY E-601 see MNH000
BAY E-605 see PAK000
BAY-E 6975 see CJN900
BAY-E 9736 see NDY700
BAY ENE 11183 B see EAT600
BAYER 73 see DFV400
BAYER 73 see DFV600
BAYER-186 see CMW700
BAYER 205 see BAT000
BAYER 693 see SLP600
BAYER 1219 see AFL500
BAYER 1355 see PDC850
BAYER 1362 see BSZ000
BAYER 1420 see PMM000
BAYER 2353 see DFV400
BAYER 2502 see NGG000
BAYER 3231 see TND000
BAYER 4245 see DET000
BAYER 4964 see ORU000
BAYER 5072 see DOU600
BAYER 5080 see DOR400
BAYER 5081 see EPY000
BAYER 5312 see EPQ000
BAYER 5360 see MMN250
BAYER 8169 see DAO500
BAYER 8169 see DAO600
BAYER 9007 see FAQ900
BAYER 9013 see DST200
BAYER 9015 see DFD000
BAYER 9051 see TDX750
BAYER 15080 see BDD000
BAYER 15922 see TIQ250
BAYER 16259 see EKN000
BAYER 17147 see ASH500
BAYER 18510 see DRR400
BAYER 19564 see DKB170
BAYER 19639 see DXH325
BAYER 20315 see DAP600
BAYER 23655 see DSK600
BAYER 25142 see PHM000
BAYER 25198 see EAV000
BAYER 25 634 see EAT600
BAYER 25648 see DFV600
BAYER 25820 see HMV000
BAYER 29492 see LIN400
BAYER 29952 see MOB599
BAYER 32394 see TNH750
BAYER 33172 see FQK000
BAYER 34727 see COQ399
BAYER 36205 see ORU000
BAYER 36743 see EHL670
BAYER 37289 see EPY000
BAYER 37341 see DJR800
BAYER 37342 see DST200
BAYER 37344 see DST000
BAYER 38819 see BIM000
BAYER 38920 see HCK000
BAYER 39007 see PMY300
BAYER 39731 see MIA250
BAYER 41637 see MOV000
BAYER 41831 see DSQ000

BAYER 42903 see EOO000
BAYER 44646 see DOR400
BAYER 45,432 see DNX800
BAYER 46131 see ZMA000
BAYER 47531 see DFL200
BAYER 52553 see PHN000
BAYER 6159H see MQR275
BAYER 62863 see DLS800
BAYER 6443H see MQR275
BAYER 70533 see CFC750
BAYER 71628 see DTQ400
BAYER 78418 see EIM000
BAYER 94337 see MQR275
BAYER 96610 see TBT175
BAYER 21/116 see MIW100
BAYER 21/199 see CNU750
BAYER 25/154 see DAP400
BAYER 41367C see MOV000
BAYER A-128 see PAF550
BAYER A 139 see BDC750
BAYER B-186 see CMW700
BAYER-E 393 see SOD100
BAYER E-605 see PAK000
BAYER G4073 see BGW750
BAYERITIAN see TGG760
BAYER 1440 L see DNA800
BAYER 1440 L see MJE780
BAYER L 13/59 see TIQ250
BAYER R39 SOLUBLE see BDC750
BAYER S767 see FAQ800
BAYER S 4400 see EPY000
BAYER S 5660 see DSQ000
BAYERTITAN see TGG760
BAY 6681 F see CJO250
BAY FCR 1272 see REF250
BAYFIDAN see CJN300
BAYFRDAN EW see CJN300
BAY G 2821 see EAE675
BAY-G 5421 see AAD900
BAYGON see PMY300
BAYGON MEB see TIL800
BAYGON, NITROSO derivative see PMY310
BAY H 4502 see BGA825
BAY HOL 0574 see EON500
BAY-HOX-1901 see EPR000
BAY-HWG 1608 see TAN050
BAY KWG 0519 see CJN300
BAY KWG 0599 see SCF525
BAY LEAF OIL see BAT500
BAY LEAF OIL see LBK000
BAYLECTROL 4200 see PHW585
BAYLETON see CJO250
BAYLUSCID see DFV400
BAYLUSCID see DFV600
BAYLUSCIDE see DFV600
BAYMAT-SPRAY see SCF525
BAY MEB 6046 see TIL800
BAY-MEB-6447 see CJO250
BAYMIX 50 see CNU750
BAY NTN80 see AMX300
BAY-NTN 5006 see AMX400
BAY NTN6867 see AMX300
BAY NTN 8629 see DGC800
BAY-NTN-9306 see SOU625
BAY NTN 19701 see UTA400
BAY-NTN 33893 see CKW400
BAY OIL see BAT500
BAY OIL see LBK000
BAYOL F see MQV750
BAYONOX see HKV100
BAYPIVAL see CJN900
BAYPRESOL see DNA800
BAYPRESOL see MJE780
BAYRE 77488 see BAT750
BAYRITES see BAP000
BAYROGEL see HKK000
BAYRUSIL see DJY200
BAY S 2758 see MIB250
BAYSAN see CJN900
BAY SLJ 0312 see FDA890
BAY SMY 1500 see ENJ600
BAY-SRA 7660 see MOC275

BAY-SRA-12869 see IMF300
BAYTAN see CJN300
BAYTAN see MEP250
BAYTAN 15 see CJN300
BAYTAN TF 3479B see CJN300
BAYTEC 110 see DJP100
BAYTEC E 505 see DJP100
BAYTENAL see SFG700
BAYTEX see FAQ900
BAYTHION see BAT750
BAYTHROID see REF250
BAYTHROID H see REF250
BAYTINAL see SFG700
BAYTITAN see TGG760
BAY VA 1470 see DMW000
BAY-VA 4059 see BOL325
BAY Y 3118 see DKV250
BAY-Y 7432 see SDY625
BAZUDEN see DCM750
BB-8 see OAH000
BBAL see NJO200
BBC see BAV575
BBC see BMW250
BBC 12 see DDL800
BBCE see BIQ500
BBH see BBQ500
BB-K8 see APT000
"B" BLASTING POWDER see ERF500
BBN see BMW250
BBN see HJQ350
BBNOH see HJQ350
BBOT see BHK600
BBOT 150 see BHK600
BBP see BEC500
BBR 3464 see HBY100
BC 7 see PJT300
BC 10 see PJT300
BC 20 see PJT300
BC 20 see UTU500
BC 27 see MCB050
BC 40 see UTU500
BC 48 see DAM625
BC 71 see MCB050
BC 77 see UTU500
BC 336 see MCB050
BCF-BUSHKILLER see TAA100
BCG see BNA940
BCG see TBJ505
B-CHLORO-N,N-DIMETHYLAMINODIBORANE see CGD399
BCM see MHC750
BCME see BIK000
BCM (NH) see BCD325
BCNA see COJ800
BCNU see BIF750
BiCNU see BIF750
BCP see BQW825
BCPE see BIN000
BCPE mixed with SPAS see CKL500
BCPN see BQQ250
BC 20 (POLYMER) see UTU500
BCS COPPER FUNGICIDE see CNP250
BCTB see DIG400
BC 20 TX see PJT300
BC 30 TX see PJT300
BD 40A see FNE100
4-BDAF see TKK025
BDCM see BND500
(BDH) see EQL000
BDH 312 see GGS000
BDH 1298 see VTF000
BDH 2700 see BAT795
BDH 6140 see BAT800
BDH 6146 see TKH050
BDH 29-790 see SMQ500
BDH 29-801 see TAI250
B.D.H. 200 HYDROCHLORIDE see EEQ000
6-Bz-1-DIBROMBENZANTHRON (CZECH) see DDK875
B-DIETHYLAMINOETHYLAMINOPHENYLACETIC ACID ISOAMYL ESTER see NOC000

BDMA see DQP800
BDM-CHLORIDE (RUSSIAN) see AFP075
BDNPF see MJO800
BD(a,h)P see DCY200
BDU see BNC750
5-BDU see BNC750
BE see BNI500
BE 100 see IAB000
BE 1293 see CLF325
BEACH APPLE see MAO875
BEACILLIN see BFC750
BEACON see TCI150
BEACON RED see CMG750
BEAD TREE see CDM325
BEAM see MQC000
BEAMETTE see PMP500
BEAN SEED PROTECTANT see CBG000
BEAN TREE see GIW195
BEAUTYLEAF see MBU780
BEAUVERIA BASSIANA see BAT830
BEAUVERIN see BAT830
BEAVER POISON see WAT325
BEBUXINE see CQH325
BEC 001 see AQP800
BECAMPICILLIN see BAB250
BECANTAL see DEE600
BECANTEX see DEE600
BECANTYL see DEE600
BECAPTAN see AJT250
BECAPTAN DISULFURE (FRENCH) see MCN500
BECILAN see PPK500
BECKAMINE 21-511 see UTU500
BECKAMINE APH see MCB050
BECKAMINE APM see MCB050
BECKAMINE G 82 see MCB050
BECKAMINE J 101 see MCB050
BECKAMINE J 820 see MCB050
BECKAMINE J 1012 see MCB050
BECKAMINE J 820-60 see MCB050
BECKAMINE L 105-60 see MCB050
BECKAMINE MA-S see MCB050
BECKAMINE NF 5 see UTU500
BECKAMINE P 136 see UTU500
BECKAMINE P 138 see UTU500
BECKAMINE P 196M see UTU500
BECKAMINE P 138-60 see UTU500
BECKAMINE PM see MCB050
BECKAMINE PM-N see MCB050
BECLACIN see AFJ625
BECLAMID see BEG000
BECLAMIDE see BEG000
BECLOFORTE see AFJ625
BECLOMETASONE DIPROPIONATE see AFJ625
BECLOMETASONE-17,21-DIPROPIONATE see AFJ625
BECLOMETHASONE DIPROPIONATE see AFJ625
BECLOVAL see AFJ625
BECLOVENT see AFJ625
BECOREL see PAN100
BECOTIDE see AFJ625
BEECHWOOD CRESOATE see BAT850
BEESIX see PPK250
BEESWAX see BAU000
BEESWAX, WHITE see BAU000
BEESWAX, YELLOW see BAU000
BEET-KLEEN see CBM000
BEET-KLEEN see CKC000
BEET-KLEEN see DTP400
BEETLE 55 see UTU500
BEETLE 60 see UTU500
BEETLE 65 see UTU500
BEETLE 80 see UTU500
BEETLE 336 see MCB050
BEETLE 338 see MCB050
"BEETLE" see BAU250
BEETLE 212-9 see UTU500
BEETLE 3735 see MCB050
BEETLE BC 27 see MCB050
BEETLE BC 71 see MCB050

BEETLE BC 309 see MCB050
BEETLE BC 371 see MCB050
BEETLE BE 336 see MCB050
BEETLE BE 645 see MCB050
BEETLE BE 669 see MCB050
BEETLE BE 670 see MCB050
BEETLE BE 681 see MCB050
BEETLE BE 683 see MCB050
BEETLE BE 685 see UTU500
BEETLE BE 687 see MCB050
BEETLE BE 3021 see MCB050
BEETLE BE 3735 see MCB050
BEETLE BE 3747 see MCB050
BEETLE BT 309 see MCB050
BEETLE BT 323 see MCB050
BEETLE BT 336 see MCB050
BEETLE BT 370 see MCB050
BEETLE BT 670 see MCB050
BEETLE BU 700 see UTU500
BEETLE RESIN 323 see MCB050
BEETLE XB 1050 see UTU500
BEET SUGAR see SNH000
BEFLAVINE see RIK000
BEFRAN see GLQ100
BEFUNOLOL see HLK600
BEFUNOLOL HYDROCHLORIDE see BAU255
BEHA see AEO000
BEHEN see MBU800
N⁴-BEHENOYL-1-β-d-ARABINOFURANOSYLCYTOSINE see EAU075
BEHENOYLCYTOSINE ARABINOSIDE see EAU075
N⁴-BEHENOYLCYTOSINE ARABINOSIDE see EAU075
BEHP see DVL700
BEI-1293 see CLF325
BEIENO (MEXICO) see HAQ100
BEIVON see TES750
BEJUCO AHOJA VACA (DOMINICAN REPUBLIC) see YAK300
BEJUCO DO PEO (PUERTO RICO) see CDH125
BEJUCO de LOMBRIZ (CUBA) see PGQ285
BEK see BJU000
BEKADID see OOI000
BEKANAMYCIN see BAU270
BEKANAMYCIN SULFATE see KBA100
BEKAPTAN see MCN750
BEKLAMID see BEG000
BELACHROME FAST ORANGE GR see CMP882
BELACID MILLING RED G see CMM325
BELACID MILLING YELLOW R see CMM759
BELACID PHLOXINE G see CMM300
BELACYL FAST YELLOW W see KDA075
BELAMINE BLACK GX see AQP000
BELAMINE BLUE 2B see CMO000
BELAMINE BORDEAUX B see CMO872
BELAMINE DIAZO BLACK BH see CMN800
BELAMINE FAST BROWN BP see CMO820
BELAMINE FAST BROWN M see CMO800
BELAMINE FAST RED 8 BL see CMO885
BELAMINE FAST RED FC see CMO870
BELAMINE FAST TURQUOISE LGL see COF420
BELAMINE GREEN BX see CMO840
BELAMINE SKY BLUE A see CMO500
BELAMINE SKY BLUE FF see CMN750
BELCOMYCIN see PKD300
BELDAVRIN see HOT500
BELFENE see LJR000
BELGANYL see BAT000
BELGENINE see BAU325
BELGRAN see IRA050
BELLADONA (HAWAII) see AOO825
BELLADONNA see BAU500
BELLADONNA see DAD880
BELLADONNA LILY see AHI635
BELLASTHMAN see DMV600
BELL CML(E) see CAU500

BELLECHEM ESTRACYL YELLOW W see KDA075
BELL MINE see CAT225
BELL MINE PULVERIZED LIMESTONE see CAO000
BELLUAINE (CANADA) see SED550
BELLYACHE BUSH see CNR135
BELMARK see FAR100
BELOC see SBV500
BELOPHAR KLA see OOI200
BELOPHOR OD see DXB450
BELORAN see BDJ600
BELOSIN see NOC000
BELOTEX PAD see BIU600
BELSEREN see CDQ250
BELT see CDR750
BELUSTINE see CGV250
BEMACO see CLD000
BEMAPHATE see CLD000
BEMAPHATE see CLD250
BEMASULPH see CLD000
BEMEGRIDE see MKA250
BEN-30 see BAV000
BENA see BAU750
BENA see BBV500
BENACHLOR see BBV500
BENACTIZINA (ITALIAN) see DHU900
BENACTIZINE HYDROCHLORIDE see BCA000
BENACTYZIN see DHU900
BENACTYZIN (CZECH) see BCA000
BENACTYZINE see DHU900
BENACTYZINE CHLORIDE see BCA000
BENACTYZINE HYDROCHLORIDE see BCA000
BENADON see BBV500
BENADON see PPK500
BENADRIN see BBV500
BENADRYL see BAU750
BENADRYL see BBV500
BENADRYL HYDROCHLORIDE see BAU750
BEN-A-HIST see PDC000
BENAKTIN see BCA000
BENALGIN see BBW500
BEN-ALLERGIN see BBV500
BENANSERIN HYDROCHLORIDE see BEM750
BENAPON see BBV500
BENASPIR see ADA725
BENAZALOX see BAV000
BENAZIDE see IKC000
BENAZOLIN see BAV000
BENAZOLIN-ETHYL see EHL600
BENAZOLIN ETHYL ESTER see EHL600
BENAZOL P see HML500
BENAZYL see RDK000
BENCAINAMIDE see DHU700
BENCARBATE see DQM600
BENCHINOX see BDD000
BENCHMARK see MGJ775
BENCICLANE see BAV250
BENCIDAL BLACK E see AQP000
BENCIDAL BLUE 2B see CMO000
BENCIDAL BLUE 3B see CMO250
BENCIDAL DARK GREEN B see CMO830
BENCIDAL FAST BLACK G see CMN240
BENCIDAL FAST RED F see CMO870
BENCIDAL FAST VIOLET N see CMP000
BENCIDAL GREEN B see CMO840
BENCIDAL NAVY BLUE BH see CMN800
BENCIDAL PURPLE 4B see DXO850
BENCONASE see AFJ625
BEN-CORNOX see BAV000
BENCYCLANE see BAV250
BENCYCLANE FUMARATE see BAV250
BENDACORT see BAV275
BENDAZAC see BAV325
BENDAZOL see BEA825
BENDAZOLE see BEA825
BENDAZOLIC ACID see BAV325
BENDECTIN see BAV350

BENDEX see BLU000
BENDIGON see RDK000
BENDIOCARB see DQM600
BENDIOCARB see MHZ000
BENDIOXIDE see MJY500
BENDOPA see DNA200
B-ENDOSULFAN-β see EAQ800
BENDRALAN see PDD350
BENDROFLUAZIDE see BEQ625
BENDROFLUMETHIAZIDE see BEQ625
BENDYLATE see BAU750
BENECARDIN see AHK750
BENECID see DWW000
BENEIDAL FAST BROWN M see CMO800
BENEMID see DWW000
BENEPEN see PAQ100
BENESAL see SAH000
BENETACIL see PAQ100
BENETHAMINE PENICILLIN see PAQ100
BENETHAMINE PENICILLIN G see PAQ100
BENETOLIN see PAQ100
BENFOS see DGP900
BENFURACARB see AKC570
BENGAL GELATIN see AEX250
BENGAL ISINGLASS see AEX250
BENGUINOX see BDD000
BEN-HEX see BBQ500
BENHEXAL see DPH000
BENICOT see NCR000
BENIDIPINE HYDROCHLORIDE see BAV400
(+−)-BENIDIPINE HYDROCHLORIDE see BAV400
BENIHINAL see FNK150
BENIROL see BBA500
BENIT see EDW100
BENIT see VFA100
BENKFURAN see NGE000
BENLATE 50 see BAV575
BENLATE and SODIUM NITRITE see BAV500
BENMOXINE see NCQ100
BENNIE see AOB250
BENOCTEN see BAU750
BENODAINE HYDROCHLORIDE see BCI500
BENODIN see BBV500
BENODINE see BBV500
BENOMYL see BAV575
BENOMYL 50W see BAV575
BENOPAN see BBV500
BENOQUIN see AEY000
BENORAL see SAN600
BENORILATE see SAN600
BENORTAN see SAN600
BENORTERONE see MNC150
BENORYLATE see SAN600
BENOTERONE see MNC150
BENOVOCYLIN see EDP000
BENOXACOR see DEN500
BENOXAPROFEN see OJI750
BENOXIL see OPI300
BENOXINATE HYDROCHLORIDE see OPI300
BENOXYL see BDS000
BENOZIL see DAB800
BEN-P see BFC750
BENPERIDOL see FGU000
BENPERIDOL see FLK100
BENPROPERINE PHOSPHATE see PJA130
BENQUINOX see BDD000
BENSECAL see BAV000
BENSERAZIDE HYDROCHLORIDE see SCA400
BENSULFOID see SOD500
BENSULFURON METHYL see BAV600
BENSULIDE see DNO800
BENSULTAP see NCN650
BENSYLYT see DDG850
BENSYLYTE see PDT250
BENT see MDQ250
BENTANEX see EEO500

BENTANIDOL see BFW250
BENTAZEPAM see BAV625
BENTAZON see MJY500
BENTHIOCARB see SAZ000
BENTHIOCARB SULFOXIDE see FMY050
BENTHIOZONE see FNF000
BENTIROMIDE see CML870
BENTONE see KBB600
BENTONITE see BAV750
BENTONITE 2073 see BAV750
BENTONITE MAGMA see BAV750
BENTONYL see TJL250
BENTOX see BAU255
BENTOX 10 see BBQ500
BENTRIDE see BEQ625
BENTROL see DNF400
BENTROL see HKB500
BENURIDE see PFB350
BENURON see BEQ625
BEN-U-RON see HIM000
BENURYL see DWW000
BENVIL see MOV500
(5R,6R)-BENXYLPENICILLIN see BDY669
BENYLAN see BBV500
BENYLATE see BCM000
BENZABAR see TIK500
BENZAC see BDS000
BENZAC see PJQ000
BENZAC see TIK500
BENZAC 1281 see DOR800
BENZ(1)ACEANTHRENE see BAW000
BENZ(1)ACEANTHRYLENE see BAW125
BENZ(a)ACEANTHRYLENE see BAW130
BENZ(j)ACEANTHRYLENE see CMC000
BENZ(j)ACEANTHRYLENE see NAH900
11H-BENZ(bc)ACEANTHRYLENE see MJL300
BENZ(j)ACEANTHRYLENE, 1,2-DIHYDRO-3,6-DIMETHYL-(9CI) see DRD850
BENZ(j)ACEANTHRYLENE, 1,2-DIHYDRO-5-METHYL- see MIK000
BENZ(j)ACEANTHRYLENE, 1,2,6,7,8,9,10,12b-OCTAHYDRO-3-METHYL- see HDR500
BENZ(j)ACEANTHRYLEN-10-OL see BAW150
1,2-BENZACENAPHTHENE see FDF000
BENZ(k)ACEPHENANTHRENE see AAF000
BENZ(e)ACEPHENANTHRYLENE see BAW250
3,4-BENZ(e)ACEPHENANTHRYLENE see BAW250
BENZACETONITRILE see FMR600
BENZACILLIN see BFC750
BENZACIN see BAW500
BENZACINE see BAW500
BENZACINE HYDROCHLORIDE see BAW500
BENZACIN HYDROCHLORIDE see BAW500
BENZACONINE see PIC250
BENZ(a)ACRIDIN-10-AMINE see AIC750
BENZ(c)ACRIDINE see BAW750
3,4-BENZACRIDINE see BAW750
7,8-BENZACRIDINE (FRENCH) see BAW750
3,4-BENZACRIDINE-9-ALDEHYDE see BAX250
BENZ(c)ACRIDINE-7-CARBONITRILE see BAX000
BENZ(c)ACRIDINE-7-CARBOXALDEHYDE see BAX250
BENZ(a)ACRIDINE-5,6-DIOL, 5,6-DIHYDRO-12-METHYL-, (Z)- see DLE500
BENZ(c)ACRIDINE 3,4-DIOL-1,2-EPOXIDE-1 see EBH875
BENZ(c)ACRIDINE 3,4-DIOL-1,2-EPOXIDE-2 see EBH850
BENZ(c)ACRIDINE, 3-METHOXY-7-METHYL- see MET875
BENZ(c)ACRIDIN-4-OL, 7-METHYL-, ACETATE (ESTER) see ABQ600
N'-BENZ(c)ACRIDIN-7-YL-N-(2-CHLOROETHYL)-N-ETHYL-1,2-

ETHANEDIAMINE DIHYDROCHLORIDE
see EHI500
α-(BENZ(c)ACRIDIN-7-YL)-N-(p-
(DIMETHYLAMINO)PHENYL)NITRONE
see BAY250
α-(9-(3,4-BENZACRIDYL))-N-(μ-
DIMETHYLAMINO-PHENYL)-NITRONE
see BAY250
BENZADONE BLUE RC see DFN300
BENZADONE BLUE RS see IBV050
BENZADONE BRILLIANT PURPLE 2R see
DFN450
BENZADONE BRILLIANT PURPLE 4R see
DFN450
BENZADONE BROWN BR see CMU770
BENZADONE BROWN R see CMU780
BENZADONE GOLDEN YELLOW see
DCZ000
BENZADONE GREY M see CMU475
BENZADONE OLIVE R see DUP100
BENZADOX see ANB000
BENZAHEX see BBQ750
BENZAIDIN see BFW250
BENZAKNEW see BDS000
BENZALACETON (GERMAN) see SMS500
BENZALACETONE see SMS500
trans-BENZALACETONE see BAY275
2-BENZALACETOPHENONE see CDH000
BENZALACETYLACETONE see DBH900
BENZAL ALCOHOL see BDX500
BENZAL-(BENZYL-CYANID) (GERMAN)
see DVX600
BENZAL CHLORIDE see BAY300
BENZALDEHYDE see BAY500
BENZALDEHYDE, 4-(ACETYLOXY)-3-
ETHOXY- see EQF100
BENZALDEHYDE, 4-AMINO- see AIC825
BENZALDEHYDE, p-BROMO- see BMT700
BENZALDEHYDE, 4-BROMO-(9CI) see
BMT700
BENZALDEHYDE, p-CHLORO- see CEI600
BENZALDEHYDE CYANOHYDRIN see
MAP250
BENZALDEHYDE, 3,5-DIBROMO-4-
HYDROXY-, (2,4-DINITROPHENYL)OXIME
see BNK400
BENZALDEHYDE, 4-(DIETHYLAMINO)-,
DIPHENYLHYDRAZONE see DHL850
BENZALDEHYDE, 4-(DIETHYLAMINO)-2-
HYDROXY- see DIJ230
BENZALDEHYDE, 2,4-DIHYDROXY- see
REF100
BENZALDEHYDE, DIMETHYL ACETAL
see DOG700
BENZALDEHYDE, 3-ETHOXY-2-
HYDROXY- see NOC100
BENZALDEHYDE, 4-ETHOXY-3-
HYDROXY- see IKO100
BENZALDEHYDE FFC see BBM500
BENZALDEHYDE, 2-FLUORO- see FFY100
BENZALDEHYDE GLYCERYL ACETAL
(FCC) see BBA000
BENZALDEHYDE GREEN see AFG500
BENZALDEHYDE GREEN see BAY750
BENZALDEHYDE, m-HYDROXY- see
FOE100
BENZALDEHYDE, 3-HYDROXY-(9CI) see
FOE100
BENZALDEHYDE, 4-(p-
HYDROXYANILINO)- see HJB260
BENZALDEHYDE, 2-HYDROXY-4-
METHOXY-(9CI) see HLR600
BENZALDEHYDE, 2-HYDROXY-5-NITRO-
see HMX700
BENZALDEHYDE, p-HYDROXY-,
THIOSEMICARBAZONE see HJG050
BENZALDEHYDE, 2-METHOXY-(9CI) see
AOT525
BENZALDEHYDE, 3-METHOXY-(9CI) see
AOT300
BENZALDEHYDE, 4-METHYL- see FOJ025

BENZALDEHYDE, PENTACHLORO-,
OXIME see PAV300
BENZALDEHYDE, 3-PHENOXY- see
PDR600
BENZALDEHYDE THIOSEMICARBAZONE
see BAZ000
BENZALDEHYDE, 2,3,4-TRIHYDROXY- see
TKN800
BENZALDEHYDE, 2,4,6-TRIMETHYL- see
MDJ745
BENZALDEHYDKYANHYDRIN (CZECH)
see MAP250
BENZALETAS see BEL900
BENZAL GLYCERYL ACETAL see BBA000
BENZALIN see DLY000
BENZALKONIUM BROMIDE see BEO000
BENZALKONIUM CHLORIDE see AFP250
BENZALKONIUM CHLORIDE see BBA500
BENZALKONIUM SACCHARINATE see
BBA625
BENZALMALONONITRILE see BBA750
BENZAL-m-NITROANILINE see BBA800
BENZAMIDE see BBB000
BENZAMIDE, N-(ACETYLOXY)-N-
(OCTYLOXY)- see ABU100
BENZAMIDE, m-(ALLYLOXY)-N-(BIS(1-
AZIRIDINYL)PHOSPHINYL)- see AGM800
BENZAMIDE, o-AMINO- see AID620
BENZAMIDE, 2-AMINO-(9CI) see AID620
BENZAMIDE, p-AMINO-N-ANTIPYRINYL-
N-METHYL- see AIN100
BENZAMIDE, 4-AMINO-5-BROMO-N-(2-
(DIETHYLAMINO)ETHYL)-2-METHOXY-
(9CI) see VCK100
BENZAMIDE, N-
(((AMINOCARBONYL)AMINO)THIOXOME
THYL)-2-CHLORO- see CEO120
BENZAMIDE, N-
(((AMINOCARBONYL)AMINO)THIOXOME
THYL)-4-NITRO- see NFL600
BENZAMIDE, 4-AMINO-5-CHLORO-2-
ETHOXY-N-((4-((4-
FLUOROPHENYL)METHYL)-2-
MORPHOLINYL)MET HYL)- see MRU253
BENZAMIDE, 4-AMINO-5-CHLORO-N-(1-
(3-(4-FLUOROPHENOXY)PROPYL)-3-
METHOXY-4-PIPERIDINYL)-2-METHOXY-,
MONOHYDRATE, cis- see CMS232
BENZAMIDE, 4-AMINO-N-(2,3-DIHYDRO-
1,5-DIMETHYL-3-OXO-2-PHENYL-1H-
PYRAZOL-4-YL)-N-METHYL- see AIN100
BENZAMIDE, 5-(2-((3-(1,3-BENZODIOXOL-
5-YL)-1-METHYLPROPYL)AMINO)-1-
HYDROXYETHYL)-2-HYDROXY-,
MONOHYDROCHLORIDE see MBZ120
BENZAMIDE, N-(BIS(1-
AZIRIDINYL)PHOSPHINYL)- see BGY050
BENZAMIDE, N-(BIS(1-
AZIRIDINYL)PHOSPHINYL)-p-BROMO- see
BGY056
BENZAMIDE, N-(BIS(1-
AZIRIDINYL)PHOSPHINYL)-m-IODO- see
BGY130
BENZAMIDE, N-(BIS(1-
AZIRIDINYL)PHOSPHINYL)-o-IODO- see
BGY135
BENZAMIDE, 5-BROMO-N-(4-
BROMOPHENYL)-2-HYDROXY- see
BOD600
BENZAMIDE, N-(1-BUTOXY-2,2,2-
TRICHLOROETHYL)-2-HYDROXY- see
BPT300
BENZAMIDE, N-(5-CHLORO-4-((4-
CHLOROPHENYL)CYANOMETHYL)-2-
METHYLPHENYL)-2-HYDROXY-3,5-
DIIODO- see CFC100
BENZAMIDE, 3-CHLORO-5-ETHYL-6-
HYDROXY-2-METHOXY-N-((1-METHYL-2-
PYRROLIDINYL)METHYL)-,
MONOHYDROCHLORIDE, (S)- see CHC100

BENZAMIDE, 5-(2-CHLORO-6-FLUORO-4-
(TRIFLUOROMETHYL)PHENOXY)-N-
(ETHYLSULFONYL)-2-NITRO- see CHJ400
BENZAMIDE, N-(((4-(2-CHLORO-4-
(TRIFLUOROMETHYL)PHENOXY) 2
FLUOROPHENYL)AMINO)CARBONYL)-
2,6-DIFLUORO- see FDB400
BENZAMIDE, 5-(2-CHLORO-4-
(TRIFLUOROMETHYL)PHENOXY)-N-
(METHYLSULFONYL)-2-NITRO- see CLS050
BENZAMIDE, 2-CYANO- see COK300
BENZAMIDE, 4-CYANO- see COK400
BENZAMIDE, o-CYANO-(8CI) see COK300
BENZAMIDE, p-CYANO-(8CI) see COK400
BENZAMIDE, 3,5-DIBROMO-N-(4-
BROMOPHENYL)-2-HYDROXY- see
THW750
BENZAMIDE, 3,5-DICHLORO-N-(3-
CHLORO-1-ETHYL-1-METHYL-2-
OXOPROPYL)-4-METHYL- see ZUJ500
BENZAMIDE, N-(((3,5-DICHLORO-4-((3-
CHLORO-5-(TRIFLUOROMETHYL)-2-
PYRIDINYL)OXY)PHENYL) AMINO)
CARBONYL)-2,6-DIFLUORO- see CDS800
BENZAMIDE, N-(((3,5-DICHLORO-4-(1,1,2,2-
TETRAFLUOROETHOXY)PHENYL)AMINO
)CARBONYL)-2,6-DIFLUORO- see HCY600
BENZAMIDE, N-(2-
(DIETHYLAMINO)ETHYL)- see DHU700
BENZAMIDE, N-(2-
(DIETHYLAMINO)ETHYL)-4-((1-OXO-2-
PROPENYL)AMINO)- see PMD850
BENZAMIDE, 2,6-DIMETHOXY-N-(3-(1-
ETHYL-1-METHYLPROPYL)-5-
ISOXAZOLYL)- see ENF100
BENZAMIDE, 3-((2-((2-(3,4-
DIMETHOXYPHENYL)ETHYL)AMINO)-2-
OXOETHYL)AMINO)-N-METHYL- see
DOK500
BENZAMIDE, N,N'-(DITHIODI-2,1-
PHENYLENE)BIS- see BDK800
BENZAMIDE, N-((1-ETHYL-2-
PYRROLIDINYL)METHYL)-5-
(ETHYLSULFONYL)-2-METHOXY- see
EPD100
BENZAMIDE, 2-HYDROXY-N-PHENYL- see
SAH500
BENZAMIDE, N-(3-(1-
METHYLETHOXY)PHENYL)-2-
(TRIFLUOROMETHYL)- see TKB286
BENZAMIDE, 2-METHYL-N-(3-(1-
METHYLETHOXY)PHENYL)- see INE050
BENZAMIDE, o-NITRO- see NEV523
BENZAMIDE, 2-NITRO- (9CI) see NEV523
BENZAMIDEPHENYLHYDRAZONE
HYDROCHLORIDE see PEK675
BENZAMIDE, N-(2-
PIPERIDINYLMETHYL)-2,5-BIS(2,2,2-
TRIFLUOROETHOXY)- see FCC065
BENZAMIDE, N,N'-(10,15,16,17-
TETRAHYDRO-5,10,15,17-TETRAOXO-5H-
DINAPHTHO(2,3-A:2',3'-I) CARBAZOLE-4,9-
DIYL)BIS- see CMU780
BENZAMIDINE, HYDROCHLORIDE see
BBM750
BENZAMIDINE, 4,4'-
(PENTAMETHYLENEDIOXY)DI-
,DIHYDROCHLORIDE see PBJ800
BENZAMIDOACETIC ACID see HGB300
(2-BENZAMIDO)ACETOHYDROXAMIC
ACID see BBB250
N-(4-BENZAMIDO-4-CARBAMOYLBUTYL)-
N-NITROSOCYANAMIDE see CBK250
1-BENZAMIDO-5-CHLORO-
ANTHRAQUINONE see BDK750
dl-α-BENZAMIDO-p-(2-
(DIETHYLAMINO)ETHOXY)-N,N-
DIPROPYLHYDROCINNAMAMIDE see
TGF175
dl-4-BENZAMIDO-N,N-
DIPROPYLGLUTARAMIC ACID see BGC625

dl-4-BENZAMIDO-N,N-
DIPROPYLGLUTARAMIC ACID SODIUM
SALT see PMH575
(S)-p-(α-BENZAMIDO-p-
HYDROXYHYDROCINNAMAMIDO)BENZ
OIC ACID see CML870
BENZAMIDOOXY ACETIC ACID,
AMMONIUM SALT see ANB000
1-BENZAMIDO-1-PHENYL-3-
PIPERIDINOPROPANE HYDROCHLORIDE
see DKK800
N-(3-BENZAMIDO-3-PHENYL)PROPYL
PIPERIDINE HYDROCHLORIDE see
DKK800
BENZAMIL BLACK E see AQP000
BENZAMIL SUPRA BROWN BRLL see
CMO750
BENZAMIN BLACK DS see CMN230
BENZAMINE BLUE see CMO250
BENZAMINE, 4,4'-(1,4-PHENYLENEBIS(1-
METHYLETHYLIDENE))BIS- see BGV800
BENZAMIZOLE see ENF100
BENZAMPHETAMINE see AOB250
BENZANIDINE see BFW250
BENZANIL BLACK BH see CMN800
BENZANIL BLUE 2B see CMO000
BENZANIL BLUE RW see CMO600
BENZANIL BORDEAUX B see CMO872
BENZANIL BROWN BS see CMO820
BENZANIL BROWN M see CMO800
BENZANIL DARK GREEN BW see CMO830
BENZANIL FAST BLACK D see CMN230
BENZANIL FAST BLACK G see CMN240
BENZANIL FAST RED F see CMO870
BENZANIL FAST RED K see CMO885
BENZANIL FAST SCARLET 4BSN see
CMO870
BENZANIL GREEN B see CMO840
BENZANIL GREEN BN see CMO840
BENZANILIDE, 3-AMINO-4-METHOXY- see
AKM500
BENZANILIDE, 2'-CHLORO-2-(2-
(DIETHYLAMINO)ETHOXY)- see DHP450
BENZANILIDE, 4'-CHLORO-2-(2-
(DIETHYLAMINO)ETHOXY)- see DHP550
BENZANILIDE, 2',2'''-DITHIOBIS- see
BDK800
BENZANIL PURPURINE 4B see DXO850
BENZANIL SCARLET 3B see CMO875
BENZANIL SKY BLUE see CMO500
BENZANIL VIOLET N see CMP000
BENZANOL BLUE RW see CMO600
BENZANOL BRILLIANT BORDEAUX BN
see CMO872
BENZANOL BRILLIANT SCARLET 3B see
CMO875
BENZANOL BROWN M see CMO800
BENZANOL FAST RED F see CMO870
BENZ(a)ANTHRACEN-7-ACETIC ACID,
METHYL ESTER see BBC500
BENZ(a)ANTHRACEN-7-ACETONITRILE
see BBB500
BENZ(a)ANTHRACEN-7-AMINE see BBB750
BENZ(a)ANTHRACEN-8-AMINE see BBC000
BENZANTHRACENE see BBC250
BENZ(a)ANTHRACENE see BBC250
BENZ(b)ANTHRACENE see NAI000
1,2-BENZANTHRACENE see BBC250
2,3-BENZANTHRACENE see NAI000
1,2-BENZ(a)ANTHRACENE see BBC250
1,2:5,6-BENZANTHRACENE see DCT400
1,2-BENZANTHRACENE-10-ACETIC ACID,
METHYL ESTER see BBC500
1,2-BENZANTHRACENE-10-ALDEHYDE
see BBC750
BENZ(a)ANTHRACENE, 8-BROMO-7,12-
DIMETHYL- see BNF315
BENZ(a)ANTHRACENE-7-
CARBOXALDEHYDE see BBC750
BENZ(a)ANTHRACENE-7,12-
DICARBOXALDEHYDE see BBD000

BENZ(a)ANTHRACENE-1,2-
DIHYDRODIOL see BBD250
BENZ(a)ANTHRACENE-3,4-
DIHYDRODIOL see BBD500
BENZ(a)ANTHRACENE-5,6-
DIHYDRODIOL see BBD980
BENZ(a)ANTHRACENE-10,11-
DIHYDRODIOL see BBF000
BENZ(a)ANTHRACENE-5,6-cis-
DIHYDRODIOL see BBE000
BENZ(a)ANTHRACENE-5,6-trans-
DIHYDRODIOL see BBD980
trans-BENZ(a)ANTHRACENE-8,9-
DIHYDRODIOL see BBE750
(+)-(3S,4S)trans-BENZ(a)ANTHRACENE-3,4-
DIHYDRODIOL see BBD750
BENZ(a)ANTHRACENE, 3,4-DIHYDROXY-
1,2-EPOXY-1,2,3,4-TETRAHYDRO-, (Z), (+)-
see DMO500
BENZ(a)ANTHRACENE-7,12-
DIMETHANOL see BBF500
BENZ(a)ANTHRACENE-7,12-
DIMETHANOLDIACETATE see BBF750
BENZ(a)ANTHRACENE-3,9-DIOL see
BBG200
(−)(3R,4R)-trans-BENZ(a)ANTHRACENE-3,4-
DIOL see BBG000
BENZ(a)ANTHRACENE 3,4-DIOL-1,2-
EPOXIDE-2 see DLE000
BENZ(a)ANTHRACENE-5,6-EPOXIDE see
BDJ500
BENZ(a)ANTHRACENE-7-ETHANOL see
BBG500
BENZ(a)ANTHRACENE-7-
METHANEDIOLDIACETATE (ester) see
BBG750
BENZ(a)ANTHRACENE-7-
METHANETHIOL see BBH000
BENZ(a)ANTHRACENE-7-METHANOL see
BBH250
BENZ(a)ANTHRACENE-7-METHANOL
ACETATE see BBH500
BENZ(a)ANTHRACENE, 1,12-METHYLENE-
see MJL300
1,2-BENZANTHRACENE, 1',9-
METHYLENE- see MJL300
BENZ(a)ANTHRACENE, 7-NITRO- see
NEW600
7H-BENZ(de)ANTHRACENE-7-ONE see
BBI250
BENZ(a)ANTHRACENE-5,6-OXIDE see
BDJ500
BENZ(a)ANTHRACENE-5,6-OXIDE see
EBP000
BENZ(a)ANTHRACENE, 8-PHENYL see
PEK750
BENZ(a)ANTHRACENE-7-THIOL see
BBH750
BENZ(a)ANTHRACEN-5-OL see BBI000
7H-BENZ(de)ANTHRACEN-7-ONE see
BBI250
N-(BENZ(a)ANTHRACEN-5-
YLCARBAMOYL)GLYCINE see BBI750
N-(BENZ(a)ANTHRACEN-7-
YLCARBAMOYL)GLYCINE see BBJ000
BENZ(a)ANTHRACEN-7-YL-OXIRANE see
ONC000
1-(BENZ(a)ANTHRACEN-7-YL)-2,2,2-
TRICHLOROETHANONE see TIJ000
BENZ(a)ANTHRACEN-7-YL
TRICHLOROMETHYL KETONE see TIJ000
BENZ(a)ANTHRA-5,6-OXIDE see EBP000
BENZ(3,4)ANTHRA(1,2-6)OXIRENE see
BDJ500
1,2-BENZANTHRAZEN (GERMAN) see
BBC250
BENZANTHRENE see BBC250
1,2-BENZANTHRENE see BBC250
2,3-BENZANTHRENE see NAI000
BENZANTHRENONE see BBI250
BENZANTHRONE see BBI250

1,2-BENZANTHRYL-3-
CARBAMIDOACETIC ACID see BBI750
1,2-BENZANTHRYL-10-
CARBAMIDOACETIC ACID see BBJ000
1,2-BENZANTHRYL-10-ISOCYANATE see
BBJ250
1,2-BENZANTHRYL-10-MERCAPTAN see
BBH750
1,2-BENZANTHRYL-10-
METHYLMERCAPTAN see BBH000
7-BENZANTHRYLOXIRANE see ONC000
BENZANTINE see BBV500
BENZAR see BAV000
BENZARONE see BBJ500
5H-BENZ(B)ARSINDOLE, 5-HYDROXY-, 5-
OXIDE see ARA100
BENZATHINE see BHB300
BENZATHINE BENZYLPENICILLIN see
BFC750
BENZATHINE PENICILLIN see BFC750
BENZATHINE PENICILLIN G see BFC750
BENZATIN see BHB300
BENZATROPINE METHANESULFONATE
see TNU000
1H-3-BENZAZEPINE, 2,3,4,5-
TETRAHYDRO-2-(NITROMETHYLENE)-
see NHN600
BENZAZIDE see BDL750
BENZAZIMIDE see BDH000
BENZAZIMIDOL HYDRATE see HJN650
BENZAZIMIDONE see BDH000
1-BENZAZINE see QMJ000
2-BENZAZINE see IRX000
1-BENZAZOLE see ICM000
BENZAZOLINE see BBW750
BENZAZOLINE HYDROCHLORIDE see
BBJ750
BENZAZON VII see NGC400
BENZBROMARON see DDP200
BENZBROMARONE see DDP200
1,2-BENZCARBAZOLE see BCG250
BENZ-o-CHLOR see DER000
BENZCHLOROPROPAMIDE see BEG000
BENZCHLORPROPAMID see BEG000
3,4-BENZCHRYSENE see PIB750
BENZCURINE IODIDE see PDD300
7H-BENZ(DE)ANTHRACEN-ONE, 3-
NITRO- see NEW700
15,16-BENZDEHYDROCHOLANTHRENE
see DCR400
1H-BENZ(DE)ISOQUINOLINE-1,3(2H)-
DIONE, 5-AMINO-2-
(DIMETHYLAMINO)ETHYL- see NAC600
1,4-BENZ3NEDIAMINE, N-PHENYL-(9CI)
see PFU500
BENZEDREX (SKF) see PNN300
BENZEDRINE see BBK000
(±)-BENZEDRINE see BBK000
dl-BENZEDRINE see BBK000
BENZEDRINE SULFATE see BBK250
d-BENZEDRINE SULFATE see BBK500
l-BENZEDRINE SULFATE see BBK750
BENZEDRYNA see AOB250
BENZEEN (DUTCH) see BBL250
BENZEHIST see BAU750
BENZEN (POLISH) see BBL250
BENZENACETIC ACID see PDY850
BENZENAMINE see AOQ000
BENZENAMINE, 3-(2-(4-
AMINOPHENYL)ETHENYL)- see DCD100
BENZENAMINE, 4-((4-AMINOPHENYL)(4-
IMINO-2,5-CYCLOHEXADIEN-1-
YLIDENE)METHYL)-2-METHYL- see
MAC500
BENZENAMINE, 4-((4-AMINOPHENYL)(4-
IMINO-3-METHYL-2,5-CYCLOHEXADIEN-
1-YLIDENE)METHYL)-2-METHYL-,
MONOHYDROCHLORIDE see DSB300
BENZENAMINE, 4,4'-AZOBIS-(9CI) see
ASK925
BENZENAMINE, 2,4-BIS((4-
AMINOPHENYL)METHYL)- see BGU100

BENZENAMINE, 2,6-BIS(1-METHYLETHYL)- see DNN630

BENZENAMINE, 4-((4-BROMOPHENYL)AZO)-N,N-DIMETHYL- see BNV760

BENZENAMINE, 4-((3-BROMOPHENYL)AZO)-N,N-DIMETHYL-(9CI) see BNE600

BENZENAMINE, 4-BUTYL-(9CI) see AJA550

BENZENAMINE, N-BUTYL-(9CI) see BQH850

BENZENAMINE, 4,4'-CARBONIMIDOYLBIS(N,N-DIETHYL-, MONONITRATE see EGM200

BENZENAMINE, 2-CHLORO-4,6-DINITRO- see CGL325

(BENZENAMINE) CHLORO((1,2,5,6-ETA)-1,5-CYCLOOCTADIENE)RHODIUM see BBK800

BENZENAMINE, N-(2-CHLOROETHYL)-N-ETHYL-(9CI) see EHJ600

BENZENAMINE, 3-CHLORO-, HYDROCHLORIDE see CEH690

BENZENAMINE, 4-CHLORO-, HYDROCHLORIDE see CJR200

BENZENAMINE, 2-CHLORO-4-METHYL- see CLK210

BENZENAMINE, 2-CHLORO-4-(METHYLSULFONYL)- see CIX200

BENZENAMINE, 4-CHLORO-2-NITRO-(9CI) see KDA050

BENZENAMINE, 2-CHLORO-6-NITRO-3-PHENOXY- see CJD300

BENZENAMINE, N-(2-CHLORO-5-(TRIFLUOROMETHYL)PHENYL)-2,4-DINITRO-6-(TRIFLUOROMETHYL)-(9CI) see FDB300

BENZENAMINE, 2,3-DICHLORO- see DEO210

BENZENAMINE, 2,4-DICHLORO-(9CI) see DEO290

BENZENAMINE, 2,6-DICHLORO-N-2-IMIDAZOLIDINYLIDENE- (9CI) see DGB500

BENZENAMINE, 4-((3,4-DICHLOROPHENYL)AZO)-N,N-DIMETHYL-(9CI) see DFD400

BENZENAMINE, 2,6-DIETHYL-(9CI) see DIS650

BENZENAMINE, N,N-DIETHYL-(9CI) see DIS700

BENZENAMINE, N,N-DIETHYL-, HYDROCHLORIDE see DIS720

BENZENAMINE, N,N-DIETHYL-4-NITROSO-(9CI) see NJW600

BENZENAMINE, N,N-DIETHYL-4-((5-NITRO-2-THIAZOLYL)AZO)- see DJT050

BENZENAMINE, 4-((3,4-DIETHYLPHENYL)AZO)-N,N-DIMETHYL-(9CI) see DJB400

BENZENAMINE, 2,4-DIFLUORO-(9CI) see DKF700

BENZENAMINE, 2,5-DIMETHOXY-(9CI) see AKD925

BENZENAMINE, 2,4-DIMETHOXY-, HYDROCHLORIDE see MEA625

BENZENAMINE, N,N-DIMETHYL-3'-BROMO-4'-ETHYL-4-(PHENYLAZO)- see BNK275

BENZENAMINE, N,N-DIMETHYL-4'-BROMO-3'-ETHYL-4-(PHENYLAZO)- see BNK100

BENZENAMINE, N,N-DIMETHYL-3'-BROMO-4'-METHYL-4-(PHENYLAZO)- see BNQ100

BENZENAMINE, N,N-DIMETHYL-4'-BROMO-3'-METHYL-4-(PHENYLAZO)- see BNQ110

BENZENAMINE, N,N-DIMETHYL-3'-CHLORO-4'-ETHYL-4-(PHENYLAZO)- see CGW100

BENZENAMINE, N,N-DIMETHYL-4'-CHLORO-3'-ETHYL-4-(PHENYLAZO)- see CGW105

BENZENAMINE, 4-(1,1-DIMETHYLETHYL)-N-(1-METHYLPROPYL)-2,6-DINITRO-(9CI) see BQN600

BENZENAMINE, N,N-DIMETHYL-2'-ETHYL-4-(PHENYLAZO)- see EIF450

BENZENAMINE, N,N-DIMETHYL-2-NITRO-(9CI) see NFW600

BENZENAMINE, 3,5-DINITRO-4-METHYL- see MJG600

BENZENAMINE, 2,4-DINITRO-N-METHYL-N-(2,4,6-TRIBROMOPHENYL)-6-(TRIFLUOROMETHYL)- see BMO300

BENZENAMINE, 2,4-DINITRO-N-(2,4,6-TRIBROMOPHENYL)-6-(TRIFLUOROMETHYL)- see DBA520

BENZENAMINE, 4,4'-DITHIOBIS-(9CI) see ALW100

BENZENAMINE, 2,2'-(1,2-ETHANEDIYLBIS(THIO))BIS- see EJC050

BENZENAMINE, ar-ETHOXY- see EEL100

BENZENAMINE, 3-ETHOXY-(9CI) see PDK800

BENZENAMINE, 4-ETHOXY-(9CI) see PDK790

BENZENAMINE, 4-ETHOXY-, HYDROCHLORIDE (9CI) see PDL750

BENZENAMINE, 2-ETHOXY-5-NITRO- see NIC200

BENZENAMINE, 2-ETHYL-(9CI) see EGK500

BENZENAMINE, 3-ETHYL-(9CI) see EOH100

BENZENAMINE, 4-ETHYL-N-HYDROXY- see EOL100

BENZENAMINE, N-ETHYL-N-(2-METHOXYETHYL)-3-METHYL-4-NITROSO- see EMI510

BENZENAMINE, 2-ETHYL-6-METHYL- see MJY000

BENZENAMINE, N-ETHYL-3-METHYL- see EPT100

BENZENAMINE, 4-((2-ETHYLPHENYL)AZO)-N,N-DIMETHYL- see EIF450

BENZENAMINE, 4-ETHYL-N-(PHENYLMETHYLENE)-, N-OXIDE see EOL700

BENZENAMINE, 4-FLUORO-(9CI) see FFY000

BENZENAMINE, 4-FLUORO-2-METHYL- see FJR900

BENZENAMINE, 4-FLUORO-3-NITRO- see FKK100

BENZENAMINE HYDROCHLORIDE see BBL000

BENZENAMINE, N-HYDROXY-2-METHYL-3-NITRO- see HLN135

BENZENAMINE, N-HYDROXY-2-METHYL-5-NITRO- see HLN125

BENZENAMINE, N-HYDROXY-4-METHYL-3-NITRO- see HMI100

BENZENAMINE, N-HYDROXY-4-NITRO- see NIR050

BENZENAMINE, N,N'-METHANETETRAYLBIS(2,6-BIS(1-METHYLETHYL)- see BJE550

BENZENAMINE, 2-METHOXY-(9CI) see AOV900

BENZENAMINE, 3-METHOXY-(9CI) see AOV890

BENZENAMINE, 2-METHOXY-, HYDROCHLORIDE (9CI) see AOX250

BENZENAMINE, 4-METHOXY-N-(4-METHOXYPHENYL)- see BKO600

BENZENAMINE, 2-METHOXY-4-NITRO-(9CI) see NEQ000

BENZENAMINE, 3-METHOXY-4-(PHENYLAZO)- see MDY400

BENZENAMINE, N-METHYL-(9CI) see MGN750

BENZENAMINE, 4-(6-METHYL-2-BENZOTHIAZOLYL)-(9CI) see MHJ300

BENZENAMINE, 4,4'-METHYLENEBIS- see MJQ000

BENZENAMINE, 4,4'-METHYLENEBIS-, DIHYDROCHLORIDE see MJQ100

BENZENAMINE, 4,4'-METHYLENEBIS(2-ETHYL)- (9CI) see MJN100

BENZENAMINE, 4,4'-METHYLENEBIS(N-METHYL)-(9CI) see MJO000

BENZENAMINE, 2-(1-METHYLETHYL)-(9CI) see INW100

BENZENAMINE, 2-METHYL-4-NITRO- see MMF780

BENZENAMINE, 2-METHYL-5-NITRO- see NMP500

BENZENAMINE, 3-METHYL-6-NITRO- see NMP550

BENZENAMINE, 2-METHYL-5-NITRO-, MONOHYDROCHLORIDE see NMP600

BENZENAMINE, N-(2-METHYL-2-NITROPROPYL)-p-NITROSO-(9CI) see NHK800

BENZENAMINE, 2-(2-METHYLPHENOXY)-(9CI) see THB100

BENZENAMINE, 2-METHYL-4-(9H-PYRIDO(3,4-B)INDOL-9-YL)- see AKZ400

BENZENAMINE, N-METHYL-N,2,4,6-TETRANITRO-(9CI) see TEG250

BENZENAMINE, 3-(METHYLTHIO)-(9CI) see ALX100

BENZENAMINE, 4-(METHYLTHIO)-(9CI) see AMS675

BENZENAMINE, N,N,-DIMETHYL-(9CI) see DQF800

BENZENAMINE, N-NITRO- see NEO510

BENZENAMINE, 4-NITRO-(9CI) see NEO500

BENZENAMINE, 4-(2-(4-NITROPHENYL)ETHENYL)- see NIM575

BENZENAMINE, 3-NITRO-N-(PHENYLMETHYLENE)-(9CI) see BBA800

BENZENAMINE, N,N,4-TRIMETHYL- see TLG150

BENZENAMINE, N,N,3-TRIMETHYL-(9CI) see TLG700

BENZENAMINE, 2,3,4,5,6-PENTANITRO- see PBM100

BENZENAMINE, 2-PHENOXY-(9CI) see PDR490

BENZENAMINE, 4-PHENOXY-(9CI) see PDR500

BENZENAMINE, N-(3-(PHENYLAMINO)-2-PROPENYLIDENE)-, MONOHYDROCHLORIDE see PFJ500

BENZENAMINE, N-(3-PHENYL-4,5-BIS((TRIFLUOROMETHYL)IMINO)-2-THIAZOLIDINYLIDENE)- see FDA890

BENZENAMINE, 4,4'-(1,3-PHENYLENEBIS(OXY))BIS- see REF070

BENZENAMINE, N-PHENYL-, STYRENATED see SMP700

BENZENAMINE, 4-(9H-PYRIDO(3,4-B)INDOL-9-YL)- see ALX120

BENZENAMINE, 4-STIBONO-(9CI) see SLP600

BENZENAMINE, 4,4'-(SULFONYLBIS(4,1-PHENYLENEOXY))BIS-(9CI) see SNZ000

BENZENAMINE, N,2,4,6-TETRANITRO- see TDY075

BENZENAMINE, 2,3,4-TRIFLUORO- see TJX900

BENZENAMINE, 4,4'-((2,2,2-TRIFLUORO-1-(TRIFLUOROMETHYL)ETHYLIDENE)BIS(4,1-PHENYLENEOXY))BIS- see TKK025

BENZENAMINE, 4-((1,7,7-TRIMETHYLBICYCLO(2.2.1)HEPT-2-YL)OXY)- see IHY500

BENZENAMINE, compounded with 1,3,5-TRINITROBENZENE (1:1) see BBL100

BENZENAMINIUM, 3,3'-(1,10-DECANEDIYLBIS((METHYLIMINO)CARBONYLOXY))BIS(N,N,N-TRIMETHYL-, 2BR see DAM625

BENZENAMINIUM, N,N-DIETHYL-3-HYDROXY-N-METHYL-, CHLORIDE see DJN430

BENZENAMINIUM, 3-(((DIMETHYLAMINO)CARBONYL)OXY)-N,N,N-TRIMETHYL-, BROMIDE (9CI) see POD000

BENZENAMINIUM N,N,N-TRIMETHYL-, BROMIDE (9CI) see TMB000

BENZENE see BBL250

BENZENEACETALDEHYDE see BBL500

BENZENEACETALDEHYDE, α-(HYDROXYIMINO)-, OXIME see HLI400

BENZENEACETALDEHYDE, 4-(1-METHYLETHYL) (9CI) see IRA000

BENZENEACETAMIDE (9CI) see PDX750

BENZENEACETAMIDE, N-(AMINOCARBONYL)-α-ETHYL-, (−)- see PDM100

BENZENEACETAMIDE, N,N-DIETHYL-(9CI) see PEU100

BENZENEACETAMIDE, N,N-DIETHYL-α-HYDROXY- see DJL700

BENZENEACETAMIDE, α-ETHYL- see NMV300

BENZENEACETAMIDE, 4-HYDROXY- see HNG550

BENZENEACETAMIDE, N-(2-(HYDROXYAMINO)-2-OXOETHYL)- see HIY300

BENZENEACETAMIDE, α-(METHOXYIMINO)-N-METHYL-2-PHENOXY-, (α-E)- see MQQ800

BENZENEACETIC ACID see PDY850

BENZENEACETIC ACID, 4-((5-ACETYL-2-FURANYL)OXY)-α-ETHYL-, METHYL ESTER see MFX650

BENZENEACETIC ACID, 4-AMINO-(9CI) see AID700

BENZENEACETIC ACID, 4-AMINO-3,5-DICHLORO-α-HYDROXY- see AJM575

BENZENEACETIC ACID, α-AMINO-3-HYDROXY-4-(HYDROXYMETHYL)-, (S)- see FOJ100

BENZENEACETIC ACID, BUTYL ESTER (9CI) see BQJ350

BENZENEACETIC ACID, 4-CHLORO-α-ETHYL-, (5-(PHENYLMETHYL)-3-FURANYL)METHYL ESTER see PFR150

BENZENEACETIC ACID, 4-CHLORO-α-(1-METHYLETHENYL)-, (3-PHENOXYPHENYL)METHYL ESTER see PDR670

BENZENEACETIC ACID, 4-CHLORO-α-(1-METHYLETHYL)-, CYANO (3-PHENOXYPHENYL)METHYL ESTER, (S-(R*,R*))- see FAR150

BENZENEACETIC ACID, 4-CHLORO-α-(1-METHYLETHYL)-, (3-PHENOXYPHENYL)METHYL ESTER see PDV330

BENZENEACETIC ACID, 2-((6-(2-CYANOPHENOXY)-4-PYRIMIDINYL)OXY)-α-(METHOXYMETHYLENE)-, METHYL ESTER, (E)- see ASP525

BENZENEACETIC ACID, α-CYCLOHEXYL-α-HYDROXY-,(4-DIETHYLAMINO)-1,1-DIMETHYL-2-BUTEYNYL ESTER,HYDROCHLORIDE see UTU700

BENZENEACETIC ACID, 2,4-DICHLORO-, ETHYL ESTER (9CI) see EHY600

BENZENEACETIC ACID, 4-(DIFLUOROMETHOXY)-α-(1-METHYLETHYL)-, CYANO(3-PHENOXYPHENYL) METHYL ESTER see COQ385

BENZENEACETIC ACID, 3,4-DIHYDROXY-see HGI980

BENZENEACETIC ACID, 3,4-DIMETHOXY-(9CI) see HGK600

BENZENEACETIC ACID, α-((DIMETHOXYPHOSPHINYL)THIO)-, ETHYL ESTER (9CI) see PDW800

BENZENEACETIC ACID, 1,5-DIMETHYL-1-ETHENYL-4-HEXENYL ESTER see LGC050

BENZENEACETIC ACID, 3,7-DIMETHYL-2,6-OCTADIENYL ESTER, (E)- see GDM400

BENZENEACETIC ACID, 3,7-DIMETHYL-6-OCTENYL ESTER (9CI) see CMU050

BENZENEACETIC ACID, ETHYL ESTER (9CI) see EOH000

BENZENEACETIC ACID, α-ETHYL-4-METHOXY-, (3-PHENOXYPHENYL)METHYL ESTER see PDR660

BENZENEACETIC ACID, α-ETHYL-4-METHYL-, (3-PHENOXYPHENYL)METHYL ESTER see PDV360

BENZENEACETIC ACID, α-ETHYL-, (3-PHENOXYPHENYL)METHYL ESTER see PDR664

BENZENEACETIC ACID, α-ETHYL-, (5-(PHENYLMETHYL)-3-FURANYL)METHYL ESTER see BEP600

BENZENEACETIC ACID, α-ETHYL-, 1-(5-(2-PROPYNYL)-3-FURANYL)-2-PROPYNYL ESTER see PMN600

BENZENEACETIC ACID, 3-HEXENYL ESTER, (Z)- see HFE625

BENZENEACETIC ACID, 4-HYDROXY-(9CI) see HNG600

BENZENEACETIC ACID, α-HYDROXY-, 2-(2-ETHOXYETHOXY)ETHYL ESTER see HJG100

BENZENEACETIC ACID, 4-HYDROXY-3-METHOXY-, (3A,3B,6,6A,9A,10,11,11A-OCTAHYDRO-6A-HYDROXY-8,10-DIMETHYL-11A-(1-METHYLETHENYL)-7-OXO-2-(PHENYLMETHYL)-7H-2,9B-EPOXYAZULENO (5,4-E)-1,3-BENZODIOXOL-5-YL)METHYL ESTER, (2S,3AR,3BS,6AR,9AR,9BR,10R,11AR)- see RDP100

BENZENEACETIC ACID, α-HYDROXY-α-(1-METHYLETHYL)-, 3-(DIETHYLAMINO)PROPYL ESTER HCL see DIQ200

BENZENEACETIC ACID, α-HYDROXY-α-PHENYL-(9CI) see BBY990

BENZENEACETIC ACID, α-(METHOXYIMINO)-2-((2-METHYLPHENOXY)METHYL)-, METHYL ESTER, (α-E)- see MLI900

BENZENEACETIC ACID, 4-METHOXY-α-(1-METHYLETHYL)-, (3-PHENOXYPHENYL)METHYL ESTER see PDR666

BENZENEACETIC ACID, 4-METHOXY-α-(1-METHYLETHYL)-, (5-PHENOXY-2-FURANYL)METHYL ESTER see PDS925

BENZENEACETIC ACID, 4-METHOXY-α-(1-METHYLETHYL)-, (3-PHENOXYPHENYL)METHYL ESTER see PDR675

BENZENEACETIC ACID, 4-METHOXY-α-(1-METHYLETHYL)-, (5-(PHENYLMETHYL)-3-FURANYL)METHYL ESTER see BEP650

BENZENEACETIC ACID-2-METHOXY-4-(2-PROPENYL)PHENYL ESTER see AGL000

BENZENEACETIC ACID, 3-METHYLBUTYL ESTER (9CI) see IHV000

BENZENEACETIC ACID, METHYL ESTER see MHA500

BENZENEACETIC ACID, 4-METHYL-α-(1-METHYLETHENYL)-, (3-PHENOXYPHENYL)METHYL ESTER see PDR668

BENZENEACETIC ACID, 5-METHYL-2-(1-METHYLETHYL)CYCLOHEXYL ESTER, (1R-(1-α-2-β,5α-))- see MCG910

BENZENEACETIC ACID, 4-METHYL-α-(1-METHYLETHYL)-, (3-PHENOXYPHENYL)METHYL ESTER see PDR680

BENZENEACETIC ACID, α-METHYL-4-((2-METHYL-2-PROPENYL)AMINO)- see MDP850

BENZENEACETIC ACID, α-METHYL-4-((2-METHYL-2-PROPENYL)AMINO)-, (±)- see MGC200

BENZENEACETIC ACID, α-METHYL-4-(2-METHYLPROPYL)-, SODIUM SALT (9CI) see IAB100

BENZENEACETIC ACID, 4-NITRO-(9CI) see NII510

BENZENEACETIC ACID, α-OXO-(9CI) see OOK150

BENZENEACETIC ACID, α-PHENYL-, 2-(DIETHYLAMINO)ETHYL ESTER, (9CI) see DHX800

BENZENEACETIC ACID, 2-PHENYLETHYL ESTER see PDI000

BENZENEACETIC ACID, 2-PROPENYL ESTER see PMS500

BENZENEACETIC ACID, SODIUM SALT (9CI) see SFA200

BENZENEACETONITRILE see PEA750

BENZENEACETONITRILE, α-ACETYL-(9CI) see PEA500

BENZENEACETONITRILE, 4-CHLORO- see CEP300

BENZENEACETONITRILE, α-((CYANOMETHOXY)IMINO)- see COP700

BENZENEACETONITRILE, α-((DIMETHOXYPHOSPHINOTHIOYL)OXY)IMINO- see MOC275

BENZENEACETONITRILE, α-((1,3-DIOXOLAN-2-YLMETHOXY)IMINO)- see OKS200

BENZENEACETONITRILE, α-((1-METHYL-2-PHENYLETHYL)AMINO)-(9CI) see AOB300

BENZENEACETONITRILE, 2-NITRO-(9CI) see NFN600

BENZENEACETONITRILE, 3-NITRO-(9CI) see NFN500

BENZENEACETONITRILE, 2,3,4,5,6-PENTACHLORO-α-HYDROXY- see PAX775

BENZENE, (ACETOXYMERCURI)- see ABU500

BENZENE, (ACETOXYMERCURIO)- see ABU500

BENZENE ACID, (2-THIENYLMETHYLENE)HYDRAZIDE (9CI) see TEO100

BENZENE, C₁₄₋₃₀-ALKYL DERIVATIVES see BBM100

BENZENEAMINE, 4-BUTYL-N-((4-METHOXYPHENYL)METHYLENE)- see MEE100

BENZENEAMINE, 2-METHYL-4-((2-METHYLPHENYL)AZO)-, MONOHYDROCHLORIDE see MPY750

BENZENEAMINE, 1,5-METHYL-4-(9H-PYRIDO(3,4-B)INDOL-9-YL)- see AKZ400

BENZENEARSONIC ACID see BBL750

BENZENEARSONIC ACID, p-ACETONYLOXY-, SODIUM SALT see SJK475

BENZENEARSONIC ACID, 4-AMINO-2-HYDROXY- see HJE400

BENZENEARSONIC ACID, 3,4-DIFLUORO-see DKG100

BENZENEARSONIC ACID, 4-(p-DIMETHYLAMINOPHENYLAZO)-, HYDROCHLORIDE see DPO275

BENZENEARSONIC ACID, p-FLUORO- see FGA100

BENZENE AZIMIDE see BDH250

4-BENZENEAZOANILINE see PEI000
BENZENEAZO-2-ANTHROL see PEI750
BENZENEAZOBENZENE see ASL250
BENZENE-1-AZOBENZENE-4-AZO-o-CRESOL see CMP090
BENZENEAZOBENZENEAZO-β-NAPHTHOL see OHI200
BENZENEAZODIMETHYLANILINE see DOT300
BENZENE-1-AZO-2-NAPHTHOL see PEJ500
BENZENEAZO-β-NAPHTHOL see PEJ500
1-BENZENEAZO-2-NAPHTHYLAMINE see FAG130
1-BENZENE-AZO-β-NAPHTHYLAMINE see FAG130
p-BENZENEAZOPHENOL see HJF000
BENZENEAZORESORCINOL see CMP600
BENZENE, BIS(1,1-DIMETHYLETHYL)- see DDV450
BENZENE, 1,1'-(1,2-BIS(1,1-DIMETHYLETHYL)-1,2-ETHANEDIYL)BIS(4-CHLORO-, (R*,S*)- see BJK525
BENZENE, 1,3-BIS(ISOCYANATOMETHYL)-(9CI) see XIJ000
BENZENE, 1,3-BIS(1-ISOCYANATO-1-METHYLETHYL)- see TDX300
BENZENE, 1,4-BIS(1-ISOCYANATO-1-METHYLETHYL)- see TDX400
BENZENE, 1,3-BIS(1-METHYLETHENYL)-see DNL700
BENZENE, 1,4-BIS(1-METHYLETHENYL)-see IMX100
BENZENE, 1,4-BIS(1-METHYLETHYL)-(9CI) see DNN830
BENZENE, 1,3-BIS(1-METHYLETHYL)-2-ISOCYANATO- see DNS100
BENZENE-1,3-BIS(SULFONYL AZIDE) see BBL825
BENZENE, 1,3-BIS(TRIFLUOROMETHYL)-5-ISOCYANO- see BLO390
BENZENEBORONIC ACID see BBM000
BENZENEBORONIC ACID, m-CARBOXY-see CCI590
BENZENEBORONIC ACID, p-CARBOXY-see CCI600
BENZENE, 1-BROMO-2,4-DINITRO- see DVB820
BENZENE, 1-(2-BROMO-1,2-DIPHENYLETHENYL)-4-ETHYL-, (Z)- see BOL315
BENZENE, (2-BROMOETHYL)- see PFB770
BENZENE, 1-BROMO-2-METHOXY-(9CI) see BMT400
BENZENE, 1-BROMO-4-METHOXY-(9CI) see AOY450
BENZENE, 1-BROMO-3-METHYL- see BOG300
BENZENE, 1-BROMO-2-METHYL-(9CI) see BOG260
BENZENE, 1-(BROMOMETHYL)-3-PHENOXY- see PDR610
BENZENE, 1-BROMO-2-NITRO- see NFQ080
BENZENE, 1-BROMO-3-NITRO- see NFQ090
BENZENE, 1-BROMO-4-NITRO- see NFQ100
BENZENE, 1,1',1'',1'''-(1,3-BUTADIENE-1,4-DIYLIDENE)TETRAKIS- see TEA400
BENZENEBUTANOIC ACID, 3-CHLORO-4-CYCLOHEXYL-α-OXO- see CPJ000
BENZENE, BUTOXYPENTACHLORO- see BPM690
BENZENE, 1-tert-BUTYL-3,5-DIMETHYL-2,4,6-TRINITRO- see TML750
BENZENE, 1-tert-BUTYL-2,6-DINITRO-3,4,5-TRIMETHYL- see MRW272
BENZENE, C10-16 ALKYL DERIV. see LGE200
BENZENECARBALDEHYDE see BAY500
BENZENECARBINOL see BDX500
BENZENECARBONAL see BAY500
BENZENECARBONYL CHLORIDE see BDM500

BENZENECARBOPEROXOIC ACID (9CI) see PCM000
BENZENECARBOTHIOAMIDE see BBM250
BENZENECARBOTHIOIC ACID see TFC550
BENZENECARBOXALDEHYDE see BBM500
BENZENECARBOXIMIDAMIDE, 4-AMINO-(9CI) see AID650
BENZENECARBOXIMIDAMIDE HYDROCHLORIDE see BBM750
BENZENECARBOXIMIDAMIDE, 4,4'-METHYLENEBIS- see MJT050
BENZENECARBOXYLIC ACID see BCL750
1,2-BENZENECARBOXYLIC ACID, MONOETHYL ESTER see MRI275
BENZENE CHLORIDE see CEJ125
BENZENE, 1-CHLORO-2-(CHLOROMETHYL)- see CEO200
BENZENE, 4-CHLORO-1-(4-CHLOROPHENOXY)-2-NITRO- see CJD600
BENZENE, 1-CHLORO-2-(2,2-DICHLORO-1-(4-CHLOROPHENYL)ETHENYL)- see DEV900
BENZENE, 1-CHLORO-4-(DICHLOROMETHYL)-(9CI) see TJD650
BENZENE, CHLORODIMETHYL(1-PHENYLETHYL)- see MRG060
BENZENE, 2-CHLORO-1,3-DINITRO-5-(TRIFLUOROMETHYL)- see CGM225
BENZENE, 2-CHLORO-1,5-DINITRO-3-(TRIFLUOROMETHYL)- see CGM200
BENZENE, (((2-CHLOROETHYL)THIO)METHYL) see BFL100
BENZENE, 2-CHLORO-1-FLUORO-4-NITRO- see CHI950
BENZENE, 1-CHLORO-4-METHYL- see TGY075
BENZENE, CHLOROMETHYL-(9CI) see CLK130
BENZENE, 1-(CHLOROMETHYL)-4-METHYL-(9CI) see MHN300
BENZENE, 1-CHLORO-2-METHYL-3-NITRO- see CJG825
BENZENE, 1-(CHLOROMETHYL)-3-NITRO-(9CI) see NFN010
BENZENE, 1-(CHLOROMETHYL)-4-NITRO-(9CI) see NFN400
BENZENE, 1-(CHLOROMETHYL)-3-PHENOXY- see PDR650
BENZENE, CHLOROMETHYL(1-PHENYLETHYL)- see PFR600
BENZENE, 5-CHLORO-1-NITRO-2,4-DIMETHOXY- see CGC100
BENZENE, 1-CHLORO-4-(4-NITROPHENOXY)-2-(PROPYLTHIO)- see CJD400
BENZENE, 1-CHLORO-3-(2-(4-NITROPHENYL)ETHENYL)-, (E)- see CJG600
BENZENE, 1-CHLORO-4-(2-(4-NITROPHENYL)ETHENYL)-, (E)- see CJG610
BENZENE, CHLOROPENTAFLUORO- see PBE100
BENZENE, 1-((2-(4-CHLOROPHENYL)-2-METHYLPROPOXY)METHYL)-3-(3-METHYLPHENOXY)- see MNQ600
BENZENE, 1-CHLORO-3-(PHENYLTHIO)-see CKI612
BENZENE, 1-CHLORO-4-(TRICHLOROMETHYL)-(9CI) see TIR900
BENZENE, 2-CHLORO-1,3,5-TRINITRO- see PIE530
BENZENE, 3-CYCLOHEXEN-1-YL- see PES500
n-BENZENE-n-CYCLOPENTADIENYL IRON(II)PERCHLORATE see BBN000
m-BENZENEDIAMINE see PEY000
o-BENZENEDIAMINE see PEY250
p-BENZENEDIAMINE see PEY500
1,2-BENZENEDIAMINE see PEY250

1,3-BENZENEDIAMINE see PEY000
1,4-BENZENEDIAMINE see PEY500
1,4-BENZENEDIAMINE, N-(2-AMINO-4-IMINO-5-METHOXY-2,5-CYCLOHEXADIEN-1-YLIDENE)-ar-METHYL- see AKN800
1,3-BENZENEDIAMINE, 5-CHLORO- see CEJ200
1,4-BENZENEDIAMINE, 2-CHLORO-5-NITRO- see ALM100
1,2-BENZENEDIAMINE, N^2-((4-CHLOROPHENYL)METHYLENE)-4-NITRO-N^1)-PHENYL- see NFS800
1,4-BENZENEDIAMINE, N,N'-DICYCLOHEXYL- see BBN100
m-BENZENEDIAMINE DIHYDROCHLORIDE see PEY750
p-BENZENEDIAMINE DIHYDROCHLORIDE see PEY650
1,4-BENZENEDIAMINE DIHYDROCHLORIDE see PEY650
1,4-BENZENEDIAMINE, 2,3-DIMETHYL-5-NITRO- see ALM140
1,4-BENZENEDIAMINE, 2,5-DIMETHYL-3-NITRO- see ALM120
1,4-BENZENEDIAMINE, N-(1,4-DIMETHYLPENTYL)-N'-PHENYL- see DTI800
1,2-BENZENEDIAMINE, N^2-((4-(DIPHENYLAMINO)PHENYL)METHYLENE)-4-NITRO-N^1)-PHENYL- see NFY100
1,3-BENZENEDIAMINE, 4-ETHOXY- see EFC300
1,3-BENZENEDIAMINE, 4-((4-ETHOXYPHENYL)AZO)- see EEQ600
1,4-BENZENEDIAMINE, 2-FLUORO-5-NITRO- see FKL500
1,3-BENZENEDIAMINE HYDROCHLORIDE see PEY750
1,4-BENZENEDIAMINE, 2-METHOXY-(9CI) see MEB800
1,3-BENZENEDIAMINE, 4-METHOXY-, DIHYDROCHLORIDE see DBO100
1,4-BENZENEDIAMINE, 2-METHOXY-5-NITRO- see MFB300
1,4-BENZENEDIAMINE, 2-METHOXY-, SULFATE (1:1) see MEB820
1,3-BENZENEDIAMINE, 4-METHOXY, SULFATE (1:1) (9CI) see DBO400
1,3-BENZENEDIAMINE, 4-METHOXY-, SULFATE (1:1), HYDRATE see MEB900
BENZENEDIAMINE, ar-METHYL- see TGL500
1,2-BENZENEDIAMINE, 3-METHYL-(9CI) see TGY800
1,3-BENZENEDIAMINE, 2-METHYL-4-((2-METHYLPHENYL)AZO)- see MIX100
1,3-BENZENEDIAMINE, 4-METHYL-6-((2-METHYLPHENYL)AZO)- see MLP770
1,3-BENZENEDIAMINE, 2-METHYL-5-NITRO- see MMG100
1,3-BENZENEDIAMINE, 4-METHYL-5-NITRO- see NFV540
1,4-BENZENEDIAMINE, 2-METHYL-5-NITRO- see MMG200
1,4-BENZENEDIAMINE, 2-METHYL-6-NITRO- see MMG210
1,3-BENZENEDIAMINE, 2-METHYL-4-(PHENYLAZO)- see MNS100
1,3-BENZENEDIAMINE, 4-METHYL-6-(PHENYLAZO)- see CMM800
1,3-BENZENEDIAMINE, 4-NITRO- see NIM550
1,2-BENZENEDIAMINE, 4-NITRO-N^2)-((4-NITROPHENYL)METHYLENE)-N^1)-PHENYL- see NHR600
1,4-BENZENEDIAMINE, 2-NITRO-N^1)-PHENYL- see NEM350
1,3-BENZENEDIAMINE, 2,4(OR 4,6)-DIETHYL-6(OR 2)-METHYL- see DJP100
1,4-BENZENEDIAMINE, N,N'- mixed PHENYL and TOLYL derivs. see DCJ700

BENZENE, 1,1'-(1,2-DIETHYL-1,3-PROPANEDIYL)BIS(4-METHOXY- see MEE150

BENZENE, 2,4-DIFLUORO-1-NITRO- see DKH900

BENZENE, p-DIHYDROXY- see HIH000

BENZENE, 1-((DIIODOMETHYL)SULFONYL)-4-METHYL- see DNF850

BENZENE-1,3-DIISOCYANATE see BBP000

BENZENE-1,3-DIISOCYANATE see BBP000

BENZENE, 1,3-DIISOCYANATO- see BBP000

BENZENE, 1,4-DIISOCYANATO- see PFA300

BENZENE-, 1,3-DIISOCYANATOMETHYL- see TGM740

BENZENE, 1,3-DIISOCYANATOMETHYL-, POLYMER WITH NIAX E 488 see DNK250

BENZENE, 2,4-DIISOCYANATO-1,3,5-TRIS(1-METHYLETHYL)-, POLYMER WITH 1,3-DIISOCYANATO-2,4-BIS(1-METHYLETHYL)BENZENE see SLI250

BENZENE, m-DIISOPROPENYL- see DNL700

BENZENE, p-DIISOPROPENYL-(6CI,7CI,8CI) see IMX100

BENZENE, p-DIISOPROPYL- see DNN830

1,3-BENZENEDIMETHANOL, 2-HYDROXY-5-METHYL- see HLX950

1,3-BENZENEDIMETHANOL, 4-HYDROXY-α1-(((6-(4-PHENYLBUTOXY)HEXYL)AMINO)METHYL)-, (+−)- see HNI600

BENZENE, m-DIMETHOXY- see REF025

BENZENE, 1,3-DIMETHOXY- see REF025

BENZENE, 1,2-DIMETHOXY-4-ETHYL- see EQF150

BENZENE, 1,2-DIMETHOXY-4-(2-FLUOROETHYL)- see DOF430

BENZENE, 1,2-DIMETHOXY-4-(2-FLUORO-2-PROPENYL)- see DOF440

BENZENE, 1-(DIMETHOXYMETHYL)-2-NITRO- see NEV510

BENZENE, (3,3-DIMETHOXY-1-PROPENYL)- (9CI) see PGA775

BENZENE, 1,3-DIMETHYL-, BENZYLATED see DQL820

BENZENE, 1-(1,1-DIMETHYLETHYL)-2,6-DINITRO-3,4,5-TRIMETHYL- see MRW272

BENZENE, 1,3-DIMETHYL-, HEXACHLORO DERIV. see HCM100

BENZENE, 1,4-DIMETHYL-2-NITRO- see NMS520

p-BENZENEDINITRILE see BBP250

BENZENE, p-DINITROSO- see DVE260

BENZENE, 1,4-DINITROSO-(9CI) see DVE260

m-BENZENEDIOL see REA000

o-BENZENEDIOL see CCP850

p-BENZENEDIOL see HIH000

1,2-BENZENEDIOL see CCP850

1,3-BENZENEDIOL see REA000

1,4-BENZENEDIOL see HIH000

1,2-BENZENEDIOL, 4-(2-AMINO-1-HYDROXYETHYL)-, (+−)-, (R-(R*,R*))-2,3-DIHYDROXYBUTANEDIOATE (1:1) (SALT) see NNE525

(1,2-BENZENEDIOLATO-O)PHENYLMERCURY see PFO750

1,4-BENZENEDIOL, 2,6-BIS(1-METHYLETHYL)-, 4-(METHYLCARBAMATE) see MIA800

1,2-BENZENEDIOL, BORON COMPLEX see DEK500

1,4-BENZENEDIOL, 2-BROMO- see BNL260

1,3-BENZENEDIOL, DIACETATE see REA100

1,3-BENZENEDIOL, DIBENZOATE see BHB100

1,3-BENZENEDIOL, 5-(2-((1,1-DIMETHYLETHYL)AMINO)-1-HYDROXYETHYL)- (9CI) see TAN100

BENZENE-1,3-DIOL, 2,4-DINITROSO- see DVF300

1,2-BENZENEDIOL, 4-(1-HYDROXY-2-(METHYLAMINO)ETHYL)-, HYDROCHLORIDE, (R)- (9CI) see AES500

1,3-BENZENEDIOL, 5-(2-(4-HYDROXYPHENYL)ETHENYL)-, (E)- see TKP200

1,2-BENZENEDIOL, 3-METHOXY- see PPQ550

1,3-BENZENEDIOL, 2-METHYL-(9CI) see MPH400

1,3-BENZENEDIOL, MONOACETATE see RDZ900

1,2-BENZENEDIOL, MONO(METHYLCARBAMATE) (9CI) see HNK900

1,3-BENZENEDIOL, 4-(PHENYLAZO)- see CMP600

1,2-BENZENEDIOL, 3,4,6-TRICHLORO-(9CI) see TIL600

1,4-BENZENEDIOL, 2,3,5-TRIMETHYL-(9CI) see POG300

1,3-BENZENEDIOL, 2,4,6-TRINITRO-, BARIUM SALT, HYDRATE (2:1:1) see BAO900

m-BENZENEDISULFONIC ACID, 4,5-DIHYDROXY- see PPQ100

1,3-BENZENEDISULFONIC ACID, 4,5-DIHYDROXY-(9CI) see PPQ100

BENZENE-1,3-DISULFONYL CHLORIDE FLUORIDE see FLY200

BENZENE, DIVINYL- see DXQ740

BENZENE, (1,2-EPOXYPROPYL)-, (1R,2R)- see ECH600

BENZENEETHANAMINE, 2,5-DIMETHOXY-α,4-DIMETHYL- see SLU600

BENZENEETHANAMINE, N,α-DIMETHYL-, (S)- see PFP850

BENZENEETHANAMINE, 4-ETHOXY-α-METHYL-, HYDROCHLORIDE see EEL050

BENZENEETHANAMINE, β-METHYL- see PGB760

BENZENEETHANAMINE, α-METHYL-N-(1-METHYLETHYL)-, HYDROCHLORIDE see IQH500

BENZENEETHANAMINE, N,N,α-TRIMETHYL- see TMA600

BENZENEETHANAMINIUM, 3-HYDROXY-N,N,N-TRIMETHYL-, HYDROXIDE (9CI) see HNF100

BENZENE, 1,1'-(1,2-ETHANEDIYLBIS(OXY))BIS(2,3,4,5,6-PENTABROMO- see EEB230

BENZENE, 1,1'-(1,2-ETHANEDIYLBIS(OXY))BIS(2,4,6-TRIBROMO- see EEB250

BENZENEETHANOL, β-((4-(2-(DIMETHYLAMINO)ETHOXY)PHENYL)PHENYLMETHYLENE)-α-METHYL-, (Z)- see HOH200

BENZENEETHANOL, α-(2-(DIMETHYLAMINO)-1-METHYLETHYL)-α-PHENYL-, (R*,S*)-(+−)- see PNA300

BENZENEETHANOLE, 2,5-DIAMINO-3-NITRO- see ALN800

BENZENEETHANOL, β-((4-IODOPHENYL)(4-(2-(1-PYRROLIDINYL)ETHOXY)PHENYL)METHYLENE)-α-METHYL-, (β-(Z))- see HLI200

BENZENEETHANOL, α-METHYL-(9CI) see PGA600

BENZENEETHANOL, β-METHYL-(9CI) see HGR600

BENZENEETHANOL, 2-NITRO- see NIM560

BENZENE, 1,1'-(1,2-ETHENEDIYL)BIS-(9CI) see SLR000

BENZENE, 1,1'-(1,2-ETHENEDIYL)BIS-, (E)-(9CI) see SLR100

BENZENE, 1,1'-(1E)-1,2-ETHENEDIYLBIS(3,4-DIMETHOXY- see TDJ200

BENZENE, 1,1'-(1,2-ETHENEDIYL)BIS(4-NITRO-, (E)- see EEE300

BENZENE, ETHENYL-, HOMOPOLYMER (9CI) see SMQ500

BENZENE ETHENYL-, HOMOPOLYMER, BROMINATED see EEE650

BENZENE, 1-ETHENYL-2-METHYL- (9CI) see VQK660

BENZENE, 1-ETHENYL-3-METHYL-(9CI) see VQK670

BENZENE, 1-ETHENYL-4-METHYL-(9CI) see VQK700

BENZENE, 1-(ETHENYLOXY)-4-NITRO- see VPZ333

BENZENE, ETHENYL-, TRIBROMO DERIV., HOMOPOLYMER see EEE650

BENZENE, ETHOXY- see PDM000

BENZENE, (2-(1-ETHOXYETHOXY)ETHYL)- see EES380

BENZENE, 1-(ETHOXYMETHYL)-2-METHOXY- see EMF070

BENZENE, 1-ETHOXY-4-NITROSO- see EEY550

BENZENE, 1-((2-(4-ETHOXYPHENYL)-2-METHYLPROPOXY)METHYL)-3-PHEN OXY- see EQN725

BENZENE, 1,1'-(2-ETHYL-2-((4-ISOCYANATOPHENOXY)METHYL)-1,3-PROPANEDIYLBIS(OXY))BIS(4-ISOCYANATO- see TNJ400

BENZENE, 1-ETHYL-3-METHYL- see EPS600

BENZENE, 1-(ETHYLTHIO)-2-NITRO- see ENQ600

BENZENE, 1-FLUORO-2-METHYL-(9CI) see FLZ100

BENZENEFORMIC ACID see BCL750

BENZENEFORMIC ACID see OOK150

BENZENEGLYOXYLIC ACID see OOK150

BENZENE HEXACHLORIDE see BBP750

α-BENZENEHEXACHLORIDE see BBQ000

β-BENZENEHEXACHLORIDE see BBR000

Δ-BENZENEHEXACHLORIDE see BFW500

γ-BENZENE HEXACHLORIDE see BBQ500

BENZENEHEXACHLORIDE (mixed isomers) see BBQ750

trans-α-BENZENEHEXACHLORIDE see BBR000

BENZENE HEXACHLORIDE-α-isomer see BBQ000

BENZENE HEXACHLORIDE-γ-isomer see BBQ500

BENZENEIODIDE see IEC500

BENZENE, 1-IODO-3-METHYL- see IFG100

BENZENE, 1-ISOCYANATO-2-METHYL- see IKG725

BENZENE, 1-ISOCYANATO-4-METHYL- see IKG735

BENZENE, 1-(1-ISOCYANATO-1-METHYLETHYL)-3-(1-METHYLETHENYL)- see IKG800

BENZENE, 1-ISOCYANATO-4-NITRO- see NIR100

BENZENE, 1-ISOCYANO-2-METHOXY-4-NITRO- see IKH700

BENZENE, 2-ISOCYANO-1-METHOXY-4-NITRO- see IKH720

BENZENE, 1-ISOCYANO-4-METHYL-3-NITRO- see IKH740

BENZENE, 1-ISOCYANO-3-NITRO- see NIR200

BENZENE, 1-ISOCYANO-4-NITRO- see IKH780

BENZENE ISOPROPYL see COE750

BENZENE-1-ISOTHIOCYANATE see ISQ000

BENZENE, 1-ISOTHIOCYANATO-4-(4-NITROPHENOXY)-(9CI) see LIG100

BENZENEMETHANAMIDE, N-NITROSO-N-(PHENYLMETHYL)- (9CI) see NJV600

BIS(PHENYLAMINO)- 1,3,5-TRIAZIN-2-YL)AMINO- see BIU600
BENZENESULFONIC ACID, 2,2'-(1,2-ETHENEDIYL)BIS(5-((4-(ETHYLAMINO)-6-(PHENYLAMINO)-1,3,5-TRIAZIN-2-YL)AMINO-, DISODIUM SALT see DXB500
BENZENESULFONIC ACID, ETHENYL-, HOMOPOLYMER, SODIUMSALT see SFO100
BENZENESULFONIC ACID, 2,2'-(1,2-ETHYLENEDIYL)BIS(5-AMINO)-(9CI) see FCA100
BENZENESULFONIC ACID, HYDRAZIDE see BBS300
BENZENESULFONIC ACID, p-HYDROXY- see HNL600
BENZENESULFONIC ACID, 4-((2-HYDROXY-1-NAPHTHALENYL)AZO)-, MONOSODIUM SALT see CMM220
BENZENESULFONIC ACID, 4-HYDROXY-, ZINC SALT (2:1) see ZIJ300
BENZENESULFONIC ACID, p-HYDROXY-, ZINC SALT (2:1) see ZIJ300
BENZENESULFONIC ACID, 4-((METHOXYCARBONYL)AMINO)-, HYDRAZIDE see MQJ300
BENZENESULFONIC ACID, METHYL ESTER see MHB300
BENZENESULFONIC ACID, METHYL-, SODIUM SALT see SIK460
BENZENESULFONIC ACID, NITRO-, SODIUM SALT see SIU100
BENZENESULFONIC ACID, OXYBIS-, DIHYDRAZIDE (9CI) see OPE000
BENZENESULFONIC ACID, SODIUM SALT see SJH050
BENZENESULFONIC ACID, TETRAPROPYLENE- see TED550
BENZENESULFONIC ACID, TETRAPROPYLENE-, SODIUM SALT see SKF600
BENZENESULFONIC ACID, THIO-, S,S'-(2-(DIMETHYLAMINO)TRIMETHYLENE) ESTER see NCN650
BENZENESULFONIC ACID, 2,4,6-TRINITRO- see TMK800
BENZENESULFONIC HYDRAZIDE see BBS300
BENZENESULFONOHYDRAZIDE see BBS300
BENZENE SULFONYL AZIDE see BBS500
BENZENE, 1,1'-SULFONYLBIS(4-FLUORO-3-NITRO)- see DKH250
BENZENESULFONYL CHLORIDE see BBS750
BENZENESULFONYL CHLORIDE, 4-(ACETYLAMINO)- see AAM600
BENZENESULFONYL CHLORIDE, 2-CHLORO- see CEK370
BENZENESULFONYL CHLORIDE, o-CHLORO-(6CI,7CI,8CI) see CEK370
BENZENESULFONYL CHLORIDE, m-(FLUOROSULFONYL)- see FLY200
2-(BENZENESULFONYL)ETHANOL see BBT000
BENZENESULFONYL FLUORIDE, 4-METHYL-(9CI) see TGO500
BENZENESULFONYL HYDRAZIDE see BBS300
BENZENESULFONYL HYDRAZINE see BBS300
BENZENESULFONYL ISOCYANATE, 4-METHYL- see IKG925
BENZENESULPHONAMIDE see BBR500
BENZENE SULPHONOHYDRAZIDE see BBS300
BENZENE SULPHONYL CHLORIDE (DOT) see BBS750
BENZENESULPHONYL FLUORIDE see BBT250
1,2,4,5-BENZENETETRACARBOXYLIC ACID see PPQ630

1,2,4,5 BENZENETETRACARBOXYLIC 1,2:4,5 DIANHYDRIDE see PPQ635
BENZENE TETRACHLORIDE see TBN220
BENZENE, 1,2,3,5-TETRACHLORO- see TBN745
BENZENE, 1,2,4,5-TETRACHLORO- see TBN220
BENZENETETRAHYDRIDE see CPC579
BENZENE, 1,1'-(THIOBIS(METHYLENE))BIS- see BEY900
BENZENETHIOL, 2-AMINO-, HYDROCHLORIDE see AIF800
BENZENETHIOL, p-tert-BUTYL- see BSO100
BENZENETHIOL, 2,4-DINITRO- see DUR425
BENZENETHIOL, o-METHOXY- see MFQ300
BENZENETHIOL, 2-METHOXY-(9CI) see MFQ300
BENZENETHIOL, 3-METHYL-(9CI) see TGO800
BENZENETHIOL, p-NITRO-, ACETATE see NIW450
1,3,5-BENZENETRIAMINE see THN775
1,3,5-BENZENETRIAMINE, 2-METHYL- see TGO800
1,3,5-BENZENETRIAMINE, 2,4,6-TRINITRO- see TMK900
1,2,4-BENZENETRICARBOXYLIC ACID see TKU700
1,2,4-BENZENETRICARBOXYLIC ACID ANHYDRIDE see TKV000
1,2,4-BENZENETRICARBOXYLIC ACID, CYCLIC 1,2-ANHYDRIDE see TKV000
1,2,4-BENZENETRICARBOXYLIC ACID, TRIS(2-ETHYLHEXYL)ESTER see TJR600
1,2,4-BENZENETRICARBOXYLIC ANHYDRIDE see TKV000
BENZENE, 1,2,3-TRICHLORO- see TIK100
BENZENE, 1,3,5-TRICHLORO- see TIK300
BENZENE, 1,2,4-TRICHLORO-5-(2,6-DICHLOROPHENOXY)- see TIL700
BENZENE, 1,1'-(2,2,2-TRICHLOROETHYLIDENE)BIS(4-METHYL- (9CI) see MIF763
BENZENE, 1,2,4-TRICHLORO-3-METHYL- see TJD600
BENZENE, 1,3,5-TRICHLORO-2-(2,4,5-TRICHLOROPHENOXY)- see HCH495
BENZENE, 1,3,5-TRICHLORO-2,4,6-TRINITRO- see TJE200
BENZENE, (TRIFLUOROETHENYL)-(9CI) see TKH310
BENZENETRIFUROXAN see BBU125
cis-BENZENE TRIIMINE see THQ600
BENZENE, 1,2,4-TRIMETHOXY-5-PROPENYL-, (E)- see IHX400
BENZENE, 1,2,4-TRIMETHOXY-5-PROPENYL-, (Z)- see ARM500
BENZENE, 1,2,4-TRIMETHOXY-5-(1-PROPENYL)-, (Z)-(9CI) see ARM100
BENZENE, 1,2,4-TRIMETHOXY-5-PROPENYL-, trans- see IHX400
BENZENE, 1,3,5-TRIMETHYL- see TLM050
BENZENE, 1,3,5-TRIMETHYL-2,4-DINITRO- see DUW505
BENZENE, 1,3,5-TRIMETHYL-2,4,6-TRINITRO- see TMI800
BENZENE-s-TRIOL see PGR000
1,2,3-BENZENETRIOL see PPQ500
1,2,4-BENZENETRIOL see BBU250
BENZENE-1,3,5-TRIOL see PGR000
1,3,5-BENZENETRIOL see PGR000
1,2,4-BENZENETRIOL, TRIACETATE see HLF600
BENZENE TRIOZONIDE see BBU500
BENZENOL see PDN750
BENZENOSULFOCHLOREK (POLISH) see BBS750
BENZENOSULPHOCHLORIDE see BBS750
BENZENSULFINAN SODNY see SJH025
BENZENYL CHLORIDE see BFL250

BENZENYL FLUORIDE see BDH500
BENZENYL TRICHLORIDE see BFL250
BENZETAMOPHYLLINE HYDROCHLORIDE see THL750
BENZETHACIL see BFC750
BENZETHIDIN see BBU625
BENZETHIDINE see BBU625
BENZETHONIUM CHLORIDE see BEN000
BENZETHONIUM CHLORIDE MONOHYDRATE see BBU750
BENZETIMIDE see BBU800
BENZETONIUM CHLORIDE see BEN000
BENZEX see BBQ750
1,2-BENZFLUORANTHENE see BAW130
2,3-BENZFLUORANTHENE see BAW250
3,4-BENZFLUORANTHENE see BAW250
10,11-BENZFLUORANTHENE see BCJ250
trans-BENZ(a,e)FLUORANTHENE-3,4-DIHYDRODIOL see BBU810
trans-BENZ(a,e)FLUORANTHENE-12,13-DIHYDRODIOL see BBU825
1,2-BENZFLUORANTHRENE see BAW130
BENZ(j)FLUOROANTHRENE see BCJ250
BENZHEXOL see PAL500
BENZHEXOL CHLORIDE see BBV000
BENZHEXOL HYDROCHLORIDE see BBV000
BENZHORMOVARINE see EDP000
BENZHYDRAMINE see BBV500
BENZHYDRAMINE HYDROCHLORIDE see BAU750
BENZHYDRAMINUM see BBV500
BENZHYDRAZIDE see BBV250
BENZHYDRIL see BBV500
BENZHYDROL see HKF300
BENZHYDRYL see BBV500
BENZHYDRYL ALCOHOL see HKF300
o-BENZHYDRYLDIMETHYLAMINOETHANOL see BBV500
o-BENZHYDRYLDIMETHYLAMINOETHANOL-8-CHLOROTHEOPHYLLINATE see DYE600
1-BENZHYDRYL-4-(2-(2-HYDROXYETHOXY)ETHYL)PIPERAZINE see BBV750
2-BENZHYDRYL-3-HYDROXY-N-METHYLPIPERIDINE HYDROCHLORIDE see BBW000
BENZHYDRYL METHYL KETONE see MJH910
N-BENZHYDRYL-N-METHYL PIPERAZINE see EAN600
N-BENZHYDRYL-N'-METHYLPIPERAZINE HYDROCHLORIDE see MAX275
N-BENZHYDRYL-N'-METHYLPIPERAZINE MONOHYDROCHLORIDE see MAX275
(N-BENZHYDRYL)(N'-METHYL)DIETHYLENEDIAMINE see EAN600
2-(BENZHYDRYLOXY)-N,N-DIMETHYLETHYLAMINE see BBV500
2-(BENZHYDRYLOXY)-N,N-DIMETHYLETHYLAMINE with 8-CHLOROTHEOPHYLLINE see DYE600
2-(BENZHYDRYLOXY)-N,N-DIMETHYLETHYLAMINEHYDROCHLORIDE see BAU750
2-(BENZHYDRYLOXYETHYL)GUANIDINE see BBW250
4-(BENZHYDRYLOXY)-1-METHYLPIPERIDINE see LJR000
BENZHYDRYL PHENYL KETONE see TMS100
BENZIDAMINE HYDROCHLORIDE see BBW500
BENZIDAZOL see BBW750
BENZIDIN (CZECH) see BBX000
BENZIDINA (ITALIAN) see BBX000
BENZIDINE see BBX000

3,3'-BENZIDINEDICARBOXYLIC ACID see BFX250

3,3'-BENZIDINE DICARBOXYLIC ACID, DISODIUM SALT see BBX250

γ,γ'-,3,3'-BENZIDINE DIOXYDIBUTYRIC ACID see DEK400

3,3'-BENZIDINE-γ,γ'-DIOXYDIBUTYRIC ACID see DEK400

2,2'-BENZIDINEDISULFONIC ACID see BBX500

BENZIDINE HYDROCHLORIDE see BBX750

BENZIDINE LACQUER YELLOW G see DEU000

BENZIDINE ORANGE see CMS145

BENZIDINE ORANGE 45-2850 see CMS145

BENZIDINE ORANGE 45-2880 see CMS145

BENZIDINE ORANGE TONER see CMS145

BENZIDINE ORANGE WD 265 see CMS145

BENZIDINE SULFATE see BBY000

BENZIDINE-3-SULFURIC ACID see BBY250

BENZIDINE SULPHATE and HYDRAZINE-BENZENE see BBY300

BENZIDINE-3-SULPHURIC ACID see BBY250

BENZIDINE, N,N,N',N'-TETRAMETHYL-see BJF600

BENZIDINE YELLOW see DEU000

BENZIDINE YELLOW 20544 see CMS208

BENZIDINE YELLOW AAOT see CMS210

BENZIDINE YELLOW ABZ 249 see CMS210

BENZIDINE YELLOW G see CMS210

BENZIDINE YELLOW GE see CMS208

BENZIDINE YELLOW GGT see CMS210

BENZIDINE YELLOW GR see CMS208

BENZIDINE YELLOW L see CMS210

BENZIDINE YELLOW LEMON 12221 see CMS208

BENZIDINE YELLOW OT (6CI) see CMS210

BENZIDINE YELLOW OTYA 8055 see CMS210

BENZIDINE YELLOW TONER YT-378 see DEU000

BENZIDIN-3-YL ESTER SULFURIC ACID see BBY500

BENZIDIN-3-YL HYDROGEN SULFATE see BBY500

BENZIES see AOB250

BENZIL see BBY750

BENZILAN see DER000

BENZILATE DU DIETHYLAMINO-ETHANOL CHLORHYDRATE (FRENCH) see BCA000

1-BENZILBIGUANIDE CLORIDRATO (ITALIAN) see BEA850

BENZILE (CLORURO di) (ITALIAN) see BEE375

N-β-(BENZILFENILAMINO)ETILPIPERIDINA CLORIDRATO (ITALIAN) see BEA275

1-(2-(2-BENZILFENOSSI)-1-METILETIL)-PIPERIDINA see MOA600

1-(2-(2-BENZILFENOSSI)-1-METILETIL)-PIPERIDINA FOSFATO (ITALIAN) see PJA130

BENZILIC ACID see BBY990

BENZILIC ACID-β-DIETHYLAMINOETHYL ESTER see DHU900

BENZILIC ACID-β-DIETHYLAMINOETHYL ESTER HYDROCHLORIDE see BCA000

BENZILIC ACID,-3-(2,5-DIMETHYL-1-PYRROLIDINYL)PROPYL ESTER, HYDROCHLORIDE see BCA250

BENZILIC ACID ESTER with 1-ETHYL-3-HYDROXY-1-METHYLPIPERIDINIUM BROMIDE see PJA000

BENZILIC ACID, ester with ETHYL (2-HYDROXYETHYL)DIMETHYLAMMONIUM CHLORIDE see ELF500

BENZILIC ACID ester with 3-HYDROXY-1,1-DIMETHYLPIPERIDINIUM BROMIDE see CBF000

BENZILIC ACID ester with 2-(HYDROXYMETHYL)-1,1-DIMETHYLPIPERIDINIUM METHYL SULFATE see CDG250

BENZILIC ACID-1-METHYL-3-PIPERIDYL ESTER see MON250

BENZILIC ACID, 3-QUINUCLIDINYL ESTER, HYDROCHLORIDE see QWJ000

BENZIL, MONOOXIME see BCA300

α-BENZIL MONOOXIME see BCA300

BENZIL, MONOXIME see BCA300

α-BENZIL MONOXIME see BCA300

BENZIL, β-MONOXIME see BCA300

BENZIL, OXIME see BCA300

(2-BENZILOXYETHYL)DIMETHYLOCTYLAMMONIUM BROMIDE see DSH000

8-α-BENZILOYLOXY-6,10-ETHANO-5-AZONIASPIRO(4.5)DECANE CHLORIDE see KEA300

8-BENZILOYLOXY-6,10-ETHANO-5-AZONIASPIRO(4.5)DECDANE CHLORIDE see BCA375

4-BENZILOYLOXY-1,1,2,2,6-PENTAMETHYLPIPERIDINIUM CHLORIDE (β FORM) see BCB000

4-BENZILOYLOXY-1,1,2,6,6-PENTAMETHYLPIPERIDINIUM CHLORIDE (α FORM) see BCB250

BENZILSAEURE-(N,N-DIMETHYL-2-HYDROXYMETHYL-PIPERIDINIUM)-ESTER-METHYLSULFAT (GERMAN) see CDG250

BENZILSAEURE-DIMETHYL-OCTYL-AMMONIUM-AETHYLESTER BROMIDE (GERMAN) see DSH000

BENZILSAEURE-DIMETHYL-PENTYL-AMMONIUM-AETHYLESTER BROMIDE (GERMAN) see DSH200

BENZILYLOXYETHYLDIMETHYLETHYLAMMONIUM CHLORIDE see ELF500

4-BENZILYLOXY-1,2,2,6-TETRAMETHYLPIPERIDINE METHOCHLORIDE (α FORM) see BCB250

4-BENZILYLOXY-1,2,2,6-TETRAMETHYLPIPERIDINE METHOCHLORIDE (β FORM) see BCB000

1H-BENZIMIDAZOL-5-AMINE, 2-(4-AMINOPHENYL)- see ALV050

1H-BENZIMIDAZOL-1-AMINE, N-((5-NITRO-2-FURANYL)METHYLENE)- see NGE100

BENZIMIDAZOLE see BCB750

o-BENZIMIDAZOLE see BCB750

1H-BENZIMIDAZOLE (9CI) see BCB750

2-BENZIMIDAZOLEACETONITRILE see BCC000

1H-BENZIMIDAZOLE-4-AMINE see BCC050

BENZIMIDAZOLE, 4-AMINO- see BCC050

BENZIMIDAZOLE, 5-AMINO-2-(p-AMINOPHENYL)- see ALV050

BENZIMIDAZOLE, 2-AMINO-5-BENZOYL-see AIH000

BENZIMIDAZOLE, 2-(BIS-(2-CHLOROETHYL)AMINOMETHYL)- see BHQ300

BENZIMIDAZOLE, 2-(o-BROMOPHENYL)-see BNV765

1H-BENZIMIDAZOLE, 2-(2-BROMOPHENYL)- see BNV765

BENZIMIDAZOLE CARBAMATE see BCC100

2-BENZIMIDAZOLECARBAMIC ACID see BCC100

2-BENZIMIDAZOLECARBAMIC ACID, 5-BENZOYL-, METHYL ESTER see MHL000

2-BENZIMIDAZOLECARBAMIC ACID, 5-(p-FLUOROBENZOYL)-, METHYL ESTER see FDA887

BENZIMIDAZOLE-2-CARBAMIC ACID, METHYL ESTER see MHC750

2-BENZIMIDAZOLECARBAMIC ACID, METHYL ESTER, HYDROCHLORIDE see CBN200

2-BENZIMIDAZOLECARBAMIC ACID, 5-(PROPYLSULFONYL)-, METHYL ESTER see PNV900

1H-BENZIMIDAZOLE, 5-CHLORO-6-(2,3-DICHLOROPHENOXY)-2-(METHYLTHIO)-see CFL200

1H-BENZIMIDAZOLE, 2-(2-CHLOROPHENYL)- see CJR550

BENZIMIDAZOLE, 1-(2-DIETHYLAMINOETHYL)-2-(p-ETHOXYBENZYL)-5-NITRO-, HYDROCHLORIDE see EQN750

BENZIMIDAZOLE, 5,6-DIMETHYL- see DQM100

1H-BENZIMIDAZOLE, 1-(2-ETHOXYETHYL)-2-(HEXAHYDRO-4-METHYL-1H-1,4-DIAZEPIN-1-YL)-,(E)-2-BUTENEDIOATE (1:2) see EAL050

BENZIMIDAZOLE,-1-(2-ETHOXYETHYL)-2-(4-METHYLHEXAHYDRO-1H-1,4-DIAZEPIN-1-YL)-, FUMARATE (1:2) see EAL050

BENZIMIDAZOLE, 2-ISOPROPYL- see IOB500

BENZIMIDAZOLE METHYLENE MUSTARD see BCC250

1H-BENZIMIDAZOLE, 2-(1-METHYLETHYL)- see IOB500

BENZIMIDAZOLE MUSTARD see BCC250

BENZIMIDAZOLE, 5-NITRO-, MONONITRATE see NFD600

1H-BENZIMIDAZOLE, 5-NITRO-, MONONITRATE see NFD600

BENZIMIDAZOLE, 4(OR 7)-AMINO- see BCC050

1H-BENZIMIDAZOLE, 2-PHENYL-(9CI) see PEL250

2-BENZIMIDAZOLETHIOL see BCC500

2-BENZIMIDAZOLETHIOL, 1-(2-(4-PYRIDYL)ETHYL)- see MCQ800

2H-BENZIMIDAZOLE-2-THIONE, 1,3-DIHYDRO-4(OR5)-METHYL- see MCO400

2H-BENZIMIDAZOLE-2-THIONE, 1,3-DIHYDRO-1-(2-(4-PYRIDINYL)ETHYL)- see MCQ800

BENZIMIDAZOLE, 1-(TRIBUTYLPLUMBYL)- see TIA800

1H-BENZIMIDAZOLE, 1-(TRIBUTYLPLUMBYL)- see TIA800

BENZIMIDAZOLE, 2-TRIFLUOROMETHYL- see TKB800

1H-BENZIMIDAZOLE, 2-(TRIFLUOROMETHYL)- see TKB800

2-BENZIMIDAZOLINONE see ONI100

1H-BENZIMIDAZOLIUM HEXAKIS-1-DODECYL-3-METHYL-2-PHENYL-(CYANO-C)FERRATE(1−) see TNH750

BENZIMIDAZOLIUM-1-NITROIMIDATE see BCD125

2-BENZIMIDAZOLYLACETONITRILE see BCC000

N-2-(BENZIMIDAZOLYL) CARBAMATE see MHC750

2-BENZIMIDAZOLYLCARBAMIC ACID see BCC100

1H-BENZIMIDAZOL-2-YLCARBAMIC ACID METHYL ESTER see MHC750

1-(2-BENZIMIDAZOLYL)-3-METHYLUREA see BCD325

4-(2-BENZIMIDAZOLYL)THIAZOLE see TEX000

1-BENZIMIDAZOLYLTRI-N-BUTYLLEAD see TIA800

(1-BENZIMIDAZOLYL)TRIBUTYLPLUMBANE see TIA800

BENZIMINAZOLE see BCB750

BENZIN B70 see NAH600
BENZINDAMINE see BCD750
BENZINDAMINE HYDROCHLORIDE see BBW500
1H-BENZ(6,7)INDAZOLO(2,3,4-fgh)NAPHTH(2'',3'':6',7')INDOLO(3',2':5,6)ANT HR A(2,1,9-mna) ACRIDINE-5,8,13,25-TETRAONE see CMU475
BENZ(e)INDENO(1,2-b)INDOLE see BCE000
BENZ(b)INDENO(1,2-d)PYRAN-9(6H)-ONE, 6a,7-DIHYDRO-3,4,6a,10-TETRAHYDROXY- see HAO600
BENZ(b)INDENO(1,2-d)PYRAN-3,6a,9,10(6H)-TETROL, 7,11b-DIHYDRO- see LFT800
BENZ(3,4)INDOLO(1,2-a)BENZIMIDAZOLE see NAK700
BENZ(cd)INDOL-2(1H)-ONE see NAX100
BENZINDOPYRINE HYDROCHLORIDE see BCE250
1-BENZINE see QMJ000
BENZINE BR-1 see GCC200
BENZINE BR-2 see GCC200
BENZINE (LIGHT PETROLEUM DISTILLATE) see PCT250
BENZINE (MOTOR FUEL) see GCC200
BENZINE (OBS.) see BBL250
BENZIN (OBS.) see BBL250
BENZINOFORM see CBY000
BENZINOL see TIO750
BENZIODARON see EID200
BENZIODARONE see EID200
BENZISOQUINOLINEDIONE see NAC600
1H-BENZ(de)ISOQUINOLINE-1,3(2H)-DIONE, 2-(2-(DIMETHYLAMINO)ETHYL)-5-NITRO- see MQX775
1H-BENZ(de)ISOQUINOLINE-1,3(2H)-DIONE, 2-PHENYL- see PEL650
1,2-BENZISOTHIAZOLIN-3-ONE see BCE475
3-BENZISOTHIAZOLINONE-1,1-DIOXIDE see BCE500
1,2-BENZISOTHIAZOLIN-3-ONE 1,1-DIOXIDE AMMONIUM SALT see ANT500
1,2-BENZISOTHIAZOLIN-3-ONE 1,1-DIOXIDE CALCIUM SALT see CAW600
1,2-BENZISOTHIAZOL-3(2H)-ONE see BCE475
1,2-BENZISOTHIAZOL-3(2H)-ONE-1,1-DIOXIDE see BCE500
1,2-BENZISOTHIAZOL-3(2H)-ONE-1,1-DIOXIDE, CALCIUM SALT see CAM750
BENZISOTRIAZOLE see BDH250
1,2-BENZISOXAZOLE-3-METHANESULFONAMIDE see BCE750
BENZITRAMIDE see BCE825
BENZ(J)ACEANTHRYLENE-9,10-OXIDE see EBO990
3,4-BENZOACRIDINE see BAW750
BENZOANTHRACENE see BBC250
BENZO(a)ANTHRACENE see BBC250
1,2-BENZOANTHRACENE see BBC250
BENZO(a)ANTHRACENE-5,6-OXIDE see EBP000
7H-BENZO(de)ANTHRACEN-7-ONE see BBI250
BENZOANTHRONE see BBI250
BENZOARIC ACID see EAI200
BENZOATE see BCL750
17-BENZOATE-3-n-BUTYRATE d'OESTRADIOL (FRENCH) see EDP500
BENZOATE of MONOETHANOLAMINE see MRH300
BENZOATE d'OESTRADIOL (FRENCH) see EDP000
BENZOATE d'OESTRONE (FRENCH) see EDV500
BENZOATE of SODA see SFB000
BENZOATE SODIUM see SFB000
1-BENZOAZO-2-NAPHTHOL see PEJ500
BENZOBARBITAL see BDS300

BENZO(g)(1,3)BENZODIOXOLO(5,6-a)QUINOLIZINIUM,5,6-DIHYDRO-9,10-DIMETHOXY-, CHLORIDE (9CI) see BFN600
BENZO(f)(1)BENZOTHIENO(3,2-b)QUINOLINE see BCF500
BENZO(h)(1)BENZOTHIENO(3,2-b)QUINOLINE see BCF750
BENZO(e)(1)BENZOTHIOPYRANO(4,3-b)INDOLE see BCG000
BENZO(g)(1)BENZOTHIOPYRANO(4,3-b)INDOLE, 3-FLUORO- see FGD100
BENZO BLACK BLUE BH see CMN800
BENZO BLACK BLUE FBH see CMN800
BENZO BLUE see CMO250
BENZO BLUE GS see CMO000
BENZO BLUE RWA see CMO600
BENZO BLUE RWS see CMO600
BENZO BORDEAUX B see CMO872
BENZO BRILLIANT RED 8BS see CMO880
BENZO BROWN D 3GA-CF see CMO810
BENZO BROWN M see CMO800
BENZO BROWN MC see CMO800
BENZOCAINE see EFX000
11H-BENZO(a)CARBAZOLE see BCG250
8,9-BENZO-γ-CARBOLINE see BDB500
BENZOCHINAMIDE see BCL250
BENZO-CHINON (GERMAN) see QQS200
BENZOCHLORPROPAMID see BEG000
BENZO(a)CHRYSENE see PIB750
BENZO(b)CHRYSENE see BCG650
BENZO(c)CHRYSENE see BCG750
BENZO(g)CHRYSENE see BCH000
2,3-BENZOCHRYSENE see BCG500
BENZO(d,e,f)CHRYSENE see BCS750
BENZO(g)CHRYSENE-9,10-OXIDE see BCH100
BENZO(11,12)CHRYSENO(5,6-b)OXIRENE, 1A,13c-DIHYDRO- see BCH100
BENZO(10,11)CHRYSENO(1,2-b)OXIRENE-6-β,7-α-DIHYDRO see BCV750
BENZO(10,11)CHRYSENO(3,4-B)OXIRENE-7,8-DIOL, 7,8,8A,9A-TETRAHYDRO-3-NITRO-, (7-α,8-β, 8A-α,9A-α)- see DMS530
BENZO(10,11)CHRYSENO(3,4-B)OXIRENE-7,8-DIOL, 7,8,8A,9A-TETRAHYDRO-1-NITRO-, (7-α,8-β, 8A-α,9A-α)- see DMS550
BENZO CONGO RED see SGQ500
5,6-BENZOCOUMARIN-3-CARBONIC ACID ETHYL ETHER see OOI200
5,6-BENZOCOUMARIN-3-CARBOXYLIC ACID ETHYL ESTER see OOI200
N-6-(3,4-BENZOCOUMARINYL)ACETAMIDE see BCH250
BENZOCTAMINE see BCH300
BENZOCTAMINE HYDROCHLORIDE see BCH750
1-BENZOCYCLOBUTENYL n-BUTYL KETONE see BCH800
BENZO(def)CYCLOPENTA(hi)CHRYSENE see BCH900
BENZO(de)CYCLOPENT(a)ANTHRACENE see BCI000
1H-BENZO(a)CYCLOPENT(b)ANTHRACENE see BCI250
BENZO(l)CYCLOPENTA(cd)PYRENE see BCI265
BENZO DARK GREEN B see CMO830
BENZO DARK GREEN BA-CF see CMO830
BENZO DEEP BLACK E see AQP000
BENZO DEEP BROWN NZ see CMO820
BENZODIAPIN see MDQ250
2H-1,4-BENZODIAZEPIN-2-ONE, 7-AMINO-1,3-DIHYDRO-5-PHENYL- see AJO280
2H-1,4-BENZODIAZEPIN-2-ONE, 7-CHLORO-5-(4-CHLOROPHENYL)-1,3-DIHYDRO-1-METHYL- see CFK150
2H-1,4-BENZODIAZEPIN-2-ONE, 7-CHLORO-5-(p-CHLOROPHENYL)-1,3-DIHYDRO-1-METHYL- see CFK150

2H-1,4-BENZODIAZEPIN-2-ONE, 7-CHLORO-1,3-DIHYDRO-3-HEMISUCCINYLOXY-5-PHENYL-, DIMETHYLAMINOETHANOL SALT see BCI300
1,4-BENZODIAZINE see QRJ000
1,3-BENZODIAZOLE see BCB750
1H,3H-BENZO(1,2-c:4,5-c')DIFURAN-1,3,5,7-TETRONE see PPQ635
1,2-BENZODIHYDROPYRONE (FCC) see HHR500
1,4-BENZODIOXAN-2-CARBOXYLIC ACID, PROPYL ESTER see PNE800
BENZODIOXANE HYDROCHLORIDE see BCI500
2-((1,4-BENZODIOXAN-2-YLMETHYL)AMINO)ETHANOL HYDROCHLORIDE see HKN500
3-(((1,4-BENZODIOXAN-2-YL)METHYL)AMINO)-N-METHYLPROPIONAMIDE see MHD300
3-(((1,4-BENZODIOXAN-2-YL)METHYL)AMINO)-1-MORPHOLINO-1-PROPANONE see MRR125
1-(1,4-BENZODIOXAN-2-YLMETHYL-1-BENZYL)-HYDRAZINE TARTRATE see BCI750
1-(1,4-BENZODIOXAN-2-YLMETHYL)PIPERIDINEHYDROCHLORIDE see BCI500
4H-1,3,2-BENZODIOXAPHOSPHORIN, 2-PHENOXY-, 2-OXIDE see SAN300
1,4-BENZODIOXIN-2-CARBOXYLIC ACID, 2,3-DIHYDRO-, PROPYL ESTER see PNE800
2,3-BENZODIOXIN-1,4-DIONE see PHY500
1,4-BENZODIOXINE, 2,3-DIHYDRO-, 6-(4-FLUOROPHENYL)- see DKT500
1,3-BENZODIOXOLE-4-CARBOXALDEHYDE see BCI800
3,4-BENZODIOXOLE-5-CARBOXALDEHYDE see PIW250
1,3-BENZODIOXOLE-2,2-DICARBOXYLIC ACID, 5-((2R)-2-((2R)-2-(3-CHLOROPHENYL)-2-HYDROXYETHYL) AMINO)PROPYL)-, DISODIUM SALT see SJN675
1,3-BENZODIOXOLE, 5-(2-(3,4-DIMETHOXYPHENYL)ETHENYL)-, (E)- see DOG800
1,3-BENZODIOXOLE-5-ETHANAMINE, N-ETHYL-α-METHYL- see EMS100
1,3-BENZODIOXOLE-5-ETHANAMINE, N-ETHYL-α-METHYL-, (+−)- see EMS100
1,3-BENZODIOXOLE, 5,5'-1,2-ETHENEDIYLBIS-, (1E)- see DRS700
1,3-BENZODIOXOLE, 5-(1-(2-(2-ETHOXYETHOXY)ETHOXY)ETHOXY)- see MJS550
1,3-BENZODIOXOLE-5-METHANOL, α-(OXIRANYL)-, ACETATE (ester) see ABW250
1,3-BENZODIOXOLE, 5-NITROVINYL- see MJR800
1,3-BENZODIOXOLE-5-(2-PROPEN-1-OL) see BCJ000
(1,3)BENZODIOXOLO(5,6-C)-1,3-DIOXOLO(4,5-I)PHENANTHRIDINIUM, 13-METHYL-, CHLORIDE see SAU100
1,3-BENZODIOXOL-4-OL, ACETYLMETHYLCARBAMATE see BCJ025
1,3-BENZODIOXOL-4-OL, 2,2-DIMETHYL-, ACETYLMETHYLCARBAMATE see BCJ005
1,3-BENZODIOXOL-4-OL, METHYLCARBAMATE see BCJ100
1,3-BENZODIOXOL-4-YL ACETYLMETHYLCARBAMATE see BCJ025
1,3-BENZODIOXOL-4-YL N-ACETYL-N-METHYLCARBAMATE see BCJ025
4-(1,3-BENZODIOXOL-5-YL)-3-BUTEN-2-ONE see MJR250
3-(1,3-BENZODIOXOL-5-YL)-1-(1,1-DIMETHYLETHYL)-2-

PROPENYLTETRADECANOIC ACID ESTER see BCJ090
1,3-BENZODIOXOL-4-YL METHYLCARBAMATE see BCJ100
2-(4-(1,3-BENZODIOXOL-5-YLMETHYL)-1-PIPERAZINYL)PYRIMIDINE see TNR485
5-(2-((3-(1,3-BENZODIOXOL-5-YL)-1-METHYLPROPYL)AMINO)-1-HYDROXYETHYL)-2-HYDROXYBENZAMIDE HCL see MBZ120
1-(5-(1,3-BENZODIOXOL-5-YL)-1-OXO-2,4-PENTADIENYL)PIPERIDINE (E,E)- (9CI) see PIV600
1,3-BENZODIOXOL-5-YL-OXO-2,4-PENTADIENYL-PIPERINE see PIV600
1,3-BENZODITHIOLIUM PERCHLORATE see BCJ125
BENZODODECINIUM BROMIDE see BEO000
BENZODOL see BCJ150
BENZOE-DIAETHYL (GERMAN) see BQA020
BENZOEPIN see EAQ750
α-BENZOEPIN see EAQ810
β-BENZOEPIN see EAQ800
BENZOESAEURE (GERMAN) see BCL750
BENZOESAEURE (NA-SALZ) (GERMAN) see SFB000
BENZOESTROFOL see EDP000
BENZO FAST BLACK G see CMN240
BENZO FAST RED F see CMO870
BENZO FAST RED 8BL see CMO885
BENZO FAST SCARLET 4BS see CMO870
BENZO FAST SCARLET 4BSA-CF see CMO870
BENZOFENAP see DGM730
BENZO(h)FLAVONE see NBI100
7,8-BENZOFLAVONE (7CI) see NBI100
BENZOFLEX 9-88 see DWS800
BENZOFLEX 9-98 see DWS800
BENZOFLEX P-600 see PKE750
BENZOFLEX S-404 see GGU000
BENZOFLEX S-552 see PBC000
BENZOFLEX P 200 see PKE750
BENZOFLEX 9-88 SG see DWS800
BENZO(1)FLUORANTHENE see BCJ250
BENZO(a)FLUORANTHENE see BAW130
BENZO(b)FLUORANTHENE see BAW250
BENZO(e)FLUORANTHENE see BAW250
BENZO(j)FLUORANTHENE see BCJ250
BENZO(k)FLUORANTHENE see BCJ280
1,2-BENZOFLUORANTHENE see BAW130
2,3-BENZOFLUORANTHENE see BAW250
3,4-BENZOFLUORANTHENE see BAW250
7,8-BENZOFLUORANTHENE see BCJ250
8,9-BENZOFLUORANTHENE see BCJ280
2,13-BENZOFLUORANTHENE see BCJ200
7,10-BENZOFLUORANTHENE see BCJ200
BENZO(ghi)FLUORANTHENE see BCJ200
BENZO(mno)FLUORANTHENE see BCJ200
11,12-BENZOFLUORANTHENE see BCJ280
11,12-BENZO(k)FLUORANTHENE see BCJ280
BENZO(j)FLUORANTHENE-4,5-DIOL, 4,5-DIHYDRO- see DKT600
BENZO(j)FLUORANTHENE, 4-FLUORO- see BCP530
BENZO(j)FLUORANTHENE, 10-FLUORO- see FGG900
11H-BENZO(7,8)FLUORANTHENO(1,10B-B)OXIRENE-11,12-DIOL , 12,12A-DIHYDRO-, (11S-(1AR*,11-α-12-β,12Aα-)- see DMP310
11H-BENZO(7,8)FLUORANTHENO(1,10B-B)OXIRENE-11,12-DIOL , 12,12A-DIHYDRO-, (11S-(1AS*,11-α-12-β,12A-β))- see DMP300
2,3-BENZOFLUORANTHRENE see BAW250
BENZO(a)FLUORENE see NAU600
BENZO(b)FLUORENE see BCJ300
1,2-BENZOFLUORENE see NAU600
2,3-BENZOFLUORENE see BCJ800

3,4-BENZOFLUORENE see BCJ900
BENZO(jk)FLUORENE see FDF000
7H-BENZO(c)FLUORENE see BCJ900
11H-BENZO(a)FLUORENE see NAU600
11H-BENZO(b)FLUORENE see BCJ800
BENZOFOLINE see EDP000
BENZOFORM BLACK BCN-CF see AQP000
BENZOFORM BLACK RRA-CF see CMN240
BENZOFURAN see BCK250
BENZO(b)FURAN see BCK250
2,3-BENZOFURAN see BCK250
4-BENZOFURANACETAMIDE, N-METHYL-N-(7-(1-PYRROLIDINYL)-1-OXOSPIRO(4.5)DEC-8-Y L)-, MONOHYDROCHLORIDE, (5R-(5-α-7-α-8-β))- see EAN900
2-BENZOFURANCARBOXYLIC ACID see CNU875
2-BENZO-FURANCETONITRILE see BCK750
BENZOFURAN, 3-(3,5-DIBROMO-4-HYDROXYBENZOYL)-2-MESITYL- see DDP300
BENZOFURAN, 3-(p-(2-(DIETHYLAMINO)ETHOXY)BENZOYL)-2-ETHYL- see EDV700
BENZOFURAN, 3-(p-(2-(DIETHYLAMINO)ETHOXY)BENZOYL)-2-MESITYL-, HYDROCHLORIDE see DHR900
BENZOFURAN, 2,3-DIHYDRO-2,2-DIMETHYL-7-(N-(N-METHYL-N-BUTOXYCARBONYLAMINOTHIO)-N-METHYLCARBAMOYLOXY)- see FPO200
BENZOFURAN, 3-(3,5-DIIODO-4-HYDROXYBENZOYL)-2-ETHYL- see EID200
BENZOFURAN, 3-(3,5-DIIODO-4-HYDROXYBENZOYL)-2-MESITYL- see DNF550
BENZOFURAN, (2-ETHYL-3-(4'-HYDROXYBENZOYL)) see BBJ500
BENZOFURAN, 3-(p-HYDROXYBENZOYL)-2-MESITYL- see HNK700
BENZOFURAN, 4-(p-HYDROXYBENZOYL)-2-MESITYL- see HNK800
BENZOFURAN, 2-MESITYL-3-(p-ANISOYL)- see AOY300
BENZOFURAN, 2-MESITYL-3-(p-METHOXYBENZOYL)- see AOY300
1-(2-BENZOFURANYL)ETHANONE see ACC100
2-BENZOFURANYL METHYL KETONE see ACC100
BENZO(b)FURAN-2-YL METHYL KETONE see ACC100
BENZOFURAZAN, 4-NITRO- see NFD800
BENZOFUR D see PEY500
BENZOFURFURAN see BCK250
BENZOFUR GG see ALT000
BENZOFUR MT see TGL750
6H-BENZOFURO(3,2-c)(1)BENZOPYRAN-6-ONE, 3,9-DIHDYROXY- see COF350
2H-BENZOFURO(3,2-G)-1-BENZOPYRAN-2-ONE, 4-(HYDROXYMETHYL)- see HLV960
22H-BENZOFURO(3A,3-H)(1,5,10)TRIAZACYCLOEICOSINE-3,14,22-TRIONE,4,5,6,7,8,9,10,11,12,13,20A,21,23,24-TETRADECAHYDRO-17,19-ETHENO-, HYDROCHLORIDE see LIP000
BENZOFUROLINE see BEP500
BENZOFUR P see ALT250
p-(5-BENZOFURYLAZO)-N,N-DIMETHYLANILINE see BCK800
p-(7-BENZOFURYLAZO)-N,N-DIMETHYLANILINE see BCL100
BENZO GREEN B see CMO840
BENZO GREEN BG-CF see CMO840
BENZO GREEN CA-CF see CMO840
BENZO GREEN GA-CF see CMO840
BENZO GREY LBGV see CMN230

BENZOGUANAMINE see BCL250
BENZO-GYNOESTRYL see EDP000
BENZO(a)HEPTALEN-9(5H)-ONE, 7-AMINO-6,7-DIHYDRO-1,2,3-TRIMETHOXY-10-(METHYLTHIO)-, (S)- see DBA200
BENZO(a)HEPTALEN-9(5H)-ONE, 6,7-DIHYDRO-1,2,3,10-TETRAMETHOXY-7-(METHYLAMINO)-, (S)- see MIW500
BENZOHEXONIUM see HEG100
BENZOHYDRAZIDE see BBV250
BENZOHYDRAZINE see BBV250
BENZOHYDROL see HKF300
BENZOHYDROQUINONE see HIH000
BENZOHYDROXAMATE see BCL500
BENZOHYDROXAMIC ACID see BCL500
2-(BENZOHYDRYLOXY)-N,N-DIMETHYLETHYLAMINE see BBV500
BENZOIC ACID see BCL750
BENZOIC ACID (DOT) see BCL750
BENZOIC ACID, 2-(ACETYLAMINO)-(9CI) see AAJ150
BENZOIC ACID, 3-(ACETYLAMINO)-2,4,6-TRIIODO-5-((METHYLAMINO)CARBONYL)-, MONOSODIUM SALT see IGC100
BENZOIC ACID, 2-(ACETYLOXY)-, COMPD. WITH dl-LYSINE (1:1) see ARP125
BENZOIC ACID, 2-(ACETYLOXY)-, COMPD. WITH 4-(3-(TRIFLUOROMETHYL)PHENYL)-1-PIPERAZINEETHANOL see TKF723
BENZOIC ACID, 2-(ACETYLSELENO)- see SBU100
BENZOIC ACID, p-ACRYLAMIDO-, 2-(DIETHYLAMINO)ETHYL ESTER see PME100
BENZOIC ACID AMIDE see BBB000
BENZOIC ACID, 4-AMINO- see AIH600
BENZOIC ACID, 3-AMINO-5-CHLORO- see AJE333
BENZOIC ACID, 4-AMINO-, 2-(DIETHYLAMINO)ETHYL ESTER, MONOHYDROCHLORIDE, MIXT. WITH 3,7-DIHYDRO-1,3,7-TRIMETHYL-1H-PURINE-2,6-DIONE see CNG827
BENZOIC ACID, 2-AMINO-3,5-DIIODO-(9CI) see API800
BENZOIC ACID, compd. with 2-AMINOETHANOL (1:1) see MRH300
BENZOIC ACID, 4-AMINO-2-HYDROXY-, COMPD. WITH 8-QUINOLINOL (1:2) see BKJ300
BENZOIC ACID, 5-((4'-((2-AMINO-8-HYDROXY-6-SULFO-1-NAPHTHALENYL)AZO)(1,1'-BIPHENYL)-4-YL)AZO)-2-HYDROXY-, DISODIUM SALT see CMO870
BENZOIC ACID, 5-((4'-((7-AMINO-1-HYDROXY-3-SULFO-2-NAPHTHALENYL)AZO)(1,1'-BIPHENYL)-4-YL)AZO)-2-HYDROXY-, DISODIUM SALT see CMO800
BENZOIC ACID, 4-((AMINOIMINOMETHYL)AMINO)-, 6-(AMINOIMINOMETHYL)-2-NAPHTHALENYL ESTER, DIMETHANESULFONATE see AHN700
BENZOIC ACID, 4-((AMINOIMINOMETHYL)AMINO)-, 2-(METHOXYCARBONYL)PHENYL ESTER see CBT175
BENZOIC ACID, 4-((AMINOIMINOMETHYL)AMINO)-, 2-METHOXY-4-(2-PROPENYL)PHENYL ESTER see MDY300
BENZOIC ACID, 2-(6-AMINO-3-IMINO-3H-XANTHEN-9-YL)-, METHYL ESTER, MONOHYDROCHLORIDE see RGP600
BENZOIC ACID, p-AMINO-, METHYL ESTER see AIN150
BENZOIC ACID, 2-AMINO-, 2-PHENYLETHYL ESTER see APJ500

BENZOIC ACID, 4-(AMINOSULFONYL)-(9CI) see SNL890

BENZOIC ACID, 5-(AMINOSULFONYL)-2,4-DICHLORO- see DGK900

BENZOIC ACID, 5-AMINO-2-(SULFOOXY)-, DISODIUM SALT see AMN275

BENZOIC ACID, AMMONIUM SALT see ANB100

BENZOIC ACID, 4-ARSENOSO- see CCI550

BENZOIC ACID AZIDE see BDL750

BENZOIC ACID, p-BENZAMIDO-, ETHYL ESTER see EGN600

BENZOIC ACID, 2-BENZOYL- see BDL850

BENZOIC ACID, 4-BENZOYL- see BDL860

BENZOIC ACID, o-BENZOYL- see BDL850

BENZOIC ACID, BENZYL ESTER see BCM000

BENZOIC ACID, p-(BIS(2-BROMOETHYL)AMINO)- see BHD300

BENZOIC ACID, 2-(((((4,6-BIS(DIFLUOROMETHOXY)-2-PYRIMIDINYL)AMINO)CARBONYL)AMINO)SULFONYL)-, METHYL ESTER see TCI150

BENZOIC ACID, 3,5-BIS(1,1-DIMETHYLETHYL)-4-(((METHYLAMINO)CARBONYL)OXY)- see CCJ390

BENZOIC ACID, 3-BORONO- (9CI) see CCI590

BENZOIC ACID, 4-BORONO- (9CI) see CCI600

BENZOIC ACID, m-BORONO- (6CI,7CI,8CI) see CCI590

BENZOIC ACID, p-BORONO- (6CI,7CI,8CI) see CCI600

BENZOIC ACID, 2-BROMO- see BMT800

BENZOIC ACID, o-BROMO- see BMT800

BENZOIC ACID, p-BROMO- see BMU100

BENZOIC ACID, 4-BROMO-(9CI) see BMU100

BENZOIC ACID, p-BROMO-, 2-PHENYLHYDRAZIDE see BMU150

BENZOIC ACID-n-BUTYL ESTER see BQK250

BENZOIC ACID, 3-((4-CARBOXY-4-METHYLPENTYL)OXY)-4-METHYL- see GCK400

BENZOIC ACID, 2-((2-CARBOXYPHENYL)AMINO)-4-CHLORO-, DISODIUM SALT see LHZ600

BENZOIC ACID, CHLORIDE see BDM500

BENZOIC ACID, 3-CHLORO- see CEL290

BENZOIC ACID, m-CHLORO- see CEL290

BENZOIC ACID, p-CHLORO- see CEL300

BENZOIC ACID, 4-CHLORO-(9CI) see CEL300

BENZOIC ACID, 5-CHLORO-2-HYDROXY- see CLD825

BENZOIC ACID, 3-CHLORO-2-HYDROXYPROPYL ESTER see CKQ500

BENZOIC ACID, o-CHLORO-5-NITRO-, METHYL ESTER see CJC515

BENZOIC ACID, p-CHLORO-, 2-PHENYLHYDRAZIDE see CEL600

BENZOIC ACID, 2-((7-CHLORO-4-QUINOLINYL)AMINO)-, 2,3-DIHYDROXYPROPYL ESTER see GGQ050

BENZOIC ACID,3-(2-CHLORO-4-(TRIFLUOROMETHYL)PHENOXY)- see CLS025

BENZOIC ACID, 5-(2-CHLORO-4-(TRIFLUOROMETHYL)PHENOXY)-2-NITRO- see CLS075

BENZOIC ACID, 5-(2-CHLORO-4-(TRIFLUOROMETHYL)PHENOXY)-2-NITRO-, ESTER WITH ETHYL GLYCOLATE see FIW100

BENZOIC ACID, 5-(2-CHLORO-4-(TRIFLUOROMETHYL)PHENOXY)-2-NITRO-, 2-ETHOXY-2-OXOETHYL ESTER see FIW100

BENZOIC ACID,5-(2-CHLORO-4-(TRIFLUOROMETHYL)PHENOXY)-2-NITRO-,SODIUM SALT see SFV650

BENZOIC ACID, CINNAMYL ESTER see CMQ750

BENZOIC ACID, 3,5-DIACETAMIDO-2,4,6-TRIIODO-, compd. with 1-DEOXY-1-(METHYLAMINO)-d-GLUCITOL see AOO875

BENZOIC ACID, 3,5-DIAMINO- see DBQ190

BENZOIC ACID, 3,5-DIAMINO-, DIHYDROCHLORIDE see DBQ200

BENZOIC ACID, 5-((4'-((2,6-DIAMINO-3-((8-HYDROXY-3,6-DISULFO-7-((4-SULFO-1-NAPHTHALENYL) AZO)-2-NAPHTHALENYL)AZO)-5-METHYLPHENYL)AZO)(1,1'-BIPHENYL)-4-YL)AZO)-2-HYDROXY-, TETRASODIUM SALT see CMO820

BENZOIC ACID,5-((4'-((2,6-DIAMINO-3-METHYL-5-SULFOPHENYL)AZO)-3,3'-DIMETHYL(1,1'-BIPHENYL)-4-YL)AZO)-2-HYDROXY-, DISODIUM SALT see CMO860

BENZOIC ACID, 5-((4'-((2,6-DIAMINO-3-METHYL-5-((4-SULFOPHENYL)AZO)PHENYL)AZO)(1,1'-BIPHE NYL)-4-YL)AZO)-2-HYDROXY-, DISODIUM SALT see CMO810

BENZOIC ACID, 5-((4'-((2,6-DIAMINO-3-METHYL-5-((4-SULFOPHENYL)AZO)PHENYL)AZO)(1,1'-BIPHE NYL)-4-YL)AZO)-2-HYDROXY-3-METHYL-, DISODIUM SALT see CMO825

BENZOIC ACID, 3,5-DI-tert-BUTYL-4-HYDROXY- see DEC900

BENZOIC ACID, 2,4-DICHLORO- see DER100

BENZOIC ACID, 3,6-DICHLORO-2-HYDROXY-(9CI) see DGK250

BENZOIC ACID, 3,6-DICHLORO-2-METHOXY-, COMPD. WITH N-METHYLMETHANAMINE (1:1), MIXT. WITH N-METHYLMETHANAMINE 2-(4-CHLORO-2-METHYLPHENOXY)PROPANOATE AND N-METHYLMETHANAMINE (2,4-DICHLOROPHENOXY)ACETATE see TKU680

BENZOIC ACID, 5-(2,4-DICHLOROPHENOXY)-2-NITRO-, METHYL ESTER see MJB600

BENZOIC ACID, 5-(2,4-DICHLOROPHENOXY)-2-NITRO-, POTASSIUM SALT see PLD050

BENZOIC ACID, 2,4-DICHLORO-5-SULFAMOYL- see DGK900

BENZOIC ACID, DIESTER with DIETHYLENE GLYCOL see DJE000

BENZOIC ACID DIESTER with DIPROPYLENE GLYCOL see DWS800

BENZOIC ACID DIESTER with POLYETHYLENE GLYCOL 600 see PKE750

BENZOIC ACID DIETHYLAMIDE see BCM250

BENZOIC ACID-N,N-DIETHYLAMIDE see BCM250

BENZOIC ACID, 3,4-DIHYDROXY- see POE200

BENZOIC ACID, 2,4-DIHYDROXY- (9CI) see HOE600

BENZOIC ACID, 3,5-DIHYDROXY-(9CI) see REF200

BENZOIC ACID, 2,5-DIHYDROXY-, MONOSODIUM SALT (9CI) see GCU050

BENZOIC ACID, 3,5-DIIODO-4-HYDROXY- see DNF300

BENZOIC ACID, 3,4-DIMETHOXY- see VHP600

BENZOIC ACID, 2-((((((4,6-DIMETHOXY-2-PYRIMIDINYL)AMINO)CARBONYL)AMINO)SULFONYL)METHYL)-, METHYL ESTER see BAV600

BENZOIC ACID, 3,4-DIMETHYL- see DQM850

BENZOIC ACID, 3,5-DIMETHYL- see MDJ748

BENZOIC ACID, p-(DIMETHYLAMINO)- see DOU650

BENZOIC ACID, 3,5-DIMETHYL-, 1-(1,1-DIMETHYLETHYL)-2-(4-ETHYLBENZOYL)HYDRAZIDE see TAI175

BENZOIC ACID,2-((3-(4-(1,1-DIMETHYLETHYL)PHENYL)-2-METHYLPROPYLIDENE)AMINO)-, METHYLESTER see LFT100

BENZOIC ACID, 2,6-DIMETHYL-4-PROPOXY-, 2-METHYL-2-(1-PYRROLIDINYL)PROPYL ESTER, HYDROCHLORIDE see DTS625

BENZOIC ACID, o-((3-(4,6-DIMETHYL-2-PYRIMIDINYL)UREIDO)SULFONYL)-, METHYL ESTER see SNW550

BENZOIC ACID, 3,4-DINITRO- see DUR550

BENZOIC ACID, 3,5-DINITRO- see DUR600

BENZOIC ACID-n-DIPROPYLENE GLYCOL DIESTER see DWS800

BENZOIC ACID, 2,2'-DISELENOBIS- see SBU150

BENZOIC ACID, 2,2'-DITHIOBIS-(9CI) see BHM300

BENZOIC ACID, 3,3'-DITHIOBIS(6-NITRO- see DUV700

BENZOIC ACID, 2,2'-DITHIODI- see BHM300

BENZOIC ACID ESTRADIOL see EDP000

BENZOIC ACID, 2-((((((4-ETHOXY-6-(METHYLAMINO)-1,3,5-TRIAZIN-2-YL)AMINO)CARBONYL)AMINO)SULFONYL)-, METHYL ESTER see MJW900

BENZOIC ACID, 3-((3-(ETHOXYMETHYL)-5-FLUORO-3,6-DIHYDRO-2,6-DIOXO-1(2H)-PYRIMIDINYL)CARB ONYL)-, see BDN600

BENZOIC ACID, 2-ETHYLHEXANEDIOL DIESTER see HEQ120

BENZOIC ACID-1-ETHYL-4-PIPERIDYL ESTER, HYDROCHLORIDE see EOW000

BENZOIC ACID-2-(1-ETHYL-2-PIPERIDYL)ETHYL ESTER HYDROCHLORIDE see EOY100

BENZOIC ACID, p-FLUORO-, 2-PHENYLHYDRAZIDE see FGH050

BENZOIC ACID, 2-FLUORO-, SODIUM SALT see SHG600

BENZOIC ACID, o-FLUORO-, SODIUM SALT see SHG600

BENZOIC ACID,4-(1-FLUORO-2-(5,6,7,8-TETRAHYDRO-5,5,8,8-TETRAMETHYL-2-NAPHTHALENYL)-1-PROPENYL)-, (Z)- see FLZ080

BENZOIC ACID, 2-FORMYL-(9CI) see FNK010

BENZOIC ACID, 3-FORMYL-2,4-DIHYDROXY-6-METHYL-, 3-HYDROXY-4-(METHOXYCARBONYL)-2,5-DIMETHYLPHENYL ESTER see ARQ600

BENZOIC ACID, (2-FURANYLMETHYLENE)HYDRAZIDE (9CI) see BDM700

BENZOIC ACID, FURFURYLIDENE HYDRAZONE see BDM700

BENZOIC ACID, p-GUANIDINO-, 4-ALLYL-2-METHOXYPHENYL ESTER see MDY300

BENZOIC ACID, p-GUANIDINO-, 4-METHYL-2-OXO-2H-1-BENZOPYRAN-7-YL ESTER see GLC100

BENZOIC ACID, HEXYL ESTER see HFL500

BENZOIC ACID HYDRAZIDE, 3-HYDRAZONE with DAUNORUBICIN, MONOHYDROCHLORIDE see ROZ000

BENZOIC ACID, p-HYDRAZINO- see HHB100

BENZOIC ACID, 4-HYDRAZINO-(9CI) see HHB100

BENZOIC ACID, 4-HYDROXY-3,5-DIMETHOXY- see SPE700
BENZOIC ACID, 2-HYDROXY-, 4-(1,1-DIMETHYLETHYL)PHENYL ESTER see BSH100
BENZOIC ACID, 2-HYDROXY-3,5-DINITRO-(9CI) see HKE600
BENZOIC ACID, 2-((2-HYDROXY-3,6-DISULFO-1-NAPHTHALENYL)AZO)-, TRISODIUM SALT (9CI) see CMG750
BENZOIC ACID, p-HYDROXY-, HEPTYL ESTER see HBO650
BENZOIC ACID, 4-HYDROXY-, HEPTYL ESTER (9CI) see HBO650
BENZOIC ACID, 2-HYDROXY-, HYDRAZIDE see HJL100
BENZOIC ACID(4-(HYDROXYIMINO)-2,5-CYCLOHEXADIEN-1-YLIDENE) HYDRAZIDE see BDD000
BENZOIC ACID, o-(HYDROXYMERCURI)- see HLO450
BENZOIC ACID, 4-HYDROXY-3-METHOXY-, METHYL ESTER see VFF100
BENZOIC ACID, 2-HYDROXY-, 3-METHYL-2-BUTENYL ESTER see PMB600
BENZOIC ACID, 2-HYDROXY-, 4-METHYLPHENYL ESTER (9CI) see THD850
BENZOIC ACID, 2-HYDROXY-, compounded with MORPHOLINE (1:1) see SAI100
BENZOIC ACID, 2-HYDROXY-4-NITRO- see NJF100
BENZOIC ACID, 4-HYDROXY-3-NITRO-, METHYL ESTER see HMY075
BENZOIC ACID, 2-HYDROXY-5-((3-NITROPHENYL)AZO)-, MONOSODIUM SALT see SIT850
BENZOIC ACID-m-HYDROXYPHENYL ESTER see HNH500
BENZOIC ACID, 2-HYDROXY-4-SULFO-, COMPD. WITH1,3,5,7-TETRAAZATRICYCLO(3.3.1.1 ^{3,7})DECANE(1:1) see UVS600
BENZOIC ACID, 2-HYDROXY-5-((4-((4-SULFOPHENYL)AZO)PHENYL)AZO)-, DISODIUM SALT see CMP882
BENZOIC ACID, p-IODO- see IEE025
BENZOIC ACID, p-ISOBUTOXY-, 3-(DIETHYLAMINO)-1,2-DIMETHYLPROPYL ESTER see GBU700
BENZOIC ACID, 4-ISOCYANO-, ETHYL ESTER see ELT100
BENZOIC ACID, p-ISOCYANO-, ETHYL ESTER see ELT100
BENZOIC ACID, ISOPROPYL ESTER see IOD000
BENZOIC ACID, LITHIUM SALT see LGW000
BENZOIC ACID, 3-METHOXY-2-METHYL-, 2-(3,5-DIMETHYLBENZOYL)-2-(1,1-DIMETHYLETHYL)HYDRAZIDE see MEQ100
BENZOIC ACID, 2-METHOXY-, METHYL ESTER see MLH800
BENZOIC ACID, 2-(((((4-METHOXY-6-METHYL-1,3,5-TRIAZIN-2-YL)AMINO)CARBONYL)AMINO)SULFONYL)-, METHYL ESTER see MQR400
BENZOIC ACID, 2-(((((4-METHOXY-6-METHYL-1,3,5-TRIAZIN-2-YL)METHYLAMINO)CARBONYL)AMINO)SULFONYL)-, METHYL ESTER see THT800
BENZOIC ACID, 4-METHOXY-, 2-PHENYLHYDRAZIDE see MEC332
BENZOIC ACID, p-METHOXY-, 2-PHENYLHYDRAZIDE see MEC332
BENZOIC ACID, 2-(6-(METHYLAMINO)-3-(METHYLIMINO)-3H-XANTHEN-9-YL)-, MONOPERCHLORATE see RGW100
BENZOIC ACID, 3-METHYL-2-BUTENYL ESTER see MHU150
BENZOIC ACID, 1-(3-METHYL)BUTYL ESTER see IHP100

BENZOIC ACID, 2-((2-METHYLPHENYL)AMINO)- see MNR300
BENZOIC ACID, 4-METHYLPHENYL ESTER see TGX100
BENZOIC ACID-2-(4-METHYLPIPERIDINO)ETHYL ESTER HYDROCHLORIDE see MOI250
BENZOIC ACID-1-METHYL-4-PIPERIDYL ESTER HYDROCHLORIDE see MON500
BENZOIC ACID, 4-(2-METHYLPROPOXY)-, 3-(DIETHYLAMINO)-1,2-DIMETHYLPROPYL ESTER see GBU700
BENZOIC ACID, 2-(METHYLSELENO)- see MPI200
BENZOIC ACID, o-(METHYLSELENO)- see MPI200
BENZOIC ACID, o-(METHYLSELENO)-, see MPI205
BENZOIC ACID, o-(METHYLTELLURO)- see ADF600
BENZOIC ACID, o-(METHYLTELLURO)-, SODIUM SALT see MPN275
BENZOIC ACID, 2-(2-NAPHTHALENYLAMINO)-(9CI) see NBF600
BENZOIC ACID NITRILE see BCQ250
BENZOIC ACID, 3-NITRO-(9CI) see NFG000
BENZOIC ACID, 2-((2-NITROPHENYL)AMINO)-(9CI) see NIJ450
BENZOIC ACID, p-NITRO-, 2-PHENYLHYDRAZIDE see NFH100
BENZOIC ACID, 4-((1-OXO-2-PROPENYL)AMINO)-, 2-(DIETHYLAMINO)ETHYL ESTER see PME100
BENZOIC ACID, PENTAFLUORO- see PBD275
BENZOIC ACID, PENTYL ESTER see PBV800
BENZOIC ACID, PEROXIDE see BDS000
BENZOIC ACID-1-PHENETHYL-4-PIPERIDYL ESTER HYDROCHLORIDE see PDJ250
BENZOIC ACID, 2-PHENOXY- see PDR603
BENZOIC ACID, 4-PHENOXY- see PDR605
BENZOIC ACID, 2-(2-PHENYLETHYLPIPERIDINO) ETHYL ESTER, HYDROCHLORIDE see PFD500
BENZOIC ACID-3-(2-PHENYLETHYLPIPERIDINO) PROPYL ESTER, HYDROCHLORIDE see PFD750
BENZOIC ACID, o-(PHENYLHYDROXYARSINO)- see PFI600
BENZOIC ACID, PHENYLMETHYL ESTER see BCM000
BENZOIC ACID, o-(PHENYLTHIO)- see PGL270
BENZOIC ACID, 2-(PHENYLTHIO)-(9CI) see PGL270
BENZOIC ACID, 2-(PHOSPHONOOXY)-(9CI) see PHA575
BENZOIC ACID, 1-PROPYL-1-PIPERIDYL ESTER HYDROCHLORIDE see PNT750
BENZOIC ACID, 8-QUINOLYL ESTER see HOE100
BENZOIC ACID, SODIUM SALT see SFB000
BENZOIC ACID p-SULFAMIDE see SNL890
BENZOIC ACID, p-SULFAMOYL- see SNL890
BENZOIC ACID, 2,3,4,5-TETRACHLORO-6-(2,4,5,7-TETRABROMO-6-HYDROXY-3-OXO-3H-XANTHEN-9-YL)- see SKS400
BENZOIC ACID, TETRAESTER with PENTAERYTHRITOL see PBC000
BENZOIC ACID, 2-((TETRAHYDRO-4,6-DIOXO-2-THIOXO-5(2H)-PYRIMIDINYLIDENE)METHYL)- see TCQ550
BENZOIC ACID, p-((E)-2-(5,6,7,8-TETRAHYDRO-5,5,8,8-TETRAMETHYL-2-NAPHTHYL)-1-PROPENYL)-, ETHYL ESTER see AQZ400
BENZOIC ACID, THIO- see TFC550

BENZOIC ACID, p-TOLYL ESTER see TGX100
BENZOIC ACID TRIESTER with GLYCERIN see GGU000
BENZOIC ACID, 3,4,5-TRIHYDROXY , ETHYL ESTER (9CI) see EKM100
BENZOIC ACID, 2,3,5-TRIIODO-, SODIUM SALT see SKL700
BENZOIC ACID, 3,4,5-TRIMETHOXY-, METHYL ESTER see MQF300
BENZOIC ACID, 2,4,6-TRIMETHYL- see IKN300
BENZOIC ACID, VINYL ESTER see VMK000
BENZOIC ALDEHYDE see BAY500
BENZOIC-3-CHLORO-N-ETHOXY-2,6-DIMETHOXYBENZIMIDIC ANHYDRIDE see BCP000
BENZOIC ETHER see EGR000
BENZOIC HYDRAZIDE see BBV250
o-BENZOIC SULPHIMIDE see BCE500
BENZOIC TRICHLORIDE see BFL250
BENZOIMIDAZOLE see BCB750
BENZOIN see BCP250
4,5-BENZO-2,3-1',2'-INDENOINDOLE (FRENCH) see BCE000
BENZOIN METHYL ETHER see MHE000
BENZOINOXIM (CZECH) see BCP500
α-BENZOIN OXIME see BCP500
3,4-BENZOISOQUINOLINE see PCX300
BENZOKETCTRIAZINE see BDH000
BENZOKTAMINA see BCH300
BENZOL (DOT) see BBL250
BENZOLAMIDE see PGJ000
BENZOLE see BBL250
BENZO LEATHER BLACK E see AQP000
BENZOLENE see BBL250
BENZOLIN see NOF500
BENZOLINE see PCT250
BENZOLIN HYDROCHLORIDE see NOF500
BENZOLO (ITALIAN) see BBL250
BENZOMATE see BCP000
BENZOMETAN see BCP650
BENZOMETHAMINE BROMIDE see BCP685
BENZON 00 see HND100
BENZONAL see BDS300
BENZO(B)NAPHTHACENE see PAV000
5H-BENZO(d)NAPHTH(2,1-B)AZEPIN-12-OL, 11-CHLORO-6,6A,7,8,9,13b-HEXAHYDRO-7-METHYL-, HYDROCHLORIDE, (6as,13br)- see BCP690
BENZO(a)NAPHTHO(8,1,2-cde)NAPHTHACENE see BCP750
BENZO(h)NAPHTHO(1,2-f,s-3)QUINOLINE see BCQ250
BENZONE see BRF500
BENZONEROL ABS see CMN300
BENZONITRILE see BCQ250
BENZONITRILE (DOT) see BCQ250
BENZONITRILE, 3-AMINO-(9CI) see AIR125
BENZONITRILE, p-FLUORO- see FGH100
BENZONITRILE, p-ISOPROPYL- see IOD050
BENZONITRILE, 3-METHYL-(9CI) see TGT250
BENZONITRILE, 4-(1-METHYLETHYL)- see IOD050
BENZONITRILE, p-NITRO- see NFI010
BENZONITRILE, 4-NITRO-(9CI) see NFI010
BENZONITRILE, 3-(2-(4-NITROPHENYL)ETHENYL)-, (E)- see COQ328
BENZONITRILE, 3-(2-(4-NITROPHENYL)ETHENYL)-, (Z)- see NIM610
BENZONITRILE, 4-(2-(4-NITROPHENYL)ETHENYL)-, (E)- see NIM605
BENZONITRILE, 4-(2-(4-NITROPHENYL)ETHENYL)-, (Z)- see NIM620
BENZONITRILE, 4-OXIRANYL- see ONC100
BENZO ORANGE PG see CMO860

BENZOPARADIAZINE see QRJ000
BENZOPENICILLIN see BDY669
BENZO(rst)PENTAPHENE see BCQ500
BENZO(rst)PENTAPHENE-5-
CARBOXALDEHYDE see BCQ750
BENZOPERIDOL see FGU000
BENZOPERIDOL see FLK100
BENZOPEROXIDE see BDS000
BENZO(a)PERYLENE see BCQ800
BENZO(b)PERYLENE see BCQ810
1,2-BENZOPERYLENE see BCQ800
2,3-BENZOPERYLENE see BCQ810
1,12-BENZOPERYLENE see BCR000
BENZO(ghi)PERYLENE see BCR000
BENZO(a)PHENALENO(1,9-h,i)ACRIDINE
see BCR250
BENZO(a)PHENALENO(1,9-i,j)ACRIDINE
see BCR500
BENZO(h)PHENALENO(1,9-bc)ACRIDINE
see BCR500
BENZO(c)PHENALENO(1,9-I,j)ACRIDINE
see BCR250
BENZO(1)PHENANTHRENE see TMS000
BENZO(a)PHENANTHRENE see BBC250
BENZO(a)PHENANTHRENE see CML810
BENZO(b)PHENANTHRENE see BBC250
BENZO(c)PHENANTHRENE see BCR750
1,2-BENZOPHENANTHRENE see CML810
2,3-BENZOPHENANTHRENE see BBC250
3,4-BENZOPHENANTHRENE see BCR750
9,10-BENZOPHENANTHRENE see TMS000
BENZO(def)PHENANTHRENE see PON250
(±)-BENZO(c)PHENANTHRENE-3,4-
DIHYDRODIOL see BCS100
(−)-BENZO(c)PHENANTHRENE-3,4-DIOL-
1,2-EPOXIDE-2 see BCS110
(+)-BENZO(c)PHENANTHRENE-3,4-DIOL-
1,2-EPOXIDE-1 see BCS103
(+)-BENZO(c)PHENANTHRENE-3,4-DIOL-
1,2-EPOXIDE-2 see BCS105
(±)-BENZO(c)PHENANTHRENE-3,4-DIOL-
1,2-EPOXIDE-1 see BMK634
BENZO(c)PHENANTHRENE-3-α-4-β-DIOL,
1,2,3,4-TETRAHYDRO-1-β,2-β-EPOXY-, (±)-
see BMK634
BENZO(c)PHENANTHRENE-3,4-DIOL,
1,2,3,4-TETRAHYDRO-1,2-EPOXY-1, (E)-(+)-
(1S,2R,3R,4S)- see BCS105
BENZO(c)PHENANTHRENE-3,4-DIOL,
1,2,3,4-TETRAHYDRO-1,2-EPOXY-, (E)-(−)-
(1R,2S,3S,4R)- see BCS110
BENZO(c)PHENANTHRENE-3,4-DIOL,
1,2,3,4-TETRAHYDRO-1,2-EPOXY-, (Z)-(+)-
(1R,2S,3R,4S)- see BCS103
BENZO(rst)PHENANTHRO(10,1,2-
cde)PENTAPHENE-9,18-DIONE,
DICHLORO- see DFN450
BENZO(c)PHENANTHRENE-8-
CARBOXALDEHYDE see BCS000
BENZO(a)PHENAZINE DI-N-OXIDE see
BCS100
BENZO(a)PHENAZINE, 7,12-DIOXIDE see
BCS100
BENZOPHENONE see BCS250
BENZOPHENONE-2 see BCS325
BENZOPHENONE-3 see MES000
BENZOPHENONE 4 see HLR700
BENZOPHENONE-6 see BJY800
BENZOPHENONE-8 see DMW250
BENZOPHENONE 12 see HND100
BENZOPHENONE, 4-BROMO- see BMU170
BENZOPHENONE-2-CARBOXYLIC ACID
see BDL850
BENZOPHENONE, 2,2'-DIHYDROXY-4,4'-
DIMETHOXY- see BJY800
BENZOPHENONE, 4,4'-DIHYDROXY-, (2,4-
DINITROPHENYL)HYDRAZONE see
BKI300
BENZOPHENONE, 2,2'-DIHYDROXY-4-
METHOXY- see DMW250
BENZOPHENONE, 4-HYDROXY-
(6CI,7CI,8CI) see HJN100

BENZOPHENONE, 2-HYDROXY-4-
(OCTYLOXY)- see HND100
BENZOPHENONE, 4-METHOXY-(6CI,8CI)
see MEC333
p-BENZOPHENONE, METHYL- see MHF750
BENZOPHENONE, OXIME see BCS400
BENZOPHENONE PINACOL see TEA600
12H-BENZO(B)PHENOSELENAZINE see
BCS450
BENZOPHENOXIME see BCS400
BENZOPHOSPHATE see BDJ250
BENZOPINACOL see TEA600
BENZOPINACONE see TEA600
BENZOPIPERILONE (ITALIAN) see BCP650
3,4-BENZOPIRENE (ITALIAN) see BCS750
BENZO(PQR)PICENE see BCS460
BENZOPROPYL see AHI250
BENZOPURPURIN 4B see DXO850
BENZOPURPURINE 4B see DXO850
BENZOPURPURINE 4BKX see DXO850
BENZOPURPURINE 4BX see DXO850
2H-1-BENZOPYRAN, 3-ACETYL- see BCS500
2H-1-BENZOPYRAN, 3-ACETYL-5-
METHOXY- see MEC340
2H-1-BENZOPYRAN-3-CARBONITRILE, 6-
METHOXY-2-METHYL- see MEU275
2H-1-BENZOPYRAN-3-
CARBOXALDEHYDE, 6-CHLORO- see
CEM300
3H-2-BENZOPYRAN-7-CARBOXYLIC ACID,
4,6-DIHYDRO-8-HYDROXY-3,4,5-
TRIMETHYL-6-OXO-, (3R-trans)- see CMS775
2H-1-BENZOPYRAN-3-CARBOXYLIC ACID,
6-METHYL-2-OXO- see MNE275
2H-1-BENZOPYRAN, 6,7-DIMETHOXY-2,2-
DIMETHYL- see AEX850
2H-1-BENZOPYRAN, 2,2-DIMETHYL-7-
METHOXY- see DAM830
(1)BENZOPYRANO(5,4,3-
CDE)(1)BENZOPYRAN-5,10-DIONE,2,3,7,8-
TETRAHYDROXY- see EAI200
2H-1-BENZOPYRAN-7-OL, 3,4-DIHYDRO-3-
(4-HYDROXYPHENYL)-, (3S)- see ECW700
2H-1-BENZOPYRAN-6-OL, 3,4-DIHYDRO-
2,5,7,8-TETRAMETHYL-2-(4,8,12-
TRIMETHYLTRIDECYL)-, ACETATE see
TGJ055
2H-1-BENZOPYRAN-5-OL, 2-METHYL-2-(4-
METHYL-3-PENTENYL)-7-PENTYL- see
PBW400
2H-1-BENZOPYRAN-2-ONE see CNV000
4H-1-BENZOPYRAN-4-ONE, 3-AMINO-2-(3-
AMINO-4-METHOXYPHENYL)- see DBX100
2H-1-BENZOPYRAN-2-ONE, 3-(3-(1,1'-
BIPHENYL)-4-YL-1,2,3,4-TETRAHYDRO-1-
NAPHTHALENYL)-4-HYDROXY- see
BGO100
4H-1-BENZOPYRAN-4-ONE, 5,7-
BIS(ACETYLOXY)-2-(4-
ACETYLOXY)PHENYL- see THM300
4H-1-BENZOPYRAN-4-ONE,5,6-
BIS(ACETYLOXY)-2-(3,4-
BIS(ACETYLOXY)PHENYL)-7-METHOXY-
see PAO160
4H-1-BENZOPYRAN-4-ONE, 7-((6-O-(6-
DEOXY-α-l-MANNOPYRANOSYL)-β-D-
GLUCOPYRANOSYL)OXY)-2,3-DIHYDRO-
5-HYDROXY-2-(3-HYDROXY-4-
METHOXYPHENYL)-, MONOMETHYL
ETHER see MKK600
4H-BENZOPYRAN-4-ONE, 7-((6-o-(6-
DEOXY-α-l-MANNOPYRANOSYL)-β-d-
GLUCOPYRANOSYL)OXY)-2-(3,4-
DIHYDROXYPHENYL)-5-HYDROXY- see
LIU450
4H-1-BENZOPYRAN-4-ONE, 2,3-DIHYDRO-
5,7-DIHYDROXY-2-(4-HYDROXYPHENYL)-
, (S)- see NBP350
4H-1-BENZOPYRAN-4-ONE, 2,3-DIHYDRO-
2-PHENYL-(9CI) see FBW150
4H-1-BENZOPYRAN-4-ONE, 2,3-DIHYDRO-
3,5,7-TRIHYDROXY-2-(3,4,5-

TRIHYDROXYPHENYL)-, (2R-trans)- see
HDW125
4H-1-BENZOPYRAN-4-ONE, 5,7-
DIHYDROXY-8-β-d-GLUCOPYRANOSYL-2-
(4-HYDROXYPHENYL)- see GFC050
4H-1-BENZOPYRAN-4-ONE, 5,7-
DIHYDROXY-3-(4-HYDROXYPHENYL)-6-
METHOXY- see TKP050
4H-1-BENZOPYRAN-4-ONE, 5,8-
DIHYDROXY-7-METHOXY-2-PHENYL- see
ITD020
2H-1-BENZOPYRAN-2-ONE, 6,7-
DIHYDROXY-4-METHYL-(9CI) see MJV800
4H-1-BENZOPYRAN-4-ONE, 5,7-
DIHYDROXY-2-PHENYL- see DMS900
4H-1-BENZOPYRAN-4-ONE, 2-(3,4-
DIHYDROXYPHENYL)-5,7-DIHYDROXY-
see TDD550
2H-1-BENZOPYRAN-2-ONE, 7-
(DIMETHYLAMINO)-4-METHYL-(9CI) see
DPJ800
4H-1-BENZOPYRAN-4-ONE, 8-
((DIMETHYLAMINO)METHYL)-7-
METHOXY-2,3-DIMETHYL-,
HYDROCHLORIDE see DRL600
2H-1-BENZOPYRAN-2-ONE, 4-ETHYL-7-
HYDROXY-3-(4-METHOXYPHENYL)- see
ELH800
4H-1-BENZOPYRAN-4-ONE, 2-(3-(β-d-
GLUCOPYRANOSYLOXY)-4,5-
DIHYDROXYPHENYL)-3,5,7-
TRIHYDROXY- see GFC200
2H-1-BENZOPYRAN-2-ONE, 7-HYDROXY-
see HJY100
2H-1-BENZOPYRAN-2-ONE, 4-HYDROXY-
3-(1-(4-NITROPHENYL)-3-OXOBUTYL)- see
ABF750
2H-1-BENZOPYRAN-2-ONE, 7-METHOXY-
see MEK300
4H-1-BENZOPYRAN-4-ONE, 7-METHOXY-
2,3-DIMETHYL-8-(4-
MORPHOLINYLMETHYL)-,
HYDROCHLORIDE see DSM600
4H-1-BENZOPYRAN-4-ONE, 2-(4-
METHOXY-3-NITROPHENYL)-3-NITRO-
see MFB430
4H-1-BENZOPYRAN-4-ONE, 2-PHENYL- see
PER700
4H-1-BENZOPYRAN-4-ONE, 5,6,7,8-
TETRAHYDRO-6-ETHYL-5-HYDROXY-3-
(HYDROXYMETHYL)-, (5S-trans)- see
ELH650
2H-1-BENZOPYRAN-2-ONE, 4,5-7-
TRIHYDROXY- see TKO300
4H-1-BENZOPYRAN-4-ONE, 3,5,7-
TRIHYDROXY-2-(4-HYDROXYPHENYL)-
see ICE000
4H-1-BENZOPYRAN-4-ONE, 3,5,7-
TRIHYDROXY-2-(3,4,5-
TRIHYDROXYPHENYL)- see HDW150
4H-1-BENZOPYRAN-4-ONE, 3,5,7-
TRIHYDROXY-2-(3,4,5-
TRIHYDROXYPHENYL)-, HEXAACETATE
see MRZ200
4H-1-BENZOPYRAN-4-ONE, 3,5,7-
TRIS(ACETYLOXY)-2-(3,4,5-
TRIS(ACETYLOXY)PHENYL)- see MRZ200
2H,5H-(1)BENZOPYRANO(4,3-B)-1,4-
OXAZIN-9-OL, 3,4,4A,10B-TETRAHYDRO-4-
PROPYL-, (4AR,10BR)- see TDB250
2-(5H-(1)BENZOPYRANO(2,3-b)PYRIDIN-7-
YL)PROPIONIC ACID see PLX400
1-(2H-1-BENZOPYRAN-3-YL)ETHANONE
see BCS500
2H-1-BENZOPYRAN-3-YL METHYL
KETONE see BCS500
1-(2H-1-BENZOPYRAN-8-YLOXY)-3-
ISOPROPYLAMINO-2-PROPANOL see
HLK800
BENZO(A)PYRAZINE see QRJ000
BENZO(a)PYREN-3-AMINE see BCS550
BENZO(a)PYRENE see BCS750

BENZO(e)PYRENE see BCT000
1,2-BENZOPYRENE see BCT000
3,4-BENZOPYRENE see BCS750
4,5-BENZOPYRENE see BCT000
6,7-BENZOPYRENE see BCS750
BENZO(a)PYRENE, 3-AMINO- see BCS550
BENZO(a)PYRENE-6-CARBOXYALDEHYDE see BCT250
BENZO(a)PYRENE-6-CARBOXYALDEHYDE THIOSEMICARBAZONE see BCT500
BENZO(e)PYRENE-4,5-DIHYDRODIOL see DMK400
BENZO(a)PYRENE, 9,10-DIHYDRO-9,10-DIHYDROXY-, (Z)- see BMK550
BENZO(A)PYRENE, 7,8-DIHYDRO-7,8-DIHYDROXY-9,10-EPOXY, syn- see BMK560
BENZO(a)PYRENE-7,8-DIHYDRODIOL see BCT750
(E)-BENZO(a)PYRENE-4,5-DIHYDRODIOL see DLB850
(E)-BENZO(a)PYRENE-7,8-DIHYDRODIOL see DLC800
cis-BENZO(a)PYRENE-9,10-DIHYDRODIOL see BMK550
BENZO(a)PYRENE-7,8-DIHYDRODIOL-9,10-EPOXIDE (anti) see BCU000
syn-BENZO(A)PYRENE-7,8-DIHYDRODIOL-9,10-OXIDE see BMK560
anti-BENZO(a)PYRENE-7,8-DIHYDRODIOL-9,10-OXIDE see BCU000
BENZO(a)PYRENE-7,8-DIHYDRO-7,8-EPOXY see BCV750
BENZO(a)PYRENE, 4,5-DIHYDROXY-4,5-DIHYDRO- see DLB800
BENZO(a)PYRENE-9,10-DIOL, 9,10-DIHYDRO-(Z)- see BMK550
BENZO(a)PYRENE-9,10-DIOL, 9,10-DIHYDRO-3-NITRO-, trans- see NFI230
anti-BENZO(a)PYRENE-DIOLEPOXIDE see DMQ000
anti(±)BENZO(a)PYRENE-DIOL-EPOXIDE see BCU250
BENZO(e)PYRENE, 9,10-DIOL-11,12-EPOXIDE 1 (cis) see DMR150
BENZO(a)PYRENE DIOL EPOXIDE ANTI see BCU250
1,6-BENZO(a)PYRENEDIONE see BCU500
3,6-BENZO(a)PYRENEDIONE see BCU750
BENZO(a)PYRENE-1,6-DIONE see BCU500
BENZO(a)PYRENE-3,6-DIONE see BCU750
6,12-BENZO(a)PYRENEDIONE see BCV000
BENZO(a)PYRENE-6,12-DIONE see BCV000
BENZO(a)PYRENE-4,5-EPOXIDE see BCV500
BENZO(a)PYRENE-7,8-EPOXIDE see BCV750
BENZO(a)PYRENE-4,5-IMINE see BCV125
BENZO(a)PYRENE-6-METHANOL see BCV250
BENZO(A)PYRENE-6-METHANOL, HYDROGEN SULFATE see SOB550
BENZO(a)PYRENE MONOPICRATE see BDV750
BENZO(a)PYRENE, 2-NITRO- see NFI210
BENZO(e)PYRENE, 1-NITRO- see NFI200
BENZO(e)PYRENE, 3-NITRO- see NFI220
BENZO(a)PYRENE-4,5-OXIDE see BCV500
BENZO(a)PYRENE-7,8-OXIDE see BCV750
BENZO(a)PYRENE-9,10-OXIDE see BCW000
BENZO(a)PYRENE-11,12-OXIDE see BCW250
6,12-BENZOPYRENE QUINONE see BCV000
BENZO(a)PYRENE-1,6-QUINONE see BCU500
BENZO(a)PYRENE-3,6-QUINONE see BCU750
BENZO(a)PYRENE-6,12-QUINONE see BCV000

BENZO(A)PYRENE, 7,8,9,10-TETRAHYDRO-trans-7,8-DIHYDROXY-9,10-EPOXY-, syn- see BMK560
BENZO(a)PYRENE, 7,8,9,10-TETRAHYDRO-7-β,8-α-9-α-10-α-TETRAHYDROXY- see TDD750
BENZO(a)PYRENE-7-β,8-α-9-α-10-α-TETRAOL see TDD750
BENZO(a)PYRENE-6-YL ACETATE see ABL750
BENZO(a)PYREN-1-OL see BCW750
BENZO(a)PYREN-2-OL see BCX000
BENZO(a)PYREN-3-OL see BCX250
BENZO(a)PYREN-4-OL see HJN500
BENZO(a)PYREN-5-OL see BCX500
BENZO(a)PYREN-6-OL see BCX750
BENZO(a)PYREN-7-OL see BCY000
BENZO(a)PYREN-9-OL see BCY250
BENZO(a)PYREN-10-OL see BCY500
BENZO(a)PYREN-11-OL see BCY750
BENZO(a)PYREN-12-OL see BCZ000
6H-BENZO(cd)PYREN-6-ONE see BCZ100
BENZO(1,2)PYRENO(4,5-b)OXIRENE-3b,4b-DIHYDRO see BCV500
BENZO(b)PYRIDINE see QMJ000
BENZO(c)PYRIDINE see IRX000
7H-BENZO(a)PYRIDO(3,2-g)CARBAZOLE see BDA000
7H-BENZO(c)PYRIDO(2,3-g)CARBAZOLE see BDA250
7H-BENZO(c)PYRIDO(3,2-g)CARBAZOLE see BDA500
13H-BENZO(a)PYRIDO(3,2-i)CARBAZOLE see BDA750
13H-BENZO(g)PYRIDO(2,3-a)CARBAZOLE see BDB000
13H-BENZO(g)PYRIDO(3,2-a)CARBAZOLE see BDB250
1,2-BENZOPYRIDO(3',2':5,6)CARBAZOLE see BDA000
3,4-BENZOPYRIDO(3',2':5,6)CARBAZOLE see BDA500
5,6-BENZOPYRIDO(2',3':1,2)CARBAZOLE see BDB000
5,6-BENZOPYRIDO(3',2':1,2)CARBAZOLE see BDB250
5,6-BENZOPYRIDO(3',2':3,4)CARBAZOLE see BDA250
7,8-BENZOPYRIDO(2',3':1,2)CARBAZOLE see BDA750
11H-BENZO(g)PYRIDO(4,3-b)INDOLE see BDB500
1,2-BENZOPYRONE see CNV000
BENZOPYRROLE see ICM000
2,3-BENZOPYRROLE see ICM000
1-BENZOPYRYIUM, 3,5,7-TRIHYDROXY-2-(3,4,5-TRIHYDROXYPHENYL)- see HDW200
1-BENZOPYRYLIUM, 2-(3,4-DIHYDROXYPHENYL)-3,5,7-TRIHYDROXY-, CHLORIDE see COI750
BENZOQUIN see AEY000
BENZOQUINAMIDE see BCL250
1,4-BENZOQUINE see QQS200
BENZOQUINOL see HIH000
BENZO(b)QUINOLINE see ADJ500
BENZO(c)QUINOLINE see PCX300
BENZO(f)QUINOLINE see BDB750
BENZO(h)QUINOLINE see BDC000
2,3-BENZOQUINOLINE see ADJ500
3,4-BENZOQUINOLINE see PCX300
5,6-BENZOQUINOLINE see BDB750
7,8-BENZOQUINOLINE see BDC000
α-BENZOQUINOLINE see BDC000
2H-BENZO(A)QUINOLIZIN-2-OL, 2-ETHYL-1,3,4,6,7,11B-HEXAHYDRO-9,10-DIMETHOXY-3-(2-METHYLPROPYL)- see HKS900
2H-BENZO(A)QUINOLIZIN-2-OL, 2-ETHYL-1,3,4,6,7,11B-HEXAHYDRO-3-ISOBUTYL-9,10-DIMETHOXY- see HKS900

2H-BENZO(a)QUINOLIZIN-2-ONE, 1,3,4,6,7,11b-HEXAHYDRO-3-ISOBUTYL-9,10-DIMETHOXY- see TBJ275
o-BENZOQUINONE see BDC250
p-BENZOQUINONE see QQS200
1,2-BENZOQUINONE see BDC250
1,4-BENZOQUINONE see QQS200
BENZOQUINONE (DOT) see BDC250
BENZOQUINONE (DOT) see QQS200
p-BENZOQUINONE AMIDINOHYDRAZONE THIOSEMICARBAZONE see AHI875
BENZOQUINONE AZIRIDINE see BDC750
1,4-BENZOQUINONE-N'-BENZOYLHYDRAZONE OXIME see BDD000
BENZOQUINONE-1,4-BIS(CHLOROIMINE)(1,4-BIS(CHLORIMIDO)-2,5-CYCLOHEXADIENE) see BDD125
1,4-BENZOQUINONE, 2-(N,N-BIS(2-HYDROXYETHYL)AMINO)- see BKB100
p-BENZOQUINONE, 2,5-BIS(PHENYLTHIO)- see BLF600
p-BENZOQUINONE, 2,6-DI-tert-BUTYL- see DDV500
p-BENZOQUINONE DIIMINE see BDD250
1,4-BENZOQUINONE DIIMINE see BDD200
1,4-BENZOQUINONE DIOXINE see DVR200
BENZOQUINONE GUANYLHYDRAZONE THIOSEMICARBAZONE see AHI875
p-BENZOQUINONE, compounded with HYDROQUINONE see QFJ000
p-BENZOQUINONE IMINE see BDD500
p-BENZOQUINONE MONOIMINE see BDD500
1,2-BENZOQUINONE MONOXIME see NLF300
p-BENZOQUINONE OXIME BENZOYLHYDRAZONE see BDD000
p-BENZOQUINONE, 2-PHENYL- see PEL750
p-BENZOQUINONE, TETRAMETHYL- see TDM810
p-BENZOQUINONE, 2,3,5,6-TETRAMETHYL- see TDM810
p-BENZOQUINONE, TETRAMETHYL-, SEMIQUINONE see TDM810
p-BENZOQUINONE, 2,3,5-TRIMETHYL-(8CI) see POG400
BENZOQUINONE, TRIMETHYL-(6CI) see POG400
p-BENZOQUINONIMINE see BDD500
BENZO RED 3B see CMO875
BENZO SCARLET 4BS see CMO870
2,1,3-BENZOSELENADIAZOLE, 5,6-DIMETHYL- see DQO650
2,1,3-BENZOSELENADIAZOLE, 5-METHYL- see MHI300
BENZOSELENAZOLIUM, 3-ETHYL-2-(3-(3-ETHYL-2-BENZOSELENAZOLINYLIDENE)-2-METHYLPROPENYL)-, IODIDE see DJQ300
BENZO SKY BLUE A-CF see CMO500
BENZO SKY BLUE S see CMO500
o-BENZOSULFIMIDE see BCE500
BENZOSULFONAMIDE see BBR500
BENZOSULFONAZOLE see BDE500
BENZOSULPHIMIDE see BCE500
BENZO-2-SULPHIMIDE see BCE500
BENZOTEF see BGY050
BENZO-TEPA see BGY050
BENZOTEPHE see BGY050
3,4-BENZOTETRACENE see BCG500
BENZO(c)TETRAPHENE see BCG500
3,4-BENZOTETRAPHENE see BCG500
BENZO(h)THEBENIDINE see BDD600
1H-2,1,4-BENZOTHIADIAZIN-3-YL-CARBAMIC ACID METHYL ESTER (9CI) see MHI600

BENZO-1,2,3-THIADIAZOLE-1,1-DIOXIDE see BDE000
BENZOTHIAMIDE see BBM250
BENZOTHIAZIDE see BDE250
2H-1,2-BENZOTHIAZINE-3-CARBOXYLIC ACID, 4-HYDROXY-2-METHYL-, ETHYL ESTER, 1,1-DIOXIDE see ELI550
2-BENZOTHIAZOLAMINE, 6-NITRO- see NFI240
BENZOTHIAZOLE see BDE500
3(2H)-BENZOTHIAZOLEACETIC ACID, 4-CHLORO-2-OXO-, ETHYL ESTER see EHL600
BENZOTHIAZOLE, 6-AMINO-2-MERCAPTO- see AIS550
BENZOTHIAZOLE, 2-(p-AMINOPHENYL)-6-METHYL- see MHJ300
BENZOTHIAZOLE, 5-CHLORO-2-METHYL- see CIH100
BENZOTHIAZOLE DISULFIDE see BDE750
BENZOTHIAZOLE, 2-MERCAPTO-6-AMINO- see AIS550
2-BENZOTHIAZOLESULFENAMIDE, N,N-DIISOPROPYL- see DNN900
7-BENZOTHIAZOLESULFONIC ACID see ALX000
7-BENZOTHIAZOLESULFONIC ACID, 2,2'-(1-TRIAZENE-1,3-DIYLDI-4,1-PHENYLENE)BIS(6-METHYL-), DISODIUM SALT see CMP050
2-BENZOTHIAZOLETHIOL see BDF000
2-BENZOTHIAZOLETHIOL, ZINC SALT (2:1) see BHA750
BENZOTHIAZOLE-2-THIONE see BDF000
2(3H)-BENZOTHIAZOLETHIONE see BDF000
2(3H)-BENZOTHIAZOLETHIONE, COMPD. WITH 2-AMINOETHANOL (1:1) see HKO012
3-BENZOTHIAZOLINEACETIC ACID, 4-CHLORO-2-OXO-, ETHYL ESTER see EHL600
2-BENZOTHIAZOLINONE, 3-METHYL-, HYDRAZONE, HYDROCHLORIDE see MHJ275
BENZOTHIAZOLIUM, 3-METHYL-2-METHYLTHIO-, p-TOLUENESULFONATE see MLX250
2(3H)-BENZOTHIAZOLONE, 3-METHYL-, HYDRAZONE, HYDROCHLORIDE see MHJ275
BENZOTHIAZOL-2-ONE, 3-TRICHLOROMETHYLTHIO- see TIT050
2(3H)-BENZOTHIAZOLONE, 3-((TRICHLOROMETHYL)THIO)- see TIT050
N-(2-BENZOTHIAZOLYL)-ACETOACETAMIDE see BDF100
S-2-BENZOTHIAZOLYL-(Z)-2-AMINO-α-(METHOXYIMINO)-4-THIAZOLEETHANETHIOATE see MCK800
S-(2-(2-BENZOTHIAZOLYLAMINO)-2-OXOETHYL) ETHANETHIOATE see ADC400
S-BENZOTHIAZOL-2-YL-(Z)-2-(2-AMINO-1,3-THIAZOL-4-YL)-2-(METHOXYIMINO)THIOACETATE see MCK800
(p-(2-BENZOTHIAZOLYL)BENZYL)PHOSPHONIC ACID DIETHYL ESTER see DIU500
2-BENZOTHIAZOLYL-N,N-DIETHYLTHIOCARBAMYL SULFIDE see BDF250
BENZOTHIAZOLYL DISULFIDE see BDE750
2-BENZOTHIAZOLYL DISULFIDE see BDE750
2-BENZOTHIAZOLYL MERCAPTAN see BDF000
N-(2-BENZOTHIAZOLYL)-N'-METHYL-N'NITROSOUREA see NKR000
N-(2-BENZOTHIAZOLYL)-N'-METHYLUREA see MHM500

2-BENZOTHIAZOLYL MORPHOLINODISULFIDE see BDF750
2-BENZOTHIAZOLYL-N-MORPHOLINOSULFIDE see BDG000
2-(1,3-BENZOTHIAZOL-2-YLOXY)-N-METHYLACETANILIDE see BDG100
2-(2-BENZOTHIAZOLYLOXY)-N-METHYL-N-PHENYLACETAMIDE see BDG100
2-BENZOTHIAZOLYLSULFENYL MORPHOLINE see BDG000
S-2-BENZOTHIAZOLYLTHIOGLYCOLIC ACID see CCH750
4-(2-BENZOTHIAZOLYLTHIO)MORPHOLINE see BDG000
BENZOTHIAZYL-2-CYCLOHEXYLSULFENAMIDE see CPI250
4-BENZOTHIENYL METHYLCARBAMATE see BDG250
BENZO(b)THIEN-4-YL METHYLCARBAMATE see BDG250
BENZOTHIOAMIDE see BBM250
BENZO(b)THIOPHENE-2-CARBOXAMIDE, 5-METHOXY-3-(1-METHYLETHOXY)-N-1H-TETRAZOL-5-YL-, MONOSODIUM SALT see BDG275
1-BENZOTHIOPHENE-4-OL see HJN550
BENZO(B)THIOPHENE-4-OL see HJN550
BENZO(b)THIOPHENE-4-OL METHYLCARBAMATE see BDG250
6-BENZOTHIOPURINE see BDG325
2H-(1)BENZOTHIOPYRANO(4,3,2-CD)INDAZOLE-5-METHANOL, 2-(2-(DIETHYLAMINO)ETHYL)- see DHU910
BENZOTHIOZANE see FNF000
BENZOTHIOZON see FNF000
BENZOTRIAZINEDITHIOPHOSPHORIC ACID DIMETHOXY ESTER see ASH500
BENZOTRIAZINE derivative of an ETHYL DITHIOPHOSPHATE see EKN000
BENZOTRIAZINE derivative of a METHYL DITHIOPHOSPHATE see ASH500
1,2,3-BENZOTRIAZIN-4(1H)-ONE see BDH000
3H-1,2,3-BENZOTRIAZIN-4-ONE see BDH000
BENZOTRIAZOLE see ABX833
1H-BENZOTRIAZOLE see BDH250
2H-BENZOTRIAZOLE see ABX836
1,2,3-BENZOTRIAZOLE see BDH250
1H-BENZOTRIAZOLE, 1-HYDROXY-, AMMONIUM SALT see HJN600
1H-BENZOTRIAZOLE, 6-NITRO- see NFJ000
2-(2H-BENZOTRIAZOL-2-YL)-4-METHYLPHENOL see HML500
BENZOTRICHLORIDE (DOT, MAK) see BFL000
BENZOTRIFLUORIDE see BDH500
BENZOTRIFLUORIDE, 4-CHLORO-3,5-DINITRO- see CGM225
3,5-BENZOTRIFLUORODIAMINE see TKB775
BENZOTRIFUROXAN see BBU125
2,5,8-BENZOTRIOXACYCLOUNDECIN-1,9-DIONE, 3,4,6,7-TETRAHYDRO-(9CI) see DJD700
BENZO(b)TRIPHENYLENE see BDH750
BENZOTRIS(c)FURAZAN-2-OXIDE see BBU125
BENZOTROPINE see BDI000
BENZOTROPINE MESYLATE see TNU000
BENZOTROPINE METHANESULFONATE see TNU000
BENZOURACIL see QEJ800
BENZO VIOLET N see CMP000
2,1,3-BENZOXADIAZOLE, 4-NITRO- see NFD800
BENZOXALE see TMP750
BENZOXAMATE see BCP000
2-BENZOXAXOLOL see BDJ000

2H-1,4-BENZOXAZINE, 4-(DICHLOROACETYL)-3,4-DIHYDRO-3-METHYL-, (+-)- see DEN500
2H-3,1-BENZOXAZINE-2,4(1H)-DIONE see IHN200
2H-1,3-BENZOXAZINE-2,4(3H)-DIONE, 6-BROMO-2-THIO- see BOG255
2H-1,3-BENZOXAZINE-2,4(3H)-DIONE, 6-CHLORO-2-THIO- see CLH800
2H-1,3-BENZOXAZINE-2,4(3H)-DIONE, 6,8-DIBROMO-2-THIO- see DDM820
2H-1,3-BENZOXAZINE-2,4(3H)-DIONE, 6,8-DICHLORO-2-THIO- see DFC300
2H-3,1-BENZOXAZINE-2,4(1H)-DIONE, 1-METHYL- see MKW300
2H-1,3-BENZOXAZINE-2,4(3H)-DIONE, 6-METHYL-2-THIO- see MPS600
2H-3,1-BENZOXAZINE-2,4(1H)-DIONE, 6-NITRO- see NHK600
2H-1,3-BENZOXAZINE-2,4(3H)-DIONE, 8-NITRO-2-THIO- see NFW210
2H-1,3-BENZOXAZINE-2,4(3H)-DIONE, 2-THIO- see TFC570
2H-1,4-BENZOXAZIN-3(4H)-ONE, 2,4-DIHYDROXY- see BDI100
2H-1,4-BENZOXAZIN-3(4H)-ONE, 2,4-DIHYDROXY-7-METHOXY- see DMW300
BENZOXAZOLE see BDI500
3(2H)-BENZOXAZOLECARBOXYLIC ACID, 6-CHLORO-2-THIOXO-, ETHYL ESTER see EHH700
BENZOXAZOLE, 2-CHLORO- see CEM850
BENZOXAZOLE, 5-CHLORO-2-DIMETHYLAMINO- see CGD270
3(2H)-BENZOXAZOLEPROPANESULFONIC ACID, 2-(4-(1,3-DIBUTYLTETRAHYDRO-4,6-DIOXO-2-THIOXO-5(2H)-PYRIMIDINYLIDENE)-2-BUTENYLIDENE)-, SODIUM SALT see EDD500
2-BENZOXAZOLETHIOL see MCK900
BENZOXAZOLE, 2,2'-(2,5-THIOPHENEDIYL)BIS(5-tert-BUTYL- see BHK600
BENZOXAZOLE, 2,2'-(2,5-THIOPHENEDIYL)BIS(5-(1,1-DIMETHYLETHYL)- see BHK600
BENZOXAZOLINONE see BDJ000
2-BENZOXAZOLINONE see BDJ000
2-BENZOXAZOLINONE, 6-METHOXY- see MEC550
2-BENZOXAZOLINONE, 3-((TRICHLOROMETHYL)THIO)- see TIT100
S-((3-BENZOXAZOLINYL-6-CHLORO-2-OXO)METHYL) O,O-DIETHYLPHOSPHORODITHIOATE see BDJ250
BENZOXAZOLIUM, 3-ETHYL-2-METHYL-, IODIDE see MJY525
BENZOXAZOLIUM, 3-PENTYL-2-((3-PENTYL-2(3H)-BENZOXAZOLYLIDENE)-1-PROPENYL)-, IODIDE see DVU700
BENZOXAZOLONE see BDJ000
2(3H)-BENZOXAZOLONE see BDJ000
2(3H)-BENZOXAZOLONE, 6-METHOXY-(9CI) see MEC550
BENZOXINE see BFW250
BENZOXIQUINE see HOE100
BENZ(a)OXIRENO(c)ANTHRACENE see BDJ500
BENZ(h)OXIRENO(5,6)BENZ(1,2-A)ACRIDINE-2,3-DIOL, 1A,2,3,13C-TETRAHYDRO-, (1A-α,2-β,3-α,13C-α)-(+−)- see DCS821
BENZ(A)OXIRENO(5,6)BENZ(1,2-H)ACRIDINE-2,3-DIOL, 1A,2,3,13C-TETRAHYDRO-, (1A-α-2-α-3-β,13Cα)-(+)- see DMN460
BENZ(A)OXIRENO(5,6)BENZ(1,2-H)ACRIDINE-2,3-DIOL, 1A,2,3,13C-

BENZOYLOXYTRIPHENYLSTANNANE see TMV000
BENZOYLPEROXID (GERMAN) see BDS000
BENZOYL PEROXIDE see BDS000
BENZOYL PEROXIDE, WET see BDS250
BENZOYLPEROXYDE (DUTCH) see BDS000
BENZOYLPHENOBARBITAL see BDS300
4-BENZOYLPHENOL see HJN100
p-BENZOYLPHENOL see HJN100
2-BENZOYLPHENYLACETIC ACID see BDS500
o-BENZOYL PHENYLACETIC ACID see BDS500
BENZOYLPHENYLACETONITRILE see OOK200
α-BENZOYLPHENYLACETONITRILE see OOK200
(4-BENZOYL-4-PHENYLBUTYL)TRIETHYLAMMONIUM IODIDE see BDS750
BENZOYLPHENYLCARBINOL see BCP250
1,1-BENZOYL PHENYL DIAZOMETHANE see BDT000
2-(3-BENZOYLPHENYL)-N,N-DIMETHYLACETAMIDE see BDT500
2-(3-BENZOYLPHENYL)-N,N-DIMETHYLPROPIONAMIDE see BDT750
(4-BENZOYL-o-PHENYLENEDIAMINE)DICHLOROPLATINUM(II) see BDU000
1-BENZOYL-1-PHENYLETHENE see CDH000
(5-BENZOYL-5-PHENYLPENTYL)TRIETHYLAMMONIUM IODIDE see BDU250
2-(3-BENZOYLPHENYL)PROPIONATE SODIUM see KGK100
2-(3-BENZOYLPHENYL)PROPIONIC ACID see BDU500
2-(m-BENZOYLPHENYL)PROPIONIC ACID see BDU500
2-BENZOYL-2-PROPANOL see HMQ100
BENZOYLPROP ETHYL see EGS000
1-(2-BENZOYLPROPYL)-2-(2-ETHOXY-2-PHENYLETHYL)PIPERAZINE DIHYDROCHLORIDE see ECU550
BENZOYLPSEUDOTROPINE HYDROCHLORIDE see TNS200
2-BENZOYLPYRIDINE see PGE760
3-BENZOYLPYRIDINE see PGE765
4-BENZOYLPYRIDINE see PGE768
2-BENZOYLPYRROLE see PGF900
β-BENZOYLSTYRENE see CDH000
N¹-BENZOYLSULFANILAMIDE see SNH800
o-BENZOYL SULFIMIDE see BCE500
o-BENZOYL SULPHIMIDE see BCE500
BENZOYL SUPEROXIDE see BDS000
BENZOYL THIOL see TFC550
N-BENZOYL TMCA METHYL ETHER see BDV250
α-BENZOYLTRIETHYLAMINE HYDROCHLORIDE see DIP600
α-BENZOYLTRIETHYLAMMONIUM CHLORIDE see DIP600
N-BENZOYL TRIMETHYL COLCHICINIC ACID METHYL ETHER see BDV250
O-BENZOYLTROPINE HYDROCHLORIDE see TNS200
p-((N-BENZOYL-l-TYROSIN)AMIDO)BENZOIC ACID see CML870
BENZOYLTYROSYL-p-AMINOBENZOIC ACID see CML870
N-BENZOYL-l-TYROSYL-p-AMINOBENZOIC ACID see CML870
BENZOYLUMINAL see BDS300
BENZPERIDOL see FGU000
BENZPERIDOL see FLK100
1,12-BENZPERYLENE see BCR000
BENZ(a)PHENANTHRENE see CML810
1,2-BENZPHENANTHRENE see CML810
2,3-BENZPHENANTHRENE see BBC250

3,4-BENZPHENANTHRENE see BCR750
9,10-BENZPHENANTHRENE see TMS000
BENZPHETAMINE HYDROCHLORIDE see BDV500
BENZPHOS see BDJ250
BENZPIPERILONE see BCP650
BENZPIPERYLON see BCP650
3,4-BENZPYREN (GERMAN) see BCS750
BENZ(a)PYRENE see BCS750
1,2-BENZPYRENE see BCT000
3,4-BENZ(a)PYRENE see BCS750
3,4-BENZPYRENE-5-ALDEHYDE see BCT250
3,4-BENZPYRENE-5-ALDEHYDE THIOSEMICARBAZONE see BCT500
BENZ(a)PYRENE 4,5-OXIDE see BCV500
1:2-BENZPYRENE PICRATE see BDV750
BENZQUINAMIDE see BCL250
BENZQUINAMIDE HYDROCHLORIDE see BDW000
BENZQUINAMIDU (POLISH) see BCL250
BENZSULFOHYDROXAMIC ACID see BDW100
N-(2-BENZTHIAZOLYL)-N'-METHYLHARNSTOFF (GERMAN) see MHM500
BENZTHIAZURON see MHM500
BENZTROPINE see BDI000
BENZTROPINE MESYLATE see TNU000
BENZTROPINE METHANESULFONATE see TNU000
BENZULFIDE see DNO800
BENZVALENE see BDW650
BENZYDAMINE see BCD750
BENZYDAMINE HYDROCHLORIDE see BBW500
BENZYDROFLUMETHIAZIDE see BEQ625
BENZYDYNA (POLISH) see BBX000
BENZYENEMETHAMAMINE HYDROCHLORIDE see BDY000
BENZYHYDRYLCYANIDE see DVX200
BENZYLACETALDEHYDE see HHP000
(3-(N-BENZYLACETAMIDO)-2,4,6-TRIIODOPHENYL)ACETIC ACID see BDW750
BENZYL ACETATE see BDX000
BENZYLACETONE see PDF800
N-BENZYL-N-(α-ACETOXYBENZYL)NITROSAMINE see ABL875
10-BENZYLACRIDINIUM ORANGE see BDX033
BENZYLADENINE see BDX090
6-BENZYLADENINE see BDX090
9-BENZYLADENINE see BDX100
N-BENZYLADENINE see BDX090
N⁶-BENZYLADENINE see BDX090
N⁹-BENZYLADENINE see BDX100
BENZYL ALCOHOL see BDX500
BENZYL ALCOHOL-α-(1-AMINOETHYL)-2,4-DIMETHOXY HYDROCHLORIDE see AJV850
BENZYL ALCOHOL, α-(AMINOMETHYL)-3,4-DIHYDROXY-, (+−)-, TARTRATE (1:1) (SALT) see NNE525
BENZYL ALCOHOL BENZOIC ESTER see BCM000
BENZYL ALCOHOL, α-((t-BUTYLAMINO)METHYL)-3,5-DIHYDROXY- see TAN100
BENZYL ALCOHOL, o-CHLORO-α-((ISOPROPYLAMINO)METHYL)-, HYDROCHLORIDE see IMX150
BENZYL ALCOHOL CINNAMIC ESTER see BEG750
BENZYL ALCOHOL, CYCLOHEXYL- see CPM300
BENZYL ALCOHOL, 3,4-DICHLORO-α-(TRICHLOROMETHYL)-, ACETATE (8CI) see TIL800
BENZYL ALCOHOL, 2,4-DIMETHYL-, ACETATE see DQP500

BENZYL ALCOHOL ETHER with ISOEUGENOL see BES750
BENZYL ALCOHOL FORMATE see BEP250
BENZYL ALCOHOL, HEXAHYDRO- see HDH200
BENZYL ALCOHOL, o-HYDROXY- see HMK100
BENZYL ALCOHOL, p-HYDROXY-α-METHYL-, 4-ACETATE see ABU600
BENZYL ALCOHOL, α-(1-(ISOPROPYLAMINO)ETHYL)-2,5-DIMETHOXY- see IPY500
BENZYL ALCOHOL, α-METHYL-, ACETATE see SMP600
BENZYL ALCOHOL, p-METHYL-, ACETATE (6CI,7CI,8CI) see XQJ700
BENZYL ALCOHOL, p-NITRO-, ACETATE (ester) see NFM100
BENZYL ALCOHOL, m-PHENOXY- see HMB625
BENZYL ALCOHOL, SULFATE (ESTER), SODIUM SALT see SFB150
BENZYL ALCOHOL, α-(TRICHLOROMETHYL)- see TIR800
BENZYLAMIDE see BEG000
BENZYLAMINE see BDX750
BENZYLAMINE, N,N-BIS(2-CHLOROETHYL)-p-FLUORO-, HYDROCHLORIDE see FHP100
BENZYLAMINE, N,N-BIS(2-CHLOROETHYL)-p-METHYL-, HYDROCHLORIDE see EAK100
BENZYLAMINE, N-(2-CHLOROETHYL)-2-METHOXY-5-NITRO- see CGQ400
BENZYLAMINE, 3,4-DIMETHOXY- see VIK050
BENZYLAMINE, N,N-DIMETHYL-, HEXAFLUOROARSENATE (1-) see BEL550
BENZYLAMINE HYDROCHLORIDE see BDY000
BENZYLAMINE, o-METHYL- see MHM510
2-(BENZYLAMINO)-6'-CHLORO-o-ACETOTOLUIDIDE HYDROCHLORIDE see BDY250
2-(2-(BENZYLAMINO)ETHYL)-2-METHYL-1,3-BENZODIOXOLE HYDROCHLORIDE see BDY500
4-(BENZYLAMINO)-2-METHYL-7H-PYRROLO(2,3-d)PYRIMIDINE see RLZ000
BENZYL-6-AMINOPENICILLINIC ACID see BDY669
4-(BENZYLAMINO)PHENOL see BDY750
BENZYLAMINOPURINE see BDX090
6-BENZYLAMINOPURINE see AKY250
6-(BENZYLAMINO)PURINE see BDX090
9-BENZYLAMINOPURINE see BDX100
6-(N-BENZYLAMINO)PURINE see BDX090
9-BENZYL-6-AMINOPURINE see BDX100
N⁶-(BENZYLAMINO)PURINE see BDX090
2-BENZYLAMINOPYRIDINE HYDROCHLORIDE see BEA000
6-BENZYLAMINO-9-TETRAHYDROPYRAN-2-YL-9H-PURINE see BEA100
BENZYLAMMONIUM CHLORIDE see BDY000
BENZYL AMMONIUM TETRACHLOROIODATE see BEA250
1-(2-(N-BENZYLANILINO)ETHYL)PIPERIDINE HYDROCHLORIDE see BEA275
2-(N-BENZYLANILINOMETHYL)-2-IMIDAZOLINE see PDC000
BENZYL ANTISEROTONIN see BEM750
BENZYL AZIDE see BEA325
BENZYLBARBITAL see BEA500
BENZYL BENZENECARBOXYLATE see BCM000
2-BENZYLBENZIMIDAZOLE see BEA825
2-BENZYLBENZIMINAZOLE see BEA825
BENZYL BENZOATE (FCC) see BCM000

BENZYL DIMETHYLCARBINYL
BUTYRATE see BEL850
BENZYL DIMETHYLCARBINYL n-
BUTYRATE see BEL850
BENZYLDIMETHYLCETYLAMMONIUM
CHLORIDE see BEL900
BENZYLDIMETHYLDODECYLAMMONIU
M BROMIDE see BEO000
BENZYLDIMETHYLDODECYLAMMONIU
M CHLORIDE see BEM000
BENZYLDIMETHYLEICOSANYLAMMONI
UM CHLORIDE see BEM250
1-BENZYL-2,3-DIMETHYL-GUANIDINE
SULFATE (1:1/2) see BFW250
BENZYLDIMETHYLHEXADECYLAMMON
IUM CHLORIDE see BEL900
N-BENZYL-N',N'-DIMETHYL-N-1-
NAPHTHYLETHYLENEDIAMINE see
BEM325
N-BENZYL-N',N'-DIMETHYL-N-2-
NAPHTHYLETHYLENEDIAMINE see
BEM330
BENZYLDIMETHYLOCTADECYL
AMMONIUM 3-
NITROBENZENESULFONATE see BEM400
BENZYLDIMETHYLOCTYLAMMONIUM
CHLORIDE see OEW000
BENZYLDIMETHYLOLEYLAMMONIUM
CHLORIDE see OHK200
(+)-N-BENZYL-N,α-
DIMETHYLPHENETHYLAMINE
HYDROCHLORIDE see BDV500
N-BENZYL-N',N'-DIMETHYL-N-
PHENYLETHYLENEDIAMINE see BEM500
S-BENZYL 1,2-
DIMETHYLPROPYL(ETHYL)THIOCARBAM
ATE (IUPAC) see PFR120
N-BENZYL-N',N'-DIMETHYL-N-2-
PYRIDYLETHYLENE DIAMINE see TMP750
N-BENZYL-N',N'-DIMETHYL-N-2-
PYRIDYL-ETHYLENEDIAMINE
HYDROCHLORIDE see POO750
1-BENZYL-2,5-DIMETHYL SEROTONIN
HYDROCHLORIDE see BEM750
BENZYLDIMETHYLSTEARYLAMMONIUM
CHLORIDE see DTC600
BENZYLDIMETHYL-p-(1,1,3,3-
TETRAMETHYLBUTYL)PHENOXYETHOX
Y-ETHYLAMMONIUM CHLORIDE see
BEN000
BENZYLDIMETHYL(2-(2-(p-(1,1,3,3-
TETRAMETHYLBUTYL)PHENOXY)ETHOX
Y)ETHYL) AMMONIUM CHLORIDE see
BEN000
2-BENZYLDIOXOLAN see BEN250
(+)-1-BENZYL-4-(2,6-DIOXO-3-PHENYL-3-
PIPERIDYL)PIPERIDINE
HYDROCHLORIDE see DBE000
dl-1-BENZYL-4-(2,6-DIOXO-3-PHENYL-3-
PIPERIDYL)PIPERIDINE
HYDROCHLORIDE see BBU800
4-(BENZYL-(2-((2,5-DIPHENYLOXAZOLE-
4-
CARBONYL)AMINO)ETHYL)CARBAMOYL)
-2-DECANOYLAMINOBUTYRIC ACID see
BEN260
1-BENZYL-5-(3-
(DIPROPYLAMINO)PROPOXY)-3-
METHYLPYRAZOLE see BEN750
1-BENZYL DIPROPYL KETONE see
BEN800
S-BENZYL DIPROPYLTHIOCARBAMATE
(IUPAC) see PFR130
BENZYL DIPROPYLTHIOLCARBAMATE
see PFR130
S-BENZYL DIPROPYL THIOLCARBAMATE
see PFR130
BENZYL DODECANOATE see BEU750
BENZYLDODECYLBIS(2-
HYDROXYETHYL)AMMONIUM
CHLORIDE (6CI,7CI) see BDJ600

BENZYLDODECYLDIMETHYL
AMMONIUM BROMIDE see BEO000
BENZYLE (CHLORURE de) (FRENCH) see
BEE375
BENZYLENE CHLORIDE see BAY300
BENZYL ETHANOATE see BDX000
BENZYL ETHER see BEO250
6-BENZYL-1-(ETHOXYMETHYL)-5-
ISOPROPYLURACIL see EAN525
α-BENZYLETHYLAMINE see PGB760
5-BENZYL-5-ETHYLBARBITURIC ACID see
BEA500
8-BENZYL-7-(2-(ETHYL(2-
HYDROXYETHYL)AMINO)ETHYL)THEOP
HYLLINE, HYDROCHLORIDE see THL750
1-BENZYL-4-ETHYNYL-3-(1-(3-
INDOLYL)ETHYL)-4-PIPERIDINOL see
BEO500
BENZYLETS see BCM000
BENZYL FAST BLUE R see ADE750
BENZYL FAST ORANGE 2RN see ADG000
BENZYL FAST RED BG see CMM330
BENZYL FAST RED 2BG see NAO600
BENZYL FAST RED GRG see CMM320
BENZYL FAST YELLOW RS see CMM759
BENZYL FORMATE see BEP250
BENZYLFUROLINE see BEP500
5-BENZYL-3-FURYLMETHYL(+)-trans-
CHRYSANTHEMATE see BEP750
5-BENZYL-3-FURYLMETHYL (+)-cis-
CHRYSANTHEMATE see RDZ875
5-BENZYL-3-FURYL METHYL(±)-cis,trans-
CHRYSANTHEMATE see BEP500
(5-BENZYL-3-FURYL) METHYL-2,2-
DIMETHYL-3-(2-METHYLPROPENYL)-
CYCLOPROPANECARBOXYLATE see
BEP500
5'-BENZYL-3'-FURYLMETHYL α-ETHYL-
PHENYLACETATE see BEP600
5'-BENZYL-3'-FURYLMETHYL α-
ISOPROPYL-4-METHOXYPHENYL
ACETATE see BEP650
(R)-o-BENZYLGLYCIDOL see BEP670
(S)-o-BENZYLGLYCIDOL see BFC225
(+)-BENZYL GLYCIDYL ETHER see BFC225
3-BENZYL-4-HEPTANONE see BEN800
BENZYLHYDRAZINE see BEQ000
BENZYLHYDRAZINE
DIHYDROCHLORIDE see BEQ250
BENZYLHYDRAZINE HYDROCHLORIDE
see BEQ500
BENZYLHYDROFLUMETHIAZIDE see
BEQ625
BENZYL HYDROGEN PHTHALATE see
MRK100
BENZYL HYDROQUINONE see AEY000
BENZYLHYDROSULFIDE see TGO750
1-BENZYL-2-(HYDROXYACETYL)INDOLE
see BES300
BENZYL-o-HYDROXYBENZOATE see
BFJ750
2-BENZYL-4-HYDROXYMETHYL-1,3-
DIOXANE see PDX250
2-BENZYL-4-HYDROXYMETHYL-1,3-
DIOXOLANE see PDX250
4-BENZYL-α-(p-HYDROXYPHENOL)-β-
METHYL-1-PIPERIDINEETHANOL see
IAG600
4-BENZYL-α-(p-HYDROXYPHENYL)-β-
METHYL-1-PIPERIDINE-ETHANOL-(L)-(+)-
TARTRATE see IAG625
4-BENZYL-α-(p-HYDROXYPHENYL)-β-
METHYL-1-PIPERIDINEETHANOL
TARTRATE see IAG625
4,6-o-BENZYLIDE-β-d-
GLUCOPYRANOSIDEPODOPHYLLOTOXI
N see PJJ250
trans-BENZYLIDENACETONE see BAY275
BENZYLIDENEACETALDEHYDE see
CMP969
2-BENZYLIDENEACETAMIDE see CMP973
BENZYLIDENE ACETONE see SMS500

trans-BENZYLIDENEACETONE see BAY275
2-BENZYLIDENEACETOPHENONE see
CDH000
BENZYLIDENE ACETYLACETONE see
DBH900
3-BENZYLIDENE ACETYLACETONE see
DBH900
S,S'-BENZYLIDENE BIS(O,O-DIMETHYL
PHOSPHORODITHIOATE) see BES250
1,1'-(BENZYLIDENEBIS((2-METHOXY-p-
PHENYLENE))(AZO))DI-2-NAPHTHOL see
DOO400
3-BENZYLIDENE-2-BUTANONE see
MNS600
BENZYLIDENE CHLORIDE see BAY300
BENZYLIDENE CHLORIDE (DOT) see
BAY300
BENZYLIDENE-4-
ETHYLBENZENEAMINE N-OXIDE see
EOL700
4,6-o-BENZYLIDENE-β-d-
GLUCOPYRANOSIDE
PODOPHYLLOTOXIN see BER500
BENZYLIDENE GLYCEROL see BBA000
2-BENZYLIDENE-1-HEPTANOL see
AOH000
BENZYLIDENEMETHYLPHOSPHORODIT
HIOATE see BES250
N-BENZYLIDENE-m-NITROANILINE see
BBA800
3-BENZYLIDENE-2,4-PENTANEDIONE see
DBH900
BENZYLIDENEPHENYLACETONITRILE
see DVX600
BENZYLIDYNE CHLORIDE see BFL250
BENZYLIDYNE FLUORIDE see BDH500
2-BENZYL-2-IMIDAZOLINE see BBW750
2-BENZYL-4,5-IMIDAZOLINE see BBW750
BENZYLIMIDAZOLINE
HYDROCHLORIDE see BBJ750
2-BENZYL-4,5-IMIDAZOLINE
HYDROCHLORIDE see BBW750
2-BENZYL-2-IMIDAZOLINE
MONOHYDROCHLORIDE see BBJ750
((1-BENZYL-1H-INDAZOL-3-
YL)OXY)ACETIC ACID see BAV325
4-(1-BENZYL-3-INDOLETHYL)PYRIDINE
HYDROCHLORIDE see BCE250
N-BENZYL 3-(α-(3'-INDOLYL)ETHYL)-4-
HYDROXY-4-ETHINYLPIPERIDINE see
BEO500
N-BENZYL-3-(α-(3'-INDOLYL)ETHYL)-4-
PIPERIDONE see ICW150
1-BENZYL-2-INDOLYL
HYDROXYMETHYL KETONE see BES300
BENZYL ISOAMYL ETHER see BES500
BENZYL ISOBUTYRATE (FCC) see IJV000
BENZYL ISOEUGENOL see BES750
BENZYL ISOEUGENOL ETHER see BES750
N-BENZYL-β-
(ISONICOTINOYLHYDRAZINE)PROPION
AMIDE see BET000
N-BENZYL-β-
(ISONICOTINYLHYDRAZINO)PROPIONA
MIDE see BET000
BENZYL ISOPENTYL ETHER see BES500
4-(3-BENZYLISOPROPYLAMINO-2-
HYDROXYPROPOXY)-9-METHOXY-7-
METHYL-FURO(3,2-g)CHROMONE,
HYDROCHLORIDE see BET750
BENZYLISOPROPYL PROPIONATE see
DQQ400
BENZYL-ISOTHIOCYANATE see BEU250
BENZYLISOTHIOUREA
HYDROCHLORIDE see BEU500
BENZYLISOTHIOURONIUM CHLORIDE
see BEU500
2-BENZYLISOTHIOURONIUM CHLORIDE
see BEU500
BENZYL ISOVALERATE (FCC) see ISW000
BENZYLKYANID see PEA750
BENZYL LAURATE see BEU750

BENZYL MERCAPTAN see TGO750
6-BENZYLMERCAPTOPURINE see BDG325
BENZYL METHANOATE see BEP250
2-BENZYL-4-METHANOL-1,3-DIOXANE see PDX250
4-BENZYL-α-(4-METHOXYPHENYL)-β-METHYL-1-PIPERIDINEETHANOL see BEU800
BENZYL-2-METHOXY-4-PROPENYLPHENYL ETHER see BES750
2-BENZYL-1-(4-(METHYLAMINO)BUTOXY)BENZENE HYDROCHLORIDE see MGE200
1-BENZYL-2-METHYL-3-(2-AMINOETHYL)-5-METHOXYINDOLE HYDROCHLORIDE see BEM750
BENZYL-3-METHYLBUTANOATE see ISW000
BENZYL-3-METHYL BUTYRATE see ISW000
BENZYL METHYL CARBINOL see PGA600
BENZYLMETHYLCARBINYL ACETATE see ABU800
1-BENZYL-2-METHYLHYDRAZINE see MHN750
N-BENZYL-3-(2'-METHYL-3'-INDOLYL)METHYL-4-PIPERIDONE see MKW150
1-BENZYL-1-(5-METHYL-3-ISOXAZOIYLCARBONYL)HYDRAZINE see IKC000
1-BENZYL-2-(5-METHYL-3-ISOXAZOIYL-CARBONYL)HYDRAZINE see IKC000
N'-BENZYL N-METHYL-5-ISOXAZOLECARBOXYLHYDRAZIDE-3 see IKC000
1-BENZYL-2-(3-METHYLISOXAZOL-5-YL)CARBONYL HYDRAZINE see BEW000
BENZYL METHYL KETONE see MHO100
1-BENZYL-2-METHYL-5-METHOXYTRYPTAMINE HYDROCHLORIDE see BEM750
1-BENZYL-3-METHYL-5-(2-(4-METHYL-1-PIPERAZINYL)ETHOXY)PYRAZOLE see BEW500
1-BENZYL-3-METHYL-5-(2-(2-METHYLPIPERIDINO)ETHOXY)PYRAZOLE see BEW750
4-BENZYL-1-(1-METHYL-4-PIPERIDYL)-3-PHENYL-3-PYRAZOLIN-5-ONE see BCP650
BENZYL-2-METHYL PROPIONATE see IJV000
N-BENZYL-N-METHYL-2-PROPYNYLAMINE see MOS250
BENZYLMETHYLPROPYNYLAMINE HYDROCHLORIDE see BEX500
N-BENZYL-N-METHYL-2-PROPYNYLAMINE HYDROCHLORIDE see BEX500
(1-BENZYL-3-METHYL-5-PYRAZOLYLOXYETHYL)TRIMETHYLAMMONIUM IODIDE see BEX750
N-BENZYL-α-METHYL-3-TRIFLUOROMETHYLPHENETHYLAMINE see BEY800
N-BENZYL-α-METHYL-m-TRIFLUOROMETHYLPHENETHYLAMINE see BEY800
BENZYL MONOCHLORACETATE see BEE500
BENZYL MONOSULFIDE see BEY900
8-BENZYL-7-(1'-MORPHOLINO-2'-AMINO)ETHYLTHEOPHYLLINE HYDROCHLORIDE see BFA000
6-BENZYL-MP see BDG325
BENZYL MUSTARD OIL see BEU250
BENZYL MUSTARD OIL see BFL000
BENZYL NICOTINATE see NCR040
BENZYL NITRATE see BFA250
BENZYL NITRILE see PEA750
α-(BENZYLNITROSAMINO)BENZYL ALCOHOL ACETATE (ester) see ABL875
1-BENZYL-1-NITROSOUREA see NJM000

N-BENZYL-N',N"-DIMETHYLGUANIDINE SULFATE see BFW250
BENZYL NORMECHLORETHAMINE see BIA750
BENZYL ORANGE 2R see ADG000
BENZYL OXIDE (CZECH) see BEO250
BENZYLOXY ACETYLENE see BFA899
7-(BENZYLOXY)-6-N-BUTYL-1,4-DIHYDRO-4-OXO-3-QUINOLINECARBOXYLIC ACID METHYL ESTER see NCN600
7-(BENZYLOXY)-6-N-BUTYL-4-HYDROXY-3-QUINOLINECARBOXYLIC ACID METHYL see NCN600
BENZYLOXYCARBONYL CHLORIDE see BEF500
BENZYLOXYCARBONYLGLYCINE see CBR125
N-BENZYLOXYCARBONYLGLYCINE see CBR125
(BENZYLOXYCARBONYL)PHENYLALANINE see CBR220
l-N-BENZYLOXYCARBONYL-3-PHENYLALANINE-1,2-DIBROMOETHYL ESTER see CBR225
N-BENZYLOXYCARBONYL-l-PHENYLALANINE VINYL ESTER see CBR235
5-BENZYLOXY-8-CHLORO-N,N-DIMETHYL-1,2,3,4-TETRAHYDRO-1-NAPHTHYLAMINE HYDROCHLORIDE see BFA930
(R)-1-(BENZYLOXY)-2,3-EPOXYPROPANE see BFC225
(S)-1-(BENZYLOXY)-2,3-EPOXYPROPANE see BEP670
2-BENZYLOXYETHANOL see EJI500
S-((N-(2-BENZYLOXYETHYL)AMIDINO)METHYL) HYDROGEN THIOSULFATE see BFC000
5-BENZYLOXY-3-ISONIPECOTOYLINDOLE see BFC200
(+)-(BENZYLOXYMETHYL)OXIRANE see BFC225
(R)-(−)-BENZYLOXYMETHYLOXIRANE see BEP670
(S)-(BENZYLOXYMETHYL)OXIRANE see BFC225
(S)-(+)-BENZYLOXYMETHYLOXIRANE see BFC225
5-BENZYLOXY-3-(1-METHYL-2-PYRROLIDINYL)INDOLE see BFC250
p-BENZYLOXYPHENOL see AEY000
p-(BENZYLOXY)PHENYL BIS(1-AZIRIDINYL)PHOSPHINATE see BFC300
1-(p-(BENZYLOXY)PHENYL)-2-(o-FLUOROPHENYL)-1-PHENYLETHYLENE see BFC400
2-(m-(BENZYLOXY)PHENYL)PYRAZOLO(1,5-a)QUINOLINE see BFC450
N,N'-(3-BENZYLOXY-1,2-PROPANEDIOXYSULFINYL)BIS(3-METHYLPHENYLMETHYLCARBAMATE) see BFC460
N,N'-(3-BENZYLOXY-1,2-PROPANEDIOXYSULFINYL)BIS(1-NAPHTHYLMETHYLCARBAMATE) see BFC470
S-((N-(3-BENZYLOXYPROPYL)AMIDINO)METHYL) HYDROGEN THIOSULFATE see BFC500
BENZYLPENCILLINDIBENZYLETHYLENEDIAMINE SALT see BFC750
BENZYLPENICILLIN see BDY669
BENZYLPENICILLIN BENZATHINE see BFC750
BENZYLPENICILLIN G see BDY669
BENZYLPENICILLINIC ACID see BDY669
BENZYLPENICILLINIC ACID POTASSIUM SALT see BFD000

BENZYL PENICILLINIC ACID SODIUM SALT see BFD250
BENZYLPENICILLIN NOVOCAINE SALT see PAQ200
BENZYLPENICILLIN POTASSIUM see BFD000
BENZYLPENICILLIN POTASSIUM SALT see BFD000
BENZYLPENICILLIN PROCAINE see PAQ200
BENZYLPENICILLIN SODIUM see BFD250
N-BENZYL-N-PHENOXYISOPROPYL-β-CHLORETHYLAMINE HYDROCHLORIDE see DDG800
4-(o-BENZYLPHENOXY)-N-METHYLBUTYLAMINE HYDROCHLORIDE see MGE200
1-(2-BENZYLPHENOXY)-2-PIPERIDINOPROPANE PHOSPHATE see PJA130
BENZYL PHENYLACETATE see BFD400
N-BENZYL-N'-PHENYLACETYLHYDRAZIDE see PDY870
BENZYL γ-PHENYLACRYLATE see BEG750
N-β-(BENZYL-PHENYLAMINO)ETHYLPIPERIDINE HYDROCHLORIDE see BEA275
p-BENZYLPHENYL CARBAMATE see PFR325
BENZYL PHENYLFORMATE see BCM000
BENZYL PHENYL KETONE see PEB000
BENZYLPHENYL NITROSAMINE see BFD750
BENZYLPHOSPHONIC ACID DIBUTYL ESTER see BFD760
4-BENZYLPIPERAZINYL β-(p-CHLOROPHENYL)PHENETHYL KETONE see BFE770
2-(4-BENZYL-PIPERIDINO)-1-(4-HYDROXYPHENYL)-1-PROPANOL TARTRATE (2:1) see IAG625
(+)-2-(1-BENZYL-4-PIPERIDYL)-2-PHENYLGLUTARIMIDE HYDROCHLORIDE see DBE000
(+)-3-(1-BENZYL-4-PIPERIDYL)-3-PHENYLPIPERIDINE-2,6-DIONE HYDROCHLORIDE see DBE000
BENZYL PROPIONATE see BFD800
N-BENZYL-4-PROTOADAMANTANEMETHANAMINE MALEATE see BFG500
2-BENZYLPYRIDINE see BFG600
4-BENZYLPYRIDINE see BFG750
BENZYL PYRIDINE-3-CARBOXYLATE see NCR040
1-BENZYLPYRIDINIUM CHLORIDE see BFH000
BENZYL-(α-PYRIDYL)-DIMETHYLAETHYLENDIAMIN (GERMAN) see TMP750
N-BENZYL-N-α-PYRIDYL-N',N'-DIMETHYL-AETHYLENDIAMIN-HYDROCHLORID (GERMAN) see POO750
1-BENZYL-3-(2-(4-PYRIDYL)ETHYL)INDOLE HYDROCHLORIDE see BCE250
BENZYL 4-PYRIDYL KETONE THIOSEMICARBAZONE see BFH100
2-(BENZYL(2-(PYRROLIDINYL)ETHYL)AMINO)-2'-CHLOROACETANILIDE DIHYDROCHLORIDE see BFI250
BENZYL RED BR see CMM330
BENZYL RED GR see CMM320
BENZYL RED GS see CMM325
BENZYL RED MG see CMM325
4-BENZYL RESORCINOL see BFI400
BENZYLRODIURAN see BEQ625
BENZYL SALICYLATE see BFJ750
BENZYL SCARLET 3BS see CMO875
BENZYLSENFOEL (GERMAN) see BEU250
BENZYL SILANE see BFJ825

BENZYL SODIUM see BFJ850
BENZYL SODIUM SULFATE see SFB150
BENZYLSTEARYLDIMETHYLAMMONIUM CHLORIDE see DTC600
BENZYL SULFIDE (8CI) see BEY900
BENZYL SULFITE see BFK000
BENZYLSULFONYL FLUORIDE see TGO300
BENZYL SULFOXIDE see DDH800
BENZYLSULPHONYL FLUORIDE see TGO300
3-BENZYLSYDNONE-4-ACETAMIDE see BED750
BENZYLT see PDT250
1-BENZYL-2-(3-(4,5,6,7-TETRAHYDROBENZISOXAZOYLYL)CARBONYL)HYDRAZINE HYDROCHLORIDE see BFK325
N-BENZYL-9-(TETRAHYDRO-2H-PYRAN-2-YL)ADENINE see BEA100
2-(4-BENZYL-1,2,3,6-TETRAHYDROPYRIDINO)-1-(4'-METHOXYPHENYL)-1-PROPANOL see RCA450
D-BENZYL TG see EDC560
7-BENZYL-3-THIA-7-AZABICYCLO(3.3.1)NONANE PERCHLORATE see BFK370
S-BENZYL THIOBENZOATE see BFK750
BENZYL THIOCYANATE see BFL000
BENZYL THIOETHER see BEY900
2-BENZYLTHIOETHYL CHLORIDE see BFL100
BENZYLTHIOGUANINE see BFL125
6-BENZYLTHIOGUANINE see BFL125
BENZYLTHIOL see TGO750
3-((BENZYLTHIO)METHYL)-6-CHLORO-1,2,4-BENZOTHIADIAZINE-7-SULFONAMIDE-1,1-DIOXIDE see BDE250
3-BENZYLTHIOMETHYL-6-CHLORO-2H-1,2,4-BENZOTHIADIAZINE-7-SULFONAMIDE-1,1-DIOXIDE see BDE250
3-BENZYLTHIOMETHYL-6-CHLORO-7-SULFAMOYL-1,2,4-BENZOTHIADIAZINE-1,1-DIOXIDE see BDE250
3-BENZYLTHIOMETHYL-6-CHLORO-7-SULFAMYL-1,2,4-BENZOTHIADIAZINE-1,1-DIOXIDE see BDE250
3-BENZYLTHIOMETHYL-6-CHLORO-7-SULFAMYL-2H-1,2,4-BENZOTHIADIAZINE-1,1-DIOXIDE see BDE250
BENZYL THIOPSEUDOUREA HYDROCHLORIDE see BEU500
2-BENZYL-2-THIO-PSEUDOUREA HYDROCHLORIDE see BEU500
6-(BENZYLTHIO)PURINE see BDG325
BENZYLTHIURONIUM CHLORIDE see BEU500
S-BENZYLTHIURONIUM CHLORIDE see BEU500
BENZYL TRICHLORIDE see BFL250
BENZYLTRIETHYLAMMONIUM CHLORIDE see BFL300
3-BENZYL-6-TRIFLUOROMETHYL-7-SULFAMOYL-3,4-DIHYDRO-1,2,4-BENZOTHIADIAZINE-1,1-DIOXIDE see BEQ625
1-BENZYL-2-TRIMETHYLACETYLHYDRAZINE HYDROCHLORIDE see BFM000
BENZYLTRIMETHYLAMMONIUM CHLORIDE see BFM250
BENZYLTRIMETHYLAMMONIUM HYDROXIDE see BFM500
BENZYL TRIMETHYL AMMONIUM IODIDE see BFM750
BENZYLTRIMETHYLAMMONIUM METHOXIDE see TLM100
BENZYLUREA see BFN125
1-BENZYLUREA see BFN125
N-BENZYLUREA see BFN125

BENZYL VIOLET see FAG120
BENZYL VIOLET 3B see FAG120
BENZYLYT see DDG800
BENZYNA see GCC200
3,4-BENZYPYRENE see BCS750
BENZYRIN see BBW500
BENZYTOL see CLW000
BEOCID-ISOPTAL see SNQ000
BEOSIT see EAQ750
BEP see BJP899
BEP see BKH625
BEPANTHEN see PAG200
BEPANTHENE see PAG200
BEPANTOL see PAG200
BEPARON see TCC000
BEPERIDEN see BGD500
BEPRIDIL HYDROCHLORIDE MONOHYDRATE see IIG600
BEPROCHINE see RHZ000
BERBERIN see BFN500
BERBERINE see BFN500
BERBERINE CHLORIDE see BFN600
BERBERINE CHLORIDE DIHYDRATE see BFN550
BERBERINE HYDROCHLORIDE see BFN600
BERBERINE HYDROCHLORIDE BIHYDRATE see BFN550
BERBERINE SULFATE see BFN625
BERBERINE SULFATE (2:1) see BFN625
BERBERINE SULFATE TRIHYDRATE see BFN750
BERBERINIUM CHLORIDE see BFN600
BERBERIN SULFATE see BFN625
BERBINIUM, 7,8,13,13a-TETRADEHYDRO-9,10-DIMETHOXY-2,3-(METHYLENEDIOXY)-, CHLORIDE see BFN600
BERCEMA see EIR000
BERCEMA FERTAM 50 see FAS000
BERCEMA NMC50 see CBM750
BERCULON A see FNF000
BERELEX see GEM000
BERGAMIOL see LFY600
BERGAMOT MINT OIL see BFN990
BERGAMOT OIL rectified see BFO000
BERGAMOTTE OEL (GERMAN) see BFO000
BERGAPTEN see MFN275
BERGENIN HYDRATE see BFO100
BERGIUS COAL HYDROGENATION PRODUCTS FRACTION 1 see HHW509
BERGIUS COAL HYDROGENATION PRODUCTS FRACTION 3 see HHW519
BERGIUS COAL HYDROGENATION PRODUCTS FRACTION 4 see HHW529
BERGIUS COAL HYDROGENATION PRODUCTS FRACTION 4 see HHW549
BERGIUS COAL HYDROGENATION PRODUCTS FRACTION 7 see HHW539
BERKAZON see FNF000
BERKENDYL see CDP000
BERKFLAM B 10 see PCC480
BERKFLAM B 10E see PAU500
BERKFURIN see NGE000
BERKMYCEN see HOH500
BERKOLOL see ICC000
BERKOMINE see DLH600
BERKOMINE see DLH630
BERKOZIDE see BEQ625
BERLINER see BAC040
BERLISON F see HHQ800
BERMAT see CJJ250
BERNARENIN see VGP000
BERNICE see CNE750
BERNIES see CNE750
BERNOCAINE see AIT250
BERNSTEINSAEURE (GERMAN) see SMY000
BERNSTEINSAEURE-ANHYDRID (GERMAN) see SNC000

BERNSTEINSAEURE-2,2-DIMETHYLHYDRAZID (GERMAN) see DQD400
BEROCILLIN see AOD000
BEROL 28 see PJT300
BEROL 478 see DJL000
BEROL VISCO 31 see VRP780
BEROL VISCO 34 see VRP790
BEROMYCIN see PDT500
BEROMYCIN see PDT750
BEROMYCIN 400 see PDT750
BEROMYCIN (penicillin) see PDT750
BERONALD see CHJ750
BEROTEC see FAQ100
BEROTEC HYDROBROMIDE see FAQ100
BERRY ALDER see MBU825
BERSAMA ABYSSINICA Fres. ssp. ABYSSINICA, leaf extract see BFO125
BERSEN see DDT300
BERTHOLITE see CDV750
BERTHOLLET SALT see PLA250
BERTRANDITE see BFO250
BERUBIGEN see VSZ000
BERYL see BFO500
BERYLLATE(2-), TETRAFLUORO-, DIAMMONIUM see ANH300
BERYLLATE(2-), TETRAFLUORO-, DIAMMONIUM, (T-4)- see ANH300
BERYLLIA see BFT250
BERYLLIUM see BFO750
BERYLLIUM-9 see BFO750
BERYLLIUM ACETATE see BFP000
BERYLLIUM ACETATE, BASIC see BFT500
BERYLLIUM ACETATE, NORMAL see BFP000
BERYLLIUM ALUMINOSILICATE see BFO500
BERYLLIUM ALUMINUM ALLOY see BFP250
BERYLLIUM ALUMINUM SILICATE see BFO500
BERYLLIUM CARBONATE see BFP500
BERYLLIUM CARBONATE (1:1) see BFP750
BERYLLIUM CARBONATE, BASIC see BFP500
BERYLLIUM CARBONATE BASIC see BFP755
BERYLLIUM CHLORIDE see BFQ000
BERYLLIUM COMPOUND with NIOBIUM (12:1) see BFQ750
BERYLLIUM COMPOUNDS see BFQ500
BERYLLIUM COMPOUNDS, n.o.s. (UN 1566) (DOT) see BFO750
BERYLLIUM COMPOUND with TITANIUM (12:1) see BFR000
BERYLLIUM COMPOUND with VANADIUM (12:1) see BFR250
BERYLLIUM-COPPER ALLOY see CNI600
BERYLLIUM-COPPER-COBALT ALLOY see CNK700
BERYLLIUM DICHLORIDE see BFQ000
BERYLLIUM DIFLUORIDE see BFR500
BERYLLIUM DIHYDROXIDE see BFS250
BERYLLIUM DINITRATE see BFT000
BERYLLIUM DINITRATE TETRAHYDRATE see BFR300
BERYLLIUM FLUORIDE see BFR500
BERYLLIUM HYDRATE see BFS250
BERYLLIUM HYDRIDE see BFR750
BERYLLIUM HYDROGEN PHOSPHATE (1:1) see BFS000
BERYLLIUM HYDROXIDE see BFS250
BERYLLIUM LACTATE see LAH000
BERYLLIUM MANGANESE ZINC SILICATE see BFS750
BERYLLIUM MONOXIDE see BFT250
BERYLLIUM-NICKEL ALLOY see NCX510
BERYLLIUM NITRATE see BFT000
BERYLLIUM NITRATE TETRAHYDRATE see BFR300
BERYLLIUM NITRATE TRIHYDRATE see BFT100

BERYLLIUM ORTHOSILICATE see SCN500
BERYLLIUM OXIDE see BFT250
BERYLLIUM OXIDE ACETATE see BFT500
BERYLLIUMOXIDE CARBONATE see BFP500
BERYLLIUM OXYACETATE see BFT500
BERYLLIUM OXYFLUORIDE see BFT750
BERYLLIUM PERCHLORATE see BFU000
BERYLLIUM PHOSPHATE see BFS000
BERYLLIUM SILICATE see PCV400
BERYLLIUM SILICATE see SCN500
BERYLLIUM SILICATE HYDRATE see BFO250
BERYLLIUM SILICIC ACID see SCN500
BERYLLIUM SULFATE (1:1) see BFU250
BERYLLIUM SULFATE TETRAHYDRATE (1:1:4) see BFU500
BERYLLIUM SULPHATE TETRAHYDRATE see BFU500
BERYLLIUM TETRAHYDROBORATE see BFU750
BERYLLIUM TETRAHYDROBORATETRIMETHYLAMINE see BFV000
BERYLLIUM, powder (UN 1567) (DOT) see BFO750
BERYLLIUM ZINC SILICATE see BFV250
BERYL ORE see BFO500
BESAN see POH000
BESANTIN see MHL000
BESTATIN see BFV300
BESTHORN'S HYDRAZONE HYDROCHLORIDE see MHJ275
BE-STILL TREE see YAK350
BESTRABUCIL see BFV325
BETABION see TES750
BETACIDE P see HNU500
BETADID see HJS850
BETADINE see PKE250
BETADRENOL see BQB250
BETADRENOL HYDROCHLORIDE see BQB250
BETAFEDRINA see BBK500
BETAFEDRINE see BBK500
BETAFEN see AOB250
BETAGAN see LFA100
BETAHISTINE see HGE820
BETAHISTINE MESILATE see BFV350
BETAHISTINE MESYLATE see BFV350
BETAINE see GHA050
BETAINE CEPHALORIDINE see TEY000
BETAINE HYDRAZIDE HYDROCHLORIDE see GEQ500
BETAINE LAURYLDIMETHYLAMINOACETATE see LBU200
BETAINE NICOTINATE see TKL890
BETAIN NICOTINATE see TKL890
BETAISODONA see PKE250
BETALGIL see DTL200
BETALIN 12 CRYSTALLINE see VSZ000
BETALING see BFW250
BETALIN S see TES750
BETALOC see SBV500
BETAMEC see DNO800
BETAMETHASONE see BFV750
BETAMETHASONE ACETATE and BETAMETHASONE PHOSPHATE see BFV755
BETAMETHASONE ACETATE mixed with BETAMETHASONE SODIUM PHOSPHATE see CCS675
BETAMETHASONE BENZOATE see BFV760
BETAMETHASONE 17-BENZOATE see BFV760
BETAMETHASONE DIPROPIONATE see BFV765
BETAMETHASONE 17,21-DIPROPIONATE see BFV765
BETAMETHASONE DISODIUM PHOSPHATE see BFV770

BETAMETHASONE-21-DISODIUM PHOSPHATE see BFV770
BETAMETHASONE PHOSPHATE and BETAMETHASONE ACETATE see BFV755
BETAMETHASONE SODIUM PHOSPHATE see BFV770
BETAMETHASONE SODIUM PHOSPHATE mixed with BETAMETHASONE ACETATE see CCS675
BETAMETHASONE VALERATE see VCA000
BETAMETHASONE 17-VALERATE see VCA000
BETA METHYL DIGOXIN see MJD300
BETA-NAFTYLOAMINA (POLISH) see NBE500
BETANAL see MEG250
BETANAL AM see EEO500
BETANAPHTHOL ORANGE see CMM220
BETA-NEG see ICC000
BETANEX see EEO500
BETANIDINE SULFATE see BFV900
BETANIDIN SULFATE see BFV900
BETANIDOLE see BFW250
BETAPAL see NBJ700
BETAPEN see PAQ100
BETAPEN-VK see PDT750
d-BETAPHEDRINE see BBK500
BETAPRONE see PMT100
BETAPTIN see AGW000
BETAPYRIMIDUM see DJS200
BETAQUIL see CBI000
BETARUNDUM see SCQ000
BETARUNDUM ST-S see SCQ000
BETARUNDUM UF see SCQ000
BETARUNDUM ULTRAFINE see SCQ000
BETASAN see DNO800
BETATRON see DXO300
BETAXIN see TES750
BETAXINA see EID000
BETAXOLOL HYDROCHLORIDE see KEA350
BETAZED see BRF500
BETEL LEAVES see BFV975
BETEL NUT see AQT650
BETEL NUT see BFW000
BETEL NUT, polyphenol fraction see BFW010
BETEL NUT TANNIN see BFW050
BETEL QUID see BFW120
BETEL QUID EXTRACT see BFW125
BETEL TOBACCO EXTRACT see BFW135
BETHABARRA WOOD see HLY500
BETHAINE CHOLINE CHLORIDE see HOA500
BETHAMETHASONE 17-BENZOATE see BFV760
BETHANECHOL CHLORIDE see HOA500
BETHANID see BFW250
BETHANIDINE, HEMISULFATE see BFW250
BETHANIDINE SULFATE see BFW250
BETHIAMIN see TES750
BETNELAN see BFV750
BETNELAN PHOSPHATE see BFW325
BETNESOL see BFV770
BETNOVATE see VCA000
BETNOVATEAT see VCA000
BETOXON see NBJ700
BETRAMIN see BBV500
BETRILOL see BON400
BETSOLAN see BFV750
BETULA OIL see MPI000
BEVALOID 35 see BLX000
BEVANTOLOL HYDROCHLORIDE see BFW400
BEVATINE-12 see VSZ000
BEVIDOX see VSZ000
BEVONIUM METHYL SULFATE see CDG250
BEVONIUM METILSULFATE see CDG250
BEWON see TES750
BEXANE see CLN750
BEXIDE see BJU000

BEXOL see BBQ500
BEXON see MMN250
BEXONE see CLN750
BEXT see BJU000
BEXT see DKE400
BEXTON see CHS500
BEXTRENE XL 750 see SMQ500
BEZITRAMIDE see BCE825
B(b)F see BAW250
B(j)F see BCJ250
BF 121 see FOJ100
BF 200 see CAT775
BF 5930 see OJW000
BFE 60 see BAU255
B 169-FERRICYANIDE see TNH750
BFH see BRK100
BFP see BJE750
BFPO see BJE750
BFV see FMV000
BG-356 see EPW600
BG 5930 see OJW000
BG 6080 see CBT500
B. GABONICA VENOM see BLV080
BGE see BRK750
t-BGE see BRK800
BGE (OSHA) see BRK750
B"-GUTTIFERIN see CBA125
BH 6 see BGS250
3-tert-BHA see BQI010
BHA (FCC) see BQI000
BHANG see CBD750
(d-BHA⁶)-PRO-NHET⁹))-LHRH ACETATE see LIU347
BHB see BPI125
BHBN see HJQ350
BHC see BBQ500
BHC see BJZ000
α-BHC see BBQ000
β-BHC see BBR000
Δ-BHC see BFW500
γ-BHC see BBQ500
BHC (USDA) see BBP750
BHD see BDN125
BH 2,4-D see DAA800
BH DALAPON see DGI400
BHEN see BRO000
B-HERBATOX see SFS000
BHFT see BEQ625
BHIMSAIM CAMPHOR see BMD000
BH MCPA see CIR250
BH MECOPROP see CIR500
BHP see DNB200
BHPBN see HIE600
BHT (food grade) see BFW750
BI-58 see DSP400
BI 6013 see AQN550
4',4'''-BIACETANILIDE see BFX000
BIACETYL see BOT500
BIACETYLMONOXIME see OMY910
BIALAMICOL HYDROCHLORIDE see BFX125
BIALAPHOS SODIUM see SEO550
BIALCOL see BDJ600
BIALFLAVINA see DBX400
BIALLYL see HCR500
BIALLYLAMICOL DIHYDROCHLORIDE see BFX125
BIALLYLAMICOL HYDROCHLORIDE see BFX125
BIALMINAL see EOK000
BIALPIRINIA see ADA725
BIALZEPAM see DCK759
p,p-BIANILINE see BBX000
4,4'-BIANILINE see BBX000
N,N'-BIANILINE see HHG000
o,p'-BIANILINE see BGF109
BIANISIDINE see TGJ750
BItertANOL see SCF525
(2,9'-BIANTHRACENE-4',8(1'H,5H)-DIONE, 2',3',6,7-TETRAHYDRO-2',6-DIMETHYL-1,2',5',6,9,10'-HEXAHYDROXY- see PCL300

1,1'-BIANTHRACENE-9,9',10,10'-
TETRAONE, 2,2'-DIMETHYL- see DQR350
5,5'-BIANTHRANILIC ACID see BFX250
(3,3'-BIANTHRA(1,9-cd)PYRAZOLE)-
6,6'(1H,1'H)-DIONE, 1,1'-DIETHYL- see
CMU825
BIARISON see POB300
BIARSAN see POB300
1,1'-BIAZIRIDINYL see BFX325
BIAZOLINA see CCS250
BIBENZAL see SLR000
BIBENZENE see BGE000
α-α'-BIBENZHYDROL see TEA600
2,5'-BIBENZIMIDAZOLE, 2'-(p-
ETHOXYPHENYL)-5-(4-METHYL-1-
PIPERAZINYL)- see BGZ100
2,5'-BI-1H-BENZIMIDAZOLE, 2'-(4-
ETHOXYPHENYL)-5-(4-METHYL-1-
PIPERAZINYL)- see BGZ100
((Δ²,²'(3H,3'H))-BIBENZO(b)THIOPHENE)-
3,3'-DIONE, see CMU815
(Δ²,²'(3H,3'H)-BIBENZO(b)THIOPHENE)-3,3'-
DIONE see DNT300
BIBENZYL see BFX500
BIBENZYLIDENE see SLR000
BIBENZYLIDINE see SLR000
BIBESOL see DGP900
BIC see BRQ500
BIC see IAN000
BICAM ULV see DQM600
BICA-PENICILLIN see BFC750
BICARBAMAMIDE see HGU050
BICARBAMIMIDIC ACID see HGU050
BICARBONATE of SODA see SFC500
BICARBURET of HYDROGEN see BBL250
BICARBURETTED HYDROGEN see EIO000
BICARNESINE see CCK655
BICEP see MQQ450
BICETONIUM see BEL900
2,2'-BICHAVICOL see BFX520
BICHE PRIETO (MEXICO) see CNG825
BICHLORACETIC ACID see DEL000
BICHLORENDO see MQW500
BICHLORIDE of MERCURY see MCY475
BICHLORURE d'ETHYLENE (FRENCH) see
EIY600
BICHLORURE de MERCURE (FRENCH) see
MCY475
BICHLORURE de PROPYLENE (FRENCH)
see PNJ400
BICHOL see DYE700
BICHROMATE d'AMMONIUM (FRENCH)
see ANB500
BICHROMATE OF POTASH see PKX250
BICHROMATE of SODA see SGI000
BICHROMATE de SODIUM (FRENCH) see
SGI000
BICILLIN see BFC750
BICIRON see RGP450
BICKIE-MOL see HIM000
BICOL see PPN100
BICOLASTIC A 75 see SMQ500
BICOLENE C see PJS750
BICOLENE H see SMQ500
BICOLENE P see PMP500
BICORTONE see PLZ000
BICP see CEX250
BICUCULLINE see BFX530
BI-CURVAL see ZJS300
Δ-1,1'-BICYCLOBUTYL see BFX545
BICYCLOBUTYLIDINE see BFX545
BICYCLO(4.4.0)DECANE see DAE800
BICYCLO(5.3.0)DECAPENTAENE see
ASP600
BICYCLO(0.3.5)DECA-1,3,5,7,9-PENTAENE
see ASP600
BICYCLO(5.3.0)-DECA-2,4,6,8,10-PENTAENE
see ASP600
BICYCLO(2,2,2)-1,4-DIAZAOCTANE see
DCK400
BICYCLO(2.2.1)HEPTADIENE see NNG000

BICYCLO(2.2.1)HEPTAN-2-AMINE, N,2,3,3-
TETRAMETHYL-, (1R,2S,4S)- see MBW778
BICYCLO(2.2.1)HEPTANE, 2,5-DIOL,
DIALLYL ETHER see BFX650
BICYCLO(2.2.1)HEPTANE, 2-CHLORO-1,7,7-
TRIMETHYL-, endo- see BMD300
BICYCLO(2.2.1)HEPTANE, 2,2-DIMETHYL-
3-METHYLENE-(9CI) see CBA500
BICYCLO(2.2.1)HEPTANE, 2,2-DIMETHYL-
3-METHYLENE-, OCTACHLORO DERIV.
see OAH100
BICYCLO(2.2.1)HEPTANE-1-
METHANESULFONIC ACID, 7,7-
DIMETHYL-2-OXO-, (1S)-(9CI) see RFU100
BICYCLO(2.2.1)HEPTANE, 2-METHOXY-
1,7,7-TRIMETHYL-, exo- see IHX500
BICYCLO(2.2.1)HEPTANE, 2,2,5,6-
TETRACHLORO-1,7,7-
TRIS(CHLOROMETHYL)-, (5-endo,6-exo)- see
THH575
BICYCLO(2.2.1)HEPTAN-2-OL, 1,7,7-
TRIMETHYL-, endo-(9CI) see BMD000
BICYCLO(3.1.1)HEPTAN-2-OL, 2,6,6-
TRIMETHYL-, (1-α-2-α-5α-- see PIH050
BICYCLO(2.2.1)HEPTAN-2-OL, 1,7,7-
TRIMETHYL-, ACETATE, EXO- see IHX600
BICYCLO(4.1.0)HEPTAN-3-ONE, 1-
METHYL-4-(1-METHYLETHYLIDENE)-7-(3-
OXOBUTYL)-, (1S,6R,7R)- see COF855
BICYCLO(2.2.1)HEPTAN-2-ONE, 1,3,3-
TRIMETHYL-, (1R)-(9CI) see FAM300
BICYCLO(2.2.1)HEPTAN-2-ONE, 1,7,7-
TRIMETHYL-, MIXT. WITH PHENOL see
PDO275
BICYCLO(2.2.1)HEPT-5-ENE-2-
CARBONITRILE see NNH600
BICYCLO(3.1.1)HEPT-2-ENE-2-
CARBOXALDEHYDE, 6,6-DIMETHYL- see
FNK150
BICYCLO(2.2.1)HEPT-5-ENE-2-
CARBOXALDEHYDE, 3-PROPYL- see
CML620
BICYCLO(2.2.1)HEPT-5-ENE, 2-(4-
CYCLOHEXENYL)- see CPD300
cis-BICYCLO(2.2.1)HEPT-5-ENE-2,3-
DICARBOXYLIC ACID, DIMETHYL ESTER
see DRB400
(endo,endo)-BICYCLO(2.2.1)HEPT-5-ENE-2,3-
DICARBOXYLIC ACID DIMETHYL ESTER
see DRB400
BICYCLO(2.2.1)HEPTENE-2-
DICARBOXYLIC ACID, 2-
ETHYLHEXYLIMIDE see OES000
BICYCLO(2.2.1)HEPT-5-ENE-2,3-
DICARBOXYLIC ACID, 1,4,5,6,7,7-
HEXACHLORO-, DIBUTYL ESTER see
CDS025
cis-BICYCLO(2,2,1-HEPTENE-2,3-
DICARBOXYLIC ACID) METHYL ESTER see
DRB400
BICYCLO(2.2.1)-HEPT-5-ENE-2,3-
DICARBOXYLIC ANHYDRIDE see BFY000
BICYCLO(2.2.1)HEPT-5-ENE-2,3-
DIMETHANOL, 1,4,5,6,7,7-HEXACHLORO-
see HCC550
BICYCLO(2.2.1)HEPT-2-ENE, 1,2,3,4,7,7-
HEXACHLORO-5,6-BIS(CHLOROMETHYL)-
(9CI) see BFY100
BICYCLO(3.1.1)HEPT-2-ENE-2-METHANOL,
6,6-DIMETHYL-, ACETATE, (1S)- see MSC050
BICYCLO(2.2.1)HEPT-5-ENE-2-METHYLOL
ACRYLATE see BFY250
BICYCLO(4.1.0)HEPT-3-ENE, 3,7,7(OR 4,7,7)-
TRIMETHYL- see CCK510
trans-BICYCLO(2.2.1)HEPT-5-ENYL-2,3-
DICARBONYLCHLORIDE see NNI000
α-BICYCLO(2.2.1)HEPT-5-EN-1-YL-α-
PHENYL-PIPERIDINEPROPANOL
HYDROCHLORIDE see BGD750
α-(BICYCLO(2.2.1)HEPT-5-EN-2-YL)-α-
PHENYL-1-PIPERIDINEPROPANOL
HYDROCHLORIDE see BGD750

1-BICYCLOHEPTENYL-1-PHENYL-3-
PIPERIDINO-PROPANOL-1 see BGD500
α-(BICYCLO(2.2.1)HEPT-5-EN-2-YL)-α-
PHENYL-1-PIPERIDINO PROPANOL see
BGD500
1-BICYCLOHEPTENYL-1-PHENYL-3-
PIPERIDINOPROPANOL-1
HYDROCHLORIDE see BGD750
N'-BICYCLO(2.2.1)HEPT-2-YL-N-(2-
CHLOROETHYL)-N-NITROSOUREA see
CHE500
BICYCLO(2.2.1)HEPT-2,5-YLENE, BISALLYL
ETHER see BFX650
2-(BICYCLO(2.2.1)HEPT-2-
YLIDENEMETHYL)-1-METHYL-5-NITRO-
1H-IMIDAZOLE see BFY400
(BICYCLOHEXYL)-1-CARBOXYLIC ACID,
2-PIPERIDINOETHYL ESTER,
HYDROCHLORIDE see NCK100
(1,1'-BICYCLOHEXYL)-1-CARBOXYLIC
ACID, 2-(1-PIPERIDINYL)ETHYL ESTER,
HYDROCHLORIDE (9CI) see NCK100
(1,1'-BICYCLOHEXYL)-2-ONE see LBX100
BICYCLO(4,3,0)NONA-3,7-DIENE see
TCU250
BICYCLONONADIENE DIEPOXIDE see
BFY750
BICYCLONONALACTONE see OBV200
3,3'-(BICYCLO(2.2.2)OCTANE-1,4-
DIYLBIS(CARBONYLIMINO-4,1-
PHENYLENECARBONYL IMINO))BIS(1-
ETHYLPYRIDINIUM) SALT with 4-
METHYLBENZENESULFONIC ACID (1:2)
see BFZ000
BICYCLO(4.2.0)OCTA-1,3,5-TRIENE, 7-
ACETYL- see BFZ120
BICYCLO(4.2.0)OCTA-1,3,5-TRIENE, 7-
BENZOYL- see BFZ180
BICYCLO(4.2.0)OCTA-1,3,5-TRIENE, 7-
VALERYL- see BCH800
BICYCLO(4.2.0)OCTA-1,3,5-TRIEN-7-YL
BENZYL KETONE see BFZ100
BICYCLO(4.2.0)OCTA-1,3,5-TRIEN-7-YL
BENZYL KETONE OXIME see BFZ110
BICYCLO(4.2.0)OCTA-1,3,5-TRIEN-7-YL
METHYL KETONE see BFZ120
BICYCLO(4.2.0)OCTA-1,3,5-TRIEN-7-YL
METHYL KETONE O-ACETYLOXIME see
BFZ130
BICYCLO(4.2.0)OCTA-1,3,5-TRIEN-7-YL
METHYL KETONE O-ALLYLOXIME see
BFZ140
BICYCLO(4.2.0)OCTA-1,3,5-TRIEN-7-YL
METHYL KETONE O-BUTYLOXIME see
BFZ150
BICYCLO(4.2.0)OCTA-1,3,5-TRIEN-7-YL
METHYL KETONE O-(2-
HYDROXYPROPYL)OXIME see HNT550
BICYCLO(4.2.0)OCTA-1,3,5-TRIEN-7-YL
METHYL KETONE O-METHYLOXIME see
MFX550
BICYCLO(4.2.0)OCTA-1,3,5-TRIEN-7-YL
METHYL KETONE OXIME see BFZ160
1-BICYCLO(4.2.0)OCTA-1,3,5-TRIEN-7-YL-1-
PENTANONE see BCH800
1-BICYCLO(4.2.0)OCTA-1,3,5-TRIEN-7-YL-1-
PENTANONE OXIME see BFZ170
BICYCLO(4.2.0)OCTA-1,3,5-TRIEN-7-YL
PENTYL KETONE OXIME see BFZ170
BICYCLO(4.2.0)OCTA-1,3,5-TRIEN-7-YL
PHENYL KETONE see BFZ180
BICYCLOPENTADIENE see DGW000
BICYCLOPENTADIENE DIOXIDE see
BGA250
BICYCLOPENTADIENYLBIS(TRICARBONY
LIRON) see BGA500
BICYCLO(2.1.0)PENT-2-ENE see BGA650
BIDERON see DGC800
BIDIPHEN see TFD250
BIDIRL see DGQ875
BIDISIN see CFC750
BIDOCEF see DYF700

BIDRIN see DGQ875
BIEBRICH SCARLET BPC see SBC500
BIEBRICH SCARLET RED see SBC500
BIEBRICH SCARLET R MEDICINAL see SBC500
BIETASERPINE see DIG800
BIETASERPINE BITARTRATE see DIH000
BIETHYLENE see BOP500
1,1'-BI(ETHYLENE OXIDE) see BGA750
BIETHYLXANTHOGENTRISULFIDE see BJU000
BIFEMELANE HYDROCHLORIDE see MGE200
BIFENOX see MJB600
BIFENTHRIN see TAC850
BIFEX see PMY300
BIFLORINE see FOW000
BIFLUORIDEN (DUTCH) see FEZ000
BIFLUORURE de POTASSIUM (FRENCH) see PKU250
BIFONAZOL see BGA825
BIFONAZOLE see BGA825
BIFORMYCHLORAZIN see TKL100
BIFORMYLCHLORAZIN see TKL100
BIFORON see BQL000
BIFURON see NGG500
BIG DIPPER see DVX800
BIGITALIN see GEU000
BIG LEAF IVY see MRU359
BIGUANIDE, 1,1'-HEXAMETHYLENEBIS(5-(p-CHLOROPHENYL))-, DIGLUCONATE see CDT250
BIGUANIDE, 1-o-TOLYL- see TGX550
BIGUANIDINE, CHROMATE see BJW825
BIGUMAL see CKB250
BIGUNAL see BQL000
BIHEXYL see DXT200
BIHOROMYCIN (crystalline) see BGB250
10204-BII see FBP300
2,4'-BI-1H-IMIDAZOLE, 1,1'-DIMETHYL-2'-NITRO- see DSV240
(2,2'-BIINDOLINE)-3,3'-DIONE see BGB275
((Δ²'²')-BIINDOLINE)-3,3'-DIONE see BGB275
(Δ²'²'-(BIINDOLINE)-3,3'-DIONE, 5,5',7,7'-TETRABROMO-(7CI,8CI) see ICU135
BIKAVERIN see LJB700
BI-KELLINA see AHK750
BIKLIN see APT000
BILAGEN see TGZ000
BILANAFOS-SODIUM see SEO550
BILARCIL see TIQ250
BILCOLIC see MKP500
BILENE see DYE700
BILETAN see DXN800
BILEVON see HCL000
BILEVON M see DFD000
BILHARCID see PIJ600
BILICANTE see MKP500
BILIDREN see DAL000
BILIGRAFIN FORTE see BGB315
BILIGRAFIN NATRIUM (GERMAN) see BGB325
BILIGRAFIN SODIUM see BGB325
BILIMIN see SKM000
BILIMIRO see IGA000
BILIMIRON see IGA000
BILINE-8,12-DIPROPIONIC ACID, 1,10,19,22,23,24-HEXAHYDRO-2,7,13,17-TETRAMETHYL-1,19-DIOXO-3,18-DIVINYL- see HAO900
BILINEURINE see CMF000
BILIOGNOST see PDM750
BILI-ORAL see EQC000
BILIPHORIN see HAQ570
BILIPHORINE see HAQ570
BILIRON see SGD500
BILIRUBIN see HAO900
BILIRUBIN IX-α see HAO900
BILISCOPIN see IGD100
BILISELECTAN see PDM750
BILITRAST see TDE750

BILIVISTAN NATRIUM (GERMAN) see BGB350
BILIVISTAN SODIUM see BGB350
BILOBORN see MRH209
BILOBRAN see MRH209
BILOCOL see DYE700
BI-LOFT, combustion products see ADX750
BILOPAC see SKO500
BILOPAQUE see SKO500
BILOPTIN see SKM000
BILOPTINON see SKM000
BILOSTAT see DAL000
BILOXAZOL see SCF525
BILTRICIDE see BGB400
BIM see MQC000
BIM 21009 see ACS100
BIM-23052 see PED800
BIM-23056 see PED850
Δ²'²'-BIMALONONITRILE see EEE500
6,6'-BIMETANILIC ACID see BBX500
BIMETHYL see EDZ000
BIMOX M see MJM700
BINAPACRYL see BGB500
(1,1'-BINAPHTHALENE)-2,2'-DIAMINE see BGB750
(1,2'-BINAPHTHALENE)-1,2'-DIAMINE see BGC000
(2,2-BINAPHTHALENE)-8,8'-DICARBOXALDEHYDE, 1,1',6,6',7,7'-HEXAHYDROXY-3,3'-DIMETHYL-5,5'-BIS(1-METHYLETHYL)-, (±)- see GJM030
(1,1'-BINAPHTHALENE)-2,2'-DIOL see BGC100
1,1'-BI-2-NAPHTHOL see BGC100
(±)-1,1-BI-2-NAPHTHOL see CDM625
β-BINAPHTHOL see BGC100
(8,8'-BI-1H-NAPHTHO(2,3-c)PYRAN)-3,3'-DIACETIC ACID, 3,3',4,4'-TETRAHYDRO-9,9',10,10'-TETRAHYDRO-7,7'-DIMETHOXY-1,1'-DIOXO-, DIMETHYL ESTER see VRP200
2,3,1',8'-BINAPHTHYLENE see BCJ280
β,β-BINAPHTHYLENEETHENE see PIB750
BINAZIN see TGJ150
BINAZINE see TGJ150
BINDAN see PFJ750
BINDAR OS-0704 see BGC125
BINDAZAC see BAV325
BINDON see ONY000
BINDON ATHYLATHER see BGC250
BINDON ETHYL ETHER see BGC250
BINITROBENZENE see DUQ200
BINOCTAL see AMX750
BINODALINE HYDROCHLORIDE see BGC500
BINODALIN HYDROCHLORID (GERMAN) see BGC500
BINOSIDE see BGC625
BINOTAL see AIV500
BINOTAL SODIUM see SEQ000
BINOVA see DXO300
BINOX M see MJM700
BINOX-M see MJM700
BIO 1,137 see BIT250
BIO 5,462 see EAQ750
BIOACRIDIN see DBX400
BIOALETRINA (PORTUGUESE) see BGC750
BIOALLETHRIN see AFR250
BIOALLETHRIN see BGC750
S-BIOALLETHRIN see AFR750
S-trans-BIOALLETHRIN see AFR750
BIOALTRINA see AFR250
BIOBAMAT see MQU750
BIOBAN-C see CAW400
BIO-BEADS S-S 2 see SMQ500
BIOCALC see CAT225
BIOCAMYCIN see HGP550
BIOCETIN see CDP250
BIOCIDE see ADR000
N-1386 BIOCIDE see BLM500
BIO-CLAVE see CJU250
BIOCOLINA see CMF750
BIOCORT ACETATE see CNS825

BIOCORTAR see HHQ800
BIO-DAC 50-22 see DGX200
BIO-DES see DKA600
BIODIASTASE 1000 see BGC825
BIODOPA see DNA200
BIOEPIDERM see BGD100
BIOEPIDERM see VSU100
BIOFANAL see NOH500
BIOFERMIN see DWX600
BIOFLAVONOID see RSU000
BIOFUREA see NGE500
BIOGASTRONE see BGD000
BIOGASTRONE see CBO500
BIOGRISIN-FP see GKE000
BIO (JAPANESE) see SDY600
BIOMET 204 see TMV850
BIOMET TBTO see BLL750
BIOMINE 1651 see MCB050
BIOMIORAN see CDQ750
BIOMITSIN see CMA750
BIOMYCIN see CMA750
BIOMYDRIN see SPC500
BIONIC see NCQ900
BIOPAL CVL-10 see NND000
BIOPAL NR-20 see NND000
BIOPAL VRO 10 see NND000
BIOPAL VRO 20 see NND000
BIO-PERGE see ISD066
BIOPERMETHRIN see PCK075
BIOPHEDRIN see EAW000
BIOPHENICOL see CDP250
BIO-PHYLLINE see HLC000
BIOPOLYENE see BGD050
BIOPRASE see BAC000
BIOPTERIN see BGD075
6-BIOPTERIN see BGD075
l-BIOPTERIN see BGD075
BIOQUAT 80 see QAT520
BIOQUAT 501 see QAT520
BIO-QUAT 50-24 see AFP250
BIOQUIN see BLC250
BIOQUIN see QPA000
BIOQUIN 1 see BLC250
BIORAL see BGD000
BIORAL see CBO500
BIORENINE see VGP000
BIORESMETHRIN see BEP750
BIORESMETHRINE see BEP750
BIORESMETRINA (PORTUGUESE) see BEP750
BIOREX see BGD088
BIORPHEN see MJH900
BIOSCLERAN see ARQ750
BIOSECHS see PII100
BIOSEDAN see NBU000
BIOSEPT see CCX000
BIOSERPINE see RDK000
BIOSHIK see TGD250
BIOS II see BGD100
BIOS II see VSU100
BIO-SOFT D-40 see DXW200
BIO-SOFT S 100 see LBU100
BIOSOL VETERINARY see NCG000
BIOSONE see GIE000
BIOSORB 130 see HND100
BIOSTAT see HOH500
BIOSTAT PA see HOH500
BIOSTEROL see VSK600
BIOSUPRESSIN see HOO500
BIOTERTUSSIN see CMW500
BIO-TESTICULINA see TBG000
BIO-TETRA see TBX000
BIOTHION see TAL250
BIOTIN see BGD100
BIOTIN see VSU100
(+)-BIOTIN see BGD100
(+)-BIOTIN see VSU100
d-BIOTIN see BGD100
D-BIOTIN see BGD100
d-BIOTIN see VSU100
D-(+)-BIOTIN see BGD100
d-(+)-BIOTIN see VSU100

BIOTIRMONE see SKJ300
BIOTROL see BAC040
BIOXIRANE see BGA750
2,2'-BIOXIRANE see BGA750
(R*,S*)-2,2'-BIOXIRANE see DHB800
(S-(R*,R*))-2,2'-BIOXIRANE see BOP750
BIOXONE see BGD250
BIOXYDE d'AZOTE (FRENCH) see NEG100
BIOXYDE de PLOMB (FRENCH) see LCX000
BIPANAL see SBN000
BIPC (the herbicide) see CEX250
BIPERIDEN see BGD500
BIPERIDEN HYDROCHLORIDE see
BGD750
BIPERIDINE HYDROCHLORIDE see
BGD750
2,2'-BIPHENOL see BGG000
o,o'-BIPHENOL see BGG000
p,p'-BIPHENOL see BGG500
BIPHENOL AF see HCZ100
BIPHENTHRIN see TAC850
BIPHENYL see BGE000
1,1'-BIPHENYL see BGE000
4-BIPHENYLACETAMIDE see PDY500
N-4-BIPHENYLACETAMIDE see PDY500
4-BIPHENYLACETHYDROXAMIC ACID see
ACD000
4-BIPHENYLACETIC ACID see BGE125
(4-BIPHENYL) ACETIC ACID see BGE125
4-BIPHENYLACETIC ACID see BGE125
p-BIPHENYLACETIC ACID see BGE125
(1,1'-BIPHENYL)-4-ACETIC ACID see
BGE125
4-BIPHENYLACETIC ACID, 2-
FLUOROETHYL ESTER see FDB200
(1,1'-BIPHENYL)-4-ACETIC ACID, 2-
FLUOROETHYL ESTER see FDB200
4-BIPHENYLACETIC ACID, 2-FLUORO-α-
METHYL- see FLG100
(1,1'-BIPHENYL)-4-ACETIC ACID, 2-
FLUORO-α-METHYL- (9CI) see FLG100
4-BIPHENYLACETIC ACID, 2-FLUORO-α-
METHYL-, 1-ACETOXYETHYL ESTER see
FJT100
(1,1'-BIPHENYL)-4-ACETIC ACID, 2-
FLUORO-α-METHYL-, 1-
(ACETYLOXY)ETHYL ESTER see FJT100
BIPHENYLAMINE see AJS100
2-BIPHENYLAMINE see BGE250
4-BIPHENYLAMINE see AJS100
o-BIPHENYLAMINE see BGE250
p-BIPHENYLAMINE see AJS100
(1,1'-BIPHENYL)-4-AMINE see AJS100
(1,1'-BIPHENYL)-2-AMINE (9CI) see BGE250
(1,1'-BIPHENYL)-4-AMINE, 4'-BUTYL- see
BQA100
4-BIPHENYLAMINE, DIHYDROCHLORIDE
see BGE300
4-BIPHENYLAMINE, 4,4'-DIMETHOXY- see
BKO600
(1,1'-BIPHENYL)-4-AMINE, 2',3-DIMETHYL-
N-HYDROXY- see HKB650
(1,1'-BIPHENYL)-2-AMINE, 4,4'-DINITRO-
see DUS100
(1,1'-BIPHENYL)-4-AMINE, 4'-ETHYL- see
EFX600
2-BIPHENYLAMINE, HYDROCHLORIDE
see BGE325
(1,1'-BIPHENYL)-2-AMINE, N-HYDROXY-
see HIT600
(1,1'-BIPHENYL)-4-AMINE, N-HYDROXY-
see BGI250
(1,1'-BIPHENYL)-4-AMINE, 4'-IODO- see
AKL600
N-4-BIPHENYLBENZAMIDE see BGF000
BIPHENYL, mixed with BIPHENYL OXIDE
(3:7) see PFA860
1,1'-BIPHENYL, 3,5-BIS(1-METHYLETHYL)-
4-NITRO- see DNQ890
BIPHENYL, 3-BROMO- see BMW290
(1,1'-BIPHENYL)-4-CARBONITRILE, 4'-
AMINO- see AJJ770

(4-
BIPHENYLCARBONYLMETHYL)DIMETHY
L(2-HYDROXYETHYL)AMMONIUM
BROMIDE see BGF050
(1,1'-BIPHENYL)-4-CARBOXAMIDE, N-(4-(8-
CYANO-1,3A,4,9B-
TETRAHYDRO(1)BENZOPYRANO(3,4-
C)PYRROLE-2(3H)-YL)BU TYL)-, (3AR,9BS)-
see COS780
4-BIPHENYLCARBOXYLIC ACID see
PEL600
(1,1'-BIPHENYL)-4-CARBOXYLIC ACID
(9CI) see PEL600
2-BIPHENYLCARBOXYLIC ACID, 2'-
HYDROXY-5'-NITRO-, Δ-LACTON see
NFV600
BIPHENYL, CHLORO- see CGM550
BIPHENYL, 3-CHLORO- see CGM770
BIPHENYL, 4-CHLORO- see CGM780
1,1'-BIPHENYL, CHLORO- see CGM550
1,1'-BIPHENYL, 3-CHLORO- see CGM770
BIPHENYL, DECABROMO- see PCC480
1,1'-BIPHENYL, 2,2',3,3',4,4',5,5',6,6'-
DECABROMO- see PCC480
2,4'-BIPHENYLDIAMINE see BGF109
4,4'-BIPHENYLDIAMINE see BBX000
(1,1'-BIPHENYL)-2,4'-DIAMINE see BGF109
(1,1'-BIPHENYL)-4,4'-DIAMINE (9CI) see
BBX000
(1,1'-BIPHENYL)-4,4'-DIAMINE,
DIHYDROCHLORIDE see BBX750
(1,1'-BIPHENYL)-4,4'-DIAMINE SULFATE
(1:1) see BBY000
(1,1'-BIPHENYL)-4,4'-DIAMINE, N,N,N',N'-
TETRAMETHYL- see BJF600
4,4'-BIPHENYLDICARBONITRILE see
BGF250
2,2-BIPHENYL DICARBONYL PEROXIDE
see BGF500
2,2'-BIPHENYLDICARBOXALDEHYDE see
DVV800
BIPHENYL, DICHLORO- see DET700
BIPHENYL, 2,4'-DICHLORO- see DET900
BIPHENYL, 4,4'-DICHLORO- see DET850
1,1'-BIPHENYL, 3,3'-DICHLORO- see DET825
1,1'-BIPHENYL, 4,4'-DICHLORO- see DET850
1,1'-BIPHENYL, DICHLORO-(9CI) see
DET700
1,1'-BIPHENYL, 2,4'-DICHLORO-(9CI) see
DET900
1,1'-BIPHENYL, 3,5-DIETHYL-4-NITRO- see
DJP600
4-BIPHENYLDIMETHYLAMINE see BGF899
1,1'-BIPHENYL, 3-(1,1-DIMETHYLETHYL)-
4-NITROSO- see BRY300
2,4-BIPHENYLDIOL see BGF910
2,5-BIPHENYLDIOL see BGG250
2,2'-BIPHENYLDIOL see BGG000
4,4'-BIPHENYLDIOL see BGG500
(1,1'-BIPHENYL)-2,4-DIOL see BGF910
(1,1'-BIPHENYL)-2,5-DIOL see BGG250
2,2'-BIPHENYLDIOL, 5,5'-DIALLYL- see
BFX520
(1,1'-BIPHENYL)-2,2'-DIOL, 5,5'-DI-2-
PROPENYL- see BFX520
2,2'-BIPHENYLDIOL, 3,3',5,5'-
TETRABROMO- see DAZ110
(1,1'-BIPHENYL)-2,2'-DIOL, 3,3',5,5'-
TETRABROMO- see DAZ110
2,2'-BIPHENYLDIOL, 3,3',5,5'-
TETRABROMO-, MONO(DIHYDROGEN
PHOSPHATE) see BNV500
(1,1'-BIPHENYL)-2,2'-DIOL, 3,3',5,5'-
TETRABROMO-, MONO(DIHYDROGEN
PHOSPHATE) (9CI) see BNV500
BIPHENYL-DIPHENYL ETHER mixture see
PFA860
(1,1'-BIPHENYL)-2,2'-DISULFONIC ACID,
4,4'-BIS((4,5-DIHYDRO-3-METHYL-5-OXO-
1-PHENYL-1H-PYRAZOL-4-YL)AZO)-,
DISODIUM SALT see CMM759

(1,1'-BIPHENYL)-2,2'-DISULFONIC ACID,
4,4'-BIS((2-HYDROXY-1-
NAPHTHALENYL)AZO)-, DISODIUM SALT
see CMM325
N,N'-(1,1'-BIPHENYL)-4,4'-DIYLBIS-
ACETAMIDE 4',4'''-BIACETANILIDE see
BFX000
2,2'-(BIPHENYL)-4,4'-DIYLBIS(2-HYDROXY-
4,4-DIMETHYLMORPHOLINIUM) see
HAP100
2,2'-(1,1'-BIPHENYL)-4,4'-DIYLBIS(2-
HYDROXY-4,4-DIMETHYL)-
MORPHOLINIUM DIBROMIDE see HAQ000
2,2'-((1,1'-BIPHENYL)-4,4'-DIYLDI-2,1-
ETHENEDIYL)BIS-BENZENESULFONIC
ACID DISODIUM SALT see TGE150
1,1'-(p'p'-
BIPHENYLENEBIS(CARBONYLMETHYL))
DI-2-PICOLINIUM DIBROMIDE see BGH000
2,2'-(4,4'-BIPHENYLENE)BIS(2-HYDROXY-
4,4-DIMETHYLMORPHOLINIUM) see
HAP100
4,4'-BIPHENYLENEBIS(2-
OXOETHYLENE)BIS(DIMETHYL(2-
HYDROXYETHYL)AMMONIUM)
DIBROMIDE see BGH250
4,4'-BIPHENYLENEBIS(3-
OXOPROPYLENE)BIS(DIMETHYL(2-
HYDROXYETHYL)AMMONIUM)DIBROMI
DE see BGH500
4,4'-BIPHENYLENEDIAMINE see BBX000
o-BIPHENYLENEMETHANE see FDI100
2,2'-BIPHENYLENE OXIDE see DDB500
1,1'-BIPHENYL, 3-ETHYL-4-NITRO- see
ENO300
BIPHENYL, 2-FLUORO- see FGI050
1,1'-BIPHENYL, 2-FLUORO-(9CI) see FGI050
BIPHENYL, 2,2',4,4',6,6'-HEXACHLORO- see
HCC800
1,1'-BIPHENYL, 2,3,3',4,4',5-HEXACHLORO-
see HCC600
1,1'-BIPHENYL, 2,2',3,3',4,4'-HEXACHLORO-
see HCC700
1,1'-BIPHENYL, 2,2',4,4',6,6'-HEXACHLORO-
see HCC800
1,1'-BIPHENYL, 2,3',4,4',5',6-HEXACHLORO-
see HCC830
4-BIPHENYLHYDROXYLAMINE see BGI250
o-BIPHENYLHYDROXYLAMINE see HIT600
BIPHENYL, 4-ISOPROPYL- see IOF250
o-BIPHENYLMETHANE see FDI100
4,4'-BIPHENYLMETHANEBISMALEIMIDE
see BKL800
4-BIPHENYLMETHANOL see HLY400
(1,1'-BIPHENYL)-4-METHANOL see HLY400
BIPHENYL, 4-METHYL- see MJH905
N-(p-BIPHENYLMETHYL)-ATROPINIUM
BROMIDE see PEM750
1,1'-BIPHENYL, 4-(1-METHYLETHYL)- see
IOF250
1,1'-BIPHENYL, 3-(1-METHYLETHYL)-4-
NITRO- see IQQ500
N,4-BIPHENYL-METHYL-dl-TROPEYL-α-
TROPINIUMBROMID (GERMAN) see
PEM750
p-BIPHENYLMETHYL-(dl-TROPYL-α-
TROPINIUM)BROMIDE see PEM750
BIPHENYL, 2-NITROSO- see NJM100
1,1'-BIPHENYL, 2-NITROSO- see NJM100
1,1'-BIPHENYL, 2-NITROSO- see NJM100
1,1'-BIPHENYL, 3-NITROSO- see NJM395
1,1'-BIPHENYL, NONABROMO- see NMV735
2-BIPHENYLOL see BGJ250
4-BIPHENYLOL see BGJ500
o-BIPHENYLOL see BGJ250
(1,1'-BIPHENYL)-2-OL see BGJ250
(1,1'-BIPHENYL)-4-OL, 3-(1-
PYRROLIDINYLMETHYL)- see PPT400
2-BIPHENYLOL, SODIUM SALT see BGJ750
(1,1'-BIPHENYL)-2-OL, SODIUM SALT see
BGJ750

(1,1'-BIPHENYL)-4-OL, 3-((4-(2-(SULFOOXY)ETHOXY)PHENYL)AZO)-, MONOSODIUM SALT see FAL050
BIPHENYL OXIDE see PFA850
1,1'-BIPHENYL, mixed with 1,1'-OXYBIS(BENZENE) see PFA860
(2-BIPHENYLOXY)TRIBUTYLTIN see BGK000
2-BIPHENYLPENICILLIN SODIUM see BGK250
BIPHENYL, PENTACHLORO- see PAV600
1,1'-BIPHENYL, PENTACHLORO-(9CI) see PAV600
BIPHENYL, 2,3',4,4',5-PENTACHLORO- see PAV620
BIPHENYL, 2,3,3',4,4'-PENTACHLORO- see PAV630
1,1'-BIPHENYL, 2,2',3,4,5-PENTACHLORO- see PAV610
1,1'-BIPHENYL, 2,3',4,4',5-PENTACHLORO- see PAV620
1,1'-BIPHENYL, 3,3',4,4',5-PENTACHLORO- see PAV700
1,1'-BIPHENYL, 2,3,3'4,4'-PENTACHLORO-(9CI) see PAV630
BIPHENYL, POLYCHLORO- see PJL750
BIPHENYL SELENIUM see PGH250
4-BIPHENYLSULFONIC ACID, SODIUM SALT see SFC600
BIPHENYL, 2,2',6,6'-TETRACHLORO- see TBO600
1,1'-BIPHENYL, 2,2',3,3'-TETRACHLORO- see TBO520
1,1'-BIPHENYL, 2,2',5,5'-TETRACHLORO- see TBO530
1,1'-BIPHENYL, 2,2',6,6'-TETRACHLORO- see TBO600
3,3',4,4'-BIPHENYLTETRAMINE see BGK500
3,3',4,4'-BIPHENYLTETRAMINE TETRAHYDROCHLORIDE see BGK750
BIPHENYL, TRICHLORO- see TIM100
BIPHENYL, 2',3,4-TRICHLORO- see TIL260
BIPHENYL, 2,4',5-TRICHLORO- see TIL275
BIPHENYL, 2,4,4'-TRICHLORO- see TIL270
1,1'-BIPHENYL, TRICHLORO-(9CI) see TIM100
1,1'-BIPHENYL, 2,2',5-TRICHLORO- see TIL258
1,1'-BIPHENYL, 2,3',4-TRICHLORO- see TIL260
1,1'-BIPHENYL, 2,3',5-TRICHLORO- see TIL262
1,1'-BIPHENYL, 2,4,4'-TRICHLORO- see TIL270
1,1'-BIPHENYL, 2,4',5-TRICHLORO-(9CI) see TIL275
N-(2-BIPHENYLYL)ACETAMIDE see PDY000
N-(3-BIPHENYLYL)ACETAMIDE see PDY250
N-(4-BIPHENYLYL)ACETAMIDE see PDY500
N-(4-BIPHENYLYL)ACETOHYDROXAMIC ACETATE see ABJ750
N-(4-BIPHENYLYL)ACETOHYDROXAMIC ACID ACETATE see ABJ750
N-4-BIPHENYLYLBENZAMIDE see BGF000
N-4-BIPHENYLYLBENZENESULFONAMIDE see BGL000
N-4-BIPHENYLYL BENZENESULFONAMIDE see BGL000
N-4-BIPHENYLYLBENZOHYDROXAMIC ACID see HJP500
1-(α-(4-BIPHENYLYL)BENZYL)IMIDAZOLE see BGA825
3-(4-BIPHENYLYLCARBONYL)PROPIONIC ACID see BGL250
1-(1,1'-BIPHENYL)-4-YL-2-((4-(DICHLOROACETYL)PHENYL)AMINO)-2-HYDROXYETHANONE see BGL400

2-(4-BIPHENYLYL)-5,6-DIHYDRO-S-TRIAZOLO(5,1-A)ISOQUINOLINE see BGL450
1-BIPHENYLYL-3,3-DIMETHYLTRIAZENE see BGL500
2-BIPHENYLYL DIPHENYL PHOSPHATE see XFS300
N,N'-4,4'-BIPHENYLYLENEBISACETAMIDE see BFX000
7,7'-(p,p'-BIPHENYLYLENEBIS(CARBONYLIMINO)) BIS(2-ETHYLQUINOLINIUM) DITOSYLATE see BGM000
7,7'-(4,4'-BIPHENYLYLENEBIS(CARBONYLIMINO)) BIS(1-ETHYLQUINOLINIUM)DI-p-TOLUENESULFONATE see BGM000
(4,4'-BIPHENYLYLENEBIS(2-OXOETHYLENE))BIS(3-IODOPYRIDINIUM) DIBROMIDE see BGM050
(4,4'-BIPHENYLYLENEBIS(2-OXOETHYLENE))-2-PICOLINIUM DIBROMIDE see BGH000
(4,4'-BIPHENYLYLENEBIS(2-OXOETHYLENE))-3-PICOLINIUM DIBROMIDE see BGP750
2,2'-BIPHENYLYLENE SULFIDE see TES300
1-(1,1'-BIPHENYL)-4-YLETHANONE see MHP500
4-(2-(1,1'-BIPHENYL)-4-YLETHOXY)QUINAZOLINE see BGM070
4-BIPHENYLYL ETHYLKETONE see BGM100
N-(4-BIPHENYLYL)FORMOHYDROXAMIC ACID see HLE650
4-BIPHENYLYL GLYCIDYL ETHER see PFU600
N-4-BIPHENYLYL-N-HYDROXYBENZENESULFONAMIDE see BGN000
N-(2-BIPHENYLYL)HYDROXYLAMINE see HIT600
N-4-BIPHENYLYLHYDROXYLAMINE see BGI250
4-BIPHENYLYL METHYL KETONE see MHP500
4-(4-BIPHENYLYL)-4-OXOBUTYRIC ACID see BGL250
β-((1,1'-BIPHENYL)-4-YLOXY)-α-(1,1-DIMETHYLETHYL)-1H-1,2,4-TRIAZOLE-1-ETHANOL see SCF525
α-(4-BIPHENYLYLOXY)PROPIONIC ACID see BGN100
((2-BIPHENYLYLOXY)TRIBUTYL)STANNANE see BGK000
((1,1'-BIPHENYL)-2-YLOXY)TRIBUTYL-(9CI) STANNANE see BGK000
2-(2-BIPHENYLYLOXY)TRIETHYLAMINE HYDROCHLORIDE see BGO000
(2-BIPHENYLYL)PENICILLIN see BGK250
1-((4-BIPHENYLYL)PHENYLMETHYL)-1H-IMIDAZOLE see BGA825
3-(3-BIPHENYL-4-YL-1,2,3,4-TETRAHYDRO-1-NAPHTHYL)-4-HYDROXYCOUMARIN see BGO100
3-(3-(4-BIPHENYLYL)-1,2,3,4-TETRAHYDRO-1-NAPHTHYL)-4-HYDROXYCOUMARIN3-(3-(1,1'-BIPHENYL)-4-YL-1,2,3,4-TETRAHYDRO-1-NAPHTHALENYL)-4-HYDROXY-2H-1-BENZOPYRAN-2-ONE see BGO100
2-(1,1'-BIPHENYL-4-YL)-s-TRIAZOLE(5,1-a)ISOQUINOLINE see BGO325
2-(4-BIPHENYLYL)-5H-s-TRIAZOLO(5,1-A)ISOINDOLE see BGO200
2-(4-BIPHENYLYL)-s-TRIAZOLO(5,1-a)ISOQUINOLINE see BGO325
2-(1,1'-BIPHENYL)-4-YL-(1,2,4)TRIAZOLO(5,1-a)ISOQUINOLINE (9CI) see BGO325

BIPINAL SODIUM see SBN000
BIPINDOGENIN-l-RHAMNOSID (GERMAN) see RFZ000
(1,4'-BIPIPERIDINE)-1'-CARBOXYLIC ACID, 4,11-DIETHYL-3,4,12,14-TETRAHYDRO-4-HYDROXY-3,14-DIOXO-1H-PYRANO(3',4':6,7)INDOLIZINO(1,2-B)QUINOLIN-9-YL ESTER, MONOHYDROCHLORIDE,TRIHYDRATE, (S)- see IGJ550
BIPOTASSIUM CHLORAZEPATE see CDQ250
BIPOTASSIUM CHROMATE see PLB250
(Δ2,2')-BIPSEUDOINDOXYL see BGB275
BIPYRIDINE see BGO500
2,2'-BIPYRIDINE see BGO500
4,4'-BIPYRIDINE see BGO600
α,α'-BIPYRIDINE see BGO500
(3,4'-BIPYRIDINE)-5-CARBONITRILE,1,6-DIHYDRO-2-METHYL-6-OXO- see MQU600
(2,4'-BIPYRIDINE)-3',5'-DICARBOXYLIC ACID, 1',4'-DIHYDRO-2',6'-DIMETHYL-,-DIETHYL ESTER see DJB460
2,2'-BIPYRIDINE, 4,4'-DIMETHYL- see DQS100
2,3'-BIPYRIDINE, 3,4,5,6-TETRAHYDRO- see TCJ825
2,3'-BIPYRIDINE, 1,2,3,6-TETRAHYDRO-1-NITROSO- see NJK300
4,4'-BIPYRIDINIUM, 1,1'-DIMETHYL-, DIBROMIDE see PAI995
BIPYRIDINIUM, 1,1'-DIMETHYL-4,4'-, DICHLORIDE see PAJ000
4,4-BIPYRIDYL see BGO600
2,2'-BIPYRIDYL see BGO500
4,4'-BIPYRIDYL see BGO600
α,α'-BIPYRIDYL see BGO500
γ,γ'-BIPYRIDYL see BGO600
BIPYROMUCYL see FPZ000
BIQUIN DURULES see QFS100
BIRCH TAR OIL see BGO750
BIRCH TAR OIL, RECTIFIED (FCC) see BGO750
BIRD of PARADISE see CAK325
BIRD PEPPER see PCB275
BIRMI, LEAF EXTRACT see KCA100
BIRNENOEL see AOD725
BIRTHWORT see AQY250
BIRUTAN see RSU000
BISABOLOL see BGO775
(−)-α-BISABOLOL see BGO775
2,7-BIS(ACETAMIDO)FLUORENE see BGP250
BIS(4-ACETAMIDOPHENYL)SULFONE see SNY500
BIS(p-ACETAMIDOPHENYL) SULFONE see SNY500
BIS-4-ACETAMINO PHENYL SELENIUMDIHYDROXIDE see BGP500
BIS(ACETATO-O)(ω-(2-(ACETYLAMINO)-5-NITRO-1,3-PHENYLENE)DI)-MERCURY see BGQ250
BIS(ACETATO-O,O')OXOZIRCONIUM see ZTS000
BIS(ACETATO)PALLADIUM see PAD300
BIS(ACETATO)TETRAHYDROXYTRILEAD see LCH000
BIS(ACETATO)TRIHYDROXYTRILEAD see LCJ000
BIS(ACETO)DIHYDROXYTRILEAD see LCH000
BIS(ACETO)DIOXOURANIUM DIHYDRATE see UQT700
BIS(ACETO-O)DIOXOURANIUM DIHYDRATE see UQT700
4,4'-BISACETOPHENONE-α,α'-DI-(3-METHYLPYRIDINIUM) DIBROMIDE see BGP750
BIS-(ACETOXYAETHYL)NITROSAMIN (GERMAN) see NKM500
BIS(ACETOXY)CADMIUM see CAD250

BIS(ACETOXYDIBUTYLSTANNANE) OXIDE see BGQ000
N,N-BIS(ACETOXYETHYL)ACETAMIDE see BGQ050
BIS(2-ACETOXYETHYL)SULFONE see BGQ100
2,6-BIS(ACETOXYMERCURI)-4-NITROACETANILIDE see BGQ250
2,6-BIS(ACETOXYMERCURI)-4-NITROANILINE see BGQ300
9,10-BISACETOXYMETHYL-1,2-BENZANTHRACENE see BBF750
2,3-BIS(ACETOXYMETHYL)QUINOXALINE see QRJ100
2,3-BIS(ACETOXYMETHYL)QUINOXALINE DI-N-OXIDE see QTS100
BIS-(p-ACETOXYPHENYL)-CYCLOHEXYLIDENEMETHANE see FBP100
BIS(p-ACETOXYPHENYL)-2-METHYLCYCLOHEXYLIDENEMETHANE see BGQ325
BIS(p-ACETOXYPHENYL)-2-PYRIDYLMETHANE see PPN100
BIS(ACETYLACETONATO) TITANIUM OXIDE see BGQ750
BIS(ACETYL ACETONE)COPPER see BGR000
2,5-BIS(ACETYLAMINO)FLUORENE see BGR250
3,5-BIS(ACETYLAMINO)-2,4,6-TRIIODOBENZOIC ACID see DCK000
4,4'-BIS(N-ACETYL-N-METHYLAMINO)AZOBENZENE see DQH200
1,1'-(3,17-BIS(ACETYLOXY)ANDROSTANE-2,16-DIYL)BIS(1-METHYLPIPERIDINIUM) DIBROMIDE see PAF625
1,1'-((2-β,3-α,5-α,16-β,17-β)-3,17-BIS(ACETYLOXY)ANDROSTANE-2,16-DIYL)BIS(1-METHYLPIPERIDINIUM) DIBROMIDE see BGR325
17,21-BIS(ACETYLOXY)-2-BROMO-6-β,9-DIFLUORO-11-β-HYROXYPREGNA-1,4-DIEN-3,20-DIONE see HAG325
BIS(ACETYLOXY)DIBUTYLSTANNANE see DBF800
17,21-BIS(ACETYLOXY)-6,9-DIFLUORO-11-HYDROXY-16-METHYLPREGNA-1,4-DIENE-3,20-DIONE (6-α,11-β,16-β)- see DKF125
BIS(ACETYLOXY)MERCURY see MCS750
2,2-BIS((ACETYLOXY)METHYL)-1,3-PROPANEDIOL DIACETATE see NNR400
BISACETYLPALLADIUM see PAD300
BIS(ACETYL-N-PHENYLCARBAMYLMETHYL)-4,4'-DISAZO-3,3'-DICHLOROBIPHENYL see DEU000
BISACODYL see PPN100
BIS(ACRYLONITRILE) NICKEL (O) see BGR500
S-(1,2-BIS(AETHOXY-CARBONYL)-AETHYL)-O,O-DIMETHYL-DITHIOPHOSPHAT (GERMAN) see MAK700
2,4-BIS(AETHYLAMINO)-6-CHLOR-1,3,5-TRIAZIN (GERMAN) see BJP000
BISAKLOFEN BP see MJO500
1,4-BIS(4-ALDOXIMINOPYRIDINIUM)BUTANEDIOL-2,3-BIBROMIDE see BGR750
1,5-BIS(4-ALDOXIMINOPYRIDINIUM)DIETHYLETHER BIBROMIDE see OPO100
1,3-BIS(4-ALDOXIMINOPYRIDINIUM) DIMETHYL ETHER BICHLORIDE see BGS250
BIS-3-ALLOXY-2-HYDROXYPROPYL-1-ESTER KYSELINY FUMAROVE (CZECH) see BGS750

2,4-BIS(ALLYLAMINO)-6-(4-(BIS-(p-FLUOROPHENYL)METHYL)-1-PIPERAZINYL)-s-TRIAZINE see BGS500
1-(4,6-BISALLYLAMINO-s-TRIAZINYL)-4-(p,p'-DIFLUOROBENZHYDRYL)-PIPERAZINE see BGS500
BIS(ALLYLDITHIOCARBAMATO)ZINC see ZCS000
BIS(3-ALLYLOXY-2-HYDROXYPROPYL) FUMARATE see BGS750
4,5-BIS(ALLYLOXY)-2-IMIDAZOLINDINONE see BGS825
BISAMIDINE see APL250
BIS AMINE see MJM200
BIS(4-AMINO-1-ANTHRAQUINONYL)AMINE see IBD000
BIS(AMINOBUTYL)TETRAMETHYLDISILOXANE see OKU300
1,3-BIS(4-AMINOBUTYL)TETRAMETHYLDISILOXANE see OKU300
1,3-BIS(4-AMINOBUTYL)-1,1,3,3-TETRAMETHYLDISILOXANE see OKU300
1,3-BIS(Δ-AMINOBUTYL)TETRAMETHYLDISILOXANE see OKU300
BIS(4-AMINO-3-CHLOROPHENYL) ETHER see BGT000
BIS(4-AMINOCYCLOHEXYL)METHANE see MJQ260
BIS(p-AMINOCYCLOHEXYL)METHANE see MJQ260
BIS(2-AMINOETHYL)AMINE see DJG600
BIS(β-AMINOETHYL)AMINE see DJG600
BIS(2-AMINOETHYL)AMINE COBALT(III) AZIDE see BGT125
BIS(2-AMINOETHYL)AMINEDIPEROXOCHROMIUM(IV) see BGT150
N,N'-BIS(2-AMINOETHYL)-1,2-DIAMINOETHANE see TJR000
BIS(β-AMINOETHYL)DISULFIDE see MCN500
N,N-BIS(2-AMINOETHYL)-1,2-ETHANEDIAMINE see NEI800
N,N'-BIS(2-AMINOETHYL)ETHYLENEDIAMINE see TJR000
N,N'-BIS(2-AMINOETHYL)-1,2-ETHYLENEDIAMINE see TJR000
BIS-p-AMINOFENYLMETHAN see MJQ000
4,4'-BIS(1-AMINO-8-HYDROXY-2,4-DISULFO-7-NAPHTHYLAZO)-3,3'-BITOLYL, TETRASODIUM SALT see BGT250
4,4'-BIS(7-(1-AMINO-8-HYDROXY-2,4-DISULFO)NAPHTHYLAZO)-3,3' -BITOLYL, TETRASODIUM SALT see BGT250
4,4'-BIS(1-AMINO-8-HYDROXY-2,4-DISULPHO-7-NAPHTHYLAZO)-3,3' -BITOLYL, TETRASODIUM SALT see BGT250
1,3-BIS-AMINOMETHYLBENZEN (CZECH) see XHS800
1,4-BIS-AMINOMETHYLBENZEN (CZECH) see PEX250
1,3-BIS(AMINOMETHYL)CYCLOHEXANE see BGT500
1,4-BIS(AMINOMETHYL)CYCLOHEXANE see BGT750
BIS(4-AMINO-3-METHYLCYCLOHEXYL)METHANE see BGT800
BIS-4-AMINO-3-METHYLFENYLMETHAN (CZECH) see MJO250
BIS(2-AMINO-1-NAPHTHYL)SODIUM PHOSPHATE see BGU000
1,3-BIS(4-AMINOPHENOXY)BENZENE see REF070
2,2-BIS(4-(4-AMINOPHENOXY)PHENYL)HEXAFLUOROPROPANE see TKK025

BIS(4-AMINOPHENOXYPHENYL)SULFONE see SNZ000
BIS(2-AMINOPHENYL)DISULFIDE see DXJ800
BIS(o-AMINOPHENYL)DISULFIDE see DXJ800
1,1'-BIS(2-AMINOPHENYL)DISULFIDE see DXJ800
BIS(4-AMINOPHENYL)ETHER see OPM000
BIS(p-AMINOPHENYL)ETHER see OPM000
BIS(4-AMINOPHENYL)METHANE see MJQ000
BIS(p-AMINOPHENYL)METHANE see MJQ000
2',4-BIS(AMINOPHENYL)METHANE see MJP750
2,4-BIS((4-AMINOPHENYL)METHYL)BENZENAMINE see BGU100
BIS(4-AMINOPHENYL) SULFIDE see TFI000
BIS(p-AMINOPHENYL)SULFIDE see TFI000
BIS(4-AMINOPHENYL) SULFONE see SOA500
BIS(m-AMINOPHENYL) SULFONE see SOA000
BIS(p-AMINOPHENYL) SULFONE see SOA500
BIS(4-AMINOPHENYL) SULPHIDE see TFI000
BIS(p-AMINOPHENYL)SULPHIDE see TFI000
BIS(4-AMINOPHENYL)SULPHONE see SOA500
BIS(p-AMINOPHENYL)SULPHONE see SOA500
BIS(o-AMINOPHENYLTHIO)ETHANE see EJC050
BIS(2-AMINOPHENYLTHIO)ZINC see BGV500
2,2-BIS(p-AMINOPHENYL)-1,1,1-TRICHLOROETHANE see BGU500
1,4-BIS(3-AMINOPROPOXY)BUTANE see BGU600
1,4-BIS(γ-AMINOPROPOXY)BUTANE see BGU600
BIS-(3-AMINOPROPYL)AMINE see AIX250
1,4-BIS(AMINOPROPYL) BUTANEDIAMINE see DCC400
N,N'-BIS(3-AMINOPROPYL)-1,4-BUTANEDIAMINE see DCC400
1,4-BIS(AMINOPROPYL)BUTANEDIAMINE TETRAHYDROCHLORIDE see GEK000
N,N'-BIS(3-AMINOPROPYL)-1,4-DIAMINOBUTANE see DCC400
N,N'-BIS(3-AMINOPROPYL)DIAMINOETHANE see TBI700
N,N'-BIS(3-AMINOPROPYL)ETHYLENEDIAMINE see TBI700
BIS(3-AMINOPROPYL)METHYLAMINE see BGU750
N,N-BIS(3-AMINOPROPYL)METHYLAMINE see BGU750
BIS(γ-AMINOPROPYL)METHYLAMINE see BGU750
BIS(ω-AMINOPROPYL)METHYLAMINE see BGU750
N,N-BIS(γ-AMINOPROPYL)METHYLAMINE see BGU750
BIS(3-AMINOPROPYL)METHYLAMINE-EPICHLOROHYDRIN COPOLYMER see EAZ600
N,N-BIS(3-AMINOPROPYL)METHYLAMINE-EPICHLOROHYDRIN COPOLYMER see EAZ600
1,4-BIS(AMINOPROPYL)PIPERAZINE see BGV000

BIS(AMINOPROPYL)PIPERAZINE (DOT) see BGV000

1,3-BIS(3-AMINOPROPYL)-1,1,3,3-TETRAMETHYLDISILOXANE see OPC100

BIS(2-AMINOTHIOPHENOL), ZINC SALT see BGV500

1,3-BIS(5-AMINO-1,3,4-TRIAZOL-2-YL)TRIAZENE see BGV750

BISANILINE-P see BGV800

2,2,-BIS(p-ANILINE)-1,1,1-TRICHLOROETHANE see BGU500

4,4'-BIS((4-ANILINO-6-BIS(2-HYDROXYETHYL)AMINO-w-TRIAZIN-2-YL)AMINO)-2,2'-STILBENEDISULFONIC ACID DISODIUM SALT see BGW000

4,4'-BIS(((4-ANILINO-6-METHOXY-s-TRIAZIN-2-YL)AMINO)-2,2'-STILBENEDISULFONIC ACID) DISODIUM SALT see BGW100

4,4'-BIS((4-ANILINO-6-MORPHOLINO-1,3,5-TRIAZINE-2-YL)AMINO)STILBENE 2,2-DISULFONATE 2Na see CMP200

BIS(p-ANISYLAMINE) see BKO600

2,2-BIS(p-ANISYL)-1,1,1-TRICHLOROETHANE see MEI450

1,4-BIS-1'-ANTHRACHINONYLAMINO-ANTHRACHINON (CZECH) see APL750

1,5-BIS-1'-ANTHRACHINONYLAMINO-ANTHRACHINON (CZECH) see APM000

BISANTRENE HYDROCHLORIDE see BGW325

BIS(AQUO)-N,N'-(2-IMINO-5-PHENYL-4-OXAZOLIDINONE)MAGNESIUM(II) see PAP000

cis-BISASCORBATO(RACEMIC-1,2-DIAMINOCYCLOHEXANE)PLATINUM(II) HYDRATE see BGW400

BIS-A-TDA see MJL750

2,5-BIS-ATHYLENIMINOBENZOCHINON-1,4 (GERMAN) see BGW750

BIS-o-AZIDO BENZOYL PEROXIDE see BGW500

1,2-BIS(AZIDOCARBONYL)CYCLOPROPANE see BGW650

BIS(2-AZIDOETHOXYMETHYL)NITRAMINE see BGW700

1,5-BIS-(4-AZIDOFENYL)-1,4-PENTADIEN-3-ON see BGW720

N,N-BIS(AZIDOMETHYL)NITRIC AMIDE see DCM499

3,3-BIS(AZIDOMETHYL)OXETANE see BGW710

1,5-BIS(p-AZIDOPHENYL)-1,4-PENTADIEN-3-ONE see BGW720

BIS(AZIDOTHIOCARBONYL)DISULFIDE see BGW739

N,N'-BIS(AZIRIDINEACETYL)-1,8-OCTAMETHYLENEDIAMINE see BGX500

2,5-BIS(AZIRIDINO)BENZOQUINONE see BGW750

N,N'-BIS-AZIRIDINYLACETYL-1,4-CYCLOHEXYLDIMETHYLENEDIAMINE see CPL100

N,N'-BIS(AZIRIDINYLACETYL)-1,8-OCTAMETHYLENE DIAMINE see BGX500

P,P-BIS(1-AZIRIDINYL)-p-(1-ADAMANTYL)-PHOSPHINE OXIDE see BJP325

BIS(1-AZIRIDINYL)AMINOPHOSPINE SULFIDE see DNU850

2,5-BIS(1-AZIRIDINYL)-3,6-BIS(2-METHOXYETHOXY)-p-BENZOQUINONE see BDC750

2,5-BIS(1-AZIRIDINYL)-3,6-BIS(2-METHOXYETHOXY)-2,5-CYCLOHEXADIENE-1,4-DIONE see BDC750

2,5-BIS(1-AZIRIDINYL)-3-(2-CARBAMOYLOXY-1-METHOXYETHYL)-6-METHYL-1,4-BENZOQUINONE see BGX750

2,4-BIS(1-AZIRIDINYL)-6-CHLOROPYRIMIDINE see EQI600

2,6-BIS(1-AZIRIDINYL)-4-CHLOROPYRIMIDINE see EQI600

p,p-BIS(1-AZIRIDINYL)-2,5-DIIODOBENZOYLPHOSPINIC AMIDE see DNE800

p,p-BIS(1-AZIRIDINYL)-N,N-DIMETHYLAMINOPHOSPHINE OXIDE see DOV600

p,p-BIS(1-AZIRIDINYL)-N,N-DIMETHYLPHOSPHINIC AMIDE see DOV600

2,5-BIS(1-AZIRIDINYL)-3,6-DIOXO-1,4-CYCLOHEXADIENE-1,4-DICARBAMIC ACID DIETHYL ESTER see ASK875

2,5-BIS(1-AZIRIDINYL)3,6-DIPROPOXY-2,5-CYCLOHEXADIENE-1,4-DIONE (9CI) see DCN500

P,P-BIS(1-AZIRIDINYL)-N-ETHYLAMINOPHOSPHINE OXIDE see BGX775

P,P-BIS(1-AZIRIDINYL)-N-ETHYLPHOSPHINIC AMIDE see BGX775

N,N'-BIS-AZIRIDINYLFORMYL-1,4-XYLENEDIAMINE see XSJ000

2,5-BIS(1-AZIRIDINYL)-3-(2-HYDROXY-1-METHOXYETHYL)-6-METHYL-p-BENZOQUINONE CARBAMATE (ESTER) see BGX750

p,p-BIS(1-AZIRIDINYL)-N-ISOPROPYLAMINOPHOSPHINE OXIDE see BGX850

p,p-BIS(1-AZIRIDINYL)-N-ISOPROPYLPHOSPHINIC AMIDE see BGX850

BIS(1-AZIRIDINYL)KETONE see BJP450

P,P-BIS(1-AZIRIDINYL)-N-METHYLPHOSPHINIC AMIDE see MGE100

BIS(1-AZIRIDINYL)(2-METHYL-3-THIAZOLIDINYL)PHOSPHINE OXIDE see BGY000

p,p-BIS(1-AZIRIDINYL)PHOSPHINOTHIOIC AMIDE see DNU850

N-(BIS(1-AZIRIDINYL)PHOSPHINYL)BENZAMIDE see BGY050

N-(BIS(1-AZIRIDINYL)PHOSPHINYL)-p-BROMOBENZAMIDE see BGY056

(BIS(1-AZIRIDINYL)PHOSPHINYL)CARBAMIC ACID, ETHYL ESTER see EHV500

N-(BIS(1-AZIRIDINYL)PHOSPHINYL)-p-CHLOROBENZAMIDE see BGY125

N-(BIS(1-AZIRIDINYL)PHOSPHINYL)-2,5-DIIODOBENZAMIDE see DNE800

N-(BIS(1-AZIRIDINYL)PHOSPHINYL)-m-IODOBENZAMIDE see BGY130

N-(BIS(1-AZIRIDINYL)PHOSPHINYL)-o-IODOBENZAMIDE see BGY135

N-(BIS(1-AZIRIDINYL)PHOSPHINYL)-p-IODOBENZAMIDE see BGY140

p,p-BIS(1-AZIRIDINYL)-N-PROPYLAMINOPHOSPHINE OXIDE see BGY500

p,p-BIS(1-AZIRIDINYL)-N-PROPYLPHOSPHINIC AMIDE see BGY500

2,5-BIS(1-AZIRIDYNYL)BENZOQUINONE see BGW750

BIS(2-BENZAMIDOPHENYL) DISULFIDE see BDK800

BIS(o-BENZAMIDOPHENYL) DISULFIDE see BDK800

BIS-BENZENE CHROMIUM see BGY700

BIS(BENZENE)CHROMIUM IODIDE see BGY720

BIS(BENZENE)CHROMIUM(1+)IODIDE see BGY720

BIS BENZENE DIAZO OXIDE see BGY750

BIS(n-BENZENE)IRON(O) see BGZ000

BISBENZIMIDAZO(2,1-b:2',1'-i)BENZO(lmn)(3,8)PHENANTHROLINE-8,17-DIONE see CMU820

BISBENZIMIDAZOLE see MOD500

BISBENZIMIDE see BGZ100

BIS(BENZOATO)DIOXOCHROMIUM TRIHYDRATE see BHA000

BIS(1,3)BENZODIOXOLO(5,6-A:4',5'-G)QUINOLIZINIUM, 6,7-DIHYDRO- see CNR100

4H-BIS(1,3)BENZODIOXOLO(5,6-A:4',5'-G)QUINOLIZIN-4-ONE, 6,7-DIHYDRO- see ONO500

BIS(BENZOTHIAZOLYL)DISULFIDE see BDE750

N,N'-BIS(2-BENZOTHIAZOLYLTHIOMETHYLENE)UREA see BHA500

1,3-BIS((2-BENZOTHIAZOLYLTHIO)METHYL)UREA see BHA500

BIS(2-BENZOTHIAZOLYLTHIO)ZINC see BHA750

BIS(2-BENZOTHIAZYL) DISULFIDE see BDE750

4,4'-BIS-BENZOYLAMINO-1,1'-DIANTHRIMID (CZECH) see IBJ000

4,5'-BIS-BENZOYLAMINO-1,1'-DIANTHRIMID (CZECH) see IBE000

BIS-o-BENZOYLAMINOFENYL-DISULFID see BDK800

5,4'-BIS-BENZOYLAMINO-8-METHOXY-1,1'- DIANTHRIMID (CZECH) see MES500

BIS(2-BENZOYLBENZOATO)BIS(3-(1-METHYL-2-PYRROLIDINYL)PYRIDINE) NICKEL TRIHYDRATE see BHB000

1,3-BIS(BENZOYLOXY)BENZENE see BHB100

2,2-BIS((BENZOYLOXY)METHYL)-1,3-PROPANEDIOL, DIBENZOATE see PBC000

1,1-BIS(BENZOYLPEROXY)CYCLOHEXANE see BHB250

1,2-BIS(BENZYLAMINO)ETHANE see BHB300

cis-BIS(2,2'-BIPYRIDINE)DICHLORORHODIUM CHLORIDE see BHB400

(OC-6-22)-BIS(2,2'-BIPYRIDINE-N,N')DICHLORORHODIUM CHLORIDE see BHB400

1,4-BIS(BIS(1-AZIRIDINYL)PHOSPHINYL)PIPERAZINE see BJC250

(4,6-BIS(BIS(BUTOXYMETHYL)AMINO)-s-TRIAZIN-2-YLIMINO)DIMETHANOL see BHB500

(N,N-BIS(2-(BIS(CARBOXYMETHYL)AMINO)ETHYL)GLYCINATO(5-))TECHNETATE(1-)-9-9TC SODIUM see TAI200

BIS((4-(BIS(2-CHLOROETHYL)AMINO)BENZENE)ACETATE)ESTRA-1,3,5(10)-TRIENE-3,17-DIOL(17-β) see EDR500

BIS((4-(BIS(2-CHLOROETHYL)AMINO)BENZENE)ACETATE)OESTRA-1,3,5(10)-TRIENE-3,17-DIOL(17-β) see EDR500

2,5-BIS(BIS-(2-CHLOROETHYL)AMINOMETHYL)HYDROQUINONE see BHB750

BIS((p-(BIS(2-CHLOROETHYL)AMINO)PHENYL)ACETATE)ESTRADIOL see EDR500

BIS((p-(BIS(2-CHLOROETHYL)AMINO)PHENYL)ACETATE)ESTRA-1,3,5(10)-TRIENE-3,17-β-DIOL see EDR500

BIS((p-(BIS(2-CHLOROETHYL)AMINO)PHENYL)ACETATE)OESTRADIOL see EDR500

BIS((p-BIS(2-
CHLOROETHYL)AMINOPHENYL)ACETAT
E)OESTRA-1,3,5(10)-TRIENE-3,17-β-DIOL
see EDR500
BIS(BISDIMETHYLAMINOPHOSPHONOUS
)ANHYDRIDE see OCM000
4,4'-BIS((4-BIS((2-
HYDROXYETHYL)AMINO)-6-CHLORO-s-
TRIAZIN-2-YL)AMINO)-2,2'-
STILBENEDISULFONIC ACID, DISODIUM
SALT see BHB950
2,6-BIS(BIS(2-HYDROXYETHYL)AMINO)-
4,8-DIPIPERIDINOPYRIMIDO(5,4-
d)PYRIMIDINE see PCP250
4,4'-BIS((4-BIS(2-HYDROXYETHYL)AMINO-
6-METHOXY-s-TRIAZIN-2-YL)AMINO)-2,2'-
STILBENEDISULFONIC ACID DISODIUM
SALT see BHC500
4,4'-BIS((4-BIS((2-
HYDROXYETHYL)AMINO)-6-(m-
SULFOANILINO)-s-TRIAZIN-2-YL)AMINO)-
2,2'-STILBENEDISULFONIC ACID
TETRASODIUM SALT see BHC750
BIS(BIS(β-
HYDROXYETHYL)SULFONIUMMETHYL)SU
LFIDE DICHLORIDE see BHD000
BIS-(β-o-METHOXYPHENYL-ISOPROPYL)-
AMINE LACTATE see BKP200
BIS-BP see BIV825
1,4-BIS(BROMOACETOXY)-2-BUTENE see
BHD150
1,2-BIS(BROMOACETOXY)ETHANE see
BHD250
p-(BIS(2-BROMOETHYL)AMINO)BENZOIC
ACID see BHD300
p-(BIS(2-BROMOETHYL)AMINO)PHENOL-
m-(α,α,α-TRIFLUOROMETHYL)BENZOATE
see BHG000
N,N-BIS(2-BROMOETHYL)ANILINE see
BHG100
2,2-BIS(BROMOMETHYL)-1,3-
PROPANEDIOL see DDQ400
2,3-BIS(BROMOMETHYL)QUINOXALINE
see BHI750
3,12-BIS(3-BROMO-1-OXOPROPYL)-3,12-
DIAZA-6,9-
DIAZONIADISPIRO(5.2.5.2)HEXADECANE
DICHLORIDE see SLD600
BIS(p-BROMOPHENYL) ETHER see BHJ000
1,4-BIS(3-BROMOPROPIONYL)-
PIPERAZINE see BHJ250
BISBUTENYLENETETRAHYDROFURFURA
L see BHJ500
2,3,4,5-BIS(2-
BUTENYLENE)TETRAHYDROFURFURAL
see BHJ500
2,3,4,5-BIS(Δ²-
BUTENYLENE)TETRAHYDROFURFURAL
see BHJ500
4,5-BIS(2-BUTENYLOXY)-2-
IMIDAZOLIDINONE see BHJ625
BIS(2-BUTOXYETHYL) ADIPATE see
BHJ750
BIS(BUTOXYETHYL) ETHER see DDW200
BIS(2-BUTOXYETHYL) ETHER see DDW200
BIS(2-BUTOXYETHYL)PHTHALATE see
BHK000
BIS(BUTOXYMALEOYLOXY)DIBUTYLSTA
NNANE see BHK250
BIS(BUTOXYMALEOYLOXY)DIOCTYLSTA
NNANE see BHK500
2,5-BIS(5-tert-BUTYLBENZOXAZOL-2-
YL)THIOPHENE see BHK600
BIS(BUTYLCARBITOL)FORMAL see BHK750
BIS(4-tert-BUTYL-m-CRESYL)SULFIDE see
BHL100
2,3,4,5-BIS(2-BUTYLENE)TETRAHYDRO-2-
FURALDEHYDE see BHJ500
2,3:4,5-BIS(2-BUTYLENE)TETRAHYDRO-2-
FURFURAL see BHJ500

BIS-Δ²-
BUTYLENETETRAHYDROFURFURAL see
BHJ500
2,3,4,5-BIS(Δ²-
BUTYLENE)TETRAHYDROFURFURAL see
BHJ500
BIS(2-BUTYL)ETHER see BRH760
BIS(2-BUTYL)ETHER see OPE030
BIS(3-tert-BUTYL-4-HYDROXY-6-
METHYLPHENYL) SULFIDE see TFC600
2,2'-BIS-6-terc.BUTYL-p-KRESYLMETHAN
(CZECH) see MJO500
α-α'-BIS(tert-
BUTYLPEROXY)DIISOPROPYLBENZENE
see BHL100
2,6-BIS(tert-BUTYL)PHENOL see DEG100
BIS-(4-tert-BUTYLPYRIDINE)-1-
METHYLETHER DICHLORIDE see SAB800
BIS(n-BUTYL)SEBACATE see DEH000
BIS(BUTYLTHIO)DIMETHYL STANNANE
see BHL500
BIS(BUTYLTHIO)DIMETHYLTIN see
BHL500
BIS-BUTYLXANTHOGEN see BSS550
BIS(10-
CAMPHORSULFONATO)DIETHYLSTANN
ANE see DKC600
N,N'-BISCARBAMOYLHYDRAZINE see
HGU050
1,3-BIS(CARBAMOYLTHIO)-2-(N,N-
DIMETHYLAMINO)PROPANE
HYDROCHLORIDE see BHL750
S-(1,2-BIS(CARBETHOXY)ETHYL)-O,O-
DIMETHYL DITHIOPHOSPHATE see
MAK700
BIS(1-(CARBO-β-
DIETHYLAMINOETHOXY)-1-
PHENYLCYCLOPENTANE)ETHANE
DISULFONATE see CBG250
2,3-
BIS(CARBOMETHOXYMERCAPTO)QUINO
XALINE see BHL800
BIS(CARBONATO(2-
))DIHYDROXYTRIBERYLLIUM see BFP500
(Z,Z)-BIS((3-
CARBOXYACRYLOYL)OXY)DIOCTYL-
STANNANE DIISOOCTYL ESTER (8CI) see
BKL000
BIS-β-CARBOXYETHYLGERMANIUM
SESQUIOXIDE see CCF125
BIS(2-CARBOXYETHYL) SULFIDE see
BHM000
3,6-BIS(CARBOXYMETHYL)-3,5-
DIAZOOCTANEDIOIC ACID see EIX000
N,N'-
BIS(CARBOXYMETHYL)DITHIOOXAMIDE
see BHM250
BIS(CARBOXYMETHYL)ETHER see
ONQ100
N,N-BIS(CARBOXYMETHYL)GLYCINE see
AMT500
N,N-BIS(CARBOXYMETHYL)GLYCINE
DISODIUM SALT see DXF000
N,N-BIS(CARBOXYMETHYL)GLYCINE
MONOSODIUM SALT see NEH800
N,N-BIS(CARBOXYMETHYL)GLYCINE
SODIUM SALT see SEP500
N,N-BIS(CARBOXYMETHYL)GLYCINE
TRISODIUM SALT MONOHYDRATE see
NEI000
p-
(BIS(CARBOXYMETHYLMERCAPTO)ARSIN
O)BENZAMIDE see TFA350
BIS(CARBOXYMETHYLMERCAPTO)(p-
CARBAMYLPHENYL)-ARSINE see TFA350
BIS(CARBOXYMETHYLMERCAPTO)(p-
UREIDOPHENYL)ARSINE see CBI250
BIS(CARBOXYMETHYLTHIO)(p-
UREIDOPHENYL)ARSINE see CBI250
BIS(2-CARBOXYPHENYL) DISULFIDE see
BHM300

BIS(o-CARBOXYPHENYL) DISULFIDE see
BHM300
p-(BIS(o-CARBOXYPHENYLMERCAPTO)-
ARSINO)-PHENYLUREA see TFD750
BIS(3-CARBOXYPROPIONYL)PEROXIDE
see BHM500
BIS(3-CARBOXYPROPIONYL) PEROXIDE
see SNC500
N,N-BIS-(β-CHLORAETHYL)-AMIN
(GERMAN) see BHN750
3-(BIS-(2-
CHLORAETHYL)AMINOMETHYL)BENZO
XAZOLON-(2) (GERMAN) see BHQ750
N-(BIS-(2-
CHLORAETHYL)AMINOMETHYLBENZYL
URETHAN) (GERMAN) see BEB000
4-(BIS-(2-
CHLORAETHYL)AMINOMETHYL)-1-
PHENYL-2,3-DIMETHYLPYRAZOLON
HYDROCHLORID (GERMAN) see BHR500
N,N-BIS-(β-CHLORAETHYL)-N',O-
PROPYLEN-PHOSPHORSAEURE-ESTER-
DIAMID (GERMAN) see CQC650
BIS-(4-CHLORBUT-1-YL)-ETHER see
OPE040
BIS-(2-CHLORETHYL)VINYLFOSFONAT
(CZECH) see VQF000
BIS(5-CHLOR-2-HYDROXYPHENYL)-
METHAN see MJM500
BIS(p-CHLOROBENZOYL) PEROXIDE see
BHM750
trans-N,N'-BIS(2-CHLOROBENZYL)-1,4-
CYCLOHEXANEBIS(METHYLAMINE)
DIHYDROCHLORIDE see BHN000
1,3-BIS((p-
CHLOROBENZYLIDENE)AMINO)GUANID
INE see RLK890
4,4'-BIS(4-CHLORO-6-BIS(2-
HYDROXYETHYLAMINO))-s-TRIAZIN-2-
YL-AMINO-2,2'-STILBENEDISULFONIC
ACID see BHN250
BIS(4-CHLOROBUTYL) ETHER see OPE040
1,2-
BIS((CHLOROCARBONYL)OXY)ETHANE
see EIQ000
1,2-BIS(2-CHLOROETHOXY)ETHANE see
TKL500
BIS(2-CHLOROETHOXY)METHANE see
BID750
N,N-BIS(2-CHLOROETHYL)ACETAMIDE
see BHN300
N,N-BIS(β-CHLOROETHYL)ACETAMIDE
see BHN300
N,N-BIS(β-CHLOROETHYL)-dl-ALANINE
HYDROCHLORIDE see BHN500
BIS(2-CHLOROETHYL)AMINE see BHO000
BIS-β-CHLOROETHYLAMINE see BHN750
BIS(2-CHLOROETHYL)AMINE
HYDROCHLORIDE see BHO250
N,N-BIS(2-CHLOROETHYL)AMINE
HYDROCHLORIDE see BHO250
BIS(β-CHLOROETHYL)AMINE
HYDROCHLORIDE see BHO250
4'-(BIS(2-
CHLOROETHYL)AMINO)ACETANILIDE
see BHO500
4-(BIS(2-CHLOROETHYL)AMINO)-
BENZENEACETIC ACID (9CI) see PCU425
4-(BIS(2-
CHLOROETHYL)AMINO)BENZENEBUTA
NOIC ACID see CDO500
4-(BIS(2-CHLOROETHYL)AMINO)-
BENZENEPENTANOIC ACID (9CI) see
BHY625
4-(BIS(2-
CHLOROETHYL)AMINO)BENZOIC ACID
see BHP125
p-(BIS(2-
CHLOROETHYL)AMINO)BENZOIC ACID
see BHP125

5-(4-BIS(2-CHLOROETHYL)AMINOPHENYL)PENTA NOIC ACID see BHY625

o-(4-(BIS(2-CHLOROETHYL)AMINO)PHENYL)-dl-TYROSINE see BHY500

5-(p-(BIS(2-CHLOROETHYL)AMINO)PHENYL)VALERI C ACID see BHY625

β-(BIS(2-CHLOROETHYLAMINO))PROPIONITRILE see BHY750

9-((3-(BIS(2-CHLOROETHYL)AMINO)PROPYL)AMINO) ACRIDINE DIHYDROCHLORIDE see BHZ000

N,N-BIS(β-CHLOROETHYL)-AMINO-N'-O-PROPYLENE-PHOSPHORIC ACID ESTER DIAMIDE see IMH000

N,N-BIS(2-CHLOROETHYL)-o-(3-AMINOPROPYL)PHOSPHORAMIDATE, ZWITTERION see CQN125

5-(BIS(2-CHLOROETHYL)AMINO)-2,4(1H,3H)PYRIMIDINEDIONE see BIA250

2-(BIS(2-CHLOROETHYL)AMINO)TETRAHYDROO XAZAPHOSPHORINE CYCLOHEXYLAMINE SALT see CQN000

(BIS(CHLORO-2-ETHYL)AMINO)-2-TETRAHYDRO-3,4,5,6-OXAZAPHOSPHORINE-1,3,2-OXIDE-2-MONOHYDRATE see CQC500

3-BIS(2-CHLOROETHYL)AMINO-p-TOLUIC ACID see AFQ575

1-(3-(BIS(2-CHLOROETHYL)AMINO)-p-TOLUOYL)PIPERIDINE see BIA100

o-(4-BIS(β-CHLOROETHYL)AMINO-o-TOLYLAZO)BENZOIC ACID see BIA000

3-(BIS(2-CHLOROETHYL)AMINO)-p-TOLYL PIPERIDYL KETONE see BIA100

2-(BIS(2-CHLOROETHYL)AMINO)-6-TRIFLUOROMETHYLTETRAHYDRO-2H-1,3,2-OXAZAPHOSPHORINE 2-OXIDE see TKD000

5-(BIS(2-CHLOROETHYL)AMINO)URACIL see BIA250

5-N,N-BIS(2-CHLOROETHYL)AMINOURACIL see BIA250

BIS(2-CHLOROETHYL)AMMONIUM CHLORIDE see BHO250

N,N-BIS(2-CHLOROETHYL)ANILINE see AOQ875

N,N-BIS(2-CHLOROETHYL)-p-ARSANILIC ACID see BIA300

N,N-BIS(2-CHLOROETHYL)BENZENAMINE see AOQ875

BIS(2-CHLOROETHYL) 1,2-BENZENEDICARBOXYLATE see BIG600

N,N-BIS(2-CHLOROETHYL)BENZENEMETHANAMI NE see BIA750

N,N-BIS(2-CHLOROETHYL)BENZENEMETHANAMI NE HYDROCHLORIDE see EAK000

BIS(2-CHLOROETHYL)BENZYLAMINE see BIA750

N,N-BIS(2-CHLOROETHYL)BENZYLAMINE see BIA750

N,N-BIS(2-CHLOROETHYL)BENZYLAMINE HYDROCHLORIDE see EAK000

N-N-BIS(2-CHLOROETHYL)BUTYLAMINE see DEV300

N,N-BIS(2-CHLOROETHYL)BUTYLAMINE HYDROCHLORIDE see BIB250

N,N-BIS(2-CHLOROETHYL)-p-CHLOROBENZYLAMINE HYDROCHLORIDE see BIB750

BIS-CHLOROETHYL 2-CHLOROETHANEPHOSPHONATE see BIB800

BIS(2-CHLOROETHYL) (2-CHLOROETHYL)PHOSPHONATE see BIB800

BIS(β-CHLOROETHYL) β-CHLOROETHYLPHOSPHONATE see BIB800

O,O-BIS(2-CHLOROETHYL)-O-(3-CHLORO-4-METHYL-7-COUMARINYL) PHOSPHATE see DFH600

N,N-BIS(2-CHLOROETHYL)-2-CHLOROPROPYLAMINE HYDROCHLORIDE see NOB700

BIS(β-CHLOROETHYL)-β-CHLOROPROPYLAMINE HYDROCHLORIDE see NOB700

2,3-(N,N(₁)-BIS(2-CHLOROETHYL)DIAMIDO)-1,3,2-OXAZAPHOSPHORIDINOXY see IMH000

1,4-BIS(2-CHLOROETHYL)-1,4-DIAZONIABICYCLO(2.2.1)HEPTANE (Z)-2-BUTENEDIOATE (1:2) see BIC325

N,N'-BIS(2-CHLOROETHYL)-N,N'-DIETHYLETHYLENEDIAMINE DIHYDROCHLORIDE see BIC300

N,N-BIS(2-CHLOROETHYL)-2,3-DIMETHOXYANILINE see BIC600

BIS(2-CHLOROETHYL) ETHER see DFJ050

BIS(α-CHLOROETHYL) ETHER see BID000

BIS(β-CHLOROETHYL) ETHER see DFJ050

BIS(2-CHLOROETHYL)ETHYLAMINE see BID250

N,N-BIS(2-CHLOROETHYL)-p-FLUOROBENZYLAMINE HYDROCHLORIDE see FHP100

BIS(2-CHLOROETHYL)FORMAL see BID750

BIS(β-CHLOROETHYL)FORMAL see BID750

N,N-BIS(2-CHLOROETHYL)FURFURYLAMINE HYDROCHLORIDE see FPX000

N,N-BIS(β-CHLOROETHYL)GLYCINE HYDROCHLORIDE see GHE000

N,N-BIS(β-CHLOROETHYL)GLYCINE HYDROCHLORIDE see GHE000

9-(2',2'-BIS-β-CHLORO-ETHYL-HYDRAZINO)ACRIDINE HYDROCHLORIDE see BID800

9-(2,2-BIS(2-CHLOROETHYL)HYDRAZINO)ACRIDINE MONOHYDROCHLORIDE see BID800

N,N-BIS(2-CHLOROETHYL)-N'-(3-HYDROXYPROPYL)PHOSPHORODIAMID ATE, CYCLOHEXYLAMMONIUM SALT see CQN000

N,N-BIS(2-CHLOROETHYL)-N'-(3-HYDROXYPROPYL)PHOSPHORODIAMIDI C ACID intramol. ESTER see CQC650

N,N-BIS(2-CHLOROETHYL)ISOBUTYLAMINE HYDROCHLORIDE see BIE750

N,N-BIS(2-CHLOROETHYL)ISOPROPYLAMINE HYDROCHLORIDE see IPG000

1,2-BIS(2-CHLOROETHYLMERCAPTO)ETHANE see SCB500

2,2'-BIS(2-CHLOROETHYLMERCAPTO)-N-METHYLDIETHYLAMINE HYDROCHLORIDE see MHQ500

BIS (β-CHLOROETHYL)-α-METHOXYETHYLAMINE HYDROCHLORIDE see MEN775

BIS(2-CHLOROETHYL)METHYLAMINE see BIE250

N,N-BIS(2-CHLOROETHYL)METHYLAMINE see BIE250

BIS(β-CHLOROETHYL)METHYLAMINE see BIE250

BIS(2-CHLOROETHYL)METHYLAMINE HYDROCHLORIDE see BIE500

N,N-BIS(2-CHLOROETHYL)-4-(6-(5-(4-METHYL-1-PIPERAZINYL)(2,5'-BI-1H-BENZIMI DAZOL)-2'-YL)HEXYL)-BENZENAMINE, HYDROCHLORIDE, HYDRATE (2:6:3) see BIE600

N,N-BIS(2-CHLOROETHYL)-4-(3-(5-(4-METHYL-1-PIPERAZINYL)(2,5'-BI-1H-BENZIMI DAZOL)-2'-YL)PROPYL)BENZENAMIDE see BIE610

N,N-BIS(2-CHLOROETHYL)-2-METHYL-1-PROPANAMINE HYDROCHLORIDE see BIE750

(N,N-BIS(2-CHLOROETHYL))-2-METHYLPROPYLAMINE HYDROCHLORIDE see BIE750

N,N-BIS(2-CHLOROETHYL)-2-NAPHTHYLAMINE see BIF250

BIS(2-CHLOROETHYL)-β-NAPHTHYLAMINE see BIF250

BIS(2-CHLOROETHYL)NITROSOAMINE see BIF500

N,N'-BIS((2-CHLOROETHYL)-N-NITROSOCARBAMOYL)CYSTAMINE see BIF625

BISCHLOROETHYLNITROSOUREA see BIF750

BIS(2-CHLOROETHYL)NITROSOUREA see BIF750

1,3-BIS-(2-CHLOROETHYL)-1-NITROSOUREA see BIF750

N,N'-BIS(2-CHLOROETHYL)-N-NITROSOUREA see BIF750

1,3-BIS(β-CHLOROETHYL)-1-NITROSOUREA see BIF750

1,3-BIS(2-CHLOROETHYL)-1-NITROSOUREA-DIPHENYLMETHANE see BIG000

N,N-BIS-(β-CHLOROETHYL)-N',N''-PROPYLENEPHOSPHORICACIDTRIAMID E see BHQ000

N,N-BIS(β-CHLOROETHYL)-N',O-PROPYLENEPHOSPHORIC ACID ESTER AMINE MONOHYDRATE see CQC500

N,N-BIS(2-CHLOROETHYL)-N',O-PROPYLENEPHOSPHORIC ACID ESTER DIAMIDE see CQC650

N,N-BIS(β-CHLOROETHYL)-N',O-TRIMETHYLENEPHOSPHORIC ACID ESTER DIAMIDE MONOHYDRATE see CQC500

N,N-BIS(β-CHLOROETHYL)-N',O-TRIMETHYLENEPHOSPHORIC ACID ESTER DIAMIDE see CQC650

N,N-BIS(2-CHLOROETHYL)-p-PHENYLENEDIAMINE see BIG250

BIS(2-CHLOROETHYL)PHOSPHITE see BIG500

BIS(2-CHLOROETHYL)PHOSPHORAMIDE CYCLIC PROPANOLAMIDE ESTER MONOHYDRATE see CQC500

BIS(2-CHLOROETHYL)PHOSPHORAMIDE-CYCLIC PROPANOLAMIDE ESTER see CQC650

N,N-BIS(2-CHLOROETHYL)PHOSPHORODIAMIDAT E HYDRACRYLIC ACID see CCE250

N,N-BIS(2-CHLOROETHYL)PHOSPHORODIAMIDIC ACID, CYCLOHEXYL AMMONIUM SALT see PHA750

N,N-BIS(2-CHLOROETHYL)-N'-3-PHOSPHORODIAMIDIC ACID HYDROXYLPROPYLCYCLOHEXYLAMINE SALT see CQN000

BIS(2-CHLOROETHYL) PHTHALATE see BIG600

N,N'-BIS(2-CHLOROETHYL)-1,4-PIPERAZINE HYDROCHLORIDE see BIG750

N⁴,N⁴-BIS(2-CHLOROETHYL)SULFANILAMIDE see BIH000

BIS(2-CHLOROETHYL)SULFIDE see BIH250

BIS(β-CHLOROETHYL)SULFIDE see BIH250

1,1-BIS(2-CHLOROETHYL)-2-SULFINYLHYDRAZINE see BIH325

BIS(2-CHLOROETHYL)SULFONE see BIH500

BIS(β-CHLOROETHYL)SULFONE see BIH500

BIS(2-CHLOROETHYLSULFONYLMETHYL)ETHER see OPG100

BIS(2-CHLOROETHYL)SULPHIDE see BIH250

N,N-BIS(2-CHLOROETHYL)TETRAHYDRO-2H-1,3,2-OXAPHOSPHORIN-2-AMINE-2-OXIDE see CQC650

N,N-BIS(2-CHLOROETHYL)TETRAHYDRO-2H-1,3,2-OXAPHOSPHORIN-2-AMINE-2-OXIDE MONOHYDRATE see CQC500

N,3-BIS(2-CHLOROETHYL)TETRAHYDRO-2H-1,3,2-OXAZAPHOSPHORIN-2-AMINE 2-OXIDE see IMH000

BIS(1-CHLOROETHYL THALLIUM CHLORIDE) OXIDE see BIH750

N,N-BIS(2-CHLOROETHYL)-2-THENYLAMINE HYDROCHLORIDE see BII000

BIS(2-CHLOROETHYLTHIO)ETHANE see SCB500

1,2-BIS(β-CHLOROETHYLTHIO)ETHANE see SCB500

BIS(2-CHLOROETHYLTHIOETHYL) ETHER see DFK200

BIS(β-CHLOROETHYLTHIOETHYL) ETHER see DFK200

5-(3,3-BIS(2-CHLOROETHYL)-1-TRIAZENO)IMIDAZOLE-4-CARBOXAMIDE see IAN000

BIS(5-CHLORO-2-HYDROXYPHENYL)METHANE see MJM500

3,12-BIS(3-CHLORO-2-HYDROXYPROPYL)-3,12-DIAZA-6,9-DIAZONIADISPIRO(5.2.5.2)HEXADECANE, DICHLORIDE see POC000

BIS(2-CHLOROISOPROPYL) ETHER see BII250

BIS(β-CHLOROISOPROPYL)ETHER see BII250

2,5-BIS(CHLOROMERCURI)FURAN see BII500

N,N-BIS(CHLOROMERCURI)HYDRAZINE see BII750

4,5-BIS(CHLOROMERCURI)-2-THIAZOLECARBAMIC ACID BENZYL ESTER see BIJ000

BIS-1,2-(CHLOROMETHOXY)ETHANE see BIJ250

1,4-BIS(CHLOROMETHOXYMETHYL)BENZENE see BIJ500

BIS-1,4-(CHLOROMETHOXY)-p-XYLENE see BIJ500

9,10-BIS(CHLOROMETHYL)ANTHRACENE see BIJ750

BIS(CHLOROMETHYL)BENZENE see XSS250

1,2-BIS(CHLOROMETHYL)BENZENE see DGP400

1,3-BIS(CHLOROMETHYL)BENZENE see DGP200

1,4-BIS(CHLOROMETHYL)BENZENE see DGP600

BIS(CHLOROMETHYL) ETHER see BIK000

BIS(2-CHLORO-1-METHYLETHYL) ETHER see BII250

BIS(2-CHLORO-1-METHYLETHYL)ETHER mixed with 2-CHLORO-1-METHYLETHYL-(2-CHLOROPROPYL) ETHER see BIK100

BIS(2-CHLORO-1-METHYLETHYL)ETHER mixed with 2-CHLORO-1-METHYLETHYL-(2-CHLOROPROPYL) ETHER see EEG100

2,3-BIS(CHLOROMETHYL)-1,4,5,6,7,7-HEXACHLOROBICYCLO(2.2.1)HEPT-4-ENE see BFY100

BIS(CHLOROMETHYL)KETONE see BIK250

3,3-BIS(CHLOROMETHYL)OXETANE see BIK325

2,2-BIS(CHLOROMETHYL)-1,3-PROPANEDIOL see BIK500

2,2-BIS(CHLOROMETHYL)-1,3-PROPANEDIOL SULFATE see BIK750

BIS(2-CHLOROMETHYL-2-PROPYL)SULFIDE see BIL000

1,3-BIS(CHLOROMETHYL)-1,1,3,3-TETRAMETHYLDISILAZANE see BIL250

BIS(p-CHLOROPHENOXY)METHANE see NCM700

BIS(2-(p-CHLOROPHENOXY)-2-METHYLPROPIONATO)HYDROXYALUMINUM see AHA150

BIS(4-CHLOROPHENYL)ACETIC ACID see BIL500

BIS(p-CHLOROPHENYL)ACETIC ACID see BIL500

O,O-BIS(p-CHLOROPHENYL)ACETIMIDOYLPHOSPHORAMIDOTHIOATE see BIM000

2,7-BIS(4-CHLOROPHENYL)BENZO(lmn)(3,8)PHENANTHROLINE-1,3,6,8(2H,7H)-TETRONE see BIM100

1,6-BIS(5-(p-CHLOROPHENYL)BIGUANDINO)HEXANE DIGLUCONATE see CDT250

1,6-BIS(5-(p-CHLOROPHENYL)BIGUANIDINO)HEXANE see BIM250

1,6-BIS(5-(p-CHLOROPHENYL)BIGUANIDINO)HEXANE DIACETATE see CDT125

1,1-BIS(p-CHLOROPHENYL)-2-CHLOROETHYLENE see BIM300

3,3-BIS(4-CHLOROPHENYL)CYCLOPENTANAMINE HYDROCHLORIDE see AJM550

BIS(4-CHLOROPHENYL)DIAZENE 1-OXIDE see DEP450

1,1-BIS(4-CHLOROPHENYL)-2,2-DICHLOROETHANE see BIM500

1,1-BIS(p-CHLOROPHENYL)-2,2-DICHLOROETHANE see BIM500

2,2-BIS(4-CHLOROPHENYL)-1,1-DICHLOROETHANE see BIM500

2,2-BIS(p-CHLOROPHENYL)-1,1-DICHLOROETHANE see BIM500

2,2-BIS(p-CHLOROPHENYL)-1,1-DICHLOROETHYLENE see BIM750

1,6-BIS(p-CHLOROPHENYLDIGUANIDO)HEXANE see BIM250

2,2-BIS(p-CHLOROPHENYL)ETHANE see BIM775

1,1-BIS(4-CHLOROPHENYL)ETHANOL see BIN000

1,1-BIS(p-CHLOROPHENYL)ETHANOL see BIN000

BIS(CHLOROPHENYL) ETHER see DFQ000

BIS(p-CHLOROPHENYL)METHANE see BIM800

BIS(p-CHLOROPHENYL)METHYL CARBINOL see BIN000

1,1-BIS(p-CHLOROPHENYL)METHYLCARBINOL see BIN000

1,1-BIS(p-CHLOROPHENYL)-2-NITROPROPANE see BIN500

1,1-BIS(p-CHLOROPHENYL)-2-NITROPROPANE mixed with 1,1-BIS(p-CHLOROPHENYL)-2-NITROBUTANE(1:2) see BON250

BIS(p-CHLOROPHENYL)SULFONE see BIN750

BIS(p-CHLOROPHENYL)SULFOXIDE see BIN900

BIS(p-CHLOROPHENYLTHIO)DIMETHYL STANNANE see BIO500

BIS(p-CHLOROPHENYLTHIO)DIMETHYLTIN see BIO500

1,1-BIS-(p-CHLOROPHENYL)-2,2,2-TRICHLOROETHANE see DAD200

2,2-BIS(p-CHLOROPHENYL)-1,1,1-TRICHLOROETHANE see DAD200

2,2-BIS(o,p-CHLOROPHENYL)-1,1,1-TRICHLOROETHANE see BIO625

α,α-BIS(p-CHLOROPHENYL)-β,β,β-TRICHLOROETHANE see DAD200

1,1-BIS(CHLOROPHENYL)-2,2,2-TRICHLOROETHANOL see BIO750

1,1-BIS(4-CHLOROPHENYL)-2,2,2-TRICHLOROETHANOL see BIO750

1,1-BIS(p-CHLOROPHENYL)-2,2,2-TRICHLOROETHANOL see BIO750

3,6-BIS(5-CHLORO-2-PIPERIDYL)-2,5-PIPERAZINEDIONE DIHYDROCHLORIDE see PIK075

BIS(1-CHLORO-2-PROPYL) ETHER see BII250

BIS-5-CHLORO TOLUENE DIAZONIUM ZINC TETRACHLORIDE see BIP250

1,3-BIS(6-CHLORO-o-TOLYL)-1-(2-(DIETHYLAMINO)ETHYL)UREA HYDROCHLORIDE see BIP500

1,3-BIS(6-CHLORO-o-TOLYL)-1-(2-PYRROLIDINYLETHYL)UREA HYDROCHLORIDE see BIP750

BIS(2-CHLOROVINYL)CHLOROARSINE see BIQ250

1,1-BIS(4-CHLORPHENYL)-AETHANOL (GERMAN) see BIN000

BIS(p-CHLORPHENYL)ESSIGSAEURE (GERMAN) see BIL500

5-(BIS(2-CLOROETIL)-AMINO)-6-METILURACILE (ITALIAN) see DYC700

BIS-CME see BIK000

BISCOMATE see PFA500

BIS-CONRAY see IDJ500

BIS(CUMENE)CHROMIUM see DGR200

BIS(pi-CUMENE)CHROMIUM see DGR200

BIS-(2-CYANOETHYL)AMINE see BIQ500

N,N-BIS(2-CYANOETHYL)AMINE see BIQ500

BIS(β-CYANOETHYL)AMINE see BIQ500

BIS(2-CYANOETHYL) N,N'-(DITHIOBIS((METHYLIMINO)CARBONYLOXY))DIETHANIMIDOTHIOATE see BIQ630

N,N-BIS(2-CYANOETHYL)-N-4-HYDROXY-1-ANTHRAQUINONYLSULFONILAMIDE see BIQ600

N,N'-BIS-(1-(2-CYANOETHYL)THIOACETALDEHYDE o-(N-METHYLCARBAMOYL)OXIME)DISULFIDE see BIQ630

N,N'-BIS(2-CYANO-2-METHYLPROPIONALDEHYDEO-(N-METHYLCARBAMOYL)OXIME)SULFIDE see BIQ660

cis-BIS(CYCLOBUTYLAMMINE)DICHLOROPLATINUM(II) see DGT200

BISCYCLOHEXANONE OXALDIHYDRAZONE see COF675

BIS(CYCLOHEXYL)CARBODIIMIDE see MDR800

BIS(CYCLOHEXYL)CARBOXYLIC ACID DIETHYLAMINOETHYL ESTER HYDROCHLORIDE see ARR750

BIS(CYCLOHEXYL) DISULFIDE see CPK700

BIS-O,O-DIETHYLPHOSPHOROTHIONIC ANHYDRIDE see SOD100

BIS(DIETHYLTHIOCARBAMOYL) DISULFIDE see DXH250

BIS(N,N-DIETHYLTHIOCARBAMOYL) DISULFIDE see DXH250

BIS(DIETHYLTHIOCARBAMOYL)DISULFIDE mixed with SODIUM NITRITE see SIS200

BIS(N,N-DIETHYLTHIOCARBAMOYL) DISULPHIDE see DXH250

BIS(DIETHYLTHIO)CHLORO METHYL PHOSPHONATE see BJD000

BIS(2,2-DIFLUOROAMINO)-1,3-BIS(DINITRO-FLUOROETHOXY)PROPANE see SPA500

BIS(DIFLUOROAMINO)DIFLUOROMETHANE see BJD250

1,1-BIS(DIFLUOROAMINO)-2,2-DIFLUORO-2-NITROETHYL METHYL ETHER see BJD375

1,2-BIS(DIFLUOROAMINO)ETHANOL see BJD500

1,2-BIS(DIFLUOROAMINO)ETHYL VINYL ETHER see BJD750

4,4-BIS(DIFLUOROAMINO)-3-FLUOROIMINO-1-PENTENE see BJE000

1,2-BIS(DIFLUOROAMINO)-N-NITROETHYLAMINE see BJE250

BIS(DIFLUOROBORYL)METHANE see BJE325

3-(4,6-BIS(DIFLUOROMETHOXY)PYRIMIDIN-2-YL)-1-(2-METHOXYCARBONYLPHENYLSULFONYL)UREA see TCI150

(T-4-(R)(R)-BIS(N-(2-DIHYDROXY-3,3-DIMETHYL)-1-OXOBUTYL-β-ALANIN-ATO)ZINC see ZKS000

2,2-BISDIHYDROXYMETHYL-1,3-PROPANEDIOL TETRANITRATE see PBC250

1,4-BIS(3,4-DIHYDROXYPHENYL)-2,3-DIMETHYLBUTANE see NBR000

N,N'-BIS(2-(3',4'-DIHYDROXYPHENYL)-2-HYDROXYETHYL)HEXAMETHYLENEDIAMINE DIHYDROCHLORIDE see HFG600

N,N'-BIS(2-(3',4'-DIHYDROXYPHENYL)-2-HYDROXYETHYL)HEXAMETHYLENEDIAMINE SULFATE see HFG650

BIS(DIHYDROXYPHENYL)SULFIDE see BJE500

BIS(2,6-DIISOPROPYLPHENYL)CARBODIIMIDE see BJE550

2,5-BIS(3,4-DIMETHOXYPHENYL)-1,3,4-THIADIAZOLE see BJE600

3,5-BIS(3,4-DIMETHOXYPHENYL)-1H-1,2,4-TRIAZOLE see BJE700

BIS(DIMETHYLAMIDO)FLUORO PHOSPHATE see BJE750

BIS(DIMETHYLAMIDO)PHOSPHORYL FLUORIDE see BJE750

2,8-BISDIMETHYLAMINOACRIDINE see BJF000

3,6-BIS(DIMETHYLAMINO)ACRIDINE see BJF000

BIS(DIMETHYLAMINO)-3-AMINO-5-PHENYLTRIAZOLYL PHOSPHINE OXIDE see AIX000

p-BIS(DIMETHYLAMINO)BENZENE see BJF500

1,4-BIS(DIMETHYLAMINO)BENZENE see BJF500

4,4'-BIS(DIMETHYLAMINO)BENZHYDRYLIDENIMINE HYDROCHLORIDE see IBA000

4,4'-BIS(DIMETHYLAMINO)BENZOHYDROL see TDO750

4,4'-BIS(DIMETHYLAMINO)BENZOPHENONE see MQS500

p,p'-BIS(N,N-DIMETHYLAMINO)BENZOPHENONE see MQS500

4,4'-BIS(DIMETHYLAMINO)BENZOPHENONE-IMINE HYDROCHLORIDE see IBA000

3,6-BIS(DIMETHYLAMINO)-10-BENZYLACRIDINIUM CHLORIDE see BDX033

4,4'-BIS(N,N-DIMETHYLAMINO)BIPHENYL see BJF600

BIS(DIMETHYLAMINOBORANE)ALUMINUM TETRAHYDROBORATE see BJG000

BIS((DIMETHYLAMINO)CARBONOTHIOYL) DISULPHIDE see TFS350

2,2'-BIS(DIMETHYLAMINO) DIETHYLSULPHIDE DIHYDROCHLORIDE see BJG100

BIS(DIMETHYLAMINO)DIMETHYLSTANNANE see BJG125

4,4'-BIS(DIMETHYLAMINO)DIPHENYLMETHANE see MJN000

p,p'-BIS(DIMETHYLAMINO)DIPHENYLMETHANE see MJN000

3,5-BIS-DIMETHYLAMINO-1,2,4-DITHIAZOLIUM CHLORIDE see BJG150

1,2-BIS-(DIMETHYLAMINO)-ETHANE (DOT) see TDQ750

BIS(2-DIMETHYLAMINOETHOXY)ETHANE see BJG250

3,6-BIS(2-(DIMETHYLAMINO)ETHOXY)-9H-XANTHEN-9-ONE DIHYDROCHLORIDE see BJG500

1,3-BIS(2-DIMETHYLAMINOETHYL)ADAMANTANE DIHYDROCHLORIDE see BJG750

1-(BIS(2-(DIMETHYLAMINO)ETHYL)AMINO)-5-METHYL-3-PHENYLINDOLE DIHYDROCHLORIDE see BJH000

1-(BIS(2-(DIMETHYLAMINO)ETHYL)AMINO)-3-PHENYLINDOLE DIHYDROCHLORIDEHYDRATE see BJH500

BIS(2-DIMETHYLAMINOETHYL) ETHER see BJH750

BIS(2-DIMETHYLAMINOETHYL)SUCCINATE BIS(METHOCHLORIDE) see HLC500

BIS(β-DIMETHYLAMINOETHYL)SUCCINATE BIS(METHYLIODIDE) see BJI000

BIS(DIMETHYLAMINO)FLUOROPHOSPHATE see BJE750

BISDIMETHYLAMINOFLUOROPHOSPHINE OXIDE see BJE750

BIS(DIMETHYLAMINO)ISOPROPYLMETHACRYLATE see BJI125

3,7-BIS(DIMETHYLAMINO)-4-NITRO-PHENOTHIAZIN-5-IUM, CHLORIDE see MJU250

3,7-BIS(DIMETHYL AMINO)PHENAZA THIONIUM CHLORIDE see BJI250

3,7-BIS(DIMETHYLAMINO)PHENOTHIAZIN-5-IUM CHLORIDE see BJI250

BIS(p-(N,N-DIMETHYLAMINO)PHENYL)KETONE see MQS500

BIS(p-DIMETHYLAMINOPHENYL)METHANE see MJN000

BIS(p-(N,N-DIMETHYLAMINO)PHENYL)METHANE see MJN000

p,p'-BIS(N,N-DIMETHYLAMINOPHENYL)METHANE see MJN000

α,α-BIS(p-DIMETHYLAMINOPHENYL)METHANOL see TDO750

BIS(4-(DIMETHYLAMINO)PHENYL)METHANONE see MQS500

BIS(p-DIMETHYLAMINOPHENYL)METHYLENE IMINE see IBB000

1,1-BIS(p-DIMETHYLAMINOPHENYL)METHYLENIMINEHYDROCHLORIDE see IBA000

BIS(p-(DIMETHYLAMINO)PHENYL)PHENYLMETHANOL see MAK500

BIS(DIMETHYLAMINO)PHOSPHONOUS ANYHYDRIDE see OCM000

BIS(DIMETHYLAMINO)PHOSPHORIC ANHYDRIDE see OCM000

1,3-BIS(2-DIMETHYLAMINOPROPYL)ADAMANTANE DIHYDROCHLORIDE see BJI750

N,N'-BIS(3-DIMETHYLAMINOPROPYL)DITHIOOXAMIDE see BJJ000

BIS(DIMETHYLAMINO)SULFOXIDE see BJJ125

BIS(DIMETHYLARSINYLDIAZOMETHYL) MERCURY see BJJ200

BIS-DIMETHYL ARSINYL OXIDE see BJJ250

BIS-DIMETHYL ARSINYL SULFIDE see BJJ500

BIS(α,α-DIMETHYLBENZYL)PEROXIDE see DGR600

1,2-BIS(3,7-DIMETHYL-5-n-BUTOXY-1-AZA-5-BORA-4,6-DIOXOCYCLOOCTYL)ETHANE see BJJ750

BIS(1,3-DIMETHYL)BUTYLAMINE see TDN250

BIS(DIMETHYLCARBAMODITHIOATO)((1,2-ETHANEDIYLBIS(CARBAMODITHIOATO))(2-)) DIZINC see BJK000

BIS(DIMETHYLCARBAMODITHIOATO-S,S')LEAD see LCW000

BIS(DIMETHYLCARBAMODITHIOATO-S,S')ZINC see BJK500

BIS(DIMETHYLDITHIOCARBAMATE de ZINC) (FRENCH) see BJK500

BIS(DIMETHYLDITHIOCARBAMATO)NICKEL see BJK250

BIS(DIMETHYLDITHIOCARBAMATO)ZINC see BJK500

BIS(DIMETHYLDITHIOCARBAMIATO)LEAD see LCW000

2,5-BIS(1,1-DIMETHYLETHYL)-2,5-CYCLOHEXADIENE-1,4-DIONE see DDV550

N,N'-BIS(1,1-DIMETHYLETHYL)-1,2-ETHANEDIAMINE see DEC100

(R*,S*)-1,1'-(1,2-BIS(1,1-DIMETHYLETHYL)-1,2-ETHANEDIYL)BIS(4-CHLOROBENZENE) see BJK525

3,5-BIS(1,1-DIMETHYLETHYL)-4-HYDROXY-BENZENEMETHANOL see IFX200

3,5-BIS(1,1-DIMETHYLETHYL)-4-HYDROXY-, 1,2-ETHANEDIYLBIS (OXY-2,1-ETHANEDIYL) ESTER BENZENEPROPANOIC ACID see BJK550

3-((3,5-BIS(1,1-DIMETHYLETHYL)-4-HYDROXYPHENYL)METHYLENE)-1-METHOXY-2-PYRROLIDINONE see MEL100

((3,5-BIS(1,1-DIMETHYLETHYL)-4-HYDROXYPHENYL)METHYLENE)PROPANEDINITRILE see DED000

((3,5-BIS(1,1-DIMETHYLETHYL)-4-HYDROXYPHENYL)METHYL)PHOSPHONIC ACID, MONOETHYL ESTER, NICKEL(2+) SALT (2:1) see BJK560

(((3,5-BIS(1,1-DIMETHYLETHYL)-4-HYDROXYPHENYL)METHYL)THIO)ACETIC ACID 2-ETHYLHEXYL ESTER see BJK600

2,4-BIS(1,1-DIMETHYLETHYL)-6-(1-(4-METHOXYPHENYL)ETHYL)PHENOL see BJK580

2,6-BIS(1,1-DIMETHYLETHYL)-4-METHYLPHENOL see BFW750

2,4-BIS(1,1-DIMETHYLETHYL)-6-(1-PHENYLETHYL)PHENOL see BJK650

BIS((4-(1,1-DIMETHYLETHYL)PHENYL)METHYL) 3-PYRIDINYLCARBONIMIDODITHIOATE see BJK660

2-((3,5-BIS(1,1-DIMETHYL)-4-HYDROXYPHENYL)METHYLENE)PROPANEDINITRILE see DED000

BIS((5,5-DIMETHYL-2-ISOPROPYLIMINO-4-(o-(N-METHYLCARBAMOYL)OXIMINO)-1,3-DITHIOLANE))SULFIDE see BJK670

1,1'-BIS(3,5-DIMETHYLMORPHOLINOCARBONYLMETHYL)-4,4'-BIPYRIDINIUM-DICHLORID (GERMAN) see BJK750

1,1'-BIS(3,5-DIMETHYLMORPHOLINOCARBONYLMETHYL)-4,4'-BIPYRIDYNIUM DICHLORIDE see BJK750

1,1'-BIS(2-(3,5-DIMETHYL-4-MORPHOLINYL)-2-OXOETHYL)-4,4'-BIPYRIDINIUM DICHLORIDE see BJK750

11,11-BIS-((3,7-DIMETHYL-2,6-OCTADIENYL)OXY)-1-UNDECENE see UNA100

1,1'-BIS(DIMETHYLOCTOXYSILYL)FERROCENE see BJK780

1,1'-BIS(DIMETHYL(OCTYLOXY)SILYL)FERROCENE see BJK780

N,N'-BIS(1,4-DIMETHYLPENTYL)-p-PHENYLENEDIAMINE see BJL000

2,5-BIS(1,1-DIMETHYLPROPYL)HYDROQUINONE see DCH400

cis-BIS(3,5-DIMETHYLPYRIDINE)DICHLOROPLATINUM see BJL100

1,3-BIS(N,N-DIMETHYL-2-PYRROLIDINIUM)PROPANE DIBENZENESULFONATE see COG500

1,3-BIS(N,N-DIMETHYL-2-PYRROLIDINIUM)PROPANONE DIIODIDE see COG750

1,4-BIS(DIMETHYLSILYL)BENZENE see PEW725

BIS(DIMETHYLSILYL) ETHER see TDP775

BIS(DIMETHYLSILYL) OXIDE see TDP775

BISDIMETHYL STIBINYL OXIDE see BJL250

BIS(DIMETHYL THALLIUM)ACETYLIDE see BJL500

BIS(DIMETHYL-THIOCARBAMOYL)-DISULFID (GERMAN) see TFS350

BIS(DIMETHYLTHIOCARBAMOYL) DISULFIDE see TFS350

BIS(DIMETHYLTHIOCARBAMOYL)DISULFIDE and NITROSOTRIS(DIMETHYLDITHIOCARBAMATO)IRON see FAZ000

BIS(DIMETHYLTHIOCARBAMOYL)SULFIDE see BJL600

BIS(DIMETHYLTHIOCARBAMOYLTHIO)METHYL-ARSINE see USJ075

BIS(DIMETHYLTHIOCARBAMYL)MONOSULFIDE see BJL600

BIS(DIMETHYL(VINYL)SILYL)AMINE see TDQ050

BIS(N,N-DIMETIL-DITIOCARBAMMATO) DI ZINCO (ITALIAN) see BJK500

BIS(β-(N,N-DIMORPHOLINO)ETHYL)SELENIDE DIHYDROCHLORIDE see MRU255

1,3-BIS(2,2,-DINITRO-2-FLUOROETHOXY)-N,N,N'N'- TETRAFLUORO-2,2-PROPANEDIAMINE see SPA500

BIS(DINITROPROPYL)FORMAL see MJO800

BIS(2,2-DINITROPROPYL)FORMAL see MJO800

(±)-1,2-BIS(3,5-DIOXOPIPERAZINE-1-YL)PROPANE see PIK250

(±)-1,2-BIS(3,5-DIOXOPIPERAZINYL)PROPANE see PIK250

2,6-BIS(DIPHENYLHYDROXYMETHYL)PIPERIDINE see BJM250

2,6-BIS(DIPHENYLHYDROXYMETHYL)PYRIDINE see BJM500

3-(BIS(3,3-DIPHENYLPROPYL)AMINO)PROPANE-1-OL see BJM625

3-(BIS(3,3-DIPHENYLPROPYL)AMINO)-1-PROPANOL see BJM625

BIS-O,O-DI-n-PROPYLPHOSPHOROTHIONIC ANHYDRIDE see TED500

BISDITHANE see BJK000

N,N'-BIS-(1,4-DITHIANE-2-o-(N-METHYLCARBAMOYL)OXIMINO)SULFIDE see BJM650

BIS(4-(DITHIOCARBOXY)-1-PIPERAZINEACETATO(2−))-MERCURY(2−), DISODIUM see SJR500

BIS(1,3-DITHIOCYANATO-1,1,3,3-TETRABUTYLDISTANNOXANE) see BJM700

BIS(DITHIOPHOSPHATE de O,O-DIETHYLE) de S,S'-(1,4-DIOXANNE-2,3-DIYLE) (FRENCH) see DVQ709

BIS(DODECANOLOXY)DIOCTYLSTANNANE see DVJ800

BIS(DODECANOYLOXY)DI-n-BUTYLSTANNANE see DDV600

BIS(DODECYLOXYCARBONYLETHYL)SULFIDE see TFD500

2,6-BIS-(DWUBENZYLOHYDROKSYMETYLO)-PIPERYDYNA (POLISH) see BIV000

BISECURIN II see DAP810

BIS(2,5-ENDOMETHYLENECYCLOHEXYLMETHYL)AMINE see BJM750

BIS(3,4-EPOXYBUTYL) ETHER see BJN000

BIS((3,4-EPOXYCYCLOHEXYL)METHYL)ADIPATE see BJN100

BIS(2,3-EPOXYCYCLOPENTYL) ETHER see BJN250

BIS(2,3-EPOXYCYCLOPENTYL) ETHER mixed with DIGLYCIDYL ETHER of BISPHENOL A (1:1) see OPI200

BIS(3,4-EPOXY-6-METHYLCYCLOHEXYLMETHYL)ADIPATE see AEN750

BIS(2,3-EPOXY-2-METHYLPROPYL)ETHER see BJN500

m-BIS(2,3-EPOXYPROPOXY)BENZENE see REF000

1,3-BIS(2,3-EPOXYPROPOXY)BENZENE see REF000

1,4-BIS(2,3-EPOXYPROPOXY)BUTANE see BOS100

2,2'-BIS(2,3-EPOXYPROPOXY)-N-tert-BUTYLDIPROPYLAMINE see DKM400

1,3-BIS(2,3-EPOXYPROPOXY)-2,2-DIMETHYLPROPANE see NCI300

2,3-BIS(2,3-EPOXYPROPOXY)-1,4-DIOXANE see BJN750

N,N-BIS(2-(2,3-EPOXYPROPOXY)ETHOXY)ANILINE see BJN850

1,2-BIS((2,3-EPOXYPROPOXY)ETHOXY)ETHANE see TJQ333

N,N-BIS(2-(2,3-EPOXYPROPOXY)ETHYL)ANILINE see BJN875

BIS(2-(2,3-EPOXYPROPOXY)ETHYL)ETHER see DJE200

2,2-BIS(p-(2,3-EPOXYPROPOXY)PHENYL)PROPANE mixed with 2,2'-OXYBIS(6-OXABICYCLO(3.1.0)HEXANE) see OPI200

BIS(2,3-EPOXYPROPYL)ANILINE see DKM120

N,N-BIS(2,3-EPOXYPROPYL)ANILINE see DKM120

1,3-BIS(2,3-EPOXYPROPYL)-5,5-DIMETHYLHYDANTOIN see DKM130

BIS(2,3-EPOXYPROPYL)ETHER see DKM200

BIS(2,3-EPOXYPROPYL)-5-ETHYL-5-METHYLHYDANTOIN see ECI200

2,2-BIS(4-(2,3-EPOXYPROPYLOXY)PHENYL)PROPANE see BLD750

BIS(EPOXYPROPYL)PHENYLAMINE see DKM120

2,6-BIS(2,3-EPOXYPROPYL)PHENYL-2,3-EPOXYPROPYLETHER see BJO000

BIS(2,6-(2,3-EPOXYPROPYL))PHENYL GLYCIDYL ETHER see BJO000

N,N'-BIS(2,3-EPOXYPROPYL)PIPERAZINE see DHE100

N,N-BIS(2,3-EPOXYPROPYL)-p-TOLUENESULFONAMIDE see DKM800

BISEPTOL see TKX000

BIS(ETHANE-1,2-DIAMINE)COPPER(2+) DIPERCHLORATE see BJO050

(SP-4-1)-BIS(1,2-ETHANEDIAMINE-KAPPAN,KAPPAN')COPPER(2+) see BJO050

BIS(ETHENYLDIMETHYLSILYL) ETHER see TDQ075

BIS(2-(2-ETHOXYBUTOXY)ETHYL) SUCCINIC ACID ESTER see BJO075

BIS(ETHOXYCARBONYLDIAZOMETHYL) MERCURY see BJO125

S-(1,2-BIS(ETHOXY-CARBONYL-ETHYL)-O,O-DIMETHYL-DITHIOFOSFAAT (DUTCH) see MAK700

S-(1,2-BIS(ETHOXYCARBONYL)ETHYL)-O,O-DIMETHYL PHOSPHORODITHIOATE see MAK700

S-1,2-BIS(ETHOXYCARBONYL)ETHYL-O,O-DIMETHYL THIOPHOSPHATE see MAK700

1,2-BIS-(3-ETHOXYCARBONYLTHIOUREIDO)BENZENE see DJV000

1,2-BIS(3-ETHOXYCARBONYL-2-THIOUREIDO) BENZENE see DJV000

BIS(2-ETHOXYETHYL) ADIPATE see BJO225

BIS(2-ETHOXYETHYL)ETHER see DIW800

BIS(2-ETHOXYETHYL)NITROSOAMINE see BJO250

N,N'-BIS(p-ETHOXYPHENYL)ACETAMIDINE see BJO500

N',N2-BIS(p-ETHOXYPHENYL)ACETAMIDINE see BJO500

N,N'-BIS(4-ETHOXYPHENYL)ETHANIMIDAMIDE see BJO500

1,4-BIS-(2,3-ETHOXYPROPYL)PIPERAZINE see DHE100

BIS(ETHOXYTHIOCARBONYL)TRISULFIDE see DKE400

N,N'-BIS(4-(ETHYLAMINO)BUTYL)-1,4-BUTANEDIAMINE see BOS050

2,4-BIS(ETHYLAMINO)-6-CHLORO-s-TRIAZINE see BJP000

2,4-BIS(ETHYLAMINO)-6-METHOXY-s-TRIAZINE see BJP250

4,6-BIS(ETHYLAMINO)-2-METHOXY-s-TRIAZINE see BJP250

1,2-BIS(ETHYLAMMONIO)ETHANE PERCHLORATE see BJP300

BIS(N-ETHYLDITHIOCARBANILATO)ZINC see ZHA000

N,N'-BIS(ETHYLENE)-p-(1-ADAMANTYL)PHOSPHONIC DIAMIDE see BJP325

BIS(ETHYLENEDIAMINE)COPPER(2+) DIPERCHLORATE see BJO050

BIS(ETHYLENEDIAMINE)(MERCURICTETRATHIOCYANATO)COPPER see BJP425

2,5-BIS(ETHYLENEIMINO)-3,6-DIPROPOXY-1,4-BENZOQUINONE see DCN000

ω,ω'-BIS-(ETHYLENEIMINOSULPHONYL)PROPANE see BJP899

BISETHYLENEUREA see BJP450

BIS(ETHYLENIMIDO)PHOSPHORYLURETHAN see EHV500

2,6-BIS(ETHYLEN-IMINO)-4-AMINO-s-TRIAZINE see BJP500

2,5-BIS-ETHYLENIMINOBENZOQUINONE see BGW750

1,3-BIS(ETHYLENIMINOSULFONYL)PROPANE see BJP899

BIS(ETHYLHEXANOATE)TRIETHYLENE GLYCOL see FCD525

BIS(2-ETHYLHEXANOYLOXY)DIBUTYL STANNANE see BJQ250

BIS(2-ETHYLHEXYL) ADIPATE see AEO000

BIS-2-ETHYLHEXYLAMIN see DJA800

2-(BIS(2-ETHYLHEXYL)AMINO)ETHANOL see DJK200

BIS(2-ETHYLHEXYL) AZELATE see BJQ500

BIS(2-ETHYLHEXYL)-1,2-BENZENEDICARBOXYLATE see DVL700

BIS(2-ETHYLHEXYL) ESTER PHOSPHOROUS ACID CADMIUM SALT see CAD500

BIS(ETHYLHEXYL) ESTER of SODIUM SULFOSUCCINIC ACID see DJL000

BIS(2-ETHYLHEXYL)ETHER see DJK600

BIS(2-ETHYLHEXYL) FUMARATE see DVK600

BIS(2-ETHYLHEXYL)HYDROGEN PHOSPHATE see BJR750

BIS(2-ETHYLHEXYL) HYDROGEN PHOSPHITE see BJQ709

BIS(2-ETHYLHEXYL) ISOPHTHALATE see BJQ750

BIS(2-ETHYLHEXYL) MALEATE see BJR000

BIS(2-ETHYLHEXYL)ORTHOPHOSPHORIC ACID see BJR750

BIS(2-ETHYLHEXYLOXYCARBONYLMETHYLTHIO)DIBUTYLSTANNANE see BKK250

BIS(2-ETHYLHEXYLOXYCARBONYLMETHYLTHIO)DIBUTYLSTANNANE see DDY600

BIS(2-ETHYLHEXYLOXYCARBONYLMETHYLTHIO)DIMETHYLSTANNANE see BKK500

BIS((2-(ETHYL)HEXYLOXY)MALEOYLOXY) DI(n-BUTYL)STANNANE see BJR250

BIS(2-ETHYLHEXYL) PHENYL PHOSPHATE see BJR625

BIS(2-ETHYLHEXYL) PHOSPHATE see BJR750

BIS(2-ETHYLHEXYL)PHOSPHORIC ACID see BJR750

BIS(2-ETHYLHEXYL)PHTHALATE see DVL700

BIS(2-ETHYLHEXYL) SEBACATE see BJS250

BIS(2-ETHYLHEXYL)SODIUM PHOSPHATE see TBA750

BIS(2-ETHYLHEXYL)SODIUM SULFOSUCCINATE see DJL000

BIS(2-ETHYLHEXYL)-S-SODIUM SULFOSUCCINATE see DJL000

1,4-BIS(2-ETHYLHEXYL) SODIUM SULFOSUCCINATE see DJL000

1,4-BIS(2-ETHYLHEXYL)SULFOBUTANEDIOIC ACID ESTER, SODIUM SALT see DJL000

BIS(2-ETHYLHEXYL) TEREPHTHALATE see BJS300

BIS(2-ETHYLHEXYL) TEREPHTHALATE see BJS500

BIS(2-ETHYLHEXYL)THIODIPROPIONATE see BJS550

BIS(2-ETHYLHEXYL) 3,3'-THIODIPROPIONATE see BJS550

BIS(2-ETHYLHEXYLTHIOGLYCOLATE)DIBUTYLTIN see DDY600

BIS(2-ETHYLHEXYLTHIOGLYCOLATE)DIOCTYLTIN see DVM800

N^1,N^{14}-BIS(ETHYL)HOMO-SPRINE see BOS050

BIS(2-ETHYL-2-HYDROXYBUTANOATO(2-)-(O^1,O^2)-OXOCHROMATE(1-) SODIUM see BJS600

BISETHYLIDENEHYDRAZINE see AAG100

BIS(ETHYLMERCURI) PHOSPHATE see BJT250

N,N'-BIS(1-ETHYL-3-METHYLPENTYL)-p-PHENYLENEDIAMINE see BJT500

BIS(ETHYLPHENYLCARBAMODITHIOATO-S,S')-(T-4)-ZINC see ZHA000

1,1-BIS(p-ETHYLPHENYL)-2,2-DICHLOROETHANE see DJC000

2,2-BIS(p-ETHYLPHENYL)-1,1-DICHLOROETHANE see DJC000

BIS(N-ETHYL-N-PHENYL)UREA see DJC400

2,2-BIS(ETHYLSULFONYL)BUTANE see BJT750

2,2-BIS(ETHYLSULFONYL)PROPANE see ABD500

BIS(ETHYLTHIO)METHYLENE MALONONITRILE see BJT800

(BIS(ETHYLTHIO)METHYLENE)PROPANE DINITRILE see BJT800

BIS(ETHYLXANTHIC)DISULFIDE see BJU000

BIS(ETHYLXANTHOGEN) DISULFIDE see BJU000

BIS(ETHYLXANTHOGEN) TETRASULFIDE see BJU250

BIS(ETHYLXANTHOGEN) TRISULFIDE see DKE400

S-(1,2-BIS(ETOSSI-CARBONIL)-ETIL)-O,O-DIMETIL-DITIOFOSFATO (ITALIAN) see MAK700

BISEXOVIS see AOO410

BISEXOVISTER see AOO410

BISFEROL A (GERMAN) see BLD500

1,6-BIS(9 FLUORENYLDIMETHYL-AMMONIUM)HEXANE BROMIDE see HEG000

BIS(2-FLUORO-2,2-DINITROETHOXY)DIMETHYLSILANE see BJU350

BIS(2-FLUORO-2,2-DINITROETHOXY)METHANE see FMV050

BIS(2-FLUORO-2,2-DINITROETHYL)AMINE see BJU500

BIS(2-FLUORO-2,2-DINITROETHYL) FORMAL see FMV050

BIS(4-FLUORO-3-NITROPHENYL)SULFONE see DKH250

1,1-BIS(FLUOROOXY)HEXAFLUOROPROPANE see BJV625

2,2-BIS(FLUOROOXY)HEXAFLUOROPROPANE see BJV630

1,1-BIS(FLUOROOXY)TETRAFLUOROETHANE see BJV635

1-(4,4-BIS(p-FLUOROPHENYL)BUTYL)-4-(2-OXO-1-BENZIMIDAZOLINYL)PIPERIDINE see PIH000

trans-4-(4,4-BIS(p-FLUOROPHENYL)BUTYL)-1-(2-(4'-PHENYLCYCLOHEXYLAMINO)ETHYL)PIPERAZINE TRIHYDROCHLORIDE see BJV750

8-(4,4-BIS(p-FLUOROPHENYL)BUTYL)-1-PHENYL-1,3,8-TRIAZASPIRO(4,5)DECAN-4-ONE see PFU250

trans-2-(4-(4,4-BIS(p-FLUOROPHENYL)BUTYL)PIPERAZINYL)-N-(4'-PHENYLCYCLOHEXYL)ACETAMIDE DIHYDROCHLORIDE see BJW000

1-(1-(4,4-BIS(p-FLUOROPHENYL)BUTYL)-4-PIPERIDYL)-2-BENZIMIDAZOLINONE see PIH000

1-(2-(BIS(4-FLUOROPHENYL)METHOXY)ETHYL)-4-(3-PHENYLPROPYL)PIPERAZINE DIHYDROCHLORIDE see BJW100

(3)-1-(BIS(p-FLUOROPHENYL)METHYL)-4-CINNAMYLPIPERAZINE DIHYDROCHLORIDE see FDD080

(E)-1-(BIS(4-FLUOROPHENYL)METHYL)-4-(3-PHENYL-2-PROPENYL)PIPERAZINE DIHYDROCHLORIDE see FDD080

1-((BIS(4-FLUOROPHENYL)METHYLSILYL)METHYL)-1H-1,2,4-TRIAZOLE see THS860

1,1-BIS(4-FLUOROPHENYL)-2-PROPYNYL-N-CYCLOHEPTYLCARBAMATE see BJW250

1,1-BIS(4-FLUOROPHENYL)-2-PROPYNYL-N-CYCLOOCTYL CARBAMATE see BJW500

α-α-BIS(4-FLUOROPHENYL)-1H-1,2,4-TRIAZOLE-1-ETHANOL see BJW600

1,1-BIS(p-FLUOROPHENYL)-2,2,2-TRICHLOROETHANE see FHJ000

2,2-BIS(p-FLUOROPHENYL)-1,1,1-TRICHLOROETHANE see FHJ000

BIS(3-FLUOROSALICYLALDEHYDE)-ETHYLENEDIIMINE-COBALT see EIS000

1,4-BIS(1-FORMAMIDO-2,2,2-TRICHLOROETHYL)PIPERAZINE see TKL100

N,N'-BIS(1-FORMAMIDO-2,2,2-TRICHLOROETHYL)PIPERAZINE see TKL100

BIS(N-FORMYL-p-AMINOPHENYL)SULFONE see BJW750

BIS(FORMYLMETHYL) MERCURY see BJW800

1,3-BIS(4-FORMYLPYRIDINIUM)-PROPANE BISOXIDE DIBROMIDE see TLQ500

1,3-BIS(4-FORMYLPYRIDINIUM)-PROPANE BISOXIME DICHLORIDE see TLQ750

2-(BIS(FURFURYLIDENAMINO))METHYLFURAN see FPS000

BIS(2-FURYL)GLYOXIME see BJW810

3,5-BIS(2-FURYL)-1H-1,2,4-TRIAZOLE see AMX000

3,5-BIS-d-GLUCONAMIDO-2,4,6-TRIIODO-N-METHYLBENZAMIDE see IFS400

m-BIS(GLYCIDYLOXY)BENZENE see REF000

1,4-BIS(GLYCIDYLOXY)BUTANE see BOS100

2,3-BIS(GLYCIDYLOXY)-1,4-DIOXANE see BJN750

1,2-BIS(GLYCIDYLOXY)ETHANE see EEA600

BIS(4-GLYCIDYLOXYPHENYL)DIMETHYLAMETHANE see BLD750

2,2-BIS(p-GLYCIDYLOXYPHENYL)PROPANE see BLD750

BIS-GMA see BLD810

BISGUANIDINIUM CARBONATE see GKW100

BIS(GUANIDINIUM) CHROMATE see BJW825

BIS(2-GUANIDOETHYL)DISULPHIDE see GLC200

(BIS-(HEPTYLOXY)PHOSPHINYL)MERCURY ACETATE see AAR750

BIS(HEXANOYLOXY)DI-n-BUTYLSTANNANE see BJX750

BIS(HEXANOYLOXY)DI-n-BUTYL-TIN see BJX750

BIS(2-(HEXYLOXY)ETHYL)ADIPATE see AEQ000

BIS(2-(HEXYLOXY)ETHYL) SUCCINATE see SMZ000

(BIS(HEXYLOXY)PHOSPHINYL)MERCURY ACETATE see AAS000

BIS(l-HISTIDINATO)COBALT see BJY000

BIS(l-HISTIDINATO)MANGANESE TETRAHYDRATE see BJX800

BIS(l-HISTIDINE)COBALT see BJY000

BIS-HM-A-TDA see BJY125

BISHYDRAZINE NICKEL(II)PERCHLORATE see BJY250

BISHYDRAZINE TIN(II)CHLORIDE see BJY500

BIS(HYDROGEN MALEATO)DIBUTYL-TIN BIS(2-ETHYLHEXYL) ESTER see BJR250

BIS(HYDROGEN MALEATO)DIOCTYLTIN BIS(2-ETHYLHEXYL) ESTER see DVM600

BIS(1-HYDROPEROXY CYCLOHEXYL)PEROXIDE see BJY750

2,2-BIS(HYDROPEROXY)PROPANE see BJY825

1,2-BIS(HYDROXOMERCURIO)-1,1,2,2-BIS(OXYDIMERCURIO)ETHANE see BKH125

BIS(HYDROXYAETHYL)-AETHER-DINITRAT (GERMAN) see DJE400

BIS-HYDROXYAETHYLAMINOPROPYL-HYDROXYAETHYL-OCTADECYLAMINDIHYDROFLUORID see BJY800

BIS(β-HYDROXYAETHYL)NITROSAMIN (GERMAN) see NKM000

BIS(2-HYDROXYBENZOATO-O¹O²-), (T-4)-CADMIUM (9CI) see CAI400

BIS(2-HYDROXY-3-tert-BUTYL-5-ETHYLPHENYL)METHANE see MJN250

BIS(4-HYDROXY-5-tert-BUTYL-2-METHYLPHENYL) SULFIDE see TFC600

BIS-2-HYDROXY-5-CHLORFENYLMETHAN see MJM500

BIS(2-HYDROXY-5-CHLOROPHENYL)METHANE see MJM500

BISHYDROXYCOUMARIN see BJZ000

BIS(4-HYDROXY-3-COUMARIN) ACETIC ACID ETHYL ESTER see BKA000

BIS-3,3'-(4-HYDROXYCOUMARINYL)ACETIC ACID ETHYL ESTER see BKA000

BIS-(4-HYDROXY-3-COUMARINYL)ETHYL ACETATE see BKA000

BIS(4-HYDROXYCOUMARIN-3-YL)METHANE see BJZ000

BIS(1-HYDROXYCYCLOHEXYL)PEROXIDE see BKA250

2,2-BIS(4-HYDROXY-3,5-DICHLOROPHENYL)PROPANE see TBO750

BIS(2-HYDROXY-3,5-DICHLOROPHENYL) SULFIDE see TFD250

N,N-BIS(2-HYDROXYETHYL)ACETAMIDE DIACETATE see BGQ050

BIS(2-HYDROXYETHYL)AMINE see DHF000

2-BIS-HYDROXYETHYLAMINO-4-ACETAMINOFENETOL (CZECH) see BKB000

3'-(BIS(2-HYDROXYETHYL)AMINO)-p-ACETOPHENETIDIDE see BKB000

2-(N,N-BIS(2-HYDROXYETHYL)AMINO)-1,4-BENZOQUINONE see BKB100

2-(BIS(β-HYDROXYETHYL)AMINO)-4,5-DIPHENYLOXAZOLE MONOHYDRATE see BKB250

1,4-BIS(2-((2-HYDROXYETHYL)AMINO)ETHYLAMINO)-9,10-ANTHRACENEDIONE see BKB325

1,4-BIS((2-((2-HYDROXYETHYL)AMINO)ETHYL)AMINO)-9,10-ANTHRACENEDIONE DIACETATE see BKB300

1,4-BIS((2-((HYDROXYETHYL)AMINO)ETHYL)AMINO)-9,10-ANTHRACENEDIONE DIACETATE (SALT) (9CI) see BKB300

1,4-BIS(2-((2-HYDROXYETHYL)AMINO)ETHYL)AMINO)ANTHRAQUINONE see BKB325

5,8-BIS((2-((2-HYDROXYETHYL)AMINO)ETHYL)AMINO)-1,4-DIHYDROXY-9,10-ANTHRACENEDIONE DIACETATE see DBE875

5,8-BIS((2-((HYDROXYETHYL)AMINO)ETHYL)AMINO)-1,4-DIHYDROXYANTHRAQUINONE see MQY090

5,8-BIS((2-((2-HYDROXYETHYL)AMINO)ETHYL)AMINO)-1,4-DIHYDROXYANTHRAQUINONE 1,4-DIACETATE see DBE875

3-(BIS(2-HYDROXYETHYL)AMINO)-6-HYDRAZINOPYRIDAZINEDIHYDROCHLORIDE see BKB500

4-(N,N-BIS(2-HYDROXYETHYL)AMINO)-7-NITROBENZOFURAZAN-1-OXIDE see NFF000

4-BIS(2-HYDROXYETHYL)AMINO-2-(5-NITRO-2-FURYL)QUINAZOLINE see BKB750

2-(BIS(2-HYDROXYETHYL)AMINO)-5-NITROPHENOL see BKC000

4-BIS(2-HYDROXYETHYL)AMINO-2-(5-NITRO-2-THIENYL)QUINAZOLINE see BKC250

N-(3-BIS(2-HYDROXYETHYL)AMINO)PROPYL)BENZAMIDE HYDROCHLORIDE see BKC500

10-(3-(BIS(2-HYDROXYETHYL)AMINO)PROPYL)-7-CHLOROISOALLOXAZINE SULFATE see BKC750

N-(3-(BIS(2-HYDROXYETHYL)AMINO)PROPYL)-o-PROPOXYBENZAMIDE HYDROCHLORIDE see BKD000

4,4'-BIS((4-(2-HYDROXYETHYL)AMINO-6-(p-SULFOANILINO)-s-TRIAZIN-2-YL)AMINO)-2,2'-STILBENEDISULFONIC ACID TETRASODIUM SALT see BKD250

N,N-BIS(2-HYDROXYETHYL)ANILINE see BKD500

N,N-BIS(2-HYDROXYETHYL)-p-ARSANILIC ACID see BKD600

1,1-BIS(2-HYDROXYETHYL)AZIRIDINIUM CHLORIDE see BKE750

BIS(2-HYDROXYETHYL)CARBAMODITHIOIC ACID, MONOPOTASSIUM SALT see PKX500

N,N-BIS(2-HYDROXYETHYL)CHLOROANILIDE see CJU400

BIS(2-HYDROXYETHYL)-2-((2-CHLORO ETHYL THIO)ETHYL SULFONIUM) CHLORIDE see BKD750

N,N-BIS(2-HYDROXYETHYL)COCOAMIDE see CNF330

N,N-BIS(2-HYDROXYETHYL)COCONUT FATTY ACID AMIDE see CNF330

N,N-BIS(2-HYDROXYETHYL)COCONUT OIL AMIDE see CNF330

BIS(2-HYDROXYETHYL)DITHIOCARBAMIC ACID, MONOPOTASSIUM SALT see PKX500

BIS(2-HYDROXYETHYL)DITHIOCARBAMIC ACID, POTASSIUM SALT see PKX500

N,N'-BIS(2-HYDROXYETHYL)-DITHIOOXAMIDE see BKD800

N,N-BIS(2-HYDROXYETHYL)DODECANAMIDE see BKE500

BIS(2-HYDROXYETHYL) ETHER see DJD600

1,1-BIS(β-HYDROXYETHYL)ETHYLENINONIUM CHLORIDE see BKE750

N,N-BIS-2-HYDROXYETHYL-p-FENYLENDIAMIN STRAN (CZECH) see BKF750

1,1-BIS(2-HYDROXYETHYL)HYDRAZINE see HHH000

BIS(2-HYDROXYETHYL)LAURAMIDE see BKE500

N,N-BIS(HYDROXYETHYL)LAURAMIDE see BKE500

N,N-BIS(2-HYDROXYETHYL)LAURAMIDE see BKE500

N,N-BIS(β-HYDROXYETHYL)LAURAMIDE see BKE500

BIS(2-HYDROXYETHYL)METHYLAMINE see MKU250

N,N-BIS(2-HYDROXYETHYL)-3-METHYLANILINE see DHF400

N,N-BIS(β-HYDROXYETHYL)-3-METHYLANILINE see DHF400

N',N'-BIS(2-HYDROXYETHYL)-N-METHYL-2-NITRO-p-PHENYLENEDIAMINE see BKF250

BIS(β-HYDROXYETHYL)NITROSAMINE see NKM000

N,N-BIS(2-HYDROXYETHYL)-9,12-OCTADECADIENAMIDE see BKF500

(Z)-N,N-BIS(2-HYDROXYETHYL)-9-OCTADECENAMIDE see OHU200

N,N-BIS(2-HYDROXYETHYL)-p-PHENYLENEDIAMINE SULFATE (1:1) see BKF750

N,N-BIS(2-HYDROXYETHYL)-1,4-PHENYLENEDIAMINE SULFATE MONOHYDRATE see BKF800

1,4-BIS(2-HYDROXYETHYL)PIPERAZINE see PIJ750

N,N'-BIS(β-HYDROXYETHYL)PIPERAZINE see PIJ750

BIS[HYDROXYETHYLPOLY(ETHYLENEOXY)ETHYLPROPYLENEGLYCOL see PJK150

BIS[HYDROXYETHYLPOLY(ETHYLENEOXY)ETHYLPROPYLENEGLYCOL see PJK151

BIS(2-HYDROXYETHYL)SULFIDE see TFI500

BIS(β-HYDROXYETHYL)SULFIDE see TFI500

N,N-BIS(2-HYDROXYETHYL)-m-TOLUIDINE see DHF400

2,2-BIS-4'-HYDROXYFENYLPROPAN (CZECH) see BLD500

l-(+)-N,N'BIS((2-HYDROXY)-1-HYDROXYMETHYLETHYL)-2,4,6-TRIIODO-5-LACTAMIDE ISOPHTHALAMIDE see IFY000

1,3-BIS(4-HYDROXYIMINOMETHYL-1-PYRIDINIO)-2-OXAPROPANE DICHLORIDE see BGS250

1,3-BIS(4-HYDROXYIMINOMETHYL-1-PYRIDINIO)PROPANE DICHLORIDE see TLQ750

1,4-BIS(4-HYDROXYIMINOMETHYL-PYRIDINIUM-(1))-BUTANEDIOL-2,3 DIBROMID (GERMAN) see BGR750

1,4-BIS(4-HYDROXYIMINOMETHYL-PYRIDINIUM-(1))BUTANEDIOL(2,3)-DIJODID see BKF900

1,4-BIS(4-HYDROXYIMINOMETHYL-PYRIDINIUM-(1))BUTANEDIOL(2,3)-DIPERCHLORAT (GERMAN) see DND400

BIS((HYDROXYIMINO-METHYL)-PYRIDINIUM)-(1)-METHYL-AETHER-DICHLORID (GERMAN) see PPC000

BIS(2-HYDROXYIMINOMETHYLPYRIDINIUM-1-METHYL)ETHER DICHLORIDE see PPC000

BIS(4-HYDROXYIMINOMETHYLPYRIDINIUM-1-METHYL)ETHER DICHLORIDE see BGS250

BIS(HYDROXYLAMINE) SULFATE see OLS000

BISHYDROXYL AMINE ZINC(II)CHLORIDE see BKG250

3,5-BIS(3-HYDROXYMERCURI-2-METHOXYPROPYL)BARBITURIC ACID SODIUM SALT see BKG500

5,5-BIS(3-HYDROXYMERCURI-2-METHOXYPROPYL)BARBITURIC ACID SODIUM SALT see BKG750

2,6-BIS(HYDROXYMERCURI)-4-NITROANILINE see BKH000

BIS(2-HYDROXY-4-METHOXYPHENYL)METHANONE see BJY800

3-BIS(HYDROXYMETHYL)AMINO-6-(5-NITRO-2-FURYLETHENYL)-1,2,4-TRIAZINE see BKH500

2-N,N-BIS(HYDROXYMETHYL)AMINO-1,3,4-THIADIAZOLE see BJY125

9:10-BISHYDROXYMETHYL-1:2-BENZANTHRACENE see BBF500

BIS(HYDROXYMETHYL)-m-CARBORANE see CCC120

BIS(HYDROXYMETHYL)-o-CARBORANE see CCC130

1,2-BIS(HYDROXYMETHYL)CARBORANE see CCC130

1,7-BIS(HYDROXYMETHYL)CARBORANE see CCC120

1,4-BIS(HYDROXYMETHYL)CYCLOHEXANE see BKH325

4,4-BIS(HYDROXYMETHYL)-1-CYCLOHEXENE see BKH400

N-(1,3-BIS(HYDROXYMETHYL)-2,5-DIOXO-4-IMIDAZOLIDINYL)-N,N'-BIS(HYDROXYME THYL)UREA see IAS100

BIS(HYDROXYMETHYL)FURATRIZINE see BKH500

3,3-BIS(HYDROXYMETHYL)HEPTANE see BKH625

1,3-BIS(HYDROXYMETHYL)IMIDAZOLIDIN-2-THIONE see BKH650

1,2-BISHYDROXYMETHYL-1-METHYLPYRROLE see MPB175

2,3-BISHYDROXYMETHYL-1-METHYLPYRROLE see MPB175

3,3-BIS(HYDROXYMETHYL)PENTANE see PMB250

BIS HYDROXYMETHYL PEROXIDE see BKH750

BIS(HYDROXYMETHYL)PEROXIDE see DMN200

BIS-(1-HYDROXYMETHYL)PEROXIDE see DMN200

meso-3,4-BIS(4-HYDROXY-3-METHYLPHENYL)HEXANE see DJI350

2,2-BIS(4-HYDROXY-3-METHYLPHENYL)PROPANE see IPK000

BIS(HYDROXYMETHYL)PHOSPHINE OXIDE see DSG600

2,2-BIS(HYDROXYMETHYL)-1,3-PROPANEDIOL see PBB750

2,2-BIS(HYDROXYMETHYL)-1,3-PROPANEDIOL ALLYL ETHER see AGQ050

2,2-BIS(HYDROXYMETHYL)-1,3-PROPANEDIOL TETRANITRATE see PBC250

d,N,N'-BIS(1-HYDROXYMETHYLPROPYL)ETHYLENED IAMINE see TGA500

2,3-BIS(HYDROXYMETHYL)QUINOXALINE DI-N-OXIDE see DVQ800

BIS(HYDROXYMETHYL)TRICYCLO(5.2.1.0(2,6))DECANE see TJG550

1,3-BIS(HYDROXYMETHYL)UREA see DTG700

N,N'-BIS(HYDROXYMETHYL)UREA see DTG700

BIS(4-HYDROXY-3-NITROPHENYL)MERCURY see MCS600

BIS(4-HYDROXY-2-OXO-2H-1-BENZOPYRAN-3-YL)ACETIC ACID ETHYL ESTER see BKA000

N,N'-BIS(2-HYDROXY-3-PHENOXYPROPYL)ETHYLENEDIAMINE DIHYDROCHLORIDE see IAG700

BIS(4-HYDROXYPHENYL)AMINE see IBJ100

BIS(p-HYDROXYPHENYL)AMINE see IBJ100

BIS-(p-HYDROXYPHENYL)-CYCLOHEXYLIDENEMETHANE DIACETATE see FBP100

BIS(4-HYDROXYPHENYL)DIMETHYLMETHANE see BLD500

BIS(4-HYDROXYPHENYL)DIMETHYLMETHANE DIGLYCIDYL ETHER see BLD750

3,4-BIS(4-HYDROXYPHENYL)-2,4-HEXADIENE see DAL600

3,4-BIS(p-HYDROXYPHENYL)-2,4-HEXADIENE see DAL600

3,4-BIS(p-HYDROXYPHENYL)HEXANE see DLB400

meso-3,4-BIS(p-HYDROXYPHENYL)-n-HEXANE see DLB400

3,4-BIS(p-HYDROXYPHENYL)-2-HEXANONE see BKH800

3,4-BIS(p-HYDROXYPHENYL)-3-HEXENE see DKA600

BIS(4-HYDROXYPHENYL)METHANE see BKI250

BIS(p-HYDROXYPHENYL)METHANE see BKI250

p,p'-BIS(HYDROXYPHENYL)METHANE see BKI250

BIS(4-HYDROXYPHENYL)METHANONE (2,4-DINITROPHENYL)HYDRAZONE see BKI300

3,3-BIS(p-HYDROXYPHENYL)PHTHALIDE see PDO750

BIS(4-HYDROXYPHENYL)PROPANE see BLD500

2,2-BIS(4-HYDROXYPHENYL)PROPANE see BLD500

2,2-BIS(p-HYDROXYPHENYL)PROPANE see BLD500

2,2-BIS(4-HYDROXYPHENYL)PROPANE, DIGLYCIDYL ETHER see BLD750

2,2-BIS(p-HYDROXYPHENYL)PROPANE, DIGLYCIDYL ETHER see BLD750

BIS(4-HYDROXYPHENYL) SULFIDE see TFJ000

2,2-BIS(p-HYDROXYPHENYL)-1,1,1-TRICHLOROETHANE see BKI500

2,3-BIS(p-HYDROXYPHENYL)VALERONITRILE see BKI750

BIS(2-HYDROXYPROPANOATO-O¹,O²)COBALT see CNC242

BIS(2-HYDROXYPROPYL)AMINE see DNL600

N,N'-(BIS-ω-HYDROXYPROPYL)HOMOPIPERAZINE 3,4,5-TRIMETHOXYBENZOATE DIHYDROCHLORIDE see CNR750

N-BIS(2-HYDROXYPROPYL)NITROSAMINE see DNB200

2,2'-BISHYDROXYPROPYLNITROSAMINE see DNB200

BIS(3-HYDROXY-1-PROPYNYL)MERCURY see BKJ250

BIS(1-HYDROXY-2(1H)-PYRIDINETHIONATO)ZINC see ZMJ000

BIS(8-HYDROXYQUINOLINE-5-SULFONIC ACID) COBALT(II) see BKJ260

BIS(8-HYDROXYQUINOLINE-5-SULFONIC ACID) MANGANESE(II) see BKJ275

(BIS(8-HYDROXYQUINOLYL)AMINO)SALICYLIC ACID see BKJ300

BIS-(5-HYDROXY-7-SULFO-2NAFTYL)AMIN (CZECH) see IBF000

1,3-BIS(1-HYDROXY-2,2,2-TRICHLOROETHYL)UREA see DGQ200

BIS(2-HYDROXY-3,5,6-TRICHLOROPHENYL)METHANE see HCL000

BISISIBUTIAMINE see BKJ325

BISINA see CKF500

BIS(3-INDOLEMETHYLENEMORPHOLINIUM)HEXACHLOROSTANNATE see BKJ500

N,N-BIS(2-IODOETHYL)ANILINE see BKJ600

N₄,N₄-BIS(2-IODOETHYL)SULFANILAMIDE see BKJ650

BIS(ISOBUTYL)ALUMINUM CHLORIDE see CGB500

BIS(ISOBUTYL)HYDROALUMINUM see DNI600

BIS(4-ISOCYANATOCYCLOHEXYL)METHANE see MJM600

BIS(2-ISOCYANATOETHYL)CARBONATE see CBV600

BIS(2-ISOCYANATOETHYL)CARBONATE see IKG400

BIS(2-ISOCYANATOETHYL)-4-CYCLOHEXENE-1,2-DICARBOXYLATE see BKJ700

BIS(2-ISOCYANATOETHYL)-5-NORBORNENE-2,3-DICARBOXYLATE see BKJ800

1,3-BIS(ISOCYANATOMETHYL)BENZENE see XIJ000

BIS(ISOCYANATOMETHYL)DURENE see TDX400

1,3-BIS(1-ISOCYANATO-1-METHYLETHYL)BENZENE see TDX300

1,4-BIS(1-ISOCYANATO-1-METHYLETHYL)BENZENE see TDX400

BIS(4-ISOCYANATOPHENYL)METHANE see MJP400

BIS(p-ISOCYANATOPHENYL)METHANE see MJP400

BIS(1,4-ISOCYANATOPHENYL)METHANE see MJP400

BIS(ISODECYL)PHTHALATE see PHW575

1,3-BIS-(ISOKYANATOMETHYL)BENZEN see XIJ000

BIS(ISONICOTINALDOXIME 1-METHYL) ETHER DICHLORIDE see BGS250

BIS(ISOOCTYLOXYCARBONYLMETHYLTHIO)DIBUTYL STANNANE see BKK250

2,6-BIS(1-METHYLBUTYL)PHENOL see BKQ750

N,N'-BIS-(2-(o-(N-METHYLCARBAMOYL)OXIMINO)-1,4-DITHIAN)-DISULFID see BKQ780

N,N'-BIS-(2-(o-(N-METHYLCARBAMOYL)OXIMINO)-1,4-DITHIANE)DISULFIDE see BKQ780

1,4-BIS(METHYLCARBAMYLOXY)-2-ISOPROPYL-5-METHYLBENZENE see BKR000

BIS((6-METHYL-3-CYCLOHEXEN-1-YL)METHYL) ESTER HEXANEDIOIC ACID see BKR100

BIS(3-METHYLCYCLOHEXYL PEROXIDE) see BKR250

1,1-BIS((3,4-METHYLENEDIOXYPHENOXY)METHYL)-N,N-DIMETHYL-1-BUTANOL CITRATE see BKR500

α,α-BIS((3,4-(METHYLENEDIOXY)PHENOXY)METHYL)-1-PIPERIDINEBUTANOLACETATE CITRATE see BKR750

1-(1,3-BIS(3,4-(METHYLENEDIOXY)PHENOXY)-2-PROPYL)PYRROLIDINE CITRATE see BKS500

1,3-BIS(1-METHYLETHENYL)BENZENE see DNL700

1,4-BIS(1-METHYLETHENYL)BENZENE see IMX100

α-((2-BIS(1-METHYLETHYL)AMINO)ETHYL)-α-PHENYL-2-PYRIDINEACETAMIDE PHOSPHATE (9CI) see RSZ600

1,4-BIS(1-METHYLETHYL)BENZENE see DNN830

BIS(1-METHYLETHYL)DIAZENE 1-OXIDE see ASP510

BIS(1-METHYLETHYL)ESTERTHIOPEROXYDICARBONIC ACID see IRS500

2,3:4,5-BIS-o-(1-METHYLETHYLIDENE)-β-d-FRUCTOPYRANOSE, METHYL((2-(1-METHYLETHOXY)PHENOXY)CARBONYL)AMIDOSULFITE see BKS600

1,2:4,5-BIS-o-(1-METHYLETHYLIDENE)-β-d-FRUCTOPYRANOSE, METHYL((3-METHYLPHENOXY)CARBONYL)AMIDOSULFITE see BKS610

2,3:4,5-BIS-o-(1-METHYLETHYLIDENE)-β-d-FRUCTOPYRANOSE, METHYL((3-METHYLPHENOXY)CARBONYL)AMIDOSULFITE see BKS620

1,2:4,5-BIS-o-(1-METHYLETHYLIDENE)-β-d-FRUCTOPYRANOSE, METHYL((1-NAPHTHALENYLOXY)CARBONYL)AMIDOSULFITE see BKS630

1,2:5,6-BIS-o-(1-METHYLETHYLIDENE)-α-d-GLUCOFURANOSE, ((((2-(DIMETHYLAMINO)-2-OXO-1-(METHYLTHIO)ETHYLIDENE)AMINO)OXY)CARBONYL)METHYLAMIDOSULFITE see BKS640

1,3-BIS(1-METHYLETHYL)-2-ISOCYANATOBENZENE see DNS100

N,N'-BIS(1-METHYLETHYL)-6-METHYL-THIO-1,3,5-TRIAZINE-2,4-DIAMINE see BKL250

2,6-BIS(1-METHYLETHYL)PHENOL see DNR800

3,5-BIS(1-METHYLETHYL)PHENOL METHYLCARBAMATE see DNS200

N-(2,6-BIS(1-METHYLETHYL)-4-PHENOXYPHENYL)-N'-(1,1-DIMETHYLETHYL)THIOUREA see BRB100

3,5-BIS(1-METHYLETHYL)PHENYL ESTER METHYL CARBAMIC ACID see DNS200

3,5-BIS(1-METHYLETHYL)PHENYL METHYLCARBAMATE see DNS200

O,O-BIS(1-METHYLETHYL)-S-(PHENYLMETHYL)PHOSPHOROTHIOATE see BKS750

BIS((4-(1-METHYLETHYL)PHENYL)METHYL) 3-PYRIDINYLCARBONIMIDODITHIOATE see BKS770

O,O-BIS(1-METHYLETHYL)-S-(2-((PHENYLSULFONYL)AMINO)ETHYL)PHEOSPHORODITHIOATE see DNO800

(T-4)-(o,o-BIS(1-METHYLETHYL)PHOSPHORODITHIOATO-S,S')DIPHENYLANTIMONY see BKS780

N,N'-BIS(1-METHYLETHYL)THIOUREA see DNS800

2,4-BIS(1-METHYLETHYL)-6-(((TRIPHENYLSTANNYL)OXY)CARBONYL)PHENOL see TMV830

BIS(2-METHYLGLYCIDYL) ETHER see BJN500

BIS(6-METHYLHEPTYL)ESTER of PHTHALIC ACID see ILR100

N,N'-BIS(5-METHYL-3-HEPTYL)-p-PHENYLENEDIAMINE see BJT500

BIS(3-METHYLHEXYL)PHTHALIC ACID ESTER see DSF200

1,1-BIS(2-METHYL-4-HYDROXY-5-tert-BUTYLPHENYL)BUTANE see BRP750

BIS(2-METHYL-3-HYDROXY-4-METHOXYMETHYL-5-METHYLPYRIDYL)DISULFIDE DIHYDROCHLORIDE see BKS800

3,5-BIS-METHYLKARBOXY-BENZENSULFONAN SODNY (CZECH) see BKN000

BIS(METHYLMERCURIC)SULFATE see BKS810

BIS-(METHYLMERCURY)-SULFATE see BKS810

BIS-(METHYLMERKURI)SULFAT see BKS810

BIS(2-METHYL-3-NITROPHENYL)DIAZENE 1-OXIDE see BKS812

BIS(2-METHYL-5-NITROPHENYL)DIAZENE 1-OXIDE see BKS813

BIS(4-METHYL-3-NITROPHENYL)DIAZENE 1-OXIDE see BKS814

2,7-BIS(4-METHYLPHENYL)BENZO(lmn)(3,8)PHENANTHROLINE-1,3,6,8(2H,7H)-TETRONE see BKS815

N,N-BIS(N-METHYL-N-PHENYL-tert-BUTYLACETAMIDO)-β-HYDROXYETHYLAMINE see DTL200

BIS(4-METHYLPHENYL) ETHER see THC600

N,N'-BIS((1-METHYL-4-PHENYL-4-PIPERIDINYL)METHYL)-DECANEDIAMIDE (9CI) see BKS825

N,N'-BIS(1-METHYL-4-PHENYL-4-PIPERIDYLMETHYL)SEBACAMIDE see BKS825

N,N'-BIS(2-METHYLPHENYL)THIOUREA see DXP600

BIS(2-METHYLPROPYL)CARBAMOTHIOIC ACID-S-ETHYL ESTER see EID500

N-BISMETHYLPTEROYLGLUTAMIC ACID see MDV500

BIS(2-METHYL PYRIDINE)SODIUM see BKT250

N,N-BIS(METHYLQUECKSILBER)-p-TOLUOL-SULFAMID see MLH100

N,N'-BIS(2-METHYL-4-QUINOLINYL)-1,10-DECANEDIAMINE DIACETATE see SAP600

N,N-BIS(METHYLSULFONEPROPOXY)AMINE HYDROCHLORIDE see YCJ000

1,6-BIS-o-METHYLSULFONYL-d-MANNITOL see BKM500

N,N'-BIS(2-METHYLSULFONYL-2-METHYLPROPIONALDEHYDE-o-(N-METHYLCARBAMOYL)OXIME)SULFIDE see BKT300

BIS((METHYLSULFONYL)OXY)DIBUTYLSTANNANE see DEI400

BIS((METHYLSULFONYL)OXY)DIPROPYLSTANNANE see DWW400

BIS(3-METHYLSULFONYLOXYPROPYL)AMINE p-TOLUENESULFONATE see IBQ100

N,N'-BIS(3-METHYL-2-THIAZOLIDINYLIDENE)UREA see BKU000

N,N'-BIS-(1-METHYLTHIOACETALDEHYDE o-(N-METHYLCARBAMOYL)OXIME)DISULFIDE see BKU100

N,N'-BIS-(1-METHYLTHIOACETALDEHYD-o-(N-METHYLCARBAMOYL)OXIM)-DISULFID see BKU100

N,N'-BIS(1-METHYLTHIO-1-(N,N-DIMETHYLCARBONYL)FORMALDEHYDE-o-(N-METHYLCARBAMOYL)OXIME)SULFIDE see BKU120

3,5-BIS(METHYLTHIO)-N-((5-NITRO-2-FURANYL)METHYLENE)-4H-1,2,4-TRIAZOL-4-AMINE see BKU150

BIS(METHYLXANTHOGEN) DISULFIDE see DUN600

1,3-BIS(3,4-METILENDIOSSIFENOSSI)-2-AMINOPROPANO (ITALIAN) see MJS250

1,3-BIS-(3,4-METILENDIOSSIFENOSSI)-2-(3-DIMETILAMINOPROPIL)PROPAN-2-OLO CITRATO (ITALIAN) see BKR500

1,3-BIS-(3,4-METILENDIOSSIFENOSSI)-2-PIRROLIDINOPROPANO CITRATO (ITALIAN) see BKS500

BIS(MONOISOPROPYLAMINO)FLUOROPHOSPHATE see PHF750

BIS(MONOISOPROPYLAMINO)FLUOROPHOSPHINE OXIDE see PHF750

BIS(4-MORPHOLINECARBODITHIOATO)MERCURY see BKU250

N,N'-BISMORPHOLINE DISULFIDE see BKU500

BISMORPHOLINO DISULFIDE see BKU500

BIS(MORPHOLINO-)METHAN (GERMAN) see MJQ750

BISMORPHOLINO METHANE see MJQ750

5,7-BIS(MORPHOLINOMETHYL)-2-HYDROXY-3-ISOPROPYL-2,4,6-CYCLOHEPTATRIEN-1-ONE DIHYDROCHLORIDE see IOW500

BIS(MORPHOLINOTHIOCARBONYL)DISULFIDE see MRR090

BISMUTH see BKU750

BISMUTH-209 see BKU750

BISMUTH AMIDE OXIDE see BKV000

BISMUTH ARSPHENAMINE SULFONATE see BKV250

BISMUTH CHROMATE see DDT250

BISMUTH COMPOUNDS see BKV750

BISMUTH DIMETHYL DITHIOCARBAMATE see BKW000

BISMUTH EMETINE IODIDE see EAM500

BISMUTH HYDROXIDE NITRATE OXIDE see BKW100

BISMUTH, (2-HYDROXYBENZOATO-O1,O2)OXO- see SAI600

BISMUTH MAGISTERY see BKW100

BISMUTH NITRATE see BKW250

BISMUTH NITRIDE see BKW500

BISMUTHOUS OXIDE see BKW600

BISMUTH OXIDE see BKW600

BISMUTH(3+) OXIDE see BKW600

BISMUTH, OXO(SALICYLATO)- see SAI600

BISMUTH OXYSALICYLATE see SAI600

BISMUTH PENTAFLUORIDE see BKW750

BISMUTH PERCHLORATE see BKW850
BISMUTH PLUTONIDE see BKX000
BISMUTH POTASSIUM SODIUM TARTRATE (SOLUBLE) see BKX250
BISMUTH SALICYLATE, BASIC see SAI600
BISMUTH SESQUIOXIDE see BKW600
BISMUTH SESQUISULFIDE see DDI200
BISMUTH SESQUITELLURIDE see BKY000
BISMUTH SODIUM-p-AMINOPHENYLARSONATE see BKX500
BISMUTH SODIUM THIOGLYCOLLATE see BKX750
BISMUTH STANNATE PENTAHYDRATE see BKY250
BISMUTH SUBCITRATE see BKX800
BISMUTH SUBNITRATE see BKW100
BISMUTH SUBNITRICUM see BKW100
BISMUTH SUBSALICYLATE see SAI600
BISMUTH SULFATE see BKY600
BISMUTH(3+) SULFIDE see DDI200
BISMUTH TELLURIDE see BKY000
BISMUTH TELLURIDE, UNDOPED see BKY000
BISMUTH TIN OXIDE see BKY250
BISMUTH TITANATE(IV) see BKY300
BISMUTH TITANIUM OXIDE see BKY300
BISMUTH TRIOXIDE see BKW600
BISMUTH TRISODIUM THIOGLYCOLLATE see BKY500
BISMUTH TRISULFATE see BKY600
BISMUTH VIOLET see AOR500
BISMUTH WHITE see BKW100
BISMUTH YELLOW see BKW600
BISMUTHYL NITRATE see BKW100
BIS-β-NAPHTHOL see BGC100
BIS(NITRATO)DIOXOURANIUM HEXAHYDRATE see URS000
2,5-BIS-(NITRATOMERCURIMETHYL)-1,4-DIOXANE see BKZ000
BIS(NITRATO-O,O')DIOXO URANIUM (solid) see URA200
BIS(NITRATO-O)OXOZIRCONIUM see BLA000
BIS-p-NITRO BENZENE DIAZO SULFIDE see BLA250
1,5-BIS(5-NITRO-2-FURANYL)-1,4-PENTADIEN-3-ONE, (AMINOIMINOMETHYL)HYDRAZONE see PAF500
BIS(5-NITROFURFURYLIDENE)ACETONE GUANYLHYDRAZONE see PAF500
sym-BIS(5-NITRO-2-FURFURYLIDENE) ACETONE GUANYLHYDRAZONE see PAF500
1,5-BIS(5-NITRO-2-FURYL)-1,4-PENTADIENE-3-AMINOHYDRAZONE HYDROCHLORIDE see DKK100
1,5-BIS(5-NITRO-2-FURYL)-3-PENTADIENONE AMIDINONHYDRAZONE see PAF500
1,5-BIS(5-NITRO-2-FURYL)-3-PENTADIENONE GUANYLHYDRAZONE see PAF500
BIS(4-NITROPHENYL)DISULFIDE see BLA500
BIS(p-NITROPHENYL)DISULFIDE see BLA500
BIS(4-NITROPHENYL)ETHER see NIM600
BIS(p-NITROPHENYL)ETHER see NIM600
BIS(4-NITROPHENYL) PHOSPHATE see BLA600
BIS(p-NITROPHENYL) PHOSPHATE see BLA600
BIS(p-NITROPHENYL)SULFIDE see BLA750
N,N'-BIS(4-NITROPHENYL)UREA compd. with 4,6-DIMETHYL-2-PYRIMIDINOL (1:1) see NCW100
BIS(2-NITRO-4-TRIFLUOROMETHYLPHENYL) DISULFIDE see BLA800
BIS(NONYLOXYMALEOYLOXY)DIOCTYL STANNANE see DEI800

BIS(OCTANOYLOXY)DI-n-BUTYL STANNANE see BLB250
BIS(OCTANOYLOXY)DI-n-BUTYLTIN see BLB250
2,2-BIS(3'-tert-OCTYL)-4'-HYDROXYPHENYLPROPANE see BLB500
BIS(1-OCTYL) MALEATE see DVK800
BIS(2-OCTYL)PHTHALATE see BLB750
BIS(2-OCTYL)PHTHALATE see BLB750
BISODIUM TARTRATE see BLC000
BISOFLEX 81 see DVL700
BISOFLEX 91 see DVJ000
BISOFLEX DOA see AEO000
BISOFLEX DOP see DVL700
BISOFLEX DOS see BJS250
BISOFLEX L79P see BBO635
BISOFLEX L911P see DEK550
BIS-(2-OKTYL)ESTER KYSELINY FTALOVE see BLB750
2,2-BIS-3'-terc.OKTYL-4'-HYDROXYFENYLPROPAN (CZECH) see BLB500
BISOL 2 see TDR800
BIS(OLEOYLOXY)DIBUTYLSTANNANE see DEJ000
BISOLVOMYCIN see HOI000
BISOLVON see BMO325
BISOLVON HYDROCHLORIDE see BMO325
BISOMEL see IQN000
BISOMER 2HEA see ADV250
2,2'-(2,2-BIS((OXIRANYLMETHOXY)METHYL)-1,3-PROPANEDIYLBIS(OXYMETHYLENE)B ISOXIRANE) see PBB850
BIS(OXIRANYLMETHYL)BENZENAMINE (9CI) see DKM100
BISOXIRENO(4B,5:8A,9)PHENANTHRO(1,2-C)FURAN-4(2H)-ONE, 11-CHLORO-1B,3,6,6B,7,7A,9,10,11,11A-DECAHYDRO-9,10-DIHYDROXY-1B-METHYL-10-(1-METHYLETHYL)-, (1AS,1BS,6BR,7AS,8AS,9R,10R,11R,11AR)- see TNC200
BISOXIRENO(f,h)QUINOLINE, 1A,1B,2A,6B-TETRAHYDRO-, (1A-α-1B-β,2A-β,6Bα)-(+-)- see QNA100
N,N'-BIS(4-OXO-4H-1-BENZOPYRAN-2-YL)-5-OXO-1,2-PYRROLIDINEDICARBOXAMIDE see CBT600
BIS(2-OXO-9-BORNANESULFONIC ACID) DIETHYLSTANNYL ESTER see DKC600
BIS(1-OXODODECYL)PEROXIDE see LBR000
(8S-(8R*,16R*))-8,16-BIS(2-OXOPROPYL)-1,9-DIOXACYCLOHEXADECANE-2,5,10-TRIONE see BLC100
BIS(2-OXOPROPYL)-N-NITROSAMINE see NJN000
BIS(1-OXOPROPYL)PEROXIDE see DWQ800
BIS(8-OXYQUINOLINE)COPPER see BLC250
BIS(PANTOTHENAMIDOETHYL) DISULFIDE see PAG150
1,2-BIS(PENTABROMOPHENOXY)ETHANE see EEB230
1,2-BIS(2,3,4,5,6-PENTABROMOPHENOXY)ETHANE see EEB230
BIS(PENTACHLOR-2,4-CYCLOPENTADIEN-1-YL) see DAE425
BIS(PENTACHLOROCYCLOPENTADIENY L) see DAE425
BIS(PENTACHLORO-2,4-CYCLOPENTADIEN-1-YL) see DAE425
BIS(PENTACHLOROPHENOL), ZINC SALT see BLC500
BIS(PENTAFLUOROETHYL)ETHER see PCG760

BIS(PENTA FLUORO PHENYL)ALUMINUM BROMIDE see BLC750
BISPENTAFLUOROSULFUR OXIDE see BLD000
BIS(PENTAMETHYLENETHIURAM)-TETRASULFIDE see TEF750
BIS(2,4-PENTANEDIONATO)CHROMIUM see BLD250
BIS(2,4-PENTANEDIONATO)COPPER see BGR000
BIS(2,4-PENTANEDIONATO-O,O')ZINC see ZCJ000
BIS(2,4-PENTANEDIONATO)TITANIUM OXIDE see BGQ750
4,5-BIS(4-PENTENYLOXY)-2-IMIDAZOLIDINONE see BLD325
BISPHENOL A see BLD500
BISPHENOL A DIGLYCIDYL ETHER see BLD750
BISPHENOL A DISODIUM SALT see BLD800
BISPHENOL AF see HCZ100
BISPHENOL A GLYCIDYLMETHACRYLATE see BLD810
BISPHENOL A SODIUM SALT see BLD800
BISPHENOL C see IPK000
BISPHENOL DIGLYCIDYL ETHER, MODIFIED see BLE000
BIS(PHENOXARSIN-10-YL) ETHER see OMY850
BIS(10-PHENOXARSYL) OXIDE see OMY850
BIS(10-PHENOXYARSINYL) OXIDE see OMY850
10,10'-BIS(PHENOXYARSINYL) OXIDE see OMY850
BIS(p-PHENOXYPHENYL)DIPHENYLSTANNAN E see BLE250
BIS(p-PHENOXYPHENYL)DIPHENYLTIN see BLE250
1,4-BIS(PHENYL AMINO)BENZENE see BLE500
BISPHENYL-(2-CHLORPHENYL)-1-IMIDAZOLYL-METHAN (GERMAN) see MRX500
5,8-BIS((2-PHENYLETHYL)AMINO)-1,4-DIHYDROXY-9,10-ANTHRACENEDIONE see BLE550
2,7-BIS(2-PHENYLETHYL)BENZO(lmn)(3,8)PHENAN THROLINE-1,3,6,8(2H,7H)-TETRONE see BLE600
2,6-BIS(1-PHENYLETHYL)-4-METHYLPHENOL see MHR050
N,N'-BIS(PHENYLISOPROPYL)PIPERAZINE DIHYDROCHLORIDE see BLF250
BIS(PHENYLMERCURI)METHYLENEDINA PHTHALENESULFONATE see PFN000
BIS(PHENYLMERCURYLAURYL)SULFIDE see PFN750
2,7-BIS(PHENYLMETHYL)BENZO(lmn)(3,8)PHE NANTHROLINE-1,3,6,8(2H,7H)-TETRONE see BLE650
BIS-N,N'-(3-PHENYLPROPYL-2)-PIPERAZINE DIHYDROCHLORIDE see BLF250
BIS(PHENYLSELENIDE) see BLF500
2,5-BIS(PHENYLTHIO)BENZOQUINONE see BLF600
2,5-BIS(PHENYLTHIO)-p-BENZOQUINONE see BLF600
BIS(PHENYLTHIO)DIMETHYLTIN see BLF750
1,3-BIS((PHENYL)TRIAZENO)BENZENE see BLG000
4,4'-BIS(4-PHENYL-2H-1,2,3-TRIAZOL-2-YL)-2,2'-STILBENEDISULFONIC ACID DIPOTASSIUM SALT see BLG100
N,N-BIS(PHOSPHONOMETHYL)GLYCINE see BLG250

2,6-BIS(PICRYLAMINO)-3,5-DINITROPYRIDINE see PQC525

1,3-BIS(2-PIPERIDINOETHYL)-5-PHENYL-5-PIPERIDINOBARBITURIC ACID CITRATE (1:2) see BLG280

BIS(PIPERIDINOTHIOCARBONYL) TETRASULFIDE see TEF750

3,8-BIS(1-PIPERIDINYLMETHYL)-2,7-DIOXASPIRO(4.4)NONANE-1,6-DIONE see BLG325

2,6-BIS(1-PIPERIDYLMETHYL)-4-(α,α-DIMETHYLBENZYL)PHENOL DIHYDROBROMIDE see BHR750

BIS(2-PROPANOL)AMINE see DNL600

BIS(2-PROPOXYETHYL)-1,4-DIHYDRO-2,6-DIMETHYL-4-(3-NITROPHENYL)-3,5-PYRIDINEDICARBOXYLATE see NDY600

2,4-BIS(PROPYLAMINO)-6-CHLOR-1,3,5-TRIAZIN (GERMAN) see PMN850

BIS(2-PROPYL) DIXANTHOGEN see IRS500

N,N'-BISPROPYLENEISOPHTHALAMIDE see BLG400

BIS(2-PROPYLOXY)DIAZENE see DNQ700

trans-1,2-BIS(n-PROPYLSULFONYL)ETHYLENE see BLG500

BIS(2-PYRIDINETHIOL 1-OXIDE)COPPER see BLG600

2,2-BIS(((3-PYRIDINYLCARBONYL)OXY)METHYL)-1,3-PROPANEDIYL ESTER of 3-PYRIDINECARBOXYLIC ACID see NCW300

BIS(2-PYRIDYLTHIO)ZINC, 1,1'-DIOXIDE see ZMJ000

BIS(8-QUINOLINATO)COPPER see BLC250

BIS(8-QUINOLINOLATO)COPPER see BLC250

BIS(8-QUINOLINOLATO-N¹,O⁸)-COPPER see BLC250

BIS(SALICYLALDEHYDE)ETHYLENEDIIMINE COBALT(II) see BLH250

BIS(SUCCINYLDICHLOROCHOLINE) see HLC000

BIS(5-SULFO-8-QUINOLINOLATO-N¹,O⁸) MANGANESE(II) see BKJ275

BISTERIL see PDC250

BISTER ML see LBU200

BIS(2,3,3,3-TETRACHLOROPROPYL) ETHER see OAL000

BIS(TETRADECANOYLOXY)DIBUTYLSTANNANE see BLH309

BIS(TETRAETHYLAMMONIUM) TETRACHLOROCOBALTATE(II) see BLH315

1,3-BIS(TETRAHYDRO-2-FURANYL)-5-FLUORO-2,4-PYRIMIDINEDIONE see BLH325

1,3-BIS(TETRAHYDRO-2-FURYL)-5-FLUOROURACIL see BLH325

7-(3,5-BIS((TETRAHYDRO-2H-PYRAN-2-YL)OXY)-2-(4-PHENOXY-3-((TETRAHYDRO-2H-PYRAN-2-YL)OXY)-1-BUTENYL)CYCLOPENTYL)-2-(PHENYLSELENO)-5-HEPTENOIC ACID, METHYL ESTER see BLH400

BIS(TETRAKIS(HYDROXYMETHYL)PHOSPHONIUM)SULFATE (salt) see TDI000

BIS-N,N,N',N'-TETRAMETHYLPHOSPHORODIAMIDIC ANHYDRIDE see OCM000

1,6-BIS(5-TETRAZOLYL)HEXAAZ-1,5-DIENE see BLI000

3,4-BIS(1,2,3,4-THIATRIAZOL-5-YL THIO) MALEIMIDE see BLI250

BIS(1,2,3,4-THIATRIAZOL-5-YL THIO)METHANE see BLI500

l-3,3-BIS(3'-THIENYL)-2-PROPENYL-(3-HYDROXY-3-PHENYLPROPYL-2)AMINE see DXI800

BIS(THIOCARBAMOYL)DISULFIDE see TFS500

BISTHIOCARBAMYL HYDRAZINE see BLJ250

1,2-BIS(THIOCYANATO)ETHANE see EJC035

BIS(2-THIOCYANATOETHYL) ETHER see TFF200

BIS(THIOCYANATO)-MERCURY see MCU250

BIS(β-THIOETHYL ACETATE) see TFI100

p,p'-BIS(α-THIOL CARBAMYLACETAMIDO)BIPHENYL see MCL500

BISTHIOSEMI see MJN300

BIS(THIOUREA) see BLJ250

BISTOLUENE DIAZO OXIDE see BLJ500

1,4-BIS(p-TOLYLAMINO)ANTHRAQUINONE see BLK000

BIS-1,4-p-TOLYLAMINOANTHRCHINON (CZECH) see BLK000

N-BIS(p-TOLYLSULFONYL)AMIDOMETHYL MERCURY see BLK250

1,3-BIS(o-TOLYL)-2-THIOUREA see DXP600

3,5-BIS(o-TOLYL)-s-TRIAZOLE see BLK500

BISTON see DCV200

BISTRAMIDE A see BLK600

BIS(TRIBENZYLSTANNYL)SULFIDE see BLK750

BIS(TRIBENZYLTIN) SULFIDE see BLK750

BIS-(TRI-N-BUTYLCIN)OXID (CZECH) see BLL750

BIS(TRIBUTYLOXIDE) of TIN see BLL750

BIS(TRI-N-BUTYLPHOSPHINE)DICHLORONICKEL see BLS250

BIS(TRIBUTYL(SEBACOYLDIOXY))TIN see BLL000

BIS((TRI-n-BUTYLSTANNYL)CYCLOPENTADIENYL)IRON see BLL250

1,1'-BIS(TRIBUTYLSTANNYL)FERROCENE see BLL250

BIS(TRIBUTYLSTANNYL)OXIDE see BLL750

BIS(TRIBUTYLTIN) ITACONATE see BLL500

BIS(TRIBUTYL TIN)OXIDE see BLL750

BIS(TRIBUTYLTIN)SULFIDE see HCA700

BIS(TRI-N-BUTYLZINN)-OXYD (GERMAN) see BLL750

BIS-2,3,5-TRICHLOR-6-HYDROXYFENYLMETHAN (CZECH) see HCL000

m-BIS(TRICHLORMETHYL)BENZENE see BLL825

BIS(TRICHLOROACETYL)PEROXIDE see BLM000

BIS-2,4,5-TRICHLORO BENZENE DIAZO OXIDE see BLM250

1,3-BIS(2,2,2-TRICHLORO-1-HYDROXYETHYL)UREA see DGQ200

BIS(3,5,6-TRICHLORO-2-HYDROXYPHENYL)METHANE see HCL000

m-BIS(TRICHLOROMETHYL)BENZENE see BLL825

1,3-BIS(TRICHLOROMETHYL)BENZENE see BLL825

1:4-BIS-TRICHLOROMETHYL BENZENE see HCM500

BIS(TRICHLOROMETHYL)SULFONE see BLM500

BISTRICHLOROMETHYLTRISULFID (CZECH) see BLM750

BIS(TRICHLORO METHYL)TRISULFIDE see BLM750

1,3-BIS(2,4,5-TRICHLOROPHENOXY)-1,1,3,3-TETRABUTYLDISTANNOXANE see OPE100

BIS(2,3,5-TRICHLOROPHENYLTHIO)ZINC see BLN000

BIS(TRIETHYLENETETRAMINE)TUNGSTATONICKEL see BLN100

BIS(TRIETHYLLEAD)SILICONHEXAFLUORIDE see BLN200

BIS(TRIETHYL TIN)ACETYLENE see BLN250

BIS(TRIETHYL TIN) SULFATE see BLN500

BIS(TRIFLUOROACETIC) ANHYDRIDE see TJX000

BIS(TRIFLUOROACETOXY)DIBUTYLTIN see BLN750

BIS(TRIFLUOROACETYL)PEROXIDE see BLO000

BIS(TRIFLUOROETHYL)ETHER see HDC000

BIS(2,2,2-TRIFLUOROETHYL)ETHER see HDC000

3,5-BIS(TRIFLUOROMETHYL)ANILINE see BLO250

1,3-BIS(TRIFLUOROMETHYL)BENZENE see BLO270

BIS(TRIFLUOROMETHYL)CHLOROPHOSPHINE see BLO280

BIS(TRIFLUOROMETHYL)CYANOPHOSPHINE see BLO300

BIS(TRIFLUOROMETHYL)DISULFIDE see BLO325

1,1-BIS(TRIFLUOROMETHYL)ETHENE see HDC450

1,3-BIS(TRIFLUOROMETHYL)-5-ISOCYANOBENZENE see BLO390

BISTRIFLUOROMETHYLMETHANE see HDE100

2,2-BIS(TRIFLUOROMETHYL)-4-METHYL-5-PHENYLOXAZOLIDINE HYDRATE see BLP250

BIS(TRIFLUOROMETHYL)NITROXIDE see BLP300

2-(3,5-BIS(TRIFLUOROMETHYL)PHENYL)-N-METHYL-HYDRAZINECARBOTHIOAMIDE (9CI) see BLP325

1-((3,5-BIS-TRIFLUOROMETHYL)PHENYL)-4-METHYL-THIOSEMICARBAZIDE see BLP325

BIS(TRIFLUOROMETHYL)PHOSPHORUS(III) AZIDE see BLP500

α,α-BIS(TRIFLUOROMETHYL)-1-PIPERIDINEMETHANOL HYDRATE see BLQ250

BIS(TRIFLUOROMETHYL)SULFIDE see BLQ325

2,2-BIS(TRIFLUOROMETHYL)THIAZOLIDINE HYDRATE see BLQ500

BIS(TRIFLUOROMETHYLTHIO)MERCURY see BLQ525

2,2'-BIS(1,6,7-TRIHYDROXY-3-METHYL-5-ISOPROPYL)-8-ALDEHYDONAPHTHALENE see GJM000

BIS(TRIISOBUTYLSTANNANE) see HDY100

BISTRIMATE see BKX750

N,N'-BIS(3-(3,4,5-TRIMETHOXYBENZOYLOXY)PROPYL)HOMOPIPERAZINE DIHYDROCHLORIDE see CNR750

1,4-BIS(3-(3,4,5-TRIMETHOXYBENZOYLOXY)-PROPYL)PERHYDRO-1,4-DIAZEPINE DIHYDROCHLORIDE see CNR750

BIS-3,4,5-TRIMETHOXY-β-PHENETHYLAMMONIUM TETRACHLOROMANGANATE(II) see BKM100

1,10-BIS-TRIMETHOXYSILYLDECANE see BLQ550

1,2-BIS(TRIMETHOXYSILYL)ETHANE see EIT200

2,3-BISTRIMETHYLACETOXYMETHYL-1-METHYLPYRROLE see BLQ600

α,ω-BIS(TRIMETHYL AMMONIUM)HEXANE DIBROMIDE see HEA000

α,ω-BIS(TRIMETHYLAMMONIUM)HEXANE DICHLORIDE see HEA500

BIS(TRIMETHYLHEXYL)TIN DICHLORIDE see BLQ750

BIS(TRIMETHYLSILYL)ACETAMIDE see TMF000

N,o-BIS(TRIMETHYLSILYL)ACETAMIDE see TMF000

BIS(TRIMETHYLSILYL)AMINE see HED500

N,N'-BIS(TRIMETHYLSILYL)AMINOBORANE see BLQ850

cis-BIS(TRIMETHYLSILYLAMINO)TELLURIUM TETRAFLUORIDE see BLQ900

BIS(TRIMETHYLSILYL)CARBODIIMIDE see BLQ950

BIS(TRIMETHYLSILYL)CHROMATE see BLR000

1,2-BIS(TRIMETHYLSILYL)HYDRAZINE see BLR125

BIS(TRIMETHYLSILYL)MERCURY see BLR140

BISTRIMETHYL SILYL OXIDE see BLR250

BIS(TRIMETHYLSILYL)PEROXOMONOSULFATE see BLR500

BIS(1,3,7-TRIMETHYL-8-XANTHINYL)MERCURY see MCT000

N,N'-BIS(2,2,2-TRINITROETHYL)UREA see BLR625

BIS(2,4,6-TRINITRO-PHENYL)-AMIN (GERMAN) see HET500

BIS(TRINITROPHENYL)SULFIDE see BLR750

BISTRIPERCHLORATO SILICON OXIDE see BLS000

BIS(TRIPHENYLPHOSPHINE)DICHLORONICKEL see BLS250

BIS(TRIPHENYL PHOSPHINE)NICKEL DITHIOCYANATE see BLS500

BIS(TRIPHENYL SILYL)CHROMATE see BLS750

BIS(TRIPHENYLTIN)ACETYLENEDICARBOXYLATE see BLS900

BIS(TRIPHENYLTIN)SULFATE see BLT000

BIS(TRIPHENYLTIN)SULFIDE see BLT250

BIS(TRIPROPYLTIN)OXIDE see BLT300

BIS(TRIS(p-CHLOROPHENYL)PHOSPHINE)MERCURIC CHLORIDE COMPLEX see BLT500

BIS(TRIS(p-DIMETHYLAMINOPHENYL)PHOSPHINE)MERCURIC CHLORIDE COMPLEX see BLT750

BIS(TRIS(p-DIMETHYLAMINOPHENYL)PHOSPHINE OXIDE)STANNIC CHLORIDE COMPLEX see BLT775

BIS(TRIS(β,β-DIMETHYLPHENETHYL)TIN)OXIDE see BLU000

BIS(TRIS(p-METHOXYPHENYL)PHOSPHINE)MERCURIC CHLORIDE COMPLEX see BLU250

BIS(TRIS(2-METHYL-2-PHENYLPROPYL)TIN)OXIDE see BLU000

BIS(TRIS(p-METHYLTHIOPHENYL)PHOSPHINE)MERCURIC CHLORIDE COMPLEX see BLU500

BISTRIUM CHLORIDE see HEA500

BISULFAN see BOT250

BISULFITE see HIC600

BISULFITE see SOH500

BISULFITE de SODIUM (FRENCH) see SFE000

BISULPHANE see BOT250

BISULPHITE see HIC600

BIS(VINYLDIMETHYLSILYL)AMINE see TDQ050

BIS(2-VINYLOXYETHYL)ETHER see DJE600

1,4-BIS((VINYLOXY)METHYL)CYCLOHEXANE see BLU600

N,N-BIS(2,4-XYLYLIMINOMETHYL)METHYLAMINE see MJL250

BITEMOL see BJP000

BITEMOL S 50 see BJP000

BITHALLIUM TRISULFATE see TEM100

BITHIODINE see ARP875

BITHION see TAL250

BITHIONOL see TFD250

BITHIONOL SULFIDE see TFD250

BITIN see TFD250

BITIODIN see BLV000

BITIRAZINE see DIW000

BITIS ARIETANS VENOM see BLV075

BITIS GABONICA VENOM see BLV080

BITOKSYBACILLIN see BAC040

BITOLTEROL MESILATE see BLV125

BITOLTEROL MESYLATE see BLV125

4,4'-BI-o-TOLUIDINE see TGJ750

5,5'-BI-p-TOLUQUINONE see BLV200

(m,o'-BITOLYL)-4-AMINE see BLV250

BITOSCANATE see PFA500

BITREX see DAP812

β-BITTER ACID see LIU000

BITTER ALMOND OIL see BLV500

BITTER ALMOND OIL CAMPHOR see BCP250

BITTER CUCUMBER see FPD100

BITTER FENNEL OIL see FAP000

BITTER GOURD see FPD100

BITTER ORANGE OIL see BLV750

BITTER ORANGE OIL see OJK330

BITTER SALTS see MAJ500

BITTERSWEET see AHJ875

BITUMEN (MAK) see ARO500

Δ(1,1')-BIUREA see ASM270

BIUREA (6CI,7CI,8CI) see HGU050

BIVERM see PDP250

BIVINYL see BOP500

BIXA ORELLANA see APE100

BIZMUTHIOL II (CZECH) see MCP500

BK see BML500

BK8 see TOC000

BK 15 see TOB750

BKF see MJO500

B-K LIQUID see SHU500

B-K POWDER see HOV500

BL 9 see DXY000

BL 15 see HKQ100

BL 25 see MCB050

BL 35 see MCB050

BL 139 see DOY400

BL 191 see PBU100

BL-225 see AFJ172

BL 434 see MCB050

BL 6341 see GKW050

BL 6341A see GKW050

γ-BL see BOV000

BLA see LCH000

BLACAR 1716 see PKQ059

1743 BLACK see BMA000

11557 BLACK see IHC550

BLACK ACACIA see GJU475

BLACK ALG AETRINE see QAT520

BLACK AND WHITE BLEACHING CREAM see HIH000

BLACK ANTIMONY see AQL500

BLACK BIRCH OIL see SOY100

BLACK BLASTING POWDER see ERF500

BLACK CALLA see ITD050

BLACK COPPER OXIDE see CNO250

BLACK DOGWOOD see MBU825

BLACK EYED SUSAN see RMK250

BLACK GOLD F 89 see IHC550

BLACK IRON BM see IHC550

BLACK LEAD see CBT500

BLACK LEAF see NDN000

BLACK LOCUST see GJU475

BLACK MANGANESE OXIDE see MAS000

BLACK MAX see EHP700

BLACK 2EMBL see AQP000

BLACK 4EMBL see AQP000

BLACK NIGHTSHADE see DAD880

BLACK OXIDE of IRON see IHC450

BLACK PEARLS see CBT750

BLACK PEPPER OIL see BLW250

BLACK PN see BMA000

BLACK POWDER, compressed (DOT) see ERF500

BLACK POWDER, compressed (DOT) see PLL750

BLACK POWDER, granular or as a meal (UN 0027) (DOT) see ERF500

BLACK POWDER, granular or as a meal (UN 0027) (DOT) see PLL750

BLACK POWDER, in pellets (UN 0028) (DOT) see ERF500

BLACK POWDER, in pellets (UN 0028) (DOT) see PLL750

BLACK WIDOW SPIDER VENOM see BLW500

BLACOSOLV see TIO750

BLADAFUM see SOD100

BLADAFUME see SOD100

BLADAFUN see SOD100

BLADAN see EEH600

BLADAN see HCY000

BLADAN see PAK000

BLADAN see TCF250

BLADAN BASE see HCY000

BLADAN-M see MNH000

BLADDERON see FCB100

BLADDERPOD LOBELIA see CCJ825

BLADEX see BLW750

BLADEX G see DFY800

BLADEX H see TAH900

BLADEX 80WP see BLW750

BLAETTERALKOHOL see HFE000

BLANC de FARD see BKW100

BLANC FIXE see BAP000

BLANCOFOR BBU see CMP120

BLANCOL see BLX000

BLANCOL DISPERSANT see BLX000

BLANDLUBE see MQV750

BLANKOPHOR BBH see CMP200

BLANKOPHOR BBU see CMP120

BLANKOPHOR HZPA see DXB450

BLANKOPHOR MBBH see CMP200

BLANOSE BWM see SFO500

BLA-S see BLX500

BLASCORID see MOA600

BLASCORID see PJA130

BLASTICIDEN-S-LAURYLSULFONATE see BLX250

BLASTICIDIN see BLX500

BLASTICIDIN S see BLX500

BLASTING GELATIN (DOT) see NGY000

BLASTING OIL see NGY000

BLASTING POWDER see ERF500

BLASTING POWDER see PLL750

BLASTMYCIN see BLX750

BLASTOESTIMULINA see ARN500

BLASTOMYCIN see BLX750

BLATTANEX see PMY300

L-BLAU 1 see IBV050

L-BLAU 3 see ADE500

L-BLAU 3 see CMM062

L-BLAU 2 (GERMAN) see FAE100

BLAUES PYOKTANIN see AOR500

BLAUSAEURE (GERMAN) see HHS000

BLAUWZUUR (DUTCH) see HHS000

BLAZER see SFV650

BLAZER 2S see SFV650

BLAZING RED see CJD500

BLEACHED-DEODORIZED LARD see LBE300

BLEACHED-DEODORIZED TALLOW see TAC100

BLEACHED LARD see LBE300

BLEACHING POWDER see HOV500

BLEACHING POWDER, containing 39% or less chlorine (DOT) see HOV500
BLEDO CARBONERO (CUBA) see PJJ315
BLEIACETAT (GERMAN) see LCV000
BLEIAZETAT (GERMAN) see LCJ000
BLEIPHOSPHAT (GERMAN) see LDU000
BLEISTEARAT (GERMAN) see LDX000
BLEISTIFTBAUMS (GERMAN) see EQY000
BLEISULFAT (GERMAN) see LDY000
BLEKIT EVANSA (POLISH) see BGT250
BLEKIT TURKUSOWY A see ERG100
BLEMINOL see ZVJ000
BLENDED RED OXIDES of IRON see IHC450
BLENOXANE see BLY000
BLENOXANE see BLY780
BLEO see BLY000
BLEOCIN see BLY000
BLEOMYCETIN see BLY500
BLEOMYCIN see BLY000
BLEOMYCIN A2 see BLY250
BLEOMYCIN A5 see BLY500
BLEOMYCIN A COMPLEX see BLY750
BLEOMYCINAMIDE, N1-(3-((3-(BIS((3,4-BIS(PHENYLMETHOXY)PHENYL)METHYL AMINO)PROPYL)METHYLAMIN O)PROPYL)- see LFJ500
BLEOMYCINAMIDE, N1-(3-(DIMETHYLSULFONIO)PROPYL)-, IRON COMPLEX see IGO500
BLEOMYCIN B2 see BLY760
BLEOMYCIN PEP see BLY770
BLEOMYCIN SULFATE see BLY780
BLEOMYCIN, SULFATE (salt) (9CI) see BLY780
BLEPH 10 see SNQ000
BLEPH-10 see SNQ710
BLEU BRILLIANT FCF see FMU059
BLEU DIAMINE see CMO250
BLEU PATENTE V see ADE500
BLEU PATENTE V see CMM062
BLEU SOLANTHRENE see IBV050
BLEX see DIN800
BLEXANE see BLY780
BL H368 see BEQ625
BLIGHIA SAPIDA see ADG400
BLIGHTOX see EIR000
BLISTER FLOWER see FBS100
BLISTERING BEETLES see CBE250
BLISTERING FLIES see CBE250
BLISTER WORT see FBS100
BLITEX see EIR000
BLITOX see CNK559
BLITOX 50 see CNK559
BLIZENE see EIR000
BLM see BLY000
BLM-PEP see BLY770
BLO see BOV000
BLOAT GUARD see PJK150
BLOAT GUARD see PJK151
BLOC see FAK100
BLOCADREN see DDG800
BLOCAN see SBH500
BLOCKADE see DWS200
BLON see BOV000
BLOODBERRY see ROA300
BLOODSTONE see HAO875
BLOTIC see MKA000
BLO-TROL see THX100
BLOX see LIH000
BLOXANTH see ZVJ000
BL P 152 see PDD350
BLP-1011 see DGE200
BLP 1322 see HMK000
BL-S 578 see DYF700
BLS 640 see APT250
BLUE 2B see CMO000
BLUE 1084 see ADE500
1085 BLUE see ADE500
1206 BLUE see FAE000
1311 BLUE see FAE100
11388 BLUE see FMU059

11669 BLUE see BGB275
12070 BLUE see FAE100
BLUE ANTHRAQUINONE PIGMENT see IBV050
BLUE 'APE (HAWAII) see XCS800
BLUE ASBESTOS (DOT) see ARM275
BLUEBELL see CMV390
BLUEBERRY ROOT see BMA150
BLUE BH see CMN800
BLUE BLACK 12B see FAB830
BLUE BLACK BN see BMA000
BLUE BLACK SX see FAB830
BLUE BN BALSE see DCJ200
BLUECAIN see BMA125
BLUE CARDINAL FLOWER see CCJ825
BLUE CHAMOMILE OIL see CDH500
BLUE COHOSH see BMA150
BLUE COPPER see CNK559
BLUE COPPER see CNP250
BLUE COPPER-50 see CNK559
BLUE COPPERRAS see CNP500
BLUE CROSS see CGN000
BLUE DEVIL WEED see VQZ675
BLUE EMB see CMO250
BLUE GINSENG see BMA150
BLUE JESSAMINE see CMV390
BLUE K see DFN300
BLUE NO. 201 see BGB275
BLUE O see IBV050
BLUE OIL see AOQ000
BLUE OIL see COD750
BLUE-OX see ZLS000
BLUE POWDER see ZBJ000
BLUE STAR see AQF000
BLUE STONE see CNP250
BLUESTONE see CNP500
BLUE TARO (HAWAII) see XCS800
BLUE URS see ADE500
BLUE VITRIOL see CNP250
BLUE VITRIOL see CNP500
BLUE VRS see ADE500
BLUE ZN 3 see CMM062
BLUTENE see AJP250
BLUTENE CHLORIDE see AJP250
BLUTON see IIU000
BM 1 see HNI500
BM 3055 see IBP200
BM 06011 see SFX730
7-BMBA see BNO750
BMC see BRS750
BMC see MHC750
BMD see BAC260
BMF 1 see MCB050
BMF 1 (AMINOPLAST) see MCB050
BMIH see IKC000
BMII see DQT100
BMOO see BRT000
BMY 27857 see SLJ800
BMY 28100 see CCS527
BMY 28488 see AMS650
BN see BFW000
BN 30 see MCB050
(d-BNA6))-LHRH ACETATE see LIU309
B-NINE see DQD400
BNM see BAV575
B. N. MEXICANUS VENOM see BMJ500
BNP see NIY500
BNP 30 see BRE500
BNPP see BLA600
BNS see NMC100
BNU see BSA250
BO 714 see TCZ000
BO-ANA see FAB600
BOB see NKL300
BOCEP VITI see GJU050
BOC-(SER(BZ1)1)-DES-HIS2)-d-TRP6))-LHRH HYDROCHLORIDE HYDRATE (2:3:8) see LIU317
BOC-(SER(BZL)1-DES-HIS2-d-TRP6)-LHRH ACETATE TRIHYDRATE see LIU315
4-BOC-STYRENE see BPI400
BOEA see BKA000

B.O.E.A. see BKA000
BOF-A2 see BDN600
BOG MANGANESE see MAS000
BOG ONION see JAJ000
BOH see HHC000
BOISAMBRENE see FMV200
BOISAMBRENE FORTE see FMV100
BOIS BLEUDE HONQRIE see FBW000
BOIS D'ARC (FRENCH) see MRN500
BOIS D'INDE see LBK000
BOIS GENTIL (CANADA) see LAR500
BOIS d'INDE see BAT500
BOIS JAMBETTE (HAITI) see GIW200
BOIS JOLI (CANADA) see LAR500
BOIS de PLOMB (CANADA) see LEF100
BOIS de ROSE OIL see BMA550
BOL see BNM250
BOL-148 see BNM250
BOLAFFININ (9CI) see BMA575
BOLAFFININE see BMA575
BOLATRON see PKQ059
BOLDIN see DNZ100
BOLDINE see DNZ100
(+)-BOLDINE see DNZ100
(S)-BOLDINE see DNZ100
(+)-(S)-BOLDINE see DNZ100
BOLDINE DIMETHYL ETHER see TDI475
BOLDO LEAF OIL see BMA600
BOLERO see SAZ000
BOLETIC ACID see FOU000
BOLETIC ACID DIMETHYL ESTER see DSB600
BOLINAN see PKQ250
BOLLS-EYE see HKC000
BOLLS-EYE see HKC500
BOLSTAR see SOU625
BOLTAGE see POK500
BOLVIDON see BMA625
BOMBITA see DBA800
BOMT see BMA650
BOMYL see SOY000
BON see HMX520
BONA see HMX520
BONABOL see XQS000
BON ACID see HMX520
BONADETTES see HGC500
BONADOXIN see HGC500
BONAID see BOO632
BONAMID see PJY500
BONAMINE see HGC500
BONAPAR see PGG350
BONAPHTHON see BNS750
BONAPICILLIN see AIV500
BONARE see CFZ000
BONAZEN see ZNA000
BONBONNIER (HAITI) see LAU600
BONBRAIN see TEH500
BOND CH 18 see AAX250
BONDELANE A see SNW500
BONDERITE 40 see ZJS400
BONDERITE 880 see ZJS400
BONDING AGENT M 3 see OMM300
BONDOLANE A see SNW500
BONE OIL see BMA750
BONGAY see HGL575
BONIBAL see DXH250
BONICOR see HNY500
BONIDE BLUE DEATH RAT KILLER see PHP010
BONIDE KRAB CRABGRASS KILLER see PLC250
BONIDE RYATOX see RSZ000
BONIDE TOPZOL RAT BAITS and KILLING SYRUP see RCF000
BONIFEN see BMB000
BONINE see MBX500
BONITON see AEH750
BONJELA see BEL900
BONLAM see CBA100
BONLOID see PKQ059
BONNECOR see BMB125
BONOFORM see TBQ100

BONOMOLD OE see HJL000
BONOMOLD OP see HNU500
BON RED YELLOW SHADE see CMS148
BONZI see TMK125
BOOKSAVER see AAX250
BOOMER-RID see SMN500
BOOTS BTS 27419 see MJL250
BOP see NJN000
BORACIC ACID see BMC000
BORACSU see SFF000
BORANE-AMMONIA see BMB150
BORANECARBOXYLIC ACID, AMMONIUM
SALT see ANB700
BORANE, COMPD. WITH N2H4 see HGU025
BORANE, COMPOUND with N,N-
DIMETHYLMETHANAMINE (1:1) see
BMB250
BORANE, COMPOUND with
TRIMETHYLAMINE (1:1) see BMB250
BORANE with DIMETHYLAMINE (1:1) see
DOR200
BORANE, compound with
DIMETHYLSULFIDE see MPL250
BORANE-HYDRAZINE see BMB260
BORANE, compounded with MORPHOLINE
see MRQ250
BORANE-PHOSPHORUS TRIFLUORIDE see
BMB270
BORANE-PYRIDINE see POQ250
BORANES see BMB280
BORANE-TETRAHYDROFURAN see
BMB300
BORANE, TRIFLUORO-, DIHYDRATE see
BMG800
BORASSUS FLABELLIFER Linn., extract see
BMB325
BORATE(1-), BIS(1,2-BENZENEDIOLATO(2-
)-KAPPA-O,KAPPA-O')-, (T-4)-,
HYDROGEN,COMPD. WITH N,N'-BIS(2-
METHYLPHENYL)GUANIDINE (1:1) see
DEK500
BORATE(1-), BIS(PYROCATECHOLATO(2-
))-, HYDROGEN, COMPD. WITH 1,3-DI-o-
TOLYLGUANIDINE (1:1) see DEK500
BORATES, TETRA, SODIUM SALT,
anhydrous (OSHA) see DXG035
BORATES, TETRA, SODIUM SALT,
anhydrous (OSHA, ACGIH) see SFE500
BORATES, TETRA, SODIUM SALT,
anhydrous (OSHA, ACGIH) see SFF000
BORATES, TETRA, SODIUM SALT,
PENTAHYDRATE see SKC550
BORATES, TETRA, SODIUM SALTS
(PENTAHYDRATE) see SKC550
BORATE(1-), TETRAFLUORO-, CADMIUM
(2:1) (9CI) see CAG000
BORATE(1-), TETRAFLUORO-, COBALT(2+)
(8CI,9CI) see CNC050
BORATE(1-), TETRAFLUORO-,
HYDROGEN see FDD125
BORATE(1-), TETRAFLUORO-,
HYDROGEN see HHS600
BORATE(1-), TETRAHYDRO-, ALUMINUM
(3:1) (9CI) see AHG875
BORAX (8CI) see SFF000
BORAX DECAHYDRATE see SFF000
BORAX GLASS see DXG035
BORAZINE see BMB500
BORAZOLE see BMB500
BORDEAUX see FAG020
BORDEAUX ARSENITE see BMB750
BORDEAUX BG see CMO872
BORDEAUX DIRECT see CMO872
BORDEAUX EMBL see CMO885
BORDEAUX RRN see CMU820
BORDEAUX SALT CIBA II see MPY750
BORDEAUX SALT IRGA II see MPY750
BORDEN 2123 see AAX250
BORDERMASTER see CIR250
BOREA see BMM650
BORER SOL see EIY600
BORESTER 2 see THX750

BORESTER O see TLN000
BORIC ACID see BMC000
BORIC ACID, DISODIUM SALT see DXG035
BORIC ACID, ETHYL ESTER see BMC250
BORIC ACID (H2B4O7), CALCIUM SALT (1:1)
(8CI) see CAN250
BORIC ACID(H2B4O7), DISODIUM SALT,
PENTAHYDRATE see SKC550
BORIC ACID, MONOSODIUM SALT see
SII100
BORIC ACID, PHENYLMERCURY SILVER
derivative see PFP250
BORIC ACID, SODIUM SALT see SJD000
BORIC ACID, TRI-n-AMYL ESTER see
TMQ000
BORIC ACID, TRIBUTYL ESTER see THX500
BORIC ACID, TRI-sec-BUTYL ESTER see
THX750
BORIC ACID, TRI-sec-BUTYL ESTER see
THY000
BORIC ACID, TRI-o-CHLOROPHENYL
ESTER see TIY750
BORIC ACID, TRI-o-CRESYL ESTER see
TJF500
BORIC ACID, TRIETHYL ESTER see TJK250
BORIC ACID, TRIETHYL ESTER see TJP500
BORIC ACID, TRIHEXYL ESTER see
TKM000
BORIC ACID, TRIISOBUTYL ESTER see
TKR750
BORIC ACID, TRIISOPROPYL ESTER see
IOI000
BORIC ACID, TRIOCTADECYL ESTER see
TMN750
BORIC ACID, TRI-n-OCTYL ESTER see
TMO550
BORIC ACID, TRIOLEYL ESTER see
BMC500
BORIC ACID, TRI-n-PENTYL ESTER see
TMQ000
BORIC ACID, TRIS(2-AMINOETHYL)
ESTER see TJK750
BORIC ACID, TRIS(1-AMINO-2-PROPYL)
ESTER see TKT200
BORIC ACID, TRIS(2-ETHYLHEXYL)
ESTER see TJR500
BORIC ACID, TRIS(1-METHYLHEPTYL)
ESTER see TMO500
BORIC ACID, TRIS(4-METHYL-2-PENTYL)
ESTER see BMC750
BORIC ACID, TRIS(PHENYLCYCLOHEXYL)
ESTER see TMR750
BORIC ACID, TRISTEARYL ESTER see
TMN750
BORIC ANHYDRIDE see BMG000
BORICID see BMC800
BORICIDE see BMC800
BORICIN see SFF000
BOR-IND see IDA400
BORINE, TRIPHENYL see TMR300
BORNANE, 2-CHLORO-, endo- see BMD300
BORNANE, 2,2,5-endo,6-exo,8,9,10-
HEPTACHLORO- see THH575
BORNANE, 2-METHOXY-, exo-(8CI) see
IHX500
10-BORNANESULFONIC ACID, 2-OXO-,
(1S,4R)-(+)- see RFU750
1-2-BORNANOL see NCQ820
2-BORNANOL, endo- see BMD000
2-BORNANONE see CBA750
(+)-2-BORNANONE see CBB250
d-2-BORNANONE see CBB250
BORNATE see IHZ000
BORNEO CAMPHOR see BMD000
BORNEOL see BMD000
(−)-BORNEOL see NCQ820
BORNEOL (DOT) see BMD000
trans-BORNEOL see BMD000
(1S,2R,4S)-(−)-1-BORNEOL see NCQ820
BORNYL ACETATE see BMD100
l-BORNYL ACETATE see BMD100
BORNYL ALCOHOL see BMD000

1-BORNYL ALCOHOL see NCQ820
S-((N-BORNYLAMIDIN)METHYL)
HYDROGEN THIOSULFATE see BMD250
BORNYL CHLORIDE see BMD300
2-BORNYL CHLORIDE see BMD300
BORNYL ISOVALERATE see HOX100
BORNYVAL see HOX100
BOROETHANE see DDI450
BOROFAX see BMC000
BOROFLUORIC ACID see FDD125
BOROFLUORIC ACID see HHS600
BOROHYDRURE de POTASSIUM (FRENCH)
see PKY250
BOROHYDRURE de SODIUM (FRENCH) see
SFF500
BOROLIN see PIB900
BORON see BMD500
BORON AZIDE DICHLORIDE see BMD750
BORON AZIDE DIIODIDE see BMD825
BORON BROMIDE see BMG400
BORON BROMIDE DIIODIDE see BME250
BORON CALCIUM OXIDE see CAN250
BORON CHLORIDE see BMG500
BORON COMPOUNDS see BME500
BORON DIBROMIDE IODIDE see BME750
BORON, (N,N-DIMETHYL-1-
OCTADECANAMINE)TRIHYDRO-,(T-4)- see
DTC300
BORON FLUORIDE see BMG700
BORON FLUORIDE, compd. with ACETIC
ACID see BMG750
BORON FLUORIDE DIHYDRATE see
BMG800
BORON, (HYDRAZINE-
KAPPAN)TRIHYDRO-, (T-4)- see HGU025
BORON, (HYDRAZINE-N)TRIHYDRO-, (T-
4)- see HGU025
BORON HYDRIDE see DDI450
p-BORONOBENZOIC ACID see CCI600
BORON OXIDE see BMG000
BORON PHOSPHIDE see BMG250
BORON SESQUIOXIDE see BMG000
BORON SODIUM OXIDE(B4NA2O7),
PENTAHYDRATE see SKC550
BORON TRIAZIDE see BMG325
BORON TRIBROMIDE see BMG400
BORON TRICHLORIDE see BMG500
BORON TRIFLUORIDE see BMG700
BORON TRIFLUORIDE−ACETIC ACID
COMPLEX see BMG750
BORON TRIFLUORIDE DIETHYL
ETHERATE see BMH250
BORON TRIFLUORIDE DIHYDRATE see
BMG800
BORON TRIFLUORIDE DIHYDRATE
(DOT) see BMG800
BORON TRIFLUORIDE-DIMETHYL
ETHER see BMH000
BORON TRIFLUORIDE DIMETHYL
ETHERATE (DOT) see BMH000
BORON TRIFLUORIDE ETHERATE see
BMH250
BORONTRIFLUORIDE
MONOETHYLAMINE see EFU500
BORON TRIIODIDE see BMH500
BORON TRIOXIDE see BMG000
BORON TRISULFIDE see BMH659
BOROPHENYLIC ACID see BBM000
BOROXIN, TRIMETHOXY- see TKZ100
BORRELIDIN see BMH750
BORSAEURE (GERMAN) see BMC000
BORSIL P see SCK600
BORTRAN see RDP300
BORTRYSAN see DEV800
BORUTA BLACK A see FAB830
BOSAN SUPRA see DAD200
BOSENTAN see BMH800
BOSMIN see VGP000
BOTHROPS ASPER VENOM see BMI000
BOTHROPS ATROX VENOM see BMI125
BOTHROPS COLOMIBIENSIS VENOM see
BMI250

BOTHROPS GODMANI VENOM see BMI500
BOTHROPS LATERALIS VENOM see BMI750
BOTHROPS NASUTUS VENOM see BMJ000
BOTHROPS NIGROVIRIDIS NEGROVIRIDIS VENOM see BMJ250
BOTHROPS NUMMIFER MEXICANUS VENOM see BMJ500
BOTHROPS OPHYOMEGA VENOM see BMJ750
BOTHROPS PICADOI VENOM see BMK000
BOTHROPS SCHLEGLII VENOM see BMK250
BOTHROPS VENOM PROTEINASE see RDA350
BOTRAN see RDP300
BOTROPASE see RDA350
BOTRYODIPLODIN see BMK290
(−)-BOTRYODIPLODIN see BMK290
BOTULINUM NEUROTOXIN see BMM292
BOTULINUSTOXIN see CMY030
BOURBONAL see EQF000
BOURGEONAL see BMK300
BOURREAU DES ARBRES (CANADA) see AHJ875
BOUTON d'OR (CANADA) see FBS100
BOUVARDIN see BMK325
BOV see SOI500
BOVERIN see BAT830
BOVERINE see BAT830
BOVIDAM see CBA100
BOVIDERMOL see DAD200
BOVILENE see FAQ500
BOVINE LACTOGENIC HORMONE see PMH625
BOVINE PINEAL GLAND EXTRACT see BMK400
BOVINE PROLACTIN see PMH625
BOVINE RENNET see RCZ100
BOVINOCIDIN see NIY500
BOVINOX see TIQ250
BOVITROL see DAM300
BOVIZOLE see TEX000
BOVOFLAVIN see DBX400
BOVOLIDE see BMK500
BOXER see PFR130
BOY see HBT500
BOYGON see PMY300
B P 1 see MCB050
BP2 see AFJ625
B(a)P see BCS750
B(e)P see BCT000
BP 1.02 see SDY625
BP 400 see MOO750
BP-400 see MOP000
BPA see BEA100
BPDE see BCU250
BPDE-syn see DMR000
anti-BPDE see BCU250
anti-BPDE see DMQ600
BP-4,5-DIHYDRODIOL see DLB800
BP-7,8-DIHYDRODIOL see BCT750
BP-7,8-DIHYDRODIOL see DML000
BP-7,8-DIHYDRODIOL see DML200
BP-9,10-DIHYDRODIOL see DLC400
B(e)P-4,5-DIHYDRODIOL see DMK400
B(E)P 9,10-DIHYDRODIOL see DMK600
BP cis-9,10-DIHYDRODIOL see BMK550
trans-BP-7,8-DIHYDRODIOL DIACETATE see DBG200
BP-7,8-DIHYDRODIOL-9,10-EPOXIDE (anti) see BCU000
syn-BP-7,8-DIHYDRODIOL-9,10-OXIDE see BMK560
anti-BP-7,8-DIHYDRODIOL-9,10-OXIDE see BCU000
B(e)P DIOL EPOXIDE-1 see DMR150
B(e)P DIOL EPOXIDE-1 see DMR400
B(E)P DIOL EPOXIDE-2 see DMR200
anti-BP-DIOLEPOXIDE see DMQ000
BP 7,8-DIOL-9,10-EPOXIDE 2 see DMP900

B(e)P 9,10-DIOL-11,12-EPOXIDE-1 see DMR150
(+)-BP-7,α,8-β-DIOL-9,α,10,α-EPOXIDE 1 see DMP600
(+)-BP-7-β,8-α-DIOL-9-α,10-α-EPOXIDE 2 see BMK620
(−)-BP 7,α,8-β-DIOL-9-β,10-β-EPOXIDE 2 see DVO175
(−)-BP-7,β,8,α-DIOL-9,β,10,β-EPOXIDE 1 see DMP800
BP DIOL EPOXIDE ANTI see BCU250
BPE-I see PJS750
BP-4,5-EPOXIDE see BCV500
BP 7,8-EPOXIDE see BCV750
B(a)P EPOXIDE I see DMR000
B(a)P EPOXIDE II see BMK630
BPG 400 see PKK500
BPG 800 see PKK750
B(E)P H4-9,10-DIOL see DNC400
B(c)PH DIOL EPOXIDE-1 see BMK634
B(c)PH DIOL EPOXIDE-2 see BMK635
B(e)P H4-9,10-EPOXIDE see ECQ150
BP-3-HYDROXY see BCX250
BP-KLP see SMQ500
BPL see PMT100
BPMC see MOV000
BP 4,5-OXIDE see BCV500
BP 7,8-OXIDE see BCV750
BP-9,10-OXIDE see BCW000
BP-11,12-OXIDE see BCW250
BPPS see SOP000
BP-3,6-QUINONE see BCU750
BP-6,12-QUINONE see BCV000
BPZ see FLL000
BR 33N see HCA550
BR 55N see PAU500
BR 700 see CKI750
BR 750 see GKO750
BR-931 see CLW500
BRACE see PHK000
BRACKEN FERN, CHLOROFORM FRACTION see BMK750
BRACKEN FERN, DRIED see BML000
BRACKEN FERN TANNIN see BML250
BRACKEN FERN, TANNIN-FREE see TAE250
BRACKEN FERN TOXIC COMPONENT see SCE000
BRADILAN see TDX860
BRADOPHEN see BDJ600
BRADYKININ see BML500
BRADYKININ (synthetic) see BML500
BRADYL see NAC500
BRALEN KB 2-11 see PJS750
BRALEN RB 03-23 see PJS750
BRAMYCIN see BML750
BRAN ABSOLUTE see WBJ700
BRANCHED MONO-OLEFINS see VRU300
BRASILACA RED R see NAP100
BRASILAMINA BLACK GN see AQP000
BRASILAMINA BLUE 2B see CMO000
BRASILAMINA BLUE 3B see CMO250
BRASILAMINA BLUE RW see CMO600
BRASILAMINA CONGO 4B see SGQ500
BRASILAMINA FAST RED F see CMO870
BRASILAMINA GREEN B see CMO830
BRASILAMINA GREEN G see CMO840
BRASILAMINA RED 4B see DXO850
BRASILAMINA VIOLET 3R see CMP000
BRASILAN AZO RUBINE 2NS see HJF500
BRASILAN BLACK BS see FAB830
BRASILAN CHROME VIOLET B see HLI000
BRASILAN FUCHSINE D see CMS228
BRASILAN METANIL YELLOW see MDM775
BRASILAN ORANGE 2G see HGC000
BRASILAN ORANGE A see CMM220
BRASILAZET BLUE GR see TBG700
BRASILAZINA OIL RED B see SBC500
BRASILAZINA OIL SCARLET see OHI200
BRASILAZINA OIL SCARLET 6G see XRA000
BRASILAZINA OIL YELLOW G see PEI000

BRASILAZINA OIL YELLOW R see AIC250
BRASILAZINA ORANGE Y see PEK000
BRASILAZOL BLACK BH see CMN800
BRASIL (CUBA) see CAK325
BRASILETTO (BAHAMAS) see CAK325
BRASILIAN CHROME ORANGE R see NEY000
BRASILIN see LFT800
BRASORAN see ASG250
BRASSICOL see PAX000
BRAUNOSAN H see PKE250
BRAUNSTEIN (GERMAN) see MAS000
BRAVO see TBQ750
BRAVO 6F see TBQ750
BRAVO-W-75 see TBQ750
BRAXIN C see POI100
BRAXORONE see BML825
BRAZILAMINA FAST BROWN 3RA see CMO800
BRAZILETTO see LFT800
BRAZILIAN PEPPER TREE see PCB300
BRAZILIN see LFT800
BRAZIL WAX see CCK640
BRB-I-28 PERCHLORATE see BFK370
BREADFRUIT VINE see SLE890
BRECHWEINSTEIN see AQJ500
BRECOLANE NDG see DJD600
BREDININ see BMM000
BREDININE see BMM000
BREITHAUPTITE see BMM075
BREK see LIH000
BRELLIN see GEM000
BREMFOL see BMM125
BREMIL see CFY000
BRENAL see AAE500
BRENDIL see VGK000
BRENOL see ART250
BRENTAMINE FAST BLUE B BASE see DCJ200
BRENTAMINE FAST GARNET GBC SALT see MPY750
BRENTAMINE FAST ORANGE GC BASE see CEH690
BRENTAMINE FAST ORANGE GR BASE see NEO000
BRENTAMINE FAST RED B BASE see NEQ000
BRENTAMINE FAST RED TR BASE see CLK220
BRENTAMINE FAST RED TR SALT see CLK235
BRENTHOL AS see CMM760
BREON see PKQ059
BREON 202 see CGW300
BREON 351 see AAX175
BREON CS 100/30 see CGW300
BRESIT see ARQ750
BRESTAN see ABX250
BRESTANOL see CLU000
BRETHINE see TAN250
BRETOL see EKN500
BRETYLAN see BMV750
BRETYLATE see BMV750
BRETYLIUM-p-TOLUENESULFONATE see BMV750
BRETYLIUM TOSYLATE see BMV750
BRETYLOL see BMV750
BREVETOXIN see BMM150
BREVIMYTAL see MDU500
BREVINYL see DGP900
BREVIRENIN see VGP000
BREVITAL SODIUM see MDU500
BRIANIL see IPU000
BRICAN see TAN100
BRICANYL see TAN100
BRICANYL see TAN250
BRICAR see TAN100
BRICARIL see TAN100
BRICK OIL see CMY825
BRICYN see TAN100
BRIDAL see BEM500
BRIER (BAHAMAS) see CAK325

BRIETAL SODIUM see MDU500
BRIGADE see TAC850
BRIGHT RED see CHP500
BRIGHT RED see CMS150
BRIGHT RED G TONER see CMS148
BRIJ 30 see DXY000
BRIJ 38 see PJT300
BRIJ 52 see PJT300
BRIJ 56 see PJT300
BRIJ 58 see PJT300
BRIJ 92 see OIG000
BRIJ 98 see PJW500
BRIJ 92((2)-OLEYL) see OIG000
BRIJ 96((10) OLEYL) see OIG040
BRIJ W1 see PJT300
BRILLIANT 15 see CAT775
BRILLIANT ACID BLACK BNA EXPORT see BMA000
BRILLIANT ACID BLACK BN EXTRA PURE A see BMA000
BRILLIANT ACID BLUE A EXPORT see ADE500
BRILLIANT ACID BLUE AS see ERG100
BRILLIANT ACID BLUE N EXTRA see ERG100
BRILLIANT ACID BLUE V EXTRA see ADE500
BRILLIANT ACID BLUE VS see ADE500
BRILLIANT ACID RED 6A see CMM400
BRILLIANT ACID RED G see CMM300
BRILLIANT ACID ROSAMINE 2G see CMM300
BRILLIANT ACID ROSAMINE 6B see CMM400
BRILLIANT ACRIDINE ORANGE E see BJF000
BRILLIANT ALIZARINE CYANINE R see CMM080
BRILLIANT ALIZARINE LIGHT BLUE 3FR see CMM080
BRILLIANT ALIZARINE SKY BLUE BS see CMM100
BRILLIANT BLACK see BMA000
BRILLIANT BLACK A see BMA000
BRILLIANT BLACK BN see BMA000
BRILLIANT BLACK NAF see BMA000
BRILLIANT BLACK N.FQ see BMA000
BRILLIANT BLUE see FMU059
BRILLIANT BLUE FCD No. 1 see FAE000
BRILLIANT BLUE FCF see FAE000
BRILLIANT BLUE GS see ADE500
BRILLIANT BLUE R see BMM500
BRILLIANT CHROME LEATHER BLACK H see AQP000
BRILLIANT COLACID RED 6A see CMM400
BRILLIANT COLACID RED G see CMM300
BRILLIANT CRESYL BLUE see BMM550
BRILLIANT CRESYL BLUE BB see BMM550
BRILLIANT CRIMSON RED see HJF500
BRILLIANT FAST YELLOW see DOT300
BRILLIANT FAT SCARLET R see CMS238
BRILLIANT GREEN 3EMBL see FAE950
BRILLIANT GREEN PHTHALOCYANINE see PJQ100
BRILLIANT GREEN SULFATE see BAY750
BRILLIANT INDIGO 4B see ICU135
BRILLIANT INDIGO 4BJD see ICU135
BRILLIANT INDIGO 4BR see ICU135
BRILLIANT INDIGO 4BV see ICU135
BRILLIANT LAKE B see CMG750
BRILLIANT LAKE M see CMS160
BRILLIANT LAKE PBB see CMG750
BRILLIANT LAKERED R see CMS160
BRILLIANT MILLING RED see NAO600
BRILLIANT OIL ORANGE R see PEJ500
BRILLIANT OIL ORANGE R BASE see CMM800
BRILLIANT OIL ORANGE Y BASE see PEK000
BRILLIANT OIL SCARLET B see XRA000
BRILLIANT OIL YELLOW see IBB000
BRILLIANT ORANGE see CMP882

BRILLIANT ORANGE GR see CMU820
BRILLIANT ORANGE (INDICATOR) see CMP882
BRILLIANT PINK AS see CMM840
BRILLIANT PINK B see FAG070
BRILLIANT PONCEAU 3R see FMU080
BRILLIANT PONCEAU G see FMU070
BRILLIANT RED see CHP500
BRILLIANT RED 5SKH see PMF540
BRILLIANT RED TONER RA see CMS160
BRILLIANTSAEURE GRUEN BS see ADF000
BRILLIANT SAFRANINE BR see GJI400
BRILLIANT SAFRANINE G see GJI400
BRILLIANT SAFRANINE GR see GJI400
BRILLIANT SCARLET see CHP500
BRILLIANT SCARLET see FMU080
BRILLIANT SCARLET G see CMS160
BRILLIANTSCHWARZ BN (GERMAN) see BMA000
BRILLIANT SULFAFLAVINE see CMM750
BRILLIANT TANGERINE 13030 see DVB800
BRILLIANT TONER Z see CHP500
BRILLIANT TONING RED AMINE see AJJ250
BRILLIANT VIOLET 5B see AOR500
BRILLIANT VIOLET K see DFN450
BRILLIANT YELLOW SLURRY see DEU000
BRIMONIDINE see BMM575
BRIMSTONE see SOD500
BRINDERDIN see RDK000
BRIPADON see FLG000
BRIPHOS L 2D see PKE850
BRIQUEST 543-33S see DJG700
BRISERINE see RDK000
BRISPEN see DGE200
BRISTACICLIN α see TBX000
BRISTACIN see PPY250
BRISTACYCLINE see TBX000
BRISTACYCLINE see TBX250
BRISTAMIN HYDROCHLORIDE see DTO800
BRISTAMYCIN see EDJ500
BRISTOL A-649 see OIU499
BRISTOL LABORATORIES BC 2605 see CQF079
BRISTOPHEN see MNV250
BRISTURIC see BEQ625
BRISTURON see BEQ625
BRITACIL see AIV500
BRITAI see CFH825
BRITAI see CMV500
BRITESIL see SCN700
BRITISH ALUMINUM AF 260 see AHC000
BRITISH ANTILEWISITE see BAD750
BRITISH EAST INDIAN LEMONGRASS OIL see LEG000
BRITOMYA M see CAT775
BRITON see TIQ250
BRITONE RED Y see NAP100
BRITTEN see TIQ250
BRITTOX see DDP000
BRL see AIV500
BRL 152 see PDD350
BRL 556 see IBP200
BRL 1341 see AIV500
BRL 1383 see SGS500
BRL 1400 see DSQ800
BRL 1400 see MNV250
BRL-1621 see SLJ000
BRL 1621 see SPD600
BRL-1702 see DGE200
BRL-2064 see CBO250
BRL 3475 see CBO000
BRL 25000 see ARS125
BRL 51308 see CBO800
BRL 14151K see PLB775
BRL 147777 see MFA300
BRL 39123A see POK100
BRL-1621 SODIUM SALT see SLJ050
BRL 2333 TRIHYDRATE see AOA100
BRN 4729620 see SKS150
BROBAMATE see MQU750

BROCADISIPAL see MJH900
BROCADISIPAL see OJW000
BROCADOPA see DNA200
BROCASIPAL see MJH900
BROCASIPAL see OJW000
BROCIDE see EIY600
BROCKMANN, ALUMINUM OXIDE see AHE250
BROCRESIN see BMM600
BROCRESINE see BMM600
BROCSIL see PDD350
BRODAN see CMA100
BRODIAR see DDS600
BRODIFACOUM see TAC800
BROFAREMINE HYDROCHLORIDE see BMM625
BROFENE see BNL250
BROGDEX 555 see SGM500
BROM (GERMAN) see BMP000
BROMACETOCARBAMIDE see BNK000
BROMACETYLENE see BMS500
BROMACIL see BMM650
BROMADAL see BNK000
BROMADEL see BNK000
BROMADIALONE see BMN000
BROMADIOLONE see BMN000
BROMADRYL see BMN250
BROMAL HYDRATE see THU500
BROMALLYLENE see AFY000
5-(2'-BROMALLYL)-5-ISOPROPYLBARBITURIC ACID see QCS000
BROMAMID see BMN350
BROMAMIDE see BMN350
BROMAMIDE see BMT250
BROMAMIDE (pharmaceutical) see BMN350
p-BROMANILID KYSELINY 5-BROMSALICYLOVE see BOD600
4-BROMANILINU (CZECH) see BMT325
p-BROMANISOLE see AOY450
BROMANMINAN SODNY (CZECH) see DKR000
BROMANYLPROMIDE see BMN350
BROMARAL see BNP750
BROMAT see HCQ500
BROMATES see BMN500
BROMATE de SODIUM (FRENCH) see SFG000
BROMAZEPAM see BMN750
BROMAZIL see BMM650
10-BROM-1,2-BENZANTHRACEN (GERMAN) see BMT750
3-BROMBENZANTHRONE see BMU000
2-BROMBENZOTRIFLUORID (CZECH) see BOJ750
3-BROMBENZOTRIFLUORID (CZECH) see BOJ500
BROMBENZYL CYANIDE see BMW250
N-p-BROMBENZYL-N-α-PYRIDYL-N',N'-DIMETHYL-AETHYLENDIAMIN-HYDROCHLORIDE (GERMAN) see HGA500
N-p-BROMBENZYL-N-α-PYRIDYL-N'-METHYL-N'-AETHYL-AETHYLENDIAMIN-MALEINAT (GERMAN) see BMW000
BROMCARBAMIDE see BNP750
BROMCHLOPHOS see NAG400
BROMCHLORENONE see BMZ000
BROMCHOLITIN see TDI475
BROMDEFENURON see MHS375
BROMDIAN see MKA270
O-(4-BROM-2,5-DICHLOR-PHENYL)-O,O-DIMETHYL-MONOTHIOPHOSPHAT (GERMAN) see BNL250
d-2-BROM-DIETHYLAMIDE of LYSERGIC ACID see BNM250
BROME (FRENCH) see BMP000
BROMEK DWUMETYLOLAURYLOBENZYLOAMONIOWY see BEO000
BROMELAIN see BMO000
BROMELAINS see BMO000
BROMELIA see EEY500
BROMELIN see BMO000

BROMEOSIN see BMO250
BROMETHALIN see BMO300
BROMETHALINE see BMO300
BROMETHOL see ARW250
BROMETHOL see THV000
BROMEX see CES750
BROMEX see DFK600
BROMEX see NAG400
BROMFENOFOS see BNV500
BROMFENPHOS see BNV500
BROMFENVINFOS see BMO310
BROMFENVINPHOS see BMO310
BROMFENVINPHOS-METHYL see MHR150
BROMFENWINFOS see BMO310
1-p-BROMFENYL-3,3-DIMETHYLTRIAZEN (CZECH) see BNW250
2-BROMFLUORBENZEN (CZECH) see FGX000
3-BROMFLUORBENZEN (CZECH) see FGY000
BROMHEXINE CHLORIDE see BMO325
BROMHEXINE HYDROCHLORIDE see BMO325
BROMIC ACID, AMMONIUM SALT see ANC000
BROMIC ACID, POTASSIUM SALT see PKY300
BROMIC ACID, SODIUM SALT see SFG000
BROMIC ETHER see EGV400
BROMIDES see BMO750
BROMIDE SALT OF POTASSIUM see PKY500
BROMIDE SALT of SODIUM see SFG500
BROMID UHLICITY see CBX750
BROMINAL see DDP000
BROMINAL M & PLUS see CIR250
BROMINATED VEGETABLE (SOYBEAN) OIL see BMO825
BROMINDIGO see ICU135
BROMINDIGO 2BD see ICU135
BROMINE see BMP000
BROMINE, solution (DOT) see BMP000
BROMINE AZIDE see BMP250
BROMINE CYANIDE see COO500
BROMINE DIOXIDE see BMP500
BROMINE FLUORIDE see BMP750
BROMINE NITRIDE see BMP250
BROMINE PENTAFLUORIDE see BMQ000
BROMINE PERCHLORATE see BMQ250
BROMINE TRIFLUORIDE see BMQ325
BROMINE(1) TRIFLUOROMETHANESULFONATE see BMQ500
BROMINE TRIOXIDE see BMQ750
BROMINEX see DDP000
BROMINIL see DDP000
5-BROMISATIN (CZECH) see BNL750
5-BROM-3-ISOPROPYL-6-METHYL-URACIL (GERMAN) see BNM000
BROMISOVAL see BNP750
BROMISOVALERYLUREA see BNP750
α-BROMISOVALERYLUREA see BNP750
BROMISOVALUM see BNP750
BROMIZOVAL see BNP750
BROMKAL 80 see BMQ800
BROMKAL 80 see OAH000
BROMKAL 80-9D see NMV735
BROMKAL 79-8DE see OAF200
BROMKAL 83-10DE see PAU500
BROMKAL G1 see PAT830
BROMKAL 82-ODE see PAU500
BROMKAL P 67-6HP see TNC500
BROM LSD see BNM250
BROMLYSERGAMIDE see BNM250
2-BROM-d-LYSERGIC ACID DIETHYLAMINE see BNM250
BROM-METHAN (GERMAN) see MHR200
BROMNATRIUM (GERMAN) see SFG500
5-BROM-5-NITRO-1,3-DIOXAN (GERMAN) see BNT000
BROMO (ITALIAN) see BMP000
BROMOACETALDEHYDE see BMR000

2-BROMOACETALDEHYDE see BMR000
α-BROMOACETALDEHYDE see BMR000
BROMOACETAMIDE see BMR025
2-BROMOACETAMIDE see BMR025
β-BROMOACETAMIDE see BMR025
3-BROMOACETANILIDE see BMR050
4-BROMOACETANILIDE see BMR100
m-BROMOACETANILIDE see BMR050
p-BROMOACETANILIDE see BMR100
3'-BROMOACETANILIDE see BMR050
4'-BROMOACETANILIDE see BMR100
p-BROMO-N-ACETANILIDE see BMR100
BROMOACETIC ACID see BMR750
α-BROMOACETIC ACID see BMR750
BROMOACETIC ACID, solid or solution (DOT) see BMR750
BROMOACETIC ACID ETHYLENE ESTER see BHD250
BROMOACETIC ACID, ETHYL ESTER see EGV000
BROMOACETIC ACID METHYL ESTER see MHR250
BROMOACETONE see BNZ000
BROMOACETONE (DOT) see BNZ000
BROMOACETONE, liquid (DOT) see BNZ000
BROMOACETONE OXIME see BMS000
BROMOACETONITRILE see BMS100
1-BROMOACETOXY-2-PROPANOL see BMS250
1-BROMOACETYL-α-α-DIPHENYL-4-PIPERIDINEMETHANOL see BMS300
BROMOACETYLENE see BMS500
BROMOACETYLENYLETHYLMETHYLCARBINOL see BNK350
BROMO ACID see BNH500
BROMO ACID see BNK700
4'-(3-BROMO-9-ACRIDINYLAMINO)METHANESULFONANILIDE see BMS750
2-BROMOACROLEIN see BMT000
3-BROMOADAMANTYL DIAZOMETHYL KETONE see EEE025
1-BROMO-3-ADAMANTYL ETHOXYMETHYL KETONE see BMT100
1-BROMO-3-ADAMANTYL HYDROXYMETHYL KETONE see BMT130
1-(2-BROMO-1-ADAMANTYL)-N-METHYL-2-PROPYLAMINE HYDROCHLORIDE see BNO000
1-(3-BROMO-1-ADAMANTYL)-N-METHYL-2-PROPYLAMINE HYDROCHLORIDE see BNO250
5-(2-BROMOALLYL)-5-sec-BUTYLBARBITURIC ACID see BOR000
γ-BROMOALLYLENE see PMN500
3-BROMOALLYL ISOCYANATE see BMT150
5-(2'-BROMOALLYL)-5-(1'-METHYL-N-PROPYL)BARBITURIC ACID see BOR000
BROMOAMINE see BMT250
3'-BROMO-trans-ANETHOLE see BMT300
BROMOANILIDE see BMR100
4-BROMOANILINE see BMT325
p-BROMOANILINE see BMT325
4-BROMO-2-(((2-ANILINO-5-NITROPHENYL)IMINO)METHYL)PHENOL see NFQ300
2-BROMOANISOLE see BMT400
4-BROMOANISOLE see AOY450
o-BROMOANISOLE see BMT400
p-BROMOANISOLE see AOY450
BROMOANTIFEBRIN see BMR100
BROMOAPROBARBITAL see QCS000
1-BROMOAZIRIDINE see BMT500
BROMO B see BNK700
4-BROMOBENZALDEHYDE see BMT700
p-BROMOBENZALDEHYDE see BMT700
10-BROMO-1,2-BENZANTHRACENE see BMT750
3-BROMOBENZ(d,e)ANTHRONE see BMU000
3-BROMO-7H-BENZ(DE)ANTHRACEN-7-ONE see BMU000

4-BROMO-BENZENAMINE (9CI) see BMT325
BROMOBENZENE (DOT) see PEO500
4-BROMOBENZENEACETONITRILE see BNV750
2-BROMO-1,4-BENZENEDIOL see BNL260
β-(p-BROMOBENZHYDRYLOXY)ETHYLDIMETHYLAMINE HYDROCHLORIDE see BNW500
2-(4-BROMOBENZOHYDRYLOXY)ETHYLDIMETHYLAMINE HYDROCHLORIDE see BNW500
2-BROMOBENZOIC ACID see BMT800
4-BROMOBENZOIC ACID see BMU100
o-BROMOBENZOIC ACID see BMT800
p-BROMOBENZOIC ACID see BMU100
p-BROMOBENZOIC ACID 2-PHENYLHYDRAZIDE see BMU150
4-BROMOBENZOPHENONE see BMU170
p-BROMOBENZOPHENONE see BMU170
6-BROMOBENZO(a)PYRENE see BMU500
2-BROMOBENZOTRIFLUORIDE see BOJ750
3-BROMOBENZOTRIFLUORIDE see BOJ500
m-BROMOBENZOTRIFLUORIDE see BOJ500
o-BROMOBENZOTRIFLUORIDE see BOJ750
5-BROMO-2-BENZOXAZOLINONE see BMU750
6-BROMO-2-BENZOXAZOLINONE see BMV000
p-BROMOBENZOYL AZIDE see BMV250
p-BROMOBENZOYLTHIOHYDROXIMIC ACID-5-DIETHYLAMINOETHYL ESTER HYDROCHLORIDE see DHZ000
4-BROMOBENZYLCYANIDE see BNV750
p-BROMOBENZYL CYANIDE see BNV750
α-BROMOBENZYL CYANIDE see BMW250
2-((p-BROMOBENZYL)(2-(DIMETHYLAMINO)ETHYL)AMINO)PYRIDINE HYDROCHLORIDE see HGA500
N-p-BROMOBENZYL-N',N'-DIMETHYL-N-2-PYRIDYLETHYLENE-DIAMINE HYDROCHLORIDE see HGA500
(o-BROMOBENZYL)ETHYLDIMETHYLAMMONIUM-p-TOLUENESULFONATE see BMV750
N-p-BROMOBENZYL-N'-ETHYL-N'-METHYL-N-2-PYRIDYLETHYLENEDIAMINE MALEATE see BMW000
BROMOBENZYLNITRILE see BMW250
α-BROMOBENZYLNITRILE see BMW250
3-BROMOBENZYLTRIFLUORIDE see BOJ500
o-BROMOBENZYLTRIFLUORIDE see BOJ750
3-BROMOBIPHENYL see BMW290
4-BROMOBIPHENYL see BMW300
3-(3-(4'-BROMO(1,1'-BIPHENYL)-4-YL)3-HYDROXY-1-PHENYLPROPYL)-4-HYDROXY-2H-1-BENZOPYRAN-2-ONE see BMN000
3-(3-(4'-BROMOBIPHENYL-4-YL)-1,2,3,4-TETRAHYDRONAPHTH-1-YL)-4-HYDROXYCOUMARIN see TAC800
3-(3-(4'-BROMO-1,1'-BIPHENYL-4-YL)-1,2,3,4-TETRAHYDRO-1-NAPHTHYL)-4-HYDROXYCOUMARIN see TAC800
α-BROMO-β,β-BIS(p-ETHOXYPHENYL)STYRENE see BMX000
4-BROMO-7-BROMOMETHYLBENZ(a)ANTHRACENE see BMX250
2-BROMO-2-(BROMOMETHYL)GLUTARONITRILE see DDM500
2-BROMO-S-(2-BROMO-5-NITROETHENYL)FURAN see BMX300

4-BROMO-α-(4-BROMOPHENYL)-α-HYDROXYBENZENEACETIC ACID-1-METHYLETHYL ESTER see IOS000

2-BROMO-6-(N-(p-BROMOPHENYL)THIOCARBAMOYL)-4-CHLORO-BENZOIC ACID see BOL325

1-BROMOBUTANE see BMX500

2-BROMOBUTANE see BMX750

4-BROMO-1-BUTENE see BMX825

3-BROMO-3-BUTEN-2-ONE see MHS400

5-BROMO-3-sec-BUTYL-6-METHYLURACIL see BMM650

2-BROMOBUTYRIC ACID see BMY250

α-BROMOBUTYRIC ACID see BMY250

4-BROMOBUTYRONITRILE see BMY500

BROMOCARBAMIDE see BNP750

BROMOCET see HCP800

BROMOCHLOROACETONITRILE see BMY800

BROMOCHLOROACETYLENE see BMY825

6-BROMO-5-CHLORO-2-BENZOXAZOLINONE see BMZ000

6-BROMO-5-CHLOROBENZOXAZOLONE see BMZ000

1-BROMO-1-CHLORO-2,2-DIFLUOROETHENE see BNA000

1-BROMO-1-CHLORO-2,2-DIFLUOROETHYLENE see BNA000

2-BROMO-2-CHLORO-1,1-DIFLUOROETHYLENE see BNA000

BROMOCHLORODIFLUOROMETHANE see BNA250

1-BROMO-3-CHLORO-5,5-DIMETHYLHYDANTOIN see BNA300

3-BROMO-1-CHLORO-5,5-DIMETHYLHYDANTOIN see BNA325

N-BROMO-N'-CHLORO-5,5-DIMETHYLHYDANTOIN see BNA300

1-BROMO-3-CHLORO-5,5-DIMETHYL-2,4-IMIDAZOLIDINEDIONE see BNA300

3-BROMO-1-CHLORO-5,5-DIMETHYL-2,4-IMIDAZOLIDINEDIONE see BNA325

7-BROMO-6-CHLOROFEBRIFUGINE HYDROBROMIDE see HAF600

7-BROMO-6-CHLORO-3-[3-(3-HYDROXY-2-PIPERDINYL)-2-OXOPROPYL]-4(3H)-QUINAZOLINONE HYDROBROMIDE see HAF600

3-BROMO-N-(2-CHLOROMERCURICYCLOHEXYL)PROPIONAMIDE see CET000

BROMOCHLOROMETHANE see CES650

BROMOCHLOROMETHYL CYANIDE see BMY800

2-(3-BROMO-4-CHLOROPHENYL)-4-CHLORO-5-((6-CHLORO-3-PYRIDINYL)METHOXY)-3(2H)-PYRIDAZINONE see BNA350

1-BROMO-1-(p-CHLOROPHENYL)-2,2-DIPHENYLETHYLENE see BNA500

4-BROMO-2-(4-CHLOROPHENYL)-1-ETHOXYMETHYL-5-TRIFLUOROMETHYLPYRROLE-3-CARBONITRILE see BNA600

O-(4-BROMO-2-CHLOROPHENYL)-O-ETHYL-S-PROPYL PHOSPHOROTHIOATE see BNA750

3-(4-BROMO-3-CHLOROPHENYL)-1-METHOXY-1-METHYLUREA see CES750

N'-(4-BROMO-3-CHLOROPHENYL)-N-METHOXY-N-METHYLUREA see CES750

N-(4-BROMO-3-CHLOROPHENYL)-N'-METHOXY-N'-METHYLUREA see CES750

2-BROMO-4-(2-CHLOROPHENYL)-9-METHYL-6H-THIENO(3,2-f)(1,2,4)TRIAZOLO(4,3-a)(1,4)DIAZEPINE see LEJ600

1-BROMO-3-CHLOROPROPANE see BNA825

3-BROMO-1-CHLOROPROPENE see BNA880

7-BROMO-5-CHLOROQUINOLIN-8-YL ACRYLATE see BNA900

BROMOCHLOROTRIFLUOROETHANE see HAG500

2-BROMO-2-CHLORO-1,1,1-TRIFLUOROETHANE see HAG500

BROMOCONAZOLE see BNA920

BROMOCRESOL GREEN see BNA940

BROMOCRESOL GREEN see TBJ505

BROMOCRIPTIN see BNB250

BROMOCRIPTINE see BNB250

BROMOCRIPTINE MESILATE see BNB325

BROMOCYAN see COO500

BROMOCYANOGEN see COO500

BROMOCYCLEN see BNQ600

BROMOCYCLENE see BNQ600

1-BROMO-12-CYCLOTRIDECADIEN-4,8,10-TRIYNE see BNB750

BROMODAN see BNQ600

1-BROMODECANE see BNB800

BROMODEOXYGLYCEROL see MRF275

BROMODEOXYURIDINE see BNC750

5-BROMODEOXYURIDINE see BNC750

5-BROMO-2-DEOXYURIDINE see BNC750

5-BROMO-2'-DEOXYURIDINE see BNC750

5-BROMODESOXYURIDINE see BNC750

2-BROMO-1,5-DIAMINO-4,8-DIHYDROXYANTHRAQUINONE see BNC800

2-BROMO-1,8-DIAMINO-4,5-DIHYDROXYANTHRAQUINONE see BND250

BROMODIBORANE see BND325

BROMODICHLOROMETHANE see BND500

2-BROMO-1-(3,4-DICHLORO-5-NITRO-2-FURANYL)ETHANONE see BND600

4-BROMO-2,5-DICHLOROPHENOL see LEN050

4-BROMO-2,5-DICHLOROPHENOL-o-ESTER with O,O-DIETHYL PHOSPHOROTHIOATE see EGV500

O-(4-BROMO-2,5-DICHLOROPHENYL)-O,O-DIETHYL PHOSPHOROTHIOATE see EGV500

O-(4-BROMO-2,5-DICHLOROPHENYL)-O,O-DIETHYLPHOSPHOROTHIONATE see EGV500

4-BROMO-2,5-DICHLOROPHENYL DIMETHYL PHOSPHOROTHIONATE see BNL250

o-(4-BROMO-2,5-DICHLOROPHENYL)-o-ETHYL PHENYLPHOSPHONOTHIOATE see BND750

4-BROMO-2-(3,4-DICHLOROPHENYL)-5-((6-IODO-3-PYRIDINYL)METHOXY)-3(2H)-PYRIDAZINONE see BND775

O-(4-BROMO-2,5-DICHLOROPHENYL)-O-METHYL PHENYLPHOSPHONOTHIOATE see LEN000

1-((4-BROMO-2-(2,4-DICHLOROPHENYL)TETRAHYDRO-2-FURANYL)METHYL)-1H-1,2,4-TRIAZOLE see BNA920

1-((2RS,4RS,2RS,4SR)-4-BROMO-2-(2,4-DICHLOROPHENYL)TETRAHYDROFURFURYL)-1 H-1,2,4-TRIAZOLE see BNA920

O-(4-BROMO-2,5-DICLORO-FENIL)-O,O-DIMETIL-MONOTIOFOSFATO (ITALIAN) see BNL250

2-BROMO-9,10-DIDEHYDRO-N,N-DIETHYL-6-METHYLERGOLINE-8-β-CARBOXAMIDE see BNM250

BROMODIETHYLACETYLCARBAMIDE see BNK000

BROMODIETHYLACETYLUREA see BNK000

5-BROMO-2-(2-(DIETHYLAMINO)ETHOXY)BENZANILIDE see DHP400

BROMODIETHYLGOLD see DJJ850

BROMODIFLUOROMETHANE see BND800

7-BROMO-1,3-DIHYDRO-5-(2-PYRIDYL)-2H-1,4-BENZDIAZEPIN-2-ONE see BMN750

dl-4-BROMO-2,5-DIMETHOXYAMPHETAMINE HYDROBROMIDE see BNE250

2-BROMO-3,5-DIMETHOXYANILINE see BNE325

dl-4-BROMO-2,5-DIMETHOXY-α-METHYLPHENETHYLAMINE HYDROBROMIDE see BNE250

21-BROMO-3,17-DIMETHOXY-19-NOR-17-α-PREGNA-1,3,5(10)-TRIEN-20-YNE see BAT800

2-BROMO-N,N-DIMETHYL-1-ADAMANATANEMETHANAMINE HYDROCHLORIDEHEMIHYDRATE see BNE500

2-BROMO-N,N-DIMETHYL-1-ADAMANTANEPROPANAMINE HYDROCHLORIDE see BNF000

3'-BROMO-4-DIMETHYLAMINOAZOBENZENE see BNE600

4'-BROMO-4-DIMETHYLAMINOAZOBENZENE see BNV760

2-((p-BROMO-α-(2-DIMETHYLAMINO)ETHYL)BENZYL)PYRIDINE BIMALEATE see BNE750

(+)-2-(p-BROMO-α-(2-(DIMETHYLAMINO)ETHYL)BENZYL)PYRIDINE MALEATE see DXG100

(±)-2-(p-BROMO-α-(2-(DIMETHYLAMINO)ETHYL)BENZYL)PYRIDINE MALEATE see DNW759

2-(p-BROMO-α-(2-(DIMETHYLAMINO)ETHYL)BENZYL)PYRIDINE MALEATE (1:1) see BNE750

2-BROMO-1-(N,N-DIMETHYLAMINOMETHYL)ADAMANTANE HYDROCHLORIDEHEMIHYDRATE see BNE500

2-BROMO-1-(3-DIMETHYLAMINOPROPYL)ADAMANTANE HYDROCHLORIDE see BNF000

4-BROMODIMETHYLANILINE see BNF250

4-BROMO-N,N-DIMETHYL ANILINE see BNF250

3-BROMO-7,12-DIMETHYLBENZ(a)ANTHRACENE see BNF300

4-BROMO-7,12-DIMETHYLBENZ(a)ANTHRACENE see BNF310

5-BROMO-9,10-DIMETHYL-1,2-BENZANTHRACENE see BNF315

1-BROMO-3,3-DIMETHYL-2-BUTANONE see PJB100

p-BROMO-α,α-DIMETHYLPHENETHYLAMINE HYDROCHLORIDE see BNF750

4-BROMO-2,6-DIMETHYLPHENOL see BOL303

α-BROMO-β-DIMETHYLPROPANOYLUREA see BNP750

3-BROMO-5,7-DIMETHYL PYRAZOLYL-2-PYRIMIDINEPHOSPHOROTHIOIC ACID-O,O-DIETHYL ESTER see BAS000

N¹-(5-BROMO-4,6-DIMETHYL-2-PYRIMIDINYL)BENZENESULFONAMIDE see SNH875

2-BROMO-4,6-DINITROANILINE see DUS200

6-BROMO-2,4-DINITROANILINE see DUS200

1-BROMO-2,4-DINITROBENZENE see DVB820

6-BROMO-2,4-DINITROBENZENEDIAZONIUM HYDROGEN SULFATE see BNG125

9-(BROMOMETHYL)-10-CHLOROANTHRACENE see BNP850

7-BROMOMETHYL-4-CHLOROBENZ(a)ANTHRACENE see BNQ000

3-BROMOMETHYL-4-CHLOROMALEIMIDE see BNQ050

BROMOMETHYL p-CHLOROPHENYL KETONE see CJJ100

BROMOMETHYL CYANIDE see BMS100

3'-BROMO-4'-METHYL-4-DIMETHYLAMINOAZOBENZENE see BNQ100

4'-BROMO-3'-METHYL-4-DIMETHYLAMINOAZOBENZENE see BNQ110

7-BROMOMETHYL-6-FLUOROBENZ(a)ANTHRACENE see BNQ250

2-BROMO METHYL FURAN see BNQ500

BROMOMETHYLHEXACHLOROBICYCLOHEPTENE see BNQ600

5-(BROMOMETHYL)-1,2,3,4,7,7-HEXACHLOROBICYCLO(2.2.1)HEPT-2-ENE see BNQ600

5-(BROMOMETHYL)-1,2,3,4,7,7-HEXACHLORO-2-NORBORNENE see BNQ600

7-BROMOMETHYL-1-METHYLBENZ(a)ANTHRACENE see BNQ750

12-BROMOMETHYL-7-METHYLBENZ(a)ANTHRACENE see BNR250

7-BROMOMETHYL-12-METHYLBENZ(a)ANTHRACENE see BNR000

2-BROMOMETHYL-5-METHYLFURAN see BNR325

BROMOMETHYL METHYL KETONE see BNZ000

5-BROMO-6-METHYL-3-(1-METHYLPROPYL)-2,4(1H,3H)-PYRIMIDINEDIONE see BMM650

5-BROMO-6-METHYL-3-(1-METHYLPROPYL)URACIL see BMM650

p-(BROMOMETHYL)NITROBENZENE see BEC000

1-(BROMOMETHYL)-4-NITROBENZENE see NFN000

6-α-BROMO-17-β-METHYL-4-OXA-5-α-ANDROSTAN-3-ONE see BMA650

1-BROMO-3-METHYLPENTIN-3-OL see BNK350

1-BROMO-3-METHYL-1-PENTYN-3-OL see BNK350

1-(BROMOMETHYL)-3-PHENOXYBENZENE see PDR610

2-((p-BROMO-α-METHYL-α-PHENYLBENZYL)OXY)-N,N-DIMETHYLETHYLAMINE HYDROCHLORIDE see BMN250

m-(BROMOMETHYL)PHENYL PHENYL ETHER see PDR610

1-BROMO-2-METHYLPROPANE see BNR750

2-BROMO-2-METHYLPROPANE see BNS000

2-BROMO-2-METHYLPROPANE (DOT) see BQM250

2-(BROMOMETHYL)-TETRAHYDROFURAN see TCS550

1-BROMONAPHTHALENE see BNS200

α-BROMONAPHTHALENE see BNS200

6-BROMO-1,2-NAPHTHOQUINONE see BNS750

(6-(3-(1-BROMO-2-NAPHTHYL)-d-ALANINE))-LHRH ACETATE see LIU309

BROMONE see BMN000

8-β-((5-BROMONICOTINOYLOXY)METHYL)-1,6-DIMETHYL-10-α-METHOXYERGOLINE see NDM000

2-BROMO-4-NITROANILINE see BNS800

2-BROMONITROBENZENE see NFQ080

3-BROMONITROBENZENE see NFQ090

4-BROMONITROBENZENE see NFQ100

m-BROMONITROBENZENE see NFQ090

o-BROMONITROBENZENE see NFQ080

p-BROMONITROBENZENE see NFQ100

5-BROMO-5-NITRO-m-DIOXANE see BNT000

5-BROMO-5-NITRO-1,3-DIOXANE see BNT000

2-BROMO-2-NITROPANE-1,3-DIOL see BNT250

4-BROMO-2-NITROPHENOL see NFQ200

2-BROMO-2-NITROPROPAN-1,3-DIOL see BNT250

2-BROMO-2-NITRO-1,3-PROPANEDIOL see BNT250

3-BROMO-8-NITROQUINOLINE see BNT300

3-BROMO-4-NITROQUINOLINE-1-OXIDE see BNT500

β-BROMO-β-NITROSOSTYRENE see BNT600

α-BROMO-p-NITROTOLUENE see NFN000

β-BROMO-β-NITROTRIMETHYLENEGLYCOL see BNT250

1-BROMOOCTANE see BNU000

1-BROMO-2-OXIMINOPROPANE see BMS000

3-BROMO-2-OXOPROPANOIC ACID see BOD550

α-BROMOPARANITROTOLUENE see NFN000

4-BROMO-PDMT see BNW250

1-BROMOPENTABORANE (9) see BNU125

3-BROMO-1,1,1,2,2-PENTAFLUOROPROPANE see PBF600

4-BROMO-1,2,2,6,6-PENTAMETHYLPIPERIDINE see BNU250

1-BROMOPENTANE see AOF800

2-BROMOPENTANE see BNU500

4-(2-(5-BROMO-2-PENTYLOXYBENZYLOXY)ETHYL)MORPHOLINE see BNU660

2-(5-BROMO-2-PENTYLOXYBENZYLOXY)TRIETHYLAMINE see BNU700

BROMOPERIDOL see BNU725

4-BROMOPHENACYL BROMIDE see DDJ600

p-BROMOPHENACYL BROMIDE see DDJ600

BROMOPHENIRAMINE MALEATE see BNE750

dl-BROMOPHENIRAMINE MALEATE see DNW759

BROMOPHENOL see BNU800

4-BROMOPHENOL see BNV010

o-BROMOPHENOL see BNV000

p-BROMOPHENOL see BNV010

BROMO PHENOLS see BNV250

BROMOPHENOPHOS see BNV500

BROMOPHENOXIM see BNK400

4-BROMOPHENYLACETONITRILE see BNV750

p-BROMOPHENYLACETONITRILE see BNV750

2-(4-BROMOPHENYL)ACETONITRILE see BNV750

α-BROMOPHENYLACETONITRILE see BMW250

p-BROMOPHENYLAMINE see BMT325

4-((3-((4-BROMOPHENYL)AMINO)-4,5-DIHYDRO-2H-BENZ(g)NDAZOL-2-YL)ACETYL)MORPHOLINE see BNV752

3-((4-BROMOPHENYL)AMINO)-N,N-DIMETHYL-PROPANAMIDE (9CI) see BMN350

3-((4-BROMOPHENYL)AMINO)-N-(2-ETHOXYETHYL)-4,5-DIHYDRO-2H-BENZ(g)INDAZOLE-2-ACETAMIDE see BNV754

p-(m-BROMOPHENYLAZO)-N,N-DIMETHYLANILINE see BNE600

p-(p-BROMOPHENYLAZO)-N,N-DIMETHYLANILINE see BNV760

2-(2-BROMOPHENYL)-1H-BENZIMIDAZOLE see BNV765

p-BROMOPHENYL BROMIDE see BNV775

2-(4-BROMOPHENYL)-4-CHLORO-5-((4-CHLOROPHENYL)METHOXY)-3(2H)-PYRIDAZINONE see BNV800

4-BROMOPHENYL CHLOROMETHYL SULFONE see BNV850

(S)-γ-(4-BROMOPHENYL)-N,N-DIMETHYL-2-PYRIDINEPROPANAMINE (Z)-2-BUTENEDIOATE (1:1) see DXG100

(±)-(Z)-γ-(4-BROMOPHENYL)-N,N-DIMETHYL-2-PYRIDINEPROPANAMINE 2-BUTENEDIOATE (1:1) see DNW759

3-(4-BROMOPHENYL)-N,N-DIMETHYL-3-(3-PYRIDINYL)-2-PROPEN-1-AMINE see ZBA500

(Z)-3-(4-BROMOPHENYL)-N,N-DIMETHYL-3-(3-PYRIDINYL)-2-PROPEN-1-AMINE DIHYDROCHLORIDE see ZBA525

3-(p-BROMOPHENYL)-N,N-DIMETHYL-3-(3-PYRIDYL)ALLYLAMINE see ZBA500

1-(4-BROMOPHENYL)-3,3-DIMETHYLTRIAZENE see BNW250

p-BROMOPHENYL ESTER ISOTHIOCYANIC ACID see BNW825

α-BROMO-β-PHENYLETHYLENE see BOF000

3-(3-(4-(2-(4-BROMOPHENYL)ETHYL)PHENYL)-1,2,3,4-TETRAHYDRO-1-NAPHTHALENYL)-4-HYDROXY2H-1-BENZOPYRAN-2-ONE see BNW300

BROMOPHENYL HYDRAMINE HYDROCHLORIDE see BNW500

4-BROMOPHENYL HYDRAZINE HYDROCHLORIDE see BNW500

3-(α-(p-(p-BROMOPHENYL)-β-HYDROXYPHENETHYL)BENZYL)-4-HYDROXYCOUMARIN see BMN000

4-(4-BROMOPHENYL)-4-HYDROXYPIPERIDINO)-4'-FLUOROBUTYROPHENONE see BNU725

4-(4-(p-BROMOPHENYL)-4-HYDROXYPIPERIDINO)-4'-FLUOROBUTYROPHENONE see BNU725

4-(4-(p-BROMOPHENYL)-4-HYDROXYPIPERIDINOL)-4'-FLUOROBUTYROPHENONE see BNU725

4-(4-(4-BROMOPHENYL)-4-HYDROXY-1-PIPERIDINYL)-1-(4-FLUOROPHENYL)-1-BUTANONE see BNU725

2-(p-BROMOPHENYL)IMIDAZO(2,1-A)ISOQUINOLINE see BNW600

2-(p-BROMOPHENYL)IMIDAZO(2,1-a)ISOQUINOLINE see BNW625

4-BROMO-2-PHENYL-1,3-INDANDIONE see BNW750

5-BROMO-2-PHENYLINDAN-1,3-DIONE see UVJ400

5-BROMO-2-PHENYL-1,3-INDANDIONE see UVJ400

5-BROMO-2-PHENYL-1H-INDENE-1,3(2H)-DIONE see UVJ400

p-BROMOPHENYL ISOTHIOCYANATE see BNW825

p-BROMO PHENYL LITHIUM see BNX000

BROMOPHENYLMETHANE see BEC000

5-((4-BROMOPHENYL)METHOXY)-4-CHLORO-2-(4-CHLORO-2-FLUOROPHENYL)-3(2H)-PYRIDAZINONE see BNX035

5-((4-BROMOPHENYL)METHOXY)-4-CHLORO-2-(4-CHLOROPHENYL)-3(2H)-PYRIDAZINONE see BNX040

3-(p-BROMOPHENYL)-1-METHOXY-1-METHYLUREA see PAM785

N'-(4-BROMOPHENYL)-N-METHOXY-N-METHYLUREA see PAM785

3-(3-(4-((4-BROMOPHENYL)METHOXY)PHENYL)-1,2,3,4-TETRAHYDRO-1-NAPHTHALENYL)-4-HYDROXY2H-1-BENZOPYRAN-2-ONE see BNX045

(4-BROMOPHENYL)METHYL BUTYL 3-PYRIDINYLCARBONIMIDODITHIOATE see BNX050

S-((4-BROMOPHENYL)METHYL) o-BUTYL 3-PYRIDINYLCARBONIMIDOTHIOATE see BNX055

o-BROMOPHENYL METHYL ETHER see BMT400

p-BROMOPHENYL METHYL ETHER see AOY450

3-(p-BROMOPHENYL)-1-METHYL-1-METHOXYUREA see PAM785

(4-BROMOPHENYL)METHYL 1-METHYLETHYL 3-PYRIDINYLCARBONIMIDODITHIOATE see BNX060

3-(p-BROMOPHENYL)-1-METHYL-1-NITROSOUREA see BNX125

1-(p-BROMOPHENYL)-3-METHYLUREA see MHS375

1-(p-BROMOPHENYL)-3-METHYLUREA mixed with SODIUM NITRITE see SIQ675

2-(m-BROMOPHENYL)-N-(4-MORPHOLINOMETHYL)SUCCINIMIDE see BNX250

(p-BROMOPHENYL)OXIRANE see BOF250

(4-BROMOPHENYLOXIRANE) (9CI) see BOF250

3-(4-BROMOPHENYL)OXIRANEMETHANOL (2R-trans)- see EBH900

3-(4-BROMOPHENYL)OXIRANEMETHANOL (2S-trans)- see EBH905

1-(p-BROMOPHENYL)-1-PHENYL-1-(2-DIMETHYLAMINOETHOXY)ETHANE HYDROCHLORIDE see BMN250

2-(1-(4-BROMOPHENYL)-1-PHENYLETHOXY)-N,N-DIMETHYLETHANAMINE HYDROCHLORIDE see BMN250

(2-(1-p-BROMOPHENYL-1-PHENYLETHOXY)ETHYL)DIMETHYLETHYLAMINE HYDROCHLORIDE see BMN250

4-BROMOPHENYL PHENYL KETONE see BMU170

N-(p-BROMOPHENYL)PHTHALIMIDE see BNX300

3-(4-BROMOPHENYL)-N-(4-PROPYLCYCLOHEXYL)-2-PROPENAMIDE see BNX330

(Z)-3-(4'-BROMOPHENYL)-3-(3"-PYRIDYL)DIMETHYLALLYLAMINE see ZBA500

4-(p-BROMOPHENYL)SEMICARBAZONE 1-METHYL-1H-PYRROLE-2-CARBOXALDEHYDE see BNX400

4-(p-BROMOPHENYL)SEMICARBAZONE-1H-PYRROLE-2-CARBOXALDEHYDE see BNX420

4-BROMOPHENYL TRIFLUOROACETATE see BNX500

p-BROMOPHENYL TRIFLUOROACETATE see BNX500

BROMOPHOS see BNL250

BROMOPHOSETHYL see EGV500

BROMOPHTHAL see TBJ700

BROMOPINACOLIN see PJB100

1-BROMOPINACOLIN see PJB100

BROMOPINACOLONE see PJB100

1-BROMOPINACOLONE see PJB100

α-BROMOPINACOLONE see PJB100

9-BROMOPREGN-4-ENE-3,11,20-TRIONE see BML825

9-α-BROMOPREGN-4-ENE-3,11,20-TRIONE see BML825

BROMOPRIDA see VCK100

BROMOPRIDE see VCK100

3-BROMO-1-PROPANAMINE HYDROBROMIDE see AJA000

1-BROMOPROPANE see BNX750

2-BROMOPROPANE see BNY000

1-BROMOPROPANE (DOT) see BNX750

3-BROMO-1,2-PROPANEDIOL see MRF275

3-BROMOPROPANOL see BNY750

3-BROMO-1-PROPANOL see BNY750

BROMO-2-PROPANONE see BNZ000

1-BROMO-2-PROPANONE see BNZ000

2-BROMOPROPENALDEHYDE see BMT000

1-BROMOPROPENE see BOA000

2-BROMOPROPENE see BOA250

3-BROMOPROPENE see AFY000

1-BROMO-1-PROPENE see BOA000

1-BROMO-2-PROPENE see AFY000

(E)-p-(3-BROMOPROPENYL)ANISOLE see BMT300

5-(3-BROMO-1-PROPENYL)-1,3-BENZODIOXOLE see BOA750

5-(2-BROMO-2-PROPENYL)-5-(1-METHYLETHYL)-2,4,6(1H,3H,5H)-PYRIMIDINETRIONE see QCS000

3-BROMOPROPIONIC ACID see BOB250

α-BROMOPROPIONIC ACID see BOB000

β-BROMOPROPIONIC ACID see BOB250

3-BROMOPROPIONITRILE see BOB500

2-BROMOPROPIOPHENONE see BOB550

α-BROMOPROPIOPHENONE see BOB550

3-BROMOPROPYLAMINE HYDROBROMIDE see AJA000

BROMOPROPYLATE see IOS000

3-BROMOPROPYL CHLORIDE see BNA825

4-BROMO-N-(4-PROPYLCYCLOHEXYL)BENZAMIDE see BOB570

2-BROMOPROPYLENE see BOA250

3-BROMOPROPYLENE see AFY000

3-BROMO-1-PROPYNE see PMN500

3-BROMOPROPYNE (DOT) see PMN500

2-BROMOPYRIDINE see BOB600

3-BROMOPYRIDINE see BOC510

5-((6-BROMO-3-PYRIDINYL)METHOXY)-4-CHLORO-2-(4-CHLOROPHENYL)-3(2H)-PYRIDAZINONE see BOC000

5-((6-BROMO-3-PYRIDINYL)METHOXY)-4-CHLORO-2-(3,4-DICHLOROPHENYL)-3(2H)-PYRIDAZINONE see BOC650

7-BROMO-5-(2-PYRIDYL)-3H-1,4-BENZODIAZEPIN-2(1H)-ONE see BMN750

3-(2-(5-BROMO-2-PYRIDYLOXY)ETHYL)THIAZOLIDINE HYDROCHLORIDE see BOD000

2-(6-(5-BROMO-2-PYRIDYLOXY)HEXYL)AMINOETHANE THIOL HYDROCHLORIDE see BOD500

1-BROMO-2,5-PYRROLIDINEDIONE see BOF500

3-BROMOPYRUVATE see BOD550

BROMOPYRUVIC ACID see BOD550

3-BROMOPYRUVIC ACID see BOD550

β-BROMOPYRUVIC ACID see BOD550

BROMOQUIN see QJJ100

2-BROMOQUINOL see BNL260

5-BROMOSALICYL-4-BROMOANILIDE see BOD600

5-BROMOSALICYLIC ACID see BOE500

BROMO SELTZER see ABG750

BROMOSILANE see BOE750

β-BROMOSTYRENE see BOF000

ω-BROMOSTYRENE see BOF000

4-BROMOSTYRENE OXIDE see BOF250

p-BROMOSTYRENE OXIDE see BOF250

4'-BROMOSTYRENE OXIDE see BOF250

p-BROMOSTYRENE-7,8-OXIDE see BOF250

BROMOSTYROL see BOF000

BROMOSTYROLENE see BOF000

N-BROMOSUCCIMIDE see BOF500

N-BROMOSUCCINIMIDE see BOF500

BROMOSULFALEIN see HAQ600

5-BROMOSULFAMETHAZINE see SNH875

BROMOSULFOPHTHALEIN see HAQ600

BROMOSULPHALEIN see HAQ600

BROMOSULPHTHALEIN see HAQ600

BROMOTALEINA see HAQ600

3-BROMO-1,1,2,2-TETRAFLUOROPROPANE see BOF750

3-BROMOTETRAHYDROTHIOPHENE-1,1-DIOXIDE see BOG000

N-BROMOTETRAMETHYL GUANIDINE see BOG250

5-BROMO-2-(2-(3-THIAZOLIDINYL)ETHOXY)PYRIDINE HYDROCHLORIDE see BOD000

p-BROMOTHIOBENZOHYDROXIMIC ACID-S-DIETHYLAMINOETHYL ESTER HYDROCHLORIDE see DHZ000

6-BROMO-2-THIO-2H-1,3-BENZOXAZINE-2,4(3H)-DIONE see BOG255

2-BROMOTOLUENE see BOG260

3-BROMOTOLUENE see BOG300

5-BROMOTOLUENE see BOG300

m-BROMOTOLUENE see BOG300

o-BROMOTOLUENE see BOG260

p-BROMOTOLUENE see BOG255

ω-BROMOTOLUENE see BEC000

α-BROMOTOLUENE (DOT) see BEC000

2-BROMOTOLUIDINE see BOG500

α-BROMO-α-TOLUNITRILE see BMW250

p-((3-BROMO-p-TOLYL)AZO)-N,N-DIMETHYLANILINE see BNQ100

p-((4-BROMO-m-TOLYL)AZO)-N,N-DIMETHYLANILINE see BNQ110

BROMOTRIBUTYLSTANNANE see TIC250

BROMOTRICHLOROMETHANE see BOH750

3-BROMO-1,1,1-TRICHLORO PROPANE see BOI000

3-BROMOTRICYCLOQUINAZOLINE see BOI250

1-BROMOTRIDECANE see BOI500

BROMOTRIETHYLSTANNANE see BOI750

BROMOTRIETHYLSTANNANE compounded with 2-PIPECOLINE (1:1) see TJU850

BROMOTRIFLUOROETHENE see BOJ000

BROMO TRIFLUOROETHYLENE see BOJ000

BROMOTRIFLUOROMETHANE see TJY100

3-BROMOTRIFLUOROMETHYLBENZENE see BOJ500

m-BROMO(TRIFLUOROMETHYL)BENZENE see BOJ500

m-BROMO-α,α,α-TRIFLUOROTOLUENE see BOJ500

o-BROMO-α,α,α-TRIFLUOROTOLUENE see BOJ750

BROMOTRIPENTYLSTANNANE see BOK250

BROMOTRIPHENYLETHYLENE see BOK500

BROMOTRIPHENYLMETHANE see TNP600

BROMOTRIPROPYLSTANNANE see BOK750

5-BROMOURACIL see BOL000

BROMOURACIL DEOXYRIBOSIDE see BNC750

5-BROMOURACIL DEOXYRIBOSIDE see BNC750

5-BROMOURACIL-2-DEOXYRIBOSIDE see BNC750

BROMOVAL see BNP750

α-BROMOVALERIC ACID see BOL250

BROMOVALEROCARBAMIDE see BNP750

BROMOVALERYLUREA see BNP750

(E)-5-(2-BROMOVINYL)-2'-DEOXYURIDINE see BOL300

trans-5-(2-BROMOVINYL)-2'-DEOXYURIDINE see BOL300
BROMOWODOR (POLISH) see HHJ000
BROMOXIL see BNP750
4-BROMO-2,6-XYLENOL see BOL303
BROMOXYNIL see DDP000
BROMOXYNIL OCTANOATE see DDM200
BROMPERIDOL see BNU725
p-BROMPHENACYL-8 see DDJ600
d-BROMPHENIRAMINE MALEATE see DXG100
(±)-BROMPHENIRAMINE MALEATE see DNW759
dl-BROMPHENIRAMINE MALEATE see DNW759
BROMPHENPHOS see BNV500
BROMPHENVINPHOS see BMO310
3-(4-BROMPHENYL)-1-METHOXYHARNSTOFF (GERMAN) see PAM785
omega-BROMPINAKOLIN see PJB100
β-BROMSTYROL see BOF000
BROMSULFALEIN see HAQ600
BROMSULFAN see HAQ600
BROMSULFOPHTHALEIN see HAQ600
BROMSULFTHALEIN see HAQ600
BROMSULPHALEIN see HAQ600
BROMSULPHTHALEIN see HAQ600
BROM-TETRAGNOST see HAQ600
BROMTHALEIN see HAQ600
BROMUCONAZOLE see BNA920
BROMURAL see BNP750
BROMURE de CYANOGEN (FRENCH) see COO500
BROMURE d'ETHYLE see EGV400
BROMURE de METHYLE (FRENCH) see MHR200
BROMURE de VINYLE (FRENCH) see VMP000
BROMURE de XYLYLE (FRENCH) see XRS000
BROMURO di ETILE (ITALIAN) see EIY500
BROMURO di METILE (ITALIAN) see MHR200
BROMURO de OXITROPIO(SPANISH) see ONI000
BROMUVAN see BNP750
BROMVALERYLUREA see BNP750
BROMVALETONE see BNP750
BROMVALETONUM see BNP750
BROMVALUREA see BNP750
BROMWASSERSTOFF (GERMAN) see HHJ000
BROMYL see BNP750
BROMYL FLUORIDE see BOL310
BRONCHIOCAIN see BQH250
BRONCHOCAIN see BQH250
BRONCHOCAINE see BQH250
BRONCHODIL see DNA600
BRONCHOLYSIN see ACH000
BRONCHOSELECTAN see AAN000
BRONCHOSPASMIN see DNA600
BRONCOVALEAS see BQF500
BRONKAID MIST see VGP000
BRONKEPHRINE see DMV600
BRONKEPHRINE HYDROCHLORIDE see ENX500
BRONOCOT see BNT250
BRONOPOL see BNT250
BRONOSOL see BNT250
BRONOX see TJL500
BRONTIN see EAG100
BRONTINA see EAG100
BRONTINE see EAG100
BRONTISOL see EAG100
BRONTYL see HOA000
BRONZE BROMO see BNH500
BRONZE BROMO see BNK700
BRONZE GREEN TONER A-8002 see AFG500
BRONZE ORANGE see CMS150
BRONZE ORANGE TONER see CMS150

BRONZE POWDER see CNI000
BRONZE RED RO see CHP500
BRONZE SCARLET see CHP500
BROOM (DUTCH) see BMP000
BROOM ABSOLUTE see GCM000
O-(4-BROOM-2,5-DICHLOOR-FENYL)-O,O-DIMETHYL-MONOTHIOFOSFAAT (DUTCH) see BNL250
5-BROOM-3-ISOPROPYL-6-METHYL-URACIL DUTCH) see BNM000
BROOMMETHAAN (DUTCH) see MHR200
BROOMWATERSTOF (DUTCH) see HHJ000
cis-BROPARESTROL see BOL315
BROPIRAMINE see AIY850
BROPIRIMINE see AIY850
B ROSE LIQUID see CCK590
BROSERPINE see RDK000
BROTIANIDE see BOL325
BROTIZOLAM see LEJ600
BROTOPON see CLY500
BROVALIN see BNP750
BROVALUREA see BNP750
BROVARIN see BNP750
BROVEL see ECU600
BROWN 5R see ADG000
1545 BROWN see CMP250
11460 BROWN see XMA000
11660 BROWN see CMB750
BROWN ACETATE see CAL750
BROWN ALGAE see AFK920
BROWN ASBESTOS (DOT) see ARM275
BROWN COPPER OXIDE see CNO000
BROWN DRAGON see JAJ000
BROWN FK see CMP250
BROWN HEMATITE see LFW000
BROWNIINE (7CI) see BOL400
BROWN IRON ORE see LFW000
BROWN IRONSTONE CLAY see LFW000
BROWN M see CMO800
BROWN SALT NV see MIH500
BROWN SK see CMU770
BROXALAX see PPN100
BROXIL see PDD350
BROXURIDINE see BNC750
BROXYKINOLIN see DDS600
BROXYNIL see DDP000
BROXYQUINOLINE see DDS600
BRS 640 see BML500
BRUCEANTIN see BOL500
BRUCELLA MELITENSIS ENDOTOXIN see BOL600
BRUCINA (ITALIAN) see BOL750
BRUCINE see BOL750
(−)-BRUCINE see BOL750
BRUCINE (DOT) see BOL750
BRUCINE IODOMETHYLATE see BOM000
BRUCINE IODOMETHYLE (FRENCH) see BOM000
BRUCINE METHIODIDE see BOM000
BRUCITE see NBT000
BRUDR see BNC750
BRUFANEUXOL see DOT000
BRUFANIC see IIU000
BRUFEN see IIU000
BRUGMANSIA ARBOREA see AOO825
BRUGMANSIA SANGUINEA see AOO825
BRUGMANSIA SUAVEOLENS see AOO825
BRUGMANSIA X CANDIDA see AOO825
BRUINSTEEN (DUTCH) see MAS000
BRULAN see BSN000
BRUMIN see WAT200
BRUNEOMYCIN see SMA000
BRUOMOPHOS (RUSSIAN) see BNL250
BRUSATOL see YAG500
(+)-BRUSATOL see YAG500
BRUSH BUSTER see MEL500
BRUSH-OFF see MQR400
BRUSH-OFF 445 MLD VOLATILE BRUSH KILLER see TAA100
BRUSH-RHAP see DAA800
BRUSH RHAP see TAA100
BRUSHTOX see TAA100

BRYAMYCIN see TFQ275
BS see BSL600
BS 572 see DNU100
BS 666 see AHR600
BS 4231 see PMG000
BS 5930 see OJW000
BS 5933 see DRR500
BS 6825 see TGX500
BS 6987 see EAG100
BS 7029 see CQH625
BS 7051 see HBT000
BS 7331 see TGJ250
BS 7020a see DAZ140
BS100-141 see GKU300
BSA see BBR500
BSB-S-E see SMQ500
BSB-S 40 see SMQ500
BSC-REFINE D see BBS750
B-SELEKTONON see DAA800
B-SELEKTONON M see CIR250
BSF see HAQ600
BSF SIMES see HAQ600
BSP see HAQ600
BSP SODIUM see HAQ600
BT see BPG325
BT see BPU000
BT 31 see DEJ100
BT 93 see EIT150
BT 201 see BAP502
BT 204 see BAP502
BT 303 see BAP502
BT 621 see CBS000
BT-93D see EIT150
BT 93W see EIT150
BTB see BAC040
BTB 202 see BAC040
BTC see AFP250
BTC 471 see BBA500
BTC 835 see QAT520
BTC 1010 see DGX200
BTF see BBU125
BTFMEA see EFU500
621-BT HYDROCHLORIDE HYDRATE see TGJ150
BTKH see BEU500
BTM see BFM250
BTO see BLL750
BTPABA see CML870
B-1,3,5-TRICHLOROBORAZINE see TIL300
B-TRIMETHYLBORAZINE see TLN100
BTS see PQC100
BTS 18322 see FLG100
BTS 27,419 see MJL250
BTS 40542 see IAL200
BTS 40542-7877 see IAL200
BTT see BSO750
BTZ see DKU875
BU-6 see BGS250
BU 533 see PIN100
BU 700 see UTU500
BU 2231A see TAC500
BUBAN 37 see DUV400
BUBBIE BLOSSOMS see CCK675
BUBBY BUSH see CCK675
BUBURONE see IIU000
BUCACID ALIZARINE LIGHT BLUE SE see APG700
BUCACID AZURE BLUE see FMU059
BUCACID BLUE BLACK see FAB830
BUCACID BRILLIANT SCARLET 3R see FMU080
BUCACID FAST CRIMSON see CMM300
BUCACID FAST CRIMSON 6B see CMM400
BUCACID FAST ORANGE G see HGC000
BUCACID FAST WOOL BLUE R see ADE750
BUCACID GUINEA GREEN BA see FAE950
BUCACID INDIGOTINE B see FAE100
BUCACID METANIL YELLOW see MDM775
BUCACID ORANGE A see CMM220
BUCACID ORANGE R see ADG000
BUCACID PATENT BLUE AF see ERG100
BUCACID PATENT BLUE VF see ADE500

BUCACID TARTRAZINE see FAG140
BUCACID WOOL GREEN see ADF000
BUCARBAN see BSM000
BUCB see DJF200
BUCCALSONE see HHR000
BUCETIN see HJS850
BUCK BRUSH see SED550
BUCKEYE see HGL575
BUCKTHORN see BOM125
BUCKTHORN see MBU825
BUCLADESINE see DEJ300
BUCLIFEN see VJZ050
BUCLIZINE see VJZ050
BUCLIZINE DIHYDROCHLORIDE see BOM250
BUCLODIN see BOM250
BUCLOSINSAEURE (GERMAN) see CPJ000
BUCLOXIC ACID see CPJ000
BUCLOXIC ACID CALCIUM see CPJ250
BUCLOXIC ACID CALCIUM SALT see CPJ250
BUCLOXINSAEURE KALZIUM (FERMAN) see CPJ250
BUCLOXONIC ACID see CPJ000
BUCLOXONIC ACID CALCIUM SALT see CPJ250
BUCOLOM see BQW825
BUCOLOME see BQW825
BUCROL see BSM000
BUCS see BPJ850
BUCTRIL see DDP000
BUCTRIL INDUSTRIAL see DDP000
BUCUMOLOL HYDROCHLORIDE see BOM510
dl-BUCUMOLOL HYDROCHLORIDE see BOM510
BUDAMIN MF 55I see MCB050
BUDAMIN MF 60I see MCB050
BUDESONIDE see BOM520
BUDIPIN (GERMAN) see BOM530
BUDIPINE see BOM530
BUD-NIP see CKC000
BUDOFORM see CHR500
BUDORM see BPF500
BUDR see BNC750
5-BUDR see BNC750
BUDRALAZINE see DQU400
BU2AE see DDU600
BUENO see MRL750
BUFA-20,22-DIENOLIDE, 3-(ACETYLOXY)-5,14-DIHYDROXY-19-OXO-, (3-β,5-β)- see HAN550
BUFA-20,22-DIENOLIDE, 3,16-DIHYDROXY-14,15-EPOXY-, 3,16-DIACETATE see CMS126
5-β-BUFA-20,22-DIENOLIDE, 14,15-β-EPOXY-3-β,16-β-DIHYDROXY-, 3,16-DIACETATE see CMS126
BUFA-20,22-DIENOLIDE, 11-OXO-3,12,14-TRIHYDROXY-, (3-β,5-β,12-α)- see POG275
5-β-BUFA-20,22-DIENOLIDE, 3-β,12-α,14-TRIHYDROXY-11-OXO- see POG275
5-β-BUFA-20,22-DIENOLIDE, 3-β,5,14-TRIHYDROXY-19-OXO-, 3-ACETATE see HAN550
BUFAPTO METHALOSE see MIF760
BUFA-4,20,22-TRIENOLIDE, 3-(β-d-GLUCOPYRANOSYLOXY)-8,14-DIHYDROXY-, (3-β)- see SBF510
BUFEDIL see BOM600
BUFEMID see BGL250
BUFEN see ABU500
BUFENCARB see BTA250
BUFETOLOL HYDROCHLORIDE see AER500
BUFEXAMIC ACID see BPP750
BUFF-A-COMP see ABG750
BUFLOMEDIL see BOM600
BUFLOMEDIL HYDROCHLORIDE see BOM600
BUFOGENIN see BOM650
BUFOGENIN B see BOM655

BUFON see DKA600
BUFONAMIN see BOM750
BUFONAMIN see BQL000
BUFOPTO ZINC SULFATE see ZNA000
BUFORMIN see BRA625
BUFORMINE see DRA625
BUFORMIN HYDROCHLORIDE see BOM750
BUFORMIN HYDROCHLORIDE see BQL000
BUFOTALIN see BON825
BUFOTALINE see BON000
BUFOTENIN see DPG109
BUFURALOL see BQD000
BUG MASTER see CBM750
BUHACH see POO250
BUIS de SAPIA (CANADA) see YAK500
BUKARBAN see BSM000
BUKS see BFW750
BUKSAMIN see AKF375
BULANA, combustion products see ADX750
BULAN and PROLAN MIXTURE (2:1) see BON250
BULBOCAPNINE see HLT000
d-BULBOCAPNINE see HLT000
BULBONIN see BQL000
BULBOSAN see TJE200
BULEN A see PJS750
BULEN A 30 see PJS750
BULGARIAN see RNF000
BULKALOID see MIF760
BULKOSOL see BON300
BULL FLOWER see MBU550
BULPUR see PLC250
BUMADIZON CALCIUM SALT HEMIHYDRATE see CCD750
BUMETANIDE see BON325
BUMEX see BON325
BUNAIOD see EQC000
BUNAMIODYL see EQC000
BUNAZOCINE HYDROCHLORIDE see BON350
BUNAZOSIN HYDROCHLORIDE see BON350
BUNDOLIN CORRECTOR ZN 14R see ZGW200
BUNGARUS CAERULEUS VENOM see BON365
BUNGARUS FASCIATUS TOXIN VI see THI300
BUNGARUS FASCIATUS VENOM see BON367
BUNGARUS MULTICINCTUS VENOM see BON370
BUNIODYL see EQC000
BUNITROLOL HYDROCHLORIDE see BON400
BUNK see PJJ300
l-BUNOLOL HYDROCHLORIDE see LFA100
BUNSENITE see NDF500
BUNT-CURE see HCC500
BUNT-NO-MORE see HCC500
BUPATOL see BOV825
BUPHENINE HYDROCHLORIDE see DNU200
BUPICAINE HYDROCHLORIDE (+) see BON750
BUPICAINE HYDROCHLORIDE (±) see BOO000
BUPIVACAINE see BSI250
l(−)-BUPIVACAINE see BOO500
d(+)-BUPIVACAINE see BOO250
dl-BUPIVACAINE see BSI250
BUPIVACAINE HYDROCHLORIDE see BOO000
BUPLEURUM FALCATUM L see SAF300
BUPLEURUM MARGINATUM WALL. EX. DC., EXTRACT see BOO625
BUPRANOLOL HYDROCHLORIDE see BQB250
BUPRENORPHINE see TAL325
BUPRENORPHINE HYDROCHLORIDE see BOO630

BUPROFEZIN see BOO631
BUPROFEZINE see BOO631
BUPROPION HYDROCHLORIDE see WBJ500
BUQUINOLATE see BOO632
BUR see EHA100
BURACYL see CPR600
BURCOL see BSM000
BURCO TME see PJV100
BURESE see CNE750
BUREX (CZECH) see PEE750
BURFOR see NCW300
BURGODIN see BCE825
BURINE see BON325
BURINEX see BON325
BURITAL SODIUM see SOX500
BURLEY TOBACCO see TGH725
BURLEY TOBACCO see TGH750
BURMA GREEN B see AFG500
BURN BEAN see NBR800
BURNISH GOLD see GIS000
BURNOL see DBX400
BURNOL see XAK000
BURNT ALUM see AHF100
BURNT ALUM see PKU725
BURNTISLAND RED see IHC450
BURNT LIME see CAU500
BURNT SIENNA see IHC450
BURNT UMBER see IHC450
BUROFLAVIN see DBX400
BURONIL see FKI000
BURPLEURUM FALCATUM LINN. VAR. MARGINATUM (WALL. EX. DC.) CL., EXTRACT see BOO625
BURSINE see CMF800
BURTOLIN see DMC600
BURTONITE 44 see FPQ000
BURTONITE V-7-E see GLU000
BURTONITE-V-40-E see CCL250
BUSAN 72A see BOO635
BUSCAPINA see SBG500
BUSCAPINE see SBG500
BUSCOL see SBG500
BUSCOLAMIN see SBG500
BUSCOLYSINE see SBG500
BUSCOPAN see SBG500
BUSCOPAN COMPOSITUM see BOO650
BUSERELIN see LIU420
BUSHMAN'S POISON see BOO700
BUSONE see BRF500
BUSPIRONE see PPP250
BUSTREN see SMQ500
BUSTREN K 500 see SMQ500
BUSTREN K 525-19 see SMQ500
BUSTREN U 825 see SMQ500
BUSTREN U 825E11 see SMQ500
BUSTREN Y 825 see SMQ500
BUSTREN Y 3532 see SMQ500
BUTABARB see BPF000
BUTABARBITAL see BPF000
BUTABARBITAL SODIUM see BPF250
BUTABARBITONE see BPF000
BUTABITAL see AFY500
BUTACAINE see BOO750
BUTACAINE SULFATE see BOP000
BUTACARB see DEG400
BUTACARBE (FRENCH) see DEG400
BUTACHLOR see CFW750
BUTACIDE see PIX250
BUTACOMPREN see BRF500
BUTACOTE see BRF500
BUTADIEEN (DUTCH) see BOP500
BUTA-1,3-DIEEN (DUTCH) see BOP500
BUTADIEN (POLISH) see BOP500
BUTA-1,3-DIEN (GERMAN) see BOP500
BUTADIENDIOXYD (GERMAN) see BGA750
BUTADIENE see BOP100
1,2-BUTADIENE see BOP250
1,3-BUTADIENE see BOP500
BUTA-1,3-DIENE see BOP500
α-γ-BUTADIENE see BOP500

BUTANEDIOIC ACID, SULFO-, 1,4-DIHEXYL ESTER, SODIUM SALT (9CI) see DKP800
BUTANEDIOIC ACID, SULFO-, 1,4-DIOCTYL ESTER, SODIUM SALT (9CI) see SOD050
BUTANEDIOIC ANHYDRIDE see SNC000
1,2-BUTANEDIOL see BOS250
1,3-BUTANEDIOL see BOS500
BUTANE-1,3-DIOL see BOS500
1,4-BUTANEDIOL see BOS750
BUTANE-1,4-DIOL see BOS750
2,3-BUTANEDIOL see BOT000
1,4-BUTANEDIOL BIS(3-AMINOPROPYL) ETHER see BGU600
1,4-BUTANEDIOL, BISSULFAMATE (ester) see BOU100
1,2-BUTANEDIOL, CYCLIC CARBONATE see BOT200
1,3-BUTANEDIOL, CYCLIC SULFITE see MQG500
1,3-BUTANEDIOL DIACRYLATE see BRG500
1,4-BUTANEDIOL DIACRYLATE see TDQ100
BUTANEDIOL DIGLYCIDYL ETHER see BOS100
BUTANE-1:4-DIOL DIGLYCIDYL ETHER see BOS100
1,4-BUTANEDIOL DIGLYCIDYL ETHER see BOS100
2,3-BUTANEDIOL, 1,4-DIMERCAPTO-, (R*,S*)-(9CI) see DXN350
1,3-BUTANEDIOL, DIMETHACRYLATE see BRG100
1,4-BUTANEDIOL DIMETHANESULPHONATE see BOT250
1,4-BUTANEDIOL DIMETHYL SULFONATE see BOT250
1,2-BUTANEDIOL, 3,4-EPOXY- see ONE050
1,4-BUTANEDIOL, POLYMER with 1,6-DIISOCYANATOHEXANE see PKO750
1,4-BUTANEDIOL POLYMER with 1,1'-METHYLENEBIS(4-ISOCYANATOBENZENE) see PKP000
2,3-BUTANEDIONE see BOT500
BUTANEDIONE (DOT) see BOT500
2,3-BUTANEDIONE, MONOOXIME see OMY910
2,3-BUTANEDIONE 2-OXIME see OMY910
1,3-BUTANEDIONE, 1-PHENYL- see BDJ800
BUTANEDIPEROXOIC ACID, BIS(1,1-DIMETHYLETHYL) ESTER see BOT600
BUTANEDIPEROXOIC ACID DI-tert-BUTYL ESTER see BOT600
N,N'-(1,4-BUTANEDIYL)BIS-1-AZIRIDIENCARBOXAMIDE see TDQ225
2,2'-(1,4-BUTANEDIYL.)BISOXIRANE see DHD800
2,2'-(1,4-BUTANEDIYLBIS(OXYMETHYLENE))BISOXIRANE see BOS100
1,4-BUTANEDIYL SULFAMATE see BOU100
BUTANE, 2,3-EPOXY-, cis- see EBJ200
BUTANEFRINE HYDROCHLORIDE see ENX500
BUTANE, 1-IODO-3-METHYL- see IHU200
BUTANE, 1-ISOTHIOCYANATO-(9CI) see ISD100
BUTANE, 1-METHOXY-(9CI) see BRU780
BUTANE, 2-METHOXY-2-METHYL- see PBX400
BUTANE MIXTURES (DOT) see BOR500
BUTANEN (DUTCH) see BOR500
BUTANENITRILE see BSX250
n-BUTANENITRILE see BSX250
BUTANENITRILE, 2,2'-AZOBIS(2-METHYL- see ASM025
BUTANENITRILE, 4-CHLORO-(9CI) see CEU300
BUTANENITRILE, 2-HYDROXY-4-(METHYLTHIO)- see HMR550

BUTANE, 2,2'-OXYBIS-(9CI) see BRH760
BUTANE, 2,2'-OXYBIS-(9CI) see OPE030
BUTANE, 1,1'-OXYBIS(4-CHLORO-(9CI) see OPE040
BUTANESULFONE see BOU250
1-BUTANESULFONIC ACID, 3-HYDROXY-, γ-SULTONE see BOU300
1-BUTANESULFONIC ACID, 1,1,2,2,3,3,4,4,4-NONAFLUORO- see PCG625
1-BUTANESULFONIC ACID, NONAFLUORO- (6CI,7CI,8CI) see PCG625
BUTANE SULTONE see BOU250
1,3-BUTANESULTONE see BOU300
Δ-BUTANE SULTONE see BOU250
γ-BUTANE SULTONE see BOU300
1,4-BUTANESULTONE (MAK) see BOU250
BUTANETETRACARBOXYLIC ACID see BOU500
1,2,3,4-BUTANETETRACARBOXYLIC ACID see BOU500
BUTANE, 1,2,3,4-TETRACHLORO- see TBO762
5H,6H-6,5A,13A,14-(1,2,3,4)BUTANETRACYCLOOCTA(1,2-B:5,6-B')DINAPHTHALENE see LIV000
tert-BUTANETHIOL see MOS000
2-BUTANETHIOL, 2-METHYL- see MHS550
BUTANETHIOL (OSHA) see BRR900
1-BUTANETHIOL, TIN(2+) SALT see BOU550
1,2,4-BUTANETRICARBOXYLIC ACID, 2-(DIMETHOXYPHOSPHINYL)-, TRIMETHYL ESTER see PHA300
1,2,4-BUTANETRICARBOXYLIC ACID, 2-PHOSPHONO-, TETRASODIUM SALT see TEE400
BUTANE, 1,1,3-TRIETHOXY- see TJL700
1,2,3-BUTANETRIOL, TRINITRATE see MKI300
1,2,4-BUTANETRIOL TRINITRATE see BOU700
1,2,4-BUTANETRIOL, TRINITRATE see BOU700
BUTANEX see CFW750
BUTANI (ITALIAN) see BOR500
BUTANILICAINE HYDROCHLORIDE see BQA750
BUTANIMIDE see SND000
BUTANOIC ACID see BSW000
BUTANOIC ACID, 2-AMINO-4-(S-BUTYLSULFONIMIDOYL)-, (2S)- see BPE800
BUTANOIC ACID, 4-((AMINOCARBONYL)NITROSOAMINO)-, METHYL ESTER see MEI300
BUTANOIC ACID, 2-AMINO-4-(HYDROXYMETHYLPHOSPHINYL)-, (+−)- see GFM222
BUTANOIC ACID, 2-AMINO-4-(HYDROXYMETHYLPHOSPHINYL)-, MONOAMMONIUM SALT see ANI800
BUTANOIC ACID, 2-AMINO-4-(METHYLSELENO)-, (+−)- see SBU800
BUTANOIC ACID, 2-AMINO-4-(METHYLSULFINYL)-(9CI) see ALF600
BUTANOIC ACID, ANHYDRIDE (9CI) see BSW550
BUTANOIC ACID, 2-BROMO-3-METHYL-(9CI) see BNM100
BUTANOIC ACID-2-BUTOXY-1-METHYL-2-OXOETHYL ESTER (9CI) see BQP000
BUTANOIC ACID, 4-CHLORO-3-OXO-, ETHYL ESTER (9CI) see EHH100
BUTANOIC ACID, 4-CHLORO-3-OXO-, METHYL ESTER (9CI) see MIH600
BUTANOIC ACID, 3-((5-(2-CHLORO-4-(TRIFLUOROMETHYL)PHENOXY)-2-NITROPHENYL)AMINO)-, METHYL ESTER see MIJ275
BUTANOIC ACID, CYCLOHEXYL ESTER (9CI) see CPI300

BUTANOIC ACID, 2-((DIETHOXYPHOSPHINYL)OXY)-3-OXO-, ETHYL ESTER see EHY700
BUTANOIC ACID, 2,3-DIHYDROXY-2-(1-METHYLETHYL)-, (2,3,5,7A-TETRAHYDRO-1H-PYRROLIZIN-7-YL) METHYL ESTER, HYDROCHLORIDE, (7ar-(7(2S*,3S*),7AR*))- see CQI525
BUTANOIC ACID, 3,7-DIMETHYL-2,6-OCTADIENYL ESTER, (E)-(9CI) see GDE810
BUTANOIC ACID, 4-((2-((2,6-DIMETHYLPHENYL)AMINO)-2-OXOETHYL)AMINO)- see DTK400
BUTANOIC ACID, 4-((2-((2,6-DIMETHYLPHENYL)AMINO)-2-OXOETHYL)AMINO)-4-OXO- see DTK420
BUTANOIC ACID, 4,4'-DISELENOBIS(2-AMINO)-(9CI) see SBU710
BUTANOIC ACID ETHYL ESTER see EHE000
BUTANOIC ACID, 2-((2-FURANYLMETHYL)(1H-IMIDAZOL-1-YLCARBONYL)AMINO)-, 4-PENTENYL ESTER see PAO170
BUTANOIC ACID, HEPTYL ESTER see HBN150
BUTANOIC ACID, 2,4-HEXADIENYL ESTER see HCS600
BUTANOIC ACID, 2-HYDROXY-4-(METHYLTHIO)-(9CI) see HMR600
BUTANOIC ACID, 4-HYDROXY-, MONOLITHIUM SALT see LHM800
BUTANOIC ACID, 2-(IMINO(5-NITRO-2-FURANYL)METHYL)HYDRAZIDE see BSX600
BUTANOIC ACID, 4-ISOTHIOCYANATO-, ETHYL ESTER see ELX527
BUTANOIC ACID, 4-((4-METHOXYBENZOYL)AMINO)- see AOY270
BUTANOIC ACID, 2-METHYL- see MHS600
BUTANOIC ACID, 3-METHYL-, 6-(ACETYLOXY)-6,7A-DIHYDRO-4-((3-METHYL-1-OXOBUTOXY)METHYL)SPIRO(CYCLOPENTA(C)PYRAN-7-(1H),2'-OXIRAN)-1-YL ESTER, (1S,2'R,6S,7AS)- see ITD018
BUTANOIC ACID, 3-METHYL-, 1-((ACETYLOXY)METHYL)-6,7A-DIHYDROSPIRO(CYCLOPENTA(C)PYRAN-7(1H),2'-OXIRAN E)-1,6-DIYL ESTER, (1S-(1-α,6-α,7-β,7A-α))- see VAD200
BUTANOIC ACID, 3-METHYLBUTYL ESTER (9CI) see IHP400
BUTANOIC ACID, 3-METHYL-, HEXYL ESTER see HFQ575
BUTANOIC ACID, 3-METHYL-, 5-METHYL-2-(1-METHYLETHYL)CYCLOHEXYL ESTER see MCG900
BUTANOIC ACID, 2-METHYL-, 2-PHENYLETHYL ESTER see PDF780
BUTANOIC ACID, 3-METHYL-, 3-PHENYL-2-PROPENYL ESTER see PEE200
BUTANOIC ACID, 3-METHYL-, 1,7,7-TRIMETHYLBICYCLO(2.2.1)HEPT-2-YL ESTER, endo- see HOX100
BUTANOIC ACID, 3-OXO-, CYCLOHEXYL ESTER (9CI) see CPP100
BUTANOIC ACID, 3-OXO-, 2-((1,3-DIOXOBUTOXY)METHYL)-2-ETHYL-1,3-PROPANEDIYL ESTER see ELJ600
BUTANOIC ACID, 3-OXO-, 5-METHYL-2-(1-METHYLETHYL)CYCLOHEXYL ESTER, (1R-(1-α-2-β, 5α-)- see MCG850
BUTANOIC ACID, 3-OXO-, 2-((1-OXO-2-PROPENYL)OXY)ETHYL ESTER (9CI) see AAY750
BUTANOIC ACID, 2-OXO-, SODIUM SALT (9CI) see SIY600
BUTANOIC ACID PENTYL ESTER see AOG000
BUTANOIC ACID, 1,2,3-PROPANETRIYL ESTER see TIG750

BUTANOIC ACID, PROPYL ESTER (9CI) see PNF100
BUTANOIC ACID 2,2,2-TRICHLORO-1-(DIMETHOXYPHOSPHINYL)ETHYL ESTER see BPG000
BUTANOIC ACID, 4,4,4-TRINITRO- see TML100
BUTANOIC ANHYDRIDE see BSW550
BUTAN-1-OL see BPW500
1-BUTANOL see BPW500
BUTAN-2-OL see BPW750
2-BUTANOL see BPW750
n-BUTANOL see BPW500
BUTANOL (DOT) see BPW500
tert-BUTANOL see BPX000
BUTANOL (FRENCH) see BPW500
sec-BUTANOL (DOT) see BPW750
2-BUTANOL ACETATE see BPV000
3-BUTANOLAL see AAH750
BUTANOL-2-AMINE see AJA250
2-BUTANOL, 4-(BUTYLBIS((1-OXODODECYL)OXY)STANNYL)- see BRN600
BUTANOL (4)-BUTYL-NITROSAMINE see HJQ350
2-BUTANOL, 4-(DIETHYLAMINO)-3-METHYL-, p-ISOBUTOXYBENZOATE (ESTER) see GBU700
2-BUTANOL, 3,3-DIMETHYL- see BRU300
2-BUTANOL, 2-((1,1-DIMETHYLETHYL)AZO)- see TAH950
BUTANOLEN (DUTCH) see BPW500
1-BUTANOL, 2,2,3,3,4,4,4-HEPTAFLUORO- see HAW100
4-BUTANOLIDE see BOV000
1-BUTANOL, 3-METHYL-, NITRATE (9CI) see ILW100
1-BUTANOL, 2-METHYL-4-(PURIN-6-YLAMINO)- see RAF400
1-BUTANOL, 2-METHYL-4-(1H-PURIN-6-YLAMINO)- see RAF400
1-BUTANOL, 4-(NITROSO-2-PROPENYLAMINO)- see HJS400
BUTANOLO (ITALIAN) see BPW500
2-BUTANOL-3-ONE see ABB500
2,3-BUTANOLONE see ABB500
BUTANOL SECONDAIRE (FRENCH) see BPW750
BUTANOL TERTIAIRE (FRENCH) see BPX000
BUTANONE 2 (FRENCH) see MKA400
1-BUTANONE, 1-(4-AMINOPHENYL)-(9CI) see AJC750
2-BUTANONE, AZINE see EMT600
2-BUTANONE, 1-BROMO-3,3-DIMETHYL- see PJB100
2-BUTANONE, 3-CHLORO- see CEU800
2-BUTANONE, 1-CHLORO-3,3-DIMETHYL- see MRG070
2-BUTANONE, 1-(p-CHLOROPHENOXY)-3,3-DIMETHYL-1-(1-IMIDAZOLYL)- see CJN900
2-BUTANONE, 1,1-DICHLORO-3,3-DIMETHYL- see DGF300
2-BUTANONE, 4-DIETHYLAMINO- see DHM100
2-BUTANONE, 3,3-DIMETHYL-1-(METHYLTHIO)-, OXIME see DSS900
2-BUTANONE, O,O'-(ETHENYLMETHYLSILYLENE)DIOXIME, (E,Z)- see MQM150
2-BUTANONE, O,O'-(ETHENYLMETHYLSILYLENE)DIOXIME, (2E,2'Z)- see MQM150
2-BUTANONE, 3,3'-ETHYLENEDINITRILODI-, DIOXIME, IRON(II) COMPLEX see EJA150
1-BUTANONE, 1-(4-FLUOROPHENYL)-4-(1,3,4,4A,5,9B-HEXAHYDRO-8-METHYL-2H-PYRIDO(4,3-B)INDOL-2-YL)-, DIHYDROCHLORIDE see MKD800

2-BUTANONE, 4-(4-HYDROXY-3-METHOXYPHENYL)- see VFP100
2-BUTANONE, 4-(4-HYDROXYPHENYL)- see RBU000
2-BUTANONE, 4-(6-METHOXY-2-NAPHTHALENYL)- see MFA300
2-BUTANONE, 4-(p-METHOXYPHENYL)-(6CI,7CI,8CI) see MFF580
2-BUTANONE, (1-METHYLPROPYLIDENE)HYDRAZONE see EMT600
2-BUTANONE (OSHA) see MKA400
2-BUTANONE, OXIME see EMU500
2-BUTANONE OXIME HYDROCHLORIDE see BOV625
2-BUTANONE, 4-PHENYL-, 3-THIOSEMICARBAZONE see PEO600
1-BUTANONE, 1-(4-PYRIDYL)- see PNV755
2-BUTANONE, SEMICARBAZONE see MKA750
BUTAN-3-ONE-2-YL BUTYRATE see BOV700
BUTANOVA see HNI500
BUTANOX LPT see MKA500
BUTANOX M 50 see MKA500
BUTANOX M 105 see MKA500
BUTANTRONE see BSY400
BUTAPERAZINE DIMALEATE see BSZ000
BUTAPHENE see BRE500
BUTAPIRAZOL see BRF500
BUTAPIRONE see HNI500
BUTAPYRAZOLE see BRF500
BUTARECBON see BRF500
n-BUTARSAMIDE see CBK750
BUTARTRINA see BRF500
BUTATAB see BPF000
BUTATAL see BPF000
BUTATENSIN see MBW750
BUTAZATE see BIX000
BUTAZATE 50-D see BIX000
BUTAZINA see BRF500
BUTAZOLIDINE SODIUM see BOV750
BUTAZONA see BRF500
BUTAZONE see BRF500
BUTEA FRONDOSA, seed extract see BOV800
BUTEA FRONDOSA, FLOWER PETALS, ALCOHOLIC EXTRACT see BOV810
BUTEA MONO-SPRA (LAM.) KUNTZE, FLOWER EXTRACT see BOV810
BUTEDRIN see BOV825
BUTEDRINE see BQF250
BUTELLINE see BOP000
2-BUTENAL see COB250
(E)-2-BUTENAL see COB260
trans-2-BUTENAL see COB260
2-BUTENAMIDE, 3-METHYL- see SBW965
1-BUTENE see BOW250
2-BUTENE see BOW255
cis-2-BUTENE see BOW500
trans-2-BUTENE see BOW750
1-BUTENE, 4-CHLORO- see CEV300
2-BUTENEDIAL, (Z)- see BOW800
cis-2-BUTENE-1,4-DIAL see BOW800
2-BUTENE, 1,4-DIBROMO-2-METHYL- see IMS100
BUTENE, DICHLORO- see DEV200
1-BUTENE, 3,4-DICHLORO- see DEV100
2-BUTENE, 1,3-DICHLORO- see DEU650
1-BUTENE, 3,4-DICHLORO-(RACEMIC MIXTURE) see DEU200
2-BUTENE, 1,4-DIHYDROXY- see BOX300
2-BUTENEDINITRILE, (E)- see FOX000
(±)-(Z)-2-BUTENEDIOATE-1H-INDEN-5-OL, 2,3-DIHYDRO-6-((2-METHYL-1-PIPERIDINYL)METHYL)- see PJI600
(Z)-2-BUTENEDIOATE (1:1) 1,2,3,4-TETRAHYDRO-2-METHYL-4-PHENYL-8-ISOQUINOLINAMINE see NMV725
(E)-BUTENEDIOIC ACID see FOU000
(Z)-BUTENEDIOIC ACID see MAK900
cis-BUTENEDIOIC ACID see MAK900
trans-BUTENEDIOIC ACID see FOU000

2-BUTENEDIOIC ACID BIS(1,3-DIMETHYLBUTYL) ESTER see DKP400
2-BUTENEDIOIC ACID BIS(2-ETHYLHEXYL) ESTER see DVK600
2-BUTENEDIOIC ACID BIS(1-METHYLETHYL) ESTER see BOX250
2-BUTENEDIOIC ACID, DIBUTYL ESTER see DED600
(Z)-2-BUTENEDIOIC ACID DIETHYL ESTER see DJO200
trans-BUTENEDIOIC ACID DIMETHYL ESTER see DSB600
2-BUTENEDIOIC ACID (Z)-, DIOCTYL ESTER (9CI) see DVK800
BUTENEDIOIC ACID, METHYL-, (E)- see MDI250
2-BUTENEDIOIC ACID, MONO(2-(2-HYDROXYETHOXY)ETHYL ESTER see MAL500
2-BUTENEDIOIC ACID, MONO(2-HYDROXYPROPYL) ESTER see MAL750
cis-BUTENEDIOIC ANHYDRIDE see MAM000
2-BUTENE-1,4-DIOL see BOX300
2-BUTENE-1,4-DIOL BIS(BROMOACETATE) see BHD150
2-BUTENE-1,4-DIOL, 2,3-DIBROMO- see DDL700
2-BUTENE-1,4-DIOL, DIMETHANESULFONATE, (E)- see DNX000
2-BUTENE-1,4-DIONE, 1,4-DIPHENYL- see DDE100
cis-2-BUTENE EPOXIDE see EBJ200
2-BUTENE EXPOXIDE see EBJ100
2-BUTENENITRILE see COQ750
3-BUTENE NITRILE see BOX500
1-BUTENE-4-NITRILE see BOX500
2-BUTENENITRILE, (Z)- see BOX600
cis-2-BUTENENITRILE see BOX600
1-BUTENE, 2-NITRO- see NFQ550
3-BUTENE-2-ONE see BOY500
1-BUTENE OXIDE see BOX750
2-BUTENE OXIDE see EBJ100
cis-2-BUTENE OXIDE see EBJ200
trans-2-BUTENE OZONIDE see BOX825
BUTENE, POLYMERS see PJL400
1-BUTENE, 2,3,4-TRICHLORO- see TIL360
2-BUTENOIC ACID see COB500
trans-2-BUTENOIC ACID see ERE000
α-BUTENOIC ACID see COB500
2-BUTENOIC ACID, 3-AMINO-, 1-METHYLETHYL ESTER see INM100
2-BUTENOIC ACID, ANHYDRIDE (9CI) see COB900
2-BUTENOIC ACID, 3-BROMO-4-(4-METHOXYPHENYL)-4-OXO-, SODIUM SALT, (E)- (9CI) see CQK600
2-BUTENOIC ACID, 4-BROMO-, METHYL ESTER (9CI) see MHR790
2-BUTENOIC ACID, 2-BUTENYLIDENE ESTER, (E,E,E)-(8CI,9CI) see COD100
2-BUTENOIC ACID, 2-CHLORO-3-(CHLOROMETHYL)-4-OXO-, (2Z)- see CFB720
2-BUTENOIC ACID, 3-(CHLOROMETHYL)-4,4-DICHLORO- see CIK850
2-BUTENOIC ACID, 2-CHLORO-3-METHYL-4-OXO-, (2Z)- see CIQ650
2-BUTENOIC ACID, 3-(CHLOROMETHYL)-2,4,4-TRICHLORO- see TIL610
2-BUTENOIC ACID, 4,4-DICHLORO-3-(DICHLOROMETHYL)- see DFP550
2-BUTENOIC ACID, 3,6-DICHLORO-2,4-DINITROPHENYL ESTER see DFE560
2-BUTENOIC ACID, 2,3-DICHLOR-4-OXO-, (Z)-(9CI) see MRU900
2-BUTENOIC ACID-3-(DIETHOXY PHOSPHINOTHIOYL)ETHYL ESTER see EIB000
2-BUTENOIC ACID, 3-((DIETHYLPHOSPHINOTHIOYL)OXY)-, ETHYL ESTER see EIB600

2-BUTENOIC ACID-3,7-DIMETHYL-6-OCTENYL ESTER see CMT500

2-BUTENOIC ACID, ETHENYL ESTER see VNU000

2-BUTENOIC ACID, 3-(((ETHYLAMINO)METHOXYPHOSPHINOTHIOYL)OXY)-, 1-METHYLETHYL ESTER, (E)-, mixt. with 2,2-DICHLOROETHENYL DIMETHYL PHOSPHATE see SAD100

2-BUTENOIC ACID, ETHYL ESTER see EHO200

2-BUTENOIC ACID, ETHYL ESTER, (E)-(9CI) see COB750

trans-2-BUTENOIC ACID ETHYL ESTER see COB750

2-BUTENOIC ACID, 3-((ETHYL(PROPYLAMINO)PHOSPHINOTHIOYL)OXY)-, METHYL ESTER see MKB320

2-BUTENOIC ACID, 3-FORMYL-2,4,4-TRICHLORO-, (Z)- see FOJ033

2-BUTENOIC ACID, HEXYL ESTER see HFM600

2-BUTENOIC ACID, 2-METHYL-, (E)-(9CI) see TGA700

2-BUTENOIC ACID, 2-METHYL-, 2,3,3A,4,5,6,7,8,9,11A-DECAHYDRO-7,9-DIHYDROXY-6,10-DIMETHYL-3-METHYLENE-2-OXO-6,9-EPOXYCYCLODECA(B)FURAN-4-YL ESTER, (3ar-(3ar*,4R*(Z),6R*,7S*,9R*,10Z,11ar*))- see NMV200

trans-2-BUTENOIC ACID METHYL ESTER see COB825

2-BUTENOIC ACID, 2-METHYL-, ETHYL ESTER, (E)- see TGA800

2-BUTENOIC ACID, 2-METHYL-, 3-HEXENYL ESTER, (E,Z)- see HFE710

2-BUTENOIC ACID, 2-METHYL, 1-ISOPROPYL ESTER (E)- see IRN100

2-BUTENOIC ACID, 2-METHYL-, METHYL ESTER, (E)- see MPW700

2-BUTENOIC ACID, 2-METHYL-, 2,3,3A,4,5,8,9,11A-OCTAHYDRO-5,9-DIHYDROXY-6,10-DIMETHYL-3-METHYLENE-2-OXOCYCLODECA(B)FURAN-4-YL ESTER, (3as-(3AR*,4S*(Z),5S*,6E,9S*,10Z,11AS*))- see ZTS300

BUTENOIC ACID, 2-((3-NITROPHENYL)METHYLENE)-3-OXO-, ETHYL ESTER see ENO200

BUTENOIC ACID, 2-((3-NITROPHENYL)METHYLENE)-3-OXO-, 2-METHOXYETHYL ESTER see MEP600

2-BUTENOL see BOY000

2-BUTEN-1-OL see BOY000

3-BUTEN-2-OL see MQL250

3-BUTENO-β-LACTONE see KFA000

2-BUTEN-1-OL, 3-METHYL-, ACETATE see DOQ350

2-BUTEN-1-OL, 3-METHYL-, BENZOATE see MHU150

2-BUTEN-1-OL, 3-METHYL-, SALICYLATE see PMB600

3-BUTEN-2-OL, 4-(2,6,6-TRIMETHYL-1-CYCLOHEXEN-1-YL)- see IFT500

3-BUTEN-2-OL, 4-(2,6,6-TRIMETHYL-2-CYCLOHEXEN-1-YL)- see IFT400

3-BUTEN-2-OL, 4-(2,6,6-TRIMETHYL-2-CYCLOHEXEN-1-YL)-, ACETATE see IFX300

1-BUTEN-3-ONE see BOY250

1-BUTEN-3-ONE see MQM100

3-BUTEN-2-ONE see BOY500

3-BUTEN-2-ONE, 3,4-BIS(p-METHOXYPHENYL)- see BKO800

3-BUTEN-2-ONE, 3,4-BIS(4-METHOXYPHENYL)-(9CI) see BKO800

3-BUTEN-2-ONE, 4-BROMO- see MHS400

3-BUTEN-2-ONE, 4-(2-FURANYL)- see FPX030

3-BUTEN-2-ONE, 4-(2-FURYL)-(6CI,8CI) see FPX030

3-BUTEN-2-ONE, 4-(4-HYDROXY-3-METHOXYPHENYL)- see MLI800

3-BUTEN-2-ONE, 4-(4-METHOXYPHENYL)- see MLI400

3-BUTEN-2-ONE, 4-(3,4-(METHYLENEDIOXY)PHENYL)-(6CI,7CI,8CI) see MJR250

3-BUTEN-2-ONE, 3-METHYL-4-PHENYL- see MNS600

3-BUTEN-2-ONE, 4-PHENYL-, (E)- see BAY275

3-BUTEN-2-ONE, 4-(2,5,6,6-TETRAMETHYL-2-CYCLOHEXEN-1-YL)-(9CI) see IGW500

3-BUTEN-2-ONE, 4-(2,4,6-TRIMETHYL-3-CYCLOHEXEN-1-YL)- see IGJ600

3-BUTEN-2-ONE,4-(2,6,6-TRIMETHYL-1-CYCLOHEXEN-1-YL)-, (E)- see IFX100

β-BUTENONITRILE see BOX500

2-BUTENYL ALCOHOL see BOY000

2-BUTENYL CHLORIDE see CEU825

cis-1-BUTENYL CYANIDE see PBQ275

2-BUTEN-1-YL DIAZOACETATE see BPA250

5-(1-BUTENYL)-5-ETHYLBARBITURIC ACID see BPA500

5-(2-BUTENYL)-5-ETHYL-2-THIOBARBITURIC ACID SODIUM SALT see BPA750

BUTENYL(3-HYDROXYPROPYL)NITROSAMINE see BPC600

(2-BUTENYLIDENE)ACETIC ACID see SKU000

2-BUTENYLIDENE CROTONATE see COD100

5-(1-BUTENYL)-5-ISOPROPYLBARBITURIC ACID see BPB500

5-(2-BUTENYL)-5-(1-METHYLBUTYL)-2-THIOBARBITURIC ACID SODIUM SALT see BPC500

3-(3-BUTENYLNITROSAMINO)-1-PROPANOL see BPC600

2-BUTENYLPHENOL (mixed isomers) see BPC750

3-BUTENYL-(2-PROPENYL)-N-NITROSAMINE see BPD000

1-(2'-BUTENYL)THEOBROMINE see BSM825

BUTEN-3-YNE see BPE109

3-BUTEN-1-YNYL DIETHYL ALUMINUM see BPE250

3-BUTEN-1-YNYL DIISOBUTYL ALUMINUM see BPE500

2-BUTEN-1-YNYL TRIETHYL LEAD see BPE750

BUTERAZINE see DQU400

BUTETHAL see BPF500

BUTETHAL SODIUM see BPF750

BUTETHAMINE HYDROCHLORIDE see IAC000

BUTETHANOL see TBN000

BUTFORMIN see BRA625

BUTHALITAL SODIUM see SFG700

BUTHALITONE SODIUM see SFG700

BUTHIDAZOLE see RCA300

l-BUTHIONINE-(S,R)-SULFOXIMINE see BPE800

l-BUTHIONINE SULFOXIMINE see BPE800

BUTIBATOL see BOV825

BUTIBUFEN see IJH000

BUTICAPS see BPF000

BUTIDIONA see BRF500

BUTIFOS see BSH250

n-BUTILAMINA (ITALIAN) see BPX750

BUTILATE see EID500

BUTILCHLOROFOS see BPG000

BUTILCHLOROFOS see DDP000

o-(4-terz.-BUTIL-2-CLORO-FENIL)-o-METIL-FOSFORAMMIDE (ITALIAN) see COD850

BUTILE (ACETATI di) (ITALIAN) see BPU750

BUTILENE see HNI500

BUTIL METACRILATO (ITALIAN) see MHU750

BUTILOPAN see IJH000

2-BUTIN HEXAMETHYL-DEWAR-BENZOL (GERMAN) see HEC500

BUTINOX see BLL750

BUTIPHOS see BSH250

BUTIROSIN A see BSX325

BUTISAN see CDS275

BUTISANE see CDS275

BUTISERPAZIDE-25 see RDK000

BUTISERPAZIDE-50 see RDK000

BUTISERPINE see RDK000

BUTISOL see BPF000

BUTISOL SODIUM see BPF250

BUTISULFINA see BSM000

BUTOBARBITAL see BPF500

BUTOBARBITAL SODIUM see BPF750

BUTOBARBITONE SODIUM see BPF750

BUTOBARBITONE see BPF500

BUTOBARBITURAL see BPF500

BUTOBEN see BSC000

BUTOBEN see DTC800

BUTOBENDINE DIHYDROCHLORIDE see CNW125

BUTOCARBOXIM (GERMAN) see MPU250

BUTOCIDE see PIX250

BUTOCTAMIDE HYDROGEN SUCCINATE see BPF825

BUTOCTAMIDE SEMISUCCINATE see BPF825

BUTOFLIN see DAF300

BUTOKSYETYLOWY ALKOHOL (POLISH) see BPJ850

BUTOLAN see PFR325

BUTOLEN see PFR325

BUTONATE see BPG000

BUTONE see BRF500

BUTOPHEN see BPG250

BUTOPYRONOXYL see BRT000

BUTORPHANOL TARTRATE see BPG325

2-BUTOSSI-ETANOLO (ITALIAN) see BPJ850

BUTOX see DAF300

BUTOXICARBOXIM (GERMAN) see SOB500

BUTOXIDE see PIX250

BUTOXON see DGA000

BUTOXONE see DGA000

BUTOXONE AMINE see DGA000

BUTOXONE ESTER see DGA000

BUTOXONE SB see EAK500

BUTOXY ACETYLENE see BPG500

2-BUTOXY-AETHANOL (GERMAN) see BPJ850

2-BUTOXY-3-AMINOBENZOIC ACID β-DIETHYLAMINOETHYL ESTER HYDROCHLORIDE see AJA500

o-BUTOXYBENZAMIDE see BPG750

2-N-BUTOXYBENZAMIDE see BPG750

2-BUTOXYBENZOESAEURE-4'-DIAETHYLAMINO-L'-METHYL-BUTYLAMID (1') HYDROCHLORID (GERMAN) see BPJ500

2-BUTOXYBENZOESAEURE-3'-DIAETHYLAMINOPROPYLAMID-(1') HYDROCHLORID (GERMAN) see BPJ750

p-BUTOXYBENZOIC ACID-3-(2-METHYLPIPERIDINO)PROPYL ESTER HYDROCHLORIDE see BPH750

1-(2-(4-BUTOXYBENZOYL)ETHYL)PIPERIDINE HYDROCHLORIDE see BPR500

(−)-8-(p-BUTOXYBENZYL)-3-α-HYDROXY-1-α-H,5-α-H-TROPANIUM BROMIDE TROPATE (ester) see BPI125

BUTOXYBENZYL HYOSCYAMINE BROMIDE see BPI125

p-BUTOXYBENZYL HYOSCYAMINIUM BROMIDE see BPI125

1-(1-(p-n-BUTOXYBENZYL)HYOSCYAMINIUM) BROMIDE see BPI125

1-BUTOXYBUTANE see BRH750
t-BUTOXYCARBONYL AZIDE see BQI250
tert-BUTOXYCARBONYL AZIDE (DOT) see BQI250
2-BUTOXYCARBONYLMETHYLENE-4-OXOTHIAZOLIDONE see BPI300
p-tert-BUTOXYCARBONYLOXYSTYRENE MONOMER see BPI400
N-(2-BUTOXY-7-CHLOROBENZO)(b)-1,5-NAPHTHYRIDIN-10-(YL)-N'-(2-CHLOROETHYL-1,3-PROPANEDIAMINE-N'-ETHYL-) see BPI500
1-(4-BUTOXY-3-CHLORO-5-METHYLPHENYL)-3-(1-PIPERIDINYL)1-PROPANONE HYDROCHLORIDE (9CI) see BPI625
4'-BUTOXY-3'-CHLORO-5'-METHYL-3-PIPERIDINO-PROPIOPHENONE HYDROCHLORIDE see BPI625
4'-BUTOXY-2'-CHLORO-2-PYRROLIDINYL ACETANILIDE HYDROCHLORIDE see BPI750
BUTOXYCINCHONINIC ACID DIETHYLETHYLENEDIAMIDE HYDROCHLORIDE see NOF500
2-BUTOXY-CYCLOPROPANECARBOXYLIC ACID BUTYL ESTER see BQM750
4'-BUTOXY-2-(DIETHYLAMINO)ACETANILIDE HYDROCHLORIDE see BPJ000
2-BUTOXY-N-(2-(DIETHYLAMINO)ETHYL)CINCHONINAMIDE see DDT200
2-BUTOXY-N-(β-DIETHYLAMINOETHYL)CINCHONINAMIDE see DDT200
2-BUTOXY-N-(2-DIETHYLAMINOETHYL)CINCHONINAMIDE HYDROCHLORIDE see NOF500
2-N-BUTOXY-N-(2-DIETHYLAMINOETHYL)CINCHONINAMIDE HYDROCHLORIDE see NOF500
2-BUTOXY-N-(2-DIETHYLAMINOETHYL)CINCHONINIC ACID AMIDE HYDROCHLORIDE see NOF500
2-BUTOXY-N-(2-(DIETHYLAMINO)ETHYL)-N-(2,6-XYLYL)CINCHONINAMIDE HYDROCHLORIDE see BPJ250
2-BUTOXY-N-((2-DIETHYLAMINO)ETHYL)-N-(2,6-XYLYL)-4-QUINOLINECARBOXAMIDE HYDROCHLORIDE see BPJ250
o-BUTOXY-N-(5-(DIETHYLAMINO)-2-PENTYL)BENZAMIDE HYDROCHLORIDE see BPJ500
o-BUTOXY-N-(3-(DIETHYLAMINO)PROPYL)BENZAMIDE HYDROCHLORIDE see BPJ750
BUTOXYDIETHYLENE GLYCOL see DJF200
BUTOXYDIGLYCOL see DJF200
BUTOXYETHANOL see BPJ850
2-BUTOXYETHANOL see BPJ850
n-BUTOXYETHANOL see BPJ850
2-BUTOXY-1-ETHANOL see BPJ850
2-sec-BUTOXYETHANOL see EJJ000
2-BUTOXYETHANOL ACETATE see BPM000
2-BUTOXYETHANOL PHOSPHATE see BPK250
2-BUTOXYETHANOL PHTHALATE (2:1) see BHK000
BUTOXYETHENE see VMZ000
2-BUTOXYETHOXY ACRYLATE see BPK500
1-BUTOXY-2-ETHOXYETHANE see BPK750
2-(2-BUTOXYETHOXY)ETHANOL see DJF200
2-(2-BUTOXYETHOXY)ETHANOL ACETATE see BQP500

2-(2-(2-BUTOXYETHOXY)ETHOXY)ETHANOL see TKL750
α-(2-(2-BUTOXYETHOXY)ETHOXY)-4,5-METHYLENEDIOXY-2-PROPYLTOLUENE see PIX250
α-(2-(2-n-BUTOXYETHOXY)-ETHOXY)-4,5-METHYLENEDIOXY-2-PROPYLTOLUENE see PIX250
5-((2-(2-BUTOXYETHOXY)ETHOXY)METHYL)-6-PROPYL-1,3-BENZODIOXOLE see PIX250
2-(2-BUTOXYETHOXY)ETHYL ACETATE see BQP500
2-(2-BUTOXY ETHOXY)ETHYL THIOCYANATE see BPL250
2-(2-(BUTOXY)ETHOXY)ETHYL THIOCYANIC ACID ESTER see BPL250
1-BUTOXY ETHOXY-2-PROPANOL see BPL500
3-(2-BUTOXYETHOXY)PROPANOL see BPL750
1-(2-BUTOXYETHOXY)-2-PROPANOL see BPL500
2-BUTOXYETHYL ACETATE see BPM000
BUTOXYETHYL-2,4-DICHLOROPHENOXYACETATE see DFY709
2-BUTOXYETHYL ESTER ACETIC ACID see BPM000
β-BUTOXYETHYL PHTHALATE see BHK000
BUTOXYETHYL 2,4,5-T see TAH900
2-BUTOXYETHYLVINYL ETHER see VMU000
(3-n-BUTOXY-2-HYDROXYPROPYL)PHENYL ETHER see BPP250
BUTOXYL see MHV750
tert-BUTOXYLITHIUM see LGY100
N-(BUTOXYMETHYL)ACRYLAMIDE see BPM660
N-(BUTOXYMETHYLAKRYLAMID see BPM660
N-BUTOXYMETHYL-2-CHLORO-2',6'-DIETHYLACETANILIDE see CFW750
N-(BUTOXYMETHYL)-2-CHLORO-N-(2,6-DIETHYLPHENYL)ACETAMIDE see CFW750
(BUTOXYMETHYL)NITROSOMETHYLAMINE see BPM500
N-(BUTOXYMETHYL)-2-PROPENAMIDE see BPM660
BUTOXYPENTACHLOROBENZENE see BPM690
2-(p-BUTOXYPHENOXY)-N-(2-(DIETHYLAMINO)ETHYL)-2,5'-DIETHOXYACETANILIDE MONOHYDROCHLORIDE see BPM750
2-(p-BUTOXYPHENOXY)-N-(2-(DIETHYLAMINO)ETHYL)-N-(2,4-DIMETHOXYPHENYL)ACETAMIDE HYDROCHLORIDE see BPN000
2-(p-BUTOXYPHENOXY)-N-(2-(DIETHYLAMINO)ETHYL)-N-(2,5-DIMETHOXYPHENYL)ACETAMIDE HYDROCHLORIDE see BPN250
2-(p-BUTOXYPHENOXY)-N-(2(DIMETHYLAMINO)ETHYL)-N-(2,6-DIMETHYLPHENYL)ACETAMIDE HYDROCHLORIDE see BPO250
1-BUTOXY-3-PHENOXY-2-PROPANOL see BPP250
3-n-BUTOXY-1-PHENOXY-2-PROPANOL see BPP250
BUTOXYPHENYL see BSF750
4-BUTOXYPHENYLACETOHYDROXAMIC ACID see BPP750
p-BUTOXYPHENYLACETOHYDROXAMIC ACID see BPP750

4-N-BUTOXYPHENYLACETOHYDROXAMIC ACID-o-ACETATE ESTER see ACE250
4-N-BUTOXYPHENYLACETOHYDROXAMIC ACID-o-FORMATE ESTER see BPQ250
4-N-BUTOXYPHENYLACETOHYDROXAMIC ACID-o-PROPIONATE ESTER see BPQ000
N-(p-BUTOXYPHENYL ACETYL)-o-FORMYLHYDROXYLAMINE see BPQ250
4-(p-BUTOXYPHENYL)SEMICARBAZONE 1-METHYL-1H-PYRROLE-2-CARBOXALDEHYDE see BPQ300
4-(p-BUTOXYPHENYL)SEMICARBAZONE-1H-PYRROLE-2-CARBOXALDEHYDE see BPQ330
4'-BUTOXY-2-PIPERIDINOACETANILIDE HYDROCHLORIDE see BPR000
4-BUTOXY-3-(PIPERIDINO)PROPIOPHENONE HYDROCHLORIDE see BPR250
4'-BUTOXY-3-PIPERIDINO PROPIOPHENONE HYDROCHLORIDE see BPR500
4-n-BUTOXY-β-(1-PIPERIDYL)PROPIOPHENONE HYDROCHLORIDE see BPR500
BUTOXYPOLYPROPYLENE GLYCOL see BRP250
BUTOXYPROPANEDIOL POLYMER see BRP250
3-BUTOXYPROPANENITRILE see BPT000
3-BUTOXY PROPANOIC ACID see BPS000
BUTOXYPROPANOL see PML260
1-BUTOXY-2-PROPANOL see BPS250
3-BUTOXY-1-PROPANOL see BPS500
BUTOXYPROPANOL (mixed isomers) see BPS750
n-BUTOXYPROPANOL (mixed isomers) see BPS750
3-BUTOXYPROPIONIC ACID see BPS000
3-BUTOXYPROPIONITRILE see BPT000
4'-BUTOXY-2-PYRROLIDINYLACETANILIDE HYDROCHLORIDE see BPT250
2-BUTOXYQUINOLINE-4-CARBOXYLIC ACID DIETHYLAMINOETHYLAMIDE see DDT200
BUTOXYRHODANODIETHYL ETHER see BPL250
1-BUTOXY-2-(2-THIOCYANATOETHYXY)ETHANE see BPL250
2-BUTOXY-2'-THIOCYANODIETHYL ETHER see BPL250
β-BUTOXY-β'-THIOCYANODIETHYL ETHER see BPL250
1-BUTOXY-2-(2-THIOCYANOETHOXY)ETHANE see BPL250
N-(1-BUTOXY-2,2,2-TRICHLOROETHYL)-2-HYDROXYBENZAMIDE see BPT300
N-(l-BUTOXY-2,2,2-TRICHLOROETHYL)SALICYLAMIDE see BPT300
(RS)-N-(1-BUTOXY-2,2,2-TRICHLOROETHYL)SALICYLAMIDE see BPT300
BUTOXYTRIETHYLENE GLYCOL see TKL750
BUTOXYTRIGLYCOL see TKL750
1-BUTOXY-2-(VINYLOXY)ETHANE see VMU000
BUTOZ see BRF500
BUTRALIN see BQN600
BUTRALINE see BQN600
BUTRATE see BPF000
BUTRIPTYLINE see BPT500
BUTRIPTYLINE HYDROCHLORIDE see BPT750
BUTRIZOL see BPU000

BUTROPIPAZON see FLL000
BUTROPIPAZONE see FLL000
BUTROPIUM BROMIDE see BPI125
BUTTER of ANTIMONY see AQC500
BUTTER of ANTIMONY see AQD000
BUTTER CRESS see FBS100
BUTTERCUP see FBS100
BUTTERCUP YELLOW see CMK500
BUTTERCUP YELLOW see PLW500
BUTTERCUP YELLOW see ZFJ100
BUTTER DAISY see FBS100
BUTTERSAEURE (GERMAN) see BSW000
BUTTER STARTER DISTILLATE see SLJ700
BUTTER YELLOW see AIC250
BUTTER YELLOW see DOT300
BUTTER of ZINC see ZFA000
BUTURON see CKF750
BUTVAR see PKI000
N-BUTYLACETANILIDE see BPU500
BUTYLACETAT (GERMAN) see BPU750
BUTYL ACETATE see BPU750
1-BUTYL ACETATE see BPU750
2-BUTYL ACETATE see BPV000
n-BUTYL ACETATE see BPU750
sec-BUTYL ACETATE see BPV000
sec-BUTYL ACETATE see BPV000
tert-BUTYL ACETATE see BPV100
BUTYLACETATEN (DUTCH) see BPU750
BUTYLACETIC ACID see HEU000
BUTYL ACETOACETATE see BPV250
N-BUTYL-N-(1-
ACETOXYBUTYL)NITROSAMINE see
BPV325
N-BUTYL-N-(1-
ACETOXYBUTYL)NITROSAMINE see
BRX000
BUTYL ACETOXYMETHYLNITROSAMINE
see BRX500
N-BUTYL-N-
(ACETOXYMETHYL)NITROSAMINE see
BRX500
sec-BUTYL ACETOXYMETHYL
NITROSAMINE see BPV500
N-sec-BUTYL-N-
(ACETOXYMETHYL)NITROSOAMINE see
BPV500
n-BUTYL-3,o-ACETYL-12-β-13-α-
DIHYDROJERVINE see BPW000
n-BUTYL ACID PHOSPHATE see ADF250
2-BUTYLACROLEIN see BPW025
3-BUTYLACROLEIN see HBI770
β-BUTYLACROLEIN see HBI770
N-tert-BUTYLACRYLAMIDE see BPW050
BUTYL ACRYLATE see BPW100
n-BUTYL ACRYLATE see BPW100
BUTYL ACRYLATE-1,1-
DICHLOROETHYLENE COPOLYMER see
VPK333
BUTYLACRYLATE, INHIBITED (DOT) see
BPW100
BUTYL ACRYLATE-VINYLIDINE
CHLORIDE COPOLYMER see VPK333
BUTYL ACRYLATE-VINYLIDINE
CHLORIDE POLYMER see VPK333
tert-BUTYL-1-ADAMANTANE
PEROXYCARBOXYLATE see BPW250
BUTYL ADIPATE see AEO750
BUTYLAETHYLMALONSAEURE-AETHYL-
DIAETHYLAMINOAETHYL-DI-ESTER
(GERMAN) see BRJ125
2-BUTYL ALCOHOL see BPW750
n-BUTYL ALCOHOL see BPW500
BUTYL ALCOHOL (DOT) see BPW500
sec-BUTYL ALCOHOL see BPW750
tert-BUTYL ALCOHOL see BPX000
sec-BUTYL ALCOHOL ACETATE see
BPV000
BUTYL ALCOHOL HYDROGEN
PHOSPHITE see DEG800
tert-BUTYL ALCOHOL, LITHIUM SALT see
LGY100
n-BUTYL ALDEHYDE see BSU250

sec-BUTYL ALLYL BARBITURIC ACID see
AFY500
BUTYLALYLONAL see BOR000
n-BUTYL AMIDO SULFURYL AZIDE see
BPX500
n-BUTYLAMIN (GERMAN) see BPX750
n-BUTYLAMINE see BPX750
(+)-2-BUTYLAMINE see BPY100
S-2-BUTYLAMINE see BPY100
sec-BUTYLAMINE see BPY100
sec-BUTYLAMINE, (S)- see BPY100
tert-BUTYLAMINE see BPY250
BUTYLAMINE, tertiary see BPY250
BUTYLAMINE, N,N-DIMETHYL- see
BRB300
BUTYLAMINE, N-ETHYL- see EHA050
tert-BUTYLAMINE, HYDROCHLORIDE see
MGJ800
BUTYLAMINE, N-METHYL-4-(o-
BENZYLPHENOXY)-, HYDROCHLORIDE
see MGE200
BUTYLAMINE OLEATE see OHU100
BUTYLAMINE (OSHA) see BPX750
2-(BUTYLAMINO)-p-ACETOPHENETIDIDE
HYDROCHLORIDE see BPY500
3-(tert-BUTYLAMINO)ACETYLINDOLE
HYDROCHLORIDE HYDRATE see BPY625
3-((tert-BUTYLAMINO)ACETYL)INDOLE
HYDROCHLORIDE HYDRATE see BPY625
2-tert-BUTYLAMINO-4-AETHYLAMINO-6-
CHLOR-1,3,5-TRIAZIN (GERMAN) see
BQB000
BUTYL-p-AMINOBENZOATE see BPZ000
p-(BUTYLAMINO)BENZOIC ACID-2-
(DIETHYLAMINO)ETHYL ESTER
MONOHYDROCHLORIDE see BQA020
p-(BUTYLAMINO)BENZOIC ACID-2-
(DIMETHYLAMINO)ETHYL ESTER see
BQA010
p-(BUTYLAMINO)BENZOIC ACID, 2-
(DIMETHYLAMINO)ETHYL ESTER,
HYDROCHLORIDE see TBN000
p-BUTYLAMINOBENZOYL-2-
DIMETHYLAMINOETHANOL see BQA010
p-BUTYLAMINOBENZOYL-2-
DIMETHYLAMINOETHANOL
HYDROCHLORIDE see TBN000
4'-N-BUTYL-4-AMINOBIPHENYL see
BQA100
N-((BUTYLAMINO)CARBONYL)-4-
METHYLBENZENESULFONAMIDE see
BSQ000
2-(BUTYLAMINO)-2'-
CHLOROACETANILIDE
HYDROCHLORIDE see BQA500
2-(BUTYLAMINO)-6'-CHLORO-o-
ACETOTOLUIDIDE
MONOHYDROCHLORIDE see BQA750
2-tert-BUTYLAMINO-4-CHLORO-6-
ETHYLAMINO-s-TRIAZINE see BQB000
2-(BUTYLAMINO)-6'-CHLORO-o-
HEXANOTOLUIDIDE HYDROCHLORIDE
see CAB500
1-(tert-BUTYLAMINO)-3-(2-CHLORO-5-
METHYLPHENOXY)-2-PROPANOL
HYDROCHLORIDE see BQB250
2-(BUTYLAMINO)-N-(2-CHLORO-6-
METHYLPHENYL)ACETAMIDE
HYDROCHLORIDE see BQA750
(±)-α-tert-BUTYLAMINO-3-
CHLOROPROPIOPHENONE
HYDROCHLORIDE see WBJ500
1-
(BUTYLAMINO)CYCLOHEXYLPHOSPHON
IC ACID DIBUTYL ESTER see ALZ000
(−)-1-(tert-BUTYLAMINO)-3-(o-
CYCLOPENTYLPHENOXY)-2-PROPANOL
SULFATE see PAP230
5-BUTYLAMINO-2-(2-
DIETHYLAMINOETHYL)-1H-ISOINDOLE-
1,3(2H)-DIONE HYDROCHLORIDE see
BQB825

4-BUTYLAMINO-N-(2-
(DIETHYLAMINO)ETHYL)PHTHALIMIDE
HYDROCHLORIDE see BQB825
2-BUTYLAMINOETHANOL see BQC000
2-sec-BUTYLAMINO-4-ETHYLAMINO-6-
METHOXY-s-TRIAZINE see BQC250
2-tert-BUTYLAMINO-4-ETHYLAMINO-6-
METHOXY-s-TRIAZINE see BQC500
2-sec-BUTYLAMINO-4-ETHYLAMINO-6-
METHOXY-1,3,5-TRIAZINE see BQC250
2-tert-BUTYLAMINO-4-ETHYLAMINO-6-
METHOXY-1,3,5-TRIAZINE see BQC500
2-tert-BUTYLAMINO-4-ETHYLAMINO-6-
METHYLMERCAPTO-s-TRIAZINE see
BQC750
2-tert-BUTYLAMINO-4-ETHYLAMINO-6-
METHYLTHIO-s-TRIAZINE see BQC750
2-tert-BUTYLAMINO-1-(7-ETHYL-2-
BENZOFURANYL)ETHANOL
HYDROCHLORIDE see BQD000
3-(2-(tert-BUTYLAMINO)ETHYL)-6-
HYDROXYBENZYL ALCOHOL SULFATE
(2:1) see BQD125
tert-BUTYL AMINO ETHYL
METHACRYLATE see BQD250
2-(tert-BUTYLAMINO)ETHYL
METHACRYLATE see BQD250
(5-(2-(tert-BUTYLAMINO)-1-
HYDROXYETHYL)-2-
HYDROXYPHENYL)UREA
HYDROCHLORIDE see BQD500
6-(2-(tert-BUTYLAMINO)-1-
HYDROXYETHYL)-3-HYDROXY-2-
PYRIDINEMETHANOL
DIHYDROCHLORIDE see POM800
4-(2-(tert-BUTYLAMINO)-1-
HYDROXYETHYL)-o-PHENYLENE DI-p-
TOLUATE MESILATE see BLV125
2-(tert-BUTYLAMINO)-1-(4-HYDROXY-3-
HYDROXYMETHYLPHENYL)ETHANOL
see BQF500
2-BUTYLAMINO-1-p-
HYDROXYPHENYLETHANOL see BQF250
o-(3-tert-BUTYLAMINO-2-
HYDROXYPROPOXY)BENZONITRILE
HYDROCHLORIDE see BON400
5-(3-tert-BUTYLAMINO-2-
HYDROXY)PROPOXY-3,4-
DIHYDROCARBOSTYRIL
HYDROCHLORIDE see MQT550
5-(3-tert-BUTYLAMINO-2-HYDROXY)-
PROPOXY)-3,4-DIHYDRO-(2(1H)-
CHINOLINON-HYDROCHLORID
(GERMAN) see MQT550
8-(3-tert-BUTYLAMINO-2-
HYDROXY)PROPOXY-5-
METHYLCOUMARIN HYDROCHLORIDE
see BOM510
(±)-2-(3'-tert-BUTYLAMINO-2'-
HYDROXYPROPYLTHIO)-4-(5'-
CARBAMOYL-2'-THIENYL)THIAZOLE
HYDROCHLORIDE see BQE000
2-(tert-BUTYLAMINO)-1-(3-INDOLYL)-1-
PROPANONE HYDROCHLORIDE
HYDRATE see BQG850
2-(tert-BUTYLAMINO)-1-(3-INDOLYL)-1-
PROPANONE MONOHYDROCHLORIDE,
MONOHYDRATE see BQG850
α-(tert-BUTYLAMINO)METHYL-2-
CHLOROBENZYL ALCOHOL
HYDROCHLORIDE see BQE250
α-((tert-BUTYLAMINO)METHYL)-o-
CHLOROBENZYL ALCOHOL
HYDROCHLORIDE see BQE250
α-((BUTYLAMINO)METHYL)-3,5-
DIHYDROXYBENZYL ALCOHOL,
SULFATE (2:1) see TAN250
α-((BUTYLAMINO)METHYL)-4-
HYDROXYBENZENEMETHANOL see
BQF250
α-((BUTYLAMINO)METHYL)-p-
HYDROXYBENZYL ALCOHOL see BQF250

α-((BUTYLAMINO)METHYL)-p-
HYDROXYBENZYL ALCOHOL SULFATE
see BOV825
α'-((tert-BUTYL AMINO)METHYL)-4-
HYDROXY-m-XYLENE-α,α'-DIOL see
BQF500
α-1-((tert-BUTYLAMINO)METHYL)-4-
HYDROXY-m-XYLENE-α,α-DIOL see
BQF500
1-(tert-BUTYLAMINO)3-(3-METHYL-2-
NITROPHENOXY)-2-PROPANOL see
BQF750
1-(BUTYLAMINO)-3-((4-
METHYLPHENYL)AMINO)-2-PROPANOL
(9CI) see BQH800
2-(BUTYLAMINO)-2-METHYL-1-
PROPANOL BENZOATE
HYDROCHLORIDE see BQF825
2-(BUTYLAMINO)-2-METHYL-1-
PROPANOL BENZOATE (ester)
HYDROCHLORIDE see BQF825
2-(BUTYLAMINO)-N-METHYL-N-(1-(2,6-
XYLYLOXY)-2-PROPYL) ACETAMIDE
HYDROCHLORIDE see BQG250
2-(sec-BUTYLAMINO)-N-METHYL-N-(1-(2,4-
XYLYLOXY)-2-PROPYL)ACETAMIDE
HYDROCHLORIDE see BQG500
(−)-1-(tert-BUTYLAMINO)-3-((4-
MORPHOLINO-1,2,5-THIADIAZOL-3-
YL)OXY)-2-PROPANOL MALEATE see
TGB185
1-BUTYLAMINO-3-(NAPHTHYLOXY)-2-
PROPANOL see BQG600
4-(BUTYLAMINO)PHENOL see BQG650
2-(BUTYLAMINO)-N-(1-PHENOXY-2-
PROPYL)ACETAMIDE HYDROCHLORIDE
see BQG750
3-(BUTYLAMINO)-4-PHENOXY-5-
SULFAMOYLBENZOIC ACID see BON325
3-(tert-BUTYLAMINO)PROPIONYLINDOLE
HYDROCHLORIDE HYDRATE see BQG850
4-(BUTYLAMINO)SALICYLIC ACID 2-
(DIETHYLAMINO)ETHYL ESTER
HYDROCHLORIDE see BQH250
p-BUTYLAMINO SALICYLIC ACID-2-
(DIETHYLAMINO)ETHYL ESTER
HYDROCHLORIDE see BQH250
4-(BUTYLAMINO)-SALICYLIC ACID 2-
(DIETHYLAMINO)ETHYL ESTER
MONOHYDROCHLORIDE see BQH250
p-BUTYLAMINO SALICYLIC ACID-2-
(DIMETHYLAMINO)ETHYL ESTER
HYDROCHLORIDE see BQH500
p-BUTYLAMINO SALICYLIC ACID-1-
ETHYL-4-PIPERIDYL ESTER
HYDROCHLORIDE see BQH750
1-(tert-BUTYLAMINO)-3-((5,6,7,8-
TETRAHYDRO-cis-6,7-DIHYDROXY-1-
NAPHTHYL)OXY)-2-PROPANOL see
CNR675
1-(tert-BUTYLAMINO)-3-(o-
((TETRAHYDROFURFURYL)OXY)PHENOX
Y)-2-PROPANOL HYDROCHLORIDE see
AER500
1-(BUTYLAMINO)-3-p-TOLUIDINO-2-
PROPANOL see BQH800
tert-BUTYLAMMONIUM CHLORIDE see
MGJ800
BUTYLAMMONIUM OLEATE see OHU100
BUTYLAMYLNITROSAMIN (GERMAN) see
BRY250
N-BUTYLANILINE see BQH850
N-(n-BUTYL)ANILINE see BQH850
p-n-BUTYLANILINE see AJA550
N-n-BUTYLANILINE (DOT) see BQH850
BUTYLATE see EID500
BUTYLATED HYDROXYANISOLE see
BQI000
3-tert-BUTYLATED HYDROXYANISOLE see
BQI010
BUTYLATED HYDROXYMETHYLPHENOL
see BQI050

BUTYLATED HYDROXYTOLUENE see
BFW750
BUTYLATED MELAMINE,
TOLUENESULFONAMIDE,
FORMALDEHYDE see MCB075
BUTYLATE-2,4,5-T see BSQ750
N-BUTYL-N-2-AZIDOETHYLNITRAMINE
see BQI125
tert-BUTYL AZIDOFORMATE see BQI250
(E)-1-T-BUTYLAZO-1-
HYDROXYCYCLOPENTANE see BQI270
2-t-BUTYLAZO-2-HYDROXY-5-
METHYLHEXANE see BQI300
2-tert-BUTYLAZO-2-HYDROXYPROPANE
see BQI400
8-BUTYLBENZ(a)ANTHRACENE see BQI500
5-n-BUTYL-1,2-BENZANTHRACENE see
BQI500
4-BUTYLBENZENAMINE see AJA550
N-BUTYLBENZENAMINE (9CI) see BQH850
n-BUTYLBENZENE see BQI750
sec-BUTYLBENZENE see BQJ000
tert-BUTYLBENZENE see BQJ250
BUTYLBENZENEACETATE see BQJ350
α-BUTYLBENZENEMETHANOL see BQJ500
N-BUTYLBENZENESULFONAMIDE see
BQJ650
2-tert-BUTYLBENZIMIDAZOLE see BQJ750
5-BUTYL-2-BENZIMIDAZOLECARBAMIC
ACID METHYL ESTER see BQK000
N-(BUTYL-5-BENZIMIDAZOLYL)-2-
CARBAMATE de METHYLE (FRENCH) see
BQK000
(4-BUTYL-1H-BENZIMIDAZOL-2-YL)-
CARBAMIC ACID METHYL ESTER see
BQK000
BUTYL BENZOATE see BQK250
n-BUTYL BENZOATE see BQK250
2-BUTYL-3-BENZOFURANYL p-((2-
DIETHYLAMINO)ETHOXY)-m,m-
DIIODOPHENYL KETONE see AJK750
p-tert-BUTYL BENZOIC ACID see BQK500
N-tert-BUTYL-2-
BENZOTHIAZOLESULFENAMIDE see
BQK750
α-BUTYLBENZYL ALCOHOL see BQJ500
1-(p-tert-BUTYLBENZYL)-4-(p-
CHLORODIPHENYLMETHYL)PIPERAZIN
E DIHYDROCHLORIDE see BOM250
N-(4-T-BUTYLBENZYL)-4-CHLORO-3-
ETHYL-1-METHYLPYRAZOLE-5-
CARBOXAMIDE see CGH750
1-(p-tert-BUTYLBENZYL-4-p-CHLORO-α-
PHENYLBENZYL)PIPERAZINE
DIHYDROCHLORIDE see BOM250
1-(p-tert-BUTYLBENZYL)-4-(3-(p-
CHLOROPHENYL)-3-
PHENYLPROPIONYL)PIPERAZINE see
BQK800
BUTYL BENZYL PHTHALATE see BEC500
n-BUTYL BENZYL PHTHALATE see BEC500
4-(p-tert-BUTYLBENZYL)PIPERAZINYL β-
(p-CHLOROPHENYL) KETONE see BQK800
4-(p-tert-BUTYLBENZYL)PIPERAZINYL
3,4,5-TRIMETHOXYPHENYL KETONE see
BQK830
1-(p-tert-BUTYLBENZYL)-4-(3,4,5-
TRIMETHOXYBENZOYL)PIPERAZINE see
BQK830
tert-BUTYL-BICYCLOPHOSPHATE see
BQK850
BUTYLBIGUANIDE see BRA625
1-BUTYLBIGUANIDE HYDROCHLORIDE
see BQL000
N-BUTYLBIGUANIDE HYDROCHLORIDE
see BQL000
N-BUTYL-BIS(2-CHLOROETHYLAMINE)
HYDROCHLORIDE see BIB250
sec-BUTYLBIS(2-CHLOROETHYL)AMINE
HYDROCHLORIDE see BQL500
tert-BUTYLBIS(2-CHLOROETHYL)AMINE
HYDROCHLORIDE see BQL750

BUTYLBIS(β-CHLOROETHYL)AMINE
HYDROCHLORIDE see BIB250
sec-BUTYL-BIS(β-CHLOROETHYL)AMINE
HYDROCHLORIDE see BQL500
tert-BUTYLBIS(β-CHLOROETHYL)AMINE
HYDROCHLORIDE see BQL750
N-BUTYL-N,N-BIS(HYDROXY
ETHYL)AMINE see BQM000
4-(BUTYLBIS((1-
OXODODECYL)OXY)STANNYL)-2-
BUTANOL see BRN600
BUTYL BORATE see THX750
n-BUTYL BORATE see THX750
sec-BUTYL-BROM-ALLYL BARBITURIC
ACID SODIUM SALT see BOR250
1-BUTYL BROMIDE see BNR750
BUTYL BROMIDE (DOT) see BMX500
iso-BUTYL BROMIDE see BMX750
sec-BUTYL BROMIDE see BMX750
n-BUTYL BROMIDE (DOT) see BMX500
tert-BUTYL BROMIDE see BQM250
3-sek.BUTYL-5-BROM-6-METHYLURACIL
(GERMAN) see BMM650
5-sec-BUTYL-5-(β-
BROMOALLYL)BARBITURIC ACID see
BOR000
tert-BUTYL BROMOMETHYL KETONE see
PJB100
N-BUTYL-1-BUTANAMINE see DDT800
N-tert-BUTYL-1,4-BUTANEDIAMINE
DIHYDROCHLORIDE see BQM309
BUTYL BUTANEDIOATE see SNA500
n-BUTYL n-BUTANOATE see BQM500
BUTYL-BUTANOL(4)-NITROSAMIN see
HJQ350
BUTYL-BUTANOL-NITROSAMINE see
HJQ350
BUTYL-2-BUTOXYCYCLOPROPANE-1-
CARBOXYLATE see BQM750
p-(N-BUTYL-2-
(BUTYLAMINO)ACETAMIDO)BENZOIC
ACID BUTYL ESTER HYDROCHLORIDE
see BQN250
N-BUTYL-2-(BUTYLAMINO)-2',6'-
PROPIONOXYLIDIDE HYDROCHLORIDE
see BQN500
N-sec-BUTYL-4-tert-BUTYL-2,6-
DINITROANILINE see BQN600
S-sec-BUTYL S-tert-BUTYL o-ETHYL
PHOSPHORODITHIOATE see ENF050
n-BUTYL BUTYRATE see BQM500
n-BUTYL n-BUTYRATE see BQM500
BUTYL BUTYRATE (FCC) see BQM500
BUTYL BUTYROLACTATE see BQP000
γ-n-BUTYL-γ-BUTYROLACTONE see
OCE000
N-n-BUTYL-N-4-(1,4-
BUTYROLACTONE)NITROSAMINE see
NJO200
BUTYL BUTYRYL LACTATE see BQP000
BUTYL CAPROATE see BRK900
BUTYLCAPTAX see BSN325
BUTYL CARBAMATE see BQP250
tert-BUTYLCARBAMIC ACID ESTER with 3-
(m-HYDROXYPHENYL)-1,1-
DIMETHYLUREA see DUM800
1-(BUTYLCARBAMOYL)-2-
BENZIMIDAZOLECARBAMIC ACID,
METHYL ESTER see BAV575
1-(BUTYLCARBAMOYL)-2-
BENZIMIDAZOLECARBAMIC ACID
METHYL ESTER and SODIUM NITRITE
(1:6) see BAV500
1-(BUTYLCARBAMOYL)-2-
BENZIMIDAZOL-METHYLCARBAMAT
(GERMAN) see BAV575
1-(N-BUTYLCARBAMOYL)-2-(METHOXY-
CARBOXAMIDO)-BENZIMIDAZOL
(GERMAN) see BAV575
N'-(BUTYLCARBAMOYL)SULFANILAMIDE
see BSM000

N¹-(BUTYLCARBAMOYL)SULFANILAMIDE see BSM000

N-BUTYLCARBINOL see AOE000

dl-sec-BUTYLCARBINOL see MHS750

BUTYL CARBITOL see DJF200

BUTYL CARBITOL ACETATE see BQP500

BUTYLCARBITOL FORMAL see BHK750

BUTYL CARBITOL 6-PROPYLPIPERONYL ETHER see PIX250

BUTYL CARBITOL RHODANATE see BPL250

BUTYL CARBITOL THIOCYANATE see BPL250

BUTYL-CARBITYL (6-PROPYLPIPERONYL) ETHER see PIX250

BUTYL CARBOBUTOXYMETHYL PHTHALATE see BQP750

5-BUTYL-2-(CARBOMETHOXYAMINO)BENZIMIDAZOLE see BQK000

1-((tert-BUTYLCARBONYL-4-CHLOROPHENOXY)METHYL)-1H-1,2,4-TRIAZOLE see CJO250

N-BUTYL-N-(2-CARBOXYETHYL)NITROSAMINE see BRX250

4-BUTYL-4-(β-CARBOXYPROPIONYL-OXYMETHYL)-1,2-DIPHENYL-3,5-PYRAZOLIDINEDIONE see SOX875

N-BUTYL-(3-CARBOXY PROPYL)NITROSAMINE see BQQ250

4-tert-BUTYLCATECHOL see BSK000

BUTYL CELLOSOLVE see BPJ850

BUTYL CELLOSOLVE ACETATE see BPM000

BUTYL CELLOSOLVE ACRYLATE see BPK500

BUTYL "CELLOSOLVE" ADIPATE see BHJ750

BUTYL "CELLOSOLVE" PHTHALATE see BHK000

BUTYL CHEMOSEPT see BSC000

o-(4-tert BUTYL-2-CHLOOR-FENYL)-o-METHYL-FOSFORZUUR-N-METHYL-AMIDE (DUTCH) see COD850

BUTYL-CHLORHYDRINETHER (CZECH) see BQT500

n-BUTYL CHLORIDE see BQQ750

BUTYL CHLORIDE (DOT) see BQQ750

sec-BUTYL CHLORIDE see CEU250

tert-BUTYL CHLORIDE see BQR000

3-tert-BUTYL-5-CHLOR-6-METHYLURACIL (GERMAN) see BQT750

tert-BUTYL CHLOROACETATE see BQR100

α-BUTYL-p-CHLOROBENZYL ESTER of SUCCINIC ACID see CKI000

β-sec-BUTYL-3-CHLORO-N,N-DIMETHYL-4-ETHOXYPHENETHYLAMINE see BQR250

β-sec-BUTYL-3-CHLORO-N,N-DIMETHYL-4-METHOXYPHENETHYLAMINE see BQR750

β-sec-BUTYL-5-CHLORO-N,N-DIMETHYL-2-METHOXYPHENETHYLAMINE see BQS000

β-sec-BUTYL-p-CHLORO-N,N-DIMETHYLPHENETHYLAMINE see BQS250

tert-BUTYLCHLORODIMETHYLSILANE see BQS300

β-sec-BUTYL-5-CHLORO-2-ETHOXY-N,N-DIISOPROPYLPHENETHYLAMINE see BQT000

1-(β-sec-BUTYL-5-CHLORO-2-ETHOXYPHENETHYL)PIPERIDINE see BQT250

BUTYL (3-CHLORO-2-HYDROXYPROPYL) ETHER see BQT500

t-BUTYL-CHLORO-2-METHYL-CYCLOHEXANECARBOXYLATE see BQT600

tert-BUTYL CHLOROMETHYL KETONE see MRG070

3-tert-BUTYL-5-CHLORO-6-METHYLURACIL see BQT750

tert-BUTYL CHLOROPEROXYFORMATE see BQU000

4-tert-BUTYL-2-CHLORO PHENYL METHYL METHYL PHOSPHORAMIDATE see COD850

4-tert.-BUTYL 2-CHLOROPHENYL METHYLPHOSPHORAMIDATE de METHYLE (FRENCH) see COD850

o-(4-tert-BUTYL-2-CHLOROPHENYL)-o-METHYL PHOSPHORAMIDOTHIONATE see BQU500

BUTYL(4-CHLOROPHENYL)METHYL 3-PYRIDINYLCARBONIMIDODITHIOATE see BQU600

α-BUTYL-α(4-CHLOROPHENYL)-1H-1,2,4-THIAZOLE-1-PROPANENITRILE see MRW775

o-(4-tert-BUTYL-2-CHLOR-PHENYL)-o-METHYL-PHOSPHORSAEURE-N-METHYL AMID (GERMAN) see COD850

3-tert-BUTYLCHOLANTHRENE see BQU750

tert-20-BUTYLCHOLANTHRENE see BQU750

tert-BUTYL CHROMATE see BQV000

α-BUTYLCINNAMALDEHYDE see BQV250

n-BUTYL CINNAMATE see BQV500

BUTYL CINNAMIC ALDEHYDE see BQV250

α-BUTYLCINNAMIC ALDEHYDE see BQV250

BUTYL CITRATE see THY100

2-tert-BUTYL-p-CRESOL see BQV750

6-tert-BUTYL-m-CRESOL see BQV600

4-tert-BUTYLCYCLOHEXANOL see BQW000

2-tert-BUTYLCYCLOHEXANOL ACETATE see BQW490

2-sec-BUTYLCYCLOHEXANONE see MOU800

p-tert-BUTYLCYCLOHEXANONE see BQW250

2-tert-BUTYLCYCLOHEXYL ACETATE see BQW490

4-tert-BUTYLCYCLOHEXYL ACETATE see BQW500

p-tert-BUTYLCYCLOHEXYL ACETATE see BQW500

N-BUTYL CYCLOHEXYL AMINE see BQW750

5-BUTYL-1-CYCLOHEXYLBARBITURIC ACID see BQW825

N-(4-tert-BUTYL CYCLOHEXYL)-3,3-DIPHENYL PROPYLAMINE HYDROCHLORIDE see BQX000

5-BUTYL-1-CYCLOHEXYL-2,4,6(1H,3H,5H)-PYRIMIDINETRIONE see BQW825

5-n-BUTYL-1-CYCLOHEXYL-2,4,6-TRIOXOPERHYDROPYRIMIDINE see BQW825

BUTYL 2,4-D see BQZ000

BUTYL DECYL PHTHALATE see BQX250

N-tert-BUTYL-1,4-DIAMINOBUTANE DIHYDROCHLORIDE see BQM309

tert-BUTYL DIAZOACETATE see BQX750

14-n-BUTYL DIBENZ(a,h)ACRIDINE see BQY000

10-n-BUTYL-1,2,5,6-DIBENZACRIDINE (FRENCH) see BQY000

n-BUTYL-2-DIBUTYLTHIOUREA see BQY250

sec-BUTYLDICHLORARSINE see BQY300

tert-BUTYLDICHLOROAMINE see BQY275

sec-BUTYLDICHLOROARSINE see BQY300

BUTYLDICHLOROBORANE see BQY000

N-BUTYL-2,2'-DICHLORODIETHYLAMINE see DEV300

N-sec-BUTYL-2,2'-DICHLORODIETHYLAMINE, HYDROCHLORIDE see BQL500

N-tert-BUTYL-2,2'-DICHLORO-DIETHYLAMINE HYDROCHLORIDE see BQL750

2-tert-BUTYL-4-(2,4-DICHLORO-5-ISOPROPYLOXYPHENYL)-1,3,4-OXADIAZOLIN-5-ONE see OMM200

BUTYL DICHLOROPHENOXYACETATE see BQZ000

BUTYL (2,4-DICHLOROPHENOXY)ACETATE see BQZ000

BUTYL (3,4-DICHLOROPHENYL)METHYL 3-PYRIDINYLCARBONIMIDODITHIOATE see BQZ100

o-BUTYL S-((3,4-DICHLOROPHENYL)METHYL) 3-PYRIDINYLCARBONIMIDOTHIOATE see PPK600

1-BUTYL-3-(3,4-DICHLOROPHENYL)-1-METHYLUREA see BRA250

N-BUTYL-N'-(3,4-DICHLOROPHENYL)-N-METHYLUREA see BRA250

α-BUTYL-α-(2,4-DICHLOROPHENYL)-1H-1,2,4-TRIAZOLE-1-ETHANOL (+-)- see HCN050

5-tert-BUTYL-3-(2,4-DICHLORO-5-PROPARGYLOXYPHENYL)-1,3,4-OXADIAZOL-2(3H)-ONE see BRA300

N-BUTYLDIETHANOLAMINE see BQM000

2-(BUTYL(2-(DIETHYLAMINO)ETHYL)AMINO)-6'-CHLORO-o-ACETOTOLUIDIDE HYDROCHLORIDE see BRA500

o-BUTYL DIETHYLENE GLYCOL see DJF200

n-BUTYLDIETHYLTIN IODIDE see BRA550

tert-BUTYLDIFLUOROPHOSPHINE see BRA600

BUTYL DIGLYME see DDW200

BUTYLDIGUANIDE see BRA625

1-BUTYLDIGUANIDE see BRA625

1-BUTYLDIGUANIDE HYDROCHLORIDE see BQL000

p-tert-BUTYLDIHYDROCINNAMALDEHYDE see BMK300

BUTYL 2,3-DIHYDRO-2,2-DIMETHYLBENZOFURAN-7-YL N,N-DIMETHYL-N,N-THIODICARBAMATE see FPO200

BUTYL-3,4-DIHYDRO-2,2-DIMETHYL-4-OXO-2H-PYRAN-6-CARBOXYLATE see BRT000

n-BUTYL-12-β-13-α-DIHYDROJERVINE-3-ACETATE see BPW000

9-BUTYL-1,9-DIHYDRO-6H-PURINE-6-THIONE see BRS500

N-BUTYL-N-(2,4-DIHYDROXYBUTYL)NITROSAMINE see BRB000

2-BUTYL-3-(3,5-DIIODO-4-(2-DIETHYLAMINOETHOXY)BENZOYL)BENZOFURAN see AJK750

2-N-BUTYL-3',5'-DIIODO-4'-N-DIETHYLAMINOETHOXY-3-BENZOYLBENZOFURAN see AJK750

1-tert-BUTYL-3-(2,6-DI-ISOPROPYL-4-PHENOXYPHENYL)THIOUREA see BRB100

BUTYLDIMETHYL(2-(3-ACETYL-2-METHYL-1-INDOLIZINYL)-2-METHYLETHYL)AMMONIUMIODIDE see BRB200

BUTYLDIMETHYLAMINE see BRB300

4'-n-BUTYL-4-DIMETHYLAMINOAZOBENZENE see BRB450

4'-tert-BUTYL-4-DIMETHYLAMINOAZOBENZENE see BRB460

3-BUTYL-1-(2-(DIMETHYLAMINO)ETHOXY)ISOQUINOLINE HYDROCHLORIDE see DNX400

4-(1-sec-BUTYL-2-(DIMETHYLAMINO)ETHYL)PHENOL see BRB500

2-(1-sec-BUTYL-2-(DIMETHYLAMINO)ETHYL)QUINOLINE see BRB750

2-(1-sec-BUTYL-2-(DIMETHYLAMINO)ETHYL)QUINOXALINE see BRC000

2-(1-sec-BUTYL-2-(DIMETHYLAMINO)ETHYL)THIOPHENE see BRC250

5-n-BUTYL-2-DIMETHYLAMINO-4-HYDROXY-6-METHYLPYRIMIDINE see BRD000

6-BUTYL-5-DIMETHYLAMINO-5H-INDENO(5,6-d)-1,3-DIOXOLE HYDROCHLORIDE see BRC500

2-n-BUTYL-3-DIMETHYLAMINO-5,6-METHYLENEDIOXYINDENE HYDROCHLORIDE see BRC500

BUTYL-3-((DIMETHYLAMINO)METHYL)-4-HYDROXYBENZOATE see BRC750

5-BUTYL-2-(DIMETHYLAMINO)-6-METHYL-4-PYRIMIDINOL see BRD000

5-BUTYL-2-(DIMETHYLAMINO)-6-METHYL-4(1H)-PYRIMIDINONE see BRD000

2-(4-tert-BUTYL-2,6-DIMETHYLBENZYL)-2-IMIDAZOLINE HYDROCHLORIDE see OKO500

2-(4-tert-BUTYL-2,6-DIMETHYLBENZYL)-2-IMIDAZOLINE MONOHYDROCHLORIDE see OKO500

tert-BUTYLDIMETHYLCHLOROSILANE see BQS300

8-tert-BUTYL-4,6-DIMETHYLCOUMARIN see DQV000

6-BUTYL-2,4-DIMETHYLDIHYDROPYRANE see GMG100

N-(7-BUTYL-4,9-DIMETHYL-2,6-DIOXO-8-HYDROXY-1,5-DIOXONAN-3-YL)-3-FORMAMIDOSALICYLAMIDE see DAM000

β-sec-BUTYL-N,N-DIMETHYL-2-ETHOXY-5-FLUOROPHENETHYLAMINE see BRD500

o-BUTYL S-((4-(1,1-DIMETHYLETHYL)PHENYL)METHYL)-3-PYRIDINYLCARBONIMIDOTHIOATE see BRD600

β-sec-BUTYL-N,N-DIMETHYL-5-FLUORO-2-METHOXYPHENETHYLAMINE see BRD750

2-(4-tert-BUTYL-2,6-DIMETHYL-3-HYDROXYBENZYL)-2-IMIDAZOLINIUM CHLORIDE see AEX000

1-BUTYL-3,3-DIMETHYL-1-NITROSOUREA see BRE000

β-sec-BUTYL-N,N-DIMETHYLPHENETHYLAMINE see BRE250

β-sec-BUTYL-N,N-DIMETHYLPHENETHYLAMINE HYDROCHLORIDE see BRE255

6-tert-BUTYL-2,4-DIMETHYLPHENOL see BST000

N-BUTYL-2,6-DIMETHYL-1-PIPERIDINECARBOXAMIDE see BRE300

tert-BUTYLDIMETHYLSILYL CHLORIDE see BQS300

2-sek.BUTYL-4,6-DINITROFENYLESTER KYSELINY OCTOVE (CZECH) see ACE500

2-sec-BUTYL-4,6-DINITROPHENOL see BRE500

o-tert-BUTYL-4,6-DINITROPHENOL see DRV200

2-sec-BUTYL-4,6-DINITROPHENOL ACETATE (ester) see ACE500

2-sec-BUTYL-4,6-DINITROPHENOL AMMONIUM SALT see BPG250

2-sec-BUTYL-4,5-DINITROPHENOL ISOPROPYL CARBONATE see CBW000

2-sec-BUTYL-4,6-DINITROPHENOL- 2,2',2''-NITRILOTRIETHANOL SALT see BRE750

o-sec-BUTYL-4,6-DINITROPHENOLTRIETHANOLAMINE SALT see BRE750

2-sec-BUTYL-4,6-DINITROPHENYLACETATE see ACE500

6-sec-BUTYL-2,4-DINITROPHENYLACETATE see ACE500

2-tert-BUTYL-4,6-DINITROPHENYL ACETATE see DVJ400

2-sec-BUTYL-4,6-DINITROPHENYL-3,3-DIMETHYLACRYLATE see BGB500

2-sec-BUTYL-4,6-DINITROPHENYL ISOPROPYL CARBONATE see CBW000

2-sec-BUTYL-4,6-DINITROPHENYL-3-METHYL-2-BUTENOATE see BGB500

2-sec-BUTYL-4,6-DINITROPHENYL-3-METHYLCROTONATE see BGB500

2-sec-BUTYL-4,5-DINITROPHENYL SENECIOATE see BGB500

BUTYL DIOXITOL see DJF200

4-BUTYL-1,2-DIPHENYL-3,5-DIOXO PYRAZOLIDINE see BRF500

4-BUTYL-1,2-DIPHENYLPYRAZOLIDINE-3,5-DIONE see BRF500

4-BUTYL-1,1-DIPHENYL-3,5-PYRAZOLIDINEDIONE with 4-(DIMETHYLAMINO)-1,2-DIHYDRO-1,5-DIMETHYL-2-PHENYL-3H-PYRAZOL-3-ONE see IGI000

4-BUTYL-1,2-DIPHENYL-3,5-PYRAZOLIDINEDIONE SODIUM SALT see BOV750

BUTYL DISELENIDE see BRF550

6-BUTYLDODECAHYDRO-7,14-METHANO-2H,6H-DIPYRIDO(1,2-a:1',2'-e)(1,5)DIAZOCINE see BSL450

BUTYLE (ACETATE de) (FRENCH) see BPU750

BUTYLENE see BOW250

α-BUTYLENE see BOW250

β-BUTYLENE see BOW255

γ-BUTYLENE see IIC000

1,2-BUTYLENE CARBONATE see BOT200

BUTYLENE DIACRYLATE see TDQ100

1,3-BUTYLENE DIACRYLATE see BRG500

1,4-BUTYLENE DIACRYLATE see TDQ100

BUTYLENEDIAMINE see BOS000

1,4-BUTYLENEDIAMINE see BOS000

2-BUTYLENE DICHLORIDE see BRG600

1,3-BUTYLENE DIMETHACRYLATE see BRG100

1,2-BUTYLENE GLYCOL see BOS250

1,4-BUTYLENE GLYCOL see BOS750

2,3-BUTYLENE GLYCOL see BOT000

β-BUTYLENE GLYCOL see BOS500

BUTYLENE GLYCOL BIS(MERCAPTOACETATE) see BKM000

1,3-BUTYLENE GLYCOL DIACRYLATE see BRG500

1,4-BUTYLENE GLYCOL DIACRYLATE see TDQ100

1,3-BUTYLENE GLYCOL (FCC) see BOS500

BUTYLENE HYDRATE see BPW750

BUTYLENE OXIDE see BOX750

BUTYLENE OXIDE see TCR750

1,2-BUTYLENE OXIDE see BOX750

2,3-BUTYLENE OXIDE see EBJ100

cis-2-BUTYLENE OXIDE see EBJ200

β-BUTYLENE OXIDE see EBJ100

1,2-BUTYLENE OXIDE, stabilized (DOT) see BOX750

1,4-BUTYLENE SULFONE see BOU250

1,2-BUTYLENIMINE see EGN500

BUTYLENIN see IIU000

BUTYL 2,3-EPOXYPROPYL FUMARATE see BRG700

BUTYL-9,10-EPOXYSTEARATE see BRH250

2,4,5-T-N-BUTYL ESTER see BSQ750

2,4,5-T,n-BUTYL ESTER mixed with 2,4-d,n-BUTYL ESTER see AEX750

2,4-d,n-BUTYL ESTER mixed with 2,4,5-T,n-BUTYL ESTER (1:1) see AEX750

BUTYL ESTER 2,4-D see BQZ000

n-BUTYL ESTER of 3,4-DIHYDRO-2,2-DIMETHYL-4-OXO-2H-PYRAN-6-CARBOXYLIC ACID see BRT000

N-BUTYLESTER KYSELINI-2,4,5-TRICHLORFENOXYOCTOVE (CZECH) see BSQ750

BUTYLESTER KYSELINY MRAVENCI see BRK000

terc.BUTYLESTER KYSELINY PEROXYBENZOOVE (CZECH) see BSC500

BUTYL ETHANOATE see BPU750

n-BUTYL ETHER see BRH750

BUTYL ETHER (DOT) see BRH750

sec-BUTYL ETHER see BRH760

sec-BUTYL ETHER see OPE030

BUTYL ETHYL ACETALDEHYDE see BRI000

BUTYL ETHYL ACETIC ACID see BRI250

BUTYLETHYLAMINE see EHA050

n-BUTYL-2-(ETHYLAMINO)-2',6'-ACETOXYLIDIDE HYDROCHLORIDE see BRI500

5-n-BUTYL-2-ETHYLAMINO-4-HYDROXY-6-METHYL-PYRIMIDINE see BRI750

5-BUTYL-2-(ETHYLAMINO)-6-METHYL-4(1H)-PYRIMIDINONE see BRI750

5-BUTYL-2-ETHYLAMINO-6-METHYLPYRIMIDIN-4-YL DIMETHYLSULPHAMATE see BRJ000

5-BUTYL-5-ETHYLBARBITURIC ACID see BPF500

5-sec-BUTYL-5-ETHYLBARBITURIC ACID see BPF000

5-sec-BUTYL-5-ETHYLBARBITURIC ACID SODIUM SALT see BPF250

BUTYL ETHYLENE see HFB000

o-BUTYL ETHYLENE GLYCOL see BPJ850

tert-BUTYL ETHYL ETHER see EHA550

n-BUTYL ETHYL KETONE see EHA600

BUTYLETHYLMALONIC ACID-2-(DIETHYLAMINO)ETHYL ETHYL ESTER see BRJ125

5-sec-BUTYL-5-ETHYLMALONYL UREA see BPF000

5-sec-BUTYL-5-ETHYL-1-METHYLBARBITURIC ACID see BRJ250

BUTYLETHYL-PROPANEDIOIC ACID-2-(DIETHYLAMINO)ETHYL ETHYL ESTER (9CI) see BRJ125

2-BUTYL-2-ETHYL-1,3-PROPANEDIOL see BKH625

5-BUTYL-5-ETHYL-2,4,6(1H,3H,5H)-PYRIMIDINETRIONE (9CI) see BPF500

5-BUTYL-5-ETHYL-2,4,6(1H,3H,5H)-PYRIMIDINETRIONE MONOSODIUM SALT (9CI) see BPF750

BUTYLETHYLTHIOCARBAMIC ACID S-PROPYL ESTER see PNF500

2-sec.-BUTYLFENOL (CZECH) see BSE000

p-tert-BUTYLFENOL (CZECH) see BSE500

2-sek.BUTYLFENYLESTER KYSELINY METHYLKARBAMINOVE (CZECH) see MOV000

p-terc.BUTYLFENYLESTER KYSELINY SALICYLOVE see BSH100

BUTYL FLUFENAMATE see BRJ325

BUTYL FORMAL see VAG000

tert-BUTYL FORMAMIDE see BRJ750

n-BUTYL FORMATE see BRK000

BUTYL FORMATE (DOT) see BRK000

N-n-BUTYL-N-FORMYLHYDRAZINE see BRK100

N-sek.BUTYLFTALIMID see BSH600

n-BUTYL-N'-(2-FUROYL)HYDRAZINE see BRK250

1-BUTYL-3-(2-FUROYL)UREA see BRK250

BUTYL GLYCIDYL ETHER see BRK750

n-BUTYL GLYCIDYL ETHER see BRK750

t-BUTYL GLYCIDYL ETHER see BRK800

BUTYL GLYCOL see BPJ850

BUTYLGLYCOL (FRENCH, GERMAN) see BPJ850

BUTYL GLYCOL PHTHALATE see BHK000

4-tert-BUTYLHEXAHYDROPHENYL ACETATE see BQW500

N-BUTYLHEXAMETHYLENEDIAMINE see BRK830

N-BUTYL-1,6-HEXANEDIAMINE see BRK830

BUTYL HEXANOATE see BRK900

n-BUTYL HEXANOATE see BRK900

BUTYLHYDRAZINE ETHANEDIOATE see BRL750

n-BUTYLHYDRAZINE HYDROCHLORIDE see BRL500

BUTYLHYDRAZINE OXALATE see BRL750

O,O-tert-BUTYL HYDROGEN MONOPEROXY MALEATE see BRM000

BUTYL HYDROGEN PHTHALATE see MRF525

terc.BUTYLHYDROPEROXID (CZECH) see BRM250

tert-BUTYLHYDROPEROXIDE see BRM250

N-BUTYL-N-(1-HYDROPEROXYBUTYL)NITROSAMINE see HIE600

tert-BUTYLHYDROQUINONE see BRM500

BUTYL HYDROXIDE see BPW500

tert-BUTYL HYDROXIDE see BPX000

6-BUTYL-4-HYDROXYAMINOQUINOLINE-1-OXIDE see BRM750

BUTYLHYDROXYANISOLE see BQI000

tert-BUTYLHYDROXYANISOLE see BQI000

tert-BUTYL-4-HYDROXYANISOLE see BQI000

3-tert-BUTYL-4-HYDROXYANISOLE see BRN000

2(3)-tert-BUTYL-4-HYDROXYANISOLE see BQI000

BUTYL-o-HYDROXYBENZOATE see BSL250

BUTYL-p-HYDROXYBENZOATE see BSC000

BUTYL p-HYDROXYBENZOATE see DTC800

n-BUTYL-o-HYDROXYBENZOATE see BSL250

α-n-BUTYL-β-HYDROXY-Δ^{α,β}-BUTENOLID (GERMAN) see BRO250

n-BUTYL-(4-HYDROXYBUTYL)NITROSAMINE see HJQ350

n-BUTYL-N-(2-HYDROXYBUTYL)NITROSAMINE see BRN250

n-BUTYL-N-(3-HYDROXYBUTYL)NITROSAMINE see BRN500

N-BUTYL-N-(4-HYDROXYBUTYL)NITROSAMINE see HJQ350

BUTYL(3-HYDROXYBUTYL)TIN DILAURATE see BRN600

N-(7-BUTYL-8-HYDROXY-4,9-DIMETHYL-2,6-DIOXO-1,5-DIOXONAN-3-YL)-3-FORMAMIDOSALICYLAMIDE see DAM000

BUTYL(2-HYDROXYETHYL)NITROSOAMINE see BRO000

3-BUTYL-4-HYDROXY-2(5H)FURANONE see BRO250

BUTYLHYDROXYISOPROPYLARSINE OXIDE see BRQ800

n-BUTYL-N-(2-HYDROXYL-3-CARBOXYPROPYL)NITROSAMINE see BRO500

4-BUTYL-4-HYDROXYMETHYL-1,2-DIPHENYL-3,5-PYRAZOLIDINEDIONE HYDROGEN SUCCINATE see SOX875

2-(tert-BUTYL)-2-(HYDROXYMETHYL)-1,3-PROPANEDIOL, CYCLIC PHOSPHITE (1:1) see BRO750

BUTYLHYDROXYOXOSTANNANE see BSL500

4-BUTYL-2-(4-HYDROXYPHENYL)-1-PHENYL-3,5-DIOXOPYRAZOLIDINE see HNI500

4-BUTYL-1-(4-HYDROXYPHENYL)-2-PHENYL-3,5-PYRAZOLIDINEDIONE see HNI500

4-BUTYL-1-(p-HYDROXYPHENYL)-2-PHENYL-3,5-PYRAZOLIDINEDIONE see HNI500

4-BUTYL-2-(p-HYDROXYPHENYL)-1-PHENYL-3,5-PYRAZOLIDINEDIONE see HNI500

α-BUTYL-ω-HYDROXYPOLY(OXY(METHYL-1,2-ETHANEDIYL)) see BRP250

BUTYL α-HYDROXYPROPIONATE see BRR600

BUTYL-(3-HYDROXYPROPYL)NITROSAMINE see BRX750

BUTYLHYDROXYTOLUENE see BFW750

2-(tert-BUTYL)-2-(HYEROXYMETHYL)-1,3-PROPANEDIOL, CYCLIC PHOSPHATE (1:1) see BQK850

BUTYLHYOSCINE see SBG500

N-BUTYLHYOSCINE BROMIDE see SBG500

N-BUTYLHYOSCINIUM BROMIDE see SBG500

tert-BUTYL HYPOCHLORITE see BRP500

4,4'-BUTYLIDENEBIS(6-tert-BUTYL-m-CRESOL) see BRP750

4,4'-BUTYLIDENEBIS(6-tert-BUTYL-3-METHYLPHENYL) see BRP750

4,4'-BUTYLIDENEBIS(3-METHYL-6-tert-BUTYLPHENOL) see BRP750

(11-β,16-α)-16,17-(BUTYLIDENEBIS(OXY))-11,21-DIHYDROXYPREGNA-1,4-DIENE-3,20-DIONE see BOM520

6,6'-BUTYLIDENEBIS(2,4-XYLENOL) see BRQ000

BUTYLIDENE CHLORIDE see BRQ050

16-α,17-α-BUTYLIDENEDIOXY-11-β,21-DIHYDROXY-1,4-PREGNADIENE-3,20-DIONE see BOM520

BUTYLIDENE PHTHALIDE see BRQ100

3-BUTYLIDENE PHTHALIDE see BRQ100

n-BUTYLIDENE PHTHALIDE see BRQ100

6-t-BUTYL-3-(2-IMIDAZOLIN-2-YLMETHYL)-2,4-DIMETHYLPHENOL see ORA100

6-tert-BUTYL-3-(2-IMIDAZOLIN-2-YLMETHYL)-2,4-DIMETHYLPHENOL HYDROCHLORIDE see AEX000

N-BUTYLIMIDODICARBONIMIDIC DIAMIDE MONOHYDROCHLORIDE (9CI) see BQL000

N-BUTYL-2,2'-IMINODIETHANOL see BQM000

2-tert-BUTYLIMINO-3-ISOPROPYL-5-PHENYLPERHYDRO-1,3,5-THIADIAZINAN-4-ONE see BOO631

2-tert-BUTYLIMINO-3-ISOPROPYL-5-PHENYL-3,4,5,6-TETRAHYDRO-2H-1,3,5-THIADIAZIN-4-ONE see BOO631

2-tert-BUTYLIMINO-3-ISOPROPYL-5-PHENYL-1,3,5-THIADIAZINAN-4-ONE see BOO631

n-BUTYL IODIDE see BRQ250

sec-BUTYL IODIDE see IEH000

tert-BUTYL IODIDE see TLU000

3-(1-BUTYL-3-ISOBUTYL-3-PYRROLIDINYL)PHENOL CITRATE see BRQ300

BUTYL ISOBUTYRATE see BRQ350

n-BUTYL ISOCYANATE see BRQ500

tert-BUTYL ISOCYANIDE see BRQ750

tert-BUTYLISONITRILE see BRQ750

tert-BUTYLISOPENTANAL see ILJ100

n-BUTYL ISOPENTANOATE see ISX000

BUTYL(ISOPROPYL)ARSINIC ACID see BRQ800

tert-BUTYL ISOPROPYL BENZENE HYDROPEROXIDE see BRR250

tert-BUTYL ISOPROPYL BENZENE HYDROPEROXIDE (DOT) see BRR250

2-sec-BUTYL-6-ISOPROPYLPHENOL see BRR500

2-((3-BUTYL-1-ISOQUINOLINYL)OXY)-N,N-DIMETHYLETHANAMINE MONOHYDROCHLORIDE see DNX400

n-BUTYL ISOTHIOCYANATE see ISD100

1-BUTYL ISOVALERATE see ISX000

n-BUTYL ISOVALERATE see ISX000

BUTYL ISOVALERIANATE see ISX000

3-(5-tert-BUTYLISOXAZOL-3-YL)-1,1-DIMETHYLUREA see ISR200

BUTYL KETONE see NMZ000

2-tert-BUTYL-p-KRESOL (CZECH) see BQV750

BUTYL LACTATE see BRR600

n-BUTYL LACTATE see BRR600

BUTYL LAEVULINATE see BRR700

n-BUTYL LAEVULINATE see BRR700

BUTYL LEVULINATE see BRR700

n-BUTYL LEVULINATE see BRR700

BUTYL LITHIUM see BRR739

tert-BUTYL LITHIUM see BRR750

BUTYLMALONIC ACID MONO(1,2-DIPHENYLHYDRAZIDE) CALCIUM SALT HEMIHYDRATE see CCD750

BUTYL-MALONSAEURE-MONO-(1,2-DIPHENYL-HYDRAZID)-CALCIUM-SEMIHYDRAT (German) see CCD750

N-BUTYLMELAMINE see BRR800

BUTYL MERCAPTAN see BRR900

n-BUTYL MERCAPTAN see BRR900

tert-BUTYL MERCAPTAN see MOS000

n-BUTYL MERCAPTAN (ACGIH,DOT) see BRR900

p-BUTYLMERCAPTOBENZHYDRYL-β-DIMETHYLAMINOETHYLSULPHIDE see BRS000

2-(BUTYLMERCAPTO)ETHYL VINYL ETHER see VNA000

BUTYLMERCAPTOMETHYLPENICILLIN see BRS250

9-BUTYL-6-MERCAPTOPURINE see BRS500

n-BUTYLMERCURIC CHLORIDE see BRS750

S-(BUTYLMERCURIC)-THIOGLYCOLIC ACID, SODIUM SALT see SFJ500

n-BUTYL MESITYL OXIDE OXALATE see BRT000

n-BUTYLMESITYLOXID OXALATE see BRT000

BUTYL MESYLATE see BRT250

BUTYLMETHACRYLAAT (DUTCH) see MHU750

BUTYL-2-METHACRYLATE see MHU750

N-BUTYL METHACRYLATE see MHU750

BUTYL METHANESULFONATE see BRT250

n-BUTYL METHANESULFONATE see BRT250

BUTYL METHOXYMETHYLNITROSAMINE see BRT750

sec-BUTYL METHOXYMETHYLNITROSAMINE see BRU000

2-tert-BUTYL-4-METHOXYPHENOL see BRN000

3-tert-BUTYL-4-METHOXYPHENOL see BQI010

BUTYL-p-METHYLBENZENESULFONATE see BSP750

n-BUTYL-α-METHYLBENZYLAMINE see BRU250

BUTYL 3-METHYLBUTYRATE see ISX000

tert-BUTYL METHYL CARBINOL see BRU300

6-tert-BUTYL-3-METHYL-2,4-DINITRO ANISOLE see BRU500

2-tert-BUTYL-5-METHYL-4,6-DINITROPHENYL ACETATE see BRU750

BUTYL METHYL ETHER (DOT) see BRU780

1-BUTYL-2-METHYL-HYDRAZINE DIHYDROCHLORIDE see MHW250
p-tert-BUTYL-α-METHYLHYDROCINNAMALDEHYDE see LFT000
p-tert-BUTYL-α-METHYLHYDROCINNAMIC ALDEHYDE see LFT000
BUTYL METHYL KETONE see HEV000
n-BUTYL METHYL KETONE see HEV000
tert-BUTYL METHYL KETONE see DQU000
2-tert-BUTYL-4-METHYLPHENOL see BQV750
2-tert-BUTYL-6-METHYLPHENOL see BRU790
4-tert-BUTYL-2-METHYLPHENOL see BRU800
1-BUTYL-3-(p-METHYLPHENYLSULFONYL)UREA see BSQ000
2-sec-BUTYL-2-METHYL-1,3-PROPANEDIOL DICARBAMATE see MBW750
BUTYL-2-METHYL-2-PROPENOATE see MHU750
BUTYL 2-METHYL-2-PROPENOATE HOMOPOLYMER see PJL500
N-BUTYL-2-METHYL-2-PROPYL-1,3-PROPANEDIOL DICARBAMATE see MOV500
N-N-BUTYL-2-METHYL-2-PROPYL-1,3-PROPANEDIOL DICARBAMATE see MOV500
3-BUTYL-5-METHYL-TETRAHYDRO-2H-PYRAN-4-YL ACETATE see MHX000
tert-BUTYL-N-(3-METHYL-2-THIAZOLIDINYLIDENE)CARBAMATE see BRV000
2-sec-BUTYL-2-METHYLTRIMETHYLENE DICARBAMATE see MBW750
BUTYLMIN see SBG500
BUTYL MONOSULFIDE see BSM125
4-BUTYLMORPHOLINE see BRV100
N-BUTYLMORPHOLINE see BRV100
N-(n-BUTYL)MORPHOLINE see BRV100
9-BUTYL-6-MP see BRS500
BUTYL MUSTARD OIL see ISD100
BUTYL MYRISTATE see MSA300
BUTYL NAMATE see SGF500
BUTYL NITRATE see BRV325
N-BUTYL-2-NITRATOETHYL NITRAMINE see BRW100
n-BUTYL NITRITE see BRV500
BUTYL NITRITE (DOT) see BRV500
sec-BUTYL NITRITE see BRV750
tert-BUTYL NITRITE see BRV760
tert-BUTYLNITROACETATE see DSV289
tert-BUTYL NITROACETYLENE see BRW000
2-(BUTYLNITROAMINO)ETHANOL NITRATE (ESTER) see BRW100
N-BUTYL-N'-NITRO-N-NITROSOGUANIDINE see NLC000
tert-BUTYL-p-NITRO PEROXY BENZOATE see BRW250
BUTYL-p-NITROPHENYL ESTER of ETHYLPHOSPHONIC ACID see BRW500
6-BUTYL-4-NITROQUINOLINE-1-OXIDE see BRW750
BUTYLNITROSAMINE see NJO000
4-(BUTYLNITROSAMINO)-1-BUTANOL see HJQ350
4-(n-BUTYLNITROSAMINO)-1-BUTANOL see HJQ350
4-(BUTYLNITROSAMINO)BUTYL ACETATE see BRX000
2-(BUTYLNITROSAMINO)ETHANOL see BRO000
4-(N-BUTYLNITROSAMINO)-4-HYDROXYBUTYRIC ACID LACTONE see NJO200
N-BUTYL-N-NITROSO-β-ALANINE see BRX250

4-(BUTYLNITROSOAMINO)-1,3-BUTANEDIOL see BRB000
4-(BUTYLNITROSOAMINO)BUTANOIC ACID see BQQ250
1-(BUTYLNITROSOAMINO)-2-BUTANOL see BRN250
4-(BUTYLNITROSOAMINO)-2-BUTANOL see BRN500
1-(BUTYLNITROSOAMINO)BUTYL ACETATE see BPV325
4-(BUTYLNITROSOAMINO)-3-HYDROXYBUTYRIC ACID see BRO500
BUTYLNITROSOAMINOMETHYL ACETATE see BRX500
3-(BUTYLNITROSOAMINO)-1-PROPANOL see BRX750
1-(BUTYLNITROSOAMINO)-2-PROPANONE see BRY000
N-BUTYL-N-NITROSO AMYL AMINE see BRY250
3-tert-BUTYL-4-NITROSOBIPHENYL see BRY300
n-BUTYL-N-NITROSO-1-BUTAMINE see BRY500
N-BUTYL-N-NITROSOBUTYRAMIDE see NJO150
N-BUTYL-N-NITROSOCARBAMIC ACID-1-NAPHTHYL ESTER see BRY750
N-BUTYL-N-NITROSO ETHYL CARBAMATE see BRZ000
BUTYLNITROSOHARNSTOFF (GERMAN) see BSA250
N-BUTYL-N-NITROSOPENTYLAMINE see BRY250
4-tert-BUTYL-1-NITROSOPIPERIDINE see BRZ200
4-tert-BUTYL-1-NITROSOPIPERIDINE see NJO300
N-BUTYL-N-NITROSOSUCCINAMIC ACID ETHYL ESTER see EHC800
2-BUTYL-3-NITROSOTHIAZOLIDINE see BSA000
n-BUTYLNITROSOUREA see BSA250
1-BUTYL-1-NITROSOUREA see BSA250
N-n-BUTYL-N-NITROSOUREA see BSA250
1-sec-BUTYL-1-NITROSOUREA see NJO500
1-BUTYL-1-NITROSOURETHAN see BRZ000
N-BUTYL-N-NITROSOURETHAN see BRZ000
n-BUTYLNORSYMPATHOL see BQF250
BUTYLNORSYMPATOL see BOV825
BUTYL-NOR-SYMPATOL see BQF250
n-BUTYLNORSYNEPHRINE see BQF250
BUTYLOCAINE see TBN000
BUTYL OCTADECANOATE see BSL600
n-BUTYL OCTADECANOATE see BSL600
2-BUTYL-1-OCTANOL see BSA500
2-BUTYLOCTYL ALCOHOL see BSA500
2-BUTYLOCTYL ESTER METHACRYLIC ACID see BSA750
BUTYLOHYDROKSYANIZOL (POLISH) see BQI000
BUTYL OLEATE see BSB000
BUTYLONE see NBU000
BUTYLOWY ALKOHOL (POLISH) see BPW500
7-BUTYL-2-OXEPANONE see BSB100
BUTYLOXIRANE see HFC100
2-BUTYLOXIRANE see HFC100
BUTYL OXITOL see BPJ850
N-BUTYL-N-(1-OXOBUTYL)NITROSAMINE see NJO150
N-BUTYL-N-(2-OXOBUTYL)NITROSAMINE see BSB500
N-BUTYL-N-(3-OXOBUTYL)NITROSAMINE see BSB750
BUTYL 4-OXOPENTANOATE see BRR700
4-tert-BUTYL-1-OXO-1-PHOSPHA-2,6,7-TRIOXABICYCLO(2.2.2)OCTANE see BQK850
BUTYL(2-OXOPROPYL)NITROSOAMINE see BRY000

tert-BUTYLOXYCARBONYL AZIDE see BQI250
2-(n-BUTYLOXYCARBONYLMETHYLENE)THIAZOLID-4-ONE see BPI300
α-BUTYLOXYCINCHONINIC ACID DIETHYLETHYLENEDIAMIDE see DDT200
N-(BUTYLOXY)METHYL-N-METHYLNITROSAMINE see BPM500
BUTYL PARABEN see BSC000
n-BUTYL PARAHYDROXYBENZOATE see BSC000
BUTYL PARASEPT see BSC000
N-BUTYL-N-PENTYLINITROSAMINE see BRY250
t-BUTYL PERACETATE see BSC250
tert-BUTYL PERACETATE see BSC250
terc.BUTYLPERBENZOAN (CZECH) see BSC500
t-BUTYL PERBENZOATE see BSC500
tert-BUTYL PERBENZOATE see BSC500
tert-BUTYL PERCAPRYLATE see BSD100
tert-BUTYL PERISOBUTYRATE see BSC600
tert-BUTYL PEROCTANOATE see BSD100
tert-BUTYL PEROCTOATE see BSD100
tert-BUTYL PEROXIDE see BSC750
t-BUTYL PEROXYACETATE see BSC250
tert-BUTYL PEROXYACETATE, >76% in solution (DOT) see BSC250
t-BUTYL PEROXY BENZOATE see BSC500
BUTYL PEROXYDICARBONATE see BSC800
sec-BUTYL PEROXYDICARBONATE see BSD000
n-BUTYL PEROXYDICARBONATE, >52% in solution (DOT) see BSC800
tert-BUTYL PEROXYISOBUTYRATE see BSC600
tert-BUTYL PEROXYISOBUTYRATE, >77% in solution (DOT) see BSC600
tert-BUTYL PEROXYOCTANOATE see BSD100
tert-BUTYL PEROXYOCTOATE see BSD100
t-BUTYL PEROXYPIVALATE see BSD250
tert-BUTYL PEROXYPIVALATE see BSD250
tert-BUTYL PERPIVALATE see BSD250
BUTYLPHEN see BSE500
4-tert-BUTYLPHENETHYLQUINAZOLIN-4-YL ETHER see FAK200
2-n-BUTYLPHENOL see BSE440
2-t-BUTYLPHENOL see BSE460
4-n-BUTYLPHENOL see BSE450
4-t-BUTYLPHENOL see BSE500
4-sec BUTYL PHENOL see BSE250
o-sec-BUTYLPHENOL see BSE000
p-sec-BUTYLPHENOL see BSE250
tert-BUTYLPHENOL GLYCIDYL ETHER see BSE600
p-tert-BUTYLPHENOL (MAK) see BSE500
p-tert-BUTYLPHENOL SODIUM SALT see BSE700
2-(p-tert-BUTYLPHENOXY)CYCLOHEXYL PROPARGYL SULFITE see SOP000
2-(p-tert-BUTYLPHENOXY)CYCLOHEXYL 2-PROPYNYL SULFITE see SOP000
3-(p-tert-BUTYLPHENOXY)-1,2-EPOXYPROPANE see BSE600
4'-(3-(4'-tert-BUTYLPHENOXY)-2-HYDROXYPROPOXY)BENZOIC ACID see BSE750
BUTYLPHENOXYISOPROPYL CHLOROETHYL SULFITE see SOP500
2-(p-BUTYLPHENOXY)ISOPROPYL 2-CHLOROETHYL SULFITE see SOP500
2-(4-tert-BUTYLPHENOXY)ISOPROPYL-2-CHLOROETHYL SULFITE see SOP500
2-(p-tert-BUTYLPHENOXY)ISOPROPYL 2'-CHLOROETHYL SULPHITE see SOP500
2-(p-tert-BUTYLPHENOXY)-1-METHYLETHYL 2-CHLOROETHYL ESTER of SULPHUROUS ACID see SOP500
2-(p-BUTYLPHENOXY)-1-METHYLETHYL 2-CHLOROETHYL SULFITE see SOP500

2-(p-tert-BUTYLPHENOXY)-1-METHYLETHYL-2-CHLOROETHYL SULFITE ESTER see SOP500
2-(p-tert-BUTYLPHENOXY)-1-METHYLETHYL 2'-CHLOROETHYL SULPHITE see SOP500
2-(p-tert-BUTYLPHENOXY)-1-METHYLETHYL SULPHITE of 2-CHLOROETHANOL see SOP500
1-(o-SEC-BUTYL PHENOXY)-2-PROPANOL see PNK500
1-(p-tert-BUTYLPHENOXY)-2-PROPANOL-2-CHLOROETHYL SULFITE see SOP500
BUTYL PHENYL ACETATE see BBA000
BUTYL PHENYLACETATE see BQJ350
n-BUTYL PHENYLACETATE see BQJ350
α-n-BUTYL-β-PHENYLACROLEIN see BQV250
n-BUTYL PHENYLACRYLATE see BQV500
p-((p-BUTYLPHENYL)AZO)-N,N-DIMETHYLANILINE see BRB450
p-((p-tert-BUTYLPHENYL)AZO)-N,N-DIMETHYLANILINE see BRB460
o-sec-BUTYLPHENYL CARBAMATE see BSF250
5-N-(p-N-BUTYLPHENYL)-2,4-DIAMINO-6,6-DIMETHYL-1,6-DIHYDRO-1,3,5-TRIAZINE see BSF275
4-BUTYL-1-PHENYL-3,5-DIOXOPYRAZOLIDINE see MQY400
p-tert-BUTYLPHENYL DIPHENYLPHOSPHATE see BSF300
BUTYL PHENYL ETHER see BSF750
S-p-tert-BUTYLPHENYL-o-ETHYL ETHYLPHOSPHONODITHIOATE see BSG000
tert-BUTYLPHENYL GLYCIDYL ETHER see BSE600
3-(o-BUTYLPHENYL)-5-(m-METHOXYPHENYL)-s-TRIAZOLE see BSG100
o-3-TERT-BUTYLPHENYL (6-METHOXY-2-PYRIDYL)(METHYL)THIOCARBAMATE see POO800
o-sec-BUTYLPHENYL METHYLCARBAMATE see MOV000
2-sec-BUTYLPHENYL N-METHYLCARBAMATE see MOV000
3-sec-BUTYLPHENYL-N-METHYLCARBAMATE see BSG250
m-sec-BUTYLPHENYL-N-METHYLCARBAMATE see BSG250
3-tert-BUTYLPHENYL N-METHYLCARBAMATE see BSG300
β-(4-tert-BUTYLPHENYL)-α-METHYLPROPIONALDEHYDE see LFT000
1-(3-(4-tert-BUTYLPHENYL)-2-METHYLPROPYL)PIPERIDINE see FAQ230
2-tert-BUTYL-3-PHENYL OXAZIRANE see BSH000
3-(o-BUTYLPHENYL)-5-PHENYL-s-TRIAZOLE see BSH075
α-sec-BUTYL-α-PHENYL-l-PIPERIDINEBUTYRONITRILE HYDROCHLORIDE see EQZ000
4-BUTYL-1-PHENYL-3,5-PYRAZOLIDINEDIONE see MQY400
p-tert-BUTYLPHENYL SALICYLATE see BSH100
BUTYL PHOSPHITE see MRF500
BUTYL PHOSPHORIC ACID see ADF250
BUTYL PHOSPHOROTRITHIOATE see BSH250
n-BUTYL PHTHALATE (DOT) see DEH200
BUTYL PHTHALATE BUTYL GLYCOLATE see BQP750
BUTYLPHTHALIDE see BSH500
3-BUTYLPHTHALIDE see BSH500
3-n-BUTYLPHTHALIDE see BSH500
N-sec-BUTYLPHTHALIMIDE see BSH600
BUTYL PHTHALYL BUTYL GLYCOLATE see BQP750

5-BUTYL PICOLINIC ACID see BSI000
5-BUTYLPICOLINIC ACID CALCIUM SALT HYDRATE see FQR100
1-BUTYL-2',6'-PIPECOLOXYLIDIDE see BSI250
1-BUTYL-2',6'-PIPECOLOXYLIDIDE (±) see BOO000
l-(−)-1-BUTYL-2',6'-PIPECOLOXYLIDIDE see BOO500
d-(+)-1-BUTYL-2',6'-PIPECOLOXYLIDIDE see BOO750
1-BUTYL-2',6'-PIPECOLOXYLIDIDE HYDROCHLORIDE (+) see BON750
(±)-1-BUTYL-2',6'-PIPECOLOXYLIDIDE MONOHYDROCHLORIDE, MONOHYDRATE see BOO000
p-(N-BUTYL-2-(PIPERIDINO)ACETAMIDO)BENZOIC ACID BUTYL ESTER HYDROCHLORIDE see BSI750
BUTYL POTASSIUM XANTHATE see PKY850
BUTYLPROPANEDIOIC ACID MONO(1,2-DIPHENYLHYDRAZIDE) CALCIUM SALT HEMIHYDRATE see CCD750
BUTYL PROPANOATE see BSJ500
BUTYL-2-PROPENOATE see BPW100
BUTYL PROPIONATE see BSJ500
n-BUTYL PROPIONATE see BSJ500
9-BUTYL-9H-PURINE-6-THIOL see BRS500
3-BUTYLPYRIDINE see BSJ550
3-n-BUTYLPYRIDINE see BSJ550
5-BUTYL-2-PYRIDINECARBOXYLIC ACID see BSI000
5-BUTYL-2-PYRIDINECARBOXYLIC ACID CALCIUM SALT HYDRATE see FQR100
BUTYL 4-PYRIDYL KETONE see VBA100
2-T-BUTYLPYRIMIDINE see DRW100
BUTYLPYRIN see BRF500
4-tert-BUTYLPYROCATECHOL see BSK000
p-tert-BUTYLPYROCATECHOL see BSK000
4-tert-BUTYLPYROKATECHIN (CZECH) see BSK000
n-BUTYLPYRROLIDINE see BSK250
N-BUTYL-α-PYRROLIDINE-CARBOXY-MESIDIDE HYDROCHLORIDE see PQB750
n-BUTYL RHODANATE see BSN500
BUTYL RUBBER see IIQ500
BUTYL SALICYLATE see BSL250
n-BUTYL SALICYLATE see BSL250
BUTYLSCOPOLAMINE BROMIDE see SBG500
N-BUTYLSCOPOLAMINE BROMIDE see SBG500
n-BUTYLSCOPOLAMINE TANNATE see BSL325
N-BUTYLSCOPOLAMINIUM BROMIDE see SBG500
BUTYLSCOPOLAMMONIUM BROMIDE see SBG500
N-BUTYLSCOPOLAMMONIUM BROMIDE see SBG500
N-BUTYLSCOPOLAMMONIUM BROMIDE combined with SODIUM SULPYRINE (1:25) see BOO650
BUTYL SELENOCYANOACETATE see BSL350
17-BUTYLSPARTEIN see BSL450
BUTYL STANNOIC ACID see BSL500
BUTYL STEARATE see BSL600
n-BUTYL STEARATE see BSL600
n-BUTYL-k-STROPHANTHIDIN see BSL750
N-BUTYLSULFANILYLUREA see BSM000
1-BUTYL-3-SULFANILYL UREA see BSM000
BUTYL SULFIDE see BSM125
n-BUTYL-SULFIDE see BSM125
1-BUTYLSULFONIMIDOCYCLOHEXAMETHYLENE see BSM250
BUTYLSYMPATHOL see BQF250
BUTYL-2,4,5-T see BSQ750
BUTYL TEGOSEPT see BSC000

BUTYL TETRADECANOATE see MSA300
BUTYL n-TETRADECANOATE see MSA300
N-BUTYL-1,2,3,6-TETRAHYDRONAPHTHALIMIDE see BSM400
1-BUTYL THEOBROMINE see BSM825
N-(5-tert-BUTYL-1,3,4-THIADIAZOL-2-YL)BENZENESULFONAMIDE see GFM200
1-(5-tert-BUTYL-1,3,4-THIADIAZOL-2-YL)-3-DIMETHYLHARNSTOFF (GERMAN) see BSN000
1-(5-(tert-BUTYL)-1,3,4-THIADIAZOL-2-YL)-1,3-DIMETHYLUREA see BSN000
1-(5-(tert-BUTYL-1,3,4-THIADIAZOL-2-YL)-4-HYDROXY-1-METHYL-2-IMIDAZOLIDIN ONE see RCA300
2-BUTYLTHIOBENZOTHIAZOLE see BSN325
BUTYLTHIOBUTANE see BSM125
n-BUTYL THIOCYANATE see BSN500
5-((1-(BUTYLTHIO)ETHYL)-5-ETHYLBARBITURIC ACID SODIUM SALT see SFJ875
2-(BUTYLTHIO)ETHYL VINYL ETHER see VNA000
S-((tert-BUTYLTHIO)METHYL)-O,O-DIETHYLPHOSPHORODITHIOATE see BSO000
n-BUTYLTHIOMETHYLPENICILLIN see BRS250
4-tert-BUTYLTHIOPHENOL see BSO100
2-((p-(BUTYLTHIO)-α-PHENYLBENZYL)THIO)-N,N-DIMETHYLETHYLAMINE see BRS000
(BUTYLTHIO)TRIOCTYLSTANNANE see BSO200
(BUTYLTHIO)TRIPROPYLSTANNANE see TMY850
n-BUTYL THIOUREA see BSO500
n-BUTYLTIN TRICHLORIDE see BSO750
BUTYLTIN TRI(DODECANOATE) see BSO750
BUTYLTIN TRILAURATE see BSO750
n-BUTYLTIN TRIS(DIBUTYLDITHIOCARBAMATE) see BSP000
BUTYL TITANATE see BSP250
p-tert-BUTYLTOLUENE see BSP500
BUTYL-p-TOLUENESULFONATE see BSP750
n-BUTYL-p-TOLUENESULFONATE see BSP750
n-BUTYL-N'-p-TOLUENESULFONYLUREA see BSQ000
1-BUTYL-3-(p-TOLYL SULFONYL)UREA see BSQ000
1-BUTYL-3-(p-TOLYLSULFONYL)UREA, SODIUM SALT see BSQ250
BUTYL TOSYLATE see BSP750
1-BUTYL-3-TOSYLUREA see BSQ000
N-n-BUTYL-N'-TOSYLUREA see BSQ000
4-BUTYL-s-TRIAZOLE see BPU000
4-N-BUTYL-4H-1,2,4-TRIAZOLE see BPU000
BUTYLTRICHLOROGERMANE see BSQ500
BUTYL-2,4,5-TRICHLOROPHENOXYACETATE see BSQ750
N-BUTYL (2,4,5-TRICHLOROPHENOXY)ACETATE see BSQ750
BUTYLTRICHLOROSILANE see BSR000
BUTYL TRICHLORO STANNANE see BSR250
3-tert-BUTYLTRICYCLOQUINAZOLINE see BSR500
BUTYL-2-((3-(TRIFLUOROMETHYL)PHENYL)AMINO)BENZOATE see BRJ325
BUTYL-o-((m-(TRIFLUOROMETHYL)PHENYL)AMINO)BENZOATE see BRJ325

BUTYL 2-(4-(5-TRIFLUOROMETHYL-2-PYRIDINYLOXY)PHENOXY)PROPANOATE see FDA885
BUTYLTRI(LAUROYLOXY)STANNANE see BSO750
5-tert-BUTYL-1,2,3-TRIMETHYL-4,6-DINITROBENZENE see MRW272
2-BUTYL-4,4,6-TRIMETHYL-1,3-DIOXANE see BSR600
tert-BUTYL TRIMETHYLPEROXYACETATE see BSD250
1-BUTYL-N-(2,4,6-TRIMETHYLPHENYL)-2-PYRROLIDINECARBOXAMIDE MONOHYDROCHLORIDE see PQB750
N-tert-BUTYL-N-TRIMETHYLSILYLAMINOBORANE see BSR825
5-tert-BUTYL-2,4,6-TRINITROXYLENE see TML750
5-tert-BUTYL-2,4,6-TRINITRO-m-XYLENE (DOT) see TML750
4-(tert-BUTYL)-2,6,7-TRIOXA-1-PHOSPHABICYCLO(2.2.2)OCTANE see BRO750
4-(tert-BUTYL)-2,6,7-TRIOXA-1-PHOSPHABICYCLO(2.2.2)OCTAN-1-ONE see BQK850
n-BUTYLTRIPHENYLPHOSPHONIUM BROMIDE see BSR900
BUTYLTRIPROPYLAMMONIUM IODIDE see BSR930
BUTYLTRIS(DIBUTYLDITHIOCARBAMATO)STANNANE see BSP000
BUTYLTRIS(2-ETHYLHEXYLOXYCARBONYLMETHYLTHIO)STANNANE see BSS000
BUTYLTRIS(ISOOCTYLOXYCARBONYLMETHYLTHIO)STANNANE see BSS000
BUTYL 10-UNDECENOATE see BSS100
BUTYL UNDECYLENATE see BSS100
N-BUTYLUREA see BSS250
sec-BUTYLUREA see BSS300
tert-BUTYLUREA see BSS310
1-BUTYLUREA and SODIUM NITRITE (2:1) see BSS500
1-BUTYLURETHAN see EHA100
BUTYLURETHANE see EHA100
1-BUTYLURETHANE see EHA100
N-BUTYLURETHANE see EHA100
BUTYL VINYL ETHER see VMZ000
BUTYL VINYL ETHER (inhibited) see VMZ000
BUTYL-XANTHIC ACID POTASSIUM SALT see PKY850
sec-BUTYLXANTHIC ACID SODIUM SALT see DXM000
BUTYLXANTHIC DISULFIDE see BSS550
5-tert-BUTYL-m-XYLENE see DQU800
6-tert-BUTYL-2,4-XYLENOL see BST000
BUTYL ZIMATE see BIX000
BUTYL ZINC PHOSPHORODITHIOATE (ZN((BUO)2(S)PS)2) see ZGA500
BUTYL ZIRAM see BIX000
BUTYN see BOO750
3-BUTYN-2-AMINE, 2-METHYL- see MHX200
1-BUTYNE see EFS500
2-BUTYNE see COC500
2-BUTYNE, 1-CHLORO-4-MERCAPTO-, S-ESTER WITH DIETHYL PHOSPHOROTHIOATE see CEV830
2-BUTYNE, 1-CHLORO-4-MERCAPTO-, S-ESTER WITH DIPHENYLPHOSPHINOTHIOATE see CEV840
2-BUTYNE, 1-CHLORO-4-MERCAPTO-, S-ESTER WITH ETHYL PHENYLPHOSPHONOTHIOATE see CEV850
2-BUTYNEDIAMIDE see ACJ250
2-BUTYNEDINITRILE see DGS000
2-BUTYNE-1,4-DIOL see BST500

1,4-BUTYNEDIOL (DOT) see BST500
2-BUTYNE-1-THIOL see BST750
BUTYNOIC ACID, 3-PHENYL-2-PROPENYL ESTER see CMQ800
1-BUTYN-3-OL see EQM600
BUTYN-1-OL, 4-(DIMETHYLAMINO)-, ACETATE see DOV825
BUTYN-1-OL-3-ESTER of m-CHLOROPHENYLCARBAMIC ACID see CEX250
1-BUTYN-3-OL, 3-METHYL- see MHX250
3-BUTYN-2-OL, 2-PHENYL- see EQN230
BUTYNORATE see DDV600
BUTYN SULFATE see BOP000
3-BUTYNYL-m-CHLOROCARBANILATE see CEX250
2-BUTYNYL-4-CHLORO-m-CHLOROCARBANILATE see CEW500
1-BUTYN-3-YL-m-CHLOROPHENYLCARBAMATE see CEX250
BUTYNYL-3N-3-CHLOROPHENYLCARBAMATE mixed with 3-CYCLOOCTYL-1,1-DIMETHYL UREA see AFM375
2-BUTYNYLENEDIAMINE, N,N,N'N'-TETRAETHYL- see BJA250
1,1'-(2-BUTYNYLENEDIOXY)BIS(3-CHLORO)-2-PROPANOL) see BST900
1,1'-(2-BUTYNYLENE)DIPYRROLIDINE see DWX600
1,1'-(2-BUTYNYLENE)DIPYRROLIDINE DIHYDROCHLORIDE see THL575
17-α-(1-BUTYNYL)-17-β-HYDROXYESTR-4-EN-3-ONE see EKF550
3-BUTYN-1-YL-p-TOLUENE SULFONATE see BSU000
2-BUTYOXY-N-(2-(DIETHYLAMINO)ETHYL)-4-QUINOLINECARBOXAMIDE see DDT200
BUTYRAC see DGA000
BUTYRAC see EAK500
BUTYRAC ESTER see DGA000
BUTYRAL see BSU250
BUTYRALDEHYD (GERMAN) see BSU250
n-BUTYRALDEHYDE see BSU250
BUTYRALDEHYDE (CZECH) see BSU250
BUTYRALDEHYDE, 3-ETHOXY-, DIETHYL ACETAL see TJL700
BUTYRALDEHYDE, 3-(ETHYLTHIO)- see EPN600
BUTYRALDEHYDE, 4-FLUORO- see FGY100
BUTYRALDEHYDE, 2-METHYLENE- see EFS700
n-BUTYRALDEHYDE OXIME see BSU500
N-BUTYRALDOXIME see BSU500
BUTYRALDOXIME (DOT) see BSU500
BUTYRAMIDE, 4-(4-(p-CHLOROPHENYL)-5-FLUORO-2-HYDROXYBENZYLIDENEAMINO)- see CKA125
BUTYRAMIDE, N,N-DIMETHYL- see DQV300
BUTYRAMIDE, 2-PHENYL- see NMV300
BUTYRAMIDE, α-PHENYL- see NMV300
3-BUTYRAMIDO-α-ETHYL-2,4,6-TRIIODOCINNAMIC ACID SODIUM SALT see EQC000
3-BUTYRAMIDO-α-ETHYL-2,4,6-TRIIODOHYDROCINNAMIC ACID SODIUM SALT see SKO500
5'-BUTYRAMIDO-2'-(2-HYDROXY-3-ISOPROPYLAMINOPROPOXY)ACETOPHENONE see AAE100
2-(3-BUTYRAMIDO-2,4,6-TRIIODOPHENYLMETHYLENE)BUTYRIC ACID SODIUM SALT see EQC000
2-(3-BUTYRAMIDO-2,4,6-TRIIODOPHENYL)PROPIONIC ACID see BSV250
BUTYRANHYDRID see BSW550
n-BUTYRANILIDE see BSV500

BUTYRANILIDE, 4'-CHLORO-2'-METHYL-3-OXO- see AAY300
BUTYRANILIDE, 4'-ETHOXY-3-HYDROXY- see HJS850
BUTYRANILIDE, 3'-HYDROXY- see HJS450
BUTYRATE SODIUM see SFN600
(BUTYRATO)PHENYLMERCURY see BSV750
BUTYRHODANID (GERMAN) see BSN500
n-BUTYRIC ACID see BSW000
BUTYRIC ACID, 4-ACETAMIDO- see AAJ350
BUTYRIC ACID, 2-AMINO-4-MERCAPTO-, dl- (9CI) see HGI300
BUTYRIC ACID, 2-AMINO-4-(METHYLSELENYL)-, dl- see SBU800
BUTYRIC ACID, 2-AMINO-4-(METHYLSULFINYL)- see ALF600
BUTYRIC ACID ANHYDRIDE see BSW550
n-BUTYRIC ACID ANHYDRIDE see BSW550
BUTYRIC ACID, 2-BROMO-3-METHYL- see BNM100
BUTYRIC ACID, 4-CHLORO-, TRIBUTYLSTANNYL ESTER see TID000
BUTYRIC ACID, CINNAMYL ESTER see CMQ800
BUTYRIC ACID, CYCLOHEXYL ESTER see CPI300
BUTYRIC ACID, 4-(2,4-DICHLOROPHENOXY)-, DIMETHYLAMINE SALT see DGA100
BUTYRIC ACID, 3,7-DIMETHYL-2,6-OCTADIENYL ESTER, (E)- see GDE810
BUTYRIC ACID-3,7-DIMETHYL-6-OCTENYL ESTER see DTF800
BUTYRIC ACID, α-α-DIMETHYLPHENETHYL ESTER see BEL850
BUTYRIC ACID, 4,4'-DISELENOBIS(2-AMINO)- see SBU710
BUTYRIC ACID ESTER with BUTYL LACTATE see BQP000
BUTYRIC ACID-γ-FLUORO-β-HDYROXY-THIOL-METHYL ESTER see FJG000
BUTYRIC ACID, 4-FLUORO-, ISOPROPYL ESTER see IPA100
BUTYRIC ACID, HEPTYL ESTER see HBN150
BUTYRIC ACID, HEXYL ESTER see HFM700
BUTYRIC ACID, 2-HYDROXY-4-(METHYLTHIO)- see HMR600
BUTYRIC ACID ISOBUTYL ESTER see BSW500
BUTYRIC ACID, 4-ISOTHIOCYANATO-, ETHYL ESTER see ELX527
BUTYRIC ACID LACTONE see BOV000
BUTYRIC ACID, 2-METHYL-(6CI,8CI) see MHS600
BUTYRIC ACID, 2-METHYL-, PHENETHYL ESTER (8CI) see PDF780
BUTYRIC ACID NITRILE see BSX250
BUTYRIC ACID, 2-(5-NITRO-α-IMINOFURFURYL)HYDRAZIDE see BSX600
BUTYRIC ACID, 2-OXO-, SODIUM SALT see SIY600
BUTYRIC ACID, PROPYL ESTER see PNF100
BUTYRIC ACID TRIESTER with GLYCERIN see TIG750
BUTYRIC ACID, 4,4,4-TRINITRO- see TML100
BUTYRIC ACID, VINYL ESTER see VNF000
BUTYRIC ALDEHYDE see BSU250
BUTYRIC ANHYDRIDE see BSW550
n-BUTYRIC ANHYDRIDE see BSW550
BUTYRIC ETHER see EHE000
BUTYRIC or NORMAL PRIMARY BUTYL ALCOHOL see BPW500
BUTYRINASE see GGA800
17-BUTYRLOXY-11-β-HYDROXY-21-PROPIONYLOXY-4-PREGNENE-3,20-DIONE see HHQ850
BUTYROLACTAM see PPT500
γ-BUTYROLACTAM see PPT500

α-BUTYROLACTONE see BOV000
β-BUTYROLACTONE see BSX000
(I)-β-BUTYROLACTONE see BSX100
dl-β-BUTYROLACTONE see BSX100
(RS)-β-BUTYROLACTONE see BSX100
γ-BUTYROLACTONE (FCC) see BOV000
BUTYROLACTONE I see BSX150
BUTYRON see CKF750
BUTYRONE (DOT) see DWT600
BUTYRONITRILE see BSX250
BUTYRONITRILE (DOT) see BSX250
BUTYRONITRILE, 2,2'-AZOBIS(2-METHYL-
see ASM025
BUTYRONITRILE, 4-CHLORO- see CEU300
BUTYRONITRILE, 4-
(DIETHOXYMETHYLSILYL)- see COR500
BUTYRONITRILE, 4-(TRICHLOROSILYL)-
see COR750
BUTYRONITRILE, 4-(TRIETHOXYSILYL)-
see COR800
BUTYROPHENONE, 4'-AMINO- see AJC750
BUTYROPHENONE, 4-(4-(p-
CHLOROPHENYL)-4-
HYDROXYPIPERIDINO)-4'-
(DIMETHYLAMINO)- see CKA580
BUTYROPHENONE, 3-(p-
CHLOROPHENYL)-2-PHENYL-4'-(2-(1-
PYRROLIDINYL)ETHOXY)-, erythro- see
CKI600
BUTYROSIN A see BSX325
4-BUTYROTHIOLACTONE see TDC800
N-(1-
BUTYROXYMETHYL)METHYLNITROSAMI
NE see BSX500
N-(1-BUTYROXYMETHYL)-N-
NITROSOMETHYLAMINE see BSX500
N'-BUTYROYL-5-NITRO-2-
FUROHYDRAZIDE see NGC550
N²)-BUTYROYL-5-NITRO-2-
FUROHYDRAZIDE IMIDE see BSX600
12-o-BUTYROYL-
PHORBOLDODECANOATE see BSX750
3-(3-BUTYRYLAMINO-2,4,6-
TRIIODOPHENYL)-2-ETHYLACRYLIC
ACID SODIUM SALT see EQC000
1-BUTYRYLAZIRIDINE see BSY000
1-n-BUTYRYLAZIRIDINE see BSY000
BUTYRYL CHLORIDE see BSY250
BUTYRYLCHOLINE IODIDE see BSY300
1-BUTYRYL-4-CINNAMYLPIPERAZINE
HYDROCHLORIDE see BTA000
1-N-BUTYRYL-4-CINNAMYL PIPERAZINE
HYDROCHLORIDE see BTA000
2-BUTYRYL-β-(N,N-
DIISOPROPYL)PHENOXYETHYLAMINE
HYDROCHLORIDE see DNN000
2-BUTYRYL-10-(3-
DIMETHYLAMINOPROPYL)PHENOTHIAZ
INE MALEATE see BTA125
BUTYRYL DITHRANOL see BSY400
10-BUTYRYLDITHRANOL see BSY400
10-BUTYRYL DITHRANOL see BSY400
BUTYRYLETHYLENEIMINE see BSY000
BUTYRYLETHYLENIMINE see BSY000
BUTYRYL LACTONE see BOV000
BUTYRYL NITRATE see BSY750
BUTYRYL OXIDE see BSW550
BUTYRYLPERAZINE DIMALEATE see
BSZ000
1-BUTYRYL-4-
(PHENYLALLYL)PIPERAZINE
HYDROCHLORIDE see BTA000
BUTYRYLPROMAZINE MALEATE see
BTA125
4-BUTYRYLPYRIDINE see PNV755
BUTYRYL TRIGLYCERIDE see TIG750
BUVETZONE see BRF500
BUX see BTA250
BUX-TEN see BTA250
2-n-BUYTLAMINOETHANOL see BQC000
BUZEPIDE METHIODIDE see BTA325
BUZON see BRF500

BUZULFAN see BOT250
BVU see BNP750
B-W see SJU000
BW-197U see MQR100
BW 248U see AEC700
BW 283U see DBX875
BW 47-83 see EAN600
BW 5071 see AMH250
BW 56-72 see TKZ000
BW 57-43 see BSF275
BW 61-43 see IPY500
BW 33-T-57 see MKW250
BW 50-197 see MQR100
BW 56-158 see ZVJ000
B.W. 57-233 see SBE500
BW 57-322 see ASB250
BW 57-323 see AKY250
BW 58-271 see RLZ000
BW 58-283 see DBX875
BW 57-323H see AKY250
BW 58-283b see DBX875
BW-A 43U see BSF275
BW-A 509U see ASE900
B.W. 356-C-61 see KFA100
BW 467-C-60 see BFW250
BW-21-Z see AHJ750
BX 112 see CPB065
BY 935 see NDY550
B-2847-Y see THA600
BYKOMYCIN see NCD550
BYLADOCE see VSZ000
BYLADOCE see VSZ000
2,2'-BYPYRIDIN see BGO500
BZ see BBU800
BZ see QVA000
B-3-Zh see CMS212
BZ 55 see BSQ000
BZCF see BEF500
BZF-60 see BDS000
BZI see BCB750
BZL see BTA500
BZQ see BCL250
BZT see BEN000
BZT see CBF825
C 2 see DXY725
C 2 see LBX050
C 6 see HEA000
C-10 see BPH750
C 45 see CGG500
C-56 see HCE500
C 78 see BQE250
C-076 see ARW150
C 172 see PAQ060
C-257 see NFW525
C-272 see BLG500
C 283 see LEF300
C 283 see NFW500
C-299 see OHJ875
C-410 see NFW200
4-C-32 see TGA525
C-492 see NFW460
C-516 see NFW100
C-541 see NFW430
C 570 see FAB400
C 600 see MJE793
C 609 see NFW450
C 661 see MNX260
C-666 see CAB125
C-684 see NFW470
C-702 see NFW400
C 702 see NFW425
C 709 see DGQ875
C-776 see DTP600
C-829 see NFW435
C-835 see NFW350
C-847 see CEW500
C-854 see CJT750
C 857 see NHE550
C-908 see ABW550
C-909 see ABU600
C 1,006 see CJT750
C 1006 see HFK500
C 1120 see DLS800

C-1228 see OIU499
C 1414 see MRH209
C-1566 see BPI400
C 1686 see MDK000
C 1739 see DOK600
C 1863 see DIN000
C 1983 see CJQ000
C 2018 see CCU250
C 2039 see DII800
C 2046 see DIB200
C 2047 see DIK200
C 2048 see MDK750
C 2052 see DIE600
C 2053 see DIE400
C 2054 see EOG500
C 2057 see DIJ000
C 2059 see DUK800
C 2060 see DIM000
C 2061 see MDK250
C 2085 see MRT250
C 2094 see MNX250
C 2095 see CJQ750
C 2096 see CJQ500
C 2097 see DIK400
C 2098 see DIB800
C 2102 see EOU500
C 2103 see END500
C 2126 see EOG000
C 2127 see EQI500
C 2136 see MDK500
C 2137 see EOF500
C 2138 see EPV000
C 2140 see MQO500
C 2141 see MQO750
C 2142 see EQI000
C 2242 see CIS250
C 2446 see AHO750
C 3037 see CFM250
C 3039 see DDM600
C 3049 see CFU250
C 3053 see DEX600
C 3054 see DHO800
C 3057 see DFF200
C 3058 see CKT750
C 3059 see CHS750
C 3061 see CFM000
C 3062 see PIN000
C 3063 see CGP380
C 3065 see DHK800
C 3067 see CIK500
C 3068 see CFO000
C 3069 see CIJ250
C 3070 see CFL500
C 3071 see CGN250
C 3072 see CFL000
C 3073 see CLC125
C 3074 see CFW250
C 3078 see CFW000
C 3078 see CLC100
C 3080 see DHK200
C 3085 see PIM750
C 3087 see PPU750
C 3089 see DHL400
C 3094 see DHK000
C 3095 see DHJ800
C 3101 see CFM500
C 3102 see DHL200
C 3103 see DDU200
C 3104 see AJE250
C 3115 see CFH000
C 3117 see BDY250
C 3120 see CFI000
C 3121 see BPJ000
C 3124 see AFW250
C 3125 see BPR000
C-3126 see PAM785
C 3127 see DIP800
C 3130 see BPT250
C 3133 see CGE750
C 3134 see MQO250
C 3135 see DHN800
C 3136 see BEE750

CADMIUM, BIS(PENTYLDITHIOCARBAMATO)- see CAD550
CADMIUM, BIS(SALICYLATO)- see CAI400
CADMIUM BROMIDE see CAD600
CADMIUM CAPRYLATE see CAD750
CADMIUM CARBONATE see CAD800
CADMIUM CATION see CAG600
CADMIUM CDTA see CAD900
CADMIUM CHLORATE see CAE000
CADMIUM CHLORIDE see CAE250
CADMIUM CHLORIDE, DIHYDRATE see CAE375
CADMIUM CHLORIDE, HYDRATE (2:5) see CAE425
CADMIUM CHLORIDE, MONOHYDRATE see CAE500
CADMIUM COMPOUNDS see CAE750
CADMIUM-COPPER ALLOY see CNI610
CADMIUM DIACETATE see CAD250
CADMIUM DIACETATE DIHYDRATE see CAD275
CADMIUM DIACETATE MONOHYDRATE see CAE800
CADMIUM DIAMIDE see CAD325
CADMIUM DIAMYL DITHIOCARBAMATE see CAD550
CADMIUM DIAZIDE see CAD350
CADMIUM DIBROMIDE see CAD600
CADMIUM DICHLORIDE see CAE250
CADMIUM DICYANIDE see CAF500
CADMIUM DIETHYL DITHIOCARBAMATE see BJB500
CADMIUM DIHYDROXIDE see CAG525
CADMIUM DIIODIDE see CAG550
CADMIUM DILAURATE see CAG775
CADMIUM DINITRATE see CAH000
CADMIUM DODECANOATE see CAG775
CADMIUM(II) EDTA COMPLEX see CAF750
CADMIUM FLUOBORATE see CAG000
CADMIUM FLUORIDE see CAG250
CADMIUM FLUOROBORATE see CAG000
CADMIUM FLUOROSILICATE see CAG500
CADMIUM FLUORURE (FRENCH) see CAG250
CADMIUM FLUOSILICATE see CAG500
CADMIUM FUME see CAH750
CADMIUM GOLDEN see CMS212
CADMIUM GOLDEN 366 see CAJ750
CADMIUM HEXAFLUOROSILICATE (7CI) see CAG500
CADMIUM HYDROXIDE see CAG525
CADMIUM IODIDE see CAG550
CADMIUM ION see CAG600
CADMIUM, ION (Cd²⁺) see CAG600
CADMIUM LACTATE see CAG750
CADMIUM LAURATE see CAG775
CADMIUM LEMON see CMS212
CADMIUM LEMON YELLOW 527 see CAJ750
CADMIUM MONOCARBONATE see CAD800
CADMIUM MONOSULFIDE see CAJ750
CADMIUM MONOTELLURIDE see CAJ800
CADMIUM MONOXIDE see CAH500
CADMIUM NITRATE see CAH000
CADMIUM(II) NITRATE see CAH000
CADMIUM(II) NITRATE TETRAHYDRATE (1:2:4) see CAH250
CADMIUM NITRIDE see TIH000
CADMIUM NONBASE, Cd,Cu see CAD290
CADMIUM OCTADECANOATE see OAT000
CADMIUM ORANGE see CAJ750
CADMIUM OXIDE see CAH500
CADMIUM OXIDE FUME see CAH750
CADMIUM PHOSPHATE see CAI000
CADMIUM PHOSPHIDE see CAI125
CADMIUM PRIMROSE see CMS212
CADMIUM PRIMROSE 819 see CAJ750
CADMIUM PROPIONATE see CAI250
CADMIUM PT see CAI350
CADMIUM 2-PYRIDINETHIONE see CAI350

CADMIUM SALICYLATE see CAI400
CADMIUM SELENIDE see CAI500
CADMIUM SELENIDE SULFIDE see CAI600
CADMIUM SILICON FLUORIDE see CAG500
CADMIUM STEARATE see OAT000
CADMIUM(II) STEARATE see OAT000
CADMIUM SUCCINATE see CAI750
CADMIUM SULFATE see CAJ000
CADMIUM SULFATE (1:1) see CAJ000
CADMIUM SULFATE, HYDRATE see CAJ100
CADMIUM SULFATE (1:1) HYDRATE (3:8) see CAJ250
CADMIUM SULFATE OCTAHYDRATE see CAJ250
CADMIUM SULFATE TETRAHYDRATE see CAJ500
CADMIUM SULFIDE see CAJ750
CADMIUM SULFIDE (AMORPHOUS) see CAJ760
CADMIUM SULFIDE SELENIDE see CAI600
CADMIUM SULFIDE mixed with ZINC SULFIDE (1:1) see CMS212
CADMIUM SULFIDE mixed with ZINC SULFIDE (5:95) see CAJ770
CADMIUM SULFIDE mixed with ZINC SULFIDE (8:92) see CAJ772
CADMIUM SULFOSELENIDE see CAI600
CADMIUM SULPHATE see CAJ000
CADMIUM SULPHIDE see CAJ750
CADMIUM SULPHOSELENIDE see CAI600
CADMIUM TELLURIDE see CAJ800
CADMIUM TETRAFLUOROBORATE (7CI) see CAG000
CADMIUM THERMOVACUUM AEROSOL see CAK000
CADMIUM-THIONEINE see CAK250
CADMIUM salt of 2,4,5-TRIBROMOIMIDAZOLE see THV500
CADMIUM YELLOW see CAJ750
CADMIUM YELLOW 000 see CAJ750
CADMIUM YELLOW 892 see CAJ750
CADMIUM YELLOW 10G CONC. see CAJ750
CADMIUM YELLOW CONC. DEEP see CAJ750
CADMIUM YELLOW CONC. GOLDEN see CAJ750
CADMIUM YELLOW CONC. LEMON see CAJ750
CADMIUM YELLOW CONC. PRIMROSE see CAJ750
CADMIUM YELLOW OZ DARK see CAJ750
CADMIUM YELLOW PRIMROSE 47-4100 see CAJ750
CADMOPUR GOLDEN YELLOW N see CAJ750
CADMOPUR YELLOW see CAJ750
CADOX see BDS000
CADOX see BSC750
CADOX see MKA500
CADOX TBH see BRM250
CADOX TS see BIX750
CADOX TS 40,50 see BIX750
CADPX PS see BHM750
CADRALAZINE see CAK275
CADUCID see FLG000
CAERULEIN see CAK285
CAESALPINIA (various species) see CAK325
CAESIUM ARSENATE see CAK350
CAESIUM HYDROXIDE, solid (UN 2682) (DOT) see CDD750
CAESIUM HYDROXIDE, solution (UN 2681) (DOT) see CDD750
CAF see CDP250
CAF see CEA750
CAFFEARINE see TKL890
CAFFEIC ACID see CAK375
CAFFEIC ACID PHENETHYL ESTER see CAK400
CAFFEIN see CAK500
CAFFEINE see CAK500
CAFFEINE BROMIDE see CAK750

CAFFEINE HYDROBROMIDE see CAK750
CAFFEINE, 8-METHOXY- see MEF800
CAFFEINE and SODIUM BENZOATE see CAK800
3-CAFFEOYLQUINIC ACID see CHK175
3-o-CAFFEOYLQUINIC ACID see CHK175
CAFRON see BCA000
CAID see CJJ000
CAIMONICILLO (DOMINICAN REPUBLIC) see ROA300
CAIN'S QUINOLINIUM see QOJ250
CAIROX see PLP000
CAJEPUTENE see MCC250
CAJEPUTOL see CAL000
CAJRAB see BAC040
CAKE ALUM see AHG750
CAKE ALUM see AHG800
CALACIDOL see CAL075
CALADIO (PUERTO RICO) see CAL125
CALADIUM see CAL125
CALAMINE (spray) see ZKA000
CALAMUS OIL see OGK000
CALAN see VHA450
CALAR see CAM000
CALCALOID PRINTING ORANGE RYW see CMU815
CALCAMINE see DME300
CALCENE CO see CAT775
CAL CHEM 5655 see MOU750
CALCIA see CAU500
CALCIATE(2-), ((ETHYLENEDINITRILO)TETRAACETATO)-, DIHYDROGEN (8CI) see CAR800
CALCIATE(2⁻), ((ETHYLENEDINITRILO)TETRAACETATO)-, DISODIUM see CAR780
CALCICAT see CAL250
CALCIC LIVER of SULFUR see CAY000
CALCICOL see CAS750
CALCICOLL see CAT775
CALCID see CAQ500
CALCIDAR 40 see CAT775
CALCIFEROL see VSZ100
CALCIFERON 2 see VSZ100
CALCILIT 8 see CAT775
CALCINED BARYTA see BAO000
CALCINED BRUCITE see MAH500
CALCINED DIATOMITE see SCJ000
CALCINED MAGNESIA see MAH500
CALCINED MAGNESITE see MAH500
CALCINED SODA see SIN500
CALCIOFON see CAS750
CALCIOPEN K see PDT750
CALCIPUR see CAS750
CALCIRETARD see CAM675
CALCITE see CAO000
CALCITETRACEMATE DISODIUM see CAR780
CALCITRIOL see DMJ400
CALCIUM see CAL250
CALCIUM ACETARSONE see CAL500
CALCIUM ACETATE see CAL750
CALCIUM ACETYLIDE see CAN750
CALCIUM ACID METHANEARSONATE see CAM000
CALCIUM ACID METHYL ARSONATE see CAM000
CALCIUM ACRYLATE, MONOHYDRATE see CAM100
CALCIUM ALGINATE see CAM200
CALCIUM ALUMINUM SILICATE see AGY100
CALCIUMARSENAT see ARB750
CALCIUM ARSENATE see CAM222
CALCIUM ARSENATE (DOT) see ARB750
CALCIUM ARSENITE see CAM300
CALCIUM ARSENITE see CAM500
CALCIUM ARSENITE, solid (DOT) see CAM300
CALCIUM ARSENITE, solid (DOT) see CAM500
CALCIUM ARSONATE (1:1) see CAM520

CALCIUM ASCORBATE see CAM600
CALCIUM ASPARTATE see CAM675
CALCIUM-l-ASPARTATE see CAM675
CALCIUM BARIUM TITANATE see BAI810
CALCIUM BENZOATE see CAM680
CALCIUM-o-BENZOSULFIMIDE see CAM750
CALCIUM-2-BENZOSULPHIMIDE see CAM750
CALCIUM-o-BENZOSULPHIMIDE see CAM750
CALCIUM BIPHOSPHATE see CAW110
CALCIUM BIS(DODECYLBENZENESULFONATE) see CAR790
CALCIUM BISULFITE see CAN000
CALCIUM BORATE see CAN250
CALCIUM BROMATE see CAN400
CALCIUM BROMIDE see CAR375
CALCIUM BUCLOXATE see CPJ250
CALCIUM 5-BUTYLPICOLINATE HYDRATE see FQR100
CALCIUM CARAGEENIN see CAO250
CALCIUM CARBIDE see CAN750
CALCIUM CARBIMIDE see CAQ250
CALCIUM CARBONATE see CAO000
CALCIUM CARBONATE (1:1) see CAT775
CALCIUM CARRAGEENAN see CAO250
CALCIUM CARRAGHEENATE see CAO250
CALCIUM CHEL-330 see CAY500
CALCIUM CHLORATE see CAO500
CALCIUM CHLORATE, aqueous solution (DOT) see CAO500
CALCIUM CHLORATE, MIXED WITH CALCIUM CHLORIDE, HEXAHYDRATE see CAO600
CALCIUM CHLORIDE see CAO750
CALCIUM CHLORIDE, anhydrous see CAO750
CALCIUM CHLORIDE with STREPTOMYCIN (1:1) see SLY000
CALCIUM CHLORITE see CAP000
CALCIUM CHLOROHYDROCHLORITE see HOV500
CALCIUM-4-(p-CHLOROPHENYL)-2-PHENYL-5-THIAZOLEACETATE see CAP250
CALCIUM CHROMATE see CAP500
CALCIUM CHROMATE (VI) see CAP500
CALCIUM CHROMATE(VI) DIHYDRATE see CAP750
CALCIUM CHROME YELLOW see CAP500
CALCIUM CHROME YELLOW see CAP750
CALCIUM CHROMIUM OXIDE (CaCrO$_4$) see CAP500
CALCIUM CITRATE see CAP850
CALCIUM COMPOUNDS see CAQ000
CALCIUM CYANAMID see CAQ250
CALCIUM CYANAMIDE see CAQ250
CALCIUM CYANIDE see CAQ500
CALCIUM CYANIDE MIXTURE, solid (DOT) see CAQ500
CALCIUM CYCLAMATE see CAR000
CALCIUM CYCLAMATE DIHYDRATE see CAQ600
CALCIUM CYCLOHEXANESULFAMATE see CAR000
CALCIUM CYCLOHEXANESULFAMATE DIHYDRATE see CAQ600
CALCIUM CYCLOHEXANE SULPHAMATE see CAR000
CALCIUM CYCLOHEXYLSULFAMATE see CAR000
CALCIUM CYCLOHEXYLSULFAMATE DIHYDRATE see CAQ600
CALCIUM CYCLOHEXYLSULPHAMATE see CAR000
CALCIUM DIACETATE see CAL750
CALCIUM DIBROMIDE see CAR375
CALCIUM DICARBIDE see CAN750
CALCIUM DICHROMATE(VI) see CAR400
CALCIUM DIFLUORIDE see CAS000

CALCIUM DIHYDRIDE see CAT200
CALCIUM DIHYDROXIDE see CAT225
CALCIUM-d-(+)-4-(2,4-DIHYDROXY-3,3-DIMETHYLBUTYRAMIDE)BUTYRATE HEMIHDYRATE see CAT175
CALCIUM d(+)-N-(α,γ-DIHYDROXY-β,β-DIMETHYLBUTYRYL)-β-ALANINATE see CAU750
CALCIUM DINITRATE see CAU000
CALCIUM DIOXIDE see CAV500
CALCIUM DIPROPIONATE see CAW400
CALCIUM DIPROPYLACETATE see VCK200
CALCIUM α,α-DIPROPYLACETATE see VCK200
CALCIUM DISILICIDE see CAR750
CALCIUM DISODIUM EDATHAMIL see CAR780
CALCIUM DISODIUM EDETATE see CAR775
CALCIUM DISODIUM EDETATE see CAR780
CALCIUM DISODIUM EDTA see CAR775
CALCIUM DISODIUM EDTA see CAR780
CALCIUM DISODIUM ETHYLENEDIAMINETETRAACETATE see CAR775
CALCIUM DISODIUM ETHYLENEDIAMINETETRAACETATE see CAR780
CALCIUM DISODIUM (ETHYLENEDINITRILO)TETRAACETATE see CAR775
CALCIUM DISODIUM (ETHYLENEDINITRILO)TETRAACETATE see CAR780
CALCIUM DISODIUM VERSENATE see CAR780
CALCIUM DISTEARATE see CAX350
CALCIUM DITHIOCYANATE see CAY250
CALCIUM DITHIONITE (DOT) see CAN000
CALCIUM DOBESILATE see DMI300
CALCIUM DODECYLBENZENESULFONATE see CAR790
CALCIUM-DTPA see CAY500
CALCIUM EDTA see CAR780
CALCIUM EDTA COMPLEX see CAR800
CALCIUM (−)-(1R,2S)-(1,2-EPOXYPROPYL)PHOSPHONATE HYDRATE see CAW376
CALCIUM ESFAR see CPJ250
CALCIUM-N-2-ETHYLHEXYL-β-OXYBUTYRAMIDE SEMISUCCINATE see CAR875
CALCIUM FLUORIDE see CAS000
CALCIUM FLUOROSILICATE see CAX250
CALCIUM FLUOSILICATE see CAX250
CALCIUM FORMATE see CAS250
CALCIUM FOSFOMYCIN HYDRATE see CAW376
CALCIUM FUSARATE see FQR100
CALCIUM 4-(β-d-GALACTOSIDO)-d-GLUCONATE see CAT650
CALCIUM GLUCONATE see CAS750
CALCIUM d-GLUCONATE see CAS750
CALCIUM GLYCEROPHOSPHATE see CAS800
CALCIUM HEXAFLUOROSILICATE see CAX250
CALCIUM HEXAGLUCONATE see CAS750
CALCIUM HEXAMETAPHOSPHATE see CAS825
CALCIUM HOMOPANTOTHENATE see CAT125
CALCIUM-d-HOMOPANTOTHENATE see CAT125
CALCIUM HOPANTENATE see CAT125
CALCIUM HOPANTENATE HEMIHYDRATE see CAT175
CALCIUM HYDRATE see CAT225
CALCIUM HYDRIDE see CAT200

CALCIUM HYDROGEN METHANEARSONATE see CAM000
CALCIUM HYDROGEN PHOSPHATE see CAT210
CALCIUM HYDROSILICATE see CAW850
CALCIUM HYDROSULFITE (DOT) see CAN000
CALCIUM HYDROXIDE see CAT225
CALCIUM HYDROXIDE (ACGIH, OSHA) see CAT225
CALCIUM HYDROXIDE HYPOCHLORITE see CAT235
CALCIUM HYPOCHLORIDE see HOV500
CALCIUM HYPOCHLORITE see HOV500
CALCIUM HYPOCHLORITE DIHYDRATE see HOV503
CALCIUM HYPOPHOSPHITE see CAT250
CALCIUM IODATE see CAT500
CALCIUM LACTATE see CAT600
CALCIUM LACTOBIONATE see CAT650
CALCIUM LIGNOSULFONATE see CAT675
CALCIUM MAGNESIUM ACETATE see CAT685
CALCIUM METHANEARSONATE see CAM000
CALCIUM METHIONATE see CAT700
CALCIUM MOLYBDATE see CAT750
CALCIUM MOLYBDENUM OXIDE (CaMoO$_4$) see CAT750
CALCIUM MONOCARBONATE see CAT775
CALCIUM MONOCHROMATE see CAP500
CALCIUM MONOSILICATE see CAW850
CALCIUM NAPHTHENATE see NAR200
CALCIUM NEMBUTAL see CAV000
CALCIUM NITRATE (DOT) see CAU000
CALCIUM(II) NITRATE (1:2) see CAU000
CALCIUM(II) NITRATE TETRAHYDRATE (1:2:4) see CAU250
CALCIUM NITRIDE see TIH250
CALCIUM OLEATE see CAU300
CALCIUM ORTHOARSENATE see ARB750
CALCIUM OXIDE see CAU500
CALCIUM 3-OXIDO-5-OXO-4-PROPIONYLCYCLOHEX-3-ENECARBOXYLATE (IUPAC) see CPB065
CALCIUM OXYCHLORIDE see HOV500
CALCIUM PANTHOTHENATE (FCC) see CAU750
CALCIUM PANTOTHENATE see CAU750
CALCIUM-d-PANTOTHENATE see CAU750
d-CALCIUM PANTOTHENATE see CAU750
CALCIUM PANTOTHENATE, CALCIUM CHLORIDE DOUBLE SALT see CAU780
CALCIUM PENTOBARBITAL see CAV000
CALCIUM PERMANGANATE see CAV250
CALCIUM PEROXIDE see CAV500
CALCIUM PEROXOCHROMATE see CAV750
CALCIUM PEROXODISULPHATE see CAW000
CALCIUM-2-(m-PHENOXYPHENYL)PROPIONATE DIHYDRATE see FAP100
CALCIUM-2-PHENYL-4-(p-CHLOROPHENYL)-5-THIAZOLEACETATE see CAP250
CALCIUM PHOSPHATE, DIBASIC see CAW100
CALCIUM PHOSPHATE, MONOBASIC see CAW110
CALCIUM PHOSPHATE, TRIBASIC see CAW120
CALCIUM PHOSPHIDE see CAW250
CALCIUM PHOSPHINATE see CAT250
CALCIUM PHOSPHONOMYCIN HYDRATE see CAW376
CALCIUM PHOTOPHOR see CAW250
CALCIUM POLYSILICATE see CAW850
CALCIUM POLYSULFIDE see CAX800
CALCIUM PROPIONATE see CAW400
CALCIUM PROPIONATE see CAW400
CALCIUM PYROPHOSPHATE see CAW450

CALCIUM RESINATE see CAW500
CALCIUM RESINATE (UN 1313) (DOT) see CAW500
CALCIUM RESINATE, fused (UN 1314) (DOT) see CAW500
CALCIUMRHODANID see CAY250
CALCIUM RHODANID (GERMAN) see CAY250
CALCIUM RICINOLEATE see CAW525
CALCIUM SACCHARIN see CAM750
CALCIUM SACCHARIN see CAW600
CALCIUM SACCHARINA see CAM750
CALCIUM SACCHARINATE see CAM750
CALCIUM SALTPETER see CAU000
CALCIUM SILICATE see CAW850
CALCIUM SILICATE, synthetic nonfibrous (ACGIH) see CAW850
CALCIUM SILICIDE see CAX000
CALCIUM SILICOFLUORIDE see CAX250
CALCIUM SODIUM HYPOCHLORITE see CAX255
CALCIUM SODIUM METAPHOSPHATE see CAX260
CALCIUM SORBATE see CAX275
CALCIUM STEARATE see CAX350
CALCIUM STEAROYL LACTATE see CAX375
CALCIUM STEAROYL-2-LACTATE see CAX375
CALCIUM STRONTIUM ACETATE (4:1:1) see CAX400
CALCIUM SULFATE see CAX500
CALCIUM(II) SULFATE DIHYDRATE (1:1:2) see CAX750
CALCIUM SULFIDE see CAX800
CALCIUM SULFIDE see CAY000
CALCIUM SULFOCYANATE see CAY250
CALCIUM SUPEROXIDE see CAV500
CALCIUM SUPERPHOSPHATE see CAY100
CALCIUM TETRABORATE see CAN250
CALCIUM THIOCYANATE see CAY250
CALCIUM TITANATE see CAY300
CALCIUM TITANIUM OXIDE see CAY300
CALCIUM TITANIUM TRIOXIDE see CAY300
CALCIUM TITRIPLEX see CAR780
CALCIUM TRISODIUM CHEL 330 see CAY500
CALCIUM TRISODIUM DIETHYLENE TRIAMINE PENTAACETATE see CAY500
CALCIUM TRISODIUM DTPA see CAY500
CALCIUM TRISODIUM PENTETATE see CAY500
CALCIUM TRISODIUM SALT of DIETHYLENETRIAMINEPENTAACETIC ACID see CAY500
CALCIUM VALPROATE see CAY675
CALCIUM VALPROATE see VCK200
CALCO 2246 see MJO500
CALCOCHROME ALIZARINE RED SC see SEH475
CALCOCHROME BLUE BLACK BC see CMP880
CALCOCHROME ORANGE GR see CMP882
CALCOCHROME ORANGE R see NEY000
CALCOCHROME YELLOW 2G see SIT850
CALCOCID ALIZARINE BLUE SE see APG700
CALCOCID ALIZARINE BLUE SKY see CMM090
CALCOCID AMARANTH see FAG020
CALCOCID BLUE AX see ERG100
CALCOCID BLUE BLACK see FAB830
CALCOCID BLUE BLACK 2R see FAB830
CALCOCID BLUE EG see FMU059
CALCOCID BRILLIANT SCARLET 3RN see FMU080
CALCOCID ERYTHROSINE N see FAG040
CALCOCID FAST BLUE SR see ADE750
CALCOCID FAST LIGHT ORANGE 2G see HGC000
CALCOCID FUCHSINE 6B see CMM400

CALCOCID GREEN G see FAE950
CALCOCID 2RIL see FMU070
CALCOCID MILLING RED G see CMM325
CALCOCID MILLING RED RC see NAO600
CALCOCID MILLING YELLOW R see CMM759
CALCOCID ORANGE Y see CMM220
CALCOCID PHLOXINE 2G see CMM300
CALCOCID URANINE B4315 see FEW000
CALCOCID VIOLET 4BNS see FAG120
CALCOCID YELLOW MCG see FAG140
CALCOCID YELLOW MXXX see MDM775
CALCOCID YELLOW XX see FAG140
CALCODUR BROWN BRL see CMO750
CALCODUR RED 8BL see CMO885
CALCODUR RESIN FAST BLUE see CMN750
CALCODUR TURQUOISE GL see COF420
CALCOFLUOR WHITE MR see DXB450
CALCOGAS ORANGE NC see PEJ500
CALCOGENE BLACK GX-CF see CMS250
CALCOGENE RED BROWN EU-CF see CMS257
C 10 ALCOHOL see DAI600
CALCOLAKE SCARLET 2R see FMU070
CALCOLOID BLUE BLC see DFN300
CALCOLOID BLUE BLD see DFN300
CALCOLOID BLUE BLFD see DFN300
CALCOLOID BLUE BLR see DFN300
CALCOLOID BLUE RS see IBV050
CALCOLOID BROWN BR see CMU770
CALCOLOID BROWN R see CMU780
CALCOLOID BROWN RK see CMU780
CALCOLOID BROWN RNB see CMU780
CALCOLOID BROWN RNBC see CMU780
CALCOLOID DIAZO BLACK BHL see CMN800
CALCOLOID GOLDEN YELLOW see DCZ000
CALCOLOID NAVY BLUE see VGP100
CALCOLOID NAVY BLUE 2GC see CMU500
CALCOLOID NAVY BLUE NTC see VGP100
CALCOLOID OLIVE R see DUP100
CALCOLOID OLIVE RC see DUP100
CALCOLOID OLIVE RL see DUP100
CALCOLOID PRINTING ORANGE RE see CMU815
CALCOLOID VIOLET 4RD see DFN450
CALCOLOID VIOLET 4RP see DFN450
CALCOMINE BLACK see AQP000
CALCOMINE BLACK EXL see AQP000
CALCOMINE BLUE 2B see CMO000
CALCOMINE BROWN B see CMO820
CALCOMINE BROWN MCW see CMO800
CALCOMINE CATECHU 2B see CMO820
CALCOMINE DARK GREEN BG see CMO830
CALCOMINE DIAZO BLACK BHD see CMN800
CALCOMINE DIAZO BLACK BTCW see CMN800
CALCOMINE GREEN BY see CMO840
CALCOMINE RED FC see CMO870
CALCOMINE RED 4BX see DXO850
CALCOMINE SCARLET 3B see CMO875
CALCOMINE SCARLET 4BSY see CMO870
CALCOMINE VIOLET N see CMP000
CALCON see HLI100
CALCO OIL ORANGE 7078 see PEJ500
CALCO OIL RED D see SBC500
CALCO OIL SCARLET BL see XRA000
CALCOSYN BRILLIANT SCARLET BN see ENP100
CALCOSYN PINK B see AKE250
CALCOSYN SAPPHIRE BLUE R see MGG250
CALCOSYN YELLOW GC see AAQ250
CALCOSYN YELLOW GCN see AAQ250
CALCOTONE GREEN G see PJQ100
CALCOTONE ORANGE 2R see DVB800
CALCOTONE ORANGE R see CMS145
CALCOTONE RED see IHC450
CALCOTONE RED 3B see NAY000

CALCOTONE TOLUIDINE RED YP see MMP100
CALCOTONE WHITE T see TGG760
CALCOTONE YELLOW GP see CMS210
CALCOZINE BLUE ZF see BJI250
CALCOZINE BRILLIANT GREEN G see BAY750
CALCOZINE CHRYSOIDINE Y see PEK000
CALCOZINE FUCHSINE HO see MAC250
CALCOZINE MAGENTA N see RMK020
CALCOZINE MAGENTA RTN see MAC250
CALCOZINE MAGENTA XX see MAC250
CALCOZINE ORANGE YS see PEK000
CALCOZINE RED 6G see RGW000
CALCOZINE RED BG Liquid see CMM840
CALCOZINE RED BX see FAG070
CALCOZINE RED Y see GJI400
CALCOZINE RHODAMINE BX see FAG070
CALCOZINE RHODAMINE 6GX see RGW000
CALCOZINE VIOLET C see AOR500
CALCOZINE VIOLET 6BN see AOR500
CALCOZINE YELLOW OX see IBA000
CALCYAN see CAQ500
CALCYANIDE see CAQ500
CALDAN see BHL750
CALDEDON NAVY BLUE AR see VGP100
C-8 ALDEHYDE see OCO000
C-9 ALDEHYDE see NMW500
C-10 ALDEHYDE see DAG000
C-16 ALDEHYDE see ENC000
C-12 ALDEHYDE, LAURIC see DXT000
C-14 ALDEHYDE, MYRISTIC see TBX500
CALDON see BRE500
CALDOPENTAMINE see TED600
CALEDON BLUE RN see IBV050
CALEDON BLUE XRC see DFN300
CALEDON BLUE XRN see IBV050
CALEDON BRILLIANT BLUE RN see IBV050
CALEDON BRILLIANT PURPLE 4R see DFN450
CALEDON BRILLIANT PURPLE 4RP see DFN450
CALEDON BROWN R see CMU780
CALEDON BROWN R 300 see CMU780
CALEDON DARK BLUE G see CMU500
CALEDON DARK BROWN 3R see CMU770
CALEDONE OLIVE RP see DUP100
CALEDON GOLDEN YELLOW see DCZ000
CALEDON GREY M see CMU475
CALEDON JADE GREEN 2G see APK500
CALEDON NAVY BLUE 2R see VGP100
CALEDON NAVY BLUE ART see VGP100
CALEDON OLIVE R see DUP100
CALEDON PAPER BLUE RN see IBV050
CALEDON PRINTING BLUE 3G see DLO900
CALEDON PRINTING BLUE RN see IBV050
CALEDON PRINTING BLUE XRN see IBV050
CALEDON PRINTING NAVY G see CMU500
CALEDON PRINTING PURPLE 4R see DFN450
CALEDON PRINTING YELLOW see DCZ000
CALF KILL see MRU359
CALFLO E see CAW850
CALF THYMUS DNA see DAZ000
CALGINATE see CAM200
CALGLUCOL see CAS750
CALGLUCON see CAS750
CALGON see SHM500
CALGON 261 see DTS500
CALGON POLYMER 261 see DTS500
CALGON 261LV see DTS500
CALIBENE see SOX875
CALIBRITE see CAT775
CALICHEAMICIN γ1 see CAY700
CALICO BUSH see MRU359
CALICO YELLOW see MRN500
CALIDRIA RG 100 see ARM268

CALIDRIA RG 144 see ARM268
CALIDRIA RG 600 see ARM268
CALIFORNIA CHEMICAL COMPANY
RE5305 see BSG250
CALIFORNIA FERN see PJJ300
CALIFORNIA PEPPER TREE see PCB300
CALIGRAN M see MAP300
CALIXIN see TJJ500
C12-14-tert-ALKYL AMINES see CAY710
CALLA see CAY800
CALLA LILY see CAY800
CALLA PALUSTRIS see WAT300
CAL-LIGHT SA see CAT775
CALMADIN see MQU750
CALMATHION see MAK700
CALMAX see MQU750
CALMDAY see CGA500
CALMINAL see EOK000
CALMINOL see MNM500
CALMIREN see MQU750
CALMIXENE see MOO750
CALMOCITENE see DCK759
CALMODEN see MDQ250
CALMONAL see HGC500
CALMORE see TEH500
CALMOREX see TEH500
CALMOS see CAT775
CALMOSINE see EJA379
CALMOTE see CAT775
CALMOTIN see BNP750
CALNEGYT see CAY875
CALOCAIN see MNQ000
CALOCHLOR see MCY475
CALOCID GREEN S see ADF000
CALOCID GREEN SB see ADF000
CALO-CLOR see CAY950
CALOFIL A 4 see CAT775
CALOFORT S see CAT775
CALOFORT U see CAT775
CALOFOR U 50 see CAT775
CALOGREEN see MCW000
CALOGREEN see MCY300
CALOMEL see MCW000
CALOMEL see MCY300
CALOMELANO (ITALIAN) see MCW000
CALOMEL and MAGNESIUM SULFATE (5:8)
see CAZ000
CALOPAKE F see CAT775
CALOPAKE HIGH OPACITY see CAT775
CALOPHYLLUM INOPHYLLUM see MBU780
CALOSAN see MCW000
CALOTAB see MCY300
CALOTROPAGENIN see CAZ050
CALOTROPIS PROCERA (Ait.) R.Br., flower
extract see CAZ075
CALOTROPIS (VARIOUS SPECIES) see
COD675
CALOXOL CP2 see CAU500
CALOXOL W3 see CAU500
CALPANATE see CAU750
CALPHOSAN see CAT600
CALPLUS see CAO750
CALPOL see HIM000
CALPURNINE see CAZ125
CALSEEDS see CAT775
CALSIL see CAW850
CALSMIN see DLY000
CALSOFT F-90 see DXW200
CALSOFT LAS 99 see LBU100
CALSOL see EIV000
CALSTAR see CAX350
CALTAC see CAO750
CALTEC see CAT775
CALTHA (VARIOUS SPECIES) see MBU550
CALTHOR see AJJ875
CALVACIN see CBA000
CALVISKEN see VSA000
CALVIT see CAT225
CALX see CAU500
CALXYL see CAU500
CALYCANTH see CCK675

CALYCANTHINE, HYDROCHLORIDE see
CBA075
CALYCANTHUS (VARIOUS SPECIES) see
CCK675
CALYSTIGINE see PAE100
CAM see CDP250
CAMA see CAM000
CAMATROPINE see MDL000
CAMAZEPAM see CFY250
CAMBAXIN see BAB250
CAMBENDAZOLE see CBA100
CAMBENZOLE see CBA100
CAMBET see CBA100
CAMBOGIC ACID see CBA125
CAMCOLIT see LGZ000
CAMEL-CARB see CAT775
CAMELIA OIL see CBA200
CAMELLIA SINENSIS see ARP500
CAMEL-TEX see CAT775
CAMEL-WITE see CAT775
CAMFOSULFONATO del d-3-4-(1'DIBENZIL-
2-CHETO-IMIDAZOLIDO)-1,2-
TRIMETILTHIOPHANIUM (ITALIAN) see
TKW500
CAMILAN see SNY500
CAMILICHIGUI (MEXICO) see DHB309
C-27-AMINE FLUORIDE see BJY800
CAMITE see BMW250
CAMIVERINE see CBA375
CAMOFORM HYDROCHLORIDE see
BFX125
CAMOMILE OIL, ENGLISH TYPE (FCC) see
CDH750
CAMOMILE OIL GERMAN see CDH500
CAMPANA (CUBA, PUERTO RICO) see
AOO825
CAMPAPRIM A 1544 see AMY050
CAMPBELLINE OIL ORANGE see PEJ500
CAMPECHE (PUERTO RICO) see LED500
CAMPHANE, 2-HYDROXY- see BMD000
2-CAMPHANOL see BMD000
1-2-CAMPHANOL see NCQ820
2-CAMPHANONE see CBA750
d-2-CAMPHANONE see CBB250
CAMPHECHLOR see CDV100
CAMPHENE see CBA750
CAMPHEROL see ICE000
CAMPHERSULFOSAEURE see RFU100
CAMPHIDONIUM see TLG000
CAMPHOCHLOR see CDV100
CAMPHOCLOR see CDV100
CAMPHOFENE HUILEUX see CDV100
CAMPHOGEN see CQI000
CAMPHOL see BMD000
CAMPHO-PHENIQUE see PDO275
CAMPHOPHYLINE see CNR125
CAMPHOR see CBA750
(+)-CAMPHOR see CBB250
d-CAMPHOR see CBB250
l-(−)-CAMPHOR see CBB000
l-CAMPHOR see CBB000
(±)-CAMPHOR see CBA800
d-(+)-CAMPHOR see CBB250
dl-CAMPHOR see CBA800
(1R,4R)-(+)-CAMPHOR see CBB250
CAMPHOR-natural see CBC500
CAMPHOR, synthetic (ACGIH, DOT) see
CBA750
CAMPHORATED OIL see CBB375
CAMPHORIC ACID, P,α-
DIMETHYLBENZYL ESTER, COMPD.
WITH 2,2'-IMINODIETHANOL see HAQ570
CAMPHORIC ACID, 1-(P,α-
DIMETHYLBENZYL) ESTER, COMPD.
WITH 2,2'-IMINODIETHANOL (1:1) see
HAQ570
CAMPHOR LINIMENT see CBB375
CAMPHOR OIL see CBB500
CAMPHOR OIL, RECTIFIED see CBB500
CAMPHOR OIL WHITE see CBB500
CAMPHOR OIL YELLOW see CBB500
CAMPHORSULFONIC ACID see RFU100

(+)-CAMPHORSULFONIC ACID see RFU100
d-CAMPHORSULFONIC ACID see RFU100
d-10-CAMPHORSULFONIC ACID see
RFU100
(+)-β-CAMPHORSULFONIC ACID see
RFU100
CAMPHOR TAR see NAJ500
CAMPHOR USP see CBB250
CAMPHOZONE see DJS200
CAMPILIT see COO500
CAMPOSAN see CDS125
CAMPOVITON 6 see PPK500
CAMPTOTHECIN see CBB870
CAMPTOTHECINE see CBB870
20(S)-CAMPTOTHECINE see CBB870
CAMPTOTHECIN, SODIUM SALT see
CBB875
CAMUZULENE see DRV000
CAMYLOFINE see NOC000
CAMYLOFINE DIHYDROCHLORIDE see
AAD875
CAMYLOFINE HYDROCHLORIDE see
AAD875
CAMYLOFIN HYDROCHLORIDE see
AAD875
CAMYLOPIN see NOC000
CANACERT AMARANTH see FAG020
CANACERT BRILLIANT BLUE FCF see
FAE000
CANACERT ERYTHROSINE BS see FAG040
CANACERT INDIGO CARMINE see FAE100
CANACERT SUNSET YELLOW FCF see
FAG150
CANACERT TARTRAZINE see FAG140
CANADA MOONSEED see MRN100
CANADIAN BALSAM see FBS200
CANADIAN FIR NEEDLE OIL see CBB900
CANADIEN 2000 see BMN000
CANADINE see TCJ800
(−)-CANADINE see TCJ800
α-CANADINE see TCJ800
CANADOL see PCT250
CANAFISTOLA (CUBA) see GIW300
CANANGA see CAL125
CANANGA OIL see CBC100
CANARIO (PUERTO RICO) see AFQ625
CANARY CHROME YELLOW 40-2250 see
LCR000
CANARY IVY see AFK950
CANAVANIN see AKD500
l-CANAVANINE see AKD500
CANCARB see CBT750
CANCER JALAP see PJJ315
CANDAMIDE see LGZ000
CANDASEPTIC see CFD990
CANDELILLA (MEXICO) see SDZ475
CANDELILLA WAX see CBC175
CANDEPTIN see LFF000
CANDEREL see ARN825
CANDEX see ARQ725
CANDEX see NOH500
CANDIDA ALBICANS GLYCOPROTEINS
see CBC375
CANDIDA LIPOLYTICA see CBC400
CANDIDIA GUILLIERMONDII see CBC425
CANDIDIN see CBC500
CANDIDIN B see CBC750
CANDIDINE see CBC500
CANDIMON see LFF000
CANDIO-HERMAL see NOH500
CANDLEBERRY see TOA275
CANDLENUT see TOA275
CANDLE SCARLET 2B see SBC500
CANDLE SCARLET B see SBC500
CANDLE SCARLET G see SBC500
CANDLETOXIN A see CBD250
CANDLETOXIN B see CBD500
CANESCEGENIN see NDY000
CANESCEGENINE see NDY565
CANESCINE see RDF000
CANESCINE 10-METHOXYDERIVATIVE
see MEK700

CARBAMALDEHYDE see FMY000
CARBAMAMIDINE see GKW000
CARBAMATE see FAS000
CARBAMATE de l'ETHINYLCYCLOHEXANOL (FRENCH) see EEH000
CARBAMATE (ISOAMYL) see IHQ000
CARBAMATE de METHYLPENTINOL (FRENCH) see MNM500
CARBAMATE du PROPINYLCYCLOHEXANOL (FRENCH) see POA250
CARBAMATES see CBH750
CARBAMAZEPEN see DCV200
CARBAMAZEPINE see DCV200
CARBAMAZINE see DIW000
CARBAMEZEPINE see DCV200
CARBAMIC ACID, ACETYLMETHYL-, 1,3-BENZODIOXOL-4-YL ESTER see BCJ025
CARBAMIC ACID, ACETYLMETHYL-, 2,2-DIMETHYL-1,3-BENZODIOXOL-4-YL ESTER see BCJ005
CARBAMIC ACID, ACETYLMETHYL-, 2,3-(ISOPROPYLIDENEDIOXY)PHENYL ESTER see BCJ005
CARBAMIC ACID, ALLYL ESTER see AGA750
CARBAMIC ACID, (AMINOCARBONYL)-, 1-ETHYNYLCYCLOHEXYL ESTER (9CI) see EQL600
CARBAMIC ACID, ((4-AMINOPHENYL)SULFONYL)-, METHYL ESTER, MONOPOTASSIUM SALT see PKV700
CARBAMIC ACID, 1H-BENZIMIDAZOL-2-YL- see BCC100
CARBAMIC ACID, 1H-BENZIMIDAZOL-2-YL-, METHYL ESTER, HYDROCHLORIDE see CBN200
CARBAMIC ACID, (BIS(1-AZIRIDINYL)PHOSPHINYL)-, 1,2,3-PROPANETRIYL ESTER see CBH770
CARBAMIC ACID, ((p-BROMOPHENYL)THIO)METHYL-, 2,3-DIHYDRO-2,2-DIMETHYL-7-BENZOFURANYL ESTER see DRL100
CARBAMIC ACID, (BUTOXYSULFINYL)METHYL-, 2,3-DIHYDRO-2,2-DIMETHYL-7-BENZOFURANYL ESTER see DLH820
CARBAMIC ACID, BUTYL ESTER see BQP250
CARBAMIC ACID-2-sec-BUTYL-2-METHYLTRIMETHYLENE ESTER see MBW750
CARBAMIC ACID, (BUTYLTHIO)METHYL-, 2,3-DIHYDRO-2,2-DIMETHYL-7-BENZOFURANYL ESTER see DLH830
CARBAMIC ACID, (CHLOROCARBONYL)(4-(TRIFLUOROMETHOXY)PHENYL)-, METHYL ESTER see MIG100
CARBAMIC ACID, N-(2-CHLOROETHYL)-N-ETHYL-, 4-NITROSO-3,5-XYLYL ESTER see CBH800
CARBAMIC ACID, (4-CHLOROPHENYL)-, 1-METHYLETHYL ESTER (9CI) see IOK100
CARBAMIC ACID, 1-(4-CHLOROPHENYL)-1-PHENYL-2-PROPYNYL ESTER see CKI500
CARBAMIC ACID, ((DIBUTYLAMINO)THIO)METHYL-, 2,2-DIMETHYL-2,3-DIHYDRO-7-BENZOFURANYL ESTER see CCC280
CARBAMIC ACID, DIBUTYLDITHIO-, COPPER(II) SALT see CNJ600
CARBAMIC ACID, DI-sec-BUTYLTHIO-, S-BENZYL ESTER see TGF030
CARBAMIC ACID, ((1R,4Z,8S,13E)-8-((4,6-DIDEOXY-4-(((2,6-DIDEOXY-4-S-(4-((6-DEOXY-3-o-METH YL-α-l-manNOPYRANOSYL)OXY)-3-IODO-5,6-DIMETHOXY-2-METHYLBENZOYL)-4-THIO-B ETA-d-RIBO-

HEXOPYRANOSYL)OXY)AMINO)-2-o-(2,4-DIDEOXY-4-(ETHYLAMINO)-3-o-METHYL-α-l-THREO-PENTOPYRANOSYL)-β-d-GLUCOPYRANOSYL)OXY)-1-HYDROXY-13-(2-(METHYLTRITHIO) ETHYLIDENE)-11-OXOBICYCLO(7.3.1)TRIDECA-4,9-DIENE-2,6-DIYN-10-YL)-, METHYL ESTER see CAY700
CARBAMIC ACID, (3,4-DIETHOXYPHENYL)-, 1-METHYLETHYL ESTER see IOS100
CARBAMIC ACID, DIETHYLDITHIO-, COPPER(II) SALT see CNL300
CARBAMIC ACID, DIETHYLDITHIO-, IRON(III) SALT see IGT100
CARBAMIC ACID, DIETHYLDITHIO-, TIN(II) SALT see TGB600
CARBAMIC ACID, DIMETHYL-(9CI) see DQY950
CARBAMIC ACID-3-DIMETHYLAMINOPHENYL ESTER, METHOSULFATE see HNP500
CARBAMIC ACID, DIMETHYL-, 5-(2-((1,1-DIMETHYLETHYL)AMINO)-1-HYDROXYETHYL)-1,3-PHENYLENE ESTER see BAF100
CARBAMIC ACID, (((2,2-DIMETHYL-1,3-DIOXOLAN-4-YL)METHOXY)SULFINYL)METHYL-,3-METHYLPHENYL ESTER see MNT250
CARBAMIC ACID, DIMETHYLDITHIO-, ANHYDROSULFIDE see BJL600
CARBAMIC ACID, DIMETHYLDITHIO-, 2,4-DINITROPHENYL ESTER see DVB850
CARBAMIC ACID, DIMETHYL-, 2-(1,3-DITHIOLAN-2-YL)PHENYL ESTER see DXN830
CARBAMIC ACID, DIMETHYL-, o-(1,3-DITHIOLAN-2-YL)PHENYL ESTER see DXN830
CARBAMIC ACID, DIMETHYLDITHIO-, ZINC SALT (2:1) see BJK500
CARBAMIC ACID, ((1,1-DIMETHYLETHOXY)SULFINYL)METHYL-, 2,3-DIHYDRO-2,2-DIMETHYL-7-BENZOFURANYL ESTER see DLH860
CARBAMIC ACID, (((1,1-DIMETHYLETHOXY)SULFINYL)METHYL)-, 2-(1-METHYLETHOXY)PHENYL ESTER see INE062
CARBAMIC ACID, DIMETHYL-, ester with (m-HYDROXYPHENYL)TRIMETHYLAMMONIUM BROMIDE see POD000
CARBAMIC ACID, N,N-DIMETHYL-, m-ISOPROPYLPHENYL ESTER see DQX300
CARBAMIC ACID, ((2,6-DIMETHYLPHENOXY)SULFINYL)METHYL-, 2,3-DIHYDRO-2,2-DIMETHYL-7-BENZOFURANYL ESTER see XWS100
CARBAMIC ACID, DIPROPYLTHIO-, S-BENZYL ESTER (7CI) see PFR130
CARBAMIC ACID, (((((1,4-DITHIAN-2-YLIDENEAMINO)OXY)CARBONYL)METHYLAMINO)THIO)METHYL-,ETHYL ESTER see MLL100
CARBAMIC ACID, DITHIO-, N,N-DIMETHYL-, DIMETHYLAMINOMETHYL ESTER see DRQ650
CARBAMIC ACID, DITHIO-, MONOSODIUM SALT (8CI) see SGR600
CARBAMIC ACID, DITHIO-, SODIUM SALT see SGR600
CARBAMIC ACID, (DODECYLTHIO)METHYL-, 2,3-DIHYDRO-2,2-DIMETHYL-7-BENZOFURANYL ESTER see DLH840
CARBAMIC ACID, ESTER with CHOLINE CHLORIDE see CBH250
CARBAMIC ACID, ESTER with 2-(HDYROXYMETHYL)-1-

METHYLPENTYLISOPROPYLCARBAMATE see IPU000
CARBAMIC ACID, ESTER with 2-(HYDROXYMETHYL)-2-METHYLPENTYL BUTYLCARBAMATE see MOV500
CARBAMIC ACID, ESTER with 2-METHYL-2-PROPYL-1,3-PROPANEDIOL BUTYLCARBAMATE see MOV500
CARBAMIC ACID, ESTER with 2-METHYL-2-PROPYL-1,3-PROPANEDIOL ISOPROPYLCARBAMATE see IPU000
CARBAMIC ACID, ESTER WITH SALICYLALDEHYDE DIMETHYL ACETAL see SAG100
CARBAMIC ACID, ETHYLENEBIS(DITHIO)-, MANGANESE SALT see MAS500
CARBAMIC ACID, ETHYL ESTER see UVA000
CARBAMIC ACID, 2-ETHYLHEXYL ESTER see EKT600
CARBAMIC ACID, ETHYL-, 2-(((MERCAPTOMETHYL)THIO)ETHYL) ESTER, S-ESTER WITH o-ISOPROPYL o-METHYL PHOSPHORODITHIOATE see MEV600
CARBAMIC ACID, ETHYL-, 6-METHOXY-8-METHYL-7-OXA-3,5-DITHIA-6-PHOSPHANON-1-YL ESTER, p-SULFIDE see MEV600
CARBAMIC ACID, ETHYL-, 6-METHOXY-8-METHYL-6-SULFIDO-7-OXA-3,5-DITHIA-6-PHOSPHANON-1-YL ESTER see MEV600
CARBAMIC ACID-1-ETHYL-1-METHYL-2-PROPYNYL ESTER see MNM500
CARBAMIC ACID-2-ETHYNYL-2-BUTYL ESTER see MNM500
CARBAMIC ACID, (5-(4-FLUOROBENZOYL)-1H-BENZIMIDAZOL-2-YL)-, METHYL ESTER (9CI) see FDA887
CARBAMIC ACID, ((4-FLUOROPHENYL)THIO)METHYL-, 2,3-DIHYDRO-2,2-DIMETHYL-7-BENZOFURANYL ESTER see DLH850
CARBAMIC ACID, HEPTYL-, 1,2,3,3A,8,8A-HEXAHYDRO-1,3A,8-TRIMETHYLPYRROLO(2,3-B)INDOL-5-YL ESTER, (3AS-CIS)- see HBP410
CARBAMIC ACID, HEXAMETHYLENEBIS(DITHIO-, DISODIUM SALT see HEF400
CARBAMIC ACID, ((HEXYLOXY)SULFINYL)METHYL-, 2,3-DIHYDRO-2,2-DIMETHYL-7-BENZOFURANYL ESTER see HFU550
CARBAMIC ACID, ((HEXYLOXY)SULFINYL)METHYL-, 2-(1-METHYLETHOXY)PHENYL ESTER see INE065
CARBAMIC ACID, ((HEXYLOXY)SULFINYL)METHYL-, 3-(1-METHYLETHOXY)PHENYL ESTER see IRC060
CARBAMIC ACID, ((HEXYLOXY)SULFINYL)METHYL-, 1-NAPHTHALENYL ESTER see NAP525
CARBAMIC ACID HYDRAZIDE see HGU000
CARBAMIC ACID-2-HYDROXYETHYL ESTER see HKQ000
CARBAMIC ACID, (2-HYDROXYETHYL)-, 2-HYDROXYETHYL ESTER see HKS550
CARBAMIC ACID, 2-HYDROXY-3-(o-METHOXYPHENOXY)PROPYL ESTER see GKK000
CARBAMIC ACID-β-HYDROXYPHENETHYL ESTER see PFJ000
CARBAMIC ACID, (3-HYDROXYPHENYL)-, ETHYL ESTER (9CI) see ELE600
CARBAMIC ACID (2-ISOBUTOXYETHYL) ESTER see IIE000
CARBAMIC ACID, ISOBUTYRYLMETHYL-, m-CUMENYL ESTER see IRC050

CARBAMIC ACID, (3-ISOCYANATOMETHYLPHENYL)-, 2-ETHYLHEXYL ESTER see EKW200
CARBAMIC ACID, ISOPROPYL ESTER see IOJ000
CARBAMIC ACID, METHYL-, 3-tert-BUTYLPHENYL ESTER see BSG300
CARBAMIC ACID, METHYL-, 2,6-DI-tert-BUTYL-p-TOLYL ESTER see TAN300
CARBAMIC ACID, N-METHYL-, 3-DIETHYLARSINOPHENYL ESTER, METHIODIDE see MID900
CARBAMIC ACID-N-METHYL-3-DIMETHYLAMINOPHENYL ESTER METHIODIDE see HNO500
CARBAMIC ACID-1-METHYLETHYL ESTER see IOJ000
CARBAMIC ACID, METHYL-, 4-HYDROXY-m-CUMENYL ESTER see IPI100
CARBAMIC ACID, METHYL-, α-HYDROXY-m-CUMENYL ESTER see HLK900
CARBAMIC ACID, METHYL-, 4-HYDROXY-3,5-DIISOPROPYLPHENYL ESTER see MIA800
CARBAMIC ACID, METHYL-, 4-HYDROXY-2-ISOPROPOXYPHENYL ESTER see INE025
CARBAMIC ACID, METHYL-, 4-HYDROXY-3-ISOPROPYLPHENYL ESTER see IPI100
CARBAMIC ACID, METHYL-, o-HYDROXYPHENYL ESTER see HNK900
CARBAMIC ACID, METHYL-, o-ISOPROPYLPHENYL ESTER see MIA250
CARBAMIC ACID, METHYL-, 2,3-(METHYLENEDIOXY)PHENYL ESTER see BCJ100
CARBAMIC ACID, METHYL-, 2-(1-METHYLETHYL)PHENYL ESTER see MIA250
CARBAMIC ACID, METHYL(2-METHYL-1-OXOPROPYL)-, 3-(1-METHYLETHYL)PHENYL ESTER see IRC050
CARBAMIC ACID, METHYL-, 3-METHYLPHENYL ESTER (9CI) see MIB750
CARBAMIC ACID, N-METHYL-N-(MORPHOLINOTHIO)-, 2,3-DIHYDRO-2,2-DIMETHYL-7-BENZOFURANYL ESTER see MRU050
CARBAMIC ACID, METHYL-, NAPHTHALENYL ESTER see MME800
CARBAMIC ACID, METHYL((4-NITROPHENYL)THIO)-, 2,3-DIHYDRO-2,2-DIMETHYL-7-BENZOFURANYL ESTER see DLH870
CARBAMIC ACID, METHYLNITROSO-, o-ISOPROPOXYPHENYL ESTER see PMY310
CARBAMIC ACID, METHYLNITROSO-, 2-(1-METHYLETHOXY)PHENYL ESTER (9CI) see PMY310
CARBAMIC ACID, METHYLNITROSO-, 3-METHYLPHENYL ESTER (9CI) see MNV500
CARBAMIC ACID, METHYLNITROSO-, m-TOLYL ESTER see MNV500
CARBAMIC ACID, METHYL(1-OXOBUTYL)-, 2,2-DIMETHYL-1,3-BENZODIOXOL-4-YL ESTER see MNE300
CARBAMIC ACID, METHYL(1-OXOPENTYL)-, 2,2-DIMETHYL-1,3-BENZODIOXOL-4-YL ESTER see MNF300
CARBAMIC ACID, METHYL(1-OXOPROPYL)-, 2,2-DIMETHYL-1,3-BENZODIOXOL-4-YL ESTER see DQM650
CARBAMIC ACID-α-METHYLPHENETHYL ESTER see CBI000
CARBAMIC ACID, METHYL(PHENOXYSULFINYL)-, 2,3-DIHYDRO-2,2-DIMETHYL-7-BENZOFURANYL ESTER see DLH880
CARBAMIC ACID, METHYL(((2-PHENYL-1,3-DIOXAN-5-YL)METHOXY)SULFINYL)-, 3-METHYLPHENYL ESTER see MNV600

CARBAMIC ACID, (3-METHYLPHENYL)-, 3-((METHOXYCARBONYL)AMINO)PHENYL ESTER (9CI) see MEG250
CARBAMIC ACID, METHYLPHENYL-, 1-METHYLETHYL ESTER (9CI) see MOS400
CARBAMIC ACID, METHYL-, SPIRO(1,3-BENZODIOXOLE-2,1'-CYCLOHEXAN)-4-YL ESTER see SLD550
CARBAMIC ACID, METHYL((1,1,2,2-TETRACHLOROETHYL)THIO)-, 2,3-DIHYDRO-2,2-DIMETHYL-7-BENZOFURANYL ESTER see DLH890
CARBAMIC ACID, METHYL-, 3-TOLYL ESTER see MIB750
CARBAMIC ACID, METHYL((TRICHLOROMETHYL)THIO)-, 2,3-DIHYDRO-2,2-DIMETHYL-7-BENZOFURANYL ESTER see DLH900
CARBAMIC ACID, (2-(4-PHENOXYPHENOXY)ETHYL)-, ETHYL ESTER see EOE200
CARBAMIC ACID-3-PHENYLPROPYL ESTER see PGA750
CARBAMIC ACID, 1-PHENYL-2-PROPYNYL ESTER see PGE000
CARBAMIC ACID, 2-PHENYLTRIMETHYLENE ESTER see FAG200
CARBAMIC ACID, PROPYL ESTER see PNG250
CARBAMIC ACID, (5-(PROPYLSULFONYL)-1H-BENZIMIDAZOL-2-YL)-, METHYL ESTER see PNV900
CARBAMIC ACID, SULFANILYL-, METHYL ESTER, POTASSIUM SALT see PKV700
CARBAMIC ACID, THIO, S-ESTER with 2-MERCAPTOACETANILIDE see MCL500
CARBAMIC ACID, 2,2,2-TRICHLOROETHYL ESTER see TIO500
CARBAMIC ACID, (m-TRIMETHYLAMMONIO)PHENYL ESTER, METHYLSULFATE see HNP500
CARBAMIC ACID, VINYL ESTER see VNK000
CARBAMIC ESTER of 3-OXYPHENYLTRIMETHYLAMMONIUM METHYLSULFATE see HNP500
CARBAMIDAL see DJS200
CARBAMIDE see USS000
CARBAMIDE PEROXIDE see HIB500
CARBAMIDE PHENYLACETATE see PEC250
CARBAMIDE RESIN see USS000
CARBAMIDINE see GKW000
CARBAMIDINE HYDROCHLORIDE see GKY000
p-CARBAMIDOBENZENEARSONIC ACID see CBJ000
β-CARBAMIDOCARBOMETHYL-O,O-DIETHYLDITHIOPHOSPHATE see AAT000
p-CARBAMIDOPHENYL ARSENOUS ACID see CBI500
p-CARBAMIDOPHENYL ARSENOUS OXIDE see CBI500
4-CARBAMIDOPHENYL BIS(CARBOXYMETHYLTHIO)ARSENITE see CBI250
p-CARBAMIDOPHENYL-BIS(2-CARBOXYPHENYLMERCAPTO)ARSINE see TFD750
4-CARBAMIDOPHENYL BIS(o-CARBOXYPHENYLTHIO)ARSENITE see TFD750
p-CARBAMIDOPHENYL-DI(1'-CARBOXYPHENYL-2')THIOARSENITE see TFD750
4-CARBAMIDOPHENYLOXOARSINE see CBI500
CARBAMIDSAEURE-AETHYLESTER (GERMAN) see UVA000
CARBAMIMIDIC ACID see USS000

CARBAMIMIDOTHIOIC ACID, DECYL ESTER, MONOHYDROCHLORIDE see DAJ480
CARBAMIMIDOTHIOIC ACID-2-(DIMETHYLAMINO)ETHYL ESTER DIHYDROCHLORIDE see NNL400
CARBAMIMIDOTHIOIC ACID, ETHYL ESTER with METAPHOSPHORIC ACID (1:1) see ELY575
CARBAMIMIDOTHIOIC ACID, ETHYL ESTER, MONO(DIETHYL PHOSPHATE) see CBI675
CARBAMIMIDOTHIOIC ACID, IMINODI-2,1-ETHANEDIYL ESTER,DIHYDROBROMIDE see IBJ050
CARBAMIMIDOTHIOIC ACID, METHYL ESTER, SULFATE (1:1) see MPV750
CARBAMIMIDOTHIOIC ACID, METHYL ESTER, SULFATE (9CI) see MPV790
CARBAMINE see CBM750
CARBAMINOCHOLINE CHLORIDE see CBH250
CARBAMINOPHENYL-p-ARSONIC ACID see CBJ000
p-CARBAMINO PHENYL ARSONIC ACID see CBJ000
CARBAMINOTHIOGLYCOLIC ACID ANILIDE see MCL500
CARBAMINOYLCHOLINE CHLORIDE see CBH250
CARBAMIOTIN see CBH250
CARBAMMATO di FENILETINILCARBINOLO see PGE000
CARBAMODITHIOIC ACID, DIETHYL-(9CI) see DJC800
CARBAMODITHIOIC ACID, DIETHYL-, compd. with N-ETHYLETHANAMINE (1:1) (9CI) see DJD000
CARBAMODITHIOIC ACID, METHYL-, MONOSODIUM SALT, DIHYDRATE (9CI) see SIL550
CARBAMODITHIOIC ACID, MONOSODIUM SALT (9CI) see SGR600
CARBAMODITHIOIC ACID, PHENYL-, ETHYL ESTER see EOK550
CARBAMOHYDROXAMIC ACID see HOO500
CARBAMOHYDROXIMIC ACID see HOO500
CARBAMOHYDROXYAMIC ACID see HOO500
CARBAMOL see UTU500
CARBAMONITRILE see COH500
CARBAMOTHIOIC ACID-S,S'-(2-(DIMETHYLAMINO)-1,3-PROPANEDIYL) ESTER, MONOHYDROCHLORIDE (9CI) see BHL750
CARBAMOTHIOIC ACID, (1,2-DIMETHYLPROPYL)ETHYL-, S-(PHENYLMETHYL) ESTER see PFR120
CARBAMOTHIOIC ACID, DIPROPYL-, S-(PHENYLMETHYL) ESTER see PFR130
CARBAMOTHIOIC ACID, (6-METHOXY-2-PYRIDINYL)METHYL-, o-(3-(1,1-DIMETHYLETHYL)PHENYL) ESTER see POO800
4'-(4-CARBAMOYL-9-ACRIDINYLAMINO)-3'-METHOXY-1-PROPANESULFONANILIDE HYDROCHLORIDE see MFN000
(p-CARBAMOYLAMINO)PHENYLARSINOBIS(2-THIO-ACETIC ACID) see CBI250
2-CARBAMOYLANILINE see AID620
N-CARBAMOYLARSANILIC ACID see CBJ000
p-CARBAMOYLBENZAMIDE see TAN600
8-CARBAMOYL-3-(2-CHLOROETHYL)IMIDAZO(5,1-d)-1,2,3,5-TETRAZIN-4(3H)-ONE see MQY110
CARBAMOYLCHOLINE CHLORIDE see CBH250

γ-CARBAMOYL CHOLINE CHLORIDE see CBH250

5-CARBAMOYL-5H-DIBENZ(b,f)AZEPINE see DCV200

5-CARBAMOYLDIBENZO(b,f)AZEPINE see DCV200

5-CARBAMOYL-5H-DIBENS(b,f)AZEPINE see DCV200

N-CARBAMOYL-2-(2,6-DICHLOROPHENYL)ACETAMIDINE HYDROCHLORIDE see PFV000

N-(2-CARBAMOYLETHYL)ARSANILIC ACID SODIUM SALT see SJG000

CARBAMOYLFORMIC ACID HYDRAZIDE see OLM310

1-CARBAMOYLFORMIMIDIC ACID see OLO000

4'-CARBAMOYL-2-FORMYL-1,1'-(OXYDIMETHYLENE)DI-PYRIDINIUM-DICHLORIDE-2-OXIME see HGE900

1-CARBAMOYLGUANIDINE see GLS900

N-CARBAMOYLGUANIDINE see GLS900

CARBAMOYLHYDRAZINE see HGU000

N-CARBAMOYLHYDROXYLAMINE see HOO500

p-((CARBAMOYLMETHYL)AMINO)-BENZENEARSONIC ACID see CBJ750

N-(CARBAMOYLMETHYL)ARSANILIC ACID see CBJ750

N-(CARBAMOYLMETHYL)-2-DIAZOACETAMIDE see CBK000

p-(CARBAMOYLMETHYL)PHENOL see HNG550

1-p-CARBAMOYLMETHYLPHENOXY-3-ISOPROPYLAMINO-2-PROPANOL see TAL475

2-CARBAMOYL-2-NITROACETONITRILE see CBK125

2-CARBAMOYLNITROBENZENE see NEV523

N-(1-CARBAMOYL-4-(NITROSOCYANAMIDO)BUTYL)BENZAMIDE see CBK250

N-CARBAMOYL-N-NITROSOGLYCINE see NKI500

l-N⁵-CARBAMOYL-N⁵-NITROSOORNITHINE see NJT000

dl-N⁵-CARBAMOYL-N⁵-NITROSOORNITHINE see NJS500

CARBAMOYL OXIME see HOO500

1-CARBAMOYLOXY-2-HYDROXY-3-(o-METHYLPHENOXY)PROPANE see CBK500

3-CARBAMOYLOXY-3-METHYL-4-PENTYNE see MNM500

(3-(N-CARBAMOYLOXY)PHENYL)DIETHYLMETHYL-AMMONIUM METHOSULFATE see HNK550

1-CARBAMOYLOXY-3-PHENYLPROPANE see PGA750

((m-CARBAMOYLOXY)PHENYL)TRIMETHYLAMMONIUM METHYLSULFATE see HNP500

2-CARBAMOYLOXYPROPYLTRIMETHYLAMMONIUM CHLORIDE see HOA500

1-CARBAMOYLOXY-1-(2-PROPYNYL)CYCLOHEXANE see POA250

4-CARBAMOYL-4-PHTHALYLBUTYRIC ACID see PIA250

10-(3-(4-CARBAMOYLPIPERIDINE)PROPYL)-2-(METHANESULFONYL)PHENOTHIAZINE see MQR000

N-(1-CARBAMOYLPROPYL)ARSANILIC ACID see CBK750

4-CARBAMOYLPYRIDINE see ILC100

4-CARBAMOYL-1-β-d-RIBOFURANOSYL-IMIDAZOLIUM-5-OLATE see BMM000

CARBAMULT see CQI500

4-CARBAMYLAMINOPHENYLARSONIC ACID see CBJ000

N-CARBAMYL ARSANILIC ACID see CBJ000

CARBAMYL CHLORIDE, N,N-DIMETHYL- see DQY950

CARBAMYLCHOLINE CHLORIDE see CBH250

5-CARBAMYLDIBENZO(b,f)AZEPINE see DCV200

5-CARBAMYL-5H-DIBENZO(b,f)AZEPINE see DCV200

CARBAMYLHYDRAZINE see HGU000

CARBAMYLHYDRAZINE HYDROCHLORIDE see SBW500

CARBAMYL HYDROXAMATE see HOO500

CARBAMYLMETHYLCHOLINE CHLORIDE see HOA500

4-CARBAMYLPHENYL BIS(CARBOXYMETHYLTHIO)ARSENITE see TFA350

1-CARBAMYL-2-PHENYLHYDRAZINE see CBL000

2-CARBAMYL PYRAZINE see POL500

CARBANIL see PFK250

CARBANILALDEHYDE see FNJ000

CARBANILIC ACID, p-CHLORO-, ISOPROPYL ESTER see IOK100

CARBANILIC ACID, DITHIO-, ETHYL ESTER see EOK550

d-(−)-CARBANILIC ACID (1-ETHYLCARBAMOYL)ETHYL ESTER see CBL500

CARBANILIC ACID ETHYL ESTER see CBL750

CARBANILIC ACID, m-HYDROXY-, ETHYL ESTER see ELE600

CARBANILIC ACID ISOPROPYL ESTER see CBM000

CARBANILIC ACID, N-METHYL-, ISOPROPYL ESTER see MOS400

CARBANILIC ACID, p-SULFO-, C-METHYL ESTER, HYDRAZIDE see MQJ300

CARBANILIC ACID, p-SULFO-, N-METHYL ESTER, S-HYDRAZIDE see MQJ300

CARBANILIDE see CBM250

CARBANILIDE, 3,3'-BIS((5-((4,6,8-TRISULFO-1-NAPHTHYL)CARBAMOYL)-o-TOLYL)CARBAMOYL)- see BAT000

m-CARBANILOYLOXYCARBANILIC ACID ETHYL ESTER see EEO500

CARBANOCHLORIDIC ACID, NAPHTHYL ESTER see NBH200

CARBANOLATE see CBM500

CARBANOLATE see CGI500

CARBANTHRENE BLUE 2R see IBV050

CARBANTHRENE BLUE BCF see DFN300

CARBANTHRENE BLUE BCS see DFN300

CARBANTHRENE BLUE RBCF see DFN300

CARBANTHRENE BLUE RCS see DFN300

CARBANTHRENE BLUE RS see IBV050

CARBANTHRENE BLUE RSP see IBV050

CARBANTHRENE BRILLIANT VIOLET 4R see DFN450

CARBANTHRENE BROWN AR see CMU780

CARBANTHRENE BROWN ARP see CMU780

CARBANTHRENE BROWN BR see CMU770

CARBANTHRENE GOLDEN YELLOW see DCZ000

CARBANTHRENE NAVY BLUE G see CMU500

CARBANTHRENE NAVY BLUE RA see VGP100

CARBANTHRENE OLIVE R see DUP100

CARBANTHRENE RED BROWN 5R see CMU800

CARBANTHRENE RED G 2B see CMU825

CARBANTHRENE RED G 2BP see CMU825

CARBANTHRENE VIOLET 2R see DFN450

CARBANTHRENE VIOLET 2RP see DFN450

CARBARSONE OXIDE see CBI500

CARBARSONE (USDA) see CBJ000

CARBARYL see CBM750

CARBARYL (ACGIH,DOT,OSHA) see CBM750

CARBASED see ACE000

CARBASONE see CBJ000

CARBATENE see MQQ250

CARBATHIONE see VFU000

CARBATOX see CBM750

CARBATOX-60 see CBM750

CARBATOX-75 see CBM750

CARBAURINE see PFR325

CARBAVINE see CBM875

CARBAVUR see CBM750

CARBAX see BIO750

CARBAZALDEHYDE see FNN000

CARBAZAMIDE see HGU000

CARBAZEPINE see DCV200

CARBAZIC ACID, ETHYL ESTER see EHG000

CARBAZIC ACID HYDRAZIDE see CBS500

CARBAZIDE see CBS500

CARBAZILQUINONE see BGX750

CARBAZINC see BJK500

CARBAZINE see ADI775

CARBAZOCHROME SODIUM SULFONATE TRIHYDRATE see AER666

9H-CARBAZOL-3-AMINE, 1,4-DIMETHYL-6-METHOXY- see DSM480

9H-CARBAZOL-3-AMINE, 9-METHYL- see MMI700

CARBAZOLE see CBN000

9H-CARBAZOLE see CBN000

9H-CARBAZOLE, 1,4-DIMETHYL-3,6-DINITRO- see DRM900

9H-CARBAZOLE, 1,4-DIMETHYL-3-NITRO- see DSV300

9H-CARBAZOLE, 1,4-DIMETHYL-6-NITRO- see DSV310

9H-CARBAZOLE, 9-ETHENYL-(9CI) see VNK100

9H-CARBAZOLE, 9-ETHENYL-, HOMOPOLYMER see VNK200

CARBAZOLE, 9-ETHYL- see EHG025

9H-CARBAZOLE, 9-ETHYL-(9CI) see EHG025

CARBAZOLE, 3-(p-HYDROXYANILINO)- see CBN100

CARBAZOLE, 9-METHYL- see MIE300

9H-CARBAZOLE, 9-METHYL- see MIE300

9H-CARBAZOLE, 9-METHYL-3-NITRO- see MMI710

9H-CARBAZOLE, 2-NITRO- see NFR700

9H-CARBAZOLE, 1,3,6,8-TETRANITRO- see TDY120

CARBAZOLE, 1,4,9-TRIMETHYL- see TLN133

9H-CARBAZOLE, 1,4,9-TRIMETHYL- see TLN133

CARBAZOLE, 9-VINYL- see VNK100

CARBAZOLE, 9-VINYL-, POLYMER (8CI) see VNK200

9H-CARBAZOL-3-OL, 6-AMINO-5,8-DIMETHYL- see AJQ300

9H-CARBAZOL-3-OL, 5,8-DIMETHYL-6-NITRO- see DSH500

4H-CARBAZOL-4-ONE, 1,2,3,9-TETRAHYDRO-9-METHYL-3-((2-METHYL-1H-IMIDAZOL-1-YL)METHYL)- see OJD150

4-(3-CARBAZOLYLAMINO)PHENOL see CBN100

CARBAZONE, DIPHENYLTHIO- see DWN200

CARBAZOTIC ACID see PID000

CARBECIN see CBO250

CARBENDAZIM see MHC750

CARBENDAZIME see MHC750

CARBENDAZIME and SODIUM NITRITE (1:1) see SIQ700

CARBENDAZIM HYDROCHLORIDE see CBN200

CARBENDAZIM and SODIUM NITRITE (5:1) see CBN375

CARBENDAZOL see MHC750

CARBENDAZOLE see MHC750

CARBENDAZYM see MHC750

CARBENICILLIN DISODIUM SALT see CBO250
CARBENICILLIN PHENYL see CBN750
CARBENICILLIN PHENYL ESTER see CBN750
CARBENICILLIN PHENYL SODIUM see CBO000
CARBENICILLIN SODIUM see CBO250
CARBENOXALONE, DISODIUM SALT see CBO500
CARBENOXOLONE see BGD000
CARBENOXOLONE, DISODIUM SALT see CBO500
CARBENOXOLONE SODIUM see CBO500
CARBESTROL see CBO625
CARBETAMEX see CBL500
CARBETAMID (GERMAN) see CBL500
CARBETAMIDE see CBL500
CARBETHOXYACETIC ESTER see EMA500
3-CARBETHOXYAMINO-5-DIMETHYLAMINOACETYL-10,11-DIHYDRODIBENZ(b,f)AZEPINE HYDROCHLORIDE see BMB125
3-CARBETHOXYAMINOPHENOL see ELE600
α-CARBETHOXY-β,β-BISCYCLOPROPYL ACRYLONITRILE see DGX000
1-CARBETHOXY-1,2-DIHYDROQUINOLINE see CBO750
4-CARBETHOXY-5-(3,3-DIMETHYL-1-TRIAZENO)-2-METHYLIMIDAZOLE see CBO800
N-CARBETHOXYETHYLENIMINE see ASH750
CARBETHOXYHYDRAZINE see EHG000
1-CARBETHOXY HYDRAZINE see EHG000
N-(CARBETHOXY)HYDRAZINE see EHG000
CARBETHOXY MALAOXON see OPK250
CARBETHOXY MALATHION see MAK700
2-CARBETHOXYMETHYLENE-3-METHYL-5-PIPERIDINO-4-THIAZOLIDONE see MNG000
CARBETHOXYMETHYL ISOCYANATE see ELS600
CARBETHOXYMETHYL ISOTHIOCYANATE see ELX530
p-CARBETHOXYPHENOL see HJL000
1(4-CARBETHOXYPHENYL)-3,3-DIMETHYLTRIAZENE see CBP250
2-(N-(4-CARBETHOXY-4-PHENYL)PIPERIDINO)PROPIOPHENONE HYDROCHLORIDE see CBP325
CARBETHOXYSYRINGOYL METHYLRESERPATE see RCA200
S-CARBETHOXYTHIAMINE HYDROCHLORIDE see EEQ500
CARBETOVUR see MAK700
CARBETOX see MAK700
CARBICRON see DGQ875
CARBIDE 6-12 see EKV000
CARBIDE BLACK D see CMN230
CARBIDE BLACK DU see CMN230
CARBIDE BLACK E see AQP000
CARBIDIUM ETHANESULFONATE see CBQ125
CARBIDIUM ETHANESULPHONATE see CBQ125
CARBIDOPA see CBQ500
CARBIDOPA MONOHYDRATE see CBQ529
CARBILAZINE see DIW000
CARBIMIDE see COH500
CARBIN see CEW500
CARBINAMINE see MGC250
CARBINOL see MGB150
CARBINOLBASE DES KRISTALLVIOLETT (GERMAN) see TJK000
CARBINOLBASE des MALACHITGRUEN (GERMAN) see MAK500
CARBINOLBASE des METHYLVIOLETT (GERMAN) see MQN250
CARBINOXAMIDE MALEATE see TAI500

CARBINOXAMINE DIPHENYLDISULFONATE see CBQ575
CARBINOXAMINE MALEATE see CGD500
p-CARBINOXAMINE MALEATE see CGD500
CARBIPHENE HYDROCHLORIDE see CBQ625
CARBITAL 90 see CAT775
CARBITOL see CBR000
CARBITOL see DJD600
CARBITOL ACETATE see CBQ750
CARBITOL ACRYLATE see EHG030
CARBITOL CELLOSOLVE see CBR000
CARBITOL SOLVENT see CBR000
CARBIUM see CAT775
CARBIUM MM see CAT775
CARBOBENZOXY CHLORIDE see BEF500
N-CARBOBENZOXYGLYCINE-1,2-DIBROMOETHYL ESTER see CBR175
N-CARBOBENZOXYGLYCINE VINYL ESTER see CBR200
N-CARBOBENZOXY-l-LEUCINE-1,2-DIBROMOETHYL ESTER see CBR210
N-CARBOBENZOXY-l-LEUCINE VINYL ESTER see CBR215
CARBOBENZOXYLGLYCINE see CBR125
CARBOBENZOXYPHENYLALANINE see CBR220
CARBOBENZOXY-l-PHENYLALANINE see CBR220
N-CARBOBENZOXY-l-PHENYLALANINE see CBR220
N-CARBOBENZOXY-l-PHENYLALANINE-1,2-DIBROMOETHYL ESTER see CBR225
N-CARBOBENZOXY-l-PHENYLALANINE VINYL ESTER see CBR235
N-CARBOBENZOXY-l-PROLINE-1,2-DIBROMOETHYL ESTER see CBR245
N-CARBOBENZOXY-l-PROLINE VINYL ESTER see CBR247
CARBOBENZOYL GLYCINE see CBR125
N-CARBOBENZOYLGLYCINE see CBR125
CARBOBENZYLOXY CHLORIDE see BEF500
CARBOBENZYLOXYGLYCINE see CBR125
N-CARBOBENZYLOXYGLYCINE see CBR125
2-CARBO-n-BUTOXY-6,6-DIMETHYL-5,6-DIHYDRO-1,4-PYRONE see BRT000
CARBOCAINE see SBB000
CARBOCAINE HYDROCHLORIDE see CBR250
CARBOCALCITONIN see CBR300
CARBOCHOL see CBH250
CARBOCHOLIN see CBH250
CARBOCHROMENE HYDROCHLORIDE see CBR500
CARBOCISTEINE see CBR675
CARBOCIT see CBR675
CARBO-CORT see CMY800
CARBOCROMENE see CBR500
CARBOCYSTEINE see CBR675
CARBO D see BJE550
CARBODIHYDRAZIDE see CBS500
CARBODIIMIDE, BIS(2,6-DIISOPROPYLPHENYL)- see BJE550
CARBODIIMIDE, BIS(TRIMETHYLSILYL)-(7CI,8CI) see BLQ950
CARBODIIMIDE, DICYCLOHEXYL- see MDR800
CARBODIIMIDE, (3-DIMETHYLAMINOPROPYL)ETHYL-, MONOHYDROCHLORIDE see EAE100
CARBODIMEDON see ZTS200
CARBODIS see CBT750
CARBOETHOXYHYDRAZINE see EHG000
CARBOETHOXYMETHYL ISOTHIOCYANATE see ELX530
2-CARBOETHOXY-1-METHYLVINYL-DIETHYLPHOSPHATE see CBR750
N¹-CARBOETHOXY-N²-PHTHALAZINO HYDRAZINE see CBS000

CARBOETHOXYPHTHALAZINO HYDRAZINE see CBS000
CARBOFENOTHION (DUTCH) see TNP250
CARBOFLUORENE AMINO ESTER see CBS250
CARBOFOS see MAK700
CARBOFRAX M see SCQ000
CARBOFURAN see CBS275
CARBOGEN (8CI) see CBV250
CARBOHYDRASE see AOM125
CARBOHYDRASE, ASPERGILLUS see CBS400
CARBOHYDRASE and CELLILASE see CBS405
CARBOHYDRASE and PROTEASE, mixed see CBS410
CARBOHYDRASE, RHIZOPUS see CBS415
CARBOHYDRAZIDE see CBS500
CARBOHYDRAZIDE, 1,5-DIPHENYL- see DVY925
N-CARBOISOPROPOXY-o-ACETYL-N-PHENYL CARBAMATE see ING400
CARBOLAC see CBT750
CARBOLAC 1 see CBT750
CARBOLATED CAMPHOR see PDO275
CARBOLIC ACID see PDN750
CARBOLITH see LGZ000
CARBOLON see SCQ000
CARBOLSAEURE (GERMAN) see PDN750
CARBOMAL see BNK000
CARBOMATE see CBM750
CARBOMER 934 see ADV950
CARBOMER 940 see ADV900
CARBOMER 940 see PIB300
CARBOMER 941 see CBS600
CARBOMER 934P see ADV900
CARBOMET see CBT750
CARBOMETHENE see KEU000
2-CARBOMETHOXYANILINE see APJ250
o-CARBOMETHOXYANILINE see APJ250
2-β-CARBOMETHOXY-3-β-BENZOXYTROPANE see CNE750
4'-CARBOMETHOXY-2,3'-DIMETHYLAZOBENZENE see CBS750
4'-CARBOMETHOXY-2,3'-DIMETHYLAZOBENZOL see CBS750
CARBOMETHOXY MALATHION see CBS800
1-(CARBOMETHOXYMETHYL)AZIRIDINE see MGR900
N-(CARBOMETHOXYMETHYL)AZIRIDINE see MGR900
N-CARBOMETHOXYMETHYLIMINOPHOSPHORYL CHLORIDE see CBT125
α-2-CARBOMETHOXY-1-METHYLVINYL DIMETHYL PHOSPHATE see MQR750
2-CARBOMETHOXY-1-METHYLVINYL-N-PROPYL ETHYLPHOSPHONOAMIDOTHIOATE see MKB320
2'-CARBOMETHOXYPHENYL 4-GUANIDINOBENZOATE see CBT175
2-CARBOMETHOXY-1-PROPEN-2-YL DIMETHYL PHOSPHATE see MQR750
N-(3-CARBOMETHOXYPROPYL)-N-(1-ACETOXYBUTYL)NITROSAMINE see MEI000
p-CARBOMETHOXYTOLUENE see MPX850
CARBOMYCIN see CBT250
CARBOMYCIN A see CBT250
CARBON see CBT500
CARBON-12 see CBT500
CARBON, activated (DOT) see CBT500
CARBON, animal or vegetable origin (DOT) see CBT500
CARBONA see CBY000
CARBON, ACTIVATED see CDI000
N-(CARBONAMIDO-2 CHROMONE)-1-((CHROMONYL AMINO)-2 CARBONYL)-5-PYRROLIDONE-2 see CBT600
CARBONATE MAGNESIUM see MAC650

(CARBONATO)DIHYDROXYDICOBALT see BAP800

(CARBONATO)DIHYDROXYDICOPPER see CNJ750

CARBONAZIDIC ACID, 1,1-DIMETHYLETHYL ESTER see BQI250

CARBONAZIDODITHIOIC ACID see ASF750

CARBON BICHLORIDE see PCF275

CARBON BISULFIDE (DOT) see CBV500

CARBON BISULPHIDE see CBV500

CARBON BLACK see CBT750

CARBON BLACK, ACETYLENE see CBT750

CARBON BLACK BV and V see CBT750

CARBON BLACK, CHANNEL see CBT750

CARBON BLACK, FURNACE see CBT750

CARBON BLACK, LAMP see CBT750

CARBON BLACK, THERMAL see CBT750

CARBON BROMIDE see CBX750

CARBON CHLORIDE see CBY000

CARBON CHLOROSULFIDE see TFN500

CARBON D see DXD200

CARBON DICHLORIDE see PCF275

CARBON DIFLUORIDE OXIDE see CCA500

CARBON DIOXIDE see CBU250

CARBON DIOXIDE, mixture with NITROGEN OXIDE (N₂O) see CBV000

CARBON DIOXIDE mixed with NITROUS OXIDE see CBV000

CARBON DIOXIDE–NITROUS OXIDE mixture (DOT) see CBV000

CARBON DIOXIDE mixed with OXYGEN see CBV250

CARBON DIOXIDE-OXYGEN mixture (DOT) see CBV250

CARBON DIOXIDE, solid (UN 1845) (DOT) see CBU250

CARBON DIOXIDE, refrigerated liquid (UN 2187) (DOT) see CBU250

CARBON DISULFIDE see CBV500

CARBON DISULPHIDE see CBV500

CARBONE (OXYCHLORURE de) (FRENCH) see PGX000

CARBONE (OXYDE de) (FRENCH) see CBW750

CARBONE (SUFURE de) (FRENCH) see CBV500

CARBON FERROCHROMIUM see FBD000

CARBON FLUORIDE see CBY250

CARBON FLUORIDE HYDRIDE see MJQ300

CARBON FLUORIDE OXIDE see CCA500

CARBON HEXACHLORIDE see HCI000

CARBON HYDRIDE NITRIDE (CHN) see HHS000

CARBONIC ACID, ALLYL ESTER, DIESTER with DIETHYLENE GLYCOL see AGD250

CARBONIC ACID, AMMONIUM SALT see ANE000

CARBONIC ACID ANHYDRIDE see CBU250

CARBONIC ACID, BARIUM SALT (1:1) see BAJ250

CARBONIC ACID BERYLLIUM SALT (1:1) see BFP750

CARBONIC ACID, BERYLLIUM SALT, BASIC see BFP755

CARBONIC ACID, BERYLLIUM SALT, MIXTURE WITH BERYLLIUM HYDROXIDE see BFP755

CARBONIC ACID, BIS(2-ISOCYANATOETHYL) ESTER see CBV600

CARBONIC ACID, BIS(2-ISOCYANATOETHYL) ESTER see IKG400

CARBONIC ACID BIS(2-METHYLALLYL) ESTER see CBV750

CARBONIC ACID-2-sec-BUTYL-4,6-DINITROPHENYL-2,4-DINITROPHENYL ESTER (8CI) see DVC200

CARBONIC ACID-2-sec-BUTYL-4,6-DINITROPHENYL ISOPROPYL ESTER see CBW000

CARBONIC ACID, BUTYL ESTER, ESTER WITH 2-(p-IODOBENZYL)BUTANOL see CBW100

CARBONIC ACID, BUTYL 3-(p-IODOPHENYL)PROPYL ESTER see CBW100

CARBONIC ACID, CADMIUM SALT see CAD800

CARBONIC ACID, CALCIUM SALT (1:1) see CAO000

CARBONIC ACID, CALCIUM SALT (1:1) see CAT775

CARBONIC ACID, CHROMIUM SALT see CMJ100

CARBONIC ACID, COBALT COMPLEX see BAP800

CARBONIC ACID, COBALT(2+) SALT (1:1) see CNB475

CARBONIC ACID, COBALT(2+) SALT, BASIC see BAP802

CARBONIC ACID, COPPER(2+) SALT (1:1) see CBW200

CARBONIC ACID, CYCLIC 3-CHLOROPROPYLENE ESTER see CBW400

CARBONIC ACID, CYCLIC ETHYLENE ESTER see GHM000

CARBONIC ACID, CYCLIC ETHYLETHYLENE ESTER see BOT200

CARBONIC ACID CYCLIC PROPYLENE ESTER see CBW500

CARBONIC ACID, DIAMMONIUM SALT see ANE000

CARBONIC ACID, DICESIUM SALT see CDC750

CARBONIC ACID, DIESTER with 1-(2,3-DIMETHYLPHENYL)-3-(2-HYDROXYETHYL)UREA see PLK580

CARBONIC ACID DIHYDRAZIDE see CBS500

CARBONIC ACID, DILITHIUM SALT see LGZ000

CARBONIC ACID, 1,3-DIMETHYLBUTYL p-IODOBENZYL ESTER see IEF200

CARBONIC ACID, 1,3-DIMETHYLBUTYL (4-IODOPHENYL)METHYL ESTER see IEF200

CARBONIC ACID, 1,1-DIMETHYLETHYL 4-ETHENYLPHENYL ESTER see BPI400

CARBONIC ACID, DIPHENYL ESTER see DVZ000

CARBONIC ACID, DIPOTASSIUM SALT see PLA000

CARBONIC ACID, DIRUBIDIUM SALT see RPB200

CARBONIC ACID, DISILVER(1+) SALT see SDN200

CARBONIC ACID, DISODIUM SALT see SFO000

CARBONIC ACID, DITHALLIUM(1+) SALT see TEJ000

CARBONIC ACID, DITHIODI-, o,o'-DIMETHYL S,S'-(2,3-QUINOXALINEDIYL) ESTER see BHL800

CARBONIC ACID, DITHIO-, O-ETHYL ESTER see EQG600

CARBONIC ACID, DITHIO-, O-ISOPROPYL ESTER, POTASSIUM SALT see IRG050

CARBONIC ACID, DITHIO-, O-PENTYL ESTER, POTASSIUM SALT see PKV100

CARBONIC ACID, 1-ETHYLBUTYL p-IODOBENZYL ESTER see HFQ510

CARBONIC ACID, 1-ETHYLBUTYL (4-IODOPHENYL)METHYL ESTER see HFQ510

CARBONIC ACID, 2-ETHYLHEXYL p-IODOBENZYL ESTER see EKW100

CARBONIC ACID, 2-ETHYLHEXYL (4-IODOPHENYL)METHYL ESTER see EKW100

CARBONIC ACID, ETHYL p-IODOBENZYL ESTER see ELQ050

CARBONIC ACID, ETHYL p-IODOPHENETHYL ESTER see ELQ200

CARBONIC ACID, ETHYL (4-IODOPHENYL)METHYL ESTER see ELQ050

CARBONIC ACID GAS see CBU250

CARBONIC ACID, compd. with GUANIDINE (1:2) see GKW100

CARBONIC ACID, HEXYL p-IODOBENZYL ESTER see HFQ520

CARBONIC ACID, HEXYL (4-IODOPHENYL)METHYL ESTER see HFQ520

CARBONIC ACID, p-IODOBENZYL ISOBUTYL ESTER see IEF100

CARBONIC ACID, p-IODOBENZYL PENTYL ESTER see AOJ950

CARBONIC ACID, (4-IODOPHENYL)METHYL 2-METHYLPROPYL ESTER see IEF100

CARBONIC ACID, (4-IODOPHENYL)METHYL PENTYL ESTER see AOJ950

CARBONIC ACID 3-(p-IODOPHENYL)-3-METHYLPROPYL ISOPROPYL ESTER see CBW600

CARBONIC ACID, ISOPROPYL ESTER, ESTER WITH 3-(p-IODOPHENYL)BUTANOL see CBW600

CARBONIC ACID, LEAD(2+) SALT (1:1) see LCP000

CARBONIC ACID LITHIUM SALT see LGZ000

CARBONIC ACID, MAGNESIUM SALT see MAC650

CARBONIC ACID METHYL-4-(o-TOLYLAZO)-o-TOLYL ESTER see CBS750

CARBONIC ACID, MONOAMMONIUM SALT see ANB250

CARBONIC ACID MONOSODIUM SALT see SFC500

CARBONIC ACID, NICKEL SALT (1:1) see NCY500

CARBONIC ACID, NICKEL SALT, BASIC see NCY600

CARBONIC ACID, THIO-, o,o'-DIMETHYL S,S'-2,3-QUINOXALINEDIYL ESTER see BHL800

CARBONIC ACID, THIO-, o-METHYL ESTER, S,S-DIESTER WITH 2,3-QUINOXALINEDITHIOL see BHL800

CARBONIC ACID, TRITHIO-, BIS(2-CHLOROETHYL)ESTER see CBW650

CARBONIC ACID, TRITHIO-, CYCLIC ETHYLENE ESTER see EJQ100

CARBONIC ACID, cyclic VINYLENE ESTER see VOK000

CARBONIC ACID, ZINC SALT (1:1) see ZEJ050

CARBONIC ANHYDRASE INHIBITOR NO. 6063 see AAI250

CARBONIC ANHYDRIDE see CBU250

CARBONIC DIAZIDE see CCA000

CARBONIC DIFLUORIDE see CCA500

CARBONIC DIHYDRAZIDE see CBS500

CARBONIC DIHYDRAZIDE, 2,2'-DIPHENYL-(9CI) see DVY925

CARBONIC OXIDE see CBW750

CARBONIMIDIC DICHLORIDE, PHENYL-(9CI) see PFJ400

CARBONIMIDIC DIHYDRAZIDE, BIS((4-CHLOROPHENYL)METHYLENE)- see RLK890

CARBONIMIDODITHIOIC ACID, 3-PYRIDINYL-, BIS((4-(1,1-DIMETHYLETHYL)PHENYL)METHYL) ESTER see BJK660

CARBONIMIDODITHIOIC ACID, 3-PYRIDINYL-, BIS((4-(1-METHYLETHYL)PHENYL)METHYL) ESTER see BKS770

CARBONIMIDODITHIOIC ACID, 3-PYRIDINYL-, (4-BROMOPHENYL)METHYL BUTYL ESTER see BNX050

CARBONYL DIAMIDE see USS000
CARBONYLDIAMINE see USS000
CARBONYL DIAZIDE see CCA000
CARBONYL DICYANIDE see OOM400
CARBONYL DIFLUORIDE see CCA500
CARBONYLDIHYDRAZINE see CBS500
CARBONYL DIISOTHIOCYANATE see
CCA125
CARBONYL FLUORIDE see CCA500
CARBONYL IRON see IGK800
CARBONYL LITHIUM see CCB250
CARBONYL POTASSIUM see CCB500
CARBONYLS see CCB609
CARBONYL SODIUM see CCB750
CARBONYL SULFIDE see CCC000
CARBONYL SULFIDE-³²S see CCC000
CARBOPHOS see MAK700
CARBOPLATIN see CCC075
CARBOPOL 934 see ADV900
CARBOPOL 934 see ADV950
CARBOPOL 940 see ADV900
CARBOPOL 940 see PIB300
CARBOPOL 941 see ADV900
CARBOPOL 941 see CBS600
CARBOPOL 960 see ADV900
CARBOPOL 961 see ADV900
CARBOPOL 934P see ADV900
CARBOPOL CV 940 see PIB300
CARBOPOL EXTRA see CBT500
CARBOPOL M see CBT500
CARBOPOL Z 4 see CBT500
CARBOPOL Z EXTRA see CBT500
CARBOPROST see CCC100
CARBOPROST TROMETHAMINE see
CCC110
CARBOQUONE see BGX750
CARBORAFFIN see CDI000
CARBORAFINE see CDI000
m-CARBORANE see NBV100
o-CARBORANE see DEJ500
o-CARBORANE(12) see DEJ500
m-CARBORANEDIMETHANOL see CCC120
o-CARBORANEDIMETHANOL see CCC130
CARBORANYLMETHYLETHYL SULFIDE
see DEJ600
CARBORANYLMETHYLPROPYL SULFIDE
see DEJ800
CARBOREX 2 see CAT775
CARBORUNDEUM see SCQ000
CARBORUNDUM see SCQ000
CARBOSET see ADV900
CARBOSET 515 see ADV900
CARBOSET RESIN NO. 515 see ADV900
CARBOSIEVE see CBT500
CARBOSORBIT R see CBT500
CARBOSPOL see AGJ250
CARBOSTESIN see BOO000
CARBOSTYRIL see CCC250
CARBOSTYRIL, THIO- see QOJ100
CARBOSULFAN see CCC280
2-CARBOSYPYRIDINE see PIB930
CARBOTHIALDIN see DSB200
CARBOTHIALDIN D47 see RCK730
CARBOTHIALDINE see DSB200
CARBOTHIALDINE see RCK730
CARBOTHRONE see APM500
CARBOWAX see PJW000
CARBOWAX see PJT200
CARBOWAX 1000 see PJT250
CARBOWAX 1500 see PJT500
CARBOWAX 4000 see PJT750
CARBOWAX 6000 see PJU000
CARBOWAX 1000 DISTEARATE see PJU750
CARBOWAX 1000 MONOSTEARATE see
PJW000
16-β-CARBOXAMIDE-3-β-ACETOXY-Δ⁵-(17-
α)-ISOPREGNENE-20-ONE see HND800
CARBOXAMIDE HYDROCHLORIDE see
CCC290
CARBOXAMIDOACETAMIDE see MAO000
1-p-(CARBOXAMIDOPHENYL)-3,3-
DIMETHYLTRIAZINE see CCC325

5-CARBOXANILIDO-2,3-DIHYDRO-6-
METHYL-1,4-OXATHIIN see CCC500
CARBOXIDE see CAT225
CARBOXINE see CCC500
CARBOXIN (USDA) see CCC500
2-CARBOXYACETANILIDE see AAJ150
CARBOXYACETIC ACID see CCC750
CARBOXYACETYLENE see PMT275
1-(p-CARBOXYAETHYLPHENYL)-3,3-
DIMETHYLTRIAZEN (GERMAN) see
CBP250
CARBOXYANILINE see API500
2-CARBOXYANILINE see API500
3-CARBOXYANILINE see AIH500
4-CARBOXYANILINE see AIH600
o-CARBOXYANILINE see API500
p-CARBOXYANILINE see AIH600
(3-(α-CARBOXY-o-ANISAMIDO)-2-(2-
HYDROXYETHOXY)PROPYL)HYDROXY-
MERCURY MONOSODIUM SALT see
HLO300
(3-(α-CARBOXY-p-ANISAMIDO)-2-
HYDROXYPROPYL)HYDROXYMERCURY
see HLP500
3-(α-CARBOXY-o-ANISAMIDO)-2-
METHOXYPROPYL HYDROXYMERCURY,
MONOSODIUM SALT see SIH500
9-CARBOXYANTHRACENE see APG550
3-CARBOXYAZETIDINE see ASC350
2-CARBOXYBENZALDEHYDE see FNK010
o-CARBOXYBENZALDEHYDE see FNK010
CARBOXYBENZENE see BCL750
p-CARBOXYBENZENEBORONIC ACID see
CCI600
4-CARBOXYBENZENESULFONAMIDE see
SNL890
p-CARBOXYBENZENESULFONAMIDE see
SNL890
p-
CARBOXYBENZENESULFONDICHLOROA
MIDE see HAF000
CARBOXYBENZENESULFONYL AZIDE see
CCD625
(o-CARBOXYBENZOYL)-p-
AMINOPHENYLSULFONAMIDOTHIAZOL
E see PHY750
N⁴-(o-CARBOXYBENZOYL)-N'-2-
THIAZOLYL-SULFANILAMIDE
SULFAPHIHALAZOLE see PHY750
CARBOXYBENZYLPENICILLIN PHENYL
ESTER SODIUM SALT see CBO000
CARBOXYBENZYLPENICILLIN SODIUM
see CBO250
4-CARBOXYBIPHENYL see PEL600
p-CARBOXYBROMOBENZENE see BMU100
α-CARBOXYCAPROYL-N,N'-
DIPHENYLHYDRAZINE CALCIUM SALT
HEMIHYDRATE see CCD750
N-(2-
CARBOXYCAPROYL)HYDRAZOBENZENE
CALCIUM SALT HEMIHYDRATE see
CCD750
p-CARBOXYCARBANILIC ACID-4-BIS(2-
CHLOROETHYLAMINO)PHENYL ESTER
see CCE000
(6R-(6-α,7-β(R*)))-1-((2-CARBOXY-7-(((((5-
CARBOXY-1H-IMIDAZOL-4-
YL)CARBONYL)AMINO)PHENYLACETYLE
)AMINO)-8-OXO-5-THIA-1-
AZABICYCLO(4.2.0)OCT-2-EN-3-
YL)METHYL)-4-(2-SULFOETHYL)-
PYRIDINIUM HYDROXODIE, inner salt,
MONOSODIUM SALT see CCS525
p-CARBOXYCHLOROBENZENE see CEL300
CARBOXYCYCLOHEXANE see HDH100
CARBOXYCYCLOPHOSPHAMIDE see
CCE250
N-(1-((1-CARBOXY-5-DIAZO-4-
OXOPENTYL)CARBAMOYL)-5-DIAZO-4-
OXOPENTYL)-GLUTAMINE SODIUM SALT
see ASO510

2-CARBOXY-1-((2,6-DICARBOXY-2,3-
DIHYDRO-4(1H)-
PYRIDINYLIDENE)ETHYLIDENE)5,6-
DIHYDROXY-1H-INDOLIUM,
HYDROXIDE, INNER SALT, SULFATE
(SALT) see BFV900
2-CARBOXY-4'-
(DIMETHYLAMINO)AZOBENZENE see
CCE500
3'-CARBOXY-4-
(DIMETHYLAMINO)AZOBENZENE see
CCE750
N-(2-CARBOXY-3,3-DIMETHYL-7-OXO-4-
THIA-1-AZABICYCLO(3.2.0)HEPT-6-YL)-2-
PHENYL-MALONAMIC ACID DISODIUM
SALT see CBO250
2-CARBOXYDIPHENYLAMINE see PEG500
2-CARBOXYDIPHENYLARSINOUS ACID
see PFI600
CARBOXYETHANE see PMU750
2-CARBOXYETHANEPHOSPHONIC ACID
see PHA560
1-CARBOXYETHANE-2-PHOSPHONIC
ACID see PHA560
2-CARBOXYETHANE-1-PHOSPHONIC
ACID see PHA560
2-CARBOXYETHYL ACRYLATE see
HGO700
β-CARBOXYETHYL ACRYLATE see
HGO700
S-2-CARBOXYETHYL-l-CYSTEINE see
CCF250
(2-
CARBOXYETHYL)DIETHYLMETHYLAMM
ONIUM IODIDE ETHYL ESTER see CCE800
(2-
CARBOXYETHYL)DIMETHYLETHYLAMM
ONIUM IODIDE ETHYL ESTER see CCE820
CARBOXYETHYLGERMANIUM
SESQUIOXIDE see CCF125
2-CARBOXYETHYLGERMASESQUIOXANE
see CCF125
3-CARBOXY-1-ETHYL-7-METHYL-1,8-
NAPHTHIDIN-4-ONE see EID000
N-(2-CARBOXYETHYL)-N-NITRO-β-
ALANINE see NHI600
4'-(2-CARBOXYETHYL)PHENYL-trans-4-
AMINOMETHYLCYCLOHEXANE
CARBOXYLATE HYDROCHLORIDE see
CDF380
2-(4-(1-CARBOXYETHYL)PHENYL)-1-
ISOINDOLINONE see IDA400
CARBOXYETHYLPHOSPHONIC ACID see
PHA560
2-CARBOXYETHYLPHOSPHONIC ACID see
PHA560
2-CARBOXYETHYL 2-PROPENOATE see
HGO700
3-((2-CARBOXYETHYL)THIO)ALANINE see
CCF250
6-(2-CARBOXYETHYLTHIO)PURINE see
CCF270
(2-
CARBOXYETHYL)TRIMETHYLAMMONIU
M IODIDE N-BUTYL ESTER see CCF300
(2-
CARBOXYETHYL)TRIMETHYLAMMONIU
M IODIDE ETHYL ESTER see CCF330
12-CARBOXYEUDESMA-3,11(13)-DIENE see
EQR000
2-CARBOXYFURAN see FQF000
3-CARBOXY-4-
HYDROXYBENZENESULFONIC ACID see
SOC500
2'-CARBOXY-2-HYDROXY-4-
METHOXYBENZOPHENONE(o-(2-
HYDROXY-p-ANISOYL)BENZOIC ACID)
see HLS500
3-(3-CARBOXY-4-HYDROXYPHENYL)-2-
PHENYL-4,5-DIHYDRO-3H-
BENZ(e)INDOLE see FAO100

(3-CARBOXY-2-HYDROXYPROPYL)TRIMETHYLAMMONIUM CHLORIDE see CCK650

(−)-(3-CARBOXY-2-HYDROXYPROPYL)TRIMETHYLAMMONIUM CHLORIDE see CCK660

l-(3-CARBOXY-2-HYDROXYPROPYL)TRIMETHYLAMMONIUM CHLORIDE see CCK660

(±)-(3-CARBOXY-2-HYDROXYPROPYL)TRIMETHYLAMMONIUM CHLORIDE see CCK655

3-CARBOXY-5-HYDROXY-1-p-SULFOPHENYL-4-p-SULFOPHENYLAZOPYRAZOLE TRISODIUM SALT see FAG140

(R)-3-CARBOXY-2-HYDROXY-N,N,N-TRIMETHYL-1-PROPANAMINIUM CHLORIDE see CCK660

3-CARBOXY-2-HYDROXY-N,N,N-TRIMETHYL-1-PROPANAMINIUM CHLORIDE (9CI) see CCK650

1-(4'-CARBOXYLAMIDOPHENYL)-3,3-DIMETHYLTRIAZINE see CCC325

(3-(4-(CARBOXYLATOMETHOXY)PHENYL)-2-HYDROXYPROPYL)HYDROXY-MERCURATE(1-), SODIUM see CCF500

(4-CARBOXYLATOPHENYL)HYDROXYMERCURATE(1-) HYDROGEN see HLN800

l-N-CARBOXYLEUCINE-N-BENZYL-1-(1,2-DIBROMOETHYL) ESTER see CBR210

l-N-CARBOXYLEUCINE N-BENZYL 1-VINYL ESTER see CBR215

6-CARBOXYL-4-HYDROXYLAMINOQUINOLINE-1-OXIDE see CCF750

CARBOXYLIC ACID HYDROCHLORIDE see FLT200

6-CARBOXYL-4-NITROQUINOLINE-1-OXIDE see CCG000

CARBOXYMETHOCEL A see AGW700

(CARBOXYMETHOXY)AMINE see ALQ650

2-(CARBOXY-METHOXY)BENZALDEHYDE SODIUM SALT see CCI250

(3-(o-(CARBOXYMETHOXY)BENZAMIDO)-2-METHOXYPROPYL)HYDROXY MERCURY, MONOSODIUM SALT compounded with THEOPHYLLINE see SAQ000

2'-CARBOXYMETHOXY-4,4'-BIS(3-METHYL-2-BUTENYLOXY)CHALCONE see IMS300

(4-(CARBOXY METHOXY)-3-CHLOROPHENYL)(5,5-DIETHYL-2,4,6(1H,3H,5H)-PYRIMIDINETRIONATO)-O²-MERCURY, MONOSODIUM SALT see CCG500

4-CARBOXYMETHYLANILINE see AID700

4-CARBOXYMETHYLBIPHENYL see BGE125

CARBOXYMETHYL CARBAMIMONIOTHIOATE CHLORIDE see CCH199

1-CARBOXYMETHYL-1-CARBOXYETHOXYETHYL-2-COCO-IMIDAZOLINIUM BETAINE see AOC250

CARBOXYMETHYL CELLULOSE see SFO500

CARBOXYMETHYL CELLULOSE, AMMONIUM SALT see CCH000

CARBOXYMETHYLCELLULOSE NORDIC see CCH000

CARBOXYMETHYL CELLULOSE, SODIUM see SFO500

CARBOXYMETHYL CELLULOSE, SODIUM SALT see SFO500

l-CARBOXYMETHYLCYSTEINE see CBR675

S-(CARBOXYMETHYL)CYSTEINE see CBR675

S-CARBOXYMETHYLCYSTEINE see CCH125

4-o-(CARBOXYMETHYL)-1-DEOXY-1,4-DIHYDRO-4-HYDROXY-1-OXO-RIFAMYCIN γ-LACTONE see RKP400

N-(CARBOXYMETHYL)-N,N-DIMETHYL-1-HEXADECANAMINIUM HYDROXIDE inner salt see CDF450

N-(CARBOXYMETHYL)-N-((2-(2,6-DIMETHYLPHENYL)AMINO)-2-OXOETHYL)-GLYCINE (9CI) see LFO300

3,3'-(CARBOXYMETHYLENE)BIS(4-HYDROXYCOUMARIN) ETHYL ESTER see BKA000

3-CARBOXYMETHYLENEPHTHALIDE see CCH150

N-(CARBOXYMETHYL)GLYCINE see IBH000

(CARBOXYMETHYL)HEXADECYLDIMETHYLAMMONIUM HYDROXIDE, inner salt (7CI) see CDF450

N-(CARBOXYMETHYL)-N'-(2-HYDROXYETHYL)-N,N'-ETHYLENEDIGLYCINE see HKS000

(o-CARBOXYMETHYL)HYDROXYLAMINE see ALQ650

((CARBOXYMETHYLIMINO)BIS(ETHYLENENITRILO))TETRAACETIC ACID see DJG800

2-CARBOXYMETHYLISOTHIOURONIUM CHLORIDE see CCH199

N-(γ-CARBOXYMETHYL-MERCAPTO-MERCURI-β METHOXY)PROPYL CAMPHORAMIC ACID see TFK260

N-(γ-CARBOXYMETHYLMERCAPTOMERCURI-β-METHOXY)PROPYLCAMPHORAMIC ACID DISODIUM SALT see TFK270

2-(CARBOXYMETHYLMERCAPTO)PHENYLSTIBONIC ACID see CCH250

2-(CARBOXYMETHYLMERCAPTO)PHENYL-STIBONSAEURE (GERMAN) see CCH250

1-CARBOXYMETHYL-1-METHYLPYRROLIDINIUM IODIDE METHYL ESTER see CCH300

5-CARBOXYMETHYL-3-METHYL-2H-1,3,5-THIADIAZINE-2-THIONE see MPP750

CARBOXYMETHYLNITROSOUREA see CCH500

CARBOXYMETHYLNITROSOUREA see NKI500

1-(CARBOXYMETHYL)-1-NITROSOUREA see NKI500

6-(5-CARBOXY-3-METHYL-2-PENTENYL)-7-HYDROXY-5-METHOXY-4-METHYLPHTHALIDE see MRX000

3-CARBOXY-1-METHYLPYRIDINIUM HYDROXIDE INNER SALT see TKL890

4-o-(CARBOXYMETHYL)RIFAMYCIN see RKK000

(CARBOXYMETHYLTHIO)ACETIC ACID see MCM750

3-(CARBOXYMETHYLTHIO)ALANINE see CBR675

3-((CARBOXYMETHYL)THIO)ALANINE see CCH125

l-3-((CARBOXYMETHYL)THIO)ALANINE see CBR675

2-CARBOXYMETHYLTHIOBENZOTHIAZOLE see CCH750

5-CARBOXYMETHYL-3-p-TOLYL-THIAZOLIDINE-2,4-DIONE-2-ACETOPHENONE HYDRAZONE see CCH800

(CARBOXYMETHYL)TRIMETHYLAMMONIUM CHLORIDE see CCH850

(CARBOXYMETHYL)TRIMETHYLAMMONIUM CHLORIDE HYDRAZIDE see GEQ500

(CARBOXYMETHYL)TRIMETHYLAMMONIUM HYDROXIDE, inner salt see GHA050

(CARBOXYMETHYL)TRIMETHYLAMMONIUM IODIDE-4-(DIMETHYLAMINO)-3-ISOPROPYLPHENYL ESTER see IOU200

(CARBOXYMETHYL)TRIMETHYLAMMONIUM IODIDE-5-(DIMETHYLAMINO)-4-ISOPROPYL-o-TOLYL ESTER see DOW875

(CARBOXYMETHYL)TRIMETHYLAMMONIUM IODIDE-6-(DIMETHYLAMINO)-4-ISOPROPYL-m-TOLYL ESTER see DQD500

(CARBOXYMETHYL)TRIMETHYLAMMONIUM IODIDE METHYL ESTER see CCH890

N-CARBOXY-3-MORPHOLINOSYDNONIMINE ETHYL ESTER see MRN275

1-CARBOXYNAPHTHALENE see NAV490

1-CARBOXY-4-NITROBENZENE see CCI250

6-CARBOXY-4-NITROQUINOLINE-1-OXIDE see CCG000

3-(3-CARBOXY-1-OXOPROPOXY)-11-OXOOLEAN-12-EN-29-OIC ACID, DISODIUM SALT (3-β,20-β) see CBO500

21-(3-CARBOXY-1-OXOPROPOXY)-5-β-PREGNANE-3,20-DIONE SODIUM SALT see VJZ000

α-(3-CARBOXY-1-OXOSULFOPROPYL)-ω-HYDROXY-POLY(OXY-1,2-ETHANEDIYL), C10-C16 ALKYL ETHERS, DISODIUM SALTS see CCI300

3-CARBOXY-2,4-PENTADIENALLACTOL see APV750

2-(5-CARBOXYPENTYL)-4-THIAZOLIDONE see CCI500

3-CARBOXYPHENOL see HJI100

4-CARBOXYPHENOL see SAI500

(2-((2-CARBOXYPHENOXY)CARBONYL)PHENYL)-1-(4-CHLOROBENZOYL)-5-METHOXY-2-METHYLINDOLE-3-ACETATE see CCI525

o-CARBOXYPHENYL ACETATE see ADA725

l-N-CARBOXY-3-PHENYLALANINE-N-BENZYL ESTER see CBR220

l-N-CARBOXY-3-PHENYLALANINE N-BENZYL 1-VINYL ESTER see CBR235

p-CARBOXYPHENYLAMINE see AIH600

2-((2-CARBOXYPHENYL)AMINO)-4-CHLOROBENZOIC ACID DISODIUM SALT see LHZ600

4-((2-CARBOXYPHENYL)AMINO)-7-CHLOROQUINOLINE α-MONOGLYCERIDE see GGQ050

p-CARBOXY PHENYLARSENOXIDE see CCI550

9-(2-CARBOXYPHENYL)-3,6-BIS(METHYLAMINO)XANTHYLIUM PERCHLORATE see RGW100

4-CARBOXYPHENYLBORONIC ACID see CCI600

m-CARBOXYPHENYLBORONIC ACID see CCI590

p-CARBOXYPHENYLBORONIC ACID see CCI600

(p-CARBOXYPHENYL)CHLOROMERCURY see CHU500

9-o-CARBOXYPHENYL-6-DIETHYLAMINO-3-ETHYLIMINO-3-ISOXANTHENE, 3-ETHOCHLORIDE see FAG070

(9-(o-CARBOXYPHENYL)-6-(DIETHYLAMINO)-3H-XANTHEN-3-YLIDENE) DIETHYLAMMONIUM CHLORIDE see FAG070

(4-CARBOXYPHENYL)HYDRAZINE see HHB100

p-CARBOXYPHENYLHYDRAZINE see HHB100

9-(o-CARBOXYPHENYL)-6-HYDROXY-3-ISOXANTHENONE see FEV000

9-o-CARBOXYPHENYL-6-HYDROXY-3-ISOXANTHONE, DISODIUM SALT see FEW000

(2-CARBOXYPHENYL)HYDROXYMERCURY see HLO450

(o-CARBOXYPHENYL)HYDROXYMERCURY see HLO450

(p-CARBOXYPHENYL)HYDROXYMERCURY see HLN800

(o-CARBOXYPHENYL)HYDROXY-MERCURY SODIUM SALT see SHT500

9-(o-CARBOXYPHENYL)-6-HYDROXY-2,4,5,7-TETRAIODO-3-ISOXANTHONE see FAG040

9-(o-CARBOXYPHENYL)-6-HYDROXY-3H-XANTHEN-3-ONE see FEV000

4'-CARBOXYPHENYLMETHANESULFONANILIDE, SODIUM SALT see CCJ000

o-CARBOXYPHENYL PHOSPHATE see PHA575

(S)-1-(N²)-(1-CARBOXY-3-PHENYLPROPYL)-l-LYSYL)-l-PROLINE see LGM400

(S)-1-(N²)-(1-CARBOXY-3-PHENYLPROPYL)-l-LYSYL)-l-PROLINE DIHYDRATE see CCJ100

((o-CARBOXYPHENYL)THIO)ETHYLMERCURY SODIUM SALT see MDI000

CARBOXYPHOSPHAMIDE see CCE250

4-CARBOXYPHTHALATO(1,2-DIAMINOCYCLOHEXANE)PLATINUM(II) see CCJ350

4-CARBOXYPHTHALIC ANHYDRIDE see TKV000

3-o-(β-CARBOXYPROPIONYL)-11-OXO-18-β-OLEAN-12-EN-30-OIC ACID, DISODIUM SALT see CBO500

3-β-(3-CARBOXYPROPIONYLOXY)-11-OXO-OLEAN-12-EN-30-OIC ACID see BGD000

3-CARBOXYPROPYL(2-PROPENYL)NITROSAMINE see CCJ375

(3-CARBOXYPROPYL)TRIMETHYLAMMONIUM IODIDE ETHYL ESTER see CCJ385

(3-CARBOXYPROPYL)TRIMETHYLAMMONIUM IODIDE METHYL ESTER see CCJ388

3-CARBOXYPYRIDINE see NCQ900

4-CARBOXYPYRIDINE see ILC000

4-CARBOXYRESORCINOL see HOE600

5-CARBOXYRESORCINOL see REF200

trans-β-CARBOXYSTYRENE see CMP980

4-CARBOXYTERBUTOL see CCJ390

4-CARBOXYTHIAZOLIDINE see TEV000

2-CARBOXYTHIOPHENE see TFM600

6-CARBOXYURACIL see OJV500

CARBOXY VINYL POLYMER see CCJ400

CARBRITAL see NBU000

CARBUTAMID see BSM000

CARBUTAMIDE see BSM000

CARBUTEN see MBW750

CARBUTEROL HYDROCHLORIDE see BQD500

CARBYL see CBH250

CARBYL SULFATE see DXI500

CARBYNE see CEW500

CARCHOLIN see CBH250

CARCINOCIDIN see CCN250

CARCINOLIPIN see CCJ500

5-β-CARDA-16,20(22)-DIENOLIDE, 3-β-((6-DEOXY-3-o-METHYL-α-l-MANNOPYRANOSYL)OXY)-14-HYDROXY- see DBA100

CARDAMINE see DJS200

CARDAMIST see NGY000

CARDAMON see CCJ625

CARDAMON OIL see CCJ625

5-β-CARDANOLIDE, 3-β-((o-2,6-DIDEOXY-β-d-RIBO-HEXOPYRANOSYL)-(1-4)-o-2,6-DIDEOXY-β-d-HEXOPYRANOSYL)-(1-4)-2,6-DIDEOXY-β-d-RIBO-HEXOPYRANOSYL)OXY)-14-HYDROXY- see DLB650

CARDELMYCIN see SMB000

CARDENAL de MACETA (MEXICO) see CCJ825

CARD-20(22)-ENOLIDE, 16-(ACETYLOXY)-3-((6-DEOXY-3-o-METHYL-α-l-ALTROPYRANOSYL)OXY)-14-HYDROXY-, (3-β,5-β,16-β)- see VCA100

CARD-20(22)-ENOLIDE, 3-((6-DEOXY-α-l-MANNOPYRANOSYL)OXY)-11,14-DIHYDROXY- see MAN400

CARD-20(22)-ENOLIDE, 3-((6-DEOXY-3-o-METHYL-α-l-MANNOPYRANOSYL)OXY)-14-HYDROXY-, (3-β, 5-β)- see SKS150

CARD-20(22)-ENOLIDE, 3-((o-2,6-DIDEOXY-3,4-DI-o-FORMYL-β-d-RIBO-HEXOPYRANOSYL)-(1-4)-o-2, 6-DIDEOXY-3-o-FORMYL-β-d-RIBO-HEXOPYRANOSYL-(1-4)- 2,6-DIDEOXY-3-o-FORMYl-β-d-RIBO-HEXOPYRANOSYL)OXY)-16-(FORMYLOXY)-14-HYDROXY-, (3-β,5-β,16-β)- see PBG200

CARD-20(22)-ENOLIDE, 3-((2,6-DIDEOXY-3-o-METHYL-β-d-LYXO-HEXOPYRANOSYL)OXY)-14-HYDROXY-16-(3-METHYL-1-OXOBUTOXY)-, (3-β,5-β,16-β)- see AEM800

CARD-20(22)-ENOLIDE, 3-((2,6-DIDEOXY-3-o-METHYL-β-d-RIBO-HEXOPYRANOSYL)OXY)-5,14,19-TRIHYDROXY-, (3-β,5-β)- see SMM100

CARD-20(22)-ENOLIDE, 3-((2,6-DIDEOXY-β-d-RIBOHEXOPYRANOSYL)OXY)-5,14-DIHYDROXY-19-OXO-, (3-β,5-β,17-α)- see HAN900

CARD-20(22)-ENOLIDE, 3,14-DIHYDROXY-, (3-β,5-α,17-α)- see UWJ200

5-α,17-α-CARD-20(22)-ENOLIDE, 3-β,14-DIHYDROXY- see UWJ200

5-β-CARD-20(22)-ENOLIDE, 5,14-DIHYDROXY-3-β-((d-GLUCOPYRANOSYL)OXY)-19-OXO- see SMN200

CARD-20(22)-ENOLIDE, 19-OXO-3,14,15-TRIHYDROXY-, (3-β,5-α,15-β)- see AFS800

CARD-20(22)-ENOLIDE, 3,5,11,14-TETRAHYDROXY-19-OXO-, (3-β,5-β,11-α)- see NDY565

5-β-CARD-20(22)-ENOLIDE, 3-β,5,11-α,14-TETRAHYDROXY-19-OXO- see NDY565

CARD-20(22)-ENOLIDE, 3,11,14-TRIHYDROXY-, (3-α,5-β,11-α)- see EBB800

5-β-CARD-20(22)-ENOLIDE, 3-α,11-α,14-TRIHYDROXY- see EBB800

CARD-20(22)-ENOLIDE, 2,3,14-TRIHYDROXY-19-OXO-, (2-α,3-β,5-α)- see CAZ050

5-α-CARD-20(22)-ENOLIDE, 2-α,3-β,14-TRIHYDROXY-19-OXO- see CAZ050

5-α-CARD-20(22)-ENOLIDE, 3-β,14,15-β-TRIHYDROXY-19-OXO- see AFS800

CARDIAGEN see DJS200

CARDIAMID see DJS200

CARDIAMINA see DJS200

CARDIAMINE see DJS200

neo-CARDIAMINE see GCE600

CARDIDIGIN see DKL800

CARDIEM see DNU600

CARDIGIN see DKL800

CARDILATE see VSA000

CARDIMON see DJS200

CARDINAL FLOWER see CCJ825

CARDINOPHILLIN see CCN500

CARDINOPHYLLIN see CCN500

CARDIO see CCK125

CARDIOFILINA see TEP500

CARDIOGRAFIN see AOO875

CARDIO-GREEN see CCK000

CARDIO-KHELLIN see AHK750

CARDIOLANATA see LAU400

CARDIOLIPOL see NCW300

CARDIOMIN see TEP500

CARDIOMONE see AOA125

CARDION see POB500

CARDIOQUIN see GAX000

CARDIORYTHMINE see AFH250

CARDIOSERPIN see RDK000

CARDIOTRAST see DNG400

CARDIOVITAL see GCE600

CARDIOVITE see POB500

CARDIS see CCK125

CARDITIN see PEV750

CARDITIVO see RDK000

CARDITOXIN see DKL800

CARDIVIX see EID200

CARDOGENEN-(20:22)-DIOL-(3-β,14) (GERMAN) see DMJ000

CARDOGENEN-(20:22)-TRIOL-(3-β,12,14) (GERMAN) see DKN300

β-CARDONE see CCK250

CARDOPHYLIN see TEP500

CARDOPHYLLIN see TEP500

CARDOVERINA see PAH250

CARDOXIN see PCP250

CARENA see TEP500

CARENE see CCK510

3-CARENE see CCK500

(+)CAR-3-ENE see CCK510

S-3-CARENE see CCK500

Δ³-CARENE see CCK500

CARFECILLIN see CBN750

CARFECILLIN SODIUM see CBO000

CARFENE see ASH500

CARFENTRAZONE-ETHYL see CCK520

CARFIMAT see PGE000

CARFIMATE see PGE000

CARFLOC D 1000 see CNH125

CARGUTOCIN see CCK550

CARIAQUILLO (PUERTO RICO) see LAU600

CARICIDE see DIW000

CARICIDE see DIW200

CARIDOROL see ICC000

CARINA see PKQ059

CARINEX GP see SMQ500

CARINEX HR see SMQ500

CARINEX HRM see SMQ500

CARINEX SB 59 see SMQ500

CARINEX SB 61 see SMQ500

CARINEX SL 273 see SMQ500

CARINEX TGX/MF see SMQ500

CARIOMIN see TEP500

CARIOTA (CUBA) see FBW100

CARISOL see IPU000

CARISOMA see IPU000

CARISOPRODATE see IPU000

CARISOPRODATUM see IPU000

CARISOPRODOL see IPU000

CARITROL see DIW200

CARLONA 900 see PJS750

CARLONA 58-030 see PJS750

CARLONA 18020 FA see PJS750

CARLONA P see PMP500

CARLONA PXB see PJS750

CARLSODAL see IPU000

CARLSOMA see IPU000

(+)-CARLUMINE see CNR745

CARLYTENE see TFY000

CARMAZINE see DXI400

CARMAZON see POB500

CARMETHIZOLE see MLX275

CARMETHOSE see SFO500

CARMETIZIDE see CCK575

CARMINAPH see PEJ500

CARMIN BLUE VS see ADE500

CARMINE see CCK590

CARMINE 3B see CMG750

CARMINE BLUE AF see ERG100

CARMINE BLUE (BIOLOGICAL STAIN) see FAE100

CARMINE BLUE V see CMM062
CARMINE BLUE VF see ADE500
CARMINIC ACID see CCK590
CARMINOMICIN I see CCK625
CARMINOMYCIN see KBU000
CARMINOMYCIN HYDROCHLORIDE see KCA000
CARMINOMYCIN I see CCK625
CARMIN (PUERTO RICO) see ROA300
CARMOFUR see CCK630
CARMOISIN (GERMAN) see HJF500
CARMOISINE ALUMINUM LAKE see HJF500
CARMOISINE SUPRA see HJF500
CARMOL HC see HHQ800
CARMUBRIS see BIF750
CARMURIT see EEQ600
CARMUSTIN see BIF750
CARMUSTINE see BIF750
CARNACID-COR see GEW000
CARNATION RED TONER B see NAY000
CARNAUBA WAX see CCK640
CARNELIO HELIO RED see MMP100
CARNELIO ORANGE G see CMS145
CARNELIO PALE LITHOL RED see NAP100
CARNELIO RED 2G see DVB800
CARNELIO RED R see CJD500
CARNELIO RUBINE LAKE see SEH475
CARNELIO YELLOW GX see DEU000
CARNITINE CHLORIDE see CCK650
l-CARNITINE CHLORIDE see CCK660
(±)-CARNITINE CHLORIDE see CCK655
dl-CARNITINE CHLORIDE see CCK655
l-CARNITINE HYDROCHLORIDE see CCK660
(R)-CARNITINE HYDROCHLORIDE see CCK660
(±)-CARNITINE HYDROCHLORIDE see CCK655
d,l-CARNITINE HYDROCHLORIDE see CCK655
dl-CARNITINE HYDROCHLORIDE see CCK655
CARNOSINE see CCK665
l-CARNOSINE see CCK665
CAROB BEAN GUM see LIA000
CAROB FLOUR see LIA000
CAROFAM see EID200
CAROID see PAG500
CAROLINA ALLSPICE see CCK675
CAROLINA JASMINE see YAK100
CAROLINA PINK see PIH800
CAROLINA TEA see HGF100
CAROLINA WILD WOODBINE see YAK100
CAROLINA YELLOW JASMINE see YAK100
CAROLYSINE see BIE500
CAROTENE see CCK685
β-CAROTENE see CCK685
CAROTENE COCHINEAL see CCK691
β-CAROTENE-4,4'-DIONE see CBE800
CARPENE see DXX400
CARPERITIDE see HGL680
CARPHENAZINE DIMALEATE see CCK700
CARPHENAZINE MALEATE see CCK700
CARPHENOL see PFR325
CARPIPRAMINE see CBH500
CARPIPRAMINE DIHYDROCHLORIDE see CCK775
CARPIPRAMINE DIHYDROCHLORIDE MONOHYDRATE see CCK780
CARPIPRAMINE HYDROCHLORIDE see CCK775
CARPIPRAMINE MALEATE see CCK790
CARPOLENE see ADV900
CARPOLIN see CBM750
CARPROFEN see CCK800
CARQUEJOL see CCL109
CARRAGEEN see CCL250
κ-CARRAGEEN see CCL350
κ-CARRAGEENAN see CCL350
CARRAGEENAN, CALCIUM(II) SALT see CAO250

CARRAGEENAN, DEGRADED see CCL500
CARRAGEENAN (FCC) see CCL250
CARRAGEENAN GUM see CCL250
CARRAGEENAN, SODIUM SALT see SFP000
κ-CARRAGEENIN see CCL350
CARRAGHEANIN see CCL250
CARRAGHEEN see CCL250
CARRAGHEENAN see CCL250
CARREL-DAKIN SOLUTION see SHU500
CARRIOMYCIN, SODIUM SALT see SFP500
CARROT, SEED EXTRACT see GAP500
CARROT SEED OIL see CCL750
CARRSERP see RDK000
CARRTIME see BBK500
CARSIL see SDX625
CARSIL (SILICATE) see SCN700
CARSODOL see IPU000
CARSONOL SLS see SIB600
CARSONON N-9 see PKF000
CARSONON PEG-4000 see PJT750
CARSOQUAT SDQ-25 see DTC600
CARSORON see DER800
CARSTAB 700 see HND100
CARSTAB DLTDP see TFD500
CARTAGYL see ARQ750
CARTAP HYDROCHLORIDE see BHL750
CARTEOLOL see CCL800
CARTEOLOL HYDROCHLORIDE see MQT550
CARTOSE see GFG000
CARTWHEELS see AOB250
CARUBICIN see CCK625
CARUSIS P see CAT775
CARVACROL see CCM000
CARVACRON see HII500
CARVACRYL 2-PROPYLVALERATE see CCM050
CARVANIL see CCK125
CARVASEPT see CEX275
CARVASIN see CCK125
1-CARVEOL see MKY250
CARVIL see MOV000
4-CARVOMENTHENOL see TBD825
3-CARVOMENTHENONE see MCF250
(−)-CARVONE see CCM120
CARVONE see MCD250
(+)-CARVONE see CCM100
1-CARVONE see CCM120
d-CARVONE see CCM100
l(−)-CARVONE see CCM120
(R)-CARVONE see CCM120
(S)-CARVONE see CCM100
d(+)-CARVONE see CCM100
(S)-(+)-CARVONE see CCM100
1-CARVYL ACETATE see CCM750
l-CARVYL PROPIONATE see MCD000
CARYLDERM see CBM750
CARYNE see CEW500
CARYOLYSIN see BIE250
CARYOLYSINE see BIE500
CARYOLYSINE HYDROCHLORIDE see BIE500
CARYOPHYLLENE see CCN000
α-CARYOPHYLLENE see HGM550
CARYOPHYLLENE ACETATE see CCN050
CARYOPHYLLENE EPOXIDE see CCN100
β-CARYOPHYLLENE EPOXIDE see CCN100
β-CARYOPHYLLENE (FCC) see CCN000
CARYOPHYLLENE OXIDE see CCN100
(−)-CARYOPHYLLENE OXIDE see CCN100
β-CARYOPHYLLENE OXIDE see CCN100
CARYOPHYLLIC ACID see EQR500
CARYOTA (VARIOUS SPECIES) see FBW100
CARZAZO (DOMINICAN REPUBLIC) see CAK325
CARZENID see SNL890
CARZENIDE see SNL890
CARZINOCIDIN see CCN250
CARZINOPHILIN see CCN500
CARZINOPHILIN A see CCN750
CARZINOSTATININ see CCO000
CARZOL see CJJ250

CARZOL SP see DSO200
CARZONAL see FLZ050
CARZONAL see FMM000
CASALIS GREEN see CMJ900
CASANTIN see DHF600
CASANTIN see DII200
CASATE see GCU050
CASATE SODIUM see GCU050
CASCABELILLO (PUERTO RICO) see RBZ400
CASCADE see FDB400
CASCAMITE see UTU500
CASCARA see MBU825
CASCARILLA OIL see CCO500
CASCARITA see CMV390
CASCO 5H see UTU500
CASCO PR 335 see UTU500
CASCO RESIN see UTU500
CASCO UL 30 see UTU500
CASCO WS 114-79 see UTU500
CASCO WS 138-43 see UTU500
CASCO WS 138-44 see UTU500
CASEIN and CASEINATE SALTS (FCC) see SFQ000
CASEIN-SODIUM see SFQ000
CASEIN, SODIUM COMPLEX see SFQ000
CASEINS, SODIUM COMPLEXES see SFQ000
CASHES see PJJ300
CASHMILON, combustion products see ADX750
CASIFLUX VP 413-004 see WCJ000
CASIMAN (MEXICO, PUERTO RICO) see SLE890
CASORON 133 see DER800
CASPAN see MDD750
CASSAINE HYDROCHLORIDE see CCO675
CASSAPPRET SR see PKF750
CASSAVA see CCO680
CASSAVA see CCO700
CASSAVA, MANIHOT UTILISSIMA see CCO700
CASSAVA MEAL see CCO700
CASSAVA POWDER see CCO700
CASSE (HAITI) see GIW300
CASSEL BROWN see MAT500
CASSEL GREEN see MAT250
CASSELLA 532 see OJD300
CASSELLA 4489 see CBR500
CASSENA see HGF100
CASSIA ALDEHYDE see CMP969
CASSIA FISTULA see GIW300
CASSIA OCCIDENTALIS see CNG825
CASSIA OIL see CCO750
CASSIAR AK see ARM268
CASSIA TORA Linn., leaf extract see CCO800
CASSIC ACID see RHZ700
CASSURIT HML see MCB050
CASSURIT LR see DTG000
CASSURIT MLP see MCB050
CASSURIT MLS see MCB050
CASSURIT MT see MCB050
CASTANEA SATIVA MILL TANNIN see CDM250
CASTOR BEAN see CCP000
CASTOR BEANS (DOT) see CCP000
CASTOR FLAKE (DOT) see CCP000
CASTOR MEAL (DOT) see CCP000
CASTOR OIL see CCP250
CASTOR OIL AROMATIC see CCP250
CASTOR OIL, ETHOXYLATED see CCP330
CASTOR OIL, HYDROGENATED see HHW502
CASTOR OIL, HYDROGENATED, ETHOXYLATED, HCO 40 see CCP300
CASTOR OIL, HYDROGENATED, ETHOXYLATED, HCO 50 see CCP305
CASTOR OIL, HYDROGENATED, ETHOXYLATED, HCO 60 see CCP310
CASTOR OIL PLANT see CCP000
CASTOR OIL POLYOXYETHYLENE ETHER see CCP330

CASTOR POMACE (DOT) see CCP000
CASTORWAX see HHW502
CASTORWAX MP-70 see HHW502
CASTORWAX MP-80 see HHW502
CASTORWAX NF see HHW502
CASTRIX see CCP500
CASTRON see PDN000
CASUL 70HF see CAR790
CASWELL NO. 481G see CDT250
CAT (herbicide) see BJP000
CATACIDE see DIW000
CATALASE from MICROCOCCUS
LYSODEIKTICUS see CCP525
CATALIN CAO-3 see BFW750
CATALOID see SCH002
CATALYTIC CRACKED CLARIFIED OIL see
CMU890
CATALYTIC-DEWAXED HEAVY
NAPHTHENIC DISTILLATE see MQV776
CATALYTIC-DEWAXED HEAVY
PARAFFINIC DISTILLATE see MQV778
CATALYTIC-DEWAXED LIGHT
NAPHTHENIC DISTILLATE see MQV777
CATALYTIC-DEWAXED LIGHT
PARAFFINIC DISTILLATE see MQV779
CATAMINE AB see AFP250
CATANAC SP see CCP675
CATANAC SP ANTISTATIC AGENT see
CCP675
CATANIL see CKK000
CATAPRES see CMX760
CATAPRESAN see CMX760
CATAPYRIN see AFW500
CATATOXIC STEROID No. 1 see CCP750
CAT CRACKED CLARIFIED OIL-
DECANTED OIL see CMU890
CATECHIN see CCP800
CATECHIN see CCP850
CATECHIN see CCP875
(+)-CATECHIN see CCP875
d-CATECHIN see CCP875
d-(+)-CATECHIN see CCP875
CATECHIN (FLAVAN) see CCP875
CATECHIN HYDRATE see DMD000
CATECHINIC ACID see CCP875
CATECHOL see CCP850
CATECHOL see CCP875
(+)-CATECHOL see CCP875
d-CATECHOL see CCP875
CATECHOL DIETHYL ETHER see CCP900
CATECHOL-3,5-DISULFONIC ACID see
PPQ100
CATECHOL (FLAVAN) see CCP875
CATECHU see CCP800
CATECHUIC ACID see CCP875
CATENULIN see NCF500
CATERGEN see CCP875
CATESBY'S VINE (BAHAMAS) see YAK300
CATEUDYL see QAK000
CAT-FLOC see DTS500
CATHA EDULIS Forsk, leaf extract see
KGK350
CATHARANTHIN see CCP925
CATHARANTHINE see CCP925
(+)-CATHARANTHINE see CCP925
CATHILON PINK FGH see CMM850
CATHINE see NNM510
CATHOCIN see SMB000
CATHOMYCIN see SMB000
CATHOMYCIN SODIUM see NOB000
CATHOMYCIN SODIUM LYOVAC see
NOB000
CATIGENE T80 see QAT520
CATILAN see CDP250
CATINAL LTC 35A see LBX075
CATIOGEN L see LBX075
CATION AB see TLW500
CATION BB see LBX075
CATION FB see LBX075
CATIONIC ORANGE ZH see CMM820
CATIONIC PINK 2S see CMM850
CATIONIC RED VIOLET see EFC650

CATIONIC ROSE 2S see CMM850
CATIONIC SP see CCP675
CATIONIC YELLOW 6Z see CCP950
CATOLIN 14 see MJO500
CATOVITAN see PNS000
CATRAL see PDN000
CATRAN see PDN000
CATRON HYDROCHLORIDE see PDN250
CATRONIACID PDN250
CATRONIAZIDE see PDN000
C. ATROX VENOM see WBJ600
CAT'S BLOOD see ROA300
CATS EYES see PAM780
CAULOPHYLLUM THALICTROIDES see
BMA150
CAULOPHYLLUM THALICTROIDES,
glycoside extract see CCQ125
CAURITE see DTG700
CAUSOIN see DKQ000
CAUSTIC BARLEY see VHZ000
CAUSTIC BARYTA see BAM500
CAUSTIC POTASH see PLJ500
CAUSTIC POTASH, dry, solid, flake, bead, or
granular (DOT) see PLJ500
CAUSTIC POTASH, liquid or solution (DOT)
see PLJ500
CAUSTIC SODA see SHS000
CAUSTIC SODA, dry (DOT) see SHS000
CAUSTIC SODA, bead (DOT) see SHS000
CAUSTIC SODA, flake (DOT) see SHS000
CAUSTIC SODA, solid (DOT) see SHS000
CAUSTIC SODA, solution see SHS500
CAUSTIC SODA, liquid (DOT) see SHS000
CAUSTIC SODA, granular (DOT) see SHS000
CAUSTIC SODA, solution (DOT) see SHS000
CAUTIVA (PUERTO RICO) see AFQ625
CAVALITE BRILLIANT BLUE R see BMM500
CAV-ECOL see WBJ700
CAVINTON see EGM100
CAVI-TROL see SHF500
CAVODIL see PDN000
CAVONLY see TDA500
CAVUMBREN see BGB315
CAYENNE PEPPER see PCB275
CAZ PENTAHYDRATE see CCQ200
4-CB see TBO700
CB-154 see BNB250
CB-154 see BNB325
CB 304 see THM750
CB-337 see HMC000
804 CB see CPJ000
CB 1331 see PCU425
CB 1348 see CDO500
CB 1356 see BHY625
CB-1385 see AJE000
CB 1506 see CHC750
1522 CB see ABH500
1613-CB see BTA125
CB 1639 see AJK250
1664 CB see AAF800
1678 CB see IDA500
1678 CB see PMX500
CB 1689 see CQA000
CB 1707 see PMH905
CB 1729 see BHT250
C.B. 2041 see BOT250
CB2058 see DNX510
CB 2095 see DNX000
CB 2511 see BKM500
CB 2562 see TFU500
CB 3008 see BHV000
CB 3025 see BHV250
CB 3025 see PED750
3026 C.B. see SAX200
CB-3026 see SAX200
CB-3307 see BHT750
CB 3717 see PMN550
CB 4261 see CFG750
CB 4306 see CDQ250
4311 CB see DKV700
4361 CB see CFG750
CB-4564 see CQC500

CB 4564 see CQC650
CB-4835 see BIA250
CB 8000 see DEQ200
8002 CB see IIY100
CB 8019 see IPU000
8022 C.B. see ENX600
8065 C.B. see MPN000
8089 CB see FGU000
8089 C.B. see FLK100
CB 10286 see CCC325
C-39089-Ba see THK750
2-CBA see CEL250
CBA (PESTICIDE) see PAV300
CBBP see THY500
CBC 806495 see TFQ750
CBC 900139 see SBE500
CBC 906288 see TND250
CB 804 CALCIUM see CPJ250
CBD see CBD599
CBD 90 see TIQ750
CBDCA see CCC075
CBDZ see CBA100
C. BICOLOR see CAL125
CBN see CBD625
CBN see CEW500
C. BONDUC see CAK325
CBPC see CBO250
CBS see CPI250
CBS-1114 see PEK675
CB 1348 SODIUM SALT see CDO625
CBSP see HAQ600
(CBZ)GLY see CBR125
CC 914 see CBI250
CC-1065 see APT375
CC 11511 see DYC800
CCA see LHZ600
(C6-C12) ALKYL ALCOHOL see AFJ100
"C" CARRIE see CNE750
CCC see CAQ250
CCC G-WHITE see CAT775
CCC No. AA OOLITIC see CAT775
CCCP see CKA550
CCC PLANT GROWTH REGULANT see
CMF400
C. CERASTES VENOM see CCX620
CCH see HOV500
CCHO see CPD000
C 7337 CIBA see PDW400
CCK 179 see DLL400
CCL see PAG075
CCN52 see RLF350
C.C. No. 914 see CBI250
CCNU see CGV250
CCP see CKA550
CCR see CAT775
CCRG 81010 see MQY110
CCS see CJT750
CCS 203 see BPW500
CCS 301 see BPW750
CCUCOL see ASB250
CCW see CAT775
CD see TGD000
CD 2 see LFK000
CD 3 see MDQ850
CD 4 see AKZ100
CD 68 see CDR750
CD 79 see OAF200
CD 336 see TDB765
CD 437 see HOL145
CD 2019 see MEU400
CD-3400 see MKR100
CDA see CAK350
CDA 101 see CNI000
CDA 102 see CNI000
CDA 110 see CNI000
CDA 122 see CNI000
CD 15006 A see BOS100
CDAA see CFK000
CDAAT see CFK000
3',4'-Cl2-DAB see DFD400
CDA: CETYLCIDE see EKN500
CDB 60 see DGN200

CDB 63 see SGG500
CDBAC see BEL900
CDBM see CFK500
CDC see CDL325
CDCA see CDL325
CDDP see PJD000
CDEA BR see CDF500
CDEC see CDO250
2-CDF see PBT100
CD III see MDQ850
C. DIURNUM see DAC500
CDM see CJJ250
CDM see SMP450
CDNA see RDP300
CDP see LFK000
CDP-CHOLIN see CMF350
CDP-CHOLINE see CMF350
CDP-COLINA see CMF350
67/20CDRI see CCW750
CDRI 77/735 see MEE150
C. DRUMMONDII see CAK325
CDT see BJP000
CDT see COW935
CDTA see CPB120
CDX see DYF700
CE see TEG250
C6E3 see HFT550
114 C.E. see HNM000
746 CE see ECU550
CE 3624 see TGA600
C-14919 E-2 see MAB400
C 1215AE30 see AFJ160
CEBESINE see OPI300
CEBETOX see MIW250
CEBITATE see ARN125
CEBOLLA see WBS850
CEBOLLEJA (MEXICO) see GJU460
CEBROGEN see GFO050
CEBRUM see MDQ250
CECA see COM830
CECALGINE TBV see SEH000
CECARBON see CBT500
CE CE CE see CMF400
CECENU see CGV250
CECIL see CNE750
CECLOR see CCR850
CECOLENE see TIO750
CEDAD see BCA000
CEDAR LEAF OIL see CCQ500
CEDARWOOD OIL ATLAS see CCQ750
CEDARWOOD OIL MOROCCAN see
CCQ750
CEDARWOOD OIL (VIRGINIA) see CCR000
CEDILANID see LAU000
CEDOCARD see CCK125
CEDRAMBER see CCR525
CEDRANE, 8,9-EPOXIDE see CCR510
8-β-H-CEDRAN-8-OL ACETATE see CCR250
CEDRANYL ACETATE see CCR250
CEDR-8-ENE see CCR500
α-CEDRENE see CCR500
CEDR-8-ENE EPOXIDE see CCR510
CEDROL FORMATE see CCR524
CEDROL METHYL ETHER see CCR525
CEDRO OIL see LEI000
CEDRUS ATLANTICA OIL see CCQ750
CEDRYL ACETATE see CCR250
CEDRYL FORMATE see CCR524
CEE see PMB000
CEE DEE see HCQ500
CEENU see CGV250
CEEPRYN see CCX000
CEEPRYN see CDF750
CEEPRYN CHLORIDE see CCX000
CEFACETRILE SODIUM see SGB500
CEFACIDAL see CCS250
CEFACLOR see CCR850
CEFACLOR see PAG075
CEFACLOR HYDRATE see CCR850
CEFADOL see CCR875
CEFADOLE see CCX300
CEFADROXIL see DYF700

CEFADYL see HMK000
CEFA-ISKIA see ALV000
CEFALOGLYCIN see CCR890
CEFALOJECT see HMK000
CEFALORIDIN see TEY000
CEFALORIDINE see TEY000
CEFALORIZIN see TEY000
CEFALOTHINE SODIUM see SFQ500
CEFALOTIN see CCX250
CEFALOTINA SODICA (SPANISH) see
SFQ500
CEFALOTO see ALV000
CEFAMANDOL see CCX300
CEFAMANDOLE see CCX300
l-CEFAMANDOLE see CCX300
l-CEFAMANDOLE NAFATE see FOD000
CEFAMANDOLE SODIUM see CCR925
CEFAMANDOL NAFATO see FOD000
CEFAMEDIN see CCS250
CEFAMEZIN see CCS250
CEFAPIRIN (GERMAN) see CCX500
CEFAPIRIN SODIUM see HMK000
CEFAPRIN SODIUM see HMK000
CEFATIN see OAV000
CEFATOXIME SODIUM see CCR950
CEFATREXYL see HMK000
CEFATRIAXONE HYDRATE see CCS588
CEFATRIZINE see APT250
CEFAZEDONE SODIUM SALT see RCK000
CEFAZIL see CCS250
CEFAZINA see CCS250
CEFAZOLIN see CCS250
CEFAZOLINE SODIUM see CCS250
CEFAZOLIN SODIUM SALT see CCS250
CEFBUPERAZONE SODIUM see TAA400
CEFDINIR see AMS650
CEFDINYL see AMS650
CEFEDRIN see CCX600
CEFLORIN see TEY000
CEFMENOXIME HEMIHYDROCHLORIDE
see CCS300
CEFMETAZOLE see CCS350
CEFMETAZOLE SODIUM see CCS360
CEFMINOX see CCS365
CEFODIZIME DISODIUM see CCS367
CEFODIZIME SODIUM see CCS367
CEFOPERAZONE SODIUM see CCS369
CEFOTAN see CCS371
CEFOTAXIME see CCS372
CEFOTAXIME SODIUM see CCR950
CEFOTETAN see CCS373
CEFOTETAN DISODIUM SALT see CCS371
CEFOTIAM DIHYDROCHLORIDE see
CCS375
CEFOTIAM HYDROCHLORIDE see CCS375
CEFOXITIN see CCS500
CEFOXITIN SODIUM SALT see CCS510
CEFOXOTIN SODIUM see CCS510
CEFPIMIZOLE SODIUM see CCS525
CEFPROZIL see CCS527
CEFRACYCLINE SUSPENSION see TBX000
CEFRACYCLINE TABLETS see TBX250
CEFRADINE see SBN440
CEFROXADIN see CCS530
CEFROXADIN DIHYDRATE see CCS535
CEFROXADINE see CCS530
CEFSULODIN SODIUM see CCS550
CEFSULODIN SODIUM see CCS550
CEFTAZIDIME see CCQ200
CEFTAZIDIME PENTAHYDRATE see
CCQ200
CEFTEZOLE SODIUM see CCS560
CEFTIOFUR see CCS575
CEFTIZOXIME SODIUM see EBE100
CEFTIZOXIM-NATRIUM (GERMAN) see
EBE100
CEFTRIAXONE SODIUM HYDRATE see
CCS588
CEFUROXIM see CCS600
CEFUROXIME see CCS600
CEFUROXIME AXETIL see CCS625
CEFUROXIME SODIUM see SFQ300

CEFUROXIME SODIUM SALT see SFQ300
CEFZONAME SODIUM see CCS635
CEGLUTION see LGZ000
CEKIURON see DXQ500
CEKUBARYL see CBM750
CEKU C.B. see HCC500
CEKUDAZIM see MHC750
CEKUDIFOL see BIO750
CEKUFON see TIQ250
CEKUGIB see GEM000
CEKUMETA see TDW500
CEKUMETHION see MNH000
CEKUQUAT see PAJ000
CEKUSAN see BJP000
CEKUSAN see DGP900
CEKUSIL see ABU500
CEKUSIL UNIVERSAL A see MEO750
CEKUSIL UNIVERSAL C see MEP250
CEKUTHOATE see DSP400
CEKUTROTHION see DSQ000
CEKUZINA-S see BJP000
CEKUZINA-T see ARQ725
CELA 50 see TKL100
CELA S-2225 see EGV500
CELA A-36 see DAE600
CELACOL M see MIF760
CELACOL M20 see MIF760
CELACOL M 20P see MIF760
CELACOL M450 see MIF760
CELACOL MM see MIF760
CELACOL MM 10P see MIF760
CELANAR see PKF750
CELANDINE see CCS650
CELANEX see BBQ500
CELANOL 252 see SNY100
CELANOL DOS 75 see DJL000
CELANTHRENE BRILLIANT BLUE see
MGG250
CELANTHRENE FAST BLUE 2G see
DMM400
CELANTHRENE FAST PINK 3B see DBX000
CELANTHRENE PURE BLUE BRS see
TBG700
CELANTHRENE RED 3BN see AKE250
CELANTHRENE RED VIOLET R see
DBP000
CELA S 1942 see BNL250
CELASTROL-METHYLETHER see PMD525
CELASTRUS SCANDENS see AHJ875
CELATOX see TNX375
CELA W 524 see TKL100
CELCOT RF see CMM760
CELEPORT see MGE200
CELERY OIL see OGL100
CELERY SEED OIL see CCS660
CELERY SEED OIL see OGL100
CELESTAN-DEPOT see CCS675
CELESTODERM see VCA000
CELESTONE see BFV750
CELESTONE CHRONODOSE see CCS675
CELESTONE SOLOSPAN see CCS675
CELESTONE SOLUSPAN see BFV755
CELESTONE SOLUSPAN see CCS675
CELEX see CCU250
CELGARD 2500 see PMP500
CELINHOL -A see OAV000
CELIOMYCIN see VQZ000
CELIPROLOL HYDROCHLORID
(GERMAN) see SBN475
CELIPROLOL HYDROCHLORIDE see
SBN475
CELITE see DCJ800
CELLACETATE see CCU050
CELLAPRET see MIF760
CELLATIVE see AAE500
CELLEX MX see CCU150
CELLIDRIN see ZVJ000
CELLILASE and CARBOHYDRASE see
CBS405
CELLITAZOL B see DCJ200
CELLITON BLUE FFR see MGG250
CELLITON BLUE G see TBG700

CELLITON BLUE GREEN B see DMM400
CELLITON BLUE RN see IBV050
CELLITON BRILLIANT YELLOW 8G see MEB750
CELLITON DISCHARGE SCARLET B see ENP100
CELLITON DISCHARGE YELLOW GL see AAQ250
CELLITON DISCHARGE YELLOW 5RL see CMP090
CELLITON DISCHARGING RUBINE BL see CMP080
CELLITON FAST BLUE GREEN B see DMM400
CELLITON FAST BLUE GREEN BA-CF see DMM400
CELLITON FAST PINK BA-CF see AKE250
CELLITON FAST PINK BN see AKE250
CELLITON FAST PINK FF3B see DBX000
CELLITON FAST PINK FF3BA-CF see DBX000
CELLITON FAST PINK RF see AKO350
CELLITON FAST PINK RFA-CF see AKO350
CELLITON FAST RED VIOLET see DBP000
CELLITON FAST RUBINE B see CMP080
CELLITON FAST RUBINE BA-CF see CMP080
CELLITON FAST VIOLET 6B see AKP250
CELLITON FAST VIOLET 6BA-CF see AKP250
CELLITON FAST VIOLET B see DBY700
CELLITON FAST VIOLET BA-CF see DBY700
CELLITON FAST YELLOW 5R see CMP090
CELLITON FAST YELLOW G see AAQ250
CELLITON FAST YELLOW GA see AAQ250
CELLITON FAST YELLOW GA-CF see AAQ250
CELLITON FAST YELLOW GGLL-CF see KDA075
CELLITON FAST YELLOW RR see DUW500
CELLITON ORANGE R see AKP750
CELLITON ROSE FF3B see DBX000
CELLITON RUBINE B see CMP080
CELLITON RUBY B see CMP080
CELLITON SCARLET BA-CF see ENP100
CELLITON SCARLET B (6CI) see ENP100
CELLITON VIOLET B see DBY700
CELLITON YELLOW 5R see CMP090
CELLITON YELLOW G see AAQ250
CELLMIC S see OPE000
CELLOCIDIN see ACJ250
CELLOFAS see SFO500
CELLOFOR (CZECH) see DWO800
CELLOGEL C see SFO500
CELLOGRAN see MIF760
CELLOIDIN see CCU250
CELLON see TBQ100
CELLOPHANE see CCT250
CELLOSIZE 4400H16 see HKQ100
CELLOSIZE QP see HKQ100
CELLOSIZE QP3 see HKQ100
CELLOSIZE QP 1500 see HKQ100
CELLOSIZE QP 4400 see HKQ100
CELLOSIZE QP 30000 see HKQ100
CELLOSIZE UT 40 see HKQ100
CELLOSIZE WP see HKQ100
CELLOSIZE WP 300 see HKQ100
CELLOSIZE WP 300H see HKQ100
CELLOSIZE WP 400H see HKQ100
CELLOSIZE WP 4400 see HKQ100
CELLOSIZE WPO 9H17 see HKQ100
CELLOSOLVE (DOT) see EES350
CELLOSOLVE ACETATE (DOT) see EES400
CELLOSOLVE ACRYLATE see ADT500
CELLOSOLVE SOLVENT see EES350
CELLOTHYL see MIF760
CELLPRO see SFO500
CELLRYL see CCU150
CELLU-BRITE see DXB450
CELLUFIX FF 100 see SFO500
CELLUFLEX see CGO500

CELLUFLEX 179C see TNP500
CELLUFLEX DOP see DVL600
CELLUFLEX DPB see DEH200
CELLUFLEX FR-2 see TNG750
CELLUFLEX TPP see TMT750
CELLUGEL see SFO500
CELLULASE AP3 see CCT900
CELLULOSE 248 see CCU150
α-CELLULOSE see CCU150
CELLULOSE, ACETATE HYDROGEN 1,2-BENZENEDICARBOXYLATE (9CI) see CCU050
CELLULOSE ACETATE MONOPHTHALATE see CCU050
CELLULOSE, ACETATE PHTHALATE see CCU050
CELLULOSE ACETOPHTHALATE see CCU050
CELLULOSE ACETYLPHTHALATE see CCU050
CELLULOSE (ACGIH,OSHA) see CCU150
CELLULOSE, CARBOXYMETHYL ETHER, ALUMINIUM SALT see AGW700
CELLULOSE CRYSTALLINE see CCU150
CELLULOSE ETHYL see EHG100
CELLULOSE ETHYLATE see EHG100
CELLULOSE GEL see CCU100
CELLULOSE GLYCOLIC ACID SODIUM SALT see CCU075
CELLULOSE GLYCOLIC ACID, SODIUM SALT see SFO500
CELLULOSE GUM see SFO500
CELLULOSE HYDROXYETHYLATE see HKQ100
CELLULOSE HYDROXYETHYL ETHER see HKQ100
CELLULOSE, 2-HYDROXYETHYL ETHER see HKQ100
CELLULOSE, 2-HYDROXYETHYL 2-(2-HYDROXY-3-(TRIMETHYLAMMONIO)PROPOXY)ETHYL 2-HYDROXY-3-(TRIMETHYLAMMONIO)PROPYL ETHER, CHLORIDE see QAT600
CELLULOSE METHYL see MIF760
CELLULOSE METHYLATE see MIF760
CELLULOSE, MICROCRYSTALLINE see CCU100
CELLULOSE NITRATE see CCU250
CELLULOSE, NITRATE (9CI) see CCU250
CELLULOSE, POWDERED see CCU150
CELLULOSE SODIUM GLYCOLATE see SFO500
CELLULOSE TETRANITRATE see CCU250
CELLUMETH see MIF760
CELLUPHOS 4 see TIA250
CELLU-QUIN see BLC250
CELLUTATE RED VIOLET RH see DBP000
CELMER see ABU500
CELMER see MEP250
CELMIDE see EIY500
CELMIDOL see CCR875
CELMONE see NAK500
CELOCURINE see BJI000
CELOGEN BSH see BBS300
CELOGEN OT see OPE000
CELON A see EIX000
CELON ATH see EIX000
CELON E see EIV000
CELON H see EIV000
CELON IS see EIV000
CELONTIN see MLP800
CELOSEN AZ see ASM270
CELOSPOR see SGB500
CELPHIDE see AHE750
CELPHOS see AHE750
CELPHOS see PGY000
CELTHIGN see MAK700
CELUFI see CCU150
CELUTATE BLUE BLT see MGG250
CELUTATE GREEN BLUE BGH see DMM400

CELUTATE PINK B see AKE250
CELUTATE PINK BN see AKE250
CELUTATE PINK BY see AKE250
CELUTATE SCARLET BH see ENP100
CELUTATE YELLOW GH see AAQ250
CEMEDINE 3000RP see EHP700
CEMEDINE 3000RP TYPE-II see EHP700
CEMEDINE 3000RS see EHP700
CEMEDINE 3000RS TYPE-II see EHP700
CEMENT (rubber) see CCW250
CEMENT BLACK see MAS000
CEMENT, PORTLAND see PKS750
CEMENT, RUBBER see CCW250
CEMIDON see ILD000
CEMULSOL D-8 see PJY100
CEMULSOL 1050 see PJY100
CEMULSOL A see PJY100
CEMULSOL C 105 see PJY100
CENESTIL see PMS825
CENITRON OB see OPE000
CENOL GARDEN DUST see RNZ000
CENOLATE see ARN125
CENESTIM see DLH630
CENSTIN see DLH600
CENSTIN see DLH630
CENTAUREPENSIN see HOW800
CENTBUCRIDINE HYDROCHLORIDE see CCW375
CENTBUTINDOLE see CCW500
CENTCHROMAN see CCW725
CENTCHROMAN HYDROCHLORIDE see CCW750
CENTEDEIN see MNQ000
CENTEDRIN see RLK000
CENTELASE see ARN500
CENTIMIDE see HCQ500
CENTPHENAQUIN see CCW800
CENTRALGIN see DAM700
CENTRALINE BLUE 3B see CMO250
CENTRALITE II see DRB200
CENTRAX see DAP700
CENTREDIN see MNQ000
CENTRINE see DOY400
CENTROFENOXINA see DPE000
CENTROPHENOXINE see AAE500
CENTRUROIDES SUFFUSUS SUFFUSUS VENOM see CCW925
CENTURINA see AOD000
CENTURY 1240 see SLK000
CENTURY CD FATTY ACID see OHU000
CENTYL see BEQ625
CEP see CDS125
2-CEPA see CDS125
CEPA CABALLERO (CUBA) see MQW525
CEPACILINA see BFC750
CEPACILLINA see BFC750
CEPACOL see CDF750
CEPACOL CHLORIDE see CCX000
CEPALORIDIN see TEY000
CEPALORIN see TEY000
CEPAVERIN see PAH250
CEPH 87/4 see TEY000
CEPHA see CDS125
CEPHACETRILE SODIUM see SGB500
CEPHADOLE see CCX300
CEPHADROXIL see DYF700
(−)-CEPHAELINE DIHYDROCHLORIDE see CCX125
CEPHAELINE HYDROCHLORIDE see CCX125
CEPHAELINE METHYL ETHER see EAL500
CEPHALEXIN see ALV000
CEPHALOGLYCIN see CCR890
CEPHALOGLYCINE see CCR890
d-CEPHALOGLYCINE see CCR890
CEPHALOMYCIN see CCX175
CEPHALORIDIN see TEY000
CEPHALORIDINE see TEY000
(3(R)-CEPHALOTAXINE-4-METHYL-2-HYDROXY-2-(4-HYDROXY-4-

METHYLPENTYL)BUTANEDIOATE
(ESTER) see HGI575
CEPHALOTHIN see CCX250
CEPHALOTHIN SODIUM see HMK000
CEPHALOTHIN SODIUM see SFQ500
CEPHALOTIN see CCX250
CEPHAMANDOLE see CCX300
CEPHAMANDOLE NAFATE see FOD000
CEPHAMYCIN see CCS365
CEPHAOGLYCIN ACID see CCR890
CEPHAPIRIN see CCX500
CEPHARANTHIN see CCX550
CEPHARANTHINE see CCX550
CEPHA 10LS see CDS125
CEPHATREXYL see HMK000
CEPHEDRINE see CCX600
CEPHOTAXIME see CCS372
CEPHOXITIN see CCS500
CEPHRADIN see SBN440
CEPHRADINE see SBN440
CEPHROL see CMT250
CEPHROL see DTF410
CEPHTRIAXONE see CCS588
CEPHUROXIME see CCS600
CEPO see CCU150
CEPO CFM see CCU150
CEPORAN see TEY000
CEPOREX see ALV000
CEPOREXIN see ALV000
CEPOREXINE see ALV000
CEPORINE see TEY000
CEPO S 20 see CCU150
CEPO S 40 see CCU150
CEPOVENIN see SFQ500
CEPRIM see CCX000
CEQUARTYL see BBA500
CERAMIC FIBRE see AHF500
CERAPHYL 230 see DNL800
CERAPHYL 368 see OFG100
CERAPHYL 375 see ISC550
CERASINE YELLOW GG see DOT300
CERASIN RED see OHI200
CERASINROT see OHI200
CERASTES CERASTES VENOM see CCX620
CERASYNT see HKJ000
CERASYNT 1000-D see OAV000
CERASYNT PA see SLL000
CERASYNT PN see SLL000
CERASYNT S see OAV000
CERASYNT SD see OAV000
CERASYNT SE see OAV000
CERASYNT WM see OAV000
CERAZOL (suspension) see TEX250
CERBERIGENIN see DMJ000
CERBEROSID (GERMAN) see CCX625
CERBEROSIDE see CCX625
CERBROSIDE see CCX625
CER-o-CILLIN see AGK250
CERCINE see DCK759
CERCOBIN see DJV000
CERCOBIN METHYL see PEX500
CEREB see CMF350
CEREBON see DPE000
CEREBROFORTE see NNE400
CERECHLOR 54 see PAH780
CERECLOR see PAH780
CERECLOR 30 see PAH780
CERECLOR 42 see PAH780
CERECLOR 48 see PAH780
CERECLOR 52 see PAH780
CERECLOR 54 see PAH780
CERECLOR 70 see PAH780
CERECLOR 51L see PAH780
CERECLOR 56L see PAH780
CERECLOR 63L see PAH780
CERECLOR 65L see PAH780
CERECLOR 70L see PAH780
CERECLOR S 42 see PAH780
CERECLOR S52 see PAH780
CERECLOR S70 see PAH780
CERECLOR 50LV see PAH780
CEREDON see BDD000

CERELINE see BDD000
CERELOSE see GFG000
CERENOX see BDD000
CEREPAP see DNA200
CEREPAX see CFY750
CERESAN see ABU500
CERESAN see CHC500
CERESAN M see EME500
CERESAN UNIVERSAL see ABU500
CERESAN UNIVERSAL-FEUCHTBEIZE see
BKS810
CERESAN-UNIVERSAL NASSBEIZE see
MEP250
CERESAN UNIVERSAL NAZBEIZE see
MEP250
CERES GREEN 3B see PJQ100
CERESOL see ABU500
CERES ORANGE G see CMP600
CERES ORANGE GN see CMP600
CERES ORANGE R see PEJ500
CERES ORANGE RR see XRA000
CERESPAN see PAH250
CERES RED 7B see EOJ500
CERES RED BB see SBC500
CERES RED G see CMS238
CERES RED G 102 see CMS238
CERES YELLOW GGN see OHJ875
CERES YELLOW R see PEI000
CEREWET see BKS810
CEREXIN A see CCX725
CEREZA (SPANISH) see AQP890
CERFA 114 see HNM000
CERIA see CCY000
CERIC DIOXIDE see CCY000
CERIC DISULFATE see CDB400
CERIC OXIDE see CCY000
CERIC SULFATE see CDB400
CERIC SULPHATE see CDB400
CERIMAN see SLE890
CERIMAN de MEJICO (CUBA) see SLE890
CERISE B see MAC250
CERISE TONER X1127 see FAG070
CERISOL SCARLET G see XRA000
CERISOL YELLOW AB see FAG130
CERISOL YELLOW GR see CMP600
CERISOL YELLOW TB see FAG135
CERIT FAC 3 see HOG000
CERIUM see CCY250
CERIUM ACETATE see CCY500
CERIUM AZIDE see CCY699
CERIUM CHLORIDE see CCY750
CERIUM(III) CHLORIDE see CCY750
CERIUM CHLORIDE, HYDRATE see
CCY800
CERIUM CITRATE see CCZ000
CERIUM(III) CITRATE see CCZ000
CERIUM COMPOUNDS see CDA250
CERIUM DIOXIDE see CCY000
CERIUM DISULFATE see CDB400
CERIUM EDETATE see CDA500
CERIUM FLUORIDE see CDA750
CERIUM FLUORURE (FRENCH) see CDA750
CERIUM NITRATE see CDB000
CERIUM(3+) NITRATE see CDB000
CERIUM(III) NITRATE see CDB000
CERIUM NITRATE, HEXAHYDRATE see
CDB250
CERIUM(III) NITRATE, HEXAHYDRATE
(1:3:6) see CDB250
CERIUM NITRIDE see CDB325
CERIUM(4+) OXIDE see CCY000
CERIUM SULFATE see CDB400
CERIUM(4+) SULFATE see CDB400
CERIUM(IV) SULFATE see CDB400
CERIUM(III) TETRAHYDROALUMINATE
see CDB500
CERIUM TRIACETATE see CCY500
CERIUM TRICHLORIDE see CCY750
CERIUM TRIFLUORIDE see CDA750
CERIUM TRIHYDRIDE see CDB750
CERIUM TRINITRATE see CDB000

CERIUM TRINITRATE HEXAHYDRATE see
CDB250
CERIUM TRISULFIDE see DEK600
CERIVASTATIN SODIUM see RLK750
CERM-1766 see IRP000
CERM-1841 see TKJ500
3024 CERM see MFG250
CERM 10,137 see TGJ885
CERNILTON see CDB760
CERNITIN GBX see CDB770
CERNITIN T-60 see CDB772
CERN KYPOVA 8 see CMU475
CERN KYPOVA 27 see DUP100
CERN KYSELA 1 see FAB830
CERN OSTAZINOVA H-N (CZECH) see
DGN600
CERN PRIMA 17 see CMN230
CERN PRIMA 19 see CMN240
CERN PRIMA 38 see AQP000
CERN REAKTIVNI 8 see CMS220
CEROTINE PONCEAU 3B see SBC500
CEROTINORANGE G see PEJ500
CEROTINSCHARLACH G see XRA000
CEROTINSCHARLACH R see OHI200
CEROUS ACETATE see CCY500
CEROUS CHLORIDE see CCY750
CEROUS CITRATE see CCZ000
CEROUS FLUORIDE see CDA750
CEROUS NITRATE see CDB000
CEROUS NITRATE HEXAHYDRATE see
CDB250
CEROXIN GL see HOG000
CEROXONE see BJK750
CERTICOL BLACK PNW see BMA000
CERTICOL CARMOISINE S see HJF500
CERTICOL ORANGE GS see HGC000
CERTICOL PONCEAU MXS see FMU070
CERTICOL PONCEAU 4RS see FMU080
CERTICOL PONCEAU SXS see FAG050
CERTICOL RED B see CMS228
CERTICOL SUNSET YELLOW CFS see
FAG150
CERTICOL TARTRAZOL YELLOW S see
FAG140
CERTINAL see ALT250
CERTIQUAL ALIZARINE see DMG800
CERTIQUAL EOSINE see BNH500
CERTIQUAL EOSINE see BNK700
CERTIQUAL FLUORESCEINE see FEW000
CERTIQUAL LITHOL RED see NAP100
CERTIQUAL OIL RED see OHI200
CERTIQUAL ORANGE I see FAG010
CERTIQUAL ORANGE II see CMM220
CERTIQUAL RHODAMIEN see FAG070
CERTOL see DNF400
CERTOLAKE SUNSET YELLOW see FAG150
CERTOMYCIN see NCP550
CERTOX see SMN500
CERTROL see HKB500
CERUBIDIN see DAC000
CERUBIDINE see DAC200
CERULEIN see CAK285
CERULENIN see ECE500
CERULIGNOL see MFM750
CERUSSETE see LCP000
CERUTIL see AAE500
CERVAGEM see CDB775
CERVEN 2G see CMM300
CERVEN BRILANTNI OSTACETOVA F-LB
(CZECH) see AKI750
CERVEN BRILANTNI OSTAZINOVA H-3B
(CZECH) see CHG250
CERVEN BRILANTNI OSTAZINOVA S-5B
(CZECH) see DGN800
CERVEN DISPERZNI 4 see AKO350
CERVEN DISPERZNI 11 see DBX000
CERVEN DISPERZNI 15 see AKE250
CERVEN DISPERZNI 60 see AKI750
CERVEN KOSENILOVA A see FMU080
CERVEN KUMIDINOVA see FAG018
CERVEN KYPOVA 13 see CMU825
CERVEN KYSELA 1 see CMM300

CERVEN KYSELA 26 see FMU070
CERVEN KYSELA 27 see FAG020
CERVEN KYSELA 114 see CMM330
CERVEN PIGMENT 3 see MMP100
CERVEN PIGMENT 53 see CMS150
CERVEN PIGMENT 57 see CMS155
CERVEN POTRAVINARSKA 1 see FAG050
CERVEN POTRAVINARSKA 2 see CMP620
CERVEN POTRAVINARSKA 9 see FAG020
CERVEN POTRAVINARSKA 10 see CMM300
CERVEN PRIMA 2 see DXO850
CERVEN PRIRODNI 20 see HLJ650
CERVEN ROZPOUSTEDLOVA 23 see
OHI200
CERVEN ROZPOUSTEDLOVA 24 see
SBC500
CERVEN ZASADITA 1 see RGW000
CERVEN ZASADITA 2 see GJI400
CERVICUNDIN see EQJ500
CERVILAXIN see RCK740
CERVOLIDE see OKW110
CERVOXAN see DOZ000
CES see ECU750
CES see SOP500
CESALIN see CDB800
CESAR see CJU275
CESIUM see CDC000
CESIUM-133 see CDC000
CESIUM ACETYLIDE see CDC125
CESIUM AMIDE see CDC250
CESIUM ARSENATE see CDC375
CESIUM BROMIDE see CDC500
CESIUM BROMOXENATE see CDC699
CESIUM CARBONATE see CDC750
CESIUM CHLORIDE see CDD000
CESIUM CHLOROXENATE see CDD250
CESIUM
CYANOTRIDECAHYDRODECABORATE
(2-) see CDD325
CESIUM DIHYDROGEN ARSENATE see
CAK350
CESIUM FLUORIDE see CDD500
CESIUM GRAPHITE see CDD625
CESIUM HYDRATE see CDD750
CESIUM HYDROXIDE see CDD750
CESIUM HYDROXIDE (ACGIH, OSHA) see
CDD750
CESIUM HYDROXIDE DIMER see CDD750
CESIUM IODIDE see CDE000
CESIUM LITHIUM
TRIDECAHYDRONONABORATE see
CDE125
CESIUM MONOCHLORIDE see CDD000
CESIUM MONOFLUORIDE see CDD500
CESIUM MONOIODIDE see CDE000
CESIUM NITRATE (DOT) see CDE250
CESIUM(I) NITRATE (1:1) see CDE250
CESIUM NITRIDE see TIH750
CESIUM OXIDE see CDE325
CESIUM OZONIDE see CDF000
CESIUM PENTACARBONYLVANADATE (3-
) see CDE400
CESIUM SELENIDE see DEJ400
CESIUM SULFATE see CDE500
CESIUM TRIOXIDE ("OZONATE") see
CDF000
CESOL see BGB400
CESTRUM (VARIOUS SPECIES) see DAC500
CET see BJP000
CET see CCX250
C12BET see LBU200
C16BET see CDF450
CETAB see HCQ500
CETAC see HCQ525
CETACORT see CNS750
CETADOL see HIM000
CETAFFINE see HCP000
CETAIN see AIT250
CETAL see HCP000
CETALKONIUM CHLORIDE see BEL900
CETALOL CA see HCP000
CETAMIUM see CCX000

CETANE see HCO600
n-CETANE see HCO600
CETAPHARM see HCP800
CETARIN see MKR250
CETAROL see HCQ500
CETASOL see HCP800
CETAVLON see HCQ500
CETAZOL see HCP800
CETETH see PJT300
CETETH 1 see PJT300
CETETH 2 see PJT300
CETHYLOSE see MIF760
CETHYTIN see MIF760
CETIL CHROMINE YELLOW GR see
MRL100
CETIL LIGHT ORANGE GG see HGC000
CETIL LIGHT RED 6B see CMM400
CETIL LIGHT RED GG see CMM300
CETIN see HCP700
CETOBEMIDON see KFK000
CETOBEMIDONE see KFK000
CETOCIRE see PJT300
CETOCYLINE see CDF250
CETOL see BEL900
CETOMACROGOL 1000 see PJT300
CETONAL see TDO255
CETRAMIN see CNH125
CETRAXATE see CDF375
CETRAXATE HYDROCHLORIDE see
CDF380
CETRIMIDE see HCQ500
CETRIMONIUM BROMIDE see HCQ500
CETYL ACETATE see HCP100
CETYL ALCOHOL see HCP000
CETYL ALCOHOL ETHOXYLATE see
PJT300
CETYLAMIN (GERMAN) see HCO500
CETYLAMINE see HCO500
CETYLAMINE see HCQ500
CETYLAMINE-HF see CDF400
CETYLAMINE HYDROFLUORIDE see
CDF400
CETYLAMINHYDROFLUORID (GERMAN)
see CDF400
CETYL-γ-AMINOBUTYRATE see AJC500
CETYL BETAINE see CDF450
CETYLDIETHYLETHYLAMMONIUM
BROMIDE see CDF500
CETYLDIMETHYLAMINE see HCP525
CETYL DIMETHYL ETHYL AMMONIUM
BROMIDE see EKN500
CETYL ETHYL DIMETHYLAMMONIUM
BROMIDE see EKN500
CETYL 2-ETHYLHEXANOATE see HCP550
CETYLETHYLMORPHOLINIUM
ETHOSULFATE see EKN550
N-CETYL-N-ETHYLMORPHOLINIUM
ETHYLSULFATE see EKN550
CETYL GABA see AJC500
CETYLIC ACID see PAE250
CETYLIC ALCOHOL see HCP000
CETYLOL see HCP000
CETYLON see BEL900
CETYL PALMITATE see HCP700
CETYL POLY(OXYETHYLENE) ETHER see
PJT300
CETYLPYRIDINIUM BROMIDE see HCP800
1-CETYLPYRIDINIUM BROMIDE see
HCP800
N-CETYLPYRIDINIUM BROMIDE see
HCP800
CETYLPYRIDINIUM CHLORIDE see
CCX000
1-CETYLPYRIDINIUM CHLORIDE see
CCX000
N-CETYLPYRIDINIUM CHLORIDE see
CCX000
CETYLPYRIDINIUM CHLORIDE
MONOHYDRATE see CDF750
CETYL SODIUM SULFATE see HCP900
CETYL SULFATE SODIUM SALT see
HCP900

CETYLTRIMETHYLAMMONIUM
BROMIDE see HCQ500
N-CETYLTRIMETHYLAMMONIUM
BROMIDE see HCQ500
CETYLTRIMETHYLAMMONIUM
CHLORIDE see HCQ525
CETYLUREUM see PEC250
CETYL ZEPHIRAN see BEL900
CEVADENE see CDG000
CEVADIC ACID see TGA700
CEVADILLA see VHZ000
CEVADIN see CDG000
CEVADINE see CDG000
CEVADINE see VHZ000
CEVANE-3-β,4-β,7-α,14,15-α,16-β,20-
HEPTOL,4,9-EPOXY-, 15-((+)-2-HYDROXY-
2-METHYLBUTYRATE) 3-((−)-2-
METHYLBUTYRATE) see VHF000
CEVANOL see BCA000
CEVIAN A 678 see AAX250
CEVIAN HL see ADY500
CEVIN see EBL000
CEVINE see EBL000
CEVITAMIC ACID see ARN000
CEVITAMIN see ARN000
CEX see ALV000
CEYLON CINNAMON BARK OIL see
CMQ510
CEYLON ISINGLASS see AEX250
CEYLON-ZIMT OEL see CMQ510
CEZ SODIUM see CCS250
CF 8 see CBT500
CF 125 see CDT000
CF 1142 see PJQ790
C1-F-ARA-A see CFJ200
CFC see PGE000
CFC 22 see CFX500
CFC 31 see CHI900
CFC-112 see TBP050
CFC-112a see TBP000
CFC 133a see TJY175
CFC 142b see CFX250
CF 8 (CARBON) see CBT500
C. FERTILIS see CCK675
C. FLORIDUS see CCK675
CFNP see FKM100
CFNU see CPL750
CFNU see FIJ000
C. FRUTESCENS see PCB275
C.F.S. see TEM000
CFT 1201 see AGR125
CFV see CDS750
CFX see CCS500
CG 113 see PMB850
CG 117 see MDM100
CG-120 see SFX725
CG 201 see CDG250
CG 315 see THJ500
CG 601 see MRU080
CG-1283 see MQW500
CG 3117 see MFA500
CGA see CHI200
CGA 10832 see CQG250
CGA-12223 see PHK000
CGA 13586 see CHI200
CGA 15324 see BNA750
CGA-18731 see IRA050
CGA-18762 see PMF600
CGA 18809 see ARY800
CGA 20168 see MDN150
CGA-24705 see MQQ450
CGA 26351 see CDS750
CGA 26423 see PMB850
CGA 29170 see PCA400
CGA 30599 see BNA900
CGA-41065 see FDD078
CGA-43089 see COP700
CGA 45156 see LBF100
CGA 48988 see MDM100
CGA-64250 see PMS930
CGA 64251 see EDW100
CGA 64251 see VFA100

CHI 38 see MQD800
CHICAGO ACID S see AKH750
CHICAGO BLUE 6B see CMN750
CHICAGO BLUE RW see CMO600
CHICK ANTIDERMATITIS FACTOR see PAG300
CHICLIDA see HGC500
1,4-CHIDM see BKH325
CHILDREN'S BANE see WAT325
CHILE PEPPER see PCB275
CHILE SALTPETER see SIO900
CHIMASSORB 81 see HND100
CHIMASSORB 944 see PKL150
CHIMCOCCIDE see RLK890
CHIMIPAL AE 3 see DXY000
CHIMOREPTIN see DLH630
CHINABERRY see CDM325
CHINACRIN HYDROCHLORIDE see CFU750
CHINA GREEN (BIOLOGICAL STAIN) see AFG500
CHINALDINE see QEJ000
CHINALPHOS see DJY200
CHINA TREE see CDM325
CHINAWOOD OIL see TOA510
CHINAWOOD OIL TREE see TOA275
CHINE APE see EAI600
CHINESE INKBERRY see DAC500
CHINESE ISINGLASS see AEX250
CHINESE LANTERN PLANT see JBS100
CHINESE RED see LCS000
CHINESE SEASONING see MRL500
CHINESE WHITE see ZKA000
CHINGAMIN see CLD000
CHINGAMIN see CLD250
CHINIDIN (GERMAN) see QFS000
CHINIDIN DURULES see QFS100
CHINIDINE SULFATE see QHA000
CHINIDIN VUFB see QFS100
CHININ (GERMAN) see QIJ000
CHININDIHYDROCHLORID (GERMAN) see QIJ000
CHININ HYDROBROMID (GERMAN) see QJJ100
CHINIOFON see IEP200
CHINOFER see IGS000
CHINOFORM see CHR500
CHINOFUNGIN see TGB475
CHINOGELB see CMM510
CHINOGELB EXTRA see CMM510
CHINOGELB WASSERLOESLICH see CMM510
CHINOIN see EID000
CHINOIN 103 see CPG500
CHINOIN-127 see CDM500
CHINOIN-170 see CDM575
CHINOLEINE see QMJ000
CHINOLIN see QMJ000
CHINOLINE see QMJ000
CHINOLINE YELLOW D SOL. IN SPIRITS see CMS240
CHINOLINE YELLOW ZSS see CMS240
CHINOMETHIONATE see ORU000
p-CHINON (GERMAN) see QQS200
CHINON (DUTCH, GERMAN) see QQS200
CHINONE see QQS200
CHINON I (GERMAN) see BGW750
CHINONOXIM-BENZOYLHYDRAZON (GERMAN) see BDD000
CHINONOXIME-BENZOYLHYDRAZONE see BDD000
CHINORTA see NIM500
CHINOTILIN see KGK400
CHINOXIDIN see QTS100
CHINOXONE see OPK300
3-CHINUCLIDYLBENZILATE see QVA000
CHIP see IGG775
CHIP-CAL see ARB750
CHIP-CAL GRANULAR see ARB750
CHIPCO 26019 see GIA000
CHIPCO BUCTRIL see DDP000
CHIPCO CRAB-KLEEN see DDP000

CHIPCO CRAB KLEEN see DXE600
CHIPCO THIRAM 75 see TFS350
CHIPCO TURF HERBICIDE "D" see DAA800
CHIPCO TURF HERBICIDE MCPP see CIR500
CHIPMAN 3,142 see TBR750
CHIPMAN 6199 see AMX825
CHIPMAN 6200 see DJA400
CHIPMAN 11974 see BDJ250
CHIPMAN R-6, 199 see AMX825
CHIPTOX see CIR250
CHIRAL BINAPHTHOL see CDM625
d-CHIRO-INOSITOL, 3-O-(2-AMINO-4-((CARBOXYIMINOMETHYL)AMINO)-2,3,4,6-TETRADEOXY-α-d-ARABINO-HEXOPYRANOSYL)- see KCA200
l-CHIRO-INOSITOL, 4-AMINO-1,4-DIDEOXY-3-o-(2,6-DIAMINO-2,3,4,6,7-PENTADEOXY-β-l-LYXO-HEPTOPYRANOSYL)-6-o-METHYL-1-(2-(FORMIMIDOYLAMINO)-N-METHYLACETAMIDO)-, SULFATE (1:2),HYDRATE see DAB630
CHIRONEX FLECKERI TOXIN see CDM700
CHISSO 507B see PMP500
CHISSONOX 201 see ECB000
CHISSONOX 206 see PKQ070
CHISSONOX 206 see VOA000
CHITA ROOT EXTRACT see PJH615
CHITIN see CDM750
CHITINA (ITALIAN) see CDM750
CHITRAKA ROOT EXTRACT see PJH615
CHKHZ 5 see MQJ300
CHKHZ 18 see DVF400
CHL see CKN250
CHLODITAN see CDN000
CHLODITHANE see CDN000
CHLODRONATE SODIUM see SFX730
CHLOFENVINPHOS see CDS750
CHLOFENAMIDE see CJN750
CHLOFUCID see DEL300
CHLOMAPHENE see CMX500
CHLOMETHOXYFEN see DGA850
CHLOMETHOXYNIL see DGA850
CHLOMIN see CDP250
CHLOMYCOL see CDP250
CHLONIXIN see CMX770
CHLOOR (DUTCH) see CDV750
3-CHLOORANILINEN (DUTCH) see CEH675
2-CHLOORBENZALDEHYDE (DUTCH) see CEI500
o-CHLOORBENZALDEHYDE (DUTCH) see CEI500
CHLOORBENZEEN (DUTCH) see CEJ125
CHLOORBENZIDE (DUTCH) see CEP000
(4-CHLOOR-BENZYL)-(4-CHLOOR-FENYL)-SULFIDE (DUTCH) see CEP000
2-CHLOOR-1,3-BUTADIEEN (DUTCH) see NCI500
(4-CHLOOR-BUT-2-YN-YL)-N-(3-CHLOOR-FENYL)-CARBAMAAT (DUTCH) see CEW500
CHLOORDAAN (DUTCH) see CDR750
O-2-CHLOOR-1-(2,4-DICHLOOR-FENYL)-VINYL-O,O-DIETHYLFOSFAAT (DUTCH) see CDS750
(2-CHLOOR-3-DIETHYLAMINO-1-METHYL-3-OXO-PROP-1-EN-YL)-DIMETHYL-FOSFAAT see FAB400
2-CHLOOR-4-DIMETHYLAMINO-6-METHYL-PYRIMIDINE (DUTCH) see CCP500
1-CHLOOR-2,4-DINITROBENZEEN (DUTCH) see CGM000
1-CHLOOR-2,3-EPOXY-PROPAAN (DUTCH) see EAZ500
CHLOORETHAAN (DUTCH) see EHH000
2-CHLOORETHANOL (DUTCH) see EIU800
CHLOORFACINON (DUTCH) see CJJ000
3-((4-(4-CHLOOR-FENOXY)-FENOXY)-FENYL)-1,1-DIMETHYLUREUM (DUTCH) see CJQ000

CHLOORFENSON (DUTCH) see CJT750
(4-CHLOOR-FENYL)-BENZEEN-SULFONAAT (DUTCH) see CJR500
(4-CHLOOR-FENYL)-4-CHLOOR-BENZEEN-SULFONAAT (DUTCH) see CJT750
3-(4-CHLOOR-FENYL)-1,1-DIMETHYLUREUM (DUTCH) see CJX750
2(2-(4-CHLOOR-FENYL-2-FENYL)-ACETYL)-INDAAN-1,3-DION (DUTCH) see CJJ000
N-(3-CHLOOR-FENYL)-ISOPROPYL CARBAMAAT (DUTCH) see CKC000
CHLOOR-HEXAVIET see HEA500
CHLOOR-METHAAN (DUTCH) see MIF765
2-(4-CHLOOR-2-METHYL-FENOXY)-PROPIONZUUR (DUTCH) see CIR500
1-CHLOOR-4-NITROBENZEEN (DUTCH) see NFS525
O-(3-CHLOOR-4-NITRO-FENYL)-O,O-DIMETHYL-MONOTHIOFOSFAAT (DUTCH) see MIJ250
O-(4-CHLOOR-3-NITRO-FENYL)-O,O-DIMETHYLMONOTHIOFOSFAAT (DUTCH) see NFT000
CHLOORPIKRINE (DUTCH) see CKN500
CHLOORTHION (DUTCH) see MIJ250
CHLOORWATERSTOF (DUTCH) see HHL000
CHLOPHEDIANOL HYDROCHLORIDE see CMW700
CHLOPHEN see PJL750
CHLOR (GERMAN) see CDV750
CHLORACETAMID (GERMAN) see CDY850
CHLORACETAMIDE-N-METHOLOL see CHO775
α-CHLORACETESSIGSAEUREAETHYLAMID see CEA100
CHLORACETIC ACID see CEA000
CHLORACETOFON see DUI900
CHLORACETONE see CDN200
CHLORACETONE see CDN200
CHLORACETONITRILE see CDN500
CHLORACETOPHENONE see CDN505
CHLORACETOPHENONE see CEA750
CHLORACETOPHON see DUI900
CHLORACETOPHONE see DUI900
CHLORACETOPHOS see CDN510
CHLORACETYL CHLORIDE see CEC250
5-CHLOR-2-ACETYL THIOPHEN see CDN525
CHLORACON see BEG000
CHLORACTIL see CKP500
2-CHLORAETHANOL (GERMAN) see EIU800
N-(2-CHLORAETHYL)-N'-(2 CHLOROETHYL)-N'-o-PROPYLEN-PHOSPHORSAEUREESTER-DIAMID (GERMAN) see IMH000
α-CHLOR-6'-AETHYL-n-(2-METHOXY-1-METHYLAETHYL)-ACET-o-TOLUIDIN (GERMAN) see MQQ450
2-CHLORAETHYL-PHOSPHONSAEURE (GERMAN) see CDS125
2-CHLORAETHYL-TRIMETHYLAMMONIUMCHLORID see CMF400
CHLORAK see TIQ250
CHLORAKON see BEG000
3-CHLORAKRYLAN SODNY see SFV250
CHLORAL see CDN550
CHLORAL, anhydrous, inhibited (DOT) see CDN550
CHLORAL ALCOHOLATE see TIO000
CHLORALDEHYDE see DEM200
CHLORALDURAT see CDO000
CHLORAL ETHYLALCOHOLATE see TIO000
CHLORAL, ETHYL HEMIACETAL see TIO000
CHLORAL HYDRATE see CDO000

CHLORALLYL
DIETHYLDITHIOCARBAMATE see CDO250
2-CHLORALLYL
DIETHYLDITHIOCARBAMATE see CDO250
CHLORALLYLENE see AGB250
CHLORALONE see CDP000
CHLORALOSANE see GFA000
α-CHLORALOSE see GFA000
CHLORALUREA see HOL100
CHLORAMBEN see AJM000
CHLORAMBUCIL see CDO500
CHLORAMBUCIL SODIUM SALT see
CDO625
CHLORAMEISENSAEUREAETHYLESTER
(GERMAN) see EHK500
CHLORAMEISENSAEURE METHYLESTER
(GERMAN) see MIG000
CHLORAMEX see CDP250
CHLORAMFICIN see CDP250
CHLORAMFILIN see CDP250
CHLORAMIDE see CDO750
CHLORAMIFENE see CMX500
CHLORAMIN see BIE500
CHLORAMINE see BIE500
CHLORAMINE see CDO750
CHLORAMINE (inorganic compound) see
CDO750
CHLORAMINE B see SFV275
CHLORAMINE BLACK BH see CMN800
CHLORAMINE BLACK C see AQP000
CHLORAMINE BLACK EC see AQP000
CHLORAMINE BLACK ERT see AQP000
CHLORAMINE BLACK EX see AQP000
CHLORAMINE BLACK EXR see AQP000
CHLORAMINE BLACK SD see CMN230
CHLORAMINE BLACK XO see AQP000
CHLORAMINE BLUE see CMO250
CHLORAMINE BLUE 2B see CMO000
CHLORAMINE BRILLIANT RED 8B see
CMO880
CHLORAMINE BROWN 2R see CMO800
CHLORAMINE BROWN 2ME see CMO800
CHLORAMINE BROWN M see CMO800
CHLORAMINE BROWN MR see CMO800
CHLORAMINE CARBON BLACK S see
AQP000
CHLORAMINE CARBON BLACK SJ see
AQP000
CHLORAMINE CARBON BLACK SN see
AQP000
CHLORAMINE FAST BROWN BRL see
CMO750
CHLORAMINE FAST RED F see CMO870
CHLORAMINE FAST RED FS see CMO870
CHLORAMINE FAST RED K see CMO885
CHLORAMINE FAST RED 5BL see CMO885
CHLORAMINE FAST SCARLET 4B see
CMO870
CHLORAMINE FAST SCARLET SE see
CMO870
CHLORAMINE GREEN 2B see CMO840
CHLORAMINE GREEN 3G see CMO840
CHLORAMINE GREEN B see CMO840
CHLORAMINE GREEN BC see CMO840
CHLORAMINE RED 3B see CMO875
CHLORAMINE RED 8B see CMO880
CHLORAMINE SKY BLUE 4B see CMO500
CHLORAMINE SKY BLUE A see CMO500
CHLORAMINE T see CDP000
CHLORAMINE-T see SFV550
CHLORAMIN HYDROCHLORIDE see
BIE500
1-CHLOR-5-AMINOANTHRACHINON
(CZECH) see AJE325
CHLORAMINOPHEN see CDO500
CHLORAMINOPHENE see CDO500
CHLORAMIPHENE see CMX500
CHLORAMIPHENE see CMX700
CHLORAMIPHENE CITRATE see CMX700
CHLORAMP see PLQ760
CHLORAMP (RUSSIAN) see PIB900
CHLORAMPHENICOL see CDP250

d-CHLORAMPHENICOL see CDP250
d-threo-CHLORAMPHENICOL see CDP250
l(+)-threo-CHLORAMPHENICOL see CDP325
CHLORAMPHENICOL ACID SUCCINATE
see CDP725
CHLORAMPHENICOL ALCOHOL see
CDP350
CHLORAMPHENICOL HEMISUCCINATE
see CDP725
CHLORAMPHENICOL HYDROGEN
SUCCINATE see CDP725
CHLORAMPHENICOL MONOPALMITATE
see CDP700
CHLORAMPHENICOL MONOSUCCINATE
see CDP725
CHLORAMPHENICOL MONOSUCCINATE
SODIUM SALT see CDP500
CHLORAMPHENICOL PALMITATE see
CDP700
CHLORAMPHENICOL SODIUM
MONOSUCCINATE see CDP500
CHLORAMPHENICOL SODIUM
SUCCINATE see CDP500
CHLORAMPHENICOL SUCCINATE see
CDP725
CHLORAMPHENICOL SUCCINATE
SODIUM see CDP500
CHLORAMPHENICOL-SUKZINAT-
NATRIUM (GERMAN) see CDP500
CHLORAMSAAR see CDP250
CHLORANAUTINE see DYE600
CHLORANIFORMETHAN see CDP750
CHLORANIFORMETHANE see CDP750
CHLORANIL see TBO500
4-CHLORANILIN (CZECH) see CEH680
2-(2-CHLORANILIN)-4,6-DICHLOR-1,3,5-
TRIAZIN (GERMAN) see DEV800
m-CHLORANILINE see CEH675
o-CHLORANILINE see CEH670
p-CHLORANILINE see CEH680
CHLORANOCRYL see DFO800
1-CHLORANTHRACHINON (CZECH) see
CEI000
CHLORANTINE FAST RED see CMO885
CHLORANTINE FAST RED 5B (6CI) see
CMO885
CHLORANTINE FAST TURQUOISE VLL see
COF420
CHLORAQUINE see CLD000
CHLORARSENOL see CEI250
CHLORARSOL see DFX400
CHLORASAN see CDP000
CHLORASEN see DFX400
CHLORASEPTINE see CDP000
CHLORASOL see CDP250
CHLORA-TABS see CDP250
CHLORATE de CALCIUM (FRENCH) see
CAO500
CHLORATE of POTASH (DOT) see PLA250
CHLORATE de POTASSIUM (FRENCH) see
PLA250
CHLORATES see CDQ000
CHLORATE SALT of MAGNESIUM see
MAE000
CHLORATE SALT of SODIUM see SFS000
CHLORATE of SODA (DOT) see SFS000
2-(3-(2-CHLORATHYL)-3-
NITROSOUREIDO)ATHYLMETHANSULFO
NAT (GERMAN) see CHF250
CHLORAX see SFS000
CHLORAZAN see CDP000
CHLORAZENE see CDP000
CHLORAZEPAM see CDQ250
CHLORAZEPATE DIPOTASSIUM see
CDQ250
CHLORAZIN see CKP500
CHLORAZINE see CDQ325
CHLORAZOL BLACK BH see CMN800
CHLORAZOL BLACK E see AQP000
CHLORAZOL BLACK EA see AQP000
CHLORAZOL BLACK E (BIOLOGICAL
STAIN) see AQP000

CHLORAZOL BLACK EN see AQP000
CHLORAZOL BLUE 3B see CMO250
CHLORAZOL BLUE B see CMO000
CHLORAZOL BLUE RW see CMO600
CHLORAZOL BORDEAUX B see CMO872
CHLORAZOL BORDEAUX BP see CMO872
CHLORAZOL BRILLIANT PURPURINE 8B
see CMO880
CHLORAZOL BROWN LF see CMO820
CHLORAZOL BROWN MP see CMO800
CHLORAZOL BURL BLACK E see AQP000
CHLORAZOL DARK GREEN PL see
CMO830
CHLORAZOL DIAZO BLACK SD see
CMN230
CHLORAZOL FAST RED FP see CMO870
CHLORAZOL FAST RED FS see CMO870
CHLORAZOL FAST SCARLET 4B see
CMO870
CHLORAZOL FAST SCARLET 4BSP see
CMO870
CHLORAZOL GREEN BN see CMO840
CHLORAZOL GREEN BNP see CMO840
CHLORAZOL LEATHER BLACK BH see
CMN800
CHLORAZOL LEATHER BLACK ENP see
AQP000
CHLORAZOL ORANGE BROWN X see
CMO810
CHLORAZOL PAPER GREEN BN see
CMO840
CHLORAZOL SILK BLACK G see AQP000
CHLORAZOL SKY BLUE FF see BGT250
CHLORAZOL SKY BLUE FF see CMN750
CHLORAZOL VIOLET N see CMP000
CHLORAZOL VISCOSE BLACK B see
CMN240
CHLORAZOL YELLOW 2G see CMP050
CHLORAZOL YELLOW DP see CMP050
CHLORAZONE see CDP000
CHLORBENSID (GERMAN) see CEP000
CHLORBENSIDE see CEP000
CHLORBENXIDE see CEP000
2-CHLORBENZALDEHYD (GERMAN) see
CEI500
CHLORBENZENE see CEJ125
p-CHLORBENZENESULFOCHLORID
(CZECH) see CEK375
p-CHLORBENZENSULFONAN SODNY
(CZECH) see CEK250
1-p-CHLORBENZHYDRYL-m-
METHYLBENZYLPIPERAZINE
DIHYDROCHLORIDE see MBX250
CHLORBENZIDE see CEP000
CHLORBENZILATE see DER000
p-CHLORBENZOIC ACID see CEL300
CHLORBENZOL see CEJ125
o-CHLORBENZONITRIL (CZECH) see
CEM000
CHLORBENZOSAMINE
DIHYDROCHLORIDE see CDQ500
m-CHLORBENZOTRIFLUORIDE see
CDQ330
CHLORBENZOXAMINE
DIHYDROCHLORIDE see CDQ500
5-CHLORBENZOXAZOLIN-2-ON see
CDQ750
CHLORBENZOXYETHAMINE
DIHYDROCHLORIDE see CDQ500
1-CHLOR-5-
BENZOYLAMINOANTHRACHINON
(CZECH) see BDK750
N-(p-CHLORBENZOYL)-γ-(2,6-
DIMETHYLANILINO)-BUTTERSAEURE
(GERMAN) see CLW625
(1-(p-CHLORBENZOYL)-5-METHOXY-2-
METHYLINDOL-3-
ACETOXY)ESSIGSAEURE (GERMAN) see
AAE625
N-p-CHLORBENZOYL-5-METHOXY-2-
METHYLINDOLE-3-ACETIC ACID see
IDA000

5-CHLORBENZOZAZOLIN-2-ON see CDQ750

(4-CHLOR-BENZYL)-(4-CHLOR-PHENYL)-SULFID (GERMAN) see CEP000

4-CHLOR-BENZYL-CYANID see CEP300

1-p-CHLORBENZYL-2-METHYL-BENZIMIDAZOL (GERMAN) see CDY325

p-CHLORBENZYL-α-PYRIDYL-DIMETHYL-AETHYLENDIAMIN (GERMAN) see CKV625

CHLORBICYCLEN see BFY100

CHLORBICYCLENE (FRENCH) see DAM700

CHLORBISAN see DMN000

CHLORBROMURON see CES750

CHLORBUFAM see CEX250

CHLORBUFAN mixed with CYCEURON see AFM375

CHLORBUPHAM see CEX250

2-CHLOR-1,3-BUTADIEN (GERMAN) see NCI500

CHLORBUTANOL see ABD000

4-CHLORBUTAN-1-OL (GERMAN) see CEU500

(4-CHLOR-BUT-2-IN-YL)-N-(3-CHLOR-PHENYL)-CARBAMAT (GERMAN) see CEW500

CHLORBUTOL see ABD000

CHLORCARVACROL see CEX275

CHLORCHOLINCHLORID see CMF400

CHLORCHOLINE CHLORIDE see CMF400

CHLORCOSANE see PAH780

p-CHLOR-m-CRESOL see CFD990

CHLORCYAN see COO750

CHLORCYCLINE see CFF500

CHLORCYCLIZINE see CFF500

CHLORCYCLIZINE DIHYDROCHLORIDE see CDR000

CHLORCYCLIZINE HYDROCHLORIDE see CDR250

CHLORCYCLIZINE HYDROCHLORIDE A see CDR500

CHLORCYCLIZINIUM CHLORIDE see CDR250

CHLORCYCLOHEXAMIDE see CDR550

4-(3-CHLOR-4-CYCLOHEXYL-PHENYL)-4-OXO-BUTTERSAEURE KALZIUM (GERMAN) see CPJ250

6-CHLOR-N-CYCLOPROPYL-N'-(1-METHYLETHYL)-1,3,5-TRIZAINE-2,4-DIAMINE see CQI750

CHLORDAN see CDR750

cis-CHLORDAN see CDR675

trans-CHLORDAN see CDR575

trans-CHLORDAN see CDR800

α-CHLORDAN see CDR675

β-CHLORDAN see CDR800

γ-CHLORDAN see CDR575

γ-CHLORDAN see CDR750

CHLORDANE see CDR750

CHLORDANE see CDR760

cis-CHLORDANE see CDR675

trans-CHLORDANE see CDR800

α-CHLORDANE see CDR675

β-CHLORDANE see CDR800

CHLORDANE, liquid (DOT) see CDR750

α(cis)-CHLORDANE see CDR675

γ(trans)-CHLORDANE see CDR575

CHLORDANE, TECHNICAL see CDR760

CHLORDECONE see KEA000

CHLORDENE see HCN000

α-CHLORDENE see CDR820

γ-CHLORDENE see CDR860

7-CHLOR-4-(4-(DIAETHYLAMINO)-1-METHYLBUTYLAMINO)-CHINOLINDIPHOSPHAT (GERMAN) see CLD250

(2-CHLOR-3-DIAETHYLAMINO-1-METHYL-3-OXO-PROP-1-EN-YL)-DIMETHYL-PHOSPHAT see FAB400

CHLORDIAZACHEL see MDQ250

CHLORDIAZEPOXIDE see LFK000

CHLORDIAZEPOXIDE HYDROCHLORIDE see MDQ250

CHLORDIAZEPOXIDE MONOHYDROCHLORIDE see MDQ250

CHLORDIAZEPOXIDE, NITROSATED see SIS000

CHLORDIAZEPOXIDE mixed with SODIUM NITRITE (1:1) see SIS000

O-2-CHLOR-1-(2,4-DICHLOR-PHENYL)-VINYL-O,O-DIAETHYLPHOSPHAT (GERMAN) see CDS750

7-CHLOR-2,3-DIHYDRO-1-METHYL-5-PHENYL-1H-1,4-BENZODIAZEPIN HYDROCHLORID (GERMAN) see MBY000

CHLORDIMEFORM see CJJ250

CHLORDIMEFORM HYDROCHLORIDE see CJJ500

2-CHLOR-11-(2-DIMETHYAMINOAETHOXY)-DIBENZO(b,f)-THIEPIN (GERMAN) see ZUJ000

2-CHLOR-4-DIMETHYLAMINO-6-METHYLPYRIMIDIN (GERMAN) see CCP500

CHLORDIMETHYLETHER (CZECH) see CIO250

1-CHLOR-2,4-DINITROBENZENE see CGM000

CHLORE (FRENCH) see CDV750

CHLOREFENIZON (FRENCH) see CJT750

CHLORENDIC ACID see CDS000

CHLORENDIC ACID DIBUTYL ESTER see CDS025

CHLORENDIC ANHYDRIDE see CDS050

CHLORENDIC DIOL see HCC550

CHLORENDIC IMIDE see CDS100

CHLOREPIN see CIR750

1-CHLOR-2,3-EPOXY-PROPAN (GERMAN) see EAZ500

CHLORESENE see BBQ500

CHLORESSIGSAEURE-N-ISOBUTINYLANILID (GERMAN) see CDS275

CHLORESSIGSAEURE-N-ISOPROPYLANILID (GERMAN) see CHS500

CHLORESSIGSAEURE-N-(METHOXYMETHYL)-2,6-DIAETHYLANILID (GERMAN) see CFX000

CHLORESTROLO see CLO750

CHLORETHAMINACIL see BIA250

CHLORETHAMINE see BIE500

CHLOR-ETHAMINE see EIW000

2-CHLORETHANOL (GERMAN) see EIU800

CHLORETHAZINE see BIE500

CHLORETHENE see VNP000

CHLORETHEPHON see CDS125

CHLORETHIAZOL see CHD750

2-(2-CHLORETHOXY)ETHYL 2'-CHLORETHYL ETHER see TKL500

CHLORETHYL see EHH000

2-CHLORETHYLACETAT see CGO600

CHLORETHYLBENZMETHOXAZONE see CDS250

CHLORETHYLENE see VNP000

2-CHLORETHYLESTER KYSELINY CHLORMRAVENCI see CGU199

2-CHLORETHYLISOKYANAT see IKH000

2-CHLORETHYLPHOSPHONIC ACID see CDS125

2-CHLORETHYL VINYL ETHER see CHI250

CHLORETIN see CDS275

CHLORETONE see ABD000

CHLOREX see DFJ050

CHLOREXTOL see PJL750

CHLOREZ 700 see PAH780

CHLOREZ 700HMP see PAH780

CHLORFACINON (GERMAN) see CJJ000

CHLORFENAC see TIY500

CHLORFENAMIDINE see CJJ250

CHLORFENAPYR see BNA600

CHLORFENETHOL see BIN000

CHLORFENIDIM see CJX750

p-CHLORFENOL (CZECH) see CJK750

2-(4'-CHLORFENOXY)ETHANOL (CZECH) see CJO500

CHLORFENPROP-METHYL see CFC750

CHLORFENSON see CJT750

CHLORFENSONE see CJT750

CHLORFENSULFID (GERMAN) see CDS500

CHLORFENSULFIDE see CDS500

CHLORFENVINFOS see CDS750

CHLORFENVINFOS see CDS750

CHLORFENVINPHOS see CDS750

1-p-CHLORFENYL-3,3-DIMETHYLTRIAZEN (CZECH) see CJI100

3-CHLOR-p-FENYLENDIAMIN (CZECH) see CEG600

m-CHLORFENYLISOKYANAT see CKA750

p-CHLORFENYLISOKYANAT (CZECH) see CKB000

p-CHLORFENYLMERKAPTOMETHYLCHLORID (CZECH) see CFB750

p-CHLORFENYLMONOGLYKOLETHER (CZECH) see CJO500

p-CHLORFENYLSILATRAN (CZECH) see CKM750

CHLORFLUAZURON see CDS800

CHLORFLURAZOLE see DGO400

CHLORFLURECOL see CDT000

CHLORFLURECOL-METHYL see CDT000

CHLORFLURECOL-METHYL ESTER see CDT000

CHLORFLURENOL see CDT000

CHLORFLURENOL METHYL ESTER see CDT000

CHLORFONIUM see THY500

CHLORFOS see TIQ250

CHLOR-N-(2-FURYLMETHYL)-5-SULFAMYLANTHRANILSAEURE (GERMAN) see CHJ750

CHLORGUANIDE see CKB250

CHLORGUANIDE HYDROCHLORIDE see CKB500

CHLORGUANIDE TRIAZINE see COX400

CHLORHEXAMIDE see CDR550

CHLORHEXIDIN (CZECH) see BIM250

CHLORHEXIDINE see BIM250

CHLORHEXIDINE ACETATE see CDT125

CHLORHEXIDINE DIACETATE see CDT125

CHLORHEXIDINE DIGLUCONATE see CDT250

CHLORHEXIDINE GLUCONATE see CDT250

CHLORHEXIDINE GLUCONATE see CDT500

CHLORHEXIDIN GLUKONATU see CDT250

CHLORHEXIDIN GLUKONATU (CZECH) see CDT500

6-CHLORHEXYLISOKYANAT see CHL250

CHLORHYDRATE de ACETOXY-THYMOXY-ETHYL-DIMETHYLAMINE (FRENCH) see TFY000

CHLORHYDRATE d'AMIKHELLINE (FRENCH) see AHP375

CHLORHYDRATE d'ANILINE (FRENCH) see BBL000

CHLORHYDRATE de 4-CHLOROORTHOTOLUIDINE (FRENCH) see CLK235

CHLORHYDRATE de CONESSINE (FRENCH) see CNH660

CHLORHYDRATE de N-(DIETHOXY-2,5-PHENYL)-N-DIETHYLAMINO-2-ETHYL BUTOXY-4-PHENOXYACETAMIDE see BPM750

CHLORHYDRATE de DIETHYLAMINOETHYLTHEOPHYLLINE (FRENCH) see DIH600

CHLORHYDRATE de (N-ETHYL,N,β-CHLORETHYL)AMINO-METHYLBENZODIOXANE (FRENCH) see CGX625

CHLORHYDRATE d'HISTAMINE (FRENCH) see HGE500
CHLORHYDRATE de (NAPHTHYLOXY-1)-4 HYDROXY 3 BUTYRAMIDOXIME (FRENCH) see NAC500
CHLORHYDRATE de NICOTINE (FRENCH) see NDP400
CHLORHYDRATE de PAPAVERINE (FRENCH) see PAH250
CHLORHYDRATE de PHENETHYL-8-OXA-1-DIAZA-3,8-SPIRO(4,5)DECANONE-2 (FRENCH) see DAI200
CHLORHYDRATE de α-PHENYL-α-(β'-DIETHYLAMINOETHYL)GLUTARIMIDE (FRENCH) see ARR875
CHLORHYDRATE de PIPERIDINOMETHYLCYCLOHEXANE (FRENCH) see PIR000
CHLORHYDRATE de (PIPERONYL-4-PIPERAZINO)-1)-(PHENYL-1-PYRROLIDONE-2-CARBOXAMIDE-4) (FRENCH) see PGA250
CHLORHYDRATE de RAUGALLINE (FRENCH) see AFH280
CHLORHYDRATE de TETRACYCLINE (FRENCH) see TBX250
CHLORHYDRATE de (TRIMETHOXY-2-4-6) PHENYL-(PYRROLIDINE-3) PROPYLACETONE (FRENCH) see BOM600
CHLORHYDRIN see CDT750
α-CHLORHYDRIN see CDT750
CHLORHYDROL see AHA000
CHLORHYDROL, GRANULAR see AHA000
CHLORHYDROL, IMPALPABLE see AHA000
CHLORHYDROXYALUMINUM ALLANTOINATE see AHA135
3-CHLOR-4-HYDROXYBIFENYL (CZECH) see CHN500
2-CHLOR-9-HYDROXYFLUOREN-CARBONSAEURE-(9)-METHYLESTER (GERMAN) see CDT000
CHLORIAZID see CLH750
CHLORIC ACID see CDU000
CHLORIC ACID, solution, containing not more than 10% acid (DOT) see CDU000
CHLORIC ACID, AMMONIUM SALT see ANE250
CHLORIC ACID, BARIUM SALT see BAJ500
CHLORIC ACID, COPPER SALT see CNJ900
CHLORIC ACID, SODIUM SALT, MIXT. WITH UREA see CDU100
CHLORIC ACID, STRONTIUM SALT see SMF500
CHLORIC ACID, THALLIUM(1+) SALT see TEJ100
CHLORICOL see CDP250
CHLORID AMONNY (CZECH) see ANE500
CHLORID ANILINU (CZECH) see BBL000
CHLORID ANTIMONITY see AQC500
CHLORIDAZON see PEE750
CHLORID-N-BUTYLCINICITY (CZECH) see BSR250
CHLORID CHROMITY HEXAHYDRAT see CMK450
CHLORID DI-n-BUTYLCINICITY (CZECH) see DDY200
CHLORID DRASELNY (CZECH) see PLA500
CHLORIDEAZEPOXIDE HYDROCHLORIDE see MDQ250
CHLORIDE de CHOLINE (FRENCH) see CMF750
CHLORIDE of DIAMINOMETHYLPHENYLDIMETHYL-p-BENZOQUINONE-DIIMINE see DCE800
CHLORIDE of LIME (DOT) see HOV500
CHLORIDE of PHOSPHORUS see PHT275
CHLORIDES see CDU250
CHLORIDE of SULFUR see SOG500
CHLORIDE of SULFUR (DOT) see SON510
CHLORID FENYLRTUTNATY (CZECH) see PFM500
CHLORIDIAZEPIDE see LFK000

CHLORIDIAZEPOXIDE see LFK000
CHLORIDIN see TGD000
CHLORIDINE see TGD000
CHLORID KREMICITY (CZECH) see SCQ500
CHLORID KYSELINY-p-CHLORBENSULFONOVE (CZECH) see CEK375
CHLORID KYSELINY CHLORMETHANSULFONOVE (CZECH) see CHY000
CHLORID KYSELINY CHLOROCTOVE see CEC250
CHLORID KYSELINY DICHLOROCTOVE see DEN400
CHLORID KYSELINY DIMETHYLKARBAMINOVE see DQY950
CHLORID MEDNY (CZECH) see CNK250
CHLORID RTUTNATY (CZECH) see MCY475
CHLORID TRIBENZYLCINICITY (CZECH) see CLP000
CHLORID TRI-n-BUTYLCINICITY (CZECH) see CLP500
CHLORIDUM see EHH000
CHLORIERTE BIPHENYLE, CHLORGEHALT 42% (GERMAN) see PJM500
CHLORIERTE BIPHENYLE, CHLORGEHALT 54% (GERMAN) see PJN000
CHLORIERTES CAMPHEN see CDU325
CHLOR-IFC see CKC000
CHLORIMINE MUSTARD see MIG800
CHLORIMIPRAMINE see CDU750
CHLORIMIPRAMINE HYDROCHLORIDE see CDV000
CHLORINAT see CEW500
CHLORINATED BIPHENYL see PJL750
CHLORINATED CAMPHENE see CDV100
CHLORINATED DIBENZO DIOXINS see CDV125
CHLORINATED DIPHENYL see PJL750
CHLORINATED DIPHENYLENE see PJL750
CHLORINATED DIPHENYL OXIDE see CDV175
CHLORINATED HC, ALIPHATIC see CDV250
CHLORINATED HC AROMATIC see CDV500
CHLORINATED HYDROCARBONS, ALIPHATIC see CDV250
CHLORINATED HYDROCARBONS, AROMATIC see CDV500
CHLORINATED HYDROCHLORIC ETHER see DFF809
CHLORINATED LIME (DOT) see HOV500
CHLORINATED NAPHTHALENES see CDV575
CHLORINATED PARAFFINS see PAH780
CHLORINATED PARAFFINS (C$_{12}$, 60% CHLORINE) see PAH800
CHLORINATED PARAFFINS (C$_{23}$, 43% CHLORINE) see PAH810
CHLORINATED POLYETHER POLYURETHAN see CDV625
CHLORINATED TRISODIUM PHOSPHATE see SJH095
CHLORINDAN see CDR750
CHLORINDANOL see CDV700
CHLORINE see CDV750
CHLORINE AZIDE see CDW000
CHLORINE CYANIDE see COO750
CHLORINE DIOXIDE see CDW450
CHLORINE DIOXIDE, not hydrated (DOT) see CDW450
CHLORINE DIOXYGEN TRIFLUORIDE see CDW500
CHLORINE FLUORIDE see CDX750
CHLORINE FLUORIDE (ClF$_5$) see CDX250
CHLORINE FLUORIDE OXIDE see PCF750
CHLORINE MOL. see CDV750
CHLORINE NITRATE see CDX000
CHLORINE NITRIDE see NGQ500

CHLORINE OXIDE see CDW450
CHLORINE(IV) OXIDE see CDW450
CHLORINE OXYFLUORIDE see PCF750
CHLORINE PENTAFLUORIDE see CDX250
CHLORINE PENTAFLUORIDE (DOT) see CDX250
CHLORINE PERCHLORATE see CDX500
CHLORINE PEROXIDE see CDW450
CHLORINE SULFIDE see SOG500
CHLORINE TETROXYFLUORIDE see FFD000
CHLORINE TRIFLUORIDE see CDX750
CHLORINE(1)TRIFLUOROMETHANESULFONATE see CDX800
CHLOR-IPC see CKC000
CHLORISONDAMINE see CDY000
CHLORISONDAMINE CHLORIDE see CDY000
CHLORISONDAMINE DIMETHOCHLORIDE see CDY000
CHLORISOPROPAMIDE see CDY100
CHLORITES see CDY250
5-CHLOR-7-JOD-8-HYDROXY-CHINOLIN (GERMAN) see CHR500
CHLORKETONE see CDY260
CHLOR KIL see CDR750
CHLORKU LITU (POLISH) see LHB000
CHLORMADINON see CDY275
CHLORMADINON ACETATE see CBF250
CHLORMADINONE see CDY275
CHLORMADINONE ACETATE see CBF250
CHLORMADINONE ACETATE mixed with MESTRANOL see CNV750
CHLORMADINONU (POLISH) see CBF250
CHLORMENE see TAI500
CHLORMEPHOS see CDY299
CHLORMEQUAT see CMF400
CHLORMEQUAT CHLORIDE see CMF400
CHLORMEROPRIN see CHX250
CHLOR-METHAN (GERMAN) see MIF765
CHLORMETHANSULFOCHLORID (CZECH) see CHY000
CHLORMETHAZANONE see CKF500
CHLORMETHAZONE see CKF500
CHLORMETHIAZOLE see CHD750
CHLORMETHINE see BIE250
CHLORMETHINE HYDROCHLORIDE see BIE500
CHLORMETHINE-N-OXIDE HYDROCHLORIDE see CFA750
CHLORMETHINUM see BIE500
3-(3-CHLOR-4-METHOXYPHENYL)-1,1-DIMETHYLHARNSTOFF (GERMAN) see MQR225
N'-(3-CHLOR-4-METHOXY-PHENYL)-N,N-DIMETHYLHARNSTOFF (GERMAN) see MQR225
CHLORMETHYL-METHYL-DIETHOXYSILAN (CZECH) see CIO000
α-(CHLORMETHYL)-2-METHYL-5-NITRO-IMIDAZOL-1-AETHANOL (GERMAN) see OJS000
4-(4-CHLOR-2-METHYLPHENOXY)-BUETTERSAEURE (GERMAN) see CLN750
4-(4-CHLOR-2-METHYLPHENOXY)-BUTTERSAEURE (GERMAN) see CLN750
4-(4-CHLOR-2-METHYL-PHENOXY)-BUTTERSAEURE NATRIUMSALZ (GERMAN) see CLO000
2-(4-CHLOR-2-METHYL-PHENOXY)-PROPIONSAEURE (GERMAN) see CIR500
3-(3-CHLOR-4-METHYLPHENYL)-1,1-DIMETHYLHARNSTOFF (GERMAN) see CIS250
N-(3-CHLOR-METHYLPHENYL)-2-METHYLPENTANAMID (GERMAN) see SKQ400
3-CHLOR-2-METHYL-PROP-1-EN (GERMAN) see CIU750
CHLORMETHYL-TRIETHOXYSILAN (CZECH) see CIY500
CHLORMEZANONE see CKF500

CHLORMIDAZOLE see CDY325
CHLORMITE see PNH750
CHLORNAFTINA see BIF250
CHLORNAPHAZIN see BIF250
α-CHLORNAPHTHALENE see CIZ000
CHLORNAPHTHIN see BIF250
1-CHLOR-5-NITROANTHRACHINON (CZECH) see CJA250
1-CHLOR-4-NITROBENZOL (GERMAN) see NFS525
CHLORNITROFEN see NIW500
CHLORNITROMYCIN see CDP250
O-(3-CHLOR-4-NITRO-PHENYL)-O,O-DIMETHYL-MONOTHIOPHOSPHAT (GERMAN) see MIJ250
O-(4-CHLOR-3-NITRO-PHENYL)-O,O-DIMETHYL-MONOTHIOPHOSPHAT (GERMAN) see NFT000
CHLOROACETALDEHYDE see CDY500
2-CHLOROACETALDEHYDE see CDY500
CHLOROACETALDEHYDE HYDRATE see CDY600
CHLOROACETALDEHYDE MONOMER see CDY500
CHLOROACETAMIDE see CDY850
2-CHLOROACETAMIDE see CDY850
N-CHLOROACETAMIDE see CDY825
α-CHLOROACETAMIDE see CDY850
CHLOROACETAMIDE OXIME see CDZ000
CHLOROACETANILIDE see PER300
2-CHLOROACETANILIDE see PER300
m-CHLOROACETANILIDE see CDZ050
3'-CHLOROACETANILIDE see CDZ050
4'-CHLOROACETANILIDE see CDZ100
α-CHLOROACETANILIDE see PER300
CHLOROACETIC ACID see CEA000
α-CHLOROACETIC ACID see CEA000
CHLOROACETIC ACID BENZYL ESTER see BEE500
CHLOROACETIC ACID tert-BUTYL ESTER see BQR100
CHLOROACETIC ACID CHLORIDE see CEC250
CHLOROACETIC ACID, ETHYL ESTER see EHG500
CHLOROACETIC ACID METHYL ESTER see MIF775
CHLORO-ACETIC ACID, PHENETHYL ESTER see PDF500
CHLOROACETIC ACID SODIUM SALT see SFU500
CHLOROACETIC ACID 2,2,2-TRICHLORO-1-(DIMETHOXYPHOSPHINYL)ETHYL ESTER see DUI900
CHLOROACETIC ACID, solid (UN 1751) (DOT) see CEA000
CHLOROACETIC ACID, liquid (UN 1750) (DOT) see CEA000
CHLOROACETIC CHLORIDE see CEC250
4-CHLOROACETOACETANILIDE see CEA050
o-CHLOROACETOACETANILIDE see AAY600
p-CHLOROACETOACETANILIDE see AAY250
2'-CHLOROACETOACETANILIDE see AAY600
4'-CHLOROACETOACETANILIDE see AAY250
γ-CHLOROACETOACETANILIDE see CEA050
γ-CHLOROACETO ACETIC ACID ANILIDE see CEA050
α-CHLOROACETOACETIC ACID MONOETHYLAMIDE see CEA100
CHLOROACETONE see CDN200
CHLOROACETONE, stabilized (DOT) see CDN200
2-CHLOROACETONITRILE see CDN500
CHLOROACETONITRILE (DOT) see CDN500
α-CHLOROACETONITRILE see CDN500

1-CHLOROACETOPHENONE see CEA750
4-CHLOROACETOPHENONE see CEB250
p-CHLOROACETOPHENONE see CEB250
2'-CHLOROACETOPHENONE see CEB000
4'-CHLOROACETOPHENONE see CEB250
α-CHLOROACETOPHENONE see CEA750
ω-CHLOROACETOPHENONE see CEA750
CHLOROACETOPHENONE, liquid or solid (DOT) see CEA750
2-CHLORO-4-ACETOTOLUIDIDE see CEB500
2-CHLOROACETO-p-TOLUIDIDE see CEB500
3'-CHLORO-p-ACETOTOLUIDIDE see CEB750
2-CHLORO-10-(3-(4-(2-ACETOXYETHYL)PIPERAZINYL)PROPYL)PHENOTHIAZINE see TFP250
8-CHLOROACETOXY-9-HYDROXY-8,9-DIHYDRO-AFLATOXIN B1 see CEB800
6-CHLORO-17-α-ACETOXY-4,6-PREGNADIENE-3,20-DIONE see CBF250
6-CHLORO-Δ⁶-17-ACETOXYPROGESTERONE see CBF250
Δ⁶-6-CHLORO-17-α-ACETOXYPROGESTERONE see CBF250
6-α-CHLORO-17-α-ACETOXYPROGESTERONE see CEB875
6-CHLORO-Δ⁶-(17-α)ACETOXYPROGESTERONE see CBF250
(CHLOROACETOXY)TRIBUTYLSTANNANE see TIC750
4'-CHLOROACETYL ACETANILIDE see CEC000
4'-(CHLOROACETYL)ACETANILIDE see CEC000
N-(CHLOROACETYL)-3-AZABICYCLO(3.2.1)NONANE see CEC100
CHLOROACETYL CHLORIDE see CEC250
N-CHLOROACETYLDIETHYLAMINE see DIX400
N-CHLOROACETYL-3,5-DIETHYL-4-AMINOPHENOL see CEC260
N-CHLOROACETYL-3,5-DIETHYL-p-BENZOQUINONE-4-IMINE see CEC270
1-CHLOROACETYL-α-α-DIPHENYL-4-PIPERIDINEMETHANOL see CEC300
CHLOROACETYLENE see CEC500
2-CHLOROACETYLFLUORENE see CEC700
l-N-(α-(CHLOROACETYL)PHENETHYL)-p-TOLUENESULFONAMIDE see THH450
p-(CHLOROACETYL)PHENOL see HNE500
N-(CHLOROACETYL)-3-PHENYL-N-(p-TOLYLSULFONYL)ALANINE see THH550
5-CHLORO-2-ACETYL THIOPHEN see CDN525
N-(CHLOROACETYL)-p-TOLUIDINE see CEB500
4'-(2-CHLORO-9-ACRIDINYLAMINO)METHANESULFONANILIDE see CED500
4'-(3-CHLORO-9-ACRIDINYLAMINO)METHANESULFONANILIDE see CED750
N-(4-((2-CHLORO-9-ACRIDINYL)AMINO)PHENYL)METHANESULFONAMIDE see CED500
2-CHLOROACROLEIN see CED800
α-CHLOROACROLEIN see CED800
CHLOROACRYLIC ACID see CEE500
2-CHLOROACRYLIC ACID see CEE500
α-CHLOROACRYLIC ACID see CEE500
2-CHLOROACRYLIC ACID ETHYL ESTER see EHH200
2-CHLOROACRYLIC ACID, METHYL ESTER see MIF800
cis-β-CHLOROACRYLIC ACID SODIUM SALT see SFV250
CHLOROACRYLONITRILE see CEE750
2-CHLOROACRYLONITRILE see CEE750
α-CHLOROACRYLONITRILE see CEE750

(3-CHLORO-1-ADAMANTANOYL)DIAZOMETHANE see CEE800
3-CHLOROADAMANTYL DIAZOMETHYL KETONE see CEE800
1-CHLORO-3-ADAMANTYL ETHOXYMETHYL KETONE see CEE825
1-CHLORO-3-ADAMANTYL HYDROXYMETHYL KETONE see CEE850
2-CHLOROADENOSINE see CEF100
2-CHLOROADENOSINE-5'-SULFAMATE see CEF125
CHLOROAETHAN (GERMAN) see EHH000
pi-CHLORO ALLYL ALCOHOL see CEF500
β-CHLORO ALLYL ALCOHOL see CEF250
α-CHLOROALLYL CHLORIDE see DGG950
γ-CHLOROALLYL CHLORIDE see DGG950
2-CHLOROALLYL DIETHYLDITHIOCARBAMATE see CDO250
2-CHLOROALLYL-N,N-DIETHYLDITHIOCARBAMATE see CDO250
CHLOROALLYLENE see AGB250
1-(3-CHLOROALLYL)-3,5,7-TRIAZA-1-AZONIAADAMANTANE CHLORIDE see CEG550
CHLOROALONIL see TBQ750
CHLOROALOSANE see GFA000
CHLOROAMBUCIL see CDO500
CHLOROAMINE see CDO750
3-CHLORO-4-AMINOANILINE see CEG600
3-CHLORO-4-AMINOANILINE SULFATE see CEG625
5-CHLORO-1-AMINOANTHRAQUINONE see AJE325
2-CHLORO-4-AMINOBENZOIC ACID see CEG750
4-CHLORO-3-AMINOBENZOTRIFLUORIDE see CEG800
4'-CHLORO-4-AMINOBIPHENYL ETHER see CEH125
3-CHLORO-4-AMINODIPHENYL see CEH000
4-CHLORO-4'-AMINODIPHENYL ETHER see CEH125
2-CHLORO-3-AMINO-1,4-NAPHTHOQUINONE see AJI250
2-CHLORO-5-AMINOPHENOL see AJI260
p-CHLORO-o-AMINOPHENOL see CEH250
3-CHLORO-4-AMINOSTILBENE see CLE500
2-CHLORO-4-AMINOTOLUENE see CLK215
3-CHLORO-2-AMINOTOLUENE see CLK227
4-CHLORO-2-AMINOTOLUENE see CLK225
5-CHLORO-2-AMINOTOLUENE see CLK220
5-CHLORO-2-AMINOTOLUENE HYDROCHLORIDE see CLK235
2-CHLORO-4-AMINOTOLUENE-5-SULFONIC ACID see AJJ250
CHLOROAMITRIPTYLINE HYDROCHLORIDE see CEH500
2-CHLOROANILINE see CEH670
3-CHLOROANILINE see CEH675
4-CHLOROANILINE see CEH680
m-CHLOROANILINE see CEH675
o-CHLOROANILINE see CEH670
p-CHLOROANILINE see CEH680
m-CHLOROANILINE, solid see CEH675
o-CHLOROANILINE, solid see CEH670
p-CHLOROANILINE, solid see CEH680
m-CHLOROANILINE, liquid see CEH675
o-CHLOROANILINE, liquid see CEH670
p-CHLOROANILINE, liquid see CEH680
3-CHLOROANILINE (ITALIAN) see CEH675
3-CHLOROANILINE HYDROCHLORIDE see CEH690
4-CHLOROANILINE HYDROCHLORIDE see CJR200
m-CHLOROANILINE HYDROCHLORIDE see CEH690
p-CHLOROANILINE HYDROCHLORIDE see CJR200
p-CHLOROANILINIUM CHLORIDE see CJR200

(o-CHLOROANILINO)DICHLOROTRIAZINE see DEV800
1-((p-(2-(CHLORO-o-ANISAMIDO)ETHYL)PHENYL)SULFONYL)-3-CYCLOHEXYL UREA see CEH700
3-CHLOROANISIDINE see CEH750
1-CHLORO-9,10-ANTHRACENEDIONE see CEI000
2-CHLORO-9,10-ANTHRACENEDIONE see CEI100
1-CHLOROANTHRAQUINONE see CEI000
2-CHLOROANTHRAQUINONE see CEI100
1-CHLORO-9,10-ANTHRAQUINONE see CEI000
α-CHLOROANTHRAQUINONE see CEI000
CHLOROARSENOL see CEI250
1-CHLOROAZIRIDINE see CEI325
N-CHLOROAZIRIDINE see CEI340
CHLOROBEN see DEP600
CHLOROBENZAL see BAY300
4-CHLOROBENZALCHLORIDE see TJD650
p-CHLOROBENZALCHLORIDE see TJD650
2-CHLOROBENZALDEHYDE see CEI500
4-CHLOROBENZALDEHYDE see CEI600
o-CHLOROBENZALDEHYDE see CEI500
p-CHLOROBENZALDEHYDE see CEI600
α-CHLOROBENZALDEHYDE see BDM500
2-CHLOROBENZAL MALONONITRILE see CEQ600
o-CHLOROBENZAL MALONONITRILE see CEQ600
1-CHLORO-5-BENZAMIDO-ANTHRAQUINONE see BDK750
7-CHLOROBENZ(a)ANTHRACENE see CEJ000
10-CHLORO-1,2-BENZANTHRACENE see CEJ000
CHLOROBENZEN (POLISH) see CEJ125
3-CHLOROBENZENAMINE see CEH675
4-CHLOROBENZENAMINE see CEH680
2-CHLORO-BENZENAMINE (9CI) see CEH670
4-CHLOROBENZENAMINE HYDROCHLORIDE see CJR200
CHLOROBENZENE see CEJ125
4-CHLOROBENZENEACETONITRILE see CEP300
4-CHLOROBENZENEAMINE see CEH680
3-CHLORO-BENZENECARBOPEROXOIC ACID (9CI) see CJI750
o-CHLOROBENZENECARBOXALDEHYDE see CEI500
p-CHLOROBENZENECARBOXALDEHYDE see CEI600
2-CHLORO-1,4-BENZENEDIAMINE see CEG600
4-CHLORO-1,3-BENZENEDIAMINE see CJY120
5-CHLORO-1,3-BENZENEDIAMINE see CEJ200
2-CHLORO-1,4-BENZENEDIAMINE SULFATE see CEG625
2-CHLOROBENZENEDIAZONIUM SALTS see CEJ500
m-CHLOROBENZENEDIAZONIUM SALTS see CEJ250
o-CHLOROBENZENEDIAZONIUM SALTS see CEJ500
p-CHLOROBENZENESULFINIC ACID see CEJ600
o-CHLOROBENZENESULFOCHLORIDE see CEK370
p-CHLOROBENZENESULFONAMIDE see CEK000
4-CHLOROBENZENESULFONATE de 4-CHLOROPHENYLE (FRENCH) see CJT750
4-CHLOROBENZENESULFONIC ACID see CEK100
p-CHLOROBENZENESULFONIC ACID see CEK100

p-CHLOROBENZENESULFONIC ACID-p-CHLOROPHENYL ESTER see CJT750
p-CHLOROBENZENESULFONIC ACID, SODIUM SALT see CEK250
2-CHLOROBENZENESULFONYL CHLORIDE see CEK370
o-CHLOROBENZENESULFONYL CHLORIDE see CEK370
p-CHLOROBENZENESULFONYL CHLORIDE see CEK375
1-(4-CHLOROBENZENESULFONYL)FESTUCLAVINE see CEK400
1-(p-CHLOROBENZENESULFONYL)FESTUCLAVINE see CEK400
1-(p-CHLOROBENZENESULFONYL)-3-PROPYLUREA see CKK000
N-(p-CHLOROBENZENESULFONYL)-N'-PROPYLUREA see CKK000
4-CHLOROBENZENETHIOL see CEK425
6-CHLOROBENZENO(a)PYRENE see CEK500
1-(p-CHLOROBENZHYDRYL)-4-(p-tert-BUTYLBENZYL)DIETHYLENEDIAMINE DIHYDROCHLORIDE see BOM250
1-p-CHLOROBENZHYDRYL-4-p-(tert)-BUTYLBENZYLPIPERAZINE DIHYDROCHLORIDE see BOM250
1-(p-CHLOROBENZHYDRYL)-4-(2-(2-HYDROXYETHOXY)ETHYL)DIETHYLENE DIAMINE see CJR909
1-(p-CHLOROBENZHYDRYL)-4-(2-(2-HYDROXYETHOXY)ETHYL)DIETHYLENE DIAMINE HYDROCHLORIDE see VSF000
N-(4-CHLOROBENZHYDRYL)-N'-(HYDROXYETHOXYETHYL)PIPERAZINE see CJR909
1-(p-CHLOROBENZHYDRYL)-4-(2-(2-HYDROXYETHOXY)ETHYL)PIPERAZINE see CJR909
1-(p-CHLOROBENZHYDRYL)-4-(m-METHYLBENZYL)DIETHYLENEDIAMINE see HGC500
1-p-CHLOROBENZHYDRYL-4-m-METHYLBENZYLPIPERAZINE see HGC500
1-(4-CHLOROBENZHYDRYL)-4-METHYLPIPERAZINE see CFF500
1-(4-CHLOROBENZHYDRYL)-4-METHYLPIPERAZINE DIHYDROCHLORIDE see CDR000
1-(p-CHLOROBENZHYDRYL)-4-METHYLPIPERAZINE HYDROCHLORIDE see CDR250
N¹-(4'-CHLOROBENZHYDRYL)-N⁴-SPIROMORPHOLINO-PIPERAZINIUM CHLORIDE HYDROCHLORIDE see CEK875
1-(4-CHLOROBENZHYDRYL)PIPERAZINE see NNK500
4-(4-CHLOROBENZHYDRYL)PIPERAZINE see NNK500
N-(p-CHLOROBENZHYDRYL)PIPERAZINE see NNK500
1-(p-CHLOROBENZIL)-2-PIRROLIDIL-METIL-BENZIMIDAZOLO CLORIDATO (ITALIAN) see CMV400
2-CHLOROBENZO(e)(1)BENZOTHIOPYRANO(4,3-b)INDOLE see CEL000
2-CHLOROBENZOIC ACID see CEL250
3-CHLOROBENZOIC ACID see CEL290
4-CHLOROBENZOIC ACID see CEL300
m-CHLOROBENZOIC ACID see CEL290
o-CHLOROBENZOIC ACID see CEL250
p-CHLOROBENZOIC ACID see CEL300
4-CHLOROBENZOIC ACID-3-ETHYL-7-METHYL-3,7-DIAZABICYCLO(3.3.1)NON-9-YL ESTER HYDROCHLORIDE see YGA700
o-CHLOROBENZOIC ACID NICKEL(II) SALT see CEL500
p-CHLOROBENZOIC ACID 2-PHENYLHYDRAZIDE see CEL600

CHLOROBENZOL (DOT) see CEJ125
CHLOROBENZONE see CEL750
o-CHLOROBENZONITRILE see CEM000
p-CHLOROBENZONITRILE see CEM250
6-CHLORO-2H-1-BENZOPYRAN-3-CARBOXALDEHYDE see CEM300
6-CHLORO-2H-1,2,4-BENZOTHIADIAZINE-7-SULFONAMIDE-1,1-DIOXIDE see CLH750
2-CHLOROBENZOTHIAZOLE see CEM500
1-CHLOROBENZOTRIAZOL see CEM625
4-CHLOROBENZOTRICHLORIDE see TIR900
p-CHLOROBENZOTRICHLORIDE see TIR900
p-CHLOROBENZOTRIFLUORIDE see CEM825
5-CHLORO-2-BENZOXAZOLAMINE see AJF500
2-CHLOROBENZOXAZOLE see CEM850
5-CHLOROBENZOXAZOLIDONE see CDQ750
5-CHLORO-2-BENZOXAZOLINONE see CDQ750
6-CHLORO-2-BENZOXAZOLINONE see CDQ750
5-CHLOROBENZOXAZOL-2-ONE see CDQ750
p-CHLOROBENZOYL AZIDE see CEO000
5-(4-CHLOROBENZOYL)-1,4-DIMETHYL-1H-PYRROLE-2-ACETIC ACID SODIUM SALT DIHYDRATE see ZUA300
2-(3'-CHLOROBENZOYL)FLUORENE see CKA100
m-CHLOROBENZOYL HYDROPEROXIDE see CJI750
1-(4-CHLOROBENZOYL)-N-HYDROXY-5-METHOXY-2-METHYL-1H-INDOLE-3-ACETAMIDE see OLM300
1-(p-CHLOROBENZOYL)-5-METHOXY-2-METHYLINDOLE-3-ACETIC ACID see IDA000
1-(4-CHLOROBENZOYL)-5-METHOXY-2-METHYL-1H-INDOLE-3-ACETIC ACID CARBOXYMETHYL ESTER see AAE625
1-(p-CHLOROBENZOYL)-5-METHOXY-2-METHYLINDOLE-3-ACETOHYDROXAMIC ACID see OLM300
((1-(4-CHLOROBENZOYL)-5-METHOXY-2-METHYLINDOLE-3-YL)ACETOXY)ACETIC ACID see AAE625
1-(p-CHLOROBENZOYL)-5-METHOXY-2-METHYL-3-INDOLYLACETOHYDROXAMIC ACID see OLM300
1-(p-CHLOROBENZOYL)-2-METHYL-5-METHOXYINDOLE-3-ACETIC ACID see IDA000
1-(p-CHLOROBENZOYL)-2-METHYL-5-METHOXY-3-INDOLE-ACETIC ACID see IDA000
α-(1-(p-CHLOROBENZOYL)-2-METHYL-5-METHOXY-3-INDOLYL)ACETIC ACID see IDA000
2-(p-CHLOROBENZOYL)-1-(2-MORPHOLINOETHYL)PYRROLE MONOHYDROCHLORIDE see CEO100
1-(2-(2-CHLOROBENZOYL)-4-NITROPHENYL)-2-(DIETHYLAMINOMETHYL)IMIDAZOLE FUMARATE see NMV400
8-(3-CHLOROBENZOYLOXY)-9-HYDROXY-8,9-DIHYDRO-AFLATOXIN B1 see CEO110
p-CHLOROBENZOYL PEROXIDE (DOT) see BHM750
1-(o-CHLOROBENZOYL)-2-THIOBIURET see CEO120
3-(2-(4-CHLOROBENZYLAMINO)ETHYL)INDOLE MONOHYDROCHLORIDE see CEO125
CHLOROBENZYLATE see DER000

2-CHLOROBENZYL CHLORIDE see CEO200

o-CHLOROBENZYL CHLORIDE see CEO200

p-CHLOROBENZYL-p-CHLOROPHENYL SULFIDE see CEP000

4-CHLOROBENZYL-4-CHLOROPHENYL SULPHIDE see CEP000

p-CHLOROBENZYL-p-CHLOROPHENYL SULPHIDE see CEP000

1-(4-CHLOROBENZYL)-3-(6-CHLORO-o-TOLYL)-1-(2-PYRROLIDINYLETHYL) UREA HYDROCHLORIDE see CEP250

4-CHLOROBENZYL CYANIDE see CEP300

p-CHLOROBENZYL CYANIDE see CEP300

1-(p-CHLOROBENZYL)-1-CYCLOPENTYL-3-PHENYLUREA see UTA400

α-(p-CHLOROBENZYL)-4-DIETHYLAMINOETHOXY-4'-METHYLBENZHYDROL see TMP500

S-(4-CHLOROBENZYL)-N,N-DIETHYLTHIOCARBAMATE see SAZ000

2-(p-CHLOROBENZYL)(2-(DIMETHYLAMINO)ETHYL)AMINO)PYRIDINE see CKV625

4-(p-CHLOROBENZYL)-2-((2-DIMETHYLAMINO)ETHYL)-1(2H)-PHTHALAZINONE HYDROCHLORIDE see CEP675

(2-(p-CHLOROBENZYL)-3-DIMETHYLAMINOMETHYL-2-BUTANOL HYDROCHLORIDE see CMW500

2-(2-CHLOROBENZYL)-4,4-DIMETHYL-1,2-OXAZOLIDIN-3-ONE see CEP800

N-(p-CHLOROBENZYL)-N',N'-DIMETHYL-N-(2-PYRIDYL)ETHYLENEDIAMINE see CKV625

N-3'-CHLOROBENZYL-N'-ETHYLUREA see LII400

4-(p-CHLOROBENZYL)-2-(HEXAHYDRO-1-METHYL-1H-AZEPIN-4-YL)-1-(2H)-PHTHALAZINONE see ASC125

4-(p-CHLOROBENZYL)-2-(HEXAHYDRO-1-METHYL-1H-AZEPIN-4-YL)-1-(2H)-PHTHALAZINONE HCl see ASC130

p-CHLOROBENZYL-3-HYDROXYCROTONATE DIMETHYL PHOSPHATE see CEQ500

2-CHLOROBENZYLIDENEACETOPHENONE see CLF150

(4-CHLOROBENZYLIDENE)ACETOPHENONE see CLF100

4-CHLOROBENZYLIDENE CHLORIDE see TJD650

o-CHLOROBENZYLIDENE MALONITRILE see CEQ600

2-CHLOROBENZYLIDENE MALONONITRILE see CEQ600

o-CHLOROBENZYLIDENE MALONONITRILE see CEQ600

1-(4-CHLOROBENZYL)-1H-INDAZOLE-3-CARBOXYLIC ACID see CEQ625

1-p-CHLOROBENZYL-1H-INDAZOLE-3-CARBOXYLIC ACID see CEQ625

1-(p-CHLOROBENZYL)-1H-INDAZOLE-3-CARBOXYLIC ACID 1,3-DIHYDROXY-2-PROPYL ESTER see GGR100

4-CHLOROBENZYL ISOTHIOCYANATE see CEQ750

3-(p-CHLOROBENZYL)OCTAHYDRO-QUINOLIZINE TARTRATE (1:1) see CMX920

1-(p-CHLOROBENZYLOXYCARBONYL)-1-PROPEN-2-YL-DIMETHYLPHOSPHATE see CEQ500

1-(1-(2-((3-CHLOROBENZYL)OXY)PHENYL)VINYL)-1H-IMIDAZOLE HYDROCHLORIDE see CMW550

α-CHLOROBENZYL PHENYL KETONE see CFJ100

p-CHLOROBENZYLPSEUDOTHIURONIUM CHLORIDE see CEQ800

4-(4-CHLOROBENZYL)PYRIDINE see CEQ900

N-(p-CHLOROBENZYL)PYROGLUTAMATE d'AMMONIUM see CKF810

N-(p-CHLOROBENZYL)PYROGLUTAMATE DE METHYLE see CKF850

N-(p-CHLOROBENZYL)PYROGLUTAMATE de DIISOPROPYLAMINE see PMH915

N-(p-CHLOROBENZYL)PYROGLUTAMATE de N'-ISOPROPYLBENZYLAMINE see PMH912

2-(p-CHLOROBENZYL(2-(PYRROLIDINYL)ETHYL)AMINO)-o-ACETOTOLUIDIDE DIHYDROCHLORIDE see CER250

1-(p-CHLOROBENZYL)-2-(1-PYRROLIDINYLMETHYL)BENZIMIDAZOLE HYDROCHLORIDE see CMV400

1-p-CHLOROBENZYL-PYRROLIDYL-METHYLENE-BENZIMIDAZOLE HYDROCHLORIDE see CMV400

3-(p-CHLOROBENZYL)QUINOLIZIDINE TARTRATE see CMX920

5-(o-CHLOROBENZYL)-4,5,6,7-TETRAHYDROTHIENO(3,2-c)PYRIDINE HYDROCHLORIDE see TGA525

2-(4-CHLOROBENZYL)-2-THIOPSEUDOUREA HYDROCHLORIDE see CEQ800

7-CHLOROBICYCLO(3.2.0)HEPTA-2,6-DIEN-6-YL DIMETHYL PHOSPHATE see HBK700

CHLOROBIPHENYL see CGM550

CHLORO BIPHENYL see PJL750

3-CHLOROBIPHENYL see CGM770

4-CHLOROBIPHENYL see CGM780

m-CHLOROBIPHENYL see CGM770

CHLORO 1,1-BIPHENYL see PJL750

2-CHLORO-1,1'-BIPHENYL see CGM750

4-CHLORO-1,1'-BIPHENYL see CGM780

3-CHLOROBIPHENYLAMINE see CEH000

2-CHLORO-N,N-BIS(2-CHLOROETHYL)-1-PROPYLAMINE HYDROCHLORIDE see NOB700

2-CHLORO-4,6-BIS(DIETHYLAMINO)-s-TRIAZINE see CDQ325

2-CHLORO-4,6-BIS(ETHYLAMINO)-s-TRIAZINE see BJP000

1-CHLORO-3,5-BISETHYLAMINO-2,4,6-TRIAZINE see BJP000

2-CHLORO-4,6-BIS(ETHYLAMINO)-1,3,5-TRIAZINE see BJP000

4-CHLORO-2,6-BIS-ETHYLENEIMINOPYRIMIDINE see EQI600

2-CHLORO-1,1-BIS(FLUOROOXY)TRIFLUOROETHANE see CER825

CHLOROBIS(2-METHYLPROPYL)ALUMINUM see CGB500

CHLOROBLE M see MAS500

2-CHLOROBMN see CEQ600

1-CHLORO-3-BROMO-BUTENE-1 see CES250

1-CHLORO-2-BROMOETHANE see CES500

sym-CHLOROBROMOETHANE see CES500

CHLOROBROMOMETHANE see CES650

4-CHLORO-7-BROMOMETHYLBENZ(a)ANTHRACENE see BNQ000

1-(3-CHLORO-4-BROMOPHENYL)-3-METHYL-3-METHOXYUREA see CES750

1-CHLORO-3-BROMOPROPANE (DOT) see BNA825

ω-CHLOROBROMOPROPANE see BNA825

trans-CHLORO(2-(3-BROMOPROPIONAMIDO)CYCLOHEXYL) MERCURY see CET000

CHLOROBRUMURON see CES750

CHLOROBUFAM see CEX250

CHLOROBUTADIENE see NCI500

1-CHLOROBUTADIENE see CET250

1-CHLORO-1,3-BUTADIENE see CET250

2-CHLOROBUTA-1,3-DIENE see NCI500

2-CHLORO-1,3-BUTADIENE see NCI500

2-CHLORO-1,3-BUTADIENE HOMOPOLYMER (9CI) see PJQ050

CHLOROBUTADIENE POLYMER see PJQ050

2-CHLORO-1,3-BUTADIENE POLYMER see PJQ050

1-CHLOROBUTANE see CEU000

2-CHLOROBUTANE see CEU250

1-CHLOROBUTANE (DOT) see BQQ750

4-CHLOROBUTANENITRILE see CEU300

4-CHLORO-1-BUTANE-OL see CEU500

3-CHLOROBUTANOIC ACID see CEW000

CHLOROBUTANOL see ABD000

4-CHLOROBUTANOL see CEU500

4-CHLORO-1-BUTANOL see CEU500

3-CHLOROBUTANONE see CEU800

1-CHLORO-2-BUTANONE see CEU750

3-CHLORO-2-BUTANONE see CEU800

1-CHLORO-2-BUTENE see CEU825

2-CHLORO-2-BUTENE see CEV000

3-CHLORO-1-BUTENE see CEV250

3-CHLORO-1-BUTENE see CEV250

4-CHLORO-1-BUTENE see CEV300

1-CHLORO-1-BUTEN-3-ONE see CEV500

CHLOROBUTIN see CDO500

CHLOROBUTINE see CDO500

N-(3-CHLORO-1-sec-BUTYLACETONYL)-p-TOLUENESULFONAMIDE see THH470

o-CHLORO-α-((tert-BUTYLAMINO)METHYL)BENZYLALCOHOL HYDROCHLORIDE see BQE250

4-CHLORO-2-(tert-BUTYLAMINO)-6-(4-METHYLPIPERAZINO)-5-METHYLTHIOPYRIMIDINE see CEV750

5-CHLORO-3-tert-BUTYL-6-METHYLURACIL see BQT750

4-CHLORO-2-BUTYNOL see CEV800

CHLORO-2-BUTYNYL-m-CHLOROCARBAMATE see CEW500

4-CHLOROBUT-2-YNYL-m-CHLOROCARBANILATE see CEW500

4-CHLORO-2-BUTYNYL-m-CHLOROCARBANILATE see CEW500

4-CHLOROBUT-2-YNYL-3-CHLOROPHENYLCARBAMATE see CEW500

4-CHLORO-2-BUTYNYL-N-(3-CHLOROPHENYL)CARBAMATE see CEW500

S-(4-CHLORO-2-BUTYNYL) o,o-DIETHYL PHOSPHOROTHIOATE see CEV830

S-(4-CHLORO-2-BUTYNYL) DIPHENYLPHOSPHINOTHIOATE see CEV840

S-(4-CHLORO-2-BUTYNYL) o-ETHYL PHENYLPHOSPHONOTHIOATE see CEV850

3-CHLOROBUTYRIC ACID see CEW000

β-CHLOROBUTYRIC ACID see CEW000

4-CHLOROBUTYRIC ACID TRIBUTYLSTANNYL ESTER see TID000

4-CHLOROBUTYRONITRILE see CEU300

γ-CHLOROBUTYRONITRILE see CEU300

CHLOROCAIN see CBR250

CHLOROCAINE see AIT250

2-CHLOROCAMPHANE see BMD300

CHLOROCAMPHENE see CDV100

CHLOROCAPS see CDP250

m-CHLORO CARBANILIC ACID-4-CHLORO-2-BUTYNYL ESTER see CEW500

3-CHLOROCARBANILIC ACID, ISOPROPYL ESTER see CKC000

m-CHLOROCARBANILIC ACID, ISOPROPYL ESTER see CKC000

m-CHLOROCARBANILIC ACID-1-METHYL-2-PROPYNYL ESTER see CEX250

CHLOROCARBONATE D'ETHYLE (FRENCH) see EHK500

CHLOROCARBONATE de METHYLE (FRENCH) see MIG000

CHLOROCARBONIC ACID METHYL ESTER see MIG000

N-(CHLOROCARBONYLOXY)TRIMETHYLUREA see CEX255

3-CHLOROCARPIPRAMINE DIHYDROCHLORIDE see DKV309

5-CHLOROCARVACROL see CEX275

2-CHLOROCHALCONE see CLF150

4-CHLOROCHALCONE see CLF100

p-CHLOROCHALCONE see CLF100

CHLOROCHIN see CLD000

3-CHLOROCHLORDENE see HAR000

6'-CHLORO-2-(p-CHLOROBENZYL(2-(DIETHYLAMINO)ETHYL)AMINO)-o-ACETOLUIDIDE DIHYDROCHLORIDE see CEX500

6'-CHLORO-2-(p-CHLOROBENZYL(2-(PYRROLIDINYL)ETHYL)AMINO)-o-ACETOTOLUIDIDE DIHYDROCHLORIDE see CEX750

3-CHLORO-N-(5-CHLORO-2,6-DINITRO-4-TRIFLUOROMETHYLPHENYL)-5-TRIFLUOROMETHYL-2-PYRIDINAMINE see CEX800

1-CHLORO-2-(β-CHLOROETHOXY)ETHANE see DFJ050

6-CHLORO-9-((2-((2-CHLOROETHYL)AMINO)ETHYL)AMINO)-2-METHOXYACRIDINE 2-HYDROCHLORIDE SESQUIHYDRATE see CEY250

2-CHLORO-N-(2-CHLOROETHYL)-N,N-DIMETHYLETHANAMINIUM CHLORIDE see DQS600

13-CHLORO-N-(2-CHLOROETHYL)-N,11-DINITROSO-10-OXO-5,6-DITHIA-2,9,11-TRIAZATRIDECANAMIDE see BIF625

2-CHLORO-N-(2-CHLOROETHYL)ETHANAMINE HYDROCHLORIDE see BHO250

7-CHLORO-10-(3-(N-(2-CHLOROETHYL)-N-ETHYL)AMINOPROPYLAMINO)-2-METHOXY-BENZO(B)(1,5)NAPHTHYRIDINE DIHYDROCHLORIDE see IAE000

6-CHLORO-9-(3-(2-CHLOROETHYL)MERCAPTOPROPYLAMINO)-2-METHOXYACRIDINE HYDROCHLORIDE see CFA250

2-CHLORO-N-(2-CHLOROETHYL)-N-METHYLETHANAMINE HYDROCHLORIDE see BIE500

2-CHLORO-N-(2-CHLOROETHYL)-N-METHYLETHANAMINE-N-OXIDE see CFA500

2-CHLORO-N-(2-CHLOROETHYL)-N-METHYLETHANAMINE-N-OXIDE HYDROCHLORIDE see CFA750

1-CHLORO-2-(β-CHLOROETHYLTHIO)ETHANE see BIH250

4-CHLORO-2-(4-CHLORO-2-FLUOROPHENYL)-5-((4-CHLOROPHENYL)METHOXY)-3(2H)-PYRIDAZINONE see CFA800

CHLORO(CHLOROMETHOXY)METHANE see BIK000

9-CHLORO-10-CHLOROMETHYL ANTHRACENE see CFB500

1-CHLORO-2-(CHLOROMETHYL)BENZENE see CEO200

1-CHLORO-4-CHLOROMETHYLBENZENE see CFB600

3-CHLORO-4-(CHLOROMETHYL)-5-HYDROXY-2(5H)-FURANONE see CFB700

cis-2-CHLORO-3-(CHLOROMETHYL)OXIRANE see DAC975

trans-2-CHLORO-3-(CHLOROMETHYL)OXIRANE see DGH500

(Z)-2-CHLORO-3-(CHLOROMETHYL)-4-OXOBUTENOIC ACID see CFB720

3-CHLORO-4-CHLOROMETHYL-1H-PYRROLE-2,5-DIONE see CFB730

1-CHLORO-4-(CHLOROMETHYLTHIO)BENZENE see CFB750

2-CHLORO-5-CHLOROMETHYLTHIOPHENE see CFB825

5-CHLORO-N-(2-CHLORO-4-NITROPHENYL)-2-HYDROXYBENZAMIDE see DFV400

5-CHLORO-N-(2-CHLORO-4-NITROPHENYL)-2-HYDROXYBENZAMIDE with 2-AMINOETHANOL (1:1) see DFV600

3-CHLORO-4-(3-CHLORO-2-NITROPHENYL)PYRROLE see CFC000

5-CHLORO-2'-CHLORO-4'-NITROSALICYLANILIDE see DFV400

4-CHLORO-2-(4-CHLOROPHENYL)-5-((4-CHLOROPHENYL)METHOXY)-3(2H)-PYRIDAZINONE see CFC050

N-(5-CHLORO-4-((4-CHLOROPHENYL)CYANOMETHYL)-2-METHYLPHENYL)-2-HYDROXY-3, 5-DIIODOBENZAMIDE see CFC100

4-CHLORO-α-(4-CHLOROPHENYL)-α-CYCLOPROPYLBENZENEMETHANOL (9CI) see PMF550

7-CHLORO-5-(2-CHLOROPHENYL)-1,3-DIHYDRO-3-HYDROXY-2H-1,4-BENZODIAZEPIN-2-ONE see CFC250

7-CHLORO-5-(o-CHLOROPHENYL)-1,3-DIHYDRO-3-HYDROXY-2H-1,4-BENZODIAZEPIN-2-ONE see CFC250

7-CHLORO-5-(2-CHLOROPHENYL)-1,3-DIHYDRO-3-HYDROXY-1-METHYL-2H-1,4-BENZODIAZEPIN-2-ONE see MLD100

5-CHLORO-3-(4-CHLOROPHENYL)-4'-FLUORO-2'-METHYLSALICYLANILIDE see CFC500

7-CHLORO-5-(2-CHLOROPHENYL)-3-HYDROXY-1H-1,4-BENZODIAZEPIN-2(3H)-ONE see CFC250

7-CHLORO-5-(2-CHLOROPHENYL)-3-HYDROXY-1-METHYL-2,3-DIHYDRO-1H-1,4-BENZODIAZEPIN-2-ONE see MLD100

4-CHLORO-2-(4-CHLOROPHENYL)-5-((6-IODO-3-PYRIDINYL)METHOXY)-3(2H)-PYRIDAZINONE see CFC600

2-CHLORO-3-(4-CHLOROPHENYL)METHYLPROPIONATE see CFC750

1-CHLORO-4-(((4-CHLOROPHENYL)METHYL)THIO)BENZENE see CEP000

8-CHLORO-6-(o-CHLOROPHENYL)-1-METHYL-4H-s-TRIAZOLO(4,3-a)(1,4)BENZODIAZEPINE see THS800

8-CHLORO-6-(2-CHLOROPHENYL)-1-METHYL-4H-(1,2,4)TRIAZOLO(4,3-a)(1,4)BENZODIAZEPINE see THS800

2-CHLORO-3-(4-CHLOROPHENYL)PROPIONIC ACID METHYL ESTER see CFC750

4-CHLORO-α-(4-CHLOROPHENYL)-α-(TRICHLOROMETHYL)BENZENEMETHANOL see BIO750

3-CHLORO-N-(3-CHLORO-5-TRIFLUOROMETHYL-2-PYRIDYL)-α,α,α-TRIFLUORO-2,6-DINITRO-p- see CEX800

CHLORO(2-CHLOROVINYL)MERCURY see CFD250

CHLOROCHOLINE CHLORIDE see CMF400

CHLOROCID see CDP250

CHLOROCIDE see CEP000

CHLOROCIDIN C TETRAN see CDP250

CHLOROCOL see CDP250

CHLOROCRESOL see CFD990

p-CHLOROCRESOL see CFD990

4-CHLORO-m-CRESOL see CFD990

4-CHLORO-o-CRESOL see CFE000

6-CHLORO-m-CRESOL see CFD990

6-CHLORO-m-CRESOL see CFE500

p-CHLORO-m-CRESOL see CFD990

4-CHLORO o CRESOXYACETIC ACID see CIR250

CHLOROCTAN SODNY (CZECH) see SFU500

CHLOROCYAN see COO750

CHLOROCYANIDE see COO750

CHLOROCYANOACETYLENE see CFE750

2-CHLORO-1-CYANOETHANOL see CHU000

2-CHLORO-N-(2-CYANOETHYL)ACETAMIDE see COM830

CHLOROCYANOGEN see COO750

2-CHLORO-α-CYANO-6-METHYLERGOLINE-8-PROPIONAMIDE see CFF100

2-CHLORO-4-(1-CYANO-1-METHYLETHYLAMINO)-6-ETHYLAMINO-1,3,5-TRIAZINE see BLW750

3-CHLORO-6-CYANO-2-NORBORNANONE-o-(METHYLCARBAMOYL)OXIME see CFF250

endo-3-CHLORO-exo-6-CYANO-2-NORBORNANONE-o-(METHYLCARBAMOYL)OXIME see CFF250

2-exo-CHLORO-6-endo-CYANO-2-NORBORNANONE-o-(METHYLCARBAMOYL)OXIME2-CARBONITRILE see CFF250

3-CHLORO-6-CYANONORBORNANONE-2-OXIME-o,N-METHYLCARBAMATE see CFF250

CHLOROCYCLAMIDE-R see CDR550

CHLOROCYCLINE see CFF500

CHLOROCYCLIZINE see CFF500

CHLOROCYCLIZINE HYDROCHLORIDE see CDR500

2-CHLOROCYCLOHEPTANONE see CFF600

α-CHLOROCYCLOHEPTANONE see CFF600

CHLOROCYCLOHEXANE see CPI400

2-CHLOROCYCLOHEXANONE see CFG250

α-CHLOROCYCLOHEXANONE see CFG250

3-CHLORO-1-CYCLOHEXENE see CFG300

4-CHLORO-4-CYCLOHEXENE-1,2-DICARBOXYLIC ANHYDRIDE see CFG500

7-CHLORO-5-(CYCLOHEXEN-1-YL)-1,3-DIHYDRO-1-METHYL-2H-1,4-BENZODIAZEPIN-2-ONE see CFG750

7-CHLORO-5-(1-CYCLOHEXENYL)-1-METHYL-2-OXO-2,3-DIHYDRO-1H-(1,4)-BENZO(f)DIAZEPINE see CFG750

6'-CHLORO-2-(CYCLOHEXYLAMINO)-o-ACETOTOLUIDIDE HYDROCHLORIDE see CFH000

5-CHLORO-N-(2-(4-((((CYCLOHEXYLAMINO) CARBONYL)AMINO)SULFONYL)PHENYL) ETHYL)-2-METHOXYBENZAMIDE see CEH700

3-(3-CHLORO-4-CYCLOHEXYLBENZOYL)PROPIONIC ACID see CPJ000

3-(3-CHLORO-4-CYCLOHEXYLBENZOYL)PROPIONIC ACID CALCIUM SALT see CPJ250

(Z)-3-(2-CHLOROCYCLOHEXYL)-1-(2-CHLOROETHYL)-1-NITROSOUREA see CFH500

cis-3-(2-CHLOROCYCLOHEXYL)-1-(2-CHLOROETHYL)-1-NITROSOUREA see CFH500

cis-N'-(2-CHLOROCYCLOHEXYL)-N-(2-CHLOROETHYL)-N-NITROSOUREA see CFH500

trans-3-(2-CHLOROCYCLOHEXYL)-1-(2-CHLOROETHYL)-1-NITROSOUREA see CFH750 trans-N'-(2-CHLOROCYCLOHEXYL)-N-(2-CHLOROETHYL)-N-NITROSOUREA see CFH750

6-CHLORO-5-CYCLOHEXYL-2,3-DIHYDRO-1H-INDENE-1-CARBOXYLIC ACID (9CI) see CMV500

(±)-6-CHLORO-5-CYCLOHEXYL-2,3-DIHYDRO-1H-INDENE-1-CARBOXYLIC ACID (9CI) see CFH825

6-CHLORO-5-CYCLOHEXYL-1-INDANCARBOXYLIC ACID see CFH825

6-CHLORO-5-CYCLOHEXYL-1-INDANCARBOXYLIC ACID see CMV500

(±)-6-CHLORO-5-CYCLOHEXYLINDAN-1-CARBOXYLIC ACID see CFH825

(±)-6-CHLORO-5-CYCLOHEXYL-1-INDANCARBOXYLIC ACID see CFH825

6'-CHLORO-2-(N-CYCLOHEXYL-N-METHYLAMINO)-o-ACETOTOLUIDIDE HYDROCHLORIDE see CFI000

3-CHLORO-4-CYCLOHEXYL-α-OXOBENZENEBUTANOIC ACID see CPJ000

3-CHLORO-4-CYCLOHEXYL-α-OXO-BENZENEBUTANOIC ACID see CPJ250

3-CHLORO-4-CYCLOHEXYL-α-OXOBENZENEBUTANOIC ACID CALCIUM SALT see CPJ250

4-(3-CHLORO-4-CYCLOHEXYLPHENYL)-4-OXO-BUTYRIC ACID see CPJ000

4-(3-CHLORO-4-CYCLOHEXYLPHENYL)-4-OXOBUTYRIC ACID CALCIUM SALT see CPJ250

CHLOROCYCLOPENTANE see CFI250

2-CHLOROCYCLOPENTANONE see CFI500

α-CHLOROCYCLOPENTANONE see CFI500

3-CHLOROCYCLOPENTENE see CFI625

4-CHLORO-2-CYCLOPENTYL PHENOL see CFI750

2-CHLORO-4-CYCLOPROPYLAMINO-6-ISOPROPYLAMINO-1,3,5-TRIAZINE see CQI750

2-CHLORO-4-CYCLOPROPYLAMINO-6-ISOPROPYLAMINO-sec-TRIAZINE see CQI750

2-((4-CHLORO-6-(CYCLOPROPYLAMINO)-1,3,5-TRIAZIN-2-YL)AMINO)-2-METHYLPROPANENITRILE see PMF600

2-(4-CHLORO-6-(CYCLOPROPYLAMINO)-s-TRIAZIN-2-YL)AMINO-2-METHYLPROPIONITRILE see PMF600

3-CHLORO-4-CYCLO-PROPYLMETHOXYPHENYLACETIC ACID LYSINE SALT (d,l) see CFJ000

2-(3-CHLORO-4-CYCLOPROPYLMETHOXYPHENYL)ACETIC ACID LYSINE SALT (d,l) see CFJ000

7-CHLORO-1-CYCLOPROPYLMETHYL-1,3-DIHYDRO-5-(2-FLUOROPHENYL)-2H-1,4-BENZODIAZEPIN-2-ONE see FMR100

7-CHLORO-1-(CYCLOPROPYLMETHYL)-1,3-DIHYDRO-5-PHENYL-2H-1,4-BENZODIAZEPIN-2-ONE see DAP700

7-CHLORO-1-CYCLOPROPYLMETHYL-5-PHENYL-1H-1,4-BENZODIAZEPIN-2(3H)-ONE see DAP700

6-CHLORO-1-(CYCLOPROPYLMETHYL)-4-PHENYL-2(1H)-QUINAZOLINONE see CQF125

CHLORODANE see CDR750

CHLORODECANE see DAJ200

6-CHLORO-Δ⁶-DEHYDRO-17-ACETOXYPROGESTERONE see CBF250

6-CHLORO-6-DEHYDRO-17-α-ACETOXYPROGESTERONE see CBF250

6-CHLORO-6-DEHYDRO-17-α-ACETOXYPROGESTERONE mixed with MESTRENOL see CNV750

6-CHLORO-6-DEHYDRO-17-α-HYDROXYPROGESTERONE ACETATE see CBF250

7-CHLORO-6-DEMETHYLTETRACYCLINE see MIJ500

7-CHLORO-6-DEMETHYLTETRACYCLINE HYDROCHLORIDE see DAI485

CHLORODEN see DEP600

α-CHLORODEOXYBENZOIN see CFJ100

2-CHLORO-9-(2-DEOXY-2-FLUORO-β-d-ARABINOFURANOSYL)ADENINE see CFJ200

6-CHLORO-6-DEOXYGLUCOSE see CFJ375

6-CHLORO-6-DEOXY-d-GLUCOSE see CFJ375

CHLORODEOXYGLYCEROL see CDT750

CHLORODEOXYGLYCEROL DIACETATE see CIL900

7(S)-CHLORO-7-DEOXYLINCOMYCIN see CMV675

7(S)-CHLORO-7-DEOXYLINCOMYCIN-2-PHOSPHATE see CMV690

5-CHLORODEOXYURIDINE see CFJ750

5-CHLORO-2'-DEOXYURIDINE see CFJ750

CHLORODIABINA see CKK000

2-CHLORO-N,N-DIALLYLACETAMIDE see CFK000

α-CHLORO-N,N-DIALLYLACETAMIDE see CFK000

1-CHLORO-2,4-DIAMINOBENZENE see CJY120

4-CHLORO-1,2-DIAMINOBENZENE see CFK125

α-(2-CHLORO-4-(4,6-DIAMINO-2,2-DIMETHYL-S-TRIAZINE-1(2H)-YL)PHENOXY)-N,N-DIMETHYL-m-TOLUAMIDE ETHANESULFONATE see THR500

CHLORODIAZEPAM see CFK150

4'-CHLORODIAZEPAM see CFK150

CHLORODIAZEPOXIDE see LFK000

1-CHLORODIBENZO-p-DIOXIN see CDV125

3-CHLORODIBENZOFURAN see MRF575

2-((8-CHLORODIBENZO(b,f)THIEPIN)-10-YL)OXY-N,N-DIMETHYLETHANAMINE see ZUJ000

CHLORODIBORANE see CFK325

CHLORODIBROMOMETHANE see CFK500

1-CHLORO-2,3-DIBROMOPROPANE see DDL800

3-CHLORO-1,2-DIBROMOPROPANE see DDL800

CHLORODIBROMOTRIFLUOROETHANE see MRF600

CHLORO(DIBUTOXYPHOSPHINYL)MERCURY see CFK750

6'-CHLORO-2-(DIBUTYLAMINO)-o-ACETOTOLUIDIDE, HYDROCHLORIDE see CFL000

1-CHLORO-2-(2,2-DICHLORO-1-(4-CHLOROPHENYL)ETHYL)BENZENE see CDN000

p-CHLORO-DI-(2-CHLOROETHYL)BENZYLAMINE HYDROCHLORIDE see BIB750

1-CHLORO-2,2-DICHLOROETHYLENE see TIO750

6-CHLORO-3-(DICHLOROMETHYL)-3,4-DIHYDRO-2H-1,2,4-BENZOTHIADIAZINE-7-SULFONAMIDE-1,1-DIOXIDE see HII500

6-CHLORO-3-(DICHLOROMETHYL)3,4-DIHYDRO-7-SULFAMYL-1,2,4-BENZOTHIADIAZINE-1,1-DIOXIDE see HII500

CHLORO(DICHLOROMETHYL)-5-HYDROXY-2(5H)-FURANONE see CFL100

3-CHLORO-4-DICHLOROMETHYL-5-HYDROXY-2(5H)-FURANONE see CFL100

(Z)-2-CHLORO-3-(DICHLOROMETHYL)-4-OXOBUTENOIC ACID see FOJ033

3-CHLORO-4-DICHLOROMETHYL-1H-PYRROLE-2,5-DIONE see CFL120

6-CHLORO-5-(2,3-DICHLOROPHENOXY)-2-METHYLTHIO-BENZIMIDAZOLE see CFL200

5-CHLORO-6-(2,3-DICHLOROPHENOXY)-2-(METHYLTHIO)-1H-BENZIMIDAZOLE see CFL200

5-CHLORO-2-(2,4-DICHLOROPHENOXYPHENOL) see TIQ000

O-(2-CHLORO-1-(2,5-DICHLOROPHENYL))-O,O-DIETHYL ESTER PHOSPHOROTHIOIC ACID see DIX600

4-CHLORO-2-(3,4-DICHLOROPHENYL)-5-((6-IODO-3-PYRIDINYL)METHOXY)-3(2H)-PYRIDAZINONE see CFL225

2-CHLORO-3-(3,4-DICHLOROPHENYL)-1-ISOPROPYL-4-METHYL-IMIDAZOLIUM CHLORIDE see CFL300

2-CHLORO-1-(2,4-DICHLOROPHENYL)VINYL DIETHYL PHOSPHATE see CDS750

β-2-CHLORO-1-(2',4'-DICHLOROPHENYL)VINYL DIETHYLPHOSPHATE see CDS750

O-(2-CHLORO-1-(2,5-DICHLOROPHENYL)VINYL)-O,O-DIETHYL PHOSPHOROTHIOATE see DIX600

CHLORODIETHOXYSILANE see DHF800

CHLORODIETHYLALUMINUM see DHI885

2'-CHLORO-2-(DIETHYLAMINO)ACETANILIDE HYDROCHLORIDE see CFL500

3'-CHLORO-2-(DIETHYLAMINO)ACETANILIDE HYDROCHLORIDE see CFL750

4'-CHLORO-2-(DIETHYLAMINO)ACETANILIDE HYDROCHLORIDE see CFM000

3'-CHLORO-2-(DIETHYLAMINO)-o-ACETOTOLUIDIDE HYDROCHLORIDE see CFM250

4'-CHLORO-2-(DIETHYLAMINO)-o-ACETOTOLUIDIDE HYDROCHLORIDE see CFM500

5'-CHLORO-2-(DIETHYLAMINO)-o-ACETOTOLUIDIDE HYDROCHLORIDE see CFM750

6'-CHLORO-2-(DIETHYLAMINO)-m-ACETOTOLUIDIDE HYDROCHLORIDE see CFN000

6'-CHLORO-3-(DIETHYLAMINO)-o-BUTYROTOLUIDIDE HYDROCHLORIDE see CFN500

(6'-CHLORO-2-(2-DIETHYLAMINO)ETHOXY)ACETANILIDE HYDROCHLORIDE see CFN750

6'-CHLORO-2-(2-(DIETHYLAMINO)ETHOXY)-o-ACETOTOLUIDIDE HYDROCHLORIDE see CFO000

5-CHLORO-2-(2-(DIETHYLAMINO)ETHOXY)BENZANILIDE see CFO250

5-CHLORO-2-(2-(DIETHYLAMINO)ETHOXY)ETHYL)-2-METHYL-1,3-BENZODIOXOLE see CFO750

5-CHLORO-2-(4-(2-(DIETHYLAMINO)ETHOXY)PHENYL)BENZOTHIAZOLE TARTRATE see EBD000

2-CHLORO-1-(p-(β-DIETHYLAMINOETHOXY)PHENYL)-1,2-DIPHENYLETHYLENE see CMX700

6'-CHLORO-2-(2-(DIETHYLAMINO)ETHYL)AMINO-o-ACETOTOLUIDIDE HYDROCHLORIDE see CFP000

5-CHLORO-2-(2-(2-(DIETHYLAMINO)ETHYLAMINO)ETHYL)-2-METHYL-1,3-BENZODIOXOLE DIHYDROCHLORIDE see CFP250

2-CHLORO-4-(DIETHYLAMINO)-6-(ETHYLAMINO)-s-TRIAZINE see TJL500

8-CHLORO-2-(2-(DIETHYLAMINO)ETHYL)-2H-(1)-BENZOTHIOPYRANO(4,3,2-cd)INDAZOLE-5-METHANOL MONOMETHANE SULFONATE see CFP750

8-CHLORO-2-(2-(DIETHYLAMINO)ETHYL)-2H-(1)BENZOTHIOPYRANO(4,3,2-cd)INDAZOLE-5-METHANOL-N-OXIDE see CFQ000

2'-CHLORO-2-((2-(DIETHYLAMINO)ETHYL)ETHYLAMINO)ACETANILIDE DIHYDROCHLORIDE see CFQ250

6'-CHLORO-2-((2-(DIETHYLAMINO)ETHYL)ETHYLAMINO)-o-ACETOTOLUIDIDE HYDROCHLORIDE see CFQ500

7-CHLORO-1-(2-(DIETHYLAMINO)ETHYL)-5-(2-FLUOROPHENYL)-1H-1,4-BENZODIAZEPIN-2(3H)-ONE see FMQ000

6'-CHLORO-2-((2-(DIETHYLAMINO)ETHYL)ISOPROPYLAMINO)-o-ACETOTOLUIDIDE HYDROCHLORIDE see CFQ750

2'-CHLORO-2-(2-(DIETHYLAMINO)ETHYL)METHYLAMINOACETANILIDE DIHYDROCHLORIDE see CFR000

4'-CHLORO-2-(2-(DIETHYLAMINO)ETHYL)METHYLAMINOACETANILIDE DIHYDROCHLORIDE see CFR250

3'-CHLORO-2-(2-(DIETHYLAMINO)ETHYL)METHYLAMINO-o-ACETOTOLUIDIDE DIHYDROCHLORIDE see CFR500

4'-CHLORO-2-(2-(DIETHYLAMINO)ETHYL)METHYLAMINO-o-ACETOTOLUIDIDE DIHYDROCHLORIDE see CFR750

5'-CHLORO-2-(2-(DIETHYLAMINO)ETHYL)METHYLAMINO-o-ACETOTOLUIDIDE DIHYDROCHLORIDE see CFS000

6'-CHLORO-2-(2-(DIETHYLAMINO)ETHYL)METHYLAMINO-m-ACETOTOLUIDIDE DIHYDROCHLORIDE see CFS250

6'-CHLORO-2-((2-(DIETHYLAMINO)ETHYL)METHYLAMINO)-o-ACETOTOLUIDIDE HYDROCHLORIDE see CFS500

6'-CHLORO-2-((2-(DIETHYLAMINO)ETHYL)OCTYLAMINO)-o-ACETOTOLUIDIDE HYDROCHLORIDE see CFS750

6'-CHLORO-2-((2-(DIETHYLAMINO)ETHYL)(2-PHENOXYETHYL)AMINO)-o-ACETOTOLUIDIDE HYDROCHLORIDE see CFT000

(4-CHLORO-N-(2-DIETHYLAMINO)ETHYL)-2-(2-PROPENYLOXY)BENZAMIDE see AFS625

4-CHLORO-N-(2-(DIETHYLAMINO)ETHYL)-2-(2-PROPENYLOXY)-BENZAMIDE (9CI) see AFS625

6'-CHLORO-2-((2-(DIETHYLAMINO)ETHYL)PROPYLAMINO)-o-ACETOTOLUIDIDE HYDROCHLORIDE see CFT250

2'-CHLORO-2-(2-(DIETHYLAMINO)ETHYLTHIO)ACETANILIDE HYDROCHLORIDE see CFT500

6'-CHLORO-2-(2-(DIETHYLAMINO)ETHYLTHIO)-o-ACETOTOLUIDIDE HYDROCHLORIDE see CFT750

7-CHLORO-10-(3-(DIETHYLAMINO)-2-HYDROXYPROPYL)ISOALLOXAZINE SULFATE see CFU000

2-CHLORO-4-(DIETHYLAMINO)-6-(ISOPROPYLAMINO)-s-TRIAZINE see IKM050

4'-CHLORO-2-(DIETHYLAMINO)-N-METHYLACETANILIDE HYDROCHLORIDE see CFU250

2'-CHLORO-2-(DIETHYLAMINO)-2'-METHYLACETANILIDE HYDROCHLORIDE see CFM250

2'-CHLORO-2-(DIETHYLAMINO)-5'-METHYLACETANILIDE HYDROCHLORIDE see CFN000

3'-CHLORO-2-(DIETHYLAMINO)-6'-METHYLACETANILIDE HYDROCHLORIDE see CFM750

4'-CHLORO-2-(DIETHYLAMINO)-2'-METHYLACETANILIDE HYDROCHLORIDE see CFM500

6'-CHLORO-2-(DIETHYLAMINO)-N-METHYL-o-ACETOTOLUIDIDE HYDROCHLORIDE see CFU500

6-CHLORO-9-((4-(DIETHYL AMINO)-1-METHYL BUTYL)AMINO)-2-METHOXYACRIDINE see ARQ250

6-CHLORO-9-((4-(DIETHYLAMINO)-1-METHYLBUTYL)AMINO)-2-METHOXYACRIDINE DIHYDROCHLORIDE see CFU750

3-CHLORO-9-(4'-DIETHYLAMINO-1'-METHYLBUTYLAMINO)-7-METHOXYACRIDINE DIHYDROCHLORIDE see CFU750

2-CHLORO-5-(ω-DIETHYLAMINO-α-METHYLBUTYLAMINO)-7-METHOXYACRIDINE DIHYDROCHLORIDE see CFU750

7-CHLORO-4-(4-DIETHYLAMINO-1-METHYLBUTYLAMINO)QUINOLINE see CLD000

7-CHLORO-4-((4'-DIETHYLAMINO-1-METHYLBUTYL)AMINO)QUINOLINE DIPHOSPHATE see CLD250

7-CHLORO-4-((4-(DIETHYLAMINO)-1-METHYLBUTYL)AMINO)-QUINOLINE PHOSPHATE (1:1) see AQT250

4-CHLORO-2-DIETHYLAMINO-6-(4-METHYLPIPERAZINO)-5-METHYLTHIOPYRIMIDINE see CFV250

2-CHLORO-10-(3-DIETHYLAMINOPROPYL)PHENOTHIAZINE see CLY750

2-CHLORO-10-(3'-DIETHYLAMINOPROPYL)PHENOTHIAZINE HYDROCHLORIDE see CLZ000

2'-CHLORO-2-(DIETHYLAMINO)-5'-TRIFLUOROMETHYLACETANILIDE HYDROCHLORIDE see CFW000

4'-CHLORO-2-(DIETHYLAMINO)-3'-TRIFLUOROMETHYLACETANILIDE HYDROCHLORIDE see CFW250

CHLORODIETHYLBORANE see CFW625

2-CHLORO-2',6'-DIETHYL-N-(BUTOXYMETHYL)ACETANILIDE see CFW750

2-CHLORO-2-DIETHYLCARBAMOYL-1-METHYLVINYL DIMETHYLPHOSPHATE see FAB400

1-CHLORO-DIETHYLCARBAMOYL-1-PROPEN-2-YL DIMETHYL PHOSPHATE see FAB400

CHLORODIETHYLENETRIAMINE PLATINUM(II) CHLORIDE see CFW800

(Z)-2-CHLORO-N,N-DIETHYL-3-HYDROXYCROTONAMIDE DIMETHYL PHOSPHORATE see PGX275

2-CHLORO-N-(2,6-DIETHYL-4-HYDROXYPHENYL)ACETAMIDE see CEC260

2-CHLORO-2',6'-DIETHYL-N-(METHOXYMETHYL)ACETANILIDE see CFX000

2-CHLORO-N-(2,6-DIETHYL-4-OXO-2,5-CYCLOHEXADIEN-1-YLIDENE)ACETAMIDE see CEC270

2-CHLORO-N-(2,6-DIETHYLPHENYL)-N-(METHOXYMETHYL)ACETAMIDE see CFX000

2-CHLORO-N-(2,6-DIETHYLPHENYL)-N-(2-PROPOXYETHYL)ACETAMIDE see PMB850

2-CHLORO-2',6'-DIETHYL-N-(2-PROPOXYETHYL)ACETANILIDE see PMB850

CHLORODIFLUOROACETYL HYPOCHLORITE see CFX125

CHLORODIFLUOROBROMOMETHANE (DOT) see BNA250

1-CHLORO-1,1-DIFLUOROETHANE see CFX250

CHLORODIFLUOROETHANES (DOT) see CFX250

m-CHLORO-N-(2,2-DIFLUOROETHYL)ANILINE see CFX300

CHLORODIFLUOROMETHANE see CFX500

CHLORODIFLUOROMETHANE see CFX500

CHLORODIFLUOROMETHANE (ACGIH,DOT,OSHA) see CFX500

CHLORODIFLUOROMETHANE and CHLOROPENTAFLUOROETHANE MIXTURE (DOT) see FOO560

1-CHLORO-3,3-DIFLUORO-2-METHOXYCYCLOPROPENE see CFX625

2-CHLORO-1-(DIFLUOROMETHOXY)-1,1,2-TRIFLUOROETHANE see EAT900

CHLORODIFLUOROMONOBROMOMETHANE see BNA250

10-CHLORO-5,10-DIHYDROARSACRIDINE see PDB000

6-CHLORO-3,4-DIHYDRO-2H-1,2,4-BENZOTHIADIAZINE-7-SULFONAMIDE-1,1-DIOXIDE see CFY000

S-(6-CHLORO-3,4-DIHYDRO-2H-1-BENZOTHIOPYRAN-4-YL) o,o-DIETHYLPHOSPHORODITHIOATE see CLH810

S-(6-CHLORO-3,4-DIHYDRO-2H-1-BENZOTHIOPYRAN-4-YL) o,o-DIMETHYL PHOSPHORODITHIOATE see CLH820

4'-CHLORO-2-((3-(10,11-DIHYDRO-5H-DIBENZ(b,f)AZEPIN-5-YL)PROPYL)METHYLAMINO)ACETOPHENONE HCl see IFZ900

1-(8-CHLORO-10,11-DIHYDRODIBENZO(b,f)THIEPIN-10-YL)-4-METHYLPIPERAZINE see ODY100

12-CHLORO-7,12-DIHYDRO-8,11-DIMETHYLBENZO(a)PHENARSAZINE see CGH500

7-CHLORO-1,3-DIHYDRO-3-(N,N-DIMETHYLCARBAMOYL-1-METHYL-5-PHENYL-2H-1,4-BENZODIAZEPIN-2-ONE see CFY250

3-CHLORO-10,11-DIHYDRO-N,N-DIMETHYL-5H-DIBENZ(b,f)AZEPINE-5-PROPANAMINE MONOHYDROCHLORIDE see CDV000

5-CHLORO-2,3-DIHYDRO-1,1-DIMETHYL-1H-INDOLIUM BROMIDE (9CI) see MIH300

N-(5-CHLORO-9,10-DIHYDRO-9,10-DIOXO-1-ANTHRACENYL)-BENZAMIDE see BDK750

N-(4-CHLORO-9,10-DIHYDRO-9,10-DIOXO-1-ANTHRACENYL) BENZAMIDE (9CI) see CEL750

S-(2-CHLORO-1-(1,3-DIHYDRO-1,3-DIOXO-2H-ISOINDOL-2-YL)ETHYL)-O,O-DIETHYL PHOSPHORODITHIOATE see DBI099

2-CHLORO-4,5-DIHYDRO-1,3,2-DITHIARSENOLE see EIU900

7-CHLORO-1,3-DIHYDRO-3-HEMISUCCINYLOXY-2H-1,4-BENZODIAZEPIN-2-ONE see CFY500

(R)N-((5-CHLORO-3,4-DIHYDRO-8-HYDROXY-3-METHYL-1-OXO-1H-2-

BENZOPYRAN-7-YL))PHENYLALANINE see CHP250

7-CHLORO-1,3-DIHYDRO-3-HYDROXY-1-METHYL-5-PHENYL-2H-1,4-BENZODIAZEPIN-2-ONE see CFY750

7-CHLORO-1,3-DIHYDRO-3-HYDROXY-1-METHYL-5-PHENYL-1,4-BENZODIAZEPIN-2-ONE DIMETHYLCARBAMATE see CFY250

7-CHLORO-1,3-DIHYDRO-3-HYDROXY-5-PHENYL-2H-1,4-BENZODIAZEPINE-2-ONE see CFZ000

7-CHLORO-1,3-DIHYDRO-1-METHYL-5-(p-CHLOROPHENYL)-2H-1,4-BENZODIAZEPIN-2-ONE see CFK150

7-CHLORO-2,3-DIHYDRO-1-METHYL-5-PHENYL-1H-1,4-BENZODIAZEPINE see CGA000

7-CHLORO-2,3-DIHYDRO-1-METHYL-5-PHENYL-1H-1,4-BENZODIAZEPINE, HYDROCHLORIDE see MBY000

7-CHLORO-1,3-DIHYDRO-1-METHYL-5-PHENYL-2H-1,4-BENZODIAZEPIN-2-ONE see DCK759

6-CHLORO-3,4-DIHYDRO-2-METHYL-3-(((2,2,2-TRIFLUOROETHYL)THIO)METHYL)2H-1,2,4-BENZOTHIADIAZINE-7-SULFONAMIDE, 1,1-DIOXIDE see PKL250

10-CHLORO-5,10-DIHYDROPHENARSAZINE see PDB000

7-CHLORO-1,3-DIHYDRO-5-PHENYL-2H-1,4-BENZODIAZEPIN-2-ONE see CGA500

7-CHLORO-1,3-DIHYDRO-5-PHENYL-1-(2-PROPYNYL)-2H-1,4-BENZODIAZEPIN-2-ONE see PIH100

7-CHLORO-1,3-DIHYDRO-5-PHENYL-1-TRIMETHYLSILYL-2H-1,4-BENZODIAZEPIN-2-ONE see CGB000

8-CHLORO-3,4-DIHYDROSPIRO(NAPHTHALENE-2(1H),4'(5'H)-OXAZOL)-2'-AMINE see CGB100

6-CHLORO-3,4-DIHYDRO-7-SULFAMOYL-2H-1,2,4-BENZOTHIADIAZINE-1,1-DIOXIDE see CFY000

2-CHLORO-1,1-DIHYDROXYETHANE see CDY600

(S-(e,e))-3-CHLORO-4,6-DIHYDROXY-2-METHYL-5-(3-METHYL-7-(TETRAHYDRO-5,5-DIMETHYL-4-OXO-2-FURANYL)-2,6-OCTADIENYL)-BENZALDEHYDE see ARM750

1-CHLORO-2,3-DIHYDROXYPROPANE see CDT750

3-CHLORO-1,2-DIHYDROXYPROPANE see CDT750

4'-CHLORO-3,5-DIIODOSALICYLANILIDE ACETATE see CGB250

CHLORO DIISOBUTYL ALUMINUM see CGB500

CHLORO(DIISOPROPOXYPHOSPHINYL)MERCURY see CGB750

6'-CHLORO-2-(DIISOPROPYLAMINO)-o-ACETOTOLUIDIDE HYDROCHLORIDE see CGC000

4-CHLORO-2,5-DIMETHOXYANILINE see CGC050

1-CHLORO-2,4-DIMETHOXY-5-NITROBENZENE see CGC100

21-CHLORO-3,17-DIMETHOXY-19-NOR-17-α-PREGN-1,3,5(10)-TRIEN-20-YNE see BAT795

3-CHLORO-5-(((((4,6-DIMETHOXY-2-PYRIMIDINYL)AMINO)CARBONYL)AMINO)SULFONYL)-1-METHYL-1H-PYRAZOLE-4-CARBOXYLIC ACID, METHYL ESTER see CGC150

N-CHLORODIMETHYLAMINE see CGC200

6'-CHLORO-2-(DIMETHYLAMINO)-o-ACETOTOLUIDIDE HYDROCHLORIDE see CGD000

p-CHLORO DIMETHYLAMINOAZOBENZENE see CGD250

2'-CHLORO-4-DIMETHYLAMINOAZOBENZENE see DRD000

3'-CHLORO-4-DIMETHYLAMINOAZOBENZENE see DRC800

4'-CHLORO-4-DIMETHYLAMINOAZOBENZENE see CGD250

5-CHLORO-2-DIMETHYLAMINOBENZOXAZOLE see CGD270

β-CHLORODIMETHYLAMINO DIBORANE see CGD399

CHLORO(DIMETHYLAMINO)ETHANE see CGW000

2-(p-CHLORO-α-(2-(DIMETHYLAMINO)ETHOXY)BENZYL)PYRIDINE BIMALEATE see CGD500

2-(p-CHLORO-α-(2-(DIMETHYLAMINO)ETHOXY)BENZYL)-PYRIDINE DIPHENYLDISULFONATE see CBQ575

2-(p-CHLORO-α-(2-(DIMETHYLAMINO)ETHOXY)BENZYL)PYRIDINE MALEATE see CGD500

2-CHLORO-11-(2-(DIMETHYLAMINO)ETHOXY)DIBENZO(b,f)THIEPIN see ZUJ000

2-CHLORO-11-2-DIMETHYLAMINOETHOXY)DIBENZO(b,f)THIEPINE see ZUJ000

2-CHLORO-α-(2-(DIMETHYLAMINO)-ETHYL)BENZHYDROL HYDROCHLORIDE see CMW700

2-(p-CHLORO-α-(2-(DIMETHYLAMINO)ETHYL)BENZYL)PYRIDINE see CLX300

dl-2-(-p-CHLORO-α-2-(DIMETHYLAMINO)ETHYLBENZYL)PYRIDINE BIMALEATE see TAI500

(+)-2-(p-CHLORO-α-(2-(DIMETHYLAMINO)ETHYL)BENZYL)PYRIDINE MALEATE see PJJ325

2-(p-CHLORO-α-(2-(DIMETHYLAMINO)ETHYL)BENZYL)PYRIDINE MALEATE (1:1) see CLD250

2'-CHLORO-2-((2-(DIMETHYLAMINO)ETHYL)ETHYLAMINO)ACETANILIDE DIHYDROCHLORIDE see CGD750

7-CHLORO-10-(2-(DIMETHYLAMINO)ETHYL)ISOALLOXAZINE SULFATE see CGE000

2-CHLORO-α-(2-(DIMETHYLAMINO)ETHYL)-α-PHENYL-BENZENEMETHANOL HYDROCHLORIDE (9CI) see CMW700

2'-CHLORO-2-(2-(DIMETHYLAMINO)ETHYLTHIO)ACETANILIDE HYDROCHLORIDE see CGE250

6'-CHLORO-2-(2-(DIMETHYLAMINO)ETHYLTHIO)-o-ACETOTOLUIDIDE see CGE500

6'-CHLORO-2-(DIMETHYLAMINO)-N-METHYL-o-ACETOTOLUIDIDE HYDROCHLORIDE see CGE750

5-CHLORO-3-(DIMETHYLAMINOMETHYL)-2-BENZOXAZOLINONE see CGF000

7-CHLORO-10-(4-(DIMETHYLAMINO)-1-METHYLBUTYL)ISOALLOXAZINE SULFATE see CGF250

4-CHLORO-α-(2-(DIMETHYLAMINO)-1-METHYLETHYL)-α-METHYLBENZENEETHANOL HYDROCHLORIDE see CMW500

(p-CHLORO-α-(2-(DIMETHYLAMINO)-1-METHYLETHYL)-α-METHYL-PHENETHYL ALCOHOL see CMW459

p-CHLORO-α-(2-DIMETHYLAMINO-1-METHYLETHYL)-α-METHYLPHENETHYL ALCOHOL HYDROCHLORIDE see CMW500

2-CHLORO-4-DIMETHYLAMINO-6-METHYL-PYRIMIDINE see CCP500

7-CHLORO-4-(DIMETHYLAMINO)-1,4,4a,5,5a,6,11,12a-OCTAHYDRO-2-NAPHTHACENECARBOXAMIDE see CMA750

p-CHLORO-α,α-DIMETHYL-2-AMINOPROPIONATE-PHENETHYL ALCOHOL HYDROCHLORIDE see CJX000

6'-CHLORO-3-(DIMETHYLAMINO)-o-PROPIONOTOLUIDIDE HYDROCHLORIDE see CGF500

(α-2-CHLORO-9-ω-DIMETHYLAMINO-PROPYLAMINE)THIOXANTHENE see TAF675

7-CHLORO-10-(3-DIMETHYLAMINOPROPYL)-BENZO-(b)(1,8)-5(10H)-NAPHTHAPYRIDONE HYDROCHLORIDE see CGG500

3-CHLORO-5-(3-(DIMETHYLAMINO)PROPYL)-10,11-DIHYDRO-5H-DIBENZ(b,f)AZEPINE see CDU750

3-CHLORO-5-(3-(DIMETHYLAMINO)PROPYL)-10,11-DIHYDRO-5H-DIBENZ(b,f)AZEPINE HYDROCHLORIDE see CGG600

3-CHLORO-5-(3-(DIMETHYLAMINO)PROPYL)-10,11-DIHYDRO-5H-DIBENZ(b,f)AZEPINE MONOHYDROCHLORIDE see CDV000

2-CHLORO-9-(3-(DIMETHYLAMINO)PROPYLIDENE)-THIOXANTHENE see TAF675

2-CHLORO-9-(ω-DI-METHYLAMINOPROPYLIDENE)THIOXANTHENE see TAF675

7-CHLORO-10-(3-(DIMETHYLAMINO)PROPYL)ISOALLOXAZINE HYDROCHLORIDE see CGG750

2-CHLORO-10-(3-(DIMETHYLAMINO)PROPYL)PHENOTHIAZINE see CKP250

CHLORO-3-(DIMETHYLAMINO-3-PROPYL)-10 PHENOTHIAZINE (FRENCH) see CKP250

2-CHLORO-10-(3-(DIMETHYLAMINO)PROPYL)PHENOTHIAZINE 5,5-DIOXIDE see ONI300

2-CHLORO-10-(3-DIMETHYLAMINOPROPYL)PHENOTHIAZINE MONOHYDROCHLORIDE see CKP500

3'-CHLORO-N,N-DIMETHYLAMINOSTIBEN (GERMAN) see CGK750

4'-CHLORO-N,N-DIMETHYLAMINOSTIBEN (GERMAN) see CGL000

2'-CHLORO-4-DIMETHYLAMINOSTILBENE see CGK500

3'-CHLORO-4-DIMETHYLAMINOSTILBENE see CGK750

4'-CHLORO-4-DIMETHYLAMINOSTILBENE see CGL000

9-CHLORO-8,12-DIMETHYLBENZ(a)ACRIDINE see CGH250

2-CHLORO-1,10-DIMETHYL-5,6-BENZACRIDINE (FRENCH) see CGH250

2-CHLORO-1,10-DIMETHYL-7,8-BENZACRIDINE (FRENCH) see DRB800

2-CHLORO-α,α-DIMETHYLBENZENEETHANIAMINE HYDROCHLORIDE see DRC600

1-CHLORO-3,3-DIMETHYLBUTAN-2-ONE see MRG070

10-CHLORO-6,9-DIMETHYL-5,10-DIHYDRO-3,4-BENZOPHENARSAZINE see CGH500

p-CHLORO-5,10-DIMETHYL-2,4-DIOXA-p-THIONO-3-PHOSPHABICYCLO(4.4.0)DECANE see CGH675

CHLORO((3-(5,5-DIMETHYL-2,4-DIOXO-3-IMIDAZOLIDINYL)-2-METHOXY)PROPYL)MERCURY see CHV250

2-CHLORO-N,N-DIMETHYLETHYLAMINE HYDROCHLORIDE see DRC000

5-CHLORO-3-(1,1-DIMETHYLETHYL)-6-METHYL-2,4(1H,3H)-PYRIMIDINEDIONE see BQT750

4-CHLORO-N-((4-(1,1-DIMETHYLETHYL)PHENYL)METHYL)-3-ETHYL-1-METHYL-1H-PYRAZOLE-5-CARBOXAMIDE see CGH750

2-CHLORO-9-(2,2-DIMETHYLHYDRAZINO)ACRIDINE see CGH800

1-(2-CHLORO-5,7-DIMETHYL-3-METHYL-1-ADAMANTYL)-N-METHYL-2-PROPYL AMINE HYDROCHLORIDE see CIF000

2-CHLORO-N,N-DIMETHYL-3-OXOBUTANAMIDE see CGH810

2-CHLORO-N-(2,6-DIMETHYL-4-OXO-2,5-CYCLOHEXADIEN-1-YLIDENE)ACETAMIDE see CGH820

2-CHLORO-2',6'-DIMETHYL-N-(2-OXOTETRAHYDRO-3-FURYL)ACETANILIDE see CGI550

p-CHLORO-N-α-DIMETHYLPHENETHYLAMINE see CIF250

p-CHLORO-α,α-DIMETHYLPHENETHYLAMINE see CLY250

4-CHLORO-α,α-DIMETHYLPHENETHYLAMINE HYDROCHLORIDE see ARW750

O-CHLORO-α,α-DIMETHYLPHENETHYLAMINE HYDROCHLORIDE see DRC600

p-CHLORO-α,α-DIMETHYLPHENETHYLAMINE HYDROCHLORIDE see ARW750

N-(p-CHLORO-α,α-DIMETHYLPHENETHYL)-2-(DIETHLAMINO)PROPIONAMIDE HYDROCHLORIDE see CGI125

4-CHLORO-3,5-DIMETHYLPHENOL see CLW000

2-CHLORO-4,5-DIMETHYLPHENOL, METHYL CARBAMATE see CGI500

((4-CHLORO-6-((2,3-DIMETHYLPHENYL)AMINO)-2-PYRIMIDINYL)THIO)ACETIC ACID see CLW250

(2-CHLORO-4,5-DIMETHYL)PHENYL ESTER, CARBAMIC ACID see CGI500

CHLORODIMETHYL(1-PHENYLETHYL)BENZENE see MRG060

2-CHLORO-N-(2,6-DIMETHYL)PHENYL-N-ISOPROPOXYMETHYLACETAMIDE see CGI200

2-CHLORO-N-(2,6-DIMETHYLPHENYL)-N-(2-METHOXYETHYL)ACETAMIDE see DSM500

2-CHLORO-4,5-DIMETHYLPHENYL METHYLCARBAMATE see CGI500

2-CHLORO-N-(2,6-DIMETHYLPHENYL)-N-((2-METHYLPROPOXY)METHYL)ACETAMIDE see IIE200

2-CHLORO-N-(2,6-DIMETHYLPHENYL)-N-(TETRAHYDRO-2-OXO-3-FURANYL)ACETAMIDE see CGI550

CHLORODIMETHYLPHOSPHINE see CGI625

7-CHLORO-1-((DIMETHYLPHOSPHINYL)METHYL)-1,3-

DIHYDRO-5-PHENYL-2H-1,4-BENZODIAZEPINE-2-ONE see FOL100

6'-CHLORO-2-(2,6-DIMETHYLPIPERIDINO)-o-ACETOTOLUIDIDE HYDROCHLORIDE see CGI750

6'-CHLORO-3-(2,6-DIMETHYLPIPERIDINO)-o-PROPIONOTOLUIDIDE HYDROCHLORIDE see CGJ000

2-CHLORO-5-(3,5-DIMETHYLPIPERIDINO SULPHONYL)BENZOIC ACID see CGJ250

1-CHLORO-N,N-DIMETHYL-2-PROPANAMINE HYDROCHLORIDE see CGJ280

2-CHLORO-N,N-DIMETHYL PROPYLAMINE see CGJ290

3-CHLORO-N,N-DIMETHYL-1-PROPYLAMINE HYDROCHLORIDE see CGJ300

2-CHLORO-4-(1,1-DIMETHYLPROPYL)PHENYL METHYL METHYLPHOSPHORAMIDATE see DYE200

2'-CHLORO-N,N-DIMETHYL-4-STILBENAMINE see CGK500

3'-CHLORO-N,N-DIMETHYL-4-STILBENAMINE see CGK750

4'-CHLORO-N,N-DIMETHYL-4-STILBENAMINE see CGL000

2-CHLORO-N,N-DIMETHYLTHIOXANTHENE-Δ⁹-γ-PROPYLAMINE see TAF675

N-CHLORO-4,5-DIMETHYLTRIAZOLE see CGL125

4'-CHLORO-2,2-DIMETHYLVALERANILIDE see CGL250

2-CHLORO-4,6-DINITROANILINE see CGL325

4-CHLORO-2,6-DINITROANILINE see CGL500

6-CHLORO-2,4-DINITROANILINE see CGL325

CHLORODINITROBENZENE see CGL750

1-CHLORO-2,4-DINITROBENZENE see CGM000

4-CHLORO-1,3-DINITROBENZENE see CGM000

6-CHLORO-1,3-DINITROBENZENE see CGM000

CHLORODINITROBENZENE (mixed isomers) (DOT) see CGL750

4-CHLORO-2,5-DINITROBENZENE DIAZONIUM-6-OXIDE see CGM199

1-CHLORO-2,4-DINITROBENZOL (GERMAN) see CGM000

4-CHLORO-3,5-DINITROBENZOTRIFLUORIDE see CGM225

1-CHLORO-2,4-DINITRONAPHTHALENE see DUS600

2-CHLORO-1,5-DINITRO-3-(TRIFLUOROMETHYL)BENZENE see CGM200

2-CHLORO-N-(2,6-DINITRO-4-(TRIFLUOROMETHYL)PHENYL)-N-ETHYL-6-FLUOROBENZE NEMETHANAMINE see FDD078

4-CHLORO-3,5-DINITRO-α-α-α-TRIFLUOROTOLUENE see CGM225

p-CHLORO-2,4-DIOXA-5-ETHYL-p-THIONO-3-PHOSPHABICYCLO(4.4.0)DECANE see CGM375

p-CHLORO-2,4-DIOXA-5-METHYL-p-THIONO-3-PHOSPHABICYCLO(4.4.0)DECANE see CGM400

CHLORO((3-(2,4-DIOXO-3-IMIDAZOLIDINYL)-2-METHOXY)PROPYL)MERCURY see CHV750

CHLORO((3-(2,4-DIOXO-5-IMIDAZOLIDINYL)-2-METHOXY)PROPYL) MERCURY see CGM450

6-CHLORO-1,3-DIOXO-5-ISOINDOLINESULFONAMIDE see CGM500

6-CHLORO-1,3-DIOXO-5-ISOINDOLINESULFONAMIDE see CLG825

CHLORO((3-(2,4-DIOXO-1-METHYL-3-IMIDAZOLIDINYL)-2-METHOXY)PROPYL)MERCURY see CHW250

CHLORO((3-(2,4-DIOXO-3-METHYL-1-IMIDAZOLIDINYL)-2-METHOXY)PROPYL)MERCURY see CHW000

CHLORO((3-(2,4-DIOXO-3-METHYL-5-IMIDAZOLIDINYL)-2-METHOXY)PROPYL)MERCURY see CHW500

CHLORODIPHENYL see CGM550

2-CHLORODIPHENYL see CGM750

3-CHLORODIPHENYL see CGM770

4-CHLORODIPHENYL see CGM780

o-CHLORODIPHENYL see CGM750

p-CHLORODIPHENYL see CGM780

2-CHLORO-2,2-DIPHENYLACETIC ACID-2-(DIETHYLAMINO)ETHYL ESTER HYDROCHLORIDE see DHW200

CHLORODIPHENYLARSINE see CGN000

1-(o-CHLORO-α,α-DIPHENYLBENZYL)IMIDAZOLE see MRX500

CHLORODIPHENYL (21% Cl) see PJM000

CHLORODIPHENYL (32% Cl) see PJM250

CHLORODIPHENYL (41% Cl) see PJL800

CHLORODIPHENYL (48% Cl) see PJM750

CHLORODIPHENYL (60% Cl) see PJN250

CHLORODIPHENYL (62% Cl) see PJN500

CHLORODIPHENYL (68% Cl) see PJN750

CHLORODIPHENYL (42% Cl) (OSHA) see PJM500

CHLORODIPHENYL (54% Cl) (OSHA) see PJN000

2-CHLORO-1,2-DIPHENYLETHANONE see CFJ100

2-(4-(2-CHLORO-1,2-DIPHENYLETHENYL)PHENOXY)-N,N-DIETHYLETHANAMINE see CMX500

1-(p-CHLORODIPHENYLMETHYL)-4-(2-(2-HYDROXYETHOXY)ETHYL)PIPERAZINE see CJR909

4-CHLORODIPHENYL SULFONE see CKI625

(E)-2-(p-(2-CHLORO-1,2-DIPHENYLVINYL)PHENOXY)TRIETHYLAMINE, CITRATE see CMX750

2-(p-(2-CHLORO-1,2-DIPHENYL VINYL)PHENOXY)TRIETHYLAMINE CITRATE (1:1) see CMX700

2-CHLORO-N,N-DI-2-PROPENYLACETAMIDE see CFK000

6'-CHLORO-2-(DIPROPYLAMINO)-o-ACETOTOLUIDIDE HYDROCHLORIDE see CGN250

CHLORODIPROPYLBORANE see CGN325

CHLORODRACYLIC ACID see CEL300

CHLORODWUFENOL see CGM550

CHLOROEPOXYETHANE see CGX000

1-CHLORO-2,3-EPOXYPROPANE see EAZ500

3-CHLORO-1,2-EPOXYPROPANE see EAZ500

cis-1-CHLORO-1,2-EPOXYPROPANE see CKS099

trans-1-CHLORO-1,2-EPOXYPROPANE see CKS100

CHLOROETENE see MIH275

2-CHLOROETHANAL see CDY500

2-CHLORO-1-ETHANAL see CDY500

2-CHLOROETHANAMIDE see CDY850

CHLOROETHANE see EHH000

CHLORO(1,2-ETHANEDIAMINE-N,N')(SULFINYLBIS(METHANE)-O)PLATINUM(1+) CHLORIDE (SP-4-3) see CGN400

2-CHLORO-1,1-ETHANEDIOL see CDY600

2-CHLOROETHANEPHOSPHONIC ACID see CDS125

2-CHLOROETHANE SULFOCHLORIDE see CGO125

2-CHLOROETHANESULFONYL CHLORIDE see CGO125

β-CHLOROETHANESULFONYL CHLORIDE see CGO125

2-CHLOROETHANESULFONYL FLUORIDE see CGO200

CHLOROETHANOIC ACID see CEA000

Δ-CHLOROETHANOL see EIU800

2-CHLOROETHANOL ACETATE see CGO600

2-CHLOROETHANOL ACRYLATE see ADT000

2-CHLOROETHANOL-2-(p-tert-BUTYLPHENOXY)-1-METHYLETHYL SULFITE see SOP500

2-CHLOROETHANOL ESTER with 2-(p-tert-BUTYLPHENOXY)-1-METHYLETHYL SULFITE see SOP500

2-CHLOROETHANOL HYDROGEN PHOSPHATE ESTER with 3-CHLORO-7-HYDROXY-4-METHYLCOUMARIN see DFH600

2-CHLOROETHANOL (MAK) see EIU800

2-CHLORO-ETHANOL-4-METHYLBENZENESULFONATE (9CI) see CHI125

2-CHLOROETHANOL PHOSPHATE see CGO500

2-CHLOROETHANOL PHOSPHATE DIESTER ESTER with 3-CHLORO-7-HYDROXY-4-METHYLCOUMARIN see DFH600

2-CHLOROETHANOL PHOSPHITE (3:1) see PHO000

2-CHLORO-ETHANOL, PHOSPHOROTHIOATE (3:1) see PHN500

2-CHLORO-ETHANOL-p-TOLUENESULFONATE (8CI) see CHI125

CHLOROETHENE see MIH275

CHLOROETHENE see VNP000

CHLOROETHENE HOMOPOLYMER see PKQ059

(2-CHLOROETHENYL) ARSONOUS DICHLORIDE see CLV000

1,1',1"-(1-CHLORO-1-ETHENYL-2-YLIDENE)-TRIS(4-METHOXYBENZENE) see CLO750

17-α-CHLOROETHINYL-17-β-HYDROXYESTRA-4,9-DIEN-3-ONE see CHP750

(1-(((2-CHLOROETHOXY)(2-CHLOROETHYL)PHOSPHINYL)OXY)ETHYL)-PHOSPHONIC ACID, 1-(BIS(2-CHLOROETHOXY)PHOSPHINYL)ETHYL 2-CHLOROETHYL ESTER see CGO525

2-(2-CHLOROETHOXY)ETHANOL see DKN000

(2-CHLOROETHOXY)ETHENE see CHI250

2-(2-CHLOROETHOXY)ETHYL 2'-CHLOROETHYL ETHER see TKL500

2-CHLORO-N-(ETHOXYMETHYL)-6'-ETHYL-O-ACETOTOLUIDIDE see CGO550

2-CHLORO-N-(ETHOXYMETHYL)-N-(2-ETHYL-6-METHYLPHENYL)ACETAMIDE see CGO550

2-CHLORO-1-(3-ETHOXY-4-NITROPHENOXY)-4-(TRIFLUOROMETHYL)BENZENE see OQU100

1-(2-(2-CHLOROETHOXY)PHENYLSULFONYL)-3-(4-METHOXY-6-METHYL-1,3,5-TRIAZIN-2-YL)UREA (IUPAC) see THQ550

2-CHLOROETHYL ACETATE see CGO600

β-CHLOROETHYL ACETATE see CGO600

CHLOROETHYL ACRYLATE see ADT000

2-CHLOROETHYL ACRYLATE see ADT000

β-CHLOROETHYL ACRYLATE see ADT000

2-CHLOROETHYL ALCOHOL see EIU800

β-CHLOROETHYL ALCOHOL see EIU800

2-CHLOROETHYLAMINE see CGP125

2-CHLOROETHYLAMINE HYDROCHLORIDE see CGP250

β-CHLOROETHYLAMINE HYDROCHLORIDE see CGP250

2-CHLORO-4-ETHYLAMINEISOPROPYLAMINE-s-TRIAZINE see ARQ725

3'-CHLORO-2-ETHYLAMINO-o-ACETOTOLUIDIDE HYDROCHLORIDE see CGP375

6'-CHLORO-2-(ETHYLAMINO)-o-ACETOTOLUIDIDE HYDROCHLORIDE see CGP380

2-CHLORO-4-ETHYLAMINO-6-(1-CYANO-1-METHYL)ETHYLAMINO-s-TRIAZINE see BLW750

7-((2-((2-CHLOROETHYL)AMINO)ETHYL)AMINO)BENZ(c)ACRIDINE DIHYDROCHLORIDE HYDRATE see CGP500

9-(2-((2-CHLOROETHYL)AMINO)ETHYLAMINO)-6-CHLORO-2-METHOXYACRIDINE, DIHYDROCHLORIDE see QCS875

2-CHLOROETHYLAMINOETHYL DEHYDROABIETATE HYDROCHLORIDE see CGQ250

N-(2-CHLOROETHYL)AMINOETHYL-4-ETHOXYNITROBENZENE see CGX325

6'-CHLORO-2-(ETHYLAMINO)-o-HEXANOTOLUIDIDE HYDROCHLORIDE see CAB750

1-CHLORO-3-ETHYLAMINO-5-ISOPROPYLAMINO-s-TRIAZINE see ARQ725

2-CHLORO-4-ETHYLAMINO-6-ISOPROPYLAMINO-s-TRIAZINE see ARQ725

1-CHLORO-3-ETHYLAMINO-5-ISOPROPYLAMINO-2,4,6-TRIAZINE see ARQ725

2-CHLORO-4-ETHYLAMINO-6-ISOPROPYLAMINO-1,3,5-TRIAZINE see ARQ725

2'-CHLORO-2-(ETHYLAMINO)-6'-METHYLACETANILIDE, HYDROCHLORIDE see CGP380

N-(2-CHLOROETHYL)AMINOMETHYL-4-HYDROXYNITROBENZENE see CGQ280

N-(2-CHLOROETHYL)AMINOMETHYL-4-METHOXYNITROBENZENE see CGQ400

2-(2-CHLOROETHYL)AMINO)METHYL)-4-NITROPHENOL see CGQ280

2-CHLORO-3-(ETHYLAMINO)-1-METHYL-3-OXO-1-PROPENYL DIMETHYL ESTER PHOSPHORIC ACID see DTP600

2-CHLORO-3-(ETHYLAMINO)-1-METHYL-3-OXO-1-PROPENYL DIMETHYL PHOSPHATE see DTP600

6-CHLORO-2-ETHYLAMINO-4-METHYL-4-PHENYL-4H-3,1-BENZOXAZINE see CGQ500

9-((3-((2-CHLOROETHYL)AMINO)PROPYL)AMINO)ACRIDINE DIHYDROCHLORIDE HYDRATE see CGR000

7-((3-((2-CHLOROETHYL)AMINO)PROPYL)AMINO)BENZ(c)ACRIDINE DIHYDROCHLORIDE SESQUIHYDRATE see CGR250

7-((3-((2-CHLOROETHYL)AMINO)PROPYL)AMINO)BENZO(b)(1,8)PHENANTHROLINE DIHYDROCHLORIDE HYDRATE see CGR750

7-((3-((2-CHLOROETHYL)AMINO)PROPYL)AMINO)BENZO(b)(1,10)PHENANTHROLINE DIHYDROCHLORIDE see CGR500

10-((2-CHLOROETHYLAMINO)PROPYLAMINO)-2-METHOXY-7-CHLOROBENZO(b)-(1,5)-NAPHTHYRIDINE see CGS500

4-((3-((2-CHLOROETHYL)AMINO)PROPYL)AMINO)-6-METHOXYQUINOLINE HYDROCHLORIDE see CGS750

2-(4-CHLORO-6-ETHYLAMINO-s-TRIAZINE-2-YLAMINO)-2-METHYL-PROPIONITRILE see BLW750

2-(4-CHLORO-6-ETHYLAMINO-1,3,5-TRIAZINE-2-YLAMINO)-2-METHYLPROPIONITRILE see BLW750

2-((4-CHLORO-6-(ETHYLAMINO)-1,3,5-TRIAZIN-2-YL)AMINO)-2-METHYL-PROPANENITRILE see BLW750

2-((4-CHLORO-6-(ETHYLAMINO)-s-TRIAZIN-2-YL)AMINO)-2-METHYLPROPIONITRILE see BLW750

CHLOROETHYLAMINOURACIL see DYC700

6'-CHLORO-2-(ETHYLAMINO)-o-VALEROTOLUIDIDE HYDROCHLORIDE see CGT000

2-CHLOROETHYLAMMONIUM CHLORIDE see CGP250

2-(2-CHLOROETHYL)-3-AZA-4-CHROMANONE see CDS250

CHLOROETHYLBENZENE see EHH500

2-CHLOROETHYL BENZYL SULFIDE see BFL100

3-(2-CHLOROETHYL)-2-(BIS(2-CHLOROETHYL)AMINO)PERHYDRO-2H-1,3,2-OXAZAPHOSPHORINE-2-OXIDE see TNT500

β-CHLOROETHYL-β-(BIS(β-HYDROXYETHYL)SULFONIUM)ETHYL SULFIDE CHLORIDE see BKD750

β-CHLOROETHYL-β'-(p-tert-BUTYLPHENOXY)-α'- METHYLETHYL SULFITE see SOP500

β-CHLOROETHYL-β-(p-tert-BUTYLPHENOXY)-α-METHYLETHYL SULPHITE see SOP500

β-CHLOROETHYL CARBAMATE see CGU000

S-((2-CHLOROETHYL)CARBAMOYL)GLUTATHIONE see CGU100

α-(2-CHLOROETHYL)-ω-(2-CHLOROETHOXY)POLY(OXY-1,2-ETHANEDIYL) see PJU300

3-(2-CHLOROETHYL)-2-(2-CHLOROETHYL)AMINO-4-HYDROPEROXYTETRAHYDRO-2H-1,3,2-OXAZAPHOSPHORINE see HIE550

3-(2-CHLOROETHYL)-2-(2-CHLOROETHYL)AMINO)PERHYDRO-2H-1,3,2-OXAZAPHOSPHORINE OXIDE see IMH000

(6-CHLORO-9-(3-ETHYL-2-CHLOROETHYL)AMINOPROPYLAMINO)-2-METHOXYACRIDINE DIHYDROCHLORIDE see ADJ875

3-(2-CHLOROETHYL)-2-(2-CHLOROETHYL)AMINO)TETRAHYDRO-2H-1,3,2-OXAZAPHOSPHORINE-2-OXIDE see IMH000

N-(2-CHLOROETHYL)-N'-(2-CHLOROETHYL)-N',O-PROPYLENEPHOSPHORIC ACID DIAMIDE see IMH000

N-(2-CHLOROETHYL)-N'-(2-CHLOROETHYL)-N',O-

2-CHLOROETHYL TOSYLATE see CHI125
N-(2-CHLOROETHYL)-α,α,α-TRIFLUORO-2,6-DINITRO-N-PROPYL-p-TOLUIDINE see FDA900
(2-CHLOROETHYL)TRIMETHYLAMMONIUM CHLORIDE see CMF400
(β-CHLOROETHYL)TRIMETHYLAMMONIUM CHLORIDE see CMF400
2-CHLOROETHYLTRIS(2-METHOXYETHOXY)SILANE see CHI200
2-CHLOROETHYL VINYL ETHER see CHI250
CHLOROETHYNE see ACJ000
17-α-CHLOROETHYNLY-19-NOR-4,9-ANDROSTADIEN-17-β-OL-3-ONE see CHP750
17-α-CHLOROETHYNYL-3,17-β-DIMETHOXY-OESTRA-1,3,5(10)-TRIENE see BAT795
17-α-CHLOROETHYNYL-17-β-HYDROXY-19-NOR-4,9-ANDROSTADIEN-3-ONE see CHP750
CHLOROETHYNYL NORGESTREL mixed with MESTRANOL (20:1) see CHI750
4-CHLORO-α-ETHYNYL-α-PHENYLBENZENEMETHANOL CARBAMATE see CKI250
4-CHLORO-α-ETHYNYL-α-PHENYLBENZENEMETHANOL CARBAMATE see CKI500
(CHLORO-2-ETIL)-1-(RIBOFURANOSILISOPROPILIDENE-2',3'-PARANITROBENZOATO)-3-NITROSOUREA see RFU600
CHLOROFENIZON see CJT750
3-(4-(4-CHLORQ-FENOSSIL))-1,1-DIMETIL-UREA (ITALIAN) see CJQ000
CHLOROFENSULPHIDE see CKL500
CHLOROFENVINPHOS see CDS750
p-CHLOROFENYLESTER KYSELINY BENZENSULFONOVE (CZECH) see CJR500
N-2-(7-CHLORO)FLUORENYLACETAMIDE see CHI825
N-(7-CHLORO-2-FLUORENYL)ACETAMIDE see CHI825
2-CHLORO-N-(9-FLUORENYL)DIETHYLAMINE HYDROCHLORIDE see FEE100
1-CHLORO-10-FLUORO-DECANE see FHN000
21-CHLORO-9-FLUORO-11-β,17-DIHYDROXY-16-β-METHYLPREGNA-1,4-DIENE-3,20-DIONE-17-PROPIONATE see CMW300
1-CHLOROFLUOROETHANE see FOO553
21-CHLORO-9-FLUORO-17-HYDROXY-16-β-METHYLPREGNA-1,4-DIENE-3,11,20-TRIONE BUTYRATE see CMW400
CHLOROFLUOROMETHANE see CHI900
3-CHLORO-4-FLUORONITROBENZENE see CHI950
2-CHLORO-1-FLUORO-4-NITROBENZENE see CHI950
1-CHLORO-8-FLUOROOCTANE see FKT000
1-CHLORO-5-FLUOROPENTANE see FFW000
7-CHLORO-5-(o-FLUOROPHENYL)-1,3-DIHDYRO-1-METHYL-2H-1,4-BENZODIAZEPIN-2-ONE see FDB100
7-CHLORO-5-(2-FLUOROPHENYL)-1-METHYL-1H-1,4-BENZODIAZEPIN-2(3H)-ONE see FDB100
8-CHLORO-6-(2-FLUOROPHENYL)-1-METHYL-4H-IMIDAZO(1,5-a)(1,4)BENZODIAZEPINE see MQT525
6-(3-(2-CHLORO-6-FLUOROPHENYL)-5-METHYL-4-ISOXAZOLECARBOXAMIDO)PENICILLANIC ACID SODIUM SALT see CHJ000

3-(2-CHLORO-6-FLUOROPHENYL)-5-METHYL-4-ISOXAZOLYLPENICILLIN SODIUM MONOHYDRATE see CHJ000
1-CHLORO-3-FLUORO-2-PROPANOL MIXT. WITH 1,3-DIFLUORO-2-PROPANOL see GEW725
3-CHLORO-2-FLUOROPROPENE see CHJ250
3-CHLORO-2-FLUORO-1-PROPENE see CHJ250
2-CHLORO-4'-FLUORO-α-(PYRIMIDIN-5-YL)BENZHYDRYL ALCOHOL see CHJ300
(+−)-2-CHLORO-4'-FLUORO-α-(PYRIMIDIN-5-YL)BENZHYDRYL ALCOHOL see CHJ300
2-CHLORO-4'-FLUORO-α-(PYRIMIDIN-5-YL)DIPHENYLMETHANOL see CHJ300
CHLORO FLUORO SULFONE see SOT500
5-(2-CHLORO-6-FLUORO-4-(TRIFLUOROMETHYL)PHENOXY)-N-(ETHYLSULFONYL)-2-NITROBENZAMIDE see CHJ400
5-(2-CHLORO-6-FLUORO-4-TRIFLUOROMETHYLPHENOXY)-N-ETHYLSULFONYL-2-NITROBENZAMIDE see CHJ400
21-CHLORO-9-FLUORO-11-β,16-α-17-TRIHYDROXYPREGN-4-ENE-3,20-DIONE cyclic 16,17-ACETAL with ACETONE see HAF300
CHLOROFLURAZOLE see DGO400
CHLOROFLURENOL-METHYL ESTER see CDT000
CHLOROFOLIN see CKN000
CHLOROFORM see CHJ500
CHLOROFORMAMIDINIUM CHLORIDE see CHJ599
CHLOROFORMAMIDINIUM NITRATE see CHJ625
CHLOROFORME (FRENCH) see CHJ500
CHLOROFORMIC ACID ALLYL ESTER see AGB500
CHLOROFORMIC ACID BENZYL ESTER see BEF500
CHLOROFORMIC ACID 2-CHLOROETHYL ESTER see CGU199
CHLOROFORMIC ACID DIMETHYLAMIDE see DQY950
CHLOROFORMIC ACID, ESTER with 1-NAPHTHOL see NBH200
CHLOROFORMIC ACID ETHYL ESTER see EHK500
CHLOROFORMIC ACID 2-FLUOROETHYL ESTER see FIH100
CHLOROFORMIC ACID ISOPROPYL ESTER see IOL000
CHLOROFORMIC ACID METHYL ESTER see MIG000
CHLOROFORMIC ACID 1-NAPHTHYL ESTER see NBH200
CHLOROFORMIC ACID PHENYL ESTER see CBX109
CHLOROFORMIC ACID PROPYL ESTER see PNH000
CHLOROFORMIC DIGITALIN see DKN400
CHLOROFORMYL CHLORIDE see PGX000
4-CHLORO-N-FORMYL-o-TOLUIDINE see CHJ700
CHLOROFOS see TIQ250
CHLOROFTALM see TIQ250
CHLORO-2-FURANYL MERCURY see CHK000
4-CHLORO-N-FURFURYL-5-SULFAMOYLANTHRANILIC ACID see CHJ750
CHLORO(2-FURYL)MERCURY see CHK000
6'-CHLORO-2-(2-FURYLMETHYL)AMINO-o-ACETOTOLUIDIDE HYDROCHLORIDE see CHK125
4-CHLORO-N-(2-FURYLMETHYL)-5-SULFAMOYLANTHRANILIC ACID see CHJ750

CHLOROFYL see CKN000
CHLOROGENIC ACID see CHK175
CHLOROGERMANE see CHK250
CHLOROGUANIDE see CKB250
CHLOROGUANIDE HYDROCHLORIDE see CKB500
CHLOROGUANIDINE HYDROCHLORIDE see CKB500
6-CHLOROGUANINE see CHK300
5-CHLORO-3(H)-2-BENZOXAZOLONE see CDQ750
(2-CHLORO-1-HEPTENYL)ARSONIC ACID MONOAMMONIUM SALT see CEI250
2-CHLORO-1,1,1,4,4,4-HEXAFLUOROBUTENE-2 see CHK750
1-CHLORO-1,1,2,2,3,3-HEXAFLUOROPROPANE see CHK800
endo-4-CHLORO-N-(HEXAHYDRO-4,7-METHANOISOINDOL-2-YL)-3-SULFAMOYLBENZAMIDE see CHK825
2-CHLORO-10-(3-(HEXAHYDROPYRROLO(1,2-a)PYRAZIN-2(1H)-YL)-PROPIONYL)PHENOTHIAZINE 2HCl see NMV750
CHLORO(2-HEXANAMIDOCYCLOHEXYL)MERCURY, (E)- see CHL000
trans-CHLORO(2-HEXANAMIDOCYCLOHEXYL)MERCURY see CHL000
CHLOROHEXYL ISOCYANATE see CHL250
4-CHLORO-2-HEXYLPHENOL see CHL500
CHLOROHYDRIC ACID see HHL000
epi-CHLOROHYDRIN see EAZ500
α-CHLOROHYDRIN see CDT750
dl-α-CHLOROHYDRIN see CHL875
α-CHLOROHYDRIN DIACETATE see CIL900
CHLOROHYDROL see AHA000
CHLOROHYDROQUINONE see CHM000
5-CHLORO-4-(HYDROXYAMINO)QUINOLINE-1-OXIDE see CHM500
6-CHLORO-4-(HYDROXYAMINO)QUINOLINE-1-OXIDE see CHM750
7-CHLORO-4-(HYDROXYAMINO)QUINOLINE-1-OXIDE see CHN000
4'-CHLORO-2'-(α-HYDROXYBENZYL)ISONICOTINANILIDE see CHN100
3-CHLORO-4-HYDROXYBIPHENYL see CHN500
CHLORO(2-HYDROXY-3,5-DINITROPHENYL)MERCURY see CHN750
3-CHLORO-4-HYDROXYDIPHENYL see CHN500
5-CHLORO-2-HYDROXYDIPHENYLMETHANE see CJU250
2-CHLORO-N-(2-HYDROXYETHYL)ANILINE see CHO125
4-CHLORO-6-(2-HYDROXYETHYLPIPERAZINO-2-METHYLAMINO-5)-METHYLTHIOPYRIMIDINE see CHO250
5-CHLORO-3-(4-(2-HYDROXYETHYL)-1-PIPERAZINYL)CARBONYLMETHYL-2-BENZOTHIAZOLINONE see CJH750
2-CHLORO-10-3-(1-(2-HYDROXYETHYL)-4-PIPERAZINYL)PROPYL PHENOTHIAZINE see CJM250
5-CHLORO-8-HYDROXY-7-IODOQUINOLINE see CHR500
2-CHLORO-4-(HYDROXY MERCURI)PHENOL see CHO750
5-CHLORO-4-HYDROXYMETANILIC ACID see AJH500
2-CHLORO-N-HYDROXYMETHYLACETAMIDE see CHO775

2-CHLORO-9-HYDROXY-9-METHYLCARBOXYLATEFLUORENE see CDT000

3-CHLORO-7-HYDROXY-4-METHYLCOUMARIN BIS(4-CHLOROBUTYL)PHOSPHATE see CHO800

3-CHLORO-7-HYDROXY-4-METHYLCOUMARIN BIS(2-CHLOROETHYL)PHOSPHATE see DFH600

3-CHLORO-7-HYDROXY-4-METHYLCOUMARIN BIS(2,3-DICHLOROPROPYL)PHOSPHATE see CHO850

3-CHLORO-7-HYDROXY-4-METHYL-COUMARIN-O,O-DIETHYL PHOSPHOROTHIOATE see CNU750

3-CHLORO-7-HYDROXY-4-METHYL-COUMARIN-O-ESTER with O,O-DIETHYL PHOSPHOROTHIOATE see CNU750

4-CHLORO-17-β-HYDROXY-17-METHYLESTR-4-EN-3-ONE see CIQ600

(−)-N-((5-CHLORO-8-HYDROXY-3-METHYL-1-OXO-7-ISOCHROMANYL)CARBONYL)-3-PHENYLALANINE see CHP250

6-CHLORO-17-α-HYDROXY-16-α-METHYLPREGNA-4,6-DIENE-3,20-DIONE see CHP375

5-CHLORO-2-((2-HYDROXY-1-NAPHTHALENYL)AZO)-4-METHYLBENZENESULFONIC ACID SODIUM SALT see CMS150

5-CHLORO-2-((2-HYDROXY-1-NAPHTHALENYL)AZO)-4-METHYLBENZENE SULFONIC ACID, BARIUM SALT (2:1) see CHP500

5-CHLORO-2-((2-HYDROXY-1-NAPHTHALENYL)AZO)-4-METHYLBENZENE SULPHONIC ACID, BARIUM SALT see CHP500

5-CHLORO-2-((2-HYDROXY-1-NAPHTHYL)AZO)-p-TOLUENE SULFONIC ACID, BARIUM SALT see CHP500

21-CHLORO-17-HYDROXY-19-NOR-17-α-PREGNA-4,9-DIEN-20-YN-3-ONE see CHP750

7-CHLORO-3-HYDROXY-5-PHENYL-1,3-DIHYDRO-2H-1,4-BENZODIAZEPIN-2-ONE see CFZ000

2-CHLORO-1-(4-HYDROXYPHENYL)ETHANONE see HNE500

(3-CHLORO-4-HYDROXYPHENYL)HYDROXYMERCURY see CHO750

CHLORO(o-HYDROXYPHENYL)MERCURY see CHW675

CHLORO(p-HYDROXYPHENYL)MERCURY see CHW750

N-(4-CHLORO-2-(HYDROXYPHENYLMETHYL)PHENYL)-4-PYRIDINECARBOXAMIDE see CHN100

7-CHLORO-10-(2-HYDROXY-3-PIPERIDINOPROPYL)ISOALLOXAZINE SULFATE see CHQ000

6-CHLORO-17-HYDROXYPREGNA-4,6-DIENE-3,20-DIONE see CDY275

6-CHLORO-17-α-HYDROXYPREGNA-4,6-DIENE-3,20-DIONE ACETATE see CBF250

6-α-CHLORO-17-α-HYDROXYPREGN-4-ENE-3,20-DIONE ACETATE see CEB875

6-CHLORO-17-α-HYDROXY-Δ⁶-PROGESTERONE ACETATE see CBF250

1-(3-CHLORO-2-HYDROXYPROPYL)-5-IODO-2-METHYL-4-NITROIMIDAZOLE see CIN000

1-(3-CHLORO-2-HYDROXYPROPYL)-2-METHYL-5-NITROIMIDAZOLE see OJS000

1-(3-CHLORO-2-HYDROXYPROPYL)-2-NITROIMIDAZOLE see CIQ250

3-CHLORO-2-HYDROXYPROPYL PERCHLORATE see CHQ250

(3-CHLORO-2-HYDROXYPROPYL)TRIMETHYLAMMONIUM CHLORIDE see CHQ300

5-CHLORO-8-HYDROXYQUINOLINE see CLD600

2-CHLORO-HYDROXYTOLUENE see CFD990

6-CHLORO-3-HYDROXYTOLUENE see CFD990

CHLOROHYSSOPIFOLIN A see HOW800

CHLOROHYSSOPIFOLIN C see CHQ400

5-CHLORO-4-(2-IMIDAZOLIN-2-YLAMINO)-2,1,3-BENZOTHIADIAZOLE HYDROCHLORIDE see TGH600

4-CHLOROIMINO-2,5-CYCLOHEXADIENE-1-ONE see CHQ500

4-CHLOROIMINO-2,6-DIBROMO-2,5-CYCLOHEXADIENE-1-ONE see CHQ750

4-CHLOROIMINO-2,6-DICHLORO-2,5-CYCLOHEXADIENE-1-ONE see CHR000

3-CHLOROIMIPRAMINE see CDU750

3-CHLOROIMIPRAMINE HYDROCHLORIDE see CDV000

CHLOROIMIPRAMINE MONOHYDROCHLORIDE see CDV000

CHLOROIN see CLD250

7-CHLORO-4-INDANOL see CDV700

2-CHLORO-6H-INDOLO(2,3-B)QUINOXALINE-6-ACETIC ACID ((3,4-DIMETHOXYPHENYL)METHYLENE)HYDRAZIDE see CHR200

CHLOROIODOACETYLENE see CHR325

CHLOROIODOETHYNE see CHR325

5-CHLORO-7-IODO-8-HYDROXYQUINOLINE see CHR500

3-CHLORO-1-IODOPROPYNE see CHR400

CHLOROIODOQUINE see CHR500

5-CHLORO-7-IODO-8-QUINOLINOL see CHR500

2-CHLOROISOBUTANE see BQR000

2-CHLORO-N-(ISOBUTOXYMETHYL)-2',6'-ACETOXYLIDIDE see IIE200

3'-CHLORO-2-ISOBUTYLAMINO-p-ACETOTOLUIDIDE HYDROCHLORIDE see CHR700

6'-CHLORO-2-(ISOBUTYLAMINO)-o-ACETOTOLUIDIDE HYDROCHLORIDE see CHR750

3'-CHLORO-3-ISOBUTYLAMINO-o-PROPIONOTOLUIDIDE HYDROCHLORIDE see CHR850

α-CHLOROISOBUTYLENE see IKE000

γ-CHLOROISOBUTYLENE see CIU750

p-CHLOROISOBUTYROPHENONE see IOL100

4'-CHLOROISOBUTYROPHENONE see IOL100

2-CHLORO-N-(ISOPROPOXYMETHYL)-2',6'-ACETOXYLIDIDE see CGI200

1-CHLORO-2-ISOPROPOXY-2-PROPANOL see CHS250

2-CHLORO-N-ISOPROPYLACETANILIDE see CHS500

α-CHLORO-N-ISOPROPYLACETANILIDE see CHS500

N-(3-CHLORO-1-ISOPROPYLACETONYL)-p-TOLUENESULFONAMIDE see THH555

2-(α-CHLORO-β-ISOPROPYLAMINE)ETHYLNAPHTHALENE HYDROCHLORIDE see CHT500

6'-CHLORO-2-(ISOPROPYLAMINO)-o-ACETOTOLUIDIDE HYDROCHLORIDE see CHS750

2'-CHLORO-2-(ISOPROPYLAMINO)-6'-METHYLACETANILIDE HYDROCHLORIDE see CHS750

S-4-CHLORO-N-ISOPROPYLCARBANILOLYLMETHYL o,o-DIMETHYL PHOSPHORODITHIOATE see AOT255

S-4-CHLORO-N-ISOPROPYLCARBANILOYLMETHYL-o,o-DIMETHYLPHOSPHORODITHIOATE see AOT255

(β-CHLOROISOPROPYL)DIMETHYLAMINE HYDROCHLORIDE see CGJ280

CHLORO ISOPROPYL KETONE see IJV100

β-CHLORO-N-ISOPROPYL-2-NAPHTHALENEETHYLAMINE HYDROCHLORIDE see CHT500

2-CHLORO-N-ISOPROPYL-N-PHENYLACETAMIDE see CHS500

CHLOROJECT L see CDP250

3-CHLORO-LACTONITRILE see CHU000

7-CHLOROLINCOMYCIN HYDROCHLORIDE see CHU050

CHLOROMADINONE ACETATE see CBF250

CHLOROMAX see CDP250

CHLOROMELAMINE see TNE775

5-CHLORO-2-MERCAPTOANILINE HYDROCHLORIDE see CHU100

S-(6-CHLORO-3-(MERCAPTOMETHYL)-2-BENZOXAZOLINONE)-O,O-DIETHYL PHOSPHORODITHIOATE see BDJ250

(CHLOROMERCURI)BENZENE see PFM500

p-(CHLOROMERCURI)BENZOIC ACID see CHU500

p-CHLOROMERCURIC BENZOIC ACID see CHU500

N-(2-CHLOROMERCURICYCLOHEXYL) HEXANAMIDE, (E)- see CHL000

N-(2-CHLOROMERCURICYCLOHEXYL)PROPIONAMIDE see CIC000

2-(CHLOROMERCURI)-4,6-DINITROPHENOL see CHN750

N-(CHLOROMERCURI)FORMANILIDE see CHU750

2-CHLOROMERCURIFURAN see CHK000

3-(3-CHLOROMERCURI-2-METHOXY-1-PROPYL)-5,5-DIMETHYLHYDANTOIN see CHV250

1-(3-CHLOROMERCURI-2-METHOXY)PROPYLHYDANTOIN see CHV500

1-(3-CHLOROMERCURI-2-METHOXY-1-PROPYL)-HYDANTOIN see CHV500

3-(3-CHLOROMERCURI-2-METHOXY-1-PROPYL)HYDANTOIN see CHV750

3-(3-(CHLOROMERCURI)-2-METHOXYPROPYL)-1-METHYLHYDANTOIN see CHW250

1-(3-CHLOROMERCURI-2-METHOXY-1-PROPYL)-3-METHYLHYDANTOIN see CHW000

3-(3-CHLOROMERCURI-2-METHOXY-1-PROPYL)-1-METHYLHYDANTOIN see CHW250

5-(3-CHLOROMERCURI-2-METHOXY-1-PROPYL)-3-METHYLHYDANTOIN see CHW500

(3-(CHLOROMERCURI)-2-METHOXYPROPYL)UREA see CHX250

1-(3-(CHLOROMERCURI)-2-METHOXYPROPYL)UREA see CHX250

o-CHLOROMERCURIPHENOL see CHW675

p-CHLOROMERCURIPHENOL see CHW750

p-(CHLOROMERCURI)PHENOL see CHW750

3-(CHLOROMERCURI)PYRIDINE see CKW500

1-(3-(CHLOROMERCURY)-2-METHOXYPROPYL)HYDANTOIN see CHV500

CHLOROMERIDIN see CHX250

CHLOROMERODRIN see CHX250

CHLOROMERODRIN see CHX250

CHLOROMETHANE see MIF765

CHLOROMETHANE mixed with DICHLOROMETHANE see CHX750

CHLOROMETHANE SULFONATE d'ETHYLE (FRENCH) see CHC750

CHLOROMETHANE SULFONYL CHLORIDE see CHY000

CHLOROMETHAPYRILENE see CHY250

2-((3-((6-CHLORO-2-METHOXY-9-ACRIDINYL)AMINO))PROPYL)ETHYLAMI NOETHANOL DIHYDROCHLORIDE see CHY750

1-((6-CHLORO-2-METHOXY-9-ACRIDYL)-AMINO)-3-(DIETHYLAMINO)-2-PROPANOL DIHYDROCHLORIDE see ADI750

N-(5-CHLORO-4-METHOXYANTHRAQUINONYL)BENZAMI DE see CIA000

N-(4-(2-(5-CHLORO-2-METHOXYBENZAMIDO)ETHYL)PHENYLS ULFONYL)-N'-CYCLOHEXYLUREA see CEH700

3-CHLORO-4-METHOXY-BENZENAMINE (9CI) see CEH750

CHLORO(trans-2-METHOXYCYCLOOCTYL)MERCURY see CIB500

3-CHLORO-3-METHOXYDIAZIRINE see CIB625

2-CHLORO-N-(2-METHOXY-3,6-DIMETHYLPHENYL)-N-((1-METHYLETHOXY)METHYL)ACETAMIDE see CIB650

5-CHLORO-4-METHOXYDIPHENYLAMINE-2-CARBOXYLIC ACID see CIB700

CHLOROMETHOXY ETHANE see CIM000

2-(2-CHLORO-1-METHOXYETHOXY)PHENOL METHYLCARBAMATE see LAU500

2-CHLORO-N-(2-METHOXYETHYL)ACET-2',6'-XYLIDIDE see DSM500

CHLORO(2-METHOXYETHYL)MERCURY see MEP250

3-CHLORO-7-METHOXY-9-(1-METHYL-4-DIETHYLAMINOBUTYLAMINO)ACRIDIN E see ARQ250

3-CHLORO-7-METHOXY-9-(1-METHYL-4-DIETHYLAMINOBUTYLAMINO)ACRIDIN E DIHYDROCHLORIDE see CFU750

2-CHLORO-N-((4-METHOXY-6-METHYL-1,3,5-TRIAZIN-2-YL)AMINOCARBONYL)-BENZENESULFONAMIDE see CMA700

CHLOROMETHOXYNIL see DGA850

2-CHLORO-1-METHOXY-4-NITROBENZENE (9CI) see CJA200

4-CHLORO-N-(p-METHOXYPHENYL)ANTHRANILIC ACID see CIB700

CHLORO-(4-METHOXYPHENYL)DIAZIRINE see CIB725

3-(3-CHLORO-4-METHOXYPHENYL)-1,1-DIMETHYL-3-NITROSOUREA see NKY500

3-(3-CHLORO-4-METHOXYPHENYL)-1,1-DIMETHYL-UREA see MQR225

N-(3-CHLORO-4-METHOXYPHENYL)-N',N'-DIMETHYLUREA see MQR225

5-CHLORO-2-METHOXYPROCAINAMIDE see AJH000

CHLORO(2-(3-METHOXYPROPIONAMIDO)CYCLOHEXY L)MERCURY see CIC000

7-CHLORO-8-METHOXY-10-(2-PYRROLIDINYLETHYL)ISOALLOXAZINE ACETATE see CIC500

2-CHLORO-6-METHOXY-4-(TRICHLOROMETHYL)PYRIDINE see CIC600

2-CHLORO-4'-METHYLACETANILIDE see CEB500

l-N-(3-CHLORO-1-METHYLACETONYL)-p-TOLUENESULFONAMIDE see THH360

2-CHLORO-N-METHYL-1-ADAMANTANE METHANAMINE HYDROCHLORIDE see CID000

6'-CHLORO-2-(METHYLAMINO)-o-ACETOTOLUIDIDE HYDROCHLORIDE see CID250

5-CHLORO-6-(((((METHYLAMINO)CARBONYL)OXY)IMIN O)BICYCLO(2.2.1)HEPTANE see CFF250

4'-CHLORO-2-((METHYLAMINO)METHYL)BENZHYDRO L HYDROCHLORIDE see CID825

7-CHLORO-2-METHYLAMINO-5-PHENYL-3H-1,4-BENZODIAZEPINE 4-OXIDE see LFK000

7-CHLORO-2-METHYLAMINO-5-PHENYL-3H-1,4-BENZODIAZEPINE-4-OXIDE, mixed with SODIUM NITRITE (1:1) see SIS000

7-CHLORO-2-METHYLAMINO-5-PHENYL-3H-1,4-BENZODIAZEPIN 4-OXIDE see LFK000

7-CHLORO-2-METHYLAMINO-5-PHENYL-3H-1,4-BENZODIAZEPIN, 4-OXIDE, HYDROCHLORIDE see MDQ250

6'-CHLORO-2-(METHYLAMINO)-o-PROPIONOTOLUIDIDE HYDROCHLORIDE see CIE250

6'-CHLORO-3-(METHYLAMINO)-o-PROPIONOTOLUIDIDE HYDROCHLORIDE see CIE500

2-CHLORO-1-(2-METHYLAMINOPROPYL)-3,5,7-TRIMETHYLADAMANTANE HYDROCHLORIDE see CIF000

4-CHLORO-5-(METHYLAMINO)-2-(α,α,α-TRIFLUORO-m-TOLYL)-3(2H)-PYRIDAZINONE see NNQ100

p-CHLORO-N-METHYLAMPHETAMINE see CIF250

d-1-p-CHLORO-METHYLAMPHETAMINE (FRENCH) see CIF250

2-CHLORO-4-METHYLANILINE see CLK210

3-CHLORO-2-METHYLANILINE see CLK200

3-CHLORO-4-METHYLANILINE see CLK215

3-CHLORO-6-METHYLANILINE see CLK225

4-CHLORO-2-METHYLANILINE see CLK220

4-CHLORO-6-METHYLANILINE see CLK220

5-CHLORO-2-METHYLANILINE see CLK225

6-CHLORO-2-METHYLANILINE see CLK227

4-CHLORO-2-METHYLANILINE HYDROCHLORIDE see CLK235

4-CHLORO-6-METHYLANILINE HYDROCHLORIDE see CLK235

o-CHLOROMETHYLANISALDEHYDE see CIF750

2-CHLOROMETHYL-p-ANISALDEHYDE see CIF750

CHLOROMETHYLATED AMINATED STRYENE-DIVINYLBENZENE RESIN see CIF775

7-CHLOROMETHYL BENZ(a)ANTHRACENE see CIG250

8-CHLORO-7-METHYLBENZ(a)ANTHRACENE see CIG000

10-CHLORO-7-METHYLBENZ(a)ANTHRACENE see CIG500

5-CHLORO-10-METHYL-1,2-BENZANTHRACENE see CIG000

7-CHLORO-10-METHYL-1,2-BENZANTHRACENE see CIG500

CHLOROMETHYLBENZENE see BEE375

CHLOROMETHYLBENZENE see CLK130

4-CHLORO-1-METHYLBENZENE see TGY075

2-CHLORO-1-METHYLBENZENE (9CI) see CLK100

4-CHLORO-2-METHYLBENZENEAMINE see CLK220

4-CHLORO-2-METHYLBENZENEAMINE HYDROCHLORIDE see CLK235

4-CHLORO-2-METHYLBENZENEDIAZONIUM SALTS see CIG750

7-CHLORO-1-METHYL-5-3H-1,4-BENZODIAZEPIN-2(1H)-ONE see DCK759

2-(2-(5-CHLORO-2-METHYL-1,3-BENZODIOXOL-2-YL)ETHOXY)-N,N-DIETHYLETHANAMINE see CFO750

6-CHLOROMETHYL BENZO(a)PYRENE see CIH000

7-CHLORO-3-METHYL-2H-1,2,4-BENZOTHIADIAZINE-1,1-DIOXIDE see DCQ700

5-CHLORO-2-METHYLBENZOTHIAZOLE see CIH100

1-(4-CHLORO-2-METHYLBENZYL)-1H-INDAZOLE-3-CARBOXYLIC ACID see TGJ875

4-CHLOROMETHYLBIPHENYL see CIH825

CHLOROMETHYL BISMUTHINE see CIH900

1-CHLORO-3-METHYLBUTANE see CII000

2-CHLORO-2-METHYLBUTANE see CII250

CHLOROMETHYL tert-BUTYL KETONE see MRG070

2-CHLORO-6-METHYLCARBANILIC ACID-2-(DIETHYLAMINO)ETHYL ESTER, HYDROCHLORIDE see CIJ250

2-CHLORO-6-METHYLCARBANILIC ACID-N-METHYL-4-PIPERIDINYL ESTER see CIK250

2-CHLORO-6-METHYLCARBANILIC ACID-2-(PYRROLIDINYL)ETHYL ESTER HYDROCHLORIDE see CIK500

dl-6-CHLORO-α-METHYLCARBAZOLE-2-ACETIC ACID see CCK800

10-CHLOROMETHYL-9-CHLOROANTHRACENE see CFB500

1-(CHLOROMETHYL)-N-((CHLOROMETHYL)DIMETHYLSILYL)-1,1-DIMETHYL-SILANAMINE see BIL250

3-CHLORO-4-METHYL-7-COUMARINYL DIETHYLPHOSPHATE see CIK750

3-CHLORO-4-METHYL-7-COUMARINYL DIETHYL PHOSPHOROTHIOATE see CNU750

O-3-CHLORO-4-METHYL-7-COUMARINYL-O,O-DIETHYL PHOSPHOROTHIOATE see CNU750

CHLOROMETHYL CYANIDE see CDN500

6-CHLORO-16-α-METHYL-Δ⁶-DEHYDRO-17-α-ACETOXYPROGESTERONE see CHP375

3-CHLORO-3-METHYLDIAZIRINE see CIK825

3-(CHLOROMETHYL)-4,4-DICHLORO-2-BUTENOIC ACID see CIK850

o-CHLORO-2-(METHYL(2-(DIETHYLAMINO)ETHYL)AMINO)PROPIO NANILIDE DIHYDROCHLORIDE see CIL000

S-(CHLOROMETHYL)-O,O-DIETHYL PHOSPHORODITHIOATE see CDY299

S-CHLOROMETHYL-O,O-DIETHYL PHOSPHOROTHIOLOTHIOATE see CDY299

S-CHLOROMETHYL-O,O-DIETHYL PHOSPHOROTHIOLOTHIONATE see CDY299

7-CHLORO-2-METHYL-3,3a-DIHYDRO-2H,9H-ISOXAZOLO(3,2-b)(1,3)BENZOXAZIN-9-ONE see CIL500

12-(CHLOROMETHYL)-2-β,3-β-DIHYDROXY-27-NORO-13-ENE-23,28-DIOIC ACID see SBY000

3'-CHLORO-4'-METHYL-4-DIMETHYLAMINOAZOBENZENE see CIL700

4'-CHLORO-3'-METHYL-4-DIMETHYLAMINOAZOBENZENE see CIL710

o-CHLORO-2-(METHYL(2-(DIMETHYLAMINO)ETHYL)AMINO)ACET ANILIDE DIHYDROCHLORIDE see CIL750

2-CHLORO-4-METHYL-6-DIMETHYLAMINOPYRIMIDINE see CCP500

4-(CHLOROMETHYL)-2,2-DIMETHYL-1,3-DIOXA-2-SILACYCLOPENTANE see CIL775

4-(CHLOROMETHYL)-2,2-DIMETHYL-1,3-DIOXOLANE see CIL800

S-(CHLOROMETHYL)-O,O-DIMETHYL PHOSPHORODITHIOIC ACID, ESTER see CDY299

1-(4-CHLOROMETHYL-1,3-DIOXOLAN-2-YL)-2-PROPANONE see CIL850

6-CHLORO-1,2-α-METHYLENE-6-DEHYDRO-17-α-HYDROXYPROGESTERONE ACETATE see CQJ500

6-CHLORO-Δ⁶-1,2-α-METHYLENE-17-α-HYDROXYPROGESTERONE ACETATE see CQJ500

CHLORO-1,2-α-METHYLENE-17-α-HYDROXY-Δ⁶-PROGESTERONE ACETATE see CQJ500

d-2-CHLORO-6-METHYLERGOLINE-8-β-ACETONITRILE METHANESULFONIC ACID SALT see LEP000

1-CHLOROMETHYL-1,2-ETHANEDIOL DIACETATE see CIL900

2-CHLORO-N-METHYL-ETHYLAMINE HYDROCHLORIDE see MIG250

2-CHLORO-1-METHYL-10-ETHYL-7,8-BENZACRIDINE (FRENCH) see EHL000

4-CHLORO-α-(1-METHYLETHYL)BENZENEACETIC ACID, (1,3-DIHYDRO-1,3-DIOXO-2H-ISOINDOL-2-YL) METHYL ESTER see DELETED

4-CHLORO-α-(1-METHYLETHYL)BENZENEACETIC ACID, (2,6-DIMETHYL-4-(2-PROPYNYL)PHENYL) METHYL ESTER see CIL920

1-CHLOROMETHYLETHYLENE GLYCOL CYCLIC SULFITE see CKQ750

(CHLOROMETHYL)ETHYLENE OXIDE see EAZ500

CHLOROMETHYL ETHYL ETHER see CIM000

(2-CHLORO-1-METHYLETHYL) ETHER see BII250

2-CHLORO-2'-METHYL-6'-ETHYL-N-ETHOXYMETHYLACETANILIDE see CIM100

4-CHLORO-β-(1-METHYLETHYL)-α-(3-PHENOXYPHENYL)BENZENEPROPANENITRI LE see DAI100

2-CHLORO-N-(1-METHYLETHYL)-N-PHENYLACETAMIDE see CHS500

O-(5-CHLORO-1-(1-METHYLETHYL)-1H-1,2,4-TRIAZOL-3-YL) O,O-DIETHYL PHOSPHOROTHIOATE see PHK000

7-(CHLOROMETHYL)-5-FLUORO-12-METHYLBENZ(a)ANTHRACENE see FHH025

3-CHLOROMETHYLFURAN see CIM300

3-CHLOROMETHYLHEPTANE see EKU000

3-CHLORO-4-METHYL-7-HYDROXYCOUMARIN DIETHYL THIOPHOSPHORIC ACID ESTER see CNU750

3-CHLORO-4-METHYL-5-HYDROXY-2(5H)-FURANONE see CIM325

N-CHLORO-4-METHYL-2-IMIDAZOLINONE see CIM399

4-CHLORO-1-METHYLIMIDAZOLIUM NITRATE see CIM500

4-CHLORO-N-(2-METHYL-1-INDOLINYL)-3-SULFAMOYLBENZAMIDE see IBV100

α-(CHLOROMETHYL)-5-IODO-2-METHYL-4-NITROIMIDAZOLE-2-ETHANOL see CIN000

5-CHLORO-2-METHYL-4-ISOTHIAZOLIN-3-ONE see CIN100

CHLOROMETHYLMERCURY see MDD750

2'-CHLORO-6'-METHYL-2-(METHYLAMINO)ACETANILIDE HYDROCHLORIDE see CID250

4-CHLORO-N-METHYL-3-((METHYLAMINO)SULFONYL)BENZAMIDE see CIP500

10-CHLOROMETHYL-9-METHYLANTHRACENE see CIN500

7-CHLOROMETHYL-12-METHYL BENZ(a)ANTHRACENE see CIN750

CHLOROMETHYLMETHYLDIETHOXY SILANE see CIO000

CHLOROMETHYL METHYL ETHER see CIO250

4-CHLORO-5-METHYL-2-(1-METHYLETHYL)PHENOL see CLJ800

2-CHLOROMETHYL-5-METHYLFURAN see CIO275

α-(CHLOROMETHYL)-2-METHYL-5-NITRO-1H-IMIDAZOLE-1-ETHANOL see OJS000

4-(CHLOROMETHYL)-2-METHYL-2-PENTYL-1,3-DIOXOLANE see CIO375

2-CHLORO-10-((2-METHYL-3-(4-METHYL-1-PIPERAZINYL)PROPYL)-PHENOTHIAZINE see CIO500

7-CHLORO-1-METHYL-4-((p-((1-METHYLPYRIDINIUM-4-YL)AMINO)PHENYL)CARBAMOYL)ANILINO)QUINOLINIUM DIBROMIDE see CIO750

6-CHLORO-1-METHYL-4-(p-((p-((1-METHYLPYRIDINIUM-4-YL)AMINO)PHENYL)CARBAMOYL)ANILINO)QUINOLINIUM, DI-p-TOLUENESULFONATE see CIP000

8-CHLORO-1-METHYL-4-((p-((p-((1-METHYLPYRIDINIUM-4-YL)AMINO)PHENYL)CARBAMOYL)ANILINO)QUINOLINIUM) DI-p-TOLUENESULFONATE see CIP250

4-CHLORO-N-METHYL-3-(METHYLSULFAMOYL)BENZAMIDE see CIP500

3'-CHLORO-α-(METHYL((MORPHOLINOCARBONYL)METHYL)AMINO)-o-BENZOTOLUIDIDE HYDROCHLORIDE see FMU000

3'-CHLORO-2'-(N-METHYL-N-((MORPHOLINOCARBONYL)METHYL)AMINOMETHYL)BENZANILIDE HYDROCHLORIDE see FMU000

3'-CHLORO-α-(N-METHYL-N-((MORPHOLINOCARBONYL)METHYL)AMINOMETHYL)BENZANILIDE HYDROCHLORIDE see FMU000

1-CHLOROMETHYL NAPHTHALENE see CIP750

α-CHLOROMETHYLNAPHTHALENE see CIP750

2-(8-CHLOROMETHYL-1-NAPHTHYLTHIO)ACETIC ACID see CIQ000

4-(CHLOROMETHYL)NITROBENZENE see NFN400

p-(CHLOROMETHYL)NITROBENZENE see NFN400

1-(CHLOROMETHYL)-2-NITROBENZENE see NFN500

1-(CHLOROMETHYL)-3-NITROBENZENE see NFN010

1-(CHLOROMETHYL)-4-NITROBENZENE see NFN400

5-CHLORO-3-METHYL-2-NITROBENZOFURAN see NHN550

5-CHLORO-N-METHYL-4-NITROIMIDAZOLE see CIQ100

α-(CHLOROMETHYL)-2-NITROIMIDAZOLE-2-ETHANOL see CIQ250

4-(CHLOROMETHYL)-2-(o-NITROPHENYL)-1,3-DIOXOLANE see CIQ400

2-CHLORO-2-METHYL-N-NITROSOETHANAMINE see CIQ500

2-CHLORO-N-METHYL-N-NITROSOETHYLAMINE see CIQ500

4-CHLORO-17-α-METHYL-19-NORTESTOSTERONE see CIQ600

N-CHLORO-5-METHYL-2-OXAZOLIDINONE see CIQ625

CHLOROMETHYLOXIRANE see EAZ500

2-(CHLOROMETHYL)OXIRANE see EAZ500

cis-2-CHLORO-3-METHYLOXIRANE see CKS099

trans-2-CHLORO-3-METHYLOXIRANE see CKS100

2-CHLORO-N-METHYL-3-OXOBUTANAMIDE see CEA100

(Z)-2-CHLORO-3-METHYL-4-OXOBUTENOIC ACID see CIQ650

7-CHLORO-1-METHYL-2-OXO-5-PHENYL-3H-1,4-BENZODIAZEPINE see DCK759

2-CHLORO-5-METHYLPHENOL see CFE500

4-CHLORO-2-METHYLPHENOL see CIR000

4-CHLORO-3-METHYLPHENOL see CFD990

(4-CHLORO-2-METHYLPHENOXY)ACETIC ACID see CIR250

4-CHLORO-2-METHYLPHENOXYACETIC ACID, ETHYL ESTER see EMR000

(4-CHLORO-2-METHYLPHENOXY)ACETIC ACID POTASSIUM SALT see MIH800

4-CHLORO-2-METHYLPHENOXYACETIC ACID SODIUM SALT see SIL500

1-(CHLOROMETHYL)-3-PHENOXYBENZENE see PDR650

4-(CHLORO-2-METHYLPHENOXY)BUTANOIC ACID see CLN750

4-(4-CHLORO-2-METHYLPHENOXY)BUTANOIC ACID, SODIUM SALT see CLO000

4-(4-CHLORO-2-METHYLPHENOXY)BUTYRIC ACID see CLN750

γ-(4-CHLORO-2-METHYLPHENOXY)BUTYRIC ACID see CLN750

CHLOROMETHYLPHENOXYBUTYRIC ACID SODIUM SALT see CLO000

4-(4-CHLORO-2-METHYLPHENOXY)BUTYRIC ACID SODIUM SALT see CLO000

1-(2-CHLORO-5-METHYLPHENOXY)-3-((1,1-DIMETHYLETHYL)AMINO)-2-PROPANOL HYDROCHLORIDE see BQB250

2-(4-CHLORO-2-METHYLPHENOXY)-N,N-DIMETHYLPROPIONAMIDE see CIR275

2-(4-CHLORO-2-METHYLPHENOXY)PROPANOIC ACID (R) (9CI) see CIR325

2-(4-CHLORO-2-METHYLPHENOXY)PROPIONIC ACID see CIR500

4-CHLORO-2-METHYLPHENOXY-α-PROPIONIC ACID see CIR500

(+)-α-(4-CHLORO-2-METHYLPHENOXY) PROPIONIC ACID see CIR500

8-CHLORO-4-(2-METHYLPHENOXY)QUINOLINE see CIR600

N-(3-CHLORO-2-METHYLPHENYL)ANTHRANILIC ACID see CLK325

7-CHLORO-N-METHYL-5-PHENYL-3H-1,4-BENZODIAZEPIN-2-AMINE-4-OXIDE see LFK000

7-CHLORO-1-METHYL-5-PHENYL-1H-1,5-BENZODIAZEPINE-2,4(3H,5H)-DIONE see CIR750

7-CHLORO-1-METHYL-5-PHENYL-2H-1,4-BENZODIAZEPIN-2-ONE see DCK759

2-((p-CHLORO-α-METHYL-α-PHENYLBENZYL)OXY)-N,N-

m-CHLORONITROBENZENE see CJB250
CHLORO-o-NITROBENZENE see CJB750
o-CHLORONITROBENZENE see CJB750
p-CHLORONITROBENZENE see NFS525
1-CHLORO-2-NITROBENZENE see CJB750
1-CHLORO-3-NITROBENZENE see CJB250
1-CHLORO-4-NITROBENZENE see NFS525
2-CHLORO-1-NITROBENZENE see CJB750
4-CHLORO-1-NITROBENZENE see NFS525
m-CHLORONITROBENZENE (DOT) see CJB250
o-CHLORONITROBENZENE, liquid (DOT) see CJB750
CHLORONITROBENZENE, ortho, liquid (DOT) see CJA950
2-CHLORO-4-NITROBENZENEAZO-2'-AMINO-4'-METHOXY-5'-METHYLBENZENE see MIH500
2-CHLORO-5-NITRO-1,4-BENZENEDIAMINE see ALM100
2-CHLORO-5-NITROBENZENESULFONIC ACID see CJB825
2-CHLORO-3,5-NITROBENZENESULFONIC ACID, SODIUM SALT see CJC000
2-CHLORO-4-NITROBENZOIC ACID see CJC250
4-CHLORO-3-NITROBENZOIC ACID see CJC500
2-CHLORO-5-NITROBENZOIC ACID METHYL ESTER see CJC515
4-CHLORO-7-NITROBENZO-2-OXA-1,3-DIAZOLE see NFS550
2-CHLORO-5-NITROBENZYL ALCOHOL see CJC549
6-CHLORO-2-NITROBENZYL BROMIDE see CJC600
2-CHLORO-4-NITROBENZYL CHLORIDE see CJC610
4-CHLORO-2-NITROBENZYL CHLORIDE see CJC625
2-CHLORO-2-NITROBUTANE see CJC750
3-CHLORO-4-(2'-NITRO-3'-CHLOROPHENYL)PYRROLE see CFC000
1-CHLORO-1-NITROETHANE see CJC800
CHLORONITROMETHANE see CJC900
2-((o-CHLORO-α-(NITROMETHYL)BENZYL)THIO)ETHYLAMINE HYDROCHLORIDE see NEC000
2-CHLORO-4-NITROPHENOL see CJD250
2-CHLORO-6-NITRO-3-PHENOXYANILINE see CJD300
2-CHLORO-6-NITRO-3-PHENOXYBENZENAMINE see CJD300
1-CHLORO-4-(4-NITROPHENOXY)-2-(PROPYLTHIO)BENZENE see CJD400
2-CHLORO-1-(4-NITROPHENOXY)-4-TRIFLUOROMETHYLBENZENE see NGB600
2-CHLORO-4-NITROPHENYLAMIDE-6-CHLOROSALICYLIC ACID see DFV400
1-((2-CHLORO-4-NITROPHENYL)AZO)-2-NAPHTHOL see CJD500
4-CHLORO-2-NITROPHENYL p-CHLOROPHENYL ETHER see CJD600
N-(2-CHLORO-4-NITROPHENYL)-5-CHLOROSALICYLAMIDE see DFV400
O-(2-CHLORO-4-NITROPHENYL) O,O-DIMETHYL PHOSPHOROTHIOATE see NFT000
O-(3-CHLORO-4-NITROPHENYL) O,O-DIMETHYL PHOSPHOROTHIOATE see MIJ250
2-CHLORO-5-NITROPHENYL ESTER ACETIC ACID see CJD625
(E)-1-CHLORO-3-(2-(4-NITROPHENYL)ETHENYL)BENZENE see CJG600
(E)-1-CHLORO-4-(2-(4-NITROPHENYL)ETHENYL)BENZENE see CJG610

1-(4-CHLORO-3-NITROPHENYL)-2-ETHOXY-2-((4-(METHYLTHIO)PHENYL)AMINO)ETHANONE see CJD630
o-(2-CHLORO-4-NITROPHENYL)-o-ISOPROPYL ETHYLPHOSPHONOTHIOATE see CJD650
CHLORONITROPROPAN (POLISH) see CJD750
CHLORONITROPROPANE see CJD750
CHLORONITROPROPANE see CJE000
1-CHLORO-1-NITROPROPANE see CJE000
1-CHLORO-2-NITROPROPANE see CJD750
2-CHLORO-2-NITROPROPANE see CJE250
3-CHLORO-4-NITROQUINOLINE-1-OXIDE see CJE500
5-CHLORO-4-NITROQUINOLINE-1-OXIDE see CJE750
6-CHLORO-4-NITROQUINOLINE-1-OXIDE see CJF000
7-CHLORO-4-NITROQUINOLINE-1-OXIDE see CJF250
3'-CHLORO-5-NITROSALICYLANILIDE see CJF500
1-CHLORO-1-NITROSOCYCLOHEXANE see CJF825
2-CHLORO-1-NITROSO-2-PHENYLPROPANE see CJG000
3-CHLORONITROSOPIPERIDINE see CJG375
4-CHLORONITROSOPIPERIDINE see CJG500
3-CHLORO-1-NITROSOPIPERIDINE see CJG375
4-CHLORO-1-NITROSOPIPERIDINE see CJG500
(E)-3-CHLORO-4'-NITROSTILBENE see CJG600
(E)-4-CHLORO-4'-NITROSTILBENE see CJG610
α-CHLORO-p-NITROSTYRENE see CJG750
2-CHLORO-4-NITROTOLUENE see CJG800
2-CHLORO-6-NITROTOLUENE see CJG825
6-CHLORO-2-NITROTOLUENE see CJG825
α-CHLORO-m-NITROTOLUENE see NFN010
α-CHLORO-o-NITROTOLUENE see NFN500
α-CHLORO-p-NITROTOLUENE see NFN400
4-CHLORO-3-NITRO-α,α,α-TRIFLUOROTOLUENE see NFS700
CHLORONIUM PERCHLORATE see CJH250
6-CHLORO-N(sup 2),N(sup 2),N(sup 4)-TRIETHYL-1,3,5-TRIAZINE-2,4-DIAMINE (IUPAC) see TJL500
9-CHLORONONANOIC ACID see CJH500
CHLOROOXIRANE see CGX000
(3-CHLORO-4-(OXIRANYLMETHOXY)PHENYL)ACETIC ACID see AFI950
5-CHLORO-1-(1-(3-(2-OXO-1-BENZIMIDAZOLINYL)PROPYL)-4-PIPERIDYL)-2-BENZIMIDAZOLINONE see DYB875
4-CHLORO-2-OXO-3(2H)-BENZOTHIAZOLEACETIC ACID see BAV000
4-CHLORO-2-OXOBENZOTHIAZOLIN-3-YL ACETIC ACID see BAV000
4-((5-CHLORO-2-OXO-3(2H)-BENZOTHIAZOLYL)ACETYL)-1-PIPERAZINEETHANOL see CJH750
3-(6-CHLORO-2-OXOBENZOXAZOLIN-3-YL)METHYL-O,O-DIETHYL PHOSPHOROTHIOLOTHIONATE see BDJ250
N-(4-CHLORO-3-OXOBUTYL)-p-TOLUENESULFONAMIDE see THH355
exo-5-CHLORO-6-OXO-endo-2-NORBORNANECARBONITRILE-o-(METHYLCARBAMOYL)OXIME see CFF250
S-((6-CHLORO-2-OXOOXAZOLO(4,5-B)PYRIDIN-3(2H)-YL)METHYL) O,O-

DIMETHYLPHOSPHOROTHIOATE see ARY800
4-CHLORO-3-OXO-N-PHENYLBUTANAMIDE see CEA050
9-CHLORO-5-OXO-7-(1H-TETRAZOL-5-YL)-5H-1-BENZOPYRANO(2,3-b)PYRIDINE SODIUM SALT PENTAHYDRATE see THK850
CHLOROPARACIDE see CEP000
CHLOROPARAFFINE 40G see PAH780
CHLOROPARAFFIN XP-470 see CJI000
CHLORO-PDMT see CJI100
CHLOROPENTABROMOCYCLOHEXANE see CPB550
CHLOROPENTAFLUOROACETONE HYDRATE see CJI250
CHLOROPENTAFLUOROBENZENE see PBE100
CHLOROPENTAFLUOROETHANE see CJI500
3-CHLORO-1,1,1,2,2-PENTAFLUOROPROPANE see CJI525
3-CHLOROPENTAFLUOROPROPENE see CJI550
3-CHLORO-1,1,2,3,3-PENTAFLUORO-1-PROPENE see CJI550
CHLOROPENTAHYDROXYDIALUMINUM see AHA000
1-CHLOROPENTANE see PBW500
1-CHLORO-3-(PENTYLOXY)-2-PROPANOL see CHS250
CHLOROPEPTIDE see CJI609
CHLOROPERALGONIC ACID see CJH500
3-CHLOROPERBENZOIC ACID see CJI750
m-CHLOROPERBENZOIC ACID see CJI750
CHLOROPERFLUOROBENZENE see PBE100
3-CHLOROPEROXYBENZOIC ACID see CJI750
m-CHLOROPEROXYBENZOIC ACID see CJI750
CHLOROPEROXYL see CDW450
CHLOROPEROXYTRIFLUOROMETHANE see CJI809
CHLOROPHACINONE see CJJ000
CHLOROPHEN see PAX250
p-CHLOROPHENACYL BROMIDE see CJJ100
4'-CHLOROPHENACYL BROMIDE see CJJ100
CHLOROPHENACYLE see HNE500
CHLOROPHENAMADIN see CJJ250
CHLOROPHENAMIDINE see CJJ250
CHLOROPHENAMIDINE HYDROCHLORIDE see CJJ500
CHLOROPHENE see CJU250
4-CHLOROPHENE-1,3-DIAMINE see CJY120
1-(p-CHLOROPHENETHYL)-6,7-DIMETHOXY-2-METHYL-1,2,3,4-TETRAHYDROISOQUINOLINE see MDV000
1-(p-CHLOROPHENETHYL)HYDRAZINE HYDROGEN SULFATE see CJK000
1-(p-CHLOROPHENETHYL)-2-METHYL-6,7-DIMETHOXY-1,2,3,4-TETRAHYDROISOQUINOLINE see MDV000
1-(p-CHLOROPHENETHYL)-1, 2,3,4-TETRAHYDRO-6,7-DIMETHOXY-2-METHYLISOQUINOLINE see MDV000
2-(o-CHLOROPHENETHYL)-3-THIOSEMICARBAZIDE see CJK100
4-CHLOROPHENIRAMINE see CLX300
2-CHLOROPHENOL see CJK250
3-CHLOROPHENOL see CJK500
4-CHLOROPHENOL see CJK750
m-CHLOROPHENOL see CJK500
o-CHLOROPHENOL see CJK250
p-CHLOROPHENOL see CJK750
o-CHLOROPHENOL, solid see CJK250
o-CHLOROPHENOL, liquid see CJK250
CHLOROPHENOLS see CJL000

CHLOROPHENOTHAN see DAD200
CHLOROPHENOTHANE see DAD200
2-CHLOROPHENOTHIAZINE see CJL100
2-(2-(4-(2-((2-CHLORO-10-
PHENOTHIAZINYL)METHYL)PROPYL)-1-
PIPERAZINYL)ETHOXY)ETHANOL see
CJL409
1-(3-(3-CHLOROPHENOTHIAZIN-10-
YL)PROPYL)-ISONIPECOTAMIDE see
CJL500
4-(3-(2-CHLOROPHENOTHIAZIN-10-
YL)PROPYL)-1-PIPERAZINEETHANOL see
CJM250
4-(3-(2-CHLOROPHENOTHIAZIN-10-
YL)PROPYL)-1-PIPERAZINEETHANOL
DIHYDROCHLORIDE see PCO500
4-(3-(2-CHLOROPHENOTHIAZIN-10-
YL)PROPYL)-1-PIPERAZINEETHANOL
MALEATE see PCO850
8-(3-(2-CHLOROPHENOTHIAZIN-10-
YL))PROPYL-1-THIA-4,8-
DIAZASPIRO(4.5)DECAN-3-ONE
HYDROCHLORIDE see SLB000
(3-(2-CHLOROPHENOTHIAZIN-10-
YL)PROPYL)TRIMETHYLAMMONIUM
IODIDE see MIJ300
CHLOROPHENOTOXUM see DAD200
10-CHLOROPHENOXARSINE see CJM750
10-CHLORO-10H-PHENOXARSINE see
CJM750
(4-CHLOROPHENOXY)ACETIC ACID see
CJN000
p-CHLOROPHENOXYACETIC ACID see
CJN000
p-CHLOROPHENOXYACETIC ACID-β-
DIMETHYLAMINOETHYL ESTER see
DPE000
(p-CHLOROPHENOXY)ACETIC ACID 2-
(DIMETHYLAMINO)ETHYL ESTER
HYDROCHLORIDE see AAE500
p-CHLOROPHENOXYACETIC ACID-2-
ISOPROPYLHYDRAZIDE see CJN250
1-(p-CHLOROPHENOXYACETYL)-2-
ISOPROPYL HYDRAZINE see CJN250
4-(4-CHLOROPHENOXY)ANILINE see
CEH125
p-CHLOROPHENOXYANILINE see
CEH125
10-CHLOROPHENOXYARSINE see CJM750
4-(4-CHLOROPHENOXY)-BENZENAMINE
(9CI) see CEH125
2-(4-CHLOROPHENOXY)-1-tert-BUTYL-2-
(1H-1,2,4-TRIAZOLE-1-YL)ETHANOL see
CJN300
2-(4-CHLOROPHENOXY)-N-(2-
(DIETHYLAMINO)ETHYL)ACETAMIDE see
CJN750
2-(p-CHLOROPHENOXY)-N-(2-
(DIETHYLAMINO)ETHYL)ACETAMIDE see
CJN750
2-(p-CHLOROPHENOXY)-N-(2-
(DIETHYLAMINO)ETHYL ACETAMIDE
COMPOUND with 4-BUTYL-1,2-DIPHENYL-
3,5-PYRAZOLIDINEDIONE (1:1) see
CMW750
(p-CHLOROPHENOXY)DIMETHYL-
ACETIC ACID see CMX000
2-(p-CHLOROPHENOXY)-N-(2-
(DIMETHYLAMINO)ETHYL)ACETAMIDE
see CJN800
β-(4-CHLOROPHENOXY)-α-(1,1-
DIMETHYLETHYL)-1H-1,2,4-TRIAZOLE-1-
ETHANOL see CJN300
1-(p-CHLOROPHENOXY)-3,3-DIMETHYL-1-
(1-IMIDAZOLYL)-2-BUTANONE see CJN900
(1RS,2RS,1RS,2SR)-1-(4-CHLOROPHENOXY)-
3,3-DIMETHYL-1-(1H-1,2,4-TRIAZOL-1-
YL)BUTAN-2-OL (IUPAC) see CJN300
1-(4-CHLOROPHENOXY)-3,3-DIMETHYL-1-
(1,2,4-TRIAZOL-1-YL)-2-BUTAN-2-ONE see
CJO250

1-(4-CHLOROPHENOXY)-3,3-DIMETHYL-1-
(1H-1,2,4-TRIAZOL-1-YL)-2-BUTANONE see
CJO250
2-(4-CHLOROPHENOXY)ETHANOL see
CJO500
2-(p-CHLOROPHENOXY)ETHANOL see
CJO500
(2-(p-
CHLOROPHENOXY)ETHYL)HYDRAZINE
HYDROCHLORIDE see CJP250
1-(2-(o-
CHLOROPHENOXY)ETHYL)HYDRAZINE
HYDROGEN SULFATE see CJP500
3-(p-CHLOROPHENOXY)-2-
HYDROXYPROPYL CARBAMATE see
CJQ250
1-(4-CHLOROPHENOXY)-1-(IMIDAZOL-1-
YL)-3,3-DIMETHYLBUTANONE see CJN900
CHLORO-4 PHENOXYISOBUTYRATE
D'HYDROXY-4 N-
DIMETHYLBUTYRAMIDE (FRENCH) see
LGK200
α-p-CHLOROPHENOXYISOBUTYRIC
ACID see CMX000
α-p-CHLOROPHENOXYISOBUTYRYL
ETHYL ESTER see ARQ750
N-2(p-CHLOROPHENOXY)ISOBUTYRYL-
N'-MORPHOLINOMETHYLUREA see PJB500
((4-
CHLOROPHENOXY)METHYL)OXIRANE
see CKA200
2-((4-
CHLOROPHENOXY)METHYL)OXIRANE
see CKA200
2-(4-CHLOROPHENOXY)-2-
METHYLPROPANOIC ACID see CMX000
2-(4-CHLOROPHENOXY)-2-
METHYLPROPANOIC ACID, 3,4-
DIHYDRO-2,5,7,8-TETRAMETHYL-2-(4,8,12-
TRIMETHYLTRIDECYL)-2H-1-
BENZOPYRAN-6-YL-ESTER, (2r(4r,8r)) see
TGJ000
2-(4-CHLOROPHENOXY)-2-
METHYLPROPANOIC ACID ETHYL ESTER
see ARQ750
2-(4-CHLOROPHENOXY)-2-
METHYLPROPANOIC ACID-1,3-
PROPANEDIYL ESTER see SDY500
2-(4-CHLOROPHENOXY-2-
METHYL)PROPIONIC ACID see CIR500
2-(p-CHLOROPHENOXY)-2-
METHYLPROPIONIC ACID see CMX000
2-(4-CHLOROPHENOXY)-2-METHYL-
PROPIONIC ACID 4-(DIMETHYLAMINO)-
4-OXOBUTYL ESTER (9CI) see LGK200
2-(p-CHLOROPHENOXY)-2-
METHYLPROPIONIC ACID ETHYL ESTER
see ARQ750
2-(p-CHLOROPHENOXY)-2-
METHYLPROPIONIC ACID
TRIMETHYLENE ESTER see SDY500
N-2(p-CHLOROPHENOXY)-2-
METHYLPROPIONYL-N'-
MORPHOLINOMETHYLUREA see PJB500
2-(4-(4-
CHLOROPHENOXY)PHENOXY)PROPIONI
C ACID see CJP750
3-(p-(p-CHLOROPHENOXY)PHENYL)-1,1-
DIMETHYLUREA see CJQ000
N'-4-(4-CHLOROPHENOXY)PHENYL-N,N-
DIMETHYLUREA see CJQ000
1-(4-(4-CHLORO-PHENOXY)PHENYL)-3,3-
D'METHYLUREE (FRENCH) see CJQ000
3-(4-CHLOROPHENOXY)-1,2-
PROPANEDIOL-1-CARBAMATE see CJQ250
3-(p-CHLOROPHENOXY)-1,2-
PROPANEDIOL-1-CARBAMATE see CJQ250
2-(3-CHLOROPHENOXY)PROPANOIC
ACID see CJQ300
2-(3-CHLOROPHENOXY)PROPIONIC ACID
see CJQ300

2-(m-CHLOROPHENOXY)PROPIONIC
ACID see CJQ300
N-(1-(o-CHLOROPHENOXY)-2-PROPYL)-2-
(DIETHYLAMINO)-N-ETHYLACETAMIDE
HYDROCHLORIDE see CJQ500
N-(1-(o-CHLOROPHENOXY)-2-PROPYL)-2-
(DIETHYLAMINO)-N-
METHYLACETAMIDE HYDROCHLORIDE
see CJQ750
7-(2-(4-(3-CHLOROPHENOXY)-3-
((TETRAHYDRO-2H-PYRAN-2-YL)OXY)-1-
BUTENYL) see CJQ800
CHLOROPHENTERMINE see CLY250
CHLOROPHENTERMINE
HYDROCHLORIDE see ARW750
2-CHLORO-N-PHENYLACETAMIDE see
PER300
N-(4-CHLOROPHENYL)ACETAMIDE see
CDZ100
N-(2-
CHLOROPHENYL)ACETOACETAMIDE see
AAY600
(4-CHLOROPHENYL)ACETONITRILE see
CEP300
p-CHLOROPHENYLACETONITRILE see
CEP300
2-(4-CHLOROPHENYL)ACETONITRILE see
CEP300
2-CHLORO-2-PHENYLACETOPHENONE
see CFJ100
2-(α-p-CHLOROPHENYLACETYL)INDANE-
1,3-DIONE see CJJ000
4-CHLOROPHENYLALANINE see CJR125
p-CHLOROPHENYLALANINE see CJR125
3-(p-CHLOROPHENYL)ALANINE see
CJR125
dl-4-CHLOROPHENYLALANINE see
FAM100
dl-p-CHLOROPHENYLALANINE see
FAM100
p-CHLORO-dl-PHENYLALANINE see
CJR125
3-CHLOROPHENYLAMINE see CEH675
4-CHLOROPHENYLAMINE see CEH680
m-CHLOROPHENYLAMINE see CEH675
4-CHLOROPHENYLAMINE
HYDROCHLORIDE see CJR200
p-CHLOROPHENYLAMINE
HYDROCHLORIDE see CJR200
β-(p-CHLOROPHENYL)-γ-AMINOBUTYRIC
ACID see BAC275
N-(((4-
CHLOROPHENYL)AMINO)CARBONYL)-
2,6-DIFLUOROBENZAMIDE see CJV250
4-((3-((4-CHLOROPHENYL)AMINO)-4,5-
DIHYDRO-2H-BENZ(G)INDAZOL-2-
YL)ACETYL)MORPHOLINE see CJR210
3-((4-CHLOROPHENYL)AMINO)-4,5-
DIHYDRO-N-(1-METHYLETHYL)-2H-
BENZ(g)INDAZOLE-2-ACETAMIDE see
CJR215
3-((4-CHLOROPHENYL)AMINO)-4,5-
DIHYDRO-N-(PHENYLMETHYL)-2H-
BENZ(g)INDAZOLE-2-ACETAMIDE see
CJR220
3-((4-CHLOROPHENYL)AMINO)-N-(2-
ETHOXYETHYL)-4,5-DIHYDRO-2H-
BENZ(G)INDAZOLE-2-ACETAMIDE see
CJR230
N-(3-CHLOROPHENYL)-1-
AZIRIDINECARBOXAMIDE see CJR250
O-(p-(p-CHLOROPHENYLAZO)PHENYL)
O,O-DIMETHYL PHOSPHOROTHIOATE
see CJR300
4-CHLOROPHENYL
BENZENESULFONATE see CJR500
p-CHLOROPHENYL
BENZENESULFONATE see CJR500
4-CHLOROPHENYL
BENZENESULPHONATE see CJR500
p-CHLOROPHENYL
BENZENESULPHONATE see CJR500

1-(α-(2-CHLOROPHENYL)BENZHYDRYL)IMIDAZOLE see MRX500

2-(o-CHLOROPHENYL)BENZIMIDAZOLE see CJR550

2-(2-CHLOROPHENYL)-1H-BENZIMIDAZOLE see CJR550

2-(4-CHLOROPHENYL)-1H-BENZ(de)ISOQUINOLINE-1,3(2H)-DIONE see CJR580

1-(p-CHLORO-α-PHENYLBENZYL)HEXAHYDRO-4-METHYL-1H-1,4-DIAZEPINE DIHYDROCHLORIDE see CJR809

1-((p-CHLORO-α-PHENYLBENZYL)HEXAHYDRO-4-METHYL)-1H-1,4-DIAZEPINE HYDROCHLORIDE see HGI200

1-(p-CHLORO-α-PHENYLBENZYL)-4-(2-((2-HYDROXYETHOXY)ETHYL)PIPERAZINE) see CJR909

1-(p-CHLORO-α-PHENYLBENZYL)-4-(m-METHYLBENZYL)PIPERAZINE see HGC500

1-(p-CHLORO-α-PHENYLBENZYL)-4-(m-METHYLBENZYL)PIPERAZINE HYDROCHLORIDE see MBX500

1-(p-CHLORO-α-PHENYLBENZYL)-4-METHYLPIPERAZINE see CFF500

1-(p-CHLORO-α-PHENYLBENZYL)-4-METHYL-PIPERAZINE DIHYDROCHLORIDE see CDR000

1-(p-CHLORO-α-PHENYLBENZYL)-4-METHYLPIPERAZINE HYDROCHLORIDE see CDR500

2-(α-(p-CHLOROPHENYL)BENZYLOXY)-N,N-DIMETHYLETHYLAMINE HYDROCHLORIDE see CJR959

3-α-((p-CHLORO-α-PHENYLBENZYL)OXY)-1-α-H,5-α-H-TROPANE HYDROCHLORIDE see CMB125

1-(4-CHLORO-α-PHENYLBENZYL)PIPERAZINE see NNK500

1-(α-(4-CHLOROPHENYL)BENZYL)-PIPERAZINE HYDROCHLORIDE see NNL000

1-(p-CHLORO-α-PHENYLBENZYL)-PIPERAZINE HYDROCHLORIDE see NNL000

2-(2-(4-(p-CHLORO-α-PHENYLBENZYL)-1-PIPERAZINYL)ETHOXY)ETHANOL see CJR909

2-(2-(4-(p-CHLORO-α-PHENYLBENZYL)-1-PIPERAZINYL)ETHOXY)ETHANOLDIHYDROCHLORIDE see HOR470

2-(2-(2-(4-(p-CHLORO-α-PHENYLBENZYL)-1-PIPERAZINYL)ETHOXY)ETHOXY)ETHANOL DIMALEATE see HHK100

1-(4-CHLOROPHENYL)BIGUANIDINIUM HYDROGEN DICHROMATE see CJT125

1-(o-CHLOROPHENYL)-2-tert-BUTYLAMINO ETHANOL HYDROCHLORIDE see BQE250

N-(3-CHLORO PHENYL) CARBAMATE de 4-CHLORO 2-BUTYNYLE (FRENCH) see CEW500

N-(3-CHLORO PHENYL) CARBAMATE D'ISOPROPYLE (FRENCH) see CKC000

(3-CHLOROPHENYL)CARBAMIC ACID 4-CHLORO-2-BUTYNYL ESTER see CEW500

N-(3-CHLOROPHENYL)CARBAMIC ACID, ISOPROPYL ESTER see CKC000

(3-CHLOROPHENYL)CARBAMIC ACID, 1-METHYLETHYL ESTER see CKC000

3-CHLOROPHENYLCARBAMIC ACID-1-METHYLPROPYNYL ESTER see CEX250

3-CHLOROPHENYL-N-CARBAMOYLAZIRIDINE see CJR250

p-CHLOROPHENYL CHLORIDE see DEP800

4-CHLOROPHENYL-4-CHLOROBENZENESULFONATE see CJT750

p-CHLOROPHENYL-p-CHLOROBENZENE SULFONATE see CJT750

4-CHLOROPHENYL-4-CHLOROBENZENESULPHONATE see CJT750

4-CHLOROPHENYL-4'-CHLOROBENZYL SULFIDE see CEP000

2-(o-CHLOROPHENYL)-2-(p-CHLOROPHENYL)-1,1-DICHLOROETHANE see CDN000

2-(4-CHLOROPHENYL)-5-((4-CHLOROPHENYL)METHOXY)-4-IODO-3(2H)-PYRIDAZINONE see CJT800

1-(o-CHLOROPHENYL)-4-(3-(p-CHLOROPHENYL)-3-PHENYLPROPIONYL)PIPERAZINE see CKI020

(2-CHLOROPHENYL)-α-(4-CHLOROPHENYL)-5-PYRIMIDINEMETHANOL see FAK100

α-(2-CHLOROPHENYL)-α-(4-CHLOROPHENYL)-5-PYRIMIDINEMETHANOL see FAK100

p-CHLOROPHENYL-N-(4'-CHLOROPHENYL)THIOCARBAMATE see CJU125

2-(4-CHLOROPHENYL)-5-((6-CHLORO-3-PYRIDINYL)METHOXY)-4-IODO-3(2H)-PYRIDAZINONE see CJU150

4-CHLORO-α-PHENYLCRESOL see BEF750

4-CHLORO-α-PHENYL-o-CRESOL see CJU250

trans-5-(4-CHLOROPHENYL)-N-CYCLOHEXYL-4-METHYL-2-OXO-3-THIAZOLIDINECARBOXAMIDE see CJU275

α-(4-CHLOROPHENYL)-α-(1-CYCLOPROPYLETHYL)-1H-1,2,4-TRIAZOLE-1-ETHANOL see CQJ150

(2RS,3RS. 2RS,3SR)-2-(4-CHLORO-PHENYL)-3-CYCLOPROPYL-1-(1H-1,2,4-TRIAZOL-1-YL)BUTAN-2-OL see CQJ150

1-(p-CHLOROPHENYL)-4,6-DIAMINO-2,2-DIMETHYL-1,2-DIHYDRO-s-TRIAZINE see COX400

5-(4'-CHLOROPHENYL)-2,4-DIAMINO-6-ETHYLPYRIMIDINE see TGD000

1-(3-CHLOROPHENYL)-2-((4-(DICHLOROACETYL)PHENYL)AMINO)-2-HYDROXYETHANONE see CJU300

(o-CHLOROPHENYL)(3-(2,4-DICHLOROPHENOXY)-2-HYDROXYPROPYL)HYDROXYARSINEOXIDE see DGA425

N-(4-CHLOROPHENYL)-N'-(3,4-DICHLOROPHENYL)UREA see TIL500

3-(4-CHLOROPHENYL)-4',5-DICHLOROSALICYLANILIDE see TIP750

N-(3-CHLOROPHENYL)DIETHANOLAMINE see CJU400

N-(m-CHLOROPHENYL)DIETHANOLAMINE see CJU400

2-p-CHLOROPHENYL-1-(p-(2-DIETHYLAMINOETHOXY)PHENYL)-1-p-TOLYLETHANOL see TMP500

2-(p-CHLOROPHENYL)-1-(p-(β-DIETHYLAMINOETHOXY)PHENYL)-1-(p-TOLYL)ETHANOL see TMP500

1-(4-CHLOROPHENYL)-3-(2,6-DIFLUOROBENZOYL)UREA see CJV250

1-p-CHLOROPHENYL-1,2-DIHYDRO-2,2-DIMETHYL-4,6-DIAMINO-s-TRIAZINE see COX400

1-(4-CHLOROPHENYL)-1,6-DIHYDRO-6,6-DIMETHYL-1,3,5-TRIAZINE-2,4-DIAMINE MONOHYDROCHLORIDE see COX325

5-p-CHLOROPHENYL-2,3-DIHYDRO-5H-IMIDAZO(2,1-A)ISOINDOL-5-OL see MBV250

5-(o-CHLOROPHENYL)-1,3-DIHYDRO-7-NITRO-2H-1,4-BENZODIAZEPIN-2-ONE see CMW000

(4-CHLOROPHENYL)(DIMETHOXYPHOSPHINYL)METHYL PHOSPHORIC ACID DIMETHYL ESTER see CMU875

1-(3-CHLOROPHENYL)-3,N,N-DIMETHYLCARBAMOYL-5-METHOXYPYRAZOLE see CJW500

1-(m-CHLOROPHENYL)-3-N,N-DIMETHYLCARBAMOYL-5-METHOXYPYRAZOLE see CJW500

1-(4-CHLOROPHENYL)-2,3-DIMETHYL-4-DIMETHYLAMINO-2-BUTANOL see CMW459

1-p-CHLOROPHENYL-2,3-DIMETHYL-4-DIMETHYLAMINO-2-BUTANOL HYDROCHLORIDE see CMW500

β-(p-CHLOROPHENYL)-α,α-DIMETHYLETHYLAMINE see CLY250

2-(4-CHLOROPHENYL)-1,1-DIMETHYLETHYL 2-AMINOPROPANOATE HYDROCHLORIDE see CJX000

(+)-1-(3-CHLOROPHENYL)-2-((1,1-DIMETHYLETHYL)AMINO)1-PROPANONE HYDROCHLORIDE (9CI) see WBJ500

N-(2-(4-CHLOROPHENYL)-1,1-DIMETHYLETHYL)-2-(DIETHYLAMINO)-PROPANAMIDE HYDROCHLORIDE see CGI125

N-(4-CHLOROPHENYL)-2,2-DIMETHYLPENTANAMIDE see CGL250

S-(p-CHLOROPHENYL)-O,O-DIMETHYL PHOSPHOTHIOATE see FOR000

1-(4-CHLOROPHENYL)-3,3-DIMETHYLTRIAZENE see CJI100

1-(p-CHLOROPHENYL)-3,3-DIMETHYL-TRIAZENE see CJI100

(RS)-1-(4-CHLOROPHENYL)-4,4-DIMETHYL-3-(1H-1,2,4-TRIAZOL-1-YLMETHYL)PENTAN-3-OL see TAN050

(RS)-1-p-CHLOROPHENYL-4,4-DIMETHYL-3-(1H-1,2,4-TRIAZOL-1-YLMETHYL)PENTAN-3-OL see TAN050

(E)-1-(4-CHLOROPHENYL)-4,4-DIMETHYL-2-(1,2,4-TRIAZOL-1-YL)-1-PENTEN-3-OL see SOU800

1-(p-CHLOROPHENYL)-3,3-DIMETHYLUREA see CJX750

3-(4-CHLOROPHENYL)-1,1-DIMETHYLUREA see CJX750

3-(p-CHLOROPHENYL)-1,1-DIMETHYLUREA see CJX750

N'-(4-CHLOROPHENYL)-N,N-DIMETHYLUREA see CJX750

N-(p-CHLOROPHENYL)-N',N'-DIMETHYLUREA see CJX750

3-(p-CHLOROPHENYL)-1,1-DIMETHYLUREA TRICHLOROACETATE see CJY000

3-(p-CHLOROPHENYL)-1,1-DIMETHYLUREA compounded with TRICHLOROACETIC ACID (1:1) see CJY000

1-(4-CHLORO PHENYL)-3,3-DIMETHYLUREE (FRENCH) see CJX750

N-(4-CHLOROPHENYL)-2,2-DIMETHYLVALEROAMIDE see CGL250

1-((2-CHLOROPHENYL)DIPHENYLMETHYL)-1H-IMIDAZOLE see MRX500

2-CHLORO-p-PHENYLENEDIAMINE see CEG600

4-CHLORO-m-PHENYLENEDIAMINE see CJY120

4-CHLORO-o-PHENYLENEDIAMINE see CFK125

(4-CHLOROPHENYL)METHYL DODECYL 3-PYRIDINYLCARBONIMIDODITHIOATE see CKF040

N²)-((4-CHLOROPHENYL)METHYLENE)-4-NITRO-N¹)-PHENYL-1,2-BENZENEDIAMINE see NFS800

(4-CHLOROPHENYL)METHYL 1-ETHYL-1-METHYLPROPYL-3-PYRIDINYLCARBONIMIDODITHIOATE see CKF043

S-((4-CHLOROPHENYL)METHYL) o-ETHYL 3-PYRIDINYLCARBONIMIDOTHIOATE see CKF057

(4-CHLOROPHENYL)METHYL HEPTYL 3-PYRIDINYLCARBONIMIDODITHIOATE see CKF063

(4-CHLOROPHENYL)METHYL HEXADECYL 3-PYRIDINYLCARBONIMIDODITHIOATE see CKF100

(4-CHLOROPHENYL)METHYL HEXYL 3-PYRIDINYLCARBONIMIDODITHIOATE see CKF150

1-((4-CHLOROPHENYL)METHYL)-1H-INDOLE-3-CARBOXYLIC ACID see CEQ625

6-(3-(o-CHLOROPHENYL)-5-METHYL-4-ISOXAZOLECARBOXAMIDEO)-3,3-DIMETHYL-7-OXO-4-THIA-1-AZABICYCLO(3.2.0)HEPTANE-2-CARBOXYLIC ACID, SODIUM SALT, MONOHYDRATE see SLJ000

6-(3-(o-CHLOROPHENYL)-5-METHYL-4-ISOXAZOLECARBOXAMIDEO)-3,3-DIMETHYL-7-OXO-4-THIA-1-AZABICYCLO(3.2.0)HEPTANE-2-CARBOXYLIC ACID, MONOSODIUM SALT see SLJ050

3-o-CHLOROPHENYL-5-METHYL-4-ISOXAZOLYLPENICILLIN SODIUM see SLJ050

2-(4-CHLOROPHENYL)-3-METHYL-4-METATHIAZANONE-1,1-DIOXIDE see CKF500

1-((4-CHLOROPHENYL)METHYL)-2-METHYL-1H-BENZIMIDAZOLE (9CI) see CDY325

(4-CHLOROPHENYL)METHYL 1-METHYLETHYL 3-PYRIDINYLCARBONIMIDODITHIOATE see CKF530

(4-CHLOROPHENYL)METHYL 1-METHYLPROPYL 3-PYRIDINYLCARBONIMIDODITHIOATE see CKF535

o-((4-CHLOROPHENYL)METHYL) S-(2-METHYLPROPYL)-3-PYRIDINYLCARBONIMIDOTHIOATE see CKF540

S-((4-CHLOROPHENYL)METHYL) o-(2-METHYLPROPYL)-3-PYRIDINYLCARBONIMIDOTHIOATE see CKF545

3-(p-CHLOROPHENYL)-1-METHYL-1-(1-METHYL-2-PROPYNYL)UREA see CKF750

N'-(4-CHLOROPHENYL)-N-METHYL-N-(1-METHYL-2-PROPYNYL)-UREA see CKF750

(2-CHLOROPHENYL)METHYL METHYL 3-PYRIDINYLCARBONIMIDODITHIOATE see CKF760

1-((4-CHLOROPHENYL)METHYL)-2-(NITROMETHYLENE)IMIDAZOLIDINE see CKF765

2-CHLOROPHENYL METHYLNITROSOCARBAMATE see MMV250

1-(p-CHLOROPHENYL)-3-METHYL-3-NITROSOUREA see MMW775

3-(p-CHLOROPHENYL)-1-METHYL-1-NITROSOUREA see MMW775

(4-CHLOROPHENYL)METHYL OCTADECYL 3-

PYRIDINYLCARBONIMIDODITHIOATE see CKF767

(4-CHLOROPHENYL)METHYL OCTYL 3-PYRIDINYLCARBONIMIDODITHIOATE see CKF770

1-((4-CHLOROPHENYL)METHYL)-5-OXO-l-PROLINE see CKF800

1-((4-CHLOROPHENYL)METHYL)-5-OXO-l-PROLINE AMMONIUM SALT see CKF810

1-((4-CHLOROPHENYL)METHYL)-5-OXO-L-PROLINE METHYL ESTER see CKF850

2-(p-CHLOROPHENYL)-4-METHYLPENTANE-2,4-DIOL see CKG000

2-(p-CHLOROPHENYL)-4-METHYL-2,4-PENTANEDIOL see CKG000

4-CHLORO-2-(PHENYLMETHYL)PHENOL see CJU250

4-CHLORO-2-(PHENYLMETHYL)PHENOL SODIUM SALT see SFB200

3-CHLORO-N-(PHENYLMETHYL)PROPANAMIDE see BEG000

1-(4-CHLOROPHENYL)-2-METHYL-1-PROPANONE see IOL100

1-((2-(4-CHLOROPHENYL)-2-METHYLPROPOXY)METHYL)-3-(3-METHYLPHENOXY)BENZENE see MNQ600

1-(3-CHLOROPHENYL)-4-(2-(5-METHYL-1H-PYRAZOL-3-YL)ETHYL)PIPERAZINE DIHYDROCHLORIDE see MCH535

3-(o-CHLOROPHENYL)-2-METHYL-4(3H)-QUINAZOLINONE see MIH925

3-(o-CHLOROPHENYL)-2-METHYL-4-QUINAZOLONE see MIH925

p-CHLOROPHENYL METHYL SULFIDE see CKG500

1-(((4-CHLOROPHENYL)METHYL)SULFINYL)-N,N-DIETHYLFORMAMIDE see FMY050

4-CHLOROPHENYL METHYL SULFONE see CKG750

p-CHLOROPHENYL METHYL SULFONE see CKG750

4-CHLOROPHENYL METHYL SULFOXIDE see CKH000

p-CHLOROPHENYL METHYL SULFOXIDE see CKH000

(4-CHLOROPHENYL)METHYL TETRADECYL 3-PYRIDINYLCARBONIMIDODITHIOATE see CKH100

(±)-1-(2-(((4-CHLOROPHENYL)METHYL)THIO)-2-(2,4-DICHLOROPHENYL)ETHYL)-1H-IMIDAZOLE MONONITRATE see SNH480

6-CHLORO-3-(((PHENYLMETHYL)THIO)METHYL)-2H-1,2,4,-BENZOTHIADIAZINE-7-SULFONAMIDE DIOXIDE see BDE250

N-4-CHLOROPHENYL-N⁵-ISOPROPYLDIGUANIDE HYDROCHLORIDE see CKB500

5-(o-CHLOROPHENYL)-7-NITRO-1H-1,4-BENZODIAZEPIN-2(3H)-ONE see CMW000

4-(4-CHLOROPHENYL)-6H-1,3,5-OXATHIAZINE see CKH120

(m-CHLOROPHENYL)OXIRANE see CLF000

(3-CHLOROPHENYL)OXIRANE (9CI) see CLF000

N-(4-CHLOROPHENYL)-3-OXOBUTANAMIDE see AAY250

1-p-CHLOROPHENYL PENTYL SUCCINATE see CKI000

β-(p-CHLOROPHENYL)PHENETHYL 4-(o-CHLOROPHENYL)PIPERAZINYL KETONE see CKI020

β-(p-CHLOROPHENYL)PHENETHYL 4-(2-HYDROXYPROPYL)PIPERAZINYL KETONE see CKI180

β-(p-CHLOROPHENYL)PHENETHYL 4-(o-METHOXYPHENYL)PIPERAZINYL KETONE see CKI185

β-(p-CHLOROPHENYL)PHENETHYL 4-(m-METHYLBENZYL)PIPERAZINYL KETONE see CKI030

β-(p-CHLOROPHENYL)PHENETHYL 4-PHENETHYLPIPERAZINYL KETONE see CKI040

β-(p-CHLOROPHENYL)PHENETHYL 4-(2-PYRIDYL)PIPERAZINYL KETONE see CKI050

β-(p-CHLOROPHENYL)PHENETHYL 4-(2-PYRIMIDYL)PIPERAZINYL KETONE see CKI060

β-(p-CHLOROPHENYL)PHENETHYL 4-(2-THIAZOLYL)PIPERAZINYL KETONE see CKI070

β-(p-CHLOROPHENYL)PHENETHYL 4-(m-TOLYL)PIPERAZINYL KETONE see CKI080

β-(p-CHLOROPHENYL)PHENETHYL 4-(o-TOLYL)PIPERAZINYL KETONE see CKI090

β-(p-CHLOROPHENYL)PHENETHYL 4-(p-TOLYL)PIPERAZINYL KETONE see CKI190

2-CHLORO-4-PHENYLPHENOL see CHN500

6-CHLORO-2-PHENYLPHENOL, SODIUM SALT see SFV500

2-((p-CHLOROPHENYL)PHENYLACETYL)-1,3-INDANDIONE see CJJ000

2(2-(4-CHLOROPHENYL)-2-PHENYLACETYL)INDAN-1,3-DIONE see CJJ000

2-((4-CHLOROPHENYL)PHENYLACETYL)-1H-INDENE-1,3(2H)-DIONE see CJJ000

1-o-CHLOROPHENYL-1-PHENYL-3-DIMETHYLAMINO-1-PROPANOL HYDROCHLORIDE see CMW700

2-(1-(4-CHLOROPHENYL)-1-PHENYLETHOXY)-N,N-DIMETHYLETHANAMINE HYDROCHLORIDE see CIS000

(1-(p-CHLOROPHENYL)-1-PHENYL)ETHYL (β-DIMETHYLAMINOETHYL) ETHER HYDROCHLORIDE see CIS000

1-(2-((4-CHLOROPHENYL)PHENYLMETHOXY)ETHYL)PIPERIDINE (9CI) see CMX800

1-((4-CHLOROPHENYL)PHENYLMETHYL)HEXAHYDRO-4-METHYL-1H-1,4-DIAZEPINE HYDROCHLORIDE (9CI) see HGI200

9-((4-CHLOROPHENYL)PHENYLMETHYL)3-OXA-9-AZA-6-AZONIASPIRO(5.5)UNDECANE CHLORIDE HCl see CEK875

8-(4-(4-CHLOROPHENYLPHENYLMETHYL)PIPERAZINYL)-3,6-DIOXAOCTANOL see HHK050

α-(p-CHLOROPHENYL)-α-PHENYL-2-PIPERIDINEMETHANOL HYDROCHLORIDE see CKI175

3-(2-CHLOROPHENYL)-1-PHENYL-2-PROPEN-1-ONE see CLF150

3-(4-CHLOROPHENYL)-1-PHENYL-2-PROPEN-1-ONE see CLF100

1-(3-(p-CHLOROPHENYL)-3-PHENYLPROPIONYL)-4-(2-HYDROXYPROPYL)PIPERAZINE see CKI180

1-(3-(p-CHLOROPHENYL)-3-PHENYLPROPIONYL)-4-(o-METHOXYPHENYL)PIPERAZINE see CKI185

1-(3-(p-CHLOROPHENYL)-3-PHENYLPROPIONYL)-4-(m-METHYLBENZYL)PIPERAZINE see CKI030

1-(3-(p-CHLOROPHENYL)-3-PHENYLPROPIONYL)-4-PHENETHYLPIPERAZINE see CKI040

p-CHLOROPHENYL-2,4,5-TRICHLOROPHENYL SULFONE see CKM000
p-CHLOROPHENYL-2,4,5-TRICHLOROPHENYL SULPHONE see CKM000
CHLOROPHENYLTRICHLOROSILANE see CKM250
1-(4-CHLOROPHENYL)-2,2,2-TRIFLUOROETHANONE o-(1,3-DIOXOLAN-2-YLMETHYL)OXIME see FMR600
(p-CHLOROPHENYL)TRIFLUOROMETHANE see CEM825
1-(p-CHLOROPHENYL)-4-(3,4,5-TRIMETHOXYBENZOYL)PIPERAZINE see CKJ100
1-(p-CHLOROPHENYL)-2,8,9-TRIOXA-5-AZA-1-SILABICYCLO(3.3.3) UNDECANE see CKM750
3-β-(4-CHLOROPHENYL)TROPANE-2-β-CARBOXYLIC ACID PHENYL ESTER HYDROCHLORIDE see CKM800
CHLOROPHIBRINIC ACID see CMX000
CHLOROPHOS see TIQ250
CHLORO-PHOSPHONOTHIOIC ACID-O,O-DIETHYL ESTER see DJW600
CHLOROPHOSPHORIC ACID DIETHYL ESTER see DIY000
S-(2-CHLORO-1-PHTHALIMIDOETHYL)-O,O-DIETHYL PHOSPHORODITHIOATE see DBI099
CHLOROPHTHALM see TIQ250
CHLOROPHYLL see CKN000
CHLOROPHYLLIN see CKN250
CHLOROPHYLLIN A see CKN250
CHLOROPHYLLS see CKN000
6-CHLOROPICOLINIC ACID see CKN375
CHLOROPICRIN see CKN500
CHLOROPICRIN, liquid (DOT) see CKN500
CHLOROPICRIN, ABSORBED (DOT) see CKN500
CHLOROPICRINE (FRENCH) see CKN500
α-CHLOROPINACOLIN see MRG070
α-CHLOROPINACOLINE see MRG070
1-CHLOROPINACOLONE see MRG070
2-CHLORO-11-(1-PIPERAZINYL)DIBENZ(b,f)(1,4)OXAZEPINE see AOA095
1-CHLOROPIPERIDINE see CKN675
N-CHLOROPIPERIDINE see CKN675
6'-CHLORO-2-PIPERIDINO-o-ACETOTOLUIDIDE HYDROCHLORIDE see CKN750
6'-CHLORO-3-(PIPERIDINO)-o-PROPIONOTOLUIDIDE HYDROCHLORIDE see CKO000
CHLOROPIRIL see CLX300
CHLOROPLATINIC ACID see CKO750
CHLOROPLATINIC(IV) ACID see CKO750
CHLOROPOTASSURIL see PLA500
CHLOROPREEN (DUTCH) see NCI500
6-CHLORO-Δ⁴·⁶-PREGNADIENE-17-α-OL-3,20-DIONE-17-ACETATE see CBF250
6-CHLORO-PREGNA-4,6-DIEN-17-α-OL-3,20-DIONE ACETATE see CBF250
CHLOROPREN (GERMAN, POLISH) see NCI500
CHLOROPRENE see NCI500
3-CHLOROPRENE see AGB250
CHLOROPRENE, inhibited (DOT) see NCI500
CHLOROPRENE, uninhibited (DOT) see NCI500
CHLOROPRENE MONOEPOXIDE see CKO800
β-CHLOROPRENE (OSHA, MAK) see NCI500
CHLOROPRENE POLYMER see PJQ050
CHLOROPROCAINE see DHT300
CHLOROPROMAZINE see CKP250
CHLOROPROMAZINE HYDROCHLORIDE see CKP500

CHLOROPROMAZINE METHOIODIDE see MIJ300
CHLOROPROMAZINE MONOHYDROCHLORIDE see CKP500
CHLOROPROMURITE see DEQ000
CHLOROPROPAMIDE see CKK000
2-CHLOROPROPANAL see CKP700
2-CHLOROPROPANAL see CKP720
3-CHLOROPROPANAL see CKP700
α-CHLOROPROPANAL see CKP700
1-CHLOROPROPANE see CKP750
2-CHLOROPROPANE see CKQ000
CHLOROPROPANE DIOL-1,3 see CKQ250
CHLORO-1,3-PROPANEDIOL see CKQ250
1-CHLOROPROPANE-2,3-DIOL see CDT750
1-CHLORO-2,3-PROPANEDIOL see CDT750
3-CHLOROPROPANE-1,2-DIOL see CDT750
3-CHLORO-1,2-PROPANEDIOL see CDT750
(±)-3-CHLORO-1,2-PROPANEDIOL see CHL875
dl-3-CHLORO-1,2-PROPANEDIOL see CHL875
3-CHLORO-1,2-PROPANEDIOL 1-BENZOATE see CKQ500
CHLOROPROPANEDIOL CYCLIC SULFITE see CKQ750
1-CHLORO-2,3-PROPANEDIOL DIACETATE see CIL900
3-CHLOROPROPANENITRILE see CKT250
3-CHLORO-1-PROPANESULFONIC ACID, MONOSODIUM SALT see CKP720
2-CHLOROPROPANOIC ACID SODIUM SALT see CKT100
2-CHLOROPROPANOL see CKR500
3-CHLOROPROPANOL see CKP600
2-CHLORO-1-PROPANOL see CKR500
1-CHLORO-2-PROPANOL with 2-CHLORO-1-PROPANOL see CKR750
CHLOROPROPANONE see CDN200
1-CHLORO-2-PROPANONE see CDN200
3-CHLOROPROPANONITRILE see CKT250
7-CHLORO-1-PROPARGYL-5-PHENYL-2H-1,4-BENZODIAZEPIN-2-ONE see PIH100
2-CHLORO-2-PROPENAL see CED800
2-CHLOROPROPENALDEHYDE see CED800
1-CHLOROPROPENE see PMR750
3-CHLOROPROPENE see AGB250
1-CHLORO-1-PROPENE see PMR750
1-CHLORO PROPENE-2 see AGB250
1-CHLORO-2-PROPENE see AGB250
2-CHLORO-1-PROPENE see CKS000
3-CHLORO-1-PROPENE see AGB250
2-CHLOROPROPENE (DOT) see CKS000
cis-1-CHLOROPROPENE OXIDE see CKS099
trans-1-CHLOROPROPENE OXIDE see CKS100
2-CHLORO-2-PROPENE-1-THIOL DIETHYLDITHIOCARBAMATE see CDO250
2-CHLORO-2-PROPENOIC ACID METHYL ESTER (9CI) see MIF800
cis-3-CHLOROPROPENOIC ACID, SODIUM SALT see SFV250
2-CHLORO-2-PROPEN-1-OL see CEF250
3-CHLORO-2-PROPEN-1-OL see CEF500
2-CHLORO-2-PROPENYL DIETHYLCARBAMODITHIOATE see CDO250
2-CHLORO-4-(2-PROPENYLOXY)BENZENEACETIC ACID see AGN000
2-CHLORO-2-PROPENYL TRIFLUOROMETHANE SULFONATE see CKS325
CHLOROPROPHAM see CKC000
CHLOROPROPHENPYRIDAMINE see CLX300
CHLOROPROPHENYLPYRIDAMINE MALEATE see TAI500
2-CHLOROPROPIONALDEHYDE see CKP700

α-CHLOROPROPIONALDEHYDE see CKP700
2-CHLOROPROPIONATE SODIUM SALT see CKT100
3-CHLOROPROPIONIC ACID see CKS500
α-CHLOROPROPIONIC ACID see CKS750
β-CHLOROPROPIONIC ACID see CKS500
2-CHLOROPROPIONIC ACID METHYL ESTER see CKT000
2-CHLOROPROPIONIC ACID SODIUM SALT see CKT100
α-CHLOROPROPIONIC ACID SODIUM SALT see CKT100
3-CHLOROPROPIONITRILE see CKT250
β-CHLOROPROPIONITRILE see CKT250
N-(3-CHLOROPROPIONYL)BENZYLAMINE see BEG000
p-CHLOROPROPIOPHENONE see CKT500
2-CHLOROPROPYL ALCOHOL see CKR500
α-CHLOROPROPYLALDEHYDE see CKP700
6'-CHLORO-2-(PROPYLAMINO)-o-ACETOTOLUIDIDE HYDROCHLORIDE see CKT750
6'-CHLORO-2-(PROPYLAMINO)-o-BUTYROTOLUIDIDE HYDROCHLORIDE see CKU000
4-CHLORO-4-((PROPYLAMINO)CARBONYL)BENZENESULFONAMIDE see CKK000
9-((2-((2-((2-CHLOROPROPYL)AMINO)ETHYL)AMINO)-2-METHOXYACRIDINE DIHYDROCHLORIDE HEMIHYDRATE see CKU250
2-CHLORO-4-(2-PROPYLAMINO)-6-ETHYLAMINO-s-TRIAZINE see ARQ725
CHLOROPROPYLATE see PNH750
3-CHLOROPROPYL BROMIDE see BNA825
4-CHLORO-N-(4-PROPYLCYCLOHEXYL)BENZAMIDE see CKU300
3-CHLOROPROPYLENE see AGB250
3-CHLORO-1-PROPYLENE see AGB250
α-CHLOROPROPYLENE see AGB250
1-CHLORO-2,3-PROPYLENE DINITRATE see CKU625
3-CHLOROPROPYLENE GYLCOL see CDT750
CHLOROPROPYLENE OXIDE see EAZ500
3-CHLORO-1,2-PROPYLENE OXIDE see EAZ500
γ-CHLOROPROPYLENE OXIDE see EAZ500
N-(3-CHLOROPROPYL)-α-METHYLPHENETHYLAMINE HYDROCHLORIDE see PKS500
3-CHLOROPROPYL-n-OCTYLSULFOXIDE see CKU750
(3-CHLOROPROPYL)TRIMETHOXYSILANE see TLC300
Δ-CHLOROPROPYLTRIMETHOXYSILANE see TLC300
(γ-CHLOROPROPYL)TRIMETHOXYSILANE see TLC300
3-CHLOROPROPYNE see CKV275
1-CHLORO-2-PROPYNE see CKV250
CHLOROPROPYNENITRILE see CFE750
CHLOROPROTHIXENE see TAF675
CHLOROPTIC see CDP250
6-CHLORO-1H-PURIN-2-AMINE see CHK300
6-CHLOROPURINE see CKV500
6-CHLORO-9H-PURINE see CKV500
6-CHLORO-1H-PURINE (9CI) see CKV500
CHLOROPYRAMINE see CKV625
CHLOROPYRIBENZAMINE see CKV625
2-CHLOROPYRIDINE see CKW000
3-CHLOROPYRIDINE see CKW250
m-CHLOROPYRIDINE see CKW250
o-CHLOROPYRIDINE see CKW000

α-CHLOROPYRIDINE see CKW000
6-CHLORO-2-PYRIDINECARBOXYLIC ACID see CKN375
2-CHLOROPYRIDINE-N-OXIDE see CKW325
5-((6-CHLORO-3-PYRIDINYL)METHOXY)-2-(3,4-DICHLOROPHENYL)-4-IODO-3(2H)-PYRIDAZINONE see CKW330
(1-((6-CHLORO-3-PYRIDINYL)METHYL)-4,5-DIHYDRO-1H-IMIDAZOL-2-YL)CYANAMIDE see CKW335
N-((6-CHLORO-3-PYRIDINYL)METHYL)-1,2-ETHANEDIAMINE see CKW340
((6-CHLORO-3-PYRIDINYL)METHYL)GUANIDINE see CKW345
1-((6-CHLORO-3-PYRIDINYL)METHYL)HEXAHYDRO-2-(NITROMETHYLENE)PYRIMIDINE see CKW350
1-((6-CHLORO-3-PYRIDINYL)METHYL)-2-IMIDAZOLIDINONE see CKW355
1-((6-CHLORO-3-PYRIDINYL)METHYL)-2-IMIDAZOLIDINONE HYDRAZONE see CKW360
(1-((6-CHLORO-3-PYRIDINYL)METHYL)-2-IMIDAZOLIDINYLIDENE)ACETONITRILE see CKW365
1-((6-CHLORO-3-PYRIDINYL)METHYL)-3-METHYL-N-NITRO-2-IMIDAZOLIDINIMINE see CKW370
N-((6-CHLORO-3-PYRIDINYL)METHYL)-N'-NITROGUANIDINE see CKW380
1-((6-CHLORO-3-PYRIDINYL)METHYL)-N-NITRO-1H-IMIDAZOL-2-AMINE see CKW385
1-((6-CHLORO-3-PYRIDINYL)METHYL)-N-NITRO-2-IMIDAZOLIDINIMINE see CKW400
(3-((6-CHLORO-3-PYRIDINYL)METHYL)-2-THIAZOLIDINYLIDENE)CYANAMIDE see CKW410
((6-CHLORO-3-PYRIDINYL)METHYL)UREA see CKW435
2-(4-(3-(3-CHLORO-10H-PYRIDO(3,2-b)-1,4-BENZOTHIAZINE-10-YL)PROPYL))-1-PIPERAZINYLETHANOL see CMY535
2-(4-(3-(3-CHLORO-10H-PYRIDO(3,2-b)(1,4)BENZOTHIAZIN-1-OYL)PROPYL)-1-PIPERAZINYL) ETHANOL see CMY535
4-(3-(3-CHLORO-10H-PYRIDO(3,2-b)(1,4)-BENZOTHIAZIN-10-YL)PROPYL)-1-PIPERAZINE ETHANOL see CMY535
CHLORO-3-PYRIDYLMERCURY see CKW500
1-(6-CHLORO-3-PYRIDYLMETHYL)-N-NITROIMIDAZOLIDIN-3-YLIDENEAMINE see CKW400
4-(2-(6-CHLORO-2-PYRIDYL)THIO)ETHYL)MORPHOLINE MONOHYDROCHLORIDE see FMU225
CHLOROPYRILENE see CHY250
1-CHLORO-2,5-PYRROLIDINEDIONE see SND500
4-CHLORO-N-((1-PYRROLIDINYLAMINO)CARBONYL)BENZENESULFONAMIDE (9CI) see GHR609
6'-CHLORO-2-(PYRROLIDINYL)-o-DIACETOTOLUIDIDE HYDROCHLORIDE see CLB250
6'-CHLORO-2-PYRROLIDINYL-o-HEXANOTOLUIDIDE HYDROCHLORIDE see CAB250
4'-CHLORO-2-PYRROLIDINYL-α,α,α-TRIFLUORO-m-ACETOTOLUIDIDE, HYDROCHLORIDE see CLC125
6'-CHLORO-2-PYRROLIDINYL-α,α,α-TRIFLUORO-m-ACETOTOLUIDINE, HYDROCHLORIDE see CLC100
2'-CHLORO-2-PYRROLIDINYL-5'-TRIFLUOROMETHYLACETANILIDE HYDROCHLORIDE see CLC100

4'-CHLORO-2-PYRROLIDINYL-3'-TRIFLUOROMETHYLACETANILIDE HYDROCHLORIDE see CLC125
3-CHLORO-4-(3-PYRROLIN-1-YL)HYDRATROPIC ACID see PJA220
CHLOROQUINALDOL see CLC500
6-CHLORO-4-QUINAZOLINONE see CLC750
6-CHLORO-4(3H)-QUINAZOLINONE see CLC750
CHLOROQUINE see CLD000
CHLOROQUINE DIHYDROCHLORIDE see CLD100
CHLOROQUINE DIPHOSPHATE see CLD250
CHLOROQUINE MUSTARD see CLD500
CHLOROQUINE PHOSPHATE see AQT250
CHLOROQUINE PHOSPHATE see CLD250
CHLOROQUINIUM see CLD000
5-CHLORO-8-QUINOLINOL see CLD600
4-((7-CHLORO-4-QUINOLINYL)AMINO)-2-(1-PYRROLIDINYLMETHYL)PHENOL DIHYDROCHLORIDE see PMY000
N⁴-(7-CHLORO-4-QUINOLINYL)-N¹,N¹-DIETHYL-1,4-PENTANEDIAMINE see CLD000
2-(((4-(7-CHLORO-4-QUINOLYL)AMINO)PENTYL)-ETHYLAMINO)ETHANOL see PJB750
4-((7-CHLORO-4-QUINOLYL)AMINO)-α-1-PYRROLIDINYL-o-CRESOL DIHYDROCHLORIDE see PMY000
2-(4-((6-CHLORO-2-QUINOXALINYL)OXY)PHENOXY)PROPANOIC ACID ETHYL ESTER see QMA100
4-CHLORORESORCINOL see CLD750
7-CHLORO-3-β-d-RIBOFURANOSYL-3H-IMIDAZO(4,5-b)PYRIDINE see AEB500
CHLOROS see SHU500
5-CHLOROSALICYLALDEHYDE see CLD800
5-CHLOROSALICYLIC ACID see CLD825
CHLORO-S.C.T.Z. see CHD750
CHLOROSILANE see DGK300
CHLOROSILANES see CLE250
3-CHLORO-4-STILBENAMINE see CLE500
2'-CHLORO-4-STILBENYL-N,N-DIMETHYLAMINE see CGK500
3'-CHLORO-4-STILBENYL-N,N-DIMETHYLAMINE see CGK750
4'-CHLORO-4-STILBENYL-N,N-DIMETHYLAMINE see CGL000
CHLOROSTOP see PKQ059
CHLOROSTYRENE see CLE600
o-CHLOROSTYRENE see CLE750
3-CHLOROSTYRENE OXIDE see CLF000
m-CHLOROSTYRENE OXIDE see CLF000
4-CHLOROSTYRYL PHENYL KETONE see CLF100
o-CHLOROSTYRYL PHENYL KETONE see CLF150
p-CHLOROSTYRYL PHENYL KETONE see CLF100
CHLOROSULFACIDE see CEP000
N-(4'-CHLORO-3'-SULFAMOYLBENZENESULFONYL)-N-METHYL-2-AMINOMETHYL-2-METHYLTETRAHYDROFURAN see MCA100
6-CHLORO-7-SULFAMOYL-2H-1,2,4-BENZOTHIADIAZINE-1,1-DIOXIDE see CLH750
6-CHLORO-7-SULFAMOYL-3,4-DIHYDRO-2H-1,2,4-BENZOTHIADIAZINE-1,1-DIOXIDE see CFY000
4-CHLORO-5-SULFAMOYL-2',6'-SALICYLOXYLIDIDE see CLF325
3-CHLORO-6-SULFANILAMIDOPYRIDAZINE see SNH900
N-CHLOROSULFINYLIMIDE see CLF500
4-CHLORO-4'-(6-SULFO-2H-NAPHTHO(1,2-d)TRIAZOL-2-YL)-2,2'-

STILBENEDISULFONIC ACID TRISODIUM SALT see CLG000
CHLOROSULFONIC ACID see CLG500
CHLOROSULFONIC ACID (with or without sulfur trioxide) (UN 1754) (DOT) see CLG500
CHLOROSULFONIC ANHYDRIDE see PPR500
4-CHLOROSULFONYLACETANILIDE see AAM600
p-(CHLOROSULFONYL)ACETANILIDE see AAM600
4'-(CHLOROSULFONYL)ACETANILIDE see AAM600
3-(CHLOROSULFONYL)BENZOYL CHLORIDE see CLG100
m-(CHLOROSULFONYL)BENZOYL CHLORIDE see CLG100
5-(CHLOROSULFONYL)-2,4-DICHLOROBENZOIC ACID see CLG200
1-CHLOROSULFONYL-5-DIMETHYLAMINONAPHTHALENE see DPN200
CHLOROSULFONYL FLUORIDE see SOT500
CHLOROSULFONYLISOCYANATE see CLG250
1-(4-CHLORO-o-SULFO-5-TOLYLAZO)-2-NAPHTHOL,BARIUM SALT see CHP500
CHLOROSULFURIC ACID see CLG500
4-CHLORO-5-SULPHAMOYLPHTHALIMIDE see CGM500
4-CHLORO-5-SULPHAMOYLPHTHALIMIDE see CLG825
CHLOROSULTHIADIL see CFY000
CHLOROTESTOSTERONE ACETATE see CLG900
4-CHLOROTESTOSTERONE ACETATE see CLG900
4-CHLOROTESTOSTERONE 17-ACETATE see CLG900
7-CHLOROTETRACYCLINE see CMA750
CHLOROTETRACYCLINE HYDROCHLORIDE see CMB000
6-CHLORO-N,N,N',N'-TETRAETHYL-1,3,5-TRIAZINE-2,4-DIAMINE see CDQ325
CHLOROTETRAFLUOROETHANE see CLH000
1-CHLORO-1,1,2,2-TETRAFLUOROETHANE see TCH150
6-CHLORO-2,3,4,5-TETRAHYDRO-3-METHYL-1-(3-METHYLPHENYL)-1H-3-BENZAZEPINE-7,8-DIOL, HYDROBROMIDE see CLH050
7-CHLORO-1,2,3,4-TETRAHYDRO-2-METHYL-3-(2-METHYLPHENYL)-4-OXO-6-QUINAZOLINESULFONAMIDE see ZAK300
5-CHLORO-1,2,3,4-TETRAHYDRO-9-MORPHOLINOACRIDINE HYDROCHLORIDE see CLH100
6-CHLORO-2,3,4,5-TETRAHYDRO-1-PHENYL-1H-3-BENZAZEPINE-7,8-DIOL, HYDROBROMIDE see CLH150
6-CHLORO-2,3,4,5-TETRAHYDRO-1-PHENYL-3-(2-PROPENYL)-1H-3-BENZAZEPINE-7,8-DIOL, HYDROBROMIDE see CLH160
CHLOROTETRAHYDROXY((2-HYDROXY-5-OXO-2-IMIDAZOLIN-4-YL)UREATO)DIALUMINUM see AHA135
N-CHLOROTETRAMETHYLGUANIDINE see CLH500
1-CHLORO-2,2,5,5-TETRAMETHYL-4-IMIDAZOLIDINONE see CLH550
9-CHLORO-7-(1H-TETRAZOL-5-YL)-5H-1-BENZOPYRANO(2,3-b)PYRIDIN-5-ONE SODIUM PENTAHYDRATE see THK850
2-CHLORO-5-(1H-TETRAZOL-5-YL)-N⁴-2-THENYLSULFANILAMIDE see ASO375
CHLOROTHALIDONE see CLY600
CHLOROTHALONIL see TBQ750
CHLOROTHANE NU see MIH275
CHLOROTHEN see CHY250

CHLOROTHENE see MIH275
CHLOROTHENE (inhibited) see MIH275
CHLOROTHENE NU see MIH275
CHLOROTHENE VG see MIH275
2-((5-CHLORO-2-THENYL)(2-DIMETHYLAMINOETHYL)AMINO)PYRIDINE see CHY250
CHLOROTHENYLPYRAMINE see CHY250
8-CHLORO-THEOPHYLLINE compounded with 4-(DIPHENYLMETHOXY)-1-METHYLPIPERIDINE (1:1) see PIZ250
5-CHLORO-1,2,3-THIADIAZOLE see CLH625
CHLOROTHIAMIDE see DGM600
CHLOROTHIAZID see CLH750
CHLOROTHIAZIDE see CLH750
4-CHLOROTHIOANISOLE see CKG500
p-CHLOROTHIOANISOLE see CKG500
6-CHLORO-2-THIO-2H-1,3-BENZOXAZINE-2,4(3H)-DIONE see CLH800
p-CHLOROTHIOCARBANILIC ACID-o-(p-CHLOROPHENYL) ESTER see CJU125
6-CHLORO-4-THIOCHROMANYL-o,o-DIETHYL DITHIOPHOSPHATE see CLH810
6-CHLORO-4-THIOCHROMANYL o,o-DIMETHYL DITHIOPHOSPHATE see CLH820
CHLOROTHIOFORMIC ACID ETHYL ESTER see CLJ750
2,3,4,5-CHLOROTHIOPHENE see TBV750
4-CHLOROTHIOPHENOL see CEK425
p-CHLOROTHIOPHENOL see CEK425
(Z)-3-(2-CHLORO-9H-THIOXANTHEN-9-YLIDENE)-N,N-DIMETHYL-1-PROPANAMINE see TAF675
4-(3-(2-CHLOROTHIOXANTHEN-9-YLIDENE)PROPYL)-1-PIPERAZINEETHANOL DIHYDROCHLORIDE see CLX250
CHLOROTHYMOL see CLJ800
6-CHLOROTHYMOL see CLJ800
4-CHLORO-o-TOLOXYACETIC ACID see CIR250
CHLOROTOLUENE see CLK130
2-CHLOROTOLUENE see CLK100
4-CHLOROTOLUENE see TGY075
o-CHLOROTOLUENE see CLK100
p-CHLOROTOLUENE see TGY075
ar-CHLOROTOLUENE see CLK130
p-CHLOROTOLUENE (DOT) see TGY075
α-CHLOROTOLUENE see BEE375
ω-CHLOROTOLUENE see BEE375
CHLOROTOLUENES see CLK130
2-CHLORO-p-TOLUIDINE see CLK210
3-CHLORO-o-TOLUIDINE see CLK200
3-CHLORO-p-TOLUIDINE see CLK215
4-CHLORO-2-TOLUIDINE see CLK220
4-CHLORO-o-TOLUIDINE see CLK220
5-CHLORO-o-TOLUIDINE see CLK225
6-CHLORO-2-TOLUIDINE see CLK227
6-CHLORO-o-TOLUIDINE see CLK227
3-CHLORO-p-TOLUIDINE HYDROCHLORIDE see CLK230
4-CHLORO-2-TOLUIDINE HYDROCHLORIDE see CLK235
4-CHLORO-o-TOLUIDINE HYDROCHLORIDE see CLK235
4-CHLORO-o-TOLUIDINE HYDROCHLORIDE (DOT) see CLK235
2-(2-CHLORO-p-TOLUIDINO)-2-IMIDAZOLINE NITRATE see TGJ885
CHLOROTOLURON see CIS250
N-(3-CHLORO-o-TOLYL)ANTHRANILIC ACID see CLK325
p-((3-CHLORO-p-TOLYL)AZO)-N,N-DIMETHYLANILINE see CIL700
p-((4-CHLORO-m-TOLYL)AZO)-N,N-DIMETHYLANILINE see CIL710
1-(6-CHLORO-o-TOLYL)-3-CYCLOHEXYL-3-(2-(DIETHYLAMINO)ETHYL)UREA HYDROCHLORIDE see CLK500

1-(6-CHLORO-o-TOLYL)-3-(3-(DIBUTYLAMINO)PROPYL)UREA HYDROCHLORIDE see CLK750
1-(6-CHLORO-o-TOLYL)-3-(2-(DIETHYLAMINO)ETHYL)-3-METHYLUREA see CLL000
1-(6-CHLORO-o-TOLYL)-3-(2-(DIETHYLAMINO)ETHYL)UREA HYDROCHLORIDE see CLL250
1-(6-CHLORO-o-TOLYL)-1-(2-(DIETHYLAMINO)ETHYL)-3-(2,6-XYLYL)UREA HYDROCHLORIDE see CLL750
1-(6-CHLORO-o-TOLYL)-3-(2-(DIETHYLAMINO)ETHYL)-3-(2,6-XYLYL)UREA HYDROCHLORIDE see CLL500
1-(6-CHLORO-o-TOLYL)-3-(3-(DIETHYLAMINO)PROPYL)UREA see CLM000
1-(6-CHLORO-o-TOLYL)-3-(2-(DIMETHYLAMINO)ETHYL)-3-ISOPROPYLUREA HYDROCHLORIDE see CLM250
1-(6-CHLORO-o-TOLYL)-3-(2-(DIMETHYLAMINO)ETHYL)UREA HYDROCHLORIDE see CLM500
1-(6-CHLORO-o-TOLYL)-3-(3-(DIMETHYLAMINO)PROPYL)UREA HYDROCHLORIDE see CLM750
N'-(4-CHLORO-o-TOLYL)-N,N-DIMETHYLFORMAMIDINE see CJJ250
N'-(4-CHLORO-o-TOLYL)-N,N-DIMETHYLFORMAMIDINE HYDROCHLORIDE see CJJ500
1-(6-CHLORO-o-TOLYL)-3-(4-METHOXYBENZYL)-3-(2-PIPERIDINOETHYL)UREA see CLN000
1-(6-CHLORO-o-TOLYL)-3-(4-METHOXYBENZYL)-3-(2-(PYRROLIDINYL)ETHYL)UREA HYDROCHLORIDE see CLN250
3-(4-CHLORO-o-TOLYL)-5-(m-METHOXYPHENYL)-s-TRIAZOLE see CLN325
1-(4-CHLORO-o-TOLYL)-3-(p-METHYLBENZYL)-3-(2-PYRROLIDINYLETHYL)UREA HYDROCHLORIDE see CLN500
((4-CHLORO-o-TOLYL)OXY)ACETIC ACID see CIR250
((4-CHLORO-o-TOLYL)OXY)ACETIC ACID, ETHYL ESTER see EMR000
((4-CHLORO-O-TOLYL)OXY)-ACETIC ACID with 2,2'-IMINODIETHANOL (1:1) see MIH750
((4-CHLORO-o-TOLYL)OXY)ACETIC ACID POTASSIUM SALT see MIH800
(p-CHLORO-o-TOLYLOXY)ACETIC ACID SODIUM SALT see SIL500
(4-CHLORO-o-TOLYLOXY)BUTYRIC ACID see CLN750
4-((4-CHLORO-o-TOLYL)OXY)BUTYRIC ACID see CLN750
(4-CHLORO-o-TOLYLOXY)BUTYRIC ACID SODIUM SALT see CLO000
2-(p-CHLORO-o-TOLYLOXY)PROPIONIC ACID see CIR500
2-((4-CHLORO-o-TOLYL)OXY)PROPIONIC ACID POTASSIUM SALT see CLO200
1-(6-CHLORO-o-TOLYL)-3-(2-PYRROLIDINYLETHYL)UREA HYDROCHLORIDE see CLO500
CHLOROTOLYLTHIOGLYCOLIC ACID see CLO600
CHLOROTRIAMMINEPLATINUM TRICHLOROPLATINATE(1-) see CLO700
CHLOROTRIANISENE see CLO750
CHLOROTRIANIZEN see CLO750
CHLOROTRIAZINE see TJD750
CHLOROTRIBENZYLSTANNANE see CLP000

CHLOROTRIBUTYLGERMANIUM see CLP250
CHLOROTRIBUTYLSTANNANE see CLP500
1-CHLORO-4-(TRICHLOROMETHYL)BENZENE see TIR900
3-CHLORO-3-TRICHLOROMETHYLDIAZIRINE see CLP625
2-CHLORO-6-(TRICHLOROMETHYL)PYRIDINE see CLP750
2-CHLORO-1-(2,4,5-TRICHLOROPHENYL)VINYL DIMETHYL PHOSPHATE see TBW100
(Z)-2-CHLORO-1-(2,4,5-TRICHLOROPHENYL)VINYL DIMETHYL PHOSPHATE see RAF100
2-CHLORO-1-(2,4,5-TRICHLOROPHENYL)VINYL PHOSPHORIC ACID DIMETHYL ESTER see TBW100
2-CHLORO-1,1,3-TRIETHOXY PROPANE see CLP800
2-CHLOROTRIETHYLAMINE see CGV500
β-CHLOROTRIETHYLAMINE see CGV500
2-CHLOROTRIETHYLAMINE HYDROCHLORIDE see CLQ250
CHLORO(TRIETHYLPHOSPHINE)GOLD see CLQ500
CHLOROTRIETHYLSTANNANE see TJV000
CHLOROTRIETHYLTIN see TJV000
6-CHLORO-N,N,N'-TRIETHYL-1,3,5-TRIAZINE-2,4-DIAMINE see TJL500
CHLOROTRIFLUORIDE see CDX750
1-CHLORO-2,2,2-TRIFLUOROETHANE see TJY175
2-CHLORO-1,1,1-TRIFLUOROETHANE see TJY175
CHLOROTRIFLUOROETHENE HOMOPOLYMER see KDK000
2-CHLORO-1,1,2-TRIFLUOROETHYL DIFLUOROMETHYL ETHER see EAT900
CHLOROTRIFLUOROETHYLENE see CLQ750
1-CHLORO-1,2,2-TRIFLUOROETHYLENE see CLQ750
2-CHLORO-1,1,2-TRIFLUOROETHYLENE see CLQ750
CHLOROTRIFLUOROETHYLENE POLYMER see KDK000
CHLOROTRIFLUOROETHYLENE POLYMERS see KDK000
2-CHLORO-1,1,2-TRIFLUOROETHYL METHYL ETHER see CLR000
CHLOROTRIFLUOROMETHANE see CLR250
CHLOROTRIFLUOROMETHANE mixed with TRIFLUOROMETHANE see FOO562
CHLOROTRIFLUOROMETHANE and TRIFLUOROMETHANE AZEOTROPIC MIXTURE (DOT) see FOO562
2-CHLORO-4-TRIFLUOROMETHYL-3'-ACETOXYDIPHENYL ETHER see CLR300
2-CHLORO-5-(TRIFLUOROMETHYL)ANILINE see CEG800
4-CHLOROTRIFLUOROMETHYLBENZENE see CEM825
p-CHLOROTRIFLUOROMETHYLBENZENE see CEM825
3-CHLORO-3-TRIFLUOROMETHYLDIAZIRINE see CLR825
4-CHLORO-3-TRIFLUOROMETHYLPHENOL see CLS000
p-CHLORO-m-TRIFLUOROMETHYLPHENOL see CLS000

N-(4-CHLORPHENYL)-2,2-DIMETHYLPENTAMID (GERMAN) see CGL250

1-(p-CHLOR-PHENYL)-3,3-DIMETHYL-TRIAZEN (GERMAN) see CJI100

N-(4-CHLOR-PHENYL)-2,2-DIMETHYL-VALERIANSAEUREAMID (GERMAN) see CGL250

4-(2-CHLORPHENYL)-2-ETHYL-9-METHYL-6H-THIENO(3,2-f)(1,2,4)TRIAZOLO(4,3-a)(1,4)DIAZEPINE see EQN600

γ-(4-(p-CHLORPHENYL)-4-HYDROXPIPERIDINO)-p-FLUORBUTYROPHENONE see CLY500

N-(3-CHLOR-PHENYL)-ISOPROPYL-CARBAMAT (GERMAN) see CKC000

4-CHLOR-PHENYL-ISOTHIOCYANAT (GERMAN) see ISH000

3-(4-CHLORPHENYL)-1-METHOXY-1-METHYLHARNSTOFF (GERMAN) see CKD500

3-(4-CHLORPHENYL)-1-METHYL-1-ISOBUTINYLHARNSTOFF (GERMAN) see CKF750

N-(4-CHLORPHENYL)-N'-METHYL-N'-ISOBUTINYLHARNSTOFF (GERMAN) see CKF750

2-(p-CHLORPHENYL)-3-METHYL-1,3-PERHYDROTHIAZIN-4-ON-1,1-DIOXIDE see CKF500

((4-CHLORPHENYL)-1-PHENYL)-ACETYL-1,3-INDANDION (GERMAN) see CJJ000

1-(4-CHLORPHENYL)-1-PHENYL-ACETYL-INDAN-1,3-DION (GERMAN) see CJJ000

2(2-(4-CHLOR-PHENYL-2-PHENYL)ACETYL)INDAN-1,3-DION (GERMAN) see CJJ000

4-(p-CHLORPHENYLTHIO)-BUTANOL (GERMAN) see CKK500

4-CHLORPHENYL-2',4',5'-TRICHLORPHENYLAZOSULFID (GERMAN) see CDS500

CHLORPHONIUM CHLORIDE see THY500

CHLORPHTHALIDOLONE see CLY600

CHLORPHTHALIDONE see CLY600

CHLOR-O-PIC see CKN500

CHLORPIKRIN (GERMAN) see CKN500

CHLORPINAKOLIN see MRG070

CHLORPROETHAZINE see CLY750

CHLORPROETHAZINE HYDROCHLORIDE see CLZ000

CHLORPROHEPTADIEN see CEH500

CHLORPROHEPTADIENE HYDROCHLORIDE see CEH500

CHLORPROHEPTATRIEN see CMA000

CHLORPROMAZIN see CKP250

CHLORPROMAZINE see CKP250

CHLORPROMAZINE METHIODIDE see MIJ300

CHLORPROMAZINE SULFONE see ONI300

CHLORPROPAMID see CKK000

CHLORPROPAMIDE see CKK000

3-CHLORPROPAN-1-OL see CKP600

3-CHLORPROPEN (GERMAN) see AGB250

CHLORPROPHAM see CKC000

CHLORPROPHAME (FRENCH) see CKC000

CHLORPROPHENPYRIDAMINE see CLX300

N-(3-CHLORPROPYL)-1-METHYL-2-PHENYL-AETHYLAMIN-HYDROCHLORID (GERMAN) see PKS000

CHLORPROTHIXEN see TAF675

CHLORPROTHIXENE see TAF675

cis-CHLORPROTHIXENE see TAF675

α-CHLORPROTHIXENE see TAF675

CHLORPROTIXEN see TAF675

CHLORPROTIXENE see TAF675

CHLORPROTIXINE see TAF675

3-(2-CHLOR-5-PYRIDYLMETHYL)-2-CYANIMINOTHIAZOLIDIN see CKW410

CHLORPYRIFOS see CMA100

CHLORPYRIFOS-ETHYL see CMA100

CHLORPYRIFOS-METHYL see CMA250

CHLORPYRIPHOS see CMA100

CHLORPYRIPHOS-ETHYL see CMA100

CHLOR-PZ see CKP250

CHLORQUINALDOL see CLC500

CHLORQUINOL see DEL300

CHLORQUINOX see CMA500

CHLORSAEURE (GERMAN) see SFS000

CHLORSAL see CLH750

CHLORSEPTOL see CDP000

2-(4"-CHLOR-4'-STILBYL)NAFTOTRIAZOL-6,2',2"-TRISULFONAN SODNY (CZECH) see CLG000

N-CHLORSUCCINIMIDE see SND500

CHLORSUCCINYLCHOLIN (GERMAN) see HLC500

4-CHLOR-5-SULFAMOYL-2',6'-SALICYLOXYLIDID (GERMAN) see CLF325

CHLORSULFAQUINOXALINE see CMA600

CHLORSULFON see CMA700

CHLORSULFONAMIDO DIHYDROBENZOTHIADIAZINE DIOXIDE see CFY000

CHLORSULFURON see CMA700

CHLORSULPHACIDE see CEP000

CHLORTALIDONE see CLY600

CHLORTEN see MIH275

CHLORTETRACYCLINE see CMA750

CHLORTETRACYCLINE, 6-DEMETHYL- see MIJ500

CHLORTETRACYCLINE HYDROCHLORIDE see CMB000

4-CHLORTETRAHYDROFTALANHYDRID (CZECH) see CFG500

CHLORTETRIN see DAI485

CHLORTHAL-DIMETHYL see TBV250

CHLORTHALIDON see CLY600

CHLORTHALIDONE see CLY600

CHLORTHAL-METHYL see TBV250

CHLORTHALONIL (GERMAN) see TBQ750

CHLORTHIAZIDE see CLH750

CHLORTHIEPIN see EAQ750

p-CHLORTHIOFENOL (CZECH) see CEK425

CHLORTHION METHYL see MIJ250

CHLORTHIOPHOS see CLB022

CHLORTHYMOL see CLJ800

CHLORTION (CZECH) see MIJ250

2-CHLOR-4-TOLUIDIN (CZECH) see CLK210

3-CHLOR-2-TOLUIDIN (CZECH) see CLK200

α-CHLORTOLUOL (GERMAN) see BEE375

CHLORTOLURON see CIS250

N'-(4-CHLOR-o-TOLYL)-N,N-DIMETHYLFORMAMIDIN (GERMAN) see CJJ250

CHLORTOX see CDR750

CHLORTRIANISEN see CLO750

CHLORTRIFLUORAETHYLEN (GERMAN) see CLQ750

CHLOR-TRIMETON see CLD250

CHLOR-TRIMETON see CLX300

CHLOR-TRIMETON see TAI500

CHLOR-TRIMETON MALEATE see TAI500

CHLOR-TRIPOLON see CLX300

CHLOR-TRIPOLON see TAI500

CHLORTROPBENZYL see CMB125

CHLORURE d'ALUMINUM (FRENCH) see AGY750

CHLORURE ANTIMONIEUX see AQC500

CHLORURE d'ARSENIC (FRENCH) see ARF500

CHLORURE ARSENIEUX (FRENCH) see ARF500

CHLORURE de BENZENYLE (FRENCH) see BFL250

CHLORURE de 1-BENZYL-3-BENZYL-CARBOXY-PYRIDINIUM (FRENCH) see SAA000

CHLORURE de BENZYLE (FRENCH) see BEE375

CHLORURE de BENZYLIDENE see BAY300

CHLORURE de BORE (FRENCH) see BMG500

CHLORURE de BUTYLE (FRENCH) see BQQ750

CHLORURE de CHLORACETYLE (FRENCH) see CEC250

CHLORURE de CHROMYLE (FRENCH) see CML125

CHLORURE de CYANOGENE see COO750

CHLORURE de DICHLORACETYLE (FRENCH) see DEN400

CHLORURE de l'ETHYLAL TRIMETHYLAMMONIUM PROPANEDIOL (FRENCH) see MJH800

CHLORURE d'ETHYLE (FRENCH) see EHH000

CHLORURE d'ETHYLENE (FRENCH) see EIY600

CHLORURE d'ETHYLIDENE (FRENCH) see DFF809

CHLORURE de FUMARYLE (FRENCH) see FOY000

CHLORURE de LITHIUM (FRENCH) see LHB000

CHLORURE de MAGNESIUM HYDRATE (FRENCH) see MAE500

CHLORURE MERCUREUX see MCY300

CHLORURE MERCUREUX (FRENCH) see MCW000

CHLORURE MERCURIQUE (FRENCH) see MCY475

CHLORURE de METHALLYLE (FRENCH) see CIU750

CHLORURE de METHYLE (FRENCH) see MIF765

CHLORURE de METHYLENE (FRENCH) see MJP450

CHLORURE PERRIQUE see FAU000

CHLORURE de SUCCINILCOLINE (FRENCH) see HLC500

CHLORURE de VINYLE (FRENCH) see VNP000

CHLORURE de VINYLIDENE (FRENCH) see VPK000

CHLORURE de ZINC (FRENCH) see ZFA000

CHLORURIT see CLH750

CHLORVINPHOS see DGP900

CHLORWASSERSTOFF (GERMAN) see HHL000

CHLORXYLAM see DET600

CHLORYL see EHH000

CHLORYL ANESTHETIC see EHH000

CHLORYLEA see TIO750

CHLORYL HYPOFLUORITE see CMB250

CHLORYL PERCHLORATE see CMB500

CHLORYL RADICAL see CDW450

CHLORZIDE see CFY000

CHLORZOXAZONE see CDQ750

CHLOTAZOLE see CMB675

CHLOTHIXEN see TAF675

CHLOTRIDE see CLH750

CHLOTRIMAZOLE see MRX500

CHLZ see CLX000

CHNU-I see NKJ050

CHOB-i-QUT see CNT350

CHOCOLA A see VSK600

CHOCOLATE BROWN FB see CMB750

CHOCOLATE EMBL see CMO820

CHOKE CHERRY see AQP890

CHOKEGARD see HCP050

CHOLAGON see DAL000

CHOLAIC ACID see TAH250

CHOLALIN see CME750

CHOLAN DH see DAL000

CHOLANORM see CDL325

CHOLANTHRENE see CMC000

CHOLANTHRYLENE see NAH900

CHOLAXINE see SKV200

CHOLEBRINE see IDJ550

CHOLECALCIFEROL see CMC750

CHOLECYSTOKININ TETRAPEPTIDE see GCE200

CHOLEDYL see CMG300

CHOLEGYL see CMG300

CHOLEIC ACID see DAQ400
CHOLEPULVIS see TDE750
CHOLERA ENDOTOXIN see VJZ100
CHOLERA ENTERO-EXOTOXIN see CMC800
CHOLERA ENTERO-EXOTOXIN see VJZ200
CHOLERA ENTEROTOXIN see CMC800
CHOLERA ENTEROTOXIN see VJZ200
CHOLERA EXOTOXIN see VJZ200
CHOLERAGEN see CMC800
CHOLERAGEN see VJZ200
CHOLERA TOXIN see VJZ200
CHOLEREBIC see DAQ400
CHOLESOLVIN see SDY500
5,7-CHOLESTADIEN-3-β-OL see DAK600
(3-β)CHOLESTA-5,7-DIEN-3-OL see DAK600
CHOLESTA-5,7-DIEN-3-β-OL ACETATE see DAK800
CHOLESTANE-3,7,12,26,27-PENTOL, (3-α,5-α,7-α,12-α)- see CQJ100
epi-CHOLESTANOL see EBA100
3-β-CHOLESTANOL see DKW000
(3-β,5-β)-CHOLESTAN-3-OL see DKW000
CHOLESTAN-3-OL, (3-α-5α-)-(9CI) see EBA100
5-α-CHOLESTAN-3-α-OL (8CI) see EBA100
α-CHOLESTANOL (7CI) see EBA100
Δ⁷-CHOLESTENOL see CMD000
CHOLEST-5-EN-3-β-OL see CMD750
5-CHOLESTEN-3-β-OL see CMD750
7-CHOLESTEN-3-β-OL see CMD000
5:6-CHOLESTEN-3-β-OL see CMD750
Δ⁵-CHOLESTEN-3-β-OL see CMD750
5-α-CHOLEST-7-EN-3-β-OL see CMD000
5-CHOLESTEN-3-β-OL 3-(p-(BIS(2-CHLOROETHYL)AMINO)PHENYL)ACETATE see CME250
CHOLEST-5-EN-3-β-OL, 24-β-ETHYL-, SULFATE SALT (1:1) see SDZ370
Δ⁶-CHOLESTEN-3-β-OL-5-α-HYDROPEROXIDE see CMD250
CHOLEST-6-EN-3-β-OL-5-α-HYDROPEROXIDE see CMD250
Δ⁶-CHOLESTEN-3-β-OL-5-α-HYDROPEROXYD (GERMAN) see CMD250
CHOLESTENONE see CMD500
CHOLEST-5-EN-3-ONE see CMD500
5-CHOLESTEN-3-ONE see CMD500
Δ⁵-CHOLESTENONE see CMD500
5-β-CHOLEST-7-EN-6-ONE, 2-β,3-β,14,22,25-PENTAHYDROXY-, (20S,22R)- see EAB600
CHOLEST-7-EN-6-ONE, 2,3,14,22,25-PENTAHYDROXY-, (2-β,3-β,5-β,22R)-(9CI) see EAB600
CHOLESTERIN see CMD750
CHOLESTERIN (GERMAN) see CMD000
CHOLESTEROL see CMD750
Δ⁷-CHOLESTEROL see DAK600
Δ⁵,⁷-CHOLESTEROL see DAK600
CHOLESTEROL BASE H see CMD750
CHOLESTEROL-α-EPOXIDE see EBM000
CHOLESTEROL-5-α,6-α-EPOXIDE see EBM000
CHOLESTEROL-5-α-HYDROPEROXIDE see CMD250
CHOLESTEROL ISOHEPTYLATE see CME000
CHOLESTEROL-5-METHYL-1-HEXANOATE see CME000
CHOLESTEROL mixed with OROTIC ACID mixed with CHOLIC ACID (2:2:1) see OJV525
CHOLESTEROL OXIDE see EBM000
CHOLESTEROL-α-OXIDE see EBM000
CHOLESTERONE see CMD500
CHOLESTERYL ALCOHOL see CMD750
CHOLESTERYL-p-BIS(2-CHLOROETHYL)AMINO PHENYLACETATE see CME250
CHOLESTERYL-14-METHYLHEXADECANOATE see CCJ500
CHOLESTRIN see CMD750

CHOLESTROL see CMD750
CHOLESTYRAMINE see CME400
CHOLESTYRAMINE CHLORIDE see CME400
CHOLESTYRAMINE RESIN see CME400
CHOLETH 24 see PKE700
CHOLEXAMIN see CME675
CHOLEXAMINE see CME675
CHOLIBIL see CNG980
CHOLIC ACID see CME750
CHOLIC ACID mixed with CHOLESTEROL mixed with OROTIC ACID (1:2:2) see OJV525
CHOLIC ACID, MONOSODIUM SALT see SFW000
CHOLIFLAVIN see DBX400
CHOLIMED see DAL000
CHOLIMIL see IDJ550
CHOLINE see CMF000
CHOLINE ACETATE see CMF250
CHOLINE ACETATE (ESTER) see CMF250
CHOLINE ACETATE (ESTER), BROMIDE see CMF260
CHOLINE, ACETYL-, BROMIDE see CMF260
CHOLINE, ACETYL-β-METHYL- see MFX560
CHOLINE, S-ACETYLTHIO-, IODIDE see ADC300
CHOLINE BITARTRATE see CMF300
CHOLINE CARBAMATE CHLORIDE see CBH250
CHOLINE CHLORHYDRATE see CMF750
CHOLINE CHLORIDE ACETATE see ABO000
CHOLINE, CHLORIDE CARBAMATE(ESTER) see CBH250
CHOLINE CHLORIDE (FCC) see CMF750
CHOLINE CHLORINE CARBAMATE see CBH250
CHOLINE CYTIDINE DIPHOSPHATE see CMF350
CHOLINE 5'-CYTIDINE DIPHOSPHATE see CMF350
CHOLINE DICHLORIDE see CMF400
CHOLINE HYDROCHLORIDE see CMF750
CHOLINE HYDROXIDE see CMF800
CHOLINE, HYDROXIDE, DIHYDROGEN PHOSPHATE, INNER SALT, 3-ESTER WITH 1-MONOSTEARIN see SLM100
CHOLINE, HYDROXIDE, 5'-ESTER with CYTIDINE 5'-(TRIHYDROGEN PYROPHOSPHATE), inner salt see CMF350
CHOLINE, INNER SALT, METHYLPHOSPHONOFLUORIDATE see MKF250
CHOLINE IODIDE SUCCINATE (2:1) see BJI000
CHOLINE ION see CMF000
CHOLINE 1,5-NAPHTHALENEDISULFONATE (2:1), DILACTATE, DIACETATE see AAC875
CHOLINE, PHOSPHATE, 3-ESTER WITH 1-MONOSTEARIN see SLM100
CHOLINE PHOSPHATE, HEXADECYL ESTER, HYDROXIDE, INNER SALT (6CI) see HCP750
CHOLINE, PROPIONATE, IODIDE see PMW750
CHOLINE SALICYLATE see CMG000
CHOLINE SALICYLATE B see CMG000
CHOLINE, SALICYLATE (SALT) see CMG000
CHOLINE SALICYLIC ACID SALT see CMG000
CHOLINE SUCCINATE (ester) see CMG250
CHOLINE SUCCINATE DICHLORIDE see HLC500
CHOLINE SUCCINATE (2:1) (ESTER) see CMG250
CHOLINE THEOPHYLLINATE see CMG300
CHOLINE, with THEOPHYLLINE (1:1) see CMG300
CHOLINE THEOPHYLLINE SALT see CMG300

CHOLINE-2,6-XYLYL ETHER BROMIDE see XSS900
CHOLINIUM CHLORIDE see CMF750
CHOLINOPHYLLINE see CMG300
CHOLIT-URSAN see DMJ200
CHOLLY see CNE750
CHOLOGON see DAL000
CHOLOGRAF-N-METHYLGLUCAMINE see BGB315
CHOLOLIN see DAL000
CHOLOREBIC see DAQ400
CHOLOVUE see IFP800
CHOLOXIN see SKJ300
CHOLSAEURE (GERMAN) see CME750
CHOLUMBRIN see TDE750
CHOLYLTAURINE see TAH250
CHONDROITIN, HYDROGEN SULFATE (9CI) see CMF600
CHONDROITIN POLYSULFATE see CMF600
CHONDROITIN POLYSULFATE SODIUM see SFW300
CHONDROITIN SULFATE see CMF600
CHONDROITIN SULFURIC ACID see CMF600
CHONDROITIN SULFURIC ACIDS see CMF600
CHONDROITIN SULPHATE see CMF600
CHONDRON see SFW300
CHONDRUS see CCL250
CHONDRUS EXTRACT see CCL250
CHONGRASS see PJJ315
CHONSURID see CMF600
CHOPSUI POTATO see YAG000
CHORAFURONE see AHK750
CHORIGON see CMG675
CHORIGONADOTROPIN see CMG675
CHORIGONIN see CMG675
CHORIONIC GONADOTROPHIN see CMG675
CHORIONIC GONADOTROPIC HORMONE see CMG675
CHORIONIC GONADOTROPIN see CMG675
C. HORRIDUS HORRIDUS VENOM see TGB150
CHORULON see CMG675
CHORYLEN see TIO750
CHOT see PBC250
CHP see PNN300
CHP-PHENOBARBITALAT (GERMAN) see CPO500
CHQ see DEL300
CHR 9 see HFT550
CHRISTENSENITE see SCK000
CHRISTMAS BERRY TREE see PCB300
CHRISTMAS CANDLE see SDZ475
CHRISTMAS FLOWER see EQX000
CHRISTMAS ROSE see CMG700
CHROMACID FAST RED 3B see CMG750
CHROMACID VIOLET R see HLI000
CHROMALUM HEXAHYDRATE see CMG800
CHROMAL VIOLET B see HLI000
CHROMAL YELLOW M see SIT850
6-CHROMANOL, 2,5,7,8-TETRAMETHYL-2-(4,8,12-TRIMETHYLTRIDECYL)-, ACETATE see TGJ055
2-CHROMANONE see HHR500
CHROMAR see XHS000
CHROMARGYRE see MCV000
CHROMATE(1-), BIS(2-ETHYL-2-HYDROXYBUTANOATO(2-)-(O¹,O²)OXO-, SODIUM see BJS600
CHROMATE(1-), DIAMMINETETRAKIS(ISOTHIOCYANATO)-, AMMONIUM, HYDRATE see ANT300
CHROMATE(1-), DIAMMINETETRAKIS(THIOCYANATO-N)-, BARIUM, (OC-6-11)- see BAO400
CHROMATE OF POTASSIUM see PLB250

CHROMATE de PLOMB (FRENCH) see LCR000
CHROMATE of SODA see DXC200
CHROMATEX GREEN G see PJQ100
CHROMATEX ORANGE R see DVB800
CHROMATEX RED J see MMP100
CHROMAVEN MILLING ORANGE G see CMP882
CHROMAVEN VIOLET B see HLI000
CHROMAZINE BLUE BLACK B see CMP880
CHROME see CMI750
CHROME ALUM see CMG850
CHROME ALUM see PLB500
CHROME ALUM (DODECAHYDRATE) see CMG850
CHROME BLACK PB see EDC625
CHROME BLACK SPECIAL see EDC625
CHROME BLACK T see EDC625
CHROME BLUE BLACK BF see CMP880
CHROMEDIA CC 31 see CCU150
CHROMEDIA CF 11 see CCU150
CHROMEDOL see CMJ565
CHROME FAST BLUE 2R see HJF500
CHROME FAST BROWN FC see TKN500
CHROME FAST CYANIDE G see CMP880
CHROME FAST CYANINE BP see CMP880
CHROME FAST CYANINE GN see CMP880
CHROME FAST CYANINE GNN see CMP880
CHROME FAST CYANINE GP see CMP880
CHROME FAST CYANINE GSS see CMP880
CHROME FAST ORANGE G see CMP882
CHROME FAST ORANGE GR see CMP882
CHROME FAST ORANGE RW see NEY000
CHROME FAST RED 3B see CMG750
CHROME FAST RED F see CMO870
CHROME FAST RED FB see CMO870
CHROME FAST RED FW see CMO870
CHROME FAST RED P see CMG750
CHROME FAST VIOLET B see HLI000
CHROME FERROALLOY see FBD000
CHROME FLUORURE see CMJ560
CHROME GREEN see CMJ900
CHROME GREEN see LCR000
CHROME IRON NICKEL BLACK SPINEL see CMS135
CHROMEL C see NCX515
CHROME LEATHER BLACK BH see CMN800
CHROME LEATHER BLACK CR see CMN800
CHROME LEATHER BLACK D see CMN230
CHROME LEATHER BLACK DS see CMN800
CHROME LEATHER BLACK E see AQP000
CHROME LEATHER BLACK EC see AQP000
CHROME LEATHER BLACK EM see AQP000
CHROME LEATHER BLACK G see AQP000
CHROME LEATHER BLACK GNA see CMN240
CHROME LEATHER BLUE 2B see CMO000
CHROME LEATHER BLUE 3B see CMO250
CHROME LEATHER BORDEAUX BC see CMO872
CHROME LEATHER BRILLIANT BLACK ER see AQP000
CHROME LEATHER BROWN BRLL see CMO750
CHROME LEATHER BROWN BS see CMO820
CHROME LEATHER BROWN M see CMO800
CHROME LEATHER DARK BLUE BHM see CMN800
CHROME LEATHER DARK GREEN N see CMO830
CHROME LEATHER DARK GREEN S see CMO830
CHROME LEATHER FAST RED N see CMO870
CHROME LEATHER GREEN B see CMO830

CHROME LEATHER PURE BLUE see CMO500
CHROME LEATHER RED 4B see DXO850
CHROME LEATHER RED 5B see CMO885
CHROME LEATHER RED F see CMO870
CHROME LEATHER RED F EXTRA see CMO870
CHROME LEATHER SCARLET 3BS see CMO875
CHROME LEATHER SCARLET SE see CMO870
CHROME LEATHER SKY BLUE see CMN750
CHROME LEMON see LCR000
CHROMELIN see OLW100
CHROME OCHER see CMJ900
CHROME ORANGE see CMP882
CHROME ORANGE see LCS000
CHROME ORANGE MR see NEY000
CHROME ORANGE R see NEY000
CHROME ORANGE RLE see NEY000
CHROME ORE see CMI500
CHROME OXIDE see CMJ900
CHROME OXIDE GREEN see CMJ900
CHROME POTASH ALUM see PLB500
CHROME RED ALIZARINE see SEH475
CHROME TAN see CMJ565
CHROME (TRIOXYDE de) (FRENCH) see CMK000
CHROME VERMILION see MRC000
CHROME VIOLET B see HLI000
CHROME VIOLET K see HLI000
CHROME VIOLET R see HLI000
CHROME YELLOW see LCR000
CHROME YELLOW 2G see SIT850
CHROME YELLOW 3RN see NEY000
CHROME YELLOW 2GR see SIT850
CHROMIA see CMJ900
CHROMIC ACETATE see CMH000
CHROMIC ACETATE(III) see CMH000
CHROMIC ACETYLACETONATE see TNN250
CHROMIC ACID see CMH250
CHROMIC ACID see CMJ900
CHROMIC ACID see CMK000
CHROMIC(VI) ACID see CMH250
CHROMIC(VI) ACID see CMK000
CHROMIC(III) ACID see CMH260
CHROMIC ACID, BARIUM SALT (1:1) see BAK250
CHROMIC ACID, BIS(TRIPHENYLSILYL) ESTER see BLS750
CHROMIC ACID, CALCIUM SALT (1:1) see CAP500
CHROMIC ACID, CALCIUM SALT (1:1) (9CI) see CAR400
CHROMIC ACID, CALCIUM SALT (1:1), DIHYDRATE see CAP750
CHROMIC ACID, CHROMIUM(3+) SALT (3:2) see CMI250
CHROMIC ACID, COPPER-ZINC-COMPLEX see CNQ750
CHROMIC ACID, DIAMMONIUM SALT see ANF500
CHROMIC ACID, DIAMMONIUM SALT see NCQ550
CHROMIC ACID, DI-tert-BUTYL ESTER see BQV000
CHROMIC ACID, DILITHIUM SALT see LHD000
CHROMIC ACID, DIPOTASSIUM SALT see PKX250
CHROMIC ACID, DISODIUM SALT see SGI000
CHROMIC ACID, DISODIUM SALT, DECAHYDRATE see SFW500
CHROMIC ACID GREEN see CMJ900
CHROMIC ACID, LEAD and MOLYBDENUM SALT see LDM000
CHROMIC ACID, LEAD(2+) SALT (1:1) see LCR000
CHROMIC ACID LEAD SALT with LEAD MOLYBDATE see LDM000

CHROMIC ACID, MERCURY ZINC COMPLEX see ZJA000
CHROMIC ACID, solid (NA 1463) (DOT) see CMK000
CHROMIC ACID, NICKEL(2+) SALT (1:1) see CMH270
CHROMIC ACID, POTASSIUM ZINC SALT (2:2:1) see PLW500
CHROMIC ACID, STRONTIUM SALT (1:1) see SMH000
CHROMIC ACID, solution (UN 1755) (DOT) see CMK000
CHROMIC ACID, ZINC SALT see ZFJ100
CHROMIC ACID, ZINC SALT (1:2) see CMK500
CHROMIC ACID, ZINC SALT, BASIC see ZFJ130
CHROMIC AMMONIUM SULFATE see ANF625
CHROMIC ANHYDRIDE see CMK000
CHROMIC CHLORIDE see CMJ250
CHROMIC CHLORIDE HEXAHYDRATE see CMK450
CHROMIC CHLORIDE STEARATE see CMH300
CHROMIC CHROMATE see CMI250
CHROMIC FLUORIDE see CMJ560
CHROMIC FLUORIDE, solid (UN1756) (DOT) see CMJ560
CHROMIC FLUORIDE, solution (UN1757) (DOT) see CMJ560
CHROMIC (III) HYDROXIDE see CMH260
CHROMIC ION see CMJ580
CHROMIC NITRATE see CMJ600
CHROMIC NITRATE NONAHYDRATE see CMJ610
CHROMIC OXIDE see CMJ900
CHROMIC OXYCHLORIDE see CML125
CHROMIC PERCHLORATE see CMI300
CHROMIC PHOSPHATE see CMK300
CHROMIC POTASSIUM SULFATE see PLB500
CHROMIC POTASSIUM SULPHATE see PLB500
CHROMIC SULFATE see CMK415
CHROMIC SULPHATE see CMK415
CHROMIC TRIFLUORIDE see CMJ560
CHROMIC TRIOXIDE see CMK000
CHROMIS ACID ($H_2Cr_2O_7$), DISODIUM SALT, DIHYDRATE (9CI) see SGI500
CHROMITAN B see CMK415
CHROMITAN MS see CMK415
CHROMITAN NA see CMK415
CHROMITE see CMI500
CHROMITE (mineral) see CMI500
CHROMITE ORE see CMI500
CHROMIUM see CMI750
CHROMIUM(+) see CMJ582
CHROMIUM (3+) see CMJ580
CHROMIUM(VI) see CMJ582
CHROMIUM (III) see CMJ580
CHROMIUM ACETATE see CMH000
CHROMIUM(2+) ACETATE see CMJ000
CHROMIUM(II) ACETATE see CMJ000
CHROMIUM(III) ACETATE see CMH000
CHROMIUM ACETATE HYDRATE see CMJ000
CHROMIUM ACETYLACETONATE see TNN250
CHROMIUM(3+) ACETYLACETONATE see TNN250
CHROMIUM(III) ACETYLACETONATE see TNN250
CHROMIUM ALLOY, BASE, Cr,C,Fe,N,Si (FERROCHROMIUM) see FBD000
CHROMIUM ALLOY, Cr,C,Fe,N,Si see FBD000
CHROMIUM, BIS(BENZENE)-(8CI) see BGY700
CHROMIUM(1+), BIS(BENZENE)-, IODIDE (8CI) see BGY720

HROMIUM(1+), BIS(eta^6)-BENZENE)-, IODIDE (9CI) see BGY720
CHROMIUM, BIS(BENZENE)IODO- see BGY720
CHROMIUM CARBONATE see CMJ100
CHROMIUM CARBONYL (MAK) see HCB000
CHROMIUM CARBONYL (OC-6-11) (9CI) see HCB000
CHROMIUM CHLORIDE see CMJ250
CHROMIUM CHLORIDE see CMJ300
CHROMIUM CHLORIDE see CMJ355
CHROMIUM(II) CHLORIDE see CMJ300
CHROMIUM(VI) CHLORIDE see CMJ355
CHROMIUM(II) CHLORIDE (1:2) see CMJ300
CHROMIUM(III) CHLORIDE (1:3) see CMJ250
CHROMIUM CHLORIDE, anhydrous see CMJ250
CHROMIUM(III) CHLORIDE, HEXAHYDRATE (1:3:6) see CMK450
CHROMIUM CHLORIDE, HEXAHYDRATE (8CI,9CI) see CMK450
CHROMIUM CHLORIDE, HEXAUREA see HEZ800
CHROMIUM CHLORIDE OXIDE see CML125
CHROMIUM CHROMATE (MAK) see CMI250
CHROMIUM-COBALT-MOLYBDENUM ALLOY see VSK000
CHROMIUM COMPOUNDS see CMJ500
CHROMIUM (CR^{6+} see CMJ582
CHROMIUM DIACETATE see CMJ000
CHROMIUM DICHLORIDE see CMJ300
CHROMIUM DICHLORIDE DIOXIDE see CML125
CHROMIUM DIOXIDE DICHLORIDE see CML125
CHROMIUM(VI) DIOXYCHLORIDE see CML125
CHROMIUM(II), DIPHENYL- see BGY700
CHROMIUM(III), DIPHENYL-, IODIDE see BGY720
CHROMIUM DISODIUM OXIDE see DXC200
CHROMIUM EDTA COMPLEX see EJA410
CHROMIUM(III) FLUORIDE see CMJ560
CHROMIUM HEXACARBONYL see HCB000
CHROMIUM HEXACHLORIDE see CMJ355
CHROMIUM(3+), HEXAKIS(UREA-O)-, TRICHLORIDE, (OC-6-11)-(9CI) see HEZ800
CHROMIUM(3+), HEXAKIS(UREA)-, TRICHLORIDE (8CI) see HEZ800
CHROMIUM(III) HEXA-UREA CHLORIDE see HEZ800
CHROMIUM HEXAVALENT ION see CMJ582
CHROMIUM(III) HYDROXIDE see CMH260
CHROMIUM HYDROXIDE SULFATE see CMJ565
CHROMIUM HYDROXIDE SULFATE see NBW000
CHROMIUM III SULFATE see CMK415
CHROMIUM ION (3+) see CMJ580
CHROMIUM(6+) ION see CMJ582
CHROMIUM (III) ION see CMJ580
CHROMIUM, ION (CR^{3+}) see CMJ580
CHROMIUM, ION (CR^{6+}) see CMJ582
CHROMIUM, ION (CR^{6+}) (8CI,9CI) see CMJ582
CHROMIUM LEAD OXIDE see LCS000
CHROMIUM LITHIUM OXIDE see LHD000
CHROMIUM METAL (OSHA) see CMI750
CHROMIUM MONOPHOSPHATE see CMK300
CHROMIUM NICKEL OXIDE see CMH270
CHROMIUM NICKEL OXIDE see NDA100
CHROMIUM NITRATE see CMJ600
CHROMIUM (3+) NITRATE see CMJ600
CHROMIUM NITRATE (DOT) see CMJ600
CHROMIUM(III) NITRATE see CMJ600
CHROMIUM NITRATE NONAHYDRATE see CMJ610

CHROMIUM(III) NITRATE, NONAHYDRATE (1:3:9) see CMJ610
CHROMIUM NITRIDE see CMJ850
CHROMIUM ORTHOPHOSPHATE see CMK300
CHROMIUM OXIDE see CMJ900
CHROMIUM OXIDE see CMK000
CHROMIUM(3+) OXIDE see CMJ900
CHROMIUM(VI) OXIDE see CMK000
CHROMIUM(III) OXIDE see CMJ900
CHROMIUM(VI) OXIDE (1:3) see CMK000
CHROMIUM(III) OXIDE (2:3) see CMJ900
CHROMIUM OXIDE, aerosols see CMJ910
CHROMIUM OXIDE, NICKEL OXIDE, and IRON OXIDE FUME see IHE000
CHROMIUM OXYCHLORIDE see CML125
CHROMIUM, PENTACARBONYL(PIPERIDINE)- see PAU700
CHROMIUM, PENTACARBONYL(PIPERIDINE)-, (OC-6-22)- see PAU700
CHROMIUM PENTAFLUORIDE see CMK275
CHROMIUM PERCHLORATE see CMI300
CHROMIUM PHOSPHATE see CMK300
CHROMIUM POTASSIUM SULFATE (1:1:2) see PLB500
CHROMIUM POTASSIUM SULPHATE see PLB500
CHROMIUM POTASSIUM ZINC OXIDE see CMK400
CHROMIUM SESQUICHLORIDE see CMK450
CHROMIUM SESQUIOXIDE see CMJ900
CHROMIUM SODIUM OXIDE see DXC200
CHROMIUM SODIUM OXIDE see SGI000
CHROMIUM SULFATE see CMJ565
CHROMIUM SULFATE see CMK405
CHROMIUM SULFATE see NBW000
CHROMIUM SULFATE (2:3) see CMK415
CHROMIUM (III) SULFATE (2:3) see CMK415
CHROMIUM SULFATE, BASIC see NBW000
CHROMIUM(III) SULFATE, HEXAHYDRATE (2:3:6) see CMG800
CHROMIUM SULFATE, PENTADECAHYDRATE see CMK425
CHROMIUM SULPHATE see CMK415
CHROMIUM SULPHATE see NBW000
CHROMIUM SULPHATE (2:3) see CMK415
CHROMIUM, TETRACHLORO-μ-HYDROXY(μ-(OCTADECANOATO-O:O))DI- see CMH300
CHROMIUM, TETRACHLORO-μ-HYDROXY(μ-STEARATO)DI- see CMH300
CHROMIUM TRIACETATE see CMH000
CHROMIUM TRIACETYLACETONATE see TNN250
CHROMIUM TRICHLORIDE see CMJ250
CHROMIUM TRICHLORIDE HEXAHYDRATE see CMK450
CHROMIUM TRIFLUORIDE see CMJ560
CHROMIUM TRIHYDROXIDE see CMH260
CHROMIUM TRINITRATE see CMJ600
CHROMIUM TRINITRATE NONAHYDRATE see CMJ610
CHROMIUM TRIOXIDE see CMK000
CHROMIUM(3+) TRIOXIDE see CMJ900
CHROMIUM(6+) TRIOXIDE see CMK000
CHROMIUM TRIOXIDE, anhydrous (DOT) see CMK000
CHROMIUM TRIOXIDE, anhydrous (UN 1463) (DOT) see CMK000
CHROMIUM TRIPERCHLORATE see CMI300
CHROMIUM TRIS(ACETYLACETONATE) see TNN250
CHROMIUM TRIS(BENZOYLACETONATE) see TNN500
CHROMIUM TRIS(2,4-PENTANEDIONATE) see TNN250
CHROMIUM YELLOW see LCR000
CHROMIUM ZINC OXIDE see ZFA100

CHROMIUM ZINC OXIDE see ZFJ100
CHROMIUM(6+)ZINC OXIDE HYDRATE (1:2:6:1) see CMK500
CHROM LEATHER TURQUOISE BLUE GLL see COF420
CHROMOCARD BLUE BLACK B see CMP880
CHROMOCARD ORANGE 3R see CMP882
CHROMOCOR see PER700
CHROMOFLAVINE see DBX400
CHROMOFLAVINE see XAK000
CHROMOGENE BLACK ET00 SPECIAL see EDC625
CHROMOGENE BLACK T 160 see EDC625
CHROMOL ORANGE R. EXTRA see NEY000
CHROMOL YELLOW G see SIT850
CHROMOL YELLOW N see SIT850
CHROMOMYCIN see OIS000
CHROMOMYCIN A3 see CMK650
CHROMOMYCIN SODIUM see CMK750
CHROMOMYSIN A$_3$ see CMK650
CHROMONAR HYDROCHLORIDE see CBR500
CHROMONE, 5,6,7,8-TETRAHYDRO-6-ETHYL-5-HYDROXY-3-(HYDROXYMETHYL)-, (E)- see ELH650
CHROMOSMON see BJI250
CHROMOSORB T see TAI250
CHROMOSULFURIC ACID (UN 2240) (DOT) see CLG500
CHROMOSULFURIC ACID (UN2240) (DOT) see CMK405
CHROMOTRICHIA FACTOR see AIH600
CHROMO (TRIOSSIDO di) (ITALIAN) see CMK000
CHROMOUS ACETATE see CMJ000
CHROMOUS ACETATE MONOHYDRATE see CMJ000
CHROMOUS CHLORIDE see CMJ300
CHROMOXYCHLORID (GERMAN) see CML125
CHROMSAEUREANHYDRID (GERMAN) see CMK000
CHROMTRIOXID (GERMAN) see CMK000
CHROMYL AZIDE CHLORIDE see CML000
CHROMYLCHLORID (GERMAN) see CML125
CHROMYL CHLORIDE see CML125
[(CHROMYLDIOXY)IODO]BENZENE see PFJ775
CHROMYL NITRATE see CML325
CHROMYL PERCHLORATE see CML500
CHRONICIN FOAM see CDP725
CHRONISULFAT see CMK405
CHRONIUM ZINCATE see ZFJ130
CHRONIUM ZINC OXIDE (9CI) see ZFJ130
CHRONOGYN see DAB830
CHROOMOXYLCHLORIDE (DUTCH) see CML125
CHROOMTRIOXYDE (DUTCH) see CMK000
CHROOMZUURANHYDRIDE (DUTCH) see CMK000
CHRYSAMMIC ACID see CML600
CHRYSAMMINIC ACID see CML600
CHRYSANTHAL see CML620
CHRYSANTHEMIC ACID see CML650
CHRYSANTHEMUM CINERAREAEFOLIUM see POO250
CHRYSANTHEMUMDICARBOXYLIC ACID MONOMETHYL ESTER PYRETHROLONE ESTER see POO100
CHRYSANTHEMUMIC ACID see CML650
CHRYSANTHEMUMIC ACID, 2,4-DIMETHYLBENZYL ESTER see DQQ500
(+)-trans-CHRYSANTHEMUMIC ACID ESTER of (+−)-ALLETHROLONE see BGC750
CHRYSANTHEMUMMONOCARBOXYLIC ACID see CML650
CHRYSANTHEMUMMONOCARBOXYLIC ACID, 2,4-DIMETHYLBENZYL ESTER see DQQ500

CHRYSANTHEMUM MONOCARBOXYLIC ACID PYRETHROLONE ESTER see POO050
CHRYSAROBIN see CML750
CHRYSAZIN see DMH400
CHRYSAZIN-3-CARBOXYLIC ACID see RHZ700
6-CHRYSENAMINE see CML800
CHRYSENE see CML810
1,2-CHRYSENEDIOL, 1,2-DIHYDRO-5-METHYL-, trans-(+−)- see CML812
1,2-CHRYSENEDIOL, 1,2-DIHYDRO-11-METHYL-, trans-(+−)- see CML814
1,2-CHRYSENEDIOL, 1,2-DIHYDRO-11-METHYL-, trans-(+−)- see DLT640
syn-CHRYSENE-3,4-DIOL 1,2-OXIDE see CML820
anti-CHRYSENE-3,4-DIOL 1,2-OXIDE see TCN300
CHRYSENE-5,6-EPOXIDE see CML830
CHRYSENE, 1,2,3,4,5,6-HEXAHYDRO-8-NITRO- (9CI) see NHB600
CHRYSENE, 7,8,9,10,11,12-HEXAHYDRO-2-NITRO- see NHB600
CHRYSENE-K-REGION EPOXIDE see CML830
CHRYSENE-5,6-OXIDE see CML830
CHRYSENEX see CML800
CHRYSENO(1,2-B)OXIRENE-2,3-DIOL, 1A,2,3,11B-TETRAHYDRO-, (1A-α-2-β,3-α-11Bα)- see TCN300
CHRYSENO(1,2-B)OXIRENE-2,3-DIOL,1A,2,3,11B-TETRAHYDRO-, (1ar-(1a-α,2-α,3-β,11B-α))- see CML820
CHRYSENO(3,4-B)OXIRENE-1,2-DIOL, 1,2,2A,3A-TETRAHYDRO-10-METHYL-, (1-α,2-β,2A-β,3A-β)- see CML835
CHRYSENO(3,4-B)OXIRENE-1,2-DIOL, 1,2,2A,3A-TETRAHYDRO-4-METHYL-, (1-α,2-β,2A-α,3A-α)-(+−)- see CML831
CHRYSENO(3,4-B)OXIRENE-1,2-DIOL, 1,2,2A,3A-TETRAHYDRO-4-METHYL-, (1-α,2-β,2A-β,3A-β)- see CML832
CHRYSENO(3,4-B)OXIRENE-1,2-DIOL, 1,2,2A,3A-TETRAHYDRO-4-METHYL-, (1-α,2-β,2A-β-3A-β)-(+−)- see CML834
CHRYSENO(3,4-B)OXIRENE-1,2-DIOL, 1,2,2A,3A-TETRAHYDRO-4-METHYL-, (1-α,2-β,2A-α,3A-α)- see MIN800
CHRYSENO(3,4-B)OXIRENE-1,2-DIOL, 1,2,2A,3A-TETRAHYDRO-4-METHYL-, (1-α,2-β,2A-β,3A-β)- see TCU750
α-CHRYSIDINE see BAW750
CHRYSIN see DMS900
CHRYSODERMOL see DMG900
CHRYSOFLUORENE see NAU600
CHRYSOIDIN see PEK000
CHRYSOIDIN A see DBP999
CHRYSOIDINE see PEK000
CHRYSOIDINE(II) see PEK000
CHRYSOIDINE A see PEK000
CHRYSOIDINE B see PEK000
CHRYSOIDINE C CRYSTALS see PEK000
CHRYSOIDINE G see PEK000
CHRYSOIDINE GN see PEK000
CHRYSOIDINE HR see PEK000
CHRYSOIDINE J see PEK000
CHRYSOIDINE M see PEK000
CHRYSOIDINE 3RN BASE see CMM800
CHRYSOIDINE ORANGE see PEK000
CHRYSOIDINE PRL see PEK000
CHRYSOIDINE PRR see PEK000
CHRYSOIDINE R BASE see CMM800
CHRYSOIDINE SL see PEK000
CHRYSOIDINE SPECIAL (biological stain and indicator) see PEK000
CHRYSOIDINE SS see PEK000
CHRYSOIDINE Y see PEK000
CHRYSOIDINE Y BASE NEW see PEK000
CHRYSOIDINE Y CRYSTALS see PEK000
CHRYSOIDINE Y EX see PEK000
CHRYSOIDINE YGH see PEK000

CHRYSOIDINE YL see PEK000
CHRYSOIDINE YN see PEK000
CHRYSOIDINE Y SPECIAL see PEK000
CHRYSOIDIN FB see PEK000
CHRYSOIDIN Y see PEK000
CHRYSOIDIN YN see PEK000
CHRYSOINE see MRL100
CHRYSOINE EXTRA see MRL100
CHRYSOINE EXTRA PURE A see MRL100
CHRYSOINE N see MRL100
CHRYSOINE S see MRL100
CHRYSOINE S EXTRA PURE see MRL100
CHRYSOIN G see MRL100
CHRYSOIN S (6CI) see MRL100
CHRYSOIN S SPECIALLY PURE see MRL100
CHRYSOMYKINE see CMA750
CHRYSON see BEP500
CHRYSONEX see CML800
CHRYSONINE S see MRL100
CHRYSOPHANIC ACID ANTHRANOL see CML750
CHRYSOTILE ASBESTOS see ARM268
CHRYSRON see BEP500
CHRYTEMIN see DLH630
CHRYZOIDYNA F.B. (POLISH) see PEK000
CHS see CPF750
CHUANGHSINMYCIN see CML847
CHUANGHSINMYCIN SODIUM see CML850
CHUANGXIMYCIN SODIUM see CML850
CHUANGXINMYCIN see CML847
CHUANLIANSU see CML860
CHWASTOKS see SIL500
CHWASTOX see CIR250
CHWASTOX see SIL500
CHYMAR see CML880
CHYMEX see CML870
CHYMOTEST see CML880
α-CHYMOTRYPSIN see CML880
CHYMOTRYPSIN A see CML880
CHYMOTRYPSIN B see CML880
CI-2 see MAV750
C.I. 23 see OHI200
C.I. 27 see HGC000
C.I. 79 see FMU070
CI 159 see BDG275
C.I. 184 see FAG020
C.I. 185 see FMU080
C.I. 258 see SBC500
CI-337 see ASA500
CI 366 see ENG500
CI395 see AOO500
CI-406 see PAN100
C.I. 440 see TKH750
C.I. 456 see CIP500
CI-473 see XQS000
CI-505 see BQM309
C.I. 515 see GLS700
C.I. 556 see SNY500
CI 581 see CKD750
CI-588 see BGO500
CI 624 see MGD200
CI-628 see NHP500
C.I. 633 see CGB250
C.I. 640 see FAG140
C.I. 671 see FMU059
CI-683 see POL475
CI-705 see QAK000
C.I. 712 see ADE500
C.I. 712 see CMM062
CI-719 see GCK300
CI 720 see TDO260
C.I. 749 see FAG070
C.I. 766 see FEW000
CI 775 see BFW400
C.I. 801 see CMM510
CI 881 see BKB300
CI-914 see CML890
CI 918 see CMM510
C.I. 925 see AJP250
CI 925 see MRA260
CI-937 see XDJ025
CI-977 see EAN900

CI-981 see DMS950
C.I. 1106 see IBV050
C.I. 1956 see CKN000
C.I. 7581 see FAE100
C.I. 10000 see NLB000
C.I. 10020 see NAX500
C.I. 10305 see PID000
C.I. 10315 see DUX800
C.I. 10338 see KDA075
C.I. 10345 see DUW500
C.I. 10355 see DVX800
C.I. 10360 see HET500
C.I. 10385 see SGP500
C.I. 11000 see PEI000
C.I. 11020 see DOT300
C.I. 11021 see OHJ875
C.I. 11025 see DPO200
C.I. 11050 see DHM500
C.I. 11085 see BAQ750
C.I. 11110 see ENP100
C.I. 11115 see CMP080
C.I. 11160 see AIC250
C.I. 11270 see DBP999
C.I. 11270 see PEK000
C.I. 11285 see NBG500
C.I. 11350 see PEJ600
C.I. 11380 see FAG130
C.I. 11390 see FAG135
C.I. 11855 see AAQ250
C.I. 11860 see OHK000
C.I. 11920 see CMP600
C.I. 12055 see PEJ500
C.I. 12075 see DVB800
C.I. 12085 see CJD500
C.I. 12100 see TGW000
C.I. 12120 see MMP100
C.I. 12140 see XRA000
C.I. 12150 see CMS238
C.I. 12156 see DOK200
C.I. 12210 see CMM770
C.I. 12355 see NAY000
C.I. 13015 see CMM758
C.I. 13020 see CCE500
C.I. 13025 see MND600
C.I. 13065 see MDM775
C.I. 13390 see ADE750
C.I. 14025 see SIT850
C.I. 14030 see NEY000
C.I. 14270 see MRL100
C.I. 14600 see FAG010
C.I. 14640 see CMP880
C.I. 14700 see FAG050
C.I. 14720 see HJF500
C.I. 14815 see CMP620
C.I. 15510 see CMM220
C.I. 15585 see CMS150
C.I. 15630 see NAP100
C.I. 15670 see HLI000
C.I. 15850 see CMS155
C.I. 15985 see FAG150
C.I. 16035 see FAG100
C.I. 16105 see CMG750
C.I. 16150 see FMU070
C.I. 16155 see FAG018
C.I. 16185 see FAG020
C.I. 16255 see FMU080
C.I. 18050 see CMM300
C.I. 18055 see CMM400
C.I. 19140 see FAG140
C.I. 19540 see CMP050
C.I. 20285 see CMP500
C.I. 20470 see FAB830
C.I. 21010 see CMM780
C.I. 21090 see DEU000
C.I. 21095 see CMS210
C.I. 21100 see CMS208
C.I. 21110 see CMS145
C.I. 22120 see SGQ500
C.I. 22155 see CMO872
C.I. 22195 see ADG000
C.I. 22245 see CMM320
C.I. 22310 see CMO870

C.I. 77061 see AQF500
C.I. 77086 see ARI000
C.I. 77099 see BAJ250
C.I. 77103 see BAK250
C.I. 77120 see BAP000
C.I. 77160 see BKW600
C.I. 77169 see BKW100
C.I. 77172 see DDI200
C.I. 77180 see CAD000
C.I. 77185 see CAD250
C.I. 77199 see CAJ750
C.I. 77205 see CMS212
C.I. 77223 see CAP500
C.I. 77223 see CAP750
C.I. 77231 see CAX750
C.I. 77265 see CBT500
C.I. 77266 see CBT750
C.I. 77288 see CMJ900
C.I. 77295 see CMJ250
C.I. 77305 see CMK415
C.I. 77320 see CNA250
C.I. 77322 see CND125
C.I. 77323 see CND825
C.I. 77353 see CNB475
C.I. 77402 see CNI000
C.I. 77402 see CNO000
C.I. 77403 see CNO250
C.I. 77410 see COF500
C.I. 77450 see CNQ000
C.I. 77480 see GIS000
C.I. 77491 see IHC450
C.I. 77504 see CMS135
C.I. 77575 see LCF000
C.I. 77577 see LDN000
C.I. 77578 see LDS000
C.I. 77580 see LCX000
C.I. 77600 see LCR000
C.I. 77601 see LCS000
C.I. 77605 see MRC000
C.I. 77610 see LCU000
C.I. 77620 see LCV100
C.I. 77622 see LDU000
C.I. 77630 see LDY000
C.I. 77640 see LDZ000
C.I. 77713 see MAC650
C.I. 77718 see TAB750
C.I. 77726 see MAT250
C.I. 77727 see MAT500
C.I. 77728 see MAS000
C.I. 77755 see PLP000
C.I. 77760 see MCT500
C.I. 77764 see MCW000
C.I. 77775 see NCW500
C.I. 77777 see NDF500
C.I. 77779 see NCY500
C.I. 77795 see PJD500
C.I. 77805 see SBO500
C.I. 77820 see SDI500
C.I. 77847 see SMM000
C.I. 77864 see TGC000
C.I. 77891 see TGG760
C.I. 77901 see TOC750
C.I. 77938 see VDU000
C.I. 77940 see VEZ000
C.I. 77945 see ZBJ000
C.I. 77947 see ZKA000
C.I. 77955 see ZFJ100
C.I. 77964 see ZJS400
C.I. 11160B see AIC250
C.I. 11320B see CMM800
C.I. 406225 see CMP100
C.I. 45370:1 see DDO200
C.I. 45380:2 see BMO250
C.I. 45410:1 see SKS400
C.I. 45410A see SKS400
C.I. 3/11855 see AAQ250
C.I. 52015 (CZECH) see BJI250
C.I. 59815 (CZECH) see CMU750
C.I. 61 570 (CZECH) see APL500
C.I. ACID BLACK 1, DISODIUM SALT (8CI) see FAB830
C.I. ACID BLACK 1 (7CI) see FAB830

C.I. ACID BLUE 1 see ADE500
C.I. ACID BLUE 3 see ADE500
C.I. ACID BLUE 3 see CMM062
C.I. ACID BLUE 41 see CMM070
C.I. ACID BLUE 43 see APG700
C.I. ACID BLUE 71 see CMM070
C.I. ACID BLUE 74 see FAE100
C.I. ACID BLUE 78 see CMM090
C.I. ACID BLUE 80 see CMM092
C.I. ACID BLUE 87 see COF420
C.I. ACID BLUE 92 see ADE750
C.I. ACID BLUE 129 see CMM100
C.I. ACID BLUE 185 see CMM120
C.I. ACID BLUE 62 (8CI) see CMM080
C.I. ACID BLUE 3, CALCIUM SALT (2:1) (8CI) see CMM062
C.I. ACID BLUE 9, DIAMMONIUM SALT see FMU059
C.I. ACID BLUE 9, DISODIUM SALT see FAE000
C.I. ACID BLUE 7 (7CI) see ERG100
C.I. ACID BLUE 1, SODIUM SALT see ADE500
C.I. ACID BLUE 7, SODIUM SALT (8CI) see ERG100
C.I. ACID BLUE 92, TRISODIUM SALT see ADE750
C.I. ACID GREEN 1 see NAX500
C.I. ACID GREEN 3 see FAE950
C.I. ACID GREEN 5 see FAF000
C.I. ACID GREEN 16 see CMM100
C.I. ACID GREEN 40 see CMM200
C.I. ACID GREEN 5, DISODIUM SALT see FAF000
C.I. ACID GREEN 3, MONOSODIUM SALT see FAE950
C.I. ACID GREEN 50, MONOSODIUM SALT see ADF000
C.I. ACID GREEN 3, SODIUM SALT see FAE950
C.I. ACID ORANGE 3 see SGP500
C.I. ACID ORANGE 7 see CMM220
C.I. ACID ORANGE 10 see HGC000
C.I. ACID ORANGE 20 see FAG010
C.I. ACID ORANGE 45 see ADG000
C.I. ACID ORANGE 52 see MND600
C.I. ACID ORANGE 45, DISODIUM SALT see ADG000
C.I. ACID ORANGE 6 (7CI) see MRL100
C.I. ACID ORANGE 7, MONOSODIUM SALT see CMM220
C.I. ACID ORANGE 20, MONOSODIUM SALT see FAG010
C.I. ACID ORANGE 6, MONOSODIUM SALT (8CI) see MRL100
C.I. ACID RED 1 see CMM300
C.I. ACID RED 2 see CCE500
C.I. ACID RED 18 see FMU080
C.I. ACID RED 26 see FMU070
C.I. ACID RED 27 see FAG020
C.I. ACID RED 33 see CMS228
C.I. ACID RED 51 see FAG040
C.I. ACID RED 85 see CMM320
C.I. ACID RED 87 see BNK700
C.I. ACID RED 92 see ADG250
C.I. ACID RED 94 see CMM325
C.I. ACID RED 98 see CMM000
C.I. ACID RED 99 see NAO600
C.I. ACID RED 114 see CMM330
C.I. ACID RED 1, DISODIUM SALT see CMM300
C.I. ACID RED 14, DISODIUM SALT see HJF500
C.I. ACID RED 26, DISODIUM SALT see FMU070
C.I. ACID RED 85, DISODIUM SALT see CMM320
C.I. ACID RED 99, DISODIUM SALT see NAO600
C.I. ACID RED 114, DISODIUM SALT see CMM330

C.I. ACID RED 97, DISODIUM SALT (8CI) see CMM325
C.I. ACID VIOLET 7 see CMM400
C.I. ACID VIOLET 7, DISODIUM SALT see CMM400
C.I. ACID YELLOW 3 see CMM510
C.I. ACID YELLOW 7 see CMM750
C.I. ACID YELLOW 9 see CMM758
C.I. ACID YELLOW 23 see FAG140
C.I. ACID YELLOW 36 see MDM775
C.I. ACID YELLOW 42 see CMM759
C.I. ACID YELLOW 73 see FEW000
C.I. ACID YELLOW 135 see FAL050
C.I. ACID YELLOW 42 DISODIUM SALT see CMM759
C.I. ACID YELLOW 9, DISODIUM SALT (8CI) see CMM758
C.I. ACID YELLOW 36 MONOSODIUM SALT see MDM775
C.I. ACID YELLOW 23, TRISODIUM SALT see FAG140
CIAFOS see COQ399
CIANATIL MALEATE see COS899
CIANAZIL see COH250
CIANIDANOL see CCP875
CIANURO di SODIO (ITALIAN) see SGA500
CIANURO di VINILE (ITALIAN) see ADX500
CIATYL see CLX250
C.I. AZOIC COUPLING COMPONENT 1 see NAX000
C.I. AZOIC COUPLING COMPONENT 2 see CMM760
C.I. AZOIC COUPLING COMPONENT 107 see TGR000
C.I. AZOIC DIAZO COMPONENT 3 see DEO295
C.I. AZOIC DIAZO COMPONENT 5 see NEQ000
C.I. AZOIC DIAZO COMPONENT 6 see NEO000
C.I. AZOIC DIAZO COMPONENT 7 see NEN500
C.I. AZOIC DIAZO COMPONENT 9 see KDA050
C.I. AZOIC DIAZO COMPONENT 11 see CLK235
C.I. AZOIC DIAZO COMPONENT 12 see NMP500
C.I. AZOIC DIAZO COMPONENT 13 see NEQ500
C.I. AZOIC DIAZO COMPONENT 21 see MIH500
C.I. AZOIC DIAZO COMPONENT 22 see PFU500
C.I. AZOIC DIAZO COMPONENT 37 see BIN500
C.I. AZOIC DIAZO COMPONENT 37 see NEO500
C.I. AZOIC DIAZO COMPONENT 48 see DCJ200
C.I. AZOIC DIAZO COMPONENT 112 see BBX000
C.I. AZOIC DIAZO COMPONENT 113 see TGJ750
C.I. AZOIC DIAZO COMPONENT 114 see NBE700
C.I. AZOIC RED 83 see MGO750
C.I. 41000B see IBB000
C.I. 42555B see TJK000
CIBA 34 see MAW850
CIBA 570 see FAB400
CIBA 709 see DGQ875
CIBA 1414 see MRH209
CIBA 1983 see CJQ000
CIBA 2059 see DUK800
CIBA 2446 see AHO750
CIBA-3126 see PAM785
CIBA 5968 see HGP495
CIBA 5968 see HGP500
CIBA 6313 see CES750
CIBA 7115 see KFK000
CIBA 8353 see DVS000

CIBA 8514 see CJJ250
CIBA 8514 see KEA000
CIBA 9295 see MKU750
CIBA 9491 see IEN000
CIBA 11925 see PCY300
CIBA 12223 see PHK000
CIBA 32644 see NML000
CIBA 12669A see MIW500
CIBA 34,647-Ba see BAC275
CIBA 39089-Ba see THK750
CIBA 17309 BA see DAL300
CIBA 32644-BA see NML000
CIBA 34276 BA see MAW850
CIBA 36278-BA see SGB500
CIBA 42155-BA see AGO750
CIBA 42244-BA see AGP000
CIBA BLUE 2B see ICU135
CIBA BLUE 2BD see ICU135
CIBA BLUE 2BDG see ICU135
CIBA BLUE 2BN see ICU135
CIBA BLUE 2BPF see ICU135
CIBA BRILLIANT BLUE BS see ICU135
CIBA C-768 see DTP800
CIBA C-776 see DTP600
CIBA C-2307 see DOL800
CIBA C 7019 see ASG250
CIBA C-7824 see MPG250
CIBA C-9491 see IEN000
CIBACET BLUE GREEN C see DMM400
CIBACET BLUE GREEN CB see DMM400
CIBACET BRILLIANT BLUE BG NEW see MGG250
CIBACET BRILLIANT PINK 4BN see DBX000
CIBACET BRILLIANT VIOLET 3B see DBY700
CIBACETE BRILLIANT PINK 4BN see DBX000
CIBACETE DIAZO NAVY BLUE 2B see DCJ200
CIBACETE RED 3B see AKE250
CIBACETE YELLOW GBA see AAQ250
CIBACET RED 3B see AKE250
CIBACET RED E3B see AKE250
CIBACET RUBINE BS see CMP080
CIBACET RUBINE R see CMP080
CIBACET SAPPHIRE BLUE G see TBG700
CIBACET SCARLET 2B see ENP100
CIBACET SCARLET BRN see ENP100
CIBACET SCARLET BS see ENP100
CIBACET TURQUOISE BLUE 2G see DMM400
CIBACET TURQUOISE BLUE 4G see DMM400
CIBACET TURQUOISE BLUE G see DMM400
CIBACET VIOLET 2R see DBP000
CIBACET YELLOW 2GC see AAQ250
CIBACET YELLOW GBA see AAQ250
CIBACET YELLOW GWL see KDA075
CIBACET YELLOW GWN see KDA075
CIBA CO. 2825 see PIT250
CIBACROLAN BLUE 8G see CMM120
CIBACRON BLACK B-D see CMS220
CIBACTHEN see AES650
CIBA-GEIGY 18809 see ARY800
CIBA-GEIGY C-9491 see IEN000
CIBA-GEIGY C-10015 see DRP600
CIBA-GEIGY GS 13005 see DSO000
CIBA-GEIGY GS 19851 see IOS000
CIBA GO.4350 see CMV375
CIBA 9333 GO see AOA050
CIBA 34276 HYDROCHLORIDE see MAW850
CIBAMIN M 84 see MCB050
CIBAMIN M 100 see MCB050
CIBAMIN ML 100GB see MCB050
CIBANAPHTHOL RF see CMM760
CIBANONE BLUE FG see DFN300
CIBANONE BLUE FGF see DFN300
CIBANONE BLUE FGL see DFN300
CIBANONE BLUE FRS see IBV050
CIBANONE BLUE FRSN see IBV050

CIBANONE BLUE GF see DFN300
CIBANONE BLUE RS see IBV050
CIBANONE BRILLIANT BLUE FR see IBV050
CIBANONE BRILLIANT ORANGE GR see CMU820
CIBANONE BROWN BR see CMU770
CIBANONE BROWN FBR see CMU770
CIBANONE BROWN FGR see CMU780
CIBANONE BROWN GR see CMU780
CIBANONE GOLDEN YELLOW see DCZ000
CIBANONE NAVY BLUE FRA see VGP100
CIBANONE NAVY BLUE RA see VGP100
CIBANONE OLIVE 2R see DUP100
CIBANONE OLIVE F2R see DUP100
CIBANONE RED 6B see CMU825
CIBANONE RED F 6B see CMU825
CIBANONE VIOLET 2R see DFN450
CIBANONE VIOLET 4R see DFN450
CIBANONE VIOLET F 4R see DFN450
CIBANONE VIOLET F 2RB see DFN450
CIBA 2696GO see BLP325
CIBA ORANGE R see CMU815
CIBA ORANGE RDL see CMU815
CIBA ORANGE RP see CMU815
CIBA PINK B see DNT300
C.I. BASIC BLUE 9 see BJI250
C.I. BASIC BLUE 12 see AJN250
C.I. BASIC BLUE 16 see CMM770
C.I. BASIC BLUE 17 see AJP250
C.I. BASIC BROWN 4 see CMM780
C.I. BASIC GREEN 4 see AFG500
C.I. BASIC GREEN 1, SULFATE (1:1) see BAY750
C.I. BASIC ORANGE 1 see CMM800
C.I. BASIC ORANGE 2 see PEK000
C.I. BASIC ORANGE 3 see PEK000
C.I. BASIC ORANGE 14 see BAQ250
C.I. BASIC ORANGE 14 see BJF000
C.I. BASIC ORANGE 21 see CMM820
C.I. BASIC ORANGE 28 see CMM830
C.I. BASIC ORANGE 2, MONOHYDROCHLORIDE see PEK000
C.I. BASIC RED 1 see RGW000
C.I. BASIC RED 2 see GJI400
C.I. BASIC RED 5 see AJQ250
C.I. BASIC RED 12 see CMM840
C.I. BASIC RED 13 see CMM850
C.I. BASIC RED 18 see BAQ750
C.I. BASIC RED 29 see CMM870
C.I. BASIC RED 1, MONOHYDROCHLORIDE see RGW000
C.I. BASIC RED 5, MONOHYDROCHLORIDE see AJQ250
C.I. BASIC RED 9, MONOHYDROCHLORIDE see RMK020
C.I. BASIC VIOLET 1 see MQN025
C.I. BASIC VIOLET 3 see AOR500
C.I. BASIC VIOLET 5 see AJP300
C.I. BASIC VIOLET 10 see FAG070
C.I. BASIC VIOLET 14 see MAC250
C.I. BASIC VIOLET 4 (8CI) see EQG550
C.I. BASIC VIOLET 14, FREE BASE see MAC500
C.I. BASIC VIOLET 14, MONOHYDROCHLORIDE (8CI) see MAC250
C.I. BASIC YELLOW 2 see IBA000
C.I. BASIC YELLOW 11 see CMM890
C.I. BASIC YELLOW 18 see CMM890
C.I. BASIC YELLOW 28 see CMM895
C.I. BASIC YELLOW 2, FREE BASE see IBB000
C.I. BASIC YELLOW 2, MONOHYDROCHLORIDE see IBA000
CIBA THIOCRON see AHO750
CIC see BAR800
CIC see IKH000
C.I. 15800 CA SALT see CMS160
CICHORIUM INTYBUS, ETHANOL EXTRACT see CMM900
CI-628 CITRATE see NHP500

CICLACILLIN see AJJ875
CICLACILLUM see AJJ875
CICLIZINA see EAN600
CICLOBIOTIC see MDO250
CICLOESANO (ITALIAN) see CPB000
CICLOESANOLO (ITALIAN) see CPB750
CICLOESANONE (ITALIAN) see CPC000
6-CICLOESIL-2,4-DINITRO-FENOLO (ITALIAN) see CPK500
1-CICLOESIL-3-p-TOLILSOLFONILUREA (ITALIAN) see CPR000
CICLONIUM IODIDE see OLW400
CICLOPIROX ETHANOLAMINE SALT (1:1) see BAR800
CICLOPIROXOLAMIN see BAR800
CICLOPIROXOLAMINE see BAR800
CICLORAL see BSM000
CICLOSERINA (ITALIAN) see CQH000
CICLOSOM see TIQ250
CICLOSPASMOL see DNU100
CICLOSPORIN see CQH100
CICP see CKC000
CICUTA BULBIFERA L. see WAT325
CICUTA DOUGLASII see WAT325
CICUTAIRE (CANADA) see WAT325
CICUTA MACULATA see WAT325
CICUTIN see PNT000
CICUTINE see PNT000
CICUTOXIN see CMN000
CIDAL see SAH000
CIDALON see IHZ000
CIDAMEX see AAI250
CIDANCHIN see CLD000
CIDANDOPA see DNA200
CIDEFERRON see CMN125
CIDEMUL see DRR400
C.I. DEVELOPER 1 see NNT000
C.I. DEVELOPER 4 see REA000
C.I. DEVELOPER 5 see NAX000
C.I. DEVELOPER 8 see HMX520
C.I. DEVELOPER 13 see PEY500
C.I. DEVELOPER 15 see PFU500
C.I. DEVELOPER 17 see NEO500
C.I. DEVELOPER 20 (obs.) see HMX520
CIDEX see GFQ000
CIDIAL see DRR400
C.I. DIRECT BLACK 17 see CMN230
C.I. DIRECT BLACK 19 see CMN240
C.I. DIRECT BLACK 32 see CMN300
C.I. DIRECT BLACK 38 see AQP000
C.I. DIRECT BLACK 19:1 see CMN150
C.I. DIRECT BLACK 19, DISODIUM SALT see CMN240
C.I. DIRECT BLACK 38, DISODIUM SALT see AQP000
C.I. DIRECT BLACK 17, MONOSODIUM SALT see CMN230
C.I. DIRECT BLACK 32, TRISODIUM SALT see CMN300
C.I. DIRECT BLUE 1 see CMN750
C.I. DIRECT BLUE 2 see CMN800
C.I. DIRECT BLUE 14 see CMO250
C.I. DIRECT BLUE 15 see CMO500
C.I. DIRECT BLUE 53 see BGT250
C.I. DIRECT BLUE 218 see CMO650
C.I. DIRECT BLUE 22, DISODIUM SALT (8CI) see CMO600
C.I. DIRECT BLUE 22 (7CI) see CMO600
C.I. DIRECT BLUE 86 (6CI) see COF420
C.I. DIRECT BLUE 1, TETRASODIUM SALT see CMN750
C.I. DIRECT BLUE 6, TETRASODIUM SALT see CMO000
C.I. DIRECT BLUE 14, TETRASODIUM SALT see CMO250
C.I. DIRECT BLUE 15, TETRASODIUM SALT see CMO500
C.I. DIRECT BLUE 2, TRISODIUM SALT see CMN800
C.I. DIRECT BROWN see CMO750
C.I. DIRECT BROWN 2 see CMO800
C.I. DIRECT BROWN 1:2 see CMO810

C.I. DIRECT BROWN 31 see CMO820
C.I. DIRECT BROWN 154 see CMO825
C.I. DIRECT BROWN 78, DIAMMONIUM
SALT see FMU059
C.I. DIRECT BROWN 2, DISODIUM SALT
see CMO800
C.I. DIRECT BROWN 1A, DISODIUM SALT
see CMO810
C.I. DIRECT BROWN 154, DISODIUM SALT
(8CI) see CMO825
C.I. DIRECT BROWN 31, TETRASODIUM
SALT see CMO820
C.I. DIRECT GREEN 1 see CMO830
C.I. DIRECT GREEN 6, DISODIUM SALT see
CMO840
C.I. DIRECT GREEN 6 (7CI) see CMO840
C.I. DIRECT ORANGE 6 see CMO860
C.I. DIRECT ORANGE 6, DISODIUM SALT
see CMO860
C.I. DIRECT RED 2 see DXO850
C.I. DIRECT RED 13 see CMO872
C.I. DIRECT RED 23 see CMO870
C.I. DIRECT RED 28 see SGQ500
C.I. DIRECT RED 39 see CMO875
C.I. DIRECT RED 46 see CMO880
C.I. DIRECT RED 81 see CMO885
C.I. DIRECT RED 13, DISODIUM SALT see
CMO872
C.I. DIRECT RED 23, DISODIUM SALT see
CMO870
C.I. DIRECT RED 28, DISODIUM SALT see
SGQ500
C.I. DIRECT RED 1, DISODIUM SALT (8CI)
see CMO870
C.I. DIRECT RED 1 (6CI,7CI) see CMO870
C.I. DIRECT RED 46, TETRASODIUM SALT
see CMO880
C.I. DIRECT VIOLET 28 see ASM100
C.I. DIRECT VIOLET 1, DISODIUM SALT
see CMP000
C.I. DIRECT VIOLET 28, DISODIUM SALT
see ASM100
C.I. DIRECT YELLOW 9 see CMP050
C.I. DIRECT YELLOW 9, DISODIUM SALT
see CMP050
C.I. 45350 DISODIUM SALT see FEW000
C.I. DISPERSE BLACK 3 see DPO200
C.I. DISPERSE BLACK 6 see DCJ200
C.I. DISPERSE BLACK 6
DIHYDROCHLORIDE see DOA800
C.I. DISPERSE BLUE 1 see TBG700
C.I. DISPERSE BLUE 3 see MGG250
C.I. DISPERSE BLUE 7 see DMM400
C.I. DISPERSE BLUE 27 see CMP060
C.I. D ISPERSE BLUE 56 see CMP070
C.I. DISPERSE BLUE 56 see DBQ220
C.I. DISPERSE BLUE 59 see CMP070
C.I. DISPERSE BLUE 59 see DBQ220
C.I. DISPERSE BLUE 71 see CMP070
C.I. DISPERSE BLUE 71 see DBQ220
C.I. DISPERSE BLUE 73 see CMP075
C.I. DISPERSE BLUE 78 see BKP500
C.I. DISPERSE BLUE 113 see CMP075
C.I. DISPERSE ORANGE 11 see AKP750
C.I. DISPERSE RED 4 see AKO350
C.I. DISPERSE RED 11 see DBX000
C.I. DISPERSE RED 13 see CMP080
C.I. DISPERSE RED 15 see AKE250
C.I. DISPERSE RED 71 see AKI750
C.I. DISPERSE RED 83 see AKI750
C.I. DISPERSE RED 60 (8CI) see AKI750
C.I. DISPERSE VIOLET 1 see DBP000
C.I. DISPERSE VIOLET 4 see AKP250
C.I. DISPERSE VIOLET 8 see DBY700
C.I. DISPERSE YELLOW 1 see DUW500
C.I. DISPERSE YELLOW 3 see AAQ250
C.I. DISPERSE YELLOW 7 see CMP090
C.I. DISPERSE YELLOW 13 see MEB750
C.I. DISPERSE YELLOW 37 see KDA075
C.I. DISPERSE YELLOW 42 see KDA075
CIDOCETINE see CDP250
CIDOFOVIR see HNS550

CIDOXEPIN HYDROCHLORIDE see
AEG750
CIDREX see CFY000
C.I. FLUORESCENT BRIGHTENER 9 see
DXB450
C.I. FLUORESCENT BRIGHTENER 46 see
TGE155
C.I. FLUORESCENT BRIGHTENER 85 see
CMP100
C.I. FLUORESCENT BRIGHTENER 208 see
DXB500
C.I. FLUORESCENT BRIGHTENER 220 see
CMP120
C.I. FLUORESCENT BRIGHTENER 260 see
CMP200
C.I. FLUORESCENT BRIGHTENING
AGENT 46, SODIUM SALT see TGE155
C.I. FOOD BLACK 1, TETRASODIUM SALT
see BMA000
C.I. FOOD BLUE 1 see FAE100
C.I. FOOD BLUE 2 see FAE000
C.I. FOOD BLUE 2 see FMU059
C.I. FOOD BLUE 3 see ADE500
C.I. FOOD BLUE 5 see CMM062
C.I. FOOD BROWN 1 see CMP250
C.I. FOOD BROWN 2 see CMB750
C.I. FOOD BROWN 3, DISODIUM SALT see
CMP500
C.I. FOOD GREEN 1 see FAE950
C.I. FOOD GREEN 2 see FAF000
C.I. FOOD GREEN 3 see FAG000
C.I. FOOD GREEN 4 see ADF000
C.I. FOOD ORANGE 3 see CMP600
C.I. FOOD ORANGE 4 see HGC000
C.I. FOOD RED 1 see FAG050
C.I. FOOD RED 2 see CMP620
C.I. FOOD RED 3 see HJF500
C.I. FOOD RED 5 see FMU059
C.I. FOOD RED 6 see FAG018
C.I. FOOD RED 7 see FMU080
C.I. FOOD RED 9 see FAG020
C.I. FOOD RED 10 see CMM300
C.I. FOOD RED 11 see CMM400
C.I. FOOD RED 12 see CMS228
C.I. FOOD RED 15 see FAG070
C.I. FOOD RED 16 see CMS238
C.I. FOOD RED 17 see FAG100
C.I. FOOD RED 1, DISODIUM SALT see
FAG050
C.I. FOOD RED 2, DISODIUM SALT see
CMP620
C.I. FOOD RED 6, DISODIUM SALT see
FAG018
C.I. FOOD VIOLET 2 see FAG120
C.I. FOOD VIOLET 3 see VQU500
C.I. FOOD YELLOW 2 see CMM758
C.I. FOOD YELLOW 3 see CMM510
C.I. FOOD YELLOW 3 see FAG150
C.I. FOOD YELLOW 4 see FAG140
C.I. FOOD YELLOW 6 see AIB350
C.I. FOOD YELLOW 8 see MRL100
C.I. FOOD YELLOW 10 see FAG130
C.I. FOOD YELLOW 11 see FAG135
C.I. FOOD YELLOW 13 see CMM510
C.I. FOOD YELLOW 3, DISODIUM SALT see
FAG150
C.I. 45350 (FREE ACID) see FEV000
CIGARETTE REFINED TAR see CMP800
CIGARETTE SMOKE CONDENSATE see
SEC000
CIGARETTE TAR see CMP800
CIGNOLIN see DMG900
CIGTHRANOL see DMG900
CIGUATOXIN 2 see CMP805
CIGUATOXIN CTX 2 see CMP805
CIGUE (CANADA) see PJJ300
CI 624 HYDROCHLORIDE see MGD210
CI-IPC see CKC000
CIKHAT see CDU100
CILAG 61 see HDY000
CILAZAPRIL MONOHYDRATE see CMP810
CILEFA BLACK B see BMA000

CILEFA ORANGE S see FAG150
CILEFA PINK B see FAG040
CILEFA PONCEAU 4R see FMU080
CILEFA RED G see CMP620
CILEFA RUBINE 2B see FAG020
CILEFA YELLOW R see CMM758
CILEFA YELLOW T see FAG140
CILIFOR see TEY000
CILLA BLUE EXTRA see TBG700
CILLA FAST BLUE FFR see MGG250
CILLA FAST BLUE GREEN B see DMM400
CILLA FAST PINK BN see AKE250
CILLA FAST PINK FF3B see DBX000
CILLA FAST PINK RF see AKO350
CILLA FAST RED VIOLET RN see DBP000
CILLA FAST RUBINE B see CMP080
CILLA FAST VIOLET 6B see AKP250
CILLA FAST VIOLET B see DBY700
CILLA FAST YELLOW 5R see CMP090
CILLA FAST YELLOW G see AAQ250
CILLA FAST YELLOW RR see DUW500
CILLA ORANGE R see AKP750
CILLA SCARLET B see ENP100
CILLENTA see BFC750
CILLORAL see BDY669
CILLORAL see BFD000
CILOPEN see BDY669
CILOSTAZOL see CMP825
CIM see CDU750
CIMAGEL see DXY000
CIMCOOL WAFERS see HMJ500
CIMETIDINE see TAB250
CIMETIDINE mixed with SODIUM NITRITE
(4:1) see CMP875
CIMEXAN see MAK700
C.I. MORDANT BLACK 3 see CMP880
C.I. MORDANT BLACK 3, MONOSODIUM
SALT see CMP880
C.I. MORDANT BLACK 11, MONOSODIUM
SALT see EDC625
C.I. MORDANT BROWN 42 see TKN500
C.I. MORDANT ORANGE 1 see NEY000
C.I. MORDANT ORANGE 6 see CMP882
C.I. MORDANT ORANGE 44 see CMP882
C.I. MORDANT ORANGE 6, DISODIUM
SALT see CMP882
C.I. MORDANT ORANGE 1,
MONOSODIUM SALT see SIU000
C.I. MORDANT RED 3 see SEH475
C.I. MORDANT RED 9 see CMG750
C.I. MORDANT RED 11 see DMG800
C.I. MORDANT RED 57 see CMO870
C.I. MORDANT RED 9, TRISODIUM SALT
see CMG750
C.I. MORDANT VIOLET 5 see HLI000
C.I. MORDANT VIOLET 26 see TDD000
C.I. MORDANT VIOLET 5, MONOSODIUM
SALT see HLI000
C.I. MORDANT VIOLET 39,
TRIAMMONIUM SALT (8CI) see AGW750
C.I. MORDANT YELLOW 1 see SIT850
C.I. MORDANT YELLOW 1, MONOSODIUM
SALT see SIT850
C.I. No. 77278 see CMJ900
C.I. No. 46005:1 see BJF000
CI-583 NA see SIF425
CINAMINE see TLN500
CINAMONIN see CMP885
CINANSERIN HYDROCHLORIDE see
CMP900
CINATABS see TLN500
C.I. NATURAL BROWN 1 see FBW000
C.I. NATURAL BROWN 3 see CCP800
C.I. NATURAL BROWN 7 see WAT000
C.I. NATURAL BROWN 8 see MAT500
C.I. NATURAL ORANGE 6 see HMX600
C.I. NATURAL RED 1 see QCA000
C.I. NATURAL RED 20 see HLJ650
C.I. NATURAL RED 25 see CMP905
C.I. NATURAL YELLOW 1 see CDH250
C.I. NATURAL YELLOW 2 see TDD550
C.I. NATURAL YELLOW 3 see ICC800

C.I. NATURAL YELLOW 8 see MRN500
C.I. NATURAL YELLOW 10 see QCA000
C.I. NATURAL YELLOW 11 see MRN500
C.I. NATURAL YELLOW 14 see MQF250
C.I. NATURAL YELLOW 16 see HLY500
CINCAINE HYDROCHLORIDE see NOF500
CINCH see CMP955
CINCHOCAINE see DDT200
CINCHOCAINE HYDROCHLORIDE see NOF500
CINCHOCAINIUM CHLORIDE see NOF500
CINCHOL see SDZ350
CINCHOLEPIDINE see LEL000
CINCHOMERONIC ACID IMIDE see POR000
CINCHONAN-9-OL, (8-α-9R)-(9CI) see CMP910
CINCHONAN-9-OL, 6'-METHOXY-, DIHYDROCHLORIDE, (8-α-9R)-, mixt. with 2-(ACETYLOXY) BENZOIC ACID and 2-HYDROXY-1,2,3-PROPANETRICARBOXYLIC ACID, TRILITHIUM SALT see TGJ350
CINCHONAN-9-OL, 6'-METHOXY-, (9S)-, SULFATE (1:1) (SALT) (9CI) see QFS100
CINCHONAN-9-OL, 6'-METHOXY-, (9S)-, SULFATE (2:1) (SALT) (9CI) see QHA000
CINCHONIDINE see CMP910
(−)-CINCHONIDINE see CMP910
(8S,9R)-CINCHONIDINE see CMP910
d-CINCHONINE see CMP925
CINCHONINIC ACID, 3-ETHYL-, ETHYL ESTER see EHM200
CINCHOPHENE see PGG000
CINCHOPHENIC ACID see PGG000
CINCHOPHEN SODIUM see SJH000
CINCHOPHEN, SODIUM SALT see SJH000
CINCHOVATINE see CMP910
CINCO NEGRITOS (MEXICO) see LAU600
CINCOPHEN see PGG000
CINDOMET see CMP950
CINEB see EIR000
CINENE see MCC250
1,4-CINEOL see IKC100
1,8-CINEOL see CAL000
CINEOLE see CAL000
1,4-CINEOLE see IKC100
1,8-CINEOLE see CAL000
CINEPAZIDE MALEATE see VGK000
CINERIN I ALLYL HOMOLOG see AFR250
CINERIN I or II see POO250
CINERUBIN A see TAH675
CINERUBIN B see TAH650
CINERUBINE A see TAH675
CINERUBINE B see TAH650
CINMETACIN see CMP950
CINMETHACIN see CMP950
CINMETHYLIN see CMP955
CINNAMAL see CMP969
CINNAMALDEHYDE see CMP969
CINNAMALDEHYDE, (E)- see CMP971
(E)-CINNAMALDEHYDE see CMP971
trans-CINNAMALDEHYDE see CMP971
CINNAMALDEHYDE DIMETHYL ACETYL see PGA775
CINNAMAMIDE see CMP973
CINNAMEIN see BEG750
CINNAMENE see SMQ000
CINNAMENOL see SMQ000
CINNAMIC ACID see CMP975
CINNAMIC ACID, (E)- see CMP980
(E)-CINNAMIC ACID see CMP980
trans-CINNAMIC ACID see CMP980
CINNAMIC ACID, α-ACETYL-m-NITRO-, ETHYL ESTER see ENO200
trans-CINNAMIC ACID BENZYL ESTER see BEG750
CINNAMIC ACID-n-BUTYL ESTER see BQV500
CINNAMIC ACID, CINNAMYL ESTER see CMQ850

CINNAMIC ACID-3-(DIETHYLAMINO) PROPYL ESTER see AQP750
CINNAMIC ACID-1,5-DIMETHYL-1-VINYL-4-HEXENYL ESTER see LGA000
CINNAMIC ACID-1,5-DIMETHYL-1-VINYL-4-HEXEN-1-YL ESTER see LGA000
CINNAMIC ACID, o-HYDROXY-, (E)- see CNU850
CINNAMIC ACID, p-(1H-IMIDAZOL-1-YLMETHYL)-, SODIUM SALT, (E)- see SHV100
CINNAMIC ACID, ISOBUTYL ESTER see IIQ000
CINNAMIC ACID, LINALYL ESTER see LGA000
CINNAMIC ACID, NICKEL(II) SALT see CMQ000
CINNAMIC ACID, p-NITRO- see NFT440
CINNAMIC ACID, 3-PHENYLPROPYL ESTER (7CI,8CI) see PGB800
CINNAMIC ACID, 3,4,5-TRIMETHOXY- see CMQ100
CINNAMIC ALCOHOL see CMQ740
trans-CINNAMIC ALDEHYDE see CMP971
CINNAMIC ALDEHYDE DIMETHYL ACETAL see PGA775
CINNAMIN see AQN750
CINNAMOHYDROXAMIC ACID see CMQ475
CINNAMON BARK OIL see CCO750
CINNAMON BARK OIL, CEYLON TYPE (FCC) see CCO750
CINNAMONIN see CCI500
CINNAMONITRILE see CMQ500
CINNAMON LEAF OIL see CMQ510
CINNAMON LEAF OIL, CEYLON see CMQ510
CINNAMON OIL see CCO750
CINNAMON OIL, CEYLON see CMQ510
CINNAMOPHENONE see CDH000
CINNAMOYLHYDROXAMIC ACID see CMQ475
1-CINNAMOYL-2-METHOXY-5-METHOXY-3-INDOLYLACETIC ACID see CMP950
1-CINNAMOYL-5-METHOXY-2-METHYLINDOLE-3-ACETIC ACID see CMP950
14-CINNAMOYLOXYCODEINONE see CMQ625
CINNAMYCIN see CMQ725
CINNAMYL ACETATE see CMQ730
CINNAMYL ALCOHOL see CMQ740
CINNAMYL ALCOHOL ANTHRANILATE see API750
CINNAMYL ALCOHOL, BENZOATE see CMQ750
CINNAMYL ALCOHOL, CINNAMATE see CMQ850
CINNAMYL ALCOHOL, FORMATE see CMR500
CINNAMYL ALCOHOL, SYNTHETIC see CMQ740
CINNAMYL ALDEHYDE see CMP969
trans-CINNAMYLALDEHYDE see CMP971
CINNAMYL-2-AMINOBENZOATE see API750
CINNAMYL-o-AMINOBENZOATE see API750
CINNAMYL ANTHRANILATE (FCC) see API750
CINNAMYL BENZOATE see CMQ750
CINNAMYL BUTYRATE see CMQ800
CINNAMYL CINNAMATE see CMQ850
α-CINNAMYL-N-CYCLOPROPYLMETHYL-α-ETHYL-N-METHYL-2-FURFURYLAMINE see CMQ855
1-CINNAMYL-4-(DIPHENYLMETHYL)PIPERAZINE see CMR100
trans-1-CINNAMYL-(4-DIPHENYLMETHYL)PIPERAZINE see CMR100

d-CINNAMYLEPHEDRINE HYDROCHLORIDE see CMR250
CINNAMYLEPHEDRINE HYDROCHLORIDE, DEXTRO see CMR250
CINNAMYLESTER KYSELINY SKORICOVE see CMQ850
CINNAMYL FORMATE see CMR500
CINNAMYL ISOBUTYRATE see CMR750
CINNAMYL ISOVALERATE see CMR800
CINNAMYL ISOVALERATE see PEE200
CINNAMYL METHANOATE see CMR500
CINNAMYL 3-METHYL BUTYRATE see PEE200
CINNAMYL NITRILE see CMQ500
CINNAMYL PROPIONATE see CMR850
CINNARIZIN see ARQ750
CINNARIZINE see CMR100
CINNARIZINE CLOFIBRATE see CMS125
CINNIMIC ALDEHYDE see CMP969
CINNOLINE, 4-METHYL- see MIO770
CINNOLINE, 4-METHYL- see MIO800
CINNOPROPAZONE see AQN750
CINOBAC see CMS130
CINOBUFAGIN ACETATE see CMS126
CINOBUFAGIN, 3-ACETYL- see CMS126
CINOPAL see BGL250
CINOPOP see BGL250
CINOSULFURON see CMS127
CINOXACIN see CMS130
CIN-QUIN see QFS000
CIN-QUIN see QHA000
CINU see CGV250
CINX see CMS130
CIODRIN see COD000
CIODRIN VINYL PHOSPHATE see COD000
CIOVAP see COD000
C.I. OXIDATION BASE see TGL750
C.I. OXIDATION BASE 2 see PFU500
C.I. OXIDATION BASE 4 see TGM400
C.I. OXIDATION BASE 7 see ALS990
C.I. OXIDATION BASE 10 see PEY500
C.I. OXIDATION BASE 12 see DBO000
C.I. OXIDATION BASE 16 see PEY250
C.I. OXIDATION BASE 17 see ALT000
C.I. OXIDATION BASE 19 see AHQ250
C.I. OXIDATION BASE 20 see TGL750
C.I. OXIDATION BASE 21 see DUP400
C.I. OXIDATION BASE 22 see ALL750
C.I. OXIDATION BASE 26 see CCP850
C.I. OXIDATION BASE 31 see REA000
C.I. OXIDATION BASE 32 see PPQ500
C.I. OXIDATION BASE 33 see NAW500
C.I. OXIDATION BASE 35 see TGL750
C.I. OXIDATION BASE 6A see ALT250
C.I. OXIDATION BASE 10A see PEY650
C.I. OXIDATION BASE 12A see DBO400
C.I. OXIDATION BASE 13A see CEG625
C.I. OXIDATION BASE 200 see TGL750
CIPC see CKC000
4-CIPC see IOK100
CIPE see PJS750
C.I. PIGMENT BLACK 6 see CBT750
C.I. PIGMENT BLACK 7 see CBT750
C.I. PIGMENT BLACK 10 see CBT500
C.I. PIGMENT BLACK 13 see CND125
C.I. PIGMENT BLACK 14 see MAS000
C.I. PIGMENT BLACK 15 see CNO250
C.I. PIGMENT BLACK 16 see ZBJ000
C.I. PIGMENT BLACK 30 see CMS135
C.I. PIGMENT BLUE 34 see CNQ000
C.I. PIGMENT BLUE 60 see IBV050
C.I. PIGMENT BROWN 8 see MAS000
C.I. PIGMENT BROWN 28 see CMU780
C.I. PIGMENT BROWN 34 see CMS137
C.I. PIGMENT GREEN 7 see PJQ100
C.I. PIGMENT GREEN 12 see NAX500
C.I. PIGMENT GREEN 17 see CMJ900
C.I. PIGMENT GREEN 36 see CMS140
C.I. PIGMENT GREEN 38 see CMS140
C.I. PIGMENT GREEN 41 see CMS140
C.I. PIGMENT GREEN 42 see PJQ100
C.I. PIGMENT GREEN 21 (9CI) see COF500

C.I. PIGMENT METAL 2 see CNI000
C.I. PIGMENT METAL 3 see GIS000
C.I. PIGMENT METAL 4 see LCF000
C.I. PIGMENT METAL 6 see ZBJ000
C.I. PIGMENT ORANGE 5 see DVB800
C.I. PIGMENT ORANGE 13 see CMS145
C.I. PIGMENT ORANGE 20 see CAJ750
C.I. PIGMENT ORANGE 21 see LCS000
C.I. PIGMENT ORANGE 43 see CMU820
C.I. PIGMENT RED see CHP500
C.I. PIGMENT RED see LCS000
C.I. PIGMENT RED 3 see MMP100
C.I. PIGMENT RED 4 see CJD500
C.I. PIGMENT RED 23 see NAY000
C.I. PIGMENT RED 49 see NAP100
C.I. PIGMENT RED 53 see CMS150
C.I. PIGMENT RED 83 see DMG800
C.I. PIGMENT RED 101 see IHC450
C.I. PIGMENT RED 104 see LDM000
C.I. PIGMENT RED 104 see MRC000
C.I. PIGMENT RED 105 see LDS000
C.I. PIGMENT RED 107 see AQL500
C.I. PIGMENT RED 122 see DTV360
C.I. PIGMENT RED 195 see CMU825
C.I. PIGMENT RED 48:1 see CMS148
C.I. PIGMENT RED 64:1 see CMS160
C.I. PIGMENT RED 48, BARIUM SALT (1:1)
(8CI) see CMS148
C.I. PIGMENT RED 64, CALCIUM SALT (2:1)
(8CI) see CMS160
C.I. PIGMENT RED 57, DISODIUM SALT
(8CI) see CMS155
C.I. PIGMENT RED 57 (7CI) see CMS155
C.I. PIGMENT RED 49, MONOSODIUM
SALT see NAP100
C.I. PIGMENT VIOLET 31 see DFN450
C.I. PIGMENT WHITE 3 see LDY000
C.I. PIGMENT WHITE 4 see ZKA000
C.I. PIGMENT WHITE 6 see TGG760
C.I. PIGMENT WHITE 7 see ZNJ100
C.I. PIGMENT WHITE 10 see BAJ250
C.I. PIGMENT WHITE 11 see AQF000
C.I. PIGMENT WHITE 17 see BKW100
C.I. PIGMENT WHITE 18 see CAT775
C.I. PIGMENT WHITE 21 see BAP000
C.I. PIGMENT WHITE 25 see CAX750
C.I. PIGMENT WHITE 32 see ZJS400
C.I. PIGMENT YELLOW see ARI000
C.I. PIGMENT YELLOW 12 see DEU000
C.I. PIGMENT YELLOW 13 see CMS208
C.I. PIGMENT YELLOW 14 see CMS210
C.I. PIGMENT YELLOW 31 see BAK250
C.I. PIGMENT YELLOW 32 see SMH000
C.I. PIGMENT YELLOW 33 see CAP500
C.I. PIGMENT YELLOW 33 see CAP750
C.I. PIGMENT YELLOW 34 see LCR000
C.I. PIGMENT YELLOW 35 see CMS212
C.I. PIGMENT YELLOW 36 see ZFJ100
C.I. PIGMENT YELLOW 37 see CAJ750
C.I. PIGMENT YELLOW 40 see PLI750
C.I. PIGMENT YELLOW 46 see LDN000
C.I. PIGMENT YELLOW 48 see LCU000
C.I. PIGMENT YELLOW 97 (8CI) see CMS214
CIPLAMYCETIN see CDP250
CIPOVIOL W 72 see PKP750
CIPRIL (PUERTO RICO) see SLJ650
CIPROFIBRATE see CMS216
CIPROMID see CQJ250
CIRAM see BJK500
CIRANTIN see HBU000
CIRCAIN see DNC000
CIRCAIR see DNC000
CIRCANOL see DLL400
CIRCOSOLV see TIO750
CIRCULIN see CMS218
C.I. REACTIVE BLACK 8 see CMS220
C.I. REACTIVE BLUE 4 see CMS222
C.I. REACTIVE BLUE 19 see BMM500
C.I. REACTIVE BLUE 21 see CMS224
C.I. REACTIVE BLUE 19, DISODIUM SALT
see BMM500
C.I. REACTIVE RED 2 see PMF540

C.I. REACTIVE YELLOW 14 see RCZ000
C.I. REACTIVE YELLOW 73 see RCM226
C.I. RED 33, DISODIUM SALT see CMS228
C.I. REDUCING AGENT 6 see CMS228
CIRENE BRILLIANT BLUE R see ADE750
CIROLEMYCIN see CMS230
CIRPONYL see MQU750
CIRRASOL 185A see NMY000
CIRRASOL ALN-WF see PJT300
CIRRASOL-OD see HCQ500
CISAPRIDE see CMS232
CISCLOMIPHENE see CMX500
CISMETHRIN see RDZ875
CISOBITAN see CMS234
C.I. SOLVENT BLACK 3 see PCJ200
C.I. SOLVENT BLACK 5 see CMS236
C.I. SOLVENT BLUE 7 see PEI000
C.I. SOLVENT BLUE 18 see TBG700
C.I. SOLVENT BLUE 69 see DMM400
C.I. SOLVENT BLUE 78 see BKP500
C.I. SOLVENT BLUE 93 see BKP500
C.I. SOLVENT BROWN 1 see NBG500
C.I. SOLVENT BROWN PR see NBG500
C.I. SOLVENT GREEN 3 see BLK000
C.I. SOLVENT GREEN 7 see TNM000
C.I. SOLVENT ORANGE 2 see TGW000
C.I. SOLVENT ORANGE 3 see PEK000
C.I. SOLVENT ORANGE 7 see XRA000
C.I. SOLVENT ORANGE 15 see BJF000
C.I. SOLVENT ORANGE 1 (8CI) see CMP600
C.I. SOLVENT RED see CMS238
C.I. SOLVENT RED 14 see ENP100
C.I. SOLVENT RED 19 see EOJ500
C.I. SOLVENT RED 23 see OHI200
C.I. SOLVENT RED 24 see SBC500
C.I. SOLVENT RED 41 see MAC500
C.I. SOLVENT RED 43 see BMO250
C.I. SOLVENT RED 48 see SKS400
C.I. SOLVENT RED 53 see AKE250
C.I. SOLVENT RED 72 see DDO200
C.I. SOLVENT RED 80 see DOK200
C.I. SOLVENT VIOLET 9 see TJK000
C.I. SOLVENT VIOLET 11 see DBP000
C.I. SOLVENT VIOLET 12 see AKP250
C.I. SOLVENT VIOLET 13 see HOK000
C.I. SOLVENT VIOLET 26 see DBX000
C.I. SOLVENT YELLOW 1 see PEI000
C.I. SOLVENT YELLOW 2 see DOT300
C.I. SOLVENT YELLOW 3 see AIC250
C.I. SOLVENT YELLOW 4 see PEJ600
C.I. SOLVENT YELLOW 5 see FAG130
C.I. SOLVENT YELLOW 7 see HJF000
C.I. SOLVENT YELLOW 12 see OHK000
C.I. SOLVENT YELLOW 14 see PEJ500
C.I. SOLVENT YELLOW 33 see CMS240
C.I. SOLVENT YELLOW 34 see IBB000
C.I. SOLVENT YELLOW 52 see DUW500
C.I. SOLVENT YELLOW 56 see OHJ875
C.I. SOLVENT YELLOW 92 see AAQ250
C.I. SOLVENT YELLOW 94 see FEV000
C.I. SOLVENT YELLOW 99 see AAQ250
C.I. SOLVENT YELLOW 1,
MONOHYDROCHLORIDE see PEI250
C.I. SOLVENT YELLOW 3
MONOHYDROCHLORIDE see MPY750
CISPLATINO (SPANISH) see PJD000
CISPLATYL see PJD000
CISTANCHE TUBULOSA Wight (extract) see
CMS245
CISTAPHOS see AKB500
CISTEAMINA (ITALIAN) see AJT250
C.I. SULFUR BLACK 1 see CMS250
C.I. SULPHUR BLACK 1 see CMS250
C.I. SULPHUR RED 6 see CMS257
CITALOPRAM HYDROBROMIDE see
CMS258
CITANEST see CMS260
CITANEST HYDROCHLORIDE see CMS260
CITARABINA see AQQ750
CITARIN see TDX750
CITARIN L see LFA020
CITEXAL see QAK000

CITEX BCL 462 see DDM300
CITEX BT 93 see EIT150
CITGRENILE see MNB500
CITHROL PO see PJY100
CITICHOLINE see CMF350
CITICOLINE see CMF350
CITIDIN DIFOSFATO de COLINA see
CMF350
CITIDOLINE see CMF350
CITIFLUS see ARQ750
CITILAT see AEC750
CITIOLASE see TCZ000
CITIOLONE see TCZ000
CITIREUMA see SOU550
CITOBARYUM see BAP000
CITOCOR see DJS200
CITODON see ERD500
CITOFUR see FLZ050
CITOL see ALT250
CITOMULGAN M see OAV000
CITOPAN see ERD500
CITOSARIN see AJJ875
CITOSULFAN see BOT250
CITOX see DAD200
CITRACETAL see CMS324
CITRACONIC ACID see CMS320
CITRACONIC ACID ANHYDRIDE see
CMS322
CITRACONIC ANHYDRIDE see CMS322
CITRA-FORT see ABG750
(E)-CITRAL see GCU100
trans-CITRAL see GCU100
CITRAL α see GCU100
α-CITRAL see GCU100
CITRAL DIETHYL ACETAL see CMS323
CITRAL DIMETHYL ACETAL see DOE000
CITRAL ETHYLENE GLYCOL ACETAL see
CMS324
CITRAL (FCC) see DTC800
CITRAL METHYLANTHRANILATE,
SCHIFF'S BASE see CMS325
CITRAM see AMX825
CITRAM see DJA400
CITRAMON see ARP250
CITRATE de ACETOXY-THYMOXY-
ETHYL-DIMETHYLAMINE (FRENCH) see
MRN600
CITRAZINIC ACID see DMV400
CITRAZON see BCP000
CITREOVIRIDIN see CMS500
CITREOVIRIDINE see CMS500
CITRETTEN see CMS750
CITRIC ACID see CMS750
CITRIC ACID, anhydrous see CMS750
CITRIC ACID, ACETYL TRIETHYL ESTER
see ADD750
CITRIC ACID, AMMONIUM IRON(3+) SALT
see FAS700
CITRIC ACID, AMMONIUM SALT see
ANF800
CITRIC ACID, COPPER(2+) SALT (8CI) see
CNK625
CITRIC ACID, DYSPROSIUM(3+) salt (1:1) see
DYG800
CITRIC ACID, GALLIUM SALT (1:1) see
GBO000
CITRIC ACID, MONOSODIUM SALT see
MRL000
CITRIC ACID, SAMARIUM SALT see SAS000
CITRIC ACID, SODIUM SALT see MRL000
CITRIC ACID, TRIBUTYL ESTER see
THY100
CITRIC ACID, TRIBUTYL ESTER, ACETATE
see THX100
CITRIC ACID, TRILITHIUM SALT see
TKU500
CITRIC ACID, TRILITHIUM SALT,
HYDRATE see LHD150
CITRIC ACID, TRIPOTASSIUM SALT see
PLB750
CITRIC ACID, YTTRIUM SALT (3:1) see
YFA000

CITRIC ACID, ZINC SALT (2:3) see ZFJ250
CITRIDIC ACID see ADH000
CITRININ see CMS775
CITRO see CMS750
CITRODYLE see OHJ100
CITROFLEX 2 see TJP750
CITROFLEX 4 see THY100
CITROFLEX A see THX100
CITROFLEX A 2 see ADD750
CITROFLEX A 4 see THX100
CITROFLUYL see MRL000
CITRONELLAL see CMS845
CITRONELLAL HYDRATE see CMS850
CITRONELLA OIL see CMT000
CITRONELLENE see CMT050
CITRONELLIC ACID see CMT125
CITRONELLOL see CMT250
CITRONELLOL see DTF410
α-CITRONELLOL see DTF400
CITRONELLOXYACETALDEHYDE see
CMT300
α-CITRONELLYL ACETATE see RHA000
CITRONELLYL ACETATE (FCC) see AAU000
CITRONELLYL-2-BUTENOATE see CMT500
CITRONELLYL BUTYRATE see CMT600
CITRONELLYL-α-CROTONATE see CMT500
CITRONELLYL ETHYL ETHER see EES100
CITRONELLYL FORMATE see CMT750
CITRONELLYL ISOBUTYRATE see CMT900
CITRONELLYL NITRILE see CMU000
CITRONELLYLOXYACETALDEHYDE see
CMT300
CITRONELLYL PHENYLACETATE see
CMU050
CITRONELLYL PROPIONATE see CMU100
CITRON YELLOW see PLW500
CITRON YELLOW see ZFJ100
CITROSODINE see TNL000
CITROVIOL see DTC000
CITRULLAMON see DKQ000
CITRULLAMON see DNU000
l-CITRULLINE and SODIUM NITRITE (2:1)
see SIS100
CITRUS AURANTIUM see LBE000
CITRUS FIX see DAA800
CITRUS HYSTRIX DC., fruit peel extract see
CMU300
CITRUS RED No. 2 see DOK200
CITRYLIDENEACETONE see POH525
CITTERAL see SEQ000
C.I. VAT BLACK 1 see CMU320
C.I. VAT BLACK 8 see CMU475
C.I. VAT BLACK 27 see DUP100
C.I. VAT BLACK 28 see IBJ000
C.I. VAT BLUE 1 see BGB275
C.I. VAT BLUE 4 see IBV050
C.I. VAT BLUE 5 see ICU135
C.I. VAT BLUE 6 see DFN300
C.I. VAT BLUE 12 see DLO900
C.I. VAT BLUE 16 see CMU500
C.I. VAT BLUE 18 see VGP100
C.I. VAT BLUE 22 see CMU750
C.I. VAT BROWN 1 see CMU770
C.I. VAT BROWN 3 see CMU780
C.I. VAT BROWN 25 see CMU800
C.I. VAT BROWN 44 see CMU770
C.I. VAT GREEN 2 see APK500
C.I. VAT GREEN 3 see CMU810
C.I. VAT ORANGE 5 see CMU815
C.I. VAT ORANGE 7 seé CMU820
C.I. VAT RED 13 see CMU825
C.I. VAT RED 41 see DNT300
C.I. VAT VIOLET 1 (8CI) see DFN450
C.I. VAT YELLOW see DCZ000
C.I. VAT YELLOW 2 see VGP200
CIVET see OGM100
CIVET ABSOLUTE see OGM100
CIVETONE see CMU850
cis-CIVETONE see CMU850
CIZARON see TMP750
CK3 see CBT750
ChKhZ 9 see BBS300

ChKhZ 21 see ASM270
ChKhZ 21R see ASM270
CL 337 see ASA500
CL 369 see CKD750
CL-395 see PDC890
CL-845 see PJA170
CL-911C see DVP400
CL 912C see LFG100
CL 9140 see QOJ200
CL 10304 see AJD000
CL 11344 see FAQ600
CL 11366 see PGJ000
CL 12150 see IRA050
CL 12503 see POP000
CL 12,625 see PIF750
CL 12880 see DSP400
CL 13494 see AKO500
CL 13,900 see AEI000
CL-14377 see MDV500
CL 16,536 see POK300
CL 18133 see EPC500
CL 24055 see DUI000
CL 26691 see CQL250
CL 27,319 see MFD500
CL 34433 see AQY375
CL-34699 see COW825
CL-38023 see FAB600
CL 40881 see EDW875
CL-43,064 see DXN600
CL 47300 see DSQ000
CL-47,470 see DHH400
CL 52160 see TAL250
CL 54998 see BMM600
CL 59806 see MQW250
CL-62362 see DCS200
CL 64475 see DHH200
CL 65336 see AJV500
CL 67772 see AOA095
CL 69049 see PMF550
CL-71563 see DCS200
CL82204 see BGL250
CL 88236 see AJI550
CL 216942 see BGW325
CL 217300 see HGP525
CL 227193 see SJJ200
CL 232315 see MQY100
CL316,243 see SJN675
CL 67310465 see PBT100
CL 1950675526 see TES000
CL 19217 4090L 7-5525 see EGI000
Cl-ADO see CEF100
CLAFEN see CQC500
CLAFEN see CQC650
CLAIRFORMIN see CMV000
CLAIRSIT see PCF300
CLAM POISON DIHYDROCHLORIDE see
SBA500
CLANDILON see DNU100
CLANICLOR see CMU875
CLAODICAL see CCK125
CLAPHENE see CQC650
CLARIFIED OILS (PETROLEUM),
CATALYTIC CRACKED see CMU890
CLARIFIED SLURRY OIL see CMU890
CLARIPEX see ARQ750
CLARITHROMYCIN see MJV775
CLARK I see CGN000
CLARO 5591 see SLJ500
CLARY OIL see CMU900
CLARY SAGE OIL see CMU900
CLARY SAGE OIL see OGQ200
CLATHROMYCIN see MJV775
CLAUDELITE see ARI750
CLAUDETITE see ARI750
CLAVACIN see CMV000
CLAVELLINA (PUERTO RICO) see CAK325
CLAVULANIC ACID SODIUM SALT see
CMV250
CLAYTON YELLOW see CMP050
CLD see CNH285
CLEARASIL BENZOYL PEROXIDE
LOTION see BDS000

CLEARASIL BP ACNE TREATMENT see
BDS000
CLEAREL see SLJ500
CLEARJEL see SLJ500
CLEARTRAN see ZNJ100
CLEARTUF see PKF750
CLEARY 3336 see DJV000
CLEBOPRIDE HYDROGEN MALATE see
CMV325
CLEBOPRIDE MALATE see CMV325
CLEFNON see CAT775
CLEISTANTHIN A see CMV375
CLELAND'S REAGENT see DXO775
CLELAND'S REAGENT see DXO800
CLEMANIL see FOS100
CLEMASTINE FUMARATE see FOS100
CLEMASTINE HYDROGEN FUMARATE see
FOS100
CLEMATIS see CMV390
CLEMATITE AUX GEAUX (CANADA) see
CMV390
CLEMIZOLE HDYROCHLORIDE see
CMV400
CLENBUTEROL HYDROCHLORIDE see
VHA350
CLENIL-A see AFJ625
CLENOLIXIMAB see CMV425
CLEOCIN see CMV675
CLEOCIN PHOSPHATE see CMV690
CLEP see CMV475
C. LEPTOSEPALA see MBU550
CLERA see NCW000
CLERIDIUM 150 see PCP250
CLESTOL see DJL000
CLETHODIM see CMV485
CLEVE'S ACID-1,6 see ALI250
CLEVE'S BETA-ACID see ALI250
CLEXANE see HAQ500
CLF II see CBT500
CLIACIL see PDT750
CLIDANAC see CFH825
CLIDANAC see CMV500
CLIFT see MCI750
CLIFTON CMPP 60 see RBF500
CLIMATERINE see DKA600
CLIMBAZOL see CJN900
CLIMBAZOLE see CJN900
CLIMBING BITTERSWEET see AHJ875
CLIMBING LILY see GEW800
CLIMBING ORANGE ROOT see AHJ875
CLIMESTRONE see ECU750
CLIN see SJO000
CLINDAMYCIN see CMV675
CLINDAMYCINE (FRENCH) see CMV675
CLINDAMYCIN HYDROCHLORIDE see
CHU050
CLINDAMYCIN-2-PALMITATE
MONOHYDROCHLORIDE see CMV680
CLINDAMYCIN PHOSPHATE see CMV690
CLINDAMYCIN-2-PHOSPHATE see CMV690
CLINDROL 2000 see OHU200
CLINDROL 2020 see OHU200
CLINDROL 101CG see BKE500
CLINDROL 200CGN see CNF330
CLINDROL 202CGN see CNF330
CLINDROL LT 15-73-1 see BKF500
CLINDROL SDG see HKJ000
CLINDROL SEG see EJM500
CLINDROL SUPERAMIDE 100L see BKE500
CLINDROL SUPERAMIDE 100CG see
CNF330
CLINESTROL see DKB000
CLINOFIBRATE see CMV700
CLINOMEL see AHR600
CLINOMIX see AHR600
CLINOPTILOLITE see CMV850
CLINOPTILOLITE (8CI) see CMV850
CLINORIL see SOU550
CLINOXAN see CFG750
CLIOQUINOL see CHR500
CLIOXANIDE see CGB250
CLIPPER see TMK125

CLIQUINOL see CHR500
CLIRADON see KFK000
CLIRADONE see KFK000
CLISTIN see CGD500
CLISTINE MALEATE see CGD500
CLISTIN MALEATE see CGD500
CLIVIA (VARIOUS SPECIES) see KAJ100
CLIVORINE see CMV950
CLIXODYNE see HIM000
CLM see SFY500
CLOAZEPAM see CMW000
CLOAZEPAM see CMW000
CLOBAZAM see CIR750
CLOBBER see CQJ250
CLOBENZEPAM HYDROCHLORIDE see CMW250
CLOBERAT see ARQ750
CLOBETASOL PROPIONATE see CMW300
CLOBETASOL-17-PROPIONATE see CMW300
CLOBETASONE BUTYRATE see CMW400
CLOBETASONE-17-BUTYRATE see CMW400
CLOBRAT see ARQ750
CLOBREN-SF see ARQ750
CLOBUTINOL see CMW459
CLOBUTINOL HYDROCHLORIDE see CMW500
CLOCAPRAMINE DIHYDROCHLORIDE see DKV309
CLOCARPRAMINE DIHYDROCHLORIDE see DKV309
CLOCETE see AAE500
CLOCONAZOLE HYDROCHLORIDE see CMW550
CLODAZONE see CMW600
CLODRONATE SODIUM see SFX730
CLOETHOCARB see LAU500
CLOETOCARB see LAU500
CLOFAR see ARQ750
CLOFEDANOL HYDROCHLORIDE see CMW700
CLOFENOTANE see DAD200
CLOFENOXIN see DPE000
CLOFEXAMIDE see CJN750
CLOFEXAMIDE PHENYLBUTAZONE see CMW750
CLOFEXAMIDE-PHENYLBUTAZONE MIXTURE see CMW750
CLOFEZON see CMW750
CLOFEZONE see CMW750
CLOFIBRAM see ARQ750
CLOFIBRAT see ARQ750
CLOFIBRATO (SPANISH) see ARQ750
CLOFIBRATO de CINARIZINA (SPANISH) see CMS125
CLOFIBRIC ACID see CMX000
CLOFIBRINIC ACID see CMX000
CLOFIBRINSAEURE (GERMAN) see CMX000
CLOFINIT see ARQ750
CLOFIPRONT see ARQ750
CLOFLUPEROL HYDROCHLORIDE see CLS250
CLOFUZID see DEL300
CLOMAZONE see CEP800
CLOMEPHENE B see CMX500
CLOMETHIAZOLE see CHD750
CLOMETHIAZOLUM see CHD750
CLOMID see CMX700
CLOMIDAZOLE see CDY325
CLOMIFEN CITRATE see CMX700
CLOMIFENE see CMX500
trans-CLOMIFENE CITRATE see CMX750
CLOMIFENO see CMX700
CLOMIPHENE see CMX500
CLOMIPHENE CITRATE see CMX700
trans-CLOMIPHENE CITRATE see CMX750
racemic-CLOMIPHENE CITRATE see CMX700
CLOMIPHENE DIHYDROGEN CITRATE see CMX700
CLOMIPHENE-R see CMX700

CLOMIPHINE see CMX700
CLOMIPRAMINE see CDU750
CLOMIPRAMINE HYDROCHLORIDE see CDV000
CLOMIVID see CMX700
CLOMPHID see CMX700
CLONAZEPAM see CMW000
CLONIDIN see DGB500
CLONIDINE see DGB500
CLONIDINE HYDROCHLORIDE see CMX760
CLONITARLID see DFV600
CLONITRALID see DFV400
CLONIXIC ACID see CMX770
CLONIXIN see CMX770
CLONIXINE see CMX770
CLONT see MMN250
CLOPANE HYDROCHLORIDE see CPV609
CLOPENTHIXOL DIHYDROCHLORIDE see CLX250
CLOPEPRAMINE HYDROCHLORIDE see IFZ900
CLOPERASTINA CLORIDRATO (ITALIAN) see CMX820
CLOPERASTINE see CMX800
CLOPERASTINE HYDROCHLORIDE see CMX820
CLOPERIDONE HYDROCHLORIDE see CMX840
CLOPHEDIANOL HYDROCHLORIDE see CMW700
CLOPHEN see PJL750
CLOPHEN A-30 see CMX845
CLOPHEN A-50 see CMX847
CLOPHEN A60 see PJN250
CLOPHENOXATE see DPE000
CLOPIDOL see CMX850
CLOPIPAZAN MESYLATE see CMX860
CLOPIXOL see CLX250
CLOPOXIDE see LFK000
CLOPRADONE see EQO000
CLOPROMAZINA (ITALIAN) see CKP250
CLOPROP see CJQ300
CLOPROSTENOL see CMX880
CLOPYRALID see CMX900
CLOPYRALID see DGJ100
CLOQUINOZINE TARTRATE see CMX920
CLORAFIN see PAH780
CLORALIO see CDN550
CLORAMIDINA see CDP250
CLORAMIN see BIE250
CLORARSEN see DFX400
CLORAZEPATE DIPOTASSIUM see CDQ250
CLOR CHEM T-590 see CDV100
CLORDAN (ITALIAN) see CDR750
CLORDIAZEPOSSIDO (ITALIAN) see LFK000
CLORDION see CBF250
CLOREPIN see CIR750
CLORESTROLO see CLO750
CLOREX see DFJ050
CLORFENIRAMINA see CLX300
CLORGYLINE HYDROCHLORIDE see CMY000
CLORHIDRATO de CELIPROLOL (SPANISH) see SBN475
CLORIDRATO DI-2-BENZIL-4,5-IMIDAZOLINA (ITALIAN) see BBW750
CLORILAX see CKF500
CLORINA see CDP000
CLORINDANOL see CDV700
CLORMETAZANONE see CKF500
CLORMETHAZON see CKF500
CLORNAPHAZINE see BIF250
CLORO (ITALIAN) see CDV750
CLOROAMFENICOLO (ITALIAN) see CDP250
4-CLORO-3-AMINOBENZOATO di DIMETILAMINOETILE CLORIDRATO (ITALIAN) see AJE350
CLOROBEN see DEP600

2-CLOROBENZALDEIDE (ITALIAN) see CEI500
CLOROBENZENE (ITALIAN) see CEJ125
(4-CLORO-BENZIL)-(4-CLORO-FENIL)-SOLFURO (ITALIAN) see CEP000
1-p-CLORO-BENZOIL-5-METOXI-2-METILINDOL-3-ACIDO ACETICO (SPANISH) see IDA000
2-CLORO-1,3-BUTADIENE (ITALIAN) see NCI500
(4-CLORO-BUT-2-IN-IL)-N-(3-CLORO-FENIL)-CARBAMMATO (ITALIAN) see CEW500
CLOROCHINA see CLD000
CLORODANE see CDR750
O-2-CLORO-1-(2,4-DICLORO-FENIL)-VINYL-O,O-DIETILFOSFATO (ITALIAN) see CDS750
(2-CLORO-3-DIETILAMINO-1-METIL-3-OXO-PROP-1-EN-IL)-DIMETIL-FOSFATO see FAB400
CLORODIFENILI, CLORO 42% (ITALIAN) see PJM500
CLORODIFENILI, CLORO 54% (ITALIAN) see PJN000
p-CLORO-α-(2-DIMETILAMINO)-1-METILETIL)-α-METIL FENETIL ALCOOL (ITALIAN) see CMW459
p-CLORO-α-(2-DIMETILAMINO)-1-METILETIL)-α-METIL FENETIL ALCOOL CLORIDRATO (ITALIAN) see CMW500
2-CLORO-4-DIMETILAMINO-6-METIL-PIRIMIDINA (ITALIAN) see CCP500
2-CLORO-10-(3-DIMETILAMINOPROPIL)FENOTIAZINA (ITALIAN) see CKP250
1-CLORO-2,4-DINITROBENZENE (ITALIAN) see CGM000
1-CLORO-2,3-EPOSSIPROPANO (ITALIAN) see EAZ500
CLOROETANO (ITALIAN) see EHH000
2-CLOROETANOLO (ITALIAN) see EIU800
(CLORO-2-ETIL)-1-CICLOESIL-3-NITROSOUREA (ITALIAN) see CGV250
(CLORO-2-ETIL)-1-(RIBOPIRANOSILTRIACETATO-2',3',4')-3-NITROSOUREA (ITALIAN) see ROF200
1-(2-(p-CLORO-α-FENILBENZILOSSI)ETIL)PIPERIDINA CLORIDRATO (ITALIAN) see CMX820
(4-CLORO-FENIL)-BENZOL-SOLFONATO (ITALIAN) see CJR500
(4-CLORO-FENIL)-4-CLORO-VENZOL-SOLFONATO (ITALIAN) see CJT750
3-(4-CLORO-FENIL)-1,1-DIMETIL-UREA (ITALIAN) see CJX750
2(2-(4-CLORO-FENIL-2-FENIL)-ACETIL)INDAN-1,3-DIONE (ITALIAN) see CJJ000
α-(p-CLOROFENIL)-α-FENIL-2-PIPERIDILMETANOLO CLORIDRATO (ITALIAN) see CKI175
N-(3-CLORO-FENIL)-ISOPROPIL-CARBAMMATO (ITALIAN) see CKC000
CLOROFORMIO (ITALIAN) see CHJ500
CLOROFOS (RUSSIAN) see TIQ250
CLOROMETANO (ITALIAN) see MIF765
7-CLORO-2-METILAMINO-5-FENIL-3H-1,4-BENZOIDIAZEPINA 4-OSSIDO (ITALIAN) see LFK000
7-CLORO-3-METIL-2H-1,2,4-BENZOTIODIAZINE-1,1-DIOSSIDO (ITALIAN) see DCQ700
3-CLORO-2-METIL-PROP-1-ENE (ITALIAN) see CIU750
CLOROMISAN see CDP250
1-CLORO-4-NITROBENZENE (ITALIAN) see NFS525
O-(4-CLORO-3-NITRO-FENIL)-O,O-DIMETIL-MONOIIOFOSFATO (ITALIAN) see NFT000

O-(3-CLORO-4-NITRO-FENIL)-O,O-DIMETIL-MONOTIOFOSFATO (ITALIAN) see MIJ250
CLOROPHENE see CJU250
CLOROPICRINA (ITALIAN) see CKN500
CLOROPIRIL see CLX300
CLOROPIRIL see TAI500
CLOROPRENE (ITALIAN) see NCI500
CLOROSAN see CDP000
CLOROSINTEX see CDP250
CLOROTEPINE see ODY100
CLOROTETRACICLINA CLORIDRATO (ITALIAN) see CMB000
CLOROTRISIN see CLO750
CLOROX see SHU500
CLORPRENALINE HYDROCHLORIDE see IMX150
CLORPROPAMIDE (ITALIAN) see CKK000
CLORSULON see CMX010
CLORTERMINE HYDROCHLORIDE see DRC600
CLORTOKEM see CIS250
CLORTRAN see ABD000
CLORURO DI ETILE (ITALIAN) see EHH000
CLORURO di ETHENE (ITALIAN) see EIY600
CLORURO di ETILIDENE (ITALIAN) see DFF809
CLORURO di MERCURIO (ITALIAN) see MCY475
CLORURO MERCUROSO (ITALIAN) see MCW000
CLORURO di METALLILE (ITALIAN) see CIU750
CLORURO di METILE (ITALIAN) see MIF765
CLORURO di SUCCINILCOLINA (ITALIAN) see HLC500
CLORURO di VINILE (ITALIAN) see VNP000
CLOSANTEL see CFC100
SYMCLOSEN see TIQ750
SYMCLOSENE see TIQ750
CLOSTEBOL ACETATE see CLG900
CLOSTRIDIUM BOTULINUM NEUROTOXIN see BMM292
CLOSTRIDIUM BOTULINUM TOXIN see CMY030
CLOSTRIDIUM BUTYRICUM MIYAIRI, powder see CMY050
CLOSTRIDIUM DIFFICILE TOXIN see CMY070
CLOSTRIDIUM DIFFICILE TOXIN B see CMY090
CLOSTRIDIUM NOVYI α-TOXIN see CMY130
CLOSTRIDIUM OEDEMATIENS TYPE A TOXIN see CMY150
CLOSTRIDIUM PERFRINGENS EXOTOXIN see CMY170
CLOSTRIDIUM PERFRINGENS β-TOXIN see CMY190
CLOSTRIDIUM PERFRINGENS (Welchii) TYPE A ENTEROTOXIN see CMY220
CLOSTRIDIUM SORDELLII TOXIN see CMY240
CLOSTRIDIUM TETANI TOXIN see CMY260
CLOSTRIDIUM TETANI, TYPE BE TOXIN see CMY280
CLOSTRIDIUM TETANI, TYPE S TOXIN see CMY300
CLOTAM see CLK325
CLOTEPIN see ODY100
CLOTHEPIN see ODY100
CLOTIAZEPAM see CKA000
CLOTRIDE see CLH750
CLOTRIMAZOL see MRX500
CLOUT see DXE600
CLOUT see FBP350
CLOVE BUD OIL see CMY475
CLOVE LEAF OIL see CMY500
CLOVE LEAF OIL MADAGASCAR see CMY500
CLOVE OIL, stem see CMY510

CLOWN TREACLE see WBS850
CLOXACILLIN see SPD600
CLOXACILLIN SODIUM MONOHYDRATE see SLJ000
CLOXACILLIN SODIUM SALT see SLJ050
CLOXAPEN see SLJ000
CLOXAPEN see SLJ050
CLOXAZEPINE see DCS200
CLOXAZOLAM see CMY525
CLOXAZOLAZEPAM see CMY525
CLOXYPEN see SLJ050
CLOXYPENDYL see CMY535
CLOZAPIN see CMY650
CLOZAPINE see CMY650
ClP see CKV500
CL 13,850 SODIUM see DRU875
CL 5343 SODIUM SALT see AMR750
CL 251931 SODIUM SALT see CCS635
CLUDR see CFJ750
CLUSIA ROSEA see BAE325
CLY-503 see SDY500
CLYSAR see PMP500
CM 6912 see EKF600
CMA see CBF250
CMA see CIF250
pCMA see CIF250
4-CMB see CIH825
CMB 50 see CBT500
CMB 200 see CBT500
CMC see SFO500
CMC 7H see SFO500
CM-CELLULOSE Na SALT see SFO500
CMC SODIUM SALT see SFO500
CMDP see MQR750
CME see CBD750
CME 127 see CJD300
CME 74770 see TKL100
C-METON see TAI500
CMF see EHF500
CMH see MAE500
C. MITIS see FBW100
CML 21 see CAU500
CML 31 see CAU500
CMME see CIO250
CMMP see SKQ400
CMP see CQL300
5'-CMP see CQL300
CMP (nucleotide) see CQL300
CMPABN see MEI000
CMPF see MIT600
CMPP see CIR500
CMPP see RBF500
CMPT see CIU800
CM-S 2957 see CLB022
CMU see CJX750
CMW BONE CEMENT see PKB500
CMZ SODIUM see CCS360
CN see CEA750
CN 2 see EHP700
CN 4 see EHP700
CN 009 see CDB760
CN 447 see DEJ000
CN 3123 see SNL850
CN 8676 see EEI000
CN-15,757 see ASA500
CN-27,554 see TKH750
CN-35355 see XQS000
CN-36337 see CIP500
CN 38703 see QAK000
CN 59,567 see CGB250
CN-11-2936 see DWS200
CN-25,253-2 see AOO500
CN-52,372-2 see CKD750
CN-55945-27 see NHP500
CNA see RDP300
CNC see NAS000
CNCC see BIF625
CNO see CNT645
C. NOCTURNUM see DAC500
C'-NORVINCALEUKOBLASTINE, 3',4'-DIDEHYDRO-4'-DIOXY-, (R-(R*,R*))-2,3-

DIHYDROXYBUTANEDIOATE (1:2) see VLF400
CNP see NIW500
CNP 1032 see NIW500
CNU see CHE750
CNUEMS see CHF250
CNU-ETHANOL see CHB750
CO 12 see DXV600
CO 433 see SNY100
CO 436 see ANO600
CO-1214 see DXV600
CO-1670 see HCP000
CO-1895 see OAX000
CO-1897 see OAX000
COAGULASE see CMY725
COAKUM see PJJ315
COAL ASH see CMY765
COAL CONVERSION MATERIALS, SRC-II HEAVY DISTILLATE see CMY750
COAL DUST see CMY760
COAL DUST, EXTRACT, NITROSATED see NJH750
COAL FACINGS see CMY760
COAL FLY ASH see CMY765
COAL GAS (UN 1023) (DOT) see HHJ500
COAL, GROUND BITUMINOUS (DOT) see CMY760
COALITE NTP see XLS100
COAL LIQUID see PCR250
COAL-MILLED see CMY760
COAL NAPHTHA see BBL250
COAL OIL see KEK000
COAL OIL see PCR250
COAL SLAG-MILLED see CMY760
COAL TAR see CMY800
COAL TAR, AEROSOL see CMY800
COAL TAR, AEROSOL see CMY805
COAL TAR CREOSOTE see CMY825
COAL TAR DISTILLATE see CMY900
COAL TAR DISTILLATES see CMY900
COAL TAR DISTILLATES, flammable (DOT) see CMY900
COAL TAR DYE see CMY920
COAL TAR OIL see CMY825
COAL TAR OIL (DOT) see CMY825
COAL TAR PITCH VOLATILES see CMZ100
COAL TAR PITCH VOLATILES: PHENANTHRENE see PCW250
COAL TAR SOLUTION USP see CMY800
COAPT see MIQ075
COATHYLENE HA 1671 see PJS750
COATHYLENE PF 0548 see PMP500
COBADEX see CNS750
COBADOCE FORTE see VSZ000
COBALIN see VSZ000
COBALT see CNA250
COBALT-59 see CNA250
COBALT ACETATE see CNA300
COBALT ACETATE see CNC000
COBALT(2+) ACETATE see CNC000
COBALT(3+) ACETATE see CNA300
COBALT(II) ACETATE see CNC000
COBALT(III) ACETATE see CNA300
COBALT ACETATE TETRAHYDRATE see CNA500
COBALT ALLOY see CNA600
COBALT ALLOY, BASE, Co 31-47, Cr 20-24, Ni 20-24, W 13-16, Fe 0-3, Mn 0-1.2, Si 0.2-0.5, C 0-0.2, La 0-0.2 see CNA600
COBALT ALLOY, BASE, Co 48-58, CR 24-26, Ni 9.5-12, W 7-8, Fe 2, Mn 0-1, Si 0-1, C 0.4-0.6 (ASTM A567-2) see CNB825
COBALT ALLOY, Co,Cr see CNA750
COBALT(III) AMIDE see CNB000
COBALTATE(3-), HEXAKIS(NITRITO-N)-, TRIPOTASSIUM, (OC-6-11)-(9CI) see PLI750
COBALTATE(3-), HEXAKIS(NITRITO-N)-, TRISODIUM (OC-6-11)- (9CI) see SFX750
COBALTATE(3-), HEXANITRO-, TRISODIUM see SFX750
COBALTATE(1-), (29H,31H-PHTHALOCYANINE-C-SULFONATO(3-)-

N29,N30,N31,N32)-, HYDROGEN see CND940

COBALTATE(2-), TETRACHLORO-, BIS(TETRAETHYLAMMONIUM) see BLH315

COBALT(II) AZIDE see CNB099

COBALT, BIS(CARBONATO(2-))HEXAHYDROXYPENTA- see CNB495

COBALT, BIS(CARBONATO(2-))HEXAHYDROXYPENTA-, MONOHYDRATE see BAP802

COBALT, BIS(N-9H-FLUOREN-2-YL-N-HYDROXYACETAMIDATO-O,O')-(9CI) see FDU875

COBALT, BIS(2-HYDROXYPROPANOATO-OO¹,O²)- see CNC242

COBALT, BIS(LACTATO)-(8CI) see CNC242

COBALT BIS(NITRATE) see CNC500

COBALT, BIS(5-SULFO-8-QUINOLINOLATO)- see BKJ260

COBALT(II), BIS(5-SULFO-8-QUINOLINOLATO-N¹,O⁸)- see BKJ260

COBALT BLACK see CND125

COBALT BOROFLUORIDE see CNC050

COBALT BORON TETRAFLUORIDE see CNC050

COBALT(II) BROMIDE see CNB250

COBALT CAPRYLATE see CNB450

COBALT CARBONATE see CNB475

COBALT CARBONATE (1:1) see CNB475

COBALT(2+) CARBONATE see CNB475

COBALT CARBONATE, COBALT DIHYDROXIDE (2:3) see CNB495

COBALT CARBONATE HYDROXIDE see CNB495

COBALT(II) CARBONATE HYDROXIDE (1:1) see BAP800

COBALT CARBONATE HYDROXIDE (CO₂CO₃(OH)₂) see BAP800

COBALT(II)CARBONATE HYDROXIDE (2:3) MONOHYDRATE see BAP802

COBALT, (CARBONATO)DIHYDROXYDI-(8CI) see BAP800

COBALT CARBONYL see CNB500

COBALT CARBONYL (CO₄(CO)₁₂) see CNB510

COBALT(II) CHLORIDE see CNB599

COBALT(III) CHLORIDE see CNB750

COBALT(2+) CHLORIDE HEXAHYDRATE see CNB800

COBALT(II) CHLORIDE HEXAHYDRATE see CNB800

COBALT CHLORIDE, HEXAHYDRATE (8CI, 9CI) see CNB800

COBALT-CHROMIUM ALLOY see CNA750

COBALT-CHROMIUM-MOLYBDENUM ALLOY see VSK000

COBALT-CHROMIUM-NICKEL-TUNGSTEN ALLOY see CNB825

COBALT COMPOUNDS see CNB850

COBALT DIACETATE see CNC000

COBALT DIACETATE TETRAHYDRATE see CNA500

COBALT DIBROMIDE see CNB250

COBALT DICHLORIDE see CNB599

COBALT DICHLORIDE HEXAHYDRATE see CNB800

COBALT DIFLUORIDE see CNC100

COBALT DIHYDROXIDE see CNC238

COBALT, DI-MU-CARBONYLHEXACARBONYLDI-, (CO-CO) see CNB500

COBALT DINITRATE see CNC500

COBALT DINITRATE HEXAHYDRATE see CND010

COBALT DIPERCHLORATE HEXAHYDRATE see CND900

COBALT(2)-EDATHAMIL see DGQ400

COBALT(II)FLUOBORATE see CNC050

COBALT N-FLUOREN-2-YLACETOHYDROXAMATE see FDU875

COBALT(II) FLUORIDE see CNC100

COBALT(3⁺), HEXAAMMINE-, (OC-6-11)-, TRIACETATE see HBU425

COBALT-HISTIDINE see BJY000

COBALT HYDROCARBONYL see CNC230

COBALT, (HYDROGEN PHTHALOCYANINESULFONATO(2-))- see CND940

COBALT HYDROXIDE see CNC233

COBALT(2+) HYDROXIDE see CNC238

COBALT(II) HYDROXIDE see CNC238

COBALT(III) HYDROXIDE see CNC233

COBALT HYDROXIDE OXIDE see CNC240

COBALT HYDROXIDE OXIDE (CoO(OH)) see CNC240

COBALTIC ACETATE see CNA300

COBALTIC-COBALTOUS OXIDE see CND020

COBALTIC HYDROXIDE see CNC233

COBALTIC OXIDE see CND825

COBALTIC POTASSIUM NITRITE see PLI750

COBALT LACTATE see CNC242

COBALT LINOLEATE see CNC245

COBALT-METHYLCOBALAMIN see VSZ050

COBALT MOLYBDATE see CNC250

COBALT(2+) MOLYBDATE see CNC250

COBALT MOLYBDENUM OXIDE see CNC250

COBALT MONOCARBONATE see CNB475

COBALT MONOOXIDE see CND125

COBALT MONOSULFATE HEPTAHYDRATE see CNE150

COBALT MONOSULFIDE see CNE200

COBALT MONOXIDE see CND125

COBALT, (MU-(CARBONATO(2-)-O:O'))DIHYDROXYDI see BAP800

COBALT MURIATE see CNB599

COBALT NAPHTHENATE, POWDER (DOT) see NAR500

COBALT NITRATE see CNC500

COBALT(2+) NITRATE see CNC500

COBALT(II) NITRATE see CNC500

COBALT NITRATE HEXAHYDRATE see CND010

COBALT(2+) NITRATE HEXAHYDRATE see CND010

COBALT(II) NITRATE HEXAHYDRATE see CND010

COBALT(II) NITRIDE see CNC750

COBALT NITROPRUSSIDE see CND000

COBALT NITROSOPENTACYANOFERRATE(3) see CND000

COBALT NITROSYLPENTACYANOFERRATE see PAY610

COBALTOCENE see BIR529

COBALTO-COBALTIC OXIDE see CND020

COBALTO-COBALTIC TETROXIDE see CND020

COBALT OCTACARBONYL see CNB500

COBALTOSIC OXIDE see CND020

COBALTOUS ACETATE TETRAHYDRATE see CNA500

COBALTOUS BROMIDE see CNB250

COBALTOUS CARBONATE see CNB475

COBALTOUS CARBONATE, BASIC see BAP802

COBALTOUS CHLORIDE see CNB599

COBALTOUS CHLORIDE, HEXAHYDRATE see CNB800

COBALTOUS DIACETATE see CNC000

COBALTOUS DICHLORIDE see CNB599

COBALTOUS FLUORIDE see CNC100

COBALTOUS HYDROXIDE see CNC238

COBALTOUS MOLYBDATE see CNC250

COBALTOUS NITRATE see CNC500

COBALTOUS NITRATE HEXAHYDRATE see CND010

COBALTOUS OXIDE see CND125

COBALTOUS PERCHLORATE, HEXAHYDRATE see CND900

COBALTOUS PHOSPHATE see CND920

COBALTOUS SULFATE see CNE125

COBALTOUS SULFATE HEPTAHYDRATE see CNE150

COBALTOUS SULFIDE see CNE200

COBALTOUS TETRAFLUOROBORATE see CNC050

COBALT OXIDE see CND020

COBALT OXIDE see CND125

COBALT(2+) OXIDE see CND125

COBALT(3+) OXIDE see CND825

COBALT(II) OXIDE see CND125

COBALT(III) OXIDE see CND825

COBALT OXIDE HYDROXIDE (CoOOH) see CNC240

COBALT OXIDE (8CI,9CI) see CND825

COBALT OXYHYDROXIDE see CNC240

COBALT PERCHLORATE HEXAHYDRATE see CND900

COBALT(II) PERCHLORATE, HEXAHYDRATE see CND900

COBALT PEROXIDE see CND825

COBALT PHOSPHATE see CND920

COBALT(II) PHOSPHATE see CND920

COBALT PHTHALOCYANINESULFONATE see CND940

COBALT RESINATE, precipitated see CNE000

COBALT SALTS see FDU875

COBALT SESQIOXIDE see CND825

COBALT SESQUIOXIDE see CND825

COBALT SULFATE see CNE125

COBALT SULFATE (1:1) see CNE125

COBALT (2+) SULFATE see CNE125

COBALT(II) SULFATE (1:1) see CNE125

COBALT(II) SULFATE (1:1), HEPTAHYDRATE see CNE150

COBALT SULFIDE see CNE200

COBALT(II) SULFIDE see CNE200

COBALT SULFIDE (amorphous) see CNE200

COBALT(II) SULPHATE see CNE125

COBALT(II) SULPHATE HEPTAHYDRATE see CNE150

COBALT TALLATE see CNE240

COBALT TETRACARBONYL see CNB500

COBALT TETRACARBONYL DIMER see CNB500

COBALT TETRAOXIDE see CND020

COBALT TRIACETATE see CNA300

COBALT TRIFLUORIDE see CNE250

COBALT TRIHYDROXIDE see CNC233

COBALT, TRI-MU-CARBONYLNONACARBONYLTETRA-, TETRAHYDRO- see CNB510

COBALT TRIOXIDE see CND825

COBALT YELLOW see PLI750

COBAMIN see VSZ000

COBAN see MRE230

COBEN see CNE375

COBEN P see PIC100

COBEX see CNE500

COBEX (polymer) see PKQ059

COBEXO see CNE500

COBH see BDD000

COBINAMIDE, COBALT-METHYL derivative, HYDROXIDE, DIHYDROGEN PHOSPHATE (ester), inner salt, 3'-ESTER with 5,6-DIMETHYL-1-α-D-RIBOFURANOSYLBENZIMIDAZOLE see VSZ050

COBIONE see VSZ000

COBOX see CNK559

COBOX BLUE see CNK559

COBRATEC #99 see BDH250

COBRATEC TT 100 see MHK000

COCAFURIN see NGE500

COCAIN-CHLORHYDRAT (GERMAN) see CNF000

COCAINE see CNE750

(−)-COCAINE see CNE750

l-COCAINE see CNE750

β-COCAINE see CNE750

COCAINE CHLORIDE see CNF000

COCAINE HYDROCHLORIDE see CNF000
(−)-COCAINE HYDROCHLORIDE see CNF000
l-COCAINE HYDROCHLORIDE see CNF000
COCAINE MURIATE see CNF000
COCAMIDE DEA see CNF330
CO CAP IMIPRAMINE 25 see DLH630
COCARTRIT see CLD000
C. OCCIDENTALIS see CCK675
COCCIDINE A see DUP300
COCCIDIOSTAT C see CMX850
COCCIDOT see DUP300
COCCINE see FMU080
COCCOCLASE see PPO000
COCCULIN see PIE500
COCCULUS see PIE500
COCCULUS solid (DOT) see PIE500
COCHENILLE DYE see CNF050
COCHENILLEROT A see FMU080
COCHIN see LEG000
COCHINEAL see CNF050
COCHINEAL (dye) see CNF050
COCHINEAL RED A see FMU080
COCHINEAL TINCTURE see CNF050
COCHLIOBOLIN see CNF109
COCHLIOBOLIN A see CNF109
COCHLIODINOL see CNF159
COCOA FATTY ACIDS, POTASSIUM SALTS see CNF175
COCO AMIDO BETAINE see CNF185
COCO-DIAZINE see PPP500
COCO DIETHANOLAMIDE see BKE500
COCONUT ALDEHYDE see CNF250
COCONUT AMINE OIL CONDENSATE see CNF270
COCONUT BUTTER see CNR000
COCONUT DIETHANOLAMIDE see CNF330
COCONUT DIETHANOLAMINE see CNF330
COCONUT DIMETHYL AMINE OXIDE see CNF325
COCONUT MEAL PELLETS, containing 6−13% moisture and no more than 10% residual fat (DOT) see CNR000
COCONUT OIL ACID DIETHANOLAMINE see CNF330
COCONUT OIL ACID DIETHANOLAMINE see CNF330
COCONUT OIL ACID DIETHANOLAMINE CONDENSATE see CNF330
COCONUT OIL AMIDE of DIETHANOLAMINE see BKE500
COCONUT OIL, ESTERS WITH POLYETHYLENE GLYCOL NONYLPHENYL ETHER see CNF335
COCONUT OIL, ESTER WITH POLYOXYETHYLENE NONYLPHENYL ESTER see CNF335
COCONUT OIL (FCC) see CNR000
COCONUT PALM OIL see CNR000
N-COCOPYRROLIDINONE see CNF340
COCUM see PJJ315
COD see CPR825
CODAL see MQQ450
CODECARBOXYLASE see PII100
CODECHINE see BBQ500
CODE H 133 see DER800
CODEHYDRASE I see CNF390
CODEHYDRASE II see CNF400
CODEHYDROGENASE I see CNF390
CODEHYDROGENASE II see CNF400
CODEINE see CNF500
CODEINE HYDROCHLORIDE see CNF750
CODEINE METHOCHLORIDE see CNG000
CODEINE NICOTINATE (ESTER) see CNG250
CODEINE PHOSPHATE see CNG500
CODEINE PHOSPHATE SESQUIHYDRATE see CNG675
CODEINE SULFATE see CNG750

CODEINONE, DIHYDRO-, TARTRATE see DKX050
CODELCORTONE see PMA000
CODELSOL see PLY275
CODEMPIRAL see ABG750
CO-DERGOCRINE MESYLATE see DLL400
CODETHYLINE see ENK000
CODETHYLINE HYDROCHLORIDE see DVO700
CODHYDRINE see DKW800
CODIAZINE see PPP500
CODIBARBITA see EOK000
CODLELURE see CNG760
CODLEMONE see CNG760
CODUSAFOS see EHY100
CODYLIN see TCY750
COE 536 see HKC575
COENZYME I see CNF390
COENZYME II see CNF400
COENZYME Q$_{10}$ see UAH000
COENZYME R see BGD100
COENZYME R see VSU100
CO-ESTRO see ECU750
COFFEARIN see TKL890
COFFEARINE see TKL890
COFFEBERRY see MBU825
COFFEE see CNG775
COFFEE BEAN EXTRACT see CNG800
COFFEE ESSENCE see CNG800
COFFEE EXTRACT see CNG800
COFFEE SENNA see CNG825
COFFEIN (GERMAN) see CAK500
COFFEINE see CAK500
COFOCAIN see CNG827
COFPLATON see CNG850
CO-FRAM see SNL850
COGESIC see DTO200
COGILOR BLUE 512.12 see FAE000
COGILOR RED 321.10 see FAG070
COGLA see ARP125
COGNAC OIL see CNG920
COGNAC OIL see EKN050
COGNAC OIL, GREEN see CNG920
COGNAC OIL, WHITE see CNG920
COGOMYCIN see FPC000
COHASAL-1H see SEH000
COHEDUR A see MCB050
COHORTAN see BDJ600
COHOSH see BAF325
CO-HYDELTRA see PMA000
COHYDRIN see DKW800
COIR DEEP BLACK C see AQP000
COIXOL see MEC550
COKAN see PJJ315
COKE see CNE750
COKE OVEN EMISSIONS see CNG929
COKE OVEN EMISSIONS (OSHA) see CNG929
COKE POWDER see CBT500
COLACE see DJL000
COLACID BLACK 10A see FAB830
COLACID BLUE A see ADE750
COLACID ORANGE see CMM220
COLACID PONCEAU 4R see FMU080
COLACID PONCEAU SPECIAL see FMU070
COLACID RED 2A see CMS228
COLALIN see CME500
COLAMINE see EEC600
COLAMINE HYDROCHLORIDE see EEC700
COLANYL GREEN GG see PJQ100
COLARYL YELLOW FGL 30 see CMS214
COLATRON see HID100
COLCEMID see MIW500
COLCEMIDE see MIW500
COLCHAMIN see MIW500
COLCHAMINE see MIW500
COLCHICIN (GERMAN) see CNG938
COLCHICINA (ITALIAN) see CNG938
COLCHICINE see CNG938
7-α-H-COLCHICINE see CNG938
COLCHICINE, 7-DEACETAMIDO-7-(METHYLAMINO)- see MIW500

COLCHICINE, N-DEACETYL-10-DEMETHOXY-10-METHYLTHIO- see DBA200
COLCHICINE, DEACETYL-N-METHYL- see MIW500
COLCHICINE, 17-ETHOXY-10-THIO- see EEK600
COLCHICOSIDE see DAN375
COLCHICUM AUTUMNALE see CNX800
COLCHICUM SPECIOSUM see CNX800
COLCHICUM VERNUM see CNX800
COLCHINEOS see CNG938
COLCHINIC ACID TRIMETHYL see TLO000
COLCHISOL see CNG938
COLCIN see CNG938
COLCOTHAR see IHC450
COLDAN see NCW000
COLDRIN see CMW700
COLEBENZ see BCM000
COLECALCIFEROL see CMC750
COLEMANITE see CAN250
COLEMID see MIW500
COLEP see MOB699
COLEPAX see IFY100
COLEPUR see DDS600
COLESTERINEX see PPH050
COLESTYRAMIN see CME400
COL-EVAC see SFC500
COLEYTL see CBH250
COLFARIT see ADA725
COLIBIL see CNG980
COLIMYCIN see PKD250
COLIMYCIN M see SFY500
COLIMYCIN SULFATE see PKD300
COLIOPAN see BPI125
COLIPAR see DDS600
COLISONE see PLZ000
COLISTICINA see PKD250
COLISTIMETHATE SODIUM see SFY500
COLISTIN see PKD250
COLISTINASE see BAC000
COLISTIN SODIUM METHANESULFONATE see SFY500
COLISTIN SULFAT see PKD300
COLISTIN SULFATE see PKD300
COLISTIN, SULFATE (SALT) see PKD300
COLISTIN SULFOMETHAT see SFY500
COLISTIN SULFOMETHATE SODIUM see SFY500
COLISTRIMETHATE SODIUM see SFY500
COLITE see CMF350
COLLAGENS, HYDROLYZATES see HID100
COLLARGOL see SDI750
s-COLLIDINE see TME272
2,4,6-COLLIDINE see TME272
sym-COLLIDINE see TME272
γ-COLLIDINE see TME272
α-γ,α'-COLLIDINE see TME272
COLLIDINE, ALDEHYDECOLLIDINE see EOS000
COLLIRON I.V. see IHG000
COLLOCARB see CBT750
COLLODION see CCU250
COLLODION COTTON see CCU250
COLLODION WOOL see CCU250
COLLOID 775 see CCL250
COLLOIDAL ARSENIC see ARA750
COLLOIDAL CADMIUM see CAD000
COLLOIDAL FERRIC OXIDE see IHC450
COLLOIDAL GOLD see GIS000
COLLOIDAL MANGANESE see MAP750
COLLOIDAL MERCURY see MCW250
COLLOIDAL SELENIUM see SBO500
COLLOIDAL SILICA see SCH002
COLLOIDAL SILICON DIOXIDE see SCH002
COLLOIDAL SULFUR see SOD500
COLLOIDOX see CNK559
COLLOKIT see SOD500
COLLOMIDE see SNM500
COLLONE AC see PJT300
COLLOWELL see SFO500

COLLOXYLIN see CCU250
COLLUNOSOL see TIV750
COLLUNOVAR see NCJ500
COLLUNOVER see NCJ500
COLLUSUL-HC see HHQ800
COLOCASIA ESCULENTA see EAI600
COLOCASIA GIGANTEA see EAI600
COLOGNE EARTH see MAT500
COLOGNE SPIRIT see EFU000
COLOGNE UMBER see MAT500
COLOGNE YELLOW see LCR000
COLOMBIAN BLACK TOBACCO
CIGARETTE REFINED TAR see CMP800
COLOMYCIN SYRUP see PKD300
COLONATRAST see BAP000
COLONIAL SPIRIT see MGB150
COLORADO RIVER HEMP see SBC550
COLORFIX see CNH125
COLORINES (MEXICO) see NBR800
COLOR-SET see TIX500
COLPOVISTER see EDU500
COLPRO see MBZ100
COLPRONE see MBZ100
COLSALOID see CNG938
COLSUL see SOD500
COLSULANYDE see SNM500
COLTIROT see TOG500
COLTSFOOT see CNH250
COLTS FOOT see PCR000
COLUMBIA BLACK EP see AQP000
COLUMBIA BROWN M see CMO800
COLUMBIA CARBON see CBT750
COLUMBIA FAST BLACK G see CMN240
COLUMBIA FAST BLACK GB see CMN240
COLUMBIA FAST RED F see CMO870
COLUMBIA FAST SCARLET 4BS see
CMO870
COLUMBIA LCK see CBT500
COLUMBIAN SPIRITS (DOT) see MGB150
COLUMBIUM see NDZ000
COLUMBIUM PENTACHLORIDE see
NEA000
COLUMBIUM POTASSIUM FLUORIDE see
PLN500
COLUTOID see GEK500
COLYER PECTIN see PAO150
COLY-MYCIN see PKD250
COLY-MYCIN INJECTABLE see SFY500
COLYMYSIN S see PKD250
COLYONAL see DBD750
COLYSTINMETHANSULFONAT
(GERMAN) see SFY500
COLZA OIL see RBK200
COMAC see CNM500
COMACID BLUE BLACK B see FAB830
COMBANTRIN see POK575
COMBAT see HGP525
COMBETIN see SMN000
COMBINACE see CAM200
COMBINAL E see TGJ050
COMBINAL K1 see VTA000
COMBOT EQUINE see TIQ250
COMBRETODENDRON AFRICANUM
(Welw), extract see CNH275
COMBUSTION IMPROVER-2 see MAV750
COMELIAN see CNR750
COMESA see EQL000
COMESA see MNM500
COMESTROL see DKA600
COMESTROL ESTROBENE see DKA600
COMFREY, RUSSIAN see RRK000
COMFREY, RUSSIAN see RRP000
COMITAL see DKQ000
COMITE see SOP000
COMITIADONE see PEC250
COMMAND see CEP800
COMMANDO see FBW135
COMMERCIAL DIESEL FUEL NO. 2 see
DHE850
COMMERCIAL LIGHT DUTY LIQUID
DETERGENT see CNH285
COMMISTERONE see HKG500

COMMON GROUNDSEL see RBA400
COMMON SALT see SFT000
COMMON SENSE COCKROACH and RAT
PREPARATIONS see PHP010
COMMOTIONAL see ABG750
COMPALOX see AHE250
COMPAZINE see PMF250
COMPAZINE see PMF500
COMPENDIUM see BMN750
COMPERLAN LD see BKE500
COMPERLAN OD see OHU200
COMPITOX see CIR500
COMPITOX see RBF500
COMPITOX PLUS see RBF500
COMPLAMEX see XCS000
COMPLAMIN see XCS000
COMPLEMIX see DJL000
COMPLEXAMINE see AHR600
COMPLEXONE see EIV000
COMPLEXON I see AMT500
COMPLEXON II see EIX000
COMPLEXON III see EIX500
COMPLEXON IV see CPB120
COMPOCILLIN G see BDY669
COMPOCILLIN-VK see PDT750
COMPOSE 134 P (FRENCH) see MLK800
COMPOUND 42 see WAT200
COMPOUND 118 see AFK250
COMPOUND 269 see EAT500
COMPOUND 338 see DER000
COMPOUND 347 see EAT900
COMPOUND 497 see DHB400
COMPOUND 604 see DFT000
COMPOUND-666 see BBP750
COMPOUND 711 see IKO000
COMPOUND 864 see YCJ000
COMPOUND 889 see DVL700
COMPOUND 88R see SOP500
COMPOUND 923 see DFY400
COMPOUND M-81 see PHI500
COMPOUND 1081 see FFF000
COMPOUND 1189 see KEA000
COMPOUND 1275 see PDV700
COMPOUND 13-61 see OOI200
COMPOUND 1836 see CLV375
COMPOUND 19-28 see DXB450
COMPOUND 2046 see MQR750
COMPOUND 3422 see PAK000
COMPOUND-3916 see DRB400
COMPOUND 3956 see CDV100
COMPOUND-4018 see CNL500
COMPOUND 4049 see MAK700
COMPOUND 4072 see CDS750
COMPOUND 47-83 see EAN600
COMPOUND-4992 see BKS810
COMPOUND 593A see PIK075
COMPOUND 6515 see DJA300
COMPOUND 6890 see TLQ000
COMPOUND 7215 see FBP300
COMPOUND 7744 see CBM750
COMPOUND 8958 see CQH500
COMPOUND 01748 see DJT800
COMPOUND 10854 see COF250
COMPOUND 17309 see DAL300
COMPOUND 20-438 see CNH300
COMPOUND 33355 see MKB750
COMPOUND 33,828 see MLJ500
COMPOUND 33T57 see MKW250
COMPOUND 38,174 see INS000
COMPOUND 42339 see ADR750
COMPOUND 48/80 see CNH375
COMPOUND 64716 see CMS130
COMPOUND 67/20 see CCW725
COMPOUND 72500 see IRN500
COMPOUND 74-637 see ALU875
COMPOUND 90459 see OJI750
COMPOUND S-6,999 see NNF000
COMPOUND 69/183 see CNH500
COMPOUND 78/702 see CNH525
COMPOUND B see CNS625
COMPOUND B DICAMBA see MEL500
COMPOUND C-9491 see IEN000

COMPOUND E see CNS800
COMPOUND E ACETATE see CNS825
COMPOUND F see CNS750
COMPOUND F-2 see ZAT000
COMPOUND-1452-F see EME500
COMPOUND 1571 F see PFB850
COMPOUND F ACETATE see HHQ800
COMPOUND G-11 see HCL000
COMPOUND HP1275 see PDV700
COMPOUND 26539 HYDROCHLORIDE see
EPL600
COMPOUND 6-12 INSECT REPELLENT see
EKV000
COMPOUND 14045 METHIODIDE see
CNH550
COMPOUND 14045 METHOCHLORIDE see
EAI875
COMPOUND 14045 METHSULFATE see
TJG225
COMPOUND No. 1080 see SHG500
COMPOUND R-242 see CKI625
COMPOUND R-25788 see DBI300
COMPOUND 2339 RP see PEN000
COMPOUND SN see SAY000
COMPOUND UC-20047 A see CFF250
COMPOUND W see TKL100
COMPOUNE 732 see BQT750
COMPTIE see CNH789
COMYCETIN see CDP250
CON A see CNH625
CONAC A see CPI250
CONAC S see CPI250
CONCANAVALIN A see CNH625
CONCEP see COP700
CONCEP II see OKS200
CONCEP III see FMR600
CONCHININ see QFS000
CONCILIUM see FGU000
CONCLYTE CALCIUM see CAT600
CONCO AAS-35 see DXW200
CONCOGEL 2 CONCENTRATE see SIY000
CONCO NI-90 see PKF000
CONCO NIX-100 see PKF500
CONCORD see RCZ050
CONCO SULFATE C see HCP900
CONCO SULFATE WA see SIB600
CONCO SXS see XJJ010
CONCO XAL see DRS200
CONCTASE C see CNH650
CONDACAPS see VSZ100
CONDENSATE PL see BKE500
CONDENSATES (PETROLEUM), VACUUM
TOWER (9CI) see MQV755
CONDITION see DCK759
CONDITIONER 1 see OBA000
CONDOCAPS see VSZ100
CONDOL see VSZ100
CONDUCTEX see CBT500
CONDUCTEX see CBT750
CONDUCTIVE POLYMER 261 see DTS500
CONDYLON see CNG938
CONDY'S CRYSTALS see PLP000
CONESSINE DIHYDROBROMIDE see
DOX000
CONESSINE HYDROCHLORIDE see
CNH660
CONEST see ECU750
CONESTORAL see EDV600
CONESTRON see ECU750
CONFECTIONER'S SUGAR see SNH000
CONFIDOR see CKW400
CONFIDOR 200 SL see CKW400
CONFORTID see IDA000
CONGOBLAU 3B see CMO250
CONGO BLUE see CMO250
CONGOCIDIN see NCP875
CONGOCIDINE see NCP875
CONGOCIDINE DIHYDROCHLORIDE see
NCQ000
CONGO RED see SGQ500
CONGO RED R-138 see NAY000
γ-CONICEIN see CNH730

γ-CONICEINE see CNH730
d-CONICINE see PNT000
CONIGON BC see EIV000
CONIIN see PNT000
CONIINE see PNT000
(+)-CONIINE see PNT000
CONINE see PNT000
α-CONINE see PNT000
CONIUM MACULATUM see CNH750
CONIUM MACULATUM see PJJ300
CONJES see ECU750
CONJUGATED EQUINE ESTROGEN see PMB000
CONJUGATED ESTROGENS see ECU750
CONJUNCAIN see OPI300
CONJUTABS see ECU750
7-CON-o-METHYLNOGAROL see MCB600
CONOCO C-50 see DXW200
CONOCO DBCL see DXW600
CONOTRANE see PFN000
CONOVA 30 see DAP810
CONOVID see EAP000
CONOVID E see EAP000
CONQUERORS see HGL575
CONQUININE see QFS000
CONRAXIN H see IDJ600
CONRAY see IGC000
CONRAY 30 see IGC000
CONRAY 60 see IGC000
CONRAY 80 see IGC100
CONRAY 280 see IGC000
CONRAY 300 see IGC100
CONRAY-400 see IGC100
CONRAY MEGLUMIN see IGC000
CONRAY MEGLUMINE 282 see IGC000
CONSDRIN see SPC500
CONSDRIN HYDROCHLORIDE see SPC500
CONSTAPHYL see DGE200
CONSTONATE see DJL000
CONSULFA see SNH900
CONT see MMN250
CONTALAX see PPN100
CONTAVERM see PDP250
CONTEBEN see FNF000
CONTERGAN see TEH500
CONTIMET 30 see TGF250
CONTINAL see NBU000
CONTINENTAL see CBT750
CONTINENTAL see KBB600
CONTINEX see CBT750
CONTIZELL see PKQ059
CONTOPHERON see TGJ050
CONTRAC see BMN000
CONTRA CREME see ABU500
CONTRADOL see ABG750
CONTRALGIN see BQA010
CONTRALIN see DXH250
CONTRAMINE see DJD000
CONTRAPAR see KFA100
CONTRAPOT see DXH250
CONTRATHION see PLX250
CONTRHEUMA RETARD see ADA725
CONTRISTAMINE HYDROCHLORIDE see CIS000
CONTRIX 28 see IGC000
CONTROL see MFD500
CONTROVLAR see EEH520
CONTUREX see SHX000
CONVALLAOTOXIN see CNH780
CONVALLARIA MAJALIS see LFT700
CONVALLARIN see CNH775
CONVALLATON see CNH780
CONVALLATOXIGENIN see SMM500
CONVALLATOXIN see CNH780
CONVALLATOXOL see CNH785
CONVALLATOXOSIDE see CNH780
CONVAL LILY see LFT700
CONVALLOTOXIN see CNH780
CONVALLOTOXOL see CNH785
CONVALOTOXOL see CNH785
CONVENIXA see TLP750
CONVUL see DKQ000

CONVULEX see PNX750
COOLSPAN see EPD500
COOMASSIE BLUE see ADE750
COOMASSIE BLUE B see CMM092
COOMASSIE BLUE MEDICINAL see ADE750
COOMASSIE BLUE RL see ADE750
COOMASSIE MILLING SCARLET G see CMM325
COOMASSIE MILLING SCARLET GP see CMM325
COOMASSIE RED PG see CMM320
COOMASSIE RED PGP see CMM320
COOMASSIE RED R see NAO600
COOMASSIE TURQUOISE BLUE 3G see CMM120
COOMASSIE VIOLET see FAG120
COOMASSIE YELLOW R see CMM759
COOMASSIE YELLOW RP see CMM759
COONTIE see CNH789
CO-OP HEXA see HCC500
COPAGEL PB 25 see SFO500
COPAIBA BALSAM see CNH792
COPAIBA OIL see CNH792
COPAIBA OLEORESIN see CNH792
COPAL (CUBA) see PCB300
COPAL Z see SMQ500
COPANOIC see IFY100
COPAROGIN see FLZ050
COPELLIDIN see END000
COPEY see BAE325
COPHARCILIN see AIV500
COPIAMYCIN see CNH800
COPIRENE see KGK000
COPOLYVIDON see AAU300
COPOX see CNO000
COPPER see CNI000
COPPER-8 see BLC250
COPPER ACETATE see CNI250
COPPER(2+) ACETATE see CNI250
COPPER(II) ACETATE see CNI250
COPPER(2+) ACETATE, MONOHYDRATE see CNI325
COPPER(II) ACETATE MONOHYDRATE see CNI325
COPPER ACETOARSENITE (DOT) see COF500
COPPER ACETOARSENITE, solid (DOT) see COF500
COPPER(II) ACETYLACETONATE see BGR000
COPPER ACETYLIDE see CNI500
COPPER-AIRBORNE see CNI000
COPPER ALLOY, Cu, Be see CNI600
COPPER ALLOY, Cu, Be, Co see CNK700
COPPER ALLOY, Cu,Cd see CNI610
COPPER ALLOY, Cu 99.60-100, Cd 0.10-0.30 see CNI800
COPPER ALLOY, Cu 99.75-100, Cd 0.05-0.15 see CNI700
COPPER ARSENATE (BASIC) see CNI900
COPPER ARSENATE HYDRATE see CNI750
COPPER ARSENATE HYDROXIDE see CNI900
COPPER ARSENITE, solid (DOT) see CNN500
COPPERAS see FBN100
COPPERAS see FBO000
COPPER ASCORBATE see CNJ325
COPPER(II) AZIDE see CNJ500
COPPER-BERYLLIUM ALLOY see CNI600
COPPER-BICHLORIDE see CNK500
COPPER BIS(ACETYLACETONATE) see BGR000
COPPER BIS(ACETYLACETONE) see BGR000
COPPER BIS(DIBUTYLCARBAMODITHIOATO-S,S')-, (SP-4-1)- see CNJ600
COPPER BIS(DIBUTYLDITHIOCARBAMATE) see CNJ600

COPPER, BIS(DIBUTYLDITHIOCARBAMATO)- see CNJ600
COPPER, BIS(DIETHYLCARBAMODITHIOATO-S,S')-, (SP-4-1)- see CNL300
COPPER BIS(DIETHYLDITHIOCARBAMATE) see CNL300
COPPER, BIS(DIETHYLDITHIOCARBAMATO)- see CNL300
COPPER(2+), BIS(1,2-ETHANEDIAMINE-KAPPAN,KAPPAN')-, (SP-4-1)-, DIPERCHLORATE see BJO050
COPPER(2+), BIS(1,2-ETHANEDIAMINE-N,N')-, (SP-4-1)-, DIPERCHLORATE see BJO050
COPPER, BIS(ETHYLENEDIAMINE)(MERCURICTETRATHIOCYANATO)- see BJP425
COPPER(2+), BIS(ETHYLENEDIAMINE)-, TETRAKIS(THIOCYANATO)MERCURATE(2-), POLYMERS see BJP425
COPPER, BIS(HYDRAZINE)BIS(HYDROGENSULFATO)- see CNL310
COPPER, BIS(1-(HYDROXY-KAPPAO)-2(1H)-PYRIDINETHIONATO-KAPPAS2)- see BLG600
COPPER, BIS(1-HYDROXY-2(1H)-PYRIDINETHIONATO)- see BLG600
COPPER, BIS(1-HYDROXY-2(1H)-PYRIDINETHIONATO-O,S)- see BLG600
COPPER BIS(2,4-PENTANEDIONATE) see BGR000
COPPER BLUE see CNQ000
COPPER, (4-BROMO-3-HYDROXY-2-NAPHTHOATO)(8-QUINOLINOLATO)- see BNL300
COPPER BRONZE see CNI000
COPPER BROWN see CNO250
COPPER CARBIDE see CNI500
COPPER CARBONATE see CBW200
COPPER CARBONATE (1:1) see CBW200
COPPER(II) CARBONATE see CBW200
COPPER CARBONATE HYDROXIDE see CNJ750
COPPER(II) CARBONATE HYDROXIDE (2:1:2) see CNJ750
COPPER CHELATE of N-HYDROXY-2-ACETYLAMINOFLUORENE see HIP500
COPPER CHLORATE see CNJ900
COPPER CHLORATE (DOT) see CNJ900
COPPER CHLORIDE see CNJ950
COPPER(I) CHLORIDE see CNK250
COPPER(2+) CHLORIDE see CNK500
COPPER(II) CHLORIDE see CNK500
COPPER CHLORIDE (DOT) see CNJ950
COPPER(II) CHLORIDE (1:2) see CNK500
COPPER CHLORIDE, BASIC see CNK559
COPPER CHLORIDE, mixed with COPPER OXIDE, HYDRATE see CNK559
COPPER CHLORIDE OXIDE see CNK559
COPPER CHLORIDE OXIDE, HYDRATE (9CI) see CNK559
COPPER(I) CHLOROACETYLIDE see CNK599
COPPER CHLOROXIDE see CNK559
COPPER CHROMATE see CNK609
COPPER CITRATE see CNK625
COPPER(I) CITRATE see CNK625
COPPER-COBALT-BERYLLIUM see CNK700
COPPER COMPOUNDS see CNK750
COPPER (CUPRIC) DIHYDRAZINIUM SULFATE see CNL310
COPPER CYANAMIDE see CNL250
COPPER CYANIDE see CNL000
COPPER(I) CYANIDE see CNL000
COPPER(II) CYANIDE see CNL250
COPPER CYANIDE (DOT) see CNL250
COPPER DIACETATE see CNI250

COPPER(2+) DIACETATE see CNI250
COPPER DIACETATE MONOHYDRATE see CNI325
COPPER DIACETYLACETONATE see BGR000
COPPER(II) DIBUTYLDITHIOCARBAMATE see CNJ600
COPPER DIETHYL DITHIOCARBAMATE see CNL300
COPPER(2+) DIETHYLDITHIOCARBAMATE see CNL300
COPPER(II) DIETHYLDITHIOCARBAMATE see CNL300
COPPER DIHYDRAZINE SULFATE see CNL310
COPPER, (DIHYDROGEN PHTHALOCYANINEDISULFONATO(2-))-, DISODIUM SALT (7CI,8CI) see COF420
COPPER DIHYDROXIDE see CNM500
COPPER DIMETHYLDITHIOCARBAMATE see CNL500
COPPER DINITRATE see CNM750
COPPER DINITRATE TRIHYDRATE see CNN000
COPPER(II)-1,3-DI(5-TETRAZOLYL)TRIAZENIDE see CNL625
COPPER EDTA COMPLEX see CNL750
COPPER-ETHYLENEDIAMINE COMPLEX see DBU800
COPPERFINE-ZINC see CNP500
COPPER (II) FLUOROACETATE see CNL800
COPPER FUME see CNM000
COPPER GLUCONATE see CNM100
COPPER, (1,3,8,16,18,24-HEXABROMO-2,4,9,10,11,15,17,22,23,25-DECACHLOROPHTHALOCYAN INATO(2-))- see CMS140
COPPER(I) HYDRIDE see CNM250
COPPER HYDROXIDE see CNM500
COPPER(2+) HYDROXIDE see CNM500
COPPER HYDROXIDE SULFATE see CNM600
COPPER HYDROXYQUINOLATE see BLC250
COPPER-8-HYDROXYQUINOLATE see BLC250
COPPER-8-HYDROXYQUINOLINATE see BLC250
COPPER-8-HYDROXYQUINOLINE see BLC250
COPPER INDIUM DISELENIDE see CNM650
COPPER INDIUM SELENIDE see CNM650
COPPER INDIUM SELENIDE see CNM650
COPPER IODIDE see COF680
COPPER(I) IODIDE see COF680
COPPER LONACOL see ZJS300
COPPER METHANE ARSONATE see CNM660
COPPER-MILLED see CNI000
COPPER MONOCARBONATE see CBW200
COPPER MONOCHLORIDE see CNK250
COPPER MONOOXIDE see CNO250
COPPER MONOSULFATE see CNP250
COPPER MONOSULFIDE see CNQ000
COPPER MONOXIDE see CNO250
COPPER NAPHTHENATE see NAS000
COPPER(2+) NITRATE see CNM750
COPPER(II) NITRATE see CNM750
COPPER(II) NITRATE, TRIHYDRATE (1:2:3) see CNN000
COPPER(I) NITRIDE see CNN250
COPPER NORDOX see CNO000
COPPER,(N,N',N'',N'''-TETRAOCTADECYL-29H,31H-PHTHALOCYANINETETRASULFONAMID ATO(2-)-N(29),N(30),N(31),N(32)- see CNQ300
COPPER OC FUNGICIDE see CNK559
COPPER 1,3,5-OCTATRIEN-7-YNIDE see CNN399
COPPER ORTHOARSENITE see CNN500
COPPER OXALATE see CNN755

COPPER(I) OXALATE see CNN750
COPPER(II) OXALATE see CNN755
COPPER(I) OXIDE see CNO000
COPPER(2+) OXIDE see CNO250
COPPER(II) OXIDE see CNO250
COPPER(II) OXIDE see CNO250
COPPER OXINATE see BLC250
COPPER (2+) OXINATE see BLC250
COPPER OXINE see BLC250
COPPER OXYCHLORIDE see CNK559
COPPER OXYCHLORIDE-ZINEB mixture see ZJS300
COPPER OXYQUINOLATE see BLC250
COPPER OXYQUINOLINE see BLC250
COPPER OXYSULFATE see CNM600
COPPER(I) PERCHLORATE see CNO325
COPPER(II) PERCHLORATE see CNO350
COPPER(II) PERCHLORATE, DIHYDRATE see CNO500
COPPER(II) PHOSPHINATE see CNO750
COPPER, (PHTHALOCYANINATO(2-))- see CNO800
COPPER, (29H,31H-PHTHALOCYANINATO(2-)-(N29,N30,N31),N32)-, (SP-4-1)- see CNO800
COPPER PHTHALOCYANINE see CNO800
COPPER PHTHALOCYANINE GREEN see PJQ100
COPPER(I) POTASSIUM CYANIDE see PLC175
COPPER QUINOLATE see BLC250
COPPER-8-QUINOLATE see BLC250
COPPER-8-QUINOLINOL see BLC250
COPPER QUINOLINOLATE see BLC250
COPPER-8-QUINOLINOLATE see BLC250
COPPER SALT 2-HYDROXY-1,2,3-PROPANETRICARBOXYLIC ACID (1:2) see CNK625
COPPERSAN see CNK559
COPPER SARDEX see CNO000
COPPER SLAG-AIRBORNE see CNI000
COPPER SLAG-MILLED see CNI000
COPPER SODIUM CYANIDE see SFZ100
COPPER SORBATE see CNP000
COPPER complex with trans-N-(p-STYRYLPHENYL)ACETOHYDROXAMIC ACID see SMU000
COPPER SULFATE see CNP250
COPPER(II) SULFATE (1:1) see CNP250
COPPER(II) SULFATE PENTAHYDRATE (1:1:5) see CNP500
COPPER SULFIDE see CNP750
COPPER SULFIDE see CNQ000
COPPER(I) SULFIDE see CNP750
COPPER(2+) SULFIDE see CNQ000
COPPER (II) SULFIDE see CNQ000
COPPER(I) TETRAHYDROALUMINATE see CNQ250
COPPER TETRA-4-(OCTADECYLSULPHONAMIDE)PHTHALO CYANINE see CNQ300
COPPER TRICHLOROPHENOLATE see CNQ375
COPPER-2,4,5-TRICHLOROPHENOLATE see CNQ375
COPPER UVERSOL see NAS000
COPPER-ZINC ALLOYS see CNQ500
COPPER-ZINC CHROMATE COMPLEX see CNQ750
COPPESAN see CNK559
COPPESAN BLUE see CNK559
COPRA (DOT) see CNR000
COPRAMAT see ZJS300
COPRANTOL see CNK559
COPRA (OIL) see CNR000
COPRA PELLETS (DOT) see CNR000
COPREN see GFW000
COPREX see CNK559
COPROL see DJL000
COPROSAN BLUE see CNK559
COPROSTANOL see DKW000
COPROSTAN-3-β-OL see DKW000

COPROSTEROL see DKW000
COPSAMINE see WAK000
COPTICIDE see SNM500
COPTISINE see CNR100
COP-TOX see CNK559
COQUE MOLLE (HAITI) see JBS100
COQUERET (CANADA) see JBS100
COQUES DU LEVANT (FRENCH) see PIE500
CORACON see DJS200
CORADON see WAK000
CORAETHAMIDE see DJS200
CORAETHAMIDUM see DJS200
CORAFIL see CNR125
CORAL BEAD PLANT see RMK250
CORAL BEAN see NBR800
CORAL BERRY see ROA300
CORALEPT see DJS200
CORALITOS (CUBA) see ROA300
CORAL PLANT see CNR135
CORAL SNAKE VENOM see CNR150
CORAL VEGETAL (CUBA) see CNR135
CORALYNE SULFOACETATE see CNR250
CORAMINE see DJS200
CORANIL BROWN H EPS see SGP500
CORANIL DIRECT BLACK B see CMN240
CORATOL see POB500
CORAVITA see DJS200
CORAX see CBT750
CORAX see MDQ250
CO-RAX see WAT200
CORAX P see CBT750
CORAZON de CABRITO (CUBA) see CAL125
CORAZONE see DJS200
CORBEL see MRQ300
CORCAT see PJX000
CORCHORGENIN see SMM500
CORCHORIN see SMM500
CORCHOSIDE A AGLYCON see SMM500
CORCHSULARIN see SMM500
CORDABROMIN see HNY500
CORDALEROMIN see HNY500
CORDALIN see HLC000
CORDIAMID see DJS200
CORDIAMIN see DJS200
CORDIAMINE see DJS200
CORDILAN see LAU400
CORDILOX see VHA450
CORDIPIN see AEC750
CORDITON see DJS200
CORDOVAL see GEW000
CORDULAN see CMD750
CORDYCEPIN see DAQ225
CORDYCEPINE see DAQ225
9-CORDYCEPOSIDOADENINE see DAQ225
CORDYNIL see DJS200
COREDIOL see DJS200
COREINE see CCL250
CORESPIN see DJS200
CORETAL see CNR500
CORETAL see THK750
CORETHAMIDE see DJS200
CORETONE see DJS200
CORETONIN see GCE600
COREXIT 9527 see CNR550
CORFLEX 880 see ILR100
CORGARD see CNR675
CORGLYCON see CNH780
CORGLYCONE see CNH780
CORGLYKON see CNH780
CORIACID SCARLET R see CMM325
CORIAL EM FINISH F see CCU250
CORIAMYRTIN see CNR725
CORIAMYRTINE see CNR725
CORIAMYRTIONE see CNR725
CORIANDER OIL see CNR735
CORIANTIN see CMG675
CORIARI MYRTIFOLIA see CNR740
CORICIDIN see ABG750
CORID see AOD175
CORIFORTE see ABG750
CORIL see ELH600

CORINE see CNE750
CORINTH FLOUR see AIB250
CORISOL see VGP000
CORIZIUM see NGG500
CORLAN see HHR000
CORLIN see CNS800
CORLIN see ELH600
Δ-CORLIN see PLZ100
CORLUMINE see CNR745
(+)-CORLUMINE see CNR745
CORLUTIN see PMH500
CORLUTIN L.A. see HNT500
CORLUVITE see PMH500
CORMALONE see SOV100
CORMED see DJS200
CORMELIAN see CNR750
CORMELIAN-DIGOTAB see CNR825
CORMID see DJS200
CORMOGRIZIN see GJU800
CORMOTYL see DJS200
CORNE CABRITE (HAITI) see YAK300
CORN ENDOSPERM OIL see CNR850
CORN GLUTEN see CNR980
CORN GLUTEN MEAL see CNR980
CORN LILY see FAB100
CORNMINT OIL see CNR990
CORNMINT OIL, PARTIALLY
DEMENTHOLIZED see MCB625
CORNOCENTIN see EDB500
CORNOCENTIN see LJL000
CORN OIL see CNS000
CORNOTONE see DJS200
CORNOX CWK see BAV000
CORNOX-M see CIR250
CORNOX RD see DGB000
CORNOX RK see DGB000
CORN PRODUCTS see SLJ500
CORN SILK and CORN SILK EXTRACT see
CNS100
CORN SUGAR see GFG000
CORN SYRUP, HIGH-FRUCTOSE see
HGB100
CORNUCOPIA see AOO825
CORODANE see OPC000
CORODIL see HNY500
CORODILAN see DHS200
CORODINOC see DUU600
CORONA COROZATE see BJK500
CORONAL see DNC000
CORONAL-CRINOS see EID200
CORONARIDINE HYDROCHLORIDE see
CNS200
(−)-CORONARIDINE
MONOHYDROCHLORIDE see CNS200
CORONARIN see DNC000
CORONARINE see PCP250
CORONATE EH see HEG300
CORONATE MR 200 see PKB100
CORONENE see CNS250
CORONIN see AHK750
COROPHOS see MRH209
COROPHYLLIN-N see HLC000
COROSANIN see ELH600
COROSORBIDE see CCK125
COROSUL D AND S see SOD500
COROTHION see PAK000
COROTONIN see DJS200
COROTOXIGENIN-RHAMNOSE see
CNS300
COROTRAN see CJT750
COROVIT see DJS200
COROVLISS see CCK125
COROXON see CIK750
COROZATE see BJK500
CORPAX see PEV750
CORPHOS see HHQ875
CORPHYLLIN see DNC000
CORPORIN see PMH500
CORPS 1571 F see PFB850
CORPS PRALINE see MAO350
CORPS R. 261 see DBA200
CORPS 2339 R P (FRENCH) see PEN000

CORPUS LUTEUM HORMONE see PMH500
CORRIGAST see HKR500
CORRIGEN see OHQ000
CORRONAROBETIN see TEH500
CORROSIVE MERCURY CHLORIDE see
MCY475
CORROSIVE SUBLIMATE see MCY475
CORRY'S SLUG DEATH see TDW500
CORSODYL see CDT250
CORSONE see SOW000
CORSTILINE see AES650
CORT A see CNS650
CORTACET see DAQ800
CORTACREAM see HHQ800
CORTADREN see CNS800
CORTADREN see CNS825
CORTAID see HHQ800
CORTAN see PLZ000
CORTANCYL see PLZ000
CORTANCYL see PLZ100
CORTATE see DAQ800
CORT-DOME see CNS750
CORTEF ACETATE see HHQ800
CORTELAN see CNS825
Δ-CORTELAN see PLZ000
CORTELL see HHQ800
CORTENIL see DAQ800
CORTES see HHQ800
CORTESAN see DAQ800
CORTEXAL see IRA000
CORTEX ALDEHYDE see PDR000
CORTEXILAR see FDD075
CORTEXONE see DAQ600
CORTEXONE ACETATE see DAQ800
COR-THEOPHYLLINE see DNC000
CORTHION see PAK000
CORTHIONE see PAK000
CORTICOBISS see HGL700
CORTICOSTERON see CNS625
CORTICOSTERONE see CNS625
CORTICOSTERONE ACETATE see CNS650
CORTICOTROPHIN see AES650
CORTICOTROPIN see AES650
CORTICOTROPIN-LIKE SUBSTANCES see
AES650
CORTICOTROPIN-RELEASING FACTOR
(HORSE) see HGL700
CORTICOTROPIN-RELEASING FACTOR
(HUMAN) see HGL700
CORTICOTROPIN-RELEASING FACTOR
(RAT) see HGL700
CORTICOTROPIN-RELEASING FACTOR
(SHEEP), 2-l-GLUTAMIC ACID-22-l-
ALANINE-23-l-ARGININE-25-l-GLUTAMIC
ACID-38-l-METHIONINE-39-l-GLUTAMIC
ACID-41-l-ISOLEUCINAMIDE- see HGL700
CORTICOTROPIN-RELEASING
HORMONE (HUMAN) see HGL700
CORTIDELT see PLZ000
CORTIFAR see DAQ800
CORTIFOAM see HHQ800
CORTIGEN see DAQ800
CORTILAN-NEU see CDR750
CORTINAQ see DAQ800
CORTINAZINE see ILD000
CORTINELLUS SHIITAKE EXTRACT
(JAPANESE) see JDJ100
CORTIPHATE INJECTABLE see HHQ875
CORTIPHYSON see AES650
CORTIPRED see SOV100
CORTIRON see DAQ800
CORTISAL see CNS800
CORTISAL see CNS825
CORTISATE see CNS800
CORTISATE see CNS825
CORTISOL see CNS750
Δ¹-CORTISOL see PMA000
CORTISOL ACETATE see HHQ800
CORTISOL ALCOHOL see CNS750
CORTISOL HEMISUCCINATE SODIUM
SALT see HHR000
CORTISOL PHOSPHATE see HHQ875

CORTISOL-21-PHOSPHATE see HHQ875
CORTISOL SODIUM HEMISUCCINATE see
HHR000
CORTISOL SODIUM SUCCINATE see
HHR000
CORTISOL-21-SODIUM SUCCINATE see
HHR000
CORTISOL SUCCINATE, SODIUM SALT see
HHR000
CORTISONE see CNS800
Δ-CORTISONE see PLZ000
Δ¹-CORTISONE see PLZ000
CORTISONE ACETATE see CNS825
CORTISONE-21-ACETATE see CNS825
CORTISONE MONOACETATE see CNS825
CORTISPRAY see CNS750
CORTISTAB see CNS825
CORTISTAL see CNS800
CORTISYL see CNS825
CORTIVIS see DAQ800
CORTIVITE see CNS800
CORTIVITE see CNS825
CORTIXYL see DAQ800
CORTOCIN-F see FDB000
CORTOGEN see CNS800
CORTOGEN see CNS825
CORTOGEN ACETATE see CNS825
CORTONE see CNS800
CORTONE see CNS825
Δ-CORTONE see PLZ000
CORTONE ACETATE see CNS825
CORTRIL ACETATE see HHQ800
CORTRIL ACETATE-AS see HHQ800
CORTROPHIN see AES650
CORTROPHYSON see AES650
CORTUSSIN see RLU000
CORUNDUM see EAL100
CORUNDUM FUME see CNT250
CORVASAL see MRN275
CORVASYMTON see SPD000
CORVATON see MRN275
CORVIC 55/9 see PKQ059
CORVIC 236581 see AAX175
CORVITAN see DJS200
CORVITIN see DBA800
CORVITOL see DJS200
CORVITONE see DJS200
CORWIN see XAH000
CORYBAN-D see ABG750
CORYDALOID see CNT325
CORYDININE see FOW000
CORYLON see HMB500
CORYLONE see HMB500
CORYLOPHYLINE see GFG100
CORYNINE see YBJ000
CORYSTIBIN see AQH500
CORYWAS see DJS200
CORYZOL see DPJ400
COSAN see SOD500
COSBIOL see SLG700
COSCOPIN see NBP275
COSCOPIN see NOA000
COSCOPIN HYDROCHLORIDE see NOA500
COSCOTABS see NOA000
COSCOTABS HYDROCHLORIDE see
NOA500
COSDEN 550 see SMQ500
COSDEN 945E see SMQ500
COSLAN see XQS000
COSMEGEN see AEB000
COSMETIC BLUE LAKE see FAE000
COSMETIC BRILLIANT PINK BLUISH D
CONC see FAG070
COSMETIC CORAL RED KO BLUISH see
CHP500
COSMETIC GREEN BLUE R25396 see
ADE500
COSMETIC WHITE see BKW100
COSMETIC WHITE C47-5175 see TGG760
COSMETOL see CCP250
COSMIC see TJJ500
COSMOPEN see BDY669

COSMOPEN see BFD000
COSMOPHLOXINE F see CMM840
COSPANON see TKP100
COSTUS OIL see CNT350
COSTUS ROOT see CNT350
COSTUS ROOT OIL see CNT400
COSULID see SNH900
COTALMON see AMK250
COTARNIN see CNT625
COTARNINE see CNT625
COTEL see VSZ000
COTINAZIN see ILD000
COTININ see FBW000
COTININE-N-OXIDE see CNT645
COTNION-ETHYL see EKN000
COTNION METHYL see ASH500
COTOFILM see HCL000
COTOFOR see EPN500
COTONE see PLZ000
COTONERAL ABS see CMN300
COTONEROL AB see CMN300
COTORAN see DUK800
COTORAN MULTI see MQQ450
COTORAN MULTI 50WP see DUK800
CO-TRIFAMOLE see SNL850
CO-TRIMOXAZOLE see SNK000
CO-TRIMOXAZOLE see TKX000
COTTON AIDE HC see HKC000
COTTON BORDEAUX B see CMO872
COTTON DUST see CNT750
COTTONEX see DUK800
COTTON GREEN B see CMO840
COTTON RED 4B see DXO850
COTTON RED 10B see CMO872
COTTON RED L see SGQ500
COTTONSEED, MODIFIED PRODUCTS see CNT950
COTTONSEED OIL (unhydrogenated) see CNU000
COTTON VIOLET R see CMP000
COUER SAIGNANT (HAITI) see CAL125
COUMADIN see WAT200
COUMADIN SODIUM see WAT220
COUMAFENE see WAT200
COUMAFURYL see ABF500
COUMAMYCIN see CNV500
COUMAPHOS see CNU750
COUMAPHOS-O-ANALOG see CIK750
COUMAPHOS OXYGEN ANALOG (USDA) see CIK750
4-COUMARIC ACID see CNU825
p-COUMARIC ACID see CNU825
trans-o-COUMARIC ACID see CNU850
COUMARILIC ACID see CNU875
COUMARIN see CNV000
COUMARIN 1 see DIL400
COUMARIN 4 see MKP500
COUMARIN 311 see DPJ800
COUMARIN, 3-(3-(4-BIPHENYLYL)-1,2,3,4-TETRAHYDRO-1-NAPHTHYL)-4-HYDROXY- see BGO100
COUMARIN, 3-CHLORO-7-HYDROXY-4-METHYL-, BIS(4-CHLOROBUTYL)PHOSPHATE see CHO800
COUMARIN, 3-CHLORO-7-HYDROXY-4-METHYL-, BIS(3-CHLOROPROPYL)PHOSPHATE see CNV050
COUMARIN, 3-CHLORO-7-HYDROXY-4-METHYL-, BIS(2,3-DICHLOROPROPYL)PHOSPHATE see CHO850
COUMARIN, 6,7-DIHYDROXY-4-METHYL-see MJV800
COUMARIN, 4-ETHYL-7-HYDROXY-3-(p-METHOXYPHENYL)- see ELH800
COUMARIN, 7-HYDROXY- see HJY100
COUMARIN, 7-HYDROXY-3-(p-HYDROXYPHENYL)-4-PHENYL- see HNK600
cis-o-COUMARINIC ACID LACTONE see CNV000
COUMARINIC ANHYDRIDE see CNV000

COUMARIN, 7-METHOXY-(8CI) see MEK300
COUMARIN, 4-METHYL-7-HYDROXY-, p-GUANIDINOBENZOATE see GLC100
COUMARIN, 4,5,7-TRIHYDROXY- see TKO300
COUMARONE see BCK250
COUMARONE-INDENE RESIN see CNV100
COUMATETRALYL see EAT600
COUMERMYCIN AL see CNV500
COUMESTROL see COF350
COUNTER see BSO000
COUNTER 15G SOIL INSECTICIDE see BSO000
COUNTER 15G SOIL INSECTICIDE-NEMATICIDE see BSO000
COUNTRY WALNUT see TOA275
COURLENE-X3 see PJS750
COURLOSE A 590 see SFO500
COVALLATOXOL see CNH785
COVATIN see BRS000
COVATIX see BRS000
CO-VIDARABINE see PBT100
COVI-OX see VSZ450
COVIT see VSZ000
COVOL see PKP750
COVOL 971 see PKP750
COWBUSH (BAHAMAS) see LED500
COW GARLIC see WBS850
COWSLIP see MBU550
CO X-40 see CNB825
COXIGON see OJI750
COXISTAC see SAN500
COXISTAT see NGE500
COXYSAN see CNK559
COYDEN see CMX850
COYOTILLO see BOM125
COZYMASE see CNF390
COZYMASE I see CNF390
COZYMASE II see CNF400
COZYME see PAG200
CP see CCJ400
CP see CQC650
3CP see CJQ300
4-CP see CJN000
CP20 see DSI250
CP 34 see TFM250
CP 105 see ECI000
CP 205 see BPM690
CP 261 see DTS500
CP 293 see PAX800
CP 767 see BIG600
CP 877 see BIB800
CP 3438 see TFD250
CP 4517 see HMR550
CP 4572 see CDO500
CP 556S see SOU600
CP 6,343 see CFK000
CP 8574 see PGZ900
CP 10,188 see FAM100
CP 10502 see PGZ915
CP 12574 see TGD250
CP 13842 see ENQ600
CP 14,957 see OAN000
CP 15,336 see DBI200
CP 15575 see ZGA500
CP 16171 see FAJ100
CP 18851 see CGO525
CP 18928 see CKI612
CP 19699 see CON300
CP 23426 see DNS600
CP 25017 see NHK800
CP 26890 see BJT800
CP 31393 see CHS500
CP 34089 see SOU650
CP 40294 see MOB699
CP-40507 see MOB750
CP 43858 see TIP750
CP 47114 see DSQ000
CP 48985 see CFC500
CP 49527 see DWX100
CP 49674 see DOP200
CP 50144 see CFX000

CP 50296 see DNE900
CP 52223 see IIE200
CP 52665 see CGI200
CP-53619 see CFW750
CP-53619 see IIE200
CP 53926 see DRR200
CP-67,015 see DLB500
CP-69637 see SFO700
CP 78601 see TBV300
CP 80288 see MIT615
CP 89141 see CIB650
CP 93520 see MDT800
CP 96320 see ACU600
CP 98576 see LGE200
CP 99109 see BQY275
CP 99213 see TBL600
CP-12,252-1 see NBP500
CP-15-639-2 see CBO250
CP 15749-3 see THD875
CP-16533-1 see IRV000
CP 10423-18 see TCW750
CP 1402304 see CKJ850
CP 1044 J3 see BPP750
CP-15467-61 see LGZ000
CPA see CJN000
CPA see CQC650
CPA see CQJ500
3-CPA see CJQ300
6-CPA see CKN375
C-PAL see FAM100
C. PALUSTRIS see MBU550
CPAS see CDS500
CPAS mixed with BCPE see CKL500
CPB see BSS550
CPB see CJR500
CP BASIC SULFATE see CNP250
CPBS see CJR500
CPBU 7 see GHR609
CPC 3005 see SLJ500
CPC 6448 see SLJ500
CPCA see BIO750
CPCBS see CJT750
CP20 (CHELATING AGENT) see DSI250
C.P. CHROME LIGHT 2010 see LCS000
C.P. CHROME ORANGE DARK 2030 see LCS000
C.P. CHROME ORANGE MEDIUM 2020 see LCS000
C.P. CHROME YELLOW LIGHT see LCR000
2-Cl-P-PD see CEG625
4-Cl-m-PD see CJY120
4-Cl-o-PD see CFK125
CPDC see PJD000
CPDD see PJD000
CPE see PJS750
CPE 16 see PJS750
CPE 25 see PJS750
CPH see CBL000
CPH see CDP250
CPIB see ARQ750
CPIRON see FBJ100
CP 1044 J3 see BPP750
CPMC see CKF000
CPNU-I see NKJ100
cis-CPO see CKS099
trans-CPO see CKS100
CPP-3,4-OXIDE see CPX625
C15-PREDIOXIN see TIL750
C. PROCERA see COD675
CPS see MIG850
CPSA see CDP725
CPS-M see TLC300
CP 45899 SODIUM SALT see PAP600
CPT see CLK215
CPT see TAF675
CdPT see CAI350
C.P. TITANIUM see TGF250
CP 49952 p-TOLUENESULFONATE see SOU675
C.P. TOLUIDINE TONER A-2989 see MMP100
C.P. TOLUIDINE TONER A-2990 see MMP100

C.P. TOLUIDINE TONER DARK RS-3340 see MMP100
C.P. TOLUIDINE TONER DEEP X-1865 see MMP100
C.P. TOLUIDINE TONER LIGHT RS-3140 see MMP100
C.P. TOLUIDINE TONER RT-6101 see MMP100
C.P. TOLUIDINE TONER RT-6104 see MMP100
C. PULCHERRIMA see CAK325
CP 261LV see DTS500
CPX see TAF675
C3-PYRIDINYLARBONIMIDOTHIOIC ACID, o-(3,4-DIMETHYLCYCLOHEXYL) S-((4-(1,1-DIMETHYLETHYL) PHENYL)METHYL) ESTER see CNV600
CPZ see CCS369
CPZ see CKP250
CPZ see CKP500
C.P. ZINC YELLOW X-883 see ZFJ100
CQ see CLD250
CoQ$_{10}$ see UAH000
C-QUENS see CBF250
C-QUENS see CNV750
CR see DDE200
CR 39 see AGD250
CR-144 see PKL150
CR 200 see PKB100
CR 242 see BGC625
CR 409 see BJE750
CR-604 see POF550
CR-605 see TGF175
CR 733 see NMV460
CR/662 see BLV000
CR 1505 see LII100
CR 2024 see MCB050
CR 3029 see MAS500
CRAB-E-RAD see DXE600
CRAB'S EYES see AAD000
CRAB'S EYES see RMK250
CRADEX see BBK500
CRAG 341 see GII000
CRAG 85W see DSB200
CRAG 974 see DSB200
CRAG DCU-73w see DGQ200
CRAG EXPERIMENTAL HERBICIDE 2 see DGQ200
CRAG FLY REPELLENT see BRP250
CRAG FRUIT FUNGICIDE 34 see GIO000
CRAG FRUIT FUNGICIDE 341 see GII000
CRAG FUNGICIDE 658 see CNQ750
CRAG FUNGICIDE 974 see DSB200
CRAG HERBICIDE see CNW000
CRAG HERBICIDE 1 see CNW000
CRAG HERBICIDE 2 see DGQ200
CRAG NEMACIDE see DSB200
CRAG SESONE see CNW000
CRAG SEVIN see CBM750
CRAIN see FBS100
CRALO-E-RAD see DXE600
CRAMCILLIN-S see PDD350
CRAMPOL see ACX500
CRAMPOLE see ACX500
CRANOMYCIN see CNW100
CRANOMYCIN HYDROCHLORIDE see CNW105
CRASTIN S 330 see PKF750
CRATAEGUS, EXTRACT see EDK600
CRATECIL see EOK000
CRAVITEN see CNW125
CRAVITEN see CNW125
CRAWHASPOL see TIO750
CRC 7001 see CEK875
CRD-401 see DNU600
CREAM of TARTAR see PKU600
CREATININE SULFATE compounded with 3-(2-AMINOETHYL)INDOLE-5-OL (1:1:1), MONOHYDRATE see AJX750
CREATININE SULFATE compounded with 3-(2-AMINOETHYL)INDOL-5-OL (1:1:1) see AJX750

CRECHLOR S 45 see PAH780
C RED 2 see CMS238
CREDO see SHF500
CREIN see DAL300
CREMODIAZINE see PPP500
CREMOMERAZINE see ALF250
CREMOMETHAZINE see SNJ000
CREMOPHOR EL see CCP330
CREMOPHOR RH 40 see CCP305
CREMOPHOR RH 40 see CCP310
CREMOPHOR RH 40/60 see CCP305
CREMOPHOR RH 40/60 see CCP310
CREMOR TARTARI see PKU600
p-CREOSOL see MEK325
CREOSOTE see CMY825
CREOSOTE, from COAL TAR see CMY825
CREOSOTE OIL see CMY825
CREOSOTE P1 see CMY825
CREOSOTUM see CMY825
CRESAN UNIVERSAL TROCKENBEIZE see MEP000
CRESIDINE see MGO750
m-CRESIDINE see MGO500
p-CRESIDINE see MGO750
CRESOATE, WOOD see BAT850
CRESODIOL see GGS000
CRESOL see CNW500
2-CRESOL see CNX000
3-CRESOL see CNW750
4-CRESOL see CNX250
m-CRESOL see CNW750
o-CRESOL see CNX000
p-CRESOL see CNX250
p-CRESOL ACETATE see MNR250
m-CRESOL, α-(AMINOOXY)-6-BROMO- see BMM600
m-CRESOL, 4,4'-(3H-2,1-BENZOXATHIOL-3-YLIDENE)BIS(2,6-DIBROMO-,S,S-DIOXIDE see BNA940
m-CRESOL, 4,4'-(3H-2,1-BENZOXATHIOL-3-YLIDENE)BIS(2,6-DIBROMO-,S,S-DIOXIDE see TBJ505
p-CRESOL, 2,6-BIS(α-METHYLBENZYL)- see MHR050
p-CRESOL, α-CYCLOHEXYLIDENE-α-(p-HYDROXYPHENYL)- see CPM770
p-CRESOL, 2,6-DI-tert-BUTYL-α-(DIMETHYLAMINO)- see DEA100
p-CRESOL, 2,6-DI-tert-BUTYL-α-(DIMETHYLAMINO)- see FAB000
p-CRESOL, 2,6-DI-tert-BUTYL-α-METHOXY- see DEE300
o-CRESOL, 4,6-DI-tert-BUTYL-α-PHENYL- see DEG150
m-CRESOL, 4,4'-(1,2-DIETHYLETHYLENE)DI- see DJI300
o-CRESOL, 4,4'-(1,2-DIETHYLETHYLENE)DI- see DJI350
m-CRESOL, 4,6-DINITRO- see DUT700
o-CRESOL, 4,6-DINITRO-, BARIUM DERIVATIVE see DUT900
o-CRESOL, DINITRO-, SODIUM SALT see SGP500
CRESOL DIPHENYL PHOSPHATE see TGY750
CRESOL FAST VIOLET see AJN250
CRESOL FLYCIDYL ETHER see TGZ100
o-CRESOL GLYCERYL ETHER see GGS000
p-CRESOL GLYCIDYL ETHER see GGY175
CRESOLI (ITALIAN) see CNW500
p-CRESOL METHYL ETHER see MGP000
m-CRESOL, 4,4',4"-(1-METHYL-1-PROPANYL-3-YLIDENE)TRIS(6-tert-BUTYL)-(7CI,8CI) see MOS100
o-CRESOL, 4-NITRO- see NFV010
m-CRESOL, 4-NITRO-, DIMETHYL PHOSPHATE see PHD750
m-CRESOL, 4-NITRO-α,α,α-TRIFLUORO- see TKD400
p-CRESOL, 2-NITRO-α-α-α-TRIFLUORO- see NMQ100

p-CRESOL, α-PHENYL-, CARBAMATE see PFR325
o-CRESOLPHTHALEIN see CNX400
CRESOL SODIUM SALT see SFZ050
m-CRESOL, α,α,α-TRIFLUORO-4-NITRO- see TKD400
CRESON see RLU000
CRESOPUR see BAV000
CRESORCINOL DIISOCYANATE see TGM750
CRESOSSIDIOLO see GGS000
CRESOSSIPROPANDIOLO see GGS000
CRESOTIC ACID see CNX625
o-CRESOTIC ACID see CNX625
2,3-CRESOTIC ACID see CNX625
CRESOTINE BLUE 2B see CMO000
CRESOTINE BLUE 3B see CMO250
CRESOTINE BORDEAUX BG see CMO872
CRESOTINE DARK GREEN B see CMO830
CRESOTINE FAST RED F see CMO870
CRESOTINE GREEN B see CMO840
CRESOTINE PURE BLUE see CMO500
CRESOTINIC ACID see CNX625
o-CRESOTINIC ACID see CNX625
2,3-CRESOTINIC ACID see CNX625
β-CRESOTINIC ACID see CNX625
CRESOTOL see DUU600
CRESOXYDIOL see GGS000
CRESOXYPROPANEDIOL see GGS000
CRESSA CRETICA Linn., extract see CNX700
CRESTABOLIC see AOO475
CRESTANIL see MQU750
CRESTINE BROWN RC see CMO800
CRESTOMYCIN see NCF500
CRESTOXO see CDV100
p-CRESYL ACETATE (FCC) see MNR250
p-CRESYL BENZOATE see TGX100
p-CRESYL CAPRYLATE see THB000
CRESYL DIPHENYL PHOSPHATE see TGY750
CRESYL FAST VIOLET see AJN250
o-CRESYL-α-GLYCERYL ETHER see GGS000
CRESYLGLYCIDE ETHER see TGZ100
CRESYL GLYCIDYL ETHER see TGZ100
p-CRESYL GLYCIDYL ETHER see GGY175
CRESYLIC ACID see CNW500
m-CRESYLIC ACID see CNW750
o-CRESYLIC ACID see CNX000
p-CRESYLIC ACID see CNX250
CRESYLIC CREOSOTE see CMY825
p-CRESYL ISOBUTYRATE see THA250
CRESYLITE see TML500
m-CRESYL METHYLCARBAMATE see MIB750
p-CRESYL METHYL ETHER see MGP000
p-CRESYL OCTANOATE see THB000
CRESYL PHOSPHATE see TNP500
o-CRESYL PHOSPHATE see TMO600
p-CRESYL SALICYLATE see THD850
CRF 41 see HGL700
CRILL 2 see MRJ800
CRILL 3 see SKV150
CRILL 10 see PKL100
CRILL 16 see SKV170
CRILL 26 see SLL000
CRILL K 3 see SKV150
CRILL K 16 see SKV170
CRILLON L.D.E. see BKE500
CRIMIDIN (GERMAN) see CCP500
CRIMIDINA (ITALIAN) see CCP500
CRIMIDINE see CCP500
CRIMSON ANTIMONY see AQL500
CRIMSON EMBL see HJF500
CRIMSON SX see FMU080
CRINODORA see CFU750
CRINURYL see DFP600
CRINUM (VARIOUS SPECIES) see SLB250
CRISALBINE see GJG000
CRISALIN see DUV600
CRISAPON see DGI400

CRISATRINA see ARQ725
CRISAZINE see ARQ725
CRISEOCICLINE see TBX000
CRISEOCIL see MCH525
CRISFOLATAN see TBQ280
CRISODIN see MRH209
CRISODRIN see MRH209
CRISONAR see CFY750
CRISPATINE see FOT000
CRISPIN see THJ750
CRISPIN RED GM see CMM325
CRISQUAT see PAJ000
CRISTALLOSE see SJN700
CRISTALLOVAR see EDV000
CRISTALOMICINA see KAM000
CRISTALOMICINA see KAV000
CRISTAPEN see BFD000
CRISTAPURAT see DKL800
CRISTERONA MB see DME500
CRISTERONE T see TBF500
CRISTOBALITE see SCI500
CRISTOBALITE see SCJ000
CRISTOXO 90 see CDV100
CRISULFAN see EAQ750
CRISURON see DXQ500
CRITTOX see EIR000
CRIXIVAN see ICE100
CROCEOMYCIN see HAL000
CROCIDOLITE (DOT) see ARM275
CROCIDOLITE ASBESTOS see ARM275
CROCOITE see LCR000
CROCUS see CNX800
CROCUS MARTIS ADSTRINGENS see IHC450
CRODACID see MSA250
CRODACOL A.10 see OBA000
CRODACOL-CAS see HCP000
CRODACOL-O see OBA260
CRODACOL-S see OAX000
CRODAMINE 1.18D see OBC000
CRODAMOL IPM see IQN000
CRODAMOL IPP see IQW000
CRODET O 6 see PJY100
CROFLEX see CBT750
CROLAC see CBT750
CROLEAN see ADR000
CROMAL ORANGE G see CMP882
CROMAL ORANGE P see CMP882
CROMARIL see PER700
CROMILE, CLORURO di (ITALIAN) see CML125
CROMOCI see CQM325
CROMOGLYCATE see CNX825
CROMOGLYCATE DISODIUM see CNX825
CROMOLYN SODIUM see CNX825
CROMOLYN SODIUM SALT see CNX825
CROMO, OSSICLORURO di (ITALIAN) see CML125
CROMOPHTAL BLUE A 3R see IBV050
CROMOPHTHAL GREEN GF see PJQ100
CROMOSAN see MAD000
CROMO SOLFATO BASIFICATO see CMJ565
CRONETAL see DXH250
CRONETON see EPR000
CRONIL see EHP000
CROPOTEX see FDA890
CROP RIDER see DAA800
CROSPOVIDONE see PKQ150
CROSSLINKER CX 100 see EML600
CROTALARIA BERTEROANA see RBZ400
CROTALARIA INCANA see RBZ400
CROTALARIA JUNCEA see RBZ400
CROTALARIA JUNCEA Linn., seed extract see CNX827
CROTALARIA RETUSA see RBZ400
CROTALARIA SPECTABILIS see RBZ400
CROTALINE see MRH000
CROTALUS ADAMANTEUS VENOM see EAB225
CROTALUS ATROX VENOM see WBJ600
CROTALUS CERASTES VENOM see CNX830

CROTALUS DURISSUS DURISSUS VENOM see CNX835
CROTALUS DURISSUS TERRIFICUS VENOM see CNY000
CROTALUS HORRIDUS HORRIDUS VENOM see TGB150
CROTALUS HORRIDUS VENOM see CNY300
CROTALUS RUBER RUBER VENOM see CNY325
CROTALUS SCUTULATUS SCUTULATUS (CALIF) VENOM see CNY339
CROTALUS SCUTULATUS SCUTULATUS VENOM see CNY350
CROTALUS SCUTULATUS VENOM see CNY375
CROTALUS VIRIDIS CERERUS VENOM see CNY390
CROTALUS VIRIDIS CONCOLOR VENOM see CNY750
CROTALUS VIRIDIS HELLERI VENOM see COA000
CROTALUS VIRIDIS LUTOSUS VENOM see COA250
CROTALUS VIRIDIS OREGANUS VENOM see COA500
CROTALUS VIRIDIS VENOM see COA750
CROTALUS VIRIDIS VIRIDIS TOXIN see VRP200
CROTALUS VIRIDIS VIRIDIS VENOM see PLX100
CROTEIN HKP see AHR600
CROTEIN SPO see HID100
CROTILIN see DAA800
CROTILIN see KHU025
CROTILINE see KHU025
CROTOCIN see COB000
CROTONAL see COB260
CROTONALDEHYDE see COB250
CROTONALDEHYDE see COB260
(E)-CROTONALDEHYDE see COB260
CROTONALDEHYDE, stabilized (DOT) see COB250
CROTONAMIDE, 2-CHLORO-N,N-DIETHYL-3-HYDROXY-, DIMETHYL PHOSPHATE see FAB400
CROTONAMIDE, 3-METHYL- see SBW965
CROTONATE DE 2,5-DICHLORO-4,6-DINITROPHENYLE see DFE560
CROTONATE de 2,4-DINITRO 6-(1-METHYL-HEPTYL)-PHENYLE (FRENCH) see AQT500
CROTONATE d'ETHYLE (FRENCH) see COB750
CROTONIC ACID see COB500
CROTONIC ACID, solid see COB500
α-CROTONIC ACID see COB500
CROTONIC ACID, 3-AMINO-, ISOPROPYL ESTER see INM100
CROTONIC ACID ANHYDRIDE see COB900
CROTONIC ACID, 4-BROMO-, METHYL ESTER see MHR790
CROTONIC ACID, 2-BUTENYLIDENE ESTER see COD100
CROTONIC ACID, 3,6-DICHLORO-2,4-DINITROPHENYL ESTER see DFE560
CROTONIC ACID, ETHYL ESTER see EHO200
α-CROTONIC ACID ETHYL ESTER see COB750
CROTONIC ACID GERNAIOL ESTER see DTE000
CROTONIC ACID, 2-METHYL-, (E)- see TGA700
(E)-CROTONIC ACID METHYL ESTER see COB825
CROTONIC ACID, 2-METHYL-, ETHYL ESTER, (E)-(8CI) see TGA800
CROTONIC ACID, 2-METHYL-, METHYL ESTER, (E)-(8CI) see MPW700
CROTONIC ACID, VINYL ESTER see VNU000

CROTONIC ALDEHYDE see COB250
CROTONIC ALDEHYDE see COB260
CROTONIC ANHYDRIDE see COB900
CROTONIC NITRILE see COC300
CROTONIQUE NITRILE see COC300
CROTONITRILE see COC300
CROTONOEL (GERMAN) see COC250
CROTON OIL see COC250
CROTONONITRILE see COC300
(Z)-CROTONONITRILE see BOX600
cis-CROTONONITRILE see BOX600
CROTONONITRILE, (Z)-(8CI) see BOX600
CROTONOSID see IKS450
CROTONOSIDE see IKS450
CROTON RESIN see COC250
CROTON TIGLIUM L. OIL see COC250
CROTONYL ALCOHOL see BOY000
CROTONYLENE see COC500
α-(N-CROTONYL-N-ETHYL)AMINO-BUTYRIC ACID mixed with BUTRIC ACID, α-(N-CROTONYL-N-PROPYL)AMINO see COC750
CROTONYL-N-ETHYL-o-TOLUIDINE see EHO500
CROTONYLOXYMETHYL-4,5,6-TRIHYDROOXYCYCLOHEX-2-ENONE see COC825
CROTOXIN see COC875
CROTOXYPHOS see COD000
CROTURAL see EHP000
CROTYL ALCOHOL see BOY000
CROTYL CHLORIDE see CEU825
CROTYL-2,4-DICHLOROPHENOXYACETATE see DAD000
CROTYLIDENE ACETIC ACID see SKU000
CROTYLIDENE DICROTONATE see COD100
CROTYLIN see KHU025
1-CROTYL THEOBROMINE see BSM825
CROVARIL see HNI500
CROW BERRY see PJJ315
CROWFOOD see FBS100
CROWN 18 see COD575
12-CROWN-4 see COD475
15-CROWN-5 see PBO000
18-CROWN-6 see COD500
CROWN BEAUTY see BAR325
CROWN FLOWER see COD675
CROYSULFONE see SOA500
CRP-401 see DNU600
CRR-733S see REA050
CRTRON see VSZ100
CRUDE ARSENIC see ARI750
CRUDE COAL TAR see CMY800
CRUDE ERGOT see EDB500
CRUDE OIL see PCR250
CRUDE OIL, synthetic see COD725
CRUDE PETROLEUM see PCR250
CRUDE PROPYLENE POLYMERS see VRU300
CRUDE SAIKOSIDE see SAF300
CRUDE SHALE OILS see COD750
CRUFOMATE see COD850
CRUFOMATE A see COD850
CRUFORMATE see COD850
CRUMERON, combustion products see ADX750
CRUNCH see CBM750
CRUSTECDYSON see HKG500
CRYOFLEX see BHK750
CRYOFLUORAN see FOO509
CRYOFLUORANE see FOO509
CRYOGENINE see CBL000
CRYOLITE see SHF000
CRYOPOLYTHENE see PJS750
CRY-O-VAC L see PJS750
CRYPTOAESCIN see COD900
CRYPTOCILLIN see MNV250
CRYPTOCRYSTALLINE QUARTZ see SCK600
CRYPTOCYANINE IODIDE see KHU050

CRYPTOCYANINE O. A. 2 see KHU050
CRYPTOESCIN see COD900
CRYPTOGIL OL see PAX250
CRYPTOGYL NA (ITALIAN) see PAX750
CRYPTOHALITE see COE000
CRYPTOSTEGIA GRANDIFLORA see
ROU450
CRYPTOSTEGIA MADAGASCARIENSIS see
ROU450
CRYPUR GOLDEN YELLOW GL see
CMM895
CRYSALBA see CAX500
CRYSTALBUMINS see AFI780
CRYSTAL CHROME ALUM see PLB500
CRYSTALETS see VSK900
CRYSTALLINA see VSZ100
CRYSTALLINE DEHYDROXY SODIUM
ALUMINUM, CARBONATE see DAC450
CRYSTALLINE DIGITALIN see DKL800
CRYSTALLINE LIENOMYCIN see LFP000
CRYSTALLINIC ACID see SMB000
CRYSTALLIZED VERDIGRIS see CNI250
CRYSTALLOMYCIN see COE100
CRYSTALLOSE see SJN700
CRYSTAL O see CCP250
CRYSTAL ORANGE 2G see HGC000
CRYSTAL PROPANIL-4 see DGI000
CRYSTAL STRUCTURE TYPES, ZEOLITES
see ZAT300
CRYSTALS of VENUS see CNI250
CRYSTAL VIOLET see AOR500
CRYSTAL VIOLET 6B see AOR500
CRYSTAL VIOLET 10B see AOR500
CRYSTAL VIOLET AO see AOR500
CRYSTAL VIOLET AON see AOR500
CRYSTAL VIOLET BASE see AOR500
CRYSTAL VIOLET BP see AOR500
CRYSTAL VIOLET BPC see AOR500
CRYSTAL VIOLET CHLORIDE see AOR500
CRYSTAL VIOLET EXTRA PURE see
AOR500
CRYSTAL VIOLET EXTRA PURE APN see
AOR500
CRYSTAL VIOLET EXTRA PURE APNX see
AOR500
CRYSTAL VIOLET FN see AOR500
CRYSTAL VIOLET HL2 see AOR500
CRYSTAL VIOLET O see AOR500
CRYSTAL VIOLET 5BO see AOR500
CRYSTAL VIOLET 6BO see AOR500
CRYSTAL VIOLET PURE DSC see AOR500
CRYSTAL VIOLET PURE DSC BRILLIANT
see AOR500
CRYSTAL VIOLET SS see AOR500
CRYSTAL VIOLET TECHNICAL see AOR500
CRYSTAL VIOLET USP see AOR500
CRYSTAMET see SJU000
CRYSTAMIN see VSZ000
CRYSTAPEN see BFD000
CRYSTAPEN see BFD250
CRYSTAR see ADA725
CRYSTAR see SCQ000
CRYSTEX see SOD500
CRYSTHION 2L see ASH500
CRYSTHYON see ASH500
CRYSTIC PREFIL S see CAT775
CRYSTODIGIN see DKL800
CRYSTOGEN see EDV000
CRYSTOIDS see HFV500
CRYSTOL CARBONATE see SFO000
CRYSTOLON 37 see SCQ000
CRYSTOLON 39 see SCQ000
CRYSTOSERPINE see RDK000
CRYSTOSOL see MQV750
CRYSTWEL see VSZ000
CS see CEQ600
CS-1 see CCP750
CS-61 see DON700
C-Sn-9 see BLL750
CS 359 see BOM510
CS 370 see CMY525
CS 386 see MQR760

CS-430 see HAG800
CS-439 see ALF500
CS-514 see LGK500
CS-600 see LII300
CS 708 see BON250
CS-847 see CEW500
CS 1170 see CCS350
CS 4030 see QVA000
50-CS-46 see EME050
60-CS-16 see CMF400
CS 645A see BIN500
CS 12602 see TCJ075
CSAC see EES400
CSC see SEC000
CSF-GIFTWEIZEN see TEM000
CSI PASTE see DTG700
CS LAFARGE see CAW850
C-3 SODIUM SALT see CBF625
CSP see CNP500
ChS-RR2 see BOS100
CS 1170 SODIUM see CCS360
C. SUFFUSUS SUFFUSUS VENOM see
CCW925
CT see CAY300
CT see CCX250
CT see CLH750
CT-1341 see AFK500
CT 3318 see COE125
4365 CT see MLI750
CT 4436 see MFO250
CTA see CLO750
CTAB see HCQ500
CTAP-III see HGL900
CTC see CMA750
CTC 236 see DHL850
CTCP see CNQ375
C-3,T-4-DIHYDROXY-R-1,2-EPOXY-1,2,3,4-
TETRAHYDRODIBENZ(A,H)ANTHRACEN
E see NAY200
CTFE see CLQ750
CTH see CLK230
CTMA see HCQ525
C-TOXIFERINE 1 see THI000
CTP see CQL400
5'-CTP see CQL400
CTR 6669 see MHC750
CTT see CCS373
CTX see CCR950
CTX see CQC650
CTX 2 see CMP805
CU-56 see CNK559
CUBAN LILLY see SLH200
CUBE see RNZ000
CUBEB OIL see COE175
CUBE EXTRACT see RNZ000
CUBE-PULVER see RNZ000
CUBE ROOT see RNZ000
CUBES see DJO000
CUBIC NITER see SIO900
CUBISOL see DNM400
CUBOR see RNZ000
CUCKOOPINT see ITD050
CUCKOO PLANT see JAJ000
CUCUMBER ALCOHOL see NMV780
CUCUMBER ALDEHYDE see NMV760
CUCUMBER DUST see ARB750
CUCURBITACIN A see COE180
CUCURBITACIN B see COE190
CUCURBITACIN E see COE250
CUCURBITACINE (D) see EAH100
CUCURBITACINE B see COE190
CUCURBITACINE-E see COE250
CUEMID see CME400
CUENTAS de ORO (PUERTO RICO) see
GIW200
CUIPU (MEXICO) see CNR135
CULLEN EARTH see MAT500
CULMINAL K 42 see MIF760
CULPEN see CHJ000
CULTAR see TMK125
CULVERAM CDG see LBU200
CUM see COE750

CUMA see BJZ000
CUMAFOS (DUTCH) see CNU750
CUMAFURYL (GERMAN) see ABF500
CUMALDEHYDE see COE500
CUMAN see BJK500
CUMAN L see BJK500
p-CUMARIC ACID see CNU825
CUMATE see CNL500
CUMATETRALYL (GERMAN, DUTCH) see
EAT600
CUMEEN (DUTCH) see COE750
CUMEENHYDROPEROXYDE (DUTCH) see
IOB000
CUMENE see COE750
psi-CUMENE see TLL750
CUMENE ALDEHYDE see COF000
CUMENE HYDROPEROXIDE (DOT) see
IOB000
CUMENE HYDROPEROXIDE,
TECHNICALLY PURE (DOT) see IOB000
CUMENE PEROXIDE see DGR600
m-CUMENOL see IQX090
p-CUMENOL see IQZ000
m-CUMENOL METHYLCARBAMATE see
COF250
CUMENT HYDROPEROXIDE see IOB000
p-(p-CUMENYLAZO)-N,N-
DIMETHYLANILINE see IOT875
3-p-CUMENYL-1,1-DIMETHYLUREA see
IRA050
CUMENYL HYDROPEROXIDE see IOB000
m-CUMENYL METHYLCARBAMATE see
COF250
o-CUMENYL METHYLCARBAMATE see
MIA250
p-CUMENYL PHENYL PHOSPHATE (7CI)
see IRA200
CUMERTILIN SODIUM see MCS000
CUMIC ALCOHOL see CQI250
p-CUMIC ALDEHYDE see COE500
CUMID see BJZ000
o-CUMIDINE see INW100
psi-CUMIDINE see TLG250
psi-CUMIDINE HYDROCHLORIDE see
TLG750
CUMINALDEHYDE see COE500
CUMINIC ACETALDEHYDE see IRA000
CUMINIC ALCOHOL see CQI250
CUMINIC ALDEHYDE (FCC) see COE500
CUMIN OIL see COF325
CUMINOL see CQI250
CUMINYL ACETATE see IOF050
CUMINYL ALCOHOL see CQI250
CUMINYL ALDEHYDE see COE500
CUMINYL NITRILE see IOD050
CUMMIN see COF325
CUMMINUM CYMINUM LINN., EXTRACT
see COF335
CUMOESTEROL see COF350
ψ-CUMOHYDROQUINONE see POG300
CUMOLHYDROPEROXID (GERMAN) see
IOB000
CUMOQUINONE see POG400
psi-CUMOQUINONE see POG400
CUMOSTROL see COF350
CUMYL ACETALDEHYDE see IRA000
CUMYL ALCOHOL see CQI250
α-CUMYL ALCOHOL see DTN100
CUMYL HYDROPEROXIDE see IOB000
α-CUMYL HYDROPEROXIDE see IOB000
CUMYL HYDROPEROXIDE, TECHNICAL
PURE (DOT) see IOB000
CUMYL PEROXIDE see DGR600
p-CUMYLPHENOL see COF400
p-(α-CUMYL)PHENOL see COF400
CUNAPSOL see NAS000
CUNDEAMOR (CUBA, PUERTO RICO) see
FPD100
CUNILATE see BLC250
CUNILATE 2472 see BLC250
CUPEY see BAE325
CUPFERRON see ANO500

CUP-OF-GOLD see CDH125
CUPPER OXIDE (RUSSIAN) see CNO000
CUPRAL see SGJ000
CUPRAL 45 see CNK559
CUPRAMAR see CNK559
CUPRAMER see CNK559
CUPRANIL BROWN BCW see CMO820
CUPRANIL BROWN BCWR see CMO820
CUPRANTOL see CNK559
CUPRASULFIDE see CNP750
CUPRATE(1-), DICYANO-, POTASSIUM see PLC175
CUPRATE(4-), (mu-((3,3'-((3,3'-DIHYDROXY(1,1'-BIPHENYL)-4,4'-DIYL)BIS(AZO))BIS(5-AMINO-4-HYDROXY-2,7-NAPHTHALENEDISULFONATO))(8-)))DI-, TETRASODIUM see CMO650
CUPRATE(2-), (29H,31H-PHTHALOCYANINE-C,C-DISULPHONATO(4-)-N^{29},N^{30},N^{31},N^{32}-, DISODIUM see COF420
CUPRATE(2-), TRIS(CYANO-C)-, DISODIUM see SFZ100
CUPRAVET see CNK559
CUPRAVIT see CNK559
CUPRAVIT BLAU see CNM500
CUPRAVIT BLUE see CNM500
CUPRAVIT FORTE see CNK559
CUPRAVIT GREEN see CNK559
CUPRENIL see MCR750
CUPREOL see SDZ350
CUPRIC ACETATE see CNI250
CUPRIC ACETATE MONOHYDRATE see CNI325
CUPRIC ACETOARSENITE see COF500
CUPRIC ACETYLACETONATE see BGR000
CUPRIC ARSENITE see CNN500
CUPRIC CARBONATE see CBW200
CUPRIC CARBONATE see CNJ750
CUPRIC CARBONATE (1:1) see CBW200
CUPRIC CHELATE of 2-N-HYDROXYFLUORENYL ACETAMIDE see HIP500
CUPRIC CHLORIDE see CNK500
CUPRIC CITRATE see CNK625
CUPRIC CYANIDE (DOT) see CNL250
CUPRIC DIACETATE see CNI250
CUPRIC DICHLORIDE see CNK500
CUPRIC DIETHYLDITHIOCARBAMATE see CNL300
CUPRIC DINITRATE see CNM750
CUPRIC DIPERCHLORATE TETRAHYDRATE see CNO500
CUPRICELLULOSE see CCU150
CUPRIC GREEN see CNN500
CUPRIC HYDROXIDE see CNM500
CUPRIC-8-HYDROXYQUINOLATE see BLC250
CUPRICIN see CNL000
CUPRIC NITRATE (DOT) see CNM750
CUPRIC NITRATE TRIHYDRATE see CNN000
CUPRICOL see CNK559
CUPRIC OXALATE see CNN750
CUPRIC OXALATE see CNN755
CUPRIC OXIDE see CNO000
CUPRIC OXIDE CHLORIDE see CNK559
CUPRIC-8-QUINOLINOLATE see BLC250
CUPRIC SULFATE see CNP250
CUPRIC SULFATE PENTAHYDRATE see CNP500
CUPRIC SULFIDE see CNQ000
CUPRIETHYLENE DIAMINE see DBU800
CUPRIETHYLENEDIAMINE, solution (DOT) see DBU800
CUPRIMINE see MCR750
CUPRINOL see NAS000
CUPRITOX see CNK559
CUPRIZANE see COF675
CUPRIZONE see COF675
CUPROCIN see ZJS300
CUPROCITROL see CNK625

CUPROFIX BLUE GREEN B see COF420
CUPROFIX BLUE GREEN FB see COF420
CUPROKYLT see CNK559
CUPROL see CNK559
CUPRON (CZECH) see BCP500
CUPRONE see BCP500
CUPROSAN see CNK559
CUPROSANA see CNK559
CUPROSAN BLUE see CNK559
CUPROUS ARSENATE, BASIC see CNI900
CUPROUS CHLORIDE see CNK250
CUPROUS CYANIDE see CNL000
CUPROUS DICHLORIDE see CNK250
CUPROUS IODIDE see COF680
CUPROUS OXIDE see CNO000
CUPROUS POTASSIUM CYANIDE see PLC175
CUPROUS SULFIDE see CNP750
CUPROVINOL see CNK559
CUPROX see CNK559
CUPROXAT see CNM600
CUPROXAT FLOWABLE see CNM600
CUPROXOL see CNK559
CUPROZAN see ZJS300
CURACIT see BJI000
CURACRON see BNA750
CURALIN M see MJM200
CURAMAGUEY (CUBA) see YAK300
CURANTYL see PCP250
CURARE see COF750
CURARIL see GGS000
CURARIN see COF825
CURARINE see COF825
(+)-CURARINE see COF825
CURARIN-HAF see TOA000
CURARYTHAN see GGS000
CURATERR see CBS275
CURATIN see AEG750
CURAX see CPI250
CURBISET see CDT000
CURCUMA see ICC800
CURCUMA LONGA Linn., rhizome extract see COF850
CURCUMA OIL see COG000
CURCUMENONE see COF855
CURCUMIN see COG000
CURCUMIN see ICC800
CURCUMINE see COG000
CURDIONE see GDO100
(+)-CURDIONE see GDO100
CURENE see PKL500
CURENE 442 see MJM200
C. URENS see FBW100
CURE-RITE 18 see OPQ100
CURESAN see MEP250
CURETAN see MEP250
CURETARD see COG250
CURETARD A see DWI000
CUREX FLEA DUSTER see RNZ000
CURITAN see DXX400
CURITHANE see MJQ000
CURITHANE 103 see MGA500
CURITHANE C126 see DEQ600
CURL FLOWER see CMV390
CURLY HEADS see CMV390
CUROL BRIGHT RED 4R see FMU080
CUROL ORANGE see CMM220
CUROL ORANGE G see MRL100
CUROL PURE BLUE B see CMM090
CUROL RED 6B see CMM400
CURON FAST YELLOW 5G see FAG140
CUROSAJIN see HGI100
CURRAL see AFS500
CURTACAIN see TBN000
CURZATE see COM300
CURZATE M see MAS500
CUSCOHYGRIN DIMETHIODIDE see COG750
CUSCOHYGRINE BIS(METHYL BENZENESULFONATE) see COG500
CUSCOHYGRINE DIMETHYLIODIDE see COG750

CUSCUTA REFLEXA Roxb., extract excluding roots see AHI630
CUSITER see TIL500
CUSTOS see MHC750
CUTAMIN BLACK CG see CMN240
CUTAMIN BLUE CR see CMO600
CUTAMIN BRILLIANT RED CG see CMM320
CUTAMIN BROWN CM see CMO800
CUTAMIN DARK BLUE CB see CMN800
CUTAMIN RED CF see CMO870
CUTCH (DYE) see CCP800
CUTICURA ACNE CREAM see BDS000
CUTISAN see TIL500
CUTISTEROL see FDB000
CUT LEAF PHILODENDRON see SLE890
CUTLESS see IRN500
CUTTING OILS see COH000
CUVALIT see LJE500
CUZ 3 see CBT500
CV 399 see MGH800
CV 1006 see CDF380
CV 3317 see DAM315
CV205-502 see DJT150
C. VESICARIA see CAK325
C. VIRIDIS VIRIDIS TOXIN see VRP200
C. VIRIDIS VIRIDIS VENOM see PLX100
CVK see PDD350
CVMP see RAF100
CVP see CDS750
CW 524 see TKL100
C-WEISS 7 (GERMAN) see TGG760
CWN 2 see CBT500
C-WR BLUE 10 see CMM092
C-WR VIOLET 8 see MAC250
CX-59 see HKE000
CX 100 see EML600
CX 206 see PKQ070
CXA-DPS see CBQ575
CXD see CCS530
CXM see CCS600
CXM see SFQ300
CXM-AX see CCS625
CY see CQC650
CY-39 see PHU500
CY 116 see AJD000
CY 216 see HAQ500
CY 222 see HAQ500
CYAANWATERSTOF (DUTCH) see HHS000
CYACETACID see COH250
CYACETACIDE see COH250
CYACETAZID see COH250
CYACETAZIDE see COH250
CYALANE see DXN600
CYAMELUROTRIAMIDE see HAQ700
CYAMEMAZINE MALEATE see COS899
CYAMEPROMAZINE MALEATE see COS899
CYAMOPSIS GUM see GLU000
CYANACETAMIDE see COJ250
CYANACETATE ETHYLE (GERMAN) see EHP500
CYANACETHYDRAZIDE see COH250
CYANACETIC ACID HYDRAZIDE see COH250
CYANACETIC ACID HYDRAZIDE see COH250
CYANACETOHYDRAZIDE see COH250
CYANACETYLHYDRAZIDE see COH250
CYANACURE see EJC050
CYANAMID 24055 see DUI000
CYANAMIDE see CAQ250
CYANAMIDE see COH500
CYANAMIDE, (4,6-BIS((1-METHYLETHYL)AMINO)-1,3,5-TRIAZIN-2-YL)METHYL- see MDM900
CYANAMIDE CALCIQUE (FRENCH) see CAQ250
CYANAMIDE, CALCIUM SALT (1:1) see CAQ250
CYANAMIDE, (3-((6-CHLORO-3-PYRIDINYL)METHYL)-2-THIAZOLIDINYLIDENE)- see CKW410
CYANAMIDE, DIETHYL- see DIY100

CYANAMIDE, DIETHYL- see DIY150
CYANAMID GRANULAR see CAQ250
CYANAMID SPECIAL GRADE see CAQ250
CYANASET see MJM200
CYANATES see COH750
CYANATOTRIBUTYLSTANNANE see COI000
CYANATRYN see COI050
CYANAZIDE see COH250
CYANAZINE see BLW750
CYANAZINE-METOLACHLOR mixture see MQQ475
CYANEA CAPILLATA TOXIN see COI125
CYANESSIGSAEURE (GERMAN) see COJ500
CYANEX 921 see TMO575
CYAN GREEN 15-3100 see PJQ100
CYANHYDRINE d'ACETONE (FRENCH) see MLC750
CYANIC ACID, m-PHENYLENE ESTER see REA200
CYANIC ACID, 1,3-PHENYLENE ESTER see REA200
CYANIC ACID, POTASSIUM SALT see PLC250
CYANIC ACID, SODIUM SALT see COI250
CYANIC ACID, TRIMETHYLSTANNYL ESTER see TMI100
CYANIDE see COI500
CYANIDE ANION see COI500
CYANIDE(CN1-) see COI500
CYANIDE(CN1-) ION see COI500
CYANIDE ION see COI500
CYANIDELONON 1522 see QCA000
CYANIDENON 1470 see TDD550
CYANIDE of POTASSIUM see PLC500
CYANIDES see HHT000
CYANIDE of SODIUM see SGA500
CYANIDE SOLUTIONS (DOT) see COI500
CYANIDES (OSHA) see PLC500
CYANIDE, dry (UN 1588) see COI500
CYANIDIN see COI750
CYANIDIN CHLORIDE see COI750
CYANIDINE see THR525
CYANIDINE 3-RUTINOSIDE see KEA325
CYANIDIN 3-RHAMNOGLUCOSIDE see KEA325
CYANIDIN-3-RHAMNOGLUCOSIDE CHLORIDE see KEA325
CYANIDIN 3-RHAMNOSYLGLUCOSIDE see KEA325
CYANIDIN 3-RUTINOSIDE see KEA325
CYANIDOL see COI750
CYANIDOL CHLORIDE see COI750
CYANIDOL 3-RHAMNOGLUCOSIDE see KEA325
CYANINE ACID BLUE R see ADE750
CYANINE ACID BLUE R NEW see ADE750
CYANINE FAST SCARLET G see CMM325
CYANINE FAST YELLOW M see CMM759
CYANINE GREEN G BASE see BLK000
CYANINE GREEN GP see PJQ100
CYANINE GREEN NB see PJQ100
CYANINE GREEN T see PJQ100
CYANINE GREEN TONER see PJQ100
CYANINOSIDE see KEA325
CYANITE see AHF500
CYANIZIDE see COH250
CYANOACETAMIDE see COJ250
2-CYANOACETAMIDE see COJ250
CYANOACETHYDRAZIDE see COH250
CYANOACETIC ACID see COJ500
CYANOACETIC ACID, ALLYL ESTER see AGC150
CYANOACETIC ACID ETHYL ESTER see EHP500
CYANOACETIC ACID HYDRAZIDE see COH250
CYANOACETIC ACID, ISOBUTYL ESTER see IIQ100
CYANOACETIC ACID METHYL ESTER see MIQ000
CYANOACETIC ESTER see EHP500

CYANOACETOHYDRAZIDE see COH250
α-CYANOACETOHYDRAZIDE see COH250
CYANOACETONITRILE see MAO250
CYANOACETYL CHLORIDE see COJ625
N-CYANOACETYL ETHYL CARBAMATE see COJ750
CYANOACETYLHYDRAZIDE see COH250
1-(CYANOACETYL)MORPHOLINE see MRR750
4-CYANOACETYLMORPHOLINE see MRR750
2-CYANOACRYLIC ACID, METHYL ESTER see MIQ075
α-CYANOACRYLIC ACID METHYL ESTER see MIQ075
3-CYANOALANINE see COJ800
β-CYANOALANINE see COJ800
CYANOAMINE see COH500
2-CYANO-4-AMINOSTILBENE see COS750
2-CYANOANILINE see APJ750
3-CYANOANILINE see AIR125
4-CYANOANILINE see COK125
m-CYANOANILINE see AIR125
o-CYANOANILINE see APJ750
p-CYANOANILINE see COK125
CYANO-B12 see VSZ000
7-CYANOBENZ(c)ACRIDINE see BAX000
4-CYANOBENZALDEHYDE see COK250
p-CYANOBENZALDEHYDE see COK250
2-CYANOBENZAMIDE see COK300
4-CYANOBENZAMIDE see COK400
o-CYANOBENZAMIDE see COK300
p-CYANOBENZAMIDE see COK400
7-CYANOBENZ(a) ANTHRACENE see COK500
10-CYANO-1,2-BENZANTHRACENE see COK500
CYANOBENZENE see BCQ250
p-CYANOBENZENECARBOXALDEHYDE see COK250
7-CYANOBENZO(c)ACRIDINE see BAX000
4-CYANOBENZONITRILE see BBP250
α-CYANOBENZYL PHENYL KETONE see OOK200
5-o-CYANOBENZYL-4,5,6,7-TETRAHYDROTHIENO(3,2-c)PYRIDINE METHANESULFONATE see TDC725
CYANOBOND W100 see EHP700
CYANOBOND W300 see EHP700
CYANOBORANE OLIGOMER see COK659
CYANOBRIK see SGA500
CYANOBROMIDE see COO500
N-CYANO-2-BROMOETHYLBUTYLAMINE see COK750
N-CYANO-2-BROMOETHYLCYCLOHEXYLAMINE see COL000
1-CYANOBUTANE see VAV300
1-CYANO-1-BUTENE see PBQ275
CYANO-CARBAMIC ACID METHYL ESTER, DIMER see MIQ350
dl-CYANOCARNITINE CHLORIDE see COL050
2-CYANOCETYLCOUMARONE see BCK750
CYANOCOBALAMIN see VSZ000
p-CYANOCUMENE see IOD050
2-CYANO-N-(2-CYANOETHYL)ETHANAMINE see BIQ500
CYANOCYCLINE A see COL125
1-(1-CYANOCYCLOHEXYL)PIPERIDINE see PIN225
α-CYANODEOXYBENZOIN see OOK200
N-CYANODIALLYLAMINE see DBJ200
5-CYANO-10,11-DIHYDRO-5-(3-DIMETHYLAMINOPROPYL)-5H-DIBENZO (a,d)CYCLOHEPTENE HYDROCHLORIDE see COL200
4-CYANO-2,6-DIIODOPHENOL see HKB500
4-CYANO-2,6-DIJODPHENOL (GERMAN) see HKB500

4-CYANO-2,6-DIJODPHENOL CAPRYSAEUREESTER (GERMAN) see DNG200
4-CYANO-2,6-DIJODPHENOL LITHIUMSALZ (GERMAN) see DNF400
CYANO-3-(DIMETHYLAMINO-3-METHYL-2-PROPYL)-10-PHENOTHIAZINE MALEATE see COS899
CYANODIMETHYLARSINE see COL750
5-CYANO-9,10-DIMETHYL-1,2-BENZANTHRACENE see COM000
α-CYANO-2,6-DIMETHYLERGOLINE-8-PROPIONAMIDE see COM075
N'-CYANO-N,N-DIMETHYLGUANIDINE see COM085
2-CYANO-3,3-DIPHENYLACRYLIC ACID, ETHYL ESTER see DGX000
α-CYANODIPHENYLMETHANE see DVX200
2-CYANO-3,3-DIPHENYL-2-PROPENOIC ACID, ETHYL ESTER see DGX000
1'-(3-CYANO-3,3-DIPHENYLPROPYL)(1,4'-BIPIPERIDINE)-4'-CARBOXAMIDE see PJA140
1-(3-CYANO-3,3-DIPHENYLPROPYL)-4-(2-OXO-3-PROPIONYL-1-BENZIMIDAZOLINYL)PIPERIDINE see BCE825
1-(3-CYANO-3,3-DIPHENYLPROPYL)-4-PHENYLISONIPECOTIC ACID ETHYL ESTER HYDROCHLORIDE see LIB000
1-(1-(3-CYANO-3,3-DIPHENYLPROPYL)-4-PIPERIDYL)-3-PROPIONYL-2-BENZIMIDAZOLINONE see BCE825
1-CYANO-3,4-EPITHIOBUTANE see EBD600
CYANOETHANE see PMV750
2-CYANOETHANEPHOSPHONIC ACID DIETHYL ESTER see COM100
2-CYANOETHANOL see HGP000
2-(2-CYANOETHOXY)ETHYL ESTER ACRYLIC ACID see COM125
4-CYANOETHOXY-2-METHYL-2-PENTANOL see COM250
CYANOETHYDRAZIDE see COH250
CYANOETHYL ACRYLATE see ADT111
2-CYANOETHYL ACRYLATE see ADT111
2-CYANOETHYL ALCOHOL see HGP000
β-CYANOETHYLAMINE see AMB500
2-CYANO-N-((ETHYLAMINO)CARBONYL)-2-(METHOXYIMINO)-ACETAMIDE see COM300
N-(CYANOETHYL)ANILINE see AOT100
N-(2-CYANOETHYL)ANILINE see AOT100
N-(β-CYANOETHYL)ANILINE see AOT100
1-(2-CYANOETHYL)AZIRIDINE see ASI250
N-(2-CYANOETHYL)AZIRIDINE see ASI250
(2-CYANOETHYL)BENZENE see HHP100
N-(β-CYANOETHYL)CHLOROACETAMIDE see COM830
2-(N-(2-CYANOETHYL)-N-CYCLOHEXYL)AMINO-ETHANOL see CPJ500
N-(CYANOETHYL)DIETHYLENETRIAMINE see COM500
CYANOETHYLENE see ADX500
1-(2-CYANOETHYL)ETHYLENIMINE see ASI250
N-(β-CYANOETHYL)ETHYLENIMINE see ASI250
2-CYANOETHYL-2'-FLUOROETHYLETHER see CON500
N-(β-CYANOETHYL)-N-(β-HYDROXYETHY)-ANILINE see CPJ500
β-CYANOETHYLMERCAPTAN see COM750
(2-CYANOETHYL)METHYLAMINE see COM800
N-(β-CYANOETHYL)MONOCHLOROACETAMIDE see COM830
CYANOETHYL-p-PHENETIDINE see COM900

α-CYANOSTILBENE see DVX600
p-CYANOSTYRENE OXIDE see ONC100
(3AR,9BS)-N-(4-(8-CYANO-1,3A,4,9B-
TETRAHYDRO-3H-BENZOPYRANO(3,4-
C)PYRROL-E-2-YL)BUTYL)-4- see COS780
2-(5-
CYANOTETRAZOLE)PENTAMMINECOBA
LT(III) PERCHLORATE see COS825
2-CYANOTOLUENE see TGT500
3-CYANOTOLUENE see TGT250
4-CYANOTOLUENE see TGT750
m-CYANOTOLUENE see TGT250
o-CYANOTOLUENE see TGT500
p-CYANOTOLUENE see TGT750
α-CYANOTOLUENE see PEA750
ω-CYANOTOLUENE see PEA750
CYANOTOXIN see SMM500
CYANOTRICHLOROMETHANE see TII750
CYANOTRIMEPRAZINE MALEATE see
COS899
CYANOTRIMETHYLANDROSTENOLONE
see COS909
2-α-CYANO-4,4,17-α-
TRIMETHYLANDROST-5-EN-17-β-OL-3-
ONE see COS909
2-CYANO-1,2,3-
TRIS(DIFLUOROAMINO)PROPANE see
COT000
CYANOTUBERICIDIN see VGZ000
α-CYANOVINYL ACETATE see ABL500
CYANOX see COQ399
CYANOX 425 see MJN250
CYANOX LTDP see TFD500
CYANOX-STDP see DXG700
CYANSAN see COI250
CYANTIN see NGE000
CYANURAMIDE see MCB000
CYANURCHLORIDE see TJD750
CYANURE see COI500
CYANURE d'ARGENT (FRENCH) see
SDP000
CYANURE de CALCIUM (FRENCH) see
CAQ500
CYANURE de CUIVRE (FRENCH) see
CNL250
CYANURE de MERCURE (FRENCH) see
MDA250
CYANURE de METHYL (FRENCH) see
ABE500
CYANURE de PLOMB (FRENCH) see LCU000
CYANURE de POTASSIUM (FRENCH) see
PLC500
CYANURE de SODIUM (FRENCH) see
SGA500
CYANURE de VINYLE (FRENCH) see
ADX500
CYANURE de ZINC (FRENCH) see ZGA000
CYANURIC ACID see THS000
CYANURIC ACID CHLORIDE see TJD750
CYANURIC ACID TRIGLYCIDYL ESTER see
TKL250
CYANURIC CHLORIDE (DOT) see TJD750
CYANURIC FLUORIDE see TKK000
CYANURIC TRIAMIDE see MCB000
CYANURIC TRIAZIDE see THR250
CYANURIC TRIAZIDE (DOT) see THR250
CYANURIC TRICHLORIDE (DOT) see
TJD750
CYANUROTRIAMIDE see MCB000
CYANUROTRIAMINE see MCB000
CYANURYL CHLORIDE see TJD750
CYANUSTINE see CQI525
(+)-CYANUSTINE see CQI525
CYANWASSERSTOFF (GERMAN) see
HHS000
CYAP see COQ399
CYASORB UV 9 see MES000
CYASORB UV 12 see BJY800
CYASORB UV 24 see DMW250
CYASORB UV 531 see HND100
CYASORB UV 24 LIGHT ABSORBER see
DMW250

CYAZID see COH250
CYAZIDE see COH250
CYAZIN see ARQ725
CYBIS see EID000
CYBOLT see COQ385
CYCAD HUSK see COT500
CYCAD MEAL see COT750
CYCAD NUT, aqueous extract see COT750
CYCAS CIRCINALIS HUSK see COT500
CYCASIN see COU000
CYCASIN ACETATE see MGS750
CYCAS REVOLUTA GLUCOSIDE see
COU000
CYCEURON plus CHLORBUFAN see
AFM375
CYCHLORAL see CPR000
CYCLACILLIN see AJJ875
CYCLADIENE see DAL600
CYCLAIN see OKK100
CYCLAINE see COU250
CYCLAINE HYDROCHLORIDE see COU250
CYCLAL CETYL ALCOHOL see HCP000
CYCLALIA see CMS850
CYCLAMAL see COU500
CYCLAMATE see CPQ625
CYCLAMATE see SGC000
CYCLAMATE CALCIUM see CAR000
CYCLAMATE CALCIUM DIHYDRATE see
CAQ600
CYCLAMATE, CALCIUM SALT see CAR000
CYCLAMATE SODIUM see SGC000
CYCLAMEN ALDEHYDE see COU500
CYCLAMEN ALDEHYDE DIETHYL
ACETAL see COU510
CYCLAMEN ALDEHYDE DIMETHYL
ACETAL see COU525
CYCLAMIC ACID see CPQ625
CYCLAMIC ACID SODIUM SALT see
SGC000
CYCLAMID see CPR000
CYCLAMIDE see ABB000
CYCLAMIDE see CPR000
CYCLAMIDOMYCIN see COV125
CYCLAN see CAR000
CYCLANDELATE see DNU100
CYCLANILIDE see COV133
CYCLAPEN see AJJ875
CYCLAPROP see TJG600
CYCLATE see BOV825
CYCLAZOCINE see COV500
CYCLE see PMF600
CYCLERGINE see DNU100
CYCLIC ADENOSINE-3',5'-PHOSPHATE see
AOA130
CYCLIC(l-ALANYL-2-MERCAPTO-l-
TRYPTOPHYL-4,5-DIHYDROXY-l-LEUCYL-
l-VALYL-ERYTHRO-3-HYDROXY-d-α-
ASPARTYL-l-CYSTEINYL-cis-4-HYDROXY-l-
PROLYL) CYCLIC (2-6)-SULFIDE see
COV525
CYCLIC AMP see AOA130
CYCLIC-3',5'-AMP see AOA130
CYCLIC AMP DIBUTYRATE see COV625
3',5'-CYCLIC AMP DIBUTYRATE see COV625
CYCLIC AMP N⁶,2'-DIBUTYRYL cAMP see
COV625
CYCLIC DIBUTYRYL AMP see COV625
CYCLIC DIMETHYLSILOXANE
PENTAMER see DAF350
CYCLIC DIMETHYLSILOXANE
TETRAMER see OCE100
CYCLIC-2,6-cis-
DIPHENYLHEXAMETHYLCYCLOTETRASI
LOXANE see CMS234
CYCLIC ETHYLENE ACETAL-2-
BUTANONE see EIO500
CYCLIC ETHYLENE CARBONATE see
GHM000
CYCLIC ETHYLENE
(DIETHOXYPHOSPHINOTHIOYL)DITHIOI
MIDOCARBONATE see DXN600

CYCLIC
ETHYLENE(DIETHOXYPHOSPHINOTHIO
YL)DITHIOIMIDOCARBONATE see
PGW750
CYCLIC ETHYLENE P,P-DIETHYL
PHOSPHONODITHIOIMIDOCARBONATE
see PGW750
CYCLIC ETHYLENE ESTER of
(DIETHOXYPHOSPHINOTHIOYL)DITHIOI
MIDOCARBONIC ACID see DXN600
CYCLIC ETHYLENE SULFITE see COV750
CYCLIC ETHYLENE
TRITHIOCARBONATE see EJQ100
CYCLIC (HYDROXYMETHYL)ETHYLENE
ACETAL ACETONE see DVR600
CYCLIC METHYLETHYLENE
CARBONATE see CBW500
CYCLIC-S,S-(6-METHYL-2,3-
QUINOXALINEDIYL)
DITHIOCARBONATE see ORU000
CYCLIC NEOPENTANETETRAYL BIS(2,4-
DI-tert-BUTYLPHENYL)ESTER
PHOSPHOROUS ACID see COV800
CYCLIC N',O-PROPYLENE ESTER of N,N-
BIS(2-
CHLOROETHYL)PHOSPHORODIAMIDIC
ACID MONOHYDRATE see CQC500
CYCLIC PROPYLENE CARBONATE see
CBW500
CYCLIC-1,2-PROPYLENE CARBONATE see
CBW500
CYCLIC PROPYLENE
(DIETHOXYPHOSPHINYL)DITHIOIMIDO
CARBONATE see DHH400
CYCLIC SALIGENIN PHENYL
PHOSPHATE see SAN300
CYCLIC SODIUM TRIMETAPHOSPHATE
see TKP750
CYCLIC SOSO see DVO920
CYCLIC-SOSO see DVO920
CYCLIC TETRAMETHYLENE SULFONE see
SNW500
CYCLIC N,N'-TRIMETHYLENE-N"-BIS(2-
CHLOROETHYL)-PHOSPHORIC
TRIAMIDE see BHQ000
CYCLISCHES
TRINATRIUMMETAPHOSPHAT (GERMAN)
see TKP750
CYCLIZINE see EAN600
CYCLIZINE CHLORIDE see MAX275
CYCLIZINE HYDROCHLORIDE see
MAX275
α-CYCLOAMYLOSE see COW925
CYCLOATE see EHT500
CYCLOBARBITAL see TDA500
CYCLOBARBITAL-SALICYLAMIDE
COMPLEX see COV825
CYCLOBARBITOL see TDA500
CYCLOBARBITONE see TDA500
CYCLOBENDAZOLE see CQE325
CYCLOBENZAPRINE see PMH600
CYCLOBENZAPRINE HYDROCHLORIDE
see DPX800
CYCLOBRAL see DNU100
CYCLOBUTANE see COW000
CYCLOBUTANE, 1-ACETYL-2,2-
DIMETHYL-3-ETHYL-, (cis)- see EII700
CYCLOBUTANECARBOXAMIDE-N-(2-
FLUORENYL) see COW500
CYCLOBUTANE, CYCLOBUTYLIDINE- see
BFX545
cis-(1,1-
CYCLOBUTANEDICARBOSYLATO)DIAMM
INEPLATINUM(II) see CCC075
1,1-CYCLOBUTANEDICARBOXYLATE
DIAMMINE PLATINUM(II) see CCC075
cis-(1,1-
CYCLOBUTANEDICARBOXYLATO)DIAM
MINEPLATINUM(II) see CCC075
CYCLOBUTANE, 1,2-
DICHLOROHEXAFLUORO- see DFM025

CYCLOBUTANE, 1,2-DICHLORO-1,2,3,3,4,4-HEXAFLUORO-(9CI) see DFM025
CYCLOBUTENE see COW250
3-CYCLOBUTENE-1,2-DIONE, 3-AMINO-4-((4-((4-(1-PIPERIDINYLMETHYL)-2-PYRIDINYL)OXY)-2-BUTENYL) AMINO)-, (Z)-, MONOHYDROCHLORIDE see PIB650
CYCLOBUTYLENE see COW250
CYCLOBUTYL-N-(2-FLUORENYL)FORMAMIDE see COW500
CYCLOBUTYLIDINECYCLOBUTANE see BFX545
17-CYCLOBUTYLMETHYL-3-HYDROXY-6-METHYLENE-8-β-METHYLMORPHINAN see COW675
(8-β)-17-(CYCLOBUTYLMETHYL)-6-METHYLENEMORPHINAN-3-OL METHANESULFONATE see COW675
CYCLOBUTYROL see COW700
CYCLOCAPRON see AJV500
CYCLOCEL see CMF400
CYCLOCHEM GMS see OAV000
CYCLOCHEM INEO see ISC550
CYCLOCHLOROTINE see COW750
α-CYCLOCITRYLIDENEACETONE see IFW000
β-CYCLOCITRYLIDENEACETONE see IFX000
α-CYCLOCITRYLIDENE-4-METHYLBUTAN-3-ONE see COW780
CYCLO-CMP HYDROCHLORIDE see COW900
CYCLOCORT see COW825
CYCLOCYTIDINE see COW875
CYCLOCYTIDINE see COW900
2,2'-CYCLOCYTIDINE see COW875
2,2'-o-CYCLOCYTIDINE see COW875
o-2,2'-CYCLOCYTIDINE see COW875
CYCLOCYTIDINE HYDROCHLORIDE see COW900
2,2'-CYCLOCYTIDINE HYDROCHLORIDE see COW900
2,2'-o-CYCLOCYTIDINE HYDROCHLORIDE see COW900
o-2,2'-CYCLOCYTIDINE MONOHYDROCHLORIDE see COW900
CYCLODAN see BFY100
CYCLODAN see EAQ750
3,7-CYCLODECADIEN-1-ONE, 3,7-DIMETHYL-10-(2-HYDROXY-1-METHYLETHYLIDENE)-, (3E,7E,10E)- see HLE680
3,7-CYCLODECADIEN-1-ONE, 3,7-DIMETHYL-10-(1-METHYLETHYLIDENE)-, (3E,7E)- see GDO200
CYCLODECA(B)FURAN, 4,7,8,11-TETRAHYDRO-3,6,10-TRIMETHYL-, (5E,9E)- see FPM150
CYCLODECENE-1,4-DIONE, 6,10-DIMETHYL-3-(1-METHYLETHYL)-, see GDO100
6-CYCLODECENE-1,4-DIONE, 6,10-DIMETHYL-3-(1-METHYLETHYL)-, (3R,6E,10S)- see NBW100
6-CYCLODECENE-1,4-DIONE, 6,10-DIMETHYL-3-(1-METHYLETHYLIDENE)-, (6E,10S)- see DAL010
β-CYCLODEXTRIN see COW925
CYCLODIOL see MEX285
CYCLODISONE see DVO920
CYCLODODECALACTAM see COW930
CYCLODODECANE, (ETHOXYMETHOXY)- see FMV100
CYCLODODECANE, (METHOXYMETHOXY)- see FMV200
CYCLODODECANE, 1-METHOXY-1-METHYL- see MEU300
cis,trans,trans-CYCLODODECA-1,5,9-TRIENE see COW935
1,5,9-CYCLODODECATRIENE (Z,E,E) see COW935

CYCLODODECYL-2,6-DIMETHYLMORPHOLINE ACETATE see COX000
N-CYCLODODECYL-2,6-DIMETHYLMORPHOLINIUM ACETATE see COX000
CYCLODOL see BBV000
CYCLODORM see TDA500
CYCLOESTROL see DLB400
4-N-CYCLOETHYLENEUREIDOAZOBENZENE see COX250
p-N-CYCLO-ETHYLENEUREIDOAZOBENZENE see COX250
CYCLOFEM see EDQ600
CYCLOFENIL see FBP100
CYCLOFENIL DIPHENOL see CPM770
CYCLOFENIL see FBP100
CYCLOFOS see SKE550
CYCLOGEST see PMH500
CYCLOGUANIL see COX400
CYCLOGUANIL HYDROCHLORIDE see COX325
CYCLOGUANYL see COX400
CYCLOGYL see CPZ125
3-CYCLOHENENYL CYANIDE see CPC625
CYCLOHEPTAAMYLOSE see COW925
β-CYCLOHEPTAAMYLOSE see COW925
9-CYCLOHEPTADECEN-1-ONE see CMU850
9-CYCLOHEPTADECEN-1-ONE, (Z)-(8CI,9CI) see CMU850
CYCLOHEPTAGLUCOSAN see COW925
CYCLOHEPTANE see COX500
CYCLOHEPTANECARBAMIC ACID-1,1-BIS(p-FLUOROPHENYL)-2-PROPYNYL ESTER see BJW250
CYCLOHEPTANONE see SMV000
CYCLOHEPTANONE, 2-CHLORO- see CFF600
6H-CYCLOHEPTA(b)QUINOLINE-11-CARBOXAMIDE, N,N-DIETHYL-7,8,9,10-TETRAHYDRO- see TCO100
6H-CYCLOHEPTA(b)QUINOLINE, 7,8,9,10-TETRAHYDRO-11-(MORPHOLINOCARBONYL)- see MRU077
1,3,5-CYCLOHEPTATRIENE see COY000
CYCLOHEPTATRIENE (DOT) see COY000
CYCLOHEPTATRIENE MOLYBDENUM TRICARBONYL see COY100
2,4,6-CYCLOHEPTATRIEN-1-ONE, 2-HYDROXY-5-ISOPROPYL-(8CI) see TFV750
2,4,6-CYCLOHEPTATRIEN-1-ONE, 2-HYDROXY-4-(1-METHYLETHENYL)- see DYB100
2,4,6-CYCLOHEPTATRIEN-1-ONE, 2-HYDROXY-3-(1-METHYLETHYL)- see TFU900
CYCLOHEPTENE see COY250
CYCLOHEPTENYL ETHYLBARBITURIC ACID see COY500
5-(1-CYCLOHEPTEN-1-YL)-5-ETHYLBARBITURIC ACID see COY500
CYCLOHEPTENYLETHYLMALONYLUREA see COY500
5-(1-CYCLOHEPTEN-1-YL)-5-ETHYL-2,4,6(1H,3H,5H)-PYRIMIDINETRIONE (9CI) see COY500
CYCLOHEXAAMINE, N,N'-METHANETETRAYLBIS- see MDR800
CYCLOHEXAAN (DUTCH) see CPB000
CYCLOHEXADECANOLIDE see OKU000
2,5-CYCLOHEXADIEN-1,4-DIONE, 2,3,5-TRIMETHYL- see POG400
1,3-CYCLOHEXADIENE see CPA500
1,4-CYCLOHEXADIENE see CPA750
3,5-CYCLOHEXADIENE-1,2-DIOL, trans-(+−)- see CPA760
3,5-CYCLOHEXADIENE-1,2-DIOL, (1S-trans)- see DMJ900
CYCLOHEXADIENEDIONE see QQS200
1,4-CYCLOHEXADIENEDIONE see QQS200

2,5-CYCLOHEXADIENE-1,4-DIONE see QQS200
3,5-CYCLOHEXADIENE-1,2-DIONE see BDC250
2,5-CYCLOHEXADIENE-1,4-DIONE, 2,5-BIS(1,1-DIMETHYLETHYL)- see DDV550
2,5-CYCLOHEXADIENE-1,4-DIONE, 2-(BIS(2-HYDROXYETHYL)AMINO)- see BKB100
2,5-CYCLOHEXADIENE-1,4-DIONE, 2,5-BIS(PHENYLTHIO)- see BLF600
2,5-CYCLOHEXADIENE-1,4-DIONE DIOXIME see DVR200
2,5-CYCLOHEXADIENE-1,4-DIONE, 2-PHENYL-(9CI) see PEL750
2,5-CYCLOHEXADIENE-1,4-DIONE, 2,3,5,6-TETRAAZIDO- see TBJ250
2,5-CYCLOHEXADIENE-1,4-DIONE, 2,3,5,6-TETRAMETHYL- see TDM810
CYCLOHEXA-1,3-DIENE DIOXIDE see DHB820
1,4-CYCLOHEXADIENE DIOXIDE see QQS200
CYCLOHEXADIENE-1-ETHANOL, 4-(1-METHYLETHYL)-, FORMATE see IHX450
CYCLOHEXADIENOL-4-ONE-1-SULFONATE de DIETHYLAMINE (FRENCH) see DIS600
2,5-CYCLOHEXADIEN-1-ONE, 5-AMINO-4-((4-AMINOMETHYLPHENYL)IMINO)-2-METHOXY- see AKO300
2,5-CYCLOHEXADIEN-1-ONE, 2,6-DICHLORO-4-((p-HYDROXYPHENYL)IMINO)-, SODIUM SALT see SGG650
2,5-CYCLOHEXADIEN-1-ONE, 3,5-DIETHYL-4-IMINO- see DIU433
2,5-CYCLOHEXADIEN-1-ONE, 3,5-DIMETHYL-4-IMINO- see DTY650
2,5-CYCLOHEXADIEN-1-ONE, 3-ETHYL-4-IMINO-5-METHYL- see ENF300
2,5-CYCLOHEXADIEN-1-ONE, 4-IMINO- see BDD500
CYCLOHEXAMETHYLENE CARBAMIDE see CPB050
CYCLOHEXAMETHYLENIMINE see HDG000
CYCLOHEXAMINE SULFATE see CPF750
CYCLOHEXAN (GERMAN) see CPB000
CYCLOHEXANAMIDE see CPB050
CYCLOHEXANAMINE see CPF500
CYCLOHEXANAMINE HYDROCHLORIDE see CPA775
CYCLOHEXANAMINE, N-METHYL-(9CI) see MIT000
CYCLOHEXANAMINE, 4,4'-METHYLENEBIS-(9CI) see MJQ260
CYCLOHEXANAMINE, 4,4'-METHYLENEBIS(2-METHYL- see BGT800
CYCLOHEXANE see CPB000
CYCLOHEXANEACETIC ACID, 1-(HYDROXYMETHYL)-, MONOSODIUM SALT (9CI) see SHL500
CYCLOHEXANEACETIC ACID, α-METHYL-, ETHYL ESTER see EHT600
1,3-CYCLOHEXANEBIS(METHYLAMINE) (8CI) see BGT500
CYCLOHEXANECARBAMIC ACID, 1,1-DIPHENYL-2-BUTYNYL ESTER see DVY900
CYCLOHEXANECARBAMIC ACID, 1-METHYL-1-PHENYL-2-PROPYNYL ESTER see MNX850
CYCLOHEXANECARBINOL see HDH200
CYCLOHEXANECARBONITRILE, 3,4-EPOXY- see EBM100
CYCLOHEXANECARBONITRILE, 5-OXO-1,3,3-TRIMETHYL- see IMG500
CYCLOHEXANECARBOXAMIDE see CPB050
CYCLOHEXANECARBOXAMIDE, N-(2,3-DICHLORO-4-HYDROXYPHENYL)-1-METHYL- see FAO250

CYCLOHEXANECARBOXAMIDE, N-(2-(4-(2-METHOXYPHENYL)-1-PIPERAZINYL)ETHYL)-N-(2-PYRIDINYL)-, TRIHYDROCHLORIDE see TKN100
CYCLOHEXANECARBOXYLIC ACID see HDH100
CYCLOHEXANECARBOXYLIC ACID, 3-(1-(ALLYLOXYAMINO)BUTYLIDENE)-6,6-DIMETHYL-2,4-DIOXO-, METHYL ESTER, SODIUM SALT see FBP350
CYCLOHEXANECARBOXYLIC ACID, CHLORO-2-METHYL-, tert-BUTYL ESTER see BQT600
CYCLOHEXANECARBOXYLIC ACID, 4(or 5)-CHLORO-2-METHYL-, tert-BUTYL ESTER (8CI) see BQT600
CYCLOHEXANECARBOXYLIC ACID, 4(or 5)-CHLORO-2-METHYL-, 1,1-DIMETHYLETHYL ESTER (9CI) see BQT600
CYCLOHEXANECARBOXYLIC ACID, 4-(CYCLOPROPYLHYDROXYMETHYLENE)-3,5-DIOXO-, ETHYL ESTER see EHU550
CYCLOHEXANECARBOXYLIC ACID,2,2-DIMETHYL-4,6-DIOXO-5-(1-((2-PROPENYLOXY)AMINO)BUTYLIDENE)-, METHYL ESTER see ZTS200
CYCLOHEXANECARBOXYLIC ACID, 3,5-DIOXO-4-(1-OXOPROPYL)-, ION(1-), CALCIUM, CALCIUM SALT see CPB065
CYCLOHEXANECARBOXYLIC ACID, ETHYL ESTER see CPB075
CYCLOHEXANECARBOXYLIC ACID-(2-HYDROXYETHYL) ESTER see HKQ500
CYCLOHEXANECARBOXYLIC ACID, LEAD SALT see NAS500
CYCLOHEXANECARBOXYLIC ACID, TRIBUTYLSTANNYL ESTER see TID100
CYCLOHEXANE, CHLORO- see CPI400
1,2-CYCLOHEXANEDIAMINE see CPB100
1,3-CYCLOHEXANEDIAMINE see DBQ800
1,2-CYCLOHEXANEDIAMINETETRAACETIC ACID see CPB120
1,2-CYCLOHEXANEDIAMINE-N,N,N',N'-TETRAACETIC ACID see CPB120
(CYCLOHEXANE-1,2-DIAMMINE)(4-CARBOXYPHTHLATO)PLATINUM(II) see CCJ350
(Z)-(CYCLOHEXANE-1,2-DIAMMINE)ISOCITRATOPLATINUM(II) see PGQ275
CYCLOHEXANE, 1,2-DIBROMO-4-(1,2-DIBROMOETHYL)- see DDM300
1,2-CYCLOHEXANEDICARBOXYLIC ACID, BIS(2,3-EPOXYPROPYL) ESTER see DKM500
1,2-CYCLOHEXANEDICARBOXYLIC ACID, BIS(OXIRANYLMETHYL)ESTER (9CI) see DKM500
CYCLOHEXANE-1,2-DICARBOXYLIC ACID, DIETHYL ESTER see HDS250
1,2-CYCLOHEXANEDICARBOXYLIC ACID, DIETHYL ESTER (9CI) see HDS250
CYCLOHEXANE, DICHLOROPHENYL- see CPJ525
trans-1,4-CYCLOHEXANEDIISOCYANATE see CPB150
CYCLOHEXANE, 1,4-DIISOCYANATO-, trans- see CPB150
$\Delta^{1-\alpha-4-\alpha'}$-CYCLOHEXANEDIMALONONITRILE see BIY600
CYCLOHEXANEDIMETHANAMINE (9CI) see BGT500
CYCLOHEXANE, 1,1-DINITRO- see DUU700
1,2-CYCLOHEXANEDIONE see CPB200
1,3-CYCLOHEXANEDIONE, 5-(2-(ETHYLTHIO)PROPYL)-2-(1-OXOPROPYL)- see EPR100
2,2'-CYCLOHEXANE-1,1-DIYLBIS(p-PHENYLENEOXY)BIS(2-METHYLBUTYRIC ACID) see CMV700
CYCLOHEXANEETHANOL, ACETATE see EHS000
CYCLOHEXANE ETHYL ACETATE see EHS000
CYCLOHEXANEETHYLAMINE see CPB500
CYCLOHEXANEFORMAMIDE see CPB050
cis-1,2,3,5-trans-4,6-CYCLOHEXANEHEXOL see IDE300
CYCLOHEXANE, ISOCYANATO-(9CI) see CPN500

CYCLOHEXANE, 5-ISOCYANATO-1-(ISOCYANATOMETHYL)-1,3,3-TRIMETHYL-(9CI) see IMG000
Δ^{1},α)-CYCLOHEXANEMALONONITRILE see CPM800
CYCLOHEXANEMETHANOL see HDH200
CYCLOHEXANEMETHANOL, α,4-DIMETHYL- see MIT615
1,4-CYCLOHEXANEMETHANOL, DIVINYL ESTER see BLU600
CYCLOHEXANEMETHANOL, α-METHYL- see CPL125
CYCLOHEXANEMETHANOL, α-α-4-TRIMETHYL-(9CI) see MCE100
CYCLOHEXANEMETHYLAMINE, N-METHYL- see MIT610
CYCLOHEXANE OXIDE see CPD000
CYCLOHEXANE, 1,2,3,4,5-PENTABROMO-6-CHLORO- see CPB550
CYCLOHEXANE, PIPERIDINOMETHYL-, CAMPHOSULFATE see PIQ750
CYCLOHEXANE, PIPERIDINOMETHYL-, HYDROCHLORIDE see PIR000
CYCLOHEXANESPIRO-5'-HYDANTOIN see DVO600
CYCLOHEXANESULFAMIC ACID, BARIUM SALT see BAL100
CYCLOHEXANESULFAMIC ACID, CALCIUM SALT see CAR000
CYCLOHEXANESULFAMIC ACID, CALCIUM SALT, DIHYDRATE see CAQ600
CYCLOHEXANESULFAMIC ACID, MONOSODIUM SALT see SGC000
CYCLOHEXANESULPHAMIC ACID see CPQ625
CYCLOHEXANESULPHAMIC ACID, MONOSODIUM SALT see SGC000
CYCLOHEXANETHIOL see CPB625
CYCLOHEXANE, 1-(TRICHLOROSILYL)- see CPR250
1,2,3-CYCLOHEXANETRIONE TRIOXIME see CPB650
CYCLOHEXANOIC ACID see HDH100
CYCLOHEXANOL see CPB750
CYCLOHEXANOL ACETATE see CPF000
CYCLOHEXANOL, 1-AMINO-3-AMINOMETHYL-3,5,5-TRIMETHYL- see AHT875
CYCLOHEXANOLAZETAT (GERMAN) see CPF000
1-CYCLOHEXANOL-α-BUTYRIC ACID see COW700
trans-($\pm$)-CYCLOHEXANOL-2-((DIMETHYLAMINO)METHYL)-1-(3-METHOXYPHENYL) see THJ600
CYCLOHEXANOL, 2-(1,1-DIMETHYLETHYL)-, ACETATE see BQW490
CYCLOHEXANOL, 1-((1,1-DIMETHYLETHYL)AZO)- see DRU900
CYCLOHEXANOL, 1,1'-(DIOXYBIS(METHYL- see DVT600
CYCLOHEXANOL, 1-ETHYNYL-, ACETATE see EQL550
CYCLOHEXANOL, 1-ETHYNYL-, ALLOPHANATE see EQL600
CYCLOHEXANOL, 1-ETHYNYL-2-(1-METHYLPROPYL)-, ACETATE see EQN300
CYCLOHEXANOL, 4-(4-(4-FLUOROPHENYL)-5-(2-METHOXY-4-PYRIMIDINYL)-1H-IMIDAZOL-1-YL)-, trans- see HKA123
CYCLOHEXANOL, 1-((1-HYDROPEROXYCYCLOHEXYL)DIOXY)- see CPC300
CYCLOHEXANOL, p-ISOPROPYL- see IOO300
CYCLOHEXANOL, 2-ISOPROPYL-5-METHYL- see MCF750
CYCLOHEXANOL, 1-METHYL-4-(1-METHYLETHENYL)-(9CI) see TBD775

CYCLOHEXANOL, 2-METHYL-5-(1-METHYLETHENYL)-, ACETATE,(1-α-2-β,5α)- (9CI) see DKV160
CYCLOHEXANOL, 3,3,5-TRIMETHYL-, cis- see TLO510
CYCLOHEXANON (DUTCH) see CPC000
CYCLOHEXANONE see CPC000
CYCLOHEXANONE-Δ see CPC250
CYCLOHEXANONE, 2-sec-BUTYL-(7CI,8CI) see MOU800
CYCLOHEXANONE, 4-(1-ETHOXYETHENYL)-3,3,5,5-TETRAMETHYL- see EES370
CYCLOHEXANONE ISO-OXIME see CBF700
CYCLOHEXANONE, 5-METHYL-2-(1-METHYLETHYLIDENE)-, (R)- see POI615
CYCLOHEXANONE, 2-(1-METHYLPROPYL)- see MOU800
CYCLOHEXANONE OXIME see HLI500
CYCLOHEXANONE PEROXIDE see CPC300
CYCLOHEXANONE, PIPERONYL- see PIX300
CYCLOHEXANYL ACETATE see CPF000
CYCLOHEXATRIENE see BBL250
CYCLOHEXENE see CPC579
CYCLOHEXENEBUTANAL, α-2,2,6-TETRAMETHYL- see TDO255
3-CYCLOHEXENE-1-CARBONITRILE see CPC625
CYCLOHEXENECARBOXALDEHYDE see TCJ100
3-CYCLOHEXENE-1-CARBOXALDEHYDE see FNK025
3-CYCLOHEXENE-1-CARBOXALDEHYDE, DIMETHYL- see DRG700
3-(CYCLOHEXENE-1-CARBOXALDEHYDE 4-(4-HYDROXY-4-METHYLPENTYL)-METHYLPHENYL)-3-CYCLOHEXEN-1-CARBOXALDEHYDE see LJE200
3-CYCLOHEXENE-1-CARBOXALDEHYDE, 1-METHYL-4-(4-METHYLPENTYL)- see VIZ150
3-CYCLOHEXENE-1-CARBOXALDEHYDE, 4-(4-METHYL-3-PENTENYL)- see IKT100
3-CYCLOHEXENE-1-CARBOXYLIC ACID see CPC650
1-CYCLOHEXENE-1-CARBOXYLIC ACID, 3,4,5 see BML000
CYCLOHEXENE, 3-CHLORO- see CFG300
4-CYCLOHEXENE-1,2-DICARBOXIMIDE see TDB100
4-CYCLOHEXENE-1,2-DICARBOXIMIDE, N-BUTYL- see BSM400
4-CYCLOHEXENE-1,2-DICARBOXIMIDE, N-(2,6-DIOXO-3-PIPERIDYL)- see TDB200
4-CYCLOHEXENE-1,2-DICARBOXIMIDE,N-((1,1,2,2-TETRACHLOROETHYL)THIO)-, cis- see TBQ280
CYCLOHEXENE-1-DICARBOXIMIDOMETHYLCHRYSANTHE MATE see NCI400
4-CYCLOHEXENE-1,2-DICARBOXYLIC ACID, BIS(2-ISOCYANATOETHYL) ESTER see BKJ700
1-CYCLOHEXENE-1,2-DICARBOXYLIC ACID DIMETHYL ESTER see DUF400
CYCLOHEXENE, 6-(3,7-DIMETHYL-2,4,6,8-NONATETRAENYLIDENE)-1,5,5-TRIMETHYL-, (ALL-E)- see AFQ700
CYCLOHEXENE EPOXIDE see CPD000
3-CYCLOHEXENE-1-METHANOL, α-HYDROXYMETHYL- see BKH400
CYCLOHEX-1-ENE-1-METHANOL, 4-(1-METHYLETHENYL)- see PCI550
3-CYCLOHEXENE-1-METHANOL, 1,2,4(or 1,3,5)-TRIMETHYL- see ISR100
CYCLOHEXENE OXIDE see CPD000
CYCLOHEXENE-1-OXIDE see CPD000
1,2-CYCLOHEXENE OXIDE see CPD000

CYCLOHEXENE, 1,3,4,5,6-PENTACHLORO-, γ- see PAV780
CYCLOHEXENE, 1,3,4,5,6-PENTACHLORO-, (3-α-4-β,5-β,6α- (9CI) see PAV780
CYCLOHEXENE, 4-(TRICHLOROSILYL)- see CPE500
CYCLOHEXENE, 1-VINYL- see VNZ990
CYCLOHEXENONE see CPD250
2-CYCLOHEXEN-1-ONE see CPD250
2-CYCLOHEXEN-1-ONE,2-(1-(((3-CHLORO-2-PROPENYL)OXY)IMINO)PROPYL)-5-(2-(ETHYLTHIO)PROPYL)-3-HYDROXY- see CMV485
2-CYCLOHEXEN-1-ONE, 2-(1-(ETHOXYIMINO)BUTYL)-5-(2-(ETHYLTHIO)PROPYL)-3-HYDROXY- see CDK800
2-CYCLOHEXEN-1-ONE, 2-(1-(ETHOXYIMINO)PROPYL)-3-HYDROXY-5-(2,4,6-TRIMETHYLPHENYL)- see GJU200
5-Δ²·³-CYCLOHEXENYL-5-ALLYL-2-THIOBARBITURIC ACID see TES500
3-CYCLOHEXEN-1-YLBENZENE see PES500
2-(4-CYCLOHEXENYL)BICYCLO(2.2.1)HEPT-5-ENE see CPD300
2-((4-CYCLOHEXEN-3-YLBUTYL)AMINO)ETHANETHIOL HYDROGEN SULFATE (ESTER) see CPD500
S-2-((4-CYCLOHEXEN-3-YLBUTYL)AMINO)ETHYL THIOSULFATE see CPD500
1-(2-CYCLOHEXEN-1-YLCARBONYL)-2-METHYLPIPERIDINE see CPD625
1-(3-CYCLOHEXEN-1-YLCARBONYL)-2-METHYLPIPERIDINE see CPD630
3-CYCLOHEXENYL CHLORIDE see CFG300
5-(2-CYCLOHEXEN-1-YL)DIHYDRO-5-(2-PROPENYL)-2-THIOXO-4,6(1H,5H)-PYRIMIDINEDIONE (9CI) see TES500
5-(1-CYCLOHEXEN-1-YL)-1,5-DIMETHYLBARBITURIC ACID see ERD500
5-(1-CYCLOHEXEN-1-YL)-1,5-DIMETHYLBARBITURIC ACID SODIUM SALT see ERE000
5-(1-CYCLOHEXEN-1-YL)-1,5-DIMETHYL-2,4,6(1H,3H,5H)-PYRIMIDINETRIONE see ERD500
5-(1-CYCLOHEXEN-1-YL)-1,5-DIMETHYL-2,4,6(1H,3H,5H,)-PYRIMIDINETRIONE MONOSODIUM SALT see ERE000
5-(1-CYCLOHEXEN-1-YL)-1,5-DIMETHYL-2,4,6(1H,3H,5H)-PYRIMIDINETRIONE SODIUM SALT (9CI) see ERE000
3-(3-CYCLOHEXENYL)-2,4-DIOXASPIRO(5.5)UNDEC-8-ENE see CPD650
CYCLOHEXENYL-ETHYL BARBITURIC ACID see TDA500
5-(1-CYCLOHEXENYL)-5-ETHYLBARBITURIC ACID see TDA500
5-(1-CYCLOHEXEN-1-YL)-5-ETHYLBARBITURIC ACID see TDA500
CYCLOHEXENYLETHYLENE see CPD750
5-(1-CYCLOHEXEN-1-YL)-5-ETHYL-2,4,6(1H,3H,5H)-PYRIMIDINETRIONE TDA500
2-CYCLOHEXENYL HYDROPEROXIDE see CPE125
5-(1-CYCLOHEXENYL-1)-1-METHYL-5-METHYLBARBITURIC ACID see ERD500
5-(Δ-1,2-CYCLOHEXENYL)-5-METHYL-N-METHYL-BARBITURSAEURE (GERMAN) see ERD500
p-(2-CYCLOHEXEN-1-YLOXY)BENZOIC ACID, 3-(2-METHYL-1-PYRROLIDINYL)PROPYL ESTER see UAG025
1-(2-(1-CYCLOHEXEN-1-YL)PHENOXY)-3-((1-METHYLETHYL)AMINO)-2-PROPANOL HYDROCHLORIDE see ERE100

CYCLOHEXENYL TRICHLOROSILANE see CPE500
CYCLOHEXENYLTRICHLOROSILANE (DOT) see CPE500
CYCLOHEXIMIDE see CPE750
CYCLOHEXYL ACETATE see CPF000
CYCLOHEXYLACETIC ACID ALLYL ESTER see AGC250
CYCLOHEXYL ACETOACETATE see CPP100
CYCLOHEXYLACETONACETATE see CPP100
CYCLOHEXYL ACRYLATE see PMP800
(6-(3-CYCLOHEXYL-d-ALANINE))-LHRH ACETATE see LIU311
6-(3-CYCLOHEXYL-d-ALANINE)LUTEINIZING HORMONE-RELEASING FACTOR(PIG) MONOACETATE (SALT) see LIU311
CYCLOHEXYL ALCOHOL see CPB750
CYCLOHEXYLALLYL-ESSIGSAEUREESTER DES 3-DIAETHYLAMINO-2,2-DIMETHYL-1-PROPANOL (GERMAN) see AGC750
CYCLOHEXYLAMIDOSULPHURIC ACID see CPQ625
CYCLOHEXYLAMINE see CPF500
CYCLOHEXYLAMINE, 4,4'-METHYLENEBIS(2-METHYL- see BGT800
CYCLOHEXYLAMINE SULFATE see CPF750
CYCLOHEXYLAMINESULPHONIC ACID see CPQ625
CYCLOHEXYLAMINO ACETIC ACID see CPG000
2-(CYCLOHEXYLAMINO)ETHANOL see CPG125
2-(2-(CYCLOHEXYLAMINO))ETHYL-2-METHYL-1,3-BENZODIOXOLE HYDROCHLORIDE see CPG250
3-(2-(CYCLOHEXYLAMINO)ETHYL)-2-(3,4-METHYLENEDIOXYPHENYL)-4-THIAZOLIDINONE HYDROCHLORIDE see WBS860
N-CYCLOHEXYL-N'-(3-AMINO-4-METHYLBENZENESULFONYL)UREA see AKQ250
4-(CYCLOHEXYLAMINO)-1-(NAPHTHALENYLOXY)-2-BUTANOL see CPG500
dl-1-CYCLOHEXYL-2-AMINOPROPANE HYDROCHLORIDE see CPG625
1-CYCLOHEXYLAMINO-2-PROPANOL see CPG700
1-(CYCLOHEXYLAMINO)-2-PROPANOL BENZOATE (ESTER) HYDROCHLORIDE see COU250
CYCLOHEXYLAMMONIUM FORMATE see CPH250
CYCLOHEXYLAMMONIUM STEARATE see CPH500
3-(5-CYCLOHEXYL-o-ANISOYL)-PROPIONIC ACID SODIUM SALT see MEK350
N-CYCLOHEXYL-1-AZIRIDINECARBOXAMIDE see CPI000
CYCLOHEXYLBENZENE see PER750
ar-CYCLOHEXYLBENZENEMETHANOL see CPM300
N-CYCLOHEXYL-2-BENZOTHIAZOLESULFENAMIDE see CPI250
N-CYCLOHEXYL-2-BENZOTHIAZOLESULFENAMIDE see CPI250
N-CYCLOHEXYL-2-BENZOTHIAZYLSULFENAMIDE see CPI250
2-(α-CYCLOHEXYLBENZYL)-N,N,N',N'-TETRAETHYL-1,3-PROPANEDIAMINE HYDROCHLORIDE see LFK200
N-CYCLOHEXYL-l-BUTANESULFONAMIDE see BSM250

CYCLOHEXYL BUTANOATE see CPI300
CYCLOHEXYL BUTYRATE see CPI300
N-CYCLOHEXYLCARBAMIC ACID 1-PHENYL-1-(3,4-XYLYL)-2-PROPYNYL ESTER see PGQ000
CYCLOHEXYL-N-CARBAMOYLAZIRIDINE see CPI000
N-CYCLOHEXYL-N-CARBAMOYLAZIRIDINE see CPI000
CYCLOHEXYLCARBINOL see HDH200
2-CYCLOHEXYLCARBONYL-1,2,3,6,7,11b-HEXAHYDRO-4H-PYRAZINO(2,1-a)ISOQUINOLIN-4-ONE see BGB400
1-(CYCLOHEXYLCARBONYL)-3-METHYLPIPERIDINE see CPI350
((CYCLOHEXYLCARBONYL)OXY)TRIBUTYLSTANNANE see TID100
4-(CYCLOHEXYLCARBONYL)PYRIDINE see CPI375
1-(CYCLOHEXYLCARBONYL)-1,2,3,6-TETRAHYDROPYRIDINE see CPI380
CYCLOHEXYLCARBOXAMIDE see CPB050
CYCLOHEXYL CARBOXYAMIDE see CPB050
CYCLOHEXYLCARBOXYLIC ACID see HDH100
CYCLOHEXYL CHLORIDE see CPI400
4-(4-CYCLOHEXYL-3-CHLOROPHENYL)-4-OXOBUTYRIC ACID see CPJ000
4-(4-CYCLOHEXYL-3-CHLOROPHENYL)-4-OXOBUTYRIC ACID CALCIUM SALT see CPJ250
2-CYCLOHEXYL-p-CRESOL see CPP050
CYCLOHEXYLCYANOETHYLETHANOLAMINE see CPJ500
N-CYCLOHEXYLCYCLOHEXANAMINE see DGT600
2-CYCLOHEXYLCYCLOHEXANONE see LBX100
CYCLOHEXYLDICHLOROBENZENE see CPJ525
2-CYCLOHEXYL-4,5-DICHLORO-4-ISOTHIAZOLIN-3-ONE see CPJ600
N-CYCLOHEXYL-N-DIETHYLTHIOCARBONYL SULFONAMIDE see CPK000
CYCLOHEXYLDIMETHYLAMINE see DRF709
N-CYCLOHEXYLDIMETHYLAMINE see DRF709
3-CYCLOHEXYL-6-(DIMETHYLAMINO)-1-METHYL-s-TRIAZINE-2,4(1H,3H)-DIONE see HFA300
3-CYCLOHEXYL-6-(DIMETHYLAMINO)-1-METHYL-1,3,5-TRIAZINE-2,4(1H,3H)-DIONE see HFA300
CYCLOHEXYL 4-(1,1-(DIMETHYLETHYL)PHENYL)METHYL-3-PYRIDINYLCARBONIMIDODITHIOATE see CPK100
2-CYCLOHEXYL-4,6-DINITROFENOL (DUTCH) see CPK500
2-CYCLOHEXYL-4,6-DINITROPHENOL see CPK500
6-CYCLOHEXYL-2,4-DINITROPHENOL see CPK500
2-CYCLOHEXYL-4,6-DINITROPHENOL DICYCLOHEXYLAMINE see CPK550
(+)-1-CYCLOHEXYL-4-(1,2-DIPHENYLETHYL)PIPERAZINE DIHYDROCHLORIDE see CPK625
(S)-1-CYCLOHEXYL-4-(1,2-DIPHENYLETHYL)-PIPERAZINE DIHYDROCHLORIDE see CPK625
(±)-1-CYCLOHEXYL-4-(1,2-DIPHENYLETHYL)PIPERAZINE DIHYDROCHLORIDE see MRU757
CYCLOHEXYL DIPHENYL PHOSPHATE see CPP800
CYCLOHEXYL DISULFIDE (6CI,7CI,8CI) see CPK700

1,2-CYCLOHEXYLENEDIAMINETETRAACETIC ACID see CPB120

N,N'-(1,4-CYCLOHEXYLENEDIMETHYLENE)BIS(2-(1-AZIRIDINYL)ACETAMIDE) see CPL100

trans-N,N'-(1,4-CYCLOHEXYLENEDIMETHYLENE)BIS(2-CHLOROBENZYLAMINE) DIHYDROCHLORIDE see BHN000

(1,2-CYCLOHEXYLENEDINITRILO)TETRAACETIC ACID see CPB120

trans-1,4-CYCLOHEXYLENEISOCYANATE see CPB150

CYCLOHEXYLENE OXIDE see CPD000

CYCLOHEXYLESTER KYSELINY OCTOVE see CPF000

1-CYCLOHEXYLETHANOL see CPL125

2-CYCLOHEXYLETHANOL see CPL250

CYCLOHEXYLETHYL ACETATE see EHS000

CYCLOHEXYLETHYL ALCOHOL see CPL250

CYCLOHEXYLETHYLCARBAMOTHIOIC ACID-S-ETHYL ESTER see EHT500

1-CYCLOHEXYL-4-((ETHYL-p-METHOXY-α-METHYLPHENETHYL)AMINO)-1-BUTANONE HYDROCHLORIDE see SBN300

CYCLOHEXYLETHYLTHIOCARBAMIC ACID-S-ETHYL ESTER see EHT500

CYCLOHEXYL FLUOROETHYL NITROSOUREA see CPL750

CYCLOHEXYL FLUOROETHYL NITROSOUREA see FIJ000

3-CYCLOHEXYL-1-(2-FLUOROETHYL)-1-NITROSOUREA see CPL750

N'-CYCLOHEXYL-N-(2-FLUOROETHYL)-N-NITROSOUREA see CPL750

(+)-α-CYCLOHEXYL-α-HYDROXY-BENZENEACETIC ACID-1-METHYL-3-PIPERIDINYL ESTER, HCl see PMS800

α-CYCLOHEXYL-β-HYDROXY-Δ^{α,β}-BUTENOLID (GERMAN) see CPM250

3-CYCLOHEXYL-4-HYDROXY-2(5H)FURANONE see CPM250

N-CYCLOHEXYLHYDROXYLAMINE see HKA109

CYCLOHEXYLHYDROXYMETHYLBENZENE see CPM300

6-CYCLOHEXYL-1-HYDROXY-4-METHYL-2(1H)-PYRIDINONE compounded with 2-AMINOETHANOL (1:1) see BAR800

6-CYCLOHEXYL-1-HYDROXY-4-METHYL-2(1H)-PYRIDON, 2-AMINOETHANOL-SALZ (GERMAN) see BAR800

6-CYCLOHEXYL-1-HYDROXY-4-METHYL-2(1H)-PYRIDONE, 2-AMINOETHANOL-SALT see BAR800

6-CYCLOHEXYL-1-HYDROXY-4-METHYL-2(1H)-PYRIDONE ETHANOLAMINE SALT see BAR800

4-(β-CYCLOHEXYL-β-HYDROXYPHENETHYL)-1,1-DIMETHYLPIPERAZINIUM METHYLSULFATE see HFG400

4-(β-CYCLOHEXYL-β-HYDROXYPHENETHYL)-1,1-DIMETHYL PIPERAZINIUM SULFATE see HFG000

N-(β-CYCLOHEXYL-β-HYDROXY-β-PHENYLETHYL)-N'-METHYLPIPERAZINE DIMETHYLSULFATE see HFG400

1-(3-CYCLOHEXYL-3-HYDROXY-3-PHENYLPROPYL)-1-METHYL-PIPERIDINIUM IODIDE see CPM750

1-(3-CYCLOHEXYL-3-HYDROXY-3-PHENYLPROPYL)-1-METHYL-PYRROLIDINIUM CHLORIDE see EAI875

(±)-N-((3-CYCLOHEXYL-3-HYDROXY-3-PHENYL)PROPYL)-N-

METHYLPYRROLIDINIUM CHLORIDE see EAI875

1-(3-CYCLOHEXYL-3-HYDROXY-3-PHENYLPROPYL.)-1-METHYL-PYRROLIDINIUM IODIDE see CNH550

1-(3-CYCLOHEXYL-3-HYDROXY-3-PHENYLPROPYL)-1-METHYL-PYRROLIDINIUM METHYL SULFATE see TJG225

2,2'-(CYCLOHEXYLIDENEBIS(4,1-PHENYLENEOXY)BIS(2-METHYLBUTANOIC ACID) see CMV700

2,2'-(4,4'-CYCLOHEXYLIDENEDIPHENOXY)-2,2'-DIMETHYLDIBUTYRIC ACID see CMV700

α-CYCLOHEXYLIDENE-α-(p-HYDROXYPHENYL)-p-CRESOL see CPM770

4-(CYCLOHEXYLIDENE(4-HYDROXYPHENYL)METHYL)PHENOL see CPM770

CYCLOHEXYLIDENEMALONONITRILE see CPM800

CYCLOHEXYLIDENEPROPANEDINITRILE see CPM800

2,2'-CYCLOHEXYLIMINODIETHANOL see DMT400

CYCLOHEXYL ISOCYANATE see CPN500

2-(N-CYCLOHEXYL-N-ISOPROPYLAMINOMETHYL)-1,3,4-OXADIAZOLE see CPN750

CYCLOHEXYLISOPROPYLMETHYLAMINE HYDROCHLORIDE see PNN300

CYCLOHEXYL-ISOTHIOCYANAT (GERMAN) see ISJ000

CYCLOHEXYL MERCAPTAN (DOT) see CPB625

CYCLOHEXYLMETHANE see MIQ740

CYCLOHEXYLMETHANOL see HDH200

CYCLOHEXYLMETHYLAMINE see MIT000

1-CYCLOHEXYL-2-METHYLAMINOPROPAN (GERMAN) see PNN400

1-CYCLOHEXYL-2-METHYLAMINOPROPANE HYDROCHLORIDE see PNN300

1,1-CYCLOHEXYL-2-METHYLAMINOPROPANE-5,5-PHENYLETHYLBARBITURATE see CPO500

CYCLOHEXYLMETHYLCARBINOL see CPL125

N-(2-CYCLOHEXYL-1-METHYLETHYL)-3,3-DIPHENYLPROPYLAMINE HYDROCHLORIDE see CPP000

1-CYCLOHEXYL-3-((p-(2-(5-METHYL-3-ISOXAZOLECARBOXAMIDO)ETHYL)PHENYL)SULFONYL)UREA see DBE885

4-(CYCLOHEXYLMETHYL)-α-(4-METHOXYPHENYL)-β-METHYL-1-PIPERIDINEETHANOL see RCA435

2-CYCLOHEXYL-4-METHYLPHENOL see CPP050

CYCLOHEXYL METHYLPHOSPHONOFLUORIDATE see MIT600

1-(CYCLOHEXYLMETHYL)PIPERIDINE HYDROCHLORIDE see PIR000

2-(4-CYCLOHEXYLMETHYLPIPERIDINO)-1-(4'-METHOXYPHENYL)-1-PROPANOL see RCA435

1-CYCLOHEXYL-N-METHYL-2-PROPANAMINE see PNN400

1-CYCLOHEXYL-1-NITROSOUREA see NJV000

CYCLOHEXYL 3-OXOBUTANOATE see CPP100

1-(2-CYCLOHEXYLPHENOXY)-1-(2-IMIDAZOLINYL)ETHANE HYDROCHLORIDE see CPP750

CYCLOHEXYL PHENYL PHOSPHATE see CPP800

CYCLOHEXYL PHENYL PHOSPHATE ((C6H11O)(PHO)2PO) see CPP800

α-CYCLOHEXYL-α-PHENYL-1-PIPERIDINEPROPANOL see PAL500

α-CYCLOHEXYL-α-PHENYL-1-PIPERIDINEPROPANOL HYDROCHLORIDE see BBV000

2-CYCLOHEXYL-2-PHENYL-4-PIPERIDINOMETHYL-DIOXOLANE-1,3 METHIODIDE see OLW400

2-CYCLOHEXYL-2-PHENYL-1-PIPERIDINO-1-PROPANOL see PAL500

1-CYCLOHEXYL-1-PHENYL-3-PIPERIDINO-PROPANOL, METHYLIODIDE see CPM750

1-CYCLOHEXYL-1-PHENYL-3-PYRROLIDINO-1-PROPANOL see CPQ250

1-CYCLOHEXYL-1-PHENYL-3-PYRROLIDINO-1-PROPANOL METHSULFATE see TJG225

1-CYCLOHEXYL-1-PHENYL-3-PYRROLIDINO-1-PROPANOL METHYL CHLORIDE see EAI875

1-CYCLOHEXYL-1-PHENYL-3-(1-PYRROLIDINYL)-1-PROPANOL see CPQ250

α-CYCLOHEXYL-α-(2-(PIPERIDINO)ETHYL)-BENZYLALCOHOL METHYLIODIDE see CPM750

CYCLOHEXYL 4-PYRIDYL KETONE see CPI375

N-CYCLOHEXYLPYRROLIDINONE see CPQ275

1-CYCLOHEXYL-2-PYRROLIDINONE see CPQ275

N-CYCLOHEXYLPYRROLIDONE see CPQ275

CYCLOHEXYLSULFAMIC ACID (9CI) see CPQ625

N-CYCLOHEXYLSULFENYLPHTHALIMIDE see CPQ700

CYCLOHEXYL SULPHAMATE SODIUM see SGC000

CYCLOHEXYLSULPHAMIC ACID see CPQ625

N-CYCLOHEXYLSULPHAMIC ACID see CPQ625

CYCLOHEXYLSULPHAMIC ACID, CALCIUM SALT see CAR000

3-CYCLOHEXYLSYDNONE IMINE MONOHYDROCHLORIDE see CPQ650

6-(4-(1-CYCLOHEXYL-1H-TETRAZOL-5-YL)BUTOXY)-3,4-DIHYDRO-2(1H)-QUINOLINONE see CMP825

N-(CYCLOHEXYLTHIO)PHTHALIMIDE see CPQ700

N-CYCLOHEXYL-p-TOLUENESULFONAMIDE see CPQ800

1-CYCLOHEXYL-3-p-TOLUENESULFONYLUREA see CPR000

1-CYCLOHEXYL-3-p-TOLYSULFONYLUREA see CPR000

CYCLOHEXYLTRICHLOROSILANE see CPR250

1-CYCLOHEXYLTRIMETHYLAMINE see CPR500

3-CYCLOHEXYL-5,6-TRIMETHYLENEURACIL see CPR600

3-CYCLOHEXYL-5,6-TRIMETHYLENURACIL see CPR600

CYCLOL ACRYLATE see BFY250

CYCLOLEUCINE see AJK250

CYCLOLYT see DNU100

CYCLOMALTOHEPTAOSE see COW925

CYCLOMANDOL see DNU100

CYCLOMEN see DAB830

CYCLOMETHIAZIDE see CPR750

CYCLOMETHICONE see PJR300

CYCLOMIDE DIN 295/S see BKF500

CYCLOMORPH see COX000

CYCLOMYCIN see CQH000

CYCLOMYCIN see TBX000

CYCLON see HHS000

CYCLONAL see ERD500

CYCLONAL SODIUM see ERE000
CYCLONAMINE see DIS600
CYCLONE B see HHS000
CYCLONITE see CPR800
CYCLONITE, wetted (UN 0072) (DOT) see CPR800
CYCLONITE, desensitized (UN 0483) (DOT) see CPR800
CYCLONIUM IODIDE see OLW400
CYCLONOL see TLO500
14,21-CYCLO-19-NORPREGNA-1,3,5(10)-TRIENE-3,16,17-TRIOL, (16-α,17-α)- see CQH275
(16-α,17-α)-14,21-CYCLO-19-NORPREGNA-1,3,5(10)-TRIENE-3,16,17-TRIOL see CQH275
cis,cis-CYCLOOCTA-1,5-DIENE see CPR825
1,5-CYCLOOCTADIENE DIEPOXIDE see CPR835
CYCLOOCTA-1,5-DIENE DIOXIDE see CPR835
(1,5-CYCLOOCTADIENE)(2,4-PENTANEDIONATO)RHODIUM see CPR840
1,5-CYCLOOCTADIENE (Z,Z) see CPR825
CYCLOOCTAFLUOROBUTANE see CPS000
CYCLOOCTANECARBAMIC ACID-1,1-BIS(p-FLUOROPHENYL)-2-PROPYNYL ESTER see BJW500
CYCLOOCTANE, 1,2:5,6-DIEPOXY- see CPR835
CYCLOOCTANONE see CPS250
1,3,5,7-CYCLOOCTATETRAENE see CPS500
3-CYCLOOCTYL-1,1-DIMETHYLHARNSTOFF (GERMAN) see CPT000
3-CYCLOOCTYL-1,1-DIMETHYLUREA see CPT000
N-CYCLOOCTYL-N',N'-DIMETHYLUREA see CPT000
3-CYCLOOCTYL-1,1-DIMETHYL UREA mixed with BUTYNYL-3N-3-CHLOROPHENYLCARBAMATE see AFM375
CYCLOOXABUTANE see OMW000
CYCLOPAMINE see CPT750
CYCLOPAN see ERD500
CYCLOPAR see TBX250
CYCLOPENIL see FBP100
CYCLOPENTABENZO(e)PYRENE see BCI265
CYCLOPENTA(ij)BENZO(a)PYRENE see BCH900
4H-CYCLOPENTA(def)CHRYSENE see CPU000
CYCLOPENTACYCLOHEPTENE see ASP600
CYCLOPENTADECANONE see CPU250
CYCLOPENTADECANONE, 3-METHYL- see MIT625
CYCLOPENTADIENE see CPU500
1,3-CYCLOPENTADIENE see CPU500
1,3-CYCLOPENTADIENE, DIMER see DGW000
1,3-CYCLOPENTADIENE, 2-METHYL- see MIT700
1,3-CYCLOPENTADIENE, 1,2,3,4,5-PENTACHLORO- see PAV800
1,3-CYCLOPENTADIENE, 1,2,3,4-TETRACHLORO- see TBO768
pi-CYCLOPENTADIENYL COMPOUND with NICKEL see NDA500
CYCLOPENTADIENYL GOLD(1) see CPU750
CYCLOPENTADIENYLMANGANESE TRICARBONYL see CPV000
CYCLOPENTADIENYL SILVER PERCHLORATE see CPV250
CYCLOPENTADIENYL SODIUM see CPV500
CYCLOPENTA(c)FURO(3',2':4,5)FURO(2,3-h)(1)BENZOPYRAN-1,11-DIONE, 2,3,6a,8,9,9a-HEXAHYDRO-8,9-DICHLORO-4-METHOXY-, (6aS-(6a-α-8-β,9-α-9aα-))- see AEU500

CYCLOPENTA(C)FURO(3',2':4,5)FURO(2,3-H)(1)BENZOPYRAN-1,11-DIONE, 2,3,6A,8,9,9A-HEXAHYDRO-8,9-DIHYDROXY-4-METHOXY-, CHLOROACETATE see CEB800
CYCLOPENTA(C)FURO(3',2':4,5)FURO(2,3-H)(1)BENZOPYRAN-1,11-DIONE, 2,3,6A,8,9,9A-HEXAHYDRO-8,9-DIHYDROXY-4-METHOXY-, 8-m-CHLOROBENZOATE see CEO110
CYCLOPENTA(C)FURO(3',2':4,5)FURO(2,3-H)(1)BENZOPYRAN-1,11-DIONE, 2,3,6A,8,9,9A-HEXAHYDRO-8,9-DIHYDROXY-4-METHOXY-, 8-p-NITROBENZOATE see NFL200
CYCLOPENTA(C)FURO(3',2':4,5)FURO(2,3-H)(1)BENZOPYRAN-11(1H)-ONE, 2,3,6A,9A-TETRAHYDRO-1-HYDROXY-4-METHOXY-, (1R-(1-α,6A-α,9A-α))- see AET630
α,β-CYCLOPENTAMETHYLENETETRAZOLE see PBI500
CYCLOPENTAMINE HYDROCHLORIDE see CPV609
CYCLOPENTANAMINE see CQA000
CYCLOPENTANAMINE, 3,3-BIS(4-CHLOROPHENYL)-, HYDROCHLORIDE see AJM550
CYCLOPENTA(de)NAPHTHALENE see AAF500
CYCLOPENTANE see CPV750
CYCLOPENTANEACETIC ACID, 3-OXO-2-PENTYL-, METHYL ESTER see HAK100
1,3-CYCLOPENTANEDISULFONYL DIFLUORIDE see CPW250
CYCLOPENTANE EPOXIDE see EBO100
CYCLOPENTANE, 1,2-EPOXY- see EBO100
CYCLOPENTANE, NITRO- see NFV530
CYCLOPENTANE OXIDE see EBO100
CYCLOPENTANEOXIDE see EBO100
CYCLOPENTANE, TETRACHLORO- see TBO770
CYCLOPENTANETHIOL see CPW300
4,5-CYCLOPENTANOFURAZAN-N-OXIDE see CPW325
CYCLOPENTANOL, 5-((4-CHLOROPHENYL)METHYLENE)-2,2-DIMETHYL-1-(1H-1,2,4-TRIAZOLE-1-YLMETHYL)- see TNP260
CYCLOPENTANOL, 1-((1,1-DIMETHYLETHYL)AZO)-, (E)- see BQI270
CYCLOPENTANONE see CPW500
CYCLOPENTANONE, 2-(4-(DIMETHYLAMINO)-2-BUTYNYL)-, HYDROCHLORIDE see DOV850
CYCLOPENTANONE-2-α,3-α-EPITHIO-5-α-ANDROSTAN-17-β-YL METHYL ACETAL see MCH600
CYCLOPENTANONE, 2-(2-HEXENYL)- see HFE513
CYCLOPENTANONE, 2-HEXYL- see HFO600
CYCLOPENTANONE OXIME see CPW750
CYCLOPENTANONE, 3-(2-OXOPROPYL)-2-PENTYL- see OOO100
1H-CYCLOPENTA(cd)PHENALENE-5,7(2H,6H)-DIONE,6,6-DICHLORO- see CDY260
CYCLOPENTAPHENANTHRENE see CPX250
3H-CYCLOPENTA(c)PHENANTHRENE see CPW800
4H-CYCLOPENTA(def)PHENANTHRENE see CPX250
1H-CYCLOPENTA(c)PHENANTHRENE, 2,3-DIHYDRO- see DKX150
17H-CYCLOPENTA(A)PHENANTHREN-17-ONE, 15,16-DIHYDRO-11-BUTOXY- see DKV140
CYCLOPENTA(cd)PYRENE see CPX500
CYCLOPENTA(cd)PYRENE-3,4-OXIDE see CPX625

CYCLOPENTA(cd)PYREN-3(4H)-ONE see CPX650
1H-CYCLOPENTAPYRIMIDINE-2,4,(3H,5H)-DIONE,6,7-DIHYDRO-3-CYCLOHEXYL- see CPR600
CYCLOPENTENE see CPX750
CYCLOPENTENE, 1,2-DICHLOROHEXAFLUORO- see DFM050
CYCLOPENTENE, 1,2-DICHLORO-3,3,4,4,5,5-HEXAFLUORO-(9CI) see DFM050
2-CYCLOPENTENE-1-OL see CPY000
1-CYCLOPENTENE-1-PROPANOL, 5-(1-METHYLETHENYL)-β,β,2-TRIMETHYL-, PROPANOATE see MJW300
2-CYCLOPENTENE-1-TRIDECANOIC ACID, SODIUM SALT see SFR000
1,2-CYCLOPENTENO-5,10-ACEANTHRENE see CPY500
5:6-CYCLOPENTENO-1:2-BENZANTHRACENE see CPY750
6,7-CYCLOPENTENO-1,2-BENZANTHRACENE see BCI250
1-CYCLOPENTEN-3-OL see CPY000
2-CYCLOPENTEN-1-ONE, 2,5-DIHYDROXY-5-METHYL-3-PIPERIDINOAMINO- see PIO900
2-CYCLOPENTEN-1-ONE, 3-DIMETHYLAMINO-2-HYDROXY-5-METHYLENE- see AOP510
2-CYCLOPENTEN-1-ONE, HEXACHLORO- see HCE600
2-CYCLOPENTEN-1-ONE, 2,3,4,4,5,5-HEXACHLORO- see HCE600
2-CYCLOPENTEN-1-ONE, 2-HEXYL- see HFO700
2-CYCLOPENTEN-1-ONE, 2-HYDROXY-5-METHYLENE-3-PIPERIDINOAMINO- see AOP800
2-CYCLOPENTEN-1-ONE, 3-METHYL-2-(2-PENTENYL)-, (Z)- see JCA100
2-CYCLOPENTEN-1-ONE, 2-PENTYL- see PBW600
CYCLOPENTENO(c,d)PYRENE see CPX500
2-CYCLOPENTENYL-4-HYDROXY-3-METHYL-2-CYCLOPENTEN-1-ONE CHRYSANTHEMATE see POO000
3-(2-CYCLOPENTEN-1-YL)-2-METHYL-4-OXO-2-CYCLOPENTEN-1-YL CHRYSANTHEMUMATE see POO000
3-(2-CYCLOPENTENYL)-2-METHYL-4-OXO-2-CYCLOPENTENYL CHRYSANTHEMUMMONOCARBOXYLATE see POO000
N-(1-CYCLOPENTEN-1-YL)-MORPHOLINE see CPY800
(+−)-1-(2-(1-CYCLOPENTEN-1-YL)PHENOXY)-3-((1,1-DIMETHYLETHYL)AMINO)-2-PROPANOL see CPY850
CYCLOPENTENYL PROPIONATE MUSK see MJW300
CYCLOPENTENYLRETHONYL CHRYSANTHEMATE see POO000
CYCLOPENTHIAZIDE see CPR750
CYCLOPENTIMINE see PIL500
CYCLOPENT(B)INDOL-3(2H)-ONE, 1,4-DIHYDRO-, OXIME see KFK150
CYCLOPENT(B)INDOL-3-ONE, 1,2,3,4-TETRAHYDRO-, OXIME see KFK150
CYCLOPENTOLATE HYDROCHLORIDE see CPZ125
CYCLOPENT(b)OXIRENO(c)PYRIDINE, 7-ETHYLIDENE-,HEXAHYDRO DERIV. see ELM600
CYCLOPENTYLAMINE see CQA000
CYCLOPENTYL 3,4-DIHYDROXYPHENYL KETONE see CQA100
3-(α-CYCLOPENTYL-4,6-DIMETHOXY-m-TOLUOYL)-PROPIONIC ACID SODIUM SALT see DOB325
(8-β)-1-CYCLOPENTYL-6,8-DIMETHYLERGOLINE see CQB300

CYCLOPROPANECARBOXYLIC ACID, 2,2-DIMETHYL-3-(2-METHYL-1-PROPENYL)-, (3-(PHENYLMETHYL)PHENYL) METHYL ESTER, (1S-trans)- see CQD930
CYCLOPROPANECARBOXYLIC ACID, 2,2-DIMETHYL-3-(1,2,2,2-TETRABROMOETHYL)-, CYANO(3-PHENOXY PHENYL)METHYL ESTER see THJ300
CYCLOPROPANECARBOXYLIC ACID, HEXADECYL ESTER see HCP500
CYCLOPROPANE, METHOXY-(9CI) see CQE750
CYCLOPROPANE, PENTACHLORO- see PAV810
CYCLOPROPYLAMINE see CQE250
N-CYCLOPROPYL-4-AMINO-3,5-DICHLOROBENZAMIDE see AJK500
5-(CYCLOPROPYLCARBONYL)-2-BENZIMIDAZOLECARBAMIC ACID METHYL ESTER see CQE325
N-CYCLOPROPYL-3,5-DICHLORO-4-AMINOBENZAMIDE see AJK500
N-CYCLOPROPYL-N'-(2,5-DIFLUOROPHENYL)UREA see CQE350
CYCLOPROPYLMELAMINE see CQE400
1-(4-(2-(CYCLOPROPYLMETHOXY)ETHYL)PHENOXY)-3-ISOPROPYLAMINOPROPAN-2-OL HYDROCHLORIDE see KEA350
1-N-CYCLOPROPYLMETHYL-3,14-DIHYDROXYMORPHINAN see CQF079
2-CYCLOPROPYLMETHYL-5,9-DIMETHYL-2'-HYDROXY-6,7-BENEOMORPHAN see COV500
3-CYCLOPROPYLMETHYL-6(eq),11(ax)-DIMETHYL-2,6-METHANO-3-BENZAZOCIN-8-OL see COV500
(5-α)-17-(CYCLOPROPYLMETHYL)-4,5-EPOXY-3,14-DIHYDROXY-MORPHINAN-6-ONE) (9CI) see CQF099
17-(CYCLOPROPYLMETHYL)-4,5-α-EPOXY-3,14-DIHYDROXY-MORPHINAN-6-ONEHYDROCHLORIDE see NAH100
CYCLOPROPYL METHYL ETHER see CQE750
CYCLOPROPYL METHYL ETHER see CQE750
3-(CYCLOPROPYLMETHYL)1-1,2,3,4,5,6-HEXAHYDRO-6,11-DIMETHYL-2,6-METHANO-3-BENZAZOCIN-8-OL see COV500
N-CYCLOPROPYLMETHYL-14-HYDROXYDIHYDROMORPHINONE see CQF099
2-CYCLOPROPYLMETHYL-2'-HYDROXY-5,9-DIMETHYL-6,7-BENZOMORPHAN see COV500
CYCLOPROPYL METHYL KETONE see CQF059
(−)-17-CYCLOPROPYLMETHYLMORPHINAN-3,4-DIOL see CQF079
17-(CYCLOPROPYLMETHYL)MORPHINAN-3-OL see CQG750
N-CYCLOPROPYLMETHYLNOROXYMORPHONE see CQF099
N-CYCLOPROPYLMETHYL-NOROXYMORPHONE HYDROCHLORIDE see NAH100
1-CYCLOPROPYLMETHYL-4-PHENYL-6-CHLORO-2(1H)-QUINAZOLINONE see CQF125
N-(CYCLOPROPYLMETHYL)-α,α,α-TRIFLUORO-2,6-DINITRO-N-PROPYL-p-TOLUIDINE see CQG250
1-(o-CYCLOPROPYLPHENOXY)-3-(ISOPROPYLAMINO)-2-PROPANOL HYDROCHLORIDE see PMF525

dl-1-(o-CYCLOPROPYLPHENOXY)-3-ISOPROPYLAMINO-2-PROPANOL HYDROCHLORIDE see PMF535
N-CYCLOPROPYL-1,3,5-TRIAZINE-2,4,6-TRIAMINE see CQE400
CYCLOPROVERA see EDQ600
CYCLORPHAN see CQG750
CYCLORYL 21 see SIB600
CYCLORYL OS see OFU200
CYCLORYL TAWF see SON000
CYCLORYL WAT see SON000
CYCLOSAN see MCW000
CYCLOSAN see MCY300
CYCLOSERINE see CQH000
CYCLO-d-SERINE see CQH000
CYCLOSIA see CMS850
CYCLOSILOXANES, DI-ME see PJR300
CYCLOSPASMOL see DNU100
CYCLOSPORIN see CQH100
CYCLOSPORIN A see CQH100
CYCLOSPORINE see CQH100
CYCLOSPORINE A see CQH100
CYCLOSTIN see CQC650
CYCLOTEN see HMB500
CYCLOTETRAMETHYLENE OXIDE see TCR750
CYCLOTETRAMETHYLENE SULFONE see SNW500
CYCLOTETRAMETHYLENE TETRANITRAMINE see CQH250
CYCLOTETRAMETHYLENETETRANITRAMINE (dry or unphlegmatized) (DOT) see CQH250
CYCLOTETRAMETHYLENETETRANITRAMINE, wetted (UN 0226) (DOT) see CQH250
CYCLOTETRAMETHYLENETETRANITRAMINE, desensitized (UN 0483) (DOT) see CQH250
CYCLOTETRASILOXANE, 2,6-DIPHENYL-2,4,4,6,8,8-HEXAMETHYL- see DWC650
CYCLOTETRASILOXANE, ETHENYLHEPTAMETHYL- see VPF150
CYCLOTETRASILOXANE, 2,4,6,8-TETRAETHENYL-2,4,6,8-TETRAMETHYL- see TCB600
CYCLOTETRASILOXANE, 2,4,6,8-TETRAMETHYL-2,4,6,8-TETRAVINYL- see TCB600
CYCLOTON 7LUF see OHK200
CYCLOTON V see HCQ500
CYCLOTRIMETHYLENENITRAMINE see CPR800
CYCLOTRIMETHYLENETRINITRAMINE see CPR800
CYCLOTRIMETHYLENETRINITRAMINE, wetted (UN 0072) (DOT) see CPR800
CYCLOTRIMETHYLENETRINITRAMINE, desensitized (UN 0483) (DOT) see CPR800
CYCLOTRIOL see CQH275
CYCLOTRISILAZANE, 2,2,4,4,6,6-HEXAMETHYL- see HEC600
CYCLOTRISILOXANE, 2,4-DIPHENYL-2,4,6,6-TETRAMETHYL-, (E)- see DWN100
CYCLOTRISILOXANE, PENTAMETHYLPHENYL- see PBI700
CYCLOTRISILOXANE, 2,4,4,6,6-PENTAMETHYL-2-PHENYL- see PBI700
CYCLOTRISILOXANE, 2,4,6-TRIMETHYL-2,4,6-TRIS(3,3,3-TRIFLUOROPROPYL)- see FLV100
1,4,8-CYCLOUNDECATRIENE, 2,6,6,9-TETRAMETHYL-, (E,E,E)- see HGM550
5-CYCLOUNDECENE-1-CARBOXYLIC ACID, 1,5-DIMETHYL-9-METHYLENE-2-((2-OXO-2H-1-BENZOPYRAN-7-YL) OXY)- see KBB800
CYCLOURON see CPT000
7,20-CYCLOVEATCHANE-1,12,15-TRIOL, 21-ETHYL-4-METHYL-16-METHYLENE-, (1-α,12-α, 15-β)- see LIN050
CYCLOVIROBUXIN D see CQH325
CYCLOVIROBUXINE see CQH325

CYCLOVIROBUXINE D see CQH325
CYCLURON see CPT000
CYCOCEL see CMF400
CYCOCEL-EXTRA see CMF400
CYCOGAN see CMF400
CYCOGAN EXTRA see CMF400
CYCOLAMIN see VSZ000
CYCRIMINE HYDROCHLORIDE see CQH500
CYCTEINAMINE see AJT250
CYDRIN see CQK500
CYDTA see CPB120
C YELLOW 12 see MRL100
CYETHOXYDIM see CDK800
CYFEN see DSQ000
CYFLEE see CQL250
CYFLEE see FAB600
CYFLUTHIN see REF250
CYFLUTHRIN see REF250
CYFLUTHRINE see REF250
CYFOS see IMH000
CYFOXYLATE see REF250
CYGON see DSP400
CYGON INSECTICIDE see DSP400
CYHALOTHRIN see GJU600
CYHALOTHRINE see GJU600
CYHALOTHRIN K see LAS200
CYHEPTAMIDE see CQH625
CYHEPTAMINE see CQH625
CYHEXATIN see CQH650
3-CYJANOPIRYDYNA see NDW515
CYJANOWODOR (POLISH) see HHS000
CYKAZINE see COU000
CYKLOHEKSAN (POLISH) see CPB000
CYKLOHEKSANOL (POLISH) see CPB750
CYKLOHEKSANON (POLISH) see CPC000
CYKLOHEKSEN (POLISH) see CPC579
CYKLOHEXANTHIOL see CPB625
CYKLOHEXYLAMINACETAT (CZECH) see CPG000
CYKLOHEXYLESTER KYSELINY THIOKYANOOCTOVE (CZECH) see TFF000
CYKLOHEXYLMERKAPTAN (CZECH) see CPB625
CYKLOHEXYLTHIOKYANOACETAT (CZECH) see TFF000
CYKLONIT see CPR800
CYKOBEMINET see VSZ000
CY-L 500 see CAQ250
CYLAN see CAR000
CYLAN see DXN600
CYLAN see PGW750
CYLERT see PAP000
CYLINDRO-SPROPSIN see CQH700
CYLOCIDE see AQQ750
CYLOCIDE see AQR000
CYLPHENICOL see CDP250
CYMAG see SGA500
CYMARIGENIN see SMM500
CYMARIN see CQH750
CYMARINE see CQH750
CYMAROL see SMM100
3-β-(β-d-CYMAROSYLOXY)-5,14-DIHYDROXY-19-OXO-5-β-CARD-20(22)-ENOLIDE see CQH750
3-β-(β-d-CYMAROSYLOXY)-5,14,19-TRIHYDROXY-5-β-CARD-20(22)-ENOLIDE see SMM100
CYMATE see BJK500
CYMBI see AIV500
CYMBI see AOD125
CYMBUSH see RLF350
CYMEL see MCB000
CYMEL see MCB050
CYMEL 200 see MCB050
CYMEL 202 see MCB050
CYMEL 235 see MCB050
CYMEL 245 see MCB050
CYMEL 255 see MCB050
CYMEL 285 see MCB050
CYMEL 300 see MCB050
CYMEL 301 see MCB050

CYMEL 303 see HDY500
CYMEL 303 see MCB050
CYMEL 305 see MCB050
CYMEL 323 see MCB050
CYMEL 325 see MCB050
CYMEL 327 see MCB050
CYMEL 350 see MCB050
CYMEL 370 see MCB050
CYMEL 373 see MCB050
CYMEL 380 see MCB050
CYMEL 385 see MCB050
CYMEL 412 see MCB050
CYMEL 428 see MCB050
CYMEL 481 see MCB050
CYMEL 482 see MCB050
CYMEL 1080 see MCB050
CYMEL 1116 see MCB050
CYMEL 1130 see MCB050
CYMEL 1133 see MCB050
CYMEL 1135 see MCB050
CYMEL 1156 see MCB050
CYMEL 1158 see MCB050
CYMEL 1161 see MCB050
CYMEL 1168 see MCB050
CYMEL 1370 see MCB050
CYMEL 243-3 see DBB200
CYMEL 265J see MCB050
CYMEL 266J see MCB050
CYMEL 247-10 see MCB050
CYMEL 7273-7 see MCB050
CYMEL 1130-235J see MCB050
CYMEL 1130-254J see MCB050
CYMEL 1130-285J see MCB050
CYMEL C 1156 see MCB050
CYMEL HM 6 see MCB050
CYMEL 401 RESIN see MCB050
CYMEL 481 RESIN see THR790
CYMEL XM 1116 see MCB050
CYMENE see CQI000
p-CYMENE see CQI000
m-CYMENE (8CI) see IRN400
o-CYMENE (8CI) see IRN300
β-CYMENE see IRN400
p-CYMENE-7-CARBOXALDEHYDE see IRA000
p-CYMENE-2-CARBOXYLIC ACID, 3-(3-(DIETHYLAMINO)PROPYLCARBAMOYLM ETHOXY)-, METHYL ESTER see MJD200
2-p-CYMENOL see CCM000
3-p-CYMENOL see TFX810
m-CYMEN-4-OL see IQJ000
p-CYMEN-3-OL see TFX810
p-CYMEN-7-OL see CQI250
CYMETHION see MDT750
CYMETOX see MIW250
CYMIDON see KFK000
CYMOL see CQI000
m-CYMOL see IRN400
o-CYMOL see IRN300
CYMONIC ACID see FIC000
CYMOXANIL see COM300
m-CYM-5-YL METHYLCARBAMATE see CQI500
CYNARON see MDT740
CYNAUSTINE HYDROCHLORIDE see CQI525
CYNCAL 80 see QAT520
CYNEM see EPC500
CYNKOMIEDZIAN see ZJS300
CYNKOTOX see EIR000
CYNKU TLENEK (POLISH) see ZKA000
CYNOGAN see BMM650
CYNOTOXIN see SMM500
CYOCEL see CMF400
CYODRIN see COD000
CYOLAN see DXN600
CYOLANE see PGW750
CYOLANE INSECTICIDE see DXN600
CYOLANE INSECTICIDE see PGW750
CYOMETRINIL see COP700
CYP see CON300
CYPENTIL see PIL500

CYPERKILL see RLF350
CYPERMETHRIN see RLF350
CYPERQUAT see CQI550
CYPERUS SCARIOSUS OIL see NAE505
CYPIP see DIW000
CYPONA see DGP900
CYPRAZINE see CQI750
CYPRESS OIL see CQJ000
CYPREX see DXX400
CYPREX 65W see DXX400
5-α-CYPRINOL see CQJ100
CYPROCONAZOLE see CQJ150
CYPROHEPTADIENE HYDROCHLORIDE see PCI250
CYPROHEPTADINE see PCI250
CYPROHEPTADINE see PCI500
CYPROHEPTADINE HYDROCHLORIDE see PCI250
CYPROMAZINE see CQE400
CYPROME ETHER see CQE750
CYPROMID see CQJ250
CYPRON see MQU750
CYPROSTERONE ACETATE see CQJ500
CYPROTERONE ACETATE see CQJ500
CYPROTERON-R ACETATE see CQJ500
CYRAL see DBB200
CYREDIN see VSZ000
CYREN see DKA600
CYREN B see DKB000
CYREZ see MCB050
CYREZ 933 see MCB050
CYREZ 933 see UTU500
CYREZ 963 see MCB050
CYREZ 966 see MCB050
CYREZ 963P-A see MCB050
CYREZ 963 P see MCB050
CYREZ 963 RESIN see HDY500
CYRSTHION see EKN000
CYSTAMIN see HEI500
CYSTAMINE see MCN500
CYSTAMINE DIHYDROCHLORIDE see CQJ750
CYSTAMINE "MCCLUNG" see PDC250
CYSTAPHOS see AKB500
CYSTAPHOS SODIUM SALT see AKB500
CYSTEAMIDE see AJT250
CYSTEAMINE see AJT250
CYSTEAMINE HYDROCHLORIDE see MCN750
CYSTEAMINE S-PHOSPHATE see AKA900
CYSTEAMINHYDROCHLORID (GERMAN) see MCN750
CYSTEIN see CQK000
l-CYSTEINAMIDE, d-PHENYLALANYL-l-CYSTEINYL-l-PHENYLALANYL-d-TRYPTOPHYL-l-LYSYL-l-THREONYL-N-(2-HYDROXY-1-(HYDROXYMETHYL)PROPYL)-, CYCLIC(2-7)-DISULFIDE, (R-(R*,R*))-, ACETATE (SALT) see ODY200
CYSTEINAMINE DISULFIDE see MCN500
(l-CYSTEINATO(2-)-S)METHYLMERCURATE(1-) HYDROGEN see MCS100
CYSTEINE see CQK000
l-CYSTEINE see CQK000
l-(+)-CYSTEINE see CQK000
l-CYSTEINE, N-ACETYL-ETHYL ESTER, ACETATE (ESTER) see DBG600
l-CYSTEINE, N-ACETYL-S-(2-HYDROXYPHENYLETHYL)- see ACN600
l-CYSTEINE, N-ACETYL-S-(2-HYDROXY-1-PHENYLETHYL)- see ACN700
CYSTEINE CHLORHYDRATE see CQK250
CYSTEINE DISULFIDE see CQK325
CYSTEINE ETHYL ESTER HYDROCHLORIDE see EHU600
CYSTEINE-GERMANIC ACID see CQK100
CYSTEINE HYDRAZIDE see CQK125
CYSTEINE HYDROCHLORIDE see CQK250
l-CYSTEINE HYDROCHLORIDE see CQK250

l-CYSTEINE HYDROCHLORIDE see CQK250
l-CYSTEINE, S-(3-HYDROXYPROPYL)- see HNV050
CYSTEINE, l-, MERCURY COMPLEX see BIS300
l-CYSTEINE MONOHYDROCHLORIDE (FCC) see CQK250
(l-CYSTEINE)TETRAHYDROXYGERMANIU M see CQK100
l-CYSTEINYLGLYCINE see CQK300
CYSTIN see CQK325
CYSTINAMIN (GERMAN) see MCN500
(−)-CYSTINE see CQK325
l-CYSTINE see CQK325
CYSTINE ACID see CQK325
CYSTINEAMINE see MCN500
l-CYSTINE-BIS(N,N-β-CHLOROETHYL)HYDRAZIDEHYDROBRO MIDE see CQK500
CYSTISINE see CQL500
CYSTOCEVA see BGB275
CYSTO-CONRAY see IGC000
CYSTOGEN see HEI500
CYSTOGRAFIN see AOO875
CYSTOIDS ANTHELMINTIC see HFV500
CYSTOKON see AAN000
CYSTOPYRIN see PDC250
CYSTORELIN see LIU305
CYSTORELIN see LIU370
CYSTURAL see EEQ600
CYSTURAL see PDC250
CYSTURAL B see EEQ600
CYTACON see VSZ000
CYTADREN see AKC600
CYTAMEN see VSZ000
CYTARABIN see AQQ750
CYTARABINA see AQQ750
CYTARABINE see AQQ750
CYTARABINE HYDROCHLORIDE see AQR000
CYTARABINOSIDE see AQQ750
CYTEL see DSQ000
CYTEMBENA see CQK600
CYTEMBENA see SIK000
CYTEN see DSQ000
CYTHIOATE see CQL250
CYTHION see MAK700
CYTHRIN see COQ385
CYTIDINDIPHOSPHOCHOLIN see CMF350
CYTIDINE see CQM500
CYTIDINE CHOLINE DIPHOSPHATE see CMF350
CYTIDINE 5'-(CHOLINE DIPHOSPHATE) see CMF350
CYTIDINE, 2'-DEOXY-2',2'-DIFLUORO- see GCK500
CYTIDINE, 2'-DEOXY-2',2'-DIFLUORO-, MONOHYDROCHLORIDE see GCK100
CYTIDINE, 2'-DEOXY-5-HYDROXY- see HKA760
CYTIDINE, 2',3'-DIDEOXY- see DHA350
CYTIDINE 5'-(DIHYDROGENPHOSPHATE) see CQL300
CYTIDINE DIPHOSPHATE CHOLINE see CMF350
CYTIDINE 5'-DIPHOSPHATE CHOLINE see CMF350
CYTIDINE DIPHOSPHATE CHOLINE ESTER see CMF350
CYTIDINE DIPHOSPHATE CHOLIN ESTER see CMF350
CYTIDINE DIPHOSPHOCHOLINE see CMF350
CYTIDINE 5'-DIPHOSPHOCHOLINE see CMF350
CYTIDINE DIPHOSPHORYLCHOLINE see CMF350
CYTIDINE, 5-HYDROXY- see HKA250
CYTIDINE, N-HYDROXY- see HKA270
CYTIDINE MONOPHOSPHATE see CQL300

CYTIDINE 5'-MONOPHOSPHATE see CQL300

CYTIDINE 5'-MONOPHOSPHORIC ACID see CQL300

CYTIDINE 5'-PHOSPHATE see CQL300

CYTIDINE 5'-PHOSPHORIC ACID see CQL300

CYTIDINE, 5'-(TETRAHYDROGEN TRIPHOSPHATE) see CQL400

CYTIDINE-5'-TRIPHOSPHATE see CQL400

CYTIDINE 5'-TRIPHOSPHORIC ACID see CQL400

CYTIDOLINE see CMF350

CYTIDYLIC ACID see CQL300

5'-CYTIDYLIC ACID see CQL300

CYTISINE see CQL500

CYTISINE HYDROCHLORIDE see CQL750

(−)7R:9S-CYTISINE HYDROCHLORIDE see CQL750

CYTITONE see CQL500

CYTOBION see VSZ000

CYTOCHALASIN B see CQM125

CYTOCHALASIN D see ZUS000

CYTOCHALASIN E see CQM250

CYTOCHALASIN-H see PAM775

CYTOCHROME C see CQM325

CYTOPHOSPHAN see CQC500

CYTOPHOSPHAN see CQC650

CYTOREST see CQM325

CYTOSAR see AQQ750

CYTOSAR HYDROCHLORIDE see AQR000

CYTOSAR-U see AQQ750

CYTOSINE (8CI) see CQM600

CYTOSINE, 1-β-d-ARABINOFURANOSYL-, 5'-PALMITATE see PAE260

CYTOSINEARABINOSIDE see AQQ750

CYTOSINE-β-ARABINOSIDE see AQQ750

CYTOSINE β-D-ARABINOSIDE see AQQ750

CYTOSINE ARABINOSIDE HYDROCHLORIDE see AQR000

CYTOSINE ARABINOSIDE PALMITATE see AQS875

CYTOSINE, 1-β-D-ARABINOSYL- see AQQ750

CYTOSINE,1-(2-DEOXY-2,2-DIFLUORO-β-d-ERYTHRO-PENTOFURANOSYL)- see PPP150

CYTOSINE DEOXYRIBOSIDE see DAQ850

CYTOSINE PROPENAL see CQM400

CYTOSINE RIBOSIDE see CQM500

CYTOSINIMINE see CQM600

CYTOSTASAN see CQM750

CYTOTEC see MJE775

CYTOVENE see GBU200

CYTOVIRIN see BLX500

CYTOXAL ALCOHOL see CQN000

CYTOXAN see CQC500

CYTOXAN see CQC650

CYTOXYL ALCOHOL CYCLOHEXYLAMMONIUM SALT see CQN000

CYTOXYL AMINE see CQN125

CYTROL see AMY050

CYTROL AMITROLE-T see AMY050

CYTROLANE see DHH400

CYTROLE see AMY050

CYURAM DS see TFS350

CYZINE PREMIX see ABY900

CYZONE see PFL000

CZERN HELIONOWA GF see CMN150

CZERN KWASOWO-CHROMOWA ETN see EDC625

CZERWIEN KWASOWA TRWALA E6B see CMM400

CZON see CCS635

CZT see CLX000

CZTEROCHLOREK WEGLA (POLISH) see CBY000

2,3,7,8-CZTEROCHLORODWUBENZO-p-DWUOKSYNY (POLISH) see TAI000

1,1,2,2-CZTEROCHLOROETAN (POLISH) see TBQ100

CZTEROCHLOROETYLEN (POLISH) see PCF275

CZTEROETHLEK OLOWIU (POLISH) see TCF000

D₂ see DBB800

D-D see DGG000

P-D see BBO500

D-13 see AHL000

2,4-D see DAA800

3,4-D see DFY500

D-40 see AFO250

D 47 see RCK730

D 4T see SLJ800

D-50 see POL500

D 50 see AAX250

D 50 see DAA800

D 109 see COU250

12FD see KDK000

D 201 see OGI075

D 206 see DYB600

D 212 see DIR100

D 268 see DRW000

D301 see BDJ600

D-365 see IRV000

D 442 see BCJ090

D 514 see TFM625

D-638 see BPJ750

D-649 see BPJ500

D-695 see BKC500

D-701 see BKD000

D-703 see EEY000

D 704 see ORI400

D 705 see BGC125

D 735 see CCC500

838-D see MHQ775

D 854 see CJT750

D 860 see BSQ000

D 100-2 see MCB050

D-1126 see FMU225

D 1221 see CBS275

D 1308 see EOS100

D-1410 see DSP600

D 1593 see CIP500

30-D-11 see MHX200

D-10,242 see DAB200

D-13,312 see TAL560

D 18506 see HCP750

20ND3-5 see MHB300

D17-1242 see PJQ800

DA see CGN000

DA see DNA200

3,4-DA see DFY500

DA-241 see PEO750

DA 339 see DWA500

DA-398 see MCH550

DA-688 see GDG200

DA79P see LGF875

DA-1773 see SJJ175

DA 2370 see PEW000

DA 737S see EHP700

2,4-DAA see DBO000

1,2-DAA (RUSSIAN) see DBO800

DAAB see DWO800

DAAE see DCN800

2,4-DAA SULFATE see DBO400

DAB see BIU750

DAB see DOT300

DABCO see DCK400

DABCO S-25 see DCK400

DABCO CRYSTAL see DCK400

DABCO EG see DCK400

DABCO R-8020 see DCK400

DABCO 33LV see DCK400

DABI see DOT600

DABICYCLINE see HOH500

DAB-O-LITE P 4 see WCJ000

DABOIA RUSSELLI SIAMENSIS TOXIN see DAB250

DABOIATOXIN see DAB250

DAB-N-OXIDE see DTK600

DABROSIN see ZVJ000

DABYLEN see BAU750

DABYLEN see BBV500

DAC 649 see TBV300

DAC 2797 see TBQ750

DACAMINE see DAA800

DACAMINE see TAA100

DACAMOX see DAB400

DACARBAZINE see DAB600

DACE see DBG600

2,4-D ACETATE see DFY800

2,4-D ACID see DAA800

DACONATE 6 see MRL750

DACONIL see TBQ750

DACONIL 2787 FLOWABLE FUNGICIDE see TBQ750

DACORENE HYDROCHLORIDE see BGO000

DACORTIN see PLZ000

DACOSOIL see TBQ750

DACOTE see CAT775

DACOVIN see PKQ059

DACPLAT see OKY100

DAC PRO see DGL200

DACTHAL see TBV250

DACTIL see EOY000

DACTIL HYDROCHLORIDE see EOY000

DACTIMICIN SULFATE see DAB630

DACTIN see DFE200

DACTINOL see RNZ000

DACTINOMYCIN see AEB000

DACTINOMYCIN (10%), ACTINOMYCIN C2 (45%), and ACTINOMYCIN C3 (45%) mixture see AEA750

DAD see DCI600

DADA see DNM400

DADDS see SNY500

DADEX see BBK500

DADIBUTOL see TGA500

DA 1257-O-(2,6-DIDEOXY-2-FLUORO-α-TALOPYRANOSYL)ADRIAMYCINONE-14-β-ALANIATE HYDROCHLORIDE see AFH650

DADOX d-CITRAMINE see BBK500

DADPE see OPM000

DADPM see MJQ000

DADPS see SOA500

DAEP see DOP200

DAEP-ES see AEF000

DAF 68 see DVL700

DAFEN see LJR000

DAFF see BJR625

DAFFODIL see DAB700

DAFTAZOL HYDROCHLORIDE see DCA600

DAG see DCI600

DAGADIP see TNP250

DAGC see AGD250

DAGENAN see PPO000

DAGENAN CHLORIDE see AAM600

DAGUTAN see SJN700

1,1-DAH see DBK100

1,2-DAH HYDROCHLORIDE see DBK120

DAHLIA see HGE925

DAI CARI XBN see BQK250

DAICEL 1150 see SFO500

DAI-EI ACID PURE BLUE VX see CMM062

DAIFLOIL 3 see KDK000

DAIFLOIL 10 see KDK000

DAIFLOIL 20 see KDK000

DAIFLOIL 50 see KDK000

DAIFLOIL 100 see KDK000

DAIFLON see CLQ750

DAIFLON 22 see CFX500

DAIFLON CTF3-D 55P see KDK000

DAIFLON CTFE see KDK000

DAIFLON D 45S see KDK000

DAIFLON M 300 see KDK000

DAIFLON M 300P see KDK000

DAIFLON S 3 see FOO000

DAILON see DXQ500

DAIMETON see SNL800

DAINICHI BENZIDINE YELLOW GRT see DEU000

DAINICHI BENZIDINE YELLOW 2GR see CMS208
DAINICHI BRILLIANT SCARLET G see CMS160
DAINICHI BRILLIANT SCARLET RG see CMS160
DAINICHI CHROME ORANGE R see LCS000
DAINICHI CHROME YELLOW G see LCR000
DAINICHI CYANINE GREEN FG see PJQ100
DAINICHI CYANINE GREEN FGH see PJQ100
DAINICHI FAST ORANGE RR see CMS145
DAINICHI FAST RED B BASE see NEQ000
DAINICHI FAST SCARLET G BASE see NMP500
DAINICHI LAKE RED C see CHP500
DAINICHI PERMANENT RED GG see DVB800
DAINICHI PERMANENT RED 4 R see MMP100
DAINICHI PERMANENT RED RX see CJD500
DAINICHI PIGMENT SCARLET 3B see CMG750
DAIOMIN see TES800
DAIPIN see DAB750
DAIRYLIDE YELLOW AAA see DEU000
DAISAZIN see TES800
DAISEN see EIR000
DAISHIKI AMARANTH see FAG020
DAISHIKI BRILLIANT SCARLET 3R see FMU080
DAISOLAC see PJS750
DAITO ORANGE BASE GC see CEH690
DAITO ORANGE BASE R see NEN500
DAITO ORANGE SALT RD see CEG800
DAITO RED BASE B see NEQ000
DAITO RED BASE 3GL see KDA050
DAITO RED BASE RL see MMF780
DAITO RED BASE TR see CLK220
DAITO RED SALT TR see CLK235
DAITO SCARLET BASE G see NMP500
DAIVOUGEN BLUE BF see COF420
DAIYA FOIL see PKF750
DAKINS SOLUTION see SHU500
DAKTARIN see MQS550
DAKTIN see DFE200
DAKTOSE B see DIB300
DAKURON see SKQ400
DALACIN C see CMV675
DALAPON see DGI600
DALAPON 85 see DGI400
DALAPON SODIUM see DGI600
DALAPON SODIUM SALT see DGI600
DALAPON (USDA) see DGI400
DAL-E-RAD see MRL750
DAL-E-RAD 100 see DXE600
DALF see MNH000
DALGOL see EQL000
DALMADORM see DAB800
DALMADORM HYDROCHLORIDE see DAB800
DALMANE see DAB800
DALMATE see DAB800
DALMATIAN SAGE OIL see SAE500
DALMATION INSECT FLOWERS see POO250
DALTOGEN see TKP500
DALTOLITE FAST GREEN GN see PJQ100
DALTOLITE FAST ORANGE G see CMS145
DALTOLITE FAST YELLOW GT see DEU000
DALYSEP see MFN500
DALZIC see ECX100
DAM see OMY910
DAM-57 see LJH000
DAMA de DIA (PUERTO RICO) see DAC500
DAMA de NOCHE (PUERTO RICO) see DAC500

DAMANTOYLDIAZOMETHANE see DAB807
DAMC see DPJ800
DAMILAN see EAH500
DAMILEN HYDROCHLORIDE see EAI000
2,4-D AMINE SALT see DFY800
DAMINOZIDE (USDA) see DQD400
2,4-D AMMONIUM SALT see DAB815
1,4-DA-2-MOA see DBX000
DAMPA D see DAB820
DAMP-ES see AEF250
DAN see DSU600
(DA)N see PJQ400
DA-2-N see DSU800
DANA see NJW500
DANABOL see DAL300
DANAMID see PJY500
DANAMINE see DJS200
DANANTIZOL see MCO500
DANAZOL see DAB830
D AND C RED NO. 27 see SKS400
DANDELION (JAMAICA) see CNG825
DANERAL see TMK000
DANEX see TIQ250
DANFIRM see AAX250
DANIFOS see DIX800
DANILON see SOX875
DANILONE see PFJ750
DANINON see CKL500
DANITOL see DAB825
DANIZOL see MMN250
DANOCRINE see DAB830
DANOL see DAB830
DANSYL see DPN200
DANSYL CHLORIDE see DPN200
DANTAFUR see NGE000
DANTEN see DKQ000
DANTEN see DNU000
DANTHION see PAK000
DANTHRON see DMH400
DANTINAL see DKQ000
DANTOIN see DFE200
DANTOIN see DNU000
DANTOINAL KLINOS see DKQ000
DANTOINE see DKQ000
DANTOROLENE SODIUM see DAB840
DANTRIUM see DAB840
DANTRIUM HEMIHEPTAHYDRATE see DAB840
DANTROLENE see DAB845
DANTROLENE SODIUM see DAB840
DANTRON see DMH400
DANU see DPN400
DANUVIL 70 see PKQ059
DAONIL see CEH700
DAP see DOT000
DAP see POJ500
DAP-1 see BGW400
DAPA see DNM400
DAPA see DOU600
DAPACRYL see BGB500
DAPAZ see MQU750
DAPHENE see DSP400
DAPHNE MEZEREUM see LAR500
DAPHNETOXIN see DAB850
DAPLEN see PJS750
DAPLEN AD see PMP500
DAPLEN 1810 H see PJS750
DAPM see MJQ000
DAPOCEL see DNM400
DAPON 35 see DBL200
DAPON R see DBL200
DAPPU 100 see IMK000
DAPRISAL see ABG750
DAPSONE see SOA500
DAPTAZILE HYDROCHLORIDE see DCA600
DAPTAZOLE HYDROCHLORIDE see DCA600
DARACLOR see TGD000
DARAL see VSZ100
DARAMIN see ANT500

DARAMIN see CAM750
DARAMMON see ANE500
DARAN see CGW300
DARAN 212 see VPK333
DARAN CR 6795H see CGW300
DARANIDE see DEQ200
DARAN X 66919M see VPK333
DARAPRAM see TGD000
DARAPRIM see TGD000
DARAPRIME see TGD000
DARATAK see AAX250
DARBID see DAB875
DAR-CHEM 14 see SLK000
DARCIL see PDD350
DARCO see CBT500
DAREBON see RDK000
DARENTHIN see BMV750
DARID QH see SEH000
DARILOID QH see SEH000
DARK GREEN EMBL see CMO830
DAROCUR 1173 see HMQ100
DAROLON see ACE000
DAROPERVAMIN see DBA800
DARVAN 1 see BLX000
DARVAN No. 1 see BLX000
DARVIC 110 see PKQ059
DARVIS CLEAR 025 see PKQ059
DARVON see DAB879
DARVON COMPOUND see ABG750
DARVON HYDROCHLORIDE see PNA500
DARVON-N see DAB880
DARVON-N see DYB400
DAS see AOP250
DAS see DOU600
DASANIDE see DEQ200
DASANIT see FAQ800
DASD see FCA100
DASEN see SCA625
DASERD see GGS000
DASEROL see GGS000
DASHEEN see EAI600
DASIKON see ABG750
DASKIL see NCQ900
DATC see DBI200
DATHROID see LFG050
DATRIL see HIM000
DATURALACTONE see DAB925
DATURA STRAMONIUM see SLV500
DATURINE see HOU000
DAUCUS CAROTA LINN., SEED EXTRACT see GAP500
DAUCUS OIL see CCL750
DAUNAMYCIN see DAC000
DAUNOBLASTIN see DAC200
DAUNOBLASTINA see DAC200
DAUNOMYCIN see DAC000
DAUNOMYCIN BENZOYLHYDRAZONE see ROU800
DAUNOMYCIN CHLOROHYDRATE see DAC200
DAUNOMYCIN, 4-DEMETHOXY-, HYDROCHLORIDE see DAN100
DAUNOMYCIN HYDROCHLORIDE see DAC200
DAUNOMYCINOL see DAC300
DAUNORUBICIN see DAC000
DAUNORUBICIN, BENZOYLHYDRAZONE, MONOHYDROCHLORIDE see ROZ000
DAUNORUBICINE see DAC000
DAUNORUBICIN HYDROCHLORIDE see DAC200
DAUNORUBICINOL see DAC300
DAURAN see AFJ400
DAVA see VGU750
DAVANA OIL see DAC400
DAVISON SG-67 see SCH002
DAVITAMON D see VSZ100
DAVITAMON-K see VTA100
DAVITAMON-K-ORAL see VTA100
DAVITAMON PP see NCQ900
DAVITIN see VSZ100

DAVOSIN see AKO500
DAWE'S DESTROL see DKA600
DAWSON 100 see MHR200
DAWSONITE see DAC450
DAXAD 11 see BLX000
DAXAD 15 see BLX000
DAXAD 18 see BLX000
DAXAD No. 11 see BLX000
DAY BLOOMING JESSAMINE see DAC500
DAYFEN see DOZ000
DAYFEN see LJR000
DAZOMET see DSB200
DAZZEL see DCM750
DB see CQK100
DB 1 see MAD100
2,4-DB see DGA000
2NDB see ALL750
4NDB see ALL500
DB 133 see FQL200
DB 134 see DNF500
DB 135 see DSC100
DB 136 see DNF450
DB 138 see EID250
DB-905 see TBS000
DB 2041 see IDJ500
DBA see DCT400
DBA see DQJ200
DB(a,c)A see BDH750
DB(a,h)A see DCT400
1,2,5,6-DBA see DCT400
DB(a,h)AC see DCS400
DB(a,j)AC see DCS600
DBA-1,2-DIHYDRODIOL see DMK200
trans-DBA-3,4-DIHYDRODIOL see DLD400
DBA-5,6-EPOXIDE see EBP500
DBB see DDL000
7H-DB(c,g)C see DCY000
DBCP see DDL800
DBD see ASH500
DBD see DDJ000
2,4-DB-DIMETHYLAMMONIUM see
DGA100
DBDPO see PAU500
DBE see EIY500
DBED see BHB300
DBED DIACETATE see DDF800
DBED DIHYDROCHLORIDE see DDG400
DBED DIPENCILLIN G see BFC750
DBED PENICILLIN see BFC750
DBF see DEC400
DBF see DJY100
DBH see BBP750
DBH see BBQ500
DBH see DDO800
1,1-DBH see DEC725
DBHMD see DEC699
DB 2182 HYDROCHLORIDE see IFZ900
DBI see PDF000
DBI-TD see PDF250
DBM see DDP600
DBM see DED600
DBMP see BFW750
DBN see BRY500
2,6-DBN see DER800
DBN (the herbicide) see DER800
DBNA see BRY500
DBNE see DDQ470
DBNPA see DDM000
DBOT see DEF400
DBP see DEH200
DBP see DES000
DB(a,e)P see NAT500
DB(a,i)P see BCQ500
DB(a,l)P see DCY400
DBPC (technical grade) see BFW750
DBQ see DDV500
2,4-DB SODIUM SALT see EAK500
D.B.T.C. see DDY200
DBTL see DDV600
2,4-D BUTOXYETHANOL ESTER see
DFY709
2,4-D BUTOXYETHYL ESTER see DFY709

2,4-D 2-BUTOXYETHYL ESTER see DFY709
2,4-D BUTYL ESTER see BQZ000
2,4-D BUTYRIC see DGA000
DBV see BRA625
DC-11 see TEF725
DC 360 see SCR400
DC 0572 see AHI875
DC 1173 see HMQ100
DCA see DAQ800
DCA see DEL000
DCA see DEO300
DCA see DFE200
3,4-DCA see DEO300
DC-38-A see GEO200
DCA 70 see AAX250
DCAA see AFH275
DCA-ETHER (1:9) see DEN800
d-(DCA⁶))LHRH ACETATE see LIU313
DCAOB see DEP450
D,l-CARNITINENITRILE CHLORIDE see
COL050
D,l-CARNITINNITRILCHLORID see COL050
DCB see COC750
DCB see DEP600
DCB see DEQ600
DCB see DER800
DCB see DES000
DCB see DEV000
1,4-DCB see DEV000
DC-45-B2 see TMO775
DCBA see BIA750
D&C BLUE No. 4 see FAE000
D&C BLUE No. 4 see FMU059
D and C BLUE No. 3 see ERG100
D and C BLUE No. 9 see DFN300
D&C BLUE NO. 6 see BGB275
D and C BLUE NO. 6 see BGB275
D&C BLUE NUMBER 1 see BJI250
DCBN see DGM600
p,p-DCBP see DET850
DCC see MDR800
DCCD see MDR800
DCCI see MDR800
DCDB see DEU375
DCDD see DAC800
2,3-DCDT see DBI200
1-1-DCE see VPK000
1,2-DCE see EIY600
DCEE see DFJ050
D and C GREEN 1 see NAX500
D&C GREEN No. 4 see FAF000
D&C GREEN No. 6 see BLK000
D&C GREEN No. 8 see TNM000
DCH 21 see ERE200
DCHA see DGT600
DCHFB see DFM000
2,4-D, α-CHLOROCROTYL ESTER see
KHU025
DCI see DFN400
DCI LIGHT MAGNESIUM CARBONATE see
MAC650
DCIP see BII250
DCIP (nematocide) see BII250
DCL see DFN500
DCM see DFO000
DCM see DFO800
DCM see DGQ200
DCM see MJP450
3DCM see BGT800
DCMA see DFO800
DCMA-13-35-7 see CMS137
DCMA-13-50-9 see CMS135
DCMC see DAI000
DCMO see CCC500
DCMOD see DLV200
DCMU see DXQ500
DCNA see RDP300
DCNA (fungicide) see RDP300
DCNB see DFT600
DCNU see CLX000
d-CON see WAT200
D&C ORANGE No. 2 see TGW000

D&C ORANGE No. 3 see FAG010
D&C ORANGE No. 3 see HGC000
D&C ORANGE No. 17 see DVB800
D and C ORANGE No. 4 see CMM220
D&C ORANGE NO. 5 see DDO200
D and C ORANGE NUMBER 15 see DMG800
DCP see DFX800
DCP see DGG500
2,4-DCP see DFX800
DCPA see DGI000
DCPA see DGW400
DCPC see BIN000
DCPE see BIN000
DCPM see NCM700
cis-DCPO see DAC975
trans-DCPO see DGH500
D&C RED 2 see FAG020
D&C RED No. 3 see FAG040
D&C RED No. 5 see FMU070
D&C RED No. 9 see CHP500
D&C RED No. 14 see TGX000
D&C RED No. 19 see FAG070
D&C RED No. 21 see BMO250
D&C RED No. 22 see BNH500
D&C RED No. 36 see CJD500
D and C RED No. 8 see CMS150
D and C RED No. 10 see NAP100
D and C RED No. 33 see CMS228
D & C RED NO. 17 see OHI200
D and C RED NO. 6 see CMS155
D and C RED NO. 28 see ADG250
D and C RED NO.31 see CMS160
D and C RED NO. 35 see MMP100
2,4-D CROTYL ESTER see DAD000
D.C.S. see BGJ750
DCTA see CPB120
DCU see DGQ200
DCUK-OME see MJV900
DC-38-V see GEO200
D+C VIOLET No. 2 see HOK000
D,l-(3-CYANO-2-
HYDROXYPROPYL)TRIMETHYLAMMONI
UM CHLORIDE see COL050
dCYD see DAQ850
D&C YELLOW No. 7 see FEV000
D&C YELLOW No. 8 see FEW000
D and C YELLOW No. 10 see CMM510
D and C YELLOW NO. 5 see FAG140
D and C YELLOW NO. 11 see CMS240
DD 234 see DAB750
DDA see DRR800
DDC see DQY950
DDC see SGJ000
cis-DDCP see DAD040
trans(−)-DDCP see DAD075
trans(+)-DDCP see DAD050
DDD see BIM500
2,4'-DDD see CDN000
o,p'-DDD see CDN000
p,p'-DDD see BIM500
DDDM see MJM500
DDE see BIM750
o,'-DDE see DEV900
2,4'-DDE see DEV900
p,p'-DDE see BIM750
DDETA see HMQ500
DDFC see GCK500
2,4-D DIMETHYLAMINE SALT see DFY800
D,4-DINITRO-1-NAPHTHOL-7-SULFONIC
ACID see FBZ100
DDM see DSU000
DDM see MJM500
DDM see MJQ000
DD-METHYL ISOTHIOCYANATE
MIXTURE see MLC000
DD MIXTURE see DGG000
DDMP see MQR100
DDN see LBU200
DDNO see DRS200
DDNP see DUR800
DDNS see BIM775
DDOA see ABC250

DDP see PJD000
cis-DDP see PJD000
DDS see DJC875
DDS see DXV000
DDS see SOA500
DDS A see DXV000
DD SOIL FUMIGANT see DGG000
DDT see DAD200
o,p'-DDT see BIO625
p,p'-DDT see DAD200
DDT DEHYDROCHLORIDE see BIM750
DDT WARF ANTIRESISTANT see DDW500
DDVF see DGP900
DDVP see DGP900
D.E. see DCJ800
DE 79 see OAF200
DE-79 see OAF200
DE 83R see PAU500
DEA see DHF000
DEA see DIS700
DEACETYLANDROMEDOTOXIN see GJU310
DEACETYLANGUIDIN see ECT700
4-DEACETYLANGUIDIN see ECT700
DEACETYLANHYDROANDROMEDOTOXIN see GJU300
DEACETYLCHOLCHICEINE see TLN750
N-DEACETYLCHOLCHICEINE see TLN750
DEACETYLCOLCHICINE see TLO000
N-DEACETYLCOLCHICINE see TLO000
DEACETYLCOLCHICINE l-TARTRATE see DBA175
N-DEACETYLCOLCHICINE l-TARTRATE(1:1), HYDRATE see DBA175
N-DEACETYL-10-DEMETHOXY-10-METHYLTHIOCOLCHICINE see DBA200
DEACETYLDEMETHYLTHYMOXAMINE see DAD500
DEACETYL-HT-2 TOXIN see DAD600
DEACETYLLANATOSIDE B see DAD650
DEACETYL-LANATOSIDE B (8CI) see DAD650
DEACETYLLANATOSIDE C see DBH200
DEACETYLLYONIATOXIN see DAD700
3-DE(2-(ACETYLMETHYLAMINO)PROPIONYLOXY)-3-HYDROXYMAYTANSINE ISOVALERATE (ESTER) see APE529
DEACETYLMETHYLCOLCHICINE see MIW500
DEACETYL-N-METHYLCOLCHICINE see MIW500
N-DEACETYL-N-METHYLCOLCHICINE see MIW500
N-DEACETYLMETHYLTHIOCOLCHICINE see DBA200
DEACETYLMULDAMINE see DAD800
N-DEACETYLTHIOCOLCHICINE see DBA200
DEACETYLTHYMOXAMINE see DAD850
DEACTIVATOR E see DJD600
DEACTIVATOR H see DJD600
DEACYLASEBOTOXIN I see GJU310
DEADLY NIGHTSHADE see BAU500
DEADLY NIGHTSHADE see DAD880
DEAD MEN'S FINGERS see WAT315
DEADOPA see DNA200
DEAE see DHO500
DEAE-D see DHW600
DEALCA TP1 see GLU000
DEALCA TP2 see GLU000
DEALKYLPRAZEPAM see CGA500
DEAMELIN S see GHR609
3'-DEAMINO-3'-(3-CYANO-4-MORPHOLINYL)DOXORUBICIN see COP765
DEAMINO-DICARBA-(GLY⁷)-OXYTOCIN see CCK550
DEAMINOHYDROXYTUBERCINDIN see DAE200
DEAMINOISOCYTOSINE see ORS050
DEAMINOISOCYTOSINE see PPP140

3'-DEAMINO-3'-MORPHOLINO-ADRIAMYCIN see MRT100
3'-DEAMINO-3'-(4-MORPHOLINYL)DAUNORUBICIN see MRT100
DEANER see DOZ000
DEA/NO see DHJ300
DEANOL see DOY800
DEANOL ACETAMIDOBENZOATE see DOZ000
DEANOL-p-ACETAMIDOBENZOATE see DOZ000
DEANOL-p-CHLOROPHENOXYACETATE see DPE000
DEANOLESTERE see DPE000
DEANOX see IHC450
DEA OXO-5 see DBA800
DEAPASIL see AMM250
DEASERPYL see MEK700
DEATH CAMAS see DAE100
DEATH-OF-MAN see WAT325
7-DEAZAADENOSINE see TNY500
7-DEAZAADENOSINE-7-CARBOXAMIDE see SAU000
7-DEAZAINOSINE see DAE200
DEB see BGA750
DEB see DKA600
2,4-DEB see SCB200
DEBA see BAG000
DEBANTIC see RAF100
DEBECACIN see DCQ800
DEBECACIN SULFATE see PAG050
DEBECILLIN see BFC750
DEBECYLINA see BFC750
DEBENAL see PPP500
DEBENAL-M see ALF250
DEBENDOX see BAV350
DEBENDRIN see BBV500
DEBETROL see SKJ300
DEBLASTON see PIZ000
DEBRICIN see FBS000
DEBRIDAT see TKU675
DEBRISOQUIN HYDROBROMIDE see DAI475
DEBRISOQUIN SULFATE see IKB000
DEBROUSSAILLANT 600 see DAA800
DEBROUSSAILLANT CONCENTRE see TAA100
DEBROXIDE see BDS000
DEC see DAE800
DEC see DIX200
DECABANE see DER800
DECABORANE see DAE400
DECABORANE(14) see DAE400
DECABROMOBIPHENYL see PCC480
2,2',3,3',4,4',5,5',6,6'-DECABROMOBIPHENYL see PCC480
DECABROMOBIPHENYL ETHER see PAU500
DECABROMOBIPHENYL OXIDE see PAU500
DECABROMODIPHENYL see PCC480
DECABROMODIPHENYL OXIDE see PAU500
DECABROMOPHENYL ETHER see PAU500
DECACHLOR see DAE425
DECACHLOROBI-2,4-CYCLOPENTADIEN-1-YL see DAE425
1,1',2,2',3,3',4,4',5,5'-DECACHLOROBI-2,4-CYCLOPENTADIEN-1-YL see DAE425
1,2,3,5,6,7,8,9,10,10-DECACHLORO(5.2.1.0²·⁶.0³·⁹.0⁵·⁸)DECANO-4-ONE see KEA000
DECACHLOROKETONE see KEA000
DECACHLORO-1,3,4-METHENO-2H-CYCLOBUTA(cd)PENTALEN-2-ONE see KEA000
DECACHLOROOCTAHYDROKEPONE-2-ONE see KEA000
DECACHLOROOCTAHYDRO-1,3,4-METHENO-2H-

CYCLOBUTA(cd)PENTALEN-2-ONE see KEA000
1,1a,3,3a,4,5,5,5a,5b,6-DECACHLOROOCTAHYDRO-1,3,4-METHENO-2H-CYCLOBUTA(cd)PENTALEN-2-ONE see KEA000
DECACHLOROPENTACYCLO(5.2.1.0²·⁶.0³·⁹.0⁵·⁸)DECAN-4-ONE see KEA000
DECACHLOROPENTACYCLO(5.3.0.0²·⁶.0⁴·¹⁰.0⁵·⁹)DECAN-3-ONE see KEA000
DECACHLOROTETRACYCLODECANONE see KEA000
DECACHLOROTETRAHYDRO-4,7-METHANOINDENEONE see KEA000
DECACIL see LFK000
DECACURAN see DAF600
DECADERM see SOW000
(2E,4E)-DECADIENAL see DAE450
(E,E)-2,4-DECADIENAL see DAE450
(2E,4E)-2,4-DECADIENAL see DAE450
trans,trans-2,4-DECADIENAL see DAE450
2,4-DECADIENOIC ACID, ETHYL ESTER, (E,Z)- see EHV100
1,6-DECADIEN-3-OL, 3,7,9-TRIMETHYL- see IIW100
DECADONIUM DIIODIDE see DAE500
DECADRON see SOW000
DECADRON-LA see DBC400
DECADRON PHOSPHATE see DAE525
DECA-DURABOL see NNE550
DECA-DURABOLIN see NNE550
DECAETHOXY OLEYL ETHER see OIG040
DECAFENTIN see DAE600
DECAFLUOROBUTYRAMIDINE see DAE625
1,1,1,2,2,3,4,5,5,5-DECAFLUOROPENTANE see DLY900
DECAHYDRO-4-α-HYDROXY-2,8,8-TRIMETHYL-2-NAPHTHOIC ACID, γ-LACTONE see LAQ100
cis-N-(DECAHYDRO-2-METHYL-5-ISOQUINOLYL)-3,4,5-TRIMETHOXYBENZAMIDE see DAE695
trans-N-(DECAHYDRO-2-METHYL-5-ISOQUINOLYL)-3,4,5-TRIMETHOXYBENZAMIDE see DAE700
DECAHYDRONAPHTHALENE see DAE800
DECAHYDRO-2-NAPHTHALENOL see DAF000
DECAHYDRONAPHTHALEN-2-OL see DAF000
DECAHYDRO-β-NAPHTHOL see DAF000
trans-DECAHYDRO-β-NAPHTHOL see DAF000
DECAHYDRO-β-NAPHTHYL ACETATE see DAF100
DECAHYDRO-β-NAPHTHYL FORMATE see DAF150
DECAHYDRONAPTHOL-2 see DAF000
DECAHYDRO-4a,7,9-TRIHYDROXY-2-METHYL-6,8-BIS(METHYLAMINO)-4H-PYRANO(2,3-b)(1,4)BENZODIOXIN-4-ONE DIHYDROCHLORIDE, (2R-(2-α,4a-β,5a-β,6-β,7-β,8-β,9-α,9a-α,10a-β))- see SLI325
Δ-DECALACTONE see DAF200
ε-DECALACTONE see BSB100
γ-N-DECALACTONE see HKA500
DECALDEHYDE see DAG000
n-DECALDEHYDE see DAG000
DECALIN see DAE800
DECALIN (DOT) see DAE800
2-DECALINOL see DAF000
DECALIN SOLVENT see DAE800
DECALINYL FORMATE see DAF150
2-DECALOL see DAF000
DECAMETHIONIUM IODIDE see DAF800
DECAMETHONIUM see DAF600
DECAMETHONIUM BROMIDE see DAF600
DECAMETHONIUM DIBROMIDE see DAF600
DECAMETHONIUM DIIODIDE see DAF800

DECAMETHONIUM IODIDE see DAF800
DECAMETHRIN see DAF300
DECAMETHRINE see DAF300
DECAMETHYLCYCLOPENTASILOXANE see DAF350
N,N'-DECAMETHYLENEBIS((1-ADAMANTYL)DIMETHYLAMMONIUM, DIIODIDE see DAE500
N⁴,N⁴-DECAMETHYLENE-BIS-(4-AMINOQUINALDINIUM)-N,N-DIHYDROACETATE see SAP600
1,1'-DECAMETHYLENEBIS(1-METHYLPIPERIDINIUM IODIDE) see DAF450
DECAMETHYLENEBIS(TRIMETHYLAMMONIUM BROMIDE) see DAF600
DECAMETHYLENE-1,10-BISTRIMETHYLAMMONIUM DIBROMIDE see DAF600
DECAMETHYLENEBIS(TRIMETHYLAMMONIUM DIIODIDE) see DAF800
DECAMETHYLENEBIS(TRIMETHYLAMMONIUM IODIDE) see DAF800
DECAMINE see DAA800
DECAMINE 4T see TAA100
1-DECAMINIUM, N-OCTYL-N,N-DIMETHYL-, CHLORIDE see OES400
DECANAL see DAG000
1-DECANAL see DAG000
n-DECANAL see DAG000
1-DECANAL (mixed isomers) see DAG200
DECANALDEHYDE see DAG000
DECANAL, DIMETHYLACETAL see AFJ700
DECANAL DIMETHYL ACETAL see DAI600
1-DECANAMINIUM, N,N-DIMETHYL-N-OCTYL-, CHLORIDE see OES400
DECANE see DAG400
n-DECANE (DOT) see DAG400
1-DECANEAMINE see DAG600
DECANE, 1-BROMO- see BNB800
1-DECANECARBOXYLIC ACID see UKA000
1,10-DECANEDIAMINE see DAG650
1,10-DECANEDIAMINE, N,N'-BIS(2-METHYL-4-QUINOLINYL)-, DIACETATE see SAP600
DECANEDINITRILE see SBK500
DECANEDIOIC ACID see SBJ500
DECANEDIOIC ACID, BIS(2-ETHYLHEXYL) ESTER see BJS250
DECANEDIOIC ACID, BIS(2-METHOXYETHYL) ESTER see DAG700
DECANEDIOIC ACID, DIBUTYL ESTER see DEH600
DECANEDIOIC ACID, compd. with 1,6-HEXANEDIAMINE (1:1) (9CI) see HEG130
1,3-DECANEDIOL see DAG750
1,10-DECANEDIOL, 2,2,9,9-TETRAMETHYL- see TDO260
DECANE, 1-(ETHENYLOXY)- see EEE700
DECANE, HENEICOSAFLUORO-1-IODO- see IEU075
DECANE, 1-IODO-1,1,2,2,3,3,4,4,5,5,6,6,7,7,8,8,9,9,10,10,10-HENEICOSAFLUORO- see IEU075
DECANE, 1-METHOXY- see MIW075
DECANOIC ACID see DAH400
n-DECANOIC ACID see DAH400
tert-DECANOIC ACID see DAH425
DECANOIC ACID-4-(4-CHLOROPHENYL)-1-(4-(4-FLUOROPHENYL)-4-OXYBUTYL)-4-PIPERIDINYL ESTER see HAG300
DECANOIC ACID, DIESTER with TRIETHYLENE GLYCOL (mixed isomers) see DAH450
DECANOIC ACID, ETHYL ESTER see EHE500
DECANOIC ACID, 4-HYDROXY-4-METHYL-, γ-LACTONE see MIW050
DECANOIC ACID, METHYL ESTER see MHY650
DECANOIC ACID, SODIUM SALT see SGC100

DECANOL see DAI600
n-DECANOL see DAI600
1-DECANOL (FCC) see DAI600
DECANOLIDE-1,4 see HKA500
DECANOLIDE-1,5 see DAF200
DECANOL (MIXED ISOMERS) see DAI800
2-DECANONE see OFE050
DECANOPHENONE, 10-FLUORO- see FKQ100
4-DECANOYLMORPHOLINE see CBF725
9-DECAOCTENOIC ACID, TRIBUTYLSTANNYL ESTER see TIA000
DECAPRYN see PGE775
DECAPRYN SUCCINATE see PGE775
DECAPS see VSZ100
DECARBAMOYLMITOMYCIN C see DAI000
10-DECARBAMOYLMITOMYCIN C see DAI000
DECARBAMYLMITOMYCIN C see DAI000
DECARBOFURAN see DLS800
DECARBORINENE see DEJ500
m-DECARBOROCARBORANE see NBV100
2-DECARBOXAMIDO-2-ACETYL-4-DESDIMETHYLAMINO-4-AMINO-9-METHYL-5A,6-ANHYDROTETRACYCLINE see CDF250
DECARBOXYCYSTEINE see AJT250
DECARBOXYCYSTINE see MCN500
DECARBOXYFENVALERATE see DAI100
DECARIS see LFA020
DECARPYN SUCCINATE (1:1) see PGE775
DECASERPIL see MEK700
DECASERPINE see MEK700
DECASERPYL see MEK700
DECASERPYL PLUS see MEK700
DECASONE see SOW000
DECASPIRIDE HYDROCHLORIDE see DAI200
DECASPRAY see SOW000
DECATOL see IOO310
n-DECATYL ALCOHOL see DAI600
DECCO SALT NO 5 see TNE775
DECCOTANE see BPY000
DECCOX see DAI495
DECELITH H see PKQ059
DECEMTHION P-6 see PHX250
2-DECENAL see DAI350
cis-4-DECENAL see DAI360
trans-2-DECEN-1-AL see DAI350
DECENALDEHYDE see DAI350
cis-4-DECEN-1-AL (FCC) see DAI360
9-DECEN-1-OL see DAI400
1-DECEN-10-OL see DAI400
ω-DECENOL see DAI400
9-DECEN-1-OL, ACETATE see DAI450
DECENTAN see CJM250
DECENYL ACETATE see DAI450
9-DECENYL ACETATE see DAI450
DECHAN see DGU200
DECHLORANE 605 see DAI460
DECHLORANE 4070 see MQW500
DECHLORANE A-O see AQF000
DECHLORANE PLUS see DAI460
DECHLORANE PLUS 515 see DAI460
DECHLORANE PLUS 2520 see DAI460
DECHLOROETHYLCYCLOPHOSPHAMIDE see MRG025
N-DECHLOROETHYLCYCLOPHOSPHAMIDE see MRG025
3-DECHLOROETHYLIFOSFAMIDE see MRG025
DECHOLIN see DAL000
DECHOLIN SODIUM SALT see SGD500
DECICAINE see TBN000
DECIMEMIDE see DAJ400
DECINCAN see VLF000
DECIS see DAF300
DECLINAX see DAI475
DECLINAX see IKB000
DECLOMYCIN see MIJ500

DECLOMYCIN HYDROCHLORIDE see DAI485
DECLOXIZINE see BBV750
DECOFOL see BIO750
n-DECOIC ACID see DAH400
DECONTRACTIL see GGS000
DECOQUINATE see DAI495
DECORPA see GLU000
DECORTANCYL see PLZ000
DECORTIN see DAQ800
DECORTIN see PLZ000
DECORTIN H see PMA000
DECORTISYL see PLZ000
DECORTON see DAQ800
DECOSERPYL see MEK700
DECOSTERONE see DAQ800
DECOSTRATE see DAQ800
DECROTOX see COD000
DECTAN see DBC400
DECTANCYL see SOW000
DECURVON see IDF300
DE-CUT see DMC600
DECYL ACRYLATE see DAI500
n-DECYL ACRYLATE see DAI500
DECYL ALCOHOL see DAI600
n-DECYL ALCOHOL see DAI600
DECYL ALCOHOL (mixed isomers) see DAI800
DECYL ALDEHYDE see DAG000
1-DECYL ALDEHYDE see DAG000
1-DECYL ALDEHYDE see UJJ000
n-DECYL ALDEHYDE see DAG000
DECYLALDEHYDE DMA see AFJ700
DECYLAMINE see DAG600
DECYL BENZENE SODIUM SULFONATE see DAJ000
DECYL BROMIDE see BNB800
1-DECYL BROMIDE see BNB800
n-DECYL BROMIDE see BNB800
DECYL BUTYL PHTHALATE see BQX250
DECYL CHLORIDE (mixed isomers) see DAJ200
N-DECYL-N,N-DIMETHYL-1-DECANAMINIUM CHLORIDE (CI) see DGX200
DECYL DIMETHYL OCTYL AMMONIUM CHLORIDE see OES400
DECYLENIC ALCOHOL see DAI400
1-DECYL-1-ETHYLPIPERIDINIUM BROMIDE see DAJ300
α-DECYL-ω-HYDROXYPOLY(OXY-1,2-ETHANEDIYL) see PKE390
α-DECYL-ω-HYDROXYPOLY(OXY-1,2-ETHANEDIYL) PHOSPHATE POTASSIUM SALT see PKE800
DECYLIC ACID see DAH400
n-DECYLIC ACID see DAH400
DECYLIC ALCOHOL see DAI600
DECYLIC ALDEHYDE see DAG000
DECYL METHYL ETHER see MIW075
cis-2-DECYL-3-(5-METHYLHEXYL)OXIRANE see ECB200
DECYL OCTYL ALCOHOL see OAX000
DECYLOCTYLDIMETHYLAMMONIUM CHLORIDE see OES400
DECYL OCTYL PHTHALATE see OEU000
n-DECYL n-OCTYL PHTHALATE see OEU000
4-(DECYLOXY)-3,5-DIMETHOXYBENZAMIDE see DAJ400
4-n-DECYLOXY-3,5-DIMETHOXYBENZOIC ACID AMIDE see DAJ400
6-DECYLOXY-7-ETHOXY-4-HYDROXY-3-QUINOLINECARBOXYLID ACID ETHYL ESTER see DAI495
4-DECYLOXY-2-HYDROXYPHENYL 4-DECYLOXYPHENYL KETONE see DAJ450
n-DECYLSUCCINIC ANHYDRIDE see DAJ475
2-DECYL-2-THIOPSEUDOUREA, MONOHYDROCHLORIDE see DAJ480

S-DECYLTHIOUREA CHLORIDE see DAJ480

DECYLTHIOURONIUM CHLORIDE see DAJ480

DECYLTHIURONIUM CHLORIDE see DAJ480

S-DECYLTHIURONIUM CHLORIDE see DAJ480

DECYLTRIMETHYLAMMONIUM BROMIDE see DAJ500

DECYLTRIPHENYLPHOSPHONIUM BROMOCHLOROTRIPHENYLSTANNATE see DAE600

(DECYL-TRIPHENYL-PHOSPHONIUM)-TRIPHENYL-BROM-CHLOR-STANNAT (GERMAN) see DAE600

DECYL VINYL ETHER see EEE700

DEDC see SGJ000

DEDELO see DAD200

DEDEVAP see DGP900

DEDK see SGJ000

DEDORAN see MCI500

DED-WEED see CIR250

DED-WEED see DAA800

DED-WEED see DGI400

DED-WEED see TIX500

DED-WEED BRUSH KILLER see TAA100

DED-WEED CRABGRASS KILLER see PLC250

DED-WEED LV-69 see DAA800

DED-WEED LV-6 BRUSH KIL and T-5 BRUSH KIL see TAA100

DEDYL see DNM400

DEE-OSTEROL see VSZ100

DEEP CRIMSON MADDER 10821 see DMG800

DEEP FASTONA RED see MMP100

9,10-DEEPITHIO-9,10-DIDEHYDROACANTHIFOLICIN see OHK100

DEEP LEMON YELLOW see SMH000

DEER BERRY see HGF100

DEE-RON see VSZ100

DEE-RONAL see VSZ100

DEE-ROUAL see VSZ100

DEER'S TONGUE see DAJ800

DEERTONGUE INCOLORE see DAJ800

DEET see DKC800

16-DEETHYL-3-o-DEMETHYL-16-METHYL-3-o-(1-OXOPROPYL)MONENSIN see DAK000

N-DEETHYLDORZOLAMIDE HYDROCHLORIDE see DAK100

DEETHYL HYCANTHONE see EGA600

DEETILATO METOCLOPRAMIDE (ITALIAN) see AJH125

DEETILMETOCLOPRAMIDE (ITALIAN) see AJH125

DEF see BSH250

DEF DEFOLIANT see BSH250

DEFEKTON see CCK775

DEFEKTON see CCK780

DE-FEND see DSP400

DEFERIPRONE see DSI250

DEFEROXAMINE see DAK200

DEFEROXAMINE MESILATE see DAK300

DEFEROXAMINE MESYLATE see DAK300

DEFEROXAMINE METHANESULFONATE see DAK300

DEFEROXAMINUM see DAK200

DEFERRIOXAMINE see DAK200

DEFERRIOXAMINE B see DAK200

DEFIBRASE see RDA350

DEFIBRASE R see RDA350

DEFICOL see PPN100

DEFILIN see DJL000

DEFILTRAN see AAI250

DEFLAMENE see FDB000

DEFLAMON-WIRKSTOFF see MMN250

DEFLEXOL see AJF500

DEFLOGIN see HNI500

DEFLORIN see TEY000

DEFOAMER S-10 see SCP000

DE-FOL-ATE see MAE000

DE-FOL-ATE see SFS000

DEFOLIANT 713 see DKE400

DEFOLIANT MN see CDU100

DEFOLIANT 2929 RP see TFH500

DEFOLIT see TEX600

DEFONIN see ILD000

DEFRADIN HYDRATE see SBN450

DEFTOR see MQR225

DEFY see DFY800

DEG see DJD600

DEGALAN S 85 see PKB500

DEGALOL see DAQ400

DEGLARESIN see MCB050

DEGLARESIN N 12 see MCB050

DEGLYCOSYLATED HCG see DAK325

DEGLYCOSYLATED HUMAN CHORIONIC GONADOTROPIN see DAK325

DEGMVE (RUSSIAN) see DJG400

DEGRANOL see MAW500

DEGRASSAN see DUS700

DE-GREEN see BSH250

DEGUELIA ROOT see DBA000

DEGUSSA see CBT750

DEH see DHL850

DEH see HHH000

D.E.H. 20 see DJG600

DEH 24 see TJR000

D.E.H. 26 see TCE500

DEHA see AEO000

DEHA see DJN000

DEHACODIN see DKW800

DEHERBAN see DAA800

DEHIDROBENZPERIDOL see DYF200

DEHISTIN see TMP750

DEHISTIN HYDROCHLORIDE see POO750

DEHP see DVL700

DEHPA EXTRACTANT see BJR750

DEHYCHOL see DAL000

DEHYDOL LS 4 see DXY000

DEHYDRACETIC ACID see MFW500

DEHYDRATIN see AAI250

DEHYDRITE see PCE000

DEHYDROABIETIC ACID see DAK400

DEHYDRO-ABIETIC ACID-2-(2-(CHLOROETHYL)AMINO)ETHYL ESTER see CGQ250

DEHYDROACETIC ACID (FCC) see MFW500

DEHYDROACETIC ACID, SODIUM SALT see SGD000

Δ^6-DEHYDRO-17-ACETOXYPROGESTERONE see MCB375

Δ^6-DEHYDRO-17-α-ACETOXYPROGESTERONE see MCB375

5,6-DEHYDRO-N-ACETYLLOLINE see DAK450

trans-DEHYDROANDROSTERONE see AOO450

trans-DEHYDROANDROSTERONE ACETATE see HJB250

DEHYDROANDROSTERONE SULFATE see SNV100

DEHYDROBENZPERIDOL see DYF200

DEHYDROBROWNIINE see DAK500

14-DEHYDROBROWNIINE see DAK500

DEHYDROCHLORAMPHENICOL see AAI110

6-DEHYDRO-6-CHLORO-17-α-ACETOXYPROGESTERONE see CBF250

DEHYDROCHOLATE SODIUM see SGD500

7-DEHYDROCHOLESTERIN see DAK600

7-DEHYDROCHOLESTERIN (GERMAN) see DAK600

DEHYDROCHOLESTEROL see DAK600

7-DEHYDROCHOLESTEROL see DAK600

7-DEHYDROCHOLESTEROL ACETATE see DAK800

7-DEHYDROCHOLESTERYL ACETATE see DAK800

7-DEHYDROCHOLESTROL, ACTIVATED see CMC750

DEHYDROCHOLIC ACID see DAL000

DEHYDROCHOLIC ACID, SODIUM SALT see SGD500

DEHYDROCHOLSAEURE (GERMAN) see DAL000

Δ^1-DEHYDROCORTISOL see PMA000

1-DEHYDROCORTISONE see PLZ000

Δ-1-DEHYDROCORTISONE see PLZ000

Δ^1-DEHYDROCORTISONE ACETATE see PLZ100

DEHYDROCURDIONE see DAL010

DEHYDROEPIANDROSTERONE see AOO450

5-DEHYDROEPIANDROSTERONE see AOO450

DEHYDROEPIANDROSTERONE ACETATE see HJB225

DEHYDROEPIANDROSTERONE ACETATE see HJB250

DEHYDROEPIANDROSTERONE MONOSULFATE see SNV100

DEHYDROEPIANDROSTERONE SODIUM SULFATE DIHYDRATE see DAL030

DEHYDROEPIANDROSTERONE-SULFATE see SNV100

DEHYDROEPIANDROSTERONE 3-SULFATE see SNV100

DEHYDROEPIANDROSTERONE 3-β-SULFATE see SNV100

DEHYDROEPIANDROSTERONE SULFATE SODIUM see DAL040

DEHYDROERGOTAMINE see DLK800

DEHYDROFOLLICULINIC ACID see BIT000

DEHYDROHELIOTRIDINE see DAL060

DEHYDROHELIOTRINE see DAL100

1-DEHYDROHYDROCORTISONE see PMA000

Δ^1-DEHYDROHYDROCORTISONE see PMA000

11-DEHYDRO-17-HYDROXYCORTICOSTERONE see CNS800

11-DEHYDRO-17-HYDROXYCORTICOSTERONE ACETATE see CNS825

11-DEHYDRO-17-HYDROXYCORTICOSTERONE-21-ACETATE see CNS825

DEHYDROISOANDROSTERONE see AOO450

5,6-DEHYDROISOANDROSTERONE see AOO450

DEHYDROISOANDROSTERONE ACETATE see HJB250

DEHYDROISOANDROSTERONE 3-ACETATE see HJB250

DEHYDROISOANDROSTERONE SULFATE see SNV100

DEHYDROLINALOOL see LFY333

DEHYDRO-β-LINALOOL see LFY333

6-DEHYDRO-6-METHYL-17-α-ACETOXYPROGESTERONE see VTF000

DEHYDRO-3-METHYLCHOLANTHRENE see DAL200

1,2-DEHYDRO-3-METHYLCHOLANTHRENE see DAL200

6-DEHYDRO-16-METHYLENE-6-METHYL-17-ACETOXYPROGESTERONE see MCB380

1-DEHYDRO-16-α-METHYL-9-α-FLUOROHYDROCORTISONE see SOW000

DEHYDROMETHYLTESTERONE see DAL300

1-DEHYDRO-17-α-METHYLTESTOSTERONE see DAL300

DEHYDROMONOCROTALINE see DAL350

DEHYDRONIVALENOL see VTF500

DEHYDRONIVALENOL MONOACETATE see ACH075

DEHYDROPHELOMYCIN D1 see BLY760

l-3,4-DEHYDROPROLINE see DAL375

11-DEHYDROPROSTAGLANDIN F2-α see POC275

DEHYDRORETRONECINE see DAL400

6-DEHYDRO-RETRO-PROGESTERONE see DYF759

DEHYDROSTILBESTROL see DAL600

DEHYDRO-p-TOLUIDINE see MHJ300

Δ³-DEHYDRO-3,4-TRIMETHYLENE-ISOBENZANTHRENE-2 see BCI000

DEHYQUART A see HCQ525

DEHYQUART C see DXY725

DEHYQUART C see LBX050

DEHYQUART CBB see BEL900

DEHYQUART CDB see BEL900

DEHYQUART LT see LBX075

DEHYQUART STC-25 see DTC600

DEHYSTOLIN see DAL600

DEIDROBENZPERIDOLO see DYF200

DEIDROCOLICO VITA see DAL000

DEINAIT see CJR909

DEIQUAT see DWX800

DEISOVALERYL BLASTMYCIN see DAM000

DEJO see DJT800

DEK see DJN750

DE-KALIN see DAE800

DEKALINA (POLISH) see DAE800

DEKAMETHYLCYKLOPENTASILOXAN see DAF350

N⁴,N⁴-DEKAMETHYLEN-BIS-(4-AMINOCHINALDINIUM)-N,N-DIHYDROACETAT see SAP600

DEKAMETRIN (HUNGARIAN) see DAF300

DEKENE see DEJ500

DEKORTIN see PLZ000

DEKRYSIL see DUS700

DEKSONAL see DOU600

DEL see AJH000

DELACHLOR see IIE200

DELACHLORE see IIE200

DELAC J see DWI000

DELAC S see CPI250

DELACURARINE see TOA000

DELADIOL see EDS100

DELADROXONE see DAM300

DELAGIL see CLD000

DELAGIL see CLD250

DELAHORMONE UNIMATIC see EDS100

DELALUTIN see HNT500

DELAN see DLK200

DELAN-COL see DLK200

DELAPHOS see ZJS400

DELAPHOS 2M see ZJS400

DELAPRIL HYDROCHLORIDE see DAM315

DELARTINE see MLD140

DELATESTRYL see MPN500

DELATESTRYL see TBF750

DELATESTRYL and DEPO-MEDROXYPROGESTERONE ACETATE see DAM325

DELATESTRYL and DEPO-PROVERA see DAM325

DELAVAN see MHB500

DELCORTIN see PLZ100

DELCORTOL see PMA000

DELEAF DEFOLIANT see TIG250

DELESTROGEN see EDS100

DELESTROGEN 4X see EDS100

DELGESIC see ADA725

DELICIA see AHE750

DELICIA see PGY000

DELIVA see ARQ750

DELLIPSOIDS see BBK500

DELMOFULVINA see GKE000

DELMONEURINA see BET000

DELNAV see DVQ709

DELONIN AMIDE see NCR000

DELOWAS S see TBD000

DELOWAX OM see TBD000

DELPET 50M see PKB500

DELPHELATINE see EAI050

DELPHENE see DKC800

m-DELPHENE see DKC800

DELPHINIC ACID see ISU000

DELPHINIDOL see DAM400

DELPHINIUM see LBF000

DELSEMIDINE see MLD140

DELSEMINE see DAM410

DELSENE see MHC750

DELSENE M see MAS500

DELSINE (6CI,7CI) see LJB100

DELSTEROL see CMC750

DELTA see CJJ000

DELTA-CORTEF see PMA000

DELTACORTELAN see PLZ000

DELTACORTENOL see PMA000

DELTACORTISONE see PLZ000

DELTACORTONE see PLZ000

DELTACORTRIL see PMA000

DELTA-DOME see PLZ000

DELTA F see PMA000

DELTAFLUORENE see SOW000

DELTAGLUCONOLACTONE see GFA200

DELTALIN see VSZ100

DELTALINE see EAI050

DELTALONE see PLZ100

DELTAMETHRIN see DAF300

trans-DELTAMETHRIN see DAM430

DELTAMINE 6-ACETATE see EAI050

DELTA-MVE see MKB750

DELTAMYCIN A see CBT250

DELTAN see DUD800

DELTANET see FPO200

DELTA-STAB see PMA000

DELTATHIONE see GFW000

DELTAZINA see PPP500

DELTILEN see SOV100

DELTISONE see PLZ000

DELTOIN see ENC500

DELTOSIDE see AHK750

DELTRATE-20 see PBC250

DELTYL see IQW000

DELTYLEXTRA see IQN000

DELTYL PRIME see IQW000

DELURSAN see DMJ200

DELUSSA BLACK FW see CBT750

DELVEX see DJT800

DELVINAL SODIUM see VKP000

DELYSID see DJO000

DEM see AJH125

DEMA see BIE500

DEMAROL see DAM600

DEMASORB see DUD800

DEMAVET see DUD800

DEMECARIUM BROMIDE see DAM625

DEMECLOCYCLINE see MIJ500

DEMECLOCYCLINE HYDROCHLORIDE see DAI485

DEMECOLCIN see MIW500

DEMECOLCINE see MIW500

DEMEFLINE see DNV200

DEMEKARIUM BROMIDE see DAM625

DEMEKASTIGMINE BROMIDE see DAM625

DEMEPHION see MIW250

DEMEROL see DAM600

DEMEROL see DAM700

DEMEROL HYDROCHLORIDE see DAM700

DEMESO see DUD800

DEMETHON-METHYL (MAK) see MIW100

4-DEMETHOXYADRIAMYCIN see DAM800

DEMETHOXYAGERATOCHROMENE see DAM830

6-DEMETHOXYAGERATOCHROMENE see DAM830

4-DEMETHOXYDAUNOMYCIN see DAN000

4-DEMETHOXYDAUNORUBICIN see DAN000

4-DEMETHOXYDAUNORUBICIN HYDROCHLORIDE see DAN100

11-DEMETHOXYRESERPINE see RDF000

N-DEMETHYLACLACINOMYCIN A see DAN200

DEMETHYLAMITRIPTYLENE see NNY000

5-O-DEMETHYLAVERMECTIN ALA see ARW150

DEMETHYLCHLOROTETRACYCLINE see MIJ500

6-DEMETHYLCHLOROTETRACYCLINE see MIJ500

6-DEMETHYL-7-CHLOROTETRACYCLINE see MIJ500

DEMETHYLCHLOROTETRACYCLINE HYDROCHLORIDE see DAI485

DEMETHYLCHLORTETRACYCLIN see MIJ500

DEMETHYLCHLORTETRACYCLINE see MIJ500

6-DEMETHYLCHLORTETRACYCLINE see MIJ500

6-DEMETHYL-7-CHLORTETRACYCLINE see MIJ500

DEMETHYLCHLORTETRACYCLINE BASE see MIJ500

DEMETHYLCHLORTETRACYCLINE HYDROCHLORIDE see DAI485

2-DEMETHYLCOLCHICINE see DAN300

O²-DEMETHYLCOLCHICINE see DAN300

O¹⁰-DEMETHYLCOLCHICINE see ADE000

3-DEMETHYLCOLCHICINE GLUCOSIDE see DAN375

o-DEMETHYLDAUNOMYCIN see KBU000

(6R,25)-5-o-DEMETHYL-28-DEOXY-6,28-EPOXY-25-(1-METHYLETHYL)MILBEMYCIN B see MQT600

4'-DEMETHYLDEOXYPODOPHYLLOTOXIN see DAN388

4'-DEMETHYLDESOXYPODOPHYLLOTOXIN see DAN388

(11R(2R,5S,6R),12R)-10-DEMETHYL-19-DE((TETRAHYDRO-5-METHOXY-6-METHYL-2H-PYRAN-2-YL)OXY)-12-METHYL-11-o-(TETRAHYDRO-5-METHOXY-6-METHYL-2H-PYRAN-2-YL)-DIANEMYCIN see LEJ700

DEMETHYLDIAZEPAM see CGA500

1-DEMETHYLDIAZEPAM see CGA500

N-DEMETHYLDIAZEPAM see CGA500

DEMETHYLDOPAN see BIA250

DEMETHYL-EPIODOPHYLLOTOXIN ETHYLIDENE GLUCOSIDE see EAV500

4-DEMETHYLEPIODOPHYLLOTOXIN-β,d-ETHYLIDENEGLUCOSIDE see EAV500

4'-DEMETHYLEPIPODOPHYLLOTOXIN-9-(4,6-O-ETHYLIDENE-β-d-GLUCOPYRANOSIDE see EAV500

4'-DEMETHYLEPIPODOPHYLLOTOXIN ETHYLIDENE-β,d-GLUCOSIDE see EAV500

4-DEMETHYL-EPIPODOPHYLLOTOXIN-β,d-ETHYLIDEN-GLUCOSIDE see EAV500

4'-DEMETHYLEPIPODOPHYLLOTOXIN-9-(4,6-O-2-THENYLIDENE-β-d-GLUCOPYRANOSIDE see EQP000

4'-DEMETHYL-EPIPODOPHYLLOTOXIN-β-d-THENYLIDENE-GLUCOSIDE see EQP000

4'-O-DEMETHYL-1-O-(4,6-O-ETHYLIDENE-β,d-GLUCOPYRANOSYL)EPIPODOPHYLLOTOXIN see EAV500

5-o-DEMETHYL-26-HYDROXYAVERMECTIN A1A see HJE525

5-o-DEMETHYL-26-HYDROXY-4"-o-((2-METHOXYETHOXY)METHYL)AVERMECTIN A1A see DAN400

DEMETHYLIMIPRAMINE see DSI709

DEMETHYLMISONIDAZOLE see NHI000

1'-DEMETHYL-(s)-NICOTINE see NNR500

N-DEMETHYLORPHENADRINE HYDROCHLORIDE see TGJ250

4'-DEMETHYL 1-O-(4,6-O,O-(2-THENYLIDENE)-β-d-GLUCOPYRANOSYL)EPIPODOPHYLLOTOXIN see EQP000

1-DEMETHYLTOXOFLAVINE see DAN450

6-DEMETIL-7-CLOROTETRACICLINA see MIJ500

DEMETON see DAO500

DEMETON see DAO600

DEMETON METHYL see MIW100

DEMETON-O-METHYL see DAO800

DEMETON-S-METHYL see DAP400

DEMETON-S-METHYLSULFON (GERMAN) see DAP600

DEMETON-S-METHYLSULFONE see DAP600

DEMETON-S-METHYL-SULFOXID (GERMAN) see DAP000

DEMETON-O-METHYL SULFOXIDE see DAP000

DEMETON-S-METHYL SULFOXIDE see DAP000

DEMETON-S-METHYL-SULPHONE see DAP600

DEMETON-METHYL SULPHOXIDE see DAP000

DEMETON-O-METILE (ITALIAN) see DAO800

DEMETON-S-METILE (ITALIAN) see DAP400

DEMETON-O see DAO500

DEMETON-O + DEMETON-S see DAO600

DEMETON-S see DAP200

DEMETRACICLINA see DAI485

DEMETRIN see DAP700

DEMISE see DFY800

DEMOCRACIN see TBX000

DEMOLOX see AOA095

DEMORPHAN see DBE200

DEMOSAN see CJA100

DEMOS-L40 see DSP400

DEMOTIL see DAP800

DEMOX see DAO600

DEMSODROX see DUD800

DEMULEN see DAP810

DEN see NJW500

DENA see NJW500

DENAMONE see VSZ450

DENAPONE see CBM750

DENAPON, NITROSATED (JAPANESE) see NBJ500

DENATONIUM BENZOATE see DAP812

DENATURED ALCOHOL (DOT) see AFJ000

DENATURED SPIRITS see AFJ000

DENDREPAR see PEM750

DENDRID see DAS000

DENDRITIS see SFT000

DENDROASPIS ANGUSTICEPS VENOM see DAP815

DENDROASPIS JAMESONI VENOM see DAP820

DENDROASPIS POLYLEPIS POLYLEPIS TOXIN I see THI252

DENDROASPIS VIRIDIS VENOM see GJU500

DENDROBACILLIN see DAP822

DENDROBAN-12-ONE HYDROCHLORIDE see DAP825

DENDROBINE HYDROCHLORIDE see DAP825

DENDROCALAMUS MEMBRANACEUS Munro, extract excluding roots see DAP840

DENEGYT see DAJ400

DENKA F 90 see SCK600

DENKA FB 44 see SCK600

DENKALAC 61 see AAX175

DENKA QP3 see SMQ500

DENKA VINYL SS 80 see PKQ059

DE-NOL see BKX800

DE-NOLTAB see BKX800

DENOPAMINE see DAP850

DENSIC C 500 see SCQ000

DENSINFLUAT see TIO750

DENUDATINE see DAP875

DENVER RESEARCH CENTER No. DRC-4575 see AMU125

DENYL see DKQ000

DENYL see DNU000

DENYLSODIUM see DNU000

DENZIMOL HYDROCHLORIDE see DAP880

DEOBASE see DAP900

DEOBASE see KEK000

DEODOPHYLL see CKN000

DEODORIZED KEROSENE see DAP900

DEODORIZED KEROSINE see DAP900

DEODORIZED WINTERIZED COTTONSEED OIL see CNU000

DEOFED see DBA800

DEOFED see MDT600

DEORLENE GREEN JJO see BAY750

DEORLENE YELLOW 2G see CMM890

DEOSAN see SHU500

DEOVAL see DAD200

DEOXIN-1 see GFG100

11-DEOXO-12-β,13-α-DIHYDRO-11-α-HYDROXYJERVINE see DAQ000

11-DEOXO-12-β,13-α-DIHYDRO-11-β-HYDROXYJERVINE see DAQ002

9-DEOXO-16,16-DIMETHYL-9-METHYLENE-PGE2 see DAQ100

9-DEOXO-16,16-DIMETHYL-9-METHYLENEPROSTAGLANDIN E2 see DAQ100

11-DEOXOGLYCYRRHETINIC ACID HYDROGEN MALEATE SODIUM SALT see DAQ110

11-DEOXOJERVINE see CPT750

11-DEOXOJERVINE-4-EN-3-ONE see DAQ125

DEOXYADENOSINE see DAQ200

2'-DEOXYADENOSINE see DAQ200

3'-DEOXYADENOSINE see DAQ225

DEOXYADENYLIC ACID POLYMER see PJQ400

2-DEOXY-3-ARABINO-HEXOSE see DAR600

2-DEOXY-d-ARABINO-HEXOSE see DAR600

6-DEOXY-6-AZIDODIHYDROISOCODEINE see ASE875

6-DEOXY-6-AZIDODIHYDROISOMORPHINE see ASG675

DEOXYBENZOIN see PEB000

7-DEOXY-7(S)-CHLOROLINCOMYCIN HYDROCHLORIDE see CHU050

DEOXYCHOLATE SODIUM see SGE000

DEOXYCHOLATIC ACID see DAQ400

7-α-DEOXYCHOLIC ACID see DAQ400

DEOXYCHOLIC ACID (FCC) see DAQ400

DEOXYCHOLIC ACID SODIUM SALT see SGE000

4'-DEOXYCIRRAMYCIN A¹ see RMF000

o-3-DEOXY-4-C-METHYL-3-(METHYLAMINO)-β-l-ARABINOPYRANOSYL-(1-6)-o-(2,b-DIAMINO-2,3,4,6-TETRADEOXY-α-d-glycero-HEX-4-ENOPYRANOSYL-(1-4))-2-DEOXY-N¹-ETHYL-d-STREPTAMINE SULFATE (2:5) (salt) see NCP550

o-3-DEOXY-4-C-METHYL-3-(METHYLAMINO)-β-l-ARABINOPYRANOSYL-(1-6)-o-(2,6-DIAMINO)-2,3,4,6-TETRADEOXY-α-d-glycero-HEX-4-ENOPYRANOSYL-(1-4)-2-DEOXY-d-STREPTAMINE see SDY750

o-3-DEOXY-4-C-METHYL-3-(METHYLAMINO)-β-l-ARABINOPYRANOSYL-(1-6)-o-(2,6-DIAMINO)-2,3,4,6-TETRADEOXY-α-d-glycero-HEX-4-ENDOPYRANOSYL-(1-4)-2-DEOXY-d-STREPTAMINE HYDROCHLORIDE see SDY755

DEOXYCOFORMYCIN see PBT100

2'-DEOXYCOFORMYCIN see PBT100

11-DEOXYCORTICOSTERONE see DAQ600

11-DEOXYCORTICOSTERONE ACETATE see DAQ800

11-DEOXYCORTICOSTERONE-21-ACETATE see DAQ800

17-DEOXYCORTISOL see CNS625

DEOXYCORTONE ACETATE see DAQ800

DEOXYCYTIDINE see DAQ850

2'-DEOXYCYTIDINE see DAQ850

N-DEOXYDEMOXAPAM see CGA500

2-((2-DEOXY-2-(3,6-DIAMINOHEXANAMIDO)-α-d-GLUOPYRANOSYL)AMINO)-3,3a,5,6,7,7a-HEXAHYDRO-7-HYDROXY-4H-IMIDAZO(4,5-C)PYRIDIN-4-ONE-6'-CARBAMATE see RAG300

14-DEOXY-14-((2-DIETHYLAMINOETHYL)MERCAPTOACETOXY)-MUTILIN HYDROGEN FUMARATE see DAR000

14-DEOXY-14-((2-DIETHYLAMINOETHYL-THIO)-ACETOXY)MUTILINE see TET800

2'-DEOXY-2',2'-DIFLUOROCYTIDINE see GCK500

2'-DEOXY-2',2'-DIFLUOROCYTIDINE MONOHYDROCHLORIDE see GCK100

6-DEOXY-7,8-DIHYDROMORPHINE see DKX600

DEOXYEPHEDRINE see DBA800

DEOXYEPHEDRINE see DBB000

d-DEOXYEPHEDRINE see PFP850

(+)-(S)-DEOXYEPHEDRINE see PFP850

d-DEOXYEPHEDRINE HYDROCHLORIDE see MDT600

dl-DEOXYEPHEDRINE HYDROCHLORIDE see DAR100

(12R,13S)-9-DEOXY-12,13-EPOXY-12,13-DIHYDRO-9-OXO-3-ACETATE-4B-(3-METHYLBUTANOATE) LEUCOMYCIN V HYDROCHLORIDE see MAC600

9-DEOXY-12,13-EPOXY-9-OXOLEUCOMYCIN V 3-ACETATE 4ᴮ-(3-METHYLBUTANOATE) see CBT250

3-DEOXY-d-ERYTHRO-HEXOS-2-ULOSE see DAR700

2'-DEOXY-5-ETHYLURIDINE see EHV200

6-DEOXY-6-FLUOROGLUCOSE see DAR150

3'-DEOXY-3'-FLUOROTHYMIDINE see DAR200

DEOXYFLUOROURIDINE see DAR400

2'-DEOXY-5-FLUOROURIDINE see DAR400

5'-DEOXY-5-FLUOROURIDINE see DYE415

2-DEOXYGLUCOSE see DAR600

2-DEOXY-d-GLUCOSE see DAR600

d-2-DEOXYGLUCOSE see DAR600

2-DEOXY-d-GLUCOSE (FRENCH) see DAR600

3-DEOXYGLUCOSONE see DAR700

d-3-DEOXYGLUCOSONE see DAR700

2-DEOXYGLYCEROL see PML250

DEOXY-GTP see GLS100

DEOXYGUANOSINE see DAR800

2'-DEOXYGUANOSINE see DAR800

(DEOXYGUANOSINE)PENTAAMMINERUTHENIUM(3+) TRICHLORIDE see DAR900

2'-DEOXYGUANOSINE 5'-(TETRAHYDROGEN TRIPHOSPHATE) see GLS100

DEOXYGUANOSINE TRIPHOSPHATE see GLS100

DEOXYGUANOSINE 5'-TRIPHOSPHATE see GLS100

2'-DEOXY-5-HYDROXYCYTIDINE see HKA760

2'-DEOXY-5-(HYDROXYMETHYL)URIDINE see HMB550

α-6-DEOXY-5-HYDROXYTETRACYLINE see DYE425

7"-DEOXY-7"-HYDROXY-TRIOXACARCIN A see TMO775

2'-DEOXY-5'-INOSINIC ACID, HOMOPOLYMER, COMPLEX WITH 2'-DEOXY-5'-CYTIDYLIC ACID HOMOPOLYMER (1:1) see DAR930

2'-DEOXY-5-IODOURIDINE see DAS000

DEOXYISOGUANOSINE see HKA750
2'-DEOXYISOGUANOSINE see HKA750
7-(6-o-(6-DEOXY-α-l-MANNOPYRANOSYL)-β-d-GLUCOPYRANOSIDE)HESPERETIN see HBU000
(3-β)-2-((6-DEOXY-α-l-MANNOPYRANOSYL)OXY)-14-HYDROXYBUFA-4,20,22-TRIENOLIDE see POB500
3-β-((6-DEOXY-α-l-MANNOSYL)OXY)-14-HYDROXYBUFA-4,20,22-TRIENOLIDE see POB500
17-DEOXYMETHANSONE see DBA875
DEOXYMETHASONE see DBA875
1-DEOXY-1-(METHYLAMINO)-d-GLUCITOL 5-ACETAMIDO-2,4,6-TRIIODO-N-METHYLISOPHTHALAMATE (SALT) see IGC000
3-((6-DEOXY-3-O-METHYL-α-l-GLUCOPYRANOSYL)OXY)-14-HYDROXY-19-OXO-CARD-20(22)-ENOLIDE see EAQ050
1-DEOXY-1-(METHYLNITROSAMINO)-d-GLUCITOL see DAS400
2-DEOXY-2-(((METHYLNITROSOAMINO)CARBONYL)AMINO)-d-GLUCOPYRANOSE see SMD000
1-DEOXY-1-(3-METHYL-3-NITROSOUREIDO)-d-GALACTOPYRANOSE see MNB000
2-DEOXY-2-(3-METHYL-3-NITROSOUREIDO)-d-GLUCOPYRANOSE see SMD000
2-DEOXY-2-(3-METHYL-3-NITROSOUREIDO)-α-(and β)-d-GLUCOPYRANOSE see SMD000
1-DEOXY-1-(N-NITROSOMETHYLAMINO)-d-GLUCITOL see DAS400
DEOXYNIVALENOL see VTF500
4-DEOXYNIVALENOL see VTF500
DEOXYNIVALENOL MONOACETATE see ACH075
7-DEOXYNOGALAROL see DAS500
DEOXYNOREPHEDRINE see AOB250
DEOXYNOREPHEDRINE see BBK000
α-6-DEOXYOXYTETRACYCLINE see DYE425
6-α-DEOXY-5-OXYTETRACYCLINE see DYE425
3'-DEOXYPAROMOMYCIN I see DAS600
(R)-3-(2-DEOXY-β-d-PENTOFURANOSYL)-3,6,7,8-TETRAHYDROIMIDAZO(4,5-d)(1,3)DIAZEPIN-8-OL see PBT100
2-DEOXYPHENOBARBITAL see DBB200
12-DEOXY-PHORBOL-20-ACETATE-13-DECDIENOATE see DAU400
12-DEOXY-PHORBOL-20-ACETATE-13-DODECANOATE see DAS800
12-DEOXYPHORBOL-20-ACETATE-13-ISOBUTYRATE see DAT000
12-DEOXY-PHORBOL-20-ACETATE-13-(2-METHYLBUTYRATE) see DAT200
12-DEOXY-PHORBOL-20-ACETATE-13-OCTENOATE see DAT400
12-DEOXY-PHORBOL-20-ACETATE-13-TIGLATE see DAT600
12-DEOXYPHORBOL-13-(4-ACETOXYPHENYLACETATE)-20-ACETATE see DAT800
12-DEOXYPHORBOL-13-ANGELATE see DAU000
12-DEOXYPHORBOL-13-ANGELATE-20-ACETATE see DAU200
12-DEOXY-PHORBOL-13-DECDIENOATE-20-ACETATE see DAU400
12-DEOXY-PHORBOL-13-DODECANOATE see DAU600
12-DEOXYPHORBOL-13-DODECANOATE-20-ACETATE see DAS800
12-DEOXYPHORBOL-13-DODECENOATE see DXU600
12-DEOXYPHORBOL-13-ISOBUTYRATE-20-ACETATE see DAT000

12-DEOXY-PHORBOL-13-α-METHYLBUTYRATE see DAW200
12-DEOXY-PHORBOL-13-α-METHYLBUTYRATE-20-ACETATE see DAT200
12-DEOXY-PHORBOL-13-OCTENOATE-20-ACETATE see DAT400
12-DEOXYPHORBOL-13-PHENYLACETATE see DAW300
12-DEOXYPHORBOL-13-PHENYLACETATE-20-ACETATE see DAW350
12-DEOXY-PHORBOL-13-TIGLATE see DAY600
12-DEOXYPHORBOL-13-TIGLATE-20-ACETATE see DAT600
4-DEOXYPYRIDOXAL see DAY800
DEOXYPYRIDOXINE see DAY800
4-DEOXYPYRIDOXINE see DAY800
4-DEOXYPYRIDOXOL see DAY800
4-DEOXYPYRIDOXOL HYDROCHLORIDE see DAY825
1-DEOXYPYRROMYCIN see DAY835
1-β-d-2'-DEOXYRIBOFURANOSYL-5-FLUOROURACIL see DAR400
1-(2-DEOXY-β-d-RIBOFURANOSYL)-5-IODOURACIL see DAS000
1-β-d-2'-DEOXYRIBOFURANOSYL-5-IODOURACIL see DAS000
1-(2-DEOXY-β-d-RIBOFURANOSYL)-5-(TRIFLUOROMETHYL)-2,4(1H,3H)-PYRIMIDINEDIONE see TKH325
DEOXYRIBONUCLEASE see EDK650
DEOXYRIBONUCLEASE (pancreatic) see EDK650
DEOXYRIBONUCLEASE A see EDK650
DEOXYRIBONUCLEASE I see EDK650
DEOXYRIBONUCLEIC ACID see DAZ000
DEOXYRIBONUCLEIC ACID, D(G-SP-T-G-G-T-G-G-G-T-G-G-G-T-G-G-G-SP-T) see ZNS400
DEOXYRIBONUCLEIC ACID, D(P-THIO)(G-T-G-G-T-G-G-G-T-G-G-G-T-G-G-G-T) see ZNS400
DEOXYRIBONUCLEIC PHOSPHATASE see EDK650
DEOXYRIBONUCLEOSIDE CYTOSINE see DAQ850
DEOXYRIBOSE CYTIDINE see DAQ850
DEOXYRIBOSE URACIL see DAZ050
1-(4-DEOXY-4-(SARCOSYL-d-SERYL)AMINO-β-d-GLUCOPYRANURONAMIDE) CYTOSINE see ARP000
DEOXYTETRARIC ACID see MAN000
4-DEOXYTETRONIC ACID see BOV000
2'-DEOXYTHIOGUANOSINE see TFJ200
β-DEOXYTHIOGUANOSINE see TFJ250
β-2'-DEOXYTHIOGUANOSINE see TFJ200
β-2'-DEOXY-6-THIOGUANOSINE MONOHYDRATE see TFJ250
DEOXYTHYMIDINE see TFX790
2'-DEOXYTHYMIDINE see TFX790
DEOXY(α-(4-TRIETHYLAMMONIO)BUTYL)BENZOIN IODIDE see BDU250
DEOXY(α-(3-TRIETHYLAMMONIO)PROPYL)BENZOIN, IODIDE see BDS750
2'-DEOXY-5-(TRIFLUOROMETHYL)URIDINE see TKH325
DEOXYURIDINE see DAZ050
2'-DEOXYURIDINE see DAZ050
6-DEOXYVERSICOLORIN A see DAZ100
2'-DEOXY-5-VINYLURIDINE see VOA550
DEOXY-V-TRYPT E see VKZ100
DEP see PMB250
DEP (pesticide) see TIQ250
DEPAKENE see PNR750
DEPAKENE see PNX750
DEPAKINE see PNR750

DEPAKINE see PNX750
DEPALLETHRIN see AFR250
DEPAMID see PNX600
DEPAMIDE see PNX600
DEPARAL see CMC750
DEPARKIN see DHF600
DEPAS see EQN600
DEPC see DIZ100
DEPC and AMMONIA see DJY050
DEPEN see MCR750
DEPHADREN see AOA500
DEPHADREN see BBK500
DEPHENIDOL HYDROCHLORIDE see CCR875
DEPHOSPHATE BROMOFENOFOS see DAZ110
DEPHRADINE HYDRATE see SBN450
DEPIGMAN see AEY000
DEPINAR see VSZ000
DEPOCID see AIF000
DEPOCILLIN see PAQ200
DEPOESTRADIOL see DAZ115
DEPOESTRADIOL CYPIONATE see DAZ115
DEPOFEMIN see DAZ115
DEPO-HEPARIN see HAQ550
DEPO-INSULIN see IDF325
DEPO-LUPRON see LEZ300
DEPO-MEDRATE see DAZ117
DEPO-MEDROL see DAZ117
DEPO-MEDRONE see DAZ117
DEPO-MEDROXYPROGESTERONE ACETATE and DELATESTRYL see DAM325
DEPO-MEDROXYPROGESTERONE ACETATE and TESTOSTERONE ENANTHATE see DAM325
DEPO-METHYLPREDNISOLONE see DAZ117
DEPO-METHYLPREDNISOLONE ACETATE see DAZ117
DEPOMIDE see AIE750
DEPON see FAQ200
DEPO-PROLUTON see HNT500
DEPO-PROVERA see MCA000
DEPO-PROVERA and DELATESTRYL see DAM325
DEPO-PROVERA and TESTOSTERONE ENANTHATE see DAM325
DEPOSTAT see GEK510
DEPOSUL see SNN300
DEPOSULIN see IDF325
DEPO-TESTOSTERONE see TBF600
DEPOT-MEDROL see DAZ117
DEPOT-OESTROMENINE see DJB200
DEPOT-OESTROMON see DJB200
DEPOVERNIL see AKO500
DEPOVIRIN see TBF600
DEPOXIN see DBA800
DEPP see BJR625
DEPPT see DJV700
DEPRANCOL see PNA500
DEPRELIN see SNE000
DEPREMOL G see DTG000
DEPREMOL M see UTU500
(+)-DEPRENIL HYDROCHLORIDE see DAZ120
1-DEPRENIL HYDROCHLORIDE see DAZ125
(±)-DEPRENIL HYDROCHLORIDE see DAZ118
(−)-DEPRENYL HYDROCHLORIDE see DAZ125
(+)-DEPRENYL HYDROCHLORIDE see DAZ120
DEPRESSIN see HEA500
DEPREX see EAI000
DEPRIDOL see MDP750
DEPRINOL see DLH630
DEPROMEL see FMR575
DEPROMIC see PNA500
DEPT see DAZ135
DEPTHON see TIQ250

DEPTOSULFONAMIDE see AIF000
DEPTRIN see EAG100
DEPTROPINE CITRATE see EAG100
DEPTROPINE METHOBROMIDE see DAZ140
DEPUALONE see AOB500
DEQUEST 2000 see NEI100
DEQUEST 2006 see PBP200
DEQUEST 2010 see HKS780
DEQUEST 2015 see HKS780
DEQUEST 2051 see HEQ600
DEQUEST 2054 see HEQ610
DEQUEST 2066 see DJG700
DEQUEST 2051 DEFLOCCULANT and SEQUESTRANT see HEQ600
DEQUEST 2054 DEFLOCCULANT and SEQUESTRANT see HEQ610
DEQUEST 2066 DEFLOCCULANT and SEQUESTRANT see DJG700
DEQUEST Z 010 see HKS780
D.E.R. 332 see BLD750
DER 438 see ECC750
DER 736 see PKK100
DERACIL see TFR250
DERALIN see ICC000
DERATOL see VSZ100
DERESPERINE see RDF000
DEREUMA see DOT000
DERGRAMIN see SOW000
DERIBAN see DGP900
DERICON 700 see MCB050
DERIL see RNZ000
DERIZENE see DNU000
DERIZENE see SPC500
DERMACAINE see DDT200
DERMACORT see CNS750
DERMADEX see HCL000
DERMA FAST BROWN W-GL see CMO750
DERMA FAST TURQUOISE W 2GL see COF420
DERMAFFINE see OBA000
DERMAFOSU (POLISH) see RMA500
DERMA FUR YELLOW RT see SGP500
DERMAGAN see ACR300
DERMAGEN see ACR300
DERMAGINE see OAV000
DERMALAR see SPD500
DERMAPHOS see RMA500
DERMASORB see DUD800
DERMATOLOGICO see ARN500
DERMATON see CDS750
DERMA YELLOW P see SGP500
DERMINA YELLOW G see MRL100
DERMISTINE see BBV500
DERMODRIN see BBV500
DERMOFURAL see NGE500
DERMOGLANCIN see MEY750
DERMOXIN see TGB475
DERONIL see SOW000
DEROSAL see MHC750
DERRIBANTE see DGP900
DERRIN see RNZ000
DERRIS see RNZ000
DERRIS ELLIPTICA, root see DBA000
DERRIS RESINS see DBA000
DERRIS ROOT see DBA000
DESdp see DKA200
DES (synthetic estrogen) see DKA600
DESACCHROMIN DISPERSION see PJA200
DESACE see DBH200
16-DESACETYL-16-ANHYDROACOSCHIMPEROSIDE P see DBA100
DESACETYLBUFOTALIN see BOM655
DESACETYLCHOLCHICEINE see TLN750
DESACETYLCOLCHICINE see TLO000
N-DESACETYLCOLCHICINE see TLO000
DESACETYLCOLCHICINE-d-TARTRATE see DBA175
DESACETYLLANATOSIDE B see DAD650
DESACETYLLANATOSIDE C see DBH200

DESACETYL-LANTOSID B (GERMAN) see DAD650
DESACETYLMETHYLCOLCHICINE see MIW500
N-DESACETYLMETHYLCOLCHICINE see MIW500
N-DESACETYL-N-METHYLCOLCHICINE see MIW500
N-DESACETYLTHIOCOLCHICINE see DBA200
N-DESACETYLTHIOCOLCHICINE see DBA200
DESACETYLTHYMOXAMINE see DAD850
DESACETYLVINBLASTINE AMIDE see VLF350
DESACETYLVINBLASTINE AMIDE SULFATE see VGU750
DESACI see DBH200
DESADRENE see SOW000
DESAGLYBUZOLE see GFM200
DESALKYLPRAZEPAM see CGA500
DESAMETASONE see SOW000
DESAMINE see DBA800
DESBENZYL CLEBOPRIDE see DBA250
DESBENZYL IBP see IRF500
DESCETYLDIGILANIDE C see DBH200
DESCHLOROBIOMYCIN see TBX000
DESCORTERONE see DAQ800
DESCOTONE see DAQ800
N-DESCYCLOPROPYLMETHYLPRAZEPAM see CGA500
DESD see DKB000
DESDANINE see COV125
DESDEMIN see CHJ750
DESDIMETHYLTAMOXIFEN see DHA250
DES DISODIUM SALT see DKA400
DESENEX see ULS000
DESENTOL see BBV500
DESERIL see MLD250
DESERNYL see MLD250
DESERPIDINE, 10-METHOXY- see MEK700
DESERPINE see RDF000
DESERPINE see RDK000
DESERTALC 57 see TAB750
DESERTOMYCIN see DBA400
DESERT RED see CHP500
DESERT ROSE see DBA450
DESERYL see MLD250
DESETHYLAPRINDINE HYDROCHLORIDE see DBA475
DESETHYLHYCANTHONE see EGA600
DESETHYLLUCANTHONE see EGA650
DESFEDRIN see DBA800
DESFERAL see DAK200
DESFERAL METHANESULFONATE see DAK200
DESFERRAL see DAK200
DESFERRIN see DAK200
DESFERRIOXAMINE see DAK200
DESFERRIOXAMINE B see DAK200
DESFERRIOXAMINE B MESYLATE see DAK300
DESFERRIOXAMINE B METHANESULFONATE see DAK300
DESFERRIOXAMINE METHANESULFONATE see DAK300
DESGLUCO-DIGITALINUM VERUM (GERMAN) see SMN275
DESGLUCODIGITONIN see DBA500
DESGLUCO-TRANSVAALIN see POB500
(DES-GLY10(d-TRP6))-LH-RH ETHYLAMIDE see LIU353
DESICAL P see CAU500
DES-I-CATE see DXD000
DESICCANT L-10 see ARB250
DESIMEX i see LBU200
DESIMIPRAMINE see DSI709
DESINFECT see CDP000
DESIPRAMIN see DSI709
DESIPRAMINE (D4) see DSI709
DESIPRAMINE HYDROCHLORIDE see DLS600

DESLANATOSIDE see DBH200
DESLANOSIDE see DBH200
DESLORELIN see LIU353
DESMA see DKA600
DESMECOLCHINE see MIW500
DESMECOLCINE see MIW500
DESMEDIPHAM see EEO500
DESMEL see PMS930
DESMETHOXYRESERPINE see RDF000
11-DESMETHOXYRESERPINE see RDF000
DESMETHYLAMITRIPTYLINE see NNY000
DESMETHYLBROMETHALIN see DBA520
DESMETHYLCYPROHEPTADINE see DBA550
DESMETHYLDIAZEPAM see CGA500
N-DESMETHYLDIAZEPAM see CGA500
DESMETHYLDOPAN see BIA250
DESMETHYLDOXEPIN see DBA600
DESMETHYLIMIPRAMINE see DSI709
DESMETHYLIMIPRAMINE HYDROCHLORIDE see DLS600
DESMETHYLMISONIDAZOLE see DBA700
DESMETHYLMISONIDAZOLE see NHI000
DESMETHYLTAMOXIFEN see DBA710
N-DESMETHYLTAMOXIFEN see DBA710
S-DESMETHYLZOPICLONE see DBA720
DESMETRYN (GERMAN, DUTCH) see INR000
DESMETRYNE see INR000
DESMODUR 44 see MJP400
DESMODUR 44V20 see PKB100
DESMODUR H see DNJ800
DESMODUR N see DNJ800
DESMODUR PU 1520A20 see PKB100
DESMODUR T80 see TGM750
DESMODUR T100 see TGM740
DES-N see TNJ300
2,4-DES-Na see CNW000
2,4-DES-NATRIUM (GERMAN) see CNW000
DESO see EPI500
DESOGESTREL see DBA750
DESOLET see SFS000
DESOMORPHINE see DKX600
DESORMONE see DAA800
DESORMONE see DGB000
14-DESOSSI-14-((2-DIETILAMINOETIL)MERCAPTO-ACETOSSI)MUTILIN IDROGENO FUMARATO (ITALIAN) see TET800
DESOSSIEFEDRINA see DBA800
DES-OXA-D see DBA800
DESOXEDRINE see DBA800
DES-3,4-OXIDE see DKB100
DES-α,β-OXIDE see DKB100
DESOXIMETASONE see DBA875
17-DESOXIMETHASONE see DBA875
DESOXIN see DBA800
DESOXO-5 see DBA800
DESOXO-5 see MDT600
DESOXON 1 see PCL500
DESOXYADENOSINE see DAQ200
DESOXYBENZOIN see PEB000
DESOXYCHOLATE AMPHOTERICIN B see FPC200
DESOXYCHOLIC ACID see DAQ400
DESOXYCHOLSAEURE (GERMAN) see DAQ400
DESOXYCORTICOSTERONE see DAQ600
DESOXYCORTICOSTERONE ACETATE see DAQ800
DESOXYCORTONE see DAQ600
DESOXYCORTONE ACETATE see DAQ800
DESOXYCYTIDIN (GERMAN) see DAQ850
14-DESOXY-14-((DIETHYLAMINOETHYL)-MERCAPTO ACETOXYL)-MUTILIN HYDROGEN FUMARATE see TET800
DESOXYEPHEDRINE see DBB000
d-DESOXYEPHEDRINE see PFP850
DESOXYEPHEDRINE HYDROCHLORIDE see DBA800
d-DESOXYEPHEDRINE HYDROCHLORIDE see MDT600

l-DESOXYEPHEDRINE HYDROCHLORIDE see MDQ500
dl-DESOXYEPHEDRINE HYDROCHLORIDE see DAR100
DESOXYFED see DBA800
DESOXYFED see MDT600
DESOXYMETASONE see DBA875
DESOXYMETHASONE see DBA875
DESOXYN see BBK500
DESOXYN see DBA800
DESOXYN see DBB000
DESOXYN see MDT600
DESOXYNE see MDT600
DESOXYNIVALENOL see VTF500
DESOXYNOREPHEDRINE see AOA250
DESOXYNOREPHEDRINE see AOB250
(±)-DESOXYNOREPHEDRINE see BBK000
racemic-DESOXYNOREPHEDRINE see BBK000
3-DESOXYNORLUTIN see NNV000
DESOXYPHED see DBA800
2-DESOXYPHENOBARBITAL see DBB200
DESOXYPHENOBARBITONE see DBB200
DESOXYPYRIDOXIME HYDROCHLORIDE see DAY825
DESOXYPYRIDOXINE see DAY800
5-DESOXYQUERCETIN see FBW000
DESOXYRIBONUCLEASE see EDK650
2'-DESOXYURIDINE see DAZ050
DESPHEN see CDP250
DESPIROL see MDT625
DE-SPROUT see DMC600
DESSIN see CBW000
DESSON see CLW000
DESTENDO see BCA000
DESTIM see DBA800
DESTIM see MDT600
DESTOLIT see DMJ200
DESTOMYCIN A see DBB400
DESTONATE 20 see DBB400
DESTRIOL see EDU500
DESTROL see DKA600
DESTRONE see EDV000
DESTRUXOL APPLEX see DXE000
DESTRUXOL BORER-SOL see EIY600
DESTRUXOL ORCHID SPRAY see NDN000
DESTUN see TKF750
DESURIC see DDP200
DESYL CHLORIDE see CFJ100
DESYPHED see MDT600
DESYREL see CKJ000
DESYREL see THK875
DESYREL see THK880
DET see DKC800
m-DET see DKC800
DETA see DJG600
m-DETA see DKC800
DETAL see DUS700
DETALUP see VSZ100
DETAMIDE see DKC800
DETA/NO see DHJ333
DETA/NONOATE see DHJ333
DETARIL see DWK400
DETDA see DJP100
DETERGENT 66 see SIB600
DETERGENT ALKYLATE see PEW500
DETERGENT HD-90 see DXW200
DETERGENTS, LIQUID containing AES see DBB450
DETERGENTS, LIQUID containing LAS see DBB460
DETF see TIQ250
DETHMORE see WAT200
DETHYLANDIAMINE see TEO000
2,4-D ETHYL ESTER see EHY600
2,4-D 2-ETHYLHEXYL ESTER see DBB480
DETHYRONA see SKJ300
DETIA GAS EX-B see AHE750
DETIA GAS EX-B see PGY000
DETICENE see DAB600
DETIGON see CMW700
DETIGON-BAYER see CMW700

DE60FTM see PAT830
DETMOL-EXTRAKT see BBQ500
DETMOL MA see MAK700
DETMOL MA 96% see MAK700
DETMOL U.A. see CMA100
DETOX see DAD200
DETOX 25 see BBQ500
DETOXAN see DAD200
DETOXARGIN see AQW000
DETRALFATE see DBB500
DETRAVIS see DAI485
DETREOMYCINE see CDP250
DETREOPAL see CDP700
DETREX see DBA800
DETTOL see CLW000
DETYROXIN see SKJ300
DEURSIL see DMJ200
DEUSLON-A see EDU500
DEUTERIOMORPHINE see DBB600
DEUTERIUM see DBB800
DEUTERIUM FLUORIDE see DBC000
DEUTERIUM OXIDE see HAK000
2-DEUTERO-2-NITROPROPANE see DBC100
DEVAL RED K see CLK220
DEVAL RED TR see CLK220
DEVEGAN see ABX500
DEVELIN see PNA500
DEVELOPER 11 see PEY000
DEVELOPER 13 see PEY500
DEVELOPER 14 see TGL750
DEVELOPER A see NAX000
DEVELOPER AMS see NAX000
DEVELOPER B see TGL750
DEVELOPER BN see NAX000
DEVELOPER BON see HMX520
DEVELOPER DB see TGL750
DEVELOPER DBJ see TGL750
DEVELOPER H see TGL750
DEVELOPER MC see TGL750
DEVELOPER MT see TGL750
DEVELOPER MT-CF see TGL750
DEVELOPER MTD see TGL750
DEVELOPER P see NEO500
DEVELOPER PF see PEY500
DEVELOPER R see REA000
DEVELOPER SODIUM see NAX000
DEVELOPER T see TGL750
DEVELOPER Z see NNT000
DEVICARB see CBM750
DEVICOPPER see CNK559
DEVICORAN see HID350
DEVIGON see DSP400
DEVIKOL see DGP900
DEVIL'S APPLE see MBU800
DEVIL'S APPLE see SLV500
DEVIL'S BACKBONE see SDZ475
DEVIL'S CLAWS see FBS100
DEVILS HAIR see CMV390
DEVIL'S IVY see PLW800
DEVILS THREAD see CMV390
DEVINCAN see VLF000
DEVIPON see DGI400
DEVISULPHAN see EAQ750
DEVITHION see MNH000
DEVOL GARNET GB SALT see MPY750
DEVOL ORANGE B see NEO000
DEVOL ORANGE C see CEH690
DEVOL ORANGE GC see CEH690
DEVOL ORANGE R see NEN500
DEVOL RED E see NEQ000
DEVOL RED F see KDA050
DEVOL RED GG see NEO500
DEVOL RED K see CLK235
DEVOL RED SALT E see MMF780
DEVOL RED TA SALT see CLK235
DEVOL RED TR see CLK235
DEVOL SCARLET A (FREE BASE) see DEO295
DEVOL SCARLET B see NMP500
DEVOL SCARLET G SALT see NMP500
DEVONIUM see AOD000

DEVORAN see BBQ500
DEVOTON see MFW100
DEX see BJU000
DEXA see SOW000
DEXACORT see DAE525
DEXACORT see SOW000
DEXA-CORTIDELT see SOW000
DEXA-CORTIDELT HOSTACORTIN H see PMA000
DEXADELTONE see SOW000
DEXADRESON see DAE525
DEXAGRO see DAE525
DEXAIME see BBK500
DEXALINE see BBK500
DEXALME see BBK500
DEXAMBUTOL see EDW875
DEXAMED see BBK500
DEXAMETH see SOW000
DEXAMETHASONE ACETATE see DBC400
DEXAMETHASONE ALCOHOL see SOW000
DEXAMETHASONE DIPROPIONATE see DBC500
DEXAMETHASONE 17,21-DIPROPIONATE see DBC500
DEXAMETHASONE DISODIUM PHOSPHATE see DAE525
DEXAMETHASONE ISONICOTINATE see DBC510
DEXAMETHASONE-21-ISONICOTINATE see DBC510
DEXAMETHASONE-21-ORTHOPHOSPHATE see BFW325
DEXAMETHASONE PALMITATE see DBC525
DEXAMETHASONE-21-PALMITATE see DBC525
DEXAMETHASONE PHOSPHATE see BFW325
DEXAMETHASONE-21-PHOSPHATE see BFW325
DEXAMETHASONE SODIUM HEMISULFATE see DBC550
DEXAMETHASONE SODIUM PHOSPHATE see DAE525
DEXAMETHASONE SODIUM SULFATE see DBC550
DEXAMETHASONE VALERATE see DBC575
DEXAMETHASONE-17-VALERATE see DBC575
DEXAMETHAZONE SODIUM PHOSPHATE see DAE525
DEXAMINE see BBK500
DEXAMPHAMINE see BBK500
DEXAMPHETAMINE see AOA500
DEXAMPHETAMINE see BBK500
DEXAMPHETAMINE SULFATE see BBK500
DEXAMYL see BBK500
DEXA-SCHEROSON (INJECTABLE) see DBC550
DEXBENZETIMIDE HYDROCHLORIDE see DBE000
DEXBROMPHENIRAMINE MALEATE see DXG100
DEXCHLOROPHENIRAMINE MALEATE see PJJ325
DEXCHLORPHENIRAMINE MALEATE see PJJ325
DEXEDRINA see BBK500
DEXEDRINE see AOA500
DEXEDRINE SULFATE see BBK500
DEXETIMIDE HYDROCHLORIDE see DBE000
DEXIES see BBK500
DEXIUM see DMI300
DEXNIGULDIPINE HYDROCHLORIDE see NDY550
DEXON see DOU600
DEXONE see SOW000
DEXON E 117 see PMP500
DEXOPHRINE see DBA800
DEXOVAL see DBA800

DEXOVAL see MDT600
DEXOXADROL HYDROCHLORIDE see DVP400
DEXPANTHENOL (FCC) see PAG200
DEXTELAN see SOW000
DEXTIM see MDT600
DEXTRAN see DBD700
DEXTRAN 1 see DBC800
DEXTRAN 2 see DBD000
DEXTRAN 5 see DBD200
DEXTRAN 10 see DBD400
DEXTRAN 11 see DBD600
DEXTRAN 70 see DBD700
DEXTRAN ION COMPLEX see IGS000
DEXTRANS see DBD800
DEXTRAN SULFATE SODIUM see DBD750
DEXTRAN SULFATE SODIUM ALUMINIUM see DBB500
DEXTRARINE see DBD750
DEXTRAVEN see DBD700
DEXTRIFERRON see IGT000
DEXTRIFERRON INJECTION see IGT000
β-DEXTRIN see COW925
DEXTRINS see DBD800
DEXTROAMPHETAMINE SULFATE see BBK500
DEXTROBENZETIMIDE HYDROCHLORIDE see DBE000
DEXTRO CALCIUM PANTOTHENATE see CAU750
DEXTROCHLORPHENIRAMINE MALEATE see PJJ325
DEXTROFER 75 see IGS000
DEXTROID see SKJ300
DEXTROMETHADONE see DBE100
DEXTROMETHORPHAN see DBE150
DEXTROMETHORPHAN BROMIDE see DBE200
DEXTROMETHORPHAN HYDROBROMIDE see DBE200
DEXTRO-α-METHYLPHENETHYLAMINE SULFATE see BBK500
DEXTROMETORPHAN HYDROBROMIDE see DBE200
DEXTROMORAMIDE see AFJ400
DEXTROMYCETIN see CDP250
DEXTROMYCIN HYDROCHLORIDE see DBE600
DEXTRONE see DWX800
DEXTRONE see PAJ000
DEXTRONE-X see PAJ000
DEXTRO-1-PHENYL-2-AMINOPROPANE SULFATE see BBK500
DEXTRO-β-PHENYLISOPROPYLAMINE SULFATE see BBK500
DEXTROPROPOXYPHENE see DAB879
DEXTROPROPOXYPHENE HYDROCHLORIDE see PNA500
DEXTROPROPOXYPHENE NAPSYLATE see DBE625
DEXTROPROXYPHEN HYDROCHLORIDE see PNA500
DEXTROPUR see GFG000
DEXTRORPHAN see DBE800
DEXTRORPHAN TARTRATE see DBE825
DEXTROSE, anhydrous see GFG000
DEXTROSE (FCC) see GFG000
DEXTROSOL see GFG000
DEXTROSULPHENIDOL see MPN000
DEXTROTHYROXINE SODIUM see SKJ300
DEXTROTUBOCURARINE CHLORIDE see TOA000
DEXTROXIN see SKJ300
DEXULATE see DBD750
DEXURON see PAJ000
DEZIBARBITUR see EOK000
DEZONE see SOW000
DF 118 see DKW800
DF 118 see DKX000
DF 125 see GLQ100
DF 468 see PIM500
DF 469 see AAJ350

DF 493 see NOB800
DF-521 see RDA350
DFA see DVX800
Df B see DAK200
DFD 0173 see PJS750
DFD 0188 see PJS750
DFD 2005 see PJS750
DFD 6005 see PJS750
DFD 6032 see PJS750
DFD 6040 see PJS750
DFDJ 5505 see PJS750
DFM see DHE750
α-DFMO see DBE835
α-DFMO see DKH875
DFO see DAK200
DFOA see DAK200
DFOM see DAK200
DFP see IRF000
DFT see DWN800
5'-DFUR see DYE415
2-DG see DAR600
6CDG see CFJ375
DGE see DKM200
DG-HCG see DAK325
DGNB 3825 see PJS750
DGTP see GLS100
DH 245 see AOO300
DH-524 see DGA800
DH 700 see VNK200
DHA see AOO450
DHA see DAK400
DHA see MFW500
DHAQ see MQY090
DHAQ DIACETATE see DBE875
DHA-SODIUM see SGD000
DHA-S SODIUM see DAL040
2,4-DHBA see HOE600
2,5-DHBA see GCU000
DHBP see DYF200
DHC see DAL000
β-DHC see DLO880
DHEA SULFATE see SNV100
D"-HEXENOLLACTONE see PAJ500
DHK see DLR000
DHMS see DME000
DHNT see BKH500
D.H.O. 180 see DLK780
(S)-DHPA see AMH800
(RS)-DHPA see AMH850
DHPN see DNB200
DHPT see MHJ300
DHPTA see DCB100
DHS see MFW500
DHT see DME500
DHT2 see DME300
5-β-DHT see HJB100
DHTP see DME525
DHUTRA see SLV500
D 201 HYDROCHLORIDE see AEG625
D 58SI see DLU900
DI-μ-(THIOCYANATODI-n-BUTYLSTANNYLOXO)BIS(THIOCYANATO DI-n-BUTYLTIN) see BJM700
DIABARIL see CKK000
DIABASE ORANGE GC BASE see CEH690
DIABASE RED B see NEQ000
DIABASE RED RL see MMF780
DIABASE SCARLET G see NMP500
DIABASIC MAGENTA see MAC250
DIABASIC MALACHITE GREEN see AFG500
DIABASIC RHODAMINE B see FAG070
DIABECHLOR see CKK000
DIABEFAGOS see DQR800
DIABEN see BSQ000
DIABENAL see CKK000
DIABENESE see CKK000
DIABENEZA see CKK000
DIABENOR see DBE885
DIABENYL see BBV500
DIABETA see CEH700

DIABETAMID see BSQ000
DIABETOL see BSQ000
DIABETORAL see CKK000
DIABET-PAGES see CKK000
DIABINESE see CKK000
DIABORAL see BSM000
DIABORAL see CPR000
DIABRIN see BOM750
DIABRIN see BQL000
DIABUTAL see NBU000
DIABUTON see BSQ000
DIABYLEN see BBV500
DIACARB see AAI250
DIACELLITON FAST BLUE BORDEAUZ B see CMP080
DIACELLITON FAST BLUE GREEN B see DMM400
DIACELLITON FAST BLUE R see TBG700
DIACELLITON FAST BORDEAUX B see CMP080
DIACELLITON FAST BRILLIANT BLUE B see MGG250
DIACELLITON FAST GREY G see DCJ200
DIACELLITON FAST PINK B see AKE250
DIACELLITON FAST SCARLET B see ENP100
DIACELLITON FAST VIOLET 5R see DBP000
DIACELLITON FAST VIOLET B see DBY700
DIACELLITON FAST VIOLET BF see AKP250
DIACELLITON FAST YELLOW G see AAQ250
DIACELLITON FAST YELLOW YLP see KDA075
DIACELLITON SCARLET B see ENP100
DIACEL NAVY DC see DCJ200
DIACEPAN see DCK759
DIACEPHIN see HBT500
DIACESAL see SAN000
DIACETAMIDE, 2,2'-DIIODO- see DNE900
4,4'-DIACETAMIDODIPHENYL SULFONE see SNY500
2-DIACETAMIDOFLUORENE see DBF200
2,7-DIACETAMIDOFLUORENE see BGP250
N-1-DIACETAMIDOFLUORENE see DBF000
2,4-DIACETAMIDO-6-(5-NITRO-2-FURYL)-s-TRIAZINE see DBF400
2,6-DIACETAMIDOPYRIDINE see POR300
3,5-DIACETAMIDO-2,4,6-TRIIODOBENZOIC ACID see DCK000
3,5-DIACETAMIDO-2,4,6-TRIIODOBENZOIC ACID, SODIUM SALT see SEN500
α-5-DIACETAMIDO-2,4,6-TRIIODO-m-TOLUIC ACID see AAI750
10,040 DIACETATE see CDT125
2',4'-DIACETATE-DIS-NOGAMYCIN see DBF500
1,3-DIACETATE GLYCEROL see DBF600
1,2-DIACETATE 1,2,3-PROPANETRIOL see DBF600
DIACETATOPALLADIUM see PAD300
DIACETATOZIRCONIC ACID see ZTS000
DIACETAZOTOL see ACR300
DIACETIC ETHER see EFS000
DIACETIN see DBF600
DIACETIN see GGA100
1,2-DI-ACETIN see DBF600
1,3-DIACETIN see DBF600
2,3-DIACETIN see DBF600
DIACETONALCOHOL (DUTCH) see DBF750
DIACETONALCOOL (ITALIAN) see DBF750
DIACETONALKOHOL (GERMAN) see DBF750
DIACETONE see DBF750
DIACETONE ACRYLAMIDE see DTH200
DIACETONE ALCOHOL see DBF750
DIACETONE ALCOHOL PEROXIDES, >57% in solution with >9% hydrogen peroxide (DOT) see HMK050

DIACETONE-ALCOOL (FRENCH) see DBF750

DIACETONE PEROXIDES, solid, or >25% in solution (DOT) see ACV500

DIACETOTOLUIDE see ACR300

o-DIACETOTOLUIDIDE, 4"-(o-TOLYLAZO)-(8CI) see ACR300

1,3-DIACETOXYBENZENE see REA100

DIACETOXYBUTYLTIN see DBF800

1,2-DIACETOXY-3-CHLOROPROPANE see CIL900

DIACETOXYDIBUTYLPLUMBANE see DED400

DIACETOXYDIBUTYL STANNANE see DBF800

DIACETOXYDIBUTYLTIN see DBF800

1,1-DIACETOXY-2,3-DICHLOROPROPANE see DBF875

trans-7,8-DIACETOXY-7,8-DIHYDROBENZO(a)PYRENE see DBG200

4-β,15-DIACETOXY-3-α,8-α-DIHYDROXY-12,13-EPOXYTRICHOTHEC-9-ENE see NCK000

DIACETOXYDIMETHYLPLUMBANE see DSL400

2-(4,4'-DIACETOXYDIPHENYLMETHYL)PYRIDINE see PPN100

4,4'-DIACETOXYDIPHENYLPYRID-2-YLMETHANE see PPN100

(4,4'-DIACETOXYDIPHENYL)(2-PYRIDYL)METHANE see PPN100

3-α,17-β-DIACETOXY-2-β,16-β-DIPIPERIDINO-5-α-ANDROSTANE DIMETHOBROMIDE see PAF625

3-β,17-β-DIACETOXY-17-α-ETHINYL-19-NOR-Δ³,⁵-ANDROSTADIENE see DBG300

3-β,17-β-DIACETOXY-17-α-ETHYNYL-4-OESTRENE see EQJ500

4-β,15-DIACETOXY-3-α-HYDROXY-12,13-EPOXYTRICHOTHEC-9-ENE see AOP250

DIACETOXYMERCURIPHENOL see HNK575

DIACETOXYMERCURY see MCS750

7-DIACETOXYMETHYLBENZ(a)ANTHRACENE see BBG750

4,15-DIACETOXY-8-(3-METHYLBUTYRYLOXY)-12,13-EPOXY-Δ-9-TRICHOTHECEN-3-OL see FQS000

4-β,15-DIACETOXY-8-α-(3-METHYLBUTYRYLOXY)-3-α-HYDROXY-12,13-EPOXYTRICHOTHEC-9-ENE see FQS000

1,4-DIACETOXY-2-METHYLNAPHTHALENE see VTA100

3-β,17-β-DIACETOXY-19-NOR-17-α-PREGN-4-EN-20-YNE see EQJ500

DIACETOXYPALLADIUM see PAD300

DI-(4-ACETOXYPHENYL)-2-PYRIDYLMETHANE see PPN100

DI-(p-ACETOXYPHENYL)-2-PYRIDYLMETHANE see PPN100

DIACETOXYPROPENE see ADR250

1,3-DIACETOXYPROPENE see PMO800

3,3-DIACETOXYPROPENE see ADR250

1,1-DIACETOXYPROPENE-2 see ADR250

DIACETOXYSCIRPENOL see AOP250

4,15-DIACETOXYSCIRPEN-3-OL see AOP250

DIACETOXYTETRABUTYLDISTANNOXANE see BGQ000

DIACETYLAMINOAZOTOLUENE see ACR300

4,4'-DIACETYLAMINOBIPHENYL see BFX000

2-DIACETYLAMINOFLUORENE see DBF200

2,7-DIACETYLAMINOFLUORENE see BGP250

N-DIACETYL-2-AMINOFLUORENE see DBF200

N,N-DIACETYL-2-AMINOFLUORENE see DBF200

DI-(p-ACETYLAMINOPHENYL)SULFONE see SNY500

3,5-DIACETYLAMINO-2,4,6-TRIJODBENZOSAEURE NATRIUM (GERMAN) see SEN500

4,4'-DIACETYLBENZIDINE see BFX000

N,N'-DIACETYL BENZIDINE see BFX000

2-((2,4-DIACETYL-5-BENZOFURANYL)OXY)TRIETHYLAMINE HYDROCHLORIDE see DBG499

N,O-DIACETYL-N-(4-BIPHENYLYL)HYDROXYLAMINE see ABJ750

N,o-DIACETYL-N-CARBOXYHYDROXYLAMINE ETHYL ESTER see DBH100

DIACETYLCHOLINE see CMG250

DIACETYLCHOLINE CHLORIDE see HLC500

DIACETYLCHOLINE DICHLORIDE see HLC500

DIACETYLCHOLINE DIIODIDE see BJI000

DIACETYLDAPSONE see SNY500

N,N'-DIACETYLDAPSONE see SNY500

4,4'-DIACETYLDIAMINODIPHENYL SULFONE see SNY500

N,N'-DIACETYL-4,4'-DIAMINODIPHENYL SULFONE see SNY500

2,6-N,N-DIACETYLDIAMINOPYRIDINE see POR300

2,6-DIACETYL-7,9-DIHYDROXY-8,9b-DIMETHYL-1,3(2H,9bH)-DIBENZOFURANDIONE see UWJ000

N,N'-DIACETYL-3,3'-DIMETHYLBENZIDINE see DRI400

N,N'-DIACETYL-N,N'-DINITRO-1,2-DIAMINOETHANE see DBG900

DIACETYL DIOXIME see DBH000

1,2-DIACETYLETHANE see HEQ500

α,β-DIACETYLETHANE see HEQ500

DIACETYL (FCC) see BOT500

N,N-DIACETYL-2-FLUORENAMINE see DBF200

DIACETYL GLYCERINE see DBF600

DIACETYLGLYCEROL see GGA100

O,O'-DIACETYL 4-HYDROXYAMINOQUINOLINE-1-OXIDE see ABN500

N,o-DIACETYL-N-HYDROXYURETHAN see DBH100

DIACETYLLANATOSID C (GERMAN) see DBH200

DIACETYLLANATOSIDE see DBH200

DIACETYLMANGANESE see MAQ000

DIACETYLMETHANE see ABX750

DIACETYLMONOOXIME see OMY910

DIACETYLMONOXIME see OMY910

DIACETYLMORFIN see HBT500

DIACETYLMORPHINE see HBT500

DIACETYLMORPHINE HYDROCHLORIDE see DBH400

N,O-DIACETYL-N-(1-NAPHTHYL)HYDROXYLAMINE see ABS250

1,3-DIACETYLOXYPROPENE see PMO800

DIACETYL PEROXIDE (MAK) see ACV500

N,O-DIACETYL-N-(p-STYRYLPHENYL)HYDROXYLAMINE see ABW500

trans-N,o-DIACETYL-N-(p-STYRYLPHENYL)HYDROXYLAMINE see ABL250

DIACETYL TARTARIC ACID ESTERS of MONO- and DIGLYCERIDES see DBH700

3,4-DI(ACETYLTHIOMETHYL)-5-HYDROXY-6-METHYLPYRIDINE HYDROBROMIDE see DBH800

N,N-DIACETYL-o-TOLYLAZO-o-TOLUIDINE see ACR300

(2,2-DIACETYLVINYL)BENZENE see DBH900

DIACID see BNK000

DIACID ALIZARINE SKY BLUE B see CMM090

DIACID ALIZARIN SKY BLUE B see CMM090

DIACID BLUE BLACK 10B see FAB830

DIACID FAST RED 6B see CMM400

DIACID LIGHT BLUE BR see CMM070

DIACID METANIL YELLOW see MDM775

DIACID ORANGE II see CMM220

DIACOTTON BENZOPURPURINE 4B see DXO850

DIACOTTON BLACK BH see CMN800

DIACOTTON BLUE BB see CMO000

DIACOTTON BORDEAUX GS see CMO872

DIACOTTON BORDEAUX KS see CMO872

DIACOTTON BRILLIANT BLUE RW see CMO600

DIACOTTON BROWN M see CMO800

DIACOTTON CONGO RED see SGQ500

DIACOTTON DARK GREEN see CMO830

DIACOTTON DEEP BLACK see AQP000

DIACOTTON DEEP BLACK RX see AQP000

DIACOTTON FAST BLACK D see CMN230

DIACOTTON FAST RED F see CMO870

DIACOTTON FAST SCARLET 4BS see CMO870

DIACOTTON GREEN B see CMO840

DIACOTTON SKY BLUE 5B see CMO500

DIACOTTON SKY BLUE 6B see CMN750

DIACRID see DBX400

DIACROMO BLACK PSS see EDC625

DIACROMO BLUE G see CMP880

DIACROMO ORANGE GR see CMP882

DIACROMO VIOLET N see HLI000

DIACRYALTE DIETHYLENE GLYCOL see ADT250

DIACRYL GOLDEN YELLOW GL-N see CMM895

DIACRYL SUPRA RED GTL see BAQ750

DIACTOL see VSZ100

DIACYCINE see TBX250

DIADEM CHROME BLUE R see HJF500

DIADENOSINE 5',5'''-P¹),P²)-TETRAPHOSPHATE see AEM300

5',5'''-DIADENOSINE TETRAPHOSPHATE see AEM300

DIADILAN see PFJ750

DIADOL see AFS500

DI-ADRESON F see PMA000

DI-ADRESON-F-AQUOSUM see PMA100

DIAETHANOLAMIN (GERMAN) see DHF000

DIAETHANOLAMIN-3,5-DIJODPYRIDON-(4)-ESSIGSAEURE (GERMAN) see DNG400

DIAETHANOLNITROSAMIN (GERMAN) see NKM000

1,1-DIAETHOXY-AETHAN (GERMAN) see AAG000

1,2-DIAETHOXY-AETHEN see DHG100

2-DIAETHOXYPHOSPHINYL-THIOAETHYL-TRIMETHYL-AMMONIUM-JODID (GERMAN) see TLF500

DIAETHYLACETAL (GERMAN) see AAG000

DIAETHYLAETHER (GERMAN) see EJU000

DIAETHYLAETHER-β,β'-(4-HYDROXYLIMINOFORMYLPYRIDINIUM)-DIBROMID see OPO100

O,O-DIAETHYL-S-(2-AETHYLTHIO-AETHYL)-DITHIOPHOSPHAT (GERMAN) see DXH325

O,O-DIAETHYL-S-(2-AETHYLTHIO-AETHYL)-MONOTHIOPHOSPHAT (GERMAN) see DAP200

O,O-DIAETHYL-S-(AETHYLTHIO-METHYL)-DITHIOPHOSPHAT (GERMAN) see PGS000

DIAETHYLALLYLACETAMIDE (GERMAN) see DJU200

DIAETHYLAMIN (GERMAN) see DHJ200

DIAETHYLAMINOAETHANOL (GERMAN) see DHO500

o-DIAETHYLAMINOAETHOXY-BENZANILID (GERMAN) see DHP200

o-DIAETHYLAMINOAETHOXY-5-CHLOR-BENZANILID (GERMAN) see CFO250

1-DIAETHYLAMINO-AETHYLAMINO-4-METHYL-THIOXANTHONHYDROCHLORID (GERMAN) see SBE500

DIAETHYLAMINOAETHYL DIPHENYLCARBAMAT HYDROCHLORID (GERMAN) see DHY200

N-DIAETHYLAMINOAETHYL-RESERPIN (GERMAN) see DIG800

DIAETHYLAMINOAETHYL-THEOPHYLLIN (GERMAN) see CNR125

7-(β-DIAETHYLAMINO-AETHYL)-THEOPHYLLIN-HYDROCHLORID (GERMAN) see DIH600

DIAETHYLAMMONIUM-DIAETHYLDITHIOCARBAMAT see DJD000

DIAETHYLANILIN (GERMAN) see DIS700

O,O-DIAETHYL-O-(4-BROM-2,5-DICHLOR)-PHENYL-MONOTHIOPHOSPHAT (GERMAN) see EGV500

DIAETHYLCARBONAT (GERMAN) see DIX200

O,O-DIAETHYL-O-(CHINOXALYL-(2))-MONOTHIOPHOSPHAT (GERMAN) see DJY200

O,O-DIAETHYL-O-(3-CHLOR-4-METHYL-CUMARIN-7-YL)-MONOTHIOPHOSPHAT (GERMAN) see CNU750

O,O-DIAETHYL-S-(6-CHLOR-2-OXO-BEN(b)-1,3-OXALIN-3-YL)-METHYL-DITHIOPHOSPHAT (GERMAN) see BDJ250

O,O-DIAETHYL-S-((4-CHLOR-PHENYL-THIO)-METHYL)DITHIOPHOSPHAT (GERMAN) see TNP250

O,O-DIAETHYL-o-(α-CYANBENZYLIDEN-AMINO)-THIONPHOSPHAT (GERMAN) see BAT750

O,O-DIAETHYL-o-(α-CYANO-BENZYLIDENAMINO)-MONOTHIOPHOSPHAT (GERMAN) see BAT750

N-DIAETHYL CYSTEAMIN (GERMAN) see DIY000

O,O-DIAETHYL-O-(2,5-DICHLOR-4-BROMPHENYL)-THIONOPHOSPHAT (GERMAN) see EGV500

O,O-DIAETHYL-O-1-(4,5-DICHLORPHENYL)-2-CHLOR-VINYL-PHOSPHAT (GERMAN) see CDS750

O,O-DIAETHYL-O-2,4-DICHLOR-PHENYL-MONOTHIOPHOSPHAT (GERMAN) see DFK600

O,O-DIAETHYL-S-((2,5-DICHLOR-PHENYL-THIO)-METHYL)-DITHIOPHOSPHAT (GERMAN) see PDC750

O,O-DIAETHYL-O-2,4-DICHLORPHENYL-THIONOPHOSPHAT (GERMAN) see DFK600

1,2-DIAETHYLHYDRAZINE (GERMAN) see DJL400

O,O-DIAETHYL-O-(2-ISOPROPYL-4-METHYL-PYRIMIDIN-6-YL)-MONOTHIOPHOSPHAT (GERMAN) see DCM750

O,O-DIAETHYL-O-(2-ISOPROPYL-4-METHYL)-6-PYRIMIDYL-THIONOPHOSPHAT (GERMAN) see DCM750

O,O-DIAETHYL-S-(1-METHYLAETHYL)-CARBAMOYL-METHYL-MONOTHIOPHOSPHAT (GERMAN) see PHK250

O,O-DIAETHYL-O-(4-METHYL-COUMARIN-7-YL)-MONOTHIOPHOSPHAT (GERMAN) see PKT000

O,O-DIAETHYL-S-(3-METHYL-2,4-DIOXO-5-OXA-3-AZA-HEPTYL)-DITHIOPHOSPHAT (GERMAN) see DJI000

O,O-DIAETHYL-O-(3-METHYL-1H-PYRAZOL-5-YL)-PHOSPHAT (GERMAN) see MOX250

O,O-DIAETHYL-O-4-METHYLSULFINYL-PHENYL-MONOTHIOPHOSPHAT (GERMAN) see FAQ800

DIAETHYL-NICOTINAMID (GERMAN) see DJS200

O,O-DIAETHYL-O-(4-NITROPHENYL)-MONOTHIOPHOSPHAT (GERMAN) see PAK000

O,O'-DIAETHYL-p-NITROPHENYLPHOSPHAT (GERMAN) see NIM500

DIAETHYL-p-NITROPHENYLPHOSPHORSAEUREESTER (GERMAN) see NIM500

DIAETHYLNITROSAMIN (GERMAN) see NJW500

O,O-DIAETHYL-S-(4-OXOBENZOTRIAZIN-3-METHYL)-DITHIOPHOSPHAT (GERMAN) see EKN000

O,O-DIAETHYL-S-((4-OXO-3H-1,2,3-BENZOTRIAZIN-3-YL)-METHYL)-DITHIOPHOSPHAT (GERMAN) see EKN000

O,O-DIAETHYL-N-PHTALIMIDOTHIOPHOSPHAT (GERMAN) see DJX200

DI-AETHYL-PROPANEDIOL (GERMAN) see PMB250

O,O-DIAETHYL-O-(PYRAZIN-2YL)-MONOTHIOPHOSPHAT (GERMAN) see EPC500

O,O-DIAETHYL-O-(2-PYRAZINYL)-THIONOPHOSPHAT (GERMAN) see EPC500

DIAETHYLSULFAT (GERMAN) see DKB110

O,O-DIAETHYL-S-(3-THIA-PENTYL)-DITHIOPHOSPHAT (GERMAN) see DXH325

DIAETHYLTHIOPHOSPHORSAEUREESTER des AETHYLTHIOGLYKOL (GERMAN) see DAO500

DIAETHYLTHIOPHOSPHORSAEUREESTER des AETHYLTHIOGLYKOL (GERMAN) see DAP200

O,O-DIAETHYL-O-3,5,6-TRICHLOR-2-PYRIDYLMONOTHIOPHOSPHAT see CMA100

DIAETHYLZINNDICHLORID (GERMAN) see DEZ000

DIAFEN see DVW700

DIAFEN see LJR000

DIAFEN (antihistamine) see DVW700

DIAFEN HYDROCHLORIDE see DVW700

DIAFENTHIURON see BRB100

DIAFORM UR see UTU500

DIAFURON see NGG500

DIAGINOL see AAN000

DIAGNORENOL see SHX000

DIAGRABROMYL see BNP750

DIAK 7 see THS100

DIAKARB see AAI250

DIAKARMON see SKV200

DIAKOL DM see UTU500

DIAKOL F see UTU500

DIAKOL M see UTU500

DIAKON see MLH750

DIAKON see PKB500

DIAL see AFS500

DIAL see CFU750

DIAL-A-GESIC see HIM000

DIALFERIN see DBK400

DIALICOR see DHS200

DIALIFOR see DBI099

DIALKYL 79 PHTHALATE see LGF875

DIALKYLPHTHALATE C7-C9 see LGF875

DIALKYLZINCS see DBI159

DIALLAAT (DUTCH) see DBI200

DIALLAT (GERMAN) see DBI200

DIALLATE see DBI200

DIALLYINITROSAMIN (GERMAN) see NJV500

DIALLYL see HCR500

DIALLYLAMINE see DBI600

4-DIALLYLAMINO-3,5-DIMETHYLPHENYL-N-METHYLCARBAMATE see DBI800

4-(DIALLYLAMINO)-3,5-XYLENOL METHYLCARBAMATE (ester) see DBI800

4-DIALLYL-AMINO-3,5-XYLYL N-METHYLCARBAMATE see DBI800

DIALLYLBARBITAL see AFS500

DIALLYLBARBITURIC ACID see AFS500

5,5-DIALLYLBARBITURIC ACID see AFS500

1,1-DIALLYL-3-(1,4-BENZODIOXAN-2-YLMETHYL)-3-METHYLUREA see DBJ100

5,5'-DIALLYL-2,2'-BIPHENYLDIOL see BFX520

2-(N,N-DIALLYLCARBAMYLMETHYL)AMINOMETHYL-1,4-BENZODIOXAN see DBJ100

DIALLYLCHLOROACETAMIDE see CFK000

N,N-DIALLYLCHLOROACETAMIDE see CFK000

N,N-DIALLYL-2-CHLOROACETAMIDE see CFK000

N,N-DIALLYL-α-CHLOROACETAMIDE see CFK000

DIALLYLCYANAMIDE see DBJ200

DIALLYLDIBROMO STANNANE see DBJ400

N,N-DIALLYLDICHLOROACETAMIDE see DBI300

N,N-DIALLYL-2,2-DICHLOROACETAMIDE see DBI300

DIALLYL DIGLYCOLATE see DKN100

DIALLYL DIGLYCOL CARBONATE see AGD250

DIALLYL DISULFIDE see AGF300

DIALLYL DISULPHIDE see AGF300

DIALLYL ETHER see DBK000

DIALLYLETHER ETHYLENGLYKOLU (CZECH) see EJE000

DIALLYLHYDRAZINE see DBK100

1,1-DIALLYLHYDRAZINE see DBK100

1,2-DIALLYLHYDRAZINE DIHYDROCHLORIDE see DBK120

DIALLYLMAL see AFS500

DIALLYL MALEATE see DBK200

DIALLYL MONOSULFIDE see AGS250

DIALLYLNITROSAMINE see NJV500

DIALLYLNORTOXIFERINE DICHLORIDE see DBK400

N,N'-DIALLYLNORTOXIFERINIUM DICHLORIDE see DBK400

DIALLYL PEROXYDICARBONATE see DBK600

2,6-DIALLYLPHENOL see DBK800

DIALLYL PHOSPHITE see DBL000

DIALLYL PHTHALATE see DBL200

DIALLYL SELENIDE see DBL300

DIALLYL SULFATE see DBL400

DIALLYL SULFIDE see AGS250

DIALLYL THIOETHER see AGS250

DIALLYL THIOUREA see DBL600

DIALLYLTIN DIBROMIDE see DBJ400

DIALUMINOUS BLACK AB see CMN300

DIALUMINOUS RED 4B see CMO885

DIALUMINUM DIPOTASSIUM SULFATE see AHF100

DIALUMINUM DIPOTASSIUM SULFATE see PKU725

DIALUMINUM OCTAVANADIUM TRIDECASILICIDE see DBL649

DIALUMINUM SULFATE OCTADECAHYDRATE see AHG800

DIALUMINUM SULPHATE see AHG750

DIALUMINUM TRIOXIDE see AHE250

DIALUMINUM TRISULFATE see AHG750

DIALUX see ADY500

DIAMARIN see DYE600

DIAMAZO see ACR300

DIAMAZOL see EEQ600

DIAMELKOL see MCB050

DIAMET see SIL500

DIAMFEN HYDROCHLORIDE see DHY200

DIAMICRON see DBL700
DIAMIDAFOS see PEV500
DIAMIDE see DRP800
DIAMIDE see HGS000
DIAMIDFOS see PEV500
DIAMIDINE see DBL800
N,N'-DIAMIDINO-9-AZA-1,17-HEPTADECANEDIAMINE HYDROGEN SULFATE see GLQ000
4,4'-DIAMIDINODIPHENOXYPENTANE see DBM000
4,4'-DIAMIDINO-α,ω-DIPHENOXYPENTANE see DBM000
4,4'-DIAMIDINO-1,5-DIPHENOXYPENTANE DIHYDROCHLORIDE see PBJ800
4,4'-DIAMIDINODIPHENOXYPENTANE DI(β-HYDROXYETHANESULFONATE see DBL800
4,4'-DIAMIDINO-α,ω-DIPHENOXYPENTANE ISETHIONATE see DBL800
4,4'-DIAMIDINO-1,3-DIPHENOXYPROPANE DIHYDROCHLORIDE see DBM400
4,4'-DIAMIDINO-α,γ-DIPHENOXYPROPANE DIHYDROCHLORIDE see DBM400
p,p'-DIAMIDINODIPHENYLMETHANE see MJT050
DIAMIDINO STILBENE see SLS000
4,4'-DIAMIDINOSTILBENE see SLS000
4,4'-DIAMIDINOSTILBENE DIHYDROCHLORIDE see SLS500
DIAMINE see HGS000
2,4-DIAMINEANISOLE see DBO000
DIAMINE BLACK BH see CMN800
DIAMINE BLACK BHM see CMN800
DIAMINE BLUE 2B see CMO000
DIAMINE BLUE 3B see CMO250
DIAMINE BORDEAUX B see CMO872
DIAMINE BORDEAUX BA-CF see CMO872
DIAMINE BORDEAUX BC see CMO872
DIAMINE BROWN M see CMO800
DIAMINE BROWN MBA-CF see CMO800
DIAMINE BROWN MR see CMO800
DIAMINE BROWN MRC see CMO800
DIAMINE BROWN 3GN-CF see CMO825
DIAMINE DARK GREEN B see CMO830
DIAMINE DARK GREEN N see CMO830
DIAMINE DEEP BLACK EC see AQP000
(4,4'-DIAMINE)-3,3'-DIMETHYL(1,1'-BIPHENYL) see TGJ750
DIAMINE DIPENICILLIN G see BFC750
DIAMINE DIRECT BLACK E see AQP000
DIAMINE FAST BLACK B see CMN240
DIAMINE FAST RED F see CMO870
DIAMINE FAST RED FA-CF see CMO870
DIAMINE FAST RED N see CMO870
DIAMINE FAST RED OJCD see CMO870
DIAMINE FAST SCARLET 4BA see CMO870
DIAMINE FAST SCARLET 4BS see CMO870
DIAMINE GREEN B see CMO840
DIAMINE PURPURINE 4B see DXO850
DIAMINE SCARLET 3BA-CF see CMO875
DIAMINE SKY BLUE CI see CMO500
DIAMINE SKY BLUE FF see BGT250
DIAMINE VIOLET N see CMP000
DIAMINIDE MALEATE see DBM800
2,6-DIAMINOACRIDINE see DBN000
2,8-DIAMINOACRIDINE see DBN600
3,6-DIAMINOACRIDINE see DBN600
3,7-DIAMINOACRIDINE see DBN000
3,9-DIAMINOACRIDINE see ADJ625
3,6-DIAMINOACRIDINE BISULPHATE see DBN400
3,6-DIAMINOACRIDINE mixture with 3,6-DIAMINO-10-METHYLACRIDINIUM CHLORIDE see DBX400
2,5-DIAMINOACRIDINE (EUROPEAN) see ADJ625

3,6-DIAMINOACRIDINE HEMISULFATE see PMH100
3,6-DIAMINOACRIDINE HYDROCHLORIDE HEMIHYDRATE see DBN200
3,6-DIAMINOACRIDINE MONOHYDROCHLORIDE see PMH250
3,6-DIAMINOACRIDINE SULFATE (1:1) see DBN400
3,6-DIAMINOACRIDINE SULPHATE (1:1) see DBN400
2,8-DIAMINOACRIDINIUM see DBN600
3,6-DIAMINOACRIDINIUM see DBN600
3,6-DIAMINOACRIDINIUM CHLORIDE see PMH250
3,6-DIAMINOACRIDINIUM CHLORIDE HYDROCHLORIDE see PMH250
2,8-DIAMINOACRIDINIUM CHLORIDE MONOHYDROCHLORIDE see PMH250
3,6-DIAMINOACRIDINIUM MONOHYDROGEN SULPHATE see DBN400
2,8-DIAMINOACRIDINIUM SULPHATE see DBN400
1,2-DIAMINOAETHAN (GERMAN) see EEA500
2,6-DIAMINO-4-(((AMINO-CARBONYL)OXY)METHYL)-3a,4,8,9-TETRAHYDRO-1H,10H-PYRROLO(1,2-c)PURINE-10,10-DIOL (3aS-(3a-α-4-α,10aR*)) see SBA600
4,4'-DIAMINO-1,1'-ANIHRIMIDE see IBD000
2,4-DIAMINOANISOL see DBO000
2,4-DIAMINOANISOLE see DBO000
2,5-DIAMINOANISOLE see MEB800
3,4-DIAMINOANISOLE see MFG000
2,4-DIAMINOANISOLE BASE see DBO000
m-DIAMINOANISOLE 1,3-DIAMINO-4-METHOXYBENZENE see DBO000
2,4-DIAMINOANISOLE DIHYDROCHLORIDE see DBO100
2,5-DIAMINOANISOLE SULFATE see MEB820
2,4-DIAMINOANISOLE SULFATE HYDRATE see MEB900
2,4-DIAMINOANISOLE SULPHATE see DBO400
2,4-DIAMINOANISOLE SULPHATE see DBO400
2,4-DIAMINO-ANISOL SULPHATE see DBO400
1,4-DIAMINOANTHRACENE-9,10-DIOL see DBO600
1,4-DIAMINO-9,10-ANTHRACENEDIOL see DBO600
1,4-DIAMINO-9,10-ANTHRACENEDIONE see DBP000
2,6-DIAMINOANTHRACHINON see APK850
2,7-DIAMINOANTHRACHINON see DBP420
1,4-DIAMINOANTHRACHINON (CZECH) see DBP000
1,5-DIAMINOANTHRACHINON (CZECH) see DBP200
1,8-DIAMINOANTHRACHINON (CZECH) see DBP400
1,2-DIAMINOANTHRAQUINONE see DBO800
1,4-DIAMINOANTHRAQUINONE see DBP000
1,5-DIAMINOANTHRAQUINONE see DBP200
1,5-DIAMINOANTHRAQUINONE see DBP400
1,8-DIAMINOANTHRAQUINONE see DBP400
2,6-DIAMINOANTHRAQUINONE see APK850
2,7-DIAMINOANTHRAQUINONE see DBP420
1,5-DIAMINO-9,10-ANTHRAQUINONE see DBP200

2,6-DIAMINO-9,10-ANTHRAQUINONE see APK850
1,5-DIAMINOANTHRARUFIN see DBP909
4,8-DIAMINOANTHRARUFIN see DBP909
3,7-DIAMINO-5-AZAANTHRACENE see DBN600
2,4-DIAMINOAZOBENZEN (CZECH) see DBP999
DIAMINOAZOBENZENE see DBP999
p-DIAMINOAZOBENZENE see ASK925
4,4'-DIAMINOAZOBENZENE see ASK925
2,4-DIAMINOAZOBENZENE HYDROCHLORIDE see PEK000
2',4-DIAMINOBENZANILIDE see DBQ125
m-DIAMINOBENZENE see PEY000
o-DIAMINOBENZENE see PEY250
p-DIAMINOBENZENE see PEY500
1,2-DIAMINOBENZENE see PEY250
1,3-DIAMINOBENZENE see PEY000
1,4-DIAMINOBENZENE see PEY500
m-DIAMINOBENZENE DIHYDROCHLORIDE see PEY750
p-DIAMINOBENZENE DIHYDROCHLORIDE see PEY650
1,3-DIAMINOBENZENE DIHYDROCHLORIDE see PEY750
1,4-DIAMINOBENZENE DIHYDROCHLORIDE see PEY650
1,3-DIAMINOBENZENESULFONIC ACID see PFA250
2,4-DIAMINOBENZENESULFONIC ACID see PFA250
2,5-DIAMINOBENZENE SULFONIC ACID see PEY800
1,3-DIAMINOBENZENE-4-SULFONIC ACID see PFA250
1,3-DIAMINOBENZENE-6-SULFONIC ACID see PFA250
3,3'-DIAMINOBENZIDENE see BGK500
3,3'-DIAMINOBENZIDINE TETRAHYDROCHLORIDE see BGK750
3,5-DIAMINOBENZOIC ACID see DBQ190
3,5-DIAMINOBENZOIC ACID DIHYDROCHLORIDE see DBQ200
3,5-DIAMINOBENZOTRIFLUORIDE see TKB775
6,6'-DIAMINO-m,m'-BIPHENOL see DMI400
4,4'-DIAMINOBIPHENYL see BBX000
o,p'-DIAMINOBIPHENYL see BGF109
p,p'-DIAMINOBIPHENYL see BBX000
4,4'-DIAMINO-1,1'-BIPHENYL see BBX000
4,4'-DIAMINOBIPHENYL-3,3'-DICARBOXYLIC ACID see BFX250
4,4'-DIAMINO-3,3'-BIPHENYLDICARBOXYLIC ACID see BFX250
4,4'-DIAMINO-3,3'-BIPHENYLDICARBOXYLIC ACID DISODIUM SALT see BBX250
4,4'-DIAMINO-3,3'-BIPHENYLDIOL see DMI400
4,4'-DIAMINOBIPHENYL-2,2'-DISULFONIC ACID see BBX500
4,4'-DIAMINO-2,2'-BIPHENYLDISULFONIC ACID see BBX500
4,4'-((4,4'-DIAMINO-(1,1'-BIPHENYL)-3,3'-DIYL)BIS(OXY))BISBUTANOIC ACID, DIHYDROCHLORIDE see DEK000
4,4'-(3,3'-DIAMINO-p,p'-BIPHENYLENEDIOXY)DIBUTYRIC ACID see DEK400
4,4'-DIAMINOBIPHENYLOXIDE see OPM000
4,4'-DIAMINO-3-BIPHENYL-3-SULFONIC ACID see BBY250
DIAMINOBIURET see IBC000
1,5-DIAMINOBIURET see IBC000
1,5-DIAMINOBIURET DIHYDRAZIDE see IBC000
1,5-DIAMINOBROMO-4,8-DIHYDROXY-9,10-ANTHRACENEDIONE see DBQ220
1,3-DIAMINOBUTANE see BOR750

1,4-DIAMINOBUTANE see BOS000
1,4-DIAMINOBUTANE
DIHYDROCHLORIDE see POK325
2,6-DIAMINO-4-BUTYLAMINO-s-TRIAZINE
see BRR800
α,ε-DIAMINOCAPROIC ACID see LJM700
3,4-DIAMINOCHLOROBENZENE see
CFK125
3,5-DIAMINOCHLOROBENZENE see
CEJ200
1,2-DIAMINO-4-CHLOROBENZENE see
CFK125
3,4-DIAMINO-1-CHLOROBENZENE see
CFK125
4,6-DIAMINO-1-(p-CHLOROPHENYL)-1,2-
DIHYDRO-2,2-DIMETHYL-s-TRIAZINE
MONOHYDROCHLORIDE see COX325
2,4-DIAMINO-5-(4-CHLOROPHENYL)-6-
ETHYLPYRIMIDINE see TGD000
2,4-DIAMINO-5-p-CHLOROPHENYL-6-
ETHYLPYRIMIDINE see TGD000
DI-(4-AMINO-3-
CHLOROPHENYL)METHANE see MJM200
4,5-DIAMINOCHRYSAZIN see DBQ250
1,8-DIAMINOCHRYSAZINE see DBQ250
DIAMINOCILLIAN see BFC750
DI-(4-AMINO-3-CLOROFENIL)METANO
(ITALIAN) see MJM200
2,5-DIAMINO-2,4,6-CYCLOHEPTATRIEN-1-
ONE HYDROCHLORIDE see DCF710
1,2-DIAMINOCYCLOHEXANE see CPB100
1,3-DIAMINOCYCLOHEXANE see DBQ800
trans-l-DIAMINOCYCLOHEXANE
OXALATOPLATINUM see OKY100
1,2-
DIAMINOCYCLOHEXANEPLATINUM(II)
CHLORIDE see DAD040
1,2-
DIAMINOCYCLOHEXANETETRAACETIC
ACID see CPB120
1,2-DIAMINOCYCLOHEXANE-N,N'-
TETRAACETIC ACID see CPB120
DI(p-AMINOCYCLOHEXYL)METHANE see
MJQ260
2,5-DIAMINO-2,4,6-CYCLOHEPTATRIEN-1-
ONE see DCF700
1,12'-DIAMINODECANE see DXW800
2,4-DIAMINO-5-(p-(p-((p-(2,4-DIAMINO-1-
ETHYLPYRIMIDINIUM-5-
YL)PHENYL)CARBAMOYLCINNAMAMIDO
)PHENYL)-1-ETHYLPYRIMIDINIUM), DI-p-
TOLUENE SULFONATE see DBR000
2,4-DIAMINO-5-(p-(p-((-(2,4-DIAMINO-1-
METHYLPYRIMIDINIUM-5-
YL)PHENYL)CARBAMOYL)CINNAMAMID
O)PHENYL-1-METHYLPYRIMIDINIUM-DI-
p-TOLUENE SULFONATE see DBR200
4,4'-DIAMINO-1,1'-
DIANTHRAQUINONYLAMINE see IBD000
4,4'-DIAMINO-1,1'-
DIANTHRAQUINONYLIMINE see IBD000
4,4'-DIAMINO-1,1'-DIANTHRIMID (CZECH)
see IBD000
4,4'-DIAMINO-1,1'-DIANTHRIMIDE see
IBD000
1,4-DIAMINO-2,6-DICHLOROBENZENE see
DBR400
4,4'-DIAMINO-3,3'-DICHLOROBIPHENYL
see DEQ600
4,4'-DIAMINO-3,3'-DICHLORODIPHENYL
see DEQ600
4,4'-DIAMINO-3,3'-
DICHLORODIPHENYLMETHANE see
MJM200
2,4-DIAMINO-5-(3,4-DICHLOROPHENYL)-
6-METHYLPYRIMIDINE see MQR100
2,4-DIAMINO-5-(3',4'-DICHLOROPHENYL)-
6-METHYLPYRIMIDINE see MQR100
2,4-DIAMINO-6-(2,5-DICHLOROPHENYL)-s-
TRIAZINE MALEATE see MQY300
4:4'-DIAMINO-2:2'-DICHLOROSTILBENE
see DGK400

4:4'-DIAMINO-3:3'-DICHLOROSTILBENE
see DGK600
4,4'-DIAMINODICYCLOHEXYLMETHANE
see MJQ260
p,p'-DIAMINODICYCLOHEXYLMETHANE
see MJQ260
o-2,6-DIAMINO-2,6-DIDEOXY-α-d-
GLUCOPYRANOSYL-(1-4)-O-(β-d-
RIBOFURANOSYL-(1-5))-2-DEOXY-d-
STREPTAMINE see XQJ650
o-2,6-DIAMINO-2,6-DIDEOXY-α-d-
GLUCOPYRANOSYL-(1-4)-o-(β-d-
XYLOFURANOSYL-(1-5))-N¹-(4-AMINO-2-
HYDROXY-1-OXOBUTYL)-2-DEOXY-d-
STREPTAMINE see BSX325
o-2,6-DIAMINO-2,6-DIDEOXY-β-l-
IDOPYRANOSYL-(1-3)-o-(β-d-
RIBOFURANOSYL-(1-5)-o-(2,6)-DIAMINO-
2,6-DIDEOXY-α-d-GLUCOPYRANOSYL-(1-
4))-2-DEOXYD-STREPTAMINE SULFATE
(SALT) see NCD550
2,2'-DIAMINODIETHYLAMINE see DJG600
1,3-DIAMINO-7-(DIETHYLAMINO)-8-
METHYLPHENOXAZIN-5-IUM CHLORIDE
see BMM550
β,β'-DIAMINODIETHYL DISULFIDE see
MCN500
3,8-DIAMINO-5-(3-
(DIETHYLMETHYLAMMONIO)PROPYL)-6-
PHENYLPHENANTHRIDINIUM DIIODIDE
see PMT000
DIAMINODIFENILSULFONA (SPANISH)
see SOA500
p,p'-DIAMINODIFENYLMETHAN see
MJQ000
3,3'-DIAMINODIFENYLSULFON (CZECH)
see SOA000
1,4-DIAMINO-2,3-
DIHYDROANTHRAQUINONE see DBR450
1,4-DIAMINO-9,10-
DIHYDROXYANTHRACEN (CZECH) see
DBO600
1,5-DIAMINO-4,8-DIHYDROXY-9,10-
ANTHRACENEDIONE see DBP909
1,8-DIAMINO-4,5-
DIHYDROXYANTHRACHINON see
DBQ250
1,5-DIAMINO-4,8-
DIHYDROXYANTHRAQUINONE see
DBP909
1,8-DIAMINO-4,5-
DIHYDROXYANTHRAQUINONE see
DBQ250
4,5-DIAMINO-1,8-
DIHYDROXYANTHRAQUINONE see
DBQ250
4,8-DIAMINO-1,5-
DIHYDROXYANTHRAQUINONE see
DBP909
1,8-DIAMINO-4,5-DIHYDROXY-9,10-
ANTHRAQUINONE see DBQ250
leuco-1,5-DIAMINO-4,8-
DIHYDROXYANTHRAQUINONE see
DBP909
3,3'-DIAMINO-4,4'-
DIHYDROXYARSENOBENZENE
DIHYDROCHLORIDE see SAP500
3,3'-DIAMINO-4,4'-DIHYDROXY
ARSENOBENZENE
METHYLENESULFOXYLATE SODIUM see
NCJ500
1,5-DIAMINO-4,8-DIHYDROXY-3-(p-
METHOXYPHENYL)ANTHRAQUINONE
see DBT000
2,4-DIAMINO-6-
DIMETHOXYPHOSPHINOTHIONYLTHIO
METHYL-s-TRIAZINE see ASD000
2,8-DIAMINO-3,7-DIMETHYLACRIDINE see
DBT200
3,6-DIAMINO-2,7-DIMETHYLACRIDINE see
DBT200

3,6-DIAMINO-2,7-DIMETHYLACRIDINE
HYDROCHLORIDE see DBT400
4,4'-DIAMINO-3,3'-DIMETHYLBIPHENYL
see TGJ750
4,4'-DIAMINO-3,3'-DIMETHYLBIPHENYL
DIHYDROCHLORIDE see DQM000
4,4'-DIAMINO-3,3'-
DIMETHYLDICYCLOHEXYLMETHANE see
BGT800
4,4'-DIAMINO-3,3'-DIMETHYLDIPHENYL
see TGJ750
2,6-DIAMINO-3,4-DIMETHYL-7-
OXOPYRANO(4,3-G)BENZIMIDAZOLE see
DBT500
2,7-DIAMINO-3,8-DIMETHYLPHENAZINE
see DBT550
2,6-DIAMINO-3,4-DIMETHYLPYRANO(3,4-
E)BENZIMIDAZOL-7(3H)-ONE see DBT500
1:2'-DIAMINO-1':2-DINAPHTHYL see
BGC000
2,2'-DIAMINO-1,1'-DINAPHTHYL see
BGB750
1,12-DIAMINO-4,9-DIOXADODECANE see
BGU600
p-DIAMINODIPHENYL see BBX000
2,4'-DIAMINODIPHENYL see BGF109
4,4'-DIAMINODIPHENYL see BBX000
O,O'-DIAMINO DIPHENYL DISULFIDE see
DXJ800
4,4'-DIAMINODIPHENYL-2,2'-DISULFONIC
ACID see BBX500
DIAMINODIPHENYL ETHER see OPM000
4,4-DIAMINODIPHENYL ETHER see
OPM000
p,p'-DIAMINODIPHENYL ETHER see
OPM000
4,4'-DIAMINODIPHENYLMETHAN see
MJQ000
2,4'-DIAMINODIPHENYLMETHAN
(GERMAN) see MJP750
DIAMINODIPHENYLMETHANE see
MJQ000
2,4'-DIAMINODIPHENYLMETHANE see
MJP750
4,4'-DIAMINODIPHENYLMETHANE see
MJQ000
o,p'-DIAMINODIPHENYLMETHANE see
MJP750
p,p'-DIAMINODIPHENYLMETHANE see
MJQ000
4,4'-DIAMINODIPHENYLMETHANE (DOT)
see MJQ000
4,4'-DIAMINODIPHENYL OXIDE see
OPM000
4,4'-DIAMINODIPHENYL SULFIDE see
TFI000
p,p'-DIAMINODIPHENYL SULFIDE see
TFI000
3,3'-DIAMINODIPHENYL SULFONE see
SOA000
DIAMINO-4,4'-DIPHENYL SULFONE see
SOA500
4,4'-DIAMINODIPHENYL SULFONE see
SOA500
p,p'-DIAMINODIPHENYL SULFONE see
SOA500
p,p'-DIAMINODIPHENYLSULFONE-N,N'-
DI(DEXTROSE SODIUM SULFONATE) see
AOO800
p,p'-DIAMINODIPHENYL SULPHIDE see
TFI000
p,p-DIAMINODIPHENYL SULPHONE see
SOA500
DIAMINO-4,4'-DIPHENYL SULPHONE see
SOA500
p,p'-
DIAMINODIPHENYLTRICHLOROETHAN
E see BGU500
4:4'-DIAMINO-3-DIPHENYLYL
HYDROGEN SULFATE see BBY250
4,4'-DIAMINO-3-DIPHENYLYL
HYDROGEN SULFATE see BBY500

DI-β-AMINOPROPIONITRILE FUMARATE see AMB750

DI(3-AMINOPROPOXY)ETHANE see DCB200

m-DI-(2-AMINOPROPYL)BENZENE DIHYDROCHLORIDE see DCC100

p-DI-(2-AMINOPROPYL)BENZENE DIHYDROCHLORIDE see DCC125

DI(3-AMINOPROPYL) ETHER of DIETHYLENE GLYCOL see DJD800

N,N'-DIAMINOPROPYLETHYLENEDIAMINE see TBI700

4-((4-(((4-(2,4-DIAMINO-1-PROPYLPYRIMIDINIUM-5-YL)PHENYL)AMINO)CARBONYL)PHENYL)AMINO)-1-PROPYLQUINOLINIUM) DIIODIDE see DCC200

DIAMINOPROPYLTETRAMETHYLENEDI AMINE see DCC400

l-(+)-N-(p-(((2,4-DIAMINO-6-PTERIDINYL)METHYL)METHYLAMINO)B ENZOYL)GLUTAMIC ACID see MDV500

dl-N-(p-((2,4-DIAMINO-6-PTERIDYLMETHYL)AMINO)BENZOYL)AL ANINE SODIUM SALT see DCC500

2,6-DIAMINOPURINE see POJ500

2,6-DIAMINOPURINE SULFATE see DCC600

2,6-DIAMINOPYRIDINE see DCC800

3,4-DIAMINOPYRIDINE see DCD000

DIAMINO-3,4-PYRIDINE see DCD000

2,6-DIAMINO-4-(2-PYRIDYL)-s-TRIAZINE see DCD050

DIAMINOPYRITAMIN see TGD000

2,4-DIAMINOSOLE SULPHATE see DBO400

3,4'-DIAMINOSTILBENE see DCD100

4:4'-DIAMINOSTILBENE see SLR500

4,4'-DIAMINOSTILBENE-2,2'-STILBENEDISULFONIC ACID see FCA100

p,p'-DIAMINOSTILBENE-o,o'-DISULFONIC ACID DISODIUM SALT see FCA200

DIAMINOSTILBENE DISULPHONATE DISODIUM SALT see FCA200

2,4-DIAMINOTOLUEN (CZECH) see TGL750

DIAMINOTOLUENE see TGL500

DIAMINOTOLUENE see TGL750

2,3-DIAMINOTOLUENE see TGY800

2,4-DIAMINOTOLUENE see TGL750

2,5-DIAMINOTOLUENE see TGM000

2,6-DIAMINOTOLUENE see TGM100

3,4-DIAMINOTOLUENE see TGM250

2,4-DIAMINO-1-TOLUENE see TGL750

2,4-DIAMINOTOLUENE DIHYDROCHLORIDE see DCE000

2,5-DIAMINOTOLUENE DIHYDROCHLORIDE see DCE200

2,6-DIAMINOTOLUENE DIHYDROCHLORIDE see DCE400

p-DIAMINOTOLUENE SULFATE see DCE600

2,5-DIAMINOTOLUENE SULFATE see DCE600

2,5-DIAMINOTOLUENE SULPHATE see DCE600

2,4-DIAMINOTOLUOL see TGL750

(4-((4-DIAMINO-m-TOLYL)IMINO)-2,5-CYCLOHEXADIEN-1-YLIDENE) DIMETHYLAMMONIUM CHLORIDE see DCE800

(4-((4-DIAMINO-m-TOLYL)IMINO)-2,5-CYCLOHEXADIEN-1-YLIDENE)DIMETHYLAMMONIUM CHLORIDE H₂O see TGU500

4,6-DIAMINO-s-TRIAZINE-2-METHANETHIOL S-ESTER with O,O-DIMETHYLPHOSPHORODITHIOATE see ASD000

4,6-DIAMINO-s-TRIAZINE-2-THIONE see DCF000

4,6-DIAMINO-1,3,5-TRIAZINE-2-THIONE see DCF000

4,6-DIAMINO-1,3,5-TRIAZINE-2(1H)-THIONE see DCF000

N-(4,6-DIAMINO-s-TRIAZIN-2-YL)-p-ARSANILIC ACID SODIUM SALT (2:3) see SIF450

S-((4,6-DIAMINO-1,3,5-TRIAZIN-2-YL)-METHYL)-O,O-DIMETHYL-DITHIOFOSFAAT (DUTCH) see ASD000

S-((4,6-DIAMINO-1,3,5-TRIAZIN-2-YL)-METHYL)-O,O-DIMETHYL-DITHIOPHOSPHAT (GERMAN) see ASD000

S-((4,6-DIAMINO-s-TRIAZIN-2-YL)METHYL)-O,O-DIMETHYL PHOSPHORODITHIOATE see ASD000

4,6-DIAMINO-1,3,5-TRIAZIN-2-YL-METHYL-O,O-DIMETHYL PHOSPHORODITHIOATE see ASD000

S-(4,6-DIAMINO-1,3,5-TRIAZIN-2-YLMETHYL)-O,O-DIMETHYL PHOSPHORODITHIOATE see ASD000

S-(4,6-DIAMINO-1,3,5-TRIAZIN-2-YLMETHYL) DIMETHYL PHOSPHOROTHIOLOTHIONATE see ASD000

3,5-DIAMINO-s-TRIAZOLE see DCF200

3,10-DIAMINOTRICYCLO(5.2.1.0²,⁶)DECANE see DCF600

2,4-DIAMINO-5-(3,4,5-TRIMETHOXYBENZYL)PYRIMIDINE see TKZ000

2,5-DIAMINOTROPONE see DCF700

2,5-DIAMINOTROPONE HYDROCHLORIDE see DCF710

1,3-DIAMINOUREA see CBS500

(S)-α,Δ-DIAMINOVALERIC ACID see OJS100

DIAMIRA TURQUOISE BLUE see CMS224

DIAMMIDE SEBACICA della 4-FENIL-4-AMMINOMETIL-N-METILPIPERIDINA (ITALIAN) see BKS825

DIAMMINE(BENZYLMALONATO)PLATIN UM (II) see DCF720

DIAMMINEBORONIUM HEPTAHYDROTETRABORATE see DCF725

DIAMMINEBORONIUM TETRAHYDROBORATE see DCF750

DIAMMINE(1,1-CYCLOBUTANEDICARBOXYLATO)PLATI NUM (II) see CCC075

cis-DIAMMINE(1,1-CYCLOBUTANEDICARBOXYLATO)PLATI NUM(II) see CCC075

cis-DIAMMINEDIBROMOPLATINUM see DCF760

cis-DIAMMINEDIBROMOPLATINUM(II) see DCF760

cis-DIAMMINEDICHLOROPLATINUM see PJD000

trans-DIAMMINEDICHLOROPLATINUM(II) see DEX000

cis-DIAMMINEDINITRATO PLATINUM (II) see DCF800

DIAMMINEMALONATO PLATINUM (II) see DCG000

DIAMMINEPALLADIUM (II) NITRATE see DCG600

cis-DIAMMINEPLATINUM DIBROMIDE see DCF760

DIAMMINETETRACHLOROPLATINUM (OC-6-22) (9CI) see TBO776

cis-DIAMMINOTETRACHLOROPLATINUM see TBO776

1,2-DIAMMONIOETHANE NITRATE see EIV700

DIAMMONIUM ARSENATE see DCG800

DIAMMONIUM BERYLLIUMTETRAFLUORIDE see ANH300

DIAMMONIUM CARBONATE see ANE000

DIAMMONIUM CHROMATE see ANF500

DIAMMONIUM CHROMATE see NCQ550

DIAMMONIUM CITRATE see ANF800

DIAMMONIUM FLUOSILICATE see COE000

DIAMMONIUM HEXACHLOROPALLADATE see ANF000

DIAMMONIUM HEXACHLOROPLATINATE (2-) see ANF250

DIAMMONIUM HEXAFLUOROSILICATE(2-) see COE000

DIAMMONIUM HYDROGEN ARSENATE see DCG800

DIAMMONIUM HYDROGEN PHOSPHATE see ANR500

DIAMMONIUM MOLYBDATE see ANM750

DIAMMONIUM MONOHYDROGEN ARSENATE see DCG800

DIAMMONIUM NICKEL DISULFATE HEXAHYDRATE see NCY060

DIAMMONIUM SILICON HEXAFLUORIDE see COE000

DIAMMONIUM SULFATE see ANU750

DIAMMONIUM TARTRATE see DCH000

DIAMMONIUM TETRACHLOROPALLADATE see ANE750

DIAMMONIUM TETRACHLOROPALLADATE(2-) see ANE750

DIAMMONIUM TETRAFLUOROBERYLLATE see ANH300

DIAMMONIUM THIOSULFATE see ANK600

DIAMMONIUM TRISULFIDE see ANT000

DIAMOND BLUE BLACK AE see CMP880

DIAMOND BLUE BLACK EB see CMP880

DIAMOND BLUE BLACK EBS-CF see CMP880

DIAMOND CHROME ORANGE GR see CMP882

DIAMOND CHROME YELLOW 3R see NEY000

DIAMOND CORINTH N see HLI000

DIAMOND FAST RED BT see CMG750

DIAMOND FUCHSINE see MAC250

DIAMOND GREEN B see AFG500

DIAMOND GREEN G see BAY750

DIAMOND RED BHA see CMG750

DIAMOND RED W see SEH475

DIAMOND SHAMROCK 40 see PKQ059

DIAMOND SHAMROCK 744 see AAX175

DIAMOND SHAMROCK DS-15647 see DAB400

DIAMORFINA see HBT500

DIAMORPHINE see HBT500

DIAMORPHINE HYDROCHLORIDE see DBH400

DIAMOX see AAI250

DIAMPHETAMINE SULFATE see BBK250

DIAMYCELINE see CDY325

DIAMYL AMINE see DCH200

DI-n-AMYLAMINE (DOT) see DCH200

α-(DIAMYLAMINOMETHYL)-1,2,3,4-TETRAHYDRO-9-PHENATHRENEMETHANOL see DCH300

DIAMYLESTER KYSELINY MALEINOVE (CZECH) see MAL000

DIAMYL ETHER see PBX000

DI-n-AMYL ETHER see PBX000

2,5-DI-tert-AMYLHYDROQUINONE see DCH400

DIAMYL KETONE see ULA000

DIAMYL MALEATE see MAL000

DIAMYLNITROSAMIN (GERMAN) see DCH600

DI-n-AMYLNITROSAMINE see DCH600

DIAMYLPHENOL see DCH800

DI-tert-AMYLPHENOL see DCI000

2,4-DI-tert-AMYLPHENOL see DCI000

2,4-DI-2-AMYLPHENOXYACETYL CHLORIDE see DCI100

DIAMYL PHTHALATE see AON300

DIAN see BLD500

DIANABOL see DAL300

DIANABOL see MPN500

DIANABOLE see DAL300

DIANAT (RUSSIAN) see MEL500

DIANATE see MEL500

DIANCINA see AOD000
DIANDRON see AOO450
DIANDRONE see AOO450
DIANEMYCIN see DCI400
DIANHYDROCULCITOL see DCI600
1,2:5,6-DIANHYDRODULCITOL see DCI600
1,2:3,4-DIANHYDROERYTHRITOL see DHB800
DIANHYDROGALACTITOL see DCI600
1,2:5,6-DIANHYDROGALACTITOL see DCI600
d-1,4:3,6-DIANHYDROGLUCITOL see HID350
DIANHYDROMANNITOL see DCI800
1:2:5:6-DIANHYDRO-d-MANNITOL see DCI800
1,4:3,6-DIANHYDROSORBITOL see HID350
1,2-5,6-DIANHYDRO-d-SORBITOL see DCI900
1,4:3,6-DIANHYDROSORBITOL-2,5-DINITRATE see CCK125
1,2:3,4-DIANHYDRO-dl-THREITOL see DHB600
DIANILBLAU see CMO250
DIANIL BLUE see CMO250
DIANIL DARK BLUE H see CMN800
o,p'-DIANILINE see BGF109
p,p'-DIANILINE see BBX000
DIANILINOMERCURY see DCJ000
DIANILINOMETHANE see MJQ000
o-DIANISIDIN (CZECH, GERMAN) see DCJ200
o-DIANISIDINA (ITALIAN) see DCJ200
o-DIANISIDINE see DCJ200
3,3'-DIANISIDINE see DCJ200
o-DIANISIDINE DIHYDROCHLORIDE see DOA800
DIANISIDINE DIISOCYANATE see DCJ400
DI-p-ANISYLAMINE see BKO600
α,γ-DI-p-ANISYL-β-(ETHOXYPHENYL)GUANIDINE HYDROCHLORIDE see PDN500
3,4-DIANISYL-3-HEXENE see DJB200
DIANISYL-MONOPHENETHYLGUANIDINE HYDROCHLORIDE see PDN500
DIANISYLTRICHLORETHANE see MEI450
2,2-DI-p-ANISYL-1,1,1-TRICHLOROETHANE see MEI450
DIANIX BLUE BG-FS see CMP075
DIANIX FAST BLUE BG-FS see CMP075
DIANIX FAST PINK R see AKO350
DIANIX FAST VIOLET B see DBY700
DIANIX FAST YELLOW 5R see CMP090
DIANIX YELLOW 5R-E see CMP090
DIANIX YELLOW YL see KDA075
DIANOL see AHP000
DIANON see DCM750
DI-1,1'-ANTHRACHINONYLAMIN (CZECH) see IBI000
DIANTHRAQUINONYLAMINE see IBI000
1,1'-DIANTHRAQUINONYLAMINE see IBI000
1,1-DIANTHRIMID (CZECH) see IBI000
DIANTHRIMIDE see IBI000
1,1'-DIANTHRIMIDE see IBI000
DIANTHUS SUPERBUS L., extract see DCJ450
DIANTIMONY PENTOXIDE see AQF750
DIANTIMONY TRIOXIDE see AQF000
DIANTIMONY TRISULFATE see AQJ250
DIANTIMONY TRISULFIDE see AQL500
DIAPADRIN see DGQ875
DIAPAM see DCK759
DIAPAMIDE see CIP500
DIAPARENE see MHB500
DIAPARENE CHLORIDE see MHB500
DIAPHEN see DVW700
DIAPHEN (NEUROPLEGIC) see DHW200
DIAPHENYL see EEQ600
DIAPHENYLSULFONE see SOA500
DIAPHENYLSULPHON see SOA500
DIAPHENYLSULPHONE see SOA500

DIAPHORM see HBT500
DIAPHTAMINE BLACK BH see CMN800
DIAPHTAMINE BLACK V see AQP000
DIAPHTAMINE BLUE BB see CMO000
DIAPHTAMINE BLUE RW see CMO600
DIAPHTAMINE BRILLIANT YELLOW 6GS see CMP050
DIAPHTAMINE BROWN 3GC see CMO825
DIAPHTAMINE BROWN M see CMO800
DIAPHTAMINE FAST BLACK KG see CMN240
DIAPHTAMINE FAST BROWN TB see CMO820
DIAPHTAMINE FAST PURPURINE 8B see CMO880
DIAPHTAMINE FAST RED FC see CMO870
DIAPHTAMINE FAST SCARLET see CMO875
DIAPHTAMINE FAST SCARLET 4BS see CMO870
DIAPHTAMINE GREEN B see CMO840
DIAPHTAMINE LIGHT RED 4B see CMO885
DIAPHTAMINE LIGHT TURQUOISE G see COF420
DIAPHTAMINE PURE BLUE see CMO500
DIAPHTAMINE PURPURINE see DXO850
DIAPHTAMINE VIOLET N see CMP000
DIAPHTHAMINE FAST BLACK FE see CMO830
DIAPHTOGENE BLACK D see CMN230
DIAPHYLLINE see TEP500
DIAPP see BEN000
DIAQUODIAMMINEPLATINUM DINITRATE see DCJ600
cis-DIAQUODIAMMINEPLATINUM(II) DINITRATE see DCJ600
DIAQUONE see DMH400
DIARESIN BLUE K see BKP500
DIAREX 43G see SMQ500
DIAREX 600 see SMR000
DIAREX HF 55 see SMQ500
DIAREX HF 77 see SMQ000
DIAREX HF 77 see SMQ500
DIAREX HF 55-247 see SMQ500
DIAREX HS 77 see SMQ500
DIAREX HT 88 see SMQ500
DIAREX HT 90 see SMQ500
DIAREX HT 190 see SMQ500
DIAREX HT 500 see SMQ500
DIAREX HT 88A see SMQ500
DIAREX YH 476 see SMQ500
DIARLIDAN see DGQ500
DIARSEN see DXE600
(1,2-DIARSENEDIYLBIS((6-HYDROXY-3,1-PHENYLENE)IMINO))BISMETHANESULFONIC ACID DISODIUM SALT see SNR000
DIARSENIC ACID, TETRASODIUM SALT see SEY100
DIARSENIC PENTOXIDE see ARH500
DIARSENIC TRIOXIDE see ARI750
DIARSENIC TRISELENIDE see ARH800
DIARSENIC TRISELENIDE (ASSE3) see ARH800
DIARSENIC TRISULFIDE see ARI000
DIARSENIC TRISULPHIDE see ARI000
DIARYLANILIDE YELLOW see DEU000
DIARYLIDE ORANGE see CMS145
DIARYLIDE YELLOW see CMS208
DIARYLIDE YELLOW AAOT see CMS210
N,N'-DIARYL-p-PHENYLENE DIAMINE see DCJ700
DIASAN see FNF000
DIASETIELMORFIEN see HBT500
DIASETILMORFIN see HBT500
DIASETYLMORFIIMI see HBT500
DIASONE HYDROCHLORIDE see MDP000
DIASTATIN see NOH500
DIAST T-514 see PCL300
DIASTYL see DKA600
DIASULFA see SNN300
DIASULFYL see SNN300
DIAT (GERMAN) see DCK000

DIATER see DXQ500
DIATERR-FOS see DCM750
DIATHESIN see HMK100
DIATHOL AS see CMM760
DIATHOL ASF see CMM760
DIATO BLUE BASE B see DCJ200
DIATOMACEOUS EARTH see DCJ800
DIATOMACEOUS EARTH, NATURAL see DCJ800
DIATOMACEOUS SILICA see DCJ800
DIATOMITE see DCJ800
DIATRAST see DNG400
DIATRIN HYDROCHLORIDE see DCJ850
DIATRIZOATE MEGLUMINE see AOO875
DIATRIZOATE METHYLGLUCAMINE see AOO875
DIATRIZOATE SODIUM SALT see SEN500
DIATRIZOESAURE (GERMAN) see DCK000
DIATRIZOIC ACID see DCK000
DIA-TUSS see TCY750
6,12-DIAZAANTHANTHRENE see ADK250
6,12-DIAZAANTHANTHRENE SULFATE see DCK200
6,12-DIAZAANTHANTHRENE SULPHATE see DCK200
1,2-DIAZABENZENE see OJW200
1,3-DIAZABENZENE see PPO750
1,4-DIAZABENZENE see POL490
10'-(β-(1,4-DIAZABICYCLO(4.3.0)NONANYL-4)-PROPIONYL)-2'-CHLOROPHENTHIAZIN DICHLORIDE see NMV750
1,4-DIAZABICYCLO(2,2,2)OCTANE see DCK400
1,4-DIAZABICYCLO(2.2.2)OCTANE HYDROGEN PEROXIDATE see DCK500
DIAZACHEL (OBS.) see MDQ250
DIAZACOSTEROL HYDROCHLORIDE see ARX800
1,4-DIAZACYCLOHEPTANE see HGI900
1,3-DIAZA-2,4-CYCLOPENTADIENE see IAL000
7,14-DIAZADIBENZ(a,h)ANTHRACENE see DDC800
1,12-DIAZADIBENZO(a,i)PYRENE see NAU000
4,11-DIAZADIBENZO(a,h)PYRENE see NAU500
1,2-DIAZA-3,4:9,10-DIBENZPYRENE see APF750
1,3,2,4-DIAZADIPHOSPHETIDINE-2,4-DIAMINE, N,N',1,3-TETRAPHENYL- see TEA550
1,3,2,4-DIAZADIPHOSPHETIDINE, 2,4-DIANILINO-1,3-DIPHENYL-(8CI) see TEA550
2,5-DIAZAHEXANE see DRI600
1,6-DIAZA-3,4,8,9,12,13-HEXAOXABICYCLO(4.4.4)TETRADECANE see DCK700
1,3-DIAZAINDENE see BCB750
2,3-DIAZAINDOLE see BDH250
DIAZAJET see DCM750
DIAZALE see THE500
DIAZAMINE BLACK D see CMN230
DIAZAMINE FAST SCARLET 4BS see CMO870
DIAZAMINE GOLDEN YELLOW T see CMP050
DIAZAMINE PURPURINE 4B see DXO850
DIAZAN see FNF000
1,4-DIAZANAPHTHALENE see QRJ000
3,6-DIAZAOCTANE-1,8-DIAMINE see TJR000
4,5-DIAZAPHENANTHRENE see PCY250
1,3,2,4-DIAZAPHOSPHETIDINE , 2,4-DIANILINO-1,3-DIPHENYL -(7CI) see TEA550
3,5-DIAZAPHTHALIMID (GERMAN) see PPP000
3,6-DIAZAPHTHALIMID (GERMAN) see POL750

DIAZATOL see DCM750
DIAZENE 42 see EHY000
DIAZENE, BIS(4-CHLORPHENYL)-, 1-OXIDE (9CI) see DEP450
DIAZENE, BIS(2-METHYL-3,5-DINITROPHENYL)-, 1-OXIDE see TDY090
DIAZENE, BIS(4-METHYL-3,5-DINITROPHENYL)-, 1-OXIDE see TDY085
DIAZENE, BIS(2-METHYL-3-NITROPHENYL)-, 1-OXIDE see BKS812
DIAZENE, BIS(2-METHYL-5-NITROPHENYL)-, 1-OXIDE see BKS813
DIAZENE, BIS(4-METHYL-3-NITROPHENYL)-, 1-OXIDE see BKS814
DIAZENECARBOXYLIC ACID, PHENYL-, 2-PHENYLHYDRAZIDE (9CI) see DVY950
DIAZENEDICARBOXAMIDE see ASM270
DIAZENE, ETHYLHYDROXY-, POTASSIUM SALT, (E)- see ELE700
DIAZENE, ETHYLHYDROXY-, POTASSIUM SALT, (Z)- see ELE720
DIAZENE-15N2, HYDROXYMETHYL-, (E)- see HMB560
DIAZENE, HYDROXYMETHYL-, POTASSIUM SALT, (E)- see HMB565
DIAZENE, HYDROXYMETHYL-, POTASSIUM SALT, (Z)- see HMB570
DIAZENE, HYDROXYPROPYL-, POTASSIUM SALT, (E)- see HNV100
DIAZENE, HYDROXYPROPYL-, POTASSIUM SALT, (Z)- see HNV150
DIAZENE, (2-METHYL-3,5-DINITROPHENYL)(4-METHYL-3,5-DINITROPHENYL)-, 1-OXIDE see TDY080
DIAZEPAM see DCK759
1H-1,4-DIAZEPINE, HEXAHYDRO- see HGI900
1H-1,4-DIAZEPINE, HEXAHYDRO-1-(p-CHLORO-α-PHENYLBENZYL)-4-METHYL- see HGI100
DIAZETARD see DCK759
DIAZETOXYSKIRPENOL (GERMAN) see AOP250
DIAZETYLMORPHINE see HBT500
DIAZIDE see DCM750
1,3-DIAZIDOBENZENE see DCL100
1,4-DIAZIDOBENZENE see DCL125
1,4-DIAZIDOBENZENE see DCL125
p-DIAZIDOBENZENE (DOT) see DCL125
2,2-DIAZIDOBUTANE see DCL159
1,2-DIAZIDOCARBONYL HYDRAZINE see DCL200
DIAZIDODIBENZALACETON see BGW720
DIAZIDODIBENZALACETON (CZECH) see BJA000
2,5-DIAZIDO-3,6-DICHLOROBENZOQUINONE see DCL300
DIAZIDODICYANOMETHANE see DCM000
DIAZIDODIMETHYLSILANE see DCL350
1,1-DIAZIDOETHANE see DCL400
1,2-DIAZIDOETHANE see DCL600
DIAZIDO ETHIDIUM see DCL800
3,8-DIAZIDO-5-ETHYL-6-PHENYLPHENANTHRIDINIUM BROMIDE see DCL800
DIAZIDOGLYCEROL see DCM550
DIAZIDOMALONONITRILE see DCM000
DIAZIDOMETHYLENEAZINE see DCM200
DIAZIDOMETHYLENECYANAMIDE see DCM400
1,3-DIAZIDO-2-NITROAZAPROPANE see DCM499
1,3-DIAZIDO-2-PROPANOL see DCM550
1,3-DIAZIDOPROPENE see DCM600
2,6-DIAZIDOPYRAZINE see DCM700
DIAZIL see DHU900
m-DIAZINE see PPO750
p-DIAZINE see POL490
1,2-DIAZINE see OJW200
1,4-DIAZINE see POL490
DIAZINE BLACK BHC see CMN800
DIAZINE BLACK DR see CMN230

DIAZINE BLACK E see AQP000
DIAZINE BLACK H see CMN800
DIAZINE BLACK HDW see CMN800
DIAZINE BLACK HNJ see CMN800
DIAZINE BLUE 2B see CMO000
DIAZINE BLUE 3B see CMO250
DIAZINE BROWN M see CMO800
DIAZINE BROWN MWR see CMO800
DIAZINE DARK GREEN BO see CMO830
DIAZINE DARK GREEN P see CMO830
DIAZINE DIRECT BLACK E see AQP000
DIAZINE DIRECT BLACK G see AQP000
DIAZINE FAST RED F see CMO870
DIAZINE FAST RED 8BK see CMO885
DIAZINE GREEN B see CMO840
DIAZINE GREEN DB see CMO840
DIAZINE RED 4B see DXO850
DIAZINE VIOLET N see CMP000
DIAZINON see DCM750
DIAZINONE see DCM750
DIAZINON mixed with METHOXYCHLOR see MEI500
DIAZIQUONE see ASK875
3,6-DIAZIRIDINYL-2,5-BIS(CARBOETHOXYAMINO)-1,4-BENZOQUINONE see ASK875
5-(4,6-DIAZIRIDINYL-2-s-TRIAZINYLAMINO)-2-FURYL-m-DIOXANE-5-METHANOL see FQB100
DIAZIRINE see DCM800
DIAZIRINE see DCP800
DIAZIRINE-3,3-DICARBOXYLIC ACID see DCM875
2,5-DIAZIRINO-3,6-DIPROPOXY-p-BENZOQUINONE see DCN000
DIAZITOL see DCM750
DIAZO see DUR800
DI-AZO see PDC250
DIAZO A see DCP300
DIAZOACETALDEHYDE see DCN200
2-(DIAZOACETAMINO)-N-ETHYLACETAMIDE see DCN600
DIAZOACETATE (ESTER)-l-SERINE see ASA500
l-DIAZOACETATE (ESTER) SERINE see ASA500
DIAZO-ACETIC ACID ESTER with SERINE see ASA500
DIAZOACETIC ACID, ETHYL ESTER see DCN800
DIAZOACETIC ESTER see DCN800
N-DIAZOACETILGLICINA-AMIDE (ITALIAN) see CBK000
N-DIAZOACETILGLICINA-IDRAZIDE (ITALIAN) see DCO800
DIAZOACETONITRILE see DCN875
DIAZOACETOPHENONE see PET800
2-DIAZOACETOPHENONE see PET800
omega-DIAZOACETOPHENONE see PET800
α-DIAZOACETOPHENONE see PET800
2-((DIAZOACETYL)AMINO)-N-ETHYLACETAMIDE see DCN600
2-((DIAZOACETYL)AMINO)-N-METHYLACETAMIDE see DCO200
DIAZOACETYL AZIDE see DCO509
DIAZOACETYLGLYCINAMIDE see CBK000
N-(DIAZOACETYL)GLYCINAMIDE see CBK000
DIAZOACETYLGLYCINE AMIDE see CBK000
N-DIAZOACETYLGLYCINE AMIDE see CBK000
N-DIAZOACETYLGLYCINEETHYLAMIDE see DCN600
DIAZOACETYLGLYCINE ETHYL ESTER see DCO600
N-DIAZOACETYLGLYCINE ETHYL ESTER see DCO600
DIAZOACETYLGLYCINE HYDRAZIDE see DCO800
N-(DIAZOACETYL)GLYCINE HYDRAZINE see DCO800

N-DIAZOACETYLGLYCINE METHYLAMIDE see DCO200
N-DIAZOACETYL GLYCYLHYDRAZIDE see DCO800
N-(1-(DIAZOACETYL)-2-METHYLBUTYL)-p-TOLUENESULFONAMIDE see THH480
o-DIAZOACETYL-l-SERINE see ASA500
DIAZOAMINOBENZEN (CZECH) see DWO800
DIAZOAMINOBENZENE see DWO800
p-DIAZOAMINOBENZENE see DWO800
DIAZOAMINOBENZOL (GERMAN) see DWO800
6-DIAZO-2-(2-(4-AMINO-4-CARBOXYBUTYRAMIDO)-6-DIAZO-5-OXOHEXANAMIDO)-HEXANOIC ACID SODIUM see ASO510
DIAZO A-SS see DOU100
DIAZOBEN see MND600
DIAZOBENZENE see ASL250
DIAZO BLACK BH see CMN800
DIAZO BLACK BHN-CF see CMN800
DIAZO BLACK BHSW see CMN800
DIAZO BLACK BHSWK see CMN800
DIAZO BLACK CR see CMN800
DIAZO BLACK D see CMN230
DIAZOBLEU see BGT250
DIAZO BROWN MC see CMO800
DIAZOCARD CHRYSOIDINE G see PEK000
20,25-DIAZOCHOLESTEROL DIHYDROCHLORIDE see ARX800
DIAZOCYCLOPENTADIENE see DCP200
DIAZODICYANOIMIDAZOLE see DCP880
DIAZODICYANOMETHANE see DCP775
4-DIAZO-N,N-DIMETHYLANILIN CHLOROZINCATE see DCP300
2-DIAZO-4,6-DINITROBENZENE-1-OXIDE see DUR800
DIAZODINITROPHENOL (dry) (DOT) see DUR800
DIAZODINITROPHENOL, wetted with not <40% H2O or mixture of alcohol & H2O (UN 0074) (DOT) see DUR800
DIAZODIPHENYLMETHANE (DOT) see DVZ100
DIAZO DIRECT BLACK N see CMN800
DIAZOESSIGSAEURE-AETHYLESTER (GERMAN) see DCN800
DIAZO FAST BLACK BH see CMN800
DIAZO FAST BLACK D see CMN230
DIAZO FAST BLACK MBH see CMN800
DIAZO FAST BLACK MD see CMN230
DIAZO FAST GARNET GBC see MPY750
DIAZO FAST ORANGE GR see NEO000
DIAZO FAST ORANGE R see NEN500
DIAZO FAST ORANGE RD see CEG800
DIAZO FAST RED AL see AIA750
DIAZO FAST RED B see NEQ000
DIAZO FAST RED GG see NEO500
DIAZO FAST RED 3GL see KDA050
DIAZO FAST RED RL see MMF780
DIAZO FAST RED TR see CLK235
DIAZO FAST RED TRA see CLK220
DIAZO FAST RED TRA see CLK235
DIAZO FAST SCARLET G see NMP500
DIAZO-ICA see DCP400
DIAZOIMIDAZOLE-4-CARBOXAMIDE see DCP400
5-DIAZOIMIDAZOLE-4-CARBOXAMIDE see DCP400
5-DIAZOIMIDAZOLE-4-CARBOXAMIDE HYDROCHLORIDE see DCP600
DIAZOIMIDE see HHG500
N-(3-DIAZO-1-ISOBUTYLACETONYL)-p-TOLUENESULFONAMIDE see THH490
DIAZOL see DCM750
DIAZOL BLACK 2V see AQP000
DIAZOL BLACK BH see CMN800
DIAZOL BLACK SD see CMN230
DIAZOL BLUE 2B see CMO000
DIAZOL BLUE RW see CMO600
DIAZOL BORDEAUX B see CMO872

DIAZOL BROWN M see CMO800
DIAZOL-C see AKK250
DIAZOL CUTCH F see CMO820
DIAZOL CUTCH FB see CMO820
1,2-DIAZOLE see POM500
1,3-DIAZOLE see IAL000
DIAZOLE ORANGE O see DCP650
DIAZOLE RED 4S see NMO700
DIAZOL FAST BLACK JRA see CMN300
DIAZOL FAST PURPURINE 8B see CMO880
DIAZOL FAST RED F see CMO870
DIAZOL FAST RED FS see CMO870
DIAZOL FAST SCARLET 4BS see CMO870
DIAZOL GREEN B see CMO840
DIAZOL GREEN BJ see CMO840
DIAZOL GREEN BLACK N see CMO830
DIAZOLIDINYL UREA see IAS100
DIAZO LIGHT RED 8B see CMO885
DIAZO LIGHT RED 8BD see CMO885
DIAZOL LIGHT TURQUOISE JL see COF420
DIAZOLONE see PPP500
DIAZOL PURE BLUE 4B see CMO500
DIAZOL PURE BLUE FF see BGT250
DIAZOL PURPURINE 4B see DXO850
DIAZOL RED S see NMO700
DIAZOL SCARLET 3B see CMO875
DIAZOL VIOLET N see CMP000
DIAZOL YELLOW J see CMP050
DIAZOMALONIC ACID see DCP700
DIAZOMALONONITRILE see DCP775
DIAZOMETHANE see DCP800
N-(3-DIAZO-1-METHYLACETONYL)-p-TOLUENESULFONAMIDE see THH380
DIAZOMETHYL PHENYL KETONE see PET800
DIAZOMYCIN B see ASO510
DIAZO NAVY BLUE BH see CMN800
DIAZO NERO MICROSETILE G see DPO200
2-DIAZONIO-4,5-DICYANOIMIDAZOLIDE see DCP880
5-DIAZONIOTETRAZOLIDE see DCQ200
N-(4-DIAZO-3-OXOBUTYL)-p-TOLUENESULFONAMIDE see THH375
N-(7-DIAZO-6-OXOHEPTYL)-p-TOLUENESULONAMIDE see THH460
DIAZO-OXO-NORLEUCINE see DCQ400
6-DIAZO-5-OXONORLEUCINE see DCQ400
6-DIAZO-5-OXO-l-NORLEUCINE see DCQ400
DIAZOPHENYL BLACK BH see CMN800
DIAZOPHENYL BLACK D see CMN230
DIAZOPHENYL BLACK DC see CMN230
2-DIAZO-1-PHENYLETHANONE see PET800
4-DIAZO-5-PHENYL-1,2,3-TRIAZOLE see DCQ500
1,3-DIAZOPROPANE see DCQ525
3-DIAZOPROPENE see DCQ550
5-DIAZOPYRIMIDINE-2,4(3H)-DIONE see DCQ600
5-DIAZO-2,4(1H,3H)-PYRIMIDINEDIONE see DCQ600
DIAZO RED RD see DCQ560
DIAZO RESIN V see DCQ650
DIAZORESORCINOL see HNG500
DIAZO 8 SS see DOU100
DIAZOSSIDO (ITALIAN) see DCQ700
DIAZOTIZING SALTS see SIQ500
3-DIAZOTYRAMINE HYDROCHLORIDE see DCQ575
DIAZOURACIL see DCQ600
5-DIAZOURACIL see DCQ600
DIAZO V see DCQ650
DIAZOXIDE see DCQ700
DIAZO Y see MPC275
DIAZO 4VZS see DCP300
DIAZYL see PPP500
DIAZYL see SNJ000
DIBA see DNH125
DIBAM see SGM500
DI-BAPN FUMARATE see AMB750
DIBAR see DTP400

DIBASIC AMMONIUM ARSENATE see DCG800
DIBASIC AMMONIUM PHOSPHATE see ANR500
DIBASIC LEAD ACETATE see LCV000
DIBASIC LEAD ARSENATE see LCK000
DIBASIC LEAD CARBONATE see LCP000
DIBASIC LEAD METAPHOSPHATE see LCV100
DIBASIC LEAD PHOSPHITE see LCV100
DIBASIC SODIUM PHOSPHATE see SJH090
DIBASIC ZINC STEARATE see ZMS000
DIBASOL see BEA825
DIBAVIT see IAL200
DIBAZOL see BEA825
DIBAZOLE see BEA825
DIBEKACIN see DCQ800
DIBEKACIN SULFATE see PAG050
DIBENAMINE see DCR200
DIBENAMINE HYDROCHLORIDE see DCR200
DIBENCIL see BFC750
DIBENCILLIN see BFC750
DIBENYLIN see PDT250
DIBENYLINE see PDT250
DIBENZ(a,e)ACEANTHRYLENE see DCR300
DIBENZ(a,j)ACEANTHRYLENE see DCR400
13H-DIBENZ(bc,j)ACEANTHRYLENE see DCR600
4H-DIBENZ(f,g,j)ACEANTHRYLENE, 5,5a,6,7-TETRAHYDRO- see DCR800
DIBENZ(1,2:4,5)ACEANTHRYLENO(9,10-B)OXIRENE-12,13-DIOL, 1A,12,13,13A-TETRAHYDRO-(1A-α,12-α,13-β,13A-α)- see DMS430
DIBENZ(2,3:4,5)ACEPHENANTHRYLENO(9,10-B)OXIRENE-7,8-DIOL, 5C,6A,7,8-TETRAHYDRO-,(5C-α,6A-α,7-β,8-α)- see DMS480
DIBENZACEPIN see DCS200
DIBENZ(a,d)ACRIDINE see DCS400
DIBENZ(a,f)ACRIDINE see DCS600
DIBENZ(a,h)ACRIDINE see DCS400
DIBENZ(a,j)ACRIDINE see DCS600
DIBENZ(c,h)ACRIDINE see DCS800
1,2,5,6-DIBENZACRIDINE see DCS400
1,2,7,8-DIBENZACRIDINE see DCS600
3,4,5,6-DIBENZACRIDINE see DCS600
3,4:5,6-DIBENZACRIDINE see DCS800
1,2,7,8-DIBENZACRIDINE (FRENCH) see DCS800
DIBENZ(a,h)ACRIDINE 3,4-DIOL-1,2-EPOXIDE see DCS821
DIBENZ(a,j)ACRIDINE METHOSULFATE see DCT000
3,4:5,6-DIBENZACRIDINE METHOSULFATE see DCT000
4,4'-DIBENZAMIDO-1,1'-DIANTHRIMIDE see IBJ000
o,o'-DIBENZAMIDODIPHENYL DISULFIDE see BDK800
DI-o-BENZAMIDOPHENYL DISULPHIDE see BDK800
DIBENZAMINE see DCT050
1,2,5,6-DIBENZANTHRACEEN (DUTCH) see DCT400
DIBENZ(a,c)ANTHRACENE see BDH750
DIBENZ(a,h)ANTHRACENE see DCT400
DIBENZ(a,j)ANTHRACENE see DCT600
1,2:3,4-DIBENZANTHRACENE see BDH750
1,2:5,6-DIBENZANTHRACENE see DCT400
1,2:7,8-DIBENZANTHRACENE see DCT600
2,3:6,7-DIBENZANTHRACENE see PAV000
DIBENZ(de,kl)ANTHRACENE see PCQ250
1,2:5,6-DIBENZ(a)ANTHRACENE see DCT400
1:2:5:6-DIBENZANTHRACENE-9-CARBAMIDO-ACETIC ACID see DCU600
1,2,5,6-DIBENZANTHRACENECHOLEIC ACID see DCT800
DIBENZ(a,h)ANTHRACENE-3,4-DIOL , 3,4-DIHYDRO-, trans-(+−)- see DKX930

DIBENZ(a,h)ANTHRACENE-5,6-OXIDE see EBP500
1,2:5,6-DIBENZANTHRACENE-9,10-endo-α,β-SUCCINIC ACID see DCU200
1,2,5,6-DIBENZANTHRACENE-9,10-endo-α,β-SUCCINIC ACID, SODIUM SALT see SGF000
DIBENZ(a,h)ANTHRACEN-5-OL see DCU400
4,4'-DI-7H-BENZ(de)ANTHRACEN-7-ONE see DCV000
N-(DIBENZ(a,h)ANTHRACEN-7-YLCARBAMOYL)GLYCINE see DCU600
DIBENZANTHRANYL GLYCINE COMPLEX see DCU600
DIBENZANTHRONE see DCU800
4,4'-DIBENZANTHRONIL see DCV000
4,4'-DIBENZANTHRONYL see DCV000
DIBENZARSENOLE, 5-HYDROXY-, 5-OXIDE see ARA100
DIBENZARSENOLIC ACID see ARA100
5H-DIBENZARSOLE, 5-HYDROXY-, 5-OXIDE see ARA100
5H-DIBENZ(b,f)AZEPINE-5-CARBOXAMIDE see DCV200
5H-DIBENZ(b,f)AZEPINE, 5-(3-((p-CHLOROBENZOYLMETHYL)-N-METHYLAMINO)PROPYL)-, HYDROCHLORIDE see IFZ900
5H-DIBENZ(b,f)AZEPINE, 3-CHLORO-5-(3-(4-CARBAMOYL-4-PIPERIDINOPIPERIDINO)PROPYL)-10,11-DIHYDRO-, DIHYDROCHLORIDE, MONOHYDRATE see DCV400
5H-DIBENZ(b,f)AZEPINE, 3,7-DICHLORO-10,11-DIHYDRO-5-(3-(DIMETHYLAMINO)PROPYL)- see DEL400
5H-DIBENZ(b,f)AZEPINE,-3,7-DICHLORO-5-(3-(DIMETHYLAMINO)PROPYL)-10,11-DIHYDRO- see DEL400
5H-DIBENZ(b,f)AZEPINE,-10,11-DIHYDRO-3,7-DICHLORO-5-(3-(DIMETHYLAMINO)PROPYL)- see DEL400
5H-DIBENZ(b,f)AZEPINE, 10,11-DIHYDRO-5-(2'-(N,N-DIMETHYLAMINO)-2'-METHYL)ETHYL- see DPK500
4-(3-(5H-DIBENZ(b,f)AZEPIN-5-YL)PROPYL)-1-PIPERAZINEETHANOL see DCV800
4-(3-(5H-DIBENZ(b,f)AZEPIN-5-YL)PROPYL)-1-PIPERAZINEETHANOL DIHYDROCHLORIDE see IDF000
5H-DIBENZ(B,F)AZEPINE, 10,11-DIHYDRO-5-(3-(DIMETHYLAMINO)-2-METHYLPROPYL)-, (+)- see TMK100
5H-DIBENZ(B,F)AZEPINE, 5-(3-(DIMETHYLAMINO)-2-METHYLPROPYL)-10,11-DIHYDRO-, (+)- see TMK100
5H-DIBENZ(B,F)AZEPINE-5-PROPANAMINE, 10,11-DIHYDRO-N,N,β-TRIMETHYL-, (+)- see TMK100
3,4,5,6-DIBENZCARBAZOL see DCY000
1,2,5,6-DIBENZCARBAZOLE see DCX600
1,2,7,8-DIBENZCARBAZOLE see DCX800
3,4,5,6-DIBENZCARBAZOLE see DCY000
2,4-DIBENZENEAZORESORCINOL see CMP600
DIBENZENECHROMIUM see BGY700
DIBENZENECHROMIUM IODIDE see BGY720
DI(BENZENEDIAZONIUM)ZINC TETRACHLORIDE see DCW000
DI(BENZENEDIAZO)SULFIDE see DCW200
DIBENZENESULFONYL PEROXIDE see DCW400
DIBENZEPIN see DCW600
DIBENZEPINE see DCW600
DIBENZEPINE HYDROCHLORIDE see DCW800
DIBENZEPIN HYDROCHLORIDE see DCW800
1,2,3,4-DIBENZFLUORENE see DCX000
1,2,7,8-DIBENZFLUORENE see DDB200

DI-(BISTRIFLUOROMETHYLPHOSFIDO)MERCURY see DDI400
DIBOA see BDI100
DIBONDRIN see BBV500
DIBORANE see DDI450
DIBORANE MIXTURES (NA 1911) see DDI450
DIBORON HEXAHYDRIDE see DDI450
DIBORON OXIDE see DDI500
DIBORON TETRACHLORIDE see DDI600
DIBORON TETRAFLUORIDE see DDI800
DIBOVAN see DAD200
DIBP see DNJ400
DIBROLUUR see BNP750
DIBROM see NAG400
1,2-DIBROMAETHAN (GERMAN) see EIY500
DIBROMANNIT see DDP600
DIBROMANNITOL see DDP600
d-DIBROMANNITOL see DDP600
DIBROMANTIN see DDI900
DIBROMANTINE see DDI900
3,9-DIBROMBENZANTHRONE see DDK875
1,4-DIBROMBUTAN (GERMAN) see DDL000
DIBROMCHLORPROPAN (GERMAN) see DDL800
1,2-DIBROM-3-CHLOR-PROPAN (GERMAN) see DDL800
O-(1,2-DIBROM-2,2-DICHLORAETHYL)-O,O-DIMETHYL-PHOSPHAT (GERMAN) see NAG400
3,5-DIBROM-4-HYDROXYLBENZALDOXIM-O-(2',4'-DINITROPHENYL)-AETHER see BNK400
DIBROMOACETIC ACID see DDJ100
DIBROMOACETONITRILE see DDJ400
2,4'-DIBROMOACETOPHENONE see DDJ600
α,p-DIBROMOACETOPHENONE see DDJ600
DIBROMOACETYLENE see DDJ800
2-((4-(DIBROMOACETYL)PHENYL)AMINO)-2-ETHOXY-1-(4-NITROPHENYL)ETHANONE see DDJ850
2,4-DIBROMO-1-ANTHRAQUINONYLAMINE see AJL500
3,9-DIBROMO-7H-BENZ(de)ANTHRACEN-7-ONE see DDK875
DIBROMOBENZENE see DDJ900
m-DIBROMOBENZENE see DDK050
p-DIBROMOBENZENE see BNV775
1,3-DIBROMOBENZENE see DDK050
1,4-DIBROMOBENZENE see BNV775
4,4'-DIBROMOBENZILIC ACID ISOPROPYL ESTER see IOS000
2,2'-DIBROMOBIACETYL see DDK600
α,α'-DIBROMOBIACETYL see DDK600
DIBROMOBICYCLOHEPTANE see DDK800
DIBROMOBICYCLOHEPTANE (mixed isomers) see DDK800
3,5-DIBROMO-N-(4-BROMOPHENYL)-2-HYDROXYBENZAMIDE see THW750
1,4-DIBROMO-1,3-BUTADIYNE see DDK875
1,4-DIBROMOBUTANE see DDL000
DIBROMOBUTENE see DDL600
1,4-DIBROMO-2-BUTENE see DDL400
1,4-trans-DIBROMOBUTENE-2 see DDL600
trans-1,4-DIBROMOBUT-2-ENE see DDL600
2,3-DIBROMO-2-BUTENE-1,4-DIOL see DDL700
DIBROMOCHLOROMETHANE see CFK500
1,2-DIBROMO-3-CHLORO-2-METHYLPROPANE see MBV720
DIBROMOCHLOROPROPANE see DDL800
1,2-DIBROMO-3-CHLOROPROPANE see DDL800
3,4'-DIBROMO-5-CHLOROTHIOSALICYLANILIDE ACETATE (ESTER) see BOL325

1,2-DIBROMO-3-CLORO-PROPANO (ITALIAN) see DDL800
DIBROMOCYANOACETAMIDE see DDM000
α,α-DIBROMO-α-CYANOACETAMIDE see DDM000
2,6-DIBROMO-4-CYANOPHENOL see DDP000
2,6-DIBROMO-4-CYANOPHENYL OCTANOATE see DDM200
cis-DIBROMODIAMMINEPLATINUM see DCF760
cis-DIBROMODIAMMINEPLATINUM(II) see DCF760
1,2-DIBROMO-4-(1,2-DIBROMOETHYL)CYCLOHEXANE see DDM300
DIBROMODIBUTYLSTANNANE see DDM400
DIBROMODIBUTYLTIN see DDM400
1,2-DIBROMO-1,1-DICHLOROETHANE see DDM425
1,2-DIBROMO-2,2-DICHLOROETHYL DIMETHYL PHOSPHATE see NAG400
O-(1,2-DIBROMO-2,2-DICLORO-ETIL)-O,O-DIMETIL-FOSTATO (ITALIAN) see NAG400
1,2-DIBROMO-2,4-DICYANOBUTANE see DDM500
1,6-DIBROMODIDEOXYDULCITOL see DDJ000
1,6-DIBROMO-1,6-DIDEOXYDULCITOL see DDJ000
1,6-DIBROMO-1,6-DIDEOXYGALACTITOL see DDJ000
1,6-DIBROMO-1,6-DIDEOXY-d-GALACTITOL see DDJ000
1,6-DIBROMO-1,6-DIDEOXY-d-MANNITOL see DDP600
1,6-DIBROMO-1,6-d-DIDESOXYMANNITOL see DDP600
2',6'-DIBROMO-2-(DIETHYLAMINO)-p-ACETOTOLUIDIDE HYDROCHLORIDE see DDM600
5,6-DIBROMO-2-(2-(2-(DIETHYLAMINO)ETHYLAMINO)ETHYL)-2-METHYL-1,3-BENZODIOXOLE DIHYDROCHLORIDE see DDM800
2',6'-DIBROMO-2-(DIETHYLAMINO)-4'-METHYLACETANILIDE HYDROCHLORIDE see DDM600
DIBROMODIFLUOROMETHANE see DKG850
6,8-DIBROMO-DIHYDRO-1,3-BENZOXAZINE-2-THIONE-4-ONE see DDM820
1,2-DIBROMO-1,2-DIISOCYANATOETHANE POLYMERS see DDN100
2,2-DIBROMO-1,3-DIMETHYLCYCLOPROPANOIC ACID see DDN150
N,N'-DIBROMODIMETHYLHYDANTOIN see DDI900
1,3-DIBROMO-5,5-DIMETHYL-2,4-IMIDAZOLIDINEDIONE see DDI900
DIBROMODIMETHYL STANNANE see DUG800
DIBROMODIPHENYLSTANNANE see DDN200
1,12-DIBROMODODECANE see DDN300
α,ω-DIBROMODODECANE see DDN300
DIBROMODULCITOL see DDJ000
DIBROMODULCITOL see DDJ000
1,6-DIBROMODULCITOL see DDJ000
2,3-DIBROMO-5,6-EPOXY-7,8-DIOXABICYCLO(2.2.2)OCTANE see DDN700
1,2-DIBROMOETANO (ITALIAN) see EIY500
1,1-DIBROMOETHANE see DDN800
sym-DIBROMOETHANE see EIY500
α,β-DIBROMOETHANE see EIY500
1,2-DIBROMOETHANE (MAK) see EIY500

1,1-DIBROMOETHENE see VPF200
(E)-1,2-DIBROMOETHENE see DDN820
trans-1,2-DIBROMOETHENE see DDN820
(1,2-DIBROMOETHYL)BENZENE see DDN900
α-β-DIBROMOETHYLBENZENE see DDN900
1-(1,2-DIBROMOETHYL)-3,4-DIBROMOCYCLOHEXANE see DDM300
1,1-DIBROMOETHYLENE see VPF200
1,2-DIBROMOETHYLENE see DDN950
(E)-1,2-DIBROMOETHYLENE see DDN820
trans-DIBROMOETHYLENE see DDN820
trans-1,2-DIBROMOETHYLENE see DDN820
DIBROMOFLUORESCEIN see DDO200
DIBROMOFLUOROPHOSPHINE SULFIDE see TFO250
4',5'-DIBROMOFLUORORESCEIN see DDO200
1,2-DIBROMOHEPTAFLUOROISOBUTYL METHYL ETHER see DDO400
1,2-DIBROMOHEXAFLUOROPROPANE see DDO450
1,2-DIBROMO-1,1,2,3,3,3-HEXAFLUOROPROPANE see DDO450
1,6-DIBROMOHEXAN (GERMAN) see DDO800
1,6-DIBROMOHEXANE see DDO800
3,5-DIBROMO-4-HYDROXYBENZALDEHYDE 2,4-DINITROPHENYL OXIME see BNK400
3,5-DIBROMO-4-HYDROXYBENZALDEHYDE (2',4'-DINITROPHENYL)OXIME see BNK400
3,5-DIBROMO-4-HYDROXYBENZALDEHYDE-O-(2',4'-DINITROPHENYL)OXIME see BNK400
3,5-DIBROMO-4-HYDROXYBENZONITRILE see DDP000
3-(3,5-DIBROMO-4-HYDROXYBENZOYL)-2-ETHYLBENZOFURAN see DDP200
(DIBROMO-3,5 HYDROXY-4 BENZOYL)-3 MESITYL-2 BENZOFURANNE see DDP300
2,7-DIBROMO-4-HYDROXYMERCURIFLUORESCEINE DISODIUM SALT see MCV000
3,5-DIBROMO-4-HYDROXYPHENYLCYANIDE see DDP000
3,5-DIBROMO-4-HYDROXYPHENYL-2-ETHYL-3-BENZOFURANYL KETONE see DDP200
(3,5-DIBROMO-4-HYDROXYPHENYL)(2-ETHYL-3-BENZOFURANYL)METHANONE see DDP200
3,5-DIBROMO-4-HYDROXYPHENYL 2-MESITYL-3-BENZOFURANYL KETONE see DDP300
(3,5-DIBROMO-4-HYDROXYPHENYL)(2-(2,4,6-TRIMETHYLPHENYL)-3-BENZOFURANYL)M ETHANONE see DDP300
5,7-DIBROMO-8-HYDROXYQUINOLINE see DDS600
1,4-DIBROMOISOPRENE see IMS100
DIBROMOMALEINIMIDE see DDP400
1,6-DIBROMOMANNITOL see DDP600
2,7-DIBROMOMESOBENZANTHRONE see DDK875
DIBROMOMETHANE see DDP800
N,N-DIBROMOMETHYLAMINE see DDQ100
DIBROMOMETHYLBORANE see DDQ125
1,4-DIBROMO-2-METHYL-2-BUTENE see IMS100
1,2-DIBROMO-2-METHYLPROPANE see DDQ150
DIBROMONEOPENTYL GLYCOL see DDQ400
DIBROMONEOPENTYL GLYCOL see DDQ400
2,2-DIBROMO-3-NITRILOPROPIONAMIDE see DDM000

2,6-DIBROMO-4-NITROANILINE see DDQ450
2,2-DIBROMO-2-NITROETHANOL see DDQ470
2,6-DIBROMO-4-NITROPHENOL see DDQ500
3,4-DIBROMONITROSOPIPERIDINE see DDQ800
DIBROMONORBORNANE see DDK800
3,5-DIBROMO-4-OCTANOYLOXYBENZONITRILE see DDM200
DIBROMOPENTAERYTHRITOL see DDQ400
1,5-DIBROMOPENTANE see DDR000
2-(3,5-DIBROMO-2-PENTYLOXYBENZYLOXY)TRIETHYLAMINE see DDR100
1,2-DIBROMOPERFLUOROETHANE see FOO525
2,4-DIBROMOPHENOL see DDR150
DIBROMOPHENYLARSINE see DDR200
1,2-DIBROMO-1-PHENYLETHANE see DDN900
DIBROMOPROPANAL see DDS400
1,2-DIBROMOPROPANE see DDR400
1,3-DIBROMOPROPANE see TLR500
2,3-DIBROMOPROPANE see DDR600
α,γ-DIBROMOPROPANE see TLR000
ω,ω'-DIBROMOPROPANE see TLR000
1,3-DIBROMOPROPANE polymer with N,N,N',N'-TETRAMETHYL-1,6-HEXANEDIAMINE see HCV500
2,3-DIBROMOPROPANOIC ACID see DDS500
2,3-DIBROMOPROPANOL see DDS000
1,3-DIBROMO-2-PROPANOL see DDR800
2,3-DIBROMO-1-PROPANOL see DDS000
2,3-DIBROMO-1-PROPANOL DIHYDROGEN PHOSPHATE see MRH217
2,3-DIBROMO-1-PROPANOL HYDROGEN PHOSPHATE see BIV825
2,3-DIBROMO-1-PROPANOL HYDROGEN PHOSPHATE AMMONIUM SALT see MRH214
2,3-DIBROMO-1-PROPANOL PHOSPHATE see TNC500
2,3-DIBROMO-1-PROPANOL, PHOSPHATE (3:1) see TNC500
2,3-DIBROMOPROPANOYL CHLORIDE see DDS100
2,3-DIBROMOPROPENE see DDS200
2,3-DIBROMOPROPIONALDEHYDE see DDS400
2,3-DIBROMOPROPIONIC ACID see DDS500
α-β-DIBROMOPROPIONIC ACID see DDS500
2,3-DIBROMOPROPIONYL CHLORIDE see DDS100
α,β-DIBROMOPROPIONYL CHLORIDE see DDS100
N,N³-DI(β-BROMOPROPIONYL)-N¹,N²-DISPIROTRIPIPERAZINIUM DICHLORIDE see SLD600
2,3-DIBROMOPROPYL ACRYLATE see DDS550
2,3-DIBROMOPROPYLPHOSPHATE see MRH217
(2,3-DIBROMOPROPYL) PHOSPHATE see TNC500
5,7-DIBROMO-8-QUINOLINOL see DDS600
4',5-DIBROMOSALICYLANILIDE see BOD600
3,4-DIBROMOSULFOLANE see DDT000
1,2-DIBROMOTETRAFLUOROETHANE see FOO525
sym-DIBROMOTETRAFLUOROETHANE see FOO525
1,2-DIBROMO-1,1,2,2-TETRAFLUOROETHANE see FOO525

3,4-DIBROMOTETRAHYDROTHIOPHENE-1,1-DIOXIDE see DDT000
trans-DIBROMOTETRAKIS(PYRIDINE)RHODIUM BROMIDE see DDT050
6,8-DIBROMO-2-THIO-2H-1,3-BENZOXAZINE-2,4(3H)-DIONE see DDM820
1,2-DIBROMO-1,1,5-TRICHLOROPENTANE see DDT100
DIBROMOTRIFLUOROMONOCHLOROETHANE see MRF600
DIBROMOXYQUINOLINE see DDS600
DIBROMSALAN see BOD600
DIBROMURE d'ETHYLENE (FRENCH) see EIY500
1,2-DIBROOM-3-CHLOORPROPAAN (DUTCH) see DDL800
O-(1,2-DIBROOM-2,2-DICHLOOR-ETHYL)-O,O-DIMETHYL-FOSFAAT (DUTCH) see NAG400
1,2-DIBROOMETHAAN (DUTCH) see EIY500
DIBUCAIN see NOF500
DIBUCAINE see DDT200
DIBUCAINE HYDROCHLORIDE see NOF500
DIBULINESULFAT see DDW000
DIBULINE SULFATE see DDW000
DIBUSMUTH DICHROMIUM NONAOXIDE see DDT250
DIBUTADIAMIN DIHYDROCHLORIDE see BQM309
DIBUTALIN see BQN600
DIBUTAMID (GERMAN) see DDT300
DIBUTAMIDE see DDT300
2,3:4,5-DI(2-BUTENYL)TETRAHYDROFURFURAL see BHJ500
DIBUTIL see DIR000
DIBUTIN see ILD000
DIBUTOLINE see DDW000
DIBUTOLINE SULFATE see DDW000
1,1-DIBUTOXYETHANE see DDT400
1,2-DIBUTOXYETHANE see DDW400
DIBUTOXYETHOXYETHYL ADIPATE see DDT500
DIBUTOXYETHYL ADIPATE see BHJ750
DI(2-BUTOXYETHYL) ADIPATE see BHJ750
2,2'-DIBUTOXYETHYL ETHER see DDW200
DI(BUTOXYETHYL)PHTHALATE see BHK000
(DIBUTOXYPHOSPHINYL)MERCURY CHLORIDE see CFK750
DI(BUTOXYTHIOCARBONYL) DISULFIDE see BSS550
DIBUTYL ACETAL see DDT400
DIBUTYL ACID PHOSPHATE see DEG700
DIBUTYL ADIPATE see AEO750
DI-N-BUTYL ADIPATE see AEO750
DIBUTYL ADIPINATE see AEO750
DIBUTYLAMID KYSELINY MRAVENCI see DEC400
n-DIBUTYLAMINE see DDT800
DI-n-BUTYLAMINE see DDT800
DI-sec-BUTYLAMINE see DDT600
DI(n-BUTYL)AMINE (DOT) see DDT800
DIBUTYLAMINE, HEXACHLOROSTANNANE (2:1) see BIV900
DIBUTYLAMINE, HEXAFLUOROARSENATE(1-) see DDV225
DIBUTYLAMINE, 4-HYDROXY-N-NITROSO- see HJQ350
DIBUTYLAMINE TETRAFLUOROBORATE see DDU000
2-(DIBUTYLAMINO)-2',6'-ACETOXYLIDIDE HYDROCHLORIDE see DDU200
1-(((DIBUTYLAMINO)CARBONYL)OXY)-N-ETHYL-N,N-DIMETHYLETHANAMINIUM SULFATE (2:1) see DDW000
DIBUTYLAMINOETHANOL see DDU600
2-DIBUTYLAMINOETHANOL see DDU600

2-N-DIBUTYLAMINOETHANOL see DDU600
2-DI-n-BUTYLAMINOETHANOL see DDU600
N,N-DI-n-BUTYLAMINOETHANOL (DOT) see DDU600
β-N-DIBUTYLAMINOETHYL ALCOHOL see DDU600
5-((2-(DIBUTYLAMINO)-ETHYL)AMINO)-3-PHENYL-1,2,4-OXADIAZOLE HYDROCHLORIDE see PEU000
DI-(N-BUTYLAMINO)FLUOROPHOSPHINE OXIDE see DDU700
α-DIBUTYL-AMINO-4-METHOXYBENZENEACETAMIDE (9CI) see DDT300
2-DIBUTYLAMINO-2-(p-METHOXYPHENYL)ACETAMIDE see DDT300
α-DIBUTYL-AMINO-p-METHOXYPHENYLACETAMIDE see DDT300
α-DIBUTYLAMINO-α-(p-METHOXYPHENYL)ACETAMIDE see DDT300
3-((DIBUTYLAMINO)METHYL)-4,5,6-TRIHYDROXYPHTHALIDE HYDROCHLORIDE see APE625
3-(DIBUTYLAMINO)-1-PROPANOL-p-AMINOBENZOATE see BOO750
3-(DIBUTYLAMINO)-1-PROPANOL-p-AMINOBENZOATE (ESTER) SULFATE (2:1) see BOP000
3-DIBUTYLAMINO-1-PROPANOL-4-AMINOBENZOATE (ESTER) SULFATE (SALT) (2:1) see BOP000
N,N-DIBUTYL-3-AMINOPROPIONITRILE see DDU800
3-(DIBUTYLAMINO)PROPYLAMINE see DDV200
3-DIBUTYLAMINOPROPYL-p-AMINOBENZOATE see BOO750
DIBUTYLAMINOPROPYL-p-AMINOBENZOATE SULFATE see BOP000
3'-DIBUTYLAMINOPROPYL-4-AMINOBENZOATE SULFATE see BOP000
((DIBUTYLAMINO)THIO)METHYLCARBAMIC ACID, 2,2-DIMETHYL-2,3-DIHYDRO-7-BENZOFURANYL ESTER see CCC280
DI-N-BUTYLAMMONIUM HEXAFLUOROARSENATE see DDV225
DI-n-BUTYLAMMONIUM TETRAFLUOROBORATE see DDU000
DIBUTYLARSINIC ACID see DDV250
DIBUTYLATED HYDROXYTOLUENE see BFW750
DIBUTYLBARBITURIC ACID see DDV400
5,5-DIBUTYLBARBITURIC ACID see DDV400
p-DI-tert-BUTYLBENZENE see DDV450
DIBUTYL-1,2-BENZENEDICARBOXYLATE see DEH200
2,5-DI-tert-BUTYLBENZENE-1,4-DIOL see DEC800
2,6-DI-tert-BUTYL-p-BENZOQUINONE see DDV500
2,5-DI-tert-BUTYL-1,4-BENZOQUINONE see DDV550
DI-N-BUTYL BENZYLPHOSPHONATE see BFD760
N,N-DI-sec-BUTYL-S-BENZYLTHIOCARBAMATE see TGF030
DIBUTYLBIS((3-CARBOXYACRYLOYL)OXY)-STANNANE DIMETHYL ESTER (Z,Z) (8CI) see BKO250
DIBUTYLBIS((2-ETHYLHEXANOYL)OXY)-STANNANE see BJQ250
DIBUTYLBIS((2-ETHYL-1-OXOHEXYL)OXY)-STANNANE (9CI) see BJQ250
DIBUTYLBIS(LAUROYLOXY)STANNANE see DDV600

DIBUTYLBIS(LAUROYLOXY)TIN see DDV600

DIBUTYLBIS(OCTANOYLOXY)STANNANE see BLB250

DIBUTYLBIS(OLEOYLOXY)STANNANE see DEJ000

DIBUTYLBIS((1-OXO-9-OCTADECENYL)OXY)STANNANE (Z,Z) see DEJ000

DIBUTYLBIS((1-OXOOCTADECYL)OXY)STANNANE see DEJ250

DIBUTYLBIS((1-OXOOCTYL)OXY)STANNANE see BLB250

DIBUTYLBIS(STEAROYLOXY)STANNANE see DEJ250

DIBUTYLBIS(TRIFLUOROACETOXY)STANNANE see BLN750

DIBUTYL BUTANEPHOSPHONATE see DDV800

O,O-DIBUTYL-1-BUTYLAMINO-CYCLOHEXYLPHOSPHONATE see ALZ000

DIBUTYL BUTYLPHOSPHONATE see DDV800

DI-n-BUTYL-CARBAMYLCHOLINE SULPHATE see DDW000

(2-DIBUTYLCARBAMYLOXYETHYL)-DIMETHYLETHYLAMMONIUM SULFATE see DDW000

DIBUTYL CARBITOL see DDW200

DIBUTYLCARBITOLFORMAL see BHK750

DIBUTYL-o-(o-CARBOXYBENZOYL) GLYCOLATE see BQP750

DIBUTYL-o-CARBOXYBENZOYLOXYACETATE see BQP750

DIBUTYL CELLOSOLVE see DDW400

DIBUTYL CELLOSOLVE ADIPATE see BHJ750

DIBUTYL CELLOSOLVE PHTHALATE see BHK000

DIBUTYL CHLORENDATE see CDS025

N,N-DI-N-BUTYL-p-CHLOROBENZENESULFONAMIDE see DDW500

4,6-DI-tert-BUTYL-m-CRESOL see DDX000

2,6-DI-tert-BUTYL-p-CRESOL (OSHA, ACGIH) see BFW750

DI-(4-tert-BUTYL-m-CRESOL)SULFIDE see BHL000

9,10-DI-n-BUTYL-1,2,5,6-DIBENZANTHRACENE see DDX200

DI-n-BUTYL(DIBUTYRYLOXY)STANNANE see DDX600

N,N'-DIBUTYL-N,N'-DICARBOXYETHYLENE DIAMINEMORPHOLIDE see DUO400

N,N'-DIBUTYL-N,N'-DICARBOXYMORPHOLIDE-ETHYLENEDIAMINE see DUO400

DIBUTYLDICHLOROGERMANE see DDY000

DIBUTYLDICHLOROSTANNANE see DDY200

DIBUTYLDICHLOROTIN see DDY200

DIBUTYL (DIETHYLENE GLYCOL BISPHTHALATE) see DDY400

DIBUTYLDI(2-ETHYLHEXYLOXYCARBONYLMETHYLTHIO)STANNANE see DDY600

DIBUTYLDIFLUOROSTANNANE see DDY800

DIBUTYL(DIFORMYLOXY)STANNANE see DDZ000

2,2-DIBUTYLDIHYDRO-6H-1,3,2-OXATHIASTANNIN-6-ONE see DEJ200

DIBUTYLDIHYDROXYSTANNANE-3,3'-THIODIPROPIONATE see DEA400

DIBUTYLDIIODOSTANNANE see DEA000

2,6-DI-tert-BUTYL-α-(DIMETHYLAMINO)-p-CRESOL see DEA100

2,6-DI-tert-BUTYL-α-(DIMETHYLAMINO)-p-CRESOL see FAB000

2,2-DIBUTYL-1,3-DIOXA-7,9-DITHIA-2-STANNACYCLODODECAN see DEA200

2,2-DIBUTYL-1,3-DIOXA-2-STANNA-7,9-DITHIACYCLODODECAN-4,12-DIONE see DEA200

2,2-DIBUTYL-1,3-DIOXA-2-STANNA-7-THIACYCLODECAN-4,10-DIONE see DEA400

2,2-DIBUTYL-1,3,2-DIOXASTANNEPIN-4,7-DIONE see DEJ100

2,2-DIBUTYL-1,3,7,2-DIOXATHIASTANNECANE-4,10-DIONE see DEA400

DIBUTYLDIPENTANOYLOXYSTANNANE see DEA600

DI-tert-BUTYL DIPEROXYCARBONATE see DEA800

DI-tert-BUTYL DIPEROXYOXALATE see DEB000

DI-tert-BUTYL DIPEROXYPHTHALATE see DEB200

DIBUTYLDIPROPIONYLOXYSTANNANE see DEB400

DIBUTYL DISELENIDE see BRF550

DI-n-BUTYL-DISELENIDE see BRF550

DIBUTYLDISELENIUM see BRF550

DIBUTYLDITHIOCARBAMIC ACID, NICKEL SALT see BIW750

DIBUTYLDITHIOCARBAMIC ACID SODIUM SALT see SGF500

DIBUTYLDITHIOCARBAMIC ACID-S-TRIBUTYLSTANNYL ESTER see DEB600

DIBUTYLDITHIO-CARBAMIC ACID ZINC COMPLEX see BIX000

DIBUTYLDITHIOCARBAMIC ACID ZINC SALT see BIX000

((DIBUTYLDITHIOCARBAMOYL)OXY)TRI BUTYLSTANNANE see DEB600

N,N-DI-sec-BUTYL DITHIOOXAMIDE see DEB800

DIBUTYL DIXANTHOGEN see BSS550

DIBUTYLDIXANTOGENATE see BSS550

DIBUTYLESTER KYSELINY FUMAROVE see DEC600

DI-n-BUTYLESTER KYSELINY JANTAROVE see SNA500

DI-sec-BUTYL ESTER PHOSPHOROFLUORIDIC ACID see DEC200

DIBUTYL ESTER SULFURIC ACID see DEC000

N,N-DIBUTYLETHANOLAMINE see DDU600

DI-sec-BUTYL ETHER see BRH760

DI-sec-BUTYL ETHER see OPE030

DI-n-BUTYL ETHER (DOT) see BRH750

N,N-DI-tert-BUTYLETHYLENEDIAMINE see DEC100

N,N'-DI-n-BUTYLETHYLENEDIAMINE-N,N'-DICARBOXYBISMORPHOLIDE see DUO400

2,6-DI-sec-BUTYLFENOL (CZECH) see DEF800

DI-sec-BUTYLFLUOROPHOSPHATE see DEC200

DI-sec-BUTYL FLUOROPHOSPHONATE see DEC200

N,N-DI-n-BUTYLFORMAMIDE see DEC400

N,N-DI-n-BUTYLFORMAMIDE see DEC400

DIBUTYL FUMARATE see DEC600

DI-n-BUTYLGERMANEDICHLORIDE see DDY000

DIBUTYLGLYCOL PHTHALATE see BHK000

DIBUTYL 1,4,5,6,7,7-HEXACHLOROBICYCLO(2.2.1)HEPT-5-ENE-2,3-DICARBOXYLATE see CDS025

DIBUTYLHEXAMETHYLENEDIAMINE see DEC699

N,N'-DIBUTYLHEXAMETHYLENEDIAMINE see DEC699

1,6-N,N'-DIBUTYLHEXANEDIAMINE see DEC699

N,N'-DIBUTYL-1,6-HEXANEDIAMINE see DEC699

DIBUTYL HEXANEDIOATE see AEO750

1,1-DIBUTYLHYDRAZINE see DEC725

N,N-DIBUTYLHYDRAZINE see DEC725

1,1-DI-n-BUTYLHYDRAZINE see DEC725

1,2-DI-n-BUTYLHYDRAZINE DIHYDROCHLORIDE see DEC775

1,1-DIBUTYLHYDRAZINE ETHANEDIOATE (1:1) see DEC785

1,2-DIBUTYLHYDRAZINE ETHANEDIOATE (1:1) see DEC795

1,1-DIBUTYLHYDRAZINE OXALATE see DEC785

1,2-DIBUTYLHYDRAZINE OXALATE see DEC795

9-(2,2-DIBUTYLHYDRAZINO)ACRIDINE MONOHYDROCHLORIDE see DEC797

DIBUTYL HYDROGEN PHOSPHATE see DEG700

DIBUTYL HYDROGEN PHOSPHITE see DEG800

2,5-DI-t-BUTYLHYDROQUINONE see DEC800

3,5-DI-tert-BUTYL-4-HYDROXYBENZOIC ACID see DEC900

3,5-DI-tert-BUTYL-4-HYDROXYBENZYL ALCOHOL see IFX200

(3,5-DI-tert-BUTYL-4-HYDROXYBENZYLIDENE)MALONONITRILE see DED000

DIBUTYL (+-)-HYDROXYBUTANEDIOATE see DED500

N,N-DIBUTYL-N-(2-HYDROXYETHYL)AMINE see DDU600

DIBUTYL 2-HYDROXYETHYL PHOSPHATE see HKQ700

3,5-DI-TERT-BUTYL-4-HYDROXY-HYDROCINNAMIC ACID, NEOPENTANETETRAYL ESTER see DED100

2,6-DI-tert-BUTYL-1-HYDROXY-4-METHYLBENZENE see BFW750

2,6-DI-tert-BUTYL-4-HYDROXYMETHYLPHENOL see IFX200

N,N-DIBUTYL(2-HYDROXYPROPYL)AMINE see DED200

3,5-DI-tert-BUTYL-4-HYDROXYTOLUENE see BFW750

DIBUTYL ISOPHTHALATE see DED300

DIBUTYL KETONE see NMZ000

2,6-DI-terc.BUTYL-p-KRESOL (CZECH) see BFW750

DIBUTYL LEAD DIACETATE see DED400

dl-DIBUTYL MALATE see DED500

DIBUTYL MALEATE see DED600

DIBUTYL(MALEOYLDIOXY)TIN see DEJ100

DIBUTYLMALOYLOXYSTANNANE see DED800

DIBUTYL(3-MERCAPTOPROPIONATO(2-))TIN see DEJ200

DIBUTYLMERCURY see DEE000

DI-sec-BUTYLMERCURY see DEE200

2,6-DI-tert-BUTYL-4-METHOXYMETHYLPHENOL see DEE300

N,N-DIBUTYLMETHYLAMINE see DEE400

4,6-DI-T-BUTYL-2-(α-METHYL-4-METHOXYBENZYL)PHENOL see BJK580

2,4-DI-tert-BUTYL-5-METHYLPHENOL see DDX000

2,6-DI-tert-BUTYL-4-METHYLPHENOL see BFW750

2,6-DI-tert-BUTYL-p-METHYLPHENOL see BFW750

2,6-DI-tert-BUTYL-4-METHYLPHENYL-N-METHYLCARBAMATE see TAN300

2,6-DI-tert-BUTYL NAPHTALENE SULFONATE SODIQUE (FRENCH) see DEE600
DIBUTYL-NAPHTHALENE SULFATE, SODIUM SALT see NBS700
2,6-DI-tert-BUTYLNAPHTHALENESULFONIC ACID SODIUM SALT see DEE600
2,6-DI-tert-BUTYL-4-NITROPHENOL see DEE800
DI-n-BUTYLNITROSAMIN (GERMAN) see BRY500
DI-n-BUTYLNITROSAMINE see BRY500
N,N-DI-n-BUTYLNITROSAMINE see BRY500
DIBUTYLNITROSOAMINE see BRY500
N,N-DIBUTYLNITROSOAMINE see BRY500
1,3-DIBUTYL-3-NITROSOUREA see DEF000
N,N'-DIBUTYL-N-NITROSOUREA see DEF000
DI-tert-BUTYL NITROXIDE see DEF090
2,2-DIBUTYL-1-OXA-2-STANNA-3-THIACYCLOHEXAN-6-ONE see DEJ200
2,2-DIBUTYL-1,3,2-OXATHIASTANNOLANE see DEF150
2,2-DIBUTYL-1,3,2-OXATHIASTANNOLANE-5-OXIDE see DEF200
DIBUTYL OXIDE see BRH750
DIBUTYLOXIDE of TIN see DEF400
DIBUTYLOXOSTANNANE see DEF400
DIBUTYLOXOTIN see DEF400
DI-tert-BUTYLPEROXID (GERMAN) see BSC750
DI-tert-BUTYL PEROXIDE (MAK) see BSC750
2,2-DI(tert-BUTYLPEROXY)BUTANE see DEF600
DI-tert-BUTYL PEROXYDE (DUTCH) see BSC750
DIBUTYL PEROXYDICARBONATE see BSC800
DI-sec-BUTYL PEROXYDICARBONATE see BSD000
DI-n-BUTYL PEROXYDICARBONATE, >52% in solution (DOT) see BSC800
DI-sec-BUTYL PEROXYDICARBONATE, not more than 52% in solution (DOT) see BSD000
DI-sec-BUTYL PEROXYDICARBONATE, technically pure (DOT) see BSD000
2,6-DI-sec-BUTYLPHENOL see DEF800
2,4-DI-tert-BUTYLPHENOL see DEG000
2,6-DI-tert-BUTYLPHENOL see DEG100
4,6-DI-tert-BUTYL-α-PHENYL-o-CRESOL see DEG150
N,N'-DI-sec-BUTYL-p-PHENYLENEDIAMINE see DEG200
3,5-DI-tert-BUTYLPHENYLMETHYLCARBAMATE see DEG400
N,N-DIBUTYL-N'-(3-PHENYL-1,2,4-OXADIAZOL-5-YL)-1,2-ETHANEDIAMINE HYDROCHLORIDE see PEU000
DIBUTYLPHENYL-PHENOL SODIUM DISULFONATE see AQV000
DIBUTYL PHENYL PHOSPHATE see DEG600
DIBUTYL PHOSPHATE see DEG700
DIBUTYL PHOSPHATE see DEG700
DI-n-BUTYL PHOSPHATE see DEG700
DIBUTYL-PHOSPHINIC ACID, 4-NITROPHENYL ESTER see NIM000
DIBUTYL PHOSPHITE see DEG800
DIBUTYL PHTHALATE see DEH200
DI-n-BUTYL PHTHALATE see DEH200
N,N-DIBUTYLPROPIONAMIDE see DEH300
5,5-DIBUTYL-2,4,6(1H,3H,5H)-PYRIMIDINETRIONE see DDV400
DIBUTYLRTUT see DEE000
DIBUTYL SEBACATE see DEH600
DI-n-BUTYL SEBACATE see DEH600
DIBUTYLSTANNANE see DEI700
DIBUTYLSTANNANE OXIDE see DEF400

1,1-DIBUTYL-1H-STANNOLE-2,5-DIONE see SLI330
2,2'-((DIBUTYLSTANNYLENE)BIS(THIO))BISACETIC ACID DINONYL ESTER see DEH650
DIBUTYLSTANNYLENE MALEATE see DEJ100
DIBUTYL SUCCINATE see SNA500
DI-n-BUTYLSUCCINATE see SNA500
DI-n-BUTYLSULFAT (GERMAN) see DEC000
DIBUTYL SULFATE see DEC000
n-DIBUTYL SULFIDE see BSM125
DI-n-BUTYLSULFIDE see BSM125
DIBUTYL SULPHIDE see BSM125
DIBUTYL TEREPHTHALATE see DEH700
3,3-DIBUTYL-6,7,8,9-TETRACHLORO-2,4,3-BENZODIOXASTANNEPIN-1,5-DIONE see DEH800
DIBUTYL(TETRACHLOROPHTHALATO)STANNANE see DEH800
DIBUTYL(THIOACETOXY)STANNANE see DEF200
DIBUTYL THIOETHER see BSM125
1,3-DIBUTYLTHIOUREA see DEI000
1,3-DIBUTYL-2-THIOUREA see DEI000
N,N'-DIBUTYLTHIOUREA see DEI000
1,3-DI-n-BUTYL-2-THIOUREA see DEI000
DIBUTYLTHIOXOSTANNANE see DEI200
DIBUTYLTIN see DEI700
DIBUTYLTIN BIS(2-ETHYLHEXANOATE) see BJQ250
DIBUTYLTIN BIS(α-ETHYLHEXANOATE) see BJQ250
DIBUTYL-TIN BIS(ISOOCTYLTHIOGLYCOLLATE) see BKK250
DI-n-BUTYLTIN BISMETHANESULFONATE see DEI400
DIBUTYLTIN BIS(METHYL MALEATE) see BKO250
DIBUTYLTIN BIS(MONOMETHYL MALEATE) see BKO250
DIBUTYLTIN BIS(TRIFLUOROACETATE) see BLN750
DIBUTYLTIN CHLORIDE see DDY200
DIBUTYL TIN DIACETATE see DBF800
DIBUTYL TIN DIBROMIDE see DDM400
DI-n-BUTYLTIN DIBUTYRATE see DDX600
DIBUTYLTIN DICAPRYLATE see BLB250
DIBUTYLTIN DICHLORIDE see DDY200
DI-n-BUTYLTIN DICHLORIDE see DDY200
DI-n-BUTYLTIN DI(DODECANOATE) see DDV600
DIBUTYLTIN DI(2-ETHYLHEXANOATE) see BJQ250
DI-n-BUTYLTIN DI-2-ETHYLHEXANOATE see BJQ250
DIBUTYLTIN DI(2-ETHYLHEXOATE) see BJQ250
DI-n-BUTYLTIN DI-2-ETHYLHEXYLTHIOGLYCOLATE see DDY600
DIBUTYLTIN DIFLUORIDE see DDY800
DI-n-BUTYLTIN DIFORMATE see DDZ000
DI-n-BUTYL TIN DI(HEXADECYLMALEATE) see DEI600
DIBUTYLTIN DIHYDRIDE see DEI700
DIBUTYLTIN DIIODIDE see DEA000
DIBUTYLTIN DILAURATE (USDA) see DDV600
DI-n-BUTYLTIN DI(MONOBUTYL)MALEATE see BHK250
DI-n-BUTYLTIN DI(MONONONYL)MALEATE see DEI800
DIBUTYLTIN DIOCTANOATE see BLB250
DIBUTYLTIN DIOCTATE see BLB250
DIBUTYLTIN DIOLEATE see DEJ000
DI-n-BUTYLTIN DIPENTANOATE see DEA500
DI-n-BUTYLTIN DIPROPIONATE see DEB400
DIBUTYLTIN DISTEARATE see DEJ250

DI-n-BUTYL-TIN DI(TETRADECANOATE) see BLH309
DIBUTYLTIN HYDRIDE see DEI700
DIBUTYLTIN LAURATE see DDV600
DIBUTYLTIN MALATE see DED800
DIBUTYLTIN MALEATE see DEJ100
DI-N-BUTYLTIN MALEATE see SLI330
DIBUTYLTIN MERCAPTOPROPIONATE see DEJ200
DIBUTYLTIN-O,S-MERCAPTOPROPIONATE see DEJ200
DIBUTYLTIN-S,O-3-MERCAPTOPROPIONATE see DEJ200
DIBUTYLTIN-S,O-β-MERCAPTOPROPIONATE see DEJ200
DIBUTYLTIN METHYL MALEATE see BKO250
DIBUTYLTIN OCTANOATE see BLB250
DIBUTYLTIN OXIDE see DEF400
DI-n-BUTYLTIN OXIDE see DEF400
DIBUTYLTIN STEARATE see DEJ250
DIBUTYLTIN SULFIDE see DEI200
DIBUTYLTIN TETRACHLOROPHTHALATE see DEH800
DIBUTYLTIN 3,3'-THIODIPROPIONATE see DEA400
2,6-DI-T-BUTYL-p-TOLYL METHYLCARBAMATE see TAN300
6,6-DIBUTYL-4,8,11-TRIOXO-5,7,12-TRIOXA-6-STANNATRIDECA-2,9-DIENOIC ACID METHYL ESTER see BKO250
DIBUTYL XANTHOGEN DISULFIDE see BSS550
DIBUTYL-ZINN see DEI700
DIBUTYLZINN-S,S'-BIS(ISOOCTYLTHIOGLYCOLAT) (GERMAN) see BKK250
DI-n-BUTYL-ZINN DI-2-AETHYLHEXYL THIOGLYKOLAT (GERMAN) see DDY600
DI-n-BUTYL-ZINN-DICHLORID (GERMAN) see DDY200
DIBUTYL-ZINN-DILAURAT (GERMAN) see DDV600
DI-N-BUTYL-ZINN-DI(MONOBUTYL)MALEINAT (GERMAN) see BHK250
DI-n-BUTYLZINN-DIMONOMETHYLMALEINAT (GERMAN) see BKO250
DI-n-BUTYL-ZINN-DI(MONONONYL)MALEINAT (GERMAN) see DEI800
DI-n-BUTYL-ZINN-OXYD (GERMAN) see DEF400
DI-N-BUTYLZINN THIOGLYKOLAT (GERMAN) see DEF200
DIBUTYRYL cAMP see COV625
N^6,2'-o-DIBUTYRYL cAMP see COV625
N^6,O$^{2'}$-DIBUTYRYL cAMP see COV625
DIBUTYRYL CYCLIC AMP see COV625
DIBUTYRYL CYCLIC AMP see DEJ300
DIBUTYRYL-3',5'-CYCLIC AMP see COV625
DIBUTYRYL CYCLIC-3',5'-AMP see COV625
N^6,2'-o-DIBUTYRYL CYCLIC AMP see COV625
N^6,O$^{2'}$-DIBUYTYRL CYCLIC AMP see COV625
DIC see DAB600
DIC 1468 see MQR275
DICA see DEL200
DICACODYL SULFIDE see CAC250
DICAESIUM SELENIDE see DEJ400
DICAIN see BQA010
DICAINE see BQA010
DICALCIUM PHOSPHATE see CAW100
DICALITE see SCH002
DICAMBA (DOT) see MEL500
DICAMOYLMETHTANE see MBW750
DICANDIOL see MQU750
DICAPRYL 1,2-BENZENEDICARBOXYLATE see BLB750
DICAPRYL PHTHALATE see BLB750

DICAPRYLYL PEROXIDE see CBF705
DICAPTOL see BAD750
1,7-DICARBACALCITONIN (sal), 1-BUTANOIC ACID-26-l-ASPARTIC ACID-27-l-VALINE-29-l-ALANINE- see CBR300
1,7-DICARBACALCITONIN (EEL), 1-BUTANOIC ACID- see CBR300
m-DICARBADODECABORANE (12) see NBV100
o-DICARBADODECABORANE(12) see DEJ500
1,2-DICARBADODECABORANE(12) see DEJ500
1,7-DICARBADODECABORANE (12) see NBV100
1,2-DICARBADODECABORANE(12)-1,2-DIMETHANOL see CCC130
1,7-DICARBADODECABORANE(12)-1,7-DIMETHANOL see CCC120
DICARBADODECABORANYLMETHYLET HYL SULFIDE see DEJ600
DICARBADODECABORANYLMETHγLPRO PYL SULFIDE see DEJ800
DICARBAM see CBM750
2,2-DI(CARBAMOYLOXYMETHYL)PENTANE see MQU750
2,2-DICARBAMYLOXYMETHYL-3-METHYLPENTANE see MBW750
DICARBAZAMIDE see IBC000
S-(1,2-DICARBETHOXYETHYL)-O,O-DIMETHYLDITHIOPHOSPHATE see MAK700
DICARBETHOXYMETHANE see EMA500
DICARBOETHOXYETHYL-O,O-DIMETHYL PHOSPHORODITHIOATE see MAK700
DI(CARBOMETHOXY)ACETYLENE see DOP400
DICARBOMETHOXYZINC see ZBS000
DICARBONIC ACID DIETHYL ESTER see DIZ100
2,3-DICARBONITRILO-1,4-DIATHIAANTHRACHINON (GERMAN) see DLK200
DICARBONYL MOLYBDENUM DIAZIDE see DEJ849
DICARBONYLPYRAZINE RHODIUM(1) PERCHLORATE see DEJ859
DICARBONYLTUNGSTEN DIAZIDE see DEJ880
DICARBOSULF see LBF100
DICARBOXIDINE HYDROCHLORIDE see DEK000
o-DICARBOXYBENZENE see PHW250
3,5-DICARBOXYBENZENESULFONIC ACID, SODIUM SALT see DEK200
3,3'-DICARBOXYBENZIDINE see BFX250
DICARBOXYDINE see DEK400
((1,2-DICARBOXYETHYL)THIO)GOLD DISODIUM SALT see GJC000
4,5-DICARBOXYIMIDAZOLE see IAM000
DICARBOXYLATE see SJN675
DICARBOXYMETHANE see CCC750
DICAROCIDE see DIW200
DICARZOL see DSO200
DICATECHOL BORATE, DI-ORTHO-TOLYLGUANIDINE SALT see DEK500
DICATECHOL BORATE 1,3-DI(o-TOLYL)GUANIDINE SALT see DEK500
DICATRON see PDN000
DI-(C9-C11 ALKYL) PHTHALATE see DEK550
DICENTRINE see ERE150
(+)-DICENTRINE see ERE150
d-DICENTRINE see ERE150
DICERIUM TRISULFIDE see DEK600
DICESIUM CARBONATE see CDC750
DICESIUM DICHLORIDE see CDD000
DICESIUM DIFLUORIDE see CDD500
DICESIUM DIIODIDE see CDE000
DICESIUM SELENIDE see DEJ400

DICESIUM SULFATE see CDE500
DICESTAL see MJM500
DICETYLDIMETHYLAMMONIUM CHLORIDE see DRK200
DICHA see DGT600
DICHA (CUBA) see DHB309
DICHAN (CZECH) see DGU200
DICHAPETULUM CYMOSUM (HOOK) ENGL see PLG000
DICHINALEX see CLD000
DICHLOBENIL (DOT) see DER800
DICHLOFENAMIDE see DEQ200
DICHLOFENTHION see DFK600
DICHLOFENTION see DFK600
DICHLOFLUANID see DFL200
DICHLOFLUANIDE see DFL200
DICHLONE (DOT) see DFT000
p-DICHLOORBENZEEN (DUTCH) see DEP800
1,4-DICHLOORBENZEEN (DUTCH) see DEP800
1,1-DICHLOOR-2,2-BIS(4-CHLOOR FENYL)-ETHAAN (DUTCH) see BIM500
1,1-DICHLOORETHAAN (DUTCH) see DFF809
1,2-DICHLOORETHAAN (DUTCH) see EIY600
2,2'-DICHLOORETHYLETHER (DUTCH) see DFJ050
DICHLOORFEEN see MJM500
(2,4-DICHLOOR-FENOXY)-AZIJNZUUR (DUTCH) see DAA800
2-(2,4-DICHLOOR-FENOXY)-PROPIONZUUR (DUTCH) see DGB000
(3,4-DICHLOOR-FENYL-AZO)-THIOUREUM (DUTCH) see DEQ000
3-(3,4-DICHLOOR-FENYL)-1,1-DIMETHYLUREUM (DUTCH) see DXQ500
3-(3,4-DICHLOOR-FENYL)-1-METHOXY-1-METHYLUREUM (DUTCH) see DGD600
3,6-DICHLOOR-2-METHOXY-BENZOEIZUUR (DUTCH) see MEL500
1,1-DICHLOOR-1-NITROETHAAN (DUTCH) see DFU000
(2,2-DICHLOOR-VINYL)-DIMETHYL-FOSFAAT (DUTCH) see DGP900
DICHLOORVO (DUTCH) see DGP900
2,6-DICHLOQUINONE see DES400
DICHLOR see DGG800
2,5-DICHLORACETANILID see DGB480
DICHLORACETIC ACID see DEL000
DICHLORACETYL CHLORIDE see DEN400
1,1-DICHLORAETHAN (GERMAN) see DFF809
1,2-DICHLOR-AETHAN (GERMAN) see EIY600
1,2-DICHLOR-AETHEN (GERMAN) see DFI210
p-DI-(2-CHLORAETHYL)-AMINO-dl-PHENYL-ALANIN (GERMAN) see BHT750
o-(p-DI(2-CHLORAETHYL)-AMINOPHENYL)-dl-TYROSIN-DIHYDROCHLORID (GERMAN) see DFH100
DICHLORALANTIPYRIN see SKS700
DICHLORALANTIPYRINE see SKS700
S-(2,3-DICHLOR-ALLYL)-N,N-DIISOPROPYL-MONOTHIOCARBAMAAT (DUTCH) see DBI200
2,3-DICHLORALLYL-N,N-(DIISOPROPYL)-THIOCARBAMAT (GERMAN) see DBI200
DICHLORALPHENAZONE see SKS700
DICHLORAL UREA see DGQ200
DICHLORAMINE see BIE250
DICHLORAN see AFP300
DICHLORAN see RDP300
DICHLORAN (amine fungicide) see RDP300
DICHLORAN (FLAME RETARDANT) see AFP300
2,4-DICHLORANILIN see DEO290
3,4-DICHLORANILIN see DEO300
3,4-DICHLORANILINE see DEO300

1-(3,4-DICHLORANILINO)-1-FORMYLAMINO-2,2,2-TRICHLORAETHAN (GERMAN) see CDP750
1,5-DICHLORANTHRACHINON see DEO700
1,8-DICHLORANTHRACHINON see DEO750
DICHLORANTIN see DFE200
3,3'-DICHLORBENZIDIN (CZECH) see DEQ600
4,4'-DICHLORBENZILSAEUREAETHYLESTER (GERMAN) see DER000
o-DICHLOR BENZOL see DEP600
p-DICHLORBENZOL (GERMAN) see DEP800
1,4-DICHLOR-BENZOL (GERMAN) see DEP800
2,6-DICHLORBENZONITRIL (GERMAN) see DER800
1-(2,4-DICHLORBENZYL)INDAZOLE-3-CARBOXYLIC ACID see DEL200
2,2'-DICHLORBIPHENYL (GERMAN) see DET800
1,1-DICHLOR-2,2-BIS(4-CHLOR-PHENYL)-AETHAN (GERMAN) see BIM500
2,3-DICHLOR-1,3-BUTADIEN (CZECH) see DEU400
2,2'-DICHLOR-DIAETHYLAETHER (GERMAN) see DFJ050
3,3'-DICHLOR-4,4'-DIAMINO-DIPHENYLAETHER (GERMAN) see BGT000
3,3'-DICHLOR-4,4'-DIAMINODIPHENYLMETHAN (GERMAN) see MJM200
DICHLOR-DIFENYLSILAN see DFF000
DICHLORDIMETHYLAETHER (GERMAN) see BIK000
DICHLOREMULSION see EIY600
DICHLOREN see BIE500
DICHLOREN (GERMAN) see BIE250
DICHLOREN HYDROCHLORIDE see BIE500
DICHLORETHANOIC ACID see DEL000
2,2'-DICHLORETHYL ETHER see DFJ050
2-(α,β-DICHLORETHYL)PYRIDINE HYDROCHLORIDE see CMV475
β,β-DICHLOR-ETHYL-SULPHIDE see BIH250
DICHLORFENIDIM see DXQ500
2,6-DICHLORFENOL (CZECH) see DFY000
3,4-DICHLORFENYLAMID KYSELINY 3,5-DICHLORSALICYLOVE see TBV000
DICHLOR-FENYLARSIN see DGB600
3,4-DICHLORFENYLISOKYANAT see IKH099
N-DICHLORFLUORMETHYLTHIO-N',N'-DIMETHYLAMINOSULFONSAEUREANILI D (GERMAN) see DFL200
N-(DICHLOR-FLUOR-METHYL-THIO)-N',N'-DIMETHYL-N-PHENYL-SCHWEFEL-SAEUREDIAMID (GERMAN) see DFL200
DICHLORFOS (POLISH) see DGP900
DICHLORHYDRATE de1-p.CHLORBENZHYDRYL-4-(2-(2-HYDROXYETHOXY)ETHYL)PIPERAZINE see HOR470
DICHLORHYDRATE de DIMETHOXY-3,4 BENZYL PIPERAZINE (FRENCH) see VIK200
5,7-DICHLOR-8-HYDROXYCHINOLIN see DEL300
DI-CHLORICIDE see DEP800
DICHLORICIDE MOTHPROOFER see TBC500
DICHLORID DIMETHYLCINICITY see DUG825
DICHLORID KYSELINY FUMAROVE (CZECH) see FOY000
DICHLORIMIPRAMINE see DEL400
DICHLORINE OXIDE see DEL600
DICHLORINE TRIOXIDE see DEL800

DICHLORISOPRENALINE (GERMAN) see DFN400

DICHLORISOPROTERENOL see DFN400

3,4-DICHLOR-ISOPROTERENOL (GERMAN) see DFN400

DICHLORMETHAZANONE see DEM000

3,6-DICHLOR-3-METHOXY-BENZOESAEURE (GERMAN) see MEL500

DICHLORMEZANONE see DEM000

DI-CHLOR-MULSION see EIY600

2,3-DICHLOR-1,4-NAPHTHOCHINON (GERMAN) see DFT000

1,1-DICHLOR-1-NITROAETHAN (GERMAN) see DFU000

2,6-DICHLOR-4-NITROANILIN (CZECH) see RDP300

2,5-DICHLORNITROBENZEN (CZECH) see DFT400

3,4-DICHLORNITROBENZEN (CZECH) see DFT600

4,4'-DICHLOR-2-NITRODIFENYLETHER see CJD600

2,4-DICHLOR-6-NITROFENOL (CZECH) see DFU600

2,4-DICHLOR-6-NITROFENYLESTER KYSELINY OCTIVE (CZECH) see DFU800

2',5-DICHLOR-4'-NITRO-SALIZYLSAEUREANILID (GERMAN) see DFV400

DICHLOROACETALDEHYDE see DEM200

2,2-DICHLOROACETALDEHYDE see DEM200

α,α-DICHLOROACETALDEHYDE see DEM200

DICHLOROACETAMIDE see DEM300

2,2-DICHLOROACETAMIDE see DEM300

d-(−)-threo-2-DICHLOROACETAMIDO-1-p-NITROPHENYL-1,3-PROPANEDIOL see CDP250

2',5'-DICHLOROACETANILIDE see DGB480

DICHLOROACETATE SODIUM SALT see SGG000

DICHLOROACETATO di DIISOPROPILAMMONIO see DNM400

2,2-DICHLOROACETIC ACID see DEL000

DICHLOROACETIC ACID ANHYDRIDE WITH DIETHYL HYDROGEN PHOSPHATE see DEM400

DICHLOROACETIC ACID, DIISOPROPYLAMINE SALT see DNM400

DICHLOROACETIC ACID METHYL ESTER see DEM800

DICHLOROACETIC ACID SODIUM SALT see SGG000

DICHLOROACETIC ANHYDRIDE see DEM825

1,1-DICHLOROACETONE see DGG500

1,3-DICHLOROACETONE see BIK250

sym-DICHLOROACETONE see BIK250

1,3-DICHLOROACETONE (DOT) see BIK250

α-α-DICHLOROACETONE see DGG500

α,γ-DICHLOROACETONE see BIK250

α,α'-DICHLOROACETONE see BIK250

DICHLOROACETONITRILE see DEN000

2,2-DICHLOROACETOPHENONE see DEN200

α,α-DICHLOROACETOPHENONE see DEN200

ω,ω-DICHLOROACETOPHENONE see DEN200

8-DICHLOROACETOXY-9-HYDROXY-8,9-DIHYDRO-AFLATOXIN B1 see DEN300

DICHLOROACETYL CHLORIDE see DEN400

2,2-DICHLOROACETYL CHLORIDE see DEN400

DICHLOROACETYL CHLORIDE (DOT) see DEN400

α,α-DICHLOROACETYL CHLORIDE see DEN400

(+-)-4-(DICHLOROACETYL)-3,4-DIHYDRO-3-METHYL-2H-1,4-BENZOXAZINE see DEN500

DICHLOROACETYLENE see DEN600

DICHLOROACETYLENE mixed with ETHER (1:9) see DEN800

d-threo-N-DICHLOROACETYL-1-p-NITROPHENYL-2-AMINO-1,3-PROPANEDIOL see CDP250

2-((4-(DICHLOROACETYL)PHENYL)AMINO)-2-ETHOXY-1-(4-NITROPHENYL)ETHANONE see DEN820

2-((4-(DICHLOROACETYL)PHENYL)AMINO)-2-HYDROXY-1-(4-METHOXYPHENYL)ETHANONE see DEN840

2-((4-(DICHLOROACETYL)PHENYL)AMINO)-2-HYDROXY-1-(4-METHYLPHENYL)ETHANONE see DEN860

2-((4-(DICHLOROACETYL)PHENYL)AMINO)-2-HYDROXY-1-(4-PHENOXYPHENYL)ETHANONE see DEN880

2-((4-(DICHLOROACETYL)PHENYL)AMINO)-2-HYDROXY-1-PHENYLETHANONE see DEN900

2-((4-(DICHLOROACETYL)PHENYL)AMINO)-2-HYDROXY-1-(4-(PHENYLTHIO)PHENYL)ET HANONE see DEN910

2,3-DICHLOROAFLATOXIN B1 see AEU500

3,3-DICHLOROALLYL CHLORIDE see TJB800

DICHLOROALLYL DIISOPROPYLTHIOCARBAMATE see DBI200

S-2,3-DICHLOROALLYL DIISOPROPYLTHIOCARBAMATE see DBI200

2,3-DICHLOROALLYL-N,N-DIISOPROPYLTHIOLCARBAMATE see DBI200

DICHLOROALLYL LAWSONE see DFN500

DICHLOROAMETHOPTERIN see DFO000

3',5'-DICHLOROAMETHOPTERIN see DFO000

2,5-DICHLORO-3-AMINOBENZOIC ACID see AJM000

3',5'-DICHLORO-4-AMINO-4-DEOXY-N10-METHYLPTEROGLUTAMIC ACID see DFO000

2,4-DICHLORO-6-AMINOPHENOL see AJM525

1-DICHLOROAMINOTETRAZOLE see DEO200

3,4-DICHLOROANILIDE-α-METHYLACRYLIC ACID see DFO800

2,5-DICHLOROANILIN (CZECH) see DEO295

2,3-DICHLOROANILINE see DEO210

2,4-DICHLOROANILINE see DEO290

2,5-DICHLOROANILINE see DEO295

3,4-DICHLOROANILINE see DEO300

4,5-DICHLOROANILINE see DEO300

N,N-DICHLOROANILINE see DEO500

2-(2,6-DICHLOROANILINO)-2-IMIDAZOLINE HYDROCHLORIDE see CMX760

(o-(2,6-DICHLOROANILINO)PHENYL)ACETIC ACID MONOSODIUM SALT see DEO600

(o-((2,6-DICHLOROANILINO)PHENYL))ACETIC ACID SODIUM SALT see DEO600

3-(2,4-DICHLOROANILINO)-1-(2,4,6-TRICHLOROPHENYL)-2-PYRAZOLINE-5-ONE see DEO625

3,6-DICHLORO-o-ANISIC ACID see MEL500

1,5-DICHLOROANTHRAQUINONE see DEO700

1,8-DICHLOROANTHRAQUINONE see DEO750

1,5-DICHLORO-9,10-ANTHRAQUINONE see DEO700

1,8-DICHLORO-9,10-ANTHRAQUINONE see DEO750

DICHLOROANTHRARUFIN see DFC600

p-DICHLOROARSINOANILINE HYDROCHLORIDE see AOR640

2-DICHLOROARSINOPHENOXATHIIN see DEP400

4,4'-DICHLOROAZOXYBENZENE see DEP450

p,p'-DICHLOROAZOXYBENZENE see DEP450

DICHLOROBENZALKONIUM CHLORIDE see AFP750

2,3-DICHLOROBENZENAMINE see DEO210

2,4-DICHLOROBENZENAMINE see DEO290

3,4-DICHLOROBENZENAMINE (9CI) see DEO300

m-DICHLOROBENZENE see DEP599

o-DICHLOROBENZENE see DEP600

o-DICHLOROBENZENE see DEP600

p-DICHLOROBENZENE see DEP800

1,2-DICHLOROBENZENE see DEP600

1,3-DICHLOROBENZENE see DEP599

o-DICHLOROBENZENE (DOT) see DEP600

o-DICHLOROBENZENE (ACGIH,OSHA) see DEP600

2,5-DICHLOROBENZENEAMINE see DEO295

2,6-DICHLOROBENZENECARBOTHIOAMIDE see DGM600

1-(3',4'-DICHLOROBENZENEDIAZOL)-2-THIOUREA see DEQ000

3,4-DICHLOROBENZENE DIAZOTHIOCARBAMID see DEQ000

3,4-DICHLOROBENZENE DIAZOTHIOUREA see DEQ000

4,5-DICHLORO-1,2-BENZENEDIOL see DGJ200

4,5-DICHLORO-m-BENZENEDISULFONAMIDE see DEQ200

4,5-DICHLORO-1,3-BENZENEDISULFONAMIDE see DEQ200

1,4-DICHLOROBENZENE (MAK) see DEP800

3,4-DICHLOROBENZENEMETHANOL METHYLCARBAMATE see DET400

2,3(or 3,4)-DICHLOROBENZENEMETHANOL METHYL CARBAMATE see DET600

DICHLOROBENZENE, PARA, solid (DOT) see DEP800

3,3'-DICHLOROBENZIDINA (SPANISH) see DEQ600

DICHLOROBENZIDINE see DEQ600

2,2'-DICHLOROBENZIDINE see DEQ400

3,3'-DICHLOROBENZIDINE see DEQ600

o,o'-DICHLOROBENZIDINE see DEQ600

3',3'-DICHLOROBENZIDINE see DEQ600

DICHLOROBENZIDINE BASE see DEQ600

3,3'-DICHLOROBENZIDINE DIHYDROCHLORIDE see DEQ800

4,4'-DICHLOROBENZILATE see DER000

4,4'-DICHLOROBENZILIC ACID ETHYL ESTER see DER000

2,4-DICHLOROBENZOIC ACID see DER100

2,5-DICHLOROBENZOIC ACID see DER400

3,4-DICHLOROBENZOIC ACID see DER600

p-DICHLOROBENZOL see DEP800

2,6-DICHLOROBENZONITRILE see DER800

1,1-DICHLORO-N-((DIMETHYLAMINO)SULFONYL)-1-FLUORO-N-PHENYLMETHANE SULFENAMIDE see DFL200

2,2-DICHLORO-3,3-DIMETHYLBUTANE see DFE100

1,1-DICHLORO-3,3-DIMETHYL-2-BUTANONE see DGF300

sym-DICHLORODIMETHYL ETHER (DOT) see BIK000

N,N-DICHLORO-1,1-DIMETHYLETHYLAMINE see BQY275

DICHLORODIMETHYLHYDANTOIN see DFE200

1,3-DICHLORO-5,5-DIMETHYL HYDANTOIN see DFE200

1,3-DICHLORO-5,5-DIMETHYL-2,4-IMIDAZOLIDINEDIONE see DFE200

DICHLORO(4,5-DIMETHYL-o-PHENYLENEDIAMMINE)PLATINUM(II) see DFE229

cis-DICHLORO(4,5-DIMETHYL-O-PHENYLENEDIAMMINE)PLATINUM(II) see DFE229

3,5-DICHLORO-N-(1,1-DIMETHYL-2-PROPYNYL)BENZAMIDE see DTT600

3,5-DICHLORO-2,6-DIMETHYL-4-PYRIDINOL see CMX850

2,3-DICHLORO-N,N-DIMETHYL-6-QUINOXALINESULFONAMIDE see DFE235

DICHLORODIMETHYLSILANE see DFE259

DICHLORODIMETHYLSTANNANE see DUG825

3,6-DICHLORO-3,6-DIMETHYLTETRAOXANE see DFE300

DICHLORODIMETHYLTIN see DUG825

(trans-4)-DICHLORO(4,4-DIMETHYLZINC 5((((METHYLAMINO)CARBONYL)OXY)IMINO)PENTANENITRILE) see DFE469

4,4'-DICHLORO-6,6'-DINITRO-O,O'-BIPHENOL see DFD000

3,3'-DICHLORO-5,5'-DINITRO-O,O'-BIPHENOL (FRENCH) see DFD000

5,5'-DICHLORO-3,3'-DINITRO(1,1'-BIPHENYL)-2,2'-DIOL see DFD000

DICHLORODINITROMETHANE see DFE550

2,5-DICHLORO-4,6-DINITROPHENYL CROTONATE see DFE560

3,6-DICHLORO-2,4-DINITROPHENYL CROTONATE see DFE560

3,6-DICHLORO-2,4-DINITROPHENYL METHACRYLATE see DFE570

DICHLORODIOCTYLSTANNANE see DVN300

trans-2,3-DICHLORO-p-DIOXANE see DFE600

trans-2,3-DICHLORO-1,4-DIOXANE see DFE600

DICHLORODIOXOCHROMIUM see CML125

4,5-DICHLORO-3,6-DIOXO-1,4-CYCLOHEXADIENE-1,2-DICARBONITRILE see DEX400

DICHLORODIPENTYLSTANNANE see DVV200

DI-p-CHLORODIPHENOXYMETHANE see NCM700

DICHLORODIPHENYL see DET700

DICHLORODIPHENYLACETIC ACID see BIL500

p,p'-DICHLORODIPHENYLACETIC ACID see BIL500

2,3-DICHLORO-6,12-DIPHENYL-DIBENZO(b,f)(1,5)DIAZOCINE see DFE700

2,8-DICHLORO-6,12-DIPHENYL-DIBENZO(b,f)(1,5)DIAZOCINE see DFE700

DICHLORODIPHENYL DICHLOROETHANE see BIM500

o,p'-DICHLORODIPHENYLDICHLOROETHANE see CDN000

p,p'-DICHLORODIPHENYLDICHLOROETHANE see BIM500

p,p'-DICHLORODIPHENYLDICHLOROETHENE see BIM750

2,2'-((3,3'-DICHLORO(1,1'-DIPHENYL)-4,4'-DIYL)BIS(AZO)BIS(3-OXO-N)-PHENYLBUTANAMIDE see DEU000

p,p'-DICHLORODIPHENYL ETHANE see BIM775

DICHLORODIPHENYLETHANOL see BIN000

p,p'-DICHLORODIPHENYLMETHYLCARBINOL see BIN000

DICHLORODIPHENYL OXIDE see DFE800

DICHLORO DIPHENYLSILANE see DFF000

DICHLORODIPHENYLSTANNANE see DWO400

DICHLORODIPHENYLSTANNANE complex with PYRIDINE (1:2) see DWA400

p,p'-DICHLORODIPHENYL SULFIDE see CEP000

4,4-DICHLORODIPHENYL SULFOXIDE see BIN900

DICHLORODIPHENYLTRICHLOROETHANE see DAD200

4,4'-DICHLORODIPHENYLTRICHLOROETHANE see DAD200

DICHLORODIPHENYLTRICHLOROETHANE (DOT) see DAD200

p,p'-DICHLORODIPHENYLTRICHLOROETHANE see DAD200

2,2-DICHLORO-N,N-DI-2-PROPENYLACETAMIDE see DBI300

2',6'-DICHLORO-2-(DIPROPYLAMINO)ACETANILIDE HYDROCHLORIDE see DFF200

DICHLORODIPROPYLSTANNANE see DFF400

DICHLORODIPROPYLTIN see DFF400

DICHLORODIPYRIDINEPLATINUM(II) (Z) see DFF500

cis-DICHLORO(DIPYRIDINE)PLATINUM(II) see DFF500

4,5-DICHLORO-1,3-DISULFAMOYLBENZENE see DEQ200

DICHLORODIVINYLSTANNANE see DXR450

DICHLORODIVINYLTIN see DXR450

1,4-DICHLORO-2,3-EPOXYBUTANE see DFF600

cis-1,3-DICHLORO-1,2-EPOXYPROPANE see DAC975

trans-1,3-DICHLORO-1,2-EPOXYPROPANE see DGH500

DICHLOROETHANE see DFF800

1,1-DICHLOROETHANE see DFF809

1,2-DICHLOROETHANE see EIY600

sym-DICHLOROETHANE see EIY600

DICHLORO-1,2-ETHANE (FRENCH) see EIY600

α,β-DICHLOROETHANE see EIY600

DICHLOROETHANOIC ACID see DEL000

2,2-DICHLOROETHANOL see DFG000

1,2-DICHLOROETHANOL ACETATE see DFG159

DICHLOROETHANOYL CHLORIDE see DEN400

1,1-DICHLOROETHENE see VPK000

1,1-DICHLOROETHENE POLYMER with CHLOROETHENE see CGW300

2,2-DICHLOROETHENOL DIMETHYL PHOSPHATE see DGP900

(E)-S-(1,2-DICHLOROETHENYL)-l-CYSTEINE (9CI) see DGP125

2,2-DICHLOROETHENYL DIETHYL PHOSPHATE see DFG200

3-(2,2-DICHLOROETHENYL)-2,2-DIMETHYLCYCLOPROPANECARBOXYLIC ACID, CYANO(3-PHENOXYPHENYL) METHYL ESTER, (1-α(S*),3-β)-(+−)- see DFG300

2,2-DICHLOROETHENYL DIMETHYL PHOSPHATE see DGP900

1,1'-(DICHLOROETHENYLIDENE)BIS(4-CHLOROBENZENE) see BIM750

2,2-DICHLOROETHENYL PHOSPHORIC ACID DIMETHYL ESTER see DGP900

DICHLOROETHER see DFJ050

1,2-DICHLORO-1-ETHOXYETHANE see DFG333

DICHLORO(4-ETHOXY-O-PHENYLENEDIAMMINE)PLATINUM(II) see DFG400

DI(2-CHLOROETHYL) ACETAL see DFG600

1,2-DICHLOROETHYL ACETATE see DFG159

DICHLOROETHYLALUMINUM see EFU050

2,2-DICHLOROETHYLAMINE see DFG700

DI-2-CHLOROETHYLAMINE HYDROCHLORIDE see BHO250

p-(DI-2-CHLOROETHYLAMINE)PHENYL BUTYRIC ACID SODIUM SALT see CDO625

9-(2-(DI(2-CHLOROETHYL)AMINO)ETHYLAMINO)-6-CHLORO-2-METHOXYACRIDINE see DFH000

2-(DI(2-CHLOROETHYL)AMINOMETHYL-5,6-DIMETHYLBENZIMIDAZOLE see BCC250

2-(DI(2-CHLOROETHYL)AMINO)-1-OXA-3-AZA-2-PHOSPHACYCLOHEXANE-2-OXIDE MONOHYDRATE see CQC500

p-N,N-DI-(2-CHLOROETHYL)AMINOPHENYL ACETIC ACID see PCU425

p-N-DI(CHLOROETHYL)AMINOPHENYLALANINE see PED750

p-DI-(2-CHLOROETHYL)-AMINO-d-PHENYLALANINE see SAX200

p-DI(2-CHLOROETHYL)AMINO-d-PHENYLALANINE see SAX200

p-DI-(2-CHLOROETHYL)AMINO-l-PHENYLALANINE see PED750

3-p-(DI(2-CHLOROETHYL)AMINO)-PHENYL-l-ALANINE see PED750

o-DI-2-CHLOROETHYLAMINO-dl-PHENYLALANINE see BHT250

p-DI(2-CHLOROETHYL)AMINO-dl-PHENYLALANINE see BHT750

p-(N,N-DI-2-CHLOROETHYL)AMINOPHENYL BUTYRIC ACID see CDO500

γ-(p-DI(2-CHLOROETHYL)AMINOPHENYL)BUTYRIC ACID see CDO500

p-N,N-DI-(β-CHLOROETHYL)AMINOPHENYL BUTYRIC ACID see CDO500

N,N-DI-2-CHLOROETHYL-γ-p-AMINOPHENYLBUTYRIC ACID see CDO500

p-(N,N-DI-2-CHLOROETHYLAMINO)PHENYL-N-(p-CARBOXYPHENYL)CARBAMATE see CCE000

o-(p-DI-(2-CHLOROETHYL)AMINOPHENYL)-dl-TYROSINE DIHYDROCHLORIDE see DFH100

p-N,N-DI-(2-CHLOROETHYL)AMINOPHENYLVALERIC ACID see BHY625

N,N-DI(2-CHLOROETHYL)AMINO-N,O-PROPYLENE PHOSPHORIC ACID ESTER DIAMIDE MONOHYDRATE see CQC500

5-(DI-2-CHLOROETHYL)AMINOURACIL see BIA250

5-(DI-(β-CHLOROETHYL)AMINO)URACIL see BIA250

N,N-DI(2-CHLOROETHYL)ANILINE see AOQ875

DICHLOROETHYLARSINE see DFH200

DICHLOROETHYLBENZENE see EHY500

DI-(2-CHLOROETHYL)BENZYLAMINE see BIA750

DICHLOROETHYLBORANE see DFH300

DI-(2-CHLOROETHYL)-3-CHLORO-4-METHYLCOUMARIN-7-YL PHOSPHATE see DFH600

DI-(2-CHLOROETHYL)-3-CHLORO-4-METHYL-7-COUMARINYL PHOSPHATE see DFH600

O,O-DI(2-CHLOROETHYL)-7-(3-CHLORO-4-METHYLCOUMARINYL)PHOSPHATE see CIK750

O,O-DI(2-CHLOROETHYL)-O-(3-CHLORO-4-METHYLCOUMARIN-7-YL) PHOSPHATE see DFH600

DICHLOROETHYLENE see DFH800

DICHLOROETHYLENE see EIY600

1,1-DICHLOROETHYLENE see VPK000

1,2-DICHLOROETHYLENE see DFI200

1,2-DICHLOROETHYLENE see DFI210

cis-DICHLOROETHYLENE see DFI200

sym-DICHLOROETHYLENE see DFI210

trans-DICHLOROETHYLENE see ACK000

DICHLORO-1,2-ETHYLENE (FRENCH) see DFI210

1,2-DICHLOROETHYLENE CARBONATE see DFI800

DICHLORO(ETHYLENEDIAMMINE)PLATINUM(II) see DFJ000

trans-1,2-DICHLOROETHYLENE (MAK) see ACK000

1,1-DICHLOROETHYLENE-MONOCHLOROETHYLENE POLYMER see CGW300

1,1-DICHLOROETHYLENE POLYMER with CHLOROETHYLENE see CGW300

S-(1,2-DICHLOROETHYLENEYL)-l-CYSTEINE see DGP000

DI(2-CHLOROETHYL) ESTER, MALEIC ACID see DFJ200

DICHLOROETHYL ETHER see DFJ050

sym-DICHLOROETHYL ETHER see DFJ050

DI(β-CHLOROETHYL)ETHER see DFJ050

β,β'-DICHLOROETHYL ETHER see DFJ050

2,2'-DICHLOROETHYL ETHER (MAK) see DFJ050

DICHLOROETHYL FORMAL see BID750

DI-2-CHLOROETHYL FORMAL see BID750

1,2-DICHLOROETHYL HYDROPEROXIDE see DFJ100

DI-2-CHLOROETHYL MALEATE see DFJ200

2,3-DICHLORO-N-ETHYLMALEINIMIDE see DFJ400

DI(2-CHLOROETHYL)METHYLAMINE see BIE250

DI(2-CHLOROETHYL)METHYLAMINE HYDROCHLORIDE see BIE500

2-(1,2-DICHLOROETHYL)-4-METHYL-1,3-DIOXOLANE see DFJ500

2-N,N-DI(2-CHLOROETHYL)NAPHTHYLAMINE see BIF250

DICHLOROETHYL-β-NAPHTHYLAMINE see BIF250

DI(2-CHLOROETHYL)-β-NAPHTHYLAMINE see BIF250

N,N-DI(2-CHLOROETHYL)-β-NAPHTHYLAMINE see BIF250

DI((CHLORO-2-ETHYL)-2-N-NITROSO-N-CARBAMOYL)-N,N-CYSTAMINE see BIF625

DICHLOROETHYL OXIDE see DFJ050

DICHLOROETHYLPHENYLSILANE see DFJ800

DICHLOROETHYLPHOSPHINE see EOQ000

DICHLOROETHYLPHOSPHINE SULFIDE see EOP600

DI-2-CHLOROETHYL PHTHALATE see BIG600

N,N-DI(2-CHLOROETHYL)-N,o-PROPYLENE-PHOSPHORIC ACID ESTER DIAMIDE see CQC650

2-(1,2-DICHLOROETHYL)PYRIDINE HYDROCHLORIDE see CMV475

DICHLOROETHYLSILANE see DFK000

DI-2-CHLOROETHYL SULFIDE see BIH250

β,β'-DICHLOROETHYL SULFIDE see BIH250

2,2'-DICHLOROETHYL SULPHIDE (MAK) see BIH250

DI-(2-CHLOROETHYL)THENYLAMINE HYDROCHLORIDE see BII000

2-2'-DI(3-CHLOROETHYLTHIO)DIETHYL ETHER see DFK200

DICHLOROETHYLVINYLSILANE see DFK400

DICHLOROETHYNE see DEN600

DICHLOROFEN see MJM500

3-(3,4-DICHLORO-FENIL)-1-METOSSI-1-METIL-UREA (ITALIAN) see DGD600

DICHLOROFENTHION see DFK600

1,1-DICHLORO-1-FLUOROETHANE see FOO550

DICHLOROFLUOROMETHANE see DFL000

DICHLOROFLUOROMETHANE-TRICHLOROFLUOROMETHANE (DOT) see DFB800

(DICHLOROFLUOROMETHYL)BENZENE see DFL100

4',5-DICHLORO-N-(4-FLUORO-2-METHYLPHENYL)-2-HYDROXY-(1,1'-BIPHENYL)-3-CARBOXAMIDE see CFC500

N-((DICHLOROFLUOROMETHYL)THIO)-N-((DIMETHYLAMINO)SULFONYL)ANILINE see DFL200

N-(DICHLOROFLUOROMETHYLTHIO)-N',N'-DIMETHYL-N-PHENYLSULFAMIDE see DFL200

N-(DICHLOROFLUOROMETHYLTHIO)-N-(DIMETHYLSULFAMOYL)ANILINE see DFL200

N'-DICHLOROFLUOROMETHYLTHIO-N,N-DIMETHYL-N'-(4-TOLYL)SULFAMIDE see DFL400

2,4-DICHLORO-6-FLUOROPHENYL p-NITROPHENYL ETHER see FKM100

2,4-DICHLORO-6-FLUOROPHENYL-4'-NITROPHENYL ETHER see FKM100

α-DICHLORO-α-FLUOROTOLUENE see DFL100

α-β-DICHLORO-β-FORMYL ACRYLIC ACID see MRU900

2,2'-DICHLORO-N-FURFURYLDIETHYLAMINE HYDROCHLORIDE see FPX000

DICHLOROGERMANE see DFL600

N,N-DICHLOROGLYCINE see DFL709

4,5-DICHLOROGUAIACOL see DFL720

1,6-DICHLORO-2,4-HEXADIYNE see DFL800

2,3-DICHLOROHEXAFLUOROBUTENE-2 see DFM000

2,3-DICHLOROHEXAFLUORO-2-BUTENE see DFM000

2,3-DICHLORO-1,1,1,4,4,4-HEXAFLUOROBUTENE-2 see DFM000

1,2-DICHLOROHEXAFLUOROCYCLOBUTANE see DFM025

1,2-DICHLORO-1,2,3,3,4,4-HEXAFLUOROCYCLOBUTANE see DFM025

1,2-DICHLOROHEXAFLUOROCYCLOPENTENE see DFM050

1,2-DICHLORO-3,3,4,4,5,5-HEXAFLUOROCYCLOPENTENE see DFM050

4,5-DICHLORO-3,3,4,5,6,6-HEXAFLUORO-1,2-DIOXANE see DFM099

1,3-DICHLORO-1,1,2,2,3,3-HEXAFLUOROPROPANE see DFM110

DICHLOROHYDRIN see DGG400

α-DICHLOROHYDRIN see DGG400

6,7-DICHLORO-4-(HYDROXYAMINO)QUINOLINE-1-OXIDE see DFM200

3,6-DICHLORO-2-HYDROXYBENZOIC ACID see DGK250

3,4-DICHLORO-2-HYDROXYCROTONOLACTONE see MRU900

3,4-DICHLORO-2-HYDROXYCROTONOLACTONIC ACID see MRU900

6,7-DICHLORO-10-(3-(N-(2-HYDROXYETHYL)ETHYLAMINO))ISOALLOXAZINE SULFATE see DFM600

6,7-DICHLORO-10-(3-(N-(2-HYDROXYETHYL)METHYLAMINO)PROPYL) ISOALLOXAZINE SULFATE see DFM800

d-(−)-threo-2,2-DICHLORO-N-(β-HYDROXY-α-(HYDROXYMETHYL))-p-NITROPHENETHYLACETAMIDE see CDP250

d-(−)-2,2-DICHLORO-N-(β-HYDROXY-α-(HYDROXYMETHYL)-p-NITROPHENYLETHYL)ACETAMIDE see CDP250

(+−)-2,2-DICHLORO-N-(α-(HYDROXYMETHYL)-p-NITROPHENACYL)ACETAMIDE see AAI110

2,6-DICHLORO-4-((p-HYDROXYPHENYL)IMINO)-2,5-CYCLOHEXADIEN-1-ONE SODIUM SALT see SGG650

DI-(5-CHLORO-2-HYDROXYPHENYL)METHANE see MJM500

5,7-DICHLORO-8-HYDROXYQUINALDINE see CLC500

DICHLOROHYDROXYQUINOLINE see DEL300

5,7-DICHLORO-8-HYDROXYQUINOLINE see DEL300

2,6-DICHLORO-N-2-IMIDAZOLIDINYLIDENE-BENZENAMINE HYDROCHLORIDE see CMX760

1-(2,5-DICHLORO-6-(1-(1H-IMIDAZOL-1-YL)VINYL)PHENOXY)-3-(ISOPROPYLAMINO)-2-PROPANOL HYDROCHLORIDE see DFM875

DICHLOROINDANTHRONE see DFN300

3,3'-DICHLOROINDANTHRONE see DFN300

7,16-DICHLOROINDANTHRONE see DFN300

2,6-DICHLOROINDOPHENOL, SODIUM SALT see SGG650

O-(2,5-DICHLORO-4-IODOPHENYL) O,O-DIMETHYL PHOSPHOROTHIOATE see IEN000

2,2'-DICHLORO-N-ISOBUTYL-DIETHYLAMINE HYDROCHLORIDE see BIE750

1,3-DICHLOROISOBUTYLENE see DFN330

1,3-DICHLORO-5-ISOCYANOBENZENE see DFN350

DICHLOROISOCYANURATE see DGN200

DICHLOROISOCYANURIC ACID see DGN200

DICHLOROISOCYANURIC ACID, dry or dichloroisocyanuric acid salts (DOT) see DGN200

DICHLOROISOCYANURIC ACID POTASSIUM SALT see PLD000

DICHLOROISOCYANURIC ACID SODIUM SALT (DOT) see SGG500

sym-DICHLOROISOPROPYL ALCOHOL see DGG400

3,4-DICHLORO-α-((ISOPROPYLAMINO)METHYL)BENZYL ALCOHOL see DFN400

2,2'-DICHLORO-N-ISOPROPYLDIETHYLAMINE HYDROCHLORIDE see IPG000

DICHLOROISOPROPYL ETHER see BII250

2,2'-DICHLOROISOPROPYL ETHER see BII250

DICHLOROISOPROPYL ETHER (DOT) see BII250

3-(2,4-DICHLORO-5-ISOPROPYLOXY-PHENYL)-Δ⁴-5-(tert-BUTYL)-1,3,4-OXADIAZOLINE-2-ONE see OMM200

DICHLOROISOVIOLANTHRONE see DFN450

DICHLOROKELTHANE see BIO750

DICHLOROLAWSONE see DFN500

DICHLOROMALEALDEHYDIC ACID see MRU900

2,3-DICHLOROMALEIC ALDEHYDE ACID see MRU900

DICHLOROMALEIC ANHYDRIDE see DFN700

DICHLOROMALEIMIDE see DFN800

DICHLOROMALEINIMIDE see DFN800

DICHLOROMALONONITRILE see DFN850

DICHLOROMAPHARSEN see DFX400

3',4'-DICHLORO-2-METHACRYLANILIDE see DFO800

DICHLOROMETHANE DIPHOSPHONATE see SFX730

DICHLOROMETHANE (MAK, DOT) see MJP450

DICHLOROMETHANETHIOSULFONIC ACID-S-TRICHLOROMETHYL ESTER see DFS600

DICHLOROMETHAZANONE see CKF500

DICHLORO(l-METHIONINATO-N,S)PLATINATE(1-) HYDROGEN (SP-4-3)- see MDT900

DICHLORO-l-METHIONINEPLATINUM(II) see MDT900

DICHLOROMETHOTREXATE see DFO000

3'5'-DICHLOROMETHOTREXATE see DFO000

2,5-DICHLORO-6-METHOXYBENZOIC ACID see MEL500

3,6-DICHLORO-2-METHOXYBENZOIC ACID see MEL500

DICHLORO(4-METHOXYCARBONYL-O-PHENYLENEDIAMMINE)PLATINUM(II) see DFO200

4,5-DICHLORO-2-METHOXYPHENOL see DFL720

DICHLORO(4-METHOXY-O-PHENYLENEDIAMMINE)PLATINUM(II) see DFO400

((2,3-DICHLORO-4-METHOXYPHENYL)-2-FURANYLMETHANONE)-O-(2-(DIETHYLAMINO)ETHYL) OXIME,MONOMETHANE SULFONATE see DFO600

(DICHLORO-2,3-METHOXY-4) PHENYL FURYL-2-O-(DIETHYLAMINOETHYL)-CETONE-OXIME (FRENCH) see DFO600

3',4'-DICHLORO-2-METHYLACRYLANILIDE see DFO800

N,N-DICHLOROMETHYLAMINE see DFO900

9,10-DI(CHLOROMETHYL)ANTHRACENE see BIJ750

DICHLOROMETHYLARSINE see DFP200

1,5-DICHLORO-3-METHYL-3-AZAPENTANE HYDROCHLORIDE see BIE500

(DICHLOROMETHYL)BENZENE see BAY300

2,4-DICHLORO-1-METHYLBENZENE see DGM700

4,4'-DICHLORO(METHYL BENZHYDROL) see BIN000

4,4'-DICHLORO-α-METHYLBENZHYDROL see BIN000

4,4'-DICHLORO-α-METHYLBENZOHYDROL see BIN000

p-(DICHLOROMETHYL)BENZYL CHLORIDE see TJD650

1-(2,4-DICHLORO-β-(p-METHYLBENZYLOXY)PHENETHYL)IMIDAZOLE NITRATE see DFP500

DICHLOROMETHYL tert-BUTYL KETONE see DGF300

3-DICHLOROMETHYL-6-CHLORO-7-SULFAMOYL-3,4-DIHYDRO-1,2,4-BENZOTHIADIAZINE-1,1-DIOXIDE see HII500

3-DICHLOROMETHYL-6-CHLORO-7-SULFAMYL-3,4-DIHYDRO-1,2,4-BENZOTHIADIAZINE-1,1-DIOXIDE see HII500

DICHLOROMETHYL CYANIDE see DEN000

3-(DICHLOROMETHYL)-4,4-DICHLORO-2-BUTENOIC ACID see DFP550

2,2'-DICHLORO-N-METHYLDIETHYLAMINE see BIE250

2,2'-DICHLORO-N-METHYLDIETHYLAMINE HYDROCHLORIDE see BIE500

2,2'-DICHLORO-N-METHYLDIETHYLAMINE-N-OXIDE see CFA500

2,2'-DICHLORO-N-METHYLDIETHYLAMINE N-OXIDE HYDROCHLORIDE see CFA750

1-(DICHLOROMETHYLDIMETHYLSILYL)-1-HEXYN-3-OL see HFZ000

N-(DICHLOROMETHYLENE)ANILINE see PFJ400

2,3-DICHLORO-4-(2-METHYLENEBUTYRL)PHENOXY ACETIC ACID see DFP600

(2,3-DICHLORO-4-(2-METHYLENEBUTYRYL)PHENOXY)ACETIC ACID see DFP600

4,4'-DICHLORO-2,2'-METHYLENEDIPHENOL see MJM500

DICHLOROMETHYLENEDIPHOSPHONIC ACID DISODIUM SALT see SFX730

(2,3-DICHLORO-4-(2-METHYLENE-1-OXOBUTYL)PHENOXY)ACETIC ACID see DFP600

sym-DICHLOROMETHYL ETHER see BIK000

α,α-DICHLOROMETHYL ETHER see DFQ000

3,4-DICHLORO-α-(((1-METHYLETHYL)AMINO)METHYL)BENZENEMETHANOL see DFN400

1,3-DICHLORO-5,5'-METHYLHYDANTOIN see DFE200

5,7-DICHLORO-2-METHYL-8-HYDROXYQUINOLINE see CLC500

DICHLORO-N-METHYLMALEIMIDE see DFP800

2,3-DICHLORO-N-METHYLMALEIMIDE see DFP800

α,α-DICHLOROMETHYL METHYL ETHER see DFQ000

DICHLOROMETHYL METHYL KETONE see DGG500

d-threo-2-(DICHLOROMETHYL)-α-(p-NITROPHENYL)-2-OXAZOLINE-4-METHANOL see DFQ100

6,7-DICHLORO-2-METHYL-1-OXO-2-PHENYL-5-INDANYLOXYACETIC ACID see IBQ400

2,5-DICHLORO-4-(3-METHYL-5-OXO-2-PYRAZOLIN-1-YL) BENZENESULFONIC ACID see DFQ200

3,3-DICHLOROMETHYLOXYCYCLOBUTANE see BIK325

2-((2,6-DICHLORO-3-METHYLPHENYL)AMINO)-BENZOIC ACID (9CI) see DGM875

2-((2,6-DICHLORO-3-METHYLPHENYL)AMINO)BENZOIC ACID ETHOXYMETHYL ESTER see DGN000

2-((2,6-DICHLORO)-3-METHYLPHENYL)AMINO-BENZOIC ACID MONOSODIUM SALT see SIF425

DICHLORO(4-METHYL-o-PHENYLENEDIAMMINE)PLATINUM(II) see DFQ400

6-(3,5-DICHLORO-4-METHYLPHENYL)-3(2H)-PYRIDAZINONE see DFQ500

DICHLOROMETHYLPHENYLSILANE see DFQ800

DICHLOROMETHYL PHOSPHINE see EOQ000

1,2-DICHLORO-2-METHYLPROPENE see IIQ200

1,3-DICHLORO-2-METHYLPROPENE see DFN330

1,3-DICHLORO-2-METHYL-1-PROPENE see DFN330

2,3-DICHLORO-2-METHYLPROPIONALDEHYDE see DFR400

DICHLORO(1-METHYLPROPYL)ARSINE see BQY300

5,7-DICHLORO-2-METHYL-8-QUINOLINOL see CLC500

DICHLOROMETHYLSILANE see DFS000

4-(DICHLOROMETHYLSILYL)BUTYRONITRILE see COR325

1,2-DICHLORO-1-(METHYLSULFONYL)ETHYLENE see DFS200

DICHLOROMETHYL TRICHLOROMETHYLTHIOSULFONE see DFS600

2,2'-DICHLORO-1"-METHYLTRIETHYLAMINE see IOF300

2,2'-DICHLORO-1"-METHYLTRIETHYLAMINE HYDROCHLORIDE see IPG000

DICHLOROMETHYL-3,3,3-TRIFLUOROPROPYLSILANE see DFS700

DICHLOROMETHYLVINYLSILANE see DFS800

DICHLOROMONOETHYLALUMINUM see EFU050

DICHLOROMONOFLUOROMETHANE (OSHA, DOT) see DFL000

2,3-DICHLORO-1,4-NAPHTHALENEDIONE see DFT000

2,3-DICHLORO-1,4-NAPHTHAQUINONE see DFT000

DICHLORONAPHTHOQUINONE see DFT000

2,3-DICHLORONAPHTHOQUINONE see DFT000

2,3-DICHLORO-1,4-NAPHTHOQUINONE see DFT000

2,3-DICHLORONAPHTHOQUINONE-1,4 see DFT000

2,3-DICHLORO-α-NAPHTHOQUINONE see DFT000

DICHLORO(2,3-NAPHTHYLENEDIAMMINE)PLATINUM(II) see DFT033

3,4-DICHLORO-5-NITRO-2-ACETYLFURAN see DFT053

2,5-DICHLORO-4-NITROANILINE see DFT100

2,6-DICHLORO-4-NITROANILINE see RDP300

2,6-DICHLORO-4-NITROBENZENAMINE (9CI) see RDP300
2,3-DICHLORONITROBENZENE see DFT200
2,5-DICHLORONITROBENZENE see DFT400
3,4-DICHLORONITROBENZENE see DFT600
1,2-DICHLORO-4-NITROBENZENE see DFT600
1,4-DICHLORO-2-NITROBENZENE see DFT400
2',4'-DICHLORO-4-NITROBIPHENYL ETHER see DFT800
3,4-DICHLORO-5-NITRO-2-BROMOACETYLFURAN see BND600
2,4-DICHLORO-4'-NITRODIPHENYL ETHER see DFT800
DICHLORONITROETHANE see DFU000
1,1-DICHLORO-1-NITROETHANE see DFU000
1-(3,4-DICHLORO-5-NITRO-2-FURANYL)ETHANONE see DFT053
DICHLORONITROMETHANE see DFT250
1,2-DICHLORO-3-NITRONAPHTHALENE see DFU400
2,4-DICHLORO-6-NITROPHENOL see DFU600
2,4-DICHLORO-6-NITROPHENOL ACETATE see DFU800
2,4-DICHLORO-1-(4-NITROPHENOXY)BENZENE see DFT800
DICHLORO(4-NITRO-o-PHENYLENEDIAMMINE)PLATINUM(II) see DFV000
6,7-DICHLORO-4-NITROQUINOLINE-1-OXIDE see DFV200
2',5-DICHLORO-4'-NITROSALICYLANILIDE see DFV400
2',5-DICHLORO-4'-NITROSALICYLANILIDE-2-AMINOETHANOL SALT see DFV600
5,2'-DICHLORO-4'-NITROSALICYLANILIDE ETHANOLAMINE SALT see DFV600
5,2-DICHLORO-4-NITROSALICYLIC ANILIDE-2-AMINOETHANOL SALT see DFV600
2',5-DICHLORO-4'-NITROSALICYLOYLANILIDE ETHANOLAMINE SALT see DFV600
3,4-DICHLORO-N-NITROSOCARBANILIC ACID METHYL ESTER see DFV800
2,2'-DICHLORO-N-NITROSODIPROPYLAMINE see DFW000
3,4-DICHLORONITROSOPIPERIDINE see DFW200
3,4-DICHLORO-N-NITROSOPYRROLIDINE see DFW600
2,6-DICHLORO-4-OCTYLPHENOL- see DFW700
5,7-DICHLOROOXINE see DEL300
2,3-DICHLORO-4-OXO-2-BUTENOIC ACID see MRU900
4,5-DICHLORO-2-OXO-1,3-DIOXOLANE see DFI800
4,6-DICHLORO-3-((1E)-3-OXO-3-(PHENYLAMINO)-1-PROPENYL)-1H-INDOLE-2-CARBOXYLIC ACID, see DFW730
DICHLOROOXOVANADIUM see DFW800
DICHLOROOXOZIRCONIUM see ZSJ000
1,3-DICHLORO-1,1,2,2,3-PENTAFLUOROPROPANE see DFW830
3,3-DICHLORO-1,1,1,2,2-PENTAFLUOROPROPANE see DFW850
DICHLOROPENTANE see DFX000
1,5-DICHLOROPENTANE see DFX200
DICHLOROPENTANES (DOT) see DFX000
DICHLOROPENTYLARSINE see AOI200
1,2-DICHLOROPERFLUOROCYCLOBUTANE see DFM025

1,2-DICHLOROPERFLUORO CYCLOPENTENE see DFM050
DICHLOROPHEN see MJM500
DICHLOROPHENAMIDE see DEQ200
DICHLOROPHENARSINE HYDROCHLORIDE see DFX400
DICHLOROPHEN B see MJM500
DICHLOROPHENE see MJM500
2,3-DICHLOROPHENOL see DFX500
2,4-DICHLOROPHENOL see DFX800
2,5-DICHLOROPHENOL see DFX850
2,6-DICHLOROPHENOL see DFY000
3,4-DICHLOROPHENOL see DFY425
3,5-DICHLOROPHENOL see DFY450
2,4-DICHLOROPHENOL BENZENESULFONATE see DFY400
3-(3,4-DICHLOROPHENOL)-1,1-DIMETHYLUREA see DXQ500
2,4-DICHLORO-PHENOL-O-ESTER with O,O-DIETHYL PHOSPHOROTHIOATE see DFK600
3,4-DICHLOROPHENOL, O-ESTER with O-METHYL METHYLPHOSPHORAMIDOTHIOATE see IEN000
(2,4-DICHLOROPHENOXY)ACETATE DIMETHYLAMINE see DFY500
DICHLOROPHENOXYACETIC ACID see DAA800
3,4-DICHLOROPHENOXYACETIC ACID see DFY500
2,4-DICHLOROPHENOXYACETIC ACID (DOT) see DAA800
(2,4-DICHLOROPHENOXY)ACETIC ACID BUTOXYETHYL ESTER see DFY709
(2,4-DICHLOROPHENOXY)ACETIC ACID, BUTYL ESTER see BQZ000
2,4-DICHLOROPHENOXYACETIC ACID BUTYL ESTER and 2,4,5-TRICHLOROPHENOXYACETIC ACID (45.5%:48.2%) see AEX750
2,4-DICHLOROPHENOXYACETIC ACID, 4-CHLOROCROTONYL ESTER see KHU025
(2,4-DICHLOROPHENOXY)ACETIC ACID DIMETHYLAMINE see DFY800
(2,4-DICHLOROPHENOXY)ACETIC ACID ETHYL ESTER see EHY600
(2,4-DICHLOROPHENOXY)ACETIC ACID 2-ETHYLHEXYL ESTER see DBB480
2,4-DICHLOROPHENOXYACETIC ACID ISOOCTYL ESTER see ILO000
(2,4-DICHLOROPHENOXY)ACETIC ACID, ISOPROPYL ESTER see IOY000
(2-4-DICHLOROPHENOXY)ACETIC ACID-1-METHYLETHYL ESTER (9CI) see IOY000
2,4-DICHLOROPHENOXYACETIC PROPYLENE GLYCOL BUTYL ETHER ESTER see DFZ000
2,4-DICHLOROPHENOXYACETIC ACID, SODIUM SALT see SGH500
dl-N-(2,4-DICHLORO-PHENOXYACETYL)-3-PHENYLALANINE see DFZ100
4-(2,4-DICHLOROPHENOXY)BUTYRIC ACID see DGA000
γ-(2,4-DICHLOROPHENOXY)BUTYRIC ACID see DGA000
4-(2,4-DICHLOROPHENOXY)BUTYRIC ACID DIMETHYLAMINE SALT see DGA100
2,4-DICHLOROPHENOXYBUTYRIC ACID, SODIUM SALT see EAK500
γ-(2,4-DICHLOROPHENOXY)BUTYRIC ACID, SODIUM SALT see EAK500
5,6-DICHLORO-1-PHENOXYCARBONYL-2-TRIFLUOROMETHYLBENZIMIDAZOLE see DGA200
2,4-DICHLOROPHENOXY ETHANEDIOL see DGA400
2,4-DICHLOROPHENOXY-1,2-ETHANEDIOL see DGA400
2-(2,4-DICHLOROPHENOXY)ETHANOL see DGP800

2-(2,4-DICHLOROPHENOXY)ETHANOL HYDROGEN SULFATE SODIUM SALT see CNW000
2-(2,4-DICHLOROPHENOXY)ETHYL BENZOATE see SCB200
2-(1-(2,6-DICHLOROPHENOXY)ETHYL)-4,5-DIHYDRO-1H-IMIDAZOLE MONOHYDROCHLORIDE see LIA400
2-(1-(2,6-DICHLOROPHENOXY)ETHYL)-2-IMIDAZOLINE HYDROCHLORIDE see LIA400
2,4-DICHLOROPHENOXYETHYL SULFATE, SODIUM SALT see CNW000
3-(2,4-DICHLOROPHENOXY)-2-HYDROXYPROPYL-o-CHLOROPHENYL ARSINIC ACID see DGA425
DI-(4-CHLOROPHENOXY)METHANE see NCM700
DI-(p-CHLOROPHENOXY)METHANE see NCM700
4-(2,4-DICHLOROPHENOXY)-2-METHOXY-1-NITROBENZENE see DGA850
2-((3,4-DICHLOROPHENOXY)METHYL)-2-IMIDAZOLINE HYDROCHLORIDE see DGA800
2-((3,4-DICHLOROPHENOXY)METHYL)-2-IMIDAZOLINE MONOHYDROCHLORIDE see DGA800
5-(2,4-DICHLOROPHENOXY)-2-NITROANISOLE see DGA850
4-(2,4-DICHLOROPHENOXY)NITROBENZENE see DFT800
5-(2,4-DICHLOROPHENOXY)-2-NITROBENZOIC ACID METHYL ESTER see MJB600
2-(4-(2,4-DICHLOROPHENOXY)PHENOXY)-METHYL-PROPIONATE see IAH050
DICHLOROPHENOXYPHOSPHINE see PHE850
DICHLOROPHENOXYPHOSPHINE OXIDE see PHE800
(R)-2-(2,4-DICHLOROPHENOXY)PROPANOIC ACID see DGA880
2-(2,4-DICHLOROPHENOXY) PROPIONIC ACID see DGB000
2-(2,5-DICHLOROPHENOXY)PROPIONIC ACID see DGB200
(+)-2-(2,4-DICHLOROPHENOXY)PROPIONIC ACID see DGB100
α-(2,4-DICHLOROPHENOXY) PROPIONIC ACID see DGB000
α-(2,5-DICHLOROPHENOXY)PROPIONIC ACID see DGB200
(2,4-DICHLOROPHENOXY)TRIBUTYLSTANNANE see DGB400
2-(2,4-DICHLOROPHENOXY)-4,5,6-TRICHLOROPHENOL see TIL750
6-(2,4-DICHLOROPHENOXY)-2,3,4-TRICHLOROPHENOL see TIL750
N-(2,5-DICHLOROPHENYL)ACETAMIDE see DGB480
DI(p-CHLOROPHENYL)ACETIC ACID see BIL500
2-((2,6-DICHLOROPHENYL)AMINO)BENZENEACETIC ACID MONOSODIUM SALT see DEO600
2-(2,6-DICHLOROPHENYLAMINO)-2-IMIDAZOLINE see DGB500
2-(2,6-DICHLOROPHENYLAMINO)-2-IMIDAZOLIN HYDROCHLORID (GERMAN) see CMX760
DICHLOROPHENYLARSINE see DGB600
N-(3,4-DICHLOROPHENYL)-1-AZIRIDINECARBOXAMIDE see DGB800
p-((3,4-DICHLOROPHENYL)AZO)-N,N-DIMETHYLANILINE see DFD400

3,4-DICHLOROPHENYLAZOTHIOUREA see DEQ000

3,4-DICHLOROPHENYL-AZOTHIOUREE (FRENCH) see DEQ000

2,4-DICHLOROPHENYL BENZENESULFONATE see DFY400

2,4-DICHLOROPHENYL BENZENESULPHONATE see DFY400

2-(3,4-DICHLOROPHENYL)-1H-BENZ(de)ISOQUINOLINE-1,3(2H)-DIONE see DGB810

DICHLOROPHENYLBORANE see DGB875

O-1-(2,4-DICHLOROPHENYL)-2-BROMOVINYL O,O-DIETHYL PHOSPHATE see BMO310

O-1-(2,4-DICHLOROPHENYL)-2-BROMOVINYL-O,O-DIMETHYL PHOSPHATE see MHR150

(3,4-DICHLOROPHENYL)CARBAMIC ACID METHYL ESTER see DEV600

3,4-DICHLOROPHENYL-N-CARBAMOYLAZIRIDINE see DGB800

2,4-DICHLOROPHENYL "CELLOSOLVE" see DGC000

N-(3,4-DICHLOROPHENYL)-N'-(4-CHLOROPHENYL)UREA see TIL500

N-(3,4-DICHLOROPHENYL)CYCLOPROPANECARBOXAMIDE see CQJ250

2,4'-DICHLOROPHENYLDICHLOROETHANE see CDN000

1-((2-(2,4-DICHLOROPHENYL)-2-(2,6-DICHLOROPHENYL)METHOXY)ETHYL)-1H-IMIDAZOLE MONONITRATE see IKN200

O-2,4-DICHLOROPHENYL-O,O-DIETHYL PHOSPHOROTHIOATE see DFK600

2,4-DICHLORO-PHENYL DIETHYL PHOSPHOROTHIONATE see DFK600

1,6-DI(4'-CHLOROPHENYLDIGUANIDINO)HEXANE DIACETATE see CDT125

1,6-DI(4'-CHLOROPHENYLDIGUANIDO)HEXANE see BIM250

N-(2,4-DICHLOROPHENYL)-N-(4,5-DIHYDRO-2-THIAZOLYL)-3-PYRIDINEMETHANAMINE see DGC050

3-(3,5-DICHLOROPHENYL)-1,5-DIMETHYL-3-AZABICYCLO(3.1.0)HEXANE-2,4-DIONE see PMF750

N-(3',5'-DICHLOROPHENYL)-1,2-DIMETHYLCYCLOPROPANE-1,2-DICARBOXIMIDE see PMF750

3-(3,4-DICHLOROPHENYL)-1,1-DIMETHYL-3-NITROSOUREA see NKO500

(E)-1-(2,4-DICHLOROPHENYL)-4,4-DIMETHYL-2-(1,2,4-TRIAZOL-1-YL)PENTEN-3-OL see DGC100

N'-(3,4-DICHLOROPHENYL)-N,N-DIMETHYLUREA see DXQ500

1-(3,4-DICHLOROPHENYL)-3,3-DIMETHYLUREE (FRENCH) see DXQ500

2,6-DICHLORO-p-PHENYLENEDIAMINE see DBR400

cis-DICHLORO(o-PHENYLENEDIAMINE)PLATINUM(II) see DGC600

DICHLORO(1,2-PHENYLENEDIAMMINE)PLATINUM(II) see DGC600

2,4-DICHLOROPHENYL ESTER of BENZENESULFONIC ACID see DFY400

2,4-DICHLOROPHENYL ESTER BENZENESULPHONIC ACID see DFY400

DI-(p-CHLOROPHENYL)ETHANOL see BIN000

3-(3,5-DICHLOROPHENYL)-5-ETHENYL-5-METHYL-2,4-OXAZOLIDINEDIONE see RMA000

DICHLOROPHENYL ETHER see DFE800

1-((2-(2,4-DICHLOROPHENYL)-4-ETHYL-1,3-DIOXOLAN-2-YL)METHYL)-1H-1,2,4-TRIAZOLE see EDW100

1-((2-(2,4-DICHLOROPHENYL)-4-ETHYL-1,3-DIOXOLAN-2-YL)METHYL)-1H-1,2,4-TRIAZOLE see VFA100

N-(3,4-DICHLOROPHENYL)-N'-(4-((1-ETHYL-3-PIPERIDYL)AMINO)-6-METHYL-2-PYRIMIDINYL)GUANIDINE see WCJ750

O-(2,4-DICHLOROPHENYL)-O-ETHYL-S-PROPYLPHOSPHORODITHIOATE see DGC800

(2,5-DICHLOROPHENYL)HYDRAZINE see DGC850

N-(β-(3,4-DICHLOROPHENYL)-β-HYDROXYETHYL)ISOPROPYLAMINE see DFN400

3,4-DICHLOROPHENYL HYDROXYLAMINE see DGD075

N-(3,4-DICHLOROPHENYL)-N'-HYDROXYUREA see DGD085

2-(3,4-DICHLOROPHENYL)IMIDAZO(2,1-A)ISOQUINOLINE see DGD090

2-((2,6-DICHLOROPHENYL)IMINO)IMIDAZOLIDINE MONOHYDROCHLORIDE see CMX760

3,4-DICHLOROPHENYL ISOCYANATE see IKH099

3,5-DICHLOROPHENYL ISOCYANIDE see DFN350

1-(3,4-DICHLOROPHENYL)-2-ISOPROPYLAMINOETHANOL see DFN400

1-(3,4-DICHLOROPHENYL)-5-ISOPROPYLBIGUANIDE HYDROCHLORIDE see DGD100

N-(3,4-DICHLOROPHENYL)METHACRYLAMIDE see DFO800

DI-(4-CHLOROPHENYL)METHANE see BIM800

DI-(p-CHLOROPHENYL)METHANE see BIM800

2,4-DICHLOROPHENYLMETHANESULFONATE see DGD400

3-(3,4-DICHLOROPHENYL)-1-METHOXYMETHYLUREA see DGD600

3-(3,4-DICHLOROPHENYL)-1-METHOXY-1-METHYLUREA see DGD600

N'-(3,4-DICHLOROPHENYL)-N-METHOXY-N-METHYLUREA see DGD600

1-(3,4-DICHLOROPHENYL)3-METHOXY-3-METHYLUREE (FRENCH) see DGD600

2,4-DICHLOROPHENYL 3'-METHOXY-4'-NITROPHENYL ETHER see DGA850

7-(((3,4-DICHLOROPHENYL)METHYL)AMINO)-ACTINOMYCIN D see DET125

3-(3,4-DICHLOROPHENYL)-1-METHYL-1-BUTYLUREA see BRA250

DI(p-CHLOROPHENYL) METHYLCARBINOL see BIN000

2-((2,6-DICHLOROPHENYL)METHYLENE)HYDRAZINECARBOXIMIDAMIDE MONOACETATE (9CI) see GKO750

3-(3,5-DICHLOROPHENYL)-N-(1-METHYLETHYL)-2,4-DIOXO-1-IMIDAZOLIDINECARBOXAMIDE see GIA000

1-((2,4-DICHLOROPHENYL)METHYL)-1H-INDAZOLE-3-CARBOXYLIC ACID see DEL200

O-(2,4-DICHLOROPHENYL)-O-METHYLISOPROPYLPHOSPHORAMIDOTHIOATE see DGD800

O-(2,4-DICHLOROPHENYL)-O-METHYL-N-ISOPROPYLPHOSPHORAMIDOTHIOATE see DGD800

3-(2,6-DICHLOROPHENYL)-5-METHYL-4-ISOXAZOLYL PENICILLIN SODIUM MONOHYDRATE see DGE200

2-(3,4-DICHLOROPHENYL)-3-METHYL-4-METATHIAZANONE-1,1-DIOXIDE see DEM000

N-(3,4-DICHLOROPHENYL)-N'-METHYL-N'-METHOXYUREA see DGD600

(2,6-DICHLOROPHENYL)METHYL METHYL 3-PYRIDINYLCARBONIMIDODITHIOATE see DGE230

2-(3,4-DICHLOROPHENYL)-4-METHYL-1,2,4-OXADIAZOLIDINE-3,5-DIONE see BGD250

1-((2,4-DICHLOROPHENYL)METHYL)-5-OXO-l-PROLINE see DGE300

1-((2,6-DICHLOROPHENYL)METHYL)-5-OXO-l-PROLINE see DGE305

1-((3,4-DICHLOROPHENYL)METHYL)-5-OXO-l-PROLINE see DGE310

1-((2,6-DICHLOROPHENYL)METHYL)-5-OXO-l-PROLINE METHYL ESTER see DGE315

1-((3,4-DICHLOROPHENYL)METHYL)-5-OXO-l-PROLINE METHYL ESTER see DGE320

1-((2,6-DICHLOROPHENYL)METHYL)-5-OXO-l-PROLINE compounded with 2-PROPANAMINE (1:1) see PMH923

1-((3,4-DICHLOROPHENYL)METHYL)-5-OXO-l-PROLINE compounded with 2-PROPANAMINE (1:1) DGE325

N-(3,4-DICHLOROPHENYL)-2-METHYL-2-PROPENAMIDE see DFO800

5-(3,4-DICHLOROPHENYL)-6-METHYL-2,4-PYRIMIDINEDIAMINE see MQR100

3-(3,5-DICHLOROPHENYL)-5-METHYL-5-VINYL-2,4-OXAZOLIDINEDIONE see RMA000

N¹-3,4-DICHLOROPHENYL-N⁵-ISOPROPYLDIGUANIDE HYDROCHLORIDE see DGD100

2,4-DICHLOROPHENYL-4-NITROPHENYL ETHER see DFT800

2,4-DICHLOROPHENYL-p-NITROPHENYL ETHER see DFT800

1-(2-(2,4-DICHLOROPHENYL)PENTYL)-1H-1,2,4-TRIAZOLE see PAP240

α-(2,4-DICHLOROPHENYL)-α-PHENYL-5-PYRIMIDINEMETHANOL see THQ525

DICHLOROPHENYLPHOSPHINE see DGE400

DICHLOROPHENYLPHOSPHINE SULFIDE see PFW200

DICHLOROPHENYLPHOSPHINE SULFIDE see PFW210

6,8-DICHLORO-2-PHENYL-α-(2-PIPERIDYL)-4-QUINOLINEMETHANOL MONOHYDROCHLORIDE see PIU200

N-(3,4-DICHLOROPHENYL)PROPANAMIDE see DGI000

1-(2-(2,4-DICHLOROPHENYL)-2-(2-PROPENYLOXY)ETHYL)-1H-IMIDAZOLE see FPB875

N-(3,4-DICHLOROPHENYL)PROPIONAMIDE see DGI000

3-(2,4-DICHLOROPHENYL)-N-(4-PROPYLCYCLOHEXYL)-2-PROPENAMIDE see DGE500

1-(2-(2,4-DICHLOROPHENYL)-4-PROPYL-1,3-DIOXOLAN-2-YLMETHYL)-1H-1,2,4-TRIAZOLE see PMS930

4,5-DICHLORO-2-PHENYL-3(2H)-PYRIDAZINONE see DGE800

1-(2,4-DICHLOROPHENYL)-2-(3-PYRIDINYL)ETHANONE o-METHYLOXIME see DGJ160

4-(2,3-DICHLOROPHENYL)-1H-PYRROLE-3-CARBONITRILE see FAQ220

1-(3,5-DICHLOROPHENYL)-2,5-PYRROLIDINEDIONE see DGF000

N-(3,5-DICHLOROPHENYL)SUCCINIMIDE see DGF000

2-(3,4-DICHLOROPHENYL)TETRAHYDRO-3-METHYL-4H-1,3-THIAZIN-4-ONE-1,1-DIOXIDE see DEM000

((2,5-DICHLOROPHENYLTHIO)METHANETHIOL)-S-ESTER with O,O-DIMETHYL PHOSPHORODITHIOATE see MNO750

((2,5-DICHLOROPHENYL)THIO)METHYLCARBAMIC ACID, 2,3-DIHYDRO-2,2-DIMETHYL-7-BENZOFURANYL ESTER see DGF100

((3,4-DICHLOROPHENYL)THIO)METHYLCARBAMIC ACID, 2,3-DIHYDRO-2,2-DIMETHYL-7-BENZOFURANYL ESTER see DGF130

2,5-DICHLOROPHENYLTHIOMETHYL O,O-DIETHYL PHOSPHORODITHIOATE see PDC750

S-(2,5-DICHLOROPHENYLTHIOMETHYL) O,O-DIETHYL PHOSPHORODITHIOATE see PDC750

S-((3,4-DICHLOROPHENYLTHIO)METHYL)-O,O-DIETHYL PHOSPHORODITHIOATE see DJA200

S-(2,5-DICHLOROPHENYLTHIOMETHYL) DIETHYL PHOSPHOROTHIOLOTHIONATE see PDC750

S-(((2,5-DICHLOROPHENYL)THIO)METHYL) O,O-DIMETHYL PHOSPHORODITHIOATE see MNO750

(RS)-2-(2,4-DICHLOROPHENYL)-1-(1H-1,2,4-TRIAZOL-1-YL)HEXAN-2-OL see HCN050

(RS)-2-(3,5-DICHLOROPHENYL)-2-(2,2,2-TRICHLOROETHYL)OXIRANE see TJK100

2,4-DICHLOROPHENYLTRICHLOROMETHANE see PAY550

DI-(p-CHLOROPHENYL)TRICHLOROMETHYLCARBINOL see BIO750

(DICHLOROPHENYL)TRICHLOROSILANE see DGF200

DICHLOROPHENYLTRICHLOROSILANE (DOT) see DGF200

DICHLOROPHOS see DGP900

DICHLOROPHOSPHORIC ACID, ETHYL ESTER see EOR000

3,6-DICHLOROPICOLINIC ACID see DGJ100

DICHLOROPICRIN see DFT250

DICHLOROPINACOLIN see DGF300

α-α-DICHLOROPINACOLIN see DGF300

DICHLOROPINAKOLIN see DGF300

α,ω-DICHLOROPOLYETHYLENE GLYCOL see PJU300

DICHLOROPROP see DGB000

DICHLOROPROPANE see DGF350

1,1-DICHLOROPROPANE see DGF400

1,2-DICHLOROPROPANE see PNJ400

1,3-DICHLOROPROPANE see DGF800

2,2-DICHLOROPROPANE see DGF900

α,β-DICHLOROPROPANE see PNJ400

1,2-DICHLOROPROPANE mixed with 1,3-DICHLOROPROPENE and ISOTHIOCYANATOMETHANE see MLC000

DICHLOROPROPANE-DICHLOROPROPENE MIXTURE see DGG000

2,3-DICHLOROPROPANOL see DGG450

1,2-DICHLOROPROPANOL-3 see DGG450

1,2-DICHLORO-3-PROPANOL see DGG450

1,3-DICHLORO-2-PROPANOL see DGG400

2,3-DICHLORO-1-PROPANOL see DGG450

1,3-DICHLOROPROPANOL-2 (DOT) see DGG400

2,3-DICHLOROPROPANOL PHOSPHATE (3:1) see TNG750

1,3-DICHLORO-2-PROPANOL PHOSPHATE (3:1) see FQU875

1,1-DICHLOROPROPANONE see DGG500

1,3-DICHLORO-2-PROPANONE see BIK250

DICHLOROPROPENE see DGG700

1,1-DICHLOROPROPENE see DGG750

1,2-DICHLOROPROPENE see DGG800

1,3-DICHLOROPROPENE see DGG950

2,3-DICHLOROPROPENE see DGH400

1,1-DICHLORO-1-PROPENE see DGG750

2,3-DICHLORO-1-PROPENE see DGH400

DICHLOROPROPENE (DOT) see DGG950

(E)-1,3-DICHLOROPROPENE see DGH225

(Z)-1,3-DICHLOROPROPENE see DGH200

cis-1,3-DICHLOROPROPENE see DGH200

trans-1,3-DICHLOROPROPENE see DGH225

1,3-DICHLOROPROPENE and 1,2-DICHLOROPROPANE MIXTURE see DGG000

cis-1,3-DICHLOROPROPENE OXIDE see DAC975

trans-1,3-DICHLOROPROPENE OXIDE see DGH500

2,3-DICHLORO-2-PROPENE-1-THIOL DIISOPROPYLCARBAMATE see DBI200

2,3-DICHLORO-2-PROPEN-1-OL see DGH600

S-(2,3-DICHLORO-2-PROPENYL)ESTER, BIS(1-METHYLETHYL) CARBAMOTHIOIC ACID see DBI200

1,2-DICHLORO-3-PROPIONAL see DGH800

2,3-DICHLORO PROPIONALDEHYDE see DGH800

α,β-DICHLOROPROPIONALDEHYDE see DGH800

DICHLOROPROPIONANILIDE see DGI000

3,4-DICHLOROPROPIONANILIDE see DGI000

3',4'-DICHLOROPROPIONANILIDE see DGI000

2,2-DICHLOROPROPIONIC ACID see DGI400

α-DICHLOROPROPIONIC ACID see DGI400

α,α-DICHLOROPROPIONIC ACID see DGI400

2,2-DICHLOROPROPIONIC ACID, SODIUM SALT see DGI600

α,α-DICHLOROPROPIONIC ACID SODIUM SALT see DGI600

2,2-DICHLOROPROPIONIC ACID, 2-(2,4,5-TRICHLOROPHENOXY)ETHYL ESTER see PBK000

1,3-DICHLOROPROPYENE-1 see DGG950

2,4-DICHLORO-N-(4-PROPYLCYCLOHEXYL)BENZAMIDE see DGI630

DICHLOROPROPYLENE see DGG700

DICHLOROPROPYLENE see DGG950

DICHLOROPROPYLENE see DGI700

1,1-DICHLOROPROPYLENE see DGG750

1,2-DICHLOROPROPYLENE see DGG800

1,3-DICHLOROPROPYLENE see DGG950

2,3-DICHLOROPROPYLENE see DGH400

cis-1,3-DICHLOROPROPYLENE see DGH200

trans-1,3-DICHLOROPROPYLENE see DGH225

α,γ-DICHLOROPROPYLENE see DGG950

3-(2,4-DICHLORO-5-(2-PROPYNYLOXY)PHENYL)-5-(1,1-DIMETHYLETHYL)-1,3,4-OXADIAZOL-2(3H)-ONE see BRA300

2,6-DICHLORO PYRIDINE see DGI800

3,6-DICHLORO-2-PYRIDINECARBOXYLIC ACID see CMX900

3,6-DICHLORO-2-PYRIDINECARBOXYLIC ACID see DGJ100

3,5-DICHLORO-2-PYRIDONE see DGJ150

2',4'-DICHLORO-2-(3-PYRIDYL)ACETOPHENONE o-METHYLOXIME see DGJ160

4,6-DICHLORO-2-PYRIMIDINAMINE see DGJ175

4,5-DICHLOROPYROCATECHOL see DGJ200

3,4-DICHLORO-2,5-PYRROLIDINEDIONE see DFN800

5,7-DICHLORO-8-QUINALDINOL see CLC500

4,7-DICHLOROQUINOLINE see DGJ250

3,7-DICHLOROQUINOLINE-8-CARBOXYLIC ACID see QMS100

DICHLOROQUINOLINOL see DEL300

2,6-DICHLORO-p-QUINONE see DES400

2,6-DICHLOROQUINONE CHLOROIMIDE see CHR000

2,3-DICHLOROQUINOXALINE see DGJ950

2,3-DICHLOROQUINOXALINE-6-CARBONYLCHLORIDE see DGK000

2,3-DICHLOROQUINOXALINE-6-SULFON-DIMETHYLAMIDE see DFE235

5,6-DICHLORO-1-β-d-RIBOFURANOSYLBENZIMIDAZOLE see DGK100

DICHLOROSAL see CFY000

3,5-DICHLOROSALICYLIC ACID see DGK200

3,6-DICHLOROSALICYLIC ACID see DGK250

DICHLOROSILANE see DGK300

DICHLOROSILANE see DGK300

2,2'-DICHLORO-4,4'-STILBENAMINE see DGK400

2,2'-DICHLORO-4,4'-STILBENEDIAMINE see DGK400

3,3'-DICHLORO-4,4'-STILBENEDIAMINE see DGK600

DICHLOROSTYRENE see DGK800

α,β-DICHLOROSTYRENE see DGK800

p-DICHLOROSULFAMOYLBENZOIC ACID see HAF000

2,4-DICHLORO-5-SULFAMOYLBENZOIC ACID see DGK900

3,4-DICHLORO-5-SULFAMYLBENZENESULFONAMIDE see DEQ200

p-(N,N-DICHLOROSULFAMYL)BENZOIC ACID see HAF000

DICHLOROSULFANE see SOG500

3,4-DICHLOROSULFOLANE see DGL200

2,4-DICHLORO-5-SULPHAMOYLBENZOIC ACID see DGK900

4.6-DICHLORO-2',4',5',7'-TETRABROMOFLUORESCEIN DIPOTASSIUM SALT see CMM000

DICHLOROTETRAFLUOROACETONE see DGL400

sym-DICHLOROTETRAFLUOROACETONE see DGL400

DICHLOROTETRAFLUOROETHANE see DGL600

sym-DICHLOROTETRAFLUOROETHANE see FOO509

1,2-DICHLORO-1,1,2,2-TETRAFLUOROETHANE (MAK) see FOO509

DICHLOROTETRAFLUOROETHANE (OSHA, ACGIH) see FOO509

1-(3,5-DICHLORO-4-(1,1,2,2-TETRAFLUOROETHOXY)PHENYL)-3-(2,6-DIFLUOROBENZOYL)UREA (IUPAC) see HCY600

1,3-DICHLORO-1,1,3,3-TETRAFLUORO-2-PROPANONE see DGL400

DICHLORO(2,3,6,7-TETRAHYDRO-1H,5H-BENZO(i,j)QUINOLIZINE)PLATINUM see DGL700

2,3-DICHLOROTETRAHYDROFURAN see DGL800

3,4-DICHLOROTETRAHYDROTHIOPHENE-1,1-DIOXIDE see DGL200

trans-DICHLOROTETRAKIS(3-PICOLINE)RHODIUM CHLORIDE see DGL825

trans-DICHLOROTETRAKIS(PYRIDINE)RHODIUM CHLORIDE see DGL835
(2,3-DICHLORO-4-(2-THENOYL)PHENOXY)ACETIC ACID see TGA600
endo-2,5-DICHLORO-7-THIABICYCLO(2.2.1)HEPTANE see DGL875
(2,3-DICHLORO-4-(2-THIENYLCARBONYL)PHENOXY)ACETIC ACID see TGA600
2,6-DICHLOROTHIOBENZAMIDE see DGM600
6,8-DICHLORO-2-THIO-2H-1,3-BENZOXAZINE-2,4(3H)-DIONE see DFC300
DICHLOROTHIOCARBONYL see TFN500
DICHLOROTHIOLANE DIOXIDE see DGL200
(2,3-DICHLORO-4-(2-THIOPHENECARBONYL)PHENOXY)ACETIC ACID see TGA600
DICHLOROTITANOCENE see DGW200
2,4-DICHLOROTOLUENE see DGM700
α-2-DICHLOROTOLUENE see CEO200
α-o-DICHLOROTOLUENE see CEO200
α-α-DICHLOROTOLUENE see BAY300
2-(4-(2,4-DICHLORO-m-TOLUOYL)-1,3-DIMETHYLPYRAZOL-5-YLOXY)-4'-METHYLACETOPHENONE see DGM730
N-(2,6-DICHLORO-m-TOLYL)ANTHRANILIC ACID see DGM875
N-(2,6-DICHLORO-m-TOLYL)ANTHRANILIC ACID ETHOXYMETHYL ESTER see DGN000
1,3-DICHLORO-s-TRIAZINE-2,4,6(1H,3H,5H)-TRIONE see DGN200
DICHLORO-S-TRIAZINE-2,4,6(1H,3H,5H)-TRIONE POTASSIUM DERIV see PLD000
1,3-DICHLORO-s-TRIAZINE-2,4,6(1H,3H,5H)TRIONE POTASSIUM SALT see PLD000
2-(4,6-DICHLORO-s-TRIAZIN-2-YLAMINO)-4-(4-AMINO-3-SULFO-1-ANTHRAQUINONYLAMINO)BENZENESULFONIC ACID, DISODIUM SALT see DGN400
4-(4,6-DICHLORO-s-TRIAZIN-2-YLAMINO)-5-HYDROXY-6-(2-HYDROXY-5-NITROPHENYLAZO)-2,7-NAPHTHALENEDISULFONIC ACID see DGN600
5-(3,5-DICHLORO-s-TRIAZINYLAMINO)-4-HYDROXY-3-PHENYLAZO-2,7-NAPHTHALENEDISULFONIC ACID see DGN800
2-(6-(4,6-DICHLORO-s-TRIAZINYL)METHYLAMINO-1-HYDROXY-3-SULFONAPHTHYLAZO)-1,5-NAPHTHALENEDISULFONIC ACID see DGO000
4,4'-DICHLORO-α-(TRICHLOROMETHYL)BENZHYDROL see BIO750
DICHLORO(4,5,6-TRICHLORO-o-PHENYLENEDIAMMINE)PLATINUM(II) see DGO200
2,2'-DICHLOROTRIETHYLAMINE see BID250
cis-DICHLORO(TRIETHYLENETETRAMINE)RHODIUM(1+) CHLORIDE see DGO250
2,2-DICHLORO-1,1,1-TRIFLUOROETHANE see TJY500
1,2-DICHLORO-4-(TRIFLUOROMETHYL)BENZENE see DGO300
4,5-DICHLORO-2-TRIFLUOROMETHYLBENZIMIDAZOLE see DGO400
5,6-DICHLORO-2-TRIFLUOROMETHYLBENZIMIDAZOLE-1-CARBOXYLATE see DGA200

5,6-DICHLORO-2-(TRIFLUOROMETHYL)-1H-BENZIMIDAZOLE-1-CARBOXYLIC ACID PHENYL ESTER see DGA200
1,3-DICHLORO-6-TRIFLUOROMETHYL-9-(3-(DIBUTYLAMINO)-1-HYDROXYPROPYL)PHENANTHRENE HCl see HAF500
1-(1,3-DICHLORO-6-TRIFLUOROMETHYL-9-PHENANTHRYL)-3-(DI-N-BUTYLAMINO)PROPANOL HYDROCHLORIDE see HAF500
DICHLORO(m-TRIFLUOROMETHYLPHENYL)ARSINE see DGO600
DICHLOROTRIPHENYLANTIMONY see DGO800
DICHLOROTRIPHENYLSTIBINE see DGO800
DICHLOROVAS see DGP900
2,2-DICHLOROVINYL ALCOHOL, DIMETHYL PHOSPHATE see DGP900
DICHLOROVINYLARSINE CHLORIDE see BIQ250
DICHLOROVINYLCHLOROARSINE (DOT) see BIQ250
S-DICHLOROVINYL-l-CYSTEINE see DGP000
S-(trans-1,2-DICHLOROVINYL)-l-CYSTEINE see DGP125
2,2-DICHLOROVINYL DIETHYL PHOSPHATE see DFG200
2,2-DICHLOROVINYL DIMETHYL PHOSPHATE see DGP900
2,2-DICHLOROVINYL DIMETHYL PHOSPHORIC ACID ESTER see DGP900
l-3-((1,2-DICHLOROVINYL)THIO)ALANINE see DGP000
DICHLOROVOS see DGP900
DICHLOROXIN see DEL300
5,7-DICHLOROXINE see DEL300
DICHLOROXYLENE see XSS250
α,α'-DICHLORO-m-XYLENE see DGP200
α,α'-DICHLORO-o-XYLENE see DGP400
α,α'-DICHLORO-p-XYLENE see DGP600
α,α'-DICHLOROXYLENE see XSS250
DICHLORPHEN see MJM500
DICHLORPHENAMIDE see DEQ200
2,4-DICHLORPHENOXYACETIC ACID see DAA800
2,4-DICHLORPHENOXYACETIC ACID OCTYL ESTER see OEU050
2-(1-(2,6-DICHLORPHENOXY)ATHYL)-2-IMIDAZOLIN-HYDROCHLORID (GERMAN) see LIA400
(2,4-DICHLOR-PHENOXY)-ESSIGSAEURE (GERMAN) see DAA800
2-(2,4-DICHLOR-PHENOXY)-PROPIONSAEURE (GERMAN) see DGB000
(+)-2-(2,4-DICHLORPHENOXY)PROPIONSAFEURE (GERMAN) see DGB100
(3,4-DICHLOR-PHENYL-AZO)-THIOHARNSTOFF (GERMAN) see DEQ000
3-(3,4-DICHLORPHENYL)-1-N-BUTYL-HARNSTOFF (GERMAN) see BRA250
2,4-DICHLORPHENYL "CELLOSOLVE" see DGP800
3-(3,4-DICHLOR-PHENYL)-1,1-DIMETHYL-HARNSTOFF (GERMAN) see DXQ500
3-(3,4-DICHLOR-PHENYL)-1-METHOXY-1-METHYL-HARNSTOFF (GERMAN) see DGD600
3-(4,5-DICHLORPHENYL)-1-METHOXY-1-METHYLHARNSTOFF (GERMAN) see DGD600
2,4,-DICHLORPHENYL-4-NITROPHENYLAETHER (GERMAN) see DFT800
DICHLORPHENYLOXYUREA see DGD085
1-(2-(2,4-DICHLORPHENYL)-2-(PROPENYLOXY)AETHYL)-1H-IMIDAZOLE see FPB875

DICHLORPHOS see DGP900
ω,ω-DICHLORPINAKOLIN see DGF300
DICHLORPROP see DGB000
(+)-DICHLORPROP see DGA880
DICHLORPROPAN see DGF350
DICHLORPROPAN-DICHLORPROPENGEMISCH (GERMAN) see DGG000
DICHLORPROPEN-GEMISCH (GERMAN) see DGG800
DICHLORPROP-P see DGA880
DICHLOR STAPENOR see DGE200
DICHLORSULFOFENYL-METHYLPYRAZOLON (CZECH) see DFQ200
DICHLOR-s-TRIAZIN-2,4,6(1H,3H,5H)TRIONE POTASSIUM see PLD000
DICHLORURE de TRIMETHYLAMMONIUM-1-(β-N-METHYLINDOYL-3") ETHYL-4'-PYRIDINIUM)-3 PROPANE see MKW100
O-(2,2-DICHLORVINYL)-O,O-DIETHYLPHOSPHAT (GERMAN) see DFG200
(2,2-DICHLOR-VINYL)-DIMETHYL-PHOSPHAT (GERMAN) see DGP900
O-(2,2-DICHLORVINYL)-O,O-DIMETHYLPHOSPHAT (GERMAN) see DGP900
DICHLORVOS see DGP900
DICHLORVOS-ETHYL see DFG200
DICHLOSALE see DFV400
DICHLOTIAZID see CFY000
DICHLOTRANE K see TDM820
DICHLOTRAN K see TDM820
DICHLOTRIDE see CFY000
DICHOLINE SUCCINATE see CMG250
DICHROMIC ACID, CALCIUM SALT (1:1) see CAR400
DICHROMIC ACID, ZINC SALT (1:1) see ZFA102
DICHROMIUM SULFATE see CMK415
DICHROMIUM SULPHATE see CMK415
DICHROMIUM TRIOXIDE see CMJ900
DICHROMIUM TRISULFATE see CMK415
DICHROMIUM TRISULPHATE see CMK415
DICHRONIC see DEO600
DICHYSTROLUM see DME300
DI(2-CIANOETIL)AMMINA (ITALIAN) see BIQ500
DI(2-CIANOETIL)METILAMMINA see MHQ750
DI(CIANOMETIL)AMMINA see IBB100
DICIDOL see TJG550
DICK (GERMAN) see DFH200
DICLOCIL see DGE200
DICLOFENAC SODIUM see DEO600
DICLOFOP-METHYL see IAH050
DICLOMEZINE see DFQ500
DICLONDAZOLIC ACID see DEL200
DICLONIA see BPR500
DICLOPHENAC SODIUM see DEO600
DICLORALUREA see DGQ200
DICLORAN see RDP300
S-(2,3-DICLORO-ALLIL)-N,N-DIISOPROPIL-MONOTIOCARBAMMATO (ITALIAN) see DBI200
p-DICLOROBENZENE (ITALIAN) see DEP800
1,4-DICLOROBENZENE (ITALIAN) see DEP800
1,1-DICLORO-2,2-BIS(4-CLORO-FENIL)-ETANO (ITALIAN) see BIM500
3,3'-DICLORO-4,4'-DIAMINODIFENILMETANO (ITALIAN) see MJM200
1,1-DICLOROETANO (ITALIAN) see DFF809
1,2-DICLOROETANO (ITALIAN) see EIY600
2,2'-DICLOROETILETERE (ITALIAN) see DFJ050

(3,4-DICLORO-FENIL-AZO)-TIOUREA
(ITALIAN) see DEQ000
3-(3,4-DICLORO-FENYL)-1,1-DIMETIL-
UREA (ITALIAN) see DXQ500
d-threo-2-DICLOROMETIL-4-((4'-
NITROFENIL)-OSSIMETIL)-2-OSSAZOLINA
(ITALIAN) see DFQ100
1,1-DICLORO-1-NITROETANO (ITALIAN)
see DFU000
(2,2-DICLORO-VINIL)DIMETILFOSFATO
(ITALIAN) see DGP900
DICLOTRIDE see CFY000
DICLOXACILLIN SODIUM
MONOHYDRATE see DGE200
DICLOXACILLIN SODIUM SALT see
DGE200
DICO see OOI000
DICOBALT BORIDE see DGQ300
DICOBALT CARBONYL see CNB500
DICOBALT EDETATE see DGQ400
DICOBALT EDTA see DGQ400
DICOBALT OCTACARBONYL see CNB500
DICOBALT OXIDE see CND825
DICOBALT TRIOXIDE see CND825
DICODID see OOI000
DICOFERIN see DGQ500
DICOFOL see BIO750
DICOL see DJD600
DICONIRT D see SGH500
DICOPHANE see DAD200
DICOPPER(I) ACETYLIDE see DGQ600
DICOPPER DICHLORIDE see CNK250
DICOPPER DIHYDROXYCARBONATE see
CNJ750
DICOPPER(I)-1,5-HEXADIYNIDE see
DGQ625
DICOPPER(I) KETENIDE see DGQ650
DICOPPER MONOSULFIDE see CNP750
DICOPPER MONOXIDE see CNO000
DICOPPER SULFIDE see CNP750
DICOPUR see DAA800
DICOPUR-M see CIR250
DICORANTIL see DNN600
DICORTOL see PMA000
DICORVIN see DKA600
DICOTEX see CIR250
DICOTEX 80 see SIL500
DICOTOX see DAA800
DICOTOX see EHY600
DICOUMARIN see BJZ000
DICOUMAROL see BJZ000
DICRESOL see DGQ700
DICRESYL see MIB750
DICRESYL N-METHYLCARBAMATE see
MIB750
DICRESYLOLPROPANE see IPK000
DICRODEN see MQH250
DICROTALIC ACID see HMC000
DICROTOFOS (DUTCH) see DGQ875
DICROTONYL PEROXIDE see DGQ859
DICROTOPHOS see DGQ875
DICRYL see DFO800
DICTYCIDE see COH250
DICTYOCARPINE see DGQ900
DICTYOCARPINE 14-ACETATE see ACH300
DICTYOCARPININE 6-ACETATE see
DGQ900
DICTYOCARPININE 6,14-DIACETATE see
ACH300
DICTYZIDE see COH250
DICUMACYL see BKA000
DICUMAN see BJZ000
DICUMARINE see BJZ000
DICUMENE CHROMIUM see DGR200
DICUMENYLCHROMIUM see DGR200
DICUMYLMETHANE see DGR400
DICUMYL PEROXIDE (DOT) see DGR600
DI-α-CUMYL PEROXIDE see DGR600
DI-CUP see DGR600
DI-CUP 40 KF see DGR600
DI-CUPR see DGR600
DICUPRAL see DXH250

DICURAN see CIS250
DICURONE see GFM000
DICYANDIAMIDE-FORMALDEHYDE
ADDUCT see CNH125
DICYANDIAMIDE-FORMALDEHYDE
POLYMER see CNH125
DICYANDIAMIDE-FORMALDEHYDE
RESIN see CNH125
DICYANDIAMIDINE see GLS900
DICYANDIAMIN see COP125
DICYANOACETYLENE see DGS000
2,2'-DICYANO-2,2'-AZOPROPANE see
ASL750
m-DICYANOBENZENE see PHX550
o-DICYANOBENZENE see PHY000
p-DICYANOBENZENE see BBP250
1,2-DICYANOBENZENE see PHY000
1,3-DICYANOBENZENE see PHX550
1,4-DICYANOBENZENE see BBP250
1,4-DICYANOBUTANE see AER250
1,4-DICYANO-2-BUTENE see DGS200
β,β-DICYANO-o-CHLOROSTYRENE see
CEQ600
DICYANODIAMIDE see COP125
DICYANODIAZENE see DGS300
2,2'-DICYANODIETHYLAMINE see BIQ500
β,β'-DICYANODIETHYL ETHER see
OQQ000
β,β'-DICYANODIETHYL SULFIDE see
DGS600
2,3-DICYANO-1,4-DITHIA-
ANTHRAQUINONE see DLK200
s-DICYANOETHANE see SNE000
1,2-DICYANOETHANE see SNE000
trans-1,2-DICYANOETHENE see FOX000
DI-(2-CYANOETHYL)AMINE see BIQ500
(E)-1,2-DICYANOETHYLENE see FOX000
DI(2-CYANOETHYL)SULFIDE see DGS600
DICYANOFURAZAN see DGS700
DICYANOFURAZAN-N-OXIDE see DGS800
DICYANOFUROXAN see DGS800
DICYANOGEN see COO000
DICYANOGEN-N,N-DIOXIDE see DGT000
1,6-DICYANOHEXANE see OCW050
DICYANOMETHANE see MAO250
1,8-DICYANOOCTANE see SBK500
1,5-DICYANOPENTANE see HBD000
1,3-DICYANOPROPANE see TLR500
1,3-DICYANOTETRACHLOROBENZENE
see TBQ750
DICYANOTRISULFIDE see SOR200
cis-
DICYCLOBUTYLAMMINEDICHLOROPLAT
INUM(II) see DGT200
DICYCLOBUTYLIDENE see BFX545
DICYCLOHEXANO-18-CROWN-6 see
DGV100
DICYCLOHEXANO-24-CROWN-8 see
DGT300
3,9-DI-(3-CYCLOHEXENYL)-2,4,8,10-
TETRAOXASPIRO(5,5)UNDECANE see
DGT400
DICYCLOHEXYL ADIPATE see DGT500
N,N-DICYCLOHEXYLAMINE see DGT600
DICYCLOHEXYLAMINE (DOT) see DGT600
DICYCLOHEXYLAMINE, COMPD. WITH 2-
CYCLOHEXYL-4,6-DINITROPHENOL (1:1)
see CPK550
DICYCLOHEXYLAMINE NITRITE see
DGU200
DICYCLOHEXYLAMINE, N-NITROSO- see
NJW100
DICYCLOHEXYLAMINE PENTANOATE
see DGU400
DICYCLOHEXYLAMINE SALT OF DINEX
see CPK550
DICYCLOHEXYLAMINE SALT OF 4,6-
DINITRO-o-CYCLOHEXYLPHENOL see
CPK550
DICYCLOHEXYLAMINE, SALT OF
DNOCHP see CPK550

1-(2-(DICYCLOHEXYLAMINO)ETHYL)-1-
METHYL-PIPERIDINIUM BROMIDE see
MJC775
1-(2-(DICYCLOHEXYLAMINO)ETHYL)-1-
METHYL-PIPERIDINIUM CHLORIDE see
CAL075
DICYCLOHEXYLAMINONITRITE see
DGU200
cis-
DICYCLOHEXYLAMMINEDICHLOROPLA
TINUM(II) see DGU709
DICYCLOHEXYLAMMONIUM 2-
CYCLOHEXYL-4,6-DINITROPHENATE see
CPK550
DICYCLOHEXYLAMMONIUM 4,6-
DINITRO-o-CYCLOHEXYLPHENATE see
CPK550
DICYCLOHEXYLAMMONIUM NITRITE see
DGU200
N,N-DICYCLOHEXYL-2-
BENZOTHIAZOLESULFENAMIDE see
DGU800
DICYCLOHEXYLCARBODIIMIDE see
MDR800
1,3-DICYCLOHEXYLCARBODIIMIDE see
MDR800
N,N'-DICYCLOHEXYLCARBODIIMIDE see
MDR800
DICYCLOHEXYLCARBONYL PEROXIDE
see DGV000
DICYCLOHEXYL-18-CROWN-6 see DGV100
DICYCLOHEXYL DISULFIDE see CPK700
2-(2,2-
DICYCLOHEXYLETHYL)PIPERIDINE
MALEATE see PCH800
DICYCLOHEXYLFLUOROPHOSPHATE see
DGV200
DICYCLOHEXYL
FLUOROPHOSPHONATE see DGV200
DICYCLOHEXYLHYDROXYPHENYLSTAN
NANE see DGV670
DICYCLOHEXYL KETONE see DGV600
DICYCLOHEXYLMETHANE-4,4'-
DIISOCYANATE see MJM600
(6-(3-(DICYCLOHEXYLMETHYL)-d-
ALANINE))-LHRH ACETATE see LIU313
DICYCLOHEXYLNITROSAMIN see NJW100
DICYCLOHEXYLNITROSAMINE see
NJW100
DICYCLOHEXYL PEROXIDE CARBONATE
see DGV650
DICYCLOHEXYL PEROXYDICARBONATE,
not >91% with water (UN 2153) (DOT) see
DGV650
DICYCLOHEXYL PEROXYDICARBONATE,
technically pure (UN 2152) (DOT) see DGV650
N,N'-DICYCLOHEXYL-p-
PHENYLENEDIAMINE see BBN100
DICYCLOHEXYLPHENYLTIN
HYDROXIDE see DGV670
DICYCLOHEXYL PHTHALATE see DGV700
DICYCLOHEXYL THIOUREA see DGV800
N,N'-DICYCLOHEXYLTHIOUREA see
DGV800
DICYCLOHEXYLTIN OXIDE see DGV900
DICYCLOMINE HYDROCHLORIDE see
ARR750
DICYCLOPENTADIENE see DGW000
DICYCLOPENTADIENE DIEPOXIDE see
BGA250
DICYCLOPENTADIENE DIOXIDE see
BGA250
DICYCLOPENTADIENYLCOBALT see
BIR529
DICYCLOPENTADIENYLDICHLOROTITA
NIUM see DGW200
DICYCLOPENTADIENYLHAFNIUM
DICHLORIDE see HAE500
DI-2,4-CYCLOPENTADIEN-1-YL IRON see
FBC000
DICYCLOPENTADIENYL IRON (OSHA,
ACGIH) see FBC000

DI-pi-CYCLOPENTADIENYLNICKEL see NDA500

DI-pi-CYCLOPENTADIENYLTITANIUM see TGH500

DICYCLOPENTADIENYLTITANIUMDICH LORIDE see DGW200

DICYCLOPENTA(c,lmn)PHENANTHREN-1(9H)-ONE, 2,3-DIHYDRO- see DGW300

DICYCLOPENTENYL ACRYLATE see DGW400

DICYCLOPENTENYL ACRYLATE see DGW400

DICYCLOPENTENYLOXYETHYL METHACRYLATE see DGW450

α,α-DICYCLOPENTYL-ACETIC ACID-DIETHYLAMINO-ETHYLESTER BROMOCTYLATE see PAR600

DICYCLOPENTYLACETIC ACID-β-DIETHYLAMINOETHYL ESTER ETHOBROMIDE see DGW600

(2-(DICYCLOPENTYLACETOXY)ETHYL)TRIE THYLAMMONIUM BROMIDE see DGW600

N-(2-((DICYCLOPENTYLACETYL)OXY)ETHYL)-N,N-DIETHYL-1-OCTANAMINIUM BROMIDE (9CI) see PAR600

2-((DICYCLOPENTYLACETYL)OXY)-N,N,N-TRIETHYL-ETHANAMINIUM BROMIDE see DGW600

cis-DICYCLOPENTYLAMMINEDICHLOROPL ATINUM(II) see BIS250

α,α-DICYCLOPENTYLESSIGSAURE-DIAETHYLAMINO-AETHYLESTER-BROMOCTYLAT (GERMAN) see PAR600

DICYCLOPROPYLDIAZOMETHANE see DGW875

3,3-DICYCLOPROPYL-2-(ETHOXYCARBONYL)ACRYLONITRILE see DGX000

DICYKLOHEXYLAMIN (CZECH) see DGT600

DICYKLOHEXYLAMINKAPRONAT (CZECH) see DGU400

DICYKLOHEXYLAMIN NITRIT (CZECH) see DGU200

N,N-DICYKLOHEXYLBENZTHIAZOLSULFENA MID (CZECH) see DGU800

DICYKLOPENTADIEN (CZECH) see DGW000

DICYNENE see DIS600

DICYNIT (CZECH) see DGU200

DICYNONE see DIS600

DICYSTEINE see CQK325

DID 47 see OMY850

DID 95 see CJM750

DIDAKENE see PCF275

DIDANDIN see DVV600

DIDAN-TDC-250 see DKQ000

DIDECANOYLTRIETHYLENE GLYCOL ESTER (mixed isomers) see DAH450

DIDECYL DIMETHYL AMMONIUM CHLORIDE see DGX200

DIDECYL PHTHALATE see DGX600

DI-N-DECYL PHTHALATE see DGX600

7,8-DIDEHYDROCHOLESTEROL see DAK600

(+)-3,4-DIDEHYDROCORONARIDINE see CCP925

2',3'-DIDEHYDRO-3'-DEOXYTHYMIDINE see SLJ800

7,8-DIDEHYDRO-3-(2-(DIETHYLAMINO)ETHOXY)-4,5-α-EPOXY-17-METHYLMORPHINAN-6-α-OL see DIE300

9,10-DIDEHYDRO-N,N-DIETHYL-2-BROMO-6-METHYLERGOLINE-8-β-CARBOXAMIDE see BNM250

9,10-DIDEHYDRO-N,N-DIETHYL-6-METHYL-ERGOLINE-8-β-CARBOXAMIDE see DJO000

9,10-DIDEHYDRO-N,N-DIETHYL-6-METHYL-ERGOLINE-8-β-CARBOXAMIDE-d-TARTRATE with METHANOL (1:2) see LJG000

12,13-DIDEHYDRO-13,14-DIHYDRO-α-ERYTHROIDINE see EDG500

8,9-DIDEHYDRO-6,8-DIMETHYLERGOLINE see AEY375

(5-α,6-α)-7,8-DIDEHYDRO-4,5-EPOXY-3-ETHOXY-17-METHYLMORPHINAN-6-OL see ENK000

7,8-DIDEHYDRO-4,5-α-EPOXY-3-ETHOXY-17-METHYLMORPHINAN-6-α-OL HYDROCHLORIDE DIHYDRATE see ENK500

7,8-DIDEHYDRO-4,5-α-EPOXY-14-HYDROXY-3-METHOXY-17-METHYLMORPHINAN-6-ONE see HJX500

7,8-DIDEHYDRO-4,5-EPOXY-14-HYDROXY-3-METHOXY-17-METHYLMORPHINAN-5-α-6-ONE see HJX500

7,8-DIDEHYDRO-4,5-α-EPOXY-14-HYDROXY-3-METHOXY-17-METHYLMORPHINAN-6-ONE-N-OXIDE see DGY000

7,8-DIDEHYDRO-4,5-α-EPOXY-3-METHOXY-17-METHYL-MORPHINAN-6-α-OL PHOSPHATE SESQUIHYDRATE (3:3:2) see CNG675

7,8-DIDEHYDRO-4,5-α-EPOXY-17-METHYLMORPHINAN-3,6-α-DIOL see MRP000

7,8-DIDEHYDRO-4,5-α-EPOXY-17-METHYLMORPHINAN-3,6-α-DIOL HYDROCHLORIDE see MRO750

7,8-DIDEHYDRO-4,5-α-EPOXY-17-METHYLMORPHINE HYDROCHLORIDE see MRO750

7,8-DIDEHYDRO-4,5-α-EPOXY-17-METHYL-3-(2-MORPHOLINOETHOXY)MORPHINAN-6-α-OL see TCY750

7,8-DIDEHYDRO-4,5-α-EPOXY-17-METHYL-3-(2-PIPERIDINOETHOXY)MORPHINAN-6-α-OL see PIT600

9,10-DIDEHYDRO-N-ETHYL-1,6-DIMETHYLERGOLINE-8-β-CARBOXAMIDE see MLD500

16,17-DIDEHYDRO-21-ETHYL-4-METHYL-7,20-CYCLOATIDANE-11-β,15-β-DIOL see DAP875

9,10-DIDEHYDRO-N-ETHYL-6-METHYLERGOLINE-8-β-CARBOXAMIDE, N-ETHYLLYSERGAMIDE see LJI000

3,8-DIDEHYDRO-HELIOTRIDINE see DAL060

9,10-DIDEHYDRO-N-(α-(HYDROXYMETHYL)ETHYL)-6-METHYLERGOLINE-8-β-CARBOXAMIDE see LJL000

9,10-(DIDEHYDRO-N-(1-HYDROXYMETHYL)PROPYL)-1,6-DIMETHYLERGOLINE-8-β-CARBOXAMIDE see MLD250

9,10-DIDEHYDRO-N-(α-(HYDROXYMETHYL)PROPYL)-6-METHYL-ERGOLINE-8-β-CARBOXAMIDE see PAM000

13,19-DIDEHYDRO-12-HYDROXY-SENECIONAN-11,16-DIONE see SBX500

13,19-DIDEHYDRO-12-HYDROXY-SENECIONAN-11,16-DIONE HYDROCHLORIDE see SBX525

5,6-DIDEHYDROISOANDROSTERONE see AOO450

13,14-DIDEHYDROMATRIDIN-15-ONE HYDROBROMIDE see SKS800

13,14-DIDEHYDRO-MATRIDIN-15-ONE MONOHYDROBROMIDE see SKS800

9,10-DIDEHYDRO-6-METHYL ERGOLINE-8-β-CARBOXYLIC ACID MORPHOLIDE, TARTARIC ACID SALT see LJK000

N'-((8-α)-9,10-DIDEHYDRO-6-METHYLERGOLIN-8-YL)-N,N-DIETHYL-UREA (Z)-2-BETENEDIOATE see LJE500

3-(9,10-DIDEHYDRO-6-METHYLERGOLIN-8-YL)-1,1-DIETHYLUREA HYDROGEN MALEATE see LJE500

3-(9,10-DIDEHYDRO-6-METHYLERGOLIN-8-α-YL)-1,1-DIETHYLUREA MALEATE (1:1) see LJE500

16,17-DIDEHYDRO-19-METHYLOXAYOHIMBAN-16-CARBOXYLIC ACID METHYL ESTER see AFG750

9,10-DIDEHYDRO-N,N,6-TRIMETHYLERGOLINE-8-β-CARBOXAMIDE see LJH000

7,8-DIDEHYDRORETINOIC ACID see DHA200

3,8-DIDEHYDRORETRONECINE see DAL400

4,4'-DIDEMETHYL-4,4'-DI-2-PROPENYLTOXIFERINE I DICHLORIDE see DBK400

DI-N-DEMETHYLTAMOXIFEN see DHA250

N,N-DIDEMETHYLTAMOXIFEN see DHA250

DIDEMNIN B see DHA300

2',3'-DIDEOXYADENOSINE see DHA325

1,4-DIDEOXY-1,4-BIS((2-HYDROXYETHYL)AMINO)ERYTHRITOL 1,4-DIMETHANESULFONATE (ESTER) see LJD500

DIDEOXYCYTIDINE see DHA350

2',3'-DIDEOXYCYTIDINE see DHA350

1,6-DIDEOXY-1,6-DI(2-CHLOROETHYLAMINO)-d-MANNITOLDIHYDROCHLORIDE see MAW750

2',3'-DIDEOXY-2',3'-DIDEHYDROTHYMIDINE see SLJ800

1,4-DIDEOXY-1,4-DIHYDRO-1,4-DIOXORIFAMYCIN see RKU000

1,6-DIDEOXY-1,6-DIIODO-d-MANNITOL see DHA375

2',3'-DIDEOXY-2'-FLUOROADENOSINE see DHA385

2,3-DIDEOXY-2-((2-HYDROXY-1-(HYDROXYMETHYL)ETHYL)AMINO)-4-C-(HYDROXYMETHYL)-EPI-INOSITOL see DNC100

DIDEOXYKANAMYCIN B see DCQ800

3',4'-DIDEOXYKANAMYCIN B see DCQ800

3',4'-DIDEOXYKANAMYCIN B SULFATE see DHA400

7-((4,6-DIDEOXY-3-(METHYL-4-(METHYLAMINO)-β-ALTROPYRANOSYL)OXY)-10,11-DIHYDRO-9,11-DIHYDROXY-8-METHYL-2-(1-PROPENYL)-1-PYRROLO(2,1-C)(1,4)BENZODIAZEPIN-5-ONE see SCF500

N,N-DIDESMETHYLTAMOXIFEN see DHA250

1,6-DIDESOXY-1,6-DIIODO-d-MANNITOL see DHA375

2,4-DIDEUTERIOESTRADIOL see DHA425

DIDEUTERIUM OXIDE see HAK000

DIDEUTERODIAZOMETHANE see DHA450

DI-2,4-DICHLOROBENZOYL PEROXIDE, >75% with water (DOT) see BIX750

4,4'-DI(DIETHYLAMINO)-4',6'-DISULPHOTRIPHENYLMETHANOL ANHYDRIDE, SODIUM SALT see ADE500

DIDIGAM see DAD200

DIDIMAC see DAD200

DI-N,N'-DIMETHYLAMIDE ETHYLENEIMIDO PHOSPHATE see ASK500

3,6-DI(DIMETHYLAMINO)ACRIDINE see BJF000

1,2-DI-(DIMETHYLAMINO)ETHANE (DOT) see TDQ750

p-DI-3-DIMETHYLAMINOPROPOXYBENZENE DIETHIODIDE see DHA500

N,N-DI(1,4-DIMETHYLPENTYL)-p-PHENYLDIAMINE see BJL000

1,5-DI(2,4-DIMETHYLPHENYL)-3-METHYL-1,3,5-TRIAZAPENTA-1,4-DIENE see MJL250

DIDOC see AAI250

DIDOCOL see DAL000

DIDODECANOYLOXYDIOCTYLSTANNA NE see DVJ800

DIDODECYL PHTHALATE see PHW550

DI-n-DODECYL PHTHALATE see PHW550

DIDODECYL 3,3'-THIODIPROPIONATE see TFD500

DIDODECYLTIN DICHLORIDE see DHA600

DIDP see DEK550

DIDP (PLASTICIZER) see PHW575

DIDRATE see DKW800

DIDROCOLO see DAL000

DIDRONEL R see DXD400

DIDROXAN see MJM500

DIDROXANE see MJM500

DIECA see DJC800

DIEDI see DNM400

DIEFFENBACHIA MACUALTA see DHB309

DIEFFENBACHIA SEQUINE see DHB309

DIEFFENBACHIA (VARIOUS SPECIES) see DHB309

DIELDREX see DHB400

DIELDRIN see DHB400

DIELDRINE (FRENCH) see DHB400

DIELDRITE see DHB400

DIELTAMID see DKC800

DIEMAL see BAG000

DIENESTROL see DAL600

(E,E)-DIENESTROL see DHB550

DIENESTROL DIACETATE see DHB500

α-DIENESTROLPHENOL, 4,4'-(DIETHYLIDENEETHYLENE)DI-, trans-, (E,E)- see DHB550

DIENOCHLOR see DAE425

DIENOESTROL see DAL600

β-DIENOESTROL see DAL600

DIENOESTROL DIACETATE see DHB500

DIENOL see DAL600

DIENOL S see SMR000

DIENONE see DDA100

DIENPAX see DCK759

DIEPIN see CGA000

3,6:10,13-DIEPOXY-8,1,15-BENZOTHIADIAZACYCLOHEPTADECINE -2,14(7H,9H)-DIONE, 1,15-DIHYDRO- see DKY900

DIEPOXYBUTANE see BGA750

l-DIEPOXYBUTANE see BOP750

2,4-DIEPOXYBUTANE see BGA750

dl-DIEPOXYBUTANE see DHB600

1,2:3,4-DIEPOXYBUTANE see BGA750

(2S,3S)-DIEPOXYBUTANE see BOP750

meso-DIEPOXYBUTANE see DHB800

(R*,S*)-DIEPOXYBUTANE see DHB800

l-1,2:3,4-DIEPOXYBUTANE see BOP750

(±)-1,2,3,4-DIEPOXYBUTANE see DHB600

dl-1,2,3,4-DIEPOXYBUTANE see DHB600

(2S,3S)-1,2:3,4-DIEPOXYBUTANE see BOP750

meso-1,2,3,4-DIEPOXYBUTANE see DHB800

1,2:3,4-DIEPOXYCYCLOHEXANE see DHB820

trans-1,2,3,4-DIEPOXYCYCLOHEXANE see DHB875

1,2:5,6-DIEPOXYCYCLOOCTANE see CPR835

1,2,9,10-DIEPOXYDECANE see DHC000

DIEPOXYDIHYDRO-7-METHYL-3-METHYLENE-1,6-OCTADIENE see DHC200

DIEPOXYDIHYDROMYRCENE see DHC200

2,3:5,6-DIEPOXY-7,8-DIOXABICYCLO[2.2.2]OCTANE see DHC309

1,2:5,6-DIEPOXYDULCITOL see DCI600

2,5-DI(1,2-EPOXYETHYL)TETRAHYDRO-2H-PYRAN see DHC400

1,2,6,7-DIEPOXYHEPTANE see DHC600

1,2:5,6-DIEPOXYHEXAHYDROINDAN see BFY750

1,2:5,6-DIEPOXYHEXAHYDRO-4,7-METHANOINDAN see BGA250

1,2:5,6-DIEPOXY-3a,4,5,6,7,7a-HEXAHYDRO-4,7-METHANOINDAN see BGA250

1,2:5,6-DIEPOXYHEXANE see DHC800

1,2,8,9-DIEPOXYLIMONENE see LFV000

1,2:8,9-DIEPOXYMENTHANE see LFV000

1,2:8,9-DIEPOXY-p-MENTHANE see LFV000

1,2,3,4-DIEPOXY-2-METHYLBUTANE see DHD200

DI(3,4-EPOXY-6-METHYLCYCLOHEXYLMETHYL)ADIPATE see AEN750

1,2:8,9-DIEPOXYNONANE see DHD400

9,10,12,13-DIEPOXYOCTADECANOIC ACID see DHD600

1,2:7,8-DIEPOXYOCTANE see DHD800

1,2,4,5-DIEPOXYPENTANE see DHE000

DIEPOXYPIPERAZINE see DHE100

DI(2,3-EPOXYPROPYL) ETHER see DKM200

DIEPOXY-SORBITOL see DCI900

9,10:12,13-DIEPOXYSTEARIC ACID see DHD600

1,2,15,16-DIEPOXY-4,7,10,13-TETRAOXAHEXADECANE see TJQ333

7-β,8-β:12,13-DIEPOXY-TRICHOTHEC-9-EN-6-β-OL see COB000

(4-β(Z),7-β,8-β)-7;8:12,13-DIEPOXY-TRICHOTHEC-9-EN-4-OL 2-BUTENOATE see COB000

DIESEL EXHAUST see DHE485

DIESEL EXHAUST EXTRACT see DHE500

DIESEL EXHAUST PARTICLES see DHE700

DIESEL FUEL (DOT) see DHE900

DIESEL FUEL (DOT) see FOP000

DIESEL FUEL MARINE see DHE750

DIESEL FUEL MARINE see DHE800

DIESEL FUEL NO. 2 see DHE850

DIESEL FUEL NO. 4 see DHE750

DIESEL FUELS see DHE900

DIESEL OIL (PETROLEUM) see DHE900

DIESEL OIL (PETROLEUM) see FOP000

DIESEL OILS see DHE900

DIESEL OILS see FOP000

DIESEL TEST FUEL see DHE900

DIESEL TEST FUEL see FOP000

S,S-DIESTER with DITHIO-p-UREIDOBENZENEARSONOUS ACID o-MERCAPTOBENZOIC ACID see TFD750

DIESTER with o-MERCAPTOBENZOIC ACID DITHIO-p-UREIDOBENZENEARSONOUS ACID see TFD750

DI-ESTRYL see DKA600

DIETADIONE (ITALIAN) see DJT400

DIETAMINE see AOB500

DIETELMIN see HEP000

DIETHADION see DJT400

DIETHADIONE see DJT400

DIETHAMPHENAZOL MONOHYDRATE see BKB250

DIETHANOLAMIN (CZECH) see DHF000

DIETHANOLAMINE see DHF000

DIETHANOLAMINE-3,5-DIIODO-4-PYRIDONE-N-ACETATE see DNG400

DIETHANOLAMINE SALT OF TRIIODOTHYROACETIC ACID see TKR100

DIETHANOLAMINE p-TOLYLMETHYLCARBINOL CAMPHORATE see HAQ570

DIETHANOLAMINE TRIIODOTHYROACETATE see TKR100

DIETHANOLAMINOBENZENE see BKD500

DIETHANOLAMINOCHLOROBENZENE see CJU400

DIETHANOLAMMONIUM MALEIC HYDRAZIDE see DHF200

N,N-DIETHANOLANILIDE, 3-CHLORO- see CJU400

DIETHANOLANILINE see BKD500

N,N-DIETHANOLANILINE see BKD500

DIETHANOLCHLOROANILIDE see CJU400

DIETHANOLETHYLAMINE see ELP000

1,1-DIETHANOLHYDRAZINE see HHH000

DIETHANOLLAURAMIDE see BKE500

N,N-DIETHANOLLAURAMIDE see BKE500

N,N-DIETHANOLLAURIC ACID AMIDE see BKE500

DIETHANOLMETHYLAMINE see MKU250

DIETHANOLNITRAMINE DINITRATE see NFW000

DIETHANOL-N-NITRAMINE DINITRATE see NFW000

DIETHANOLNITROSOAMINE see NKM000

DIETHANOLOLEAMIDE see OHU200

N,N-DIETHANOLOLEAMIDE see OHU200

DIETHANOL-m-TOLUIDINE see DHF400

DIETHAZIN see DII200

DIETHAZINE see DII200

DIETHAZINE HYDROCHLORIDE see DHF600

1,3-DIETHENYL-1,1,3,3-TETRAMETHYLDISILOXANE see TDQ075

DIETHIBUTIN HYDROCHLORIDE see DJP500

DIETHION see EEH600

DIETHOFENCARB see IOS100

2,2-DIETHOXYACETOPHENONE see PFH260

α-α-DIETHOXYACETOPHENONE see PFH260

o-DIETHOXYBENZENE see CCP900

1,2-DIETHOXYBENZENE see CCP900

1-(3,4-DIETHOXYBENZYL)-6,7-DIETHOXYISOQUINOLINE HYDROCHLORIDE see PAH260

1,2-DIETHOXYCARBONYLDIAZENE see DIT300

1,2-DI(ETHOXYCARBONYL)ETHYL-O,O-DIMETHYL PHOSPHORODITHIOATE see MAK700

S-(1,2-DIETHOXYCARBONYL)ETHYL O,O-DIMETHYL PHOSPHOROTHIOATE see OPK250

S-(1,2-DI(ETHOXYCARBONYL)ETHYL) DIMETHYL PHOSPHOROTHIOLOTHIONATE see MAK700

DIETHOXYCHLOROSILANE see DHF800

1,1-DIETHOXY-3,7-DIMETHYL-2,6-OCTADIENE see CMS323

DIETHOXYDIMETHYLSILANE see DHG000

1,1-DIETHOXY-ETHAAN (DUTCH) see AAG000

1,1-DIETHOXYETHANE see AAG000

1,2-DIETHOXYETHANE see EJE500

1,2-DIETHOXY ETHENE see DHG100

DIETHOXY ETHYL ADIPATE see BJO225

1,2-DIETHOXY ETHYLENE see DHG100

1,1-DIETHOXY-trans-2-HEXENE see HFA600

DIETHOXY-3-KYANPROPYL-METHYLSILAN see COR500

DIETHOXYMETHANE (DOT) see EFT500

α,α-DI(p-ETHOXYPHENYL)-β-BROMO-β-PHENYLETHYLENE see BMX000

α-(((DIETHOXYPHOSPHINOTHIOYL)OXY)I MINO)BENZENEACETONITRILE see BAT750

((DIETHOXYPHOSPHINOTHIOYL)THIO)A CETIC ACID, ETHYL ESTER see DIX000

(2-((DIETHOXYPHOSPHINOTHIOYL)THIO)ETHYL)CARBAMIC ACID, ETHYL ESTER see EMC000

(DIETHOXYPHOSPHINYL)DITHIOIMIDO CARBONIC ACID CYCLIC ETHYLENE ESTER see PGW750

2-(DIETHOXYPHOSPHINYL)ETHYLTRIETHOXYSILANE see DKD500

(DIETHOXYPHOSPHINYLIMINO)-1,3-DITHIETANE see DHH200

DIETHOXYPHOSPHINYLIMINO-2-DITHIETANNE-1,3 (FRENCH) see DHH200

2-(DIETHOXYPHOSPHINYLIMINO)-1,3-DITHIOLANE see DXN600

2-(DIETHOXYPHOSPHINYLIMINO)-1,3-DITHIOLANE see PGW750

2-(DIETHOXYPHOSPHINYLIMINO)-4-METHYL-1,3-DITHIOLANE see DHH400

(DIETHOXY-PHOSPHINYL)MERCURY ACETATE see AAS250

(α-(DIETHOXYPHOSPHINYL)-p-METHOXYBENZYL)BIS(2-CHLOROPROPYL)ANTIMONITE see DHH600

2-((DIETHOXYPHOSPHINYL)OXY)-1H-BENZ(de)ISOQUINOLINE-1,3(2H)-DIONE see HMV000

2-DIETHOXY-PHOSPHINYLTHIOETHYL-TRIMETHYLAMMONIUM IODIDE see TLF500

N-(2-(DIETHOXYPHOSPHINYLTHIO)ETHYL)TRIMETHYLAMMONIUM IODIDE see TLF500

DIETHOXYPHOSPHORUS OXYCHLORIDE see DIY000

DIETHOXYPHOSPHORYL CYANIDE see DJW800

DIETHOXYPHOSPHORYL-THIOCHOLINE IODIDE see TLF500

2,2-DIETHOXYPROPANE see ABD250

3,3-DIETHOXYPROPENE see DHH800

3,3-DIETHOXY-1-PROPENE see DHH800

5-((3,5-DIETHOXY-4-(1H-PYRROL-1-YL)PHENYL)METHYL)-2,4-PYRIMIDINEDIAMINE see EBB700

DIETHOXYTETRAETHYLENE GLYCOL see PBO250

6,6'-DIETHOXYTHIOINDIGO see CMU815

DI-ETHOXYTHIOKARBONYL-TRISULFID see DKE400

DIETHOXY THIOPHOSPHORIC ACID ESTER of 2-ETHYLMERCAPTOETHANOL see DAO600

DIETHOXY THIOPHOSPHORIC ACID ESTER OF 7-HYDROXY-4-METHYL COUMARIN see PKT000

(DIETHOXY-THIOPHOSPHORYLOXYIMINO)-PHENYL ACETONITRILE see BAT750

DIETHQUINALPHION see DJY200

DIETHQUINALPHIONE see DJY200

7,12-DIETHYENYL-3,8,13,17-TETRAMETHYL-21H,23H-PORPHINE-2,18-DIPROPANOIC ACID DISODIUM SALT see DXF700

DIETHYL see BOR500

DIETHYL ACETAL see AAG000

DIETHYL ACETALDEHYDE see DHI000

N,N-DIETHYLACETAMIDE see DHI200

DIETHYLACETIC ACID see DHI400

DIETHYLACETOACETAMIDE see DHI600

N,N-DIETHYLACETOACETAMIDE see DHI600

1-DIETHYLACETYLAZIRIDINE see DHI800

DIETHYL ACETYLENE DICARBOXYLATE see DHI850

DIETHYLACETYLETHYLENEIMINE see DHI800

O,O-DIETHYL-O-ACETYL PHOSPHATE see MJD100

DIETHYL ACID PHOSPHATE see DJW500

O,O-DIETHYL-S-N-(A-CYANOISOPROPYL)CARBOMOYLMETHYL PHOSPHOROTHIOATE see PHK250

DIETHYL ADIPATE see AEP750

l-N,N-DIETHYLALANINE-6-CHLORO-o-TOLYL ESTER HYDROCHLORIDE see DHI875

l-N,N-DIETHYLALANINE-2,6-DIETHYLPHENYL ESTER HYDROCHLORIDE see FAC157

l-N,N-DIETHYLALANINE-2,6-DIMETHOXYPHENYL ESTER HYDROCHLORIDE see FAC163

l-N,N-DIETHYLALANINE MESITYL ESTER HYDROCHLORIDE see FAC179

l-N,N-DIETHYLALANINE-2,6-XYLYL ESTER HYDROCHLORIDE see FAC100

4'-(N,N-DIETHYLALANYL)METHANESULFONANILIDE HYDROCHLORIDE see DHI876

DIETHYLALUMINUM BROMIDE see DHI880

DIETHYLALUMINUM CHLORIDE see DHI885

DIETHYLALUMINUM MONOCHLORIDE see DHI885

N,N-DIETHYLAMIDE OF PHENOXYACETIC ACID see RCZ200

DIETHYLAMIDE of PROPIONIC ACID see DJX250

DIETHYLAMIDE de VANILLIQUE see DKE200

DIETHYLAMINE see DHJ200

N,N-DIETHYLAMINE see DHJ200

DIETHYLAMINE ACETARSONE see ACN250

DIETHYLAMINE-3-ACETYLAMINO-4-HYDROXYPHENYLARSONATE see ACN250

DIETHYLAMINE, 2-CHLORO-N-(9-FLUORENYL)-, HYDROCHLORIDE see FEE100

DIETHYLAMINE DIAZENIUMDIOLATE see DHJ333

DIETHYLAMINE, 2,2'-DICHLORO-N-METHYL-, OXIDE see CFA500

DIETHYLAMINE, 2,2'-DICHLORO-N-PROPYL-, HYDROCHLORIDE see PNF050

DIETHYLAMINE, 2,2'-DICHLORO-N-TETRAHYDROFURFURYL- see FPW100

DIETHYLAMINE/NO see DHJ300

DIETHYLAMINE NONOATE see DHJ333

2-(DIETHYLAMINO)ACETANILIDE see DHJ400

(DIETHYLAMINO)ACETONITRILE see DHJ600

N,N-DIETHYLAMINOACETONITRILE see DHJ600

2-(DIETHYLAMINO)-o-ACETOPHENETIDIDE, HYDROCHLORIDE see DHJ800

2-(DIETHYLAMINO)-p-ACETOPHENETIDIDE, HYDROCHLORIDE see DHK000

2-(DIETHYLAMINO)-o-ACETOTOLUIDIDE HYDROCHLORIDE see DHK200

DIETHYLAMINOACETO-2,6-XYLIDIDE see DHK400

2-(DIETHYLAMINO)-2',6'-ACETOXYLIDIDE see DHK400

α-DIETHYLAMINOACETO-2,6-XYLIDIDE see DHK400

α-DIETHYLAMINO-2,6-ACETOXYLIDIDE see DHK400

2-(DIETHYLAMINO)-2',6'-ACETOXYLIDIDE HYDROCHLORIDE see DHK600

2-(DIETHYLAMINO)-3',5'-ACETOXYLIDIDE HYDROCHLORIDE see DHK800

2-(DIETHYLAMINO)-2',6'-ACETOXYLIDIDE MONOHYDROCHLORIDE see DHK600

α-DIETHYLAMINO-2,5-ACETOXYLIDINE HYDROCHLORIDE see DHK600

DIETHYLAMINOACET-2,6-XYLIDIDE see DHK400

2-(((DIETHYLAMINO)ACETYL)AMINO)-3-METHYL-BENZOIC ACID METHYL ESTER, MONOHYDROCHLORIDE see BMA125

N-((DIETHYLAMINO)ACETYL)ANTHRANILIC ACID, ETHYL ESTER, HYDROCHLORIDE see DHL200

N-(2-DIETHYLAMINO)ACETYLANTHRANILIC ACID, ETHYL ESTER HYDROCHLORIDE see DHL200

N-((DIETHYLAMINO)ACETYL)ANTHRANILIC ACID, METHYL ESTER, HYDROCHLORIDE see DHL400

N-(2-DIETHYLAMINO)ACETYLANTHRANILIC ACID, METHYL ESTER, HYDROCHLORIDE see DHL400

β-4-(1-DIETHYLAMINOACETYL-2-PIPERIDYL)-2,2-DIPHENYL-1,3-DIOXOLANE HYDROCHLORIDE see DHL600

DIETHYLAMINOACETYL-2,4,6-TRIMETHYLANILINE HYDROCHLORIDE see DHL800

(DIETHYLAMINO)ACYLANILIDE see DHJ400

4-(DIETHYLAMINO)ANILINE see DJV200

p-(DIETHYLAMINO)ANILINE see DJV200

4-(DIETHYLAMINO)AZOBENZENE see OHJ875

p-(DIETHYLAMINO)AZOBENZENE see OHJ875

N,N-DIETHYL-4-AMINOAZOBENZENE see OHJ875

4-(DIETHYLAMINO)BENZALDEHYDE DIPHENYLHYDRAZONE see DHL850

4-DIETHYLAMINOBENZALDEHYDE-1,1-DIPHENYLHYDRAZONE see DHL850

N,N-DIETHYLAMINOBENZENE see DIS700

3-(DIETHYLAMINO)BENZENESULFONIC ACID, SODIUM SALT see DHM000

4-(DIETHYLAMINO)-2-BUTANONE see DHM100

1-DIETHYLAMINO-1-BUTEN-3-YNE see DHM200

4-DIETHYLAMINO-2-BUTYNYLPHENYL(CYCLOHEXYL)GLYCOLATE HYDROCHLORIDE see OPK000

2-(DIETHYLAMINO)BUTYRIC ACID-2,6-XYLYL ESTER HYDROCHLORIDE see DHM309

3-(DIETHYLAMINO)-2',6'-BUTYROXYLIDIDE HYDROCHLORIDE see DHM400

DIETHYLAMINOCARBETHOXYBICYCLOHEXYL HYDROCHLORIDE see ARR750

2-(DIETHYLAMINO)CHLOROETHANE see CGV500

α-DIETHYLAMINO-2,6-DIMETHYLACETANILIDE see DHK400

ω-DIETHYLAMINO-2,6-DIMETHYLACETANILIDE see DHK400

2-(DIETHYLAMINO)-3',5'-DIMETHYLACETANILIDE HYDROCHLORIDE see DHK800

ω-DIETHYLAMINO-2,6-DIMETHYLACETANILIDE HYDROCHLORIDE see DHK600

3-(DIETHYLAMINO)-7-((p-(DIMETHYLAMINO)PHENYL)AZO)-5-PHENYLPHENAZINIUM CHLORIDE see DHM500

(+−)-4-DIETHYLAMINO-1,1-DIMETHYLBUT-2-YN-1-YL2-CYCLOHEXYL-2-HYDROXY-2-PHENYLACETATE HCL H2O see UTU700

2-(DIETHYLAMINO)-N-(1,3-DIMETHYL-4-(o-FLUOROBENZOYL)-5-PYRAZOLYL)ACETAMI de HYDROCHLORIDE see AAI118

2-(DIETHYLAMINO)-N-(2,6-DIMETHYLPHENYL)ACETAMIDE MONOHYDROCHLORIDE see DHK600

3-(DIETHYLAMINO)-2,2-DIMETHYL-1-PROPANOL-p-AMINOBENZOATE see DNY000

3-DIETHYLAMINO-2,2-DIMETHYLPROPYL TROPATE PHOSPHATE see AOD250

2-(DIETHYLAMINO)-N,N-DIPHENYLACETAMIDE HYDROCHLORIDE see DHN800

2-(DIETHYLAMINO)-N-(DIPHENYLMETHYL)ACETAMIDE HYDROCHLORIDE see DHO000

3-DIETHYLAMINO-1,1-DI(2'-THIENYL)BUT-1-ENE HYDROCHLORIDE see DJP500

3-DIETHYLAMINO-1,2-EPOXYPROPANE see GGW800

(DIETHYLAMINO)ETHANE see TJO000

DIETHYLAMINOETHANETHIOL see DIY600

2-(DIETHYLAMINO)ETHANETHIOL see DIY600

DIETHYLAMINOETHANETHIOL HYDROCHLORIDE see DHO400

DIETHYLAMINOETHANOL see DHO500

2-DIETHYLAMINOETHANOL see DHO500

2-(DIETHYLAMINO)ETHANOL see DHO500

N-DIETHYLAMINOETHANOL see DHO500

2-N-DIETHYLAMINOETHANOL see DHO500

DIETHYLAMINOETHANOL (DOT) see DHO500

β-DIETHYLAMINOETHANOL see DHO500

DIETHYLAMINOETHANOL-4-AMINOBENZOATE HYDROCHLORIDE see AIT250

DIETHYLAMINOETHANOL-p-AMINOSALICYLATE see DHO600

DIETHYLAMINOETHANOL ESTER OF DIPHENYLPROPYLACETIC ACID HYDROCHLORIDE see PBM500

2-DIETHYLAMINOETHANOL HYDROCHLORIDE see DHO700

2-(DIETHYLAMINO)-2'-ETHOXYACETANILIDE, HYDROCHLORIDE see DHJ800

2-(2-(DIETHYLAMINO)ETHOXY)-2',6'-ACETOXYLIDIDE HYDROCHLORIDE see DHO800

2-(2-(DIETHYLAMINO)ETHOXY)ADAMANTANE ETHYL IODIDE see DHP000

o-(DIETHYLAMINOETHOXY)BENZANILIDE see DHP200

2-(2-(DIETHYLAMINO)ETHOXY)BENZANILID E see DHP200

2-(2-(DIETHYLAMINO)ETHOXY)-5-BROMOBENZANILIDE see DHP400

2-(2-(DIETHYLAMINO)ETHOXY)-2'-CHLORO-BENZANILIDE see DHP450

2-(2-(DIETHYLAMINO)ETHOXY)-3'-CHLORO-BENZANILIDE see DHP500

2-(2-(DIETHYLAMINO)ETHOXY)-4'-CHLORO-BENZANILIDE see DHP550

6-(2-(DIETHYLAMINOETHOXY))-4,7-DIMETHOXY-5-CINNAMOYLBENZOFURANMALEATE see DHP600

6-(2-DIETHYLAMINOETHOXY)-4,7-DIMETHOXY-5-(p-

METHOXYCINNAMOYL)BENZOFURAN OXALATE see DHQ000

2-(DIETHYLAMINOETHOXY)DIPHENYL HCl see BGO000

DIETHYLAMINOETHOXYETHANOL see DHQ100

2-(2-(DIETHYLAMINO)ETHOXY)ETHANOL see DHQ100

2-β-(DIETHYLAMINO)ETHOXY)ETHANOL see DHQ100

2-(2-(DIETHYLAMINO)ETHOXY)ETHANOL-2-ETHYL-2-PHENYLBUTYRATE see DHQ200

2-(2-DIETHYLAMINOETHOXY)ETHYL α,α-DIETHYLPHENYLACETATE see DHQ200

2-(2-DIETHYLAMINOETHOXY)ETHYL-2-ETHYL-2-PHENYLBUTYRATE see DHQ200

2-((2-(DIETHYLAMINO)ETHOXY)ETHYL-2-PHENYLBUTYRATE CITRATE see BOR350

DIETHYLAMINOETHOXYETHYL-1-PHENYL-1-CYCLOPENTANE CARBOXYLATE HYDROCHLORIDE see DHQ500

6-(2-DIETHYLAMINOETHOXY)-N-(o-METHOXYPHENYL)NICOTINAMIDE HYDROCHLORIDE see DHQ600

2-(2-(DIETHYLAMINO)ETHOXY)-3-METHYLBENZANILIDE see DHQ800

5-(2-(DIETHYLAMINO)ETHOXY)-3-METHYL-1-PHENYLPYRAZOLE see DHR000

1-((p-(2-DIETHYLAMINO)ETHOXY)PHENYL)-3,4-DIHYDRO-6-METHOXY-2-PHENYLNAPHTHALENE HYDROCHLORIDE see MFF625

1-(p-(β-DIETHYLAMINOETHOXY)PHENYL)-1,2-DIPHENYLCHLOROETHYLENE see CMX500

1-(p-(β-DIETHYLAMINO ETHOXY)PHENYL)-1,2-DIPHENYL-2-CHLOROETHYLENE CITRATE see CMX700

p-(2-(DIETHYLAMINO)ETHOXY)PHENYL 2-ETHYL-3-BENZOFURANYL KETONE see EDV700

1-DIETHYLAMINO-3-(p-ETHOXYPHENYL)INDAN CITRATE see DHR800

p-(2-(DIETHYLAMINO)ETHOXY)PHENYL 2-MESITYL-3-BENZOFURANYL KETONE HYDROCHLORIDE see DHR900

1-(p-(β-DIETHYLAMINOETHOXY)PHENYL)-2-NITRO-1,2-DIPHENYLETHYLENE CITRATE see EAF100

(p-2-DIETHYLAMINOETHOXYPHENYL)-1-PHENYL-2-p-ANISYLETHANOL see DHS000

1-(4-(2-DIETHYLAMINOETHOXY)PHENYL)-1-PHENYL-2-(p-ANISYL)ETHANOL see DHS000

1-(p-2-DIETHYLAMINOETHOXYPHENYL)-1-PHENYL-2-p-ANISYLETHANOL see DHS000

1-(p-(β-DIETHYLAMINOETHOXY)PHENYL)-1-PHENYL-2-(p-METHOXYPHENYL)-ETHANE HYDROCHLORIDE see MFG530

1-(p-(2-DIETHYLAMINO)ETHOXY)PHENYL)-1-PHENYL-2-(p-METHOXYPHENYL)ETHANOL see DHS000

1-(2-(2-DIETHYLAMINO)ETHOXY)PHENYL)-3-PHENYL-1-PROPANONE HYDROCHLORIDE see DHS200

(o-β-DIETHYLAMINOETHOXY)-PHENYL PROPIOPHENONE HYDROCHLORIDE see DHS200

1-(4-(2-(DIETHYLAMINO)ETHOXY)PHENYL)-1-(p-TOLYL)-2-(p-CHLOROPHENYL)ETHANOL see TMP500

1-(p-(β-DIETHYLAMINOETHOXY)PHENYL)-1-(p-TOLYL)-2-(p-CHLOROPHENYL)ETHANOL see TMP500

2-(2-(DIETHYLAMINO)ETHOXY)-N-(2,6-XYLYL)CINCHONINAMIDE, HYDROCHLORIDE see DHS400

2-(2-(DIETHYLAMINO)ETHOXY)-N-(2,6-XYLYL)-4-QUINOLINECARBOXAMIDE, HYDROCHLORIDE see DHS400

2-(N'-(2-(DIETHYLAMINO)ETHYL)ACETAMIDO)-2',6'-ACETOXYLIDIDE HYDROCHLORIDE see DHS600

2-(DIETHYLAMINO)-N-ETHYL-o-ACETOTOLUIDIDE HYDROCHLORIDE see DHS800

N-(2-(DIETHYLAMINO)ETHYL)-2',6'-ACETOXYLIDIDE HYDROCHLORIDE see DHT000

DIETHYLAMINOETHYL ACRYLATE see DHT125

2-(DIETHYLAMINO)ETHYL ACRYLATE see DHT125

N,N-DIETHYLAMINOETHYL ACRYLATE see DHT125

β-DIETHYLAMINOETHYL ACRYLATE see DHT125

N-(2-(DIETHYLAMINO)ETHYL)-1-ADAMANTANEACETAMIDE ETHYL IODIDE see DHT200

β-DIETHYLAMINOETHYL ALCOHOL see DHO500

α-(2-(DIETHYLAMINOETHYL)AMINO)-BENZENEACETIC ACID-3-METHYLBUTYL ESTER see NOC000

DIETHYLAMINOETHYL-p-AMINOBENZOATE see AIL750

2-DIETHYLAMINOETHYL-p-AMINOBENZOATE see AIL750

β-DIETHYLAMINOETHYL-4-AMINOBENZOATE see AIL750

2-DIETHYLAMINOETHYL-p-AMINOBENZOATE HYDROCHLORIDE see AIT250

2'-DIETHYLAMINOETHYL-3-AMINO-2-BUTOXYBENZOATE HYDROCHLORIDE see AJA500

2-(DIETHYLAMINO)ETHYL-4-AMINO-2-CHLOROBENZOATE see DHT300

2-(2-(DIETHYLAMINO)ETHYLAMINO)ETHYL)-1,3-BENZODIOXOLE, DIMALEATE see DJU800

2-(2-(2-DIETHYLAMINO)ETHYLAMINO)ETHYL-2,5-DIMETHYL-1,3-BENZODIOXOLE DIMALEATE see DHT400

2-(2-(2-(DIETHYLAMINO)ETHYLAMINO)ETHYL)-2-ETHYL-1,3-BENZODIOXOLE DIMALEATE see DJI200

2-(2-(2-(DIETHYLAMINO)ETHYLAMINO)ETHYL)-2-METHYL-1,3-BENZODIOXOLE, DIMALEATE see DHT600

1-((2-(DIETHYLAMINO)ETHYL)AMINO)-4-(HYDROXYMETHYL)THIOXANTHEN-9-ONE see LIM000

1-((2-(DIETHYLAMINO)ETHYL)AMINO)-4-(HYDROXYMETHYL)9H-THIOXANTHEN-9-ONE see LIM000

1-((2-(DIETHYLAMINO)ETHYL)AMINO)-4-(HYDROXYMETHYL)-9H-THIOXANTHEN-9-ONE MONOMETHANE-SULFONATE (SALT) see HGO500

2-(2-(DIETHYLAMINO)ETHYLAMINOMETHYL

)-2-METHYL-1,3-BENZODIOXOLE DIMALEATE see DHT800

1-(2-DIETHYLAMINOETHYLAMINO)-4-METHYLTHIAXANTHONE HYDROCHLORIDE see SBE500

1-(β-DIETHYLAMINOETHYLAMINO)-4-METHYLTHIAXANTHONE HYDROCHLORIDE see SBE500

1-(2'-DIETHYLAMINO)ETHYLAMINO-4-METHYLTHIOXANTHENONE see DHU000

1-((2-DIETHYLAMINOETHYL)AMINO)-4-METHYL-9H-THIOXANTHEN-9-ONE see DHU000

1-((2-DIETHYLAMINOETHYL)AMINO))-4-METHYLTHIOXANTHEN-9-ONE HYDROCHLORIDE see SBE500

1-((2-DIETHYLAMINO)ETHYL)AMINO-4-METHYL-9H-THIOXANTHEN-9-ONE, MONOHYDROCHLORIDE see SBE500

α-(N,β-DIETHYLAMINOETHYL) AMINOPHENYLACETIC ACID ISOAMYL ESTER see NOC000

2-(3-(2-(DIETHYLAMINO)ETHYLAMINO)PROPYL)-2-METHYL-1,3-BENZODIOXOLE DIMALEATE see DHU400

2-(3-(2-(DIETHYLAMINO)ETHYLAMINO)PROPYL)-2-METHYL-1,3-NAPHTHOL(2,3-d)DIOXOLE DIHYDROCHLORIDE see DHU600

DIETHYLAMINOETHYL-p-AMINOSALICYLATE see DHO600

2-DIETHYLAMINOETHYL-4-AMINOSALICYLATE see DHO600

2-DIETHYLAMINOETHYL-p-AMINOSALICYLATE see DHO600

N-(2-(DIETHYLAMINO)ETHYL)BENZAMIDE see DHU700

DIETHYLAMINOETHYL BENZILATE see DHU900

2-(DIETHYLAMINO)ETHYL BENZILATE see DHU900

β-DIETHYLAMINOETHYL BENZILATE see DHU900

2-DIETHYLAMINOETHYL BENZILATE HYDROCHLORIDE see BCA000

β-DIETHYLAMINOETHYL BENZILATE HYDROCHLORIDE see BCA000

2-(2-(DIETHYLAMINO)ETHYL)-2H(1)BENZOTHIOPYRANO(4,3,2-CD)INDAZOLE-5-METHANOL see DHU910

o-(DIETHYLAMINOETHYL)-N-BENZOYL-dl-TYROSIL-DI-n-PROPYLAMIDE see TGF175

β-DIETHYLAMINOETHYL-2-BUTOXY-3-AMINOBENZOATE HYDROCHLORIDE see AJA500

N-(2-(DIETHYLAMINO)ETHYL)-2-BUTOXYCINCHONINAMIDE see DDT200

N-((2-DIETHYLAMINO)ETHYL)CARBAMIC ACID-6-CHLORO-o-TOLYL ESTER HYDROCHLORIDE see DHV200

N-(2-(DIETHYLAMINO)ETHYLCARBAMIC ACID) MESITYL ESTER HYDROCHLORIDE see DHV400

N-(2-(DIETHYLAMINO)ETHYL)CARBAMIC ACID-2,6-XYLYL ESTER HYDROCHLORIDE see DHV600

DIETHYLAMINOETHYL CHLORIDE see CGV500

2-(DIETHYLAMINO)ETHYL CHLORIDE see CGV500

N-DIETHYLAMINOETHYL CHLORIDE see CGV500

β-(DIETHYLAMINO)ETHYL CHLORIDE see CGV500

DIETHYLAMINOETHYL CHLORIDE HYDROCHLORIDE see CLQ250

β-DIETHYLAMINOETHYL CHLORIDE HYDROCHLORIDE see CLQ250

2-(DIETHYLAMINO)ETHYLCHLORODIPHENYLACETATE HYDROCHLORIDE see DHW200

N-(2-(DIETHYLAMINO)ETHYL)-2',6'-CROTONXYLILIDE HYDROCHLORIDE see DHX600

β-DIETHYLAMINOETHYL CUMATE HYDROCHLORIDE see IOI600

β-DIETHYLAMINOETHYL-1-CYCLOHEXYLCYCLOHEXANECARBOXYLATE HYDROCHLORIDE see ARR750

β-DIETHYLAMINOETHYL-1-CYCLOHEXYLHEXAHYDROBENZOATE HYDROCHLORIDE see ARR750

2-DIETHYLAMINOETHYL CYCLOPENTYL(2-THIENYL)GLYCOLATE HYDROCHLORIDE see DHW400

2-DIETHYLAMINOETHYL-2-CYCLOPENTYL-2-(2-THIENYL)HYDROXYACETATE METHOBROMIDE see PBS000

2-DIETHYLAMINOETHYL-α-CYCLOPENTYL-2-THIOPHENEGLYCOLATE METHOBROMIDE see PBS000

DIETHYLAMINOETHYL-DEXTRAN see DHW600

DIETHYLAMINOETHYLDEXTRAN POLYMER see DHW600

N-(2'-DIETHYLAMINOETHYL)DIBENZOPARATHIAZINE see DII200

DIETHYLAMINOETHYL-α,α-DICYCLOPENTYLACETATE ETHOBROMIDE see DGW600

(2-(2-DIETHYLAMINO)ETHYL-O,O-DIETHYL ESTER, OXALATE (1:1) see AMX825

S-(DIETHYLAMINOETHYL)-O,O-DIETHYL PHOSPHOROTHIOATE see DJA400

S-(2-DIETHYLAMINOETHYL)-O,O-DIETHYLPHOSPHOROTHIOATE HYDROGEN OXALATE see AMX825

7-(2-(DIETHYLAMINO)ETHYL)-3,7-DIHYDRO-1,3-DIMETHYL-1H-PURINE-2,6-DIONE (9CI) see CNR125

7-(2-(DIETHYLAMINO)ETHYL)-3,7-DIHYDRO-1,3-DIMETHYL-1H-PURINE-2,6-DIONE HYDROCHLORIDE see DIH600

N-(2-(DIETHYLAMINO)ETHYL)-4-(DIMETHYLAMINO)-2-METHOXY-5-NITROBENZAMIDE MONOHYDROCHLORIDE see DUO300

N-(2-(DIETHYLAMINO)ETHYL)-2',6'-DIMETHYLCROTONANILIDE HYDROCHLORIDE see DHX600

2-DIETHYLAMINOETHYL DIPHENYLACETATE see DHX800

2-DIETHYLAMINOETHYL DIPHENYL ACETATE see THK000

DIETHYLAMINOETHYLDIPHENYL CARBAMATE HYDROCHLORIDE see DHY200

2-DIETHYLAMINOETHYLDIPHENYLCARBAMATE HYDROCHLORIDE see DHY200

2-(DIETHYLAMINO)ETHYL DIPHENYLGLYCOLATE see DHU900

2-DIETHYLAMINOETHYL DIPHENYLGLYCOLATE HYDROCHLORIDE see BCA000

3-(2-(DIETHYLAMINO)ETHYL)-5,5-DIPHENYLHYDANTOIN see DHY300

DIETHYLAMINOETHYLDIPHENYLHYDANTOINE (FRENCH) see DHY300

2'-DIETHYLAMINOETHYL-2,2-DIPHENYLPENTANOATE HYDROCHLORIDE see PBM500

β-DIETHYLAMINOETHYL DIPHENYLPROPYLACETATE HYDROCHLORIDE see PBM500

2-(DIETHYLAMINO)ETHYL DIPHENYLTHIOACETATE HYDROCHLORIDE see DHY400

β-DIETHYLAMINOETHYL DIPHENYLTHIOACETATE HYDROCHLORIDE see DHY400

S-(2-(DIETHYLAMINO)ETHYL)DIPHENYLTHIOACETIC ACID HYDROCHLORIDE see DHY400

S-(2-(DIETHYLAMINO)ETHYL)DIPHENYLTHIOACETIC ACID HYDROCHLORIDE see DHY400

S-(2-(DIETHYLAMINO)ETHYL) DIPHENYLTHIOCARBAMATE HYDROCHLORIDE see PDC875

2-DIETHYLAMINOETHYL DIPHENYLTHIOLACETATE HYDROCHLORIDE see DHY400

2-DIETHYLAMINOETHYL-2,2-DIPHENYLVALERATE see DIG400

2-DIETHYLAMINOETHYL-2,2-DIPHENYLVALERATE HYDROCHLORIDE see PBM500

S-DIETHYLAMINOETHYL ESTER-p-BROMOTHIOBENZO HYDROXIMIC ACID HYDROCHLORIDE see DHZ000

2-DIETHYLAMINOETHYLESTER KYSELINY DIFENYLOCTOVE see DHX800

2-DIETHYLAMINOETHYL ESTER KYSELINY DIFENYLOCTOVE (CZECH) see THK000

2-DIETHYLAMINOETHYLESTER KYSELINY METHAKRYLOVE see DIB300

2-(DIETHYLAMINO)ETHYL ESTER-2-PHENYL BENZENE ACETIC ACID see THK000

1-(2-(DIETHYLAMINO)ETHYL)-2-(p-ETHOXYBENZYL)-5-BENZIMIDAZOLYL METHYL KETONE see DHZ050

1-(2-DIETHYLAMINOETHYL)-2-(p-ETHOXYBENZYL)-5-NITROBENZIMIDAZOLE HYDROCHLORIDE see EQN750

1-(2-(DIETHYLAMINO)ETHYL)-2-((p-ETHOXYPHENYL)THIO)BENZIMIDAZOLE HYDROCHLORIDE see DHZ100

N-(2-(DIETHYLAMINO)ETHYL)-2-ETHOXY-N-(2,6-XYLYL)CINCHONINAMIDE HYDROCHLORIDE see DIA000

N-(2-(DIETHYLAMINO)ETHYL)-2-ETHOXY-N-(2,6-XYLYL)-4-QUINOLINECARBOXAMIDE HYDROCHLORIDE see DIA000

DIETHYLAMINOETHYL ETHYLANILINE see PFB850

5-(2-(DIETHYLAMINO)ETHYL)-3-(α-ETHYLBENZYL)-1,2,4-OXADIAZOLECITRATE see POF500

3-(2-(DIETHYLAMINO)ETHYL)-5-ETHYL-5-PHENYLBARBITURIC ACID HYDROCHLORIDE see HEL500

N-(2-(DIETHYLAMINO)ETHYL)-2-ETHYL-2-PHENYLMALONAMIC ACID ETHYL ESTER see FAJ150

3-(2-(DIETHYLAMINO)ETHYL)-5-(2-FURYL)-1-PHENYL-1H-PYRAZOLINE HYDROCHLORIDE see DIA400

N-(2-(DIETHYLAMINO)ETHYL)-2,2',4,4',6,6'-HEXAMETHYLBENZANILIDE HYDROCHLORIDE see DIA800

DIETHYLAMINOETHYL-3-HYDROXY-4-AMINOBENZOATE see DHO600

7-(β-DIETHYLAMINOETHYL)-8-(α-HYDROXYBENZYL) THEOPHYLLINE HYDROCHLORIDE see BTA500

3-((1-DIETHYLAMINOETHYLIDENE)AZINOMETHYL)RIFAMYCIN SV see RJZ100

5-((γ-DIETHYLAMINO-β-HYDROXYPROPYL)AMINO)-3-METHOXY-8-CHLOROACRIDINE DIHYDROCHLORIDE see ADI750

4-(DIETHYLAMINO)-2-ISOPROPYL-2-PHENYLVALERONITRILE see DIK000

2-(DIETHYLAMINO)-N-(1-MESITYLOXY-2-PROPYL)-N-METHYLACETAMIDE HYDROCHLORIDE see DIK200

2-(DIETHYLAMINO)-N-(1-(p-METHOXYPHENOXY)-2-PROPYL)-N-METHYLACETAMIDE HYDROCHLORIDE see DIK400

2-(DIETHYLAMINO)-N-METHYL-o-ACETOTOLUIDIDE HYDROCHLORIDE see DIK600

2-(DIETHYLAMINO)-N-METHYL-2',6'-ACETOXYLIDIDE HYDROCHLORIDE see DIK800

7-(DIETHYLAMINO)-4-METHYL-2H-1-BENZOPYRAN-2-ONE see DIL400

p-(2-((DIETHYLAMINO)METHYL)BUTOXY)BENZOIC ACID, p-METHOXYPHENYL ESTER HYDROCHLORIDE see UAG000

8-((4-(DIETHYLAMINO)-1-METHYLBUTYL)AMINO)-6-METHOXYQUINOLINE see RHZ000

6-((4-(DIETHYLAMINO)-1-METHYLBUTYL)AMINO)-2-METHYL-4,5,8-TRIMETHOXYQUINOLINE-1,5-NAPHTHALENE DISULFONATE see DIL000

6((-4-(DIETHYLAMINO)-1-METHYLBUTYL))-5,8-DIMETHOXY-2,4-DIMETHYLQUINOLINE-1,5-NAPHTHALENE DISULFONATE see DIL200

7-DIETHYLAMINO-4-METHYLCOUMARIN see DIL400

7-DIETHYLAMINO-4-METHYLCOUMARIN, HYDROGEN SULFATE see DIL600

10-(2-DIETHYLAMINO-2-METHYLETHYL)PHENOTHIAZINE see DIR000

3-((DIETHYLAMINO)METHYL)-4-HYDROXYBENZOIC ACID ETHYL ESTER see EIA000

2-(DIETHYLAMINO)-N-METHYL-N-(2-MESITYLOXYETHYL)ACETAMIDE HYDROCHLORIDE see DIM000

5-(DIETHYLAMINO)METHYL-3-(1-METHYL-5-NITROIMIDAZOL-2-YLMETHYLENE AMINO)-2-OXAZOLIDINONE HYDROCHLORIDE see DIM200

2-(DIETHYLAMINO)-N-(2-METHYL-1-NAPHTHYL)ACETAMIDE HYDROCHLORIDE see DIM400

4-DIETHYLAMINOMETHYL-2-(5-NITRO-2-THIENYL)THIAZOLE HYDROCHLORIDE see DIM600

2-(DIETHYLAMINO)-4-METHYL-1-PENTANOL-p-AMINOBENZOATE (ESTER) see LET000

2-(DIETHYLAMINO)-4-METHYL-1-PENTANOL, p-AMINOBENZOATE ESTER, METHANESULFONATE see LEU000

2-(DIETHYLAMINO)-N-METHYL-N-(2-PHENETHYL)ACETAMIDE HYDROCHLORIDE see DIM800

3-(DIETHYLAMINO)-N-METHYL-N-(1-PHENOXY-2-PROPYL)PROPIONAMIDE HYDROCHLORIDE see DIN000

2-(DIETHYLAMINO)-N-METHYL-N-(3-PHENYLPROPYL)ACETAMIDE HYDROCHLORIDE see DIN200

2-(DIETHYLAMINO)-N-(4-METHYL-2-PYRIDYL)ACETAMIDE DIHYDROCHLORIDE see DIN400

O-(2-(DIETHYLAMINO)-6-METHYL-4-PYRIMIDINYL)-O,O-DIETHYL PHOSPHOROTHIOATE see DIN600

2-DIETHYLAMINO-6-METHYLPYRIMIDIN-4-YL DIETHYLPHOSPHOROTHIONATE see DIN600

2-DIETHYLAMINO-6-METHYLPYRIMIDIN-4-YL DIMETHYL PHOSPHOROTHIONATE see DIN800

O-(2-DIETHYLAMINO-6-METHYLPYRIMIDIN-4-YL)-O,O-DIMETHYL PHOSPHOROTHIOATE see DIN800

O-(2-(DIETHYLAMINO)-6-METHYL-4-PYRIMIDINYL)-O,O-DIMETHYL PHOSPHOROTHIOATE see DIN800

7-DIETHYLAMINO-5-METHYL-s-TRIAZOLO(1,5-a)PYRIMIDINE see DIO200

4-DIETHYLAMINONITROSOBENZENE see NJW600

5-(DIETHYLAMINO)-2-NITROSOPHENOL HYDROCHLORIDE see DIO300

4-(2-(DIETHYLAMINO)-2-OXOETHOXY)-3-METHOXYBENZENEACETIC ACID, PROPYL ESTER see PMM000

4-o-(2-(DIETHYLAMINO)-2-OXOETHYL)RIFAMYCIN see RKA000

N-(4-(2-(DIETHYLAMINO)-1-OXOPROPYL)PHENYL)METHANESULFONAMIDE MONOHYDROCHLORIDE see DHI876

1-(DIETHYLAMINO)-4-PENTANONE see NOB800

5-DIETHYLAMINO-2-PENTANONE see NOB800

3-(DIETHYLAMINO)PHENOL see DIO400

m-(DIETHYLAMINO)PHENOL see DIO400

DIETHYL-m-AMINO-PHENOLPHTHALEIN HYDROCHLORIDE see FAG070

2-(DIETHYLAMINO)-N-PHENYLACETAMIDE see DHJ400

4-((4-(DIETHYLAMINO)PHENYL)AZO)PYRIDINE-1-OXIDE see DIP000

2-(p-(DIETHYLAMINOPHENYL))-1,3,2-DITHIARSENOLANE see DIP100

DIETHYLAMINOPHENYLMERCURIC ACETATE see AAS300

2-(DIETHYLAMINO)-1-PHENYL-1-PROPANONE HYDROCHLORIDE see DIP600

10-DIETHYLAMINOPROPIONYL-3-TRIFLUOROMETHYL PHENOTHIAZINE HYDROCHLORIDE see FDE000

2-(DIETHYLAMINO)PROPIOPHENONE see DIP400

α-DIETHYLAMINOPROPIOPHENONE see DIP400

2-DIETHYLAMINOPROPIOPHENONE HYDROCHLORIDE see DIP600

2-(DIETHYLAMINO)-2'-PROPOXYACETANILIDE HYDROCHLORIDE see DIP800

3-DIETHYLAMINOPROPYLAMINE see DIQ100

N-(3-DIETHYLAMINOPROPYL)AMINE see DIY800

N,N-DIETHYLAMINOPROPYLAMINE see DIY800

3-(DIETHYLAMINO)PROPYLAMINE (DOT) see DIY800

1-(3-DIETHYLAMINOPROPYLAMINO)-9-METHOXYELLIPTICINE BIMALEATE see DIQ125

2-DIETHYLAMINO-1-PROPYL-N-DIBENZOPARATHIAZINE see DIR000

3-(DIETHYLAMINO)PROPYL ISOPROPYL(PHENYL)GLYCOLATE HYDROCHLORIDE see DIQ200

10-(2-DIETHYLAMINOPROPYL)PHENOTHIAZINE see DIR000

10-(2-(DIETHYLAMINO)PROPYL)-10H-PYRIDO(3,2-B)(1,4)BENZOTHIAZINE see DIR100

3-DIETHYLAMINO-5H-PYRIDO(4,3-b)INDOLE see DIR800

4-(DIETHYLAMINO)SALICYLALDEHYDE see DIJ230

p-(DIETHYLAMINO)SALICYLALDEHYDE see DIJ230

DIETHYLAMINO STILBENE see DKA000

DIETHYLAMINOSULFUR TRIFLUORIDE see DIR875

2-(DIETHYLAMINO)-2',4',6'-TRICHLOROACETANILIDE HYDROCHLORIDE see DIS000

2-DIETHYLAMINO-2',4',6'-TRIMETHYLACETANILIDE HYDROCHLORIDE see DHL800

2-(DIETHYLAMINO)-2',4',6'-TRIMETHYLACETANILIDE MONOHYDROCHLORIDE see DHL800

DIETHYLAMINOTRIMETHYLENAMINE see DIY800

2-(DIETHYLAMINO)-N-(2,4,6-TRIMETHYLPHENYL)ACETAMIDE MONOHYDROCHLORIDE see DHL800

(6-(DIETHYLAMINO)-3H-XANTEN-3-YLIDENE)DIETHYLAMMONIUM CHLORIDE see DIS200

N-(6-(DIETHYLAMINO)-3H-XANTHEN-3-YLIDINE)-N-ETHYLETHANAMINIUM CHLORIDE see DIS200

2-DIETHYLAMMONIOETHYL NITRATE see DIS400

DIETHYLAMMONIUM CHLORIDE see DIS500

DIETHYLAMMONIUM CYCLOHEXADIEN-4-OL-1-ONE-4-SULFONATE see DIS600

DIETHYLAMMONIUM DIETHYLDITHIOCARBAMATE see DJD000

DIETHYLAMMONIUM-2,5-DIHYDROXYBENZENE SULFONATE see DIS600

N,N-DIETHYLANILIN (CZECH) see DIS700

DIETHYLANILINE see DIS700

2,6-DIETHYLANILINE see DIS650

N,N-DIETHYLANILINE see DIS700

N,N-DIETHYLANILINE HYDROCHLORIDE see DIS720

2-α-17-α-DIETHYL-A-NOR-5-α-ANDROSTANE-2-β,17-β-DIOL see EQN320

N,N-DIETHYL-p-ARSANILIC ACID see DIS775

DIETHYL ARSINE see DIS800

DIETHYL ARSINIC ACID see DIS850

DIETHYL AZOFORMATE see DIT300

DIETHYL AZOMALONATE see DIT350

DIETHYLBARBITONE see BAG000

DIETHYLBARBITURATE MONOSODIUM see BAG250

DIETHYL-BARBITURIC ACID see BAG000

5,5-DIETHYLBARBITURIC ACID see BAG000

5,5-DIETHYLBARBITURIC ACID SODIUM deriv. see BAG250

N,N-DIETHYLBENZAMIDE see BCM250

6,8-DIETHYLBENZ(a)ANTHRACENE see DIT400

7,12-DIETHYLBENZ(a)ANTHRACENE see DIT800

8,12-DIETHYLBENZ(a)ANTHRACENE see DIT600

9,10-DIETHYL-1,2-BENZANTHRACENE see DIT800

2,6-DIETHYLBENZENAMINE see DIS650

N,N-DIETHYLBENZENAMINE see DIS700

DIETHYL BENZENE see DIU000

m-DIETHYLBENZENE see DIU200

o-DIETHYLBENZENE see DIU300

DIETHYLBENZENE (DOT) see DIU000

N,N-DIETHYLBENZENEACETAMIDE see PEU100

α,α-DIETHYLBENZENEACETIC ACID 2-(2-(DIETHYLAMINO)ETHOXY)ETHYL ESTER see DHQ200

N,N-DIETHYLBENZENESULFONAMIDE see DIU400

DIETHYLBENZOL see DIU000

3,5-DIETHYL-p-BENZOQUINONE-4-IMINE see DIU433

DIETHYL-4-(BENZOTHIAZOL-2-YL)BENZYLPHOSPHONATE see DIU500

DIETHYL BENZYLPHOSPHONATE see DIU600

O,O-DIETHYL-S-BENZYL THIOPHOSPHATE see DIU800

DIETHYLBERYLLIUM see DIV000

meso-α,α'-DIETHYLBIBENZYL-4,4'-DISULFONIC ACID DIPOTASSIUM SALT see SPA650

sym-DIETHYL BIS(DIMETHYLAMIDO)PYROPHOSPHATE see DIV200

unsym-DIETHYL BIS(DIMETHYLAMIDO)PYROPHOSPHATE see DJA300

DIETHYL BIS-DIMETHYLPYROPHOSPHORADIAMIDE (symmetrical) see DIV200

DIETHYL BIS-DIMETHYL PYROPHOSPHORDIAMIDE asym see DJA300

DIETHYL ((BIS(2-HYDROXYETHYL)AMINO)METHYL)PHOSPHONATE see DIV300

1,2-DIETHYL-1,3-BIS-(p-METHOXYPHENYL)-1-PROPENE see CNH525

DIETHYLBISMUTH CHLORIDE see DIV400

DIETHYLBIS(OCTANOYLOXY)STANNANE see DIV600

DIETHYLBIS(1-OXOOCTYL)OXY)STANNANE see DIV600

O,O-DIETHYL-O-(4-BROOM-2,5-DICHLOOR-FENYL)-MONOTHIOFOSFAAT (DUTCH) see EGV500

DI-2-(2-ETHYLBUTOXY)ETHYL ADIPATE see AEP250

DI(2-ETHYLBUTYL) ADIPATE see AEP500

o,o-DIETHYL o-(4-(1-((((BUTYLAMINO)CARBONYL)OXY)IMINO)ETHYL)PHENYL) PHOSPHOROTHIOIC ACID ESTER see DIV700

DIETHYLCADMIUM see DIV800

DIETHYLCARBAMAZANE CITRATE see DIW200

DIETHYLCARBAMAZINE see DIW000

DIETHYLCARBAMAZINE ACID CITRATE see DIW200

DIETHYLCARBAMAZINE CITRATE see DIW200

DIETHYLCARBAMAZINE HYDROCHLORIDE see DIW300

DIETHYLCARBAMAZINE HYDROGEN CITRATE see DIW200

DIETHYLCARBAMIC ACID ESTER with (m-HYDROXYPHENYL)TRIMETHYLAMMONIUM METHYLSULFATE see HNQ000

N,N-DIETHYLCARBAMIC ACID-3-(TRIMETHYLAMMONIO)PHENYL ESTER, METHYLSULFATE see HNQ000

DIETHYLCARBAMIC CHLORIDE see DIW400

DIETHYLCARBAMIC ESTER of 3-OXYPHENYLTRIMETHYLAMMONIUM METHYLSULFATE see HNQ000

DIETHYLCARBAMIDOYL CHLORIDE see DIW400

DIETHYLCARBAMODITHIOIC ACID 2-CHLORO-2-PROPENYL ESTER see CDO250

DIETHYLCARBAMODITHIOIC ACID, SODIUM SALT see SGJ000

DIETHYL-CARBAMOTHIOIC ACID S-((4-CHLOROPHENYL)METHYL) ESTER see SAZ000

DIETHYLCARBAMOYL CHLORIDE see DIW400

N,N-DIETHYLCARBAMOYL CHLORIDE see DIW400

(4-((DIETHYLCARBAMOYL)METHOXY)-3-METHOXYPHENYL)ACETIC ACID PROPYL ESTER see PMM000

(p-((DIETHYLCARBAMOYL)METHOXY)-3-METHOXYPHENYL)ACETIC ACID PROPYL ESTER see PMM000

17-β-N,N-DIETHYLCARBAMOYL-4-METHYL-4-AZA-5-α-ANDROSTAN-3-ONE see DJP700

1-DIETHYLCARBAMOYL-4-METHYLPIPERAZINE see DIW000

1-DIETHYLCARBAMOYL-4-METHYLPIPERAZINE DIHYDROGEN CITRATE see DIW200

N-((DIETHYLCARBAMOYL)METHYL)-3,4,5-TRIMETHOXYBENZAMIDE see TIH800

(3-(N',N'-DIETHYLCARBAMOYLOXY)PHENYL)TRIMETHYLAMMONIUM METHYLSULFATE see HNQ000

DIETHYLCARBAMYL CHLORIDE see DIW400

1-DIETHYLCARBAMYL-4-METHYLPIPERAZINE HYDROCHLORIDE see DIW300

1-DIETHYLCARBAMYL-4-METHYLPIPERZINE see DIW000

N,N-DIETHYLCARBANILIDE see DJC400

O,O-DIETHYL-S-(CARBETHOXY)METHYL PHOSPHOROTHIOLATE see DIW600

DIETHYL CARBINOL see IHP010

DIETHYLCARBINOL see IHP010

DIETHYL CARBITOL see DIW800

1,1'-DIETHYL-4,4'-CARBOCYANINE IODIDE see KHU050

O,O-DIETHYL-S-CARBOETHOXYMETHYL DITHIOPHOSPHATE see DIX000

O,O-DIETHYL-S-CARBOETHOXYMETHYL PHOSPHORODITHIOATE see DIX000

O,O-DIETHYL-S-CARBOETHOXYMETHYL PHOSPHOROTHIOATE see DIW600

O,O-DIETHYL-S-CARBOETHOXYMETHYL THIOPHOSPHATE see DIW600

DIETHYL CARBONATE see DIX200

DIETHYL CARBONATE (DOT) see DIX200

DIETHYL CELLOSOLVE (DOT) see EJE500

DIETHYLCETONE (FRENCH) see DJN750

O,O-DIETHYL-O-(2-CHINOXALYL)PHOSPHOROTHIOATE see DJY200

O,O-DIETHYL-S-((4-CHLOOR-FENYL-THIO)-METHYL)-DITHIOFOSFAAT (DUTCH) see TNP250

O,O-DIETHYL-O-(3-CHLOOR-4-METHYL-CUMARIN-7-YL)MONOTHIOFOSFAAT (DUTCH) see CNU750

O,O-DIETHYL-S-((6-CHLOOR-2-OXO-BENZOXAZOLIN-3-YL)-METHYL)-DITHIOFOSFAAT (DUTCH) see BDJ250

N,N-DIETHYLCHLORACETAMIDE see DIX400

O,O-DIETHYL-S-p-CHLORFENYLTHIOMETHYLESTER KYSELINY DITHIOFOSFORECNE (CZECH) see TNP250

DIETHYLCHLOROALUMINUM see DHI885

O,O-DIETHYL-S-(6-CHLOROBENZOXAZOLINYL-3-METHYL)DITHIOPHOSPHATE see BDJ250

O,O-DIETHYL-O-(2-CHLORO-1-(2',4'-DICHLOROPHENYL)VINYL) PHOSPHATE see CDS750

O,O-DIETHYL-O-(2-CHLORO-1,2,5-DICHLOROPHENYLVINYL) PHOSPHOROTHIOATE see DIX600

DIETHYL(2-CHLOROETHYL)AMINE see CGV500

DIETHYL-β-CHLOROETHYLAMINEHYDROCHLORIDE see CLQ250

O,O-DIETHYL-S-p-CHLOROLPHENYLTHIOMETHYL DITHIOPHOSPHATE see TNP250

DIETHYL-3-CHLORO-4-METHYL-7-COUMARINYL PHOSPHATE see CIK750

O,O-DIETHYL-O-(3-CHLORO-4-METHYLCOUMARIN-7-YL)PHOSPHATE see CIK750

O,O-DIETHYL-O-(3-CHLORO-4-METHYL-7-COUMARINYL)PHOSPHOROTHIOATE see CNU750

O,O-DIETHYL-O-(3-CHLORO-4-METHYLCOUMARINYL-7)THIOPHOSPHATE see CNU750

O,O-DIETHYL-O-(3-CHLORO-4-METHYL-2-OXO-2H-BENZOPYRAN-7-YL)PHOSPHOROTHIOATE see CNU750

S,S-DIETHYL(CHLOROMETHYL)PHOSPHONODITHIOATE see BJD000

O,O-DIETHYL-3-CHLORO-4-METHYL-7-UMBELLIFERONE THIOPHOSPHATE see CNU750

O,O-DIETHYL-O-(3-CHLORO-4-METHYLUMBELLIFERYL)PHOSPHOROTHIOATE see CNU750

DIETHYL-3-CHLORO-4-METHYLUMBELLIFERYL THIONOPHOSPHATE see CNU750

O,O-DIETHYL-S-((6-CHLORO-2-OXOBENZOXAZOLIN-3-YL)METHYL) PHOSPHORODITHIOATE see BDJ250

O,O-DIETHYL-S-(6-CHLORO-2-OXO-BENZOXAZOLIN-3-YL)METHYL-PHOSPHORO THIOLOTHIONATE see BDJ250

O,O-DIETHYL-P-CHLOROPHENYLMERCAPTOMETHYL DITHIOPHOSPHATE see TNP250

O,O-DIETHYL-S-(4-CHLOROPHENYLTHIOMETHYL) DITHIOPHOSPHATE see TNP250

O,O-DIETHYL-S-(p-CHLOROPHENYLTHIOMETHYL) PHOSPHORODITHIOATE see TNP250

O,O-DIETHYL-S-p-CHLOROPHENYL THIOMETHYLPHOSPHOROTHIOATE see DIX800

DIETHYL CHLOROPHOSPHATE see DIY000

O,O-DIETHYL-S-(2-CHLORO-1-PHTHALIMIDOETHYL)PHOSPHORODITHIOATE see DBI099

DIETHYLCHLOROTHIOPHOSPHATE see DJW600

DIETHYL-2-CHLOROVINYL PHOSPHATE see CLV375

O,O-DIETHYL-O-(2-CHLOROVINYL) PHOSPHATE see CLV375

DIETHYLCHLORTHIOFOSFAT (CZECH) see DJW600

O,O-DIETHYL-S-((2-CYAAN-2-METHYL-ETHYL)-CARBAMOYL)-METHYL-MONOTHIOFOSFAAT (DUTCH) see PHK250

DIETHYLCYANAMIDE see DIY100

DIETHYLCYANAMIDE see DIY150

N,N-DIETHYLCYANAMIDE see DIY100

N,N-DIETHYLCYANAMIDE see DIY150

DIETHYLCYANOPHOSPHATE see DJW800

DIETHYL CYANOPHOSPHONATE see DJW800

P,P-DIETHYL CYCLIC ETHYLENE ESTER OF PHOSPHONODITHIOIMIDOCARBONIC ACID see PGW750 p,p-DIETHYL CYCLIC PROPYLENE ESTER of PHOSPHONODITHIOIMIDOCARBONIC ACID see DHH400

DIETHYLCYCLOHEXANE (mixed isomers) see DIY200

DIETHYLCYSTEAMIN see DIY600

DIETHYLCYSTEAMINE see DIY600

N-DIETHYL CYSTEAMINE see DIY600

N,N-DIETHYL CYSTEAMINE see DIY600

DIETHYL DECANEDIOATE see DJY600

DIETHYL-1,10-DECANEDIOATE see DJY600

N,N-DIETHYL-1,3-DIAMINOPROPANE see DIY800

DIETHYLDIAZENE-1-OXIDE see ASP000

DIETHYL DICARBONATE see DIZ100

O,O-DIETHYL-O-(2,4-DICHLOOR-FENYL)-MONOTHIOFOSFAAT (DUTCH) see DFK600

O,O-DIETHYL O-2,5-DICHLORO-4-BROMOPHENYL-PHOSPHOROTHIOATE see EGV500

O,O-DIETHYL O-(2,5-DICHLORO-4-BROMOPHENYL) THIOPHOSPHATE see EGV500

O,O-DIETHYL O-(DICHLORO(METHYLTHIO)PHENYL) PHOSPHOROTHIOATE see CLB022

O,O-DIETHYL O-DICHLORO(METHYLTHIO)PHENYL THIOPHOSPHATE see CLB022

O,O-DIETHYL-O-(2,4-DICHLOROPHENYL) PHOSPHOROTHIOATE see DFK600

DIETHYL 2,4-DICHLOROPHENYL PHOSPHOROTHIONATE see DFK600

O,O-DIETHYL-S-(2,5-DICHLOROPHENYLTHIOMETHYL) DITHIOPHOSPHATE see PDC750

O,O-DIETHYL-S-(2,5-DICHLOROPHENYLTHIOMETHYL) DITHIOPHOSPHORAN see PDC750

O,O-DIETHYL-S-(2,5-DICHLOROPHENYLTHIOMETHYL) PHOSPHORODITHIOATE see PDC750

O,O-DIETHYL-S-(3,4-DICHLOROPHENYL-THIO)METHYL PHOSPHOROTHIOATE see DJA200

O,O-DIETHYL-S-(2,5-DICHLOROPHENYLTHIOMETHYL) PHOSPHOROTHIOLOTHIONATE see PDC750

O,O-DIETHYL-O-2,4-DICHLOROPHENYL THIOPHOSPHATE see DFK600

DIETHYLDICHLOROSILANE (DOT) see DEY800

DIETHYLDICHLOROSTANNANE see DEZ000

DIETHYL DI(DIMETHYLAMIDO)PYROPHOSPHATE (symmetrical) see DIV200

DIETHYL DI(DIMETHYLAMIDO)PYROPHOSPHATE (unsymmetrical) see DJA300

DIETHYL (2-(DIETHOXYMETHYLSILYL)ETHYL)PHOSPHONATE see DJA330

N,N-DIETHYL-N'-(2,5-DIETHOXYPHENYL)-N'-(4-BUTOXYPHENOXYACETYL) ETHYLENEDIAMINE HCl see BPM750

DIETHYL-S-2-DIETHYLAMINOETHYL PHOSPHOROTHIOATE see DJA400

O,O-DIETHYL-S-2-DIETHYLAMINOETHYL PHOSPHOROTHIOATE see DJA400

O,O-DIETHYL-S-(2-DIETHYLAMINO) ETHYLPHOSPHOROTHIOATE HYDROGEN OXALATE see AMX825

O,O-DIETHYL-S-DIETHYLAMINOETHYL PHOSPHOROTHIOLATE see DJA400

O,O-DIETHYL-S-2-DIETHYLAMINOETHYL PHOSPHOROTHIOLATE see DJA400

O,O-DIETHYL-S-(β-DIETHYLAMINO)ETHYL PHOSPHOROTHIOLATE see DJA400

O,O-DIETHYL-S-(β-DIETHYLAMINO)ETHYL PHOSPHOROTHIOLATE HYDROGEN OXALATE see AMX825

O,O-DIETHYL-S-(2-DIETHYLAMINOETHYL) THIOPHOSPHATE see DJA400

O,O-DIETHYL O-(2-DIETHYLAMINO-6-METHYL-4-PYRIMIDINYL)PHOSPHOROTHIOATE see DIN600

1-(2-(2-(2,6-DIETHYL-α-(2,6-DIETHYLPHENYL)BENZYLOXY)ETHOXY)ETHYL)-4-METHYLPIPERAZINE see DJA600

2,2'-DIETHYLDIHEXYLAMINE see DJA800

O,O-DIETHYL-(1,2-DIHYDRO-1,3-DIOXO-2H-ISOINDOL-2-YL)PHOSPHONOTHIOATE see DJX200

5,5-DIETHYLDIHYDRO-2H-1,3-OXAZINE-2,4(3H)-DIONE see DJT400

O,O-DIETHYL O-(2,3-DIHYDRO-3-OXO-2-PHENYL-6-PYRIDAZINYL)PHOSPHOROTHIOATE see POP000

DIETHYLDIIODOSTANNANE see DJB000

DIETHYL DIMERCAPTOSUCCINATE see DJB100

DIETHYL 2,3-DIMERCAPTOSUCCINATE see DJB100

DIETHYL (DIMETHOXYPHOSPHINOTHIOYLTHIO) BUTANEDIOATE see MAK700

DIETHYL (DIMETHOXYPHOSPHINOTHIOYLTHIO)SUCCINATE see MAK700

α,α'-DIETHYL-4,4'-DIMETHOXYSTILBENE see DJB200

trans-α,α'-DIETHYL-4,4'-DIMETHOXYSTILBENE see DJB200

3',4'-DIETHYL-4-DIMETHYLAMINOAZOBENZENE see DJB400

o,o-DIETHYL o-(4-(1-((((DIMETHYLAMINO)CARBONYL)OXY)IMINO)ETHYL)PHENYL) PHOSPHOROTHIOIC ACID ESTER see DJB420

8,8-DIETHYL-N,N-DIMETHYL-2-AZA-8-GERMASPIRO(4.5)DECANE-2-PROPANAMINE DIHYDROCHLORIDE see SLD800

DIETHYLDIMETHYLMETHANE see DTI000

DIETHYL 2,6-DIMETHYL-4(2-PYRIDYL)-1,4-DIHYDRO-3,5-PYRIDINE-DICARBOXYLATE see DJB460

2,3-DIETHYL-N,5-DIMETHYL-6-QUINOXALINAMINE see DJP550

O,O-DIETHYL-O-(4-DIMETHYLSULFAMONYLPHENYL)PHOSPHOROTHIOATE see DJB500

p'-DIETHYL-p-DIMETHYL THIOPYROPHOSPHATE see DJB600

p-DIETHYL-p'-DIMETHYLTHIOPYROPHOSPHATE see DJB600

N,N'-DIETHYL-N,N'-DINITROSOETHYLENEDIAMINE see DJB800

N³,N³-DIETHYL-2,4-DINITRO-6-(TRIFLUOROMETHYL)-1,3-BENZENEDIAMINE see CNE500

3,3-DIETHYL-2,4-DIOXO-5-METHYLPIPERIDINE see DNW400

DIETHYLDIPHENYL DICHLOROETHANE see DJC000

DIETHYLDIPHENYLTHIURAM DISULFIDE see DJC200

N,N-DIETHYL-N,N-DIPHENYLTHIURAMDISULFIDE see DJC200

sym-DIETHYLDIPHENYLUREA see DJC400

1,3-DIETHYL-1,3-DIPHENYLUREA see DJC400

N,N'-DIETHYL-N,N'-DIPHENYLUREA see DJC400

DIETHYL DISELENIDE see EIN550

DIETHYLDISULFID (CZECH) see DJC600

DIETHYLDISULFIDE see DJC600

N,N-DIETHYL-4,4-DI-2-THIENYL-3-BUTEN-2-AMINE HYDROCHLORIDE see DJP500

N,N-DIETHYL-3,3-DI-2-THIENYL-1-METHYLALLYLAMINE HYDROCHLORIDE see DJP500

DIETHYLDITHIO BIS(THIONOFORMATE) see BJU000

DIETHYL DITHIOCARBAMATE see DJC800

DIETHYLDITHIOCARBAMATE SODIUM see SGJ000

DIETHYLDITHIOCARBAMIC ACID see DJC800

DIETHYLDITHIOCARBAMIC ACID ANHYDROSULFIDE with DIMETHYLTHIOCARBAMIC ACID see DJC875

DIETHYLDITHIOCARBAMIC ACID-2-CHLOROALLYL ESTER see CDO250

DIETHYLDITHIOCARBAMIC ACID DIETHYLAMINE SALT see DJD000

DIETHYLDITHIOCARBAMIC ACID LEAD(II) SALT see DJD200

DIETHYLDITHIOCARBAMIC ACID SELENIUM(II) SALT see DJD400

DIETHYLDITHIOCARBAMIC ACID SODIUM see SGJ000

DIETHYLDITHIOCARBAMIC ACID, SODIUM SALT see SGJ000

DIETHYLDITHIOCARBAMIC ACID SODIUM SALT TRIHYDRATE see SGJ500

DIETHYLDITHIOCARBAMIC ACID TELLURIUM SALT see EPJ000

DIETHYLDITHIOCARBAMIC ACID ZINC SALT see BJC000

DIETHYLDITHIOCARBAMIC ANHYDRIDE of O,O-DIISOPROPYL THIONOPHOSPHORIC ACID see DKB600

DIETHYLDITHIOCARBAMIC ANHYDROSULFIDE see DKB600

DIETHYLDITHIOCARBAMIC SODIUM TRIHYDRATE see SGJ500

DIETHYLDITHIOCARBAMINIC ACID see DJC800

2-(N,N-DIETHYLDITHIOCARBAMYL)BENZOATHIAZOLE see BDF250

O,O-DIETHYLDITHIOFOSFORECNAN SODNY (CZECH) see PHG750

O,O-DIETHYL 1,3-DITHIOLAN-2-YLIDENEPHOSPHORAMIDOTHIOATE see DXN600

DIETHYL-N-1,3-DITHIOLANYL-2-IMINO PHOSPHATE see DXN600

DIETHYLDITHIONE see DJC800

O,O-DIETHYL-DITHIOPHOSPHORIC ACID, p-CHLOROPHENYLTHIOMETHYL ESTER see TNP250

O,O-DIETHYLDITHIOPHOSPHORYLACETIC ACID-N-MONOISOPROPYLAMIDE see IOT000

3-DIETHYLDITHIOPHOSPHORYLMETHYL-6-CHLOROBENZOXAZOLONE-2 see BDJ250

DIETHYL DIXANTHOGEN see BJU000

N,N-DIETHYLDODECANAMIDE see DJD500

DIETHYL EMME see EEV200

DIETHYLENDIAMINE see DPJ200

1,4-DIETHYLENEDIAMINE see PIJ000
N,N-DIETHYLENE DIAMINE (DOT) see PIJ000
DIETHYLENE DIOXIDE see DVQ000
1,4-DIETHYLENE DIOXIDE see DVQ000
DIETHYLENE ETHER see DVQ000
DIETHYLENE GLYCOL see DJD600
DIETHYLENE GLYCOL, BIS(ALLYL CARBONATE)- see AGD250
DIETHYLENE GLYCOL, BISCHLOROFORMATE see OPO000
DIETHYLENE GLYCOL BISPHTHALATE see DJD700
DIETHYLENE GLYCOL-n-BUTYL ETHER see DJF200
DIETHYLENE GLYCOL BUTYL ETHER ACETATE see BQP500
DIETHYLENE GLYCOL DIACETATE see DJD750
DIETHYLENE GLYCOL DIACRYLATE see ADT250
DIETHYLENE GLYCOL DI(3-AMINOPROPYL) ETHER see DJD800
DIETHYLENE GLYCOL DIBENZOATE see DJE000
DIETHYLENEGLYCOL DIBUTYL ETHER see DDW200
DIETHYLENEGLYCOL DI-n-BUTYL ETHER see DDW200
DIETHYLENE GLYCOL, DIESTER with BUTYLPHTHALATE see DDY400
DIETHYLENE GLYCOL DIETHYL ETHER see DIW800
DIETHYLENE GLYCOL DIGLYCIDYL ETHER see DJE200
DIETHYLENE GLYCOL DIMETHYL ETHER see BKN750
DIETHYLENE GLYCOL DINITRATE see DJE400
DIETHYLENE GLYCOL DINITRATE, containing at least 25% phlegmatizer (DOT) see DJE400
DIETHYLENE GLYCOL DIVINYL ETHER see DJE600
DIETHYLENE GLYCOL ETHYL ETHER see CBR000
DIETHYLENE GLYCOL ETHYL ETHER ACRYLATE see EHG030
DIETHYLENE GLYCOL ETHYL METHYL ETHER see DJE800
DIETHYLENE GLYCOL ETHYLVINYL ETHER see DJF000
DIETHYLENE GLYCOL-n-HEXYL ETHER see HFN000
DIETHYLENE GLYCOL ISOPROPYL ETHER see IOJ500
DIETHYLENE GLYCOL METHYL ETHER see DJG000
DIETHYLENE GLYCOL MONOBUTYL ETHER see DJF200
DI(ETHYLENE GLYCOL MONOBUTYL ETHER)PHTHALATE see DJF400
DIETHYLENE GLYCOL MONOCHLOROACETATE see HKI600
DIETHYLENE GLYCOL MONO-2-CYANOETHYL ETHER see DJF600
DIETHYLENE GLYCOL, MONOESTER with STEARIC ACID see HKJ000
DIETHYLENE GLYCOL MONOETHYL ETHER see CBR000
DIETHYLENE GLYCOL MONOETHYL ETHER ACETATE see CBQ750
DIETHYLENE GLYCOL MONOHEPTYL ETHER see HBP275
DIETHYLENE GLYCOL MONOHEXYL ETHER see HFN000
DIETHYLENE GLYCOL, MONO(HYDROGEN MALEATE) see MAL500
DIETHYLENE GLYCOL MONOISOBUTYL ETHER see DJF800

DIETHYLENE GLYCOL MONOMETHYL ETHER see DJG000
DIETHYLENE GLYCOL MONOMETHYL ETHER ACETATE see MIE750
DIETHYLENE GLYCOL MONOMETHYLPENTYL ETHER see DJG200
DIETHYLENE GLYCOL-MONO-2-METHYLPENTYL ETHER see DJG200
DIETHYLENE GLYCOL MONOPHENYL ETHER see PEQ750
DIETHYLENE GLYCOL MONOSTEARATE see HKJ000
DIETHYLENE GLYCOL MONOVINYL ETHER see DJG400
DIETHYLENE GLYCOL PHENYL ETHER see PEQ750
DIETHYLENE GLYCOL STEARATE see HKJ000
DIETHYLENE GLYCOL VINYL ETHER see DJG400
DIETHYLENE IMIDE OXIDE see MRP750
DIETHYLENE IMIDOXIDE see MRP750
DIETHYLENE IMINEAMIDOTHIOPHOSPHORIC ACID see DNU850
2,6-DIETHYLENEIMINO-4-CHLOROPYRIMIDINE see EQI600
DI(ETHYLENE OXIDE) see DVQ000
DIETHYLENE OXIDE see TCR750
DIETHYLENE OXIMIDE see MRP750
1,3-DI(ETHYLENESULPHAMOYL)PROPANE see BJP899
DIETHYLENETRIAMINE see DJG600
DIETHYLENETRIAMINE, 4-(2-AMINOETHYL)- see NEI800
DIETHYLENETRIAMINEPENTAACETIC ACID see DJG800
1,1,4,7,7-DIETHYLENETRIAMINEPENTAACETIC ACID see DJG800
DIETHYLENETRIAMINE PENTAACETIC ACID, CALCIUM TRISODIUM SALT see CAY500
DIETHYLENETRIAMINEPENTA(METHYL ENEPHOSPHONIC ACID), SODIUM SALT see DJG700
DIETHYLENETRIAMINOMETHYLPHENOL see AJU700
(DIETHYLENETRINITRILO)PENTAACETIC ACID see DJG800
DIETHYLENEUREA see BJP450
N,N'-DIETHYLENEUREA see BJP450
DIETHYLEN-GLYCOL MONOVINYL ESTER see DJG400
DIETHYLENGLYKOLDINITRATE (CZECH) see DJE400
DIETHYLENIMIDE OXIDE see MRP750
α,α'-DIETHYL-α,α'-EPOXYBIBENZYL-4,4'-DIOL see DKB100
DIETHYL-β,γ-EPOXYPROPYLPHOSPHONATE see DJH200
DIETHYLESTER KYSELINY ACETYLAMINOMALONOVE (CZECH) see AAK750
DIETHYLESTER KYSELINY HEXAHYDRFTALOVE see HDS250
DIETHYLESTER KYSELINY SIROVE see DKB110
DIETHYL ESTER of PYROCARBONIC ACID see DIZ100
DIETHYL ESTER SULFURIC ACID see DKB110
N,N-DIETHYLETHANAMINE see TJO000
N,N-DIETHYL-1,2-ETHANEDIAMINE see DJI400
DIETHYL ETHANEDIOATE see DJT200
(R*,S*)-4,4'-(1,2-DIETHYL-1,2-ETHANEDIYL)BIS-BENZENESULFONIC ACID DIPOTASSIUM SALT see SPA650

DIETHYL ETHANE PHOSPHONITE see DJH500
DIETHYLETHANOLAMINE see DHO500
N,N-DIETHYLETHANOLAMINE see DHO500
(E)-1,1'-(1,2-DIETHYL-1,2-ETHENE-DIYL)BIS(4-METHOXYBENZENE) see DJB200
4,4'-(1,2-DIETHYL-1,2-ETHENEDIYL)BIS-PHENOL see DKA600
trans-4,4'-(1,2-DIETHYL-1,2-ETHENEDIYL)BISPHENOL see DKA600
4,4'-(1,2-DIETHYL-1,2-ETHENEDIYL)BISPHENOL-(E)-BIS(DIHYDROGEN PHOSPHATE) see DKA200
trans-4,4'-(1,2-DIETHYL-1,2-ETHENEDIYL)BISPHENOL DIPROPIONATE see DKB000
DIETHYL ETHER (DOT) see EJU000
DIETHYLETHEROXODIPEROXOCHROMIUM(VI) see DJH800
DIETHYL-S-(2-ETHIOETHYL)THIOPHOSPHATE see DAP200
DIETHYLETHOXYALUMINUM see EER000
O,O-DIETHYL-S-(N-ETHOXYCARBONYL-N-METHYLCARBAMOYLMETHYL) PHOSPHORODITHIOATE see DJI000
O,O-DIETHYL S-(N-ETHOXYCARBONYL-N-METHYLCARBAMOYLMETHYL) PHOSPHOROTHIOLOTHIONATE see DJI000
DIETHYL (ETHOXYMETHYLENE)MALONATE see EEV200
DIETHYLETHOXYMETHYLENEOXALACETATE see DJI100
O,O-DIETHYL S-ETHSULFONYLMETHYL THIOTHIONOPHOSPHATE see TEZ100
O,O-DIETHYL-O-(2-ETHTHIOETHYL)PHOSPHOROTHIOATE see DAO500
O,O-DIETHYL-S-(2-ETHTHIOETHYL)PHOSPHOROTHIOATE see DAP200
DIETHYL 2-ETHTHIOETHYL THIONOPHOSPHATE see DAO500
O,O-DIETHYL-S-(2-ETHTHIONYLETHYL) PHOSPHOROTHIOATE see ISD000
DIETHYL-S-(2-ETHTHIONYLETHYL) THIOPHOSPHATE see ISD000
O,O-DIETHYL S-ETHTHIONYLMETHYL THIOTHIONOPHOSPHATE see TEZ200
N,N-DIETHYL-N'-(2-(2-ETHYL-1,3-BENZODIOXOL-2-YL)ETHYL) ETHYLENEDIAMINE DIMALEATE see DJI200
O,O-DIETHYL-S-(2-ETHYL-N,N-DIETHYLAMINO) PHOSPHOROTHIOATE HYDROGEN OXALATE see AMX825
4,4'-(1,2-DIETHYLETHYLENE)BIS(2-AMINOPHENOL) see DJI250
N,N-DIETHYLETHYLENEDIAMINE see DJI400
4,4'-(1,2-DIETHYLETHYLENE)DI-m-CRESOL see DJI300
4,4'-(1,2-DIETHYLETHYLENE)DI-o-CRESOL see DJI350
4,4'-(1,2-DIETHYLETHYLENE)DIPHENOL see DLB400
4,4'-(1,2-DIETHYLETHYLENE)DIRESORCINOL see DJJ390
O,O-DIETHYL-S-ETHYL-2-ETHYLMERCAPTOPHOSPHOROTHIOLATE see DAP200
O,O-DIETHYL-S-ETHYL-2-ETHYLMERCAPTO PHOSPHOROTHIOLATE SULFOXIDE see ISD000

O,O-DIETHYL-S-(2-ETHYLMERCAPTOETHYL) DITHIOPHOSPHATE see DXH325

O,O-DIETHYL 2-ETHYLMERCAPTOETHYL THIOPHOSPHATE see DAO600

O,O-DIETHYL-2-ETHYLMERCAPTOETHYL THIOPHOSPHATE, THIONO ISOMER see SPF000

O,O-DIETHYL-S-ETHYLMERCAPTOMETHYL DITHIOPHOSPHONATE see PGS000

O,O-DIETHYL-S-ETHYL PHOSPHOROTHIOATE see TJU800

(T-4)-DIETHYL(2-(1-ETHYL-1-(2H-PYRROL-5-YL)PROPYL)-1H-PYRROLATO-N¹,N²)-BORON see MRW275

O,O-DIETHYL-S-((ETHYLSULFINYL)ETHYL)PHOSPHORODITHIOATE see OQS000

O,O-DIETHYL S-(2-(ETHYLSULFINYL)ETHYL) PHOSPHORODITHIOATE see OQS000

O,O-DIETHYL-O-(2-ETHYLSULFONYLETHYL)PHOSPHOROTHIOATE see SPF000

DIETHYL-2-ETHYLSULFONYLETHYL THIONOPHOSPHATE see SPF000

O,O-DIETHYL-S-ETHYLSULFONYL METHYLPHOSPHORODITHIOATE see TEZ100

O,O-DIETHYL S-ETHYLSULFONYLMETHYL THIONOPHOSPHATE see TEZ100

O,O-DIETHYL-S-(2-ETHYLTHIO-ETHYL)-DITHIOFOSFAAT (DUTCH) see DXH325

O,O-DIETHYL-S-(2-ETHYLTHIO-ETHYL)-MONOTHIOFOSFAAT (DUTCH) see DAP200

O,O-DIETHYL-2-ETHYLTHIOETHYL PHOSPHORODITHIOATE see DXH325

O,O-DIETHYL-S-(2-ETHYLTHIOETHYL) PHOSPHORODITHIOATE see DXH325

O,O-DIETHYL-S-2-(ETHYLTHIO)ETHYL PHOSPHORODITHIOATE see DXH325

O,O-DIETHYL-2-ETHYLTHIO ETHYL PHOSPHOROTHIOATE see DAO500

O,O-DIETHYL-O-2-(ETHYLTHIO)ETHYL PHOSPHOROTHIOATE see DAO500

O,O-DIETHYL-S-2-(ETHYLTHIO)ETHYL PHOSPHOROTHIOATE see DAP200

O,O-DIETHYL-S-(2-(ETHYLTHIO)ETHYL) PHOSPHOROTHIOLATE (USDA) see DAP200

DIETHYL-2-(ETHYLTHIO(ETHYL PHOSPHOROTHIONATE)) see DAO500

O,O-DIETHYL-S-(2-ETHYLTHIOETHYL) THIONOPHOSPHATE see DXH325

O,O-DIETHYL-S-ETHYLTHIOMETHYL DITHIOPHOSPHONATE see PGS000

O,O-DIETHYL-ETHYLTHIOMETHYL PHOSPHORODITHIOATE see PGS000

o,o-DIETHYL S-(2-(ETHYLTHIO)-6-METHYL-4-PYRIMIDINYL) PHOSPHORODITHIOATE see DJJ393

O,O-DIETHYL-S-ETHYLTHIOMETHYL THIOTHIONOPHOSPHATE see PGS000

O,O-DIETHYL-S-ETHYLTHIONYLMETHYLPHOSPHORODITHIOATE see TEZ200

O,O-DIETHYL-S-ETHYLTHIONYLMETHYL THIONOPHOSPHATE see TEZ200

N,N'-DIETHYL-p-FENYLENDIAMIN see DJV200

DIETHYL FLUOROPHOSPHATE see DJJ400

DIETHYL FORMAMIDE see DJJ600

DIETHYL FUMARATE see DJJ800

DIETHYL GALLIUM HYDRIDE see DJJ829

N,N-DIETHYLGLYCINE-2,6-DIMETHOXYPHENYL ESTER HYDROCHLORIDE see FAC166

N,N-DIETHYLGLYCINE MESITYL ESTER HYDROCHLORIDE see FAC170

N,N-DIETHYLGLYCINE-2,6-XYLYL ESTER HYDROCHLORIDE see FAC050

N,N-DIETHYLGLYCINONITRILE see DHJ600

DIETHYL GLYCOL DIMETHYL ETHER see BKN750

DIETHYL GOLD BROMIDE see DJJ850

DIETHYLGOLD BROMIDE (DOT) see DJJ850

DIETHYLGUANIDINE HYDROCHLORIDE DIHYDRATE see DJJ875

3,3'-DIETHYLHEPTAMETHINETHIACYANINE IODIDE see DJK000

DIETHYL HEXAFLUOROGLUTARATE see DJK100

DIETHYL HEXAHYDROPHTHALATE see HDS250

DIETHYL HEXANEDIOATE see AEP750

DI-2-ETHYLHEXYL ADIPATE see AEO000

DI-(2-ETHYLHEXYL)AMINE see DJA800

2-DI-(2-ETHYLHEXYL)AMINOETHANOL see DJK200

O,O'-DI-(2-ETHYLHEXYL) DITHIOPHOSPHORIC ACID see DJK400

DI-(2-ETHYLHEXYL) ETHER see DJK600

DI-(2-ETHYLHEXYL) FUMARATE see DVK600

DI-2-ETHYLHEXYL ISOPHTHALATE see BJQ750

DI-(2-ETHYLHEXYL)MALEATE see BJR000

DI-(2-ETHYLHEXYL)ORTHOPHTHALATE see DVL700

DI(2-ETHYLHEXYL) PEROXYDICARBONATE see DJK800

DI(2-ETHYLHEXYL)PHENYL PHOSPHATE see BJR625

DI(2-ETHYLHEXYL)PHOSPHATE see BJR750

DI-2(ETHYLHEXYL)PHOSPHORIC ACID see BJR750

DI-(2-ETHYLHEXYL)PHOSPHORIC ACID (DOT) see BJR750

DI(2-ETHYLHEXYL)PHTHALATE see DVL700

DI-(2-ETHYLHEXYL)SEBACATE see BJS250

DI-(2-ETHYLHEXYL) SODIUM SULFOSUCCINATE see DJL000

DI(2-ETHYLHEXYL) TEREPHTHALATE see BJS300

DI-2-ETHYLHEXYLTIN DICHLORIDE see DJL200

DI(2-ETHYLHEXYL)TIN DICHLORIDE see DJL200

N¹,N¹⁴-DIETHYLHOMO-SPRINE see BOS050

1,2-DIETHYLHYDRAZINE see DJL400

N-N'-DIETHYLHYDRAZINE see DJL400

sym-DIETHYLHYDRAZINE see DJL400

1,2-DIETHYLHYDRAZINE DIHYDROCHLORIDE see DJL600

DIETHYL HYDROGEN PHOSPHATE see DJW500

O,O-DIETHYL HYDROGEN PHOSPHATE see DJW500

DIETHYL HYDROGEN PHOSPHITE see DJW400

DIETHYLHYDROXY ARSINE OXIDE see DIS850

N,N-DIETHYL-2-HYDROXYBENZAMIDE see DJY400

N,N-DIETHYL-α-HYDROXYBENZENEACETAMIDE see DJL700

N,N-DIETHYL-2-((HYDROXYDIPHENYLACETYL)METHYLAMINO)-N-METHYL-ETHANAMINIUM BROMIDE (9CI) see BCP685

N,N-DIETHYL-N-(β-HYDROXYETHYL)AMINE see DHO500

DIETHYL(2-HYDROXYETHYL)AMMONIUM CHLORIDE DIPHENYLCARBAMATE see DHY200

DIETHYL(2-HYDROXYETHYL)METHYLAMMONIUM BROMIDE α-PHENYLCYCLOHEXANEGLYCOLATE see ORQ000

DIETHYL(2-HYDROXYETHYL)METHYLAMMONIUM BROMIDE XANTHENE-9-CARBOXYLATE see DJM800

DIETHYL(2-HYDROXYETHYL)METHYLAMMONIUMBROMIDE XANTHENE-9-CARBOXYLATE see XCJ000

DIETHYL(2-HYDROXYETHYL)METHYLAMMONIUM-3-METHYL-2-PHENYLVALERATE BROMIDE see VBK000

DIETHYLHYDROXYLAMINE see DJN000

N,N-DIETHYLHYDROXYLAMINE see DJN000

O,O-DIETHYL N-HYDROXYNAPHTHALIMIDE PHOSPHATE see HMV000

DIETHYL(m-HYDROXYPHENYL)ARSINE METHIODIDE see DJN400

DIETHYL(m-HYDROXYPHENYL)METHYLAMMONIUM BROMIDE see HNK000

DIETHYL(m-HYDROXYPHENYL)METHYLAMMONIUM CHLORIDE see DJN430

DIETHYL(m-HYDROXYPHENYL)METHYLAMMONIUM IODIDE DIMETHYLCARBAMATE see HNK500

DIETHYL(m-HYDROXYPHENYL)METHYLARSONIUM IODIDE see DJN400

O,O-DIETHYL-7-HYDROXY-3,4-TETRAMETHYLENE COUMARINYL PHOSPHOROTHIOATE see DXO000

DIETHYL HYDROXYTIN HYDROPEROXIDE see DJN489

4,4'-(1,2-DIETHYLIDENE-1,2-ETHANEDIYL)BISPHENOL see DAL600

4,4'-(1,2-DIETHYLIDENE-1,2-ETHANEDIYL)BIS(PHENOL) DIACETATE see DHB500

4,4'-(DIETHYLIDENEETHYLENE)DIPHENOL see DAL600

p,p'-(DIETHYLIDENEETHYLENE)DIPHENOL see DAL600

4,4'-(DIETHYLIDENEETHYLENE)DIPHENOL DIACETATE see DHB500

DIETHYLIDINEHYDRAZINE see AAG100

3,5-DIETHYL-4-IMINO-2,5-CYCLOHEXADIEN-1-ONE see DIU433

N,N-DIETHYL-N'-2-INDANYL-N'-PHENYL-1,3-PROPANEDIAMINE HYDROCHLORIDE see FBP850

DIETHYL ISOPHTHALATE see IMK100

O,O-DIETHYL-S-(N-ISOPROPYLCARBAMOYLMETHYL) DITHIOPHOSPHATE see IOT000

O,O-DIETHYL-S-ISOPROPYLCARBAMOYLMETHYL PHOSPHORODITHIOATE see IOT000

O,O-DIETHYL-S-(N-ISOPROPYLCARBAMOYLMETHYL) PHOSPHORODITHIOATE see IOT000

O,O-DIETHYL-S-2-ISOPROPYLMERCAPTOMETHYLDITHIOPHOSPHATE see DJN600

O,O-DIETHYL-S-(ISOPROPYLMERCAPTOMETHYL) PHOSPHORODITHIOATE see DJN600

O,O-DIETHYL-O-(2-ISOPROPYL-4-METHYL-PYRIMIDIN-6-YL)MONOTHIOFOSFAAT (DUTCH) see DCM750

O,O-DIETHYL-O-(2-ISOPROPYL-4-METHYL-6-PYRIMIDINYL)PHOSPHOROTHIOATE see DCM750

O,O-DIETHYL-O-(2-ISOPROPYL-6-METHYL-4-PYRIMIDINYL) PHOSPHOROTHIOATE see DCM750

DIETHYL 4-(2-ISOPROPYL-6-METHYLPYRIMIDINYL)PHOSPHOROTHIONATE see DCM750

O,O-DIETHYL-O-(2-ISOPROPYL-4-METHYL-6-PYRIMIDYL)PHOSPHOROTHIOATE see DCM750

O,O-DIETHYL-O-(2-ISOPROPYL-4-METHYL-6-PYRIMIDYL) THIONOPHOSPHATE see DCM750

O,O-DIETHYL-2-ISOPROPYL-4-METHYLPYRIMIDYL-6-THIOPHOSPHATE see DCM750

O,O-DIETHYL-S-(ISOPROPYLTHIOMETHYL) PHOSPHORODITHIOATE see DJN600

DIETHYLKETENE see DJN700

O,O-DIETHYL-O-(2-KETO-4-METHYL-7-α',β'-BENZO-α'-PYRANYL) THIOPHOSPHATE see PKT000

DIETHYL KETONE see DJN750

DIETHYLKYANAMID see DIY100

DIETHYLKYANAMID see DIY150

DIETHYLLAURAMIDE see DJD500

N,N-DIETHYLLAURAMIDE see DJD500

N,N-DIETHYLLAURYLAMIDE see DJD500

DIETHYL LEAD DIACETATE see DJN800

DIETHYL LEAD DINITRATE see DJN875

N,N-DIETHYLLEUCINON-p-AMINOBENZOIC ACID METHANESULFONATE see LEU000

N,N-DIETHYLLYSERGAMIDE see DJO000

DIETHYL MAGNESIUM see DJO100

DIETHYL MALEATE see DJO200

DIETHYL MALONATE (FCC) see EMA500

DIETHYLMALONYLUREA see BAG000

DIETHYLMALONYLUREA SODIUM see BAG250

N,N-DIETHYLMANDELAMIDE see DJL700

DIETHYL(2-MERCAPTOETHYL)AMINE see DIY600

DIETHYL MERCAPTOSUCCINATE-O,O-DIMETHYL DITHIOPHOSPHATE, S-ESTER see MAK700

DIETHYL MERCAPTOSUCCINATE-O,O-DIMETHYL PHOSPHORODITHIOATE see MAK700

DIETHYL MERCAPTOSUCCINATE-O,O-DIMETHYL THIOPHOSPHATE see MAK700

DIETHYL MERCAPTOSUCCINATE-S-ESTER with O,O-DIMETHYLPHOSPHORODITHIOATE see MAK700

DIETHYL MERCAPTOSUCCINIC ACID-O,O-DIMETHYL PHOSPHORODITHIOATE see MAK700

DIETHYL MERCURY see DJO400

N,N-DIETHYLMETANILAN SODNY (CZECH) see DHM000

N,N-DIETHYL-2-(4-(6-METHOXY-2-PHENYL-1H-INDEN-3-YL)PHENOXY)-ETHANAMINE HYDROCHLORIDE see MFG260

N,N-DIETHYL-3-(4-METHOXYPHENYL)-1,2,4-OXADIAZOLE-5-ETHANAMINE see CPN750

N,N-DIETHYL-2-(4-(2-(4-METHOXYPHENYL)-1-PHENYLETHYL)PHENOXY)-ETHANAMINE (9CI) see MFG525

o,o-DIETHYL S-((5-METHOXY-1,3,4-THIADIAZOL-2-YL)METHYL) PHOSPHOROTHIOATE see DJO410

o,o-DIETHYL o-(4-(1-((((METHYLAMINO)CARBONYL)OXY)IMINO)ETHYL)PHENYL) PHOSPHOROTHIOIC ACID ESTER see DJO420

(((2-(DIETHYLMETHYLAMMONIO)-1-METHYL)ETHOXY)ETHYL)TRIMETHYLAMMONIUM DIIODIDE see MQF750

5,5-DIETHYL-1-METHYLBARBITURIC ACID see DJO800

N,N-DIETHYL-3-METHYLBENZAMIDE see DKC800

1,3-DIETHYL-5-METHYLBENZENE see DJP000

N,N-DIETHYL-5-METHYL-2H-(1)BENZOTHIOPYRANO(4,3,2-CD)INDAZOLE-2-ETHANAMINE, MONOMETHANESULFONATE see DJP100

DIETHYL METHYL CARBINOLURETHAN see ENF000

O,O-DIETHYL S-(N-METHYL-N-CARBOETHOXYCARBAMOYLMETHYL) DITHIOPHOSPHATE see DJI000

O,O-DIETHYL-O-(4-METHYLCOUMARIN-7-YL)-MONOTHIOFOSFAAT (DUTCH) see PKT000

O,O-DIETHYL-O-(4-METHYL-7-COUMARINYL) PHOSPHOROTHIOATE see PKT000

O,O-DIETHYL-O-(4-METHYL-7-COUMARINYL) THIONOPHOSPHATE see PKT000

O,O-DIETHYL-O-(4-METHYLCOUMARINYL-7) THIOPHOSPHATE see PKT000

DIETHYL (2-(METHYLDIETHOXYSILYL)ETHYL)PHOSPHONATE see DJA330

O,O-DIETHYL-S-(3-METHYL-2,4-DIOXO-5-OXA-3-AZA-HEPTYL)-DITHIOFOSFAAT (DUTCH) see DJI000

1,1-DIETHYL-2-METHYL-3-DIPHENYLMETHYLENEPYRROLIDINIUM BROMIDE see PAB750

N,N-DIETHYL-1-METHYL-3,3-DI-2-THIENYLALLYLAMINE HYDROCHLORIDE see DJP500

DIETHYL (4-METHYL-1,3-DITHIOLAN-2-YLIDENE)PHOSPHOROAMIDATE see DHH400

N,N-DIETHYL-N'-((8-α)-6-METHYLERGOLIN-8-YL)UREA (Z)-2-BUTENEDIOATE see DLR150

o,o-DIETHYL o-(4-(1-((((1-METHYLETHYL)AMINO)CARBONYL)OXY)IMINO)ETHYL) PHENYL) PHOSPHOROTHIOIC ACID ESTER see DJP520

O,O-DIETHYL-O-6-METHYL-2-ISOPROPYL-4-PYRIMIDINYL PHOSPHOROTHIOATE see DCM750

O,O-DIETHYL-O-(4-METHYL-7-KUMARINYL) ESTER KYSELINY THIOFOSFORESCNE (CZECH) see PKT000

DIETHYLMETHYL METHANE see MNI500

2,3-DIETHYL-5-METHYL-6-METHYLAMINOQUINOXALINE see DJP550

DIETHYLMETHYL(2-(N-METHYLBENZILAMIDO)ETHYL)AMMONIUM BROMIDE see BCP685

O,O-DIETHYL-O-(3-METHYL-4-(METHYLTHIO)PHENYL)PHOSPHOROTHIOATE see LIN400

1,1-DIETHYL-3-METHYL-3-NITROSOUREA see DJP600

N,N-DIETHYL-4-METHYL-3-OXO-5-α-4-AZAANDROSTANE-17-β-CARBOXAMIDE see DJP700

O,O-DIETHYL-O-(4-METHYL-2-OXO-2H-1-PHOSPHOROTHIOIC ACID BENZOPYRAN-7-YL)ESTER (9CI) see PKT000

N,N'-DI(1-ETHYL-3-METHYLPENTYL)-p-PHENYLENEDIAMINE see BJT500

N,N-DIETHYL-α-METHYL-10H-PHENOTHIAZINE-10-ETHANAMINE see DIR000

N,N-DIETHYL-2-(2-(2-METHYL-5-PHENYL-1H-PYRROL-1-YL)PHENOXY)-ETHANAMINE see LEF400

DIETHYLMETHYLPHOSPHINE see DJQ200

O,S-DIETHYL METHYLPHOSPHONOTHIOATE see DJR700

N,N-DIETHYL-4-METHYL-1-PIPERAZINECARBOXAMIDE see DIW000

N,N-DIETHYL-4-METHYL-1-PIPERAZINE CARBOXAMIDE CITRATE see DIW200

N,N-DIETHYL-4-METHYL-1-PIPERAZINECARBOXAMIDE DIHYDROGEN CITRATE see DIW200

N,N-DIETHYL-4-METHYL-1-PIPERAZINECARBOXAMIDE-2-HYDROXY-1,2,3-PROPANETRICARBOXYLATE scc DIW200

3,3-DIETHYL-5-METHYL-2,4-PIPERIDINEDIONE see DNW400

3,3-DIETHYL-5-METHYLPIPERIDINE-2,4-DIONE see DNW400

O,O-DIETHYL-O-(3-METHYL-1H-PYRAZOL-5-YL)-FOSFAAT (DUTCH) see MOX250

DIETHYL-3-METHYL-5-PYRAZOLYL PHOSPHATE see MOX250

O,O-DIETHYL-O-(3-METHYL-5-PYRAZOLYL) PHOSPHATE see MOX250

5,5-DIETHYL-1-METHYL-2,4,6(1H,3H,5H)-PYRIMIDINETRIONE see DJO800

3,3'-DIETHYL-9-METHYLSELENOCARBOCYANINE IODIDE see DJQ300

O,O-DIETHYL-O-(p-(METHYLSULFINYL)PHENYL) PHOSPHOROTHIOATE see FAQ800

O,O-DIETHYL-O-p-(METHYLSULFINYL)PHENYL THIOPHOSPHATE see FAQ800

DIETHYLMETHYLSULFONIUM IODIDEMERCURIC IODIDE (ADDITION COMPOUND) see DJQ800

DIETHYLMETHYL SULFONIUM IODINE with MERCURY IODIDE (1:1) see DJQ800

N,N-DIETHYL-4-METHYLTETRAMETHYLENEDIAMINE see DJQ850

3,3'-DIETHYL-9-METHYLTHIACARBOCYANINE IODIDE see DJR200

O,S-DIETHYL METHYLTHIOPHOSPHONATE see DJR700

O,O-DIETHYL-O-(4-(METHYLTHIO)-3,5-XYLYL)PHOSPHOROTHIOATE see DJR800

N,N-DIETHYL-5-METHYL-(1,2,4)TRIAZOLO(1,5-a)PYRIMIDINE-7-AMINE see DIO200

N,N-DIETHYL-N-METHYL-2-(2-(TRIMETHYLAMMONIO)ETHOXY)-1-PROPANAMINIUM DIIODIDE see MQF750

O,O-DIETHYL-O-(4-METHYLUMBELLIFERONE) ESTER OF THIOPHOSPHORIC ACID see PKT000

O,O-DIETHYL-O-(4-METHYLUMBELLIFERONE) PHOSPHOROTHIOATE see PKT000

DIETHYL (4-METHYLUMBELLIFERYL) THIONOPHOSPHATE see PKT000

N,N-DIETHYL-N-METHYL-2-((9H-XANTHEN-9-YLCARBONYL)OXY)ETHANAMINIUM BROMIDE see DJM800

α,α-DIETHYL-1-NAPHTHALENEACETIC ACID SODIUM SALT see DJS100
O,O-DIETHYL-o-NAPHTHALIMIDE PHOSPHOROTHIOATE see NAQ500
O,O-DIETHYL-o-NAPHTHALOXIMIDO PHOSPHOROTHIOATE see NAQ500
O,O-DIETHYL-o-NAPHTHALOXIMIDOPHOSPHOROTHIONATE see NAQ500
O,O-DIETHYL-o-NAPHTHYLAMIDOPHOSPHOROTHIOATE see NAQ500
DIETHYL-NICOTAMIDE see DJS200
N,N-DIETHYLNICOTINAMIDE see DJS200
DIETHYLNITRAMINE see DJS500
N,N-DIETHYL-N'-(1-NITRO-9-ACRIDINYL)-1,2-ETHANEDIAMINE DIHYDROCHLORIDE (9CI) see NFW100
N,N-DIETHYL-N'-(1-NITRO-9-ACRIDINYL)-1,3-PROPANEDIAMINE DIHYDROCHLORIDE (9CI) see NFW200
3,5-DIETHYL-4-NITROBIPHENYL see DJP600
N,N-DIETHYL-2-(4-(2-NITRO-1,2-DIPHENYLETHENYL)PHENOXY)ETHANAMINE CITRATE see EAF100
O,O-DIETHYL-O-(4-NITRO-FENIL)-MONOTHIOFOSFAAT (DUTCH) see PAK000
DIETHYL-p-NITROFENYL ESTER KYSELINY FOSFORECNE (CZECH) see NIM500
O,O-DIETHYL-O-p-NITROFENYLESTER KYSELINYTHIOFOSFORECNE (CZECH) see PAK000
O,O-DIETHYL-S-p-NITROFENYLESTER KYSELINY THIOFOSFORECNE (CZECH) see DJS800
O,O-DIETHYL-o-p-NITROFENYLFOSFAT (CZECH) see NIM500
O,O-DIETHYL-O-p-NITROFENYLTIOFOSFAT (CZECH) see PAK000
DIETHYL p-NITROPHENYL PHOSPHATE see NIM500
O,O-DIETHYL O-p-NITROPHENYL PHOSPHATE see NIM500
O,O-DIETHYL-O-(4-NITROPHENYL) PHOSPHOROTHIOATE see PAK000
O,O-DIETHYL-O-4-NITROPHENYLPHOSPHOROTHIOATE see PAK000
O,O-DIETHYL-O-(p-NITROPHENYL) PHOSPHOROTHIOATE see PAK000
O,O-DIETHYL-S-(4-NITROPHENYL) PHOSPHOROTHIOATE see DJS800
O,S-DIETHYL-O-(4-NITROPHENYL)PHOSPHOROTHIOATE see DJT000
O,S-DIETHYL-O-(p-NITROPHENYL) PHOSPHOROTHIOATE see DJT000
O,O-DIETHYL-S-(4-NITROPHENYL)PHOSPHOROTHIOIC ACID ESTER see DJS800
O,S-DIETHYL-O-(4-NITROPHENYL)PHOSPHOROTHIOIC ACID ESTER see DJT000
O,S-DIETHYL-O-(p-NITROPHENYL)PHOSPHOROTHIOIC ACID ESTER see DJT000
DIETHYL-4-NITROPHENYL PHOSPHOROTHIONATE see PAK000
DIETHYL-p-NITROPHENYLTHIONOPHOSPHATE see PAK000
O,O-DIETHYL-O-(p-NITROPHENYL)THIONOPHOSPHATE see PAK000
DIETHYL-p-NITROPHENYLTHIOPHOSPHATE see PAK000

O,O-DIETHYL-O-4-NITROPHENYL THIOPHOSPHATE see PAK000
O,O-DIETHYL-O-p-NITROPHENYL THIOPHOSPHATE see PAK000
O,O-DIETHYL-S-(4-NITROPHENYL)THIOPHOSPHATE see DJS800
O,S-DIETHYL-O-(4-NITROPHENYL)THIOPHOSPHATE see DJT000
DIETHYLNITROSAMINE see NJW500
N,N-DIETHYLNITROSAMINE see NJW500
DIETHYLNITROSOAMINE see NJW500
N,N-DIETHYL-4-NITROSOBENZENAMINE see NJW600
1,3-DIETHYL-1-NITROSOUREA see NJW700
1,1'-DIETHYL-1-NITROSOUREA see NJW700
N,N'-DIETHYL-N-NITROSOUREA see NJW700
N,N-DIETHYL-4-((5-NITRO-2-THIAZOLYL)AZO)BENZENAMINE see DJT050
DIETHYL OCTAFLUOROADIPATE see DJT100
DIETHYL OCTAFLUOROHEXANEDIOATE see DJT100
N,N-DIETHYL-N-(1,2,3,4,4A,5,10,10A-OCTAHYDRO-6-HYDROXY-1-PROPYLBENZO(G)QUI NOLIN-3-YL)-SULFAMIDE, (3-α,4A-α,10A-β)-(+−)- see DJT150
DIETHYLOLAMINE see DHF000
O,O-DIETHYL-O,2-PYRAZINYL PHOSPHOROTHIOATE see EPC500
O,O-DIETHYL O(and S)-2-(ETHYLTHIO)ETHYL PHOSPHOROTHIOATE MIXTURE see DAO600
DIETHYL OXALACETATE SODIUM SALT see SGJ550
DIETHYL OXALATE see DJT200
5,5-DIETHYL-1,3-OXAZIN-2,4-DIONE see DJT400
5,5-DIETHYL-1,3-OXAZINE-2,4-DIONE see DJT400
DIETHYL OXIDE see EJU000
O,O-DIETHYL-S-(4-OXO-3H-1,2,3-BENZOTRIAZINE-3-YL)-METHYL-DITHIOPHOSPHATE see EKN000
O,O-DIETHYL-S-(4-OXOBENZOTRIAZINO-3-METHYL)PHOSPHORODITHIOATE see EKN000
O,O-DIETHYL-S-((4-OXO-3H-1,2,3-BENZOTRIAZIN-3-YL)-METHYL)-DITHIO FOSFAAT (DUTCH) see EKN000
N,N-DIETHYL-3-OXO-BUTANAMIDE (9CI) see DHI600
DIETHYL OXOBUTANEDIOATE ION(1-) SODIUM see SGJ550
DIETHYL OXYDIFORMATE see DIZ100
DIETHYL PARAOXON see NIM500
DIETHYL-PARA-PHENYLENEDIAMINE see DJV200
DIETHYLPARATHION see PAK000
3,3'-DIETHYLPENTAMETHINETHIACYANINE IODIDE see DJT800
3,3-DIETHYLPENTANE see DJU000
2,2-DIETHYL-4-PENTENAMIDE see DJU200
DIETHYL PERFLUOROADIPATE see DJT100
DIETHYL PERFLUOROGLUTARATE see DJK100
DIETHYL PEROXIDE see DJU400
DIETHYL PEROXYDICARBONATE see DJU600
DIETHYL PEROXYDICARBONATE, >27% in solution (DOT) see DJU600
DIETHYL PEROXYDIFORMATE see DJU600
2,6-DIETHYLPHENOL see DJU700

N,N-DIETHYL-2-PHENOXYACETAMIDE see RCZ200
N,N-DIETHYLPHENYLACETAMIDE see PEU100
DIETHYLPHENYLAMINE see DIS700
N,N-DIETHYL-p-(PHENYLAZO)ANILINE see OHJ875
N,N-DIETHYL-4-(PHENYLAZO)BENZENAMINE see OHJ875
p-((3,4-DIETHYLPHENYL)AZO)-N,N-DIMETHYLANILINE see DJB400
N,N-DIETHYL-N'-(2-(2-PHENYL-1,3-BENZODIOXOL-2-YL)ETHYL)ETHYLENEDIAMINE DIMALEATE see DJU800
DI(p-ETHYLPHENYL)DICHLOROETHANE see DJC000
DIETHYL-4,4'-o-PHENYLENEBIS(3-THIOALLOPHANATE) see DJV000
DIETHYL-p-PHENYLENEDIAMINE see DJV200
N,N'-DIETHYL-p-PHENYLENEDIAMINE see DJV200
N,N-DIETHYL-p-PHENYLENEDIAMINE HYDROCHLORIDE see AJO250
N,N'-DIETHYL-p-PHENYLENEDIAMINE SULFATE see DJV250
N,N-DIETHYL-N'-PHENYLETHYLENEDIAMINE see DJV300
N,N-DIETHYL-2-PHENYL-GLYCINE-2,6-XYLYL ESTER HYDROCHLORIDE see FAC150
O,O-DIETHYL-O-(5-PHENYL-3-ISOXAZOLYL) PHOSPHOROTHIOATE see DJV600
O,O-DIETHYL-O-(3-(5-PHENYL)-1,2-ISOXAZOLYL)PHOSPHOROTHIOATE see DJV600
O,O-DIETHYL-O-(5-PHENYL-3-ISOXAZOLYL)PHOSPHOROTHIOIC ACID ESTER see DJV600
N,N-DIETHYL-3-PHENYL-1,2,4-OXADIAZOLE-5-ETHANAMINE see OOC000
O,O-DIETHYL-O-PHENYLPHOSPHOROTHIOATE see DJV700
N,N-DIETHYL-3-(1-PHENYLPROPYL)-1,2,4-OXADIAZOLE-5-ETHANAMINE CITRATE see POF500
DIETHYL PHENYLTIN ACETATE see DJV800
3,3-DIETHYL-1-PHENYLTRIAZENE see PEU500
O,O-DIETHYL O-(1-PHENYL-1H-1,2,4-TRIAZOL-3-YL)PHOSPHOROTHIOATE see THT750
DIETHYL PHOSPHATE see DJW500
DIETHYL PHOSPHINE see DJW000
DIETHYLPHOSPHINIC ACID ANHYDRIDE WITH DIETHYL PHOSPHOROTHIONATE see DJW100
DIETHYLPHOSPHINIC ACID-p-NITROPHENYL ESTER see DJW200
DIETHYL PHOSPHITE see DJW400
DIETHYL PHOSPHORIC ACID see DJW500
O,O-DIETHYL PHOSPHORIC ACID O-p-NITROPHENYL ESTER see NIM500
O,O-DIETHYLPHOSPHOROCHLORIDOTHIOATE see DJW600
DIETHYL PHOSPHOROCYANIDATE see DJW800
O,O-DIETHYL PHOSPHORODITHIOATE AMMONIUM see DJW875
O,O-DIETHYL PHOSPHORODITHIOATE S-ester with 3-(MERCAPTOMETHYL)-1,2,3-BENZOTRIAZIN-4(3H)-ONE see EKN000
O,O-DIETHYLPHOSPHOROTHIOATE, O-ESTER with 6-HYDROXY-2-PHENYL-3(2H)-PYRIDAZINONE see POP000

OXAPENTAMETHYLENE)BIS(AMMONIUM IODIDE) see MQF750
DIETHYL TETRAOXOSULFATE see DKB110
DIETHYL THALLIUM PERCHLORATE see DKB175
DIETHYLTHIADICARBOCYANINE IODIDE see DJT800
3,3'-DIETHYLTHIADICARBOCYANINE IODIDE see DJT800
DIETHYLTHIAMBUTENE HYDROCHLORIDE see DJP500
N,N-DIETHYLTHIOCARBAMIDE see DKC400
N,N-DIETHYLTHIOCARBAMYL-O,O-DIISOPROPYLDITHIOPHOSPHATE see DKB600
DIETHYLTHIOETHER see EPH000
2,2-DIETHYL-3-THIOMORPHOLINONE see DKC200
DIETHYL THIOPHOSPHORIC ACIDESTER of 3-CHLORO-4-METHYL-7-HYDROXYCOUMARIN see CNU750
DIETHYLTHIOPHOSPHORYL CHLORIDE (DOT) see DJW600
1,3-DIETHYLTHIOUREA see DKC400
1,3-DIETHYL-2-THIOUREA see DKC400
N,N'-DIETHYLTHIOUREA see DKC400
DIETHYLTIN CHLORIDE see DEZ000
DIETHYLTIN DI(10-CAMPHORSULFONATE) see DKC600
DIETHYLTIN DICAPRYLATE see DIV600
DIETHYLTIN DICHLORIDE see DEZ000
DIETHYLTIN DIIODIDE see DJB000
DIETHYLTIN DIOCTANOATE see DIV600
DIETHYLTOLUAMIDE see DKC800
DIETHYL-m-TOLUAMIDE see DKC800
N,N-DIETHYL-m-TOLUAMIDE see DKC800
N,N-DIETHYL-o-TOLUAMIDE see DKD000
DIETHYLTOLUENEDIAMINE see DJP100
DIETHYL TRIAZENE see DKD200
DIETHYL-TRIAZENE see DKD200
1,3-DIETHYLTRIAZENE see DKD200
1,3-DIETHYL-1-TRIAZENE see DKD200
N,N'-DIETHYLTRIAZENE see DKD200
1,3-DIETHYLTRIAZINE see DKD200
O,O-DIETHYL O-3,5,6-TRICHLORO-2-PYRIDYL PHOSPHOROTHIOATE see CMA100
3,9-DIETHYLTRIDECYL-6-SULFATE see DKD400
DIETHYL (2-(TRIETHOXYSILYL)ETHYL)PHOSPHONIC ACID see DKD500
N⁴,N⁴-DIETHYL-α,α,α-TRIFLUORO-3,5-DINITRO-TOLUENE-2,4-DIAMINE see CNE500
N,N-DIETHYL-4-(α-(α,α,α-TRIFLUORO-o-TOLYL) BENZYLOXY)PENTYLAMINE CITRATE see DKD600
O,O-DIETHYL-S-2-TRIMETHYLAMMONIUM ETHYLPHOSPHONOTHIOLATE IODIDE see TLF500
1,1-DIETHYLUREA see DKD650
N,N-DIETHYLUREA see DKD650
asym-DIETHYLUREA see DKD650
N,N-DIETHYLVANILLAMIDE see DKE200
DIETHYL XANTHOGENATE see BJU000
DIETHYLXANTHOGEN DISULFIDE see BJU000
DI(ETHYLXANTHOGEN)TRISULFIDE see DKE400
DIETHYL YELLOW see OHJ875
DIETHYLZINC see DKE600
17-β-2-ε,17-α-DIETHYNYL, A-NOR-ANDROSTANE-2-ε, DIHYDROXYDIPROPINATE see AOO150
2-α,17-α-DIETHYNYL-A-NOR-5-α-ANDROSTANE-2-β,17-β-DIOL DIHEMISUCCINATE see SDY675

2-α-17-α-DIETHYNYL-A-NOR-5-α-ANDROSTANE-2-β,17-β-DIOL DIPROPIONATE see AOO150
DIETHYXIME see DHZ000
DIETIL see CAR000
DIETILAMIDE-CARBOPIRIDINA see DJS200
DIETILAMINA (ITALIAN) see DHJ200
α-DIETILAMINO-2,6-DIMETILACETANILIDE (ITALIAN) see DHK400
5-(2-DIETILAMINOETIL)-3-FENIL-1,2,4-OXADIEZOLO CLORIDRATO (ITALIAN) see OOE100
O,O-DIETIL-O-(4-BROMO-2,5-DICLORO-FENIL)-MONOTIOFOSFATO (ITALIAN) see EGV500
O,O-DIETIL-S-((2-CIAN-2-METIL-ETIL)-CARBAMOIL)-METIL-MONOTIOFOSFATO (ITALIAN) see PHK250
O,O-DIETIL-S-((4-CLORO-FENIL-TIO)-METILE)-DITIOFOSFATO (ITALIAN) see TNP250
O,O-DIETIL-O-(3-CLORO-4-METIL-CUMARIN-7-IL-MONOTIOFOSFATO) (ITALIAN) see CNU750
O,O-DIETIL-S-((6-CLORO-2-OXO-BENZOSSAZOLIN-3-IL)-METIL)-DITIOFOSFATO (ITALIAN) see BDJ250
O,O-DIETIL-O-(2,4-DICLORO-FENIL)-MONOTIOFOSFATO (ITALIAN) see DFK600
5,5-DIETILDIIDRO-1,3-OSSAZIN-2,4-DIONE (ITALIAN) see DJT400
DIETILESTILBESTROL (SPANISH) see DKA600
O,O-DIETIL-S-(2-ETILTIO-ETIL)-DITIOFOSFATO (ITALIAN) see DXH325
O,O-DIETIL-S-(2-ETILTIO-ETIL)-MONOTIOFOSFATO (ITALIAN) see DAP200
O,O-DIETIL-S-(ETILTIO-METIL)-DITIOFOSFATO (ITALIAN) see PGS000
O,O-DIETIL-S-(N-ETOSSI-CARBONIL-N-METIL-CARBAMOIL-METIL)-DITIOFOSFATO (ITALIAN) see DJI000
O,O-DIETIL-O-(2-ISOPROPIL-4-METIL-PIRIMIDIN-6-IL)-MONOTIOFOSFATO (ITALIAN) see DCM750
O,O-DIETIL-O-(4-METILCUMARIN-7-IL)-MONOTIOFOSFATO (ITALIAN) see PKT000
O,O-DIETIL-O-(3-METIL-1H-PIRAZOL-5-IL)-FOSFATO (ITALIAN) see MOX250
O,O-DIETIL-O-(4-NITRO-FENIL)-MONOTIOFOSFATO (ITALIAN) see PAK000
O,O-DIETIL-S-((4-OXO-3H-1,2,3-BENZOTRIAZIN-3-IL)-METIL)-DITIOFOSFATO (ITALIAN) see EKN000
DIETILPROPANDIOLO see PMB250
1,1-DIETOSSIETANO (ITALIAN) see AAG000
DIETREEN see RAF100
DIETROL see DKE800
DIETROXINE see DJT400
O,O-DIETYL-S-2-ETYLMERKAPTOETYLTIOFOSFAT (CZECH) see DAP200
O,O-DIETYL-O-4-METHYLKUMARINYL(7)TIOFOSFAT (CZECH) see PKT000
DIF 4 see DRP800
DIFACIL see DHX800
DIFACIL see THK000
DIFEDRYL see BBV500
DIFENACOUM see BGO100
DIFENAKUM see BGO100
DIFENAMIZOLE see PAM500
DIFENHYDRAMIN see BBV500
DIFENHYDRAMINE HYDROCHLORIDE see BAU750
DIFENIDOL see DWK200
DIFENIDOL HYDROCHLORIDE see CCR875
DIFENIDOLIN see CCR875
DIFENIDRAMINA (ITALIAN) see BBV500

DIFENILDICHETOPIRAZOLIDINA (ITALIAN) see DWA500
DIFENILHIDANTOINA (SPANISH) see DKQ000
DIFENIL-METAN-DIISOCIANATO (ITALIAN) see MJP400
DIFENIN see DKQ000
DIFENIN see DNU000
DIFENOCONAZOLE see THS900
DIFENOXURON see LGM300
1,5-DIFENOXYANTHRACHINON see DVW100
DIFENSON see CJT750
DIFENTHOS see TAL250
DIFENYL-DIHYDROXYSILAN see DMN450
N,N-DIFENYLETHYLENDIAMIN see DWB400
N,N'-DIFENYL-p-FENYLENDIAMIN (CZECH) see BLE500
2-(DIFENYL-HYDROXYACETOXY)ETHYL-DIETHYLAMMONIUMCHLORID (CZECH) see BCA000
DIFENYLIN see BGF109
DIFENYLMETHAAN-DIISSOCYANAAT (DUTCH) see MJP400
DIFENYLSULFON (CZECH) see PGI750
DIFENZOQUAT METHYL SULFATE see ARW000
DIFERULOYLMETHANE see ICC800
DIFETOIN see DNU000
DIFEXAMIDE METHIODIDE see BTA325
DIFFLAM see BBW500
DIFFOLLISTEROL see EDP000
DIFHYDAN see DKQ000
DIFHYDAN see DNU000
DIFLAVINE (ACRIDINE) see DBN000
DIFLORASONE DIACETATE see DKF125
DIFLUBENZURON see CJV250
DIFLUCORTOLONE VALERATE see DKF130
DIFLUCORTOLONE 21-VALERATE see DKF130
DIFLUCORTOLONVALERIANAT (GERMAN) see DKF130
DIFLUFENICAN see DKJ210
DIFLUFENICANIL see DKJ210
DIFLUFENZOPYR, 98.1% see DKI300
DIFLUNISAL see DKI600
DIFLUORO see MJD275
2,4'-DIFLUOROACETANILIDE see DKF170
DIFLUOROACETIC ACID see DKF200
DIFLUOROAMINE see DKF400
3-DIFLUOROAMINO-1,2,3-TRIFLUORODIAZIRIDINE see DKF600
DIFLUOROAMMONIUM HEXAFLUOROARSENATE see DKF620
2,4-DIFLUOROANILINE see DKF700
2,4-DIFLUOROBENZENAMINE see DKF700
m-DIFLUOROBENZENE see DKF800
p-DIFLUOROBENZENE see DKG000
3,4-DIFLUOROBENZENEARSONIC ACID see DKG100
2,10-DIFLUOROBENZO(rst)PENTAPHENE see DKG400
1,1-DIFLUORO-1-CHLOROETHANE see CFX250
DIFLUOROCHLOROETHANES (DOT) see CFX250
DIFLUOROCHLOROMETHANE see CFX500
2',2'-DIFLUORODEOXYCYTIDINE see PPP150
2',2'-DIFLUORO-2'-DEOXYCYTIDINE see GCK500
2',2'-DIFLUORODEOXYCYTIDINE MONOHYDROCHLORIDE see GCK100
DIFLUORODIAZENE see DKG600
DIFLUORODIAZIRINE see DKG700
2,10-DIFLUORODIBENZO(a,i)PYRENE see DKG400
DIFLUORODIBROMOETHANE see DKG800

1,1-DIFLUORO-1,2-DIBROMOETHANE see DKG800
DIFLUORODIBROMOMETHANE see DKG850
1,1-DIFLUORO-2,2-DICHLOROETHYLENE see DFA300
DIFLUORODICHLOROMETHANE see DFA600
2',4'-DIFLUORO-4-DIMETHYLAMINOAZOBENZENE see DKG980
2',5'-DIFLUORO-4-DIMETHYLAMINOAZOBENZENE see DKH000
3',4'-DIFLUORO-4-DIMETHYLAMINOAZOBENZENE see DRL000
3',5'-DIFLUORO-4-DIMETHYLAMINOAZOBENZENE see DKH100
DIFLUORODIMETHYLSTANNANE see DKH200
4,4'-DIFLUORO-3,3-DINITRODIPHENYL SULFONE see DKH250
p,p'-DIFLUORO-m,m'-DINITRODIPHENYL SULFONE see DKH250
DIFLUORODIPHENYLTRICHLOROETHANE see FHJ000
DIFLUOROETHANE see ELN500
1,1-DIFLUOROETHANE see ELN500
1,2-DIFLUOROETHANE see DKH300
1,1-DIFLUOROETHENE see VPP000
1,1-DIFLUOROETHYLENE (DOT, MAK) see VPP000
1,1-DIFLUOROETHYLENE POLYMERS (PYROLYSIS) see DKH600
DIFLUORO-N-FLUOROMETHANIMINE see DKH825
DIFLUOROFORMALDEHYDE see CCA500
2,2-DIFLUORO-1,1,1,3,3,3-HEXACHLOROPROPANE see HCH475
2',4'-DIFLUORO-4-HYDROXY-3-BIPHENYLCARBOXYLIC ACID see DKI600
2',4'-DIFLUORO-4-HYDROXY-(1,1'-BIPHENYL)-3-CARBOXYLIC ACID see DKI600
2',4'-DIFLUORO-4-HYDROXY-(1',1-DIPHENYL)-3-CARBOXYLIC ACID see DKI600
6-α,9-DIFLUORO-11-β-HYDROXY-16-α-METHYL-21-VALERYLOXY-1,4-PREGNADIENE-3,20-DIONE see DKF130
6-α,9-α-DIFLUORO-16-α-HYDROXYPREDNISOLONE-16,17-ACETONIDE see SPD500
6-α,2-DIFLUORO-11-β-HYDROXY-21-VALERYLOXY-16-α-METHYL-1,4-PREGNADIENE-3,20-DIONE see DKF130
DIFLUOROMETHANE see MJQ300
DIFLUOROMETHYLENE DIHYPOFLUORITE see DKH830
2-(DIFLUOROMETHYL)ORNITHINE see DBE835
α-DIFLUOROMETHYLORNITHINE see DBE835
dl-α-DIFLUOROMETHYLORNITHINE see DKH875
2-(DIFLUOROMETHYL)-dl-ORNITHINE HYDROCHLORIDE see EAE775
α-DIFLUOROMETHYLORNITHINE HYDROCHLORIDE see EAE775
DIFLUOROMETHYLPHOSPHINE OXIDE see MJD275
DIFLUOROMONOCHLOROMETHANE see CFX500
2,4-DIFLUORONITROBENZENE see DKH900
3,4-DIFLUORO-2-NITROBENZENEDIAZONIUM-6-OXIDE see DKI200

3,6-DIFLUORO-2-NITROBENZENEDIAZONIUM-4-OXIDE see DKI289
2-(1-((((3,5-DIFLUOROPHENYL)AMINO)CARBONYL)HYDRAZONO)ETHYL)-3-PYRIDINECARBOXYLIC ACID, 98.1% see DKI300
DIFLUOROPHENYLARSINE see DKI400
p-((3,5-DIFLUOROPHENYL)AZO)-N,N-DIMETHYLANILINE see DKH100
5-(2,4-DIFLUOROPHENYL)SALICYLIC ACID see DKI600
1,1-DI(4-FLUOROPHENYL)-2-(1,2,4-TRIAZOLE-1-YL)-ETHANOL see BJW600
N-(2,4-DIFLUOROPHENYL)-2-(3-(TRIFLUOROMETHYL)PHENOXY)-3-PYRIDINECARBOXAMIDE see DKJ210
6-α,9-α-DIFLUOROPREDNISOLONE 17-BUTYRATE 21-ACETATE see DKJ300
1,3-DIFLUORO-2-PROPANOL see DKI800
5,7-DIFLUOROQUINOLINE see DKI850
1,1-DIFLUORO-1,2,2,2-TETRACHLOROETHANE see TBP000
1,2-DIFLUORO-1,1,2,2-TETRACHLOROETHANE see TBP050
1,2-DIFLUOROTRICHLOROETHANE see DKI900
1,1-DIFLUORO-1,2,2-TRICHLOROETHANE see TIM000
1,2-DIFLUORO-1,1,2-TRICHLOROETHANE see DKI900
1,2-DIFLUORO-1,2,2-TRICHLOROETHANE see DKI900
3,8-DIFLUOROTRICYCLOQUINAZOLINE see DKJ200
3,3-DIFLUORO-2-(TRIFLUOROMETHYL)ACRYLIC ACID, METHYL ESTER see MNN000
3,3-DIFLUORO-2-(TRIFLUOROMETHYL)-2-PROPENOIC ACID, METHYL ESTER see MNN000
2',4'-DIFLUORO-2-(α-α-α-TRIFLUORO-m-TOLYLOXY)NICOTINANILIDE see DKJ210
6-α,9-DIFLUORO-11-β,17,21-TRIHYDROXYPREGNA-1,4-DIENE-3,20-DIONE-21-ACETATE-17-BUTYRATE see DKJ300
1,1-DIFLUOROUREA see DKJ225
DIFLUPREDNATE see DKJ300
DIFLUPYL see IRF000
DIFLUREX see TGA600
DIFLURON see CJV250
DIFLUROPHATE see IRF000
DIFO see BJE750
DIFOLATAN see CBF800
DIFOLLICULINE see EDP000
DIFONATE see FMU045
DIFORENE see DOZ000
7,12-DIFORMYLBENZ(a)ANTHRACENE see BBD000
1,4-DIFORMYLBENZENE see TAN500
2,2'-DIFORMYLBIPHENYL see DVV800
N,N'-DIFORMYL-p,p'-DIAMINODIPHENYLSULFONE see BJW750
1,2-DIFORMYLHYDRAZIN (GERMAN) see DKJ600
1,2-DIFORMYLHYDRAZINE see DKJ600
2,4'-DIFORMYL-1,1'-(OXYDIMETHYLENE)DIPYRIDINIUM DICHLORIDE, DIOXIME see HGL650
DIFOSAN see CBF800
DIFOSGEN see TIR920
DIFURAN see DKK100
DIFURAN see PAF500
DI-2-FURANYLETHANEDIONE see FPZ000
DIFURAZONE see PAF500
N,N-DIFURFURAL-n-PHENYLENEDIAMINE see DKK600
2,2'-DIFURFURYL ETHER see FPX025
DIFURFURYL ETHER (7CI) see FPX025
DI-2-FUROYL PEROXIDE see DKK400

DI-2-FURYLGLYOXAL see FPZ000
DI-2-FURYLGLYOXAL MONOXIME see FQB000
DI-2-FURYLGLYOXIME see BJW810
DIFURYLMETHANE see FPX028
DI-2-FURYLMETHANE see FPX028
2,2'-DIFURYLMETHANE see FPX028
DI-α-FURYLMETHANE see FPX028
DIGACIN see DKN400
DIGADOLINIUM TRIOXIDE see GAP000
DIGALLIUM TRIOXIDE see GBS050
DIGALLIUM TRISULFATE see GBS100
DIGAMMACAINE see DKK800
DIGENEA SIMPLEX MUCILAGE see AEX250
DIGENIC ACID see KAJ200
DIGENIN see KAJ200
11,11-DIGERANYLOXY-1-UNDECENE see UNA100
DIGERMANE see DKL000
DIGERMIN see DUV600
DIGIBUTINA see BRF500
DIGILANID A see LAT000
DIGILANID B see LAT500
DIGILANID C see LAU000
DIGILANIDE B see LAT500
DIGILANIDES see LAU400
DIGILONG see DKL800
DIGIMED see DKL800
DIGIMED see LAU400
DIGIMERCK see DKL800
DIGISIDIN see DKL800
DIGITALIN see DKL325
DIGITALIN see DKL800
DIGITALINE (FRENCH) see DKL800
DIGITALINE CRISTALLISEE see DKL800
DIGITALINE NATIVELLE see DKL800
DIGITALIN (GLYCOSIDE) see DKL325
DIGITALINUM VERUM see DKL325
DIGITALINUM VERUM see DKL800
DIGITALIS see DKL200
DIGITALIS see FOM100
DIGITALIS GLYCOSIDE see DKN400
DIGITALIS LANATA STANDARD see DKL300
DIGITALIS PURPUREA see FOM100
DIGITALIS PURPUREA, LEAF see DKL200
DIGITALUM VERUM see DKL325
DIGITANNOID see DKL200
DIGITIN see DKL400
DIGITOFLAVONE see TDD550
DIGITONIN see DKL400
DIGITOPHYLLIN see DKL800
DIGITOXIGENIN see DMJ000
DIGITOXIGENIN + 2 DIGITOXOSE + ACETYL-DIGILANIDOBOSE (GERMAN) see LAT000
DIGITOXIGENIN + 2-DIGITOXOSE + 1-ACETYL-(4)-DIGITOSE (GERMAN) see ACH750
DIGITOXIGENIN + 2-DIGITOXOSE + ACETYL-(3)-DIGITOXOSE (GERMAN) see ACH750
DIGITOXIGENINE see DMJ000
DIGITOXIGENIN-TRIDIGITOXOSID (GERMAN) see DKL800
DIGITOXIGENIN TRIDIGITOXOSIDE see DKL800
DIGITOXIN see DKL800
DIGITOXIN, 20,22-DIHYDRO- see DLB650
DIGITOXOSIDE see DKL875
DI-l-GLUTAMIC ACID, MONOCALCIUM SALT, TETRAHYDRATE, l- see MRF550
DIGLYCERIDE ACETIC ACID see DBF600
DIGLYCEROL TETRANITRATE see TDY100
1,4-DIGLYCIDLOXYBUTANE see BOS100
N,N-DIGLYCIDYLANILIN (CZECH) see DKM120
DIGLYCIDYLANILINE see DKM100
N-N-DIGLYCIDYLANILINE see DKM120
DIGLYCIDYL BISPHENOL A ETHER see BLD750

N,N'-DIGLYCIDYL-5,5-DIMETHYLHYDANTOIN see DKM130
DIGLYCIDYLESTER KYSELINY HEXAHYDROFTALOVE (CZECH) see DKM500
DIGLYCIDYL ETHER see DKM200
DIGLYCIDYL ETHER of N,N-BIS(2-HYDROXYETHOXYETHYL)ANILINE see BJN850
DIGLYCIDYL ETHER of 2,2-BIS(4-HYDROXYPHENYL)PROPANE see BLD750
DIGLYCIDYL ETHER of 2,2-BIS(p-HYDROXYPHENYL)PROPANE see BLD750
DIGLYCIDYL ETHER of N,N-BIS(2-HYDROXYPROPYL)-tert-BUTYLAMINE see DKM400
DIGLYCIDYL ETHER of BISPHENOL A see BLD750
DIGLYCIDYL ETHER of BISPHENOL A mixed with BIS(2,3-EPOXYCYCLOPENTYL) ETHER (1:1) see OPI200
DIGLYCIDYL ETHER of 4,4'-ISOPROPYLIDENEDIPHENOL BLD750
DIGLYCIDYL ETHER of NEOPENTYL GLYCOL see NCI300
DIGLYCIDYL ETHER of PHENYLDIETHANOLAMINE see BJN875
DIGLYCIDYLETHYLENE GLYCOL see EEA600
DIGLYCIDYL HEXAHYDROPHTHALATE see DKM500
1,3-DIGLYCIDYLOXYBENZENE see REF000
1,2-DIGLYCIDYLOXYETHANE see EEA600
N-N-DIGLYCIDYLPHENYLAMINE see DKM120
DIGLYCIDYL PHTHALATE see DKM600
DIGLYCIDYL PIPERAZINE see DHE100
DIGLYCIDYL RESORCINOL ETHER see REF000
N,N-DIGLYCIDYL-p-TOLUENESULFONAMIDE see DKM800
N,N-DIGLYCIDYL-p-TOLUENESULPHONAMIDE see DKM800
DIGLYCIDYLTRIETHYLENE GLYCOL see TJQ333
DIGLYCIN see IBH000
DIGLYCINE see IBH000
DIGLYCOALDEHYDE see IDE050
DIGLYCOL see DJD600
DIGLYCOLAMINE see AJU250
DIGLYCOL CHLORHYDRIN see DKN000
DIGLYCOL DIMETHACRYLATE see BKM250
DIGLYCOLDINITRAAT (DUTCH) see DJE400
DIGLYCOL (DINITRATE de) (FRENCH) see DJE400
DIGLYCOLIC ACID, BIS(2-METHYLALLYL) ESTER see BKP400
DIGLYCOLIC ACID, DIALLYL ESTER see DKN100
DIGLYCOLIC ACID DI-(3-CARBOXY-2,4,6-TRIIODOANILIDE) DISODIUM see BGB350
DIGLYCOLIC ACID, DIMETHYL ESTER see DTH500
DIGLYCOLIC ACID (6CI) see ONQ100
DIGLYCOL MONOBUTYL ETHER see DJF200
DIGLYCOL MONOBUTYL ETHER ACETATE see BQP500
DIGLYCOL MONOETHYL ETHER see CBR000
DIGLYCOL MONOETHYL ETHER ACETATE see CBQ750
DIGLYCOL MONOMETHYL ETHER see DJG000
DIGLYCOL MONOSTEARATE see HKJ000
DIGLYCOLSAEURE-DI-(3-CARBOXY-2,4,6-TRIJOD-ANILID) DINATRIUM (GERMAN) see BGB350
DIGLYCOL STEARATE see HKJ000

DIGLYKOKOLL see IBH000
DIGLYKOLDINITRAT (GERMAN) see DJE400
DIGLYME see BKN750
DIGOLD(I) KETENIDE see DKN250
DIGORID A see ACI000
DIGORID B see ACI250
DIGOXIGENIN see DKN300
DIGOXIGENINE see DKN300
DIGOXIGENIN-TRIDIGITOXOSID (GERMAN) see DKN400
DIGOXIGENIN + ZUCKERKETTE WIE BEI ACETYL-DIGITOXIN-α (GERMAN) see ACI250
DIGOXIGENIN + ZUCKERKETTE WIE BEI ACETYL-DIGITOXIN A (GERMAN) see ACI000
DIGOXIN see DKN400
ε-DIGOXIN ACETATE see DKN600
DIGOXINE see DKN400
DIGOXIN PENTAFORMATE see DKN875
DIGUANIDINIUM CARBONATE see GKW100
DI-(8-GUANIDINO-OCTYL)AMINE SULFATE see GLQ000
DIGUANYL see CKB500
DIHDYROPYRONE see BRT000
((DIHDYROXYPROPYL)THIO)METHYLMERCURY see MLG750
DIHEPTYL ETHER see HBO000
DIHEPTYL KETONE see HBO790
DIHEPTYLMERCURY see DKO000
DIHEPTYL PHTHALATE see HBP400
DI-n-HEPTYL PHTHALATE see HBP400
DIHEXANOYL PEROXIDE see DKO400
DIHEXYL see DXT200
DIHEXYL ADIPATE see HEO200
DIHEXYLAMINE see DKO600
DI-N-HEXYLAMINE see DKO600
DI-N-HEXYL AZELATE see ASC000
DIHEXYL trans-BUTENEDIOATE see DKP000
DIHEXYL ETHER see DKO800
DIHEXYL FUMARATE see DKP000
DIHEXYL HEXANEDIOATE see HEO200
DIHEXYL KETONE see TJJ300
DI-n-HEXYL KETONE see TJJ300
DIHEXYL LEAD DIACETATE see DKP200
DIHEXYL MALEATE see DKP400
DIHEXYLOXYETHYL ADIPATE see AEQ000
DI-(2-(2-HEXYLOXY)ETHYL)ESTER KYSELINY ADIPOVE see AEQ000
DI-(2-HEXYLOXYETHYL)ESTER KYSELINY JANTAROVE see SMZ000
DI-2-HEXYLOXYETHYL SUCCINATE see SMZ000
DIHEXYL PHTHALATE see DKP600
DI-n-HEXYL PHTHALATE see DKP600
DIHEXYL SODIOSULFOSUCCINATE see DKP800
DIHEXYL SODIUM SULFOSUCCINATE see DKP800
DIHEXYL SULFOSUCCINATE SODIUM SALT see DKP800
DIHEXYLTIN DICHLORIDE see DFC200
DIHEXYVERIN HYDROCHLORIDE see NCK100
DIHIDRAL see BBV500
DIHIDROBENZPERIDOL see DYF200
DIHIDROCLORURO de BENZIDINA (SPANISH) see BBX750
DIHYCON see DKQ000
DI-HYDAN see DKQ000
DI-HYDAN see DNU000
DIHYDANTOIN see DKQ000
DIHYDANTOIN see DNU000
DIHYDRALAZIN see OJD300
DIHYDRALAZINE see OJD300
DIHYDRALAZINE HYDROCHLORIDE see DKQ200
DIHYDRALAZINE SULFATE see DKQ600

DIHYDRALLAZIN see OJD300
DIHYDRAZINECOBALT(II) CHLORATE see DKQ400
1,4-DIHYDRAZINONAPHTHALAZINE see OJD300
DIHYDRAZINOPHTHALAZINE see OJD300
1,4-DIHYDRAZINOPHTHALAZINE see OJD300
1,4-DIHYDRAZINOPHTHALAZINE HYDROCHLORIDE see DKQ200
1,4-DIHYDRAZINOPHTHALAZINE SULFATE see DKQ600
DIHYDREL see DKQ650
DIHYDRIN see DKW800
DIHYDROABIKOVIROMYCIN see ELM600
1,2-DIHYDRO-5-ACENAPHTHYLENAMINE see AAF250
9,10-DIHYDROACRIDINE see ADI775
DIHYDROAFLATOXIN B1 see AEU750
DIHYDROAFLATOXIN G1 see AEV500
DIHYDROAMBRETTOLIDE see OKU000
9,10-DIHYDRO-1-AMINO-4-BROMO-9,10-DIOXO-2-ANTHRACENE SULFONIC ACID SODIUM SALT see DKR000
2,3-DIHYDRO-6-AMINO-2-(2-CHLOROETHYL)-4H-1,3-BENZOXAZIN-4-ONE see DKR200
4,5-DIHYDRO-2-AMINO-1-((6-CHLORO-3-PYRIDINYL)METHYL)-1H-IMIDAZOLE-4,5-DIOL see DKR300
9,10-DIHYDRO-1-AMINO-4-(CYCLOHEXYLAMINO)-9,10-DIOXO-2-ANTHRACENESULFONIC ACID SODIUM SALT see CMM080
2,4-DIHYDRO-5-AMINO-2-(1-(3,4-DICHLOROPHENYL)ETHYL)-3H-PYRAZOL-3-ONE see EAE675
1,9-DIHYDRO-2-AMINO-9-(3,4-DIHYDROXYBUTYL)-6H-PURIN-6-ONE see DMI700
2,3-DIHYDRO-6-AMINO-1,3-DIOXO-2-(p-TOLYL) 1H-BENZ(de)ISOQUINOLINE-5-SULFONIC ACID, MONOSODIUM SALT see CMM750
1,9-DIHYDRO-2-AMINO-9-((2-HYDROXYETHOXY)METHYL)-6H-PURIN-6-ONE SODIUM SALT see AEC725
9,10-DIHYDRO-1-AMINO-4-(3-(2-HYDROXYETHYL)AMINOSULFONYL-4-METHYLPHENYLAMINO)-9,10-DIOXO-2-ANTHRACENE SULFONIC ACID SODIUM SALT see DKR400
1,9-DIHYDRO-2-AMINO-9-(4-HYDROXY-3-(HYDROXYMETHYL)BUTYL)-6H-PURIN-6-ONE MONOSODIUM SALT see POK100
2,4-DIHYDRO-5-AMINO-2-PHENYL-3H-PYRAZOL-3-ONE see DKR600
DIHYDROANETHOLE see PNE250
N,N-DIHYDRO-1,1,1',2'-ANTHRAQUINONE-AZINE see IBV050
DIHYDROAVERMECTIN B1A see DKR700
22,23-DIHYDROAVERMECTIN B1 see ITD875
22,23-DIHYDROAVERMECTIN B1A see DKR700
DIHYDROAZIRENE see EJM900
DIHYDRO-1H-AZIRINE see EJM900
DIHYDROBAIKIANE see HDS300
1,2-DIHYDROBENZ(1)ACEANTHRYLENE see BAW000
1,2-DIHYDROBENZ(e)ACEANTHRYLENE see AAE000
1,2-DIHYDRO-BENZ(j)ACEANTHRYLENE see CMC000
4,5-DIHYDROBENZ(k)ACEPHENANTHRYLENE see AAF000
5,6-DIHYDROBENZ(c)ACRIDINE-7-CARBOXYLIC ACID see TEF775
1,2-DIHYDROBENZ(a)ANTHRACENE see DKS400

3,4-DIHYDROBENZ(a)ANTHRACENE see DKS600

1a,11b-DIHYDROBENZ(3,4)ANTHRA(1,2-b)OXIRENE see EBP000

5,6-DIHYDROBENZENE(e)ACEANTHRYLENE see AAE000

7,11b-DIHYDROBENZ(b)INDENO(1,2-d)PYRAN-3,6a,9,10(6H)-TETROL see LFT800

1,2-DIHYDROBENZO(a)ANTHRACENE see DKS400

3,4-DIHYDROBENZO(a)ANTHRACENE see DKS600

6,13-DIHYDROBENZO(e)(1)BENZOTHIOPYRANO(4,3-b)INDOLE see DKS800

5,6-DIHYDRO-7H-BENZO(c)CARBAZOLE-8-CARBOXYLIC ACID-2-(DIETHYLAMINO)ETHYL ESTER HYDROCHLORIDE see DKS909

5,6-DIHYDRO-7H-BENZO(c)CARBAZOLE-9-CARBOXYLIC ACID-2-(DIETHYLAMINO)ETHYL ESTER HYDROCHLORIDE see DKT000

5,6-DIHYDRO-7H-BENZO(c)CARBAZOLE-10-CARBOXYLIC ACID-2-(DIETHYLAMINO)ETHYL ESTER HYDROCHLORIDE see DKT200

2,3-DIHYDRO-1H-BENZO(h,i)CHRYSENE see DKT400

1A,13c-DIHYDROBENZO(11,12)CHRYSENO(5,6-b)OXIRENE see BCH100

6-β,7-α-DIHYDROBENZO(10,11)CHRYSENO(1,2-b)OXIRENE see BCV750

2,3-DIHYDRO-1H-BENZO(a)CYCLOPENT(b)ANTHRACENE see CPY750

(2,3-DIHYDRO-1,4-BENZODIOXIN-6-YL)(4-FLUOROPHENYL)METHANONE see DKT500

N-((2,3-DIHYDRO-1,4-BENZODIOXIN-2-YL)METHYL)-N-METHYL-N',N'-DI-2-PROPENYL-UREA (9CI) see DBJ100

4,5-DIHYDROBENZO(j)FLUORANTHENE-4,5-DIOL see DKT600

7,8-DIHYDROBENZO(a)PYRENE see DKU000

9,10-DIHYDROBENZO(e)PYRENE see DKU400

(E)-7,8-DIHYDROBENZO(a)PYRENE-7,8-DIOL see DLC800

9,10-DIHYDROBENZO(a)PYRENE-9,10-DIOL see DLC000

(Z)-9,10-DIHYDROBENZO(a)PYRENE-9,10-DIOL see BMK550

trans-4,5-DIHYDROBENZO(a)PYRENE-4,5-DIOL see DLB850

3b,4a-DIHYDRO-4H-BENZO(1,2)PYRENO(4,5-b)AZIRINE see BCV125

3,4-DIHYDRO-2H-1,4-BENZOTHIAZINE HYDROCHLORIDE see DKU875

S-(3,4-DIHYDRO-2H-1-BENZOTHIOPYRAN-4-YL) o,o-DIMETHYL PHOSPHORODITHIOATE see TFE300

(+−)-1-((3,4-DIHYDRO-2H-1-BENZOTHIOPYRAN-8-YL)OXY)-3-((1,1-DIMETHYLETHYL)AMINO)-2-PROPANOL see TBF400

2,3-DIHYDRO-1,3-BENZOXAZINE-4H-2-THIONE-4-ONE see TFC570

7,8-DIHYDRO-N-BENZYLADENINE see DKV125

1,8-DIHYDRO-8-(4-BROMOPHENYL)-4H-PYRAZOLO-(3,4-D)(1,2,4)TRIAZOLO(1,5-A)PYRIMIDIN-4-ONE see DKV135

1,3-DIHYDRO-7-BROMO-5-(2-PYRIDYL)-2H-1,4-BENZODIAZEPIN-2-ONE see BMN750

DIHYDROBUTADIENE SULPHONE see SNW500

15,16-DIHYDRO-11-N-BUTOXYCYCLOPENTA(A)PHENANTHREN-17-ONE see DKV140

(−)-3,4-DIHYDRO-5-(3-(tert-BUTYLAMINO)-2-HYDROXYPROPOXY)-1(2H)-NAPHTHALENONE HYDROCHLORIDE see LFA100

2,3-DIHYDRO-5-CARBOXANILIDO-6-METHYL-1,4-OXATHIIN see CCC500

2,3-DIHYDRO-5-CARBOXANILIDO-6-METHYL-1,4-OXATHIIN-4,4-DIOXIDE see DLV200

DIHYDROCARVEOL see DKV150

1,6-DIHYDROCARVEOL see DKV150

DIHYDROCARVEOL ACETATE see DKV160

DIHYDROCARVEYL ACETATE see DKV160

d-DIHYDROCARVONE see DKV175

DIHYDROCARVYL ACETATE see DKV160

5,10-DIHYDRO-7-CHLOR-10-(2-(DIMETHYLAMINO)ETHYL)-11H-DIBENZO(b,e)(1,4)DIAZEPIN-11-ONE HCl see CMW250

DIHYDROCHLORIDE-1-NITRO-9-((2-DIMETHYLAMINO)-1-METHYLETHYLAMINO)-ACRIDINE see NFW435

DIHYDROCHLORIDE SALT OF DIETHYLENEDIAMINE see PIK000

2,3-DIHYDRO-6-CHLORO-2-(2-CHLOROETHYL)-4H-1,3-BENZOXAZIN-4-ONE see DKV200

(±)-2,3-DIHYDRO-6-CHLORO-5-CYCLOHEXYL-1H-INDENE-1-CARBOXYLIC ACID see CFH825

1,4-DIHYDRO-8-CHLORO-1-CYCLOPROPYL-6-FLUORO-7-((4AS,7AS)-OCTAHYDRO-6H-PYRROLO(3,4-B)PYRIDIN-6-YL)-4-OXO-3-QUINOLINECARBOXYLIC ACID, MONOHYDROCHLORIDE see DKV250

1'-(3-(10,11-DIHYDRO-3-CHLORO-5H-DIBENZ(b,f)AZEPIN-5-YL)PROPYL)-(1,4'-BIPIPERIDINE)-4'-CARBOXAMIDE DIHYDROCHLORIDE see DKV309

1,3-DIHYDRO-7-CHLORO-5-(o-FLUOROPHENYL)-3-HYDROXY-1-(2-HYDROXYETHYL)-2H-1,4-BENZODIAZEPIN-2-ONE see DKV400

1,3-DIHYDRO-7-CHLORO-3-HYDROXY-1-METHYL-5-PHENYL-2H-1,4-BENZODIAZEPIN-2-ONE see CFY750

2,3-DIHYDRO-7-CHLORO-1H-INDEN-4-OL (9CI) see CDV700

3,3a-DIHYDRO-7-CHLORO-2-METHYL-2H,9H-ISOXAZOLO(3,2-b)(1,3)BENZOXAZIN-9-ONE see CIL500

2,3-DIHYDRO-7-CHLORO-1-METHYL-5-PHENYL-1H-1,4-BENZODIAZEPINE see CGA000

4,5-DIHYDRO-N-(2-CHLORO-4-METHYLPHENYL)-1H-IMIDAZOL-2-AMINE MONONITRATE see TGJ885

2,3-DIHYDRO-7-CHLORO-2-OXO-5-PHENYL-1H-1,4-BENZODIAZEPINE-3-CARBOXYLIC ACID MONOPOTASSIUM SALT see DKV700

4,5-DIHYDRO-3-(4-CHLOROPHENYL)-4-METHYL-(((4-(TRIFLUOROMETHYL)PHENYL)AMINO)CARBONYL)-1H-PYRAZOLE-4-CARBOXYLIC ACID METHYL ESTER see DKV710

4,5-DIHYDRO-5-((p-CHLOROPHENYL)THIOMETHYL)OXAZOLAMINE see AJI500

4,5-DIHYDRO-((6-CHLORO-3-PYRIDINYL)METHYL)-1H-IMIDAZOL-2-AMINE, see DKV720

4,5-DIHYDRO-1-((6-CHLORO-3-PYRIDINYL)METHYL)-2-(NITROAMINO)-1H-IMIDAZOLE-4,5-DIOL see DKV730

3,4-DIHYDRO-6-CHLORO-7-SULFAMYL-1,2,4-BENZOTHIADIAZINE-1,1-DIOXIDE see CFY000

DIHYDROCHLOROTHIAZID see CFY000

DIHYDROCHLOROTHIAZIDE see CFY000

3,4-DIHYDROCHLOROTHIAZIDE see CFY000

6,12,b-DIHYDROCHOLANTHRENE see DKV800

meso-DIHYDROCHOLANTHRENE see DKV800

DIHYDROCHOLESTEROL see DKW000

DIHYDROCHRYSENE see DKW200

1,2-DIHYDROCHRYSENE see DKW200

3,4-DIHYDROCHRYSENE see DKW400

(E)-1,2-DIHYDRO-1,2-CHRYSENEDIOL see DLD200

trans-1,2-DIHYDROCHRYSENE-1,2-DIOL see DLD200

DIHYDROCINNAMALDEHYDE see HHP000

DIHYDROCITRONELLAL see TCY300

6,7-DIHYDROCITRONELLAL see TCY300

DIHYDROCITRONELLOL see DTE600

DIHYDROCITRONELLYL ACETATE see DTE800

DIHYDROCODEINE see DKW800

7,8-DIHYDROCODEINE see DKW800

DIHYDROCODEINE ACID TARTRATE see DKX000

DIHYDROCODEINE BITARTRATE see DKX000

DIHYDROCODEINE TARTRATE see DKX000

DIHYDROCODEINE TARTRATE (1:1) see DKX000

DIHYDROCODEINONE see OOI000

DIHYDROCODEINONE BITARTRATE see DKX050

DIHYDROCORTISOL see PMA500

5-β-DIHYDROCORTISOL see PMA500

DIHYDROCOUMARIN see HHR500

3,4-DIHYDROCOUMARIN see HHR500

DIHYDROCUMINYL ALCOHOL see PCI550

DIHYDROCUMINYL ALDEHYDE see DKX100

10,11-DIHYDRO-5-CYANO-N,N-DIMETHYL-5H-DIBENZO(a,d)CYCLOHEPTENE-5-PROPYLAMINE HCl see COL250

6,7-DIHYDRO-3-CYCLOHEXYL-1H-CYCLOPENTAPYRIMIDINE-2,4(3H,5H)-DIONE see CPR600

3,4-DIHYDRO-6-(4-(1-CYCLOHEXYL-1H-TETRAZOL-5-YL)BUTOXY)-2(1H)-QUINOLINONE see CMP825

2,3-DIHYDRO-1H-CYCLOPENTA(c)PHENANTHRENE see DKX150

DIHYDRODAUNOMYCIN see DAC300

13-DIHYDRODAUNOMYCIN see DAC300

13-DIHYDRODAUNORUBICIN see DAC300

22,23-DIHYDRO-5-o-DEMETHYLAVERMECTIN A1A see DKR700

22,23-DIHYDRO-5-o-DEMETHYL-26-((2-METHOXYETHOXY)METHOXY)AVERMECTIN A1A see DKX200

12-β,13-α-DIHYDRO-11-DEOXO-11-β-HYDROXYJERVINE see DAQ002

DIHYDRODEOXYMORPHINE see DKX600

1,2-DIHYDRO-2'-DEOXY-2-OXOADENOSINE see HKA750

7,8-DIHYDRO-2'-DEOXY-8-OXOGUANOSINE see HKA770

DIHYDRODESOXYMORPHINE-D see DKX600

9,10-DIHYDRO-4,5-DIAMINO-1-HYDROXY-2,7-ANTHRACENE DISULFONIC ACID DISODIUM SALT see DKX800

9,10-DIHYDRO-8a,10,-DIAZONIAPHENANTHRENE DIBROMIDE see DWX800

9,10-DIHYDRO-8A,10,-
DIAZONIAPHENANTHRENE DIBROMIDE
see EJC025
9,10-DIHYDRO-8a,10a-
DIAZONIAPHENANTHRENE(1,1'-
ETHYLENE-2,2'-
BIPYRIDYLIUM)DIBROMIDE see DWX800
9,10-DIHYDRO-8A,10A-
DIAZONIAPHENANTHRENE(1,1'-
ETHYLENE-2,2'-
BIPYRIDYLIUM)DIBROMIDE see EJC025
trans-1,2-
DIHYDRODIBENZ(a,e)ACEANTHRYLENE-
1,2-DIOL see DKX875
trans-10,11-
DIHYDRODIBENZ(a,e)ACEANTHRYLENE-
10,11-DIOL see DKX900
5,6-DIHYDRODIBENZ(a,h)ANTHRACENE
see DKY000
5,6-DIHYDRODIBENZ(a,j)ANTHRACENE
see DKY200
7,14-DIHYDRODIBENZ(a,h)ANTHRACENE
see DKY400
9,10-DIHYDRO-1,2,5,6-
DIBENZANTHRACENE see DKY400
3,4-DIHYDRODIBENZ(a,h)ANTHRACENE-
3,4-DIOL trans-(+−)- see DKX930
10,11-DIHYDRO-5-DIBENZ(b,f)AZEPINE
see DKY800
3,4-DIHYDRO-1,2,5,6-DIBENZCARBAZOLE
see DLA000
12,13-DIHYDRO-7H-
DIBENZO(a,g)CARBAZOLE see DLA000
5,8-DIHYDRODIBENZO(a,def)CHRYSENE
see DLA100
7,14-DIHYDRODIBENZO(b,def)CHRYSENE
see DLA120
10,11-DIHYDRO-5H-
DIBENZO(a,d)CYCLOHEPTENE-5-
CARBOXAMIDE see CQH625
10,11-DIHYDRO-5H-
DIBENZO(a,d)CYCLOHEPTEN-5-ONE see
DDA100
10,11-DIHYDRO-5H-
DIBENZO(a,d)CYCLOHEPTEN-5-ONE
OXIME see DDD000
3,10-DIHYDRO-5H-
DIBENZO(a,d)CYCLOHEPTEN-5-
YLIDENE-N,N-DIMETHYL-1-
PROPANAMINE see EAH500
3-(10,11-DIHYDRO-5H-
DIBENZO(a,d)CYCLOHEPTEN-5-
YLIDENE)-1-ETHYL-2-
METHYLPYRROLIDINE see PJA190
3-(10,11-DIHYDRO-5H-
DIBENZO(a,d)CYCLOHEPTEN-5-
YLIDENE)-1-ETHYL-2-METHYL-
PYRROLIDINE HYDROCHLORIDE see
TMK150
4-((10,11-DIHYDRO-5H-
DIBENZO(a,d)CYCLOHEPTEN-5-YL)OXY)-
1-METHYLPIPERIDINE HYDROGEN
MALEATE see HBT000
4-((10,11-DIHYDRO-5H-
DIBENZO(a,d)CYCLOHEPTEN-5-YL)OXY)-
1-METHYLPIPERIDINE, MALEATE (1:1) see
HBT000
3-α-((10,11-DIHYDRO-5H-
DIBENZO(a,d)CYCLOHEPTEN-5-YL)OXY)-
8-METHYLTROPANIUM BROMIDE see
DAZ140
3-α-((10,11-DIHYDRO-5H-
DIBENZO(A,D)CYCLOHEPTEN-5-
YL)OXY)TROPAN DIHYDROGEN
CITRATE see EAG100
3-α-(10,11-DIHYDRO-5H-
DIBENZO(A,D)CYCLOHEPTEN-5-
YLOXY)TROPANE CITRATE see EAG100
5,10-DIHYDRO-3,4:8,9-DIBENZOPYRENE
see DLA120
5,8-DIHYDRO-3,4:9,10-DIBENZOPYRENE
see DLA100

1,2-DIHYDRO-5,7-DIBROMO-2-(5,7-
DIBROMO-1,3-DIHYDRO-3-OXO-2H-
INDOL-2-YLIDENE)-3H-INDOL-3-ONE see
ICU135
2,3-
DIHYDRODICYCLOPENTA(c,lmn)PHENAN
THREN-1(9H)-ONE see DGW300
1,15-DIHYDRO-3,6:10,13-DIEPOXY-8,1,15-
BENZOTHIADIAZACYCLOHEPTADECINE
-2,14(7H,9H)-DIONE see DKY900
2,3-DIHYDRO-1-DIETHYLAMINOETHYL-
2-METHYL-3-PHENYL-4(3H)-
QUINAZOLINONE OXALATE see DID800
2,3-DIHYDRO-1-DIETHYLAMINOETHYL-
2-METHYL-3-(o-TOLYL)-4(3H)-
QUINAZOLINONE OXALATE see DIE200
DIHYDRO-5,5-DIETHYL-2H-1,3-OXAZINE-
2,4(3H)-DIONE see DJT400
DIHYDRODIETHYLSTILBESTROL see
DLB400
1,4-DIHYDRO-6,8-DIFLUORO-1-ETHYL-4-
OXO-7-(4-PYRIDINYL)-3-
QUINOLINECARBOXYLIC ACID see
DLB500
1,4-DIHYDRO-4-(2-
(DIFLUOROMETHOXY)PHENYL)-2,6-
DIMETHYL-3,5-PYRIDINEDICARBOXYLIC
ACID DIMETHYL ESTER see RSZ375
N-((3S)-2,3-DIHYDRO-6-(2,6-
DIFLUOROPHENYL)METHOXY)-3-
BENZOFURANYL)-N-HYDROUREA, XY-
see DLB600
DIHYDRODIGITOXIN see DLB650
20,22-DIHYDRODIGITOXIN see DLB650
cis-5,6-DIHYDRO-5,6-
DIHYDROXYBENZ(a)ANTHRACENE see
BBE000
(−)(3R,4R)-trans-3,4-DIHYDRO-3,4-
DIHYDROXYBENZ(a)ANTHRACENE see
BBG000
(+)-(3S,4S)-trans-3,4-DIHYDRO-3,4-
DIHYDROXYBENZ(a)ANTHRACENE see
BBD750
trans-3,4-DIHYDRO-3,4-
DIHYDROXYBENZO(a)ANTHRACENE see
BBD500
(−)(3R,4R)trans-3,4-DIHYDRO-3,4-
DIHYDROXYBENZO(a)ANTHRACENE see
BBG000
(+)-(3S,4S)-trans-3,4-DIHYDRO-3,4-
DIHYDROXYBENZO(a)ANTHRACENE see
BBD750
4,5-DIHYDRO-4,5-
DIHYDROXYBENZO(J)FLUORANTHENE
see DKT600
4,5-DIHYDRO-4,5-
DIHYDROXYBENZO(a)PYRENE see
DLB800
4,5-DIHYDRO-4,5-
DIHYDROXYBENZO(e)PYRENE see
DMK400
9,10-DIHYDRO-9,10-
DIHYDROXYBENZO(a)PYRENE see
DLC000
trans-4,5-DIHYDRO-4,5-
DIHYDROXYBENZO(a)PYRENE see
DLB850
trans-7,8-DIHYDRO-7,8-
DIHYDROXYBENZO(a)PYRENE see
DLC800
trans-9,10-DIHYDRO-9,10-
DIHYDROXYBENZO(a)PYRENE see
DLC200
(±)-trans-9,10-DIHYDRO-9,10-
DIHYDROXYBENZO(a)PYRENE see
DLC400
7,8-DIHYDRO-7,8-
DIHYDROXYBENZO(a)PYRENE-9,10-
OXIDE see DLD000
trans-1,2-DIHYDRO-1,2-
DIHYDROXYCHRYSENE see DLD200

(E)-1,2-DIHYDRO-1,2-
DIHYDROXYDIBENZ(a,h)ANTHRACENE
see DMK200
trans-3,4-DIHYDRO-3,4-
DIHYDROXYDIBENZ(a,h)ANTHRACENE
see DLD400
trans-3,4-DIHYDRO-3,4-
DIHYDROXYDIBENZO(a,h)ANTHRACENE
see DLD400
trans-3,4-DIHYDRO-3,4-
DIHYDROXYDIBENZO(a,e)FLUORANTHE
NE see BBU810
trans-3,4-DIHYDRO-3,4-
DIHYDROXYDIBENZO(a,e)FLUORANTHE
NE see DKX900
trans-12,13-DIHYDRO-12,13-
DIHYDROXYDIBENZO(a,e)FLUORANTHE
NE see BBU825
trans-12,13-DIHYDRO-12,13-
DIHYDROXYDIBENZO(a,e)FLUORANTHE
NE see DKX875
(E)-8,9-DIHYDRO-8,9-DIHYDROXY-7,12-
DIMETHYLBENZ(a)ANTHRACENE see
DLD800
trans-3,4-DIHYDRO-3,4-DIHYDROXY-7,12-
DIMETHYLBENZ(a)ANTHRACENE see
DLD800
trans-8,9-DIHYDRO-8,9-DIHYDROXY-7,12-
DIMETHYLBENZ(a)ANTHRACENE see
DLD800
trans-10,11-DIHYDRO-10,11-DIHYDROXY-
7,12-DIMETHYLBENZ(a)ANTHRACENE see
DLD875
9,10-DIHYDRO-4,5-DIHYDROXY-9,10-
DIOXO-2-ANTHRACENECARBOXYLIC
ACID see RHZ700
9,10-DIHYDRO-3,4-DIHYDROXY-9,10-
DIOXO-2-ANTHRACENESULFONIC ACID
MONOSODIUM SALT see SEH475
9,10-DIHYDRO-9,10-DIHYDROXY-9,10-DI-
n-PROPYL-1,2:5,6-DIBENZANTHRACENE
see DLK600
trans-3,4-DIHYDRO-3,4-DIHYDROXY DMBA
see DLD600
trans-8,9-DIHYDRO-8,9-DIHYDROXY DMBA
see DLD800
(±)-(1R,2S,3R,4R)-3,4-DIHYDRO-3,4-
DIHYDROXY-1,2-
EPOXYBENZ(a)ANTHRACENE see DLE000
(±)-(1S,2R,3R,4R)-3,4-DIHYDRO-3,4-
DIHYDROXY-1,2-
EPOXYBENZ(a)ANTHRACENE see DLE200
(S)-2,3-DIHYDRO-5,7-DIHYDROXY-2-(4-
HYDROXYPHENYL)-4H-1-BENZOPYRAN-
4-ONE see NBP350
trans-1,2-DIHYDRO-1,2-
DIHYDROXYINDENO(1,2,3-cd)PYRENE see
DLE400
cis-5,6-DIHYDRO-5,6-DIHYDROXY-12-
METHYLBENZ(a)ACRIDINE see DLE500
trans-3,4-DIHYDRO-3,4-DIHYDROXY-7-
METHYLBENZ(c)ACRIDINE see MGU550
(E)-1,2-DIHYDRO-1,2-DIHYDROXY-7-
METHYLBENZ(a)ANTHRACENE see
DLE600
(E)-5,6-DIHYDRO-5,6-DIHYDROXY-7-
METHYL-BENZ(a)ANTHRACENE see
DLF000
trans-1,2-DIHYDRO-1,2-DIHYDROXY-7-
METHYLBENZ(a)ANTHRACENE see
DLE600
trans-3,4-DIHYDRO-3,4-DIHYDROXY-7-
METHYLBENZ(a)ANTHRACENE see
MBV500
trans-5,6-DIHYDRO-5,6-DIHYDROXY-7-
METHYLBENZ(a)ANTHRACENE see
DLF000
trans-8,9-DIHYDRO-8,9-DIHYDROXY-7-
METHYLBENZ(a)ANTHRACENE see
DLF200

(+)-3,4-DIHYDRO-3,8-DIHYDROXY-3-METHYLBENZ(a)ANTHRACENE-1,7,12(2H)-TRIONE see TDX840

trans-7,8-DIHYDRO-7,8-DIHYDROXY-3-METHYLCHOLANTHRENE see MJD600

trans-9,10-DIHYDRO-9,10-DIHYDROXY-3-METHYLCHOLANTHRENE see MJD610

cis-2-α,3-DIHYDRODIHYDROXY-3-METHYLCHOLANTHRENE see DLF300

trans-11,12-DIHYDRO-11,12-DIHYDROXY-3-METHYLCHOLANTHRENE see DML800

1,2-DIHYDRO-1,2-DIHYDROXY-5-METHYLCHRYSENE see DLF400

7,8-DIHYDRO-7,8-DIHYDROXY-5-METHYLCHRYSENE see DLF600

(+−)-trans-1,2-DIHYDRO-1,2-DIHYDROXY-5-METHYLCHRYSENE see CML812

(+−)-trans-7,8-DIHYDRO-7,8-DIHYDROXY-5-METHYLCHRYSENE see CML814

(+−)-trans-7,8-DIHYDRO-7,8-DIHYDROXY-5-METHYLCHRYSENE see DLT640

14,19-DIHYDRO-12,13-DIHYDROXY(13-α,14-α)-20-NORCROTALANAN-11,15-DIONE see MRH000

(E)-1,2-DIHYDRO-1,2-DIHYDROXYTRIPHENYLENE see DMM200

1,4-DIHYDRO-1,4-DIKETONAPHTHALENE see NBA500

5,6-DIHYDRO-9,10-DIMETHOXYBENZO(g)-1,3-BENZODIOXOLO(5,6-a)QUINOLIZINIUM CHLORIDE DIHYDRATE see BFN550

5,6-DIHYDRO-9,10-DIMETHOXYBENZO(g)-1,3-BENZODIOXOLO(5,6-a)QUINOLIZINIUM SULFATE TRIHYDRATE see BFN750

5,6-DIHYDRO-9,10-DIMETHOXY-BENZO(g)-1,3-BENZODIOXOLO(5,6-a)QUINOLIZINIUM SULFATE (2:1) see BFN625

3,4-DIHYDRO-6-(4-(3,4-DIMETHOXYBENZOYL)-1-PIPERAZINYL)-2(1H)-QUINOLINONE see DLF700

3,4-DIHYDRO-6,7-DIMETHOXY-N-(2-(2-METHOXYPHENOXY)ETHYL)-2(1H)-ISOQUINOLINECARBOXIMIDAMIDE see SBA875

5,10-DIHYDRO-10-(2-(DIMETHYLAMINO)ETHYL)-8-ETHYLSULFONYL-5-METHYL-11H-DIBENZO(b,e)(1,4)DIAZEPIN-11-ONE see DLG000

5,10-DIHYDRO-10-(2-(DIMETHYLAMINO)ETHYL)-5-METHYL-11H-DIBENZO(b,e)(1,4)DIAZEPIN-11-ONE see DCW600

10,11-DIHYDRO-5-(3-DIMETHYLAMINO-2-METHYLPROPYL)-5H-DIBENZ(b,f)AZEPINE see DLH200

10,11-DIHYDRO-5-(3-(DIMETHYLAMINO)-2-METHYLPROPYL)-5H-DIBENZ(b,f)AZEPINE MALEATE (1:1) see SOX550

1,2-DIHYDRO-3-DIMETHYLAMINO-7-METHYL-1,2-(PROPYLMALONYL)-1,2,4-BENZOTRIAZINE see AQN750

1,2-DIHYDRO-3-DIMETHYLAMINO-7-METHYL-1,2-(PROPYLMALONYL)-1,2,4-BENZOTRIAZINE DIHYDRATE see ASA000

1,2-DIHYDRO-3-DIMETHYLAMINO-7-METHYL-1,2-(PROPYLMALONYL)-1,2,4-BENZOTRIAZINE SODIUM SALT see ASA250

3,7-DIHYDRO-8-(3-(DIMETHYLAMINO)PROPOXY)-1,3,7-TRIMETHYL-1H-PURINE-2,6-DIONE HYDROCHLORIDE see POF525

5,6-DIHYDRO-N-(3-(DIMETHYLAMINO)PROPYL)-11H-DIBENZ(b,e)AZEPINE see DLH600

10,11-DIHYDRO-5-(3-(DIMETHYLAMINO)PROPYL)-5H-DIBENZ(b,f)AZEPINE see DLH600

10,11-DIHYDRO-5-(3-(DIMETHYLAMINO)PROPYL)-5H-DIBENZ(b,f)AZEPINE HYDROCHLORIDE see DLH630

10,11-DIHYDRO-5-(γ-DIMETHYLAMINOPROPYLIDENE)-5H-DIBENZO(a,d)CYCLOHEPTENE see EAH500

5,6-DIHYDRO-7,12-DIMETHYLBENZ(a)ANTHRACENE see DLH800

2,3-DIHYDRO-2,2-DIMETHYL-7-BENZOFURANYL (BUTOXYSULFINYL)METHYLCARBAMATE see DLH820

2,3-DIHYDRO-2,2-DIMETHYL-7-BENZOFURANYL (BUTYLTHIO)METHYLCARBAMATE see DLH830

2,3-DIHYDRO-2,2-DIMETHYL-7-BENZOFURANYL (DI-N-BUTYLAMINOSULFENYL)METHYLCARBAMATE see CCC280

2,3-DIHYDRO-2,2-DIMETHYL-7-BENZOFURANYL((DIBUTYLAMINO)THIO) METHYL CARBAMATE see CCC280

2,3-DIHYDRO-2,2-DIMETHYL-7-BENZOFURANYL (DODECYLTHIO)METHYLCARBAMATE see DLH840

2,3-DIHYDRO-2,2-DIMETHYL-7-BENZOFURANYL((4-FLUOROPHENYL)THIO)METHYLCARBAMATE see DLH850

2,3-DIHYDRO-2,2-DIMETHYLBENZOFURANYL-7-(METHYL)(T-BUTOXYFULFINYL)CARBAMATE see DLH860

2,3-DIHYDRO-2,2-DIMETHYLBENZOFURANYL-7-(METHYL)(N-BUTOXYSULFINYL)CARBAMATE see DLH820

2,3-DIHYDRO-2,2-DIMETHYL-7-BENZOFURANYL METHYLCARBAMATE see CBS275

2,3-DIHYDRO-2,2-DIMETHYLBENZOFURANYL-7-N-METHYLCARBAMATE see CBS275

2,3-DIHYDRO-2,2-DIMETHYLBENZOFURANYL-7-(METHYL)(2,6-DIMETHYLPHENOXYSULFINYL)CARBAMATE see XWS100

2,3-DIHYDRO-2,2-DIMETHYLBENZOFURANYL 7-(METHYL)(HEXOXYSULFINYL)CARBAMATE see HFU550

2,3-DIHYDRO-2,2-DIMETHYL-7-BENZOFURANYLMETHYL((4-NITROPHENYL)THIO)CARBAMATE see DLH870

2,3-DIHYDRO-2,2-DIMETHYL-7-BENZOFURANYL METHYL(PHENOXYSULFINYL)CARBAMATE see DLH880

2,3-DIHYDRO-2,2-DIMETHYL-7-BENZOFURANYLMETHYL(1,1,2,2-TETRACHLOROETHYL)THIOCARBAMATE see DLH890

2,3-DIHYDRO-2,2-DIMETHYL-7-BENZOFURANYLMETHYL((TRICHLOROMETHYL)THIO)CARBAMATE see DLH900

2,3-DIHYDRO-2,2-DIMETHYL-7-BENZOFURANYL NITROSO-METHYL CARBAMATE see NJQ500

3,4-DIHYDRO-1,11-DIMETHYLCHRYSENE see DLI000

16,17-DIHYDRO-11,17-DIMETHYLCYCLOPENTA(a)PHENANTHRENE see DLI200

16,17-DIHYDRO-11,12-DIMETHYL-15H-CYCLOPENTA(a)PHENANTHRENE see DRH800

15,16-DIHYDRO-11,12-DIMETHYLCYCLOPENTA(a)PHENANTHREN-17-ONE see DLI400

15,16-DIHYDRO-7,11-DIMETHYL-17H-CYCLOPENTA(a)PHENANTHREN-17-ONE see DLI300

10,11-DIHYDRO-N,N-DIMETHYL-5H-DIBENZ(b,f)AZEPINE-5-PROPANAMINE MONOHYDROCHLORIDE see DLH630

10,11-DIHYDRO-N,N-DIMETHYL-5H-DIBENZO(a,d)-CYCLOHEPTENE-Δ^{5,γ}-PROPYLAMINE HCL see EAI000

10,11-DIHYDRO-N,N-DIMETHYL-5H-DIBENZO(a,d)HEPTALENE-Δ^5-γ-PROPYLAMINE see EAH500

2,3-DIHYDRO-5,6-DIMETHYL-1,4-DITHIIN 1,1,4,4-TETROXIDE see OMY825

5,6-DIHYDRO-7,12-DIMETHYL-5,6-EPOXYBENZ(a)ANTHRACENE see DQL600

2,3-DIHYDRO-2,5-DIMETHYL-2-FORMYL-1,4-PYRAN see DLI600

3,7-DIHYDRO-1,3-DIMETHYL-7-(2-HYDROXYPROPYL)-1H-PURINE-2,6-DIONE see HOA000

10,11-DIHYDRO-N,N-DIMETHYL-5,10-METHANO-5H-DIBENZO(a,d)CYCLOHEPTENE-12-METHANAMINE HCl see DPK000

1,2-DIHYDRO-1,5-DIMETHYL-4-((1-METHYLETHYL)AMINO)-2-PHENYL-3H-PYRAZOL-3-ONE see INM000

1,2-DIHYDRO-1,5-DIMETHYL-4-((1-METHYLETHYL)AMINO)-2-PHENYL-3H-PYRAZOL-3-ONE see INY000

3,7-DIHYDRO-1,3-DIMETHYL-7-((5-METHYL-1,2,4-OXADIAZOL-3-YL)METHYL)-1H-PURINE-2,6-DIONE see CDM575

3,7-DIHYDRO-1,3-DIMETHYL-7-(3-(METHYLPHENYLAMINO)PROPYL)-1H-PURINE-2,6-DIONE see DLI625

3,7-DIHYDRO-1,3-DIMETHYL-7-(2-(METHYLTHIO)ETHYL)-1H-PURINE-2,6-DIONE (9CI) see MDN100

2,5-DIHYDRO-1,2-DIMETHYL-3-(2-NAPHTHALENYL)-1H-PYRROLE, REL-(2R,3R)-2,3-DIHYDROXYBUTANEDIOATE (1:1) see DLI630

3,4-DIHYDRO-2,2-DIMETHYL-2H-NAPHTHO(1,2-B)PYRAN-5,6-DIONE see LBC500

1,4-DIHYDRO-2,6-DIMETHYL-4-(2-NITROPHENYL)-3,5-PYRIDINEDICARBOXYLIC ACID DIMETHYL ESTER see AEC750

1,4-DIHYDRO-2,6-DIMETHYL-4-(3-NITROPHENYL)-3,5-PYRIDINEDICARBOXYLIC ACID ETHYL METHYL ESTER see EMR600

1,4-DIHYDRO-2,6-DIMETHYL-4-(m-NITROPHENYL)-3,5-PYRIDINEDICARBOXYLIC ACID BIS(2-PROPOXYETHYL) ESTER see NDY600

1,4-DIHYDRO-2,6-DIMETHYL-4-(m-NITROPHENYL)-3,5-PYRIDINEDICARBOXYLIC ACID 2-(BENZYLMETHYL AMINO)ETHYL METHYL ESTER, MONOHYDROCHLORIDE see PCG550

1,4-DIHYDRO-2,6-DIMETHYL-4-(3-NITROPHENYL)-3,5-PYRIDINEDICARBOXYLIC ACID 2-METHOXYETHYL-1-METHYLETHYL ESTER see NDY700

10,11-DIHYDRO-α-8-DIMETHYL-11-OXO-DIBENZ(b,f)OXEPIN-2-ACETIC ACID see DLI650

3,7-DIHYDRO-3,7-DIMETHYL-1-(5-OXOHEXYL)-1H-PURINE-2,6-DIONE see PBU100

N-(2,3-DIHYDRO-1,5-DIMETHYL-3-OXO-2-PHENYL-1H-PYRAZOL-4-YL)-2-(DIMETHYLAMINO)PROPANAMIDE see AMF375

3,4-DIHYDRO-2,2-DIMETHYL-4-OXO-2H-PYRAN-6-CARBOXYLIC ACID-n-BUTYL ESTER see BRT000

2,3-DIHYDRO-2,2-DIMETHYL-6-((4-(PHENYLAZO)-1-NAPHTHYL)AZO)PERIMIDINE see PCJ200

2,3-DIHYDRO-N,N-DIMETHYL-3-PHENYL-1H-INDEN-1-AMINE see DRX400

7,8-DIHYDRO-1,3-DIMETHYL-8-(2-PROPENYL)-1H-IMIDAZO(2,1-f)PURINE-2,4(3H,6H)-DIONE see KHU100

3,7-DIHYDRO-1,3-DIMETHYL-1H-PURINE-2,6-DIONE see TEP000

3,7-DIHYDRO-3,7-DIMETHYL-1H-PURINE-2,6-DIONE see TEO500

3,7-DIHYDRO-1,3-DIMETHYL-1H-PURINE-2,6-DIONE compounded with 1,2-ETHANEDIAMINE (2:1) see TEP500

3,4-DIHYDRO-2,5-DIMETHYL-2H-PYRAN-2-CARBOXALDEHYDE see DLI600

4,5-DIHYDRO-3,3-DIMETHYL-1H-(PYRAZOLO(3,4-b)((1,4)-BENZOXAZEPINE)) HYDROCHLORIDE see DLJ000

6-(2,5-DIHYDRO-1,2-DIMETHYL-1H-PYRROL-3-YL)-2-NAPHTHALENOL HYDROCHLORIDE see DLJ050

1,4-DIHYDRO-2,6-DIMETHYL-4-(α,α,α-TRIFLUORO-o-TOLYL)-3,5-PYRIDINEDICARBOXYLIC ACID DIETHYL ESTER see FOO875

4,5-DIHYDRO-2,7-DINITROPYRENE see DLJ100

endo,endo-DIHYDRODI(NORBORNADIENE) see DLJ500

DIHYDRO-2,2-DIOCTYL-6H-1,3,2-OXATHIASTANNIN-6-ONE see DVN800

9,10-DIHYDRO-9,10-DIOXO-1,8-ANTHRACENE DISULFONIC ACID DIPOTASSIUM SALT see DLJ600

9,10-DIHYDRO-9,10-DIOXO-1,5-ANTHRACENE DISULFONIC ACID DISODIUM SALT see DLJ700

9,10-DIHYDRO-9,10-DIOXOANTHRACENE-1,8-DISULFONIC ACID, DISODIUM SALT see DLJ730

N',N'''-(9,10-DIHYDRO-9,10-DI-OXO-2,6-ANTHRACENEDIYL)BIS(N,N-DIETHYLETHANIMIDAMIDE) see APL250

9,10-DIHYDRO-9,10-DIOXO-1-ANTHRACENE SULFONIC ACID SODIUM SALT see DLJ800

1,3-DIHYDRO-1,3-DIOXO-5-ISOBENZOFURANCARBOXYLIC ACID see TKV000

3-(1,3-DIHYDRO-1,3-DIOXO-2H-ISOINDOL-2-YL)-2-OXOPIPERIDINE see OOM300

3,7-DIHYDRO-7-(1,3-DIOXOLAN-2-YLMETHYL)-1,3-DIMETHYL-1H-PURINE-2,6-DIONE see TEQ175

3,4-DIHYDRO-3,4-DIOXO-1-NAPHTHALENE SULFONIC ACID SODIUM SALT see DLK000

5,10-DIHYDRO-5,10-DIOXONAPHTHO(2,3-b)-p-DITHIIN-2,3-DICARBONITRILE see DLK200

N-(4,5-DIHYDRO-4,5-DIOXO-1-PYRENYL)ACETAMIDE see AAM680

7,14-DIHYDRO-7,14-DIPROPYLDIBENZ(a,h)ANTHRACENE-7,14-DIOL see DLK600

5,6-DIHYDRO-DIPYRIDO(1,2a;2,1c)PYRAZINIUM DIBROMIDE see DWX800

5,6-DIHYDRO-DIPYRIDO(1,2A.2,1C)PYRAZINIUM DIBROMIDE see EJC025

5,6-DIHYDRO-p-DITHIIN-2,3-DICARBOXIMIDE see DLK700

3,4-DIHYDRO-3,4-EPOXY-1,2-BENZANTHRACENE see EBP000

DIHYDROEPOXYBENZENE see OKS250

5,6-DIHYDRO-5,6-EPOXYDIBENZ(a,h)ANTHRACENE see EBP500

1,2-DIHYDRO-1,2-EPOXYINDENO('₁,2,3-cd)PYRENE see DLK750

11,12-DIHYDRO-11,12-EPOXY-3-METHYLCHOLANTHRENE see MIL500

7,8-DIHYDRO-4,5-α-EPOXY-17-METHYLMORPHINAN-3,6-α-DIOL DIACETATE see HBT500

α-DIHYDROEQUILENIN see EDT500

β-DIHYDROEQUILENIN see EDU000

α-DIHYDROEQUILENINA (SPANISH) see EDT500

β-DIHYDROEQUILENINA (SPANISH) see EDU000

DIHYDROERGOCORNINE see DLK780

9,10-DIHYDROERGOCORNINE see DLK780

DIHYDROERGOTAMINE see DLK800

DIHYDROERGOTAMINE TARTRATE (2:1) see DLL000

DIHYDROERGOTOXIN see DLL100

DIHYDROERGOTOXINE MESYLATE see DLL400

DIHYDROERGOTOXINE METHANE SULFONATE see DLL400

DIHYDROERGOTOXINE METHANESULPHONATE see DLL400

DIHYDROERGOTOXINE MONOMETHANESULFONATE (SALT) see DLL400

DIHYDROERGOTOXIN MESYLATE see DLL400

DIHYDROERGOTOXIN METHANESULFONATE see DLL400

DIHYDRO-β-ERYTHROIDINE HYDROBROMIDE see DLL600

DIHYDROESTRIN BENZOATE see EDP000

3-(9,10-DIHYDRO-9,10-ETHANOANTHRACEN-9-YL)PROPYLMETHYLAMINE see LIN800

7,14-DIHYDRO-7,14-ETHANODIBENZ(a,b)ANTHRACENE-15,16-DICARBOXYLIC ACID see DCU200

1,2-DIHYDRO-6-ETHOXY-2,2,4-TRIMETHYLQUINOLINE see SAV000

15,16-DIHYDRO-11-ETHYLCYCLOPENTA(a)PHENANTHREN-17-ONE see DLM000

1,4-DIHYDRO-1-ETHYL-6-FLUORO-4-OXO-7-(1-PIPERAZINYL)-3-QUINOLINECARBOXYLIC ACID see BAB625

1,4-DIHYDRO-1-ETHYL-6-FLUORO-4-OXO-7-(1H-PYRROL-1-YL)-3-QUINOLINECARBOXYLIC ACID see FLV050

2,3-DIHYDRO-3-ETHYL-6-METHYL-1H-CYCLOPENTA(a)ANTHRACENE see DLM600

1,4-DIHYDRO-1-ETHYL-7-METHYL-4-OXO-1,8-NAPHTHYRIDINE-3-CARBOXYLIC ACID see EID000

5,8-DIHYDRO-8-ETHYL-5-OXO-2-(1-PIPERAZINYL)-PYRIDO(2,3-d)PYRAMIDINE-6-CARBOXYLIC ACID TRIHYDRATE see PII350

5,8-DIHYDRO-8-ETHYL-5-OXO-2-(1-PIPERAZINYL)PYRIDO(2,3-d)PYRIMIDINE-6-CAR BOXYLIC ACID see PIZ000

5,8-DIHYDRO-8-ETHYL-5-OXO-2-(1-PYRROLIDINYL) PYRIDO(2,3-D)PYRIMIDINE-6-CARBOXYLIC ACID see PAF250

DIHYDROETORPHINE HYDROCHLORIDE see DLM700

DIHYDROEUGENOL see MFM750

6,13-DIHYDRO-2-FLUOROBENZO(g)(1)BENZOTHIOPYRAN O(4,3-b)INDOLE see DLN000

6,13-DIHYDRO-3-FLUOROBENZO(e)(1)BENZOTHIOPYRAN O(4,3-b)INDOLE see DLN200

6,13-DIHYDRO-4-FLUOROBENZO(e)(1)BENZOTHIOPYRAN O(4,3-b)INDOLE see DLN400

6,11-DIHYDRO-2-FLUORO(1)BENZOTHIOPYRANO(4,3-b)INDOLE see DLO000

6,11-DIHYDRO-4-FLUORO(1)BENZOTHIOPYRANO(4,3-b)INDOLE see DLO200

2,3-DIHYDRO-9-FLUORO-3-METHYL-10-(4-METHYL-1-PIPERAZINYL)-7-OXO-7H-PYRIDO(1,2,3-de)-1,4-BENZOXAZINE-6-CARBOXYLIC ACID see OGI300

1,3-DIHYDRO-5-(o-FLUOROPHENYL)-1-METHYL-7-NITRO-2H-1,4-BENZODIAZEPIN-2-ONE see FDD100

6,11-DIHYDRO-2-FLUORO-THIOPYRANO(4,3-b)BENZ(e)INDOLE see DLO000

DIHYDROFOLLICULAR HORMONE see EDO000

DIHYDROFOLLICULIN see EDO000

DIHYDROFOLLICULIN BENZOATE see EDP000

α-DIHYDROFUCOSTEROL see SDZ350

DIHYDRO-2,5-FURANDIONE see SNC000

2,5-DIHYDROFURAN-2,5-DIONE see MAM000

DIHYDRO-2(3H)-FURANONE see BOV000

DIHYDROGEN DIOXIDE see HIB050

DIHYDROGEN (ETHYLENEDIAMINETETRAACETATO(4-))NICKELATE (2-) see EJA500

DIHYDROGEN HEXACHLOROPLATINATE see CKO750

DIHYDROGEN HEXACHLOROPLATINATE(2-) see CKO750

DIHYDROGEN HEXACHLOROPLATINATE HEXAHYDRATE see DLO400

(DIHYDROGEN MERCAPTOSUCCINATO)GOLD DISODIUM SALT see GJC000

DIHYDROGEN METHYL PHOSPHATE see DLO800

DIHYDROGEN OXIDE see WAT259

(2-(10',11'-DIHYDRO-5'H-AZEPIN-5'-YL)-1-METHYL) ETHYLTRIMETHYLAMMONIUMIODIDE see DKR800

DIHYDROHELENALIN see DLO875

11,13-DIHYDROHELENALIN see DLO875

β-DIHYDROHEPTACHLOR see DLO880

(Z)-DIHYDRO-5-(3-HEXENYL)-5-METHYL-2(3H)-FURANONE see MIW060

6,15-DIHYDROHYDROXY-5,9,14,18-ANTHRAZINETETRONE see DLO900

DIHYDROHYDROXYCODEINONE see PCG500

DIHYDRO-14-HYDROXYCODEINONE see PCG500

15,16-DIHYDRO-11-HYDROXYCYCLOPENTA(a)PHENANTHREN-17-ONE see DLO950

3,4-DIHYDRO-7-METHOXY-3-PHENYL-4-(4-(2-(1-PYRROLIDINYL)ETHOXY)PHENYL)-2H-1-BENZOTHIOPYRAN-4-OL HYDROCHLORIDE see DLS275

(R)-5,6-DIHYDRO-4-METHOXY-6-STYRYL-2H-PYRAN-2-ONE see GJI250

3,12-DIHYDRO-6-METHOXY-3,3,12-TRIMETHYL-7H-PYRANO(2,3-C)ACRIDIN-7-ONE see ADR750

10,11-DIHYDRO-5-(3-(METHYLAMINO)PROPYL)-5H-DIBENZ(b,f)AZEPINE HYDROCHLORIDE see DLS600

1,2-DIHYDRO-3-METHYL-BENZ(j)ACEANTHRYLENE see MIJ750

1,2-DIHYDRO-3-METHYLBENZ(j)ACEANTHRYLENE COMPOUND with 2,4,6-TRINITROPHENOL (1:1) see MIL750

(Z)-1,2-DIHYDRO-3-METHYL-BENZ(j)ACEANTHRYLENE-2a,3(3H)-DIOL see DLF300

2,3-DIHYDRO-2-METHYLBENZOPYRANYL-7,N-METHYLCARBAMATE see DLS800

9,10-DIHYDRO-7-METHYLBENZO(a)PYRENE see DLT000

4,5-DIHYDRO-2-((2-METHYLBENZO(b)THIEN-3-YL)METHYL)-1H-IMIDAZOLE HYDROCHLORIDE see MHJ500

1':2'-DIHYDRO-4'-METHYL-3:4-BENZPYRENE see DLT000

DIHYDRO-5-(1-METHYLBUTYL)-5-(2-PROPENYL)-2-THIOXO-4,6(1H,5H)-PYRIMIDINEDIONE (9CI) see AGL375

5,6-DIHYDRO-2-METHYL-3-CARBOXANILIDO-1,4-OXATHIIN (GERMAN) see CCC500

5,6-DIHYDRO-2-METHYL-3-CARBOXANILIDO-1,4-OXATHIIN-4,4-DIOXID (GERMAN) see DLV200

11,12-DIHYDRO-3-METHYLCHOLANTHRENE see DLT400

6,12b-DIHYDRO-3-METHYLCHOLANTHRENE see DLT200

meso-DIHYDRO-3-METHYLCHOLANTHRENE see DLT200

(E)-11,12-DIHYDRO-3-METHYLCHOLANTHRENE-11,12-DIOL see DML800

9,10-DIHYDRO-3-METHYLCHOLANTHRENE-1,9,10-TRIOL see DLT600

9,10-DIHYDRO-3-METHYL-CHOLANTHRENE-1,9,10-TRIOL see TKO750

1,2-DIHYDRO-5-METHYL-1,2-CHRYSENEDIOL see DLF400

7,8-DIHYDRO-5-METHYL-7,8-CHRYSENEDIOL see DLF600

(+−)-trans-1,2-DIHYDRO-5-METHYL-1,2-CHRYSENEDIOL see CML812

(+−)-trans-1,2-DIHYDRO-11-METHYL-1,2-CHRYSENEDIOL see CML814

(+−)-trans-1,2-DIHYDRO-11-METHYL-1,2-CHRYSENEDIOL see DLT640

16,17-DIHYDRO-7-METHYL-15H-CYCLOPENTA(a)PHENANTHRENE see MIV250

16,17-DIHYDRO-11-METHYL-15H-CYCLOPENTA(a)PHENANTHRENE see MJD750

15,16-DIHYDRO-11-METHYL-17H-CYCLOPENTA(a)PHENANTHREN-17-OL see DLT800

15,16-DIHYDRO-7-METHYLCYCLOPENTA(a)PHENANTHREN-17-ONE see DLT900

15,16-DIHYDRO-11-METHYLCYCLOPENTA(a)PHENANTHREN-17-ONE see MJE500

16,17-DIHYDRO-11-METHYLCYCLOPENTA(a)PHENANTHREN-15-ONE see DLU200

15,16-DIHYDRO-7-METHYL-17H-CYCLOPENTA(a)PHENANTHREN-17-ONE see DLT900

15,16-DIHYDRO-11-METHYL-17H-CYCLOPENTA(a)PHENANTHREN-17-ONE see MJE500

10,11-DIHYDRO-N-METHYL-5H-DIBENZO(a,d)CYCLOHEPTANE-Δ,γ-PROPYLAMINE see NNY000

16,17-DIHYDRO-17-METHYLENE-15H-CYCLOPENTA(a)PHENANTHRENE see DLU600

4,5-DIHYDRO-N-(1-METHYLETHYL)-3-(PHENYLAMINO)-2H-BENZ(g)INDAZOLE-2-ACETAMIDE see DLU650

1,3-DIHYDRO-3-METHYL-2H-IMIDAZO(4,5-F)QUINOLIN-2-ONE OXIME see HIU600

DIHYDROMETHYL-α-IONONE see TLO530

15,16-DIHYDRO-11-METHYL-15-METHOXYCYCLOPENTA(a)PHENANTHREN-17-ONE see DLU700

15,16-DIHYDRO-11-METHYL-6-METHOXY-17H-CYCLOPENTA(a)PHENANTHREN-17-ONE see MEV250

2-(4,5-DIHYDRO-4-METHYL-4-(1-METHYLETHYL)-5-OXO-1H-IMIDAZOL-2-YL)-3-QUINOLINECARBOXYLIC ACID see IAH100

2,4-DIHYDRO-2-METHYL-5-(METHYLTHIO)-4-(((5-NITRO-2-FURANYL)METHYLENE)AMINO)-3H-1,2,4-TRIAZOLE-3-THIONE see DLU750

2,3-DIHYDRO-2-METHYL-1-(MORPHOLINOACETYL)-3-PHENYL-4(1H)-QUINAZOLINONE HYDROCHLORIDE see HGL630

R-(+)-(2,3-DIHYDRO-5-METHYL-3-((4-MORPHOLINYL)METHYL)PYRROLO(1,2,3-DE)-1,4-BENZOXAZINYL)(1- see NAP545

1,4-DIHYDRO-1-METHYL-7-(2-(5-NITRO-2-FURYL)VINYL)-4-OXO-1,8-NAPHTHYRIDINE-3-CARBOXYLIC ACID, POTASSIUM SALT see DLU800

2,3-DIHYDRO-N-METHYL-7-NITRO-2-OXO-5-PHENYL-1H-1,4-BENZODIAZEPINE-1-CARBOXAMIDE see DLU900

1,3-DIHYDRO-1-METHYL-7-NITRO-5-PHENYL-2H-1,4-BENZODIAZEPIN-2-ONE see DLV000

DIHYDRO-5-METHYL-1-NITROSO-2,4(1H,3H)-PYRIMIDINEDIONE see NJX500

2,3-DIHYDRO-6-METHYL-1,4-OXATHIIN-5-CARBOXANILIDE see CCC500

5,6-DIHYDRO-2-METHYL-1,4-OXATHIIN-3-CARBOXANILIDE see CCC500

5,6-DIHYDRO-2-METHYL-1,4-OXATHIIN-3-CARBOXANILIDE-4,4-DIOXIDE see DLV200

cis-(+−)-1A,7B-DIHYDRO-4-METHYLOXIRENO(F)QUINOLINIUM see MPF900

1,6-DIHYDRO-2-METHYL-6-OXO-(3,4'-BIPYRIDINE)-5-CARBONITRILE see MQU600

N-(4,5-DIHYDRO-4-METHYL-5-OXO-1,2-DITHIOLO(4,3-B)PYRROL-6-YL)ACETAMIDE see ABI250

2-(1,2-DIHYDRO-1-METHYL-2-OXO-3H-INDOLE-3-YLIDENE)HYDRZAINECARBOTHIOAMIDE see MKW250

3,4-DIHYDRO-2-METHYL-4-OXO-3-o-TOLYLQUINAZOLINE see QAK000

(±)-DIHYDRO-5-(1-METHYL-2-PENTYNYL)-5-(2-PROPENYL)-2-THIOXO-4,6(1H,5H)-PYRIMIDINEDIONE see AGL875

3,6-DIHYDRO-α-((2-METHYLPHENOXY)METHYL)-1-(2H)-

PYRIDINEETHANOL HYDROCHLORIDE see TGK225

2,3-DIHYDRO-2-METHYL-9-PHENYL-1H-INDENO(2,1-c)PYRIDINE HYDROBROMIDE see NOE525

5,6-DIHYDRO-2-METHYL-N-PHENYL-1,4-OXATHIIN-3-CARBOXAMIDE see CCC500

5,6-DIHYDRO-2-METHYL-N-PHENYL-1,4-OXATHIIN-3-CARBOXAMIDE-4,4-DIOXIDE see DLV200

5,11-DIHYDRO-11-((4-METHYL-1-PIPERAZINYL)ACETYL)-6H-PYRIDO(2,3-b)(1,4)BENZODIAZEPIN-6-ONE DIHYDROCHLORIDE see GCE500

10,11-DIHYDRO-2-(4-METHYL-1-PIPERAZINYL)-11-(2-ATHIAZOLYL)-PYRIDAZINO(3,4-b)(1,4)BENZOXAZEPINE see DLV400

10,11-DIHYDRO-2-(4-METHYL-1-PIPERAZINYL)-11-(3,4-XYLYL)PYRIDAZINO(3,4-b)(1,4)BENZOXAZEPINE MALEATE see DLV600

9,10-DIHYDRO-10-(1-METHYL-4-PIPERIDINYLIDENE)-9-ANTHRACENOL HYDROCHLORIDE see WAJ000

3,4-DIHYDRO-4-METHYL-2-PIPERIDINYL-1(2H)-NAPTHALENONE HYDROCHLORIDE see PIR100

1,2-DIHYDRO-2-(1-METHYL-4-PIPERIDINYL)-5-PHENYL-4-(PHENYLMETHYL)-3H-PYRAZOL-3-ONE (9CI) see BCP650

6,11-DIHYDRO-11-(1-METHYL-4-PIPERIDYLIDENE)-5H-BENZO(5,6)CYCLOHEPTA(1,2-b)PYRIDINE DIMALEATE see DLV800

3,4-DIHYDROMETHYL-2H-PYRAN see MJE750

(S)-(+)-5,6-DIHYDRO-6-METHYL-2H-PYRAN-2-ONE see PAJ500

4,5-DIHYDRO-6-METHYL-4-((3-PYRIDINYLMETHYLENE)AMINO)-1,2,4-TRIAZIN-3(2H)-ONE, (E)- see DLV850

4,9-DIHYDRO-1-METHYL-3H-PYRIDO(3,4-b)INDOL-7-OL MONOHYDROCHLORIDE see HAI300

1,5-DIHYDRO-5-METHYL-1-β-d-RIBOFURANOSYL-1,4,5,6,8-PENTAAZAACENAPHTHYLEN-3-AMINE see TJE870

4,5-DIHYDRO-2-METHYLTHIAZOLE see DLV900

cis-(−)-3,5-DIHYDRO-3-METHYL-2H-THIOPYRANO(4,3,2-cd)INDOLE-2-CARBOXYLIC ACID see CML847

2,3-DIHYDRO-5-METHYL-2-THIOXO-4(1H)-PYRIMIDINONE see TFQ650

2,3-DIHYDRO-6-METHYL-2-THIOXO-4(1H)-PYRIMIDINONE see MPW500

DIHYDROMORPHINE HYDROCHLORIDE see DNU310

DIHYDROMORPHINONE see DLW600

DIHYDROMORPHINONE HYDROCHLORIDE see DNU300

DIHYDROMORPHINON-N-OXYD-DITARTARAT (GERMAN) see EBY500

3,4-DIHYDROMORPHOL see PCW500

2,3-DIHYDRO-1-(MORPHOLINOACETYL)-3-PHENYL-4(1H)-QUINAZOLINONE see MRN250

2,3-DIHYDRO-1-(MORPHOLINOACETYL)-3-PHENYL-4(1H)-QUINAZOLINONE HYDROCHLORIDE see PCJ350

DIHYDROMUCODINITRILE see HFB600

DIHYDROMUCONSAEUREDINITRIL see HFB600

DIHYDROMYRCENE see CMT050

DIHYDROMYRCENOL see DLX000

DIHYDROMYRCENYL ACETATE see DLX100

DIHYDROMYRICETIN see HDW125

(+)-DIHYDROMYRICETIN see HDW125

4,5-DIHYDRONAPHTH(1,2-k)ACEPHENANTHRYLENE see PCW000

3,4-DIHYDRO-1(2H)-NAPHTHALENONE see DLX200

2-((3,4-DIHYDRO-1(2H)-NAPHTHALENYLIDENE)METHYL)-1-METHYL-5-NITRO-1H-IMIDAZOLE see DLX225

4,5-DIHYDRO-2-(1-NAPHTHALENYLMETHYL)-1H-IMIDAZOLE MONOHYDROCHLORIDE see NCW000

4,5-DIHYDRO-2-(1-NAPHTHALENYLMETHYL)-1H-IMIDAZOLE MONONITRATE (9CI) see NAH550

2,3-DIHYDRO-2-(1-NAPHTHALENYL)-4(1H)-QUINAZOLINONE see DLX300

2,3-DIHYDRO-2-(1-NAPHTHYL)-4(1H)-QUINAZOLINONE see DLX300

DIHYDRONE HYDROCHLORIDE see DLX400

DIHYDRONEOPINE see DKW800

1,2-DIHYDRO-5-NITRO-ACENAPHTHYLENE see NEJ500

trans-9,10-DIHYDRO-3-NITROBENZO(a)PYRENE 9,10 DIOL sec NFI230

9,10-DIHYDRO-1-NITRO-9,10-DIOXO-2-ANTHROIC ACID see NFS502

4H-2,3-DIHYDRO-2-(5'-NITRO-2'-FURYL)-3-HYDROXY-1,3-BENZOXALINE-4-ONE see HMY100

1,2-DIHYDRO-2-(5'-NITROFURYL)-4-HYDROXY-CHINAZOLIN-3-OXID (GERMAN) see DLX800

1,2-DIHYDRO-2-(5'-NITROFURYL)-4-HYDROXYQUINAZOLINE-3-OXIDE see DLX800

4,5-DIHYDRO-N-NITRO-1-NITROSO-1H-IMIDAZOL-2-AMINE see NLB700

1,3-DIHYDRO-7-NITRO-5-PHENYL-2H-1,4-BENZODIAZEPIN-2-ONE see DLY000

4,5-DIHYDRO-2-NITROPYRENE see NFW220

4,5-DIHYDRO-N-NITRO-1-(3-PYRIDINYLMETHYL)-1H-IMIDAZOL-2-AMINE see DLY100

3,6-DIHYDRO-2-NITROSO-2H-1,2-OXAZINE see NJX000

DIHYDRO-1-NITROSO-2,4(1H,3H)-PYRIMIDINEDIONE see NJY000

2,5-DIHYDRO-1-NITROSO-1H-PYRROLE see NLQ000

5,6-DIHYDRO-1-NITROSOTHYMINE see NJX500

DIHYDRO-5-NITROSO-2,4,6-TRIMETHYL-4H-1,3,5-DITHIAZINE see NLU600

5,6-DIHYDRO-1-NITROSOURACIL see NJY000

1,2-DIHYDRO-2-(5-NITRO-2-THIENYL)QUINAZOLIN-4(3H)-ONE see DLY200

1,2-DIHYDRO-2-(5-NITRO-2-THIENYL)-4(3H)-QUINAZOLINONE see DLY200

DIHYDRONIVALENOL see DLY300

DIHYDRONORDICYCLOPENTADIENYL ACETATE see DLY400

DIHYDRONORGUAIARETIC ACID see NBR000

4,5-DIHYDRO-7-NORTALL OIL-1H-IMIDAZOLE-1-ETHANOL see TAC050

10,11-DIHYDRO-N,N,β-TRIMETHYL-5H-DIBENZ(b,f)AZEPINE-5-PROPANAMINE see DLH200

(±)-10,11-DIHYDRO-N,N,β-TRIMETHYL-5H-DIBENZO(a,d)-CYCLOHEPTENE-5-PROPANAMINE HCl see BPT750

d,l-10,11-DIHYDRO-N,N,β-TRIMETHYL-5H-DIBENZO(a,d)-CYCLOHEPTENE-5-PROPYLAMINE see BPT500

(±)-10,11-DIHYDRO-N,N,β-TRIMETHYL-5H-DIBENZO(a,d)-CYCLOHEPTENE-5-PROPYLAMINE HCl see BPT750

(+−)-9,10-DIHYDRO-N,N,10-TRIMETHYL-2-(TRIFLUORMETHYL)-9-ANTHRACENPROPANAMIN (GERMAN) see DMF800

cis-(±)-9,10-DIHYDRO-N,N,10-TRIMETHYL-2-(TRIFLUOROMETHYL)-9-ANTHRACENE PROPANAMINE see DMF800

3,6-DIHYDRO-1,2,2H-OXAZINE see DLY700

DIHYDROOXIRENE see EJN500

1A,7B-DIHYDROOXIRENO(H)QUINOLINE (+−)- see QNJ200

1,2-DIHYDRO-2-OXOADENOSINE see IKS450

9,10-DIHYDRO-9-OXOANTHRACENE see APM500

2,3-DIHYDRO-3-OXOBENZISOSULFONAZOLE see BCE500

2,3-DIHYDRO-3-OXOBENZISOSULPHONAZOLE see BCE500

5,13-DIHYDRO-5-OXOBENZO(e)(2)BENZOPYRANO(4,3-b)INDOLE see DLY800

3,4-DIHYDRO-4-OXO-3-BENZOTRIAZINYLMETHYL O,O-DIETHYL PHOSPHORODITHIOATE see EKN000

S-(3,4-DIHYDRO-4-OXO-1,2,3-BENZOTRIAZIN-3-YLMETHYL) O,O-DIETHYL PHOSPHORODITHIOATE see EKN000

S-(3,4-DIHYDRO-4-OXO-1,2,3-BENZOTRIAZIN-3-YLMETHYL)-O,O-DIMETHYL PHOSPHORODITHIOATE see ASH500

S-(3,4-DIHYDRO-4-OXO-BENZO(α)(1,2,3)TRIAZIN-3-YLMETHYL)-O,O-DIMETHYL PHOSPHORODITHIOATE see ASH500

6,11-DIHYDRO-11-OXO-DIBENZ(b,e)OXEPIN-3-ACETIC ACID see OMU000

N-(10,12-DIHYDRO-10-OXOISOINDOLO(1,2-B)QUINAZOLIN-2-YL)ACETAMIDE see ACB300

3-(1,3-DIHYDRO-1-OXO-2H-ISOINDOL-2-YL)-2,6-DIOXOPIPERIDINE see DVS600

4-(1,3-DIHYDRO-1-OXO-2H-ISOINDOL-2-YL)-α-METHYLBENZENEACETIC ACID see IDA400

3-(1,3-DIHYDRO-1-OXO-2H-ISOINDOL-2-YL)-2-OXOPIPERIDINE see EAJ100

O-(1,6)-(DIHYDRO-6-OXO-1-PHENYLPYRIDAZIN-3-YL), O,O-DIETHYL PHOSPHOROTHIOATE see POP000

10-((3,6-DIHYDRO-6-OXO-2H-PYRAN-2-YL)HYDROXYMETHYL)-5,9,10,11-tert-TRAHYDRO-4-HYDROXY-5-(1-HYDROXYHEPTYL)-1H-CYCLONONA(1,2-C :5,6-C')DIFURAN-1,3,6,8(4H)-TETRONE see RQP000

DIHYDROOXYCODEINONE HYDROCHLORIDE see DLX400

DIHYDROOXYCODEINON-N-OXYD (GERMAN) see DGY000

2,3-DIHYDRO-3,3',4',5,7-PENTAHYDROXYFLAVONE see DMD000

1,1-DIHYDROPERFLUOROBUTANOL see HAW100

α-α-DIHYDROPERFLUOROBUTANOL see HAW100

1,1-DIHYDROPERFLUOROHEPTYL ACRYLATE see DLY850

N-(1,1-DIHYDROPERFLUOROOCTYL)PYRIDINIUM TRIFLUOROMETHANESULFONATE see MBV710

2,3-DIHYDROPERFLUOROPENTANE see DLY900

2,2-DIHYDROPERFLUOROPROPANE see HDE100

1,2-DIHYDROPHENANTHRENE see DLZ000

1,2-DIHYDRO-1,2-PHENANTHRENEDIOL see DMA000

3,4-DIHYDRO-3,4-PHENANTHRENEDIOL see PCW500

1a,9b-DIHYDROPHENANTHRO(9,10-B)OXIRENE (9CI) see PCX000

N-(9,10-DIHYDRO-2-PHENANTHRYL)ACETAMIDE see DMA400

4-((4,5-DIHYDRO-3-(PHENYLAMINO)-2H-BENZ(G)INDAZOL-2-YL)ACETYL)MORPHOLINE see DMA500

4,5-DIHYDRO-3-(PHENYLAMINO)-N-(PHENYLMETHYL)-2H-BENZ(G)INDAZOLE-2-ACETAMIDE see DMA550

5-(4,5-DIHYDRO-2-PHENYL-3H-BENZ(e)INDOL-3-YL)-2-HYDROXYBENZOIC ACID see FAO100

5-(4,5-DIHYDRO-2-PHENYL-3H-BENZ(e)INDOL-3-YL)SALICYLIC ACID see FAO100

2,3-DIHYDRO-2-PHENYL-4H-1-BENZOPYRAN-4-ONE see FBW150

1,3-DIHYDRO-3-(((2-PHENYLETHYL)AMINO)METHYLENE)-2H-INDOL-2-ONE (E)- see ICU145

(4,5-DIHYDRO-3-PHENYL-5-ISOXAZOLYL)PHOSPHONIC ACID see PFL100

N-((3S)-2,3-DIHYDRO-6-(PHENYLMETHOXY)-3-BNENZOFURANYL)-N-HYDROXYUREA see DMA600

4,5-DIHYDRO-N-PHENYL-N-PHENYLMETHYL-1H-IMIDAZOLE-2-METHANAMINE see PDC000

5,6-DIHYDRO-2-PHENYLPYRAZOLO(5,1-a)ISOQUINOLINE see PEU600

5-(2-(3,6-DIHYDRO-4-PHENYL-1(2H)-PYRIDYL)ETHYL)-3-METHYL-2-OXAZOLIDINONE see DMB000

5,6-DIHYDRO-2-PHENYL-(1,2,4)TRIAZOLO(5,1-a)ISOQUINOLINE (9CI) see PEU650

2,3-DIHYDROPHORBOL ACETATE MYRISTATE see DMB200

2,3-DIHYDROPHORBOL MYRISTATE ACETATE see DMB200

DIHYDROPICROTOXININ see DMB300

α-DIHYDROPICROTOXININ see DMB300

10,11-DIHYDRO-5-(3-(4-PIPERIDINO-4-CARBAMOYLPIPERIDINO))PROPYL-(b,f)AZEPINE see CBH500

6,7-DIHYDRO-6-(2-PROPENYL)-5H-DIBENZ(c,e)AZEPINE PHOSPHATE see AGD500

4,5-DIHYDRO-2-(2-PROPENYLTHIO)THIAZOLE (9CI) see AGT250

(S)-2-(DIHYDRO-5-PROPYL-2(3H)-FURYLIDENE)-1,3-CYCLOPENTANEDIONE see OKS100

2,3-DIHYDRO-6-PROPYL-2-THIOXO-4(1H)-PYRIMIDINONE see PNX000

3,4-DIHYDRO-1-PROPYL-4,4,6-TRIMETHYL-2(1H)-PYRIMIDINETHIONE see DMB330

DIHYDROPSEUDOIONONE see GDE400

α-β-DIHYDROPSEUDOIONONE see GDE400

3,7-DIHYDRO-1H-PURINE-2,6-DIONE see XCA000

1,7-DIHYDRO-6H-PURINE-6-THIONE see POK000

1,7-DIHYDRO-6H-PURINE-6-THIONE MONOHYDRATE see MCQ100

1,7-DIHYDRO-6H-PURIN-6-ONE see DMC000

1,7-DIHYDRO-6H-PURIN-6-THION, MONOHYDRAT see MCQ100
1,2-DIHYDRO-3,6-PYRADIZINEDIONE see DMC600
DIHYDROPYRAN see DMC200
3,4-DIHYDROPYRAN see DMC200
2H-3,4-DIHYDROPYRAN see DMC200
Δ²-DIHYDROPYRAN see DMC200
3,4-DIHYDRO-2H-PYRAN-2-CARBOXALDEHYDE see ADR500
2,3-DIHYDRO-1,4-PYRAN-2-KARBOXALDEHYD see ADR500
2,3-DIHYDRO-1H-PYRAZOLO(2,3-a)IMIDAZOLE see IAN100
1,5-DIHYDRO-4H-PYRAZOLO(3,4-d)PYRIMIDIN-4-ONE see ZVJ000
1,2-DIHYDROPYRIDAZINE-3,6-DIONE see DMC600
1,2-DIHYDRO-3,6-PYRIDAZINEDIONE see DMC600
6,7-DIHYDROPYRIDO(1,2a;2',1'-c)PYRAZINEDIUM DIBROMIDE see DWX800
6,7-DIHYDROPYRIDO(1,2-A. 2',1'-C)PYRAZINEDIUM DIBROMIDE see EJC025
1,2-DIHYDROPYRIDO(2,1,e)TETRAZOLE see DMC800
(S)-2,5-DIHYDRO-1H-PYRROLE-2-CARBOXYLIC ACID see DAL375
3,4-DIHYDROPYRROLIDINE see SND000
DIHYDROQUERCETIN see DMD000
(+)-DIHYDROQUERCETIN see DMD000
2,3-DIHYDROQUERCETIN see DMD000
(2R,3R)-DIHYDROQUERCETIN see DMD000
DIHYDROQUINIDINE see HIG500
10,11-DIHYDROQUINIDINE see HIG500
7,8-DIHYDRORETINOIC ACID see DMD100
4a,5-DIHYDRO-RIBOFLAVIN-5'-PHOSPHATE SODIUM SALT see DMD200
DIHYDRORUBRATOXIN B see RQP050
DIHYDROSAFROLE see DMD600
DIHYDROSAMIDIN see DUO350
1,4-DIHYDRO-4-(3-((((3-(SPIRO(INDENE-4,1'-PIPERIDIN-1-YL))PROPYL)AMINO)C-ARBONYL)AMINO)PHENYL)- see DTU852
DIHYDROSTERIGMATOCYSTIN see HKB300
22,23-DIHYDROSTIGMASTEROL see SDZ350
DIHYDROSTILBESTROL see DLB400
DIHYDROSTREPTOMYCIN see DME000
DIHYDROSTREPTOMYCIN and STREPTOMYCIN see SLY200
DIHYDROSTREPTOMYCIN SULFATE see DME200
2,3-DIHYDROSUCCINIC ACID see TAF750
DIHYDROTACHYSTEROL see DME300
DIHYDROTACHY STEROL see IGF000
DIHYDROTACHYSTEROL₂ see DME300
DIHYDRO-α-TERPINEOL see MCE100
DIHYDROTERPINYL ACETATE see DME400
DIHYDROTESTOSTERONE see DME500
4-DIHYDROTESTOSTERONE see DME500
5-α-DIHYDROTESTOSTERONE see DME500
5-β-DIHYDROTESTOSTERONE see HJB100
4,5-α-DIHYDROTESTOSTERONE see DME500
DIHYDROTESTOSTERONE PROPIONATE see DME525
5-α-DIHYDROTESTOSTERONE PROPIONATE see DME525
DIHYDROTESTOSTERONE-17-β-PROPIONATE see DME525
DIHYDRO-5-TETRADECYL-2(3H)-FURANONE see SLL400
4,5-DIHYDRO-N-(4,6,7,8-TETRAHYDRO-1-NAPHTHALENYL)-1H-IMIDAZOL-2-AMINE HYDROCHLORIDE HYDRATE see RGP450

4,5-DIHYDRO-N-(4,6,7,8-TETRAHYDRO-1-NAPHTHALENYL)-1H-IMIDAZOL-2-AMINE MONOHYDROCHLORIDE see THJ825
4,5-DIHYDRO-2-(1,2,3,4-TETRAHYDRO-1-NAPHTHALENYL)-1H-IMIDAZOLE MONOHYDROCHLORIDE see VRZ000
5,6-DIHYDRO-2,3,9,10-TETRAMETHOXYDIBENZO(a,g)QUINOLIZINIUM HYDROXIDE see PAE100
6,7-DIHYDRO-1,2,3,10-TETRAMETHOXY-7-(METHYLAMINO)-BENZO(α)HEPTALEN-9(5H)-ONE see MIW500
1,3-DIHYDROTETRAMETHYLDISILOXANE see TDP775
1,2-DIHYDRO-2,2,4,6-TETRAMETHYLPYRIDINE see DME600
6,7-DIHYDRO-3,5,5,7-TETRAMETHYL-5H-THIAZOLO(3,2-a)PYRIMIDIN-7-OL HYDROCHLORIDE see DME700
DIHYDRO-3-(TETRAPROPENYL)-2,5-FURANDIONE see TEC600
DIHYDROTHEELIN see EDO000
N-(5,6-DIHYDRO-4H-1,3-THIAZINYL)-2,6-XYLIDINE see DMW000
4,5-DIHYDRO-2-THIAZOLAMINE see TEV600
2-((DIHYDRO-2(3H)-THIENYLIDENE)METHYL)-1-METHYL-5-NITRO-1H-IMIDAZOLE S,S-DIOXIDE see DME750
2,3-DIHYDROTHIIRENE see EJP500
2,5-DIHYDROTHIOPHENE DIOXIDE see DMF000
2,5-DIHYDROTHIOPHENE-1,1-DIOXIDE see DMF000
2,5-DIHYDROTHIOPHENE SULFONE see DMF000
DIHYDRO-2(3H)-THIOPHENONE see TDC800
4,5-DIHYDRO-3(2H)-THIOPHENONE see TDC800
DIHYDRO-2-THIOXO-4,6(1H,5H)-PYRIMIDINEDIONE see MCK500
2,3-DIHYDRO-2-THIOXO-4(1H)-PYRIMIDINONE see TFR250
1,2-DIHYDRO-s-TRIAZINE-4,6-DIAMINO-2,2-DIMETHYL-1-PHENYL-2,4,5-TRICHLOROPHENOXYACETATE see DMF400
(2R-trans)-2,3-DIHYDRO-3,5,7-TRIHYDROXY-2-(3,4,5-TRIHYDROXYPHENYL)-4H-1-BENZOPYRAN-4-ONE see HDW125
16,17-DIHYDRO-11,12,17-TRIMETHYLCYCLOPENTA(a)PHENANTHRENE see DMF600
1,2-DIHYDRO-2,2,4-TRIMETHYL-6-ETHOXYQUINOLINE see SAV000
(1,3-DIHYDRO-1,3,3-TRIMETHYL-2H-INDOL-2-YLIDENE)ACETALDEHYDE see FBV050
3,7-DIHYDRO-1,3,7-TRIMETHYL-1H-PURINE-2,6-DIONE see CAK500
3,7-DIHYDRO-1,3,7-TRIMETHYL-1H-PURINE-2,6-DIONE MONOHYDROBROMIDE see CAK750
1,2-DIHYDRO-2,2,4-TRIMETHYLQUINOLINE see TLP500
1,2-DIHYDROTRIPHENYLENE see DMG000
DIHYDROXO(1,2-DIAMINOCYCLOHEXANE)PLATINUM(II) DIHYDRATE see DMG100
DIHYDROXYACETONE see OLW100
1,3-DIHYDROXYACETONE see OLW100
2,4-DIHYDROXYACETOPHENONE see DMG400
2,5-DIHYDROXYACETOPHENONE see DMG600
2',4'-DIHYDROXYACETOPHENONE see DMG400

2',5'-DIHYDROXYACETOPHENONE see DMG600
11-β,17-α-DIHYDROXY-21-ACETOXYPREGESTERONE see HHQ800
1,8-DIHYDROXY-10-ACETYL-9-ANTHRONE see ACA750
3-α,17-β-DIHYDROXY-5-α-ANDROSTANE see DMG700
3-β,17-β-DIHYDROXY-5-α-ANDROSTANE see AOO405
2,3-DIHYDROXYANISOLE see PPQ550
1,8-DIHYDROXY-9,10-ANTHRACENEDIONE see DMH400
1,2-DIHYDROXYANTHRACHINON see DMG800
1,4-DIHYDROXYANTHRACHINON (CZECH) see DMH000
1,5-DIHYDROXYANTHRACHINON (CZECH) see DMH200
1,8-DIHYDROXYANTHRACHINON (CZECH) see DMH400
1,2-DIHYDROXY-ANTHRACHINON-3-METHYLEN-IMINODIESSIGSAEURE see AFM400
DIHYDROXYANTHRANOL see APH250
1,8-DIHYDROXYANTHRANOL see APH250
1,8-DIHYDROXY-9-ANTHRANOL see APH250
DIHYDROXYANTHRAQUINONE see MQY090
1,2-DIHYDROXYANTHRAQUINONE see DMG800
1,4-DIHYDROXYANTHRAQUINONE see DMH000
1,5-DIHYDROXYANTHRAQUINONE see DMH200
1,8-DIHYDROXYANTHRAQUINONE see DMH400
2,6-DIHYDROXYANTHRAQUINONE see DMH600
1,2-DIHYDROXY-9,10-ANTHRAQUINONE see DMG800
1,4-DIHYDROXY-9,10-ANTHRAQUINONE see DMH000
1,5-DIHYDROXY-9,10-ANTHRAQUINONE see DMH200
6,6'-((4,8-DIHYDROXY-1,5-ANTHRAQUINONYLENE)DIIMINO) DI-m-TOLUENE SULFONIC ACID DISODIUM SALT see TGS000
(((3,4-DIHYDROXY-2-ANTHRAQUINONYL)METHYL)IMINO)DIACETIC ACID see AFM400
1,8-DIHYDROXY-ANTHRONE see DMG900
1,8-DIHYDROXYANTHRONE see DMG900
1,8-DIHYDROXY-9-ANTHRONE see APH250
1,8-DIHYDROXY-9-ANTHRONE see DMG900
2,4-DIHYDROXYAZOBENZENE see CMP600
5,5-DIHYDROXY BARBITURIC ACID see DMG950
2,4-DIHYDROXYBENZALDEHYDE see REF100
3,4-DIHYDROXYBENZALDEHYDE METHYLENE KETAL see PIW250
N,2-DIHYDROXYBENZAMIDE see SAL500
3,9-DIHYDROXYBENZ(a)ANTHRACENE see BBG200
1,4-DIHYDROXY-BENZEEN (DUTCH) see HIH000
1,4-DIHYDROXYBENZEN (CZECH) see HIH000
DIHYDROXYBENZENE see HIH000
m-DIHYDROXYBENZENE see REA000
o-DIHYDROXYBENZENE see CCP850
p-DIHYDROXYBENZENE see HIH000
1,2-DIHYDROXYBENZENE see CCP850
1,3-DIHYDROXYBENZENE see REA000
1,4-DIHYDROXYBENZENE see HIH000
3,4-DIHYDROXYBENZENEACETIC ACID see HGI980

3,4-DIHYDROXYBENZENEACRYLIC ACID see CAK375

2,4-DIHYDROXYBENZENECARBONAL see REF100

1,3-DIHYDROXYBENZENE DIACETATE see REA100

4,5-DIHYDROXY-1,3-BENZENEDISULFONIC ACID see PPQ100

4,5-DIHYDROXY-1,3-BENZENEDISULFONIC ACID DISODIUM SALT see DXH300

DIHYDROXYBENZENE (OSHA) see HIH000

2,5-DIHYDROXYBENZENESULFONIC ACID CALCIUM SALT see DMI300

2,5-DIHYDROXYBENZENESULFONIC ACID with N-ETHYLETHANAMINE see DIS600

3,3'-DIHYDROXYBENZIDINE see DMI400

2,4-DIHYDROXYBENZOFENON (CZECH) see DMI600

2,4-DIHYDROXYBENZOIC ACID see HOE600

2,5-DIHYDROXYBENZOIC ACID see GCU000

3,4-DIHYDROXYBENZOIC ACID see POE200

3,5-DIHYDROXYBENZOIC ACID see REF200

1,4-DIHYDROXY-BENZOL (GERMAN) see HIH000

2,4-DIHYDROXYBENZOPHENONE see DMI600

2,4-DIHYDROXY-1,4-BENZOXAZINONE see BDI100

2,4-DIHYDROXY-2H-1,4-BENZOXAZIN-3(4H)-ONE see BDI100

2,2'-DIHYDROXYBINAPHTHALENE see BGC100

2,5-DIHYDROXYBIPHENYL see BGG250

2,2'-DIHYDROXYBIPHENYL see BGG000

2,6-DIHYDROXY-5-BIS(2-CHLOROETHYL)AMINOPYRAMIDINE see BIA250

1,4-DIHYDROXY-5,8-BIS((2-HYDROXYETHYL)AMINO)-9,10-ANTHRACENEDIONE see DMM400

1,4-DIHYDROXY-5,8-BIS(2-((2-HYDROXYETHYL)AMINO)ETHYLAMINO)-9,10-ANTHRACENEDIONE DIACETATE see DBE875

1,4-DIHYDROXY-5,8-BIS((2-((HYDROXYETHYL)AMINO)ETHYL)AMINO)-9,10-ANTHRACENEDIONE (9CI) see MQY090

(4,5-DIHYDROXY-1,3-BIS(HYDROXYMETHYL))-2-IMIDAZOLIDINONE see DTG000

2,5-DIHYDROXY-3,6-BIS(5-(3-METHYL-2-BUTENYL)-1H-INDOL-3-YL)-2,5-CYCLOHEXADIENE-1,4-DIONE see CNF159

3-β,14-β-DIHYDROXYBUFA-4,20,22-TRIENOLIDE 3-RHAMNOSIDE see POB500

1,3-DIHYDROXYBUTANE see BOS500

1,4-DIHYDROXYBUTANE see BOS750

2,3-DIHYDROXYBUTANE see BOT000

2,3-DIHYDROXYBUTANEDIOIC ACID see TAF750

2,3-DIHYDROXYBUTANEDIOIC ACID, DIAMMONIUM SALT see DCH000

(R*,S*)-2,3-DIHYDROXY-BUTANEDIOIC ACID ION(2−) (9CI) see TAF775

1,4-DIHYDROXY-1,4-BUTANEDISULFONIC ACID, DISODIUM SALT see SMX000

1,4-DIHYDROXY-2-BUTENE see BOX300

DIHYDROXYBUTENEDIOIC ACID see DMW200

9-(3,4-DIHYDROXYBUTYL)GUANINE see DMI700

2,6-DIHYDROXY-4-CARBOXYPYRIDINE see DMV400

3,β,14-DIHYDROXY-5,β-CARD-20(22)ENOLIDE see DMJ000

(3-β,5-α,17-α)-3,14-DIHYDROXYCARD-20(22)-ENOLIDE see UWJ200

(3-β,5-β)-3,14-DIHYDROXY-CARD-20(22)-ENOLIDE see DMJ000

3-β,14-DIHYDROXY-5-β-CARD-20(22)-ENOLIDE-3-FORMATE see FNK050

R-4,T-5-DIHYDROXY-C-6,6A-EPOXY-4,5,6,6A-TETRAHYDROBENZO(J)FLUORANTHENE see DMP300

(±)-2,3-DIHYDROXYCHLOROPROPANE see CHL875

DIHYDROXYCHLOROTHIAZIDUM see CFY000

DIHYDROXY 3-12 CHOLANATE de Na (FRENCH) see SGE000

3,12-DIHYDROXYCHOLANIC ACID see DAQ400

3-α,12-α-DIHYDROXYCHOLANIC ACID see DAQ400

3-α,7-α-DIHYDROXYCHOLANIC ACID see CDL325

3-α,7-β-DIHYDROXYCHOLANIC ACID see DMJ200

3,7-DIHYDROXYCHOLAN-24-OIC ACID see DMJ200

3-α,12-α-DIHYDROXY-5-β-CHOLANOIC ACID see DAQ400

3-α,12-α-DIHYDROXY-5-β-CHOLAN-24-OIC ACID see DAQ400

(3-α,5-β,7-β)-3,7-DIHYDROXYCHOLAN-24-OIC ACID see DMJ200

3-α,7-α-DIHYDROXY-5-β-CHOLAN-24-OIC ACID see CDL325

3-α,7-β-DIHYDROXY-6-β-CHOLAN-24-OIC ACID see DMJ200

3-α,7-β-DIHYDROXY-5-β-CHOLANOIC ACID see DMJ200

3-α-12-α-DIHYDROXY-5-β-CHOLAN-24-OIC ACID with DIBENZ(a,h)ANTHRACENE see DCT800

(3-α,5-β,12-α)-3,12-DIHYDROXY-CHOLAN-24-OIC ACID MONOSODIUM SALT see SGE000

3-α,12-α-DIHYDROXY-5-β-CHOLAN-24-OIC ACID SODIUM SALT see SGE000

3-α,12-α-DIHYDROXYCHOLANSAEURE (GERMAN) see DAQ400

3-α,7-α-DIHYDROXYCHOLANSAEURE (GERMAN) see DMJ200

1,25-DIHYDROXYCHOLECALCIFEROL see DMJ400

1a,25-DIHYDROXYCHOLECALCIFEROL see DMJ400

24(R),25-DIHYDROXYCHOLECALCIFEROL see HJS900

1-α,25-DIHYDROXYCHOLECALCIFEROL see DMJ400

3-α,12-α-DIHYDROXY-5-β-CHOL-8(14)-EN-24-OIC ACID see AQO500

3,4-DIHYDROXYCINNAMIC ACID see CAK375

DIHYDROXYCODEINONE HYDROCHLORIDE see DLX400

DI-(4-HYDROXY-3-COUMARINYL)METHANE see BJZ000

R-4,T-3-DIHYDROXY-C-1,2-OXY-1,2,3,4-TETRAHYDROCHRYSENE see CML820

DIHYDROXYCYCLOBUTENEDIONE see DMJ600

3,4-DIHYDROXYCYCLOBUTENE-1,2-DIONE see DMJ600

3,4-DIHYDROXY-3-CYCLOBUTENE-1,2-DIONE see DMJ600

(−)-3,14-DIHYDROXY-N-(CYCLOBUTYLMETHYL)MORPHINAN see CQF079

2,3-DIHYDROXY-2,4,6-CYCLOHEPTATRIEN-1-ONE see HON500

1,8-DIHYDROXY-4,5-DIAMINOANTHRACHINON see DBQ250

1,5-DIHYDROXY-4,8-DIAMINOANTHRACHINON (CZECH) see DBP909

1,5-DIHYDROXY-4,8-DIAMINOANTHRAQUINONE see DBP909

DIHYDROXYDIBENZANTHRONE see DMJ800

16,17-DIHYDROXYDIBENZANTHRONE see DMJ800

DI(4-HYDROXY-3,5-DI-tert-BUTYLPHENYL)METHANE see MJM700

2,2'-DIHYDROXY-5,5'-DICHLORODIPHENYLMETHANE see MJM500

2,2'-DIHYDROXYDIETHYLAMINE see DHF000

4,4'-DIHYDROXY-α,β-DIETHYLDIPHENYLETHANE see DLB400

DIHYDROXYDIETHYL ETHER see DJD600

β,β'-DIHYDROXYDIETHYL ETHER see DJD600

4,4'-DIHYDROXY-α,α'-DIETHYL-STILBEN-DIPHOSPHATE TETRASODIUM see TEE300

4,4'-DIHYDROXYDIETHYLSTILBENE see DKA600

4,4'-DIHYDROXY-α,β-DIETHYLSTILBENE see DKA600

DIHYDROXYDIETHYLSTILBENE DIPROPIONATE see DKB000

4,4'-DIHYDROXY-α,β-DIETHYLSTILBENE DIPROPIONATE see DKB000

4,4'-DIHYDROXY-α,β-DIETHYLSTILBENE PALMITATE see DKA800

β,β'-DIHYDROXYDIETHYL SULFIDE see TFI500

trans-1,2-DIHYDROXY-1,2-DIHYDROBENZ(a)ANTHRACENE see BBD250

trans-3,4-DIHYDROXY-3,4-DIHYDROBENZ(a)ANTHRACENE see BBD500

trans-5,6-DIHYDROXY-5,6-DIHYDROBENZ(a)ANTHRACENE see BBD980

trans-8,9-DIHYDROXY-8,9-DIHYDROBENZ(a)ANTHRACENE see BBE750

trans-1,2-DIHYDROXY-1,2-DIHYDROBENZ(a,h)ANTHRACENE see DMK200

trans-10,11-DIHYDROXY-10,11-DIHYDROBENZ(a)ANTHRACENE see BBF000

(E)-3,4-DIHYDROXY-3,4-DIHYDROBENZ(a)ANTHRACENE-7,12-DIMETHANOL see DML775

(+)-(1S,2S)-1,2-DIHYDROXY-1,2-DIHYDROBENZENE see DMJ900

(+)-1-α,2-β-DIHYDROXY-1,2-DIHYDROBENZENE see DMJ900

1,2-DIHYDROXY-1,2-DIHYDROBENZENE, racemic mixture of (+)-and (−)- see CPA760

trans-1,2-DIHYDROXY-1,2-DIHYDROBENZO(a,h)ANTHRACENE see DMK200

trans-4,5-DIHYDROXY-4,5-DIHYDROBENZO(a)PYRENE see DLB850

trans-4,5-DIHYDROXY-4,5-DIHYDROBENZO(e)PYRENE see DMK400

trans-7,8-DIHYDROXY-7,8-DIHYDRO-BENZO(a)PYRENE see DLC800

(−)-trans-7,8-DIHYDROXY-7,8-DIHYDROBENZO(a)PYRENE see DML200

(+,−)-trans-7,8-DIHYDROXY-7,8-DIHYDROBENZO(a)PYRENE see DML000

(+)-trans-7,8-DIHYDROXY-7,8-DIHYDROBENZO(a)PYRENE see DML400

trans-9,10-DIHYDROXY-9,10-DIHYDROBENZO(a)PYRENE see DLC200

2,2'-DIHYDROXY-4-METHOXYBENZOPHENONE see DMW250

2,4-DIHYDROXY-7-METHOXY-1,4-BENZOXAZINONE see DMW300

2,4-DIHYDROXY-7-METHOXY-2H,1,4-BENZOXAZIN-3(4H)ONE see DMW300

5,8-DIHYDROXY-7-METHOXYFLAVONE see ITD020

3,7-DIHYDROXY-9-METHOXY-1-METHYL-6H-DIBENZO(b,d)PYRAN-6-ONE see AGW500

1,2-DIHYDROXY-3-(2-METHOXYPHENOXY)PROPANE see RLU000

2-(3,4-DIHYDROXY-5-METHOXYPHENYL)-3,5,7-TRIHYDROXYBENZOPYRYLIUM, ACID ANION see PCU000

5,8-DIHYDROXY-3-METHOXYXANTHONE-1-o-GLUCOSIDE see SOZ100

3,4'-DIHYDROXY-2-(METHYLAMINO)ACETOPHENONE see MGC350

3,4-DIHYDROXY-α-METHYLAMINOACETOPHENONE see MGC350

3,4-DIHYDROXY-α-((METHYLAMINO)METHYL)BENZYL ALCOHOL see VGP000

(±)-3,4-DIHYDROXY-α-((METHYLAMINO)METHYL)BENZYL ALCOHOL HYDROCHLORIDE see AES625

(−)-3,4-DIHYDROXY-α-(((METHYLAMINO)METHYL)BENZYL) ALCOHOL (+)-TARTRATE (1:1) SALT see AES000

3-DI(HYDROXYMETHYL)AMINO-6-(5-NITRO-2-FURYLETHENYL)-1,2,4-TRIAZINE see BKH500

3-DI(HYDROXYMETHYL)AMINO-6-(2-(5-NITRO-2-FURYL)VINYL)-1,2,4-TRIAZINE see BKH500

7:12-DIHYDROXYMETHYLBENZ(a)ANTHRACENE see BBF500

1,3-DIHYDROXY-5-METHYLBENZENE see MPH500

6,7-DIHYDROXY-4-METHYL-2H-1-BENZOPYRAN-2-ONE see MJV800

2,6-DIHYDROXY-4-METHYL-5-BIS(2-CHLOROETHYL)AMINOPYRIMIDINE see DYC700

2,2-(DIHYDROXYMETHYL)-1-BUTANOL, MONOALLYL ETHER see TLX110

cis-1,2-DIHYDROXY-3-METHYLCHOLANTHRENE see MIK750

6,7-DIHYDROXY-4-METHYLCOUMARIN see MJV800

4,4-DIHYDROXYMETHYL-1-CYCLOHEXENE see BKH400

(E)-3,4-DIHYDROXY-7-METHYL-3,4-DIHYDROBENZ(a)ANTHRACENE-12-METHANOL see DMX000

2,12-DIHYDROXY-4-METHYL-11,16-DIOXOSENECIONANIUM see DMX200

DI-4-HYDROXY-3,3'-METHYLENEDICOUMARIN see BJZ000

2,4-DIHYDROXYMETHYLENE-3-(2,2-DIMETHOXYETHYL)GLUTARALDEHYDE see DMX300

DIHYDROXYMETHYL FURATRIZINE see BKH500

11-α,17-β-DIHYDROXY-17-METHYL-3-OXOANDROSTA-1,4-DIENE-2-CARBOXALDEHYDE see FNK040

14,16-DIHYDROXY-3-METHYL-7-OXO-trans-BENZOXACYCLOTETRADEC-11-EN-1-ONE see ZAT000

(5Z,11-α,13E,15S,17Z)-11,15-DIHYDROXY-15-METHYL-9-OXO-PROSTA-5,13-DIEN-1-OIC ACID see MOV800

(11-α-13E)-(±)-11,16-DIHYDROXY-16-METHYL-9-OXOPROST-13-EN-1-OIC ACID METHYL ESTER see MJE775

2,4-DIHYDROXY-2-METHYLPENTANE see HFP875

DIHYDROXYMETHYL PEROXIDE see DMN200

1,2-DIHYDROXY-3-(2-METHYLPHENOXY)PROPANE see GGS000

α,β-DIHYDROXY-γ-(2-METHYLPHENOXY)PROPANE see GGS000

3,4-DIHYDROXY-3-METHYL-4-PHENYL-1-BUTYNE see DMX800

d-N,N'-DI(1-HYDROXYMETHYLPROPYL)ETHYLENEDIAMINE DIHYDROCHLORIDE mixed with SODIUM NITRITE see SIS500

N,N'-DIHYDROXYMETHYLUREA see DTG700

1,8-DIHYDROXY-10-MYRISTOYL-9-ANTHRONE see MSA750

1,5-DIHYDROXYNAPHTHALENE see NAN505

1,6-DIHYDROXYNAPHTHALENE see NAN507

1,7-DIHYDROXYNAPHTHALENE see NAN508

2,5-DIHYDROXYNAPHTHALENE see NAN507

2,7-DIHYDROXYNAPHTHALENE see NAO500

N,3-DIHYDROXY-4-(1-NAPHTHALENYLOXY)BUTANINIDAMIDE HYDROCHLORIDE see NAC500

1,5-DIHYDROXYNAPTHALENE see NAN505

d-threo-N-(1,1'-DIHYDROXY-1-p-NITROPHENYLISOPROPYL)DICHLOROACETAMIDE see CDP250

2,2'-DIHYDROXY-N-NITROSODIETHYLAMINE see NKM000

3,4-DIHYDROXYNOREPHEDRINE HYDROCHLORIDE see AMB000

3,17-β-DIHYDROXYOESTRA-1,3,5-TRIENE see EDO000

3,17-α-DIHYDROXYOESTRA-1,3,5(10)-TRIENE see EDO500

3,17-β-DIHYDROXY-1,3,5(10)-OESTRATRIENE see EDO000

DIHYDROXYOESTRIN see EDO000

(11-β)-11,21-DIHYDROXY-17-(1-OXOBUTOXY)-PREGN-4-ENE-3,20-DIONE see HHQ825

1,8-DIHYDROXY-10-(1-OXOPENTYL)-9(10H)-ANTHRACENONE see DMX900

(5Z,11-α,13E,15S)-11,15-DIHYDROXY-9-OXOPROSTA-5,13-DIEN-1-OIC ACID see DVJ200

9,15-DIHYDROXY-11-OXO-PROSTA-5,13-DIEN-1-OIC ACID, (5Z,9-α,13E,15S)- see POC275

(5Z,11-α,13E,15S)-11,15-DIHYDROXY-9-OXOPROSTA-5,13-DIEN-1-OIC ACID MONOSODIUM SALT see POC360

(11-α,13E,15S)-11,15-DIHYDROXY-9-OXO-PROST-13-EN-1-OIC ACID (9CI) see POC350

11,15-DIHYDROXY-9-OXO-PROST-13-EN-1-OIC ACID, (11-α,13E,15S)-, and α-CYCLODEXTRIN see AGW275

1,8-DIHYDROXY-10-(1-OXOTETRADECYL)-9(10H)-ANTHRACENONE see MSA750

1,2-DIHYDROXY-2-OXO-N-(2,6-XYLYL)-3-PYRIDINECARBOXAMIDE see XLS300

R-7,t-8-DIHYDROXY-t-9,10-OXY-7,8,9,10-TETRAHYDROBENZO(a)PYRENE see DMQ800

anti-r-7,trans-8-DIHYDROXY-trans-9,10-OXY-7,8,9,10-TETRAHYDROBENZO(a)PYRENE see BCU250

trans-7,8-DIHYDROXY-9,10-OXY-7,8,9,10-TETRAHYDROBENZO(a)PYRENE see BCU250

R-4,T-3-DIHYDROXY-T-1,2-OXY-1,2,3,4-TETRAHYDROCHRYSENE see TCN300

2,2'-DIHYDROXY-3,3',5,5',6-PENTACHLOROBENZANILIDE see DMZ000

1,5-DIHYDROXYPENTANE see PBK750

3,5-DIHYDROXYPHENOL see PGR000

DIHYDROXYPHENYLACETIC ACID see HGI980

3,4-DIHYDROXY-PHENYLACETIC ACID see HGI980

DIHYDROXY-l-PHENYLALANINE see DNA200

3,4-DIHYDROXYPHENYLALANINE see DNA200

(−)-3,4-DIHYDROXYPHENYLALANINE see DNA200

l-DIHYDROXYPHENYL-l-ALANINE see DNA200

3,4-DIHYDROXYPHENYL-l-ALANINE see DNA200

3,4-DIHYDROXY-l-PHENYLALANINE see DNA200

l-3,4-DIHYDROXYPHENYLALANINE see DNA200

(−)-3-(3,4-DIHYDROXYPHENYL)-l-ALANINE see DNA200

3-(3,4-DIHYDROXYPHENYL)-l-ALANINE see DNA200

l-3-(3,4-DIHYDROXYPHENYL)ALANINE see DYC200

l-α-DIHYDROXYPHENYLALANINE see DNA200

β-(3,4-DIHYDROXYPHENYL)-l-ALANINE see DNA200

l-β-(3,4-DIHYDROXYPHENYL)ALANINE see DNA200

l-3,4-DIHYDROXYPHENYL-α-ALANINE see DNA200

β-(3,4-DIHYDROXYPHENYL)-α-ALANINE see DNA200

l-3-(3,4-DIHYDROXYPHENYL)ALANINE METHYL ESTER see DYC300

1-(3,4-DIHYDROXYPHENYL)-2-AMINO-1-BUTANOL HYDROCHLORIDE see ENX500

l-1-(3,4-DIHYDROXYPHENYL)-2-AMINOETHANOL see NNO500

3,4-DIHYDROXYPHENYL)-1-AMINO-2-ETHANOL-1-HYDROCHLORIDE see NNP050

3,4-DIHYDROXYPHENYLAMINOPROPANOL HYDROCHLORIDE see AMB000

5,7-DIHYDROXY-2-PHENYL-4H-1-BENZOPYRAN-4-ONE see DMS900

1-(3,5-DIHYDROXYPHENYL)-2-tert-BUTYLAMINOETHANOL SULPHATE see TAN250

2-(3,4-DIHYDROXYPHENYL)-2,3-DIHYDRO-3,5,7-TRIHYDROXY-4H-1-BENZOPYRAN-4-ONE see DMD000

(2R-trans)-2-(3,4-DIHYDROXYPHENYL)-2,3-DIHYDRO-3,5,7-TRIHYDROXY-4H-1-BENZOPYRAN-4-ONE see DMD000

α-(3,4-DIHYDROXYPHENYL)-β-DIMETHYLAMINOETHANOL see MJV000

l-3,4-DIHYDROXYPHENYLETHANOLAMINE see NNO500

DIHYDROXYPHENYLETHANOLISOPROPYLAMINE see DMV600

1-(2,4-DIHYDROXYPHENYL)ETHANONE see DMG400

3,4-DIHYDROXYPHENYLETHYLMETHYLAMINE HYDROCHLORIDE see EAZ000

3,4-DIHYDROXYPHENYLGLYOXIME see DNA300

meso-3,4-DI(p-HYDROXYPHENYL)-n-HEXANE see DLB400

γ,Δ-DI(p-HYDROXYPHENYL)-HEXANE see DLB400

3,4'(4,4'-DIHYDROXYPHENYL)HEX-3-ENE see DKA600

1-(3,4-DIHYDROXYPHENYL)-1-HYDROXY-2-AMINOBUTANE HYDROCHLORIDE see ENX500

1-3,5-DIHYDROXY-PHENYL-2-((1-(4-HYDROXYBENZYL)ETHYL)AMINO)-ETHANOL) HYDROBROMIDE see FAQ100

α-(3,4-DIHYDROXYPHENYL)-α-HYDROXY-β-DIMETHYLAMINOETHANE see MJV000

7-(3-(2-(3,5-DIHYDROXYPHENYL-2-HYDROXY-ETHYLAMINO)PROPYL))THEOPHYLLINE HYDROCHLORIDE see DNA600

1-(3,4-DIHYDROXYPHENYL)-2-ISOPROPYLAMINOETHANOL see DMV600

(±)1-(3,4-DIHYDROXYPHENYL)-2-ISOPROPYLAMINOETHANOL HYDROCHLORIDE see IQS500

1-(3,5-DIHYDROXYPHENYL)-2-(ISOPROPYLAMINO)ETHANOL SULFATE see MDM800

dl-α-3,4-DIHYDROXYPHENYL-β-ISOPROPYLAMINOETHANOL SULFATE see IRU000

l-(−)-3-(3,4-DIHYDROXYPHENYL)-2-METHYLALANINE see DNA800

l(−)-β-(3,4-DIHYDROXYPHENYL)-α-METHYLALANINE see DNA800

3,4-DIHYDROXYPHENYL-1-METHYLAMINO-2-ETHANE HYDROCHLORIDE see EAZ000

1-1-(3,4-DIHYDROXYPHENYL)-2-METHYLAMINOETHANOL see VGP000

1-1-(3,4-DIHYDROXYPHENYL)-2-METHYLAMINO-1-ETHANOL HYDROCHLORIDE see AES500

1-(3,4-DIHYDROXYPHENYL)-2-(METHYLAMINO)-ETHANONE (9CI) see MGC350

11-β,17-DIHYDROXY-21-((((1-PHENYLMETHYL)-1H-INDAZOL-3-YL)OXY)ACETYLOXY)PREGN-4-ENE-3,20-DIONE see BAV275

2-(3,4-DIHYDROXYPHENYL)-2,3,4,5,7-PENTAHYDROXY-1-BENZOPYRAN see HBA259

(DIHYDROXYPHENYL)PHENYL MERCURY see PFO250

α-DI(p-HYDROXYPHENYL)PHTHALIDE see PDO750

2,2-DI(4-HYDROXYPHENYL)PROPANE see BLD500

β-DI-p-HYDROXYPHENYLPROPANE see BLD500

3,4-DIHYDROXYPHENYLPROPANOLAMINE HYDROCHLORIDE see AMB000

3-(3,4-DIHYDROXYPHENYL)-2-PROPENOIC ACID (9CI) see CAK375

2-(3,4-DIHYDROXYPHENYL)-3,5,7-TRIHYDROXY-4H-1-BENZOPYRAN-4-ONE see QCA000

2-(3,4-DIHYDROXYPHENYL)-3,5,7-TRIHYDROXY-1-BENZOPYRYLIUM CHLORIDE see COI750

(11-β)-11,17-DIHYDROXY-21-(PHOSPHONOOXY)-PREGN-4-ENE-3,20-DIONE (9CI) see HHQ875

DIHYDROXYPHTHALOPHENONE see PDO750

17,21-DIHYDROXYPREGNA-1,4-DIENE-3,11,20-TRIONE see PLZ000

11-β,21-DIHYDROXYPREGN-3,20-DIONE see CNS625

(11-β)-11,21-DIHYDROXY-PREGN-4-ENE-3,20-DIONE (9CI) see CNS625

11-β,21-DIHYDROXY-PREGN-4-ENE-3,20-DIONE see CNS625

11-β,21-DIHYDROXY-PREGN-4-ENE-3,20-DIONE-21-ACETATE see CNS650

11-β,21-DIHYDROXYPREGN-4-ENE-3,20-DIONE ACETATE see CNS650

16-α,17-DIHYDROXYPREGN-4-ENE-3,20-DIONE CYCLIC ACETAL with ACETOPHENONE see DAM300

17-α,17-DIHYDROXY-14-α-PREGN-4-ENE-3,20-DIONE 21-IODOACETATE see DNA850

17α,21-DIHYDROXY-4-PREGNENE-3,11,20-TRIONE see CNS800

17,21-DIHYDROXYPREGN-4-ENE-3,11,20-TRIONE ACETATE see CNS825

17,21-DIHYDROXY-PREGN-4-ENE-3,11,20-TRIONE 21-ACETATE see CNS825

11,12-DIHYDROXYPROGESTERONE see CNS625

11-β,21-DIHYDROXYPROGESTERONE see CNS625

DIHYDROXYPROGESTERONE ACETOPHENIDE see DAM300

1,2-DIHYDROXYPROPANE see PML000

1,3-DIHYDROXYPROPANE see PML250

1,3-DIHYDROXYPROPANONE see OLW100

4-(2'-(β,γ-DIHYDROXYPROPOXYCARBONYL)PHENYLAMINO)-7-CHLOROQUINOLEINE see GGQ050

4-(2'-β,γ-DIHYDROXYPROPOXYCARBONYLPHENYLAMINO)-7-CHLOROQUINOLINE see GGQ050

2,3-DIHYDROXYPROPYL ACETATE see GGO000

(S)-9-(2,3-DIHYDROXYPROPYL)ADENINE see AMH800

(R,S)-9-(2,3-DIHYDROXYPROPYL)ADENINE see AMH850

(17R,21-α)-17,21-DIHYDROXY-4-PROPYLAJMALANIUM see PNC875

17R,21-α-DIHYDROXY-4-PROPYLAJMALANIUM HYDROGEN TARTRATE see DNB000

(17R,21-α)-17,21-DIHYDROXY-4-PROPYLAJMALINIUM BROMIDE see PNC925

2,3-DIHYDROXYPROPYLAMINE see AMA250

DI(2-HYDROXY-n-PROPYL)AMINE see DNB200

4-(2,3-DIHYDROXYPROPYLAMINO)-2-(5-NITRO-2-THIENYL)QUINAZOLINE see DNB600

2,3-DIHYDROXYPROPYL CHLORIDE see CDT750

(R)-2',3'-DIHYDROXYPROPYL-5-DEOXY-5-DIMETHYLARSINOYL-β-d-RIBOSIDE see DNB700

(R)-2,3-DIHYDROXYPROPYL-5-DEOXY-5-(DIMETHYLARSINYL)β-d-RIBOFURANOSIDE see DNB700

7-(2,3-DIHYDROXYPROPYL)-3,7-DIHYDRO-1,3-DIMETHYL-1H-PURINE-2,5-DIONE see DNC000

N,N-DI-(2-HYDROXYPROPYL)NITROSAMINE see DNB200

1-((2,3-DIHYDROXYPROPYL)NITROSAMINO)-2-PROPANONE see NJY550

4-(o-(2',3'-DIHYDROXYPROPYLOXYCARBONYL)PHENYL)-AMINO-8-TRIFLUOROMETHYLQUINOLINE see TKG000

DIHYDROXYPROPYL THEOPHYLLINE see DNC000

7-(2,3-DIHYDROXYPROPYL)THEOPHYLLINE see DNC000

DIHYDROXYPROPYL THEOPYLIN (GERMAN) see DNC000

(1,2-DIHYDROXY-3-PROPYL)THIOPHYLLIN see DNC000

2,3-DIHYDROXYPROPYL-N-(8-(TRIFLUOROMETHYL)-4-QUINOLYL)ANTHRANILATE see TKG000

N-(1,3-DIHYDROXY-2-PROPYL)VALIOLAMINE see DNC100

2,4-DIHYDROXY-2H-PYRAN-Δ-3(6H),α-ACETIC ACID-3,4-LACTONE see CMV000

(2,4-DIHYDROXY-2H-PYRAN-3(6H)-YLIDENE)ACETIC ACID-3,4-LACTONE see CMV000

5,7-DIHYDROXY-PYRIDOTETRAZOLE-6-CARBONITRILE see DND900

4,6-DIHYDROXYPYRIMIDIN see PPP100

2,4-DIHYDROXYPYRIMIDINE see UNJ800

5,5-DIHYDROXY-2,4,6(1H,3H,5H)-PYRIMIDINETRIONE see DMG950

4,8-DIHYDROXYQUINALDIC ACID see DNC200

4,8-DIHYDROXYQUINALDINIC ACID see DNC200

2,4-DIHYDROXYQUINAZOLINE see QEJ800

4,8-DIHYDROXYQUINOLINE-2-CARBOXYLIC ACID see DNC200

2,3-DIHYDROXYQUINOXALINE see QRS000

2,3-DIHYDROXY-(R-(R*,R*))-BUTANEDIOIC ACID DISODIUM SALT (9CI) see BLC000

8,8'-DIHYDROXY-RUGULOSIN see LIV000

12,18-DIHYDROXY-SENECIONAN-11,16-DIONE see RFP000

3',6'-DIHYDROXYSPIRO(ISOBENZOFURAN-1(3H),9'(9H)-XANTHEN)-3-ONE see FEV000

4,4'-((1,8-DIHYDROXY-4-SULFONAPHTHALENE-2,7-DIYL)BIS(AZO-4,1-PHENYLENEAZO))BIS(1-HYDROXY-2-NAPHTHALENECARBOXYLIC ACID see DNC300

(+−)-trans-1,2-DIHYDROXY- SYN-3,4-EPOXY-1,2,3,4-TETRAHYDRO-5-METHYLCHRYSENE see CML831

2,2'-DIHYDROXY-3,3',5,5'-TETRACHLORODIPHENYLSULFIDE see TFD250

trans-9,10-DIHYDROXY-9,10,11,12-TETRAHYDROBENZO(e)PYRENE see DNC400

trans-1,2-DIHYDROXY-1,2,3,4-TETRAHYDROCHRYSENE see DNC600

trans-3,4-DIHYDROXY-1,2,3,4-TETRAHYDRODIBENZ(a,h)ANTHRACENE see DNC800

trans-3,4-DIHYDROXY-1,2,3,4-TETRAHYDRODIBENZO(a,h)ANTHRACENE see DNC800

trans-1,2-DIHYDROXY-1,2,3,4-TETRAHYDROTRIPHENYLENE see DND000

1,1'-(2,3-DIHYDROXYTETRAMETHYLENE)BIS(4-FORMYLPYRIDINIUM) DIPERCHLORATE, DIOXIME see DND400

6,6'-DIHYDROXY-3,3,3',3'-TETRAMETHYL-1,1'-SPIROBIINDANE see SLD570

1,8-DIHYDROXY-2,4,5,7-TETRANITROANTHRAQUINONE (chrysamminic acid) (DOT) see CML600

5,7-DIHYDROXYTETRAZOLO(1,5-a)PYRIDINE-6-CARBONITRILE see DND900

2,3-DIHYDROXYTOLUENE see DNE200

2,5-DIHYDROXYTOLUENE see MKO250

2,6-DIHYDROXYTOLUENE see MPH400

3,4-DIHYDROXYTOLUENE see DNE200

3,5-DIHYDROXYTOLUENE see MPH500
α-2-DIHYDROXYTOLUENE see HMK100
1,3-DIHYDROXY-2,4,6-
TRINITROBENZENE see SMP500
2,4-DIHYDROXY-1,3,5-
TRINITROBENZENE see SMP500
3,5-DIHYDROXY-2,6,6-TRIS(3-METHYL-2-
BUTENYL)-4-(3-METHYL-1-OXOBUTYL)-
2,4-CYCLOHEXADIEN-1-ONE see LIU000
2,5-DIHYDROXY-3-UNDECYL-1,4-
BENZOQUINONE see EAJ600
2,5-DIHYDROXY-3-UNDECYL-2,5-
CYCLOHEXADIENE-1,4-DIONE (9CI) see
EAJ600
6-(6,10-DIHYDROXYUNDECYL)-β-
RESORCYLIC ACID-μ-LACTONE see
RBF100
DIHYDROXYVIOLANTHRON (CZECH) see
DMJ800
16,17-DIHYDROXYVIOLANTHRONE see
DMJ800
DIHYDROXYVITAMIN D3 see DMJ400
24(R),25-DIHYDROXYVITAMIN D3 see
HJS900
1-α-DIHYDROXYVITAMIN D3 see HJV000
1-α,25-DIHYDROXYVITAMIN D3 see
DMJ400
1-α,24(R)-DIHYDROXYVITAMIN D3 see
SBL600
DIHYDROZEATIN see RAF400
(+−)-DIHYDROZEATIN see RAF400
DIHYXAL see OLW100
1,4-DIIDROBENZENE (ITALIAN) see
HIH000
DIIDROBENZO(1-4)TIAZINA
CLORIDRATO (ITALIAN) see DKU875
DIIDRO-5,5-DIETIL-2H-1,3-OSSAZIN-
2,4(3H)-DIONE (ITALIAN) see DJT400
DIIDROXI-1,4-BENZENESULFONATO-3-
DI-ETILAMMONIUM (ITALIAN) see DIS600
1,4-DIIMIDO-2,5-CYCLOHEXADIENE see
BDD200
1,3-DIIMINOISOINDOLIN (CZECH) see
DNE400
1,3-DIIMINOISOINDOLINE see DNE400
DIINDIUM TRIOXIDE see ICI100
DIINDOGEN see BGB275
DIIODAMINE see DNE600
DIIODBENZOTEPH see DNE800
DIIODOACETYLENE see DNE500
DIIODOAMINE see DNE600
3,5-DIIODOANTHRANILIC ACID see
API800
1,2-DIIODOBENZENE see DNE700
2,6-DIIODO-1,4-BENZENEDIOL see
DNF200
2,6-DIIODO-p-BENZOQUINONE see
DNG800
DIIODOBENZOTEF see DNE800
N-2,5-DIIODOBENZOYL-N',N',N"N"-
DIETHYLENEPHOSPHORTRIAMIDE see
DNE800
1,4-DIIODO-1,3-BUTADIYNE see DNE875
1,4-DIIODOBUTANE see TDQ400
2,2'-DIIODODIACETAMIDE see DNE900
α,α'-DIIODODIACETAMIDE see DNE900
cis-DIIODODIAMMINEPLATIUM (II) see
DNF000
DIIODOETHYNE see DNE500
1,6-DIIODOHEXANE see HEG200
2,6-DIIODOHYDROQUINONE see DNF200
3,5-DIIODO-4-HYDROXYBENZOIC ACID
see DNF300
3,5-DIIODO-4-HYDROXYBENZONITRILE
see HKB500
3,5-DIIODO-4-HYDROXYBENZONITRILE,
LITHIUM SALT see DNF400
3,5-DIIODO-4-HYDROXYBENZONITRILE
OCTANOATE see DNG200
DIIODO-3,3 HYDROXY-4 BENZOYL 2
FURANNE see DNF500

(DIIODO-3,5 HYDROXY-4 BENZOYL)-3
MESITYL-2 BENZOFURANNE see DNF550
3,5-DIIODO-4-HYDROXYPHENYL 2,5-
DIMETHYL-3-FURYL KETONE see DNF450
3,5-DIIODO-4-HYDROXYPHENYL 2-
ETHYL-3-BENZOFURANYL KETONE see
EID200
3,5-DIIODO-4-HYDROXYPHENYL 5-
ETHYL-2-FURYL KETONE see EID250
3,5-DIIODO-4-HYDROXYPHENYL 2-
FURYL KETONE see DNF500
3,5-DIIODO-4-HYDROXYPHENYL 2-
MESITYL-3-BENZOFURANYL KETONE see
DNF550
β-(3,5-DIIODO-4-HYDROXYPHENYL)-α-
PHENYLPROPIONIC ACID see PDM750
DIIODOHYDROXYQUIN see DNF600
DIIODOHYDROXYQUIN see DNF600
DIIODOHYDROXYQUINOLINE see
DNF600
5,7-DIIODO-8-HYDROXYQUINOLINE see
DNF600
3,5-DIIODO-4-(3'-IODO-4'-
ACETOXYPHENOXY)BENZOIC ACID see
TKP850
DIIODOMETHANE see DNF800
DIIODOMETHYLARSINE see MGQ775
DIIODOMETHYLATE de la
BIS(PIPERIDINOMETHYL-COUMARANYL-
5)CETONE (FRENCH) see COE125
1-((DIIODOMETHYL)SULFONYL)-4-
METHYLBENZENE see DNF850
DIIODOMETHYL p-TOLYL SULFONE see
DNF850
DIIODOMETILATO del
BISPIPERIDINOMETILCUMARANIL-5-
CHETONE (ITALIAN) see COE125
2,6-DIIODO-4-NITROPHENOL see DNG000
3,5-DIIODO-4-
OCTANOYLOXYBENZONITRILE see
DNG200
5,7-DIIODO-OXINE see DNF600
3,5-DIIODO-4-OXO-1(4H)-
PYRIDINEACETIC ACID COMPD. WITH 1-
DEOXY-1-(METHYLAMINO)GLUCITOL see
IFA100
3,5-DIIODO-4-OXO-
1(4H)PYRIDINEACETIC ACID-2,2'-
IMINODIETHANOL SALT see DNG400
1,5-DIIODOPENTANE see PBH125
3,5-DIIODO-α-PHENYLPHLORETIC ACID
see PDM750
1,3-DIIODOPROPANE see TLR050
3,5-DIIODO-4-PYRIDONE-N-ACETATE
BIS(HYDROXYETHYL)AMMONIUM see
DNG400
3,5-DIIODO-4-PYRIDONE-N-ACETIC ACID,
DIETHANOLAMINE SALT see DNG400
DIIODOQUIN see DNF600
2,6-DIIODOQUINOL see DNF200
5,7-DIIODO-8-QUINOLINOL see DNF600
DIIODOQUINONE see DNG800
3,5-DIIODOSALICYLIC ACID see DNH000
DIIRON TRISULFATE see FBA000
DIISOAMYL ADIPATE see AEQ500
DIISOAMYLMERCURY see DNL200
DIISOBUTENE see TMA250
DIISOBUTILCHETONE (ITALIAN) see
DNI800
DIISOBUTYL ADIPATE see DNH125
DIISOBUTYLALUMINIUM HYDRIDE see
DNI600
DIISOBUTYLALUMINUM CHLORIDE see
CGB500
DIISOBUTYLALUMINUM HYDRIDE see
DNI600
DIISOBUTYLALUMINUM
MONOCHLORIDE see CGB500
DIISOBUTYLAMINE see DNH400
DIISOBUTYLAMINOBENZOYLOXYPROPY
L THEOPHYLLINE see DNH500

α-
((DIISOBUTYLAMINO)METHYL)THEOPHY
LLINE-8-ETHANOL BENZOATE (ester) see
DNH500
DIISOBUTYL CARBINOL see DNH800
DI-ISOBUTYLCETONE (FRENCH) see
DNI800
DIISOBUTYLCHLOROALUMINUM see
CGB500
p-
DIISOBUTYLCRESOXYETHYLDIMETHYL
BENZYLAMMONIUM CHLORIDE
MONOHYDRATE see MHB500
DIISOBUTYLENE see TMA250
DIISOBUTYLENE OXIDE see DNI200
DIISOBUTYLESTER KYSELINY FTALOVE
see DNJ400
DIISOBUTYL FUMARATE see DNI400
DIISOBUTYLHYDROALUMINUM see
DNI600
DIISOBUTYLKETON (DUTCH, GERMAN)
see DNI800
DIISOBUTYL KETONE see DNI800
DI-ISO-BUTYLNITROSAMINE see DRQ200
DIISOBUTYLOXOSTANNANE see DNJ000
DIISOBUTYLPHENOXYETHOXYETHYLDI
METHYL BENZYL AMMONIUM
CHLORIDE see BEN000
p-
DIISOBUTYLPHENOXYETHOXYETHYLDI
METHYLBENZYLAMMONIUM CHLORIDE
MONOHYDRATE see BBU750
DIISOBUTYL PHTHALATE see DNJ400
DIISOBUTYLSULFIDE HYDRATE see
IJO000
DIISOBUTYLTHIOCARBAMIC ACID-S-
ETHYL ESTER see EID500
DIISOBUTYLTIN OXIDE see DNJ000
DIISOBUTYRYL PEROXIDE see DNJ600
o,o'-DIISOBUTYRYLTHIAMINE DISULFIDE
see BKJ325
DIISOCARB see EID500
4-4'-DIISOCYANATE de
DIPHENYLMETHANE (FRENCH) see
MJP400
DI-ISOCYANATE de TOLUYLENE see
TGM750
1,3-DIISOCYANATOBENZENE see BBP000
1,4-DIISOCYANATOBENZENE see PFA300
trans-1,4-DIISOCYANATOCYCLOHEXANE
see CPB150
4,4'-DIISOCYANATO-3,3'-DIMETHOXY-1,1'-
BIPHENYL see DCJ400
4,4'-DIISOCYANATO-3,3'-DIMETHYL-1,1'-
BIPHENYL see DQS000
4,4'-DIISOCYANATODIPHENYLMETHANE
see MJP400
1,6-DIISOCYANATOHEXANE see DNJ800
1,6-DIISOCYANATOHEXANE
HOMOPOLYMER see HEG300
DI-ISO-CYANATOLUENE see TGM750
DIISOCYANATOMETHANE see DNK100
DIISOCYANATOMETHYLBENZENE see
DNK200
DIISOCYANATOMETHYLBENZENE see
TGM740
2,6-DIISOCYANATO-1-METHYLBENZENE
see TGM800
2,4-DIISOCYANATO-1-METHYLBENZENE
(9CI) see TGM750
1,3-DIISOCYANATOMETHYLBENZENE
POLYMER WITH NIAX E 488 see DNK250
1,5-DIISOCYANATONAPHTHALENE see
NAM500
DIISOCYANATOTOLUENE see TGM740
2,4-DIISOCYANATOTOLUENE see TGM750
2,6-DIISOCYANATOTOLUENE see TGM800
DIISOCYANAT-TOLUOL see TGM750
4,4'-(2,3-DIISOCYANO-1,3-BUTADIENE-1,4-
DIYL)BIS-1,2-BENZENEDIOL see XCS700
DIISODECYLISOPHTHALATE see BBO700
DI-ISODECYL PHTHALATE see DEK550

DIISODECYL PHTHALATE see DEK550
DIISODECYL PHTHALATE see PHW575
DIISODECYL TETRAHYDRO-4,5-EPOXYPHTHALATE see FCD570
DIISONITROSOACETONE see DNK300
2,3-DIISONITROSOBUTANE see DBH000
DIISONONYL PHTHALATE see DNK400
DIISONONYL PHTHALATE see PHW585
DIISONONYLTIN DICHLORIDE see BLQ750
DIISOOCTYL ACID PHOSPHATE see DNK800
DIISOOCTYL ((DIOCTYLSTANNYLENE)DITHIO)DIACETATE see BKK750
DIISOOCTYL PHOSPHATE (DOT) see DNK800
DIISOOCTYL PHTHALATE see ILR100
DIISOPENTYLMERCURY see DNL200
DIISOPENTYLOXOSTANNANE see DNL400
DIISOPENTYLPHTHALATE see BBO725
DIISOPENTYLRTUT see DNL200
DIISOPENTYLTIN OXIDE see DNL400
DIISOPHENOL see DNG000
DIISOPROPANOLAMINE see DNL600
DIISOPROPANOLNITROSAMINE see DNB200
m-DIISOPROPENYLBENZENE see DNL700
p-DIISOPROPENYLBENZENE see IMX100
1,3-DIISOPROPENYLBENZENE see DNL700
1,4-DIISOPROPENYLBENZENE see IMX100
N,N'-DIISOPROPIL-FOSFORODIAMMIDO-FLUORURO (ITALIAN) see PHF750
DIISOPROPOXYPHOSPHORYL FLUORIDE see IRF000
((DIISOPROPROXYPHOSPHINOTHIOYL)THIO)TRICYCLOHEXYL STANNANE see DNT200
s-DIISOPROPYLACETONE see DNI800
DIISOPROPYL ADIPATE see DNL800
DI(ISOPROPYLAMIDO)PHOSPHORYLFLUORIDE see PHF750
DIISOPROPYLAMINE see DNM200
DIISOPROPYLAMINE DICHLORACETATE see DNM400
DIISOPROPYLAMINE, compd. with DICHLOROACETIC ACID (1:1) see DNM400
DIISOPROPYLAMINE DICHLOROETHANOATE see DNM400
2-(DIISOPROPYLAMINO)-2',6'-ACETOXYLIDIDE HYDROCHLORIDE see DNM600
2-DIISOPROPYLAMINOETHANOL see DNP000
2-(2-(DIISOPROPYLAMINO)ETHOXY)BUTYROPHENONE HYDROCHLORIDE see DNN000
2-(DIISOPROPYLAMINO)ETHYL CHLORIDE HYDROCHLORIDE see CGV600
S-(2-DIISOPROPYLAMINOETHYL)-O-ETHYL METHYL PHOSPHONOTHIOLATE see EIG000
α-(2-(DIISOPROPYLAMINO)ETHYL)-α-PHENYL-2-PYRIDINEACETAMIDE see DNN600
α-(2-DIISOPROPYLAMINOETHYL)-α-PHENYL-2-PYRIDINEACETAMIDE PHOSPHATE see RSZ600
β-DIISOPROPYLAMINOETHYL-9-XANTHENECARBOXYLATE METHOBROMIDE see HKR500
2,6-DIISOPROPYLAMINO-4-METHOXYTRIAZINE see MFL250
γ-DIISOPROPYLAMINO-α-PHENYL-α-(2-PYRIDYL)BUTYRAMIDE see DNN600
DIISOPROPYLAMMINE-trans-DIHYDROXYMALONATOPLATINUM(IV) see IGG775
DIISOPROPYLAMMONIUM DICHLOROACETATE see DNM400

DIISOPROPYLAMMONIUM DICHLOROETHANOATE see DNM400
2,6-DIISOPROPYL ANILINE see DNN630
DIISOPROPYLBENZENE see DNN709
m-DIISOPROPYLBENZENE see DNN829
o-DIISOPROPYLBENZENE see DNN800
p-DIISOPROPYLBENZENE see DNN830
1,3-DIISOPROPYLBENZENE see DNN829
1,4-DIISOPROPYLBENZENE see DNN830
DIISOPROPYLBENZENE HYDROPEROXIDE, not more than 72% in solution (DOT) see DNS000
DIISOPROPYLBENZENE PEROXIDE see DGR600
1,3-DIISOPROPYLBENZENE SODIUM SALT, DIHYDROPEROXIDE see DNN840
1,4-DIISOPROPYLBENZENE SODIUM SALT, DIISOPEROXIDE see DNN850
p-DIISOPROPYLBENZOL see DNN830
N,N-DIISOPROPYL-2-BENZOTHIAZOLESULFENAMIDE see DNN900
O,O-DIISOPROPYL-S-BENZYL PHOSPHOROTHIOLATE see BKS750
O,O-DIISOPROPYL-S-BENZYL THIOPHOSPHATE see BKS750
DIISOPROPYLBERYLLIUM see DNO200
DIISOPROPYLCARBAMIC ACID, ETHYL ESTER see DNP600
DIISOPROPYLCARBODIIMIDE see DNO400
N,N'-DIISOPROPYL-DIAMIDO-FOSFORZUUR-FLUORIDE (DUTCH) see PHF750
N,N-DIISOPROPYL-DIAMIDO-PHOSPHORSAEURE-FLUORID (GERMAN) see PHF750
N,N'-DIISOPROPYLDIAMIDOPHOSPHORYL FLUORIDE see PHF750
O,O-DIISOPROPYL-S-DIETHYLDITHIOCARBAMOYLPHOSPHORODITHIOATE see DKB600
DI-ISOPROPYL 1,3-DITHIOLANE-2-YLIDENEMALONATE see IRU500
DIISOPROPYL-1,3-DITHIOL-2-YLIDENEMALONATE see MAO275
O,O-DIISOPROPYL DITHIOPHOSPHORIC ACID ESTER of-N,N-S-DIETHYLTHIOCARBAMOYL-O,O-DIISOPROPYL PHOSPHOROTHIOATE see DKB600
N-(2-(O,O-DIISOPROPYLDITHIOPHOSPHORYL)ETHYL)BENZENESULFONAMIDE see DNO800
N-(β-O,O-DIISOPROPYLDITHIOPHOSPHORYLETHYL)BEZENESULFONAMIDE see DNO800
DIISOPROPYL DIXANTHOGEN see IRS500
DIISOPROPYL ESTER of DITHIOCARBAMYL PHOSPHOROTHIOIC ACID see DKB600
DIISOPROPYL ESTER SULFURIC ACID see DNO900
DIISOPROPYL ETHANOLAMINE see DNP000
N,N-DIISOPROPYL ETHANOLAMINE see DNP000
DIISOPROPYL ETHER see IOZ750
DIISOPROPYL ETHYL CARBAMATE see DNP600
N,N-DIISOPROPYL ETHYL CARBAMATE see DNP600
N,N-DIISOPROPYL ETHYLENEDIAMINE see DNP700
O,O-DIISOPROPYL-S-ETHYLSULFINYLMETHYLDITHIOPHOSPHATE see EPH500
O,O-DIISOPROPYL-S-ETHYLSULFINYLMETHYL PHOSPHORODITHIOATE see EPH500
DIISOPROPYL FLUOROPHOSPHATE see IRF000

O,O-DIISOPROPYL FLUOROPHOSPHATE see IRF000
DIISOPROPYL FLUOROPHOSPHONATE see IRF000
DIISOPROPYLFLUOROPHOSPHORIC ACID ESTER see IRF000
DIISOPROPYLFLUORPHOSPHORSAEURE ESTER (GERMAN) see IRF000
DIISOPROPYL FUMARATE see DNQ200
DIISOPROPYL HYDROGEN PHOSPHITE see DNQ600
DIISOPROPYL(2-HYDROXYETHYL)METHYLAMMONIUMBROMIDE with XANTHENE-9-CARBOXYLATE see HKR500
3,5-DIISOPROPYL-4-HYDROXYPHENYL METHYLCARBAMATE see MIA800
DIISOPROPYL HYPONITRITE see DNQ700
DIISOPROPYLIDENE ACETONE see PGW250
sym-DIISOPROPYLIDENE ACETONE see PGW250
1:2,5:6-DI-O-ISOPROPYLIDENE-α-D-GLUCOFURANOSE see DVO100
1:2,5:6-DI-O-ISOPROPYLIDEN-α-D-GLUCOFURANOSE see DVO100
DIISOPROPYL KETONE see DTI600
DIISOPROPYLMERCURY see DNQ800
DIISOPROPYL METHANEPHOSPHONATE see DNQ875
N,N'-DIISOPROPYL-6-METHOXY-1,3,5-TRIAZINE-2,4-DIYLDIAMINE see MFL250
DIISOPROPYL METHYLPHOSPHONATE see DNQ875
DIISOPROPYLNAPHTHALENE see BKL600
3,5-DIISOPROPYL-4-NITROBIPHENYL see DNQ890
DIISOPROPYL-p-NITROPHENYL PHOSPHATE see DNR309
O,O-DIISOPROPYL-o,p-NITROPHENYL PHOSPHATE see DNR309
DIISOPROPYLNITROSAMIN (GERMAN) see NKA000
DIISOPROPYL OXIDE see IOZ750
DIISOPROPYLOXOSTANNANE see DNR200
DIISOPROPYL PARAOXON see DNR309
DIISOPROPYL PERDICARBONATE see DNR400
DIISOPROPYL PEROXYDICARBONATE see DNR400
2,6-DIISOPROPYLPHENOL see DNR800
3,5-DIISOPROPYLPHENOL METHYLCARBAMATE see DNS200
DIISOPROPYLPHENYLHYDROPEROXIDE (solution) see DNS000
2,6-DIISOPROPYLPHENYL ISOCYANATE see DNS100
3,5-DIISOPROPYLPHENYL METHYLCARBAMATE see DNS200
3,5-DIISOPROPYLPHENYL-N-METHYLCARBAMATE see DNS200
DIISOPROPYL PHOSPHITE see DNQ600
DIISOPROPYL PHOSPHOFLUORIDATE see IRF000
DIISOPROPYLPHOSPHONATE see DNQ600
O,O-DIISOPROPYL PHOSPHONATE see DNQ600
N,N'-DIISOPROPYLPHOSPHORODIAMIDIC FLUORIDE see PHF750
S-(O,O-DIISOPROPYL PHOSPHORODITHIOATE) ESTER of N-(2-MERCAPTOETHYL)BENZENESULFONAMIDE see DNO800
DIISOPROPYL PHOSPHOROFLUORIDATE see IRF000
O,O'-DIISOPROPYL PHOSPHORYL FLUORIDE see IRF000
DIISOPROPYL PHTHALATE see PHW600
DIISOPROPYL PYRIDINE-2,5-DICARBOXYLATE see IKC070

DI-ISOPROPYLSULFAT (GERMAN) see DNO900

DI-ISOPROPYLSULFATE see DNO900

DI-ISOPROPYL TARTRATE see TAF760

N-DIISOPROPYLTHIOCARBAMIC ACID-S-2,3,3-TRICHLOROALLYL ESTER see DNS600

N-DIISOPROPYLTHIOCARBAMIC ACID S-2,3,3-TRICHLORO-2-PROPENYL ESTER see DNS600

DI-ISOPROPYLTHIOLOCARBAMATE de S-(2,3-DICHLOROALLYLE) (FRENCH) see DBI200

o,o-DIISOPROPYLTHIOLPHOSPHORIC ACID see IRF500

o,o-DIISOPROPYL THIOPHOSPHATE see IRF500

DIISOPROPYL THIOUREA see DNS800

1,3-DIISOPROPYLTHIOUREA see DNS800

N,N'-DIISOPROPYLTHIOUREA see DNS800

DIISOPROPYLTIN DICHLORIDE see DNT000

DIISOPROPYLTIN OXIDE see DNR200

N,N-DIISOPROPYL-2,3,3-TRICHLORALLYL-THIOLCARBAMAT (GERMAN) see DNS600

DIISOPROPYLTRICHLOROALLYLTHIOCARBAMATE see DNS600

O,O-DIISOPROPYL-S-TRICYCLOHEXYLTIN PHOSPHORODITHIOATE see DNT200

DIISOPROPYL XANTHOGENATE DISULFIDE see IRS500

DIISOPROPYL XANTHOGEN DISULFIDE see IRS500

DIISOPYRAMIDE PHOSPHATE see RSZ600

1,4-DIISOTHIOCYANATOBENZENE see PFA500

1,2-DIISOTHIOCYANATOETHANE see ISK000

4,4'-DIISOTHIOINDIGO see DNT300

1,6-DIJOD-1,6-DIDESOXY-d-MANNIT see DHA375

3,5-DIJOD-4-HYDROXY-BENZONITRIL (GERMAN) see HKB500

3,5-DIJOD-4-HYDROXY-BENZONITRIL CAPRYSAEUREESTER (GERMAN) see DNG200

3,5-DIJOD-4-HYDROXY-BENZONITRILE LITHIUMSALZ (GERMAN) see DNF400

3,5-DIJOD-4-PYRIDON-N-ESSIGSAEURE METHYLGLUKAMIN see IFA100

DIKAIN see BQA010

DIKAIN HYDROCHLORIDE see TBN000

DIKAR see MAO880

3,5-DIKARBOXYBENZENSULFONAN SODNY (CZECH) see DEK200

DIKETENE see KFA000

DIKETENE, inhibited (DOT) see KFA000

2,3-DIKETOBUTANE see BOT500

4,4'-DIKETO-β-CAROTENE see CBE800

2,5-DIKETOHEXANE see HEQ500

1,3-DIKETOHYDRINDENE see IBS000

2,3-DIKETOINDOLINE see ICR000

DIKETONE ALCOHOL see DBF750

3,20-DIKETO-11-β,18-OXIDO-4-PREGNENE-18,21-DIOL see ECA100

2,5-DIKETOPYRROLIDINE see SND000

2,5-DIKETOTETRAHYDROFURAN see SNC000

DIKOL see PDM750

DIKONIT see SGG500

DIKOTEKS see SIL500

DIKOTEX 30 see SIL500

DI-KU-SHUANG see MJL750

DILABIL see DAL000

DILABIL SODIUM see SGD500

DILACORAN see IRV000

DILACTONE ACTINOMYCINDIOIC D ACID see AEB000

DILAFURANE see EID200

DILAHIL see DAL000

DILAN see BON250

DILANGIL see MAW250

DILANGIO see POD750

DILANTHANUM OXIDE see LBA100

DILANTHANUM TRIOXIDE see LBA100

DILANTIN see DKQ000

DILANTIN see DNU000

DILANTIN DB see DEP600

DILANTINE see DKQ000

DILANTIN SODIUM see DNU000

DILAPHYLLIN see HLC000

DILATAN KORE see ELH600

DILATIN see DNU100

DILATIN DB see DEP600

DILATOL HYDROCHLORIDE see DNU200

DILATYL see DNU200

DILAUDID see DNU300

DILAUDID HYDROCHLORIDE see DNU300

DILAUDID HYDROCHLORIDE see DNU310

DILAUROYL PEROXIDE see LBR000

DILAUROYL PEROXIDE, TECHNICAL PURE (DOT) see LBR000

DILAURYLESTER KYSELINY β',β'-THIODIPROPIONOVE see TFD500

DILAURYL PHTHALATE see PHW550

DILAURYL THIODIPROPIONATE see TFD500

DILAURYL 3,3'-THIODIPROPIONATE see TFD500

DILAURYL β-THIODIPROPIONATE see TFD500

DILAURYL β',β'-THIODIPROPIONATE see TFD500

DILAURYLTIN DICHLORIDE see DHA600

DILA-VASAL see EID200

DILAVASE see VGA300

DILAVASE see VGF000

DILAZEP/β-ACETYLDIGOXIN see CNR825

DILAZEP DIHYDROCHLORIDE see CNR750

DILCIT see HFG550

DILEAD(II) LEAD(IV) OXIDE see LDS000

DI-LEN see DNU000

DILENE see BIM500

DILEXPAL see HFG550

DILIC see HKC000

1,3-DILITHIOBENZENE see DNU325

DILITHIUM N-ACETYL-l-ASPARTATE see DNU330

DILITHIUM-1,1-BIS(TRIMETHYLSILYL)HYDRAZIDE see DNU350

DILITHIUM CARBONATE see LGZ000

DILITHIUM CHROMATE see LHD000

DILITURIC ACID see NET550

DILL FRUIT OIL see DNU400

DILL HERB OIL see DNU400

DILL HERB OIL, AMERICAN TYPE see DNU390

DILL OIL see DNU390

DILL OIL see DNU400

DILL OIL, INDIAN TYPE see DNU392

DILL SEED OIL see DNU400

DILL SEED OIL, AMERICAN TYPE see DNU390

DILL SEED OIL, EUROPEAN TYPE see DNU400

DILL SEED OIL, INDIAN see DNU392

DILL SEED OIL, INDIAN TYPE see DNU392

DILL WEED OIL see DNU400

DILOMBRIN see DJT800

DILOR see DLO880

DILOR see DNC000

DILOSPAN S see PGR000

DILOSYN see MDT500

DILOSYN see MPE250

DILOXOL see GGS000

DILTIAZEM HYDROCHLORIDE see DNU600

DILUEX see PAE750

DILURAN see AAI250

DILURGEN see SIG000

DILVASENE see FMX000

DILYN see RLU000

DILZEM see DNU600

DIMACIDE YELLOW N-5RL see SGP500

DIMAGNESIUM PHOSPHATE see MAH775

1,3-DIMALEIMIDOBENZENE see BKL750

1,6-DIMALEIMIDOHEXANE see HEG050

4,4'-DIMALEIMIDOPHENYLMETHANE see BKL800

p,p'-DIMALEIMIDOPHENYLMETHANE see BKL800

2,4-DIMALEIMIDOTOLUENE see TGY770

DIMALONE see DRB400

DIMANGANESE TRIOXIDE see MAT500

DIMANIN C see SGG500

DIMANTINE see DTC400

DIMAPP see DQA400

DIMAPYRIN see DOT000

DIMAS see DQD400

DIMATE 267 see DSP400

DIMATIF see DNU850

DIMAVAL see DNU860

DIMAYAL see DNU860

DIMAZINE see DSF400

DIMAZON see ACR300

DIMBOA see DMW300

DIMEBOLIN see TCQ260

DIMEBOLINE see TCQ260

DIMEBON see TCQ260

DIMEBON DIHYDROCHLORIDE see DNU875

DIMEBONE see TCQ260

DIMECRON see FAB400

DIMECRON 100 see FAB400

DIMECROTIC ACID MAGNESIUM SALT see DOK400

DIMEDROL see BBV500

DIMEDRYL see BBV500

DIMEFADANE see DRX400

DIMEFLINE see DNV000

DIMEFLINE HYDROCHLORIDE see DNV200

DIMEFOX see BJE750

DIMEFURON see UTA300

DIMEGLUMINE IOCARMATE see IDJ500

DIMEHYPO see DXC900

DIMEHYPO JUMBO see DXC900

DIMELIN see ABB000

DIMELONE see DRB400

DIMELOR see ABB000

DIMELQX see TLT768

DIMEMORFAN PHOSPHATE see MLP250

DIMENFORMON see EDO000

DIMENFORMON BENZOATE see EDP000

DIMENFORMON DIPROPIONATE see EDR000

DIMENFORMONE see EDP000

DIMENFORMON PROLONGATUM see EDO000

DIMENHYDRINATE see DYE600

DIMENOXADOL HYDROCHLORIDE see DPE200

DIMEPHENTHIOATE see DRR400

DIMEPHENTHOATE see DRR400

DIMEPIPERATE see MNU300

DIMERAY see IDJ500

DIMERCAPROL PROPANOL see BAD750

2,3-DIMERCAPTOBUTANEDIOIC ACID see DNV610

(R*,S*)-2,3-DIMERCAPTOBUTANEDIOIC ACID see DNV800

2,3-DIMERCAPTOBUTANEDIOIC ACID, DIETHYL ESTER see DJB100

(R*,S*)-1,4-DIMERCAPTO-2,3-BUTANEDIOL see DXN350

dl-threo-DIMERCAPTO-2,3-BUTANEDIOL see DXO775

d-threo-1,4-DIMERCAPTO-2,3-BUTANEDIOL see DXO800

(R*,R*)-(±)-1,4-DIMERCAPTO-2,3-BUTANEDIOL (9CI) see DXO775

1,2-DIMERCAPTOETHANE see EEB000

DIMERCAPTOL see BAD750

2,3-DIMERCAPTOL-1-PROPANOL see BAD750

1,2-DIMERCAPTO-4-METHYLBENZENE see TGN000

4,5-DI(MERCAPTOMETHYL)-2-METHYL-3-PYRIDINOL DITHIOACETATE HYDROBROMIDE see DBH800

1,2-DIMERCAPTOPROPANE see PML300

1,3-DIMERCAPTOPROPANE see PML350

2,3-DIMERCAPTOPROPANE see PML300

2,3-DIMERCAPTOPROPANE SODIUM SULPHONATE see DNU860

2,3-DIMERCAPTOPROPANESULFONIC ACID SODIUM SALT see DNU860

2,3-DIMERCAPTO-1-PROPANESULFONIC ACID SODIUM SALT see DNU860

DIMERCAPTOPROPANOL see BAD750

2,3-DIMERCAPTOPROPANOL see BAD750

2,3-DIMERCAPTOPROPAN-1-OL see BAD750

2,3-DIMERCAPTOPROPYL-p-TOLYSULFIDE see DNV000

4,5-DIMERCAPTOPYRIDOXINDI-THIOACETAT HYDROBROMID (GERMAN) see DBH800

DIMERCAPTOSUCCINIC ACID see DNV610

2,3-DIMERCAPTOSUCCINIC ACID see DNV610

meso-DIMERCAPTOSUCCINIC ACID see DNV800

meso-2,3-DIMERCAPTOSUCCINIC ACID see DNV800

α-β-DIMERCAPTOSUCCINIC ACID see DNV610

2,3-DIMERCAPTOSUCCINIC ACID, DIETHYL ESTER see DJB100

meso-DIMERCAPTOSUCCINIC ACID SODIUM SALT see DNU860

2,5-DIMERCAPTO-1,3,4-THIADIAZOLE see TES250

DIMERCUROUS METHANE ARSONATE see DNW000

DIMERCURY DICHLORIDE see MCY300

DIMERCURY IMIDE OXIDE see DNW200

DIMER CYKLOPENTADIENU (CZECH) see DGW000

DIMERIN see DNW400

DIMER X see IDJ500

DIMESTROL see DJB200

1,6-DIMESYL-d-MANNITOL see BKM500

1,4-DIMESYLOXYBUTANE see BOT250

1,4-DI(MESYLOXYETHYLAMINO)ERYTHRITOL see LJD500

DIMET see DXE600

DIMETACRINE see DNW700

DIMETACRINE BITARTRATE see DRM000

DIMETACRIN HYDROGENTARTRATE see DRM000

DIMETAN see DRL200

dl-DIMETANE MALEATE see DNW759

DIMETATE see DSP400

DIMETAZINA see SNN300

DIMETHACHLON see DGF000

DIMETHACHLOR see DSM500

DIMETHACHLORE see DSM500

DIMETHACIN see DNW700

DIMETHACINE see DNW700

DIMETHACRINE TARTRATE see DRM000

DIMETHADIONE see PMO250

DIMETHAEN see DPA000

DIMETHAMETRYN see ARW775

DIMETHAMETRYN see DTS700

DIMETHAMETRYNE see ARW775

DIMETHAMETRYNE see DTS700

2,5-DIMETHANESULFOMYLOXYHEXANE see DSU000

(R*,S*)-DIMETHANESULFONATE-meso-2,5-HEXANEDIOL (9CI) see DSU100

1,6-DIMETHANESULFONATE-d-MANNITOL see BKM500

1,4-DIMETHANESULFONATE THREITOL see TFU500

(2s,3s)-1,4-DIMETHANESULFONATE TREITOL see TFU500

1,4-DIMETHANESULFONOXYBUTANE see BOT250

cis-1,4-DIMETHANE SULFONOXY-2-BUTENE see DNW800

trans-1,4-DIMETHANE SULFONOXY-2-BUTENE see DNX000

1,4-DIMETHANE SULFONOXY-2-BUTYNE see DNX200

1,4-DIMETHANESULFONOXY-1,4-DIMETHYLBUTANE see DSU000

1,6-DIMETHANE-SULFONOXY-d-MANNITOL see BKM500

1:3-DIMETHANESULFONOXYPROPANE see TLR250

1,4-DI(METHANESULFONYLOXY)BUTANE see BOT250

DIMETHANESULFONYL PEROXIDE see DNX300

1,6-DIMETHANESULPHONOXY-1,6-DIDEOXY-d-MANNITOL see BKM500

1,3-DIMETHANESULPHONOXYPROPANE see TLR250

1,4-DIMETHANESULPHONYLOXYBUTANE see BOT250

1,4-DIMETHANOLCYCLOHEXANE DIVINYL ETHER see BLU600

DIMETHAZONE see CEP800

DIMETHESTERONE see DRT200

DIMETHICONE 350 see PJR000

DIMETHINDENE MALEATE see FMU409

DIMETHINDEN MALEATE see FMU409

DIMETHIOTAZINE see FMU039

DIMETHIPIN see OMY825

DIMETHIRIMOL see BRD000

DIMETHISOQUIN HYDROCHLORIDE see DNX400

DIMETHISTERON see DRT200

DIMETHISTERONE see DRT200

DIMETHISTERONE and ETHINYL ESTRADIOL see DNX500

DIMETHOAAT (DUTCH) see DSP400

DIMETHOAT (GERMAN) see DSP400

DIMETHOATE O-ANALOG see DNX800

DIMETHOATE-ETHYL see DNX600

DIMETHOATE OXYGEN ANALOG see DNX800

DIMETHOATE PO ISOLOGUE see DNX800

DIMETHOATE (USDA) see DSP400

DIMETHOAT TECHNISCH 95% see DSP400

DIMETHOCAINE see DNY000

DIMETHOGEN see DSP400

DIMETHOTHIAZINE see DUC400

DIMETHOTHIAZINE MESYLATE see FMU039

DIMETHOTHIAZINE METHANESULFONATE see FMU039

DIMETHOXANE see ABC250

DIMETHOXON see DNX800

2,11-DIMETHOXY-6aA-α-APORPHIN-1-OL HYDROCHLORIDE see ISD033

1,2-DIMETHOXY-4-ALLYLBENZENE see AGE250

3,4'-DIMETHOXY-4-AMINOAZOBENZENE see DNY400

2,6-DIMETHOXY-4-(p-AMINOBENZENESULFONAMIDO)PYRIMIDINE see SNN300

(trans)-2,5-DIMETHOXY-4'-AMINOSTILBENE see DON400

2,5-DIMETHOXYAMPHETAMINE HYDROCHLORIDE see DOJ800

3,4-DIMETHOXYAMPHETAMINE HYDROCHLORIDE see DOK000

2,4-DIMETHOXYANILINE see DNY500

2,5-DIMETHOXYANILINE see AKD925

2,4-DIMETHOXYANILINE HYDROCHLORIDE see MEA625

2,3-DIMETHOXYANILINE MUSTARD see BIC600

2-(2-(2,4-DIMETHOXYANILINO)VINYL)-1,3,3-TRIMETHYL-3H-INDOLIUM CHLORIDE see CMM890

1,5-DIMETHOXY-9,10-ANTHRACENEDIONE see DNY800

1,5-DIMETHOXYANTHRACHINON (CZECH) see DNY800

1,5-DIMETHOXYANTHRAQUINONE see DNY800

1-5,6-DIMETHOXYAPORPHINE see NOE500

(R)-1,2-DIMETHOXYAPORPHINE see NOE500

1,2-DIMETHOXY-6a-β-APORPHINE see NOE500

1,10-DIMETHOXY-6a-α-APORPHINE-2,9-DIOL see DNZ100

2,3-DIMETHOXYBENZALDEHYDE see DNZ200

3,4-DIMETHOXYBENZALDEHYDE see VHK000

7,12-DIMETHOXYBENZ(a)ANTHRACENE see DOA000

2,5-DIMETHOXYBENZENAMINE see AKD925

2,4-DIMETHOXYBENZENAMINE HYDROCHLORIDE see MEA625

m-DIMETHOXYBENZENE see REF025

o-DIMETHOXYBENZENE see DOA200

p-DIMETHOXYBENZENE see DOA400

1,2-DIMETHOXYBENZENE see DOA200

1,3-DIMETHOXYBENZENE see REF025

3,4-DIMETHOXYBENZENEACETIC ACID see HGK600

3,4-DIMETHOXY-BENZENEACETONITRILE (9CI) see VIK100

2,5-DIMETHOXYBENZENEAZO-β-NAPHTHOL see DOK200

3,4-DIMETHOXYBENZENECARBONAL see VHK000

2,6-DIMETHOXY-1,4-BENZENEDIOL see DON200

3,3'-DIMETHOXYBENZIDIN (CZECH) see DCJ200

3,3'-DIMETHOXYBENZIDINE see DCJ200

3,3'-DIMETHOXYBENZIDINE DIHYDROCHLORIDE see DOA800

3,3'-DIMETHOXYBENZIDINE-4,4'-DIISOCYANATE see DCJ400

1-(4,7-DIMETHOXY-2-BENZOFURANYL)ETHANONE see DOA810

1-(6,7-DIMETHOXY-2-BENZOFURANYL)ETHANONE see DOA815

4,7-DIMETHOXY-2-BENZOFURANYL METHYL KETONE see DOA810

6,7-DIMETHOXY-2-BENZOFURANYL METHYL KETONE see DOA815

3,4-DIMETHOXYBENZOIC ACID see VHP600

6,7-DIMETHOXYBENZOPYRAN-2-ONE see DRS800

1-(5,8-DIMETHOXY-2H-1-BENZOPYRAN-3-YL)ETHANONE see DOA820

1-(7,8-DIMETHOXY-2H-1-BENZOPYRAN-3-YL)ETHANONE see DOA830

5,8-DIMETHOXY-2H-1-BENZOPYRAN-3-YL METHYL KETONE see DOA820

7,8-DIMETHOXY-2H-1-BENZOPYRAN-3-YL METHYL KETONE see DOA830

1-(3,4-DIMETHOXYBENZOYL)-4-(1,2,3,4-TETRAHYDRO-2-OXO-6-QUINOLINYL)PIPERA ZINE see DLF700

3,4-DIMETHOXYBENZYLAMINE see VIK050

β-(2,4-DIMETHOXY-5-BENZYLBENZOYL)PROPIONIC ACID SODIUM SALT see DOA875

3,4-DIMETHOXYBENZYL CHLORIDE see BKM750

3-(3',5'-DIMETHOXYPHENOXY)PROPANEDIOL-(1,2) see DOJ700

3,4-DIMETHOXYPHENYL ACETIC ACID see HGK600

(3,4-DIMETHOXYPHENYL)ACETONE see VIK300

3,4-DIMETHOXYPHENYLACETONITRILE see VIK100

DI-p-METHOXYPHENYLAMINE see BKO600

1-(2,5-DIMETHOXYPHENYL)-2-AMINOPROPANE see DOJ800

1-(3,4-DIMETHOXYPHENYL)-2-AMINOPROPANE see DOK000

1-((2,5-DIMETHOXYPHENYL)AZO)-2-NAPHTHALENOL see DOK200

1-((2,5-DIMETHOXYPHENYL)AZO)-2-NAPHTHOL see DOK200

2,5-DIMETHOXY-1-(PHENYLAZO)-2-NAPHTHOL see DOK200

1-(1-(2,5-DIMETHOXYPHENYL)AZO)-2-NAPHTHOL see DOK200

3-(2,4-DIMETHOXYPHENYL)CROTONIC ACID MAGNESIUM SALT see DOK400

3,4-DIMETHOXYPHENYLETHANE see EQF150

1,1-DIMETHOXY-2-PHENYLETHANE see PDX000

DIMETHOXYPHENYLETHYLAMINE see DOE200

3,4-DIMETHOXYPHENYLETHYLAMINE see DOE200

2-(3,4-DIMETHOXYPHENYL)ETHYLAMINE see DOE200

3,4-DIMETHOXYPHENYLETHYLAMINE (base) see DOE200

β-(3,4-DIMETHOXYPHENYL)ETHYLAMINE see DOE200

3,4-DIMETHOXY-β-PHENYLETHYLAMINE see DOE200

3,4-DIMETHOXY-β-PHENYLETHYLAMINE HYDROCHLORIDE see DOI400

3-((2-((2-(3,4-DIMETHOXYPHENYL)ETHYL)AMINO)-2-OXOETHYL)AMINO)-N-METHYLBENZAMIDE see DOK500

4-(2-(2,5-DIMETHOXYPHENYL)ETHYL)BENZENAMINE see DON400

3-(2-(3,4-DIMETHOXYPHENYL)ETHYLCARBAMOYLMETHYL)AMINO-N-METHYLBENZAMIDE see DOK500

1-(3,4-DIMETHOXYPHENYL)-5-ETHYL-7,8-DIMETHOXY-4-METHYL-5H-2,3-BENZODIAZEPINE see GJS200

3,4-DIMETHOXYPHENYL ETHYL KETONE see PMX600

2-(3,4-DIMETHOXYPHENYL)-5-ETHYLTHIAZOLIDIN-4-ONE see KGU100

1-(2',5'-DIMETHOXYPHENYL)-2-GLYCINAMIDOETHANOL HYDROCHLORIDE see MQT530

β-(2,5-DIMETHOXYPHENYL)-β-HYDROXYISOPROPYLAMINE HYDROCHLORIDE see MDW000

2-(2,5-DIMETHOXYPHENYL)ISOPROPYLAMINE see DOK600

β-(2,5-DIMETHOXYPHENYL)ISOPROPYLAMINE HYDROCHLORIDE see DOJ800

DIMETHOXYPHENYLMETHANE see DOG700

1-((3,4-DIMETHOXYPHENYL)METHYL)-6,7-DIMETHOXYISOQUINOLINE see PAH000

((3,4-DIMETHOXYPHENYL)METHYL)HYDRAZINE see VIK150

DIMETHOXYPHENYLMETHYLSILANE see DOH400

1,1-DIMETHOXY-2-PHENYLPROPANE see HII600

1-(3,4-DIMETHOXYPHENYL)-1-PROPANONE see PMX600

1-(3,4-DIMETHOXYPHENYL)-2-PROPANONE see VIK300

1-(3,4-DIMETHOXYPHENYL)-2-PROPENE see AGE250

3-(4,6-DIMETHOXY-α-PHENYL-m-TOLUOYL)-PROPIONIC ACID SODIUM SALT see DOA875

2,2-DI-(p-METHOXYPHENYL)-1,1,1-TRICHLOROETHANE see MEI450

DI(p-METHOXYPHENYL)-TRICHLOROMETHYL METHANE see MEI450

DIMETHOXYPHOSPHINE OXIDE see DSG600

3-((DIMETHOXYPHOSPHINOTHIOYL)OXY)-2-METHYL-2-PROPENOIC ACID METHYL ESTER, (E)- see MDN150

((DIMETHOXYPHOSPHINOTHIOYL)THIO)BUTANEDIOIC ACID DIETHYL ESTER see MAK700

2-DIMETHOXYPHOSPHINOTHIOYLTHIOMETHYL-4,6-DIAMINO-s-TRIAZINE see ASD000

2-((DIMETHOXYPHOSPHINYL)OXY)-1H-BENZ(d,e)ISOQUINOLINE-1,3(2H)-DIONE see DOL400

3-((DIMETHOXYPHOSPHINYL)OXY)-2-BUTENOIC ACID METHYL ESTER see MQR750

(E)-3-((DIMETHOXYPHOSPHINYL)OXY)-2-BUTENOIC ACID 1-PHENYLETHYL ESTER (9CI) see COD000

3-(DIMETHOXYPHOSPHINYLOXY)-N,N-DIMETHYL-cis-CROTONAMIDE see DGQ875

3-(DIMETHOXYPHOSPHINYLOXY)-N,N-DIMETHYLISOCROTONAMIDE see DGQ875

3-(DIMETHOXYPHOSPHINYLOXY)-N-METHYL-N-METHOXY-cis-CROTONAMIDE see DOL800

3-(DIMETHOXYPHOSPHINYLOXY)N-METHYL-cis-CROTONAMIDE see MRH209

((DIMETHOXYPHOSPHINYL)THIO)ACETIC ACID ETHYL ESTER see DRB600

((DIMETHOXYPHOSPHINYL)THIO)-BUTANEDIOIC ACID DIETHYL ESTER (9CI) see OPK250

DIMETHOXY POLYETHYLENE GLYCOL see DOM100

1,1-DIMETHOXYPROPANE see DOM200

2,2-DIMETHOXYPROPANE see DOM400

3,3-DIMETHOXYPROPENE see DOM600

1,2-DIMETHOXY-4-PROPENYLBENZENE see IKR000

3,3-DIMETHOXYPROPIONALDEHYDE see DOM625

3,4-DIMETHOXYPROPIOPHENONE see PMX600

3',4'-DIMETHOXYPROPIOPHENONE see PMX600

N-(3,6-DIMETHOXY-4-PYRIDAZINYL)SULFANILAMIDE see DON700

N-(((4,6-DIMETHOXY-2-PYRIMIDINYL)AMINO)CARBONYL)-3-(ETHYLSULFONYL)-2-PYRIDINESULFONAMIDE see DOM700

N-(((4,6-DIMETHOXY-2-PYRIMIDINYL)AMINO)CARBONYL)-3-(TRIFLUOROMETHYL)-2-PYRIDINESULFONAMIDE see FCC050

1-(4,6-DIMETHOXYPYRIMIDIN-2-YL)-3-(3-ETHYLSULFONYL-2-PYRIDYLSULFONYL)UREA see DOM700

N¹-(2,6-DIMETHOXY-4-PYRIMIDINYL)SULFANILAMIDE see SNN300

N¹-(4,6-DIMETHOXYPYRIMIDIN-2-YL)SULFANILAMIDE see SNI500

1-(4,6-DIMETHOXYPYRIMIDIN-2-YL)-3-(3-TRIFLUOROMETHYL-2-PYRIDYLSULFONYL)UREA see FCC050

N'-(5,6-DIMETHOXY-4-PYRIMIDYL)SULFANILAMIDE see AIE500

4,7-DIMETHOXY-6-(2-PYRROLIDINYLETHOXY)-5-CINNAMOYLBENZOFURAN MALEATE see DON000

2,6-DIMETHOXYQUINOL see DON200

5,6-DIMETHOXYSTERIGMATOCYSTIN see HOL200

4-(2,5-DIMETHOXY)STILBENAMINE see DON400

2',5'-DIMETHOXYSTILBENAMINE see DON400

2,5-DIMETHOXY-4'-STILBENAMINE see DON400

2,3-DIMETHOXYSTRYCHNIDIN-10-ONE see BOL750

2,3-DIMETHOXYSTRYCHNINE see BOL750

DIMETHOXY STRYCHNINE (DOT) see BOL750

DIMETHOXYSULFADIAZINE see SNN300

2,4-DIMETHOXY-6-SULFANILAMIDO-1,3-DIAZINE see SNN300

3,6-DIMETHOXY-4-SULFANILAMIDOPYRIDAZINE see DON700

2,6-DIMETHOXY-4-SULFANILAMIDOPYRIMIDINE see SNN300

DIMETHOXYTETRAETHYLENE GLYCOL see PBO500

DIMETHOXYTETRAGLYCOL see PBO500

2,5-DIMETHOXYTETRAHYDROFURAN see DON800

(s-(4*,S*))-6,7-DIMETHOXY-3-(5,6,7,8-TETRAHYDRO-4-METHOXY-6-METHYL-1,3-DIOXOLO(4,5-g)ISOQUINOLIN-5-YL)-1(3H)-ISOBENZOFURANONE, N-OXIDE, HYDROCHLORIDE see NBP300

3',5'-DIMETHOXY-3,4',5,7-TETRAHYDROXYFLAVYLIUM ACID ANION see MAO750

DI(METHOXYTHIOCARBONYL)DISULFIDE see DUN600

1-(4,6-DIMETHOXY-1,3,5-TRIAZIN-2-YL)-3-(2-(2-METHOXYETHOXY)PHENYLSULFONYL)UREA (IUPAC) see CMS127

DIMETHOXY-2,2,2-TRICHLORO-1-N-BUTYRYLOXY-ETHYLPHOSPHINE OXIDE see BPG000

DIMETHOXY-2,2,2-TRICHLORO-1-HYDROXY-ETHYL-PHOSPHINE OXIDE see TIQ250

3,3'-DIMETHOXYTRIPHENYLMETHANE-4,4'-BIS(1"-AZO-2"-NAPHTHOL) see DOO400

6,7-DIMETHOXY-1-VERATRYLISOQUINOLINE see PAH000

6,7-DIMETHOXY-1-VERATRYLISOQUINOLINE-3-CARBOXYLIC ACID SODIUM SALT see PAG750

6,7-DIMETHOXY-1-VERATRYLISOQUINOLINE HYDROCHLORIDE see PAH250

DIMETHOXYVIOLANTHRONE see JAT000

16,17-DIMETHOXYVIOLANTHRONE see JAT000

DIMETHPRAMIDE see DUO300

DIMETHPYRINDENE MALEATE see FMU409

DIMETHRIN see DQQ500

DIMETHULENE see DRV000
DIMETHWLEN see DRV000
DIMETHYL see EDZ000
DIMETHYL see SDF000
DIMETHYLACETAL see DOO600
DIMETHYLACETAMIDE see DOO800
N,N-DIMETHYLACETAMIDE see DOO800
O,O-DIMETHYL-S-(2-ACETAMIDOETHYL) ESTER PHOSPHORODITHIOIC ACID see DOP200
1,1-DIMETHYL-3-(p-ACETAMIDOPHENYL)TRIAZENE see DUI000
3,5-DIMETHYLACETAMINOPHEN see DOO900
2,4-DIMETHYLACETANILIDE see ABO758
2,6-DIMETHYLACETANILIDE see ABP760
3,4-DIMETHYLACETANILIDE see ABP770
2',4'-DIMETHYLACETANILIDE see ABO758
3',4'-DIMETHYLACETANILIDE see ABP770
DIMETHYLACETIC ACID see IJU000
N,N-DIMETHYLACETOACETAMIDE see DOP000
2',4'-DIMETHYLACETOACETANILIDE see OOI100
DIMETHYLACETONE see DJN750
DIMETHYLACETONE AMIDE see DOO800
DIMETHYLACETONITRILE see IJX000
N,N-DIMETHYL-β-ACETOXY β-PHENYLETHYLAMINE see ABN700
O,O-DIMETHYL-S-(2-(ACETYLAMINO)ETHYL) DITHIOPHOSPHATE see DOP200
O,O-DIMETHYL-S-(2-ACETYLAMINOETHYL) PHOSPHORODITHIOATE see DOP200
DIMETHYLACETYL CHLORIDE see IJV100
DIMETHYLACETYLENE see COC500
DIMETHYLACETYLENECARBINOL see MHX250
DIMETHYL ACETYLENEDICARBOXYLIC ACID see DOP400
DIMETHYLACETYLENYLCARBINOL see MHX250
2,5-DIMETHYL-3-ACETYLFURAN see ACI400
O,S-DIMETHYLACETYLPHOSPHOROAMIDOTHIOATE see DOP600
DIMETHYL ACID PHOSPHITE see DSG600
2,7-DIMETHYL-3,6-ACRIDINEDIAMINE MONOHYDROCHLORIDE see DBT400
3,3-DIMETHYLACRYLAMIDE see SBW965
N,N-DIMETHYLACRYLAMIDE see DOP800
β,β-DIMETHYLACRYLAMIDE see SBW965
3,3-DIMETHYL-ACRYLATE de 2,4-DINITRO-6-(1-METHYLPROPYLE) PHENYLE (FRENCH) see BGB500
3,3-DIMETHYLACRYLIC ACID see MHT500
(E)-2,3-DIMETHYLACRYLIC ACID see TGA700
trans-2,3-DIMETHYLACRYLIC ACID see TGA700
β,β-DIMETHYLACRYLIC ACID see MHT500
trans-α-β-DIMETHYLACRYLIC ACID see TGA700
3,3-DIMETHYLACRYLIC ACID 2-sec-BUTYL-4,5-DINITROPHENYL ESTER see BGB500
cis-α,β-DIMETHYL ACRYLIC ACID, GERANIOL ESTER see GDO000
DIMETHYL ADIPATE see DOQ300
DIMETHYLAETHANOLAMIN (GERMAN) see DOY800
O,O-DIMETHYL-S-(2-AETHYLSULFINYL-AETHYL)-THIOLPHOSPHAT (GERMAN) see DAP000
O,O-DIMETHYL-S-(2-AETHYLSULFONYL-AETHYL)-THIOLPHOSPHAT (GERMAN) see DAP600
N-(5-(1,1-DIMETHYLAETHYL)-1,3,4-THIADIAZOL-2-YL)-N,N'-

DIMETHYLHARNSTOFF (GERMAN) see BSN000
O,O-DIMETHYL-S-(2-AETHYLTHIO-AETHYL)-DITHIO PHOSPHAT (GERMAN) see PHI500
O,O-DIMETHYL-O-(2-AETHYLTHIO-AETHYL) MONOTHIOPHOSPHAT (GERMAN) see DAO800
O,O-DIMETHYL-S-(2-AETHYLTHIO-AETHYL)-MONOTHIOPHOSPHAT (GERMAN) see DAP400
DIMETHYL ALDEHYDE see DOO600
4-(N,4-DIMETHYL-l-ALLOISOLEUCINE)-8-(N,4-DIMETHYL-l-ALLOISOLEUCINE)-QUINOMYCIN A see QQS075
DIMETHYLALLYL ACETATE see DOQ350
3,3-DIMETHYLALLYL ACETATE see DOQ350
γ,γ-DIMETHYLALLYL ACETATE see DOQ350
DIMETHYLALLYL ALCOHOL see MHU110
3,3-DIMETHYLALLYL ALCOHOL see MHU110
γ,γ-DIMETHYLALLYL ALCOHOL see MHU110
2-(3,3-DIMETHYLALLYL)CYCLAZOCINE see DOQ400
2-DIMETHYLALLYL-5,9-DIMETHYL-2'-HYDROXYBENZOMORPHAN see DOQ400
2-(3,3-DIMETHYLALLYL)-5-ETHYL-2'-HYDROXY-9-METHYL-6,7-BENZOMORPHAN see DOQ600
6-(1,1-DIMETHYLALLYL)-7H-FURO(3,2-G)(1)BENZOPYRAN-7-ONE see XPJ100
2-(3,3-DIMETHYLALLYL)-2',2'-HYDROXY-5,9-DIMETHYL-6,7-BENZOMORPHAN see DOQ400
DIMETHYLALUMINUM CHLORIDE see DOQ700
DIMETHYLALUMINUM HYDRIDE see DOQ750
DIMETHYLAMIDE ACETATE see DOO800
DIMETHYLAMIDE DIETHYLENEIMIDE PHOSPHORIC ACID see DOV600
DIMETHYLAMID KYSELINY CHLORMRAVENCI see DQY950
DIMETHYLAMIDOETHOXYPHOSPHORYL CYANIDE see EIF000
DIMETHYLAMINE see DOQ800
DIMETHYLAMINE, solution (DOT) see DOQ800
DIMETHYLAMINE, anhydrous (DOT) see DOQ800
DIMETHYLAMINE, aqueous solution (DOT) see DOQ800
DIMETHYLAMINE BENZHYDRYL ESTER HYDROCHLORIDE see BAU750
DIMETHYLAMINE BORANE see DOR200
4-DIMETHYLAMINE m-CRESYL METHYLCARBAMATE see DOR400
DIMETHYLAMINE with DIBORANE (1:1) see DOX200
DIMETHYLAMINE-EPICHLOROHYDRIN COPOLYMER see DOR500
DIMETHYLAMINE HYDROCHLORIDE see DOR600
4-DIMETHYLAMINEPYRIDINE see DQB600
DIMETHYLAMINE SALT of 2,4-D see DFY800
DIMETHYLAMINE SALTS of mixed POLYCHLOROBENZOIC ACIDS see PJQ000
DIMETHYLAMINE-2,3,6-TRICHLOROBENZOATE see DOR800
4-(DIMETHYLAMINE)-3,5-XYLYL-N-METHYLCARBAMATE see DOS000
(DIMETHYLAMINO)ACETIC ACID HYDROCHLORIDE see MPI100
DIMETHYLAMINOACETONITRILE see DOS200
2-DIMETHYLAMINOACETONITRILE (DOT) see DOS200

(DIMETHYLAMINO)ACETYLENE see DOS300
N',N'-DIMETHYL-4'-AMINO-N-ACETYL-N-MONOMETHYL-4-AMINOAZOBENZENE see DPQ200
N'-DIMETHYLAMINOACETYLPARTRICIN A DIASPARTAT see DOS400
DIMETHYLAMINOAETHANOL (GERMAN) see DOY800
β-DIMETHYLAMINO-AETHYL-BENZHYDRYL-AETHER (GERMAN) see BBV500
N-(2'-DIMETHYLAMINOAETHYL)-(o-BENZYLPHENOL)-AETHER HYDROCHLORID (GERMAN) see DPD400
N-(4-((1-(DIMETHYLAMINO)-AETHYLIDEN)AMINO)PHENYL)-2-METHOXYACETAMID-HYDROCHLORID (GERMAN) see DPF200
N,N-DIMETHYL-β-AMINOAETHYL-ISOTHIURONIUM DIHYDROCHLORID (GERMAN) see NNL400
N-DIMETHYLAMINO-AETHYL-N-p-METHOXY-BENZYL-α-AMINO-PYRIDIN-MALEAT (GERMAN) see WAK000
5-(DIMETHYLAMINOAETHYL-OXYIMINO)-5H-DIBENZO(a,d)CYCLOHEPTA-1,4-DIENHYDROCHLORID (GERMAN) see DPH600
3-(DIMETHYLAMINO)-ALANINE (9CI) see ARY625
DIMETHYLAMINO-ANALGESINE see DOT000
p-DIMETHYLAMINOANILINE DIHYDROCHLORIDE see DTM000
9-(p-DIMETHYLAMINOANILINO)ACRIDINE see DOS800
DIMETHYLAMINOANTIPYRINE see DOT000
4-(DIMETHYLAMINO)ANTIPYRINE see DOT000
4-(DIMETHYLAMINO)ANTIPYRINE mixed with SODIUM NITRITE (1:1) see DOT200
p-DIMETHYLAMINOAZOBENZEN (CZECH) see DOT300
DIMETHYLAMINOAZOBENZENE see DOT300
4-DIMETHYLAMINOAZOBENZENE see DOT300
p-DIMETHYLAMINOAZOBENZENE see DOT300
4-(N,N-DIMETHYLAMINO)AZOBENZENE see DOT300
N,N-DIMETHYL-4-AMINOAZOBENZENE see DOT300
N,N-DIMETHYL-p-AMINOAZOBENZENE see DOT300
2',3-DIMETHYL-4-AMINOAZOBENZENE see AIC250
4-DIMETHYLAMINOAZOBENZENE AMINE-N-OXIDE see DTK600
p-(DIMETHYLAMINO)AZOBENZENE-o-CARBOXYLIC ACID see CCE500
4'-DIMETHYLAMINOAZOBENZENE-2-CARBOXYLIC ACID see CCE500
N,N-DIMETHYLAMINOAZOBENZENE-N-OXIDE see DTK600
4-DIMETHYLAMINOAZOBENZENE-4'-SULPHONIC ACID SODIUM SALT see MND600
DIMETHYLAMINOAZOBENZOL see DOT300
4-DIMETHYLAMINOAZOBENZOL see DOT300
p-DIMETHYLAMINO-AZOBENZOL (GERMAN) see DOT300
DIMETHYLAMINOAZOPHENE see DOT000
4-(DIMETHYLAMINO) BENZALDEHYDE see DOT400

p,p-DIMETHYLAMINODIPHENYLMETHANE see MJN000

α-4-DIMETHYLAMINO-1,2-DIPHENYL-3-METHYL-2-BUTANOL PROPIONATE see PNA250

α-(+)-4-DIMETHYLAMINO-1,2-DIPHENYL-3-METHYL-2-BUTANOL PROPIONATE ESTER see DAB879

2-(DIMETHYLAMINO)-N-(1,3-DIPHENYL-1H-PYRAZOL-5-YL) PROPANAMIDE see PAM500

4-(DIMETHYLAMINO)-2,2-DIPHENYLVALERAMIDE see DOY400

3-(DIMETHYLAMINO)-1-(2,2-DIPHOSPHONOETHYL)PYRAZINIUM INNER SALT see DOY500

2-DIMETHYLAMINO ETHANETHIOL HYDROCHLORIDE see DOY600

DIMETHYLAMINOETHANOL see DOY800

2-(DIMETHYLAMINO)ETHANOL see DOY800

N-DIMETHYLAMINOETHANOL see DOY800

N,N-DIMETHYLAMINOETHANOL see DOY800

β-DIMETHYLAMINOETHANOL see DOY800

2-DIMETHYLAMINOETHANOL-p-ACETAMIDOBENZOATE see DOZ000

DIMETHYLAMINOETHANOL ACETATE see DOZ100

2-DIMETHYLAMINOETHANOL ACETATE see DOZ100

2-(DIMETHYLAMINO)ETHANOL BITARTRATE see DPA000

2-DIMETHYLAMINOETHANOL-4-N-BUTYLAMINOBENZOATE HYDROCHLORIDE see TBN000

β-DIMETHYLAMINOETHANOL DIPHENYLMETHYL ETHER see BBV500

2-(DIMETHYLAMINO)ETHANOL METHACRYLATE see DPG600

2-(DIMETHYLAMINO)ETHANOL TARTRATE see DPA200

1-(DIMETHYLAMINOETHOXYACETAMIDO)ADAMANTANE HYDROCHLORIDE see AEF500

N-(p-(2-(DIMETHYLAMINO)ETHOXY)BENZYL)-3,4,5-TRIMETHOXYBENZAMIDE HYDROCHLORIDE see TKW750

N-(p-(2-(DIMETHYLAMINO)ETHOXY)-BENZYL)-3,4,5-TRIMETHOXYBENZAMIDE MONOHYDROCHLORIDE see TKW750

1-(β-DIMETHYLAMINOETHOXY)-3-N-BUTYLISOQUINOLINE HYDROCHLORIDE see DNX400

1-(β-DIMETHYLAMINOETHOXY)-3-N-BUTYLISOQUINOLINE MONOHYDROCHLORIDE see DNX400

5-(2-(N,N-DIMETHYLAMINO)ETHOXY)CARVACROL ACETATE CITRATE see MRN600

5-(2-(N,N-DIMETHYLAMINO)ETHOXY)CARVACROL ACETATE HYDROCHLORIDE see TFY000

2-(2-DIMETHYLAMINOETHOXY)CHALCONE CITRATE see DPA500

2-(2-DIMETHYLAMINOETHOXY)-N,N-DIETHYLPROPYLAMINE DIMETHIODIDE see MQF750

α-(2-DIMETHYLAMINOETHOXY)DIPHENYLMETHANE see BBV500

2-(2-DIMETHYLAMINOETHOXY)ETHANOL see DPA600

2-(2-DIMETHYLAMINOETHOXY)ETHANOL-1-

PHENYL-CYCLOPENTYLCARBOXYLATE see DPA800

2-(2-(DIMETHYLAMINO)ETHOXY)ETHYL-1-PHENYLCYCLOPENTANECARBOXYLATE see DPA800

4-(2-(DIMETHYLAMINO)ETHOXY)-5-ISOPROPYL-2-METHYLPHENOL see DAD850

2-(α-(2-(DIMETHYLAMINO)ETHOXY)-α-METHYLBENZYL)PYRIDINE see DYE500

DIMETHYLAMINOETHOXY-METHYL-BENZYL-PYRIDINE SUCCINATE see PGE775

2-(α-(2-DIMETHYLAMINOETHOXY)-α-METHYLBENZYL)PYRIDINE SUCCINATE see PGE775

4-(2-(DIMETHYLAMINO)ETHOXY)-2-METHYL-5-(1-METHYLETHYL)PHENOL see DAD850

6-(2-DIMETHYLAMINOETHOXY)-2-((5-NITRO-1-METHYL-2-IMIDAZOLYL)-METHYLENE)-1-TETRALON SULFATE see DPB200

cis-1-(p-(2-(N,N-DIMETHYLAMINO)ETHOXY)PHENYL)-1,2-DIPHENYLBUT-1-ENE see NOA600

trans-1-(p-β-DIMETHYLAMINOETHOXYPHENYL)-1,2-DIPHENYLBUT-1-ENE CITRATE see TAD175

2-DIMETHYLAMINOETHOXYPHENYLMETHYL-2-PICOLINE see DYE500

2-DIMETHYLAMINOETHOXYPHENYLMETHYL-2-PICOLINE SUCCINATE see PGE775

trans-4-(1-(4-(2-(DIMETHYLAMINO)ETHOXY)PHENYL)-2-PHENYL-1-BUTENYL)PHENOL see HOH100

(Z)-β-((4-(2-(DIMETHYLAMINO)ETHOXY)PHENYL)PHENYLMETHYLENE)-α-METHYLBENZENEETHANOL see HOH200

4-(2-(DIMETHYLAMINO)ETHOXY)-1(2H)-PHTHALAZINONE OXIME MONOHYDROCHLORIDE see TAC825

4-(2-DIMETHYLAMINOETHOXY)-N-(3,4,5-TRIMETHOXYBENZOYL)BENZYLAMINE HYDROCHLORIDE see TKW750

DIMETHYLAMINOETHYL ACETATE see DOZ100

2-(DIMETHYLAMINO)ETHYL ACETATE see DOZ100

N-(2-(DIMETHYLAMINO)ETHYL)-1-ACRIDINECARBOXAMIDE see DPB220

DIMETHYLAMINOETHYL ACRYLATE see DPB300

2-(DIMETHYLAMINO)ETHYL ACRYLATE METHOCHLORIDE see OOM500

N-(2-(DIMETHYLAMINO)ETHYL)-1-ADAMANTANEACETAMIDE ETHYL IODIDE see DPB400

β-DIMETHYLAMINOETHYL ALCOHOL see DOY800

2-DIMETHYLAMINOETHYLAMINE see DPC000

2-(DIMETHYLAMINO)ETHYL-p-AMINOBENZOATE HYDROCHLORIDE see DPC200

2-(2-(DIMETHYLAMINO)ETHYLAMINO)ETHYL)-2-METHYL-1,3-BENZODIOXOLEDI HYDROCHLORIDE see DPC400

2-(DIMETHYLAMINO)ETHYL-p-AMINOSALICYLATE see AMN000

β-DIMETHYLAMINOETHYL BENZHYDRYL ESTER HYDROCHLORIDE see BAU750

β-DIMETHYLAMINOETHYLBENZHYDRYLE THER see BBV500

DIMETHYLAMINOETHYL BENZILATE, HYDROCHLORIDE see BAW500

2-(DIMETHYLAMINO)ETHYL BENZILATE HYDROCHLORIDE see BAW500

β-DIMETHYLAMINOETHYL BENZILATE HYDROCHLORIDE see BAW500

N,N-DIMETHYL-β-3-AMINOETHYLBENZOTHIOPHENE HYDROCHLORIDE see DQP000

α-(1-(DIMETHYLAMINO)ETHYL)BENZYL ALCOHOL HYDROCHLORIDE see DPD200

N-DIMETHYLAMINOETHYLBENZYLANILINE HYDROCHLORIDE see PEN000

DIMETHYLAMINOETHYL BENZYLATE HYDROCHLORIDE see BAW500

(2-(DIMETHYLAMINO)ETHYL)(o-BENZYLPHENOXY)ETHERHYDROCHLORIDE see DPD400

2-(α-(2-DIMETHYLAMINOETHYL)BENZYL)PYRIDINE see TMJ750

2-(α-(2-(DIMETHYLAMINO)ETHYL)BENZYL)PYRIDINE, BIMALEATE see TMK000

2-(α-(2-(DIMETHYLAMINO)ETHYL)BENZYL)PYRIDINE, MALEATE see TMK000

β-DIMETHYLAMINOETHYL-p-BROMO-α-METHYLBENZHYDRYL ETHER HYDROCHLORIDE see BMN250

DIMETHYLAMINOETHYL-p-BUTYL-AMINOBENZOATE see BQA010

2-DIMETHYLAMINOETHYL-p-BUTYLAMINOBENZOATE see BQA010

2-(DIMETHYLAMINO)ETHYL-p-(BUTYLAMINO)BENZOATE HYDROCHLORIDE see TBN000

DIMETHYLAMINOETHYL-p-N-BUTYLAMINOBENZOATE HYDROCHLORIDE see TBN000

DIMETHYLAMINOETHYL CHLORIDE see CGW000

2-DIMETHYLAMINOETHYLCHLORIDE see CGW000

β-(DIMETHYLAMINO)ETHYL CHLORIDE see CGW000

α-(2-DIMETHYLAMINOETHYL)-o-CHLOROBENZHYDROL HYDROCHLORIDE see CMW700

β-DIMETHYLAMINOETHYL (p-CHLORO-α-METHYLBENZHYDRYL) ETHER HYDROCHLORIDE see CIS000

DIMETHYLAMINOETHYL-p-CHLOROPHENOXYACETATE see DPE000

DIMETHYLAMINOETHYL 4-CHLOROPHENOXYACETATE HYDROCHLORIDE see AAE500

DIMETHYLAMINOETHYL p-CHLOROPHENOXYACETATE HYDROCHLORIDE see AAE500

DIMETHYLAMINOETHYL-4-CHLOROPHENOXYACETIC ACID see DPE000

2-DIMETHYLAMINOETHYL 2'-DIETHYLAMINOISOPROPYL ETHER BISMETHIODIDE see MQF750

β-DIMETHYLAMINOETHYL β'-DIETHYLAMINO-α'-METHYLETHYL ETHER DIMETHIODIDE see MQF750

S-(2-DIMETHYLAMINOETHYL)-O,O-DIETHYLPHOSPHORITHIOATE METHIODIDE see TLF500

5-(2-(DIMETHYLAMINO)ETHYL)-2,3-DIHYDRO-3-HYDROXY-2-(p-METHOXYPHENYL)1,5-BENZOTHIAZEPIN-4(5H)-ONE-ACETATE (ESTER) HYDROCHLORIDE see DPE100

2-DIMETHYLAMINO-4-HYDROXY-5-n-BUTYL-6-METHYLPYRIMIDINE see BRD000

4-DIMETHYLAMINO-3-HYDROXYDIPHENYL see DOV400

β-DIMETHYLAMINOISOPROPYL CHLORIDE see CGJ290

10-(3-DIMETHYLAMINOISOPROPYL)PHENOTHIAZINE HYDROCHLORIDE see PMI750

DIMETHYLAMINO-ISOPROPYL-PHENTHIAZIN (GERMAN) see DQA400

10-DIMETHYLAMINOISOPROPYL-2-PROPIONYLPHENOTHIAZINE MALEATE see IDA500

DIMETHYLAMINO-ISOPROPYL-THIOPHENYL-PYRIDYL-AMIN see OGI075

N-DIMETHYLAMINOISOPROPYLTHIOPHENYLPYRIDYLAMINE HYDROCHLORIDE see AEG625

6-DIMETHYLAMINO-9-(3'-(p-METHOXY-l-PHENYLALANYLAMINO)-β-d-RIBOFURANOSYL)-PURINE see AEI000

2-(DIMETHYLAMINO)-2'-METHYLACETANILIDE HYDROCHLORIDE see DHK200

2-(DIMETHYLAMINO)-N-(((METHYLAMINO)CARBONYL)OXY)-2-OXOETHANIMIDOTHIOIC ACID METHYL ESTER see DSP600

4-(N,N-DIMETHYLAMINO)-3'-METHYLAZOBENZENE see DUH600

7-(DIMETHYLAMINO)-4-METHYL-2H-1-BENZOPYRAN-2-ONE see DPJ800

α-(DIMETHYLAMINOMETHYL)BENZYL ALCOHOL see DSH700

4-(DIMETHYLAMINO)-3-METHYL-2-BUTANOL p-AMINOBENZOATE (ester) see TOE150

4-(DIMETHYLAMINO)-3-METHYL-2-BUTANOL 4-AMINOBENZOATE (ester) HYDROCHLORIDE see AIT750

1-(DIMETHYLAMINO)-2-METHYL-2-BUTANOL BENZOATE (ESTER) see AOM000

2-DIMETHYLAMINO-4-METHYL-5-n-BUTYL-6-HYDROXYPYRIMIDINE see BRD000

7-DIMETHYLAMINO-4-METHYLCOUMARIN see DPJ800

anti-8-(N,N-DIMETHYLAMINOMETHYL)DIBENZOBICYCLO(3.2.1)OCTADIENE HYDROCHLORIDE see DPK000

4-(DIMETHYLAMINO)-3-METHYL-1,2-DIPHENYL-2-BUTANOL PROPIONATE see PNA250

α-4-DIMETHYLAMINO-3-METHYL-1,2-DIPHENYL-2-BUTANOLPROPIONATE see PNA250

d-4-DIMETHYLAMINO-3-METHYL-1,2-DIPHENYL-2-BUTANOL PROPIONATE HYDROCHLORIDE see PNA500

4-(DIMETHYLAMINO)-3-METHYL-1,2-DIPHENYL-2-BUTANOL PROPIONATE-2-NAPHTHALENESULFONATE see DYB400

6-DIMETHYLAMINO-5-METHYL-4,4-DIPHENYL-3-HEXANONE see IKZ000

d-4-DIMETHYLAMINO-3-METHYL-1,2-DIPHENYL-2-PROPIONOXYBUTANENAPHTHALENE-2-SULPHONATE HYDRATE see DAB880

4-DIMETHYLAMINO-3-METHYL-1,2-DIPHENYL-2-PROPOXYBUTANE see PNA250

m-(((DIMETHYLAMINO)METHYLENE)AMINO)PHENYLMETHYL CARBAMATE,HYDROCHLORIDE see DSO200

3-((DIMETHYLAMINOMETHYLENE)AMINO)-2,4,6-TRIIODOHYDROCINNAMIC ACID SODIUM SALT see SKM000

3-DIMETHYLAMINOMETHYLENEIMINOPHENYL-N-METHYLCARBAMATE, HYDROCHLORIDE see DSO200

2-DIMETHYLAMINO-N-METHYLETHYLAMINE see DPK400

9-((2-(DIMETHYLAMINO)-1-METHYLETHYL)AMINO)-1-NITROACRIDINE DIHYDROCHLORIDE see NFW435

9-((2-(DIMETHYLAMINO)-1-METHYLETHYL)AMINO)-1-NITROACRIDINE HYDROCHLORIDE see NFW435

(2-DIMETHYLAMINO-2-METHYL)ETHYL-N-DIBENZOPARATHIAZINE see DQA400

5-(2'-(N,N-DIMETHYLAMINO)-2'-METHYL)ETHYL-10,11-DIHYDRO-5H-DIBENZ(b,f)AZEPINE see DPK500

α-(2-(DIMETHYLAMINO)-1-METHYLETHYL)-2-METHYL-α-PHENYLBENZENEETHANOL, HCl see DPM200

10-(2-(DIMETHYLAMINO)-2-METHYLETHYL)PHENOTHIAZINE see DQA400

N-(2'-DIMETHYLAMINO-2'-METHYL)ETHYLPHENOTHIAZINE see DQA400

N-(2'-DIMETHYLAMINO-2'-METHYL)ETHYLPHENOTHIAZINE HYDROCHLORIDE see PMI750

1-(10-(2-(DIMETHYLAMINO)-1-METHYLETHYL)PHENOTHIAZIN-2-YL)-1-PROPANONE HYDROCHLORIDE see PMT500

s-α-(2-(DIMETHYLAMINO)-1-METHYLETHYL)-α-PHENYLBENZENEETHANOL PROPIOATE HYDROCHLORIDE see PNA500

10-(2-DIMETHYLAMINO-2-METHYLETHYL)-10H-PYRIDO(3,2-b)(1,4)BENZOTHIAZINE HYDROCHLORIDE see AEG625

N-DIMETHYLAMINO-2-METHYLETHYL THIODIPHENYLAMINE see DQA400

trans-2-((DIMETHYLAMINO)METHYLIMINO)-5-(2-(5-NITRO-2-FURYL)VINYL)-1,3,4-OXADIAZOLE see DPL000

3-(DIMETHYLAMINOMETHYL)INDOLE see DYC000

β-DIMETHYLAMINOMETHYLINDOLE see DYC000

8-(DIMETHYLAMINOMETHYL)-7-METHOXY-3-METHYLFLAVONE see DNV000

8-((DIMETHYLAMINO)METHYL)-7-METHOXY-3-METHYLFLAVONE HYDROCHLORIDE see DNV200

8-((DIMETHYLAMINO)METHYL)-7-METHOXY-3-METHYL-2-PHENYL-4H-1-BENZOPYRAN-4-ONE HYDROCHLORIDE see DNV200

8-((DIMETHYLAMINO)METHYL)-7-METHOXY-3-METHYL-2-PHENYLFLAVONE see DNV000

2-DIMETHYLAMINOMETHYL-1-(m-METHOXYPHENYL)CYCLOHEXANOL see DPL200

(+)-(E)-2-(DIMETHYLAMINOMETHYL)-1-(m-METHOXYPHENYL)CYCLOHEXANOL see THJ600

(+)-trans-2-(DIMETHYLAMINOMETHYL)-1-(m-METHOXYPHENYL)CYCLOHEXANOL see THJ600

(±)-trans-2-((DIMETHYLAMINO)METHYL-1-(m-METHOXYPHENYL))CYCLOHEXANOL see THJ500

(E)-2-((DIMETHYLAMINO)METHYL)-1-(m-METHOXYPHENYL)-1-CYCLOHEXANOL HYDROCHLORIDE see THJ750

trans-2-(DIMETHYLAMINOMETHYL)-1-(m-METHOXYPHENYL)CYCLOHEXANOL HYDROCHLORIDE see THJ750

3-(DIMETHYLAMINO)-1-METHYL-3-OXO-1-PROPENYL DIMETHYL PHOSPHATE see DGQ875

4-(DIMETHYLAMINO)-3-METHYLPHENOL METHYL CARBAMATE (ester) see DOR400

2-(DIMETHYLAMINO)-N-(((METHYL(((2-PHENYL-1,3-DIOXAN-5-YL)METHOXY)SULFINYL)AMINO)CARBONYL)OXY)-2-OXO-ETHANIMIDOTHIOIC ACID, METHYL ESTER see DPL300

3-DIMETHYLAMINO-4-METHYLPHENYL ESTER-N-METHYLCARBAMIC ACID HYDROCHLORIDE see DPL900

(4-DIMETHYLAMINO-3-METHYL-PHENYL)N-METHYL-CARBAMAAT (DUTCH) see DOR400

(4-DIMETHYLAMINO-3-METHYL-PHENYL)N-METHYL-CARBAMAT (GERMAN) see DOR400

(4-DIMETHYLAMINO-3-METHYL-PHENYL)N-METHYL-CARBAMATE see DOR400

1-(4-(DIMETHYLAMINO)-2-METHYL-5-PHENYL-1H-PYRROL-3-YL)ETHANONE see DPL950

3-DIMETHYLAMINO-2-METHYL-1-PHENYL-o-TOLYPROPANOL HYDROCHLORIDE see DPM200

2-DIMETHYLAMINO-2-METHYL-1-PROPANOL see DPM400

1-(3-DIMETHYLAMINO-2-METHYLPROPYL)-4,5-DIHYDRO-2,3:6,7-DIBENZAZEPINE see DLH200

5-(3-(DIMETHYLAMINO)-2-METHYLPROPYL)-10,11-DIHYDRO-5H-DIBENZ(b,f)AZEPINE see DLH200

5-(3-DIMETHYLAMINO-2-METHYLPROPYL)-10,11-DIHYDRO-5H-DIBENZ(b,f)AZEPINE HYDROCHLORIDE see TAL000

5-(γ-DIMETHYLAMINO-β-METHYLPROPYL)-10,11-DIHYDRO-5H-DIBENZO(b,f)AZEPINE see DLH200

3-DIMETHYLAMINO-7-METHYL-1,2-(PROPYLMALONYL)-1,2-DIHYDRO-1,2,4-BENZOTRIAZINE SODIUM SALT see ASA250

3-DIMETHYLAMINO-7-METHYL-1,2-(n-PROPYLMALONYL)-1,2-DIHYDRO-1,2,4-BENZOTRIAZINE see AQN750

3-DIMETHYLAMINO-7-METHYL-1,2-(N-PROPYLMALONYL)-1,2-DIHYDRO-1,2,4-BENZOTRIAZINE DIHYDRATE see ASA000

10-(3-(DIMETHYLAMINO)-2-METHYLPROPYL)-2-METHOXYPHENOTHIAZINE see MCI500

10-(3-(DIMETHYLAMINO)-2-METHYLPROPYL)-3-METHOXYPHENOTHIAZINE see PDP600

10-(3'-DIMETHYLAMINO-2'-METHYL-1'-PROPYL)-3-METHOXYPHENOTHIAZINE see PDP600

10-(3-(DIMETHYLAMINO)-2-METHYLPROPYL)-2-(METHYLTHIO)PHENOTHIAZINE HYDROCHLORIDE see MDT650

10-(3-(DIMETHYLAMINO)-2-METHYLPROPYL)PHENOTHIAZINE see AFL500

10-(3-(DIMETHYLAMINO)-2-METHYLPROPYL)PHENOTHIAZINE-2-CARBONITRILE MALEATE see COS899

10-(3-(DIMETHYLAMINO)-2-METHYLPROPYL)PHENOTHIAZINE-5,5-DIOXIDE see AFL750

5-DIMETHYLAMINO-9-METHYL-2-PROPYL-1H-PYRAZOLO(1,2-a)(1,2,4)BENZOTRIAZINE-1,3(2H)-DIONE see AQN750

α-(DIMETHYLAMINOMETHYL)PROTOCATECHUYL ALCOHOL see MJV000

4-DIMETHYLAMINO-2'-METHYLSTILBENE see TMF750

2-DIMETHYLAMINO-1-(METHYLTHIO)GLYOXAL-o-METHYLCARBAMOYLMONOXIME see DSP600

1-DIMETHYLAMINONAPHTHALENE see DSU400

DIMETHYLAMINONAPHTHALENESULFONYL CHLORIDE see DPN200

1-DIMETHYLAMINONAPHTHALENE-5-SULFONYL CHLORIDE see DPN200

1-(DIMETHYLAMINO)-5-NAPHTHALENESULFONYLCHLORIDE see DPN200

5-(DIMETHYLAMINO)-1-NAPHTHALENESULFONYL CHLORIDE see DPN200

2-DIMETHYLAMINO-1,4-NAPHTHOQUINONE see DPN300

5-DIMETHYLAMINONAPHTHYL-5-SULFONYL CHLORIDE see DPN200

o-(DIMETHYLAMINO)NITROBENZENE see NFW600

4-DIMETHYLAMINO-5-NITRO-2-METHOXY-N-(2-DIETHYLAMINOETHYL)BENZAMIDE HYDROCHLORIDE see DUO300

4-(DIMETHYLAMINO)NITROSOBENZENE see DSY600

p-(DIMETHYLAMINO)NITROSOBENZENE see DSY600

1,3-DIMETHYL-4-AMINO-5-NITROSOURACIL see DPN400

1-(((((4-DIMETHYLAMINO)-1,4,4A,5,5A,6,11,12A-OCTAHYDRO-3,6,10,12,12A-PENTAHYDROXY-6-METHYL-1,11-DIOXO-2-NAPHTHACENYL)CARBONYL)AMINO)METHYL)-l-PROLINE see PMI000

(DIMETHYLAMINO)OXO-ACETIC ACID 2-ACETYL-2-METHYL-1-PHENYLHYDRAZIDE (9CI) see DVU100

2-(DIMETHYLAMINO)-2-OXOETHYL-N-(((METHYLAMINO)CARBONYL)OXY)ETHANIMIDOTHIOATE see DPN500

5-(DIMETHYLAMINOOXYIMINO)-5H-DIBENZO(a,b)CYCLOHEPTA-1,4-DIENE HYDROCHLORIDE see DPH600

DIMETHYLAMINOPHENAZON (GERMAN) see DOT000

DIMETHYLAMINOPHENAZONE see DOT000

4-DIMETHYLAMINOPHENAZONE see DOT000

3-(DIMETHYLAMINO)PHENOL see HNK560

m-(DIMETHYLAMINO)PHENOL see HNK560

4-DIMETHYLAMINOPHENOL HYDROCHLORIDE see DPN800

p-DIMETHYLAMINOPHENOL HYDROCHLORIDE see DPN800

m-(DIMETHYLAMINO)PHENOL METHIODIDE see HNN500

(7-(DIMETHYLAMINO)-3H-PHENOXAZIN-3-YLIDENE)DIMETHYLAMMONIUM CHLORIDE see DPN900

3-DIMETHYLAMINO-2-PHENOXYPROPIOPHENONE HYDROCHLORIDE see DPO100

p-DIMETHYLAMINOPHENYLAMINE see DTL800

9-((p-DIMETHYLAMINO)PHENYL)AMINO))ACRIDINE see DOS800

4-(p-DIMETHYLAMINOPHENYLAZO)ANILINE see DPO200

3-((4-(DIMETHYLAMINO)PHENYL)AZO)BENZALDEHYDE see FNK100

3-(((p-DIMETHYLAMINO)PHENYL)AZO)BENZALDEHYDE see FNK100

4-DIMETHYLAMINOPHENYLAZOBENZENE see DOT300

4-(p-DIMETHYLAMINOPHENYLAZO)-BENZENEARSONIC ACID HYDROCHLORIDE see DPO275

p-((DIMETHYLAMINO)PHENYL)AZO)BENZENESULFONIC ACID SODIUM SALT see MND600

4-((DIMETHYLAMINO)PHENYL)AZO)BENZIMIDAZOLE see DQM200

2-((4-DIMETHYLAMINO)PHENYLAZO)BENZOIC ACID see CCE500

3-((p-(DIMETHYLAMINO)PHENYL)AZO)BENZOIC ACID see CCE750

o-((p-(DIMETHYLAMINO)PHENYL)AZO)BENZOIC ACID see CCE500

6-DIMETHYLAMINOPHENYLAZOBENZOTHIAZOLE see DPO400

6-((p-(DIMETHYLAMINO)PHENYL)AZO)BENZOTHIAZOLE see DPO400

7-((p-(DIMETHYLAMINO)PHENYL)AZO)BENZOTHIAZOLE see DPO600

6-DIMETHYLAMINOPHENYLAZOBENZTHIAZOLE see DPO400

3-((p-DIMETHYLAMINO)PHENYLAZO)BENZYL ALCOHOL see HMB600

m-((p-DIMETHYLAMINO)PHENYL)AZO)BENZYL ALCOHOL see HMB600

o-((p-DIMETHYLAMINO)PHENYL)AZO)BENZYL ALCOHOL see HMB595

6-((p-(DIMETHYLAMINO)PHENYL)AZO)-1H-INDAZOLE see DSI800

4-(p-DIMETHYLAMINO)PHENYL)AZO)ISOQUINOLINE see DPO800

5-((p-(DIMETHYLAMINO)PHENYL)AZO)ISOQUINOLINE see DPP000

7-((p-(DIMETHYLAMINO)PHENYL)AZO)ISOQUINOLINE see DPP200

5-((p-(DIMETHYLAMINO)PHENYL)AZO)ISOQUINOLINE-2-OXIDE see DPP400

4-((4-(DIMETHYLAMINO)PHENYL)AZO)-2,6-LUTIDINE-1-OXIDE see DPP800

4-((p-(DIMETHYLAMINO)PHENYL)AZO)-2,5-LUTIDINE 1-OXIDE see DPP600

4-((p-(DIMETHYLAMINO)PHENYL)AZO)-3,5-LUTIDINE-1-OXIDE see DPP709

4-((p-(DIMETHYLAMINO)PHENYL)AZO)-N-METHYLACETANILIDE see DPQ200

5-((p-(DIMETHYLAMINO)PHENYL)AZO)-3-METHYLQUINOLINE see MJF500

5-((p-(DIMETHYLAMINO)PHENYL)AZO)-6-METHYLQUINOLINE see MJF750

5-((p-(DIMETHYLAMINO)PHENYL)AZO)-7-METHYLQUINOLINE see DPQ400

5-((p-(DIMETHYLAMINO)PHENYL)AZO)-8-METHYLQUINOLINE see MJG000

2-(4-DIMETHYLAMINOPHENYLAZO)NAPHTHALENE see DSU800

N-(4-((4-(DIMETHYLAMINO)PHENYL)AZO)PHENYL)-N-METHYLACETAMIDE see DPQ200

4-((4-(DIMETHYLAMINO)PHENYL)AZO)-2-PICOLINE-1-OXIDE see DSS200

3'-(4-DIMETHYLAMINOPHENYL)AZOPYRIDINE see POP750

5-((p-(DIMETHYLAMINO)PHENYL)AZO)QUINALDINE see DPQ600

4-(DIMETHYLAMINO)PHENYL)AZO)QUINOLINE see DTY200

5-((p-(DIMETHYLAMINO)PHENYL)AZO)QUINOLINE see DPQ800

6-((p-(DIMETHYLAMINO)PHENYL)AZO)QUINOLINE see DPR000

4-((p-(DIMETHYLAMINO)PHENYL)AZO)QUINOLINE-1-OXIDE see DTY400

5-((p-(DIMETHYLAMINO)PHENYL)AZO)QUINOLINE-1-OXIDE see DPR200

6-((p-(DIMETHYLAMINO)PHENYL)AZO)QUINOLINE-1-OXIDE see DPR400

5-((p-(DIMETHYLAMINO)PHENYL)AZO)QUINOXALINE see DUA200

6-((p-(DIMETHYLAMINO)PHENYL)AZO)QUINOXALINE see DUA400

(±)-2-(DIMETHYLAMINO)-2-PHENYLBUTYL-3,4,5-TRIMETHOXYBENZOATE see TKU650

4-(DIMETHYLAMINO)PHENYL-3-(4-(4-CHLOROPHENYL)-4-HYDROXYPIPERIDINO)-PROPYL KETONE see CKA580

dl-trans-2-DIMETHYLAMINO-1-PHENYL-CYCLOHEX-3-EN-trans-1-CARBONSAEUREAETHYLESTER HCl (GERMAN) see EIH000

dl-trans-2-DIMETHYLAMINO-1-PHENYL-CYCLOHEX-3-ENE-trans-CARBONIC ACID ETHYL ESTER HCl see EIH000

(4-(DIMETHYLAMINO)PHENYL)DIAZENESULFONIC ACID, SODIUM SALT see DOU600

4-((DIMETHYLAMINO)PHENYL)DIAZENESULFONIC ACID, SODIUM SALT see DOU600

p-(DIMETHYLAMINO)-PHENYLDIAZONATRIUMSULFONAT (GERMAN) see DOU600

DIMETHYLAMINOPHENYLDIMETHYLPYRAZOLIN see DOT000

4-DIMETHYLAMINO-1-PHENYL-2,3-DIMETHYLPYRAZOLONE see DOT000

2-(p-DIMETHYLAMINOPHENYL)-1,6-DIMETHYLQUINOLINIUM CHLORIDE see DPS200

4-(DIMETHYLAMINO)PHENYL ESTER THIOCYANIC ACID see TFH500

4-(p-DIMETHYLAMINOPHENYL)IMINO-2,5-CYCLOHEXADIENE-1-ONE see DPS600

DIMETHYLAMINOPHENYLMERCURIC ACETATE see AAS310

1-DIMETHYLAMINO-2-PHENYL-3-METHYLPENTANE HYDROCHLORIDE see BRE255

1-(4'-DIMETHYLAMINOPHENYL)-2-(1'-NAPHTHYL)ETHYLENE see DSV000
4-DIMETHYLAMINOPHENYL-2-((4-PHENYL-1,2,5,6-TETRAHYDRO-1-PYRIDYL)ETHYL) KETONE see DPS700
2-DIMETHYLAMINO-1-PHENYLPROPANE see TMA600
2-(3-DIMETHYLAMINO-1-PHENYLPROPYL)PYRIDINE see TMJ750
1-(N,N-DIMETHYLAMINO)-3-(PHENYL-3-α-PYRIDYL)PROPANE MALEATE see TMK000
2-(4-DIMETHYLAMINOPHENYL)QUINOLINE see DPT200
2-(p-(DIMETHYLAMINO)PHENYL)QUINOLINE see DPT200
4'-(DIMETHYLAMINO)-3-(4-PHENYL-1,2,5,6-TETRAHYDRO-1-PYRIDYL)PROPIOPHENONE see DPS700
p-(DIMETHYLAMINO)PHENYLTHIOCYANATE see TFH500
5-DIMETHYLAMINO-3-PIPERIDINOACETYLINDOLE see DPT300
1,1-DIMETHYLAMINOPROPANOL-2 see DPT800
1,1-DIMETHYLAMINOPROPAN-2-OL see DPT800
4-(2-(DIMETHYLAMINO)PROPIONAMIDO)ANTIPYRINE see AMF375
3-(DIMETHYLAMINO)PROPIONITRILE see DPU000
β-DIMETHYLAMINOPROPIONITRILE see DPU000
3-(3-DIMETHYLAMINOPROPIONYL)INDOLE see DPT300
3-(DIMETHYLAMINO)PROPIOPHENONE HYDROCHLORIDE see DPU400
β-DIMETHYLAMINOPROPIOPHENONE HYDROCHLORIDE see DPU400
1-(3-(DIMETHYLAMINO)PROPOXY)ADAMANTANE ETHYL IODIDE see DPU600
5-(3-(DIMETHYLAMINO)PROPOXY)-3-METHYL-1-PHENYLPYRAZOLE see DPU800
3-(DIMETHYLAMINO)PROPYLAMINE see AJQ100
N,N-DIMETHYL-N-(3-AMINOPROPYL)AMINE see AJQ100
2,2'-(3-DIMETHYLAMINOPROPYLAMINO)BIBENZYL see DLH600
10-(2-DIMETHYLAMINOPROPYL)-1-AZAPHENOTHIAZINE HYDROCHLORIDE see AEG625
10-(3-DIMETHYLAMINOPROPYL)-1-AZAPHENOTHIAZINE HYDROCHLORIDE see DYB800
10-(2-DIMETHYLAMINOPROPYL-(1)-4-AZAPHENTHIAZIN HYDROCHLORID (GERMAN) see AEG625
DIMETHYLAMINOPROPYL CHLORIDE, HYDROCHLORIDE see CGJ300
10-(3-DIMETHYLAMINOPROPYL)-2-CHLOROPHENOTHIAZINE MONOHYDROCHLORIDE see CKP500
5-(3-(DIMETHYLAMINO)PROPYL)-5H-DIBENZ(b,f)AZEPINE see DPW600
1-(3-DIMETHYLAMINOPROPYL)-4,5-DIHYDRO-2,3,6,7-DIBENZAZEPINE see DLH600
5-(3-(DIMETHYLAMINOPROPYL)-10,11-DIHYDRO-5H-DIBENZ(b,f)AZEPINE HYDROCHLORIDE see DLH630
5-(3-(DIMETHYLAMINO)PROPYL)-10,11-DIHYDRO-5H-DIBENZ(b,f)AZEPINE-5-OXIDE see IBP309

5-(3-DIMETHYLAMINO) PROPYL-10,11-DIHYDRO 5H-DIBENZ(b,f)AZEPINE, 5-OXIDE MONOHYDROCHLORIDE see IBP000
5-(3-(DIMETHYLAMINOPROPYL)-10,11-DIHYDRO-5H-DIBENZO(b,f)AZEPINE see DLH600
dl-11-(3-DIMETHYLAMINOPROPYL)-6,11-DIHYDRODIBENZO(b,d)THIEPIN see HII000
10-(3-(DIMETHYLAMINO)PROPYL)-9,9-DIMETHYLACRIDAN TARTRATE (1:1) see DRM000
α-(3-(DIMETHYLAMINO)PROPYL)-5-(1,1-DIMETHYLETHYL)-1-PHENYL-1H-PYRAZOLE-4-METHANOL see DPW610
10-(2-(DIMETHYLAMINO)PROPYL)-N,N-DIMETHYLPHENOTHIAZINE-2-SULFONAMIDE see DUC400
α-(3-(DIMETHYLAMINO)PROPYL)-1,5-DIPHENYL-1H-PYRAZOLE-4-METHANOL see DPW630
1-(3-DIMETHYLAMINOPROPYL)-3-ETHYLCARBODIIMIDE HYDROCHLORIDE see EAE100
α-(3-(DIMETHYLAMINO)PROPYL)-5-ETHYL-1-PHENYL-1H-PYRAZOLE-4-METHANOL see DPW650
5-(3-(DIMETHYLAMINO)PROPYL)-6,7,8,9,10,11-HEXAHYDRO-5H-CYCLOOCT(b)INDOLE see DPX200
5-(3-(DIMETHYLAMINO)PROPYL)-2-HYDROXY-10,11-DIHYDRO-5H-DIBENZ(b,f)AZEPINE see DPX400
5-(3'-DIMETHYLAMINOPROPYLIDENE)-DIBENZO-(a,d)(1,4)-CYCLOHEPTADIENE see EAH500
3-(3-DIMETHYLAMINOPROPYLIDENE)-1:2-4:5-DIBENZOCYCLOHEPTA-1:4-DIENE see EAI000
5-(3-DIMETHYLAMINOPROPYLIDENE)DIBENZO(a,d)(1,4)CYCLOHEPTADIENE HYDROCHLORIDE see EAI000
5-(3-DIMETHYLAMINOPROPYLIDENE)-5H-DIBENZO-(a,d)CYCLOHEPTENE HYDROCHLORIDE see DPX800
5-(γ-DIMETHYLAMINOPROPYLIDENE)-5H-DIBENZO(a,d)-10,11-DIHYDROCYCLOHEPTENE see EAH500
11-DIMETHYLAMINO PROPYLIDENE-6H-DIBENZ(b,e)OXEPIN see AEG750
5-(3-DIMETHYLAMINOPROPYLIDENE)-10,11-DIHYDRO-5H-DIBENZO(a,d)CYCLOHEPTENE see EAH500
5-(γ-DIMETHYLAMINOPROPYLIDENE)-10,11-DIHYDRO-5H-DIBENZO(A,D)CYCLOHEPTENE see EAH500
11-(3-DIMETHYLAMINOPROPYLIDENE)-6,11-DIHYDRODIBENZO(b,e)THIEPIN see DYC875
11-(3-DIMETHYLAMINOPROPYLIDENE)-6,11-DIHYDRODIBENZO(b,e)THIEPINE HYDROCHLORIDE see DPY200
cis-11-(3-DIMETHYLAMINOPROPYLIDENE)-6,11-DIHYDRODIBENZO(b,e)THIEPIN 5-OXIDE HYDROGEN MALEATE see POD800
11-(3-(DIMETHYLAMINO)PROPYLIDENE)-6,11-DIHYDRODIBENZ(b,e)OXEPIN HYDROCHLORIDE see AEG750
11-(3-DIMETHYLAMINOPROPYLIDENE)-6,11-DIHYDRODIBENZ(b,e)OXIPIN see DYE409
9-(3-DIMETHYLAMINOPROPYLIDENE)-10,10-DIMETHYL-9,10-DIHYDROANTHRACENE HYDROCHLORIDE see TDL000

2,2'-(3-DIMETHYLAMINOPROPYLIMINO)DIBENZYL see DLH600
N-(γ-DIMETHYLAMINOPROPYL)IMINODIBENZYL see DLH600
N-(3-DIMETHYLAMINOPROPYL)IMINODIBENZYL HYDROCHLORIDE see DLH630
5-DIMETHYLAMINO-6-PROPYL-5H-INDENO(5,6-d)-1,3-DIOXOLE HYDROCHLORIDE see DPY600
10-(3-DIMETHYLAMINOPROPYL)-2-METHOXYPHENOTHIAZINE see MFK500
10-((3-(DIMETHYLAMINO)PROPYL)-2-METHOXY)PHENOTHIAZINE, MALEATE see MFK750
17-β-((3-(DIMETHYLAMINO)-PROPYL)METHYLAMINO)ANDROST-5-EN-3-β-OL DIHYDROCHLORIDE see ARX800
α-(3-(DIMETHYLAMINO)PROPYL)-5-(1-METHYLETHYL)-1-PHENYL-1H-PYRAZOLE-4-METHANOL see DPY700
α-(3-(DIMETHYLAMINO)PROPYL)-5-METHYL-1-PHENYL-1H-PYRAZOLE-4-METHANOL see DPY800
10-(3-(DIMETHYLAMINO)PROPYL)-1-NITRO-9-ACRIDANONE HYDROCHLORIDE see NFW460
10-(3-(DIMETHYLAMINO)PROPYL)-1-NITRO-9(10H)-ACRIDINONE MONOHYDROCHLORIDE (9CI) see NFW460
o-(2-DIMETHYLAMINO)PROPYL)OXIME-5H-DIBENZO(a,d)CYCLOHEPTEN-5-ONE MONOHYDROCHLORIDE see LHX498
10-(2-(DIMETHYLAMINO)PROPYL)PHENOTHIAZINE see DQA400
10-(3-(DIMETHYLAMINO)PROPYL)PHENOTHIAZINE see DQA600
10-(3-DIMETHYLAMINOPROPYL)PHENOTHIAZINE-3-ETHYLONE see ABH500
10-(2-DIMETHYLAMINOPROPYL)PHENOTHIAZINE HYDROCHLORIDE see PMI750
10-(3-(DIMETHYLAMINO)PROPYL)PHENOTHIAZINE HYDROCHLORIDE see PMI500
N-(2-DIMETHYLAMINOPROPYL-1)PHENOTHIAZINE HYDROCHLORIDE see PMI750
(DIMETHYLAMINO-2-PROPYL-10-PHENOTHIAZINE HYDROCHLORIDE (FRENCH) see DQA400
10-(γ-DIMETHYLAMINO-N-PROPYL)PHENOTHIAZINE HYDROCHLORIDE see PMI500
10-(2-(DIMETHYLAMINO)PROPYL)PHENOTHIAZINE MONOHYDROCHLORIDE see PMI750
1-(10-(3-(DIMETHYLAMINO)PROPYL)PHENOTHIAZIN-2-YL)-1-BUTANONE MALEATE see BTA125
1-(10-(3-(DIMETHYLAMINO)PROPYL)-10H-PHENOTHIAZIN-2-YL)ETHANONE see ABH500
10-(3-DIMETHYLAMINOPROPYL)PHENOTHIAZIN-3-YLMETHYL KETONE see ABH500
10-(3-(DIMETHYLAMINO)PROPYL)PHENOTHIAZIN-2-YL METHYL KETONE MALEATE (1:1) see AAF750
10-(2-(DIMETHYLAMINO)PROPYL)PHENOTHIAZIN-2-YL MORPHOLINOMETHYL KETONE see DQA700

10-(3-(DIMETHYLAMINO)PROPYL)PHENOTHIAZIN-2-YL MORPHOLINOMETHYL KETONE see DQA710

1-(10-(2-DIMETHYLAMINOPROPYL)-PHENOTHIAZIN-2-YL)-1-PROPANONE MALEATE see IDA500

1-(10-(3-(DIMETHYLAMINO)PROPYL)PHENOTHIAZIN-2-YL)-1-PROPANONE MALEATE see PMX500

1-(2-(DIMETHYLAMINO)PROPYL)-2-(PHENYLACETYL)PYRROLE CITRATE see BEL525

α-(2-(DIMETHYLAMINO)PROPYL)-α-PHENYLBENZENEACETAMIDE see DOY400

α-(3-(DIMETHYLAMINO)PROPYL)-1-PHENYL-5-PROPYL-1H-PYRAZOLE-4-METHANOL see DQA720

α-(3-(DIMETHYLAMINO)PROPYL)-1-PHENYL-1H-PYRAZOLE-4-METHANOL see DQA730

10-(2-DIMETHYLAMINOPROPYL)-2-PROPIONYLPHENOTHIAZINE MALEATE see IDA500

((4-DIMETHYLAMINOPROPYLPYRIDO (3,2b) BENZOTHIAZINE)) HYDROCHLORIDE see DYB800

1-(3-(DIMETHYLAMINO)PROPYL)-2-PYRROLIDINONE see DQA800

2-(3-DIMETHYLAMINOPROPYL)-3a,4,7,7a-TETRAHYDRO-4,7-ETHANOISOINDOLINE DIMETHIODIDE see DQB309

10-(2-DIMETHYLAMINOPROPYL)-9-THIA-1,10-DIAZAANTHRACENE HYDROCHLORIDE see AEG625

N-(3-DIMETHYLAMINOPROPYL)THIOCARBAMINSAEURE-S-AETHYLESTER-HYDROCHLORID (GERMAN) see EIH500

2'-((3-(DIMETHYLAMINO)PROPYL)THIO)CINNAMANILIDE HYDROCHLORIDE see CMP900

n-(2-((3-(DIMETHYLAMINO)PROPYL)THIO)PHENYL)-3-PHENYL-2-PROPENAMIDE MONOHYDROCHLORIDE see CMP900

DIMETHYLAMINO-N-PROPYL-THIOPHENYLPYRIDYLAMINE see DYB600

10-(3-(DIMETHYLAMINO)PROPYL)-2-(TRIFLUOROMETHYL) PHENOTHIAZINE see TKL000

2-DIMETHYLAMINOPYRIDINE see DQB400

DIMETHYLAMINO-2 PYRIDINE see DQB400

4-DIMETHYLAMINOPYRIDINE see DQB600

p-DIMETHYLAMINOPYRIDINE see DQB600

γ-(DIMETHYLAMINO)PYRIDINE see DQB600

2-(DIMETHYLAMINO) RESERPILINATE see DQB800

2-(DIMETHYLAMINO) RESERPILIN-24-OIC ACID ETHYL ESTER see DQB800

DIMETHYLAMINORHODANBENZOL see TFH500

4-DIMETHYLAMINOSTILBEN (GERMAN) see DUB800

N,N-DIMETHYL-4-AMINOSTILBENE see DUB800

cis-4-DIMETHYLAMINOSTILBENE see DUC200

4-DIMETHYLAMINO-trans-STILBENE see DUC000

trans-4-DIMETHYLAMINOSTILBENE see DUC000

trans-p-(DIMETHYLAMINO)STILBENE see DUC000

DIMETHYLAMINOSTOVAINE see AHI250

7-(p-(DIMETHYLAMINO)STYRYL)BENZ(c)ACRIDINE see DQC000

12-(p-(DIMETHYLAMINO)STYRYL)BENZ(a)ACRIDINE see DQC200

2-(4-DIMETHYLAMINOSTYRYL)BENZOTHIAZOLE see DQC400

2-(p-(DIMETHYLAMINO)STYRYL)BENZOTHIAZOLE see DQC400

14-(p-(DIMETHYLAMINO)STYRYL)DIBENZ(a,j)ACRIDINE see DOV000

4-(p-(DIMETHYLAMINO)STYRYL)-6,8-DIMETHYLQUINOLINE see DQC600

4-(4-(DIMETHYLAMINO)STYRYL)QUINOLINE see DQD000

4-(p-(DIMETHYLAMINO)STYRYL)QUINOLINE see DQD000

2-(4-N,N-DIMETHYLAMINOSTYRYL)QUINOLINE see DQD000

4-(p-(DIMETHYLAMINO)STYRYL)QUINOLINE MONOHYDROCHLORIDE see DQD200

DIMETHYLAMINOSUCCINAMIC ACID see DQD400

N-(DIMETHYLAMINO)SUCCINAMIC ACID see DQD400

N-DIMETHYLAMINO-SUCCINAMIDSAEURE (GERMAN) see DQD400

O-(4-((DIMETHYLAMINO)SULFONYL)PHENYL) O,O-DIMETHYL PHOSPHOROTHIOATE see FAB600

(DIMETHYLAMINO)-TERMINATED see SDF000

DIMETHYLAMINO-4-THIOCYANOBENZENE see TFH500

4-DIMETHYLAMINOTHIOCYANOBENZENE see TFH500

p-(DIMETHYLAMINO)THIOCYANOBENZENE see TFH500

p-N,N-DIMETHYLAMINOTHIOCYANOBENZENE see TFH500

p-DIMETHYLAMINOTHYMOLDIMETHYLURETHANE METHIODIDE see DQD500

4-((4-(DIMETHYLAMINO)-m-TOLYL)AZO)-2-PICOLINE-1-OXIDE see DQD600

4-((4-(DIMETHYLAMINO)-m-TOLYL)AZO)-3-PICOLINE-1-OXIDE see DQE000

4-((4-(DIMETHYLAMINO)-o-TOLYL)AZO)-2-PICOLINE-1-OXIDE see DQD800

4-((4-(DIMETHYLAMINO)-o-TOLYL)AZO)-3-PICOLINE-1-OXIDE see DQE200

5-((4-(DIMETHYLAMINO)-m-TOLYL)AZO)QUINOLINE see DQE400

5-((4-(DIMETHYLAMINO)-o-TOLYL)AZO)QUINOLINE see DQE600

4-(DIMETHYLAMINO)-m-TOLYL METHYLCARBAMATE see DOR400

5-DIMETHYLAMINO-4-TOLYL METHYLCARBAMATE see DQE800

6-DIMETHYLAMINO-2,3,5-TRIMETHYLAMINOQUINOXALINE see QQS330

S,S'-(2-(DIMETHYLAMINO)TRIMETHYLENE)BIS(BENZENETHIOSULFONATE) see NCN650

S,S'-(2-(DIMETHYLAMINO)TRIMETHYLENE)BIS(THIOCARBAMATE) HYDROCHLORIDE see BHL750

DIMETHYLAMINOTRIMETHYLSILANE see DQE900

4-DIMETHYLAMINOTRIPHENYLMETHAN (GERMAN) see DRQ000

4-DIMETHYLAMINOTRIPHENYLMETHANE see DRQ000

3-DIMETHYLAMINO-1,1,2-TRIS(4-METHOXYPHENYL)-1-PROPENE HYDROCHLORIDE see TNJ750

5-DIMETHYLAMINO-1,2,3-TRITHIANE HYDROGENOXALATE see TFH750

N-(6-(DIMETHYLAMINO-3H-XANTHEN-3-YLIDENE)-N-METHYLMETHANAMINIUM CHLORIDE see PPQ750

4-DIMETHYLAMINO-3,5-XYLENOL see DQF000

4-(DIMETHYLAMINO)-3,5-XYLENOL METHYLCARBAMATE (ESTER) see DOS000

4-((4-(DIMETHYLAMINO)-2,3-XYLYL)AZO)PYRIDINE-1-OXIDE see DQF200

4-((4-(DIMETHYLAMINO)-2,5-XYLYL)AZO)PYRIDINE-1-OXIDE see DQF400

4-((4-(DIMETHYLAMINO)-3,5-XYLYL)AZO)PYRIDINE-1-OXIDE see DQF600

4-(DIMETHYLAMINO)-3,5-XYLYL ESTER METHYLCARBAMIC ACID see DOS000

4-DIMETHYLAMINO-3,5-XYLYL METHYLCARBAMATE see DOS000

4-DIMETHYLAMINO-3,5-XYLYL-N-METHYLCARBAMATE see DOS000

4-(N,N-DIMETHYLAMINO)-3,5-XYLYL N-METHYLCARBAMATE see DOS000

DIMETHYLAMMONIUM CHLORIDE see DOR600

DIMETHYLAMMONIUM 2,4-DICHLOROPHENOXYACETATE see DFY800

N-((2-DIMETHYLAMMONIUM)ETHYL)-4,5,6,7-TETRACHLOROISOINDOLINIUM DIMETHOCHLORIDE see CDY000

DIMETHYLAMMONIUM PERCHLORATE see DQF650

DIMETHYLAMPHETAMINE see TMA600

DI(4-METHYL-2-AMYL) MALEATE see DKP400

4,5'-DIMETHYL ANGELICIN see DQF700

4,4'-DIMETHYLANGELICIN plus ULTRAVIOLET A RADIATION see FQD130

DIMETHYLANILINE see XMA000

2,3-DIMETHYLANILINE see XMJ000

2,4-DIMETHYLANILINE see XMS000

2,5-DIMETHYLANILINE see XNA000

2,6-DIMETHYLANILINE see XNJ000

3,4-DIMETHYLANILINE see XNS000

3,5-DIMETHYLANILINE see XOA000

N,N-DIMETHYLANILINE see DQF800

N,N-DIMETHYL-p-ANILINEDIAZOSULFONIC ACID SODIUM SALT see DOU600

2,4-DIMETHYLANILINE HYDROCHLORIDE see XOJ000

2,5-DIMETHYLANILINE HYDROCHLORIDE see XOS000

N,N-DIMETHYLANILINE METHIODIDE see TMB750

N-DIMETHYL-ANILINE (OSHA) see DQF800

2-(2,6-DIMETHYLANILINO)-5,6-DIHYDRO-4H-1,3-THIAZINE see DMW000

2-(3,4-DIMETHYLANILINO)-2-OXAZOLINE see XOS500

6,12-DIMETHYLANTHANTHRENE see DQG000

DIMETHYLANTHRACENE see DQG100

9,10-DIMETHYLANTHRACENE see DQG200

3-(10,10-DIMETHYL(10H)-ANTHRACENYLIDENE)-N,N-DIMETHYL-1-PROPANAMINE (9CI) see AEG875

3-(10,10-DIMETHYL-9(10H)-ANTHRACENYLIDENE)-N,N-DIMETHYL-1-PROPANAMINE HYDROCHLORIDE see TDL000
N,N-DIMETHYL-N'-(1-ANTHRACHINONYL)FORMAMIDINIUMCHLORID (GERMAN) see APK750
DIMETHYL ANTHRANILATE (FCC) see MGQ250
DIMETHYLANTIMONY CHLORIDE see DQG400
DIMETHYLARSENIC ACID see HKC000
DIMETHYLARSENIC ACID SODIUM SALT TRIHYDRATE see HKC550
DIMETHYLARSINAT SODNY see HKC500
DIMETHYLARSINE see DQG600
DIMETHYLARSINIC ACID see HKC000
DIMETHYLARSINIC ACID SODIUM SALT TRIHYDRATE see HKC550
DIMETHYL ARSINIC SULFIDE see DQG700
((DIMETHYLARSINO)OXY)SODIUM-As-OXIDE see HKC500
N,N-DIMETHYL-1-AZIRIDINECARBOXAMIDE see EJA100
N,N-DIMETHYL-p-AZOANILINE see DOT300
2,3'-DIMETHYLAZOBENZENE see DQH000
3,6'-DIMETHYLAZOBENZENE see DQH000
2,3'-DIMETHYLAZOBENZENE-4'-METHYLCARBONATE see CBS750
N,N'-DIMETHYL-4,4'-AZODIACETANILIDE see DQH200
2,2'-DIMETHYL-2,2'-AZODIBUTYRONITRILE see ASM025
DIMETHYL AZODIFORMATE see DQH509
DIMETHYL AZOFORMATE see DQH509
2,2'-DIMETHYL-BEBEERINIUM (8CI) see COF825
DIMETHYLBELLIDIFOLIN see HOL223
1,3-DIMETHYLBENZ(e)ACEPHENANTHRYLENE see DQH550
7,9-DIMETHYLBENZ(c)ACRIDINE see DQI200
5,7-DIMETHYL-1,2-BENZACRIDINE see DQI600
6,9-DIMETHYL-1,2-BENZACRIDINE see DQI800
7,10-DIMETHYLBENZ(c)ACRIDINE see DQI800
7,11-DIMETHYLBENZ(c)ACRIDINE see DQI400
8,10-DIMETHYL-BENZ(a)ACRIDINE see DQI600
8,12-DIMETHYLBENZ(a)ACRIDINE see DQH600
9,12-DIMETHYLBENZ(a)ACRIDINE see DQH800
1,10-DIMETHYL-5,6-BENZACRIDINE see DQH600
2,10-DIMETHYL-5,6-BENZACRIDINE see DQH800
1,10-DIMETHYL-7,8-BENZACRIDINE (FRENCH) see DQI400
2,10-DIMETHYL-7,8-BENZACRIDINE (FRENCH) see DQI800
3,10-DIMETHYL-7,8-BENZACRIDINE (FRENCH) see DQI200
N,N-DIMETHYLBENZAMIDE see DQJ000
DIMETHYLBENZANTHRACENE see DQJ200
DIMETHYLBENZ(a)ANTHRACENE see DQJ200
4,5-DIMETHYLBENZ(a)ANTHRACENE see DQJ600
6,7-DIMETHYLBENZ(a)ANTHRACENE see DQJ800
6,8-DIMETHYLBENZ(a)ANTHRACENE see DQK000
7,12-DIMETHYLBENZANTHRACENE see DQJ200

7,8-DIMETHYLBENZ(a)ANTHRACENE see DQL000
8,9-DIMETHYLBENZ(a)ANTHRACENE see DQK600
9,10-DIMETHYL-BENZANTHRACENE see DQJ200
1,12-DIMETHYLBENZ(a)ANTHRACENE see DQJ400
4,9-DIMETHYL-1,2-BENZANTHRACENE see DQK200
5,6-DIMETHYL-1,2-BENZANTHRACENE see DQK600
5,9-DIMETHYL-1,2-BENZANTHRACENE see DQK800
5:9-DIMETHYL-1:2-BENZANTHRACENE see DQK800
6,12-DIMETHYLBENZ(a)ANTHRACENE see DQK200
6,7-DIMETHYL-1,2-BENZANTHRACENE see DQL200
6,8-DIMETHYL-1,2-BENZANTHRACENE see DQK000
7,11-DIMETHYLBENZ(a)ANTHRACENE see DQK400
7,12-DIMETHYLBENZ(a)ANTHRACENE see DQJ200
8,12-DIMETHYLBENZ(a)ANTHRACENE see DQK800
9,10-DIMETHYLBENZ(a)ANTHRACENE see DQJ200
9,10-DIMETHYLBENZ(a)ANTHRACENE see DQL200
1',9-DIMETHYL-1,2-BENZANTHRACENE see DQJ400
3,4'-DIMETHYL-1,2-BENZANTHRACENE see DQJ600
4,10-DIMETHYL-1,2-BENZANTHRACENE see DQJ800
5,10-DIMETHYL-1,2-BENZANTHRACENE see DQL000
8,10-DIMETHYL-1,2-BENZANTHRACENE see DQK400
9,10-DIMETHYL-1,2-BENZANTHRACENE see DQJ200
7,12-DIMETHYLBENZ(a)ANTHRACENE-8-CARBONITRILE see COM000
7,12-DIMETHYLBENZ(a)ANTHRACENE-D16 see DQL400
7,12-DIMETHYLBENZ(a)ANTHRACENE, DEUTERATED see DQL400
7,12-DIMETHYLBENZ(a)ANTHRACENE-3,4-DIOL see DQK900
7,12-DIMETHYLBENZ(a)ANTHRACENE-5,6-OXIDE see DQL600
9:10-DIMETHYL-1:2-BENZANTHRACENE-9:10-OXIDE see DQL800
7,12-DIMETHYLBENZANTHRACENE-7,12-endo-α,β-SUCCINIC ACID see DRT000
9,10-DIMETHYL-1,2-BENZANTHRAZEN (GERMAN) see DQJ200
DIMETHYLBENZANTHRENE see DQJ200
O,O-DIMETHYL-S-(BENZAZIMINOMETHYL)DITHIOPHOSPHATE see ASH500
2,2-DIMETHYL-1,3-BENZDIOXOL-4-YL-N-METHYLCARBAMATE see DQM600
α,α-DIMETHYLBENZEETHANAMINE see DTJ400
2,3-DIMETHYLBENZENAMINE see XMJ000
2,4-DIMETHYLBENZENAMINE see XMS000
2,5-DIMETHYLBENZENAMINE see XNA000
2,6-DIMETHYLBENZENAMINE see XNJ000
3,5-DIMETHYLBENZENAMINE see XOA000
2,4-DIMETHYLBENZENAMINE HYDROCHLORIDE see XOJ000
2,5-DIMETHYLBENZENAMINE HYDROCHLORIDE see XOS000
DIMETHYLBENZENE see XGS000
m-DIMETHYLBENZENE see XHA000
o-DIMETHYLBENZENE see XHJ000
p-DIMETHYLBENZENE see XHS000
1,2-DIMETHYLBENZENE see XHJ000

1,3-DIMETHYLBENZENE see XHA000
1,4-DIMETHYLBENZENE see XHS000
N,N-DIMETHYLBENZENEAMINE see DQF800
1,3-DIMETHYLBENZENE, BENZYLATED see DQL820
α,α'-DIMETHYL-m-BENZENEBIS(ETHYLAMINE) DIHYDROCHLORIDE see DCC100
α,α'-DIMETHYL-p-BENZENEBIS(ETHYLAMINE) DIHYDROCHLORIDE see DCC125
N,N-DIMETHYL-1,4-BENZENEDIAMINE see DTL800
3,5-DIMETHYLBENZENEDIAZONIUM-2-CARBOXYLATE see DQL899
4,6-DIMETHYLBENZENEDIAZONIUM-2-CARBOXYLATE see DQL959
DIMETHYL-1,2-BENZENEDICARBOXYLATE see DTR200
DIMETHYL-1,4-BENZENE DICARBOXYLATE see DUE000
2,6-DIMETHYL-1,4-BENZENEDIOL (9CI) see DSG700
1,3-DIMETHYLBENZENE HEXACHLORO DERIV. see HCM100
N,N-DIMETHYLBENZENEMETHANAMINE see DQP800
α,4-DIMETHYLBENZENEMETHANOL see TGZ000
α,α-DIMETHYLBENZENEMETHANOL see DTN100
2,4-DIMETHYLBENZENEMETHANOL ACETATE see DQP500
DIMETHYL BENZENEORTHODICARBOXYLATE see DTR200
α,α-DIMETHYLBENZENEPROPANOL ACETATE see MNT000
2,4-DIMETHYLBENZENESULFONIC ACID see XJJ000
3,3'-DIMETHYLBENZIDIN see TGJ750
3,3'-DIMETHYLBENZIDINE see TGJ750
3,3'-DIMETHYLBENZIDINE DIHYDROCHLORIDE see DQM000
5,6-DIMETHYLBENZIMIDAZOLE see DQM100
N,N-DIMETHYL-p-(4-BENZIMIDAZOLYAZO)ANILINE see DQM200
DIMETHYLBENZIMIDAZOLYCOBAMIDE see VSZ000
5,6-DIMETHYLBENZIMIDAZOLYCOBAMIDE CYANIDE see VSZ000
N,N-DIMETHYL-4(4'-BENZIMIDAZOLYLAZO)ANILINE see DQM200
DIMETHYL BENZMIDE see DQJ000
7,12-DIMETHYLBENZO(a)ANTHRACENE see DQJ200
6,12-DIMETHYLBENZO(1,2-b:5,4-b')BIS(1)BENZOTHIOPHENE see DQM400
2,2-DIMETHYL-1,3-BENZODIOX-4-OL METHYLCARBAMATE see DQM600
2,2-DIMETHYL-1,3-BENZODIOXOL-4-YL N-BUTYRYL-N-METHYLCARBAMATE see MNE300
2,2-DIMETHYLBENZO-1,3-DIOXOL-4-YL METHYLCARBAMATE see DQM600
2,2-DIMETHYL-1,3-BENZODIOXOL-4-YL METHYL(1-OXOPROPYL)CARBAMATE see DQM650
2,2-DIMETHYL-1,3-BENZODIOXOL-4-YL N-PENTANOYL-N-METHYLCARBAMATE see MNF300
2,2-DIMETHYL-1,3-BENZODIOXOL-4-YL N-PROPIONYL-N-METHYLCARBAMATE see DQM650
6,12-DIMETHYLBENZO(1,2-b:4,5-b')DITHIONAPHTHENE see DQM800

1,3-DIMETHYLBENZO(b)FLUORANTHENE see DQH550

N,N-DIMETHYL-p-(5-BENZOFURYLAZO)ANILINE see BCK800

N,N-DIMETHYL-p-(7-BENZOFURYLAZO)ANILINE see BCL100

3,4-DIMETHYLBENZOIC ACID see DQM850

3,5-DIMETHYLBENZOIC ACID see MDJ748

7,13-DIMETHYLBENZO(b)PHENANTHRO(3,2-d)THIOPHENE see DRJ000

ar,ar-DIMETHYLBENZOPHENONE see PGP750

1,2-DIMETHYLBENZO(a)PYRENE see DQN000

1,3-DIMETHYLBENZO(a)PYRENE see DQN200

1,4-DIMETHYLBENZO(a)PYRENE see DQN400

1,6-DIMETHYLBENZO(a)PYRENE see DQN600

2,3-DIMETHYLBENZO(a)PYRENE see DQN800

3,6-DIMETHYLBENZO(a)PYRENE see DQO000

4,5-DIMETHYLBENZO(a)PYRENE see DQO400

3,12-DIMETHYLBENZO(a)PYRENE see DQO200

2,5-DIMETHYL-p-BENZOQUINONE see XQJ000

2,5-DIMETHYLBENZOSELENAZOLE see DQO600

5,6-DIMETHYL-2,1,3-BENZOSELENODIAZOLE see DQO650

2,5-DIMETHYLBENZOTHIAZOLE see DQO800

N,N-DIMETHYLBENZO(b)THIOPHENE-3-ETHYLAMINE HYDROCHLORIDE see DQP000

O,O-DIMETHYL-S-(1,2,3-BENZOTRIAZINYL-4-KETO)METHYL PHOSPHORODITHIOATE see ASH500

1,2-DIMETHYL-1H-BENZOTRIAZOLIUM IODIDE see DQP100

1,3-DIMETHYL-3H-BENZOTRIAZOLIUM IODIDE see DQP125

1,2-DIMETHYLBENZOTRIAZOLIUM JODID (GERMAN) see DQP100

1,3-DIMETHYLBENZOTRIAZOLIUM JODID (GERMAN) see DQP125

1,4-DIMETHYL-2,3-BENZPHENANTHRENE see DQJ200

2,5-DIMETHYLBENZSELENAZOL (CZECH) see DQO600

2,5-DIMETHYLBENZTHIAZOL (CZECH) see DQO800

N,N-DIMETHYL-p-(6-BENZTHIAZOLYLAZO)ANILINE see DPO400

N,N-DIMETHYL-p-(7-BENZTHIAZOLYLAZO)ANILINE see DPO600

N,N-DIMETHYL-4-(6'-BENZTHIAZOLYLAZO)ANILINE see DPO400

N,N-DIMETHYL-4-(7'-BENZTHIAZOLYLAZO)ANILINE see DPO600

4,9-DIMETHYL-2,3-BENZTHIOPHANTHRENE see DQP400

2,4-DIMETHYLBENZYL ACETATE see DQP500

p,α-DIMETHYLBENZYL ALCOHOL see TGZ000

α,α-DIMETHYLBENZYL ALCOHOL see DTN100

N,N-DIMETHYLBENZYLAMINE see DQP800

N,N-DIMETHYLBENZYLAMINE HEXAFLUOROARSENATE see BEL550

DIMETHYLBENZYLAMINE HYDROCHLORIDE see DQQ000

DIMETHYLBENZYLAMMONIUM CHLORIDE see DQQ000

DIMETHYL BENZYL CARBINOL see DQQ200

DIMETHYLBENZYL CARBINOLACETATE see BEL750

DIMETHYL BENZYL CARBINYL ACETATE see DQQ375

DIMETHYLBENZYLCARBINYL BUTYRATE see BEL850

DIMETHYL BENZYL CARBINYL BUTYRATE see DQQ380

DIMETHYL BENZYL CARBINYL PROPIONATE see DQQ400

2,4-DIMETHYLBENZYLCHRYSANTHEMUMATE see DQQ500

2,4-DIMETHYLBENZYL-(I)-cis-trans-CHRYSANTHEMUMATE see DQQ500

2,4-DIMETHYLBENZYL- 2,2-DIMETHYL-3-(2-METHYLPROPENYL)CYCLOPROPANECARBOXYLATE see DQQ500

4-α,α-DIMETHYLBENZYL-α,α'-DIPIPERIDINO-2,6-XYLENOL DIHYDROBROMIDE see BHR750

2,4-DIMETHYLBENZYLESTER KYSELINY CHRYSANTHEMOVE see DQQ500

2,4-DIMETHYLBENZYL ESTER OF cis,trans-CHRYSANTHEMUMIC ACID see DQQ500

N,N'-DI(α-METHYLBENZYL)ETHYLENEDIAMINE see DQQ600

α,α-DIMETHYLBENZYL HYDROPEROXIDE (MAK) see IOB000

1-(α,α-DIMETHYLBENZYL)-3-METHYL-3-PHENYLUREA see DQQ700

DIMETHYLBENZYLOCTADECYLAMMONIUM CHLORIDE see DTC600

p-(α-α-DIMETHYLBENZYL)PHENOL see COF400

N,N-DIMETHYL-N'-BENZYL-N'-(α-PYRIDYL)ETHYLENEDIAMINE see TMP750

DIMETHYL BERYLLIUM see DQR200

DIMETHYLBERYLLIUM-1,2-DIMETHOXYETHANE see DQR289

3',3'''-DIMETHYL-4',4'''-BIACETANILIDE see DRI400

2,2'-DIMETHYL-1,1'-BIANTHRACENE-9,9',10,10'-TETRONE see DQR350

2,2'-DIMETHYL-1,1'-BIANTHRAQUINONE see DQR350

6,12-DIMETHYL-BIBENZO(def,mno)CHRYSENE see DQG000

2,2-DIMETHYLBICYCLO(2.2.1)HEPTANE-3-CARBOXYLIC ACID, METHYL ESTER see MHY500

6,6-DIMETHYLBICYCLO(3.1.1)HEPT-2-ENE-2-CARBOXALDEHYDE see FNK150

DIMETHYL cis-BICYCLO(2,2,1)-5-HEPTENE-2,3-DICARBOXYLATE see DRB400

6,6-DIMETHYLBICYCLO-(3.1.1)-2-HEPTENE-2-ETHANOL see DTB800

6,6-DIMETHYLBICYCLO(3.1.1)-2-HEPTENE-2-ETHYL ACETATE see DTC000

(1S)-6,6-DIMETHYLBICYCLO(3.1.1)HEPT-2-ENE-2-METHANOL ACETATE see MSC050

1,1-DIMETHYLBIGUANIDE see DQR600

N,N-DIMETHYLBIGUANIDE see DQR600

DIMETHYLBIGUANIDE HYDROCHLORIDE see DQR800

1,1-DIMETHYLBIGUANIDE HYDROCHLORIDE see DQR800

6,6'-DIMETHYL-(2,2'-BINAPHTHALENE)-1,1',8,8'-TETROL see DVO809

3,2'-DIMETHYL-4-BIPHENYLAMINE see BLV250

3,3'-DIMETHYL-4-BIPHENYLAMINE see DOV200

3,3'-DIMETHYLBIPHENYL-4,4'-BIPHENYLDIAMINE DIHYDROCHLORIDE see DQM000

3,3'-DIMETHYL-4,4'-BIPHENYLDIAMINE see TGJ750

3,3'-DIMETHYLBIPHENYL-4,4'-DIAMINE see TGJ750

3,3'-DIMETHYL-(1,1'-BIPHENYL)-4,4'-DIAMINE see TGJ750

2,3'-DIMETHYLBIPHENYL-4,4'-DIAMINE DIHYDROCHLORIDE see DQM000

ar,ar'-DIMETHYL-(1,1'-BIPHENYL)-ar,ar'-DIOL (9CI) see DGQ700

3,3'-DIMETHYL-4,4'-BIPHENYLENE DIISOCYANATE see DQS000

4-N-(3,2'-DIMETHYLBIPHENYL)HYDROXAMINE see HKB650

4,4'-DIMETHYL-2,2'-BIPYRIDINE see DQS100

1,1'-DIMETHYL-4,4'-BIPYRIDINIUM see PAI990

N,N'-DIMETHYL-4,4'-BIPYRIDINIUM DIBROMIDE see PAI995

N,N'-DIMETHYL-4,4'-BIPYRIDINIUM DICHLORIDE see PAJ000

1,1'-DIMETHYL-4,4'-BIPYRIDINIUM DIHYDRATE see PAJ100

N,N'-DIMETHYL-4,4'-BIPYRIDYLIUM DICHLORIDE see PAJ000

1,1'-DIMETHYL-4,4'-BIPYRIDYNIUM DICHLORIDE see PAJ000

1,1'-DIMETHYL-4,4'-BIPYRIDYNIUM DIMETHYLSULFATE see PAJ250

DIMETHYL 1,3-BIS(CARBOMETHOXY)-1-PROPEN-2-YL PHOSPHATE see SOY000

DIMETHYL-BIS(β-CHLOROETHYL)AMMONIUM CHLORIDE see DQS600

β,γ-DIMETHYL-α,Δ-BIS(3,4-DIHYDROXYPHENYL)BUTANE see NBR000

O,O-DIMETHYL-S-(1,2-BIS(ETHOXYCARBONYL)ETHYL)DITHIOPHOSPHATE see MAK700

O,O-DIMETHYL-S-1,2-BIS(ETHOXYCARBONYL)ETHYL PHOSPHOROTHIOATE see OPK250

DIMETHYL BIS(p-HYDROXYPHENYL)METHANE see BLD500

DIMETHYLBISMUTH CHLORIDE see DQT000

5,5-DIMETHYL-1,3-BIS(OXIRANYLMETHYL)-2,4-IMIDAZOLIDINEDIONE see DKM130

5,5-DIMETHYL-1,3-BIS(OXIRANYLMETHYL)-2,4-IMIDAZOLIDINEDIONE (9CI) see DKM130

3,3'-DIMETHYLBISPHENOL A see IPK000

DIMETHYLBIS(PHENYLTHIO)STANNANE see BLF750

N,N'-DIMETHYL-N,N'-BIS(3-(3',4',5'-TRIMETHOXYBENZOXY)PROPYL)ETHYLENEDIAMINE DIHYDROCHLORIDE see HFF500

N,N-DIMETHYL-4-BROMOANILINE see BNF250

N,N-DIMETHYL-β-(p-BROMOANILINO)PROPIONAMIDE see BMN350

O,O-DIMETHYL-O-(4-BROMO-2,5-DICHLOROPHENYL) PHOSPHOROTHIOATE see BNL250

N,N-DIMETHYL-(2-BROMOETHYL)HYDRAZINIUM BROMIDE see DQT100

N,N-DIMETHYL-(2-BROMOETHYL)HYDRAZINIUM BROMIDE see DQT100

2,3-DIMETHYL-1,3-BUTADIENE see DQT150

N,N-DIMETHYL-1-BUTANAMINE see BRB300

1,2-DIMETHYLBUTANE see IKS600

2,2-DIMETHYLBUTANE see DQT200

2,3-DIMETHYLBUTANE see DQT400

DIMETHYL BUTANEDIOATE see SNB100

2,3-DIMETHYL-2,3-BUTANEDIOL see TDR000

4,4'-(2,3-DIMETHYL-1,4-BUTANEDIYL)BIS(1,2-DIMETHOXYBENZENE see DQT500

1,3-DIMETHYL BUTANOL see AOK750

3,3-DIMETHYL-2-BUTANOL see BRU300

3,3-DIMETHYL-2-BUTANOL METHYLPHOSPHONOFLUORIDATE see SKS500

2,2-DIMETHYL-3-BUTANONE see DQT800

3,3-DIMETHYL-2-BUTANONE see DQU000

5-(1,3-DIMETHYL-2-BUTENYL)-5-ETHYL BARBITURIC ACID see DQU200

5-(1,3-DIMETHYL-2-BUTENYL)-5-ETHYL-2,4,6(1H,3H,5H)PYRIMIDINETRIONE see DQU200

1-(2-(1,3-DIMETHYL-2-BUTENYLIDENE)HYDRAZINO)PHTHALAZINE see DQU400

1,3-DIMETHYLBUTYL ACETATE see HFJ000

DI(3-METHYLBUTYL)ADIPATE see AEQ500

1,3-DIMETHYL BUTYLAMINE see DQU600

N,N-DIMETHYLBUTYLAMINE see BRB300

1,3-DIMETHYLBUTYLAMINE (DOT) see DQU600

1,3-DIMETHYL-5-tert-BUTYLBENZENE see DQU800

1,1-DIMETHYL-3-((3-N-tert-BUTYLCARBAMYLOXY)-PHENYL)UREA see DUM800

4,6-DIMETHYL-8-tert-BUTYLCOUMARIN see DQV000

5-(1,3-DIMETHYLBUTYL)-5-ETHYLBARBITURIC ACID see DDH900

5-(1,3-DIMETHYLBUTYL)-5-ETHYL BARBITURIC ACID, SODIUM SALT see DQV200

5-(1,3-DIMETHYLBUTYL)-5-ETHYL-2,4,6(1H,3H,5H)-PYRIMIDINETRIONE (9CI) see DDH900

1,3-DIMETHYLBUTYL p-IODOBENZYL CARBONATE see IEF200

3,3-DIMETHYL-2-BUTYL METHYLPHOSPHONOFLUORIDATE see SKS500

3,3-DIMETHYL-n-BUT-2-YL METHYLPHOSPHONOFLUORIDATE see SKS500

N-(1,3-DIMETHYLBUTYL)-N'-PHENYL-p-PHENYLENEDIAMINE see DQV250

N,N-DIMETHYLBUTYRAMIDE see DQV300

O,O-DIMETHYL-(1-BUTYRYLOXY-2,2,2-TRICHLOROETHYL) PHOSPHONATE see BPG000

DI-2-METHYLBUTYRYL PEROXIDE see DQW600

DIMETHYLCADMIUM see DQW800

N,N-DIMETHYLCAPRAMIDE see DQX000

N,N-DIMETHYLCAPROAMIDE see DQX200

N,N-DIMETHYLCAPRYLAMIDE see DTE200

DIMETHYLCARBAMATE de 5,5-DIMETHYL DIHYDRORESORCINOL (FRENCH) see DRL200

DIMETHYLCARBAMATE-d'l-ISOPROPYL-3-METHYL-5-PYRAZOLYLE (FRENCH) see DSK200

DIMETHYLCARBAMIC ACID CHLORIDE see DQY950

N,N-DIMETHYLCARBAMIC ACID CHLORIDE see DQY950

N,N-DIMETHYLCARBAMIC ACID-3-DIETHYLAMINOPHENYL ESTER METHIODIDE see HNK500

N,N-DIMETHYLCARBAMIC ACID-3-DIETHYLAMINOPHENYL ESTER, METOCHLORIDE see TGH670

DIMETHYLCARBAMIC ACID-(α-(DIETHYLAMINO))-o-TOLYL ESTER, HYDROCHLORIDE see DQY000

DIMETHYLCARBAMIC ACID-m-(DIETHYLMETHYLAMINO)PHENYL ESTER IODIDE see HNK500

DIMETHYLCARBAMIC ACID-m-(DIETHYLMETHYLAMMONIO)PHENYL ESTER, CHLORIDE see TGH670

DIMETHYLCARBAMIC ACID-1-((DIMETHYLAMINO)CARBONYL)-5-METHYL-1H-PYRAZOL-3-YL ESTER see DQZ000

DIMETHYLCARBAMIC ACID 2-(DIMETHYLAMINO)-5,6-DIMETHYL-4-PYRIMIDINYL ESTER see DOX600

N,N-DIMETHYLCARBAMIC ACID-4-(β-DIMETHYLAMINOETHYL)PHENYL ESTER, HYDROCHLORIDE see DQX800

N,N-DIMETHYLCARBAMIC ACID-p-(β-DIMETHYLAMINOETHYL)PHENYL ESTER, HYDROCHLORIDE see DQX800

N,N-DIMETHYLCARBAMIC ACID-3-DIMETHYLAMINOPHENYL ESTER METHOSULFATE see DQY909

DIMETHYLCARBAMIC ACID ESTER of 3-HYDROXY-1-METHYLPYRIDINIUM BROMIDE see MDL600

DIMETHYLCARBAMIC ACID ESTER with 3-HYDROXY-N,N,5-TRIMETHYLPYRAZOLE-1-CARBOXAMIDE see DQZ000

DIMETHYLCARBAMIC ACID ESTER with (m-HYDROXYPHENYL)TRIMETHYLAMMONIUM METHYL SULFATE see DQY909

DIMETHYLCARBAMIC ACID ETHYL ESTER see EII500

DIMETHYLCARBAMIC ACID ester with 3-HYDROXY-5,5-DIMETHYL-2-CYCLOHEXEN-1-ONE see DRL200

N,N-DIMETHYLCARBAMIC ACID, m-ISOPROPYL PHENYL ESTER see DQX300

DIMETHYLCARBAMIC ACID 3-METHYL-1-(1-METHYLETHYL)-1H-PYRAZOL-5-YL ESTER see DSK200

DIMETHYL CARBAMIC ACID-3-METHYL-1-PHENYL PYRAZOL-5-YL ESTER see PPQ625

DIMETHYLCARBAMIC ACID-5-METHYL-1H-PYRAZOL-3-YL ESTER see DQZ000

N,N-DIMETHYLCARBAMIC ACID-8-QUINOLINYL ESTER METHOSULFATE see DQY400

N,N-DIMETHYLCARBAMIC ACID-3-(TRIMETHYLAMMONIO)PHENYL ESTER METHYLSULFATE see DQY909

DIMETHYLCARBAMIC CHLORIDE see DQY950

DIMETHYLCARBAMIC ESTER of HORDENINE HYDROCHLORIDE see DQX800

DIMETHYLCARBAMIC ESTER of 2-OXYBENZYLDIETHYLAMINE HYDROCHLORIDE see DQY000

DIMETHYLCARBAMIC ESTER of 8-OXYMETHYLQUINOLINIUM METHYLSULFATE see DQY400

DIMETHYLCARBAMIC ESTER of 3-OXYPHENYLTRIMETHYLAMMONIUM METHYLSULFATE see DQY909

DIMETHYLCARBAMIDOYL CHLORIDE see DQY950

N,N-DIMETHYLCARBAMIDOYL CHLORIDE see DQY950

DIMETHYLCARBAMODITHIOIC ACID, IRON COMPLEX see FAS000

DIMETHYLCARBAMODITHIOIC ACID, IRON(3+) SALT see FAS000

DIMETHYLCARBAMODITHIOIC ACID, ZINC COMPLEX see BJK500

DIMETHYLCARBAMODITHIOIC ACID, ZINC SALT see BJK500

3-(DIMETHYLCARBAMOXY)PHENYL TRIMETHYLAMMONIUM METHYL SULFATE see DQY909

N-(DIMETHYLCARBAMOYL)AZIRIDINE see EJA100

(4-DIMETHYLCARBAMOYLBUTYL)TRIMETHYLAMMONIUM IODIDE see DQY925

DIMETHYLCARBAMOYL CHLORIDE see DQY950

N,N-DIMETHYLCARBAMOYL CHLORIDE see DQY950

DIMETHYL CARBAMOYL CHLORIDE (ACGIH,DOT) see DQY950

2-DIMETHYLCARBAMOYL-3-METHYLPYRAZOLYL-(5)-N,N-DIMETHYLCARBAMAT see DQZ000

1-DIMETHYLCARBAMOYL-5-METHYL-3-PYRAZOLYL DIMETHYLCARBAMATE see DQZ000

2-DIMETHYLCARBAMOYL-3-METHYL-5-PYRAZOLYL DIMETHYLCARBAMATE see DQZ000

cis-2-DIMETHYLCARBAMOYL-1-METHYLVINYL DIMETHYLPHOSPHATE see DGQ875

((4-(N,N-DIMETHYLCARBAMOYLOXY)-2-ISOPROPYL)PHENYL)TRIMETHYLAMMONIUM IODIDE see HJZ000

(3-(DIMETHYLCARBAMOYLOXY)PHENYL)DIETHYLMETHYLAMMONIUM CHLORIDE see TGH670

(3-(DIMETHYLCARBAMOYLOXY)PHENYL)DIETHYLMETHYLAMMONIUM IODIDE see HNK500

(3-(DIMETHYLCARBAMOYLOXY)PHENYL)TRIMETHYLAMMONIUM METHYLSULFATE see DQY909

1-(DIMETHYLCARBAMYL)AZIRIDINE see EJA100

DIMETHYLCARBAMYL CHLORIDE see DQY950

N,N-DIMETHYLCARBAMYL CHLORIDE see DQY950

DIMETHYLCARBAMYL DIETHYLTHIOCARBAMYL SULFIDE see DJC875

2-(N,N-DIMETHYLCARBAMYL)-3-METHYLPYRAZOLYL-5 N,N-DIMETHYLCARBAMATE see DQZ000

DIMETHYL 2-CARBAMYL-3-METHYLPYRAZOLYLDIMETHYLCARBAMATE (GERMAN) see DQZ000

N,N'-DIMETHYL CARBANILIDE see DRB200

DIMETHYL CARBATE see DRB400

O,O-DIMETHYL-S-(CARBETHOXY)METHYL PHOSPHOROTHIOLATE see DRB600

DIMETHYLCARBINOL see INJ000

2,2-DIMETHYL-6-CARBOBUTOXY-2,3-DIHYDRO-4-PYRONE see BRT000

α,α-DIMETHYL-α'-CARBOBUTOXY-DIHYDRO-γ-PYRONE see BRT000

O,O-DIMETHYL-S-(1-CARBOETHOXYBENZYL) DITHIOPHOSPHATE see DRR400

O,O-DIMETHYL-S-CARBOETHOXYMETHYL THIOPHOSPHATE see DRB600

O,O-DIMETHYL-O-(2-CARBOMETHOXY-1-METHYLVINYL) PHOSPHATE see MQR750

DIMETHYL-1-CARBOMETHOXY-1-PROPEN-2-YL PHOSPHATE see MQR750

DIMETHYL CARBONATE see MIF000

O,O-DIMETHYL-S-(CARBONYLMETHYLMORPHOLINO) PHOSPHORODITHIOATE see PHI500
DIMETHYLCELLOSOLVE see DOE600
DIMETHYLCETYLAMINE see HCP525
N,N-DIMETHYLCETYLAMINE see HCP525
DIMETHYL-CHLORMETHYL-ETHOXYSILAN (CZECH) see DRC400
O,O-DIMETHYL-O-3-CHLOR-4-NITROFENYLTIOFOSFAT (CZECH) see MIJ250
O,O-DIMETHYL-O-(3-CHLOR-4-NITROPHENYL)-MONOTHIOPHOSPHAT (GERMAN) see MIJ250
DIMETHYLCHLOROACETOACETAMIDE see CGH810
2,6-DIMETHYLCHLOROACETYLQUINONEIMINE see CGH820
8,12-DIMETHYL-9-CHLOROBENZ(a)ACRIDINE see CGH250
1,10-DIMETHYL-2-CHLORO-5,6-BENZACRIDINE see CGH250
7,11-DIMETHYL-10-CHLOROBENZ(c)ACRIDINE see DRB800
1,10-DIMETHYL-2-CHLORO-7,8-BENZACRIDINE (FRENCH) see DRB800
O,O-DIMETHYL-O-(6-CHLOROBICYCLO(3.2.0)HEPTADIEN-1,5-YL)PHOSPHATE see HBK700
DIMETHYL 2-CHLORO-2-DIETHYLCARBAMOYL-1-METHYLVINYL PHOSPHATE see FAB400
O,O-DIMETHYL O-(2-CHLORO-2-(N,N-DIETHYLCARBAMOYL)-1-METHYLVINYL) PHOSPHATE see FAB400
DIMETHYLCHLOROETHER see CIO250
DIMETHYL(2-CHLOROETHYL)AMINE see CGW000
DIMETHYL(2-CHLOROETHYL)AMINE HYDROCHLORIDE see DRC000
DIMETHYL-β-CHLOROETHYLAMINE HYDROCHLORIDE see DRC000
DIMETHYLCHLOROFORMAMIDE see DQY950
2,2-DIMETHYL-4-(CHLOROMETHYL)-1,3-DIOXA-2-SILACYCLOPENTANE see CIL775
DIMETHYLCHLOROMETHYLETHOXYSILANE see DRC400
N,N-DIMETHYL-3'-CHLORO-4'-METHYL-4-(PHENYLAZO)-BENZENAMINE see CIL700
O,O-DIMETHYL O-2-CHLOR-4-NITROPHENYL PHOSPHOROTHIOATE see NFT000
O,O-DIMETHYL-O-(3-CHLORO-4-NITROPHENYL) PHOSPHOROTHIOATE see MIJ250
DIMETHYL-2-CHLORO-4-NITROPHENYLTHIONOPHOSPHATE see DRC500
DIMETHYL-3-CHLORO-4-NITROPHENYL THIONOPHOSPHATE see MIJ250
DIMETHYL-2-CHLORONITROPHENYL THIOPHOSPHATE see NFT000
O,O-DIMETHYL-O-(3-CHLORO-4-NITROPHENYL) THIOPHOSPHATE see MIJ250
α,α-DIMETHYL-o-CHLOROPHENETHYLAMINE HYDROCHLORIDE see DRC600
α,α-DIMETHYL-p-CHLOROPHENETHYLAMINE HYDROCHLORIDE see ARW750
N-DIMETHYL-4-(p-CHLOROPHENOXY-1,1'-DIMETHYLACETATE)BUTYRAMIDE see LGK200
N-DIMETHYL-4-(1,4'-CHLOROPHENOXY-1,1'-DIMETHYLACETATE)BUTYRAMIDE see LGK200
N-DIMETHYL-4-(p-CHLOROPHENOXYISOBUTYRATE)BUTYRAMIDE see LGK200

N-DIMETHYL-4-(1,4'-CHLOROPHENOXYISOBUTYRATE)BUTYRAMIDE see LGK200
N,N-DIMETHYL-p-((m-CHLOROPHENYL)AZO)ANILINE see DRC800
N,N-DIMETHYL-p-((o-CHLOROPHENYL)AZO)ANILINE see DRD000
N,N-DIMETHYL-p-((p-CHLOROPHENYL)AZO)ANILINE see CGD250
o,o-DIMETHYL-S-(2-((4-CHLOROPHENYL)(1-METHYLETHYL)AMINO)-2-OXOETHYL)PHOSPHORODITHIOATE see AOT255
O,O-DIMETHYL-S-p-CHLOROPHENYL PHOSPHOROTHIOATE see FOR000
DIMETHYL-p-CHLOROPHENYLTHIOMETHYL DITHIOPHOSPHATE see MQH750
O,O-DIMETHYL-S-(p-CHLOROPHENYLTHIOMETHYL)PHOSPHORODITHIOATE see MQH750
1,1-DIMETHYL-3-(p-CHLOROPHENYL)UREA see CJX750
N,N-DIMETHYL-N'-(4-CHLOROPHENYL)UREA see CJX750
N,N-DIMETHYL-2-CHLOROPROPYLAMINE see CGJ290
DIMETHYL CHLOROTHIOPHOSPHATE (DOT) see DTQ600
N,N-DIMETHYL-p-((4-CHLORO-m-TOLYL)AZO)ANILINE see CIL710
DIMETHYLCHLORTHIOFOSAT (CZECH) see DTQ600
O,O-DIMETHYL-O-2-CHLOR-1-(2,4,5-TRICHLORPHENYL)-VINYL-PHOSPHAT (GERMAN) see TBW100
1,3-DIMETHYLCHOLANTHRENE see DRE000
2,3-DIMETHYLCHOLANTHRENE see DRD800
3,6-DIMETHYLCHOLANTHRENE see DRD850
15,20-DIMETHYLCHOLANTHRENE see DRE000
16:20-DIMETHYLCHOLANTHRENE see DRD800
(+)-o,o'-DIMETHYLCHONDROCURARINE DIIODIDE see DUM000
1,2-DIMETHYLCHRYSENE see DRE200
4,5-DIMETHYLCHRYSENE see DRE600
5,6-DIMETHYLCHRYSENE see DRE800
5,7-DIMETHYLCHRYSENE see DRE400
1,11-DIMETHYLCHRYSENE see DRE400
5,11-DIMETHYLCHRYSENE see DRF000
DIMETHYL CITRACONATE see DRF200
2,2-DIMETHYL-7-COUMARANYL-N-METHYLCARBAMATE see CBS275
4,6-DIMETHYLCOUMARIN see DRF400
DIMETHYL l-CURINE DIMETHIODIDE see TDO100
o,o-DIMETHYLCURINE DIMETHIODIDE see TDO100
DIMETHYLCYANAMIDE scc DRF600
DIMETHYLCYANOARSINE see COL750
O,O-DIMETHYL-O-(4-CYANO-PHENYL)-MONOTHIOPHOSPHAT (GERMAN) see COQ399
O,O-DIMETHYL-O-4-CYANOPHENYL PHOSPHOROTHIOATE see COQ399
O,O-DIMETHYL-O-p-CYANOPHENYL PHOSPHOROTHIOATE see COQ399
O,O-DIMETHYL-O-4-CYANOPHENYL THIOPHOSPHATE see COQ399
N,N-DIMETHYLCYCLOHEXANAMINE see DRF709
o-DIMETHYLCYCLOHEXANE see DRF800
1,3-DIMETHYLCYCLOHEXANE see DRG000

1,4-DIMETHYLCYCLOHEXANE see DRG200
1,2-DIMETHYLCYCLOHEXANE (DOT) see DRF800
cis-1,2-DIMETHYLCYCLOHEXANE see DRF800
trans-1,2-DIMETHYLCYCLOHEXANE see DRG400
trans-1,4-DIMETHYLCYCLOHEXANE see DRG600
(±)N,α-DIMETHYL-CYCLOHEXANEETHANAMINE see PNN400
N,α-DIMETHYLCYCLOHEXANEETHYLAMINE see PNN400
α,N-DIMETHYLCYCLOHEXANEETHYLAMINE see PNN400
N,α-DIMETHYLCYCLOHEXANEETHYLAMINE HYDROCHLORIDE see PNN300
DIMETHYL-3-CYCLOHEXENE-1-CARBOXALDEHYDE see DRG700
DIMETHYL-3-CYCLOHEXENE-1-CARBOXALDEHYDE see LJE200
1,5-DIMETHYL-5-(1-CYCLOHEXENYL)BARBITURIC ACID see ERD500
1,5-DIMETHYL-5-CYCLOHEXENYL-1'-BARBITURIC ACID, SODIUM SALT see ERE000
DIMETHYLCYCLOHEXYLAMINE see DRF709
N,N-DIMETHYLCYCLOHEXYLAMINE (DOT) see DRF709
N,N-DIMETHYL-N-CYCLOHEXYLMETHYLAMINE see CPR500
N,α-DIMETHYLCYCLOPENTANEETHYLAMINE HYDROCHLORIDE see CPV609
11,17-DIMETHYL-15H-CYCLOPENTA(a)PHENANTHRENE see DRH200
12,17-DIMETHYL-15H-CYCLOPENTA(a)PHENANTHRENE see DRH400
1,2-DIMETHYLCYCLOPENTENE OZONIDE see DRH600
3,4-DIMETHYL-1,2-CYCLOPENTENOPHENANTHRENE see DRH800
DIMETHYLCYCLOPOLYSILOXANE see PJR300
N-DIMETHYLCYSTEAMINE HYDROCHLORIDE see DOY600
DIMETHYLCYSTEINE see MCR750
β,β-DIMETHYLCYSTEINE see MCR750
9,10-DIMETHYL-DBA see DRI800
N,N-DIMETHYLDECANAMIDE see DQX000
3,3'-DIMETHYL-N,N'-DIACETYLBENZIDINE see DRI400
DIMETHYL DIALKYL AMMONIUM CHLORIDE see DRI500
3,5-DIMETHYL-4-DIALLYLAMINOPHENYL-N-METHYLCARBAMATE see DBI800
3,3'-DIMETHYL-4,4'-DIAMINODICYCLOHEXYLMETHANE see BGT800
DIMETHYLDIAMINODIPHENYLMETHANE see MJO000
3,3'-DIMETHYL-4,4'-DIAMINODIPHENYLMETHANE see MJO250
N,N'-DIMETHYLDIAMINOETHANE see DRI600
asym-DIMETHYL-3,7-DIAMINOPHENAZATHIONIUM CHLORIDE see DUG700

N,N-DIMETHYL-1,3-DIAMINOPROPANE see AJQ100

O,O-DIMETHYL-S-(4,6-DIAMINO-1,3,5-TRIAZINYL-2-METHYL) DITHIOPHOSPHATE see ASD000

O,O-DIMETHYL- S-(4,6-DIAMINO-s-TRIAZIN-2-YLMETHYL)PHOSPHORODITHIOATE see ASD000

O,O-DIMETHYL-S-(4,6-DIAMINO-1,3,5-TRIAZIN-2-YL)METHYL PHOSPHORODITHIOATE see ASD000

O,O-DIMETHYL-S-(4,6-DIAMINO-1,3,5-TRIAZIN-2-YL)METHYL PHOSPHOROTHIOLOTHIONATE see ASD000

DIMETHYLDIAMINOXANTHENYL CHLORIDE see ADK000

3,3'-DIMETHYLDIAN see IPK000

DIMETHYLDIARSINE SULFIDE see DQG700

1,5-DIMETHYL-1,5-d-DIAZAUNDECAMETHYLENE POLYMETHOBROMIDE see HCV500

1,1-DIMETHYLDIAZENIUM PERCHLORATE see DRI700

2,3-DIMETHYL-1,4-DIAZINE see DTU400

2,5-DIMETHYL-1,4-DIAZINE see DTU600

7,14-DIMETHYLDIBENZ(a,h)ANTHRACENE see DRI800

9,10-DIMETHYL-1,2,5,6-DIBENZANTHRACENE see DRI800

5,9-DIMETHYL-7H-DIBENZO(c,g)CARBAZOLE see DRI900

N,N-DIMETHYL-5H-DIBENZO(a,d)CYCLOHEPTENE-Δ^5,γ-PROPYLAMINE see PMH600

N,N-DIMETHYL-5H-DIBENZO(a,d)CYCLOHEPTENE-$\Delta^{5,\gamma}$-PROPYLAMINE HYDROCHLORIDE see DPX800

N,N-DIMETHYLDIBENZO(b,e)THIEPIN-$\Delta^{11(6H)}$,γPROPYLAMINE see DYC875

N,N-DIMETHYLDIBENZO(b,e)THIEPIN-$\Delta^{11(6H)}$,γ-PROPYLAMINE HYDROCHLORIDE see DPY200

4,9-DIMETHYL-2,3,5,6-DIBENZOTHIOPHENTHRENE see DRJ000

N,N-DIMETHYLDIBENZ(b,e)OXEPIN-$\Delta^{11(6H)}$-γ-PROPYLAMINE see DYE409

N,N-DIMETHYLDIBENZ(b,e)OXEPIN-$\Delta^{11(6H,\gamma)}$-PROPYLAMINE HYDROCHLORIDE see AEG750

1,1-DIMETHYLDIBORANE see DRJ200

1,2-DIMETHYLDIBORANE see DRJ400

O,O-DIMETHYL-O-(1,2-DIBROMO-2,2-DICHLOROETHYL)PHOSPHATE see NAG400

DIMETHYL-1,2-DIBROMO-2,2-DICHLOROETHYL PHOSPHATE (OSHA) see NAG400

2,5-DIMETHYL-2,5-DI(t-BUTYLPEROXY)HEXANE see DRJ800

2,5-DIMETHYL-2,5-DI(tert-BUTYLPEROXY)HEXANE see DRJ800

2,5-DIMETHYL-2,5-DI(tert-BUTYLPEROXY)HEXYNE-3 see DRJ825

O,O-DIMETHYL-S-1,2-(DICARBAETHOXYAETHYL)-DITHIOPHOSPHAT (GERMAN) see MAK700

O,O-DIMETHYL-S-(1,2-DICARBETHOXYETHYL) DITHIOPHOSPHATE see MAK700

O,O-DIMETHYL-S-(1,2-DICARBETHOXYETHYL)PHOSPHORODITHIOATE see MAK700

O,O-DIMETHYL-S-(1,2-DICARBETHOXY)ETHYL PHOSPHOROTHIOATE see OPK250

O,O-DIMETHYL-S-(1,2-DICARBETHOXYETHYL) THIOTHIONOPHOSPHATE see MAK700

o,o-DIMETHYL S-(1,2-DICARBOMETHOXY)ETHYL PHOSPHORODITHIOATE see CBS800

DIMETHYL-1,3-DI(CARBOMETHOXY)-1-PROPEN-2-YL PHOSPHATE see SOY000

DIMETHYL DICARBONATE see DRJ850

DIMETHYLDICETYLAMMONIUM CHLORIDE see DRK200

O,O-DIMETHYL-O-(2,5-DICHLOR-4-BROMPHENYL)-THIONOPHOSPHAT (GERMAN) see BNL250

O,O-DIMETHYL-O-(2,5-DICHLOR-4-JODPHENYL)-MONOTHIOPHOSPHAT (GERMAN) see IEN000

O,O-DIMETHYL-O-(2,5-DICHLOR-4-JODPHENYL)-THIONOPHOSPHAT (GERMAN) see IEN000

O,O-DIMETHYL-O-(2,5-DICHLORO-4-BROMOPHENYL)PHOSPHOROTHIOATE see BNL250

O,O-DIMETHYL-O-(2,5-DICHLORO-4-BROMOPHENYL) THIOPHOSPHATE see BNL250

O,O-DIMETHYL-O-2,2-DICHLORO-1,2-DIBROMOETHYL PHOSPHATE see NAG400

DIMETHYL-2,2-DICHLOROETHENYL PHOSPHATE see DGP900

DIMETHYL-1,1'-DICHLOROETHER see BIK000

O,O-DIMETHYL-O-2,5-DICHLORO-4-IODOPHENYL THIOPHOSPHATE see IEN000

DIMETHYL DICHLOROMALONATE see DRK300

O,O-DIMETHYL S-(2,5-DICHLOROPHENYLTHIO)METHYL PHOSPHORODITHIOATE see MNO750

1,1-DIMETHYL-3-(3,4-DICHLOROPHENYL)UREA see DXQ500

DIMETHYL DICHLOROPROPANEDIOATE see DRK300

DIMETHYLDICHLOROSILANE (DOT) see DFE259

DIMETHYLDICHLOROSTANNANE see DUG825

DIMETHYLDICHLOROTIN see DUG825

DIMETHYL DICHLOROVINYL PHOSPHATE see DGP900

DIMETHYL-2,2-DICHLOROVINYL PHOSPHATE see DGP900

O,O-DIMETHYL DICHLOROVINYL PHOSPHATE see DGP900

O,O-DIMETHYL-O-2,2-DICHLOROVINYL PHOSPHATE see DGP900

DIMETHYL-DICHLORSILAN see DFE259

O,O-DIMETHYL-O-(2,2-DICHLOR-VINYL)-PHOSPHAT (GERMAN) see DGP900

DIMETHYLDIDECYLAMMONIUM CHLORIDE see DGX200

2,5-DIMETHYL-1,2,5,6-DIEPOXYHEX-3-YNE see DRK400

O,O-DIMETHYL-S-1,2-DI(ETHOXYCARBAMYL)ETHYL PHOSPHORODITHIOATE see MAK700

DIMETHYL-DIETHOXYSILAN (CZECH) see DHG000

DIMETHYLDIETHOXYSILANE (DOT) see DHG000

DIMETHYL DIETHYLAMIDO-1-CHLOROCROTONYL (2) PHOSPHATE see FAB400

1,2-DIMETHYL-3-DIETHYLAMINOPROPYL p-ISOBUTOXYBENZOATE see GBU700

N,N-DIMETHYL-p-((3,4-DIETHYLPHENYL)AZO)ANILINE see DJB400

2,6-DIMETHYL-1,1-DIETHYLPIPERIDINIUM BROMIDE see DRK600

trans-4,4'-DIMETHYL-α-α'-DIETHYLSTILBENE see DRK500

N,N-DIMETHYL-2,5-DIFLUORO-p-(2,5-DIFLUOROPHENYLAZO)ANILINE see DRK800

N,N-DIMETHYL-p-(2,5-DIFLUOROPHENYLAZO)ANILINE see DKH000

N,N-DIMETHYL-p-(3,4-DIFLUOROPHENYLAZO)ANILINE see DRL000

N,N-DIMETHYL-p-(3,5-DIFLUOROPHENYLAZO)ANILINE see DKH100

N,N-DIMETHYL-3',4'-DIFLUORO-4-(PHENYLAZO)BENZENEAMINE see DRL000

DIMETHYL DIGLYCOLATE see DTH500

N,N-DIMETHYLDIGUANIDE see DQR600

9:10-DIMETHYL-9-10-DIHYDRO-1,2-BENZANTHRACENE-9,10-OXIDE see DQL800

2,2-DIMETHYL-2,3-DIHYDROBENZOFURAN-7-YL-N-(4-BROMOPHENYLTHIO)-N-METHYLCARBAMATE see DRL100

2,2-DIMETHYL-2,3-DIHYDROBENZOFURAN-7-YL ESTER, METHYLCARBAMIC ACID see CBS275

2,2-DIMETHYL-2,3-DIHYDRO-7-BENZOFURANYL-N-METHYLCARBAMATE see CBS275

11,17-DIMETHYL-16,17-DIHYDRO-15H-CYCLOPENTA(a)PHENANTHRENE see DLI200

7,11-DIMETHYL-15,16-DIHYDROCYCLOPENTA(a)PHENANTHREN-17-ONE see DLI300

2,5-DIMETHYL-2,5-DIHYDROFURAN-2,5-ENDO PEROXIDE see DUL589

O,O-DIMETHYL-S-(3,4-DIHYDRO-4-KETO-1,2,3-BENZOTRIAZINYL-3-METHYL) DITHIOPHOSPHATE see ASH500

2,5-DIMETHYL-2,5-DIHYDROPEROXYHEXANE, >82% with water (DOT) see DSE800

5,5-DIMETHYL-DIHYDRORESORCINOL-N,N-DIMETHYLCARBAMAT (GERMAN) see DRL200

5,5-DIMETHYLDIHYDRORESORCINOL DIMETHYLCARBAMATE see DRL200

5,5-DIMETHYL-4,5-DIHYDRO-3-RESORCYL-DIMETHYL-CARBAMAT (GERMAN) see DRL200

1,3-DIMETHYL-7-(2,3-DIHYDROXYPROPYL)XANTHINE see DNC000

DIMETHYL DIISOPROPYL PYROPHOSPHATE see DRL300

O,O-DIMETHYL-S-1,2-DIKARBETOXYLETHYLDITIOFOSFAT (CZECH) see MAK700

DIMETHYL DIKETONE see BOT500

2,6-DIMETHYL-3,5-DIMETHOXYCARBONYL-4-(o-DIFLUOROMETHOXYPHENYL)-1,4-DIHYDROPYRIDINE see RSZ375

DIMETHYL-α-(DIMETHOXYPHOSPHINYL)-p-CHLOROBENZYL PHOSPHATE see CMU875

DIMETHYL 3-(DIMETHOXYPHOSPHINYLOXY)GLUTACONATE see SOY000

DIMETHYLDIMETHOXYSILANE see DOE100

2,4'-DIMETHYL-4-DIMETHYLAMINOAZOBENZENE see DRL400

2',3'-DIMETHYL-4-DIMETHYLAMINOAZOBENZENE see DUN800

N,5-DIMETHYL-4-((DIMETHYLAMINO)CARBONYL)-N-((4-(1,1-DIMETHYLETHYL)PHENYL)THIO)-2,7-DIOXA-3,6-DIAZAOCTA-3,5-DIENAMIDE see DRL425

N,N-DIMETHYL-4-DIMETHYLAMINO-3-ISOPROPYLPHENYL ESTER METHIODIDE, CARBAMIC ACID see HJZ000

3,3-DIMETHYL-4-(DIMETHYLAMINO)-4-(o-METHOXYPHENYL)BUTYL o-METHOXYPHENYL KETONE see DRL450

3,3-DIMETHYL-4-(DIMETHYLAMINO)-4-(p-METHOXYPHENYL)BUTYL p-METHOXYPHENYL KETONE see DRL460

2,3-DIMETHYL-8-(DIMETHYLAMINOMETHYL)-7-METHOXYCHROMONE HYDROCHLORIDE see DRL600

3-keto-1,5-DIMETHYL-4-DIMETHYLAMINO-2-PHENYL-2,3-DIHYDROPYRAZOLE see DOT000

1,5-DIMETHYL-4-DIMETHYLAMINO-2-PHENYL-3-PYRAZOLONE see DOT000

2,3-DIMETHYL-4-DIMETHYLAMINO-1-PHENYL-5-PYRAZOLONE see DOT000

9,9-DIMETHYL-10-(3-(DIMETHYLAMINO)PROPYL)ACRIDAN see DNW700

9,9-DIMETHYL-10-DIMETHYLAMINOPROPYLACRIDAN HYDROGEN TARTRATE see DRM000

9,9-DIMETHYL-10-(3-DIMETHYLAMINO)PROPYLACRIDINE TARTRATE see DRM000

5,6-DIMETHYL-2-DIMETHYLAMINO-4-PYRIMIDINYLDIMETHYLCARBAMATE see DOX600

6,8-DIMETHYL-(4-p-(DIMETHYLAMINO)STYRYL)QUINOLINE see DQC600

3,3-DIMETHYL-4-(DIMETHYLAMINO)-4-(m-TOLYL)BUTYL m-TOLYL KETONE see DRM100

3,3-DIMETHYL-4-(DIMETHYLAMINO)-4-(o-TOLYL)BUTYL o-TOLYL KETONE see DRM110

3,3-DIMETHYL-4-(DIMETHYLAMINO)-4-(p-TOLYL)BUTYL p-TOLYL KETONE see DRM120

3',4'-DIMETHYL-4-DIMETHYLAMINOZOBENZENE see DUO000

O,O-DIMETHYL-O-(2-DIMETHYL-CARBAMOYL-1-METHYL-VINYL)PHOSPHAT (GERMAN) see DGQ875

O,O-DIMETHYL-O-(N,N-DIMETHYLCARBAMOYL-1-METHYLVINYL) PHOSPHATE see DGQ875

N,N-DIMETHYL-p-(N',N'-DIMETHYLCARBAMOYLOXY)PHENETHYLAMINE, HYDROCHLORIDE see DQX800

N,N-DIMETHYL-N'-(5-(1,1-DIMETHYLETHYL)-3-ISOXAZOLYL)UREA see ISR200

cis-2,6-DIMETHYL-4-(3-(4-(1,1-DIMETHYLETHYL)PHENYL)-2-METHYLPROPYL)MORPHOLINE see FAQ300

3,4-DIMETHYL-4-(3,4-DIMETHYL-5-ISOXAZOLYAZO)-ISOXAZOLIN-5-ONE see DRM600

O,O-DIMETHYL-O-(3,5-DIMETHYL-4-METHYLTHIOPHENYL) PHOSPHOROTHIOATE see DST200

O,O-DIMETHYL-O-(1,4-DIMETHYL-3-OXO-4-AZA-PENT-1-ENYL)FOSFAAT (DUTCH) see DGQ875

O,O-DIMETHYL-O-(1,4-DIMETHYL-3-OXO-4-AZA-PENT-1-ENYL)PHOSPHATE see DGQ875

N,N-DIMETHYL-p-(2',3'-DIMETHYLPHENYLAZO)ANILINE see DUN800

N,N-DIMETHYL-p-(3',4'-DIMETHYLPHENYLAZO)ANILINE see DUO000

DIMETHYL (DIMETHYLPHOSPHINYL)BUTANEDIOATE see PHA565

DIMETHYL (DIMETHYLPHOSPHONO)SUCCINATE see PHA565

3-DIMETHYL-1,2-DIMETHYLPROPYL p-AMINOBENZOATE HYDROCHLORIDE see AIT750

N,N-DIMETHYL-4-(4'-(2',5'-DIMETHYLPYRIDYL-1'-OXIDE)AZO)ANILINE see DPP600

N,N-DIMETHYL-4-(4'-(2',6'-DIMETHYLPYRIDYL-1'-OXIDE)AZO)ANILINE see DPP800

N,N-DIMETHYL-4-(4'-(3',5'-DIMETHYLPYRIDYL-1'-OXIDE)AZO)ANILINE see DPP709

2,5-DIMETHYL-1-(5-(2,5-DIMETHYLPYRROLIDINO)-2,4-PENTADIENYLIDENE) PYRROLIDINIUM CHLORIDE SESQUIHYDRATE see DRM800

O,O-DIMETHYL-O-(p-(N,N-DIMETHYLSULFAMOYL)PHENYL)PHOSPHOROTHIOATE see FAB600

2,2'-DIMETHYL-3,3'-DINITROAZOXYBENZENE see BKS812

2,2'-DIMETHYL-5,5'-DINITROAZOXYBENZENE see BKS813

4,4'-DIMETHYL-3,3'-DINITROAZOXYBENZENE see BKS814

1,4-DIMETHYL-3,6-DINITROCARBAZOLE see DRM900

1,4-DIMETHYL-3,6-DINITRO-9H-CARBAZOLE see DRM900

3,4-DIMETHYL-2,6-DINITRO-N-(1-ETHYLPROPYL)ANILINE see DRN200

N,N'-DIMETHYL-N,N'-DINITROOXAMIDE see DRN300

DIMETHYL-DI-NITROSO-AETHYLENDIAMIN (GERMAN) see DVE400

N,N'-DIMETHYL-N,N'-DINITROSO-1,4-BENZENEDICARBOXAMIDE see DRO400

1,6-DIMETHYL-1,6-DINITROSOBIUREA see DRN400

DIMETHYLDINITROSOETHYLENEDIAMINE see DVE400

N,N'-DIMETHYL-N,N'-DINITROSOETHYLENEDIAMINE see DVE400

N,N'-DIMETHYL-N,N'-DINITROSO-1,2-HYDRAZINEDICARBOXAMIDE see DRN400

DIMETHYLDINITROSOOXAMID (GERMAN) see DRN600

N,N'-DIMETHYL-N,N'-DINITROSOOXAMIDE see DRN600

2,5-DIMETHYLDINITROSOPIPERAZINE see DRN800

2,6-DIMETHYLDINITROSOPIPERAZINE see DRO000

2,5-DIMETHYL-1,4-DINITROSOPIPERAZINE see DRN800

N,N'-DIMETHYL-N,N'-DINITROSO-1,3-PROPANEDIAMINE see DRO200

N,N'-DIMETHYL-N,N'-DINITROSOTEREPHTHALAMIDE see DRO400

DIMETHYLDIOCTADECYLAMMONIUM CHLORIDE see DXG625

DIMETHYLDIOCTYLAMMONIUM CHLORIDE see DTF820

2,2-DIMETHYL-1,3-DIOXA-6-AZA-2-SILACYCLOOCTANE-6-ETHANOL see SDH670

DIMETHYL-1,1-DIOXA-2,8-HYDROXYETHYL-5 SILA-1 AZA-5 CYCLOOCTANE (FRENCH) see SDH670

DIMETHYL DIOXANE see DRO800

2,6-DIMETHYL-1,4-DIOXANE see DRP000

4,4-DIMETHYLDIOXANE-1,3 see DVQ400

DIMETHYL-p-DIOXANE (DOT) see DRO800

2,2-DIMETHYL-m-DIOXANE-4,6-DIONE see DRP200

2,2-DIMETHYL-1,3-DIOXANE-4,6-DIONE see DRP200

DIMETHYLDIOXANES (DOT) see DRO800

2,6-DIMETHYL-m-DIOXAN-4-OL ACETATE see ABC250

2,6-DIMETHYL-m-DIOXAN-4-YL ACETATE see ABC250

2-((2,2-DIMETHYL-1,3-DIOXAN-5-YLIDENE)METHYL)-1-METHYL-5-NITRO-1H-IMIDAZOLE see DRP300

2,2-DIMETHYL-4,6-DIOXO-m-DIOXANE see DRP200

2,2-DIMETHYL-1,3-DIOXOLAN see DRP400

2,2-DIMETHYL-1,3-DIOXOLANE-4-METHANOL see DVR600

2-(4,5-DIMETHYL-1,3-DIOXOLAN-2-YL)PHENYL-N-METHYLCARBAMATE see DRP600

N,N-DIMETHYLDIPHENYLACETAMIDE see DRP800

N,N-DIMETHYL-2,2-DIPHENYLACETAMIDE see DRP800

N,N-DIMETHYL-α,α-DIPHENYLACETAMIDE see DRP800

2',3'-DIMETHYL-2-DIPHENYLAMINECARBOXYLIC ACID see XQS000

cis-N,N-DIMETHYL-2-(p-(1,2-DIPHENYL-1-BUTENYL)PHENOXY)ETHYLAMINE see NOA600

N,N-DIMETHYL-2-(p-(1,2-DIPHENYL-1-BUTENYL)PHENOXY)ETHYLAMINE CITRATE see DRP875

3,3'-DIMETHYL-4,4'-DIPHENYLDIAMINE see TGJ750

3,3'-DIMETHYLDIPHENYL-4,4'-DIAMINE see TGJ750

4,4'-DIMETHYLDIPHENYL ETHER see THC600

(R) (−)-N,N-DIMETHYL-1,2-DIPHENYLETHYLAMINE HYDROCHLORIDE see DWA600

3,3'-DIMETHYLDIPHENYLMETHANE-4,4'-DIISOCYANATE see MJN750

N,N-DIMETHYL-4-(DIPHENYLMETHYL)ANILINE see DRQ000

α-N,N-DIMETHYL-3,4-DIPHENYL-2-METHYL-3-PROPIONOXY-1-BUTYLAMINE see PNA250

1,2-DIMETHYL-3,5-DIPHENYL-1-H-PYRAZOLIUM METHYL SULFATE see ARW000

N,N'-DIMETHYL-N,N'-DIPHENYLUREA see DRB200

DIMETHYLDIPROPYLENETRIAMINE see AME100

N,N-DIMETHYLDIPROPYLENETRIAMINE see AME100

2,2'-DIMETHYLDIPROPYLINITROSOAMINE see DRQ200

1,1'-DIMETHYL-4,4'-DIPYRIDINIUM-DICHLORID see PAJ000

1',1'-DIMETHYL-4,4'-DIPYRIDINIUM DI(METHYLSULFATE) see PAJ250

N-(2,7-DIMETHYLDIPYRIDO(1,2-A:3',2'-D)IMIDAZOL-4-YL)ACETAMIDE see AAJ600

N-(4,6-DIMETHYLDIPYRIDO(1,2-A:3',2'-D)IMIDAZOL-3-YL)HYDROXYLAMINE see HIT610

4,4'-DIMETHYLDIPYRIDYL DICHLORIDE see PAJ000

1,1'-DIMETHYL-4,4'-DIPYRIDYLIUM CHLORIDE see PAJ000

N,N'-DIMETHYL-4,4'-DIPYRIDYLIUM DICHLORIDE see PAJ000

6,7-DIMETHYL-2,4-DI-1-PYRROLIDINYL-7H-PYRROLO(2,3-D)PYRIMIDINE SULFATE (1:1) see DRQ300

DIMETHYL DISULFIDE see DRQ400

3,3-DIMETHYL-1,4-DITHIAN-2-ONE-o-((METHYL((TRICHLOROMETHYL)THIO)A MINO)CARBONYL)OXIME see DRQ500

N,N-DIMETHYL-4,4-DI-2-THIENYL-3-BUTEN-2-AMINE HYDROCHLORIDE see TLQ250

N,1-DIMETHYL-3,3-DI-2-THIENYL-N-ETHYLALLYLAMINE HYDROCHLORIDE see EIJ000

DIMETHYL N,N'-(DITHIOBIS((METHYLIMINO)CARBONYL OXY))BISETHANIMIDOTHIOATE see BKU100

o,o-DIMETHYL DITHIOBIS(THIOFORMATE) see DUN600

DIMETHYLDITHIOCARBAMATE ZINC SALT see BJK500

DIMETHYLDITHIOCARBAMIC ACID COPPER SALT see CNL500

DIMETHYLDITHIOCARBAMIC ACID with DIMETHYLAMINE (1:1) see DRQ600

DIMETHYLDITHIOCARBAMIC ACID DIMETHYL AMINE SALT see DRQ600

N,N-DIMETHYLDITHIOCARBAMIC ACID DIMETHYLAMINOMETHYL ESTER see DRQ650

DIMETHYLDITHIOCARBAMIC ACID DIMETHYLAMMONIUM SALT see DRQ600

DIMETHYLDITHIOCARBAMIC ACID, IRON SALT see FAS000

DIMETHYLDITHIOCARBAMIC ACID, IRON(3+) SALT see FAS000

DIMETHYLDITHIOCARBAMIC ACID, LEAD SALT see LCW000

DIMETHYLDITHIOCARBAMIC ACID, SODIUM SALT see SGM500

N,N-DIMETHYLDITHIOCARBAMIC ACID S-TRIBUTYLSTANNYL ESTER see TID150

DIMETHYLDITHIOCARBAMIC ACID, ZINC SALT see BJK500

N,N-DIMETHYL-DITHIOCARBAMINSAEURE-DIMETHYLAMINOMETHYL-ESTER see DRQ650

O,O-DIMETHYLDITHIOFOSFORECNAN SODNY (CZECH) see PHI250

2,4-DIMETHYL-1,3-DITHIOLANE-2-CARBOXALDEHYDE O-((METHYLAMINO)CARBONYL)OXIME see DRR000

2,4-DIMETHYL-1,3-DITHIOLANE-2-CARBOXALDEHYDE O-(METHYLCARBAMOYL)OXIME see DRR000

O,O-DIMETHYLDITHIOPHOSPHATE see PHH500

O,O-DIMETHYLDITHIOPHOSPHATE DIETHYLMERCAPTOSUCCINATE see MAK700

DIMETHYLDITHIOPHOSPHORIC ACID see PHH500

O,O-DIMETHYL DITHIOPHOSPHORIC ACID see PHH500

DIMETHYLDITHIOPHOSPHORIC ACID N-METHYLBENZAZIMIDE ESTER see ASH500

O,O-DIMETHYL DITHIOPHOSPHORYLACETIC ACID-N-METHYL-N-FORMYLAMIDE see DRR200

O,O-DIMETHYLDITHIOPHOSPHORYLACETIC ACID-N-MONOMETHYLAMIDE SALT see DSP400

O,O-DIMETHYL-DITHIOPHOSPHORYLESSIGSAEURE MONOMETHYLAMID (GERMAN) see DSP400

(O,O DIMETHYLDITHIOPHOSPHORYLPHENYL)ACETIC ACID ETHYL ESTER see DRR400

DIMETHYL DIXANTHOGEN see DUN600

N,N-DIMETHYL-2-(DI-2,6-XYLYLMETHOXY)ETHYLAMINE HYDROCHLORIDE see DRR500

DIMETHYL DIZENEDICARBOXYLATE see DQH509

2,5-DIMETHYL-DNPZ see DRN800

2,6-DIMETHYL-DNPZ see DRO000

N,N-DIMETHYLDODECANAMIDE see DRR600

N,N-DIMETHYL-1-DODECANAMINE see DRR800

2,6-DIMETHYLDODECA-2,6,8-TRIEN-10-ONE see DRR700

7,11-DIMETHYL-4,6,10-DODECATRIEN-3-ONE see DRR700

N,N-DIMETHYLDODECYLAMINE see DRR800

DIMETHYLDODECYLAMINE ACETATE see DRS000

N,N-DIMETHYLDODECYLAMINE ACETATE see DRS000

DIMETHYLDODECYLAMINE HYDROCHLORIDE mixed with SODIUM NITRITE (7:8) see SIS150

DIMETHYLDODECYLAMINE-N-OXIDE see DRS200

N,N-DIMETHYLDODECYLAMINE OXIDE see DRS200

N,N-DIMETHYL-DODECYLAMINOXID (CZECH) see DRS200

N,N-DIMETHYLDODECYLBETAINE see LBU200

N,N-DIMETHYL-N-DODECYLGLYCINE see LBU200

N,N-DIMETHYL-n-DODECYL(2-HYDROXY-3-CHLOROPROPYL)AMMONIUM CHLORIDE see DRS400

N,N-DIMETHYL-n-DODECYL(3-HYDROXYPROPENYL) AMMONIUM CHLORIDE see DRS600

7,8-DIMETHYLENEBENZ(a)ANTHRACENE see CMC000

3:4-DIMETHYLENE-1:2-BENZANTHRACENE see AAF000

8:9-DIMETHYLENE-1:2-BENZANTHRACENE see BAW000

DIMETHYLENEDIAMINE see EEA500

DIMETHYLENE DIISOTHIOCYANATE see ISK000

3,4,3',4'-DIMETHYLENEDIOXYSTILBENE see DRS700

DIMETHYLENE GLYCOL see BOT000

DIMETHYLENEIMINE see EJM900

DIMETHYLENE OXIDE see EJN500

N,N-DIMETHYLENEOXIDEBIS(PYRIDINIUM-4-ALDOXIME) DICHLORIDE see BGS250

DIMETHYLENIMINE see EJM900

N,N-DIMETHYLENOXID-BIS-(PYRIDINIUM-4-ALDOXIM)-DICHLORID (GERMAN) see BGS250

7,12-DIMETHYL-5,6-EPOXY-5,6-DIHYDROBENZ(a)ANTHRACENE see DQL600

3,11-DIMETHYL-10,11-EPOXY-7-ETHYL-2,6-TRIDECADIENOIC ACID METHYL ESTER see MJG100

exo-1,2-cis-DIMETHYL-3,6-EPOXYHEXAHYDROPHTHALIC ANHYDRIDE see CBE750

6,7-DIMETHYLESCULETIN see DRS800

O,S-DIMETHYL ESTER AMIDE of AMIDOTHIOATE see DTQ400

O,O-DIMETHYLESTER KYSELINY CHLORTHIOFOSFORECNE (CZECH) see DTQ600

DIMETHYLESTER KYSELINY FOSFORITE (CZECH) see DSG600

DIMETHYLESTER KYSELINY MALEINOVE see DSL800

DIMETHYLESTER KYSELINY SIROVE (CZECH) see DUD100

DIMETHYLESTER KYSELINY TEREFTALOVE (CZECH) see IML000

DIMETHYL ESTER PHOSPHORIC ACID ESTER with METHYL 3-HYDROXYCROTONATE see MQR750

O,O-DIMETHYL ESTER PHOSPHOROTHIOIC ACID-S-ESTER with 1,2-BIS(METHOXYCARBONYL)ETHANETHIOL see OPK250

O,O-DIMETHYL ESTER PHOSPHOROTHIOIC ACID-S-ESTER with ETHYL MERCAPTOACETATE see DRB600

N,N'-DIMETHYLETHANEDIAMINE see DRI600

N,N'-DIMETHYL-1,2-ETHANEDIAMINE see DRI600

N,N-DIMETHYLETHANETHIOAMIDE see DUG450

7,14-DIMETHYL-7,14-ETHANODIBENZ(a,b)ANTHRACENE-15,16-DICARBOXYLIC ACID see DRT000

1,1-DIMETHYLETHANOL see BPX000

DIMETHYLETHANOLAMINE see DOY800

N,N-DIMETHYLETHANOLAMINE see DOY800

DIMETHYLETHANOLAMINE (DOT) see DOY800

(E)-4,4'-(1,2-DIMETHYL-1,2-ETHENEDIYL)BIS-PHENOL (9CI) see DUC300

DIMETHYL ETHER (DOT) see MJW500

DIMETHYL ETHER HYDROQUINONE see DOA400

(DIMETHYL ETHER)OXODIPEROXO CHROMIUM(VI) see DRT089

DIMETHYLETHER RESORCINOL see REF025

DIMETHYLETHER of d-TUBOCURARINE IODIDE see DUM000

6-α,21-DIMETHYLETHISTERONE see DRT200

O,O-DIMETHYL-S-α-ETHOXY-CARBONYLBENZYL PHOSPHORODITHIOATE see DRR400

3,7-DIMETHYL-3-(1-ETHOXYETHOXY)-1,6-OCTADIENE see ELZ050

2',6'-DIMETHYL-2-(2-ETHOXYETHYLAMINO)ACETANILIDE see DRT400

2',6'-DIMETHYL-2-(2-ETHOXYETHYLAMINO)ACETANILIDE HYDROCHLORIDE see DRT600

N,N-DIMETHYL-p-((3-ETHOXYPHENYL)AZO)ANILINE see DRU000

DIMETHYLETHOXYPHENYLSILANE see DRU200

DIMETHYLETHOXYSILANE (ACGIH) see EES200

O,O-DIMETHYL-S-(5-ETHOXY-1,3,4-THIADIAZOLINYL-3-METHYL)DITHIOPHOSPHATE see DRU400

O,O-DIMETHYL-S-(5-ETHOXY-1,3,4-THIADIAZOL-2(3H)-ONYL-(3)-METHYL)DITHIOPHOSPHATE see DRU400

O,O-DIMETHYL-S-(5-ETHOXY-1,3,4-THIADIAZOL-2(3H)-ONYL-(3)-METHYL)PHOSPHORODITHIOATE see DRU400

O,O-DIMETHYL-S-(2-ETHSULFONYLETHYL)PHOSPHOROTHIOATE see DAP600

DIMETHYL-S-(2-ETHSULFONYLETHYL)THIOPHOSPHATE see DAP600

O,O-DIMETHYL-S-(2-ETHTHIOETHYL)PHOSPHOROTHIOATE see DAP400

DIMETHYL-S-(2-ETHTHIOETHYL)THIOPHOSPHATE see DAP400

O,O-DIMETHYL-S-(2-ETHTHIONYLETHYL)PHOSPHOROTHIOATE see DAP000

DIMETHYL-S-(2-ETHTHIONYLETHYL)THIOPHOSPHATE see DAP000

DIMETHYL ETHYL ALLENOLIC ACID METHYL ETHER see DRU600

1,1-DIMETHYLETHYLAMINE see BPY250

N,N-DIMETHYL-4'-ETHYL-4-AMINOAZOBENZENE see EOI500

(±)-1-((1,1-DIMETHYLETHYL)AMINO)-3-(2,3-DIMETHYLPHENOXY)-2-PROPANOL HYDROCHLORIDE see XGA500

2-((1,1-DIMETHYLETHYL)AMINO)ETHANOL see BRH300

5-(2-((1,1-DIMETHYLETHYL)AMINO)-1-HYDROXYETHYL)-1,3-BENZENEDIOL see TAN100

5-(2-((1,1-DIMETHYLETHYL)AMINO)-1-HYDROXYETHYL)-1,3-BENZENEDIOL, SULFATE (2:1) (SALT) see TAN250

1-(4-(3-((1,1-DIMETHYLETHYL)AMINO)-2-HYDROXYPROPOXY)-2-BENZOFURANYL)ETHA NONE see ACN300

1-(7-(3-((1,1-DIMETHYLETHYL)AMINO)-2-HYDROXYPROPOXY)-2-BENZOFURANYL)ETHA NONE see ACN320

2-(3-((1,1-DIMETHYLETHYL)AMINO)-2-HYDROXYPROPOXY)-BENZONITRILE HYDROCHLORIDE see BON400

5-(3-((1,1-DIMETHYLETHYL)AMINO)-2-HYDROXYPROPOXY)-3,4-DIHYDRO-2(1H)-QUINOLINONE see CCL800

5-(3-((1,1-DIMETHYLETHYL)AMINO)-2-HYDROXYPROPOXY)-1,2,3,4-TETRAHYDRO-2,3-NAPHTHALENEDIOL see CNR675

α-1-(((1,1-DIMETHYLETHYL)AMINO)METHYL)-4-HYDROXY-1,3-BENZENEDIMETHANOL see BQF500

α⁶-(((1,1-DIMETHYLETHYL)AMINO)METHYL)-3-HYDROXY-2,6-PYRIDINEDIMETHANOL DIHYDROCHLORIDE see POM800

N-(5-(((1,1-DIMETHYLETHYL)AMINO)SULFONYL)-1,3,4-THIADIAZOL-2-YL)ACETAMIDE MONOSODIUM SALT see DRU875

2-((1,1-DIMETHYLETHYL)AZO)-2-BUTANOL see TAH950

1-((1,1-DIMETHYLETHYL)AZO)CYCLOHEXANOL see DRU900

(E)-1-((1,1-DIMETHYLETHYL)AZO)CYCLOPENTANOL see BQI270

2-((1,1-DIMETHYLETHYL)AZO)-5-METHYL-2-HEXANOL see BQI300

2-((1-DIMETHYLETHYL)AZO)-2-PROPANOL see BQI400

1,4-DIMETHYL-7-ETHYLAZULENE see DRV000

4-(1,1-DIMETHYLETHYL)-1,2-BENZENEDIOL see BSK000

4-(1,1-DIMETHYLETHYL)BENZENEPROPANAL see BMK300

DI-N-METHYL ETHYL CARBAMATE see EII500

O,O-DIMETHYL-S-(N-ETHYLCARBAMOYLMETHYL)DITHIOPHOSPHATE see DNX600

O,O-DIMETHYL-S-(N-ETHYLCARBAMOYLMETHYL)PHOSPHORODITHIOATE see DNX600

DIMETHYLETHYLCARBINOL see PBV000

1,1-DIMETHYLETHYL CHLOROACETATE see BQR100

2-(1,1-DIMETHYLETHYL)CYCLOHEXANOL ACETATE see BQW490

1-(4-(1,1-DIMETHYLETHYL)-2,6-DIMETHYL-3,5-DINITROPHENYL)ETHANONE see ACE600

2-(1,1-DIMETHYLETHYL)-4,6-DINITROPHENOL see DRV200

2-(1,1-DIMETHYLETHYL)-4,6-DINITROPHENOL ACETATE see DVJ400

1-(1,1-DIMETHYLETHYL)-4,4-DIPHENYLPIPERIDINE see BOM530

DIMETHYLETHYLENE see BOW500

N,N'-DIMETHYLETHYLENEDIAMINE see DRI600

sym-DIMETHYLETHYLENEDIAMINE see DRI600

N,N-DIMETHYLETHYLENEUREA see EJA100

1,1-DIMETHYLETHYL 4-ETHENYLPHENYL CARBONATE see BPI400

1,1-DIMETHYLETHYL ETHYL ETHER see EHA550

1,1-DIMETHYLETHYL GLYCIDYL ETHER see BRK800

DIMETHYL ETHYL HEXADECYL AMMONIUM BROMIDE see EKN500

1,1-DIMETHYLETHYL HYDROPEROXIDE see BRM250

DIMETHYL-ETHYL-β-HYDROXYETHYL-AMMONIUM-SULFATE-DI-n-BUTYLCARBAMATE see DDW000

DIMETHYLETHYL-β-HYDROXYETHYLAMMONIUM SULFATE DIBUTYLURETHAN see DDW000

DIMETHYLETHYL(m-HYDROXYPHENYL)AMMONIUM CHLORIDE see EAE600

α-(3-(1,1-DIMETHYLETHYL)-2-HYDROXYPHENYL)-1-METHYL-5-NITRO-1H-IMIDAZOLE-2-METHANOL see DRV250

5,5-DIMETHYL-2-(ETHYLIMINO)-1,3-DITHIOLAN-4-ONE-o-((METHYLAMINO)CARBONYL)OXIME see DRV300

2-((1,1-DIMETHYLETHYL)IMINO)TETRAHYDRO-3-(1-METHYLETHYL)-5-PHENYL-4H-1,3,5-THIADIAZIN-4-ONE see BOO300

N'-(5-(1,1-DIMETHYLETHYL)-3-ISOXAZOLYL)-N,N-DIMETHYLUREA see ISR200

O,O-DIMETHYL-S-(2-ETHYLMERCAPTOETHYL)DITHIOPHOSPHATE see PHI500

O,O-DIMETHYL-2-ETHYLMERCAPTOETHYL THIOPHOSPHATE see TIR250

O,O-DIMETHYL-O-ETHYLMERCAPTOETHYL THIOPHOSPHATE see DAO800

O,O-DIMETHYL-S-ETHYLMERCAPTOETHYL THIOPHOSPHATE see DAP400

O,O-DIMETHYL-S-ETHYLMERCAPTOETHYL THIOPHOSPHATE, THIOLO ISOMER see DAP400

O,O-DIMETHYL 2-ETHYLMERCAPTOETHYL THIOPHOSPHATE, THIONO ISOMER see DAO800

O,O-DIMETHYL-S-2-ETHYLMERKAPTOETHYLESTER KYSELINY DITHIOFOSFORECNE (CZECH) see PHI500

4-(1,1-DIMETHYLETHYL)-α-METHYLBENZENEPROPANAL see LFT000

6-(1,1-DIMETHYLETHYL)-3-METHYL-2,4-DINITROPHENYL ACETATE see BRU750

2-(1,1-DIMETHYLETHYL)-5-METHYLPHENOL see BQV600

N,N-DIMETHYL-p-(3'-ETHYL-4'-METHYLPHENYLAZO)ANILINE see EPT500

N,N-DIMETHYL-p-(4'-ETHYL-3'-METHYLPHENYLAZO)ANILINE see EPU000

4-(1,1-DIMETHYLETHYL)-N-(1-METHYLPROPYL)-2,6-DINITROBENZENAMINE see BQN600

N,N-DIMETHYL-N'-ETHYL-N'-1-NAPHTHYLETHYLENEDIAMINE see DRV500

N,N-DIMETHYL-N'-ETHYL-N'-2-NAPHTHYLETHYLENEDIAMINE see DRV550

α,α-DIMETHYLETHYL NITRITE see BRV760

1,1-DIMETHYL-3-ETHYL-3-NITROSOUREA see DRV600

3,5-DIMETHYL-5-ETHYLOXAZOLIDINE-2,4-DIONE see PAH500

4-(1,1-DIMETHYLETHYL)PHENOL see BSE500

4-(1,1-DIMETHYLETHYL)PHENOL SODIUM SALT see BSE700

2-(4-(1,1-DIMETHYLETHYL)PHENOXY)CYCLOHEXYL 2-PROPYNYL ESTER, SULFUROUS ACID see SOP000

2-(4-(1,1-DIMETHYLETHYL)PHENOXY)CYCLOHEXYL 2-PROPYNYL SULFITE see SOP000

4-(3-(4-(1,1-DIMETHYLETHYL)PHENOXY)-2-HYDROXYPROPOXY)BENZOIC ACID see BSE750

((4-(1,1-DIMETHYLETHYL)PHENOXY)METHYL)OXIRANE see BSE600

N,N-DIMETHYL-p-((4-ETHYLPHENYL)AZO)ANILINE see EOI500

N,N-DIMETHYL-p-((m-ETHYLPHENYL)AZO)ANILINE see EOI000

N,N-DIMETHYL-p-((o-ETHYLPHENYL)AZO)ANILINE see EIF450

N,N-DIMETHYL-p-(3'-ETHYLPHENYLAZO)ANILINE see EOI000

N,N-DIMETHYL-3'-ETHYL-4-(PHENYLAZO)BENZENAMINE see EOI000

4-(2-(4-(1,1-DIMETHYLETHYL)PHENYL)ETHENYL)-1-NITROBENZENE, (E)- see DRV700

4-(4-(1,1-DIMETHYLETHYL)PHENYL)ETHOXY)QUINAZOLINE see FAK200

N,N-DIMETHYL-N'-ETHYL-N'-PHENYLETHYLENEDIAMINE see DRV850

S-(1,1-DIMETHYLETHYL)PHENYL)-o-ETHYL ETHYLPHOSPHONODITHIOATE see BSG000

α-(4-(1,1-DIMETHYLETHYL)PHENYL)-4-(HYDROXYDIPHENYLMETHYL)-1-PIPERIDINEBUTANOL see TAI450

(4-(1,1-DIMETHYLETHYL)PHENYL)METHYL ETHYL 3-PYRIDINYLCARBONIMIDODITHIOATE see DRV870

S-((4-(1,1-DIMETHYLETHYL)PHENYL)METHYL) o-ETHYL-3-PYRIDINYLCARBONIMIDOTHIOATE see DRV875

(4-(1,1-DIMETHYLETHYL)PHENYL)METHYL HEPTYL-3-PYRIDINYLCARBONIMIDODITHIOATE see DRV880

(4-(1,1-DIMETHYLETHYL)PHENYL)METHYL HEXYL 3-PYRIDINYLCARBONIMIDODITHIOATE see DRV885

(4-(1,1-DIMETHYLETHYL)PHENYL)METHYL METHYL-3-PYRIDINYLCARBONIMIDODITHIOATE see DRV890

S-((4-(1,1-DIMETHYLETHYL)PHENYL)METHYL) o-OCTYL-3-PYRIDINYLCARBONIMIDOTHIOATE see DRV900

(4-(1,1-DIMETHYLETHYL)PHENYL)METHYL PENTYL-3-PYRIDINYLCARBONIMIDODITHIOATE see DRV910

S-((4-(1,1-DIMETHYLETHYL)PHENYL)METHYL) o-PENTYL-3-PYRIDINYLCARBONIMIDOTHIOATE see DRV920

4-(3-(4-(1,1-DIMETHYLETHYL)PHENYL)-2-METHYLPROPYL)-2,6-DIMETHYLMORPHOLINE see MRQ300

1-(3-(4-(1,1-DIMETHYLETHYL)PHENYL)-2-METHYLPROPYL)PIPERIDINE see FAQ230

S-((4-(1,1-DIMETHYLETHYL)PHENYL)METHYL) o-PROPYL-3-PYRIDINYLCARBONIMIDOTHIOATE see DRV930

2,2-DIMETHYL-3-(p-ETHYLPHENYL)PROPANAL see EIJ600

α-α-DIMETHYL-p-ETHYLPHENYLPROPANAL see EIJ600

α-(5-(1,1-DIMETHYLETHYL)-1-PHENYL-1H-PYRAZOL-4-YL)-1-PIPERIDINEBUTANOL see DRV940

N-(1,1-DIMETHYLETHYL)-2-PROPENAMIDE see BPW050

DIMETHYLETHYL(3-(10H-PYRIDO(3,2-b))(1,4)BENZOTHIAZIN-10-YL)PROPYLAMMONIUM ETHYL SULFATE see DRW000

2-(1,1-DIMETHYLETHYL)PYRIMIDINE see DRW100

O,O-DIMETHYL-S-(2-ETHYLSULFINYL-ETHYL)-MONOTHIOFOSFAAT (DUTCH) see DAP000

O,O-DIMETHYL-S-(2-(ETHYLSULFINYL)ETHYL) PHOSPHOROTHIOATE see DAP000

O,O-DIMETHYL-S-(2-ETHYLSULFINYL)ETHYL THIOPHOSPHATE see DAP000

DIMETHYLETHYL SULFONIUM IODIDE with MERCURY IODIDE (1:1) see EIL500

O,O-DIMETHYL-S-ETHYL-2-SULFONYLETHYL PHOSPHOROTHIOLATE see DAP600

O,O-DIMETHYL-S-ETHYLSULPHINYLETHYL PHOSPHOROTHIOLATE see DAP000

O,O-DIMETHYL-S-ETHYLSULPHONYLETHYL PHOSPHOROTHIOLATE see DAP600

O,O-DIMETHYL-S-(2-ETHYLTHIO-ETHYL)-DITHIOFOSFAAT (DUTCH) see PHI500

O,O-DIMETHYL-O-(2-ETHYL-THIO-ETHYL)-MONOTHIOFOSFAAT (DUTCH) see DAO800

O,O-DIMETHYL-S-(2-ETHYLTHIO-ETHYL)-MONOTHIOFOSFAAT (DUTCH) see DAP400

O,O-DIMETHYL S-(2-(ETHYLTHIO)ETHYL) PHOSPHORODITHIOATE see PHI500

O,O-DIMETHYL-O-2-(ETHYLTHIO)ETHYL PHOSPHOROTHIOATE see DAO800

O,O-DIMETHYL-S-(2-(ETHYLTHIO)ETHYL)PHOSPHOROTHIOATE see DAP400

S-(((1,1-DIMETHYLETHYL)THIO)METHYL)-O,O-DIETHYL PHOSPHORODITHIOATE see BSO000

N,N-DIMETHYL-p-((3-ETHYL-p-TOLYL)AZO)ANILINE see EPT500

N,N-DIMETHYL-p-((4-ETHYL-m-TOLYL)AZO) ANILINE see EPU000

(1,1-DIMETHYLETHYL)UREA see BSS310

DIMETHYL ETHYNEDICARBOXYLATE see DOP400

DIMETHYLETHYNYLCARBINOL see MHX250

DIMETHYLETHYNYLMETHANOL see MHX250

N,N-DIMETHYL-α-ETHYNYL-α-PHENYLBENZENEACETAMIDE see DWA700

DIMETHYL FANDANE see DRX400

DIMETHYL-FENYL-ETHOXYSILAN (CZECH) see DRU200

2,3-DIMETHYLFLUORANTHENE see DRX600

7,8-DIMETHYLFLUORANTHENE see DRX800

8,9-DIMETHYLFLUORANTHENE see DRY000

1,9-DIMETHYLFLUORENE see DRY100

DIMETHYLFLUOROARSINE see DRY289

2,10-DIMETHYL-3-FLUORO-5,6-BENZACRIDINE see FHR000

9,12-DIMETHYL-10-FLUOROBENZ(a)ACRIDINE see FHR000

7,12-DIMETHYL-1-FLUOROBENZ(a)ANTHRACENE see FHS000

7,12-DIMETHYL-4-FLUOROBENZ(a)ANTHRACENE see DRY400

7,12-DIMETHYL-5-FLUOROBENZ(a)ANTHRACENE see DRY600

7,12-DIMETHYL-8-FLUOROBENZ(a)ANTHRACENE see DRY800

7,12-DIMETHYL-11-FLUOROBENZ(a)ANTHRACENE see DRZ000

N,N-DIMETHYL-2-FLUORO-4-PHENYLAZOANILINE see FHQ000

N,N-DIMETHYL-p-(2-FLUOROPHENYLAZO)ANILINE see FHQ010

N,N-DIMETHYL-p-((p-FLUOROPHENYL)AZO)ANILINE see DSA000

3,3-DIMETHYL-1-(p-FLUOROPHENYL)TRIAZENE see DSA600

DIMETHYL FLUOROPHOSPHATE see DSA800

DIMETHYL FORMAL see MGA850

DIMETHYLFORMALDEHYDE see ABC750

DIMETHYLFORMAMID (GERMAN) see DSB000

DIMETHYLFORMAMIDE see DSB000

N,N-DIMETHYL FORMAMIDE see DSB000

N,N-DIMETHYLFORMAMIDE (DOT) see DSB000

DIMETHYLFORMOCARBOTHIALDINE see DSB200

2,4-DIMETHYL-2-FORMYL-1,3-DITHIOLANE OXIME METHYLCARBAMATE see DRR000

O,O-DIMETHYL-S-(N-FORMYL-N-METHYLCARBAMOYLMETHYL) PHOSPHORODITHIOATE see DRR200

DIMETHYLFOSFIT see DSG600

DIMETHYLFOSFONAT see DSG600

DIMETHYL FUCHSIN see DSB300

6,6-DIMETHYLFULVENE see DSB400

DIMETHYL FUMARATE see DSB600

DIMETHYL FURAN see DSB800

DIMETHYL FURANE see DSB800

2,5-DIMETHYL FURANE see DSC000

4,8-DIMETHYL-2H-FURO(2,3-h)-1-BENZOPYRAN-2-ONE see DQF700

4,9-DIMETHYL-2H-FURO(2,3-H)-1-BENZOPYRAN-2-ONE plus ULTRAVIOLET A RADIATION see FQD130

2,5-DIMETHYL-3-FURYL p-HYDROXYPHENYL KETONE see DSC100

α,α-DIMETHYLGLYCINE see MGB000

DIMETHYLGLYCINE HYDROCHLORIDE see MPI100

N,N-DIMETHYLGLYCINE HYDROCHLORIDE see MPI100

DIMETHYLGLYCINE HYDROCHLORIDE mixed with SODIUM NITRITE (3:1) see DSC200

N,N-DIMETHYLGLYCINONITRILE see DOS200

DIMETHYLGLYOXAL see BOT500

DIMETHYLGLYOXIME see DBH000

DIMETHYLGOLD SELENOCYANATE see DSC400

as-DIMETHYLGUANIDINE HYDROCHLORIDE see DSC800

N,N'-DIMETHYLHARNSTOFF (GERMAN) see DUM200

DIMETHYLHARNSTOFF and NATRIUMNITRIT (GERMAN) see DUM400

m-(3,3-DIMETHYLHARNSTOFF)-PHENYL-tert-BUTYLCARBAMAT (GERMAN) see DUM800

2,6-DIMETHYL-2,5-HEPTADIEN-4-ONE see PGW250

2,6-DIMETHYL-2,5-HEPTADIEN-4-ONE DIOZONIDE see DSD000

2,5-DIMETHYLHEPTANE see DSD200

3,5-DIMETHYLHEPTANE see DSD400

4,4-DIMETHYLHEPTANE see DSD600

2,6-DIMETHYL-2-HEPTANOL see FON200

2,6-DIMETHYL HEPTANOL-4 see DNH800

2,6-DIMETHYL-4-HEPTANOL see DNH800

2,6-DIMETHYL-HEPTAN-4-ON (DUTCH, GERMAN) see DNI800

2,6-DIMETHYLHEPTAN-4-ONE see DNI800

2,6-DIMETHYL-4-HEPTANONE see DNI800

2,6-DIMETHYL-5-HEPTENAL see DSD775

2,6-DIMETHYL-3-HEPTENE see DSD800

2,6-DIMETHYL-4-HEPTYLPHENOL, (o and p) see NNC500

N,N-DIMETHYLHEXADECANAMIDE see DTH700

N,N-DIMETHYLHEXADECYLAMINE see HCP525

N,N-DIMETHYL-N-HEXADECYLGLYCINE see CDF450

1,3-DIMETHYLHEXAHYDROPYRIMIDONE see DSE489

N,N-DIMETHYLHEXANAMIDE see DQX200

2,3-DIMETHYLHEXANE see DSE509

2,4-DIMETHYLHEXANE see DSE600

DIMETHYLHEXANE DIHYDROPEROXIDE (dry) see DSE800

DIMETHYL HEXANEDIOATE see DOQ300

3,4-DIMETHYL-2,5-HEXANEDIONE see DSF100

1,5-DIMETHYLHEXYLAMINE see ILM000

α,ε-DIMETHYLHEXYLAMINE see ILM000

N,1-DIMETHYLHEXYLAMINE HYDROCHLORIDE see NCL300

DI(3-METHYLHEXYL)PHTHALATE see DSF200

5,5-DIMETHYLHYDANTOIN see DSF300

1,2-DIMETHYLHYDRAZIN (GERMAN) see DSF600

DIMETHYLHYDRAZINE see DSF400

1,1-DIMETHYLHYDRAZINE see DSF400

1,2-DIMETHYLHYDRAZINE see DSF600

1,2-DIMETHYL-HYDRAZINE see DSF600

N,N-DIMETHYLHYDRAZINE see DSF400

N,N'-DIMETHYLHYDRAZINE see DSF600

sym-DIMETHYLHYDRAZINE see DSF600

uns-DIMETHYLHYDRAZINE see DSF400

asym-DIMETHYLHYDRAZINE see DSF400

unsym-DIMETHYLHYDRAZINE see DSF400

1,1-DIMETHYLHYDRAZINE (GERMAN) see DSF400

DIMETHYLHYDRAZINE, symmetrical (DOT) see DSF600

DIMETHYLHYDRAZINE, unsymmetrical (DOT) see DSF400

1,2-DIMETHYLHYDRAZINE DIHYDROCHLORIDE see DSF800

N,N'-DIMETHYLHYDRAZINE DIHYDROCHLORIDE see DSF800

sym-DIMETHYLHYDRAZINE DIHYDROCHLORIDE see DSF800

1,2-DIMETHYLHYDRAZINE ETHANEDIOATE (1:1) see DSG300

1,1-DIMETHYLHYDRAZINE HYDROCHLORIDE see DSG000

1,2-DIMETHYLHYDRAZINE HYDROCHLORIDE see DSG200

sym-DIMETHYLHYDRAZINE HYDROCHLORIDE see DSG200

1,2-DIMETHYLHYDRAZINE OXALATE see DSG300

DIMETHYLHYDRAZINIUM derivative of 2-CHLOROETHYLPHOSPHONIC ACID see DKQ650

9-(2,2-DIMETHYLHYDRAZINO)ACRIDINE see DSG320

9-(2,2-DIMETHYLHYDRAZINO)ACRIDINE MONOHYDROCHLORIDE see DSG330

9-(2,2-DIMETHYLHYDRAZINO)ACRIDINE MONO(METHYL SULFATE) see DSG340

2-(2,2-DIMETHYLHYDRAZINO)-4-(5-NITRO-2-FURYL)THIAZOLE see DSG400

DIMETHYL HYDROGEN PHOSPHATE see PHC800

O,O-DIMETHYL HYDROGEN PHOSPHATE see PHC800

DIMETHYL HYDROGEN PHOSPHITE see DSG600

DIMETHYLHYDROGENPHOSPHITE see DSG600

DIMETHYLHYDROQUINONE see DOA400

2,6-DIMETHYLHYDROQUINONE see DSG700

DIMETHYLHYDROQUINONE ETHER see DOA400

3,5-DIMETHYL-4-HYDROXYACETANILIDE see DOO900

1,4-DIMETHYL-6-HYDROXY-3-AMINOCARBAZOLE see AJQ300

N,N-DIMETHYL-HYDROXYBENZENESULFONAMIDE see DUC600

α-5,9-DIMETHYL-2'-HYDROXY-6,7-BENZOMORPHAN HYDROCHLORIDE see NNS600

DIMETHYL-2,5 (HYDROXY 4 BENZOYL) 3 FURANNE see DSC100

2',3-DIMETHYL-N-HYDROXY-(1,1'-BIPHENYL)-4-AMINE see HKB650

N,N-DIMETHYL-2-HYDROXYETHYLAMINE see DOY800

N,N-DIMETHYL-N-(2-HYDROXYETHYL)AMINE see DOY800

DIMETHYL(2-HYDROXYETHYL)OCTYLAMMONIUM BROMIDE BENZILATE see DSH000

DIMETHYL(2-HYDROXYETHYL)PENTYLAMMONIUM BROMIDE BENZILATE see DSH200

1,3-DIMETHYL-7-(2-HYDROXYETHYL)XANTHINE see HLC000

2,5-DIMETHYL-4-HYDROXY-3(2H)-FURANONE see HKC575

DIMETHYL 3-HYDROXYGLUTACONATE DIMETHYL PHOSPHATE see SOY000

1-1,4-DIMETHYL-10-HYDROXY-2,3,4,5,6,7-HEXAHYDRO-1,6-METHANO-1H-4-BENZAZONINE HYDROBROMIDE see SLI200

(−)-1,4-DIMETHYL-10-HYDROXY-2,3,4,5,6,7-HEXAHYDRO-1,6-METHANO-1H-4-BENZAZONINE HYDROBROMIDE see SLI200

3,5-DIMETHYL-3-HYDROXYHEXANE-4-CARBOXYLIC ACID-β-LACTONE see DSH400

7-(5,5-DIMETHYL-3-HYDROXY-2-(3-HYDROXY-1-OCTENYL)CYCLOPENTYL)-5-HEPTENOIC ACID see DAQ100

2,2-DIMETHYL-5-HYDROXYMETHYL-1,3-DIOXOLANE see DVR600

2',3-DIMETHYL-4-(2-HYDROXYNAPHTHYLAZO)AZOBENZENE see SBC500

1,4-DIMETHYL-6-HYDROXY-3-NITROCARBAZOLE see DSH500

4-(4,8-DIMETHYL-5-HYDROXY-7-NONENYL)-4-METHYL-3,8-DIOXABICYCLO(3.2.1)OCTANE-1-ACETIC ACID see OJK300

α-5,9-DIMETHYL-2'-HYDROXY-2(N)-PHENETHYL-6,7-BENZOMORPHAN HYDROBROMIDE see HKC625

3,7-DIMETHYL-7-HYDROXYOCTANAL see CMS850

3,7-DIMETHYL-7-HYDROXY-1-OCTANOL see DTE400

4-(4,4-DIMETHYL-3-HYDROXY-1-PENTENYL)-2-METHOXYPHENOL see DSH600

N,N-DIMETHYL-β-HYDROXYPHENETHYLAMINE see DSH700

3-(3,5-DIMETHYL-4-HYDROXYPHENYL)-2-METHYL-4(3H)-QUINAZOLINONE see DSH800

1,2-DIMETHYL-3-(m-HYDROXYPHENYL)-3-PROPYLPYRROLIDINE see DSI000

6-α,21-DIMETHYL-17-β-HYDROXY-17-α-PREG-4-EN-20-YN-3-ONE see DRT200

6-α,21-DIMETHYL-17-β-HYDROXY-17-α-PREGN-4-EN-20-YN-3-ONE see DRT200

DIMETHYL(2-HYDROXYPROPYL)AMINE see DPT800

(2,2-DIMETHYL-3-HYDROXYPROPYL)TRIETHYLAMMONIUM BROMIDE TROPATE (ESTER) see DSI200

4,6-DIMETHYL-5-HYDROXY-3-PYRIDINEMETHANOL see DAY800

1,2-DIMETHYL-3-HYDROXY-4(1H)-PYRIDINONE see DSI250

1,2-DIMETHYL-3-HYDROXYPYRID-4-ONE see DSI250

1,1-DIMETHYL 3-HYDROXYPYRROLIDINIUM BROMIDE-α-CYCLOPENTYLMANDELATE see GIC000

O,O-DIMETHYL-(1-HYDROXY-2,2,2-TRICHLORAETHYL)-PHOSPHAT (GERMAN) see TIQ250

O,O-DIMETHYL-(1-HYDROXY-2,2,2-TRICHLORAETHYL)PHOSPHONSAEURE ESTER (GERMAN) see TIQ250

O,O-DIMETHYL-(1-HYDROXY-2,2,2-TRICHLORO)ETHYL PHOSPHATE see TIQ250

DIMETHYL-1-HYDROXY-2,2,2-TRICHLOROETHYL PHOSPHONATE see TIQ250

O,O-DIMETHYL-(1-HYDROXY-2,2,2-TRICHLOROETHYL)PHOSPHONATE see TIQ250

N,N-DIMETHYL-5-HYDROXYTRYPTAMINE see DPG109

DIMETHYL HYPONITRILE see DSI489

5,5-DIMETHYL-2,4-IMIDAZOLIDINEDIONE see DSF300

1,3-DIMETHYLIMIDAZOLIUM IODIDE see MKU100

3,4-DIMETHYL-3H-IMIDAZO(4,5-f)QUINOLIN-2-AMINE see AJQ600

3,8-DIMETHYL-3H-IMIDAZO(4,5-f)QUINOXALIN-2-AMINE see AJQ675

N,N-DIMETHYLIMIDODICARBONIMIDIC DIAMIDE MONOHYDROCHLORIDE see DQR800

3,5-DIMETHYL-4-IMINO-2,5-CYCLOHEXADIEN-1-ONE see DTY650

DIMETHYLIMIPRAMINE see DSI709

DIMETHYLIMIPRAMINE HYDROCHLORIDE see DLS600

N,N-DIMETHYL-p-(6-INDAZYLAZO)ANILINE see DSI800

N,N-DIMETHYL-4-(6'-1H-INDAZYLAZO)ANILINE see DSI800

2,3-DIMETHYLINDOLE see DSI850

2,3-DIMETHYL-1H-INDOLE see DSI850

N,N-DIMETHYL-β-3-INDOLYLETHYLAMINE SULFOSALICYLATE see DPG000

N,N-DIMETHYL-α-INDOLYLIDENE-p-TOLUIDINE see DOT600

DIMETHYLIODOARSINE see DSI889

DIMETHYLIONONE see COW780

1,3-DIMETHYL-α-IONONE see COW780

2,8-DIMETHYL-6-ISOBUTYLNONANOL-4 see DSJ200

DIMETHYL ISOPHTHALATE see IML000

DIMETHYLISOPROPANOLAMINE see DPT800

α-α-DIMETHYL-m-ISOPROPENYL BENZYL ISOCYANATE see IKG800

α-α-DIMETHYL-p-ISOPROPENYL BENZYL ISOCYANATE see IKG850

1,4-DIMETHYL-7-ISOPROPYLAZULENE see DSJ800

DIMETHYL-5-(1-ISOPROPYL-3-METHYLPYRAZOLYL)CARBAMATE see DSK200

4,4-DIMETHYL-1-ISOPROPYL-2-NONYL-2-IMIDAZOLINE see DSK300

O,O-DIMETHYL-S-ISOPROPYL-2-SULFINYLETHYLPHOSPHOROTHIOATE see DSK600

O,O-DIMETHYL-S-2-(ISOPROPYLTHIO)ETHYLPHOSPHORODITHIOATE see DSK800

4,4'-DIMETHYLISOPSORALEN plus ULTRAVIOLET A RADIATION see FQD130

N,N-DIMETHYL-4-(4'-ISOQUINOLINYLAZO)ANILINE see DPO800

N,N-DIMETHYL-4-(7'-ISOQUINOLINYLAZO)ANILINE see DPP200

N,N'-DIMETHYL-4-(5'-ISOQUINOLINYLAZO)ANILINE see DPP000

N,N-DIMETHYL-4-(5'-ISOQUINOLYL-2'-OXIDE)AZOANILINE see DPP400

1,3-DIMETHYLISOTHIOUREA see DSK900

3,4-DIMETHYLISOXALE-5-SULFANILAMIDE see SNN500

3,4-DIMETHYLISOXALE-5-SULPHANILAMIDE see SNN500

3,5-DIMETHYLISOXAZOLE see DSK950

N^1-(3,4-DIMETHYL-5-ISOXAZOLYL)SULFANILAMIDE see SNN500

N[1]-(3,4-DIMETHYL-5-ISOXAZOLYL)SULFANILAMIDE LITHIUM SALT see DSL000
N[1]-(3,4-DIMETHYL-5-ISOXAZOLYL)SULFANILAMIDE SODIUM SALT see DSL200
N'-(3,4)DIMETHYLISOXAZOL-5-YL-SULPHANILAMIDE see SNN500
N[1]-(3,4-DIMETHYL-5-ISOXAZOLYL)SULPHANILAMIDE see SNN500
DIMETHYLKARBAMOYLCHLORID see DQY950
DIMETHYLKETAL see ABC750
DIMETHYLKETENE see DSL289
DIMETHYLKETOL see ABB500
DIMETHYL KETONE see ABC750
N,N-DIMETHYLLAURAMIDE see DRR600
N,N-DIMETHYLLAURYLAMINE see DRR800
DIMETHYL LAURYLBENZENE AMMONIUM BROMIDE see BEO000
DIMETHYLLAURYLBETAINE see LBU200
DIMETHYL LEAD DIACETATE see DSL400
o,N-DIMETHYLLITSEFERINE see ERE150
N,N-DIMETHYL-4-((3',5'-LUTIDYL-1'-OXIDE)AZO)ANILINE see DPP709
DIMETHYLMAGNESIUM see DSL600
o,o-DIMETHYL MALATHION see CBS800
DIMETHYL MALEATE see DSL800
DIMETHYLMALEIC ANHYDRIDE see DSM000
α,β-DIMETHYLMALEIC ANHYDRIDE see DSM000
DIMETHYL MALONATE see DSM200
DIMETHYL MANGANESE see DSM289
DIMETHYL MERCURY see DSM450
DIMETHYLMESCALINE see DOE000
N,N-DIMETHYLMETHANAMINE OXIDE, DIHYDRATE see TLE250
DIMETHYLMETHANE see PMJ750
1,4-DIMETHYL-6-METHOXY-3-AMINOCARBAZOLE see DSM480
2,2-DIMETHYL-7-METHOXY-2H-1-BENZOPYRAN see DAM830
N,N-DIMETHYL-N'-(4-METHOXYBENZYL)-N'-(2-PYRIDYL)ETHYLENEDIAMINE MALEATE see DBM800
N,N-DIMETHYL-N'-(p-METHOXYBENZYL)-N'-(2-PYRIMIDYL)ETHYLENEDIAMINE see NCD500
N,N-DIMETHYL-N'-(4-METHOXYBENZYL)-N'-(2-PYRIMIDYL)ETHYLENEDIAMINE HYDROCHLORIDE see RDU000
N,N-DIMETHYL-N'-(p-METHOXYBENZYL)-N'-(2-THIAZOLYL)-ETHYLENEDIAMINE MONOHYDROCHLORIDE see ZUA000
1,4-DIMETHYL-6-METHOXY-9H-CARBAZOL-3-AMINE see DSM480
O,O-DIMETHYL-O-2-METHOXYCARBONYL-1-METHYL-VINYL-PHOSPHAT (GERMAN) see MQR750
DIMETHYL 2-METHOXYCARBONYL-1-METHYLVINYL PHOSPHATE see MQR750
DIMETHYL METHOXYCARBONYLPROPENYL PHOSPHATE see MQR750
DIMETHYL (1-METHOXYCARBOXYPROPEN-2-YL)PHOSPHATE see MQR750
N,N-DIMETHYL-N'-(4-METHOXY-3-CHLOROPHENYL)UREA see MQR225
1,1-DIMETHYL-5-METHOXY-3-(DITHIEN-2-YLMETHYLENE)PIPERIDINIUM BROMIDE see TGB160
2',6'-DIMETHYL-2-(2-METHOXYETHYLAMINO)-PROPIONANILIDE see MEN250

O,O-DIMETHYL-S-(2-METHOXYETHYLCARBAMOYLMETHYL) DITHIOPHOSPHATE see AHO750
O,O-DIMETHYL-S-(2-METHOXYETHYLCARBAMOYL METHYL)PHOSPHORODITHIOATE see AHO750
2,6-DIMETHYL-N-(2-METHOXYETHYL)CHLOROACETANILIDE see DSM500
O,O-DIMETHYL-S-((METHOXYMETHYL)CARBAMOYL)METHYL PHOSPHORODITHIOATE see FNE000
2,3-DIMETHYL-7-METHOXY-8-(MORPHOLINOMETHYL)CHROMONE HYDROCHLORIDE see DSM600
α,α-DIMETHYL-2-(6-METHOXYNAPHTHYL)PROPIONIC ACID see DRU600
2,6-DIMETHYL-2-METHOXY-4-NITROSO-MORPHOLINE see NKO600
3,7-DIMETHYL-7-METHOXY-1-OCTANAL see DSM800
3,7-DIMETHYL-7-METHOXY-2-OCTANOL see DLR300
O,O-DIMETHYL-S-(5-METHOXY-4-OXO-4H-PYRAN-2-YL)PHOSPHOROTHIOATE see EAS000
1,1-DIMETHYL-3-(4-(4-METHOXYPHENOXY)PHENYL)UREA see LGM300
N,N-DIMETHYL-p-(2-METHOXYPHENYLAZO)ANILINE see DSN000
N,N-DIMETHYL-p-(3-METHOXYPHENYLAZO)ANILINE see DSN200
N,N-DIMETHYL-p-(4-METHOXYPHENYLAZO)ANILINE see DSN400
3,3-DIMETHYL-1-p-METHOXYPHENYLTRIAZENE see DSN600
2',6'-DIMETHYL-2-(2-METHOXYPROPYLAMINO)ACETANILIDE HYDROCHLORIDE see DSN800
2',6'-DIMETHYL-2-(2-METHOXYPROPYLAMINO)-PROPIONANILIDE PERCHLORATE see MEN500
α,α-DIMETHYL-6-METHOXY-β-PROPYL-3,4-DIHYDRO-2-NAPHTHALENEPROPIONIC ACID see EAB100
5,11-DIMETHYL-9-METHOXY-6H-PYRIDO(4,3-B)CARBAZOLE LACTATE see MEL780
1,9-DIMETHYL-7-METHOXY-9H-PYRIDO(3,4-b)INDOLE HYDROCHLORIDE see ICU100
O,O-DIMETHYL-S-((5-METHOXY-PYRON-2-YL)-METHYL)-THIOLPHOSPHAT (GERMAN) see EAS000
O,O-DIMETHYL-S-(5-METHOXYPYRONYL-2-METHYL) THIOPHOSPHATE see EAS000
(O,O-DIMETHYL)-S-(-2-METHOXY-Δ²-1,3,4-THIADIAZOLIN-5-ON-4-YLMETHYL)DITHIOPHOSPHATE DIMETHYL PHOSPHOROTHIOLOTHIONATE see DSO000
O,O-DIMETHYL-S-(5-METHOXY-1,3,4-THIADIAZOLINYL-3-METHYL) DITHIOPHOSPHATE see DSO000
O,O-DIMETHYL-S-(2-METHOXY-1,3,4-THIADIAZOL-5-(4H)-ONYL)-(4)-METHYL)-DITHIOPHOSPHAT (GERMAN) see DSO000
O,O-DIMETHYL-S-(2-METHOXY-1,3,4-THIADIAZOL-5(4H)-ONYL-(4)-METHYL) PHOSPHORODITHIOATE see DSO000
O,O-DIMETHYL-S-((2-METHOXY-1,3,4 (4H)-THIODIAZOL-5-ON-4-YL)-

METHYL)DITHIOFOSFAAT (DUTCH) see DSO000
3,7-DIMETHYL-9-(4-METHOXY-2,3,6-TRIMETHYLPHENYL)-2,4,6,8-NONANETETRAENOIC ACID ETHYL ESTER see EMJ500
all-trans-3,7-DIMETHYL-9-(4-METHOXY-2,3,6-TRIMETHYLPHENYL)-2,4,6,8-NONATETRAENOIC ACID see REP400
N,N-DIMETHYL-N'-(((METHYLAMINO)CARBONYL)OXY)PHENYLMETHANIMIDAMIDE MONOHYDROCHLORIDE see DSO200
2,2-DIMETHYL-4-(N-METHYLAMINOCARBOXYLATO)-1,3-BENZODIOXOLE see DQM600
O,O-DIMETHYL-S-(2-(METHYLAMINO)-2-OXOETHYL) PHOSPHORODITHIOATE see DSP400
5,7-DIMETHYL-1-(2-METHYLAMINOPROPYL)-2-PHENYLADAMANTANE HYDROCHLORIDE see DSO400
1,3-DIMETHYL-5-(METHYLAMINO)-4-PYRAZOLYL o-FLUOROPHENYL KETONE see DSO500
(1,3-DIMETHYL-5-(METHYLAMINO)-1H-PYRAZOL-4-YL)(2-FLUOROPHENYL)METHANONE see DSO500
2,5-DIMETHYL-6-METHYLAMINOQUINOXALINE see TME300
3,5-DIMETHYL-6-METHYLAMINOQUINOXALINE see TME305
N,N-DIMETHYL-α-METHYLBENZYLAMINE see DSO800
1,5-DIMETHYL-5-(1-METHYLBUTYL)BARBITURIC ACID see DSP200
2,2-DIMETHYL-4-(N-METHYLCARBAMATO)-1,3-BENZODIOXOLE see DQM600
O,O-DIMETHYL-S-2-(1-N-METHYLCARBAMOYLETHYLMERCAPTO) ETHYL THIOPHOSPHATE see MJG500
DIMETHYL-S-(2-(1-METHYLCARBAMOYLETHYLTHIO ETHYL)) PHOSPHOROTHIOLATE see MJG500
O,O-DIMETHYL-S-(2-(1-METHYLCARBAMOYLETHYLTHIO)ETHYL) PHOSPHOROTHIOATE see MJG500
O,O-DIMETHYL-S-(N-METHYL-CARBAMOYL)-METHYL-DITHIOFOSFAAT (DUTCH) see DSP400
(O,O-DIMETHYL-S-(N-METHYL-CARBAMOYL-METHYL)-DITHIOPHOSPHAT) (GERMAN) see DSP400
O,O-DIMETHYL-S-(N-METHYLCARBAMOYLMETHYL) DITHIOPHOSPHATE see DSP400
O,O-DIMETHYL-S-((N-METHYL-CARBAMOYL)-METHYL)MONOTHIOFOSFAAT (DUTCH) see DNX800
O,O-DIMETHYL-S-(N-METHYL-CARBAMOYL)-METHYL-MONOTHIOPHOSPHAT (GERMAN) see DNX800
O,O-DIMETHYL METHYLCARBAMOYLMETHYL PHOSPHORODITHIOATE see DSP400
O,O-DIMETHYL-S-(N-METHYLCARBAMOYLMETHYL) PHOSPHORODITHIOATE see DSP400
O,O-DIMETHYL-S-((METHYLCARBAMOYL)METHYL)PHOSPHOROTHIOATE see DNX800

O,O-DIMETHYL-S-(N-METHYLCARBAMOYLMETHYL)PHOSPHOROTHIOATE see DNX800

DIMETHYL-S-(N-METHYL-CARBAMOYL-METHYL)PHOSPHOROTHIOLATE see DNX800

O,O-DIMETHYL-S-(N-METHYLCARBAMOYLMETHYL) PHOSPHOROTHIOLATE see DNX800

O,O-DIMETHYL-S-(N-METHYLCARBAMOYLMETHYL) THIOPHOSPHATE see DNX800

O,O-DIMETHYL-O-(2-N-METHYLCARBAMOYL-1-METHYL-VINYL)-FOSFAAT (DUTCH) see MRH209

O,O-DIMETHYL-O-(2-N-METHYLCARBAMOYL-1-METHYL)-VINYL-PHOSPHAT (GERMAN) see MRH209

O,O-DIMETHYL-O-(2-N-METHYLCARBAMOYL-1-METHYL-VINYL) PHOSPHATE see MRH209

N,N-DIMETHYL-α-METHYLCARBAMOYLOXYIMINO-α-(METHYLTHIO)ACETAMIDE see DSP600

N',N'-DIMETHYL-N-((METHYLCARBAMOYL)OXY)-1-METHYLTHIOOXAMIMIDIC ACID see DSP600

N',N'-DIMETHYL-N-((METHYLCARBAMOYL)OXY)-1-THIOOXAMIMIDIC ACID METHYL ESTER see DSP600

O,O-DIMETHYL-S-(N-METHYLCARBAMYLMETHYL) THIOTHIONOPHOSPHATE see DSP400

O,O-DIMETHYL-O-(1-METHYL-2-CARBOXY-α-PHENYLETHYL)VINYL PHOSPHATE see COD000

O,O-DIMETHYL O-(1-METHYL-2-CARBOXYVINYL) PHOSPHATE see MQR750

O,O-DIMETHYL-O-(1-METHYL-2-CHLOR-2-N,N-DIAETHYL-CARBAMOYL)-VINYL-PHOSPHAT see FAB400

(O,O-DIMETHYL-O-(1-METHYL-2-CHLORO-2-DIETHYLCARBAMOYL-VINYL) PHOSPHATE) see FAB400

N,N-DIMETHYL-N'-(2-METHYL-4-CHLOROPHENYL)-FORMAMIDINE see CJJ250

N,N-DIMETHYL-N'-(2-METHYL-4-CHLOROPHENYL)-FORMAMIDINE HYDROCHLORIDE see CJJ500

N,N-DIMETHYL-N'-(2-METHYL-4-CHLORPHENYL)-FORMADIN (GERMAN) see CJJ250

2,6-DIMETHYL-1-((2-METHYLCYCLOHEXYL)CARBONYL)PIPERIDINE see DSP650

O,O-DIMETHYL-S-(3-METHYL-2,4-DIOXO-3-AZA-BUTYL)-DITHIOFOSFAAT (DUTCH) see DRR200

O,O-DIMETHYL-S-(3-METHYL-2,4-DIOXO-3-AZA-BUTYL)-DITHIOPHOSPHAT (GERMAN) see DRR200

6,6-DIMETHYL-2-METHYLENEBICYCLO(3.1.1)HEPTANE see POH750

2,2'-DIMETHYL-4,4'-METHYLENEBIS(CYCLOHEXYLAMINE) see BGT800

N,N'-DIMETHYL-4,4'-METHYLENEDIANILINE see MJO000

DIMETHYLMETHYLENE-p,p'-DIPHENOL see BLD500

DIMETHYLMETHYLENEHYDRAZINE see FMV300

5,5-DIMETHYL-2-((1-METHYLETHYL)IMINO)-1,3-DITHIOLAN-4-ONE-o-((METHYLAMINO)CARBONYL)OXIME see DSP700

5,5-DIMETHYL-2-((1-METHYLETHYL)IMINO)1,3-DITHIOLAN-4-ONE, o-((METHYL((TRICHLOROMETHYL)THIO)AMINO)CARBONYL)OXIME see DSP710

N,N-DIMETHYL-N'-(4-(1-METHYLETHYL)PHENYL)UREA see IRA050

O,O-DIMETHYL-S-(N-METHYL-N-FORMYL-CARBAMOYLMETHYL)-DITHIOPHOSPHAT see DRR200

O,O-DIMETHYL-S-(N-METHYL-N-FORMYLCARBAMOYLMETHYL)PHOSPHORODITHIOATE see DRR200

DIMETHYL METHYL MALEATE see DRF200

O,O-DIMETHYL-O-4-(METHYLMERCAPTO)-3-METHYLPHENYL PHOSPHOROTHIOATE see FAQ900

O,O-DIMETHYL-p-4-(METHYLMERCAPTO)-3-METHYLPHENYL THIOPHOSPHATE see FAQ900

O,O-DIMETHYL-O-(4-METHYLMERCAPTOPHENYL)PHOSPHATE see PHD250

N,N-DIMETHYL-N'-(2-METHYL-4-(((METHYLAMINO)CARBONYL)OXY)PHENYL)METHANIMIDAMIDE see FNE500

(E)-DIMETHYL 1-METHYL-3-(METHYLAMINO)-3-OXO-1-PROPENYL see MRH209

DIMETHYL-1-METHYL-2-(METHYLCARBAMOYL)VINYLPHOSPHATE, cis PHOSPHATE see MRH209

O,O-DIMETHYL-O-(3-METHYL-4-METHYLMERCAPTOPHENYL)PHOSPHOROTHIOATE see FAQ900

N,N-DIMETHYL-2-METHYL-4-(4'-(2'-METHYLPYRIDYL-1'-OXIDE)AZO)ANILINE see DQD600

N,N'DIMETHYL-3-METHYL-4-(4'-(2'-METHYLPYRIDYL-1'OXIDE)AZO)ANILINE see DQD800

O,O-DIMETHYL-O-(3-METHYL-4-METHYLTHIO-FENYL)-MONOTHIOFOSFAAT (DUTCH) see FAQ900

O,O-DIMETHYL-O-(3-METHYL-4-METHYLTHIOPHENYL)-MONOTHIOPHOSPHAT (GERMAN) see FAQ900

O,O-DIMETHYL-O-3-METHYL-4-METHYLTHIOPHENYL PHOSPHOROTHIOATE see FAQ900

O,O-DIMETHYL-O-(3-METHYL-4-METHYLTHIO-PHENYL)-THIONOPHOSPHAT (GERMAN) see FAQ900

N-DIMETHYL-1-METHYL-N'-(1-NITRO-9-ACRIDINYL)-1,2-ETHANEDIAMINE DIHYDROCHLORIDE see NFW435

O,O-DIMETHYL-O-(3-METHYL-4-NITROFENYL)-MONOTHIOFOSFAAT (DUTCH) see DSQ000

DIMETHYL ((1-METHYL-5-NITRO-1H-IMIDAZOL-2-YL)METHYLENE)PROPANEDIOATE see DSP800

O,O-DIMETHYL-O-(3-METHYL-4-NITRO-PHENYL)-MONOTHIOPHOSPHAT (GERMAN) see DSQ000

O,O-DIMETHYL O-(3-METHYL-4-NITROPHENYL)PHOSPHORATE see PHD750

O,O-DIMETHYL-O-(3-METHYL-4-NITROPHENYL) PHOSPHOROTHIOATE see DSQ000

DIMETHYL-3-METHYL-4-NITROPHENYLPHOSPHOROTHIONATE see DSQ000

O,O-DIMETHYL-O-(3-METHYL-4-NITROPHENYL) THIOPHOSPHATE see DSQ000

α-4-DIMETHYL-α-(4-METHYL-3-PENTENYL)-3-CYCLOHEXENE-1-METHANOL see BGO775

N,N-DIMETHYL-p-(2'-METHYLPHENYLAZO)ANILINE see DUH800

N,N-DIMETHYL-p-(3'-METHYLPHENYLAZO)ANILINE see DUH600

N,N-DIMETHYL-4-((2-METHYLPHENYL)AZO)BENZENAMINE see DUH800

N,N-DIMETHYL-4-((3-METHYLPHENYL)AZO)BENZENAMINE see DUH600

N,N-DIMETHYL-4-((4-METHYLPHENYL)AZO)BENZENAMINE see DUH400

N,N-DIMETHYL-2-(α-METHYL-α-PHENYLBENZYLOXY)ETHYLAMINE see DSQ600

N,N-DIMETHYL-2-((α-METHYL-α-PHENYLBENZYL)OXY)ETHYLAMINE see DSQ600

N,N-DIMETHYL-2-((o-METHYL-α-PHENYL-BENZYL)OXY)-ETHYLAMINE CITRATE see DPH000

N,N-DIMETHYL-2-(o-METHYL-α-PHENYLBENZYLOXY)ETHYLAMINE HYDROCHLORIDE see OJW000

DIMETHYL-cis-1-METHYL-2-(1-PHENYLETHOXYCARBONYL)VINYL PHOSPHATE see COD000

3,3-DIMETHYL-6-(((5-METHYL-3-PHENYL)-4-ISOXAZOLECARBOXAMIDE)-7-OXO)-4-THIA-1-AZABICYCLO(3.2.0)HEPTANE-2-CARBOXYLIC ACID see DSQ800

3,5-DIMETHYL-N-(2-METHYLPHENYL)-4-NITRO-1H-PYRAZOLE-1-ACETAMIDE see DSQ810

3,5-DIMETHYL-N-(2-METHYLPHENYL)-1H-PYRAZOLE-1-ACETAMIDE see DSQ820

3,5-DIMETHYL-N-(3-METHYLPHENYL)-1H-PYRAZOLE-1-ACETAMIDE see DSQ830

3,5-DIMETHYL-N-(4-METHYLPHENYL)-1H-PYRAZOLE-1-ACETAMIDE see DSQ840

DIMETHYL-5-(3-METHYL-1-PHENYLPYRAZOLYL) CARBAMATE see PPQ625

3,3-DIMETHYL-1-(m-METHYLPHENYL)TRIAZENE see DSR200

3,3-DIMETHYL-1-(o-METHYLPHENYL)TRIAZENE see MNT500

DIMETHYL METHYLPHOSPHONATE see DSR400

O,O-DIMETHYL-O-(3-METHYL) PHOSPHOROTHIOATE see DSQ000

N,N-DIMETHYL-9-(3-(4-METHYL-1-PIPERANIZYL)PROPYLIDENE)-9H-THIOXANTHENE-2-SULFONAMIDE, (Z)- see NBP500

N,N-DIMETHYL-9-(3-(4-METHYL-1-PIPERAZINYL)PROPYLIDENE)THIAXANTHENE-2-SULFONAMIDE see NBP500

N,N-DIMETHYL-9-(3-(4-METHYL-1-PIPERAZINYL)PROPYLIDENE)THIOXANTHENE-2-SULFONAMIDE see NBP500

DIMETHYL-3-(2-METHYL-1-PROPENYL)CYCLOPROPANECARBOXYLATE see BEP500

(+)-2,2-DIMETHYL-3-(2-METHYLPROPENYL)-CYCLOPROPANECARBOXYLIC ACID-(E)-,ESTER with (+)- see AFR750

(+)-(Z)-2,2-DIMETHYL-3-(2-METHYLPROPENYL)-CYCLOPROPANECARBOXYLIC ACID ESTER with 2-ALLYL-4-HYDROXY-3-METHYL-2-CYCLOPENTEN-ONE see AFR500

O,O-DIMETHYL-O-(4-NITROFENYL)-
MONOTHIOFOSFAAT (DUTCH) see
MNH000
4,6-DIMETHYL-2-(5-NITRO-2-
FURYL)PYRIMIDINE see DSV400
1,2-DIMETHYL-5-NITROIMIDAZOLE see
DSV800
4,5-DIMETHYL-2-NITROIMIDAZOLE see
DSW500
1,2-DIMETHYL-4-NITRO-1H-IMIDAZOLE
see DSV600
1,2-DIMETHYL-5-NITRO-1H-IMIDAZOLE
see DSV800
DIMETHYLNITROMETHANE see NIY000
O,O-DIMETHYL-O-(4-NITRO-3-
METHYLPHENYL)THIOPHOSPHATE see
DSQ000
N,N-DIMETHYL-p-((m-
NITROPHENYL)AZO)ANILINE see DSW600
N,N-DIMETHYL-p-((o-
NITROPHENYL)AZO)ANILINE see DSW800
O,O-DIMETHYL-O-(4-NITRO-PHENYL)-
MONOTHIOPHOSPHAT (GERMAN) see
MNH000
DIMETHYL p-NITROPHENYL
MONOTHIOPHOSPHATE see MNH000
DIMETHYL-4-NITROPHENYL
PHOSPHATE see PHD500
DIMETHYL-p-NITROPHENYL
PHOSPHATE see PHD500
O,O-DIMETHYL-O-(4-NITROPHENYL)
PHOSPHOROTHIOATE see MNH000
O,O-DIMETHYL-O-(p-NITROPHENYL)
PHOSPHOROTHIOATE see MNH000
O,O-DIMETHYL-S-(p-NITROPHENYL)
PHOSPHOROTHIOATE see DTH800
DIMETHYL 4-NITROPHENYL
PHOSPHOROTHIONATE see MNH000
O,O-DIMETHYL-O-(4-NITROPHENYL)-
THIONOPHOSPHAT (GERMAN) see
MNH000
O,O-DIMETHYL-O-(p-NITROPHENYL)-
THIONOPHOSPHAT (GERMAN) see
MNH000
DIMETHYL-p-NITROPHENYL
THIONPHOSPHATE see MNH000
DIMETHYL p-NITROPHENYL
THIOPHOSPHATE see MNH000
O,O-DIMETHYL-O-p-NITROPHENYL
THIOPHOSPHATE see MNH000
O,O-DIMETHYL-S-(4-
NITROPHENYL)THIOPHOSPHATE see
DTH800
3,3-DIMETHYL-1-(p-
NITROPHENYL)TRIAZENE see DSX400
2,3-DIMETHYL-4-NITROPYRIDINE-1-
OXIDE see DSX800
2,5-DIMETHYL-4-NITROPYRIDINE-1-
OXIDE see DSY000
3,5-DIMETHYL-4-NITROPYRIDINE 1-
OXIDE see DSY200
DIMETHYLNITROSAMIN (GERMAN) see
NKA600
DIMETHYLNITROSAMINE see NKA600
N,N-DIMETHYLNITROSAMINE see
NKA600
DIMETHYLNITROSOAMINE see NKA600
N,N-DIMETHYL-p-NITROSOANILINE see
DSY600
DIMETHYL-p-NITROSOANILINE (DOT) see
DSY600
N,N-DIMETHYL-4-
NITROSOBENZENAMINE see DSY600
N,m-DIMETHYL-N-
NITROSOBENZYLAMINE see NKS000
N,o-DIMETHYL-N-
NITROSOBENZYLAMINE see NKR500
N,p-DIMETHYL-N-
NITROSOBENZYLAMINE see NKS500
α,N-DIMETHYL-N-
NITROSOBENZYLAMINE see NKW000

3,2'-DIMETHYL-4-NITROSOBIPHENYL see
DSY800
DIMETHYLNITROSOHARNSTOFF
(GERMAN) see DTB200
1,2-DIMETHYLNITROSOHYDRAZINE see
DSY889
N,O-DIMETHYL-N-
NITROSOHYDROXYLAMINE see DSZ000
DIMETHYLNITROSOMORPHOLINE see
DTA000
2,6-DIMETHYLNITROSOMORPHOLINE see
DTA000
2,6-DIMETHYL-N-NITROSOMORPHOLINE
see DTA000
2,6-DIMETHYL-4-NITROSOMORPHOLINE
cis and trans mixture (2:1) see DTA050
(+−)-1'-DIMETHYL-1'-NITROSONICOTINE
see NLP800
DIMETHYL(p-NITROSOPHENYL)AMINE
see DSY600
N,N'-DIMETHYL-N-NITROSO-N'-
PHENYLUREA see DTN875
3,5-DIMETHYL-1-NITROSOPIPERAZINE
see NKA850
2,6-DIMETHYLNITROSOPIPERIDINE see
DTA400
3,5-DIMETHYLNITROSOPIPERIDINE see
DTA600
3,5-DIMETHYL-1-NITROSOPIPERIDINE see
DTA600
cis-3,5-DIMETHYL-1-NITROSOPIPERIDINE
see DTA690
trans-3,5-DIMETHYL-1-
NITROSOPIPERIDINE see DTA700
2,5-DIMETHYL-N-NITROSOPYRROLIDINE
see DTA800
1,3-DIMETHYLNITROSOUREA see DTB200
N-DIMETHYL-N-NITROSOUREA see
PDK750
1,3-DIMETHYL-N-NITROSOUREA see
DTB200
N,N'-DIMETHYLNITROSOUREA see
DTB200
O,O-DIMETHYL-O-4-NITRO-m-TOLYL
PHOSPHOROTHIOATE see DSQ000
N¹,N¹-DIMETHYL-N²-(1-NITRO-9-
ACRIDINYL)-1,2-PROPANEDIAMINE
DIHYDROCHLORIDE see NFW435
3,7-DIMETHYL-2,6-NONADIEN-1-AL see
EHN500
3,7-DIMETHYL-2,6-NONADIENENITRILE
see LEH100
3,7-DIMETHYL-16-NONADIEN-3-OL see
ELZ000
3,7-DIMETHYL-1,6-NONADIEN-3-YL
ACETATE see HGI585
6-(3,7-DIMETHYL-2,4,6,8-
NONATETRAENYLIDENE)-1,5,5-
TRIMETHYLCYCLOHEXENE (ALL-E)- see
AFQ700
6,6-DIMETHYL-2-NORPINENE-2-
ETHANOL see DTB800
6,6-DIMETHYL-2-NORPINENE-2-
ETHANOL ACETATE see DTC000
3,7-DIMETHYL-2,6-OCTADADIEN-1-YL
PROPIONATE see GDM450
N,N-DIMETHYLOCTADECANAMIDE see
DTC200
(N,N-DIMETHYL-1-
OCTADECANAMINE)TRIHYDROBORON
(T-4) see DTC300
(Z)-N,N-DIMETHYL-N-9-
OCTADECENYLBENZENEMETHANAMIN
IUM CHLORIDE see OHK200
N,N-DIMETHYLOCTADECYLAMINE see
DTC400
N,N-DIMETHYL-N-
OCTADECYLBENZENEMETHANAMINIU
M SALT WITH 3-
NITROBENZENESULFONIC ACID (1:1) see
BEM400

DIMETHYLOCTADECYLBENZYLAMMONI
UM CHLORIDE see DTC600
N,N-DIMETHYL-N-OCTADECYL-1-
OCTADECANAMINIUM CHLORIDE see
DXG625
3,7-DIMETHYL-2,6-OCTADIENAL see
DTC800
trans-3,7-DIMETHYL-2,6-OCTADIENAL see
GCU100
3,7-DIMETHYL-2,6-OCTADIENAL
DIETHYL ACETAL see CMS323
3,7-DIMETHYL-1,6-OCTADIENE see
CMT050
(E)-3,7-DIMETHYL-2,6-
OCTADIENENITRILE see GDM000
2,6-DIMETHYL-2,7-OCTADIENE-6-OL see
LFX000
3,7-DIMETHYL-2,6-OCTADIENOIC ACID
see GCW000
3,7-DIMETHYL-2,7-OCTADIENOIC ACID
see GCW000
2,6-DIMETHYLOCTA-2,7-DIEN-6-OL see
LFX000
2,6-DIMETHYL-5,7-OCTADIEN-2-OL see
DTC990
3,7-DIMETHYLOCTA-1,6-DIEN-3-OL see
LFX000
3,7-DIMETHYL-1,6-OCTADIEN-3-OL see
LFX000
3,7-DIMETHYL-(−)-1,6-OCTADIEN-3-OL see
LFY000
3,7-DIMETHYL-(E)-2,6-OCTADIEN-1-OL see
DTD000
3,7-DIMETHYL-(Z)-2,6-OCTADIEN-1-OL see
DTD200
2-cis-3,7-DIMETHYL-2,6-OCTADIEN-1-OL
see DTD200
2,6-DIMETHYL-trans-2,6-OCTADIEN-8-OL
see DTD000
3,7-DIMETHYL-trans-2,6-OCTADIEN-1-OL
see DTD000
3,7-DIMETHYL-1,6-OCTADIEN-3-OL
ACETATE see LFY600
cis-3,7-DIMETHYL-2,6-OCTADIEN-1-OL
ACETATE see NCO100
trans-3,7-DIMETHYL-2,6-OCTADIEN-1-OL
ACETATE see DTD800
3,7-DIMETHYL-1,6-OCTADIEN-3-OL
BENZOATE see LFZ000
trans-3,7-DIMETHYL-2,6-OCTADIEN-1-OL-
2-BUTENOATE see DTE000
3,7-DIMETHYL-1,6-OCTADIEN-3-OL
CINNAMATE see LGA000
3,7-DIMETHYL-2,6-OCTADIEN-1-OL,
FORMATE (cis) see FNC000
trans-3,7-DIMETHYL-2,6-OCTADIEN-1-OL
FORMATE see GCY000
3,7-DIMETHYL-1,6-OCTADIEN-3-OL
ISOBUTYRATE see LGB000
trans-3,7-DIMETHYL-2,6-OCTADIEN-1-OL
ISOBUTYRATE see GDI000
4,7-DIMETHYL-1,6-OCTADIEN-3-OL
ISOVALERATE see LGC000
(E)-3,7-DIMETHYL-2,6-OCTADIEN-1-OL
PROPIONATE see GDM450
(Z)-3,7-DIMETHYL-2,6-OCTADIEN-1-OL
PROPIONATE see NCP000
cis-3,7-DIMETHYL-2,6-OCTADIEN-1-OL
PROPIONATE see NCP000
3,7-DIMETHYL-1,6-OCTADIEN-3-YL
ACETATE see LFY600
3,7-DIMETHYL-2-trans-6-OCTADIENYL
ACETATE see DTD800
trans-3,7-DIMETHYL-2,6-OCTADIEN-1-YL
ACETATE see DTD800
3,7-DIMETHYL-1,6-OCTADIEN-3-YL-o-
AMINOBENZOATE see APJ000
3,7-DIMETHYL-1,6-OCTADIEN-3-YL
BENZOATE see LFZ000
3,7-DIMETHYL-2,6-OCTADIEN-1-YL
BENZOATE see GDE800

3,7-DIMETHYL-2,6-OCTADIEN-1-YL BUTYRATE see GDE825

trans-3,7-DIMETHYL-2,6-OCTADIEN-1-YL BUTYRATE see GDE810

3,7-DIMETHYL-2-trans-6-OCTADIENYL CROTONATE see DTE000

trans-3,7-DIMETHYL-2,6-OCTADIEN-1-YL cis-α,β-DIMETHYL ACRYLATE see GDO000

3-7-DIMETHYL-2,6-OCTADIENYL ESTER-2-BUTENOIC ACID see DTE000

3,7-DIMETHYL-2,6-OCTADIENYL ESTER FORMIC ACID (E) see GCY000

(E)-3,7-DIMETHYLOCTA-2,6-DIEN-1-YL ESTER, HEXANOIC ACID see GDG000

trans-3,7-DIMETHYL-2,6-OCTADIENYL ESTER ISOBUTYRIC ACID see GDI000

(Z)-3,7-DIMETHYL-2,6-OCTADIENYL ESTER ISOVALERIC ACID see NCO500

(E,E,E)-3,7-DIMETHYL-2,6-OCTADIENYL ESTER-2-METHYL-2-BUTENOIC ACID see GDO000

trans-2,6-DIMETHYL-2,6-OCTADIEN-8-YL ETHANOATE see DTD800

3,7-DIMETHYL-1,6-OCTADIEN-3-YL FORMATE see LGA050

trans-3,7-DIMETHYL-2,6-OCTADIEN-1-YL FORMATE see GCY000

(E)-3,7-DIMETHYLOCTA-2,6-DIEN-1-YL-n-HEXANOATE see GDG000

N-(3,7-DIMETHYL-2,6-OCTADIENYLIDENE)ANTHRANILIC ACID METHYL ESTER see CMS325

3,7-DIMETHYL-1,6-OCTADIEN-3-YL ISOBUTYRATE see LGB000

cis-3,7-DIMETHYL-2,6-OCTADIEN-1-YL ISOBUTYRATE see NCO200

trans-3-7-DIMETHYL-2,6-OCTADIENYL ISOBUTYRATE see GDI000

trans-3,7-DIMETHYL-2,6-OCTADIENYL ISOPENTANOATE see GDK000

3,7-DIMETHYL-1,6-OCTADIEN-3-YL ISOVALERATE see LGC000

3,7-DIMETHYL-2-cis-6-OCTADIEN-1-YL ISOVALERATE see NCO500

5-((3,7-DIMETHYL-2,6-OCTADIENYL)OXY)-2-ETHYLPYRIDINE, (E)- see DTE100

3,7-DIMETHYL-2,6-OCTADIEN-1-YL PHENYLACETATE see GDM400

trans-3,7-DIMETHYL-2,6-OCTADIEN-1-YL PHENYLACETATE see GDM400

trans-3,7-DIMETHYL-2,6-OCTADIEN-1-YL PROPIONATE see GDM450

3,7-DIMETHYLOCTANAL see TCY300

N,N-DIMETHYLOCTANAMIDE see DTE200

3,7-DIMETHYL-1,2-OCTANEDIOL see DTE400

4,4-DIMETHYLOCTANOIC ACID, TRIBUTYLSTANNYL ESTER see TIF250

DIMETHYLOCTANOL see DTE600

2,6-DIMETHYL-8-OCTANOL see DTE600

3,7-DIMETHYLOCTANOL-3 see LFY510

3,7-DIMETHYL-3-OCTANOL see TCU600

3,7-DIMETHYL-1-OCTANOL (FCC) see DTE600

(4,4-DIMETHYLOCTANOYLOXY)TRIBUTYLSTANNANE see TIF250

3,7-DIMETHYLOCTANYL ACETATE see DTE800

3,7-DIMETHYLOCTANYL BUTYRATE see DTF000

DIMETHYLOCTATRIENE see DTF200

DIMETHYLOCTATRIENE (mixed isomer) see DTF200

3,7-DIMETHYL-6-OCTENAL see CMS845

3,7-DIMETHYL-6-OCTENENITRILE see CMU000

3,7-DIMETHYL-6-OCTENOIC ACID see CMT125

2,6-DIMETHYL-1-OCTEN-8-OL see DTF400

2,6-DIMETHYL-2-OCTEN-8-OL see CMT250

2,6-DIMETHYL-2-OCTEN-8-OL see DTF410

2,6-DIMETHYL-7-OCTEN-2-OL see DLX000

3,7-DIMETHYL-6-OCTEN-1-OL see CMT250

3,7-DIMETHYL-6-OCTEN-1-OL see DTF410

3,7-DIMETHYL-7-OCTEN-1-OL see DTF400

2,6-DIMETHYL-2-OCTEN-8-OL ACETATE see AAU000

3,7-DIMETHYL-7-OCTEN-1-OL ACETATE see RHA000

2,6-DIMETHYL-2-OCTEN-8-OL-BUTYRATE see DTF800

3,7-DIMETHYL-6-OCTEN-1-OL BUTYRATE see DTF800

3,7-DIMETHYL-6-OCTEN-1-OL CROTONATE see CMT500

3,7-DIMETHYL-6-OCTEN-1-OL FORMATE see CMT750

3,7-DIMETHYL-6-OCTEN-1-YL ACETATE see AAU000

2,6-DIMETHYL-2-OCTEN-8-YL BUTYRATE see DTF800

3.7-DIMETHYL-6-OCTEN-1-YL BUTYRATE see CMT600

2,6-DIMETHYL-2-OCTEN-8-YL FORMATE see CMT750

3,7-DIMETHYL-6-OCTEN-1-YL FORMATE see CMT750

3,7-DIMETHYL-6-OCTEN-1-YL ISOBUTYRATE see CMT900

3,7-DIMETHYL-6-OCTEN-1-YL PHENYLACETATE see CMU050

3,7-DIMETHYL-6-OCTEN-1-YN-3-OL see LFY333

3,7-DIMETHYLOCTYL ACETATE see DTE800

N,N-DIMETHYL-N-OCTYLBENZENEMETHANAMINIUM CHLORIDE see OEW000

N,N-DIMETHYL-N-OCTYL-1-DECANAMINIUM CHLORIDE see OES400

3,7-DIMETHYLOCTYL ESTER BUTANOIC ACID see DTF000

N,N-DIMETHYL-N-OCTYL-1-OCTANAMINIUM CHLORIDE see DTF820

3,6-DIMETHYL-4-OCTYN-4-DIOL-(3,6) see DTF850

N,N-DIMETHYLOKTADECYLAMIN (CZECH) see DTC400

DIMETHYLOL DIHYDROXYETHYLENE UREA see DTG000

DIMETHYLOLETHYLENETHIOUREA see BKH650

DIMETHYLOLGLYOXALUREA see DTG000

N,N-DIMETHYLOL-2-METHOXYETHYL CARBAMATE see DTG200

1,1-DIMETHYLOL-1-NITROETHANE see NHO500

DIMETHYLOLPROPANE see DTG400

DIMETHYLOLPROPANE DIACRYLATE see DUL200

DIMETHYLOL-TETRAKIS-BUTOXYMETHYLMELAMIN (CZECH) see BHB500

DIMETHYLOL THIOUREA see DTG600

DIMETHYLOLTRICYCLO(5.2.1.0(2,6))DECANE see TJG550

1,3-DIMETHYLOLUREA see DTG700

2,3-DIMETHYL-7-OXABICYCLO(2.2.1)HEPTANE-2,3-DICARBOXYLIC ANHYDRIDE see CBE750

6,10-DIMETHYL-3-OXA-9-UNDECENAL see CMT300

DIMETHYL OXAZOLIDINE see DTG750

4,4-DIMETHYLOXAZOLIDINE see DTG750

DIMETHYLOXAZOLIDINEDIONE see PMO250

5,5-DIMETHYLOXAZOLIDINE-2,4-DIONE see PMO250

5,5-DIMETHYL-2,4-OXAZOLIDINEDIONE see PMO250

N¹-(4,5-DIMETHYL-2-OXAZOLYL)-SULFANILAMIDE see AIE750

3,3-DIMETHYLOXETANE see EBQ500

3,3-DIMETHYL-2-OXETANONE see DTH000

3,3-DIMETHYL-2-OXETHANONE see DTH000

2,3-DIMETHYLOXIRANE see EBJ100

cis-2,3-DIMETHYLOXIRANE see EBJ200

5,5-DIMETHYL-3-(2-(OXIRANYLMETHOXY)PROPYL)-1-(OXIRANYLMETHYL)-2,4-IMIDAZOLIDINEDIONE see DTH100

O,O-DIMETHYL-S-(2-OXO-3-AZA-BUTYL)-DITHIOPHOSPHAT (GERMAN) see DSP400

O,O-DIMETHYL-S-(2-OXO-3-AZABUTYL)-MONOTHIOPHOSPHATE see DNX800

O,O-DIMETHYL-S-(4-OXOBENZOTRIAZINO-3-METHYL)PHOSPHORODITHIOATE see ASH500

O,O-DIMETHYL-S-(4-OXO-1,2,3-BENZOTRIAZINO(3)-METHYL) THIOTHIONOPHOSPHATE see ASH500

O,O-DIMETHYL-S-((4-OXO-3H-1,2,3-BENZOTRIAZIN-3-YL)-METHYL)-DITHIOFOSFAAT (DUTCH) see ASH500

O,O-DIMETHYL-S-((4-OXO-3H-1,2,3-BENZOTRIAZIN-3-YL)-METHYL)-DITHIOPHOSPHAT (GERMAN) see ASH500

O,O-DIMETHYL-S-4-OXO-1,2,3-BENZOTRIAZIN-3(4H)-YLMETHYL PHOSPHORODITHIOATE see ASH500

O,O-DIMETHYL-S-(4-OXO-3H-1,2,3-BENZOTRIZIANE-3-METHYL)PHOSPHORODITHIOATE see ASH500

N,N-DIMETHYL-3-OXOBUTANAMIDE see DOP000

N-(1,1-DIMETHYL-3-OXOBUTYL)ACRYLAMIDE see DTH200

N-(1,1-DIMETHYL-3-OXOBUTYL)-2-PROPENAMIDE see DTH200

(5,5-DIMETHYL-3-OXO-CYCLOHEX-1-EN-YL)-N,N-DIMETHYL-CARBAMAAT (DUTCH) see DRL200

(5,5-DIMETHYL-3-OXO-CYCLOHEX-1-EN-YL)-N,N-DIMETHYL-CARBAMAT (GERMAN) see DRL200

5,5-DIMETHYL-3-OXOCYCLOHEX-1-ENYL DIMETHYLCARBAMATE see DRL200

5,5-DIMETHYL-3-OXO-1-CYCLOHEXEN-1-YL DIMETHYLCARBAMATE see DRL200

3-(2-(3,5-DIMETHYL-2-OXOCYCLOHEXYL)-2-HYDROXYETHYL)GLUTARIMIDE see CPE750

1,6-DIMETHYL-4-OXO-1,6,7,8,9,9a-HEXAHYDRO-4H-PYRIDO(1,2-a)PYRIMIDINE-3-CARBOXAMIDE see CDM500

3,7-DIMETHYL-1-(5-OXOHEXYL)-1H,3H-PURIN-2,6-DIONE see PBU100

DIMETHYLOXOHEXYLXANTHINE see PBU100

3,7-DIMETHYL-1-(5-OXOHEXYL)XANTHINE see PBU100

7,7-DIMETHYL-3-OXO-4-OXA-8-THIA-2,5-DIAZANON-5-ENE-2-SULFINIC ACID DECYL ESTER see MLX830

3,3-DIMETHYL-7-OXO-6-(2-PHENYLACETAMIDO)-4-THIA-1-AZABICYCLO(3.2.0)HEPTANE-2-CARBOXYLIC ACID compounded with EPHEDRINE (1:1) see PAQ120

2-(2,2-DIMETHYL-1-OXOPROPYL)-1H-INDENE-1,3(2H)-DIONE see PIH175

endo-8,8-DIMETHYL-3-((1-OXO-2-PROPYLPENTYL)OXY)-8-AZONIABICYCLO(3.2.1)OCTANE BROMIDE see LJS000

DIMETHYLOXOSTANNANE see DTH400

O,O-DIMETHYL-S-(3-OXO-3-THIA-PENTYL)-MONOTHIOPHOSPHAT (GERMAN) see DAP000

2,2-DIMETHYL-3-(3-OXO-3-(2,2,2-TRIFLUORO-1-(TRIFLUOROMETHYL)ETHOXY)-1-PROPENYL)CYCLOPROPANECARBOXYLIC ACID, CYANO(3-PHENOXYPHENYL)METHYL ESTER, (1R-(1-α(S*),3-α(Z)))- see DTH450
DIMETHYL 2,2'-OXYBISACETATE see DTH500
2,2-DIMETHYL-4-OXYMETHYL-1,3-DIOXOLANE see DVR600
α,γ-DIMETHYL-α-OXYMETHYL GLUTARALDEHYDE see DTH600
1-(2,5-DIMETHYLOXYPHENYLAZO)-2-NAPHTHOL see DOK200
DIMETHYLOXYQUINAZINE see AQN000
O,O-DIMETHYL-1-OXY-2,2,2-TRICHLOROETHYL PHOSPHONATE see TIQ250
N,N-DIMETHYLPALMITAMIDE see DTH700
DIMETHYLPALMITYLAMINE see HCP525
3,5-DIMETHYLPARACETAMOL see DOO900
DIMETHYL PARANITROPHENYL THIONOPHOSPHATE see DTH800
DIMETHYL PARAOXON see PHD500
DIMETHYL PARATHION see MNH000
N,N-DIMETHYLPENTANAMIDE see DUN200
2,3-DIMETHYLPENTANE see DTI000
2,4-DIMETHYLPENTANE see DTI200
2,3-DIMETHYLPENTANOL see DTI400
2,3-DIMETHYL-1-PENTANOL see DTI400
2,4-DIMETHYL-3-PENTANONE see DTI600
S,S-DIMETHYLPENTASULFUR HEXANITRIDE see DTI709
DI(4-METHYL-2-PENTYL) MALEATE see DKP400
N-(1,4-DIMETHYLPENTYL)-N'-PHENYL-1,4-BENZENEDIAMINE see DTI800
3,5-DIMETHYLPERHYDRO-1,3,5-THIADIAZIN-2-THION (CZECH, GERMAN) see DSB200
DIMETHYL PEROXIDE see DTJ000
DIMETHYLPEROXYCARBONATE see DTJ159
16,16-DIMETHYL-trans-Δ²-PGE1 METHYL ESTER see CDB775
N,N¹-DIMETHYLPHAEANTHINE DIIODIDE see TDX835
1,4-DIMETHYLPHENANTHRENE see DTJ200
α,α-DIMETHYLPHENETHYL ACETATE see BEL750
α,α-DIMETHYLPHENETHYL ACETATE see DQQ375
α,α-DIMETHYLPHENETHYL ALCOHOL see DQQ200
α,α-DIMETHYLPHENETHYL ALCOHOL ACETATE see BEL750
α,α-DIMETHYLPHENETHYL ALCOHOL PROPIONATE see DQQ400
d-N,α-DIMETHYLPHENETHYLAMINE see PFP850
α,α-DIMETHYLPHENETHYLAMINE see DTJ400
N,α-DIMETHYLPHENETHYLAMINE HYDROCHLORIDE see DBA800
(−)-N-α-DIMETHYLPHENETHYLAMINE HYDROCHLORIDE see MDQ500
α-α-DIMETHYLPHENETHYL BUTYRATE see BEL850
3,4-DIMETHYLPHENISOPROPYLAMINE SULFATE see DTK200
DIMETHYLPHENOL see XKA000
2,3-DIMETHYLPHENOL see XKJ000
2,4-DIMETHYLPHENOL see XKJ500
2,5-DIMETHYLPHENOL see XKS000
2,6-DIMETHYLPHENOL see XLA000
3,4-DIMETHYLPHENOL see XLJ000
3,5-DIMETHYLPHENOL see XLS000

3,6-DIMETHYLPHENOL see XKS000
4,5-DIMETHYLPHENOL see XLJ000
4,6-DIMETHYLPHENOL see XKJ500
3,4-DIMETHYLPHENOL METHYLCARBAMATE see XTJ000
DIMETHYLPHENOL PHOSPHATE (3:1) see XLS100
2,6-DIMETHYLPHENOL PHOSPHATE (3:1) see TNR550
3',3''-DIMETHYLPHENOLPHTHALEIN see CNX400
2,6-DIMETHYLPHENOL-4-SULFONIC ACID see XJJ005
DIMETHYLPHENOSAFRANINE see AJP300
2,8-DIMETHYLPHENOSAFRANINE see GJI400
N,N-DIMETHYL-10H-PHENOTHIAZINE-10-PROPANAMINE see DQA600
N,N-DIMETHYL-3-PHENOTHIAZINESULFONAMIDE see DTK300
5-(2,5-DIMETHYLPHENOXY)-2,2-DIMETHYLPENTANOIC ACID (9CI) see GCK300
5-((3,5-DIMETHYLPHENOXY)METHYL)-2-OXAZOLIDINONE see XVS000
(2-(2,6-DIMETHYLPHENOXY)PROPYL)TRIMETHYLAMMONIUM CHLORIDE MONOHYDRATE see TLQ000
2-(2,6-DIMETHYLPHENOXY)-N,N,N-TRIMETHYL-ETHANAMINIUM BROMIDE (9CI) see XSS900
2-(2,6-DIMETHYLPHENOXY)-N,N,N-TRIMETHYL-1-PROPANAMINIUM HYDRATE see TLQ000
α,α-DIMETHYLPHENRTHYL BUTYRATE see DQQ380
N,N-DIMETHYL-2-(2-PHENYLACETAMIDO)ACETAMIDE see PEB775
O,O-DIMETHYL-S-(PHENYLACETIC ACID ETHYL ESTER) PHOSPHORODITHIOATE see DRR400
1-(5,7-DIMETHYL-2-PHENYL-1-ADAMANTYL)-N-METHYL-2-PROPYLAMINE HYDROCHLORIDE see DSO400
DIMETHYLPHENYLAMINE see DQF800
DIMETHYLPHENYLAMINE see XMA000
2,3-DIMETHYLPHENYLAMINE see XMJ000
2,4-DIMETHYLPHENYLAMINE see XMS000
2,5-DIMETHYLPHENYLAMINE see XNA000
3,4-DIMETHYLPHENYLAMINE see XNS000
3,5-DIMETHYLPHENYLAMINE see XOA000
N,N-DIMETHYLPHENYLAMINE see DQF800
2-((2,3-DIMETHYLPHENYL)AMINO)BENZOIC ACID see XQS000
2-(2,6-DIMETHYLPHENYLAMINO)-4H-5,6-DIHYDRO-1,3-THIAZINE see DMW000
4-((2-((2,6-DIMETHYLPHENYL)AMINO)-2-OXOETHYL)AMINO)BUTANOIC ACID see DTK400
4-((2-((2,6-DIMETHYLPHENYL)AMINO)-2-OXOETHYL)AMINO)-4-OXOBUTANOIC ACID see DTK420
1,5-DIMETHYL-2-PHENYL-4-AMINOPYRAZOLINE see AIB300
N-(2,3-DIMETHYLPHENYL)ANTHRANILIC ACID see XQS000
N-(2,6-DIMETHYLPHENYL)-2-AZABICYCLO(2.2.2)OCTANE-3-CARBOXAMIDE MONOHYDROCHLORIDE (9CI) see EAV100
2,3-DIMETHYL-4-PHENYLAZOANILINE see DTL000
N,N-DIMETHYL-p-PHENYLAZOANILINE see DOT300
N,N-DIMETHYL-p-PHENYLAZOANILINE-N-OXIDE see DTK600

N,N-DIMETHYL-4-PHENYLAZO-o-ANISIDINE see DTK800
N,N-DIMETHYL-4-(PHENYLAZO)BENZAMINE see DOT300
2,3-DIMETHYL-4-(PHENYLAZO)BENZENAMINE see DTL000
N,N-DIMETHYL-4-(PHENYLAZO)BENZENAMINE see DOT300
4-((2,4-DIMETHYLPHENYL)AZO)-3-HYDROXY-2,7-NAPHTHALENEDISULFONIC ACID, DISODIUM SALT see FMU070
4-((2,4-DIMETHYLPHENYL)AZO)-3-HYDROXY-2,7-NAPHTHALENEDISULPHONIC ACID, DISODIUM SALT see FMU070
1-((2,4-DIMETHYLPHENYL)AZO)-2-NAPHTHALENOL see XRA000
N,N-DIMETHYL-4-(PHENYLAZO)-m-TOLUIDINE see TLE750
N,N-DIMETHYL-4-(PHENYLAZO)-o-TOLUIDINE see MJF000
N,N-DIMETHYL-α-PHENYLBENZENEACETAMIDE see DRP800
(R)-N,N-DIMETHYL-α-PHENYLBENZENEETHANAMINE, HYDROCHLORIDE see DWA600
2-DI(N-METHYL-N-PHENYL-tert-BUTYL-CARBAMOYLMETHYL)AMINOETHANOL see DTL200
N-(2,6-DIMETHYLPHENYLCARBAMOYLMETHYL)-IMINODIACETIC ACID see LFO300
N-(N'-(2,6-DIMETHYLPHENYL)CARBAMOYLMETHYL)IMINODIACETIC ACID see LFO300
DIMETHYLPHENYLCARBINOL see DTN100
O,O-DIMETHYL-S-(PHENYL)(CARBOETHOXY)METHYL PHOSPHORODITHIOATE see DRR400
N-(2,6-DIMETHYLPHENYL)-5,6-DIHYDRO-4H-1,3-THIAZIN-2-AMINE see DMW000
N-(2,6-DIMETHYLPHENYL)-5,6-DIHYDRO-4H-1,3-THIAZINE-2-AMINE (9CI) see DMW000
N'-(2,4-DIMETHYLPHENYL)-N-(((2,4-DIMETHYLPHENYL)IMINO)METHYL)-N-METHYLME THANIMIDAMIDE see MJL250
1,3-DIMETHYL-3-PHENYL-2,5-DIOXOPYRROLIDINE see MLP800
DIMETHYL-4,4'-o-PHENYLENE-BIS-(3-THIOALLOPHANATE) see PEX500
DIMETHYL-p-PHENYLENEDIAMINE see DTL600
DIMETHYL-p-PHENYLENEDIAMINE see DTL800
N,N-DIMETHYL-p-PHENYLENEDIAMINE see DTL600
N,N-DIMETHYL-p-PHENYLENEDIAMINE see DTL800
N,N-DIMETHYL-p-PHENYLENEDIAMINE DIHYDROCHLORIDE see DTM000
N,N-DIMETHYL-p-PHENYLENEDIAMINE HEMISULFATE see DTM200
DIMETHYL-p-PHENYLENEDIAMINE HYDROCHLORIDE see DTM000
N,N-DIMETHYL-p-PHENYLENEDIAMINE MONOHYDROCHLORIDE see DTM400
1,1-DIMETHYL-2-PHENYLETHANAMINE see DTJ400
1,1-DIMETHYL-2-PHENYLETHANOL see DQQ200
(E)-N,N,-DIMETHYL-4-(2-PHENYLETHENYL)BENZENAMINE see DUC000
(DIMETHYL-S-(PHENYLETHOXYCARBONYLMETHYL)PHOSPHOROTHIOLOTHIONATE) see DRR400

S-(O,O-DIMETHYLPHOSPHORODITHIOATE) of N-(2-MERCAPTOETHYL)ETHYLCARBAMATE see EMC000

N-((O,O-DIMETHYLPHOSPHORODITHIOYL)ETHYL)ACETAMIDE see DOP200

O,O-DIMETHYL PHOSPHOROTHIOATE-O,O-DIESTER with 4,4'-THIODIPHENOL see TAL250

O,O-DIMETHYL PHOSPHOROTHIOATE-O-ESTER with 4-HYDROXY-m-ANISONITRILE see DTQ800

DIMETHYL PHOSPHOROUS ACID see DSG600

5-(O,O-DIMETHYLPHOSPHORYL)-6-CHLOROBICYCLO(3.2.0)HEPTA-1,5-DIEN see HBK700

DIMETHYL PHTHALATE see DTR200

(O,O-DIMETHYL-PHTHALIMIDOMETHYL-DITHIOPHOSPHATE) see PHX250

1,4-DIMETHYLPIPERAZINE see LIQ500

2,5-DIMETHYLPIPERAZINE see DTR400

N,N'-DIMETHYLPIPERAZINE see LIQ500

α,4-DIMETHYL-1-PIPERAZINEACETIC ACID-6-CHLORO-o-TOLYL ESTER DIHYDROCHLORIDE see FAC185

α,4-DIMETHYL-1-PIPERAZINEACETIC ACID-2,6-DIISOPROPYLPHENYL ESTER DIHYDROCHLORIDE see FAC160

2-β,16-β-(4'-DIMETHYL-1'-PIPERAZINO)-3-α,17-β-DIACETOXY-5-α-ANDROSTANE 2BR see PII250

2,6-DIMETHYLPIPERIDINE see LIQ550

1,1-DIMETHYLPIPERIDINIUM CHLORIDE see MCH540

N,N-DIMETHYL-PIPERIDINIUM CHLORIDE see MCH540

2-(2,6-DIMETHYLPIPERIDINO)-2',6'-ACETOXYLIDIDE HYDROCHLORIDE see DTR800

2,4'-DIMETHYL-3-PIPERIDINOPROPIOPHENONE see TGK200

2,4'-DIMETHYL-3-PIPERIDINOPROPIOPHENONE HYDROCHLORIDE see MRW125

N,N-DIMETHYL-4-PIPERIDYLIDENE-1,1-DIPHENYLMETHANE METHYLSULFATE see DAP800

DIMETHYLPOLYSILOXANE see DTR850

6,17-DIMETHYLPREGNA-4,6-DIENE-3,20-DIONE see MBZ100

1,2-DIMETHYLPROPANAMINE see AOE200

2,2-DIMETHYLPROPANE see NCH000

2,2-DIMETHYLPROPANE, other than pentane and isopentane (DOT) see NCH000

N,N-DIMETHYL-1,3-PROPANEDIAMINE see AJQ100

DIMETHYL PROPANEDIOATE see DSM200

2,2-DIMETHYL-1,3-PROPANEDIOL see DTG400

2,2-DIMETHYL-1,3-PROPANEDIOL DIACRYLATE see DUL200

2,2'-((2,2-DIMETHYL-1,3-PROPANEDIYL)BIS(OXYMETHYLENE))BIS OXIRANE see NCI300

2,2-DIMETHYLPROPANOIC ACID see PJA500

2,2-DIMETHYLPROPANOIC ACID-3-(2-(ETHYLAMINO)-1-HYDROXYETHYL)PHENYL ESTER HYDROCHLORIDE see EGC500

(±)-2,2-DIMETHYL-PROPANOIC ACID-4-(1-HYDROXY-2-(METHYLAMINO)ETHYL)-1,2-PHENYLENE ESTER, HYDROCHLORIDE see DWP559

2,2-DIMETHYLPROPANOIC ACID ISOOCTADECYL ESTER see ISC550

2,2-DIMETHYL-PROPANOIC ACID-2-OXO-2-PHENYLETHYL ESTER (9CI) see PCV350

2,2-DIMETHYLPROPANOYL CHLORIDE see DTS400

1,1-DIMETHYLPROPARGYL ALCOHOL see MHX250

α-α-DIMETHYLPROPARGYL ALCOHOL see MHX250

1,1-DIMETHYLPROPARGYLAMINE see MHX200

N,N-DIMETHYL-2-PROPENAMIDE see DOP800

6-(1-1-DIMETHYL-2-PROPENYL)-7H-FURO(3,2-G)(1)BENZOPYRAN-7-ONE see XPJ100

2,10-DIMETHYL-6-(2-PROPENYLOXY)-4,8-DIOXA-3,9-DITHIA-2,10-DIAZAUNDECANEDIOIC ACID, DI-1-NAPHTHALENYL ESTER, 3,9-DIOXIDE see DTS450

N,N-DIMETHYL-N-2-PROPENYL-2-PROPEN-1-AMINIUM CHLORIDE HOMOPOLYMER (9CI) see DTS500

DIMETHYL PROPIOLACTONE see DTH000

3,3-DIMETHYL-β-PROPIOLACTONE see DTH000

N,N-DIMETHYLPROPIONAMIDE see DTS600

2,2-DIMETHYLPROPIONIC ACID see PJA500

α,α-DIMETHYLPROPIONIC ACID see PJA500

2,2-DIMETHYLPROPIONYL CHLORIDE see DTS400

2-(2,2-DIMETHYL-3-PROPIONYL)-1-METHYL-3-(METHYLETHENYL)CYCLOPENTENE see MJW300

2,6-DIMETHYL-4-PROPOXY-BENZOIC ACID 2-METHYL-2-(1-PYRROLIDINYL)PROPYLESTER see DTS625

2,6-DIMETHYL-4-PROPOXY-BENZOIC ACID 2-METHYL-2-(1-PYRROLIDINYL)PROPYL ESTER see UAG050

2,6-DIMETHYL-4-PROPOXY-BENZOIC ACID 2-(1-PYRROLIDINYL)PROPYL ESTER HYDROCHLORIDE see UAG075

1,2-DIMETHYLPROPYLAMINE see AOE200

4-(1,2-DIMETHYL-N-PROPYLAMINO)-2-ETHYLAMINO-6-METHYLTHIO-S-TRIAZINE see ARW775

4-(1,2-DIMETHYL-N-PROPYLAMINO)-2-ETHYLAMINO-6-METHYLTHIO-S-TRIAZINE see DTS700

7,12-DIMETHYL-8-PROPYL-BENZ(a)ANTHRACENE see PNI500

4-(1,1-DIMETHYLPROPYL)CYCLOHEXANONE see AOH750

N,N-DIMETHYL-1,3-PROPYLENEDIAMINE see AJQ100

DIMETHYLPROPYLENEUREA see DSE489

1,1-DIMETHYLPROPYL HYDROPEROXIDE see PBX325

5-(3-DIMETHYLPROPYLIDENE)DIBENZO(a,d)(1,4)CYCLOHEPTADIENE see EAH500

1,1-DIMETHYLPROPYL METHYL ETHER see PBX400

p-(1,1-DIMETHYLPROPYL)PHENOL see AON000

p-(α,α-DIMETHYLPROPYL)PHENOL see AON000

N,N-DIMETHYL-p-((p-PROPYLPHENYL)AZO)ANILINE see DTT400

(±)-3-(1,3-DIMETHYL-4-PROPYL-4-PIPERIDINYL)PHENOL HYDROCHLORIDE see PIB700

trans-(±)-3-(1,3-α-DIMETHYL-4-α-PROPYL-4-β-PIPERIDINYL)PHENOL HYDROCHLORIDE see PIB700

m-(1,2-DIMETHYL-3-PROPYL-3-PYRROLIDINYL)PHENOL see DSI000

1,1-DIMETHYLPROPYNOL see MHX250

1,1-DIMETHYLPROPYNYLAMINE see MHX200

(−)-N,α-DIMETHYL-N-2-PROPYNYLBENZENEETHANAMINE HYDROCHLORIDE see DAZ125

N-(1,1-DIMETHYLPROPYNYL)-3,5-DICHLOROBENZAMIDE see DTT600

(+)-N,α-DIMETHYL-N-2-PROPYNYLPHENETHYLAMINE HYDROCHLORIDE see DAZ120

(±)-N,α-DIMETHYL-N-2-PROPYNYLPHENETHYLAMINE HYDROCHLORIDE see DAZ118

DIMETHYL-1-PROPYNYLTHALLIUM see DTT800

16,16-DIMETHYL-trans-Δ²-PROSTAGLANDIN E1 METHYL ESTER see CDB775

3,4-DIMETHYLPROTOCATECHUIC ACID see VHP600

3,5-DIMETHYL-4H-PYRAN-4-ONE-2-METHOXY-6-(TETRAHYDRO-4-(β-METHYL-p-NITROCINNAMYLIDENE)-2-FURYL) see DTU200

2,3-DIMETHYLPYRAZINE see DTU400

2,5-DIMETHYLPYRAZINE see DTU600

2,6-DIMETHYLPYRAZINE see DTU800

3,6-DIMETHYLPYRAZINE-2-THIOL see DTU825

3,6-DIMETHYL-2(1H)-PYRAZINETHIONE see DTU825

3,5-DIMETHYLPYRAZOLE see DTU850

2,4-DIMETHYLPYRIDINE see LIY990

2,5-DIMETHYLPYRIDINE see LJA000

2,6-DIMETHYLPYRIDINE see LJA010

3,4-DIMETHYLPYRIDINE see LJB000

α-γ-DIMETHYLPYRIDINE see LIY990

α-α'-DIMETHYLPYRIDINE see LJA010

2,6-DIMETHYL-3,5-PYRIDINE DICARBOXYLIC ACID, DIMETHYLESTER see DTU852

2,6-DIMETHYLPYRIDINE-N-OXIDE see DTV089

2,6-DIMETHYLPYRIDINE-1-OXIDE-4-AZO-p-DIMETHYLANILINE see DPP800

N,N-DIMETHYL-3-(1-(2-PYRIDINYL)ETHYL)-1H-INDENE-2-ETHANAMINE (Z)-2-BUTENEDIOATE (1:1) see FMU409

N¹-(4,6-DIMETHYL-2-PYRIDINYL)SULFANILAMIDE, MONOSODIUM SALT see SJW500

N,N-DIMETHYL-N'-2-PYRIDINYL-N'-(2-THIENYLMETHYL)-1,2-ETHANEDIAMIDE see TEO250

N,N-DIMETHYL-N'-2-PYRIDINYL-N'-(2-THIENYLMETHYL)-1,2-ETHANEDIAMINE MONOHYDROCHLORIDE see DPJ400

5,11-DIMETHYL-6H-PYRIDO(4,3-b)CARBAZOL-9-AMINE see AJS875

5,11-DIMETHYL-6H-PYRIDO(4,3-b)CARBAZOLE see EAI850

5,11-DIMETHYL-6H-PYRIDO(4,3-b)CARBAZYL-9-OL see HKH000

1,4-DIMETHYL-5H-PYRIDO(4,3-b)INDOL-3-AMINE see TNX275

1,4-DIMETHYL-5H-PYRIDO(4,3-b)INDOL-3-AMINE ACETATE see AJR500

1,4-DIMETHYL-5H-PYRIDO(4,3-b)INDOL-3-AMINE MONOACETATE see AJR500

N,N-DIMETHYL-p-(3-PYRIDYLAZO)ANILINE see POP750

N,N-DIMETHYL-4-(3'-PYRIDYLAZO)ANILINE see POP750

N,N-DIMETHYL-N'-(2-PYRIDYL)-N'-BENZYLETHYLENEDIAMINE HYDROCHLORIDE see POO750

N,N-DIMETHYL-N'-(2-PYRIDYL)-N'-(5-CHLORO-2-THENYL)ETHYLENEDIAMINE see CHY250

(3,3-DIMETHYL-1-(m-PYRIDYL-N-OXIDE))TRIAZENE see DTV200

N,N-DIMETHYL-N'-PYRID-2-YL-N'-2-THENYLETHYLENEDIAMINE see TEO250

N,N-DIMETHYL-N'-(2-PYRIDYL)-N'-THENYLETHYLENEDIAMINE HYDROCHLORIDE see DPJ400

S-(4,6-DIMETHYL-2-PYRIMIDINYL)-O,O-DIETHYL PHOSPHORODITHIOATE see DTV400

N^1-(2,6-DIMETHYL-4-PYRIMIDINYL)SULFANILAMIDE see SNJ350

N^1-(4,6-DIMETHYL-2-PYRIMIDINYL)SULFANILAMIDE see SNJ000

(N^1-(4,6-DIMETHYL-2-PYRIMIDINYL)SULFANILAMIDO) SODIUM see SJW500

6,8-DIMETHYLPYRIMIDO(5,4-e)-as-TRIAZINE-5,7(6H,8H)-DIONE see FBP300

6,8-DIMETHYL-PYRIMIDO(5,4-e)-1,2,4-TRIAZINE-5,7(6H,8H)-DIONE see FBP300

N-(4,6-DIMETHYL-2-PYRIMIDINYL)SULFANILAMIDE see SNJ000

1,3-DIMETHYL PYROGALLATE see DOJ200

2,5-DIMETHYLPYRROLE see DTV300

20-(2,4-DIMETHYL-1H-PYRROLE-3-CARBOXYLATE) BATRACHOTOXININ A see BAR750

20-α-(2,4-DIMETHYL-1H-PYRROLE-3-CARBOXYLATE) BETRACHOTOXININ A see BAR750

3,4-DIMETHYLPYRROLIDINE ETHANOL see HKR550

N,N-DIMETHYL-3-(PYRROLIDIN-1-YL)PROPIONAMIDE see DTV330

2,4-DIMETHYLPYRROL-3-YL METHYL KETONE see ACI500

2,5-DIMETHYLPYRROL-3-YL METHYL KETONE see ACI550

N-(2,5-DIMETHYL-1H-PYRROL-1-YL)-6-(4-MORPHOLINYL)-3-PYRIDAZINAMINE HYDROCHLORIDE see MBV735

2,9-DIMETHYLQUINACRIDONE see DTV360

N,N-DIMETHYL-4-(4'-QUINOLYLAZO)ANILINE see DTY200

N,N-DIMETHYL-4-(5'-QUINOLYLAZO)ANILINE see DPQ800

N,N-DIMETHYL-4-(6'-QUINOLYLAZO)ANILINE see DPR000

N,N-DIMETHYL-p-(5'-QUINOLYLAZO)ANILINE see DPQ800

N,N-DIMETHYL-4-(5'-QUINOLYLAZO)-m-TOLUIDINE see DQE400

DIMETHYLQUINOLYL METHYLSULFATE UREA see PJA120

N,N-DIMETHYL-4-((4'-QUINOLYL-1'-OXIDE)AZO)ANILINE see DTY400

N,N-DIMETHYL-4-((5'-QUINOLYL-1'-OXIDE)AZO)ANILINE see DPR200

N,N'-DIMETHYL-4-((6'-QUINOLYL-1'-OXIDE)AZO)ANILINE see DPR400

3,3-DIMETHYL-1-(3-QUINOLYL)TRIAZENE see DTY600

2,6-DIMETHYLQUINONEIMINE see DTY650

2,3-DIMETHYL-6-QUINOXALINAMINE see AJR600

N,5-DIMETHYL-6-QUINOXALINAMINE see MLK775

2,3-DIMETHYLQUINOXALINE see DTY700

2,3-DIMETHYLQUINOXALINE DIOXIDE see DTY800

o,o'-DIMETHYL S,S'-2,3-QUINOXALINEDIYL THIOCARBONATE see BHL800

N,N-DIMETHYL-p-(6-QUINOXALINYLAZO)ANILINE see DUA400

N,N-DIMETHYL-p-(6-QUINOXALYAZO)ANILINE see DUA400

N,N-DIMETHYL-p-(5-QUINOXALYLAZO)ANILINE see DUA200

6,7-DIMETHYL-9-d-RIBITYLISOALLOXAZINE see RIK000

7,8-DIMETHYL-10-d-RIBITYLISOALLOXAZINE see RIK000

7,8-DIMETHYL-10-(d-RIBO-2,3,4,5-TETRAHYDROXYPENTYL)-4a,5-DIHYDROISOALLOXAZINE see DUA600

7,8-DIMETHYL-10-(d-RIBO-2,3,4,5-TETRAHYDROXYPENTYL)ISOALLOXAZINE see RIK000

N,N-DIMETHYLSALICYLAMIDE see DUA800

DIMETHYL SALICYLATE see MLH800

DIMETHYL SELENATE see DUB000

DIMETHYL SELENIDE see DUB200

DIMETHYLSELENIUM see DUB200

N,N-DIMETHYLSEROTONIN see DPG109

DIMETHYLSILAZANE TRIMER see HEC600

DIMETHYLSILBOESTROL see DUC300

DIMETHYL SILICONE see DTR850

DIMETHYL SILOXANE see DUB600

DIMETHYLSILOXANE PENTAMER see DAF350

DIMETHYLSILYL ETHER see TDP775

(DIMETHYL SILYLMETHYL)TRIMETHYL LEAD see DUB689

N,N-DIMETHYLSTEARAMIDE see DTC200

DIMETHYLSTEARAMINE see DTC400

N,N-DIMETHYL-4-STILBENAMINE see DUB800

(E)-N,N-DIMETHYL-4-STILBENAMINE see DUC000

(Z)-N,N-DIMETHYL-4-STILBENAMINE see DUC200

cis-N,N-DIMETHYL-4-STILBENAMINE see DUC200

trans-N,N-DIMETHYL-4-STILBENAMINE see DUC000

(E)-α,α'-DIMETHYL-4,4'-STILBENEDIOL see DUC300

(E)-α,α'-DIMETHYL-4,4'-STILBENEDIOL DIACETATE (ester) see DXS300

DIMETHYLSTILBESTROL see DUC300

trans-DIMETHYLSTILBESTROL DIACETATE see DXS300

trans-DIMETHYLSTILBOESTROL DIACETATE see DXS300

3,4-DIMETHYL STYRENE OXIDE see EBT000

N,N-DIMETHYL-p-STYRYLANILINE see DUB800

DIMETHYL SUCCINATE see SNB100

DIMETHYLSULFAAT (DUTCH) see DUD100

DIMETHYLSULFAMIC ACID 5-BUTYL-2-(ETHYLAMINO)-6-METHYL-4-PYRIMIDINYL ESTER see BRJ000

N,N-DIMETHYLSULFAMID see DUD900

N,N'-DIMETHYLSULFAMIDE see DUD900

DIMETHYLSULFAMIDO-3-(DIMETHYLAMINO-2-PROPYL)-10-PHENOTHIAZINE see DUC400

2-(DIMETHYLSULFAMOYL)-(9-(4-METHYL-1-PIPERAZINYL)PROPYLIDENE)THIOXANTHENE see NBP500

p-(N,N-DIMETHYLSULFAMOYL)PHENOL see DUC600

O,O-DIMETHYL O-p-SULFAMOYLPHENYL PHOSPHOROTHIOATE see CQL250

3,4-DIMETHYL-5-SULFANILAMIDOISOXAZOLE see SNN500

4,5-DIMETHYL-2-SULFANILAMIDOOXAZOLE see AIE750

2,4-DIMETHYL-6-SULFANILAMIDOPYRIMIDINE see SNJ350

2,6-DIMETHYL-4-SULFANILAMIDOPYRIMIDINE see SNJ350

4,6-DIMETHYL-2-SULFANILAMIDOPYRIMIDINE see SNJ000

N,N-DIMETHYLSULFANILIC ACID see DUD000

DIMETHYLSULFAT (CZECH) see DUD100

DIMETHYL SULFATE see DUD100

DIMETHYLSULFID (CZECH) see TFP000

DIMETHYL SULFIDE (DOT) see TFP000

DI-METHYLSULFIDE BORANE see MPL250

DIMETHYLSULFIDE-α,α'-DICARBOXYLIC ACID see MCM750

2-(7-(1,1-DIMETHYL-3-(4-SULFOBUTYL)BENZ(e)INDOLIN-2-YLIDENE)-1,3,5-HEPTATRIENYL)-1,1-DIMETHYL-3-(4-SULFOBUTYL)1H-BENZ(e)INDOLIUM IODIDE, INNER SALT, SODIUM SALT see ICL000

2,4-DIMETHYL SULFOLANE see DUD400

3-DIMETHYLSULFONAMIDO-10-(2-DIMETHYLAMINOPROPYL)PHENOTHIAZINE see DUC400

1,4-DIMETHYLSULFONOXYBUTANE see BOT250

3-((2,4-DIMETHYL-5-SULFOPHENYL)AZO)-4-HYDROXY-1-NAPHTHALENESULFONIC ACID, DISODIUM SALT see FAG050

DIMETHYL SULFOXIDE see DUD800

1,3-DIMETHYLSULFURYLDIAMIDE see DUD900

3-DIMETHYLSULPHAMIDOPHENOTHIAZINE see DTK300

3,4-DIMETHYL-5-SULPHANILAMIDOISOXAZOLE see SNN500

as-DIMETHYL SULPHATE see MLH500

DIMETHYL SULPHIDE (ACGIH) see TFP000

3,4-DIMETHYL-5-SULPHONAMIDOISOXAZOLE see SNN500

3-((2,4-DIMETHYL-5-SULPHOPHENYL)AZO)-4-HYDROXY-1-NAPHTHALENESULPHONIC ACID, DISODIUM SALT see FAG050

DIMETHYL SULPHOXIDE see DUD800

DIMETHYL TEREPHTHALATE see DUE000

(2,6-DIMETHYL-4-TERTIARYBUTYL-3-HYDROXYPHENYL)METHYLIMIDAZOLINE HYDROCHLORIDE see AEX000

DIMETHYL-1,2,2,2-TETRACHLOROETHYL PHOSPHATE see DUE600

DIMETHYL TETRACHLOROTEREPHTHALATE see TBV250

DIMETHYL 2,3,5,6-TETRACHLOROTEREPHTHALATE see TBV250

N,N-DIMETHYLTETRADECANAMIDE see DSU200

N,N-DIMETHYL-N-TETRADECYLBENZENEMETHANAMINIUM, CHLORIDE (9CI) see TCA500

1,3-DIMETHYL-2-TETRADECYL-2-THIOPSEUDOUREA HYDRIODIDE see DUE700

2,2-DIMETHYL-3-(2,3,3,3-TETRAFLUORO-1-PROPENYL)CYCLOPROPANECARBOXYLIC ACID, (3-PHENOXYPHENYL)METHYL ESTER, (1-α,3-β(E))-(+−)- see DUE750

7,12-DIMETHYL-1,2,3,4-TETRAHYDROBENZ(a)ANTHRACENE see TCP600

7,12-DIMETHYL-8,9,10,11-TETRAHYDROBENZ(a)ANTHRACENE see DUF000

1,11-DIMETHYL-1,2,3,4-TETRAHYDROCHRYSENE see DUF200

1,1-DIMETHYL-3-(α,α,α-TRIFLUORO-m-TOLYL) UREA see DUK800

9-cis-3,7-DIMETHYL-9-(2,6,6-TRIMETHYL-1-CYCLOHEXEN-1-YL)-2,4,6,8-NONATETRAENAL see VSK975

3,7-DIMETHYL-9-(2,6,6-TRIMETHYL-1-CYCLOHEXEN)-1-YL-2,4,6,8-NONATETRAENOIC ACID see VSK950

3,7-DIMETHYL-9-(2,6,6-TRIMETHYL-1-CYCLOHEXEN-1-YL)-2,4,6,8-NONATETRAEN-1-OL see VSK600

trans-3,7-DIMETHYL-9-(2,6,6-TRIMETHYL-1-CYCLOHEXEN-1-YL)-2,4,6-NONATRIEN OIC ACID see DMD100

trans-3,7-DIMETHYL-9-(2,6,6-TRIMETHYL-1-CYCLOHEXEN-1-YL)-7-YNE-2,4,6-NON ATRIENOIC ACID see DHA200

2,2-DIMETHYLTRIMETHYLENE ACRYLATE see DUL200

2,2-DIMETHYLTRIMETHYLENE ESTER ACRYLIC ACID see DUL200

DIMETHYLTRIMETHYLENE GLYCOL see DTG400

3,3-DIMETHYLTRIMETHYLENE OXIDE see EBQ500

β,β-DIMETHYLTRIMETHYLENE OXIDE see EBQ500

N,N-DIMETHYL-4-(3,4,5-TRIMETHYLPHENYL)AZOANILINE see DUL400

N,N-DIMETHYL-4-((3,4,5-TRIMETHYLPHENYL)AZO)BENZENAMINE see DUL400

1,2-DIMETHYL-2-TRIMETHYLSILYLHYDRAZINE see DUL500

DIMETHYLTRIMETHYLSILYLPHOSPHINE see DUL550

1,4-DIMETHYL-2,3,7-TRIOXABICYCLO[2.2.1]HEPT-5-ENE see DUL589

(3,5-DIMETHYL-1,2,4-TRIOXOLANE) see BOX825

DIMETHYL (1,2,4-TRIS(METHOXYCARBONYL)-2-BUTYL)PHOSPHONATE see PHA300

N,N-DIMETHYL-1,2,3-TRITHIAN-5-AMINE, ETHANEDIOATE (1:1) see TFH750

N,N-DIMETHYL-1,2,3-TRITHIAN-5-AMINE HYDROGENOXALATE see TFH750

N,N-DIMETHYL-1,2,3-TRITHIAN-5-YLAMMONIUM HYDROGEN OXALATE see TFH750

N,N-DIMETHYLTRYPTAMINE see DPF600

DIMETHYL TUBOCURARINE see DUL800

o,o-DIMETHYLTUBOCURARINE see DUL800

o,o'-DIMETHYLTUBOCURARINE see DUL800

DIMETHYL TUBOCURARINE IODIDE see DUM000

α,3-DIMETHYLTYROSINE METHYL ESTER HYDROCHLORIDE see DUM100

6,10-DIMETHYL-UNDECA-5,9-DIEN-2-ONE see GDE400

2,6-DIMETHYLUNDECA-2,6,8-TRIENE-10-ONE see POH525

6,10-DIMETHYL-3,5,9-UNDECATRIEN-2-ONE see POH525

1,1-DIMETHYLUREA see DUM150

1,3-DIMETHYLUREA see DUM200

N,N'-DIMETHYLUREA see DUM200

sym-DIMETHYLUREA see DUM200

DIMETHYLUREA and SODIUM NITRITE see DUM400

p-N,N-DIMETHYLUREIDOAZOBENZENE see DUM600

m-(3,3-DIMETHYLUREIDO)PHENYL-tert-BUTYL CARBAMATE see DUM800

N,N-DIMETHYLVALERAMIDE see DUN200

6,8-o-DIMETHYLVERSICOLORIN A see DUN300

6,8-o-DIMETHYLVERSICOLORIN B see DUN310

β,β-DIMETHYLVINYL CHLORIDE see IKE000

DIMETHYLVINYLETHINYL-p-HYDROXYPHENYLMETHANE see DUN400

DIMETHYL(VINYL)ETHYNYLCARBINOL see MKM300

α-(2,2-DIMETHYLVINYL)-α-ETHYNYL-p-CRESOL see DUN400

1,5-DIMETHYL-1-VINYL-4-HEXEN-1-OL BENZOATE see LFZ000

1,5-DIMETHYL-1-VINYL-4-HEXEN-1-OL CINNAMATE see LGA000

1,5-DIMETHYL-1-VINYL-4-HEXEN-1-YL-o-AMINOBENZOATE see APJ000

1,5-DIMETHYL-1-VINYL-4-HEXEN-1-YL BENZOATE see LFZ000

1,5-DIMETHYL-1-VINYL-4-HEXEN-1-YL CINNAMATE see LGA000

1,5-DIMETHYL-1-VINYL-4-HEXENYL ESTER, ISOBUTYRIC ACID see LGB000

DIMETHYL VIOLOGEN see PAI990

DIMETHYL VIOLOGEN CHLORIDE see PAJ000

DIMETHYL XANTHIC DISULFIDE see DUN600

1,3-DIMETHYLXANTHINE see TEP000

1,7-DIMETHYLXANTHINE see PAK300

3,7-DIMETHYLXANTHINE see TEO500

3-((1,3-DIMETHYLXANTHIN-7-YL)METHYL)-5-METHYL-1,2,4-OXADIAZOLE see CDM575

DIMETHYLXANTHOGEN DISULFIDE see DUN600

3,3-DIMETHYL-1-XENYL-TRIAZENE see BGL500

N,N-DIMETHYL-p-(2,3,XYLYLAZO)ANILINE see DUN800

N,N-DIMETHYL-p-(3,4-XYLYLAZO)ANILINE see DUO000

2,2-DIMETHYL-5-(2,5-XYLYLOXY)VALERIC ACID see GCK300

DIMETHYL YELLOW see DOT300

DIMETHYL YELLOW-N,N-DIMETHYLANILINE see DOT300

DIMETHYLZINC see DUO200

DIMETHYLZINN-S,S'-BIS(ISOOCTYLTHIOGLYCOLAT) (GERMAN) see BKK500

DIMETHYOXYDOPAMINE see DOE200

2,6-DIMETHYPYRIDINE see LJA010

10,11-DIMETHYSTRYCHNINE see BOL750

5-(DIMETILAMINOETILOSIMINO-5H-DIBENZO(a,d)CICLOEPTA-1,4-DIENE) CLORIDRATO (ITALIAN) see DPH600

(4-DIMETILAMINO-3-METIL-FENIL)-N-METIL-CARBAMMATO (ITALIAN) see DOR400

5-(3-DIMETILAMINOPROPILIDEN)-5H-DIBENZO-(a,d)-CICLOPENTENE (ITALIAN) see PMH600

N-(γ-DIMETILAMINOPROPIL)-IMINODIBENZILE CLORIDRATO (ITALIAN) see DLH630

9-(3-DIMETILAMINOPROPYLIDEN)-10,10-DIMETIL-9,10-DIIDROANTHRACENE (ITALIAN) see AEG875

DIMETILAN see DQZ000

DIMETILANE see DQZ000

2,5-DIMETILBENZOCHINONE (1:4) (ITALIAN) see XQJ000

O,O-DIMETIL-O-(1,4-DIMETIL-3-OXO-4-AZA-PENT-1-ENIL)-FOSFATO (ITALIAN) see DGQ875

2,6-DIMETIL-EPTAN-4-ONE (ITALIAN) see DNI800

O,O-DIMETIL-S-(2-ETILTIO-ETIL)-MONOTIOFOSFATO (ITALIAN) see DAP400

O,O-DIMETIL-S-(2-ETIL-SOLFINIL-ETIL)-MONOTIOFOSFATO (ITALIAN) see DAP000

O,O-DIMETIL-S-(ETILTIO-ETIL)-DITIOFOSFATO (ITALIAN) see PHI500

O,O-DIMETIL-O-(2-ETILTIO-ETIL)-MONOTIOFOSFATO (ITALIAN) see DAO800

2,6-DIMETILFENILICO DELL'ACIDO α-N-METILPIPERAZINOBUTIRRICO IDOCLORIDRAT (ITALIAN) see FAC130

DIMETILFORMAMIDE (ITALIAN) see DSB000

O,O-DIMETIL-S-(N-FORMIL-N-METIL-CARBAMOIL-METIL)-DITIOFOSFATO (ITALIAN) see DRR200

O,O-DIMETIL-S-(N-METIL-CARBAMOIL-METIL)-DITIOFOSFATO (ITALIAN) see DSP400

O,O-DIMETIL-S-(N-METIL-CARBAMOIL)-METIL-MONOTIOFOSFATO (ITALIAN) see DNX800

O,O-DIMETIL-O-(2-N-METILCARBAMOIL-1-METIL-VINIL)-FOSFATO (ITALIAN) see MRH209

O,O-DIMETIL-O-(3-METIL-4-METILTIO-FENIL)-MONOTIOFOSFATO (ITALIAN) see FAQ900

O,O-DIMETIL-O-(3-METIL-4-NITRO-FENIL)-MONOTIOFOSFATO (ITALIAN) see DSQ000

O,O-DIMETIL-S-((2-METOSSI-1,3,4-(4H)-TIADIZAOL-5-ON-4-IL)-METIL)-DITIFOSFATO (ITALIAN) see DSO000

O,O-DIMETIL-S-((MORFOLINO-CARBONIL)-METIL)-DITIOFOSFATO (ITALIAN) see MRU250

O,O-DIMETIL-O-(4-NITRO-FENIL)-MONOTIOFOSFATO (ITALIAN) see MNH000

O,O-DIMETIL-S-((4-OXO-3H-1,2,3-BENZOTRIAZIN-3-IL)-METIL)-DITIOFOSFATO (ITALIAN) see ASH500

(5,5-DIMETIL-3-OXO-CICLOES-1-EN-IL)-N,N-DIMETIL-CARBAMMATO (ITALIAN) see DRL200

3,5-DIMETIL-PERIDRO-1,3,5-THIADIAZIN-2-TIONE (ITALIAN) see DSB200

DIMETILSOLFATO (ITALIAN) see DUD100

DIMETIL-m-TOLUIDINA see TLG700

DIMETIL-p-TOLUIDINA see TLG150

O,O-DIMETIL-(2,2,2-TRICLORO-1-IDROSSI-ETIL)-FOSFONATO (ITALIAN) see TIQ250

DIMETINA see BEM500

DIMETINDENE MALEATE see FMU409

DIMETOL see FON200

DIMETON see DSP400

3,3'-DIMETOSSIBENZODINA (ITALIAN) see DCJ200

DIMETOX see TIQ250

DIMETPRAMIDE see DUO300

DIMETRIDAZOLE see DSV800

DIMETRIN see DQQ500

DIMETYLFORMAMIDU (CZECH) see DSB000

O,O-DIMETYL-O-p-NITROFENYLFOSFAT (CZECH) see PHD500

DIMEVAMIDE see DOY400

DIMEVUR see DSP400

DIMEX see MLC100

DIMEXAN see DUN600

DIMEXANO see DUN600

DIMEXIDE see DUD800

DIMEZATHINE see SNJ000

DIMEZOL 14 see BHQ300

DIMIC see BIM775

DIMID see DRP800

DIMIDIN see DUO350

DIMILIN see CJV250

DIMIPRESSIN see DLH600

DIMIPRESSIN see DLH630

DIMITAN see BIE500

DIMITE see BIN000

DIMITRON see CMR100

DIMITRONAL see CMR100

DIMO see DLW600
DIMONOCLOROACETILAJMALINA
CLORIDRATO (ITALIAN) see AFH275
DIMORLIN see AFJ400
DIMORPHOLAMINE see DUO400
DIMORPHOLINE DISULFIDE see BKU500
DIMORPHOLINETHIURAM DISULFIDE see
MRR090
DIMORPHOLINIUM
HEXACHLOROSTANNATE see DUO500
DIMORPHOLINO DISULFIDE see BKU500
1,5-DIMORPHOLINO-3-(1-NAPHTHYL)-
PENTANE see DUO600
DIMORPHOLINOPHOSPHINIC ACID
PHENYL ESTER see DUO700
DIMP see DNQ875
DIMPEA see DOE200
DIMPYLATE see DCM750
DIM-SA see DNV800
DI-MU-
CARBONYLHEXACARBONYLDICOBALT
see CNB500
DIMYRCETOL see DUO800
DIN 1.4876 see IGL100
DIN 2.4602 see CNA750
DIN 2.4964 see CNA750
DINA see NFW000
DINACORYL see DJS200
DINACRIN see ILD000
DINAPACRYL see BGB500
DINAPHTAZIN (GERMAN) see DUP000
3,4,5,6-DINAPHTHACARBAZOLE see
DCY000
1,2,5,6-DINAPHTHACRIDINE see DCS400
3,4,6,7-DINAPHTHACRIDINE see DCS600
DINAPHTHAZINE see DUP000
5H-DINAPHTHO(2,3-A:2',3'-I)CARBAZOLE-
5,10,15,17(16H)-TETRONE, 4,9-
DIBENZAMIDO-(7CI) see CMU780
16H-DINAPHTHO(2,3-a:2',3'-i)CARBAZOLE-
5,10,15,17-TETRAONE, 6,9-DIBENZAMIDO-
see DUP100
16H-DINAPHTHO(2,3-a:2',3'-i)CARBAZOLE-
5,10,15,17-TETRAONE, 6,9-DIBENZAMIDO-
1-METHOXY- see CMU800
2,2'-DINAPHTHOL see BGC100
DINAPHTHO(2,3-a:2',3'-
i)NAPHTH(2',3':6,7)INDOLO(2,3-
c)CARBAZOLE-5,10,15,17,22,24-HEXAONE,
16,23-DIHYDRO- see CMU770
DINAPHTHO(1,2,3-cd:3',2',1'-lm)PERYLENE-
5,10-DIONE see DCU800
DI-(1-NAPHTHOYL)PEROXIDE see DUP200
DI-β-NAPHTHYLDIIMIDE see ASN750
DI-β-NAPHTHYL-p-PHENYLDIAMINE see
NBL000
DI-β-NAPHTHYL-p-PHENYLENEDIAMINE
see NBL000
N,N'-DI-β-NAPHTHYL-p-
PHENYLENEDIAMINE see NBL000
sym-DI-β-NAPHTHYL-p-
PHENYLENEDIAMINE see NBL000
N,N'-DI(α-(1-NAPHTHYL)PROPIONYLOXY-
2-ETHYL)PIPERAZINE
DIHYDROCHLORIDE see NAD000
DINARKON see DLX400
DINATE see DXE600
DINATRIUM-
AETHYLENBISDITHIOCARBAMAT
(GERMAN) see DXD200
DINATRIUM-(N,N'-AETHYLEN-
BIS(DITHIOCARBAMAT)) (GERMAN) see
DXD200
DINATRIUM-(3,6-EPOXY-CYCLOHEXAAN-
1,2-DICARBOXYLAAT) (DUTCH) see
DXD000
DINATRIUM-(3,6-EPOXY-CYCLOHEXAN-
1,2-DICARBOXYLAT) (GERMAN) see
DXD000
DINATRIUM-(N,N'-ETHYLEEN-
BIS(DITHIOCARBAMAAT)) (DUTCH) see
DXD200

DINATRIUMPYROPHOSPHAT (GERMAN)
see DXF800
DINDEVAN see PFJ750
DINEODYMIUM TRIOXIDE see NCC000
DINEVAL see PFJ750
DINEX see CPK500
DINEZIN see DII200
DINGSABLCH, LEAF EXTRACT see KCA100
DINICKEL TRIOXIDE see NDH500
DINICONAZOLE M see THS920
DINIL see PFA860
DINILE see SNE000
DINIOBIUM PENTAOXIDE see NEA050
DINIOBIUM PENTOXIDE see NEA050
DINITOLMID see DUP300
DINITOLMIDE see DUP300
DINITRAMINE see CNE500
2,4-DINITRANILINE see DUP600
DINITRANILINE ORANGE see DVB800
DINITRATE de DIETHYLENE-GLYCOL
(FRENCH) see DJE400
1,3-DINITRATO-2,2-
BIS(NITRATOMETHYL)PROPANE see
PBC250
DINITRATODIOXOURANIUM,
HEXAHYDRATE see URS000
2,2'-DINITRATO-N-NITRODI-
ETHYLAMINE see NFW000
DINITRILE of ISOPHTHALIC ACID see
PHX550
2,3-DINITRILO-1,4-DITHIA-
ANTHRAQUINONE see DLK200
2,3-DINITRILO-1,4-
DITHIOANTHRACHINON (GERMAN) see
DLK200
DINITRO see BRE500
DINITRO-3 see BRE500
DINITROAMINE see CNE500
4,6-DINITRO-2-AMINOPHENOL see
DUP400
2,4-DINITROANILIN (GERMAN) see
DUP600
2,4-DINITROANILINA (ITALIAN) see
DUP600
2,4-DINITROANILINE see DUP600
DINITROANILINE ORANGE ND-204 see
DVB800
DINITROANILINE RED see DVB800
2-(2,4-DINITROANILINO)ETHANOL see
DVC300
2,4-DINITROANISOL see DUP800
2,4-DINITROANISOLE see DUP800
α-DINITROANISOLE see DUP800
15,18-DINITROANTHRA(9,1,2-
cde)BENZO(rst)PENTAPHENE-5,10-DIONE
see DUP830
1,5-DINITRO-9,10-ANTHRACENEDIONE
see DUQ000
1,5-DINITROANTHRACHINON (CZECH)
see DUQ000
1,5-DINITROANTHRAQUINONE see
DUQ000
4,4'-DINITRO-2,2'-AZOXYTOLUENE see
BKS813
3,5-DINITROBENZAMIDE see DUQ150
2,4-DINITROBENZENAMIME see DUP600
DINITROBENZENE see DUQ180
m-DINITROBENZENE see DUQ200
o-DINITROBENZENE see DUQ400
p-DINITROBENZENE see DUQ600
1,2-DINITROBENZENE see DUQ400
1,3-DINITROBENZENE see DUQ200
2,4-DINITROBENZENE see DUQ200
DINITROBENZENE, solution (DOT) see
DUQ180
4,6-DINITROBENZENEDIAZONIUM-2-
OXIDE see DUQ800
2,4-DINITROBENZENESULFENYL
CHLORIDE see DUR200
2,4-DINITROBENZENESULFONIC ACID
see DUR400
2,4-DINITROBENZENETHIOL see DUR425

4,6-DINITROBENZOFURAZAN-N-OXIDE
see DUR500
3,4-DINITROBENZOIC ACID see DUR550
3,5-DINITROBENZOIC ACID see DUR600
3,5-DINITROBENZOIC ACID CHLORIDE
see DUR850
1,3-DINITROBENZOL see DUQ200
DINITROBENZOL, solid (DOT) see DUQ180
5,7-DINITRO-1,2,3-BENZOXADIAZOLE see
DUR800
3,5-DINITROBENZOYL CHLORIDE see
DUR850
4,4'-DINITROBIFENYL (CZECH) see DUS000
4,4'-DINITROBIPHENYL see DUS000
4,4'-DINITRO-2-BIPHENYLAMINE see
DUS100
4,4'-DINITRO-(1,1'-BIPHENYL)-2-AMINE see
DUS100
3,5-DINITRO-N,N'-BIS(2,4,6-
TRINITROPHENYL)-2,6-
PYRIDINEDIAMINE see PQC525
2,4-DINITRO-6-BROMANILIN (CZECH) see
DUS200
2,4-DINITRO-6-BROMOANILINE see
DUS200
2,4-DINITROBROMOBENZENE see DVB820
2,3-DINITRO-2-BUTENE see DUS400
4,6-DINITRO-2-sec.BUTYLFENOL (CZECH)
see BRE500
4,6-DINITRO-2-sec.BUTYLFENOLATE
AMMONY (CZECH) see BPG250
2,4-DINITRO-6-sec-BUTYLFENYLESTER
KYSELINY OCTOVE (CZECH) see ACE500
2,4-DINITRO-6-sek.BUTYL-
ISOPROPYLPHENYLCARBONAT
(GERMAN) see CBW000
DINITROBUTYLPHENOL see BRE500
2,4-DINITRO-6-sec-BUTYLPHENOL see
BRE500
4,6-DINITRO-2-sec-BUTYLPHENOL see
BRE500
4,6-DINITRO-o-sec-BUTYLPHENOL see
BRE500
2,4-DINITRO-6-tert-BUTYLPHENOL see
DRV200
4,6-DINITRO-2-sec-BUTYLPHENOL
AMMONIUM SALT see BPG250
4,6-DINITRO-o-sec-BUTYLPHENOL
AMMONIUM SALT see BPG250
DINITROBUTYLPHENOL-2,2',2"-
NITRILOTRIETHANOL SALT see BRE750
2,4-DINITRO-6-sek.BUTYL-
PHENYLACETAT (GERMAN) see ACE500
4,6-DINITRO-2-s-BUTYLPHENYL
ACETATE see ACE500
4,6-DINITRO-2-sec-BUTYLPHENYL β,β-
DIMETHYLACRYLATE see BGB500
2,4-DINITRO-6-sec-BUTYLPHENYL
ISOPROPYL CARBONATE see CBW000
2,4-DINITRO-6-tert-BUTYLPHENYL
METHANESULFONATE see DUS500
2,4-DINITRO-6-sec-BUTYLPHENYL-2-
METHYLCROTONATE see BGB500
4,6-DINITRO-2-CAPRYLPHENYL
CROTONATE see AQT500
4,6-DINITRO-2-(2-CAPRYL)PHENYL
CROTONATE see AQT500
2,4-DINITROCHLORBENZEN-6-
SULFONAN SODNY (CZECH) see CJC000
2,6-DINITRO-4-CHLOROANILINE see
CGL500
DINITROCHLOROBENZENE see CGL750
2,4-DINITROCHLOROBENZENE see
CGM000
1,3-DINITRO-4-CHLOROBENZENE see
CGM000
2,4-DINITRO-1-CHLOROBENZENE see
CGM000
DINITROCHLOROBENZENE (DOT) see
CGL750
DINITROCHLOROBENZOL see CGM000

2,4-DINITRO-1-CHLORO-NAPHTHALENE see DUS600

3,5-DINITRO-4-CHLORO-α-α-α-TRIFLUOROTOLUENE see CGM225

4,5-DINITROCHRYSAZIN see DMN400

DINITROCRESOL see DUS700

DINITRO-o-CRESOL see DUS700

DINITRO-p-CRESOL see DUT600

2,4-DINITRO-o-CRESOL see DUS700

2,6-DINITRO-p-CRESOL see DUT600

3,5-DINITRO-o-CRESOL see DUT000

3,5-DINITRO-p-CRESOL see DUT200

4,6-DINITRO-m-CRESOL see DUT700

4,6-DINITRO-o-CRESOL see DUS700

4,6-DINITRO-o-CRESOL AMMONIUM SALT see DUT800

4,6-DINITRO-o-CRESOL BARIUM DERIVATIVE see DUT900

4,6-DINITRO-o-CRESOL DIETHYLAMINE SALT see DUU000

4,6-DINITRO-o-CRESOL METHYLAMINE (1:1) see DUU200

4,6-DINITRO-o-CRESOL MORPHOLINE (1:1) see DUU400

4,6-DINITRO-o-CRESOLO (ITALIAN) see DUS700

DINITRO-o-CRESOL SODIUM SALT see DUU600

DINITRO-o-CRESOL SODIUM SALT see SGP550

3,5-DINITRO-o-CRESOL SODIUM SALT see DUU600

4,6-DINITRO-o-CRESOL SODIUM SALT see DUU600

1,1-DINITROCYCLOHEXANE see DUU700

DINITROCYCLOHEXYLPHENOL see CPK500

DINITRO-o-CYCLOHEXYLPHENOL see CPK500

2,4-DINITRO-6-CYCLOHEXYLPHENOL see CPK500

4,6-DINITRO-o-CYCLOHEXYLPHENOL see CPK500

DINITROCYCLOHEXYLPHENOL (DOT) see CPK500

DINITROCYCLOHEXYLPHENOL, DICYCLOHEXYLAMINE SALT see CPK550

DINITRO-o-CYCLOHEXYLPHENOL (DICYCLOHEXYLAMINE SALT) see CPK550

4,6-DINITRO-o-CYCLOHEXYLPHENOL, DICYCLOHEXYLAMINE SALT see CPK550

DINITROCYCLOPENTYLPHENOL see CQB250

DINITRODENDTROXAL see DUS700

N,N'-DINITRO-1,2-DIAMINOETHANE see DUU800

DINITRODIAZOMETHANE see DUV000

2,7-DINITRODIBENZO-p-DIOXIN see DUV020

2,8-DINITRODIBENZO-p-DIOXIN see DUV030

2,7-DINITRODIBENZO(b,e)(1,4)DIOXIN see DUV020

2,8-DINITRODIBENZO(b,e)(1,4)DIOXIN see DUV030

3,3'-DINITRO-4,4'-DIFLUORODIPHENYL SULFONE see DKH250

DINITRODIGLICOL (ITALIAN) see DJE400

DINITRODIGLYKOL (CZECH) see DJE400

2,7-DINITRO-4,5-DIHYDROPYRENE see DLJ100

5,6-DINITRO-2-DIMETHYLAMINOPYRIMIDINONE see DUV089

2,4-DINITRODIPHENYLAMINE see DUV100

4,4'-DINITRODIPHENYL DISULFIDE see BLA500

p,p'-DINITRODIPHENYL DISULFIDE see BLA500

4,4'-DINITRODIPHENYL ETHER see NIM600

p,p'-DINITRODIPHENYL ETHER see NIM600

4,4'-DINITRODIPHENYL OXIDE see NIM600

3',5'-DINITRO-4'-(DI-n-PROPYLAMINO)ACETOPHENONE see DUV400

3,5-DINITRO-N4,N4-DIPROPYLSULFANILAMIDE see OJY100

2,6-DINITRO-N,N-DIPROPYL-4-(TRIFLUOROMETHYL)BENZENAMINE see DUV600

2,6-DINITRO-N,N-DI-N-PROPYL-α,α,α-TRIFLURO-p-TOLUIDINE see DUV600

2,2'-DINITRO-5,5'-DITHIODIBENZOESAEURE see DUV700

2,2'-DINITRO-5,5'-DITHIODIBENZOIC ACID see DUV700

1,1-DINITROETHANE see DUV710

1,2-DINITROETHANE see DUV720

1,1-DINITROETHANE (dry) (DOT) see DUV710

N,N-DINITROETHANEDIAMINE see DUU800

N,N'-DINITROETHYLENEDIAMINE see DUU800

N,N'-DINITROETHYLENEDIAMINE see DUV800

2,5-DINITRO-N-(1-ETHYLPROPYL)-3,4-XYLIDINE see DRN200

2,3-DINITROFENOL see DUY900

2,6-DINITROFENOL see DVA200

3,4-DINITROFENOL see DVA400

2,4-DINITROFENOL (DUTCH) see DUZ000

DINITROFENOLO (ITALIAN) see DUZ000

2,4-DINITROFENYLHYDRAZIN (CZECH) see DVC400

3,7-DINITROFLUORANTHENE see DUW100

3,9-DINITROFLUORANTHENE see DUW120

4,12-DINITROFLUORANTHENE see DUW120

2,7-DINITROFLUORENE see DUW200

2,7-DINITROFLUORENONE see DUW300

2,7-DINITRO-9-FLUORENONE see DUW300

2,4-DINITROFLUOROBENZENE see DUW400

2,4-DINITRO-1-FLUOROBENZENE see DUW400

2,2-DINITRO-2-FLUOROETHANOL see FHW000

DINITROGEN MONOXIDE see NGU000

DINITROGEN TETRAFLUORIDE see TCI000

DINITROGEN TETROXIDE see NGU500

DINITROGEN TETROXIDE, liquefied (DOT) see NGU500

DINITROGLICOL (ITALIAN) see EJG000

1,3-DINITROGLYCERIN see GGA200

DINITROGLYCOL see EJG000

2,4-DINITRO-p-HYDROXYDIPHENYLAMINE see DUW500

3,5-DINITRO-2-HYDROXYTOLUENE see DUS700

1,3-DINITRO-2-IMIDAZOLIDINONE see DUW503

1,3-DINITRO-2-IMIDAZOLIDONE see DUW503

2,4-DINITRO-6-ISOBROPYL-m-CRESOL see DVG200

2,6-DINITRO-4-ISOPROPYLPHENOL see IOX000

4,6-DINITROKRESOL (DUTCH) see DUS700

4,6-DINITRO-o-KRESOL (CZECH) see DUS700

4,6-DINITRO-o-KRESYLESTER KYSELINY OCTOVE (CZECH) see AAU250

DINITROL see DUS700

2,4-DINITROMESITYLENE see DUW505

DINITROMETHANE see DUW507

2,6-DINITRO-3-METHOXY-4-tert-BUTYLTOLUENE see BRU500

3,3'-DINITRO-4'-METHOXYFLAVONE see MFB430

2,6-DINITRO-4-METHYLANILINE see DVI100

3,5-DINITRO-2-METHYLBENZENEDIZAONIUM-4-OXIDE see DUX509

2,5-DINITRO-3-METHYLBENZOIC ACID see DUX560

2,4-DINITRO-3-METHYL-6-tert-BUTYLPHENYLACETAT (GERMAN) see BRU750

2,4-DINITRO-3-METHYL-6-tert-BUTYLPHENYL ACETATE see BRU750

DINITROMETHYL CYCLOHEXYLTRIENOL see DUS700

N,N'-DINITRO-N-METHYL-1,2-DIAMINOETHANE see DUX600

DINITRO(1-METHYLHEPTYL)PHENYL CROTONATE see AQT500

2,4-DINITRO-6-(1-METHYLHEPTYL)PHENYL CROTONATE see AQT500

2,4-DINITRO-6-METHYLPHENOL see DUS700

2,4-DINITRO-6-METHYLPHENOL SODIUM SALT see DUU600

4,6-DINITRO-2-(1-METHYL-N-PROPYL)PHENOL see BRE500

2,4-DINITRO-6-(1-METHYL-PROPYL)PHENOL (FRENCH) see BRE500

4,6-DINITRO-N-METHYL-N-(2,4,6-TRIBROMOPHENYL)-α-α-α-TRIFLUOR O-o-TOLUIDINE see BMO300

2,4-DINITRO-1-NAFTOL see DUX800

1,3-DINITRONAPHTHALENE see DUX650

1,5-DINITRONAPHTHALENE see DUX700

1,5-DINITRONAPHTHALENE see DUX700

1,8-DINITRONAPHTHALENE see DUX710

2,4-DINITRO-1-NAPHTHOL see DUX800

2,4-DINITRONAPHTHOLSULFONIC ACID see FBZ100

2-4 DINITRO-α-NAPHTOL see DUX800

2-4 DINITRO-α-NAPHTOL (FRENCH) see DUX800

2,4-DINITRO-6-(2-OCTYL)PHENYL CROTONATE see AQT500

2,6-DINITRO-4-PERCHLORYLPHENOL see DUY200

1,6-DINITROPHENANTHRENE see DUY225

2,6-DINITROPHENANTHRENE see DUY230

2,7-DINITROPHENANTHRENE see DUY232

3,5-DINITROPHENANTHRENE see DUY235

3,6-DINITROPHENANTHRENE see DUY240

2,10-DINITROPHENANTHRENE see DUY245

3,10-DINITROPHENANTHRENE see DUY250

1,7-DINITROPHENAZINE see DUY300

2,4-DINITROPHENETOLE see DUY400

DINITROPHENOL see DUY600

DINITROPHENOL see DUY600

2,3-DINITROPHENOL see DUY900

2,3-DINITROPHENOL see DUY900

2,4-DINITROPHENOL see DUZ000

2,5-DINITROPHENOL see DVA000

2,6-DINITROPHENOL see DVA200

3,4-DINITROPHENOL see DVA400

3,5-DINITROPHENOL see DVA600

α-DINITROPHENOL see DUZ000

β-DINITROPHENOL see DVA200

γ-DINITROPHENOL see DVA000

2,4-DINITROPHENOL SODIUM SALT see DVA800

DINITROPHENOL SOLUTIONS (UN 1599) (DOT) see DUY600

DINITROPHENOL, dry or wetted with <15% water, by weight (UN 0076) (DOT) see DUY600

DINITROPHENOL, wetted with not <15% water, by weight (UN 1320) (DOT) see DUY600

2,4-DINITROPHENYLACETYL CHLORIDE see DVB200

5-((2,4-DINITROPHENYL)AMINO)-2-(PHENYLAMINO)BENZENESULFONIC ACID MONOSODIUM SALT see SGP500

1-((2,4-DINITROPHENYL)AZO)-2-NAPHTHOL see DVB800

2,4-DINITROPHENYL BROMIDE see DVB820

o,p-DINITROPHENYL BROMIDE see DVB820

4,6-DINITROPHENYL-2-sec-BUTYL-3-METHYL-2-BUTENONATE see BGB500

2,4-DINITROPHENYL-DIMETHYL-DITHIOCARBAMATE see DVB850

2,4-DINITROPHENYL-2,4-DINITRO-6-sec-BUTYLPHENYL CARBONATE see DVC200

DI-4-NITROPHENYL DISULFIDE see BLA500

N-2,4-DINITROPHENYLETHANOLAMINE see DVC300

DI-4-NITROPHENYL ETHER see NIM600

2,4-DINITROPHENYL ETHER of MORPHINE see DVC800

2,4-DINITROPHENYLHYDRAZINE see DVC400

2,4-DINITROPHENYLHYDRAZINIUMPERCHLORATE see DVC600

o-(2,4-DINITROPHENYL)HYDROXYLAMINE see DVC700

DINITROPHENYLMETHANE see DVG600

2,4-DINITROPHENYLMETHYL ETHER see DUP800

2,4-DINITROPHENYLMORPHINE HYDROCHLORIDE see DVC800

DI-p-NITROPHENYL PHOSPHATE see BLA600

2,4-DINITROPHENYL THIOCYANATE see DVF800

N-(2,4)-DINITROPHYLPEPTIDASE see GIA050

2,2-DINITRO-1,3-PROPANEDIOL see DVD000

2,2-DINITROPROPANOL see DVD200

2,2-DINITRO-1-PROPANOL see DVD200

DINITROPYRENE see DVD300

DINITROPYRENE see DVD400

DINITROPYRENE see DVD600

DINITROPYRENE see DVD800

1,3-DINITROPYRENE see DVD400

1,6-DINITROPYRENE see DVD600

1,8-DINITROPYRENE see DVD800

2,7-DINITROPYRENE see DVD900

4,6-DINITROQUINOLINE-1-OXIDE see DVE000

4,7-DINITROQUINOLINE-1-OXIDE see DVE200

2,4-DINITRORESORCINOL (heavy metal salts of) (dry) (DOT) see DVF300

2,4-DINITRO-RHODANBENZOL (GERMAN) see DVF800

3,5-DINITROSALICYLIC ACID see HKE600

p-DINITROSOBENZENE see DVE260

1,4-DINITROSOBENZENE see DVE260

1,4-DINITROSOBENZENE HOMOPOLYMER see PJR500

p-DINITROSOBENZENE POLYMERS see PJR500

DINITROSOCIMETIDINE see DVE300

N,N'-DINITROSO-N,N'-DIETHYLETHYLENEDIAMINE see DJB800

N,N'-DINITROSO-N,N'-DIMETHYLETHYLENEDIAMINE see DVE400

N,N'-DINITROSO-N,N'-DIMETHYLOXAMID (GERMAN) see DRN600

DINITROSO-2,5-DIMETHYLPIPERAZINE see DRN800

DINITROSO-2,6-DIMETHYLPIPERAZINE see DRO000

1,4-DINITROSO-2,6-DIMETHYLPIPERAZINE see DRO000

N,N'-DINITROSO-2,6-DIMETHYLPIPERAZINE see DRO000

DINITROSODIMETHYLPROPANEDIAMINE see DRO200

N,N'-DINITROSO-N,N'-DIMETHYL-1,3-PROPANEDIAMINE see DRO200

N,N'-DINITROSO-N,N'-DIMETHYLTEREPHTALSAUREAMID see DRO400

N,N'-DINITROSO-N,N'-DIMETHYLTEREPHTHALAMIDE, not >72% as a paste (DOT) see DRO400

DINITROSOHOMOPIPERAZINE see DVE600

N,4-DINITROSO-N-METHYLANILINE see MJG750

DINITROSOPENTAMETHYLENETETRAMINE see DVF400

N,N-DINITROSOPENTAMETHYLENETETRAMINE see DVF400

3,4-DI-N-NITROSOPENTAMETHYLENETETRAMINE see DVF400

3,7-DI-N-NITROSOPENTAMETHYLENETETRAMINE see DVF400

N1,N3-DINITROSOPENTAMETHYLENETETRAMINE see DVF400

DI(N-NITROSO)-PERHYDROPYRIMIDINE see DVF000

DINITROSOPIPERAZIN (GERMAN) see DVF200

DINITROSOPIPERAZINE see DVF200

1,4-DINITROSOPIPERAZINE see DVF200

N,N'-DINITROSOPIPERAZINE see DVF200

DINITROSOPRODECTIN see PPH100

N-DINITROSOPYRIDINOLCARBAMATE see PPH100

N,N'-DINITROSOPYRIDINOL CARBAMATE see PPH100

DINITROSORBIDE see CCK125

2,4-DINITROSO-m-RESORCINOL see DVF300

3,7-DINITROSO-1,3,5,7-TETRAAZABICYCLO[3.3.1]NONANE see DVF400

(E)-3,4'-DINITROSTILBENE see NHR620

(E)-4,4'-DINITROSTILBENE see EEE300

(E)-p,p'-DINITROSTILBENE see EEE300

DINITROSTILBENEDISULFONIC ACID see DVF600

4,4'-DINITRO-2,2'-STILBENEDISULFONIC ACID see DVF600

2,7-DINITRO-4,5,9,10-TETRAHYDROPYRENE see TCQ400

2,4-DINITROTHIOCYANATOBENZENE see DVF800

2,4-DINITROTHIOCYANOBENZENE see DVF800

2,4-DINITRO-1-THIOCYANOBENZENE see DVF800

2,4-DINITROTHIOPHENE see DVG000

2,6-DINITROTHYMOL see DVG200

DINITROTHYMOL 1-2-4 (FRENCH) see DVG200

3,5-DINITRO-o-TOLUAMIDE see DUP300

DINITROTOLUENE see DVG600

2,3-DINITROTOLUENE see DVG800

2,4-DINITROTOLUENE see DVH000

2,5-DINITROTOLUENE see DVH200

2,6-DINITROTOLUENE see DVH400

3,4-DINITROTOLUENE see DVH600

3,5-DINITROTOLUENE see DVH800

DINITROTOLUENES, molten (DOT) see PGN500

DINITROTOLUENES, liquid or solid (DOT) see DVG600

2,6-DINITRO-p-TOLUIDINE see DVI100

3,5-DINITRO-o-TOLUIDINE see AJR750

3,5-DINITRO-p-TOLUIDINE see MJG600

2,4-DINITROTOLUOL see DVH000

2,4-DINITRO-N-(2,4,6-TRIBROMOPHENYL)-6-(TRIFLUOROMETHYL)BENZENAMINE see DBA520

4,6-DINITRO-1,2,3-TRICHLOROBENZENE see DVI600

2,6-DINITRO-4-TRIFLUORMETHYL-N,N-DIPROPYLANILIN (GERMAN) see DUV600

2,4'-DINITRO-4-TRIFLUOROMETHYL-DIPHENYL ETHER see NIX000

2,4-DINITRO-1,3,5-TRIMETHYLBENZENE (DOT) see DUW505

sym-DINITROXYDIETHYLNITRAMINE see NFW000

α-α'-DI-(NITROXY)METHYL ETHER (DOT) see OPQ200

N-DINITROZO-PIRIDINOLKARBAMAT (HUNGARIAN) see PPH100

DINKUM OIL see EQQ000

DINOBUTON see CBW000

DINOC see DUS700

DINOC see DUU600

DINOCAP-MANCOZEB MIXTURE see MAO880

DINOCTON-6 see DVI800

DINOCTON-O see DVI800

DINOFEN see CBW000

DINOLEINE see DNF600

DINONYL-1,2-BENZENEDICARBOXYLATE see DVJ000

DI-n-NONYL PHTHALATE see DVJ000

DINOPHYSISTOXIN-1 see MND300

DINOPOL 235 see OEU000

DINOPOL NOP see DVL600

DINOPROST see POC500

DINOPROST METHYL ESTER see DVJ100

DINOPROSTONE see DVJ200

DINOPROST TROMETHAMINE (USDA) see POC750

18,19-DINORANDROSTA-1,3,5(10)-TRIEN-17-ONE, 3-METHOXY- see NNP100

18,19-DINOR-17-α-PREGN-4-EN-3-ONE, 13-ETHYL-17-HYDROXY- see NNE600

DINOSEB see BRE500

DINOSEB-ACETATE see ACE500

DINOSEB (AMINE) see BPG250

DINOSEBE (FRENCH) see BRE500

DINOSEBE ACETATE see ACE500

DINOSEB METHACRYLATE see BGB500

DINOSOL see SNN300

DINOTERB see DRV200

DINOTERB ACETATE see DVJ400

DINOVEX see DAL600

DINOXOL see DAA800

DINOXOL see TAA100

DINOZOL see DUT800

DINOZOL 50 see DUT800

DINP see PHW585

DINP2 see PHW585

DINP3 see PHW585

DINTOIN see DKQ000

DINTOINA see DNU000

DINULCID see OLM300

DINURANIA see DUS700

DINYL see PFA860

DIOCID see DVJ500

DIOCIDE see DVJ500

DIOCTADECYL THIODIPROPIONATE see DXG700

DIOCTADECYL 3,3'-THIODIPROPIONATE see DXG700

DIOCTANOL-2-PHTHALATE see BLB750

DIOCTANOYL PEROXIDE see CBF705

DIOCTLYN see DJL000

DIOCTYL ADIPATE see AEO000

DIOCTYLAL see DJL000

DIOCTYLAMINE see DVJ600

DIOCTYLAMINE, N-METHYL- see MND125

DIOCTYL AZELATE see BJQ500

DIOCTYL-o-BENZENEDICARBOXYLATE see DVL600
DIOCTYLBIS(LAUROYLOXY)STANNANE see DVJ800
DIOCTYLBIS(NONYLOXYMALEOYLOXY) STANNANE see DEI800
DIOCTYLDIDODECANOYLOXYSTANNANE see DVJ800
DIOCTYLDI(LAUROYLOXY)STANNANE see DVJ800
DIOCTYLDIMETHYLAMMONIUM CHLORIDE see DTF820
2,2-DIOCTYL-1,3-DIOXA-2-STANNA-7-THIADECAN-4,10-DIONE see DVN909
2,2-DIOCTYL-1,3,2-DIOXASTANNEPIN-4,7-DIONE see DVK200
4,4'-DIOCTYLDIPHENYLAMINE see DVK400
DIOCTYLDIPHENYLSTANNANE see DVK500
DIOCTYLDIPHENYLTIN see DVK500
DIOCTYL ESTER of SODIUM SULFOSUCCINATE see DJL000
DIOCTYL ESTER of SODIUM SULFOSUCCINIC ACID see DJL000
DIOCTYL ETHER see OEY000
DIOCTYL(ETHYLENEDIOXYBIS(CARBON YLMETHYLTHIO))STANNANE see DVN400
DIOCTYL FUMARATE see DVK600
DIOCTYLISOPENTYLPHOSPHINE OXIDE see DVK709
DIOCTYLISOPENTYLPHOSPHINE OXIDE see DVK709
DIOCTYL ISOPHTHALATE see BJQ750
DIOCTYL KETONE see OFE020
DI-n-OCTYL KETONE see OFE020
DIOCTYL MALEATE see DVK800
DIOCTYL MALEATE see DVK800
DI-N-OCTYL MALEATE see DVK800
"DIOCTYL" MALEATE see BJR000
DIOCTYL-MEDO FORTE see DJL000
DIOCTYLMETHYLAMINE see MND125
DI-N-OCTYLMETHYLAMINE see MND125
N,N-DIOCTYL-N-METHYLAMINE see MND125
N,N-DIOCTYL-1-OCTANAMINE see DVL000
2,2-DIOCTYL-1,3,2-OXATHIASTANNOLANE-5-OXIDE see DVL200
DIOCTYLOXOSTANNANE see DVL400
DIOCTYL PHTHALATE see DVL600
DIOCTYL PHTHALATE see DVL700
n-DIOCTYL PHTHALATE see DVL600
DI-sec-OCTYL PHTHALATE see DVL700
DIOCTYL(1,2-PROPYLENEDIOXYBIS(MALEOYLDIOXY))STANNANE see DVL800
DIOCTYL SEBACATE see BJS250
DI-n-OCTYL SODIUM SULFOSUCCINATE see SOD050
DIOCTYL SODIUM SULFOSUCCINATE (FCC) see DJL000
DIOCTYLSTANNIUM DICHLORIDE see DVN300
(Z,Z)-4,4'-((DIOCTYLSTANNYLENE)BIS(OXY))BIS(4-OXO-2-BUTANOIC ACID) DIISOOCTYL ESTER see BKL000
DIOCTYLSTANNYLENE MALEATE see DVK200
DIOCTYL SULFOSUCCINATE SODIUM SALT see DJL000
DIOCTYL TEREPHTHALATE see BJS300
DIOCTYLTHIOACETOXYSTANNANE see DVL200
DIOCTYLTHIOXOSTANNANE see DVM000
DI-N-OCTYLTIN BIS(BUTYL MALEATE) see BHK500
DI-n-OCTYLTIN BIS(BUTYL MERCAPTOACETATE) see DVM200

DI-n-OCTYLTIN BIS(DODECYL MERCAPTIDE) see DVM400
DI-n-OCTYLTIN BIS(2-ETHYLHEXYL MALEATE) see DVM600
DI-n-OCTYLTIN BIS(2-ETHYLHEXYL) MERCAPTOACETATE see DVM800
DIOCTYLTINBIS(ISOOCTYL MALEATE) see BKL000
DIOCTYLTIN BIS(ISOOCTYL MERCAPTOACETATE) see BKK750
DIOCTYLTIN-S,S'-BIS(ISOOCTYL MERCAPTOACETATE) see BKK750
DIOCTYLTIN BIS(ISOOCTYL THIOGLYCOLATE) see BKK750
DIOCTYL-TIN BIS(ISOOCTYLTHIOGLYCOLLATE) see BKK750
DI-n-OCTYLTIN BIS(LAURYLTHIOGLYCOLATE) see DVN000
DIOCTYLTIN BIS(MONOLAURYL MALEATE) see DVN100
DI-(N-OCTYL)TIN BIS-o,o'-(MONOLAURYL MALEATE) see DVN100
DI-n-OCTYLTIN-1,4-BUTANEDIOL-BIS-MERCAPTOACETATE see DVN200
DIOCTYLTIN DICHLORIDE see DVN300
DI-n-OCTYLTINDICHLORIDE see DVN300
DI-n-OCTYLTIN DIISOOCTYL THIOGLYCOLATE see BKK750
DIOCTYLTIN DILAURATE see DVJ800
DI-n-OCTYLTIN DILAURATE see DVJ800
DI-N-OCTYLTIN DIMONOBUTYLMALEATE see BHK500
DI-n-OCTYLTIN DI(1,2-PROPYLENEGLYCOLMALEATE) see DVL800
DI-n-OCTYLTIN ETHYLENEGLYCOL DITHIOGLYCOLATE see DVN400
DI-N-OCTYLTIN-2-ETHYLHEXYLDIMERCAPTOETHANOATE see DVM800
DIOCTYLTIN MALEATE see DVK200
DI-n-OCTYLTIN MALEATE see DVK200
DIOCTYLTIN MERCAPTIDE see DVN600
DI-n-OCTYLTIN MERCAPTIDE see DVN600
DIOCTYLTIN-β-MERCAPTOPROPIONATE see DVN800
DI-n-OCTYLTIN β-MERCAPTOPROPIONATE see DVN800
DIOCTYLTIN OXIDE see DVL400
DI-n-OCTYLTIN OXIDE see DVL400
DI-n-OCTYLTIN SULFIDE see DVM000
DIOCTYLTIN-3,3'-THIODIPROPIONATE see DVN909
DIOCTYLTIN THIOGLYCOLATE see DVL200
DI-n-OCTYLTIN THIOGLYCOLATE see DVL200
DI-N-OCTYLTIN-THIOGLYCOLIC ACID 2-ETHYLHEXYL ESTER see DVM800
DI-n-OCTYL-ZINN AETHYLENGLYKOL-DITHIOGLYKOLAT (GERMAN) see DVN400
DI-n-OCTYL-ZINN-BIS(2-AETHYLHEXYLMALEINAT) (GERMAN) see DVM600
DI-n-OCTYL-ZINN-BIS(LAURYL-THIOGLYKOLAT) (GERMAN) see DVN000
DI-n-OCTYL-ZINN-1,4-BUTANDIOL-BIS-MERCAPTOACETAT (GERMAN) see DVN200
DI-n-OCTYL-ZINN DICHLORID see DVN300
DI-n-OCTYL-ZINN-DI-ISOOCTYLTHIOGLYKOLAT (GERMAN) see BKK750
DI-n-OCTYL-ZINN DILAURAT (GERMAN) see DVJ800
DI-N-OCTYLZINN-DIMONOBUTYLMALEINAT (GERMAN) see BHK500

DI-n-OCTYLZINN-DIMONOMETHYLMALEINAT (GERMAN) see BKO500
DI-n-OCTYL-ZINN-DI-(1,2-PROPYLENGLYKOLMALEINAT)(GERMAN) see DVL800
DI-n-OCTYLZINN MALEINAT see DVK200
DI-n-OCTYL-ZINN β-MERCAPTOPROPIONAT (GERMAN) see DVN800
DI-n-OCTYL-ZINN OXYD (GERMAN) see DVL400
DI-n-OCTYL-ZINN THIOGLYKOLAT (GERMAN) see DVL200
DIOCYDE see DVO000
DIODON see DNG400
DIODONE see DNG400
DIODOQUIN see DNF600
DIODOXYLIN see DNF600
DIODRAST see DNG400
DIOFORM see DFI210
DIOGYN see EDO000
DIOGYN B see EDP000
DIOGYNETS see EDO000
DIOKAN see DVQ000
DIOKSAN (POLISH) see DVQ000
DIOKSYNY (POLISH) see TAI000
DIOKTYLESTER SULFOJANTARANU SODNEHO see SOD050
DIOLAMINE see DHF000
DIOLANDRONE see AOO475
DIOLANE see HFP875
DIOLENE see IPU000
DIOL-EPOXIDE-1 see DMO500
DIOL-EPOXIDE 2 see DMO800
anti-DIOLEPOXIDE see DVO175
DIOL EPOXIDE OF BADGE see ECE800
DIOLICE see CNU750
DIOLOSTENE see AOO475
DIOMEDICONE see DJL000
DI-ON see DXQ500
DIONE 21-ACETATE see RKP000
3,17-DIONE-19-ACETOXY-Δ(1,3)-ANDROSTADIENE see ABL625
2-4-DIONE-1,3-DIAZASPIRO(4.5)DECANE see DVO600
DIONIN see ENK000
DIONIN see ENK500
DIONINE see ENK000
DIONINE HYDROCHLORIDE see DVO700
DIONIN HYDROCHLORIDE see DVO700
DIONONE see DMH400
DIONONYL PHTHALATE see DVJ000
DIOPAL see SBH500
DIOPHYLLIN see TEP500
DIORTHOTOLYLGUANIDINE see DXP200
DIOSE see GHO100
DIOSPYROL see DVO809
DIOSPYROS VIRGINIANA see PCP500
1,4-DIOSSAN-2,3-DIYL-BIS(O,O-DIETIL-DITIOFOSFATO) (ITALIAN) see DVQ709
DIOSSANO-1,4 (ITALIAN) see DVQ000
1,4-DIOSSIBENZENE (ITALIAN) see QQS200
2,4-DIOSSI-5-DIAZOPIRIMIDINA (ITALIAN) see DCQ600
DIOSSIDONE see BRF500
DIOSUCCIN see DJL000
DIOTHANE see DVO819
DIOTHANE HYDROCHLORIDE see DVV500
DIOTHENE see PJS750
DIOTILAN see DJL000
DIOVAC see DJL000
DIOVOCYLIN see EDR000
DIOVOCYLIN see EDR000
DIOXAAN-1,4 (DUTCH) see DVQ000
1,4-DIOXAAN-2,3-DIYL-BIS(O,O-DIETHYL-DITHIOFOSFAAT) (DUTCH) see DVQ709
2,4-DIOXA-5-AZA-3-PHOPHAHEPT-5-ENE-7-NITRILE, 3-METHOXY-6-PHENYL-,3-SULFIDE see MOC275

5-(1,4-DIOXA-8-AZASPIRO(4.5)DEC-8-YLMETHYL)-3-ETHYL-6,7-DIHYDRO-2-METHYL-INDOL-4(5H)-ONE see AFH550

DIOXABENZOFOS see MEC250

2,3-DIOXABICYCLO(2.2.2)OCT-5-ENE, 1-ISOPROPYL-4-METHYL- see ARM500

DIOXACARB see DVS000

1,8-DIOXACYCLOHEPTADECAN-9-ONE see OLE100

1,6-DIOXACYCLOHEPTADECAN-17-ONE see OKW110

1,7-DIOXACYCLOHEPTADECAN-17-ONE see OKW100

1,9-DIOXACYCLOHEXADECANE-2,5,10-TRIONE, 8,16-BIS(2-OXOPROPYL)-, (8S-(8R*,16R*))- see BLC100

1,3-DIOXACYCLOHEXANE see DVP600

1,3-DIOXACYCLOPENTANE see DVR800

1,4-DIOXACYCLOTRIDECANE-5,13-DIONE see EIP050

4,9-DIOXA-1,12-DIAMINODODECANE see BGU600

6,9-DIOXA-3,12-DIAZATETRADECANEDIOIC ACID, 3,12-BIS(CARBOXYMETHYL)-(9CI) see EIT000

2,15-DIOXA-3,14-DISILAHEXADECANE, 3,3,14,14-TETRAMETHOXY- see BLQ550

2,7-DIOXA-3,6-DISILAOCTANE, 3,3,6,6-TETRAMETHOXY- see EIT200

4,7-DIOXA-3,8-DITHIA-2,9-DIAZADECANEDIOIC ACID, 2,9-DIMETHYL-5-((1-METHYLETHOXY)METHYL)-,DI-1-NAPHTHALENYL ESTER, 3,8-DIOXIDE see INE085

4,7-DIOXA-3,8-DITHIA-2,9-DIAZADECANEDIOIC ACID, 2,9-DIMETHYL-5-((1-METHYLETHOXY)METHYL)-, BIS(3-METHYLPHENYL) ESTER, 3,8-DIOXIDE see INE075

4,8-DIOXA-3,9-DITHIA-2,10-DIAZADECANEDIOIC ACID, 2,10-DIMETHYL-6-((METHYL((1-NAPHTHALENYLOXY)CARBONYL)AMINO)SULFINYL)OXY)-, DI-1-NAPHTHALENYL ESTER, 3,9-DIOXIDE see PML650

4,7-DIOXA-3,8-DITHIA-2,9-DIAZADECANEDIOIC ACID, 2,9-DIMETHYL-5-((PHENYLMETHOXY)METHYL)-,DI-1-NAPHTHALENYL ESTER, 3,8-DIOXIDE see BFC470

4,7-DIOXA-3,8-DITHIA-2,9-DIAZADECANEDIOIC ACID, 2,9-DIMETHYL-5-((PHENYLMETHOXY)METHYL)-,BIS(3-METHYLPHENYL) ESTER, 3,8-DIOXIDE see BFC460

4,7-DIOXA-3,8-DITHIA-2,9-DIAZADECANEDIOIC ACID, 2,9-DIMETHYL-5-((2-PROPENYLOXY)METHYL)-,DI-1-NAPHTHALENYL ESTER, 3,8-DIOXIDE see AGP620

4,7-DIOXA-3,8-DITHIA-2,9-DIAZADECANEDIOIC ACID, 2,9-DIMETHYL-5-((2-PROPENYLOXY)METHYL)-,BIS(3-METHYLPHENYL) ESTER, 3,8-DIOXIDE see AGP610

4,8-DIOXA-3,9-DITHIA-2,10-DIAZAUNDECANEDIOIC ACID, 2,10-DIMETHYL-, BIS(2,3-DIHYDRO-2,2-DIMETHYL-7-BENZOFURANYL) ESTER, 3,9-DIOXIDE see PML270

4,8-DIOXA-3,9-DITHIA-2,10-DIAZAUNDECANEDIOIC ACID, 2,10-DIMETHYL-6-(((METHYL((2-(1-METHYLETHOXY)PHENOXY)CARBONYL)AMINO)SULFINYL)OXY)-, BIS(2-(1-METHYLETHOXY)PHENYL) ESTER, 3,9-DIOXIDE see PML550

4,8-DIOXA-3,9-DITHIA-2,10-DIAZAUNDECANEDIOIC ACID, 2,10-DIMETHYL-6-(2-PROPENYLOXY)-,BIS(3-METHYLPHENYL) ESTER, 3,9-DIOXIDE see AGP600

(R*,S*)-3,14-DIOXA-2,15-DITHIA-6,11-DIAZEHEXADECANE-8,9-DIOL, 2,2,15,15-TETRAOXIDE (9CI) see LJD500

1,5,2,4-DIOXADITHIEPANE-2,2,4,4-TETRAOXIDE see DVO920

1,5,2,4-DIOXADITHIEPANE, 2,2,4,4-TETRAOXIDE see DVO920

4,9-DIOXADODECANE-1,12-DIAMINE see BGU600

3,6-DIOXADODECANOL-1 see HFN000

1-DIOXADROL HYDROCHLORIDE see LFG100

d-DIOXADROL HYDROCHLORIDE see DVP400

2,5-DIOXAHEXANE see DOE600

m-DIOXAN see DVP600

p-DIOXAN (CZECH) see DVQ000

DIOXAN-1,4 (GERMAN) see DVQ000

2,3-p-DIOXANDITHIOL S,S-BIS(O,O-DIETHYL PHOSPHORODITHIOATE) see DVQ709

1,4-DIOXAN-2,3-DIYL-BIS(O,O-DIETHYL-DITHIOPHOSPHAT) (GERMAN) see DVQ709

1,4-DIOXAN-2,3-DIYL-BIS(O,O-DIETHYLPHOSPHOROTHIOLOTHIONATE) see DVQ709

1,4-DIOXAN-2,3-DIYL-O,O,O',O'-TETRAETHYL DI(PHOSPHORODITHIOATE) see DVQ709

DIOXANE see DVQ000

p-DIOXANE see DVQ000

1,3-DIOXANE see DVP600

2,3-p-DIOXANE-S,S-BIS(O,O-DIETHYLPHOSPHORODITHIOATE) see DVQ709

1,3-DIOXANE, 2-BUTYL-4,4,6-TRIMETHYL-see BSR600

m-DIOXANE-4,4-DIMETHYL see DVQ400

cis-2,3-p-DIOXANEDITHIOL-S,S-BIS(O,O-DIETHYLPHOSPHORODITHIOATE) see DVQ600

p-DIOXANE-2,3-DITHIOL-S,S-DIESTER with O,O-DIETHYL PHOSPHORODITHIOATE see DVQ709

p-DIOXANE-2,3-DIYL ETHYL PHOSPHORODITHIOATE see DVQ709

1,4-DIOXANE (MAK) see DVQ000

1,3-DIOXANE-5-METHANOL,-5-((4,6-BIS(1-AZIRIDINYL)-1,3,5-TRIAZIN-2-YL)AMINO)-2-(2-FURANYL)- see FQB100

m-DIOXANE-5-METHANOL, 2-FURYL-5-(4,6-DIAZIRIDINYL-2-s-TRIAZINYLAMINO)- see FQB100

1,3-DIOXANE, 2-(PHENYLMETHYL)-4,4,6-TRIMETHYL- see PFR400

DIOXANNE (FRENCH) see DVQ000

DIOXANONE see DVQ630

p-DIOXANONE see DVQ630

2-p-DIOXANONE see DVQ630

p-DIOXAN-2-ONE see DVQ630

1,4-DIOXAN-2-ONE (9CI) see DVQ630

3,6-DIOXAOCTANE-1,8-DIOL see TJQ000

3,6-DIOXAOCTANOL, 8-(4-(4-CHLOROPHENYLPHENYLMETHYL)PIPERAZINYL)- see HHK050

2,7-DIOXAPYRENE-1,3,6,8-TETRONE see NAP300

1,3-DIOXA-2-SILACYCLOPENTANE, 4-(CHLOROMETHYL)-2,2-DIMETHYL- see CIL775

2,4-DIOXASPIRO(5.5)UNDEC-8-ENE, 3-(3-CYCLOHEXENYL)- see CPD650

1,3,2-DIOXASTANNEPIN-4,7-DIONE, 2,2-DIBUTYL- see DEJ100

4,9-DIOXATETRACYCLO(5.4.0.0^{3,5}.0^{8,10})UNDECANE see BFY750

4,10-DIOXATETRACYCLO(5.4.0^{3,5}.0^{1,7}.0^{9,11})UNDECANE see BFY750

1,3,2-DIOXATHIANE-2,2-DIOXIDE (9CI) see TLR750

5,11-DIOXA-9-THIA-4,7,9,12-TETRAAZAPENTADECA-3,12-DIENEDINITRILE, 6,10-DIOXO-2,2,7,9,14,14-HEXAMETHYL- see BIQ660

1,3,2-DIOXATHIOLANE-2-OXIDE (9CI) see COV750

DIOXATHION see DVQ709

5,10-DIOXATRICYCLO(7.1.0.0^{4,6}))DECANE see CPR835

(E)-3,8-DIOXATRICYCLO(5.1.0.0^{2,4})OCTANE see DHB875

3,8-DIOXATRICYCLO(5.1.0.0^{2,4}))OCTANE, (90% TRANS,10% CIS MIXTURE) see DHB820

DIOXATRINE see BBU800

cis-1,4-DIOXENEDIOXETANE see DVQ759

1,4-DI-N-OXIDE 2,3-BIS(OXYMETHYL)QUINOXLINE see DVQ800

1,4-DI-N-OXIDE of DIHYDROXYMETHYLQUINOXALINE see DVQ800

2,2-DIOXIDE-1,3,2-DIOXATHIOLANE see EJP000

4,4-DIOXIDE-1,4-OXATHIANE see DVR000

1,4-DIOXIDE-2,3-QUINOXALINEDIMETHANOL see DVQ800

1,1-DIOXIDETETRAHYDROTHIOFURAN see SNW500

1,1-DIOXIDETETRAHYDROTHIOPHENE see SNW500

DIOXIDE of VITAVAX see DLV200

DIOXIDIN see DVQ800

DIOXIDINE see DVQ800

DIOXIME-p-BENZOQUINONE see DVR200

DIOXIME-1,4-CYCLOHEXADIENEDIONE see DVR200

DIOXIME-2,5-CYCLOHEXADIENE-1,4-DIONE see DVR200

DIOXIN (herbicide contaminant) see TAI000

DIOXINE see TAI000

DIOXIN (bactericide) (OBS.) see ABC250

DIOXITOL see CBR000

DIOXOAMINOPYRINE see DVU100

9,10-DIOXOANTHRACENE see APK250

p-DIOXOBENZENE see HIH000

2,2'-((1,4-DIOXO-1,4-BUTANEDIYL)BIS(OXY))BIS(N,N,N-TRIMETHYLETHANAMINIUM) see CMG250

2,2'-((1,4-DIOXO-1,4-BUTANEDIYL)BIS(OXY)BIS(N,N,N-TRIMETHYLETHANAMINIUM) DICHLORIDE see HLC500

1,1'-((1,4-DIOXO-1,4-BUTANEDIYL)BIS(OXY-2,1-ETHANEDIYL))BIS-QUINOLINIUM DIIODIDE see KGK400

1,1',1''-(3,6-DIOXO-1,4-CYCLOHEXADIENE-1,2,4-TRIYL)TRISAZIRIDINE see TND000

1,2-DIOXOCYCLOHEXANE see CPB200

((2,5-DIOXO-3-CYCLOHEXEN-1-YL)THIO)ACETIC ACID see DVR250

2,6-DIOXO-5-DIAZOPYRIMIDINE see DCQ600

DIOXODICHLOROCHROMIUM see CML125

3,3'-(DIOXODIGERMOXANYLENE)DIPROPANOIC ACID see CCF125

9,10-DIOXO-9,10-DIHYDRO-1-NITRO-6-ANTHRACENESULFONIC ACID see DVR400

DIOXO-9,9-(DIMETHYLAMINO-3-METHYL-2-PROPYL)-10-PHENOTHIAZINE (FRENCH) see AFL750

5,5-DIOXO-10-(2-
(DIMETHYLAMINO)PROPYL)PHENOTHIA
ZINE HYDROCHLORIDE see DVT400
3,5-DIOXO-1,2-DIPHENYL-4-N-
BUTYLPYRAZOLIDENE see BRF500
3,5-DIOXO-1,2-DIPHENYL-4-N-
BUTYLPYRAZOLIDIN SODIUM see BOV750
DI-OXO-DI-N-PROPYLNITROSAMINE see
NJN000
2,2'-DIOXO-DI-N-PROPYLNITROSAMINE
see NJN000
2,4-DIOXO-5-FLUORO-N-HEXYL-3,4-
DIHYDRO-1(2H)-
PYRIMIDINECARBOXAMIDME see CCK630
2,4-DIOXO-5-FLUORO-N-HEXYL-1,2,3,4-
TETRAHYDRO-1-
PYRIMIDINECARBOXAMIDE see CCK630
2,3-DIOXOINDOLINE see ICR000
2-(1,3-DIOXO-2-
ISOINDOLINE)ETHYLTHIURONIUM
BROMIDE see DVR500
DIOXOLAN see DVR600
1,3-DIOXOLAN see DVR800
1,4-DIOXOLAN-2,5-
DIYLDIMETHYLENEBIS(NITROMERCURY
) see BKZ000
1,3-DIOXOLANE see DVR800
DIOXOLANE (DOT) see DVR600
1,3-DIOXOLANE-2-ACETIC ACID, 2-
METHYL-, ETHYL ESTER see EFR100
1,3-DIOXOLANE, 4-(CHLOROMETHYL)-2-
(o-NITROPHENYL)- see CIQ400
1,3-DIOXOLANE, 2-(1,2-
DICHLOROETHYL)-4-METHYL- see DFJ500
1,3-DIOXOLANE, 2-(2,6-DIMETHYL-1,5-
HEPTADIENYL)- see CMS324
1,3-DIOXOLANE-4-METHANOL see
DVR909
1,3-DIOXOLANE-4-METHANOL, 2-(1-
IODOETHYL)- see IEL800
1,3-DIOXOLANE, 4-METHYL-2-PENTYL-
see MNM450
1,3-DIOXOLANE, 4-METHYL-2-PHENYL-
see MNT600
2-(1,3-DIOXOLANE-2-YL)PHENYL N-
METHYLCARBAMATE see DVS000
1,3-DIOXOLAN-2-ONE see GHM000
1,3-DIOXOLAN-2-ONE, 4-
(CHLOROMETHYL)- see CBW400
1,3-DIOXOLAN-2-ONE, 4-ETHYL- see
BOT200
α-((1,3-DIOXOLAN-2-
YLMETHOXY)IMINO)BENZENEACETONI
TRILE see OKS200
7-(1,3-DIOXOLAN-2-
YLMETHYL)THEOPHYLLINE see TEQ175
o-(1,3-DIOXOLAN-2-YL METHYL)-2,2,2-
TRIFLUORO-4'-
CHLOROACETOPHENONEOXIME see
FMR600
((1,3-DIOXOLAN-4-
YL)METHYL)TRIMETHYLAMMONIUM
IODIDE see FMX000
2-(1,3-DIOXOLAN-2-YL)PHENYL-N-
METHYLCARBAMAT see DVS000
o-(1,3-DIOXOLAN-2-YL)PHENYL
METHYLCARBAMATE see DVS000
o-(1,3-DIOXOLAN-2-YL)PHENYL
METHYLNITROSOCARBAMATE see
MMV500
1,3-DIOXOL-4-EN-2-ONE see VOK000
1,3-DIOXOLO(4,5-G)FURO(2,3-
B)QUINOLINE, 9-METHOXY- see MEL725
1,3-DIOXOLO(4,5-H)ISOQUINOLINE see
MJR775
DIOXOLONE-2 see GHM000
1,3-DIOXOL-2-ONE see VOK000
(1,3)DIOXOLO(4,5-i)PYRROLO(3,2,1-
de)PHENANTHRIDINIUM, 4,5-DIHYDRO-2-
HYDROXY-(9CI) see LJB800
2,6-DIOXO-4-METHYL-4-
ETHYLPIPERIDINE see MKA250

3-(2-(1,3-DIOXO-2-
METHYLINDANYL))GLUTARIMIDE see
DVS100
DIOXONE see DJT400
9,10-DIOXO-1-NITRO-9,10-DIHYDRO-5-
ANTHRACENESULFONIC ACID see DVS200
2,2'-DIOXO-N-NITROSODIPROPYLAMINE
see NJN000
3,5-DIOXO-4-(1-
OXOPROPYL)CYCLOHEXANECARBOXYLI
C ACID ION(1-) CALCIUM CALCIUM SALT
see CPB065
3,5-DIOXO-1-PHENYL-2-(p-
HYDROXYPHENYL)-4-N-
BUTYLPYRAZOLIDENE see HNI500
3-(2-(1,3-DIOXO-2-
PHENYLINDANYL))GLUTARIMIDE see
DVS300
3-(2-(1,3-DIOXO-2-PHENYL-4,5,6,7-
TETRAHYDRO-4,7-
DITHIAINDANYL))GLUTARIMIDE see
DVS400
1,3-DIOXOPHTHALAN see PHW750
1,3-DIOXO-5-PHTHALANCARBOXYLIC
ACID see TKV000
2,6-DIOXO-3-PHTHALIMIDOPIPERIDINE
see TEH500
2-(2,6-DIOXOPIPERIDEN-3-YL)
PHTHALIMIDINE see DVS600
2-(2-6-DIOXO-3-PIPERIDINYL)1H-
ISOINDOLE-1,3(2H)-DIONE see TEH500
(s)-2-(2,6-DIOXO-3-PIPERIDINYL)-1H-
ISOINDOLE-1,3(2H)-DIONE see TEH520
N-(2,6-DIOXO-3-
PIPERIDYL)PHTHALIMIDE see TEH500
(±)-N-(2,6-DIOXO-3-
PIPERIDYL)PHTHALIMIDE see DVT200
DIOXOPROMETHAZINE
HYDROCHLORIDE see DVT400
N,N-DI(2-OXOPROPYL)NITROSAMINE see
NJN000
2,2'-DIOXOPROPYL-N-
PROPYLNITROSAMINE see NJN000
2,6-DIOXOPURINE see XCA000
1,3-DIOXO-2-(3-
PYRIDYLMETHYLENE)INDAN see DVT459
2,4-DIOXOPYRIMIDINE see UNJ800
2,4-DIOXOTETRAHYDROQUINAZOLINE
see QEJ800
3,5-DIOXO-2,3,4,5-TETRAHYDRO-1,2,4-
TRIAZINE RIBOSIDE see RJA000
2,4-DIOXOTHIAZOLIDINE see TEV500
DIOXOTHIOLAN see SNW500
1,1-DIOXOTHIOLAN see SNW500
DIOXYAMINOPYRINE see DVU100
DIOXYANTHRANOL see APH250
1,4-DIOXYANTHRAQUINONE (RUSSIAN)
see DMH000
2,6-DIOXY-8-AZAPURINE see THT350
m-DIOXYBENZENE see REA000
o-DIOXYBENZENE see CCP850
p-DIOXYBENZENE see HIH000
1,4-DIOXYBENZENE see QQS200
3,3'-DIOXYBENZIDINE see DMI400
1,4-DIOXY-BENZOL (GERMAN) see QQS200
DIOXYBENZON see DMW250
DIOXYBENZONE see DMW250
DIOXYBIS(2,2'-DI-tert-BUTYLBUTANE see
DVT500
DIOXYBIS METHANOL see DMN200
DIOXYBUTADIENE see BGA750
3-β,14-DIOXY-CARDEN-(20:22)-OLID
(GERMAN) see DMJ000
3-α,7-β-DIOXYCHOLANIC ACID see
DMJ200
DIOXYDE de BARYUM (FRENCH) see
BAO250
DIOXYDEMETON-S-METHYL see DAP600
3,3'-(DIOXYDICARBONYL)DIPROPIONIC
ACID see SNC500
2,4-DIOXY-3,3-DIETHYL-5-
METHYLPIPERIDINE see DNW400

3-β,14-DIOXY-DIGEN-(20:22)-OLID
(GERMAN) see DMJ000
DIOXYDIMETHANOL see DMN200
1,1'-(DIOXY)DIMETHYLCYCLOHEXANOL
see DVT600
DIOXYDINE see DVQ800
4,4'-DIOXYDIPHENYLSULFIDE see TFJ000
1,4-DIOXYETHYLAMINO-5,8-
DIOXYANTHRAQUINONE see DMM400
N,N-DIOXYETHYLANILINE see BKD500
DIOXYETHYLENE ETHER see DVQ000
DIOXYGENYL TETRAFLUOROBORATE
see DVT800
DIOXYLINE see HOE100
DIOXYMETHYLENE-PROTOCATECHUIC
ALDEHYDE see PIW250
1,3-DIOXY-2-NICOTINSAEURENITRIL-
TETRAZOL (GERMAN) see DND900
1,4-DI-p-OXYPHENYL-2,3-DI-ISONITRILO-
1,3-BUTADIENE see DVU000
DI(p-OXYPHENYL)-2,4-HEXADIENE see
DAL600
2-(3,4-DIOXYPHENYL)TETRAHYDRO-1,4-
OXAZIN (GERMAN) see MRU100
7-(2,3-DIOXYPROPYL)THEOPHYLLINE see
DNC000
DIOXYPYRAMIDON see DVU100
DI-8-OXYQUINOLINE-N-
AMINOSALICYLIC ACID see BKJ300
DIOZOL see BRF500
DIP see DNS200
DIPA see DNL600
DIPA see DNM200
DIPA see DNM400
DIPAC see DNN900
DIPALLADIUM TRIOXIDE see DVU200
DIPAM see DCK759
DIPAN see DVX200
DIPANE see WAK000
DIPANOL see MCC250
DIPAR see PDF250
DIPARAANISYL-MONOPHENETHYL-
GUANIDIN-HYDROCHLORID (GERMAN)
see PDN500
DI-PARALEN see CFF500
DIPARALENE see CFF500
DIPARALENE HYDROCHLORIDE see
CDR250
DI-PARALENE-2-HYDROCHLORIDE see
CDR000
DIPARCOL see DHF600
DIPARCOL see DII200
DIPAV see PAH250
DIPAXIN see DVV600
DIPEGYL see NCR000
DIPEL see BAC040
DIPELARGONYL PEROXIDE see NNA100
DIPENICILLINA-G-ALLUMINIO-
SULFAMETOSSIPIRIDAZINA (ITALIAN) see
DVU300
DIPENICILLIN-G-ALUMINIUM-
SULPHAMETHOXYPYRIDAZINE see
DVU300
DIPENINBROMID (GERMAN) see DGW600
DIPENTAMETHYLENETHIURAM
TETRASULFIDE see TEF750
DI(PENTANOYLOXY)DIBUTYLSTANNAN
E see DEA600
DIPENTENE see MCC250
DIPENTENE DIOXIDE see LFV000
DIPENTYLAMINE see DCH200
α-((DIPENTYLAMINO)METHYL)-1,2,3,4-
TETRAHYDRO-9-
PHENANTHRENEMETHANOL see DCH300
DIPENTYL ETHER see PBX000
2,5-DI-tert-PENTYLHYDROQUINONE see
DCH400
DIPENTYL KETONE see ULA000
DIPENTYL LEAD DIACETATE see DVU600
DIPENTYL MALEATE see MAL000
DIPENTYLNITROSAMINE see DCH600
DI-n-PENTYLNITROSAMINE see DCH600

3,3'-DIPENTYLOXACARBOCYANINE IODIDE see DVU700
DIPENTYLOXOSTANNANE see DVV000
DIPENTYL PHENOL see DCH800
2,4-DI-tert-PENTYLPHENOL see DCI000
2-(2,4-DI-tert-PENTYLPHENOXY)BUTYRIC ACID see DVV109
DIPENTYL PHTHALATE see AON300
DI-n-PENTYLPHTHALATE see AON300
DIPENTYLTIN DICHLORIDE see DVV200
DIPENTYLTIN OXIDE see DVV000
DIPEPTIDE SWEETENER see ARN825
2,6-DIPERCHLORYL-4,4'-DIPHENOQUINONE see DVV400
DIPERDON HYDROCHLORIDE see DVV500
DIPERFLUOROBUTYL ETHER see PCG755
DIPERODON HYDROCHLORIDE see DVV500
DIPEROXYTEREPHTHALIC ACID see DVV550
DIPHACIL see DHX800
DIPHACIN see DVV600
DIPHACINONE see DVV600
DIPHACYL see DHX800
DIPHANTINE see BBV500
DIPHANTOIN see DKQ000
DIPHANTOINE SODIUM see DNU000
M-DIPHAR see MAS500
DIPHASTON see DYF759
DIPHEBUZOL see BRF500
DIPHEDAL see DKQ000
DIPHEDAN see DNU000
DIPHEMANIL see DAP800
DIPHEMANIL METHYLSULFATE see DAP800
DIPHENACIN see DVV600
DIPHENACOUM see BGO100
DIPHENADIONE see DVV600
DIPHENALDEHYDE see DVV800
DIPHENAMID see DRP800
DIPHENAMIDE see DRP800
DIPHENAMIZOLE see PAM500
DIPHENAN see PFR325
DIPHENAN (pharmaceutical) see PFR325
DIPHENANE see PFR325
DIPHENATE see DNU000
DIPHENATIL see DAP800
DIPHENATRILE see DVX200
DIPHENAZINE DIHYDROCHLORIDE see BLF250
DIPHENCHLOXAZINE HYDROCHLORIDE see DVW000
DIPHENETHYLAMINE, o,o'-DIMETHOXY-α-α'-DIMETHYL-, compounded with LACTIC ACID see BKP200
DIPHENEX see DGA850
DIPHENHYDRINATE see DYE600
DIPHENICILLIN see BGK250
DIPHENIDOL see DWK200
DIPHENIN see DNU000
DIPHENINE see DKQ000
DIPHENINE SODIUM see DNU000
DIPHENMANIL METHYLSULFATE see DAP800
DIPHENMETHANIL see DAP800
DIPHENMETHANIL METHYLSULFATE see DAP800
o-DIPHENOL see CCP850
1,5-DIPHENOXYANTHRAQUINONE see DVW100
N,N'-DI(3-PHENOXY-2-HYDROXYPROPYL)ETHYLENEDIAMINE DIHYDROCHLORIDE see IAG700
DIPHENOXYLATE HYDROCHLORIDE see LIB000
1,3-DIPHENOXY-2-PROPANOL see DVW600
DIPHENPROFOS see CKK530
DIPHENPYRALINE HYDROCHLORIDE see DVW700
DIPHENTHANE 70 see MJM500
DIPHENTOIN see DKQ000

DIPHENTOIN see DNU000
DIPHENYLACETAMIDE see PDX500
N,N'-DIPHENYLACETAMIDINE see DVW750
DIPHENYLACETIC ACID see DVW800
α,α-DIPHENYLACETIC ACID see DVW800
DIPHENYLACETIC ACID 2-(BIS(2-HYDROXYETHYL)AMINO)ETHYL ESTER HYDROCHLORIDE see DVW900
DIPHENYLACETIC ACID CHLORIDE see DVX500
DIPHENYLACETIC ACID DIETHYLAMINOETHYL ESTER see DHX800
DIPHENYLACETIC ACID DIETHYLAMINOETHYL ESTER see THK000
DIPHENYLACETIC ACID, 2-(DIETHYLAMINO)ETHYL ESTER see DHX800
DIPHENYLACETIC ACID, 2-(DIETHYLAMINO)ETHYL ESTER see THK000
DIPHENYLACETIC ACID, ESTER with 3-QUINUCLIDINOL, HYDROGEN SULFATE (2:1) DIHYDRATE see QVJ000
DIPHENYLACETIC ACID-1-ETHYL-3-PIPERIDYL ESTER HYDROCHLORIDE see EOY000
1,1-DIPHENYL ACETONE see MJH910
DIPHENYLACETONITRILE see DVX200
DIPHENYLACETOPHENONE see TMS100
2,2-DIPHENYLACETOPHENONE see TMS100
ω,ω-DIPHENYLACETOPHENONE see TMS100
6-DIPHENYLACETOXY-1-AZABICYCLO(3.2.1)OCTANE HYDROCHLORIDE see ARX500
O,O-DIPHENYL (1-ACETOXY-2,2,2-TRICHLOROETHYL)PHOSPHONATE see DVX400
DIPHENYLACETYL CHLORIDE see DVX500
2,2-DIPHENYLACETYL CHLORIDE see DVX500
α-α-DIPHENYLACETYL CHLORIDE see DVX500
DIPHENYLACETYLDIETHYLAMINOETHANOL see DHX800
DIPHENYLACETYLDIETHYLAMINOETHANOL see THK000
2-DIPHENYLACETYL-1,3-DIKETOHYDRINDENE see DVV600
2-(DIPHENYLACETYL)INDAN-1,3-DIONE see DVV600
2-DIPHENYLACETYL-1,3-INDANDIONE see DVV600
2-(DIPHENYLACETYL)-1H-INDENE-1,3(2H)-DIONE see DVV600
2,3-DIPHENYLACRYLONITRILE see DVX600
α,β-DIPHENYLACRYLONITRILE see DVX600
3-(2,2-DIPHENYLAETHYL)-5-(2-PIPERIDINOAETHYL)-1,2,4-OXADIAZOL (GERMAN) see LFJ000
DIPHENYLAMIDE see DRP800
DIPHENYLAMIN-(β-DIAETHYLAMINOAETHYL)CARBAMIDTHIOESTER (GERMAN) see PDC875
DIPHENYLAMINE see DVX800
N,N-DIPHENYLAMINE see DVX800
DIPHENYLAMINE, 4-AMINO- see PFU500
DIPHENYLAMINE, p-AMINO- see PFU500
DIPHENYLAMINE, 2,5'-BIS(TRIFLUOROMETHYL)-2'-CHLORO-4,6-DINITRO- see FDB300
DIPHENYLAMINE-2-CARBOXYLIC ACID see PEG500
DIPHENYLAMINECHLORARSINE see PDB000

DIPHENYLAMINECHLOROARSINE (DOT) see PDB000
DIPHENYLAMINE, 2,4-DINITRO- see DUV100
DIPHENYLAMINE, HEXANITRO- see HET500
DIPHENYLAMINE HYDROGEN SULFATE see DVY000
DIPHENYLAMINE, 4-(PHENYLAZO)- see PEI800
DIPHENYLAMINE SULFATE see DVY000
DIPHENYLAMINOCHLOROARSINE see DVY050
N²)-((4-(DIPHENYLAMINO)PHENYL)METHYLENE)-4-NITRO-N¹)-PHENYL-1,2-BENZENEDIAMINE see NFY100
DIPHENYLAN see DKQ000
N,N-DIPHENYLANILINE see TMQ500
DIPHENYLAN SODIUM see DNU000
DIPHENYLARSINOUS ACID see DVY100
DIPHENYLARSINOUS CHLORIDE see CGN000
N,N-DIPHENYLARSONAMIDOUS CHLORIDE see DVY050
DIPHENYLBENZENE see TBD000
m-DIPHENYLBENZENE see TBC620
p-DIPHENYLBENZENE see TBC750
1,2-DIPHENYLBENZENE see TBC640
1,4-DIPHENYLBENZENE see TBC750
DIPHENYL BENZENEPHOSPHONATE see TMU300
2,7-DIPHENYLBENZO(lmn)(3,8)PHENANTHROLINE-1,3,6,8(2H,7H)-TETRONE see DVY300
DIPHENYL o-BIPHENYLYL PHOSPHATE see XFS300
DIPHENYLBIS(PHENYLTHIO)STANNANE see DVY800
DIPHENYLBIS(PHENYLTHIO)TIN see DVY800
DIPHENYL BLACK see PFU500
DIPHENYL BLUE 2B see CMO000
DIPHENYL BLUE 3B see CMO250
DIPHENYL BLUE BLACK GHS see CMN800
DIPHENYL BLUE BLACK MBH see CMN800
DIPHENYL BLUE G see CMO600
DIPHENYL BORDEAUX BX see CMO872
DIPHENYL BRILLIANT BLUE see CMO500
DIPHENYL BRILLIANT BLUE FF see CMN750
DIPHENYL BROWN 3RB see CMO800
DIPHENYL BROWN BS see CMO820
DIPHENYL BROWN BVV see CMO800
DIPHENYL BROWN PT see CMO810
DIPHENYL BROWN 3GT see CMO825
DIPHENYL BROWN TB see CMO820
DIPHENYL BROWN V see CMO800
DIPHENYLBUTAZONE see BRF500
1,3-DIPHENYL-2-BUTEN-1-ONE see MPL000
(Z)-2-(p-(1,2-DIPHENYL-1-BUTENYL)-PHENOXY)-N,N-DIMETHYLETHYLAMINE see NOA600
trans-2-(p-(1,2-DIPHENYL-1-BUTENYL)PHENOXY)-N,N-DIMETHYLETHYLAMINE CITRATE see DVY875
2-(4-((1Z)-1,2-DIPHENYL-1-BUTENYL)PHENOXY)ETHANAMINE see DHA250
cis-2-(p-(1,2-DIPHENYL-1-BUTENYL)PHENOXY)-N-METHYLETHYLAMINE see DBA710
1,2-DIPHENYL-4-BUTYL-3,5-DIKETOPYRAZOLIDINE CALCIUM SALT see PEO750
1,2-DIPHENYL-4-BUTYL-3,5-DIOXOPYRAZOLIDINE see BRF500
1,1-DIPHENYL-2-BUTYNYL-N-CYCLOHEXYLCARBAMATE see DVY900

DIPHENYL ISODECYL PHOSPHITE see IKL200
4,4-DIPHENYL-N-ISOPROPYLCYCLOHEXYLAMINE HYDROCHLORIDE see SDZ000
1,2-DIPHENYL-4-(γ-KETOBUTYL)-3,5-PYRAZOLIDINEDIONE see KGK000
DIPHENYL KETONE see BCS250
DIPHENYL KETOXIME see BCS400
8,10-DIPHENYL LOBELIONOL see LHY000
DIPHENYLMERCURY see DWD800
DIPHENYLMETHAN-4,4'-DIISOCYANAT (GERMAN) see MJP400
DIPHENYLMETHANEBISMALEIMIDE see BKL800
4,4'-DIPHENYLMETHANEBISMALEIMIDE see BKL800
2,4'-DIPHENYLMETHANEDIAMINE see MJP750
4,4'-DIPHENYLMETHANEDIAMINE see MJQ000
DIPHENYL METHANE DIISOCYANATE see MJP400
4,4'-DIPHENYLMETHANE DIISOCYANATE see MJP400
p,p'-DIPHENYLMETHANE DIISOCYANATE see MJP400
DIPHENYLMETHANE 4,4'-DIISOCYANATE (DOT) see MJP400
DIPHENYLMETHANE-4,4'-DIISOCYANATE-TRIMELLIC ANHYDRIDE-ETHOMID HT POLYMER see TKV000
4,4'-DIPHENYLMETHANEDIMALEIMIDE see BKL800
DIPHENYL METHANEPHOSPHONATE see MDQ825
DIPHENYLMETHANOL see HKF300
DIPHENYLMETHANONE see BCS250
DIPHENYLMETHANONE OXIME see BCS400
2-(DIPHENYLMETHOXY)ACETAMIDOXIME HYDROGEN MALEATE see DWE200
2-(DIPHENYLMETHOXY)-N,N-DIMETHYL-ETHANAMINE HYDROCHLORIDE see BAU750
2-(DIPHENYLMETHOXY)-N,N-DIMETHYLETHYLAMINE see BBV500
2-DIPHENYLMETHOXY-N,N-DIMETHYLETHYLAMINE HYDROCHLORIDE see BAU750
endo-3-(DIPHENYLMETHOXY)-8-ETHYL-8-AZABICYCLO(3.2.1)OCTANE see DWE800
3-(DIPHENYLMETHOXY)-8-ETHYLNORTROPANE see DWE800
3-α-(DIPHENYLMETHOXY)-8-ETHYLNORTROPANE see DWE800
3-α-(DIPHENYLMETHOXY)-1-α-H,5-α-H-TROPANE see BDI000
4-(DIPHENYLMETHOXY)-1-METHYLPIPERIDINE see LJR000
4-(DIPHENYLMETHOXY)-1-METHYLPIPERIDINE CHLOROTHEOPHYLLINE see DWF000
3-DIPHENYLMETHOXYTROPANE MESYLATE see TNU000
3-DIPHENYLMETHOXYTROPANE METHANESULFONATE see TNU000
DIPHENYLMETHYL ALCOHOL see HKF300
DIPHENYLMETHYL BROMIDE (DOT) see BNG750
DIPHENYL METHYL BROMIDE, solution (DOT) see BNH000
DIPHENYLMETHYLCYANIDE see DVX200
α-(DIPHENYLMETHYLENE)BENZENEACETIC ACID see TMQ250
1,1'-DIPHENYLMETHYLENEBIS(3-(2-CHLOROETHYL)-3-NITROSOUREA) see BIG000

2-DIPHENYLMETHYLENEBUTYLAMINE HYDROCHLORIDE see DWF200
3-(DIPHENYLMETHYLENE)-1,1-DIETHYL-2-METHYLPYRROLIDINIUM BROMIDE see PAB750
4-(DIPHENYLMETHYLENE)-1,1-DIMETHYLPIPERIDINIUM METHYLSULFATE see DAP800
4-(DIPHENYLMETHYLENE)-2-ETHYL-3-METHYLCYCLOHEXANOL ACETATE see DWF300
3-(DIPHENYLMETHYLENE)-1-ETHYL-2-METHYLPYRROLIDINE ETHYL BROMIDE see PAB750
(DIPHENYLMETHYLENE)HYDROXYLAMINE see BCS400
1-DIPHENYLMETHYLENYL-2-METHYL-3-ETHYL-4-ACETOXYCYCLOHEXANE see DWF300
1-(DIPHENYLMETHYL)-4-(2-(2-HYDROXYETHOXY)ETHYL)PIPERAZINE see BBV750
1-DIPHENYLMETHYL-4-METHYLPIPERAZINE see EAN600
(±)-1-DIPHENYLMETHYL-4-METHYLPIPERAZINE HYDROCHLORIDE see MAX275
(+)-2,2-DIPHENYL-3-METHYL-4-MORPHOLINOBUTYRYLPYRROLIDINE see AFJ400
2-(3,3-DIPHENYL-3-(5-METHYL-1,3,4-OXADIAZOL-2-YL)PROPYL)-2-AZABICYCLO(2.2.2)OCTANE see DWF700
DIPHENYL METHYLPHOSPHONATE see MDQ825
2-(2-((4-DIPHENYLMETHYL)-1-PIPERAZINYL)ETHOXY)ETHANOL HYDROXYDIETHYLPHENAMINE see BBV750
1-(3-(4-(DIPHENYLMETHYL)-1-PIPERAZINYL)PROPYL)-2-BENZIMIDAZOLINONE see OMG000
1-(3-(4-(DIPHENYLMETHYL)-1-PIPERAZINYL)PROPYL)-1,3-DIHYDRO-2H-BENZIMIDAZOL-2-ONE see OMG000
4-DIPHENYLMETHYLPIPERIDINE see DWF865
4,4-DIPHENYL-1-METHYLPIPERIDINE MALEATE see DWF869
5,5-DIPHENYL-3-(3-(2-METHYLPIPERIDINO)PROPYL)-2-THIOHYDANTOIN HYDROCHLORIDE see DWF875
5,5-DIPHENYL-3-(3-(2-METHYLPIPERIDINO)PROPYL)-2-THIOHYDANTOIN MONOHYDROCHLORIDE see DWF875
α-2,2-DIPHENYL-4-(1-METHYL-2-PIPERIDYL)-1,3-DIOXOLANE HYDROCHLORIDE see DWG600
β-2,2-DIPHENYL-4-(1-METHYL-2-PIPERIDYL)-1,3-DIOXOLANE HYDROCHLORIDE see DWH000
α-2,2-DIPHENYL-4-(1-METHYL-2-PIPERIDYL)-1,3-DIOXOLANE METHYLIODIDE see DWH200
3,3-DIPHENYL-2-METHYL-1-PYRROLINE see DWH500
DIPHENYLMETHYLSILANOL see DWH550
4-DIPHENYLMETHYLTROPYLTROPINIUM BROMIDE see PEM750
4-DIPHENYLMETHYL-dl-TROPYLTROPINIUM BROMIDE see PEM750
4,4-DIPHENYL-6-MORPHOLINO-3-HEPTANONE HYDROCHLORIDE see DWH600
4,4-DIPHENYL-6-MORPHOLINO-3-HEXANONE HYDROCHLORIDE see DWH800
2,3-DIPHENYL-3H-NAPHTHO(1,2-d)TRIAZOLIUM CHLORIDE see DWH875

4,6-DIPHENYL-1-(((5-NITRO-2-FURANYL)METHYLENE)AMINO)-2(1H)-PYRIDINONE see DWH900
DIPHENYLNITROSAMIN (GERMAN) see DWI000
DIPHENYLNITROSAMINE see DWI000
N,N-DIPHENYLNITROSOAMINE see DWI000
DIPHENYL N-NITROSOAMINE see DWI000
o-DIPHENYLOL see BGJ250
2,2-DI(4-PHENYLOL)PROPANE see BLD500
DIPHENYLOLPROPANE DISODIUM SALT see BLD800
DIPHENYL ORANGE GG see CMO860
DIPHENYL (OSHA) see BGE000
2,5-DIPHENYLOXAZOLE see DWI200
4,5-DIPHENYL-2-OXAZOLEPROPANOIC ACID see OLW600
4,5-DIPHENYL-2-OXAZOLEPROPIONIC ACID see OLW600
5,5-DIPHENYL-1-OXAZOLIDIN-2,4-DIONE see DWI300
N-(4,5-DIPHENYL-2-OXAZOLYL)DIACETAMIDE see ACI629
N-(4,5-DIPHENYLOXAZOL-2-YL)DIETHANOLAMINE MONOHYDRATE see BKB250
2,2'-((4,5-DIPHENYL-2-OXAZOLYL)IMINO)-DIETHANOLMONOHYDRATE see BKB250
3,3-DIPHENYL-2-OXETANONE see DWI400
DIPHENYL OXIDE see PFA850
DIPHENYLOXIDE-4,4'-DISULFOHYDRAZIDE (DOT) see OPE000
DIPHENYL-4-γ-OXO-γ-BUTRIC ACID see BGL250
1,2-DIPHENYL-4-(3'-OXOBUTYL)-3,5-DIOXOPYRAZOLIDINE see KGK000
DIPHENYLOXOSTANNANE, POLYMER see DWO600
DIPHENYL PENTACHLORIDE see PAV600
1,5-DIPHENYL-1,4-PENTAZDIENE see DWI800
DIPHENYL-p-PHENYLENEDIAMINE see BLE500
N,N'-DIPHENYL-p-PHENYLENEDIAMINE see BLE500
DIPHENYL PHENYLPHOSPHONATE see TMU300
1,2-DIPHENYL-4-(2'-PHENYLSULFINETHYL)-3,5-PYRAZOLIDINEDIONE see DWM000
5,5-DIPHENYL-1-PHENYLSULFONYLHYDANTOIN see DWJ300
5,5-DIPHENYL-1-(PHENYLSULFONYL)-2,4-IMIDAZOLIDINEDIONE (9CI) see DWJ300
1,2-DIPHENYL-4-PHENYLTHIOETHYL-3,5-PYRAZOLIDINEDIONE see DWJ400
DIPHENYL PHOSPHATE see PHE300
(T-4)-(DIPHENYLPHOSPHINODITHIOATO-S,S')DIPHENYLANTIMONY see DWJ600
α,α-DIPHENYL-1-PIPERIDINEBUTANOL see DWK200
α,α-DIPHENYL-1-PIPERIDINEBUTANOL HYDROCHLORIDE see CCR875
α,α-DIPHENYL-2-PIPERIDINEMETHANOL see DWK400
α,α-DIPHENYL-2-PIPERIDINEMETHANOL HYDROCHLORIDE see PII750
α,α-DIPHENYL-4-PIPERIDINEMETHANOL HYDROCHLORIDE see PIY750
α,α-DIPHENYL-1-PIPERIDINEPROPANOL HYDROCHLORIDE see PMC250
α,α-DIPHENYL-1-PIPERIDINEPROPANOL METHANESULFONATE (ester) see PMC275
DIPHENYL-PIPERIDINO-AETHYL-ACETAMID-BROMMETHYLAT (GERMAN) see RDA375
1,1-DIPHENYL-3-N-PIPERIDINOBUTANOL-1 HYDROCHLORIDE see ARN700

2,2-DIPHENYL-4-(4-PIPERIDINO-4-CARBAMOYLPIPERIDINO)BUTYRONITRILE see PJA140
1,1-DIPHENYL-3-PIPERIDINO-1-PROPANOL HYDROCHLORIDE see PMC250
5,5-DIPHENYL-3-(3-PIPERIDINOPROPYL)-2-THIOHYDANTOIN HYDROCHLORIDE see DWK500
1-2,2-DIPHENYL-4-(2-PIPERIDYL)-1,3-DIOXOLANE HYDROCHLORIDE see LFG100
d-2,2-DIPHENYL-4-(2-PIPERIDYL)-1,3-DIOXOLANE HYDROCHLORIDE see DVP400
1,1-DIPHENYL-3-(1-PIPERIDYL)-1-PROPANOL HYDROCHLORIDE see PMC250
1,3-DIPHENYL-1,3-PROPANEDIONE see PFU300
1,3-DIPHENYL-1-PROPEN-3-ONE see CDH000
α,α-DIPHENYL-β-PROPIOLACTONE see DWI400
1,2-DIPHENYL-2-PROPIONOXY-3-METHYL-4-DIMETHYLAMINOBUTANE see PNA250
(+)-1,2-DIPHENYL-2-PROPIONOXY-3-METHYL-4-DIMETHYLAMINOBUTANE HYDROCHLORIDE see PNA500
3-(3,3-DIPHENYLPROPYLAMINO)PROPYL-3',4',5'-TRIMETHOXYBENZOATE HYDROCHLORIDE see DWK700
3-(N-d,d-DIPHENYLPROPYL-N-METHYL)AMINOPROPAN-1-OL HYDROCHLORIDE see DWK900
N-(3,3-DIPHENYLPROPYL)-α-METHYLPHENETHYLAMINE see PEV750
N-(3,3-DIPHENYLPROPYL)-α-METHYLPHENETHYLAMINE LACTATE see DWL200
DIPHENYLPROPYLPHOSPHINE see PNI600
1,1-DIPHENYL-2-PROPYN-1-OL CYCLOHEXANECARBAMATE see DWL400
1,1-DIPHENYL-2-PROPYNYL-N-CYCLOHEXYLCARBAMATE see DWL400
1,1-DIPHENYL-2-PROPYNYL ESTER CYCLOHEXANECARBAMIC ACID see DWL400
1,1-DIPHENYL-2-PROPYNYL-N-ETHYLCARBAMATE see DWL500
1,1-DIPHENYL-2-PROPYNYL 1-PYRROLIDINECARBOXYLATE see DWL525
DIPHENYLPYRALIN-8-CHLORTHEOPHYLLINAT (GERMAN) see PIZ250
DIPHENYLPYRALINE see LJR000
DIPHENYLPYRALINE HYDROCHLORIDE see DVW700
DIPHENYLPYRALINE TEOCLATE see PIZ250
1,2-DIPHENYL-3,5-PYRAZOLIDINEDIONE see DWA500
1,4-DIPHENYL-3,5-PYRAZOLIDINEDIONE see DWL600
1,3-DIPHENYL-5-PYRAZOLONE see DWL800
DIPHENYLPYRAZONE see DWM000
DIPHENYLPYRILENE see LJR000
3,3-DIPHENYL-3-(PYRROLIDINE-CARBONYLOXY)-1-PROPYNE see DWL525
α,α-DIPHENYL-3-QUINUCLIDINEMETHANOL HYDROCHLORIDE see DWM400
DIPHENYL RED 4B see DXO850
DIPHENYL RED 8B see CMO880
DIPHENYL RED B see CMO870
DIPHENYL RED 3BS see CMO875
DIPHENYL RED 4BS see DXO850
DIPHENYL SCARLET 3BS see CMO875
DIPHENYL SELENIDE see PGH250
DIPHENYLSELENONE see DWM600
DIPHENYL SKY BLUE 6B see CMO500

DIPHENYLSTANNANE see DWM800
DIPHENYL SULFIDE see PGI500
DIPHENYL SULFONE see PGI750
DIPHENYL SULFOXIDE see DWN000
1-((DIPHENYLSULFOXIMIDO)METHYL)-1-METHYLPYRROLIDINIUM BROMIDE see HAK200
DIPHENYL SULPHONE see PGI750
trans-2,4-DIPHENYL-2,4,6,6-TETRAMETHYLCYCLOTRISILOXANE see DWN100
1,3-DIPHENYL-1,1,3,3-TETRAMETHYLDISILOXANE see DWN150
3,3',4,4'-DIPHENYLTETRAMINE see BGK500
DIPHENYLTHIOCARBAMIC ACID-S-(2-(DIETHYLAMINO)ETHYL) ESTER see PDC850
DIPHENYLTHIOCARBAMIC ACID-S-(2-(DIETHYLAMINO)ETHYL) ESTER HYDROCHLORIDE see PDC875
N,N'-DIPHENYLTHIOCARBAMIDE see DWN800
sym-DIPHENYLTHIOCARBAMIDE see DWN800
DIPHENYL THIOCARBAZIDE see DWN400
DIPHENYLTHIOCARBAZONE see DWN200
1,5-DIPHENYL-3-THIOCARBOHYDRAZIDE see DWN400
DIPHENYL THIOETHER see PGI500
5,5-DIPHENYL-2-THIOHYDANTOIN see DWN600
DIPHENYLTHIOLACETIC ACID-2-DIETHYLAMINOETHYL ESTER HYDROCHLORIDE see DHY400
DIPHENYLTHIOUREA see DWN800
1,3-DIPHENYLTHIOUREA see DWN800
1,1-DIPHENYL-2-THIOUREA see DWO000
1,3-DIPHENYL-2-THIOUREA see DWN800
N,N'-DIPHENYLTHIOUREA see DWN800
sym-DIPHENYLTHIOUREA see DWN800
DIPHENYLTIN see DWM800
DIPHENYLTIN DIBROMIDE see DDN200
DIPHENYLTIN DICHLORIDE see DWO400
DIPHENYLTIN DIHYDRIDE see DWM800
DIPHENYLTIN OXIDE POLYMER see DWO600
DIPHENYL TOLYL PHOSPHATE see TGY750
1,3-DIPHENYLTRIAZENE see DWO800
5,6-DIPHENYL-as-TRIAZIN-3-OL see DWO875
3,5-DIPHENYL-s-TRIAZOLE see DWO950
3,5-DIPHENYL-1,2,4-TRIAZOLE see DWO950
3,5-DIPHENYL-1H-1,2,4-TRIAZOLE see DWO950
DIPHENYLTRICHLOROETHANE see DAD200
2,2-DIPHENYL-1,1,1-TRICHLOROETHANE see DWP000
DIPHENYL (2,2,2-TRICHLORO-1-HYDROXYETHYL)PHOSPHONATE see TIQ300
1,3-DIPHENYLUREA see CBM250
N,N'-DIPHENYLUREA see CBM250
sym-DIPHENYLUREA see CBM250
2,2-DIPHENYL-VALERIC ACID-2-(DIETHYLAMINO)ETHYL ESTER see DIG400
DIPHER see EIR000
DIPHERGAN see PMI750
DI-PHETINE see DKQ000
DI-PHETINE see DNU000
DIPHEXAMIDE METHIODIDE see BTA325
DIPHONE see SOA500
DIPHOSGEN see TIR920
DIPHOSGENE see PGX000
DIPHOSGENE see TIR920
DIPHOSPHANE see DWP229
1,2-DIPHOSPHINOETHANE see DWP250
DIPHOSPHOPYRIDINE NUCLEOTIDE see CNF390

DIPHOSPHORIC ACID, DISODIUM SALT see DXF800
DIPHOSPHORIC ACID TETRAETHYL ESTER see TCF250
DIPHOSPHORIC ACID, TETRAMETHYL ESTER see TDV000
DIPHOSPHORIC ACID, TETRAPOTASSIUM SALT see TEC100
DIPHOSPHORUS PENTOXIDE see PHS250
DIPHOSPHORUS TRIOXIDE see PHT500
DIPHTHERIA TOXIN see DWP300
DIPHYL see PFA860
DIPHYLLIN see DNC000
DIPHYLLIN-3,4-o-DIMETHYL XYLOSIDE see CMV375
DIPICRYLAMINE see HET500
DIPICRYLAMINE (DOT) see HET500
DIPICRYLOXAMIDE see HET700
DIPIDOLOR see PJA140
DIPIGYL see NCR000
DIPIKRYLAMIN see HET500
DIPIN see BJC250
DIPINE see BJC250
DIPIPERAL see FHG000
2-β,16-β-DIPIPERIDINO-5-α-ANDROSTAN-3-α,16-β-DIOLDIMETHOBROMIDE see AOO403
2-β,16-β-DIPIPERIDINO-5-α-ANDROSTAN-3-α,17-β-DIOL DIPIVALATE HYDROCHLORIDE see DWP500
2,2',2'',2'''-(4,8-DIPIPERIDINOPYRIMIDO(5,4-d)PYRIMIDINE-2,6-DIYLDINITRILO)TETRAETHANOL see PCP250
DIPIPERON see FHG000
DIPIPERONE see FHG000
DIPIRARTRIL-TROPICO see DUD800
DIPIRIN see DOT000
DIPIRITRAMIDE see PJA140
DIPIVEFRINE HYDROCHLORIDE see DWP559
DIPIVEFRIN HYDROCHLORIDE see DWP559
DIPLIN see DMZ000
DIPLODIOL see ELH650
DIPLOSAL see SAN000
DIPLOSPORIN see ELH650
DIPN see DNB200
DIPOFENE see DCM750
DIPOLYOXYETHYLATEDPOLYPROPYLENEGLYCOL ETHER see PJK150
DIPOLYOXYETHYLATEDPOLYPROPYLENEGLYCOL ETHER see PJK151
DIPO-SAFT see BFC750
DIPOTASSIUM 4,4'-BIS(4-PHENYL-1,2,3-TRIAZOL-2-YL)STILBENE-2,2'-DISULFONATE see BLG100
DIPOTASSIUM 4,4'-BIS(4-PHENYL-1,2,3-TRIAZOL-2-YL)STILBENE-2,2'-SULFONATE see BLG100
DIPOTASSIUM CHLORAZEPATE see CDQ250
DIPOTASSIUM CHROMATE see PLB250
DIPOTASSIUM CLORAZEPATE see CDQ250
DIPOTASSIUM CYCLOOCTATETRAENE see DWP900
DIPOTASSIUM DIAZIRINE-3,3-DICARBOXYLATE see DWP950
DIPOTASSIUM DICHLORIDE see PLA500
DIPOTASSIUM DICHROMATE see PKX250
DIPOTASSIUM DIMAGNESIUM TRISULFATE see MAI650
DIPOTASSIUM-meso-N,N-DISULFO-3,4-DIPHENYLHEXANE see SPA650
DIPOTASSIUM ETHYLENEDIAMINETETRAACETATE see EEB100
DIPOTASSIUM ETHYLENEDIAMINETETRAACETATE see EJA250
DIPOTASSIUM MOLYBDATE see PLL125

DIPOTASSIUM MONOCHROMATE see PLB250

DIPOTASSIUM MONOPHOSPHATE see PLQ400

DIPOTASSIUM NICKEL TETRACYANIDE see NDI000

DIPOTASSIUM OXALATE see PLN300

DIPOTASSIUM PENTACHLORORHODATE(2-)- see PAY300

DIPOTASSIUM PERSULFATE see DWQ000

DIPOTASSIUM PHOSPHATE see PLQ400

DIPOTASSIUM SELENITE see SBO100

DIPOTASSIUM TETRACHLOPALLADATE see PLN750

DIPOTASSIUM TETRACYANONICKELATE see NDI000

DIPOTASSIUM TETRAOXOMOLYBDATE see PLL125

DIPOTASSIUM TETRAOXOMOLYBDATE(2-) see PLL125

DIPOTASSIUM TRICHLORONITROPLATINATE see PLN050

DIPPEL'S OIL see BMA750

DIPPING ACID see SOI500

DIPRAM see DGI000

DIPRAMID see DAB875

DIPRAMIDE see DAB875

DIPRAZINE see DQA400

DIPRIVAN see DNR800

DIPROFILLIN see DNC000

DIPROFILLINE see DNC000

DIPRON see SNM500

DIPROPAMINE see DHA500

DIPROPANEDIOL DIBENZOATE see DWS800

DIPROPANOIC ACID GERMANIUM SESQUIOXIDE see CCF125

DIPROPARGYL ETHER see POA500

DIPROPARGYLNITROSAMINE see NKB600

DI-2-PROPENYLAMINE see DBI600

DI-2-PROPENYL ESTER, 1,2-BENZENEDICARBOXYLIC ACID see DBL200

DI-2-PROPENYL ISOPHTHALATE see IMK000

DI-2-PROPENYL PHOSPHONITE see DBL000

N-N-DI-2-PROPENYL-2-PROPEN-1-AMINE see THN000

5,5-DI-2-PROPENYL-2,4,6(1H,3H,5H)-PYRIMIDINETRIONE (9CI) see AFS500

DIPROPETRYN see EPN500

DIPROPETRYNE see EPN500

DIPROPHYLLIN see DNC000

DIPROPHYLLINE see DNC000

DIPROPIONATE BECLOMETHASONE see AFJ625

DIPROPIONATE d'OESTRADIOL (FRENCH) see EDR000

DIPROPIONATO de ESTILBENE (SPANISH) see DKB000

p,p'-DIPROPIONOXY-trans-α,β-DIETHYLSTILBENE see DKB000

DIPROPIONYL PEROXIDE see DWQ800

DIPROPIONYL PEROXIDE, >28% in solution (DOT) see DWQ800

1,1-DIPROPOXYETHANE see AAG850

4,5-DIPROPOXY-2-IMIDAZOLIDINONE see DWQ850

1-(DI-N-PROPOXYPHOSPHINOTHIOYLTHIOMETHYLCARBONYL-2-METHYLPIPERIDINE) see PIX775

DIPROPYL ACETAL see AAG850

DIPROPYLACETAMIDE see PNX600

DIPROPYLACETATE DE CALCIUM see VCK200

DIPROPYLACETATE SODIUM see PNX750

DIPROPYLACETIC ACID see PNR750

N-DIPROPYLACETIC ACID see PNR750

DIPROPYLACETIC ACID CALCIUM SALT see CAY675

DIPROPYLACETIC ACID CALCIUM SALT see VCK200

DIPROPYL ADIPATE see DWQ875

DI-n-PROPYL ADIPATE see DWQ875

DIPROPYLAMINE see DWR000

DI-n-PROPYLAMINE see DWR000

n-DIPROPYLAMINE see DWR000

DIPROPYLAMINE, 3,3'-BIS((3-(AMINOPROPYL))AMINO)- see TED600

1-DIPROPYLAMINOACETYLINDOLINE see DWR200

1-(4-(DIPROPYLAMINO)-2-BUTYNYL)-2-PYRROLIDINONE see DWR300

1-(4-DIPROPYLAMINOBUT-2-YNYL)PYRROLID-2-ONE see DWR300

4-(DI-N-PROPYLAMINO)-3,5-DINITRO-1-TRIFLUOROMETHYLBENZENE see DUV600

2-(R,S)-(DI-N-PROPYLAMINO)-6-(4-METHOXYPHENYLSULFONYLMETHYL)-1,2,3,4-TETRAHYDRONAPHTHALENE see DWR350

4-((DIPROPYLAMINO)SULFONYL)BENZOIC ACID see DWW000

DIPROPYLCARBAMIC ACID ETHYL ESTER see DWT400

DIPROPYLCARBAMOTHIOIC ACID-S-ETHYL ESTER see EIN500

DIPROPYLDIAZENE 1-OXIDE see ASP500

DIPROPYL-2,2'-DIHYDROXYAMINE see DNL600

9,10-DI-n-PROPYL-9-10-DIHYDROXY-9,10-DIHYDRO-1,2,5,6-DIBENZANTHRACENE see DLK600

N,N-DI-N-PROPYL-2,6-DINITRO-4-TRIFLUOROMETHYLANILINE see DUV600

N3,N3-DIPROPYL-2,4-DINITRO-6-TRIFLUOROMETHYL-m-PHENYLENEDIAMINE see DWS200

DI-n-PROPYL DISELENIDE see PNI850

α,α'-DIPROPYLENEDINITRILODI-o-CRESOL see DWS400

DIPROPYLENE GLYCOL see DWS500

DIPROPYLENE GLYCOL see OQM000

DIPROPYLENE GLYCOL BUTYL ETHER see DWS600

DIPROPYLENE GLYCOL DIACRYLATE see DWS650

DIPROPYLENE GLYCOL DIBENZOATE see DWS800

DIPROPYLENE GLYCOL DIMETHYL ETHER see DWS900

DIPROPYLENE GLYCOL DIPELARGONATE see DWT000

DIPROPYLENE GLYCOL, ETHYL ETHER see EEW100

DIPROPYLENE GLYCOL METHYL ETHER see DWT200

DIPROPYLENE GLYCOL MONOMETHYL ETHER see DWT200

DIPROPYLENE GLYCOL, 3,3,5-TRIMETHYLCYCLOHEXYL ETHER see TLO600

DIPROPYLENETRIAMINE see AIX250

DI-n-PROPYLESSIGSAURE (GERMAN) see PNR750

DIPROPYL ETHER see PNM000

N,N-DI-n-PROPYL ETHYL CARBAMATE see DWT400

1-(N,N-DIPROPYLGLYCYL)INDOLINE see DWR200

1,1-DIPROPYLHYDRAZINE ETHANEDIOATE (1:1) see DWT430

1,2-DIPROPYLHYDRAZINE ETHANEDIOATE (1:1) see DWT450

1,1-DIPROPYLHYDRAZINE OXALATE see DWT430

1,2-DIPROPYLHYDRAZINE OXALATE see DWT450

DIPROPYL ISOCINCHOMERONATE see EAU500

DI-N-PROPYL-ISOCINCHOMERONATE (GERMAN) see EAU500

DIPROPYL KETONE see DWT600

DI-n-PROPYL MALEATE-ISOSAFROLE CONDENSATE see PNP250

DIPROPYL MERCURY see DWU000

DIPROPYL METHANE see HBC500

DI-n-PROPYL 6,7-METHYLENEDIOXY-3-METHYL-1,2,3,4-TETRAHYDRONAPHTHALENE see PNP250

DI-n-PROPYL-3-METHYL-6,7-METHYLENEDIOXY-1,2,3,4-TETRAHYDRONAPHTHALENE-1,2-DICARBOXYLATE see PNP250

S,S-DIPROPYL METHYLPHOSPHONOTRITHIOATE see DWU200

O,O-DIPROPYL S-2-METHYL-PIPERIDINOCARBONYL-METHYL PHOSPHORODITHIOATE see PIX775

O,O-DI-n-PROPYL-O-(4-METHYLTHIOPHENYL)PHOSPHATE see DWU400

DI-n-PROPYLNITROSAMINE see NKB700

α-DIPROPYLNITROSAMINE METHYL ETHER see DWU800

DIPROPYLNITROSOAMINE see NKB700

DIPROPYL OXIDE see PNM000

DIPROPYLOXOSTANNANE see DWV000

DIPROPYL PEROXIDE see DWV200

DI-n-PROPYL PEROXYDICARBONATE see DWV400

DIPROPYL PHTHALATE see DWV500

DI-n-PROPYL PHTHALATE see DWV500

N,N-DIPROPYL-1-PROPANAMINE see TMY250

DIPROPYL 2,5-PYRIDINEDICARBOXYLATE see EAU500

N,N-DIPROPYL SUCCINAMIC ACID ETHYL ESTER see DWV600

DIPROPYL SUCCINATE see DWV800

DI-N-PROPYL SUCCINATE see DWV800

4-(DIPROPYLSULFAMOYL)BENZOIC ACID see DWW000

p-(DIPROPYLSULFAMOYL)BENZOIC ACID see DWW000

p-(DIPROPYLSULFAMOYL)BENZOIC ACID SODIUM SALT see DWW200

p-(DIPROPYLSULFAMYL)BENZOIC ACID see DWW000

p-(DI-N-PROPYLSULFAMYL)BENZOIC ACID SODIUM SALT see DWW200

DIPROPYL-5,6,7,8-TETRAHYDRO-7-METHYLNAPHTHO(2,3-d)-1,3-DIOXOLE-5,6-DICARBOXYLATE see PNP250

N,N-DIPROPYLTHIOCARBAMIC ACID-S-ETHYL ESTER see EIN500

DIPROPYLTHIOCARBAMIC ACID-S-PROPYL ESTER see PNI750

DI-n-PROPYLTIN BISMETHANESULFONATE see DWW400

DIPROPYLTIN CHLORIDE see DFF400

DIPROPYLTIN DICHLORIDE see DFF400

DI-n-PROPYLTIN DICHLORIDE see DFF400

DIPROPYLTIN OXIDE see DWV000

N,N-DIPROPYL-4-TRIFLUOROMETHYL-2,6-DINITROANILINE see DUV600

DIPROPYL ZINC see DWW500

DI-2-PROPYNYLAMINE, N-NITROSO- see NKB600

DI(2-PROPYNYL) ETHER see POA500

DI-2-PROPYNYLNITROSAMINE see NKB600

DIPROSONE see BFV765

DIPROSTRON see EDR000

DIPROXID see IRS500

DIPROXIDE see IRS500

DIPROZIN see DQA400

DIPTERAX see TIQ250

DIPTEREX see TIQ250

DIPTEREX 50 see TIQ250

DIPTEREX, ACETYL- see CDN510

DIPTEVUR see TIQ250
DIPTHAL see DNS600
DIPYRIDAMINE see PCP250
DIPYRIDAMOL see PCP250
DIPYRIDAMOLE see PCP250
DIPYRIDAN see PCP250
4,4'-DIPYRIDINE see BGO600
DIPYRIDINESODIUM see DWW600
DIPYRIDO(1,2-A:3',2'-D)IMIDAZOL-2-
AMINE, N-HYDROXY-6-METHYL- see
HIU550
DIPYRIDO(1,2-A:3',2'-D)IMIDAZOLE, 2-
AMINO-4-METHYL- see AKS100
DIPYRIDO(1,2-A:3',2'-D)IMIDAZOL-3-
IMINE, N-HYDROXY- see HIT620
DIPYRIDO(1,2-A:3',2'-D)IMIDAZOL-3-
IMINE, N-HYDROXY-4,6-DIMETHYL- see
HIT610
N-(DIPYRIDO(1,2-A:3',2'-D)IMIDAZOL-3-
YL)HYDROXYLAMINE see HIT620
DIPYRIDO(1,2-A:2',1'-C)PYRAZINEDIIUM,
6,7-DIHYDRO- see DWW730
DIPYRIDO(1,2-A. 2',1'-C)PYRAZINEDIIUM,
6,7-DIHYDRO-, DIBROMIDE see EJC025
DIPYRIDO(2,3-D,2,3-K)PYRENE see NAU500
DIPYRIDO(1,2-a:3',2'-d)IMIDAZOL-2-AMINE
see DWW700
DIPYRIDO(1,2-a:3',2'-d)IMIDAZOLE, 2-
AMINO-, HYDROCHLORIDE see AJS225
DIPYRIDO(1,2-a:3',2'-d)IMIDAZOLE, 2-
AMINO-6-METHYL-, HYDROCHLORIDE
see AKS275
DIPYRIDO(2,3-d,2,3-1)PYRENE see NAU000
4,4-DIPYRIDYL see BGO600
2,2'-DIPYRIDYL see BGO500
4,4'-DIPYRIDYL see BGO600
α,α'-DIPYRIDYL see BGO500
γ,γ'-DIPYRIDYL see BGO600
DIPYRIDYLDIHYDRATE see PAJ100
DIPYRIDYL HYDROGEN PHOSPHATE see
DWX000
2,2'-DIPYRIDYL KETONE HYDRAZONE
see DWX100
DI-3-PYRIDYLMERCURY see DWX200
1,2-DI-3-PYRIDYL-2-METHYL-1-
PROPANONE see MCJ370
DIPYRIDYL PHOSPHATE see DWX000
DIPYRIN see DOT000
DIPYRONE MONOHYDRATE see MDM500
DIPYROXIME see TLQ500
cis-
DIPYRROLIDINEDICHLOROPLATINUM(II
) see DEU200
1,4-DIPYRROLIDINYL-2-BUTYNE see
DWX600
1,4-DIPYRROLIDINYL-2-BUTYNE see
DWX600
DI-1H-PYRROL-2-YL KETONE see PPY300
DIPYUDAMINE see PCP250
DIQUAT see DWX800
DIQUAT (ACGIH) see DWW730
DIQUAT DIBROMIDE see DWX800
DIQUAT DIBROMIDE see EJC025
DIQUAT DICHLORIDE see DWY000
DI-QUINOL see DNF600
1,3-DIQUINOLIN-6-YLUREA
BISMETHOSULFATE see PJA120
DIRALGAN see TKG000
DIRAM A see ANG500
DIRAME see PMX250
DIRAX see AQN635
DIRAX see IDJ500
DIRCA PALUSTRIS see LEF100
DIRECT ARTIFICIAL SILK BLACK BO see
CMN300
DIRECT ARTIFICIAL SILK BLACK G see
CMN240
DIRECT BLACK 3 see AQP000
DIRECT BLACK 17 see CMN230
DIRECT BLACK 19 see CMN240
DIRECT BLACK 2S see CMN300
DIRECT BLACK 32 see CMN300

DIRECT BLACK 38 see AQP000
DIRECT BLACK 19:1 see CMN150
DIRECT BLACK A see AQP000
DIRECT BLACK AB see CMN300
DIRECT BLACK BH see CMN800
DIRECT BLACK BRN see AQP000
DIRECT BLACK CX see AQP000
DIRECT BLACK CXR see AQP000
DIRECT BLACK E see AQP000
DIRECT BLACK EW see AQP000
DIRECT BLACK EX see AQP000
DIRECT BLACK FR see AQP000
DIRECT BLACK GAC see AQP000
DIRECT BLACK GB NB see DWY100
DIRECT BLACK GREEN see CMO830
DIRECT BLACK GW see AQP000
DIRECT BLACK GX see AQP000
DIRECT BLACK GXR see AQP000
DIRECT BLACK JET see AQP000
DIRECT BLACK META see AQP000
DIRECT BLACK METHYL see AQP000
DIRECT BLACK N see AQP000
DIRECT BLACK RX see AQP000
DIRECT BLACK SD see AQP000
DIRECT BLACK WS see AQP000
DIRECT BLACK Z see AQP000
DIRECT BLUE 2 see CMN800
DIRECT BLUE 6 see CMO000
DIRECT BLUE 14 see CMO250
DIRECT BLUE 15 see CMO500
DIRECT BLUE 22 see CMO600
DIRECT BLUE 86 see COF420
DIRECT BLUE 10G see CMO500
DIRECT BLUE BLACK BH see CMN800
DIRECT BLUE BR see CMO600
DIRECT BLUE HH see CMO500
DIRECT BLUE MRW see CMO600
DIRECT BLUE RW see CMO600
DIRECT BLUE RWN see CMO600
DIRECT BORDEAUX see CMO872
DIRECT BORDEAUX A see CMO872
DIRECT BORDEAUX AN see CMO872
DIRECT BORDEAUX B see CMO872
DIRECT BORDEAUX BG see CMO872
DIRECT BORDEAUX BN see CMO872
DIRECT BRILLIANT BLUE FF see CMN750
DIRECT BRILLIANT GREEN BB see
CMO840
DIRECT BRILLIANT GREEN C see CMO840
DIRECT BRILLIANT GREEN CBM see
CMO840
DIRECT BRILLIANT RED 4A see CMO870
DIRECT BROWN 2 see CMO800
DIRECT BROWN 1:2 see CMO810
DIRECT BROWN 1A see CMO810
DIRECT BROWN 31 see CMO820
DIRECT BROWN 3B see CMO820
DIRECT BROWN 5C see CMO810
DIRECT BROWN 5G see CMO810
DIRECT BROWN 95 see CMO750
DIRECT BROWN 154 see CMO825
DIRECT BROWN B see CMO820
DIRECT BROWN 3RB see CMO800
DIRECT BROWN BR see PEY000
DIRECT BROWN BS see CMO820
DIRECT BROWN BSB see CMO820
DIRECT BROWN CGN see CMO810
DIRECT BROWN CMD see CMO825
DIRECT BROWN D3Y see CMO825
DIRECT BROWN FS see CMO820
DIRECT BROWN KKH see CMO800
DIRECT BROWN M see CMO800
DIRECT BROWN MB see CMO800
DIRECT BROWN MR see CMO800
DIRECT BROWN 5GR see CMO825
DIRECT BROWN RC see CMO800
DIRECT BROWN RMR see CMO800
DIRECT BROWN 2GS see CMO810
DIRECT BROWN TRB see CMO820
DIRECT CLARET see CMO872
DIRECT DARK BLUE BH see CMN800
DIRECT DARK GREEN A see CMO830

DIRECT DARK GREEN B see CMO830
DIRECT DARK GREEN BF see CMO830
DIRECT DARK GREEN BG see CMO830
DIRECT DARK GREEN MB see CMO830
DIRECT DARK GREEN S see CMO830
DIRECT DARK GREEN SUPRA see CMO830
DIRECT DARK GREEN WS see CMO830
DIRECT DEEP BLACK E see AQP000
DIRECT DEEP BLACK EAC see AQP000
DIRECT DEEP BLACK EA-CF see AQP000
DIRECT DEEP BLACK E EXTRA see
AQP000
DIRECT DEEP BLACK EW see AQP000
DIRECT DEEP BLACK EX see AQP000
DIRECT DEEP GREEN A see CMO830
DIRECT DIAZO BLACK see CMN800
DIRECT DIAZO BLACK C see CMN800
DIRECT DIAZO BLACK N see CMN800
DIRECT DIAZO BLACK S see CMN800
DIRECT FAST BLACK CAB see CMN300
DIRECT FAST BLACK G see CMN240
DIRECT FAST BLACK GU see CMN240
DIRECT FAST BLACK SA see CMN240
DIRECT FAST BLUE L 7V see COF420
DIRECT FAST BROWN BP see CMO820
DIRECT FAST BROWN M see CMO800
DIRECT FAST BROWN MM see CMO800
DIRECT FAST BROWN TSN see CMO820
DIRECT FAST BROWN TWC see CMO820
DIRECT FAST BROWN V see CMO800
DIRECT FAST BROWN VR see CMO800
DIRECT FAST PURPURINE 8B see CMO880
DIRECT FAST RED 2S see CMO885
DIRECT FAST RED 5B see CMO885
DIRECT FAST RED B see CMO870
DIRECT FAST RED F see CMO870
DIRECT FAST RED FN see CMO870
DIRECT FAST RED FR see CMO870
DIRECT FAST RED G see CMO870
DIRECT FAST RED 8BL see CMO885
DIRECT FAST RED MF see CMO870
DIRECT FAST SCARLET 3B see CMO875
DIRECT FAST SCARLET 4B see CMO870
DIRECT FAST SCARLET M 4BS see CMO870
DIRECT FAST SCARLET 4BS see CMO870
DIRECT FAST SCARLET SE see CMO870
DIRECT FAST SCARLET 4BSN see CMO870
DIRECT FAST TURQUOISE see COF420
DIRECT FAST VIOLET N see CMP000
DIRECT GARNET LG see CMO872
DIRECT GREEN see CMO840
DIRECT GREEN 6 see CMO840
DIRECT GREEN 2B see CMO840
DIRECT GREEN A see CMO840
DIRECT GREEN BN see CMO840
DIRECT GREEN BP see CMO840
DIRECT GREEN BX see CMO840
DIRECT GREEN MB see CMO840
DIRECT GREEN WAC see CMO830
DIRECT LIGHTFAST RED 2S see CMO885
DIRECT LIGHT RED 4B see CMO885
DIRECT LIGHT RED 8B see CMO885
DIRECT LIGHT RED M 8BL see CMO885
DIRECT LIGHT TURQUOISE BLUE GL see
COF420
DIRECT NAVY BLUE BH see CMN800
DIRECT ORANGE 6 see CMO860
DIRECT ORANGE G see CMO860
DIRECT ORANGE T see CMO860
DIRECT PURE BLUE see CMO500
DIRECT PURE BLUE M see CMO500
DIRECT PURPURINE 4B see DXO850
DIRECT PURPURINE M4B see DXO850
DIRECT RAYON BLACK KSG see CMN240
DIRECT RED 1 see CMO870
DIRECT RED 2 see DXO850
DIRECT RED 13 see CMO872
DIRECT RED 23 see CMO870
DIRECT RED 28 see SGQ500
DIRECT RED 39 see CMO875
DIRECT RED 4A see DXO850
DIRECT RED 4B see DXO850

DIRECT RED 81 see CMO885
DIRECT RED DCB see DXO850
DIRECT RED F see CMO870
DIRECT RED FR see CMO870
DIRECT RED Kh see CMO870
DIRECT RED M see CMO870
DIRECT RED M 10B see CMO872
DIRECT RED MN see CMO870
DIRECT RED 8BS see CMO880
DIRECT ROSE MN see CMO870
DIRECT SCARLET see CMO870
DIRECT SCARLET LIGHTFAST see CMO870
DIRECT SCARLET 3BS see CMO875
DIRECT SCARLET 4BS see CMO870
DIRECT SCARLET SE see CMO870
DIRECT SKY BLUE A see CMO500
DIRECT VIOLET C see CMP000
DIRECT YELLOW MTZ see CMP050
DIRECT YELLOW TZ see CMP050
DIREKTAN see NCQ900
DIREMA see CFY000
DIREN see UVJ450
DIRESORCYL SULFIDE see BJE500
DIRESUL BLACK P see CMS250
DIRESUL BORDEAUX BS see CMS257
DIRESUL CORINTH R see CMS257
DIREX 4L see DXQ500
DIREXIODE see DNF600
DIREZ see DEV800
DIRI 2434 see PFT700
DIRI 2538 see MNV760
DIRI 2635 see CKH120
DIRI 2656 see MNV765
DIRI 2657 see MFG520
DIRIAN see BOL325
DIRIDONE see PDC250
DIRIDONE see PEK250
DIRIMAL see OJY100
DIRNATE see SNL890
DIRONYL see DLR150
DIROX see HIM000
DIRUBIDIUM CARBONATE see RPB200
DIRUBIDIUM MONOCARBONATE see RPB200
DISADINE see PKE250
DISALCID see SAN000
N,N'-DISALICYCLIDENE-1,2-PROPANEDIAMINE see DWS400
DISALICYLALPROPYLENEDIIMINE see DWS400
DISALICYLIC ACID see SAN000
N,N'-DISALICYLIDENE-1,2-DIAMINOPROPANE see DWS400
N,N'-DISALICYLIDENE ETHYLENEDIAMINE see DWY200
DISALUNIL see CFY000
DISALYL see SAN000
DISAN see DNO800
DISATABS TABS see VSK600
DISCO LIPIODOL see LGK225
DISCOLITE see FMW000
DISDOLEN see PHA575
DISELENIDE, BIS(2,2-DIETHOXYETHYL)- see BJA200
DISELENIDE, DIBUTYL-(9CI) see BRF550
DISELENIDE, DIETHYL see EIN550
DISELENIDE, DIPROPYL-(9CI) see PNI850
DISELENIUM DICHLORIDE see SBS500
α,α'-DISELENOBIS-o-ACETOTOLUIDIDE see DWY400
4,4'-DISELENOBIS(2-AMINOBUTYRIC ACID) see SBU710
2,2'-DISELENOBIS(N-PHENYLACETAMIDE) see DWY600
3,3'-DISELENODIALANINE see DWY800
3,3'-DISELENODIALANINE see SBP600
d-3,3'-DISELENODIALANINE see SBU300
l-3,3'-DISELENODIALANINE see SBU303
meso-3,3'-DISELENODIALANINE see DWY900
p,p'-DISELENODIANILINE see DWZ000

β,β'-DISELENODIPROPIONIC ACID, SODIUM SALT see DWZ100
DISELENO SALICYLIC ACID see SBU150
DI-SEPTON see DAO500
DISETIL see DXH250
DISFLAMOLL TKP see TNP500
DISFLAMOLL TOF see TNI250
1,4-DISILABUTANE, 1,1,1,4,4,4-HEXAMETHOXY- see EIT200
DISILANE see DXA000
DISILAZANE, 1,1,3,3-TETRAMETHYL-1,3-DIVINYL- see TDQ050
DISILOXANE, BIS(AMINOBUTYL)TETRAMETHYL- see OKU300
DISILOXANE, 1,3-BIS(3-AMINOPROPYL)-1,1,3,3-TETRAMETHYL- see OPC100
DISILOXANE, 1,3-DIETHENYL-1,1,3,3-TETRAMETHYL- see TDQ075
DISILOXANE, 1,3-DIPHENYL-1,1,3,3-TETRAMETHYL- see DWN150
DISILOXANE, 1,1,3,3-TETRAMETHYL-1,3-DIVINYL- see TDQ075
DISILVER ACETYLIDE SILVER NITRATE see SDJ025
DISILVER CARBONATE see SDN200
DISILVER CYANAMIDE see DXA500
DISILVER KETENIDE see DXA600
DISILVER OXALATE see SDU000
DISILVER OXIDE see SDU500
DISILVER PENTATIN UNDECAOXIDE see DXA800
DISILYN see BEN000
DISIPAL see MJH900
DISIPAL HYDROCHLORIDE see OJW000
DI-SIPIDIN see ORU500
DISNOGALAMYCINIC ACID see DXA900
DISNOGAMYCIN see NMV600
DISODIUM ADENOSINE TRIPHOSPHATE see AEM100
DISODIUM ADENOSINE 5'-TRIPHOSPHATE see AEM100
DISODIUM 4-AMINO-5-HYDROXY-2,7-NAPHTHALENEDISULFONATE see HAA400
DISODIUM ANTHRAQUINONE-1,5-DISULFONATE see DLJ700
DISODIUM ARSENATE see ARC000
DISODIUM ARSENATE, HEPTAHYDRATE see ARC250
DISODIUM ARSENIC ACID see ARC000
DISODIUM ATP see AEM100
DISODIUM AUROTHIOMALATE see GJC000
DISODIUM-4,4'-BIS((4-AMINO-6-(2-HYDROXYETHYL)AMINO-s-TRIAZIN-2-YL)AMINO)-2,2'-STILBENDISULFONIC ACID see DXB400
DISODIUM-4,4'-BIS((4-ANILINO-6-ETHYLAMINO-1,3,5-TRIAZIN-2-YL)AMINO)STILBENE-2,2'-DISULFONATE see DXB500
DISODIUM-4,4'-BIS((4-ANILINO-6-METHOXY-s-TRIAZIN-2-YL)AMINO)STILBENE-2,2'-DISULFONATE see BGW100
DISODIUM 4,4'-BIS((4-ANILINO-6-MORPHOLINO-1,3,5-TRIAZIN-2-YL)AMINO)STILBENE-2,2'-DISULFONATE see CMP200
DISODIUM-4,4'-BIS((4,6-DIANILINO-1,3,5-TRIAZIN-2-YL)AMINO)STILBENE-2,2'-DISULFONATE see DXB450
DISODIUM BISETHYLPHENYLTRIAMINOTRIAZINE STILBENEDISULFONATE see DXB500
DISODIUM-4,4'-BIS(2-SULFOSTYRYL)BIPHENYL see TGE150
DISODIUM BROMOSULFOPHTHALEIN see HAQ600
DISODIUM CALCIUM EDTA see CAR780

DISODIUM CALCIUM ETHYLENEDIAMINETETRAACETATE see CAR780
DISODIUM CARBONATE see SFO000
DISODIUM-N-(3-(CARBOXYMETHYLTHIOMERCURI)-2-METHOXYPROPYL)-α-CAMPHORAMATE see TFK270
DISODIUM CHROMATE see DXC200
DISODIUM CHROMOGLYCATE see CNX825
DISODIUM CINNAMYLIDENE BISULFITE derivative of SULFAPYRIDINE see DXF400
DISODIUM CITRATE see DXC400
DISODIUM CLODRONATE see SFX730
DISODIUM CROMOGLICATE see CNX825
DISODIUM CROMOGLYCATE see CNX825
DISODIUM DEXAMETHASONE PHOSPHATE see DAE525
DISODIUM DIACID ETHYLENEDIAMINETETRAACETATE see EIX500
DISODIUM-3,3'-DIAMINO-4,4'-DIHYDROXYARSENOBENZENE-N-DIMETHYLENESULFONATE see SNR000
DISODIUM-3,3'-DIAMINO-4,4'-DIHYDROXYARSENOBENZENE-N,N'-DIMETHYLENEBISULFITE see SNR000
DISODIUM p,p'-DIAMINODIPHENYLSULFONE-N,N'-DIGLUCOSE SULFONATE see AOO800
DISODIUM-2,7-DIBROM-4-HYDROXY-MERCURI-FLUORESCEIN see MCV000
DISODIUM-2',7'-DIBROMO-4'-(HYDROXYMERCURY)FLUORESCEIN see MCV000
DISODIUM (DICHLOROMETHYLENE)BISPHOSPHONATE see SFX730
DISODIUM DICHROMATE see SGI000
DISODIUM DIFLUORIDE see SHF500
DISODIUM DIHYDROGEN ATP see AEM100
DISODIUM DIHYDROGEN ETHYLENEDIAMINETETRAACETATE see EIX500
DISODIUM DIHYDROGEN(ETHYLENEDINITRILO)TETRAACETATE see EIX500
DISODIUM DIHYDROGEN-(1-HYDROXYETHYLIDENE)DIPHOSPHONATE see DXD400
DISODIUM DIHYDROGEN PYROPHOSPHATE see DXF800
DISODIUM-1,3-DIHYDROXY-1,3-BIS-(aci-NITROMETHYL)-2,2,4,4-TETRAMETHYLCYCLO BUTANE see DXC600
N,N'-DISODIUM N,N'-DIMETHOXYSULFONYLDIAMIDE see DXC800
DISODIUM S,S'-(2-DIMETHYLAMINO-1,3-PROPANEDIYL)BIS(THIOSULFATE) see DXC900
DISODIUM (2,4-DIMETHYLPHENYLAZO)-2-HYDROXYNAPHTHALENE-3,6-DISULFONATE see FMU070
DISODIUM (2,4-DIMETHYLPHENYLAZO)-2-HYDROXYNAPHTHALENE-3,6-DISULPHONATE see FMU070
DISODIUM DIOXIDE see SJC500
DISODIUM DIPHOSPHATE see DXF800
DISODIUM DISULFITE see SII000
DISODIUM-4,4'-DISULFOXYDIPHENYL-(2-PYRIDYL)METHANE see SJJ175
DISODIUM EDATHAMIL see EIX500
DISODIUM EDETATE see EIX500
DISODIUM EDTA (FCC) see EIX500
DISODIUM-3,6-ENDOXOHEXAHYDROPHTHALATE see DXD000
DISODIUM EOSIN see BNH500

DISODIUM-3,6-EPOXYCYCLOHEXANE-1,2-DICARBOXYLATE see DXD000
DISODIUM (−)-(1R,2S)-(1,2-EPOXYPROPYL)PHOSPHONATE HYDRATE see FOL200
DISODIUM ETHANOL-1,1-DIPHOSPHONATE see DXD400
DISODIUM ETHYDRONATE see DXD400
DISODIUM ETHYLENEBIS(DITHIOCARBAMATE) see DXD200
DISODIUM ETHYLENE-1,2-BISDITHIOCARBAMATE see DXD200
DISODIUM ETHYLENEDIAMINETETRAACETATE see EIX500
DISODIUM ETHYLENEDIAMINETETRAACETIC ACID see EIX500
DISODIUM (ETHYLENEDINITRILO)TETRAACETATE see EIX500
DISODIUM (ETHYLENEDINITRILO)TETRAACETIC ACID see EIX500
DISODIUM ETIDRONATE see DXD400
DISODIUM FLUOROPHOSPHATE see DXD600
DISODIUM FOSFOMYCIN see DXF600
DISODIUM FOSFOMYCIN HYDRATE see FOL200
DISODIUM FUMARATE see DXD800
DISODIUM GLYCYRRHIZIN see DXD875
DISODIUM GLYCYRRHIZINATE see DXD875
DISODIUM GMP see GLS800
DISODIUM-5'-GMP see GLS800
DISODIUM-5'-GUANYLATE see GLS800
DISODIUM-5'-GUANYLATE mixed with DISODIUM 5'-INOSINATE (1:1) see RJF400
DISODIUM GUANYLATE (FCC) see GLS800
DISODIUM HEXAFLUOROSILICATE see DXE000
(2-)-DISODIUM HEXAFLUOROSILICATE see DXE000
DISODIUM HYDROGEN ARSENATE see ARC000
DISODIUM HYDROGEN CITRATE see DXC400
DISODIUM HYDROGEN NITRILOTRIACETATE see DXF000
DISODIUM HYDROGEN NITRILOTRIACETATE, MONOHYDRATE see NHK850
DISODIUM HYDROGEN ORTHOARSENATE see ARC000
DISODIUM HYDROGEN PHOSPHATE see SJH090
DISODIUM-6-HYDROXY-3-OXO-9-XANTHENE-o-BENZOATE see FEW000
DISODIUM 2-HYDROXY-5-((4-((4-SULFOPHENYL)AZO)PHENYL)AZO)BENZOATE see CMP882
DISODIUM-5,5'-((2-HYDROXYTRIMETHYLENE)DIOXY)-BIS(4-OXO-4H-1-BENZOPYRAN-2-CARBOXYLATE) see CNX825
DISODIUM-3-HYDROXY-4-((2,4,5-TRIMETHYLPHENYL)AZO)-2,7-NAPHTHALENEDISULFONATE see FAG018
DISODIUM-3-HYDROXY-4-((2,4,5-TRIMETHYLPHENYL)AZO)-2,7-NAPHTHALENEDISULFONIC ACID see FAG018
DISODIUM-3-HYDROXY-4-((2,4,5-TRIMETHYLPHENYL)AZO)-2,7-NAPHTHALENEDISULPHONATE see FAG018
DISODIUM-3-HYDROXY-4-((2,4,5-TRIMETHYLPHENYL)AZO)-2,7-

NAPHTHALENEDISULPHONIC ACID see FAG018
DISODIUM IMINODIACETATE see DXE200
DISODIUM IMP see DXE500
DISODIUM INDIGO-5,5-DISULFONATE see FAE100
DISODIUM INOSINATE see DXE500
DISODIUM-5'-INOSINATE see DXE500
DISODIUM-5'-INOSINATE mixed with DISODIUM 5'-GUANYLATE (1:1) see RJF400
DISODIUM INOSINE-5'-MONOPHOSPHATE see DXE500
DISODIUM INOSINE-5'-PHOSPHATE see DXE500
DISODIUM LATAMOXEF see LBH200
DISODIUM METASILICATE see SJU000
DISODIUM METHANEARSENATE see DXE600
DISODIUM METHANEARSONATE see DXE600
DISODIUM METHOTREXATE see MDV600
DISODIUM METHYLARSENATE see DXE600
DISODIUM METHYLARSONATE see DXE600
DISODIUM MOLYBDATE see DXE800
DISODIUM MOLYBDATE DIHYDRATE see DXE875
DISODIUM MONOFLUOROPHOSPHATE see DXD600
DISODIUM MONOHYDROGEN ARSENATE see ARC000
DISODIUM MONOHYDROGEN PHOSPHATE see SJH090
DISODIUM MONOMETHYLARSONATE see DXE600
DISODIUM MONOSELENIDE see SJT000
DISODIUM MONOSILICATE see SJU000
DISODIUM MONOXIDE see SIN500
DISODIUM NITRILOTRIACETATE see DXF000
DISODIUM NITRILOTRIACETATE IRON(II) CHELATE see DXF100
DISODIUM NITRILOTRIACETIC ACID MONOHYDRATE see NHK850
DISODIUM NITROPRUSSIDE DIHYDRATE see SIW500
DISODIUM NITROSYLPENTACYANOFERRATE see SIU500
DISODIUM OCTABORATE, TETRAHYDRATE see DXF200
DISODIUM ORTHOPHOSPHATE see SJH090
DISODIUM-7-OXABICYCLO(2.2.1)HEPTANE-2,3-DICARBOXYLATE see DXD000
DISODIUM OXIDE see SIN500
DISODIUM PEROXIDE see SJC500
DISODIUM-2-(p-(γ-PHENYLPROPYLAMINO)BENZENESULFONAMIDO) PYRIDINE see DXF400
DISODIUM PHOSPHATE see SJH090
DISODIUM PHOSPHONOMYCIN see DXF600
DISODIUM PHOSPHONOMYCIN HYDRATE see FOL200
DISODIUM PHOSPHORIC ACID see SJH090
DISODIUM PHOSPHOROFLUORIDATE see DXD600
DISODIUM PREDNISOLONE 21-PHOSPHATE see PLY275
DISODIUM PROTOPORPHYRIN see DXF700
DISODIUM PYROPHOSPHATE see DXF800
DISODIUM PYROSULFITE see SII000
DISODIUM-5'-RIBONUCLEOTIDE see RJF400
DISODIUM SALT of EDTA see EIX500
DISODIUM SALT of ENDOTHALL see DXD000
DISODIUM SALT of 1-INDIGOTIN-S,S'-DISULPHONIC ACID see FAE100

DISODIUM SALT of 7-OXABICYCLO(2.2.1)HEPTANE-2,3-DICARBOXYLIC ACID see DXD000
DISODIUM SALT of 2-(4-SULPHO-1-NAPHTHYLAZO)-1-NAPHTHOL-4-SULPHONIC ACID see HJF500
DISODIUM SALT of 1-p-SULPHOPHENYLAZO-2-NAPHTHOL-6-SULPHONIC ACID see FAG150
DISODIUM SALT of 1-(2,4-XYLYLAZO)-2-NAPHTHOL-3,6-DISULFONIC ACID see FMU070
DISODIUM SALT of 1-(2,4-XYLYLAZO)-2-NAPHTHOL-3,6-DISULPHONIC ACID see FMU070
DISODIUM SELENATE see DXG000
DISODIUM SELENITE see SJT500
DISODIUM SEQUESTRENE see EIX500
DISODIUM SILICOFLUORIDE see DXE000
DISODIUM 2-(4-STYRYL-3-SULFOPHENYL)-7-SULFO-2H-NAPHTHO(1,2-d)TRIAZOLE see DXG025
DISODIUM SUCCINATE see SJW100
DISODIUM SULBENICILLIN see SNV000
DISODIUM SULFATE see SJY000
DISODIUM SULFITE see SJZ000
DISODIUM SULFOBENZYLPENICILLIN see SNV000
DISODIUM α-SULFOBENZYLPENICILLIN see SNV000
DISODIUM-2-(4-SULFO-1-NAPHTHYLAZO)-1-NAPHTHOL-4-SULFONATE see HJF500
DISODIUM-2-(4-SULPHO-1-NAPHTHYLAZO)-1-NAPHTHOL-4-SULPHONATE see HJF500
DISODIUM TARTRATE see BLC000
DISODIUM l-(+)-TARTRATE see BLC000
DISODIUM TETRABORATE see DXG035
DISODIUM TETRACEMATE see EIX500
DISODIUM TETRAOXATUNGSTATE (2-) see SKN500
DISODIUM TETRAOXOTUNGSTATE (2-) see SKN500
DISODIUM-5-TETRAZOLAZOCARBOXYLATE see DXG050
DISODIUM TUNGSTATE see SKN500
DISODIUM VERSENATE see EIX500
DISODIUM VERSENE see EIX500
DISOFEN see DNG000
DISOLFURO DI TETRAMETILTIOURAME (ITALIAN) see TFS350
DISOMAR see DXE600
DISOMER MALEATE see DXG100
2,4-D ISOOCTYL ESTER see ILO000
DISOPHENOL see DNG000
2,4-D ISOPROPYL ESTER see IOY000
DISOPYRAMIDE see DNN600
DISOQUIN see DNF600
DISORLON see CCK125
DISOTAT see DNM400
DISPADOL see DAM700
DISPAL see AHE250
DISPAMIL see PAH250
DISPARICIDA see ABX500
DISPARLURE see ECB200
DISPASOL M see PKB500
DISPERGATOR NF see BLX000
DISPERGATOR REAX see LFQ500
DISPERGATOR UFOXANE see LFQ500
DISPERMINE see PIJ000
DISPERSE BLUE 7 see DMM400
DISPERSE BLUE 56 see CMP070
DISPERSE BLUE 56 see DBQ220
DISPERSE BLUE 59 see DBQ220
DISPERSE BLUE 71 see DBQ220
DISPERSE BLUE 73 see CMP075
DISPERSE BLUE 78 see BKP500
DISPERSE BLUE 110 see BKP500
DISPERSE BLUE GREEN see DMM400
DISPERSE BLUE K see MGG250

DISPERSE BLUE NO 1 see TBG700
DISPERSE BLUE PE see DBQ220
DISPERSE BLUE POLYESTER see DBQ220
DISPERSE BORDEAUX S see CMP080
DISPERSE BRILLIANT PINK see DBX000
DISPERSE BRILLIANT ROSE see DBX000
DISPERSED BLUE 12195 see FAE000
DISPERSED ORANGE 11348 see FAG150
DISPERSED VIOLET 12197 see FAG120
DISPERSE DYE FAST YELLOW 4K see CMP090
DISPERSED YELLOW 12116 see FAG150
DISPERSE FAST PINK B see AKE250
DISPERSE FAST VIOLET B see AKP250
DISPERSE FAST YELLOW 2K see DUW500
DISPERSE FAST YELLOW 4K see CMP090
DISPERSE FAST YELLOW G see AAQ250
DISPERSE MB-61 see TFC600
DISPERSE ORANGE see AKP750
DISPERSE PINK Zh see AKO350
DISPERSE POLYESTER BLUE see DBQ220
DISPERSE POLYESTER PINK 2S see AKI750
DISPERSE RED 1 see ENP100
DISPERSE RED-4 see AKO350
DISPERSE RED 11 see DBX000
DISPERSE RED 13 see CMP080
DISPERSE RED 15 see AKE250
DISPERSE RED 25 see AKE250
DISPERSE RED 60 see AKI750
DISPERSE RED ZH see ENP100
DISPERSER NF see BLX000
DISPERSE ROSE Zh see AKO350
DISPERSE SCARLET B see ENP100
DISPERSE SCARLET ZH see ENP100
DISPERSE VIOLET 2S see DBY700
DISPERSE VIOLET 4S see AKP250
DISPERSE VIOLET K see DBP000
DISPERSE YELLOW 3 see AAQ250
DISPERSE YELLOW 7 see CMP090
DISPERSE YELLOW 6Z see MEB750
DISPERSE YELLOW G see AAQ250
DISPERSE YELLOW GWL see KDA075
DISPERSE YELLOW POLYESTER see KDA075
DISPERSE YELLOW R see DUW500
DISPERSE YELLOW STABLE 2K see DUW500
DISPERSE YELLOW Z see AAQ250
DISPERSING AGENT NF see BLX000
DISPERSIVE blue-green see DMM400
DISPERSIVE RUBY ZH see ENP100
DISPERSIVE YELLOW 3T see AAQ250
DISPERSOL ACA see BLX000
DISPERSOL BLUE B-R see CMP070
DISPERSOL BLUE B-R see DBQ220
DISPERSOL FAST CRIMSON B see CMP080
DISPERSOL FAST SCARLET B see ENP100
DISPERSOL FAST YELLOW A see DUW500
DISPERSOL FAST YELLOW G see AAQ250
DISPERSOL FAST YELLOW T see KDA075
DISPERSOL ORANGE D-G see AKE250
DISPERSOL PRINTING YELLOW A see DUW500
DISPERSOL PRINTING YELLOW G see AAQ250
DISPERSOL RED B 2B see AKI750
DISPERSOL RED B 3B see DBX000
DISPERSOL RED PP see SBC500
DISPERSOL RUBINE B see CMP080
DISPERSOL SCARLET B see ENP100
DISPERSOL VIOLET B see AKP250
DISPERSOL YELLOW A-G see AAQ250
DISPERSOL YELLOW B-A see DUW500
DISPERSOL YELLOW C-T see KDA075
DISPERSOL YELLOW PP see PEJ500
DISPERSTAT A see QAT565
DISPERSTAT W see QAT565
DISPEX C40 see ADV900
DISPHEX see PKE250
DISPHOLIDUS TYPHUS VENOM see DXG150

DISPIRO(CYCLOHEXANE-1,2'(3'H)-QUINAZOLINE-4',1"(4a'H)-CYCLOHEXANE), 5',6',7',8'-TETRAHYDRO- see TCQ600
3,6-DI(SPIROCYCLOHEXANE)TETRAOXANE see DXG200
DISPRANOL see BIK500
DISRUPT see ECB200
DISSENTEN see LIH000
DISSOLVANT APV see DJD600
DISTACLOR see CCR850
DISTAKAPS V-K see PDT750
DISTAMICINA A (ITALIAN) see SLI300
DISTAMINE see PAP550
DISTAMYCIN see DXG400
DISTAMYCIN A see SLI300
DISTAMYCIN A/4 see DXG450
DISTAMYCIN A/5 see DXG500
DISTAMYCIN A HYDROCHLORIDE see DXG600
DISTANNANE, HEXAISOPROPYL- see HDY100
DISTANNATHIANE, HEXABUTYL-(9CI) see HCA700
DISTANNATHIANE, HEXAKIS(PHENYLMETHYL)-(9CI) see BLK750
DISTANNOXANE, BIS(1,3-DITHIOCYANATO-1,1,3,3-TETRABUTYL)- see BJM700
DISTANNOXANE, 1,3-BIS(2,4,5-TRICHLOROPHENOXY)-1,1,3,3-TETRABUTYL- see OPE100
DISTANNOXANE, HEXAETHYL- see HCX050
DISTANNOXANE, HEXAOCTYL- see HEW100
DISTANNOXANE, 1,1,1,3,3,3-HEXAPROPYL- see BLT300
DISTANNTHIANE, HEXABUTYL- see HCA700
DISTANNTHIANE, HEXAETHYL- see HCX100
DISTANNTHIANE, HEXAOCTYL- see HEW150
DISTANNTHIANE, HEXAPROPYL- see HEW200
DISTAQUAINE V see PDT500
DISTAQUAINE V-K see PDT750
DISTARCH PHOSPHATE see SLJ550
DISTAVAL see TEH500
DISTAXAL see TEH500
DISTEARIN see OAV000
DISTEARYL DIMETHYLAMMONIUM CHLORIDE see DXG625
DISTEARYL THIODIPROPIONATE see DXG650
DISTEARYL 3,3'-THIODIPROPIONATE see DXG700
DISTEARYL β-THIODIPROPIONATE see DXG700
DISTEARYL β,β'-THIODIPROPIONATE see DXG700
DISTEARYL THIOPROPIONATE see DXG700
DISTERYL see CDT250
DISTESOL see EHP000
DISTESSOL see EHP000
DISTHENE see AHF500
DISTIGMINE BROMIDE see DXG800
DISTILBENE see DKA600
DISTILBENE see DKB000
DISTILLATE FUEL, MARINE, PETROLEUM DERIV. see DHE750
DISTILLATE PETROLEUM, CATALYTIC REFORMER FRACTIONATOR RESIDUE, LOW-BOILING see DXG850
DISTILLATES (COAL TAR) see CMY900
DISTILLATES (PETROLEUM), ACID-TREATED HEAVY NAPHTHENIC (9CI) see MQV760

DISTILLATES (PETROLEUM), ACID-TREATED LIGHT NAPHTHENIC (9CI) see MQV770
DISTILLATES (PETROLEUM), ACID-TREATED LIGHT PARAFFINIC (9CI) see MQV775
DISTILLATES (PETROLEUM), HEAVY CATALYTIC CRACKED see DXG810
DISTILLATES (PETROLEUM), HEAVY NAPHTHENIC (9CI) see MQV780
DISTILLATES (PETROLEUM), HEAVY PARAFFINIC (9CI) see MQV785
DISTILLATES (PETROLEUM), HYDRODESULFURIZED MIDDLE see DXG820
DISTILLATES (PETROLEUM), HYDROTREATED (severe) heavy paraffinic (9CI) see MQV796
DISTILLATES (PETROLEUM), HYDROTREATED (mild) HEAVY NAPHTHENIC (9CI) see MQV790
DISTILLATES (PETROLEUM), HYDROTREATED (mild) HEAVY PARAFFINIC (9CI) see MQV795
DISTILLATES (PETROLEUM), HYDROTREATED (mild) LIGHT NAPHTHENIC (9CI) see MQV800
DISTILLATES (PETROLEUM), HYDROTREATED (mild) LIGHT PARAFFINIC (9CI) see MQV805
DISTILLATES (PETROLEUM), HYDROTREATED MIDDLE see DXG830
DISTILLATES (PETROLEUM), LIGHT CATALYTIC CRACKED see DXG840
DISTILLATES (PETROLEUM), LIGHT NAPHTHENIC (9CI) see MQV810
DISTILLATES (PETROLEUM), LIGHT PARAFFINIC (9CI) see MQV815
DISTILLATES (PETROLEUM), SOLVENT-DEWAXED HEAVY NAPHTHENIC (9CI) see MQV820
DISTILLATES (PETROLEUM), SOLVENT-DEWAXED HEAVY PARAFFINIC (9CI) see MQV825
DISTILLATES (PETROLEUM), SOLVENT-DEWAXED LIGHT NAPHTHENIC (9CI) see MQV835
DISTILLATES (PETROLEUM), SOLVENT-DEWAXED LIGHT PARAFFINIC (9CI) see MQV840
DISTILLATES (PETROLEUM), SOLVENT-REFINED (mild) HEAVY NAPHTHENIC (9CI) see MQV845
DISTILLATES (PETROLEUM), SOLVENT-REFINED (mild) HEAVY PARAFFINIC (9CI) see MQV850
DISTILLATES (PETROLEUM), SOLVENT-REFINED (mild) LIGHT NAPHTHENIC (9CI) see MQV852
DISTILLATES (PETROLEUM), SOLVENT-REFINED (mild) LIGHT PARAFFINIC (9CI) see MQV855
DISTILLATES (PETROLEUM), STRAIGHT-RUN MIDDLE see GBW000
DISTILLATES RESIDUE, LOW-BOILING see DXG850
DISTILLED LIME OIL see OGM850
DISTILLED MUSTARD see BIH250
DISTIVIT (B12 PEPTIDE) see VSZ000
DISTOBRAM see TGI250
DISTOKAL see HCI000
DISTOL 8 see EIV000
DISTOPAN see HCI000
DISTOPIN see HCI000
DISTOVAL see TEH500
DISTRANEURIN see CHD750
DISTYLIN see DMD000
DISUL see CNW000
1,3-DI(4-SULFAMOYLPHENYL)TRIAZENE see THQ900
1,3-DISULFAMYL-4,5-DICHLOROBENZENE see DEQ200

DISULFAN see DXH250
DISULFATOZIRCONIC ACID see ZTJ000
DISULFIDE, BENZYL METHYL see MHN350
DISULFIDE, BIS(MORPHOLINOTHIOCARBONYL) see MRR090
DISULFIDE, BIS(2-NITRO-α-α-α-TRIFLUORO-p-TOLYL) see BLA800
DISULFIDE DIBENZYL see DXH200
DISULFIDE, DICYCLOHEXYL see CPK700
DISULFIDE DIPHENYL see PEW250
DISULFIDE, DI-2-PROPENYL (9CI) see AGF300
DISULFINE BLUE VN see ADE500
DISULFIRAM see DXH250
DISULFIRAM mixed with SODIUM NITRITE see SIS200
2,5-DISULFO-1-AMINOBENZENE see AIE000
2,5-DISULFOANILINE see AIE000
1,5-DISULFOANTHRAQUINONE see APK625
1,8-DISULFOANTHRAQUINONE see APK635
2,2'-DISULFOBENZIDINE see BBX500
3,5-DISULFOCATECHOL DISODIUM SALT see DXH300
DISULFO-meso-4,4-DIPHENYLHEXANE DIPOTASSIUM see SPA650
4,8-DISULFO-2-NAPHTHALAMINE see ALH250
3,5-DISULFOPYROCATECHOL see PPQ100
2,2'-DISULFO-4,4'-STILBENEDIAMINE DISODIUM SALT see FCA200
DISULFOTON see DXH325
DISULFOTON DISULIDE see OQS000
DISULFOTON SULFOXIDE see OQS000
DISULFURAM see DXH250
DISULFUR DIBROMIDE see DXH350
DISULFUR DICHLORIDE see SON510
DISULFUR DINITRIDE see DXH400
DISULFURE de TETRAMETHYLTHIOURAME (FRENCH) see TFS350
DISULFUR HEPTAOXIDE see DXH600
DISULFUROUS ACID, DISODIUM SALT see SII000
DISULFUR PENTOXYDICHLORIDE see PPR500
DISULFURYL CHLORIDE see PPR500
DISULFURYL DIAZIDE see DXH800
DISUL-Na see CNW000
DISULONE see SOA500
DISULPHINE BLUE AN see ERG100
DISULPHINE BLUE VN 150 see ADE500
DISULPHINE LAKE BLUE AN see ERG100
DISULPHINE LAKE BLUE EG see FMU059
DISULPHINE VN see ADE500
DISULPHURAM see DXH250
DISULPHURIC ACID see SOI520
DISUL-SODIUM see CNW000
DISYNCRAM see MPE250
DISYNCRAN see MDT500
DISYNCRAN see MPE250
DISYNFORMON see EDV000
DI-SYSTON see DXH325
DISYSTON SULFOXIDE see OQS000
DI-TAC see DXE600
DITAK see UVJ450
DITALLOW DIMETHYL AMMONIUM CHLORIDE see QAT550
DITAVEN see DKL800
DITAZOL MONOHYDRATE see BKB250
DITEFTIN see FQJ000
1,2-DI(5-TETRAZOLYL)HYDRAZINE see DXI000
1,3-DI(5-TETRAZOYL)TRIAZENE see DXI200
DITHALLIUM CARBONATE see TEJ000
DITHALLIUM OXIDE see TEL040
DITHALLIUM SULFATE see TEM000

DITHALLIUM(1+) SULFATE see TEM000
DITHALLIUM TRIOXIDE see TEL050
DITHANE see DXI300
DITHANE D-14 see DXD200
DITHANE M-45 see DXI400
DITHANE A-4 see DUQ600
DITHANE A-40 see DXD200
DITHANE M 22 see MAS500
DITHANE M 22 SPECIAL see MAS500
DITHANE R-24 see BPU000
DITHANE S60 see DXI400
DITHANE SPC see DXI400
DITHANE STAINLESS see ANZ000
DITHANE ULTRA see DXI400
DITHANE Z see EIR000
1,4-DITHIAANTHRAQUINONE-2,3-DICARBONITRILE see DLK200
1,4-DITHIAANTHRAQUINONE-2,3-DINITRILE see DLK200
1,4-DITHIACYCLOHEPTYLIDEN-6-IMINYL N-METHYL-CARBAMATE see DXI450
1,4-DITHIACYCLOHEXANE see DXI550
DITHIADENOXIDE see DXI480
DITHIADENOXID HYDROGEN MALEATE see DXI480
6,7-DITHIA-3,10-DIAZADODECANE, 3,10-DIETHYL- see TCD300
2,4-DITHIA-1,3-DIOXANE-2,2,4,4-TETRAOXIDE see DXI500
p-DITHIANE see DXI550
1,4-DITHIANE see DXI550
m-DITHIANE-2-CARBOXALDEHYDE, 2-METHYL-, o-(METHYLCARBAMOYL)OXIME see MJJ100
p-DITHIANE, 2,3-DEHYDRO-2,3-DIMETHYL-, TETROXIDE see OMY825
DITHIANON see DLK200
DITHIANONE see DLK200
1,4-DITHIAN-2-ONE, 3,3-DIMETHYL-, o-((METHYL((TRICHLOROMETHYL)THIO)AMINO)CARBONYL)OXIME see DRQ500
1,4-DITHIAN-2-ONE, o, o'-(DITHIOBIS((METHYLIMINO)CARBONYL))DIOXIME see BKQ780
1,4-DITHIAN-2-ONE, o-((METHYLAMINO)CARBONYL)OXIME see MID870
1,4-DITHIAN-2-ONE, o-((METHYL((TRICHLOROMETHYL)THIO)AMINO)CARBONYL)OXIME see MQC300
1,4-DITHIAN-2-ONE, o, o'-(THIOBIS((METHYLIMINO)CARBONYL))DIOXIME see BJM650
(((((1,4-DITHIAN-2-YLIDENEAMINO)OXY)CARBONYL)METHYLAMINO)THIO)METHYLCARBAMIC ACID, (1-METHYLETHYLIDENE)DI-4,1-PHENYLENE ESTER see DXI560
4,5-DITHIA-1,7-OCTADIENE see AGF300
1,3,2-DITHIARSENOLANE, 2-(p-BIS(2-CHLOROETHYL)AMINOPHENYL)- see BHW300
1,3,2-DITHIARSENOLANE, 2-(p-(DIETHYLAMINO)PHENYL)- see DIP100
1,3,2-DITHIARSENOLE, 2-CHLORODIHYDRO- see EIU900
1,3,2-DITHIARSOLANE, 2-CHLORO- see EIU900
3,6-DITHIA-3,4,5,6-TETRAHYDROPHTHALIMIDE see DLK700
2,6-DITHIA-1,3,5,7-TETRAZATRICYCLO(3.3.1.1^{3,7})DECANE-2,2,6,6-TETROXIDE see TDX500
DITHIAZANINE see DXI600
DITHIAZANINE IODIDE see DJT800
DITHIAZANIN IODIDE see DJT800
4H-1,3,5-DITHIAZINE, DIHYDRO-5-NITROSO-2,4,6-TRIMETHYL- see NLU600
DITHIAZININE see DJT800
(R,S)-α-(1-((3,3-DI-3-THIENYLALLYL)AMINO)ETHYL))BENZYL

ALCOHOL (+)-(α)-HYDROCHLORIDE see DXI800
(+)-α-(1-(93,3-DI-3-THIENYLALLYL)AMINO)ETHYL)BENZYL ALCOHOL HYDROCHLORIDE see DXI800
1-((3,3-DI-2-THIENYL-1-METHYL)ALLYL)PYRROLIDINE HYDROCHLORIDE see DXJ100
3-(DI(2-THIENYL)METHYLENE)-5-METHYLDECAHYDROQUINOLIZANIUM BROMIDE see TGF075
3-(DI-2-THIENYLMETHYLENE)-1-METHYLPIPERIDINE see BLV000
3-(DI-2-THIENYLMETHYLENE)-1-METHYLPIPERIDINE CITRATE see ARP875
3-(DI-2-THIENYLMETHYLENE)-5-METHYL-trans-QUINOLIZIDINIUM BROMIDE see TGF075
3,3-DI-2-THIENYL-N,N,1-TRIMETHYLALLYLAMINE HYDROCHLORIDE see TLQ250
1,4-DITHIEPAN-6-ONE, o-((METHYLAMINO)CARBONYL)OXIME see DXI450
1,4-DITHIEPAN-6-ONE, o-(METHYLCARBAMOYL)OXIME see DXI450
1,3-DITHIETAN-2-YLIDENE PHOSPHORAMIDIC ACID DIETHYL ESTER see DHH200
1,4-DITHIIN, 2,3-DIHYDRO-5,6-DIMETHYL-, 1,1,4,4-TETRAOXIDE see OMY825
DITHIO see SOD100
2,2'-DITHIOBIS(N-(1-ADAMANTYLMETHYL)ACETAMIDINE) DIHYDROCHLORIDE HEMIHYDRATE see DXJ400
O,O-DITHIO-BIS-ANILINE see DXJ800
2,2'-DITHIOBISANILINE see DXJ800
2',2'''-DITHIOBISBENZANILIDE see BDK800
4,4'-DITHIOBISBENZENAMINE see ALW100
2,2'-DITHIOBIS(BENZOIC ACID) see BHM300
2,2'-DITHIOBIS(BENZOTHIAZOLE) see BDE750
2,2'-DITHIOBIS(N,N-DIETHYLETHANAMINE) see TCD300
1,1'-DITHIOBIS(N,N-DIETHYLTHIOFORMAMIDE) see DXH250
2,2'-DITHIOBIS(N,N-DIMETHYLETHYLAMINE) DIHYDROCHLORIDE see BJG100
1,1'-DITHIOBIS(N,N-DIMETHYLTHIO)FORMAMIDE see TFS350
α,α'-DITHIOBIS(DIMETHYLTHIO)FORMAMIDE see TFS350
2,2'-DITHIOBIS(ETHYLAMINE) see MCN500
2,2'-DITHIO-BIS-(ETHYLAMINE) DIHYDROCHLORIDE see CQJ750
3,3'-DITHIOBIS(METHYLENE)BIS(5-HYDROXY-6-METHYL-4-PYRIDINEMETHANOL) DIHYDROCHLORIDE see BMB000
2,2'-(DITHIOBIS(METHYLENE))BIS(1-METHYL-5-NITRO-1H-IMIDAZOLE) see IAM040
DITHIOBISMORPHOLINE see BKU500
4,4'-DITHIOBIS(MORPHOLINE) see BKU500
3,3'-DITHIOBIS(6-NITROBENZOIC ACID) see DUV700
2,2'-DITHIOBIS(5-NITROPYRIDINE) see DXL200
2,2'-DITHIOBIS(PYRIDINE-1-OXIDE)MAGNESIUM SULFATE TRIHYDRATE see DXL400
DITHIOBIS(THIOFORMIC ACID) O,O-DIBUTYL ESTER see BSS550
DITHIOBIS(THIOFORMIC ACID)-o,o-DIETHYL ESTER see BJU000
2,2'''-DITHIOBISTRIETHYLAMINE see TCD300
2,5-DITHIOBIUREA see BLJ250

DITHIOBIURET see DXL800
DITHIOCARB see SGJ000
DITHIOCARB see SGJ500
DITHIOCARBAMATE see SGJ000
DITHIOCARBAMOYLHYDRAZINE see MLJ500
DITHIOCARBANILIC ACID ETHYL ESTER see EOK550
DITHIOCARBONIC ACID-o-BUTYL ESTER POTASSIUM SALT see PKY850
DITHIOCARBONIC ACID-o-sec-BUTYL ESTER SODIUM SALT see DXM000
DITHIOCARBONIC ACID O-ISOPROPYL ESTER POTASSIUM SALT see IRG050
DITHIOCARBONIC ACID O-PENTYL ESTER POTASSIUM SALT see PKV100
DITHIOCARBONIC ANHYDRIDE see CBV500
DITHIOCARBOXYMETHYL-p-CARBAMIDOPHENYLARSENOUS OXIDE see DXM100
DITHIOCARBOXYPHENYL-p-CARBAMIDOPHENYLARSENOUS OXIDE see ARK800
p,p-DITHIOCYANATODIPHENYLAMINE see DXM200
1,2-DITHIOCYANATOETHANE see EJC035
1,2-DITHIOCYANOETHANE see EJC035
DITHIODEMETON see DXH325
β,β'-DITHIODIALANINE see CQK325
2,2'-DITHIODIANILINE see DXJ800
4,4'-DITHIODIANILINE see ALW100
p,p'-DITHIODIANILINE see ALW100
2',2'''-DITHIODIBENZANILIDE see BDK800
2,2'-DITHIODIBENZOESAEURE see BHM300
2,2'-DITHIODIBENZOIC ACID see BHM300
N,N'-(DITHIODICARBONOTHIOYL)BIS(N-METHYLMETHANAMINE) see TFS350
4,4'-(DITHIODICARBONOTHIOYL)BISMORPHOLINE see MRR090
N,N'-(DITHIODI-2,1-ETHANEDIYL)BISGUANIDINE see GLC200
2,2-DITHIODIETHANOL see DXM600
1,1'-DITHIODIETHYLENEBIS(3-(2-(CHLOROETHYL)-3-NITROSOUREA)) see BIF625
DITHIODIGLYCOL see DXM600
3,3'-DITHIODIMETHYLENEBIS(5-HYDROXY-6-METHYL-4-PYRIDINEMETHANOL) DIHYDROCHLORIDE HYDRATE see BMB000
5,5'-DITHIODIMETHYLENEBIS(2-METHYL-3-HYDROXY-4-HYDROXYMETHYLPYRIDINE)DIHYDROCHLORIDE HYDRATE see BMB000
N,N-DITHIODIMORPHOLINE see BKU500
4,4'-DITHIODIMORPHOLINE see BKU500
N,N'-(DITHIODI-2,1-PHENYLENE)BISBENZAMIDE see BDK800
DITHIODIPHOSPHORIC ACID, TETRAETHYL ESTER see SOD100
2,2'-DITHIODIPYRIDINE-1,1'-DIOXIDE see DXN300
DITHIOERYTHRITOL see DXN350
1,4-DITHIOERYTHRITOL see DXN350
DITHIOETHYLENEGLYCOL see EEB000
1,1'-DITHIOFORMAMIDINE DIHYDROCHLORIDE see DXN400
DITHIOFOS see SOD100
DITHIOGLYCEROL see BAD750
1,2-DITHIOGLYCEROL see BAD750
DITHIOGLYCOL see EEB000
DITHIOGLYCOLYL p-ARSENOBENZAMIDE see TFA350
DITHIOHYDANTOIN see IAT100
DITHIOLANE see DXN600
1,3-DITHIOLANE, 2-ACETYL-2-METHYL- see ACR050

1,3-DITHIOLANE-2,4-DIONE, 5,5-DIMETHYL-, 2-(DIMETHYHYDRAZONE), 4-(o-((METHYLETHYL((TRICHLOROMETHYL)THIO)AMINO)CARBONYL)OXIME) see DXN500
DITHIOLANE IMINOPHOSPHATE see DXN600
1,2-DITHIOLANE-3-PENTANAMIDE (9CI) see DXN709
1,3-DITHIOLANE-2-THIONE see EJQ100
1,2-DITHIOLANE-3-VALERAMIDE see DXN709
1,2-DITHIOLANE-3-VALERIC ACID see DXN800
1,3-DITHIOLAN-4-ONE, 5,5-DIMETHYL-2-(1,1-(DIMETHYLETHYL)IMINO)-, o-((METHYLAMINO)CARBONYL) OXIME see DXN820
1,3-DITHIOLAN-2-ONE, o-((((1,1-DIMETHYLETHYL)DITHIO)METHYLAMINO)CARBONYL)OXIME see MHX100
1,3-DITHIOLAN-4-ONE, 5,5-DIMETHYL-2-(ETHYLIMINO)-, o-((METHYLAMINO)CARBONYL)OXIME see DRV300
1,3-DITHIOLAN-4-ONE, 5,5-DIMETHYL-2-((1-METHYLETHYL)IMINO)-, o-((METHYLAMINO)CARBONYL)OXIME see DSP700
1,3-DITHIOLAN-4-ONE, 5,5-DIMETHYL-2-((1-METHYLETHYL)IMINO)-, o, o'-(THIOBIS((METHYLIMINO) CARBONYL))DIOXIME see BJK670
1,3-DITHIOLAN-4-ONE, o-((METHYL((TRICHLOROMETHYL)THIO)AMINO)CARBONYL)OXIME see MQC320
1,3-DITHIOLAN-2-YLIDENE-PHOSPHORAMIDOTHIOIC ACID DIETHYL ESTER see DXN600
1,3-DITHIOLAN-2-YLIDENE-PHOSPHORAMIDOTHIOIC ACID-O,O-DIETHYL ESTER see DXN600
4-((5-(1,2-DITHIOLAN-3-YL)-1-OXOPENTYL)AMINO)BUTANOIC ACID see LGK100
o-(1,3-DITHIOLAN-2-YL)PHENYL DIMETHYLCARBAMATE see DXN830
5-(1,2-DITHIOLAN-3-YL)VALERIC ACID see DXN800
1,3-DITHIOLIUM PERCHLORATE see DXN850
1,2-DITHIOLPROPANE see PML300
1,3-DITHIOL-2-YLIDENE-PROPANEDIOIC ACID BIS(1-METHYLETHYL) ESTER see MAO275
DITHIO-METHANEARSONOUS ACID BIS(ANHYDROSULFIDE) with DIMETHYLDITHIOCARBAMIC ACID see USJ075
DITHIOMETON (FRENCH) see PHI500
4,4'-DITHIOMORPHOLINE see BKU500
DITHION see DXO000
DITHION see SOD100
DITHIONE see DXO000
DITHIONE see SOD100
DITHIONIC ACID see SOI250
DI(THIONOCARBOMETHOXY) DISULFIDE see DUN600
DITHIONOUS ACID, ZINC SALT (1:1) see ZGJ000
DITHIONOUS ACID, ZINC SALT (1:1) see ZIJ100
6,8-DITHIOOCTANOIC ACID see DXN800
DITHIOOXALDIIMIDIC ACID see DXO200
DITHIOOXAMIDE see DXO200
DITHIOPHOS see SOD100
DITHIOPHOSPHATE de O,O-DIETHYLE et de (4-CHLORO-PHENYL) THIOMETHYLE (FRENCH) see TNP250
DITHIOPHOSPHATE de O,O-DIETHYLE et de S-(2-ETHYLTHIO-ETHYLE) see DXH325

DITHIOPHOSPHATE de O,O-DIETHYLE et d'ETHYLTHIOMETHYLE (FRENCH) see PGS000
DITHIOPHOSPHATE de O,O-DIETHYLE et de S-N-METHYL-N-CARBOETHOXY CARBAMOYLMETHYLE (FRENCH) see DJI000
DITHIOPHOSPHATE de-O,O-DIETHYLE et de S(2,5-DICHLOROPHENYL) THIOMETHYLE (FRENCH) see PDC750
DITHIOPHOSPHATE de O,O-DIMETHYLE et de S-((4,6-DIAMINO-1,3,5-TRIAZINE-2-YL)-METHYLE) (FRENCH) see ASD000
DITHIOPHOSPHATE de O,O-DIMETHYLE et de S-(1,2-DICARBOETHOXYETHYLE) (FRENCH) see MAK700
DITHIOPHOSPHATE de O,O-DIMETHYLE et de S-(2-ETHYLTHIO-ETHYLE) (FRENCH) see PHI500
DITHIOPHOSPHATE de O,O-DIMETHYLE et de S-((MORPHOLINOCARBONYLE)-METHYLE) (FRENCH) see MRU250
DITHIOPHOSPHATE de O,O-DIMETHYLE et de S-(-N-METHYLCARBAMOYL-METHYLE) (FRENCH) see DSP400
DI(THIOPHOSPHORIC) ACID, TETRAETHYL ESTER see SOD100
DITHIOPHOSPHORSAEURE-O-AETHYL-S,S-DIPHENYLESTER (GERMAN) see EIM000
2,3-DITHIOPROPANOL see BAD750
DITHIOPROPYLTHIAMINE see DXO300
DITHIOPROPYLTHIAMINE HYDROCHLORIDE see DXO400
DITHIOPYR see POQ600
DITHIOPYROPHOSPHATE de TETRAETHYLE (FRENCH) see SOD100
DITHIOSYSTOX see DXH325
DITHIOTEP see SOD100
DITHIOTEREPHTHALIC ACID see DXO600
DITHIOTHREITOL see DXO775
1,4-DITHIOTHREITOL see DXO800
dl-DITHIOTHREITOL see DXO775
d-1,4-DITHIOTHREITOL see DXO800
rac-DITHIOTHREITOL see DXO775
dl-1,4-DITHIOTHREITOL see DXO775
DITHIOTRIMETHYLENEGLYCOL see PML350
DITHIOXAMIDE see DXO200
DITHIZON see DWN200
DITHIZONE see DWN200
DITHRANOL see DMG900
DITHRANOL, 10-ACETYL- see ACI640
DITHRANOL, 10-BUTYRYL- see BSY400
DITHRANOL, 10-PROPIONYL- see PMW760
DITIAMINA see EIR000
DITILIN see CMG250
DITILIN see HLC500
DITILINE see CMG250
DITILINE see HLC500
DITILIN IODIDE see BJI000
DITIOVIT see DXO300
DITOIN see DNU000
DITOINATE see DKQ000
4,4'-DI-o-TOLUIDINE see TGJ750
1,4-DI-p-TOLUIDINOANTHRAQUINONE see BLK000
DI-o-TOLUYLTHIOUREA see DXP600
DITOLYLBIS(AZONAPHTHIONIC ACID) see DXO850
DITOLYLETHANE see DXP000
DI-p-TOLYL ETHER see THC600
DI-o-TOLYLGUANIDINE see DXP200
1,3-DI-o-TOLYLGUANIDINE see DXP200
1,3-DI-o-TOLYLGUANIDINIUM DIPYROCATECHOL BORATE see DEK500
N,N-DI(p-TOLYL)HYDRAZINE see DXP300
N,N'-DI-o-TOLYL-p-PHENYLENE DIAMINE see DXP400
DI-o-TOLYLTHIOUREA see DXP600
DITRAN see DXP800
DITRANIL see RDP300

DI-TRAPEX see ISE000
DI-TRAPEX see MLC000
DITRAZIN see DIW200
DITRAZIN CITRATE see DIW200
DITRAZINE see DIW200
DITRAZINE BASE see DIW000
DITRAZINE CITRATE see DIW200
1,3-DI(TRICHLOROMETHYL)BENZENE see BLL825
DITRIDECYLAMINE see DXQ000
DITRIDECYL PHTHALATE see DXQ200
DI(TRI-(2,2-DIMETHYL-2-PHENYLETHYL)TIN)OXIDE see BLU000
DITRIFON see TIQ250
DITRIPENTAT see CAY500
DI[TRIS-1,2-DIAMINOETHANECHROMIUM(III)]TRIPE ROXODISULFATE see DXQ339
DI[TRIS-1,2-DIAMINOETHANECOBALT(III)]TRIPEROX ODISULFATE see DXQ369
DITRIZOATE METHYLGLUCAMINE see AOO875
DITROPAN see OPK000
DITROSOL see DUS700
DITUBIN see ILD000
DIUCARDYN SODIUM see TFK270
DIULO see ZAK300
DI-n-UNDECYL KETONE see TJF250
DIUNDECYL PHTHALATE see DXQ400
DIUPRES see RDK000
DIURAL see CHJ750
DIURAMID see AAI250
DIURAPID see ASO375
1,1-DIUREIDISOBUTANE see IIV000
DIUREIDOISOBUTANE see IIV000
DIURESAL see CLH750
DIURESE see HII500
DIURETIC SALT see PKT750
DIURETICUM-HOLZINGER see AAI250
DIURETIN see SJO000
DIUREX see DXQ500
DIUREXAN see CLF325
DIURIL see CLH750
DIURILIX see CLH750
DIURITE see CLH750
DIUROBROMINE see TEO500
DIUROL see AMY050
DIUROL see DXQ500
DIUROL 5030 see AMY050
DIURON see DXQ500
DIURON 4L see DXQ500
DIURONE see CHX250
DIUTAZOL see AAI250
DIUTENSEN-R see RDK000
DIUTRID see CLH750
DIUXANTHINE see TEP500
DIVASCOL see BBW750
DIVERCILLIN see AIV500
DIVERCILLIN see AOD125
DIVERINE see NCK100
DIVERON see MQU750
DIVINYL see BOP500
DIVINYL ACETYLENE see HCU500
DIVINYL ADIPATE see HEO150
m-DIVINYLBENZEN see DXQ745
DIVINYLBENZENE see DXQ740
DIVINYLBENZENE see DXQ745
m-DIVINYLBENZENE see DXQ745
DIVINYLBENZENE COPOLYMER see DXQ750
DIVINYLENE OXIDE see FPK000
DIVINYLENE SULFIDE see TFM250
DIVINYLENIMINE see PPS250
DIVINYL ETHER (DOT) see VOP000
DIVINYL ETHER, inhibited (DOT) see VOP000
DIVINYL ETHER of TRIETHYLENE GLYCOL see TDY800
DIVINYLETHYLENE see HEZ000
DIVINYL MAGNESIUM see DXQ850

3,9-DIVINYLSPIROBI(m-DIOXANE) see DXR000
DIVINYL SULFONE see DXR200
2,5-DIVINYLTETRAHYDROPYRAN see DXR400
2,5-DIVINYLTETRAHYDRO-2H-PYRAN see DXR400
1,3-DIVINYLTETRAMETHYLDISILAZANE see TDQ050
1,3-DIVINYL-1,1,3,3-TETRAMETHYLDISILAZANE see TDQ050
1,3-DIVINYLTETRAMETHYLDISILOXANE see TDQ075
1,1'-DIVINYLTETRAMETHYLDISILOXANE see TDQ075
sym-DIVINYLTETRAMETHYLDISILOXANE see TDQ075
3,9-DIVINYL-2,4,8,10-TETRAOXASPIRO(5.5)UNDECANE see DXR000
DIVINYLTIN DICHLORIDE see DXR450
DIVINYL ZINC see DXR500
DIVIPAN see DGP900
DIVIT URTO see VSZ100
DIVULSAN see DNU000
DIVYNYL OXIDE see VOP000
DIXANTHOGEN see BJU000
DIXARIT see CMX760
DIXERAN see AEG875
DIXERAN see TDL000
DIXIBEN see EID000
DIXIE see CBT750
DIXIE see KBB600
DIXIECELL see CBT750
DIXIEDENSED see CBT750
DIXITHERM see CBT750
DIXON see FAB400
DIXON 164 see TAI250
DIXOPAK see PJS750
N,N-DI-(2,4-XYLYLIMINOMETHYL)METHYLAMINE see MJL250
DIXYRAZINE see MKQ000
DIXYRAZINE DIHYDROCHLORIDE see DXR800
(α-DIYLENE)POLY(p-AMINOBENZALDEHYDE-N) see DXS000
DIZENE see DEP600
DIZINC BIS(DIMETHYLDITHIOCARBAMATE)ETH YLENEBIS(DITHIOCARBAMATE) see BJK000
DIZINON see DCM750
DIZOXIDE see DCQ700
DJ-1461 see DQU400
DJ-1550 see SNL800
D. JAMESONI VENOM see DAP820
DKB see DCQ800
DKB SULFATE see DHA400
DKC 1347 see CLK215
DKD see DIZ100
DKhM see DGQ200
DKM (RUSSIAN) see DGR400
DL 111 see EOL600
DL 152 see DIG800
D. L. 152 see DIH000
242 DL see BGC625
DL 473 see RKZ100
dl-832 see TDI600
DL-8280 see OGI300
D,L-N-(2,6-DIMETHYLPHENYL)-N-(2'-METHOXYACETYL)ALANINATE de METHYLE see MDM100
DL 204-IT see EFC259
DL 717-IT see CKL325
DL 1047 N see MEE100
DL NORGESTREL (FRENCH) see NNQ500
DLP-87 see PPP750
DLP 787 see PPP750
trans,D,L-2-PHENYLCYCLOPROPYLAMINE SULFATE see PET500
DLT see TFD500

DLTDP see TFD500
DLTP see TFD500
DLX-6000 see TAI250
DM see DAC000
DM see PDB000
1 DM 10 see XSS900
DMA see DOO800
DMA see DOQ800
DMA see DXE600
DMA see DXS200
DMA-4 see DAA800
DMAA see HKC000
DMAB see DOR200
DMAB see DOT300
3,2'-DMAB see BLV250
DMAC see DOO800
DMAE see DOY800
DMAE p-ACETAMIDOBENZOATE see DOZ000
DMAEE see BJH750
DMAE TARTRATE see DPA200
D 9998 MALEATE see FMP100
DMAM see DKP400
DMAMP see DPM400
DMAP see VTF000
DMASA see DQD400
DMB see DOA400
DMBA see DQJ200
7,12-DMBA see DQJ200
DMBC see DQQ200
DMBCA see BEL750
DMBEB see DDH900
DMC see BIN000
DMC see SMP450
DMCBAC see BEL900
DMCC see DQY950
DMCM see EID600
DMCT see MIJ500
DMDHEU see DTG000
DMDJ 4309 see PJS750
DMDJ 5140 see PJS750
DMDJ 7008 see PJS750
DMDK see SGM500
DMDPN see DRQ200
DMDT see MEI450
p,p'-DMDT see MEI450
DMDZ see CGA500
DMEP see DOF400
DMES see DRK500
2,4-D METHYL ESTER see DAA825
DMF see BJE750
DMF see DSB000
DMFA see DSB000
DMH see DSF300
DMH see DSF400
DMH see DSF600
DMH see DSF800
DMH see DSG200
DMHQ see DSG700
DMI see DSI709
DMI see DSK950
DMI see IAL200
DMI 50475 see DSI709
DMI HYDROCHLORIDE see DLS600
DMM see BKM500
DMMP see DSR400
DMMPA see DST800
DMN see NKA600
DMNA see NKA600
DMNM see DSV200
DMNM see DTA000
DMNO see DSV200
DMN-OAC see AAW000
DMNT see DSG400
DMO see PMO250
DMP see DTR200
DMP see DTU850
DMP see TNH000
2,5-DMP see XKS000
2,6-DMP see XLA000
DMP-30 see TNH000
3,4-DMP see XLJ000

3,5-DMP see XLS000
DMPA see DGD800
DMPD see DTL800
DMPE see DOE200
DMPEA see DOE200
DMPP see DTO000
DMPP IODIDE see DTO000
DMPS see DNU860
DMPT see DTP000
DMPTP see TFD500
DMS see DNV800
DMS see DUC300
DMS see DUD100
DMS see DUD400
DMS see TFP000
DMS-70 see DUD800
DMS-90 see DUD800
DMSA see DNV800
DMSA see DQD400
DMSC see HOL200
trans-DMS-DIACETATE see DXS300
DMS(METHYL SULFATE) see DUD100
DMSO see DUD800
DMSP see FAQ800
DMT see DPF600
DMTP see FAQ900
DMTP (JAPAN) see DSO000
DMTT see DSB200
DMU see DTG700
DMU see DXQ500
DN-111 see CPK550
DN 289 see BRE500
DNA see DAZ000
DNA see DUP600
DNAASE see EDK650
DNA DEPOLYMERASE see EDK650
DNA, D(G-SP-T-G-G-T-G-G-G-T-G-G-G-T-
G-G-G-SP-T) see ZNS400
DNA ENDONUCLEASE see EDK650
DNA NUCLEASE see EDK650
DNASE I see EDK650
DNASW see EDK650
DNBA see DUR600
DNBP see BRE500
DNBP AMMONIUM SALT see BPG250
DNBS see SNA500
DNCB see CGM000
DN CUST D-4 see CPK550
DN DRY MIX No. 1 see CPK500
DN-DRY MIX No. 2 see DUS700
DN DUST No. 12 see CPK500
2,4-DNFB see DUW400
1,3-DNG see GGA200
DN-IMI see DKV720
DNNS see FBZ100
DNOC AMMONIUM SALT see DUT800
DNOCHP see CPK500
DNOC-SODIUM see SGP550
DNOC SODIUM SALT see DUU600
DNOK (CZECH) see DUS700
DNOK-ACETAT (CZECH) see AAU250
DNOP see DVL600
DNOSBP see BRE500
DNP see BIN500
DNP see DNG000
2,4-DNP see DUZ000
2,5-DNP see DVA000
DNPC see DUT600
DNPD see NBL000
2,4-DNPH see DVC400
DNPMT see DVF400
DNPOH see DVD200
DNPT see DVF400
DNPZ see DVF200
DNRB see DVF800
DNSBP see BRE500
DNT see DVH000
2,3-DNT see DVG800
2,4-DNT see DVH000
2,5-DNT see DVH200
2,6-DNT see DVH400

3,4-DNT see DVH600
3,5-DNT see DVH800
DNTB see DVF800
DNTBP see DRV200
DNTP see PAK000
DO 9 see DXY000
DO 14 see SOP000
DOA see AEO000
DOA see AOO500
DOBANIC ACID 83 see LBU100
DOBANIC ACID JN see LBU100
DOBANOL 45 see AFJ168
DOBANOL 23-3 see AFJ155
DOBANOL 23-5 see AFJ155
DOBANOL 25-3 see AFN950
DOBANOL 25-7 see AFJ160
DOBANOL 25-9 see AFJ160
DOBANOL 45-3 see AFJ168
DOBANOL 911 see AFJ150
DOBANOL 25-11 see AFJ160
DOBANOL 45-11 see AFJ168
DOBANOL 45E4 see AFJ168
DOBANOL 45E7 see AFJ168
DOBANOL 23-4.5/6 see AFJ155
DOBANOX 25I see AFJ160
DOBENDAN see CCX000
DOBESILATE CALCIUM see DMI300
DOBETIN see VSZ000
DOBO see TEP500
DOBREN see EPD500
DOBUTAMINE HYDROCHLORIDE see
DXS375
DOBUTREX see DXS375
DOCA see DAQ800
DOCA ACETATE see DAQ800
DOC-AC see DAQ800
DOC ACETATE see DAQ800
DOCEMINE see VSZ000
DOCETAXEL see TAH800
DOCETAXOL see TAH800
DOCIBIN see VSZ000
DOCIGRAM see VSZ000
DOCITON see ICB000
DOCITON see ICC000
DOCTAMICINA see CDP250
DOCUSATE SODIUM see DJL000
DODAT see DAD200
DODDLE-DO (PUERTO RICO) see CAK325
DODECABEE see VSZ000
DODECACARBONYLDIVANADIUM see
DXS400
DODECACARBONYL TETRACOBALT see
CNB510
DODECACARBONYLTRIIRON see DXS600
DODECACHLOROOCTAHYDRO-1,3,4-
METHENO-2H-
CYCLOBUTA(c,d)PENTALENE see MQW500
1,1a,2,2,3,3a,4,5,5,5a,5b,6-
DODECACHLOROOCTAHYDRO-1,3,4-
METHENO-1H-
CYCLOBUTA(c,d)PENTALENE see MQW500
DODECACHLOROPENTACYCLODECANE
see MQW500
DODECACHLOROPENTACYCLO(3,2,2,0^{2,6},0
^{3,9},0^{5,10})DECANE see MQW500
8E,10E-DODECADIEN-1-OL see CNG760
(E,E)-8,10-DODECADIEN-1-OL see CNG760
trans-8,trans-10-DODECADIEN-1-OL see
CNG760
8,10-DODECADIEN-1-OL, (E,E)- see CNG760
DODECAHYDRODIPHENYLAMINE see
DGT600
DODECAHYDRO-7,14-METHANO-2H,6H-
DIPYRIDO(1,2-a:1',2'-e)(1,5)DIAZOCINE see
SKX500
DODECAHYDROPHENYLAMINE NITRITE
see DGU200
Δ-DODECALACTONE see DXS700
Δ-DODECALACTONE see HBP450
γ-DODECALACTONE see OES100
1,12'-DODECAMETHYLENEDIAMINE see
DXW800

DODECAMETHYLENE DIBROMIDE see
DDN300
DODECAMETHYLPENTASILOXANE see
DXS800
n-DODECAN (GERMAN) see DXT200
1-DODECANAL see DXT000
DODECANAMIDE, N,N-DIETHYL- see
DJD500
1-DODECANAMINE (9CI) see DXW000
DODECANAMINE ACETATE see DXW050
DODECANAMINE HYDROCHLORIDE see
DXW060
1-DODECANAMINE, HYDROCHLORIDE
(9CI) see DXW060
1-DODECANAMINIUM, N-
(CARBOXYMETHYL)-N,N-DIMETHYL-,
HYROXIDE, inner salt see LBU200
1-DODECANAMINIUM, N,N,N-
TRIMETHYL-, CHLORIDE see LBX075
DODECANE see DXT200
1,12-DODECANEDIAMINE see DXW800
1,12'-DODECANEDIAMINE see DXW800
DODECANE, 1,12-DIBROMO- see DDN300
DODECANENITRILE see DXT400
1-DODECANEPHOSPHONIC ACID see
DXT500
n-DODECANEPHOSPHONIC ACID see
DXT500
1-DODECANETHIOL see LBX000
tert-DODECANETHIOL see DXT800
DODECANOIC ACID see LBL000
DODECANOIC ACID BENZYL ESTER see
BEU750
DODECANOIC ACID, CADMIUM SALT
(9CI) see CAG775
DODECANOIC ACID-2,3-
DIHYDROXYPROPYL ESTER see MRJ000
DODECANOIC ACID, 1-OXO-2-(2,4,6-
TRIMETHYLPHENYL)-1H-INDEN-3-YL
ESTER see OPC045
DODECANOIC ACID, 2-
THIOCYANATOETHYL ESTER see LBO000
1-DODECANOL see DXV600
n-DODECANOL see DXV600
DODECANOL ACETATE see DXV400
1-DODECANOL ACETATE see DXV400
DODECANOL, ETHOXYLATE see DXY000
DODECANOL ETHOXYLATED see EEU000
DODECANOL-ETHYLENE OXIDE (9.5
moles) CONDENSATE see DXY000
DODECANOLIDE-1,4 see OES100
1-DODECANOL, 2-OCTYL- see OEW100
DODECANOL, POLYETHOXYLATED see
DXY000
1-((n-DODECANOYLOXY)METHYL)-3-
METHYLIMIDAZOLIUM CHLORIDE see
HMG500
3-(DODECANOYLOXYMETHY)-1-
METHYL-1H-IMIDAZOLIUM CHLORIDE
see MKR150
DODECANOYL PEROXIDE see LBR000
DODECAN-1-YL ACETATE see DXV400
1,3,5,7,9-DODECAPENTAENE, 1-(2-
HYDROXYPROPOXY)-, (ALL-E)- see
AFQ820
1,3,5,7,9-DODECAPENTAENE, 1-PROPOXY-
, (ALL-E)- see DXT900
1'-(1,3,5,7,9-
DODECAPENTAENYLOXY)PROPANE see
DXT900
2'-(1,3,5,7,9-DODECAPENTAENYLOXY)-2'-
PROPANOL see AFQ820
1,6,10-DODECATRIEN-3-OL, 3,7,11-
TRIMETHYL-, ACETATE, (S-(Z))- see
NCN800
4,6,10-DODECATRIEN-3-ONE, 7,11-
DIMETHYL- see DRR700
DODECATRIETHYLAMMONIUM
BROMIDE see DXU200
5,7,11-DODECATRIYN-1-OL see DXU250
DODECAVITE see VSZ000
1-DODECEN see DXU260

2-DODECENAL see DXU280
trans-2-DODECEN-1-AL see DXU300
DODECENE EPOXIDE see DXU400
9a-DODECENOATE see DXU600
(Z)-7-DODECEN-1-OL see DXU800
(E)-9-DODECENOL ACETATE see DXU822
7-DODECEN-1-OL, ACETATE, (Z)- see
DXU830
9-DODECEN-1-OL, ACETATE, (Z)- see
GJU050
(Z)-7-DODECEN-1-OLACETATE see
DXU830
9-DODECEN-1-OL, ACETATE, (9E)- see
DXU822
(Z)-7-DODECENYL ACETATE see DXU830
(Z)-9-DODECENYL ACETATE see GJU050
cis-7-DODECENYL ACETATE see DXU830
cis-9-DODECENYL ACETATE see GJU050
DODECENYLSUCCINIC ANHYDRIDE see
DXV000
DODECOIC ACID see LBL000
DODECONIUM see HEF200
N-DODECYL ACETAMIDE see LBR100
DODECYL ACETATE see DXV400
n-DODECYL ACETATE see DXV400
DODECYL ALCOHOL see DXV600
n-DODECYL ALCOHOL see DXV600
DODECYL ALCOHOL ACETATE see
DXV400
DODECYL ALCOHOL CONDENSED with 4
MOLES ETHYLENE OXIDE see LBS000
DODECYL ALCOHOL CONDENSED with 7
MOLES ETHYLENE OXIDE see LBT000
DODECYL ALCOHOL CONDENSED with
23 MOLES ETHYLENE OXIDE see LBU000
DODECYL ALCOHOL, ETHOXYLATED see
DXY000
DODECYL ALCOHOL, HYDROGEN
SULFATE, SODIUM SALT see SIB600
1-DODECYL ALDEHYDE see DXT000
DODECYLAMINE see DXW000
1-DODECYLAMINE see DXW000
n-DODECYLAMINE see DXW000
DODECYLAMINE, ACETATE see DXW050
1-DODECYLAMINE ACETATE see DXW050
DODECYLAMINE, HYDROCHLORIDE see
DXW060
n-DODECYLAMINE HYDROCHLORIDE see
DXW060
1-DODECYL-4-AMINOQUINALDINIUM
ACETATE see AJS750
N-DODECYL-4-AMINOQUINALDINIUM
ACETATE see AJS750
DODECYLAMMONIUM CHLORIDE see
DXW060
n-DODECYLAMMONIUM CHLORIDE see
DXW060
DODECYL AMMONIUM SULFATE see
SOM500
DODECYLBENZENE see PEW500
DODECYL BENZENE SODIUM
SULFONATE see DXW200
DODECYL BENZENESULFONATE see
DXW400
n-DODECYLBENZENESULFONIC ACID
see LBU100
DODECYLBENZENESULFONIC ACID
(DOT) see LBU100
DODECYLBENZENESULFONIC ACID
COMPD. WITH 2-PROPANAMINE (1:1) see
BBS275
DODECYLBENZENESULFONIC ACID,
compd. with 2,2',2"-NITRILOTRIS(ETHANOL)
(1:1) see TJK800
DODECYLBENZENESULFONIC ACID
SODIUM SALT see DXW200
DODECYLBENZENESULFONIC ACID
TRIETHANOLAMINE SALT see TJK800
DODECYLBENZENESULPHONATE,
SODIUM SALT see DXW200
DODECYLBENZENESULPHONIC ACID see
LBU100

DODECYLBENZENSULFONAN SODNY
(CZECH) see DXW200
DODECYLBENZYL CHLORIDE see
DXW600
DODECYLBETAINE see LBU200
DODECYLBIS(AMINOETHYL)GLYCINE
HYDROCHLORIDE see DYA850
N-DODECYL-N,N-BIS(2-
HYDROXYETHYL)BENZENEMETHANAM
INIUM CHLORIDE see BDJ600
DODECYLDIAMINE see DXW800
DODECYLDIMETHYLAMINE see DRR800
N-DODECYLDIMETHYLAMINE see
DRR800
DODECYLDIMETHYLAMINE OXIDE see
DRS200
N-DODECYLDIMETHYLAMINE OXIDE see
DRS200
(DODECYLDIMETHYLAMMONIO)ACETA
TE see LBU200
N-DODECYL-N,N-
DIMETHYLBENZENEMETHANAMINIUM
BROMIDE see BEO000
DODECYL DIMETHYL
BENZYLAMMONIUM CHLORIDE see
BEM000
DODECYLDIMETHYLBETAINE see LBU200
DODECYLDIMETHYL(3,4-
DICHLOROBENZYL)AMMONIUM
CHLORIDE see LBV100
DODECYLDIMETHYL(2-
PHENOXYETHYL)AMMONIUM BROMIDE
see DXX000
DODECYL-6,6-DIOCTYL-4,8,11-TRIOXO-
5,7,12-TRIOXA-6-STANNATETRACOSA-2,9-
DIENOATE see DVN100
DODECYL-DI(β-OXYAETHYL)-BENZYL-
AMMONIUMCHLORID see BDJ600
1,12'-DODECYLENEDIAMINE see DXW800
DODECYLESTER KYSELINY GALLOVE see
DXX200
1-DODECYL-1-ETHYLPIPERIDINIUM
BROMIDE see DXX100
DODECYL GALLATE see DXX200
N-DODECYLGUANIDINACETAT
(GERMAN) see DXX400
DODECYLGUANIDINE ACETATE see
DXX400
N-DODECYLGUANIDINE ACETATE see
DXX400
DODECYLGUANIDINE ACETATE with
SODIUM NITRITE (3:5) see DXX600
DODECYLGUANIDINE
HYDROCHLORIDE see DXX800
1-DODECYLHEXAHYDRO-1H-AZEPINE-1-
OXIDE see DXX875
1-DODECYLHEXAMETHYLENIMINE-N-
OXIDE see DXX875
α-DODECYL-ω-HYDROXYPOLY(OXY-1,2-
ETHANEDIYL) PHOSPHATE see PKE850
α-DODECYL-ω-HYDROXY-
POLYOXYETHYLENE see DXY000
2-DODECYLISOQUINOLINIUM BROMIDE
see LBW000
1-DODECYL MERCAPTAN see LBX000
m-DODECYL MERCAPTAN see LBX000
tert-DODECYLMERCAPTAN see DXT800
DODECYL MERCAPTAN (ACGIH) see
LBX000
tert. DODECYLMERKAPTAN (CZECH) see
DXT800
DODECYL METHACRYLATE see DXY200
3-DODECYL-1-METHYL-2-
PHENYLBENZIMIDAZOLIUM
FERRICYANIDE see TNH750
1-DODECYL-3-METHYL-2-PHENYL-1H-
BENZIMIDAZOLIUM,
HEXACYANOFERRATE(III) see TNH750
DODECYL-2-METHYL-2-PROPENOATE see
DXY200
4-DODECYLMORPHOLINE-4-OXIDE see
DXY300

4-DODECYLMORPHOLINE-N-OXIDE see
DXY300
DODECYL NITRATE see NEE000
2-(DODECYLOXY)ETHANOL HYDROGEN
SULFATE SODIUM SALT see DYA000
2-(2-(2-
(DODECYLOXY)ETHOXY)ETHOXY)ETHA
NESULFONIC ACID, SODIUM SALT see
SIC000
2-(2-(2-
(DODECYLOXY)ETHOXY)ETHOXY)ETHA
NOL HYDROGEN SULFATE SODIUM
SALT see SIC000
N-(2-DODECYLOXYETHYL)-N-METHYL-2-
(PYRROLIDINYL)ACETAMIDE
HYDROCHLORIDE see DXY400
DODECYLPHENOL (mixed isomers) see
DXY600
DODECYL PHENYLMERCURI SULFIDE see
DYA400
1-DODECYLPIPERIDINE-1-OXIDE see
DXY700
1-DODECYLPIPERIDINE-N-OXIDE see
DXY700
DODECYL-POLYAETHYLENOXYD-
AETHER (GERMAN) see DXY000
DODECYL POLY(OXYETHYLENE)ETHER
see DXY000
DODECYLPYRIDINIUM CHLORIDE see
DXY725
DODECYLPYRIDINIUM CHLORIDE see
LBX050
1-DODECYLPYRIDINIUM CHLORIDE see
DXY725
1-DODECYLPYRIDINIUM CHLORIDE see
LBX050
N-DODECYLPYRIDINIUM CHLORIDE see
DXY725
N-DODECYLPYRIDINIUM CHLORIDE see
LBX050
N-DODECYLPYRROLIDINONE see
DXY750
1-DODECYL-2-PYRROLIDINONE see
DXY750
N-DODECYLSARCOSINE SODIUM SALT
see DXZ000
DODECYL SODIUM ETHOXYSULFATE see
DYA000
DODECYL SODIUM SULFATE see SIB600
DODECYL SODIUM SULFOACETATE see
SIB700
DODECYL SULFATE see MRH250
DODECYL SULFATE, SODIUM SALT see
SIB600
DODECYLSULFURIC ACID see MRH250
n-DODECYL THIOCYANATE see DYA200
α-(2-(tert-DODECYLTHIO)ETHYL)-ω-
HYDROXYPOLY(OXY-1,2-ETHANEDIYL)
see PKE350
tert-DODECYLTHIOL see DXT800
(DODECYLTHIO)PHENYLMERCURY see
DYA400
DODECYL-p-TOLYL TRIMETHYL
AMMONIUM CHLORIDE see DYA600
DODECYLTRICHLOROSILANE see DYA800
DODECYLTRIMETHYLAMMONIUM
BROMIDE see DYA810
DODECYLTRIMETHYLAMMONIUM
CHLORIDE see LBX075
6-DODECYL-2,2,4-TRIMETHYL-1,2-
DIHYDROQUINOLINE see SAU480
DODEMORFE (FRANCE) see COX000
DODEX see VSZ000
DODGUADINE see DXX400
DODICIN HYDROCHLORIDE see DYA850
DODIGEN 1383 see HCQ525
DODIGEN 2617 see DTF820
DODINE see DXX400
DODINE ACETATE see DXX400
DODINE, mixture with GLYODIN see
DXX400

DODINE with SODIUM NITRITE (3:5) see DXX600
1,4-DOEA-5,8-DAPFA (RUSSIAN) see DMM400
DOF see DVK600
DOFENAPYN see PCA400
DOFSOL see VSK600
DOG HOBBLE see DYA875
DOG LAUREL see DYA875
DOGMATIL see EPD500
DOGMATYL see EPD500
DOG PARSLEY see FMU200
DOG POISON see FMU200
DOGQUADINE see DXX400
DOISYNOESTROL see BIS750
DOISYNOESTROL see BIS750
DOISYNOESTROL see BIT000
DOISYNOLIC ACID see DYB000
DOJYOPICRIN see CKN500
DOKIRIN see BLC250
DOKTACILLIN see AIV500
DOL see BBQ750
β-DOLABRIN see DYB100
DOLADENE see DJM800
DOLADENE see XCJ000
DOLAFLUX see HGM100
DOLAN, combustion products see ADX750
DOLANTAL see DAM700
DOLANTIN see DAM700
DOLANTIN HYDROCHLORIDE see DAM700
DOLANTIN-N-OXIDE HYDROCHLORIDE see DYB250
DOLANTOL see DAM700
DOLAREN see DAM700
DOLARGAN see DAM700
DOLCENTAL see EQL600
DOLCOL see PIZ000
DOLCO MOUSE CEREAL see SMN500
DOLCONTRAL see DAM600
DOLCONTRAL see DAM700
DOLCYMENE see CQI000
DOLEAN pH 8 see ADA725
DOLENAL see DAM700
DOLENE see DAB879
DOLENE see PNA500
DOLENOL see DAM700
DOLEN-PUR see HCD250
DOLESTAN see BAU750
DOLESTINE see DAM700
DOLGIN see IIU000
DOL GRANULE see BBQ500
DOLICUR see DUD800
DOLIGUR see DUD800
DOLIN see DAM700
DOLIPOL see BSQ000
DOLIPRANE see HIM000
DOLISINA see DII200
DOLKWAL AMARANTH see FAG020
DOLKWAL BRILLIANT BLUE see FAE000
DOLKWAL ERYTHROSINE see FAG040
DOLKWAL INDIGO CARMINE see FAE100
DOLKWAL ORANGE SS see TGW000
DOLKWAL PONCEAU 3R see FAG018
DOLKWAL SUNSET YELLOW see FAG150
DOLKWAL TARTRAZINE see FAG140
DOLKWAL YELLOW AB see FAG130
DOLKWAL YELLOW OB see FAG135
DOLLS EYES see BAF325
DOLMIX see BBQ750
DOLOBID see DKI600
DOLOBIL see DKI600
DOLOBIS see DKI600
DOLOCAP see PNA500
DOLOCHLOR see CKN500
DOLOCONEURASE see DYB300
DOLOGAL see DAM700
DOLOMATE see GGG050
DOLOMIDE see SAH000
DOLOMITE see CAO000
DOLONEURINE see DAM700
DOLONIL see PDC250

DOLOPETHIN see DAM700
DOLOPHINE see MDO750
DOLOPHINE see MDP000
DOLOPHINE HYDROCHLORIDE see MDP000
DOLOPHINE HYDROCHLORIDE see MDP750
d-DOLOPHINE HYDROCHLORIDE see MDP240
DOLOPHIN HYDROCHLORIDE see MDP750
DOLOSAL see DAM600
DOLOSAL see DAM700
DOLOVIN see IDA500
DOLOXENE see DAB879
DOLOXENE see DYB400
DOLOXENE see PNA500
DOLPHINE see MDP000
DOLSIMA see DII200
DOLSIN see DAM600
DOLVANOL see DAM700
DOLVIRAN see SBN400
DOM see BJR000
DOM see SLU600
DOMAGK'S T.B.1 CONTEBEN see FNF000
DOMAIN see SNJ350
DOMAKOL see FNF000
DOMALIUM see DCK759
DOMAR see CAT775
DOMAR see PIH100
DOMARAX see IPU000
DOMATOL see AMY050
DOMATOL 88 see AMY050
DOMESTROL see DKA600
DOMF see MCV000
DOMICAL see EAI000
DOMICILLIN see SEQ000
DOMINAL see DYB600
DOMINAL HYDROCHLORIDE see DYB800
(−)-DOMOIC ACID see DYB825
DOMOIC ACID see DYB825
l-DOMOIC ACID see DYB825
DOMOSO see DUD800
DOMPERIDONE see DYB875
DOMUCOR see ELH600
DON see DCQ400
DONAXINE see DYC000
DONMOX see AAI250
DONOPON see SBG500
DONOREST see CKI750
DONOVAN'S SOLUTION see ARI500
DOOJE see HBT500
DOP see DVL700
(−)-DOPA see DNA200
l-DOPA see DNA200
DOPAC see HGI980
DOPACETIC ACID see HGI980
DOPAFLEX see DNA200
l-DOPA HYDROCHLORIDE see DYC200
DOPAL see DNA200
DOPAMET see DNA800
DOPAMET see MJE780
l-DOPA METHYL ESTER see DYC300
DOPAMINE see DYC400
DOPAMINE CHLORIDE see DYC600
DOPAMINE HYDROCHLORIDE see DYC600
DOPAN see DYC700
DOPANE see DYC700
DOPARKINE see DNA200
DOPASOL see DNA200
DOPATEC see MJE780
DOPEGYT see DNA800
DOPEGYT see MJE780
DOPIDRIN see DBA800
DOPN see NJN000
DOPRAM see ENL100
DOPRAM see SLU000
DOPRIN see DNA200
DOPTAEC see DNA800
DORADO see DGJ160
DORAL see VSZ100

DORANTAMIN see WAK000
DORAPHEN see PNA500
DORBANE see DMH400
DORBANEX see DMH400
DORCOSTRIN see DAQ800
DOREVANE see IDA500
ZZ-DORICIDA see NCN650
ZZ-DORICIDE see NCN650
DORICO see ERD500
DORICO SOLUBLE see ERE000
DORIDEN see DYC800
DORIDEN-SED see DYC800
DORINAMIN see BBW500
DORISUL see SNN300
DORLOTYN see AMX750
DORLYL see PKQ059
DORM see AFS500
DORMABROL see MQU750
DORMAL see CDO000
DORMALLYL see AFS500
DORMATE see MBW750
DORME see PMI750
DORMETHAN see DBE200
DORMIDIN see EQL000
DORMIGENE see BNP750
DORMIGOA see QAK000
DORMIN see TEO250
DORMIPHEN see EQL000
DORMIRAL see EOK000
DORMITURIN see BNK000
DORMODOR see DAB800
DORMOGEN see QAK000
DORMONAL see BAG000
DORMONE see DAA800
DORMOSAN see EQL000
DORMUPHAR see SKS700
DORMUTIL see QAK000
DORMWELL see SKS700
DORMYTAL see AMX750
DORNASE see EDK650
DORNAVA see EDK650
DORNAVAC see EDK650
DORNWAL see ALY250
DORNWALL see ALY250
DORSACAINE see OPI300
DORSACAINE HYDROCHLORIDE see OPI300
DORSEDIN see QAK000
DORSIFLEX see MFD500
DORSILON see MFD500
DORSITAL see NBT500
DORSULFAN see SNN500
DORVICIDE A see BGJ750
DORVON see SMQ500
DORVON FR 100 see SMQ500
DORYL (PHARMACEUTICAL) see CBH250
DOS see BJS250
DOSAFLO see MQR225
DOSAGRAN see MQR225
DOSANEX see MQR225
DOSANEX FL see MQR225
DOSANEX MG see MQR225
DOSEGRAN see AFL750
DOSULEPIN see DYC875
DOSULEPIN CHLORIDE see DPY200
DOSULEPIN HYDROCHLORIDE see DPY200
D.O.T. see DUP300
DOTAN see CDY299
DOTC see DVN300
DOTG see BKK750
DOTG ACCELERATOR see DXP200
DOTHEIPIN HYDROCHLORIDE see DPY200
DOTHIEPIN see DYC875
DOTMENT 324 see AHE250
DOTRIACONTANE see DYC900
DOTYCIN see EDH500
DOUBLE STRENGTH see TIX500
DOUGLAS FIR OIL see OJK340
DOUXAN see NIC200
DOVENIX see HLJ500

DOVIP see FAB600
DOW 5 see XFS300
DOW 209 see SMR000
DOW 360 see SMQ500
DOW 456 see SMQ500
DOW 665 see SMQ500
DOW 860 see SMQ500
DOW 874 see CGW300
DOW 1329 see DGD800
DOW 1683 see SMQ500
DOWANOL see CBR000
DOWANOL 33B see PNL250
DOWANOL-50B see DWT200
DOWANOL 62B see TNA000
DOWANOL DB see DJF200
DOWANOL DE see CBR000
DOWANOL DM see DJG000
DOWANOL DPM see DWT200
DOWANOL EB see BPJ850
DOWANOL EE see EES350
DOWANOL EIPAT see INA500
DOWANOL EM see EJH500
DOWANOL EP see PER000
DOWANOL EPH see PER000
DOWANOL PM see PNL250
DOWANOL (R) PMA GLYCOL ETHER
ACETATE see PNL265
DOWANOL PM GLYCOL ETHER see
PNL250
DOWANOL PPH GLYCOL ETHER see
PNL300
DOWANOL TE see EFL000
DOWANOL T 4PH see TCE450
DOWANOL TMAT see TJQ750
DOWANOL TPM see MEV500
DOWANOL TPM GLYCOL ETHER see
TNA000
DOWCC 132 see COD850
DOWCHLOR see CDR750
DOWCIDE 1 see BGJ250
DOWCIDE 7 see PAX250
DOWCIDE 1 ANTIMICROBIAL see BGJ250
DOWCO 109 see BQU500
DOWCO 118 see DGD800
DOWCO 133 see DYD200
DOWCO 139 see DOS000
DOWCO 159 see DYD400
DOWCO 160 see DYD600
DOWCO 161 see EOQ500
DOWCO-163 see CLP750
DOWCO 169 see PEV500
DOWCO 177 see DYD800
DOWCO 179 see CMA100
DOWCO 183 see DYE200
DOWCO 184 see CEG550
DOWCO 186 see HON000
DOWCO 187 see AGU500
DOWCO-213 see CQH650
DOWCO 217 see CMA250
DOWCO 233 see TJE890
DOWCO 269 see CIC600
DOWCO 290 see DGJ100
DOWCO 356 see TJK100
DOWCO 453EE see HAG850
DOWCO 453ME see HAG860
DOW CORNING 200 see HEE000
DOW CORNING 344 see PJR300
DOW CORNING 345 see DAF350
DOW CORNING 346 see PJR000
DOW CORNING 344 FLUID see PJR300
DOW CORNING 345 FLUID see DAF350
DOW-CORNING 200 FLUID-LOT No. AA-
4163 see DUB600
DOW CORNING SILICONE FLUID and
FLUOROHYDROCARBON see XGA000
DOW CORNING X 2-1401 see PJR300
DOW CORNING X1-6145A see EIT200
DOW DEFOLIANT see SFU500
DOW DORMANT FUNGICIDE see SJA000
DOWELL L 37 see NEI100
DOW ET 14 see RMA500
DOW ET 57 see RMA500

DOWFLAKE see CAO750
DOWFROST see PML000
DOWFROTH 250 see MKS250
DOWFUME see MHR200
DOWFUME 40 see EIY500
DOWFUME EB-5 see DYE400
DOWFUME EDB see EIY500
DOWFUME MC-2 SOIL FUMIGANT see
MHR200
DOWFUME N see DGG000
DOWFUME W-8 see EIY500
DOW GENERAL see BRE500
DOW GENERAL WEED KILLER see BRE500
DOWICIDE see BGJ750
DOWICIDE 2 see TIV750
DOWICIDE 4 see CHN500
DOWICIDE 6 see TBT000
DOWICIDE 7 see PAX250
DOWICIDE 9 see CFI750
DOWICIDE 2S see TIW000
DOWICIDE 31 see SFV500
DOWICIDE A see BGJ750
DOWICIDE A & A FLAKES see BGJ750
DOWICIDE B see SKK500
DOWICIDE B see TIV750
DOWICIDE EC-7 see PAX250
DOWICIDE G see PAX250
DOWICIDE G-ST see SJA000
DOWICIDE Q see CEG550
DOWICIL 75 see CEG550
DOWICIL 100 see CEG550
DOWIZID A see BGJ750
DOWLAP F see TKD400
DOW LATEX 612 see SMR000
DOW LATEX 874 see CGW300
DOWLEX see PKF750
DOWLEX FILM see PJS750
DOW MCP AMINE WEED KILLER see
CIR750
DOW MX 5514 see SMQ500
DOW MX 5516 see SMQ500
DOWMYCIN E see EDJ500
DOW PENTACHLOROPHENOL DP-2
ANTIMICROBIAL see PAX250
DOW-PER see PCF275
DOW PLASTICIZER 5 see XFS300
DOWPON see DGI400
DOWPON see DGI600
DOWPON M see DGI400
DOW SEED DISINFECTANT No. 5 see
TBO500
DOW SELECTIVE see BPG250
DOW SELECTIVE WEED KILLER see
BRE500
DOW SODIUM TCA INHIBITED see TII500
DOW SODIUM TCA SOLUTION see TII250
DOWSPRAY 9 see DDN900
DOWSPRAY 17 see CPK500
DOWTHERM see PFA860
DOWTHERM 209 see PNL250
DOWTHERM A see PFA860
DOWTHERM E see DEP600
DOWTHERM SR 1 see EJC500
DOW-TRI see TIO750
DOWZENE DHC see PIK000
D-OX see SHR500
DOXAL see TMJ150
DOXAPRAM see ENL100
DOXAPRAM HYDROCHLORIDE
HYDRATE see SLU000
DOXCIDE 50 see CDW450
DOXEPHRIN see DBA800
DOXEPIN see DYE409
DOXEPIN HYDROCHLORIDE see AEG750
DOXERGAN see AFL750
DOX HYDROCHLORIDE see HKA300
DOXICICLINA (ITALIAN) see DYE425
DOXIFLURIDINE see DYE415
DOXIGALUMICINA see HGP550
DOXINATE see DJL000
DOXIUM see DMI300

DOXO see DAQ800
DOXOL see DJL000
DOXORUBICIN see AES750
DOXORUBICIN see HKA300
DOXORUBICIN HYDROCHLORIDE see
HKA300
DOXYCYCLINE see DYE425
DOXYCYCLINE HYCLATE see HGP550
DOXYCYCLINE HYDROCHLORIDE see
HGP550
DOXYFED see DBA800
DOXYFED see MDT600
DOXY-II see HGP550
DOXYLAMINE see DYE500
DOXYLAMINE SUCCINATE see PGE775
DOXYLAMINE SUCCINATE (1:1) see
PGE775
DOXYPYRROMYCIN see DAY835
DOXY-TABLINEN see HGP550
DOZAR see DPJ400
2,4-DP see DGB000
(+)-2,4-DP see DGA880
2-(2,4-DP) see DGB000
DP-1904 see IAT300
DPA see DGI000
DPA see DVX800
DPA see HET500
n-DPA see PNR750
2,2-DPA see DGI600
DPAC see DVX500
D-P-A INJECTION see PAG200
DPA SODIUM see PNX750
DPBS see DFY400
DPC see DVY925
DPC see DXY725
DPC see LBX050
DPD see DJV200
DPD 63760H see MQR400
D & P DOUBLE O CRABGRASS KILLER see
PLC250
DPDP see IKL200
DPE-HCl see DWD200
DPF see DYE550
DPF-1 see DYE550
DPG see DNA300
DPG see DWC600
DPG ACCELERATOR see DWC600
2,4-D PGBE see DFZ000
DPH see DKQ000
DPH see DNU000
DPID see DLH600
DPMA see DMB200
DPN see CNF390
DPN see NKB700
DPNA see NKB700
DPO see DWI300
DPP see AON300
DPP see PAK000
DPP see PEK250
DPPD see BLE500
2,4-D PROPYLENE GLYCOL BUTYL
ETHER ESTER see DFZ000
DPS see PGI750
DPT see DCA600
3,5-DPT see DWO950
DPTA see DCB100
DPX 47 see ASH275
DPX 1108 see ANG750
DPX 1410 see DSP600
DP X 1410 see MME809
DPX 3217 see COM300
DPX 3654 see PMF550
DPX 3674 see HFA300
DPX 4189 see CMA700
DPX 6376 see MQR400
DPX 6774 see IRA050
DPX 3217M see COM300
DPX 43898 see FOL050
DPX-A 7881 see MJW900
DPX-A 8947 see ASH275
DPX-F 5384 see BAV600
DPX-H 6573 see THS860

DPX-L 5300 see THT800
DPX-T 6376 see MQR400
DPX-Y 6202 see QMA100
DPX Y5893-9 see CJU275
DQ2511 see DOK500
DQDA 1868 see PJS750
DQUIGARD see DGP900
DQV-K see PDT750
DQWA 0355 see PJS750
DR-15771 see SBE500
DRABET see BSQ000
DRACONIC ACID see AOU600
DRACYLIC ACID see BCL750
DRAGIL-P see SLL000
DRAGOCAL see CAS750
DRAGON ARUM see JAJ000
DRAGON ROOT see JAJ000
DRAGONS HEAD see JAJ000
DRAGON TAIL see JAJ000
DRAGONTHOL A see CMM760
DRAKEOL see MQV750
DRAKEOL see MQV875
DRALZINE see HGP500
DRAMAMIN see DYE600
DRAMAMINE see DYE600
DRAMARIN see DYE600
DRAMYL see DYE600
DRAPEX 4.4 see FAB920
DRAPOLENE see AFP250
DRAPOLEX see BBA500
DRASIL 507 see CNH125
DRAT see CJJ000
DRAWIN 755 see MPU250
DRAWINOL see CBW000
DRAZA see DST000
DRAZINE see PDV700
DRAZOXOLON see MLC250
DRAZOXOLONE see MLC250
DRB see DGK100
DRB see DVF800
DRC-714 see BIM000
DRC 1201 see IDJ550
DRC 1339 see CLK215
DRC-1,339 see CLK230
DRC 3340 see DTN200
DRC 3341 see MIB750
DRC 3345 see PFS350
DRC-4575 see AMU125
DRC 6246 see AGW675
DREFT see SIB600
DRENAMIST see VGP000
DRENE see SON000
DRENOBYL see DAL000
DRENOL see CFY000
DRENUSIL-R see RDK000
DREPAMON see TGF030
DREVOGENIN A see DYE650
DREWMULSE POE-SMO see PKL100
DREWMULSE TP see OAV000
DREWMULSE V see OAV000
DREWSORB 60 see SKV150
DREXEL see DXQ500
DREXEL DEFOL see SFS000
DREXEL DIURON 4L see DXQ500
DREXEL DSMA LIQUID see DXE600
DREXEL METHYL PARATHION 4E see
MNH000
DREXEL PARATHION 8E see PAK000
DREXEL-SUPER P see DMC600
DRIBAZIL see DYE700
DRICOL see AHL500
DRI-DIE PESTICIDE 67 see SCH002
DRIDOL see DYF200
DRIED WHEY see WBL150
DRIERITE see CAX500
DRI-KIL see RNZ000
DRILL TOX-SPEZIAL AGLUKON see
BBQ500
DRINALFA see DBA800
DRINALFA see MDT600
DRINOX see AFK250
DRINOX see HAR000

DRINUPAL HYDROCHLORIDE see CKB500
DRIOL see DYE700
DRIOL-LABAZ see DYE700
DRISDOL see VSZ100
DRISTAN INHALER see PNN400
DRI-TRI see SJH200
DROCODE see DKW800
DROCTIL see ERE200
DROGENIL see FMR050
DROLEPTAN see DYF200
DROMETRIZOLE see HML500
DROMILAC see DBV400
DROMISOL see DUD800
DROMORAN see MKR250
levo-DROMORAN see LFG000
racemic DROMORAN see MKR250
DROMORAN-HYDROBROMIDE see
HMH500
DROMORAN HYDROBROMIDE see
MDV250
l-DROMORAN TARTRATE see DYF000
DROMOSTAT see TKU675
DROMYL see DYE600
DRONACTIN see PCI500
DRONCIT see BGB400
DROPCILLIN see BDY669
DROPERIDOL see DYF200
DROP LEAF see SFS000
DROPP see TEX600
DROPRENILAMINE HYDROCHLORIDE
see CPP000
DROPSPRIN see SAH000
DROTEBANOL see ORE000
DROXAROL see BPP750
DROXARYL see BPP750
DROXOL see MDQ250
DROXOLAN see DAQ400
DRP 859025 see TND500
DRUMULSE AA see OAV000
DRUPINA 90 see BJK500
DRW 1139 see ALA500
DRY AND CLEAR see BDS000
DRY ICE see CBU250
DRY ICE (UN 1845) (DOT) see CBU250
DRYISTAN see BBV500
DRYLIN see TKX000
DRYLISTAN see BBV500
DRY MIX No. 1 see CPK500
DRYOBALANOPS CAMPHOR see BMD000
DRYPTAL see CHJ750
DRYSEQ see SCN700
DRY WHEY see WBL150
DS see DKB110
DS see DXN300
DS-36 see SNL800
DS-M-1 see DBD750
DS-15647 see DAB400
DS 18302 see CBW000
DSDP see DJA400
DSE see DXD200
DSMA LIQUID see DXE600
DS-Na see SFW000
2,4-D SODIUM SALT see SGH500
DSP see SJH090
DSP4 see CGY600
DSPT see THQ900
DSS see DJL000
DSS see SOA500
DS SUBSTANDE see AJX500
DST see DME000
DST 50 see SMR000
DSTDP see DXG700
DSTP see DXG700
DT see DNC000
DT see DWP000
DT see TFX790
D 65MT see XAJ000
2,4-D and 2,4,5-T (2:1) see DAB000
DTA see DCB100
DTA see DLK200
DTAC see LBX075
DTAS see MJK750

DTB see DXL800
DTBN see DEF090
DTBP see BSC750
DTBT see BLN750
DTDP see DXQ200
DTE see DXN350
DTHYD see TFX790
DTIC see DAB600
DTIC CITRATE see DUI400
DTIC-DOME see DAB600
DTMC see BIO750
DTPA see DJG800
DTPA CALCIUM TRISODIUM SALT see
CAY500
DTP HYDROCHLORIDE see DXO400
N-D1-TRIMETHYL-3,3-DI-2-
THIENYLALLYLAMINE
HYDROCHLORIDE see TLQ250
DTS see DNV800
DTT see DXO775
DTT see DXO800
DTT see MIF763
D-DTT see DXO800
DTX1 see THI252
DU see DCQ600
dU see DAZ050
DU-717 see CIT625
DU-5747 see CCK575
DU 21220 see RLK700
DU 23000 see FMR575
DU 112307 see CJV250
DU-A 1 see SCQ000
DU-A 2 see SCQ000
DU-A 3 see SCQ000
DU-A 4 see SCQ000
DU-A 3C see SCQ000
DUAL see MQQ450
DUAMINE see MCH525
DUASYN ACID GREEN V see CMM100
DUASYN ACID YELLOW RRT see SGP500
DUATOK see TEX250
DUAZOMYCIN see DYF400
DUAZOMYCIN B see ASO501
DUAZOMYCIN B see ASO510
DUBIMAX see NAE000
DUBORIMYCIN see DAC300
DUBOS CRUDE CRYSTALS see TOG500
DUBRONAX see SOA500
DUCKALGIN see SEH000
DUCOBEE see VSZ000
O-DUE see TAG750
DUFALONE see BJZ000
DUFASTON see DYF759
DUGERASE see AIT250
DUGRO see MMN750
DUKERON see TIO750
DUKSEN see DCK759
DULCIDOR see GFA000
DULCINE see EFE000
DULCITOLDIEPOXIDE see DCI600
DULCOLAN see PPN100
DULCOLAX see PPN100
DUL-DUL (PUERTO RICO) see CAK325
DULL 704 see PJY500
DULSIVAC see DJL000
DULZOR-ETAS see SGC000
DUMASIN see CPW500
DUMBCANE see DHB309
DUMB PLANT see DHB309
DUMITONE see SOA500
DUMOCYCIN see TBX250
DUMOGRAN see MPN500
DUNCAINE see DHK400
DUNCAINE HYDROCHLORIDE see
DHK600
DUNERYL see EOK000
DUNKELGELB see PEJ500
DUO CHONG WEI see PDV330
DUODECANE see DXT200
DUODECIBIN see VSZ000
DUODECYL ALCOHOL see DXV600
DUODECYLIC ACID see LBL000

DUODECYLIC ALDEHYDE see DXT000
DUOGASTRONE see CBO500
DUO-KILL see COD000
DUO-KILL see DGP900
DUOLAX see DMH400
DUOLIP see TEQ500
DUOLUTON see NNL500
DUOMYCIN see CMA750
DUOSAN see MAP300
DUOSAN (pesticide) see MAP300
DUOSOL see DJL000
DUOSOL (ULCER TREATMENT) see
BKX800
DUO-STREPTOMYCIN see SLY200
DUOTRATE see PBC250
DUO XIAO ZUO see TMK125
DUP 942 see XDJ025
DUPHAPEN see PAQ200
DUPHAR see CKM000
DUPHASTON see DYF759
DUPICAL see OBW100
DUPLEX PERMATON RED L 20-7022 see
CJD500
DUPLEX RED LAKE CD 20-5925 see CMS150
DUPLEX TOLUIDINE RED L 20-3140 see
MMP100
DUPLOSAN DP see DGA880
DUPONOL see SIB600
DUPONOL 80 see OFU200
DU PONT 326 see DGD600
DU PONT 634 see CPR600
DU PONT 732 see BQT750
DU PONT 1991 see BAV575
DUPONT DPX 3654 see PMF550
DU PONT HERBICIDE 326 see DGD600
DU PONT HERBICIDE 732 see BQT750
DU PONT HERBICIDE 976 see BMM650
DU PONT INSECTICIDE 1179 see MDU600
DU PONT INSECTICIDE 1519 see DVS000
DU PONT INSECTICIDE 1642 see MPS250
DU PONT PC CRABGRASS KILLER see
PLC250
DU PONT WK see DXY000
DUPRENE see PJQ050
DURABIOTIC see BFC750
DURABOL see DYF450
DURABOLIN see DYF450
DURABOLIN-O see EJT600
DURABORAL see EJT600
DURA CLOFIBRAT see ARQ750
DURAD see TNP500
DURAD 220B see IIQ150
DURAD 550B see IIQ150
DURAD MP280R HYDRAULIC FLUID see
DYF500
DURA-ESTRADIOL see EDS100
DURAFUR BLACK R see PEY500
DURAFUR BLACK RC see PEY650
DURAFUR BROWN see ALL750
DURAFUR BROWN 2R see ALL750
DURAFUR BROWN MN see DBO400
DURAFUR BROWN RB see ALT250
DURAFUR DEVELOPER C see CCP850
DURAFUR DEVELOPER D see NAW500
DURAFUR DEVELOPER E see NAN505
DURAFUR DEVELOPER G see REA000
DURALUTON see HNT500
DURAMAX see ADA725
DURAMITE see CAT775
DURAMYCIN (8CI) see LEX100
DURAN see DXQ500
DURANATE EXP-D 101 see PKO750
DURANIT see SMR000
DURANITRAT see CCK125
DURAN MP280R see DYF500
DURANOL BLUE GREEN B see DMM400
DURANOL BLUE TR see CMP070
DURANOL BLUE TR see DBQ220
DURANOL BRILLIANT BLUE B see
MGG250
DURANOL BRILLIANT BLUE CB see
TBG700

DURANOL BRILLIANT BLUE VIOLET BR
see DBY700
DURANOL BRILLIANT RED T 2B see
AKI750
DURANOL BRILLIANT VIOLET B see
AKP250
DURANOL BRILLIANT VIOLET BR see
DBY700
DURANOL BRILLIANT YELLOW G see
MEB750
DURANOL ORANGE G see AKP750
DURANOL PRINTING BLUE GREEN B see
DMM400
DURANOL RED 2B see AKE250
DURANOL RED X3B see DBX000
DURANOL VIOLET WR see DBP000
DURANTA REPENS see GIW200
DURAPATITE see HJE100
DURA-PENITA see BFC750
DURAPHOS see MQR750
DURAPROST see OLW600
DURASORB see DUD800
DURA-TAB S.M. AMINOPHYLLINE see
TEP500
DURATINT GREEN 1001 see PJQ100
DURATION see AEX000
DURATOX see DAP400
DURATOX see MIW100
DURAVOS see DGP900
DURAX see CPI250
DURAZOL BLUE 5G see CMO600
DURAZOL BLUE 8G see COF420
DURAZOL FAST BLUE 8GS see COF420
DURAZOL GRAY B see CMN230
DURAZOL PAPER BLUE 8G see COF420
DURAZOL RED 2B see CMO885
DURAZOL RED 2BP see CMO885
DURCAL 10 see CAT775
DUR-EM 204 see GGR200
DURETHAN BK see PJY500
DURETTER see FBN100
DUREX see CBT750
DURFAX 80 see PKL100
DURGACET RUBINE B see CMP080
DURGACET SCARLET B see ENP100
DURGACET YELLOW G see AAQ250
DURGASOL SCARLET GG SALT see
DEO295
DURICEF see DYF700
DURINDONE BLUE 4B see ICU135
DURINDONE BLUE 4BC see ICU135
DURINDONE BLUE 4BCP see ICU135
DURINDONE ORANGE R see CMU815
DURINDONE ORANGE RP see CMU815
DURINDONE PRINTING BLACK BL see
CMU320
DURINDONE PRINTING BLUE 4BC see
ICU135
DURINDONE PRINTING ORANGE R see
CMU815
DURINDONE PRINTING RED B see
DNT300
DURINDONE RED B see DNT300
DURINDONE RED BP see DNT300
DUROCHROME BLUE OCG see CMP880
DUROCHROME CYANINE G see CMP880
DUROCHROME FAST CYANINE 6BN see
CMP880
DUROCHROME ORANGE G see CMP882
DUROCHROME VIOLET B see HLI000
DUROCHROME YELLOW 2RN see NEY000
DUROFERON see FBN100
DUROFOL P see PKQ059
DUROID 5870 see TAI250
DUROLAX see PPN100
DUROMINE see DTJ400
DURONITRIN see TJL250
DUROPENIN see BFC750
DUROPROCIN see MDU300
DUROQUINONE see TDM810
DUROSPERSE YELLOW G see AAQ250
DUROTOX see PAX250

DUROX see AKO500
DURSBAN see CMA100
DURSBAN F see CMA100
DURSBAN METHYL see CMA250
DURTAN 60 see SKV150
DUSICNAN BARNATY (CZECH) see BAN250
DUSICNAN CERITY (CZECH) see CDB000
DUSICNAN HORECHATY (CZECH) see
MAH250
DUSICNAN HORECNATY (CZECH) see NEE100
DUSICNAN KADEMNATY (CZECH) see
CAH250
DUSICNAN VAPENATY (CZECH) see
CAU250
DUSICNAN ZINECNATY (CZECH) see
NEF500
DUSICNAN ZIRKONICITY (CZECH) see
ZSA000
DUSITAN DICYKLOHEXYLAMINU
(CZECH) see DGU200
DUSITAN SODNY (CZECH) see SIQ500
DUSOLINE see CMD750
DUSORAN see CMD750
DUSPAR 125B see HKS400
DU-SPREX see DER800
DUST M see RAF100
DUS-TOP see MAE250
DUTCH LIQUID see EIY600
DUTCH OIL see EIY600
DUTCH-TREAT see HKC500
DU-TER see HON000
DUTION see DXH325
DUTOM see SKQ400
DUVADILAN see VGA300
DUVADILAN see VGF000
DUVALINE see PPH050
DUVARON see DYF759
DUVILAX BD 20 see AAX250
DUVOID see HOA500
DV see DAL600
DV-17 see DBC575
DV-79 see AJV500
D 33LV see DCK400
DV 400 see PKB500
DV 714 see LEF400
DVEDEG (RUSSIAN) see DJE600
D3-VIGANTOL see CMC750
DW-61 see FCB100
DW 62 see DNV000
DW 62 see DNV200
DW-116 see FLT200
DW3418 see BLW750
DWARF BAY see LAR500
DWARF LAUREL see MRU359
DWARF PINE NEEDLE OIL see PIH400
DWARF POINCIANA see CAK325
DWELL see EFK000
DWUBROMOETAN (POLISH) see EIY500
DWUCHLOROCZTEROFLUOROETAN
(POLISH) see DGL600
DWUCHLORODWUETYLOWY ETER
(POLISH) see DFJ050
DWUCHLORODWUFLUOROMETAN
(POLISH) see DFA600
2,4-DWUCHLOROFENOKSYOCTOSY
KWAS (POLISH) see DAA800
DWUCHLOROFLUOROMETAN (POLISH)
see DFL000
DWUCHLOROPROPAN (POLISH) see
PNJ400
DWUCHLOROSTYREN (POLISH) see
DGK800
DWUETYLOAMINA (POLISH) see DHJ200
O,O-DWUETYLO-O-1-(2,4-
DWUCHLOROFENYLO)-2-
BROMOWINYLOFOSFORAN see BMO310
DWUETYLOWY ETER (POLISH) see EJU000
DWUFENYLOGUANIDYNA (POLISH) see
DWC600
DWUMETHYLOFORMAMID (POLISH) see
DSB000

DWUMETYLOANILINA (POLISH) see DQF800
O,O-DWUMETYLO-O-1-(2,4-DWUCHLOROFENYLO)-2-BROMOWINYLOFOSFORAN see MHR150
3,5-DWUMETYLOIZOKSAZOLU see DSK950
3,5-DWUMETYLOPIRAZOLU see DTU850
DWUMETYLOSULFOTLENKU (POLISH) see MAO250
DWUMETYLOWY SIARCZAN (POLISH) see DUD100
DWU-β-NAFTYLO-p-FENYLODWUAMINA (POLISH) see NBL000
DWUNITROBENZEN (POLISH) see DUQ200
DWUNITRO-o-KREZOL (POLISH) see DUS700
3,3'-DWUOKSYBENZYDYNA (POLISH) see DMI400
DWUSIARCZEK DWUBENZOTIAZYLU (POLISH) see BDE750
DX see AES750
DX80-2 see XJJ005
DXEWMULSE POE-SML see PKG000
DXG see BSS550
DXM 100 see PJS750
DXMS see SOW000
DY032 see PKQ070
DYALL see PJS750
DYANACIDE see ABU500
DYANAP see NBL200
DYAZIDE see CFY000
DYBAR see DTP400
DYCARB see DQM600
DYCHOLIUM see SGD500
DYCILL see DGE200
DYCLOCAINUM see BPR500
DYCLONE HYDROCHLORIDE see BPR500
DYCLONINE HYDROCLORIDE see BPR500
DYCLOTHANE see BPR500
DYDELTRONE see PMA000
DYDROGESTERONE see DYF759
DYDROGESTERONE and HYDROXYPROGESTERONE (9:10) see PMA250
DYE C see DYF800
DYE EVANS BLUE see BGT250
DYE FD AND C RED No. 4 see FAG050
DYE FDC RED 2 see FAG020
DYE FD&C RED No. 3 see FAG040
DYE FD & C RED No. 4 see FAG050
DYE FDC YELLOW LAKE 6 see FAG150
DYE FD & C YELLOW LAKE 6 see FAG150
DYE FDC YELLOW NO. 6 see FAG150
DYE FD & C YELLOW NO. 6 see FAG150
DYE FD and C YELLOW NO. 5 see FAG140
DYE GS see ALL750
DYE ORANGE No. 1 see FAG010
DYE QUINOLINE YELLOW see CMM510
DYE RED RASPBERRY see FAG020
DYESTROL see DKA600
DYE SUNSET YELLOW see FAG150
DYETONE see SFG000
DYFLOS see IRF000
DYFONATE see FMU045
DYGRATYL see DME300
DYKANOL see PJL750
DYKOL see DAD200
DYLAMON see BBV500
DYLAN see PJS750
DYLAN SUPER see PJS750
DYLAN WPD 205 see PJS750
DYLARK 111 see SEA500
DYLARK 230 see SEA500
DYLARK 231 see SEA500
DYLARK 232 see SEA500
DYLARK 238 see SEA500
DYLARK 250 see SMQ500
DYLARK 332 see SEA500
DYLENE see SMQ500
DYLENE 8 see SMQ500
DYLENE 9 see SMQ500

DYLENE 8G see SMQ500
DYLEPHRIN see VGP000
DYLITE F 40 see SMQ500
DYLITE F 40L see SMQ500
DYLOX see TIQ250
DYLOX-METASYSTOX-R see TIQ250
DYMADON see HIM000
DYMANTHINE see DTC400
DYMEL 22 see CFX500
DYMELOR see ABB000
DYMEX see ABH000
DYMID see DRP800
DYNA-CARBYL see CBM750
DYNACORYL see DJS200
DYNADUR see PKQ059
DYNALIN INJECTABLE see TET800
DYNALTONE see LBV100
DYNAMICARDE see DJS200
DYNAMITE see DYG000
DYNAMONE see AES650
DYNAMUTILIN see TET800
DYNAPEN see DGE200
DYNAPHENIL see AOB500
DYNAPRIN see DLH600
DYNARSAN see ABX500
DYNASTEN see PAN500
DYNA-ZINA see DLH600
DYNA-ZINA see DLH630
DYNAZONE see NGE500
DYNEL see ADY250
DYNEL NYGL see ADY250
DYNERIC see CMX700
DYNEX see DXQ500
DYNH see PJS750
DYNIUM CHLORIDE see LBV100
DYNK 2 see PJS750
DYNOMIN MM 9 see MCB050
DYNOMIN MM 75 see MCB050
DYNOMIN MM 100 see MCB050
DYNOMIN UI 16 see UTU500
DYNOMIN UM 15 see UTU500
DYNONE see EIH500
DYNONE II see CPK550
DYNOSOL see DUU600
DYNOTHEL see SKJ300
DYNOVAS see PAH250
DYODIN see DNF600
DYPERTANE COMPOUND see RDK000
DYP-97 F see LBR000
DYPHONATE see FMU045
DYPHYLLINE see DNC000
DYPNONE see MPL000
DYPRIN see MDT740
DYREN see UVJ450
DYRENE see DEV800
DYRENE 50W see DEV800
DYRENIUM see UVJ450
DYREX see TIQ250
DYSEDON see AFL750
DYSPNE-INHAL see VGP000
DYSPROSIUM see DYG400
DYSPROSIUM CHLORIDE see DYG600
DYSPROSIUM CITRATE see DYG800
DYSPROSIUM NITRATE see DYH100
DYSPROSIUM(III) NITRATE HEXAHYDRATE (1:3:6) see DYH000
DYSPROSIUM TRINITRATE see DYH100
DYTAC see UVJ450
DYTEK A see MNI515
DYTHERM X 214 see SEA500
DYTHOL see CMD750
DYTOL M-83 see OEI000
DYTOL S-91 see DAI600
DYTOL E-46 see OAX000
DYTOL F-11 see HCP000
DYTOL J-68 see DXV600
DYTRANSIN see IJG000
DYVON see TIQ250
DYZOL see DCM750
DZ see MDH500
DZhp-4K see CMP090
E2 see PJY100

E-3 see ELF500
E6 see PKO500
E-48 see DJV600
E 62 see PKQ059
E 64 see TFK255
E 100 see ICC800
E 102 see FAG140
E-103 see BSN000
E 103 see MRL100
E 104 see CMM510
E 110 see FAG150
E-111 see PNN300
E 125 see CMP620
E 127 see FAG040
E 130 see IBV050
E 131 see ADE500
E 131 see CMM062
E 132 see FAE100
E 140 see CKN000
E 141 see DIS600
E 142 see ADF000
E 151 see BMA000
E 158 see DAP600
E-212 see VGZ000
E-236 see DUJ400
(−)-E-250 see DAZ125
E 261 see DTS500
E-298 see MRU080
E 335 see AAU300
E393 see SOD100
E-41B see ELX527
E 518 see VPK333
E 534 see PKB100
E 535 see AAU300
583E see POC750
E 600 see NIM500
E-646 see EAV700
E 66P see PKQ059
686E see SMQ500
E 702 see MJM700
E 736 see TFD000
E 785 see EAB100
(+)-E-250 see DAZ120
E-0687 see MGE200
E1001 see IPM000
E1001 see IPN000
E 1004 see IPP000
E-1059 see DAO500
E 1059 see DAO600
E 1440 see CCU250
264CE see AFS625
E-2663 see CML870
305CE see FDA875
E 3314 see HAR000
E 3432 see FLV050
E 5110 see MEL100
E 6010 see PJB810
E 7256 see LBU100
(±)-E-250 see DAZ118
20SE60 see MCB050
EA 166 see GLS700
E-733-A see CQH000
EA 2277 see QVA000
EA 3547 see DDE200
EAA see EFS000
EAB see EOH500
EACA see AJD000
EACA KABI see AJD000
E-73 ACETATE see ABN000
EACS see AJD000
EAGLE GERMANTOWN see CBT750
EA-1 HYDROCHLORIDE see NCL300
EAK see ODI000
EAMN see ENR500
α-EARLEINE see GHA050
EARTHCIDE see PAX000
EARTH GALL see FAB100
EARTHNUT OIL see PAO000
EASEPTOL see HJL000
EASTBOND M 5 see PMP500
EASTER FLOWER see EQX000

EASTERN COTTONMOUTH VENOM see EAB200
EASTERN DIAMOND-BACK RATTLESNAKE VENOM see EAB225
EASTERN STATES DUOCIDE see WAT200
EAST INDIAN COPAIBA see GME000
EAST INDIAN LEMONGRASS OIL see LEG000
EAST INDIAN SANDALWOOD OIL see OGY220
EASTMAN 910 see MIQ075
EASTMAN 1334 see KHU050
EASTMAN 7663 see DJT800
EASTMAN 910 ADHESIVE see MIQ075
EASTMAN BLUE BNN see MGG250
EASTMAN DOTP PLASTICIZER see BJS300
EASTMAN FAST BLUE B-GLF see CMP060
EASTMAN HTP VIOLET 310 see DJT050
EASTMAN INHIBITOR DHPB see DMI600
EASTMAN INHIBITOR HPT see HEK000
EASTMAN INHIBITOR RMB see HNH500
EASTMAN 910 MONOMER see MIQ075
EASTONE SCARLET BG see ENP100
EASTONE YELLOW GN see AAQ250
EASTOZONE see BJL000
EASTOZONE 31 see BJT500
EASTOZONE 33 see BJL000
EASY OFF-D see TIG250
EATAN see DLY000
EAU de BROUTS ABSOLUTE see EAB500
EAU GRISON see CAX800
EAZAMINE see DII200
EB see BGT250
EB see EGP500
EB 8 see OAF200
EB 80 see PAT850
EB-382 see MGC200
EBDC see OJV600
EBECRYL 110 see PER250
EBECRYL 605A see EAB550
EBECRYL IBOA see IHX700
E-D-BEE see EIY050
EBERPINE see RDK000
EBERSPINE see RDK000
EBI see ISK000
EBIS see ISK000
EBNA see NKD500
EBNS see EHC800
EBONTA see EIT000
EBRANTIL see USJ000
EBS see FAG040
EBSERPINE see RDK000
EBUCIN see CAS750
EBUFOS see EHY100
EBZ see EDP000
E.C. 1.1.3.4 see GFG100
E.C. 3.1.1.3. see GGA800
E.C. 3.1.4.5 see EDK650
E.C. 3.4.4.5 see CML880
E.C. 3.4.4.6 see CML880
E.C. 3.4.4.7 see EAG875
E.C. 3.5.1.2 see GFO070
E.C. 3.2.1.18 see NCQ200
E.C. 3.2.1.23 see GAV100
E.C. 3.4.21.1 see CML880
E.C. 3.4.21.5 see TFU800
E.C. 3.4.4.13 see TFU800
E.C. 3.4.4.16 see BAC000
E.C. 3.4.4.24 see BMO000
E.C. 3.1.23.21 see REF262
E.C. 3.4.2.1.11 see EAG875
E.C. 3.4.21.14 see BAC000
E.C. 3.4.21.15 see SBI860
ECABET SODIUM see EAB560
ECADOTRIL see SDY625
ECARAZINE see TGJ150
ECARAZINE HYDROCHLORIDE see TGJ150
E. CARINATUS VENOM see EAD600
ECARLATE GN see CMP620
ECATOX see PAK000
ECBOLINE see EDC565

ECBOLINE ETHANESULFONATE see EDC575
ECCOTHAL see TEM000
ECDYSONE see EAB600
α-ECDYSONE see EAB600
β-ECDYSONE see HKG500
ECDYSTERONE see HKG500
β-ECDYSTERONE see HKG500
ECF see EHK500
ECGONINE, METHYL ESTER, BENZOATE (ESTER) see CNE750
ECH see EAZ500
ECHIMIDINE see EAC500
ECHINOMYCIN see EAD500
ECHINOMYCIN A see EAD500
ECHIS CARINATUS VENOM see EAD600
ECHIS COLORATA VENOM see EAD650
ECHIS COLORATUS VENOM see EAD650
ECHIUM PLANTAGINEUM see VQZ675
ECHIUM VULGARE see VQZ675
ECHLOMEZOL see EFK000
ECHODIDE see TLF500
ECHOTHIOPHATE see TLF500
ECHOTHIOPHATE IODIDE see TLF500
ECHUJETIN see DMJ000
ECIPHIN see EAW000
ECLERIN see BOV825
ECLIPSE BLACK BG see CMS250
ECLIPSE DEEP BLACK S see CMS250
ECLIPSE RED see DXO850
ECLIPSE VIOLET BROWN X see CMS257
ECLORIL see CDO500
ECLORION see HBT500
ECM see ADA725
ECODOX see HGP550
ECOFROL see TGJ050
E. COLI 0111: B4 LPS see EDK750
E. COLI 0111:B4 LPS see LGK375
ECOLID see CDY000
ECOLID CHLORIDE see CDY000
E. COLI ENDOTOXIN see EDK700
ECON see TGJ050
ECONAZOLE NITRATE see EAE000
ECONOCHLOR see CDP250
ECOPRO see TAL250
ECOSTATIN see EAE000
ECOTHIOPATE IODIDE see TLF500
ECOTHIOPHATE IODIDE see TLF500
ECOTRIN see ADA725
ECP see DAZ115
ECP see DFK600
ECPN see NKE000
ECR see EDB125
ECT 743 see EAE050
ECTEINASCIDIN 743 see EAE050
ECTEINASCIDINE 743 see EAE050
ECTIBAN see AHJ750
ECTIDA see EHP000
ECTILURAN see TEH500
ECTILUREA see EHP000
ECTODEX see MJL250
ECTON see EHP000
ECTORAL see RMA500
ECTRIN see FAR100
ECTYDA see EHP000
ECTYLCARBAMIDE see EHP000
ECTYLUREA see EHO700
ECTYLUREA see EHP000
ECTYN see EHP000
ECUANIL see MQU750
ECYLERT see PAP000
ECZECIDIN see CHR500
ED see DFH200
ED see EQJ100
EDA see DCN800
EDA 200 see EIV750
3',4'-Et2-DAB see DJB400
EDAP see EAE100
EDATHAMIL see EIX000
EDATHAMIL CALCIUM DISODIUM see CAR780
EDATHAMIL DISODIUM see EIX500

EDATHAMIL MONOSODIUM FERRIC SALT see EJA379
EDATHANIL TETRASODIUM see EIV000
EDB see EIY500
EDB-85 see EIY500
EDBC see OJV600
EDBPHA see EIV100
EDC see EIY600
EDCO see MHR200
EDC (REAGENT) see EAE100
EDDHA see EIV100
EDDO see EAI600
EDDP see EIM000
EDECRIL see DFP600
EDECRIN see DFP600
EDECRINA see DFP600
EDEINE see EAE400
EDEMEX see BDE250
EDEMOX see AAI250
EDEN see LFK000
EDENAL see MQU750
EDETAMIN see CAR780
EDETAMINE see CAR780
EDETATE CALCIUM see CAR780
EDETATE DISODIUM see EIX500
EDETATE SODIUM see EIV000
EDETATE TRISODIUM see TNL250
EDETIC ACID see EIX000
EDETIC ACID CALCIUM DISODIUM SALT see CAR780
EDETIC ACID TETRASODIUM SALT see EIV000
EDHPA see EIV100
EDICOL AMARANTH see FAG020
EDICOL BLUE CL 2 see FAE000
EDICOL PONCEAU RS see FMU070
EDICOL SUPRA 10B see CMS228
EDICOL SUPRA BLACK BN see BMA000
EDICOL SUPRA BLUE E6 see FMU059
EDICOL SUPRA BLUE VR see ADE500
EDICOL SUPRA CARMOISINE WS see HJF500
EDICOL SUPRA ERYTHROSINE A see FAG040
EDICOL SUPRA GERANINE 2G see CMM300
EDICOL SUPRA GERANINE 2GS see CMM300
EDICOL SUPRA GREEN B see ADF000
EDICOL SUPRA PONCEAU 4R see FMU080
EDICOL SUPRA PONCEAU SX see FAG050
EDICOL SUPRA ROSE B see FAG070
EDICOL SUPRA 10BS see CMS228
EDICOL SUPRA TARTRAZINE N see FAG140
EDICOL SUPRA YELLOW FC see FAG150
EDIFENPHOS see EIM000
EDION see TLP750
EDIPHENPHOS see EIM000
EDIPOSIN see VBK000
EDISOL M see MIF760
EDISTIR RB see SMQ500
EDISTIR RB 268 see SMR000
EDIWAL see NNL500
EDMBA see DQL600
EDOXUDINE see EHV200
EDPA HYDROCHLORIDE see DWF200
EDRIZAR see MJL250
EDROFURADENE see EAE500
EDROPHONIUM BROMIDE see TAL490
EDROPHONIUM CHLORIDE see EAE600
EDRUL see EAE675
EDTA (chelating agent) see EIX000
EDTA ACID see EIX000
EDTACAL see CAR780
EDTA CALCIUM DISODIUM SALT see CAR780
d'E.D.T.A. DISODIQUE (FRENCH) see EIX500
EDTA, DISODIUM SALT see EIX500
EDTA FERRIC SODIUM SALT see IGX875
EDTA, SODIUM SALT see EIV000

EDTA TETRASODIUM SALT see EIV000
EDTA TRISODIUM SALT see TNL250
EDTA TRISODIUM SALT (TRIHYDRATE) see TNL500
EDTA-ZINC see ZGW200
EDTA-ZINC COMPLEX see ZGW200
EDTA ZINC SALT see ZGW200
EDU see EHV200
E 103 (DYE) see MRL100
E-103-E see MQQ050
EEC No. E924 see PKY300
EECPE see QFA250
EEC SERIAL No. 124 see CMP250
EEDDKK see EIJ500
EENA see ELG500
EENKAPTON (DUTCH) see PDC750
(12-β(E,E))-12-((1-OXO-5-PHENYL-2,4-PENTADIENYL)OXY)-DAPHNETOXIN see MDJ250
EEREX GRANULAR WEED KILLER see BMM650
EEREX WATER SOLUBLE CONCENTRATE WEED KILLER see BMM650
EES see DKB139
E,E-3,7,11,15-TETRAMETHYL-1,6,10,14-HEXADECATETRAEN-3-OL see GDG300
EF 10 see SCK600
EFACIN see NCQ900
EF CORLIN see CNS750
EFED see TMU250
EFEDRIN see EAW000
E-FEROL see TGJ050
EFERON see PEC250
EFEROX see LFG050
EFFEMOLL DOA see AEO000
EFFISAX see MOV500
EFFIX see FBW135
EFFLUDERM (free base) see FMM000
EFFORTIL see EGE500
EFFROXINE see DBA800
EFFUSAN see DUS700
EFH see EKL250
EFIRAN 99 see TIR800
EFK 1 see EAE700
EFK 1 (flotation agent) see EAE700
EFLORAN see MMN250
EFLORNITHINE see DKH875
EFLORNITHINE HYDROCHLORIDE see EAE775
EFLOXATE see ELH600
EFO-DINE see PKE250
EFOSITE ALUMINUM see AHH800
EFRICEL see SPC500
EFROXINE see MDT600
EFSIOMYCIN see FMR500
EFTAPAN see ECU550
EFTOLON see AIF000
EFUDEX see FMM000
EFUDIX see FMM000
EFURANOL see DLH630
EFV 250/400 see IGK800
EGACID ORANGE GG see FAG010
EGACID RED 6B see CMM400
EGACID RED G see CMM300
EGBE see BPJ850
EGDME see DOE600
EGDN see EJG000
EGF see EBA275
EGF-UROGASTRONE see UVJ475
EGG-ALBUMIN see AFI780
EGGOBESIN see PNN300
EGG YELLOW A see FAG140
EGICALM see ARP125
EGITOL see HCI000
EGLONYL see EPD500
EGM see EJH500
EGME see EJH500
EGPEA see PNA225
EGRI M 5 see CAT775
EGT see EJD600
EGTA see EIT000
EGYPTIAN RATTLEPOD see SBC550

EGYT 201 see BAV250
EGYT 341 see GJS200
EGYT 739 see CAY875
EGYT-1050 see DAJ400
EH2 see DGQ200
EH 121 see TNX000
EHBC see OJV600
EHB-M see OJV600
EHBN see ELE500
EHDP see HKS780
EHEN see ELG500
EHRLICH 5 see ARL000
EHRLICH 594 see ABX500
EHRLICH 606 see SAP500
EHRLICH'S REAGENT see DOT400
EI see EJM900
EI-103 see BSN000
EI-1642 see MPS250
EI-12880 see DSP400
EI-18706 see DNX600
EI 38,555 see CMF400
EI 47031 see PGW750
EI 47300 see DSQ000
EI-47470 see DHH400
EI 52160 see TAL250
EICOSAFLUOROUNDECANOIC ACID see EAE875
2,2,3,3,4,4,5,5,6,6,7,7,8,8,9,9,10,10,11,11-EICOSAFLUOROUNDECANOIC ACID see EAE875
11-H-EICOSAFLUORUNDEKANSAEURE (GERMAN) see EAE875
ω-H-EICOSAFLUORUNDEKANSAEURE (GERMAN) see EAE875
EICOSAHYDRO DIBENZO(b,k)(1,4,7,10,13,16)HEXAOXACYCLOOCTADECIN see DGV100
EICOSANOIC ACID see EAF000
EICOSANYL DIMETHYL BENZYLAMMONIUM CHLORIDE see BEM250
5,8,11,14-EICOSATETRAENAMIDE, N-(2-HYDROXYETHYL)-, (ALL-Z)- see HKR600
5,8,11,14-EICOSATETRAENOIC ACID see EAF050
EICOSATETRAENOIC ACID, 15-HYDROXY- see HKG680
5,8,10,14-EICOSATETRAENOIC ACID, 12-HYDROXY-, (E,Z,Z,Z)- see HKG650
6,8,11,14-EICOSATETRAENOIC ACID, 5-HYDROXY-, (S-(E,Z,Z,Z))- see HKG600
5,8,11,14-EICOSATETRAENOYLETHANOLAMIDE see HKR600
EINALON S see CLY500
EIPW 111 CITRATE see EAF100
EIRENAL see PGA750
EISENDEXTRAN (GERMAN) see IGS000
EISENDIMETHYLDITHIOCARBAMAT (GERMAN) see FAS000
EISEN-III-HYDROXID-POLYMALTOSE (GERMAN) see IHA000
EISENOXYD see IHC450
EISEN(III)-TRIS(N,N-DIMETHYLDITHIOCARBAMAT) (GERMAN) see FAS000
EITDRONATE DISODIUM see DXD400
EJIBIL see EGQ000
EK 54 see DUU600
EK 1700 see PEY250
EKAFLUVIN see KDK000
EKAGOM CBS see CPI250
EKAGOM TB see TFS350
EKAGOM TEDS see DXH250
EKAGON TE see DJC200
EKALUX see DJY200
EKATIN see PHI500
EKATIN AEROSOL see PHI500
EKATINE-25 see PHI500
EKATIN TD see DXH325
EKATIN ULV see PHI500
EKATOX see PAK000

EKAVYL SD 2 see PKQ059
EKILAN see MFD500
EKKO CAPSULES see DKQ000
EKKUSAGONI see DGA850
EKOA (HAWAII) see LED500
EKOMINE see MGR500
EKTAFOS see DGQ875
EKTASOLVE de ACETATE see CBQ750
EKTASOLVE DB see DJF200
EKTASOLVE DB ACETATE see BQP500
EKTASOLVE DIB see DJF800
EKTASOLVE EB see BPJ850
EKTASOLVE EB ACETATE see BPM000
EKTASOLVE EE see EES350
EKTASOLVE EE ACETATE SOLVENT see EES400
EKTASOLVE EIB see IIP000
EKTASOLVE EP see PNG750
EKTEBIN see PNW750
EKTYLCARBAMID see EHP000
EKVACILLIN see SLJ000
EKVACILLIN see SLJ050
EK 1108GY-A see TAI250
EL-103 see BSN000
EL 107 see ENF100
EL-119 see OJY100
EL-161 see ENE500
EL 222 see FAK100
EL 228 see CHJ300
EL-273 see THQ525
EL-291 see MQC000
EL 400 see BNL250
EL 500 see IRN500
EL 614 see BMO300
EL-620 see PKE500
EL-719 see PKE500
EL 2289 see CHJ300
EL 4049 see MAK700
ELAGOSTASINE see EAI200
ELAIDIC ACID see EAF500
ELAIOMYCIN see EAG000
ELALDEHYDE see PAI250
ELAMOL see EAG100
ELAMOL see TGJ250
ELAN see PGG350
ELANCOBAN see MRE225
ELANCOLAN see DUV600
ELANIL see EAH500
ELAOL see DEH200
ELARGIN see EAG100
ELARGYL see EAG100
ELASIOMYCIN see EAG500
ELASTASE see EAG875
ELASTOFIX ACS see MCB050
ELASTONON see AOA250
ELASTONON see BBK000
ELASTOPAR see MJG750
ELASTOPAX see MJG750
ELASTOZONE 31 see BJT500
ELASTOZONE 34 see PFL000
ELASZYM see EAG875
ELATERICIN A see EAH100
α-ELATERIN see COE250
ELAVIL see EAH500
ELAVIL see EAI000
ELAVIL HYDROCHLORIDE see EAI000
ELAYL see EIO000
ELBANIL see CKC000
ELBATAN see CPR600
ELBENYL ORANGE A-3RD see SGP500
ELBRUS see CGA000
ELCACID MILLING FAST RED RS see CMM330
ELCATONIN see CBR300
ELCEMA F 150 see CCU150
ELCEMA G 250 see CCU150
ELCEMA P 050 see CCU150
ELCEMA P 100 see CCU150
ELCIDE 75 see MDI000
ELCORIL see CDO500
EL-CORTELAN SOLUBLE see HHR000
ELCOSINE see SNJ350

ELCOZINE CHRYSOIDINE Y see PEK000
ELCOZINE RHODAMINE B see FAG070
ELCOZINE RHODAMINE 6GDN see RGW000
ELDADRYL see BAU750
ELDELIN see EAI050
ELDELINE see EAI050
ELDEPRYL see DAZ125
ELDERBERRY see EAI100
ELDERFIELD PYRIMIDINE MUSTARD see DYC700
ELDESINE see VGU750
ELDEZOL see NGE500
ELDIATRIC C see CAK000
ELDODRAM see DYE600
ELDOPAL see DNA200
ELDOPAQUE see HIH000
ELDOQUIN see HIH000
ELDRIN see RSU000
ELEAGIC ACID see EAI200
ELEAGOL see SCA750
ELECOR see PEV750
ELECTRO-CF 11 see TIP500
ELECTRO-CF 12 see DFA600
ELECTRO-CF 22 see CFX500
ELECTROCORTIN see AFJ875
ELECTROCORUNDUM see EAL100
ELECTRONIC E-2 see HIC000
ELEKTROCORTIN see AFJ875
ELEMENTAL SELENIUM see SBO500
ELEMI see EAI500
ELEMI OIL see EAI500
ELEN see IBW500
ELENIUM see LFK000
ELENIUM see MDQ250
ELEPHANT'S EAR see CAL125
ELEPHANT'S EAR see EAI600
ELEPHANT TRANQUILIZER see AOO500
ELEPSINDON see DKQ000
ELESANT see DLR300
ELESTOL see CLD000
ELEUDRON see TEX250
ELEVAN see DOZ000
ELEX 334 see EAI800
ELF see CBT750
ELFANEX see RDK000
ELFAN 4240 T see SON000
ELFAN WA SULPHONIC ACID see LBU100
ELFTEX see CBT750
ELGACID ORANGE 2G see FAG010
ELGETOL see BRE500
ELGETOL see DUS700
ELGETOL see DUU600
ELGETOL 318 see BRE500
ELIAMINA RED 8BL see CMO885
ELIAMINE LIGHT TURQUOISE G see COF420
ELICIDE see MDI000
ELIMIN see DLV000
ELIMOCLAVIN see EAJ000
ELINTAAL see ELZ050
ELIPOL see DUS700
ELIPTEN see AKC600
ELITE see TAN050
ELITE FAST ORANGE R see ADG000
ELITE FAST RED BG see NAO600
ELITE FAST RED G see CMM320
ELITE FAST RED GRS see CMM320
ELITE FAST RED R see NAO600
ELITE FAST RED RS see NAO600
ELITONE see DJS200
ELIXICON see TEP000
ELIXIR of VITRIOL see SOI510
ELIXOPHYLLIN see TEP000
ELIXOPHYLLINE see TEP000
ELJON FAST ORANGE G see CMS145
ELJON FAST SCARLET PV EXTRA see MMP100
ELJON FAST SCARLET RN see MMP100
ELJON LAKE RED C see CHP500
ELJON LITHOL RED No. 10 see NAP100
ELJON MADDER see DMG800

ELJON PINK TONER see RGW000
ELJON RUBINE BS see CMS148
ELJON YELLOW BG see DEU000
ELKAPIN see MNG000
ELKON FAST YELLOW GR see CMS208
ELKOSIL see SNJ350
ELKOSIN see SNJ350
ELKOSINE see SNJ350
ELLAGIC ACID see EAI200
ELLIPTICINE see EAI850
ELLIPTISINE see EAI850
ELLSYL see MHJ500
ELMASIL see AMY050
ELMEDAL see BRF500
ELMER'S GLUE ALL see AAX250
ELOBROMOL see DDJ000
ELOCRON see DVS000
ELOKOFAC RA-6N see PKE800
ELON see MGJ750
ELON WORT see CCS650
ELORINE see CPQ250
ELORINE CHLORIDE see EAI875
ELORINE SULFATE see TJG225
EL P.E.T.N. see PBC250
ELPI see ARQ750
ELPON see PMP500
ELRODORM see DYC800
ELSAN see DRR400
ELSERPINE see RDK000
ELSYL see MHJ500
ELTESOL ST 34 see SIK460
ELTESOL ST 90 see SIK460
ELTESOL SX 30 see XJJ010
ELTEX see PJS750
ELTEX 6037 see PJS750
ELTEX A 1050 see PJS750
ELTREN see DXY725
ELTREN see LBX050
ELTRIANYL see TKX000
ELTROXIN see LFG050
ELVACITE see PKB500
ELVANOL see PKP750
ELVANOL 50-42 see PKP750
ELVANOL 52-22 see PKP750
ELVANOL 70-05 see PKP750
ELVANOL 71-30 see PKP750
ELVANOL 90-50 see PKP750
ELVANOL 522-22 see PKP750
ELVANOL 73125G see PKP750
ELVARON see DFL200
ELYMOCLAVIN see EAJ000
ELYMOCLAVINE see EAJ000
ELYSION see AOO500
ELYZOL see MMN250
ELZOGRAM see CCS250
EM see EDH500
EM 3 see RNU100
EM 12 see DVS600
EM 136 see OOM300
EM 255 see EAJ100
910EM see EHP700
EM 923 see DFY400
EM 1173 see HMQ100
EMA 1605 see EJM950
EMAFORM see CHR500
EMAGRIN see SPC500
EMALEX 103 see PJT300
EMALEX 115 see PJT300
EMAL T see SON000
EMAMECTIN BENZOATE see EAJ150
EMANAY ATOMIZED ALUMINUM POWDER see AGX000
EMANAY ZINC DUST see ZBJ000
EMANAY ZINC OXIDE see ZKA000
EMANDIONE see PFJ750
EMANIL see DAS000
EMANON 4115 see PJY100
EMAR see ZKA000
EMASOL 41S see SKV170
EMATHLITE see KBB600
EMAZOL RED B see EAJ500
EMB see TGA500

EMBACETIN see CDP250
EMBADOL see TFK300
EMBAFUME see MHR200
EMBANOX see BQI000
EMBARIN see ZVJ000
EMBARK see DUK000
EMBARK PLANT GROWTH REGULATOR see DUK000
EMBATHION see EEH600
EMBAY 8440 see BGB400
EMBECHINE see BIE500
EMBELIC ACID see EAJ600
EMBELIN see EAJ600
EMBEQUIN see DNF600
EMBERLINE see EAJ600
EMB-FATOL see EDW875
EMBICHIN see BIE250
EMBICHIN see BIE500
EMBICHIN 7 see NOB700
EMBICHIN HYDROCHLORIDE see BIE500
EMBIKHINE see BIE500
EMBINAL see BAG250
EMBIOL see VSZ000
EMBITOL see EAK000
p-EMBITOL see EAK100
EMBONIC ACID see PAF100
EMBRAMINE HYDROCHLORIDE see BMN250
EMBUTAL see NBU000
EMBUTOX see DGA000
EMBUTOX see EAK500
EMBUTOX KLEAN-UP see DGA000
EMC see CHC500
EMCEPAN see CIR250
EMCOL 888 see EAL000
EMCOL H-2A see PJY100
EMCOL CA see OAV000
EMCOL DS-50 CAD see HKJ000
EMCOL E-607 see LBD100
EMCOL H 31A see PJY100
EMCOL-IM see IQN000
EMCOL-IP see IQW000
EMCOL MSK see OAV000
EMCOL O see GGR200
EMCOL PS-50 RHP see SLL000
EMD 9806 see SDZ000
EMD 16-795 see PGF112
EMDABOL see TFK300
EMDABOLIN see TFK300
EMEDAN see BSM000
EMEDASTINE DIFUMARATE see EAL050
EMERALD GREEN see BAY750
EMERALD GREEN see COF500
EMERESSENCE 1150 see EJQ500
EMERESSENCE 1160 see PER000
EMEREST 2301 see OHW000
EMEREST 2314 see IQN000
EMEREST 2316 see IQW000
EMEREST 2325 see BSL600
EMEREST 2350 see EJM500
EMEREST 2381 see SLL000
EMEREST 2400 see OAV000
EMEREST 2401 see OAV000
EMEREST 2646 see PJY100
EMEREST 2660 see PJY100
EMEREST 2801 see OHW000
EMERGIL see FMO129
EMERICID see LIF000
EMERSAL 6400 see SIB600
EMERSAL 6434 see SON000
EMERSAL 6465 see TAV750
EMERSIST 7232 see AFJ160
EMERSOL 120 see SLK000
EMERSOL 140 see PAE250
EMERSOL 143 see PAE250
EMERSOL 210 see OHU000
EMERSOL 213 see OHU000
EMERSOL 6321 see OHU000
EMERSOL 233LL see OHU000
EMERSOL 221 LOW TITER WHITE OLEIC ACID see OHU000

EMERSOL 220 WHITE OLEIC ACID see OHU000
EMERY see EAL100
EMERY 655 see MSA250
EMERY 2218 see MJW000
EMERY 2219 see OHW000
EMERY 2310 see OHW000
EMERY 5703 see BKD500
EMERY 5709 see DHF400
EMERY 5711 see TGT000
EMERY 5712 see DMT800
EMERY 5714 see HKS100
EMERY 5715 see CJU400
EMERY 5717 see CJU400
EMERY 5770 see EHJ600
EMERY 5791 see MCN250
EMERY 6705 see PER000
EMERY 6802 see PJW500
EMERY OLEIC ACID ESTER 2221 see GGR200
EMERY OLEIC ACID ESTER 2301 see OHW000
EMERY X-88-R see DWT000
EMESIDE see ENG500
EMETE-CON see BDW000
EMETHIBUTIN HYDROCHLORIDE see EIJ000
(all-E)-9-(4-METHOXY-2,3,6-TRIMETHYLPHENYL)-3,7-DIMETHYL-2,4,6,8-NONATETRAENOIC ACID see REP400
EMETIC HOLLY see HGF100
EMETICON see BDW000
EMETIC WEED see CCJ825
EMETINE see EAL500
(−)-EMETINE see EAL500
EMETINE ANTIMONY IODIDE see EAM000
EMETINE BISMUTH IODIDE see EAM500
EMETINE with BISMUTH(III) TRIIODIDE see EAM500
EMETINE, DIHYDROCHLORIDE see EAN000
(−)-EMETINE DIHYDROCHLORIDE see EAN000
1-EMETINE DIHYDROCHLORIDE see EAN000
EMETINE DIHYDROCHLORIDE TETRAHYDRATE see EAN500
EMETINE HYDROCHLORIDE see EAN000
EMETINE TRIIODOBISMUTH(III) see EAM500
EMETIQUE (FRENCH) see AQG250
EMETIRAL see PMF500
EMETREN see CDP250
EMFAC 1202 see NMY000
EMI-CORLIN see HHR000
EMID 6511 see BKE500
EMID 6541 see BKE500
EMID 6545 see OHU200
EMINEURINA see CHD750
EMIPHEROL see VSZ450
EMISAN 6 see MEP250
EMISOL see AMY050
EMISOL 50 see AMY050
EMISOL F see AMY050
EMITEFUR see BDN600
EMIVIRINE see EAN525
EMMATOS see MAK700
EMMATOS EXTRA see MAK700
EMMI see EME050
EMOCICLINA see VSZ000
EMODIN see IIU000
EMODIN see MQF250
EMODOL see MQF250
EMO-NIK see NDN000
EMOQUIL see EAN600
EMOREN see DTL200
EMOREN see EAN650
EMORFAZONE see EAN700
EMORHALT see AJV500
EMOTIVAL see CFC250

EMP see EDT100
EMP see EME100
EMPAL see CIR250
EMPECID see MRX500
EMPENTHRIN see EQN250
d-EMPENTHRIN see EQN250
EMPETAAL see IKT100
EMPG see ENC000
EMPILAN 2848 see EJM500
EMPILAN BP 100 see PJY100
EMPILAN BQ 100 see PJY100
EMPIRIN see ADA725
EMPIRIN COMPOUND see ABG750
EMPIRIN COMPOUND see ARP250
EMPLETS POTASSIUM CHLORIDE see PLA500
EMPP see EAV700
EMQ see SAV000
EMRITE 6009 see GGR200
EMS see EMF500
EMSORB 2500 see SKV100
EMSORB 2502 see SKV170
EMSORB 2505 see SKV150
EMSORB 2510 see MRJ800
EMSORB 2515 see SKV000
EMSORB 6900 see PKL100
EMSORB 6907 see SKV195
EMSORB 6915 see PKG000
EMT 25,299 see MDV500
EMTAL 596 see TAB750
EMTEXATE see MDV500
EMTROL 1630B see DAI800
EMTRYL see DSV800
EMTRYLVET see DSV800
EMTRYMIX see DSV800
EMTS see EME500
EMULGATOR 8972 see SKV170
EMULGEN 100 see DXY000
EMULGEN 210 see PJT300
EMULGEN 703 see AFJ155
EMUL P.7 see OAV000
EMULPHOGENE DA 530 see PKE370
EMULPHOGENE DA 630 see PKE390
EMULPHOGENE LM-710 see PKE350
EMULPHOR see PKE500
EMULPHOR A see PJY100
EMULPHOR ON-870 see PJW500
EMULPHOR ON-877 see EAN725
EMULPHOR SURFACTANTS see PKE500
EMULPHOR UN-430 see PJY100
EMULPHOR VN 430 see PJY100
EMULSAMINE BK see DAA800
EMULSAMINE E-3 see DAA800
EMULSEPT see EAN750
EMULSIFIABLE OIL see MQV796
EMULSIFIABLE OIL see MQV855
EMULSIFIER No. 104 see SIB600
EMULSIFIER WHC see OHU200
EMULSION 212 see FQU875
EMULSIPHOS 440/660 see SJH200
EMULSOV O EXTRA P see EAN800
E-MYCIN see EDH500
EMYRENIL see OOG000
18EN see FMW333
EN 237 see EAL100
EN 313 see EEI050
EN-1530 see NAH000
EN 1627 see DPE000
EN 1639 see CQF099
EN 1939 see CQF099
EN-15304 see NAH000
EN 1639A see NAH100
EN 18133 see EPC500
EN-28,450 see AGT250
ENADEL see CMY525
ENADOLINE HYDROCHLORIDE see EAN900
ENALAPRIL MALEATE see EAO100
ENALLYNYMAL SODIUM see MDU500
ENAMEL WHITE see BAP000
ENANTHAL see HBB500
ENANTHALDEHYDE see HBB500

ENANTHALDOXIME see EAO200
ENANTHIC ACID see HBE000
ENANTHIC ALCOHOL see HBL500
ENANTHOLE see HBB500
ENANTHONE see TJJ300
ENANTHOTOXIN see EAO500
ENANTHYLIC ACID see HBE000
ENANTHYLIC ETHER see EKN050
ENARMON see TBG000
ENAVID see EAP000
ENBU see ENT000
ENC see ENT500
ENCEPHALARTOS HILDEBRANDTII see EAQ000
ENCETROP see NNE400
ENCORDIN see EAQ050
ENCORTON see PLZ000
ENCYPRATE see EGT000
ENDAK see MQT550
ENDAK MITE see MQT550
ENDANIL BLUE B see CMM092
ENDAVEN see EGS000
ENDECRIL see DFP600
ENDEP see EAI000
ENDIARON see DEL300
ENDIEMALUM see DJO800
ENDISON see DCV800
ENDOBIL see IFP800
ENDOBION see NCR000
ENDOCEL see EAQ750
ENDOCID see EAS000
ENDOCIDE see EAS000
ENDOCISTOBIL see BGB315
ENDODAN see EJQ000
ENDODEOXYRIBONUCLEASE BBRI see REF262
ENDODEOXYRIBONUCLEASE CHUI see REF262
ENDODEOXYRIBONUCLEASE HINBIII see REF262
ENDODEOXYRIBONUCLEASE HINDIII see REF262
ENDODEOXYRIBONUCLEASE HINFII see REF262
ENDODEOXYRIBONUCLEASE HSUI see REF262
ENDODEOXYRIBONUCLEASE I see EDK650
ENDODEOXYRIBONUCLEASE RSAI see REF270
ENDOD, EXTRACT see EAQ100
ENDODIOL see HCC550
ENDO E see VSZ450
ENDO E DOMPE see TGJ050
6,14-ENDOETHENO-7-(2-HYDROXY-2-PENTYL)-TETRAHYDRO-ORIPAVINE HYDROCHLORIDE see EQO500
ENDOFOLLICOLINA D.P. see EDR000
ENDOFOLLICULINA see EDV000
ENDOGRAFIN see BGB315
ENDOGRAPHIN see BGB315
ENDOKOLAT see PPN100
ENDOLACTON see EAQ815
ENDOLAT see DAM700
2,5-ENDOMETHYLENE CYCLOHEXANECARBOXYLIC ACID, ETHYL ESTER (mixed formyl isomers) see NNG500
(2,5-ENDOMETHYLENECYCLOHEXYLMETHYL)AMINE see AKY750
cis-3,6-ENDOMETHYLENE-Δ⁴-TETRAHYDROPHTHALIC ACID DIMETHYL ESTER see DRB400
ENDOMETHYLENETETRAHYDROPHTHALIC ACID, N-2-ETHYLHEXYL IMIDE see OES000
ENDOMYCIN see EAQ500
3,6-ENDOOXOHEXAHYDROPHTHALIC ACID see EAR000
ENDOPANCRINE see IDF300
ENDOPITUITRINA see ORU500

β-ENDORPHIN see EAQ600
ENDOSAN see BGB500
ENDOSOL see EAQ750
ENDOSULFAN see EAQ750
ENDOSULFAN 1 see EAQ810
ENDOSULFAN 2 see EAQ800
α-ENDOSULFAN see EAQ810
β-ENDOSULFAN see EAQ800
ENDOSULFAN A see EAQ810
ENDOSULFAN ALCOHOL see HCC550
ENDOSULFAN B see EAQ800
ENDOSULFANDIOL see HCC550
ENDOSULFAN ETHER see HDK200
ENDOSULFAN HYDROXYETHER see
HDK210
ENDOSULFAN α-HYDROXY ETHER see
HDK210
ENDOSULFAN LACTON see EAQ815
ENDOSULFAN SULFATE see EAQ820
ENDOSULPHAN see EAQ750
ENDOTAL see DXD000
ENDOTHAL see DXD000
ENDOTHAL see EAR000
ENDOTHAL COMBINED with IPC (1:1) see
EAR500
ENDOTHALL see EAR000
ENDOTHAL-NATRIUM (DUTCH) see
DXD000
ENDOTHAL-SODIUM see DXD000
ENDOTHAL TECHNICAL see EAR000
ENDOTHAL WEED KILLER see DXD000
ENDOTHION see EAS000
ENDOTOXIN see EAS100
ENDOTOXIN, klp see EAS260
hmi ENDOTOXIN, phenol water extract see
HAB710
ENDOTOXIN, AEROMONAS
HYDROPHILA A₃ see AET500
Δ-ENDOTOXIN, from BACILLUS
THURINGIENSIS see EAS230
ENDOTOXIN, BACT. AERTRYCKE see
EAS200
ENDOTOXIN, BACTERIODES FRAGILIS
see BAC390
ENDOTOXIN, BRUCELLA MELITENSIS see
BOL600
ENDOTOXIN, ESCHERICHIA COLI see
EDK700
hmi ENDOTOXIN, NaCl-citrate extract see
HAB700
ENDOTOXIN, PSEUDOMONAS
AERUGINOSA see POH620
ENDOTOXIN, VIBRIO CHOLERAE see
VJZ100
ENDOX see EAT600
ENDOXAN see CQC650
ENDOXANA see CQC500
ENDOXANAL see CQC650
ENDOXAN-ASTA see CQC500
ENDOXAN MONOHYDRATE see CQC500
ENDOXAN R see CQC500
3,6-ENDOXOHEXAHYDROPHTHALIC
ACID see EAR000
3,6-ENDOXOHEXAHYDROPHTHALIC
ACID DISODIUM SALT see DXD000
ENDRATE see EIX000
ENDRATE DISODIUM see EIX500
ENDRATE TETRASODIUM see EIV000
ENDREX see EAT500
ENDRIN see EAT500
ENDRINE (FRENCH) see EAT500
ENDRIN KETONE see KFK200
ENDROCID see EAT600
ENDROCIDE see EAT600
ENDURACIDIN see EAT800
ENDURACIDIN A see EAT810
ENDURANCE see DWS200
ENDURON see MIV000
ENDURONYL see RDF000
ENDUXAN see CQC500
ENDYDOL see ADA725
ENDYL see TNP250

E.N.E. see ENX500
ENE 11183 B see EAT600
ENELFA see HIM000
ENENTHOHYDROXAMIC ACID see
HBF550
ENERADE EB-7104 see FMW343
ENERIL see HIM000
ENERZER see IKC000
ENFENEMAL see ENB500
ENFLURANE see EAT900
ENGLISH HOLLY see HGF100
ENGLISH IVY see AFK950
ENGLISH RED see IHC450
ENHEPTIN see ALQ000
ENHEPTIN A see ABY900
ENHEXYMAL see ERD500
ENHEXYMAL see ERE000
ENHEXYMAL NFN see ERE000
ENHYDRINA SCHISTOSA VENOM see
EAU000
EN-1733A HYDROCHLORIDE see MRB250
ENIACID BLACK IVS see FAB830
ENIACID BLACK SH see FAB830
ENIACID BRILLIANT RUBINE 3B see
HJF500
ENIACID FUCHSINE BN see CMS228
ENIACID LIGHT ORANGE G see HGC000
ENIACID LIGHT RED 3G see CMM300
ENIACID LIGHT RED 6B see CMM400
ENIACID METANIL YELLOW GN see
MDM775
ENIACID ORANGE I see FAG010
ENIACID SUNSET YELLOW see FAG150
ENIACID YELLOW RS see MRL100
ENIACROMO ORANGE R see NEY000
ENIACROMO RED B see CMG750
ENIACROMO YELLOW G see SIT850
ENIACYL SCARLET B see ENP100
ENIALIT LIGHT RED RL see MMP100
ENIAL ORANGE I see PEJ500
ENIAL RED IV see SBC500
ENIAL YELLOW 2G see DOT300
ENIAMETHYL ORANGE see MND600
ENIANIL BLACK CN see AQP000
ENIANIL BLUE 2BN see CMO000
ENIANIL BLUE RW see CMO600
ENIANIL BRILLIANT BLUE FF see CMN750
ENIANIL BROWN 2GS see CMO810
ENIANIL DARK GREEN BG see CMO830
ENIANIL FAST BROWN M see CMO800
ENIANIL FAST RED F see CMO870
ENIANIL FAST SCARLET 4BS see CMO870
ENIANIL GREEN B see CMO840
ENIANIL GREEN BBN see CMO840
ENIANIL PURE BLUE AN see CMO500
ENIAZOL BLUE BLACK BHN see CMN800
ENICOL see CDP250
ENIDE see DRP800
ENIDRAN see DYE700
ENIDREL see CFZ000
ENIDREL see EAP000
ENILOCONAZOL (SP) see FPB875
ENIPRESSER see RDK000
ENJ 2065 see PHW585
ENJAY CD 460 see PMP500
ENJI see CNF050
ENKALON see NOH000
ENKEFAL see DNU000
ENKELFEL see DKQ000
ENNG see ENU000
ENOCIANINA see GJU100
ENOCITABINE see EAU075
ENORDEN see MQU750
ENORDET AE 1215-30 see AFJ160
ENORDET AE 1215-9.4 see AFJ160
ENOVID see EAP000
ENOVID-E see EAP000
ENOVIT see EAP000
ENOVIT M see PEX500
ENOXACIN see EAU100
ENOXACIN HYDRATE (2;3) see EAU150
ENOXAPARIN see HAQ500

ENOXOLONE see GIE000
ENPHENEMAL see ENB500
ENPROMATE see DWL400
ENPROSTIL see EAU200
ENQUIK see UTU600
ENRADIN see EAT800
ENRADINE see EAT810
ENRAMYCIN see EAT800
ENRAMYCIN see EAT810
ENRICHED SUPERPHOSPHATE see SOV500
ENRIOCHROME ORANGE AOR see
NEY000
ENRUMAY see WAK000
E.N.S. see ENX500
ENS see EPI300
ENSEAL see PLA500
ENSIGN see CMF350
ENSODORM see EOK000
ENSTAMINE HYDROCHLORIDE see
DCJ850
ENSTAR see POB000
ENSURE see EAQ750
ENS-ZEM WEEVIL BAIT see DXE000
ENT see ENU500
ENT 5 see LBO000
ENT 6 see BPL250
ENT 9 see BRT000
17-ENT see ABU000
17-ENT see ENX600
ENT 38 see PDP250
ENT 54 see ADX500
ENT 92 see IHZ000
ENT 114 see DYA200
ENT 123 see VHZ000
ENT 133 see RNZ000
ENT 154 see DUS700
ENT 157 see CPK500
ENT 262 see DTR200
ENT-337 see DED500
ENT 375 see EKV000
ENT 666 see SNA500
ENT 884 see COF500
ENT 988 see BJK500
ENT 1025 see HCP100
ENT 1,122 see BRE500
ENT 1,501 see DXE000
ENT 1,506 see DAD200
ENT 1,656 see EIY600
ENT 1,716 see MEI450
ENT 1,860 see PCF275
ENT 2,435 see NDR500
ENT 3,424 see NDN000
ENT 3,776 see DFT000
ENT 3,797 see TBO500
ENT 4,225 see BIM500
ENT 4,504 see DFJ050
ENT 4,585 see CJR500
ENT 4,705 see CBY000
ENT 7068 see EQL550
ENT 7,543 see POO100
ENT 7,796 see BBQ500
ENT 8,184 see OES000
ENT 8286 see BRP250
ENT 8,420 see DGG000
ENT 8,538 see DAA800
ENT 8,601 see BBP750
ENT 9,232 see BBQ000
ENT 9,233 see BBR000
ENT 9,234 see BFW500
ENT 9,624 see BIN000
ENT 9,735 see CDV100
ENT 9,932 see CDR750
ENT 14,250 see PIX250
ENT 14,611 see ASL250
ENT 14,689 see FAS000
ENT 14,874 see EIR000
ENT 14,875 see MAS500
ENT 15,108 see PAK000
ENT 15,152 see HAR000
ENT 15,208 see CKE750
ENT 15,208 see NCM700
ENT 15,266 see PNP250

ENT 15,349 see EIY500
ENT 15,406 see PNJ400
ENT 15,748 see HBL100
ENT 15,949 see AFK250
ENT 16,087 see NIM500
ENT 16,225 see DHB400
ENT 16,273 see SOD100
ENT 16,275 see BGC750
ENT 16,358 see CJT750
ENT 16,391 see KEA000
ENT 16,436 see DXX400
ENT 16,519 see SOP500
ENT 16,634 see ISA000
ENT 16,894 see TED500
ENT 17,034 see MAK700
ENT 17,035 see NFT000
ENT 17,251 see EAT500
ENT 17,291 see OCM000
ENT 17,292 see MNH000
ENT 17,295 see DAO600
ENT 17,470 see DFK600
ENT 17,510 see AFR250
ENT 17,588 see PPQ625
ENT 17,591 see EAU500
ENT 17591 see IKC070
ENT 17,596 see BHJ500
ENT 17,798 see EBD700
ENT 17,941 see CKI625
ENT 17,956 see CNU750
ENT 18,060 see CKC000
ENT 18,066 see BON250
ENT 18,544 see PDK000
ENT 18,596 see DER000
ENT 18,771 see TCF250
ENT 18,861 see MIJ250
ENT 18,862 see DAO800
ENT 18,862 see MIW100
ENT 18,870 see DMC600
ENT 19,059 see PNQ250
ENT 19,060 see DSK200
ENT 19,109 see BJE750
ENT 19,244 see IKO000
ENT 19,442 see TBC500
ENT 19,507 see DCM750
ENT 19,763 see TIQ250
ENT 20,218 see DKC800
ENT 20,279 see MFF580
ENT 20,696 see CEP000
ENT 20,738 see DGP900
ENT 20,852 see BPG000
ENT 20,852 see DDP000
ENT 20871 see MJS550
ENT 20,993 see AMX825
ENT 21,040 see AGE250
ENT 21,170 see DQQ500
ENT-21170 see DQQ500
ENT 22,014 see EKN000
ENT 22,335 see TBR250
ENT 22,374 see MQR750
ENT 22,542 see DKC800
ENT 22641 see FQQ500
ENT 22,784 see BIN500
ENT 22,865 see DJN600
ENT 22,897 see DVQ709
ENT 22,952 see POO000
ENT 23,233 see ASH500
ENT 23,284 see RMA500
ENT 23393 see BNQ600
ENT 23,437 see DXH325
ENT 23,438 see DRR400
ENT 23,444 see DKB170
ENT 23,584 see BIT250
ENT 23,648 see BIO750
ENT 23,708 see TNP250
ENT 23,737 see CKM000
ENT 23,968 see POP000
ENT 23,969 see CBM750
ENT 23,970 see CGM400
ENT 23,979 see EAQ750
ENT 24,042 see PGS000
ENT 24,044 see PGZ900
ENT 24,105 see EEH600

ENT 24,415 see PGZ915
ENT 24,482 see DGQ875
ENT 24,650 see DSP400
ENT 24,652 see IOT000
ENT 24,653 see EAS000
ENT 2,4716 see CEQ500
ENT 24,717 see COD000
ENT 24,723 see MOX250
ENT 24,725 see DKB600
ENT 24,727 see AQT500
ENT 24,738 see DRL200
ENT 24,833 see SOY000
ENT 24,915 see TND250
ENT 24,944 see EAV000
ENT 24,945 see FAQ800
ENT 24,954 see PHG600
ENT 24,964 see DAP000
ENT 24,969 see CDS750
ENT 24,970 see NAQ500
ENT 24,979 see BLL750
ENT 24,984 see SHF000
ENT 24,986 see DXO000
ENT 24,988 see NAG400
ENT 25,208 see ABX250
ENT 25,294 see BIE250
ENT 25,296 see TND500
ENT 25,445 see AMY050
ENT 25,456 see OAL000
ENT 25,500 see COF250
ENT 25,506 see DNX500
ENT 25,507 see PHM000
ENT 25,515 see FAB400
ENT 25,515 see PGX300
ENT 25,516 see THW750
ENT 25,540 see FAQ900
ENT 25,543 see COF250
ENT 25,545 see OAN000
ENT 25,550 see SCH002
ENT 25,567 see HMV000
ENT 25,580 see EPC500
ENT 25,582 see HCI475
ENT 25,584 see EBW500
ENT 25,585 see PDC750
ENT 25,599 see MQH750
ENT 25,606 see ORU000
ENT 25,610 see ENI500
ENT 25,612 see MOB599
ENT 25,623 see EHY700
ENT 25,636 see LIN400
ENT 25,640 see CQL250
ENT 25,644 see FAB600
ENT 25,647 see DGD800
ENT 25,650 see DIX000
ENT 25,651 see DUI000
ENT 25,670 see MIA250
ENT 25,671 see PMY300
ENT 25,673 see DJR800
ENT 25,674 see DSK600
ENT 25,675 see COQ399
ENT 25,678 see TNH750
ENT 25,684 see DST200
ENT 25,705 see PHX250
ENT 25,707 see PHL750
ENT 25,712 see EPY000
ENT 25,715 see DSQ000
ENT 25,718 see DAE425
ENT 25,719 see MQW500
ENT 25,723 see EHL670
ENT 25,726 see DST000
ENT 25,732 see PMN250
ENT 25,734 see PHD250
ENT 25,736 see DET600
ENT 25,737 see DTV400
ENT 25,739 see BES250
ENT 25,755 see CJD650
ENT 25,760 see ASD000
ENT 25,764 see TBV750
ENT 25,765 see BSG000
ENT 25,766 see DOS000
ENT 25,776 see DNX800
ENT 25,777 see MIB250
ENT 25,780 see DNS200

ENT 25,784 see DOR400
ENT 25,787 see MOB699
ENT 25,793 see BGB500
ENT 25,796 see FMU045
ENT 25,801 see EMC000
ENT 25,809 see DXN600
ENT 25,810 see MIB500
ENT 25,823 see DFV400
ENT 25,830 see PGW750
ENT 25,832 see CON300
ENT 25,841 see RAF100
ENT 25,843 see TMD000
ENT 25,870 see MOB750
ENT 25,922 see DQZ000
ENT 25,962 see CFF250
ENT 25,977 see MOB500
ENT 25,979 see DWU200
ENT 25,991 see DHH400
ENT 26,058 see DEV800
ENT 26,079 see AMG750
ENT 26,263 see EJN500
ENT 26,316 see AQO000
ENT 26,396 see EMF500
ENT 26,398 see TCQ500
ENT 26,538 see CBG000
ENT 26,592 see BGA750
ENT 26,613 see MJG500
ENT 26,925 see TIL500
ENT 26,999 see PNH750
ENT 27,039 see BSG250
ENT 27,040 see AAI115
ENT 27,041 see BDG250
ENT 27,043 see MHR150
ENT 27,045 see EMR100
ENT 27,093 see CBM500
ENT 27,102 see DIX600
ENT 27,115 see CKL750
ENT 27,127 see AON250
ENT 27,128 see MOU750
ENT 27,129 see MRH209
ENT 27,139 see TIP750
ENT 27,160 see AHO750
ENT 27,161 see PHN250
ENT 27,162 see BNL250
ENT 27,163 see BDJ250
ENT 27,164 see CBS275
ENT 27,165 see TAL250
ENT 27,179 see MEV600
ENT 27,180 see MOB250
ENT 27,186 see PGD000
ENT 27,192 see DYE200
ENT 27,193 see DSO000
ENT 27,221 see SNT100
ENT 27,223 see AIX000
ENT 27,226 see SOP000
ENT 27,230 see DTQ800
ENT 27,238 see DRU400
ENT 27,244 see CBW000
ENT 27,248 see PHN000
ENT 27,250 see EOO000
ENT 27,253 see TCY275
ENT 27,257 see DRR200
ENT 27,258 see EGV500
ENT 27,261 see TIA000
ENT 27,267 see BJD000
ENT 27,300 see CQI500
ENT 27,305 see FNE500
ENT 27,311 see CMA100
ENT 27,313 see SCD500
ENT 27,318 see EIN000
ENT 27,320 see DBI099
ENT 27,324 see DLS800
ENT 27,335 see CJJ250
ENT 27,341 see MDU600
ENT 27,346 see DOP200
ENT 27,349 see CFC500
ENT 27,350 see MDX250
ENT 27,351 see MLX000
ENT 27,357 see DTP800
ENT 27,358 see DTP600
ENT 27,389 see DVS000
ENT 27,394 see DJY200

ENT 27,396 see DTQ400
ENT 27,407 see MPG250
ENT 27,408 see IEN000
ENT 27,410 see DRP600
ENT 27,411 see MPS250
ENT 27,438 see DGA200
ENT 27,474 see BEP500
ENT 27,488 see BAT750
ENT 27,520 see CMA250
ENT 27,521 see PHE250
ENT 27,552 see IOS000
ENT 27,566 see DSO200
ENT 27,567 see CJJ250
ENT 27,567 see CJJ500
ENT 27,572 see FAK000
ENT 27,625 see DOL800
ENT 27,696 see DRR000
ENT 27,738 see BLU000
ENT 27,766 see DOX600
ENT 27,822 see DOP600
ENT 27,851 see DAB400
ENT 27,910 see DED000
ENT 27,967 see MJL250
ENT 27,989 see MKA000
ENT 28,009 see HON000
ENT 28,344 see PIZ499
ENT-29012 see MID870
ENT 29,054 see CJV250
ENT 29,118 see EPC175
ENT 30,838 see CPK550
ENT 31,472 see MJK500
ENT 31,560 see BQT600
ENT 32,833 see AAR500
ENT 33,266 see DXU830
ENT 33,335 see HFX000
ENT 33,348 see HFE520
ENT 33478 see HCP070
ENT 34,872 see CNG760
ENT 34,886 see ECB200
ENT 35,349 see TJF400
ENT 35770 see HDR700
ENT 50,003 see TNK250
ENT 50,107 see BJC250
ENT 50,146 see RDK000
ENT 50,172 see HEF500
ENT 50,324 see EJM900
ENT 50,434 see AQG250
ENT 50,439 see BIA250
ENT 50,698 see DYC700
ENT 50,787 see BGX775
ENT 50,825 see MLY000
ENT-50,838 see TDQ225
ENT 50,852 see HEJ500
ENT 50,882 see HEK000
ENT 50,909 see AGU500
ENT 50918 see HEK050
ENT 50,990 see DOV600
ENT 50,991 see ASK500
ENT-51007 see TDT850
ENT 51,253 see BGY500
ENT 51,254 see MGE100
ENT 51256 see BGX850
ENT 51,762 see NNF000
ENT 51,799 see MJQ500
ENT 51,904 see TLR250
ENT 60229 see FQQ500
ENT 61,241 see ACM750
ENT 61,969 see DNU850
ENT 70,459 see EQD000
ENT 70,460 see KAJ000
ENT 70,531 see POB000
ENT 25,832-a see CON300
ENT 27,300-A see CQI500
ENTACYL see HEP000
ENT AI3-29261 see AFK000
ENT 27,386GC see DRR400
ENT 27,699GC see DIN800
ENTEPAS see AMM250
ENTERAMINE see AJX500
ENTERICIN see ADA725
ENTERO-BIO FORM see CHR500

ENTERO-EXOTOXIN, CHOLERA see CMC800
ENTEROMYCETIN see CDP250
ENTEROPHEN see ADA725
ENTEROQUINOL see CHR500
ENTEROSALICYL see SJO000
ENTEROSALIL see SJO000
ENTEROSARINE see ADA725
ENTEROSEDIV see TEH500
ENTEROSEPT see DNF600
ENTEROSEPTOL see CHR500
EXO-ENTEROTOXIN see VJZ200
ENTEROTOXIN B, STAPHYLOCOCCAL see EAV025
ENTEROTOXIN, CHOLERA see CMC800
ENTEROTOXIN, CLOSTRIDIUM PERFRINGENS, TYPE A see CMY220
ENTEROTOXON see NGG500
ENTERO-VIOFORM see CHR500
ENTEROZOL see CHR500
ENTERUM LOCORTEN see CHR500
ENTEX see FAQ900
ENTHOHEX see PCY300
ENTIZOL see MMN250
ENTOBEX see PCY300
ENTOMOPHTORINE see EAV050
ENTOMOXAN see BBQ500
ENTPROL see QAT000
ENTRA see TMX775
ENTRAMIN see ALQ000
ENTROKIN see CHR500
ENTRONON see PCY300
ENTROPHEN see ADA725
ENTSUFON see TMN490
ENTUSIL see SNN500
ENT 24,980-X see DJA400
ENT 25,545-X see OAN000
ENT 25,552-X see CDR750
ENT 25,554-X see MNO750
ENT 25,555-X see DJA200
ENT 25,595-X see DQZ000
ENT 25,602-X see COD850
ENT 25,700-X see HCK000
ENT 27,395-X see CQH650
ENTYDERMA see AFJ625
ENU see ENV000
ENU see NKE500
ENUCLEN see BBA500
ENVERT 171 see DAA800
ENVERT DT see DAA800
ENVERT-T see TAA100
ENVIOMYCIN SULFATE see VQZ100
ENZACTIN see THM500
ENZAMIN see BBW500
ENZAPROST see POC500
ENZAPROST F see POC500
ENZEON see CML880
ENZOPRIDE see CNF390
ENZOSE see DBI800
E.O. see EJN500
EO 5A see IGK800
EO 122 see EAV100
EOC see EEO000
EOCT see COD475
EOE 13 see EEH580
EOSIN see BMO250
EOSIN see BNK700
EOSIN BLUE see ADG250
EOSINE see BMO250
EOSINE see BNH500
EOSINE B see BNK700
EOSINE BLUE see ADG250
EOSINE BLUISH see ADG250
EOSINE FA see BNK700
EOSINE LAKE RED Y see BNK700
EOSINE SODIUM SALT see BNH500
EOSINE YELLOWISH see BNH500
EOSIN GELBLICH (GERMAN) see BNH500
EP 30 see PAX250
EP-145 see ECO500
EP 160 see PKP750
EP-185 see DNI200

EP 201 see ECB000
EP-205 see BJN250
EP-206 see PKQ070
EP-206 see VOA000
EP 316 see CQI500
EP-332 see DSO200
EP-333 see CJJ250
EP 333 see CJJ500
EP-411 see PFC750
EP-452 see MEG250
EP 453 see OMY925
EP-475 see EEO500
EP 587 see EBH420
EP-1086 see TKS000
EP 1463 see AAX250
EP-161E see ISE000
EPAL see AHH800
EPAL 6 see HFJ500
EPAL 8 see OEI000
EPAL 10 see DAI600
EPAL 12 see DXV600
EPAL 810 see DAI800
EPAL 16NF see HCP000
E-PAM see DCK759
EPAMIN see DKQ000
EPAMIN see DNU000
EPANUTIN see DKQ000
EPANUTIN see DNU000
EPAREN see DFL200
EPASMIR "5" see DKQ000
EPASTATIN SODIUM see LGK500
EPATIOL see MCI375
EPC (the plant regulator) see CBL750
EPDANTOINE SIMPLE see DKQ000
EPE see EAV500
EPELIN see DKQ000
EPELIN see DNU000
EPERISONE HYDROCHLORIDE see EAV700
E 6010 (PHARMACEUTICAL) see PJB810
EPHEDRAL see EAW000
EPHEDRATE see EAW000
EPHEDREMAL see EAW000
EPHEDRIN see EAW000
EPHEDRINE see EAW000
(+)-EPHEDRINE see EAW200
EPHEDRINE, (+)- see EAW200
d-EPHEDRINE see EAW200
l-EPHEDRINE see EAW000
l(−)-EPHEDRINE see EAW000
(±)-EPHEDRINE see EAW100
EPHEDRINE, (±)- see EAW100
dl-EPHEDRINE see EAW100
l-(+)-EPHEDRINE see EAW200
psi-EPHEDRINE see POH000
d-psi-EPHEDRINE see POH000
EPHEDRINE HYDROCHLORIDE see EAW500
EPHEDRINE HYDROCHLORIDE see EAW500
EPHEDRINE HYDROCHLORIDE see EAX000
(−)-EPHEDRINE HYDROCHLORIDE see EAX000
d-EPHEDRINE HYDROCHLORIDE see EAW995
l-EPHEDRINE, HYDROCHLORIDE see EAW500
l-EPHEDRINE HYDROCHLORIDE see EAX000
dl-EPHEDRINE HYDROCHLORIDE see EAX500
EPHEDRINE PENICILLIN see PAQ120
d-EPHEDRINE PHOSPHATE (ESTER) see EAY150
l-EPHEDRINE PHOSPHATE (ESTER) see EAY155
dl-EPHEDRINE PHOSPHATE (ESTER) see EAY175
1-EPHEDRINE SULFATE see EAY500
EPHEDRINHYDROCHLORID (GERMAN) see MGH250

EPHEDRITAL see EAW000
EPHEDROL see EAW000
EPHEDROSAN see EAW000
EPHEDROTAL see EAW000
EPHEDSOL see EAW000
EPHENDRONAL see EAW000
EPHERON see PEC250
EPHETONIN see EAX500
EPHETONINE see EAX500
EPHININE HYDROCHLORIDE see EAZ000
EPHIRSULPHONATE see CJT750
EPHORRAN see DXH250
EPHOXAMIN see EAW000
EPHYNAL see VSZ450
EPHYNAL ACETATE see TGJ050
EPIANDROSTERONE, DEHYDRO-, ACETATE see HJB225
EPIB see ARQ750
(+−)-EPIBATIDINE DIHYDROCHLORIDE see EAZ100
EPIBENZALIN see DLY000
EPIBLOC see CDT750
EPIBROMHYDRIN see BNI000
EPIBROMOHYDRIN (DOT) see BNI000
EPIBROMOHYDRINE see BNI000
EPICHLOORHYDRINE (DUTCH) see EAZ500
EPICHLORHYDRIN (GERMAN) see EAZ500
EPICHLORHYDRINE (FRENCH) see EAZ500
EPICHLOROHYDRIN see EAZ500
α-EPICHLOROHYDRIN see EAZ500
(dl)-α-EPICHLOROHYDRIN see EAZ500
EPICHLOROHYDRIN-BIS(3-AMINOPROPYL)METHYLAMINE COPOLYMER see EAZ600
EPICHLOROHYDRIN-DIMETHYLAMINE COPOLYMER see DOR500
EPICHLOROHYDRYNA (POLISH) see EAZ500
EPICHLOROPHYDRIN see EAZ500
EPICHOLESTANOL see EBA100
52-EPICIGUATOXIN 3 see CMP805
EPICLASE see PEC250
EPI-CLEAR see BDS000
EPICLORIDRINA (ITALIAN) see EAZ500
EPICUR see MQU750
EPI-CURE 113 see BGT800
EPICURE DDM see MJQ000
EPICURE NMA see NAC000
EPIDEHYDROCHOLESTERIN see EBA100
EPIDERMAL GROWTH FACTOR see EBA275
EPIDERMOL see ACR300
EPIDIAN 5 see IPO000
3,17-EPIDIHYDROXYESTRATRIENE see EDO000
3,17-EPIDIHYDROXYOESTRATRIENE see EDO000
EPIDIONE see TLP750
9,10-EPIDIOXY ANTHRACENE see EBA500
1,4-EPIDIOXY-1,4-DIHYDRO-6,6-DIMETHYLFULVENE see EBA600
EPIDONE see DWI300
EPIDONE see TLP750
EPIDORM see EOK000
EPIDOSIN see VBK000
4'-EPIDOXORUBICIN see EBB100
EPIDOZIN see VBK000
EPIDROPAL see ZVJ000
EPI-DX see EBB100
EPIFEN see FMS875
EPIFENYL see DKQ000
EPIFENYL see DNU000
EPIFLUOROHYDRIN see EBU000
EPIFOAM see HHQ800
EPIFRIN see VGP000
EPIGALLOCATECHIN GALLATE see EBB200
(−)-EPIGALLOCATECHIN GALLATE see EBB200

EPIGALLOCATECHIN 3-GALLATE see EBB200
(−)-EPIGALLOCATECHIN-3-o-GALLATE see EBB200
(−)-EPIGALLOCATECHOL GALLATE see EBB200
EPIGALLOCATECHOL, 3-GALLATE, (−)- see EBB200
EPIHYDAN see DKQ000
EPIHYDAN see DNU000
EPIHYDRIN ALCOHOL see GGW500
EPIHYDRINALDEHYDE see GGW000
EPIHYDRINAMINE, N,N-DIETHYL- see GGW800
EPIHYDRINE ALDEHYDE see GGW000
3-EPIHYDROXYETIOALLOCHOLAN-17-ONE see HJB050
EPI-INOSITOL, 2,3-DIDEOXY-2-((2-HYDROXY-1-(HYDROXYMETHYL)ETHYL)AMINO)-4-C-(HYDROXYMETHYL)- see DNC100
EPIKOTE 155 see FMW333
EPIKOTE 1001 see IPM000
EPIKOTE 1001 see IPN000
EPIKOTE 1004 see IPP000
EPIKURE DDM see MJQ000
EPILAN see DKQ000
EPILAN see MKB250
EPILAN-D see DNU000
EPILANTIN see DKQ000
EPILANTIN see DNU000
EPILEO PETIT MAL see ENG500
EPILIM see PNR750
EPILIN see PNX750
EPIMID see MNZ000
EPINAL see AGN000
EPINAT see DKQ000
EPINAT see DNU000
EPINELBON see DLY000
EPINEPHRAN see VGP000
EPINEPHRINE see VGP000
(−)-EPINEPHRINE see VGP000
1-EPINEPHRINE see VGP000
d-EPINEPHRINE see AES250
(R)-EPINEPHRINE see VGP000
dl-EPINEPHRINE see EBB500
EPINEPHRINE racemic see EBB500
1-EPINEPHRINE (synthetic) see VGP000
EPINEPHRINE BITARTRATE see AES000
(−)-EPINEPHRINE BITARTRATE see AES000
EPINEPHRINE-d-BITARTRATE see AES000
l-EPINEPHRINE BITARTRATE see AES000
1-EPINEPHRINE-d-BITARTRATE see AES000
EPINEPHRINE CHLORIDE see AES500
1-EPINEPHRINE CHLORIDE see AES500
(−)-EPINEPHRINE HYDROCHLORIDE see AES500
1-EPINEPHRINE HYDROCHLORIDE see AES500
(±)-EPINEPHRINE HYDROCHLORIDE see AES625
dl-EPINEPHRINE HYDROCHLORIDE see AES625
EPINEPHRINE HYDROGEN TARTRATE see AES000
EPINEPHRINE ISOPROPYL HOMOLOG see DMV600
1-EPINEPHRINE TARTRATE see AES000
EPINOVAL see DJU200
EPI-PEVARYL see EAE000
EPIPODOPHYLLOTOXIN see EBB600
EPIPREMNUM AUREUM see PLW800
EPIRENAMINE see VGP000
EPIRENAN see VGP000
EPI-REZ 508 see BLD750
EPI-REZ 510 see BLD750
EPI-REZ 508 mixed with ERR 4205 (1:1) see OPI200
EPIRIZOLE see MCH550
EPIROPRIM see EBB700

EPIROTIN see BBW500
EPIRUBICIN see EBB100
3-EPISARMENTOGENIN see EBB800
EPISED see DKQ000
EPISEDAL see EOK000
5-EPISISOMICIN see EBC000
EPISOL TARTRATE see EBD000
EPITELIOL see VSK600
EPITHELONE see ACR300
2,3-EPITHIOANDROSTAN-17-OL see EBD500
2-α,3-α-EPITHIO-5-α-ANDROSTAN-17-β-OL see EBD500
2,2-EPITHIO-17-((1-METHOXYCYCLOPENTYL)OXY)-ANDROSTANE (2-α,3-α,5-α,17-β) see MCH600
2,3-EPITHIOPROPYL METHOXY ETHER see EBD550
4,5-EPITHIOVALERONITRILE see EBD600
2-α,3-α-EPITHIO-17-β-YL 1-METHOXYCYCLOPENTYL ETHER see MCH600
EPITIOSTANOL see EBD500
EPITOPIC see DKJ300
EPITRATE see VGP000
EPL see EDN000
EPN see EBD700
EPOBRON see IIU000
EPOCAN see BMB000
EPOCELIN see EBE100
EPODYL see TJQ333
EPOK U 9048 see UTU500
EPOK U 9192 see MCB050
EPOLAMINE see PPS700
EPOLENE C see PJS750
EPOLENE C 10 see PJS750
EPOLENE C 11 see PJS750
EPOLENE E see PJS750
EPOLENE E 10 see PJS750
EPOLENE E 12 see PJS750
EPOLENE M 5K see PMP500
EPOLENE N see PJS750
EPON 562 see EBF000
EPON 815 see EBF200
EPON 820 see EBF500
EPON 828 see BLD750
EPON 828 see IPO000
EPON 1001 see EBG000
EPON 1007 see EBG500
EPON 828 mixed with ERR 4205 (1:1) see OPI200
EPONOC B see TGX550
EPONTHOL see PMM000
EPORAL see SOA500
(3,6-EPOSSI-CICLOESAN-1,2-DICARBOSSILATO) DISODICO (ITALIAN) see DXD000
EPOSTANE see EBH400
EPOSTAR EPS-S see MCB050
EPOXIDE 7 see EBH420
EPOXIDE 8 see EBH430
EPOXIDE-201 see ECB000
EPOXIDE 269 see LFV000
EPOXIDE A see BLD750
EPOXIDE ERLA-0510 see EBH500
EPOXIDIZED SOYBEAN OIL see EBH525
EPOXUDINE see EHV200
1,2-EPOXYAETHAN (GERMAN) see EJN500
6H-3A,6-EPOXYAZULENE-6-OL, OCTAHYDRO-3-METHYL-8-METHYLENE-5-(1-METHYLETHYLIDENE)-, (3S,3AS,6R,8AS)- see IKG100
(E)-1-α-2-α-EPOXYBENZ(c)ACRIDINE-3-α-4-β-DIOL see EBH850
(Z)-1-β,2-β-EPOXYBENZ(c)ACRIDINE-3-α-4-β-DIOL see EBH875
4,7-EPOXYBENZO(c)THIOPHENE-1,3-DIONE,HEXAHYDRO-3a-METHYL-,(3a-α,4-β,7-β,7a-α)- see MKK800
2,3-EPOXY-N-BENZYLPROPANAMIDE see EBH890

(2S,3S)-(−)-2,3-EPOXY-3-(4-BROMOPHENYL)-1-PROPANOL see EBH905

(2R,3R)-(+)-2,3-EPOXY-3-(4-BROMOPHENYL)-1-PROPANOL see EBH900

1,4-EPOXY-1,3-BUTADIENE see FPK000

EPOXYBUTANE see BOX750

1,2-EPOXYBUTANE see BOX750

1,4-EPOXYBUTANE see TCR750

2,3-EPOXYBUTANE see EBJ100

2,3-EPOXYBUTANE see EBJ100

cis-2,3-EPOXYBUTANE see EBJ200

meso-2,3-EPOXYBUTANE see EBJ200

3,4-EPOXY-1,2-BUTANEDIOL see ONE050

1,2-EPOXYBUTENE-3 see EBJ500

3,4-EPOXY-1-BUTENE see EBJ500

(2S,3S)-(−)-(2,3-EPOXYBUTYLESTER)-4-NITROBENZOATE see EBJ700

(2R,3R)-(+)-(2,3-EPOXYBUTYLESTER)-4-NITROBENZOATE see EBJ600

2,3-EPOXYBUTYRIC ACID BUTYL ESTER see EBK000

2,3-EPOXYBUTYRIC ACID, ETHYL ESTER see EJS000

1,2-EPOXYBUTYRONITRILE see EBK500

3,4-EPOXYBUTYRONITRILE see EBK500

EPOXYCARYOPHYLLENE see CCN100

4,9-EPOXYCEVANE-3-α,4-β,12,14,16-β,17,20-HEPTOL see EBL000

4,9-EPOXYCEVANE-3-β,4-β,7-α,14,15-α,16-β,20-HEPTOL see EBL500

4,9-EPOXYCEVANE-3,4,12,14,16,17,20-HEPTOL 3-(3,4-DIMETHOXYBENZOATE) see VHU000

1,2-EPOXY-3-CHLOROPHENOXYPROPANE see CKA200

1,2-EPOXY-3-CHLOROPROPANE see EAZ500

5-α,6-α-EPOXYCHOLESTANOL see EBM000

5,6-α-EPOXY-5-α-CHOLESTAN-3-β-OL see EBM000

EPOXYCHOLESTEROL see EBM000

1,2-EPOXYCYCLOHEXANE see CPD000

3,4-EPOXYCYCLOHEXANE-CARBONITRILE see EBM100

3,6-EPOXY-CYCLOHEXANE 1,2-CARBOXYLATE DISODIQUE (FRENCH) see DXD000

3,6-endo-EPOXY-1,2-CYCLOHEXANEDICARBOXYLIC ACID see EAR000

EPOXYCYCLOHEXYLETHYL TRIMETHOXY SILANE see EBO000

β-(3,4-EPOXYCYCLOHEXYL)ETHYLTRIMETHOXYSILANE see EBO000

3,4-EPOXYCYCLOHEXYLMETHYL 3,4-EPOXYCYCLOHEXANE CARBOXYLATE see EBO050

1,2-EPOXYCYCLOPENTANE see EBO100

(3-α,4-β)-12,13-EPOXY-4,15-DIACETATE-TRICHOTHEC-9-ENE-3,4,15-TRIOL see AOP250

12,13-EPOXY-4-β,15-DIACETOXY-3-α,8-α-DIHYDROXYTRICHOTHEC-9-ENE see NCK000

12,13-EPOXY-4-β,15-DIAZETOXY-3-α-HYDROXY-TRICHOTHEC-9-ENE see AOP250

9,10-EPOXY-9,10-DIHYDROBENZ(j)ACEANTHRYLENE see EBO990

5,6-EPOXY-5,6-DIHYDROBENZ(a)ANTHRACENE see EBP000

7,8-EPOXY-7,8-DIHYDROBENZO(a)PYRENE see BCV750

(−)-EPOXYDIHYDROCARYOPHYLLENE see CCN100

5,6-EPOXY-5,6-DIHYDROCHRYSENE see CML830

5,6-EPOXY-5,6-DIHYDRODIBENZ(a,h)ANTHRACENE see EBP500

5,6-EPOXY-5,6-DIHYDRO-7,12-DIMETHYLBENZ(a)ANTHRACENE see DQL600

15,20-EPOXY-15,30-DIHYDRO-12-HYDROXYSENECIONAN-11,16-DIONE see JAK000

EPOXYDIHYDROLINALOOL, mixed isomers see LFY500

5,6-EPOXY-5,6-DIHYDRO-7-METHYLBENZ(A) ANTHRACENE see MGZ000

11,12-EPOXY-11,12-DIHYDRO-3-METHYLCHOLANTHRENE see MIL500

9,10-EPOXY-9,10-DIHYDROPHENANTHRENE see PCX000

(4-α-5-α-17-β)-4,5-EPOXY-3,17-DIHYDROXY-4,17-DIMETHYLANDROST-2-ENE-2-CARBONITRILE see EBH400

4,5-α-EPOXY-3,14-DIHYDROXY-17-METHYLMORPHINAN-6-ONE HYDROCHLORIDE see ORG100

11-β-18-EPOXY-18,21-DIHYDROXYPREGN-4-ENE-3,20-DIONE see ECA100

1,3-EPOXY-2,2-DIMETHYLPROPANE see EBQ500

21,23-EPOXY-19,24-DINOR-17-α-CHOLA-1,3,5(10),7,20,22-HEXAENE-3,17-DIOL 3-ACETATE see EDU600

endo-2,3-EPOXY-7,8-DIOXABICYCLO(2.2.2)OCT-5-ONE see EBQ550

1,2-EPOXYDODECANE see DXU400

3,4-EPOXY-2,5-ENDOMETHYLENECYCLOHEXANECARBOXYLIC ACID, ETHYL ESTER see ENZ000

1,2-EPOXY-4-(EPOXYETHYL)CYCLOHEXANE see VOA000

1,2-EPOXY-7,8-EPOXYOCTANE see DHD800

4,5-EPOXY-2-(2,3-EPOXYPROPYL)VALERIC ACID, METHYL ESTER see MJD000

EPOXYETHANE see EJN500

1,2-EPOXYETHANE see EJN500

1,2-EPOXY-3-ETHOXYPROPANE see EBQ700

1,2-EPOXY-3-ETHOXYPROPANE see EKM200

1,2-EPOXY-3-ETHOXYPROPANE (DOT) see EKM200

7-(EPOXYETHYL)-BENZ(a)ANTHRACENE see ONC000

1,2-EPOXYETHYLBENZENE see EBR000

EPOXYETHYLBENZENE (8CI) see EBR000

α-EPOXYETHYL-1,3-BENZODIOXOLE-5-METHANOL see HOE500

2-(1,2-EPOXYETHYL)-5,6-EPOXYBENZENE see EBR500

2-(α,β-EPOXYETHYL)-5,6-EPOXYBENZENE see EBR500

1-EPOXYETHYL-3,4-EPOXYCYCLOHEXANE see VOA000

2,3,-EPOXY-2-ETHYLHEXANAMIDE see OLU000

α-(EPOXYETHYL)-p-METHOXYBENZYL ALCOHOL see HKI075

α-EPOXYETHYL-1,2-(METHYLENEDIOXY)BENZYL ALCOHOL ACETATE see ABW250

3-(EPOXYETHYL)-7-OXABICYCLO(4.1.0)HEPTANE see VOA000

4-(EPOXYETHYL)-7-OXABICYCLO(4.1.0)HEPTANE see VOA000

3-(1,2-EPOXYETHYL)-7-OXABICYCLO(4.1.0)HEPTANE see VOA000

4-(1,2-EPOXYETHYL)-7-OXABICYCLO(4.1.0)HEPTANE see VOA000

1-EPOXYETHYLPYRENE see ONG000

4-EPOXYETHYLPYRENE see PON750

p-(EPOXYETHYL)TOLUENE see MPK750

4-(EPOXYETHYL)-1,2-XYLENE see EBT000

2′,3′-EPOXYEUGENOL see EBT500

1,2-EPOXY-3-FLUOROPROPANE see EBU000

2-β,3-β,6-β,7-α)-2,3-EPOXY-GRAYANOTOXANE-5,6,7,10,16-PENTOL-6-ACETATE see LJE100

EPOXYGUAIENE see EBU100

EPOXY HARDENER ZZL-0814 see EBU500

EPOXY HARDENER ZZL-0816 see EBV000

EPOXY HARDENER ZZL-0822 see EBV100

EPOXY HARDENER ZZL-0854 see EBV500

EPOXY HARDENER ZZLA-0334 see EBW000

EPOXYHEPTACHLOR see EBW500

1,2-EPOXYHEXADECANE see EBX500

5-β,20-EPOXY-1,2-α,4,7-β,10-β,13-α-HEXAHYDROXY-TAX-11-EN-9-ONE 4,10-DIACETATE 2-BENZOATE 13-ESTER with (2R,3S)-N-BENZOYL-3-PHENYLISOSERINE see TAH775

EPOXY-N-HEXANE see HFC100

1,2-EPOXYHEXANE see HFC100

14,15-β-EPOXY-3-β-HYDROXY-5-β-BUFA-20,22-DIENOLIDE see BOM650

4-α,5-α-EPOXY-17-β-HYDROXY-4,17-DIMETHYL-3-OXOANDROSTANE-2-α-CARBONITRILE see EBY100

4,5-EPOXY-3-HYDROXY-N-METHYLMORPHINAN see DKX600

4,5-α-EPOXY-3-HYDROXY-17-METHYLMORPHINAN-6-ONE HYDROCHLORIDE see DNU300

4,5-α-EPOXY-3-HYDROXY-17-METHYLMORPHINAN-6-ONE-N-OXIDE TARTRATE see EBY500

2,3-EPOXY-4-HYDROXYNONANAL see EBY550

(2-α-4-α-5-α-17-β)-4,5-EPOXY-17-HYDROXY-3-OXOANDROSTANE-2-CARBONITRILE see EBY600

4-α-5-EPOXY-17-β-HYDROXY-3-OXO-5-α-ANDROSTANE-2-α-CARBONITRILE see EBY600

5,6-EPOXY-3-HYDROXY-p-TOLUQUINONE see TBF325

12,13-EPOXY-4-HYDROXYTRICHOTHEC-9-EN-8-ONE CROTONATE see TJE750

4,5-EPOXY-3-HYDROXYVALERIC ACID-β-LACTONE see ECA000

4,7-EPOXYISOBENZOFURAN-1,3-DIONE, HEXAHYDRO-3A-METHYL-, (3A-α-4-β,7-β,7Aα- see HDR800

1,2-EPOXY-3-ISOPROPOXYPROPANE see IPD000

1,4-EPOXY-p-MENTHANE see IKC100

1,8-EPOXY-p-MENTHANE see CAL000

EPOXYMETHAMINE BROMIDE see SBH500

11,13-(EPOXYMETHANO)-13H-CYCLOPENTA(A)PHENANTHRENE, PREGN-4-ENE-3,20-DERIV. see ECA100

2H-3,11C-β-(EPOXYMETHANO)PHENANTHRO(10,1-BC)PYRAN-3-α(3A-β-H)-CA RBOXYLIC ACID,1,4,5,6A-β,7,7A-α,10,11,11A,11B-α-DECAHYDRO-8,11A-β-DIM ETHYL-5,10-DIOXO-1-β,2-α,4-β,9-TETRAHYDROXY-, METHYL ESTER, 4-(3-METHYLCROTONATE) see YAG500

4,5α-EPOXY-3-METHOXY-17-METHYLMORPHINAN-6-ONE see OOI000

1,2-EPOXY-3-METHOXYPROPANE see GGW600

2,3-EPOXY-2-METHYL-N-BENZYLPROPANAMIDE see ECA200

11,12-EPOXY-3-METHYLCHOLANTHRENE see ECA500

3,4-EPOXY-6-METHYLCYCLOHEXANECARBOXYLIC ACID,ALLYL ESTER see AGF500

3,4-EPOXY-6-METHYLCYCLOHEXENECARBOXYLIC

1,2-EPOXY-3,3,3-TRICHLOROPROPANE see TJC250

12,13-EPOXY-TRICHOTHEC-9-ENE-3-α,15-TETROL 15-ACETATE, 8-ISOVALERATE see THI250

(3-α)-12,13-EPOXYTRICHOTHEC-9-ENE-3,4,15-TRIOL 15-ACETATE see ECT700

(3-α,4-β)-12,13-EPOXYTRICHOTHEC-9-ENE-3,4,15-TRIOL TRIACETATE see SBF525

6-β,7-β-EPOXY-3-α-TROPANYL S-(−)-TROPATE see SBG000

EPOXYTROPINE TROPATE see SBG000

EPOXYTROPINE TROPATE METHYLBROMIDE see SBH500

EPOXYTROPINE TROPATE METHYLNITRATE see SBH520

1,2-EPOXY-4-VINYLCYCLOHEXANE see VNZ000

EPRAZIN see POL500

EPRAZINONE DIHYDROCHLORIDE see ECU550

EPRAZINONE HYDROCHLORIDE see ECU550

EPROFIL see TEX000

EPROLIN see VSZ450

E 64 (PROTEINASE INHIBITOR) see TFK255

EPROZINOL see EQY600

EPROZINOL DIHYDROCHLORIDE see ECU600

EPSAMON see AJD000

EPSICAPRON see AJD000

EPSILAN see VSZ450

EPSILAN-M see TGJ050

EPSOM SALTS see MAJ250

EPSOM SALTS see MAJ500

EPSYLON KAPROLAKTAM (POLISH) see CBF700

EPT see EQP000

EPT 500 see IHC550

EPTAC 1 see BJK500

EPTACLORO (ITALIAN) see HAR000

1,4,5,6,7,8,8-EPTACLORO-3a,4,7,7a-TETRAIDRO-4,7-endo-METANO-INDENE (ITALIAN) see HAR000

EPTAL see DKQ000

EPTAM see EIN500

EPTAM see EPC150

EPTANI (ITALIAN) see HBC500

EPTAN-3-ONE (ITALIAN) see EHA600

EPTAPUR see CKF750

EPTC see EIN500

EPTOIN see DKQ000

EPTOIN see DNU000

E-PVC see PKQ059

EQ see SAV000

EQUAL see ARN825

EQUANIL SUSPENSION see MQU750

EQUIBEN see CBA100

EQUIBRAL see MDQ250

EQUI BUTE see BRF500

EQUIGEL see DGP900

EQUIGYNE see ECU750

EQUIGYNE see ECU750

EQUILASE see CCP525

EQUILENIN see ECV000

EQUILENINA (SPANISH) see ECV000

EQUILENIN BENZOATE see ECV500

EQUILENINE see ECV000

EQUILIN see ECW000

EQUILIN BENZOATE see ECW500

EQUILIN SODIUM SULFATE see ECW520

EQUILIN, SULFATE, SODIUM SALT (6CI) see ECW520

EQUILIUM see MQU750

EQUILIUM see PGE000

EQUIMATE see ECW550

EQUINE CYONIN see SCA750

EQUINE GONADOTROPHIN see SCA750

EQUINE GONADOTROPIN see SCA750

EQUINIL see MQU750

EQUINO-ACID see TIQ250

EQUINO-AID see TIQ250

EQUIPERTINE see ECW600

EQUIPOISE see CJR909

EQUIPROXEN see MFA500

EQUIPUR see VLF000

EQUISETIC ACID see ADH000

EQUITDAZIN see MHC750

EQUIZOLE see TEX000

EQUOL see ECW700

ER see REK330

ER 115 see CFY750

ER5461 see CQG250

ERABUTOXINA see ECX000

ERADE see ORU000

ERADEX see CMA100

ERALDIN see ECX100

ERALON see ILD000

ERAMIDE see CDR250

ERAMIN see HGD000

ERANTIN see PNA500

ERASE see HKC000

ERASOL see BIE500

ERASOL HYDROCHLORIDE see BIE500

ERASOL-IDO see BIE500

ERAZE see WAT211

ERAZIDON see ORU000

ERBAPLAST see CDP250

ERBAPRELINA see TGD000

ERBIT N see EDE700

ERBITOX see DUT800

ERBIUM CHLORIDE see ECX500

ERBIUM CITRATE see ECY000

ERBIUM(III) NITRATE (1:3) see ECY500

ERBIUM(III) NITRATE, HEXAHYDRATE (1:3:6) see ECZ000

ERBIUM TRICHLORIDE see ECX500

ERBN see PBK000

ERBOCAIN see PDU250

ERBON see PBK000

ERCEFUROL see DGQ500

ERCEFURYL see DGQ500

ERCO-FER see FBJ100

ERCOFERRO see FBJ100

ERCORAX see HKR500

ERCOTINA see HKR500

ERE 1359 see REF000

EREBILE see DAL000

E-RETINAL see VSK985

all-E-RETINAL see VSK985

EREVIT see TGJ050

ERGADENYLIC ACID see AOA125

ERGAM see EDC500

ERGAMINE see HGD000

ERGATE see EDC500

ERGENYL see PNX750

ERGOATETRINE see LJL000

ERGOBASINE see LJL000

ERGOCALCIFEROL see VSZ100

ERGOCHROME AA (2,2')-5-β,6-α,10-β-5',6'-α,10'-β see EDA500

ERGOCORNINE see EDA600

ERGOCORNINE see EDA600

ERGOCORNINE, DIHYDRO- see DLK780

ERGOCORNINE, 9,10-DIHYDRO- see DLK780

ERGOCORNINE HYDROGEN MALEATE see EDA875

ERGOCORNINE HYDROGEN MALEINATE see EDA875

ERGOCORNINE MALEATE see EDA875

ERGOCORNINE MESYLATE see EDB000

ERGOCORNINE METHANESULFONATE (SALT) see EDB000

ERGOCORNINE METHANESULPHONATE see EDB000

ERGOCORNININE METHANESULFONATE see EDB020

ERGOCRISTININE METHANESULFONATE see EDB025

ERGOCRYPTINE see EDB100

α-ERGOCRYPTINE see EDB100

ERGOCRYPTINE MESYLATE see EDB125

ERGOCRYPTINE METHANESULFONATE see EDB125

ERGOCRYPTINE METHANESULPHONATE see EDB125

ERGOCRYPTININE METHANESULFONATE see EDB150

ERGOKLININE see LJL000

α-ERGOKRYPTINE see EDB100

ERGOKRYPTINE METHANESULFONATE see EDB125

ERGOLINE-8-ACETAMIDE, 6-ETHYL-, (8-β)-, (R-(R*,R*))-2,3-DIHYDROXYBUTANEDIOATE (1:1) see EJS100

ERGOLINE-8-ACETAMIDE, 6-METHYL-, (8-β)-, (R-(R*,R*))-2,3-DIHYDROXYBUTANEDIOATE (2:1) see MJV500

ERGOLINE-8-ACETAMIDE, 6-(2-PROPENYL)-, (8-β)-, (R-(R*,R*))-2,3-DIHYDROXYBUTANEDIOATE see PMR800

ERGOLINE-8-ACETAMIDE, 6-PROPYL-, (8-β)-, (R-(R*,R*))-2,3-DIHYDROXYBUTANEDIOATE (2:1) see PNL800

ERGOLINE-8-CARBAMIC ACID, 6-PROPYL-, ETHYL ESTER, (8S)- see EPC115

ERGOLINE-8-CARBOXAMIDE, 1-ACETYL-9,10-DIDEHYDRO-N,N-DIETHYL-6-METHYL-, (8-β)- see AFJ450

ERGOLINE-8-β-CARBOXAMIDE, 9,10-DIDEHYDRO-N,N-DIETHYL-6-METHYL-, TARTRATE (2:1) see LJM600

ERGOLINE-8-CARBOXAMIDE, 9,10-DIDEHYDRO-N-(1-HYDROXYETHYL)-6-METHYL-, (8-β)- see LJI100

ERGOLINE, 1-((4-CHLOROPHENYL)SULFONYL)-6,8-DIMETHYL-, (8-β)- see CEK400

ERGOLINE, 1-CYCLOPENTYL-6,8-DIMETHYL-, (8-β)- see CQB300

ERGOLINE-8-β-PROPIONAMIDE, α-ACETYL-6-METHYL- see ACR100

ERGOLINE-8-PROPIONAMIDE, 6-ALLYL-α-CYANO- see AGC200

ERGOLINE-8-PROPIONAMIDE, 2-CHLORO-α-CYANO-6-METHYL- see CFF100

ERGOLINE-8-PROPIONAMIDE, α-CYANO-2,6-DIMETHYL- see COM075

ERGOLINE-8-PROPIONAMIDE, α-CYANO-6-ISOBUTYL- see COP600

ERGOLINE-8-β-PROPIONAMIDE, N-ETHYL-6-METHYL-α-(METHYLSULFONYL)- see EMW100

ERGOLINE-8-β-PROPIONITRILE, 6-METHYL-α-(4-METHYL-1-PIPERAZINYLCARBONYL)- see MLR400

ERGOLINE-8-β-PROPIONITRILE, 6-METHYL-α-(1-PYRROLIDINYLCARBONYL)- see MPC300

ERGOMAR see EDC500

ERGOMETRINE see LJL000

ERGOMETRINE ACID MALEATE see EDB500

ERGOMETRINE MALEATE see EDB500

ERGONOVINE see LJL000

ERGONOVINE, MALEATE (1:1) (SALT) see EDB500

ERGOPLAST ADC see DGT500

ERGOPLAST AdDO see AEO000

ERGOPLAST.FDC see DGV700

ERGOPLAST FDO see DVL700

ERGORONE see VSZ100

ERGOSINE METHANESULFONATE see EDB200

α-ERGOSINE METHANESULFONATE see EDB200

ERGOSINE MONOMETHANESULFONATE see EDB200

5-β-ERGOSTA-2,24-DIEN-26-OIC ACID, 5,6-β-EPOXY-4-β,22,27-TRIHYDROXY-1-OXO-,Δ-LACTONE, (20S,22R)- see EDB300

ERGOSTAT see EDC500
ERGOSTEROL, activated see VSZ100
ERGOSTEROL, irradiated see VSZ100
ERGOT see EDB500
ERGOTAMAN-3',6',18-TRIONE, 2',5'-BIS(1-METHYLETHYL)-12'-HYDROXY-, (5'-α-8α-, MONOMETHANESULFONATE (SALT) see EDB020
ERGOTAMAN-3',6',18-TRIONE, 12'-HYDROXY-2',5'-BIS(1-METHYLETHYL)-, (5'-α)- see EDA600
ERGOTAMAN-3',6',18-TRIONE, 12'-HYDROXY-2'-(1-METHYLETHYL)-5'-(2-METHYLPROPYL)-, (5'-α-8α-, MONOMETHANESULFONATE (SALT) see EDB150
ERGOTAMAN-3',6',18-TRIONE, 12'-HYDROXY-2'-(1-METHYLETHYL)-5'-(PHENYLMETHYL)-, (5'-α-8α-, MONOMETHANESULFONATE (SALT) see EDB025
ERGOTAMAN-3',6',18-TRIONE, 12'-HYDROXY-2'-METHYL-5'-(2-METHYLPROPYL)-, (5'-α)-, MONOMETHANESULFONATE (salt) see EDB200
ERGOTAMAN-3',6',18-TRIONE, 12'-HYDROXY-2'-METHYL-5'-(PHENYLMETHYL)-, (5'-α-8α-, MONOMETHANESULFONATE (SALT) see EDC520
ERGOTAMINE see EDC000
ERGOTAMINE BITARTRATE see EDC500
ERGOTAMINE TARTRATE see EDC500
ERGOTAMININE METHANESULFONATE see EDC520
ERGOTARTRATE see EDC500
ERGOTERM OTGO see DVN600
ERGOTERM TGO see EDC560
ERGOTIDINE see HGD000
ERGOTOCINE see LJL000
ERGOTOXIN see EDC565
ERGOTOXINE see EDC565
ERGOTOXINE, DIHYDRO- see DLL100
ERGOTOXINE ETHANESULFONATE see EDC575
ERGOTOXINE ETHANESULPHONATE see EDC575
ERGOTOXINE ETHANSULFONATE see EDC575
ERGOTRATE see EDB500
ERGOTRATE see LJL000
ERGOTRATE MALEATE see EDB500
ERGOVALINE MESYLATE see EDC585
ERGOVALINE, METHANESULFONATE see EDC585
ERGOVALINE METHANESULPHONATE see EDC585
ERGOVALINE MONOMETHANESULFONATE see EDC585
ERIAMYCIN see EDC600
ERIBATE N see EDE700
ERIBUTAZONE see BRF500
ERIDAN see DCK759
ERIE BENZO 4BP see DXO850
ERIE BLACK B see AQP000
ERIE BLACK BF see AQP000
ERIE BLACK GAC see AQP000
ERIE BLACK GXOO see AQP000
ERIE BLACK JET see AQP000
ERIE BLACK NUG see AQP000
ERIE BLACK RXOO see AQP000
ERIE BORDEAUX B see CMO872
ERIE BRILLIANT BLACK S see AQP000
ERIE BROWN 3GN see CMO825
ERIE CONGO 4B see SGQ500
ERIE FAST BROWN B see CMO820
ERIE FAST BROWN 3RB see CMO800
ERIE FAST BROWN 3RBD see CMO800
ERIE FAST RED FD see CMO870
ERIE FAST SCARLET SCB see CMO870
ERIE FIBRE BLACK VP see AQP000

ERIE GREEN GPD see CMO840
ERIE GREEN MT see CMO840
ERIE GREEN TCM see CMO840
ERIE GREEN WT see CMO830
ERIE ORANGE Y see CMO860
ERIE RED 4B see DXO850
ERIE SCARLET 3B see CMO875
ERIE VIOLET 3R see CMP000
ERINA see MQU750
ERINIT see PBC250
ERINITRIT see SIQ500
ERIO ANTHRACENE BRILLIAN BLUE B see CMM090
ERIO ANTHRACENE BRILLIANT BLUE RFF see CMM080
ERIO ANTHRACENE FAST BLUE BB see CMM090
ERIOBOTRYA JAPONICA see LIH200
ERIO BRILLIANT BLUE V see ADE500
ERIOCHROMAL YELLOW G see SIT850
ERIOCHROME BLACK T see EDC625
ERIOCHROME BLUE BLACK see CMP880
ERIOCHROME BLUE BLACK 2B see CMP880
ERIOCHROME BLUE BLACK 2G see CMP880
ERIOCHROME BLUE BLACK B see CMP880
ERIOCHROME BLUE BLACK BC see CMP880
ERIOCHROME BLUE BLACK BSS see CMP880
ERIOCHROME BLUE BLACK 2BP see CMP880
ERIOCHROME ORANGE G see CMP882
ERIOCHROME ORANGE R see CMP882
ERIO CHROME RED PE see CMG750
ERIOCHROME RED PE1 see CMG750
ERIOCHROME VIOLET B see HLI000
ERIO CHROME VIOLET BA see HLI000
ERIO CHROME VIOLET BR see HLI000
ERIOCHROME YELLOW 2G see SIT850
ERIOCHROME YELLOW GS see SIT850
ERIO FAST BLUE BRL see CMM070
ERIO FAST BLUE BS see CMM090
ERIO FAST CYANINE SE see APG700
ERIO FAST ORANGE AS see HGC000
ERIO FAST YELLOW AE see SGP500
ERIO FAST YELLOW AEN see SGP500
ERIO FAST YELLOW RL see CMM759
ERIO FLOXINE 2G see CMM300
ERIO FLOXINE 6B see CMM400
ERIO FLOXINE 2GN see CMM300
ERIO FLOXINE 6BN see CMM400
ERIOGLAUCINE see ADE500
ERIOGLAUCINE see FMU059
ERIOGLAUCINE G see FAE000
ERIOGLAUCINE SUPRA see ADE500
ERIO GLAUCINE X see ERG100
ERIO GLAUCINE XS see ERG100
ERIO GREEN B see CMM100
ERIO GREEN S see ADF000
ERION see ARQ250
ERION see CFU750
ERIONITE see EDC650
ERIONITE (CAKNA (AL2SI7O18)2.14H2O) see EDC700
ERIONYL BLUE BFF see CMM100
ERIONYL BLUE E-B see CMM090
ERIONYL BLUE E-BFF see CMM100
ERIONYL BLUE E-RFF see CMM080
ERIONYL RED G see CMM320
ERIONYL RED RS see CMM330
ERIONYL YELLOW E-AEN see SGP500
ERIO ORANGE II see CMM220
ERIOSIN BLUE BLACK B see FAB830
ERIOSIN FAST BLUE B see CMM090
ERIOSIN FAST BLUE BFF see CMM100
ERIOSIN FAST BLUE G see CMM120
ERIOSIN FAST BLUE RFF see CMM080
ERIOSIN RHODAMINE B see FAG070
ERIOSKY BLUE see FMU059
ERIO TARTRAZINE see FAG140

ERIO YELLOW AEN see SGP500
ERISIMIN DIHYDRATE see HAO000
ERISPAN see FDB100
ERITADENINE see POJ300
d-ERITADENINE see POJ300
ERITRONE see VSZ000
ERITROXILINA see CNE750
ERIZIMIN see HAN800
ERIZOMYCIN see PPI775
ERL-2774 see BLD750
ERL-2795 see ECL000
ERL-4221 see EBO050
ERLA-2270 see PKQ070
ERLA-2270 see VOA000
ERLA-2271 see PKQ070
ERLA-2271 see VOA000
ERMALONE see MJE760
ERMETRINE see LJL000
EROCYANINE 540 see EDD500
EROINA see HBT500
ERR 4205 see BJN250
ERR 4205 mixed with ARALDITE 6010 (1:1) see OPI200
ERRE 0100 see FMW333
ERR 4205 mixed with EPI-REZ 508 (1:1) see OPI200
ERR 4205 mixed with EPON 828 (1:1) see OPI200
ERROLON see CHJ750
ERSERINE see PIA500
ERTALON 6SA see PJY500
ERTILEN see CDP250
ERTRON see VSZ100
ERTUBAN see ILD000
ERUNIT see CGO550
ERYCIN see EDH500
ERYCORBIN see SAA025
ERYCYTOL see VSZ000
ERYPAR see EDJ500
ERYSAN see BIF250
ERYSIMIN see HAN800
ERYSIMIN DIHYDRATE see HAO000
ERYSIMOTOXIN see HAN800
ERYSIMUPICRONE see SMM500
ERYSIMUPIKRON see SMM500
ERYSODINE HYDROCHLORIDE see EDE000
ERYSOPINE HYDROCHLORIDE see EDE500
ERYTHORBIC ACID see EDE600
ERYTHORBIC ACID see SAA025
d-ERYTHORBIC ACID see SAA025
ERYTHORBIC ACID SODIUM SALT see EDE700
ERYTHRALINE HYDROBROMIDE see EDF000
ERYTHRENE see BOP500
ERYTHRITOL ANHYDRIDE see BGA750
ERYTHRITOL ANHYDRIDE see DHB800
ERYTHRITOL, 1,4-DITHIO- see DXN350
ERYTHROCIN see EDH500
ERYTHROCIN STEARATE see EDJ500
d-ERYTHRO-d-GALACTO-OCTOPYRANOSIDE, METHYL 7-(S)-CHLORO-6,7,8-TRIDEOXY-6-(1-METHYL-4-PROPYL-l-2-PYRROLIDINECARBOXAMIDO)-1-THIO-, HYDROCHLORIDE, trans-α- see CHU050
ERYTHROGRAN see EDH500
ERYTHROGUENT see EDH500
d-ERYTHRO-HEX-2-ENONIC ACID γ-LACTONE MONOSODIUM SALT see EDE700
d-ERYTHRO-HEX-2-ENONIC ACID, γ-LACTONE, MONOSODIUM SALT see EDE700
d-ERYTHRO-HEXOSULOSE, 3-DEOXY- see DAR700
d-ERYTHRO-HEXOS-2-ULOSE, 3-DEOXY-(9CI) see DAR700
ERYTHROHYCIN GLUCEPTATE see EDI500

dl-ERYTHRO-HYDROXY(2',5'-
DIMETHOXYPHENYL)-2-
ISOPROPYLAMINOPROPANE
HYDROCHLORIDE see IPY500
β-ERYTHROIDINE see EDG500
β-ERYTHROIDINE HYDROCHLORIDE see
EDH000
ERYTHROMYCIN see EDH500
ERYTHROMYCIN A see EDH500
ERYTHROMYCIN CARBONATE see EDI000
ERYTHROMYCIN, 8,9-DIDEHYDRO-N-
DEMETHYL-9-DEOXO-4",6,12-TRIDEOXY-
6,9-EPOXY-N-ETHYL- see AOP502
ERYTHROMYCIN GLUCOHEPTONATE
(1:1) see EDI500
ERYTHROMYCIN HYDROCHLORIDE see
EDJ000
ERYTHROMYCIN, 6-o-METHYL- MJV775
ERYTHROMYCIN OCTADECANOATE (salt)
see EDJ500
ERYTHROMYCIN PALMITATE see EDJ100
ERYTHROMYCIN STEARATE see EDJ500
ERYTHROMYCIN STEARIC ACID SALT see
EDJ500
ERYTHROMYCIN SULFAMATE see EDK000
ERYTHROMYCIN SULFAMATE (SALT) see
EDK000
ERYTHROMYCIN THIOCYANATE see
EDK200
ERYTHRONIC ACID, 4-(6-AMINO-9H-
PURIN-9-YL)-4-DEOXY-, d- see POJ300
d-ERYTHRO-PENTONAMIDE, 2,3,5-
TRIDEOXY-N-(2,3-DIHYDRO-2-HYDROXY-
1H-INDEN-1-YL)-5-(2-(((1,1-
DIMETHYLETHYL)AMINO)CARBONYL)-4-
(3-PYRIDINYLMETHYL)-1-PIPERAZINYL)-
2-(PHENYLMETHYL)-,(1(1S,2R),5(S))-,
SULFATE (1:1) (SALT) see ICE100
ERYTHROSE see TKO100
ERYTHROSIN see FAG040
ERYTHROSINE B-FO (BIOLOGICAL
STAIN) see FAG040
ERYTHROTIN see VSZ000
ERYTROXYLIN see CNE750
ES 902 see PPN000
ESACHLOROBENZENE (ITALIAN) see
HCC500
ESACICLONATO see SHL500
ESAIDRO-1,3,5-TRINITRO-1,3,5-TRIAZINA
(ITALIAN) see CPR800
ESAMETILENTETRAMINA (ITALIAN) see
HEI500
ESAMETINA see HEA000
ESAMETONIO IODURO (ITALIAN) see
HEB000
ESANI (ITALIAN) see HEN000
ESANITRODIFENILAMINA (ITALIAN) see
HET500
ESANTENE see HFG550
ESAPROPIMATO see POA250
ESBATAL see BFW250
ESBECYTHRIN see DAF300
ESBERICARD see EDK600
ESBIOL see AFR750
ESBIOL CONCENTRATE 90% see AFR750
ESBRITE see SMQ500
ESBRITE 2 see SMQ500
ESBRITE 4 see SMQ500
ESBRITE 8 see SMQ500
ESBRITE 4-62 see SMQ500
ESBRITE G 10 see SMQ500
ESBRITE G-P 2 see SMQ500
ESBRITE LBL see SMQ500
ESBRITE 500HM see SMQ500
ESCAMBIA 2160 see PKQ059
ESCASPERE see RDK000
ESCHERICHIA COLI ENDONUCLEASE I
see EDK650
ESCHERICHIA COLI ENDOTOXIN see
EDK700
ESCHERICHIA COLI
LIPOPOLYSACCHARIDE see EDK750

ESCIN see EDK875
α-ESCIN see EDL000
β-ESCIN see EDL500
ESCINA (ITALIAN) see EDK875
β-ESCINIC ACID see EDL000
ESCIN, SODIUM SALT see EDM000
ESCIN TRIETHANOLAMINE SALT see
EDM500
ESCLAMA see NHH000
ESC LIPOPOLYSACCHARIDE see LGK375
ESCOPARONE see DRS800
ESCOREZ 7404 see SMQ500
ESCORPAL see PDC850
ESCORPAL see PDC875
ESCORT see MQR400
ESCULETIN DIMETHYL ETHER see
DRS800
ESDRAGOL see AFW750
ESE see EPI000
(E-β,5E,7E,10-α,22E)-9,10-SECOERGOSTA-
5,7,22-TRIEN-3-OL (9CI) see DME300
ESELIN see DIS600
ESEN see PHW750
ESERINE see PIA500
ESERINE SALICYLATE see PIA750
ESERINE SULFATE see PIB000
ESERINE SULPHATE see PIB000
ESEROLEIN, METHYLCARBAMATE
(ESTER) see PIA500
ESERPINE see RDK000
ESFAR see CPJ000
ESFAR CALCIUM see CPJ250
ESFENVALERATE see FAR150
ESGRAM see PAJ000
ESICLENE see FNK040
ESIDREX see CFY000
ESIDRIX see CFY000
ESILGAN see CKL250
ESJAYDIOL see AOO475
ESKABARB see EOK000
ESKACILLIAN V see PDT500
ESKACILLIN see BFD000
ESKADIAZINE see PPP500
ESKALIN V see VRA700
ESKALIN V see VRF000
ESKALITH see LGZ000
ESKALON 100 see CAT775
ESKASERP see RDK000
ESKAZINE see TKK250
ESKAZINE DIHYDROCHLORIDE see
TKK250
ESKEL see AHK750
ESKIMON 11 see TIP500
ESKIMON 12 see DFA600
ESKIMON 22 see CFX500
ESMAIL see CGA000
ESMARIN see HII500
ESMOLOL see MKR125
(+-)-ESMOLOL see MKR125
ESOBARBITALE (ITALIAN) see ERD500
E 39 SOLUBLE see BDC750
ESOMID CHLORIDE see HEA500
ESOPHOTRAST see BAP000
ESOPIN see MDL000
ESORB see VSZ450
ESP see DSK600
ESPADOL see CLW000
ESPARIN see DQA600
ESPASMO GASIUM see DAB750
ESPECTINOMICINA DIHYDROCHLORIDE
PENTAHYDRATE see SKY500
ESPENAL see DXH250
ESPERAL see DXH250
ESPERAN see OLW400
ESPERFOAM FR see MKA500
ESPEROX 10 see BSC500
ESPEROX 24M see BSC600
ESPEROX 31M see BSD250
ESPERSON see DBA875
ESPHYGMOGENINA see VGP000
ESPIGA de AMOR (PUERTO RICO) see
CAK325

ESPIGELIA (CUBA) see PIH800
ESPINOMYCIN A see MBY150
ESPIRAN see DAI200
ESPIRITIN see LAG010
ES POLYTAMIN see AHR600
ESPRIL see BET000
ESPROCARB see PFR120
ESQUINON see BGX750
ESSENCE of MIRBANE see NEX000
ESSENCE of MYRBANE see NEX000
ESSENCE of NIOBE see EGR000
ESSENCE OF NIOBE see MHA750
ESSENCE of ROSE see RNA000
ESSENTIAL OIL see CNT350
ESSENTIAL OIL from MYRTLE see OGU000
ESSENTIAL OIL OF ACORUS CALAMUS
Linn. see OGL020
ESSENTIAL OIL OF CYMBOPOGON
NARDUS see CMT000
ESSENTIAL PHOSPHOLIPIDS see EDN000
ESSEX see CBT750
ESSEX 1360 see HLB400
ESSEX GUM 1360 see HLB400
ESSIGESTER (GERMAN) see EFR000
ESSIGSAEURE (GERMAN) see AAT250
ESSIGSAEUREANHYDRID (GERMAN) see
AAX500
ESSO FUNGICIDE 406 see CBG000
ESSO HERBICIDE 10 see BQZ000
EST see OMY925
ESTABAL see BFW250
ESTABEX S see BOU550
ESTABEX U 18 see DVK200
ESTANE 5703 see UVA000
ESTAR see CMY800
ESTAR see PKF750
ESTASIL see MQU750
ESTAVUDINA (INN-SPANISH) see SLJ800
ESTAZOLAM see CKL250
ESTER 25 see NIM500
ESTER d'ACIDE BENZILIQUE et DU-1-
METHYLSULFATE de 1,1-DIMETHYL-(2-
HYDROXY-METHYL)PIPERIDINIUM see
CDG250
ESTER del ACIDO BENCILICO del-1,1-
DIMETIL-2-OXIMETIL-PIPERIDINIO-
METILSULFATO (SPANISH) see CDG250
ESTERASE-LIPASE see EDN100
ESTERCIDE T-2 and T-245 see TAA100
S-ESTER with O,O-DIMETHYL
PHOSPHOROTHIOATE see MAK700
ESTER DWETYLOAMINOETYLOWSKY
KWASU DWUFENYLOOCTOWEGO
(POLISH) see THK000
ESTER DWUETYLOAMINOETYLOWY
KWASU DWUFENYLOOCTOWEGO see
DHX800
ESTERE CIANOACETICO see EHP500
ESTERE ETOSSIMETILICO dell' ACIDO N-
(2,6-DICLORO-m-TOLIL)ANTRANILICO
(ITALIAN) see DGN000
ESTERE ISOAMILICO dell'ACIDO α-(N-
(PIRROLIDINOETIL))-
AMINOFENILACETICO (ITALIAN) see
CBA375
S-ESTER of (2-
MERCAPTOETHYL)TRIMETHYLAMMONI
UM IODIDE with O,O-DIETHYL
PHOSPHOROTHIOATE see TLF500
O-ESTER-p-NITROPHENOL with O-ETHYL
PHENYL PHOSPHONOTHIOATE see
EBD700
ESTERON see DAA800
ESTERON 44 see IOY000
ESTERON 99 see DAA800
ESTERON 76 BE see DAA800
ESTERON 245 BE see TAA100
ESTERON BRUSH KILLER see DAA800
ESTERON BRUSH KILLER see TAA100
ESTERON 99 CONCENTRATE see DAA800
ESTERONE see EDV000
ESTERONE FOUR see DAA800

ESTERON 44 WEED KILLER see DAA800
ESTEROQUINONE LIGHT BLUE 4JL see DMM400
ESTEROQUINONE LIGHT PINK RLL see AKO350
ESTEROQUINONE LIGHT YELLOW 4JL see AAQ250
ESTEROQUINONE LIGHT YELLOW 3JLL see KDA075
α-ESTER PALMITIC ACID with D-threo-(−)-2,2-DICHLORO-N-(β-HYDROXY-α-(HYDROXYMETHYL)-p-NITROPHENETHYL)ACETAMIDE see CDP700
ESTERS see EDN500
ESTER SULFONATE see CJT750
ESTERTRICHLOROSTANNANE see TIM500
ESTEVE see BRF500
ESTIBOGLUCONATO SODICO see AQH800
ESTIBOGLUCONATO SODICO see AQI250
ESTILBEN see DKA600
ESTILBEN see DKB000
ESTILBIN see DKB000
ESTIMULEX see DBA800
ESTINERVAL see PFC750
ESTOCINE see DPE200
ESTOL 103 see IQW000
ESTOL 603 see OAV000
ESTOL 1550 see DJX000
ESTOMYCIN see NCF500
ESTON see DSK600
ESTON see QFA250
ESTONATE see DAD200
ESTON-B see EDP000
ESTONE see DAA800
ESTONE YELLOW GN see AAQ250
ESTONMITE see CJT750
ESTONOX see CDV100
ESTOSTERIL see PCL500
ESTOTSIN see DPE200
ESTOX see DSK600
ESTRACYT see EDT100
ESTRACYT HYDRATE see EDN600
ESTRADEP see DAZ115
ESTRA-4,9-DIEN-3-ONE, 11-(4-(DIMETHYLAMINO)PHENYL)-17-HYDROXY-17-(1-PROPYNYL)-, (11-β, 17-β)-see HKB700
ESTRADIOL see EDO000
d-ESTRADIOL see EDO000
cis-ESTRADIOL see EDO000
α-ESTRADIOL see EDO000
β-ESTRADIOL see EDO000
17-α-ESTRADIOL see EDO500
ESTRADIOL-17-β see EDO000
17-β-ESTRADIOL see EDO000
3,17-β-ESTRADIOL see EDO000
d-3,17-β-ESTRADIOL see EDO000
ESTRADIOL BENZOATE see EDP000
ESTRADIOL-3-BENZOATE see EDP000
β-ESTRADIOL BENZOATE see EDP000
β-ESTRADIOL-3-BENZOATE see EDP000
ESTRADIOL-17-β-BENZOATE see EDP000
17-β-ESTRADIOL BENZOATE see EDP000
ESTRADIOL-17-β-3-BENZOATE see EDP000
17-β-ESTRADIOL-3-BENZOATE see EDP000
ESTRADIOL-17-BENZOATE-3,n-BUTYRATE see EDP500
ESTRADIOL-3-BENZOATE mixed with PROGESTERONE (1:14 moles) see EDQ000
ESTRADIOL BENZOATE mixed with PROGESTERONE (1:14 moles) see EDQ000
ESTRADIOL, 3-(BIS(2-CHLOROETHYL)CARBAMATE) see EDS200
ESTRADIOL-17-CAPRYLATE see EDQ500
ESTRADIOL-17-β 3-CYCLOPENTYL ETHER see QFA250
ESTRADIOL CYCLOPENTYLPROPIONATE see DAZ115
ESTRADIOL-17-CYCLOPENTYLPROPIONATE see DAZ115

ESTRADIOL-17-β-CYCLOPENTYLPROPIONATE see DAZ115
ESTRADIOL-CYPIONATE see DAZ115
ESTRADIOL-17-CYPIONATE see DAZ115
ESTRADIOL-17-β-CYPIONATE see DAZ115
ESTRADIOL CYPIONATE MIXED WITH MEDROXYPROGESTERONE ACETATE see EDQ600
ESTRADIOL DIPROPIONATE see EDR000
ESTRADIOL-3,17-DIPROPIONATE see EDR000
β-ESTRADIOL DIPROPIONATE see EDR000
17-β-ESTRADIOL DIPROPIONATE see EDR000
β-ESTRADIOL-3,17-DIPROPIONATE see EDR000
3,17-β-ESTRADIOL DIPROPIONATE see EDR000
ESTRADIOL, 2-METHOXY- see MEL785
ESTRADIOL 3-METHYL ETHER see MFB775
17-β-ESTRADIOL 3-METHYL ETHER see MFB775
ESTRADIOL MONOBENZOATE see EDP000
17-β-ESTRADIOL MONOBENZOATE see EDP000
ESTRADIOL MONOPAMITATE see EDR400
ESTRADIOL MUSTARD see EDR500
ESTRADIOL PHOSPHATE POLYMER see EDS000
ESTRADIOL POLYESTER with PHOSPHORIC ACID see EDS000
ESTRADIOL-TESTOSTERONE MIXT. see TBF650
ESTRADIOL VALERATE see EDS100
ESTRADIOL-17-VALERATE see EDS100
ESTRADIOL 17-β-VALERATE see EDS100
ESTRADIOL VALERIANATE see EDS100
ESTRADURIN see EDS000
ESTRADURIN see PJR750
ESTRAGARD see DAL600
ESTRAGON OIL see TAF700
ESTRALDINE see EDO000
ESTRALUTIN see HNT500
ESTRAMUSTINE see EDS200
ESTRAMUSTINE PHOSPHATE DISODIUM see EDT100
ESTRAMUSTINE PHOSPHATE DISODIUM HYDRATE see EDN600
ESTRAMUSTINE PHOSPHATE SODIUM see EDT100
ESTRAMUSTINE PHOSPHATE SODIUM HYDRATE see EDN600
α-ESTRA-1,3,5,7,9-PENTANE-3,17-DIOL see EDT500
β-ESTRA-1,3,5,7,9-PENTANE-3,17-DIOL see EDU000
ESTRATAB see ECU750
1,3,5,7-ESTRATETRAEN-3-OL-17-ONE see ECW520
ESTRA-1,3,5(10),7-TETRAEN-17-ONE, 3-HYDROXY-, HYDROGEN SULFATE SODIUM SALT (8CI) see ECW520
ESTRA-1,3,5(10),7-TETRAEN-17-ONE, 3-(SULFOOXY)-, SODIUM SALT see ECW520
(17-β)-ESTRA-1,3,5(10)-TRIEN-3,17-DIOL, 3-METHOXY-17-(3,3,3-TRIFLUORO-1-PROPYNYL) see TKH050
ESTRA-1,3,5(10)-TRIENE-2,4-D2-3,17-DIOL, (17-β)- see DHA425
1,3,5-ESTRATRIENE-3,17-α-DIOL see EDO500
1,3,5-ESTRATRIENE-3,17-β-DIOL see EDO000
ESTRA-1,3,5(10)-TRIENE-3,17-α-DIOL see EDO500
ESTRA-1,3,5(10)-TRIENE-3,17-β-DIOL see EDO000
17-β-ESTRA-1,3,5(10)-TRIENE-3,17-DIOL see EDO000
ESTRA-1,3,5(10)-TRIENE-3,17-β-DIOL, 3-BENZOATE see EDP000

1,3,5(10)-ESTRATRIENE-3,17-β-DIOL 3-BENZOATE see EDP000
ESTRA-1,3,5-TRIENE-3,17-DIOL (17-β)-3-BENZOATE see EDP000
ESTRA-1,3,5(10)-TRIENE-3,17-β-DIOL-17-BENZOATE-3-n-BUTYRATE see EDP500
ESTRA-1,3,5(10)-TRIENE-3,17-DIOL (17β)-, 3-(BIS(2-CHLOROETHYL)CARBAMATE) see EDS200
ESTRA-1,3,5(10)-TRIENE-3,17-β-DIOL 3-(BIS(2-CHLOROETHYL)CARBAMATE)17-DISODIUM PHOSPHATE see EDT100
(17-β)-ESTRA-1,3,5(10)-TRIENE-3,17-DIOL 17-CYCLOPENTANEPROPANOATE (9CI) see DAZ115
1,3,5(10)-ESTRATRIENE-3,17-β-DIOL DIPROPIONATE see EDR000
ESTRA-1,3,5(10)-TRIENE-3,17-DIOL (17-β)-DIPROPIONATE see EDR000
ESTRA-1,3,5(10)-TRIENE-3,17-β-DIOL, 11-β-ETHYL- see EJT575
ESTRA-1,3,5(10)-TRIENE-3,17-DIOL, 4-FLUORO-, (17-β)-(9CI) see FIA500
ESTRA-1,3,5(10)-TRIENE-3,17-β-DIOL, 2-METHOXY- see MEL785
ESTRA-1,3,5(10)-TRIENE-3,17-DIOL, 2-METHOXY-, (17β)- see MEL785
ESTRA-1,3,5(10)-TRIENE-3,17-β-DIOL-17-OCTANOATE see EDQ500
(17-β)-ESTRA-1,3,5(10)-TRIENE-3,17-DIOL-17-PENTANOATE (9CI) see EDS100
(17-β)-ESTRA-1,3,5(10)-TRIENE-3,17-DIOL POLYMER with PHOSPHORIC ACID see EDS000
ESTRA-1,3,5(10)-TRIENE-17-β-DIOL-17-TETRAHYDROPYRANYL ETHER see EDU100
ESTRA-1,3,5(10)-TRIENE-1,3,17-β-TRIOL see HKH550
ESTRA-1,3,5(10)-TRIENE-2,3,17-β-TRIOL see HKH600
ESTRA-1,3,5(10)-TRIENE-3,16-α,17-β-TRIOL see EDU500
1,3,5-ESTRATRIENE-3-β,16-α,17-β-TRIOL see EDU500
(16-α,17-β)-ESTRA-1,3,5(10)-TRIENE-3,16,17-TRIOL see EDU500
1,3,5-ESTRATRIEN-3-OL-17-ONE see EDV000
1,3,5(10)-ESTRATRIEN-3-OL-17-ONE see EDV000
4,9,11,-ESTRATRIEN-17β-OL-3-ONE see THL600
Δ-1,3,5-ESTRATRIEN-3-β-OL-17-ONE see EDV000
ESTRA-4,9,11-TRIEN-3-ONE, 17-(ACETYLOXY)-, (17-β)-(9CI) see HKI100
ESTRA-4,9,11-TRIEN-3-ONE, 17-α-ALLYL-17-HYDROXY- see AGW675
ESTRA-1,3,5(10)-TRIEN-17-ONE, 2,3-DIHYDROXY- see HKI200
ESTRA-1,3,5(10)-TRIEN-17-ONE, 3,4-DIHYDROXY- see HKI300
ESTRA-4,9,11-TRIEN-3-ONE, 17-β-HYDROXY-, ACETATE see HKI100
ESTRA-1,3,5(10)-TRIEN-17-ONE, 3-HYDROXY-, o-METHYLOXIME see EDV555
ESTRA-4,9,11-TRIEN-3-ONE, 17-HYDROXY-17-(2-PROPENYL)-, (17-β)-(9CI) see AGW675
ESTRA-1,3,5(10)-TRIEN-17-ONE, 3-(SULFOOXY)-, compounded with PIPERAZINE (1:1) see PIK450
ESTRA-1,3,5(10)-TRIEN-17-ONE, 3-(SULFOXY)-, SODIUM SALT (9CI) see EDV600
ESTRA-1,3,5(10)-TRIETNE-3,17-β-DIOL-3-BENZOATE mixed with PROGESTERONE (1:14 moles) see EDQ000
ESTRATRIOL see EDU500
ESTRAVEL see EDS100
ESTRELLA DEL NORTE (CUBA) see ROU450

ESTR-4-EN-3-ONE, 4-CHLORO-17-β-HYDROXY-17-METHYL- see CIQ600
ESTR-4-EN-3-ONE, 4-CHLORO-17-HYDROXY-17-METHYL , (17 β)- see CIQ600
ESTR-4-EN-3-ONE, 17-β-HYDROXY-, DECANOATE see NNE550
ESTR-4-EN-3-ONE, 17-((1-OXODECYL)OXY)-, (17-β)- (9CI) see NNE550
ESTREPTOCIDA see SNM500
ESTRIFOL see ECU750
ESTRIL see DKA600
ESTRIN see EDV000
ESTRIOL see EDU500
16-α,17-β-ESTRIOL see EDU500
3,16-α,17-β-ESTRIOL see EDU500
ESTRIOLO (ITALIAN) see EDU500
ESTROATE see ECU750
ESTROBEN see DKB000
ESTROBENE see DKA600
ESTROBENE see DKB000
ESTROCON see ECU750
ESTRODIENOL see DAL600
ESTROFOL see PKF750
ESTROFURATE see EDU600
ESTROGEN see DKA600
ESTROGEN see EEH500
ESTROGENIN see DKB000
ESTROGENS, CONJUGATES see PMB000
ESTROICI see EDR000
ESTROL see EDV000
ESTROMED see ECU750
ESTROMENIN see DKA600
ESTRON see EDV000
ESTRONA (SPANISH) see EDV000
ESTRONE see EDV000
ESTRONE-A see EDV000
ESTRONE BENZOATE see EDV500
ESTRONE, HYDROGEN SULFATE, compounded with PIPERAZINE (1:1) (8CI) see PIK450
ESTRONE, HYDROGEN SULFATE, SODIUM SALT see EDV600
ESTRONE 17-METHOXIME see EDV555
ESTRONE, o-METHYLOXIME see EDV555
ESTRONE SODIUM SULFATE see EDV600
ESTRONE SULFATE SODIUM see EDV600
ESTRONE SULFATE SODIUM SALT see EDV600
ESTRONE-3-SULFATE SODIUM SALT see EDV600
ESTRONEX see EDR000
ESTROPAN see ECU750
ESTROPIPATE see PIK450
ESTRORAL see DAL600
ESTROSEL see DGP900
ESTROSOL see DGP900
ESTROSTILBEN see DKB000
ESTROSYN see DKA600
ESTROVIS see QFA250
ESTROVIS 4000 see QFA250
ESTROVISTER see QFA250
ESTROVITE see EDO000
ESTRU BENZYLOWEGO KWASU NIKOTYNOWEGO see NCR040
ESTRUGENONE see EDV000
ESTRUMATE see CMX880
ESTRUSOL see EDV000
ESTULIC see GKU300
ESTYRENE 4-62 see SMQ500
ESTYRENE AS see ADY500
ESTYRENE G 15 see DWW000
ESTYRENE G 20 see SMQ500
ESTYRENE G-P 4 see SMQ500
ESTYRENE H 61 see SMQ500
ESTYRENE 500SH see SMQ500
ESUCOS see MKQ000
ESZ see IHL000
ETs see EHG100
ET 14 see RMA500
ET 57 see RMA500
ET 67 see CGW300
ET-394 see THW750

ET 495 see TNR485
ET 743 see EAE050
ETABENZARONE see EDV700
ETABETACIN see EDW000
ETABUS see DXH250
ETACELASIL see CHI200
ETACELSAL see CHI200
ETACONAZOLE see EDW100
ETACONAZOLE see VFA100
ETACRINIC ACID see DFP600
ETAFENONE HYDROCHLORIDE see DHS200
ETAFOS see EEE200
ETAIN (TETRACHLORURE d') (FRENCH) see TGC250
ETAKRINIC ACID see DFP600
ETAMBRO see TCC000
ETAMBUTOL see EDW875
ETAMICAN see VSZ450
ETAMIDE see EEM000
ETAMINAL SODIUM see NBU000
ETAMINOPHYLLINE see CNR125
ETAMIPHYLLIN see CNR125
ETAMIPHYLLINE see CNR125
ETAMIPHYLLINE HYDROCHLORIDE see DIH600
ETAMIPHYLLIN HYDROCHLORIDE see DIH600
ETAMON CHLORIDE see TCC250
E-TAMOXIFEN see TAC880
ETAMSYLATE see DIS600
ETAMYCIN see EDW200
ETAMYCIN A see EDW200
ETANAUTINE see BBV500
ETANOLAMINA (ITALIAN) see EEC600
ETANOLO (ITALIAN) see EFU000
ETANTIOLO (ITALIAN) see EMB100
ETAPERAZIN see CJM250
ETAPERAZINE see CJM250
ETAPHOS see EEE200
ETAVIT see VSZ450
ETAZIN see BQC250
ETAZINE see BQC250
ETC see EPY600
ETCHLORVINOLO see CHG000
ETEAI see EDW300
ETEM see EJQ000
ETERE ETILICO (ITALIAN) see EJU000
E TETRAETHYLPYRONIN see DIS200
(+)-E-7,8,9,10-TETRAHYDRO-7-α,8-β-DIHYDROXY-9-β,19-β-EPOXY-BENZO(a)PYRENE see BMK620
ETH see EEI000
ETH see EPQ000
ETHAANTHIOL (DUTCH) see EMB100
ETHACRIDINE LACTATE see EDW500
ETHACROLEIN see EFS700
ETHACRYNIC ACID see DFP600
ETHACURE 100 see DJP100
ETHAL see HCP000
ETHALFLURALIN see ENE500
ETHALFLURLIN see ENE500
ETHAL LA-X see DXY000
ETHAMBUTOL see TGA500
ETHAMBUTOL DIHYDROCHLORIDE see EDW875
ETHAMBUTOL HYDROCHLORIDE see EDW875
ETHAMBUTOL mixed with SODIUM NITRITE (1:1) see SIS500
ETHAMIDE see DWW000
ETHAMINAL see NBT500
ETHAMINAL SODIUM see NBU000
ETHAMON DS see EDX000
ETHAMOXYTRIPHETOL see DHS000
ETHAMSYLATE see DIS600
ETHAN see PII500
ETHANAL see AAG250
ETHANAL OXIME see AAH250
ETHANAMIDE see AAI000
ETHANAMIDINE HYDROCHLORIDE see AAI500

ETHANAMINE see EFU400
ETHANAMINE, 2-BROMO-(9CI) see BNI650
ETHANAMINE, 2-((3,4-DIHYDRO-2-METHYL-4-(3-(TRIFLUOROMETHYL)PHENYL)-2H-1-BENZOPYRAN-7-YL)OXY)-N,N-DIMETHYL- see EDX100
ETHANAMINE, 2-(4-((1Z)-1,2-DIPHENYL-1-BUTENYL)PHENOXY)- see DHA250
ETHANAMINE, 2-(4-(1,2-DIPHENYL-1-BUTENYL)PHENOXY)-N-METHYL-, (Z)- see DBA710
ETHANAMINE, 2,2'-DITHIOBIS(N,N-DIETHYL- see TCD300
ETHANAMINE, 2-HYDRAZINO-N,N-DIMETHYL- see HHB600
ETHANAMINE, 2,2'-(HYDROXYNITROSOHYDRAZONO)BIS- see DHJ333
ETHANAMINIUM, 2-(ACETYLOXY)-N,N,N-TRIMETHYL-, BROMIDE (9CI) see CMF260
ETHANAMINIUM, 2-(ACETYLTHIO)-N,N,N-TRIMETHYL-, IODIDE (9CI) see ADC300
ETHANAMINIUM, N-(4-(BIS(4-(DIETHYLAMINO)PHENYL)METHYLENE)-2,5-CYCLOHEXADIEN-1-YLIDENE)-N-ETHYL-, CHLORIDE see EQG550
ETHANAMINIUM, 2-CHLORO-N,N,N-TRIMETHYL-, CHLORIDE (9CI) see CMF400
ETHANAMINIUM, 2-((CYCLOHEXYLHYDROXYPHENYLACETYL)OXY)-N,N-DIETHYL-N-METHYL-, BROMIDE (9CI) see ORQ000
ETHANAMINIUM, N-(4-((4-(DIETHYLAMINO)PHENYL)(5-HYDROXY-2,4-DISULFOPHENYL)METHYLENE)-2,5-CYCLOHEXADIEN-1-YLIDENE)-N-ETHYL-, HYDROXIDE, inner salt, CALCIUM SALT (2:1) see CMM062
ETHANAMINIUM, N,N-DIETHYL-N-METHYL-2-((9H-XANTHEN-9-YLCARBONYL)OXY)-, BROMIDE (9CI) see XCJ000
ETHANAMINIUM,2-(((HEXADECYLOXY)HYDROXYPHOSPHINYL)OXY)-N,N,N-TRIMETHYL-, HYDROXIDE, INNER SALT see HCP750
ETHANAMINIUM, 2-HYDRAZINO-N,N,N-TRIMETHYL-2-OXO-, CHLORIDE (9CI) see GEQ500
ETHANAMINIUM, 2-HYDROXY-N,N,N-TRIS(2-HYDROXYETHYL)-, HYDROXIDE (9CI) see TCB500
ETHANAMINIUM, 2-((1-OXO-2-PROPENYL)OXY)-N,N,N-TRIMETHYL-, CHLORIDE see OOM500
ETHANAMINIUM, N,N,N-TRIETHYL-, IODIDE (9CI) see TCC750
ETHANAMINIUM, N,N,N-TRIETHYL-, PERCHLORATE see TCD002
ETHANAMINIUM, N,N,N-TRIETHYL-, SALT WITH 1,1,2,2,3,3,4,4,5,5,6,6,7,7,8,8,8-HEPTADECAFLUORO-1-OCTANESULFONIC ACID (1:1) see TCD100
ETHANAMINIUM, N,N,N-TRIETHYL-, TETRACHLOROCOBALTATE(2-) (2:1) see BLH315
ETHANAMINIUM, N,N,N-TRIMETHYL-, IODIDE (9CI) see EQC600
ETHANAMINIUM, N,N,N-TRIMETHYL-2-(1-OXOPROPOXY)-, IODIDE (9CI) see PMW750
ETHANDROSTATE see EDY600
ETHANE see EDZ000
ETHANE, 1,1-BIS(p-CHLOROPHENYL)- see BIM775
ETHANE, 1,2-BIS(2,3-EPOXYPROPOXY)- see EEA600
ETHANE, 1,1-BIS(p-TOLYL)-2,2,2-TRICHLORO-` see MIF763
ETHANECARBOXYLIC ACID see PMU750

ETHANE, CHLORODIBROMOTRIFLUORO- see MRF600

ETHANE, 2-CHLORO-2-(DIFLUOROMETHOXY)-1,1,1-TRIFLUORO-(9CI) see IKS400

ETHANE, 1-CHLORO-1-FLUORO- see FOO553

ETHANE, CHLOROPENTAFLUORO-, mixt. with CHLORODIFLUOROMETHANE see FOO560

ETHANE, 1-CHLORO-1,1,2,2-TETRAFLUORO- see TCH150

ETHANEDIAL DIOXIME see EEA000

ETHANEDIAMIDE see OLO000

1,2-ETHANEDIAMINE see EEA500

1,2-ETHANEDIAMINE, N,N-BIS(2-AMINOETHYL)- see NEI800

1,2-ETHANEDIAMINE, N,N'-BIS(1,1-DIMETHYLETHYL)- see DEC100

1,2-ETHANEDIAMINE, DIHYDROCHLORIDE see EIW000

1,2-ETHANEDIAMINE, N,N'-DIMETHYL-(9CI) see DRI600

ETHANEDIAMINE, N,N'-DINITRO-(9CI) see DUU800

1,2-ETHANEDIAMINE, N-METHYL-(9CI) see MJW100

1,2-ETHANEDIAMINE, POLYMER WITH FORMALDEHYDE AND PHENOL (9CI) see EEA550

ETHANE, 1,2-DIAMINO-, COPPER COMPLEX see DBU800

ETHANE, 1,2-DIAZIDO- see DCL600

ETHANE, 1,2-DIBROMO-1,1-DICHLORO- see DDM425

ETHANE, 1,2-DIBROMOTETRAFLUORO- see FOO525

ETHANE, 1,2-DIBROMO-1,1,2,2-TETRAFLUORO-(9CI) see FOO525

1,2-ETHANEDICARBAMIC ACID, TETRATHIO- see DXI300

1,2-ETHANEDICARBOXYLIC ACID see SMY000

ETHANE DICHLORIDE see EIY600

ETHANE, 1,2-DICHLORO-1-ETHOXY- see DFG333

ETHANE, 1,1-DICHLORO-1-FLUORO- see FOO550

ETHANE, 1,2-DIFLUORO- see DKH300

ETHANE, 1,2-DIFLUORO-1,1,2-TRICHLORO- see DKI900

ETHANEDINITRILE see COO000

ETHANE, 1,1-DINITRO- see DUV710

ETHANE, 1,2-DINITRO- see DUV720

ETHANEDIOIC ACID see OLA000

ETHANEDIOIC ACID BIS(CYCLOHEXYLIDENE HYDRAZIDE) see COF675

ETHANEDIOIC ACID, COPPER(2+) SALT (1:1) see CNN755

ETHANEDIOIC ACID DIAMMONIUM SALT see ANO750

ETHANEDIOIC ACID, DIPOTASSIUM SALT (9CI) see PLN300

ETHANEDIOIC ACID, DISILVER(1+) SALT see SDU000

ETHANEDIOIC ACID, TIN(2+) SALT (1:1) (9CI) see TGE250

1,2-ETHANEDIOL see EJC500

ETHANEDIOL, BIS(1-AZIRIDINEPROPIONATE) see EIP100

1,1-ETHANEDIOL, 2-CHLORO- see CDY600

1,2-ETHANEDIOL DIACETATE see EJD759

1,2-ETHANEDIOL DIGLYCIDYL ETHER see EEA600

ETHANEDIOL DIMETHACRYLATE see BKM250

1,2-ETHANEDIOL DIMETHACRYLATE see BKM250

1,2-ETHANEDIOL DIMETHANESULFONATE (9CI) see BKM125

ETHANEDIOL DINITRATE see EJG000

1,2-ETHANEDIOL DIPROPANOATE (9CI) see COB260

1,2-ETHANEDIOL, MONOACETATE see EJI000

1,2-ETHANEDIOL, 1-OXIRANYL- see ONE050

1,2-ETHANEDIOL, PHENYL- see SMQ100

1,2-ETHANEDIOL, 1-PHENYL- see SMQ100

1,2-ETHANEDIOL, 1,1,2,2-TETRAPHENYL- see TEA600

ETHANEDIONE, DI-2-FURANYL-, DIOXIME see BJW810

ETHANEDIONE, DIPHENYL-, MONOOXIME see BCA300

ETHANEDIONIC ACID see OLA000

N,N'-(1,2-ETHANEDIOXYSULFINYL)BIS(S-METHYL-N-METHYLCARBAMOYLOXYTHIOACETIMIDATE) see EEA700

ETHANEDITHIOAMIDE see DXO200

1,2-ETHANEDITHIOL see EEB000

1,2-ETHANEDITHIOL, CYCLIC ESTER with P,P-DIETHYL PHOSPHONODITHIOIMIDOCARBONATE see PGW750

1,2-ETHANEDITHIOL, CYCLIC S,S-ESTER with PHOSPHONODITHIOIMIDOCARBONIC ACID P,P-DIETHYL ESTER see PGW750

N,N'-(1,2-ETHANEDITHIOSULFINYL)BIS(2,3-DIHYDRO-2,2-DIMETHYLBENZOFURANYL-7) METHYLCARBAMATE see EEB025

N,N'-(1,2-ETHANEDITHIOSULFINYL)BIS(S-METHYL-N-METHYLCARBAMOYLOXYTHIOACETIMIDATE) see EEB050

ETHANEDIUREA see EJC100

N,N'-1,2-ETHANEDIYLBIS(N-BUTYL-4-MORPHOLINECARBOXAMIDE) see DUO400

1,2-ETHANEDIYLBIS(CARBAMODITHIOATO)(2−)-MANGANESE see MAS500

(1,2-ETHANEDIYLBIS(CARBAMODITHIOATO))(2⁻)ZINC see EIR000

1,2-ETHANEDIYLBIS(CARBAMODITHIOATO)(2⁻)-S,S'-ZINC see EIR000

1,2-ETHANEDIYLBISCARBAMODITHIOIC ACID DISODIUM SALT see DXD200

1,2-ETHANEDIYLBISCARBAMODITHIOIC ACID MANGANESE COMPLEX see MAS500

1,2-ETHANEDIYLBISCARBAMODITHIOIC ACID, MANGANESE(2+) SALT (1:1) see MAS500

1,2-ETHANEDIYLBISCARBAMODITHIOIC ACID, ZINC COMPLEX see EIR000

1,2-ETHANEDIYLBISCARBAMOTHIOIC ACID, ZINC SALT see EIR000

N,N'-1,2-ETHANEDIYLBIS(N-(CARBOXYMETHYL))GLYCINE see EIX000

N,N'-1,2-ETHANEDIYLBIS(N-(CARBOXYMETHYL)GLYCINE), DIPOTASSIUM SALT see EEB100

N,N'-1,2-ETHANEDIYLBIS(N-(CARBOXYMETHYL)GLYCINE) DISODIUM SALT see EIX500

N,N'-1,2-ETHANEDIYLBIS(N-(CARBOXYMETHYL)GLYCINE TETRASODIUM SALT see EIV000

N,N'-1,2-ETHANEDIYLBIS(N-CARBOXYMETHYL)GLYCINE, TRISODIUM SALT see TNL250

3,3'-(1,2-ETHANEDIYLBIS(DIMETHYLSILYLENE))B IS(N,N,N-TRIMETHYL-1-PROPANAMINIUM) DIIODIDE see EEB200

1,2-ETHANEDIYLBISMANEB, MANGANESE(2+) SALT (1:1) see MAS500

1,1'-(1,2-ETHANEDIYL)BIS(1-METHYLHYDRAZINE) see EEB225

1,1'-(1,2-ETHANEDIYLBIS(OXY))BIS-BUTANE see DDW400

2,2'-(1,2-ETHANEDIYLBIS(OXY))BISETHANOL see TJQ000

1,1'-(1,2-ETHANEDIYLBIS(OXY))BIS(2,3,4,5,6-PENTABROMOBENZENE) see EEB230

1,1'-(1,2-ETHANEDIYLBIS(OXY))BIS(2,4,6-TRIBROMOBENZENE) see EEB250

2,2'-(1,2-ETHANEDIYLBIS(OXYMETHYLENE))BISO XIRANE see EEA600

2,2'-(1,2-ETHANEDIYL)BIS(4,5,6,7-TETRABROMO-1H-ISOINDOLE-1,3(2H)-DIONE) see EIT150

2,2'-(1,2-ETHANEDIYLBIS(THIO))BISBENZENAMI NE see EJC050

1,2-ETHANEDIYLBIS(TRIS(2-CYANOETHYL)PHOSPHONIUM DIBROMIDE see EIU000

N,N'-1,2-ETHANEDIYLBISUREA see EJC100

(R)-2,2'-(1,2-ETHANEDIYLDIIMINO)BIS-1-BUTANOL see TGA500

1,1'-(1,2-ETHANEDIYLDIIMINO)BIS(3-(4-METHOXYPHENOXY)-2-PROPANOL, DIMETHANESULFONATE (salt) see MQY125

1,2-ETHANEDIYL DIMETHANESULFONATE see BKM125

1,2-ETHANEDIYL ESTER CARBAMIMIDOTHIOIC ACID DIHYDROBROMIDE see EJA000

1,2-ETHANEDIYL THIOCYANATE see EJC035

ETHANE HEXACHLORIDE see HCI000

ETHANE, HEXAFLUORO- see HDC100

ETHANE HEXAMERCARBIDE see BKH125

ETHANE HEXAMERCARBIDE see EEB500

ETHANE, HEXANITRO- see HET675

ETHANEHYDRAZONIC ACID see ACM750

ETHANE-1-HYDROXY-1,1-DIPHOSPHONATE see HKS780

ETHANE-1-HYDROXY-1,1-DIPHOSPHONIC ACID DISODIUM SALT see DXD400

ETHANE-1-HYDROXY-1,1-DIPHOSPHONIC ACID, TETRAPOTASSIUM SALT see TEC250

ETHANE-1-HYDROXY-1,1-DIPHOSPHONIC ACID, TRATRASODIUM SALT see TEE250

ETHANE-1-HYDROXY-1,1-DIPHOSPHONIC ACID, TRISODIUM SALT see TNL750

ETHANE, ISOTHIOCYANATO-(9CI) see ELX525

ETHANE, 1,1'-(METHYLENEBIS(OXY))BIS(2-FLUORO-2,2-DINITRO- see FMV050

ETHANENITRILE see ABE500

ETHANE, 1,1'-(OXYBIS(METHYLENESULFONYL))BIS(2-CHLORO- see OPG100

ETHANE PENTACHLORIDE see PAW500

ETHANE, PENTAFLUORO- see PBD400

ETHANEPEROXOIC ACID see PCL500

ETHANEPEROXOIC ACID, 1,1-DIMETHYLETHYL ESTER see BSC250

ETHANEPHOSPHONIC ACID, 2-CYANO-, DIETHYL ESTER see COM100

ETHANESELENOL, 2-AMINO-, HYDROCHLORIDE see AJS900

ETHANE-1,1'-SULFINYLBIS see EPI500

ETHANE, 1,1'-SULFINYLBIS(1,2-DICHLORO- see SNT100

ETHANESULFONIC ACID, 2-AMINO- see TAG750

ETHANESULFONIC ACID, 2-HYDROXY-, AMMONIUM SALT see ANL100

ETHANESULFONIC ACID, 2-HYDROXY-, compounded eith 9-((2-METHOXY-4-((METHYLSULFONYL)AMINO)PHENYL)AMINO)-N,5-DIMETHYL-4-ACRIDINECARBOXAMIDE (1:1) see EEB600

ETHANESULFONIC ACID, 2-(((3-α-5-β)-3-HYDROXY-24-OXOCHOLAN-24-YL)AMINO)- see LHW100

ETHANESULFONIC ACID, ISOPROPYL ESTER see IOX500

ETHANESULFONIC ACID, 2-(((METHOXYCARBONYL)AMINO)((2-NITRO-5-(PROPYLTHIO)PHENYL)AMINO)METHYLENE)AMINO)- see NCP600

ETHANESULFONIC ACID, 1-METHYLETHYL ESTER see IOX500

ETHANESULFONYL CHLORIDE see EEC000

ETHANESULFONYL FLUORIDE, 2-CHLORO- see CGO200

ETHANE, 1,1,1,2-TETRABROMO- see TBJ560

ETHANE, 1,1,2,2-TETRACHLORO-1,2-DIFLUORO- see TBP050

ETHANE, 1,1,1,1-TETRAFLUORO- see EEC100

ETHANETHIOAMIDE see TFA000

ETHANETHIOAMIDE, N,N-DIMETHYL-(9CI) see DUG450

ETHANETHIOIC ACID see TFA500

ETHANETHIOIC ACID, S-(2-(2-BENZOTHIAZOLYLAMINO)-2-OXOETHYL) ESTER see ADC400

ETHANETHIOIC ACID, S-(4-NITROPHENYL) ESTER see NIW450

ETHANETHIOL see EMB100

ETHANETHIOL, 2-AMINO-, DIHYDROGEN PHOSPHATE (ESTER) see AKA900

ETHANETHIOL, 2-AMINO-, S-ESTER WITH PHOSPHOROTHIOIC ACID see AKA900

ETHANETHIOLIC ACID see TFA500

ETHANE TRICHLORIDE see TIN000

ETHANE, 1,1,1-TRICHLORO-2,2-BIS(p-TOLYL)- see MIF763

ETHANE, 1,1,1-TRIFLUORO- see TJY900

ETHANE, 1,1,2-TRIFLUORO- see TJY950

1,1,1-ETHANETRIOL DIPHOSPHONATE see HKS780

ETHANE, compressed (UN 1035) (DOT) see EDZ000

ETHANE, refrigerated liquid (UN 1961) (DOT) see EDZ000

ETHANIMIDOTHIOIC ACID, 2-(DIMETHYLAMINO)-N-HYDROXY-2-OXO-, METHYL ESTER see OLT100

ETHANIMIDOTHIOIC ACID, N,N'-(DITHIOBIS((METHYLIMINO)CARBONYLOXY))BIS-, DIMETHYL ESTER see BKU100

ETHANIMIDOTHIOIC ACID, N,N'-(DITHIOBIS((METHYLIMINO)CARBONYLOXY))BIS-,BIS(2-CYANOETHYL) ESTER see BIQ630

ETHANIMIDOTHIOIC ACID, N,N'-(1,2-ETHANEDIYLBIS(OXYSULFINYL(METHYLIMINO)CARBONYLOXY)BIS-, DIMETHYL ESTER see EEA700

ETHANIMIDOTHIOIC ACID, N,N'-(1,2-ETHANEDIYLBIS(THIOSULFINYL(METHYLIMINO)CARBONYLOXY))BIS-, DIMETHYL ESTER see EEB050

ETHANIMIDOTHIOIC ACID, N-(((((HEXYLOXY)SULFINYL)METHYLAMINO)CARBONYL)OXY)-, METHYL ESTER see MKM800

ETHANIMIDOTHIOIC ACID, N-HYDROXY-, METHYL ESTER see MPS270

ETHANIMIDOTHIOIC ACID, N-(((METHYLAMINO)CARBONYL)OXY)-, 1-

(AMINOCARBONYL)PROPYL ESTER see AJD810

ETHANIMIDOTHIOIC ACID, N-(((METHYLAMINO)CARBONYL)OXY)-, 2-AMINO-1-METHYL-2-OXOETHYL ESTER see AKY880

ETHANIMIDOTHIOIC ACID, N-(((METHYLAMINO)CARBONYL)OXY)-, 2-AMINO-2-OXOETHYL ESTER see ALQ642

ETHANIMIDOTHIOIC ACID, N-(((METHYLAMINO)CARBONYL)OXY)-, 2-(DIMETHYLAMINO)-2-OXOETHYL ESTER see DPN500

ETHANIMIDOTHIOIC ACID, N-(((METHYLAMINO)CARBONYL)OXY)-, 2-OXO-2-(1-PYRROLIDINYL)ETHYL ESTER see OOO500

ETHANIMIDOTHIOIC ACID, N,N'-((1-((1-METHYLETHOXY)METHYL)-1,2-ETHANEDIYL)BIS(OXYSULFINYL(METHYLIMINO)CARBONYLOXY))BIS-, DIMETHYL ESTER see INE068

ETHANIMIDOTHIOIC ACID, N-(((METHYL(4-MORPHOLINYLTHIO)AMINO)CARBONYL)OXY)-, METHYL ESTER see MLL150

ETHANIMIDOTHIOIC ACID, N,N'-(THIOBIS((METHYLAMINO)CARBONYLOXY))BIS(2-(DIMETHYLAMINO)-2-OXO-, DIMETHYL ESTER see BKU120

ETHANIMIDOTHIOIC ACID, N,N'-(THIOBIS((METHYLIMINO)CARBONYLOXY))BIS-, DIMETHYL ESTER see LBF100

ETHANIMINIUM-N-(6-(DIETHYLAMINO)-3H-XANTHEN-3-YLIDENE)-N-ETHYL CHLORIDE see DIS200

9,10-ETHANOANTHRACENE-9(10H)-METHANAMINE, N-METHYL- see BCH300

ETHANOANTHRACENE-9(10H)-METHYLAMINE, N-METHYL- see BCH300

9,10-ETHANOANTHRACENE-9(10H)-METHYLAMINE, N-METHYL- see BCH300

6,10-ETHANO-5-AZONIASPIRO(4.5)DECAN-8-OL CHLORIDE BENZILATE see BCA375

1H-4,9a-ETHANOCYCLOHEPTA(c)PYRAN-7-CARBOXYLIC ACID, 4a-(ACETYLOXY)-3-(4-CARBOXY-1,3-PENTADIENYL)3,4,4a,5,6,9-HEXAHYDRO-3-METHYL-1-OXO-, 7-METHYL ESTER, 3-α(1E,3E),4-α-4a-α-9a-α)-(−)- (9CI) see POH600

3,8a-ETHANO-8aH-1-BENZOPYRAN-2(3H)-ONE, HEXAHYDRO-3,5,5-TRIMETHYL- see LAQ100

ETHANOIC ACID see AAT250

ETHANOIC ACID, ETHENYL ESTER see VLU250

ETHANOIC ANHYDRATE see AAX500

N-ETHANOLACETAMIDE see HKM000

ETHANOL, 2-(1-(9-ACRIDINYL)HYDRAZINO)-, MONOHYDROCHLORIDE see ADQ550

ETHANOLAMINE see EEC600

β-ETHANOLAMINE see EEC600

ETHANOLAMINE, solution (DOT) see EEC600

ETHANOLAMINE CHLORIDE see EEC700

ETHANOLAMINE-N,N-DIACETIC ACID see HKM500

ETHANOLAMINE HYDROCHLORIDE see EEC700

ETHANOLAMINE PERCHLORATE see HKM175

ETHANOLAMINE PHOSPHATE see EED000

ETHANOLAMINE SALT of 5,2'-DICHLORO-4'-NITROSALICYLICANILIDE see DFV600

ETHANOL, 2-AMINO-, COMPD. WITH 2-BENZOTHIAZOLETHIOL see HKO012

ETHANOL, 2-AMINO-, COMPD. WITH 2-BENZOTHIAZOLETHIOL (1:1) see HKO012

ETHANOL, 2-AMINO-, HYDROCHLORIDE see EEC700

ETHANOL, 1-AMINO-(8CI,9CI) see AAG500

ETHANOL, 2-((4-AMINO-3-METHYLPHENYL)ETHYLAMINO)-, SULFATE (1:1) (SALT) see AKZ100

ETHANOL, 2-((4-AMINO-2-NITROPHENYL)AMINO)- see ALO750

ETHANOL, 2-AMINO-1-PHENYL- see HNF000

ETHANOL, 2,2'-((4-AMINOPHENYL)IMINO)BIS-, SULFATE (SALT), HYDRATE (1:1:1) see BKF800

ETHANOL, 2-AZIDO-, NITRATE (ester) see ASF800

ETHANOL, 2-BROMO-, ACETATE see BNI600

ETHANOL-2-(2-BUTOXYETHOXY) THIOCYANATE see BPL250

ETHANOL, 2-BUTOXY-, MANUFACTURE OF, BY-PRODUCTS FROM see EED100

ETHANOL, 2-(BUTYLNITROAMINO)-, NITRATE (ESTER) see BRW150

ETHANOL, 2-CHLORO-, ACETATE see CGO600

ETHANOL, 2-CHLORO-, (2-CHLOROETHYL)PHOSPHONATE (2:1) see BIB800

ETHANOL, 2-CHLORO-, ETHYLENE PHOSPHATE (4:1) see TDG760

ETHANOL, 2-((4-((2-CHLORO-4-NITROPHENYL)AZO)PHENYL)ETHYLAMINO)- see CMP080

ETHANOL, 2-(4-CHLOROPHENOXY)-1-tert-BUTYL-2-(1H-1,2,4-TRIAZOLE-1-YL)- see CJN300

ETHANOL, 2-(2-(4-(p-CHLORO-α-PHENYLBENZYL)-1-PIPERAZINYL)ETHOXY)-, DIHYDROCHLORIDE see HOR470

ETHANOL, 2-(2-(2-(4-(p-CHLORO-α-PHENYLBENZYL)-1-PIPERAZINYL)ETHOXY)ETHOXY)- see HHK050

ETHANOL, 2-(2-(2-(4-(p-CHLORO-α-PHENYLBENZYL)-1-PIPERAZINYL)ETHOXY)ETHOXY)-, DIMALEATE see HHK100

ETHANOL, 2,2'-((3-CHLOROPHENYL)IMINO)BIS- see CJU400

ETHANOL, 2-(2-(4-(2-((4-CHLOROPHENYL)PHENYLMETHOXY)ETHYL)-1-PIPERAZINYL)ETHOXY)-, 2HCL see CLW650

ETHANOL,- 2-(2-(2-(4-((4-CHLOROPHENYL)PHENYLMETHYL)-1-PIPERAZINYL)ETHOXY)ETHOXY)- see HHK050

ETHANOL, 2-(2,4-DIAMINOPHENOXY)-, DIHYDROCHLORIDE see DCA250

ETHANOL, 2-(2-(4-DIBENZO(B,F)(1,4)THIAZEPIN-11-YL-1-PIPERAZINYL)ETHOXY)-, (E)-2-BUTENEDIOATE (2:1) SALT see QCJ300

ETHANOL, 2,2-DIBROMO-2-NITRO- see DDQ470

ETHANOL, 2-(2,4-DICHLOROPHENOXY)-, BENZOATE see SCB200

ETHANOL, 1-(3,4-DICHLOROPHENYL)-2,2,2-TRICHLORO-, ACETATE see TIL800

ETHANOL, 2-(2-(DIETHYLAMINO)ETHOXY)- see DHQ100

ETHANOL, 2-((1,1-DIMETHYLETHYL)AMINO)- see BRH300

ETHANOL, 2-(2,4-DINITROANILINO)- see DVC300

ETHANOL, 2-((2,4-DINITROPHENYL)AMINO)- see DVC300

ETHANOL, 2-(1,3,5,7,9-DODECAPENTAENYLOXY)-, (ALL E)- see AFQ800

ETHANOL, 2,2'-(1,2-ETHANEDIYLBIS(OXY))BIS-, DINITRATE see TJQ500

ETHANOL, 2-(ETHENYLOXY)-, NITRATE see EJK100

ETHANOLETHYLENE DIAMINE see AJW000

ETHANOL, 2,2'-(ETHYLIMINO)BIS-(9CI) see ELP000

ETHANOL, 2-(ETHYL(3-METHYL-4-NITROSOPHENYL)AMINO)- see ELG100

ETHANOL, 2-(ETHYL(3-METHYL-4-(PHENYLAZO)PHENYL)AMINO)- see ENB100

ETHANOL, 2-(ETHYLNITROAMINO)-, NITRATE (ESTER) see ENN200

ETHANOL, 2-(ETHYL(4-((4-NITROPHENYL)AZO)PHENYL)AMINO)-(9CI) see ENP100

ETHANOL, 2-(N-ETHYL-m-TOLUIDINO)- see HKS100

ETHANOL, 2-(HEPTYLOXY)- see HBN250

ETHANOL, 2-(2-(HEPTYLOXY)ETHOXY)- see HBP275

ETHANOL, 2-(2-(2-(HEXYLOXY)ETHOXY)ETHOXY)- see HFT550

ETHANOL, 2-((2-(2-HYDROXYETHOXY)-4-NITROPHENYL)AMINO)- see HKK100

ETHANOL, 2,2'-((3-(2-HYDROXYETHOXY)OCTADECYLAMINO)PROPYL)IMINO)BIS-, DIHYDROFLUORIDE see BJY800

ETHANOL, 2,2'-((3-(N-(2-HYDROXYETHYL)-N-OCTADECYLAMINO)PROPYL)IMINO)DI-, DIHYDROFLUORIDE see BJY800

ETHANOL, 2,2-IMINODI-, COMPD. WITH 2-((4-CHLORO-o-TOLYL)OXY)PROPIONIC ACID (1:1) see MIH900

ETHANOL,2,2'-IMINODI-,3,5-DIIODO-4-OXO-1(4H)-PYRIDINEACETATE (salt) see DNG400

ETHANOL,2,2'-IMINODI- with 3,5-DIIODO-4-OXO-1(4H)-PYRIDINEACETIC ACID (1:1) see DNG400

ETHANOL, 2,2'-IMINODI-, 1-(P,α-DIMETHYLBENZYL) CAMPHORATE (SALT) see HAQ570

ETHANOL, 2-(ISOBUTYL)AMINO- see IIM050

ETHANOL, 2-ISOCYANATO-, CARBONATE (2:1) (ESTER) (9CI) see CBV600

ETHANOL, 2-ISOCYANATO-, CARBONATE (2:1) (ESTER) (9CI) see IKG400

ETHANOL, 2-(2-ISOPROPOXYETHOXY)- see IOJ500

ETHANOLISOPROPYLAMINE see INN400

ETHANOL, 2-(ISOPROPYLAMINO)- see INN400

ETHANOL (MAK) see EFU000

ETHANOLMERCURY BROMIDE see EED600

ETHANOL, 2-(2-(2-METHOXYETHOXY)ETHOXY)-, ACETATE (8CI,9CI) see AAV500

ETHANOL, 2-METHOXY-, OLEATE see MEP750

ETHANOL, 2-METHOXY-, PHOSPHATE (3:1) see TLA600

ETHANOL, 2-(2-(1-METHYLETHOXY)ETHOXYL)- see IOJ500

ETHANOL, 2-((1-METHYLETHYL)AMINO)-(9CI) see INN400

ETHANOL, 2,2'-(METHYLIMINO)BIS- see MKU250

ETHANOL, 2-(METHYLNITROSOAMINO)-, 4-METHYLBENZENESULFONATE (ESTER) see NKU420

ETHANOL, 2-((2-METHYLPROPYL)AMINO)- see IIM050

ETHANOL, 2,2',2''-NITRILOTRI-, COMPD. WITH METHANEARSONIC ACID see TJK900

ETHANOL, 2,2',2''-NITRILOTRI-, HYDROCHLORIDE see NEI700

ETHANOL, 2,2',2''-NITRILOTRIS-, HYDROCHLORIDE see NEI700

ETHANOL, 2,2',2''-NITRILOTRIS-, TRINITRATE (ester), PHOSPHATE (1:2)(SALT) (9CI) see TJL250

ETHANOL, 2-((1-NITRO-9-ACRIDINYL)AMINO)-, MONOHYDROCHLORIDE see NHE550

ETHANOL, 2-(((5-NITRO-2-FURANYL)METHYLENE)AMINO)-, N-OXIDE (9CI) see HKW450

ETHANOL, 2-((5-NITROFURFURYLIDENE)AMINO)-, N-OXIDE see HKW450

ETHANOL, 2-NITRO-, NITRATE (ester) see NFY570

ETHANOL, 2-((2,3,3A,4,7,7A(OR 3A,4,5,6,7,7A)-HEXAHYDRO-4,7-METHANO-1H-INDENYL)OXY)- see EJJ100

ETHANOL, 2-PHENOXY-, ACETATE see EJL600

ETHANOL, 2-(2-(2-(2-PHENOXYETHOXY)ETHOXY)ETHOXY)- see TCE450

ETHANOL, 1-PHENYL- see PDE000

ETHANOL, 2-PHENYL- see PDD750

ETHANOL 200 PROOF see EFU000

ETHANOL, 2,2',2''-(PROPYLIDYNETRIS(METHYLENEOXY))TRI-, TRIACRYLATE see TLX100

ETHANOL, 2,2'-(PROPYLIMINO)BIS- see PNH775

ETHANOL, 2,2'-(PROPYLIMINO)DI- see PNH775

4-ETHANOLPYRIDINE see POR500

ETHANOL, 2,2',2'',2'''-SILANETETRAYLTETRAKIS- see TDH100

ETHANOL SOLUTIONS (UN 1170) (DOT) see EFU000

ETHANOLSULFONIC ACID see HKI500

ETHANOL, 2,2'-SULFONYLBIS-, DIACETATE (9CI) see BGQ100

ETHANOL, 2,2'-SULFONYLDI-, DIACETATE (7CI,8CI) see BGQ100

ETHANOL THALLIUM(1+) SALT see EEE000

ETHANOL, 2,2'-THIODI-, DIACETATE see TFI100

1-ETHANOL-2-THIOL see MCN250

ETHANOL, 2,2,2-TRICHLORO-, DIHYDROGEN PHOSPHATE see TIP300

ETHANOL-2-(2,4,5-TRICHLOROPHENOXY)-, 2,2-DICHLOROPROPIONATE see PBK000

ETHANONE, 1-(4-AMINO-5-(2-CHLOROPHENYL)-2-METHYL-1H-PYRROL-3-YL)- see AJI300

ETHANONE, 1-(4-AMINO-5-(p-CHLOROPHENYL)-2-METHYL-1H-PYRROL-3-YL)- see AJI330

ETHANONE, 1-(4-AMINO-5-(3,4-DICHLOROPHENYL)-2-METHYL-1H-PYRROL-3-YL)- see AJM600

ETHANONE, 1-(4-AMINO-5-(3,4-DIMETHOXYPHENYL)-2-METHYL-1H-PYRROL-3-YL)- see AJO800

ETHANONE, 1-(4-AMINO-1,2-DIMETHYL-5-PHENYL-1H-PYRROL-3-YL)- see AJR100

ETHANONE, 1-(4-AMINO-1-ETHYL-2-METHYL-5-PHENYL-1H-PYRROL-3-YL)- see AKA600

ETHANONE, 1-(4-AMINO-5-(3-FLUOROPHENYL)-2-METHYL-1H-PYRROL-3-YL)- see AKC550

ETHANONE, 1-(4-AMINO-5-(4-FLUOROPHENYL)-2-METHYL-1H-PYRROL-3-YL)- see AKC560

ETHANONE, 1-(4-AMINO-5-(4-METHOXY-3-METHYLPHENYL)-2-METHYL-1H-PYRROL-3-YL)- see AKO100

ETHANONE, 1-(4-AMINO-5-(2-METHOXYPHENYL)-2-METHYL-1H-PYRROL-3-YL)- see AKO450

ETHANONE, 1-(4-AMINO-5-(3-METHOXYPHENYL)-2-METHYL-1H-PYRROL-3-YL)- see AKO400

ETHANONE, 1-(4-AMINO-5-(4-METHOXYPHENYL)-2-METHYL-1H-PYRROL-3-YL)- see AKO430

ETHANONE, 1-(4-AMINO-2-METHYL-5-(2-METHYLPHENYL)-1H-PYRROL-3-YL)- see AKT800

ETHANONE, 1-(4-AMINO-2-METHYL-5-(3-METHYLPHENYL)-1H-PYRROL-3-YL)- see AKT830

ETHANONE, 1-(4-AMINO-2-METHYL-5-(4-METHYLPHENYL)-1H-PYRROL-3-YL)- see AKT850

ETHANONE, 1-(4-AMINO-2-METHYL-5-PHENYL-1H-PYRROL-3-YL)-, MONOHYDROCHLORIDE see ALA300

ETHANONE, 1-(9-AZABICYCLO(4.2.1)NON-2-EN-2-YL)-, (1R)- see AOO120

ETHANONE, 1-(2-BENZOFURANYL)-(9CI) see ACC100

ETHANONE, 1-(2H-1-BENZOPYRAN-3-YL)-(9CI) see BCS500

ETHANONE, 1-(1,1'-BIPHENYL)-4-YL-(9CI) see MHP500

ETHANONE, 1-(1,1'-BIPHENYL)-4-YL-2-((4-(DICHLOROACETYL)PHENYL)AMINO)-2-HYDROXY- see BGL400

1-ETHANONE, 1-(3-BROMOADAMANTYL)-2-DIAZO- see EEE025

ETHANONE, 1-(1-BROMO-3-ADAMANTYL)-2-ETHOXY- see BMT100

ETHANONE, 1-(1-BROMO-3-ADAMANTYL)-2-HYDROXY- see BMT130

ETHANONE, 2-BROMO-1-(4-CHLOROPHENYL)- see CJJ100

ETHANONE, 2-BROMO-1-(3,4-DICHLORO-5-NITRO-2-FURANYL)- see BND600

1-ETHANONE, 1-(3-CHLOROADAMANTYL)-2-DIAZO- see CEE800

ETHANONE, 1-(1-CHLORO-3-ADAMANTYL)-2-ETHOXY- see CEE825

ETHANONE, 1-(1-CHLORO-3-ADAMANTYL)-2-HYDROXY- see CEE850

ETHANONE, 2-CHLORO-1-(9H-FLUOREN-2-YL)- see CEC700

ETHANONE, 2-CHLORO-1-(4-HYDROXYPHENYL)- see HNE500

ETHANONE, 1-(4-CHLORO-3-NITROPHENYL)-2-ETHOXY-2-((4-(METHYLTHIO)PHENYL)AMINO)- see CJD630

ETHANONE, 2-CHLORO-1-PHENYL- see CEA750

ETHANONE, 1-(3-CHLOROPHENYL)-2-((4-(DICHLOROACETYL)PHENYL)AMINO)-2-HYDROXY- see CJU300

ETHANONE, 1-(4-CHLOROPHENYL)-2,2,2-TRIFLUORO-, o-(1,3-DIOXOLAN-2-YLMETHYL)OXIME see FMR600

ETHANONE, 2-DIAZO-1-PHENYL-(9CI) see PET800

ETHANONE, 2-((4-(DIBROMOACETYL)PHENYL)AMINO)-2-ETHOXY-1-(4-NITROPHENYL)- see DDJ850

ETHANONE, 2-((4-(DICHLOROACETYL)PHENYL)AMINO)-2-ETHOXY-1-(4-NITROPHENYL)- see DEN820

ETHANONE, 2-((4-(DICHLOROACETYL)PHENYL)AMINO)-2-HYDROXY-1-(4-METHOXYPHENYL)- see DEN840

ETHANONE, 2-((4-(DICHLOROACETYL)PHENYL)AMINO)-2-

HYDROXY-1-(4-METHYLPHENYL)- see DEN860

ETHANONE, 2-((4-(DICHLOROACETYL)PHENYL)AMINO)-2-HYDROXY-1-(4-PHENOXYPHENYL)- see DEN880

ETHANONE, 2-((4-(DICHLOROACETYL)PHENYL)AMINO)-2-HYDROXY-1-PHENYL- see DEN900

ETHANONE, 2-((4-(DICHLOROACETYL)PHENYL)AMINO)-2-HYDROXY-1-(4-(PHENYLTHIO)PHENYL)- see DEN910

ETHANONE, 2-((4-(2,4-DICHLORO-3-METHYLBENZOYL)-1,3-DIMETHYL-1H-PYRAZOL-5-YL)OXY)-1-(4-METHYLPHENYL)- see DGM730

ETHANONE, 1-(3,4-DICHLORO-5-NITRO-2-FURANYL)- see DFT053

ETHANONE, 1-(2,4-DICHLOROPHENYL)-2-(3-PYRIDINYL)-, o-METHYLOXIME see DGJ160

ETHANONE, 2,2-DIETHOXY-1-PHENYL- see PFH260

ETHANONE, 1-(3,4-DIHYDROXYPHENYL)-2-((1-METHYLETHYL)AMINO)-, HYDROCHLORIDE see DMV500

ETHANONE, 1-(4,7-DIMETHOXY-2-BENZOFURANYL)-(9CI) see DOA810

ETHANONE, 1-(6,7-DIMETHOXY-2-BENZOFURANYL)-(9CI) see DOA815

ETHANONE, 1-(5,8-DIMETHOXY-2H-1-BENZOPYRAN-3-YL)-(9CI) see DOA820

ETHANONE, 1-(7,8-DIMETHOXY-2H-1-BENZOPYRAN-3-YL)-(9CI) see DOA830

ETHANONE, 1-(4-(DIMETHYLAMINO)-2-METHYL-5-PHENYL-1H-PYRROL-3-YL)- see DPL950

ETHANONE, 1-(10-(2-(DIMETHYLAMINO)PROPYL)-10H-PHENOTHIAZIN-2-YL)-(9CI) see AAF800

ETHANONE, 1-(4-(3-((1,1-DIMETHYLETHYL)AMINO)-2-HYDROXYPROPOXY)-2-BENZOFURANYL)- see ACN300

ETHANONE, 1-(7-(3-((1,1-DIMETHYLETHYL)AMINO)-2-HYDROXYPROPOXY)-2-BENZOFURANYL)- see ACN320

ETHANONE,1-(4-(1,1-DIMETHYLETHYL)-2,6-DIMETHYL-3,5-DINITROPHENYL)- see ACE600

ETHANONE, 1-(2,4-DIMETHYL-5-PHENYL-1H-PYRROL-3-YL)- see DTO300

ETHANONE, 1-(2,5-DIMETHYL-3-THIENYL)- see DUG425

ETHANONE, 1-(2-ETHYL-7-(2-HYDROXY-3-((1-METHYLETHYL)AMINO)PROPOXY)-4-BENZOFURANYL)- see ELI600

ETHANONE, 1-(3-ETHYL-5,6,7,8-TETRAHYDRO-5,5,8,8-TETRAMETHYL-2-NAPHTHALENYL)(9CI) see ACL750

ETHANONE, 1-(9H-FLUOREN-2-YL) see FEI200

ETHANONE, 1-(2-FURANYL)-(9CI) see ACM200

ETHANONE, 2-HYDROXY-1,2-DIPHENYL- see BCP250

ETHANONE, 2-HYDROXY-1-(1-IODO-3-ADAMANTYL)- see HME100

ETHANONE, 1-(7-(2-HYDROXY-3-((1-METHYLETHYL)AMINO)PROPOXY)-2-BENZOFURANYL)-(9CI) see HLK600

ETHANONE, 1-(4-HYDROXY-2-METHYL-5-PHENYL-1H-PYRROL-3-YL)- see HMN100

ETHANONE, 1-(7-(2-HYDROXY-3-((1-METHYLPROPYL)AMINO)PROPOXY)-2-BENZOFURANYL)- see ACN310

ETHANONE, 1-(2-HYDROXYPHENYL)-(9CI) see HIN500

ETHANONE, 2-HYDROXY-1-(1-PHENYL-3-ADAMANTYL)- see HMK150

ETHANONE, 1-(10-(3-(4-HYDROXY-1-PIPERIDINYL)PROPYL)-10H-PHENOTHIAZIN-2-YL)- see HNT200

ETHANONE, 1-(8-HYDROXY-5-QUINOLINYL)-(9CI) see HOE200

ETHANONE, 1-(5-METHOXY-2-BENZOFURANYL)-(9CI) see MEC330

ETHANONE, 1-(6-METHOXY-2-BENZOFURANYL)-(9CI) see MEC300

ETHANONE, 1-(7-METHOXY-2-BENZOFURANYL)-(9CI) see MEC320

ETHANONE, 1-(5-METHOXY-2H-1-BENZOPYRAN-3-YL)-(9CI) see MEC340

ETHANONE, 1-(5-METHOXY-1H-INDOL-3-YL)-2-(4-PIPERIDINYL)-, MONOHYDROCHLORIDE (9CI) see MES900

ETHANONE, 1-(4-METHOXY-2-METHYL-5-PHENYL-1H-PYRROL-3-YL)- see MEX275

ETHANONE, 1-(7-METHOXYNAPHTHO(2,1-b)FURAN-2-YL)- see ACQ790

ETHANONE, 1-(4-METHOXYPHENYL)-(9CI) see MDW750

ETHANONE, 1-(10-(3-(4-METHOXY-1-PIPERIDINYL)PROPYL)-10H-PHENOTHIAZIN-2-YL)- see MFJ200

ETHANONE, 1-(4-METHYLPHENYL)-(9CI) see MFW250

ETHANONE, 1-(2-METHYL-5-PHENYL-1H-PYRROL-3-YL)- see MNY800

ETHANONE, 1-(5-METHYL-2-THIENYL)- see MPR300

ETHANONE, 1-(1-NAPHTHALENYL)-(9CI) see ABC475

ETHANONE, 1-(2-NITRONAPHTHO(2,1-b)FURAN-7-YL)- see NHR100

ETHANONE, 1-(4-NITROPHENYL)- see NEL600

ETHANONE, 1-(3-NITROPHENYL)-(9CI) see NEL500

ETHANONE, 1-(5-NITRO-2-THIENYL)-(9CI) see NML100

ETHANONE, 1-(4-NONYLPHENYL)-, OXIME see NND100

ETHANONE, 1-(3-PHENOXYPHENYL)-(9CI) see PDR200

ETHANONE, 1-PHENYL-, MONOCHLORO DERIV. see CDN505

ETHANONE, 1-PHENYL-, OXIME see ABH150

ETHANONE, 1-(2-PYRIDINYL)-(9CI) see PPH200

ETHANONE, 1-(5,6,7,8-TETRAHYDRO-3,5,5,6,8,8-HEXAMETHYL-2-NAPHTHALENYL)- see EEE100

ETHANONE, 1-(5,6,7,8-TETRAHYDRO-3,5,5,6,8,8-HEXAMETHYL-2-NAPHTHALENYL)- see FBW110

ETHANONE, 1-(3,7,7-TRIMETHYLBICYCLO(4.1.0)HEPTENYL)-(9CI) see ACF150

ETHANONE, 1-(1,3,5-TRINITRO-1H-PYRROL-2-YL)- see TMN100

ETHANONE, 1,2,2-TRIPHENYL- see TMS100

1H-2,10A-ETHANOPHENANTHRENE, KAURAN-17-AL DERIV. see OOS100

1,4-ETHANOPIPERIDINE see QUJ400

ETHANOX see EEH600

ETHANOX 330 see TMJ000

ETHANOX 701 see DEG100

ETHANOX 736 see TFD000

ETHANOXYTRIPHETOL see DHS000

ETHANOYL CHLORIDE see ACF750

12,3,6A-ETHANYLYLIDENE-9,11A-METHANOAZULENO(2,1-B)AZOCINE-6,8,11-TRIOL, TETRADECAHYDRO-1-ETHYL-3-METHYL-10-METHYLENE-, 6-BENZOATE, (3R-(3-α,6-β,6A-α,8-β,9-β,11A,11A-β,12A,12A-β,14R*))- see BDN150

8H-13,3,6A-ETHANYLYLIDENE-7,10-METHANOOXEPINO(3,4-I)-1-BENZAZOCIN-8-ONE,

TETRADECAHYDRO-14-(BENZOYLOXY)-1-ETHYL-12A-HYDROXY-6-METHOXY-3-METHYL-, (3R-(3-α,6-β,6A-α,7-β,7A-α,10-β,12A-α,13-α,13A-β,14S*,15R*))- see HBU410

ETHAPERAZINE see CJM250

ETHAPHENE see HLV500

ETHAPHOS see EEE200

ETHAVAN see EQF000

ETHAVERINE HYDROCHLORIDE see PAH260

ETHAZATE see BJC000

ETHAZOLE (FUNGICIDE) see EFK000

ETHBENZAMIDE see EEM000

ETHCHLOROVYNOL see CHG000

ETHCHLORVINYL see CHG000

ETHCLORVYNOL see CHG000

ETHEFON see CDS125

ETHEL see CDS125

ETHENE see EIO000

ETHENE, 1,1-DIBROMO- see VPF200

ETHENE, 1,2-DIBROMO-, (E)-(9CI) see DDN820

ETHENE, 1,1-DICHLORO-, POLYMER with CHLOROETHENE (9CI) see CGW300

ETHENE, 1,1-DICHLORO-, POLYMER WITH BUTYL 2-PROPENOATE see VPK333

ETHENE, 1,2-DIETHOXY- see DHG100

ETHENE, 1,1-DIFLUORO- see VPP000

2,3'-(1,2-ETHENEDIYL)BIS(5-AMINOBENZENESULFONIC ACID) DISODIUM SALT see FCA200

(E)-1,1'-(1,2-ETHENEDIYL)BISBENZENE see SLR100

(E)-1,1'-(1,2-ETHENEDIYL)BIS(4-NITROBENZENE) see EEE300

1,1'-(1,2-ETHENEDIYL)DIBENZENE see SLR000

ETHENE, ETHOXY-, POLYMERS see OJV600

ETHENE, FLUORO- see VPA000

ETHENE, NITRO-, HOMOPOLYMER see NFY560

ETHENE OXIDE see EJN500

ETHENE POLYMER see PJS750

ETHENE, 1,1'-SULFINYLBIS(2,2-DICHLORO- see SNT100

ETHENETETRACARBONITRILE see EEE500

ETHENE, TRIBROMO- see THV100

ETHENE, TRIFLUORO- see TKA400

ETHENE, TRIFLUORO(TRIFLUOROMETHOXY)- see TKG800

ETHENO DEOXYADENOSINE TRIPHOSPHATE see EEE550

1,N⁶)-ETHENODEOXYADENOSINE TRIPHOSPHATE see EEE550

ETHENOL HOMOPOLYMER (9CI) see PKP750

6,14-ETHENOMORPHINAN-7-METHANOL, 18,19-DIHYDRO, 4,5-EPOXY-3-HYDROXY-6-METHOXY-α,17-DIMETHYL-α-PROPYL-, HYDROCHLORIDE, (5α,7α(R))- see DLM700

6,14-ETHENOMORPHINAN-7-METHANOL, 3,6-DIMETHOXY-α-17-DIMETHYL-4,5-EPOXY-α-(2-PHENYL ETHYL)-, (5-α-7-α(R))- see EEE600

ETHENONE see KEU000

1,4-ETHENOPENTALENE, 1,2,3,5,7,8-HEXACHLORO-1,3A,4,5,6,6A-HEXAHYDRO-, (1-α-3A-α-4-β,5-α-6Aα)- see CDR820

6,14-endo-ETHENOTETRAHYDROORIPAVINE, 7-α-(1-HYDROXY-1-METHYLBUTYL)- see EQO450

1-ETHENOXY-2-ETHYLHEXANE see ELB500

ETHENYL ACETATE see VLU250

1-ETHENYLAZIRIDINE see VLZ000

α-ETHENYL-1-AZIRIDINEETHANOL (9CI) see VMA000

ETHENYLBENZENE see SMQ000
ETHENYLBENZENE HOMOPOLYMER see SMQ500
ETHENYLBENZENE POLYMER with 1,3-BUTADIENE see SMR000
4-ETHENYL-BENZENESULFONIC ACID SODIUM SALT, HOMOPOLYMER (9CI) see SJK375
ETHENYLBENZENE TRIBROMO DERIV. HOMOPOLYMER see EEE650
5-ETHENYLBICYCLO(2.2.1)HEPT-2-ENE see VQA000
N-ETHENYLCARBAZOLE see VNK100
9-ETHENYL-9H-CARBAZOLE see VNK100
1-ETHENYLCYCLOHEXENE see VNZ990
4-ETHENYL-1-CYCLOHEXENE see CPD750
α-ETHENYL-3,4-DIMETHOXYBENZENEMETHANOL see HMB700
ETHENYL ETHANOATE see VLU250
S-ETHENYL-l-HOMOCYSTEINE see VLU210
S-ETHENYL-dl-HOMOCYSTEINE see VLU200
6-ETHENYL-6-(METHOXYETHOXY)-2,5,7,10-TETRAOXA-6-SILAUNDECANE (9CI) see TNJ500
4-ETHENYL-2-METHOXYPHENOL see VPF100
1-ETHENYL-2-METHYLBENZENE see VQK660
1-ETHENYL-3-METHYLBENZENE see VQK670
1-ETHENYL-4-METHYLBENZENE see VQK700
5-ETHENYL-2-METHYLPYRIDINE see MQM500
1-(ETHENYLOXY) BUTANE see VMZ000
1-(ETHENYLOXY)DECANE see EEE700
2-(ETHENYLOXY)ETHANOL see EJL500
2-(ETHENYLOXY)ETHANOL NITRATE see EJK100
ETHENYLOXYETHENE see VOP000
1-(2-(ETHENYLOXY)ETHOXY)BUTANE see VMU000
2-(2-(ETHENYLOXY)ETHOXY)ETHANOL see DJG400
2-(ETHENYLOXY)ETHYL 2-METHYL-2-PROPENOATE see VQA150
4-ETHENYLPHENOL ACETATE see ABW550
ETHENYL 2-PROPENOATE see EEE800
1-ETHENYL PYRENE see EEF000
4-ETHENYL PYRENE see EEF500
2-ETHENYLPYRIDINE see VQK560
4-ETHENYLPYRIDINE see VQK590
ETHENYLPYRIDINE 1-OXIDE HOMOPOLYMER see PKQ100
1-ETHENYL-2-PYRROLIDINONE see EEG000
1-ETHENYL-2-PYRROLIDINONE HOMOPOLYMER see PKQ250
1-ETHENYL-2-PYRROLIDINONE HOMOPOLYMER compounded with IODINE see PKE250
1-ETHENYL-2-PYRROLIDINONE POLYMER with ETHENYLBENZENE see VQK595
1-ETHENYL-2-PYRROLIDINONE POLYMERS see PKQ250
1-ETHENYLSILATRANE see VQK600
ETHENYLTRIMETHOXYSILANE see TLD000
ETHENZAMID see EEM000
ETHENZAMIDE see EEM000
ETHEPHON see CDS125
ETHER see EJU000
ETHER, ALLYL GLYCERYL see AGP500
ETHER, ALLYL PENTABROMOPHENYL see PAU600
ETHER, BENZYL p-(o-FLUORO-α-PHENYLSTYRYL)PHENYL see BFC400

ETHER, 4-BIPHENYLYL 2,3-EPOXYPROPYL see PFU600
ETHER, BIS(4-CHLOROBUTYL) see OPE040
ETHER, BIS(2-CHLORO-1-METHYLETHYL), mixed with 2-CHLORO-1-METHYLETHYL-(2-CHLOROPROPYL)ETHER (7:3) see BIK100
ETHER, BIS(2-CHLORO-1-METHYLETHYL), mixed with 2-CHLORO-1-METHYLETHYL-(2-CHLOROPROPYL)ETHER (7:3) see EEG100
ETHER, BIS(2-CYANOETHYL) see OQQ000
ETHER, BIS(2-(2-ETHOXYETHOXY)ETHYL) see PBO250
ETHER, BIS((2-HEXYLOXY-2-ETHOXY)ETHYL) see HFT600
ETHER BIS-14-HYDROXY-IMINOMETHYLOPYRIDINE-(1)-METYLODICHLORIDE (POLISH) see BGS250
ETHER, BIS(2-(2-METHOXYETHOXY)ETHYL) see PBO500
ETHER-BIS(α-METHYLBENZYL) see MHN500
ETHER, BIS(p-NITROPHENYL) see NIM600
ETHER, BUTYL 2,3-EPOXYPROPYL see BRK750
ETHER, tert-BUTYL ETHYL see EHA550
ETHER, BUTYL GLYCIDYL see BRK750
ETHER BUTYLIQUE (FRENCH) see BRH750
ETHER, BUTYL METHYL see BRU780
ETHER, BUTYL PENTACHLOROPHENYL see BPM690
ETHER CHLORATUS see EHH000
ETHER, 4-CHLOROPHENYL (4'-CHLORO-2'-NITRO)PHENYL see CJD600
ETHER, 1-CHLORO-2,2,2-TRIFLUOROETHYL DIFLUOROMETHYL see IKS400
ETHER, 2-CHLORO-α-α-α-TRIFLUORO-p-TOLYL p-NITROPHENYL see NGB600
ETHER CYANATUS see PMV750
ETHER, DECYL METHYL see MIW075
ETHER, DECYL VINYL (6CI,7CI,8CI) see EEE700
ETHER DICHLORE (FRENCH) see DFJ050
ETHER, 2,4-DICHLORO-6-FLUOROPHENYL p-NITROPHENYL see FKM100
ETHER, (2,4-DICHLOROPHENYL) (3-METHOXY-4-NITROPHENYL) see DGA850
ETHER, DICYCLOPENTYL see CQB275
ETHER DIETHYLIQUE DE L'ETHYLENE-GLYCOL see DHG100
ETHER, DI-n-HEPTYL- see HBO000
ETHER, DI-n-PENTYL- see PBX000
ETHER, DI-n-PROPYL- see PNM000
ETHER, 2,3-EPITHIOPROPYL METHYL see EBD550
ETHER ETHYLBUTYLIQUE (FRENCH) see EHA500
ETHER ETHYLIQUE (FRENCH) see EJU000
ETHER, ETHYL PHENYL see PDM000
ETHER, ETHYL VINYL, POLYMERS see OJV600
ETHER, HEXACHLOROPHENYL see CDV175
ETHER HYDROCHLORIC see EHH000
ETHERIN see PJS750
ETHER ISOPROPYLIQUE (FRENCH) see IOZ750
ETHER METHYLIQUE MONOCHLORE (FRENCH) see CIO250
ETHER, METHYL tert-PENTYL (6CI,7CI,8CI) see PBX400
ETHER, METHYL PHENYL see AOX750
ETHER, METHYL PROPYL see MOU830
ETHER, 2-(METHYLTHIO)ETHYL VINYL see MQM200
ETHER MONOETHYLIQUE de l'ETHYLENE-GLYCOL (FRENCH) see EES350
ETHER MONOETHYLIQUE de l'HYDROQUINONE see EFA100

ETHER MONOMETHYLIQUE de l'ETHYLENE-GLYCOL (FRENCH) see EJH500
ETHER MURIATIC see EHH000
ETHER, p-NITROPHENYL VINYL see VPZ333
ETHEROL E see PJS750
ETHERON see PKL500
ETHERON SPONGE see PKL500
ETHEROPHENOL see IHX400
ETHER, PENTACHLOROPHENYL see PAW250
ETHER, PHENYL m-TOLYL- see MNV770
ETHERS see EEG500
ETHERSULFONATE see CJT750
ETHER, TRIFLUOROMETHYL TRIFLUOROVINYL see TKG800
ETHEVERSE see CDS125
ETHFAC 142W see PKE850
ETHIBI see EDW875
ETHICON PTFE see TAI250
ETHIDE see DFU000
ETHIDIUM BROMIDE see DBV400
ETHIDIUM CHLORIDE see HGI000
ETHIENOCARB see DFE469
ETHIMIDE see EPQ000
ETHINA see EPQ000
ETHINAMATE see EEH000
ETHINAMIDE see EPQ000
ETHINE see ACI750
ETHINODIOL see EQJ100
ETHINODIOL DIACETATE see EQJ500
ETHINONE see GEK500
17-α-ETHINYL-Δ⁵-ANDROSTENE-3-β,17-β-DIOL 3-CYCLOHEXYLPROPIONATE see EDY600
ETHINYLBENZENE see PEB750
2-ETHINYLBUTANOL-2 see EQL000
1-ETHINYLCYCLOHEXYL CARBAMATE see EEH000
1-ETHINYLCYCLOHEXYL CARBONATE see EEH000
17-α-ETHINYL-3,17-DIHYDROXY-Δ¹,³,⁵-ESTRATRIENE see EEH500
17-α-ETHINYL-3,17-DIHYDROXY-Δ¹,³,⁵-OESTRATRIENE see EEH500
ETHINYL ESTRADIOL see EEH500
17-ETHINYLESTRADIOL see EEH500
17-ETHINYL-3,17-ESTRADIOL see EEH500
17-α-ETHINYLESTRADIOL see EEH500
17-α-ETHINYL-17-β-ESTRADIOL see EEH500
17-α-ETHINYLESTRADIOL 3-CYCLOPENTYL ETHER see QFA250
ETHINYL ESTRADIOL and DIMETHISTERONE see DNX500
ETHINYLESTRADIOL and MEDROXYPROGESTERONE ACETATE see POF275
ETHINYLESTRADIOL-3-METHYL ETHER see MKB750
17-α-ETHINYL ESTRADIOL 3-METHYL ETHER see MKB750
ETHINYLESTRADIOL-3-METHYL ETHER and NORETHYNODRED (1:50) see EAP000
ETHINYL ESTRADIOL and NORETHINDRONE ACETATE see EEH520
ETHINYL ESTRADIOL mixed with NORGESTREL see NNL500
ETHINYLESTRADIOL mixed with dl-NORGESTREL see NNL500
17-ETHINYL-5(10)-ESTRAENEOLONE see EEH550
17-α-ETHINYL-ESTRA(5,10)ENEOLONE see EEH550
17-α-ETHINYLESTRA-4-EN-17-β-OL-3-ONE see NNP500
17-α-ETHINYLESTRA-1,3,5(10)-TRIENE-3,17-β-DIOL see EEH500
ETHINYLESTRENOL see NNV000
Δ⁴-17-α-ETHINYLESTREN-17-β-OL see NNV000

17-α-ETHINYL-5,10-ESTRENOLONE see EEH550

ETHINYLESTRIOL see EEH500

17-α-ETHINYL-13-β-ETHYL-17-β-HYDROXY-4-ESTREN-3-ONE see NNQ520

17-α-ETHINYL-17-β-HYDROXYESTR-4-ENE see NNV000

17-α-ETHINYL-17-β-HYDROXY-Δ:4-ESTREN-3-ONE see NNP500

17-α-ETHINYL-17-β-HYDROXY-Δ5(10)-ESTREN-3-ONE see EEH550

17-α-ETHINYL-17-β-HYDROXYOESTR-4-ENE see NNV000

ETHINYLMETHYLETHYLCARBINOL see EQL000

17-α-ETHINYL-19-NORTESTOSTERONE see NNP500

17-α-ETHINYL-Δ5,10-19-NORTESTOSTERONE see EEH550

17-α-ETHINYL-19-NORTESTOSTERONE ACETATE see ABU000

17-α-ETHINYL-19-NORTESTOSTERONE-17-β-ACETATE see ABU000

17-α-ETHINYL-19-NORTESTOSTERONE ENANTHATE see NNQ000

ETHINYLOESTRADIOL see EEH500

17-ETHINYL-3,17-OESTRADIOL see EEH500

ETHINYLOESTRADIOL mixed with LYNOESTRENOL see EEH575

ETHINYLOESTRADIOL-3-METHYL ETHER see MKB750

17-α-ETHINYL OESTRADIOL-3-METHYL ETHER see MKB750

ETHINYL OESTRADIOL mixed with NORETHISTERONE ACETATE see EEH520

ETHINYLOESTRADIOL mixed with NORGESTREL see NNL500

ETHINYL-OESTRANOL see EEH500

ETHINYLOESTRANOL see NNV000

17-α-ETHINYL-Δ1,3,5(10)OESTRATRIENE-3,17-β-DIOL see EEH500

17-α-ETHINYLOESTRA-1,3,5(10)-TRIENE-3,17-β-DIOL see EEH500

ETHINYL OESTRENOL see NNV000

Δ4-17-α-ETHINYLOESTREN-17-β-OL see NNV000

ETHINYLOESTRIOL see EEH500

ETHINYLTESTOSTERONE see GEK500

17-ETHINYLTESTOSTERONE see GEK500

ETHINYL TRICHLORIDE see TIO750

ETHIODAN see ELQ500

ETHIODIZED OIL (9CI) see EEH580

ETHIODOL see EEH580

ETHIOFENCARB see EPR000

ETHIOFOS see AMD000

ETHIOL see EEH600

ETHIOLACAR see MAK700

ETHION see EEH600

ETHION, dry see CMA100

ETHIONIAMIDE see EPQ000

ETHIONIN see EEI000

ETHIONINE see AKB250

ETHIONINE see EEI000

l-ETHIONINE see AKB250

(±)-ETHIONINE see EEI000

dl-ETHIONINE see EEI000

ETHIOPHENCARP see EPR000

ETHIOZIN see ENJ600

ETHIRIMOL see BRI750

ETHISTERONE see GEK500

ETHISTERONE and DIETHYLSTILBESTROL see EEI025

ETHLON see PAK000

ETHMOSINE see EEI050

ETHMOZINE see EEI050

ETHNINE see TCY750

ETHOATE METHYL see DNX600

ETHOBROM see ARW250

ETHOBROM see THV000

ETHOCAINE see AIT250

ETHOCEL see EHG100

ETHOCEL 150 see EHG100

ETHOCEL 890 see EHG100

ETHOCEL E7 see EHG100

ETHOCEL E50 see EHG100

ETHOCEL MED see EHG100

ETHOCEL N7 see EHG100

ETHOCEL N10 see EHG100

ETHOCEL N200 see EHG100

ETHOCEL STD see EHG100

ETHOCHLORVYNOL see CHG000

ETHODAN see EEH600

ETHODIN see EDW500

ETHODRYL see DIW000

ETHODRYL CITRATE see DIW200

ETHODUOMEEN see EEI060

ETHODUOMEEN, HYDROFLUORIDE see EEI100

ETHOFAT O see PJY100

ETHOFAT O 15 see PJY100

ETHOFENPROX see EQN725

ETHOGLUCID see TJQ333

ETHOGLUCIDE see TJQ333

ETHOHEPTAZINE CITRATE see MIE600

ETHOHEXADIOL see EKV000

ETHOL see HCP000

ETHOMEEN C/15 see EEJ000

ETHOMEEN S/12 see EEJ500

ETHOMEEN S/15 see EEK000

ETHOMEEN S/L5 see EEK025

ETHOMEEN T/15 see EEK050

ETHONE see ENY500

ETHONE, 2-CHLORO-1,2-DIPHENYL-(9CI) see CFJ100

ETHONIC 1214-2 see AFJ165

ETHONIC 1214-6.5 see AFJ165

ETHOPABATE see EEK100

ETHOPHENPROX see EQN725

ETHOPHYLLINE see TEP500

ETHOPIAN see EDW875

ETHOPROP see EIN000

ETHOPROPAZINE see DIR000

ETHOPROPHOS see EIN000

ETHOSALICYL see EEM000

ETHOSPERSE CL20 see PJT300

ETHOSPERSE LA-4 see DXY000

ETHOSUCCIMIDE see ENG500

ETHOSUCCINIMIDE see ENG500

ETHOSUXIDE see ENG500

ETHOSUXIMIDE see ENG500

ETHOTOIN see EOL100

ETHOVAN see EQF000

ETHOXAZENE see EEQ600

ETHOXAZOLAMIDE see EEN500

ETHOXENE see VMA000

ETHOXOL 20 see PJW500

2-ETHOXY-4-ACETAMIDOBENZOID ACID METHYL ESTER see EEK100

2-ETHOXYACETANILIDE see ABG350

3-ETHOXYACETANILIDE see ABG250

4-ETHOXYACETANILIDE see ABG750

m-ETHOXYACETANILIDE see ABG250

p-ETHOXYACETANILIDE see ABG750

2'-ETHOXYACETANILIDE see ABG350

3'-ETHOXYACETANILIDE see ABG250

ETHOXY ACETATE see EES400

ETHOXYACETIC ACID see EEK500

2-ETHOXYACETIC ACID see EEK500

4-ETHOXYACETOACETANILIDE see AAZ000

4'-ETHOXYACETOACETANILIDE see AAZ000

N-(ETHOXYACETYL)DEACETYLTHIOCOLCHICINE see EEK600

ETHOXY ACETYLENE see EEL000

3-ETHOXY ACROLEIN DIETHYL ACETAL see TJM500

4-ETHOXYAMPHETAMINE HYDROCHLORIDE see EEL050

ETHOXYANILINE see EEL100

2-ETHOXYANILINE see PDK819

3-ETHOXYANILINE see PDK800

4-ETHOXYANILINE see PDK790

m-ETHOXYANILINE see PDK800

o-ETHOXYANILINE see PDK819

p-ETHOXYANILINE see PDK790

4-ETHOXYANILINE HYDROCHLORIDE see PDL750

6-ETHOXY-m-ANOL see IRY000

ETHOXYBENZALDEHYDE see EEL500

4-ETHOXYBENZALDEHYDE see EEL500

p-ETHOXYBENZALDEHYDE see EEL500

2-ETHOXYBENZAMIDE see EEM000

o-ETHOXYBENZAMIDE see EEM000

3-ETHOXYBENZENAMINE see PDK800

4-ETHOXYBENZENAMINE see PDK790

4-ETHOXYBENZENAMINE HYDROCHLORIDE see PDL750

ETHOXYBENZENE see PDM000

o-ETHOXYBENZOIC ACID (1-CARBOXYETHYLIDENE)HYDRAZIDE see RSU450

o-ETHOXY-BENZOIL-IDRAZONE DELL'ACIDO PIRUVICO (ITALIAN) see RSU450

6-ETHOXY-2-BENZOTHIAZOLESULFONAMIDE see EEN500

o-ETHOXY-BENZOYL-HYDRAZONE of PYRUVIC ACID see RSU450

4-(p-ETHOXYBENZOYL)PYRIDINE see EEN600

1-(4-ETHOXY-2-BUTYNYL)PYRROLIDINE see EEN700

1-(4-ETHOXYBUT-2-YNYL)-PYRROLIDINE see EEN700

8-ETHOXYCAFFEINE see EEO000

ETHOXYCARBONYCYCLOHEXANE see CPB075

3-(ETHOXYCARBONYLAMINO)PHENOL see ELE600

m-(ETHOXYCARBONYLAMINO)PHENOL see ELE600

3-ETHOXYCARBONYLAMINOPHENYL-N-PHENYLCARBAMATE see EEO500

N-(ETHOXYCARBONYL)AZIRIDINE see ASH750

S-α-ETHOXYCARBONYLBENZYL-O,O-DIMETHYL PHOSPHORODITHIOATE see DRR400

S-α-ETHOXYCARBONYLBENZYL DIMETHYL PHOSPHOROTHIOLOTHIONATE see DRR400

ETHOXYCARBONYLDIAZOMETHANE see DCN800

ETHOXY CARBONYL DIGOXIN see EEP000

ETHOXYCARBONYLETHYLENE see EFT000

N-ETHOXYCARBONYLETHYLENEIMINE see ASH750

ETHOXYCARBONYL-1-ETHYLENIMINE see ASH750

ETHOXYCARBONYL HYDRAZIDE see EHG000

(ETHOXYCARBONYL) HYDRAZINE see EHG000

1-(ETHOXYCARBONYL) HYDRAZINE see EHG000

N-(ETHOXYCARBONYL) HYDRAZINE see EHG000

7-ETHOXYCARBONYL-4-HYDROXYMETHYL-6,8-DIMETHYL-1(2H)-PHTHALAZINONE see PHV725

4-ETHOXYCARBONYL-1-(2-HYDROXY-3-PHENOXYPROPYL) 4-PHENYLPIPERIDINE HYDROCHLORIDE see EEQ000

ETHOXYCARBONYLMETHYL BROMIDE see EGV000

S-(N-ETHOXYCARBONYL-N-METHYLCARBAMOYLMETHYL)-DIETHYL PHOSPHORODITHIOATE see DJI000

S-((ETHOXYCARBONYL)METHYLCARBAMOYL)METHYL-O,O-DIETHYL PHOSPHORODITHIOATE see DJI000

N-ETHOXYCARBONYL-N-METHYLCARBAMOYLMETHYL-O,O-DIETHYL PHOSPHORODITHIOATE see DJI000

(2-(ETHOXYCARBONYL)-1-METHYL)ETHYL CARBONIC ACID-p-IODOBENZYL ESTER see EEQ100

ETHOXYCARBONYLMETHYL ISOCYANATE see ELS600

2-ETHOXYCARBONYL-1-METHYLVINYL DIETHYL PHOSPHATE see CBR750

N-(ETHOXYCARBONYL)-3-(4-MORPHOLINYL)SYDNONE IMINE see MRN275

o-(4-(1-(((ETHOXYCARBONYL)OXY)IMINO)ETHYL)PHENYL) o,o-DIETHYL PHOSPHOROTHIOATE see EEQ200

N-(N-((S)-1-ETHOXYCARBONYL-3-PHENYLPROPYL)-l-ALANYL)-N-(INDAN-2-YL)GLYCINE HYDROCHLORIDE see DAM315

N-((S)-1-ETHOXYCARBONYL-3-PHENYLPROPYL)-l-ALANYL-l-PROLINE MALEATE see EAO100

(S)-1-(N-(1-(ETHOXYCARBONYL)-3-PHENYLPROPYL)-l-ALANYL)-l-PROLINE MALEATE see EAO100

S-ETHOXYCARBONYLTHIAMINE HYDROCHLORIDE see EEQ500

ETHOXY CHLOROMETHANE see CIM000

p-ETHOXYCHRYSOIDINE see EEQ600

ETHOXY CLEVE'S ACID see AJU500

2-ETHOXY-6,9-DIAMINOACRIDINE LACTATE see EDW500

2-ETHOXY-6,9-DIAMINOACRIDINE LACTATE HYDRATE see EDW500

2-ETHOXY-6,9-DIAMINOACRIDINIUM LACTATE see EDW500

ETHOXY DIETHYL ALUMINUM see EER000

ETHOXY DIGLYCOL see CBR000

11-ETHOXY-15,16-DIHYDRO-17-CYCLOPENTA(a)PHENANTHREN-17-ONE see EER400

2-ETHOXY DIHYDROPYRAN see EER500

2-ETHOXY-3,4-DIHYDRO-1,2-PYRAN see EER500

2-ETHOXY-3,4-DIHYDRO-2H-PYRAN see EER500

2-ETHOXY-2,3-DIHYDRO-γ-PYRAN see EER500

6-ETHOXY-1,2-DIHYDRO-2,2,4-TRIMETHYLQUINOLINE see SAV000

ETHOXYDIISOBUTYLALUMINUM see EES000

3'-ETHOXY-4-DIMETHYLAMINOAZOBENZENE see DRU000

1-ETHOXY-3,7-DIMETHYL-2,6-OCTADIENE see GDG100

8-ETHOXY-2,6-DIMETHYLOCTENE-2 see EES100

ETHOXYDIMETHYLSILANE see EES200

ETHOXYDIPHENYLACETIC ACID see EES300

ETHOXYETHANE see EJU000

2-ETHOXYETHANOL see EES350

2-ETHOXYETHANOL ACETATE see EES400

2-ETHOXYETHANOL, ESTER with ACETIC ACID see EES400

ETHOXY ETHENE see EQF500

4-(1-ETHOXYETHENYL)-3,3,5,5-TETRAMETHYLCYCLOHEXANONE see EES370

1-(2-ETHOXYETHOXY)-BUTANE see BPK750

2-(2-ETHOXYETHOXY)ETHANOL see CBR000

2-(2-ETHOXYETHOXY)ETHANOL ACETATE see CBQ750

1-ETHOXY-2-(β-ETHOXYETHOXY)ETHANE see DIW800

2-(2-(2-ETHOXYETHOXY)ETHOXY)ETHANOL see EFL000

5-(1-(2-(2-ETHOXYETHOXY)ETHOXY)ETHOXY)-1,3-BENZODIOXOLE see MJS550

ETHOXYETHOXYETHYL ACRYLATE see EHG030

(2-(1-ETHOXYETHOXY)ETHYL)BENZENE see EES380

2-(2-ETHOXYETHOXY)ETHYL-3,4-(METHYLENEDIOXY)PHENYLACETAL OF ACETALDEHYDE see MJS550

2-(2-ETHOXYETHOXY)ETHYL 2-PROPENOATE see EHG030

3-(2-ETHOXY)ETHOXY-2-PROPANOL see EET000

2-ETHOXY-ETHYLACETAAT (DUTCH) see EES400

ETHOXYETHYL ACETATE see EES400

2-ETHOXYETHYL ACETATE see EES400

β-ETHOXYETHYL ACETATE see EES400

ETHOXYETHYL ACRYLATE see ADT500

2-ETHOXYETHYL ACRYLATE see ADT500

2-(2-ETHOXYETHYLAMINO)-2',6'-ACETOXYLIDIDE see DRT400

2-(2-ETHOXYETHYLAMINO)-2',6'-ACETOXYLIDIDE HYDROCHLORIDE see DRT600

2-(2-ETHOXYETHYLAMINO)-2'-METHYL-PROPIONANILIDE see EES500

2-(2-ETHOXYETHYLAMINO)-o-PROPIONOTOLUIDIDE see EES500

N-(2-ETHOXYETHYL)-4,5-DIHYDRO-3-(PHENYLAMINO)-2H-BENZ(g)INDAZOLE-2-ACETAMIDE see EES600

2-ETHOXYETHYLE, ACETATE de (FRENCH) see EES400

1,1'-((1-ETHOXYETHYL)ETHANEDIYLIDENE)BIS(3-THIOSEMICARBAZIDE) see KFA100

ETHOXYETHYL ETHER of PROPYLENE GLYCOL see EET000

(1-ETHOXYETHYL)GLYOXAL BIS(THIOSEMICARBAZONE) see KFA100

(1-ETHOXYETHYLIDENE)MALONONITRILE see EET100

1-ETHOXYETHYLIDENEMALONONITRILE see EET100

(1-ETHOXYETHYLIDENE)PROPANEDINITRILE see EET100

2-ETHOXYETHYL-2-METHOXYETHYLETHER see DJE800

1-(2-ETHOXYETHYL)-2-(4-METHYLHEXAHYDRO-1H-1,4-DIAZEPIN-1-YL)BENZIMIDAZOLE FUMARATE (1:2) see EAL050

1-(2-ETHOXYETHYL)-2-(4-METHYL-1-HOMOPIPERAZINYL)BENZIMIDAZOLE DIFUMARATE see EAL050

3-((1-(2-ETHOXYETHYL)-5-NITRO-1H-IMIDAZOL-2-YL)METHYLENE)-1-METHYL-2-PYRROLIDINONE see EET200

2-ETHOXYETHYL-2-PROPENOATE see ADT500

2-ETHOXYETHYL-2-(VINYLOXY)ETHYL ETHER see DJF000

ETHOXYETHYNE see EEL000

ETHOXYFORMIC ANHYDRIDE see DIX200

3-ETHOXYHEXANAL DIETHYL ACETAL see TJM000

(Z)-1-ETHOXY-1-(3-HEXENYLOXY)ETHANE see EKS100

1-ETHOXY-1-HEXYLOXYETHANE see EKS120

3-ETHOXY-4-HYDROXYBENZALDEHYDE see EQF000

4-ETHOXY-3-HYDROXYBENZALDEHYDE see IKO100

4-ETHOXY-N-HYDROXY-BENZENAMINE see HNG100

4'-ETHOXY-3-HYDROXYBUTYRANILIDE see HJS850

N-(2-ETHOXY-3-HYDROXYMERCURIPROPYL)BARBITAL see EET500

1-ETHOXY-2-HYDROXY-4-PROPENYLBENZENE see IRY000

2-[1-(ETHOXYIMINO)BUTYL]-5-[2-(ETHYLTHIO)PROPYL]-3-HYDROXY-2-CYCLOHEXENE-1-ONE see EET550

2-(1-(ETHOXYIMINO)BUTYL)-5-(2-(ETHYLTHIO)PROPYL)-3-HYDROXY-2-CYCLOHEXEN-1-ONE see CDK800

2-(1-(ETHOXYIMINO)PROPYL)-3-HYDROXY-5-MESITYLCYCLOHEX-2-ENONE see GJU200

2-(1-(ETHOXYIMINO)PROPYL)-3-HYDROXY-5-(2,4,6-TRIMETHYLPHENYL)-2-CYCLOHEXEN-1-ONE see GJU200

2-(3-ETHOXY-1-INDANYLIDENE)-1,3-DINDANDIONE see BGC250

1-ETHOXY-3-ISOPROPOXYPROPAN-2-OL see EET600

1-ETHOXY-3-ISOPROPOXY-2-PROPANOL see EET600

ETHOXYKARBONYLMETHYL-METHYLESTER KYSELINY FTALOVE (CZECH) see MOD000

1-ETHOXY-4-(1-KETO-2-HYDROXYETHYL)-NAPHTHALENE SUCCINATE see SNB500

ETHOXYLATED FATTY ALCOHOLS (C16-18) see AFJ170

ETHOXYLATED LANOLIN see LAU560

ETHOXYLATED LAURYL ALCOHOL see DXY000

ETHOXYLATED LAURYL ALCOHOL see EEU000

ETHOXYLATED MONO- and DIGLYCERIDES see EEU100

ETHOXYLATED OCTYL PHENOL see GHS000

ETHOXYLATED PROPOXYLATED C8-10 ALCS. see AFJ172

ETHOXYLATED SORBITAN MONOOLEATE see PKL100

1-ETHOXY-1-LINALYLOXYETHANE see ELZ050

6-ETHOXY-2-MERCAPTOBENZOTHIAZOLE see EEU500

ETHOXYMETHANE see EMT000

(ETHOXYMETHOXY)CYCLODODECANE see FMV100

7-(ETHOXYMETHYL)BENZ(a)ANTHRACENE see EEV000

10-ETHOXYMETHYL-1:2-BENZANTHRACENE see EEV000

7-ETHOXY-12-METHYLBENZ(a)ANTHRACENE see MJX000

4-ETHOXY-α-METHYLBENZENEETHANAMINE HYDROCHLORIDE see EEL050

4-ETHOXY-2-METHYL-3-BUTYN-2-OL see EEV100

ETHOXY METHYL CHLORIDE see CIM000

ETHOXYMETHYL-N-(2,6-DICHLORO-m-TOLYL)ANTHRANILATE see DGN000

2-ETHOXY-4-METHYL-3,4-DIHYDROPYRAN see EEV150

2-ETHOXY-4-METHYL-2,3-DIHYDRO-4H-PYRAN see EEV150

2-(ETHOXYMETHYL)-1,3-DIHYDROXY-9,10-ANTHRACENEDIONE see DMT200

ETHOXYMETHYLENEMALONIC ACID, ETHYL ESTER see EEV200

ETHOXYMETHYLENE MALONONITRILE see EEW000

1-(2-ETHOXY-2-METHYLETHOXY)-2-PROPANOL see EEW100

2-((ETHOXY((1-METHYLETHYL)AMINO)PHOSPHINOTHIOYL)OXY)BENZOIC ACID 1-METHYLETHYL ESTER see IMF300

1-(ETHOXYMETHYL)-2-METHOXYBENZENE see EMF600

7-ETHOXY METHYL-12-METHYL BENZ(a)ANTHRACENE see EEX000

1-(ETHOXYMETHYL)-5-(1-METHYLETHYL)-6-(PHENYLMETHYL)-2,4(1H,3H)-PYRIMIDINEDIONE see EAN525

2-ETHOXY-N-METHYL-N-(2-(METHYLPHENETHYLAMINO)ETHYL)-2,2-DIPHENYLACETAMIDE HYDROCHLORIDE see CBQ625

4-ETHOXY-2-METHYL-5-MORPHOLINO-3(2H)-PYRIDAZINONE see EAN700

4-ETHOXY-2-METHYL-5-(4-MORPHOLINYL)-3(2H)-PYRIDAZINONE see EAN700

(ETHOXYMETHYL)OXIRANE see EBQ700

(ETHOXYMETHYL)OXIRANE see EKM200

2-ETHOXY-2-METHYLPROPANE see EHA550

2-ETHOXY-4-METHYL-TETRAHYDROPYRAN see EEX100

N-ETHOXYMORPHOLINO DIAZENIUM FLUOROBORATE see EEX500

O-ETHOXY-N-(3-MORPHOLINOPROPYL)BENZAMIDE see EEY000

2-ETHOXYNAPHTHALENE see EEY500

6-(2-ETHOXY-1-NAPHTHAMIDO)PENICILLIN SODIUM see SGS500

2-ETHOXY-5-NITROANILINE see NIC200

1-ETHOXY-2-NITROBENZENE see NIC990

1-ETHOXY-4-NITROBENZENE see NID000

ETHOXY-4-NITROPHENOXYPHENYLPHOSPHINE SULFIDE see EBD700

N-(4-ETHOXY-3-NITRO)PHENYLACETAMIDE see NEL000

ETHOXY-4-NITROPHENYLOXYETHYLPHOSPHINEOXIDE see ENQ000

p-ETHOXYNITROSOBENZENE see EEY550

3-ETHOXY-2-OXOBUTYRALDEHYDE BIS(THIOSEMICARBAZONE) see KFA100

1-ETHOXY-2,2,3,3,3-PENTAFLUORO-1-PROPANOL see EFA000

ETHOXYPHAS see DIX000

3-(4-(β-ETHOXYPHENETHYL)-1-PIPERAZINYL)-2-METHYL-1-PHENYL-1-PROPANONE DIHYDROCHLORIDE see ECU550

3-(4-(β-ETHOXYPHENETHYL)-1-PIPERAZINYL)-2-METHYLPROPIOPHENONE see PFB000

2-(4-(β-ETHOXYPHENETHYL)-1-PIPERAZINYLMETHYL)PROPIOPHENONE DIHYDROCHLORIDE see ECU550

4-ETHOXYPHENOL see EFA100

p-ETHOXYPHENOL see EFA100

2-((2-ETHOXYPHENOXY)METHYL)MORPHOLINE see VKA875

2-((2-ETHOXYPHENOXY)METHYL)MORPHOLINE HYDROCHLORIDE see VKF000

2-((o-ETHOXYPHENOXY)METHYL)MORPHOLINE HYDROCHLORIDE see VKF000

2-(2-ETHOXYPHENOXYMETHYL)TETRAHYDRO-1,4-OXAZINE see VKA875

2-(2-ETHOXYPHENOXYMETHYL)TETRAHYDRO-1,4-OXAZINE HYDROCHLORIDE see VKF000

N-(2-ETHOXYPHENYL)ACETAMIDE see ABG350

N-(4-ETHOXYPHENYL)ACETAMIDE see ABG750

N-p-ETHOXYPHENYLACETAMIDE see ABG750

N-(3-ETHOXYPHENYL)ACETAMIDE (9CI) see ABG250

N-(4-ETHOXYPHENYL)ACETOHYDROXAMIC ACID see HIN000

4-(p-ETHOXYPHENYLAZO)-m-PHENYLENEDIAMINE see EEQ600

α-ETHOXY-α-PHENYL-BENZENEACETIC ACID (9CI) see EES300

2-(4-ETHOXYPHENYL)-1,3-BIS(4-METHOXYPHENYL)GUANIDINE HYDROCHLORIDE see PDN500

S-2-((4-(p-ETHOXYPHENYL)BUTYL)AMINO)ETHYL THIOSULFATE see EFC000

1-(p-ETHOXYPHENYL)-1-DIETHYLAMINO-3-METHYL-3-PHENYLPROPANE HYDROCHLORIDE see PEF500

2-(3-ETHOXYPHENYL)-5,6-DIHYDRO-s-TRIAZOLO(5,1-a)ISOQUINOLINE see EFC259

4-ETHOXY-7-PHENYL-3,5-DIOXA-6-AZA-4-PHOSPHAOCT-6-ENE-8-NITRILE-4-SULFIDE see BAT750

4-ETHOXY-m-PHENYLENEDIAMINE see EFC300

5-(m-ETHOXYPHENYL)-3-(o-ETHYLPHENYL)-s-TRIAZOLE see EFC600

3-(4-(2-ETHOXY-2-PHENYLETHYL)-1-PIPERAZINYL)-2-METHYL-1-PHENYL-1-PROPANONE see PFB000

3-(4-(2-ETHOXY-2-PHENYLETHYL)-1-PIPERAZINYL)-2-METHYL-1-PHENYL-1-PROPANONE DIHYDROCHLORIDE see ECU550

N-(4-ETHOXYPHENYL)-N-HYDROXYACETAMIDE see HIN000

(p-ETHOXYPHENYL)HYDROXYLAMINE see HNG100

N-(p-ETHOXYPHENYL)HYDROXYLAMINE see HNG100

2-(m-ETHOXYPHENYL)IMIDAZO(2,1-A)ISOQUINOLINE see EFC625

2-(2-(4-((4-ETHOXYPHENYL)METHYLAMINO)PHENYL)ETHENYL)-1,3,3-TRIMETHYL-3H-INDOLIUM CHLORIDE see EFC650

1-((2-(4-ETHOXYPHENYL)-2-METHYLPROPOXY)METHYL)-3-PHENOXYBENZENE see EQN725

3-(4-ETHOXYPHENYL)-2-METHYL-4(3H)-QUINAZOLINONE see LID000

N-(4-ETHOXYPHENYL)-3'-NITROACETAMIDE see NEL000

p-ETHOXYPHENYL 2-PYRIDYLKETONE see EFC700

p-ETHOXYPHENYL 3-PYRIDYLKETONE see EFC800

p-ETHOXYPHENYL 4-PYRIDYL KETONE see EFC600

4-ETHOXYPHENYLUREA see EFE000

p-ETHOXYPHENYLUREA see EFE000

N-(4-ETHOXYPHENYL)UREA see EFE000

ETHOXYPHOS see DIX000

4-ETHOXY-β-(1-PIPERIDYL)PROPIOPHENONE HYDROCHLORIDE see EFE500

1-ETHOXYPROPANE see EPC125

3-ETHOXY-1,2-PROPANEDIOL see EFF000

ETHOXYPROPANOL see EJV000

1-ETHOXY-2-PROPANOL see EFF500

3-ETHOXY-1-PROPANOL see EFG000

ETHOXY PROPIONALDEHYDE see EFG500

ETHOXYPROPIONIC ACID see EFH000

ETHOXYPROPIONIC ACID, ETHYL ESTER see EJV500

3-ETHOXYPROPIONIC ACID, ETHYL ESTER see EJV500

β-ETHOXYPROPIONITRILE see EFH500

ETHOXYPROPYLACRYLATE see EFI000

ETHOXYPROPYL ESTER ACRYLIC ACID see EFI000

1-ETHOXY-2-PROPYNE see EFJ000

N¹-(6-ETHOXY-3-PYRIDAZINYL)SULFANILAMIDE see SNJ100

4'-ETHOXY-2-PYRROLIDINYLACETANILIDE HYDROCHLORIDE see PPU750

ETHOXYQUINE see SAV000

ETHOXYQUIN (FCC) see SAV000

2-ETHOXY-1(2H)-QUINOLINECARBOXYLIC ACID, ETHYL ESTER see EFJ500

3-ETHOXYSALICYLALDEHYDE see NOC100

ETHOXYSILATRANE see EFJ600

1-ETHOXYSILATRANE see EFJ600

6-ETHOXY-3-SULFANILAMIDOPYRIDAZINE see SNJ100

5-ETHOXY-3-TRICHLOROMETHYL-1,2,4-THIADIAZOLE see EFK000

ETHOXYTRIETHYLENE GLYCOL see EFL000

1-ETHOXY-2,2,2-TRIFLUOROETHANOL see EFK500

ETHOXYTRIGLYCOL see EFL000

6-ETHOXY-2,2,4-TRIMETHYL-1,2-DIHYDROQUINOLINE see SAV000

ETHOXYTRIMETHYLSILANE see EFL500

3-ETHOXY-1,1,1-TRIPHENYL-4-OXA-2-THIA-3-PHOSPHA-1-STANNAHEXANE 3-SULFIDE see EFL600

ETHOXYZOLAMIDE see EEN500

ETHRANE see EAT900

ETHREL see CDS125

ETHRIL see EDJ500

ETHRIOL see HDF300

ETHYBENZTROPINE see DWE800

ETHYCHLOZATE see EHK600

ETHYL 702 see MJM700

ETHYL 703 see DEA100

ETHYL 703 see FAB000

ETHYL 736 see TFD000

ETHYLAC see BDF250

ETHYLACETAAT (DUTCH) see EFR000

ETHYLACETAMIDE see EFM000

N-ETHYLACETAMIDE see EFM000

d-N-ETHYLACETAMIDE CARBANILATE see CBL500

2-(2-(3-(N-ETHYLACETAMIDO)-2,4,6-TRIIODOPHENOXY)ETHOXY)ACETIC ACID SODIUM SALT see EFM500

2-(2-(3-(N-ETHYLACETAMIDO)-2,4,6-TRIIODOPHENOXY)ETHOXY)-2-PHENYL ACETIC ACID SODIUM SALT see EFN500

2-(2-(3-(N-ETHYLACETAMIDO)-2,4,6-TRIIODOPHENOXY)ETHOXY)PROPIONIC ACID SODIUM SALT see EFO000

2-(3-(N-ETHYLACETAMIDO)-2,4,6-TRIIODOPHENYL)BUTYRIC ACID see EFP000

ETHYLACETANILIDE see EFQ500

N-ETHYLACETANILIDE see EFQ500

ETHYL ACETATE see EFR000

ETHYLACETIC ACID see BSW000

ETHYL ACETIC ESTER see EFR000

ETHYLARSONOUS DICHLORIDE see DFH200
ETHYL AURAMINE NITRATE see EGM200
ETHYL AZIDE see EGM500
ETHYL AZIDOFORMATE see EGN000
ETHYL-2-AZIDO-2-PROPENOATE see EGN100
2-ETHYLAZIRIDINE see EGN500
ETHYL AZIRIDINECARBOXYLATE see ASH750
ETHYL-1-AZIRIDINECARBOXYLATE see ASH750
ETHYL AZIRIDINOCARBOXYLATE see ASH750
ETHYL-1-AZIRIDINYLCARBOXYLATE see ASH750
ETHYL AZIRIDINYLFORMATE see ASH750
ETHYLBARBITAL see BAG000
ETHYLBENATROPINE see DWE800
7-ETHYLBENZ(c)ACRIDINE see BBM500
9-ETHYL-3,4-BENZACRIDINE see BBM500
ETHYL p-BENZAMIDOBENZOATE see EGN600
7-ETHYLBENZ(a)ANTHRACENE see EGO500
8-ETHYLBENZ(a)ANTHRACENE see EGO000
12-ETHYLBENZ(a)ANTHRACENE see EGP000
5-ETHYL-1,2-BENZANTHRACENE see EGO000
10-ETHYL-1,2-BENZANTHRACENE see EGO500
ETHYLBENZEEN (DUTCH) see EGP500
2-ETHYLBENZENAMINE see EGK500
3-ETHYLBENZENAMINE see EOH100
N-ETHYLBENZENAMINE see EGK000
N-ETHYLBENZENAMINO see EGK000
ETHYL BENZENE see EGP500
ETHYL BENZENEACETATE see EOH000
α-ETHYLBENZENEACETIC ACID-2-((2-DIETHYLAMINO)ETHOXY)ETHYL ESTER CITRATE see BOR350
α-ETHYLBENZENEMETHANOL see EGQ000
N-ETHYLBENZENESULFONAMIDE see EGQ100
ETHYL BENZOATE see EGR000
1-(7-ETHYLBENZOFURAN-2-YL)-2-tert-BUTYLAMINO-1-HYDROXYETHANE HYDROCHLORIDE see BQD000
2-ETHYL-3-BENZOFURANYL p-HYDROXYPHENYL KETONE see BBJ500
ETHYLBENZOL see EGP500
5-ETHYLBENZO(c)PHENANTHRENE see EGS500
2-ETHYLBENZOXAZOLE see EGR500
ETHYL BENZOYL ACETATE see EGR600
ETHYL-o-BENZOYL-3-CHLORO-2,6-DIMETHOXY-BENZOHYDROXIMATE see BCP000
ETHYL-N-BENZOYL-N-(3,4-DICHLOROPHENYL)-2-AMINOPROPIONATE see EGS000
(+-)-ETHYL N-BENZOYL-N-(3,4-DICHLOROPHENYL)-2-AMINOPROPIONATE see EGS000
2-ETHYL-3:4-BENZPHENANTHRENE see EGS500
ETHYLBENZTROPINE see DWE800
ETHYL BENZYL ACETOACETATE see EFS000
α-ETHYLBENZYL ALCOHOL see EGQ000
ETHYLBENZYLBARBITURIC ACID see BEA500
ETHYL-N-BENZYLCYCLOPROPANECARBAMATE see EGT000
ETHYL-N-BENZYL-N-CYCLOPROPYLCARBAMATE see EGT000

ETHYL-1-(2-BENZYLOXYETHYL)-4-PHENYLPIPERIDINE-4-CARBOXYLATE see BBU625
1,2-ETHYL BIS-AMMONIUM PERCHLORATE see EGT500
ETHYL (BIS(1-AZIRIDINYL)PHOSPHINYL)CARBAMATE see EHV500
ETHYLBIS(2-CHLOROETHYL)AMINE see BID250
ETHYLBIS(β-CHLOROETHYL)AMINE see BID250
ETHYLBIS(β-CHLOROETHYL)AMINE HYDROCHLORIDE see EGU000
ETHYL BISCOUMACETATE see BKA000
2-ETHYL-1,3-BIS(DISMETHYLAMINO)-2-PROPANOL BENZOATE see AHI250
ETHYL BIS(4-HYDROXYCOUMARINYL)ACETATE see BKA000
ETHYL BIS(4-HYDROXY-3-COUMARINYL)ACETATE see BKA000
ETHYLBIS(2-HYDROXYETHYL)AMINE see ELP000
S-ETHYL BIS(2-METHYLPROPYL)CARBAMOTHIOATE see EID500
O-ETHYL-S,S-BIS(1-METHYLPROPYL)PHOSPHORODITHIOATE see EHY100
ETHYL BORATE (DOT) see BMC250
ETHYL BROMACETATE see EGV000
ETHYL BROMIDE see EGV400
ETHYL BROMOACETATE see EGV000
ETHYL-α-BROMOACETATE see EGV000
N-ETHYL-N-o-BROMOBENZYL-N,N-DIMETHYLAMMONIUM TOSYLATE see BMV750
ETHYL α-BROMO-α-CYANOACETATE see EGV450
ETHYL BROMOPHOS see EGV500
ETHYL (2-BROMOPROPIONAMIDO)ACETATE see EGV550
2-ETHYLBUTADIENE see EGV600
2-ETHYL-1,3-BUTADIENE see EGV600
2-ETHYLBUTANAL see DHI000
2-ETHYL-1-BUTANAMINE see EHA000
ETHYL BUTANOATE see EHE000
2-ETHYL BUTANOIC ACID see DHI400
2-ETHYLBUTANOL see EGW000
2-ETHYLBUTANOL-1 see EGW000
2-ETHYL-1-BUTANOL see EGW000
2-ETHYL-1-BUTANOL, SILICATE see TCD250
2-ETHYL-2-BUTENAL see EHO000
2-ETHYL-1-BUTENE see EGW500
2-ETHYL-1-BUTENE-1-ONE see DJN700
ETHYL (E)-2-BUTENOATE see COB750
3-ETHYLBUTINOL see EQL000
2-(2-ETHYLBUTOXY)ETHANOL see EGX000
N-(2-ETHYLBUTOXYETHOXYPROPYL)BICYCLO(2.2.1)HEPTENE-2,3-DICARBOXIMIDE see EGX500
N-(2-ETHYLBUTOXYETHOXYPROPYL)-5-NORBORNENE-2,3-DICARBOXIMIDE see EGX500
2-ETHYLBUTOXYPROPANOL, MIXED ISOMERS see EGX600
3-(2-ETHYLBUTOXY)PROPIONIC ACID see EGY000
3-(2-ETHYLBUTOXY)PROPIONITRILE see EGY500
ETHYLBUTYLACETALDEHYDE see BRI000
ETHYL BUTYLACETATE (DOT) see EHF000
2-ETHYLBUTYLACRYLATE see EGZ000
2-ETHYLBUTYL ALCOHOL see EGW000
2-ETHYLBUTYLAMINE see EHA000
N-ETHYLBUTYLAMINE see EHA050

5-ETHYL-5-N-BUTYLBARBITURIC ACID see BPF500
ETHYL BUTYLCARBAMATE see EHA100
ETHYL-N,N-BUTYLCARBAMATE see EHA100
ETHYLBUTYLCETONE (FRENCH) see EHA600
2-ETHYLBUTYL ESTER, ACRYLIC ACID see EGZ000
2-ETHYLBUTYLESTER KYSELINY AKRYLOVE see EGZ000
ETHYL BUTYL ETHER see EHA500
ETHYL tert-BUTYL ETHER see EHA550
1-ETHYLBUTYL p-IODOBENZYL CARBONATE see HFQ510
ETHYLBUTYLKETON (DUTCH) see EHA600
ETHYL BUTYL KETONE see EHA600
ETHYL-N-BUTYLNITROSAMINE see EHC000
ETHYL-tert-BUTYLNITROSAMINE see NKD500
ETHYL N-BUTYL-N-NITROSOSUCCINAMATE see EHC800
ETHYL tert-BUTYL OXIDE see EHA550
2-ETHYL-2-BUTYL-1,3-PROPANEDIOL see BKH625
2-ETHYLBUTYL SILICATE see EHC900
α-ETHYL-α',sec-BUTYLSTILBENE see EHD000
α-ETHYL-β-sec-BUTYLSTILBENE see EHD000
3-ETHYLBUTYNOL see EQL000
ETHYL BUTYRALDEHYDE see DHI000
ETHYL BUTYRALDEHYDE (DOT) see DHI000
α-ETHYLBUTYRALDEHYDE see DHI000
2-ETHYLBUTYRALDEHYDE (DOT,FCC) see DHI000
ETHYL n-BUTYRATE see EHE000
ETHYL BUTYRATE (DOT,FCC) see EHE000
α-ETHYLBUTYRIC ACID see DHI400
2-ETHYLBUTYRIC ACID (FCC) see DHI400
2-ETHYLBUTYRIC ALDEHYDE see DHI000
γ-ETHYLBUTYROLACTONE see HDY600
γ-ETHYL-n-BUTYROLACTONE see HDY600
ETHYL CADMATE see BJB500
ETHYL CAPRATE see EHE500
ETHYL CAPRINATE see EHE500
α-ETHYLCAPROALDEHYDE see BRI000
ETHYL CAPROATE see EHF000
α-ETHYLCAPROIC ACID see BRI250
2-ETHYLCAPROIC ACID SODIUM SALT see SGS600
2-ETHYLCAPROYL CHLORIDE see EKO600
ETHYL CAPRYLATE see ENY000
ETHYL CARBAMATE see UVA000
N-ETHYLCARBAMIC ACID-3-DIMETHYLAMINOPHENYL ESTER, METHOSULFATE see HNQ500
ETHYLCARBAMIC ACID 1,1-DIPHENYL-2-PROPYNYL ESTER see DWL500
ETHYLCARBAMIC ACID, ETHYL ESTER see EJW500
N-ETHYLCARBAMIC ACID-3-(TRIMETHYLAMMONIO)PHENYL ESTER, METHYLSULFATE see HNQ500
ETHYLCARBAMIC ESTER of m-OXYPHENYLTRIMETHYLAMMONIUM METHYLSULFATE see HNQ500
ETHYL CARBAMMONIOTHIOATE HYDROGEN SULFATE see ELY550
d-(−)-1-(ETHYLCARBAMOYL)ETHYL PHENYLCARBAMATE see CBL500
2-(N-ETHYL CARBAMOYLHYDROXYMETHYL)FURAN see EHF500
S-(N-ETHYLCARBAMOYLMETHYL) DIMETHYL PHOSPHORODITHIOATE see DNX600
(3-(N-ETHYLCARBAMOYLOXY)PHENYL)TRIME

THYLAMMONIUM METHYLSULFATE see HNQ500
ETHYL CARBANILATE see CBL750
ETHYL CARBAZATE see EHG000
ETHYL CARBAZINATE see EHG000
9-ETHYLCARBAZOLE see EHG025
N-ETHYLCARBAZOLE see EHG025
ETHYL CARBINOL see PND000
ETHYL CARBITOL see CBR000
ETHYLCARBITOL ACRYLATE see EHG030
ETHYL-Δ-CARBOETHOXYVALERATE see AEP750
ETHYL CARBONATE see DIX200
ETHYL CARBONAZIDE see EGN000
ETHYLCARBONYL PIPERAZINE see EHG050
1-(p-ETHYLCARBOXYPHENYL)-3,3-DIMETHYLTRIAZENE see CBP250
N-ETHYL-N-(3-CARBOXYPROPYL)NITROSOAMINE see NKE000
ETHYL CELLOSOLVE see EES350
ETHYL CELLOSOLVE ACETAAT (DUTCH) see EES400
ETHYLCELLULOSE see EHG100
ETHYL CETAB see EKN500
ETHYLCHLOORFORMIAAT (DUTCH) see EHK500
ETHYL CHLORACETATE see EHG500
ETHYL CHLORIDE see EHH000
ETHYL CHLOROACETATE see EHG500
ETHYL-α-CHLOROACETATE see EHG500
ETHYL 4-CHLOROACETOACETATE see EHH100
ETHYL γ-CHLOROACETOACETATE see EHH100
ETHYL 2-CHLOROACRYLATE see EHH200
ETHYL α-CHLOROACRYLATE see EHH200
ETHYLCHLOROBENZENE see EHH500
ETHYL 2-(4-((6-CHLORO-2-BENZOTHIAZOLYL)OXY)PHENOXY)PROPANOATE see EHH600
ETHYL CHLOROBENZOXAZOLINE-2-THIONE-3-CARBOXYLATE see EHH700
ETHYL (+-)-2-(4-(6-CHLORO-2-BENZOXAZOLYLOXY)PHENOXY)PROPANOATE see FAQ200
ETHYL CHLOROCARBONATE (DOT) see EHK500
ETHYL 4-CHLORO-α-(4-CHLOROPHENYL)-α-HYDROXYBENZENEACETATE see DER000
ETHYL CHLOROETHANOATE see EHG500
ETHYL-β-CHLOROETHYLAMINE HYDROCHLORIDE see CGX250
7-(2-(ETHYL-2-CHLOROETHYL)AMINOETHYLAMINO)BENZ(c)ACRIDINE DIHYDROCHLORIDE see EHI500
9-(5-(4-(N-ETHYL-N-(2-CHLOROETHYL)AMINO)PHENOXY)PENTYLAMINO)ACRIDINE see EHI600
7-(3-(ETHYL-2-CHLOROETHYLAMINO)PROPYLAMINO))BENZ(c)ACRIDINE DIHYDROCHLORIDE see EHJ000
9-(3-(ETHYL(2-CHLOROETHYL)AMINO)PROPYLAMINO)-6-CHLORO-2-METHOXYACRIDINE DIHYDROCHLORIDE see ADJ875
9-((3-ETHYL-2-CHLOROETHYL)AMINOPROPYLAMINO)-4-METHOXYACRIDINE DIHYDROCHLORIDE see EHJ500
ETHYL(CHLOROETHYL)ANILINE see EHJ600
N,N'-ETHYL-N,N'-(β-CHLOROETHYL)ETHYLENEDIAMINE DIHYDROCHLORIDE see BIC500
ETHYL(2-CHLOROETHYL)ETHYLENIMONIUM PICRYLSULFONATE see EHK000

ETHYL-β-CHLOROETHYLETHYLENIMONIUM PICRYLSULFONATE see EHK000
1-ETHYL-1-(β-CHLOROETHYL)ETHYLENIMONIUM PICRYLSULFONATE see EHK000
ETHYL-β-CHLOROETHYL-β-HYDROXYETHYLAMINE PICRYLSULFONATE see EHK300
ETHYL-N-(β-CHLOROETHYL)-N-NITROSOCARBAMATE see CHF500
ETHYL-2-CHLOROETHYL SULFIDE see CGY750
ETHYL-β-CHLOROETHYL SULFIDE see CGY750
ETHYL-7-CHLORO-5-(o-FLUOROPHENYL)-2,3-DIHYDRO-2-OXO-1H-1,4-BENZODIAZEPINE-3-CARBOXYLATE see EKF600
ETHYL CHLOROFORMATE see EHK500
ETHYL-5-CHLORO-3(1H)-INDAZOLYLACETATE see EHK600
7-ETHYL-10-CHLORO-11-METHYLBENZ(c)ACRIDINE see EHL000
ETHYL 4-(4-CHLORO-2-METHYLPHENOXY)BUTYLATE see EHL500
ETHYL-4-(4-CHLORO-2-METHYLPHENOXY)BUTYRATE see EHL500
ETHYL 4-CHLORO-2-OXO-3(2H)-BENZOTHIAZOLEACETATE see EHL600
ETHYL 4-CHLORO-3-OXOBUTANOATE see EHH100
ETHYL CHLOROPHENOXYISOBUTYRATE see ARQ750
ETHYL-p-CHLOROPHENOXYISOBUTYRATE see ARQ750
ETHYL-2-(p-CHLOROPHENOXY)ISOBUTYRATE see ARQ750
ETHYL-α-(4-CHLOROPHENOXY)ISOBUTYRATE see ARQ750
ETHYL-α-p-CHLOROPHENOXYISOBUTYRATE see ARQ750
ETHYL 2-(4-CHLOROPHENOXY)-2-METHYLPROPANOATE CALCIUM SALT see EHL625
ETHYL 2-(4-CHLOROPHENOXY)-2-METHYLPROPIONATE see ARQ750
ETHYL 2-(p-CHLOROPHENOXY)-2-METHYLPROPIONATE see ARQ750
ETHYL-α-(4-CHLOROPHENOXY)-α-METHYLPROPIONATE see ARQ750
ETHYL-α-(p-CHLOROPHENOXY)-α-METHYLPROPIONATE see ARQ750
o-ETHYL S-4-CHLOROPHENYL ETHYLPHOSPHONODITHIOATE see EHL670
ETHYL 2-(4-(6-CHLORO-2-QUINOXALINYLOXY)PHENOXY)PROPANOATE see QMA100
ETHYL CHLOROTHIOFORMATE (DOT) see CLJ750
ETHYL 6-CHLORO-2-THIOXO-3(2H)-BENZOXAZOLECARBOXYLATE see EHH700
ETHYL-β-CHLOROVINYLETHYNYL CARBINOL see CHG000
ETHYLCHLORVYNOL see CHG000
3-ETHYLCHOLANTHRENE see EHM000
3-ETHYL-CHOLANTHRENE see EHM000
20-ETHYLCHOLANTHRENE see EHM000
24-β-ETHYLCHOLEST-5-EN-3-β-OL SULFATE see SDZ370
24-α-ETHYLCHOLESTEROL see SDZ350
ETHYL CHRYSANTHEMATE see EHM100
ETHYL CHRYSANTHEMUMATE see EHM100

3-ETHYLCINCHONINIC ACID ETHYL ESTER see EHM200
ETHYL-trans-CINNAMATE see EHN000
ETHYL CINNAMATE (FCC) see EHN000
ETHYL CITRAL see EHN500
ETHYL CITRATE see TJP750
ETHYL CLOFIBRATE see ARQ750
2-ETHYLCROTONALDEHYDE see EHO000
ETHYLCROTONATE see COB750
ETHYL CROTONATE see EHO200
ETHYL (E)-CROTONATE see COB750
ETHYL CROTONATE (DOT) see COB750
ETHYL trans-CROTONATE see COB750
N-ETHYL-o-CROTONOTOLUIDIDE see EHO500
N-ETHYL-o-CROTONOTOLUIDINE see EHO500
trans-1-(2-ETHYLCROTONOYL)UREA see EHO700
(α-ETHYL-cis-CROTONYL)CARBAMIDE see EHP000
2-ETHYLCROTONYLUREA see EHP000
2-ETHYL-cis-CROTONYLUREA see EHP000
cis-(2-ETHYLCROTONYL) UREA see EHP000
2-ETHYL-trans-CROTONYLUREA see EHO700
ETHYL CYANIDE see PMV750
ETHYL CYANOACETATE see EHP500
ETHYL CYANOACETATE see EHP500
ETHYL CYANOACRYLATE see EHP700
ETHYL 2-CYANOACRYLATE see EHP700
ETHYL α-CYANOACRYLATE see EHP700
ETHYLCYANOCYCLOHEXYL ACETATE see EHQ000
ETHYL 1-(3-CYANO-3,3-DIPHENYLPROPYL)-4-PHENYLISONIPECOTATE MONOHYDROCHLORIDE see LIB000
ETHYL CYANOETHANOATE see EHP500
N-ETHYL-N-(2-CYANOETHYL)ANILINE see EHQ500
N-ETHYL-N,β-CYANOETHYLANILINE see EHQ500
o-ETHYL-o-4-CYANOPHENYL PHENYLPHOSPHOROTHIOATE see CON300
ETHYL 2-CYANO-2-PROPENOATE see EHP700
ETHYL CYCLOBUTANE see EHR000
5-ETHYL-5-(1'-CYCLOHEPTENYL)-BARBITURIC ACID see COY500
ETHYL CYCLOHEXANECARBOXYLATE see CPB075
1-ETHYLCYCLOHEXANOL see EHR500
5-ETHYL-5-CYCLOHEXENYLBARBITURIC ACID see TDA500
ETHYLCYCLOHEXYL ACETATE see EHS000
N-ETHYL(CYCLOHEXYL)AMINE see EHT000
N-ETHYL-CYCLOHEXYLAMINE see EHT000
ETHYL CYCLOHEXYLCARBOXYLATE see CPB075
S-ETHYL CYCLOHEXYLETHYLTHIOCARBAMATE see EHT500
ETHYL 2-CYCLOHEXYLPROPIONATE see EHT600
ETHYL CYCLOPENTANE see EHU000
ETHYL CYCLOPROPANE see EHU500
ETHYL 4-(CYCLOPROPYLHYDROXYMETHYLENE)-3,5-DIOXOCYCLOHEXANECARBOXYLATE see EHU550
ETHYL CYMATE see BJC000
ETHYL CYSTEINE HYDROCHLORIDE see EHU600
3'-ETHYL-DAB see EOI000
4'-ETHYL-DAB see EOI500

ETHYL (D+)-2-(4-(6-CHLOR-2-BENZOXAZOLYLOXY)PHENOXY)PROPANOATE see FAQ200

p,p-ETHYL DDD see DJC000

p,p'-ETHYL-DDD see DJC000

ETHYL N-(O-4-DEACETYL-4'-DEOXYVINBLASTIN-23-OYL-B)-TRYPTOPHANATE see VKZ100

ETHYL DECABORANE see EHV000

ETHYL (2E,4Z)-DECADIENOATE see EHV100

ETHYL DECANOATE (FCC) see EHE500

ETHYL DECYLATE see EHE500

ETHYL 6-(N-DECYLOXY)-7-ETHOXY-4-HYDROXYQUINOLINE-3-CARBOXYLATE see DAI495

5-ETHYLDEOXYURIDINE see EHV200

5-ETHYL-2'-DEOXYURIDINE see EHV200

β-5-ETHYLDEOXYURIDINE see EHV200

ETHYL(DI-(1-AZIRIDINYL)PHOSPHINYL)CARBAMATE see EHV500

ETHYL DIAZOACETATE see DCN800

N-ETHYLDIAZOACETYLGLYCINE AMIDE see DCN600

1-ETHYLDIBENZ(a,h)ACRIDINE see EHW000

1-ETHYLDIBENZ(a,j)ACRIDINE see EHW500

8-ETHYLDIBENZ(a,h)ACRIDINE see EHX000

1"-ETHYLDIBENZ(a,h)ACRIDINE see EHX000

1'-ETHYL-1,2,5,6-DIBENZACRIDINE (FRENCH) see EHW000

1'-ETHYL-3,4,5,6-DIBENZACRIDINE (FRENCH) see EHW500

ETHYL DIBROMOBENZENE see EHY000

ETHYL 2,3-DIBROMOPROPANOATE see EHY050

ETHYL 2,3-DIBROMOPROPIONATE see EHY050

ETHYL α,β-DIBROMOPROPIONATE see EHY050

O-ETHYL-S,S-DI-sec-BUTYLPHOSPHORODITHIOATE see EHY100

ETHYLDICHLOROALUMINUM see EFU050

ETHYLDICHLOROBENZENE see EHY500

ETHYL-4,4'-DICHLOROBENZILATE see DER000

ETHYL-p,p'-DICHLOROBENZILATE see DER000

ETHYL-4,4'-DICHLORODIPHENYL GLYCOLLATE see DER000

N-ETHYL-DICHLOROMALEINIMIDE see DFJ400

ETHYL (2,4-DICHLOROPHENOXY)ACETATE see EHY600

ETHYL-4,4'-DICHLOROPHENYL GLYCOLLATE see DER000

O-ETHYL-O-(2,4-DICHLOROPHENYL)-S-n-PROPYL-DITHIOPHOSPHATE see DGC800

O-ETHYL-O-2,4-DICHLOROPHENYL THIONOBENZENEPHOSPHONATE see SCC000

ETHYL DICHLOROSILANE (DOT) see DFK000

ETHYLDICHLORTHIOFOSFAT (CZECH) see MRI000

ETHYLDICOUMAROL see BKA000

ETHYLDICOUMAROL ACETATE see BKA000

ETHYLDIETHANOLAMINE see ELP000

N-ETHYLDIETHANOLAMINE see ELP000

ETHYL ((DIETHOXYPHOSPHINOTHIOYL)THIO)ACETATE see DIX000

ETHYL (2-((DIETHOXYPHOSPHINOTHIOYL)THIO)ETHYL)CARBAMATE see EMC000

ETHYL-2-(DIETHOXYPHOSPHINYL)-3-OXOBUTANOATE see EHY700

ETHYL DIETHOXYPHOSPHORYL ACETATE see EIC000

ETHYL-3-((DIETHYLAMINO)METHYL)-4-HYDROXYBENZOATE see EIA000

ETHYL-N,N-DIETHYL CARBAMATE see EIB000

ETHYL DIETHYLENE GLYCOL see CBR000

4'-ETHYL-N,N-DIETHYL-p-(PHENYLAZO)ANILINE see EIB500

ETHYL 3-((DIETHYLPHOSPHINOTHIOYL)OXY)-2-BUTENOATE see EIB600

ETHYL (DIETHYLPHOSPHONO)ACETATE see EIC000

ETHYL (DIETHYLPHOSPHONO)ACETATE see EIC000

5-ETHYL-1,3-DIGLYCIDYL-5-METHYLHYDANTOIN see ECI200

ETHYL DIGLYME see DIW800

ETHYL DIHYDROGEN PHOSPHATE see ECI300

o-ETHYL DIHYDROGEN PHOSPHATE see ECI300

5-ETHYLDIHYDRO-5-(1-METHYLBUTYL)-2-THIOXO-4,6,(1H,5H)-PYRIMIDINEDIONE (9CI) see PBT250

5-ETHYLDIHYDRO-5-(1-METHYLBUTYL)-2-THIOXO-4,6,(1H,5H)-PYRIMIDINEDIONE, MONOSODIUM SALT see PBT500

3-ETHYL-2,3-DIHYDRO-6-METHYL-1H-CYCLOPENT(a)ANTHRACENE see DLM600

1-ETHYL-1,4-DIHYDRO-6,7-METHYLENEDIOXY-4-OXO-3-QUINOLINECARBOXYLIC ACID see OOG000

(3S-cis)-3-ETHYLDIHYDRO-4-((1-METHYL-1H-IMIDAZOL-5-YL)METHYL)-2(3H)-FURANONE see PIF000

3-ETHYL-6,7-DIHYDRO-2-METHYL-5-MORPHOLINOMETHYLINDOLE-4(5H)-ONE HYDROCHLORIDE see MRB250

3-ETHYL-6,7-DIHYDRO-2-METHYL-5-MORPHOLINOMETHYLINDOL-4(5H)-ONE HYDROCHLORIDE see MRB250

1-ETHYL-1,4-DIHYDRO-7-METHYL-4-OXO-1,8-NAPHTHYRIDINE-3-CARBOXYLIC ACID see EID000

1-ETHYL-1,4-DIHYDRO-7-METHYL-4-OXO-1,8-NAPHTHYRIDINE-3-CARBOXYLIC ACID SODIUM SALT see NAG450

1-ETHYL-4,6-DIHYDRO-3-METHYL-8-PHENYLPYRAZOLO(4,3-e)(1,4)DIAZEPINE see POL475

1-ETHYL-1,4-DIHYDRO-4-OXO(1,3)DIOXOLO(4,5-g)CINNOLINE-3-CARBOXYLIC ACID see CMS130

5-ETHYL-5,8-DIHYDRO-8-OXO-1,3-DIOXOLO(4,5-g)QUINOLINE-7-CARBOXYLIC ACID see OOG000

8-ETHYL-5,8-DIHYDRO-5-OXO-2-(1-PIPERAZINYL)PYRIDO(2,3-d)PYRAMIDINE-6-CARBOXYLIC ACID 3H₂O see PII350

5-ETHYLDIHYDRO-5-PHENYL-4,6(1H,5H)-PYRIMIDINEDIONE see DBB200

2-ETHYL-2,3-DIHYDRO-3-((4-(2-(1-PIPERIDINYL)ETHOXY)PHENYL)AMINO)-1H-ISOINDOL-1-ONE see AHP125

9-ETHYL-1,9-DIHYDRO-6H-PURINE-6-THIONE see EMC500

ETHYL-3,4-DIHYDROXYBENZENE SULFONATE see EID100

ETHYL-4,4'-DIHYDROXYDICOUMARINYL-3,3'-ACETATE see BKA000

20-ETHYL-6-β,8-DIHYDROXY-1-α-METHOXY-4-METHYLHETERATISAN-14-ONE see EID150

2-ETHYL-3-(3',5'-DIIODO-4'-HYDROXYBENZOYL)-CUMARONE see EID200

ETHYL-2-(DIIODO-3,5 HYDROXY-4 BENZOYL)5-FURANNE see EID250

ETHYL-6,7-DIISOBUTOXY-4-HYDROXYQUINOLINE-3-CARBOXYLATE see BOO632

S-ETHYLDIISOBUTYL THIOCARBAMATE see EID500

ETHYL-N,N-DIISOBUTYLTHIOCARBAMATE see EID500

S-ETHYL N,N-DIISOBUTYLTHIOCARBAMATE see EID500

ETHYL-N,N-DIISOBUTYL THIOLCARBAMATE see EID500

O-ETHYL-S-2-DIISOPROPYLAMINOETHYL METHYLPHOSPHONOTHIOTE see EIG000

ETHYL-S-DIISOPROPYLAMINOETHYL METHYLTHIOPHOSPHONATE see EIG000

ETHYL-α-((DIMETHOXYPHOSPHENOTHIOYL)THIO)BENZENEACETATE see DRR400

4-ETHYL-6,7-DIMETHOXY-9H-PYRIDO(3,4-B)INDOLE-3-CARBOXYLIC ACID, METHYL ESTER see EID600

ETHYL DIMETHYLACRYLATE see MIP800

ETHYL 3,3-DIMETHYLACRYLATE see MIP800

ETHYL β,β-DIMETHYLACRYLATE see MIP800

ETHYL DIMETHYLAMIDOCYANOPHOSPHATE see EIF000

ETHYL trans-3-DIMETHYLAMINOACRYLATE see EIF100

2'-ETHYL-4-DIMETHYLAMINOAZOBENZENE see EIF450

3'-ETHYL-4-DIMETHYLAMINOAZOBENZENE see EOI000

4'-ETHYL-4-DIMETHYLAMINOAZOBENZENE see EOI500

4'-ETHYL-N,N-DIMETHYL-4-AMINOAZOBENZENE see EIB500

ETHYL N,N-DIMETHYLAMINO CYANOPHOSPHATE see EIF000

O-ETHYL-S-(2-DIMETHYL AMINO ETHYL)-METHYLPHOSPHONOTHIOATE see EIF500

ETHYL-S-DIMETHYLAMINOETHYL METHYLPHOSPHONOTHIOLATE see EIG000

2-ETHYL-10-(3-DIMETHYLAMINO-2-METHYLPROPYL)PHENOTHIAZINE HYDROCHLORIDE see ELQ600

ETHYL-dl-trans-2-DIMETHYLAMINO-1-PHENYL-3-CYCLOHEXENE-1-CARBOXYLATE see EIH000

(±)-ETHYL-trans-2-2(DIMETHYLAMINO)-1-PHENYL-3-CYCLOHEXENE-1-CARBOXYLATE HYDROCHLORIDE see EIH000

S-ETHYL N-(3-DIMETHYLAMINOPROPYL)THIOL CARBAMATE HYDROCHLORIDE see EIH500

7-ETHYL-1,4-DIMETHYLAZULENE see DRV000

ETHYL-N,N-DIMETHYL CARBAMATE see EII500

3-ETHYL-2,2-DIMETHYLCYCLOBUTYL METHYL KETONE(E)- see EII600

3-ETHYL-2,2-DIMETHYLCYCLOBUTYL METHYL KETONE (Z)- see EII700

N-ETHYL-N-1-DIMETHYL-3,3-DI-2-THIENYLALLYLAMINE HYDROCHLORIDE see EIJ000

N-ETHYL-N-1-DIMETHYL-3,3-DI-2-THIENYL-2-PROPENAMINE HYDROCHLORIDE see EIJ000

ETHYL DIMETHYLDITHIOCARBAMATE see EIJ500

ETHYL 1,1-DIMETHYLETHYL ETHER see EHA550

ETHYL DIMETHYLHYDROCINNAMALDEHYDE see EIJ600

ETHYLDIMETHYLMETHANE see EIK000

5-ETHYL-3,5-DIMETHYLOXAZOLIDINE-2,4-DIONE see PAH500

3-ETHYL-2,3-DIMETHYL PENTANE see EIK500

4'-ETHYL-N,N-DIMETHYL-4-(PHENYLAZO)-m-TOLUIDINE see EMS000

ETHYLDIMETHYLPHOSPHINE see EIL000

ETHYL DIMETHYLPHOSPHORAMIDOCYANIDATE see EIF000

ETHYL-N,N-DIMETHYLPHOSPHORAMIDOCYANIDATE see EIF000

ETHYL-N-(2-(O,O-DIMETHYLPHOSPHORODITHIOYL)ETHYL)CARBAMATE see EMC000

ETHYL-O,O-DIMETHYL PHOSPHORODITHIOYLPHENYL ACETATE see DRR400

N-ETHYL-1,2-DIMETHYLPROPYLAMINE see EIL050

2-ETHYL-3,5(6)-DIMETHYLPYRAZINE see EIL100

4-ETHYL-3,5-DIMETHYLPYRROL-2-YL METHYL KETONE see EIL200

ETHYLDIMETHYL SULFONIUM IODIDE MERCURIC IODIDE ADDITION COMPOUND see EIL500

ETHYL 3,3-DIMETHYL-4,6,6-TRICHLORO-5-HEXENOATE see EIL600

10-ETHYL-4,4-DIOCTYL-7-OXO-8-OXA-3,5-DITHIA-4-STANNATETRADECANOIC ACID-2-ETHYLHEXYL ESTER see DVM800

ETHYLDIOL ACRILATE (RUSSIAN) see EIP000

ETHYLDIOL METACRYLATE see BKM250

1-ETHYL-2,5-DIOXO-4-PHENYLIMIDAZOLIDINE see EOL100

O-ETHYL-S,S-DIPHENYL DITHIOPHOSPHATE see EIM000

O-ETHYL-S,S-DIPHENYL PHOSPHORODITHIOATE see EIM000

2-ETHYL-3,3-DIPHENYL-2-PROPENYLAMINE HYDROCHLORIDE see DWF200

ETHYLDIPHENYLTIN ACETATE see EIM100

O-ETHYL-S,S-DIPROPYL ESTER, PHOSPHORODITHIOIC ACID see EIN000

O-ETHYL-S,S-DIPROPYLPHOSPHORODITHIOATE see EIN000

S-ETHYL-N,N-DIPROPYLTHIOCARBAMATE see EIN500

S-ETHYL-N,N-DI-N-PROPYLTHIOCARBAMATE see EIN500

ETHYL DI-N-PROPYLTHIOLCARBAMATE see EIN500

ETHYL-N,N-DIPROPYLTHIOLCARBAMATE see EIN500

ETHYL-N,N-DI-N-PROPYLTHIOLCARBAMATE see EIN500

ETHYL DISELENIDE see EIN550

O-ETHYL DITHIOCARBAMATE see EQG600

O-ETHYL DITHIOCARBONATE see EQG600

(O-ETHYL DITHIOCARBONATO)POTASSIUM see PLF000

ETHYLDITHIOURAME see DXH250

ETHYLDITHIURAME see DXH250

ETHYL DODECANOATE see ELY700

ETHYLE (ACETATE d') (FRENCH) see EFR000

ETHYLE, CHLOROFORMIAT d' (FRENCH) see EHK500

ETHYLEEN-CHLOORHYDRINE (DUTCH) see EIU800

ETHYLEENDIAMINE (DUTCH) see EEA500

ETHYLEENDICHLORIDE (DUTCH) see EIY600

ETHYLEENIMINE (DUTCH) see EJM900

ETHYLEENOXIDE (DUTCH) see EJN500

ETHYLE (FORMIATE d') (FRENCH) see EKL000

ETHYL ENANTHATE see EKN050

ETHYLENE see EIO000

ETHYLENE, compressed (DOT) see EIO000

ETHYLENE, refrigerated liquid (DOT) see EIO000

ETHYLENE ACETATE see EJD759

ETHYLENEACETIC ACID see EIO500

ETHYLENE ACRYLATE see EIP000

ETHYLENE ALCOHOL see EJC500

ETHYLENE ALDEHYDE see ADR000

ETHYLENE AZELATE see EIP050

ETHYLENE 1-AZIRIDINEPROPIONATE see EIP100

ETHYLENE, 1-(p-(BENZYLOXY)PHENYL)-2-(o-FLUOROPHENYL)-1-PHENYL- see BFC400

1,1'-ETHYLENE-2,2'-BIPYRIDYLIUM DIBROMIDE see DWX800

1,1'-ETHYLENE-2,2'-BIPYRIDYLIUM DIBROMIDE see EJC025

1,1'-ETHYLENE-2,2'-BIPYRIDYLIUM ION see DWW730

ETHYLENE BIS(BROMOACETATE) see BHD250

1,1'-ETHYLENEBIS(5-BUTOXY-3,7-DIMETHYL-1,5-AZABOROCINE-4,6-DIONE) see BJJ750

N,N-ETHYLENEBIS(N-BUTYL-4-MORPHOLINECARBOXAMIDE) see DUO400

1,1'-ETHYLENEBIS(3-(2-CHLOROETHYL)-3-NITROSOUREA) see EIP500

ETHYLENE BIS(CHLOROFORMATE) see EIQ000

ETHYLENE, 1,1-BIS(p-CHLOROPHENYL)-2-CHLORO- see BIM300

1,1'-ETHYLENEBIS-CNU see EIP500

ETHYLENEBIS-(DIPHENYLARSINE) see EIQ200

ETHYLENEBIS(DITHIOCARBAMATE) DISODIUM SALT see DXD200

ETHYLENEBISDITHIOCARBAMATE MANGANESE see MAS500

N,N-ETHYLENE BIS(DITHIOCARBAMATE MANGANEUX) (FRENCH) see MAS500

N,N-ETHYLENE BIS(DITHIOCARBAMATE de SODIUM) (FRENCH) see DXD200

ETHYLENEBIS(DITHIOCARBAMATO) MANGANESE see MAS500

ETHYLENEBIS(DITHIOCARBAMATO)MANGANESE and ZINC ACETATE (50:1) see EIQ500

ETHYLENE BIS(DITHIOCARBAMATO)ZINC see EIR000

ETHYLENEBIS(DITHIOCARBAMIC ACID) DISODIUM SALT see DXD200

ETHYLENEBIS(DITHIOCARBAMIC ACID) MANGANESE SALT see MAS500

ETHYLENEBIS(DITHIOCARBAMIC ACID MANGANESE ZINC COMPLEX (8CI) see DXI400

ETHYLENEBIS(DITHIOCARBAMIC ACID) MANGANOUS SALT see MAS500

ETHYLENEBIS(DITHIOCARBAMIC ACID) NICKEL(II) SALT see EIR500

ETHYLENEBIS(DITHIOCARBAMIC ACID), ZINC SALT see EIR000

N,N'-ETHYLENE BIS(3-FLUOROSALICYLIDENEIMINATO)COBALT(II) see EIS000

N,N'-ETHYLENEBIS(2-(o-HYDROXYPHENYL)GLYCINE) see EIV100

ETHYLENEBIS(IMINODIACETIC ACID) DISODIUM SALT see EIX500

ETHYLENEBIS(IMINODIACETIC ACID) TETRASODIUM SALT see EIV000

ETHYLENE BIS(IODOACETATE) see EIS100

ETHYLENEBISISOTHIOCYANATE see ISK000

ETHYLENE BIS(MERAPTOACETATE) see MCN000

ETHYLENE BIS(METHANESULFONATE) see BKM125

2,2'-(ETHYLENEBIS(NITROSOIMINO))BISBUTANOL see HMQ500

(ETHYLENEBIS(OXYETHYLENENITRILO))TETRAACETIC ACID see EIT000

ETHYLENE BIS(PENTABROMOPHENOXIDE) see EEB230

1,1'-ETHYLENEBIS(PYRIDINIUM)BROMIDE see EIT100

N,N'-ETHYLENEBIS(SALICYLIDENEIMINATO) COBALT(II) see BLH250

ETHYLENE BIS(TETRABROMOPHTHALIMIDE) see EIT150

ETHYLENE BIS(THIOGLYCOLATE) see MCN000

2,2'-ETHYLENE-BIS-(2-THIOPSEUDOUREA), DIHYDROBROMIDE see EJA000

ETHYLENE BISTHIURAM MONOSULFIDE see EJQ000

ETHYLENE-BIS-THIURAMMONO-SULFIDE see ISK000

1,2-ETHYLENEBIS(TRIMETHOXYSILANE) see EIT200

ETHYLENEBIS(TRIS(2-CYANOETHYL))PHOSPHONIUM BROMIDE) see EIU000

1,1'-ETHYLENEBISUREA see EJC100

ETHYLENE BRASSYLATE see EJQ500

ETHYLENE BROMIDE see EIY500

ETHYLENE BROMOACETATE see BHD250

ETHYLENE, 1-BROMO-1-(p-CHLOROPHENYL)-2,2-DIPHENYL- see BNA500

ETHYLENE, 1-BROMO-1,2-DIPHENYL-2-(p-ETHYLPHENYL)-, (Z)- see BOL315

ETHYLENEBROMOHYDRIN see BNI500

ETHYLENE, BROMOTRIPHENYL- see BOK500

ETHYLENE CARBONATE see GHM000

ETHYLENE CARBONIC ACID see GHM000

ETHYLENECARBOXAMIDE see ADS250

ETHYLENECARBOXYLIC ACID see ADS750

ETHYLENE CHLORIDE see EIY600

ETHYLENE CHLOROBROMIDE see CES500

ETHYLENE CHLOROFORMATE see EIQ000

ETHYLENE CHLOROHYDRIN see EIU800

ETHYLENE, 1-(o-CHLOROPHENYL)-1-(p-CHLOROPHENYL)-2,2-DICHLORO- see DEV900

ETHYLENE CHLOROTHIOARSENATE(III) see EIU900

ETHYLENE, CHLOROTRIFLUORO-, POLYMERS (8CI) see KDK000

ETHYLENE CYANIDE see SNE000

ETHYLENE CYANOHYDRIN see HGP000

ETHYLENE DIACRYLATE see EIP000

1,2-ETHYLENEDIAMINE see EEA500

ETHYLENE-DIAMINE (FRENCH) see EEA500

ETHYLENEDIAMINEACETIC ACID TRISODIUM SALT see TNL250

ETHYLENEDIAMINE-N,N'-BIS(2-HYDROXYPHENYLACETIC ACID) see EIV100

ETHYLENEDIAMINE, N-(5-CHLORO-2-THENYL)-N',N'-DIMETHYL-N-2-PYRIDYL- see CHY250

N,N'-ETHYLENEDIAMINEDIACETIC ACID TETRASODIUM SALT see EIV000

ETHYLENEDIAMINE, N,N'-DIBENZYL- see BHB300

ETHYLENEDIAMINE, N,N'-DIBENZYL-, DIACETATE see DDF800

ETHYLENEDIAMINEDICHLORIDE PLATINUM (II) see DFJ000

ETHYLENEDIAMINE, N,N-DIETHYL-N'-PHENYL- see DJV300

ETHYLENEDIAMINE-DI(2-HYDROXYPHENYL)ACETIC ACID see EIV100

ETHYLENEDIAMINE-DI(o-HYDROXYPHENYL)ACETIC ACID see EIV100

ETHYLENEDIAMINE, N,N-DIISOPROPYL- see DNP700

ETHYLENEDIAMINE, N,N'-DIMETHYL- see DRI600

ETHYLENEDIAMINEDINITRATE see EIV700

ETHYLENE DIAMINE DIPERCHLORATE (DOT) see EGT500

ETHYLENEDIAMINE, N,N'-DIPHENYL- see DWB400

ETHYLENEDIAMINE ETHOXYLATE see EIV750

ETHYLENEDIAMINE ETHYLENE OXIDE ADDUCT see EIV750

ETHYLENEDIAMINE-FORMALDEHYDE-PHENOL COPOLYMER see EEA550

ETHYLENEDIAMINE HYDROCHLORIDE see EIW000

ETHYLENEDIAMINE, N-METHYL- see MJW100

ETHYLENEDIAMINE (OSHA) see EEA500

ETHYLENEDIAMINE PERCHLORATE see EGT500

ETHYLENEDIAMINE-PHENOL-FORMALDEHYDE POLYMER see EEA550

ETHYLENEDIAMINE, N-PHENYL-N,N,N'-TRIETHYL-, HYDROCHLORIDE see PFB850

ETHYLENEDIAMINE, POLYMER WITH FORMALDEHYDE AND PHENOL (8CI) see EEA550

ETHYLENEDIAMINE SULFATE see EIW500

ETHYLENEDIAMINETETRAACETATE see EIX000

ETHYLENEDIAMINETETRAACETATE DISODIUM SALT see EIX500

(ETHYLENEDIAMINETETRAACETATO)ZINCATE(2-) see ZGW200

ETHYLENEDIAMINETETRAACETIC ACID see EIX000

ETHYLENEDIAMINE-N,N,N',N'-TETRAACETIC ACID see EIX000

ETHYLENEDIAMINETETRAACETIC ACID, ALUMINUM-SODIUM SALT see EJA400

ETHYLENEDIAMINETETRAACETIC ACID, CALCIUM DISODIUM CHELATE see CAR780

ETHYLENEDIAMINETETRAACETIC ACID, DISODIUM SALT see EIX500

ETHYLENEDIAMINE TETRAACETIC ACID, IRON(III) SALT see HIA000

ETHYLENEDIAMINETETRAACETIC ACID, TETRASODIUM SALT see EIV000

ETHYLENEDIAMINETETRAACETIC ACID, TRISODIUM SALT see TNL250

ETHYLENEDIAMINETETRAACETONITRILE see EJB000

ETHYLENEDIAMINE, compounded with THEOPHYLLINE (1:2) see TEP500

ETHYLENEDIAMMONIUM CHLORIDE see EIW000

ETHYLENE, 1,2-DIBENZOYL- see DDE100

1,2-ETHYLENE DIBROMIDE see EIY500

ETHYLENE, 1,2-DIBROMO- see DDN950

ETHYLENE, 1,2-DIBROMO-, (E)- see DDN820

ETHYLENE, 1,1-DIBROMO-(6CI,7CI,8CI) see VPF200

(E)1,2-ETHYLENEDICARBOXYLIC ACID see FOU000

cis-1,2-ETHYLENEDICARBOXYLIC ACID see MAK900

trans-1,2-ETHYLENEDICARBOXYLIC ACID see FOU000

trans-1,2-ETHYLENEDICARBOXYLIC ACID DIMETHYL ESTER see DSB600

ETHYLENEDICESIUM see EIY550

ETHYLENE DICHLORIDE see EIY600

1,2-ETHYLENE DICHLORIDE see EIY600

ETHYLENE, 1,1-DICHLORO-, POLYMER with CHLOROETHYLENE see CGW300

ETHYLENE DICYANIDE see SNE000

ETHYLENE, 1,2-DIETHOXY- see DHG100

ETHYLENE DIGLYCIDYL ETHER see EEA600

ETHYLENE DIGLYCOL see DJD600

ETHYLENE DIGLYCOL MONOETHYL ETHER see CBR000

ETHYLENE DIGLYCOL MONOMETHYL ETHER see DJG000

ETHYLENE DIHYDRATE see EJC500

(+)-2,2'-(ETHYLENEDIIMINO)DI-1-BUTANOL see TGA500

N,N'-ETHYLENE DIIMINO DI(o-CRESOL) see DWY200

3,3'-(ETHYLENEDIIMINODIMETHYLENE)BIS(5,5-DIPHENYLHYDANTOIN) see EIY700

ETHYLENE DIISOTHIOCYANATE see ISK000

ETHYLENE DIISOTHIOUREA DIHYDROBROMIDE see EJA000

ETHYLENE DIISOTHIOURONIUM DIBROMIDE see EJA000

ETHYLENE DIMERCAPTAN see EEB000

α-ETHYLENE DIMERCAPTAN see EEB000

ETHYLENE DIMETHANESULFONATE see BKM125

ETHYLENE DIMETHANESULPHONATE see BKM125

ETHYLENE DIMETHYL ETHER see DOE000

N,N-ETHYLENE-N',N'-DIMETHYLUREA see EJA100

ETHYLENEDINITRAMINE see DUU800

ETHYLENEDINITRAMINE see DUV800

ETHYLENE DINITRATE see EJG000

3,3'-ETHYLENEDINITRILODI-2-BUTANONE DIOXIME IRON(II) COMPLEX see EJA150

(ETHYLENEDINITRILO)TETRAACETATE DIPOTASSIUM SALT see EEB100

(ETHYLENEDINITRILO)TETRAACETATE DIPOTASSIUM SALT see EJA250

((ETHYLENEDINITRILO)TETRAACETATO(2−))-COBALTATE(2−) COBALT(2+) SALT see DGQ400

((ETHYLENEDINITRILO)TETRAACETATO)-FERATE(1-), SODIUM see EJA379

ETHYLENEDINITRILOTETRAACETIC ACID see EIX000

(ETHYLENEDINITRILO)-TETRAACETIC ACID see TNL500

(ETHYLENEDINITRILO)TETRAACETIC ACID ALUMINUM COMPLEX see EJA390

(ETHYLENEDINITRILO)TETRAACETIC ACID ALUMINUM SODIUM SALT see EJA400

(ETHYLENEDINITRILO)TETRAACETIC ACID CADMIUM(II) COMPLEX see CAF750

(ETHYLENEDINITRILO)TETRAACETIC ACID CHROMIUM COMPLEX see EJA410

(ETHYLENEDINITRILO)TETRAACETIC ACID COPPER(II) COMPLEX see CNL750

(ETHYLENEDINITRILO)-TETRAACETIC ACID DISODIUM SALT see EIX500

(ETHYLENEDINITRILO)TETRA ACETIC ACID, LEAD(II) COMPLEX see LDD000

(ETHYLENEDINITRILO)TETRA ACETIC ACID, MERCURY(II) COMPLEX see MDB250

(ETHYLENEDINITRILO)TETRAACETIC ACID NICKEL(II) COMPLEX see EJA500

(ETHYLENEDINITRILO)TETRAACETONITRILE see EJB000

1,1',1'',1'''-(ETHYLENEDINITRILO)TETRA-2-PROPANOL see QAT000

ETHYLENEDINITROAMINE see DUU800

ETHYLENEDINITROAMINE see DUV800

6,6'-(ETHYLENEDIOXY)BIS(4-AMINOQUINALDINE) DIHYDROCHLORIDE see EJB100

ETHYLENEDIOXYBIS(ETHYLENEAMINO) TETRAACETIC ACID see EIT000

2,2'-ETHYLENEDIOXYDIETHANOL see TJQ000

2,2'-ETHYLENEDIOXYDIETHANOL DIACETATE see EJB500

2,2'-(ETHYLENEDIOXY)DI(ETHYL ACETATE) see EJB500

2,2'-(ETHYLENEDIOXY)DI(ETHYL 2-ETHYLBUTYRATE) see TJQ250

2,2'-ETHYLENEDIOXYETHANOL see TJQ000

16,17-ETHYLENEDIOXYVIOLANTHRONE see CMU500

ETHYLENE DIPERCHLORATE see EJC000

1,1'-ETHYLENEDIPYRIDINIUM DIBROMIDE see EIT100

1,1'-ETHYLENE-2,2'-DIPYRIDINIUM DICHLORIDE see DWY000

ETHYLENE DIPYRIDYLIUM DIBROMIDE see DWX800

ETHYLENE DIPYRIDYLIUM DIBROMIDE see EJC025

1,1-ETHYLENE 2,2-DIPYRIDYLIUM DIBROMIDE see DWX800

1,1'-ETHYLENE-2,2'-DIPYRIDYLIUM DIBROMIDE see DWX800

1,1'-ETHYLENE-2,2'-DIPYRIDYLIUM DIBROMIDE see EJC025

ETHYLENEDITHIOCYANATE see EJC035

2,2'-(ETHYLENEDITHIO)DIANILINE see EJC050

2,2-ETHYLENEDITHIODIPSEUDOUREA DIHYDROBROMIDE see EJA000

ETHYLENE DITHIOGLYCOL see EEB000

ETHYLENEDITHIOL see EEB000

ETHYLENEDIUREA see EJC100

1,1'-ETHYLENEDIUREA see EJC100

2,2'-(1,2-ETHYLENEDIYL)BIS(5-AMINOBENZENESULFONIC ACID) see FCA100

1,2-ETHYLENEDIYLBIS(CARBAMODITHIOATO)MANGANESE see MAS500

ETHYLENE EPISULFIDE see EJP500

ETHYLENE EPISULPHIDE see EJP500

ETHYLENE FLUORIDE see ELN500

ETHYLENE, FLUORO-(8CI) see VPA000

ETHYLENE FORMATE see EJF000

ETHYLENE GLYCOL see EJC500

ETHYLENE GLYCOL ACETATE see EJD759

ETHYLENE GLYCOL ACETATE see EJI000

ETHYLENE GLYCOL ACRYLATE see ADV250

ETHYLENE GLYCOL BIS(AMINOETHYL ETHER)TETRAACETATE see EIT000

ETHYLENE GLYCOL BIS(β-AMINOETHYL ETHER)TETRAACETATE see EIT000

ETHYLENE GLYCOL BIS(2-AMINOETHYL ETHER)TETRAACETIC ACID see EIT000

ETHYLENE GLYCOL BIS(2-AMINOETHYL ETHER)-N,N,N',N'-TETRAACETIC ACID see EIT000

ETHYLENE GLYCOL BIS(β-AMINOETHYL ETHER)-N,N'-TETRAACETIC ACID see EIT000

ETHYLENE GLYCOL BIS(BROMOACETATE) see BHD250

ETHYLENE GLYCOL, BISCHLOROFORMATE see EIQ000

ETHYLENE GLYCOL BIS(CHLOROMETHYL)ETHER see BIJ250

ETHYLENE GLYCOL BIS(2,3-EPOXY-2-METHYLPROPYL) ETHER see EJD000

ETHYLENE GLYCOL-BIS-(2-HYDROXYETHYL ETHER) see TJQ000

ETHYLENE GLYCOL, BIS(IODOACETATE) see EIS100

ETHYLENE GLYCOL BIS(METHACRYLATE) see BKM250

ETHYLENE GLYCOL BIS(THIOGLYCOLATE) see MCN000

ETHYLENE GLYCOL BIS(TRICHLOROACETATE) see EJD600

ETHYLENE GLYCOL, BIS(TRICHLOROACETATE) see EJD600

ETHYLENE GLYCOL-n-BUTYL ETHER see BPJ850

ETHYLENE GLYCOL CARBONATE see GHM000

ETHYLENE GLYCOL, CHLOROHYDRIN see EIU800

ETHYLENE GLYCOL, CYCLIC CARBONATE see GHM000

ETHYLENE GLYCOL, CYCLIC SULFATE see EJP000

ETHYLENE GLYCOL DIACETATE see EJD759

ETHYLENE GLYCOL DIACRYLATE see EIP000

ETHYLENE GLYCOL DIALLYL ETHER see EJE000

ETHYLENE GLYCOL, DIBUTOXYTETRA see TCE350

ETHYLENE GLYCOL DIBUTYL see DDW400

ETHYLENE GLYCOL DI(CHLOROFORMATE) see EIQ000

ETHYLENE GLYCOL DI(2,3-EPOXY-2-METHYLPROPYL)ETHER see EJD000

ETHYLENE GLYCOL DIETHYL ETHER see DHG100

ETHYLENE GLYCOL DIETHYL ETHER see EJE500

ETHYLENE GLYCOL DIFORMATE see EJF000

ETHYLENE GLYCOL DIGLYCIDYL ETHER see EEA600

ETHYLENE GLYCOL DIHYDROXYDIETHYL ETHER see TJQ000

ETHYLENE GLYCOL DIMETHACRYLATE see BKM250

ETHYLENE GLYCOL DIMETHYL ETHER see DOE600

ETHYLENE GLYCOL DINITRATE see EJG000

ETHYLENE GLYCOL DINITRATE mixed with NITROGLYCERIN (1:1) see NGY500

ETHYLENE GLYCOL DIPROPIONATE (8CI) see COB260

ETHYLENE GLYCOL ETHYL ETHER see EES350

ETHYLENE GLYCOL ETHYL ETHER ACETATE see EES400

ETHYLENE GLYCOL-N-HEXYL ETHER see HFT500

ETHYLENE GLYCOLIDE (2,3-EPOXY-2-METHYLPROPYL)ETHER see EJD000

ETHYLENE GLYCOL ISOPROPYL ETHER see INA500

ETHYLENE GLYCOL MALEATE see EJG500

ETHYLENE GLYCOL METHACRYLATE see EJH000

ETHYLENE GLYCOL METHYL ETHER see EJH500

ETHYLENE GLYCOL METHYL ETHER ACETATE see EJJ500

ETHYLENE GLYCOL MONOACETATE see EJI000

ETHYLENE GLYCOL MONOACRYLATE see ADV250

ETHYLENE GLYCOL MONOBENZYL ETHER see EJI500

ETHYLENE GLYCOL MONO-sec-BUTYL ETHER see EJJ000

ETHYLENE GLYCOL MONOBUTYL ETHER ACETATE (MAK) see BPM000

ETHYLENE GLYCOL MONOBUTYL ETHER (MAK, DOT) see BPJ850

ETHYLENE GLYCOL MONOBUTYL ETHER, MANUFACTURE OF, BY-PRODUCTS FROM see EED100

ETHYLENE GLYCOL MONO-DICYCLOPENTENYL ETHER see EJJ100

ETHYLENE GLYCOL MONOETHYL ETHER see EES350

ETHYLENE GLYCOL MONOETHYL ETHER (DOT) see EES350

ETHYLENE GLYCOL MONOETHYL ETHER ACETATE (MAK, DOT) see EES400

ETHYLENE GLYCOL MONOETHYL ETHER ACRYLATE see ADT500

ETHYLENE GLYCOL MONOETHYL ETHER PROPENOATE see ADT500

ETHYLENE GLYCOL MONOHEPTYL ETHER see HBN250

ETHYLENE GLYCOL, MONO-2,4-HEXADIENE ETHER see HCT500

ETHYLENE GLYCOL MONOHEXYL ETHER see HFT500

ETHYLENE GLYCOL, MONO(HYDROGEN MALEATE) see EJG500

ETHYLENE GLYCOL MONOISOBUTYL ETHER see IIP000

ETHYLENE GLYCOL, MONOISOPROPYL ETHER see INA500

ETHYLENE GLYCOL, MONOMETHACRYLATE see EJH000

ETHYLENE GLYCOL MONOMETHYL ETHER ACETATE see EJJ500

ETHYLENE GLYCOL MONOMETHYL ETHER ACETYLRICINOLEATE see MIF500

ETHYLENE GLYCOL MONOMETHYL ETHER ACRYLATE see MEM250

ETHYLENE GLYCOL MONOMETHYL ETHER ACRYLATE see MIF750

ETHYLENE GLYCOL MONOMETHYL ETHER (MAK, DOT) see EJH500

ETHYLENE GLYCOL MONOMETHYL ETHER OLEATE see MEP750

ETHYLENE GLYCOL MONOMETHYLPENTYL ETHER see EJK000

ETHYLENE GLYCOL MONO-2-METHYLPENTYL ETHER see EJK000

ETHYLENE GLYCOL MONONITRATE VINYL ETHER see EJK100

ETHYLENE GLYCOL MONOPHENYL ETHER see PER000

ETHYLENEGLYCOL MONOPHENYL ETHER PROPIONATE see EJK500

ETHYLENE GLYCOL-MONO-PROPYL ETHER see PNG750

ETHYLENE GLYCOL-MONO-n-PROPYL ETHER see PNG750

ETHYLENE GLYCOL MONOPROPYL ETHER ACETATE see PNA225

ETHYLENE GLYCOL, MONOSTEARATE see EJM500

ETHYLENE GLYCOL MONO-2,6,8-TRIMETHYL-4-NONYL ETHER see EJL000

ETHYLENE GLYCOL MONOVINYL ETHER see EJL500

ETHYLENE GLYCOL PHENYL ETHER see PER000

ETHYLENE GLYCOL PHENYL ETHER ACETATE see EJL600

ETHYLENE GLYCOL SILICATE see EJM000

ETHYLENE GLYCOL STEARATE see EJM500

ETHYLENE GLYCOL VINYL ETHER see EJL500

ETHYLENE HEXACHLORIDE see HCI000

ETHYLENE HOMOPOLYMER see PJS750

ETHYLENEIMINE see EJM900

ETHYLENE IMINE, INHIBITED (DOT) see EJM900

1-ETHYLENEIMINO-2-HYDROXY-3-BUTENE see VMA000

ETHYLENE IODOHYDRIN see IEL000

ETHYLENE MALEIC ANHYDRIDE CO-POLYMER see EJM950

ETHYLENE MERCAPTOACETATE see MCN000

ETHYLENE METHACRYLATE see BKM250

N,N-ETHYLENE-N'-METHYLUREA see EJN400

ETHYLENE MONOCHLORIDE see VNP000

1,8-ETHYLENE NAPHTHALENE see AAE750

1,8-ETHYLENENAPHTHALENE see AAF275

ETHYLENE NITRATE see EJG000

ETHYLENE, NITRO-, POLYMERS see NFY560

ETHYLENENITROSOUREA see NKL000

ETHYLENE OXIDE see EJN500

ETHYLENE OXIDE, mixed with CARBON DIOXIDE see EJO000

ETHYLENE OXIDE and CARBON DIOXIDE MIXTURES (DOT) see EJO000

ETHYLENE OXIDE CYCLIC TETRAMER see COD475

ETHYLENE OXIDE, ETHYL- see BOX750

ETHYLENE OXIDE, LANOLIN ADDUCT see LAU560

ETHYLENE OXIDE POLYMER see EJO025

ETHYLENE OXIDE and PROPYLENE OXIDE BLOCK POLYMER see PJK200

ETHYLENE (OXYDE d') (FRENCH) see EJN500

1-ETHYLENEOXY-3,4-EPOXYCYCLOHEXANE see VOA000

ETHYLENE OZONIDE see EJO500

1,4-ETHYLENEPIPERAZINE see DCK400

1,4-ETHYLENEPIPERIDINE see QUJ400

ETHYLENE POLYMER see PJS750

ETHYLENE POLYMERS (8CI) see PJS750

ETHYLENE, POLYMER WITH MALEIC ANHYDRIDE see EJM950

ETHYLENE PROPIONATE see COB260

ETHYLENESUCCINIC ACID see SMY000

ETHYLENE SULFATE see EJP000

ETHYLENE SULFIDE see EJP500

ETHYLENE SULFITE see COV750

1,2-ETHYLENE SULFITE see COV750

ETHYLENE SULPHIDE see EJP500

ETHYLENE TEREPHTHALATE POLYMER see PKF750

ETHYLENE TETRACHLORIDE see PCF275

ETHYLENETHIOCARBAMYL SULFIDE see EJQ000

ETHYLENE THIOCYANATE see EJC035

ETHYLENE THIOUREA see IAQ000

1,3-ETHYLENE-2-THIOUREA see IAQ000

N,N'-ETHYLENETHIOUREA see IAQ000

ETHYLENETHIOUREA mixed with SODIUM NITRITE see IAR000

l'ETHYLENE THIOUREE (FRENCH) see IAQ000

ETHYLENE THIURAM MONOSULFIDE see EJQ000

ETHYLENE THIURAM MONOSULPHIDE see EJQ000

ETHYLENE, TRIBROMO- see THV100

ETHYLENE TRICHLORIDE see TIO750

ETHYLENE, TRICHLOROFLUORO- see TIP400

ETHYLENE TRITHIOCARBONATE see EJQ100

ETHYLENE UNDECANE DICARBOXYLATE see EJQ500

ETHYLENE UREA see IAS000

1,3-ETHYLENE UREA see IAS000

ETHYLENGLYCOL MONOVINYL ESTER (RUSSIAN) see EJL500

ETHYLENGLYKOLDIGLYCIDYLETHER see EEA600

ETHYLENGLYKOLDINITRAT (CZECH) see EJG000

ETHYLENGLYKOLTRICHLORACETAT see EJD600

ETHYLENIMINE see EJM900

ETHYLENIMINE, N-CHLORO- see CEI340

1-ETHYLENIMINO-2-HYDROXYBUTENE see VMA000

1-N-ETHYLEPHEDRINE HYDROCHLORIDE see EJR500

ETHYL-2,3-EPOXYBUTYRATE see EJS000

ETHYL-α,β-EPOXYHYDROCINNAMATE see EOK600

ETHYL α,β-EPOXY-β-METHYLHYDROCINNAMATE see ENC000

ETHYL 2,3-EPOXY-3-METHYL-3-PHENYLPROPIONATE see ENC000

ETHYL-α,β-EPOXY-α-PHENYLPROPIONATE see EOK600

(8-β)-6-ETHYLERGOLINE-8-ACETAMIDE TARTRATE see EJS100

ETHYL ESTER of N-ACETYL-dl-SARCOLYSYL-l-PHENYLALANINE see ACD250

ETHYL ESTER of N-ACETYL-dl-SARCOSYLYL-dl-VALINE see ARM000

ETHYL ESTER-1-CYSTEINE HYDROCHLORIDE (9CI) see EHU600

ETHYL ESTER l-CYSTEINE HYDROCHLORIDE (9CI) see MBX800

ETHYL ESTER of 1,2,5,6-DIBENZANTHRACENE-endo-α,β-SUCCINO GLYCINE see EJT000

ETHYL ESTER of 4,4'-DICHLOROBENZILIC ACID see DER000

ETHYL ESTER of DIMETHYLDITHIOCARBAMIC ACID see EIJ500

ETHYL ESTER of O,O-DIMETHYLDITHIOPHOSPHORYL α-PHENYL ACETATE ACID see DRR400

ETHYL ESTER of 2,3-EPOXY-3-PHENYLBUTANOIC ACID see ENC000

O-ETHYLESTER KYSELINY DICHLORTHIOFOSFORECNE (CZECH) see MRI000

ETHYLESTER KYSELINY DUSITE see ENN000

ETHYLESTER KYSELINY N-ETHYL-N-NITROSOKARBAMINOVE see NKE500

ETHYLESTER KYSELINY FLUOROCTOVE see EKG500

ETHYLESTER KYSELINY GALLOVE see EKM100

ETHYLESTER KYSELINY KROTONOVE see EHO200

ETHYLESTER KYSELINY KYANOCTOVE see EHP500

ETHYLESTER KYSELINY METHYLKARBAMINOVE see EMQ500

ETHYLESTER KYSELINY MLECNE see LAJ000

ETHYLESTER KYSELINY ORTHOMRAVENCI (CZECH) see ENY500

ETHYL ESTER of METHANESULFONIC ACID see EMF500

ETHYL ESTER of 3-METHYLCHOLANTHRENE-endo-α,β-SUCCINOGLYCINE see EJT500

ETHYL ESTER of METHYLNITROSO-CARBAMIC ACID see MMX250

ETHYL ESTER of METHYLSULFONIC ACID see EMF500

ETHYL ESTER of METHYLSULPHONIC ACID see EMF500

ETHYL ESTER of MONOACETIC ACID see MFW100

11-β-ETHYLESTRADIOL see EJT575

11-β-ETHYLESTRA-1,3,5(10)-TRIENE-3,17-β-DIOL see EJT575

ETHYLESTRENOL see EJT600

N-ETHYL-ETHANAMINE see DHJ200

N-ETHYL-ETHANAMINE HYDROCHLORIDE (9CI) see DIS500

ETHYL ETHANE SULFONATE see DKB139

ETHYL ETHANOATE see EFR000

ETHYL ETHER see EJU000

ETHYL ETHER of 10-(β-MORPHOLYLPROPIONYL)PHENTHIAZIN ECARBAMINO ACID HYDROCHLORIDE see EEI050

ETHYL ETHER of PROPYLENE GLYCOL see EJV000

11-β-ETHYL-17-α-ETHINYLESTRADIOL see EJV400

13-ETHYL-17-α-ETHINYL-17-HYDROXYGON-4,9,11-TRIEN-3-ONE see ENX575

ETHYL-β-ETHOXYPROPIONATE see EJV500

3-ETHYL-2-(5-(3-ETHYL-2-BENZOTHIAZOLINYLIDENE)-1,3-PENTADIENYL)BENZOTHIAZOLIUM IODIDE see DJT800

ETHYL-N-ETHYL CARBAMATE see EJW500

S-ETHYL-N-ETHYL-N-CYCLOHEXYLTHIOLCARBAMATE see EHT500

3-ETHYL-5-(4,4-ETHYLENEDIOXYPIPERIDINO-1-METHYL)-6,7-DIHYDRO-2-METHYLINDOL-4(5H)-ONE see AFH550

ETHYL ETHYLENE OXIDE see BOX750

2-ETHYLETHYLENIMINE see EGN500

2-ETHYL-N-(2-ETHYLHEXYL)-1-HEXANAMINE see DJA800

ETHYL-6-(ETHYL(2-HYDROXYPROPYL)AMINO)-3-PYRIDAZINECARBAZATE see CAK275

ETHYL-2-(6(ETHYL(2-HYDROXYPROPYL)AMINO)-3-PYRIDAZINYL)HYDRAZINECARBOXYLATE see CAK275

ETHYL(5-ETHYLMERCURI-3-(1,2,4-THIADIAZOLYL)THIO)MERCURY(II) see EJW600

ETHYL N-ETHYLNITROSOCARBAMATE see NKE500

ETHYL 3-ETHYL-4-OXO-5-PIPERIDINO-Δ²,ᵅ-THIAZOLIDINEACETATE see EOA500

ETHYL (Z)-(3-ETHYL-4-OXO-5-PIPERIDINOTHIAZOLIDIN-2-YLIDENE)ACETATE see EOA500

5-ETHYL-5-(1-ETHYLPROPYL)BARBITURIC ACID see EJY000

5-ETHYL-5-(1-ETHYLPROPYL)2,4,6(1H,3H,5H)-PYRIMIDINETRIONE see EJY000

1-ETHYL-4-(p-(p-((p-((1-ETHYLPYRIDINIUM-4-YL)AMINO)-2-AMINOPHENYL)CARBAMOYL)CINNAMA MIDO)ANILINO)PYRIDINIUM, DIBROMIDE see EJZ000

1-ETHYL-4-(p-(p-((1-ETHYLPYRIDINIUM-4-YL)AMINO)BENZAMIDO)ANILINO)QUIN OLINIUM DIBROMIDE see EKA000

1-ETHYL-4-(p-((p-((1-ETHYLPYRIDINIUM-4-YL)AMINO)PHENYL)CARBAMOYL)ANILIN O)QUINOLINIUM, DIBROMIDE see EKA500

1-ETHYL-4-(p-(p-((p-((1-ETHYLPYRIDINIUM-4-YL)AMINO)PHENYL)CARBAMOYL)CINNA MAMIDO)ANILINO)PYRIDINIUM, DI-p-TOLUENE SULFONATE see EKB000

1-ETHYL-4-(p-((p-(p-(1-ETHYLPYRIDINIUM-4-YL)PHENYL)CARBAMOYL)ANILINO)QUIN OLINIUM), DI-p-TOLUENE SULFONATE see EKC500

ETHYL 3-ETHYL-4-QUINOLINECARBOXYLATE see EHM200

1-ETHYL-6-((p-(p-((1-ETHYLQUINOLINIUM-6-YL)CARBAMOYL)BENZAMIDO)BENZAMI DO)QUINOLINIUM), DI-p-TOLUENE SULFONATE see EKC990

1-ETHYL-7-((p-(p-((1-ETHYLQUINOLINIUM-7-YL)CARBAMOYL)BENZAMIDO)BENZAMI DO)QUINOLINIUM), DI-p-TOLUENE SULFONATE see EKD000

O,O-ETHYL S-2(ETHYLTHIO)ETHYL PHOSPHORODITHIOATE see DXH325

ETHYLETHYNE see EFS500

13-ETHYL-17-α-ETHYNYLGON-4-EN-17-β-OL-3-ONE see NNQ500

13-ETHYL-17-α-ETHYNYLGON-4-EN-17-β-OL-3-ONE see NNQ520

13-ETHYL-17-α-ETHYNYL-17-β-HYDROXY-4-GONEN-3-ONE see NNQ500

13-ETHYL-17-α-ETHYNYL-17-β-HYDROXY-4-GONEN-3-ONE see NNQ520

dl-13-β-ETHYL-17-α-ETHYNYL-17-β-HYDROXYGON-4-EN-3-ONE see NNQ500

(±)-13-ETHYL-17-α-ETHYNYL-17-HYDROXYGON-4-EN-3-ONE see NNQ500

17-α-ETHYLETHYNYL-19-NORTESTOSTERONE see EKF550

dl-13-β-ETHYL-17-α-ETHYNYL-19-NORTESTOSTERONE see NNQ500

ETHYL ETRINOATE see EMJ500

ETHYLEX GUM 2020 see HLB400

ETHYL (E,Z)-2,4-DECADIENOATE see EHV100

2-ETHYL FENCHOL see EKF575

ETHYL FLAVONE-7-OXYACETATE see ELH600

ETHYL-7-FLAVONOXYACETATE see ELH600

ETHYL FLAVON-7-YLOXYACETATE see ELH600

ETHYL FLAVONYL-7-OXYACETATE see ELH600

ETHYL FLUCLOZEPATE see EKF600

ETHYL FLUORIDE (DOT) see FIB000

ETHYL FLUOROACETATE see EKG500

ETHYL-10-FLUORODECANOATE see EKI000

ETHYL-ω-FLUORODECANOATE see EKI000

1-ETHYL-6-FLUORO-1,4-DIHYDRO-4-OXO-7-(1-PIPERAZINYL)-1,8-NAPHTHYRIDINE-3-CARBOXYLIC ACID see EAU100

1-ETHYL-6-FLUORO-1,4-DIHYDRO-4-OXO-7-(1-PIPERAZINYL)-3-QUINOLINECARBOXYLIC ACID see BAB625

ETHYL-6-FLUOROHEXANOATE see EKJ500

ETHYL-ω-FLUOROHEXANOATE see EKJ500

ETHYL-9-FLUORONONANECARBOXYLATE see EKI000

ETHYL-8-FLUORO OCTANOATE see EKK500

ETHYL-ω-FLUOROOCTANOATE see EKK500

ETHYL-5-FLUOROPENTANECARBOXYLATE see EKJ500

ETHYL-p-FLUOROPHENYL SULFONE see FLG000

ETHYL FLUOROSULFATE see EKK550

ETHYLFORMAMIDE see EKK600

N-ETHYLFORMAMIDE see EKK600

ETHYL FORMATE see EKL000

ETHYLFORMIAAT (DUTCH) see EKL000
ETHYLFORMIC ACID see PMU750
ETHYL FORMIC ESTER see EKL000
1-ETHYL-1-FORMYLHYDRAZINE see EKL250
N-ETHYL-N-FORMYLHYDRAZINE see EKL250
6-ETHYL-7-FORMYL-1,1,4,4-TETRAMETHYL-1,2,3,4-TETRAHYDRONAPHTHALENE see FNK200
ETHYL FUMARATE see DJJ800
5-ETHYL-2(5H)-FURANONE see EKL500
4-ETHYL-3-FURAZANONE see ELK500
ETHYL FUROATE see EKM000
ETHYL GALLATE see EKM100
ETHYL GERANYL ETHER see GDG100
α-ETHYL GLYCEROL ETHER see EFF000
α-ETHYL GLYCIDYL ETHER see EBQ700
ETHYL GLYCIDYL ETHER see EKM200
ETHYL GLYCOLATE see EKM500
ETHYL GLYCOLATE ISOCYANATE see ELS600
ETHYLGLYKOLACETAT (GERMAN) see EES400
ETHYL GLYME see EJE500
ETHYL GREEN see BAY750
ETHYL-p-(6-GUANIDINOHEXANOYLOXY)BENZOATE METHANESULFONATE see GAD400
ETHYL GUSATHION see EKN000
ETHYL GUTHION see EKN000
ETHYL 10-HENDECENOATE see EQD200
ETHYL HEPTANOATE see EKN050
ETHYL HEPTANOATE see EKN050
ETHYL n-HEPTANOATE see EKN050
ETHYL HEPTOATE see EKN050
ETHYL HEPTYLATE see EKN050
(3-ETHYL-N-HEPTYL)METHYLCARBINOL see ENW500
1-ETHYL-1-HEPTYLPIPERIDINIUM BROMIDE see EKN100
ETHYLHEXABITAL see TDA500
ETHYL HEXADECYL DIMETHYL AMMONIUM BROMIDE see EKN500
4-ETHYL-4-HEXADECYL MORPHOLINIUM ETHYL SULFATE see EKN550
1-ETHYL-1-HEXADECYLPIPERIDINIUM BROMIDE see EKN600
ETHYL HEXAFLUORO-2-BROMOBUTYRATE see EKO000
sec-ETHYL HEXAHYDRO-1H-AZEPINE-1-CARBOTHIOATE see EKO500
5-ETHYLHEXAHYDRO-4,6-DIOXO-5-PHENYLPHYIMIDINE see DBB200
5-ETHYLHEXAHYDRO-5-PHENYLPYRIMIDINE-4,6-DIONE see DBB200
2-ETHYLHEXALDEHYDE see BRI000
ETHYLHEXALDEHYDE (DOT) see BRI000
ETHYL-1-HEXAMETHYLENEIMINECARBOTHIOLATE see EKO500
S-ETHYL-1-HEXAMETHYLENEIMINOTHIOCARBAMATE see EKO500
S-ETHYL-N-HEXAMETHYLENETHIOCARBAMATE see EKO500
2-ETHYLHEXANAL see BRI000
ETHYL HEXANEDIOL see EKV000
2-ETHYLHEXANEDIOL-1,3 see EKV000
2-ETHYLHEXANE-1,3-DIOL see EKV000
2-ETHYL-1,3-HEXANEDIOL see EKV000
ETHYL HEXANOATE (FCC) see EHF000
2-ETHYLHEXANOIC ACID see BRI250
2-ETHYLHEXANOIC ACID CHLORIDE see EKO600
2-ETHYLHEXANOIC ACID, 2-ETHYLHEXYL ESTER see EKW000
2-ETHYLHEXANOIC ACID SODIUM SALT see SGS600

(R)-2-ETHYLHEXANOIC ACID SODIUM SALT see EKO700
2-ETHYLHEXANOIC ACID, VINYL ESTER see VOU000
2-ETHYLHEXANOL see EKQ000
2-ETHYL-1-HEXANOL see EKQ000
2-ETHYL-1-HEXANOL ESTER with DIPHENYL PHOSPHATE see DWB800
2-ETHYL-1-HEXANOL HYDROGEN PHOSPHATE see BJR750
2-ETHYL-1-HEXANOL HYDROGEN SULFATE, SODIUM SALT see TAV750
2-ETHYL-1-HEXANOL PHOSPHATE see TNI250
2-ETHYL-1-HEXANOL SILICATE see EKQ500
2-ETHYL-1-HEXANOL SULFATE SODIUM SALT see TAV750
2-ETHYLHEXANOYL CHLORIDE see EKO600
((2-ETHYLHEXANOYL)OXY)TRIBUTYLSTANNANE see TID250
2-ETHYLHEXANYL ACETATE see OEE000
2-ETHYL HEXENAL see EKQ600
2-ETHYLHEXENAL see EKR000
2-ETHYL-2-HEXENAL see EKR000
2-ETHYL-1-HEXENE see EKR500
2-ETHYL HEXENE-1 see EKR500
2-ETHYL-2-HEXENOIC ACID see EKS000
ETHYL cis-3-HEXENYL ACETAL see EKS100
2-ETHYLHEXOIC ACID see BRI250
2-ETHYLHEXOIC ACID, VINYL ESTER see VOU000
ETHYL HEXYL ACETAL see EKS120
2-ETHYLHEXYL ACETATE see OEE000
β-ETHYLHEXYL ACETATE see OEE000
ETHYL-2-HEXYL ACETOACETATE see EKS150
ETHYL α-HEXYLACETOACETATE see EKS150
2-ETHYLHEXYL ACRYLATE see ADU250
2-ETHYLHEXYL ACRYLATE-HYDROXYETHYL ACRYLATE-METHYL ACRYLATE POLYMER see EKS200
ETHYLHEXYL ACRYLATE 50:50 MIXTURE see ADU500
2-ETHYLHEXYL ALCOHOL see EKQ000
2-ETHYL HEXYLAMINE see EKS500
2-ETHYLHEXYL-3-AMINOPROPYL ETHER see ELA000
N-(2-ETHYLHEXYL)ANILINE see EKT000
5-ETHYL-5-HEXYLBARBITURIC ACID SODIUM SALT see EKT500
N-(2-ETHYLHEXYL)BICYCLO-(2,2,1)-HEPT-5-ENE-2,3-DICARBOXIMIDE see OES000
2-ETHYLHEXYL CARBAMATE see EKT600
2-ETHYLHEXYL-1-CHLORIDE see EKU000
2-ETHYLHEXYL-6-CHLORIDE see EKU100
2-ETHYLHEXYL 2-CYANO-3,3-DIPHENYLACRYLATE see ODY150
2-ETHYLHEXYL α-CYANO-β,β'-DIPHENYLACRYLATE see ODY150
2-ETHYLHEXYL 2-CYANO-3,3-DIPHENYL-2-PROPENOATE see ODY150
2-ETHYLHEXYL 2-CYANO-3-PHENYLCINNAMATE see ODY150
N-(2-ETHYLHEXYL)CYCLOHEXYLAMINE see EKU500
2-ETHYLHEXYL DIPHENYL ESTER PHOSPHORIC ACID see DWB800
2-ETHYLHEXYL DIPHENYLPHOSPHATE see DWB800
ETHYL HEXYLENE GLYCOL see EKV000
2-ETHYLHEXYL-9,10-EPOXYOCTADECANOATE see EKV500
2-ETHYLHEXYL EPOXYSTEARATE see EKV500
2-ETHYLHEXYLESTER KYSELINY 2-ETHYLKAPRONOVE see EKW000
2-ETHYLHEXYL ETHANOATE see OEE000

2-ETHYLHEXYL-2-ETHYLHEXANOATE see EKW000
2-ETHYLHEXYL FUMARATE see DVK600
2-ETHYLHEXYL GLYCIDYL ETHER see GGY100
N-(2-ETHYLHEXYL)-3-HYDROXYBUTYRAMIDE HYDROGEN SUCCINATE see BPF825
N-2-ETHYLHEXYLIMIDEENDOMETHYLENETETRAHYDROPHTHALIC ACID see OES000
2-ETHYLHEXYL-p-IODOBENZYL CARBONATE see EKW100
2-ETHYLHEXYL (3-ISOCYANATOMETHYLPHENYL)CARBAMATE see EKW200
2-ETHYLHEXYLMALEINAN DI-N-BUTYLCINICITY (CZECH) see BJR250
2-ETHYLHEXYL MERCAPTOACETATE see EKW300
2-ETHYLHEXYL METHACRYLATE see EKW500
2-ETHYL-1-HEXYL METHACRYLATE see EKW500
ETHYLHEXYL NITRATE see EKW600
2-ETHYLHEXYL NITRATE see EKW600
N-(2-ETHYLHEXYL)-5-NORBORNENE-2,3-DICARBOXIMIDE see OES000
2-ETHYLHEXYL OCTADECANOATE see OFU300
2-ETHYLHEXYL OCTYLPHENYLPHOSPHITE see EKX000
N-2-ETHYLHEXYL-β-OXYBUTYRAMIDE SEMISUCCINATE see BPF825
2-(2-ETHYLHEXYLOXY)ETHANOL see EKX500
2-((2-ETHYLHEXYL)OXY)ETHANOL see EKX500
4-(2-ETHYLHEXYLOXY)-2-HYDROXYBENZOPHENONE see EKY000
(((2-ETHYLHEXYL)OXY)METHYL)OXIRANE see GGY100
3-((2-ETHYLHEXYL)OXY)PROPANENITRILE see EKZ000
3-(2-ETHYLHEXYLOXY)PROPIONITRILE see EKZ000
2-ETHYLHEXYLOXYPROPYLAMINE see ELA000
3-((2-ETHYLHEXYL)OXY)PROPYLAMINE see ELA000
2-ETHYLHEXYL PALMITATE see OFG100
ETHYLHEXYL PHTHALATE see DVL700
2-ETHYLHEXYL PHTHALATE see DVL700
1-ETHYL-1-HEXYLPIPERIDINIUM BROMIDE see ELA600
2-ETHYLHEXYL-2-PROPENOATE see ADU250
5-ETHYL-5-HEXYL-2,4,6-(1H,3H,5H)-PYRIMIDINETRIONE MONOSODIUM SALT see EKT500
2-ETHYLHEXYL SALICYLATE see ELB000
2-ETHYLHEXYL SEBACATE see BJS250
2-ETHYLHEXYL SODIUM SULFATE see TAV750
2-ETHYLHEXYL STEARATE see OFU300
2-ETHYLHEXYL SULFATE see ELB400
2-ETHYLHEXYL SULFOSUCCINATE SODIUM see DJL000
2-(2-ETHYLHEXYL)-3a,4,7,7a-TETRAHYDRO-4,7-METHANO-1H-ISOINDOLE-1,3(2H)-DIONE see OES000
5-ETHYL-5-HEXYL-2-THIOBARBITURIC ACID SODIUM SALT see SHN275
2-ETHYLHEXYL THIOGLYCOLATE see EKW300
2-ETHYLHEXYL THIOGLYCOLATE see EKW300
2-ETHYLHEXYL VINYL ETHER see ELB500
S-ETHYL-HOMOCYSTEINE see EEI000
S-ETHYL-l-HOMOCYSTEINE see AKB250
S-ETHYL-dl-HOMOCYSTEINE see EEI000
ETHYL HYDRATE see EFU000

ETHYLHYDRAZINE HYDROCHLORIDE see ELC000

ETHYL HYDRIDE see EDZ000

ETHYLHYDROCUPREINE see HHR700

ETHYLHYDROCUPREINE HYDROCHLORIDE see ELC500

ETHYL HYDROGEN ADIPATE see ELC600

ETHYL(HYDROGEN CYSTEINATO)MERCURY see EME000

ETHYL(HYDROGEN p-MERCAPTOBENZENESULFONATO)MERCURY SODIUM SALT see SKH150

ETHYL HYDROGEN PEROXIDE see ELD000

ETHYL HYDROPEROXIDE see ELD000

ETHYL HYDROPERSULFIDE see EEB000

ETHYL HYDROSULFIDE see EMB100

ETHYL HYDROXIDE see EFU000

1-ETHYL-3-(HYDROXYACETYL)INDOLE see ELD100

4'-ETHYL-4-HYDROXYAZOBENZENE see ELD500

ETHYL-o-HYDROXYBENZOATE see SAL000

ETHYL-p-HYDROXYBENZOATE see HJL000

2-ETHYL-3-(p-HYDROXYBENZOYL)BENZOFURAN see BBJ500

2-ETHYL-4'-HYDROXY-3-BENZOYLBENZOFURAN see BBJ500

ETHYL-2 (HYDROXY-4 BENZOYL)-3 BENZOFURANNE see BBJ500

ETHYL-2-HYDROXY-2,2-BIS(4-CHLOROPHENYL)ACETATE see DER000

N-ETHYL-N-(4-HYDROXYBUTYL)NITROSOAMINE see ELE500

ETHYL-N-HYDROXYCARBAMATE see HKQ025

ETHYL 3-HYDROXYCARBANILATE see ELE600

ETHYL m-HYDROXYCARBANILATE see ELE600

ETHYL-m-HYDROXYCARBANILATE CARBANILATE (ESTER) see EEO500

α-ETHYL-1-HYDROXYCYCLOHEXANEACETIC ACID see COW700

ETHYLHYDROXYCYCLOHEXANECARBONITRILE ACETATE (ESTER) see EHQ000

(E)-ETHYLHYDROXYDIAZENE POTASSIUM SALT see ELE700

(Z)-ETHYLHYDROXYDIAZENE POTASSIUM SALT see ELE720

N-ETHYL-3-HYDROXY-N,N-DIMETHYL-BENZENAMINIUM BROMIDE (9CI) see TAL490

N-ETHYL-3-HYDROXY-N,N-DIMETHYLBENZENAMINIUM CHLORIDE (9CI) see EAE600

13-ETHYL-17-HYDROXY-18,19-DINOR-17-α-PREGNA-4,9,11-TRIEN-20-YN-3-ONE see ENX575

(±)-13-ETHYL-17-HYDROXY-18,19-DINOR-17-α-PREGN-4-EN-20-YN-3-ONE see NNQ500

N-ETHYL-2-((HYDROXYDIPHENYLACETYL)OXY)-N,N-DIMETHYLETHANAMINIUM CHLORIDE see ELF500

16-ETHYL-17-HYDROXYESTER-4-EN-3-ONE see ELF100

16-β-ETHYL-17-β-HYDROXYESTER-4-EN-3-ONE ACETATE see ELF110

16-β-ETHYL-17-β-HYDROXYESTR-4-EN-3-ONE see ELF100

16-β-ETHYL-17-β-HYDROXY-4-ESTREN-3-ONE see ELF100

2-(N-ETHYL-N-2-HYDROXYETHYLAMINO)ETHANOL see ELP000

1-ETHYL-1-(2-HYDROXYETHYL)AZIRIDINIUM SALT with 2,4,6-TRINITROBENZENESULFONIC ACID see ELG000

1-ETHYL-1-(2-HYDROXYETHYL)AZIRIDINIUM-2,4,6-TRINITROBENZENESULFONATE see ELG000

ETHYL (2-HYDROXYETHYL)DIMETHYLAMMONIUM BENZILATE CHLORIDE see ELF500

ETHYL(2-HYDROXYETHYL)DIMETHYLAMMONIUM CHLORIDE BENZILATE see ELF500

ETHYL(2-HYDROXYETHYL)DIMETHYL-AMMONIUM SULFATE (SALT), BIS(DIBUTYLCARBAMATE) see DDW000

ETHYL(2-HYDROXYETHYL)ETHYLENIMONIUM PICRYLSULFONATE see ELG000

ETHYL-β-HYDROXYETHYLETHYLENIMONIUM PICRYLSULFONATE see ELG000

1-ETHYL-1-(β-HYDROXYETHYL)ETHYLENIMONIUM PICRYLSULFONATE see ELG000

N-ETHYL-N-(2-HYDROXYETHYL)-3-METHYL-4-NITROSOANILINE see ELG100

ETHYL-2-HYDROXYETHYLNITROSAMINE see ELG500

N-ETHYL-N-HYDROXYETHYLNITROSAMINE see ELG500

ETHYL-N-(2-HYDROXYETHYL)-N-NITROSOCARBAMATE see ELH000

ETHYL-2-HYDROXYETHYL SULFIDE see EPP500

ETHYL-2-HYDROXYETHYL THIOETHER see EPP500

N-ETHYL-N-(2-HYDROXYETHYL)-m-TOLUIDINE see HKS100

ETHYL-7-HYDROXYFLAVONE see ELH600

(E)-6-ETHYL-5-HYDROXY-3-(HYDROXYMETHYL)-5,6,7,8-TETRAHYDROCHROMONE see ELH650

ETHYL 2-HYDROXYISOBUTYRATE see ELH700

ETHYL α-HYDROXYISOBUTYRATE see ELH700

ETHYL-4-HYDROXY-3-METHOXYBENZOATE see EQE500

(8-α,9R)-1-ETHYL-9-HYDROXY-6'-METHOXYCINCHONAN-1-IUM IODIDE see QIS000

4-ETHYL-7-HYDROXY-3-(p-METHOXYPHENYL)COUMARIN see ELH800

ETHYL-2-(HYDROXYMETHYL)ACRYLATE see ELI500

ETHYL-α-(HYDROXYMETHYL)ACRYLATE see ELI500

ETHYL 4-HYDROXY-2-METHYL-2H-1,2-BENZOTHIAZINE-3-CARBOXYLATE 1,1-DIOXIDE see ELI550

1-(2-ETHYL-7-(2-HYDROXY-3-((1-METHYLETHYL)AMINO)PROPOXY)-4-BENZOFURANYL) ETHANONE see ELI600

α-ETHYL-β-(HYDROXYMETHYL)-1-METHYL-IMIDAZOLE-5-BUTYRIC ACID, γ-LACTONE see PIF000

1-ETHYL-7-HYDROXY-2-METHYL-1,2,3,4,4a,9,10,10a-OCTAHYDROPHENANTHRENE-2-CARBOXYLIC ACID see DYB000

1-ETHYL-3-HYDROXY-1-METHYL-PIPERIDINIUM BROMIDE BENZILATE see PJA000

2-ETHYL-2-(HYDROXYMETHYL)-1,3-PROPANEDIOL, CYCLIC PHOSPHATE (1:1) see ELJ500

2-ETHYL-2-(HYDROXYMETHYL)-1,3-PROPANEDIOL TRIACETOACETATE see ELJ600

2-ETHYL-2-(HYDROXYMETHYL)-1,3-PROPANEDIOL TRIACRYLATE see TLX175

2-ETHYL-2-HYDROXYMETHYL-1,3-PROPANEDIOL TRIMETHACRYLATE see TLX250

ETHYL 2-HYDROXY-2-METHYLPROPANOATE see ELH700

ETHYL-4-HYDROXY-3-MORPHOLINOMETHYLBENZOATE see ELK000

5-ETHYL-2'-HYDROXY-2(N)-(3-METHYL-2-BUTENYL)-9-METHYL-6,7-BENZOMORPHAN see DOQ600

17-α-ETHYL-17-HYDROXYNORANDROSTENONE see ENX600

17-α-ETHYL-17-HYDROXY-4-NORANDROSTEN-3-ONE see ENX600

17-α-ETHYL-17-HYDROXY-19-NORANDROST-4-EN-3-ONE see ENX600

3-ETHYL-4-HYDROXY-1,2,5-OXADIAZOLE see ELK500

β-ETHYL-β-HYDROXYPHENETHYL CARBAMATE see HNJ000

β-ETHYL-β-HYDROXYPHENETHYL CARBAMIC ACID ESTER see HNJ000

ETHYL (3-HYDROXYPHENYL)CARBAMATE see ELE600

ETHYL N-(3-HYDROXYPHENYL)CARBAMATE see ELE600

3-((10-ETHYL-11-(p-HYDROXYPHENYL)DIBENZ(B,F)OXEPIN-3-YL)OXY)-1,2-PROPANEDIOL HYDRATE (4:1) see ELK600

ETHYL(m-HYDROXYPHENYL)DIMETHYLAMMONIUM BROMIDE (8CI) see TAL490

ETHYL(m-HYDROXYPHENYL)DIMETHYLAMMONIUM CHLORIDE see EAE600

ETHYL-p-HYDROXYPHENYL KETONE see ELL500

3-ETHYL-2-(p-HYDROXYPHENYL)-1-METHYLINDEN-6-OL see ELL550

ETHYL (4-(m-HYDROXYPHENYL)-1-METHYL)-4-PIPERIDYL KETONE see KFK000

ETHYL 3-(m-HYDROXYPHENYL)-1-METHYL-3-PYROLIDINECARBOXYLATE see ELL500

ETHYL 2-HYDROXYPROPIONATE see LAJ000

ETHYL α-HYDROXYPROPIONATE see LAJ000

3-(6-(ETHYL-(2-HYDROXYPROPYL)AMINO)PYRIDAZIN-3-YL)CARBAZIC ACID ETHYL ESTER see CAK275

2-(6-ETHYL(2-HYDROXYPROPYL)AMINO)-3-PYRIDAZINYL)-HYDRAZINECARBOXYLIC ACID ETHYL ESTER see CAK275

N-ETHYL-N-(3-HYDROXYPROPYL)NITROSAMINE see ELM000

(S)-4-ETHYL-4-HYDROXY-1H-PYRANO(3',4':6,7)INDOLIZINO(1,2-b)QUINOLINE-3,14(4H,12H)-DIONE see CBB870

2-ETHYL-3-HYDROXY-4H-PYRAN-4-ONE see EMA600

α-ETHYL-3-HYDROXY-2,4,6-TRIIODOHYDROCINNAMIC ACID see IFZ800

α-ETHYL-β-(3-HYDROXY-2,4,6-TRIIODOPHENYL)PROPIONIC ACID see IFZ800

ETHYL HYPOCHLORITE see ELM500
ETHYLIC ACID see AAT250
ETHYLIDENE ACETONE see PBR500
5-ETHYLIDENEBICYCLO(2.2.1)HEPT-2-ENE see ELO500
N,N'-ETHYLIDENE-BIS(ETHYL CARBAMATE) see ELO000
1,1'-(ETHYLIDENEBIS(OXY)BISBUTANE see DDT400
1,1'-(ETHYLIDENE)BIS(OXY)BIS(2-CHLOROETHANE) see DFG600
ETHYLIDENE BROMIDE see DDN800
ETHYLIDENE CHLORIDE see DFF809
7-ETHYLIDENECYCLOPENT(b)OXIRENO(c)PYRIDINE HEXAHYDRO DERIV. see ELM600
3-ETHYLIDENE-3,4,5,6,9,11,13,14,14A,14B-DECAHYDRO-6-HYDROXY-5,6-DIMETHYL(1,6)DIOXACYCLODODECINO(2,3,4-GH)-PYRROLIZINE-2,7-DIONE see IDG500
ETHYLIDENE DIBROMIDE see DDN800
ETHYLIDENEDICARBAMIC ACID, DIETHYL ESTER see ELO000
ETHYLIDENE DICHLORIDE see DFF809
ETHYLIDENE DIETHYL ETHER see AAG000
ETHYLIDENE DIFLUORIDE see ELN500
5-ETHYLIDENEDIHYDRO-2(3H)-FURANONE see HLF000
ETHYLIDENE DIMETHYL ETHER see DOO600
ETHYLIDENE DINITRATE see ELN600
ETHYLIDENE DIURETHAN see ELO000
ETHYLIDENE FLUORIDE see ELN500
ETHYLIDENE-2(5H)-FURANONE see MOW500
ETHYLIDENE GYROMITRIN see AAH000
ETHYLIDENEHYDROXYLAMINE see AAH250
trans-15-ETHYLIDENE-12-β-HYDROXY-4,12-α,13-β-TRIMETHYL 8-OXO-4,8 SECOSENEC-1-ENINE see DMX200
ETHYLIDENELACTIC ACID see LAG000
ETHYLIDENE NORBORNENE see ELO500
5-ETHYLIDENE-2-NORBORNENE see ELO500
ETHYLIDICHLORARSINE see DFH200
ETHYLIDICHLOROARSINE (DOT) see DFH200
4,4',4''-ETHYLIDYNETRISPHENOL see ELO600
ETHYLIMINE see EJM900
2,2'-(ETHYLIMINO)BISETHANOL see ELP000
2,2'-(ETHYLIMINO)DIETHANOL see ELP000
N-ETHYL-2,2'-IMINODIETHANOL see ELP000
3-ETHYL-4-IMINO-5-METHYL-2,5-CYCLOHEXADIEN-1-ONE see ENF300
N-ETHYL-N'-2-INDANYL-N'-PHENYL-1,3-PROPANEDIAMINE HYDROCHLORIDE see DBA475
ETHYL IODIDE see ELP500
ETHYL IODIDE see ELP500
ETHYL IODOACETATE see ELQ000
ETHYL p-IODOBENZYL CARBONATE see ELQ050
ETHYLIODOMETHYLARSINE see ELQ100
ETHYL p-IODOPHENYLETHYL CARBONATE see ELQ200
ETHYL 10-(p-IODOPHENYL)UNDECANOATE see ELQ500
ETHYL-10-(p-IODOPHENYL)UNDECYLATE see ELQ500
5-ETHYL-5-ISOAMYLBARBITURIC ACID see AMX750
5-ETHYL-5-ISOAMYLMALONYL UREA see AMX750
ETHYL ISOBUTANOATE see ELS000

ETHYL ISOBUTENOATE see MIP800
ETHYLISOBUTRAZINE HYDROCHLORIDE see ELQ600
ETHYLISOBUTYLMETHANE, ISOHEPTANE see MKL250
ETHYL ISOBUTYRATE see ELS000
ETHYLISOBUTYRATE (DOT) see ELS000
ETHYL ISOCYANATE see ELS500
ETHYL ISOCYANATE (DOT) see ELS500
ETHYL ISOCYANATOACETATE see ELS600
ETHYL ISOCYANIDE see ELT000
ETHYL 4-ISOCYANOBENZOATE see ELT100
2-ETHYLISOHEXANOL see ELT500
2-ETHYLISOHEXANOL see EMZ000
ETHYL ISONICOTINATE see ELU000
2-ETHYLISONICOTINIC ACID THIOAMIDE see EPQ000
α-ETHYLISONICOTINIC ACID THIOAMIDE see EPQ000
2-ETHYLISONICOTINIC THIOAMIDE see EPQ000
α-ETHYLISONICOTINOYLTHIOAMIDE see EPQ000
ETHYL ISONITRILE see ELT000
ETHYLISOPENTYLBARBITURIC ACID see AMX750
5-ETHYL-5-ISOPENTYLBARBITURIC ACID see AMX750
5-ETHYL-5-ISOPENTYLBARBITURIC ACID SODIUM SALT see AON750
ETHYL ISOPROPENYL KETONE see IMW500
O-ETHYL-O-(2-ISOPROPOXY-CARBONYL)-PHENYL ISOPROPYLPHOSPHORAMIDOTHIOATE see IMF300
ETHYL ISOPROPYLBARBITURIC ACID see ELX000
5-ETHYL-5-ISOPROPYLBARBITURIC ACID see ELX000
ETHYL ISOPROPYL FLUOROPHOSPHONATE see ELX100
ETHYL ISOPROPYLIDENE ACETATE see MIP800
4-ETHYL-3-ISOPROPYL-4-METHYL-1-OXACYCLOBUTAN-2-ONE see DSH400
ETHYL ISOPROPYLNITROSAMINE see ELX500
5-ETHYL-5-ISOPROPYL-2-THIOBARBITURIC ACID SODIUM SALT see SHY500
ETHYLISOTHIAMIDE see EPQ000
ETHYL ISOTHIOCYANATE see ELX525
ETHYL ISOTHIOCYANATOACETATE see ELX530
ETHYL 4-ISOTHIOCYANATOBUTANOATE see ELX527
ETHYL-4-ISOTHIOCYANATOBUTANOATE see ELX527
ETHYL 4-ISOTHIOCYANATOBUTYRATE see ELX527
ETHYL ISOTHIOCYANOACETATE see ELX530
2-ETHYLISOTHIONICOTINAMIDE see EPQ000
α-ETHYLISOTHIONICOTINAMIDE see EPQ000
S-ETHYLISOTHIOURONIUM HYDROGEN SULFATE see ELY550
S-ETHYLISOTHIURONIUM DIETHYL PHOSPHATE see CBI675
S-ETHYLISOTHIURONIUM METAPHOSPHATE see ELY575
ETHYL ISOVALERATE (FCC) see ISY000
ETHYLISOVANILLIN see IKO100
ETHYLJODID see ELP500
ETHYL KETOVALERATE see EFS600
ETHYL 4-KETOVALERATE see EFS600
N-ETHYL-N-2-KYANETHYLANILIN see EHQ500

d-N-ETHYLLACTAMIDE CARBANILATE (ESTER) see CBL500
ETHYL LACTATE (DOT,FCC) see LAJ000
ETHYL LAEVULINATE see EFS600
ETHYL LAURATE see ELY700
ETHYL LEVULATE see EFS600
ETHYL LINALOOL see ELZ000
ETHYLLINALYL ACETAL see ELZ050
ETHYL LINALYL ACETATE see HGI585
ETHYLLITHIUM see ELZ100
ETHYL LOFLAZEPATE see EKF600
ETHYL MAGNESIUM IODIDE see EMA000
ETHYL MALEATE see DJO200
N-ETHYLMALEIMIDE see MAL250
ETHYL MALONATE see EMA500
ETHYL MALTOL see EMA600
ETHYL MANDELATE see EMB000
ETHYLMERCAPTAAN (DUTCH) see EMB100
ETHYL MERCAPTAN see EMB100
ETHYL MERCAPTOACETATE see EMB200
ETHYL-2-MERCAPTOACETATE see EMB200
ETHYL-α-MERCAPTOACETATE see EMB200
ETHYL(MERCAPTOACETATO(2')-O,S)-MERCURATE(1-)-POTASSIUM see PLE750
ETHYL MERCAPTOACETIC ACID see EMB200
ETHYL (2-MERCAPTOETHYL) CARBAMATE S-ESTER with O,O-DIMETHYL PHOSPHORODITHIOATE see EMC000
β-ETHYLMERCAPTOETHYL DIMETHYL THIONOPHOSPHATE see DAO800
2-ETHYLMERCAPTOMETHYLPHENYL-N-METHYLCARBAMATE see EPR000
2-ETHYLMERCAPTO-10-(3-(1-METHYL-4-PIPERAZINYL)PROPYL)PHENOTHIAZINE DIMALEATE see TEZ000
3-ETHYLMERCAPTO-10-(1'-METHYLPIPERAZINYL-4'-PROPYL)PHENOTHIAZINE DIMALEATE see TEZ000
ETHYL 2-(MERCAPTOMETHYLTHIO)ACETATE S-ESTER with O-ETHYLMETHYLPHOSPHONOTHIOATE see EMC100
ETHYL MERCAPTOPHENYLACETATE-O,O-DIMETHYL PHOSPHOROCITHIOATE see DRR400
9-ETHYL-6-MERCAPTOPURINE see EMC500
ETHYLMERCURIC ACETATE see EMD000
ETHYLMERCURIC CHLORIDE see CHC500
ETHYLMERCURIC CYSTEINE see EME000
ETHYLMERCURIC DICYANDIAMIDE see EME025
ETHYLMERCURICHLORENDIMIDE see EME050
ETHYLMERCURIC PHOSPHATE see BJT250
ETHYLMERCURIC PHOSPHATE see EME100
N-(ETHYLMERCURI)-1,4,5,6,7,7-HEXACHLOROBICYCLO(2.2.1)HEPT-5-ENE-2,3-DICARBOXIMIDE see EME050
N-ETHYLMERCURI-3,4,5,6,7,7-HEXACHLORO-3,6-ENDOMETHYLENE-1,2,3,6- TETRAHYDROPHTHALIMIDE see EME050
N-ETHYLMERCURI-N-PHENYL-p-TOLUENESULFONAMIDE see EME500
N-ETHYLMERCURI-1,2,3,6-TETRAHYDRO-3,6-ENDOMETHANO-3,4,5,6,7,7-HEXACHLOROPHTHALIMIDE see EME050
o-(ETHYLMERCURITHIO)BENZOIC ACID SODIUM SALT see MDI000
ETHYLMERCURITHIOSALICYLIC ACID SODIUM SALT see MDI000
N-(ETHYLMERCURI)-p-TOLUENESULFONANILIDE see EME500

N-(ETHYLMERCURI)-p-TOLUENESULPHONANILIDE see EME500
ETHYLMERCURY CHLORIDE see CHC500
ETHYLMERCURY DICYANDIAMIDE see EME025
ETHYLMERCURY PHOSPHATE see BJT250
ETHYLMERCURY PHOSPHATE see EME100
ETHYLMERCURY p-TOLUENESULFANILIDE see EME500
ETHYLMERCURY-p-TOLUENE SULFONAMIDE see EME500
ETHYLMERCURY-p-TOLUENESULFONANILIDE see EME500
ETHYLMERCURY TOLUENESULFONATE see EME600
ETHYLMERKAPTAN (CZECH) see EMB100
β-ETHYLMERKAPTOETHANOL (CZECH) see EPP500
β-ETHYLMERKAPTOETHYLCHLORID (CZECH) see CGY750
ETHYLMERKURIACETAT see EMD000
ETHYLMERKURIDIKYANDIAMID see EME025
ETHYL METHACRYLATE see EMF000
ETHYL METHACRYLATE, INHIBITED (DOT) see EMF000
ETHYL METHANESULFONATE see EMF500
ETHYLMETHANESULFONATO-CNU see CHF250
ETHYL METHANESULPHONATE see EMF500
ETHYL METHANOATE see EKL000
ETHYL METHANSULFONATE see EMF500
ETHYL METHANSULPHONATE see EMF500
ETHYLMETHIAMBUTENE HYDROCHLORIDE see EIJ000
7-ETHYL-5-METHOXY-BENZ(a)ANTHRACENE see MEN750
ETHYL-4-METHOXYBENZOATE see AOV000
ETHYL-p-METHOXYBENZOATE see AOV000
ETHYL 2-METHOXYBENZYL ETHER see EMF600
ETHYL o-METHOXYBENZYL ETHER see EMF600
2'-ETHYL-2-(2-METHOXY BUTYLAMINO) PROPIONANILIDE see EMG000
ETHYL O-(o-(METHOXYCARBONYL)BENZOYL)GLYCOLATE see MOD000
ETHYL o-(METHOXYCARBONYL)BENZOYLOXYACETATE see MOD000
2'-ETHYL-3-(2-METHOXYETHYL)AMINOBUTYRANILIDE HYDROCHLORIDE see EMG500
2'-ETHYL-4-(2-METHOXYETHYL)AMINOBUTYRANILIDE HYDROCHLORIDE see EMH000
2'-ETHYL-3-(2-METHOXYETHYL)AMINO-3-METHYLBUTYRANILIDE CYCLAMATE see EMH500
2'-ETHYL-3-(2-METHOXYETHYL)AMINO-3-METHYLBUTYRANILIDE CYCLOHEXANE SULFAMATE see EMH500
2'-ETHYL-2-(2-METHOXYETHYLAMINO)PROPIONANILIDE see EMI000
2'-ETHYL-2-(2-METHOXYETHYLAMINO)-PROPIONANILIDE HYDROCHLORIDE see EMI500
N-ETHYL-N-(2-METHOXYETHYL)-3-METHYL-4-NITROSOANILINE see EMI510
N-ETHYL-N-(2-METHOXYETHYL)-4-NITROSO-m-TOLUIDINE see EMI510
N-ETHYL-6-METHOXY-N'-(1-METHYLETHYL)-1,3,5-TRIAZINE-2,4-DIAMINE see EGD000

4-(1-ETHYL-2-(4-METHOXYPHENYL)-1-BUTENYL)PHENOL (E)- see EMI525
N-(2-(5-ETHYL-2-METHOXYPHENYL)ETHYL)PROPIONAMIDE see EMI530
3-ETHYL-4-(p-METHOXYPHENYL)-2-METHYL-3-CYCLOHEXENE-1-CARBOXYLIC ACID see CBO625
ETHYL 3-METHOXYPROPANOATE see EMI550
ETHYL 3-METHOXYPROPIONATE see EMI550
ETHYL (13-cis)-9-(4-METHOXY-2,3,6-TRIMETHYLPHENYL)-3,7-DIMETHYL-2,4,6,8-NONATETRAENOATE see EMJ600
ETHYL all-trans-9-(4-METHOXY-2,3,6-TRIMETHYLPHENYL)-3,7-DIMETHYL-2,4,6,8-NONATETRAENOATE see EMJ500
ETHYLMETHYLACETIC ACID see MHS600
ETHYL-2-METHYLACRYLATE see EMF000
ETHYL-α-METHYL ACRYLATE see EMF000
4-ETHYLMETHYLAMINOAZOBENZENE see ENB000
p-ETHYLMETHYLAMINOAZOBENZENE see ENB000
N-ETHYL-N-METHYL-p-AMINOAZOBENZENE see ENB000
4'-ETHYL-N-METHYL-4-AMINOAZOBENZENE see EOJ000
3-ETHYLMETHYLAMINO-1,1-DI(2'-THIENYL)BUT-1-ENE HYDROCHLORIDE see EIJ000
α-(1-(ETHYLMETHYLAMINO)ETHYL)BENZYL ALCOHOL HYDROCHLORIDE (−) see EJR500
4-((((4-ETHYL-4-METHYL)AMINO)PHENYL)AZO)PYRIDINE 1-OXIDE see MKB500
2-ETHYL-6-METHYLANILINE see MJY000
ETHYL METHYL ARSINE see EMK600
ETHYL METHYL AZIDOMETHYL PHOSPHONATE see EML500
2-ETHYL-2-((3-(2-METHYL-1-AZIRIDINYL)-1-OXOPROPYL) METHYL)-1,3-PROPANEDIYL ESTER 2-METHYL-1-AZIRIDINEPROPANOIC ACID see EML600
7-ETHYL-9-METHYLBENZ(c)ACRIDINE see EMM000
12-ETHYL-7-METHYLBENZ(a)ANTHRACENE see EMN000
7-ETHYL-12-METHYLBENZ(a)ANTHRACENE see EMM500
2-ETHYL-6-METHYL-BENZENAMINE see MJY000
N-ETHYL-3-METHYLBENZENAMINE see EPT100
o-ETHYL METHYLBENZENE see EPS500
p-ETHYLMETHYLBENZENE see EPT000
1-ETHYL-2-METHYLBENZENE see EPS500
1-ETHYL-4-METHYLBENZENE see EPT000
ETHYL-p-METHYL BENZENESULFONATE see EPW500
N-ETHYL-α-METHYL-1,3-BENZODIOXOLE-5-ETHANAMINE see EMS100
3-ETHYL-2-METHYLBENZOXAZOLIUM IODIDE (7CI) see MJY525
N-ETHYL(α-METHYLBENZYL)AMINE see EMO000
N-ETHYL-α-METHYLBENZYLAMINE see EMO000
N-ETHYL-3-METHYL-2-BUTANAMINE see EIL050
ETHYL (E)-2-METHYL-2-BUTENOATE see TGA800
5-ETHYL-5-(1-METHYL-1-BUTENYL)BARBITURATE see EMO500
5-ETHYL-5-(1-METHYL-1-BUTENYL)BARBITURIC ACID see EMO500

5-ETHYL-5-(1-METHYL-2-BUTENYL)BARBITURIC ACID see EMO875
5-ETHYL-5-(1-METHYL-1-BUTENYL)BARBITURIC ACID SODIUM SALT see VKP000
5-ETHYL-5-(3-METHYL-2-BUTENYL)BARBITURIC ACID SODIUM SALT see EMP500
5-ETHYL-5-(1-METHYL-1-BUTENYL)-2,4,6(1H,3H,5H)-PYRIMIDINETRIONE see EMO500
5-ETHYL-5-(1-METHYL-1-BUTENYL)-2,4,6(1H,3H,5H)-PYRIMIDINETRIONE SODIUM SALT see VKP000
5-ETHYL-5-(1-METHYLBUTYL)BARBITURIC ACID see NBT500
5-ETHYL-5-(3-METHYLBUTYL)BARBITURIC ACID see AMX750
5-ETHYL-5-(3-METHYLBUTYL)BARBITURIC ACID, SODIUM DERIVATIVE see AON750
5-ETHYL-5-(1-METHYLBUTYL)BARBITURIC ACID SODIUM SALT see NBU000
ETHYL 2-METHYLBUTYL KETOXINE see EMP550
5-ETHYL-5-(1-METHYLBUTYL)MALONYLUREA see NBT500
5-ETHYL-5-(1-METHYLBUTYL)-2,4,6(1H,3H,5H)-PYRIMIDINETRIONE (9CI) see NBT500
5-ETHYL-5-(1-METHYLBUTYL)-2,4,6(1H,3H,5H)-PYRIMIDINETRIONE MONOSODIUM SALT (9CI) see NBU000
5-ETHYL-5-(1-METHYLBUTYL)-2-THIOBARBITURIC ACID see PBT250
5-ETHYL-5-(1-METHYLBUTYL)-2-THIOBARBITURIC ACID MONOSODIUM see PBT500
ETHYL 2-METHYLBUTYRATE see EMP600
ETHYL METHYLCARBAMATE see EMQ500
ETHYL-N-METHYLCARBAMATE see EMQ500
ETHYLMETHYL CARBINOL see BPW750
ETHYL METHYL CETONE (FRENCH) see MKA400
3-ETHYL-7-METHYL-9-α-(4'-CHLOROBENZOYLOXY)-3,7-DIAZABICYCLO(3.3.1)NONAME HYDROCHLORIDE see YGA700
ETHYL-2-METHYL-4-CHLOROPHENOXYACETATE see EMR000
o-ETHYL-S-(3-METHYL-4-CHLOROPHENYL)ETHYL PHOSPHONODITHIOATE see EMR100
ETHYL 3-METHYLCROTONATE see MIP800
ETHYL (E)-2-METHYLCROTONATE see TGA800
ETHYL α-METHYLCROTONATE see MIP800
ETHYL α-METHYLCROTONATE see TGA800
ETHYL α-METHYLCYCLOHEXANEACETATE see EHT600
2-ETHYL-8-METHYL-2,8-DIAZASPIRO(4,5)DECANE-1,3-DIONE HYDROBROMIDE see EMR500
ETHYL METHYL 1,4-DIHYDRO-2,6-DIMETHYL-4-(m-NITROPHENYL)-3,5-PYRIDINEDICARBOXYLATE see EMR600
1-ETHYL-7-METHYL-1,4-DIHYDRO-1,8-NAPHTHYRIDINE-4-ONE-3-CARBOXYLIC ACID see EID000
1-ETHYL-7-METHYL-1,4-DIHYDRO-1,8-NAPHTHYRIDIN-4-ONE-3-CARBOXYLIC ACID see EID000

4'-ETHYL-2-METHYL-4-DIMETHYLAMINOAZOBENZENE see EMS000

2-ETHYL-2-METHYL-1,3-DIOXOLANE see EIO500

4-ETHYL-4-METHYL-2,6-DIOXOPIPERIDINE see MKA250

13-ETHYL-11-METHYLENE-18,19-DINOR -17-α-PREGN-4-EN-20-YN-17-OL see DBA750

N-ETHYL-3,4-METHYLENEDIOXYAMPHETAMINE see EMS100

1-ETHYL-6,7-METHYLENEDIOXY-4(1H)-OXOCINNOLINE-3-CARBOXYLIC ACID see CMS130

1-ETHYL-6,7-METHYLENEDIOXY-4-QUINOLONE-3-CARBOXYLIC ACID see OOG000

ETHYL METHYLENE PHOSPHORODITHIOATE see EEH600

ETHYL METHYL ETHER see EMT000

ETHYL METHYL ETHER (DOT) see EMT000

ETHYL (4-(1-METHYLETHYL)PHENYL)METHYL 3-PYRIDINYLCARBONIMIDODITHIOATE see EMT100

5-ETHYL-5-(1-METHYLETHYL)-2,4,6(1H,3H,5H)-PYRIMIDINETRIONE see ELX000

3-ETHYL-3-METHYLGLUTARIMIDE see MKA250

β-ETHYL-β-METHYLGLUTARIMIDE see MKA250

7-ETHYL-2-METHYL-4-HENDECANOL SULFATE SODIUM SALT see EMT500

7-ETHYL-2-METHYL-4-HEXADECANOL SULFATE SODIUM SALT see SIO000

ETHYLMETHYLKETON (DUTCH) see MKA400

ETHYL METHYL KETONE (DOT) see MKA400

ETHYL METHYL KETONE AZINE see EMT600

ETHYL METHYL KETONE OXIME see EMU500

ETHYL METHYL KETONE PEROXIDE see MKA500

ETHYL-METHYLKETONOXIM see EMU500

ETHYL METHYL KETOXIME see EMU500

ETHYL 2-METHYLLACTATE see ELH700

N-ETHYL-2-METHYLMALEIMIDE see EMV000

O-ETHYL-O-(4-(METHYLMERCAPTO)PHENYL)-S-N-PROPYLPHOSPHOROTHIONOTHIOLATE see SOU625

2-ETHYL-6-METHYL-1-N-(2-METHOXY-1-EMTHYLETHYL)CHLOROACETANILIDE see MQQ450

1-ETHYL-2-METHYL-7-METHOXY-1,2,3,4-TETRAHYDROPHENANTHRYL-2-CARBOXYLIC ACID see BIT000

5-ETHYL-1-METHYL-5-(1-METHYLPROPENYL)BARBITURIC ACID see EMW000

N-ETHYL-6-METHYL-α-(METHYLSULFONYL)ERGOLINE-8-β-PROPIONAMIDE see EMW100

ETHYL-3-METHYL-4-(METHYLTHIO)PHENYL(1-METHYLETHYL)PHOSPHORAMIDATE see FAK000

O-ETHYL-O-METHYL-O-p-NITROFENYLESTER KYSELINY THIOFOSFORECNE see ENI175

ETHYL (2-(2-METHYL-5-NITRO-1-IMIDAZOLYL)ETHYL)SULFONE see TGD250

ETHYLMETHYLNITROSAMINE see MKB000

2-(ETHYL(3-METHYL-4-NITROSOPHENYL)AMINO)ETHANOL see ELG100

N'-ETHYL-N-METHYL-N-NITROSOUREA see NKU370

N-ETHYL-N'-METHYL-N-NITROSOUREA see NKE100

4-ETHYL-1-METHYLOCTYLAMINE see EMY000

2-ETHYL-1-(3-METHYL-1-OXO-2-BUTENYL)PIPERIDINE see EMY100

1-ETHYL-7-METHYL-4-OXO-1,4-DIHYDRO-1,8-NAPHTHYRIDINE-3-CARBOXYLIC ACID see EID000

ETHYL (Z)-(3-METHYL-4-OXO-5-PIPERIDINO-THIAZOLIDIN-2-YLIDENE)ACETATE see MNG000

2-ETHYL-4-METHYLPENTANOL see ELT500

2-ETHYL-4-METHYLPENTANOL see EMZ000

2-ETHYL-4-METHYL-1-PENTANOL see ELT500

2-ETHYL-4-METHYL-1-PENTANOL see EMZ000

5-ETHYL-5-(1-METHYL-1-PENTENYL)BARBITURIC ACID see ENA000

ETHYL METHYL PEROXIDE see ENA500

N-ETHYL-N-METHYL-p-(PHENYLAZO)ANILINE see ENB000

2-(ETHYL(3-METHYL-4-(PHENYLAZO)PHENYL)AMINO)ETHANOL see ENB100

N-ETHYLMETHYLPHENYLBARBITURIC ACID see ENB500

5-ETHYL-1-METHYL-5-PHENYLBARBITURIC ACID see ENB500

5-ETHYL-N-METHYL-5-PHENYLBARBITURIC ACID see ENB500

3-ETHYL-2-METHYL-4-PHENYL-4-CYCLOHEXENECARBOXYLIC ACID SODIUM SALT see MBV775

5-ETHYL-6-METHYL-4-PHENYL-3-CYCLOHEXENE-1-CARBOXYLIC ACID SODIUM SALT see MBV775

ETHYL METHYLPHENYLGLYCIDATE see ENC000

5-ETHYL-1-METHYL-5-PHENYLHYDANTOIN see ENC500

5-ETHYL-3-METHYL-5-PHENYLHYDANTOIN see MKB250

5-ETHYL-3-METHYL-5-PHENYL-2,4(3H,5H)-IMIDAZOLEDIONE see MKB250

5-ETHYL-3-METHYL-5-PHENYLIMIDAZOLIDIN-2,4-DIONE see MKB250

ETHYL-1-METHYL-4-PHENYLISONIPECOTATE see DAM600

ETHYL-1-METHYL-4-PHENYLISONIPECOTATE HYDROCHLORIDE see DAM700

ETHYL-1-METHYL-4-PHENYLPIPERIDINE-4-CARBOXYLATE see DAM600

ETHYL-1-METHYL-4-PHENYLPIPERIDINE-4-CARBOXYLATE HYDROCHLORIDE see DAM700

ETHYL-1-METHYL-4-PHENYLPIPERIDYL-4-CARBOXYLATE HYDROCHLORIDE see DAM700

5-ETHYL-1-METHYL-5-PHENYL-2,4,6(1H,3H,5H)-PYRIMIDINETRIONE see ENB500

α-ETHYL-1-METHYL-α-PHENYL-3-PYRROLIDINEMETHANOL PROPIONATE FUMARATE see ENC600

α-ETHYL-4-METHYL-1-PIPERAZINEACETIC ACID-2,6-DIETHYLPHENYL ESTER DIHYDROCHLORIDE see FAC165

α-ETHYL-4-METHYL-1-PIPERAZINEACETIC ACID MESITYL ESTER DIHYDROCHLORIDE see FAC195

α-ETHYL-4-METHYL-1-PIPERAZINEACETIC ACID-2,6-XYLYL ESTER HYDROCHLORIDE see FAC130

3-ETHYL-6-METHYLPIPERIDINE see END000

5-ETHYL-2-METHYLPIPERIDINE see END000

4-ETHYL-4-METHYL-2,6-PIPERIDINEDIONE see MKA250

N-ETHYL-2-(METHYLPIPERIDINO)-N-(1-PHENOXY-2-PROPYL)ACETAMIDE HYDROCHLORIDE see EOG500

4'-ETHYL-2-METHYL-3-PIPERIDINOPROPIOPHENONE HYDROCHLORIDE see EAV700

N-ETHYL-2-(2-METHYLPIPERIDINO)-N-(1-(2,4-XYLYLOXY)-2-PROPYL) ACETAMIDE HYDROCHLORIDE see END500

ETHYL-2-METHYLPROPANOATE see ELS000

ETHYL-2-METHYL-2-PROPENOATE see EMF000

5-ETHYL-5-(1-METHYLPROPENYL)BARBITURIC ACID see ENE000

N-ETHYL-N-(2-METHYL-2-PROPENYL)-2,6-DINITRO-4-(TRIFLUOROMETHYL)BENZENAMINE see ENE500

ETHYL-2-METHYLPROPIONATE see ELS000

5-ETHYL-5-(1-METHYLPROPYL)BARBITURATE see BPF000

5-ETHYL-5-(1-METHYLPROPYL)BARBITURIC ACID see BPF000

5-ETHYL-5-(1-METHYLPROPYL)BARBITURIC ACID SODIUM SALT see BPF250

α-ETHYL-4-(2-METHYLPROPYL)BENZENEACETIC ACID see IJH000

1-ETHYL-1-METHYLPROPYL CARBAMATE see ENF000

o-ETHYL S-1-METHYLPROPYL S-1,1-DIMETHYLETHYL PHOSPHORODITHIOATE see ENF050

N-(3-(1-ETHYL-1-METHYLPROPYL)-5-ISOXAZOLYL)-2,6-DIMETHOXYBENZAMIDE see ENF100

5-ETHYL-5-(1-METHYLPROPYL)-2,4,6(1H,3H,5H)-PYRIMIDINETRIONE (9CI) see BPF000

5-ETHYL-5-(1-METHYLPROPYL)-2,4,6(1H,3H,5H)-PYRIMIDINETRIONE MONOSODIUM SALT see BPF250

1-ETHYL-1-METHYL-2-PROPYNYL CARBAMATE see MNM500

2-ETHYL-3-METHYLPYRAZINE see ENF200

3-ETHYL-6-METHYLPYRIDINE see EOS000

5-ETHYL-2-METHYLPYRIDINE see EOS000

3-ETHYL-3-METHYL-1-PYRIDYL-TRIAZENE see PPN250

3-ETHYL-3-METHYLPYRROLIDINE-2,5-DIONE see ENG500

3-ETHYL-3-METHYL-2,5-PYRROLIDINE-DIONE see ENG500

2-ETHYL-6-METHYLQUINONEIMINE see ENF300

2-ETHYL-2-METHYLSUCCINIMIDE see ENG500

α-ETHYL-α-METHYLSUCCINIMIDE see ENG500

ETHYLMETHYLTHIAMBUTENE HYDROCHLORIDE see EIJ000

N-ETHYL-N'-(3-METHYL-2-THIAZOLIDINYLIDENE)UREA see ENH000

O-ETHYL-O-(4-(METHYLTHIO)PHENYL)PHOSPHORODIT HIOIC ACID-S-PROPYL ESTER see SOU625

O-ETHYL-O-(4-(METHYLTHIO)PHENYL) S-PROPYL PHOSPHORODITHIOATE see SOU625

ETHYLMETHYLTHIOPHOS see ENI175

ETHYL-4-(METHYLTHIO)-m-TOLYL ISOPROPYL PHOSPHORAMIDATE see FAK000

O-ETHYL-O-(4-METHYLTHIO-m-TOLYL) METHYLPHOSPHORAMIDOTHIOATE see ENI500

N-ETHYL-α-METHYL-m-(TRIFLUOROMETHYL)PHENETHYLAMIN E see ENJ000

N-ETHYL-α-METHYL-m-TRIFLUOROMETHYLPHENETHYLAMINE see PDM250

N-ETHYL-α-METHYL-m-(TRIFLUOROMETHYL)PHENETHYLAMIN E HYDROCHLORIDE see PDM250

3-ETHYL-5-METHYL-1,2,4-TRIOXOLANE see PBQ300

7-ETHYL-2-METHYL-4-UNDECANOL SULFATE SODIUM SALT see EMT500

ETHYLMETHYL VALERAMIDE see ENJ500

2-ETHYL-3-METHYLVALERAMIDE see ENJ500

ETHYL METRIBUZIN see ENJ600

ETHYL MONOBROMOACETATE see EGV000

ETHYL MONOCHLORACETATE see EHG500

ETHYL MONOCHLOROACETATE see EHG500

ETHYL MONOIODOACETATE see ELQ000

ETHYL MONOSULFIDE see EPH000

ETHYLMORPHINE see ENK000

3-o-ETHYLMORPHINE see ENK000

ETHYLMORPHINE HYDROCHLORIDE see DVO700

ETHYLMORPHINE HYDROCHLORIDE see ENK500

o-ETHYLMORPHINE HYDROCHLORIDE see DVO700

ETHYL MORPHINE HYDROCHLORIDE DIHYDRATE see ENK500

4-ETHYLMORPHOLINE see ENL000

N-ETHYLMORPHOLINE see ENL000

1-(N-p-ETHYLMORPHOLINE)-5-NITROIMIDAZOLE see NHH000

1-ETHYL-4-(2-MORPHOLINOETHYL)-3,3-DIPHENYL-2-PYRROLIDINONE see ENL100

1-ETHYL-4-(2-MORPHOLINOETHYL)-3,3-DIPHENYL-2-PYRROLIDINONE HYDROCHLORIDE HYDRATE see SLU000

ETHYL 1-(2'-MORPHOLINOETHYL)-4-PHENYLPIPERIDINE-4-CARBOXYLATE DIHYDROCHLORIDE see MRN675

ETHYL-N-(4-MORPHOLINOMETHYL)CARBAMATE see ENL500

ETHYL 10-(3-MORPHOLINOPROPIONYL)PHENOTHIAZ INE-2-CARBAMATE HYDROCHLORIDE see EEI050

9-ETHYL-6-MP see EMC500

ETHYL MUSTARD OIL see ELX525

ETHYL MYRISTATE see ENL850

ETHYLNANDROL see EJT600

1-ETHYLNAPHTHALENE see ENL860

2-ETHYLNAPHTHALENE see ENL862

ETHYL 1-NAPHTHALENEACETATE see ENL900

ETHYL-β-NAPHTHOLATE see EEY500

ETHYL 1-NAPHTHYLACETATE see ENL900

ETHYL-2-NAPHTHYL ETHER see EEY500

ETHYL-β-NAPHTHYL ETHER see EEY500

1-ETHYL-1-(1-NAPHTHYL)-2-THIOUREA see ENM000

3-ETHYLNIRVANOL see MKB250

ETHYL NITRATE see ENM500

N-ETHYL-2-NITRATOETHYL NITRAMINE see ENN200

ETHYL NITRILE see ABE500

ETHYL NITRITE see ENN000

ETHYL NITRITE see ENN000

ETHYL NITRITE SOLUTIONS (DOT) see ENN000

ETHYL NITROACETATE see ENN100

2-(ETHYLNITROAMINO)ETHANOL NITRATE (ESTER) see ENN200

5-(N-ETHYL-N-NITRO)AMINO-3-(5-NITRO-2-FURYL)-s-TRIAZOLE see ENN500

ETHYL-p-NITROBENZOATE see ENO000

ETHYL NITROBENZOATE, PARA ESTER see ENO000

ETHYL (4-NITROBENZOYL)ACETATE see ENO100

ETHYL (p-NITROBENZOYL)ACETATE see ENO100

ETHYL 3-NITROBENZYLIDENEACETOACETATE see ENO200

ETHYL 2-(3-NITROBENZYLIDENE)ACETOACETATE see ENO200

ETHYL 2-(m-NITROBENZYLIDENE)ACETOACETATE see ENO200

3-ETHYL-4-NITROBIPHENYL see ENO300

N-ETHYL-N-NITROETHANAMINE (9CI) see DJS500

O-ETHYL-O-((4-NITROFENYL)-FENYL)-MONOTHIOFOSFONAAT (DUTCH) see EBD700

1-ETHYL-3-NITROGUANIDINE see ENP000

N-ETHYL-N'-NITROGUANIDINE see ENP000

ETHYLNITROLIC ACID see NHY100

N-ETHYL-N'-NITRO-N-NITROSOGUANIDINE see ENU000

ETHYL 4-NITRO-β-OXOBENZENEPROPANOATE see ENO100

2-(ETHYL(4-((4-NITROPHENYL)AZO)PHENYL)AMINO)ET HANOL see ENP100

ETHYL-p-NITROPHENYL BENZENETHIONOPHOSPHONATE see EBD700

O-ETHYL O-(4-NITROPHENYL)BENZENETHIONOPHOSP HONATE see EBD700

ETHYL-p-NITROPHENYL BENZENETHIOPHOSPHATE see EBD700

ETHYL-p-NITROPHENYL BENZENETHIOPHOSPHONATE see EBD700

ETHYL p-NITROPHENYL ETHYLPHOSPHATE see NIM500

ETHYL-4-NITROPHENYL ETHYLPHOSPHONATE see ENQ000

ETHYL-p-NITROPHENYLPENTYLPHOSPHONATE see ENQ500

ETHYL-p-NITROPHENYL PHENYLPHOSPHONOTHIOATE see EBD700

O-ETHYL-O-(4-NITROPHENYL) PHENYLPHOSPHONOTHIOATE see EBD700

O-ETHYL-O-p-NITROPHENYL PHENYLPHOSPHONOTHIOLATE see EBD700

O-ETHYL-O-p-NITROPHENYL PHENYLPHOSPHOROTHIOATE see EBD700

ETHYL 2-NITROPHENYL SULFIDE see ENQ600

ETHYL o-NITROPHENYL SULFIDE see ENQ600

ETHYL-p-NITROPHENYL THIONOBENZENEPHOSPHATE see EBD700

ETHYL-p-NITROPHENYL THIONOBENZENEPHOSPHONATE see EBD700

3-ETHYL-4-NITROPYRIDINE-1-OXIDE see ENR000

2-ETHYL-4-NITROQUINOLINE-1-OXIDE see NGA500

2-(ETHYLNITROSAMINO)ETHANOL see ELG500

1-(ETHYLNITROSAMINO)ETHYL ACETATE see ABT500

(ETHYLNITROSAMINO)METHYL ACETATE see ENR500

4-((ETHYLNITROSAMINO)METHYL)PYRIDI NE see NLH000

4-(ETHYLNITROSOAMINO)-1-BUTANOL see ELE500

4-(ETHYLNITROSOAMINO)BUTYRIC ACID see NKE000

5-(N-ETHYL-N-NITROSO)AMINO-3-(5-NITRO-2-FURYL)-s-TRIAZOLE see ENS000

ETHYLNITROSOANILINE see NKD000

N-ETHYL-N-NITROSOBENZENAMINE see NKD000

N-ETHYL-N-NITROSOBENZYLAMINE see ENS500

ETHYLNITROSOBIURET see ENT000

N-ETHYL-N-NITROSOBIURET see ENT000

N-ETHYL-N-NITROSO-tert-BUTANAMINE see NKD500

N-ETHYL-N-NITROSOBUTYLAMINE see EHC000

ETHYLNITROSOCARBAMIC ACID, ETHYL ESTER see NKE500

N-ETHYL-N-NITROSOCARBAMIC ACID ETHYL ESTER see NKE500

N-ETHYL-N-NITROSOCARBAMIC ACID 1-NAPHTHYL ESTER see NBI000

N-ETHYL-N-NITROSOCARBAMIDE see ENV000

ETHYLNITROSOCYANAMIDE see ENT500

N-ETHYL-N-NITROSO-ETHANAMINE see NJW500

N-ETHYL-N-NITROSOETHENAMINE see NKF000

N-ETHYL-N-NITROSOETHENYLAMINE see NKF000

N-ETHYL-N-NITROSO-N'-NITROGUANIDINE see ENU000

N-ETHYL-N-NITROSO-N'-(2-OXOPROPYL)UREA see NKE120

ETHYL 4-NITROSO-1-PIPERAZINECARBOXYLATE see NJQ000

2-ETHYL-3-NITROSOTHIAZOLIDINE see ENU500

2-ETHYL-N-NITROSOTHIAZOLIDINE see ENU500

ETHYLNITROSOUREA see ENV000

1-ETHYL-1-NITROSOUREA see ENV000

N-ETHYL-N-NITROSO-UREA see ENV000

ETHYL NITROSOURETHAN see NKE500

N-ETHYL-N-NITROSOURETHAN see NKE500

ETHYL NITROSOURETHANE see NKE500

N-ETHYL-N-NITROSOURETHANE see NKE500

N-ETHYL-N-NITROSOVINYLAMINE see NKF000

1-ETHYL-3-(5-NITRO-2-THIAZOLYL) UREA see ENV500

N-ETHYL-N'-(5-NITRO-2-THIAZOLYL)UREA see ENV500

ETHYL NONANOATE see ENW000

5-ETHYL-2-NONANOL see ENW500

5-ETHYL-3-NONEN-2-ONE see ENX000

ETHYL NONYLATE see ENW000

1-ETHYL-1-NONYLPIPERIDINIUM BROMIDE see ENX100

ETHYLNORADRENALINE
HYDROCHLORIDE see ENX500
ETHYL NORADRIANOL see EGE500
ETHYL NOREPINEPHRINE
HYDROCHLORIDE see ENX500
α-ETHYLNOREPINEPHRINE
HYDROCHLORIDE see ENX500
ETHYLNORGESTRIENONE see ENX575
(−)-N-ETHYL-NORHYOSCINE-
METHOBROMIDE see ONI000
N-ETHYLNORPHENYLEPHRINE see
EGE500
11-β-ETHYL-19-NOR-17-α-PREGNA-
1,3,5(10)-TRIEN-20-YNE-3,17-DIOL see
EJV400
N-ETHYL-
NORSCOPOLAMINEMETHOBROMIDE see
ONI000
ETHYLNORSUPRARENIN
HYDROCHLORIDE see ENX500
17-ETHYL-19-NORTESTOSTERONE see
ENX600
16-β-ETHYL-19-NORTESTOSTERONE see
ELF100
17-α-ETHYL-19-NORTESTOSTERONE see
ENX600
N-ETHYLNORTROPINE BENZHYDRYL
ETHER see DWE800
ETHYL OCTADECANOATE see EPF700
ETHYL n-OCTADECANOATE see EPF700
1-ETHYL-1-OCTADECYLPIPERIDINIUM
BROMIDE see ENX875
1-ETHYL-1,2,3,4,4a,9,10,10a-OCTAHYDRO-7-
HYDROXY-2-METHYL-2-
PHENANTHRENECARBOXYLIC ACID see
DYB000
ETHYL OCTANOATE see ENY000
ETHYL OCTYLATE see ENY000
1-ETHYL-1-OCTYLPIPERIDINIUM
BROMIDE see ENY100
ETHYL OENANTHATE see CNG920
ETHYL OENANTHATE see EKN050
ETHYL OENANTHYLATE see EKN050
ETHYLOLAMINE see EEC600
1-(β-ETHYLOL)-2-METHYL-5-NITRO-3-
AZAPYRROLE see MMN250
N-ETHYL-2(OR 4)-
METHYLBENZENESULFONAMIDE see
ENY200
5-ETHYL-5-(3 OR 6-OXO-1-CYCLOHEXEN-
1-YL)BARBITURIC ACID see OJV600
ETHYL ORTHOFORMATE see ENY500
ETHYL ORTHOSILICATE see EPF550
ETHYL OXALATE see DJT200
ETHYL OXALATE (DOT) see DJT200
ETHYL-3-OXATRICYCLO-
(3.2.1.0²,⁴)OCTANE-6-CARBOXYLATE see
ENZ000
1-ETHYL-3-(2-OXAZOLYL)UREA see
EOA000
ETHYLOXIRANE see BOX750
ETHYL β-OXOBENZENEPROPANOATE
see EGR600
ETHYL-3-OXOBUTANOATE see EFS000
ETHYL-3-OXOBUTYRATE see EFS000
ETHYL 3-OXOBUTYRATE ETHYLENE
KETAL see EFR100
ETHYL 4-OXOPENTANOATE see EFS600
(Z)-2-(3-ETHYL-4-OXO-5-PIPERIDINO-2-
THIAZOLIDINYLIDENE) ACETIC ACID see
EOA500
cis-2-(3-ETHYL-4-OXO-5-PIPERIDINO-2-
THIAZOLIDINYLIDENE)ACETIC ACID see
EOA500
2-(3-ETHYL-4-OXO-5-PIPERIDINO-2-
THIAZOLIDINYLIDENE)ACETIC ACID
ETHYL ESTER see EOB000
(3-ETHYL-4-OXO-5-(1-PIPERIDINYL)-2-
THIAZOLIDINYLIDENE)ACETIC ACID
ETHYL ESTER see EOA500
ETHYL 4-OXOVALERATE see EFS600
ETHYL OXYHYDRATE see EOB050

4-ETHYLOXYPHENOL see EFA100
ETHYL PABATE see EEK100
ETHYL PARABEN see HJL000
ETHYL PARAOXON see NIM500
ETHYL PARASEPT see HJL000
ETHYL PARATHION see PAK000
S-ETHYL PARATHION see DJT000
ETHYL PELARGONATE see ENW000
ETHYL PENTABORANE (9) see EOB100
1-ETHYL-1-PENTADECYLPIPERIDINIUM
BROMIDE see EOB200
3-ETHYL-2-PENTANOL see EOB300
3-ETHYL-3-PENTANOL see TJP550
5-ETHYL-5-PENTYLBARBITURIC ACID see
PBS250
2-((1-ETHYLPENTYL)OXY)ETHANOL see
EGJ000
2-(((1-
ETHYLPENTYL)OXY)ETHOXY)ETHANOL
see EGJ500
ETHYL PERCHLORATE see EOD000
ETHYL PEROXYCARBONATE see DJU600
20-ETHYL-PGF2-α see EPC200
ETHYLPHENACEMIDE see PFB350
ETHYL PHENACETATE see EOH000
ETHYLPHENACETIN see EOD500
5-ETHYLPHENAZINIUM ETHYLSULFATE
see EOE000
N-ETHYLPHENAZONIUM ETHOSULFATE
see EOE000
2-ETHYLPHENOL see PGR250
4-ETHYLPHENOL see EOE100
o-ETHYLPHENOL see PGR250
ETHYL (2-(4-
PHENOXYPHENOXY)ETHYL)CARBAMAT
E see EOE200
N-ETHYL-N-(1-PHENOXY-2-
PROPYL)CARBAMIC ACID-2-
(DIETHYLAMINO)ETHYL ESTER
HYDROCHLORIDE see EOF500
N-ETHYL-N-(1-PHENOXY-2-
PROPYL)CARBAMIC ACID-2-(2-
METHYLPIPERIDINO)ETHYL ESTER
HYDROCHLORIDE see EOG000
N-ETHYL-N-(1-PHENOXY-2-PROPYL)-2-(2-
METHYLPIPERIDINO) ACETAMIDE
HYDROCHLORIDE see EOG500
3-(2-ETHYLPHENOXY)-1-((1S)-1,2,3,4-
TETRAHYDRONAPHTH-1-YLAMINO)-(2S)-
2-PROPANOL OXALATE see EOG600
ETHYL PHENYLACETATE see EOH000
N-α-ETHYLPHENYLACETYL-N'-ACETYL
UREA see ACX500
1-((ETHYL)PHENYLACETYL)UREA see
PFB350
ETHYL-β-PHENYLACRYLATE see EHN000
ETHYLPHENYLAMINE see EGK000
3-ETHYLPHENYLAMINE see EOH100
(R)-N-ETHYL-2-
(((PHENYLAMINO)CARBONYL)OXY)PROP
ANAMIDE see CBL500
3-
(ETHYLPHENYLAMINO)PROPIONITRILE
see EHQ500
N-ETHYL-p-(PHENYLAZO)ANILINE see
EOH500
N-ETHYL-4-
(PHENYLAZO)BENZENAMINE see EOH500
p-((m-ETHYLPHENYL)AZO)-N,N-
DIMETHYLANILINE see EOI000
p-((p-ETHYLPHENYL)AZO)-N,N-
DIMETHYLANILINE see EOI500
p-(4-ETHYLPHENYLAZO)-N-
METHYLANILINE see EOJ000
N-ETHYL-1-((4-
(PHENYLAZO)PHENYL)AZO)-2-
NAPHTHALENAMINE see EOJ500
N-ETHYL-1-((p-
(PHENYLAZO)PHENYL)AZO)-2-
NAPHTHALENAMINE see EOJ500

N-ETHYL-1-((4-
(PHENYLAZO)PHENYL)AZO)-2-
NAPHTHYLAMINE see EOJ500
N-ETHYL-1-((p-
(PHENYLAZO)PHENYL)AZO)-2-
NAPHTHYLAMINE see EOJ500
N-ETHYL-N-(p-
(PHENYLAZO)PHENYL)HYDROXYLAMIN
E see HKY000
5-ETHYL-5-PHENYLBARBITURIC ACID see
EOK000
5-ETHYL-5-PHENYLBARBITURIC ACID
SODIUM see SID000
5-ETHYL-5-PHENYLBARBITURIC ACID
SODIUM SALT see SID000
S-2-((4-(p-
ETHYLPHENYL)BUTYL)AMINO)ETHYL
THIOSULFATE see EOK500
2-ETHYL-2-PHENYLBUTYRIC ACID 2-(2-
DIETHYLAMINOETHOXY)ETHYL ESTER
see DHQ200
ETHYL-N-PHENYLCARBAMATE see
CBL750
ETHYL
PHENYLCARBAMOYLOXYPHENYLCARBA
MATE see EEO500
ETHYL PHENYL CARBINOL see EGQ000
ETHYL PHENYL DICHLOROSILANE
(DOT) see DFJ800
3-ETHYL-3-PHENYL-2,6-
DIKETOPIPERIDINE see DYC800
3-ETHYL-3-PHENYL-2,6-
DIOXOPIPERIDINE see DYC800
ETHYL PHENYLDITHIOCARBAMATE see
EOK550
ETHYL N-PHENYLDITHIOCARBAMATE
see EOK550
ETHYL-2-PHENYLETHANOATE see
EOH000
ETHYL PHENYL ETHER see PDM000
O-ETHYL-S-PHENYL
ETHYLDITHIOPHOSPHONATE see
FMU045
O-ETHYL-S-PHENYL
ETHYLPHOSPHONODITHIOATE see
FMU045
2-ETHYL-2-PHENYLGLUTARIMIDE see
DYC800
α-ETHYL-α-PHENYLGLUTARIMIDE see
DYC800
ETHYL PHENYLGLYCIDATE see EOK600
ETHYL-3-PHENYLGLYCIDATE see EOK600
5-ETHYL-5-
PHENYLHEXAHYDROPYRIMIDINE-4,6-
DIONE see DBB200
ETHYLPHENYLHYDANTOIN see EOL000
3-ETHYL-5-PHENYLHYDANTOIN see
EOL100
5-ETHYL-5-PHENYLHYDANTOIN see
EOL000
(−)-5-ETHYL-5-PHENYLHYDANTOIN see
EOL050
l-5-ETHYL-5-PHENYLHYDANTOIN see
EOL050
4-ETHYLPHENYLHYDROXYLAMINE see
EOL100
N-(4-ETHYLPHENYL)HYDROXYLAMINE
see EOL100
N-(p-ETHYLPHENYL)HYDROXYLAMINE
see EOL100
3-ETHYL-5-PHENYLIMIDAZOLIDIN-2,4-
DIONE see EOL100
5-ETHYL-5-PHENYL-2,4-
IMIDAZOLIDINEDIONE see EOL000
ETHYL PHENYL KETONE see EOL500
ETHYLPHENYLMALONIC ACID 2-
(DIETHYLAMINO)ETHYL ETHYL ESTER
see PCU400
3-(o-ETHYLPHENYL)-5-(m-
METHOXYPHENYL)-s-TRIAZOLE see
EOL600

5-(2-ETHYLPHENYL)-3-(3-METHOXYPHENYL)-s-TRIAZOLE see EOL600

(3-ETHYLPHENYL)-5-(3-METHOXYPHENYL)-1,2,4-TRIAZOLE see EOL600

3-(2-ETHYLPHENYL)-5-(3-METHOXYPHENYL)-1H-1,2,4-TRIAZOLE see EOL600

5-(2-ETHYLPHENYL)-3-(3-METHOXYPHENYL)-1H-1,2,4-TRIAZOLE see EOL600

5-ETHYL-5-PHENYL-N-METHYLBARBITURIC ACID see ENB500

4-ETHYL-N-(PHENYLMETHYLENE)BENZENAMINE N-OXIDE see EOL700

5-ETHYL-5-(PHENYLMETHYL)-2,4,6(1H,3H,5H)-PYRIMIDINETRIONE (9CI) see BEA500

O-ETHYL-PHENYL-p-NITROPHENYL THIOPHOSPHONATE see EBD700

N-ETHYL-3-PHENYL-2-NORBORNANAMINE HYDROCHLORIDE see EOM000

ETHYL P-PHENYLPHOSPHONOCHLORIDOTHIOATE see EOM100

o-ETHYLPHENYLPHOSPHONOTHIOATE-o-ESTER with p-HYDROXYBENZONITRILE see CON300

3-ETHYL-3-PHENYL-2,6-PIPERIDINEDIONE see DYC800

3-(o-ETHYLPHENYL)-5-PIPERONYL-s-TRIAZOLE see EOM600

ETHYLPHENYL-PROPANEDIOIC ACID-2-(DIETHYLAMINO)ETHYL ETHYL ESTER (9CI) see PCU400

ETHYL-3-PHENYLPROPENOATE see EHN000

α-(5-ETHYL-1-PHENYL-1H-PYRAZOL-4-YL)-1-PIPERIDINEBUTANOL see EOM650

5-ETHYL-5-PHENYL-2,4,6-(1H,3H,5H)PYRIMIDINETRIONE see EOK000

5-ETHYL-5-PHENYL-2,4,6-(1H,3H,5H)PYRIMIDINETRIONE MONOSODIUM SALT see SID000

(±)-5-ETHYL-5-PHENYL-2-PYRROLIDINONE see EOM700

(±)-5-ETHYL-5-PHENYLPYRROLID-2-ONE see EOM700

2-ETHYL-2-PHENYLSUCCINIC ACID-2-(DIETHYLAMINO)ETHYL ETHYL ESTER see SBC700

ETHYL 4-PHENYL-1-(2-TETRAHYDROFURFURYLOXYETHYL)PIPERIDINE-4-CARBOXYLATE see FPQ100

ETHYL(N-PHENYL-p-TOLUENESULFONAMIDATO)MERCURY see EME500

ETHYL(N-PHENYL-p-TOLUENESULFONAMIDO)MERCURY see EME500

1-ETHYL-3-PHENYLTRIAZENE see MRI250

ETHYL PHOSPHATE see TJT750

ETHYL PHOSPHATE, DI- see DJW500

ETHYL PHOSPHATE, MONO- see ECI300

4-ETHYL-1-PHOSPHA-2,6,7-TRIOXABICYCLO(2.2.2)OCTANE see TNI750

4-ETHYL-1-PHOSPHA-2,6,7-TRIOXABICYCLO(2.2.2)OCTANE-1-OXIDE see ELJ500

ETHYL PHOSPHINE see EON000

ETHYL PHOSPHONAMIDOTHIONIC ACID-4-(METHYLTHIO)-m-TOLYL ESTER see EON500

ETHYLPHOSPHONIC ACID BUTYL-p-NITROPHENYL ESTER see BRW500

ETHYLPHOSPHONIC ACID ETHYL-p-NITROPHENYL ESTER see ENQ000

ETHYL PHOSPHONIC ACID, METHYL p-NITROPHENYL ESTER see EON600

ETHYL PHOSPHONODITHIOIC ACID-O-METHYL-S-(p-TOLYL) ESTER see EOO000

ETHYL PHOSPHONOTHIOIC ACID-(2-DIETHYLAMINOETHYL)-4,6-DICHLOROPHENYL ESTER see EOP500

ETHYL PHOSPHONOTHIOIC DICHLORIDE see EOP600

ETHYL PHOSPHONOTHIOIC DICHLORIDE, anhydrous (DOT) see EOP600

ETHYLPHOSPHONOTHIONIC DICHLORIDE see EOP600

ETHYL PHOSPHONOTHIOYL DICHLORIDE see EOP600

ETHYL PHOSPHONOUS DICHLORIDE see EOQ000

ETHYL PHOSPHONOUS DICHLORIDE, anhydrous (DOT) see EOQ000

ETHYL PHOSPHORAMIDIC ACID-2,4-DICHLOROPHENYL ESTER see EOQ500

ETHYLPHOSPHORAMIDIC ACID, METHYL-(2,4,5-TRICHLOROPHENYL) ESTER see DYD400

ETHYLPHOSPHORAMIDIC ACID-(2,4,5-TRICHLOROPHENYL) ESTER see DYD600

ETHYL PHOSPHORODICHLORIDATE see EOR000

ETHYL PHTHALATE see DJX000

N-ETHYLPHTHALIMIDE see EOR500

ETHYLPHTHALYL ETHYL GLYCOLATE see EOR525

5-ETHYL-2-PICOLINE see EOS000

5-ETHYL-α-PICOLINE see EOS000

2-(4-ETHYL-1-PIPERAZINYL)-4-PHENYLQUINOLINE DIHYDROCHLORIDE see EOS100

1-ETHYLPIPERIDINE see EOS500

1-ETHYL-3-PIPERIDINOL DIPHENYLACETATE HYDROCHLORIDE see EOY000

2-ETHYL-3-(β-PIPERIDINO-p-PHENETIDINO)PHTHALIMIDINE see AHP125

N-ETHYL-2-(PIPERIDINO)-N-(1-(2,4-XYLYLOXY)-2-PROPYL)ACETAMIDE HYDROCHLORIDE see EOU500

1-ETHYL-3-PIPERIDINYL ESTER-α-PHENYLBENZENEACETIC ACID HYDROCHLORIDE see EOY000

1-ETHYL-4-PIPERIDYL-p-AMINOBENZOATE HYDROCHLORIDE see EOV000

N-ETHYL-3-PIPERIDYLBENZILATE METHOBROMIDE see PJA000

1-ETHYL-3-PIPERIDYL BENZILATE METHYLBROMIDE see PJA000

1-ETHYL-4-PIPERIDYL BENZOATE HYDROCHLORIDE see EOW000

1-ETHYL-3-PIPERIDYL DIPHENYLACETATE HYDROCHLORIDE see EOY000

N-ETHYL-3-PIPERIDYL DIPHENYLACETATE HYDROCHLORIDE see EOY000

1-ETHYL-3-PIPERIDYL ESTER-α-PHENYLBENZENEACETIC ACID HYDROCHLORIDE see EOY000

β-(1-ETHYL-2-PIPERIDYL)ETHYL BENZOATE HYDROCHLORIDE see EOY100

β-2-(N-ETHYLPIPERIDYL)ETHYLBENZOATE HYDROCHLORIDE see EOY100

ETHYL POTASSIUM XANTHATE see PLF000

ETHYL POTASSIUM XANTHOGENATE see PLF000

2-ETHYLPROPENAL see EFS700

2-ETHYL-2-PROPENAL see EFS700

ETHYL PROPENOATE see EFT000

ETHYL-2-PROPENOATE see EFT000

ETHYL PROPENYL ETHER see EOY500

ETHYL-1-PROPENYL ETHER see EOY500

ETHYL PROPIONATE see EPB500

2-ETHYL-3-PROPYL ACROLEIN see EKR000

α-ETHYL-β-n-PROPYLACROLEIN see EKR000

5-ETHYL-5-(1-PROPYL-1-BUTENYL)BARBITURIC ACID see EPC000

ETHYL-N,N-PROPYLCARBAMATE see EPC050

N-(1-ETHYLPROPYL)-3,4-DIMETHYL-2,6-DINITROBENZENAMINE see DRN200

2-ETHYL-3-PROPYL-2,3-EPOXYPROPIONAMIDE see OLU000

ETHYL N-((5R,8S,10R)-6-PROPYL-8-ERGOLINYL)CARBAMATE see EPC115

ETHYL PROPYL ETHER see EPC125

ETHYL n-PROPYL ETHER see EPC125

2-ETHYL-3-PROPYLGLYCIDAMIDE see OLU000

ETHYL PROPYL KETONE see HEV500

3-(1-ETHYLPROPYL)PHENOL METHYLNITROSOCARBAMATE mixed with 3-(1-METHYLBUTYL)PHENYL METHYLNITROSOCARBAMATE see NJP000

ETHYL PROPYL PHOSPHOROTHIOATE see EPC130

2-ETHYL-3-PROPYL-1,3-PROPANEDIOL see EKV000

2-ETHYL-2-PROPYLTHIOBUTYRAMIDE see EPC135

O-ETHYL PROPYLTHIOCARBAMATE see EPC150

O-ETHYL-S-PROPYL-O-(2,4,6-TRICHLOROPHENYL)PHOSPHOROTHIOATE see EPC175

20-ETHYLPROSTAGLANDIN F2-α see EPC200

ETHYLPROTAL see EQF000

ETHYL PTS see EPW500

9-ETHYL-9H-PURINE-6-THIOL see EMC500

9-ETHYL-9H-PURINE-6(1H)-THIONE see EMC500

ETHYL PYRAZINYL PHOSPHOROTHIOATE see EPC500

2-ETHYL-4-PYRIDINECARBOTHIOAMIDE see EPQ000

2-(5-ETHYL-2-PYRIDYL)ETHYL ACRYLATE see ADU750

2-(5-ETHYL-2-PYRIDYL)ETHYL PROPENOATE see ADU750

ETHYL 4-PYRIDYL KETONE see PMX300

ETHYL PYROCARBONATE see DIZ100

2-ETHYL PYROMECONIC ACID see EMA600

N-ETHYLPYRROLIDINONE see EPC700

1-ETHYL-2-PYRROLIDINONE see EPC700

N-ETHYL-2-(PYRROLIDINYL)-o-ACETOTOLUIDIDE HYDROCHLORIDE see EPC875

4'-ETHYL-4-N-PYRROLIDINYLAZOBENZENE see EPC950

3(1-ETHYL-2-PYRROLIDINYL)INDOLE HYDROCHLORIDE see EPC999

N-((1-ETHYL-2-PYRROLIDINYL)METHYL)-5-(ETHYLSULFONYL)-o-ANISAMIDE see EPD100

N-((1-ETHYL-2-PYRROLIDINYL)METHYL)-2-METHOXY-5-ETHYLSULFONYLBENZAMIDE HYDROCHLORIDE see SOU725

N-((1-ETHYL-2-PYRROLIDINYL)METHYL)-2-METHOXY-5-SULFAMOYLBENZAMIDE see EPD500

N-((1-ETHYL-2-PYRROLIDINYL)METHYL)-5-SULFAMOYL-o-ANISAMIDE see EPD500

N-ETHYLPYRROLIDONE see EPC700

N-(1-ETHYL-2-PYRROLIDYLMETHYL)-2-METHOXY-5-ETHYLSULFONYL

BENZAMIDE CHLORHYDRATE see SOU725
1-ETHYLQUINALDINIUM IODIDE see EPD600
ETHYL-1(2H)-QUINOLINECARBOXYLATE see CBO750
ETHYL RETINAMIDE see REK330
N-ETHYLRETINAMIDE see REK330
ETHYL RHODANATE see EPP000
3-ETHYLRHODANINE see EPF500
3-ETHYLRODANIN see EPF500
ETHYL-S see BID250
O-ETHYL Se-(2-DIETHYLAMINOETHYL)PHOSPHONOSELENOATE see SBU900
ETHYL SEBACATE see DJY600
ETHYL SELENAC see DJD400
ETHYL SELENAC see SBP900
ETHYL SELERAM see SBP900
ETHYL SENECIOATE see MIP800
ETHYL SILICATE see EPF550
ETHYL SILICON TRICHLORIDE see EPY500
1-N-ETHYLSISOMICIN see SBD000
ETHYL SODIUM see EPF600
ETHYL STEARATE see EPF700
ETHYLSTIBAMINE see SLP600
α-ETHYL-4,4'-STILBENEDIOL see EPF800
ETHYLSTILBESTROL see EPF800
ETHYLSTILBOESTROL see EPF800
m-ETHYLSTYRENE see EPG000
ETHYL SUCCINATE see SNB000
ETHYL SULFATE see DKB250
ETHYL SULFHYDRATE see EMB100
ETHYL SULFIDE see EPH000
S-(2-(ETHYLSULFINYL)ETHYL)-O,O-DIMETHYL PHOSPHOROTHIOATE see DAP000
S-ETHYLSULFINYLMETHYL-O,O-DIISOPROPYLDITHIOFOSFAT see EPH500
S-(ETHYLSULFINYL)METHYL O,O-DIISOPROPYL PHOSPHORODITHIOATE see EPH500
S-2-ETHYL-SULFINYL-1-METHYL-ETHYL-O,O-DIMETHYL-MONOTHIOFOSFAAT see DSK600
ETHYL SULFITE see DKB119
ETHYLSULFOCHLORIDE see EEC000
ETHYL SULFOCYANATE see EPP000
ETHYLSULFONAL see BJT750
ETHYLSULFONYLETHANOL see EPI000
2-(ETHYLSULFONYL)ETHANOL see EPI000
ETHYLSULFONYLETHYL ALCOHOL see EPI000
1-(2-(ETHYLSULFONYL)-ETHYL)-2-METHYL-5-NITROIMIDAZOLE see TGD250
1-(ETHYLSULFONYL)-4-FLUOROBENZENE see FLG000
4-(ETHYLSULFONYL)-1-NAPHTHALENE SULFONAMIDE see EPI300
γ-(ETHYLSULFONYL)-N,N,α-TRIMETHYL-γ-PHENYLBENZENEPROPANAMINE HYDROCHLORIDE see DWC100
3-(3-ETHYLSULFONYL)PENTYL PIPERIDINO KETONE see EPI400
ETHYL ((p-SULFOPHENYL)THIO)MERCURY, SODIUM SALT see SKH150
ETHYL SULFOXIDE see EPI500
S-2-ETHYL-SULPHINYL-1-METHYL-ETHYL-O,O-DIMETHYL PHOSPHOROTHIOLATE see DSK600
4-ETHYLSULPHONYLNAPHTHALENE-1-SULFONAMIDE see EPI300
4-ETHYLSULPHONYLNAPHTHALENE-1-SULPHONAMIDE see EPI300
ETHYL TELLURAC see EPJ000
ETHYL TELLURIDE (8CI) see DKB150
1-ETHYL-1-TETRADECYLPIPERIDINIUM BROMIDE see EPJ600

6-ETHYL-1,2,3,4-TETRAHYDRO-9,10-ANTHRACENEDIONE see EPK000
2-ETHYL-5,6,7,8-TETRAHYDROANTHRAQUINONE see EPK000
3'-ETHYL-5',6',7',8'-TETRAHYDRO-5',5',8',8'-TETRAMETHYL-2'-ACETONAPHTHONE see ACL750
1-(3-ETHYL-5,6,7,8-TETRAHYDRO-5,5,8,8-TETRAMETHYL-2-NAPHTHALENYL)-ETHANONE see ACL750
ETHYL-p-((E)-2-(5,6,7,8-TETRAHYDRO-5,5,8,8-TETRAMETHYL-2-NAPHTHYL)-1-PROPENYL)BENZOATE see AQZ400
1-ETHYL-2,2,6,6-TETRAMETHYL-4-(N-ACETYL-N-PHENYL)PIPERIDINE see EPL000
1-ETHYL-2,2,6,6-TETRAMETHYLPIPERIDINE HYDROCHLORIDE see EPL600
1-ETHYL-2,2,6,6-TETRAMETHYLPIPERIDINE HYDROGEN TARTRATE see EPL700
1-ETHYL-2,2,6,6-TETRAMETHYL-PIPERIDINE TARTRATE see EPL700
N-(1-ETHYL-2,2,6,6-TETRAMETHYLPIPERIDIN-4-YL)ACETANILIDE see EPL000
1-ETHYL-2,2,6,6-TETRAMETHYL-4-(N-PROPIONYL-N-BENZYLAMINO)PIPERIDINE see EPM000
1-ETHYL-1,1,3,3-TETRAMETHYLTETRAZENIUM see EPM550
ETHYL TETRAPHOSPHATE see HCY000
ETHYL TETRAPHOSPHATE, HEXA- see HCY000
2-ETHYLTETRAZOLE see EPM590
5-ETHYLTETRAZOLE see EPM600
ETHYL THIOALCOHOL see EMB100
2-ETHYL-4-THIOAMIDYLPYRIDINE see EPQ000
2-ETHYLTHIO-4,6-BIS(ISOPROPYLAMINO)-s-TRIAZINE see EPN500
6-(ETHYLTHIO)N,N'-BIS(1-METHYLETHYL)-1,3,5-TRIAZINE-2,4-DIAMINE see EPN500
3-(ETHYLTHIO)BUTANAL see EPN600
3-(ETHYLTHIO)BUTYRALDEHYDE see EPN600
β-(ETHYLTHIO)BUTYRALDEHYDE see EPN600
2-ETHYL-4-THIOCARBAMOYLPYRIDINE see EPQ000
2-(ETHYLTHIO)CHLOROETHANE see CGY750
ETHYL THIOCYANATE see EPP000
ETHYL THIOCYANATOACETATE see TFF100
ETHYL THIOCYANOACETATE see TFF100
ETHYLTHIOETHANE see EPH000
2-(ETHYLTHIO)-ETHANETHIOL S-ESTER with O,O-DIETHYL PHOSPHOROTHIOATE see DAP200
2-(ETHYLTHIO)ETHANOL see EPP500
ETHYL THIOETHER see EPH000
2-ETHYLTHIOETHYL CHLORIDE see CGY750
S-2-(ETHYLTHIO)ETHYL O,O-DIETHYL ESTER OF PHOSPHORODITHIOIC ACID see DXH325
2-ETHYLTHIOETHYL O,O-DIMETHYL PHOSPHORODITHIOATE see PHI500
S-(2-(ETHYLTHIO)ETHYL) O,O-DIMETHYLPHOSPHORODITHIONATE see PHI500
O-(2-(ETHYLTHIO)ETHYL)-O,O-DIMETHYL PHOSPHOROTHIOATE see DAO800
S-(2-(ETHYLTHIO)ETHYL)-O,O-DIMETHYL PHOSPHOROTHIOATE see DAP400

S(and O)-2-(ETHYLTHIO)ETHYL-O,O-DIMETHYL PHOSPHOROTHIOATE see MIW100
S-(2-(ETHYLTHIO)ETHYL)DIMETHYL PHOSPHOROTHIOLATE see DAP400
S-(2-(ETHYLTHIO)ETHYL)DIMETHYL PHOSPHOROTHIOLOTHIONATE see PHI500
2-(ETHYLTHIO)ETHYL DIMETHYL PHOSPHOROTHIONATE see DAO800
S-(2-(ETHYLTHIO)ETHYL)-O,O-DIMETHYL THIOPHOSPHATE see DAP400
ETHYL THIOGLYCOLATE see EMB200
2-ETHYLTHIOISONICOTINAMIDE see EPQ000
α-ETHYLTHIOISONICOTINAMIDE see EPQ000
2-((ETHYLTHIO)METHYL)PHENOL METHYLCARBAMATE see EPR000
2-((ETHYLTHIO)METHYL)PHENYL METHYLCARBAMATE see EPR000
(2-ETHYLTHIOMETHYLPHENYL)-N-METHYLCARBAMATE see EPR000
2-(ETHYLTHIO)-10-(3-(4-METHYL-1-PIPERAZINYL)PROPYL)-10-PHENOTHIAZINE-(Z)-2-BUTENEDIOATE see TEZ000
2-(ETHYLTHIO)-10-(3-(4-METHYL-1-PIPERAZINYL)PROPYL)PHENOTHIAZINE DIMALEATE see TEZ000
ETHYLTHIOMETON see DAO500
ETHYL THIOMETON see DXH325
ETHYLTHIOMETON B see DXH325
ETHYLTHIOMETON SULFOXIDE see OQS000
ETHYLTHIONOPHOSPHONYL DICHLORIDE see EOP600
ETHYL THIOPHANATE see DJV000
ETHYLTHIOPHOSPHONIC DICHLORIDE see EOP600
5-(((2-ETHYLTHIO)PROPYL)-2-(1-OXOPROPYL)-1,3-CYCLOHEXANEDIONE see EPR100
2-ETHYL-2-THIO-PSEUDOUREA with METAPHOSPHORIC ACID (1:1) see ELY575
ETHYL THIOPYROPHOSPHATE see SOD100
(ETHYLTHIO)TRIOCTYLSTANNANE see EPR200
ETHYL THIOUREA see EPR600
1-ETHYLTHIOUREA see EPR600
ETHYL THIRAM see DXH250
ETHYL THIUDAD see DXH250
ETHYL THIURAD see DXH250
ETHYL TIGLATE see TGA800
ETHYLTIN TRICHLORIDE see EPS000
ETHYL-α-TOLUATE see EOH000
2-ETHYLTOLUENE see EPS500
4-ETHYLTOLUENE see EPT000
m-ETHYLTOLUENE see EPS600
o-ETHYLTOLUENE see EPS500
p-ETHYLTOLUENE see EPT000
ETHYL(p-TOLUENESULFONANILIDATO)MERCURY see EME500
ETHYL-p-TOLUENESULFONATE see EPW500
6-ETHYL-o-TOLUIDINE see MJY000
N-ETHYL-m-TOLUIDINE see EPT100
2-(N-ETHYL-m-TOLUIDINO)ETHANOL see HKS100
p-((3-ETHYL-p-TOLYL)AZO)-N,N-DIMETHYLANILINE see EPT500
p-((4-ETHYL-m-TOLYL)AZO)-N,N-DIMETHYLANILINE see EPU000
N-ETHYL-N-(1-(o-TOLYLOXY-2-PROPYL)CARBAMIC ACID)-2-(2-METHYLPIPERIDINO)ETHYL ESTER HYDROCHLORIDE see EPV000
1-ETHYL-3-p-TOLYLTRIAZENE see EPW000
ETHYL TOSYLATE see EPW500
ETHYL-p-TOSYLATE see EPW500

ETHYL-3,5,6-TRI-o-BENZYL-d-GLUCOFURANOSIDE see EPW600
ETHYL TRI-n-BUTYLAMMONIUM HYDROXIDE see EPX000
ETHYL TRICHLOROPHENYLETHYLPHOSPHONO THIOATE see EPY000
O-ETHYL-O-2,4,5-TRICHLOROPHENYL ETHYLPHOSPHONOTHIOATE see EPY000
ETHYL TRICHLOROSILANE see EPY500
ETHYLTRICHLOROSILANE (DOT) see EPY500
ETHYLTRICHLOROSTANNANE see EPS000
ETHYLTRICHLOROTIN see EPS000
ETHYLTRICHLORPHON see EPY600
3-ETHYLTRICYCLOQUINAZOLINE see EPZ000
ETHYLTRIETHOXYSILANE see EQA000
ETHYL-2,2,3-TRIFLUORO PROPIONATE see EQB500
ETHYL 3,4,5-TRIHYDROXYBENZOATE see EKM100
α-ETHYL-β-(2,4,6-TRIIODO-3-BUTYRAMIDOPHENYL)ACRYLIC ACID SODIUM SALT see EQC000
α-ETHYL-β-(2,4,6-TRIIODO-3-BUTYRAMIDOPHENYL)PROPIONIC ACID SODIUM SALT see SKO500
ETHYLTRIIODOGERMANE see EQC500
ETHYLTRIMETHYLAMMONIUM IODIDE see EQC600
ETHYL-γ-TRIMETHYLAMMONIUM PROPANEDIOL IODIDE see MJH250
ETHYL-3,7,11-TRIMETHYLDODECA-2,4-DIENOATE see EQD000
ETHYL(2E,4E)-3,7,11-TRIMETHYL-2,4-DODECADIENOATE see EQD000
ETHYL (2E,4E)-3,7,11-TRIMETHYL-DODECA-2-4-DIENOATE see EQD000
ETHYLTRIMETHYLOLMETHANE see HDF300
4-ETHYL-2,6,7-TRIOXA-1-ARSABICYCLO(2.2.2)OCTANE see EQD100
4-ETHYL-2,6,7-TRIOXA-1-PHOSPHABICYCLO(2.2.2)OCTANE see TNI750
4-ETHYL-2,6,7-TRIOXA-1-PHOSPHABICYCLO(2.2.2)OCTANE-1-OXIDE see ELJ500
ETHYLTRIPHENYLPHOSPHONIUM IODIDE see EQD150
α-ETHYLTRYPTAMINE ACETATE see AJB250
dl-α-ETHYLTRYPTAMINE ACETATE see AJB250
ETHYL TUADS see BJB500
ETHYL TUADS see DXH250
ETHYL TUEX see DXH250
ETHYL UNDECENOATE see EQD200
ETHYL 10-UNDECENOATE see EQD200
ETHYL UNDECYLENATE see EQD200
1-ETHYL-1-UNDECYLPIPERIDINIUM BROMIDE see EQD600
ETHYLUREA see EQD875
1-ETHYLUREA see EQD875
N-ETHYLUREA see EQD875
ETHYLUREA and SODIUM NITRITE (1:1) see EQD900
ETHYLUREA and SODIUM NITRITE (2:1) see EQE000
ETHYLURETHAN see UVA000
ETHYL URETHANE see UVA000
o-ETHYLURETHANE see UVA000
ETHYL VANILLATE see EQE500
ETHYL VANILLIN see EQF000
O-ETHYLVANILLIN see NOC100
ETHYL VANILLIN ACETATE see EQF100
4-ETHYLVERATROLE see EQF150
N-ETHYL-N-VINYLACETAMIDE see EQF200
m-ETHYL VINYLBENZEN (CZECH) see EPG000

m-ETHYLVINYLBENZENE see EPG000
ETHYLVINYLDICHLOROSILANE see DFK400
ETHYL VINYL ETHER see EQF500
ETHYL VINYL ETHER HOMOPOLYMER see OJV600
ETHYL VINYL ETHER POLYMER see OJV600
ETHYL VINYL KETONE see PBR250
ETHYLVINYLNITROSAMINE see NKF000
3-ETHYL-6-VINYLPYRIDINE see EQG500
5-ETHYL-2-VINYLPYRIDINE see EQG500
ETHYL VIOLET see EQG550
ETHYL VIOLET AX see EQG550
ETHYL VIOLET GGA see EQG550
ETHYL XANTHATE see EQG600
ETHYLXANTHIC ACID see EQG600
ETHYLXANTHIC ACID ANHYDROSULFIDE with O-ETHYLTHIOLCARBONATE see EQH000
ETHYLXANTHIC ACID POTASSIUM SALT see PLF000
ETHYLXANTHIC ACID SODIUM SALT see SHE500
ETHYL XANTHOGENATE see EQG600
ETHYL XANTHOGEN DISULFIDE see BJU000
N-ETHYL-N-(1-(3,5-XYLYLOXY)-2-PROPYL)CARBAMIC ACID-2-(DIETHYLAMINO)ETHYL ESTER HYDROCHLORIDE see EQI000
N-ETHYL-N-(1-(3,5-XYLYLOXY)-2-PROPYL)CARBAMIC ACID-2-(2-METHYLPIPERIDINO)ETHYL ESTER HYDROCHLORIDE see EQI500
ETHYL ZIMATE see BJC000
ETHYL ZIMATE see EIR000
ETHYL ZIRUM see BJC000
ETHYMAL see ENG500
ETHYMEMAZINE MONOHYDROCHLORIDE see ELQ600
ETHYMIDINE see EQI600
ETHYNE see ACI750
ETHYNE, DICHLORO-(9CI) see DEN600
ETHYNE, DIIODO- see DNE500
ETHYNERONE see CHP750
ETHYNERONE mixed with MESTRANOL (20:1) see EQJ000
17-α-ETHYNIL-Δ⁴-ESTRENE-17-β-OL see NNV000
ETHYNLOESTRENOL see NNV000
ETHYNODIOL see EQJ100
ETHYNODIOL ACETATE see EQJ500
ETHYNODIOL DIACETATE see EQJ500
β-ETHYNODIOL DIACETATE see EQJ500
ETHYNODIOL DIACETATE mixture with ETHYNYLESTRADIOL see DAP810
ETHYNODIOL DIACETATE mixed with MESTRANOL see EQK010
ETHYNODIOL mixed with MESTRANOL see EQK100
17-α-ETHYNYL-17-β-ACETOXY-19-NORANDROST-4-EN-3-ONE see ABU000
2-ETHYNYL-A-NOR-5-α-17-α-PREGN-20-YNE-2-β,17-DIOL see EQN320
ETHYNYLBENZENE see PEB750
α-ETHYNYLBENZENEMETHANOL CARBAMATE see PGE000
α-ETHYNYLBENZYL CARBAMATE see PGE000
2-ETHYNYL-2-BUTANOL see EQL000
2-ETHYNYL-2-BUTYL CARBAMATE see MNM500
ETHYNYLCARBINOL see PMN450
1-ETHYNYLCYCLOHEXANOL see EQL500
1-ETHYNYL-1-CYCLOHEXANOL see EQL500
1-ETHYNYLCYCLOHEXAN-1-OL see EQL500
1-ETHYNYLCYCLOHEXANOL ACETATE see EQL550

1-ETHYNYLCYCLOHEXANOL CARBAMATE see EEH000
1-ETHYNYLCYCLOHEXYL ACETATE see EQL550
1-ETHYNYLCYCLOHEXYL ALLOPHANATE see EQL600
1-ETHYNYLCYCLOHEXYL CARBAMATE see EEH000
1-ETHYNYLCYCLOHEXYL ESTER CARBAMIC ACID see EEH000
17-α-ETHYNYL-3,17-DIHYDROXY-4-ESTRENE DIACETATE see EQJ500
17-ETHYNYL-3,17-DIHYDROXY-1,3,5-OESTRATRIENE see EEH500
ETHYNYLDIMETHYLCARBINOL see MHX250
17-α-ETHYNYL-6-α,21-DIMETHYLTESTOSTERONE see DRT200
ETHYNYLENEBIS(CARBONYLOXY)BIS(TRIPHENYLSTANNANE) see BLS900
ETHYNYLESTRADIOL see EEH500
17-α-ETHYNYLESTRADIOL see EEH500
17-α-ETHYNYLESTRADIOL-17-β see EEH500
ETHYNYLESTRADIOL mixture with ETHYNODIOL DIACETATE see DAP810
ETHYNYLESTRADIOL-LYNESTRENOL mixture see EEH575
ETHYNYLESTRADIOL-3-METHYL ETHER see MKB750
17-ETHYNYLESTRADIOL-3-METHYL ETHER see MKB750
17-α-ETHYNYLESTRADIOL-3-METHYL ETHER see MKB750
ETHYNYLESTRADIOL-3-METHYL ETHER and 17-α-ETHYNYL-17-HYDROXYESTREN-3-ONE see MDL750
ETHYNYLESTRADIOL-3-METHYL ETHER and 17-α-ETHYNYL-19-NORTESTOSTERONE see MDL750
ETHYNYLESTRADIOL mixed with NORETHINDRONE see EQM500
17-α-ETHYNYL-1,3,5(10)-ESTRATRIENE-3,17-β-DIOL see EEH500
17-α-ETHYNYLESTRA-1,3,5(10)-TRIENE-3,17-β-DIOL see EEH500
17-α-ETHYNYL-ESTR-5(10)-ENE-3-α,17-β-DIOL see NNU000
17-α-ETHYNYL-ESTR-5(10)-ENE-3-β,17-β-DIOL see NNU500
17-α-ETHYNYLESTR-4-ENE-3-β,17-β-DIOL ACETATE see EQJ500
17-α-ETHYNYL-4-ESTRENE-3-β,17-DIOL DIACETATE see EQJ500
17-α-ETHYNYL-4-ESTRENE-3-β,17-β-DIOL DIACETATE see EQJ500
ETHYNYLESTRENOL see NNV000
17-α-ETHYNYLESTRENOL see NNV000
17-α-ETHYNYLESTR-4-EN-17-β-OL see NNV000
17-α-ETHYNYL-4-ESTREN-17-OL-3-ONE see NNP500
17-α-ETHYNYLESTR-5(10)-EN-17-β-OL-3-ONE see EEH550
17-α-ETHYNYL-5(10)-ESTREN-17-OL-3-ONE see EEH550
17-α-ETHYNYL-ESTR-5(10)-EN-3-ON-17-β-OL see EEH550
1-ETHYNYLETHANOL see EQM600
2-ETHYNYLFURAN see EQN000
17-α-ETHYNYL-17-β-HEPTANOYLOXY-4-ESTREN-3-ONE see NNQ000
17-α-ETHYNYL-17-HYDROXY-6-α,21-DIMETHYLANDROST-4-EN-3-ONE see DRT200
17-α-ETHYNYL-17-β-HYDROXYESTRA-4,9,11-TRIEN-3-ONE see NNR125
17-α-ETHYNYL-17-HYDROXY-4-ESTREN-3-ONE see NNP500
17-α-ETHYNYL-17-β-HYDROXY-5(10)-ESTREN-3-ONE see EEH550

17-α-ETHYNYL-17-β-HYDROXYESTR-5(10)-EN-3-ONE see EEH550

17-α-ETHYNYL-17-β-HYDROXY-Δ⁵⁽¹⁰⁾-ESTREN-3-ONE see EEH550

17-α-ETHYNYL-17-HYDROXYESTR-5(10)-EN-3-ONE see MKB750

17-α-ETHYNYL-17-HYDROXY-5(10)-ESTREN-3-ONE see EEH550

17-α-ETHYNYL-17-HYDROXYESTR-4-EN-3-ONE ACETATE see ABU000

17-α-ETHYNYL-17-HYDROXYESTREN-3-ONE and ETHYNYLESTRADIOL 3-METHYL ETHER see MDL750

(+)-17-α-ETHYNYL-17-β-HYDROXY-3-METHOXY-1,3,5(10)-ESTRATRIENE see MKB750

(+)-17-α-ETHYNYL-17-β-HYDROXY-3-METHOXY-1,3,5(10)-OESTRATRIENE see MKB750

17-α-ETHYNYL-17-β-HYDROXY-19-NORANDROST-4-EN-3-ONE see NNP500

17-α-ETHYNYL-17-β-HYDROXY-3-OXO-Δ⁵⁽¹⁰⁾-ESTRENE see EEH550

ETHYNYLMETHANOL see PMN450

α-ETHYNYL-p-METHOXYBENZYL ALCOHOL see HKA700

α-ETHYNYL-p-METHOXYBENZYL ALCOHOL ACETATE see EQN225

17-α-ETHYNYL-3-METHOXY-1,3,5(10)-ESTRATRIEN-17-β-OL see MKB750

17-ETHYNYL-3-METHOXY-1,3,5(10)-ESTRATRIEN-17-β-OL see MKB750

17-α-ETHYNYL-3-METHOXY-17-β-HYDROXY-Δ-1,3,5(10)-ESTRATRIENE see MKB750

17-α-ETHYNYL-3-METHOXY-17-β-HYDROXY-Δ-1,3,5(10)-OESTRATRIENE see MKB750

17-ETHYNYL-3-METHOXY-1,3,5(10)-OESTRATRIEN-17-β-OL see MKB750

α-ETHYNYL-α-METHYLBENZYL ALCOHOL see EQN230

ETHYNYLMETHYL ETHER see MDX000

17-ETHYNYL-18-METHYL-19-NORTESTOSTERONE see NNQ500

17-ETHYNYL-18-METHYL-19-NORTESTOSTERONE see NNQ520

(RS)-(E)-1-ETHYNYL-2-METHYL-2-PENTENYL(1R)-cis,trans-CHRYSANTHEMATE see EQN250

1-ETHYNYL-2-(1-METHYLPROPYL)CYCLOHEXANOL ACETATE see EQN300

1-ETHYNYL-2-(1-METHYLPROPYL)CYCLOHEXYL ACETATE see EQN300

17-α-ETHYNYL-19-NORANDROST-4-ENE-3-β,17-β-DIOL see EQJ100

17-α-ETHYNYL-19-NORANDROST-4-ENE-3-β,17-β-DIOL DIACETATE see EQJ500

17-α-ETHYNYL-19-NORANDROST-4-EN-17-β-OL-3-ONE see NNP500

17-α-ETHYNYL-19-NOR-4-ANDROSTEN-17-β-OL-3-ONE see NNP500

17-α-ETHYNYL-19-NOR-5(10)-ANDROSTEN-17-β-OL-3-ONE see EEH550

2-α-ETHYNYL-a-NOR-17-α-PREGN-20-YNE-2-β,17-β-DIOL see EQN320

17-α-ETHYNYL-19-NORTESTOSTERONE see NNP500

17-α-ETHYNYL-19-NORTESTOSTERONE ACETATE see ABU000

17-α-ETHYNYL-19-NORTESTOSTERONE and ETHYNYLESTRADIOL 3-METHYL ETHER see MDL750

ETHYNYLOESTRADIOL see EEH500

17-ETHYNYLOESTRADIOL see EEH500

17-α-ETHYNYLOESTRADIOL see EEH500

17-α-ETHYNYL-17-β-OESTRADIOL see EEH500

17-α-ETHYNYLOESTRADIOL-17-β see EEH500

ETHYNYLOESTRADIOL METHYL ETHER see MKB750

17-ETHYNYLOESTRADIOL-3-METHYL ETHER see MKB750

17-α-ETHYNYLOESTRADIOL METHYL ETHER see MKB750

17-α-ETHYNYLOESTRADIOL-3-METHYL ETHER see MKB750

17-α-ETHYNYL-1,3,5(10)-OESTRATRIENE-3,17-β-DIOL see EEH500

17-α-ETHYNYLOESTRA-1,3,5(10)-TRIENE-3,17-β-DIOL see EEH500

17-α-ETHYNYL-1,3,5-OESTRATRIENE-3,17-β-DIOL see EEH500

17-ETHYNYLOESTRA-1,3,5(10)-TRIENE-3,17-β-DIOL see EEH500

17-α-ETHYNYLOESTRENOL see NNV000

17-α-ETHYNYLOESTR-4-EN-17-β-OL see NNV000

ETHYNYLTESTOSTERONE see GEK500

17-α-ETHYNYLTESTOSTERONE see GEK500

ETHYNYL VINYL SELENIDE see EQN500

ETHYONOMIDE see EPQ000

ETICOL see NIM500

ETIDRONIC ACID see HKS780

ETIFELMIN HYDROCHLORIDE see DWF200

ETIFOXIN see CGQ500

ETIFOXINE see CGQ500

ETIL 702 see MJM700

ETIL ACRILATO (ITALIAN) see EFT000

ETILACRILATULUI (ROMANIAN) see EFT000

ETILAMINA (ITALIAN) see EFU400

ETILBENZENE (ITALIAN) see EGP500

ETILBUTILCHETONE (ITALIAN) see EHA000

ETIL CLOROCARBONATO (ITALIAN) see EHK500

ETIL CLOROFORMIATO (ITALIAN) see EHK500

ETILE (ACETATO di) (ITALIAN) see EFR000

ETILE (FORMIATO di) (ITALIAN) see EKL000

ETILEFRINE see EGE500

ETILEFRINE HYDROCHLORIDE see EGF000

ETILEFRINE PIVALATE HYDROCHLORIDE see EGC500

N,N'-ETILEN-BIS(DITIOCARBAMMATO) di MANGANESE (ITALIAN) see MAS500

N,N'-ETILEN-BIS(DITIOCARBAMMATO) di SODIO (ITALIAN) see DXD200

ETILENE (OSSIDO di) (ITALIAN) see EJN500

ETILENIMINA (ITALIAN) see EJM900

ETILEN-XANTISAN TABL. see TEP500

ETILFEN see EOK000

ETILMERCAPTANO (ITALIAN) see EMB100

O-ETIL-O-((4-NITRO-FENIL)-FENIL)-MONOTIOFOSFONATO (ITALIAN) see EBD700

S-2-ETIL-SULFINIL-1-METIL-ETIL-O,O-DIMETIL-MONOTIOFOSFATO see DSK600

ETIMID see EPQ000

ETIMIDIN see EQI600

ETIN see EDC500

ETINAMATE see EEH000

ETIOCHOLANE-17-β-OL-3-ONE see HJB100

ETIOCHOLAN-17-β-OL-3-ONE see HJB100

ETIOCIDAN see EPQ000

ETIOL see MAK700

ETIONAMID see EPQ000

ETIONIZINA see EPQ000

ETIZOLAM see EQN600

ETM see EJQ000

ETM (heterocycle) see EJQ000

ETMA see EQN700

ETMOZIN see EEI050

ETMT see EFK000

ETO see EJN500

ETOC see THI500

ETOCIL see EEM000

ETOCLOFENE see DGN000

ETODROXINE see HHK050

ETODROXIZINE DIMALEATE see HHK100

ETODROXYZINE see HHK050

ETODROXYZINE DIMALEATE see HHK100

ETOFEN see DGN000

ETOFENAMATE see HKK000

ETOFENPROX see EQN725

ETOFYLLINCLOFIBRAT (GERMAN) see TEQ500

ETOFYLLINE see HLC000

ETOFYLLINE CLOFIBRATE see TEQ500

ETOGLUCID see TJQ333

ETOGYN see MBZ100

ETOKSYETYLOWY ALKOHOL (POLISH) see EES350

ETOMAL see ENG500

ETOMIDATE see HOU100

ETOMIDE HYDROCHLORIDE see CBQ625

ETOMIDOLINE see AHP125

ETONITAZENE HYDROCHLORIDE see EQN750

ETOPERIDONE see EQO000

E-TOPLEX see TGJ050

ETOPOSIDE see EAV500

ETOPROPEZINA see DIR000

ETOPSIDE see EBB600

ETORPHINE see EQO450

(−)-ETORPHINE see EQO450

7-α-ETORPHINE see EQO450

ETORPHINE HYDROCHLORIDE see EQO500

ETOSALICIL see EEM000

ETOSALICYL see EEM000

2-ETOSSIETIL-ACETATO (ITALIAN) see EES400

ETOSUXIMIDA see ENG500

ETOVAL see BPF500

ETOXAZENE see EEQ600

ETOXON AF5 see SNY100

ETOZOLINE see MNG000

ETP see EPQ000

ETP see EQP000

ETs (POLYSACCHARIDE) see EHG100

SYMETRA see DKE800

ETRENOL see HGO500

ETRETIN see REP400

ETRETINATE see EMJ500

ETRIDIAZOLE see EFK000

ETRIOL see HDF300

ETROFLEX see GKK000

ETROFOL see CKF000

ETROFOLAN see MIA250

ETROLENE see RMA500

ETROPRES see RCA200

ETROZOLIDINA see HNI500

ETRUSCOMICINA see LIN000

ETRUSCOMYCIN see LIN000

SYMETRYCZNA DWUMETYLOHYDRAZYNA (POLISH) see DSF600

ETRYPTAMINE ACETATE see AJB250

ETTRIOL see HDF300

ETU see IAQ000

ETYBENZATROPINE see DWE800

ETYCHLOZATE see EHK600

ETYDION see TLP750

ETYLENU TLENEK (POLISH) see EJN500

ETYLOAMINA (POLISH) see EFU400

ETYLOBENZEN (POLISH) see EGP500

ETYLON see TCC000

ETYLOWY ALKOHOL (POLISH) see EFU000

ETYLU BROMEK (POLISH) see EGV400

ETYLU CHLOREK (POLISH) see EHH000

ETYLU KRZEMIAN (POLISH) see EPF550

E. TYPHOSA LIPOPOLYSACCHARIDE see EQP100

EU-1806 see NAE100

EU 4200 see TNR485

EU 11100 see DRV250
EUBASIN see PPO000
EUBASINUM see PPO000
EUBINE see DLX400
β-EUCAINE see EQP500
EUCAINE B see EQP500
EUCALMYL see FLU000
EUCALYPTOLE see CAL000
EUCALYPTOL (FCC) see CAL000
EUCALYPTUS CITRIODORA OIL, ACETYLATED see OHJ100
EUCALYPTUS OIL see EQQ000
EUCALYPTUS OIL, ACETYLATED see OHJ100
EUCALYPTUS REDUNCA TANNIN see MSC000
EUCANINE GB see TGL750
EUCAST see EQQ100
EUCHESSINA see PDO750
EUCHEUMA SPINOSUM GUM see CCL250
EUCHRYSINE see BJF000
EUCISTEN see EID000
EUCISTIN see PDC250
EUCLIDAN see EQQ100
EUCLORINA see CDP000
EUCODAL see DLX400
EUCORAN see DJS200
EUCTAN see TGJ885
EUCUPIN DIHYRDOCHLORIDE see EQQ500
EUDATINE see BEX500
EUDEMINE INJECTION see DCQ700
EUDESMA-3,11(13)-DIEN-12-OIC ACID see EQR000
EUDESM-4-EN-12-OIC ACID, 6-HYDROXY-1-OXO-, γ-LACTONE see GJM300
EUDRAGIT E see MDN505
EUDRAGIT E 100 see MDN505
EUDRAGIT E 12.5 see MDN505
EUDRAGIT L 30D see MDN600
EUDRAGIT L 30D55 see MDN600
EUFIN see DIX200
EUFLAVINE see DBX400
EUFLAVINE see XAK000
EUFODRIANL see MDT600
EUGENIA JAMBOS Linn., extract excluding roots see SPF200
EUGENIC ACID see EQR500
EUGENOL see EQR500
EUGENOL ACETATE see EQS000
1,3,4-EUGENOL ACETATE see EQS000
EUGENOL FORMATE see EQS100
1,3,4-EUGENOL METHYL ETHER see AGE250
EUGENOL OXIDE see EBT500
EUGENOL PHENYLACETATE see AGL000
EUGENO-2',3'-OXIDE see EBT500
EUGENYL ACETATE see EQS000
EUGENYL FORMATE see EQS100
EUGENYL METHYL ETHER see AGE250
EUGENYL PHENYLACETATE see AGL000
EUGLUCAN see CEH700
EUGLUCON see CEH700
EUGLUCON 5 see CEH700
EUGLYCIN see AKQ250
EUGLYKON see CEH700
EUGYON see NNL500
EUHAEMON see VSZ000
EUHYPNOS see CFY750
EUKAIN B see EQP500
EUKALYPTUS OEL (GERMAN) see EQQ000
EUKODAL see DLX400
EUKRATON see MKA250
EUKUPIN DIHYDROCHLORIDE see EQQ500
EUKYSTOL see CLY500
EULAN SP see REF250
EULAVA SM see MAG250
EULAXAN see PPN100
EULICIN see EQS500
EULISSIN A see DAF800
EULIXINE see DAF800

EUMICTON see AAI250
EUMIDRINA see MGR500
EUMIN see MMN250
EUMOTOL see CCD750
EUMOVATE see CMW400
EUMYCETIN see EQT000
EUMYDRIN see MGR500
EUNASIN see BAV000
EUNATROL see OIA000
EUNERPAN see FKI000
EUNOCTAL see AMX750
EUNOCTIN see DLY000
EUONYMUS EUROPAEUS see ERA309
EUPAREN see DFL200
EUPARENE see DFL200
EUPATORIOPICRIN see EQT100
EUPATORIOPICRINE see EQT100
EUPHODRIN see DBA800
EUPHORBIA ABYSSINICA LATEX see EQT500
EUPHORBIA CANARIENSIS LATEX see EQU000
EUPHORBIA CANDELABRIUM LATEX see EQU500
EUPHORBIA ESULA LATEX see EQV000
EUPHORBIA GRANDIDENS LATEX see EQV500
EUPHORBIA LATHYRIS LATEX see EQW000
EUPHORBIA OBOVALIFOLIA LATEX see EQW500
EUPHORBIA POINSETTIS BUIST see EQX000
EUPHORBIA PULCHERRIMA WILLD. see EQX000
EUPHORBIA SERRATA LATEX see EQX500
EUPHORBIA TIRUCALLI LATEX see EQY000
EUPHORBIA TIRUCALLI L., EXTRACT see EQY100
EUPHORIA WULFENII LATEX see EQY500
EUPHORIN see AHK750
EUPHORIN see CBL750
EUPHOZID see ILE000
EUPHYLLIN see TEP500
EUPHYLLINE see TEP500
EUPLACID see EHP000
EUPNERON see ECU600
EUPNERON see EQY600
EUPRACTONE see PMO250
EUPRAMIN see DLH600
EUPRAMIN see DLH630
EUPREX see TLG000
EURAZYL see EQZ000
EURECEPTOR see TAB250
EURECOR see CCK125
EUREKENE see PNX750
EURESOL see RDZ900
EUREX see EHT500
EURINOL see HII500
EUROCERT AMARANTH see FAG020
EUROCERT AZORUBINE see HJF500
EUROCERT CHRYSOINE S see MRL100
EUROCERT COCHINEAL RED A see FMU080
EUROCERT ORANGE FCF see FAG150
EUROCERT SCARLET GN see CMP620
EUROCERT TARTRAZINE see FAG140
EURODIN see CKL250
EURODOPA see DNA200
EUROGALE see MKP500
EUROPEAN HOLLY see HGF100
EUROPEAN MISTLETOE see ERA100
EUROPEAN SPINDLE TREE see ERA309
EUROPEN see MGR500
EUROPHAN see PKQ059
EUROPIC CHLORIDE see ERA500
EUROPIUM CHLORIDE see ERA500
EUROPIUM CITRATE see ERB000
EUROPIUM EDETATE see ERB500
EUROPIUM NITRATE see ERC550

EUROPIUM(III) NITRATE, HEXAHYDRATE (1:3:6) see ERC000
EUROPIUM(II) SULFIDE see ERC500
EUROPIUM TRINITRATE see ERC550
EUROTIN (A) see ERC600
EURPHYLLIN see TEP500
EUSAL see EEM000
EUSAPRIM see SNK000
EUSAPRIM see TKX000
EUSCOPOL see HOT500
EUSMANID see BFW250
EUSOLEX OCR see ODY150
EUSPIRAN see IMR000
EUSTEROL see NMV300
EUSTIDIL see DFH600
EUSTIGMIN see NCL100
EUSTIGMIN BROMIDE see POD000
EUSTIGMINE see NCL100
EUSTIGMIN METHYLSULFATE see DQY909
EUSTROPHINUM see SMN000
EUTAGEN see DLX400
EUTANOL G see OEW100
EUTENSIN see CHJ750
EUTHATAL see NBU000
EUTHYROX see LFG050
EUTIMOX see PMI250
EUTONYL see MOS250
EUTRIT see XPJ000
EUUFILIN see TEP500
EUVERNIL see SNQ550
EUVESTIN see DKB000
EUVIFOR see NNE400
EVABLIN see BGT250
EVAC-Q-KIT see PDO750
EVAC-Q-KWIK see PDO750
EVAC-U-GEN see PDO750
EVADYNE see BPT750
EVALON see ILD000
EVANOL see SJJ175
EVANS BLUE DYE see BGT250
EVASPIRINE see PGG350
EVASPRINE see PGG350
EVAU-SUPER see SFS000
EVE see EMS100
EVE see EQF500
EVE CARBAMATE see ERC700
EVENING TRUMPET FLOWER see YAK100
EVENTIN see PNN300
EVENTIN HYDROCHLORIDE see PNN300
EVERCYN see HHS000
EVERGREEN CASSENA see HGF100
EVERLASTING FLOWER OIL see HAK500
EVERNINOMICIN-D see ERC800
EVERNINOMYCIN-B see ERD000
EVEX see ECU750
EVEX see EDV600
EVION see VSZ450
EVIPAL see ERD500
EVIPAL SODIUM see ERE000
EVIPAN see ERD500
EVIPAN SODIUM see ERE000
EVIPHEROL see TGJ050
EVIPLAST 80 see DVL700
EVIPLAST 81 see DVL700
EVISECT see TFH750
EVISEKT see TFH750
EVITAMINUM see VSZ450
EVOLA see DEP800
EVONOGENIN see DMJ000
EVRONAL see SBM500
EVRONAL SODIUM see SBN000
EVRRONAL see SBN000
EV-TOXIN see ARP640
EWEISS see BAP000
EX 4355 see DLS600
EX10781 see BAV000
EX 10-781 see MHJ500
EXACTHIN see AES650
EXACT-S see TFP000
EXACYL see AJV500
EXADRIN see VGP000

EXAGAMA see BBQ500
EXAL see VLA000
EXALAMIDE see HFS759
EXAPROLOL HYDROCHLORIDE see ERE100
EXCEL see FAQ200
EXCELSIOR see CBT750
EXD see BJU000
EXDOL see HIM000
EXELL see ERE125
EXELMIN see PIJ500
EXEMESTANE see MJL275
EXHAUST GAS see CBW750
EXHORAN see DXH250
EXHORRAN see DXH250
EXIMINE see ERE150
EXIPROBEN SODIUM see ERE200
EXITELITE see AQF000
EXITLURE see OAP070
EXLAN, combustion products see ADX750
EXLUTEN see NNV000
EXLUTION see NNV000
EXLUTON see NNV000
EXLUTONA see NNV000
EXMIGRA see EDC500
EXMIN see AHJ750
EXNA see BDE250
EXOFENE see HCL000
EXOLAN see APH500
(+-)-EXO-1-METHYL-4-(1-METHYLETHYL)-2-((2-METHYLPHENYL)METHOXY)-7-OXABIC YCLO(2.2.1)HEPTANE see CMP955
EXON 450 see AAX175
EXON 454 see AAX175
EXON 605 see PKQ059
EXONAL see FLZ050
EXOSALT see BDE250
EXOTHERM see TBQ750
EXOTHERM TERMIL see TBQ750
EXOTHION see EAS000
EXOTOXIN see BAC125
EXOTOXIN A, PSEUDOMONAS AERUGINOSA see POH650
β-EXOTOXIN (BACILLUS THURINGIENSIS) see BAC125
EXOTOXIN, BACILLUS THURINGIENSIS MORRISONI see BAC140
EXOTOXIN, CLOSTRIDIUM PERFRINGENS see CMY170
EXOTOXIN, PSEUDOMONAS AERUGINOSA see POH630
EXOTOXIN, PSEUDOMONAS PSEUDOMALLEI see POH690
EXP 126 see AJU625
EXP 338 see AMQ500
EXP 999 see MQR000
EXP 105-1 see AED250
EXP-105-1 see TJG250
EXP 3864 see QMA100
EXPAND see CDK800
EXPANDEX see DBD700
EXPANSIN see CMV000
EXPERIMENTAL CHEMOTHERAPEUTANT 1,207 see NNH000
EXPERIMENTAL FUNGICIDE 341 see GII000
EXPERIMENTAL FUNGICIDE 5223 see DXX400
EXPERIMENTAL FUNGICIDE 224 (UNION CARBIDE) see ZJA000
EXPERIMENTAL HERBICIDE 2 see DGQ200
EXPERIMENTAL HERBICIDE 634 see CPR600
EXPERIMENTAL HERBICIDE 732 see BQT750
EXPERIMENTAL INSECTICIDE 711 see IKO000
EXPERIMENTAL INSECTICIDE 4049 see MAK700
EXPERIMENTAL INSECTICIDE 4124 see NFT000

EXPERIMENTAL INSECTICIDE 7744 see CBM750
EXPERIMENTAL INSECTICIDE 12008 see DJN600
EXPERIMENTAL INSECTICIDE 12,880 see DSP400
EXPERIMENTAL INSECTICIDE 52160 see TAL250
EXPERIMENTAL INSECTICIDE S-4087 see CON300
EXPERIMENTAL NEMATOCIDE 18,133 see EPC500
EXPERIMENTAL TICK REPELLENT 3 see AEO750
EXPLOSION ACETYLENE BLACK see CBT750
EXPLOSION BLACK see CBT750
EXPLOSIVE D see ANS500
EXPLOSIVES, HIGH see ERF000
EXPLOSIVES, LOW see ERF500
EXPLOSIVES, PERMITTED see ERG000
EXP. MITICIDE No. 7 see BRP250
EXPONCIT see NNW500
EXPORSAN see DNO800
EXPRESS see THT800
EXPRESS 75 DF see THT800
EXSEL see SBR000
EXSICATED SODIUM SULFITE see SJZ000
EXSICCATED FERROUS SULFATE see FBN100
EXSICCATED FERROUS SULPHATE see FBN100
EXSICCATED SODIUM PHOSPHATE see SJH090
EXTACOL see PGA750
EXT D AND C RED NO. 1 see CMM400
EXT D and C BLUE No. 3 see ERG100
EXT D and C GREEN NO. 1 see NAX500
EXT. D&C ORANGE No. 3 see FAG010
EXT. D&C RED No. 15 see FAG018
EXT. D and C RED NO. 2 see CMG750
EXT. D and C RED NO. 7 see SEH475
EXT D and C RED NO. 11 see CMM300
EXT D&C YELLOW No. 1 see MDM775
EXT. D&C YELLOW No. 9 see FAG130
EXT. D&C YELLOW No. 10 see FAG135
EXTENCILLINE see BFC750
EXTENDED ZINC INSULIN SUSPENSION see LEK000
EXTENICILLINE see BFC750
EXTERMATHION see MAK700
EXTERNAL BLUE 1 see BJI250
EXTHRIN see AFR250
EXTRACT D&C ORANGE No. 4 see TGW000
EXTRACT D&C RED No. 10 see HJF500
EXTRACT D&C RED No. 14 see XRA000
EXTRACT of JAMAICA GINGER see JBA000
EXTRACTS (PETROLEUM), HEAVY NAPHTHENIC DISTILLATE SOLVENT (9CI) see MQV857
EXTRACTS (PETROLEUM), HEAVY PARAFFINIC DISTILLATE SOLVENT (9CI) see MQV859
EXTRACTS (PETROLEUM), LIGHT NAPHTHENIC DISTILLATE SOLVENT (9CI) see MQV860
EXTRACTS (PETROLEUM), LIGHT PARAFFINIC DISTILLATE SOLVENT (9CI) see MQV862
EXTRACTS (PETROLEUM), RESIDUAL OIL SOLVENT (9CI) see MQV863
EXTRA FINE 200 SALT see SFT000
EXTRA FINE 325 SALT see SFT000
EXTRAMYCIN see APY500
EXTRANASE see BMO000
EXTRA-PLEX see DLB400
EXTRAR see DUS700
EXTRAX see RNZ000
EXTREMA see AQF000
EXTREMA see EPF550
EXTREMA see GDY000
EXTREMA see SCQ500

EXTREN see ADA725
EXTREX P 60 see PJY100
EXTRINSIC FACTOR see VSZ000
EXURATE see DDP200
EYE BRIGHT see CCJ825
EYE-CORT see HHQ800
EYEULES see ARR000
E-Z-OFF see MAE000
E-Z-OFF D see BSH250
E-Z-PAQUE see BAP000
F 1 see DEA100
F 1 see FAB000
F 1 see WCJ000
F 3 see KDK000
F 6 see NCP875
F10 see FHL000
11 F see KDK000
F 12 see DFA600
12 F see KDK000
F 13 see CLR250
F 14 see CBY250
16 F see ARM266
F 22 see CFX500
F 25 see AJU900
2KF see DOR800
F-33 see FLJ000
F 3M see KDK000
F3T see TKH325
F 44 see SCK600
4LF see KDK000
F-53 see AOO150
F-112 see TBP050
F 114 see FOO509
F-115 see CJI500
F 116 see HDC100
F 125 see PBD400
F 125 see SCK600
F-139 see BPG000
F-150 see FPI000
F 156 see HEL500
F 190 see ABX500
190 F see ABX500
F-400 see DAB840
F461 see DLV200
F 735 see CCC500
F 849 see AKR500
F-850 see DFQ500
F 933 see BCI500
F 1162 see SNM500
1167 F see DJV300
1262 F see BGO000
F 1262 see BGO000
1358F see SOA500
F 1358 see SOA500
1399 F see SNY500
2249F see FMX000
F 2387 see DVX600
F 2559 see PDD300
F 2966 see DXI400
F 5384 see BAV600
F 6060 see CPM770
F 6066 see FBP100
F 6103 see BGQ325
F 7771 see HDW100
F-114B2 see FOO525
23F203 see PJS750
F30066 see FPO000
F 1 (complexon) see EIX500
F 10 (pesticide) see MAS500
F 1 (antioxidant) see FAB000
FA see FMV000
FA see FPQ900
Fa 100 see EQR500
F 124A see TCH150
FA 142 see NAC600
4-F-3NA see FKK100
M-FA 142 see NAC600
FAA see FDR000
FAA see FFF000
FAA see FIC000
2-FAA see FDR000
2,7-FAA see BGP250

F-diAA see DBF200
FABANTOL see PMM000
FABIANOL see APT000
FABRITONE PE see PJS750
FABT (CZECH) see ALV500
FAC see FAS700
FAC see IOT000
FA-Ca see FQR100
FAC 20 see IOT000
FAC 5273 see PIX250
FACCLA see PKU600
FACCULA see PKU600
FACET see QMS100
FACTITIOUS AIR see NGU000
FACTOR II (VITAMIN) see VSZ000
FACTOR PP see NCR000
FACTOR S see BGD100
FACTOR S see VSU100
FACTOR S (vitamin) see VSU100
FACTOR S (VITAMIN) see BGD100
FAD see RIF100
FADORMIR see QAK000
FAECLA see PKU600
FAECULA see PKU600
FAFT see FQQ400
FAGINE see CMF800
FAIR 30 see DMC600
FAIR 85 see AFJ100
FAIR 85 see ODE000
FAIR PS see DMC600
FAIRY BELLS see FOM100
FAIRY CAP see FOM100
FAIRY GLOVE see FOM100
FAIRY LILY see RBA500
FAIRY THIMBLES see FOM100
FALAPEN see BFD000
FALECALCITRIOL see HDB100
FALICAIN see PNB250
FALICAINE HYDROCHLORIDE see PNB250
FALIGRUEN see CNK559
FALISAN see MEP250
FALITHION see DSQ000
FALITIRAM see TFS350
FALKITOL see HCI000
FALL see SFS000
FALL CROCUS see CNX800
FALSE ACACIA see GJU475
FALSE BITTERSWEET see AHJ875
FALSE HELLEBORE see FAB100
FALSE KOA (HAWAII) see LED500
FALSE MISTLETOE see MQW525
FALSE PARSLEY see FMU000
FALSE SYCAMORE see CDM325
FAM see MME809
FAMAFLUR see FDB300
FAMFOS see FAB400
FAMFOS see FAB600
FAMID see DVS000
FAMODIL see FAB500
FA MONOMER see FPQ900
FAMOPHOS see FAB600
FAMOPHOS WARBEX see FAB600
FAMOTIDINE see FAB500
FAMPHOS see FAB600
FAMPHUR see FAB600
FANAL PINK B see RGW000
FANAL PINK GFK see RGW000
FANAL RED 25532 see RGW000
FANASIL see AIE500
FANERON see BNK400
FANFOS see FAB600
FANFT see NGM500
FANNOFORM see FMV000
FANODORMO see TDA500
FANTERRIN see HOH500
F 1 (ANTIOXIDANT) see DEA100
FANYLINE see FFK000
FANZIL see AIE500
FAP see FPT100
FARBRUSS see CBT750
FAREDINA see TEY000
FARENAL see TMJ150

FARGAN see DQA400
FARGAN see PMI750
FAR-GO see DNS600
FARIAL see IBR200
FARINEX 100 see SLJ500
FARINGOSEPT see AHI875
FARLUTAL see MBZ150
FARLUTIN see MCA000
FARMA see BAT000
FARMA 939 see BAT000
FARMACYROL see DHB500
FARMAMID see SNQ000
FARMCO see DAA800
FARMCO ATRAZINE see ARQ725
FARMCO DIURON see DXQ500
FARMCO FENCE RIDER see TAA100
FARMCO PROPANIL see DGI000
FARMICETINA see CDP250
FARMIGLUCIN see APP500
FARMIN 80 see OBC000
FARMINOSIDIN see APP500
FARMISERINE see CQH000
FARMITALIA 204/122 see MFN500
FARMORUBICIN see EBB100
FARMOTAL see PBT250
FARMOTAL see PBT500
FARNESOL see FAB800
FARNESYL ALCOHOL see FAB800
FAROLITO (CUBA) see JBS100
FARTOX see PAX000
FAS see FAF000
FAS-CILE see MQU750
FASCIOLIN see CBY000
FASCIOLIN see HCI000
FASCO-TERPENE see CDV100
FASCO WY-HOE see CKC000
FASERTON see AHE250
FASIGIN see TGD250
FASIGYN see TGD250
FASINEX see CFL200
FASTAC see RCZ050
FASTAC 10 EC see RCZ050
FAST ACID BLUE RL see ADE750
FAST ACID LIGHT BLUE BRL see CMM100
FAST ACID MAGENTA see CMS228
FAST ACID MAGENTA B see CMS228
FASTBALLS see BBK500
FAST BENZIDENE ORANGE YB 3 see CMS145
FAST BLACK HBN see PCJ200
FAST BLUE B BASE see DCJ200
FAST BLUE R SALT see PFU500
FAST BROWN SALT RR see MIH500
FAST CHROME CYANINE 6B see CMP880
FAST CHROME CYANINE G see CMP880
FAST CORINTH BASE B see BBX000
FAST CRIMSON GR see CMM300
FAST CRIMSON 6BL see CMM400
FAST DARK BLUE BASE R see TGJ750
FAST DISPERSE YELLOW 2K see DUW500
FAST DRIMSON GR see CMM300
FAST GARNET GBC see MPY750
FAST GARNET GBC BASE see AIC250
FAST GARNET SALT GBC see MPY750
FAST GREEN see AFG500
FAST GREEN FCF see FAG000
FAST GREEN JJO see BAY750
FAST LIGHT ORANGE GA see HGC000
FAST LIGHT YELLOW E see SGP500
FASTOAN RED 2G see DVB800
FASTOGEN BLUE FP-3100 see DNE400
FASTOGEN BLUE SH-100 see DNE400
FASTOGEN GREEN 5005 see PJQ100
FASTOGEN GREEN B see PJQ100
FASTOGEN GREEN 2YK see CMS140
FASTOGEN GREEN Y see CMS140
FASTOGEN SUPER MAGNETA R see DTV360
FASTOGEN SUPER MAGNETA RE 03 see DTV360
FASTOGEN SUPER MAGNETA RG see DTV360

FASTOGEN SUPER MAGNETA RH see DTV360
FASTOGEN SUPER MAGNETA RS see DTV360
FAST OIL ORANGE see PEJ500
FAST OIL ORANGE II see XRA000
FAST OIL ORANGE T see CMP600
FAST OIL RED B see SBC500
FAST OIL SCARLET III see OHI200
FAST OIL YELLOW see AIC250
FAST OIL YELLOW 2G see CMP600
FAST OIL YELLOW B see DOT300
FAST OIL YELLOW G see CMP600
FASTOLITE BRILLIANT BORDEAUX BN see CMO872
FASTOLITE RED 8BL see CMO885
FASTOLUX GREEN see PJQ100
FASTONA ORANGE G see CMS145
FASTONA RED B see MMP100
FASTONA RED R see CJD500
FASTONA SCARLET RL see MMP100
FASTONA SCARLET YS see MMP100
FAST ORANGE see PEJ500
FAST ORANGE 3R see CJD500
FAST ORANGE BASE GC see CEH690
FAST ORANGE BASE GR see NEO000
FAST ORANGE BASE JR see NEO000
FAST ORANGE BASE JS see CEH690
FAST ORANGE G see CMS145
FAST ORANGE G BASE see CEH690
FAST ORANGE GC BASE see CEH675
FAST ORANGE GR BASE see NEO000
FAST ORANGE MC BASE see CEH690
FAST ORANGE O BASE see NEO000
FAST ORANGE RD OIL see CEG800
FAST ORANGE RD SALT see CEG800
FAST ORANGE R SALT see NEN500
FAST ORANGE SALT RD see CEG800
FAST ORANGE SALT RDA see CEG800
FAST ORANGE SALT RDN see CEG800
FAST PURPURINE see CMO880
FAST RED A see MMP100
FAST RED AB see MMP100
FAST RED 5NA BASE see NEQ000
FAST RED A (PIGMENT) see MMP100
FAST RED B see NEQ000
FAST RED 2G BASE see NEO500
FAST RED BASE 2J see NEO500
FAST RED BASE B see NEQ000
FAST RED BASE GG see NEO500
FAST RED BASE 3JL see KDA050
FAST RED BASE 3GL SPECIAL see KDA050
FAST RED BASE RL see MMF780
FAST RED BASE TR see CLK220
FAST RED BB see SBC500
FAST RED B BASE see NEQ000
FAST RED BB BASE see AOX250
FAST RED 2NC BASE see KDA050
FAST RED F see CMO870
FAST RED GG BASE see NEO500
FAST RED J see MMP100
FAST RED JE see MMP100
FAST RED KB AMINE see CLK225
FAST RED KB BASE see CLK225
FAST RED KB SALT see CLK225
FAST RED KB SALT SUPRA see CLK225
FAST RED KBS SALT see CLK225
FAST RED 3GL BASE see KDA050
FAST RED 3GL SPECIAL BASE see KDA050
FAST RED MP BASE see NEO500
FAST RED P BASE see NEO500
FAST RED R see MMP100
FAST RED R see OHI200
FAST RED RL BASE see MMF780
FAST RED 4BS see CMO870
FAST RED SALT RL see MMF780
FAST RED SALT TR see CLK235
FAST RED SALT TRA see CLK235
FAST RED SALT TRN see CLK235
FAST RED SG see NMP500
FAST RED SGG BASE see DEO295
FAST RED 5NT see MMF780

FAST RED 5CT BASE see CLK220
FAST RED TR see CLK220
FAST RED TR11 see CLK220
FAST RED TR BASE see CLK220
FAST RED TRO BASE see CLK220
FAST RED TR SALT see CLK235
FAST RED 5CT SALT see CLK235
FAST RED 5NT SALT see MMF780
FAST SCARLET see DXO850
FAST SCARLET BASE B see NBE500
FAST SCARLET BASE G see NMP500
FAST SCARLET BASE J see NMP500
FAST SCARLET G see NMP500
FAST SCARLET G BASE see NMP500
FAST SCARLET GC BASE see NMP500
FAST SCARLET G SALT see NMP500
FAST SCARLET J SALT see NMP500
FAST SCARLET M4NT BASE see NMP500
FAST SCARLET R see NEQ500
FAST SCARLET S see CMM320
FAST SCARLET 4BSA see CMO870
FAST SCARLET T BASE see NMP500
FAST SCARLET TR BASE see CLK200
FAST SILK YELLOW SH see CMM759
FAST SPIRIT YELLOW AAB see PEI000
FAST SULON BLACK BN see FAB830
FASTUM see BDU500
FASTUSOL BLUE 9GLP see CMO650
FASTUSOL RED 4BA-CF see CMO885
FASTUSOL TURQUOISE BLUE LGA see COF420
FAST WHITE see LDY000
FAST WOOL BLUE R see ADE750
FAST YELLOW AB see CMM758
FAST YELLOW AT see AIC250
FAST YELLOW B see AIC250
FAST YELLOW EXTRA SPECIALLY PURE see CMM758
FAST YELLOW GC BASE see CEH670
FAST YELLOW S see CMM758
FAST YELLOW S EXTRA SPECIALLY PURE see CMM758
FAST YELLOW Y see CMM758
FAT 92'213/A see CMM895
FATAL see TBV250
FAT BLACK HB see PCJ200
FAT BLACK HB 01 see PCJ200
FAT BROWN 2G see NBG500
FAT BROWN 2R see NBG500
FAT BROWN RR see NBG500
FATOLIAMID see EPQ000
FAT ORANGE A see CMP600
FAT ORANGE G see CMP600
FAT ORANGE GS see CMP600
FAT ORANGE II see TGW000
FAT ORANGE RG see CMP600
FAT PONCEAU R see SBC500
FAT RED 2B see SBC500
FAT RED 7B see EOJ500
FAT RED B see SBC500
FAT RED BB see SBC500
FAT RED BG see CMS238
FAT RED (BLUISH) see OHI200
FAT RED BS see SBC500
FAT RED G see CMS238
FAT RED G see OHI200
FAT RED HRR see OHI200
FAT RED R see OHI200
FAT RED RS see CMS238
FAT RED RS see OHI200
FAT RED TS see SBC500
FAT RED (YELLOWISH) see XRA000
FAT SCARLET 2G see XRA000
FAT SCARLET LB see OHI200
FATSCO ANT POISON see ARD750
FAT SOLUBLE GREEN ANTHRAQUINONE see BLK000
FAT SOLUBLE RED S see CMS238
FAT SOLUBLE RED ZH see OHI200
FATTY ACID, LANOLIN, ISO-PR ESTERS see IPS500
FATTY ACIDS see FAB850

FATTY ACIDS, COCO, 2-((2-HYDROXYETHYL)AMINO)ETHYL ESTERS see CNF270
FATTY ACID, SOAP (C12-14) see FAB880
FATTY ACID, TALL OIL, EPOXIDIZED-2-ETHYLHEXYL ESTER see FAB900
FATTY ACID, TALL OIL, EPOXIDIZED, OCTYL ESTER see FAB920
FATTY ALCOHOLS see AFJ100
FAT VICTORIA YELLOW D see CMP600
FAT YELLOW see DOT300
FAUSTAN see DCK759
FAVERIN 50 see FMR575
FAVISTAN see MCO500
FB/2 see DWX800
F-13B1 see TJY100
FB 217 see PJS750
FB 5097 see MRX500
FBA 52 see DPJ800
FBA 85 see CMP100
FBA 1420 see PMM000
FBA 1500 see MCA100
FBA 4503 see PMX250
FBC 32197 see QMA100
FBC CMPP see CIR500
FBHC see BBQ750
FB SODIUM PERCARBONATE see SJB400
FC-1 see CMB125
5-FC see FHI000
FC 12 see DFA600
FC 14 see CBY250
FC-21 see DFL000
FC 22 see CFX500
FC 31 see CHI900
FC 43 see HAS000
FC 47 see HAS000
FC 112 see TBP050
FC 113 see TJE100
FC 114 see FOO509
FC 123 see TJY500
FC 125 see PBD400
FC-143 see ANP625
FC143 see DKH300
FC 402 see FAC050
FC 403 see XSS375
FC 410 see FAC060
FC 448 see MMA525
FC 450 see DIV300
FC 455 see DHM309
FC 457 see FAC100
FC 480 see FAC130
FC 590 see FAC150
FC 591 see FAC155
FC 642 see FAC157
FC 646 see FAC160
FC 650 see FAC163
FC 651 see FAC165
FC 652 see FAC166
FC 657 see FAC170
FC 659 see FAC175
FC 660 see FAC179
FC 668 see FAC185
FC 676 see DHI875
FC 681 see FAC195
FC-1318 see OBO000
FC133a see TJE100
FC 133a see TJY175
FC142b see CFX250
FC143a see TJY900
FC 152a see ELN500
FC 3001 see SOX400
FC 4/58 see CBA375
FC 4648 see PKQ059
FC 114B2 see FOO525
FC-C 318 see CPS000
FCDR see FHO000
FCdR see FHO000
FC-MY 5450 see SMQ500
FCNU see CPL750
F-COL see FHH100
F-CORTEF see FHH100
FCR 1272 see REF250

FD-1 see BLH325
FDA see DGB600
FDA 0101 see SHF500
FDA 0109 see BHT250
FDA 0121 see ALL250
FDA 0345 see BIF750
FDA 1446 see AFR250
FDA 1541 see EIN500
FDA 1725 see TNX400
FDA 1902 see MCA100
2 FDBP see MQY400
FD&C ACID RED 32 see FAG080
FD&C BLUE No. 1 see FAE000
FD&C BLUE No. 2 see FAE100
FDC GREEN 1 see FAE950
FD&C GREEN No. 1 see FAE950
FD&C GREEN No. 2 see FAF000
FD&C GREEN No. 3 see FAG000
FD&C GREEN No. 2-ALUMINUM LAKE see FAF000
FD & C NO. 6 see FAG150
FD and C NO. 6 see FAG150
FDC ORANGE I see FAG010
FD&C ORANGE No. 1 see FAG010
FD&C RED No. 1 see FAG018
FD&C RED No. 2 see FAG020
FD&C RED No. 3 see FAG040
FD&C RED No. 4 see FAG050
FD&C RED No. 19 see FAG070
FD&C RED No. 32 see FAG080
FD&C RED No. 40 see FAG100
FD&C RED No. 2-ALUMINIUM LAKE see FAG020
FD & C RED No. 4-ALUMINIUM LAKE see FAG050
FD and C RED NO. 40 see FAG100
FD&C VIOLET No. 1 see FAG120
FD and C YELLOW 6 see FAG150
FD and C YELLOW LAKE NO. 6 see FAG150
FD&C YELLOW No. 3 see FAG130
FD&C YELLOW No. 4 see FAG135
FD&C YELLOW No. 5 see FAG140
FD&C YELLOW No. 6 see FAG150
FD and C YELLOW No. 10 see CMM510
FDC YELLOW NO. 6 see FAG150
FD and C YELLOW NO. 6 see FAG150
FD & C YELLOW NO. 6 ALUMINIUM LAKE see FAG150
FD & C YELLOW NO. 5 TARTRAZINE see FAG140
FDN see DRP800
FDUR see DAR400
FEBANTEL see BKN250
FEBRILIX see HIM000
FEBRININA see DOT000
FEBRO-GESIC see HIM000
FEBROLIN see HIM000
FEBRON see DOT000
FEBRUARY DAPHNE see LAR500
FEBUPROL see BPP250
FECAMA see DGP900
FECAPENTAENE 12 see AFS100
FECTO see CBT750
FECTRIM see SNK000
FECTRIM see TKX000
FEDACIN see NGE500
FEDAL-UN see PCF275
Fe-DEXTRAN see IGS000
FEDRIN see EAW000
FEEN-A-MINT GUM see PDO750
FEENO see PDP250
FEFO see FMV050
FEGLOX see DWX800
FEIDMIN 5 see TES800
FEINALMIN see DLH630
FEKABIT see PLA250
FELACRINOS see DAL000
FELAN see EKO500
FELAZINE see PFC750
FELBAMATE see FAG200
FELBAMATO see FAG200
FELBATOL see FAG200

FELBEN see BAU750
FELDENE see FAJ100
FELICAIN (GERMAN) see PNB250
FELICUR see EGQ000
FELISON see DAB800
FELITROPE see EGQ000
FELIXYN see CKE750
α-FELLANDRENE see MCC000
FELLING ZINC OXIDE see ZKA000
FELLOZINE see PMI750
FELMANE see FMQ000
FELONWORT see CCS650
FELSULES see CDO000
FELUREA see PEC250
FELVITEN see AOO490
FEMA 3174 see HKC575
FEMA 3186 see MJH905
FEMA 3230 see POM000
FEMA 3251 see PPH200
FEMACOID see ECU750
FEMADOL see PNA500
FEMALE WATER DRAGON see WAT300
FEMAMIDE see FAJ150
FEMA No. 2003 see AAG250
FEMA No. 2005 see MDW750
FEMA No. 2006 see AAT250
FEMA No. 2007 see THM500
FEMA No. 2008 see ABB500
FEMA No. 2009 see ABH000
FEMA No. 2011 see AEN250
FEMA No. 2026 see AGC500
FEMA No. 2031 see AGH250
FEMA No. 2032 see AGA500
FEMA No. 2033 see AGI500
FEMA No. 2034 see AGJ250
FEMA No. 2037 see AGM500
FEMA No. 2045 see ISV000
FEMA No. 2055 see IHO850
FEMA No. 2058 see IHP100
FEMA No. 2060 see IHP400
FEMA No. 2061 see AOG500
FEMA No. 2063 see AOG600
FEMA No. 2069 see IHS000
FEMA No. 2073 see AOJ900
FEMA No. 2075 see IHU100
FEMA No. 2082 see AON350
FEMA No. 2084 see IME000
FEMA No. 2085 see ITB000
FEMA No. 2086 see PMQ750
FEMA No. 2097 see AOX750
FEMA No. 2098 see AOY400
FEMA No. 2099 see MED500
FEMA No. 2109 see ARN000
FEMA No. 2127 see BAY500
FEMA No. 2134 see BCS250
FEMA No. 2135 see BDX000
FEMA No. 2137 see BDX500
FEMA No. 2138 see BCM000
FEMA No. 2140 see BED000
FEMA No. 2141 see IJV000
FEMA No. 2142 see BEG750
FEMA No. 2149 see BFD400
FEMA No. 2150 see BFD800
FEMA No. 2151 see BFJ750
FEMA No. 2152 see ISW000
FEMA No. 2159 see BMD100
FEMA No. 2160 see IHX600
FEMA No. 2170 see MKA400
FEMA No. 2174 see BPU750
FEMA No. 2175 see IIJ000
FEMA No. 2178 see BPW500
FEMA No. 2179 see IIL000
FEMA No. 2183 see BQI000
FEMA No. 2184 see BFW750
FEMA No. 2186 see BQM500
FEMA No. 2187 see BSW500
FEMA No. 2188 see BRQ350
FEMA No. 2190 see BQP000
FEMA No. 2193 see IIQ000
FEMA No. 2203 see DTC800
FEMA No. 2209 see BBA000
FEMA No. 2210 see IJF400

FEMA No. 2213 see IJN000
FEMA No. 2218 see ISX000
FEMA No. 2219 see BSU250
FEMA No. 2220 see IJS000
FEMA No. 2221 see BSW000
FEMA No. 2222 see IJU000
FEMA No. 2223 see TIG750
FEMA No. 2224 see CAK500
FEMA No. 2229 see CBA500
FEMA No. 2245 see CCM000
FEMA No. 2249 see CCM100
FEMA No. 2249 see CCM120
FEMA No. 2252 see CCN000
FEMA No. 2286 see CMP969
FEMA No. 2288 see CMP975
FEMA No. 2293 see CMQ730
FEMA No. 2294 see CMQ740
FEMA No. 2295 see API750
FEMA No. 2299 see CMR500
FEMA No. 2301 see CMR850
FEMA No. 2302 see CMR800
FEMA No. 2306 see CMS750
FEMA No. 2307 see CMS845
FEMA No. 2309 see CMT250
FEMA No. 2311 see AAU000
FEMA No. 2312 see CMT600
FEMA No. 2313 see CMT900
FEMA No. 2314 see CMT750
FEMA No. 2316 see CMU100
FEMA No. 2341 see COE500
FEMA No. 2356 see CQI000
FEMA No. 2361 see DAF200
FEMA No. 2362 see DAG000
FEMA No. 2362 see DAG200
FEMA No. 2365 see DAI600
FEMA No. 2366 see DAI350
FEMA No. 2370 see BOT500
FEMA No. 2371 see BEO250
FEMA No. 2375 see EMA500
FEMA No. 2376 see DJY600
FEMA No. 2377 see SNB000
FEMA No. 2379 see DKV150
FEMA No. 2381 see HHR500
FEMA No. 2391 see DTE600
FEMA No. 2392 see DQQ375
FEMA No. 2393 see DQQ200
FEMA No. 2394 see DQQ380
FEMA No. 2401 see DXS700
FEMA No. 2402 see DXU300
FEMA No. 2414 see EFR000
FEMA No. 2415 see EFS000
FEMA No. 2418 see EFT000
FEMA No. 2420 see AOV000
FEMA No. 2421 see EGM000
FEMA No. 2422 see EGR000
FEMA No. 2426 see DHI000
FEMA No. 2427 see EHE000
FEMA No. 2428 see ELS000
FEMA No. 2429 see DHI400
FEMA No. 2430 see EHN000
FEMA No. 2432 see EHE500
FEMA No. 2433 see EJN500
FEMA No. 2434 see EKL000
FEMA No. 2437 see EKN050
FEMA No. 2439 see EHF000
FEMA No. 2440 see LAJ000
FEMA No. 2441 see ELY700
FEMA No. 2443 see EMP600
FEMA No. 2444 see ENC000
FEMA No. 2445 see ENL850
FEMA No. 2447 see ENW000
FEMA No. 2449 see ENY000
FEMA No. 2452 see EOH000
FEMA No. 2454 see EOK600
FEMA No. 2456 see EPB500
FEMA No. 2458 see SAL000
FEMA No. 2463 see ISY000
FEMA No. 2464 see EQF000
FEMA No. 2465 see CAL000
FEMA No. 2467 see EQR500
FEMA No. 2468 see IKQ000
FEMA No. 2469 see EQS000

FEMA No. 2470 see AAX750
FEMA No. 2475 see AGE250
FEMA No. 2476 see IKR000
FEMA No. 2478 see FAB800
FEMA No. 2489 see FPQ875
FEMA No. 2497 see DSD775
FEMA No. 2497 see FQT000
FEMA No. 2507 see DTD000
FEMA No. 2509 see DTD800
FEMA No. 2511 see GDE800
FEMA No. 2512 see GDE825
FEMA No. 2514 see GCY000
FEMA No. 2516 see GDM400
FEMA No. 2517 see GDM450
FEMA No. 2539 see HBA550
FEMA No. 2540 see HBB500
FEMA No. 2544 see MGN500
FEMA No. 2545 see EHA600
FEMA No. 2548 see HBL500
FEMA No. 2557 see HEM000
FEMA No. 2559 see HEU000
FEMA No. 2560 see HFA525
FEMA No. 2562 see HFD500
FEMA No. 2563 see HFE000
FEMA No. 2565 see HFI500
FEMA No. 2567 see HFJ500
FEMA No. 2569 see HFO500
FEMA No. 2583 see CMS850
FEMA No. 2585 see HJV700
FEMA No. 2588 see RBU000
FEMA No. 2593 see ICM000
FEMA No. 2594 see IFW000
FEMA No. 2595 see IFX000
FEMA No. 2615 see DXT000
FEMA No. 2617 see DXV600
FEMA No. 2633 see LFU000
FEMA No. 2635 see LFX000
FEMA No. 2636 see LFY600
FEMA No. 2638 see LFZ000
FEMA No. 2640 see LGB000
FEMA No. 2642 see LGA050
FEMA No. 2645 see LGC100
FEMA No. 2665 see MCF500
FEMA No. 2665 see MCG000
FEMA No. 2665 see MCG250
FEMA No. 2667 see MCE250
FEMA No. 2667 see MCG275
FEMA No. 2668 see MCG500
FEMA No. 2668 see MCG750
FEMA No. 2670 see AOT530
FEMA No. 2672 see MFF580
FEMA No. 2677 see MFW250
FEMA No. 2681 see MGP000
FEMA No. 2682 see APJ250
FEMA No. 2683 see MHA750
FEMA No. 2684 see MNT075
FEMA No. 2685 see PDE000
FEMA No. 2690 see MIP750
FEMA No. 2697 see MIO000
FEMA No. 2698 see MIO500
FEMA No. 2700 see HMB500
FEMA No. 2707 see MKK000
FEMA No. 2718 see MGQ250
FEMA No. 2719 see MLL600
FEMA No. 2723 see ABC500
FEMA No. 2729 see MND275
FEMA No. 2731 see HFG500
FEMA No. 2733 see MHA500
FEMA No. 2743 see COU500
FEMA No. 2745 see MPI000
FEMA No. 2749 see MQI550
FEMA No. 2753 see MHW260
FEMA No. 2762 see MRZ150
FEMA No. 2770 see DTD200
FEMA No. 2772 see NCN700
FEMA No. 2780 see NMV780
FEMA No. 2781 see CNF250
FEMA No. 2782 see NMW500
FEMA No. 2788 see NNB400
FEMA No. 2789 see NNB500
FEMA No. 2797 see OCO000
FEMA No. 2798 see OCE000

FENAZO LIGHT BLUE RA see CMM080
FENAZO ORANGE see CMM220
FENAZO RED B see CMM080
FENAZO RED C see HJF500
FENAZO RED FG see CMM325
FENAZO SCARLET 2R see FMU070
FENAZO SCARLET 3R see FMU080
FENAZOXINE see FAL000
FENAZOXINE HYDROCHLORIDE see NBS500
FENAZO YELLOW M see MDM775
FENAZO YELLOW N3GL see FAL050
FENAZO YELLOW T see FAG140
FENAZO YELLOW XX see CMM750
FENBENDAZOL see FAL100
FENBENDAZOLE see FAL100
FENBITAL see EOK000
FENBUFEN see BGL250
FENBUTATIN OXIDE see BLU000
FENCAL see ARB750
FENCAMFAMINE HYDROCHLORIDE see EOM000
FENCAMINE HYDROCHLORIDE see FAM000
FENCARBAMIDE see PDC850
FENCARBAMIDE HYDROCHLORIDE see PDC875
FENCAROL see DWM400
FENCHEL OEL (GERMAN) see FAP000
FENCHLONINE see FAM100
FENCHLOORFOS (DUTCH) see RMA500
FENCHLORFOS see RMA500
FENCHLORFOSU (POLISH) see RMA500
FENCHLOROPHOS see RMA500
FENCHLORPHOS see RMA500
endo-FENCHOL see TLW000
α-FENCHOL see TLW000
FENCHON (GERMAN) see TLW250
FENCHONE see TLW250
(+)-FENCHONE see FAM300
d-FENCHONE see FAM300
d(+)-FENCHONE see FAM300
FENCHYL ACETATE see FAO000
α-FENCHYL ALCOHOL see TLW000
FENCLONIN see FAM100
FENCLONINE see FAM100
FENCLOR see PJL750
FENCLOR 42 see PJL750
FENCLOR 64 see FAO050
FENCLOZIC ACID see CKK250
FENDON see HIM000
FENDONA see RCZ050
FENDOSAL see FAO100
FENDOZAL see FAO100
FENELZIN see PFC750
FENERGAN see DQA400
FENERGAN see PMI750
FENESTERIN see CME250
FENESTREL see FAO200
FENESTRIN see CME250
FENETAZINA see DQA400
FENETHAZINE HYDROCHLORIDE see DPI000
β-FENETHYLALKOHOL see PDD750
FENETHYLLINE HYDROCHLORIDE see CBF825
FENETICILLINE see PDD350
p-FENETIDIN see PDK790
1-FENETILBIGUANIDE CLORIDRATO (ITALIAN) see PDF250
N'-β-FENETILFORMAMIDINILIMINOUREA (ITALIAN) see PDF000
8-N-FENETIL-1-OXA-2-OXO-3,8-DIAZASPIRO-(4,5)-DECANO CLORIDRATO (ITALIAN) see DAI200
FENFLUORAMINE HYDROCHLORIDE see PDM250
FENFLURAMINE see ENJ000
FENFLUTHRIN see FAO220
FENFORMINA see PDF000
FENHEXAMID see FAO250

FENHYDREN see PFJ750
FENIBUT see PEE500
FENIBUTAZONA see BRF500
FENIBUT HYDROCHLORIDE see GAD000
FENIBUTOL see BRF500
FENICOL see CDP250
FENICOL see EGQ000
FENIDANTOIN "S" see DKQ000
FENIDIN see DTP400
FENIDRONE see OPK300
FENIGAM see PEE500
FENIGAMA see PEE500
FENIGAM HYDROCHLORIDE see GAD000
2-FENIL-2-(p-AMINOFENIL)PROPIONAMMIDE (ITALIAN) see ALX500
2-FENIL-5-BROMO-INDANDIONE (ITALIAN) see UVJ400
FENILBUTINE see BRF500
FENILDICLOROARSINA (ITALIAN) see DGB600
FENILEP see PEC250
FENILFAR see SPC500
4-FENIL-4-FORMILPIPERIDINA (ITALIAN) see PFY100
FENILIDINA see BRF500
FENILIDRAZINA (ITALIAN) see PFI000
FENILIN see PFJ750
FENILISOPROPILIDRAZINA see PDN000
4-FENIL-α-METILFENILACETATO-γ-PROPILSOLFONATO SALE SODICO (ITALIAN) see PFR350
FENILOR see DDS600
FENILPRENAZONE see PEW000
2-FENILPROPANO (ITALIAN) see COE750
FENILPROPANOLAMINA (ITALIAN) see NNM000
FENIPENTOL see BQJ500
FENISED see PEC250
FENISTIL see FMU409
FENISTIL-RETARD see FMU409
FENITOIN see DNU000
FENITOX see DSQ000
FENITROOXON see PHD750
FENITROOXONE see PHD750
FENITROTHION see DSQ000
FENITROTHION-MALATHION MIXTURE see FAO300
FENITROTHION OXON see PHD750
FENITROTION (HUNGARIAN) see DSQ000
FENITROXON see PHD750
FENIZON (FRENCH) see CJR500
FENKAROL see DWM400
FENMEDIFAM see MEG250
FENNEL OIL see FAP000
FENNOSAN see QPA000
FENNOSAN B 100 see DSB200
FENOBARBITAL see EOK000
FENOBCARB see MOV000
FENOBOLIN see DYF450
FENOBUCARB see MOV000
FENOCICLINA see BIS750
FENOCICLINA see BIS750
FENOCYCLIN see BIS750
FENOCYCLIN see BIS750
FENOCYCLIN see BIT000
FENOCYCLINE see BIS750
FENOCYCLINE see BIS750
FENOCYCLINE see BIT000
FENOCYLIN see BIS750
FENOCYLIN see BIS750
FENOFIBRIC ACID see CJP750
FENOFLURAZOLE see DGA200
FENOL 25 see AJU900
FENOL (DUTCH, POLISH) see PDN750
FENOLFTALEIN see PDO750
FENOLIPUNA see PDO800
FENOLO (ITALIAN) see PDN750
FENOLOVO see HON000
FENOLOVO ACETATE see ABX250
FENOPHOSPHON see EPY000
FENOPON CO 436 see ANO600

FENOPON CO 433N see SNY100
FENOPON EP 110 see ANO600
FENOPON EP 120 see ANO600
FENOPRAIN see PMJ525
FENOPROFEN CALCIUM DIHYDRATE see FAP100
FENOPROFEN CALCIUM SALT DIHYDRATE see FAP100
FENOPROFEN SODIUM see FAQ000
FENOPROMIN see AOA250
FENOPRON see FAP100
FENOPROP see TIX500
FENORMONE see TIX500
FENOSMOLIN see PDP100
FENOSPEN see PDT500
FENOSTENYL see PEC250
FENOSTIL see FMU409
FENOSUCCIMIDE see MNZ000
FENOTEROL BROMIDE see FAQ100
FENOTEROL HYDROBROMIDE see FAQ100
FENOTHIAZINE (DUTCH) see PDP250
FENOTIAZINA (ITALIAN) see PDP250
FENOTONE see BRF500
FENOVERM see PDP250
FENOX see SPC500
FENOXAPROP-ETHYL see FAQ200
FENOXEDIL see BPM750
FENOXEDIL HYDROCHLORIDE see BPM750
FENOXYBENZAMIN see DDG800
FENOXYCARB see EOE200
2-FENOXYETHANOL (CZECH) see PER000
FENOXYL BLACK RD see CMS250
FENOXYL BORDEAUX B see CMS257
FENOXYL CARBON N see DUZ000
FENOXYPEN see PDT500
FENOXYPEN see PDT750
1-FENOXY-2,3-PROPANEDIOL see GGA950
FENOZAFLOR see DGA200
FENOZAN 22 see DED100
FENOZAN 23 see DED100
FENPENTADIOL see CKG000
FENPHOSPHORIN see MEC250
FENPICLONIL see FAQ220
FENPIVERIMIUM BROMIDE see RDA375
FENPROBAMATO see PGA750
FENPROPANAGE see DAB825
FENPROPATHRIN see DAB825
FENPROPAZINA see DIR000
FENPROPIDIN see FAQ230
FENPROPIDINE see FAQ230
FENPROPIMORPH see MRQ300
cis-FENPROPIMORPH see FAQ300
FENPROSTALENE see FAQ500
FENPYRATE see FAQ600
FENSON see CJR500
FENSPIRIDE see DAI200
FENSPIRIDE HYDROCHLORIDE see DAI200
FENSULFOTHION see FAQ800
FENTAL see FLZ050
FENTANEST see PDW500
FENTANEST see PDW750
FENTANIL see PDW500
FENTANYL see PDW500
FENTANYL CITRATE see PDW750
FENTAZIN see CJM250
FENTHIAPROP-ETHYL see EHH600
FENTHION see FAQ900
FENTHION SULFONE see DSS800
FENTHIURAM see FAQ930
FENTHOATE see DRR400
FENTIAPRIL see FAQ950
FENTIAZAC see CKI750
FENTIAZAC CALCIUM SALT see CAP250
FENTIAZIN see PDP250
FENTIN ACETAAT (DUTCH) see ABX250
FENTIN ACETAT (GERMAN) see ABX250
FENTIN ACETATE see ABX250
FENTIN CHLORIDE see CLU000
FENTINE ACETATE (FRENCH) see ABX250

FENTIN HYDROXIDE see HON000
FENTIURAM see FAQ930
FENTRIFANIL see FDB300
FENTRINOL see AHL500
FENUGREEK ABSOLUTE see FAR000
FENULON see DTP400
FENURAL see PEC250
FENUREA see PEC250
FENURON see DTP400
FENURONE see PEC250
FENURON TCA see FAR050
FENURON TCA SALT see FAR050
FENURON TRICHLOROACETATE see
FAR050
FENVALERATE see FAR100
FENVALERATE A ALPHA see FAR150
FENVALERATE ALPHA see FAR150
2-FENYL-5-AMINOBENZTHIAZOL
(CZECH) see ALV500
1-FENYL-4-AMINO-5-CHLOR-6-
PYRIDAZINON (CZECH) see PEE750
1-FENYL-3-AMINOPYRAZOLIN (CZECH)
see ALX750
FENYLBUTAZON see BRF500
FENYL-CELLOSOLVE (CZECH) see PER000
N-FENYL-N'-CYKLOHEXYL-p-
FENYLENDIAMIN (CZECH) see PET000
FENYLDICHLORARSIN see DGB600
1-FENYL-4,5-DICHLOR-6-PYRIDAZINON
(CZECH) see DGE800
1-FENYL-3,3-DIETHYLTRIAZEN (CZECH)
see PEU500
1-FENYL-3,3-DIMETHYLTRIAZIN see
DTP000
m-FENYLENDIAMIN (CZECH) see PEY000
FENYLENODWUAMINA (POLISH) see
PEY500
FENYLEPSIN see DKQ000
FENYLESTER KYSELINY
CHLORMRAVENCI (CZECH) see CBX109
FENYLESTER KYSELINY OCTOVE see
PDY750
FENYLESTER KYSELINY SALICYLOVE see
PGG750
1-FENYL-1,2-ETHANDIOL see SMQ100
1-FENYLETHANOL see PDE000
β-FENYLETHANOL see PDD750
4-(1-FENYLETHYL)FENOL see PFD400
FENYLETTAE see EOK000
N-FENYL-p-FENYLENDIAMIN see PFU500
FENYLFOSFIN see PFV250
FENYL-GLYCIDYLETHER (CZECH) see
PFF360
FENYLGLYCOL see SMQ100
FENYLHIST see BAU750
FENYLHYDRAZID KYSELINY OCTOVE
see ACX750
FENYLHYDRAZINE (DUTCH) see PFI000
FENYL-α-HYDROXYBENZYLKETON see
BCP250
FENYL-β-HYDROXYETHYLSULFON
(CZECH) see BBT000
N-FENYLIMID KYSELINY MALEINOVE
(CZECH) see PFL750
FENYLISOKYANAT see PFK250
FENYLKYANID see BCQ250
FENYLMERCURIACETAT (CZECH) see
ABU500
FENYLMERCURICHLORID (CZECH) see
PFM500
FENYLMERKURINITRAT see MCU750
FENYL-METHYLKARBINOL see PDE000
1-FENYL-3-METHYL-2-PYRAZOLIN-5-ON
see NNT000
2-FENYLOTIOMOCZNIK (POLISH) see
DWN800
1-FENYLPIPERAZIN see PFX000
2-FENYL-PROPAAN (DUTCH) see COE750
1-FENYL-3-PYRAZOLON (CZECH) see
PGE250
FENYLSILAN see SDX250
FENYLSILATRAN (CZECH) see PGH750

FENYL-TRIFLUORSILAN see PGO500
FENYPRIN see DBA800
FENYRAMIDOL see PGG350
FENYRAMIDOL HYDROCHLORIDE see
PGG355
FENYRIPOL see PGG350
FENYRIPOL HYDROCHLORIDE see
FAR200
FENYTAN see PEC250
FENYTOINE see DKQ000
FENYTOINE see DNU000
FENZAFLOR see DGA200
FENZEN (CZECH) see BBL250
FEOJECTIN see IHG000
FEOSOL see FBN100
FEOSOL see FBO000
FEOSPAN see FBN100
FEOSTAT see FBJ100
FEPRAZONE see PEW000
FEPRONA see FAP100
FERBAM see FAS000
FERBAM 50 see FAS000
FERBAM, IRON SALT see FAS000
FERBECK see FAS000
FERDEX 100 see IGS000
FERGON see FBK000
FERGON PREPARATIONS see FBK000
FER-IN-SOL see FBN100
FER-IN-SOL see FBO000
FERISAN see EJA379
FERKETHION see DSP400
FERLUCON see FBK000
FERMATE FERBAM FUNGICIDE see
FAS000
FERMENICIDE LIQUID see SOH500
FERMENICIDE POWDER see SOH500
FERMENTATION ALCOHOL see EFU000
FERMENTATION AMYL ALCOHOL see
IHP000
FERMENTATION BUTYL ALCOHOL see
IIL000
FERMIDE see TFS350
FERMINE see DTR200
FERMOCIDE see FAS000
FERNACOL see TFS350
FERNASAN see TFS350
FERNESTA see BQZ000
FERNESTA see DAA800
FERNEX see DIN600
FERNIDE see TFS350
FERNIMINE see DAA800
FERNIMINE 4 see MIH800
FERNISOLONE see PMA000
FERNISONE see HHQ800
FERNOS see DOX600
FERNOXENE see SGH500
FERNOXONE see DAA800
FERODIN SL see FAS100
FERO-GRADUMET see FBN100
FERO-GRADUMET see FBO000
FEROTON see FBJ100
FER PENTACARBONYLE (FRENCH) see
IHG000
FERRADOW see FAS000
FERRALYN see FBN100
FERRATE(2-), CHLORO(7,12-DIETHENYL-
3,8,13,17-TETRAMETHYL-21H,23H-
PORPHINE-2,18-DIPROPANOATO(4-)-$\textit{K}$-
N²¹),$\textit{K}$-N²²),$\textit{K}$²³),$\textit{K}$-N²⁴))-, DIHYDROGEN, (SP-
5-13)- see HAQ050
FERRATE(1-),
((ETHYLENEDINITRILO)TETRAACETATO
)- see IGX550
FERRATE (Fe₁₂-O₁₉²⁻) BARIUM (1:1) (9CI) see
BAL625
FERRATE(1-), (GLYCINATO-
N,O)(SULFATO(2-)-O',O')-, HYDROGEN, (T-
4)-(9CI) see FBD500
FERRATE(4-), HEXACYANO-, AMMONIUM
IRON(3+) see FAS800
FERRATE(4-), HEXACYANO-,
TETRAPOTASSIUM see TEC500

FERRATE(4-), HEXAKIS(CYANO-C)-,
AMMONIUM IRON(3+) (1:1:1), (OC-6-11)- see
FAS800
FERRATE(4-), HEXAKIS(CYANO-C)-,
IRON(3+) (3:4), (OC-6-11)-(9CI) see IGY000
FERRATE(4-), HEXAKIS(CYANO-C)-,
TETRAPOTASSIUM, (OC-6-11)- see TEC500
FERRATE(2-), PENTAKIS(CYANO-
C)NITROSYL-, DISODIUM, DIHYDRATE
(OC-6-22)-, (9CI) see SIW500
FERRIAMICIDE see MQW500
FERRIC ACETYLACETONATE see IGL000
FERRIC AMMONIUM CITRATE see FAS700
FERRIC AMMONIUM CITRATE, GREEN see
FAS700
FERRIC AMMONIUM FERROCYANIDE see
FAS800
FERRIC AMMONIUM OXALATE see
ANG925
FERRIC AMMONIUM OXALATE (DOT) see
ANG925
FERRIC ARSENATE, solid (DOT) see IGN000
FERRIC ARSENITE, solid (DOT) see IGO000
FERRIC ARSENITE, BASIC see IGO000
FERRIC BLEOMYCIN A2 see IGO500
FERRIC CHLORIDE see FAU000
FERRIC CHLORIDE HEXAHYDRATE see
FAW000
FERRIC CHLORIDE (UN 1733) (DOT) see
FAU000
FERRIC CHLORIDE, solution (UN 2582)
(DOT) see FAU000
FERRIC CHOLINE CITRATE see FBC100
FERRIC CITRATE see FAW100
FERRIC DEXTRAN see IGS000
FERRIC DIMETHYLDITHIOCARBAMATE
see FAS000
FERRIC FERROCYANIDE see IGY000
FERRIC FLUORIDE see FAX000
FERRIC HEXACYANOFERRATE (II) see
IGY000
FERRIC HYDROXIDE
NITRILOTRIPROPIONIC ACID COMPLEX
see FAY000
FERRIC NITRATE see FAY200
FERRIC NITRATE see IHB900
FERRIC NITRATE (DOT) see FAY200
FERRIC NITRATE (DOT) see IHB900
FERRIC NITRATE, NONAHYDRATE see
IHC000
FERRIC NITRILOTRIACETATE see IHC100
FERRIC NITROSODIMETHYL
DITHIOCARBAMATE and TETRAMETHYL
THIURAM DISULFIDE see FAZ000
FERRICON see ROF300
FERRIC ORTHOPHOSPHATE see FAZ500
FERRIC OXIDE see IHC450
FERRIC OXIDE, SACCHARATED see
IHG000
FERRIC PHOSPHATE see FAZ500
FERRIC PYROPHOSPHATE see FAZ525
FERRIC SACCHARATE IRON OXIDE (MIX.)
see IHG000
FERRIC SODIUM EDETATE see EJA379
FERRIC SODIUM EDTA see EJA379
FERRIC SODIUM GLUCONATE COMPLEX
see IHK000
FERRIC SODIUM PYROPHOSPHATE see
SHE700
FERRIC SULFATE see FBA000
FERRIC TRIACETYLACETONATE see
IGL000
FERRIC TRICHLORIDE HEXAHYDRATE
see FAW000
FERRIC VERSENATE see IGX550
FERRICYANURE de TRI(1-DODECYL-2-
PHENYL-3-METHYL)-1,3-
BENZIMIDAZOLIUM (FRENCH) see
TNH750
FERRICYTOCHROME C see CQM325
FERRIDEXTRAN see IGS000
FERRIGEN see IGT000

FERRIHEXACYANOFERRATE see IGY000
FERRITIN see FBB000
FERRIVENIN see IHG000
FERRLECIT see FBB100
FERROACTINOLITE see FBG200
FERROANTHOPHYLLITE see ARM264
FERROCENE see FBC000
FERROCENE, ACETYL- see ABA750
FERROCENE, 1,1'-
BIS(DIMETHYL(OCTYLOXY)SILYL)- see
BJK780
FERROCENE, (3,5,5-TRIMETHYL-1-
OXOHEXYL)- see TLT300
FERROCHOLINATE see FBC100
FERROCHROME see FBD000
FERROCHROME (exothermic) see FBD000
exothermic FERROCHROME see FBD000
FERROCHROMIUM see FBD000
FERROCIN see IGY000
FERROCYANIDES see FBD100
FERROCYTOCHROME C see CQM325
FERRODEXTRAN see IGS000
FERROFLUKIN 75 see IGS000
FERROFOS 509 see NEI100
FERROFOS 510 see HKS780
FERROFUME see FBJ100
FERROGLUCIN see IGS000
FERROGLUKIN 75 see IGS000
FERROGLYCINE SULFATE see FBD500
FERROGLYCINE SULFATE see FBD500
FERROGLYCINE SULFATE COMPLEX see
FBD500
FERRO-GRADUMET see FBN100
FERRO LEMON YELLOW see CAJ750
FERROLIP see FBC100
FERROMANGANESE (exothermic) see
FBE000
exothermic FERROMANGANESE (DOT) see
FBE000
FERRON see IEP200
FERRONAT see FBJ100
FERRONE see FBJ100
FERRONICKEL see IGL120
FERRONICUM see FBK000
FERRONORD see FBD500
FERRO ORANGE YELLOW see CAJ750
FERROSAN see PLZ100
FERROSANOL see FBD500
FERROSILICON see FBG000
FERROSILICON see IHJ000
FERROSILICON, containing more than 30%
but less than 90% SILICON (DOT) see FBG000
FERROSULFAT (GERMAN) see FBN100
FERROSULFATE see FBN100
FERROTEMP see FBJ100
FERRO-THERON see FBN100
FERROTREMOLITE see FBG200
FERROTSIN see IGY000
FERROUS see FBN000
FERROUS ACETATE see FBH000
FERROUS AMMONIUM SULFATE,
HEXAHYDRATE see IGL200
FERROUS ARSENATE (DOT) see IGM000
FERROUS ARSENATE, solid (DOT) see
IGM000
FERROUS ASCORBATE see FBH050
FERROUS CARBONATE see FBH100
FERROUS CHLORIDE see FBI000
FERROUS CHLORIDE, solid (NA 1759)
(DOT) see FBI000
FERROUS CHLORIDE, solution (NA 1760)
(DOT) see FBI000
FERROUS CHLORIDE TETRAHYDRATE
see FBJ000
FERROUS CITRATE see FBJ075
FERROUS FERRITE see IHG100
FERROUS FUMARATE see FBJ100
FERROUS GLUCONATE see FBK000
FERROUS GLUCONATE DIHYDRATE see
FBL000
FERROUS GLUTAMATE see FBM000
FERROUS ION see FBN000

FERROUS LACTATE see LAL000
FERROUS SULFATE see FBN100
FERROUS SULFATE (FCC) see FBO000
FERROUS SULFATE HEPTAHYDRATE see
FBO000
FERROVAC E see IGK800
FERROVANADIUM DUST see FBP000
FERROXDURE see BAL625
FERRO YELLOW see CAJ750
FERRUGO see IHC450
FERRUM see FBJ100
FERRUM see FBP050
FERSAMAL see FBJ100
FERSOLATE see FBN100
FERTENE see PJS750
FERTILVIT see TGJ050
FERTILYSIN see BIX250
FERTINORM see FMT100
FERTIRAL see LIU370
FERTODUR see FBP100
FERULA JAESCHKEANA VATKE,
EXTRACT see FBP175
FERULIC ACID see FBP200
trans-FERULIC ACID see FBP200
FERVENULIN see FBP300
FERVENULINE see FBP300
FERVIN see FBP350
FERVINAL see CDK800
FES see ZAT000
FESOFOR see FBO000
FESOTYME see FBO000
FESTID NIGHTSHADE see HAQ100
FETT see FNK200
FETTER BUSH see DYA875
FETTERBUSH see FBP520
FETTORANGE B see XRA000
FETTORANGE R see PEJ500
FETTPONCEAU G see OHI200
FETTROT see OHI200
FETTSCHARLACH see OHI200
FETTSCHARLACH LB see OHI200
FEUILLES CRABE (HAITI) see SLJ650
FEVARIN see FMR575
FEVER TWIG see AHJ875
FF see FQN000
FF see FQO000
FF 106 see IJH000
FF 680 see EEB250
FFB 32 see PHB500
F-5-FU see FLZ050
F 151 FUMARATE see MDP800
FG 400 see TBJ700
FG 834 see SMQ500
FG 2000 see MKA270
FG 4963 see FAJ200
FG 5111 see FKI000
FG 7051 see PAL650
FG7142 see MPA065
FG 8115 see TJC870
FG 8115S see TJC870
FH 099 see DTL200
FHA see FJF100
FH 122-A see NNQ500
FHCH see BBQ750
FHD-3 see BOF750
F.I 106 see AES750
FI 106 see HKA300
FI 1163 see LIN000
F.I. 58-30 see EPQ000
Fi 5853 see APP500
FI 6120 see DLH200
F.I. 6145 see PIW000
FI 6146 see BTA325
FI6339 see DAC000
FI 6341 see FDB000
FI 6714 see NDM000
FI 6804 see HKA300
FIBERGLASS see FBQ000
FIBERS, REFRACTORY CERAMIC see
RCK725
FIBER V see PKF750

FIBORAN see FBP850
FIBRALEM see ARQ750
FIBRASET TC see UTU500
FIBRE BLACK VF see AQP000
FIBRENE C 400 see TAB750
FIBRINOGENASE see TFU800
FIBROTAN see PFN000
FIBROUS CROCIDOLITE ASBESTOS see
ARM275
FIBROUS GLASS see FBQ000
FIBROUS GLASS DUST (ACGIH) see FBQ000
FIBROUS GRUNERITE see ARM250
FIBROUS TREMOLITE see ARM280
FICAM see DQM600
FICHLOR 91 see TIQ750
FICIN see FBS000
FI CLOR 71 see DGN200
FI CLOR 91 see TIQ750
FI CLOR 60S see SGG500
FICUSIN see FQD000
FICUS PROTEASE see FBS000
FICUS PROTEINASE see FBS000
FIDDLE FLOWER see SDZ475
FIDDLE-NECK see TAG250
FIELD GARLIC see WBS850
FIETIN see FBW000
FIGARON see EHK600
FIGUIER MAUDIT MARRON (HAITI) see
BAE325
FIGWORT see FBS100
F III (sugar fraction) see FAB010
FILARIOL see EGV500
FILARSEN see DFX400
FILIGRANA (CUBA) see LAU600
FILMERINE see SIQ500
FILORAL see CMG300
FILTEX WHITE BASE see CAT775
FILTRASORB see CBT500
FILTRASORB 200 see CBT500
FILTRASORB 400 see CBT500
FIMALENE see ILD000
FINA, combustion products see ADX750
FINALE see ANI800
FINAPLIX see HKI100
FINAVEN see ARW000
FINDOLAR see GGS000
FINE GUM HES see SFO500
FINEMEAL see BAP000
FINIMAL see HIM000
FINISH EN see DTG700
FINISOL BLUE GREEN G see COF420
FINLEPSIN see DCV200
FINNCARB 6002 see CAT775
FINQUEL see EFX500
FINTIN ACETATO (ITALIAN) see ABX250
FINTINE HYDROXYDE (FRENCH) see
HON000
FINTIN HYDROXID (GERMAN) see
HON000
FINTIN HYDROXYDE (DUTCH) see
HON000
FINTIN IDROSSIDO (ITALIAN) see HON000
FINTROL see AQM000
FIORINAL see ABG750
FIR BALSAM ABSOLUTE see FBS200
FIR BALSAM OREGON see OJK340
FIRE DAMP see MDQ750
FIRE GUARD 2000 see MKA270
FIREMASTER 680 see EEB250
FIREMASTER 695 see EEB230
FIREMASTER BP-6 see FBU000
FIREMASTER BP 4A see MKA270
FIREMASTER FF-1 see FBU509
FIREMASTER FF 680 see EEB250
FIREMASTER PHT 4 see TBJ700
FIREMASTER T23P-LV see TNC500
FIRMACEF see CCS250
FIRMATEX RK see DTG000
FIRMAZOLO see AIF000
FIRMOTOX see POO250
FIR NEEDLE OIL, CANADIAN TYPE see
FBU850

FIR NEEDLE OIL, SIBERIAN see FBV000
FIRON see FBJ100
FISCHER'S ALDEHYDE see FBV050
FISCHER'S SOLUTION see FBV100
FISETHOLZ see FBW000
FISETIN see FBW000
FISH BERRY see PIE500
FISH POISON see HGL575
FISHTAIL PALM see FBW100
FISH-TOX see RNZ000
FISIOQUENS see EEH575
FISONS B25 see CEW500
FISONS NC 2964 see DSO000
FISONS NC 5016 see DGA200
FISSUCAIN see BQA010
FITTIOS see DNX600
FITTIOS B/77 see DNX600
FITOHEMAGLUTYNINA (POLISH) see
PIB575
FITROL see KBB600
FITROL DESICCATE 25 see KBB600
FIXANOL BLACK E see AQP000
FIXANOL BLUE 2B see CMO000
FIXANOL BLUE BH see CMN800
FIXANOL BROWN LF see CMO820
FIXANOL BROWN M see CMO800
FIXANOL C see HCP800
FIXANOL GREEN BN see CMO840
FIXANOL ORANGE BROWN X see CMO810
FIXANOL RED FS see CMO870
FIXANOL VIOLET N see CMP000
FIXAPRET CP see DTG000
FIXATIVE IS see CNH125
FIXER IS see CNH125
FIXIERER P see THM900
FIXOL see CMS850
FIXOLIDE see FBW110
FK 101 see KGK500
FK 235 see NDY650
FK 482 see AMS650
FK 506 see TAA900
FK 749 see EBE100
FK 1160 see CJH750
F KLOT see AJP250
FL see MQR225
FL-1039 see MCB550
FLAC see ARB750
FLACAVON R see TNC500
FLACETHYLE see ELH600
FLACITRAN see TDD550
FLAGECIDIN see AOY000
FLAGEMONA see MMN250
FLAGESOL see MMN250
FLAGIL see MMN250
FLAGYL see MMN250
FLAKE WHITE see BKW100
FLAMAL 171 see FBW125
FLAMARIL see HNI500
FLAMAZINE see SNI425
FLAMENCO see TGG760
FLAME TONES see CJD500
FLAMINGO FLOWER see APM875
FLAMINGO LILY see APM875
FLAMING RED see CJD500
FLAMMEX AP see TNC500
FLAMMEX B 10 see PCC480
FLAMMEX 5AE see PAU600
FLAMOLIN MF 15711 see PJS750
(−)-FLAMPROP-ISOPROPYL see FBW135
d-FLAMPROP-ISOPROPYL see FBW135
FLAMPROP-m-ISOPROPYL see FBW135
FLAMPROP-METHYL see FDB500
FLAMRUSS see CBT750
FLAMULA (CUBA) see CMV390
FLAMYCIN see CMA750
FLANARIL see HNI500
FLANOGEN ELA see CCL250
FLAROXATE HYDROCHLORIDE see
FCB100
FLAVACRIDINE see ADQ700
FLAVACRIDINUM HYDROCHLORICUM see
DBX400

FLAVANONE see FBW150
4-FLAVANONE see FBW150
FLAVANONE, 3,3',4',5,5',7-HEXAHYDROXY-
see HDW125
FLAVANONE, 4',5,7-TRIHYDROXY- see
NBP350
FLAVASPIDIC ACID see FBY000
FLAVASPIDSAEURE (GERMAN) see FBY000
FLAVAXIN see RIK000
FLAVAZONE see NGE500
FLAVENSOMYCIN see FBZ000
FLAVIANIC ACID see FBZ100
FLAVIN see XAK000
FLAVIN ADENIN DINUCLEOTIDE see
RIF100
FLAVIN ADENINE DINUCLEOTIDE see
RIF100
FLAVINAT see RIF100
FLAVINE see DBN400
FLAVINE see DBX400
FLAVINE see XAK000
FLAVINE-ADENINE DINUCLEOTIDE see
RIF100
FLAVINE ADENOSINE DIPHOSPHATE see
RIF100
FLAVIN SULPHATE see DBN400
FLAVIOFORM see DBX400
FLAVIPIN see DBX400
FLAVISEPT see DBX400
FLAVISPIDIC ACID BB see FBY000
FLAVITAN see RIF100
FLAVITROL see EDW500
FLAVOFUNGIN (1:10) see FCA000
FLAVOMYCELIN see LIV000
FLAVOMYCIN see MRA250
FLAVONE see PER700
FLAVONE, 5,7-DIHYDROXY- see DMS900
FLAVONE, 5,8-DIHYDROXY-7-METHOXY-
see ITD020
7-FLAVONE ETHYL HYDROXYACETATE
see ELH600
FLAVONE-7-ETHYLOXYACETATE see
ELH600
FLAVONE, 3'-(GLUCOPYRANOSYLOXY)-
3,4',5,5',7-PENTAHYDROXY- see GFC200
FLAVONE, 8-d-GLUCOSYL-4',5,7-
TRIHYDROXY- see GFC050
FLAVONE, 3,3',4',5,5',7-HEXAHYDROXY- see
HDW150
FLAVONE, 3,3',4',5,5',7-HEXAHYDROXY-,
HEXAACETATE see MRZ200
FLAVONE, 3-HYDROXY- see HLC600
FLAVONE, 3',4',5,7-TETRAHYDROXY- see
TDD550
FLAVONE, 3',4',5,7-TETRAHYDROXY- see
TDD550
FLAVONE, 3',4',5,6-TETRAHYDROXY-7-
METHOXY-, TETRAACETATE see PAO160
FLAVONE, 4',5,7-TRIHYDROXY-,
TRIACETATE see THM300
FLAVONIC ACID see FCA100
FLAVONIC ACID DISODIUM SALT see
FCA200
FLAVONOID AGLUCONE see FCB000
7-FLAVONOXYACETIC ACID ETHYL
ESTER see ELH600
FLAVOPHOSPHOLIPOL see MRA250
FLAVOSAN see XAK000
FLAVOXATE HYDROCHLORIDE see
FCB100
FLAVUMYCIN B see FCC000
FLAVUROL see MCV000
FLAVYLIUM, 3,3',4',5,5',7-HEXAHYDROXY-
see HDW200
FLAVYLIUM, 3,4',5,7-TETRAHYDROXY-3',5'-
DIMETHOXY-, CHLORIDE see MAO600
FLAXEDIL see PDD300
FLAX OLIVE see LAR500
FLAZASULFURON see FCC050
FLB 524 HYDROCHLORIDE see CHC100
FLEBOCORTID see HHR000
FLECAINIDE see FCC065

(+−)-FLECAINIDE see FCC065
FLECK-FLIP see TIO750
FLECK'S EXTRAORDINARY CEMENT see
ZJS400
FLECTOL A see TLP500
FLECTOL H see TLP500
FLECTOL H, POLYMER see PJQ750
FLECTOL PASTILLES see TLP500
FLEET-X see TLM050
FLEET-X-DV-99 see TLM000
FLEUR SUREAU (CANADA, HAITI) see
EAI100
FLEX see CLS050
FLEXAL see IPU000
FLEXAMINE G see BLE500
FLEXAN 500 see SFO100
FLEXARTAL see IPU000
FLEXARTEL see IPU000
FLEXAZONE see BRF500
FLEXCHLOR see PAH780
FLEXERIL see DPX800
FLEXIBAN see DPX800
FLEXIBLE COLLODION see CCU250
FLEXICHEM see CAX350
FLEXICHEM CS see CAX350
FLEXIDOR see ENF100
FLEXILON see AJF500
FLEXIMEL see DVL700
FLEXIN see AJF500
FLEXOL 8N8 see FCN531
FLEXOL A 26 see AEO000
FLEXOL DOP see DVL700
FLEXOL EP-8 see FAB900
FLEXOL EPO see FCC100
FLEXOL GPE see PJC500
FLEXOL JPO see FCN530
FLEXOL NODP see FCN533
FLEXOL 4GO see FCD500
FLEXOL PEP see FCD570
FLEXOL PLASTICIZER 810 see FCN515
FLEXOL PLASTICIZER 8N8 see FCN531
FLEXOL PLASTICIZER CC-55 see FCD520
FLEXOL PLASTICIZER DIP see ILR100
FLEXOL PLASTICIZER DOP see DVL700
FLEXOL PLASTICIZER JPO see FCN530
FLEXOL PLASTICIZER NODP see FCN533
FLEXOL PLASTICIZER 3GO see FCD525
FLEXOL PLASTICIZER PEP see FCD570
FLEXOL PLASTICIZER TCP see TNP500
FLEXOL PLASTICIZER 3GV see FCD522
FLEXOL TOF see TNI250
FLEXOL 3GV see FCD522
FLEXON 393 see AQZ150
FLEXO RED 482 see RGW000
FLEXZONE 3C see PFL000
FL-G 5 see KDK000
FL-G 35 see KDK000
FL-G 100 see KDK000
FL-G 330 see KDK000
FLIBOL E see TIQ250
FLIEGENTELLER see TIQ250
FLINDIX see CCK125
FLINT see SCI500
FLINT see SCJ500
FLIT 406 see CBG000
FLOBACIN see OGI300
FLOCALCITRIOL see HDB100
FLOCOOL 180 see SJC500
FLOCOR see PKQ059
FLOGAR see OLM300
FLOGENE see CKI750
FLOGHENE see HNI500
FLOGICID see BPP750
FLOGINAX see MFA500
FLOGISTIN see HNI500
FLOGITOLO see HNI500
FLOGOCID N PLASTIGEL see BPP750
FLOGODIN see HNI500
FLOGORIL see HNI500
FLOGOS see SOX875
FLOGOSTOP see HNI500

FLO-MOR see PAI000
FLOMORE see BSQ750
FLOMOXEF see FCN100
FLOMOXEF SODIUM see FCN100
FLOPIRINA see HNI500
FLOPROPION see TKP100
FLOPROPIONE see TKP100
FLO PRO T SEED PROTECTANT see TFS350
FLO PRO V SEED PROTECTANT see CCC500
FLOR de ADONIS (CUBA) see PCU375
FLORALOZONE see EIJ600
FLORALTONE see GEM000
FLORALTONE see TKQ250
FLORANE see HBP425
FLORAQUIN see DNF600
FLOR de BARBERO (CUBA) see AFQ625
FLOR de CAMARON (MEXICO) see CAK325
FLOR de CULEBRA (PUERTO RICO) see APM875
FLORDIMEX see CDS125
FLOREL see CDS125
FLORES MARTIS see FAU000
FLORIDA ARROWROOT see CNH789
FLORIDA HOLLY see PCB300
FLORIDIN see TEY000
FLORIDINE see SHF500
FLORIMYCIN see VQZ000
FLORINEF see FHH100
FLORIPONDIO (PUERTO RICO) see AOO825
FLORITE see BAR900
FLORITE R see CAW850
FLOROCID see SHF500
FLOROL see PGR250
FLOROMYCIN see VQZ000
FLORONE see DKF125
FLORONE (ITALIAN) see XQJ000
FLOROPIPAMIDE see FHG000
FLOROPIPETON see FLN000
FLOROPRYL see IRF000
FLOROXENE see TKB250
FLOR del PERU see YAK350
FLOSIN see IDA400
FLOSINT see IDA400
FLOTHENE see PJS750
FLOU see POF500
FLOUVE OIL see FDA000
FLOVACIL see DKI600
FLOWER FENCE see CAK325
FLOWER of PARADISE see HMX600
FLOWERS of ANTIMONY see AQF000
FLOWERS of SULPHUR (DOT) see SOD500
FLOWERS of ZINC see ZKA000
FLOXACILLIN SODIUM see FDA100
FLOXACILLIN SODIUM MONOHYDRATE see CHJ000
FLOXAPEN see CHJ000
FLOXAPEN SODIUM see FDA100
FLOXIN see OGI300
FLOXURIDIN see DAR400
FLOXURIDINE see DAR400
FLOZENGES see SHF500
FLUALAMIDE see FDA875
FLUANISON see HAF400
FLUANISONE see HAF400
FLUANISONE HYDROCHLORIDE see FDA880
FLUANXOL see FMO129
FLUATE see TIO750
FLUAZIFOP-BUTYL see FDA885
FLUAZINAM see CEX800
FLUBE see BLX000
FLUBENDAZOLE see FDA887
FLUBENZIMINE see FDA890
FLUCARBAZONE-SODIUM see FDA895
FLUCHLORALIN see FDA900
FLUCINAR see SPD500
FLUCLOXACILLIN SODIUM see FDA100
FLUCLOXACILLIN SODIUM MONOHYDRATE see CHJ000

FLUCLOXACILLIN SODIUM SALT see FDA100
FLUCORT see FDD075
FLUCORT see SPD500
FLUCORTICIN see FDD075
FLUCORTOLONE see FDA925
FLUCYTHRINATE see COQ385
FLUCYTHRINATE see COQ390
FLUCYTOSINE see FHI000
FLUDERMA see FDB000
FLUDEX see IBV100
FLUDIAZEPAM see FDB100
FLUDILAT see BAV250
FLUDROCORTISONE see FHH100
FLUDROCORTONE see FHH100
FLUE DUST, ARSENIC CONTAINING see ARE500
FLUE GAS see CBW750
FLUENETIL see FDB200
FLUENYL see FDB200
FLUFENAMIC ACID see TKH750
FLUFENAMINE see FDB300
FLUFENAMINSAURE (GERMAN) see TKH750
FLUFENOXURON see FDB400
FLUFENPROP-METHYL see FDB500
FLUGENE 22 see CFX500
FLUGERIL see FMR050
FLUGEX 12B1 see BNA250
FLUIBIL see CDL325
FLUID-EXTRACT of JAMAICA GINGER U.S.P. see JBA000
FLUIFORT see CBR675
FLUIMUCETIN see ACH000
FLUIMUCIL see ACH000
FLUITRAN see HII500
FLUKOIDS see CBY000
FLUMAMINE see DQR600
FLUMARK see EAU100
FLUMEN see CLH750
FLUMESIL see BEQ625
FLUMETHASONE see FDD075
FLUMETHIAZIDE see TKG750
FLUMETRALIN see FDD078
FLUMETRALINE see FDD078
FLUMICIL see ACH000
FLUMIOXAZIN see FLZ075
FLUMOPERONE HYDROCHLORIDE see TKK750
FLUNARIZINE DIHYDROCHLORIDE see FDD080
FLUNARIZINE HYDROCHLORIDE see FDD080
FLUNIGET see DKI600
FLUNISOLIDE see FDD085
FLUNITRAZEPAM see FDD100
FLUOBORIC ACID see FDD125
FLUOBORIC ACID see HHS600
FLUOBORIC ACID (DOT) see FDD125
FLUOBORIC ACID (DOT) see HHS600
FLUOBRENE see FOO525
FLUOCINOLIDE see FDD150
FLUOCINOLONE ACETONIDE see SPD500
FLUOCINOLONE 16,17-ACETONIDE see SPD500
FLUOCINOLONE ACETONIDE ACETATE see FDD150
FLUOCINOLONE ACETONIDE 21-ACETATE see FDD150
FLUOCINONIDE see FDD150
FLUOCORTOLON see FDA925
FLUOCORTOLONE see FDA925
FLUODROCORTISONE see FHH100
FLUOHYDRISONE see FHH100
FLUOHYDROCORTISONE see FHH100
FLUO-KEM see TAI250
FLUOMETURON see DUK800
FLUOMINE see EIS000
FLUOMINE DUST see EIS000
FLUON see TAI250
FLUOOXENE see TKB250
FLUOPERAZINE see TKE500

FLUOPERAZINE see TKK250
FLUOPERIDOL see FLU000
FLUOPHOSGENE see CCA500
FLUOPHOSPHORIC ACID DI(DIMETHYLAMIDE) see BJE750
FLUOPHOSPHORIC ACID, DIETHYL ESTER see DJJ400
FLUOPHOSPHORIC ACID, DIISOPROPYL ESTER see IRF000
FLUOPHOSPHORIC ACID, DIMETHYL ESTER see DSA800
FLUOR (DUTCH, FRENCH, GERMAN, POLISH) see FEZ000
FLUORACETATO di SODIO (ITALIAN) see SHG500
5-FLUORACIL (GERMAN) see FMM000
FLUORACIZINE see FDE000
FLUORAKIL 100 see FFF000
FLUORAL see SHF500
FLUORAL HYDRATE see TJZ000
FLUORAMIDE see FFU000
3,6-FLUORANDIOL see FEV000
3',6'-FLUORANDIOL see FEV000
FLUORANE 114 see FOO509
4-FLUORANILIN see FFY000
1-FLUORANTHENAMINE see FDE100
8-FLUORANTHENAMINE see FDE200
9-FLUORANTHENAMINE see FDE200
FLUORANTHENE see FDF000
FLUORANTHENE, 8-METHYL- see MKC775
FLUORANTHENE, 2-NITRO- see NGA600
N-FLUORANTHEN-3-YLACETAMIDE see AAK400
N-3-FLUORANTHENYLACETAMIDE see AAK400
FLUORAPATITE see FDH000
FLUORAQUIN see DNF600
o-FLUORBENZOESAEURE (GERMAN) see FGH000
FLUORCORTOLONE see FDA925
5-FLUOR-DESOXYCYTIDIN (GERMAN) see FHO000
FLUOREN-2-AMINE see FDI000
2-FLUORENAMINE see FDI000
FLUOREN-9-AMINE, N-(2-CHLOROETHYL)-N-ETHYL-, HYDROCHLORIDE see FEE100
9H-FLUOREN-9-AMINE, N-(2-CHLOROETHYL)-N-ETHYL-, HYDROCHLORIDE (9CI) see FEE100
FLUORENE see FDI100
9H-FLUORENE see FDI100
2-FLUORENEAMINE see FDI000
9-FLUORENECARBOXYLATE-3-QUINUCLIDINOL HYDROCHLORIDE see FDK000
FLUORENE-9-CARBOXYLIC ACID-2-(DIETHYLAMINO)ETHYL ESTER see CBS250
FLUORENE-9-CARBOXYLIC ACID-3-QUINUCLIDINYL ESTER see FDK000
FLUORENE, 2-(CHLOROACETYL)- see CEC700
FLUORENE-2,7-DIAMINE see FDM000
2,7-FLUORENEDIAMINE see FDM000
9H-FLUORENE, 1,9-DIMETHYL- see DRY100
1,1'-(9H-FLUORENE-2,7-DIYL)BIS(2-(DIETHYLAMINO)ETHANONE) DIHYDROCHLORIDE TRIHYDRATE see FDN000
FLUORENE, 9-METHYL- see MKC800
9H-FLUORENE, 9-METHYL- see MKC800
FLUORENE, NITRO- see FDN100
FLUORENE, 3-NITRO- see NGB200
9H-FLUORENE, NITRO-(9CI) see FDN100
9H-FLUORENE, 3-NITRO-(9CI) see NGB200
9H-FLUORENE, 2,3,9-TRIMETHYL- see TLS600
9H-FLUOREN-9-OL, 2-NITRO- see HMY080
FLUOREN-9-ONE see FDO000
9-FLUORENONE see FDO000

9H-FLUOREN-9-ONE see FDO000
9-FLUORENONE, 2,7-DINITRO- see DUW300
9H-FLUOREN-9-ONE, 2-NITRO- see NGB300
9H-FLUOREN-9-ONE, 1,2,3,4-TETRAHYDRO- see TCR300
FLUOREN-9-ONE, 2,4,5,7-TETRANITRO- see TDY110
9H-FLUOREN-9-ONE, 2,4,5,7-TETRANITRO- see TDY110
FLUORENO(9,1-gh)QUINOLINE see FDP000
1-FLUORENYLACETAMIDE see FDQ000
2-FLUORENYLACETAMIDE see FDR000
3-FLUORENYL ACETAMIDE see FDS000
N-FLUOREN-1-YL ACETAMIDE see FDQ000
N-1-FLUORENYLACETAMIDE see FDQ000
N-FLUOREN-2-YL ACETAMIDE see FDR000
N-2-FLUORENYLACETAMIDE see FDR000
N-3-FLUORENYL ACETAMIDE see FDS000
N-FLUOREN-3-YL ACETAMIDE see FDS000
N-FLUOREN-4-YLACETAMIDE see ABY000
N-4-FLUORENYLACETAMIDE see ABY000
N-9H-FLUOREN-3-YL ACETAMIDE see FDS000
1-(N-2'-FLUORENYLACETAMIDO-2-ACETYLAMINO)FLUORENE see AAK500
1-FLUORENYL ACETHYDROXAMIC ACID see FDT000
FLUORENYL-2-ACETHYDROXAMIC ACID see HIP000
3-FLUORENYL ACETHYDROXAMIC ACID see FDU000
N-(FLUOREN-2-YL)ACETOHYDROXAMIC ACETAMIDE see ABL000
N-(FLUOREN-3-YL)ACETOHYDROXAMIC ACETATE see ABO250
N-(FLUOREN-4-YL)ACETOHYDROXAMIC ACETATE see ABO500
N-FLUOREN-1-YL ACETOHYDROXAMIC ACID see FDT000
N-FLUOREN-2-YL ACETOHYDROXAMIC ACID see HIP000
N-2-FLUORENYL ACETOHYDROXAMIC ACID see HIP000
N-FLUOREN-3-YL ACETOHYDROXAMIC ACID see FDU000
N-FLUOREN-2-YLACETOHYDROXAMIC ACID, COBALT(2+) COMPLEX see FDU875
N-FLUOREN-2-YL ACETOHYDROXAMIC ACID, COPPER(2+) COMPLEX see HIP500
N-FLUOREN-2-YL ACETOHYDROXAMIC ACID, IRON(3+) COMPLEX see HIQ000
N-FLUOREN-2-YL ACETOHYDROXAMIC ACID, MANGANESE(2+) COMPLEX see HIR000
N-FLUOREN-2-YL ACETOHYDROXAMIC ACID, NICKEL(2+) COMPLEX see HIR500
N-FLUOREN-2-YL ACETOHYDROXAMIC ACID, POTASSIUM SALT see HIS000
N-FLUOREN-2-YL ACETOHYDROXAMIC ACID SULFATE see FDV000
N-FLUOREN-2-YLACETOHYDROXAMIC ACID, ZINC COMPLEX see ZHJ000
N-(2-FLUORENYL)BENZAMIDE see FDX000
N-FLUOREN-2-YL BENZAMIDE see FDX000
N-9H-FLUOREN-2-YL-BENZAMIDE (9CI) see FDX000
N-FLUOREN-1-YL BENZOHYDROXAMIC ACID see FDY000
N-FLUOREN-2-YL BENZOHYDROXAMIC ACID see FDZ000
N-(2-FLUORENYL)BENZOHYDROXAMIC ACID see FDZ000
N-FLUOREN-2-YL BENZOHYDROXAMIC ACID ACETATE see ABO750
2,7-FLUORENYLBISACETAMIDE see BGP250
N,N'-FLUOREN-2,7-YLBISACETAMIDE see BGP250

N-FLUORENYLCYCLOBUTANECARBOXAMIDE see COW500
2-FLUORENYLDIACETAMIDE see DBF200
N-FLUOREN-1-YLDIACETAMIDE see DBF000
N-1-FLUORENYLDIACETAMIDE see DBF000
N-FLUOREN-2-YLDIACETAMIDE see DBF200
N-2-FLUORENYLDIACETAMIDE see DBF200
2-FLUORENYLDIMETHYLAMINE see DPJ600
2,5-FLUORENYLENEBISACETAMIDE see BGR250
N,N'-FLUOREN-2,5-YLENEBISACETAMIDE see BGR250
N,N'-FLUOREN-2,7-YLENEBISACETAMIDE see BGP250
N,N'-2,7-FLUORENYLENEBISACETAMIDE see BGP250
N,N'-(FLUOREN-2,7-YLENE)BIS(ACETYLAMINE) see BGP250
N,N'-FLUOREN-2,7-YLENE BIS(TRIFLUOROACETAMIDE) see FEE000
N,N'-2,7-FLUORENYLENEDIACETAMIDE see BGP250
N-(9-FLUORENYL)-N-ETHYL-β-CHLOROETHYLAMINE HYDROCHLORIDE see FEE100
N-FLUOREN-2-YL FORMAMIDE see FEF000
N,2-FLUORENYL FORMAMIDE see FEF000
N-(2-FLUORENYL)FORMOHYDROXAMIC ACID see FEG000
N-9H-FLUOREN-2-YL-N-HYDROXYBENZAMIDE see FDZ000
2-FLUORENYL HYDROXYLAMINE see HIU500
3-FLUORENYLHYDROXYLAMINE see FEH000
N-FLUOREN-2-YLHYDROXYLAMINE see HIU500
N-FLUOREN-3-YL HYDROXYLAMINE see FEH000
N-FLUOREN-2-YL-HYDROXYLAMINE-o-GLUCURONIDE see FEI000
N-2-FLUORENYLHYDROXYLAMINE-o-GLUCURONIDE see FEI000
9-FLUORENYLMETHYL CHLOROFORMATE see FEI100
2-FLUORENYL METHYL KETONE see FEI200
2-FLUORENYLMONOMETHYLAMINE see FEI500
N-(2-FLUORENYL)MYRISTOHYDROXAMIC ACID ACETATE see FEM000
N-9H-FLUOREN-2-YL-N-NITROSOACETAMIDE see FEM050
1-FLUORENYL PHENYL KETONE see FEM100
9H-FLUOREN-2-YLPHENYLMETHANONE see FEM100
N-(2-FLUORENYL)PHTHALAMIC ACID see FEN000
N-FLUORENYL-2-PHTHALIMIC ACID see FEN000
N-(2-FLUORENYL)PROPIONOHYDROXAMIC ACID see FEO000
N-2-FLUORENYL SUCCINAMIC ACID see FEP000
N-(FLUOREN-2-YL)-o-TETRADECANOYLACETOHYDROXAMIC ACID see ACS000
N-FLUOREN-2-YL-N-TETRADECANOYLHYDROXAMIC ACID see HMU000
N-FLUOREN-2-YL-2,2,2-TRIFLUOROACETAMIDE see FER000

N-(2-FLUORENYL)-2,2,2-TRIFLUOROACETAMIDE see FER000
FLUORESCEIN see FEV000
FLUORESCEIN, soluble see FEW000
FLUORESCEIN, 4',5'-DIBROMO- see DDO200
FLUORESCEINE see FEV000
FLUORESCEIN MERCURIACETATE see FEV100
FLUORESCEIN MERCURIC ACETATE see FEV100
FLUORESCEIN MERCURY ACETATE see FEV100
FLUORESCEIN SODIUM see FEW000
FLUORESCEIN SODIUM B.P see FEW000
FLUORESCEIN, 2',4',5',7'-TETRABROMO-, DISODIUM SALT see BNK700
FLUORESCEIN, 2',4',5',7'-TETRABROMO-4,5,6,7-TETRACHLORO- see SKS400
FLUORESCEIN, 2',4',5',7'-TETRABROMO-4,5,6,7-TETRACHLORO- see SKS400
FLUORESCENT BRIGHTENER 46 see TGE155
FLUORESCENT BRIGHTENER 85 see CMP100
FLUORESCENT BRIGHTENER 220 see CMP120
FLUORESONE see FLG000
FLUORESSIGSAEURE (GERMAN) see SHG500
2-FLUORETHYLESTER KYSELINY CHLORMRAVENCI see FIH100
1-p-FLUORFENYL-3,3-DIMETHYLTRIAZEN (CZECH) see DSA600
FLUORID BORITY-DIMETHYLETHER (1:1) see BMH000
FLUORIDE see FEX875
FLUORIDE(1-) see FEX875
FLUORIDE ION see FEX875
FLUORIDE ION(1-) see FEX875
FLUORIDENT see SHF500
FLUORIDES see FEY000
FLUORID HLINITY (CZECH) see AHB000
FLUORID KYSELINY FLUOROCTOVE see FFS100
FLUORID KYSELINY OCTOVE see ACM000
FLUORID SODNY (CZECH) see SHF500
FLUORIGARD see SHF500
FLUORIMIDE see DKF400
FLUORINE see FEZ000
FLUORINE, compressed (DOT) see FEZ000
FLUORINE AZIDE see FFA000
FLUORINEED see SHF500
FLUORINE FLUORO SULFATE see FFB000
FLUORINE MONOXIDE see ORA000
FLUORINE NITRATE see NMU000
FLUORINE OXIDE see ORA000
FLUORINE PERCHLORATE see FFD000
FLUORINERT FC 43 see HAS000
6-FLUORIN-7-(PYRROL-1-YL)-1-ETHYL-1,4-DIHYDRO-4-OXOCHINOLON-3-CARBONSAEUE see FLV050
FLUORINSE see SHF500
FLUORISTAN see TGD100
FLUOR-I-STRIP A.T. see FEW000
FLUORITAB see SHF500
FLUORITE see CAS000
1-FLUOR-2-JODETHAN see FIQ000
FLUOR-O-KOTE see SHF500
FLUORO (ITALIAN) see FEZ000
FLUOROACETALDEHYDE see FFE000
FLUOROACETAMIDE see FFF000
2-FLUOROACETAMIDE see FFF000
7-FLUORO-2-ACETAMIDO-FLUORENE see FFG000
FLUOROACETANILIDE see FFH000
2-FLUOROACETANILIDE see FFH000
FLUOROACETATE see FIC000
FLUOROACETIC ACID see FIC000
2-FLUOROACETIC ACID see FIC000
FLUOROACETIC ACID (DOT) see FIC000
FLUOROACETIC ACID AMIDE see FFF000

FLUOROACETIC ACID (2-ETHYLHEXYL) ESTER see FFI000
FLUOROACETIC ACID, MERCURY(II) SALT see MDB500
FLUOROACETIC ACID METHYL ESTER see MKD000
FLUOROACETIC ACID, SODIUM SALT see SHG500
FLUOROACETIC ACID, TRIETHYLLEAD SALT see TJS500
FLUOROACETONITRILE see FFJ000
FLUOROACETPHENYLHYDRAZIDE see FFK000
3-FLUORO-4-ACETYLAMINOBIPHENYL see FKX000
3'-FLUORO-4-ACETYLAMINOBIPHENYL see FKY000
4'-FLUORO-4-ACETYLAMINOBIPHENYL see FKZ000
1-FLUORO-2-ACETYLAMINOFLUORENE see FFL000
3-FLUORO-2-ACETYLAMINOFLUORENE see FFM000
4-FLUORO-2-ACETYLAMINOFLUORENE see FFN000
5-FLUORO-2-ACETYLAMINOFLUORENE see FFO000
6-FLUORO-2-ACETYLAMINOFLUORENE see FFP000
7-FLUORO-2-ACETYLAMINOFLUORENE see FFG000
8-FLUORO-2-ACETYLAMINOFLUORENE see FFQ000
p-FLUOROACETYLAMINOPHENYL DERIVATIVE of NITROGEN MUSTARD see BHP750
FLUOROACETYL CHLORIDE see FFR000
FLUORO ACETYLENE see FFS000
FLUOROACETYL FLUORIDE see FFS100
o-(FLUOROACETYL)SALICYLIC ACID see FFT000
FLUORO-β-ALANINE HYDROCHLORIDE see FFT100
α-FLUORO-β-ALANINE HYDROCHLORIDE see FFT100
FLUOROAMINE see FFU000
4'-FLUORO-4-AMINODIPHENYL see AKC500
5-FLUORO AMYLAMINE see FFV000
5-FLUOROAMYL CHLORIDE see FFW000
5-FLUOROAMYL THIOCYANATE see FFX000
4-FLUOROANILINE see FFY000
p-FLUOROANILINE see FFY000
2-FLUOROBENZALDEHYDE see FFY100
o-FLUOROBENZALDEHYDE see FFY100
4-FLUOROBENZANTHRACENE see FFZ000
4-FLUOROBENZ(a)ANTHRACENE see FFZ000
4'-FLUORO-1,2-BENZANTHRACENE see FFZ000
4-FLUOROBENZENAMINE see FFY000
FLUOROBENZENE see FGA000
4-FLUOROBENZENEACETONITRILE see FLC000
4-FLUOROBENZENEARSONIC ACID see FGA100
1-(2-(4-(6-FLUORO-1,2-BENZISOXAZOL-3-YL)-1-PIPERIDINYL)ETHYL)-3-PHENYL-2-IMIDAZOLIDINONE see FGA200
2-FLUORO-BENZO(e)(1)BENZOTHIOPYRANO(4,3-b)INDOLE see FGB000
3-FLUORO-BENZO(e)(1)BENZOTHIOPYRANO(4,3-b)INDOLE see FGD000
3-FLUORO-BENZO(g)(1)BENZOTHIOPYRANO(4,3-b)INDOLE see FGD100
4-FLUORO-BENZO(e)(1)BENZOTHIOPYRANO(4,3-b)INDOLE see FGF000

4-FLUORO-BENZO(g)(1)BENZOTHIOPYRANO(4,3-b)INDOLE see FGG000
4-FLUOROBENZO(j)FLUORANTHENE see BCP530
10-FLUOROBENZO(j)FLUORANTHENE see FGG900
o-FLUOROBENZOIC ACID see FGH000
p-FLUOROBENZOIC ACID 2-PHENYLHYDRAZIDE see FGH050
1-(4'-FLUOROBENZOIL)-3-PIRROLIDINOPROPANO MALEATO (ITALIAN) see FGW000
4-FLUOROBENZONITRILE see FGH100
p-FLUOROBENZONITRILE see FGH100
3-FLUOROBENZO(rst)PENTAPHENE see FGI000
6-FLUOROBENZO(a)PYRENE see FGI100
2-FLUORO-(1)BENZOTHIOPYRANO(4,3-b)INDOLE see FGJ000
4-FLUORO-(1)BENZOTHIOPYRANO(4,3-b)INDOLE see FGL000
4-FLUORO-6H-(1)BENZOTHIOPYRANO(4,3-b)QUINOLINE see FGO000
(5-(4-FLUOROBENZOYL)-1H-BENZIMIDAZOLE-2-YL)CARBAMIC ACID METHYL ESTER see FDA887
2-FLUOROBENZOYL CHLORIDE see FGP000
o-FLUOROBENZOYL CHLORIDE see FGP000
1'-(3-(p-FLUOROBENZOYL)PROPYL)(1,4'-BIPIPERIDINE)-4'-CARBOXAMIDE see FHG000
1-(3-p-FLUOROBENZOYLPROPYL)-4-p-CHLOROPHENYL-4-HYDROXYPIPERIDINE see CLY500
2-(3-(p-FLUOROBENZOYL)-1-PROPYL)-5-α,9-α-DIMETHYL-2'-HYDROXY-6,7-BENZOMORPHAN see FGQ000
3-(γ-(p-FLUOROBENZOYL)PROPYL)-2,3,4,4a,5,6-HEXAHYDRO-1(H)-PYRAZINO(1,2A)QUINOLINE HCl see CNH500
8-(3-p-FLUOROBENZOYL-1-PROPYL)-4-OXO-1-PHENYL-1,3,8-TRIAZASPIRO(4,5)DECANE see SLE500
8-(3-(p-FLUOROBENZOYL)PROPYL)-1-PHENYL-1,3,8-TRIAZASPIRO(4.5)DECAN-4-ONE see SLE500
4-(3-(p-FLUOROBENZOYL)PROPYL)-1-PIPERAZINOCARBOXYLIC ACID CYCLOHEXYLESTER see FLK000
1-(3-(p-FLUOROBENZOYL)PROPYL)-4-PIPERIDINOISONIPACOTAMIDE see FHG000
1-(1-(3-(p-FLUOROBENZOYL)PROPYL)-4-PIPERIDYL)-2-BENZIMIDAZOLINETHIONE see TGB175
1-(1-(3-(p-FLUOROBENZOYL)PROPYL)-4-PIPERIDYL)-2-BENZIMIDAZOLINONE see FLK000
1-(1-(3-(p-FLUOROBENZOYL)PROPYL)-4-PIPERIDYL)-2-BENZIMIDAZOLINONE, HYDROCHLORIDE MONOHYDRATE see FGU000
1-(3-(4-FLUOROBENZOYL)PROPYL)-4-PIPERIDYL-N-ISOPROPYL CARBAMATE see FGV000
1-(3-(4-FLUOROBENZOYL)PROPYL)-4-(2-PYRIDYL)PIPERAZINE see FLU000
1-(1-(3-(p-FLUOROBENZOYL)PROPYL)-1,2,3,6-TETRAHYDRO-4-PYRIDYL)-2-BENZIMIDAZOLINONE see DYF200
1-(4'-FLUOROBENZOYL)-3-PYRROLIDINYLPROPANE MALEATE see FGW000
N-p-FLUOROBENZYL((CHROMONYL-2 AMINO)-CARBONYL)-5 PYRROLIDONE-2 see FLH150
4-FLUOROBENZYLCYANIDE see FLC000

p-FLUOROBENZYL CYANIDE see FLC000
1-(p-FLUOROBENZYL)-2-((1-(2-(p-METHOXYPHENYL)ETHYL)PIPERID-4-YL)AMINO)BENZIMIDAZOLE see ARP675
N-(p-FLUOROBENZYL)PYROGLUTAMATE de CHROMONYL-2 METHYL see FLH320
N-(p-FLUOROBENZYL)PYROGLUTAMATE D'AMMONIUM see FLH300
N-(p-FLUOROBENZYL)PYROGLUTAMATE de DIETHYLAMINE see FLH325
N-(p-FLUOROBENZYL)PYROGLUTAMATE de DIISOPROPYLAMINE see PMH933
N-(p-FLUOROBENZYL)PYROGLUTAMATE de ((DIMETHYL-2,2)DIOXOLANNE-1,3-YL-4)-METHYL see FLH310
N-(p-FLUOROBENZYL)PYROGLUTAMATE de N'-ISOPROPYLBENZYLAMINE see PMH930
N-(p-FLUOROBENZYL)PYROGLUTAMATE de METHYLE see FLH315
N-(p-FLUOROBENZYL)PYROGLUTAMIDE see FGI025
N-p-FLUOROBENZYL((M-TRIFLUOROMETHYLANILINO)CARBONYL)-5-PYRROLIDONE-2 see FLH400
2-FLUOROBIPHENYL see FGI050
N-4-(4'-FLUORO)BIPHENYLACETAMIDE see FKZ000
4'-FLUORO-4-BIPHENYLAMINE see AKC500
N-(4'-FLUORO-4-BIPHENYLYL)ACETAMIDE see FKZ000
2-(2-FLUORO-4-BIPHENYLYL)PROPIONIC ACID see FLG100
FLUOROBISISOPROPYLAMINO-PHOSPHINE OXIDE see PHF750
5-FLUORO-1,3-BIS(TETRAHYDRO-2-FURANYL)-2,4(1H,3H)-PYRIMIDINEDIONE see BLH325
FLUOROBIS(TRIFLUOROMETHYL)PHOSPHINE see FGW100
FLUOROBLASTIN see FMM000
1-FLUORO-2-BROMOBENZENE see FGX000
1-FLUORO-3-BROMOBENZENE see FGY000
6-FLUORO-7-BROMOMETHYLBENZ(a)ANTHRACENE see BNQ250
4-FLUOROBUTANAL see FGY100
4-FLUOROBUTYL BROMIDE see FHA000
4-FLUOROBUTYL CHLORIDE see FHB000
4-FLUOROBUTYL IODIDE see FHC000
3-FLUOROBUTYL ISOCYANATE see FHC200
4-FLUOROBUTYL THIOCYANATE see FHD000
4-FLUORO-BUTYRIC ACID-2-CHLOROETHYL ESTER see CGZ000
2-FLUOROBUTYRIC ACID ISOPROPYL ESTER see FHD100
4-FLUOROBUTYRIC ACID METHYL ESTER see MKE000
4-FLUOROBUTYRONITRILE see FHF000
γ-FLUOROBUTYRONITRILE see FHF000
FLUOROBUTYROPHENONE see FHG000
FLUOROCARBON-12 see DFA600
FLUOROCARBON-22 see CFX500
FLUOROCARBON 113 see FOO000
FLUOROCARBON 114 see FOO509
FLUOROCARBON-115 see CJI500
FLUOROCARBON 1211 see BNA250
FLUOROCARBON FC 43 see HAS000
FLUOROCARBON FC143 see DKH300
FLUOROCARBON FC142b see CFX250
FLUOROCARBON FC143a see TJY900
FLUOROCARBON No. 11 see TIP500
FLUOROCHLOROCARBON LIQUID see FHH000
5-FLUORO-7-CHLOROMETHYL-12-METHYLBENZ(a)ANTHRACENE see FHH025
FLUOROCHROME see MCV000

FLUOROCORTISONE see FHH100
4-FLUORO-CROTONIC ACID METHYL ESTER see MKE250
2-FLUORO-2'-CYANODIETHYL ETHER see CON500
5-FLUOROCYSTOSINE see FHI000
5-FLUOROCYTOSINE see FHI000
FLUORO-DDT see FHJ000
1-FLUORODECANE see FHL000
10-FLUORODECANOL see FHM000
ω-FLUORODECANOL see FHM000
10-FLUORODECYL CHLORIDE see FHN000
5-FLUORODEOXYCYTIDINE see FHO000
5-FLUORO-2'-DEOXYCYTIDINE see FHO000
3'-FLUORO-3'-DEOXYTHYMIDINE see DAR200
FLUORODEOXYURIDINE see DAR400
5-FLUORODEOXYURIDINE see DAR400
5-FLUORO-2-DEOXYURIDINE see DAR400
5-FLUORO-2'-DEOXYURIDINE see DAR400
β-5-FLUORO-2'-DEOXYURIDINE see DAR400
6-α-FLUORODEXAMETHASONE see FDD075
6-FLUORODIBENZ(a,h)ANTHRACENE see FHP000
4-FLUORO-1,2:5,6-DIBENZANTHRACENE see FHP000
p-FLUORO-DI-(2-CHLOROETHYL)-BENZYLAMINE HYDROCHLORIDE see FHP100
FLUORODICHLOROMETHANE see DFL000
FLUORODIFEN see NIX000
9-α-FLUORO-11-β,21-DIHYDROXY-16-α-ISOPROYLIDENEDIOXY-1,4-PREGNADIENE, 3,20-DIONE see AQX500
FLUORO-9-α DIHYDROXY-11-β,17-β METHYL-17-α ANDROSTENE-4 ONE-3 (FRENCH) see AOO275
9-α-FLUORO-11-β,17-β-DIHYDROXY-17-α-METHYL-4-ANDROSTENE-3-ONE see AOO275
9-FLUORO-11-β-,17-β-DIHYDROXY-17-METHYLANDROST-4-EN-3-ONE see AOO275
6-α-FLUORO-11-β,21-DIHYDROXY-16-α-METHYLPREGNA-1,4-DIENE-3,20-DIONE see FDA925
9-FLUORO-11-β,21-DIHYDROXY-16-α-METHYLPREGNA-1,4-DIENE-3,20-DIONE see DBA875
9-α-FLUORO-11-β,17-DIHYDROXY-3-OXO-4-ANDROSTENE-17-α-PROPIONIC ACID POTASSIUM see CCP750
FLUORODIISOPROPYL PHOSPHATE see IRF000
m-FLUORODIMETHYLAMINOAZOBENZENE see FHQ100
2-FLUORO-4-DIMETHYLAMINOAZOBENZENE see FHQ000
2'-FLUORO-4-DIMETHYLAMINOAZOBENZENE see FHQ010
3'-FLUORO-4-DIMETHYLAMINOAZOBENZENE see FHQ100
4'-FLUORO-4-DIMETHYLAMINOAZOBENZENE see DSA000
4'-FLUORO-p-DIMETHYLAMINOAZOBENZENE see DSA000
4'-FLUORO-N,N-DIMETHYL-4-AMINOAZOBENZENE see DSA000
2'-FLUORO-4-DIMETHYLAMINOSTILBENE see FHU000
4'-FLUORO-4-DIMETHYLAMINOSTILBENE see FHV000

10-FLUORO-9,12-DIMETHYLBENZ(a)ACRIDINE see FHR000
3-FLUORO-2,10-DIMETHYL-5,6-BENZACRIDINE see FHR000
1-FLUORO-7,12-DIMETHYLBENZ(a)ANTHRACENE see FHS000
4-FLUORO-7,12-DIMETHYLBENZ(a)ANTHRACENE see DRY400
5-FLUORO-7,12-DIMETHYLBENZ(a)ANTHRACENE see DRY600
8-FLUORO-7,12-DIMETHYLBENZ(a)ANTHRACENE see DRY800
11-FLUORO-7,12-DIMETHYLBENZ(a)ANTHRACENE see DRZ000
4'-FLUORO-N,N-DIMETHYL-p-PHENYLAZOANILINE see DSA000
2'-FLUORO-N,N-DIMETHYL-4-STILBENAMINE see FHU000
4'-FLUORO-N,N-DIMETHYL-4-STILBENAMINE see FHV000
1-FLUORO-2,4-DINITROBENZENE see DUW400
1,2,4-FLUORODINITROBENZENE see DUW400
1-FLUORO-1,1-DINITRO-2-BUTENE see FHV300
2-FLUORO-1,1-DINITROETHANE see FHV800
2-FLUORO-2,2-DINITROETHANOL see FHW000
2-FLUORO-2,2-DINITROETHANOL see FHW000
2-FLUORO-2,2-DINITROETHYLAMINE see FHX000
FLUORO DINITROMETHANE see FHY000
FLUORO DINITROMETHYL AZIDE see FHZ000
1-FLUORO-1,1-DINITRO-2-PHENYLETHANE see FHZ200
12-FLUORO DODECANO NITRILE see FIA000
2,7-FLUOROENEDIAMINE see FDM000
4-FLUOROESTRADIOL see FIA500
4-FLUOROESTRA-1,3,5-(10)-TRIENE-3,17-β-DIOL see FIA500
FLUOROETHANE see FIB000
FLUOROETHANOIC ACID see FIC000
FLUOROETHANOL see FID000
2-FLUOROETHANOL see FIE000
β-FLUOROETHANOL see FDB000
2-FLUOROETHANOL, PHOSPHITE (3:1) see PHO250
FLUOROETHENE see VPA000
FLUOROETHYL see HDC000
β-FLUOROETHYL-N-(β-CHLOROETHYL)-N-NITROSOCARBAMATE see FIH000
2-FLUOROETHYL CHLOROFORMATE see FIH000
1-FLUOROETHYL-3-CYCLOHEXYL-1-NITROSOUREA see CPL750
1-(2-FLUOROETHYL)-3-CYCLOHEXYL-1-NITROSOUREA see FIJ000
FLUOROETHYL-O,O-DIETHYLDITHIOPHOSPHORYL-1-PHENYLACETATE see FIK000
FLUOROETHYLENE see VPA000
FLUOROETHYLENE OZONIDE see FIK875
2-FLUOROETHYL ESTER DIPHENYLACETIC ACID see FIP999
2-FLUOROETHYL FLUOROACETATE see FIM000
β-FLUOROETHYL FLUOROACETATE see FIM000
2-FLUORO ETHYL-γ-FLUORO BUTYRATE see FIN000
β-FLUOROETHYL-γ-FLUOROBUTYRATE see FIN000

2'-FLUOROETHYL-6-FLUOROHEXANOATE see EKJ500
2-FLUOROETHYL-5-FLUOROHEXOATE see FIO000
(8R)-8-(2-FLUOROETHYL)-3-α-HYDROXY-1-α-H,5-α-H-TROPANIUM BROMIDE BENZILATE H2O see FMR300
β-FLUOROETHYLIC ESTER of XENYLACETIC ACID see FIP999
2-FLUOROETHYL IODIDE see FIQ000
2-FLUOROETHYL MERCAPTOPHENYLACETATE-O,O-DIETHYL PHOSPHORODITHIOATE see FIK000
2-FLUOROETHYL-N-METHYL-N-NITROSOCARBAMATE see FIS000
1-(2-FLUOROETHYL)-1-NITROSO-UREA see NKG000
1-FLUORO-2-FAA see FFL000
3-FLUORO-2-FAA see FFM000
4-FLUORO-2-FAA see FFN000
5-FLUORO-2-FAA see FFO000
6-FLUORO-2-FAA see FFP000
7-FLUORO-2-FAA see FIT200
8-FLUORO-2-FAA see FFQ000
FLUOROFLEX see TAI250
N-(7-FLUOROFLUORENE-2-YL)ACETAMIDE see FFG000
N-(1-FLUOROFLUOREN-2-YL)ACETAMIDE see FFL000
N-(3-FLUOROFLUOREN-2-YL)ACETAMIDE see FFM000
N-(4-FLUOROFLUOREN-2-YL)ACETAMIDE see FFN000
N-(5-FLUOROFLUOREN-2-YL)ACETAMIDE see FFO000
N-(6-FLUOROFLUOREN-2-YL)ACETAMIDE see FFP000
N-(8-FLUOROFLUOREN-2-YL)ACETAMIDE see FFQ000
7-FLUORO-2-N-(FLUORENYL)ACETHYDROXAMIC ACID see FIT200
7-FLUORO-N-(FLUOREN-2-YL)ACETOHYDROXAMIC ACID see FIT200
N-(7-FLUORO-2-FLUORENYL)ACETOHYDROXAMIC ACID see FIT200
4'-FLUORO-4-(8-FLUORO-2,3,4,5-TETRAHYDRO-1H-PYRIDO(4,3-b)INDOL-2-YL)BUTYROPHENONE HYDROCHLORIDE see FIW000
FLUOROFORM see CBY750
FLUOROFORMYL FLUORIDE see CCA500
FLUOROFORMYLON see FDB000
FLUOROFUR see FLZ050
FLUOROGESAROL see FHJ000
FLUOROGLYCOFEN see FIW100
FLUOROGLYCOFEN-ETHYL see FIW100
1-FLUOROHEPTANE see FIX000
7-FLUOROHEPTANONITRILE see FIY000
7-FLUOROHEPTYLAMINE see FIZ000
FLUOROHEXANE see FJA000
1-FLUOROHEXANE see FJA000
6-FLUOROHEXANESULPHONYL FLUORIDE see FJA100
5-FLUORO-1-HEXYLCARBAMOYL-URACIL see CCK630
9-α-FLUOROHYDROCORTISONE see FHH100
7-FLUORO-N-HYDROXY-N-2-ACETYLAMINOFLUORENE see FIT200
FLUOROHYDROXYANDROSTENEDIONE see FJF100
9-α-FLUORO-11-β-HYDROXY-4-ANDROSTENE-3,17-DIONE see FJF100
4-FLUORO-3-HYDROXY-BUTANETHIOIC ACID METHYL ESTER see MKF000
γ-FLUORO-β-HYDROXY-BUTYRIC ACID METHYL ESTER see MKE750
4-FLUORO-2-HYDROXYBUTYRIC ACID SODIUM SALT see SHI000

γ-FLUORO-β-HYDROXY BUTYRIC ACID THIO METHYL ESTER see FJG000

4'-FLUORO-4-(4-HYDROXY-4-(4'-CHLOROPHENYL)PIPERIDINO)BUTYROPHENONE see CLY500

9-α-FLUORO-17-HYDROXYCORTICOSTERONE see FHH100

(1-α,3-β,4-α)-5-FLUORO-1-(3-HYDROXY-4-(HYDROXYMETHYL)CYCLOPENTYL)-2,4(1H,3H)-PYRIMIDINEDIONE, (+−) see FJF200

9-α-FLUORO-11-β-HYDROXY-17-METHYLTESTOSTERONE see AOO275

9-α-FLUORO-16-α-HYDROXYPREDNISOLONE see AQX250

9-α-FLUORO-16-HYDROXYPREDNISOLONE ACETONIDE see AQX500

4-FLUORO-2-HYDROXYTHIOBUTYRIC ACID-S-METHYL ESTER see FJG000

4'-FLUORO-4-(4-HYDROXY-4-p-TOLYLPIPERIDINO)BUTYROPHENONE, HYDROCHLORIDE see MNN250

4'-FLUORO-4-(4-HYDROXY-4-(α,α,α-TRIFLUORO-m-TOLYL)PIPERIDINO)BUTYROPHENONE see TKK500

4-FLUORO-4,4-IDROSSI-4-(m-TRIFLUOROMETIL-FENIL)-PIPERIDINO-BUTIRROFENONE (ITALIAN) see TKK500

N-FLUOROIMINO DIFLUOROMETHANE see FJI000

1-FLUOROIMINOHEXAFLUOROPROPANE see FJI500

2-FLUOROIMINOHEXAFLUOROPROPANE see FJI510

4-FLUORO-3-IODOBENZOIC ACID-3-(DIBUTYLAMINO)PROPYL ESTER, HYDROCHLORIDE see HHM500

1-FLUORO-2-IODOETHANE see FIQ000

FLUOROISOPROPOXYMETHYLPHOSPHINE OXIDE see IPX000

9-α-FLUORO-16-α-17-α-ISOPROPYLEDENE DIOXY PREDNISOLONE see AQX500

9-α-FLUORO-16-α-17-α-ISOPROPYLIDENEDIOXY-Δ-1-HYDROCORTISONE see AQX500

FLUOROLON 3 see KDK000

FLUOROLON 4 see TAI250

FLUOROLUBE 2000 see KDK000

FLUOROLUBE 300/140 see KDK000

FLUOROLUBE FS 5 see KDK000

FLUOROLUBE GR 470 see KDK000

FLUOROLUBE S 30 see KDK000

FLUOROMAR see TKB250

FLUOROMETHANE see FJK000

4'-FLUORO-4-(4-(o-METHOXYPHENYL)-1-PIPERAZINYL)BUTYROPHENONE see HAF400

4'-FLUORO-4-(4-(o-METHOXYPHENYL)-1-PIPERAZINYL)BUTYROPHENONE HYDROCHLORIDE see FDA880

10-FLUORO-12-METHYLBENZ(a)ACRIDINE see MKD750

1-FLUORO-10-METHYL-7,8-BENZACRIDINE see MKD500

3-FLUORO-10-METHYL-5,6-BENZACRIDINE see MKD750

3-FLUORO-10-METHYL-7,8-BENZACRIDINE (FRENCH) see MKD250

2-FLUORO-7-METHYLBENZ(a)ANTHRACENE see FJN000

3-FLUORO-7-METHYLBENZ(a)ANTHRACENE see FJO000

6-FLUORO-7-METHYLBENZ(a)ANTHRACENE see FJP000

9-FLUORO-7-METHYLBENZ(a)ANTHRACENE see FJQ000

10-FLUORO-7-METHYLBENZ(a)ANTHRACENE see FJR000

6-FLUORO-10-METHYL-1,2-BENZANTHRACENE see FJQ000

7-FLUORO-10-METHYL-1,2-BENZANTHRACENE see FJR000

3'-FLUORO-10-METHYL-1,2-BENZANTHRACENE see FJO000

4-FLUORO-2-METHYLBENZENAMINE see FJR900

1-FLUORO-2-METHYLBENZENE see FLZ100

4-FLUORO-7-METHYL-6H-(1)BENZOTHIOPYRANO(4,3-b)QUINOLINE see FJT000

2-FLUORO-α-METHYL-4-BIPHENYLACETIC ACID see FLG100

2-FLUORO-α-METHYL(1,1'-BIPHENYL)-4-ACETIC ACID see FLG100

2-FLUORO-α-METHYL-(1,1'-BIPHENYL)-4-ACETIC ACID 1-(ACETYLOXY)ETHYL ESTER see FJT100

2-FLUORO-3-METHYLCHOLANTHRENE see FJU000

6-FLUORO-3-METHYLCHOLANTHRENE see FJV000

9-FLUORO-3-METHYLCHOLANTHRENE see FJW000

1-FLUORO-5-METHYLCHRYSENE see FJY000

6-FLUORO-5-METHYLCHRYSENE see FKA000

7-FLUORO-5-METHYLCHRYSENE see FKB000

9-FLUORO-5-METHYLCHRYSENE see FKC000

11-FLUORO-5-METHYLCHRYSENE see FKD000

12-FLUORO-5-METHYLCHRYSENE see FKE000

Δ¹-9-α-FLUORO-16-α-METHYLCORTISOL see SOW000

FLUOROMETHYL CYANIDE see FFJ000

6-α-FLUORO-16-α-METHYL-1-DEHYDROCORTICOSTERONE see FDA925

9-α-FLUORO-17-α-METHYL-11-β,17-DIHYDROXY-4-ANDROSTEN-3-ONE see AOO275

9-α-FLUORO-17-α-METHYL-17-HYDROXY-4-ANDROSTENE-3,11-DIONE see FKF100

cis-5-FLUORO-2-METHYL-1-((4-METHYLSULFINYL)PHENYL)METHYLENE))-1H-INDENE-3-ACETIC ACID see SOU550

2-FLUORO-N-METHYL-N-1-NAPHTHALENYLACETAMIDE see MME809

2-FLUORO-N-METHYL-N-1-NAPHTHYLACETAMIDE see MME809

p-FLUORO-N-METHYL-N-NITROSOANILINE see FKF800

4-FLUORO-α-METHYLPHENETHYLAMINE see FKG000

4'-FLUORO-4-(4-METHYLPIPERIDINO)BUTYROPHENONE HYDROCHLORIDE see FKI000

9-α-FLUORO-16-α-METHYLPREDNISOLONE see SOW000

9-α-FLUORO-16-β-METHYLPREDNISOLONE see BFV750

6-α-FLUORO-16-α-METHYLPREDNISOLONE-21-ACETATE see PAL600

6-α-FLUORO-16-α-METHYLPREGNA-1,4-DIENE-11-β,21-DIOL-3,20-DIONE see FDA925

6-α-FLUORO-16-α-METHYL-Δ¹,⁴-PREGNADIENE-11-β-DIOL-3,20-DIONE see FDA925

9-α-FLUORO-16-α-METHYL-1,4-PREGNADIENE-11-β,17-α,21-TRIOL-3,20-DIONE see SOW000

9-α-FLUORO-16-β-METHYL-1,4-PREGNADIENE-11-β,17-α,21-TRIOL-3,20-DIONE see BFV750

4-α-FLUORO-16-α-METHYL-11-β,17,21-TRIHYDROXYPREGNA-1,4-DIENE-3,20-DIONE see SOW000

FLUOROMETHYL(1,2,2-TRIMETHYLPROPOXY)PHOSPHINE OXIDE see SKS500

α-FLUORONAPHTHALENE see FKK000

3-(6-FLUORO-2-NAPHTHALENYL)-2,5-DIHYDRO-1,2-DIMETHYL-1H-PYRROLE, EL-(2R,3R)-2,3-DIHYDROXYBUTANEDIOATE (1:1) see FKK035

3-(6-FLUORO-2-NAPHTHALENYL)-1,2-DIMETHYL (2R,3S)-REL-3-PYRROLIDINOL HYDROCHLORIDE see FKK045

4-FLUORO-3-NITROANILINE see FKK100

4-FLUORONITROBENZENE see FKL000

p-FLUORONITROBENZENE see FKL000

1-FLUORO-4-NITROBENZENE see FKL000

2-FLUORO-5-NITRO-1,4-BENZENEDIAMINE see FKL500

N-FLUORO-N-NITROBUTYLAMINE see FKM000

FLUORONITROFEN see FKM100

3-FLUORO-4-NITROQUINOLINE-1-OXIDE see FKO000

8-FLUORO-4-NITROQUINOLINE-1-OXIDE see FKP000

FLUORONIUM PERCHLORATE see FKQ000

9-FLUORONONYL PHENYL KETONE see FKQ100

8-FLUOROOCTANOIC ACID, ETHYL ESTER see EKK500

8-FLUOROOCTYL BROMIDE see FKS000

8-FLUOROOCTYL CHLORIDE see FKT000

8-FLUOROOCTYL PHENYL KETONE see FKT050

5-FLUOROOROTATE see TCQ500

5-FLUOROOROTIC ACID see TCQ500

4'-FLUORO-4-(4-OXO-1-PHENYL-1,3,8-TRIAZASPIRO(4,5)DECAN-8-YL)-BUTYROPHENONE see SLE500

FLUOROPAK 80 see TAI250

5-FLUOROPENTYLAMINE see FFV000

5-FLUOROPENTYL THIOCYANATE see FFX000

2-FLUOROPHENOL see FKT100

4-FLUOROPHENOL see FKV000

o-FLUOROPHENOL see FKT100

FLUOROPHENOTHIAZINE DIHYDROCHLORIDE see FKW000

7-(2-(4-(4-FLUOROPHENOXY)-3-HYDROXY-1-BUTENYL)-3,5-DIHYDROXYCYCLOPENTYL)-(1-α-(Z),2-β-(1E,3S*),3-α,5-α)-5-HEPTENOIC ACID see IAD100

2-FLUORO-N-PHENYLACETAMIDE see FFH000

4'-(m-FLUOROPHENYL)ACETANILIDE see FKY000

4'-(p-FLUOROPHENYL)ACETANILIDE see FKZ000

2'-FLUORO-4'-PHENYLACETANILIDE see FKX000

4-FLUOROPHENYLACETONITRILE see FLC000

p-FLUOROPHENYLACETONITRILE see FLC000

FLUOROPHENYLALANINE see FLD100

2-FLUOROPHENYLALANINE see FLE000

3-FLUOROPHENYLALANINE see FLD000

4-FLUOROPHENYLALANINE see FLF000

m-FLUOROPHENYLALANINE see FLD000

o-FLUOROPHENYLALANINE see FLE000

p-FLUOROPHENYLALANINE see FLF000

3-(o-FLUOROPHENYL)ALANINE see FLE000

d,l-FLUOROPHENYLALANINE see FLD100

4-FLUORO-dl-PHENYLALANINE see FLD100

dl-4-FLUOROPHENYLALANINE see FLD100

d,l-p-FLUOROPHENYLALANINE see FLD100

p-FLUOROPHENYLAMINE see FFY000

p-((p-FLUOROPHENYL)AZO)-N,N-DIMETHYLANILINE see DSA000

4-((4-FLUOROPHENYL)AZO)-N,N-DIMETHYLBENZENAMINE see DSA000

p-FLUOROPHENYL ETHYL SULFONE see FLG000

α-(2-FLUOROPHENYL)-α-(4-FLUOROPHENYL)-1H-1,2,4-TRIAZOLE-1-ETHANOL see FMR200

3-FLUORO-4-PHENYLHYDRATROPIC ACID see FLG100

1-(4-FLUOROPHENYL)-4-(4-HYDROXY-4-(4-METHYLPHENYL)-1-PIPERIDINYL)-1-BUTANONE HYDROCHLORIDE see MNN250

1-(4-FLUOROPHENYL)-4-(4-HYDROXY-4-(3-(TRIFLUOROMETHYL)PHENYL)-1-PIPERIDINYL)-1-BUTANONE see TKK500

2-(4-FLUOROPHENYL)IMIDAZO(2,1-A)ISOQUINOLINE see FLG150

2-(p-FLUOROPHENYL)IMIDAZO(2,1-A)ISOQUINOLINE see FLG150

N-(9-(p-FLUOROPHENYLIMINO)FLUOREN-2-YL)ACETAMIDE see FLG200

4-(p-FLUOROPHENYL)-1-ISOPROPYL-7-METHYL-2(1H)-QUINAZOLINONE see THH350

3-FLUOROPHENYL ISOTHIOCYANATE see ISL000

4-FLUOROPHENYL ISOTHIOCYANATE see ISM000

m-FLUOROPHENYL ISOTHIOCYANATE see ISL000

p-FLUOROPHENYL ISOTHIOCYANATE see ISM000

4-FLUOROPHENYLLITHIUM see FLH100

(±)-α-(p-FLUOROPHENYL)-4-(o-METHOXYPHENYL)-1-PIPERAZINEBUTANOL see HAH000

dl-1-(4-FLUOROPHENYL)-4-(1-(4-(2-METHOXY-PHENYL))-PIPERAZINYL)BUTANOL see HAH000

4-(4-FLUOROPHENYL)-7-METHYL-1-(1-METHYLETHYL)-2(1H)-QUINAZOLINONE see THH350

1-((4-FLUOROPHENYL)METHYL)-5-OXO-N-(4-OXO-4H-1-BENZOPYRAN-2-YL)-2-PYRROLIDINECARBOXAMIDE see FLH150

1-((4-FLUOROPHENYL)METHYL)-5-OXO-l-PROLINE see FLH200

1-((4-FLUOROPHENYL)METHYL)-5-OXO-l-PROLINE AMMONIUM SALT see FLH300

1-((4-FLUOROPHENYL)METHYL)-5-OXO-l-PROLINE (2,2-DIMETHYL-1,3-DIOXOLAN-4-YL)METHYL ESTER see FLH310

1-((4-FLUOROPHENYL)METHYL)-5-OXO-l-PROLINE compounded with N-ETHYLETHANAMINE (1:1) see FLH325

1-((4-FLUOROPHENYL)METHYL)-5-OXO-l-PROLINE METHYL ESTER see FLH315

1-((4-FLUOROPHENYL)METHYL)-5-OXO-l-PROLINE(4-OXO-4H-1-BENZOPYRAN-2-YL)METHYL ESTER see FLH320

1-((4-FLUOROPHENYL)METHYL)-5-OXO-2-PYRROLIDINECARBOXAMIDE see FGI025

1-((4-FLUOROPHENYL)METHYL)-5-OXO-N-(3-(TRIFLUOROMETHYL)PHENYL)-2-PYRROLIDINE CARBOXAMIDE see FLH400

8-(4-p-FLUORO PHENYL-4-OXOBUTYL)-2-METHYL-2,8-DIAZASPIRO(4.5)DECANE-1,3-DIONE see FLJ000

8-(4-(4-FLUOROPHENYL)-4-OXOBUTYL)-1-PHENYL-1,3,8-TRIAZASPIRO(4.5)DECAN-4-ONE see SLE500

4-(4-(4-FLUOROPHENYL)-4-OXOBUTYL)-1-PIPERAZINECARBOXYLIC ACID CYCLOHEXYL ESTER see FLK000

1-(1-(4-(4-FLUOROPHENYL)-4-OXOBUTYL)-4-PIPERIDINYL)-1,3-DIHYDRO-2H-BENZIMIDAZOL-2-ONE see FLK100

1-(1-(4-(p-FLUOROPHENYL-4-OXOBUTYL)-1,2,3,6-TETRAHYDRO-4-PYRIDYL)-2-BENZIMIDAZOLINONE see DYF200

1-(4-FLUOROPHENYL)-4-(4-PHENYL-1-PIPERAZINYL)-1-BUTANONE see FLL000

4'-FLUORO-4-(1-(4-PHENYL)PIPERAZINO)BUTYROPHENONE see FLL000

3-(4-FLUOROPHENYL)-N-(4-PROPYLCYCLOHEXYL)-2-PROPENAMIDE see FLL100

1-(4-FLUOROPHENYL)-4-(4-(2-PYRIDINYL)-1-PIPERAZINYL)-1-BUTANONE see FLU000

8-FLUORO-N-(2-(4-PHENYL-2-THIAZOLYL)ETHYL)-4-QUINOLINAMINE see FLL200

FLUOROPHOSGENE see CCA500

FLUOROPHOSPHORIC ACID, anhydrous see PHJ250

4'-FLUORO-4-(4-N-PIPERIDINO-4-CARBAMIDOPIPERIDINO)BUTYROPHENONE see FHG000

p-FLUORO-γ-(4-PIPERIDINO-4-CARBAMOYLPIPERIDINO)BUTYROPHENONE see FHG000

4'-FLUORO-4-(4-PIPERIDINO-4-PROPIONYLPIPERIDINO)BUTYROPHENONE see FLN000

FLUOROPLAST 3 see CLQ750

FLUOROPLAST 3 see KDK000

FLUOROPLAST 4 see TCH500

FLUOROPLAST 3P see KDK000

FLUOROPLAST F 3 see KDK000

FLUOROPLEX see FMM000

3-FLUOROPROPENE see AGG500

2-FLUORO-2-PROPEN-1-OL see FLQ000

3-FLUOROPROPIONIC ACID see FLR000

ω-FLUOROPROPIONIC ACID see FLR000

4-FLUORO-N-(4-PROPYLCYCLOHEXYL)BENZAMIDE see FLR050

3-FLUOROPROPYL ISOCYANATE see FLR100

FLUOROPRYL see IRF000

2-FLUOROPYRIDINE see FLT100

1-(5-FLUORO-2-PYRIDYL)-6-FLUORO-7-(4-METHYL-1-PIPERAZINYL)-1,4-DIHYDRO-4-OXOQUINOLONE-3- see FLT200

4'-FLUORO-4-(4-(2-PYRIDYL)-1-PIPERAZINYL)BUTYROPHENONE see FLU000

5-FLUORO-2,4-PYRIMIDINEDIONE see FMM000

5-FLUORO-2,4(1H,3H)-PYRIMIDINEDIONE see FMM000

5-FLUORO-4(1H)-PYRIMIDINONE see FMO100

4'-FLUORO-4-(n-(4-PYRROLIDINAMIDO)-4-m-TOLYPIPERIDINO)BUTYROPHENONE see FLV000

4'-FLUORO-4-(1-PYRROLIDINYL)BUTYROPHENONE MALEATE see FGW000

6-FLUORO-7-(1-PYRROLYL)-1-ETHYL-1,4-DIHYDRO-4-OXO-3-QUINOLINECARBOXYLIC ACID see FLV050

5-FLUOROQUINOLINE see FLV060

7-FLUOROQUINOLINE see FLV070

FLUOROSILICIC ACID see SCO500

FLUOROSILICONE TRIMER see FLV100

4'-FLUORO-4-STILBENAMINE see FLY000

2-FLUORO-4-STILBENYL-N,N-DIMETHYLAMINE see FHU000

4'-FLUORO-4-STILBENYL-N,N-DIMETHYLAMINE see FHV000

FLUOROSULFONATES see FLY100

FLUOROSULFONIC ACID (DOT) see FLZ000

m-FLUOROSULFONYLBENZENESULFONYL CHLORIDE see FLY200

FLUOROSULFONYL CHLORIDE see SOT500

FLUOROSULFURIC ACID see FLZ000

FLUOROSULFURYL HYPOFLUORITE see FFB000

FLUOROTANE see HAG500

N-(2-FLUORO-1,1,2,2-TETRACHLOROETHYLTHIO)-METHANESULFOANILIDE see FLZ025

mu-FLUOROTETRAFLUORODISTANNATE(1-), SODIUM see SJA500

5-FLUORO-1,2,3,6-TETRAHYDRO-2,6-DIOXO-4-PYRIMIDINECARBOXYLIC ACID see TCQ500

5-FLUORO-1-(TETRAHYDRO-2-FURANYL)-2,4-PYRIMIDINEDIONE see FLZ050

5-FLUORO-1-(TETRAHYDRO-2-FURANYL)-2,4(1H,3H)-PYRIMIDINEDIONE see FLZ050

5-FLUORO-1-(TETRAHYDROFURAN-2-YL)URACIL see FLZ050

5-FLUORO-1-(TETRAHYDRO-3-FURYL)URACIL see FLZ050

5-FLUORO-3-(TETRAHYDRO-2-FURYL)URACIL see FLZ065

7-FLUORO-6-(3,4,5,6-TETRAHYDROPHTHALIMIDO)-4-(2-PROPYNYL)-1,4-BENZOXAZIN-3(2H)-ONE see FLZ075

(Z)-p-(1-FLUORO-2-(5,6,7,8-TETRAHYDRO-5,5,8,8-TETRAMETHYL-2-NAPHTHYL)-1-PROPENYL)BENZOIC ACID see FLZ080

9-α-FLUORO-11-β,16-α,17,21-TETRAHYDROXYPREGNA-1,4-DIENE-3,20-DIONE see AQX250

9-α-FLUORO-11-β,16-α,17,21-TETRAHYDROXY-1,4-PREGNADIENE-3,20-DIONE see AQX250

9-α-FLUORO-11-β,16-α,17-α,21-TETRAHYDROXYPREGNA-1,4-DIENE-3,20-DIONE see AQX250

6-α-FLUORO-11-β,16-α,17,21-TETRAHYDROXYPREGNA-1,4-DIENE,-3,20-DIONE, CYCLIC 16,17-ACETAL with ACETONE see FDD085

9-FLUORO-11-β,16-α,17,21-TETRAHYDROXYPREGNA-1,4-DIENE-3,20-DIONE, CYCLIC 16,17-ACETAL with 21-(3,3-DIMETHYLBUTYRATE)ACETONE see AQY375

9-FLUORO-11-β,16-α,17,21-TETRAHYDROXYPREGNA-1,4-DIENE-3,20-DIONE-16,21-DIACETATE see AQX750

FLUOROTHENE see KDK000

8-FLUORO-N-(2-(2-THIENYL)ETHYL)-4-QUINOLINAMINE see FLZ090

2-FLUOROTHIOPYRANO(4,3-b)BENZ(e)INDOLE see FGJ000

4-FLUOROTHIOPYRANO(4,3-b)BENZ(e)INDOLE see FGL000

4'-FLUORO-4-(4-(2-THIOXOBENZIMIDAZOL-1-YL)PIPERIDINO)-BUTYROPHENONE see TGB175

FLUOROTHYL see HDC000

2-FLUOROTOLUENE see FLZ100

o-FLUOROTOLUENE see FLZ100

p-FLUOROTOLUENE see FMC000

2-(5-FLUORO-o-TOLUIDINO)-2-IMIDAZOLINE HYDROCHLORIDE see FMR080

FLUOROTRIBUTYLSTANNANE see FME000

FLUOROTRICHLOROMETHANE (OSHA) see TIP500

2-FLUOROTRICYCLOQUINAZOLINE see FMF000

3-FLUOROTRICYCLOQUINAZOLINE see FMG000

4-FLUORO-4'-TRIFLUOROMETHYLBENZOPHENONE GUANYLHYDRAZONE HYDROCHLORIDE see FMH000
9-α-FLUORO-11-β,17,21-TRIHYDROXY-16-β-METHYLPREGNA-1,4-DIENE- 3,20-DIONE see BFV750
9-α-FLUORO-11-β,17-α,21-TRIHYDROXY-16-α-METHYLPREGNA-1,4-DIENE-3,20-DIONE see SOW000
9-FLUORO-11-β,17,21-TRIHYDROXY-16-β-METHYLPREGNA-1,4-DIENE-3,20-DIONE see BFV750
9-FLUORO-11-β,17,21-TRIHYDROXY-16-α-METHYLPREGNA-1,4-DIENE-3,20-DIONE see SOW000
9-FLUORO-11-β,17,21-TRIHYDROXY-16-α-METHYLPREGNA-1,4-DIENE-3,20-DIONE ACETATE see DBC400
9-FLUORO-11-β,17,21-TRIHYDROXY-16-β-METHYLPREGNA-1,4-DIENE-3,20-DIONE, 21-(DIHYDROGEN PHOSPHATE), DISODIUM SALT see BFV770
9-FLUORO-11-β,17,21-TRIHYDROXY-16-α-METHYLPREGNA-1,4-DIENE-3,20-DIONE-21-(DIHYDROGEN PHOSPHATE) DISODIUM SALT see DAE525
9-FLUORO-11-β,17,21-TRIHYDROXY-16-β-METHYLPREGNA-1,4-DIENE-3,20-DIONE, 17,21-DIPROPIONATE see BFV765
9-FLUORO-11-β,17,21-TRIHYDROXY-16-α-METHYLPREGNA-1,4-DIENE-3,20-DIONE-17,21-DIPROPIONATE see DBC500
9-FLUORO-11-β,17,21-TRIHYDROXY-16-α-METHYLPREGNA-1,4-DIENE-3,20-DIONE-21-(HYDROGEN SULFATE), MONOSODIUM SALT see DBC550
9-FLUORO-11-β,17,21-TRIHYDROXY-16-α-METHYLPREGNA-1,4-DIENE-3,20-DIONE, 21-ISONICOTINATE see DBC510
9-FLUORO-11-β,17,21-TRIHYDROXY-16-α-METHYLPREGNA-1,4-DIENE-3,20-DIONE-21-PALMITATE see DBC525
9-FLUORO-11-β,17,21-TRIHYDROXY-16-α-METHYLPREGNA-1,4-DIENE-3,20-DIONE-17-VALERATE see DBC575
9-FLUORO-11-β,17,21-TRIHYDROXY-16-β-METHYLPREGNA-1,4-DIENE-3,20,DIONE-17-VALERATE see VCA000
9-α-FLUORO-11-β,17-α,21-TRIHYDROXY-4-PREGNENE-3,20-DIONE see FHH100
9-FLUORO-11-β,17,21-TRIHYDROXYPREGN-4-ENE-3,20-DIONE see FHH100
FLUOROTRINITROMETHANE see FMI000
3-FLUORO-1,2,4-TRIOXOLANE see FIK875
FLUOROTRIPHENYLSTANNANE see TMV850
FLUOROTROJCHLOROMETAN (POLISH) see TIP500
3-FLUOROTYROSIN see FMJ000
3-FLUOROTYROSINE see FMJ000
m-FLUOROTYROSINE see FMJ000
FLUOROURACIL see FMM000
5-FLUOROURACIL see FMM000
5-FLUOROURACIL DEOXYRIBOSIDE see DAR400
5-FLUOROURACIL-2'-DEOXYRIBOSIDE see DAR400
5-FLUOROURIDINE see FMN000
5-FLUOROVALERONITRILE see FMO000
FLUOROWODOR (POLISH) see HHU500
FLUOROXENE see TKB250
4-FLUORPHENYL-1-ISOPROPYL-7-METHYL-2(1H)-CHINAZOLINON (GERMAN) see THH350
FLUORPLAST 4 see TAI250
5-FLUORPROPYRIMIDINE-2,4-DIONE see FMM000
FLUORSPAR see CAS000
FLUORTENSID FT 248 see TCD100
FLUORTENSID FT 719 see PCH320

FLUORTHYRIN see FMJ000
3-FLUORTYROSIN (GERMAN) see FMJ000
5-FLUORURACIL (GERMAN) see FMM000
FLUORURE de BORE (FRENCH) see BMG700
FLUORURE de N,N'-DIISOPROPYLE PHOSPHORODIAMIDE (FRENCH) see PHF750
FLUORURE de POTASSIUM (FRENCH) see PLF500
FLUORURES ACIDE (FRENCH) see FEZ000
FLUORURE de SODIUM (FRENCH) see SHF500
FLUORURE de SULFURYLE (FRENCH) see SOU500
FLUORURE de N,N,N',N'-TETRAMETHYLE PHOSPHORO-DIAMIDE (FRENCH) see BJE750
FLUORURE de THIONYLE (FRENCH) see TFL250
FLUORURI ACIDI (ITALIAN) see FEZ000
FLUORURIDINE DEOXYRIBOSE see DAR400
FLUORWASSERSTOFF (GERMAN) see HHU500
FLUORWATERSTOF (DUTCH) see HHU500
FLUORXENE see TKB250
FLUORYL see CBY750
FLUOSILICATE de ALUMINUM (FRENCH) see THH000
FLUOSILICATE de AMMONIUM (FRENCH) see COE000
FLUOSILICATE de MAGNESIUM (FRENCH) see MAG250
FLUOSILICATE de SODIUM see DXE000
FLUOSILICIC ACID see SCO500
FLUOSOL 43 see HAS000
FLUOSOL-DA 20% see FMO050
FLUOSTIGMINE see IRF000
FLUOSULFONIC ACID (DOT) see FLZ000
FLUOTESTIN see AOO275
FLUOTHANE see HAG500
FLUOTITANATE de POTASSIUM (FRENCH) see PLI000
FLUOTRACEN see DMF800
FLUOVITIF see SPD500
FLUOXIDINE see FMO100
FLUOXIMESTERONE see AOO275
FLUOXYDINE see FMO100
FLUOXYMESTERONE see AOO275
FLUOXYMESTRONE see AOO275
FLUOXYPREDNISOLONE see AQX250
FLUPENTHIXOL see FMO129
cis-(Z)-FLUPENTHIXOL see FMO129
(α,β)-FLUPENTHIXOL see FMO129
FLUPENTHIXOLE see FMO129
FLUPENTIHXOL HYDROCHLORIDE see FMO150
FLUPENTIXOL see FMO129
FLUPENTIXOL DIHYDROCHLORIDE see FMO150
FLUPHENAMIC ACID see TKH750
FLUPHENAZINE see TJW500
FLUPHENAZINE DIHYDROCHLORIDE see FMP000
FLUPHENAZINE ENANTHATE see PMI250
FLUPHENAZINE HYDROCHLORIDE see FMP000
FLUPIRTINE MALEATE see FMP100
FLUPIRTIN-MALEAT (GERMAN) see FMP100
FLUPROQUAZONE see THH350
FLUPROSTENOL see ECW550
FLURACIL see FMM000
FLURA-GEL see SHF500
FLURAZEPAM see FMQ000
FLURAZEPAM DIHYDROCHLORIDE see DAB800
FLURAZEPAM HYDROCHLORIDE see DAB800
FLURAZEPAM MONOHYDROCHLORIDE see FMQ100

FLURAZOLE see BEG300
FLURBIPROFEN see FLG100
FLURBIPROFEN AXETIL see FJT100
FLURCARE see SHF500
FLURENTIXOL see FMO129
FLURI see FMM000
FLURIDONE see FMQ200
FLURIL see FMM000
4'-FLURO-4-(1,2,4,4a,5,6-HEXAHYDRO-3H-PYRANZINO(1,2-A)QUINOLIN-3-YL)-BUTYROPHENONE 2HCl see CNH500
FLUROTHYL see HDC000
FLUROXENE see TKB250
FLURPRIMIDOL see IRN500
FLURTAMONE see MGJ775
FLUSILAZOL see THS860
FLUSILAZOLE see THS860
FLUSTERON see AOO275
FLUTAMIDE see FMR050
FLUTAZOLAM see FMR075
FLUTESTOS see AOO275
FLUTOLANIL see TKB286
FLUTONE see AQX500
FLUTONIDINE HYDROCHLORIDE see FMR080
FLUTOPRAZEPAM see FMR100
FLUTRA see HII500
FLUTRIAFEN see FMR200
FLUTRIAFOL see FMR200
FLUTROPIUM BROMIDE HYDRATE see FMR300
FLUVALINATE see VBU100
FLUVET see FDD075
FLUVIN see CFY000
FLUVOMYCIN see FMR500
FLUVOXAMINE see FMR550
FLUVOXAMINE MALEATE see FMR575
FLUXANXOL see FMO129
FLUXEMA see POD750
FLUX MAAG see NDN000
FLUXOFENIM see FMR600
FLUZILAZOL see THS860
FLY BAIT GRITS see SOY000
FLY-DIE see DGP900
FLY FIGHTER see DGP900
FLYPEL see DKC800
13FM see KDK000
FM 100 see LFN000
FM 510 see PJS750
FMA see ABU500
FMA see FEV100
FMA (analytical reagent) see FEV100
FMC 249 see AFR250
FMC-1240 see EEH600
FMC 5273 see PIX250
FMC 5462 see EAQ750
FMC 5488 see CKM000
FMC 9044 see BGB500
FMC-9102 see MQQ250
FMC 10242 see CBS275
FMC 11092 see DUM800
FMC-16388 see PNV750
FMC 17370 see BEP500
FMC 30980 see RLF350
FMC 31768 see MRU050
FMC 33297 see AHJ750
FMC 35001 see CCC280
FMC 35171 see PCK060
FMC 41655 see AHJ750
FMC 45497 see RLF350
FMC 45806 see RLF350
FMC 54800 see TAC850
FMC 57020 see CEP800
FMC 58000 see TAC850
FMC 67825 see EHY100
(+−)-cis-FMC 33297 see PCK060
FM-NTS see CCU250
F-6-NDBA see NJN300
F-NORSTEARANTHRENE see TCJ500
FNT see NDY500
FO see TCQ500
FOA see TCQ500

FOBEX see BCA000
FOCUSAN see TGB475
5-FODURD see FNK033
FOE 1976 see BDG100
FOGARD see FMR700
FOGARD S see FMR700
FOIN ABSOLUTE see FMS000
FOIN COUPE see FMS000
FOLACIN see FMT000
FOLAN RED B see CMM330
FOLAN YELLOW G see CMM320
FOLATE see FMT000
FOLBEX see DER000
FOLBEX SMOKE-STRIPS see DER000
FOLCID see CBF800
FOLCODAL see CMR100
FOLCODINE see TCY750
FOLCYSTEINE see FMT000
FOLEDRIN see FMS875
FOLETHION see DSQ000
FOLEX see TIG250
FOLIANDRIN see OHQ000
FOLIC ACID see FMT000
FOLIC ACID, 4-AMINO- see AMG750
FOLIDOL see PAK000
FOLIDOL M see MNH000
FOLIGAN see ZVJ000
FOLIKRIN see EDV000
FOLIMAT see DNX800
FOLINERIN see OHQ000
FOLINEVIN see OHQ000
FOLIONE see MND275
FOLIPEX see EDV000
FOLISAN see EDV000
FOLKS GLOVE see FOM100
FOLLESTRINE see EDV000
FOLLICLE-STIMULATING HORMONE see
FMT100
FOLLICORMON see EDP000
FOLLICULAR HORMONE see EDV000
FOLLICULAR HORMONE HYDRATE see
EDU500
FOLLICULIN see EDV000
FOLLICULINE BENZOATE see EDV000
FOLLICUNODIS see EDV000
FOLLICYCLIN P see EDR000
FOLLIDIENE see DAL600
FOLLIDIENE see DKA600
FOLLIDRIN see EDP000
FOLLIDRIN see EDV000
FOLLINYL see NNL500
FOLLITROPIN see FMT100
FOLLORMON see DAL600
FOLLUTEIN see CMG675
FOLOSAN see PAX000
FOLOSAN see TBR750
FOLOSAN DB-905 FUMITE see TBS000
FOLPAN see TIT250
FOLPET see TIT250
FOLSAN see TBS000
FOMAC see HCL000
FOMAC 2 see PAX000
FOM-Ca HYDRATE see CAW376
FOMESAFEN see CLS050
FOMESAFENE see CLS050
FOMINOBEN HYDROCHLORIDE see
FMU000
FOM-Na see DXF600
FOMOCAINE see PDU250
p-FOMOCAINE see PDU250
FOMREZ SUL-3 see DBF800
FOMREZ SUL-4 see DDV600
FONATOL see DKA600
FONAZINE MESYLATE see FMU039
FONDAREN see DRP600
FONOFOS see FMU045
FONOLINE see MQV750
FONTARSAN see ARL000
FONTARSOL see DFX400
FONTEGO see BON325
FONTILIX see MQQ050
FONTILIZ see MQQ050

FONURIT see AAI250
FONZYLANE see BOM600
FOOD BLUE 1 see FMU059
FOOD BLUE 2 see FAE000
FOOD BLUE 3 see ADE500
FOOD BLUE 4 see IBV050
FOOD BLUE DYE No. 1 see FAE000
FOODCOL SUNSET YELLOW FCF see
FAG150
FOOD DYE RED No. 104 see ADG250
FOOD DYE RED No. 104 see CMM000
FOOD GREEN 1 see FAE950
FOOD GREEN 2 see FAF000
FOOD GREEN S see ADF000
FOOD RED 2 see CMP620
FOOD RED 2 see FAG020
FOOD RED 4 see FAG050
FOOD RED 5 see HJF500
FOOD RED 6 see FMU080
FOOD RED 7 see FMU080
FOOD RED 9 see FAG020
FOOD RED 14 see FAG040
FOOD RED 15 see FAG070
FOOD RED 16 see CMS238
FOOD RED COLOR No. 105, SODIUM SALT
see RMP175
FOOD RED No. 101 see FMU070
FOOD RED No. 102 see FMU080
FOOD RED No. 104 see ADG250
FOOD RED No. 105, SODIUM SALT see
RMP175
FOOD STARCH, MODIFIED see FMU100
FOOD YELLOW 2 see CMM758
FOOD YELLOW 3 see FAG150
FOOD YELLOW 4 see FAG140
FOOD YELLOW 5 see FAG140
FOOD YELLOW 6 see FAG150
FOOD YELLOW 13 see CMM510
FOOD YELLOW NO. 4 see FAG140
FOOL'S CICELY see FMU200
FOOL'S PARSLEY see FMU200
FOPIRTOLINA (SPANISH) see FMU225
FOPIRTOLINE HYDROCHLORIDE see
FMU225
FORAAT (DUTCH) see PGS000
FORALAMIN FUMARATE see MDP800
FORALMINE FUMARATE see MDP800
FORANE see IKS400
FORANE 22 see CFX500
FORAPIN see MCB525
FORBEL 750 see MRQ300
FORCE see FMU300
FORDIURAN see BON325
FORE see DXI400
FOREDEX 75 see DAA800
FORENOL see NDX500
FORHISTAL MALEATE see FMU409
FORIOD see TDE750
FORIT see ECW600
FORLEX see MLC000
FORLIN see BBQ500
FORMAGENE see PAI000
FORMAL see MAK700
FORMAL see MGA850
FORMALDEHYD (CZECH, POLISH) see
FMV000
FORMALDEHYDE see FMV000
FORMALDEHYDE, solution (DOT) see
FMV000
FORMALDEHYDE-ANILINE COPOLYMER
see FMW330
FORMALDEHYDE BIS(β-CHLOROETHYL)
ACETAL see BID750
FORMALDEHYDE BIS(2-FLUORO-2,2-
DINITROETHYL) ACETAL see FMV050
FORMALDEHYDE COPOLYMER with
UREA see UTU500
FORMALDEHYDE CYANOHYDRIN see
HIM500
FORMALDEHYDE CYCLODODECYL
ETHYL ACETAL see FMV100

FORMALDEHYDE CYCLODODECYL
METHYL ACETAL see FMV200
FORMALDEHYDE DIMETHYLACETAL see
MGA850
FORMALDEHYDE,
DIMETHYLHYDRAZONE see FMV300
FORMALDEHYDE 2,2-
DIMETHYLHYDRAZONE see FMV300
FORMALDEHYDE-ETHYLENEDIAMINE-
PHENOL COPOLYMER see EEA550
FORMALDEHYDE HYDROSULFITE see
FMW000
FORMALDEHYDE-MELAMINE
CONDENSATE see MCB050
FORMALDEHYDE-MELAMINE
COPOLYMER see MCB050
FORMALDEHYDE-MELAMINE POLYMER
see MCB050
FORMALDEHYDE-MELAMINE RESIN see
MCB050
FORMALDEHYDE OXIDE POLYMER see
FMW300
FORMALDEHYDE, OXIME see FMW400
FORMALDEHYDE, POLYMER with
BENZENAMINE see FMW330
FORMALDEHYDE, POLYMER WITH
(CHLOROMETHYL)OXIRANE AND
PHENOL see FMW333
FORMALDEHYDE, POLYMER WITH 1,2-
ETHANEDIAMINE ANDPHENOL see
EEA550
FORMALDEHYDE, POLYMER WITH 4-
NONYLPHENOL AND OXIRANE see
FMW343
FORMALDEHYDE, SELENO- see SBU650
FORMALDEHYDE SODIUM BISULFITE
ADDUCT see FMW000
FORMALDEHYDE SODIUM
SULFOXYLATE see FMW000
FORMALDEHYDESULFOXYLIC ACID
SODIUM SALT see FMW000
FORMALDEHYDE-UREA CONDENSATE
see UTU500
FORMALDEHYDE-UREA COPOLYMER see
UTU500
FORMALDEHYDE-UREA POLYMER see
UTU500
FORMALDEHYDE-UREA
PRECONDENSATE see UTU500
FORMALDEHYDE-UREA PREPOLYMER
see UTU500
FORMALDEHYDE-UREA RESIN see
UTU500
FORMALDOXIME see FMW400
FORMAL FAST BLACK 2B see CMN240
FORMAL GLYCOL see DVR800
FORMAL HYDRAZINE see FNN000
FORMALIN see FMV000
FORMALIN 40 see FMV000
FORMALIN (DOT) see FMV000
FORMALINA (ITALIAN) see FMV000
FORMALINE (GERMAN) see FMV000
FORMALINE BLACK C see AQP000
FORMALIN-LOESUNGEN (GERMAN) see
FMV000
FORMALIN-MELAMINE COPOLYMER see
MCB050
FORMALIN-UREA COPOLYMER see
UTU500
FORMALITH see FMV000
FORMAL-γ-TRIMETHYLAMMONIUM
PROPANEDIOL see FMX000
FORMAMIDE see FMY000
FORMAMIDE, N-(1,1'-BIPHENYL)-4-YL-N-
HYDROXY- see HLE650
FORMAMIDE, N-(4-CHLORO-2-
METHYLPHENYL)- see CHJ700
FORMAMIDE, 1-(((4-
CHLOROPHENYL)METHYL)SULFINYL)-
N,N-DIETHYL- see FMY050
FORMAMIDE, N,N'-(DITHIOBIS(2-(2-
HYDROXYETHYL)-1-

METHYLVINYLENE))BIS(N-((4-AMINO-2-METHYL-5-PYRIMIDINYL))METHYL)- see TES800
FORMAMIDE, N-ETHYL- see EKK600
FORMAMIDE, N,N'-(1,4-PIPERAZINEDIYLBIS(2,2,2-TRICHLOROETHYLIDENE))BIS-(8CI,9CI) see TKL100
FORMAMIDE, N-(2,2,2-TRICHLORO-1-(MORPHOLINYL)ETHYL)- see FMY100
FORMAMIDOBENZENE see FNJ000
N-(1-FORMAMIDO-2,2,2-TRICHLOROETHYL)MORFOLIN see FMY100
FORMAMINE see HEI500
FORMANILIDE see FNJ000
FORMARIN see MNM500
FORMATRIX see ECU750
3-FORMAZANTHIOL, 1,5-DIPHENYL- see DWN200
FORMEBOLONE see FNK040
FORMETANATE HYDROCHLORIDE see DSO200
FORMHYDRAZID (GERMAN) see FNN000
FORMHYDRAZIDE see FNN000
FORMHYDROXAMIC ACID see FMZ000
FORMHYDROXAMSAEURE (GERMAN) see FMZ000
FORMIATE de METHYLE (FRENCH) see MKG750
FORMIATE de PROPYLE (FRENCH) see PNM500
FORMIC ACID see FNA000
FORMIC ACID, ALLYL ESTER see AGH000
FORMIC ACID AMMONIUM SALT see ANH500
FORMIC ACID, AZIDO-, tert-BUTYL ESTER see BQI250
FORMIC ACID, AZIDODITHIO- see ASF750
FORMIC ACID, 1-BUTYLHYDRAZIDE see BRK100
FORMIC ACID, CALCIUM SALT see CAS250
FORMIC ACID, CHLORO-, 2-CHLOROETHYL ESTER see CGU199
FORMIC ACID, CHLORO-, FLUOREN-9-YLMETHYL ESTER see FEI100
FORMIC ACID, CHLORO-, 2-FLUOROETHYL ESTER see FIH100
FORMIC ACID, CHLORO-, OXYDIETHYLENE ESTER see OPO000
FORMIC ACID, CHLORO-, TRICHLOROMETHYL ESTER see TIR920
FORMIC ACID, CHLORO-, 2,4,6-TRICHLOROPHENYL ESTER see TTY800
FORMIC ACID, CINNAMYL ESTER see CMR500
FORMIC ACID, CITRONELLYL ESTER see CMT750
FORMIC ACID-3,7-DIMETHYL-6-OCTEN-1-YL ESTER see CMT750
FORMIC ACID, DITHIOBIS(THIO-, O,O-DIBUTYL ESTER see BSS550
FORMIC ACID, DITHIOBIS(THIO-, o,o-DIISOPROPYL ESTER see IRS500
FORMIC ACID, ETHYL ESTER see EKL000
FORMIC ACID, 1-ETHYLHYDRAZIDE see EKL250
FORMIC ACID, GERANIOL ESTER see GCY000
FORMIC ACID, HEPTYL ESTER see HBO500
FORMIC ACID, HEXYL ESTER see HFQ100
FORMIC ACID, HYDRAZIDE see FNN000
FORMIC ACID, ISOBUTYL ESTER see IIR000
FORMIC ACID, ISOPENTYL ESTER see IHS000
FORMIC ACID, ISOPROPYL ESTER see IPC000
FORMIC ACID, METHYLHYDRAZIDE see FNW000

FORMIC ACID, METHYLPENTYLIDENEHYDRAZIDE see PBJ875
FORMIC ACID (2-(4-METHYL-2-THIAZOLYL))HYDRAZIDE see FNB000
FORMIC ACID, NERYL ESTER see FNC000
FORMIC ACID, OCTYL ESTER see OEY100
FORMIC ACID, (PHENYLAZO)-, 2-PHENYLHYDRAZIDE see DVY950
FORMIC ACID, 1-PROPYLHYDRAZIDE see PNM650
FORMIC ACID, compounded with QUININE (1:1) see QIS300
FORMIC ALDEHYDE see FMV000
FORMIC ANAMMONIDE see HHS000
FORMIC BLACK BA see AQP000
FORMIC BLACK C see AQP000
FORMIC BLACK CW see AQP000
FORMIC BLACK MTG see AQP000
FORMIC BLACK TG see AQP000
FORMIC ETHER see EKL000
FORMIC HYDRAZIDE see FNN000
FORMIC 2-(4-(5-NITROFURYL)-2-THIAZOLYL)HYDRAZIDE see NDY500
FORMILOXIN see FND100
FORMILOXIN see PBG200
FORMILOXINE see FND100
FORMILOXINE see PBG200
FORMIN see HEI500
FORMIN K-K see MCB050
FORMOCARBAM see FNE000
FORMOCORTAL see FDB000
FORMOHYDRAZIDE see FNN000
FORMOL see FMV000
FORMOLA 40 see DAA800
FORMOMALENIC THALLIUM see TEM399
FORMONITRILE see HHS000
FORMOPAN see FMW000
FORMOSA CAMPHOR see CBA750
FORMOSA CAMPHOR OIL see CBB500
FORMOSE OIL OF CAMPHOR see CBB500
FORMOSULFACETAMIDE see SNQ710
FORMOSULFATHIAZOLE see TEX250
FORMOTEROL FUMARATE DIHYDRATE see FNE100
FORMOTHION see DRR200
o-FORMOTOLUIDIDE, 4'-CHLORO- see CHJ700
FORMOXIME see FMW400
FORMPARANATE see FNE500
FORMULA 40 see DFY800
FORMVAR 1285 see AAX250
4'-FORMYLACETANILIDE THIOSEMICARBAZONE see FNF000
2-FORMYLAMINOFLUORENE see FEF000
2-FORMYLAMINO-4-(5-NITRO-2-FURYL)THIAZOLE see NGM500
2-FORMYLAMINO-4-((2-5-NITRO-2-FURYL)VINYL)-1,3-THIAZOLE see FNG000
FORMYLANILINE see FNJ000
N-FORMYLANILINE see FNJ000
6-FORMYLANTHANTHRENE see FNK000
7-FORMYLBENZ(c)ACRIDINE see BAX250
4-FORMYLBENZALDEHYDE see TAN500
p-FORMYLBENZALDEHYDE see TAN500
7-FORMYLBENZO(c)ACRIDINE see BAX250
2-FORMYLBENZOIC ACID see FNK010
o-FORMYLBENZOIC ACID see FNK010
4-FORMYLBENZONITRILE see COK250
p-FORMYLBENZONITRILE see COK250
6-FORMYLBENZO(a)PYRENE see BCT250
2-FORMYL-3:4-BENZPHENANTHRENE see BCS000
O-FORMYLCEFAMANDOLE SODIUM see FOD000
4-FORMYLCYCLOHEXENE see FNK025
5-FORMYL-2'-DEOXYURIDINE see FNK033
5-FORMYL-1,2:3,4-DIBENZOPYRENE see DCY800
5-FORMYL-3,4:8,9-DIBENZOPYRENE see DCY600

5-FORMYL-3,4:9,10-DIBENZOPYRENE see BCQ750
N-FORMYL-N'-(3',4'-DICHLORPHENYL)-2,2,2-TRICHLORACETALDEHYDAM (GERMAN) see CDP750
FORMYLDIENOLONE see FNK040
3-FORMYL-DIGITOXIGENIN see FNK050
3-12-FORMYL-DIGOXIGENIN see FNK075
2-FORMYL-3,4-DIHYDRO-2H-PYRAN see ADR500
N-FORMYLDIMETHYLAMINE see DSB000
3'-FORMYL-N,N-DIMETHYL-4-AMINOAZOBENZENE see FNK100
p-FORMYLDIMETHYLANILINE see DOT400
2-FORMYL-6,6-DIMETHYLBICYCLO(3.1.1)HEPT-2-ENE see FNK150
N-FORMYLETHYLAMINE see EKK600
α-FORMYLETHYLBENZENE see COF000
FORMYLETHYLTETRAMETHYLTETRALIN see FNK200
N-FORMYL-N-2-FLUORENYLHYDROXYLAMINE see FEG000
N-FORMYLFORMAMIDE see FNL000
16-FORMYL-GITOXIN see GES100
5-FORMYLGUAIACOL see FNN000
6-FORMYLGUAIACOL see VFP000
FORMYLHYDRAZIDE see FNN000
FORMYLHYDRAZINE see FNN000
N-FORMYLHYDRAZINE see FNN000
2-(2-FORMYLHYDRAZINO)-4-(5-NITRO-2-FURYL)THIAZOLE see NDY500
N-FORMYL HYDROXYAMINOACETIC ACID see FNO000
(R,R)-α-(1-FORMYL-2-HYDROXYETHOXY)-1,6-DIHYDRO-6-OXO-9H-PURINE-9-ACETALDEHYDE see IDE050
N-FORMYL-N-HYDROXYGLYCINE see FNO000
N-FORMYLHYDROXYLAMINE see FMZ000
2-FORMYL-11-α-HYDROXY-Δ¹-METHYLTESTOSTERONE see FNK040
FORMYLIC ACID see FNA000
3-FORMYLINDOLE see FNO100
1-FORMYLISOQUINOLINE THIOSEMICARBAZONE see IRV300
N-FORMYLJERVINE see FNP000
S-(2-(FORMYLMETHYLAMINO)-2-OXOETHYL)-O,O-DIMETHYLPHOSPHORODITHIOATE see DRR200
2-FORMYL-17-α-METHYLANDROSTA-1,4-DIENE-11-α,17-β-DIOL-3-ONE see FNK040
6-FORMYL-12-METHYLANTHANTHRENE see FNQ000
7-FORMYL-9-METHYLBENZ(c)ACRIDINE see FNR000
7-FORMYL-11-METHYLBENZ(c)ACRIDINE see FNS000
12-FORMYL-7-METHYLBENZ(a)ANTHRACENE see MGY500
7-FORMYL-12-METHYLBENZ(a)ANTHRACENE see FNT000
N-FORMYL-N-METHYLCARBAMOYLMETHYL-O,O-DIMETHYL PHOSPHORODITHIOATE see DRR200
S-(N-FORMYL-N-METHYLCARBAMOYLMETHYL)-O,O-DIMETHYL PHOSPHORODITHIOATE see DRR200
S-(N-FORMYL-N-METHYLCARBAMOYLMETHYL) DIMETHYL PHOSPHOROTHIOLOTHIONATE see DRR200
5-FORMYL-10-METHYL-3,4:8,9-DIBENZOPYRENE see FNV000

5-FORMYL-8-METHYL-3,4:9,10-DIBENZOPYRENE see FNU000
2-FORMYL-5-METHYLFURAN see MKH550
1-FORMYL-1-METHYLHYDRAZINE see FNW000
N-FORMYL-N-METHYLHYDRAZINE see FNW000
1-FORMYL-1-METHYL-4-(4-METHYLPENTYL)-3-CYCLOHEXENE see VIZ150
2-FORMYL-2'-METHYL-1,1'-(OXYDIMETHYLENE)DIPYRIDINIUM, DICHLORIDE OXIME see FNW100
4-FORMYL-4'-METHYL-1,1'-(OXYDIMETHYLENE)DIPYRIDINIUM, DICHLORIDE OXIME see MBZ000
N-FORMYL-N-METHYL-p-(PHENYLAZO)ANILINE see FNX000
2-FORMYL-1-METHYLPYRIDINIUM CHLORIDE OXIME see FNZ000
2-FORMYL-1-METHYLPYRIDINIUM IODIDE OXIME see POS750
2-FORMYL-1-METHYLPYRIDINIUM METHANESULFONATE OXIME see PLX250
2-FORMYL-N-METHYLPYRIDINIUM OXIME CHLORIDE see FNZ000
2-FORMYL-N-METHYLPYRIDINIUM OXIME IODIDE see POS750
2-FORMYL-N-METHYLPYRIDINIUM OXIME METHANESULFONATE see PLX250
2-FORMYL-1-METHYL-PYRIDINIUM PERCHLORATE OXIME see FOA000
4-FORMYLMONOMETHYLAMINOAZOBENZENE see FNX000
N-FORMYLMORFOLIN see FOA100
4-FORMYLMORPHOLINE see FOA100
N-FORMYLMORPHOLINE see FOA100
1-FORMYLNAPHTHALENE see NAJ000
3-FORMYLNITROBENZENE see NEV000
p-FORMYLNITROBENZENE see NEV500
4A-FORMYL-1,4,4A,5A,6,9,9A,9B-OCTAHYDRODIBENZOFURAN see BHJ500
(6R-(6-α,7-β(R)))-7-(((FORMYLOXY)PHENYLACETYL)AMINO)-3-(((1-METHYL-1H-TETRAZOL-5-YL)THIO)METHYL)-8-OXO-5-THIA-1-AZABICYCLO(4.2.0)OCT-2-ENE-2-CARBOXYLIC ACID MONOSODIUM SALT see FOD000
FORMYLOXYTRIBENZYLSTANNANE see FOE000
(FORMYLOXY)TRIS(PHENYLMETHYL)STANNANE see FOE000
2-FORMYLPHENOL see SAG000
3-FORMYLPHENOL see FOE100
4-FORMYLPHENOL see FOF000
m-FORMYLPHENOL see FOE100
o-FORMYLPHENOL see SAG000
p-FORMYLPHENOL see FOF000
N-FORMYLPIPERIDIN (GERMAN) see FOH000
1-FORMYLPIPERIDINE see FOH000
3-FORMYLPYRIDINE see NDM100
4-FORMYLPYRIDINE see IKZ100
2-FORMYLPYRIDINE THIOSEMICARBAZONE see PIB925
4-FORMYLPYRIDINE THIOSEMICARBAZONE see ILB000
2-FORMYLQUINOXALINE-1,4-DIOXIDE CARBOMETHOXYHYDRAZONE see FOI000
4-FORMYLRESORCINOL see REF100
N-FORMYL-l-p-SARCOLYSIN see BHX250
3-FORMYLTETRAHYDROFURAN see FOI100
2-FORMYLTHIOPHENE see TFM500
α-FORMYLTHIOPHENE see TFM500
1-FORMYL-3-THIOSEMICARBAZIDE see FOJ000
p-FORMYLTOLUENE see FOJ025
FORMYL TRICHLORIDE see CHJ500

(Z)-3-FORMYL-2,4,4-TRICHLORO-2-BUTENOIC ACID see FOJ033
4-FORMYL-1-(3-(TRIETHYLAMMONIO)PROPYL)PYRIDINIUM DIBROMIDE OXIME see FOJ050
8-FORMYL-1,6,7-TRIHYDROXY-5-ISOPROPYL-3-METHYL-2,2'-BISNAPHTHALENE see GJM000
1-FORMYL-3,5,6-TRIMETHYL-3-CYCLOHEXENE and 1-FORMYL-2,4,6-TRIMETHYL-3-CYCLOHEXENE see IKJ000
5-FORMYLURACIL see FOJ060
FORMYL VIOLET S4BN see FAG120
FOROMACIDIN see SLC000
FORON BLUE SBGL see CMP075
FORON BRILLIANT RED E 2BL see AKI750
FORON YELLOW SE-FL see KDA075
FOROTOX see TIQ250
FORPEN see BFD000
FORPHENICINOL see FOJ100
FORRON see TAA100
FORSTAN see ORU000
FORST U 46 see TAA100
FOR-SYN see BEP500
FORTALGESIC see DOQ400
FORTALIN see DOQ400
FORTAM see CCQ200
FORTAZ see CCQ200
FORTECORTIN see SOW000
FORTEX see TAA100
FORTHION see MAK700
FORTIFICAR see SFX000
FORTIFLEX 6015 see PJS750
FORTIFLEX A 60/500 see PJS750
FORTIGRO see FOI000
FORTIMICIN A see FOK000
FORTIMICIN A SULFATE see FOL000
FORTION NM see DSP400
FORTODYL see VSZ100
FORTOMBRINE-N see AAN000
FORTOVASE see FOL025
FORTRACIN see BAC250
FORTRACIN (BACITRACIN-MD) see BAC260
FORTRAL see DOQ400
FORTRESS see FOL050
FORTROL see BLW750
FORTURF see TBQ750
FORZA see FMU300
FOSAMAX see SKL600
FOSAMINE AMMONIUM see ANG750
FOSAZEPAM see FOL100
FOSCHLOR see TIQ250
FOSCHLOREM (POLISH) see TIQ250
FOSCHLOR R-50 see TIQ250
FOSCOLIC ACID see PGW600
FOSDRIN see MQR750
FOSETYL-AL see AHH800
FOSETYL ALUMINUM see AHH800
FOSFAKOL see NIM500
FOS-FALL "A" see BSH250
FOSFAMID see DSP400
FOSFAMIDON see FAB400
FOSFAMIDONE see FAB400
FOSFAZIDE see ILE000
FOSFERMO see PAK000
FOSFESTROL see DKA200
FOSFESTROL TETRASODIUM see TEE300
FOSFEX see PAK000
FOSFIVE see PAK000
FOSFOCINA see PHA550
FOSFOMOLYBDENAN SODNY see SIM200
FOSFOMYCIN see PHA550
FOSFOMYCIN CALCIUM HYDRATE see CAW376
FOSFOMYCIN-Ca HYDRATE see CAW376
FOSFOMYCIN DISODIUM see DXF600
FOSFOMYCIN DISODIUM HYDRATE see FOL200
FOSFOMYCIN DISODIUM SALT see DXF600
FOSFOMYCIN SODIUM HYDRATE see FOL200

FOSFOMYCIN SODIUM SALT see DXF600
FOSFON D see THY500
FOSFONO 50 see EEH600
FOSFONOMYCIN see PHA550
FOSFORAN TROJ-(1,3-DWUCHLOROIZOPROPYLOWY) (POLISH) see FQU875
FOSFORO BIANCO (ITALIAN) see PHP010
FOSFORO(PENTACHLORURO di) (ITALIAN) see PHR500
FOSFORO(TRICLORURO di) (ITALIAN) see PHT275
FOSFOROWODOR (POLISH) see PGY000
FOSFOROXYCHLORID see PHQ800
FOSFORPENTACHLORIDE (DUTCH) see PHR500
FOSFORTHIOCHLORID see TFO000
FOSFORTRICHLORIDE (DUTCH) see PHT275
FOSFORYN TROJETYLOWY (CZECH) see TJT800
FOSFORYN TROJMETYLOWY (CZECH) see TMD500
FOSFORZUUROPLOSSINGEN (DUTCH) see PHB250
FOSFOSAL see PHA575
FOSFOTHION see MAK700
FOSFOTION see MAK700
FOSFOTOX see DSP400
FOSFURI di ALLUMINIO (ITALIAN) see AHE750
FOSFURI di MAGNESIO (ITALIAN) see MAI000
FOSGEEN (DUTCH) see PGX000
FOSGEN (POLISH) see PGX000
FOSGENE (ITALIAN) see PGX000
FOSOVA see PAK000
FOSPIRATE see PHE250
FOSPIRATE METHYL see PHE250
FOSSIL FLOUR see SCH002
FOSTEN see PAF550
FOSTERGE A 2523 see PKE850
FOSTER GRANT 834 see SMQ500
FOSTERN see PAK000
FOSTEX see BDS000
FOSTHIETAN see DHH200
FOSTION see IOT000
FOSTION MM see DSP400
FOSTOX see PAK000
FOSTRIL see HCL000
FOSVEL see LEN000
FOSVEX see TCF250
FOSZFAMIDON see FAB400
FOTRIN see FOM000
5-FOU see FOJ060
FOUADIN see AQH500
FOUMARIN see ABF500
FOURAMIEN 2R see ALL750
FOURAMINE see TGL750
FOURAMINE BA see DBO400
FOURAMINE BROWN AP see PPQ500
FOURAMINE D see PEY500
FOURAMINE EG see ALS990
FOURAMINE ERN see NAW500
FOURAMINE J see TGL750
FOURAMINE OP see ALT000
FOURAMINE P see ALT250
FOURAMINE PCH see CCP850
FOURAMINE RS see REA000
FOURAMINE STD see TGM400
FOURINE DS see PEY650
FOURNEAU see BAT000
FOURNEAU 190 see ABX500
FOURNEAU 309 see BAT000
FOURNEAU 933 see BCI500
FOURNEAU 1162 see SNM500
FOURNEAU 2268 see MJH250
FOURNEAU 2987 see DII200
FOURRINE 1 see PEY500
FOURRINE 36 see ALL750
FOURRINE 4R see DUP400
FOURRINE 57 see NEM480

FOURRINE 64 see PEY650
FOURRINE 65 see ALS990
FOURRINE 68 see CCP850
FOURRINE 76 see DBO400
FOURRINE 79 see REA000
FOURRINE 81 see CEG625
FOURRINE 84 see ALT250
FOURRINE 88 see AHQ250
FOURRINE 93 see DUP400
FOURRINE 94 see TGL750
FOURRINE 99 see NAW500
FOURRINE A see AHQ250
FOURRINE BROWN 2R see ALL750
FOURRINE BROWN PR see NEM480
FOURRINE BROWN PROPYL see NEM480
FOURRINE D see PEY500
FOURRINE EG see ALS990
FOURRINE ERN see NAW500
FOURRINE M see TGL750
FOURRINE P BASE see ALT250
FOURRINE PG see PPQ500
FOURRINE SLA see DBO400
FOURRINE SO see CEG625
FOUR THOUSAND FORTY-NINE see MAK700
FOVANE see BDE250
FOXGLOVE see DKL200
FOXGLOVE see FOM100
FOY see GAD400
FOZALON see BDJ250
FP 4 see PJS750
FP 70 see FLG100
FP-83 see FJT100
FPA see FHG000
FPA see FLD100
FPF 1002 see TAB250
p-FPHE see FLF000
FPL see PIV650
FPL 670 see CNX825
FPL 12924AA see RCK750
FR 2 see TNG750
FR 28 see DXG035
FR-33 see FLJ000
FR 143 see OAF200
FR 222 see MKA500
FR 300 see PAU500
FR 860 see HAQ500
FR 1138 see DDQ400
FR 1208 see OAF200
FR-1923 see NMV480
FR 3068 see TGA600
FR 651A see CPB550
FR 651P see CPB550
FR 651P see CPB550
FR-13,479 see EBE100
FR 34235 see NDY650
FRABEL see HNI500
FRACINE see NGE500
FRACTION AB see CKP250
FRADEMICINA see LGC200
FRADIOMYCIN SULFATE see NCG000
FRAESEOL see ENC000
FRAGAROL see NBJ000
FRAGIVIX see BBJ500
FRAILECILLO (CUBA) see CNR135
FRAMBINONE see RBU000
FRAMED see BJP000
FRAMYCETIN see NCF000
FRAMYCETIN SULFATE see NCD550
FRAMYCIN SULFATE see NCD550
FRANCILLADE (HAITI) see CAK325
FRANGULA EMODIN see MQF250
FRANKINCENSE GUM see OIM000
FRANKINCENSE OIL see OIM025
FRANKLIN see CAO000
FRANOCIDE see DIW200
FRANOZAN see DIW200
FRANROZE see FLZ050
FRAQUINOL see NCD550
FRATOL see SHG500
FRAXINELLONE see FOM200

FRAXINUS JAPONICA Blume, bark extract see FON100
FRAXIPARIN see HAQ500
FRAZALON see AOO300
FRE AMINE I see AHR600
FREE ACID see SLW475
FREE BENZYLPENICILLIN see BDY669
FREE COCONUT OIL see CNR000
FREE HISTAMINE see HGD000
FREEMANS WHITE LEAD see LDY000
FREESIOL see FON200
FREEURIL see BDE250
FREKAPHYLLIN see HLC000
FREKVEN see ICC000
FRENACTIL see FGU000
FRENACTIL see FLK100
FRENACTYL see FGU000
FRENACTYL see FLK100
FRENANTOL see ELL500
FRENASMA see CNX825
FRENCH GREEN see COF500
FRENCH JASMINE see COD675
FRENCH ROSE ABSOLUTE see RMP000
FRENOCK see SKE100
FRENOGASTRICO see DJM800
FRENOGASTRICO see XCJ000
FRENOHYPON see ELL500
FRENOLON DIFUMARATE see MDU750
FRENOLYSE see AJV500
FRENQUEL HYDROCHLORIDE see PIY750
FRENTIROX see MCO500
FREON see CFX500
FREON 11 see TIP500
FREON 12 see DFA600
FREON 13 see CLR250
FREON 14 see CBY250
FREON 21 see DFL000
FREON 22 see CFX500
FREON 23 see CBY750
FREON 30 see MJP450
FREON 31 see CHI900
FREON 32 see MJQ300
FREON 41 see FJK000
FREON 112 see TBP050
FREON 113 see FOO000
FREON 114 see FOO509
FREON 115 see CJI500
FREON 116 see HDC100
FREON 123 see TJY500
FREON 141 see FOO550
FREON 142 see CFX250
FREON 151 see FOO553
FREON 152 see DKH300
FREON 152 see ELN500
FREON 253 see TJY200
FREON 500 see DFB400
FREON 502 see FOO560
FREON 503 see FOO562
FREON 122A see DKI900
FREON 12B1 see BNA250
FREON 133a see TJY175
FREON 13B1 see TJY100
FREON 13T1 see IFM100
FREON 142b see CFX250
FREON 114B2 see FOO525
FREON 12-B2 see DKG850
FREON C-318 see CPS000
FREON F-12 see DFA600
FREON F-23 see CBY750
FREON FT see TJE100
FREON MF see TIP500
FREON R 112 see TBP050
FREON 113TR-T see FOO000
FRESENIUS D 6 see CCU150
FRESMIN see VSZ000
FREUND'S ADJUVANT see FOO600
FRIAR'S CAP see MRE275
FRIDERON see RBF100
FRIGEN see CFX500
FRIGEN 11 see TIP500
FRIGEN 12 see DFA600
FRIGEN 22 see CFX500

FRIGEN 114 see FOO509
FRIGEN 113a see FOO000
FRIGIDERM see FOO509
FRIJOLILLO (MEXICO) see NBR800
FRISIUM see CIR750
FROBEN see FLG100
FRON 125 see PBD400
FRON 225 see DFW850
FROWNCIDE see CEX800
FRP 53 see PAU500
FRUCOTE see BPY000
FRUCTOFURANOSE, TETRANICOTINATE see TDX860
FRUCTONE see EFR100
β-d-FRUCTOPYRANOSE, 2,3:4,5-BIS-o-(1-METHYLETHYLIDENE)-, METHYL((1-NAPHTHALENYLOXY) CARBONYL)AMIDOSULFITE see NBH300
d-FRUCTOPYRANOSE, 1-DEOXY-1-((4-METHYLPHENYL)NITROSOAMINO)- see NLW600
d-FRUCTOPYRANOSYLAMINE, N-NITROSO-N-p-TOLYL- see NLW600
FRUCTOSE (FCC) see LFI000
FRUCTUS PIPERIS LONGI see PIV650
FRUITDO see BLC250
FRUITONE see NAK000
FRUITONE see NAK500
FRUITONE A see TAA100
FRUITONE CPA see CJQ300
FRUITONE T see TIX500
FRUIT RED A EXTRA YELLOWISH GEIGY see HJF500
FRUIT RED A GEIGY see FAG020
FRUIT SALAD PLANT see SLE890
FRUIT SUGAR see LFI000
FRUMIN AL see DXH325
FRUMIN G see DXH325
FRUMTOSNIL see DAM625
FRUSEMIDE see CHJ750
FRUSEMIN see CHJ750
FRUSID see CHJ750
FRUSTAN see DCK759
FRUTABS see LFI000
FS 74 see SCK600
FS 19N see FOO700
FSH see FMT100
FSH-P see FMT100
FT see FPY000
3-FT see FLZ065
FT 8 see BEQ625
FT 248 see TCD100
F3DThd see TKH325
FTA see FQQ100
FTAALZUURANHYDRIDE (DUTCH) see PHW750
F1-TABS see SHF500
FTAFLEX DIBA see DNH125
FTALAN see TIT250
FTALOPHOS see PHX250
FTALOWY BEZWODNIK (POLISH) see PHW750
FTBG see FMH000
F3TDR see TKH325
FTHALIDE see TBT175
FTIVAZID see VEZ925
FTIVAZIDE see VEZ925
FTORAFUR see FLZ050
FTORIN see FOO875
FTORIN (PHARMACEUTICAL) see FOO875
FTORLON 3 see KDK000
FTORLON 4 see TAI250
FTORLON 3M see KDK000
FTORLON F3 see KDK000
FTOROPLAST 3 see KDK000
FTOROPLAST 4 see TAI250
FTOROPLAST 3P see KDK000
FTOROTAN (RUSSIAN) see HAG500
F-2 TOXIN see ZAT000
FT mixture with URACIL (1:4) see UNJ810
5-FU see FMM000
FUBERIDATOL see FQK000

FUBERIDAZOLE see FQK000
FUBERISAZOL see FQK000
FUBRIDAZOLE see FQK000
FUCHSIN see MAC250
p-FUCHSIN see RMK020
FUCHSIN (basic) see MAC500
FUCHSINE see MAC250
FUCHSINE A see MAC250
FUCHSINE BASE see MAC500
FUCHSINE CS see MAC250
FUCHSINE DR-001 see RMK020
FUCHSINE G see MAC250
FUCHSINE HF BASE see MAC500
FUCHSINE HO see MAC250
FUCHSINE N see MAC250
FUCHSINE RTN see MAC250
FUCHSINE SBP see MAC250
FUCHSINE SPC see RMK020
FUCHSINE Y see MAC250
FUCIDINA see SHK000
FUCIDINE see SHK000
FUCLASIN see BJK500
FUCLASIN ULTRA see BJK500
FUDIOLAN see IRU500
FUDR see DAR400
5-FUDR see DAR400
FUEL OIL see FOP000
FUEL OIL, pyrolyzate see FOP100
FUEL OIL, NO. 6 see HAJ750
FUEL OIL NO. 2 (DOT) see DHE800
FUEL OIL, RESIDUAL see FOP200
FUELS, DIESEL see DHE900
FUELS, DIESEL see FOP000
FUELS, DIESEL, NO. 2 see DHE850
FUELS, GASOLINE see GCC200
FUGACILLIN see CBO250
FUGEREL see FMR050
FUGILIN see FOZ000
FUGOA see NNW500
FUGU POISON see FOQ000
FUJI 1 see IRU500
FUJI GRASS see PFR120
FUJI HEC-BL 20 see HKQ100
FUJIONE see IRU500
FUJI-ONE see IRU500
FUJITHION see FOR000
FUKI-NO-TOH (JAPANESE) see PCR000
FUKINOTOXIN see PCQ750
FUKINOTOXIN (neutral) see PCQ750
FUKKOL AROMAX 1 see AQZ150
FUKLASIN see BJK500
FUKLASIN ULTRA see FAS000
FULAID see FLZ050
FULCIN see GKE000
FULCINE see GKE000
FULDAZIN see MQU525
FULFEEL see FLZ050
FUL-GLO see FEW000
FULL RANGE CATALYTIC REFORMED
NAPHTHA see NAQ510
FULLSAFE see TKH750
FULLY HYDROGENATED RAPESEED OIL
see RBK200
FULMINATE of MERCURY see MDC000
FULMINATE of MERCURY (dry) (DOT) see
MDC000
FULMINATES see FOS000
FULMINATING MERCURY (DOT) see
MDC000
FULMINATING SILVER (DOT) see SDR000
FULMINIC ACID see FOS050
FULSIX see CHJ750
FULUMINOL see FOS100
FULUVAMIDE see CHJ750
6-FULVENOSELONE see FOS300
FULVICAN GRISACTIN see GKE000
FULVICIN see GKE000
FULVINA see GKE000
FULVINE see FOT000
FULVISTATIN see GKE000
FUMADIL B see FOZ000
FUMAFER see FBJ100

FUMAGILLIN see FOZ000
FUMAGON see DDL800
FUMAR-F see FBJ100
FUMARIC ACID see FOU000
FUMARIC ACID, BUTYL 2,3-
EPOXYPROPYL ESTER see BRG700
FUMARIC ACID, DIBUTYL ESTER see
DEC600
FUMARIC ACID DIHEXYL ESTER see
DKP000
FUMARIC ACID, DIISOPROPYL ESTER see
DNQ200
FUMARIC ACID, DIMETHYL ESTER see
DSB600
FUMARIC ACID DINITRILE see FOX000
FUMARIC ACID ETHYL-2,3-
EPOXYPROPYL ESTER see FOV000
FUMARIC ACID, MAGNESIUM SALT see
MAF750
FUMARIC ACID, METHYL- see MDI250
FUMARINE see FOW000
FUMARONITRILE see FOX000
FUMARONITRILE see FOX000
FUMAROYL CHLORIDE see FOY000
FUMARSAEUREDINITRIL see FOX000
FUMARYLCHLORID (CZECH) see FOY000
FUMARYL CHLORIDE see FOY000
FUMAZONE see DDL800
FUMED SILICA see SCH002
FUMED SILICON DIOXIDE see SCH002
FUMETOBAC see NDN000
FUMETTE see MDR750
FUMIDIL see FOZ000
FUMIGACHLORIN see FPA000
FUMIGANT-1 (OBS.) see MHR200
FUMIGRAIN see ADX500
FUMILAT A see MJL250
FUMING LIQUID ARSENIC see ARF500
FUMING SULFURIC ACID see SOI520
FUMIRON see FBJ100
FUMITOXIN see AHE750
FUMO-GAS see EIY500
FUMONISIN B1 see MAB500
FUMONISIN B2 see FPA500
FUNBAS see MRQ300
FUNCHIONG JUJR see COQ385
FUNDAL see CJJ250
FUNDAL 500 see CJJ250
FUNDAL SP see CJJ500
FUNDASOL see BAV575
FUNDEX see CJJ250
FUNDUSCEIN see FEW000
FUNGACETIN see THM500
FUNGAFLOR see FPB875
FUNGAREST see KFK100
FUNGICHROMIN see FPC000
FUNGICHROMIN, HYDRATE see FPC000
FUNGICHTHOL see IAD000
FUNGICIDE 1991 see BAV575
FUNGICIDE FX see MJM500
FUNGICLOR see PAX000
FUNGIFEN see PAX250
FUNGIFOS see MIE250
FUNGILIN see AOC500
FUNGILON see TNH750
FUNGIMAR see CNO000
FUNGINEX see TKL100
FUNGINON see FPC000
FUNGIPLEX see RCK730
FUNGISONE see AOC500
FUNGISTOP see TGB475
FUNGITOX see PEX500
FUNGITOX OR see ABU500
FUNGIVIN see GKE000
FUNGIZONE see AOC500
FUNGIZONE INTRAVENOUS see FPC200
FUNGOCIN see BAB750
FUNGOL see ZIA000
FUNGOL B see SHF500
FUNGONIT GF 2 see ZIA000
FUNGO-POLYCID see CDY325
FUNGORAL see KFK100

FUNGOSTOP see BJK500
FUNGUS BAN TYPE II see CBG000
FUNGUS CURE see ADQ700
FUNICOLOSIN see FPD000
FUNICULOSIN (PIGMENT) see FPD050
FUQUA see FPD100
FUR see FMN000
5-FUR see FMN000
FURACILLIN see NGE500
FURACIN see SPC500
FURACINETTEN see NGE500
FURACOCCID see NGE500
FURACON see AKC570
FURACORT see NGE500
FURACYCLINE see NGE500
FURADAN see CBS275
FURADANTIN see NGE000
FURADANTIN MONOHYDRATE see
NGE780
FURADONIN see NGE000
FURADROXYL see FPE100
FURAFLUOR see FLZ050
FURAL see FPQ875
FURALAZIN see FPF000
2-FURALDEHYDE see FPQ875
2-FURALDEHYDE AZINE see FPH000
2-FURALDEHYDE, 2,3:4,5-BIS(2-
BUTENYLENE)TETRAHYDRO- see BHJ500
2-FURALDEHYDE, 5-METHYL- see MKH550
3-FURALDEHYDE, TETRAHYDRO- see
FOI100
FURALDON see NGE500
FURALE see FPQ875
FURALTADONE see FPI000
FURALTADONE HYDROCHLORIDE see
FPI100
l-FURALTADONE HYDROCHLORIDE see
FPI150
FURAMETHRIN see POD875
FURAMIZOLE see FPI200
FURAMON see FPY000
FURAMON IODIDE see FPY000
FURAN see FPK000
FURAN (DOT) see FPK000
FURANACE see NDY400
FURANACE-10 see NDY400
FURAN, 2-ACETYL- see ACM200
2-FURANACROLEIN see FPK025
3-FURANACROLEIN, 2-METHYL- see
FPX050
2-FURANACRYLIC ACID see FPK050
2-FURANALDEHYDE see FPQ875
FURAN-2-AMIDOXIME see FPK100
FURAN, 2-BROMO-S-(2-BROMO-5-
NITROETHENYL)- see BMX300
FURAN, 2-
(BROMOMETHYL)TETRAHYDRO- see
TCS550
2-FURANCARBINOL see FPU000
2-FURANCARBONAL see FPQ875
2-FURANCARBOXALDEHYDE see FPQ875
2-FURANCARBOXALDEHYDE, (2-
FURANYLMETHYLENE)HYDRAZONE
(9CI) see FPH000
3-FURANCARBOXALDEHYDE,
TETRAHYDRO- see FOI100
2-FURANCARBOXAMIDE, N-(3-((4-AMINO-
6,7-DIMETHOXY-2-
QUINAZOLINYL)METHYLAMINO)PROPY
L)TETRAHYDRO-, see XDJ050
2-FURANCARBOXIMIDAMIDE, N-
(ACETYLOXY)-5-NITRO- see NGC500
2-FURANCARBOXIMIDAMIDE, 5-NITRO-
N-(1-OXOBUTOXY)- see NGC550
2-FURANCARBOXIMIDAMIDE, 5-NITRO-
N-(1-OXOPROPOXY)- see NGC570
2-FURANCARBOXIMIDIC ACID, 5-NITRO-,
HYDRAZIDE see NGG600
α-FURANCARBOXYLIC ACID see FQF000
2-FURANCARBOXYLIC ACID, 3,4-
DICHLORO-5-NITRO-, METHYL ESTER see
MJB550

2-FURANCARBOXYLIC ACID, 2,5-DIHYDRO-4-HYDROXY-2-((4-HYDROXY-3-(3-METHYL-2-BUTENYL)PHENYL)METHYL)-3-(4-HYDROXYPHENYL)-5-OXO-, METHYL ESTER see BSX150

FURAN-α-CARBOXYLIC ACID METHYL ESTER see MKH600

2-FURANCARBOXYLIC ACID, 5-NITRO-, 2-ACETYLHYDRAZIDE see NGH600

2-FURANCARBOXYLIC ACID, TETRAHYDRO-3-HYDROXY-2,3,4-TRIMETHYL-5-OXO-, (2R-(2-α,3-β,4-α))- see MRG200

FURAN, 2,5-DIETHYLTETRAHYDRO- see DKB165

2,5-FURANDIONE see MAM000

2,5-FURANDIONE, DIHYDRO-3-(TETRAPROPENYL)-(9CI) see TEC600

2,5-FURANDIONE, 3-(DODECENYL)DIHYDRO- see DXV000

2,5-FURANDIONE, HOMOPOLYMER (9CI) see MAM500

2,5-FURANDIONE, POLYMER with ETHENYLBENZENE (9CI) see SEA500

2,5-FURANDIONE, POLYMER WITH ETHENE see EJM950

FURANEOL see HKC575

FURAN, 2-(N-ETHYLCARBAMOYLHYDROXYMETHYL)- see EHF500

FURANIDINE see TCR750

FURANIUM see FEW000

2-FURANMETHANAMINE, N-(CYCLOPROPYLMETHYL)-α-ETHYL-N-METHYL-α-(3-PHENYL-2-PROPENYL)- see CMQ855

2-FURANMETHANEDIOL, 5-NITRO-, DIACETATE see NGB800

2-FURANMETHANETHIOL see FPM000

2-FURANMETHANOL see FPU000

2-FURANMETHANOL, ACETATE (9CI) see FPM100

2-FURANMETHANOL, α-(N-ETHYLCARBAMOYL)- see EHF500

2-FURANMETHYL ACETATE see FPM100

2-FURANMETHYLAMINE see FPW000

FURAN, 2,2'-METHYLENEBIS- see FPX028

FURAN, 2,2'-METHYLENEDI- see FPX028

FURAN, 2-(2-NITROVINYL)- see FQM100

FURANODIENE see FPM150

FURAN-OFTENO see NGE500

FURANOL see FPY000

3-FURANOL, TETRAHYDRO- see HOI200

2(5H)-FURANONE, 3-CHLORO-4-(CHLOROMETHYL)-5-HYDROXY- see CFB700

2(5H)-FURANONE, 3-CHLORO-4-DICHLOROMETHYL-5-HYDROXY- see CFL100

2(5H)-FURANONE, 3-CHLORO-5-HYDROXY-4-METHYL- see CIM325

2(3H)-FURANONE, DIHYDRO-5-(3-HEXENYL)-5-METHYL-, (Z)- see MIW060

2(3H)-FURANONE, DIHYDRO-5-OCTYL- see OES100

3(2H)-FURANONE, 2,5-DIMETHYL-4-HYDROXY- see HKC575

2(3H)-FURANONE, 5-ETHYLDIHYDRO- see HDY600

2(3H)-FURANONE, 5-HEPTYLDIHYDRO- see HBN200

2(5H)-FURANONE, 5-HYDROXY-4-((1E)-2-PHENYLETHENYL)- see HNK585

3(2H)-FURANONE, 5-(METHYLAMINO)-2-PHENYL-4-(3-(TRIFLUOROMETHYL)PHENYL)-, (+-)- see MGJ775

2(5H)-FURANONE, 5-((1-METHYL-5-NITRO-1H-IMIDAZOL-2-YL)METHYLENE)- see FPM200

3(2H)-FURANONE, 2(OR 5)-ETHYL-4-HYDROXY-5(OR 2)-METHYL- see HND150

FURAN, 2,2'-(OXYBIS(METHYLENE))BIS- see FPX025

FURAN, 2,2'-(OXYDIMETHYLENE)DI-(6CI,8CI) see FPX025

FURAN, 2-PENTYL- see AOJ600

2-FURANPROPIONIC ACID, TETRAHYDRO-α-(1-NAPHTHYLMETHYL)-, 2-(DIETHYLAMINO)ETHYL ESTER, OXALATE (1:1) see NAE100

FURAN, TETRAHYDRO-2-(BROMOMETHYL)- see TCS550

FURAN, TETRAHYDRO-2-HEPTYL- see HBP425

FURANTHRIL see CHJ750

FURANTHRYL see CHJ750

FURANTOIN see NGE000

FURANTRIL see CHJ750

2-(2-FURANYL)-1H-BENZIMIDAZOLE see FQK000

4-(2-FURANYL)-3-BUTEN-2-ONE see FPX030

1-(2-FURANYL)ETHANONE see ACM200

5-(3-FURANYL)-5-HYDROXY-2-PENTANONE see IGF325

N-(2-FURANYLMETHYL)-N',N'-DIMETHYL-N-2-PYRIDINYL-1,2-ETHANEDIAMINE FUMARATE see MDP800

3-((2-FURANYLMETHYLENE)AMINO)-2-OXAZOLIDONE see FPM300

1-(3-FURANYL)-4-METHYL-1-PENTANONE see PCI750

1-(3-FURANYL)-2,4-PENTANEDIOL see IGF200

1-(3-FURANYL)-1,4-PENTANEDIONE see IGF300

3-FURANYLPHENYLMETHANONE see FQO050

FURAPLAST see NGE500

FURAPROMIDIUM see FPO000

FURAPYRIMIDONE see FPO100

FURASEPTYL see NGE500

FURATHIOCARB see FPO200

FURATOL see SHG500

FURATONE see BKH500

FURATONE-S see BKH500

FURAXONE see NGG500

FURAZABOL see AOO300

FURAZOL see NGG500

FURAZOLIDON see NGG500

FURAZOLIDONE, DENITRO- see FPM300

FURAZOLIDONE (USDA) see NGG500

FURAZOLIN see FPI000

FURAZOLINE see FPI000

FURAZON see NGG500

FURAZONAL see NGG550

FURAZONE see NGE500

FURAZOSIN see AJP000

FURAZOSIN HYDROCHLORIDE see FPP100

FUR BLACK 41867 see PEY500

FUR BROWN 41866 see PEY500

FURCELLERAN GUM see FPQ000

cis-FURCONAZOLE see FPQ050

FURESIS see CHJ750

FURESOL see NGE500

FURETHIDINE see FPQ100

FURFURAL see FPQ875

2-FURFURAL see FPQ875

FURFURALACETONE see FPX030

FURFURAL-ACETONE ADDUCT see FPQ900

FURFURAL-ACETONE MONOMER see FPQ900

1:1 FURFURAL-ACETONE MONOMER see FPQ900

FURFURAL ACETONE MONOMER FA see FPQ900

FURFURAL ALCOHOL see FPU000

FURFURALDAZINE see FPH000

FURFURALDEHYDE see FPQ875

FURFURALDEHYDE AZINE see FPH000

FURFURALE (ITALIAN) see FPQ875

FURFURAL OXIME see FPR000

FURFURAMIDE see FPS000

FURFURAN see FPK000

FURFURIN see NGE500

FURFUROL see FPQ875

FURFUROLATSETONOVYI MONOMER FA see FPQ900

FURFUROLE see FPQ875

FURFURYL ACETATE see FPM100

FURFURYLACETONE see FPT000

6-FURFURYLADENINE see FPT100

N-FURFURYLADENINE see FPT100

N6-FURFURYLADENINE see FPT100

FURFURYL ALCOHOL see FPU000

FURFURYL ALCOHOL, ACETATE see FPM100

FURFURYL ALCOHOL PHOSPHATE (3:1) see FPV000

2-FURFURYLALKOHOL (CZECH) see FPU000

FURFURYLAMINE see FPW000

FURFURYLAMINE, TETRAHYDRO-N,N-BIS(2-CHLOROETHYL)- see FPW100

6-(FURFURYLAMINO)PURINE see FPT100

N6-(FURFURYLAMINO)PURINE see FPT100

FURFURYL AZINE see FPH000

FURFURYL-BIS(2-CHLOROETHYL)AMINE HYDROCHLORIDE see FPX000

FURFURYL-BIS(β-CHLOROETHYL)AMINE HYDROCHLORIDE see FPX000

FURFURYL ETHER see FPX025

2-(2-FURFURYL)FURAN see FPX028

FURFURYLIDENEACETONE see FPX030

2-FURFURYLIDENEACETONE see FPX030

FURFURYLIDINE-2-PROPANAL see FPX050

FURFURYL MERCAPTAN see FPM000

N-(2-FURFURYL)-N-(2-PYRIDYL)-N',N'-DIMETHYLETHYLENEDIAMINE FUMARATE see MDP800

1-FURFURYLPYRROLE see FPX100

N-FURFURYL PYRROLE see FPX100

N-(2-FURFURYL)PYRROLE see FPX100

FURFURYLTRIMETHYLAMMONIUM IODIDE see FPY000

FURIDAZOL see FQK000

FURIDAZOLE see FQK000

FURIDIAZINE see NGI500

FURIDON see NGG500

FURIL see FPZ000

2,2'-FURIL see FPZ000

α-FURIL see FPZ000

FURILAZOLE see FPZ200

FURIL, DIOXIME see BJW810

α-FURIL DIOXIME see BJW810

2-FURIL-METANALE (ITALIAN) see FPQ875

α-FURILMONOXIME see FQB000

FURISYL see FQB100

FURITON see NGB700

FURLOE see CKC000

FURLOE 4EC see CKC000

FURMETHANOL see FPI000

FURMETHIDE see FPY000

FURMETHONOL see FPI000

FURMETHONOL see FPI150

FURMETONOL see FPI000

FURMITHIDE IODIDE see FPY000

FURNAL see CBT750

FURNEX see CBT750

FURNEX N 765 see CBT750

FURO(2,3-C)ACRIDIN-6(2H)-ONE, 1,11-DIHYDRO-5-HYDROXY-11-METHYL-2-(1-METHYLETHENYL)-, (−)- see RRP100

FUROBACTINA see NGE000

2H-FURO(2,3-h)(1)BENZOPYRAN-2-ONE see FQC000

7H-FURO(3,2-g)(1)BENZOPYRAN-7-ONE see FQD000

5H-FURO(3',2':6,7)(1)BENZOPYRANO(3,4-C)PYRIDIN-5-ONE, 7-METHYL plus ULTRAVIOLET A RADIATION see FQD100

5H-FURO(3',2':6,7)(1)BENZOPYRANO(3,4-C)PYRIDIN-5-ONE plus ULTRAVIOLET A RADIATION see FQD142

FUROCOUMARIN see FQD000

FURO(2',3',7,6)COUMARIN see FQD000
FURO(4',5',6,7)COUMARIN see FQD000
FURO(5',4',7,8)COUMARIN see FQC000
FURODAN see CBS275
FURO(3,4-E)-1,3-BENZODIOXOL-8(6H)-ONE, 6-(1,2,3,4-TETRAHYDRO-6,7-DIMETHOXY-2-METHYL-1-ISOQUINOLINYL)-, (R-(R*,S*))- see CNR745
FURO(3,4-E)-1,3-BENZODIOXOL-8(6H)-ONE, 6-(5,6,7,8-TETRAHYDRO-6-METHYL-1,3-DIOXOLO(4,5-G) ISOQUINOLIN-5-YL-, (S-(R*,R*))- see AER300
FURO(3,4-E)-1,3-BENZODIOXOL-8(6H)-ONE, 6-(5,6,7,8-TETRAHYDRO-6-METHYL-1,3-DIOXOLO(4,5-G) ISOQUINOLIN-5-YL)-, (R-(R*,R*))- see CBF550
7H-FURO(3',2':4,5)FURO(2,3-C)XANTHEN-7-ONE, 3A,12C-DIHYDRO-6,8-DIMETHOXY- see MPJ050
7H-FURO(3',2':4,5)FURO(2,3-C)XANTHEN-7-ONE,3A,12C-DIHYDRO-6,8-DIMETHOXY-, (3ar-CIS)- see MPJ050
7H-FURO(3',2':4,5)FURO(2,3-C)XANTHEN-7-ONE, 3A,12C-DIHYDRO-8-HYDROXY-6,10,11-TRIMETHOXY- see HOL200
7H-FURO(3',2':4,5)FURO(2,3-c)XANTHEN-7-ONE,1,2,3,12c-TETRAHYDRO-6,11-DIMETHOXY-8-HYDROXY-,(3ar-CIS)- see MEL550
7H-FURO(3',2':4,5)FURO(2,3-C)XANTHEN-7-ONE, 1,2,3A,12C-TETRAHYDRO-8-HYDROXY-6-METHOXY- see HKB300
FUROFUTRAN see FLZ050
7H-FURO(3,2-G)(1)BENZOPYRAN-7-ONE,2,3-DIHYDRO-2-(1-HYDROXY-1-METHYLETHYL)-, (S)- see MBU525
7H-FURO(3,2-G)(1)BENZOPYRAN-7-ONE, 2,3-DIHYDRO-2-(1-HYDROXY-1-METHYLETHYL)-, (S)-(+)- see MBU525
7H-FURO(3,2-G)(1)BENZOPYRAN-7-ONE, 4,9-DIMETHOXY- see IMO500
7H-FURO(3,2-G)(1)BENZOPYRAN-7-ONE,6-(1,1-DIMETHYLALLYL)- see XPJ100
7H-FURO(3,2-G)(1)BENZOPYRAN-7-ONE, 6-(1,1-DIMETHYL-2-PROPENYL)-(9CI) see XPJ100
7H-FURO(3,2-G)(1)BENZOPYRAN-7-ONE, 2,5,9-TRIMETHYL- see HOM270
2H-FURO(2,3-H)(1)BENZOPYRAN-2-ONE, 4,9-DIMETHYL-, plus ULTRAVIOLET A RADIATION see FQD130
2H-FURO(2,3-H)(1)BENZOPYRAN-2-ONE, 5-METHYL- see HLV955
2H-FURO(2,3-H)(1)BENZOPYRAN-2-ONE, 4,6,9-TRIMETHYL, plus ULTRAVIOLET A RADIATION see FQD140
FUROIC ACID see FQE000
2-FUROIC ACID see FQF000
α-FUROIC ACID see FQF000
2-FUROIC ACID, METHYL ESTER see MKH600
2-FUROIC ACID, 5-NITRO-, 2-ACETYLHYDRAZIDE see NGH600
2-FUROIC ACID, 5-NITRO-, 2-BUTYRYLHYDRAZIDE see NGC550
2-FUROIC ACID, TETRAHYDRO-3-HYDROXY-2,3,4-TRIMETHYL-5-OXO-, (2R,3R,4R)- see MRG200
FUROIN see FQI000
FUROLE see FPQ875
α-FUROLE see FPQ875
FURO(3',4':6,7)NAPHTHO(2,3-D)-1,3-DIOXOL-6(5AH)-ONE, 5,8,8A,9-TETRAHYDRO-5-(4-HYDROXY-3,5-DIMETHOXYPHENYL)-, (5R-(5-α,5A-β,8A-α))-` see DAN388
FURO(3',4':6,7)NAPHTHO(2,3-d)-1,3-DIOXOL-6(5aH)-ONE, 5,8,8a,9-TETRAHYDRO-9-HYDROXY-5-(3,4,5-TRIMETHOXYPHENYL)-, (5R-(5-α-5a-β,8a-α-9-β))- see EBB600
FURORE see FAQ200

FUROSEDON see CHJ750
FUROSEMID see CHJ750
FUROSEMIDE see CHJ750
FUROSEMIDE "MITA" see CHJ750
FUROTHIAZOLE see FQJ000
FUROVAG see NGG500
FUROX see NGG500
FUROXAL see NGG500
FUROXANE see NGG500
FUROXONE SWINE MIX see NGG500
2-FUROYL AZIDE see FQJ025
N-FUROYL-N'-n-BUTYLHARNSTOFF (GERMAN) see BRK250
FUROYL CHLORIDE see FQJ050
(FUROYLOXY)TRIETHYL PLUMBANE see TJS750
2-(4-(2-FUROYL)PIPERAZIN-1-YL)-4-AMINO-6,7-DIMETHOXYQUINAZOLINE see AJP000
2-(4-(2-FUROYL)PIPERAZIN-1-YL)-4-AMINO-6,7-DIMETHOXYQUINAZOLINE HYDROCHLORIDE see FPP100
FUROZOLIDINE see NGG500
FURPIRINOL see NDY400
FURPYRINOL see NDY400
FURRO 4R see DUP400
FURRO D see PEY500
FURRO EG see ALS990
FURRO ER see NAW500
FURRO L see DBO000
FURRO P BASE see ALT250
FURRO SLA see DBO400
FURSEMID see CHJ750
FURSEMIDE see CHJ750
FURSULTIAMIN see FQJ100
FURSULTIAMINE see FQJ100
FURTHRETHONIUM IODIDE see FPY000
FURTRETHONIUM IODIDE see FPY000
FURTRIMETHONIUM IODIDE see FPY000
FUR YELLOW see PEY500
3-(α-FURYL-β-ACETYLAETHYL)-4-HYDROXYCUMARIN (GERMAN) see ABF500
3-(1-FURYL-3-ACETYLETHYL)-4-HYDROXYCOUMARIN see ABF500
3-(2-FURYL)ACROLEIN see FPK025
β-2-FURYLACROLEIN see FPK025
FURYL ALCOHOL see FPU000
FURYLAMIDE see FQN000
2-(2-FURYL)BENZIMIDAZOLE see FQK000
2-(2'-FURYL)-BENZIMIDAZOLE see FQK000
4-(2-FURYL)-2-BUTANONE see FPT000
2-FURYLCARBINOL see FPU000
α-FURYLCARBINOL see FPU000
17-α-(3-FURYL)ESTRA-1,3,5(10),7-TETRAENE-3,17-DIOL-3-ACETATE see EDU600
FURYLFURAMIDE see FQN000
FURYLFURAMIDE see FQO000
2-FURYL α-HYDROXYFURFURYL KETONE see FQI000
1-(3-FURYL)-4-HYDROXYPENTANONE see FQL100
5-(3-FURYL)-5-HYDROXY-2-PENTANONE see IGF325
1-(β-FURYL)-4-HYDROXYPENTANONE see FQL100
2-FURYL p-HYDROXYPHENYL KETONE see FQL200
β-FURYL ISOAMYL KETONE see PCI750
2-FURYLISOPROPYLAMINE SULFATE see FQM000
β-(2-FURYL)ISOPROPYLAMINE SULFATE see FQM000
2-FURYLMERCURIC CHLORIDE see CHK000
2-FURYLMERCURY CHLORIDE see CHK000
2-FURYL-METHANAL see FPQ875
(2-FURYL)METHANOL see FPU000
1-(2-FURYL)METHYLAMINE see FPW000
2-FURYL METHYL KETONE see ACM200

1-(3-FURYL)-4-METHYL-1-PENTANONE see PCI750
2-FURYL-1-NITROETHENE see FQM100
α-2-FURYL-5-NITRO-2-FURANACYRLAMIDE see FQN000
2-(2-FURYL)-3-(5-NITRO-2-FURYL)ACRYLAMIDE see FQN000
(E)-2-(2-FURYL)-3-(5-NITRO-2-FURYL)ACRYLAMIDE see FQO000
trans-2-(2-FURYL)-3-(5-NITRO-2-FURYL)ACRYLAMIDE see FQO000
2-(2-FURYL)-3-(5-NITRO-2-FURYL)ACRYLIC ACID AMIDE see FQN000
α-(FURYL)-β-(5-NITRO-2-FURYL)ACRYLIC AMIDE see FQN000
1-(3-FURYL)-1,4-PENTANEDIONE see IGF300
3-FURYL PHENYL KETONE see FQO050
3-(α-FURYL)PROPENAL see FPK025
1-(5-(3-FURYL)TETRAHYDRO-2-METHYL-2-FURYL)-4-METHYL-2-PENTANONE, (2R,5S)-(−) see NCQ900
N-(4-(2-FURYL)-2-THIAZOLYL)ACETAMIDE see FQQ100
N-(4-(2-FURYL)-2-THIAZOLYL)FORMAMIDE see FQQ400
FURYLTRIAZINE see FQQ500
FUSARENONE X see FQR000
FUSAREX see TBR750
FUSAREX see TBS000
FUSARIC ACID see BSI000
FUSARIC ACID-Ca see FQR100
FUSARIC ACID CALCIUM SALT see FQR100
FUSARINIC ACID see BSI000
FUSARIOTOXIN T 2 see FQS000
FUSARIUM TOXIN see ZAT000
FUSED BORAX see DXG035
FUSED BORIC ACID see BMG000
FUSED QUARTZ see SCK600
FUSED SILICA see SCK600
FUSELEX see SCK600
FUSELEX RD 120 see SCK600
FUSELEX RD 40-60 see SCK600
FUSELEX ZA 30 see SCK600
FUSELOEL (GERMAN) see FQT000
FUSEL OIL see FQT000
FUSEL OIL, REFINED (FCC) see FQT000
FUSID see CHJ750
FUSIDATE SODIUM see SHK000
FUSIDIC ACID see FQU000
FUSIDIN see SHK000
FUSIDINE see FQU000
FUSILADE see FDA885
FUSIN see SHK000
FUSSOL see FFF000
FUSTEL see FBW000
FUSTET see FBW000
FUT 175 see AHN700
FUTRAFUL see FLZ050
FUTRAMINE D see PEY500
FUTRAMINE EG see ALS990
FUTRICAN see CDY325
FUVACILLIN see NGE500
FUXAL see SNN300
FUZIDIN see FQU000
45VF see CNB825
F 3 (VINYL POLYMER) see KDK000
FW 50 see WCJ000
FW 293 see BIO750
FW 734 see DGI000
FW 925 see DFT800
FW 200 (mineral) see WCJ000
FWH 114 see IIY100
FWH 399 see TNV625
FWH 429 see TNV625
FX 703 see FMO150
FX 2182 see EGS000
FYBEX see FQU200
FYCOL 8 see CNK559
FYDALIN see BNK000
FYDE see FMV000
FYFANON see MAK700

FYROL 6 see DIV300
FYROL 76 see PGZ910
FYROL CEF see CGO500
FYROLFLEX RDP see REA050
FYROL FR 2 see FQU875
FYROL FR2 see TNG750
FYROL HB32 see TNC500
FYRQUEL 150 see TNP500
FYSIOQUENS see EEH575
FYTIC ACID see PIB250
FYTOLAN see CNK559
G-0 see FQM100
G 0 see HDG000
G 1 see HIM000
G 3 see MCB050
G 4 see MJM500
G-11 see HCL000
G 50 see EHG100
G-52 see GAA100
G 130 see DTN775
G 200 see EHG100
G 251 see EKN550
G 263 see EKN550
G 301 see DCM750
G 316 see PJC000
G 338 see DER000
G 347 see DST200
G 475 see COY500
G 572 see CJR550
G 641 see BNV765
G 711 see BBS275
G 821 see MCB050
G 996 see CDS125
G-1029 see AIH000
G-2129 see PJY000
G 3063 see GAC000
G 3300 see BBS275
G 3707 see DXY000
G 3802 see PJT300
G 3816 see PJT300
G 3820 see PJT300
Ge 132 see CCF125
Go 919 see EOA500
G 14744 see DWA500
G-2130A see DXY000
G 22150 see DLH630
G 22355 see DLH600
G 22355 see DLH630
G-23350 see ABF750
G 23992 see DER000
G 24,163 see PNH750
G-24480 see DCM750
G-24622 see DJX400
G-25804 see TBO500
G 27202 see HNI500
G 27365 see DJA200
G 27901 see TJL500
G 28364 see DEL400
G-29288 see MQH750
G 30027 see ARQ725
G 30031 see IKM050
G-31435 see MFL250
G 32883 see DCV200
G 33040 see DCV800
G 33182 see CLY600
G 34161 see BKL250
G 34360 see INR000
G 34586 see CDV000
G 35020 see DLS600
G-704,650 see SKL600
GA see EIF000
GA see GEM000
GA 242 see CDF400
GA 297 see BJY800
GA-10832 see CQG250
GA 56 (enzyme) see GGA800
GABA see PIM500
GABACET see NNE400
p-GABA HYDROCHLORIDE see GAD000
GABAIL see GCU050
GABBROMICINA see APP500
GABBROMYCIN see APP500

GABBROMYCIN see NCF500
GABBROPAS see AMM250
GABBRORAL see APP500
GABBROROL see APP500
GABEXATE MESILATE see GAD400
GABEXATE MESYLATE see GAD400
GABOB see AKF375
GABOMADE see AKF375
GABRITE see UTU500
GADEXYL see MQU750
GADODIAMIDE HYDRATE see GAD500
GADOLINIA see GAP000
GADOLINIUM see GAF000
GADOLINIUM, AQUA(5,8-
BIS(CARBOXYMETHYL)-11-(2-
(METHYLAMINO)-2-OXOETHYL)-3-OXO-
2,5,8,11-TETRAAZATRIDECAN-13-OATO(3-
))-, HYDRATE see GAD500
GADOLINIUM CHLORIDE see GAH000
GADOLINIUM CITRATE see GAJ000
GADOLINIUM(III) NITRATE (1:3) see
GAL000
GADOLINIUM(III) NITRATE,
HEXAHYDRATE (1:3:6) see GAN000
GADOLINIUM OXIDE see GAP000
GADOLINIUM(3+) OXIDE see GAP000
GADOLINIUM(III) OXIDE see GAP000
GADOLINIUM SESQUIOXIDE see GAP000
GADOLINIUM TRICHLORIDE see GAH000
GADOLINIUM TRIOXIDE see GAP000
GAFAC RD 510 see PKE850
GAFAMIDE COA see CNF270
GAFCOL EB see BPJ850
GAFCOTE see PKQ059
GAF-S 630 see AAU300
GAG ROOT see CCJ825
GAIMAR see RLU000
GAINEX see PEL250
GAJAR, seed extract see GAP500
GALACTARIC ACID see GAR000
GALACTASOL see GLU000
GALACTASOL A see SLJ500
GALACTICOL see DDJ000
β-d-GALACTOPYRANOSIDE, (3-β)-
SOLANID-5-EN-3-YL O-6-DEOXY-α-l-
MANNOPYRANOSYL-(1-2)-O-(β-d-
GLUCOPYRANOSYL-(1-3))-,
HYDROCHLORIDE see SKS100
4-o-β-d-GALACTOPYRANOSYL-d-
FRUCTOFURANOSE see LAR100
4-o-β-d-GALACTOPYRANOSYL-d-
GLUCITOL DIHYDRATE see LAO500
4-o-β-d-GALACTOPYRANOSYL-d-
GLUCITOL MONOHYDRATE see LAO600
3-β-d-GALACTOPYRANOSYL-l-METHYL-l-
NITROSOUREA see MNB000
4-o-β-d-GALACTOPYRANOSYL-d-
SORBITOL DIHYDRATE see LAO500
GALACTOQUIN see GAX000
GALACTOSACCHARIC ACID see GAR000
d-GALACTOSAMINE HYDROCHLORIDE
see GAT000
GALACTOSE see GAV000
d-GALACTOSE see GAV000
α-GALACTOSIDASE see GAV050
β-GALACTOSIDASE see GAV100
4-(β-d-GALACTOSIDO)-d-GLUCOSE see
LAR000
GALACTURONIC ACID with α-(6-
METHOXY-4-QUINOLYL)-5-VINYL-2-
QUINUCLIDINEMETHANOL see GAX000
GALAN de DIA (CUBA) see DAC500
GALANGIN see GAZ000
GALAN de NOCHE (CUBA) see DAC500
GALANTHAMINE HYDROBROMIDE see
GBA000
GALANTHUS NIVALIS see SED575
GALATONE see GBB500
4-o-β-d-GALATTOPIRANOSIL-d-
FRUTTOFURANOSIO (ITALIAN) see
LAR100
GALATUR see DPX200

GALBANUM OIL see GBC000
GALCTIN see PMH625
GALECRON see CJJ250
GALECRON SP see CJJ500
GALENA see LDZ000
GALFER see FBJ100
GALIPAN see BAV000
GALITION see FAO300
GALLACETOPHENONE see TKN250
GALLALDEHYDE-3,5-DIMETHYL ETHER
see DOF600
GALLAMINE see PDD300
GALLANT see HAG850
GALLIA see GBS050
GALLIC ACID see GBE000
GALLIC ACID, DODECYL ESTER see
DXX200
GALLIC ACID, ETHYL ESTER see EKM100
GALLIC ACID, LAURYL ESTER see DXX200
GALLIC ACID, PROPYL ESTER see PNM750
GALLIMYCIN see EDJ500
GALLITO (CUBA) see SBC550
GALLIUM see GBG000
GALLIUM ARSENIDE see GBK000
GALLIUM CHLORIDE see GBM000
GALLIUM(3+) CHLORIDE see GBM000
GALLIUM CITRATE see GBO000
GALLIUM COMPOUNDS see GBO500
GALLIUM LACTATE mixed with SODIUM
LACTATE (1.6:1 moles) see GBQ000
GALLIUM MONOARSENIDE see GBK000
GALLIUM-NICKEL ALLOY see NDD500
GALLIUM NITRATE see GBS000
GALLIUM(III) NITRATE (1:3) see GBS000
GALLIUM OXIDE see GBS050
GALLIUM SESQUIOXIDE see GBS050
GALLIUM SULFATE see GBS100
GALLIUM TRIOXIDE see GBS050
GALLOCHROME see MCV000
GALLODESOXYCHOLIC ACID see CDL325
GALLOGAMA see BBQ500
GALLOGEN (ASTRINGENT) see EAI200
GALLOGEN (CHOLERETIC) see HAQ570
GALLOTANNIC ACID see TAD750
GALLOTANNIN see TAD750
GALLOTOX see ABU500
GALLOTOX see PFO550
GALLOXON see DFH600
GALOFAK see BDY669
GALOPERIDOL see CLY500
GALOXANE see DFH600
GALOXOLIDE see GBU000
GALOZONE see CCL250
GALSEPTIL see SNQ000
GALVANISONE see FLL000
GALVATOL 1-60 see PKP750
GAMACID see BBQ500
GAMAPHEX see BBQ500
GAMAQUIL see PGA750
GAMAREX see PIM500
GAMASERPIN see RDK000
GAMASOL 90 see DUD800
GAMBIER see CCP800
GAMEFAR see RHZ000
GAMENE see BBQ500
GAMIBETAL see AKF375
GAMISO see BBQ500
GAMIT see CEP800
GAMMA-COL see BBQ500
GAMMACORTEN see SOW000
GAMMAHEXA see BBQ500
GAMMAHEXANE see BBQ500
GAMMALIN see BBQ500
GAMMALON see PIM500
GAMMA OH see HJS500
GAMMAPHOS see AMD000
GAMMARIZA see OJY200
GAMMASERPINE see RDK000
GAMMEXANE see BBP750
GAMMOPAZ see BBQ500
GAMONIL see CBM750
GAMONIL see IFZ900

GAMOPHENE see HCL000
GANCICLOVIR SODIUM see GBU200
GANCIDIN (unpurified) see GBU600
GANEAKE see ECU750
GANEX E 535 see AAU300
GANEX P 804 see PKQ250
GANGESOL see HII500
GANGLEFENE see GBU700
GANGLIOSTAT see HEA000
GANIDAN see AHO250
GANJA see CBD750
GANLION see MKU750
GANOCIDE see MLC250
GANOZAN see CHC500
GANPHEN see PMI750
GANSIL see CDP000
GANTANOL see SNK000
GANTAPRIN see TKX000
GANTRIM see TKX000
GANTRISINE see SNN500
GANTRON PVP see AAU300
GANTRON S 630 see AAU300
GANTRON S 860 see AAU300
GANU see CHA000
GARAMYCIN see GCO000
GARAMYCIN see GCS000
GARANTOSE see BCE500
GARBANCILLO (CUBA) see GIW200
GARDCIDE see RAF100
GARDENAL SODIUM see SID000
GARDENIA YELLOW see GBU750
GARDENIOL II see SMP600
GARDENOL see SMP600
GARDENTOX see DCM750
GARDEPANYL see EOK000
GARDIQUAT 1450 see AFP250
GARDIQUAT 1250AF see QAT520
GARDONA see RAF100
GARDOPRIM see BQB000
GARDRIN see EAU200
GARGET see PJJ315
GARGON see TFQ275
GARI see CCO700
GARLIC see WBS850
GARLIC OIL see GBU800
GARLIC OIL see GBU825
GARLIC POWDER see GBU850
GARLON see TJE890
GARMIAN see BOV825
GARNITAN see DGD600
GAROLITE SA see CAT775
GAROX see BDS000
GARRATHION see TNP250
GARVOX see DQM600
GAS-FURNACE BLACK see CBT750
GASHOUSE TANKAGE see SKZ000
GAS OIL see DHE800
GAS OIL see GBW000
GAS OIL (DOT) see DHE800
GAS OILS (petroleum), heavy vacuum see GBW005
GAS OILS (petroleum), hydrodesulfurized heavy vacuum see GBW010
GAS OILS (petroleum), light vacuum see GBW025
GASOLINE see GBY000
GASOLINE (100–130 octane) see GCA000
GASOLINE (115–145 octane) see GCC000
GASOLINE BR-1 see GCC200
GASOLINE ENGINE EXHAUST CONDENSATE see GCE000
GASOLINE ENGINE EXHAUST "TAR" see GCE000
GASOLINE, SYNTHETIC see GCC200
GASOLINE, UNLEADED see GCE100
GASTER see FAB500
GASTEX see CBT750
GASTOMAX see CDQ500
GASTRACID see PEK250
GASTRIDENE see BGC625
GASTRIDIN see FAB500
GASTRINIDE see TEH500

GASTRIN TETRAPEPTIDE see GCE200
GASTRIN TETRAPEPTIDE AMIDE see GCE200
GASTRIPON see PEM750
GASTRODYN see GIC000
GASTROGRAFIN see AOO875
GASTROMET see TAB250
GASTRON see DJM800
GASTRON see XCJ000
GASTROPIDIL see CBF000
GASTROSEDAN see DJM800
GASTROSEDAN see XCJ000
GASTROTELOS see COI750
GASTROTEST see PEK250
GASTROTOPIC see BGC625
GASTROZEPIN see GCE500
GATALONE see GBB500
GATINON see MHM500
GAUCHO see CKW400
GAULTHERIA OIL, ARTIFICIAL see MPI000
GB see IPX000
GB 94 see BMA625
GBH see BDD000
GBL see DWT600
GBR-12909 see BJW100
GBS see SEG800
GBX see CDB770
GC 30 see DHU700
GC 928 see CJR500
GC-1106 see HCL500
GC-2466 see MRV000
GC-2603 see FAR050
GC-2996 see CJY000
GC 3707 see SOY000
GC 4072 see CDS750
GC 6936 see ABX250
GC 7787 see HDA500
GC 8993 see CLU000
GC 9160 see MDT625
GC 10000 see SCQ000
GC 3944-3-4 see PAX000
GCP-1634 see MNX900
GCP 5126 see DED000
G-CURE see ADV900
GD see SKS500
GDUE see OMM300
GEA 6414 see CLK325
GEABOL see DAL300
GEARPHOS see MNH000
GEARPHOS see PAK000
GEASTIGMOL see GCE600
GEASTIMOL see GCE600
GEBUTOX see BRE500
GECHLOREERDEDIFENYL (DUTCH) see PJM500
GEDEX see SMQ500
GEFARNATE see GDG200
GEFARNIL see GDG200
GEFARNYL see GDG200
GEIGY see IKM050
GEIGY 338 see DER000
GEIGY 2747 see PET250
GEIGY 12968 see DRU400
GEIGY 13005 see DSO000
GEIGY 19258 see DRL200
GEIGY 22870 see DQZ000
GEIGY 24480 see DCM750
GEIGY 27,692 see BJP000
GEIGY 30,027 see ARQ725
GEIGY 30,044 see BJP250
GEIGY 30494 see MNO750
GEIGY 32,293 see EGD000
GEIGY 32883 see DCV200
GEIGY-BLAU 536 see BGT250
GEIGY 444E see AKQ250
GEIGY-444E see TBO500
GEIGY G-23611 see DSK200
GEIGY G-27365 see DJA200
GEIGY G-28029 see PDC750
GEIGY G-29288 see MQH750
GEIGY G.S. 14254 see BQC250
GEIGY GS-19851 see IOS000

GEIGY HERBICIDE 444E see AKQ250
GEISSOSPERMINE see GCG300
GELACILLIN see BDY669
GELAN I see EIF000
GELATIN see PCU360
GELATINE see PCU360
GELATINE DYNAMITE see DYG000
GELATIN-EPINEPHRINE see AES500
GELATIN FOAM see PCU360
GELATINS see PCU360
L-GELB 2 see FAG140
L-GELB 3 see CMM510
GELBER PHOSPHOR (GERMAN) see PHP010
GELBIN see CAP500
GELBIN YELLOW ULTRAMARINE see CAP750
GELBORANGE-S see FAG150
GELCARIN see CCL250
GELCARIN HMR see CCL250
GELDANAMYCIN see GCI000
G-ELEVEN see HCL000
GELFOAM see PCU360
GEL II see SHF500
GELIOMYCIN see HAL000
GELITA-SOL C see HID100
GELOCATIL see HIM000
GELOSE see AEX250
GELOZONE see CCL250
GELSEMIN see GCK000
GELSEMINE see GCK000
GELSEMIUM SEMPERVIRENS see YAK100
GELSTAPH see SLJ000
GELSTAPH see SLJ050
GELTABS see VSZ100
GELUCYSTINE see CQK325
GELUTION see SHF500
GELVA CSV 16 see AAX250
GELVATOL see PKP750
GELVATOL 1-30 see PKP750
GELVATOL 1-90 see PKP750
GELVATOL 3-91 see PKP750
GELVATOL 20-30 see PKP750
GELVATOL 2090 see PKP750
GEMCADIOL see TDO260
GEMCITABINE see GCK100
GEMCITABINE HYDROCHLORIDE see GCK100
GEM-DIBROMOETHYLENE see VPF200
GEMEPROST see CDB775
GEMFIBROZIL see GCK300
GEMFIBROZIL M3 see GCK400
GEMICITABINE see GCK500
GEMONIL see DJO800
GEMONIT see DJO800
GEMZAR see GCK100
GENACORT see CNS750
GENACRON YELLOW G see AAQ250
GENACRYL ORANGE G see CMM820
GENACRYL PINK G see CMM850
GENACRYL YELLOW 3G see CMM890
GENACRYL YELLOW 4G see CMM890
GENAMIN 16R302D see HCP525
GENAMIN DSAC see DXG625
GENAPOL 24/50 see AFJ165
GENAPOL 26L80 see AFJ165
GENAPOL 24 L50 see AFJ165
GENAPOL OX 030 see AFJ160
GENAZO RED KB SOLN see CLK225
GENDRIV 162 see GLU000
GENE AFG3 METALLOPROTEINASE see GIA050
GENEP EPTC see EIN500
GENERAL CHEMICAL 9160 see MDT625
GENERAL CHEMICALS 1189 see KEA000
GENERAL CHEMICALS 3707 see SOY000
GENERAL CHEMICALS 8993 see CLU000
GENERAL CHEMICALS 9160 see MDT625
GENERAL WEED KILLER see EAQ800
GENET ABSOLUTE see GCM000
GENETRON 11 see TIP500
GENETRON 12 see DFA600

GENETRON 13 see CLR250
GENETRON 21 see DFL000
GENETRON 22 see CFX500
GENETRON 23 see CBY750
GENETRON 32 see MJQ300
GENETRON 100 see ELN500
GENETRON 101 see CFX250
GENETRON 112 see TBP050
GENETRON 113 see FOO000
GENETRON 114 see FOO509
GENETRON 115 see CJI500
GENETRON 316 see FOO509
GENETRON 1113 see CLQ750
GENETRON 133a see TJY175
GENETRON 142b see CFX250
GENETRON 1112A see DFA300
GENETRONE 1112A see DFA300
GENIPHENE see CDV100
GENIPIN see GCM300
GENISIS see ECU750
GENISTEIN see GCM350
GENISTEOL see GCM350
GENISTERIN see GCM350
GENITE see DFY400
GENITE 883 see CJT750
GENITHION see PAK000
GENITOL see DFY400
GENITOX see DAD200
GENITRON AC see ASM270
GENITRON AC 2 see ASM270
GENITRON AC 4 see ASM270
GENITRON BSH see BBS300
GENO-CRISTAUZ GREMY see TBF500
GENOFACE see TIL500
GENOGRIS see NNE400
GENOL see MGJ750
GENONOL GEV see MEL800
GENOPHYLLIN see TEP500
GENOPTIC see GCS000
GENOPTIC S.O.P. see GCS000
GENOTHERM see PKQ059
GENOXAL see CQC500
GENOXAL see CQC650
GENOZYM see CMX700
GENPROPATHRIN see DAB825
GENSALATE SODIUM see GCU050
GENTALPIN see GCU050
GENTAMICIN C2b see MQS579
GENTAMICIN C²b see MQS579
GENTAMICIN C COMPLEX see GCO200
GENTAMYCIN see GCO000
GENTAMYCIN see GCO000
GENTAMYCIN-CREME (GERMAN) see GCO000
GENTAMYCIN SULFATE see GCS000
GENTASOL see GCU050
GENTERSAL see AOR500
GENTIANAE SCABRAE RADIX (LATIN) see RSZ675
GENTIANAVIOLETT see AOR500
GENTIAN VIOLET see AOR500
GENTIAVERM see AOR500
GENTICID see AOR500
GENTIDOL see GCU050
GENTIMON see ALB000
GENTINATRE see GCU050
GENTIOLETTEN see AOR500
GENTISAN see GCU050
GENTISATE see GCU000
GENTISATE SODIUM see GCU050
GENTISIC ACID see GCU000
GENTISIC ACID, MONOSODIUM SALT see GCU050
GENTISOD see GCU050
GENTRAN see DBD700
GENTRON 142B see CFX250
GENU see CCL250
GENU see PAO150
GENUGEL see CCL250
GENUGEL CJ see CCL250
GENUGOL RLV see CCL250

GENUINE ACETATE CHROME ORANGE see LCS000
GENUINE ORANGE CHROME see LCS000
GENUINE PARIS GREEN see COF500
GENUVISCO J see CCL250
GENVIS see SLJ500
GEOCARB 50EC see MOV000
GEOFOS see DHH200
GEOMYCIN see HOH500
GEON see PJR000
GEON see PKQ059
GEON 135 see AAX175
GEON 222 see CGW300
GEON 652 see CGW300
GEON LATEX 151 see PKQ059
GEOPEN see CBO250
GEOTRICYN see MCH525
GERANALDEHYDE see GCU100
GERANIAL see GCU100
GERANIC ACID see GCW000
GERANINE 2GS see CMM300
GERANIOL ACETATE see DTD800
GERANIOL ALCOHOL see DTD000
GERANIOL BUTYRATE see GDE810
GERANIOL CROTONATE see DTE000
GERANIOL EXTRA see DTD000
GERANIOL (FCC) see DTD000
GERANIOL FORMATE see GCY000
GERANIOL TETRAHYDRIDE see DTE600
GERANIUM CRYSTALS see PFA850
GERANIUM LAKE N see FAG070
GERANIUM OIL see GDA000
GERANIUM OIL ALGERIAN TYPE see GDA000
GERANIUM OIL BOURBON see GDC000
GERANIUM OIL, EAST INDIAN TYPE see PAE000
GERANIUM OIL MOROCCAN see GDE000
GERANIUM OIL, TURKISH TYPE see PAE000
GERANIUM THUNBERGII Sieb. et Zucc., extract see GDE300
GERANONITRILE see GDM000
GERANOXY ACETALDEHYDE see GDM100
GERANYL ACETATE (FCC) see DTD800
GERANYL ACETOACETATE see AAY500
GERANYL ACETONE see GDE400
GERANYL ALCOHOL see DTD000
GERANYL BENZOATE see GDE800
GERANYL BUTANOATE see GDE810
GERANYL-2-BUTENOATE see DTE000
GERANYL BUTYRATE see GDE810
GERANYL BUTYRATE see GDE825
GERANYL n-BUTYRATE see GDE810
GERANYL CAPROATE see GDG000
GERANYL CROTONATE see DTE000
GERANYL ETHYL ETHER see GDG100
GERANYL FARNESYL ACETATE see GDG200
GERANYL FORMATE (FCC) see GCY000
GERANYL HEXANOATE see GDG000
GERANYLINALOOL see GDG300
GERANYL ISOBUTYRATE see GDI000
GERANYL ISOVALERATE see GDK000
GERANYL NITRILE see GDM000
GERANYL OXYACETALDEHYDE see GDM100
GERANYL PHENYLACETATE see GDM400
GERANYL PROPIONATE see GDM450
GERANYL TIGLATE see GDO000
GERANYL α-TOLUATE see GDM400
GERASTOP see ARQ750
GERBITOX see BAV000
GERFIL see PMP500
GERHARDITE see CNN000
GERIAPAN see NMV300
GERISON see SNM500
GERISTEROL see NMV300
GERLACH 1396 see NBJ700
GERMACRA-1(10),4,11(13)-TRIEN-12-OIC ACID, 6-β,8-α-DIHYDROXY-, 12,6-

LACTONE, 4-HYDROXY-2-(HYDROXYMETHYL)CROTONATE see EQT100
GERMACR-1(10)-ENE-5,8-DIONE see GDO100
GERMACRONE see GDO200
GERMAIN'S see CBM750
GERMALGENE see TIO750
GERMALL 11 see IAS100
GERMALL 115 see GDO800
GERMA-MEDICA see HCL000
GERMANATE(2-), BIS(2-CARBOXYLATOETHYL)TRIOXODI-, DIHYDROGEN (9CI) see CCF125
GERMAN CHAMOMILE OIL see CDH500
GERMANE (DOT) see GEI100
GERMANE, PROPYLTRIIODO- see TKR050
GERMANE, TETRAPROPYL- see TED650
GERMANE, TRIIODOISOPROPYL- see TKQ300
GERMANE, TRIIODOPROPYL- see TKR050
GERMANIA see GEC000
GERMANIC ACID see GEC000
GERMANIC OXIDE (crystalline) see GDS000
GERMANIN see BAT000
GERMANIUM see GDU000
GERMANIUM, metal powder see GDU000
GERMANIUM BROMIDE see GDW000
GERMANIUM CHLORIDE see GDY000
GERMANIUM COMPOUNDS see GEA000
GERMANIUM, (l-CYSTEINE)TETRAHYDROXY- see CQK100
GERMANIUM DIOXIDE see GEC000
GERMANIUM ELEMENT see GDU000
GERMANIUM HYDRIDE see GEI100
GERMANIUM MONOHYDRIDE see GEG000
GERMANIUM OXIDE see GEC000
GERMANIUM OXIDE (GeO₂) see GEC000
GERMANIUM(II) SULFIDE see GEI000
GERMANIUM TETRABROMIDE see GDW000
GERMANIUM TETRACHLORIDE see GDY000
GERMANIUM TETRAHYDRIDE see GEI100
GERMERIN (GERMAN) see VHF000
GERMICICLIN see MDO250
GERMIN see EBL500
GERMINE see EBL500
GERMINOL see BBA500
GERMISAN see PFO250
GERMITOL see BBA500
GERNEBCIN see TGI250
GEROBIT see DBA800
GERODYL see DWK400
GERONTINE see DCC400
GERONTINE TETRAHYDROCHLORIDE see GEK000
GEROSTOP see GEK200
GEROT-EPILAN see MKB250
GEROT-EPILAN-D see DKQ000
GEROVIT see DBA800
GEROVITAL see AIL750
GEROX see SLW500
GERSDORFFITE see GEK300
GERTLEY BORATE see SFF000
GERVOT see MDT600
GESABAL see IKM050
GESADURAL see BJP250
GESAFID see DAD200
GESAFLOC see TJL500
GESAFRAM see MFL250
GESAFRAM 50 see MFL250
GESAGARD see BKL250
GESAGRAM see MFL250
GESAKUR see PNH750
GESAMIL see PMN850
GESAPAX COMBI see HBT100
GESAPAX-H see AHK300
GESAPAX MULTI see AHK300
GESAPON see DAD200
GESAPRIM see ARQ725

GESAPRIM 1802 see HBT100
GESAPRIM FORTE see HBT100
GESARAN see BJP000
GESARAN see INQ000
GESAREX see DAD200
GESAROL see DAD200
GESATAMIN see EGD000
GESATOP see BJP000
GESFID see MQR750
GESOPRIM see ARQ725
GESTANIN see AGF750
GESTANOL see AGF750
GESTANON see AGF750
GESTANYN see AGF750
GESTATRON see DYF759
GESTEROL L.A. see HNT500
GESTID see MQR750
GESTONORONE CAPROATE see GEK510
GESTONORONE CAPRONATE see GEK510
GESTORAL see GEK500
GESTRIGONE see ENX575
GESTRINONE see ENX575
GESTRONOL CAPROATE see GEK510
GESTRONOL HEXANOATE see GEK510
GESTYL see SCA750
GETTYSOLVE-B see HEN000
GETTYSOLVE-C see HBC500
GEUM ELATUM (Royle) Hook. f., extract see
GEK600
GEVEX see TGJ050
GEVILON see GCK300
GEVILON see SHL500
GF see DVR909
G-M-F see DVR200
GF 58 see TNJ500
GFX-E see GEK875
GFX-ES see GEK880
G1V GARD DXN see ABC250
GH see MJM500
GH5 see TCU660
GH8 see MEY100
GH9 see MEV800
GH 20 see PKP750
GH32 see MFN550
GH34 see MEX400
GHA 331 see AHC000
G 3063 HYDROCHLORIDE see GAC000
G-I see AOO375
GIACOSIL HYDROCHLORIDE see LFK200
GIALLO CROMO (ITALIAN) see LCR000
GIANT MILKWOOD see COD675
GIARDIL see NGG500
GIARLAM see NGG500
GIATRICOL see MMN250
GIBBERELLIC ACID see GEM000
GIBBERELLIN see GEM000
GIBBREL see GEM000
GIBS see CAX500
GIB-SOL see GEM000
GIB-TABS see GEM000
GICHTEX see ZVJ000
GIE see DVR600
GIEGY GS-13798 see MPG250
GIESE SALT see FAS800
GIFBLAAR POISON see FIC000
GIGANTIN see CMV000
GIHITAN see DCK759
G-II see GJU300
GILDER'S WHITING see CAT775
GILEMAL see CEH700
GILOTHERM OM 2 see TBD000
GILSONITE see GEO000
GILUCARD see RDK000
GILUCOR NITRO see NGY000
GILURYTMAL see AFH250
GILUTENSIN see DWF200
GILVOCARCIN V see GEO200
GIMID see DYC800
GINANDRIN see AOO410
GINARSOL see ABX500
GINBEY (DOMINICAN REPUBLIC) see
SLJ650

GINDARINE HYDROCHLORIDE see
GEO600
GINDARIN HYDROCHLORIDE see GEO600
GINEFLAVIR see MMN250
GINGER OIL see GEQ000
GINGERONE see VFP100
GINGICAIN M see TBN000
GINGILLI OIL see SCB000
GINKGO BILOBA L., ROOT EXTRACT see
GEQ100
GINSENG see GEQ400
GINSENG, ROOT EXTRACT see GEQ425
GINSENG ROOT-NEUTRAL SAPONINS see
GEQ425
GINSENGWURZEL, EXTRACT (GERMAN)
see GEQ425
GINSENOSIDE see GEQ450
GINSENOSIDE RB1 see PAF450
GIQUEL see HKR500
GIRACID see PDC250
GIRARD REAGENT T see GEQ500
GIRARD'S REAGENT T see GEQ500
GIRARD T REAGENT see GEQ500
GIRL see CNE750
GIROSTAN see TFQ750
GITALIN see GES000
GITALOXIGENIN + DIGITALOSE
(GERMAN) see VIZ200
GITALOXIGENIN-TRIDIGITOXOSID
(GERMAN) see GES100
GITALOXIN see GES100
GITALOXIN-16-FORMATE see GES100
GITHAGENIN see GMG000
GITOFORMATE see FND100
GITOFORMATE see PBG200
GITOFORMATO see PBG200
GITOXIGENIN-GLUCODIGITALOSID see
DKL325
GITOXIGENIN-3-o-MONODIGITALOSIDE
see SMN275
GITOXIGENIN-TRIDIGITOXOSID
(GERMAN) see GEU000
GITOXIGENIN TRIDIGITOXOXIDE-16-
FORMATE see GES100
GITOXIN see GEU000
GITOXIN PENTAACETATE see GEW000
GITOXIN, PENTAFORMATE see PBG200
GITOXOGENIN-d-DIGITALOSID
(GERMAN) see SMN275
GIUBA BLACK D see CMN230
GL 02 see PKP750
GL 03 see PKP750
G.L. 102 see EIT100
G.L. 105 see PBH100
GL 2487 see DLO880
GLACIAL ACETIC ACID see AAT250
GLACIAL ACRYLIC ACID see ADS750
GLAFENIN see GGQ050
GLAFENINE see GGQ050
GLANDUBOLIN see EDV000
GLANDUCORPIN see PMH500
GLANIL see CMR100
GLAPHENIN see GGQ050
GLAPHENINE see GGQ050
GLARUBIN see GEW700
GLASS see FBQ000
GLASS FIBERS see FBQ000
GLAUCARUBIN see GEW700
GLAUCINE see TDI475
(+)-GLAUCINE see TDI475
d-GLAUCINE see TDI475
s-(+)-GLAUCINE see TDI475
GLAUCINE HYDROCHLORIDE see TDI500
dl-GLAUCINE PHOSPHATE see TDI600
dl-GLAUCINPHOSPHAT (GERMAN) see
TDI600
GLAUMEBA see GEW700
GLAUPAX see AAI250
GLAURAMINE see IBB000
GLAUVENT see TDI475
GLAXORIDIN see TEY000
GLAZD PENTA see PAX250

GLAZIDIM see CCQ200
GLEAN see CMA700
GLEAN 20DF see CMA700
GLEBOFOS see DXH325
GLEEM see SHF500
GLENTONIN-RETARD see CCK125
GLIANIMON see FGU000
GLIANIMON see FLK100
GLIANIMON MITE see FLK100
GLIBENCLAMIDE see CEH700
GLIBORNURIDE see GFY100
GLIBUTIDE see BQL000
GLICAFAN see GGQ050
GLICLAZIDE see DBL700
GLICOL MONOCLORIDRINA (ITALIAN)
see EIU800
GLICOSIL see CPR000
GLIFAN see GGQ050
GLIFANAN see GGQ050
GLIFANAR see GGQ050
GLIFOSATO ESTRELLA see GIQ100
GLIFTOR see GEW725
GLIKOCEL TA see SFO500
GLIMID see DYC800
GLIODIN see GIO000
GLIOTOXIN see ARO250
GLIPASOL see GEW750
GLIPORAL see BQL000
GLIQUIDONE see GEW780
GLIRICIDIAL SEPIUM see RBZ000
GLISEMA see CKK000
GLISOLAMIDE see DBE885
GLISOXEPID see PMG000
GLISOXEPIDE see PMG000
GLIVENOL see EPW600
GLN-ALA-THR-VAL-GLY-ASP-VAL-ASN-
THR-ASP-ARG-PRO-GLY-LEU-LEU-ASP-
LEU-LYS see OAP400
GLO 5 see PKP750
GLOBENICOL see CDP250
GLOBOCICLINA see MDO250
GLOBOID see ADA725
GLOBULARIACITRIN see RSU000
γ-GLOBULIN see IBP250
GLOBULINS, γ- see IBP250
GLOGAL see HNI500
GLOMAX see KBB600
GLONOIN see NGY000
GLONSEN see SHK800
GLORIOSA LILY see GEW800
GLORIOSA ROTHSCHILDIANA see
GEW800
GLORIOSA SUPERBA see GEW800
GLOROUS see CDP250
GLORY LILY see GEW800
GLOSSO STERANDRYL see MPN500
GLOVER see LCF000
GLOXAZON see KFA100
GLOXAZONE see KFA100
GLU-P-2 see DWW700
GLUBORID see GFY100
GLUCAGON see GEW875
GLUCAL see CAS750
d-GLUCARIC ACID, MONOPOTASSIUM
SALT see PLJ350
GLUCAZIDE see GBB500
GLUCID see BCE500
GLUCIDORAL see BSM000
GLUCINIUM see BFO750
GLUCINUM see BFO750
GLUCITOL see SKV200
d-GLUCITOL see SKV200
d-GLUCITOL, 1-DEOXY-1-
(METHYLAMINO)-, 3,5-
BIS(ACETYLAMINO)-2,4,6-
TRIIODOBENZOATE (SALT) see AOO875
d-GLUCITOL, 1-DEOXY-1-
(METHYLAMINO)-, 3,5-DIIODO-4-OXO-
1(4H)-PYRIDINEACETATE (SALT) see
IFA100
d-GLUCITOL, 1,2:5,6-DIANHYDRO- see
DCI900

GLYCODIAZINE SODIUM SALT see GHK200
GLYCODINE see TCY750
GLYCO-FLAVINE see DBX400
GLYCOGEN see GHK300
GLYCOHYDROCHLORIDE see GHK000
GLYCOL see EJC500
GLYCOL ALCOHOL see EJC500
GLYCOLALDEHYDE see GHO100
GLYCOLAMIDE, N-(β-HYDROXY-α-(HYDROXYMETHYL)-p-NITROPHENETHYL)-, d-THREO- see CDP350
GLYCOLANILIDE see GHK500
GLYCOL BIS(HYDROXYETHYL) ETHER see TJQ000
GLYCOL BIS(MERCAPTOACETATE) see MCN000
GLYCOL BROMIDE see EIY500
GLYCOL BROMOHYDRIN see BNI500
GLYCOL BUTYL ETHER see BPJ850
GLYCOL CARBONATE see GHM000
GLYCOL CHLOROHYDRIN see EIU800
GLYCOL CYANOHYDRIN see HGP000
GLYCOL DIACETATE see EJD759
GLYCOL DIBROMIDE see EIY500
GLYCOL DICHLORIDE see EIY600
GLYCOL, DIETHOXYTETRAETHYLENE see PBO250
GLYCOL DIFORMATE see EJF000
GLYCOL DIGLYCIDYL ETHER see EEA600
GLYCOL DIMERCAPTOACETATE see MCN000
GLYCOL DIMETHACRYLATE see BKM250
GLYCOL DIMETHYL ETHER see DOE600
GLYCOLDINITRAAT (DUTCH) see EJG000
GLYCOL DINITRATE see EJG000
GLYCOL (DINITRATE DE) (FRENCH) see EJG000
GLYCOL ETHER see DJD600
GLYCOL ETHER de ACETATE see CBQ750
GLYCOL ETHER DB see DJF200
GLYCOL ETHER DB ACETATE see BQP500
GLYCOL-ETHERDIAMINETETRAACETIC ACID see EIT000
GLYCOL ETHER EB see BPJ850
GLYCOL ETHER EB ACETATE see BPJ850
GLYCOL ETHER EE see EES350
GLYCOL ETHER EE ACETATE see EES400
GLYCOL ETHER EM see EJH500
GLYCOL ETHER EM ACETATE see EJJ500
GLYCOL ETHER PM see PNL250
GLYCOL ETHERS see GHN000
GLYCOL ETHYLENE ETHER see DVQ000
GLYCOL ETHYL ETHER see DJD600
GLYCOL ETHYL ETHER see EES350
GLYCOLEUCINE see AJC950
GLYCOL FORMAL see DVR800
GLYCOLIC ACID see GHO000
GLYCOLIC ACID, (4-AMINO-3,5-DICHLOROPHENYL)- see AJM575
GLYCOLIC ACID, 2,2-DI-2-THIENYL-, 6,6,9-TRIMETHYL-9-AZABICYCLO(3.3.1)NON-3-YL ESTER, HCl see MBV100
GLYCOLIC ACID, ETHYL ESTER, METHYL PHTHALATE see MOD000
GLYCOLIC ACID, PHENYL- see MAP000
GLYCOLIC ACID PHENYL ETHER see PDR100
GLYCOLIC ACID, THIO- see TFJ100
GLYCOLIC ACID, 2-THIO- see TFJ100
GLYCOLIC ALDEHYDE see GHO100
GLYCOLIC NITRILE see HIM500
GLYCOLITHOCHOLIC ACID (6CI,7CI) see HND200
GLYCOLIXIR see GHA000
GLYCOLLIC ACID PHENYL ETHER see PDR100
GLYCOL METHACRYLATE see EJH000
GLYCOL METHYL ETHER see EJH500
GLYCOL MONOACETATE see EJI000
GLYCOL-MONOACETIN see EJI000

GLYCOL MONOBUTYL ETHER see BPJ850
GLYCOL MONOBUTYL ETHER ACETATE see BPM000
GLYCOLMONOCHLOORHYDRINE (DUTCH) see EIU800
GLYCOL MONOCHLOROHYDRIN see EIU800
GLYCOL MONOETHYL ETHER see EES350
GLYCOL MONOETHYL ETHER ACETATE see EES400
GLYCOL MONOMETHACRYLATE see EJH000
GLYCOL MONOMETHYL ETHER see EJH500
GLYCOL MONOMETHYL ETHER ACETATE see EJJ500
GLYCOL MONOMETHYL ETHER ACETYLRICINOLEATE see MIF500
GLYCOL MONOMETHYL ETHER ACRYLATE see MEM250
GLYCOL MONOMETHYL ETHER ACRYLATE see MIF750
GLYCOL MONOPHENYL ETHER see PER000
GLYCOL MONOSTEARATE see EJM500
GLYCOLMUL SOC see SKV170
GLYCOLONITRILE see HIM500
GLYCOLONITRILE ACETATE see COP750
(GLYCOLOYLOXY)TRIBUTYLSTANNANE see GHQ000
GLYCOL POLYETHYLENE MONOSTEARATE 200 see PJV500
GLYCOLPYRAMIDE see GHR609
GLYCOLS, POLYETHYLENE, (ALKYLIMINO)DIETHYLENE ETHER, MONOFATTY ACID ESTER see PKE550
GLYCOLS, POLYETHYLENE, DIMETHYL ETHER see DOM100
GLYCOLS, POLYETHYLENE, MONOCHOLESTERYL ETHER see PKE700
GLYCOLS, POLYETHYLENE, condensed with 23 moles MONODODECYL ETHER see LBU050
GLYCOLS, POLYETHYLENE, MONOHEXADECYL ETHER (8CI) see PJT300
GLYCOLS, POLYETHYLENE, MONOLAURATE (8CI) see PJY000
GLYCOLS, POLYETHYLENE, MONO(NONYLPHENYL) ETHER see NND500
GLYCOLS, POLYETHYLENE, MONO((1,1,3,3-TETRAMETHYLBUTYL)PHENYL) ETHER see GHS000
GLYCOLS, POLYETHYLENE MONO(TRIMETHYLNONYL) see GHU000
GLYCOLS, POLYETHYLENE POLYPROPYLENE, MONOBUTYL ETHER (nonionic) see GHY000
GLYCOLS, POLYTETRAMETHYLENE see GHY100
GLYCOL STEARATE see EJM500
GLYCOL SULFATE see EJP000
GLYCOL SULFITE see COV750
GLYCOLYLUREA see HGO600
GLYCOMONOCHLORHYDRIN see EIU800
GLYCOMUL O see SKV100
GLYCOMUL P see MRJ800
GLYCOMUL S see SKV150
GLYCOMUL SOC see SKV170
GLYCOMUL SOC SPECIAL see SKV170
GLYCON S-70 see SLK000
GLYCON DP see SLK000
GLYCONIAZIDE see GBB500
GLYCONITRILE see HIM500
GLYCONORMAL see GHK200
GLYCON RO see OHU000
GLYCON TP see SLK000
GLYCON WO see OHU000
GLYCOPHEN see GIA000
GLYCOPHENE see GIA000

GLYCOPROTEASE see GIA050
GLYCOPROTEINS, AVIDINS see GIA100
GLYCOPYRROLATE see GIC000
GLYCOPYRROLATE BROMIDE see GIC000
GLYCOPYRRONIUM BROMIDE see GIC000
GLYCOSIDE l-E1 see HAK075
GLYCOSOLVE DIP see IOJ500
GLYCOSPERSE L-20 see PKG000
GLYCOSPERSE O-20 see PKL100
GLYCOSPERSE L-20X see PKG000
GLYCOSPERSE L-20X see PKL000
GLYCOSPERSE TO-20 see TOE250
GLYCOSPERSE TS 20 see SKV195
GLYCO STEARIN see HKJ000
GLYCOTUSS see RLU000
GLYCOXALINEDICARBOXYLIC ACID see IAM000
GLYCURONE see GFM000
GLYCYCLAMIDE see CPR000
GLYCYL ALCOHOL see GGA000
2-GLYCYLAMINOFLUORENE see AKC000
GLYCYLBETAINE see GHA050
GLYCYL FP-12 see AFQ800
GLYCYRON see GIG000
GLYCYRRHETIC ACID see GIE000
18-β-GLYCYRRHETIC ACID see GIE000
GLYCYRRHETIN see GIE000
GLYCYRRHETINIC ACID see GIE000
α-GLYCYRRHETINIC ACID see GIE000
β-GLYCYRRHETINIC ACID see GIEO5O
18-β-GLYCYRRHETINIC ACID see GIE000
GLYCYRRHETINIC ACID GLYCOSIDE see GIG000
GLYCYRRHETINIC ACID HYDROGEN SUCCINATE DISODIUM SALT see CBO500
18-β-GLYCYRRHETINIC ACID HYDROGEN SUCCINATE DISODIUM SALT see CBO500
GLYCYRRHIZA see LFN300
GLYCYRRHIZAE (LATIN) see LFN300
GLYCYRRHIZA EXTRACT see LFN300
GLYCYRRHIZIC ACID see GIG000
GLYCYRRHIZIC ACID (8CI) see GIG000
18-β-GLYCYRRHIZIC ACID see GIG000
GLYCYRRHIZIC ACID, AMMONIUM SALT see GIE100
GLYCYRRHIZIN see GIG000
β-GLYCYRRHIZIN see GIG000
GLYCYRRHIZINA see LFN300
GLYCYRRHIZINIC ACID see GIG000
GLYCYRRHIZINIC ACID DISODIUM SALT see DXD875
GLYECINE A see TFI500
GLYESTRIN see ECU750
GLYFERRO see FBD500
GLYFYLLIN see DNC000
GLYHEXYLAMIDE see AKQ250
GLYHEXYLAMINE ISODIANE see AKQ250
GLYKOKOLAN SODNY (CZECH) see SHT000
GLYKOKOLLBETAIN see GHA050
GLYKOKOLLBETAIN-CHLORID see CCH850
GLYKOLDINITRAT (GERMAN) see EJG000
GLYME see DOE600
GLYME-3 see TKL875
GLYME-23 see DOM100
GLYMIDINE SODIUM SALT see GHK200
GLYMOL see MQV750
GLYODEX 3722 see CBG000
GLYODIN see GII000
GLYODIN see GIO000
GLYODIN ACETATE see GII000
GLYOTOL see GGS000
GLYOXAL see GIK000
GLYOXAL, 29.2% see GIK050
GLYOXAL, DIMETHYL- see BOT500
GLYOXAL, DIOXIME see EEA000
GLYOXALIDIN see GIO000
GLYOXALIN see IAL000
GLYOXALINE see IAL000
GLYOXALINE-5-ALANINE see HGE700

GLYOXALINE-5-ALANINE MONOHYDROCHLORIDE see HGE800
GLYOXAL, PHENYL-, 2-(DIETHYL ACETAL) see PFH260
GLYOXAL, PHENYL-, DIOXIME see HLI400
GLYOXANILIDE OXIME see GIK100
GLYOXIDE see GII000
GLYOXIDE see GIO000
GLY-OXIDE see HIB500
GLYOXIDE DRY see GII000
GLYOXIME see EEA000
GLYOXIME, CHLOROMETHYL- see PQC200
GLYOXIME, 1-CHLORO-2-METHYL- see PQC200
GLYOXIME, PHENYL- see HLI400
GLYOXYLANILIDE, OXIME see GIK100
GLYOXYLANILIDE, 2-OXIME see GIK100
GLYOXYLIC ACID see GIQ000
GLYOXYLIC ACID, METHYL ESTER see MKI550
GLYOXYLIC ACID, 1-METHYL-4-PIPERIDYL ESTER, 2-(BIS(p-CHLOROPHENYL) ACETAL) see MON600
GLYOXYLIC ACID, PHENYL- see OOK150
GLYOXYLONITRILE, PHENYL-, OXIME, o,o-DIMETHYL PHOSPHOROTHIOATE see MOC275
GLYPASOL see GEW750
GLYPED see THM500
GLYPHOSATE see PHA500
GLYPHOSATE ISOPROPYLAMINE SALT see GIQ100
GLYPHOSINE see BLG250
GLYPHYLLIN see DNC000
GLYPHYLLINE see DNC000
GLYSANOL B see ART250
GLYSOBUZOLE see IIY100
GLYSOLETTEN see EOK000
GLYTAC see GGY200
GLYTAC A 100 see GGY200
GLYVENAL see EPW600
GLYVENOL see EPW600
GM 3 see MCB050
GM 4 see MCB050
GM 14 see PKP750
G-MAC see GGY200
GMI see DLS600
GMK 527 see PJB800
GMO 8903 see GGR200
GMP see GLS750
5'-GMP see GLS750
GMP DISODIUM SALT see GLS800
5'-GMP DISODIUM SALT see GLS800
GMP SODIUM SALT see GLS800
GM 10 (SILICATE) see SCN700
GM SULFATE see GCS000
GNATROL see BAC130
GNOSCOPINE see NBP275
GNS see GEQ425
GO-80 see FBV100
GO 186 see SHL500
GO-1261 see EIH000
GO 10213 see SAY950
GOAL see OQU100
GODALAX see PPN100
GOE 1734 see DBQ125
G 3802POE see PJT300
G 3804POE see PJT300
GOE 687 (GERMAN) see MNG000
GOHSENOL see PKP750
GOHSENOL AH 22 see PKP750
GOHSENOL GH see PKP750
GOHSENOL GH 17 see PKP750
GOHSENOL GH 20 see PKP750
GOHSENOL GH 23 see PKP750
GOHSENOL GL 03 see PKP750
GOHSENOL GL 05 see PKP750
GOHSENOL GL 08 see PKP750
GOHSENOL GM 14 see PKP750
GOHSENOL GM 94 see PKP750
GOHSENOL GM 14L see PKP750
GOHSENOL KH 17 see PKP750

GOHSENOL NH 05 see PKP750
GOHSENOL NH 17 see PKP750
GOHSENOL NH 18 see PKP750
GOHSENOL NH 20 see PKP750
GOHSENOL NH 26 see PKP750
GOHSENOL NK 114 see PKP750
GOHSENOL NL 05 see PKP750
GOHSENOL NM 14 see PKP750
GOHSENYL E 50 Y see AAX250
D-GOITRIN see VQA100
(R)-GOITRIN see VQA100
GOLAGEN K see MCB575
GOLARSYL see ACN250
GOLD see GIS000
1721 GOLD see CNI000
GOLD(I) ACETYLIDE see GIT000
GOLDBALLS see FBS100
GOLD BOND see CCP250
GOLD BRONZE see CNI000
GOLD CHLORIDE see GIW176
GOLD(III) CHLORIDE see GIW176
GOLD CHLORIDE SODIUM see GIZ100
GOLD COMPOUNDS see GIW179
GOLD CREST VENGEANCE see BMO300
GOLD(I) CYANIDE see GIW189
GOLD DUST see CNE750
GOLDEN ANTIMONY SULFIDE see AQF500
GOLDEN BROWN RK-FQ see CMP250
GOLDEN CEYLON CREEPER see PLW800
GOLDEN CHAIN see GIW195
GOLDEN DEWDROP see GIW200
GOLDEN HUNTER'S ROBE see PLW800
GOLDEN HURRICANE LILY see REK325
GOLDEN POTHOS see PLW800
GOLDEN RAIN see GIW300
GOLDEN SEAL see ICC800
GOLDEN SHOWER see GIW300
GOLDEN SPIDER LILY see REK325
GOLDEN YELLOW see DCZ000
GOLDEN YELLOW see DUX800
GOLDEN YELLOW RUAF see CMM758
GOLD FLAKE see GIS000
GOLD(III) HYDROXIDE-AMMONIA see GIX300
GOLD KERATINATE see GIX350
GOLD LEAF see GIS000
GOLD NITRIDE AMMONIA see GIY000
GOLD(I) NITRIDE-AMMONIA see GIY300
GOLD(III) NITRIDE TRIHYDRATE see GIZ000
GOLD ORANGE see MND600
GOLD ORANGE MP see DOU600
GOLD POWDER see GIS000
GOLDQUAT 276 see PAJ000
GOLD SATINOBRE see LDS000
GOLD SODIUM CHLORIDE see GIZ100
GOLD SODIUM THIOMALATE see GJC000
GOLD SODIUM THIOSULFATE see GJE000
GOLD SODIUM THIOSULFATE DIHYDRATE see GJG000
GOLD THIOGLUCOSE see ART250
GOLD TRICHLORIDE see GIW176
GOLDWEED see FBS100
GOLD YELLOW see MRL100
GOLTIX see ALA500
GOMBARDOL see SNM500
GOMELIN see BAC040
GOMISIN A see SBE450
GONABION see CMG675
GONACRINE see DBX400
GONACRINE see XAK000
GONADEX see CMG675
GONADOGRAPHON LUTEINIZING HORMONE see GJI075
GONADORELIN see LIU370
GONADORELIN ACETATE see LIU305
GONADORELIN DIACETATE see LIU380
GONADOTRAPHON FSH see SCA750
GONADOTROPIN-RELEASING FACTOR see LIU370

GONADOTROPIN RELEASING HORMONE see LIU370
GONADOTROPIN RELEASING HORMONE AGONIST see GJI100
GONADOTROPIN RELEASING HORMONE, (d-TRP^6)-PRO^9)-NET)- see LIU353
GONADYL see SCA750
GONA-1,3,5(10)-TRIENE-3,16-α-17-β-TRIOL, 13-ETHYL- see HGI700
GONA-1,3,5(10)-TRIENE-3,16,17-TRIOL, 13-ETHYL-, (16-α-17-β)-(9CI) see HGI700
GONA-1,3,5(10)-TRIEN-17-ONE, 3-METHOXY- see NNP100
GONDAFON see GHK200
GONOCRIN see XAK000
GONOSAN see GJI250
GONTOCHIN see CLD000
GONTOCHIN PHOSPHATE see CLD250
GONYAULAX TOXIC DIHYDROCHLORIDE see SBA500
GOOD-RITE see PJR000
GOODRITE 1800X73 see SMR000
GOOD-RITE GP 264 see DVL700
GOOD-RITE K 37 see ADV900
GOOD-RITE K-700 see ADV900
GOOD-RITE K 702 see ADV900
GOOD-RITE K727 see ADV900
GOOD-RITE NIX see SIA000
GOOD-RITE WS 801 see ADV900
GOOLS see MBU550
GOOSEBERRY TOMATO see JBS100
GOPHACIDE see BIM000
GOPHER BAIT see SMN500
GOPHER-GITTER see SMN500
GORDOLOBO YERBA (MEXICO) see RBA400
GORDONA see RAF100
GORDON'S MECOMEC see CLO200
GORE-TEX see TAI250
GORMAN see SCA750
GOSHTAM see CNT350
GOSLING see PAM780
GOSSYPIMINE see GJI400
GOSSYPINE see CMF800
GOSSYPLURE see GJK000
GOSSYPLURE H.F. see GJK000
GOSSYPOL see GJM000
(+)-GOSSYPOL see GJM025
(±)-GOSSYPOL see GJM030
racemic-GOSSYPOL see GJM030
GOSSYPOL ACETATE see GJM035
GOSSYPOL ACETATE see GJM259
GOSSYPOL ACETIC ACID see GJM259
GOSSYPOSE see RBA100
GOTAMINE TARTRATE see EDC500
GOTA de SANGRE (CUBA) see PCU375
GOTENSIN see LFA100
GOTHNION see ASH500
GOUGEROTIN see ARP000
GOUT STALK see CNR135
G 2 (OXIDE) see AHE250
GOYL see ABX500
GP 7I see SCK600
GP 11I see SCK600
GP-121 see AOO500
GP 130 see DTN775
2100 GP see PJS750
GP 33679 see DPX400
GP 38383 see IBP309
GP 45840 see DEO600
GP-40-66:120 see HCD250
GP-AMIN see MFB000
GPKh see HAR000
G 50 (POLYSACCHARIDE) see EHG100
GR see GLS000
GR-23 see GFG200
GR-29 see GFG205
GR 2/234 see AFK875
GR 33040 see DCV800
GR 38032 see OJD150
GR 43175 see DPH300

GR218,231 see DWR350
GR 38032X see OJD150
GR 43175X see DPH300
GRAAFINA see EDP000
de GRAAFINA see EDP000
GRACET VIOLET 2R see DBP000
GRACILIN see GJM300
GRAFESTROL see DKA600
GRAHAM'S SALT see SII500
GRAIN ALCOHOL see EFU000
GRAINES D'EGLISE (GUADELOUPE) see RMK250
GRAINS de LIN PAYS (HAITI) see LED500
GRAIN SORGHUM HARVEST-AID see SFS000
GRAMAXIN see CCS250
GRAMEVIN see DGI400
GRAMEVIN see DGI600
GRAMICIDIN see GJO000
GRAMICIDIN A see GJO025
GRAMIN see DYC000
GRAMINE see DYC000
GRAMINIC ACID see GJQ100
GRAMINON see IRA050
GRAMISAN see MEP250
GRAMIXEL see PAJ000
GRAMONOL see PAJ000
GRAMOXON see PAJ000
GRAMOXONE see PAJ000
GRAMOXONE D see PAJ000
GRAMOXONE DICHLORIDE see PAJ000
GRAMOXONE METHYL SULFATE see PAJ250
GRAMOXONE S see PAI990
GRAMOXONE S see PAJ000
GRAMOXONE W see PAJ000
GRAMPENIL see AIV500
GRAMURON see PAJ000
GRANADA GREEN LAKE GL see PJQ100
GRANALINO (DOMINICAN REPUBLIC) see LED500
GRANATICIN see GJS000
GRANATICIN A see GJS000
GRANDAXIN see GJS200
GRANEX O see SFS000
GRANMAG see MAH500
GRANODINE 80 see ZJS400
GRANODINE 16NC see ZJS400
GRANOSAN see CHC500
GRANOSAN M see EME100
GRANOSAN M see EME100
GRANOX NM see HCC500
GRANOX PPM see CBG000
GRANULAR ZINC see ZBJ000
GRANULATED SUGAR see SNH000
GRANULIN see ACR300
GRANUREX see BRA250
GRANUTOX see PGS000
GRAPE BLUE A GEIGY see FAE100
GRAPE COLOR EXTRACT see GJS300
GRAPEFRUIT OIL see GJU000
GRAPEFRUIT OIL, expressed see GJU000
GRAPEFRUIT OIL, coldpressed see GJU000
GRAPEMONE see GJU050
GRAPE SKIN EXTRACT see GJU100
GRAPE SUGAR see GFG000
GRAPHIC RED Y see NAP100
GRAPHITE see CBT500
GRAPHITE SYNTHETIC (ACGIH,OSHA) see CBT500
GRAPHLOX see HCE500
GRAPHOL see MGJ750
GRAPHTAL RED RL see CJD500
GRAPHTOL BLUE RL see IBV050
GRAPHTOL GREEN 2GLS see PJQ100
GRAPHTOL ORANGE GP see CMS145
GRAPHTOL RED A-4RL see MMP100
GRAPHTOL RED 2GL see DVB800
GRAPHTOL YELLOW A-HG see DEU000
GRAPHTOL YELLOW GXS see CMS210
GRAPHTOL YELLOW RGS see CMS208
GRASAL BRILLIANT RED B see SBC500

GRASAL BRILLIANT RED G see OHI200
GRASAL BRILLIANT YELLOW see DOT300
GRASAL YELLOW see FAG130
GRASAN BRILLIANT RED B see SBC500
GRASAN BROWN DT NEW see NBG500
GRASAN ORANGE 3R see XRA000
GRASAN ORANGE R see PEJ500
GRASCIDE see DGI000
GRASEX see CDN550
GRASIDIM see CDK800
GRASIP see FBP350
GRASIPAN see FBP350
GRASLAN see BSN000
GRASOL BLUE 2GS see TBG700
GRASOL VIOLET R see DBP000
GRASOL YELLOW RSF see CMP600
GRASP see GJU200
GRASPAZ see FBP350
GRASSLAND WEEDKILLER see BAV000
GRATIBAIN see OKS000
GRATUS STROPHANTHIN see OKS000
GRAVIDOX see PPK250
GRAVINOL see DYE600
GRAVISTAT see NNL500
GRAVOCAIN see DHK400
GRAVOCAIN HYDROCHLORIDE see DHK400
GRAVOL see DYE600
GRAY ACETATE see CAL750
GRAY AMBER see AHJ000
GRAYANOL A see GJU225
GRAYANOL B see GJU250
GRAYANOTOXANE-3,5,6,10,14,16-HEXOL, (3-β,6-β,14R)- see GJU310
GRAYANOTOXANE-3,5,6,10,14,16-HEXOL 14 ACETATE see AOO375
GRAYANOTOXANE-5,6,7,10,16-PENTOL, 2,3-EPOXY-, (2-β,3-β,6-β,7-α)- see DAD700
GRAYANOTOX-10(20)-ENE-3,5,6,14,16-PENTOL, (3-β,6-β,14R)- see GJU300
GRAYANOTOX-15-ENE-3,5,6,10,14-PENTOL, (3-β,6-β,14R)- see GJU330
GRAYANOTOXIN I see AOO375
GRAYANOTOXIN II see GJU300
GRAYANOTOXIN III see GJU310
GRAYANOTOXIN III 6,14-DIACETATE see GJU315
GRAYANOTOXIN VI see GJU330
GREAT BLUE LOBELIA see CCJ825
GREAT LAKES BA-59P see MKA270
GREEN 5 see ADF000
1724 GREEN see FAG000
11091 GREEN see BLK000
11661 GREEN see CMJ900
12078 GREEN see ADF000
GREEN BS see ADF000
GREEN CHROME OXIDE see CMJ900
GREEN CHROMIC OXIDE see CMJ900
GREEN CINNABAR see CMJ900
GREEN CROSS COUCH GRASS KILLER see TII500
GREEN CROSS CRABGRASS KILLER see PLC250
GREEN CROSS WARBLE POWDER see RNZ000
GREEN-DAISEN M see DXI400
GREEN DENSIC see SCQ000
GREEN DENSIC GC 800 see SCQ000
GREEN DRAGON see JAJ000
GREENHARTEN see HLY500
GREEN HELLEBORE see FAB100
GREEN HELLEBORE see VIZ000
GREEN HYDROQUINONE see QFJ000
GREEN LILY see GJU460
GREEN LOCUST see GJU475
GREEN MAMBA VENOM see GJU500
GREEN No. 2 see BLK000
GREEN No. 203 see FAF000
GREEN NICKEL OXIDE see NDF500
GREENOCKITE see CAJ750
GREEN OIL see APG500
GREEN OIL see COD750

GREEN ROUGE see CMJ900
GREEN S see ADF000
GREEN SEAL-8 see ZKA000
GREEN VITRIOL see FBN100
GREEN VITROL see FBO000
GRELAN see AMK250
GRELUTIN see NBL200
GRENADE see GJU600
GRENADES, empty primed (NA0349) (DOT) see GJU600
GRENADES, hand or rifle, with bursting charge (UN0284, UN0285, UN0292, UN0293) (DOT) see GJU600
GRENADES, practice, hand or rifle (UN0452, UN0110, UN0318, UN0372) (DOT) see GJU600
GREOSIN see GKE000
GRESFEED see GKE000
G 3 (RESIN) see MCB050
G-RESINS see PJS750
GREX see PJS750
GREX PP 60-002 see PJS750
GREY ARSENIC see ARA750
GREY NICKER see CAK325
Gn-RH see LIU370
GnRH-A see GJI100
GRICIN see GKE000
GRIFFEX see ARQ725
GRIFFIN MANEX see MAS500
GRIFFITH'S ZINC WHITE see LHX000
GRIFOMIN see TEP500
GRIFULVIN see GKE000
GRILON see NOH000
GRILON see PJY500
GRILONIT RV 1806 see BOS100
GRIPENIN see CBO250
GRIPPEX see TEH500
GRISACTIN see GKE000
GRISCOFULVIN see GKE000
GRISEFULINE see GKE000
GRISEIN see GJU800
GRISEMIN see CEX250
GRISEO see GKE000
(+)-GRISEOFULVIN see GKE000
GRISEOFULVIN-FORTE see GKE000
GRISEOFULVINUM see GKE000
GRISEOLIC ACID see GJU900
GRISEOLIC ACID A see GJU900
GRISEOLUTEIN B see GJW000
GRISEOMYCIN see GJY000
GRISEORUBIN COMPLEX see GJY100
GRISEORUBIN I HYDROCHLORIDE see GKA000
GRISEOVIRIDIN see GKC000
GRISETIN see GKE000
GRISIN see CEX250
GRISIN see GJU800
GRISOFULVIN see GKE000
GRISOL see TCF250
GRISOLEN see PJS750
GRISOVIN see GKE000
GRIS-PEG see GKE000
GR-M see PJQ050
GROCEL see GEM000
GROCO see LGK000
GROCO 2 see OHU000
GROCO 4 see OHU000
GROCO 54 see SLK000
GROCO 5L see OHU000
GROCO 5810 see BSL600
GROCOLENE see GGA000
GROCOR 5500 see OAV000
GROCOR 6000 see OAV000
GROFAS see QSA000
GROPPER see MQR400
GROSAFE see CBT500
GROTAN B see THR820
GROTAN BK see THR820
GROUND HEMLOCK see YAK500
GROUNDNUT OIL see PAO000
GROUND RYANIA SPECISA(VAHL) STEMWOOD (ALKOLOID RYANODINE) see RSZ000

GROUNDSEL see RBA400
GROUND VOCLE SULPHUR see SOD500
GR 20263 PENTAHYDRATE see CCQ200
L-GRUEN No. 1 see CKN000
L-GRUEN No. 1 (GERMAN) see CKN000
GRUMEX see BQW490
GRUNDIER ARBEZOL see PAX250
GRUNERITE see GKE900
GR 33343X see HNI600
GRYSIO see GKE000
GRYZBOL see DVF800
GRZYBOL see DVF800
GS 6 see IGK800
GS-95 see TEZ000
GS 015 see BMB125
G 339 S see CAX350
GS-3065 see DYE425
GS 6244 see FOI000
GS 13528 see THR600
GS 13529 see BQB000
GS-13,798 see MPG250
GS 14259 see BQC500
GS 15254 see BQC250
GS-16068 see EPN500
GS 34360 see INR000
GSH see GFW000
G-STROPHANTHIN see OKS000
G 52 SULFATE see GAA120
GSY 930 see MOS100
GT see PCU360
GT41 see BOT250
GT-1012 see DNB000
GT 2041 see BOT250
GTB see GGU000
GTG see ART250
5'-GTGGTGGGTGGGT-GGGT-3' see ZNS400
GTN see NGY000
GUABENXANE see GKG300
GUACALOTE AMARILLO (CUBA) see CAK325
GUACAMAYA (CUBA) see CAK325
GUACIS (MEXICO) see LED500
GUAIAC ACETATE see GKG400
GUAIAC GUM see GLY100
GUAIACOL see GKI000
m-GUAIACOL see REF050
GUAIACOLGLICERINETERE see RLU000
GUAIACOL GLYCERYL ETHER see RLU000
GUAIACOL GLYCERYL ETHER CARBAMATE see GKK000
GUAIACURANE see RLU000
GUAIACWOOD ACETATE see GKG400
GUAIAC WOOD OIL see GKM000
GUAIACYL GLYCERYL ETHER see RLU000
GUAIACYLPROPANE see MFM750
GUAIA-1(5),7(11)-DIENE see GKO000
GUAIAMAR see RLU000
GUAIANESIN see RLU000
s-GUAIAZULENE see DSJ800
GUAICOL see GKI000
GUAIENE see GKO000
β-GUAIENE see GKO000
GUAIFENESIN see RLU000
GUAIMERCOL see GKO500
GUAIOL ACETATE see GKG400
GUAIPHENESINE see RLU000
GUAJACOL-GLYCERINAETHER (GERMAN) see RLU000
GUAJACOL-α-GLYCERINETHER see RLU000
GUAMIDE see AHO250
GUANABENZ see GKO750
GUANABENZ ACETATE see GKO750
GUANATOL HYDROCHLORIDE see CKB500
GUANAZODINE see GKO800
GUANAZODINE SULFATE MONOHYDRATE see CAY875
GUANAZOL see AJO500
GUANAZOLE see DCF200
GUANAZOLO see AJO500

GUANERAN see AKY250
GUANETHIDINE see GKQ000
GUANETHIDINE BISULFATE see GKS000
GUANETHIDINE MONOSULFATE see GKU000
GUANETHIDINE SULFATE see GKS000
GUANFACINE HYDROCHLORIDE see GKU300
GUANICAINE see PDN500
GUANICIL see AHO250
GUANIDAN see AHO250
GUANIDINE see GKW000
GUANIDINE, AMINO-, HYDROCHLORIDE see AKC800
GUANIDINE, (4-(((2-((4-AMINO-1,2,5-THIADIAZOL-3-YL)AMINO)ETHYL)THIO)METHYL)-2-THIAZOLYL)-, S-OXIDE see GKW050
GUANIDINE CARBONATE see GKW100
GUANIDINECARBOXAMIDE see GLS900
GUANIDINE CHLORIDE see GKY000
GUANIDINE, 2-CYANO-1,1-DIMETHYL- see COM085
GUANIDINE, N'-CYANO-N,N-DIMETHYL- see COM085
GUANIDINE, CYANO-, METHYLMERCURY deriv. see MLF250
GUANIDINE, 1,3-DIPHENYL-, MONOHYDROCHLORIDE see DWC625
GUANIDINE, N,N'-(DITHIODI-2,1-ETHANEDIYL)BIS-(9CI) see GLC200
GUANIDINE, 1,1'-(DITHIODIETHYLENE)DI- see GLC200
GUANIDINE, 1-HEXYL-3-NITRO-1-NITROSO- see HFR600
GUANIDINE HYDROCHLORIDE see GKY000
GUANIDINE, 1,1'-(IMINOBIS(OCTAMETHYLENE))DI-, TRIACETATE see GLQ100
GUANIDINE, N-METHYL-N'-NITRO-N-NITROSO-(9CI) see MMP000
GUANIDINE, N-METHYL-N'-NITROSO- see MMX600
GUANIDINE, MONOHYDROCHLORIDE see GKY000
GUANIDINE MONONITRATE see GLA000
GUANIDINE MONOPHOSPHATE see GLS750
GUANIDINE, N,N'''-(IMINODI-8,1-OCTANEDIYL)BIS-, TRIACETATE see GLQ100
GUANIDINE NITRATE (DOT) see GLA000
GUANIDINE, NITRO- see NHA500
GUANIDINE, N'-NITRO-N-NITROSO-N-PENTYL- see NLC500
GUANIDINE, SULFANILYL- see AHO250
GUANIDINE THIOCYANATE see TFF250
GUANIDINIUM CARBONATE see GKW100
GUANIDINIUM CHLORIDE see GKY000
GUANIDINIUM DICHROMATE see GLB100
GUANIDINIUM HYDROCHLORIDE see GKY000
GUANIDINIUM NITRATE see GLB300
GUANIDINIUM PERCHLORATE see GLC000
GUANIDINIUM THIOCYANATE see TFF250
p-GUANIDINOBENZOIC ACID 4-ALLYL-2-METHOXYPHENYL ESTER see MDY300
p-GUANIDINOBENZOIC ACID 4-METHYL-2-OXO-2H-1-BENZOPYRAN-7-YL ESTER see GLC100
p-GUANIDINOBENZOIC ACID p-NITROPHENYL ESTER see NIQ500
γ-GUANIDINOBUTYRIC ACID AMIDE, NITROSATED see NJU000
GUANIDINOETHYL DISULFIDE see GLC200
2-GUANIDINOETHYL DISULFIDE see GLC200

1-(2-GUANIDINOETHYL)HEPTAMETHYLENIMINE see GKQ000
N-(2-GUANIDINOETHYL)HEPTAMETHYLENIMINE SULFATE see GKU000
1-(2-GUANIDINOETHYL)OCTAHYDROAZOCINE see GKQ000
1-(2-GUANIDINOETHYL)OCTAHYDROAZOCINE SULFATE (2:1) see GKS000
γ-GUANIDINO-β-HYDROXYBUTYRIC ACID AMIDE, NITROSATED see HNA000
GUANIDINO-6-METHYL-1,4-BENZODIOXANE see GKG300
6-GUANIDINOMETHYL 1,4-BENZODIOXANE SULFATE see GKG300
GUANIDINO METHYL-6-BENZODIOXANNE-1,4 (FRENCH) see GKG300
GUANIDOETHYL DISULFIDE see GLC200
GUANINE see GLI000
GUANINE DEOXYRIBOSIDE see DAR800
GUANINE HYDROCHLORIDE see GLK000
GUANINE-3-N-OXIDE see GLK100
GUANINE-7-N-OXIDE see GLM000
GUANINE-3-N-OXIDE HEMIHYDROCHLORIDE see GLO000
(GUANINE)PENTAAMMINERUTHENIUM(3+) TRICHLORIDE see GLO100
GUANINE, 9-β-D-RIBOFURANOSYL- see GLS000
GUANINE RIBOSIDE see GLS000
GUANIOL see DTD000
GUANOCTINE see GLQ000
GUANOCTINE TRIACETATE see GLQ100
GUANOSINE see GLS000
GUANOSINE, 2'-DEOXY-, 5'-(TETRAHYDROGEN TRIPHOSPHATE) see GLS100
GUANOSINE, 2'-DEOXY-6-THIO- see TFJ200
GUANOSINE 2'-DEOXY-, 5'-TRIPHOSPHATE see GLS100
GUANOSINE, 7,8-DIHYDRO-2'-DEOXY-8-OXO- see HKA770
GUANOSINE, 7,8-DIHYDRO-8-OXO- see ONW150
GUANOSINE MONOPHOSPHATE see GLS750
GUANOSINE 5'-MONOPHOSPHATE see GLS750
GUANOSINE 5'-MONOPHOSPHORIC ACID see GLS750
GUANOSINE 5'-PHOSPHATE see GLS750
GUANOSINE 5'-(TETRAHYDROGEN TRIPHOSPHATE), 2'-DEOXY-(9CI) see GLS100
GUANOTHIAZON see AHI875
GUANOXYFEN SULFATE see GLS700
GUANTLET see CHJ300
GUANYLCYSTAMINE see GLC200
GUANYL DISULFIDE DIHYDROCHLORIDE see DXN400
GUANYL HYDRAZINE see AKC750
GUANYLHYDRAZINE HYDROCHLORIDE see AKC800
GUANYLIC ACID see GLS750
5'-GUANYLIC ACID see GLS750
GUANYLIC ACID SODIUM SALT see GLS800
GUANYL NITROSAMINO GUANYL TETRAZENE see TEF500
1-GUANYL-4-NITROSAMINOGUANYLTETRAZENE see TEF500
GUANYL NITROSAMINOGUANYLTETRAZENE (dry) (DOT) see TEF500
GUANYL NITROSAMINOGUANYLTETRAZENE,

wetted or tetrazene, wetted with not <30% water
or (DOT) see TEF500
N₁-GUANYLSULFANILAMIDE see AHO250
GUANYLUREA see GLS900
GUAR see GLU000
GUARAN see GLU000
GUARANINE see CAK500
GUAR FLOUR see GLU000
GUAR GUM see GLU000
GUAR GUM, CARBOXYMETHYL ETHER,
SODIUM SALT see SFO600
GUAR GUM, SODIUM CARBOXYMETHYL
DERIVATIVE see SFO600
GUASTIL see EPD500
GUATEMALA LEMONGRASS OIL see
LEH000
GUAVA see GLW000
GUAYIGA (DOMINICAN REPUBLIC) see
CNH789
GUAZATINE TRIACETATE see GLQ100
GUAZATIN TRIACETATE see GLQ100
GUBERNAL see AGW000
GUDAKHU (INDIA) see SED400
GUESAPON see DAD200
GUESAROL see DAD200
GUIACOL-GLICERILETERE
MONOCARBAMMATO see GKK000
GUICITRINA see AIV500
GUICITRINE see AIV500
GUIDAZIDE see GBB500
GUIGNER'S GREEN see CMJ900
GUINDA (PUERTO RICO) see APM875
GUINEA GREEN see FAE950
GUINEA GREEN B see FAE950
GUINEA GREEN BA see FAE950
GULF S-15126 see DED000
GULITOL see SKV200
l-GULITOL see SKV200
GULLIOSTIN see PCP250
GUM see PJR000
GUM ARABIC see AQQ500
GUM CAMPHOR see CBA750
GUM CARRAGEENAN see CCL250
GUM CHON 2 see CCL250
GUM CHROND see CCL250
GUM CYAMOPSIS see GLU000
GUM GHATTI see GLY000
GUM GUAIAC see GLY100
GUM GUAR see GLU000
GUM NAFKACRYSTAL see ROH900
GUM OPIUM see OJG000
GUM OVALINE see AQQ500
GUM SENEGAL see AQQ500
GUM STERCULIA see KBK000
GUM STERCULIA see SLO500
GUM TARA see GMA000
GUM TRAGACANTH see THJ250
GUNACIN see GMC000
GUNCOTTON see CCU250
GUNPOWDER see ERF500
GUNPOWDER see PLL750
GUNPOWDER, compressed (UN 0028) (DOT)
see ERF500
GUNPOWDER, compressed (UN 0028) (DOT)
see PLL750
GUNPOWDER, granular or as a meal (UN
0027) (DOT) see ERF500
GUNPOWDER, granular or as a meal (UN
0027) (DOT) see PLL750
GUNPOWDER, in pellets (UN 0028) (DOT) see
ERF500
GUNPOWDER, in pellets (UN 0028) (DOT) see
PLL750
GURJUN BALSAM see GME000
GURONSAN see GFM000
GUSATHION see ASH500
GUSATHION A see EKN000
GUSERVIN see GKE000
GUSTAFSON CAPTAN 30-DD see CBG000
GUTHION (DOT) see ASH500
GUTHION (ETHYL) see EKN000
GUTRON see MQT530

GUTTAGENA see PKQ059
GUTTALAX see SJJ175
A″2-GUTTIFERIN see GME300
β-GUTTIFERIN see CBA125
α-2-GUTTIFERIN see GME300
α-GUTTIFERIN (9CI) see GME300
G-X 55 see CNB825
GYKI 11679 see RFU800
GYKI 14166 see PEE100
GYN see MAG000
GYNAESAN see EDU500
GYN-ANOVLAR see EEH520
GYNECLORINA see CDP000
GYNECORMONE see EDP000
GYNEFOLLIN see DAL600
GYNE-LOTRIMIN see MRX500
GYNE-MERFEN see MDH500
GYNERGEN see EDC500
GYNERGON see EDO000
GYNESINE see TKL890
GYNESTREL see EDO000
GYNFORMONE see EDP000
GYNOCHROME see MCV000
GYNOESTRYL see EDO000
GYNOFON see ABY900
GYNOKHELLAN see AHK750
GYNOLETT see DKB000
GYNONLAR 21 see EEH520
GYNO-PEVARYL see EAE000
GYNOPHARM see DKA600
GYNOPLIX see ABX500
GYNOREST see DYF759
GYNOVLAR see EQM500
GY-PHENE see CDV100
GYPSINE see LCK000
GYPSINE see LCK100
GYPSOGENIN see GMG000
GYPSOPHILASAPOGENIN see GMG000
GYPSOPHILASAPONIN see GMG000
GYPSUM see CAX750
GYPSUM STONE see CAX750
GYRANE see GMG100
GYROMITRIN see AAH000
GYRON see DAD200
H-22 see BSG300
H-28 see BSG250
H-33 see BPP250
H-34 see HAR000
H 3S see IHC550
H 46 see AHC000
H-69 see DTN200
H-88 see TKF250
H 95 see CKF750
H115 see CDY325
H 133 see DER800
H.17B see HHQ825
H 224 see BRA625
H241 see EQN320
H 321 see DST000
H-365 see ELL500
454H see SMQ500
H-490 see ENG500
H 520 see RDK000
H 610 see EOM000
H 774 see BKO825
H-899 see AKG500
H 940 see ENG500
H 990 see AEX500
"H" see HBT500
H 1032 see MHJ500
H-1067 see BKO840
H-1075 see HAA300
H-1286 see HAA310
H 1313 see DER800
H 1672 see PEF750
H-2053 see LIU500
2903 H see CDS275
H3 111 see DBC510
H3261 see NCX525
H 3292 see DNN600
H 3452 see FBP100
H 4007 see MCH535

H 4065 see LIU353
H4 099 see DTL200
H-4723 see CIR750
86H 60 see HMA600
H 8717 see PMN250
96H60 see DFH600
110H60 see CHF600
H-19218 see CPB150
H 22948 see IOK100
H 56/28 see AGW250
H 59/64 see DUM100
H 93/26 see MQR144
H 394/84 see DTU852
cis-H 102.09 see ZBA500
HA see HHQ800
HA 106 see TLG000
HAAS see SMT500
HAB 439 see PFL100
HABA de SAN ANTONIO (PUERTO RICO)
see CAK325
HABILLA (PUERTO RICO) see YAG000
HACHE UNO SUPER see FDA885
HACHI-SUGAR see SGC000
H ACID see AKH000
H ACID DISODIUM SALT see HAA400
cis-HAD see DGU709
HADACIDIN see FNO000
HADACIDINE see FNO000
HADACIN see FNO000
2,4-HADIYNYLENE CHLOROFORMATE see
HAB600
HAEMATEIN see HAO600
HAEMATITE see HAO875
HAEMODAN see MGC350
HAEMOFORT see FBO000
HAEMOPHILIS INFLUENZAE,
ENDOTOXIN, phenol water extract see
HAB710
HAEMOPHILIS INFLUENZAE,
ENDOTOXIN, hypertonic NaCl citrate extract
see HAB700
HAEMOSTASIN see VGP000
HAF 50 EK see DUI900
HAFFKININE see ARQ250
HAFNIUM see HAC000
HAFNIUM, wet with not less than 25% water
(DOT) see HAC000
HAFNIUM CHLORIDE see HAC800
HAFNIUM CHLORIDE OXIDE see HAD500
HAFNIUM CHLORIDE OXIDE
OCTAHYDRATE see HAD000
HAFNIUM DICYCLOPENTADIENE
DICHLORIDE see HAE500
HAFNIUM OXYCHLORIDE see HAD500
HAFNIUM OXYCHLORIDE
OCTAHYDRATE see HAD000
HAFNIUM POWDER, dry (UN 2545) (DOT)
see HAC000
HAFNIUM POWDER, wetted with not <25%
water (UN 1326) (DOT) see HAC000
HAFNIUM TETRACHLORIDE see HAC800
HAFNIUM(IV) TETRAHYDROBORATE see
HAE000
HAFNOCENE DICHLORIDE see HAE500
HAG 107 see THJ300
HAIARI see RNZ000
HAIDR see ICC800
HAIMASED see SIA500
HAIROXAL see PII100
HAIRY see HBT500
HAITIN see HON000
HAKUENKA CC see CAT775
HAKUENKA R 06 see CAT775
HALACRINATE see BNA900
HALAD see ICC800
HALAMID see CDP000
HALANE see DFE200
HALAR 200 see KDK000
HALARSOL see DFX400
HALAZONE see HAF000
HALAZUCHROME B see VAD200
HALBMOND see BAU750

HALCIDERM see HAF300
HALCIMAT see HAF300
HALCINONIDE see HAF300
HALCION see THS800
HALCORT see HAF300
HALDAR see ICC800
HALDI RHIZOME EXTRACT see COF850
HALDOL see CLY500
HALDRONE see PAL600
HALEITE see DUU800
HALF-CYSTEINE see CQK000
HALF-CYSTINE see CQK000
HALF-MUSTARD GAS see CGY750
HALF MUSTARD GAS see CHC000
HALF-MYLERAN see EMF500
HALF SULFUR MUSTARD see CHC000
HALIDO see BAV250
HALIDOR see POD750
HALITE see SFT000
HALIZAN see TDW500
HALKAN see DYF200
HALLOYSITE see HAF375
HALLTEX see SEH000
HALMARK see FAR150
HALOANISONE see HAF400
HALOANISONE COMPOSITUM see FDA880
HALOCARBON 11 see TIP500
HALOCARBON 14 see CBY250
HALOCARBON 23 see CBY750
HALOCARBON 112 see TBP050
HALOCARBON 113 see FOO000
HALOCARBON 114 see FOO509
HALOCARBON 115 see CJI500
HALOCARBON 112a see TBP000
HALOCARBON 152A see ELN500
HALOCARBON 1132A see VPP000
HALOCARBON C-138 see CPS000
HALOCARBON OIL 11-14 see KDK000
HALOCARBON OIL 11-21 see KDK000
HALOCARBON OIL 13-21 see KDK000
HALOCARBON 13/UCON 13 see CLR250
HALOFANTRINE HYDROCHLORIDE see HAF500
HALOFANTRINO (SPANISH) see HAF500
HALOFLEX 202 see ADV900
HALOFLEX 208 see ADV900
HALOFUGINONE HYDROBROMIDE see HAF600
HALOG see HAF300
HALOGABIDE see CKA125
HALOGENE T 30 see BNA300
HALOMICIN see HAF825
HALOMYCETIN see CDP250
HALON see DFA600
HALON see KDK000
HALON 14 see CBY250
HALON 1001 see MHR200
HALON 1011 see CES650
HALON 1202 see DKG850
HALON 1211 see BNA250
HALON 1301 see TJY100
HALON 2001 see EGV400
HALON 2312 see MRF600
HALON 2402 see FOO525
HALON 10001 see MKW200
HALONISON see HAF400
HALON (POLYMER) see KDK000
HALON TFEG 180 see TAI250
HALOPERIDOL see CLY500
HALOPERIDOL DECANOATE see HAG300
HALOPREDONE ACETATE see HAG325
HALOPROGIN see IFA000
HALOPROPANE see BOF750
HALOPYRAMINE see CKV625
HALOSAFEN see CHJ400
HALOSTEN see CLY500
HALOTAN see HAG500
HALOTESTIN see AOO275
HALOTEX see IFA000
HALOTHANE see HAG500
HALOWAX see TBR000
HALOWAX see TIT500

HALOWAX 1014 see HCK500
HALOWAX 1031 see MRG050
HALOXAZOLAM see HAG800
HALOXON see DFH600
HALOXYFOP-(2-ETHOXYETHYL) see HAG850
HALOXYFOP-ETOTYL see HAG850
HALOXYFOP-METHYL see HAG860
HALS 3 see PKL150
HALSAN see HAG500
HALSO 99 see CLK100
HALTRON 22 see CFX500
HALTS see BAF250
HALUD see ICC800
HALVIC 223 see PKQ059
HALVISOL see HAH000
HAMAMELIS see WCB000
HAMIDOP see DTQ400
HAMILTON RED see CHP500
HAMOVANNID see HFG550
HAMP-ENE 100 see EIV000
HAMP-ENE 215 see EIV000
HAMP-ENE 220 see EIV000
HAMP-ENE ACID see EIX000
HAMP-ENE Na4 see EIV000
HAMP-EX ACID see DJG800
HAMP-OL ACID see HKS000
HAMPSHIRE see IBH000
HAMPSHIRE GLYCINE see GHA000
HAMPSHIRE NTA see SIP500
HAMPSHIRE NTA ACID see AMT500
HAMYCIN see HAH800
HANA see HMX600
HANANE see BJE750
HANCOCK YELLOW 10010 see DEU000
HANQ see HIX200
HANSACOR see DJS200
HANSAMID see NCR000
HANSA ORANGE RN see DVB800
HANSA RED B see MMP100
HANSA RED G see MMP100
HANSA SCARLET RB see MMP100
HANSA SCARLET RN see MMP100
HANSA SCARLET RNC see MMP100
HANSOLAR see SNY500
HAOS see HLN100
HAPA-B SULFATE see SBE800
HAPASIL see NCP600
HAPLOS see EOK000
HAPPY DUST see CNE750
HAPTOCIL see CQE325
HAPTOCIL see PPO000
HAPTOGLOBINS see HAH900
HAQ see BKB300
HAQ see BKB325
HARDENER SL see BGT800
HARE-RID see SMN500
HARMALOL HYDROCHLORIDE see HAI300
HARMAN see MPA050
HARMANE see MPA050
HARMAR see PNA500
HARMINE see HAI500
HARMOGEN see PIK450
HARMONE B 79 see DFN300
HARMONIN see MQU750
HARMONYL see RDF000
HARNESS see CGO550
HAROWAX L 9 see GGR200
HARRICAL see CCK125
HARRY see HBT500
HARTOL see MQU750
HART'S HORN see MBU825
HARTSHORN PLANT see PAM780
HARVADE see OMY825
HARVADE 25F see OMY825
HARVADE F-25 see OMY825
HARVAMINE see WAK000
HARVATRATE see MGR500
HARVEN see SGD000
HARVEST-AID see SFS000
HARWAX A see HOG000
HAS (GERMAN) see HLB400

HASACH see CBD750
HA SALT see NOH100
HASETHROL see PBC250
HASHISH see CBD750
HASTELLOY C see CNA750
HASTINGS CARMINE 2G see CMM300
HATACLEAN see BPT300
HATCOL 200 see TJR600
HATCOL DIBP see DNJ400
HATCOL DOP see DVL700
HAURYMELLIN see DQR800
HAVERO-EXTRA see DAD200
HAVIDOTE see EIX000
HAVOC see TAC800
HAY ABSOLUTE see FMS000
HAYNES 25 see CNA750
HAYNES 188 see CNA600
HAYNES ALLOY NUMBER 25 see CNA750
HAYNES STELLITE 21 see CNA750
HAYNES STELLITE 31 see CNB825
HAYNON see CLX300
HAZODRIN see MRH209
HB see HEA000
HB^{17} see HHQ825
HB-218 see AKM125
HB 419 see CEH700
m-HBA see HJI100
HBB see HCA500
HBBN see HJQ350
HBC see HEF500
HBF 386 see AEA750
HBF 386 see AEB000
HBK see HAI600
HBP see HHQ850
9,10-H2 B(e)P see DKU400
HBT 3 see BAP502
HC see CNS750
3HC see HJX700
HC-3 see HAQ000
HC-58 see CBR300
HC-064 see NGB700
292 HC see PIK400
336 HC see PIK375
HC 1281 see TIK500
1352 HC see DHY200
HC 2072 see NIM500
HC7901 see DUG500
HCA see HCL500
HCA see HHQ800
HCA WEEDKILLER see HCL500
HCB see HCC500
HCB see HCC700
2,2',4,4',6,6'-HCB see HCC800
HCBD see HCD250
HC BLUE 1 see BKF250
HC BLUE No. 2 see HKN875
α-HCC see HJV000
HCCH see BBP750
HCCH see BBQ500
HCCPD see HCE500
HCDD see HAJ500
HCE see EBW500
HCFC 125 see PBD400
HCFC 124A see TCH150
HCFC 225CA see DFW850
HCFC 225CB see DFW830
HCFC 225BC see DFW830
HCFU see CCK630
HC 20-511 FUMARATE see KGK200
HCG see CMG675
HCG-004 see CJP750
HCH see BBQ500
α-HCH see BBQ000
β-HCH see BBR000
γ-HCH see BBQ500
HCL SALZ des p-AMINO-BENZOESAEURE-DIMETHYLAMINO-AETHYL-ESTER (GERMAN) see DPC200
HCN-CNCl MIXTURE see HHS500
HCO 40 see CCP300
HCO 50 see CCP305
HCO 60 see CCP310

HCP see ABD000
HCP see HCL000
HC RED NO. 1 see NEM350
HC RED NO. 3 see ALO750
HCS 3260 see CDR750
HCl SALZ des p-AMINO-SALICYLSAEURE-DIAETHYLAMINOAETHYLESTER (GERMAN) see AMM750
HCl SALZ des p-AMINO-SALICYLSAEURE-DIMETHYLAMINOAETHYL-ESTER (GERMAN) see AMN000
HCl SALZ DES p,N,N-BUTYLAMINOSALICYLSAEUREDIAETHYL AMINOAETHYLESTER (GERMAN) see BQH250
HCT see HGO500
HCTZ see CFY000
H. CYANOCINCTUS VENOM see SBI900
HC YELLOW 4 see HKK100
HC YELLOW No. 4 see BKC000
HCZ see CFY000
HD 419 see CEH700
HD AMARANTH B see FAG020
HDEHP see BJR750
H.D. EUTANOL see OBA000
1,3(2H,9bH)-DIBENZOFURANDIONE, 2,6-DIACETYL-7,9-DIHYDROXY-8,9b-DIMETHYL-, (9bR)- see UWJ100
3(9bH)-DIBENZOFURANONE, 2,6-DIACETYL-8,9b-DIMETHYL-1,7,9-TRIHYDROXY-, D- see UWJ100
3(9bH)-DIBENZOFURANONE, 2,6-DIACETYL-1,7,9-TRIHYDROXY-8,9b-DIMETHYL-, D- see UWJ100
1(H)-2,3-DIHYDROCYCLOPENTA(c)PHENANTHRENE see DKX150
HDMTX see MDV500
HDO see HEP500
HD OLEYL ALCOHOL 70/75 see OBA000
HD OLEYL ALCOHOL 80/85 see OBA000
HD OLEYL ALCOHOL 90/95 see OBA000
HD OLEYL ALCOHOL CG see OBA000
HD PONCEAU 4R see FMU080
HD SUNSET YELLOW FCF see FAG150
HD SUNSET YELLOW FCF SUPRA see FAG150
HD TARTRAZINE see FAG140
HD TARTRAZINE SUPRA see FAG140
HDTMP see HEQ600
HDU see HEF500
HE 166 see DUS500
HEADACHE WEED see CMV390
HEALON see HGN600
HEALTHIED see PAO170
HEART-OF-JESUS see CAL125
HEARTS see AOB250
HEARTS see BBK500
HEAT see HAJ700
HEAT PRE see MJG750
HEAVENLY BLUE see DJO000
HEAVY CATALYTICALLY CRACKED DISTILLATE see DXG810
HEAVY CATALYTIC CRACKED NAPHTHA see NAQ520
HEAVY CATALYTIC REFORMED NAPHTHA see NAQ530
#6 HEAVY FUEL OILS see HAJ750
HEAVY NAPHTHENIC DISTILLATE see MQV780
HEAVY NAPHTHENIC DISTILLATE SOLVENT EXTRACT see MQV857
HEAVY NAPHTHENIC DISTILLATES (PETROLEUM) see MQV780
HEAVY OIL see CMY825
HEAVY PARAFFINIC DISTILLATE see MQV785
HEAVY PARAFFINIC DISTILLATE, SOLVENT EXTRACT see MQV859
HEAVY THERMAL CRACKED NAPHTHA see NAQ580
HEAVY VACUUM DISTILLATE see GBW005

HEAVY VACUUM GAS OIL (PETROLEUM) see GBW005
HEAVY WATER see HAK000
HEAVY WATER-d2 see HAK000
HEAZLEWOODITE see HAK050
HEAZLEWOODITE see NDJ500
HEBABIONE HYDROCHLORIDE see PPK500
HEBANIL see CKP500
HEBARAL see EKT500
HEB-CORT see CNS750
HEBIN see FMT100
HEBP see HKQ700
HEBUCOL see COW700
HEC see HKQ100
HEC-AL 5000 see HKQ100
HECLOTOX see BBQ500
HECNU see CHB750
HECNU-MS see HKQ300
HECTOGRAPH VIOLET SR see AOR500
HECTO VIOLET R see AOR500
HEDAPUR M 52 see CIR250
HEDERA CANARIENSIS see AFK950
HEDERA HELIX see AFK950
α-HEDERIN see HAK075
α-HEDERINE see HAK075
HEDEROSIDE C see HAK075
HEDEX see HIM000
HEDGE PLANT see PMD550
HED-HEPARIN see HAQ550
HEDIONDA (PUERTO RICO) see CNG825
HEDIONDILLA (PUERTO RICO) see LED500
HEDIONE see HAK100
HEDOLIT see DUS700
HEDONAL see DGB000
HEDONAL see MOU500
HEDONAL DP see DGB000
HEDONAL MCPP see CIR500
HEDONAL MCPP see CLO200
HEDONAL (The herbicide) see DAA800
HEDP see HKS780
HEDTA see HKS000
HEDULIN see PFJ750
HEEDTA see HKS000
HEF-2 see JDA100
HEF-3 see JDA125
HE-HK-52 see HAK200
HEKBILIN see CDL325
HEKSAN (POLISH) see HEN000
HEKSOGEN (POLISH) see CPR800
HEKTALIN see VGP000
HELAKTYN BLACK DN see CMS220
HELAKTYN PURE BLUE FR see CMS222
HELANTHRENE BLACK BL see CMU320
HELANTHRENE BLUE BC see DFN300
HELANTHRENE BROWN GR see CMU770
HELANTHRENE YELLOW see DCZ000
HELENALIN see HAK300
HELFERGIN see AAE500
HELFERGIN see DPE000
HEL-FIRE see BRE500
HELFO DOPA see DNA200
HELFOSERPIN see RDK000
HELIACID LIGHT YELLOW 4R see SGP500
HELIANE ORANGE RF see CMU815
HELIANE RED 5B see DNT300
HELIANTHINE see MND600
HELIANTHINE B see MND600
HELIANTHRENE BLUE RS see IBV050
HELICHRYSUM OIL see HAK500
HELICOCERIN see ECE500
HELICON see ADA725
HELIC YELLOW GW see DEU000
HELINDON ORANGE R see CMU815
HELINDON RED BB see DNT300
HELIO FAST GREEN GN see CMS140
HELIO FAST GREEN GT see CMS140
HELIO FAST ORANGE RN see DVB800
HELIO FAST RED BN see MMP100
HELIO FAST RED RL see MMP100
HELIO FAST RED RN see MMP100

HELIO FAST YELLOW GRF see CMS208
HELIO FAST YELLOW GRN see CMS208
HELIOGEN see CDP000
HELIOGEN BLUE 6470 see IBV050
HELIOGEN BLUE SBL see COF420
HELIOGEN GREEN 6G see CMS140
HELIOGEN GREEN 8680 see PJQ100
HELIOGEN GREEN 8730 see PJQ100
HELIOGEN GREEN 9360 see CMS140
HELIOGEN GREEN 8681K see PJQ100
HELIOGEN GREEN 8682T see PJQ100
HELIOGEN GREEN A see PJQ100
HELIOGEN GREEN 6GA see CMS140
HELIOGEN GREEN 8GA see CMS140
HELIOGEN GREEN G see PJQ100
HELIOGEN GREEN GA see PJQ100
HELIOGEN GREEN GN see PJQ100
HELIOGEN GREEN GNA see PJQ100
HELIOGEN GREEN GTA see PJQ100
HELIOGEN GREEN GV see PJQ100
HELIOGEN GREEN GWS see PJQ100
HELIOMYCIN see HAL000
HELION BLACK GF see CMN150
HELION RED 8B see CMO885
HELION TURQUOISE BLUE FGLL see COF420
HELIOPAR see CLD000
HELIO RED RL see MMP100
HELIO RED TONER see MMP100
HELIO RED TONER LCLL see CHP500
HELIOSTABLE BRILLIANT PINK B EXTRA see RGW000
HELIOTRIDINE ESTER with LASIOCARPUM and ANGELIC ACID see LBG000
HELIOTRINE see HAL500
HELIOTRON see HAL500
HELIOTROPIN see PIW250
HELIOTROPIN ACETAL see PIZ499
HELIOTROPIUM SUPINUM L. see HAM000
HELIOTROPIUM TERNATUM see SAF500
HELIOTROPYL ACETATE see PIX000
HELIOTROPYL ACETONE see MJR250
HELIUM see HAM500
HELIUM, compressed (UN 1046) (DOT) see HAM500
HELIUM, refrigerated liquid (cryogenic liquid) (UN 1963) (DOT) see HAM500
HELIXIN see HAK075
HELLEBORE see CMG700
HELLEBOREIN see HAN500
HELLEBORUS NIGER see CMG700
HELLEBRIGENIN 3-ACETATE see HAN550
HELLEBRIGENIN-GLUCO-RHAMNOSID (GERMAN) see HAN600
HELLEBRIN see HAN600
HELLEGRIGENIN GLUCORHAMNOSIDE see HAN600
HELLIPIDYL see AMM250
HELMATAC see BQK000
HELMET FLOWER see MRE275
HELMETINA see PDP250
HELMINAL see KAJ200
HELMIRANE see DFH600
HELMIRON see DFH600
HELMIRONE see DFH600
HELMOX see COH250
HELODERMA SUSPECTUM VENOM see HAN625
HELOTHION see SOU625
HELOXY WC68 see NCI300
HELVETICOSID (GERMAN) see HAN800
HELVETICOSIDE see HAN800
17-α-HELVETICOSIDE see HAN900
HELVETICOSIDE DIHYDRATE see HAO000
HEMACHATUS HAEMACHATES VENOM see HAO500
HEMACHATUS HAEMACHATUS VENOM see HAO500
HEMATEIN see HAO600
HEMATINE see HAO600
HEMATIN-PROTEIN see CQM325

HEMATITE see HAO875
HEMATOIDIN see HAO900
HEMATOPORPHYRIN see PHV275
HEMATOPORPHYRIN MERCURY
DISODIUM SALT see HAP000
HEMATOXYLIN see HAP500
HEMEL see HEJ500
HEMETOIDIN see HAO900
HEMICHOLINE see HAQ000
HEMICHOLINIUM see HAP100
HEMICHOLINIUM-3 see HAQ000
HEMICHOLINIUM BROMIDE see HAQ000
HEMICHOLINIUM-3-BROMIDE see HAQ000
HEMICHOLINIUM DIBROMIDE see HAQ000
HEMICHOLINIUM-3-DIBROMIDE see HAQ000
HEMIMELLITENE see TLL500
HEMIN see HAQ050
HEMINEVRIN see CHD750
HEMISINE see VGP000
l'HEMISULFATE de GUANIDINO METHYL-6-BENZODIOXANNE-1,4 (FRENCH) see GKG300
HEMITON see CMX760
HEMLOCK OIL see SLG650
HEMLOCK WATER DROPWORT see WAT315
HEMO-B-DOZE see VSZ000
HEMOCAPROL see AJD000
HEMOCOAGULASE see CMY725
HEMODAL see MMD500
HEMODESIS see PKQ250
HEMODEX see DBD700
HEMODEZ see PKQ250
HEMOFURAN see NGE500
HEMOKLOT see MCB575
HEMOMIN see VSZ000
HEMOPAR see AJD000
HEMOSTASIN see VGP000
HEMOSTYPTANON see EDU500
HEMOTON see FBJ100
HEMOTROPE see PEU000
HEMPA see HEK000
HENBANE see HAQ100
HENDECANAL see UJJ000
HENDECANALDEHYDE see UJJ000
HENDECANE see UJS000
HENDECANOIC ACID see UKA000
HENDECANOIC ALCOHOL see UNA000
1-HENDECANOL see UNA000
2-HENDECANONE see UKS000
HENDECENAL see ULJ000
10-HENDECEN-1-YL ACETATE see UMS000
HENDECYL ALCOHOL see UNA000
n-HENDECYLENIC ALCOHOL see UNA000
10-HENEDECENOIC ACID see ULS000
HENEICOSAFLUORO-1-IODODECANE see IEU075
6-HENEICOSEN-11-ONE, (Z)- see HAQ200
(Z)-6-HENEICOSEN-11-ONE see HAQ200
HENINE see LIN100
HENKEL'S COMPOUND see TBC450
HENNA see HMX600
HENNOLETTEN see EOK000
HENU see HKW500
HEOD see DHB400
HEPACHOLINE see CMF750
HEPADIAL see DOK400
HEPAGON see VSZ000
HEPALIDINE see TEV000
HEPAR CALCIS see CAY000
HEPAREGENE see TEV000
HEPARIN see HAQ500
α-HEPARIN see HAQ500
HEPARINATE see HAQ500
HEPARIN COFACTOR see AQN550
HEPARIN COFACTOR B see AQN550
HEPARINIC ACID see HAQ500
HEPARINOID see SEH450
HEPARIN SODIUM see HAQ550
HEPARIN SULFATE see HAQ500

HEPARLIPON see DXN800
HEPARTEST see HAQ600
HEPARTESTABROME see HAQ600
HEPASYNTHYL see HAQ570
HEPATAMINE see AHR600
HEPATHROM see HAQ550
HEPATION see MAO275
HEPATOSAN see AHR600
HEPATOSULFALEIN see HAQ600
HEPATOXANE see HAQ570
HEPAVIS see VSZ000
HEPCOVITE see VSZ000
HEPERAL see DXG600
HEPIN see AJD000
HEPINOID see SEH450
HEPORAL see AOO490
HEPT see TCF250
1,3,4,6,7,9,9B-HEPTAAZAPHENALENE-2,5,8-TRIAMINE see HAQ700
1,3,4,6,7,9,9B-HEPTAAZAPHENALENE, 2,5,8-TRIAMINO-(6CI,7CI,8CI) see HAQ700
HEPTABARB see COY500
HEPTABARBITAL see COY500
HEPTABARBITONE see COY500
HEPTABARBUM see COY500
HEPTACAINE see PIO750
HEPTACHLOOR (DUTCH) see HAR000
1,4,5,6,7,8,8-HEPTACHLOOR-3a,4,7,7a-TETRAHYDRO-4,7-endo-METHANO-INDEEN (DUTCH) see HAR000
HEPTACHLOR see HAR000
HEPTACHLOR (technical grade) see HAR500
HEPTACHLORE (FRENCH) see HAR000
HEPTACHLOR EPOXIDE (USDA) see EBW500
2,2,5-endo,6-exo,8,9,10-HEPTACHLOROBORNANE see THH575
HEPTACHLORODIBENZO-p-DIOXIN see HAR100
1,2,3,4,6,7,8-HEPTACHLORODIBENZODIOXIN see HAR100
1,2,3,4,6,7,9-HEPTACHLORODIBENZODIOXIN see HAR150
1,2,3,4,6,7,8-HEPTACHLORODIBENZO-p-DIOXIN see HAR100
1,2,3,4,6,7,9-HEPTACHLORODIBENZO-p-DIOXIN see HAR150
3,4,5,6,7,8,8-HEPTACHLORODICYCLOPENTADIENE see HAR000
3,4,5,6,7,8,8a-HEPTACHLORODICYCLOPENTADIENE see HAR000
1,4,5,6,7,8,8-HEPTACHLORO-2,3-EPOXY-2,3,3a,4,7,7a-HEXAHYDRO-4,7-METHANOINDENE see EBW500
1,4,5,6,7,8,8-HEPTACHLORO-2,3-EPOXY-3a,4,7,7a-TETRAHYDRO-4,7-METHANOINDAN see EBW500
1,4,5,6,7,8,8-HEPTACHLORO-3a,4,5,6,7,7a-HEXAHYDRO-4,7-METHANO-1H-INDENE see SCD500
2,3,4,5,6,7,7-HEPTACHLORO-1a,1b,5,5a,6,6a-HEXAHYDRO-2,5-METHANO-2H-INDENO(1,2-b)OXIRENE see EBW500
1,4,5,6,7,8,8-HEPTACHLORO-3a,4,7,7a-TETRAHYDRO-4,7-ENDOMETHANOINDENE see HAR000
1,4,5,6,7,10,10-HEPTACHLORO-4,7,8,9,-TETRAHYDRO-4,7-ENDOMETHYLENEINDENE see HAR000
1,4,5,6,7,8,8a-HEPTACHLORO-3a,4,7,7a-TETRAHYDRO-4,7-METHANOINDANE see HAR000
1(3a),4,5,6,7,8,8-HEPTACHLORO-3a(1),4,7,7a-TETRAHYDRO-4,7-METHANOINDENE see HAR000
1,4,5,6,7,8,8-HEPTACHLORO-3a,4,7,7a-TETRAHYDRO-4,7-METHANOINDENE see HAR000

1,4,5,6,7,8,8-HEPTACHLORO-3a,4,7,7a-TETRAHYDRO 4,7-METHANOINDENE (technical grade) see HAR500
1,4,5,6,7,8,8-HEPTACHLORO-3a,4,7,7a-TETRAHYDRO-4,7-METHANOL-1H-INDENE see HAR000
1,4,5,6,7,8,8-HEPTACHLORO-3a,4,7,7,7a-TETRAHYDRO-4,7-METHYLENE INDENE see HAR000
1,4,5,6,7,8,8-HEPTACHLOR-3a,4,7,7,7a-TETRAHYDRO-4,7-endo-METHANO-INDEN (GERMAN) see HAR000
HEPTACOSAFLUOROTRIBUTYLAMINE see HAS000
HEPTADECAFLUORO-1-IODOOCTANE see PCH325
HEPTADECAFLUORONONANOIC ACID AMMONIUM SALT see HAS050
1,1,2,2,3,3,4,4,5,5,6,6,7,7,8,8,8-HEPTADECAFLUORO-1-OCTANESULFONIC ACID see HAS075
HEPTADECANE see HAS100
n-HEPTADECANE see HAS100
HEPTADECANE, 9-AZA-1,17-DIGUANIDINO-, TRIACETATE see GLQ100
1-HEPTADECANECARBOXYLIC ACID see SLK000
HEPTADECANOIC ACID see HAS500
HEPTADECANOIC ACID, 16-METHYL-, ISOPROPYL ESTER see IPS450
HEPTADECANOL (mixed primary isomers) see HAT000
9-HEPTADECANONE see OFE020
2,8,10-HEPTADECATRIENE-4,6-DIYNE-1,14-DIOL see EAO500
8,10,12-HEPTADECATRIENE-4,6-DIYNE-1,14-DIOL, (E,E,E)-(−)- see CMN000
2-(8-HEPTADECENYL)-2-IMIDAZOLINE-1-ETHANOL see AHP500
n-HEPTADECOIC ACID see HAS500
HEPTADECYL CYANIDE see SLL500
2-HEPTADECYL-4,5-DIHYDRO-1H-IMIDAZOLYL MONOACETATE see GII000
2-HEPTADECYL GLYOXALIDINE see GIO000
2-HEPTADECYL GLYOXALIDINE ACETATE see GII000
2-HEPTADECYL-1-HYDROXYETHYLIMIDAZOLINE see HAT500
n-HEPTADECYLIC ACID see HAS500
2-HEPTADECYL-2-IMIDAZOLINE see GIO000
2-HEPTADECYL-2-IMIDAZOLINE ACETATE see GII000
2-HEPTADECYL-2-IMIDAZOLINE-1-ETHANOL see HAT500
HEPTADECYL-TRIMETHYLAMMONIUM METHYLSULFAT (CZECH) see HAU500
HEPTADECYLTRIMETHYLAMMONIUM METHYLSULFATE see HAU500
2,4-HEPTADIENAL see HAV450
HEPTADIENAL-2,4 see HAV450
trans,trans-2,4-HEPTADIENAL see HAV450
1,6-HEPTADIENE-3,5-DIONE, 1,7-BIS(4-HYDROXY-3-METHOXYPHENYL)- see ICC800
4,5-HEPTADIENOIC ACID, 7-(3-HYDROXY-2-(3-HYDROXY-4-PHENOXY-1-BUTENYL)-5-OXOCYCLOPENTYL)-, METHYL ESTER, (1-α,2-β(1E,3R*),3-α)- see EAU200
1,6-HEPTADIYNE see HAV500
HEPTADONE see MDO750
HEPTADON HYDROCHLORIDE see MDP750
HEPTADORM see COY500
HEPTAFLUORJODPROPAN see HAY300
HEPTAFLUORMASELNAN STRIBRNY (CZECH) see HAW000
HEPTAFLUOROBUTANOIC ACID, SILVER SALT see HAW000

2,2,3,3,4,4,4-HEPTAFLUOROBUTANOL see HAW100

2,2,3,3,4,4,4-HEPTAFLUOROBUTYRAMIDE see HAX000

HEPTAFLUOROBUTYRIC ACID see HAX500

HEPTAFLUOROBUTYRIC ACID, ETHYL ESTER see HAY000

HEPTAFLUOROBUTYRYL HYPOCHLORITE see HAY059

HEPTAFLUOROBUTYRYL HYPOFLUORITE see HAY100

HEPTAFLUOROBUTYRYL NITRATE see HAY200

HEPTAFLUOROIODOPROPANE see HAY300

HEPTAFLUOROISOBUTYLENE METHYL ETHER see HAY500

2-HEPTAFLUOROPROPYL-1,3,4-DIOXAZOLONE see HAY600

HEPTAFLUOROPROPYL HYPOFLUORITE see HAY650

HEPTAFLUR see CDF400

HEPTAGRAN see HAR000

2,3,3',4,4'5,7-HEPTAHYDROXYFLAVAN see HBA259

HEPTAKIS (DIMETHYLAMINO)TRIALUMINUM TRIBORON PENTAHYDRIDE see HBA500

γ-HEPTALACTONE see HBA550

HEPTALDEHYDE see HBB500

n-HEPTALDEHYDE see HBB500

HEPTALDEHYDE METHYLANTHRANILATE, Schiff's base see HBO700

HEPTAMAL see COY500

HEPTAMETHYLENEIMINE mixed with SODIUM NITRITE see SIT000

HEPTAMETHYLPHENYLCYCLOTETRASIL OXANE see HBA600

HEPTAMETHYLVINYLCYCLOTETRASILO XANE see VPF150

HEPTAMINE see HBM490

HEPTAMINOL HYDROCHLORIDE see HBB000

HEPTAMUL see HAR000

HEPTAMYL HYDROCHLORIDE see HBB000

HEPTAN (POLISH) see HBC500

HEPTANAL see HBB500

HEPTANAL, CYCLIC (HYDROXYMETHYL)ETHYLENE ACETAL see HBC000

HEPTANAL-1,2-GLYCERYL ACETAL see HBC000

HEPTANAL, OXIME see EAO200

HEPTANAMIDE, N-HYDROXY- see HBF550

HEPTANAMIDE, N-PENTYL- see PBX100

1-HEPTANAMINE see HBL600

2-HEPTANAMINE see HBM490

2-HEPTANAMINE SULFATE (2:1) see AKD600

HEPTANDIOIC ACID see PIG000

HEPTANE see HBC500

n-HEPTANE see HBC500

HEPTANE, 1-BROMO- see HBN100

1-HEPTANECARBOXYLIC ACID see OCY000

HEPTANE, 1-CHLORO-5-METHYL- see EKU100

HEPTANEDICARBOXYLIC ACID see ASB750

1,7-HEPTANEDICARBOXYLIC ACID see ASB750

HEPTANEDINITRILE see HBD000

HEPTANEDIOIC ACID see PIG000

HEPTANE-1,7-DIOIC ACID see PIG000

1,7-HEPTANEDIOIC ACID see PIG000

HEPTANEN (DUTCH) see HBC500

HEPTANENITRILE, 7-AMINO- see AKD550

HEPTANE, 1,1'-OXYBIS-(9CI) see HBO000

1-HEPTANETHIOL see HBD500

1,1,1,3,5,5,5-HEPTANITROPENTANE see HBD650

HEPTANOHYDROXAMIC ACID see HBF550

HEPTANOIC ACID see HBE000

HEPTANOIC ACID, 2,7-DIAMINO-7-IMINO-, MONOHYDROCHLORIDE, (S)- see IDA525

HEPTANOIC ACID, ISOBUTYL ESTER see IIS000

HEPTANOIC ACID, METHYL ESTER see MKK100

HEPTANOIC ACID, 2-METHYLPROPYL ESTER see IIS000

HEPTANOIC ACID, ester with TESTOSTERONE see TBF750

1-HEPTANOL see HBL500

2-HEPTANOL see HBE500

HEPTANOL-2 see HBE500

3-HEPTANOL see HBF000

n-HEPTANOL see HBL500

n-HEPTANOL-1 (FRENCH) see HBL500

2-HEPTANOL, 2,6-DIMETHYL- see FON200

3-HEPTANOL, 6-(DIMETHYLAMINO)-4,4-DIPHENYL-, ACETATE (ester), (3S,6S)-(−)- see ACQ258

3-HEPTANOL, 6-(DIMETHYLAMINO)-4,4-DIPHENYL-, ACETATE (ester), HYDROCHLORIDE, (3S,6S)-(−)- see ACQ260

HEPTANOL, FORMATE see HBO500

HEPTANOLIDE-1,4 see HBA550

HEPTANOLIDE-4,1 see HBA550

3-HEPTANOL-6-METHYL-3-PHENYL-1-(N-PIPERIDYL) HYDROCHLORIDE see HBF500

HEPTANON see MDO750

HEPTAN-3-ON (DUTCH, GERMAN) see EHA600

2-HEPTANONE see MGN500

HEPTAN-3-ONE see EHA600

3-HEPTANONE see EHA600

HEPTAN-4-ONE see DWT600

4-HEPTANONE see DWT600

4-HEPTANONE, 3-BENZYL-(8CI) see BEN800

3-HEPTANONE, 6-(DIMETHYLAMINO)-4,4-DIPHENYL-, (±)- see MDO760

3-HEPTANONE, 5-METHYL- see EGI750

3-HEPTANONE, 5-METHYL-, OXIME see EMP550

4-HEPTANONE, 3-(PHENYLMETHYL)- see BEN800

1-HEPTANONE, 1-(4-PYRIDYL)- see HBF600

HEPTANOYLHYDROXAMIC ACID see HBF550

17-β-HEPTANOYLOXY-19-NOR-17-α-PREGNEN-20-YNONE see NNQ000

4-HEPTANOYLPYRIDINE see HBF600

HEPTANYL ACETATE see HBL000

3,6,9,12,15,18,21-HEPTAOXATRICOSAN-1-OL, 23-(4-NONYLPHENOXY)- see NNC700

HEPTARINOID see SEH450

HEPTA SILVER NITRATE OCTAOXIDE see HBI500

HEPTA-1,3,5-TRIYNE see HBI725

HEPTAZINE see IKM050

2,4-HEPTDAIENAL see HAV450

HEPTEDRINE see HBM490

4-HEPTENAL see HBI800

2-HEPTENAL, (E)- see HBI770

(E)-2-HEPTEN-1-AL see HBI770

cis-4-HEPTEN-1-AL see HBI800

2-HEPTENE see HBJ500

n-HEPTENE see HBJ000

1-n-HEPTENE see HBJ000

3-HEPTENE (mixed isomers) see HBK350

1-HEPTENE-4,6-DIYNE see HBK450

2-HEPTENOIC ACID see HBK500

6-HEPTENOIC ACID, 7-(2-CYCLOPROPYL-4-(4-FLUOROPHENYL)-3-QUINOLINYL)-3,5-DIHYDROXY-,CALCIUM SALT (2:1), (3R,5S,6E)- see NMV425

5-HEPTENOIC ACID, 7-(3,5-DIHYDROXY-2-(3-HYDROXY-5-PHENYLPENTYL)CYCLOPENTYL)-,1-METHYLETHYL ESTER, (1R-(1-α(Z),2-β(R*),3-α,5-α))- see XAA500

5-HEPTENOIC ACID, 7-(3,5-DIHYDROXY-2-(3-OXODECYL)CYCLOPENTYL)-, 1-METHYLETHYL ESTER,(1R-(1α(Z),2β,3α,5α))- see IRR050

6-HEPTENOIC ACID, 7-(4-(4-FLUOROPHENYL)-5-(METHOXYMETHYL)-2,6-BIS(1-METHYLETHYL)-3-PYRIDINYL)-3,5-DIHYDROXY-, MONOSODIUM SALT, (S-(R*,S*-(E)))- see RLK750

5-HEPTEN-2-OL, 6-METHYL-2-(4-METHYL-3-CYCLOHEXEN-1-YL)- see BGO775

5-HEPTEN-2-ONE, 6-METHYL- see MKK000

HEPTENOPHOS see HBK700

HEPTENYL ACROLEIN see DAE450

6-HEPTEN-4-YN-3-OL, 1-(PROPYLTHIO)-2,3,6-TRIMETHYL- see TME260

HEPTHLIC ACID see HBE000

HEPTOCOAGULASE see CMY725

n-HEPTOIC ACID see HBE000

HEPTYL ACETATE see HBL000

1-HEPTYL ACETATE see HBL000

n-HEPTYL ACETATE see HBL000

HEPTYL ACRYLATE see HBL100

HEPTYL ALCOHOL see HBL500

HEPTYLAMINE see HBL600

1-HEPTYLAMINE see HBL600

2-HEPTYLAMINE see HBM490

3-HEPTYLAMINE see HBM500

n-HEPTYLAMINE see HBL600

HEPTYLAMINE, 1-METHYL- see OEK010

2-HEPTYLAMINE SULFATE see AKD600

8-HEPTYLBENZ(a)ANTHRACENE see HBN000

5-n-HEPTYLBENZ(1:2)BENZANTHRACENE see HBN000

HEPTYL BROMIDE see HBN100

n-HEPTYL BROMIDE see HBN100

HEPTYL BUTANOATE see HBN150

HEPTYL BUTYRATE see HBN150

γ-HEPTYLBUTYROLACTONE see HBN200

γ-n-HEPTYLBUTYROLACTONE see HBN200

HEPTYL CARBINOL see OEI000

HEPTYL CELLOSOLVE see HBN250

2-n-HEPTYL CYCLOPENTANONE see HBN500

α-HEPTYL CYCLOPENTANONE see HBN500

HEPTYLDICHLORARSINE see HBN600

HEPTYL (DIETHYLAMINO)OXOACETATE see HBN700

HEPTYL N,N-DIETHYLOXAMATE see HBN700

HEPTYLENE see HBK350

1-HEPTYLENE see HBJ000

HEPTYL ETHER see HBO000

HEPTYL FORMATE see HBO500

HEPTYL HYDRAZINE see HBO600

HEPTYL HYDRIDE see HBC500

HEPTYL 4-HYDROXYBENZOATE see HBO650

HEPTYL p-HYDROXYBENZOATE see HBO650

n-HEPTYL p-HYDROXYBENZOATE see HBP300

n-HEPTYLIC ACID see HBE000

HEPTYLIDENE METHYL ANTHRANILATE see HBO700

HEPTYL KETONE see HBO790

HEPTYL MERCAPTAN see HBD500

n-HEPTYLMERCAPTAN see HBD500

n-HEPTYL METHANOATE see HBO500

HEPTYL (4-(1-METHYLETHYL)PHENYL)METHYL 3-PYRIDINYLCARBONIMIDODITHIOATE see HBO800

HEPTYL METHYL KETONE see NMY500

HEPTYLMETHYLNITROSAMINE see HBP000
1-HEPTYL-1-NITROSOUREA see HBP250
n-HEPTYL NITROSUREA see HBP250
2-HEPTYLOXYCARBANILIC ACID-2-(1-PIPERIDINYL)ETHYL ESTER HYDROCHLORIDE see PIO750
2-(HEPTYLOXY)ETHANOL see HBN250
2-(2-(HEPTYLOXY)ETHOXY)ETHANOL see HBP275
p-(HEPTYLOXY)PHENOL see HBP285
4-(HEPTYLOXY)PHENOL (9CI) see HBP285
(2-(HEPTYLOXY)PHENYL)CARBAMIC ACID-2-(1-PIPERIDINYL)ETHYL ESTER HYDROCHLORIDE see PIO750
N-(2-(HEPTYLOXYPHENYLCARBAMOYLOXY)ETHYL)PIPERIDINIUM CHLORIDE see PIO750
HEPTYL PARABEN see HBO650
HEPTYLPARABEN see HBP300
2-HEPTYLPHENOL see HBP350
o-HEPTYLPHENOL see HBP350
o-n-HEPTYLPHENOL see HBP350
HEPTYL PHTHALATE see HBP400
(−)-HEPTYLPHYSOSTIGMINE see HBP410
HEPTYL 4-PYRIDYL KETONE see HBF600
2-HEPTYL-TETRAHYDROFURAN see HBP425
α-HEPTYL-3,4,5-TRIMETHOXYPHENETHYLAMINE see HBP435
n-HEPTYL-Δ-VALEROLACTONE see HBP450
2-HEPTYN-1-OL see HBS500
HEPZIDE see ENV500
HEPZIDINE MALEATE see HBT000
HER see REK350
HERBADOX see DRN200
HERB-ALL see MRL750
HERBAN see HDP500
HERBAN M see MRL750
HERBATIM see SIL550
HERBATOX see DXQ500
HERBAXON see PAJ000
HERBAX TECHNICAL see DGI000
HERBAZIN see BJP000
HERBAZOLIN see BAV000
HERB BONNETT see PJJ300
HERB-CHRISTOPHER see BAF325
HERBE-A-BRINVILLIERS (HAITI) see PIH800
HERBE A PLOMB (HAITI) see LAU600
HERBE AUX GEAUX (CANADA) see CMV390
HERBESSER see DNU600
HERBEX see BJP000
HERBICIDE 82 see BNM000
HERBICIDE 273 see DXD000
HERBICIDE 326 see DGD600
HERBICIDE 634 see CPR600
HERBICIDE 976 see BMM650
HERBICIDE 6602 see MQR225
HERBICIDE C-2059 see DUK800
HERBICIDE ES see GCC200
HERBICIDE M see CIR250
HERBICIDES, MONURON see CJX750
HERBICIDES, SILVEX see TIX500
HERBIDAL see DAA800
HERBIDAL TOTAL see AMY050
HERBIFERT see HBT100
HERBIZOLE see AMY050
HERBOGIL see DRV200
HERBOXONE see PAJ000
HERBOXY see BJP000
HERCOFLAT 135 see PMP500
HERCOFLEX 260 see DVL700
HERCON CHEK/MATE see HCP050
HERCON DISRUPT see HCP050
HERCO PRILLS see ANN000
HERCULES 426 see BFY100
HERCULES 3956 see CDV100

HERCULES 4580 see DIX000
HERCULES 5727 see COF250
HERCULES 6937 see DJJ393
HERCULES 7531 see HDP500
HERCULES 8717 see PMN250
HERCULES 9573 see TAN300
HERCULES 9699 see MIB500
HERCULES 12402 see BFY100
HERCULES 14503 see DBI099
HERCULES N 100 see HKQ100
HERCULES P6 see PBB750
HERCULES TOXAPHENE see CDV100
HERCULON see PMP500
HERKAL see DGP900
HERMAL see TFS350
HERMAT FEDK see ZHA000
HERMAT TMT see TFS350
HERMAT ZDM see BJK500
HERMAT Zn-MBT see BHA750
HERMESETAS see BCE500
HERMOPHENYL see MCU500
HERNIARIN (6CI) see MEK300
HEROIEN see HBT500
HEROIIN see HBT500
HEROIN see HBT500
HEROIN HYDROCHLORIDE see DBH400
HEROLAN see HBT500
HEROPON see DBA800
HERPESIL see DAS000
HERPIDU see DAS000
HERPLEX see DAS000
HERPLEX LIQUIFILM see DAS000
HERYL see TFS350
HERZO see POB500
HERZO PROSCILLAN see POB500
HES see HLB400
HESOFEN see EQL000
HESPAN see HKQ100
HESPANDER see HLB400
HESPANDER INJECTION see HLB400
HESPERIDIN see HBU000
HESPERIDIN METHYLCHALCONE see HBU400
HESPERIDOSIDE see HBU000
HESPERITIN-7-RHAMNOGLUCOSIDE see HBU000
HESTRIUM CHLORIDE see HEA500
HET see HCY000
HET ACID see CDS000
2,3,3',4,4',5,7-HETAHYDROXYFLAVAN see HBA259
HETAMIDE ML see BKE500
HETAPHENONE see DHS200
HETASTARCH see HKQ100
5-HETE see HKG600
12-HETE see HKG650
15-HETE see HKG680
HETERATISAN-14-ONE, 6-(BENZOYLOXY)-2-o-ETHYL-8-HYDROXY-1-METHOXY-4-METHYL-, (1-α,6-β)- see HBU410
HETERATISAN-14-ONE, 6,8-DIHYDROXY-20-ETHYL-1-METHOXY-4-METHYL-, (1-α,6-β)- see EID150
HETERATISINE see EID150
HETERATISINE 6-BENZOATE see HBU410
HETEROAUXIN see ICN000
HETEROFOS see HBU415
HETEROPHOS see HBU415
HETOLIN see MQH250
HETP see HCY000
HETRAZAN see DIW200
HETROGE K PREMIX see MCB575
HETROGEN K see MCB575
HEUCOPHOS ZP 10 see ZJS400
HEV-4 see CNA750
HEWETEN 10 see CCU150
HEXA see BBP750
HEXAAMINE COBALT(III) ACETATE see HBU425
HEXAAMINECOBALT TRIACETATE see HBU425

HEXAAMMINECHROMIUM(III) NITRATE see HBU500
HEXAAMMINECOBALT(III) CHLORATE see HBV000
HEXAAMMINECOBALT(III) CHLORITE see HBV500
HEXAAMMINECOBALT(III) HEXANITROCOBALTATE (3-) see HBW000
HEXAAMMINECOBALT(III) IODATE see HBW500
HEXAAMMINECOBALT(III) NITRATE see HBX000
HEXAAMMINECOBALT(III) PERCHLORATE see HBX500
HEXAAMMINECOBALT(III) PERMANGANATE see HBY000
HEXAAMMINECOBALT(3+) TRIACETATE see HBU425
HEXAAMMINEDICHLOROBIS(MU-(1,5-PENTANEDIAMINE-KAPPAN:KAPPAN'))TRIPLATINUM(4+), STEREOISOMER see HBY100
HEXAAMMINERUTHENIUM(III)CHLORIDE see HBY200
HEXAAMMINERUTHENIUM TRICHLORIDE see HBY200
HEXAAMMINERUTHENIUM TRICHLORIDE, HYDRATE see HBY200
HEXAAMMINETITANIUM(III) CHLORIDE see HBY500
HEXAAMMINETRICHLORORUTHENIUM see HBY200
HEXAAQUACHROMIUM CHLORIDE see CMK450
HEXAAQUACHROMIUM (III) CHLORIDE see CMK450
HEXAAQUACOBALT(II) PERCHLORATE see HBZ000
1,4,7,10,13,16-HEXAAZACYCLOOCTADECANE SULFATE (1:3) see HCN100
1,1,3,3,5,5-HEXAAZIDO-2,4,6-TRIAZA-1,3,5-TRIPHOSPHORINE see HCA000
HEXA(1-AZIRIDINYL)TRIPHOSPHOTRIAZINE see AQO000
HEXABALM see HCL000
HEXABARBITAL see ERD500
HEXABENZOBENZENE see CNS250
HEXABETALIN see PPK500
HEXABORANE(10) see HCA275
HEXABORANE(12) see HCA285
HEXABRIX see IGD200
HEXABROMOBENZENE see HCA385
HEXABROMOBIPHENYL see HCA500
2,4,5,2',4',5'-HEXABROMOBIPHENYL see FBU509
HEXABROMOBIPHENYL (technical grade) see FBU000
HEXABROMODIPHENYL ETHER see HCA550
HEXABROMODIPHENYL OXIDE see HCA550
HEXABROMONAPHTHALENE see HCA600
1,2,3,4,6,7-HEXABROMONAPHTHALENE see HCA650
HEXABUTYLDISTANNOXANE see BLL750
HEXABUTYLDISTANNTHIANE see HCA700
1,1,1,3,3,3-HEXABUTYLDISTANNTHIANE see HCA700
HEXABUTYLDITIN see BLL750
(HEXABUTYL(mu-(SULFATO(2−)-O,O'':O',O''')))DI TIN see TIF600
HEXACAP see CBG000
HEXACARBONYLCHROMIUM see HCB000
HEXACARBONYL CHROMIUM see HCB000
HEXACARBONYLDI-PI-CYCLOPENTADIENYL-MU-MERCURIODICHROMIUM see BIR500
HEXACARBONYLMOLYBDENUM see HCB500
HEXACARBONYLTUNGSTEN see HCC000

HEXACHLOROHEXAHYDROMETHANO-2,4,3-BENZODIOXATHIEPIN-3-OXIDE see EAQ750

6,7,8,9,10,10-HEXACHLORO-1,5,5a,6,9,9a-HEXAHYDRO-6,9-METHANO-2,4,3-BENZODIOXATHIEPIN-3-OXIDE see EAQ750

6,7,8,9,10,10-HEXACHLORO-1,5,5a,6,9,9a-HEXAHYDRO-6,9-METHANO-3-METHYL-2,4-BENZODIOXEPIN see HCK000

6,7,8,9,10,10-HEXACHLORO-1,5,5a,6,9,9a-HEXAHYDRO-3-METHYL-6,9-METHANO-2,4-BENZDIOXEPIN see HCK000

HEXACHLOROIRIDATE(2-) DIHYDROGEN, (OC-6-11) see HIA500

HEXACHLOROMELAMINE see TNG275

HEXACHLORONAPHTHALENE see HCK500

1,2,3,4,6,7-HEXACHLORONAPHTHALENE see HCK550

1,2,3,5,6,7-HEXACHLORONAPHTHALENE see HCK600

1,4,5,6,7,7-HEXACHLORO-5-NORBORNENE-2,3-DICARBOXIMIDE see CDS100

1,4,5,6,7,7-HEXACHLORO-5-NORBORNENE-2,3-DICARBOXYLIC ACID see CDS000

1,4,5,6,7,7-HEXACHLORO-5-NORBORNENE-2,3-DIMETHANOL CYCLIC SULFITE see EAQ750

3,4,5,6,9,9-HEXACHLORO-1a,2,2a,3,6,6a,7,7a-OCTAHYDRO-2,7:3,6-DIMETHANONAPHTH(2,3-b)OXIRENE see DHB400

3,4,5,6,9,9-HEXACHLORO-1a,2,2a,3,6,6a,7,7a-OCTAHYDRO-2,7:3,6-DIMETHANONAPHTH(2,3-b)OXIRENE see EAT500

HEXACHLOROPHANE see HCL000

HEXACHLOROPHEN see HCL000

HEXACHLOROPHENE see HCL000

HEXACHLOROPHENE see MRM000

HEXACHLOROPHENE (DOT) see HCL000

HEXACHLOROPLATINATE(2-) DIPOTASSIUM see PLR000

HEXACHLOROPLATINATE(2-) DISODIUM HEXAHYDRATE see HCL300

HEXACHLOROPLATINIC ACID see CKO750

HEXACHLOROPLATINIC(IV) ACID see CKO750

HEXACHLOROPLATINIC(4+) ACID, HYDROGEN- see CKO750

HEXACHLORO-2-PROPANONE see HCL500

1,1,1,3,3,3-HEXACHLORO-2-PROPANONE see HCL500

HEXACHLOROPROPENE see HCM000

HEXACHLOROPROPYLENE see HCM000

HEXACHLORORHODATE(3-) TRIAMMONIUM see HCM050

4,5,6,7,8,8-HEXACHLORO-3a,4,7,7a-TETRAHYDRO-4,7-METHANOINDENE see HCN000

N,N,N',N',N'',N''-HEXACHLORO-1,3,5-TRIAZINE-2,4,6-TRIAMINE see TNG275

1,2,4,5,7,8-HEXACHLORO-9H-XANTHENE see XBA100

HEXACHLORO-m-XYLENE see HCM100

α,α,α,α',α',α'-HEXACHLOROXYLENE see BLL825

α,α,α,α',α',α'-HEXACHLORO-m-XYLENE see BLL825

α,α,α,α',α',α'-HEXACHLORO-p-XYLENE see HCM500

α,α'-HEXACHLORO-m-XYLENE see BLL825

α,α'-HEXACHLOROXYLENE see HCM500

4,5,6,7,8,8-HEXACHLOR-$\Delta^{1,5}$-TETRAHYDRO-4,7-METHANOINDEN see HCN000

HEXACID 698 see HEU000

HEXACID 898 see OCY000

HEXACID 1095 see DAH400

HEXACID C-7 see HBE000

HEXACID C-9 see NMY000

HEXACO BLUE VRS see ADE500

HEXACOL ACID YELLOW G see CMM758

HEXACOL BLACK PN see BMA000

HEXACOL BLUE VRS see ADE500

HEXACOL BRILLIANT BLUE A see FAE000

HEXACOL CARMOISINE see HJF500

HEXACOL ERYTHROSINE BS see FAG040

HEXACOL GREEN S see ADF000

HEXACOL OIL ORANGE SS see TGW000

HEXACOL OIL YELLOW GG see CMP600

HEXACOL ORANGE GG CRYSTALS see HGC000

HEXACOL PONCEAU 4R see FMU080

HEXACOL PONCEAU MX see FMU070

HEXACOL PONCEAU SX see FAG050

HEXACOL RED 2G see CMM300

HEXACOL RED 6B see CMM400

HEXACOL RED 10B see CMS228

HEXACOL RHODAMINE B EXTRA see FAG070

HEXACOL SUNSET YELLOW FCF see FAG150

HEXACOL SUNSET YELLOW FCF SUPRA see FAG150

HEXACOL SUNSET YELLOW FCP see FAG150

HEXACOL SUNSET YELLOW F & F SUPRA see FAG150

HEXACOL TARTRAZINE see FAG140

HEXACONAZOLE see HCN050

HEXACOSE see HCS500

HEXACYANOFERRATE(3-) TRIPOTASSIUM see PLF250

HEXACYANOTRIS(3-DODECYL-1-METHYL)-2-PHENYLBENZIMIIMIDAZOLINIUM FERRATE (3-) see TNH750

HEXACYCLEN TRISULFATE see HCN100

HEXACYCLONAS see SHL500

HEXACYCLONATE SODIUM see SHL500

7,11-HEXADECADIEN-1-OL, ACETATE see GJK000

HEXADECADROL see SOW000

HEXADECAFLUOROHEPTANE see PCH000

HEXADECAFLUORO-1-NONANOL see HCO000

2,2,3,3,4,4,5,5,6,6,7,7,8,8,9,9-HEXADECAFLUORONONANOL see HCO000

HEXADECANAMIDE, N,N-DIMETHYL- see DTH700

1-HEXADECANAMINE see HCO500

1-HEXADECANAMINE HYDROFLUORIDE (9CI) see CDF400

1-HEXADECANAMINIUM, N-(CARBOXYMETHYL)-N,N-DIMETHYL-, HYDROXIDE, inner salt see CDF450

1-HEXADECANAMINIUM, N-HEXADECYL-N,N-DIMETHYL-, CHLORIDE (9CI) see DRK200

1-HEXADECANAMINIUM, N,N,N-TRIETHYL-, BROMIDE (9CI) see CDF500

1-HEXADECANAMINIUM, N,N,N-TRIMETHYL-, CHLORIDE see HCQ525

HEXADECANE see HCO600

n-HEXADECANE see HCO600

HEXADECANOIC ACID see PAE250

HEXADECANOIC ACID, 2-ETHYLHEXYL ESTER (9CI) see OFG100

HEXADECANOIC ACID, HEXADECYL ESTER see HCP700

HEXADECANOIC ACID, ISOPROPYL ESTER see IQW000

HEXADECANOIC ACID,-(6-METHYL-5-((1-OXOHEXADECYL)OXY)-3,4-PYRIDINEDIYL)BIS(METHYLENE) ESTER see HMQ575

HEXADECANOL see HCP000

1-HEXADECANOL see HCP000

HEXADECAN-1-OL see HCP000

n-HEXADECANOL see HCP000

1-HEXADECANOL, ACETATE see HCP100

1,16-HEXADECANOLACTONE see OKU000

1-HEXADECANOL, HYDROGEN SULFATE, SODIUM SALT see HCP900

HEXADECANOLIDE see OKU000

1,6,10,14-HEXADECATETRAEN-3-OL, 3,7,11,15-TETRAMETHYL-, (E,E)- see GDG300

7-HEXADECENAL, (Z)- see HCP022

9-HEXADECENAL, (Z)- see HCP032

(Z)-7-HEXADECENAL see HCP022

(Z)-9-HEXADECENAL see HCP032

11-HEXADECENAL, (Z)- see HCP050

(Z)-11-HEXADECENAL see HCP050

HEXADECENE EPOXIDE see EBX500

(Z)-11-HEXADECENOL see HCP060

11-HEXADECEN-1-OL, (Z)- see HCP060

(Z)-7-HEXADECENOL ACETATE see HCP070

7-HEXADECEN-1-OL, ACETATE, (Z)- see HCP070

(Z)-7-HEXADECEN-1-OL ACETATE see HCP070

2-HEXADECEN-1-OL, 3,7,11,15-TETRAMETHYL-, (R-(R*,R*-(E)))-(9CI) see PIB600

(Z)-7-HEXADECENYL ACETATE see HCP070

cis-7-HEXADECENYL ACETATE see HCP070

n-HEXADECOIC ACID see PAE250

HEXADECYL ACETATE see HCP100

HEXADECYL ALCOHOL see HCP000

n-HEXADECYL ALCOHOL see HCP000

N-HEXADECYLAMINE see HCO500

HEXADECYLAMINE, N,N-DIMETHYL- see HCP525

HEXADECYLAMINE HYDROFLUORIDE see CDF400

HEXADECYLBETAINE see CDF450

HEXADECYL CYCLOPROPANECARBOXYLATE see HCP500

HEXADECYLDIMETHYLAMINE see HCP525

HEXADECYL 2-ETHYLHEXANOATE see HCP550

HEXADECYLIC ACID see PAE250

HEXADECYLMALEINAN DI-n-BUTYLCINICITY (CZECH) see DEI600

HEXADECYL NEODECANOATE see HCP600

2-(((HEXADECYLOXY)HYDROXYPHOSPHINYL)OXY)-N,N,N-TRIMETHYLETHANAMINIUMHYDROXIDE, INNER SALT see HCP750

HEXADECYL PALMITATE see HCP700

HEXADECYLPHOSPHOCHOLINE see HCP750

HEXADECYLPHOSPHORYLCHOLINE see HCP750

N-HEXADECYLPHOSPHORYLCHOLINE see HCP750

HEXADECYLPOLY(ETHYLENEOXY)ETHANOL see PJT300

HEXADECYLPYRIDINE BROMIDE see HCP800

HEXADECYLPYRIDINIUM BROMIDE see HCP800

1-HEXADECYLPYRIDINIUM BROMIDE see HCP800

N-HEXADECYLPYRIDINIUM BROMIDE see HCP800

1-HEXADECYL-PYRIDINIUM BROMIDE mixture with CHLORO(2-HYDROXYETHYL)MERCURY see DVJ500

HEXADECYLPYRIDINIUM CHLORIDE see CCX000

1-HEXADECYLPYRIDINIUM CHLORIDE see CCX000

n-HEXADECYLPYRIDINIUM CHLORIDE see CCX000

1-HEXADECYLPYRIDINIUM CHLORIDE MONOHYDRATE see CDF750

HEXADECYL SODIUM SULFATE see HCP900

HEXADECYLTRICHLOROSILANE see HCQ000

HEXADECYLTRIETHYLAMMONIUM BROMIDE see CDF500

HEXADECYLTRIMETHYLAMMONIUM BROMIDE see HCQ500

(1-HEXADECYL)TRIMETHYLAMMONIUM BROMIDE see HCQ500

N-HEXADECYLTRIMETHYLAMMONIUM BROMIDE see HCQ500

N-HEXADECYL-N,N,N-TRIMETHYLAMMONIUM BROMIDE see HCQ500

HEXADECYLTRIMETHYLAMMONIUM CHLORIDE see HCQ525

HEXADECYLTRIMETHYLAMMONIUM PENTACHLOROPHENOL see TLN150

ω-h-HEXADEKAFLUORNONANOL-1 (GERMAN) see HCO000

HEXADENOL see HCS500

HEXADERM RED MRG see CMM325

HEXA-2,4-DIENAL see SKT500

2,4-HEXADIENAL see SKT500

2,4-HEXADIENAL, (2E,4Z)- see HCQ535

(2E,4Z)-2,4-HEXADIENAL see HCQ535

2,4-HEXADIENAL, 6-HYDROXY-, (E,E)- see HLE800

HEXADIENE see HCQ600

1,4-HEXADIENE see HCR000

1,5-HEXADIENE see HCR500

HEXA-1,5-DIENE see HCR500

(E,E)-2,4-HEXADIENEDIAL see HCR600

2,4-HEXADIENEDIAL, (E,E)- see HCR600

1,3-HEXADIENE-5-YNE see HCS100

HEXADIENIC ACID see SKU000

HEXADIENOIC ACID see SKU000

2,4-HEXADIENOIC ACID see SKU000

trans-trans-2,4-HEXADIENOIC ACID see SKU000

2,4-HEXADIENOIC ACID, 6-OXO-, (E,E)- see ONW200

2,4-HEXADIENOIC ACID POTASSIUM SALT see PLS750

2,4-HEXADIENOL see HCS500

2,4-HEXADIEN-1-OL see HCS500

2,4-HEXADIEN-1-OL ACETATE see AAU750

2,4-HEXADIENOL BUTANOATE see HCS600

2,4-HEXADIENYL ACETATE see AAU750

2,4-HEXADIENYL BUTYRATE see HCS600

2,4-HEXADIENYL ISOBUTYRATE see HCS700

2,4-HEXADIENYL 2-METHYLPROPANOATE see HCS700

2-(2,4-HEXADIENYLOXY)ETHANOL see HCT500

1,5-HEXADIEN-3-YNE see HCU500

4,5-HEXADIEN-2-YN-1-OL see HCV000

HEXADIMETHRINE BROMIDE see HCV500

HEXADIONA see DBB200

2,4-HEXADIYN-1,6-BIS-p-TOLUENESULFONATE see HCV600

1,5-HEXADIYNE see HCV850

2,4-HEXADIYNE-1,6-DIOIC ACID see HCV875

2,4-HEXADIYNE-1,6-DIOL, DI-p-TOLUENESULFONATE see HCV600

1,5-HEXADIYNE-3-ONE see HCV880

2,4-HEXADIYNYLENE BISCHLOROFORMATE see HCW000

HEXADRIN see CQC650

HEXADRIN see EAT500

HEXADROL see SOW000

HEXAETHYLBENZENE see HCX000

HEXAETHYLDILEAD see TJS000

HEXAETHYLDISTANNOXANE see HCX050

1,1,1,3,3,3-HEXAETHYLDISTANNOXANE see HCX050

HEXAETHYLDISTANNTHIANE see HCX100

1,1,1,3,3,3-HEXAETHYLDISTANNTHIANE see HCX100

HEXAETHYLTETRAFOSFAT see HCY000

HEXAETHYL TETRAPHOSPHATE see HCY000

HEXAETHYL TETRAPHOSPHATE, liquid or solid (DOT) see HCY000

HEXAETHYLTRIALUMINUM TRITHIOCYANATE see HCY500

HEXAFEN see HCL000

HEXAFERB see FAS000

HEXAFLUMURON see HCY600

HEXAFLUORAMIN see FDB300

HEXAFLUORENIUM DIBROMIDE see HEG000

HEXAFLUOROACETIC ANHYDRIDE see TJX000

HEXAFLUOROACETONE see HCZ000

HEXAFLUOROACETONE BISPHENOL A see HCZ100

HEXAFLUOROACETONE HYDRATE see HDA000

HEXAFLUOROACETONE SESQUIHYDRATE see HDE500

HEXAFLUORO ACETONE TRIHYDRATE see HDA500

HEXAFLUOROBENZENE see HDB000

HEXAFLUOROCALCITRIOL see HDB100

1,1,1,4,5,5-HEXAFLUORO-2-CHLORO-2-BUTENE see CHK750

HEXAFLUORODICHLOROBUTENE see HDB500

HEXAFLUORODIETHYL ETHER see HDC000

HEXAFLUORODIPHENYLOLPROPANE see HCZ100

HEXAFLUOROEPOXYPROPANE see HDF050

HEXAFLUORO-1,2-EPOXYPROPANE see HDF050

HEXAFLUOROETHANE see HDC100

HEXAFLUORO FERRATE (3-) TRIAMMONIUM SALT see ANI000

HEXAFLUOROGLUTARIC ACID DIETHYL ESTER see DJK100

HEXAFLUOROGLUTARONITRILE see HDC300

HEXAFLUOROGLUTARYL DIHYPOCHLORITE see HDC425

2-(1,1,2,3,3,3-HEXAFLUORO-2-(HEPTAFLUOROPROPOXY)PROPOXY)-2,3,3,3-TETRAFLUOROPROPANOIC ACID see HDC435

HEXAFLUOROISOBUTYLENE see HDC450

3,3,3,4,4,4-HEXAFLUOROISOBUTYLENE see HDC450

HEXAFLUOROISOBUTYRIC ACID METHYL ESTER see MKK750

HEXAFLUOROISOPROPANOL see HDC500

HEXAFLUOROISOPROPYLIDENEAMINE see HDD000

HEXAFLUOROISOPROPYLIDENEAMINOLITHIUM see HDD500

3,3'-(HEXAFLUOROISOPROPYLIDENE)DIPHENOL see HCZ100

HEXAFLUOROKIESELSAEURE (GERMAN) see SCO500

HEXAFLUOROKIEZELZUUR (DUTCH) see SCO500

4,4,4,4',4',4'-HEXAFLUORO-N-NITROSODIBUTYLAMINE see NJN300

HEXAFLUOROPENTANEDIOIC ACID DIETHYL ESTER see DJK100

HEXAFLUOROPHOSPHORIC ACID see HDE000

1,1,1,2,3,3-HEXAFLUOROPROPANE see HDE050

1,1,1,3,3,3-HEXAFLUOROPROPANE see HDE100

1,1,1,3,3,3-HEXAFLUORO-2-PROPANOL see HDC500

HEXAFLUORO-2-PROPANONE HYDRATE see HDA000

HEXAFLUORO-2-PROPANONE SESQUIHYDRATE see HDE500

HEXAFLUOROPROPENE see HDF000

HEXAFLUOROPROPENE EPOXIDE see HDF050

HEXAFLUOROPROPENE OXIDE see HDF050

HEXAFLUOROPROPYLENE (DOT) see HDF000

HEXAFLUOROPROPYLENE OXIDE (DOT) see HDF050

HEXAFLUOROSILICATE(2-) ALUMINUM (3:2) see AHB400

HEXAFLUOROSILICATE(2-) DIHYDROGEN see SCO500

HEXAFLUOROSILICATE (2-1) LEAD(II) SALT DIHYDRATE see LDG000

HEXAFLUOROSILICATE(2-) MAGNESIUM (1:1) see MAF600

HEXAFLUOROSILICATE (2−), NICKEL see NDD000

HEXAFLUOROSILICATE(2-) STRONTIUM see SMJ000

HEXAFLUORO-SILICATE(2−), THALLIUM see TEK250

1,1,1,3,3,3-HEXAFLUORO-2-(TRIFLUOROMETHYL)-2-PROPANOL see HDF075

HEXAFLUORO VANADATE (3-) TRIAMMONIUM SALT see ANI500

α,α,α,α,α,α-HEXAFLUORO-3,5-XYLIDINE see BLO250

1-(α,α,α,α',α',α'-HEXAFLUORO-3,5-XYLYL)-4-METHYL-3-THIO-SEMICARBAZIDE see BLP325

HEXAFLUORURE de SOUFRE (FRENCH) see SOI000

HEXAFLUOSILICIC ACID see SCO500

HEXAFLURATE see PLH500

HEXAFLURON see HCY600

HEXAFLURONIUM BROMIDE see HEG000

HEXAFORM see HEI500

HEXAFUNGIN see HDF100

HEXAGLYCERINE see HDF300

2,3,5,6,7,8-HEXAHYDRO-9-AMINO-1H-CYCLOPENTA(b)QUINOLINE HYDROCHLORIDE HYDRATE see AKD775

HEXAHYDROANILINE see CPF500

HEXAHYDROANILINE HYDROCHLORIDE see CPA775

1,2,3,7,8,9-HEXAHYDROANTHANTHRENE see HDF500

HEXAHYDRO-1H-AZEPIN-1-AMINE see HEG400

HEXAHYDROAZEPINE see HDG000

HEXAHYDRO-1H-AZEPINE see HDG000

HEXAHYDRO-2-AZEPINONE see CBF700

HEXAHYDRO-2H-AZEPIN-2-ONE see CBF700

HEXAHYDRO-2H-AZEPIN-2-ONE HOMOPOLYMER see PJY500

N-(((HEXAHYDRO-1H-AZEPIN-1-YL)-AMINO)CARBONYL)-4-METHYLBENZENESULFONAMIDE see TGJ500

((2-HEXAHYDRO-1-AZEPINYL)ETHYL)GUANIDINE SULFATE (2:1) see HDG600

(+)-6-(((HEXAHYDRO-1H-AZEPIN-1-YL)METHYLENE)AMINO)-3,3-DIMETHYL-7-OXO-4-THIA-1-AZABICYCLO(3.2.0)HEPTANE-2-CARBOXYLIC ACID HYDROXYMETHYL

ESTER, PIVALATE (ester), MONOHYDROCHLORIDE see MCB550

3-(3-(HEXAHYDRO-1H-AZEPIN-1-YL)PROPOXY)-1-(α,α,α-TRIFLUORO-m-TOLYL)-1H-PYRAZOLO(3,4-b)PYRIDINE MONOHYDROCHLORIDE see ITD100

1-(HEXAHYDRO-1-AZEPINYL)-3-p-TOLYLSULFONYLUREA see TGJ500

1-(HEXAHYDRO-1H-AZEPIN-1-YL)-3-(p-TOLYLSULFONYL)UREA see TGJ500

(2-(HEXAHYDRO-1(2H)-AZOCINYL)ETHYL) GUANIDINE HYDROGEN SULFATE see GKU000

(2-(HEXAHYDRO-1(2H)-AZOCINYL)ETHYL)GUANIDINE SULFATE see GKS000

1,2,4,5,6,7-HEXAHYDROBENZ(e)ACEANTHRYLENE see HDH000

HEXAHYDROBENZENAMINE see CPF500

HEXAHYDROBENZENE see CPB000

HEXAHYDROBENZOIC ACID see HDH100

HEXAHYDROBENZOIC ACID AMIDE see CPB050

HEXAHYDROBENZYL ALCOHOL see HDH200

HEXAHYDROCARQUEJENE see MCE275

HEXAHYDROCRESOL see MIQ745

N-(((HEXAHYDROCYCLOPENTA(c)PYRROL-2(1H)-YL)AMINO)CARBONYL)-4-METHYL-BENZENESULFONAMIDE see DBL700

1-(HEXAHYDROCYCLOPENTA(c)PYRROL-2(1H)-YL)-3-(p-TOLYLSULFONYL)UREA see DBL700

HEXAHYDRO-p-CYMENE HYDROPEROXIDE see IQE000

HEXAHYDRO-1,8-DIAMINO-4,7-METHANOINDAN see DCF000

HEXAHYDRO-1,4-DIAZEPINE see HGI900

HEXAHYDRO-1,4-DIAZINE see PIJ000

1,2,3,4,12,13-HEXAHYDRODIBENZ(a,h)ANTHRACENE see HDK000

1,2,3,7,8,9-HEXAHYDRODIBENZO(def,mno)CHRYSENE see HDF500

1,5A,6,9,9A,9B-HEXAHYDRO-4A(4H)-DIBENZOFURANCARBOXALDEHYDE see BHJ500

HEXAHYDRO-1,3-DICYCLOHEXYL-5-((1-METHYL-5-NITRO-1H-IMIDAZOL-2-YL)METHYLENE)PYRIMIDINE see HDK100

1a-β,1b,5a6,6a7a-β-HEXAHYDRO-1b-α,6-β-DIHYDROXY-8a-ISOPROPENYL-6a-α-METHYL-SPIRO(2,5-METHANO-7H-OXIRENO(3,4)CYCLOPENT(1,2-d)OXEPIN-7,2'-OXIRAN)-3(2aH)-ONE see TOE175

HEXAHYDRO-3A,7A-DIMETHYL-4,7-EPOXYISOBENZOFURAN-1,3-DIONE see CBE750

(−)-2,3,4,5,6,7-HEXAHYDRO-1,4-DIMETHYL-1,6-METHANO-1H-4-BENZAZONIN-10-OL HYDROBROMIDE see SLI200

1,2,3,4,5,6-HEXAHYDRO-6,11-DIMETHYL-3-(3-METHYL-2-BUTENYL)-2,6-METHANO-3-BENZAZOCINE see DOQ400

1,2,3,4,5,6-HEXAHYDRO-6,11-DIMETHYL-3-(3-METHYL-2-BUTENYL)-2,6-METHANO-3-BENZAZOCIN-8-OL HCl see PBP300

HEXAHYDRO-1,4-DINITROSO-1H-1,4-DIAZEPINE see DVE600

HEXAHYDRO-1-DODECYL-1H-AZEPINE-1-OXIDE see DXX875

(4aRS,5RS,9bRS)-2,3,4,4a,5,9b-HEXAHYDRO-2-ETHYL-7-METHYL-5-p-TOLYL-1H-INDENO(1,2-c)PYRIDINE HYDROCHLORIDE see CNH300

1,3,3A,4,7,7A-HEXAHYDRO-4,5,6,7,8,8-HEXACHLORO-4,7-METHANOISOBENZOFURAN see HDK200

1,3,3A,4,7,7A-HEXAHYDRO-4,5,6,7,8,8-HEXACHLORO-4,7-METHANOISOBENZOFURAN-1-OL see HDK210

2,2,4,4,6,6-HEXAHYDRO-2,2,4,4,6,6-HEXAKIS(1-AZIRIDINYL)-1,3,5,2, 4,6-TRIAZATRIPHOSPHORINE see AQO000

1,3,4,6,7,8-HEXAHYDRO-4,6,6,7,8,8-HEXAMETHYL-CYCLOPENTA-γ-2-BENZOPYRAN see GBU000

15,16,17,18,19,20-HEXAHYDRO-18-HYDROXY-17-METHOXY-YOHIMBAN PROPIONATE METHANESULFONATE see KAH000

1,2,3a,4,5,9b-HEXAHYDRO-8-HYDROXY-3-METHYL-9b-PROPYL-3H-BENZ(e)INDOLE, HYDROCHLORIDE see HDO500

1,2,3,3a,8,8a-HEXAHYDRO-5-HYDROXY-1,3a,8-TRIMETHYL-PYRROLO(2,3-b)INCOLE METHYLCARBAMATE (ester), (3aS-cis)-, SULFATE (2:1) see PIB000

1,3,4,6,7,11b-HEXAHYDRO-3-ISOBUTYL-9,10-DIMETHOXY-2H-BENZO(a)QUINOLIZIN- 2-ONE see TBJ275

4-HEXAHYDROISONICOTINIC ACID see ILG100

exo-HEXAHYDRO-4,7-METHANOINDAN see TLR675

HEXAHYDRO-4,7-METHANOINDANDIMETHANOL see TJG550

3-(HEXAHYDRO-4,7-METHANOINDAN-5-YL)-1,1-DIMETHYLUREA see HDP500

1-(3a,4,5,6,7,7a-HEXAHYDRO-4,7-METHANO-5-INDANYL)-3,3-DIMETHYLUREA see HDP500

1-(5-(3a,4,5,6,7,7a-HEXAHYDRO-4,7-METHANOINDANYL))-3,3-DIMETHYLUREA see HDP500

3-(5-(3a,4,5,6,7,7a-HEXAHYDRO-4,6-METHANOINDANYL))-1,1-DIMETHYLUREA see HDP500

3a,4,5,6,7,7a-HEXAHYDRO-4,7-METHANO-1H-INDEN-6-OL ACETATE see DLY400

1,2,3,4,5,6-HEXAHYDRO-1,5-METHANO-8H-PYRIDO(1,2-A)(1,5)DIAZOCIN-8-ONE see CQL500

1,2,3,4,5,6-HEXAHYDRO-6-METHYLAZEPINO(4,5-b)INDOLE HYDROCHLORIDE see HDQ500

HEXAHYDRO-1-METHYL-2H-AZEPIN-2-ONE see MHY750

6,7,8,9,10,12b-HEXAHYDRO-3-METHYL CHOLANTHRENE see HDR500

HEXAHYDRO-1-((2-METHYLCYCLOHEXYL)CARBONYL)-1H-AZEPINE see HDR700

1,2,3,4,10,14b-HEXAHYDRO-2-METHYLDIBENZO(c,f)PYRAZINO(1,2-a)AZEPINE see MQS220

1,2,3,4,10,14b-HEXAHYDRO-2-METHYLDIBENZO(c,f)PYRAZINO(1,2-a)AZEPINE HYDROCHLORIDE see BMA625

(3A-α-4-β,7-β,7Aα)-HEXAHYDRO-3A-METHYL-4,7-EPOXYISOBENZOFURAN-1,3-DIONE see HDR800

(+)-cis-1,3,4,9,10,10A-HEXAHYDRO-11-METHYL-2H-10,4a-IMINOETHANOPHENANTHREN-6-OL see DBE800

dl-1,3,4,9,10,10A-HEXAHYDRO-11-METHYL-2H-10,4A-IMINOETHANOPHENANTHREN-6-OL see MKR250

1,2,3,4,5,6-HEXAHYDRO-1-(2'-METHYL-3'-(N-METHYLAMINO)PROPYL)-1-BENZAZOCINE HYDROCHLORIDE see HDS200

HEXAHYDROMETHYLPHENOL see MIQ745

1,2,3,4,10,14B-HEXAHYDRO-2-METHYLPYRAZINO(2,1-A)PYRIDO(2,3-C)(2)BENZAZEPINE see HDS210

2,3,3A,4,5,6-HEXAHYDRO-8-METHYL-1H-PYRAZINO(3,2,1J,K)CARBAZOLE HYDROCHLORIDE see HDS225

7,8,9,10,11,12-HEXAHYDRO-2-NITROCHRYSENE see NHB600

HEXAHYDRO-1-NITROSO-1H-AZEPINE see NKI000

HEXAHYDRO-2-OXO-1,4-CYCLOHEXANEDIMETHANOL see BKH325

HEXAHYDRO-3,6-endo-OXYPHTHALIC ACID see EAR000

HEXAHYDROPHENOL see CPB750

1,3,6,7,8,9-HEXAHYDRO-5-PHENYL-2H-(1)BENZOTHIENO(2,3-e)-1,4-DIAZEPIN-2-ONE see BAV625

HEXAHYDROPHENYL ETHYL ACETATE see EHS000

HEXAHYDROPHENYLETHYL ALCOHOL see CPL250

HEXAHYDROPHTHALIC ACID DIETHYL ESTER see HDS250

HEXAHYDROPICOLINIC ACID see HDS300

HEXAHYDROPYRAZINE see PIJ000

HEXAHYDROPYRIDINE see PIL500

HEXAHYDROTHYMOL see MCF750

HEXAHYDROTOLUENE see MIQ740

HEXAHYDRO-1,3,5-s-TRIAZINE see HDV500

HEXAHYDRO-1,3,5-TRIETHYL-s-TRIAZINE see HDW000

HEXAHYDRO-3,5,5-TRIMETHYL-3,8a-ETHANO-8aH-1-BENZOPYRAN-2(3H)-ONE see LAQ100

HEXAHYDRO-1,3,5-TRIMETHYL-s-TRIAZINE see HDW100

HEXAHYDRO-1,3,5-TRINITROSO-s-TRIAZINE see HDV500

HEXAHYDRO-1,3,5-TRINITROSO-1,3,5-TRIAZINE see HDV500

HEXAHYDRO-1,3,5-TRINITRO-1,3,5-TRIAZIN (GERMAN) see CPR800

HEXAHYDRO-1,3,5-TRINITRO-1,3,5-TRIAZINE see CPR800

HEXAHYDRO-1,3,5-TRIS(DIMETHYLAMINOPROPYL)-s-TRIAZINE see TNH250

HEXAHYDRO-1,3,5-TRIS(HYDROXYETHYL)TRIAZINE see THR820

HEXAHYDRO-1,3,5-TRIS(1-OXO-2-PROPENYL)-1,3,5-TRIAZINE see THM900

2,2',3,3,3',3'-HEXAHYDROXY(2,2' BI INDAN)-1,1'-DIONE see HHI000

2-β,3-β,14,20,22,25-HEXAHYDROXY-5-β-CHOLET-7-EN-6-ONE see HKG500

3,3',4',5,5',7-HEXAHYDROXY-2,3-DIHYDROFLAVANONOL see HDW125

3,3',4',5,5',7-HEXAHYDROXYFLAVONE see HDW150

3,5,7,3',4',5'-HEXAHYDROXYFLAVONE see HDW150

3,3',4',5,5',7-HEXAHYDROXYFLAVONE HEXAACETATE see MRZ200

3,3',4',5,5',7-HEXAHYDROXYFLAVYLIUM see HDW200

3,3',4',5,5',7-HEXAHYDROXYFLAVYLIUM ACID ANION see DAM400

HEXAHYDROXYLAMINECOBALT(III) NITRATE see HDX500

HEXA(HYDROXYMETHYL)MELAMINE see HDY000

HEXAISOBUTYLDITIN see HDY100

HEXAKIS(μ-ACETATO)-μ⁴-OXOTETRABERYLLIUM see BFT500

HEXAKIS(μ-ACETATO-O:O')-μ⁴-OXOTETRABERYLLIUM see BFT500

2,2,4,4,6,6-HEXAKIS(1-AZIRIDINYL)CYCLOTRIPHOSPHAZA-1,3,5-TRIENE see AQO000

2,2,4,4,6,6-HEXAKIS(1-AZIRIDINYL)-2,2,4,4,6,6-HEXAHYDRO-1,3,5,2,4,6-TRIAZATRIPHOSPHORINE see AQO000

HEXAKIS-(1-AZIRIDINYL)PHOSPHONITRILE see AQO000

HEXAKIS(AZIRIDINYL)PHOSPHOTRIAZINE see AQO000

HEXAKIS(DIHYDROGEN PHOSPHATE) MYO-INOSITOL see PIB250

HEXAKIS(β,β-DIMETHYLPHENETHYL)DISTANNOXANE see BLU000

HEXAKIS(HYDROXYMETHYL)MELAMINE see HDY000

HEXAKIS(HYDROXYMETHYL)-1,3,5-TRIAZINE-2,4,6-TRIAMINE see HDY000

HEXAKIS-METHOXYMETHYLMELAMIN (CZECH) see HDY500

HEXAKIS(METHOXYMETHYL)MELAMINE see HDY500

HEXAKIS(METHOXYMETHYL)-s-TRIAZINE-2,4,6-TRIAMINE see HDY500

N,N,N',N',N'',N''-HEXAKIS(METHOXYMETHYL)-1,3,5-TRIAZINE-2,4,6-TRIAMINE see HDY500

HEXAKIS(2-METHYL-2-PHENYLPROPYL)DISTANNOXANE see BLU000

HEXAKIS(PYRIDINE)IRON(II) TRIDECACARBONYLTETRAFERRATE(2-) see HEY000

HEXAKOSE see HCS500

γ-HEXALACTONE see HDY600

HEXALAN RED 2G see CMM300

HEXALAN RED 6B see CMM400

HEXALAN RED B see CMS228

HEXALDEHYDE (DOT) see HEM000

HEXALENE see HCP070

HEXALIN see CPB750

HEXALITHIUM DISILICIDE see HDZ000

HEXALURE see HCP070

HEXAMARIUM see DXG800

HEXAMETAPHOSPHATE, SODIUM SALT see SHM500

HEXAMETAPOL see HEK000

HEXAMETHIONIUM BROMIDE see HEA000

HEXAMETHIONIUM CHLORIDE see HEA500

HEXAMETHONIUM BROMIDE see HEA000

HEXAMETHONIUM CHLORIDE see HEA500

HEXAMETHONIUM DIBROMIDE see HEA000

HEXAMETHONIUM DICHLORIDE see HEA500

HEXAMETHONIUM DIIODIDE see HEB000

HEXAMETHONIUM IODIDE see HEB000

HEXAMETHOXYDISILYLETHANE see EIT200

HEXA(METHOXYMETHYL)MELAMINE see HDY500

HEXAMETHYLBENZENE see HEC000

HEXAMETHYL-BICYCLO(2.2.0)HEXA-2,5-DIEN (GERMAN) see HEC500

HEXAMETHYLBICYCLO(2.2.0)HEXA-2,5-DIENE see HEC500

HEXAMETHYLCYCLOTRISILAZANE see HEC600

2,2,4,4,6,6-HEXAMETHYLCYCLOTRISILAZANE see HEC600

N,N,N,N',N',N'-HEXAMETHYL-1,10-DECANEDIAMINIUM DIBROMIDE see DAF600

N,N,N,N¹,N¹,N¹-HEXAMETHYL-1,10-DECANEDIAMINIUM DIIODIDE see DAF800

cis-2,2,4,6,6,8-HEXAMETHYL-4,8-DIPHENYL-CYCLOTETRASILOXANE see CMS234

HEXAMETHYLDIPLATINUM see HED000

HEXAMETHYLDISILANE see HED425

HEXAMETHYLDISILAZANE see HED500

HEXAMETHYLDISILOXANE see HEE000

HEXAMETHYLDISTANNANE see HEE500

HEXAMETHYLDITIN see HEE500

HEXAMETHYLENAMINE see HEI500

1,6-HEXAMETHYLEN-BIS(4-HYDROXIMINOFORMYLPYRIDINIUM)-DIBROMID see HEE600

HEXAMETHYLENDIISOKYANAT see DNJ800

HEXAMETHYLEN-1,6-(N-DIMETHYLCARBODESOXYMETHYL)AMMONIUM DICHLORIDE see HEF200

HEXAMETHYLENE see CPB000

HEXAMETHYLENEAMINE see HEI500

N,N'-HEXAMETHYLENEBIS-1-AZIRIDINECARBOXAMIDE see HEF500

HEXAMETHYLENEBIS((CARBOXYMETHYL)DIMETHYLAMMONIUM), DICHLORIDE, DIDODECYL ESTER see HEF200

1,1'-HEXAMETHYLENEBIS(5-(p-CHLOROPHENYL)BIGUANIDE) see BIM250

1,1'-HEXAMETHYLENEBIS(5-(p-CHLOROPHENYL)BIGUANIDE) DIACETATE see CDT125

1,1'-HEXAMETHYLENEBIS(5-(p-CHLOROPHENYL)BIGUANIDE)DIGLUCONATE see CDT250

N,N'-HEXAMETHYLENEBIS(2,2-DICHLORO-N-ETHYLACETAMIDE) see HEF300

HEXAMETHYLENEBIS(DIMETHYL-9-FLUORENYLAMMONIUM BROMIDE) see HEG000

HEXAMETHYLENEBIS(DITHIOCARBAMIC ACID) DISODIUM SALT see HEF400

HEXAMETHYLENEBIS(ETHYLENEUREA) see HEF500

1,6-HEXAMETHYLENEBIS(ETHYLENEUREA) see HEF500

HEXAMETHYLENEBIS(FLUOREN-9-YLDIMETHYLAMMONIUM BROMIDE) see HEG000

HEXAMETHYLENE BIS(9-FLUORENYL DIMETHYLAMMONIUM)DIBROMIDE see HEG000

1,1'-HEXAMETHYLENEBIS(2-FORMYLPYRIDINIUM BROMIDE OXIME) see HEG010

N,N'-HEXAMETHYLENEBIS(2-HYDROXYIMINOMETHYLPYRIDINIUM BROMIDE) see HEG010

N,N'-HEXAMETHYLENEBIS(3-HYDROXYIMINOMETHYLPYRIDINIUM BROMIDE) see HEG020

N,N'-HEXAMETHYLENEBIS(4-HYDROXYIMINOMETHYLPYRIDINIUM BROMIDE) see HEE600

N,N'-HEXAMETHYLENEBIS(MALEIMIDE) see HEG050

HEXAMETHYLENEBIS(TRIMETHYLAMMONIUM) BROMIDE see HEA000

HEXAMETHYLENE(BISTRIMETHYLAMMONIUM)CHLORIDE see HEA500

HEXAMETHYLENEBIS(TRIMETHYLAMMONIUM) DIBENZENESULFONATE see HEG100

HEXAMETHYLENEBIS(TRIMETHYLAMMONIUM IODIDE) see HEB000

HEXAMETHYLENEDIAMINE see HEO000

1,6-HEXAMETHYLENEDIAMINE see HEO000

HEXAMETHYLENEDIAMINE ADIPATE see NOH100

HEXAMETHYLENEDIAMINE ADIPATE (1:1) see HEG120

HEXAMETHYLENEDIAMINE DIHYDROCHLORIDE see HEO100

1,6-HEXAMETHYLENEDIAMINE DIHYDROCHLORIDE see HEO100

HEXAMETHYLENEDIAMINE MONOADIPATE see HEG120

HEXAMETHYLENEDIAMINE SEBACATE see HEG130

HEXAMETHYLENEDIAMINE, solid (UN 2280) (DOT) see HEO000

HEXAMETHYLENEDIAMINE, solution (UN 1783) (DOT) see HEO000

HEXAMETHYLENEDIAMMONIUM ADIPATE see HEG120

HEXAMETHYLENEDIAMMONIUM SEBACATE see HEG130

HEXAMETHYLENEDIETHYLENEUREA see HEF500

HEXAMETHYLENE DIIODIDE see HEG200

HEXAMETHYLENE DIISOCYANATE see DNJ800

HEXAMETHYLENE-1,6-DIISOCYANATE see DNJ800

1,6-HEXAMETHYLENE DIISOCYANATE see DNJ800

HEXAMETHYLENE DIISOCYANATE (DOT) see DNJ800

HEXAMETHYLENE DIISOCYANATE POLYMER see HEG300

HEXAMETHYLENE DIISOCYANATE TRIMER see HEG300

N,N'-HEXAMETHYLENEDIMALEIMIDE see HEG050

HEXAMETHYLENEDIOL see HEP500

HEXAMETHYLENE GLYCOL see HEP500

1,1-HEXAMETHYLENEHYDRAZINE see HEG400

HEXAMETHYLENE IMINE (DOT) see HDG000

HEXAMETHYLENEIMINE-3,5-DINITROBENZOATE see ACI250

1-(2-HEXAMETHYLENEIMINOETHYL)-2-OXOCYCLOHEXANECARBOXYLIC ACID BENZYL ESTER HYDROCHLORIDE see HEI000

HEXAMETHYLENE ISOCYANATE POLYMER see HEG300

HEXAMETHYLENETETRAAMINE see HEI500

HEXAMETHYLENETETRAMINE see HEI500

HEXAMETHYLENE TETRAMINE TETRAIODIDE see HEI650

HEXAMETHYLENETETRAMMONIUM TETRAPEROXOCHROMATE(V) see HEJ000

HEXAMETHYLENETRIPEROXYDIAMINE see DCK700

HEXAMETHYLENIMINE see HDG000

2-(β-HEXAMETHYLENIMINOAETHYL)CYCLOHEXANON-2-CARBONSAUREBENZYLESTER-HYDROCHLORIDE (GERMAN) see HEI000

HEXAMETHYLENTETRAMIN (GERMAN) see HEI500

HEXAMETHYLERBIUM-HEXAMETHYLETHYLENEDIAMINE LITHIUM COMPLEX see HEJ350

2,4,6,8,9,10-HEXAMETHYLHEXAAZA-1,3,5,7-TETRAPHOSPHAADAMANTANE see HEJ375

N,N,N,N',N',N'-HEXAMETHYL-1,6-HEXANEDIAMINIUM DIBROMIDE see HEA000

N,N,N,N',N',N'-HEXAMETHYL-1,6-HEXANEDIAMINIUM DICHLORIDE see HEA500

N,N,N,N',N',N'-HEXAMETHYL-1,6-HEXANEDIAMINIUM DIIODIDE see HEB000

2,3,3,4,4,5-HEXAMETHYL-2-HEXANETHIOL see DXT800

1,1,2,3,3,6-HEXAMETHYLINDAN-5-YL METHYL KETONE see HEJ400

HEXAMETHYLMELAMINE see HEJ500

HEXAMETHYL METHYLOLMELAMINE see HDY500

HEXAMETHYLOLMELAMIN (CZECH) see HDY000

HEXAMETHYLOLMELAMINE see HDY000

HEXAMETHYLOL-MELAMIN-HEXA-METHYLAETHER see HDY500

HEXAMETHYLPARAOSANILINE CHLORIDE see AOR500

HEXAMETHYLPHOSPHORAMIDE see HEK000

HEXAMETHYLPHOSPHORIC ACID TRIAMIDE (MAK) see HEK000

HEXAMETHYLPHOSPHORIC TRIAMIDE see HEK000

N,N,N,N,N,N-HEXAMETHYLPHOSPHORIC TRIAMIDE see HEK000

HEXAMETHYLPHOSPHOROTHIOIC TRIAMIDE see HEK050

HEXAMETHYLPHOSPHOROTRIAMIDE see HEK000

HEXAMETHYLPHOSPHORUS TRIAMIDE see HEK100

HEXAMETHYLPHOSPHOTRIAMIDE see HEK000

HEXAMETHYLRHENIUM see HEK550

HEXAMETHYL-p-ROSANILINE CHLORIDE see AOR500

HEXAMETHYL-p-ROSANILINE HYDROCHLORIDE see AOR500

N,N,N',N',N'',N''-HEXAMETHYLSILANETRIAMINE see TNH300

HEXAMETHYLSILAZANE see HED500

2,6,10,15,19,23-HEXAMETHYLTETRACOSANE see SLG700

HEXAMETHYLTHIOPHOSPHORAMIDE see HEK050

HEXAMETHYLTHIOPHOSPHORIC TRIAMIDE see HEK050

N,N,N',N',N'',N''-HEXAMETHYL-1,3,5-TRIAZINE-2,4,6-TRIAMINE see HEJ500

HEXAMETHYL-s-TRITHIANE see HEL000

HEXAMETHYL-1,3,5-TRITHIANE see HEL000

2,2,4,4,6,6-HEXAMETHYLTRITHIANE see HEL000

HEXAMETHYL VIOLET see AOR500

HEXAMETON see HEA000

HEXAMETON CHLORIDE see HEA500

HEXAMIC ACID see CPQ625

HEXAMID see HEL500

HEXAMIDINE see DBB200

HEXAMIDINE (the antispasmodic) see DBB200

HEXAMINE (DOT) see HEI500

HEXAMITE see TCF250

HEXAMOL SLS see SIB600

HEXAMONE see HCP070

HEXANAL see ERE000

HEXANAL see HEM000

1-HEXANAL see HEM000

HEXANAL, 2-METHYLENE- see BPW025

HEXANAL, 3,5,5-TRIMETHYL- see ILJ100

HEXANAMIDE see HEM500

HEXANAMIDE, N-PHENYL- see HEV600

1-HEXANAMINE see HFK000

HEXANANILIDE see HEV600

HEXANAPHTHENE see CPB000

HEXANASTAB see ERE000

HEXANASTAB ORAL see ERD500

HEXANATE see TIX250

HEXANATE D see HCP600

HEXANATRIUMTETRAPOLYPHOSPHAT (GERMAN) see HEY500

n-HEXANE see HEN000

HEXANE (DOT) see HEN000

1-HEXANECARBOXYLIC ACID see HBE000

2-HEXANECARBOXYLIC ACID see HEN500

HEXANEDIAMIDE (9CI) see AEN000

1,6-HEXANEDIAMINE see HEO000

1,6-HEXANEDIAMINE, N-BUTYL- see BRK830

1,6-HEXANEDIAMINE, N,N'-DICARBOXYMETHYL-N,N'-DIMETHYL-, DIMETHOCHLORIDE, DIDODECYL ESTER see HEF200

HEXANEDIAMINE DIHYDROCHLORIDE see HEO100

1,6-HEXANEDIAMINE, DIHYDROCHLORIDE see HEO100

1,6-HEXANEDIAMINE, N,N,N',N'-TETRAMETHYL- see TDR100

1,6-HEXANEDIAMINIUM, N,N'-BIS(2-(DODECYLOXY)-2-OXOETHYL)-N,N,N',N'-TETRAMETHYL-, DICHLORIDE see HEF200

1,6-HEXANEDIAMINIUM, N,N,N,N',N',N'-HEXAMETHYL-, DIBENZENESULFONATE (9CI) see HEG100

HEXANE, 1,6-DIIODO- see HEG200

HEXANE, 1,6-DIISOCYANATO-, HOMOPOLYMER (9CI) see HEG300

HEXANE, 2,5-DIMETHYL-, 2,5-DIHYDROPEROXIDE see DSE800

HEXANEDINITRILE see AER250

1,6-HEXANEDIOIC ACID see AEN250

HEXANEDIOIC ACID, AMMONIUM SALT see HEO120

HEXANEDIOIC ACID, BIS(2-(2-BUTOXYETHOXY)ETHYL ESTER (9CI) see DDT500

HEXANEDIOIC ACID, BIS(2-BUTOXYETHYL) ESTER see BHJ750

HEXANEDIOIC ACID, BIS(2-ETHYLHEXYL) ESTER see AEO000

HEXANEDIOIC ACID, BIS(2-(HEXYLOXY)ETHYL)ESTER (9CI) see AEQ000

HEXANEDIOIC ACID, BIS(3-METHYLBUTYL) ESTER see AEQ500

HEXANEDIOIC ACID, BIS(1-METHYLETHYL) ESTER see DNL800

HEXANEDIOIC ACID BIS(4-METHYL-7-OXABICYCLO(4.1.0)HEPT-3-YL)METHYL ESTER see AEN750

HEXANEDIOIC ACID, BIS(7-OXABICYCLO(4.1.0)HEPT-3-YLMETHYL) ESTER see BJN100

HEXANEDIOIC ACID−DIBUTYL ESTER see AEO750

HEXANEDIOIC ACID, DICYCLOHEXYL ESTER (9CI) see DGT500

HEXANEDIOIC ACID, DIETHENYL ESTER (9CI) see HEO150

HEXANEDIOIC ACID, DIHEXYL ESTER (9CI) see HEO200

HEXANEDIOIC ACID DIHYDRAZIDE see AEQ250

HEXANEDIOIC ACID DINITRILE see AER250

HEXANEDIOIC ACID, DIOCTYL ESTER see AEO000

HEXANEDIOIC ACID-DI-2-PROPENYL ESTER see AEO500

HEXANEDIOIC ACID ETHENYL METHYL ESTER see MQL000

HEXANEDIOIC ACID, compd. with 1,6-HEXANEDIAMINE (1:1) (9CI) see HEG120

HEXANEDIOIC ACID, compounded with 1,6-HEXANEDIAMINE see NOH100

HEXANEDIOIC ACID, OCTAFLUORO-, DIETHYL ESTER see DJT100

HEXANEDIOIC ACID, compound with PIPERAZINE (1:1) see HEP000

HEXANEDIOIC ACID, POLYMER with 1,4-BUTANEDIOL and 1,1'-METHYLENEBIS(4-ISOCYANATOBENZENE) see PKM250

HEXANEDIOIC ACID, POLYMER with 1,3-ETHANEDIOL and 1,1'-METHYLENEBIS(4-ISOCYANATOBENZENE) see PKL750

HEXANEDIOIC ACID, TRIMETHYL- see TLT250

1,2-HEXANEDIOL see HFP875

1,6-HEXANEDIOL see HEP000

2,5-HEXANEDIOL see HEQ000

ω-HEXANEDIOL see HEP500

α-ω-HEXANEDIOL see HEP500

1,6-HEXANEDIOL DIACRYLATE see HEQ100

1,6-HEXANEDIOL DIISOCYANATE see DNJ800

2,5-HEXANEDIOL DIMETHYLSULFONATE see DSU000

HEXANEDIOL, 2-ETHYL-, DIBENZOATE see HEQ120

2,3-HEXANEDIONE see HEQ200

2,5-HEXANEDIONE see HEQ500

2,5-HEXANEDIONE, 3,4-DIMETHYL- see DSF100

2,3-HEXANEDIONE, 5-METHYL- see MKL300

N,N'-1,6-HEXANEDIYLBIS-1-AZIRIDINECARBOXAMIDE see HEF500

3,3'-(1,6-HEXANEDIYLBIS-((METHYLIMINO)CARBONYL)OXY)BIS(1-METHYLPYRIDINIUMDIBROMIDE) see DXG800

(1,6-HEXANEDIYLBIS(NITRILOBIS(METHYLENE)))TETRAKISPHOSPHONIC ACID see HEQ600

(1,6-HEXANEDIYLBIS(NITRILOBIS(METHYLENE)))TETRAKISPHOSPHONIC ACID POTASSIUM SALT see HEQ610

1,1'-(1,6-HEXANEDIYL)BIS-1H-PYRROLE-2,5-DIONE (9CI) see HEG050

HEXANE, 1,2-EPOXY- see HFC100

HEXANE, 1-(1-ETHOXYETHOXY)- see EKS120

1,2,3,4,5,6-HEXANEHEXOL see HER000

6-HEXANELACTAM see CBF700

HEXA-NEMA see DFK600

HEXANE, 3-(p-METHOXYBENZYL)-4-(p-METHOXYPHENYL)- see MEE150

HEXANEN (DUTCH) see HEN000

HEXANENITRILE see HER500

HEXANES (FCC) see HEN000

HEXANESULFONYL FLUORIDE, 6-FLUORO- see FJA100

1-HEXANETHIOL see HES000

HEXANE, 2,2,5-TRIMETHYL- see TLT200

1,2,6-HEXANETRIOL see HES500

HEXANETRIOL-1,2,6 see HES500

HEXANE-1,2,6-TRIOL see HES500

HEXANHYDROPYRIDINE HYDROCHLORIDE see HET000

HEXANICIT see HFG550

HEXANICOTINOYL INOSITOL see HFG550

HEXANICOTOL see HFG550

HEXANITROBENZENE see HET350

HEXANITROCOBALTATE(3-) TRIPOTASSIUM see PLI750

2,2',4,4',6,6'-HEXANITRODIFENYLAMIN see HET500

HEXANITRODIFENYLAMINE (DUTCH) see HET500

HEXANITRODIPHENYLAMINE see HET500

HEXANITRODIPHENYLAMINE (FRENCH) see HET500

2,2',4,4',6,6'-HEXANITRODIPHENYLAMINE see HET500

2,4,6,2',4',6'-HEXANITRODIPHENYLAMINE see HET500

2,4,6,2',4',6'-HEXANITRODIPHENYLAMINE see HET500

HEXANITRODIPHENYLSULFIDE see BLR750

HEXANITROETHANE see HET675

HEXANITROL see MAW250

HEXANITROOXANILIDE see HET700

3-HEXENYL ESTER, BENZOIC ACID (Z)- see HFE500
cis-3-HEXENYL ETHANOATE see HFE150
cis-3-HEXENYL ETHYL ACETAL see EKS100
cis-3-HEXENYL FORMATE see HFE516
cis-3-HEXENYL HEXANOATE see HFE510
3-HEXENYL ISOBUTANOATE see HFE518
β,γ-HEXENYL ISOBUTANOATE see HFE520
3-HEXENYL ISOBUTYRATE see HFE518
cis-3-HEXENYL ISOBUTYRATE see HFE520
cis-3-HEXENYL ISOVALERATE (FCC) see ISZ000
β,γ-HEXENYL METHANOATE see HFE516
cis-3-HEXENYL-2-METHYL-trans-2-BUTENOATE see HFE710
cis-3-HEXENYL 2-METHYLBUTYRATE see HFE550
3-HEXENYL 2-METHYPROPANOATE see HFE518
cis-HEXENYL OCYACETALDEHYDE see HFE600
cis-3-HEXENYL PENTANOATE see HFE800
cis-3-HEXENYL PHENYLACETATE see HFE625
2-HEXENYL PROPANOATE see HFE700
β,γ-HEXENYL PROPANOATE see HFE650
cis-3-HEXENYL PROPIONATE see HFE650
trans-2-HEXENYL PROPIONATE see HFE700
cis-3-HEXENYL SALICYLATE see SAJ000
β,γ-cis-HEXENYL SALICYLATE see SAJ000
cis-3-HEXENYL TIGLATE see HFE710
β,γ-HEXENYL α-TOLUATE see HFE625
cis-3-HEXENYL VALERATE see HFE800
4-HEXEN-1-YN-3-OL see HFF000
4-HEXEN-1-YN-3-ONE see HFF300
HEXERMIN see PPK500
HEXERMIN P see PII100
HEXESTROL see DLB400
meso-HEXESTROL see DLB400
HEXESTROL DIPHOSPHATE SODIUM see SHN150
HEXETHAL SODIUM see EKT500
HEXICIDE see BBQ500
HEXIDE see HCL000
HEXILMETHYLENAMINE see HEI500
HEXILURE see CPR600
HEXMETHYLPHOSPHORAMIDE see HEK000
HEXOBARBITAL see ERD500
HEXOBARBITAL Na see ERE000
HEXOBARBITAL SODIUM see ERE000
HEXOBARBITONE see ERD500
HEXOBARBITONE Na see ERE000
HEXOBARBITONE SODIUM see ERE000
HEXOBENDINE DIHYDROCHLORIDE see HFF500
HEXOBION see PPK500
HEXOCYCLIUM see HFG000
HEXOCYCLIUM METHYLSULFATE see HFG400
HEXOESTROL see DLB400
HEXOGEEN (DUTCH) see CPR800
HEXOGEN see CPR800
HEXOGEN 5W see CPR800
HEXOGEN (Explosive) see CPR800
HEXOGEN, wetted (UN 0072) (DOT) see CPR800
HEXOGEN, desensitized (UN 0483) (DOT) see CPR800
n-HEXOIC ACID see HEU000
1,6-HEXOLACTAM see CBF700
HEXOLITE see CPR800
HEXOLITE, dry or wetted with <15% water, by weight (UN 0118) (DOT) see CPR800
HEXON (CZECH) see HFG500
HEXON CHLORIDE see HEA500
HEXONE see HFG500
HEXONE CHLORIDE see HEA500
HEXONIUM DIBROMIDE see HEA000

HEXONIUM DIIODIDE see HEB000
HEXOPAL see HFG550
HEXOPHENE see HCL000
HEXOPRENALINE DIHYDROCHLORIDE see HFG600
HEXOPRENALINE SULFATE see HFG650
HEXOSAN see HCL000
HEXOXYACETALDEHYDE DIMETHYLACETAL see HFG700
2-HEXOXYACETALDEHYDE DIMETHYLACETAL see HFG700
β-HEXOXYACETALDEHYDE DIMETHYLACETAL see HFG700
p-HEXOXYBENZOIC ACID-3-(2'-METHYLPIPERIDINO)PROPYL ESTER see HFH500
3-HEXOXY-1-(2'-CARBOXYPHENOXY)-PROPANOL-(2) SODIUM SALT see ERE200
n-HEXOXYETHOXYETHANOL see HFN000
HEXYCLAN see BBQ750
HEXYL (GERMAN, DUTCH) see HET500
HEXYL ACETATE see HFI500
1-HEXYL ACETATE see HFI500
sec-HEXYL ACETATE see HFJ000
n-HEXYL ACETATE (FCC) see HFI500
HEXYLACETYLENE see OGI025
β-HEXYLACROLEIN see NNA300
HEXYL ACRYLATE see ADV000
N-HEXYL ACRYLATE see ADV000
HEXYL ALCOHOL see HFJ500
n-HEXYL ALCOHOL see HFJ500
sec-HEXYL ALCOHOL see EGW000
tert-HEXYL ALCOHOL see HFJ600
HEXYL ALCOHOL, ACETATE see HFI500
HEXYLAMINE see HFK000
N-HEXYLAMINE see HFK000
9-((3-(HEXYLAMINO)PROPYL)AMINO)-1-NITROACRIDINE DIHYDROCHLORIDE see HFK500
HEXYLAN see BBP750
8-HEXYL-BENZ(a)ANTHRACENE see HFL000
5-n-HEXYL-1,2-BENZANTHRACENE see HFL000
4-HEXYL-1,3-BENZENEDIOL see HFV500
HEXYL BENZOATE see HFL500
n-HEXYLBENZOATE see HFL500
HEXYL BROMIDE see HFM500
HEXYL BUTANOATE see HFM700
n-HEXYL BUTANOATE see HFM700
n-HEXYL n-BUTANOATE see HFM700
HEXYL-2-BUTENOATE see HFM600
n-HEXYL 2-BUTENOATE see HFM600
HEXYL BUTYRATE see HFM700
1-HEXYL BUTYRATE see HFM700
n-HEXYL BUTYRATE see HFM700
γ-N-HEXYL-γ-BUTYROLACTONE see HKA500
HEXYLCAINE see OKK100
HEXYLCAINE HYDROCHLORIDE see COU250
HEXYL CAPROATE see HFQ500
HEXYL CAPRYLATE see HFS600
n-HEXYL CAPRYLATE see HFS600
1-HEXYLCARBAMOYL-5-FLUOROURACIL see CCK630
HEXYL CARBITOL see HFN000
n-HEXYL CARBITOL see HFN000
n-HEXYL CARBORANE see HFN500
N-HEXYL CELLOSOLVE see HFT500
2-HEXYL-4-CHLOROPHENOL see CHL500
HEXYL CINNAMALDEHYDE see HFO500
α-HEXYLCINNAMALDEHYDE (FCC) see HFO500
HEXYL CINNAMIC ALDEHYDE see HFO500
α-HEXYLCINNAMIC ALDEHYDE see HFO500
HEXYL CROTONATE see HFM600
2-HEXYLCYCLOPENTANONE see HFO600
2-n-HEXYLCYCLOPENTANONE see HFO600

2-n-HEXYL-2-CYCLOPENTEN-1-ONE see HFO700
2-HEXYLDECANOIC ACID see HFP500
2-HEXYLDECANSAEURE (GERMAN) see HFP500
HEXYLDICARBADODECABORANE(12) see HFN500
HEXYLDICHLORARSINE see HFP600
HEXYL (DIETHYLAMINO)OXOACETATE see HFP650
HEXYL N,N-DIETHYLOXAMATE see HFP650
5-HEXYLDIHYDRO-2(3H)-FURANONE see HKA500
4-HEXYL-1,3-DIHYDROXYBENZENE see HFV500
HEXYL 2,2-DIMETHYLPROPANOATE see HFP700
HEXYLENE see HFA650
HEXYLENE see HFB000
HEXYLENE GLYCOL see HFP875
HEXYLENE GLYCOL DIACETATE see HFQ000
HEXYLENIC ALDEHYDE see HFA500
cis-β,γ-HEXYLENIC ALDEHYDE see HFA515
HEXYL ETHANOATE see HFI500
HEXYL ETHER see DKO800
N-HEXYL ETHER see DKO800
HEXYL FORMATE see HFQ100
n-HEXYL FORMATE see HFQ100
HEXYL FUMARATE see DKP000
HEXYL GLYCIDYL ETHER see GGY125
n-HEXYL HEXANOATE see HFQ500
HEXYL HEXOATE see HFQ500
3-HEXYL-p-IODOBENZYL CARBONATE see HFQ510
N-HEXYL-p-IODOBENZYL CARBONATE see HFQ520
HEXYL ISOBUTANOATE see HFQ550
n-HEXYL ISOBUTANOATE see HFQ550
HEXYL ISOBUTYRATE see HFQ550
1-HEXYL ISOBUTYRATE see HFQ550
n-HEXYL ISOBUTYRATE see HFQ550
n-HEXYL ISOPENTANOATE see HFQ575
HEXYL ISOVALERATE see HFQ575
HEXYL ISOVALERATE see HFQ600
HEXYL KETONE see TJJ300
HEXYL MANDELATE see HFR000
HEXYL MERCAPTAN see HES000
HEXYLMERCURIC BROMIDE see HFR100
n-HEXYLMERCURIC BROMIDE see HFR100
HEXYL MERCURY BROMIDE see HFR100
HEXYL 3-METHYL BUTANOATE see HFQ575
n-HEXYL trans-2-METHYL-2-BUTENOATE see HFX000
HEXYL 2-METHYLBUTYRATE see HFR200
2-HEXYL-4-METHYL-1,3-DIOXOLANE see HFR500
HEXYL NEOPENTANOATE see HFP700
1-HEXYL-3-NITRO-1-NITROSOGUANIDINE see HFR600
1-HEXYL-1-NITROSOUREA see HFS500
HEXYL OCTANOATE see HFS600
2-(HEXYLOXY)BENZAMIDE see HFS759
o-HEXYLOXYBENZAMIDE see HFS759
2-n-HEXYLOXYBENZAMIDE see HFS759
2-(HEXYLOXY)ETHANOL see HFT500
2-((2-HEXYLOXY)ETHOXY)ETHANOL see HFN000
2-(2-(2-(HEXYLOXY)ETHOXY)ETHOXY)ETHANOL see HFT550
2-HEXYLOXY-2-ETHOXYETHYL ETHER see HFT600
2-HEXYLOXY-2-ETHYLOXYETHYL ETHER see HFT600
o-(3-(HEXYLOXY)-2-HYDROXYPROPOXY)BENZOIC ACID MONOSODIUM SALT see ERE200

((HEXYLOXY)METHYL)OXIRANE see
GGY125
4'-(HEXYLOXY)-3-
PIPERIDINOPROPIOPHENONE
HYDROCHLORIDE see HFU500
4-HEXYLOXY-β-(1-
PIPERIDYL)PROPIOPHENONE
HYDROCHLORIDE see HFU500
((HEXYLOXY)SULFINYL)METHYLCARBA
MIC ACID-2,3-DIHYDRO-2,2-DIMETHYL-7-
BENZOFURANYL ESTER see HFU550
2-HEXYLPHENOL see HFU600
o-HEXYLPHENOL see HFU600
α-n-HEXYL-β-PHENYLACROLEIN see
HFO500
p-HEXYLPHENYL DIPHENYL
PHOSPHATE see HFU700
2-(1-HEXYL-3-PIPERIDYL)ETHYL ESTER,
BENZOIC ACID HYDROCHLORIDE see
HFV000
HEXYL PROPANOATE see HFV100
1-HEXYL PROPANOATE see HFV100
HEXYL-2-PROPENOATE see ADV000
HEXYL PROPIONATE see HFV100
1-HEXYL PROPIONATE see HFV100
HEXYLRESORCIN (GERMAN) see HFV500
4-HEXYLRESORCINE see HFV500
HEXYLRESORCINOL see HFV500
4-HEXYLRESORCINOL see HFV500
p-HEXYLRESORCINOL see HFV500
4-n-HEXYLRESORCINOL see HFV500
3-HEXYL-7,8,9,10-TETRAHYDRO-6,6,9-
TRIMETHYL-6H-DIBENZO(B,D)PYRAN-1-
OL see HGK500
HEXYLTHIOCARBAM see EHT500
3-(5-
(HEXYLTHIO)PENTYL)THIAZOLIDINE
HYDROCHLORIDE see HFW500
n-HEXYL TIGLINATE see HFX000
HEXYL TILGLATE see HFX000
HEXYLTRICHLOROSILANE see HFX500
N-HEXYLVALERAMIDE see VAG100
n-HEXYL VINYL SULFONE see HFY000
3-HEXYNE, 2,5-DIMETHYL-2,5-DI(t-
BUTYLPEROXY)- see DRJ825
3-HEXYNE-2,5-DIOL see HFY500
HEXYNE-3-DIOL-2,5 see HFY500
HEXYNOL see HFZ000
1-HEXYN-3-OL see HGA000
HEXYTHIAZOX see CJU275
HEYDEFLON see TAI250
HEYDEN 768 see TGC300
HF 10 see SMQ500
HF 55 see SMQ500
HF 77 see SMQ500
HF 191 see DGV700
HF 264 see BRJ325
HF-1854 see CMY650
HF 1927 see DCW600
HF 1927 see DCW800
HF-2333 see HOU059
HF3170 see DCS200
HFA see FIC000
H-35-F 87 (BVM) see DSQ000
HFC 32 see MJQ300
HFC-125 see PBD400
HFC 134A see EEC100
HFC 236EA see HDE050
HFC-236EA see HDE050
HFC 236FA see HDE100
HFC-236FA see HDE100
HFCB see HDB500
HFC-4310MEE see DLY900
HFDB 4201 see PJS750
HFE see HDC000
HF-2159 HYDROCHLORIDE see CIT000
HFIP see HDC500
81723 HFU see TET800
HG 010 see PHB550
HG-203 see CHX250
Hg 532 see PGA750
H.G. BLENDING see SFT000

HGG-12 see HGA050
HGI see BBQ500
HG 010 (PHOSPHATE) see PHB550
HH 102 see SMQ500
HH-197 see BOR350
H. HAEMACHATES VENOM see HAO500
HHDN see AFK250
Hg-HEMATOPORPHYRIN-Na see HAP000
1,1-H,H-HEPTAFLUOROBUTANOL see
HAW100
HHI 11 see SMQ500
H102/09 HYDROCHLORIDE see ZBA525
HI-6 see HGE900
HIADELON see PII100
HI-ALAZIN see TGB475
HI-A-VITA see VSK600
HIBANIL see CKP250
HIBANIL see CKP500
HIBAWOOD OIL see HGA100
HIBERNA see DQA400
HIBERNAL see CKP250
HIBERNAL see CKP500
HIBERNON HYDROCHLORIDE see
HGA500
HIBERNYL see TCY750
HIBESTROL see DKA600
HIBICLENS see CDT250
HIBICON see BEG000
HIBIDIL see CDT250
HIBISCOLIDE see OKW110
HIBISCRUB see CDT250
HIBISCUS MANIHOT Linn., extract see
HGA550
HIBISCUS ROSA-SINENSIS, flower extract see
HGA600
HIBITANE see BIM250
HIBITANE see CDT250
HIBITANE DIACETATE see CDT125
HIBROM see NAG400
HICAL-2 see HGB000
HI-CAL 3 see JDA125
HICHILLOS see KGK000
HICO CCC see CMF400
HICOFOR PR see MCB050
HICOPHOR PR see MCB000
HIDA see LFO300
HIDACHROME ORANGE R see NEY000
HIDACHROME YELLOW 2G see SIT850
HIDACIAN see COH250
HIDACIANN see COH250
HIDACID AMARANTH see FAG020
HIDACID AZO RUBINE see HJF500
HIDACID AZURE BLUE see FMU059
HIDACID BLUE A see ERG100
HIDACID BLUE AF see ERG100
HIDACID BLUE V see ADE500
HIDACID BROMO ACID REGULAR see
BNK700
HIDACID DIBROMO FLUORESCEIN see
BNH500
HIDACID DIBROMO FLUORESCEIN see
BNK700
HIDACID EMERALD GREEN see FAE950
HIDACID FAST CRIMSON see CMM300
HIDACID FAST FUCHSINE 6B see CMM400
HIDACID FAST ORANGE G see HGC000
HIDACID FAST SCARLET 3R see FMU080
HIDACID FLUORESCEIN see FEV000
HIDACID METANIL YELLOW see MDM775
HIDACID ORANGE II see CMM220
HIDACID SCARLET 2R see FMU070
HIDACID URANINE see FEW000
HIDACID WOOL GREEN see ADF000
HIDACO BRILLIANT CRYSTAL VIOLET see
AOR500
HIDACO BRILLIANT GREEN see BAY750
HIDACO CRYSTAL VIOLET see AOR500
HIDACO MALACHITAE GREEN BASE see
AFG500
HIDACO METHYLENE BLUE SALT FREE
see BJI250
HIDACO OIL ORANGE see PEJ500

HIDACO OIL RED see SBC500
HIDACO OIL YELLOW see AIC250
HIDACO SAFRANINE see GJI400
HIDACO VICTORIA BLUE R see VKA600
HIDAN see DKQ000
HIDANTILO see DKQ000
HIDANTINA SENOSIAN see DKQ000
HIDANTINA VITORIA see DKQ000
HIDANTOMIN see DKQ000
HIDAZID TARTRAZINE see FAG140
HI-DERATOL see VSZ100
HIDRALAZIN see HGP495
HIDRALAZIN see HGP500
HIDRANIZIL see ILD000
HIDRIL see CFY000
HIDRIX see HOO500
HIDROCHLORTIAZID see CFY000
HIDRO-COLISONA see CNS750
HIDROESTRON see EDP000
HIDROMEDIN see DFP600
HIDRORONOL see CFY000
HIDROTIAZIDA see CFY000
N-(trans-4-HIDROXICICLOHEXIL)-(2-
AMINO-3,5-DIBROMOBENCIL)AMINA
(SPANISH) see AHJ250
HIDROXIFENAMATO see HNJ000
HIDROXITEOFILLINA see DNC000
HIDRULTA see ILD000
HI-DRY see TCE250
HI-ENTEROL see CHR500
HIERBA de SANTIAGO (MEXICO) see
RBA400
HIESTRONE see EDV000
HI-FAX see PJS750
HI-FAX 1900 see PJS750
HI-FAX 4401 see PJS750
HI-FAX 4601 see PJS750
HI-FLASH NAPHTHA see NAH600
HIFOL see BIO750
HIGH BELIA see CCJ825
HIGH FLASH AROMATIC NAPHTHA see
SKS350
HIGH-FRUCTOSE CORN SYRUP see
HGB100
HIGHROSIN see RNU100
HIGILITE see AHC000
HIGOSAN see MEP250
HIGUERETA CIMARRONA see CNR135
HIGUERETA (CUBA, PUERTO RICO) see
CCP000
HIGUERILLA (MEXICO) see CCP000
HIGUEROXYL DELABARRE see FBS000
HI-JEL see BAV750
HIKIZIMYCIN see APF000
HILDAN see EAQ750
HILDIT see DAD200
HILITE 60 see DGN200
HILLS-OF-SNOW see HGP600
HILONG see CFZ000
HILTAMINE TURQUOISE BLUE 8GL see
COF420
HILTHION see MAK700
HILTHION 25WDP see MAK700
HILTONIL FAST BLUE B BASE see DCJ200
HILTONIL FAST ORANGE GC BASE see
CEH690
HILTONIL FAST ORANGE GR BASE see
NEO000
HILTONIL FAST ORANGE R BASE see
NEN500
HILTONIL FAST RED B BASE see NEQ000
HILTONIL FAST RED KB BASE see CLK225
HILTONIL FAST RED 3GL BASE see
KDA050
HILTONIL FAST RED RL BASE see MMF780
HILTONIL FAST SCARLET 2G BASE see
DEO295
HILTONIL FAST SCARLET G BASE see
NMP500
HILTONIL FAST SCARLET GC BASE see
NMP500

HILTONIL FAST SCARLET G SALT see NMP500
HILTONPHTHOL AS see CMM760
HILTOSAL FAST BLUE B SALT see DCJ200
HILTOSAL FAST GARNET GBC SALT see MPY750
HILTOSAL FAST ORANGE RD SALT see CEG800
HILTOSAL FAST RED RL SALT see MMF780
HILTOSAL FAST SCARLET 2G SALT see DEO295
HINDAMINE GARNET GBC see MPY750
HINDAMINE SCARLET GG see DEO295
HINDASOL BLUE B SALT see DCJ200
HINDASOL RED TR SALT see CLK235
H. INFLUENZAE ENDOTOXIN, phenol water extract see HAB710
H. INFLUENZAE ENDOTOXIN, NaCl-citrate extract see HAB700
HINOKITIOL see IRR000
HINOKITOL see IRR000
HINOSAN see EIM000
HINSALU see RQU300
HIOHEX CHLORIDE see HEA500
HIOXYL see HIB050
HIP see COF250
HIPERCILINA see BFD000
HIPHYLLIN see DNC000
HIPNAX see DLY000
HIPODERMIN see TJE885
HIPOFTALIN see HGP495
HIPOFTALIN see HGP500
HI-POINT 90 see MKA500
HI-POINT 180 see MKA500
HI-POINT PD-1 see MKA500
HIPOSERPIL see RDK000
HIPPEASTRUM (VARIOUS SPECIES) see AHI635
HIPPOBROMA LONGIFLORA see SLJ650
HIPPODIN see HGB200
HIPPOMANE MANCINELLA see MAO875
HIPPOPHAIN see AJX500
HIPPURAN see HGB200
HIPPURIC ACID see HGB300
HIPPURIC ACID SODIUM SALT see SHN500
HIPPUROHYDROXAMIC ACID see BBB250
HIPPUZON see TEH500
HIPRENE MC 532 see GHY100
HIPSAL see DLY000
HIPTAGENIC ACID see NIY500
HI-PYRIDOXIN see PII100
HIRATHIOL see IAD000
HIRSUTIC ACID N see HGB500
HISERPIA see RDK000
HISHIREX 502 see PKQ059
HISINDAMONE A see CDY000
HISMANAL see ARP675
HISPACET FAST YELLOW G see AAQ250
HISPACID BRILLIANT SCARLET 3RF see FMU080
HISPACID FAST BLUE R see ADE750
HISPACID FAST CARMOISINE 6B see CMM400
HISPACID FAST CARMOISINE G see CMM300
HISPACID FAST ORANGE 2G see HGC000
HISPACID FAST YELLOW T see FAG140
HISPACID GREEN GB see FAE950
HISPACID ORANGE 1 see FAG010
HISPACID ORANGE AF see CMM220
HISPACID YELLOW CG see MRL100
HISPACID YELLOW MG see MDM775
HISPACROM BLUE BG see CMP880
HISPACROM FAST ORANGE GR see CMP882
HISPACROM ORANGE R see NEY000
HISPACROM RED B see CMG750
HISPACROM VIOLET B see HLI000
HISPACROM YELLOW 2G see SIT850
HISPACROM YELLOW 2GR see SIT850
HISPADIAZO BLACK D see CMN230
HISPALIT FAST SCARLET RN see MMP100

HISPALUZ RED 8BL see CMO885
HISPALUZ TURQUOISE GE see COF420
HISPAMIN BLACK EF see AQP000
HISPAMIN BLUE 2B see CMO000
HISPAMIN BLUE RW see CMO600
HISPAMIN BLUE 3BX see CMO250
HISPAMIN CONGO 4B see SGQ500
HISPAMIN FAST BLACK CG see CMN240
HISPAMIN FAST BROWN 3R2B see CMO800
HISPAMIN FAST BROWN NZ see CMO820
HISPAMIN FAST RED FN see CMO870
HISPAMIN FAST SCARLET 4BS see CMO870
HISPAMIN GREEN B see CMO840
HISPAMIN GREEN WT see CMO830
HISPAMIN PURE YELLOW T2G see CMP050
HISPAMIN RED 4B see DXO850
HISPAMIN SKY BLUE 3B see CMO500
HISPAMIN SKY BLUE 6B see CMN750
HISPAMIN VIOLET 3R see CMP000
HISPAVIC 229 see PKQ059
HISPERSE YELLOW G see AAQ250
HISPRIL see LJR000
HISPRIL HYDROCHLORIDE see DVW700
HISTABID see SPC500
HISTABUTICINE see VJZ050
HISTABUTIZINE see VJZ050
HISTABUTYZINE see VJZ050
HISTABUTYZINE DIHYDROCHLORIDE see BOM250
HISTACAP see WAK000
HISTADUR see CLX300
HISTADUR see TAI500
HISTADUR DURA-TABS see TAI500
HISTADYL see TEO250
HISTADYL HYDROCHLORIDE see DPJ400
HISTAFED see DPJ400
HISTAGLOBIN see HGC400
HISTALEN see TAI500
HISTALON see WAK000
HISTAMETHINE see HGC500
HISTAMETHIZINE see HGC500
HISTAMETIZINE see HGC500
HISTAMETIZYNE see HGC500
HISTAMINE see HGD000
HISTAMINE ACID PHOSPHATE see HGE000
HISTAMINE DICHLORIDE see HGD500
HISTAMINE DIHYDROCHLORIDE see HGD500
HISTAMINE DIPHOSPHATE see HGE000
HISTAMINE HYDROCHLORIDE see HGE500
HISTAMINE PHOSPHATE (1:2) see HGE000
HISTAMINOS see ARP675
HISTAN see WAK000
HISTANTIN see CFF500
HISTANTINE see CFF500
HISTANTINE DIHYDROCHLORIDE see CDR000
HISTAPAN see TAI500
HISTAPYRAN see WAK000
HISTARGAN see DQA400
HISTASAN see WAK000
HISTASPAN see CLD250
HISTATEX see DBM800
HISTAXIN see BBV500
HISTIDINE see HGE700
l-HISTIDINE, N-β-ALANYL- see CCK665
l-HISTIDINE (FCC) see HGE700
HISTIDINE, α-(FLUOROMETHYL)-(9CI) see MRI285
l-HISTIDINE HYDROCHLORIDE see HGE750
HISTIDINE MONOHYDROCHLORIDE see HGE800
HISTIDINE, MONOHYDROCHLORIDE, l- see HGE750
HISTIDYL see DPJ400
β-HISTINE see HGE820
HISTOCARB see CBJ000
HISTOSTAB see PDC000
HISTRYL see LJR000

HISTYN see LJR000
HISTYRENE S 6F see SMR000
HI-STYROL see SMQ500
HITACERAM SC 101 see SCQ000
HITENOL N 093 see ANO600
HIVOL-44 see DAA800
HI-YIELD DESICCANT H-10 see ARB250
HI-Z see OJY200
HIZAROCIN see CPE750
HIZEX see PJS750
HIZEX 5000 see PJS750
HIZEX 5100 see PJS750
HIZEX 1091J see PJS750
HIZEX 1291J see PJS750
HIZEX 1300J see PJS750
HIZEX 2100J see PJS750
HIZEX 2200J see PJS750
HIZEX 3300F see PJS750
HIZEX 3300S see PJS750
HIZEX 6100P see PJS750
HIZEX 7300F see PJS750
HIZEX 3000B see PJS750
HIZEX 7000F see PJS750
HIZEX 2100LP see PJS750
HIZEX 5100LP see PJS750
HIZEX 3000S see PJS750
HIZEX 5000S see PJS750
HJ 6 see HGE900
HK-141 see BAW500
HK 256 see LFJ000
H.K. FORMULA No. K. 7117 see FMU059
HL 267 see DGW600
HL-331 see ABU500
HL 2153 see DPD400
HL 2197 see CEP675
HL 2447 see DFV400
HL 8700 see PMI750
HL 8727 see DIG400
HL-8731 see MQF750
HLR 4219 see MJQ260
HLR 4448 see MJQ260
HLS 831 see BSQ000
HM 100 see MCB050
HMAF see HLU600
HMAF, MGI-114 see HLU600
HMB see HFR100
7-HMBA see BBH250
HMBD see HLX925
HMD see PAN100
HMDA see HEO000
HMDI see DNJ800
HMDM see HEG050
HMDS see HED500
HMF see ORG000
HMG see HMC000
h-MG see CGY750
HMGA see HMC000
HMM see HEJ500
12-HM-7-MBA see HMF500
7-HM-12-MBA see HMF000
HMP see SHM500
HMPA see HEK000
HMPT see HEK000
HMPTS see HEK050
HMT see HEI500
HMX see CQH250
HMX (dry or unphlegmatized) (DOT) see CQH250
HMX, wetted (UN 0226) (DOT) see CQH250
beta HMY see CQH250
HN1 see BID250
HN 1 see CGW000
HN2 see BIE250
HN₂ AMINE OXIDE see CFA500
HN1 CHLOROHYDRIN see EHK300
HNED CERVENAVA OSTANTHRENOVA 5 RF see CMU800
HNED KYPOVA 1 see CMU770
HNED KYPOVA 25 see CMU800
HNED OSTANTHRENOVA BR see CMU770
HNED ROZPOUSTEDLOVA 1 see NBG500
HNED SUDAN RR see NBG500

HN2.HCl see BIE500
HN1 HYDROCHLORIDE see EGU000
HN2 HYDROCHLORIDE see BIE500
HNI•HCl see EGU000
HN₂ OXIDE HYDROCHLORIDE see CFA750
HN₂ OXIDE MUSTARD see CFA500
HNT see HHD500
HNU see HKW500
HO 342 see BGZ100
HO 11513 see TMK000
3-HO-AAF see HIO875
HOCA see DXH250
HOCH see FMV000
HODAG GMS see OAV000
HODAG PSML-20 see PKG000
HODAG SMS see SKV150
HODAG SSO see SKV170
HODAG SVO 9 see PKL100
HODOSTIN see DQY909
HODSON see DSK800
HOE 118 see PDW250
HOE 280 see OGI300
HOE 296 see BAR800
HOE 574 see AOT255
HOE 766 see LIU420
HOE 881 see FAL100
HOE 893 see PAP225
HOE 933 see CAZ125
HOE 984 see NMV725
HOE 2,671 see EAQ750
HOE 2747 see CKD500
HOE 2784 see BGB500
HOE 2810 see DGD600
HOE-2824 see ABX250
HOE 2872 see CLU000
HOE 2904 see ACE500
HOE 2982 see HBK700
HOE 766A see LIU420
HOE 837V see ECW550
HOE 893D see PAP225
HOE 893D see PAP230
HOE 00661 see ANI800
HOE 16410 see IRA050
HOE 17411 see MHC750
HOE 23408 see IAH050
HOE 30374 see AOT255
HOE 33171 see FAQ200
HOE 33258 see MOD500
HOE 33342 see BGZ100
HOE 35 609 see EHH600
HOE 36801 see CGQ500
HOE 39866 see ANI800
HOE-A 25-01 see FAQ200
HOECHST 1082 see MDP000
HOECHST 10720 see KFK000
HOECHST 10,820 see MDP750
12494 HOECHST see RDA375
33258 HOECHST see MOD500
HOECHST 33342 see BGZ100
HOECHST DYE 33258 see MOD500
HOECHST PA 190 see PJS750
HOECHST WAX PA 520 see PJS750
HOEGRASS see IAH050
HOELON see IAH050
HOELON 3EC see IAH050
HOE 2960 OJ see THT750
HOE 2982 OJ see HBK700
HOE-S 2617 see DTF820
HOE S 3837 see AMF600
HOFMAN'S VIOLET see HGE925
HOG see AOO500
HOG see PDC890
HOG APPLE see MBU800
HOG BUSH see CDM325
HOGGAR see BNK000
HOGGAR N see PGE775
HOG PHYSIC see CCJ825
HOG'S POTATO see DAE100
HOJA GRANDE (CUBA) see APM875
HOKKO-MYCIN see SLW500
HOKMATE see FAS000
HOLBAMATE see MQU750

HO LEAF OIL see HGF000
HOLIN see EDU500
HOLLICHEM HQ 3300 see BBA625
HOLLY see HGF100
HOLMIUM see HGF500
HOLMIUM CHLORIDE see HGG000
HOLMIUM CITRATE see HGG500
HOLMIUM NITRATE see HGH100
HOLMIUM(III) NITRATE, HEXAHYDRATE
(1:3:6) see HGH000
HOLMIUM TRINITRATE see HGH100
HOLOCAINE see BJO500
HOLODORM see QAK000
HOLOPAN see SBH500
HOLOXAN see IMH000
HOMANDREN see MPN500
HOMANDREN (amps) see TBG000
HOMAPIN see MDL000
HOMATROMIDE see MDL000
HOMATROPIN see HGK550
HOMATROPINE see HGK550
(±)-HOMATROPINE BROMIDE see HGH150
HOMATROPINE HYDROBROMIDE see
HGH150
HOMATROPINE METHYLBROMIDE see
MDL000
HOMBITAN see TGG760
HOME HEATING OIL No. 2 see DHE800
#2 HOME HEATING OILS see DHE800
#2 HOME HEATING OILS see HGH200
HOMEOSTAN see ACL250
HOMIDIUM BROMIDE see DBV400
HOMIDIUM CHLORIDE see HGI000
HOMOANISIC ACID see MFE250
HOMOARGININE, nitrosated see HNA500
HOMOAROMOLINE see HGI050
(+)-HOMOAROMOLINE see HGI050
HOMOATROPINE see HGK550
HOMOCAL D see CAT775
HOMOCATECHOL see DNE200
HOMOCHLORCYCLIZINE see HGI100
HOMOCHLORCYCLIZINE
DIHYDROCHLORIDE see CJR809
HOMOCHLORCYCLIZINE
HYDROCHLORIDE see HGI200
HOMOCHLOROCYCLIZINE see HGI100
HOMOCHLOROCYCLIZINE
DIHYDROCHLORIDE see CJR809
HOMOCHLOROCYCLIZINE
HYDROCHLORIDE see HGI200
HOMOCLOMINE see HGI100
HOMOCODEINE see TCY750
(+−)-HOMOCYSTEINE see HGI300
dl-HOMOCYSTEINE see HGI300
HOMODAMON see KFK000
18-HOMO-ESTRIOL see HGI700
HOMOFOLATE see HGI525
HOMOFOLIC ACID see HGI525
HOMOGUAIACOL see MEK325
HOMOHARRINGTONINE see HGI575
HOMOLINALYL ACETATE see HGI585
HOMOLLE'S DIGITALIN see DKN400
HOMOMENTHOL see TLO500
HOMOMYRETENOL see DTB800
HOMONEURINE CHLORIDE see HGI600
18-HOMO-OESTRIOL see HGI700
HOMOOLAN see HIM000
HOMOPIPERAZINE see HGI900
HOMOPIPERIDINE see HDG000
HOMOPROLINE see HDS300
HOMOPROTOCATECHUIC ACID see
HGI980
HOMOPYROCATECHOL see DNE200
HOMORESTAR see HGI100
HOMOSALICYLIC ACID see CNX625
HOMOSTERONE see TBF500
3-HOMOTETRA HYDRO CANNIBINOL see
HGK500
HOMOTHALICRINE see HGI050
HOMOTROPINE see HGK550
HOMOTRYPTAMINE HYDROCHLORIDE
see AME750

HOMOVERATRIC ACID see HGK600
HOMOVERATRONITRILE see VIK100
HOMOVERATRYLAMINE see DOE200
HON see AKG500
HONEY DIAZINE see PPP500
HONEY YELLOW 3GNT see CMO810
HONG KIEN see ABU500
HONGKONG ROSIN WW see RNU100
HONVAN see DKA200
HONVAN TETRASODIUMTETRASODIUM
SALT, (E)- see TEE300
HOOKER HRS-16 see DAE425
HOOKER HRS-1362 see AAI115
HOOKER HRS-1422 see DNS200
HOOKER HRS 1654 see DAE425
HOOKER NO. 1 CHRYSOTILE ASBESTOS
see ARM268
HOPA see CAT175
HOPANTENATE CALCIUM see CAT125
HOPANTENATE CALCIUM
HEMIHDYRATE see CAT175
HOPCIDE see CKF000
HOPCIDE, NITROSATED (JAPANESE) see
MMV250
HOPCIN see MOV000
HOP EXTRACT EI see HGK725
HOP EXTRACT, MODIFIED see HGK750
HOPFENEXTRAKTE EI see HGK725
HOPS OIL see HGK800
HORACE VERNET'S BLUE see CNQ000
HORBADOX see DRN200
HORFEMINE see DKB000
HORMALE see MPN500
HORMATOX see DGB000
HORMEX ROOTING POWDER see ICP000
HORMIN see DFY800
HORMIT see SGH500
HORMOCEL-2CCC see CMF400
HORMODIN see ICP000
HORMOESTROL see DLB400
HORMOFEMIN see DAL600
HORMOFLAVEINE see PMH500
HORMOFOLLIN see EDV000
17-HORMOFORIN see AOO450
HORMOFORT see HNT500
HORMOGYNON see EDP000
HORMOLUTON see PMH500
HORMOMED see EDU500
HORMONIN see EDU500
HORMONISENE see CLO750
HORMONIT see NAK525
HORMOSLYR 64 see DAB000
HORMOSLYR 500T see TAH900
HORMOTESTON see TBG000
HORMOTUHO see CIR250
HORMOVARINE see EDV000
HORSE see HBT500
HORSE ATRIAL NATRIURETIC PEPTIDE-
28 see HGL680
HORSE BLOB see MBU550
HORSE CHESTNUT see HGL575
HORSE-CYTOCHROME C see CQM325
HORSE GOLD see FBS100
HORSE HEAD A-410 see TGG760
HORSE HEART CYTOCHROME C see
CQM325
HORSE NICKER see CAK325
HORSE POISON (JAMAICA) see SLJ650
HORTENSIA (CUBA) see HGP600
HORTFENICOL see CDP250
HORTICULTURAL SPRAY OIL see MQV796
HORTICULTURAL SPRAY OIL see MQV855
HORTOCRITT see EJQ000
HOSALON see DSK800
HOSDON GRANULE see DSK800
HOSTACAIN see BQA750
HOSTACAINE see BQA750
HOSTACAINE HYDROCHLORIDE see
BQA750
HOSTACILLIN see PAQ200
HOSTACORTIN see PLZ000
HOSTACORTIN see PMA000

HOSTACYCLIN see TBX000
HOSTADUR see PKF750
HOSTAFLEX VP 150 see AAX175
HOSTAFLON see KDK000
HOSTAFLON see TAI250
HOSTAFLON C see KDK000
HOSTAGINAN see PEV750
HOSTALEN see PJS750
HOSTALEN GD 620 see PJS750
HOSTALEN GD 6250 see PJS750
HOSTALEN GF 4760 see PJS750
HOSTALEN GF 5750 see PJS750
HOSTALEN GM 5010 see PJS750
HOSTALEN GUR see PJS750
HOSTALEN HDPE see PJS750
HOSTALEN PP see PMP500
HOSTALIT see PKQ059
HOSTALIVAL see NMV725
HOSTANOX SE 2 see DXG700
HOSTANOX VP-SE 2 see DXG700
HOSTAPERM GREEN 8G see CMS140
HOSTAPERM GREEN GG see PJQ100
HOSTAPERM ORANGE GR see CMU820
HOSTAPERM PINK E see DTV360
HOSTAPERM PINK EB see DTV360
HOSTAPERM VAT ORANGE GR see CMU820
HOSTAPERM YELLOW GR see CMS208
HOSTAPERM YELLOW GT see CMS210
HOSTAPERM YELLOW GTT see CMS210
HOSTAPHAN see PKF750
HOSTAPON T see SIY000
HOSTAQUICK see ABU500
HOSTAQUICK see HBK700
HOSTAREX PX 324 see TMO575
HOSTATHERM PINK FBL see AKI750
HOSTATHION see THT750
HOSTAVAT BLUE 2BD see ICU135
HOSTAVAT BLUE 4BR see ICU135
HOSTAVAT BRILLIANT ORANGE GR see CMU820
HOSTAVAT GOLDEN YELLOW see DCZ000
HOSTAVAT ORANGE R see CMU815
HOSTAVIK (RUSSIAN) see HBK700
HOSTETEX L-PEC see TIK250
HOSTPERM PINK E 02 see DTV360
HOSTYREN N see SMQ500
HOSTYREN N 4000 see SMQ500
HOSTYREN N 7001 see SMQ500
HOSTYREN N 4000V see SMQ500
HOSTYREN S see SMQ500
HOT PEPPER see PCB275
HOURBESE see DKE800
HOWFLEX GBP see DDY400
HOWFLEX GBP see DJD700
HOX 1901 see EPR000
HP see PHV275
H.P. 34 see SAH000
HP 129 see FAO100
H.P. 165 see BPG750
H.P. 206 see PMY750
H.P. 209 see EEM000
H.P. 216 see HFS759
HP 1275 see PDV700
HP 1325 see HGL600
2M 4KHP see CIR500
HPA see EPI300
HPBT 1 see BAP502
HPC see HNT500
HPC see OPK300
HPMPC see HNS550
β-HPN see HGP000
HPOP see HNX500
HPP see ADF800
HPP see ZVJ000
2-HPPN see NLM500
β-HPPN see NLM500
HPT see HEK000
HQ-275 see HGL630
HQ-3,300 see AFP300
HR 376 see CIR750
HR 756 see CCR950

HR 930 see FOL100
HRS-16 see DAE425
HRS 16A see DAE425
HRS-587 see TJA200
HRS 860 see CEP000
HRS 1276 see MQW500
HRS-1362 see AAI115
HRS 1654 see DAE425
HRS 1655 see HCE500
HRW 13 see SLJ500
HS see HGW500
HS 3 see HGL650
HS 4 see PPC000
HS 21 see CNA750
HS 25 see CNA750
HS 31 see CNB825
HS 55 see AFM375
HS 103 see DTE100
HS 592 see FOS100
HS-119-1 see PEE750
2M-4KH SODIUM SALT see SIL500
HSP 2986 see SDZ000
HSR-902 see TGF075
H-STADUR see CLD250
5-HT see AJX500
HT 88 see SMQ500
HT 88A see SMQ500
HT 91-1 see SMQ500
HT 972 see MJQ000
5-HTA see AJX500
HTAC see HCQ525
HT-400 E 1/8' see CNC250
H. TERNATUM see SAF500
HT-F 76 see SMQ500
HTH see HOV500
HTP see HCY000
5-HTP see HOO100
l-5-HTP see HOA600
l-5-HTP see HOO000
1-α-H,5-α-H-TROPANE, 3-α-((10,11-DIHYDRO-5H-DIBENZO(AD)CYCLOHEPTEN-5-YL)OXY)-, CITRATE (1:1) see EAG100
1-α-H,5-α-H-TROPAN-3-α-OL, ATROPATE (ESTER) see AQO250
1-α-H,5-α-H-TROPAN-3-α-OL, 6-β,7-β-EPOXY-, (−)-TROPATE (ESTER), HYDROBROMIDE, TRIHYDRATE see SBG600
1-α-H,5-α-H-TROPAN-3-α-OL, MANDELATE (ESTER) see HGK550
1-α-H,5-α-H-TROPAN-3-α-OL, MANDELATE (ester), HYDROBROMIDE see HGH150
1-α-H,5-α-H-TROPAN-3-α-OL (±)-TROPATE (ESTER) see ARR000
1-α-H,5-α-H-TROPAN-3-α-OL (±)-TROPATE (ESTER), SULFATE (2:1) SALT see ARR500
1-α-H,5-α-H-TROPAN-3-α-OL, (−)-TROPATE (ester), SULFATE (2:1) see HOT600
HT-2 TOXIN see THI250
HUBBUCK'S WHITE see ZKA000
HUBER see CBT750
HUDSON CHROME RED B see CMG750
HUELE de NOCHE (MEXICO) see DAC500
HUEVO de GATO (CUBA) see JBS100
HUILE d'ANILINE (FRENCH) see AOQ000
HUILE de CAMPHRE (FRENCH) see CBA750
HUILE de FUSEL (FRENCH) see FQT000
HUILE H50 see TFM250
HULS P 6500 see PMP500
HUMAN ADRENOMEDULLIN see HGL670
HUMAN ANGIOTENSIN II see IKX030
HUMAN ATRIAL NATRIURETIC FACTOR (99-126) see HGL680
HUMAN ATRIAL NATRIURETIC PEPTIDE (99-126) see HGL680
HUMAN ATRIAL NATRIURETIC PEPTIDE (1-28) (99-126) see HGL680
HUMAN ATRIOPEPTIN(1-28) see HGL680
HUMAN ATRIOPEPTIN(99-126) see HGL680

HUMAN CHORIOGONADOTROPIN, DEGLYCOSYLATED see DAK325
HUMAN CHORIONIC GONADOTROPIN see CMG675
HUMAN CORTICOTROPIN-RELEASING FACTOR see HGL700
HUMAN CORTICOTROPIN-RELEASING HORMONE see HGL700
HUMAN CORTICOTROPIN-RELEASING HORMONE-41 see HGL700
HUMAN CRF see HGL700
HUMAN CRF (1-41) see HGL700
HUMAN IMMUNOGLOBULIN COG-78 see HGL800
HUMAN PLATELET EXTRACT, CTAP-III see HGL900
HUMAN RECOMBINANT TUMOR NECROSIS FACTOR-α see HGL920
HUMAN SERUM ALBUMIN see HGL930
HUMAN SPERM see HGM000
HUMATIN see APP500
HUMATIN see NCF500
HUMEDIL see BCP650
HUMENEGRO see CBT750
HUMIC ACID, SODIUM SALT see HGM100
HUMIDIN see HGM500
HUMIFEN NBL 85 see BLX000
HUMIFEN WT 27G see DJL000
HUMINSAURE NATRIUM see HGM100
HUMORSOL see DAM625
HUMULENE see HGM550
α-HUMULENE (6CI,7CI) see HGM550
HUMULIN I see IDF325
HUMYCIN see NCF500
HUMYCIN SULFATE see APP500
HUNDRED PACE SNAKE VENOM see HGM600
HUNGARIAN CHAMOMILE OIL see CDH500
HUNGAZIN see ARQ725
HUNGAZIN DT see BJP000
HUNGAZIN PK see ARQ725
HUNGER WEED see FBS100
HUNTER'S ROBE see PLW800
HURA CREPITANS see SAT875
HURRICANE PLANT see SLE890
HUSEPT EXTRA see CLW000
HUSTODIL see RLU000
HUSTOSIL see RLU000
HVA 2 see BKL750
HVA-2 CURING AGENT see BKL750
HVP see ADF800
HW 4 see CQH250
HW 920 see DXQ500
HWA 153 see HLF500
HX 3/5 see CCU250
HX-487 see EEB230
HX 752 see BLG400
HX-868 see HGN000
HXR see IDE000
HY 951 see TJR000
HYACINTH see ZSS000
HYACINTH ABSOLUTE see HGN500
HYACINTHAL see COF000
HYACINTH BASE see BOF000
HYACINTH BODY see EES380
HYACINTHIN see BBL500
HYACINTH-OF-PERU see SLH200
HYADRINE see BBV500
HYADUR see DUD800
HYALURONIC ACID, SODIUM SALT see HGN600
HYAMINE see BEN000
HYAMINE 10X see MHB500
HYAMINE 1622 see BBU750
HYAMINE 1622 see BEN000
HYAMINE 2389 see DYA600
HYAMINE 3500 see AFP250
HYAMINE 3500 see QAT520
HYANILID see SAH500
HYASORB see BFD000
HYBAR X see UNJ800

HYBERNAL see CKP500
HYCANTHON see LIM000
HYCANTHONE see LIM000
HYCANTHONE MESYLATE see HGO500
HYCANTHONE METHANESULFONATE see HGO500
HYCANTHONE METHANESULPHONATE see HGO500
HYCANTHONE MONOMETHANESULPHONATE see HGO500
HYCAR see PJR000
HYCAR LX 407 see SMR000
HY-CHLOR see HOV500
HYCHOTINE see CJR909
HYCLORATE see ARQ750
HYCLORITE see SHU500
HYCORACE see HHR000
HYCORTOLE ACETATE see HHQ800
HYCOZID see ILD000
HYDAN see DFE200
HYDAN (antiseptic) see DFE200
HYDANTAL see DKQ000
HYDANTIN SODIUM see DNU000
HYDANTOCIDIN see HGO550
HYDANTOIN see DKQ000
HYDANTOIN see HGO600
HYDANTOIN, BIS(2,3-EPOXYPROPYL)-5-ETHYL-5-METHYL- see ECI200
HYDANTOIN, 1-BROMO-3-CHLORO-5,5-DIMETHYL- see BNA300
HYDANTOIN, 1,3-DIBROMO-5,5-DIMETHYL- see DDI900
HYDANTOIN, 3,3'-(ETHYLENEDIIMINODIMETHYLENE)BIS(5,5-DIPHENYL- see EIY700
HYDANTOIN, 5-ETHYL-5-PHENYL-, (−)- see EOL050
HYDANTOIN, 5-(2-(METHYLTHIO)ETHYL- see MDT800
HYDANTOIN, 1-NITRO- see NHE100
HYDANTOIN, 1-((5-NITROFURFURYLIDENE)AMINO)-, MONOHYDRATE see NGE780
HYDANTOIN, 1-((5-(p-NITROPHENYL)FURFURYLIDENE)AMINO)- see DAB845
HYDANTOIN SODIUM see DNU000
HYDELTRA see PMA000
HYDELTRASOL see PLY275
HYDELTRONE see PMA000
HYDERGIN see DLL400
HYDERGINE see DLL400
HYDNOCARPUS OIL see CDK750
HYDOUT see DXD000
HYDOUT see EAR000
HYDOXIN see PPK250
HYDRACETIN see ACX750
HYDRACILLIN see PAQ200
HYDRACRYLIC ACID, ACRYLATE see HGO700
HYDRACRYLIC ACID β-LACTONE see PMT100
HYDRACRYLONITRILE see HGP000
HYDRACRYLONITRILE ACRYLATE see ADT111
HYDRADICARBONAMIDE see HGU050
HYDRAL see CDO000
HYDRAL 705 see AHC000
HYDRALAZINE see HGP495
HYDRALAZINE CHLORIDE see HGP500
HYDRALAZINE HYDROCHLORIDE see HGP500
HYDRALAZINE MONOHYDROCHLORIDE see HGP500
HYDRAL de CHLORAL see CDO000
HYDRALIN see CPB750
HYDRALLAZINE see HGP495
HYDRALLAZINE HYDROCHLORIDE see HGP500
HYDRAM see EKO500
HYDRAMETHYLNON see HGP525

HYDRAMYCIN see HGP550
HYDRANGEA see HGP600
HYDRANGEA MACROPHYLLA see HGP600
HYDRANGIN see HJY100
HYDRANGINE see HJY100
HYDRAPHEN see PFN000
HYDRAPRESS see HGP500
HYDRARGAPHEN see PFN000
HYDRARGYRUM BIJODATUM (GERMAN) see MDD000
HYDRASTINE HYDROCHLORIDE see HGQ500
HYDRASTININE see HGR000
HYDRASTIS see ICC800
HYDRASTIS CANADENSIS L., ROOT EXTRACT see HGR500
HYDRATED ALUMINA see AHC250
HYDRATED LIME see CAT225
HYDRATROP ALDEHYDE see COF000
HYDRATROPIC ACETATE see PGB750
HYDRATROPIC ACID, p-ISOBUTYL-, SODIUM SALT see IAB100
HYDRATROPIC ALCOHOL see HGR600
HYDRATROPIC ALDEHYDE see COF000
HYDRATROPIC ALDEHYDE DIMETHYL ACETAL see PGA800
HYDRATROPYL ACETATE see PGB750
HYDRATROPYL ALCOHOL see HGR600
HYDRAZID see ILD000
HYDRAZIDE BSG see BBS300
l-HYDRAZIDE CYSTEINE see CQK125
HYDRAZID KYSELINY MALEINOVE see DMC600
HYDRAZINE see HGS000
HYDRAZINE, anhydrous (DOT) see HGS000
HYDRAZINE AQUEOUS SOLUTIONS with >64% hydrazine, by weight (DOT) see HGS000
HYDRAZINE AQUEOUS SOLUTIONS, with not >64% hydrazine, by weight (DOT) see HGU500
HYDRAZINE AZIDE see HGS100
HYDRAZINE, AZIDO- see HGS100
HYDRAZINE-BENZENE see PFI000
HYDRAZINE-BENZENE and BENZIDINE SULFATE see BBY300
HYDRAZINE, 1-BENZOYL-2-THENYLIDENE see TEO100
HYDRAZINE BISBORANE see HGT500
HYDRAZINE, 1,1-BIS(p-METHYLPHENYL)- see DXP300
HYDRAZINE, 1,1-BIS(4-METHYLPHENYL)-(9CI) see DXP300
HYDRAZINE BORANE (1:1) see HGU025
HYDRAZINE, 1-(p-BROMOPHENYL)-, HYDROCHLORIDE see BNW550
HYDRAZINECARBOHYDRAZONOTHIOIC ACID see TFE250
HYDRAZINECARBOTHIOAMIDE see TFQ000
HYDRAZINECARBOTHIOAMIDE, 2-((3-AMINO-2-PYRIDINYL)METHYLENE)- see AMI600
HYDRAZINECARBOTHIOAMIDE, N-(4-(AMINOSULFONYL)PHENYL)-2-(1H-PYRROL-2-YLMETHYLENE)- (9CI) see SNL920
HYDRAZINECARBOTHIOAMIDE, N-(4-CHLOROPHENYL)-2-((1-METHYL-1H-PYRROL-2-YL)METHYLENE)- (9CI) see CKK600
HYDRAZINECARBOTHIOAMIDE, N-(4-CHLOROPHENYL)-2-(1H-PYRROL-2-YLMETHYLENE)- (9CI) see CKK700
HYDRAZINECARBOTHIOAMIDE, 2,2'-(1-(1-ETHOXYETHYL)-1,2-ETHANEDIYLIDENE)BIS see KFA100
HYDRAZINECARBOTHIOAMIDE, N-(3-METHOXYPHENYL)-2-((1-METHYL-1H-PYRROL-2-YL)METHYLENE)- (9CI) see MFJ010
HYDRAZINECARBOTHIOAMIDE, N-(4-METHOXYPHENYL)-2-((1-METHYL-1H-

PYRROL-2-YL)METHYLENE)- (9CI) see MFJ020
HYDRAZINECARBOTHIOAMIDE, N-(3-METHOXYPHENYL)-2-(1H-PYRROL-2-YLMETHYLENE)- (9CI) see MFJ030
HYDRAZINECARBOTHIOAMIDE, N-(4-METHOXYPHENYL)-2-(1H-PYRROL-2-YLMETHYLENE)- (9CI) see MFJ040
HYDRAZINECARBOTHIOAMIDE, 2,2'-METHYLENEBIS- see MJN300
HYDRAZINECARBOTHIOAMIDE, 2-((1-METHYL-1H-PYRROL-2-YL)METHYLENE)-N-PHENYL- (9CI) see PGN030
HYDRAZINECARBOTHIOAMIDE, 2-PHENYL-(9CI) see PGM750
HYDRAZINECARBOTHIOAMIDE, N-PHENYL-(9CI) see PGN000
HYDRAZINECARBOTHIOAMIDE, N-PHENYL-2-(1H-PYRROL-2-YLMETHYLENE)- (9CI) see PGN045
HYDRAZINECARBOTHIOAMIDE, 2-(1H-PYRROL-2-YLMETHYLENE)- (9CI) see PPS300
HYDRAZINECARBOXALDEHYDE see FNN000
HYDRAZINE CARBOXAMIDE see HGU000
HYDRAZINECARBOXAMIDE, N-(4-(AMINOSULFONYL)PHENYL)-2-((1-METHYL-1H-PYRROL-2-YL)METHYLENE)- see SNL900
HYDRAZINECARBOXAMIDE, N-(4-BROMOPHENYL)-2-((1-METHYL-1H-PYRROL-2-YL)METHYLENE)- (9CI) see BNX400
HYDRAZINECARBOXAMIDE, N-(4-BROMOPHENYL)-2-(1H-PYRROL-2-YLMETHYLENE)-(9CI) see BNX420
HYDRAZINECARBOXAMIDE, N-(4-BUTOXYPHENYL)-2-((1-METHYL-1H-PYRROL-2-YL)METHYLENE)- (9CI) see BPQ300
HYDRAZINECARBOXAMIDE, N-(4-BUTOXYPHENYL)-2-(1H-PYRROL-2-YLMETHYLENE)-(9CI) see BPQ330
HYDRAZINECARBOXAMIDE, N-(3-CHLOROPHENYL)-2-((1-METHYL-1H-PYRROL-2-YL)METHYLENE)- (9CI) see CKJ760
HYDRAZINECARBOXAMIDE, N-(4-CHLOROPHENYL)-2-((1-METHYL-1H-PYRROL-2-YL)METHYLENE)- (9CI) see CKJ765
HYDRAZINECARBOXAMIDE, N-(3-CHLOROPHENYL)-2-(1H-PYRROL-2-YLMETHYLENE)- (9CI) see CKJ770
HYDRAZINECARBOXAMIDE, N-(4-CHLOROPHENYL)-2-(1H-PYRROL-2-YLMETHYLENE)-(9CI) see CKJ775
HYDRAZINECARBOXAMIDE, N-(3-METHOXYPHENYL)-2-((1-METHYL-1H-PYRROL-2-YL)METHYLENE)- (9CI) see MFH945
HYDRAZINECARBOXAMIDE, N-(4-METHOXYPHENYL)-2-((1-METHYL-1H-PYRROL-2-YL)METHYLENE)- (9CI) see MFH950
HYDRAZINECARBOXAMIDE, N-(3-METHOXYPHENYL)-2-(1H-PYRROL-2-YLMETHYLENE)- (9CI) see MFH955
HYDRAZINECARBOXAMIDE, N-(4-METHOXYPHENYL)-2-(1H-PYRROL-2-YLMETHYLENE)- (9CI) see MFH960
HYDRAZINECARBOXAMIDE, 2-((1-METHYL-1H-PYRROL-2-YL)METHYLENE)-N-(4-NITROPHENYL)- (9CI) see NIV150
HYDRAZINECARBOXAMIDE, 2-((1-METHYL-1H-PYRROL-2-YL)METHYLENE)-N-(4-(OCTYLOXY)PHENYL)- (9CI) see OFE100
HYDRAZINECARBOXAMIDE, 2-((1-METHYL-1H-PYRROL-2-YL)METHYLENE)-N-PHENYL- (9CI) see PGH600

HYDRAZINECARBOXAMIDE
MONOHYDROCHLORIDE see SBW500
HYDRAZINECARBOXAMIDE, N-(4-
NITROPHENYL)-2-(1H-PYRROL-2-
YLMETHYLENE)- (9CI) see NIV200
HYDRAZINECARBOXAMIDE, N-(4-
(OCTYLOXY)PHENYL)-2-(1H-PYRROL-2-
YLMETHYLENE)- (9CI) see OFE200
HYDRAZINECARBOXAMIDE, N-PHENYL-
2-(1H-PYRROL-2-YLMETHYLENE)- (9CI) see
PGH650
HYDRAZINECARBOXIMIDAMIDE see
AKC750
HYDRAZINECARBOXIMIDAMIDE
HYDROCHLORIDE see AKC800
HYDRAZINECARBOXYLIC ACID, ETHYL
ESTER see EHG000
HYDRAZINE, 1-(2-(o-
CHLOROPHENOXY)ETHYL)-,
HYDROGEN SULFATE (1:1) see CJP500
HYDRAZINE, 1-(2-(o-
CHLOROPHENOXY)ETHYL)-, SULFATE
(1:1) see CJP500
HYDRAZINE, COMPD. WITH BORANE (1:1)
see HGU025
HYDRAZINE, 1,2-DIBENZOYL- see DDE250
HYDRAZINE, 1,1-DIBUTYL-,
ETHANEDIOATE (1:1) see DEC785
HYDRAZINE, 1,2-DIBUTYL-,
ETHANEDIOATE (1:1) see DEC795
HYDRAZINE DICARBONIC ACID
DIAZIDE (DOT) see DCL200
1,2-HYDRAZINEDICARBOTHIOAMIDE see
BLJ250
1,2-HYDRAZINEDICARBOXAMIDE see
HGU050
HYDRAZINEDICARBOXYLIC ACID
DIAMIDE see HGU050
HYDRAZINE, (2,5-DICHLOROPHENYL)-
see DGC850
HYDRAZINE, 1,1-DIETHYL-2-HYDROXY-2-
NITROSO-, ION (1-) see DHJ300
HYDRAZINE DIFLUORIDE see HGU100
HYDRAZINE, DIHYDROFLUORIDE see
HGU100
HYDRAZINE, (2-
(DIMETHYLAMINO)ETHYL)- see HHB600
HYDRAZINE, 1,2-DIMETHYL-,
ETHANEDIOATE (1:1) see DSG300
HYDRAZINE, 1,1-DIPHENYL- see DWD100
HYDRAZINE, 1,1-DI-2-PROPENYL-(9CI) see
DBK100
HYDRAZINE, 1,1-DIPROPYL-,
ETHANEDIOATE (1:1) see DWT430
HYDRAZINE, 1,2-DIPROPYL-,
ETHANEDIOATE (1:1) see DWT450
HYDRAZINE, 1,1-DI-p-TOLYL- see DXP300
HYDRAZINE, 1,1'-(1,2-ETHANEDIYL)BIS(1-
METHYL- see EEB225
HYDRAZINE, 1,1'-ETHYLENEBIS(1-
METHYL- see EEB225
HYDRAZINE, 1-GLUCOSYL-2-
ISONICOTINYL- see GFG070
HYDRAZINE, 2-GLUCOSYL-1-
ISONICOTINYL- see GFG070
HYDRAZINE, HEPTYL- see HBO600
HYDRAZINE HYDRATE see HGU500
HYDRAZINE HYDRATE see HGU501
HYDRAZINE HYDRATE, with not >64%
hydrazine, by weight (DOT) see HGU500
HYDRAZINE, HYDROCHLORIDE see
HGV000
HYDRAZINE HYDROGEN SULFATE see
HGW500
(T-4)-(HYDRAZINE-
KAPPAN)TRIHYDROBORON see HGU025
HYDRAZINE, 1-(o-
METHOXYPHENETHYL)-, SULFATE (1:1)
see MFG200
HYDRAZINE, N-((2-
METHYLCYCLOPROPYL)CARBONYL)- see
MIV300

HYDRAZINE, METHYL-, ETHANEDIOATE
(1:1) see MKN300
HYDRAZINE, (p-METHYLPHENETHYL)-
see MNO780
HYDRAZINE, 1-(α-METHYLPHENETHYL)-
2-PHENETHYL- see MNP400
HYDRAZINE, 1-(o-METHYLPHENETHYL)-,
SULFATE (1:1) see MNU100
HYDRAZINE, 1-(p-METHYLPHENETHYL)-,
SULFATE (1:1) see MNU150
HYDRAZINE, (1-METHYL-2-
PHENOXYETHYL)-, (Z)-2-BUTENEDIOATE
(1:1) (9CI) see PDV700
HYDRAZINE, 1-(2-(o-
METHYLPHENOXY)ETHYL)-, HYDROGEN
SULFATE (1:1) see MNR100
HYDRAZINE, (1-METHYL-2-
PHENOXYETHYL)-, MALEATE see PDV700
HYDRAZINE, (2-(4-
METHYLPHENYL)ETHYL)-, SULFATE (1:1)
(9CI) see MNU150
HYDRAZINE MONOBORANE see HGV500
HYDRAZINE MONOCHLORIDE see
HGV000
HYDRAZINE, MONOHYDRATE see
HGU501
HYDRAZINE MONOSULFATE see HGW500
HYDRAZINE, 2-(5-NITRO-α-
IMINOFURFURYL)- see NGG600
HYDRAZINE, PERCHLORATE see HGV600
HYDRAZINE PERCHLORATE (DOT) see
HGV600
HYDRAZINE, 1-(1-PHENOXY-2-PROPYL)-,
MALEATE see PDV700
HYDRAZINE PROPANEMETHANE
SULFONATE see HGW000
HYDRAZINE SELENATE see HGW100
HYDRAZINE, SELENATE see HGW100
HYDRAZINE SULFATE (1:1) see HGW500
HYDRAZINE SULPHATE see HGW500
HYDRAZINE, (THIENYL)- see TEY300
HYDRAZINE, TRIFLUOROSTANNITE see
HHA100
HYDRAZINE YELLOW see FAG140
HYDRAZINIUM CHLORATE see HGX000
HYDRAZINIUM CHLORIDE see HGV000
HYDRAZINIUM CHLORITE see HGX500
HYDRAZINIUM, 1,1-DIMETHYL-1-(2,3-
DIMETHYL-2-HYDROXY-3-BUTENYL)-2-(
1-OXOPROPYL)-, HYDROXIDE, INNER
SALT see HGX550
HYDRAZINIUM DIPERCHLORATE see
HGY000
HYDRAZINIUM HYDROGENSELENATE
see HGY500
HYDRAZINIUM MONOCHLORIDE see
HGV000
HYDRAZINIUM NITRATE see HGZ000
HYDRAZINIUM PERCHLORATE see
HHA000
HYDRAZINIUM SULFATE see HGW500
HYDRAZINIUM TRIFLUOROSTANNITE see
HHA100
2-HYDRAZINO-4-(4-
AMINOPHENYL)THIAZOLE see HHB000
2-HYDRAZINO-4-(p-
AMINOPHENYL)THIAZOLE see HHB000
HYDRAZINOBENEZENE see PFI000
4-HYDRAZINOBENZOIC ACID see HHB100
p-HYDRAZINOBENZOIC ACID see HHB100
2-HYDRAZINOBENZOTHIAZOLE see
HHB500
3-HYDRAZINO-6-(N,N-BIS-(2-
HYDROXYETHYL)AMINO) PYRIDAZINE
DIHYDROCHLORIDE see BKB500
(S)-α-HYDRAZINO-3,4-DIHYDROXY-α-
METHYL-BENZENEPROPANOIC ACID
(9CI) see CBQ500
(−)-l-α-HYDRAZINO-3,4-DIHYDROXY-α-
METHYLHYDROCINNAMIC ACID
MONOHYDRATE see CBQ529

2-HYDRAZINO-N,N-
DIMETHYLETHANAMINE see HHB600
2-HYDRAZINOETHANOL see HHC000
3-HYDRAZINO-6-((2-
HYDROXYPROPYL)METHYLAMINO)PYRI
DAZINE DIHYDROCHLORIDE see HHD000
HYDRAZINO-α-METHYLDOPA see CBQ500
p-HYDRAZINONITROBENZENE see
NIR000
2-HYDRAZINO-4-(5-NITRO-2-
FURANYL)THIAZOLE see HHD500
2-HYDRAZINO-4-(5-NITRO-2-
FURYL)THIAZOLE see HHD500
2-HYDRAZINO-4-(4-
NITROPHENYL)THIAZOLE see HHE000
1-HYDRAZINO-2-PHENYLETHANE see
PFC500
1-HYDRAZINO-2-PHENYLETHANE
HYDROGEN SULPHATE see PFC750
2-HYDRAZINO-1-PHENYLPROPANE see
PDN000
HYDRAZINOPHTHALAZINE see HGP495
1-HYDRAZINOPHTHALAZINE see HGP495
1-HYDRAZINOPHTHALAZINE ACETONE
HYDRAZONE see HHE500
1-HYDRAZINOPHTHALAZINE
HYDROCHLORIDE see HGP500
1-HYDRAZINOPHTHALAZINE
MONOHYDROCHLORIDE see HGP500
4-HYDRAZINO-2-THIOURACIL see HHF500
HYDRAZOBENZEN (CZECH) see HHG000
HYDRAZOBENZENE see HHG000
1,1'-HYDRAZOBIS(FORMAMIDE) see
HGU050
HYDRAZOCARBONAMIDE see HGU050
HYDRAZODIBENZENE see HHG000
HYDRAZODICARBONAMIDE see HGU050
HYDRAZODICARBONSAEUREABIS(METH
YLNITROSAMID) (GERMAN) see DRN400
HYDRAZODICARBOXAMIDE see HGU050
HYDRAZODICARBOXYLIC ACID
BIS(METHYLNITROSAMIDE) see DRN400
HYDRAZODIFORMIC ACID see DKJ600
HYDRAZOETHANE see DJL400
HYDRAZOIC ACID see HHG500
HYDRAZOMETHANE see DSF600
HYDRAZOMETHANE see MKN000
HYDRAZONIUM, 1-(2-BROMOETHYL)-1,1-
DIMETHYL-, BROMIDE see DQT100
HYDRAZONIUM SULFATE see HGW500
2,2'-HYDRAZONODIETHANOL see
HHH000
HYDRAZYNA (POLISH) see HGS000
HYDREA see HOO500
HYDREL see HHH100
HYDRIDES see HHH500
HYDRIN-2 see HHQ800
HYDRINDANTIN, anhydrous see HHI000
1,2-HYDRINDENE see IBR000
HYDRINDONAPHTHENE see IBR000
HYDRIODIC ACID see HHI500
HYDRIODIC ACID, solution (UN 1787)
(DOT) see HHI500
HYDRIODIC ETHER see ELP500
HYDRIODIDE-ENTROL see CHR500
HYDRITE see KBB600
HYDRO-AQUIL see CFY000
HYDROAZODICARBOXYBIS(METHYLNIT
ROSAMIDE) see DRN400
HYDROAZOETHANE see DJL400
HYDROBIS(2-
METHYLPROPYL)ALUMINUM see DNI600
HYDROBROMIC ACID see HHJ000
HYDROBROMIC ACID
MONOAMMONIATE see ANC250
HYDROBROMIC ACID SOLUTION, >49%
hydrobromic acid (UN 1788) (DOT) see HHJ000
HYDROBROMIC ACID SOLUTION, not
>49% hydrobromic acid (UN 1788) (DOT) see
HHJ000
HYDROBROMIC ETHER see EGV400
HYDROCARB 60 see CAT775

HYDROCARBON GAS see HHJ500
HYDROCARBON GASES, COMPRESSED,
N.O.S. (UN 1964) (DOT) see HHJ500
HYDROCARBON GASES, LIQUEFIED,
N.O.S. (UN 1965) (DOT) see HHJ500
HYDROCARBON GASES MIXTURES,
COMPRESSED, N.O.S. (UN 1964) (DOT) see
HHJ500
HYDROCARBON GASES MIXTURES,
LIQUEFIED, N.O.S. (UN 1965) (DOT) see
HHJ500
HYDROCELLULOSE see HHK000
HYDROCERIN see CMD750
HYDROCHINON (CZECH, POLISH) see
HIH000
HYDROCHLORBENZETHYLAMINE see
HHK050
HYDROCHLORBENZETHYLAMINE
DIMALEATE see HHK100
HYDROCHLORIC ACID see HHL000
HYDROCHLORIC ACID DICARBOXIDE see
DEK000
HYDROCHLORIC ACID DIMETHYLAMINE
see DOR600
HYDROCHLORIC ACID, mixed with NITRIC
ACID (3:1) see HHM000
HYDROCHLORIC ACID, solution (UN 1789)
(DOT) see HHL000
HYDROCHLORIC ETHER see EHH000
HYDROCHLORIDE see HHL000
L-67 HYDROCHLORIDE see CMS260
1C50 HYDROCHLORIDE see EIJ000
d-8955 HYDROCHLORIDE see DXI800
191C49 HYDROCHLORIDE see DJP500
l-HYDROCHLORIDE ARGININE see
AQW000
HYDROCHLORIDE of DI-n-
BUTYLAMINOPROPYL-3-IODO-4-
FLUOROBENZOATE see HHM500
HYDROCHLORIDE DIETHYLAMINE see
DIS500
HYDROCHLORIDE OF 1-
PHENYLCYCLOPENTANECARBOXYLIC
ACID DIETHYLAMINOETHYL ESTER see
PET250
HYDROCHLORIDE,RO 4-1284 see HKS900
HYDROCHLORID SALZ des p-N-n-
BUTYLAMINO-BENZOESAURE-
DIAETHYLAMINOAETHYLESTERS
(GERMAN) see BQA020
HYDROCHLOROAURIC ACID, SODIUM
SALT see GIZ100
HYDROCHLOROFLUOROCARBON 142b
see CFX250
HYDROCHLOROUS ACID, SODIUM SALT,
PENTAHYDRATE see SHU525
HYDROCHLORTHIAZID see CFY000
HYDROCINNAMALDEHYDE see HHP000
HYDROCINNAMALDEHYDE, p-
ISOPROPYL-α-METHYL-, DIETHYL
ACETAL see COU510
HYDROCINNAMALDEHYDE, p-
ISOPROPYL-α-METHYL-, DIMETHYL
ACETAL see COU525
HYDROCINNAMAMIDE, α-((p-
CHLOROBENZOYL)AMINO)-N,N-
DIPROPYL-4-(2-(4-METHYL-1-
PIPERAZINYL) ETHOXY)-, CITRATE, (+-)-
see NMV460
HYDROCINNAMIC ACID, α-AMINO- see
PEC750
HYDROCINNAMIC ACID, 3,5-DI-TERT-
BUTYL-4-HYDROXY-, TETRAESTER WITH
PENTAERYTHRITOL (7CI) see DED100
HYDROCINNAMIC ALCOHOL see HHP050
HYDROCINNAMIC ALDEHYDE see
HHP000
HYDROCINNAMIQUE NITRILE (FRENCH)
see HHP100
HYDROCINNAMONITRILE see HHP100
HYDROCINNAMYL ACETATE see HHP500
HYDROCINNAMYL ALCOHOL see HHP050

HYDROCINNAMYL CINNAMATE see
PGB800
HYDROCINNAMYL FORMATE see HHQ000
HYDROCINNAMYL ISOBUTYRATE see
HHQ500
HYDROCINNAMYL PROPIONATE see
HHQ550
HYDROCODIN see DKW800
HYDROCODONE see OOI000
HYDROCODONE BITARTRATE see
DKX050
HYDROCONQUININE see HIG500
Δ¹-HYDROCORTISONE see PMA000
11-β-HYDROCORTISONE see CNS750
HYDROCORTISONE-21-ACETATE see
HHQ800
HYDROCORTISONE BUTYRATE see
HHQ825
HYDROCORTISONE-17-BUTYRATE see
HHQ825
HYDROCORTISONE-17-BUTYRATE-21-
PROPIONATE see HHQ850
HYDROCORTISONE 17-α-BUYTRATE see
HHQ825
HYDROCORTISONE BUYTRATE
PROPIONATE see HHQ850
HYDROCORTISONE FREE ALCOHOL see
CNS750
HYDROCORTISONE PHOSPHATE see
HHQ875
HYDROCORTISONE-21-PHOSPHATE see
HHQ875
21-HYDROCORTISONEPHOSPHORIC
ACID see HHQ875
HYDROCORTISONE SODIUM SUCCINATE
see HHR000
HYDROCORTISONE-21-SODIUM
SUCCINATE see HHR000
HYDROCORTISYL see CNS750
HYDROCORTONE see CNS750
HYDROCOUMARIN see HHR500
HYDROCUPREINE ETHYL ESTER
HYDROCHLORIDE see ELC500
HYDROCUPREINE ETHYL ETHER see
HHR700
HYDROCYANIC ACID see HHS000
HYDROCYANIC ACID, mixed with
CYANOGEN CHLORIDE (3:2 BY WT) see
HHS500
HYDROCYANIC ACID, aqueous solutions
<5% HCN (NA 1613) (DOT) see HHS000
HYDROCYANIC ACID, ION(CN1-) see
COI500
HYDROCYANIC ACID, POTASSIUM SALT
see PLC500
HYDROCYANIC ACID (PRUSSIC),
unstabilized (DOT) see HHS000
HYDROCYANIC ACID, SALTS see HHT000
HYDROCYANIC ACID, SODIUM SALT see
SGA500
HYDROCYANIC ACID, aqueous solutions not
>20% hydrocyanic acid (UN 1613) (DOT) see
HHS000
HYDROCYANIC ETHER see PMV750
HYDROCYCLIN see HOI000
HYDROCYCLINE see MCH525
HYDRODARCO see CBT500
HYDRODELTALONE see PMA000
HYDRODELTISONE see PMA000
HYDRODESUFURIZED MIDDLE
DISTILLATE see DXG820
HYDRODESULFURIZED HEAVY VACUUM
GAS OIL see GBW010
HYDRODESULFURIZED KEROSINE
(PETROLEUM) see KEK110
HYDRODIISOBUTYLALUMINUM see
DNI600
HYDRODIURETIC see CFY000
HYDRO-DIURIL see CFY000
17-β-HYDROESTR-4-EN-3-ONE see NNX400
HYDROFLUOBORIC ACID see FDD125
HYDROFLUOBORIC ACID see HHS600

HYDROFLUORIC ACID see HHU500
HYDROFLUORIC ACID mixed with
SULFURIC ACID see HHV000
HYDROFLUORIC ACID, solution, >60%
strength (UN 1790) (DOT) see HHU500
HYDROFLUORIC ACID, solution, not >60%
strength (UN 1790) (DOT) see HHU500
HYDROFLUORIC and SULFURIC ACIDS,
MIXTURE (DOT) see HHV000
HYDROFLUORIDE see HHU500
HYDROFLUORIDE-1927 WANDER see
DCW800
HYDROFLUOSILICIC ACID see SCO500
HYDROFOL see PAE250
HYDROFOL ACID 200 see HOG000
HYDROFOL ACID 1255 see LBL000
HYDROFOL ACID 1495 see MSA250
HYDROFOL ACID 1655 see SLK000
α-HYDROFORMAMINE CYANIDE see
HHW000
HYDROFURAMIDE see FPS000
HYDROFURAN see TCR750
HYDROGEN see HHW500
HYDROGEN (DOT) see HHW500
HYDROGEN, compressed (DOT) see HHW500
HYDROGEN, refrigerated liquid (DOT) see
HHW500
HYDROGEN ANTIMONIDE see SLQ000
HYDROGEN ARSENIDE see ARK250
HYDROGENATED CASTOR OIL see
HHW502
HYDROGENATED COAL OIL FRACTION 1
see HHW509
HYDROGENATED COAL OIL FRACTION 3
see HHW519
HYDROGENATED COAL OIL FRACTION 4
see HHW529
HYDROGENATED COAL OIL FRACTION 7
see HHW539
HYDROGENATED COAL OIL FRACTION 9
see HHW549
HYDROGENATED FISH OIL see HHW560
HYDROGENATED MDI see MJM600
HYDROGENATED SPERM OIL see HHW575
HYDROGENATED TERPHENYLS see
HHW800
HYDROGEN AZIDE see HHG500
HYDROGEN BROMIDE
(ACGIH,OSHA,MAK) see HHJ000
HYDROGEN BROMIDE, anhydrous (UN
1048) (DOT) see HHJ000
HYDROGEN CARBOXYLIC ACID see
FNA000
HYDROGEN CHLORIDE see HHX000
HYDROGEN CHLORIDE (aerosol) see
HHX500
HYDROGEN CHLORIDE, anhydrous (UN
1050) (DOT) see HHL000
HYDROGEN CHLORIDE, refrigerated liquid
(UN 2186) (DOT) see HHL000
HYDROGEN CYANAMIDE see COH500
HYDROGEN CYANIDE see HHS000
HYDROGEN CYANIDE (ACGIH,OSHA) see
HHS000
HYDROGEN CYANIDE, anhydrous, stabilized
(UN 1051) (DOT) see HHS000
HYDROGEN CYANIDE, anhydrous, stabilized,
absorbed in a porous inert material (UN 1614)
(DOT) see HHS000
HYDROGEN DIMETHYL PHOSPHITE see
DSG600
HYDROGEN DIOXIDE see HIB050
HYDROGEN DISULFIDE see HHZ000
HYDROGENE SULFURE (FRENCH) see
HIC500
(HYDROGEN(ETHYLENEDINITRILO)TET
RAACETATO)IRON see HIA000
HYDROGEN FLUORIDE, anhydrous (UN
1052) (DOT) see HHU500
HYDROGEN HEXACHLOROIRIDATE (4+)
see HIA500

HYDROGEN HEXACHLOROPLATINATE(4+) see CKO750
HYDROGEN HEXAFLUOROPHOSPHATE see HDE000
HYDROGEN HEXAFLUOROSILICATE see SCO500
HYDROGEN IODIDE see HHI500
HYDROGEN IODIDE, anhydrous (UN 2197) (DOT) see HHI500
HYDROGEN ISOTHIOCYANATE see ISF100
HYDROGEN NITRATE see NED500
HYDROGEN OXALATE of AMITON see AMX825
HYDROGEN PEROXIDE, 30% see HIB010
HYDROGEN PEROXIDE, 90% see HIB050
HYDROGEN PEROXIDE, 8% to 20% see HIB005
HYDROGEN PEROXIDE, solution, 30% see HIB010
HYDROGEN PEROXIDE, solution, 8% to 20% (DOT) see HIB005
HYDROGEN PEROXIDE, stabilized with >60% hydrogen peroxide (DOT) see HIB050
HYDROGEN PEROXIDE CARBAMIDE see HIB500
HYDROGEN PEROXIDE & PEROXYACETIC ACID MIXT., with acids, H$_2$O not >5% peroxyacetic acid (DOT) see PCL500
HYDROGEN PEROXIDE, aqueous solutions with not <8% but <20% hydrogen peroxide (UN 2984) (DOT) see HIB010
HYDROGEN PEROXIDE, aqueous solutions with >40%, not >60% hydrogen peroxide (UN 2014) see HIB010
HYDROGEN PEROXIDE with UREA (1:1) see HIB500
HYDROGEN PHOSPHIDE see PGY000
HYDROGEN POTASSIUM FLUORIDE see PKU250
HYDROGEN POTASSIUM PHTHALATE see HIB600
HYDROGEN SELENIDE see HIC000
HYDROGEN SELENIDE, anhydrous (DOT) see HIC000
HYDROGEN SODIUM SELENITE see SBO200
HYDROGEN SODIUM SELENIUM OXIDE see SBO200
21-(HYDROGEN SUCCINATE)CORTISOL, MONOSODIUM SALT see HHR000
HYDROGEN SULFIDE see HIC500
HYDROGEN SULFITE see HIC600
HYDROGEN SULFITE SODIUM see SFE000
HYDROGEN SULFURIC ACID see HIC500
HYDROGEN TETRAFLUOROBORATE see FDD125
HYDROGEN TETRAFLUOROBORATE see HHS600
HYDROGEN TRISULFIDE see HID000
HYDROGESTRONE see DYF759
α-HYDRO-omega-HYDROXYPOLY(OXY-1,2-ETHANEDIYL) see PJT000
α-HYDRO-ω-HYDROXY-POLY(OXY-1,2-ETHANEDIYL) see PJT500
α-HYDRO-ω-HYDROXYPOLY(OXY(METHYL-1,2-ETHANEDIYL)), (CHLOROMETHYL)OXIRANE POLYMER see PKK100
α-HYDRO-ω-HYDROXYPOLY(OXY(METHYL-1,2-ETHANEDIYL)) ETHER WITH BIS((2-(HYDROXYETHYL)AMINO)METHYL)PHENOL (3:1) see HID050
α-HYDRO-ω-HYDROXY-POLY(OXYRTHYLENE)-POLY(OXYPROPYLENE)(51-57 MOLES)POLY(OXYETHYLENE) BLOCK POLYMER see PJK200
HYDROL see DBI800
HYDROLIN see SHR500

HYDROLIT see FMW000
HYDROLOSE see MIF760
HYDROL SW see PKF500
HYDROLYZED ANIMAL PROTEIN see HID100
HYDROLYZED COLLAGEN see HID100
HYDROLYZED MILK PROTEIN see ADF800
HYDROLYZED PLANT PROTEIN see ADF800
HYDROLYZED VEGETABLE PROTEIN see ADF800
HYDRO-MAG MA see MAG750
HYDROMAGNESITE see MAC650
HYDROMEDIN see DFP600
HYDROMIREX see MRI750
HYDROMORPHONE see DLW600
HYDROMORPHONE HYDROCHLORIDE see DNU300
HYDROMOX R see RDK000
HYDRONITRIC ACID see HHG500
HYDRONOL see HID350
HYDRONSAN see GBB500
HYDROOT see SOI500
HYDROPERIT see HIB500
HYDROPEROXIDE see HIB050
HYDROPEROXIDE, ACETYL see PCL500
HYDROPEROXIDE, ETHYL see ELD000
6-HYDROPEROXY-4-CHOLESTEN-3-ONE see HID500
6-β-HYDROPEROXYCHOLEST-4-EN-3-ONE see HID500
6-β-HYDROPEROXY-Δ⁴CHOLESTEN-3-ONE see HID500
1-HYDROPEROXYCYCLOHEX-3-ENE see HIE000
1-HYDROPEROXY-3-CYCLOHEXENE see HIE000
1-HYDROPEROXYCYCLOHEXYL-1-HYDROXYCYCLOHEXYL PEROXIDE see CPC300
4-HYDROPEROXYCYCLOPHOSPHAMIDE see HIF000
HYDROPEROXYDE de BUTYLE TERTIAIRE (FRENCH) see BRM250
HYDROPEROXYDE de CUMENE (FRENCH) see IOB000
HYDROPEROXYDE de CUMYLE (FRENCH) see IOB000
10-β-HYDROPEROXY-17-α-ETHYNYL-4-ESTREN-17-β-OL-3-ONE see HIE525
4-HYDROPEROXYIFOSFAMIDE see HIE550
N-(HYDROPEROXYMETHYL)-N-NITROSOPROPYLAMINE see HIE570
2-HYDROPEROXY-2-METHYLPROPANE see BRM250
HYDROPEROXY-N-NITROSODIBUTYLAMINE see HIE600
1-(HYDROPEROXY)-N-NITROSODIMETHYLAMINE see HIE700
4-HYDROPEROXYPHOSPHAMIDE see HIF000
2-HYDROPEROXYPROPANE see IPI000
1-HYDROPEROXY-1-VINYLCYCLOHEX-3-ENE see HIF575
4-HYDROPEROXY-4-VINYL-1-CYCLOHEXENE see HIF575
HYDROPHENOL see CPB750
HYDROPHIS CYANOCINCTUS VENOM see SBI900
HYDROPHIS ELEGANS (AUSTRALIA) VENOM see HIG000
HYDROPRENE see EQD000
HYDROPRES see RDK000
HYDROPRES KA see RDK000
HYDROQUINIDINE see HIG500
HYDROQUINOL see HIH000
HYDROQUINOLE see HIH000
HYDROQUINOL, ISOPROPYL-, 4-(METHYLCARBAMATE) see IPI100
HYDROQUINONE see HIH000
m-HYDROQUINONE see REA000
o-HYDROQUINONE see CCP850

p-HYDROQUINONE see HIH000
α-HYDROQUINONE see HIH000
HYDROQUINONE, liquid or solid (DOT) see HIH000
HYDROQUINONE, compounded with p-BENZOQUINONE see QFJ000
HYDROQUINONE BENZYL ETHER see AEY000
HYDROQUINONE, BROMO- see BNL260
HYDROQUINONE CALCIUM SULFONATE see DMI300
HYDROQUINONECARBOXYLIC ACID see GCU000
HYDROQUINONE, 2,5-DI-tert-BUTYL- see DEC800
HYDROQUINONE, 2,6-DIISOPROPYL-, 4-(METHYLCARBAMATE) see MIA800
HYDROQUINONE-β-d-GLUCOPYRANOSIDE see HIH100
HYDROQUINONE MONOBENZYL ETHER see AEY000
HYDROQUINONE MONOETHYL ETHER see EFA100
HYDROQUINONE MONOMETHYL ETHER see MFC700
HYDROQUINONE MUSTARD see BHB750
HYDROQUINONE, PHENYL- see BGG250
HYDROQUINONE, TRIMETHYL- see POG300
HYDRO-RAPID see CHJ750
HYDRORETROCORTIN see PMA000
HYDRORUBEANIC ACID see DXO200
HYDROSALURIC see CFY000
HYDROSARPAN see AFG750
HYDROSCINE HYDROBROMIDE see HOT500
HYDROSILICOFLUORIC ACID see SCO500
HYDROSORBIC ACID see HFD000
HYDROSULFITE ANION see HIC600
HYDROSULFITE AWC see FMW000
HYDROTHAL-47 see EAR000
HYDROTHIADENE see HII000
HYDROTHIDE see CFY000
HYDROTHOL see DXD000
HYDROTREATED (mild) HEAVY NAPHTHENIC DISTILLATE see MQV790
HYDROTREATED (mild) HEAVY NAPHTHENIC DISTILLATES (PETROLEUM) see MQV790
HYDROTREATED (mild) HEAVY PARAFFINIC DISTILLATE see MQV795
HYDROTREATED (severe) HEAVY PARAFFINIC DISTILLATE see MQV796
HYDROTREATED KEROSENE see KEK100
HYDROTREATED (mild) LIGHT NAPHTHENIC DISTILLATE see MQV800
HYDROTREATED (mild) LIGHT NAPHTHENIC DISTILLATES (PETROLEUM) see MQV800
HYDROTREATED (mild) LIGHT PARAFFINIC DISTILLATE see MQV805
HYDROTREATED NAPHTHA see NAH600
HYDROTRICHLOROTHIAZIDE see HII500
HYDROTRICINE see TOG500
HYDROTROPALDEHYDE DIMETHYL ACETAL see HII600
HYDROTROPE see XJJ010
HYDROTROPIC ALDEHYDE DIMETHYL ACETAL see HII600
HYDROXAMIC ACID, BENZSULFO- see BDW100
HYDROXINE see CJR909
HYDROXINE YELLOW L see FAG140
HYDROXON see HNT500
N-HYDROXY-AABP see ACD000
N-HYDROXY-AAF see HIP000
trans-4'-HYDROXY-AAS see HIK000
12-HYDROXYABIETIC ACID BIS(2-CHLOROETHYL)AMINE see HIJ000
12-HYDROXYABIETIC ACID BIS(2-CHLOROETHYL)AMINE SALT see HIJ000
HYDROXYACETALDEHYDE see GHO100

N-HYDROXY-4-ACETAMIDOBIPHENYL see ACD000

N-4-(N-HYDROXYACETAMIDO)BIPHENYL see ACD000

N-HYDROXY-4-ACETAMIDODIPHENYL see ACD000

1-HYDROXY-2-ACETAMIDOFLUORENE see HIJ400

2-(N-HYDROXYACETAMIDO)FLUORENE see HIP000

N-HYDROXY-2-ACETAMIDOFLUORENE see HIP000

2-(N-HYDROXYACETAMIDO)NAPHTHALENE see NBD000

2-(N-HYDROXYACETAMIDO)PHENANTHRENE see PCZ000

HYDROXY-2-(β-(2'-ACETAMIDOPHENYL))ETHYLNAPHTHAMIDE see HIJ500

4-(N-HYDROXYACETAMIDO)STILBENE see SMT000

trans-N-HYDROXY-4-ACETAMIDOSTILBENE see SMT500

trans-4'-HYDROXY-4-ACETAMIDOSTILBENE see HIK000

HYDROXY(o-ACETAMINOBENZOATO)MERCURY SODIUM SALT see SJP500

7-HYDROXY-2-ACETAMINOFLUORENE see HIK500

4-HYDROXYACETANILIDE see HIM000

m-HYDROXYACETANILIDE see HIL500

o-HYDROXYACETANILIDE see HIL000

p-HYDROXYACETANILIDE see HIM000

2'-HYDROXYACETANILIDE see HIL000

3'-HYDROXYACETANILIDE see HIL500

4'-HYDROXYACETANILIDE see HIM000

HYDROXYACETIC ACID see GHO000

HYDROXYACETIC ACID ETHYL ESTER see EKM500

HYDROXYACETIC ACID, MONOSODIUM SALT see SHT000

HYDROXYACETONE see ABC000

HYDROXYACETONITRILE see HIM500

2-HYDROXYACETONITRILE see HIM500

N-HYDROXY-p-ACETOPHENETIDIDE see HIN000

o-HYDROXYACETOPHENONE see HIN500

p-HYDROXYACETOPHENONE see HIO000

2'-HYDROXYACETOPHENONE see HIN500

4'-HYDROXYACETOPHENONE see HIO000

N-HYDROXY-4-ACETYLAMINOBIBENZYL see PDI250

N-HYDROXY-4-ACETYLAMINOBIPHENYL see ACD000

N-HYDROXY-2-ACETYLAMINOFLUORENE see HIP000

3-HYDROXY-N-ACETYL-2-AMINOFLUORENE see HIO875

N-HYDROXY-N-ACETYL-2-AMINOFLUORENE see HIP000

N-HYDROXY-2-ACETYLAMINOFLUORENE, COBALTOUS CHELATE see FDU875

N-HYDROXY-2-ACETYLAMINOFLUORENE, CUPRIC CHELATE see HIP500

N-HYDROXY-2-ACETYLAMINOFLUORENE, FERRIC CHELATE see HIQ000

N-HYDROXY-2-ACETYLAMINOFLUORENE-o-GLUCURONIDE see HIQ500

N-HYDROXY-2-ACETYLAMINOFLUORENE, MANGANOUS CHELATE see HIR000

N-HYDROXY-2-ACETYLAMINOFLUORENE, NICKELOUS CHELATE see HIR500

N-HYDROXY-2-ACETYLAMINOFLUORENE, POTASSIUM SALT see HIS000

N-HYDROXY-2-ACETYLAMINOFLUORENE, ZINC CHELATE see ZHJ000

N-HYDROXY-2-ACETYLAMINONAPHTHALENE see NBD000

N-HYDROXY-2-ACETYLAMINOPHENANTHRENE see PCZ000

N-HYDROXY-4-ACETYLAMINOSTILBENE see SMT000

trans-N-HYDROXY-4-ACETYLAMINOSTILBENE see SMT500

trans-N-HYDROXY-4-ACETYLAMINOSTILBENE CUPRIC CHELATE see SMU000

2-(HYDROXYACETYL)INDOLE see HIS100

5-(HYDROXYACETYL)INDOLE see HIS110

2-(HYDROXYACETYL)-1-METHYLINDOLE see HIS120

2-(HYDROXYACETYL)-3-METHYLINDOLE see HIS130

3-(HYDROXYACETYL)-1-METHYLINDOLE see HIS140

2-(2-HYDROXYACETYL)-1-METHYLINDOLE see HIS120

3-(2-HYDROXYACETYL)-2-METHYLINDOLE see HIS150

2-HYDROXYACLACINOMYCIN A see HIS300

3-(2-HYDROXY-1-ADAMANTYL)-N-METHYLPROPLYAMINE HYDROCHLORIDE see MGK750

N-HYDROXYADENINE see HIT000

2-HYDROXYADENOSINE see IKS450

6-HYDROXYADENOSINE see HIT100

N-HYDROXYADENOSINE see HIT100

6-N-HYDROXYADENOSINE see HIT100

N⁶-HYDROXYADENOSINE see HIT100

HYDROXYADIPALDEHYDE see HIT500

2-HYDROXYADIPALDEHYDE see HIT500

α-HYDROXYADIPALDEHYDE see HIT500

2-(2-HYDROXYAETHOXY)AETHYLESTER der FLUTENAMINSAEURE (GERMAN) see HKK000

4-HYDROXYAFLATOXIN B1 see AEW000

β-HYDROXYALANINE see SCA355

17-HYDROXY-17-α-ALLYL-4-ESTRENE see AGF750

(−)-3-HYDROXY-N-ALLYLMORPHINAN see AGI000

l-3-HYDROXY-N-ALLYL MORPHINAN see AGI000

1-HYDROXY-4-AMINOANTHRAQUINONE see AKE250

2-HYDROXY-4-AMINOBENZOIC ACID see AMM250

4-HYDROXYAMINOBIPHENYL see BGI250

N-HYDROXY-2-AMINOBIPHENYL see HIT600

N-HYDROXY-4-AMINOBIPHENYL see BGI250

β-HYDROXY-α-AMINOBUTYRIC ACID see AKF375

3-HYDROXYAMINO-4,6-DIMETHYLDIPYRIDO(1,2-A:3',2'-D)IMIDAZOLE see HIT610

4-HYDROXY-3-AMINODIPHENYL HYDROCHLORIDE see AIW250

3-HYDROXY-4-AMINODIPHENYL SULPHATE see AKF250

3-HYDROXYAMINODIPYRIDO(1,2-A:3',2'-D)IMIDAZOLE see HIT620

5-HYDROXY-3-(β-AMINOETHYL)INDOLE see AJX500

N-HYDROXY-2-AMINOFLUORENE see HIU500

N-HYDROXY-3-AMINOFLUORENE see FEH000

N-HYDROXY-2-AMINOFLUORENE N-GLUCURONIDE see HLE750

2-HYDROXY-4-AMINO-5-FLUOROPYRIMIDINE see FHI000

2-HYDROXYAMINO-6-METHYLDIPYRIDO(1,2-A:3',2'-D)IMIDAZOLE see HIU550

2-HYDROXYAMINO-3-METHYLIMIDAZOLO(4,5-F)QUINOLINE see HIU600

3-HYDROXY-5-AMINOMETHYLISOXAZOLE see AKT750

3-HYDROXY-5-AMINOMETHYLISOXAZOLE-AGARIN see AKT750

4-(HYDROXYAMINO)-5-METHYLQUINOLINE-1-OXIDE see HIV000

4-(HYDROXYAMINO)-6-METHYLQUINOLINE-1-OXIDE see HIV500

4-(HYDROXYAMINO)-7-METHYLQUINOLINE-1-OXIDE see HIW000

4-(HYDROXYAMINO)-8-METHYLQUINOLINE-1-OXIDE see HIW500

N-HYDROXY-1-AMINONAPHTHALENE see HIX000

N-HYDROXY-2-AMINONAPHTHALENE see NBI500

2-HYDROXYAMINO-1,4-NAPHTHOQUINONE see HIX200

3-HYDROXY-4-AMINONITROBENZENE see ALO000

4-(HYDROXYAMINO)-6-NITROQUINOLINE-1-OXIDE see HIX500

4-(HYDROXYAMINO)-7-NITROQUINOLINE-1-OXIDE see HIY000

N-HYDROXY p-AMINOOCTANOYLPHENONE see HIY200

N-(2-(HYDROXYAMINO)-2-OXOETHYL)BENZENEACETAMIDE see HIY300

N-(2-(HYDROXYAMINO)-2-OXOETHYL)-2-THIOPHENECARBOXAMIDE see TEN725

dl-α-HYDROXY-β-AMINOPROPYLBENZENE see NNM500

α-HYDROXY-β-AMINOPROPYLBENZENE HYDROCHLORIDE see PMJ500

6-HYDROXYAMINOPURINE see HIT000

4-(HYDROXYAMINO)-6-QUINOLINECARBOXYLIC ACID-1-OXIDE see CCF750

4-(HYDROXYAMINO)QUINOLINE-1-OXIDE see HIY500

4-(HYDROXYAMINO)QUINOLINE-1-OXIDE, HYDROCHLORIDE see HIZ000

4-HYDROXYAMINOQUINOLINE-1-OXIDE HYDROCHLORIDE see HIZ000

4-HYDROXYAMINOQUINOLINE-1-OXIDE MONOHYDROCHLORIDE see HIZ000

4-(HYDROXYAMINO)QUINOLINE-1-OXIDE MONOHYDROCHLORIDE see HIZ000

trans-N-HYDROXY-4-AMINOSTILBENE see HJA000

4-(HYDROXYAMINO)TOLUENE see HJA500

HYDROXYAMMONIUM CHLORIDE see HLN000

p-HYDROXYAMPHETAMINE HYDROCHLORIDE see HJA600

19-HYDROXYANDROSTA-1,4-DIENE-3,17-DIONE ACETATE see ABL625

3-α-HYDROXY-17-ANDROSTANONE see HJB050

17-β-HYDROXY-5-α-ANDROSTAN-3-ONE see DME500

17-β-HYDROXY-5-β-ANDROSTAN-3-ONE see HJB100

3-α-HYDROXY-5-α-ANDROSTAN-17-ONE see HJB050

(5-α,17-β)-17-HYDROXY-ANDROSTAN-3-ONE (9CI) see DME500

17-β-HYDROXY-5-β,14-β-ANDROSTAN-3-ONE ACRYLATE see HJB150

4-HYDROXY-4-ANDROSTENE-3,17-DIONE see HJB200

4-HYDROXY-μ₄-ANDROSTENEDIONE see HJB200

7-β-HYDROXYANDROST-4-EN 3 ONE see TBF500

17-β-HYDROXYANDROST-4-EN-3-ONE see TBF500

17-β-HYDROXY-4-ANDROSTEN-3-ONE see TBF500

3-β-HYDROXY-5-ANDROSTEN-17-ONE see AOO450

17-HYDROXY-(17-β)-ANDROST-4-EN-3-ONE see TBF500

17-β-HYDROXY-Δ⁴-ANDROSTEN-3-ONE see TBF500

3-β-HYDROXYANDROSTEN-17-ONE ACETATE see HJB225

3-β-HYDROXYANDROST-5-EN-17-ONE ACETATE see HJB250

3-β-HYDROXYANDROST-5-EN-17-ONE ESTER with SODIUM SULFATE DIHYDRATE see DAL030

3-β-HYDROXY-5-ANDROSTEN-17-ONE SODIUM SULFATE DIHYDRATE see SJK410

2-HYDROXYANILINE see ALT000

3-HYDROXYANILINE see ALS990

4-HYDROXYANILINE see ALT250

o-HYDROXYANILINE see ALT000

p-HYDROXYANILINE see ALT250

4-(p-HYDROXYANILINO)BENZALDEHYDE see HJB260

2-HYDROXY-m-ANISALDEHYDE see VFP000

3-HYDROXY-p-ANISALDEHYDE see FNM000

4-HYDROXY-m-ANISALDEHYDE see VFK000

3-HYDROXYANISIC ACID see HJC000

2-HYDROXY-m-ANISIC ACID see HJB500

3-HYDROXY-m-ANISIC ACID see HJB500

4-HYDROXY-m-ANISIC ACID see VFF000

2-HYDROXYANISOLE see GKI000

3-HYDROXYANISOLE see REF050

m-HYDROXYANISOLE see REF050

o-HYDROXYANISOLE see GKI000

1-HYDROXY-9,10-ANTHRACENEDIONE see HJE000

1-HYDROXYANTHRACHINON (CZECH) see HJE000

3-HYDROXYANTHRANILIC ACID see AKE750

3-HYDROXYANTHRANILIC ACID METHYL ESTER see HJC500

3-(3-HYDROXYANTHRANILOYL)ALANINE see HJD000

3-(3-HYDROXYANTHRANILOYL)-l-ALANINE see HJD500

3-HYDROXY-ANTHRANILSAEURE (GERMAN) see AKE750

1-HYDROXYANTHRAQUINONE see HJE000

1-HYDROXY-9,10-ANTHRAQUINONE see HJE000

4-HYDROXY-1-ANTHRAQUINONYLAMINE see AKE250

N-(4-HYDROXY-1-ANTHRAQUINONYL)-4-METHYLANILINE see HOK000

N-(4-HYDROXY-1-ANTHRAQUINONYL)-p-TOLUIDINE see HOK000

HYDROXYAPATITE see HJE100

2-HYDROXY-p-ARSANILIC ACID see HJE400

4-HYDROXY-3-ARSANILIC ACID see HJE500

4-HYDROXY-m-ARSANILIC ACID see HJE500

trans-N-HYDROXY-ASS see SMT500

HYDROXYATHYLSTARKE (GERMAN) see HLB400

4A-HYDROXYAVERMECTIN B1 see HJE525

3-HYDROXY-8-AZAXANTHINE see HJE575

14-HYDROXYAZIDOMORPHINE see HJE600

4-HYDROXYAZOBENZENE see HJF000

p-HYDROXYAZOBENZENE see HJF000

4-HYDROXY-3,4'-AZODI-1-NAPHTHALENESULFONIC ACID, DISODIUM SALT see HJF500

4-HYDROXY-3,4'-AZODI-1-NAPHTHALENESULPHONIC ACID, DISODIUM SALT see HJF500

2-HYDROXY-1,1'-AZONAPHTHALENE-3,6,4'-TRISULFONIC ACID TRISODIUM SALT see FAG020

4'-HYDROXY-2,3'-AZOTOLUENE see HJG000

4-HYDROXY BAYGON see INE025

10-HYDROXYBENZ(j)ACEANTHRYLENE see BAW150

2-HYDROXYBENZALDEHYDE see SAG000

3-HYDROXYBENZALDEHYDE see FOE100

4-HYDROXYBENZALDEHYDE see FOF000

m-HYDROXYBENZALDEHYDE see FOE100

o-HYDROXYBENZALDEHYDE see SAG000

p-HYDROXYBENZALDEHYDE see FOF000

meta-HYDROXYBENZALDEHYDE see FOE100

o-HYDROXYBENZALDEHYDE OXIME see SAG500

4-HYDROXYBENZALDEHYDE THIOSEMICARBAZONE see HJG050

2-HYDROXYBENZAMIDE see SAH000

N-HYDROXYBENZAMIDE see BCL500

o-HYDROXYBENZAMIDE see SAH000

o-HYDROXYBENZANILIDE see SAH500

5-HYDROXYBENZ(a)ANTHRACENE see BBI000

3-HYDROXY-1,2-BENZANTHRACENE see BBI000

HYDROXYBENZENE see PDN750

4-HYDROXYBENZENEACETAMIDE see HNG550

4-HYDROXYBENZENEACETIC ACID see HNG600

α-HYDROXYBENZENEACETIC ACID 2-(2-ETHOXYETHOXY)ETHYL ESTER see HJG100

α-HYDROXY-BENZENEACETIC ACID 3,3,5-TRIMETHYLCYCLOHEXYL ESTER (9CI) see DNU100

p-HYDROXYBENZENEARSONIC ACID see PDO250

4-HYDROXYBENZENEDIAZONIUM-3-CARBOXYLATE see HJH000

2-HYDROXYBENZENEMETHANOL see HMK100

N-HYDROXYBENZENESULFONANILIDE see HJH100

HYDROXYBENZENESULFONIC ACID see HJH500

p-HYDROXY-BENZENESULFONIC ACID MERCURY DERIVATIVE, DISODIUM SALT see MCU500

p-HYDROXYBENZENESULFONIC ACID ZINC SALT see ZIJ300

o-HYDROXYBENZHYDRAZIDE see HJL100

2-HYDROXYBENZHYDROXAMIC ACID see SAL500

3-HYDROXYBENZISOTHIAZOL-S,S-DIOXIDE see BCE500

2-HYDROXY-1,3,2-BENZODIOXASTIBOLE see HJI000

2-HYDROXYBENZOHYDRAZIDE see HJL100

2-HYDROXYBENZOHYDROXAMIC ACID see SAL500

o-HYDROXYBENZOHYDROXAMIC ACID see SAL500

2-HYDROXYBENZOIC ACID see SAI000

3-HYDROXYBENZOIC ACID see HJI100

4-HYDROXYBENZOIC ACID see SAI500

m-HYDROXYBENZOIC ACID see HJI100

o-HYDROXYBENZOIC ACID see SAI000

p-HYDROXYBENZOIC ACID see SAI500

2-HYDROXYBENZOIC ACID BUTYL ESTER see BSL250

p-HYDROXYBENZOIC ACID BUTYL ESTER see BSC000

p-HYDROXYBENZOIC ACID BUTYL ESTER, SODIUM SALT see HJI500

p-HYDROXYBENZOIC ACID ETHYL ESTER see HJL000

p-HYDROXYBENZOIC ACID HEPTYL ESTER see HBO650

2-HYDROXYBENZOIC ACID HYDRAZIDE see HJL100

o-HYDROXYBENZOIC ACID HYDRAZIDE see HJL100

2-HYDROXYBENZOIC ACID METHYL ESTER see MPI000

o-HYDROXYBENZOIC ACID, METHYL ESTER see MPI000

p-HYDROXYBENZOIC ACID METHYL ESTER see HJL500

2-HYDROXYBENZOIC ACID MONOAMMONIUM SALT see ANT600

2-HYDROXYBENZOIC ACID MONOSODIUM SALT see SJO000

4-HYDROXYBENZOIC ACID ((5-NITRO-2-FURANYL)METHYLENE)HYDRAZIDE see DGQ500

p-HYDROXYBENZOIC ACID (5-NITROFURFURYLIDENE)HYDRAZIDE see DGQ500

4-HYDROXYBENZOIC ACID PROPYL ESTER see HNU500

p-HYDROXYBENZOIC ACID PROPYL ESTER see HNU500

p-HYDROXYBENZOIC ACID, PROPYL ESTER, SODIUM DERIVATIVE see PNO250

p-HYDROXYBENZOIC ACID, SODIUM SALT see HJM000

2-HYDROXYBENZOIC ACID STRONTIUM SALT (2:1) see SML500

p-HYDROXYBENZOIC ETHYL ESTER see HJL000

o-HYDROXYBENZOIC SODIUM SALT see SJO000

2-HYDROXYBENZOIC-5-SULFONIC ACID see SOC500

4-HYDROXYBENZONITRILE see HJN000

p-HYDROXYBENZONITRILE see HJN000

4-HYDROXYBENZOPHENONE see HJN100

p-HYDROXYBENZOPHENONE see HJN100

4'-HYDROXYBENZOPHENONE see HJN100

4-HYDROXY-2H-1-BENZOPYRAN-2-ONE see HJY000

1-HYDROXYBENZO(a)PYRENE see BCW750

2-HYDROXYBENZO(a)PYRENE see BCX000

3-HYDROXYBENZO(a)PYRENE see BCX250

4-HYDROXYBENZO(a)PYRENE see BCX500

5-HYDROXYBENZO(a)PYRENE see BCX500

6-HYDROXYBENZO(a)PYRENE see BCX750

7-HYDROXYBENZO(a)PYRENE see BCY000

9-HYDROXYBENZO(a)PYRENE see BCY250

10-HYDROXYBENZO(a)PYRENE see BCY500

11-HYDROXYBENZO(a)PYRENE see BCY750

12-HYDROXYBENZO(a)PYRENE see BCZ000

HYDROXYBENZOPYRIDINE see QPA000

4-HYDROXYBENZOTHIOPHENE see HJN550

4-HYDROXYBENZO(B)THIOPHENE see HJN550

1-HYDROXY-1H-BENZOTRIAZOLE AMMONIUM SALT see HJN600

1-HYDROXYBENZOTRIAZOLE HYDRATE see HJN650

N-HYDROXYBENZOTRIAZOLE HYDRATE see HJN650

1-HYDROXY-1H-BENZOTRIAZOLE HYDRATE see HJN650

2-HYDROXYBENZOXAZOLE see BDJ000
N-HYDROXY-2-BENZOYLAMINOFLUORENE see FDZ000
HYDROXY-4 BENZOYL-2-FURANNE see FQL200
N-o-HYDROXYBENZOYLGLYCINE see SAN200
N-(2-HYDROXYBENZOYL)-GLYCINE (9CI) see SAN200
2-HYDROXYBENZOYLHYDRAZIDE see HJL100
o-HYDROXYBENZOYLHYDRAZIDE see HJL100
2-HYDROXYBENZOYLHYDRAZINE see HJL100
o-HYDROXYBENZOYLHYDRAZINE see HJL100
(HYDROXY-4 BENZOYL)-3 MESITYL-2 BENZOFURANNE see HNK700
(HYDROXY-4 BENZOYL)-4 MESITYL-2 BENZOFURANNE see HNK800
2-(p-HYDROXYBENZOYL)PYRIDINE see HJN700
4-HYDROXYBENZ(a)PYRENE see HJN500
8-HYDROXY-3,4-BENZPYRENE see BCX250
p-HYDROXYBENZYL ACETONE see RBU000
2-HYDROXYBENZYL ALCOHOL see HMK100
o-HYDROXYBENZYL ALCOHOL see HMK100
5-(α-HYDROXYBENZYL)-2-BENZIMIDAZOLECARBAMIC ACID METHYL ESTER see HJN875
5-((p-HYDROXYBENZYLIDENE)AMINO)-3-METHYLISOTHIAZOLO(5,4-d)PYRIMIDINE-4,6(5H,7H)-DIONE see IGE100
5-((4'-HYDROXYBENZYLIDENOIMINO)-3-METHYLISOTHIAZOLO(5,4-d)PYRIMIDINE-(7H)-4,6)-DIONE see IGE100
α-HYDROXYBENZYL PHENYL KETONE see BCP250
(p-HYDROXYBENZYL)TARTARIC ACID see HJO500
2-HYDROXYBIFENYL (CZECH) see BGJ250
2-HYDROXYBIPHENYL see BGJ250
4-HYDROXYBIPHENYL see BGJ500
o-HYDROXYBIPHENYL see BGJ250
p-HYDROXYBIPHENYL see BGJ500
N-HYDROXY-N-4-BIPHENYLACETAMIDE see ACD000
N-HYDROXY-(1,1'-BIPHENYL)-2-AMINE see HIT600
N-HYDROXY-4-BIPHENYLBENZAMIDE see HJP500
2-HYDROXYBIPHENYL SODIUM SALT see BGJ750
N-HYDROXY-4-BIPHENYLYLBENZAMIDE see HJP500
HYDROXY-4-BIPHENYLYLBENZENESULFONAMIDE see BGN000
HYDROXYBIS(2-(p-CHLOROPHENOXY)ISOBUTYRIC ACID) ALUMINUM see AHA150
12'-HYDROXY-2',5'-α-BIS(1-METHYLETHYL)ERGOTAMAN-3',6',18-TRIONE see EDA600
4-HYDROXY-6,7-BIS(2-METHYLPROPOXY)-3-QUINOLINECARBOXYLIC ACID ETHER ESTER see BOO632
2-HYDROXY-4,6-BIS(NITROAMINO)-1,3,5-TRIAZINE see HJP575
1-((1R)-2-endo-HYDROXY-3-endo-BORNYL)-3-(p-TOLYLSULFONYL)UREA see GFY100
3-HYDROXYBUTANAL see AAH750
1-HYDROXYBUTANE see BPW500
2-HYDROXYBUTANE see BPW750
HYDROXYBUTANEDIOIC ACID see MAN000

4-HYDROXYBUTANOIC ACID LACTONE see BOV000
3-HYDROXYBUTANOIC ACID-β-LACTONE see BSX000
4-HYDROXYBUTANOIC ACID MONOLITHIUM SALT see LHM800
3-HYDROXY-2-BUTANONE see ABB500
2-HYDROXY-3-BUTENENITRILE see HJQ000
1-(2-HYDROXYBUT-1-ENYL)AZIRIDINE see VMA000
o-(2-HYDROXY-3-(tert-BUTYLAMINO)PROPOXY)BENZONITRILE HYDROCHLORIDE see BON400
1-HYDROXY-4-tert-BUTYLBENZENE see BSE500
4-HYDROXYBUTYLBUTYLNITROSAMINE see HJQ350
1-HYDROXY-3-BUTYL HYDROPEROXIDE see HJQ500
3-HYDROXYBUTYL-(2-HYDROXYPROPYL)-N-NITROSAMINE see HJR000
4-HYDROXYBUTYL-(2-HYDROXYPROPYL)-N-NITROSAMINE see HJR500
4-HYDROXYBUTYL-(3-HYDROXYPROPYL)-N-NITROSAMINE see HJS000
4-((4-HYDROXYBUTYL)NITROSAMINO)BUTYRIC ACID METHYL ESTER ACETATE (ESTER) see MEI000
N-(4-HYDROXYBUTYL)-N-NITROSO-β-ALANINE, METHYL ESTER, ACETATE see MEG750
4-HYDROXYBUTYL(2-PROPENYL)NITROSAMINE see HJS400
3-HYDROXYBUTYRALDEHYDE see AAH750
β-HYDROXYBUTYRALDEHYDE see AAH750
3'-HYDROXYBUTYRANILIDE see HJS450
γ-HYDROXYBUTYRATE SODIUM SALT see HJS500
γ-HYDROXYBUTYRIC ACID CYCLIC ESTER see BOV000
3-HYDROXYBUTYRIC ACID-2-HEPTYL ESTER see MKM750
HYDROXYBUTYRIC ACID LACTONE see BSX000
3-HYDROXYBUTYRIC ACID LACTONE see BSX000
4-HYDROXYBUTYRIC ACID γ-LACTONE see BOV000
β-HYDROXYBUTYRIC ACID-p-PHENETIDIDE see HJS850
4-HYDROXYBUTYRIC ACID SODIUM SALT see HJS500
γ-HYDROXYBUTYROLACTONE see BOV000
3-HYDROXY-p-BUTYROPHENETIDIDE see HJS850
(24R)-HYDROXYCALCIDIOL see HJS900
2-HYDROXYCAMPHANE see BMD000
HYDROXYCARBAMIC ACID ETHYL ESTER see HKQ025
HYDROXYCARBAMINE see HOO500
2-HYDROXYCARBAMOYL-1-METHYLPYRIDINIUM IODIDE see HJS910
3-HYDROXYCARBAMOYL-1-METHYLPYRIDINIUM IODIDE see HJS920
4-HYDROXYCARBAMOYL-1-METHYLPYRIDINIUM IODIDE see HJS930
m-HYDROXYCARBANILIC ACID ETHYL ESTER see ELE600
m-HYDROXYCARBANILIC ACID METHYL ESTER m-METHYLCARBANILATE see MEG250
3-HYDROXY-4-CARBOXYANILINE see AMM250
cis-4-HYDROXY-CCNU see CHA500

trans-4-HYDROXY-CCNU see CHB000
HYDROXYCELLULOSE see CCU150
8-HYDROXY-CHINOLIN (GERMAN) see QPA000
8-HYDROXY-CHINOLIN-SULFAT (GERMAN) see QPS000
2-HYDROXY-5-CHLORO-N-(2-CHLORO-4-NITROPHENYL)BENZAMIDE see DFV400
4-(4-HYDROXY-4'-CHLORO-4-PHENYLPIPERIDINO)-4'-FLUOROBUTYROPHENONE see CLY500
2-(3-(4-HYDROXY-4-p-CHLOROPHENYLPIPERIDINO)-PROPYL)-3-METHYL-7-FLUOROCHROMONE see HJU000
HYDROXYCHLOROQUINE see PJB750
3-α-HYDROXYCHOLANIC ACID see LHW000
3-α-HYDROXY-5-β-CHOLANIC ACID see LHW000
(3-α,5-β)-3-HYDROXY-CHOLAN-24-OIC ACID see LHW000
3-α-HYDROXY-5-β-CHOLAN-24-OIC ACID, MONOSODIUM SALT see SIC500
(3-α,5-β)-3-HYDROXYCHOLAN-24-OIC ACID, MONOSODIUM SALT see SIC500
HYDROXYCHOLECALCIFEROL see HJV000
1-HYDROXYCHOLECALCIFEROL see HJV000
1-α-HYDROXYCHOLECALCIFEROL see HJV000
3-β-HYDROXYCHOLESTANE see DKW000
3-β-HYDROXYCHOLEST-5-ENE see CMD750
6-HYDROXYCHOLEST-4-EN-3-ONE see HJV500
3-β-HYDROXYCHOLEST-5-EN-7-ONE (8CI) see ONO000
(3-β)3-HYDROXYCHOLEST-5-EN-7-ONE (9CI) see ONO000
3-HYDROXYCINCHOPEN see OPK300
HYDROXYCINCHOPHENE see OPK300
HYDROXYCINE see CJR909
4-HYDROXYCINNAMIC ACID see CNU825
p-HYDROXYCINNAMIC ACID see CNU825
4'-HYDROXYCINNAMIC ACID see CNU825
trans-2-HYDROXYCINNAMIC ACID see CNU850
trans-o-HYDROXYCINNAMIC ACID see CNU850
o-HYDROXYCINNAMIC ACID LACTONE see CNV000
7-HYDROXYCITRONELLAL see CMS850
HYDROXYCITRONELLAL DIMETHYL ACETAL see HJV700
HYDROXYCITRONELLAL DIMETHYL ACETAL see LBO050
HYDROXYCITRONELLAL DMA see LBO050
HYDROXYCITRONELLAL (FCC) see CMS850
HYDROXYCITRONELLAL-INDOLE (SCHIFF BASE) see ICS100
HYDROXYCITRONELLA METHYL ETHER see DSM800
HYDROXYCITRONELLOL see DTE400
HYDROXYCITRONELLYLIDENE-INDOLE see ICS100
HYDROXY CNU METHANESULPHONATE see HKQ300
HYDROXYCODEINONE see HJX500
14-HYDROXYCODEINONE see HJX500
14-β-HYDROXYCODEINONE see HJX500
HYDROXYCOPPER(II) GLYOXIMATE see HJX625
11-HYDROXYCORTICOALDOSTERONE see CNS625
17-HYDROXYCORTICOSTERONE see CNS750
17-HYDROXY-CORTICOSTERONE 21-ACETATE see HHQ800
17-α-HYDROXYCORTICOSTERONE ACETATE see HHQ800

11-β-HYDROXYCORTISONE see CNS750

trans-3'-HYDROXYCOTININE see HJX700

4-HYDROXYCOUMARIN see HJY000

7-HYDROXYCOUMARIN see HJY100

(1IYDROXY-4 COUMARINYL 3)-3 PHENYL-3 (BROMO-4 BIPHENYLYL-4)-1 PROPANOL-1 (FRENCH) see BMN000

(E)-3-HYDROXY-CROTONIC ACID α-METHYLBENZYL ESTER, DIMETHYL PHOSPHATE see COD000

3-HYDROXYCROTONIC ACID METHYL ESTER DIMETHYL PHOSPHATE see MQR750

1-HYDROXYCUMENE see DTN100

(4-HYDROXY-o-CUMENYL)TRIMETHYLAMMONIUM CHLORIDE, METHYLCARBAMATE see HJY500

(4-HYDROXY-o-CUMENYL)TRIMETHYLAMMONIUM IODIDE, DIMETHYLCARBAMATE see HJZ000

1-HYDROXY-CYCLOBUT-1-ENE-3,4-DIONE see MRE250

3-HYDROXY-CYCLOBUT-3-ENE-1,2-DIONE see MRE250

1-HYDROXYCYCLOHEPTANECARBONITRILE see HKA000

3-HYDROXYCYCLOHEXADIEN-1-ONE see REA000

HYDROXYCYCLOHEXANE see CPB750

1-HYDROXY-CYCLOHEXANECARBONITRILE see HKA000

2-HYDROXYCYCLOHEXANE-1,1,3,3-TETRAMETHANOL TETRAESTER with NICOTINIC ACID see CME675

N-HYDROXYCYCLOHEXYLAMINE see HKA109

N-(trans-4-HYDROXYCYCLOHEXYL)-(2-AMINO-3,5-DIBROMOBENZYL)-AMINE see AHJ500

α-(1-HYDROXYCYCLOHEXYL)BUTYRIC ACID see COW700

trans-1-(4-HYDROXYCYCLOHEXYL)-4-(4-FLUOROPHENYL)-5-(2-METHOXYPYRIMIDIN-4-YL)IMIDAZOLE see HKA123

1-HYDROXYCYCLOPENTYL CYCLOHEXANE CARBOXYLIC ACID ESTER see HKA130

4-HYDROXYCYCLOPHOSPHAMIDE see HKA200

(−)-3-HYDROXY-N-CYCLOPROPYLMETHYLMORPHINAN see CQG750

2-HYDROXY-p-CYMENE see CCM000

3-HYDROXY-p-CYMENE see TFX810

5-HYDROXYCYTIDINE see HKA250

4-N-HYDROXYCYTIDINE see HKA270

N-4-HYDROXYCYTIDINE see HKA270

14-HYDROXYDAUNOMYCIN see AES750

14'-HYDROXYDAUNOMYCIN see AES750

14-HYDROXYDAUNORUBICINE see AES750

HYDROXYDAUNORUBICIN HYDROCHLORIDE see HKA300

2-HYDROXYDECALIN see DAF000

HYDROXYDECANOIC ACID-γ-LACTONE see HKA500

17-HYDROXY-11-DEHYDROCORTICOSTERONE see CNS800

17-α-HYDROXY-11-DEHYDROCORTICOSTERONE see CNS800

1'-HYDROXY-2',3'-DEHYDROESTRAGOLE see HKA700

2-HYDROXYDEOXYADENOSINE see HKA750

5-HYDROXYDEOXYCYTIDINE see HKA760

8-HYDROXYDEOXYGUANOSINE see HKA770

5-HYDROXY-α-6-DEOXYTETRACYCLINE see DYE425

HYDROXYDE de POTASSIUM (FRENCH) see PLJ500

HYDROXYDE de SODIUM (FRENCH) see SHS000

HYDROXYDE de TETRAMETHYLAMMONIUM (FRENCH) see TDK500

HYDROXYDE de TRIPHENYL-ETAIN (FRENCH) see HON000

3-HYDROXY-4,15-DIACETOXY-8-(3-METHYLBUTYRYLOXY)-12,13-EPOXY-Δ⁹-TRICHOTHECENE see FQS000

N-HYDROXY-N,N'-DIACETYLBENZIDINE see HKB000

HYDROXYDIAZEPAM see CFY750

3-HYDROXYDIAZEPAM see CFY750

5-HYDROXY-DIBENZ(a,h)ANTHRACENE see DCU400

4-HYDROXY-3,5-DIBROMOBENZONITRILE see DDP000

4-HYDROXY-3,5-DI-tert-BUTYLTOLUENE see BFW750

2-HYDROXY-3-(3,3-DICHLOROALLYL)-1,4-NAPHTHOQUINONE see DFN500

2-HYDROXY-3,5-DICHLOROPHENYL SULPHIDE see TFD250

HYDROXYDICHLOROQUINALDINE see CLC500

o-HYDROXY-N,N-DIETHYLBENZAMIDE see DJY400

β-HYDROXYDIETHYL SULFIDE see EPP500

HYDROXY(3-(5,5-DIETHYL-2,4,6-TRIOXO-(1H,3H,5H)-PYRIMIDINO)-2-ETHOXYPROPYL)MERCURY see EET500

HYDROXY(3-(5,5-DIETHYL-2,4,6-TRIOXO-(1H,3H,5H)PYRIMIDINO)-2-ISOPROPOXYPROPYL)MERCURY see INE000

HYDROXY(3-(5,5-DIETHYL-2,4,6-TRIOXO-(1H,3H,5H)-PYRIMIDINO)-2-METHOXYPROPYL)MERCURY see MES250

p-HYDROXYDIFENYLAMIN (CZECH) see AOT000

2-(HYDROXY)-5-(2,4-DIFLUOROPHENYL)BENZOIC ACID see DKI600

14-HYDROXYDIHYDROCODEINONE see PCG500

14-HYDROXYDIHYDROCODEINONE HYDROCHLORIDE see DLX400

HYDROXYDIHYDROCYCLOPENTADIENE see HKB200

11-HYDROXY-15,16-DIHYDROCYCLOPENTA(a)PHENANTHRACEN-17-ONE ACETATE (ESTER) see ABN250

1-HYDROXY-13-DIHYDRODAUNOMYCIN see DAC300

5-HYDROXYDIHYDROSTERIGMATOCYSTIN see HKB300

14-HYDROXYDIHYDRO-6-β-THEBAINOL 4-METHYL ETHER see ORE000

4-HYDROXY-3,5-DIIODOBENZONITRILE see HKB500

3-(4-(4-HYDROXY-3,5-DIIODOPHENOXY)-3,5-DIIODOPHENYL)ALANINE see TFZ300

o-(4-HYDROXY-3,5-DIIODOPHENYL)-3,5-DIIODO-TYROSINE see TFZ300

o-(4-HYDROXY-3,5-DIIODOPHENYL)-3,5-DIIODO-d-TYROSINE MONOSODIUM SALT see SKJ300

o-(4-HYDROXY-3,5-DIIODOPHENYL)-3,5-DIIODO-l-TYROSINE SODIUM SALT PENTAHYDRATE see SKJ325

β-(4-HYDROXY-3,5-DIIODOPHENYL)-α-PHENYLPROPIONIC ACID see PDM750

8-HYDROXY-5,7-DIIODOQUINOLINE see DNF600

4-HYDROXY-6,7-DIISOBUTOXY-3-QUINOLINECARBOXYLIC ACID ETHYL ESTER see BOO632

4-HYDROXY-3,5-DIMETHOXYBENZOIC ACID see SPE700

4-HYDROXY-3,5-DIMETHOXY-BENZOIC ACID ETHYL CARBONATE, ester with METHYL RESERPATE see RCA200

7-(4-HYDROXY-3,5-DIMETHOXYCINNAMOYLAMINO)-3-HEXYLOXY-4-HYDROXY-1-METHYL-2(1H)-QUINOLINONE see HKB550

β-HYDROXY-β-(2,5-DIMETHOXYPHENYL)-ISOPROPYLAMINE HYDROCHLORIDE see MDW000

18-α-HYDROXY-11, 17-α-DIMETHOXY-3-β, 20-α-YOHIMBAN-16-β-CARBOXYLIC ACID, METHYL ESTER, 4-HYDROXY-3,5-DIMETHOXYBENZOATE (ESTER), ETHYLCARBONATE (ESTER) see HKB600

α-HYDROXY-α,α,α-DIMETHYLACETOPHENONE see HMQ100

N-HYDROXY-3,2'-DIMETHYL-4-AMINOBIPHENYL see HKB650

17-β-HYDROXY-11-β-(4-DIMETHYLAMINOPHENYL-1)-17-α-(PROP-1-YNYL) OESTRA-4,9-DIEN-3-ONE see HKB700

HYDROXYDIMETHYLARSINE OXIDE see HKC000

HYDROXYDIMETHYLARSINE OXIDE, SODIUM SALT see HKC500

HYDROXYDIMETHYLARSINE OXIDE, SODIUM SALT see HKC500

HYDROXYDIMETHYLARSINE OXIDE, SODIUM SALT TRIHYDRATE see HKC550

1-HYDROXY-2,4-DIMETHYLBENZENE see XKJ500

4-HYDROXY-N-DIMETHYLBUTYRAMIDE-4-CHLOROPHENOXY-ISOBUTYRATE see LGK200

3-HYDROXYDIMETHYL CROTONAMIDE DIMETHYL PHOSPHATE see DGQ875

3-HYDROXY-N,N-DIMETHYL-cis-CROTONAMIDE DIMETHYL PHOSPHATE see DGQ875

3-HYDROXY-5,5-DIMETHYL-2-CYCLOHEXEN-1-ONE DIMETHYLCARBAMATE see DRL200

2'-HYDROXY-5,9-DIMETHYL-2-(3,3-DIMETHYLALLYL)-6,7-BENZOMORPHAN see DOQ400

dl-2'-HYDROXY-5,9-DIMETHYL-2-(3,3-DIMETHYLALLYL)-6,7-BENZOMORPHAN see DOQ400

17-β-HYDROXY-7-α,17-DIMETHYLESTR-4-EN-3-ONE see MQS225

(7-α,17-β)-17-HYDROXY-7,17-DIMETHYL-ESTR-4-EN-3-ONE (9CI) see MQS225

4-HYDROXY-2,5-DIMETHYL-3(2H)FURANONE see HKC575

7-HYDROXY-3,7-DIMETHYL OCTANAL see CMS850

7-HYDROXY-3,7-DIMETHYLOCTAN-1-AL see CMS850

7-HYDROXY-3,7-DIMETHYL OCTANAL:ACETAL see HJV700

6-HYDROXY-3,7-DIMETHYLOCTANOIC ACID LACTONE see HKC600

7-HYDROXY-3,7-DIMETHYLOCTAN-1-OL see DTE400

3-HYDROXY-4,5-DIMETHYLOL-α-PICOLINE see PPK250

3-HYDROXY-4,5-DIMETHYLOL-α-PICOLINE HYDROCHLORIDE see PPK500

3-HYDROXY-4,5-DIMETHYLOL-α-PICOLINE TRIS(HEXADECANOATE) see HMQ575

p-HYDROXY-N,α-DIMETHYLPHENETHYLAMINE see FMS875

2'-HYDROXY-5,9-DIMETHYL-2-PHENETHYL-6,7-BENZOMORPHAN HYDROBROMIDE see PMD325

(+−)-2'-HYDROXY-5,9-DIMETHYL-2-PHENETHYL-6,7-BENZOMORPHAN HYDROBROMIDE see HKC625

N-(3-(2-HYDROXY-4,5-DIMETHYLPHENYL)ADAMANT-1-YLMETHYL)ACETAMIDINOTHIOSULFURIC ACID see HKC700

3-HYDROXY-1,1-DIMETHYLPIPERIDINIUM BROMIDE BENZILATE see CBF000

5-HYDROXY-4,6-DIMETHYL-3-PYRIDINEMETHANOL HYDROCHLORIDE see DAY825

4-HYDROXY-3,5-DIMETHYL-1,2,4-TRIAZOLE see HKD550

4-HYDROXY-N,N-DIMETHYLTRYPTAMINE see HKE000

5-HYDROXY-N,N-DIMETHYLTRYPTAMINE see DPG109

1-HYDROXY-2,4-DINITROBENZENE see DUZ000

4-HYDROXY-3,5-DINITROBENZENEARSONIC ACID see HKE500

2-HYDROXY-3,5-DINITROBENZOIC ACID see HKE600

8-HYDROXY-5,7-DINITRO-2-NAPHTHALENESULFONIC ACID see FBZ100

β-(2-HYDROXY-3,5-DINITROPHENYL)BUTANE ACETATE see ACE500

2-HYDROXY-3,5-DINITROPYRIDINE see HKE700

HYDROXYDIONE see VJZ000

HYDROXYDIONE SODIUM see VJZ000

HYDROXYDIONE SUCCINATE see VJZ000

2-HYDROXYDIPHENYL see BGJ250

4-HYDROXYDIPHENYL see BGJ500

o-HYDROXYDIPHENYL see BGJ250

p-HYDROXYDIPHENYL see BGJ500

HYDROXYDIPHENYLACETIC ACID see BBY990

3-((HYDROXYDIPHENYLACETYL)OXY)-1,1-DIMETHYLPIPERIDINIUM BROMIDE see CBF000

2-(((HYDROXYDIPHENYLACETYL)OXY)METHYL)-1,1-DIMETHYLPIPERIDINIUM METHYL SULFATE (SALT) see CDG250

4-((HYDROXYDIPHENYLACETYL)OXY)-1,1,2,2,6-PENTAMETHYLPIPERIDINIUM CHLORIDE (α FORM) see BCB250

4-((HYDROXYDIPHENYLACETYL)OXY)-1,1,2,2,6-PENTAMETHYLPIPERIDINIUM CHLORIDE (β FORM) see BCB000

4-HYDROXYDIPHENYLAMINE see AOT000

p-HYDROXYDIPHENYLAMINE see AOT000

p-HYDROXYDIPHENYLAMINE ISOPROPYL ETHER see HKF000

4-HYDROXYDIPHENYLDIMETHYLMETHANE see COF400

4,4'-HYDROXY-γ,Δ-DIPHENYL-β,Δ-HEXADIENE see DAL600

HYDROXYDIPHENYLMETHANE see HKF300

α-HYDROXYDIPHENYLMETHANE-β-DIMETHYLAMINOETHYL ETHER HYDROCHLORIDE see BAU750

2-HYDROXYDIPHENYL SODIUM see BGJ750

2-HYDROXYDIPHENYL, SODIUM SALT see BGJ750

1-HYDROXY-1,1-DIPHOSPHONOETHANE see HKS780

7-HYDROXY-2-(DIPROPYLAMINO)TETRALINE see HKF350

8-HYDROXY-2-(DI-N-PROPYLAMINO)TETRALINE see HKF370

N-HYDROXYDITHIOCARBAMIC ACID see HKF600

5-HYDROXYDODECANOIC ACID LACTONE see HBP450

5-HYDROXYDODECANOIC ACID Δ-LACTONE see HBP450

6-HYDROXYDOPAMINE see HKF875

6-HYDROXYDOPAMINE HYDROCHLORIDE see HKG000

N-HYDROXY-EAB see HKY000

20-HYDROXYECDYSONE see HKG500

5-HYDROXYEICOSATETRAENOIC ACID see HKG600

12-HYDROXYEICOSATETRAENOIC ACID see HKG650

15-HYDROXYEICOSATETRAENOIC ACID see HKG680

5-HYDROXY-6,8,11,14-EICOSATETRAENOIC ACID see HKG600

12-HYDROXY-5,8,10,14-EICOSATETRAENOIC ACID see HKG650

9-HYDROXYELLIPTICIN see HKH000

9-HYDROXYELLIPTICINE see HKH000

HYDROXY-9 ELLIPTICINE (FRENCH) see HKH000

1-α-HYDROXYENDOSULFAN ETHER see HDK210

p-HYDROXYEPHEDRINE see HKH500

3-β-HYDROXY-14,15-β-EPOXY-5-β-BUFA-20,22-DIENOLIDE see BOM650

3-HYDROXY-1,2-EPOXYPROPANE see GGW500

1-HYDROXYESTRADIOL see HKH550

2-HYDROXYESTRADIOL see HKH600

4-HYDROXYESTRADIOL see HKH850

16-α-HYDROXYESTRADIOL see EDU500

4-HYDROXY-17-β-ESTRADIOL see HKH850

2-HYDROXYESTRADIOL, 2-METHYL ETHER see MEL785

17-β-HYDROXY-ESTRA-4-EN-3-ONE-17-PHENYLPROPIONATE see DYF450

1'-HYDROXYESTRAGOLE see HKI000

1'-HYDROXY-ESTRAGOLE-2',3'-OXIDE see HKI075

3-HYDROXYESTRA-1,3,5(10),6,8-PENATEN-17-ONE see ECV000

3-HYDROXYESTRA-1,3,5,7,9-PENTAEN-17-ONE BENZOATE see ECV500

3-HYDROXYESTRA-1,3,5(10),7-TETRAEN-17-ONE see ECW000

3-HYDROXYESTRA-1,3,5(10),7-TETRAEN-17-ONE BENZOATE see ECW500

3-HYDROXYESTRA-1,3,5(10)-TRIEN-17-ONE see EDV000

17β-HYDROXYESTRA-4,9,11-TRIEN-3-ONE see THL600

17-β-HYDROXYESTRA-4,9,11-TRIEN-3-ONE ACETATE see HKI100

3-HYDROXYESTRA-1,3,5(10)-TRIEN-17-ONE BENZOATE see EDV500

17-β-HYDROXY-ESTR-4-ENE-3-ONE-3-PHENYLPROPIONATE see DYF450

(17-β)-17-HYDROXY-ESTR-4-EN-3-ONE (9CI) see NNX400

17-β-HYDROXYESTR-4-EN-3-ONE DECANOATE see NNE550

HYDROXYESTRIN BENZOATE see EDP000

2-HYDROXYESTRONE see HKI200

4-HYDROXYESTRONE see HKI300

1-HYDROXYETHANECARBOXYLIC ACID see LAG000

HYDROXYETHANEDIPHOSPHONIC ACID see HKS780

1-HYDROXYETHANEDIPHOSPHONIC ACID see HKS780

2-HYDROXYETHANESULFONIC ACID see HKI500

2-HYDROXYETHANESULFONIC ACID AMMONIUM SALT see ANL100

2-HYDROXY-1-ETHANETHIOL see MCN250

HYDROXYETHANOIC ACID see GHO000

HYDROXY ETHER see EES350

17-β-HYDROXY-17-α-ETHINYL-5(10)-ESTREN-3-ONE see EEH550

4-HYDROXY-3-ETHOXYBENZALDEHYDE see EQF000

2-(2-HYDROXYETHOXY)ETHYL CHLOROACETATE see HKI600

2-(2-HYDROXYETHOXY)ETHYL ESTER STEARIC ACID see HKJ000

2-(2-HYDROXYETHOXY)ETHYL PERCHLORATE see HKJ500

10-(3-(4-HYDROXYETHOXYETHYL-1-PIPERAZINYL)-2-METHYLPROPYL)PHENOTHIAZINE see MKQ000

2-(2-HYDROXYETHOXY)ETHYL-N-(α,α,α-TRIFLUORO-m-TOLYL)ANTHRANILATE see HKK000

(2-HYDROXYETHOXY)METHOXYMETHANE see MET000

2-((2-(2-HYDROXYETHOXY)-4-NITROPHENYL)AMINO)ETHANOL see HKK100

HYDROXYETHYL ACETAMIDE see HKM000

N-(2-HYDROXYETHYL)ACETAMIDE see HKM000

β-HYDROXYETHYLACETAMIDE see HKM000

N-β-HYDROXYETHYLACETAMIDE see HKM000

2-HYDROXYETHYL ACETATE see EJI000

2-HYDROXYETHYL ACETOACETATE ACRYLATE see AAY750

1-(2-HYDROXYETHYL)-4-(3-(2-ACETYL-10-PHENOTHIAZYL)PROPYL)PIPERAZINE see ABG000

HYDROXYETHYL ACRYLATE see ADU500

HYDROXYETHYL ACRYLATE see ADV250

2-HYDROXYETHYL ACRYLATE see ADV250

β-HYDROXYETHYL ACRYLATE see ADV250

2-(1-HYDROXYETHYL)ACRYLONITRILE see HKY650

N-HYDROXYETHYLAMID KYSELINY 4-AMINOTOLUEN-2-SULFONOVE (CZECH) see AKG000

2-HYDROXYETHYLAMINE see EEC600

β-HYDROXYETHYLAMINE see EEC600

2-HYDROXYETHYLAMINIUM PERCHLORATE see HKM175

N-HYDROXY-N-ETHYL-4-AMINOAZOBENZENE see HKY000

2-HYDROXYETHYLAMINODIACETIC ACID see HKM500

7-(3-(N-(2-HYDROXYETHYL)AMINO)-2-HYDROXYPROPYL)THIOPHYLLINE NICOTINATE see XCS000

4-HYDROXYETHYLAMINO-1-METHYLAMINOANTHRAQUINONE see MGG250

2-(β-HYDROXYETHYLAMINOMETHYL)-1,4-BENZODIOXANE HYDROCHLORIDE see HKN500

4-(2-HYDROXYETHYL)AMINO-3-NITROANILINE see ALO750

2,2'-((4-((2-HYDROXYETHYL)AMINO-3-NITROPHENYL)IMINO))BISETHANOL see HKN875

2,2'-((4-((2-HYDROXYETHYL)AMINO)-3-NITROPHENYL)IMINO)DIETHANOL see HKN875

4-(2-HYDROXYETHYLAMINO)-2-(5-NITRO-2-THIENYL)QUINAZOLINE see HKO000

N-(2-HYDROXYETHYL)AMMONIUM BENZOTHIAZOLE-2-THIOLATE see HKO012

N-(2-HYDROXYETHYL)ANACHIDONAMIDE see HKR600

2-HYDROXY-1-ETHYLAZIRIDINE see ASI000

N-(2-HYDROXYETHYL)AZIRIDINE see ASI000

β-HYDROXY-1-ETHYLAZIRIDINE see ASI000

N-(β-HYDROXYETHYL)AZIRIDINE see ASI000

10-β-HYDROXYETHYL-1:2-BENZANTHRACENE see BBG500

β-HYDROXYETHYLBENZENE see PDD750

2-HYDROXYETHYLCARBAMATE see HKQ000

N-HYDROXY ETHYL CARBAMATE see HKQ025

β-HYDROXYETHYLCARBAMATE see HKQ000

(2-HYDROXYETHYL)CARBAMIC ACID,γ-LACTONE see OMM000

1-(((2-HYDROXYETHYL)CARBAMOYL)METHYL)PYRIDINIUM CHLORIDE LAURATE (ESTER) see LBD100

HYDROXYETHYL CELLULOSE see HKQ100

2-HYDROXYETHYL CELLULOSE see HKQ100

HYDROXYETHYL CELLULOSE ETHER see HKQ100

2-HYDROXYETHYL CELLULOSE ETHER see HKQ100

1-(2-HYDROXYETHYL)-3-(2-CHLOROETHYL)-3-NITROSOUREA see CHB750

2-HYDROXYETHYL-2-CHLOROETHYL SULFIDE see CHC000

1-(2-HYDROXYETHYL)-4-(3-(2-CHLORO-10-PHENOTHIAZINYL)PROPYL)PIPERAZINE see CJM250

HYDROXYETHYL CNU see CHB750

HYDROXYETHYL CNU METHANESULFONATE see HKQ300

2-HYDROXYETHYL CYCLOHEXANECARBOXYLATE see HKQ500

1-HYDROXY-α-ETHYLCYCLOHEXYLACETIC ACID see COW700

N-(2-HYDROXYETHYL)CYCLOHEXYLAMINE see CPG125

2-HYDROXYETHYL 2,3-DIBROMOPROPANOATE see HKQ600

2-HYDROXYETHYL-2,3-DIBROMOPROPIONATE see HKQ600

2-HYDROXYETHYL DIBUTYL PHOSPHATE see HKQ700

N-(HYDROXYETHYL)DIETHYLENETRIAMINE see HKR000

N-(2-HYDROXYETHYL)DIETHYLENETRIAMINE see HKR000

(2-HYDROXYETHYL)DIISOPROPYLMETHYL AMMONIUMBROMIDE XANTHENE-9-CARBOXYLATE see HKR500

β-HYDROXYETHYLDIMETHYLAMINE see DOY800

N-2-HYDROXYETHYL-3,4-DIMETHYLAZOLIDIN see HKR550

(2-HYDROXYETHYL)DIMETHYL(3-STEARAMIDOPROPYL)-AMMONIUM PHOSPHATE (1:1) (SALT) see CCP675

N-(2-HYDROXYETHYL)-5,8,11,14-EICOSATETRAENAMIDE (ALL-Z)- see HKR600

N-HYDROXYETHYLENEDIAMINETRIACETIC ACID see HKS000

2-HYDROXYETHYL ESTER MALEIC ACID see EJG500

2-HYDROXYETHYL ESTER METHACRYLIC ACID see EJH000

2-HYDROXYETHYL ESTER STEARIC ACID see EJM500

N-HYDROXYETHYL-1,2-ETHANEDIAMINE see AJW000

HYDROXYETHYL ETHER CELLULOSE see HKQ100

N-(2-HYDROXYETHYL)ETHYLENEDIAMINE see AJW000

N-(β-HYDROXYETHYL)ETHYLENEDIAMINE see AJW000

N-(β-HYDROXYETHYLETHYLENEDIAMINE)-N,N',N'-TRIACETIC ACID see HKS000

(N-HYDROXYETHYLETHYLENEDINITRILO)TRIACETIC ACID see HKS000

N-HYDROXYETHYL ETHYLENE IMINE see ASI000

1-(2-HYDROXYETHYL)ETHYLENIMINE see ASI000

N-(2-HYDROXYETHYL)ETHYLENIMINE see ASI000

N-HYDROXYETHYL-N-ETHYL-m-TOLUIDINE see HKS100

1-HYDROXYETHYL-2-HEPTADECENYLGLYOXALIDINE see AHP500

1-(2-HYDROXYETHYL)-2-HEPTADECENYLGLYOXALIDINE see AHP500

1-(2-HYDROXYETHYL)-2-HEPTADECENYL-2-IMIDAZOLINE see AHP500

1-(2-HYDROXYETHYL)-2-N-HEPTADECENYL-2-IMIDAZOLINE see AHP500

2-HYDROXY-3-ETHYLHEPTANOIC ACID see HKS300

2-HYDROXY-4-(2'-ETHYLHEXOXY)BENZOFENON (CZECH) see EKY000

HYDROXYETHYL HYDRAZINE see HHC000

N-(2-HYDROXYETHYL)HYDRAZINE see HHC000

β-HYDROXYETHYLHYDRAZINE see HHC000

N-HYDROXYETHYL-N-2-HYDROXYALKYLAMINE see HKS400

N-(2-HYDROXYETHYL)-N-(4-HYDROXYBUTYL)NITROSAMINE see HKS500

2-HYDROXYETHYL (2-HYDROXYETHYL)CARBAMATE see HKS550

2-(1-HYDROXYETHYL)-7-(2-HYDROXY-3-ISOPROPYLAMINOPROPOXY)BENZOFURAN see HKS600

o-(2-HYDROXYETHYL)HYDROXYLAMINE see HKS775

1-(2-HYDROXYETHYL)-2-HYDROXYMETHYL-5-NITROIMIDAZOLE see HMJ000

5-(2-HYDROXYETHYL)-3-((4-HYDROXY-2-METHYL-5-PYRIMIDINYL)METHYL)-4-METHYL-THIAZOLIUM see ORS200

1-HYDROXYETHYLIDENE-1,1-DIPHOSPHONIC ACID see HKS780

(1-HYDROXYETHYLIDENE)DIPHOSPHONIC ACID DISODIUM SALT see DXD400

(1-HYDROXYETHYLIDENE)DIPHOSPHONIC ACID, TETRAPOTASSIUM SALT see TEC250

(1-HYDROXYETHYLIDENE)DIPHOSPHONIC ACID, TETRASODIUM SALT see TEE250

(1-HYDROXYETHYLIDENE)DIPHOSPHONIC ACID, TRISODIUM SALT see TNL750

3-(1-HYDROXYETHYLIDENE)-6-METHYL-2H-PYRAN-2,4(3H)-DIONE, SODIUM SALT see SGD000

1-(2-HYDROXYETHYL)-2-IMIDAZOLINE-2-YLNORTALL OIL see TAC075

2,2'-((2-HYDROXYETHYL)IMINO BIS(N-(α,α-DIMETHYLPHENETHYL))-N-METHYL-ACETAMIDE see DTL200

2,2'-((2-HYDROXYETHYL)IMINO)BIS(N-(α,α-DIMETHYLPHENETHYL)-N-METHYLACETAMIDE) HCL see EAN650

2,2'-((2-HYDROXYETHYL)IMINO)BIS(N-(1,1-DIMETHYL-2-PHENYLETHYL))-N-METHYLACETAMIDE) see DTL200

(2-HYDROXYETHYL)IMINODIACETIC ACID see HKM500

2-HYDROXYETHYL IODOACETATE see HKS800

2-HYDROXY-2-ETHYL-3-ISOBUTYL-9,10-DIMETHOXY-1,2,3,4,6,7-HEXAHYDROBENZO(A)CHINOLIZIN see HKS900

(N-HYDROXYETHYL)ISOPROPYLAMINE see INN400

β-HYDROXYETHYL ISOPROPYL ETHER see INA500

2-(N-(2-HYDROXYETHYL)KARBAMOYL)-3-METHYLCHINOXALIN-1,4-DIOXID see HKV100

N-β-HYDROXYETHYL-N-β-KYANETHYLANILIN (CZECH) see CPJ500

N-(α-HYDROXYETHYL)LYSERGAMIDE see LJI100

2-HYDROXYETHYL MERCAPTAN see MCN250

2-HYDROXYETHYLMERCURY(II) NITRATE see HKT200

HYDROXYETHYL METHACRYLATE see EJH000

2-HYDROXYETHYL METHACRYLATE see EJH000

β-HYDROXYETHYL METHACRYLATE see EJH000

2-(N-2-HYDROXYETHYL-N-METHYLAMINO)ETHANOL see MKU250

7-(2-HYDROXYETHYL)-12-METHYLBENZ(a)ANTHRACENE see HKU000

N-HYDROXYETHYL-α-METHYLBENZYLAMINE see HKU500

1-HYDROXYETHYL METHYL KETONE see ABB500

1-HYDROXYETHYL-2-METHYL-5-NITROIMIDAZOLE see MMN250

1-(2-HYDROXYETHYL)-2-METHYL-5-NITROIMIDAZOLE see MMN250

1-(2-HYDROXY-1-ETHYL)-2-METHYL-5-NITROIMIDAZOLE see MMN250

1-(β-HYDROXYETHYL)-2-METHYL-5-NITROIMIDAZOLE see MMN250

N-(2-HYDROXYETHYL)-N-METHYL-2-NITRO-1H-IMIDAZOLE-1-ACETAMIDE see HKU600

3-(2-HYDROXYETHYL)-3-METHYL-1-PHENYLTRIAZENE see HKV000

N-(2-HYDROXYETHYL)-3-METHYL-2-QUINOXALINECARBOXAMIDE 1,4-DIOXIDE see HKV100

N-(2-HYDROXYETHYL)MORPHOLINE see MRQ500

N-β-HYDROXYETHYLMORPHOLINE see MRQ500

2-HYDROXYIMINOMETHYL-N-METHYLPYRIDINIUM METHANESULPHONATE see PLX250
2-(HYDROXYIMINO)-N-PHENYLACETAMIDE see GIK100
8-HYDROXYINDENO(1,2,3-cd)PYRENE see IBZ100
5-HYDROXYINDOLEACETIC ACID see HLJ000
5-HYDROXYINDOLE-3-ETHANOL see HLI600
5-HYDROXY-1H-INDOLE-3-ETHANOL see HLI600
5-HYDROXYINDOLYLACETIC ACID see HLJ000
2-HYDROXYINIPRAMINE see DPX400
m-HYDROXYIODOBENZENE see IEV010
4-HYDROXY-3-IODO-5-NITROBENZONITRILE see HLJ500
(4-(4-HYDROXY-3-IODOPHENOXY)-3,5-DIIODOPHENYL)ACETIC ACID DIETHANOLAMINESALT see TKR100
l-3-(4-(4-HYDROXY-3-IODOPHENOXY)-3,5-DIIODOPHENYL)ALANINE see LGK050
O-(4-HYDROXY-3-IODOPHENYL)-3,5-DIIODO-l-TYROSINE see LGK050
8-HYDROXY-7-IODOQUINOLINE SULFONATE see IEP200
8-HYDROXY-7-IODOQUINOLINESULFONIC ACID see IEP200
8-HYDROXY-7-IODO-5-QUINOLINESULFONIC ACID see IEP200
HYDROXYISOBUTYLISOPROPYLARSINE OXIDE see IPS100
2-HYDROXY-3-ISOBUTYL-6-(1-METHYLPROPYL)PYRAZINE 1-OXIDE see ARO000
α-HYDROXYISOBUTYRIC ACID ACETATE METHYL ESTER see HLJ600
α-HYDROXYISOBUTYRONITRILE see MLC750
α-HYDROXYISOBUTYROPHENONE see HMQ100
(−)-(1-HYDROXY-3-ISOHEXENYL)NAPHTHAZARINE see HLJ650
5-(1-HYDROXY-2-ISOPROPYLAMINO)BUTYL-8-HYDROXYCARBOSTYRIL HYDROCHLORIDE see PME600
4'-(1-HYDROXY-2-(ISOPROPYLAMINO)ETHYL)METHANESULFOANILIDE HYDROCHLORIDE see CCK250
4'-(1-HYDROXY-2-ISOPROPYLAMINO)ETHYL)METHANESULFONANILIDE MONOHYDROCHLORIDE see CCK250
4-HYDROXY-α-ISOPROPYLAMINOMETHYLBENZYL ALCOHOL see HLK000
p-HYDROXY-α-ISOPROPYLAMINOMETHYLBENZYL ALCOHOL see HLK000
3-HYDROXY-α-ISOPROPYLAMINOMETHYLBENZYL ALCOHOL HYDROCHLORIDE see HLK500
N-(4-(2-HYDROXY-3-(ISOPROPYLAMINO)PROPOXY)ACETAMIDE (9CI) see ECX100
4'-(2-HYDROXY-3-(ISOPROPYLAMINO)PROPOXY)ACETANILIDE see ECX100
7-(2-HYDROXY-3-(ISOPROPYLAMINO)PROPOXY)-2-BENZOFURANYL METHYL KETONE see HLK600
8-(2-HYDROXY-3-ISOPROPYLAMINO)PROPOXY-2H-1-BENZOPYRAN see HLK800

4-(2-HYDROXY-3-ISOPROPYLAMINOPROPOXY)-INDOLE see VSA000
7-(2-HYDROXY-3-(ISOPROPYLAMINO)PROPOXY)-α-METHYL-2-BENZOFURANMETHANOL see HKS600
2-(p-(2-HYDROXY-3-(ISOPROPYLAMINO)PROPOXY)PHENYL)ACETAMIDE see TAL475
2-HYDROXY-4-ISOPROPYL-2,4,6-CYCLOHEPTATRIEN-1-ONE see IRR000
2-HYDROXY-5-ISOPROPYL-2,4,6-CYCLOHEPTATRIEN-1-ONE see TFV750
β-HYDROXY-N-ISOPROPYL-3,4-DICHLOROPHENETHYLAMINE see DFN400
3-α-HYDROXY-8-ISOPROPYL-1-α-H,5-α-H-TROPANIUM BROMIDE (±)-TROPATE see IGG000
(8r)-3-α-HYDROXY-8-ISOPROPYL-1-α-H,5-α-H-TROPIUMBROMIDE-(±)-TROPATE see IGG000
HYDROXYISOPROPYLMERCURY see IPW000
(2-(4-HYDROXY-2-ISOPROPYL-5-METHYLPHENOXY)ETHYL)DIMETHYLAMINE see DAD850
(2-(4-HYDROXY-2-ISOPROPYL-5-METHYLPHENOXY)ETHYL)METHYLAMINE see DAD500
m-HYDROXYISOPROPYLPHENYL-N-METHYLCARBAMATE see HLK900
12-HYDROXY-13-ISOPROPYLPODOCARPA-7,13-DIEN-15-OIC ACID BIS(2-CHLOROETHYL)AMINE SALT see HIJ000
HYDROXYISOXAZOLE see HLM000
3-HYDROXY-17-KETO-ESTRA-1,3,5-TRIENE see EDV000
4-HYDROXY-2-KETO-4-METHYLPENTANE see DBF750
3-HYDROXY-17-KETO-OESTRA-1,3,5-TRIENE see EDV000
HYDROXYKYNURENINE see HJD000
3-HYDROXYKYNURENINE see HJD000
3-HYDROXY-l-KYNURENINE see HJD500
l-3-HYDROXYKYNURENINE see HJD500
HYDROXYLAMINE see HLM500
HYDROXYLAMINE, N-ACETYL-N-(7-IODO-2-FLUORENYL)-O-MYRISTOYL- see MSB100
HYDROXYLAMINE, o-ACETYL-N-ISONICOTINOYL- see POQ320
HYDROXYLAMINE, o-ACETYL-N-NICOTINOYL- see POQ310
HYDROXYLAMINE, o-ACETYL-N-((5-NITRO-2-FURYL)FORMIMIDOYL)- see NGC500
HYDROXYLAMINE, o-ACETYL-N-PICOLINOYL- see POQ300
HYDROXYLAMINE, N-2-BIPHENYLYL- see HIT600
HYDROXYLAMINE, o-BUTYRYL-N-((5-NITRO-2-FURYL)FORMIMIDOYL)- see NGC550
HYDROXYLAMINE CHLORIDE see HLN000
HYDROXYLAMINE CHLORIDE (1:1) see HLN000
HYDROXYLAMINE, N,o-DIACETYL-N-CARBOXY-, ETHYL ESTER see DBH100
HYDROXYLAMINE, N-(2',3-DIMETHYLBIPHENYL-4-YL)- see HKB650
HYDROXYLAMINE, N-(4,6-DIMETHYLDIPYRIDO(1,2-A:3',2'-D)IMIDAZOL-3-YL)- see HIT610
HYDROXYLAMINE, N-(DIPYRIDO(1,2-A:3',2'-D)IMIDAZOL-3-YL)- see HIT620
HYDROXYLAMINE, N-(p-DITHIAN-2-YLENE)-o-(METHYLCARBAMOYL)- see MID870

HYDROXYLAMINE, N-(p-ETHYLPHENYL)- see EOL100
HYDROXYLAMINE, HYDRIODIDE see HLM600
HYDROXYLAMINE HYDROCHLORIDE see HLN000
HYDROXYL AMINE IODIDE (DOT) see HLM600
HYDROXYLAMINE, N-(6-METHYLDIPYRIDO(1,2-A:3',2'-D)IMIDAZOL-2-YL)- see HIU550
HYDROXYLAMINE, N-METHYL-N-(4-QUINOLINYL)-, 1-OXIDE see HLX550
HYDROXYLAMINE NEUTRAL SULFATE see OLS000
HYDROXYLAMINE, N-((5-NITRO-2-FURYL)FORMIMIDOYL)-o-PROPIONYL- see NGC570
HYDROXYLAMINE, N-(p-NITROPHENYL)- see NIR050
HYDROXYLAMINE, N-(3-NITRO-o-TOLYL)- see HLN135
HYDROXYLAMINE SULFATE see OLS000
HYDROXYLAMINE SULFATE (2:1) see OLS000
HYDROXYLAMINESULFONIC ACID see HLN100
HYDROXYLAMINE-O-SULFONIC ACID see HLN100
4-HYDROXYLAMINOBIPHENYL see BGI250
4-(HYDROXYLAMINO)-2-METHYL-QUINOLINE, 1-OXIDE see MKR000
2-HYDROXYLAMINO-4-NITROTOLUENE see HLN125
2-HYDROXYLAMINO-6-NITROTOLUENE see HLN135
4-(HYDROXYLAMINO)-2-NITROTOLUENE see HMI100
6-HYDROXYLAMINOPURINE see HIT000
6-N-HYDROXYLAMINOPURINE see HIT000
6-HYDROXYLAMINOPURINE RIBOSIDE see HIT100
N6-HYDROXYLAMINOPURINE RIBOSIDE see HIT100
HYDROXYLAMMONIUM CHLORIDE see HLN000
HYDROXYLAMMONIUM PHOSPHINATE see HLN500
HYDROXYLAMMONIUM SULFATE see OLS000
O-HYDROXYLAMMONIUM SULFONATE see HLN100
HYDROXYLAPATITE (CA5(OH)(PO4)3) (9CI) see HJE100
HYDROXYLATED LECITHIN see HLN700
o-HYDROXYLBENZHYDRAZIDE see HJL100
(2-HYDROXYLIMINIOBUTANE CHLORIDE) see BOV625
HYDROXYLIZARIC ACID see TKN750
13-HYDROXYLUPANINE-2-PYRROLE CARBOXYLIC ACID ESTER see CAZ125
HYDROXYLUREA see HOO500
N-HYDROXY-MAB see HLV000
1-HYDROXY-MA-144-N1 see RHA125
1-HYDROXY MA144 S1 see APV000
(E)-1-HYDROXY-MC-9,10-DIHYDRODIOL see DLT600
6-HYDROXY-2-MERCAPTOPYRIMIDINE see TFR250
p-HYDROXYMERCURIBENZOATE see HLN800
o-HYDROXYMERCURIBENZOIC ACID see HLO450
p-HYDROXYMERCURIBENZOIC ACID see HLN800
o-(N-3-HYDROXYMERCURI-2-HYDROXYETHOXYPROPYLCARBAMYL)PHENOXY-ACETIC ACID, SODIUM SALT see HLO300

5-(3-HYDROXYMERCURI-2-METHOXYPROPYL) BARBITURIC ACID SODIUM SALT see HLU000

o-((3-HYDROXYMERCURI-2-METHOXYPROPYL)CARBAMOYL)PHENOXYACETIC ACID MONOSODIUM SALT see SIH500

N-((3-(HYDROXYMERCURI)-2-METHOXYPROPYL)-CARBAMOYL)SUCCINAMIC ACID see MFC000

8-(γ-HYDROXYMERCURI-β-METHOXYPROPYL)-3-COUMARINCARBOXYLICACID THEOPHYLLINE SODIUM see MCS000

8-(3-HYDROXYMERCURI)-2-METHOXYPROPYL)-2-OXO-2H-1-BENZOPYRAN-3-CARBOXYLIC ACID SODIUM SALT COMPOUND with THEOPHYLLINE (1:1) see MCS000

N-(γ-HYDROXYMERCURI-β-METHOXYPROPYL)SALICYLAMIDE-o-ACETIC ACID SODIUM SALT see SIH500

2-(HYDROXYMERCURI)-4-NITROANILINE see HLO350

2-HYDROXYMERCURI-3-NITROBENZOIC ACID-1,2-CYCLIC ANHYDRIDE see NHY500

HYDROXYMERCURI-o-NITROPHENOL see HLO400

o-HYDROXYMERCURIOBENZOIC ACID see HLO450

o-(HYDROXYMERCURI)PHENOL see HLO500

HYDROXYMERCURIPROPANOLAMIDE of m-CARBOXYPHENOXYACETIC ACID see HLP000

HYDROXYMERCURIPROPANOLAMIDE of o-CARBOXYPHENOXYACETIC ACID see NCM800

HYDROXYMERCURIPROPANOLAMIDE of p-CARBOXYPHENOXYACETIC ACID see HLP500

(5-(HYDROXYMERCURI)-2-THIENYL)MERCURY ACETATE see HLQ000

2-HYDROXYMESITYLENE see MDJ740

2-HYDROXYMESTRANOL see HLQ050

p-HYDROXYMETHAMPHETAMINE see FMS875

N-HYDROXY-METHANAMINE see MKQ875

HYDROXYMETHANESULFINIC ACID SODIUM SALT see FMW000

7-HYDROXYMETHOTREXATE see HLQ100

4'-HYDROXY-3'-METHOXYACETOPHENONE see HLQ500

4-HYDROXY-3-METHOXYALLYLBENZENE see EQR500

1-HYDROXY-2-METHOXY-4-ALLYLBENZENE see EQR500

4-HYDROXY-3-METHOXY-4-AMINOAZOBENZENE see HLR000

N-HYDROXY-3-METHOXY-4-AMINOAZOBENZENE see HLR000

3-HYDROXY-4'-METHOXY-4-AMINODIPHENYL see HLR500

3-HYDROXY-4'-METHOXY-4-AMINODIPHENYL HYDROCHLORIDE see AKN250

2-HYDROXY-3-METHOXYBENZALDEHYDE see VFP000

2-HYDROXY-4-METHOXYBENZALDEHYDE see HLR600

3-HYDROXY-4-METHOXYBENZALDEHYDE see FNM000

4-HYDROXY-3-METHOXYBENZALDEHYDE see VFK000

1-HYDROXY-2-METHOXYBENZENE see GKI000

6-HYDROXY-7-METHOXY-5-BENZOFURANACRYLIC ACID Δ-LACTONE see XDJ000

2-HYDROXY-3-METHOXYBENZOIC ACID see HJB500

3-HYDROXY-4-METHOXYBENZOIC ACID see HJC000

4-HYDROXY-3-METHOXYBENZOIC ACID see VFF000

4-HYDROXY-3-METHOXYBENZOIC ACID METHYL ESTER see VFF100

2-HYDROXY-4-METHOXYBENZOPHENONE see MES000

2-HYDROXY-4-METHOXYBENZOPHENONE-5-SULFONIC ACID see HLR700

o-(2-HYDROXY-4-METHOXYBENZOYL)BENZOIC ACID see HLS500

(α-HYDROXY-p-METHOXYBENZYL)-PHOSPHONIC ACID DIETHYL ESTER, ESTER with BIS(2-CHLOROPROPYL) ANTIMONATE(III) see DHH600

4-HYDROXY-3-METHOXYCINNAMIC ACID see FBP200

4-HYDROXY-3-METHOXY-1-METHYLBENZENE see MEK325

3-HYDROXY-N-METHOXY-N-METHYL-cis-CROTONAMIDE, DIMETHYL PHOSPHATE see DOL800

4-HYDROXY-3-METHOXY-5,6-METHYLENEDIOXY-APORPHIN see HLT000

6-HYDROXY-3-METHOXY-N-METHYL-4,5-EPOXYMORPHINAN see DKW800

5-HYDROXY-4-(METHOXYMETHYL)-6-METHYL-3-PYRIDINEMETHANOL see MFN600

6-(4-HYDROXY-6-METHOXY-7-METHYL-3-OXO-5-PHTHALANYL)-4-METHYL-4-HEXENOIC ACID, SODIUM SALT see SIN850

(E)-6-(4-HYDROXY-6-METHOXY-7-METHYL-3-OXO-5-PHTHALANYL)-4-METHYL-4-HEXENOIC ACID see MRX000

1-HYDROXY-6-METHOXYPHENAZINE 5,10-DIOXIDE see HLT100

2-HYDROXY-3-(o-METHOXYPHENOXY)PROPYL 1-CARBAMATE see GKK000

2-HYDROXY-3-(o-METHOXYPHENOXY)PROPYL NICOTINATE see HLT300

N-HYDROXY-2-METHOXY-4-(PHENYLAZO)BENZENAMINE see HLR000

4-(4-HYDROXY-3-METHOXYPHENYL)-2-BUTANONE see VFP100

4-(4-HYDROXY-3-METHOXYPHENYL)-3-BUTEN-2-ONE see MLI800

1-(4-HYDROXY-3-METHOXYPHENYL)ETHANONE see HLQ500

(4-HYDROXY-3-METHOXYPHENYL)ETHYL METHYL KETONE see VFP100

5-HYDROXY-3-METHOXY-7-PHENYL-2,6-HEPTADIENOIC ACID γ-LACTONE see GJI250

trans-N-((4-HYDROXY-3-METHOXYPHENYL)METHYL)-8-METHYL-6-NONEAMIDE see CBF750

N-((4-HYDROXY-3-METHOXYPHENYL)METHYL)-8-METHYL-6-NONENAMIDE see CBF750

(2-HYDROXY-4-METHOXYPHENYL)PHENYLMETHANONE see MES000

3-(4-HYDROXY-3-METHOXYPHENYL)PROPENOIC ACID see FBP200

p-HYDROXY-m-METHOXYPHENYLPROPYLENE OXIDE see EBT500

1-(1-HYDROXY-3-METHOXY-3-PHENYLPROPYL)-4-(2-METHOXY-2-PHENYLETHYL)PIPERAZINE DIHYDROCHLORIDE see MFG250

1-HYDROXY-2-METHOXY-4-PROPENYLBENZENE see IKQ000

4-HYDROXY-3-METHOXY-1-PROPENYLBENZENE see IKQ000

1-HYDROXY-2-METHOXY-4-PROP-2-ENYLBENZENE see EQR500

4-HYDROXY-3-METHOXYPROPYLBENZENE see MFM750

1-(2-HYDROXY-3-METHOXYPROPYL)-2-NITROIMIDAZOLE see NHH500

4-HYDROXY-8-METHOXYQUINALDIC ACID see HLT500

6-(HYDROXY(6-METHOXY-4-QUINOLINYL)METHYL)-1-ETHYL-3-VINYL-QUINUCLIDINIUM, IODIDE see QIS000

6-(1-HYDROXY-1-(6-METHOXY-4-QUINOLINYL)METHYL)-1-ETHYL-3-VINYLQUINUCLIDINIUM, IODIDE see QIS000

4-HYDROXY-3-METHOXYSTYRENE see VPF100

4-HYDROXY-3-METHOXYTOLUENE see MEK325

HYDROXY(2-METHOXY-3-(2,4,6-TRIOXO-(1H,3H,5H)PYRIMID-5-YL)PROPYL)MERCURY SODIUM SALT see HLU000

2-HYDROXY-3-METHYL-4-ACETYLTETRAHYDROFURANE see BMK290

N-(HYDROXYMETHYL)ACRYLAMIDE see HLU500

2-(HYDROXYMETHYL)ACRYLIC ACID, ETHYL ESTER see ELI500

6-(HYDROXYMETHYL)ACYLFULVENE see HLU600

6-HYDROXYMETHYLACYLFULVENE see HLU600

2-HYDROXY-N-METHYL-1-ADAMANTANEPROPANAMINE HYDROCHLORIDE see MGK750

17-β-HYDROXY-17-(2-METHYLALLYL)ESTR-4-EN-3-ONE see MDQ075

17-β-HYDROXY-17-α-(2-METHYLALLYL)ESTR-4-EN-3-ONE see MDQ075

17-β-HYDROXY-17-α-(1-METHYLALLYL)ESTR-4-EN-3-ONE see MDQ100

α-HYDROXY-β-METHYL AMINE PROPYLBENZENE see EAW000

N-HYDROXY-N-METHYL-4-AMINOAZOBENZENE see HLV000

(R)-4-(1-HYDROXY-2-(METHYLAMINO)ETHYL)-1,2-BENZENEDIOL (9CI) see VGP000

(−)-α-HYDROXY-β-(METHYLAMINO)ETHYL-α-(3-HYDROXYBENZENE) HYDROCHLORIDE see SPC500

3'-(1-HYDROXY-2-(METHYLAMINO)ETHYL)METHANESULFONANILIDE METHANESULFONATE see AHL500

3'-(1-HYDROXY-2-(METHYLAMINO)ETHYL)METHANESULFONANILIDE MONOMETHANESULFONATE SALT see AHL500

N-(3-(1-HYDROXY-2-(METHYLAMINO)ETHYL)PHENYL)-METHANESULFONAMIDE MONOMETHANESULFONATE SALT see AHL500

1-α-HYDROXY-β-METHYLAMINO-3-HYDROXY-1-ETHYLBENZENE see NCL500

4-HYDROXY-α-((METHYLAMINO)METHYL)BENZENEMETHANOL see HLV500

(R)-3-HYDROXY-α-((METHYLAMINO)METHYL)BENZENEMETHANOL see NCL500

(R)-3-HYDROXY-α-((METHYLAMINO)METHYL)BENZENEMETHANOL HYDROCHLORIDE see SPC500

(−)-m-HYDROXY-α-(METHYLAMINOMETHYL)BENZYL ALCOHOL see NCL500

p-HYDROXY-α-((METHYLAMINO)METHYL)BENZYL ALCOHOL see HLV500

1-m-HYDROXY-α-((METHYLAMINO)METHYL)-BENZYL ALCOHOL see NCL500

1-m-HYDROXY-α-(METHYLAMINOMETHYL)BENZYL ALCOHOL HYDROCHLORIDE see SPC500

1-HYDROXY-2-METHYL-4-AMINONAPHTHALENE see AKX500

4-HYDROXY-7-(METHYLAMINO)-2-NAPHTHALENESULFONIC ACID see HLX000

1-HYDROXY-2-METHYLAMINO-1-PHENYLPROPANE see EAW000

2-HYDROXY-1-(3-METHYLAMINOPROPYL)ADAMANTANE HYDROCHLORIDE see MGK750

4'-(1-HYDROXY-2-(METHYLAMINO)PROPYL)METHANESULFONANILIDE HYDROCHLORIDE see HLX500

4-(N-HYDROXY-N-METHYLAMINO)QUINOLINE 1-OXIDE see HLX550

17-β-HYDROXY-17-α-METHYL-5-α-ANDROSTANO(2,3-c)FURAZAN see AOO300

17-β-HYDROXY-17-METHYL-5-α-ANDROSTAN-3-ONE see MJE760

17-β-HYDROXY-17-α-METHYLANDROSTANO(3,2-c)PYRAZOLE see AOO401

17-β-HYDROXY-7-α-METHYLANDROST-5-ENE-3-ONE see HLX600

17-β-HYDROXY-17-METHYLANDROST-4-EN-3-ONE see MPN500

17-β-HYDROXY-1-METHYL-5-α-ANDROST-1-EN-3-ONE ACETATE see PMC700

17-β-HYDROXY-17-α-METHYLANDROSTRA-1,4-DIEN-3-ONE see DAL300

HYDROXY METHYL ANETHOL see IRY000

6-HYDROXYMETHYLANTHANTHRENE see HLX700

9-HYDROXYMETHYLANTHRACENE see APG600

β-d-4-HYDROXY-3-METHYL-2-ANTHRAQUINONYL-6-o-β-d-XYLOPYRANOSYLGLUCOPYRANOSIDE see ROU500

5-(HYDROXYMETHYL)-1-AZA-3,7-DIOXABICYCLO(3.3.0)OCTANE see OMM300

7-HYDROXYMETHYLBENZ(a)ANTHRACENE see BBH250

10-HYDROXYMETHYL-1,2-BENZANTHRACENE see BBH250

p-HYDROX-N-METHYLBENZEDRINE see FMS875

1-HYDROXY-2-METHYLBENZENE see CNX000

1-HYDROXY-3-METHYLBENZENE see CNW750

1-HYDROXY-4-METHYLBENZENE see CNX250

4-(HYDROXYMETHYL)BENZENEDIAZONIUM SULFATE see HLX900

4-(HYDROXYMETHYL)BENZENEDIAZONIUM TETRAFLUOROBORATE see HLX925

2-HYDROXY-5-METHYL-1,3-BENZENEDIMETHANOL see HLX950

4-HYDROXY-6-METHYL-5-BENZOFURANACRYLIC ACID γ-LACTONE see HLV955

4-(HYDROXYMETHYL)-2H-BENZOFURO(3,2-G)-1-BENZOPYRAN-2-ONE see HLV960

2-HYDROXY-3-METHYL-BENZOIC ACID (9CI) see CNX625

4-HYDROXYMETHYL-4',5'-BENZOPSORALEN see HLV960

6-HYDROXYMETHYLBENZO(a)PYRENE see BCV250

6-HYDROXYMETHYLBENZO(a)PYRENESULFATE ESTER (SODIUM SALT) see HLY000

2-HYDROXY-3-METHYL-1,4-BENZOQUINONE 5,6-EPOXIDE see TBF325

2-HYDROXYMETHYLBICYCLO(2.2.1)HEPTANE see NNH000

4-(HYDROXYMETHYL)BIPHENYL see HLY400

17-β-HYDROXY-17-METHYL-B-NORANDROST-4-EN-3-ONE see MNC150

2-HYDROXY-3-(3-METHYL-2-BUTENYL)-1,4-NAPHTHALENEDIONE see HLY500

2-HYDROXY-3-(3-METHYL-2-BUTENYL)-1,4-NAPHTHOQUINONE see HLY500

7-α-(1-(R)-HYDROXY-1-METHYLBUTYL)-6,14-ENDOETHENOTETRAHYDRO-ORIPAVINE HYDROCHLORIDE see EQO500

7-α-1-(R)-HYDROXY-1-METHYLBUTYL)-6,14-endo-ETHENOTETRAHYDROORIPAVINE see EQO450

2-HYDROXY-2-METHYL-3-BUTYNE see MHX250

4-HYDROXY-1-METHYLCARBOSTYRIL see HMR500

1-HYDROXY-3-METHYLCHOLANTHRENE see HMA000

2-HYDROXY-3-METHYLCHOLANTHRENE see HMA500

15-HYDROXY-20-METHYLCHOLANTHRENE see HMA000

(E)-1-HYDROXY-3-METHYLCHOLANTHRENE 9,10-DIHYDRODIOL see DLT600

HYDROXYMETHYLCHRYSAZIN see DMU600

7-HYDROXY-4-METHYLCOUMARIN see MKP500

7-HYDROXY-4-METHYLCOUMARIN, BIS(2-CHLOROETHYL) PHOSPHATE see HMA600

7-HYDROXY-4-METHYL COUMARIN, O-ESTER with O,O,DIETHYL PHOSPHOROTHIOATE see PKT000

7-HYDROXY-4-METHYLCOUMARIN SODIUM see HMB000

3-HYDROXY-N-METHYL-cis-CROTONAMIDE DIMETHYL PHOSPHATE see MRH209

2-HYDROXY-4-METHYL-2,4,6-CYCLOHEPTATRIEN-1-ONE see MQI000

HYDROXYMETHYLCYCLOHEXANE see HDH200

1-(HYDROXYMETHYL)CYCLOHEXANEACETIC ACID, SODIUM SALT see SHL500

2-HYDROXY-3-METHYL-2-CYCLOPENTEN-1-ONE see HMB500

4-HYDROXY-3-METHYL-2-CYCLOPENTEN-1-ONE, cis- mixed with trans-2,2-DIMETHYL-3-(2-METHYL-PROPENYL)CYCLOPROPANECARBOXYLIC ACID ESTER with 2-ALLYL-4-HYDROXY-3-METHYL-2-CYCLOPENTEN-1-ONE (1:4) see AFS000

4-HYDROXY-4-METHYL-7-cis-DECENOIC ACID LACTONE see MIW060

5-HYDROXYMETHYLDEOXYURIDINE see HMB550

5-HYDROXYMETHYL-2'-DEOXYURIDINE see HMB550

(E)-HYDROXYMETHYLDIAZENE-15N2 see HMB560

(E)-HYDROXYMETHYLDIAZENE POTASSIUM SALT see HMB565

(Z)-HYDROXYMETHYLDIAZENE POTASSIUM SALT see HMB570

4-HYDROXYMETHYL-2,6-DI-tert-BUTYLPHENOL see BQI050

N-(HYDROXYMETHYL)-N-(1,3-DIHYDROXYMETHYL-2,5-DIOXO-4-IMIDAZOLIDINYL)-N'-(HYDROXYMETHYL)UREA see IAS100

3'-HYDROXYMETHYL-4-(DIMETHYLAMINO)AZOBENZENE see HMB600

2'-HYDROXYMETHYL-N,N-DIMETHYL-4-AMINOAZOBENZENE see HMB595

3'-HYDROXYMETHYL-N,N-DIMETHYL-4-AMINOAZOBENZENE see HMB600

4-HYDROXYMETHYL-2,2-DIMETHYL-1,3-DIOXOLANE see DVR600

2-(HYDROXYMETHYL)-2,4-DIMETHYLPENTANEDIAL see DTH600

2-(HYDROXYMETHYL)-1,1-DIMETHYLPIPERIDINIUM METHYL SULFATE BENZILATE see CDG250

3-(HYDROXYMETHYL)DIPHENYL ETHER see HMB625

2-HYDROXYMETHYLENE-17-α-METHYL-5-α-ANDROSTAN-17-β-OL-3-ONE see PAN100

2-HYDROXYMETHYLENE-17-α-METHYL-DIHYDROTESTOSTERONE see PAN100

2-(HYDROXYMETHYLENE)-17-α-METHYLDIHYDROTESTOSTERONE see PAN100

2-HYDROXYMETHYLENE-17-α-METHYL-17-β-HYDROXY-3-ANDROSTANONE see PAN100

17-HYDROXY-16-METHYLENE-19-NORPREGN-4-ENE-3,20-DIONE ACETATE see HMB650

17-β-HYDROXY-17-METHYLESTR-4-EN-3-ONE see MDM350

2-(HYDROXYMETHYL)ETHANOL see PML250

4-(1-HYDROXY-2-((1-METHYLETHYL)AMINO)ETHYL)-1,2-BENZENEDIOL see DMV600

4-(1-HYDROXY-2-((1-METHYLETHYL)AMINO)ETHYL)-1,2-BENZENEDIOL HYDROCHLORIDE see IMR000

(±)-4-(1-HYDROXY-2-((1-METHYLETHYL)AMINO)ETHYL)-1,2-BENZENEDIOL HYDROCHLORIDE see IQS500

5-(1-HYDROXY-2-((1-METHYLETHYL)AMINO)ETHYL)-1,3-BENZENEDIOL SULFATE (2:1) SALT see MDM800

4-(2-HYDROXY-3-((1-METHYLETHYL)AMINO)PROPOXY)BENZENEACETAMIDE see TAL475

1-(7-(2-HYDROXY-3-((1-METHYLETHYL)AMINO)PROPOXY)-2-BENZOFURANYL)ETHANONE HYDROCHLORIDE see BAU255

1-(7-(2-HYDROXY-3-((1-METHYLETHYL)AMINO)PROPOXY)-2-BENZOFURANYL)ETHANONE see HLK600

N-(4-(2-HYDROXY-3-((1-METHYLETHYL)AMINO)PROPOXY)PHENYL)ACETAMIDE see ECX100

2-HYDROXY-4-(1-METHYLETHYL)-2,4,6-CYCLOHEPTATRIEN-1-ONE see IRR000

(HYDROXYMETHYL)ETHYLENE ACETATE see DBF600

N-(1-(HYDROXYMETHYL)ETHYL)-d-LYSERGOMIDE see LJL000

N-(α-(HYDROXYMETHYL)ETHYL)-d-LYSERGOMIDE see LJL000

12'-HYDROXY-2'-(1-METHYLETHYL)-5'-α-(2-METHYLPROPYL)ERGOTAMAN-3',6',18-TRIONE see EDB100

1-HYDROXY-1-METHYLETHYL PHENYL KETONE see HMQ100

3-(1-HYDROXY-1-METHYLETHYL)PHENYL METHYLCARBAMATE see HLK900

1'-HYDROXYMETHYLEUGENOL see HMB700

5-HYDROXYMETHYLFURALDEHYDE see ORG000

2-HYDROXYMETHYLFURAN see FPU000

5-(HYDROXYMETHYL)-2-FURANCARBOXALDEHYDE see ORG000

4-HYDROXY-5-METHYL-2(5H)-FURANONE see MPQ500

5-(HYDROXYMETHYL)FURFURAL see ORG000

HYDROXYMETHYLFURFUROLE see ORG000

3-HYDROXY-3-METHYLGLUTARIC ACID see HMC000

β-HYDROXY-β-METHYLGLUTARIC ACID see HMC000

3-HYDROXY-1-METHYLGUANINE see HMC500

3-HYDROXY-7-METHYLGUANINE see HMD000

3-HYDROXY-9-METHYLGUANINE see HMD500

3-HYDROXYMETHYL-n-HEPTAN-4-OL see EKV000

3-α-HYDROXY-8-METHYL-1-α-H,5-α-H-TROPANIUM BROMIDE MANDELATE see MDL000

3-α-HYDROXY-8-METHYL-1-α-H,5-H-TROPANIUM BROMIDE 2-PROPYLVALERATE see LJS000

3-α-HYDROXY-8-METHYL-1-α-H,5-α-H-TROPANIUM NITRATE (±)-TROPATE (ESTER) see MGR500

HYDROXYMETHYL HYDROPEROXIDE see HME000

5-HYDROXY-2-METHYL-1H-IMIDAZOLE-4-CARBOXAMIDE see MID800

N-HYDROXY-3-METHYL-3H-IMIDAZO(4,5-F)QUINOLINE-2-AMINE see HIU600

5-HYDROXY-2-METHYL-1-INDANONE SODIUM SALT see HME050

3-HYDROXYMETHYLINDOLE see ICP100

3-HYDROXY-1-METHYL-5,6-INDOLINEDIONE see AES639

HYDROXYMETHYL 2-INDOLYL KETONE see HIS100

7-(3-((2-HYDROXY-3-((2-METHYLINDOL-4-YL)OXY)PROPYL)AMINO)BUTYL)THEOPHYLLINE see TAL560

HYDROXYMETHYLINITRILE see HIM500

HYDROXYMETHYL 1-IODO-3-ADAMANTYL KETONE see HME100

6-HYDROXYMETHYL-2-ISOPROPYLAMINOMETHYL-7-NITRO-1,2,3,4-TETRAHYDROQUINOLINE see OLT000

3-HYDROXY-1-METHYL-4-ISOPROPYLBENZENE see TFX810

3-HYDROXY-5-METHYLISOXAZOLE see HLM000

2-HYDROXY-2-METHYLMALONATE see MPN250

HYDROXYMETHYLMERCURY see MLG000

12-HYDROXYMETHYL-7-METHYLBENZ(a)ANTHRACENE see HMF500

7-HYDROXYMETHYL-12-METHYLBENZ(a)ANTHRACENE see HMF000

1-HYDROXYMETHYL-2-METHYLDITMIDE-2-OXIDE see HMG000

17-HYDROXY-6-METHYL-16-METHYLENEPREGNA-4,6-DIENE-3,20-DIONE, ACETATE see MCB380

1-(HYDROXYMETHYL)-3-METHYLIMIDAZOLINIUM CHLORIDE, DODECANOATE see HMG500

3-(HYDROXYMETHYL)-1-METHYLIMIDAZOLINIUM CHLORIDE LAURATE (ESTER) see HMG600

HYDROXYMETHYL 1-METHYL-2-INDOLYL KETONE see HIS120

HYDROXYMETHYL 3-METHYL-2-INDOLYL KETONE see HIS130

2-(HYDROXYMETHYL)-2-(METHYLPENTYL) BUTYLCARBAMATE CARBAMATE see MOV500

2-(HYDROXYMETHYL)-2-METHYLPENTYL ESTER, CARBAMATE, BUTYL CARBAMIC ACID see MOV500

HYDROXYMETHYL METHYL PEROXIDE see HMH000

5-(HYDROXYMETHYL)-11-METHYL-6H-PYRIDO(4,3-B)CARBAZOLE see HMH050

5-(HYDROXYMETHYL)-11-METHYL-6H-PYRIDO(4,3-B)CARBAZOLE N-METHYLCARBAMATE see HMH100

5-HYDROXYMETHYL-6-METHYL-2,4(1H,3H)-PYRIMIDINEDIONE see HMH300

5-HYDROXYMETHYL-4-METHYLURACIL see HMH300

5-HYDROXYMETHYL-6-METHYLURACIL see HMH300

(−)-3-HYDROXY-N-METHYLMORPHINAN see LFG000

d-3-HYDROXY-N-METHYLMORPHINAN see DBE800

(±)-3-HYDROXY-N-METHYLMORPHINAN see MKR250

dl-3-HYDROXY-N-METHYLMORPHINAN see MKR250

l-3-HYDROXY-N-METHYLMORPHINAN BITARTRATE see DYF000

dl-3-HYDROXY-N-METHYLMORPHINAN HYDROBROMIDE see HMH500

dl-3-HYDROXY-N-METHYLMORPHINAN HYDROBROMIDE see MDV250

d-3-HYDROXY-N-METHYLMORPHINAN TARTRATE see DBE825

l-3-HYDROXY-N-METHYLMORPHINAN TARTRATE see DYF000

(−)-3-HYDROXY-N-METHYLMORPHINAN TARTRATE DIHYDRATE see MLZ000

(+)-3-HYDROXY-N-METHYLMORPHINAN TARTRATE HYDRATE see MMA000

5-HYDROXY-2-METHYL-1,4-NAPHTHALENEDIONE see PJH610

5-HYDROXY-7-METHYL-1,4-NAPHTHALENEDIONE see RBF200

2-HYDROXY-3-METHYL-1,4-NAPHTHOQUINONE see HMI000

3-HYDROXY-2-METHYL-1,4-NAPHTHOQUINONE see HMI000

5-HYDROXY-2-METHYL-1,4-NAPHTHOQUINONE see PJH610

5-HYDROXY-7-METHYL-1,4-NAPHTHOQUINONE see RBF200

N-HYDROXY-2-METHYL-3-NITROBENZENAMINE see HLN135

N-HYDROXY-2-METHYL-5-NITROBENZENAMINE see HLN125

N-HYDROXY-4-METHYL-3-NITROBENZENAMINE see HMI100

o-HYDROXYMETHYLNITROBENZENE see NFM550

3-HYDROXYMETHYL-1-((3-(5-NITRO-2-FURYL)ALLYLIDENE)AMINO)HYDANTOIN see HMI500

6-HYDROXYMETHYL-2-(2-(5-NITRO-2-FURYL)VINYL)-PYRIDINE see NDY400

2-HYDROXYMETHYL-5-NITROIMIDAZOLE-1-ETHANOL see HMJ000

2-HYDROXYMETHYL-2-NITROPROPANE-1,3-DIOL see HMJ500

2-HYDROXYMETHYL-2-NITROPROPANE-1,3-DIOL see HMJ500

((HYDROXYMETHYL)NITROSAMINO)-BUTYRIC ACID METHYL ESTER, ACETATE see MEI250

1-HYDROXY-3-(METHYLNITROSAMINO)-2-PROPANONE ACETATE (ESTER) see HMJ600

N-(HYDROXYMETHYL)-N-NITROSO-β-ALANINE METHYL ESTER, ACETATE see MEH000

2-HYDROXYMETHYLNORCAMPHANE see NNH000

17-β-HYDROXY-18-METHYL-19-NOR-17-α-PREGN-4-EN-20-YN-3-ONE see NNQ500

17-β-HYDROXY-18-METHYL-19-NOR-17-α-PREGN-4-EN-20-YN-3-ONE see NNQ520

3-HYDROXY-4-METHYL-7-OXABICYCLO(4.1.0)HEPT-3-ENE-2,5-DIONE see TBF325

(1R-cis)-3-HYDROXY-4-METHYL-7-OXABICYCLO(4.1.0)HEPT-3-ENE-2,5-DIONE see TBF325

7-HYDROXY-4-METHYL-2-OXO-2H-1-BENZOPYRAN see MKP500

3-(HYDROXYMETHYL)-8-OXO-7-(2-(4-PYRIDYLTHIO)ACETAMIDO)-5-THIA-1-AZABICYCLO(4.2.0)OCT-2-ENE-2-CARBOXYLIC ACID, ACETATE (ESTER) see CCX500

3-(HYDROXYMETHYL)-8-OXO-7-(2-(4-PYRIDYLTHIO)ACETAMIDO)-5-THIA-1-AZABICYCLO(4,2,0)OCT-2-ENE-2-CARBOXYLIC ACID ACETATE(ESTER), MONOSODIUM SALT see HMK000

3-HYDROXY-3-METHYLPENTANEDIOIC ACID see HMC000

4-HYDROXY-4-METHYL-PENTAN-2-ON (GERMAN, DUTCH) see DBF750

4-HYDROXY-4-METHYLPENTANONE-2 see DBF750

4-HYDROXY-4-METHYL PENTAN-2-ONE see DBF750

4-HYDROXY-4-METHYL-2-PENTANONE see DBF750

4-HYDROXY-4-METHYL-2-PENTANONE PEROXIDE see HMK050

(−)-2-(1-HYDROXY-4-METHYL-3-PENTENYL)-5,8-DIHYDROXY-1,4-NAPHTHOQUINONE see HLJ650

2-HYDROXYMETHYLPHENOL see HMK100

o-(HYDROXYMETHYL)PHENOL see HMK100

4-HYDROXY-α-(1-((1-METHYL-2-PHENOXYETHYL)AMINO)ETHYL)-BENZENEMETHANOL HYDROCHLORIDE see VGA300

p-HYDROXY-N-(1-METHYL-2-PHENOXYETHYL)NOREPHEDRINE see VGF000

HYDROXYMETHYL 1-PHENYL-3-ADAMANTYL KETONE see HMK150

1-(3-HYDROXY-4-METHYLPHENYL)-2-AMINO-1-PROPANOL HYDROCHLORIDE see MKS000

HYDROXYMETHYLPHENYLARSINE OXIDE see HMK200

N-(4-((2-HYDROXY-5-METHYLPHENYL)AZO)PHENYL)ACETAMIDE see AAQ250

2-(2-HYDROXY-5-METHYLPHENYL)BENZOTRIAZOLE see HML500

2-HYDROXY-2-METHYL-1-PHENYL-1-PROPANONE see HMQ100

p-HYDROXY-α-(1-((1-METHYL-3-PHENYLPROPYL)AMINO)ETHYL)BENZYL ALCOHOL HYDROCHLORIDE see DNU200

5-(1-HYDROXY-2-((1-METHYL-3-PHENYLPROPYL)AMINO)ETHYL)SALICYL AMIDE HYDROCHLORIDE see HMM500

2-(HYDROXYMETHYL)-6-PHENYL-3(2H)-PYRIDAZINONE see HMN000

2-HYDROXYMETHYL-6-PHENYL-3-PYRIDAZONE see HMN000

1-(4-HYDROXY-2-METHYL-5-PHENYL-1H-PYRROL-3-YL)ETHANONE see HMN100

3-HYDROXY-3-METHYL-1-PHENYLTRIAZENE see HMP000

γ-(HYDROXYMETHYLPHOSPHINYL)-l-α-AMINOBUTYRYL-l-ALANYL-L-ALANINE SODIUM SALT see SEO000

3-HYDROXY-2-METHYL-5-((PHOSPHONOOXY)METHYL)-4-PYRIDINECARBOXALDEHYDE see PII100

HYDROXYMETHYLPHTHALIMIDE see HMP100

N-(HYDROXYMETHYL)PHTHALIMIDE see HMP100

17-HYDROXY-6-METHYLPREGNA-4,6-DIENE-3,20-DIONE ACETATE see VTF000

17-HYDROXY-6-METHYLPREGNA-4,6-DIENE-3,20-DIONE ACETATE mixed with 19-NOR-17-α-PREGNA-1,3,5(10)-TRIEN-2-YNE-3,17-DIOL see MCA500

17-HYDROXY-6-METHYLPREGN-4-ENE-3,20-DIONE ACETATE see HMQ000

17-HYDROXY-6-α-METHYLPREGN-4-ENE-3,20-DIONE ACETATE see MCA000

17-α-HYDROXY-6-α-METHYLPREGN-4-ENE-3,20-DIONE ACETATE see MCA000

17-α-HYDROXY-6-α-METHYLPROGESTERONE ACETATE see MCA000

1-HYDROXYMETHYLPROPANE see IIL000

HYDROXYMETHYL-PROPANEDIOIC ACID (9CI) see MPN250

N-(HYDROXYMETHYL)-2-PROPENAMIDE see HLU500

2-HYDROXY-2-METHYLPROPIONITRILE see MLC750

α-HYDROXY-α-METHYLPROPIOPHENONE see HMQ100

1-(HYDROXYMETHYL)PROPYLAMIDE of 1-METHYL-(+)-LYSERGIC ACID HYDROGEN MALEATE see MQP500

d-N,N'-(1-HYDROXYMETHYLPROPYL)ETHYLENED INITROSAMINE see HMQ500

17-β-HYDROXY-6-α-METHYL-17-(1-PROPYNYL)ANDROST-4-EN-3-ONE see DRT200

(6-α,17-β)-17-HYDROXY-6-METHYL-17-(1-PROPYNYL)-ANDROST-4-EN-3-ONE see DRT200

3-HYDROXY-2-METHYL-4H-PYRAN-4-ONE see MAO350

4-HYDROXY-6-METHYL-2H-PYRAN-2-ONE see THL800

1-HYDROXYMETHYLPYRENE see PON400

2-HYDROXYMETHYLPYRENE see HMQ550

2-(HYDROXYMETHYL)PYRIDINE see POR800

3-(HYDROXYMETHYL)PYRIDINE see NDW510

4-(HYDROXYMETHYL)PYRIDINE see POR810

5-HYDROXY-6-METHYL-3,4-PYRIDINEDICARBINOL HYDROCHLORIDE see PPK500

5-HYDROXY-6-METHYL-3,4-PYRIDINEDICARBINOL TRIPALMITATE see HMQ575

5-HYDROXY-6-METHYL-3,4-PYRIDINEDIMETHANOL see PPK250

5-HYDROXY-6-METHYL-3,4-PYRIDINEDIMETHANOL HYDROCHLORIDE see PPK500

3-HYDROXY-1-METHYLPYRIDINIUM BROMIDE DIMETHYLCARBAMATE (ESTER) see MDL600

3-HYDROXY-1-METHYLPYRIDINIUM BROMIDE HEXAMETHYLENEBIS(METHYLCARBAMATE) see DXG800

3-HYDROXY-1-METHYL-PYRIDINIUM DIMETHYLCARBAMATE (ester) see PPI800

4-HYDROXY-2-METHYL-N-2-PYRIDINYL-2H-THIENO(2,3-e)-1,2-THIAZINE-3-CARBOXAMIDE 1,1-DIOXIDE see TAL485

4-HYDROXY-2-METHYL-N-(2-PYRIDYL)-2H-1,2-BENZOTHIAZIN-3-CARBOXYAMID-1,1-DIOXID (GERMAN) see FAJ100

4-HYDROXY-2-METHYL-N-(2-PYRIDYL)-2H-1,2-BENZOTHIAZINE-3-CARBOXAMIDE-1,1-DIOXIDE see FAJ100

3-HYDROXY-2-METHYL-4-PYRONE see MAO350

3-HYDROXY-2-METHYL-γ-PYRONE see MAO350

8-HYDROXY-2-METHYLQUINOLINE see HMQ600

8-HYDROXY-1-METHYL QUINOLINIUM IODIDE, METHYLCARBAMATE see MID500

8-HYDROXY-1-METHYLQUINOLINIUM METHYLSULFATE DIMETHYLCARBAMATE see DQY400

4-HYDROXY-1-METHYL-2-QUINOLONE see HMR500

12-HYDROXY-4-METHYL-4,8-SECOSENECIONAN-8,11,16-TRIONE see DMX200

2-HYDROXYMETHYLTETRAHYDROPYRAN see MDS500

2-HYDROXY-4-(METHYLTHIO)BUTANENITRILE see HMR550

2-HYDROXY-4-(METHYLTHIO)BUTANOIC ACID see HMR600

1-(HYDROXYMETHYL)-2,8,9-TRIOXA-5-AZA-1-SILABICYCLO(3.3.3) UNDECANEMETHACRYLATE see HMS500

3-α-HYDROXY-8-METHYL-1-α,5-α-H-TROPANIUM BROMIDE (±)TROPATE see MGR250

3-HYDROXY-α-METHYL-l-TYROSINE see DNA800

3-HYDROXY-α-METHYL-l-TYROSINE-1-(2,2-DIMETHYL-1-OXOPROPOXY)ETHYL ESTER PHOSPHATE HYDRATE see HMS875

4-HYDROXY-3-MORPHOLINOMETHYLBENZOIC ACID ETHYL ESTER see ELK000

HYDROXYMYCIN see NCF500

N-HYDROXY-N-MYRISTOYL-2-AMINOFLUORENE see HMU000

HYDROXY No. 253 see NNC500

5-HYDROXY-1,4-NAFTOCHINON see WAT000

2-HYDROXYNAPHTHALDEHYDE see HMU200

β-HYDROXYNAPHTHALDEHYDE see HMU200

1-HYDROXYNAPHTHALENE see NAW500

2-HYDROXYNAPHTHALENE see NAX000

α-HYDROXYNAPHTHALENE see NAW500

β-HYDROXYNAPHTHALENE see NAX000

2-HYDROXY-1-NAPHTHALENECARBOXALDEHYDE see HMU200

N-HYDROXY-2-NAPHTHALENECARBOXAMIDE see NAV000

3-HYDROXY-2-NAPHTHALENECARBOXYLIC ACID see HMX520

2-HYDROXY-1,4-NAPHTHALENEDIONE see HMX600

2-HYDROXY-6-NAPHTHALENESULFONIC ACID see HMU500

6-HYDROXY-2-NAPHTHALENESULFONIC ACID see HMU500

N-HYDROXY-N-2-NAPHTHALENYLACETAMIDE see NBD000

4-((4-HYDROXY-1-NAPHTHALENYL)AZO)BENZENESULPHONIC ACID, MONOSODIUM SALT see FAG010

3-(6-HYDROXY-2-NAPHTHALENYL)-1,2-DIMETHYL (2R,3S)-REL-3-PYRROLIDINOL YDROCHLORIDE see HMU600

N-HYDROXYNAPHTHALIMIDE, DIETHYL PHOSPHATE see HMV000

N-HYDROXYNAPHTHALIMIDE-O,O-DIETHYL PHOSPHOROTHIOATE see NAQ500

3-HYDROXY-2-NAPHTHAMIDE see HMW500

2-HYDROXYNAPHTHENESULFONIC ACID see HMX000

2-HYDROXY-1-NAPHTHOIC ACID see HMX500

3-HYDROXY-2-NAPHTHOIC ACID see HMX520

2-HYDROXY-3-NAPHTHOIC ACID ANILIDE see CMM760

2-HYDROXYNAPHTHOQUINONE see HMX600

2-HYDROXY-1,4-NAPHTHOQUINONE see HMX600

5-HYDROXY-1,4-NAPHTHOQUINONE see WAT000

N-HYDROXY-1-NAPHTHYLAMINE see HIX000

N-HYDROXY-2-NAPHTHYLAMINE see NBI500

1-HYDROXY-2-NAPHTHYLAMINE HYDROCHLORIDE see ALK250

2-HYDROXY-1-NAPHTHYLAMINE HYDROCHLORIDE see ALK000

p-((4-HYDROXY-1-NAPHTHYL)AZO)BENZENESULFONIC ACID, MONOSODIUM SALT see FAG010

p-((2-HYDROXY-1-NAPHTHYL)AZO)BENZENESULFONIC ACID SODIUM SALT see CMM220

p-((4-HYDROXY-1-NAPHTHYL)AZO)BENZENESULPHONIC ACID, MONOSODIUM SALT see FAG010

p-((4-HYDROXY-1-NAPHTHYL)AZO)BENZENESULPHONIC ACID, SODIUM SALT see FAG010

N-HYDROXYNAPHTHYLIMIDE DIETHYL PHOSPHATE see HMV000

3-HYDROXY-4-(1-NAPHTHYLOXY)BUTYRAMIDOXIME HYDROCHLORIDE see NAC500

p-HYDROXYNEOZONE see NBF500

2-HYDROXY-4-NITROANILINE see ALO000

2-HYDROXY-5-NITROANILINE see NEM500

4-HYDROXY-3-NITROANILINE see NEM480

2-HYDROXY-5-NITROBENZALDEHYDE see HMX700

N-HYDROXY-4-NITROBENZENAMINE see NIR050

2-HYDROXYNITROBENZENE see NIF010

3-HYDROXYNITROBENZENE see NIE600

4-HYDROXYNITROBENZENE see NIF000

m-HYDROXYNITROBENZENE see NIE600

4-HYDROXY-3-NITROBENZENEARSONIC ACID see HMY000

4-HYDROXY-3-NITROBENZENESULFONYL CHLORIDE see HMY050

4-HYDROXY-3-NITROBENZOIC ACID METHYL ESTER see HMY075

9-HYDROXY-2-NITROFLUORENE see HMY080

N-HYDROXY-5-NITRO-2-FURANCARBOXIMIDAMIDE see NGD000

3-HYDROXY-2-(5-NITRO-2-FURYL)-2H-1,3-BENZOXAZIN-4(3H)-ONE see HMY100

2-HYDROXY-5-NITROMETANILIC ACID see HMY500

4-HYDROXY-3-NITROPHENYLARSONIC ACID see HMY000

2-HYDROXY-5-((4-NITROPHENYL)AZO)BENZOIC ACID see NEY000

4-HYDROXY-3-(1-(4-NITROPHENYL)-3-OXOBUTYL)-2H-1-BENZOPYRAN-2-ONE see ABF750

3-(HYDROXYNITROSOAMINO)-l-ALANINE (9CI) see AFH750

N-HYDROXY-N-NITROSO-BENZENAMINE, AMMONIUM SALT see ANO500

3-HYDROXYNITROSOCARBOFURAN see HMZ100

3-HYDROXY-4-(NITROSOCYANAMIDO)BUTYRAMIDE see HNA000

2-HYDROXY-6-(NITROSOCYANAMIDO)HEXANAMIDE see HNA500

2-HYDROXY-3-NITROSO-2,7-NAPHTHALENEDISULFONIC ACID DISODIUM SALT see HNB000

trans-4-HYDROXY-1-NITROSO-l-PROLINE see HNB500

3-HYDROXYNITROSOPYRROLIDINE see NLP700

8-HYDROXY-5-NITROSOQUINOLINE see NLR000

4-HYDROXY-NONACHLORODIPHENYL ETHER see HNB550

N-HYDROXYNONANAMIDE see NMX600

5-HYDROXYNONANOIC ACID, LACTONE see NMV790

4-HYDROXYNONANOIC ACID, γ-LACTONE see CNF250

4-HYDROXYNONENAL see HNB600

4-HYDROXY-2-NONENAL see HNB600

trans-4-HYDROXY-2-NONENAL see HNB700

HYDROXYNOREPHEDRINE see HNB875

m-HYDROXY NOREPHEDRINE see HNB875

1-m-HYDROXYNOREPHEDRINE see HNC000

3-α-HYDROXYNORETHYNODREL see NNU000

3-β-HYDROXYNORETHYNODREL see NNU500

17-HYDROXY-19-NOR-17-α-PREGNA-4,20-DIEN-3-ONE see NCI525

17-HYDROXY-19-NOR-17-α-PREGNA-4,9,11-TRIEN-20-YN-3-ONE see NNR125

17-HYDROXY-19-NORPREGN-4-ENE-3,20-DIONE HEXANOATE see GEK510

17-HYDROXY-19-NOR-4-PREGNENE-3,20-DIONE HEXANOATE see GEK510

17-HYDROXY-19-NORPREGN-4-EN-3-ONE see ENX600

17-β-HYDROXY-19-NOR-17-α-PREGN-4-EN-3-ONE see ENX600

17-HYDROXY-19-NOR-17-α-PREGN-4-EN-3-ONE see ENX600

(17-α)-17-HYDROXY-19-NORPREGN-4-EN-20-YN-3-ONE see NNP500

(17-α)-17-HYDROXY-19-NORPREGN-5(10)-EN-20-YN-3-ONE see EEH550

17-β-HYDROXY-19-NORPREGN-4-EN-20-YN-3-ONE see NNP500

17-HYDROXY(17-α)-19-NORPREGN-5(10)-EN-20-YN-3-ONE see EEH550

17-HYDROXY-19-NOR-17-α-PREGN-4-EN-20-YN-3-ONE see NNP500

17-HYDROXY-19-NOR-17-α-PREGN-5(10)-EN-20-YN-3-ONE see EEH550

17-β-HYDROXY-19-NOR-17-α-PREGN-4-EN-20-YN-3-ONE ACETATE see ABU000

17-HYDROXY-19-NOR-17-α-PREGN-4-EN-20-YN-3-ONE ACETATE see ABU000

17-HYDROXY-19-NOR-17-α-PREGN-4-EN-20-YN-3-ONE 1-ADAMANTANECARBOXYLATE (ESTER) OXIME see HNC500

17-HYDROXY-19-NOR-17-α-PREGN-4-EN-20-YN-3-ONE OXIME see NNQ050

17-α-HYDROXY-19-NORPROGESTERONE CAPROATE see GEK510

3-HYDROXY-N,N,5-TRIMETHYLPYRAZOLE-1-CARBOXAMIDE DIMETHYLCARBAMATE (ESTER) see DQZ000

13-HYDROXY-9,11-OCTADECADIENOIC METHYL ESTER see MKR500

12-HYDROXY-cis-9-OCTADECENOIC ACID see RJP000

1-HYDROXYOCTANE see OEI000

5-HYDROXYOCTANOIC ACID LACTONE see OCE000

5-HYDROXYOCTANOIC ACID LACTONE see ODE300

2-(3-HYDROXY-1-OCTENYL)-5-OXO-3-CYCLOPENTENE-1-HEPTANOIC ACID see POC250

7-(2-(3-HYDROXY-1-OCTENYL)-5-OXO-3-CYCLOPENTEN-1-YL)-5-HEPTENOIC ACID see MCA025

4-HYDROXY-5-OCTYL-2(5H)FURANONE see HND000

2-HYDROXY-4-(OCTYLOXY)BENZOPHENONE see HND100

(2-HYDROXY-4-(OCTYLOXY)PHENYL)PHENYLMETHANONE see HND100

16-α-HYDROXYOESTRADIOL see EDU500

3-HYDROXY-OESTRA-1,3,5(10)-TRIEN-17-ONE see EDV000

3-HYDROXY-1,3,5(10)-OESTRATRIEN-17-ONE see EDV000

2-HYDROXY-4-OKTYLOXYBENZOFENON see HND100

4-HYDROXY-2(OR 5)-ETHYL-5(OR 2)-METHYL-3(2H)-FURANONE see HND150

16-HYDROXY-11-OXAHEXADECANOIC ACID, ω-LACTONE see OKW100

16-HYDROXY-12-OXAHEXADECANOIC ACID, ω-LACTONE see OKW110

γ-HYDROXY-β-OXOBUTANE see ABB500

(11-β)-11-HYDROXY-17-(1-OXOBUTOXY)-21-(1-OXOPROPOXY)-PREGN-4-ENE-3,20-DIONE see HHQ850

2-(((3-α-5-β)-3-HYDROXY-24-OXOCHOLAN-24-YL)AMINO)ETHANESULFONIC ACID see LHW100

N-((3-α-5-β)-3-HYDROXY-24-OXOCHOLAN-24-YL)GLYCINE see HND200

1-HYDROXY-4-OXO-2,5-CYCLOHEXADIENE-1-SULFONIC ACID compound with DIETHYLAMINE see DIS600

5-β-HYDROXY-19-OXODIGITOXIGENIN see SMM500

6α-HYDROXY-3-OXO-11-EPIISOEUSANTONA-1,4-DIENIC ACID, γ-LACTONE see SAU500

6α-HYDROXY-3-OXO-EUDESMA-1,4-DIEN-12-OIC ACID, γ-LACTONE (11S)-(−)- see SAU500

4-HYDROXY-3-(3-OXO-1-FENYL-BUTYL) CUMARINE (DUTCH) see WAT200

5-HYDROXY-4-OXO-NORVALINE see AKG500

5-HYDROXY-4-OXO-l-NORVALINE see HND250

Δ-HYDROXY-γ-OXO-l-NORVALINE see AKG500

3-β-HYDROXY-11-OXOOLEAN-12-EN-30-OIC ACID see GIE000

3-β-HYDROXY-23-OXO-OLEAN-12-EN-28-OIC ACID see GMG000

3-β-HYDROXY-11-OXO-18-α-OLEAN-12-EN-30-OIC ACID see GIEO5O

3-β-HYDROXY-11-OXOOLEAN-12-EN-30-OIC ACID, HYDROGEN SUCCINATE see BGD000

3-β-HYDROXY-11-OXOOLEAN-12-EN-30-OIC ACID HYDROGEN SUCCINATE DISODIUM SALT see CBO500

4-HYDROXY-3-(3-OXO-1-PHENYLBUTYL)-2H-1-BENZOPYRAN-2-ONE POTASSIUM SALT see WAT209

4-HYDROXY-3-(3-OXO-1-PHENYLBUTYL)-2H-1-BENZOPYRAN-2-ONE SODIUM SALT (9CI) see WAT220

4-HYDROXY-3-(3-OXO-1-PHENYL-BUTYL)-CUMARIN (GERMAN) see WAT200

HYDROXYOXOPHENYL IODANIUM PERCHLORATE see HND375

(HYDROXY)(OXO)(PHENYL)-LAMBDA³-IODANIUM PERCHLORATE see IFS000

3-β-HYDROXY-20-OXO-17-α-PREGN-5-ENE-16-β-CARBOXAMIDE ACETATE see HND800

(13E,15S)-15-HYDROXY-9-OXO-PROSTA-10,13-DIEN-1-OIC ACID (9CI) see POC250

15-HYDROXY-9-OXO-PROSTA-5,10-13-TRIEN-1-OIC ACID, (5Z,13E,15S)- (9CI) see MCA025

3-HYDROXY-2-OXOPURINE see HOB000

(+−)-4-HYDROXY-2-OXO-1-PYRROLIDINEACETAMIDE see HND900

trans-6-(10-HYDROXY-6-OXO-1-UNDECENYL)-μ-LACTONE, RESORCYLIC ACID see ZAT000

6-(10-HYDROXY-6-OXO-trans-1-UNDECENYL-)-β-RESORCYCLIC ACID-N-LACTONE see ZAT000

2'-HYDROXYPELARGIDENOLON 1522 see MRN500

3-HYDROXY-2-PENTANEDIOIC ACID, DIMETHYL ESTER, DIMETHYL PHOSPHATE see SOY000

4-HYDROXYPENTANOIC ACID LACTONE see VAV000

4-HYDROXYPENT-3-ENOIC ACID LACTONE see AOO750

4-HYDROXYPENT-3-ENOIC ACID LACTONE see MKH250

4-HYDROXY-2-PENTENOIC ACID γ-LACTONE see MKH500

S-3-HYDROXY-4-PENTENONITRILE see COP400

1-HYDROXY-3-n-PENTYL-Δ⁸-TETRAHYDROCANNABINOL see HNE400

1-HYDROXY-2-PENTYNE-4-ONE see HNE450

N-HYDROXYPHENACETIN see HIN000

4-HYDROXYPHENACYL CHLORIDE see HNE500

p-HYDROXYPHENACYL CHLORIDE see HNE500

HYDROXYPHENAMATE see HNJ000

N-HYDROXY-N-2-PHENANTHRENYL-ACETAMIDE (9CI) see PCZ000

N-HYDROXY-4'-PHENETHYLACETANILIDE see PDI250

β-HYDROXYPHENETHYL ALCOHOL-α-CARBAMATE see PFJ000

4-HYDROXYPHENETHYLAMINE see TOG250

p-HYDROXYPHENETHYLAMINE see TOG250

β-HYDROXYPHENETHYLAMINE see HNF000

p-HYDROXY-β-PHENETHYLAMINE see TOG250

2-(β-HYDROXYPHENETHYLAMINO)PYRIDINE see PGG350

2-(β-HYDROXY-β-PHENETHYLAMINO)-PYRIMIDINE HYDROCHLORIDE see FAR200

β-HYDROXYPHENETHYL CARBAMATE see PFJ000

(−)-2-(6-(β-HYDROXYPHENETHYL)-1-METHYL-2-PIPERIDYL)-ACETOPHENONE HYDROCHLORIDE see LHZ000

(m-HYDROXYPHENETHYL)TRIMETHYLAMMONIUM HYDROXIDE see HNF100

(m-HYDROXYPHENETHYL)TRIMETHYLAMMONIUM PICRATE see HNG000

N-HYDROXYPHENETIDINE see HNG100

N-HYDROXY-p-PHENETIDINE see HNG100

p-HYDROXYPHENETOLE see EFA100

2-HYDROXYPHENOL see CCP850

3-HYDROXYPHENOL see REA000

m-HYDROXYPHENOL see REA000

o-HYDROXYPHENOL see CCP850

p-HYDROXYPHENOL see HIH000

7-HYDROXY-3H-PHENOXAZIN-3-ONE-10 OXIDE see HNG500

1-HYDROXY-2-PHENOXYETHANE see PER000

(2-HYDROXYPHENOXY)PHENYLMERCURY (8CI) see PFO750

4-HYDROXYPHENYLACETAMIDE see HNG550

(p-HYDROXYPHENYL)ACETAMIDE see HNG550

2-HYDROXY-N-PHENYLACETAMIDE see GHK500

2-(p-HYDROXYPHENYL)ACETAMIDE see HNG550

N-(4-HYDROXYPHENYL)ACETAMIDE see HIM000

m-HYDROXYPHENYL ACETATE see RDZ900

(4-HYDROXYPHENYL)ACETIC ACID see HNG600

(p-HYDROXYPHENYL)ACETIC ACID see HNG600

p-HYDROXYPHENYLACETIC ACID see HNG700

α-HYDROXYPHENYLACETIC ACID see MAP000

HYDROXYPHENYLACETONITRILE see MAP250

2-HYDROXY-2-PHENYLACETOPHENONE see BCP250

α-HYDROXY-α-PHENYLACETOPHENONE see BCP250

p-HYDROXYPHENYLACRYLIC ACID see CNU825

β-(4-HYDROXYPHENYL)ACRYLIC ACID see CNU825

l-β-(p-HYDROXYPHENYL)ALANINE see TOG300

α-(4-HYDROXYPHENYL)-β-AMINOETHANE see TOG250

1-(m-HYDROXYPHENYL)-2-AMINOETHANOL see AKT000

1-(p-HYDROXYPHENYL)-2-AMINOETHANOL see AKT250

1-(3'-HYDROXYPHENYL)-2-AMINOETHANOL see AKT000

l-1-(m-HYDROXYPHENYL)-2-AMINO-1-PROPANOL-d-HYDROGEN TARTRATE see HNC000

4-HYDROXYPHENYLARSENOUS ACID see HNG800

p-HYDROXYPHENYLARSONIC ACID see PDO250

(4-HYDROXYPHENYL)ARSONIC ACID polymer with FORMALDEHYDE see BCJ150

7-HYDROXY-8-(PHENYLAZO)-1,3-NAPHTHALENEDISULFONIC ACID, DISODIUM SALT see HGC000

7-HYDROXY-8-(PHENYLAZO)-1,3-NAPHTHALENEDISULPHONIC ACID, DISODIUM SALT see HGC000

6-((p-HYDROXYPHENYL)AZO)URACIL see HNH475

α-HYDROXY-α-PHENYLBENZENEACETIC ACID see BBY990

α-HYDROXY-α-PHENYLBENZENEACETIC ACID-2-(DIETHYLAMINO)ETHYL ESTER see DHU900

N-HYDROXYPHENYLBENZENESULFONAMIDE see HJH100

3-HYDROXYPHENYL BENZOATE see HNH500

2-(o-HYDROXYPHENYL)BENZOXAZOLE see HNI000

p-HYDROXYPHENYL BENZYL ETHER see AEY000

3-α-HYDROXY-8-(p-PHENYLBENZYL)-1-α-H,5-α-H-TROPANIUM BROMIDE, (±)-TROPATE see PEM750

1-(p-HYDROXYPHENYL)-3-BUTANONE see RBU000

4-(4-HYDROXYPHENYL)-2-BUTANONE see RBU000

4-(p-HYDROXYPHENYL)-2-BUTANONE ACETATE see AAR500

4-(p-HYDROXYPHENYL)-2-BUTANONE (FCC) see RBU000

p-HYDROXYPHENYLBUTAZONE see HNI500

(+−)-4-HYDROXY-α1-(((6-(4-PHENYLBUTOXY)HEXYL)AMINO)METHYL)-1,3-BENZENEDIMETHANOL see HNI600

1-(p-HYDROXYPHENYL)-2-BUTYLAMINOETHANOL see BQF250

2-HYDROXY-2-PHENYLBUTYL CARBAMATE see HNJ000

3-HYDROXY-2-PHENYLCINCHONINIC ACID see OPK300

4-(p-HYDROXYPHENYL)-2,5-CYCLOHEXADIENE-1-ONE SODIUM SALT see HNJ500

(3-HYDROXYPHENYL)DIETHYLMETHYLAMMONIUM BROMIDE see HNK000

(m-HYDROXYPHENYL)DIETHYLMETHYLAMMONIUM IODIDE, DIMETHYLCARBAMATE see HNK500

(m-HYDROXYPHENYL)DIETHYLMETHYLAMMONIUM METHOSULFATE METHYLCARBAMATE see HNK550

2-(p-HYDROXYPHENYL)-5,7-DIHYDROXYCHROMONE see CDH250

4-(4-HYDROXYPHENYL)-2,3-DIISOCYANO-1,3-BUTADIENYL)-1,2-BENZENEDIOL see XCS680

(−)-(R)-1-(p-HYDROXYPHENYL)-2-((3,4-DIMETHOXYPHENETHYL)AMINO)ETHANOL see DAP850

(3-HYDROXYPHENYL)DIMETHYLAMINE see HNK560

3-HYDROXYPHENYLDIMETHYLETHYLAMMONIUM BROMIDE see TAL490

1-(4-HYDROXYPHENYL)-3,3-DIMETHYLTRIAZINE see DUI600

(4-HYDROXY-m-PHENYLENE)BIS(ACETATOMERCURY) see HNK575

m-HYDROXYPHENYLETHANOLAMINE see AKT000

p-HYDROXYPHENYLETHANOLAMINE see AKT250

m-HYDROXYPHENYLETHANOLETHYLAMINE see EGE500

1-(4-HYDROXYPHENYL)ETHANONE see HIO000

5-HYDROXY-4-(2-PHENYL-(E)-ETHENYL)-2(5H)-FURANONE see HNK585

N-HYDROXY-N-(4-(2-PHENYLETHENYL)PHENYL)ACETAMIDE see SMT000

4-HYDROXYPHENYLETHYLAMINE see TOG250

2-(p-HYDROXYPHENYL)ETHYLAMINE see TOG250

β-HYDROXYPHENYLETHYLAMINE see TOG250

β-HYDROXY-β-PHENYLETHYLAMINE see HNF000

1-(3'-HYDROXYPHENYL)-2-ETHYLAMINOETHANOL see EGE500

2-(β-(HYDROXYPHENYL)ETHYLAMINOMETHYL)TETRALONE HYDROCHLORIDE see HAJ700

2-(β-HYDROXY-β-PHENYL-ETHYL-AMINO)-PYRIMIDINE CHLORHYDRATE (FRENCH) see FAR200

2-HYDROXY-2-PHENYLETHYL CARBAMATE see PFJ000

β-HYDROXY-β-PHENYLETHYL DIMETHYLAMINE see DSH700

1-(3-HYDROXYPHENYL)-1-HYDROXY-2-AMINOETHANE see AKT000

1-(4-HYDROXYPHENYL)-1-HYDROXY-2-BUTYLAMINOETHANE see BQF250

(2-HYDROXYPHENYL)HYDROXYMERCURY see HLO500

7-HYDROXY-4-PHENYL-3-(4-HYDROXYPHENYL)COUMARIN see HNK600

1-(3-HYDROXYPHENYL)-2-ISOPROPYLAMINOETHANOL HYDROCHLORIDE see HLK500

β-(p-HYDROXYPHENYL)ISOPROPYLMETHYLAMINE see FMS875

(2R,4R)-2-(o-HYDROXYPHENYL)-3-(3-MERCAPTOPROPIONYL)-4-THIAZOLIDINECARBOX YLIC ACID see FAQ950

o-HYDROXYPHENYLMERCURIC CHLORIDE see CHW675

HYDROXYPHENYLMERCURY see PFN100

p-HYDROXYPHENYL 2-MESITYLBENZOFURAN-3-YL KETONE see HNK700

p-HYDROXYPHENYL 2-MESITYLBENZOFURAN-4-YL KETONE see HNK800

p-HYDROXYPHENYLMETHYLAMINOETHANOL see HLV500

1-(4-HYDROXYPHENYL)-2-METHYLAMINOETHANOL see HLV500

1-1-(m-HYDROXYPHENYL)-2-METHYLAMINOETHANOL see NCL500

l-1-(m-HYDROXYPHENYL)-2-METHYL-AMINOETHANOL HYDROCHLORIDE see SPC500

(+−)-1-(4-HYDROXYPHENYL)-2-METHYLAMINOETHANOL TARTRATE see SPD000

1-(p-HYDROXYPHENYL)-2-METHYLAMINOPROPANE see FMS875

α-(p-HYDROXYPHENYL)-β-METHYLAMINOPROPANE see FMS875

p-HYDROXYPHENYLMETHYLAMINOPROP ANOL see HKH500

1-(4-HYDROXYPHENYL)-2-METHYLAMINOPROPANOL see HKH500

2-HYDROXYPHENYL METHYLCARBAMATE see HNK900

o-HYDROXYPHENYL METHYLCARBAMATE see HNK900

α-(p-HYDROXYPHENYL)-α-(2-METHYLCYCLOHEXYLIDENE)-p-CRESOL DIACETATE see BGQ325

1-(3-HYDROXYPHENYL)-N-METHYLETHANOLAMINE see NCL500

1-(4-HYDROXYPHENYL)-N-METHYLETHANOLAMINE see HLV500

4-(m-HYDROXYPHENYL)-1-METHYLISONIPECOTINOYL METHYL KETONE see HNK950

o-HYDROXYPHENYL METHYL KETONE see HIN500

p-HYDROXYPHENYL METHYL KETONE see HIO000

1-(4-HYDROXYPHENYL)-2-(1-METHYL-2-PHENOXYETHYLAMINO)PROPANOL see VGF000

1-(p-HYDROXYPHENYL)-2-(1'-METHYL-2'-PHENOXY)ETHYLAMINOPROPANOL-1 HYDROCHLORIDE see VGA300

1-(p-HYDROXYPHENYL)-2-(1'-METHYL-2'-PHENOXYETHYLAMINO)PROPANOL-2-HYDROCHLORIDE see VGF000

α-(4-HYDROXYPHENYL)-β-METHYL-4-(PHENYLMETHYL)-1-PIPERIDINEETHANOL (9CI) see IAG600

1-p-HYDROXYPHENYL-2-(1'-METHYL-3'-PHENYLPROPYLAMINO)-1-PROPANOL HYDROCHLORIDE see DNU200

(±)-4-(2-((3-(4-HYDROXYPHENYL)-1-METHYLPROPYL)AMINO)ETHYL)-1,2-BENZENEDIOL HYDROCHLORIDE see DXS375

4-(2-((3-(p-HYDROXYPHENYL)-1-METHYLPROPYL)AMINO)ETHYL)PYROCA TECHOL HYDROCHLORIDE see DXS375

3-HYDROXY-N-PHENYL-2-NAPHTHALENECARBOXAMIDE see CMM760

p-HYDROXYPHENYL-2-NAPHTHYLAMINE see NBF500

N-p-HYDROXYPHENYL-2-NAPHTHYLAMINE see NBF500

p-HYDROXYPHENYL-β-NAPHTHYLAMINE see NBF500

N-p-HYDROXYPHENYL-β-NAPHTHYLAMINE see NBF500

1-HYDROXY-1-PHENYLPENTANE see BQJ500

1-(p-HYDROXYPHENYL)-2-PHENYL-4-BUTYL-3,5-PYRAZOLIDINEDIONE see HNI500

1-p-HYDROXYPHENYL-2-PHENYL-3,5-DIOXO-4-N-BUTYLPYRAZOLIDINE see HNI500

(4-HYDROXYPHENYL)PHENYLMETHANON E see HJN100

1-(p-HYDROXYPHENYL)-2-(3'-PHENYLTHIOPROPYLAMINO)-1-PROPANOL HYDROCHLORIDE see HNL100

HYDROXYPHENYLPHOSPHINE OXIDE see HNL200

p-HYDROXYPHENYL-1-PROPANONE see ELL500

1-(4-HYDROXYPHENYL)-1-PROPANONE see ELL500

N-HYDROXY-3-PHENYL-2-PROPENAMIDE (9CI) see CMQ475

3-(4-HYDROXYPHENYL)-2-PROPENOIC ACID see CNU825

(E)-3-(2-HYDROXYPHENYL)-2-PROPENOIC ACID see CNU850

l-(1-HYDROXY-1-PHENYL-2-PROPYLAMINO)-1-(m-METHOXYPHENYL)-1-PROPANONE HYDROCHLORIDE see OQU000

3-((1-HYDROXY-1-PHENYL-2-PROPYL)METHYLAMINO)PROPIONITRIL E see CCX600

(p-HYDROXYPHENYL) 2-PYRIDYL KETONE see HJN700

p-HYDROXYPHENYLPYRUVIC ACID see HNL500

3-HYDROXY-2-PHENYL-4-QUINOLINECARBOXYLIC ACID see OPK300

HYDROXYPHENYL SALICYLAMIDE see DYE700

p-HYDROXYPHENYLSALICYLAMIDE see DYE700

N-(4-HYDROXYPHENYL)SALICYLAMIDE see DYE700

N-(p-HYDROXYPHENYL)SALICYLAMIDE see DYE700

4-HYDROXYPHENYLSULFONIC ACID see HNL600

4-(3-HYDROXY-3-PHENYL-3-(2-THIENYL)PROPYL)-4-METHYLMORPHOLINIUMIODIDE see HNM000

4-HYDROXY-α-(1-((3-(PHENYLTHIO)PROPYL)AMINO)ETHYL)-BENZENEMETHANOL HYDROCHLORIDE see HNL100

p-HYDROXY-α-(1-((3-(PHENYLTHIO)PROPYL)AMINO)ETHYL)B ENZYL ALCOHOL HYDROCHLORIDE see HNL100

2-((N-(m-HYDROXYPHENYL)-p-TOLUIDINO)METHYL)-2-IMIDAZOLINE see PDW400

(m-HYDROXYPHENYL)TRIMETHYLAMMONI UM BROMIDEDECAMETHYLENEBIS(METHYL CARBAMATE) (7CI) see DAM625

3-HYDROXYPHENYLTRIMETHYLAMMONI UM BROMIDE DIMETHYLCARBAMIC ESTER see POD000

(m-HYDROXYPHENYL)TRIMETHYLAMMONI UM BROMIDE DIMETHYLCARBAMATE see POD000

(m-HYDROXYPHENYL)TRIMETHYLAMMONI UM CHLORIDE, METHYLCARBAMATE see HNN000

(m-HYDROXYPHENYL)TRIMETHYLAMMONI UM DIMETHYLCARBAMATE (ester) see NCL100

(m-HYDROXYPHENYL)TRIMETHYLAMMONI UM IODIDE see HNN500

(m-HYDROXYPHENYL)TRIMETHYLAMMONI UM IODIDE, BENZYLCARBAMATE see HNO000

(m-HYDROXYPHENYL)TRIMETHYLAMMONI UM IODIDE, METHYLCARBAMATE see HNO500

(p-HYDROXYPHENYL)TRIMETHYLAMMONI UM IODIDE, METHYLCARBAMATE see HNP000

(m-HYDROXYPHENYL)TRIMETHYLAMMONI UM METHYLSULFATE BENZYLCARBAMATE see BED500

(m-HYDROXYPHENYL)TRIMETHYLAMMONI UM METHYLSULFATE, CARBAMATE see HNP500

(m-HYDROXYPHENYL)TRIMETHYLAMMONI UM METHYLSULFATE DIETHYLCARBAMATE see HNQ000

(3-HYDROXYPHENYL)TRIMETHYLAMMONI UM METHYL SULFATE DIMETHYLCARBAMIC ESTER see DQY909

(m-HYDROXYPHENYL)TRIMETHYLAMMONI UM METHYL SULFATE DIMETHYLCARBAMATE see DQY909

(m-HYDROXYPHENYL)TRIMETHYLAMMONI UM METHYLSULFATE, ETHYLCARBAMATE see HNQ500

(m-HYDROXYPHENYL)TRIMETHYLAMMONI UM METHYLSULFATE, METHYLCARBAMATE see HNR000

(m-HYDROXYPHENYL)TRIMETHYLAMMONI UM METHYLSULFATE METHYLPHENYLCARBAMATE see HNR500

(m-HYDROXYPHENYL)TRIMETHYLAMMONI UM METHYLSULFATE, PENTAMETHYLENECARBAMATE see HNS000

(m-HYDROXYPHENYL)TRIMETHYLAMMONI UM METHYLSULFATE, PHENYLCARBAMATE see HNS500

(S)-1-(3-HYDROXY-2-PHOSPHONYLMETHOXYPROPYL)CYTOSI NE see HNS550

N-HYDROXYPHTHALIMIDE see HNS600

3-HYDROXY-2-PICOLINE-4,5-DIMETHANOL see PPK250

N-HYDROXYPIPERIDINE see PIO925

3-(1-HYDROXY-2-PIPERIDINOETHYL)-5-PHENYLISOXAZOLE see PCJ325

3-(1-HYDROXY-2-PIPERIDINOETHYL)-5-PHENYLISOXAZOLE CITRATE see HNT075

3-(1-HYDROXY-2-PIPERIDINOETHYL)-5-PHENYLISOXAZOLE CITRATE (2:1) see PCJ325

2-HYDROXY-N-(3-(m-(PIPERIDINOMETHYL)PHENOXY)PROPYL)ACETAMIDE ACETATE (ester) HYDROCHLORIDE see HNT100

10-(3-(4-HYDROXYPIPERIDINO)PROPYL)PHENOT HIAZINE-2-CARBONITRILE see PIW000

10-(3-(4-HYDROXYPIPERIDINO)PROPYL)PHENOT HIAZIN-2-YL METHYL KETONE see HNT200

10-(3-(4-HYDROXYPIPERIDINO)PROPYL)PHENOT HIAZIN-2-YL METHYL KETONE see HNT200

HYDROXYPOLYETHOXYDODECANE see DXY000

17-HYDROXYPREGNA-4,6-DIENE-3,20-DIONE ACETATE see MCB375

17-HYDROXYPREGNA-4,6-DIENE-3,20-DIONE ACETATE (ESTER) see MCB375

3-HYDROXYPREGNANE-11,20-DIONE see AFK875

3-α-HYDROXY-5-α-PREGNANE-11,20-DIONE see AFK875

3-α-HYDROXY-5-α-PREGNANE-11,20-DIONE mixed with 3-α,21-DIHYDROXY-5-α-PREGNANE-11,20-DIONE 21-ACETATE (3:1) see AFK500

21-HYDROXYPREGNANE-3,20-DIONE SODIUM HEMISUCCINATE see VJZ000

21-HYDROXY-5-β-PREGNANE-3,20-DIONE SODIUM HEMISUCCINATE see VJZ000

21-HYDROXY-5-β-PREGNANE-3,20-DIONE, SODIUM SALT, HEMISUCCINATE see VJZ000

3-β-HYDROXY-5-α-PREGNAN-20-ONE-20-ISONICOTINYLHYDRAZONE see AFT125

21-HYDROXYPREGN-4-ENE-3,20-DIONE see DAQ600

17-HYDROXYPREGN-4-ENE-3,20-DIONE ACETATE see PMG600

21-HYDROXYPREGN-4-ENE-3,20-DIONE-21-ACETATE see DAQ800

17-HYDROXYPREGN-4-ENE-3,20-DIONE HEXANOATE see HNT500

17α,21-HYDROXYPREGN-4-ENE-3,11,20-TRIONE see CNS800

17-HYDROXY-17-α-PREGN-4-EN-20-YN-3-ONE see GEK500

HYDROXYPROCAINE see DHO600

m-HYDROXYPROCAINE see DHO600

21-HYDROXYPROGESTERONE see DAQ600

17-HYDROXYPROGESTERONE ACETATE see PMG600

HYDROXYPROGESTERONE CAPROATE see HNT500

17-α-HYDROXYPROGESTERONE CAPROATE see HNT500

17-α-HYDROXY PROGESTERONE-N-CAPROATE see HNT500

17-α-HYDROXYPROGESTERONE HEXANOATE see HNT500

HYDROXYPROLINE see HNT525

4-HYDROXYPROLINE see HNT525

HYDROXY-l-PROLINE see HNT525

l-HYDROXYPROLINE see HNT525

l-4-HYDROXYPROLINE see HNT525

trans-HYDROXYPROLINE see HNT525

trans-4-HYDROXYPROLINE see HNT525

trans-l-HYDROXYPROLINE see HNT525

3ᴬ)-(4-HYDROXY-l-PROLINE)ACTINOMYCIN D see AEC185

m-HYDROXYPROPADRINE see HNB875

1-HYDROXYPROPANE see PND000

2-HYDROXY-1,2,3-PROPANECARBOXYLIC ACID TERBIUM (3+) SALT (1:1) see TAM500

((2-HYDROXY-1,3-PROPANEDIYL)BIS(NITRILOBIS(METHYLENE)))TETRAKISPHOSPHONIC ACID see DYE550

3-HYDROXYPROPANENITRILE see HGP000

3-HYDROXY-1-PROPANESULFONIC ACID γ-SULTONE see PML400

3-HYDROXY-1-PROPANESULPHONIC ACID SULTONE see PML400

2-HYDROXY-1,2,3-PROPANETRICARBOXYLIC ACID see CMS750

2-HYDROXY-1,2,3-PROPANETRICARBOXYLIC ACID COPPER(2+) SALT (1:2) (9CI) see CNK625

2-HYDROXY-1,2,3-PROPANETRICARBOXYLIC ACID EBRIUM(3+) salt (1:1) see ECY000

2-HYDROXY-1,2,3-PROPANETRICARBOXYLIC ACID MONOSODIUM SALT see MRL000

2-HYDROXY,1,2,3-PROPANETRICARBOXYLIC ACID, TRIETHYL ESTER see TJP750

2-HYDROXY-1,2,3-PROPANETRICARBOXYLIC ACID TRILITHIUM SALT HYDRATE see LHD150

2-HYDROXY-1,2,3-PROPANETRISCARBOXYLIC ACID CERIUM(3+) SALT (1:1) (9CI) see CCZ000

2-HYDROXYPROPANNITRIL see LAQ000

2-HYDROXYPROPANOIC ACID see LAG000

(S)-2-HYDROXYPROPANOIC ACID see LAG010

2-HYDROXYPROPANOIC ACID, BUTYL ESTER see BRR600

2-HYDROXYPROPANOIC ACID CALCIUM SALT see CAT600

2-HYDROXYPROPANOIC ACID MONOSODIUM SALT see LAM000

1-HYDROXY-2-PROPANONE see ABC000

3-HYDROXY-2-PROPENAL SODIUM SALT see MAN800

3-HYDROXYPROPENE see AFV500

2-HYDROXYPROPIONIC ACID see LAG000

(S)-2-HYDROXYPROPIONIC ACID see LAG010

α-HYDROXYPROPIONIC ACID see LAG000

3-HYDROXYPROPIONIC ACID LACTONE see PMT100

2-HYDROXYPROPIONITRILE see LAQ000

3-HYDROXYPROPIONITRILE see HGP000

β-HYDROXYPROPIONITRILE see HGP000

N-HYDROXY-N-2-PROPIONYLAMINO FLUORENE see FEO000

l-5α-HYDROXYPROPIONYLAMINO-2,4,6-TRIIODOISOPHTHALIC ACID DI(1,3-DIHYDROXY-2-PROPYLAMIDE) see IFY000

HYDROXYPROPIOPHENONE see ELL500

4-HYDROXYPROPIOPHENONE see ELL500

p-HYDROXYPROPIOPHENONE see ELL500

2-(2-(2-HYDROXYPROPOXY)PROPOXY)-1-PROPANOL see TMZ000

O-(2-HYDROXYPROPYL)-1-ACETYLBENZOCYCLOBUTENE OXIME see HNT550

2-HYDROXYPROPYL ACRYLATE see HNT600

β-HYDROXYPROPYL ACRYLATE see HNT600

HYDROXY PROPYL ALGINATE see PNJ750

2-HYDROXYPROPYLAMINE see AMA500

3-HYDROXYPROPYLAMINE see PMM250

2'-(2-HYDROXY-3-(PROPYLAMINO)PROPOXY)-3-PHENYLPROPIOPHENONE HYDROCHLORIDE see PMJ525

(3-HYDROXYPROPYL)BENZENE see HHP050

α-HYDROXYPROPYLBENZENE see EGQ000

p-HYDROXYPROPYL BENZOATE see HNU500

3-HYDROXYPROPYL BROMIDE see BNY750

HYDROXYPROPYL CELLULOSE see HNV000

1-(3-HYDROXYPROPYL)-CNU see CHC250

S-(3-HYDROXYPROPYL)CYSTEINE see HNV050

S-(3-HYDROXYPROPYL)-l-CYSTEINE see HNV050

(E)-HYDROXYPROPYLDIAZENE POTASSIUM SALT see HNV100

(Z)-HYDROXYPROPYLDIAZENE POTASSIUM SALT see HNV150

1-(2-HYDROXYPROPYL)-3,7-DIMETHYLXANTHINE see HNY500

1-(3-HYDROXYPROPYL)-3,7-DIMETHYLXANTHINE see HNZ000

(2-HYDROXY-1,3-PROPYLENEDIAMINE-N,N,N'₁,N'₁-TETRAMETHYLENEPHOSPHORIC ACID see DYE550

HYDROXYPROPYL ETHER of CELLULOSE see HNV000

2-HYDROXYPROPYL METHACRYLATE see HNV500

β-HYDROXYPROPYL METHACRYLATE see HNV500

HYDROXYPROPYL METHYLCELLULOSE see HNX000

2-HYDROXYPROPYLMETHYLNITROSAMINE see NKU500

2-HYDROXYPROPYL 2-METHYL-2-PROPENOATE see HNV500

3-((2-HYDROXYPROPYL)NITROSAMINO)-1,2-PROPANEDIOL see NOC400

1-(((2-HYDROXYPROPYL)NITROSO)AMINO)ACETONE see HNX500

4-(N-(2-HYDROXYPROPYL)-N-NITROSOAMINO)-1-BUTANOL see HJR500

4-(N-(2-HYDROXYPROPYL)-N-NITROSOAMINO)-2-BUTANOL see HJR000

4-(N-(3-HYDROXYPROPYL)-N-NITROSOAMINO)-1-BUTANOL see HJS000

1-((2-HYDROXYPROPYL)NITROSOAMINO)-2-PROPANONE see HNX500

1-(2-HYDROXYPROPYL)-1-NITROSOUREA see NKO400

1-(3-HYDROXYPROPYL)-1-NITROSOUREA see NKK000

2-HYDROXYPROPYL PHENYL ARSINIC ACID see HNX600

N-(3-HYDROXYPROPYL)-1,2-PROPANEDIAMINE see HNX800

β-HYDROXYPROPYLPROPYLNITROSAMINE see NLM500

(2-HYDROXYPROPYL)PROPYLNITROSOAMINE see NLM500

HYDROXYPROPYL STARCH see HNY000

1-(2-HYDROXYPROPYL)THEOBROMINE see HNY500

1-(3-HYDROXYPROPYL)THEOBROMINE see HNZ000

1-(3-HYDROXYPROPYL)THEOBROMINE see HNZ000

1-(β-HYDROXYPROPYL)THEOBROMINE see HNY500

HYDROXYPROPYLTHEOPHYLLINE see HOA000

7-(2-HYDROXYPROPYL)THEOPHYLLINE see HOA000

β-HYDROXYPROPYLTHEOPHYLLINE see HOA000

7-(β-HYDROXYPROPYL)THEOPHYLLINE see HOA000

(3-HYDROXYPROPYL)TRIMETHOXYSILANE METHACRYLATE see TLC250

(2-HYDROXYPROPYL)TRIMETHYLAMMONIUM CHLORIDE see MIM300

(2-HYDROXYPROPYL)TRIMETHYLAMMONIUMCHLORIDE ACETATE see ACR000

(2-HYDROXYPROPYL)TRIMETHYLAMMONIUM CHLORIDE CARBAMATE see HOA500

HYDROXYPROPYL UC 10854 see HLK900

17-β-HYDROXY-17-(1-PROPYNYL)ESTR-4-EN-3-ONE see MNC100

17-β-HYDROXY-17-α-(1-PROPYNYL)ESTR-4-EN-3-ONE see MNC100

4-(1-(2-HYDROXY)PROPYOXY)BENZENEARSONIC ACID see PMQ000

N-HYDROXY-1H-PURIN-6-AMINE (9CI) see HIT000

3-HYDROXYPURIN-2(3H)-ONE see HOB000

4'-HYDROXYPYRAZOLOL(3,4-d)PYRIMIDINE see ZVJ000

4-HYDROXY-3,4-PYRAZOLOPYRIMIDINE see ZVJ000

4-HYDROXYPYRAZOLO(3,4-d)PYRIMIDINE see ZVJ000

4-HYDROXY-1H-PYRAZOLO(3,4-d)PYRIMIDINE see ZVJ000

4-HYDROXYPYRAZOLYL(3,4-d)PYRIMIDINE see ZVJ000

8-HYDROXYPYRENE-1,3,6-TRISULFONIC ACID SODIUM SALT see TNM000

8-HYDROXY-1,3,6-PYRENETRISULFONIC ACID TRISODIUM SALT see TNM000

N-(3-HYDROXY-1-PYRENYL)ACETAMIDE see HOB025

N-(6-HYDROXY-1-PYRENYL)ACETAMIDE see HOB050

6-HYDROXY-3(2H)-PYRIDAZINONE see DMC600

6-HYDROXY-3-(2H)-PYRIDAZINONE DIETHANOLAMINE see DHF200

2-HYDROXYPYRIDINE see OOO200

3-HYDROXYPYRIDINE see PPH025

4-HYDROXYPYRIDINE see HOB100

β-HYDROXYPYRIDINE see PPH025

γ-HYDROXYPYRIDINE see HOB100

1-HYDROXY-2-PYRIDINETHIONE see HOB500

1-HYDROXY-2-(1H)-PYRIDINETHIONE see HOB500

1-HYDROXY-2-(1H)-PYRIDINETHIONE SODIUM SALT see HOC000

(1-HYDROXY-2-PYRIDINETHIONE), SODIUM SALT, TECH see MCQ750

β-(N-(3-HYDROXY-4-PYRIDONE))-α-AMINOPROPIONIC ACID see HOC500

5-(α-HYDROXY-α-2-PYRIDYLBENZYL)-7-(α-2-PYRIDYLBENZYLIDENE)-5-NORBORNENE-2,3-DICARBOXIMIDE see NNF000

4-HYDROXYPYRIMIDINE see ORS050

4-HYDROXYPYRIMIDINE see PPP140

6-HYDROXYPYRIMIDINE see ORS050

6-HYDROXYPYRIMIDINE see PPP140

4-HYDROXY-2(1H)-PYRIMIDINETHIONE see TFR250

6-HYDROXY-4(1H)-PYRIMIDINONE see PPP100

8-HYDROXYQUINALDIC ACID see HOE000

HYDROXYQUINALDINE see HMQ600

8-HYDROXYQUINALDINE see HMQ600

HYDROXYQUINOL see BBU250

4-HYDROXYQUINOLINAMINE-1-OXIDE MONOHYDROCHLORIDE see HIZ000

N-HYDROXY-4-QUINOLINAMINE-1-OXIDE MONOHYDROCHLORIDE see HIZ000

8-HYDROXYQUINOLINE see QPA000

8-HYDROXYQUINOLINE BENZOATE see HOE100

8-HYDROXYQUINOLINE COPPER COMPLEX see BLC250

8-HYDROXYQUINOLINE GLUCURONIDE see QPS100

8-HYDROXYQUINOLINE SUCCINATE see HOE150

8-HYDROXYQUINOLINE SULFATE see QPS000

1-(8-HYDROXY-5-QUINOLINYL)ETHANONE see HOE200

8-HYDROXY-5-QUINOLYL METHYL KETONE see HOE200

3-HYDROXYQUINUCLIDINE see QUJ990

14-HYDROXY-4,14-RETRO-RETINOL see RFK100

14(R)-HYDROXY-RETRO-VITAMIN A see RFK100

14-HYDROXY-3-β-(RHAMNOSYLOXY)BUFA-4,20,22-TRIENOLIDE see POB500

5-HYDROXY-1-β-d-RIBOFURANOSYL-1H-IMIDAZOLE-4-CARBOXAMIDE see BMM000

1'-HYDROXYSAFROLE see BCJ000

1'-HYDROXYSAFROLE-2',3'-OXIDE see HOE500

4-HYDROXYSALICYLALDEHYDE see REF100

4'-HYDROXYSALICYLANILIDE see DYE700

p'-HYDROXYSALICYLANILIDE see DYE700

5-HYDROXYSALICYLATE SODIUM see GCU050

4-HYDROXYSALICYLIC ACID see HOE600

5-HYDROXYSALICYLIC ACID see GCU000

p-HYDROXYSALICYLIC ACID see HOE600

14-β-HYDROXY-3-β-SCILLOBIOSIDOBUFA-4,20,22-TRIENOLIDE see GFC000

3-HYDROXY-16,7-SECOESTRA-1,3,5(10)-TRIEN-17-OIC ACID see DYB000

12-HYDROXYSENECIONAN-11,16-DIONE see ARS500

(15E)-12-HYDROXY-SENECIONAN-11,16-DIONE (9CI) see IDG000

HYDROXYSENKIRKINE see HOF000

5-HYDROXY-1-(N-SODIO-5-TETRAZOLYLAZO)TETRAZOLE see HOF500

3-HYDROXY-SPIRO(8-AZONIABICYCLO(3.2.1)OCTANE-8,1'-PYRROLIDINIUM CHLORIDE) BENZILATE see BCA375

3-α-HYDROXY-SPIRO(1-α-H,5-α-H-NORTROPANE-8,1'-PYRROLIDINIUM) CHLORIDE BENZILATE see KEA300

12-HYDROXYSTEARIC ACID see HOG000

12-HYDROXYSTEARIC ACID see HOG000

12-HYDROXYSTEARIC ACID, METHYL ESTER see HOG500

HYDROXYSTREPTOMYCIN see SLX500

4-HYDROXYSTYRENE see VQA200

p-HYDROXYSTYRENE see VQA200

(E)-4'-(p-HYDROXYSTYRYL) ACETANILIDE see HIK000

HYDROXYSUCCINIC ACID see MAN000

α-HYDROXYSUCCINIC ACID see MAN000

4-HYDROXY-3-((4-SULFO-1-NAPHTHALENYL)AZO)-1-NAPHTHALENESULFONIC ACID, DISODIUM SALT see HJF500

3-HYDROXY-4-((4-SULFO-1-NAPHTHALENYL)AZO)-2,7-NAPHTHLENEDISULFONIC ACID, TRISODIUM SALT see FAG020

3-HYDROXY-4-((4-SULFO-1-NAPHTHYL)AZO)-2,7-NAPHTHALENEDISULFONIC ACID, TRISODIUM SALT see FAG020

6-HYDROXY-5-((4-SULFOPHENYL)AZO)-2-NAPHTHALENESULFONIC ACID, DISODIUM SALT see FAG150

6-HYDROXY-5-((p-SULFOPHENYL)AZO)-2-NAPHTHALENESULFONIC ACID, DISODIUM SALT see FAG150

1-HYDROXY-2-(3-SULFOPROPOXY)ANTHRAQUINONE SODIUM SALT see HOH000

4-HYDROXY-3-((5-SULFO-2,4-XYLYL)AZO)-1-NAPHTHALENESULFONIC ACID, DISODIUM SALT see FAG050

3-HYDROXY-4-((4-SULPHO-1-NAPHTHALENYL)AZO)-2,7-NAPHTHALENEDISULPHONIC ACID, TRISODIUM SALT see FAG020

3-HYDROXY-4-((4-SULPHO-1-NAPHTHYL)AZO)-2,7-NAPHTHALENEDISULPHONIC ACID, TRISODIUM SALT see FAG020

6-HYDROXY-5-((4-SULPHOPHENYL)AZO)-2-NAPHTHALENESULPHONIC ACID, DISODIUM SALT see FAG150

6-HYDROXY-5-((p-SULPHOPHENYL)AZO)-2-NAPHTHALENESULPHONIC ACID, DISODIUM SALT see FAG150

4-HYDROXY-3-((5-SULPHO-2,4-XYLYL)AZO)-1-NAPHTHALENESULPHONIC ACID, DISODIUM SALT see FAG050

HYDROXYTAMOXIFEN see HOH100

4-HYDROXYTAMOXIFEN see HOH100

α-HYDROXYTAMOXIFEN see HOH200

5-HYDROXYTETRACYCLINE see HOH500

5-HYDROXYTETRACYCLINE HYDROCHLORIDE see HOI000

N-HYDROXY-N-TETRADECANOYL-2-AMINOFLUORENE see HMU000

m-HYDROXYTETRAETHYLDIAMINOTRIPHENYLCARBINOL ANHYDRIDE DISULFONIC ACID CALCIUM SALT see CMM062

3-HYDROXYTETRAHYDROFURAN see HOI200

4-HYDROXY-3-(1,2,3,4-TETRAHYDRO-1-NAFTYL)-4-CUMARINE (DUTCH) see EAT600

4-HYDROXY-3-(1,2,3,4-TETRAHYDRO-1-NAPHTHALENYL)-2H-1-BENZOPYRAN-2-ONE (9CI) see EAT600

4-HYDROXY-3-(1,2,3,4-TETRAHYDRO-1-NAPHTHYL)CUMARIN see EAT600

2-HYDROXY-1,2,3,3-TETRAHYDRO-3H-PYRANO (3,2-F)QUINOLINE-8(7H)-ONE see HOI222

4-HYDROXY-3-(1,2,3,4-TETRAHYDRO-3-(4-(4-(TRIFLUOROMETHYL)PHENOXY)PHENYL)-1-NAPHTHALENYL)-2H-1-BENZOPYRAN-2-ONE see HOI245

4-HYDROXY-3-(1,2,3,4-TETRAHYDRO-3-(4-(TRIFLUOROMETHYL)PHENYL)-1-NAPHTHALENYL)2H-BENZOPYRAN-2-ONE see HOI265

6-HYDROXY-2,5,7,8-TETRAMETHYLCHROMAN-2-CARBOXYLIC ACID see TNR625

7-HYDROXY-3,4-TETRAMETHYLENECOUMARIN-O,O-DIETHYL THIOPHOSPHATE see DXO000

4-HYDROXY-2,2,6,6-TETRAMETHYL-1-PIPERIDINYLOXY see HOI300

3-HYDROXY-2,2,5,5-TETRAMETHYLTETRAHYDRO-3-FURYL METHYL KETONE see HOI400

14-HYDROXY-6-β-THEBAINOL 4-METHYL ETHER see ORE000

7-HYDROXYTHEOPHYLLIN (GERMAN) see HOJ000

7-HYDROXYTHEOPHYLLINE see HOJ000

9-(2-HYDROXYTHEOXYMETHYL)GUANINE see AEC700

6-HYDROXY-2-THIO-8-AZAPURINE see HOJ100

HYDROXYTHIOSPASMIN see HOJ150

α-HYDROXY-THYMIDINE (9CI) see HMB550

HYDROXYTOLUENE see BDX500

4-HYDROXYTOLUENE see CNX250

m-HYDROXYTOLUENE see CNW750

o-HYDROXYTOLUENE see CNX000

p-HYDROXYTOLUENE see CNX250

α-HYDROXYTOLUENE see BDX500

α-HYDROXY-α-TOLUIC ACID see MAP000

1-HYDROXY-4-(p-TOLUIDINO)ANTHRAQUINONE see HOK000

HYDROXYTOLUOLE (GERMAN) see CNW500

2-(m-HYDROXY-N-p-TOLYLANILINOMETHYL)-2-IMIDAZOLINE see PDW400

4'-((6-HYDROXY-m-TOLYL)AZO)ACETANILIDE see AAQ250

2-HYDROXY-3-o-TOLYLOXYPROPYL-1-CARBAMATE see CBK500

(3-HYDROXY-p-TOLYL)TRIMETHYLAMMONIUM CHLORIDE,METHYLCARBAMATE see HOL000

3-HYDROXY-4β,8-α-15-TRIACETOXY-12,13-EPOXYTRICHOTHEC-9-ENE see ACS500

4-HYDROXY-3H-o-TRIAZOLO(4,5-d)PYRIMIDINE-5,7(4H-6H)-DIONE see HJE575

(HYDROXY)TRIBUTYLSTANNANE see TID500

HYDROXYTRIBUTYLSTANNANE-4,4-
DIMETHYLOCTANOATE see TIF250
HYDROXYTRIBUTYLSTANNANE,
SULFATE (2:1) see TIF600
β-HYDROXYTRICARBALLYLIC ACID see
CMS750
1-HYDROXY-2,2,2-TRICHLOROETHYL
DIETHYL PHOSPHONITE see EPY600
1-HYDROXY-2,2,2-
TRICHLOROETHYLPHOSPHONIC ACID
DIMETHYL ESTER see TIQ250
(1-HYDROXY-2,2,2-
TRICHLOROETHYL)UREA see HOL100
2'-HYDROXY-2,4,4'-TRICHLORO-
PHENYLETHER see TIQ000
2-HYDROXY-3,5,6-TRICHLOROPYRIDINE
see HOL125
6-(4-HYDROXY-3-
TRICYCLO(3.3.1.1(3,7))DEC-1-YLPHENYL)-
2-NAPHTHALENECARBOXYLIC ACID see
HOL145
2-HYDROXYTRIETHYLAMINE see DHO500
(S)-N-(3-HYDROXY-9,10,11-TRIMETHOXY-
5H-DIBENZO(a,c)CYCLOHEPTEN-5-YL)-
ACETAMIDE (9CI) see ACG250
8-HYDROXY-6,10,11-TRIMETHOXY-3A,12C-
DIHYDRO-7H-FURO(3',2':4,5)FURO(2,3-
C)XANTHEN-7-ONE see HOL200
8-HYDROXY-1,3,5-
TRIMETHOXYXANTHEN-9-ONE see
HOL223
17-β-HYDROXY-4,4,17-α-TRIMETHYL-
ANDROST-5-ENE(2,3-d)ISOXAZOLE see
HOM259
3-HYDROXY-N,N,N-
TRIMETHYLBENZENAMINIUM IODIDE
see HNN500
3-HYDROXY-N,N,N-
TRIMETHYLBENZENEETHANAMINIUM
PICRATE see HNG000
6-HYDROXY-β,2,7-TRIMETHYL-5-
BENZOFURANACRYLIC ACID γ-
LACTONE see HOM270
((2-
HYDROXYTRIMETHYLENE)DINITRILO)T
ETRAACETIC ACID see DCB100
2-HYDROXY-N,N,N-
TRIMETHYLETHANAMINIUM see CMF000
2-HYDROXY-N,N,N-
TRIMETHYLETHANAMINIUM SALT with
3,7-DIHYDRO-1,3-DIMETHYLPURINE-2,6-
DIONE see CMG300
2-HYDROXY-N,N,N-
TRIMETHYLETHANAMINIUM SALT with 2-
HYDROXYBENZOIC ACID (1:1) see
CMG000
3-HYDROXY-2,2,4-TRIMETHYL-3-
PENTENOIC ACID, β-LACTONE see IPL000
3-HYDROXY-4-((2,4,5-
TRIMETHYLPHENYL)AZO)-2,7-
NAPHTHALENEDISULFONIC ACID,
DISODIUM SALT see FAG018
3-HYDROXY-4-((2,4,5-
TRIMETHYLPHENYL)AZO)-2,7-
NAPHTHALENEDISULPHONIC ACID,
DISODIUM SALT see FAG018
2-HYDROXY-N,N,N-TRIMETHYL-1-
PROPANAMINIUM CHLORIDE see MIM300
HYDROXYTRIMETHYLSTANNANE see
TMI250
2-HYDROXY-1,3,5-TRINITROBENZENE see
PID000
3-HYDROXY-2,4,6-TRINITROPHENOL see
SMP500
HYDROXYTRIPHENYLSILANE see
HOM300
HYDROXYTRIPHENYLSTANNANE see
HON000
HYDROXYTRIPHENYLTIN see HON000
3-HYDROXYTROPOLONE see HON500
5-HYDROXYTRYPTAMINE see AJX500

5-HYDROXYTRYPTAMINE CREATININE
SULFATE see AJX750
5-HYDROXYTRYPTAMINE CREATININE
SULFATE MONOHYDRATE see AJX750
HYDROXYTRYPTOPHAN see HOO100
5-HYDROXYTRYPTOPHAN see HOA575
5-HYDROXYTRYPTOPHAN see HON800
5-HYDROXYTRYPTOPHAN see HOO100
5-HYDROXY-l-TRYPTOPHAN see HOA600
5-HYDROXY-l-TRYPTOPHAN see HOO000
dl-HYDROXYTRYPTOPHAN see HOA575
dl-HYDROXYTRYPTOPHAN see HON800
l-5-HYDROXYTRYPTOPHAN see HOA600
l-5-HYDROXYTRYPTOPHAN see HOO000
(±)-5-HYDROXYTRYPTOPHAN see HOA575
(±)-5-HYDROXYTRYPTOPHAN see HON800
dl-5-HYDROXYTRYPTOPHAN see HOA575
5-HYDROXYTRYPTOPHANE see HOO100
5-HYDROXYTRYPTOPHOL see HLI600
3-HYDROXYTYRAMINE see DYC400
3-HYDROXYTYRAMINE
HYDROCHLORIDE see DYC600
m-HYDROXYTYRAMINE
HYDROCHLORIDE see DYC600
3-HYDROXY-l-TYROSINE see DNA200
l-o-HYDROXYTYROSINE see DNA200
3-HYDROXY-l-TYROSINE
HYDROCHLORIDE see DYC200
4-HYDROXYUNDECANOIC ACID
LACTONE see HBN200
5-HYDROXYUNDECANOIC ACID
LACTONE see UKJ000
4-HYDROXYUNDECANOIC ACID, γ-
LACTONE see HBN200
HYDROXYUREA see HOO500
N-HYDROXYUREA see HOO500
N-HYDROXYURETHAN see HKQ025
N-HYDROXYURETHANE see HKQ025
3-HYDROXYURIC ACID see HOO875
4-HYDROXYVALERIC ACID LACTONE see
VAV000
17-HYDROXY-17-α-VINYL-4-ESTREN-3-
ONE see NCI525
1-α-HYDROXYVITAMIN D3 see HJV000
9-HYDROXYXANTHENE see XBJ000
3-HYDROXYXANTHINE see HOP000
3-HYDROXYXANTHINE see HOQ500
7-HYDROXYXANTHINE see HOP259
3-HYDROXYXANTHINE ACETATE see
HOP300
1-HYDROXYXANTHINE DIHYDRATE see
HOQ000
3-HYDROXYXANTHINE HYDRATE see
HOQ500
3-HYDROXY-4-(2,4-XYLYLAZO)-3,7-
NAPHTHALENEDISULFONIC ACID,
DISODIUM SALT see FMU070
3-HYDROXY-4-(2,4-XYLYLAZO)-3,7-
NAPHTHALENEDISULPHONIC ACID,
DISODIUM SALT see FMU070
3-(4-HYDROXY-3,5-XYLYL)-2-METHYL-
4(3H)-QUINAZOLINONE see DSH800
17-HYDROXYYOHIMBAN-16-
CARBOXYLIC ACID METHYL ESTER see
YBJ000
17-α-HYDROXY-20-α-YOHIMBAN-16-β-
CARBOXYLIC ACID, METHYL ESTER,
HYDROCHLORIDE see YCA000
o-HYDROXYZIMTSAEURE-LACTON
(GERMAN) see CNV000
HYDROXYZINE see CJR909
HYDROXYZINE DIHYDROCHLORIDE see
HOR470
HYDROXYZINE HYDROCHLORIDE see
HOR470
HYDROXYZINE HYDROCHLORIDE see
VSF000
HYDROXYZINE PAMOATE see HOR500
HYDRURE de LITHIUM (FRENCH) see
LHH000
HYDURA see HOO500
HYFLAVIN see RIK000

HYGROMIX-8 see AQB000
HYGROMULL see UTU500
HYGROMYCIN B (USDA) see AQB000
HYGROSTATIN see HOS500
HYGROTON see CLY600
HYGROTON-RESERPINE see RDK000
HYKINONE see MCB575
HYKOLEX see DAL000
HYLEMOX see EEH600
HYLENE M50 see MJP400
HYLENE-T see TGM740
HYLENE T see TGM750
HYLENE TCPA see TGM750
HYLENE TLC see TGM750
HYLENE TM see TGM750
HYLENE TM see TGM800
HYLENE TM-65 see TGM750
HYLENE TRF see TGM750
HYLENTA see BFD000
HYLITE LF see DTG000
HYLUTIN see HNT500
HYMECROMONE see MKP500
HYMECROMONE SODIUM see HMB000
HYMENOCALLIS (VARIOUS SPECIES) see
BAR325
HYMENOVIN see HOT000
HYMENOXON see HOT200
HYMENOXONE see HOT200
HYMEXAZOL see HLM000
HYMINAL see QAK000
HYMORPHAN see DLW600
HYMORPHAN see DNU300
HYONIC PE-250 see PKF500
HYOSAN see MJM500
HYOSCIN-N-BUTYLBROMID (GERMAN)
see SBG500
HYOSCIN-N-BUTYL BROMIDE see SBG500
HYOSCINE see SBG000
(−)-HYOSCINE see SBG000
HYOSCINE BROMIDE see HOT500
HYOSCINE BUTOBROMIDE see SBG500
HYOSCINE BUTYL BROMIDE see SBG500
HYOSCINE-N-BUTYL BROMIDE see
SBG500
HYOSCINE F HYDROBROMIDE see
HOT500
HYOSCINE HYDROBROMIDE see HOT500
(−)-HYOSCINE HYDROBROMIDE see
HOT500
1-HYOSCINE HYDROBROMIDE see
HOT500
HYOSCINE METHYL BROMIDE see SBH500
(−)-HYOSCYAMINE see HOU000
HYOSCYAMINE see HOU000
1-HYOSCYAMINE see HOU000
dl-HYOSCYAMINE see ARR000
HYOSCYAMINE METHYLBROMIDE see
MGR250
dl-HYOSCYAMINE METHYLNITRATE see
MGR500
HYOSCYAMINE SULFATE see HOT600
(−)-HYOSCYAMINE SULFATE see HOT600
HYOSCYAMUS NIGER see HAQ100
HYOSCYINE HYDROBROMIDE see
HOT500
HYOSOL see SBG000
dl-HYOSYAMINE METHYLNITRATE see
MGR500
HYOZID see ILD000
HYPAQUE see SEN500
HYPAQUE 60 see AOO875
HYPAQUE 13.4 see AOO875
HYPAQUE CYSTO see AOO875
HYPAQUE M 30 see AOO875
HYPAQUE MEGLUMINE see AOO875
HYPAQUE SODIUM see SEN500
HYPCOL see QAK000
HYPERAN see HFS759
HYPERAZIN see HGP500
HYPERBUTAL see BPF500
HYPERCAL B see RDK000
HYPERNEPHRIN see VGP000

HYPERNIC EXTRACT see LFT800
HYPEROL see HIB500
HYPERPAX see DNA800
HYPERPAX see MJE780
HYPERSIN see BFW250
HYPERSTAT see DCQ700
HYPERTANE FORTE see RDK000
HYPERTENAIN see MAW250
HYPERTENSAN see RDK000
HYPERTENSIN see AOO900
HYPERTONALUM see DCQ700
HY-PHI 1055 see OHU000
HY-PHI 1088 see OHU000
HY-PHI 1199 see SLK000
HY-PHI 2066 see OHU000
HY-PHI 2088 see OHU000
HY-PHI 2102 see OHU000
HYPHYLLINE see DNC000
HYPNODIN see HOU059
HYPNOGEN see EOK000
HYPNOGENE see BAG000
HYPNOMIDATE see HOU100
HYPNON see DLV000
HYPNONE see ABH000
HYPNOREX see LGZ000
HYPNORM see HAF400
HYPNOSTAN see PBT500
HYPNO-TABLINETTEN see EOK000
HYPO see SKI000
HYPO see SKI500
HYPOCHLORITES see HOU500
HYPOCHLORITE SOLUTIONS containing >7% available chlorine by wt. see SHU500
HYPOCHLORITE SOLUTIONS with >5% but <16% available chlorine by wt. (DOT) see SHU500
HYPOCHLORITE SOLUTIONS with 16% or more available chlorine by wt. (UN 1791) see SHU500
HYPOCHLOROUS ACID see HOV000
HYPOCHLOROUS ACID, CALCIUM SALT see HOV500
HYPOCHLOROUS ACID, CALCIUM SALT, DIHYDRATE see HOV503
HYPOCHLOROUS ACID, CALCIUM SODIUM SALT (3:1:1) see CAX255
HYPOCHLOROUS ACID, POTASSIUM SALT see PLK000
HYPOCHLOROUS ACID, SODIUM SALT, HEPTAHYDRATE see SHU510
HYPOCHLOROUS ACID, SODIUM SALT, HYDRATE (2:5) see SHU515
HYPODERMACID see TIQ250
HYPOGLYCIN see MJP500
HYPOGLYCIN A see MJP500
HYPOGLYCIN B see HOW100
HYPOGLYCINE A see MJP500
HYPOGLYCINE B see HOW100
HYPONITROUS ACID see HOW500
HYPONITROUS ACID ANHYDRIDE see NGU000
α-HYPOPHAMINE see ORU500
HYPOPHENON see ELL500
HYPOPHOSPHOROUS ACID see PGY250
HYPOPHTHALIN see HGP495
HYPOPHTHALIN see HGP500
HYPOPHYSEAL GROWTH HORMONE see PJA250
HYPOPRESOL see OJD300
HYPORENIN see VGP000
HYPOS see HGP500
HYPOSTEROL see NMV300
HYPOTHIAZIDE see CFY000
HYPOTROL see SBM500
HYPOTROL see SBN000
HYPOXANTHINE see DMC000
HYPOXANTHINE NUCLEOSIDE see IDE000
(HYPOXANTHINE)PENTAAMMINERUTHE NIUM(3+) TRICHLORIDE see HOW600
HYPOXANTHINE RIBONUCLEOSIDE see IDE000

HYPOXANTHINE RIBOSIDE see IDE000
HYPOXANTHINE-d-RIBOSIDE see IDE000
HYPOXANTHOSINE see IDE000
HYPROVAL-PA see HNT500
HYPTOR BASE see QAK000
HYRCANIN see HOW800
HYRE see RIK000
HYSCO see HOT500
HYSCYLENE P see PDX000
HYSONE-A see HHQ800
HYSSOP OIL see HOX000
HYSTEROL see HOX100
HYSTRENE 80 see SLK000
HYSTRENE 8016 see PAE250
HYSTRENE 9014 see MSA250
HYSTRENE 9512 see LBL000
HYTAKEROL see DME300
HYTANE 500L see IRA050
HYTONE LOTION see CNS750
HYTOX see MIA250
HYTRIN see TEF700
HYVAR see BMM650
HYVAR see BNM000
HYVAREX see BMM650
HYVAR X see BMM650
HYVAR X BROMACIL see BMM650
HYVAR X WEED KILLER see BMM650
HYVERMECTIN see ITD875
HYZYD see ILD000
I 535 see AAU300
I 635 see AAU300
I 677 see OLK200
I 735 see AAU300
I 1173 see HMQ100
I-1431 see CQH000
I 337A see CDP250
I 7840 see IKG900
IA see IDZ000
IA see IHN200
IA-4 see CFP750
IA-6 see DHU910
IA 887 see NCP875
IAA see ICN000
IAB see AOC500
I ACID see AKI000
IAMBOLEN see PKF750
IA-4 N-OXIDE see CFQ000
IA-PRAM see DLH630
IARACTAN see TAF675
IdB-1027 see COI750
IBA see ICP000
IBATRAN see ACE000
IBD see CCK125
IBD 78 see MGL500
IBDU see IIV000
IBENZMETHYZINE see PME250
IBENZMETHYZINE HYDROCHLORIDE see PME500
IBENZMETHYZIN HYDROCHLORIDE see PME500
IBERIN see MPN100
IBIFUR see FPI000
IBINOLO see TAL475
IBIODORM see SKS700
IBIODRAL see BBV500
IBIOFURAL see NGE500
IBIOSUC see SGC000
IBIOTON see TAI500
IBIOTYZIL see BCA000
IBIOZEDRINE see AOB250
IBN see IJD000
IBOA see IHX700
IBOGAMINE-18-CARBOXYLIC ACID, 3,4-DIDEHYDRO-, METHYL ESTER, (2-α-5-β,6-α-18-β)- see CCP925
IBOGAMINE-18-CARBOXYLIC ACID, METHYL ESTER, HYDROCHLORIDE see CNS200
IBOGAMINE-18-CARBOXYLIC ACID, METHYL ESTER, MONOHYDROCHLORIDE (9CI) see CNS200
IBOMAL see QCS000

IBOTENIC ACID see AKG250
IBOTENSAURE (GERMAN) see AKG250
IBP see BKS750
IBP see DIU800
IBUFEN see IIU000
IBUFENAC see IJG000
IBUNAC see IJG000
IBUPROCIN see IIU000
IBUPROFEN see IIU000
IBUPROFEN GUAIACOL ESTER see IJJ000
IBUPROFEN PICONOL see IAB000
IBUPROFEN SODIUM see IAB100
IBUTILIDE FUMARATE see IAB200
IBYLCAINE HYDROCHLORIDE see IAC000
IBZ see PME500
I-C 26 see DWC100
IC 6002 see PIW000
ICCSH see LIU300
ICE-NUCLEATION-ACTIVE PSEUDOMONAS SYRINGAE see IAC100
ICF 205-930 see TNU100
ICG see CCK000
ICHDEN see IAD000
ICHTAMMON see IAD000
ICHTHADONE see IAD000
ICHTHALUM see IAD000
ICHTHAMMOL see IAD000
ICHTHAMMONIUM see IAD000
ICHTHIUM see IAD000
ICHTHOSAN see IAD000
ICHTHOSAURAN see IAD000
ICHTHOSULFOL see IAD000
ICHTHYMALL see IAD000
ICHTHYNAT see IAD000
ICHTHYOL see IAD000
ICHTHYOPON see IAD000
ICHTHYSALLE see IAD000
ICI 180 see MEL780
ICI 350 see AJH129
ICI 543 see PMP500
ICI 3435 see SNI500
ICI 28257 see ARQ750
ICI 32525 see SNL800
ICI-32865 see TJQ333
I.C.I. 33,828 see MLJ500
ICI 35868 see DNR800
ICI 38174 see INT000
ICI 42464 see CHT500
ICI 43823 see BPI300
ICI 45520 see ICB000
ICI 45520 see ICC000
ICI 46,474 see NOA600
ICI 46,474 see TAD175
ICI 46638 see DMZ000
ICI 48213 see FBP100
ICI 50172 see ECX100
ICI 51426 see CQH625
ICI 54,450 see CKK250
ICI 55,548 see DBA710
ICI-58834 see VKA875
ICI 58,834 see VKF000
ICI 59118 see RCA375
ICI 66,082 see TAL475
ICI 74205 see EPC200
ICI 79,280 see HOH100
ICI 79,939 see IAD100
ICI 80996 see CMX880
ICI 81008 see ECW550
ICI 118587 see XAH000
ICI 123215 see LIU420
ICI 142268 see DHA250
ICI 146814 see GJU600
ICI 151291 see BJW600
ICI 162846 see IBM100
ICI 204636 see QCJ300
racemic-ICI 74205 see EPC200
racemic-ICI 79,939 see IAD100
racemic-ICI 80,996 see CMX880
racemic-ICI 81,008 see ECW550
ICI 47,699 CITRATE see DVY875
ICI 156834 DISODIUM see CCS371
ICI-EP 5850 see MQD750

ICIG 180 see MEL780
ICIG 770 see EAI850
ICIG 772 see MEL775
ICIG 776 see MKW150
ICIG 777 see ICW150
ICIG 778 see BEO500
ICIG 779 see MFN700
I.C.I.G. 1105 see RFU600
ICIG 1109 see CGV250
ICIG 1110 see CHD250
I.C.I.G. 1163 see ROF200
I.C.I.G. 1325 see BIF625
I.C.I. HYDROCHLORIDE see INT000
I.C.I. LTD. COMPOUND NUMBER 80996 see CMX880
ICI 54856 METHYL ESTER see MIO975
ICIPEN see PDT750
ICI-PP 333 see TMK125
ICI-PP 557 see AHJ750
ICI-PP 563 see GJU600
ICN-1229 see RJA500
ICON see LAS200
ICORAL B see HNB875
ICR 10 see QDS000
ICR-125 see QCS875
ICR 170 see ADJ875
ICR 180 see CGS750
ICR-191 see MEJ250
ICR 217 see CGX750
ICR 220 see BHZ000
ICR-25A see CLD500
ICR 290 see CGX500
ICR 292 see EHJ000
ICR 311 see EHI500
ICR 340 see IAE000
ICR 342 see CFA250
ICR 355 see BPI500
ICR 368 see CGY000
ICR 372 see CGS500
ICR 377 see EHJ500
ICR 394 see CHH000
ICR 395 see CGR500
ICR 433 see CIN500
ICR 442 see CKU250
ICR 449 see ADQ500
ICR-450 see BIJ750
ICR 451 see CIG250
ICR 486 see CFB500
ICR-48b see DFH000
ICR 498 see BNO750
ICR 502 see BNR000
ICR 506 see BNO500
ICR DIHYDROCHLORIDE see MEJ250
ICRF-159 see PIK250
ICRF 159 see RCA375
ICR 170-OH see CHY750
ICS 205930 see TNU100
ICS 205-930 see TNU100
ICSH see LIU300
ICTALIS SIMPLE see DKQ000
ID 540 see FDB100
ID-622 see IAG300
ID-1229 see FGQ000
IDA see IBH000
IDALENE see MOV500
IDANTOIL see DNU000
IDANTOIN see DKQ000
IDANTOINAL see DNU000
IDARAC see TKG000
ID 480 DIHYDROCHLORIDE see DAB800
IDEXUR see DAS000
IDMT see IOW500
IDOCYL NOVUM see SJO000
IDO-K see MCB575
IDOMETHINE see IDA000
IDONOR see PJB500
IDOSERP see RDK000
IDOXENE see DAS000
IDOXURIDIN see DAS000
IDOXURIDINE see DAS000
IDPN see BIQ500
IDRAGIN see ADA725

IDRALAZINA (ITALIAN) see HGP495
IDRAZIDE DELL'ACIDO ISONICOTINICO see ILD000
IDRAZIL see ILD000
IDRAZINA IDRATA see HGU501
IDRAZINA SOLFATO (ITALIAN) see HGW500
2-IDRAZINO-6-(N,N-BIS(2-IDROSSIETIL)-AMINO)-PIRIDAZINA CLORIDRATO (ITALIAN) see BKB500
3-IDRAZINO-6-(N-(2-IDROSSIPROPIL)METILAMINO)PIRIDAZINA DICLORIDRATO (ITALIAN) see HHD000
IDRIANOL see SPC500
IDROBUTAZINA see HNI500
IDROCHINONE (ITALIAN) see HIH000
IDROESTRIL see DKA600
IDROGENO SOLFORATO (ITALIAN) see HIC500
IDROGESTENE see HNT500
IDROPEROSSIDO di CUMENE (ITALIAN) see IOB000
IDROPEROSSIDO di CUMOLO (ITALIAN) see IOB000
IDROSSIDO DI STAGNO TRIFENILE (ITALIAN) see HON000
1',1-(2-IDROSSIETIL)4-(3-(2-CLORO-10-FENOTIAZIL)PROPILPIPERAZINA (ITALIAN) see CJM250
(+−)-1-(4-IDROSSIFENIL)-2-METILAMINOETANOLO TARTRATO (ITALIAN) see SPD000
4-IDROSSI-4-METIL-PENTAN-2-ONE (ITALIAN) see DBF750
d-N,N'-(1-IDROSSIMETIL PROPIL)-ETILENDINITROSAMINA (ITALIAN) see HMQ500
4-IDROSSI-3-(3-OXO-)-(FENIL-BUTIL)-CUMARINE (ITALIAN) see WAT200
l-5α-IDROSSIPROPIONILAMINO-2,4,6-TRIIODOISOFTAL-DI(1,3-DIIDROSSI-2-PROPILAMIDE) see IFY000
4-IDROSSI-3-(1,2,3,4-TETRAIDRO-1-NAFTIL)CUMARINA (ITALIAN) see EAT600
IDROSSIZINA see CJR909
IDROTIADENE see HII000
IDROTIAZIDE see CFY000
IDRYL see FDF000
IDSOSERP see RDK000
IDU see DAS000
IDUCHER see DAS000
IDULEA see DAS000
IDULIAN see DLV800
IDUOCULOS see DAS000
IDUR see DAS000
IDURIDIN see DAS000
I-EBU see EAN525
IEM-1-15 see CEX250
IEM 455 see CJN750
IERGIGAN see DQA400
IEROIN see HBT500
IF 2303 see PHB550
IF (fumigant) see TBV750
IFC see CBM000
IFC-45 see CGG500
IFENEC see EAE000
IFENPRODIL see IAG600
IFENPRODIL TARTRATE see IAG625
IFENPRODIL TARTRATE (2:1) see IAG625
IFENPRODIL l-(+)-TARTRATE see IAG625
IFIBRIUM see LFK000
IFOSFAMID see IMH000
IFOSFAMIDE see IMH000
IF ROM 203 see IAG700
IGE see IPD000
IGELITE F see PKQ059
IGE (OSHA) see IPD000
IGEPAL CA see GHS000
IGEPAL CA-63 see PKF500
IGEPAL CNP-10 see CNF335
IGEPAL CO-630 see PKF000

IGEPAL DA 530 see PKE370
IGEPAL GAS see IAH000
IGEPON T-33 see SIY000
IGEPON T-43 see SIY000
IGEPON T 51 see SIY000
IGEPON T-71 see SIY000
IGEPON T-73 see SIY000
IGEPON T-77 see SIY000
IGEPON TE see SIY000
IGIG 929 see HKH000
IGNAZIN see AFH250
IGNOTINE see CCK665
IGROSIN see SIH500
IGROTON see CLY600
IH 773B see FDA885
IH636 GRAPE SEED PROANTHOCYANIDIN EXTRACT see AEC300
II 17T1 see PHB550
II 17T4 see PHB550
II-C-2 see DOQ400
IIH see ILE000
IKACLOMIN see CMX700
IKADA RHODAMINE B see FAG070
IKF 1216 see CEX800
IKI 7899 see CDS800
IKhS 1 see CGW300
IKTEROSAN see PGG000
IKURIN see ANU650
IL 6001 see DLH200
ILBION see FNF000
ILETIN see IDF300
ILETIN U 40 see LEK000
ILEX AQUIFOLIUM see HGF100
ILEX OPACA see HGF100
ILEX VOMITORIA see HGF100
ILIADIN see AEX000
ILIDAR see AGD500
ILIDAR see ARZ000
ILIDAR BASE see ARZ000
ILIDAR PHOSPHATE see AGD500
ILITIA see VSZ450
ILIXATHIN see RSU000
ILLOXAN see IAH050
ILLOXOL see DHB400
ILLUDINE S see LIO600
ILLUDIN S see LIO600
IL-6302 MESYLATE see FMU039
ILOPAN see PAG200
ILOTYCIN see EDH500
ILOTYCIN HYDROCHLORIDE see EDJ000
ILOXAN see IAH050
IM see DLH600
IMADYL see CCK800
IMAGON see CLD000
IMAKOL see AFL750
IMAVATE see DLH630
IMAVEROL see FPB875
IMAZALIL see FPB875
IMAZAQUIN see IAH100
IMBARAL see SOU550
IMBENTIN C 125/85 see AFJ160
IMBENTIN C 145/100 see AFJ168
IMBENTIN C 123/1000 see AFJ155
IMBENTIN L 125/094 see AFJ160
IMBENTIN L 145/093 see AFJ168
IMBENTIN L 145/100 see AFJ168
IMBRILON see IDA000
IMC 3950 see SAZ000
IMD 760 see ARX800
IMESONAL see SBM500
IMESONAL see SBN000
IMET 3106 see IAK100
IMET 3393 see CQM750
IMETRO see IBP200
IMFERON see IGS000
IMGREITE see IAK200
IMI 115 see TGF250
IMIDACLOPRID see CKW400
IMIDA-LAB see TEH500
IMIDALIN see BBW750
IMIDALINE HYDROCHLORIDE see BBJ750

IMIDAMINE see PDC000
IMIDAN see PHX250
IMIDAN (PEYTA) see TEH500
IMIDAZO(2,1-A)ISOQUINOLINE, 2-(m-(ALLYLOXY)PHENYL)- see AGP300
IMIDAZO(2,1-A)ISOQUINOLINE, 2 (1,3 BENZODIOXOL-5-YL)- see MJT025
IMIDAZO(2,1-A)ISOQUINOLINE, 2-(4-BROMOPHENYL)- see BNW600
IMIDAZO(2,1-A)ISOQUINOLINE, 2-(p-BROMOPHENYL)- see BNW600
IMIDAZO(2,1-A)ISOQUINOLINE, 2-(m-CHLOROPHENYL)- see CKA600
IMIDAZO(2,1-A)ISOQUINOLINE, 2-(P-CHLOROPHENYL)- see CKA625
IMIDAZO(2,1-A)ISOQUINOLINE, 2-(3,4-DICHLOROPHENYL)- see DGD090
IMIDAZO(2,1-A)ISOQUINOLINE, 5,6-DIHYDRO-2-(3-ETHOXYPHENYL)- see EFC625
IMIDAZO(2,1-A)ISOQUINOLINE, 5,6-DIHYDRO-2-(2-METHOXYPHENYL)- see MFF600
IMIDAZO(2,1-A)ISOQUINOLINE, 5,6-DIHYDRO-2-(4-METHOXYPHENYL)- see MFF620
IMIDAZO(2,1-A)ISOQUINOLINE, 5,6-DIHYDRO-2-(m-METHOXYPHENYL)- see IAK300
IMIDAZO(2,1-A)ISOQUINOLINE, 5,6-DIHYDRO-2-(o-METHOXYPHENYL)- see MFF600
IMIDAZO(2,1-A)ISOQUINOLINE, 5,6-DIHYDRO-2-(p-METHOXYPHENYL)- see MFF620
IMIDAZO(2,1-A)ISOQUINOLINE, 5,6-DIHYDRO-2-(4-METHYLPHENYL)- see TGY300
IMIDAZO(2,1-A)ISOQUINOLINE, 5,6-DIHYDRO-2-p-TOLYL- see TGY300
IMIDAZO(2,1-A)ISOQUINOLINE, 2-(m-ETHOXYPHENYL)- see EFC625
IMIDAZO(2,1-A)ISOQUINOLINE, 2-(p-FLUOROPHENYL)- see FLG150
IMIDAZO(2,1-A)ISOQUINOLINE, 2-(m-METHOXYPHENYL)- see MFG252
IMIDAZO(2,1-A)ISOQUINOLINE, 2-(o-METHOXYPHENYL)- see MFG256
IMIDAZO(2,1-A)ISOQUINOLINE, 2-(p-METHOXYPHENYL)- see MFG254
IMIDAZO(2,1-A)ISOQUINOLINE, 2-(3,4-(METHYLENEDIOXY)PHENYL)- see MJT025
IMIDAZO(2,1-A)ISOQUINOLINE, 2-(4-METHYLPHENYL)- see THA100
IMIDAZO(2,1-A)ISOQUINOLINE, 3-METHYL-2-PHENYL- see MNU600
IMIDAZO(2,1-A)ISOQUINOLINE, 2-(p-NITROPHENYL)- see NIR075
IMIDAZO(2,1-A)ISOQUINOLINE, 2-(3-(2-PROPENYLOXY)PHENYL)- see AGP300
IMIDAZO(2,1-A)ISOQUINOLINE, 2-(m-PROPOXYPHENYL)- see PNA600
IMIDAZO(2,1-A)ISOQUINOLINE, 2-p-TOLYL- see THA100
IMIDAZO(1,2-A)QUINOLINE, 4,5-DIHYDRO-2-PHENYL- see PEU525
9H-IMIDAZO(1,2-a)BENZIMIDAZOLE, 9-((2-DIETHYLAMINO)ETHYL)-2-PHENYL-, DIHYDROCHLORIDE see DIF300
9H-IMIDAZO(1,2-a)BENZIMIDAZOLE-9-ETHANAMINE, N,N-DIETHYL-2-PHENYL-, DIHYDROCHLORIDE see DIF300
3H-IMIDAZO(2,1-I)PURINE, 3-(2-DEOXY-5-o-(HYDROXY((HYDROXY(PHOSPHONOOXY)PHOSPHINYL)OXY) PHOSPHINYL)-β-d-ERYTHRO-PENTOFURANOSYL)- see EEE550
IMIDAZOL see IAL000
1H-IMIDAZOL-2-AMINE, 4,5-DIHYDRO-N-(5-FLUORO-2-METHYLPHENYL)-

,MONOHYDROCHLORIDE (9CI) see FMR080
IMIDAZOLE see IAL000
IMIDAZOLE-1-ACETAMIDE, N-(2-HYDROXYETHYL)-2-NITRO- see HKW455
1H-IMIDAZOLE-1-ACETAMIDE, N-(2-HYDROXYETHYL)-2-NITRO-(9CI) see HKW455
IMIDAZOLEACRYLIC ACID see UVJ440
IMIDAZOLE-4-ACRYLIC ACID see UVJ440
5-IMIDAZOLEACRYLIC ACID see UVJ440
1H-IMIDAZOLE, 2-(BICYCLO(2.2.1)HEPT-2-YLIDENEMETHYL)-1-METHYL-5-NITRO- see BFY400
IMIDAZOLE-2-CARBOXALDEHYDE, 1-METHYL-5-NITRO- see MKG800
5-IMIDAZOLECARBOXAMIDE, 4-AMINO- see AKK250
5-IMIDAZOLECARBOXAMIDE, 4-AMINO-, HYDROCHLORIDE see IAL100
1H-IMIDAZOLE-4-CARBOXAMIDE, 5-AMINO-, MONOHYDROCHLORIDE see IAL100
1H-IMIDAZOLE-1-CARBOXAMIDE, N-PROPYL-N-(2-(2,4,6-TRICHLOROPHENOXY)ETHYL)- see IAL200
α-β-IMIDAZOLECARBOXYLIC ACID see IAM000
1H-IMIDAZOLE-4-CARBOXYLIC ACID, 5-(3,3-DIMETHYL-1-TRIAZENYL)-2-METHYL-, ETHYL ESTER see CBO800
IMIDAZOLE, 5-CHLORO-1-METHYL-4-NITRO- see CIQ100
4,5-IMIDAZOLEDICARBOXYLATE see IAM000
IMIDAZOLE-4,5-DICARBOXYLIC ACID see IAM000
IMIDAZOLE, 1-(2-(2,4-DICHLOROPHENYL)-2-((2,4-DICHLOROPHENYL)METHOXY)ETHYL)-(9CI) see MQS550
1H-IMIDAZOLE,4,5-DIHYDRO-2-(2,2-DIPHENYLCYCLOPROPYL)-, SUCCINATE(2:1) see DVZ050
1H-IMIDAZOLE, 4,5-DIHYDRO-HEPTADECENYL-, MONOHYDROCHLORIDE see IAM025
1H-IMIDAZOLE, 2-((2,3-DIHYDRO-1H-INDEN-1-YLIDENE)METHYL)-1-METHYL-5-NITRO see IAM035
1H-IMIDAZOLE, 2-((3,4-DIHYDRO-1(2H)-NAPHTHALENYLIDENE)METHYL)-1-METHYL-5-NITRO- see DLX225
1H-IMIDAZOLE, 4,5-DIHYDRO-, 2-NORTALL-OIL ALKYL DERIVS. see TAC075
1H-IMIDAZOLE, 4,5-DIHYDRO-2-PHENYL- see PFJ300
1H-IMIDAZOLE, 2-((DIHYDRO-2(3H)-THIENYLIDENE)METHYL)-1-METHYL-5-NITRO-, S,S-DIOXIDE see DME750
1H-IMIDAZOLE-4,5-DIMETHANOL, 1-METHYL-2-(METHYLTHIO)-, BIS(METHYLCARBAMATE)(ESTER) MONOHYDROCHLORIDE see MLX275
1H-IMIDAZOLE, 2-((2,2-DIMETHYL-1,3-DIOXAN-5-YLIDENE)METHYL)-1-METHYL-5-NITRO- see DRP300
1H-IMIDAZOLE, 2,2'-(DITHIOBIS(METHYLENE))BIS(1-METHYL-5-NITRO)- see IAM040
1H-IMIDAZOLE-4-ETHANAMINE see HGD000
1H-IMIDAZOLE-4-ETHANAMINE PHOSPHATE (1:2) see HGE000
1H-IMIDAZOLE-1-ETHANOL, 4,5-DIHYDRO-, 2-NORTALL OIL see TAC050
1H-IMIDAZOLE-1-ETHANOL, 4,5-DIHYDRO-, 2-NORTALL-OIL ALKYL DERIVS. see TAC050
1H-IMIDAZOLE, 1-ETHENYL-, HOMOPOLYMER see VPP100

IMIDAZOLE-4-ETHYLAMINE see HGD000
4-IMIDAZOLEETHYLAMINE see HGD000
5-IMIDAZOLEETHYLAMINE see HGD000
1H-IMIDAZOLE, 1-(2-(ETHYLSULFONYL)ETHYL)-2-METHYL-5-NITRO- see TGD250
IMIDAZOLE-2-HYDROXYBENZOATE see IAM100
1H-IMIDAZOLE-2-METHANOL, α-(3-(1,1-DIMETHYLETHYL)-2-HYDROXYPHENYL)-1-METHYL-5-NITRO- see DRV250
1H-IMIDAZOLE, 4-METHYL-5-NITRO-1-(PHENYLMETHYL)- see MMP110
1H-IMIDAZOLE, 5-METHYL-4-NITRO-1-(PHENYLMETHYL)- see MMP113
1H-IMIDAZOLE, 1-METHYL-5-NITRO-2-(2-PHENYL-1-PROPENYL)- see MMP120
1H-IMIDAZOLE, 1-METHYL-5-NITRO-2-((PHENYLSULFONYL)METHYL)- see MMP125
1H-IMIDAZOLE, 1-(2-(METHYLTHIO)-1-(2-(PENTYLOXY)PHENYL)ETHENYL)-, MONOHYDROCHLORIDE, (E)- see NCP525
IMIDAZOLE MUSTARD see IAN000
IMIDAZOLE, 4-NITRO- see NHG100
1H-IMIDAZOLE, 4-NITRO-5-(2-PHENYLETHENYL)- see NIM630
1H-IMIDAZOLE, 4-NITRO-5-(2-PHENYLETHENYL)-1-(PHENYLMETHYL)- see NIM640
IMIDAZOLE, 2-NITRO-1-β-d-RIBOFURANOSYL- see NHG200
1H-IMIDAZOLE, 2-NITRO-1-β-d-RIBOFURANOSYL- see NHG200
IMIDAZOLE-4-PROPIONIC ACID,α-AMINO-α-(FLUOROMETHYL)- see MRI285
IMIDAZOLEPYRAZOLE see IAN100
IMIDAZOLE with SALICYLIC ACID see IAM100
IMIDAZOLE-2-THIOL see IAO000
1H-IMIDAZOLE, 1-(TRIBUTYLPLUMBYL)- see IAT400
IMIDAZOLE, 1-VINYL-, POLYMERS see VPP100
IMIDAZOL-2-HYDROXYBENZOAT (GERMAN) see IAM100
2,4-IMIDAZOLIDINEDIONE (9CI) see HGO600
2,4-IMIDAZOLIDINEDIONE, 1-BROMO-3-CHLORO-5,5-DIMETHYL- see BNA300
2,4-IMIDAZOLIDINEDIONE, 1,3-DIBROMO-5,5-DIMETHYL-(9CI) see DDI900
2,4-IMIDAZOLIDINEDIONE, 5,5-DIMETHYL- see DSF300
2,4-IMIDAZOLIDINEDIONE, 5,5-DIMETHYL-3-(2-(OXIRANYLMETHOXY)PROPYL)-1-(OXIRANYLMETHYL)- see DTH100
2,4-IMIDAZOLIDINEDIONE, 5-ETHYL-5-METHYL-1,3-BIS(OXIRANYLMETHYL)- see ECI200
2,4-IMIDAZOLIDINEDIONE, 5-ETHYL-5-PHENYL-, (R)- (9CI) see EOL050
2,4-IMIDAZOLIDINEDIONE, 5-(2-(METHYLTHIO)ETHYL)- see MDT800
2,4-IMIDAZOLIDINEDIONE, 1-NITRO- see NHE100
2,4-IMIDAZOLIDINEDIONE, 1-(((5-NITRO-2-FURANYL)METHYLENE)AMINO)-, MONOHYDRATE (9CI) see NGE780
2,4-IMIDAZOLIDINEDIONE, 1-(((5-(4-NITROPHENYL)-2-FURANYL)METHYLENE)AMINO)-, SODIUM SALT see DAB840
2,4-IMIDAZOLIDINEDIONE, 3,3'-(2-(OXIRANYLMETHOXY)-1,3-(PROPANEDIYL)BIS(5,5-DIMETHYL-1-(OXIRANYLMETHYL)- see IAO100
4,5-IMIDAZOLIDINEDITHIONE see IAP000
2-IMIDAZOLIDINETHIONE see IAQ000
2-IMIDAZOLIDINETHIONE, 1,3-BIS(HYDROXYMETHYL)- see BKH650

2-IMIDAZOLIDINETHIONE mixed with SODIUM NITRITE see IAR000

2-IMIDAZOLIDINIMINE, 1-((6-CHLORO-3-PYRIDINYL)METHYL)-N-NITRO- see CKW400

2-IMIDAZOLIDINONE see IAS000

4-IMIDAZOLIDINONE, 1-ACETYL-2-THIOXO- see ADC750

2-IMIDAZOLIDINONE, 3-(5-tert-BUTYL-1,3,4-THIADIAZOL-2-YL)-4-HYDROXY-1-METHYL- see RCA300

4-IMIDAZOLIDINONE, 1-CHLORO-2,2,5,5-TETRAMETHYL- see CLH550

2-IMIDAZOLIDINONE, 3-(5-(1,1-DIMETHYLETHYL)-1,3,4-THIADIAZOL-2-YL)-4-HYDROXY-1-METHYL-(9CI) see RCA300

2-IMIDAZOLIDINONE, 1,3-DINITRO- see DUW503

2-IMIDAZOLIDINONE, 1-(1-METHYL-5-NITRO-1H-IMIDAZOL-2-YL)-3-(METHYLSULFONYL)- see SAY950

IMIDAZOLIDINYL UREA 11 see IAS100

2-IMIDAZOLIDONE see IAS000

IMIDAZOLINE see IAT000

2-IMIDAZOLINE see IAT000

2-IMIDAZOLINE, 2-(2,6-DICHLOROANILINO)- see DGB500

2-IMIDAZOLINE, 4,4-DIMETHYL-1-ISOPROPYL-2-NONYL- see DSK300

IMIDAZOLINE-2,4-DITHIONE see IAT100

2-IMIDAZOLINE, 2-(5-FLUORO-o-TOLUIDINO)-,MONOHYDROCHLORIDE see FMR080

2-IMIDAZOLINE, 1-(2-HYDROXYETHYL)-2-(TALL OIL ALKYL)- see TAC050

2-IMIDAZOLINE, 2-METHYL-1-(3,4,5-TRIMETHOXYBENZOYL)- see IAT200

2-IMIDAZOLINE, 2-PHENYL- see PFJ300

N-(2-IMIDAZOLINE-2-YL)-N-(4-INDANYL)AMINE MONOHYDROCHLORIDE see IBR200

N-(2-IMIDAZOLINE-2-YL)-N-(4-INDANYL)AMIN-MONOHYDROCHLORID (GERMAN) see IBR200

IMIDAZOLIUM, 2-CHLORO-3-(3,4-DICHLOROPHENYL)-1-ISOPROPYL-4-METHYL-, CHLORIDE see CFL300

IMIDAZOLIUM COMPOUNDS,1-(2-(CARBOXYMETHOXY)ETHYL)-1-(CARBOXYMETHYL)-4,5, HYDROXY-2-NORCOCO ALKYL, HYDROXIDES, INNER SALTS, DISODIUM SALTS see IAT250

1H-IMIDAZOLIUM, 4,5-DIHYDRO-1-(CARBOXYMETHYL)-1-(2-HYDROXYETHYL)-2-UNDECYL-, HYDROGEN SULFATE (salt), MONOSODIUM SALT see AOC275

IMIDAZOLIUM, 1,3-DIMETHYL-, IODIDE see MKU100

1H-IMIDAZOLIUM, 1,3-DIMETHYL-, IODIDE (9CI) see MKU100

4H-IMIDAZOL-4-ONE, 3,5-DIHYDRO-5-METHYL-2-(METHYLTHIO)-5-PHENYL-3-(PHENYLAMINO)-, (5S)- see FAJ300

2-(4-IMIDAZOLYL)ETHYLAMINE see HGD000

2-IMIDAZOL-4-YL-ETHYLAMINE see HGD000

β-IMIDAZOLYL-4-ETHYLAMINE see HGD000

2-(1H-IMIDAZOL-4-YLMETHYL)-8H-INDENO(1,2-D)THIAZOLE MONOFUMARATE see IAT275

6-(1-IMIDAZOLYLMETHYL)-5,6,7,8-TETRAHYDRONAPHTHALENE-2-CARBOXYLIC ACID MONOHYDROCHLORIDE see IAT300

3-(1H-IMIDAZOL-4-YL)-2-PROPENOIC ACID see UVJ440

(1-IMIDAZOLYL)TRIBUTYLPLUMBANE see IAT400

1H-IMIDAZO(4,5-b)(1,5)NAPHTHYRIDIN-2-AMINE, 1-METHYL- see MKU125

1H-IMIDAZO(4,5-b)(1,6)NAPHTHYRIDIN-2-AMINE, 1-METHYL- see MKU130

1H-IMIDAZO(4,5-B)(1,7)NAPHTHYRIDIN-2-AMINE, 1-METHYL- see MKU135

1H-IMIDAZO(4,5-b)(1,8)NAPHTHYRIDIN-2-AMINE, 1-METHYL- see MKU140

1H-IMIDAZO(2,1-f)PURINE-2,4(3H,6H)-DIONE,7,8-DIHYDRO-8-ALLYL-1,3-DIMETHYL-(8CI) see KHU100

1H-IMIDAZO(4,5-B)PYRIDIN-2-AMINE, 1,5-DIMETHYL- see AJQ510

1H-IMIDAZO(4,5-b)PYRIDIN-2-AMINE, 1,6-DIMETHYL- see AJQ520

3H-IMIDAZO(4,5-b)PYRIDIN-2-AMINE, 3,5-DIMETHYL- see AJQ530

1H-IMIDAZO(4,5-b)PYRIDIN-2-AMINE, 1-METHYL- see AKT510

3H-IMIDAZO(4,5-b)PYRIDIN-2-AMINE, 3-METHYL- see AKT520

3H-IMIDAZO(4,5-b)PYRIDIN-2-AMINE, 3-METHYL-6-PHENYL- see AKZ300

1H-IMIDAZO(4,5-B)PYRIDIN-2-AMINE, 1-METHYL-6-PHENYL-(9CI) see AKZ200

3H-IMIDAZO(4,5-b)PYRIDIN-2-AMINE, 3,5,6-TRIMETHYL- see AMV752

3H-IMIDAZO(4,5-b)PYRIDIN-2-AMINE, 3,5,6-TRIMETHYL- see AMV752

3H-IMIDAZO(4,5-b)PYRIDIN-2-AMINE, 3,5,7-TRIMETHYL- see AMV754

IMIDAZO(4,5-B)PYRIDINE, 2-AMINO-1-METHYL-6-PHENYL- see AKZ200

1H-IMIDAZO(4,5-B)PYRIDINE, 2-AMINO-1-METHYL-6-PHENYL- see AKZ200

4H-IMIDAZO(4,5-C)PYRIDIN-4-ONE, 2-((2-(3-AMINO-6-(3-AMINO-6-(3-AMINO-6-(3-AMINO-6-(3,6-DIAMINOHEXANAMIDO)HEXANAMIDO)HEXANAMIDO)HEXANAMIDO)HEXANAMIDO)-2-DEOXY-BET A-d-GULOPYRANOSYL) AMINO)-3,3A,5,6,7,7A-HEXAHYDRO-7-HYDROXY-, 6'-CARBAMATE see RAG400

4-IMIDAZO(1,2-a)PYRIDIN-2-YL-α-METHYLBENZENEACETIC ACID see IAY000

2-(4-(IMIDAZO(1,2-a)PYRIDIN-2-YL)PHENYL)PROPIONIC ACID see IAY000

2-(p-(2-IMIDAZO(1,2-a)PYRIDYL)PHENYL)PROPIONIC ACID see IAY000

7H-IMIDAZO(4,5-D)PYRIMIDINE see POJ250

IMIDAZO(1,2-c)QUINAZOLINE, 5-(4-CHLOROPHENYL)-2,3,5,6-TETRAHYDRO- see CKK050

IMIDAZO(1,2-c)QUINAZOLINE, 2,3,5,6-TETRAHYDRO-5-(p-CHLOROPHENYL)- see CKK050

1H-IMIDAZO(4,5-b)QUINOLIN-2-AMINE, 1-METHYL- see MKU145

1H-IMIDAZO(4,5-f)QUINOLIN-2-AMINE, 1-METHYL- see AKT530

3H-IMIDAZO(4,5-F)QUINOLINE-2-AMINE, N-HYDROXY-3-METHYL- see HIU600

1H-IMIDAZO(4,5-F)QUINOLINE, 2-AMINO-1-METHYL- see AKT530

2H-IMIDAZO(4,5-F)QUINOLIN-2-ONE, 1,3-DIHYDRO-3-METHYL-, OXIME see HIU600

3H-IMIDAZO(4,5-f)QUINOXALIN-2-AMINE, 3-METHYL- see AKT650

3H-IMIDAZO(4,5-f)QUINOXALIN-2-AMINE, 3,4,5,7,8-PENTAMETHYL- see PBI600

3H-IMIDAZO(4,5-F)QUINOXALIN-2-AMINE, 3,4,5,8-TETRAMETHYL- see AMQ600

3H-IMIDAZO(4,5-F)QUINOXALIN-2-AMINE, 3,4,7,8-TETRAMETHYL- see AMQ620

3H-IMIDAZO(4,5-F)QUINOXALIN-2-AMINE, 3,4,5-TRIMETHYL- see AMV760

3H-IMIDAZO(4,5-F)QUINOXALIN-2-AMINE, 3,4,8-TRIMETHYL- see TLT768

3H-IMIDAZO(4,5-F)QUINOXALIN-2-AMINE, 3,5,7-TRIMETHYL- see AMV770

3H-IMIDAZO(4,5-f)QUINOXALIN-2-AMINE, 3,7,8-TRIMETHYL- see TLT771

IMIDAZO(5,1-d)-1,2,3,5-TRIAZINE-8-CARBOXAMIDE, 3-(2-CHLOROETHYL)-3,4-DIHYDRO-4-OXO- see MQY110

IMIDAZO(2,1-β)THIAZOLE MONOHYDROCHLORIDE see LFA020

IMIDENE see TEH500

IMIDIN see NAH500

IMIDOBENZYLE see DLH600

IMIDOBENZYLE see DLH630

4,4'-(IMIDOCARBONYL)BIS(N,N-DIMETHYLAMINE) MONOHYDROCHLORIDE see IBA000

4,4'-(IMIDOCARBONYL)BIS(N,N-DIMETHYLANILINE) see IBB000

1,1'-IMIDODIACETONITRILE see IBB100

IMIDODICARBONIC DIAMIDE, N,N',2-TRIS(6-ISOCYANATOHEXYL)- see TNJ300

IMIDODICARBONIC DIHYDRAZIDE (9 CI) see IBC000

IMIDODICARBONIMIDIC DIAMIDE, N-(2-METHYLPHENYL)-(9CI) see TGX550

IMIDODICARBOXYLIC ACID, DIHYDRAZIDE see IBC000

3,3'-IMIDODI-1-PROPANOL, DIMETHANESULFONATE (ester), HYDROCHLORIDE see YCJ000

IMIDODISULFURIC ACID, AMMONIUM SALT see ANK650

IMIDOL see DLH630

IMIDOLE see PPS250

IMIGRAN see DPH300

IMILANYLE see DLH630

IMINAZOLE see IAL000

IMINOALDEHYDE see FNL000

IMINO-1,1'-BIANTHRAQUINONE see IBI000

IMINOBIS(ACETIC ACID) see IBH000

2,2'-IMINOBISACETONITRILE see IBB100

1,1'-IMINOBIS(4-AMINO-9,10-ANTHRACENEDIONE) see IBD000

1,1'-IMINOBIS(4-AMINOANTHRAQUINONE) see IBD000

1,1'-IMINOBIS-9,10-ANTHRACENEDIONE see IBI000

1,1'-IMINOBIS(4-BENZAMIDOANTHRAQUINONE) see IBJ000

4,5'-IMINOBIS(4-BENZAMIDOANTHRAQUINONE) see IBE000

2,2'-IMINOBISETHANOL see DHF000

2,2'-IMINOBISETHYLAMINE see DJG600

IMINOBISFORMALDEHYDE see FNL000

5,6'-IMINOBIS(1-HYDROXY-2-NAPHTHALENESULFONIC ACID) see IBF000

(IMINOBIS(OCTAMETHYLENE))-DIGUANIDINE SULFATE see GLQ000

1,1'-(IMINOBIS(OCTAMETHYLENE))DIGUANIDINE TRIACETATE see GLQ100

4,4'-IMINOBISPHENOL see IBJ100

3,3'-IMINOBISPROPANENITRILE see BIQ500

3,3'-IMINOBIS-1-PROPANOL DIMETHANESULFONATE (ester), 4-METHYLBENZENESULFONATE (salt) see IBQ100

IMINOBIS(PROPYLAMINE) see AIX250

3,3'-IMINOBIS(PROPYLAMINE) see AIX250

4-IMINO-2,5-CYCLOHEXADIEN-1-ONE see BDD500

4,4'-((4-IMINO-2,5-CYCLOHEXADIEN-1-YLIDENE)METHYLENE)DIANILINE MONOHYDROCHLORIDE-o-TOLUIDINE see RMK020

IMINODIACETIC ACID see IBH000

2,2'-IMINODIACETIC ACID see IBH000
IMINODIACETONITRILE see IBB100
1,1'-IMINODIANTHRAQUINONE see IBI000
N,N'-(IMINODI-4,1-
ANTHRAQUINONYLENE)BISBENZAMIDE
see IBJ000
IMINODIBENZYL see DKY800
IMINODI-2,1-ETHANEDIYL
CARBAMIMIDOTHIOATE
DIHYDROBROMIDE see IBJ050
IMINODIETHANOIC ACID see IBH000
2,2'-IMINODIETHANOL see DHF000
2,2'-IMINODI-ETHANOL with 1,2-
DIHYDRO-3,6-PYRIDAZINEDIONE (1:1) see
DHF200
2,2'-IMINODI-N-NITROSOETHANOL see
NKM000
IMINODIOCTAN SODNY (CZECH) see
DXE200
4,4'-IMINODIPHENOL see IBJ100
1,1'-IMINODI-2-PROPANOL see DNL600
3,3'-IMINODIPROPIONITRILE see BIQ500
β,β-IMINODIPROPIONITRILE see BIQ500
IMINO-β,β'-DIPROPIONITRILE see BIQ500
β,β'-IMINODIPROPIONITRILE see BIQ500
IMINODIPROPYL
DIMETHANESULFONATE 4-
TOLUENESULPHONATE see IBQ100
5-IMINO-1,2,4-DITHIAZOLIDINE-3-
THIONE see IBL000
9,18-
(IMINOETHANIMINOETHANIMINOETHA
NIMINOMETHANO)PYRROLO(1',2':8,9)(1,5,8
,11,14)THIATETRAAZACYCLOOCTADECIN
O(18,17-B)INDOLE-6-
ACETAMIDE,1,2,3,5,6,7,8,9,10,12,17,18,19,20,2
1,22,23,23A-OCTADECAHYDRO-29-sec-
BUTYL-2,14-DIHYDROXY-21-(2-HYDROXY-
1-METHYLPROPYL)-5,8,20,23,24,27,30,33-
OCTAOXO-,11-OXIDE see IBL100
l-N⁶)-(1-IMINOETHYL)LYSINE
HYDROCHLORIDE see IBL200
2-IMINOHEXAFLUOROPROPANE see
HDD000
(2-
((IMINO(METHYLAMINO)METHYL)THIO)
ETHYL)TRIETHYLAMMONIUM BROMIDE
HYDROBROMIDE see MRU775
N⁵)-(IMINO(NITROAMINO)METHYL)-l-
ORNITHINE see OIU900
(2-IMINO-5-PHENYL-4-
OXAZOLIDINONATO(2-
))DIAQUOMAGNESIUM see PAP000
2-IMINO-5-PHENYL-4-OXAZOLIDINONE
see IBM000
2-IMINO-5-PHENYL-4-OXAZOLIDINONE
MAGNESIUM CHELATE see PAP000
IMINOPHOSPHATE see DXN600
2(3H)-IMINO-9-β-D-RIBOFURANOSYL-9H-
PURIN-6(1H)-ONE see GLS000
2-IMINOTHIAZOLIDINE see TEV600
3-((IMINO((2,2,2-
TRIFLUOROETHYL)AMINO)METHYL)AMI
NO)-1H-PYRAZOLE-1-PENTAN AMIDE see
IBM100
IMINOUREA see GKW000
IMINOUREA HYDROCHLORIDE see
GKY000
IMIPHOS see BGY000
IMIPRAMINA (ITALIAN) see DLH600
IMIPRAMINA (ITALIAN) see DLH630
IMIPRAMINE see DLH600
IMIPRAMINE see DLH630
IMIPRAMINEDEMETHYL
HYDROCHLORIDE see DLS600
IMIPRAMINE HYDROCHLORIDE see
DLH630
IMIPRAMINE MONOHYDROCHLORIDE
see DLH630
IMIPRAMINE-N-OXIDE
HYDROCHLORIDE see IBP000
IMIPRIN see DLH600

IMIPRIN see DLH630
IMITREX see DPH300
IMIZIN see DLH600
IMIZINUM see DLH600
IMMEDIAL BLACK AT see CMS250
IMMEDIAL BLACK GN see CMS250
IMMEDIAL BLACK MF see CMS250
IMMEDIAL CARBON B see CMS250
IMMEDIAL CARBON BO see CMS250
IMMEDIAL CARBON BR see CMS250
IMMEDIAL CARBON CBO see CMS250
IMMEDIAL CARBON CM see CMS250
IMMEDIAL CARBON CMR see CMS250
IMMEDIAL CARBON LP see CMS250
IMMEDIAL CARBON MLB see CMS250
IMMEDIAL CARBON NGD see CMS250
IMMEDIAL CARBON RRC see CMS250
IMMEDIAL MFS GRAINS see CMS250
IMMEDIAL PRINTING BLACK B PASTE see
CMS250
IMMEDIAL RED BROWN 3B see CMS257
IMMENOCTAL see SBM500
IMMENOCTAL see SBN000
IMMENOX see SBM500
IMMETROPAN see IBP200
IMMORTELLE see HAK500
IMMUNE ENDOGLOBULIN see IBP250
IMMUNOGLOBULIN see IBP250
IMODIUM see LIH000
IMOL S 140 see TNP500
IMOTRYL see BBW500
IMOVANCE see ZUA450
IMOVANE see ZUA450
IMP see IDE200
5'-IMP see IDE200
IMPACT see FMR200
o-IMPC see PMY300
IMP DISODIUM SALT see DXE500
5'-IMP DISODIUM SALT see DXE500
IMPEDEX see SJL500
IMPERATOR see RLF350
IMPERATORIN see IHR300
IMPERIAL GREEN see COF500
IMPERON FIXER T see TND250
IMPERVOTAR see CMY800
IMPF see IPX000
IMP HYDROCHLORIDE see DLH630
IMPINGEMENT BLACK see CBT750
IMPIRAMINE-N-OXIDE see IBP309
IMPLETOL see CNG827
IMPOSIL see IGS000
IMPRAMINE see DLH600
IMPROMEN see BNU725
IMPROMIDINE HYDROCHLORIDE see
IBQ075
IMPROMIDINE TRIHYDROCHLORIDE see
IBQ075
IMPROSULFAN-p-TOLUENESULFONATE
see IBQ100
IMPROSULFAN TOSILATE see IBQ100
IMPROSULFAN TOSYLATE see IBQ100
IMPROVED WILT PRUF see PKQ059
IMPRUVOL see BFW750
IMP SODIUM SALT see DXE500
IMPY see IAN100
IMS see IPY000
IMUGAN see CDP750
IMURAN see ASB250
IMUREK see ASB250
IMUREL see ASB250
IMUTEX see IAL000
IMVITE I.G.B.A. see BAV750
IMWITOR 191 see OAV000
IMWITOR 900K see OAV000
IN-117 see HEG000
IN 399 see MKW000
IN 511 see PGG350
IN 511 see PGG355
IN 836 see FAR200
IN-A 8947 see ASH275
INABENFIDE see CHN100
INACID see IDA000

INACILIN see AOD000
INACTIVE LIMONENE see MCC250
INAKOR see ARQ725
INALONE O see AFJ625
INALONE R see AFJ625
INAMIL see AIF000
INAMYCIN see NOB000
INAMYCIN see SMB000
INAPPIN see DYF200
INAPSIN see DYF200
INBESTAN see FOS100
INBUTON see BSM000
INCASAN see IBQ300
INCAZANE see IBQ300
INCIDOL see BDS000
INCOLOY 800 see IGL100
2-(4-(3-INCOLYLMETHYL)-1-
PIPERAZINYL)-QUINOLINE DIMALEATE
see ICZ100
INCORTIN see CNS800
INCORTIN see CNS825
INCRECEL see CMF400
1H-as-INDACENO(3,2-
D)OXACYCLODODECIN-7,15-DIONE,
2,3,3A,5A,5B,6,9,10,11,12,13,14,16A,16B-
TETRADECAHYDRO-2-(((6-DEOXY-2,3,4-
TRI-o-METHYL-α-l-
manNOPYRANOSYL)OXY)-13-((5-
DIMETHYLAMINO)TETRAHYDRO-6-
METHYL-2H-PYRAN-2-YL)OXY)-9-ETHYL-
14-METHYL-,(2 R-(2R*,3AS*,5AR*,
5BS*,9S*,13S*(2R*,5S*,6R*),14R*,16AS*,16BR*)
)- see LEK500
INDACRINONE see IBQ400
INDAFLEX see IBV100
INDALCA AG see GLU000
INDALCA AG-BV see GLU000
INDALCA AG-HV see GLU000
INDALONE see BRT000
INDAMOL see IBV100
INDAN see IBR000
INDANAL see CFH825
INDANAL see CMV500
INDANAZOLIN (GERMAN) see IBR200
INDANAZOLINE see IBR200
INDANAZOLINE HYDROCHLORIDE see
IBR200
INDAN, 4,6-DINITRO-1,1,3,3,5-
PENTAMETHYL- see MRU300
1,3-INDANDIONE see IBS000
1,3-INDANDIONE, 2-ISOVALERYL- see
ITD010
1,3-INDANDIONE, 2-ISOVALERYL-, ION(1-
), CALCIUM see ITD015
INDANE see IBR000
5-INDANOL see IBU000
INDAN-5-OL see IBU000
2-INDANONE see IBV000
1-INDANONE, 5-HYDROXY-2-METHYL-,
SODIUM SALT see HME050
INDANTHREN BLUE see IBV050
INDANTHREN BLUE 3G see DLO900
INDANTHREN BLUE BC see DFN300
INDANTHREN BLUE BCA see DFN300
INDANTHREN BLUE BCS see DFN300
INDANTHREN BLUE GP see IBV050
INDANTHREN BLUE GPT see IBV050
INDANTHREN BLUE 3GN see DLO900
INDANTHREN BLUE RPT see IBV050
INDANTHREN BLUE RS see IBV050
INDANTHREN BLUE RSN see IBV050
INDANTHREN BLUE RSP see IBV050
INDANTHREN BRILLIANT BLUE 3G see
DLO900
INDANTHREN BRILLIANT BLUE R see
IBV050
INDANTHREN BRILLIANT ORANGE GR
see CMU820
INDANTHREN BRILLIANT VIOLET 4R see
DFN450
INDANTHREN BRILLIANT VIOLET RR see
DFN450

INDANTHREN BRONZE BR see CMU770
INDANTHREN BROWN BR see CMU770
INDANTHREN BROWN FFR see CMU780
INDANTHREN BROWN GR see CMU770
INDANTHREN BROWN R see CMU780
INDANTHREN BROWN RAP see CMU780
INDANTHRENE see IBV050
INDANTHRENE BLUE see IBV050
INDANTHRENE BLUE BC see DFN300
INDANTHRENE BLUE BCF see DFN300
INDANTHRENE BLUE GP see IBV050
INDANTHRENE BLUE RP see IBV050
INDANTHRENE BLUE RS see IBV050
INDANTHRENE BLUE RSA see IBV050
INDANTHRENE BLUE RSN see IBV050
INDANTHRENE BRILLIANT ORANGE GR see CMU820
INDANTHRENE BRILLIANT ORANGE GRP see CMU820
INDANTHRENE BRILLIANT VIOLET 4R see DFN450
INDANTHRENE BRILLIANT VIOLET RR see DFN450
INDANTHRENE BROWN BR see CMU770
INDANTHRENE BROWN RARWP see CMU780
INDANTHRENE BROWN R (6CI) see CMU780
INDANTHRENE BROWN RN see CMU780
INDANTHRENE BROWN RWP see CMU780
INDANTHRENE GOLDEN YELLOW see DCZ000
INDANTHRENE NAVY BLUE G see CMU500
INDANTHRENE OLIVE R see DUP100
INDANTHRENE PRINTING BLACK BL see CMU320
INDANTHRENE RED BROWN 5RF see CMU800
INDANTHRENE REDDISH BROWN 5RF see CMU800
INDANTHRENE RUBINE R see CMU825
INDANTHREN GREY M see CMU475
INDANTHREN GREY MG see CMU475
INDANTHREN NAVY BLUE TRR see CMU750
INDANTHREN OLIVE R see DUP100
INDANTHREN PRINTING BLUE FRS see IBV050
INDANTHREN PRINTING BLUE KRS see IBV050
INDANTHREN PRINTING VIOLET F 4R see DFN450
INDANTHREN RED BROWN 5RF see CMU800
INDANTHREN RUBINE R see CMU825
INDANTHREN RUBINE RS see CMU825
INDANTHRONE see IBV050
1,2,3-INDANTRIONE-2-HYDRATE see DMV200
1,2,3-INDANTRIONE MONOHYDRATE see DMV200
2-INDANYLAMINE HYDROCHLORIDE see AKL000
INDAPAMIDE see IBV100
INDAR see BPU000
2H-INDAZOL-2-AMINE, N-((5-NITRO-2-FURANYL)METHYLENE)- see NGE130
INDAZOL BLUE R see CMM770
1H-INDAZOLE-3-ACETIC ACID, 5-CHLORO-, ETHYL ESTER see EHK600
1H-INDAZOLE-3-CARBOXYLIC ACID, 1-(p-CHLOROBENZYL)-, 1,3-DIHYDROXY-2-PROPYL ESTER see GGR100
2-(p-(2H-INDAZOL-2-YL)PHENYL)PROPIONIC ACID see IBW100
INDECAINIDE HYDROCHLORIDE see IBW400
INDELOXAZINE HYDROCHLORIDE see IBW500
INDEMA see PFJ750
INDENE see IBX000

1H-INDENE-1,3(2H)-DIONE see IBS000
INDENESTROL see ELL550
INDENESTROL A see ELL550
(+-)-INDENESTROL A see ELL550
INDENE TRIPROPYLAMINE see IBY000
INDENO(1,2-E)(1,3,4)OXADIAZINE-4A(3H)-CARBOXYLIC ACID, 7-CHLORO-2,5-DIHYDRO-2-(((METHOXYCARBONYL)(4-(TRIFLUOROMETHOXY)PHENYL)AMINO)CARBONYL)-, METHYL ESTER, (4AS)- see IDA550
INDENOESTROL A see ELL550
INDEN-6-OL, 3-ETHYL-2-(p-HYDROXYPHENYL)-1-METHYL- see ELL550
1H-INDEN-6-OL, 3-ETHYL-2-(4-HYDROXYPHENYL)-1-METHYL- see ELL550
INDENOLOL CLORHIDRATO see IBY650
INDENOLOL HYDROCHLORIDE see IBY600
INDENOLOL HYDROCHLORIDE see IBY650
INDENOLOL HYDROCHLORIDE see ICA000
dl-INDENOLOL HYDROCHLORIDE see IBY650
1H-INDEN-1-ONE, 2,3-DIHYDRO-5-HYDROXY-2-METHYL-, SODIUM SALT see HME050
13H-INDENO(1,2-1)PHENANTHRENE see DCX000
13H-INDENO(1,2-c)PHENANTHRENE see IBY700
INDENO(7',1':6,7,8)PHENANTHRO(3,4-B)OXIRANE-2,3-DIOL, 1A,2,3,6,7,11C-HEXAHYDRO-8-METHYL-, (1as-(1A-α,2-β,3-α,11C-α))- see MBV715
INDENO(7',1':6,7,8)PHENANTHRO(1,2-b)OXIRENE,1A,11-DIHYDRO- see EBO990
INDENO(1,7-ab)PYRENE see BCH900
INDENO(1,2,3-cd)PYRENE see IBZ000
INDENO(1,2,3-cd)PYRENE-1,2-OXIDE see DLK750
INDENO(1,2,3-cd)PYREN-8-OL see IBZ100
INDENO(1',2',3':1,10)PYRENO(4,5-B)OXIRENE, 1A,11B-HYDRO- see DLK750
8H-INDENO(1,2-D)THIAZOLE, 2-(1H-IMIDAZOL-4-YLMETHYL)-, (E)-2-BUTENEDIOATE (1:1) see IAT275
4-(1H-INDEN-1-YLIDENEMETHYL)-N,N-DIMETHYLBENZENAMINE see DOT600
4-(1-H-INDEN-1-YLIDENEMETHYL)-N-METHYL-N-NITROSOBENZENAMINE see MMR750
(±)-1-(7-INDENYLOXY)-3-ISOPROPYLAMINOPROPAN-2-OL HYDROCHLORIDE see ICA000
1-(4(or 7)-INDENYLOXY)-3-(ISOPROPYLAMINO)-2-PROPANOL HYDROCHLORIDE see IBY600
2-(7-INDENYLOXYMETHYL)MORPHOLINE HYDROCHLORIDE see IBW500
2-(((1H-INDEN-7-YLOXY)METHYL)MORPHOLINE HYDROCHLORIDE see IBW500
INDEPENDENCE RED see MMP100
INDERAL see ICB000
INDERAL HYDROCHLORIDE see ICC000
INDEREX see ICC000
INDEROL see ICC000
INDI see ICD100
INDIA see ICI100
INDIAM ARSENIDE see ICG100
INDIAN APPLE see MBU800
INDIAN BERRY see PIE500
INDIAN BLACK DRINK see HGF100
INDIAN CANNABIS see CBD750
INDIAN COBRA VENOM see ICC700
INDIAN GUM see AQQ500
INDIAN GUM see GLY000

INDIAN HEMP see CBD750
INDIAN JACK-IN-THE-PULPIT see JAJ000
INDIAN LABURNUM see GIW300
INDIAN LAUREL see MBU780
INDIAN LICORICE SEED see AAD000
INDIAN LILAC see CDM325
INDIAN PINK see CCJ825
INDIAN PINK see PIH800
INDIAN POKE see FAB100
INDIAN POKE see VIZ000
INDIAN POLK see PJJ315
INDIAN RED see IHC450
INDIAN RED see LCS000
INDIAN SAFFRON see ICC800
INDIAN SAVIN TREE (JAMAICA) see CAK325
INDIAN TOBACCO see CCJ825
INDIAN TURMERIC see ICC800
INDIAN WALNUT see TOA275
INDIA RUBBER see ROH900
INDIA RUBBER VINE see ROU450
INDICAN (POTASSIUM SALT) see ICD000
INDICINE-N-OXIDE see ICD100
INDIGENE BLACK D see CMN230
INDIGENOUS PEANUT OIL see PAO000
INDIGO see BGB275
INDIGO 4B see ICU135
INDIGO BLUE see BGB275
INDIGO BLUE 2B see CMO000
INDIGO CARMINE see FAE100
INDIGO CARMINE (BIOLOGICAL STAIN) see FAE100
INDIGO CARMINE DISODIUM SALT see FAE100
INDIGO CIBA see BGB275
INDIGO CIBA SL see BGB275
INDIGO EXTRACT see FAE100
INDIGO J see BGB275
INDIGO-KARMIN (GERMAN) see FAE100
INDIGO N see BGB275
INDIGO NAC see BGB275
INDIGO NACCO see BGB275
INDIGO P see BGB275
INDIGO PLN see BGB275
INDIGO POWDER W see BGB275
INDIGO PURE BASF see BGB275
INDIGO PURE BASF POWDER K see BGB275
INDIGO SYNTHETIC see BGB275
INDIGOTIN see BGB275
5,5'-INDIGOTIN DISULFONIC ACID see FAE100
INDIGOTINE see FAE100
INDIGOTINE DISODIUM SALT see FAE100
INDIGO VS see BGB275
INDIGO YELLOW see ICE000
INDINAVIR SULFATE see ICE100
INDION see PFJ750
INDISAN see IKA000
INDISULFAT (GERMAN) see ICJ000
INDIUM see ICF000
INDIUM ACETYLACETONATE see ICG000
INDIUM ARSENIDE see ICG100
INDIUM CHLORIDE see ICK000
INDIUM CITRATE see ICH000
INDIUM MONOARSENIDE see ICG100
INDIUM MONOPHOSPHIDE see ICI300
INDIUM NITRATE see ICI000
INDIUM OXIDE see ICI100
INDIUM (3+) OXIDE see ICI100
INDIUM (III) OXIDE see ICI100
INDIUM PHOSPHIDE see ICI300
INDIUM SESQUIOXIDE see ICI100
INDIUM SULFATE see ICJ000
INDIUM TRICHLORIDE see ICK000
INDIUM TRIOXIDE see ICI100
INDOBLACK GR see CMN240
INDOBLOC see ICC000
INDO BLUE B-I see DFN300
INDO BLUE WD 279 see DFN300
INDOCID see IDA000
INDOCIN I.V. see IDA100

INDOCYANINE GREEN see CCK000
INDOCYANINE GREEN see ICL000
INDOCYBIN see PHU500
INDOFAST ORANGE OV 5983 see CMU820
INDOFAST VIOLET LAKE see DFN450
INDOINE BLUE see CMM770
INDOINE BLUE 3B see CMM770
INDOINE BLUE R see CMM770
INDOKLON see HDC000
INDOL (GERMAN) see ICM000
3-INDOLACETONITRILE see ICW000
INDOLACIN see CMP950
β-INDOLAETHYLAMIN-CHLORHYDRAT (GERMAN) see AJX250
α-INDOLAETHYLAMIN SALZSAEURE (GERMAN) see LIU100
INDOLAPRIL HYDROCHLORIDE see ICL500
INDOLE see ICM000
INDOLE, 3-ACETATO- see IDA600
3-INDOLEACETIC ACID see ICN000
1H-INDOLE-3-ACETIC ACID see ICN000
β-INDOLEACETIC ACID see ICN000
β-INDOLE-3-ACETIC ACID see ICN000
1H-INDOLE-3-ACETIC ACID, 1-(4-CHLOROBENZOYL)-5-METHOXY-2-METHYL-2-((2-CARBOXYPHENOXY)CARBONYL)PHENYL ESTER see CCI525
1H-INDOLE-3-ACETIC ACID, 1-(4-CHLOROBENZOYL)-5-METHOXY-2-METHYL-, SODIUM SALT, TRIHYDRATE see IDA100
INDOLEACETONITRILE see ICW000
INDOLE-3-ACETONITRILE see ICW000
1H-INDOLE-3-ACETONITRILE see ICW000
INDOLE-3-ACRYLIC ACID see ICO000
INDOLE-3-ALANINE see TNX000
INDOLE-3-ALDEHYDE see FNO100
INDOLE-3-(2-AMINOBUTYL) ACETATE see AJB250
INDOLE, 3-(2-AMINOETHYL)-5-METHOXY- see MFS400
INDOLE, 5-BENZYLOXY-3-ISONIPECOTOYL- see BFC200
1H-INDOLE-3-BUTANOIC ACID see ICP000
INDOLE BUTYRIC see ICP000
INDOLE BUTYRIC ACID see ICP000
3-INDOLEBUTYRIC ACID see ICP000
β-INDOLEBUTYRIC ACID see ICP000
γ-(INDOLE-3)-BUTYRIC ACID see ICP000
INDOLE-3-CARBALDEHYDE see FNO100
INDOLE-3-CARBINOL see ICP100
INDOLE-5-CARBONITRILE, 3-ACETYL- see COP550
INDOLE-5-CARBONITRILE, 3-(2-METHYLPROPIONYL)- see COP525
INDOLE-3-CARBOXALDEHYDE see FNO100
1H-INDOLE-3-CARBOXALDEHYDE (9CI) see FNO100
INDOLE-3-CARBOXALDEHYDE, 2-(m-AMINOPHENYL)-, 4-(m-TOLYL)-3-THIOSEMICARBAZONE see ALW900
1H-INDOLE-2-CARBOXYLIC ACID, 1-(2-((1-(ETHOXYCARBONYL)BUTYL)AMINO)-1-OXOPROPYL)OCTAHYDRO-,(2S-(1(R*(R*)),2α,3Aβ,7Aβ))-, COMPD. WITH 2-METHYL-2-PROPANAMINE (1:1) see PCJ230
1H-INDOLE-3-CARBOXYLIC ACID, 8-METHYL-8-AZABICYCLO(3.2.1)OCT-3-YL ESTER, endo- see TNU100
INDOLE, 3-(2-(4-CHLOROBENZYLAMINO)ETHYL)-, MONOHYDROCHLORIDE see CEO125
1H-INDOLE, 2,3-DIHYDRO- see ICS300
INDOLE, 2,3-DIMETHYL- see DSI850
1H-INDOLE, 2,3-DIMETHYL-(9CI) see DSI850
INDOLE, 5-(DIMETHYLAMINO)-3-(PIPERIDINOACETYL)- see DPT300
INDOLE-2,3-DIONE see ICR000

1H-INDOLE-3-ETHANAMINE see AJX000
INDOLE ETHANOL see ICS000
INDOLE-3-ETHANOL, 5-HYDROXY- see HLI600
1H-INDOLE-3-ETHANOL, 5-HYDROXY- see HLI600
INDOLE-3-ETHYLAMINE HYDROCHLORIDE see AJX250
α-INDOLEETHYLAMINE HYDROCHLORIDE see LIU100
β-INDOLE-ETHYLAMINE HYDROCHLORIDE see AJX250
INDOLE, 3-(HYDROXYACETYL)-1-METHYL see HIS140
INDOLE, 3-(HYDROXYACETYL)-2-METHYL see HIS150
1H-INDOLE-5-METHANESULFONAMIDE, 3-(2-(DIMETHYLAMINO)ETHYL)-N-METHYL- see DPH300
INDOLE-3-METHANOL see ICP100
1H-INDOLE-3-METHANOL (9CI) see ICP100
1H-INDOLE-METHANOL, 5-METHOXY-α-3-PYRIDINYL- see MFN700
3-INDOLEMETHANOL, 5-METHOXY-α-(3-PYRIDYL)- see MFN700
INDOLE, 3,3'-METHYLENEBIS(1-(PIPERIDINOMETHYL))- see MJP420
1H-INDOLE, 3,3'-METHYLENEBIS(1-(1-PIPERIDINYLMETHYL))-(9CI) see MJP420
INDOLENE see GCC200
INDOLENE see ICS100
INDOLE, 2-PHENYL- see PFJ760
INDOLEPROPIONIC ACID see ICS200
1H-INDOLE-3-PROPIONIC ACID see ICS200
β-INDOLEPROPIONIC ACID see ICS200
INDOLE, 1-PROPIONYL- see ICW100
INDOLE-3-PROPYLAMINE HYDROCHLORIDE see AME750
1H-INDOLE(3,2-C)QUINOLINE-1,4(7H)-DIONE, 8,9,10,11-TETRAHYDRO-3-METHOXY- see TCU660
1H-INDOLE-2-SULFONIC ACID, 5-((AMINOCARBONYL)HYDRAZONO)-2,3,5,6-TETRAHYDRO-1-METHYL-6-OXO-, MONOSODIUM SALT, TRIHYDRATE see AER666
INDOLE, 3-(N-2-THIAZOLYLFORMIMIDOYL)- see TEX210
INDOL-3-ETHYLAMINE see AJX000
INDOLE, 2,3,6-TRIMETHYL- see TLT800
1H-INDOLE, 2,3,6-TRIMETHYL- see TLT800
INDOLIN see BBW500
Δ²,α)-INDOLINACETALDEHYDE, 1,3,3-TRIMETHYL-(6CI) see FBV050
INDOLINE see ICS300
Δ²,α)-INDOLINEACETALDEHYDE, 1,3,3-TRIMETHYL-(7CI,8CI) see FBV050
INDOLINE, 5-ACETYL- see ACO320
2,3-INDOLINEDIONE see ICR000
INDOLINE, 2-METHYL- see MKV800
INDOLINE, 2-METHYLENE-1,3,3-TRIMETHYL- see TLU200
3-INDOLINONE, 5-BROMO-2-(9-CHLORO-3-OXONAPHTHO(1,2-b)THIEN-2(3H)-YLIDENE)- see CMU320
3H-INDOLIUM, 2-(p-((2-CHLOROETHYL)METHYLAMINO)STYRYL)-1,3,3-TRIMETHYL-, CHLORIDE see CMM850
3H-INDOLIUM, 2-(2-(2,4-DIMETHOXYANILINO)VINYL)-1,3,3-TRIMETHYL-, CHLORIDE see CMM890
3H-INDOLIUM, 2-(2-((2,4-DIMETHOXYPHENYL)AMINO)ETHENYL)-1,3,3-TRIMETHYL-, CHLORIDE see CMM890
3H-INDOLIUM, 2-(2-(4-((4-ETHOXYPHENYL)METHYLAMINO)PHENYL)ETHENYL)-1,3,3-TRIMETHYL-, CHLORIDE see EFC650
3H-INDOLIUM, 2-(((4-METHOXYPHENYL)METHYLHYDRAZON

O)METHYL)-1,3,3-TRIMETHYL-, METHYL SULFATE see CMM895
3H-INDOLIUM, 2-(2-(2-METHYLINDOL-3-YL)VINYL)-1,3,3-TRIMETHYL-, CHLORIDE see CMM820
3H-INDOLIUM, 1,3,3-TRIMETHYL-2-(3-(1,3,3-TRIMETHYL-2-INDOLINYLIDENE)PROPENYL)-, CHLORIDE see CMM840
INDOL-N-METHYLHARMINE HYDROCHLORIDE see ICU100
INDOL-3-OL, ACETATE (ester) (8CI) see IDA600
1H-INDOL-3-OL, ACETATE (ester) (9CI) see IDA600
INDOL-3-OL, HYDROGEN SULFATE (ESTER), POTASSIUM SALT see ICD000
INDOL-3-OL, POTASSIUM SULFATE see ICD000
3H-INDOL-3-ONE, 1,2-DIHYDRO-5,7-DIBROMO-2-(5,7-DIBROMO-1,3-DIHYDRO-3- OXO-2H-INDOL-2-YLIDENE)- see ICU135
3H-INDOL-3-ONE, 2(1,3-DIHYDRO-3-OXO-2H-INDOL-2-YLIDENE)-1,2-DIHYDRO-(9CI) see BGB275
2H-INDOL-2-ONE, 1,3-DIHYDRO-3-(((2-PHENYLETHYL)AMINO)METHYLENE)-, (E)- see ICU145
1H-INDOLO(3,2-C)QUINOLINE-1,4(11H)-DIONE, 3-METHOXY-6-METHYL- see MEV800
1H-INDOLO(3,2-C)QUINOLINE-1,4(7H)-DIONE, 8,9,10,11-TETRAHYDRO-3-METHOXY-6-METHYL- see MEY100
6H-INDOLO(2,3-B)QUINOXALINE-6-ACETIC ACID, 2-CHLORO-, (((3,4-DIMETHOXYPHENYL)METHYLENE)HYDRAZIDE see CHR200
Δ-INDOLYBUTYLAMINE HYDROCHLORIDE see AJB500
INDOLYLACETIC ACID see ICN000
INDOLYL-3-ACETIC ACID see ICN000
3-INDOLYLACETIC ACID see ICN000
β-INDOLYLACETIC ACID see ICN000
α-INDOL-3-YL-ACETIC ACID see ICN000
INDOLYLACETONITRILE see ICW000
3-INDOLYLACETONITRILE see ICW000
3-INDOLYLACRYLIC ACID see ICO000
1-β-3-INDOLYLALANINE see TNX000
β-INDOLYLALDEHYDE see FNO100
ω-3-INDOLYLAMYLAMINE ADIPINATE see ALR500
INDOLYL-3-BUTYRIC ACID see ICP000
4-(INDOLYL)BUTYRIC ACID see ICP000
4-(INDOL-3-YL)BUTYRIC ACID see ICP000
4-(3-INDOLYL)BUTYRIC ACID see ICP000
γ-(INDOL-3-YL)BUTYRIC ACID see ICP000
γ-(3-INDOLYL)BUTYRIC ACID see ICP000
3-INDOLYL-γ-BUTYRIC ACID see ICP000
3-INDOLYLCARBINOL see ICP100
3-INDOLYLETHANOL see ICS000
2-(3-INDOLYL)ETHYLAMINE see AJX000
β-3-INDOLYLETHYLAMINE HYDROCHLORIDE see AJX250
INDOL-1-YL ETHYL KETONE see ICW100
N-(2-(3-INDOLYL)ETHYL)-NICOTINAMIDE see NDW525
N-(2-INDOL-3-YLETHYL)NICOTINAMIDE see NDW525
3-(1-(1H-INDOL-3-YL)ETHYL)-1-(PHENYLMETHYL)-4-PIPERIDINONE see ICW150
2-INDOLYL METHOXYMETHYL KETONE see ICW200
(±)-1-(3-INDOLYL)-2-METHYLAMINOETHANOL see ICY000
1-(INDOLYL-3)-2-METHYLAMINOETHANOL-1 RACEMATE see ICY000
INDOL-3-YL METHYL KETONE see ICY100

1-(3'-INDOLYLMETHYL)-4-(2''-QUINOLYL)PIPERAZINE DIMALEATE see ICZ100
INDOLYL-3-MORPHOLINOMETHYL KETONE see ICZ150
1-(4-INDOLYLOXY)-3-(ISOPROPYLAMINO)-2-PROPANOL see VSA000
1-(1H-INDOL-4-YLOXY)-3-((1-METHYLETHYL)AMINO)-2-PROPANOL see VSA000
INDOLYL-3-PIPERIDINOMETHYL KETONE see ICZ200
INDOL-3-YL POTASSIUM SULFATE see ICD000
3-(3-INDOLYL)PROPANOIC ACID see ICS200
3-(1-H-INDOL-3-YL)-2-PROPENOIC ACID see ICO000
γ-3-INDOLYLPROPYLAMINE HYDROCHLORIDE see AME750
INDOL-3-YL SULFATE, POTASSIUM SALT see ICD000
INDO MAROON LAKE RV 6666 see CMU825
INDOMECOL see IDA000
INDOMED see IDA000
INDOMETHACIN see IDA000
INDOMETHACIN SODIUM TRIHYDRATE see IDA100
INDOMETHAZINE see IDA000
INDOMETICINA (SPANISH) see IDA000
INDON see PFJ750
INDONAPHTHENE see IBX000
INDOPAN see AME500
INDOPHENOL, 2,6-DICHLORO-, SODIUM SALT see SGG650
INDOPROFEN see IDA400
INDOPTIC see IDA000
INDO-RECTOLMIN see IDA000
INDORM see IDA500
INDOSPICINE HYDROCHLORIDE see IDA525
INDOSPICINE MONOHYDROCHLORIDE MONOHYDRATE see IDA525
INDO-TABLINEN see IDA000
INDOTONER BLUE B 79 see DFN300
INDOXACARB see IDA550
INDOXAMIC ACID see OLM300
INDOXINE KL see CMN800
INDOXYLACETATE see IDA600
INDOXYL-O-ACETATE see IDA600
INDUSTRENE 105 see OHU000
INDUSTRENE 205 see OHU000
INDUSTRENE 206 see OHU000
INDUSTRENE 4516 see PAE250
INDUSTRENE 5016 see SLK000
INERTEEN see PJL750
INETOL see INS000
INETOL see INT000
INEXIT see BBQ500
INF 1837 see TKH750
INF 3355 see XQS000
INF 4668 see DGM875
INFAMIL see HNI500
INFERNO see DJA400
INFILTRINA see DUD800
INFLAMASE see PLY275
INFLAMEN see BMO000
INFLATINE see LHY000
INFLAZON see IDA000
INFRON see VSZ100
INFUSORIAL EARTH see DCJ800
INGALAN see DFA400
INGALAN (RUSSIAN) see DFA400
INGENANE HEXADECANOATE see IDB000
INHALAN see DFA400
INH-G see GBB500
INHG-SODIUM see IDB100
INHIBINE see HIB050
INHIBISOL see MIH275
INHIBITOR OP-10 see OJD232

INHISTON see TMK000
'INIA (HAWAII) see CDM325
INICARDIO see DJS200
INIPROL see PAF550
INITIATING EXPLOSIVE DIAZODINITROPHENOL (DOT) see DUR800
INITIATING EXPLOSIVE LEAD MONONITRORESORCINATE (DOT) see LDP000
INITIATING EXPLOSIVE NITROSOGUANIDINE see NKH000
INITIATING EXPLOSIVE PENTAERYTHRITE TETRANITRATE (DOT) see PBC250
INJERTO (TEXAS, MEXICO) see MQW525
INKASAN see IBQ300
INK BERRY see PJJ315
INK ORANGE JSN see HGC000
INK RED JSN see CMM300
INKREDOL-1 see IDB200
INNOVAN see DYF200
INNOVAR see DYF200
INNOXALON see EID000
INO see IDE000
INOCOR see AOD375
INOKOSTERONE see IDD000
INOLIN see IDD100
INOPHYLLINE see TEP500
INOPSIN see DYF200
INOSIE see IDE000
INOSINE see IDE000
β-INOSINE see IDE000
INOSINE, 2-AMINO- see GLS000
INOSINE, 2-AMINO-2'-DEOXY-, OXIME see AJL200
INOSINE DIALDEHYDE see IDE050
INOSINE-5'-MONOPHOSPHATE see IDE200
INOSINE-5'-MONOPHOSPHATE DISODIUM see DXE500
INOSINE-5'-MONOPHOSPHORIC ACID see IDE200
(INOSINE)PENTAAMINERUTHENIUM(3+) TRICHLORIDE see IDE100
INOSINE-5'-PHOSPHATE see IDE200
INOSINIC ACID see IDE200
5'-INOSINIC ACID, DISODIUM SALT, mixed with DISODIUM-5'-GUANYLATE (1:1) see RJF400
5'-INOSINIC ACID, HOMOPOLYMER complex with 5'-CYTIDYLIC ACID HOMOPOLYMER (1:1) see PJY750
INOSIN-5'-MONOPHOSPHATE DISODIUM see DXE500
INOSITHEXAPHORSAEURE (GERMAN) see PIB250
INOSITOL see IDE300
i-INOSITOL see IDE300
meso-INOSITOL see IDE300
INOSITOL HEXANICOTINATE see HFG550
m-INOSITOL HEXANICOTINATE see HFG550
myo-INOSITOL HEXANICOTINATE see HFG550
meso-INOSITOL HEXANICOTINATE see HFG550
INOSITOL HEXAPHOSPHATE see PIB250
INOSITOL HEXASULFATE SODIUM SALT see IDE400
INOSITOL NIACINATE see HFG550
INOSITOL NICOTINATE see HFG550
INOSTRAL see CNX825
INOTREX see DXS375
INOVAL see DYF200
INOVITAN PP see NCR000
INOX see IDE050
INPC see PMS825
INPROQUONE see DCN000
INSANE ROOT see HAQ100
INSARIOTOXIN see FQS000
INSECTICIDE 1,179 see MDU600
INSECTICIDE No. 497 see DHB400

INSECTICIDE No. 4049 see MAK700
INSECTICIDE-NEMATICIDE 1410 see DSP600
INSECTOPHENE see EAQ750
INSECT POWDER see POO250
INSECT REPELLENT 448 see PES000
6-12-INSECT REPELLENT see EKV000
INSEGAR see EOE200
INSIDON see DCV800
INSIDON DIHYDROCHLORIDE see IDF000
INSOLUBLE GLUCOSE ISOMERASE ENZYME PREPARATIONS see GFG050
INSOLUBLE SACCHARINE see BCE500
INSOMNOL see EQL000
INSOM-RAPIDO see CAV000
IN-SONE see PLZ000
INSPIR see ACH000
INSULAMIN see BOM750
INSULAMIN see BQL000
INSULAMINA see DNA200
INSULAR see IDF300
INSULATARD see IDF325
INSULIN see IDF300
INSULIN INJECTION see IDF300
INSULIN LENTE see LEK000
INSULIN NOVO LENTE see LEK000
INSULIN PROTAMINE ZINC see IDF325
INSULIN RETARD RI see IDF325
INSULIN ZINC COMPLEX see LEK000
INSULIN ZINC PROTAMINATE see IDF325
INSULIN ZINC PROTAMINE see IDF325
INSULIN ZINC SUSPENSION see LEK000
INSULTON see MKB250
INSULYL see IDF300
INSULYL-RETARD see IDF325
INSUMIN see DAB800
INTAL see CNX825
INTALBUT see BRF500
INTALPRAM see DLH600
INTALPRAM see DLH630
INTEBAN SP see IDA000
INTEGERRIMINE see IDG000
INTEGRIN see ECW600
INTENKORDIN see CBR500
INTENSAIN see CBR500
INTENSAIN HYDROCHLORIDE see CBR500
INTENSE BLUE see FAE100
INTERCAIN see BQA010
INTERCHEM ACETATE BLUE B see MGG250
INTERCHEM ACETATE BORDEAUX B see CMP080
INTERCHEM ACETATE DEVELOPED BLACK see DPO200
INTERCHEM ACETATE FAST PINK DNA see AKO350
INTERCHEM ACETATE GREEN BLUE ALF see DMM400
INTERCHEM ACETATE PINK 3B see DBX000
INTERCHEM ACETATE PINK BLF see AKE250
INTERCHEM ACETATE SCARLET B see ENP100
INTERCHEM ACETATE VIOLET 6B see AKP250
INTERCHEM ACETATE VIOLET R see DBP000
INTERCHEM ACETATE YELLOW G see AAQ250
INTERCHEM DIRECT BLACK Z see AQP000
INTERCHEM DISPERSE YELLOW GH see AAQ250
INTERCHEM HISPERSE GREEN BLUE ALFH see DMM400
INTERCHEM HISPERSE PINK BH see AKE250
INTERCHEM HISPERSE SCARLET BH see ENP100
INTERCHEM POLYDYE YELLOW GSFD see KDA075

INTERCHEM YELLOW GSF (HDLF) see
KDA075
INTERFLO see PJS750
INTERKELLIN see AHK750
INTERLEUKIN 2 (HUMAN CLONE PTG853
PROTEIN MOIETY REDUCED) see IDG100
INTERNATIONAL ORANGE 2221 see
LCS000
INTEROX see HIB010
INTERPINA see RDK000
INTERSTITIAL CELL STIMULATING
HORMONE see LIU300
INTERTULLE FUCIDIN see SHK000
INTEXAN CTC 29 see HCQ525
INTEXAN LB-50 see AFP250
INTEXAN SB-85 see DTC600
INTEXSAN CPC see CCX000
INTEXSAN CTC 29 see HCQ525
INTEXSAN CTC 50 see HCQ525
INTEXSAN LQ75 see LBW000
INTOCOSTRIN see TOA000
INTOCOSTRINE see COF750
INTOLEX see BEQ625
INTRABILIX see EQC000
INTRABLIX see BGB315
INTRABOND LIQUID TURQUOISE LL see
COF420
INTRACID FAST ORANGE G see HGC000
INTRACID GREEN F see FAE950
INTRACID PURE BLUE L see FAE000
INTRACID PURE BLUE V see ADE500
INTRACORT see HHR000
INTRADERM TYROTHRICIN see TOG500
INTRADEX see DBD700
INTRALITE TURQUOISE GLL see COF420
INTRAMYCETIN see CDP250
INTRANARCON see SEH600
INTRANARCONE see SEH600
INTRANEFRIN see VGP000
INTRANYL ORANGE T-4R see SGP500
INTRASIL BLUE R see CMP075
INTRASPERSE YELLOW GBA see AAQ250
INTRASPERSE YELLOW GBA EXTRA see
AAQ250
INTRASPORIN see TEY000
INTRASTIGMINA see NCL100
INTRATHION see PHI500
INTRATION see PHI500
INTRAVAL see PBT250
INTRAVAL SODIUM see PBT500
INTRAVAT BLUE GF see DFN300
INTRAZONE RED BR see CMM330
INTROMENE see HII500
INTROPIN see DYC600
INULINE see LJB150
INVENOL see BSM000
INVERSAL see AHI875
INVERSINE see VIZ400
INVERSINE HYDROCHLORIDE see
MQR500
INVERTON 245 see TAA100
INVERTS OF 2,4-D AND 2,4-DP see PMM100
INVERT SUGAR see IDH200
INVERT SUGAR SYRUP see IDH200
INVIRASE see FOL025
INVISI-GARD see PMY300
IOB 82 see AFQ575
IOCARMATE MEGLUMINE see IDJ500
IOCARMIC ACID see TDQ230
IOCARMIC ACID DI-N-
METHYLGLUCAMINE SALT see IDJ500
IOCETAMIC ACID see IDJ550
IODAIRAL see HGB200
IODAMIDE see AAI750
IODAMIDE 380 see IDJ600
IODAMIDE MEGLUMINE see IDJ600
IODAMIDE METHYLGLUCAMINE see
IDJ600
IODATES see IDJ700
IODE (FRENCH) see IDM000
IODEIKON see TDE750
IODENTEROL see CHR500

IODIC ACID see IDK000
IODIC ACID, SODIUM SALT see SHV500
IODIC ACIDOIC ACID, POTASSIUM SALT
see PLK250
IODIDES see IDL000
IODIMIDE see DNE600
IODINATED CASEIN see IDL100
IODINATED GLYCEROL see IEL800
IODINE see IDM000
IODINE AZIDE see IDN000
IODINE(I) AZIDE see IDN000
IODINE AZIDE (dry) (DOT) see IDN000
IODINEBENZOL see IEC500
IODINE BROMIDE see IDN200
IODINE CHLORIDE see IDS000
IODINE CRYSTALS see IDM000
IODINE CYANIDE see COP000
IODINE DIOXIDE TRIFLUORIDE see
IDP000
IODINE DIOXYGEN TRIFLUORIDE see
IDP000
IODINE HEPTAFLUORIDE see IDQ000
IODINE ISOCYANATE see IDR000
IODINE MONOCHLORIDE see IDS000
IODINE(V) OXIDE see IDS300
IODINE PENTAFLUORIDE see IDT000
IODINE(III) PERCHLORATE see IDU000
IODINE PHOSPHIDE see PHA000
IODINE-POTASSIUM IODIDE see PLW285
IODINE and PROPYLTHIOURACIL see
PNX100
IODINE SUBLIMED see IDM000
IODINE TRIACETATE see IDV000
IODIO (ITALIAN) see IDM000
IODIPAMIDE MEGLUMINE see BGB315
IODIPAMIDE MEGLUMINE SALT see
BGB315
IODIPAMIDE METHYLGLUCAMINE SALT
see BGB315
IODOACETAMIDE see AOC500
IODOACETAMIDE see IDW000
2-IODOACETAMIDE see IDW000
α-IODOACETAMIDE see IDW000
4-IODOACETANILIDE see IDY000
p-IODOACETANILIDE see IDY000
4'-IODOACETANILIDE see IDY000
IODOACETATE see IDZ000
IODOACETATE SODIUM SALT see SIN000
IODOACETIC ACID see IDZ000
IODOACETIC ACID ETHYL ESTER see
ELQ000
IODOACETIC ACID SODIUM SALT see
SIN000
(IODOACETOXY)TRIBUTYLSTANNANE
see TID750
(IODOACETOXY)TRIPROPYLSTANNANE
see TNB500
N-(IODOACETYL)-3-
AZABICYCLO(3.2.2)NONANE see IDZ100
1-IODOACETYL-α-α-DIPHENYL-4-
PIPERIDINEMETHANOL see IDZ200
IODOACETYLENE see IDZ400
IODOALPHIONIC ACID see PDM750
3-IODOANILINE see IEB000
4-IODOANILINE see IEC000
m-IODOANILINE see IEB000
p-IODOANILINE see IEC000
IODOANTIPYRINE see IEU085
4-IODOANTIPYRINE see IEU085
IODOAZIDE see IDN000
4-IODOBENZENAMINE see IEC000
IODOBENZENE see IEC500
IODOBENZENE see IEC500
4-IODOBENZENEDIAZONIUM-2-
CARBOXYLATE see IED000
o-IODOBENZOIC ACID see IEE000
p-IODOBENZOIC ACID see IEE025
o-IODOBENZOIC ACID SODIUM SALT see
IEE050
p-IODOBENZOIC ACID SODIUM SALT see
IEE100

o-IODOBENZOIC ACID
TRIBUTYLSTANNYL ESTER see TIE000
p-IODOBENZOIC ACID
TRIBUTYLSTANNYL ESTER see TIE250
m-IODOBENZOTEF see BGY130
o-IODOBENZOTEF see BGY135
N-(2-IODOBENZOYL)-GLYCIN
MONOSODIUM SALT (9CI) see HGB200
N-p-IODOBENZOYL-N',N',N',N'-
DIETHYLENETRIAMIDE of PHOSPHORIC
ACID see BGY140
(4-
IODOBENZOYLOXY)TRIBUTYLSTANNAN
E see TIE250
(o-
IODOBENZOYLOXY)TRIPROPYLSTANNA
NE see IEF000
p-IODOBENZYL ISOBUTYL CARBONATE
see IEF100
p-IODOBENZYL-4-METHYL-2-PENTYL
CARBONATE see IEF200
IODOBIL see PDM750
4'-IODO-(1,1'-BIPHENYL)-4-AMINE see
AKL600
1-IODO-1,3-BUTADIYNE see IEG000
1-IODOBUTANE see BRQ250
2-IODOBUTANE see IEH000
IODOCHLORHYDROXYQUINOL see
CHR500
IODOCHLORHYDROXYQUINOLINE see
CHR500
7-IODO-5-CHLORO-8-
HYDROXYQUINOLINE see CHR500
7-IODO-5-CHLOROXINE see CHR500
5-IODODEOXYURIDINE see DAS000
5-IODO-2'-DEOXYURIDINE see DAS000
IODODIMETHYLARSINE see IEI000
4-IODO-3,5-DIMETHYLISOXAZOLE see
IEI600
4-IODO-3,5-DIMETHYL-N-(2-
METHYLPHENYL)-1H-PYRAZOLE-1-
ACETAMIDE see IEI700
4-IODO-3,5-DIMETHYL-N-(3-
METHYLPHENYL)-1H-PYRAZOLE-1-
ACETAMIDE see IEI730
4-IODO-3,5-DIMETHYL-N-(4-
METHYLPHENYL)-1H-PYRAZOLE-1-
ACETAMIDE see IEI740
2-IODO-3,5-DINITROBIPHENYL see IEJ000
IODOENTEROL see CHR500
IODOETHANE see ELP500
IODOETHANOL see IEL000
2-IODOETHANOL see IEL000
2-(2-IODOETHYL)-1,3-DIOXOLANE see
IEL700
2-(1-IODOETHYL)-1,3-DIOXOLANE-4-
METHANOL see IEL800
(2-IODOETHYL)TRIMETHYLAMMONIUM
IODIDE see IEM300
3-IODO-2-FAA see IEO000
IODOFENOPHOS see IEN000
N-(3-IODO-2-FLUORENYL)ACETAMIDE
see IEO000
IODOFORM see IEP000
IODOGNOST see TDE750
IODOHEPTAFLUOROPROPANE see
HAY300
IODOHIPPURA see HGB200
IODOHIPPURATE SODIUM see HGB200
o-IODOHIPPURATE SODIUM see HGB200
4-(3-IODO-4-HYDROXYPHENOXY)-3,5-
DIIODOPHENYLALANINE see LGK050
7-IODO-8-HYDROYQUINOLINE-5-
SULFONIC ACID see IEP200
IODOMETANO (ITALIAN) see MKW200
IODOMETHANE see MKW200
IODOMETHANESULFONIC ACID SODIUM
SALT see SHX000
1-IODO-3-METHYLBENZENE see IFG100
1-IODO-3-METHYLBUTANE see IHU200
IODOMETHYLMAGNESIUM see MLE250

7-IODOMETHYL-12-METHYLBENZ(a)ANTHRACENE see IER000
1-IODO-2-METHYLPROPANE see IIV509
2-IODO-2-METHYLPROPANE see TLU000
IODOMETHYLTRIMETHYLARSONIUM IODIDE see IET000
IODOMETHYLZINC see MQP250
IODONIUM, DIPHENYL-, HEXAFLUOROARSENATE (1-) see DWD300
1-IODOOCTANE see OFA100
2-IODOOCTANE see OFA200
IODOPACT see AAN000
IODOPANIC ACID see IFY100
IODOPANOIC ACID see IFY100
IODOPAQUE see AAN000
1-IODOPENTANE see IET500
1-IODO-3-PENTEN-1-YNE see IEU000
1-IODOPERFLUORODECANE see IEU075
1-IODOPERFLUOROOCTANE see PCH325
IODOPHENAZONE see IEU085
IODOPHENE see IEU100
IODOPHENE SODIUM see TDE750
2-IODOPHENOL see IEV000
3-IODOPHENOL see IEV010
4-IODOPHENOL see IEW000
m-IODOPHENOL see IEV010
o-IODOPHENOL see IEV000
p-IODOPHENOL see IEW000
1-IODO-3-PHENYL-2-PROPYNE see IEX000
IODOPHOS see IEN000
IODOPHTHALEIN SODIUM see TDE750
1-IODOPROPANE see PNO750
2-IODOPROPANE see IPS000
3-IODOPROPENE see AGI250
3-IODO-1-PROPENE see AGI250
3-IODOPROPIONIC ACID see IEY000
(IODOPROPIONYLOXY)TRIBUTYLSTANNANE see TIE500
3-IODOPROPYLENE see AGI250
IODOPROPYLIDENE GLYCEROL see IEL800
3-IODOPROPYNE see IEZ800
3-IODO-2-PROPYNYL-2,4,5-TRICHLOROPHENYL ETHER see IFA000
IODOPYRACET see DNG400
IODOPYRACET MEGLUMINE see IFA100
IODOPYRINE see IEU085
IODOQUINOL see DNF600
IODORAYORAL see TDE750
IODOSOBENZENE see IFC000
IODOSOBENZENE DIACETATE see IFD000
IODOSYLBENZENE see IFE000
4-IODOSYLTOLUENE see IFE875
2-IODOSYLVINYL CHLORIDE see IFE879
3-IODOTETRAHYDROTHIOPHENE-1,1-DIOXIDE see IFG000
5-IODO-2-THIOURACIL, SODIUM SALT see SHX500
3-IODOTOLUENE see IFG100
4-IODOTOLUENE see IFK509
m-IODOTOLUENE see IFG100
IODO(p-TOLYL)MERCURY see IFL000
IODOTRIBUTYLSTANNANE see IFM000
IODOTRIFLUOROMETHANE see IFM100
IODOTRIMETHYLSTANNANE see IFN000
IODOTRIMETHYLTIN see IFN000
IODOTRIPHENYLSTANNANE see IFO000
IODOTRIPROPYLSTANNANE see TNB250
IODO-UNDECINIC ACID see IFO700
11-IODO-10-UNDECINIC ACID see IFO700
11-IODO-10-UNDECYNOIC ACID see IFO700
5-IODOURACIL see IFP000
5-IODOURACIL DEOXYRIBOSIDE see DAS000
IODOXAMATE MEGLUMINE see IFP800
IODOXAMIC ACID MEGLUMINE SALT see IFP800
IODTETRAGNOST see TDE750
IODURE d'ETHYL-TRIMETHYL-AMMONIUM see EQC600

IODURE de MERCURE (FRENCH) see MDC750
IODURE de METHYLE (FRENCH) see MKW200
l'IODURE de α-THIENYL-1 PHENYL-1 N-METHYL MORPHOLINIUM-3 PROPANOL-1 (FRENCH) see HNM000
IODURIL see SHW000
IODURON B see DNG400
4-IODYLANISOLE see IFQ775
IODYLBENZENE see IFR000
IODYLBENZENE PERCHLORATE see IFS000
4-IODYL TOLUENE see IFS350
2-IODYLVINYL CHLORIDE see IFS385
IOFENDYLATE see ELQ500
IOGLUCOMIDE see IFS400
IOMEPROL see IFS500
IOMEPROLO see IFS500
IOMERON 350 see IFS500
IOMESAN see DFV400
IOMEX see IFT100
IOMEZAN see DFV400
IONAC PFAZ 322 see EML600
6,3-IONENE BROMIDE see IFT300
IONET S-80 see SKV100
IONET MO-400 see PJY100
IONET S 60 see SKV150
IONOL see BFW750
IONOL 4 see DEE300
IONOL 6 see MHR050
α-IONOL see IFT400
β-IONOL see IFT500
IONOL (antioxidant) see BFW750
IONOL K65 see IFT600
IONONE see IFV000
α-IONONE see IFW000
β-IONONE see IFX000
β-IONONE see IFX100
(E)-β-IONONE see IFX100
trans-β-IONONE see IFX100
IONOX 100 see IFX200
IONOX 220 see MJM700
IONOX 100 ANTIOXIDANT see IFX200
IONOX 220 ANTIOXIDANT see MJM700
α-IONYL ACETATE see IFX300
IOP see AES000
IOPAMIDOL see IFY000
IOPAMIRON see IFY000
IOPANOIC ACID see IFY100
IOPEZITE see PKX250
IOPHENDYLATE see ELQ500
IOPHENOXIC ACID see IFZ800
IOPODATE SODIUM see SKM000
IOPRAMINE HYDROCHLORIDE see IFZ900
IOPRONIC ACID see IGA000
IOPYRACIL see DNG400
IOQUIN SUSPENSION see DNF600
IOSALIDE see JDS200
IOTALAMATE de METHYLGLUCAMINE (FRENCH) see IGC000
IOTALAMATE de SODIUM see IGC100
IOTHALAMATE MEGLUMINE see IGC000
IOTHALAMATE METHYLGLUCAMINE see IGC000
IOTHALAMATE METHYLGLUCAMINE SALT see IGC000
IOTHALAMATE SODIUM see IGC100
IOTHALAMATE SODIUM (INJECTION) see IGC100
IOTHALAMATE SODIUM SALT see IGC100
IOTHALAMIC ACID see IGD000
IOTOX see HKB500
IOTROXATE MEGLUMINE see IGD075
IOTROXATE METHYLGLUCMINE SALT see IGD075
IOTROXIC ACID see IGD100
IOTROXINSAEURE (GERMAN) see IGD100
IOXAGLIC ACID see IGD200
IOXYNIL see HKB500
IOXYNIL, LITHIUM SALT see DNF400
IOXYNIL OCTANOATE see DNG200

IP-10 see IGE100
IP 50 see IRA050
IP-82 see IIU009
IPA see DMV600
IPA see IMJ000
o-IPA see INA200
IPABUTONA see HNI500
IPAMIX see IBV100
IPANER see DAA800
IPAZINE see IKM050
IPC COMBINED with ENDOTHAL (1:1) see EAR500
IPD see YCJ000
IPDI see IMG000
IP-1,2-DIOL see DLE400
IPECAC SYRUP see IGF000
IPECACUANHA see IGF000
IPE-TOBACCO WOOD see HLY500
IP-FLO see IRA050
IPHOSPHAMIDE see IMH000
IPN see ILE000
IPN see PHX550
IPNO see IBP309
IPNOFIL see QAK000
IPO 8 see TBW100
IPO 62 see BMO310
IPO 63 see MHR150
IPODATE SODIUM see SKM000
IPOFOS see BMO310
IPOGLICONE see BSQ000
IPOGNOX 89 see TFD500
IPOLAB see HMM500
IPOLINA see HGP500
1,4-IPOMEADIOL see IGF200
IPOMEANIN see IGF300
IPOMEANINE see IGF300
IPOMEANOL see FQL100
IPOMEANOL see IGF325
4-IPOMEANOL see FQL100
IPOPHOS see BMO310
IPORES see RCA200
IPOTENSIVO see MBW750
IPPC see CBM000
IPRAL see ELX000
IPRAL SODIUM see NBU000
IPRATROPIUMBROMID (GERMAN) see IGG000
IPRATROPIUM BROMIDE see IGG000
IPRAZID see ILE000
IPRINDOLE see DPX200
IPROBENFOS see BKS750
IPROCLOZIDE see CJN250
IPRODIONE see GIA000
IPROGEN see DLH630
IPROHEPTINE HYDROCHLORIDE see IGG300
IPRONIAZID see ILE000
IPRONIAZID DIHYDROCHLORIDE see IGG600
IPRONIAZID HYDROCHLORIDE see IGG600
IPRONIAZID PHOSPHATE see IGG700
IPRONID see ILE000
IPRONIDAZOLE (USDA) see IGH000
IPRONIN see ILE000
IPROPLATIN see IGG775
IPROPRAN see IGH000
IPROVERATRIL see IRV000
IPROVERATRIL HYDROCHLORIDE see VHA450
IPROX see CMY650
IPSILON see AJD000
IPSOFLAME see BRF500
IPSOTIAN see MQU750
IPSP see EPH500
IPT see IRU500
IPT see IRU500
IPT (GERMAN) see INP100
IPURON see SOZ100
IPU STEFES see IRA050
I.P.Z. see IDF325
IQ 1 see IRV300

IQ DIHYDROCHLORIDE see AKT620
IR 125 see CCK000
IRADICAV see SHF500
IRAGEN RED L-U see FAG070
IRALDEINE see IFV000
IRAMIL see DLH600
IRAMIL see DLH630
IRAX see PJS750
IRC 453 see CIN750
IRC-50 ARVIN see VGU700
IREHDIAMINE A see IGH700
IRENAL see ELX000
IRENAL see PLO500
IRENAT see PLO500
IRENE see DLS600
IRETIN see AQQ750
IRETIN see AQR000
IRG see AKY250
IRGACHROME ORANGE OS see LCS000
IRGACURE 1173 see HMQ100
IRGAFOS P-EPQFF see PCB400
IRGAFOS PEPQ see PCB400
IRGALITE 1104 see CBT500
IRGALITE BRONZE RED CL see BNH500
IRGALITE BRONZE RED CL see BNK700
IRGALITE FAST BRILLIANT GREEN GL see PJQ100
IRGALITE FAST BRILLIANT GREEN 3GL see PJQ100
IRGALITE FAST RED 2GL see DVB800
IRGALITE FAST RED P4R see MMP100
IRGALITE FAST SCARLET RND see MMP100
IRGALITE GREEN GLN see PJQ100
IRGALITE ORANGE P see CMS145
IRGALITE ORANGE PG see CMS145
IRGALITE ORANGE PX see CMS145
IRGALITE RED 4B see CMS155
IRGALITE RED C see CMS150
IRGALITE RED CBN see CHP500
IRGALITE RED PRR see CJD500
IRGALITE RED PV2 see MMP100
IRGALITE RED RL see NAP100
IRGALITE RED RNPX see MMP100
IRGALITE RUBINE PB see CMS155
IRGALITE SCARLET RB see MMP100
IRGALITE YELLOW BAW see CMS208
IRGALITE YELLOW BAWX see CMS208
IRGALITE YELLOW BO see DEU000
IRGALITE YELLOW BR see CMS210
IRGALITE YELLOW BRE see CMS210
IRGALON see EIV000
IRGANOX 1010 see DED100
IRGANOX 1040 see DED100
IRGANOX 1520 see MHR025
IRGANOX P-EPQ see PCB400
IRGANOX PS 800 see TFD500
IRGANOX PS 802 see DXG700
IRGAPHOS P-EPO see PCB400
IRGAPLAST ORANGE G see CMS145
IRGAPLAST YELLOW IRS see CMS208
IRGAPYRIN see IGI000
IRGAPYRINE see IGI000
IRGASAN see TIQ000
IRGASAN BS-200 see TBV000
IRGASAN DP300 see TIQ000
IRGASOL RED 3BNS see CMG750
IRGASTAB 2002 see BJK560
IRGASTAB 2002 HT see BJK560
IRGASTAB T 4 see DEJ100
IRGASTAB T 150 see DEJ100
IRGASTAB T 290 see DEJ100
IRIDIL see HNI500
IRIDIUM see IGJ000
IRIDIUM CHLORIDE see IGJ300
IRIDIUM(IV) CHLORIDE see IGJ499
IRIDIUM MURIATE see IGJ300
IRIDIUM TETRACHLORIDE see IGJ499
IRIDOCIN see EPQ000
IRIDOZIN see EPQ000
IRIFAN see TDA500

IRINOTECTAN HYDROCHLORIDE HYDRATE see IGJ550
IRIS see OHJ130
IRIS ABSOLUTE see OHJ130
IRISH GUM see CCL250
IRISH MOSS EXTRACT see CCL250
IRISH MOSS GELOSE see CCL250
IRISOL BASE see HOK000
IRISONE see IFV000
IRISONE ACETATE see CNS825
IRITONE see IGJ600
IRIUM see SIB600
IRLOXACIN see FLV050
IRMIN see DLH600
IROCAINE see AIT250
IROINI see HBT500
IRO-JEX see IGS000
IROMIN see FBK000
IRON see IGK800
IRON(2+) see FBN000
IRON(2+) ACETATE see FBH000
IRON(II) ACETATE see FBH000
IRON ACETYLACETONATE see IGL000
IRON ALLOY, BASE see IGL100
IRON ALLOY, BASE see IGL110
IRON ALLOY, BASE, Fe 60–69, Cr 18–21, Mn 8–10, Ni 5–7, Si 0–1, N 0.2–0.4, C 0–0.1, P 0–0.1 see IGL110
IRON ALLOY, BASE, Fe,Ni see IGL120
IRON ALLOY, BASE, Fe 39–47, Ni 30–35, Cr 19–23, Mn 0–1.5, Si 0–1, Cu 0–0.8, Al 0–0.6, Ti 0–0.6, C 0–0.1 see IGL100
IRON(III) AMMONIUM CITRATE see FAS700
IRON AMMONIUM SULFATE HYDRATE see IGL200
IRON ARSENATE (DOT) see IGM000
IRON(II) ARSENATE (3:2) see IGM000
IRON(III) ARSENATE (1:1) see IGN000
IRON(III)-o-ARSENITE PENTAHYDRATE see IGO000
IRON(II) ASCORBATE see FBH050
IRONATE see FBO000
IRON, (N,N-BIS(CARBOXYMETHYL)GLYCINATO(3-)-N,O,O',O")-, (T-4)-(9CI) see IHC100
IRON BIS(CYCLOPENTADIENE) see FBC000
IRON, BIS(MU-(METHANETHIOLATO))TETRANITROSYL DI-, (FE-FE) see BKM530
IRON BLACK see IHC550
IRON BLEOMYCIN see IGO500
IRON BLUE see IGY000
IRON(II) BROMIDE see IGP000
IRON(III) BROMIDE see IGQ000
IRON CAPRYLATE see IGQ050
IRON CARBIDE see IGQ750
IRON CARBOHYDRATE COMPLEX see IGT000
IRON(II) CARBONATE see FBH100
IRON CARBONYL see IHG500
IRON CARBONYL see NMV740
IRON, CARBONYL (FCC) see IGK800
IRON CHLORIDE see FAU000
IRON(III) CHLORIDE see FAU000
IRON(II) CHLORIDE (1:2) see FBI000
IRON(3+) CHLORIDE HEXAHYDRATE see FAW000
IRON(III), CHLORIDE HEXAHYDRATE see FAW000
IRON CHLORIDE TETRAHYDRATE see FBJ000
IRON(2+) CHLORIDE TETRAHYDRATE see FBJ000
IRON (II) CHLORIDE TETRAHYDRATE see FBJ000
IRON CHOLINE CITRATE COMPLEX see FBC100
IRON CHROMITE see CMI500
IRON(II) CITRATE see FBJ075

IRON(III) CITRATE see FAW100
IRON COMPOUNDS see IGR499
IRON CYANIDE see IGY000
IRON DEXTRAN see IGS000
IRON-DEXTRAN COMPLEX see IGS000
IRON DEXTRAN GLYCEROL GLYCOSIDE see IGS900
IRON DEXTRAN INJECTION see IGS000
IRON-DEXTRIN COMPLEX see IGT000
IRON DEXTRIN INJECTION see IGT000
IRON DIACETATE see FBH000
IRON DICHLORIDE see FBI000
IRON DICHLORIDE TETRAHYDRATE see FBJ000
IRON DICYCLOPENTADIENYL see FBC000
IRON DIETHYLDITHIOCARBAMATE see IGT100
IRON(III) DIETHYLDITHIOCARBAMATE see IGT100
IRON DIMETHYLDITHIOCARBAMATE see FAS000
IRON DISULFIDE see IGV000
IRON DUST see IGW000
α-IRONE see IGW500
IRON EDTA see IGX550
IRON(II)-EDTA see IGX550
IRON(II) EDTA COMPLEX see IGX000
IRON(III)-EDTA COMPLEX see IGX550
IRON(III)-EDTA SODIUM see IGX875
IRON, ELECTROLYTIC see IGK800
IRON, ELEMENTAL see IGK800
IRON (Fe 2+) see FBN000
IRON(3+) FERROCYANIDE see IGY000
IRON (III) FERROCYANIDE see IGY000
IRON complex with N-FLUOREN-2-YL ACETOHYDROXAMIC ACID see HIQ000
IRON FLUORIDE see FAX000
IRON FUMARATE see FBJ100
IRON GLUCONATE see FBK000
IRON(III) HEXACYANOFERRATE(4-) see IGY000
IRON HYDROGENATED DEXTRAN see IGS000
IRON(+3) HYDROXIDE COMPLEX with NITRILO-TRI-PROPIONIC ACID see FAY000
IRON(III)HYDROXIDE-POLYMALTOSE see IHA000
IRON(II,III) OXIDE see IHC550
IRON(II) ION see FBN000
IRON(2+) LACTATE see LAL000
IRON LINOLEATE see IHA050
IRON(II) MALEATE see IHB675
IRON MANGANESE ZINC OXIDE see IHB677
IRON METHANEARSONATE see IHB680
IRON MONOSULFATE see FBN100
IRON NAPHTHENATE see IHB700
IRON(3+), (N1-(3-(DIMETHYLSULFONIO)PROPYL)BLEOMY CINAMIDATO)- see IGO500
IRON NICKEL SULFIDE see NDE500
IRON NICKEL ZINC OXIDE see IHB800
IRON NITRATE see FAY200
IRON NITRATE see IHB900
IRON (III) NITRATE, ANHYDROUS see FAY200
IRON (III) NITRATE, ANHYDROUS see IHB900
IRON(III) NITRATE, NONAHYDRATE (1:3:9) see IHC000
IRON NITRILOTRIACETATE see IHC100
IRON-NITRILOTRIACETATE CHELATE see IHC100
IRON(3+) NTA see IHC100
IRON ORE see HAO875
IRONORM INJECTION see IGS000
IRON OXIDE see IHC450
IRON OXIDE see IHG100
IRON(II) OXIDE see IHC500
IRON(III) OXIDE see IHC450
IRON OXIDE, spent see IHG100

IRON OXIDE, CHROMIUM OXIDE, and NICKEL OXIDE FUME see IHE000
IRON OXIDE FUME see IHF000
IRON OXIDE RED see IHC450
IRON OXIDE RED 130B see IHG100
IRON OXIDE, SACCHARATED see IHG000
IRON PENTACARBONYL see IHG500
IRON PERSULFATE see FBA000
IRON PHOSPHATE see FAZ500
IRON-POLYSACCHARIDE COMPLEX see IHH000
IRON-POLY(SORBITOL-GLUCONIC ACID) COMPLEX see IHH300
IRON PROTOCHLORIDE see FBI000
IRON PROTOSULFATE see FBN100
IRON PYRITES see IGV000
IRON PYROPHOSPHATE see FAZ525
IRON, REDUCED (FCC) see IGK800
IRON SACCHARATE see IHG000
IRON SESQUIOXIDE see IHC450
IRON SESQUIOXIDE HYDRATED see LFW000
IRON SESQUISULFATE see FBA000
IRON-SILICON see IHJ000
IRON SODIUM GLUCONATE see IHK000
IRON SORBITEX see IHL000
IRON-SORBITOL see IHK100
IRON SORBITOL CITRATE see IHL000
IRON-SORBITOL-CITRIC ACID see IHL000
IRON SPONGE, spent obtained from coal gas purification (DOT) see IHG100
IRON SUGAR see IHG000
IRON SULFATE (2:3) see FBA000
IRON(III) SULFATE see FBA000
IRON(II) SULFATE (1:1) see FBN100
IRON(II) SULFATE (1:1), HEPTAHYDRATE see FBO000
IRON SULFIDE see IGV000
IRON(II) SULFIDE see IHN000
IRON(III) SULFIDE see IHN050
IRON TALLATE see IHN075
IRON TERSULFATE see FBA000
IRON TRICHLORIDE see FAU000
IRON TRICHLORIDE HEXAHYDRATE see FAW000
IRON TRIFLUORIDE see FAX000
IRON TRINITRATE see FAY200
IRON TRINITRATE see IHB900
IRON, TRIS(DIETHYLCARBAMODITHIOATO-S,S')-, (OC-6-11)- see IGT100
IRON TRIS(DIETHYLDITHIOCARBAMATE) see IGT100
IRON, TRIS(DIETHYLDITHIOCARBAMATO)- see IGT100
IRON, TRIS(5,6-DIHYDRO-5,6-DIOXO-2-NAPHTHALENESULFONIC ACID-5-OXIMATO)-, TRISODIUM SALT see NAX500
IRON VITRIOL see FBN100
IRON VITROL see FBO000
IROQUINE see CLD000
IROSPAN see FBN100
IROSUL see FBN100
IROSUL see FBO000
IROX (GADOR) see FBK000
IRRADIATED ERGOSTA-5,7,22-TRIEN-3-β-OL see VSZ100
IRRATHENE R see PJS750
IRTRAN 1 see MAF500
IRTRAN 2 see ZNJ100
IRTRAN 3 see CAS000
IRTRAN 6 see CAJ800
IS 401 see DNM400
1205 I.S. see MHD300
1530 I.S. see MRR125
1665 I.S. see BKS825
2341 I.S. see BIQ500
2406 I.S. see IBB100
2466 I.S. see MHQ750
2723 I.S. see PFY100

ISACONITINE see PIC250
ISADRINE see IMR000
ISADRINE-HYDROCHLORIDE see IMR000
dl-ISADRINE HYDROCHLORIDE see IQS500
ISALIZINA see BET000
ISAMID see DAB875
ISAMIN see WAK000
ISAROL see AIF000
ISATEX see ACD500
ISATIC ACID LACTAM see ICR000
ISATIDINE see RFU000
ISATIN see ICR000
ISATINIC ACID ANHYDRIDE see ICR000
ISATOIC ACID ANHYDRIDE see IHN200
ISATOIC ANHYDRIDE see IHN200
ISAZOFOS see PHK000
ISAZOPHOS see PHK000
ISCEON 22 see CFX500
ISCEON 113 see FOO000
ISCEON 122 see DFA600
ISCEON 131 see TIP500
ISCHELIUM see DLL400
ISCOBROME see MHR200
ISCOBROME D see EIY500
ISCOVESCO see DKA600
ISDIN see CCK125
I-SEDRIN see EAW000
ISEPAMICIN DISULFATE see IHN300
ISEPAMICIN SULFATE see SBE800
ISETHION see GFW000
ISETHIONIC ACID see HKI500
ISF 2001 see TNP275
ISF 2469 see CAK275
ISF 2508 see CFJ000
ISICAINA see DHK400
ISICAINE HYDROCHLORIDE see DHK600
ISIDRINA see ILD400
ISINDONE see IDA400
ISKIA-C see ARN125
ISLACTID see AES650
ISLANDICIN see FPD050
ISLANDITOXIN see COW750
ISMAZIDE see ILD000
ISMELIN see GKQ000
ISMELIN see GKS000
ISMELIN SULFATE see GKS000
ISMICETINA see CDP250
ISMIPUR see POK000
5-ISMN see ISC500
ISMOTIC see HID350
ISOACETOPHORONE see IMF400
ISOADRENALINE HYDROCHLORIDE see AMB000
ISOALLOXAZINE-ADENINE DINUCLEOTIDE see RIF100
ISOAMIDONE II see IKZ000
ISOAMINILE CITRATE see PCK639
ISOAMINILE CYCLAMATE see IHO200
ISOAMINILE CYCLAMATE see PCK669
ISOAMYCIN see BBK000
ISOAMYGDALIN see IHO700
ISOAMYL ACETATE see IHO850
ISOAMYL ALCOHOL see IHP000
ISOAMYL ALCOHOL see IHP010
ISOAMYL ALDEHYDE see MHX500
ISOAMYL ALKOHOL (CZECH) see IHP000
ISO-AMYLALKOHOL (GERMAN) see IHP000
ISOAMYL AMINOFORMATE see IHQ000
ISOAMYL BENZOATE see IHP100
ISOAMYL BENZYL ETHER see BES500
ISOAMYL BROMIDE see BNP250
ISOAMYL BUTANOATE see IHP400
ISOAMYL BUTYLATE see IHP400
ISOAMYL BUTYRATE see IHP400
ISOAMYL-n-BUTYRATE see IHP400
ISOAMYL CAPROATE see IHU100
ISOAMYL CAPRYLATE see IHP500
ISOAMYL CARBAMATE see IHQ000
ISOAMYL CINNAMATE see AOG600
ISOAMYL CYANIDE see MNJ000

ISOAMYLDICHLORARSINE see IHQ100
ISOAMYLDICHLOROARSINE see IHQ100
ISOAMYL α-(N-(β-DIETHYLAMINOETHYL))-AMINOPHENYLACETATE see NOC000
ISOAMYL N-(β-DIETHYLAMINOETHYL)-α-AMINOPHENYLACETATE see NOC000
ISOAMYL 5,6-DIHYDRO-7,8-DIMETHYL-4,5-DIOXO-4H-PYRANO(3,2-c)QUINOLINE-2-CARBOXYLATE see IHR200
ISOAMYLENE see IHR220
ISOAMYLENE ALCOHOL see PBL000
β-ISOAMYLENE OXIDE see TLY175
8-ISOAMYLENOXYPSORALEN see IHR300
4-(β-ISOAMYLENYL)-1,2-DIPHENYL-3,5-PYRAZOLIDINEDIONE see PEW000
ISOAMYL ETHANOATE see IHO850
ISOAMYLETHYLBARBITURIC ACID see AMX750
5-ISOAMYL-5-ETHYLBARBITURIC ACID see AMX750
5-ISOAMYL-5-ETHYLBARBITURIC ACID, SODIUM DERIVATIVE see AON750
ISOAMYL FORMATE see IHS000
ISOAMYL GERANATE see IHS100
1-ISOAMYL GLYCEROL ETHER see IHT000
ISOAMYL HEXANOATE see IHU100
ISOAMYL HEXANOATE see IHU100
ISOAMYLHYDRIDE see EIK000
ISOAMYL HYDROCUPREINE DIHYDROCHLORIDE see EQQ500
ISOAMYL o-HYDROXYBENZOATE see IME000
ISOAMYL IODIDE see IHU200
ISOAMYL ISOVALERATE (FCC) see ITB000
ISOAMYL METHANOATE see IHS000
ISOAMYL METHYL KETONE see MKW450
ISO-AMYL NITRATE see ILW100
ISOAMYLNITRILE see ITD000
ISOAMYL NITRITE see IMB000
ISOAMYL OCTANOATE see IHP500
ISOAMYLOL see IHP000
ISOAMYL 3-PENTYL PROPENATE see AOG600
ISOAMYL PHENYLACETATE see IHV000
ISOAMYL PHENYLAMINOACETATE HYDROCHLORIDE see PCV750
ISOAMYL PHENYLETHYL ETHER see IHV050
ISOAMYL POTASSIUM XANTHATE see PLK580
ISOAMYL PROPIONATE see AON350
ISOAMYL PROPIONATE see ILW000
ISOAMYL SALICYLATE (FCC) see IME000
1-ISOAMYL THEOBROMINE see IHX200
ISOANETHOLE see AFW750
cis-ISOASARONE see ARM100
trans-ISOASARONE see IHX400
ISOASCORBIC ACID see SAA025
d-ISOASCORBIC ACID see SAA025
ISOASCORBIC ACID, SODIUM SALT see EDE700
ISOBAC 20 see HCL000
ISOBAC 20 see MRM000
ISOBAMATE see IPU000
ISOBARB see NBU000
ISOBENZAMIC ACID see AMU750
ISOBENZAN see OAN000
5-ISOBENZOFURANCARBONITRILE, 1-(3-(DIMETHYLAMINO)PROPYL)-1-(4-FLUOROPHENYL)-1,3-DIHYDRO-, MONOHYDROBROMIDE see CMS258
1,3-ISOBENZOFURANDIONE see PHW750
1,3-ISOBENZOFURANDIONE, 4,5,6,7-TETRABROMO- see TBJ700
1,3-ISOBENZOFURANDIONE, 4,5,6,7-TETRACHLORO-(9CI) see TBT150
1,3-ISOBENZOFURANDIONE, 3a,4,7,7a-TETRAHYDROMETHYL- see MPP000
1(3H)-ISOBENZOFURANONE see PHW800
1(3H)-ISOBENZOFURANONE, 3,3-BIS(4-HYDROXYPHENYL)- see PDO750

1(3H)-ISOBENZOFURANONE, 3-
BUTYLIDENE-(9CI) see BRQ100
1(3H)-ISOBENZOFURANONE, 3-(3-
FURANYL)-3a,4,5,6-TETRAHYDRO-3a,7-
DIMETHYL-, (3R-cis)-(9CI) see FOM200
ISOBERGAMATE see IHX450
ISO-BID see CCK125
ISOBIDE see HID350
ISOBIND 100 see PKB100
ISO-BNU see IJF000
ISOBORNEOL see IHY000
dl-ISOBORNEOL see IHY000
ISOBORNEOL METHYL ETHER see IHX500
ISOBORNEOL THIOCYANATOACETATE
see IHZ000
ISOBORNYL ACETATE see IHX600
ISOBORNYL ACETATE see IHX600
ISOBORNYL ACRYLATE see IHX700
exo-ISOBORNYL ACRYLATE see IHX700
ISOBORNYL ALCOHOL see IHY000
ISOBORNYL METHYL ETHER see IHX500
p-(ISOBORNYLOXY)ANILINE see IHY500
ISOBORNYL THIOCYANATOACETATE see
IHZ000
ISOBORNYL THIOCYANOACETATE see
IHZ000
ISOBROMYL see BNP750
ISOBUTANAL see IJS000
ISOBUTANDIOL-2-AMINE see ALB000
ISOBUTANE see MOR750
ISOBUTANE (DOT) see MOR750
ISOBUTANE MIXTURES (DOT) see MOR750
ISOBUTANOL (DOT) see IIL000
ISOBUTANOLAMINE see IIA000
ISOBUTANOL-2-AMINE see IIA000
ISOBUTANOYL CHLORIDE see IJV100
ISOBUTAZINA see HNI500
ISOBUTENAL see MGA250
ISOBUTENE see IIC000
ISOBUTENYL CHLORIDE see CIU750
ISOBUTENYL METHYL KETONE see
MDJ750
ISOBUTIL see HNI500
ISOBUTINYL-N-(3-CHLORPHENYL)-
CARBAMAT (GERMAN) see CEX250
p-ISOBUTOXYBENZOIC ACID 3-
(DIETHYLAMINO)-1,2-DIMETHYLPROPYL
ESTER see GBU700
p-ISOBUTOXYBENZOIC ACID, α,β-
DIMETHYL-γ-DIETHYLAMINOPROPYL
ESTER see GBU700
2-ISOBUTOXYETHANOL see IIP000
ISOBUTOXY-2-ETHOXY-2-ETHANOL see
DJF800
2-(2-ISOBUTOXYETHOXY)ETHANOL see
DJF800
(2-ISOBUTOXYETHYL)CARBAMATE see
IIE000
N-ISOBUTOXYMETHYLACRYLAMIDE see
IIE100
N-ISOBUTOXYMETHYL-2-CHLORO-2',6'-
DIMETHYLACETANILIDE see IIE200
N-ISOBUTOXYMETHYL-2-CHLORO-N-
(2',6'-DIMETHYLPHENYL)-ACETAMIDE see
IIE200
2-ISOBUTOXYNAPHTHALENE see NBJ000
1-ISOBUTOXY-2-PROPANOL see IIG000
ISOBUTOXYPROPANOL, MIXED ISOMERS
see IIG500
3-ISOBUTOXY-2-PYRROLIDINO-N-
PHENYL-N-BENZYLPROPYLAMINE
HYDROCHLORIDE HYDRATE see IIG600
2-ISOBUTOXY TETRAHYDROPYRAN see
III000
ISOBUTYL ACETATE see IIJ000
N-ISOBUTYL-N-
(ACETOXYMETHYL)NITROSAMINE see
ABR125
ISOBUTYL ACRYLATE see IIK000
ISOBUTYL ACRYLATE, inhibited (DOT) see
IIK000
ISOBUTYL ADIPATE see DNH125

ISOBUTYL ALCOHOL see IIL000
ISOBUTYLALDEHYDE see IJS000
ISOBUTYL ALDEHYDE (DOT) see IJS000
ISOBUTYLALKOHOL (CZECH) see IIL000
ISO-BUTYLALLYLBARBITURIC ACID see
AGI750
ISOBUTYLALLYLBARTURIC ACID see
AGI750
ISOBUTYLAMINE see IIM000
2-ISOBUTYLAMINOETHANOL see IIM050
2-ISOBUTYLAMINOETHANOL
HYDROCHLORIDE ACID SALT, p-
AMINOBENZOIC ACID ESTER see IAC000
2-(ISOBUTYLAMINO)ETHYL-p-
AMINOBENZOATE HYDROCHLORIDE see
IAC000
2-(ISOBUTYLAMINO)-2-METHYL-1-
PROPANOL BENZOATE (ESTER),
HYDROCHLORIDE see IIM100
2-(ISOBUTYLAMINO)-2-METHYL-1-
PROPANOL BENZOATE
HYDROCHLORIDE see IIM100
3-(3-(ISOBUTYLAMINO)PROPYL)-2-(3,4-
METHYLENEDIOXYPHENYL)-4-
THIAZOLIDINONE HYDROCHLORIDE see
WBS855
ISOBUTYLATED UREA FORMALDEHYDE
see IIM300
ISOBUTYLBENZENE see IIN000
ISOBUTYLBIS(2-CHLOROETHYL)AMINE
HYDROCHLORIDE see BIE750
ISOBUTYLBIS(β-CHLOROETHYL)AMINE
HYDROCHLORIDE see BIE750
ISOBUTYL BROMIDE see BNR750
ISOBUTYL BUTANOATE see BSW500
ISOBUTYL-2-BUTENOATE see IIN300
3-ISOBUTYL-6-sec-BUTYL-2-
HYDROXYPYRAZINE-1-OXIDE see
ARO000
ISOBUTYL BUTYRATE (FCC) see BSW500
ISOBUTYL CAPROATE see IIT000
ISOBUTYLCARBINOL see IHP000
(2-(p-(5-(ISOBUTYLCARBOZMOYL)-2-
OCTYLOXYBENZAMIDO)BENZAMIDO)E
THYL)-TRIETHYLAMMONIUM IODIDE see
IIO750
ISOBUTYL CELLOSOLVE see IIP000
ISOBUTYL CHLORIDE see CIU500
ISOBUTYL CINNAMATE see IIQ000
ISOBUTYL CYANOACETATE see IIQ100
2-((1-ISOBUTYL-3,5-
DIMETHYLHEXYL)OXY)ETHANOL see
EJL000
1-ISOBUTYL-3,4-DIPHENYLPYRAZOLE-5-
ACETIC ACID SODIUM SALT see DWD400
ISOBUTYLDIUREA see IIV000
ISOBUTYLENATED
METHYLSTYRENATED PHENOL see
IIQ140
ISOBUTYLENATED PHENOL PHOSPHATE
(3:1) see IIQ150
ISOBUTYLENE (DOT) see IIC000
ISOBUTYLENE CHLORIDE see IIQ200
ISOBUTYLENEDIUREA see IIV000
ISOBUTYLENE-ISOPRENE COPOLYMER
see IIQ500
ISOBUTYLENEOXIDE see IIQ600
ISOBUTYLESTER KYSELINY
METHAKRYLOVE see IIY000
ISOBUTYLESTER KYSELINY MRAVENCI
see IIR000
ISOBUTYLESTER KYSELINY OCTOVE see
IIJ000
ISOBUTYL FORMATE see IIR000
ISO-BUTYL FORMATE see IIR000
ISOBUTYL-2-FURANPROPIONATE see
IIR100
ISOBUTYL FURYLPROPIONATE see IIR100
ISOBUTYLGLYCEROL, NITRO- see HMJ500
ISOBUTYL HEPTANOATE see IIS000
ISOBUTYL HEPTYLATE see IIS000
ISOBUTYL HEXANOATE see IIT000

4-ISOBUTYLHYDRATROPIC ACID see
IIU000
p-ISOBUTYLHYDRATROPIC ACID see
IIU000
p-ISOBUTYLHYDRATROPIC ACID
SODIUM SALT see IAB100
ISOBUTYL-o-HYDROXYBENZOATE see
IJN000
2,2'-ISOBUTYLIDENEBIS(4,6-
DIMETHYLPHENOL) see IIU100
1,1'-ISOBUTYLIDENEBISUREA see IIV000
ISOBUTYLIDENEDIUREA see IIV000
ISOBUTYL IODIDE see IIV509
ISOBUTYL ISOBUTYRATE see IIW000
ISOBUTYLISOBUTYRATE (DOT) see IIW000
ISOBUTYL ISOPENTANOATE see ITA000
ISOBUTYL ISOVALERATE see ITA000
ISOBUTYLJODID see IIV509
ISOBUTYL KETONE see DNI800
ISOBUTYL LINALOL see IIW100
ISOBUTYL MERCAPTAN see IIX000
ISOBUTYL METHACRYLATE see IIY000
ISOBUTYL-α-METHACRYLATE see IIY000
5-ISOBUTYL-2-p-
METHOXYBENZENESULFONAMIDO-
1,3,4-THIADIAZOLE see IIY100
2-(ISOBUTYL-3-
METHYLBUTOXY)ETHANOL see IJA000
2-(2-(1-ISOBUTYL-3-
METHYLBUTOXY)ETHOXY)ETHANOL see
IJB000
ISOBUTYL METHYL CARBINOL see
MKW600
ISOBUTYL-METHYLKETON (CZECH) see
HFG500
ISOBUTYL METHYL KETONE see HFG500
ISOBUTYL METHYL KETONE PEROXIDE
see IJB100
ISOBUTYLMETHYLMETHANOL see
MKW600
5-ISOBUTYL-5-
(METHYLTHIOMETHYL)BARBITURIC
ACID SODIUM SALT see SHY000
ISOBUTYL 2-NAPHTHYL ETHER see
NBJ000
ISOBUTYL β-NAPHTHYL ETHER see
NBJ000
ISOBUTYL NITRITE see IJD000
1-ISOBUTYL-3-NITRO-1-
NITROSOGUANIDINE see IJE000
N-ISOBUTYL-N'-NITRO-N-
NITROSOGUANIDINE see IJE000
2-ISOBUTYL-3-NITROSOTHIAZOLIDINE
see NKN500
1-ISO-BUTYL-1-NITROSOUREA see IJF000
N-ISOBUTYL-N-NITROSOUREA see IJF000
ISOBUTYL ORTHOVANADATE see VDK000
ISOBUTYL PHENYLACETATE see IJF400
4-ISOBUTYLPHENYLACETIC ACID see
IJG000
(p-ISOBUTYLPHENYL)ACETIC ACID see
IJG000
2-(4-ISOBUTYLPHENYL)BUTYRIC ACID see
IJH000
2-(4-ISOBUTYLPHENYL)PROPANOIC ACID
see IIU000
2-(p-ISOBUTYLPHENYL)PROPIONIC ACID
see IIU000
α-(4-ISOBUTYLPHENYL)PROPIONIC ACID
see IIU000
α-p-ISOBUTYLPHENYLPROPIONIC ACID
see IIU000
2-(p-ISOBUTYLPHENYL)PROPIONIC ACID-
o-METHOXYPHENYL ESTER see IJJ000
2-(p-ISOBUTYLPHENYL)PROPIONIC ACID
2-PYRIDYLMETHYL ESTER see IAB000
ISOBUTYL PHOSPHATE see IJJ200
ISOBUTYL POTASSIUM XANTHATE see
PLK600
O-ISOBUTYL POTASSIUM XANTHATE see
EQG600
ISOBUTYL PROPENOATE see IIK000

ISOBUTYL-2-PROPENOATE see IIK000
ISOBUTYL PROPIONATE (DOT) see PMV250
2-ISOBUTYLQUINOLINE see IJM000
α-ISOBUTYLQUINOLINE see IJM000
ISOBUTYL SALICYLATE see IJN000
ISOBUTYL STEARATE see IJN100
ISOBUTYLSULFHYDRATE see IJO000
p-ISOBUTYL-α-TOLUIC ACID see IJG000
ISOBUTYLTRIMETHYLETHANE see TLY500
ISOBUTYL VINYL ETHER see IJQ000
p-ISOBUTYOXYBENZOIC ACID-3-(2'-METHYLPIPERIDINO)PROPYL ESTER see IJR000
ISOBUTYRALDEHYD (CZECH) see IJS000
ISOBUTYRALDEHYDE see IJS000
ISOBUTYRALDEHYDE, OXIME see IJT000
ISOBUTYRIC ACID see IJU000
ISOBUTYRIC ACID (DOT) see IJU000
ISOBUTYRIC ACID, BENZYL ESTER see IJV000
ISOBUTYRIC ACID CHLORIDE see IJV100
ISOBUTYRIC ACID, CINNAMYL ESTER see CMR750
ISOBUTYRIC ACID, 3,7-DIMETHYL-2,6-OCTADIENYL ESTER, (Z)-(8CI) see NCO200
ISOBUTYRIC ACID, ETHYL ESTER see ELS000
ISOBUTYRIC ACID, 2,4-HEXADIENYL ESTER (8CI) see HCS700
ISOBUTYRIC ACID, HEXYL ESTER see HFQ550
ISOBUTYRIC ACID, α-HYDROXY-, ACETATE, METHYL ESTER see HLJ600
ISOBUTYRIC ACID, ISOBUTYL ESTER see IIW000
ISOBUTYRIC ACID, 1-ISOPROPYL-2,2-DIMETHYLTRIMETHYLENE ESTER see TLZ000
ISOBUTYRIC ACID, METHYL ESTER see MKX000
ISOBUTYRIC ACID, 2-PHENOXYETHYL ESTER (6CI,8CI) see PDS900
ISOBUTYRIC ACID-3-PHENYLPROPYL ESTER see HHQ500
ISOBUTYRIC ACID, p-TOLYL ESTER see THA250
ISOBUTYRIC ALDEHYDE see IJS000
ISOBUTYRIC ANHYDRIDE see IJW000
ISOBUTYRONE see DTI600
ISOBUTYRONITRILE see IJX000
ISOBUTYROYL CHLORIDE see IJV100
ISOBUTYRYL CHLORIDE see IJV100
3-ISOBUTYRYL-2-ISOPROPYLPYRAZOLO(1,5-a)PYRIDINE see KDA100
ISOBUZOLE see IIY100
ISOCAINE see IJZ000
ISOCAINE-ASID see AIT250
ISOCAINE BASE see PIV750
ISOCAINE-HEISLER see AIT250
ISOCAMPHANYL CYCLOHEXANOL (mixed isomers) see IKA000
ISOCAMPHOL see IHY000
ISOCAMPHYL CYCLOHEXANOL (mixed isomers) see IKA000
ISOCAPROALDEHYDE see MQJ500
ISOCAPRONITRILE see MNJ000
ISOCARAMIDINE SULFATE see IKB000
ISOCARB see PMY300
ISOCARBONAZID see IKC000
ISOCARBOSSAZIDE see IKC000
ISOCARBOXAZID see IKC000
ISOCARBOXAZIDE see IKC000
ISOCARBOXYZID see IKC000
ISOCARNOX see RBF500
ISOCETYL STEARATE see IKC050
ISOCHINOL see DNX400
ISOCHLOORTHION (DUTCH) see NFT000
ISOCHRYSENE see TMS000
ISOCID see ILD000

ISOCIL see BNM000
ISOCILLIN see PDT750
ISOCINCHOMERONIC ACID, 1,4-DIHYDRO-2,6-DIMETHYL-4-(2-PYRIDYL)-, DIETHYL ESTER see DJB460
ISOCINCHOMERONIC ACID, DIISOPROPYL ESTER see IKC070
ISOCINEOLE see IKC100
cis-ISOCITRATO-(1,2-DIAMINO-CYCLOHEXAN)-PLATIN(II) (GERMAN) see PGQ275
ISO-CORNOX see CIR500
ISO-CORNOX 57 see RBF500
ISOCOTIN see ILD000
ISOCROTONIC NITRILE see BOX600
ISOCROTONONITRILE (6CI,7CI) see BOX600
ISOCROTYL CHLORIDE see IKE000
ISOCUMENE see IKG000
ISOCURCUMENOL see IKG100
ISOCYANATE 580 see PKB100
ISOCYANATE de METHYLE (FRENCH) see MKX250
ISOCYANATES see IKG349
ISOCYANATOACETIC ACID ETHYL ESTER see ELS600
ISOCYANATOCYCLOHEXANE see CPN500
ISOCYANATOETHANE see ELS500
2-ISOCYANATOETHANOL CARBONATE (2:1) (ESTER) see CBV600
2-ISOCYANATOETHANOL CARBONATE (2:1) (ESTER) see IKG400
2-ISOCYANATOETHYL METHACRYLATE see IKG700
β-ISOCYANATOETHYL METHACRYLATE see IKG700
ISO-CYANATOMETHANE see MKX250
1-ISOCYANATO-2-METHYLBENZENE see IKG725
1-ISOCYANATO-4-METHYLBENZENE see IKG735
1-(1-ISOCYANATO-1-METHYLETHYL)-3-(1-METHYLETHENYL)BENZENE see IKG800
1-(1-ISOCYANATO-1-METHYLETHYL)-4-(1-METHYLETHENYL)BENZENE see IKG850
(3-ISOCYANATOMETHYLPHENYL)CARBAMIC ACID, 2-ETHYLHEXYL ESTER see EKW200
3-ISOCYANATOMETHYL-3,5,5-TRIMETHYLCYCLOHEXYLISOCYANATE see IMG000
1-ISOCYANATO-4-NITROBENZENE see NIR100
4-ISOCYANATO-N-(4-NITROPHENYLBENZENAMINE) (9CI) see IKI000
1-ISOCYANATOPROPANE see PNP000
3-ISOCYANATOPROPYLTRIETHOXYSILANE see IKG900
γ-ISOCYANATOPROPYLTRIETHOXYSILANE see IKG900
4-ISOCYANATOTOLUENE see IKG735
p-ISOCYANATOTOLUENE see IKG735
ISOCYANIC ACID, ALLYL ESTER see AGJ000
ISOCYANIC ACID, ANHYDRIDE WITH p-TOLUENESULFONIC ACID see IKG925
ISOCYANIC ACID, BENZ(a)ANTHRACEN-7-YL ESTER see BBJ250
ISOCYANIC ACID, 3-BROMOALLYL ESTER see BMT150
ISOCYANIC ACID, BUTYL ESTER see BRQ500
ISOCYANIC ACID-2-CHLOROETHYL ESTER see IKH000
ISOCYANIC ACID, 6-CHLOROHEXYL ESTER see CHL250
ISOCYANIC ACID-m-CHLOROPHENYL ESTER see CKA750

ISOCYANIC ACID-p-CHLOROPHENYL ESTER see CKB000
ISOCYANIC ACID, 1,4-CYCLOHEXYLENE ESTER, trans-(8CI) see CPB150
ISOCYANIC ACID, CYCLOHEXYL ESTER see CPN500
ISOCYANIC ACID-3,4-DICHLOROPHENYL ESTER see IKH099
ISOCYANIC ACID, DIESTER with 1,6-HEXANEDIOL see DNJ800
ISOCYANIC ACID, 2,6-DIISOPROPYLPHENYL ESTER see DNS100
ISOCYANIC ACID, 3,3'-DIMETHYL-4,4'-BIPHENYLENE ESTER see DQS000
ISOCYANIC ACID, ESTER with DI-o-TOLUENEMETHANE see MJN750
ISOCYANIC ACID, ETHYL ESTER see ELS500
ISOCYANIC ACID, 4-FLUOROBUTYL ESTER see FHC200
ISOCYANIC ACID, 3-FLUOROPROPYL ESTER see FLR100
ISOCYANIC ACID, HEXAHYDRO-4,7-METHANOINDAN-1,8-YLENE ESTER see TJG500
ISOCYANIC ACID, HEXAMETHYLENE ESTER see DNJ800
ISOCYANIC ACID, HEXAMETHYLENE ESTER, POLYMER with 1,4-BUTANEDIOL see PKO750
ISOCYANIC ACID, HEXAMETHYLENE ESTER, POLYMERS see HEG300
ISOCYANIC ACID, m-ISOPROPENYL-α-α-DIMETHYL BENZYL ESTER see IKG800
ISOCYANIC ACID, p-ISOPROPENYL-α-α-DIMETHYLBENZYL ESTER see IKG850
ISOCYANIC ACID, METHYLENEDI-p-PHENYLENE ESTER, POLYMER with 1,4-BUTANEDIOL see PKP000
ISOCYANIC ACID, METHYL ESTER see MKX250
ISOCYANIC ACID, METHYLPHENYLENE ESTER see TGM740
ISOCYANIC ACID, METHYLPHENYLENE ESTER see TGM750
ISOCYANIC ACID, 4-METHYL-m-PHENYLENE ESTER see TGM750
ISOCYANIC ACID-1,5-NAPHTHYLENE ESTER see NAM500
ISOCYANIC ACID, p-NITROPHENYL ESTER see NIR100
ISOCYANIC ACID, OCTADECYL ESTER see OBG000
ISOCYANIC ACID, p-PHENYLENEDIISOPROPYLIDENE ESTER (7CI) see TDX400
ISOCYANIC ACID, p-PHENYLENE ESTER (6CI,7CI,8CI) see PFA300
ISOCYANIC ACID, PHENYL ESTER see PFK250
ISOCYANIC ACID, PROPYL ESTER see PNP000
ISOCYANIC ACID, α-α-α',α'-TETRAMETHYL-p-XYLYLENE ESTER (8CI) see TDX400
ISOCYANIC ACID, o-TOLYL ESTER see IKG725
ISOCYANIC ACID, p-TOLYL ESTER see IKG735
ISOCYANIC ACID, TRIESTER with 1,3,5-TRIS(6-HYDROXYHEXYL)BIURET see TNJ300
ISOCYANIC ACID, 3-(TRIETHOXYSILYL)PROPYL ESTER see IKG900
ISOCYANIC ACID, (m-TRIFLUOROMETHYLPHENYL) ESTER see TKJ250
ISOCYANIC ACID, p-XYLYLENE ESTER see XSS260
ISOCYANIDE see COI500
ISOCYANIDES see IKH339

ISOCYANOAMIDE see IKH669
ISOCYANOETHANE see ELT000
1-ISOCYANO-2-METHOXY-4-NITROBENZENE see IKH700
2-ISOCYANO-1-METHOXY-4-NITROBENZENE see IKH720
1-ISOCYANO-4-METHYL-3-NITROBENZENE see IKH740
1-ISOCYANONAPHTHALENE see IKH760
α-ISOCYANONAPHTHALENE see IKH760
1-ISOCYANO-3-NITROBENZENE see NIR200
1-ISOCYANO-4-NITROBENZENE see IKH780
4-ISOCYANO-4'-NITRODIPHENYLAMINE see IKI000
ISOCYANURIC ACID see THS000
ISOCYANURIC ACID, DICHLORO- see DGN200
ISOCYANURIC ACID, DICHLORO-, POTASSIUM SALT see PLD000
ISOCYANURIC ACID TRIALLYL ESTER see THS150
ISOCYANURIC CHLORIDE see TIQ750
ISOCYANURIC DICHLORIDE see DGN200
ISOCYCLEX see IOE000
ISOCYCLOCITRAL see IKJ000
ISODECANOL see IKK000
ISODECYL ACRYLATE see IKL000
ISODECYL ALCOHOL see IKK000
ISODECYL ALCOHOL ACRYLATE see IKL000
ISODECYL DIPHENYL PHOSPHATE see IKL100
ISODECYL DIPHENYL PHOSPHITE see IKL200
α-ISODECYL-ω-HYDROXYPOLY(OXY-1,2-ETHANEDIYL) see PKE370
ISODECYL METHACRYLATE see IKM000
ISODECYL PHENYL PHOSPHITE see IKL200
ISODECYL PHOSPHITE see IKM025
ISODECYL PROPENOATE see IKL000
ISODEMETON see DAP200
ISODIAZINE see IKM050
ISODIAZOMETHANE see IKH669
ISODIENESTROL see DAL600
ISODIHYDROLAVANDULOL see IQH000
ISODIHYDRO LAVANDULYL ACETATE see AAV250
ISODIHYDROLAVANDULYL ALDEHYDE see IKM100
ISODILAN see IJG000
ISODIN see CFZ000
ISODINE see PKE250
ISODIPHENYLBENZENE see TBC620
ISODIPRENE see CCK500
ISODONAZOLE NITRATE see IKN200
ISODORMID see IQX000
ISODRIN see IKO000
ISODRINE see FMS875
ISODRINUM see FMS875
ISODUR see IIV000
ISODURENE see TDM500
β-ISODURYLIC ACID see IKN300
cis-ISOELEMICIN see ARM100
ISOENANTHIC ACID see IKN400
ISOENDOXAN see IMH000
d-ISOEPHEDRINE see POH000
ISOESTRAGOLE see PMQ750
ISOETHADIONE see PAH500
ISOETHYLVANILLIN see IKO100
ISOEUGENOL see IKQ000
ISOEUGENOL ACETATE see AAX750
ISOEUGENOL AMYL ETHER see AOK000
1,3,4-ISOEUGENOL METHYL ETHER see IKR000
ISOEUGENYL ACETATE (FCC) see AAX750
ISOEUGENYL METHYL ETHER see IKR000
ISOFEDROL see EAW000
ISOFEDROL see EAY500
ISOFENPHOS see IMF300

ISOFLAV see DBN400
ISOFLAV BASE see DBN600
ISOFLAVONE, 4',5,7-TRIHYDROXY-6-METHOXY- see TKP050
ISOFLUOROPHATE see IRF000
ISOFLURANE see IKS400
ISOFLUROPHATE see IRF000
ISOFORON see IMF400
ISOFORONE (ITALIAN) see IMF400
ISOFOSFAMIDE see IMH000
ISOFTALODINITRIL (CZECH) see PHX550
ISOGAMMA ACID see AKI000
ISOGLAUCON see CMX760
ISOGUANINE RIBOSIDE see IKS450
ISOGUANOSINE see IKS450
ISOHARRINGTONINE see IKS500
ISOHEXANAL see MQJ500
ISOHEXANE see IKS600
ISOHEXANOIC ACID (mixed isomers) see IKT000
ISOHEXENYL CYCLOHEXENYL CARBOXALDEHYDE see IKT100
ISOHEXYL ALCOHOL see AOK750
ISOHOL see INJ000
ISOHOMOGENOL see IKR000
ISOINDOLE-1,3-DIONE see PHX000
1,3-ISOINDOLEDIONE see PHX000
1H-ISOINDOLE-1,3(2H)-DIONE, 2-(CYCLOHEXYLTHIO)- see CPQ700
1H-ISOINDOLE-1,3(2H)-DIONE,2-(2,6-DIOXO-3-PIPERIDINYL)-, (R)- see TEH510
1H-ISOINDOLE-1,3(2H)-DIONE, 2,2'-(1,2-ETHANEDIYL)BIS(4,5,6,7-TETRABROMO- see EIT150
1H-ISOINDOLE-1,3(2H)-DIONE, 2-(HYDROXYMETHYL)- see HMP100
1H-ISOINDOLE-1,3(2H)-DIONE, 2-METHYL- see MOC300
1H-ISOINDOLE-1,3(2H)-DIONE, 2-METHYL-5-NITRO- see NHN700
1H-ISOINDOLE-1,3(2H)-DIONE, 5-NITRO- see NIX100
1H-ISOINDOLE-1,3(2H)-DIONE, 3-α-4,7,7-α-TETRAHYDRO-(9CI) see TDB100
1H-ISOINDOLE-1,3(2H)-DIONE, 4,5,6,7-TETRAHYDRO-2-(7-FLUORO-3,4-DIHYDRO-3-OXO-4- (2-PROPYNYL)-2H-1,4-BENZOXAZIN-6-YL)- see FLZ075
1H-ISOINDOLE-1,3(2H)-DIONE,3A,4,7,7A-TETRAHYDRO-2-((1,1,2,2-TETRACHLOROETHYL)THIO)-, cis- see TBQ280
1H-ISOINDOLE-5-SULFONAMIDE, 6-CHLORO-2,3-DIHYDRO-1,3-DIOXO- see CGM500
ISOINDOLINE, 2-(2-DIMETHYLAMINOETHYL)-4,5,6,7-TETRACHLORODIMETHO CHLORIDE see CDY000
1,3-ISOINDOLINEDIONE see PHX000
5-ISOINDOLINESULFONAMIDE, 6-CHLORO-1,3-DIOXO- see CGM500
ISOINOKOSTERONE see HKG500
ISOJASMONE see IKV000
ISO-K see BDU500
ISOKET see CCK125
ISOL see HFP875
ISOLAN see DSK200
ISOLANE (FRENCH) see DSK200
ISOLANID see LAU000
ISOLANIDE see LAU000
ISOLASALOCID A see IKW000
ISOL BENZIDINE FAST YELLOW GRX see CMS208
ISOL BENZIDINE YELLOW G see DEU000
ISOL BENZIDINE YELLOW GO see CMS210
ISOL BENZIDINE YELLOW GRX 2548 see CMS208
ISOL BONA RED N 5R BARIUM SALT see CMS148
ISOL BONA RED NR BARIUM SALT see CMS148

ISOL BONA RUBINE BK see CMS155
ISOL BONA RUBINE BKS see CMS155
ISOL BONA RUBINE KBK see CMS155
ISOL DIARYL YELLOW GRF see CMS208
ISOLEUCINE see IKX000
ISOLEUCINE see IKX000
dl-ISOLEUCINE see IKX010
5-l-ISOLEUCINEANGIOTENSIN II see IKX030
l-ISOLEUCINE (FCC) see IKX000
ISOL FAST RED 2G see DVB800
ISOL FAST RED HB see MMP100
ISOL FAST RED R see CJD500
ISOL FAST RED RN2B see MMP100
ISOL FAST RED RN2G see MMP100
ISOL FAST RED RNB see MMP100
ISOL FAST RED RNG see MMP100
ISOL LAKE RED LCL see CMS150
ISOL LAKE RED LCR see CMS150
ISOL LAKE RED LCS 12527 see CHP500
ISOL LAKE RED LCT see CMS150
ISOL RUBY BK see CMS155
ISOL RUBY BKS see CMS155
ISOL TOLUIDINE RED HB see MMP100
ISOL TOLUIDINE RED RN2B see MMP100
ISOL TOLUIDINE RED RN2G see MMP100
ISOL TOLUIDINE RED RNB see MMP100
ISOL TOLUIDINE RED RNG see MMP100
ISOLYN see ILD000
ISOMACK see CCK125
ISOMALATHION see IKX200
ISOMALIC ACID see MPN250
ISOMEBUMAL see EJY000
ISOMELAMINE see MCB000
ISOMELIN see GKS000
ISOMENTHONE see IKY000
ISOMEPROBAMATE see IPU000
β-ISOMER see BBR000
ISOMERIC CHLORTHION see NFT000
ISOMETASYSTOX see DAP400
ISOMETASYSTOX SULFONE see DAP600
ISOMETHADONE see IKZ000
ISOMETHEPTENE see ILK000
ISOMETHEPTENE HYDROCHLORIDE see ODY000
ISOMETHYLSYSTOX see DAP400
ISOMETHYLSYSTOX SULFONE see DAP600
ISOMETHYLSYSTOX SULFOXIDE see DAP000
ISOMIN see TEH500
ISOMYL see AMX750
ISOMYN see BBK000
ISOMYST see IQN000
ISOMYTAL see AMX750
ISONA see EDE700
ISONAL see AFT000
ISONAL see ENB500
ISONAL (ROUSSEL) see ENB500
ISONAPHTHOIC ACID see NAV500
ISONAPHTHOL see NAX000
ISONATE see MJP400
ISONATE 390P see PKB100
ISONEX see ILD000
ISONIACID see ILD000
ISONIAZIDE see ILD000
ISONIAZID SODIUM METHANESULFONATE see SIJ000
ISONICAZIDE see ILD000
ISONICO see ILD000
ISONICOTAN see ILD000
ISONICOTINALDEHYDE see IKZ100
ISONICOTINALDEHYDE THIOSEMICARBAZONE see ILB000
ISONICOTINAMIDE see ILC100
ISONICOTINAMIDE PENTAAMMINE RUTHENIUM(II) PERCHLORATE see ILB150
ISONICOTINHYDRAZID see ILD000
ISONICOTINIC ACID see ILC000
ISONICOTINIC ACID AMIDE see ILC100
ISONICOTINIC ACID, ETHYL ESTER see ELU000

ISONICOTINIC ACID HYDRAZIDE see ILD000
ISONICOTINIC ACID HYDRAZIDE, HYDRAZONE WITH d-GLUCOSE (7CI,8CI) see GFG070
ISONICOTINIC ACID-2-ISOPROPYLHYDRAZIDE see ILE000
ISONICOTINIC ACID, 2-ISOPROPYLHYDRAZIDE, PHOSPHATE see IGG700
ISONICOTINIC ACID, SODIUM SALT see ILF000
ISONICOTINIC ACID, VANILLYLIDENEHYDRAZIDE see VEZ925
ISONICOTINIC ALDEHYDE see IKZ100
N-ISONICOTINOYL-N'-GLUCURONID-HYDRAZIN-NATRIUMSALZ DIHYDRAT (GERMAN) see IDB100
N-ISONICOTINOYL-N'-GLUCURONSAEURE-γ-LACTON-HYDRAZON (GERMAN) see GBB500
ISONICOTINOYL HYDRAZIDE see ILD000
2-(2-ISONICOTINOYLHYDRAZINO)-d-GLUCOPYRANURONIC ACID SODIUM SALT DIHYDRATE see IDB100
4-(ISONICOTINOYLHYDRAZONE)PIMELIC ACID see ILG000
1-ISONICOTINOYL-2-ISOPROPYLHYDRAZINE see ILE000
N-ISONICOTINOYL-N'(β-N-BENZYLCARBOXAMIDOETHYL)HYDRAZINE see BET000
ISONICOTINSAEUREHYDRAZID see ILD000
1-ISONICOTINYL-2-GLUCOSYLHYDRAZINE see GFG070
ISONICOTINYL HYDRAZIDE see ILD000
1-ISONICOTINYL-2-ISOPROPYLHYDRAZINE see ILE000
ISONIDE see ILD000
ISONIKAZID see ILD000
ISONIN see ILD000
ISONIPECAINE see DAM600
ISONIPECAINE HYDROCHLORIDE see DAM700
ISONIPECOTIC ACID see ILG100
ISONIPECOTIC ACID, 1-(2-HYDROXY-3-PHENOXYPROPYL)-4-PHENYL-, ETHYL ESTER, HYDROCHLORIDE see EEQ000
ISONIPECOTIC ACID, 1-(3-HYDROXY-3-PHENYLPROPYL)-4-PHENYL-, ETHYL ESTER, HYDROCHLORIDE see PDO900
3-ISONIPECOTYLINDOLE see ILG200
ISONIRIT see ILD000
ISONITOX see MIW250
ISONITROPROPANE see NIY000
ISONITROSOACETANILIDE see GIK100
2-ISONITROSOACETANILIDE see GIK100
ISONITROSOACETOPHENONE see ILH000
ISONITROSOACETYLANILINE see GIK100
5-ISONITROSOBARBITURIC ACID see AFU000
β-ISONITROSOPROPANE see ABF000
ISONITROSOPROPIOPHENONE see ILI000
ISONIXIN see XLS300
ISONIXINE see XLS300
ISONIZIDE see ILD000
ISONONYL ACETATE see TLT500
ISONONYL ALCOHOL see ILJ000
ISONONYL ALCOHOL, PHTHALATE (2:1) see PHW585
ISONONYLALDEHYDE see ILJ100
ISONORENE see DMV600
ISONYL see ILK000
ISOOCTADECANOIC ACID, 1-METHYLETHYL ESTER see IPS450
ISOOCTANE (DOT) see TLY500
ISOOCTANOL see ILL000
ISOOCTYL ACRYLATE see ILK100
ISOOCTYL ALCOHOL see ILL000

ISOOCTYL ALCOHOL (2,4-DICHLOROPHENOXY)ACETATE see ILO000
2-ISOOCTYL AMINE see ILM000
ISOOCTYL-2,4-DICHLOROPHENOXYACETATE see ILO000
ISOOCTYL ESTER, MERCAPTOACETATE ACID see ILR000
ISOOCTYL MERCAPTOACETATE see ILR000
ISOOCTYL PHTHALATE see ILR100
p-ISOOCTYLPOLYOXYETHYLENEPHENOL FORMALDEHYDE POLYMER see TDN750
ISOOCTYL THIOGLYCOLATE see ILR000
ISOOCTYL ((TRIBUTYLSTANNYL)THIO)ACETATE see ILR110
ISOOCTYL ((TRIMETHYLSTANNYL)THIO)ACETATE see ILR120
ISOPAL see IQW000
ISOPARAFFINIC PETROLEUM HYDROCARBONS, SYNTHETIC see ILR150
ISOPARATHION see DJT000
ISOPELLETIERINE see PAO500
ISOPENTALDEHYDE see MHX500
ISOPENTANE (DOT) see EIK000
ISOPENTANOIC ACID (DOT) see ISU000
ISOPENTANOIC ACID, PHENYLMETHYL ESTER see ISW000
ISOPENTANOL see IHP000
2-ISOPENTANOYL-1,3-INDANEDIONE see ITD010
ISOPENTENE see IHR220
ISOPENTENES (DOT) see IHR220
ISOPENT-2-ENYL ACETATE see DOQ350
4-(2-ISOPENTENYL)-1,2-DIPHENYL-3,5-PYRAZOLIDINEDIONE see PEW000
5-(1-ISOPENTENYL)-5-ISOPROPYLBARBITURIC ACID see ILT000
5-(2-ISOPENTENYL)-5-ISOPROPYL-1-METHYLBARBITURIC ACID see ILU000
8-ISOPENTENYLOXYPSORALENE see IHR300
ISOPENTYL ACETATE see IHO850
ISOPENTYL ALCHOL, ESTER with N-(2-(DIETHYLAMINO)ETHYL)-2-PHENYLGLYCINE see NOC000
ISOPENTYL ALCOHOL see IHP000
ISOPENTYL ALCOHOL ACETATE see IHO850
ISOPENTYL ALCOHOL, FORMATE see IHS000
ISOPENTYL ALCOHOL, NITRATE see ILW100
ISOPENTYL ALCOHOL NITRITE see IMB000
ISOPENTYL ALCOHOL, PROPIONATE see ILW000
ISOPENTYL BENZOATE see IHP100
ISOPENTYL BROMIDE see BNP250
ISOPENTYL BUTANOATE see IHP400
ISOPENTYL BUTYRATE see IHP400
ISOPENTYLDIOCTYLPHOSPHINE OXIDE see DVK709
ISOPENTYL FORMATE see IHS000
ISOPENTYL HEXANOATE see IHU100
ISOPENTYL-n-HEXANOATE see IHU100
ISOPENTYL-2-HYDROXYPHENYL METHANOATE see IME000
ISOPENTYL IODIDE see IHU200
ISOPENTYL ISOVALERATE see ITB000
ISOPENTYL METHYL KETONE see MKW450
ISOPENTYL NITRATE see ILW100
ISOPENTYL NITRITE see IMB000
3-((ISOPENTYL)NITROSOAMINO)-2-BUTANONE see MHW350
ISOPENTYL OCTANOATE see IHP500

2-(ISOPENTYLOXY)-4-METHYL-1-PENTANOL see IJA000
ISOPENTYLPHENYLACETATE see IHV000
ISOPENTYL-2-PHENYLGLYCINATE HYDROCHLORIDE see PCV750
ISOPENTYL PROPIONATE see ILW000
ISOPENTYL SALICYLATE see IME000
1-ISOPENTYL-THEOBROMINE see IHX200
ISOPERTHIOCYANIC ACID see IBL000
ISOPESTOX see PHF750
ISOPHANE INSULIN see IDF325
ISOPHANE INSULIN INJECTION see IDF325
ISOPHANE INSULIN SUSPENSION see IDF325
ISOPHEN see CBW000
ISOPHEN see DBA800
ISOPHEN see MDT600
ISOPHEN (pesticide) see CBW000
ISOPHENERGAN see DQA400
ISOPHENICOL see CDP250
ISOPHENPHOS see IMF300
ISOPHORONE see IMF400
ISOPHORONE DIAMINE DIISOCYANATE see IMG000
ISOPHORONE DIISOCYANATE see IMG000
ISOPHORONEDIISOCYANATE, solution, 70%, by weight (DOT) see IMG000
ISOPHORONENITRILE see IMG500
ISOPHOSPHAMIDE see IMH000
ISOPHRIN see NCL500
ISOPHRINE see SPC500
ISOPHRIN HYDROCHLORIDE see SPC500
ISOPHTALDEHYDES (FRENCH) see IMI000
ISOPHTHALALDEHYDE see IMI000
ISOPHTHALAMIC ACID, 5-ACETYLAMINO-N-METHYL-2,4,6-TRIIODO-, SODIUM SALT see IGC100
ISOPHTHALIC ACID see IMJ000
ISOPHTHALIC ACID CHLORIDE see IMO000
ISOPHTHALIC ACID, DIALLYL ESTER see IMK000
ISOPHTHALIC ACID, DIALLYL ESTER (6CI,7CI,8CI) see IMK000
ISOPHTHALIC ACID, DIBUTYL ESTER see DED300
ISOPHTHALIC ACID DICHLORIDE see IMO000
ISOPHTHALIC ACID, DIETHYL ESTER see IMK100
ISOPHTHALIC ACID, DIISODECYL ESTER (6CI) see BBO700
ISOPHTHALIC ACID, DIMETHYL ESTER see IML000
ISOPHTHALODINITRILE see PHX550
ISOPHTHALONITRILE see PHX550
3,3'-(ISOPHTHALOYLBIS(IMINO-p-PHENYLENECARBONYLIMINO))BIS(1-ETHYLPYRIDINIUM), DI-p-TOLUENESULFONATE see IMM000
ISOPHTHALOYL CHLORIDE see IMO000
ISOPHTHALOYL DICHLORIDE see IMO000
ISOPHTHALYL DICHLORIDE see IMO000
ISOPIMPINELLIN see IMO500
ISOPIRINA see INM000
ISOPLAIT see VGA300
ISOPOL see AHR600
ISO PPC see PMS825
ISOPRAL see IMQ000
ISOPREGNENONE see DYF759
ISOPRENALINE see DMV600
ISOPRENALINE CHLORIDE see IMR000
ISOPRENALINE HYDROCHLORIDE see IMR000
(±)-ISOPRENALINE HYDROCHLORIDE see IQS500
dl-ISOPRENALINE HYDROCHLORIDE see IQS500
racemic ISOPRENALINE HYDROCHLORIDE see IQS500
(±)-ISOPRENALINE SULFATE see IRU000

dl-ISOPRENALINE SULFATE see IRU000
ISOPRENE see IMS000
ISOPRENE see MNH750
ISOPRENE DIBROMIDE see IMS100
ISOPRENE, INHIBITED (DOT) see IMS000
ISOPRENYL CHALCONE see IMS300
ISOPROCARB see MIA250
ISOPROCARBE see MIA250
ISOPROCIL (FRENCH) see BNM000
ISOPROMEDOL see IMT000
ISOPROMETHAZINE see DQA400
ISOPROPAMIDE see DAB875
ISOPROPAMIDE IODIDE see DAB875
ISOPROPANETHIOL see IMU000
ISOPROPANOL (DOT) see INJ000
ISOPROPANOLAMINE see AMA500
ISOPROPANOLAMINES see IMV000
ISOPROPANOLAMINES, mixed see IMV000
ISOPROPENE CYANIDE see MGA750
ISOPROPENIL-BENZOLO (ITALIAN) see MPK250
ISOPROPENYL ACETATE see MQK750
ISOPROPENYL ACETATE (DOT) see MQK750
ISOPROPENYL-BENZEEN (DUTCH) see MPK250
ISOPROPENYLBENZENE see MPK250
ISOPROPENYL-BENZOL (GERMAN) see MPK250
ISOPROPENYL CARBINOL see IMW000
4-ISOPROPENYL-1-CYCLOHEXENE-1-CARBOXALDEHYDE see DKX100
4-ISOPROPENYL-1-CYCLOHEXENE-1-CARBOXALDEHYDE, ANTI-OXIME see PCJ000
4-ISOPROPENYL-CYCLOHEX-1-ENE-1-METHANOL see PCI550
ISOPROPENYLESTER KYSELINY OCTOVE see MQK750
ISOPROPENYL ETHYL KETONE see IMW500
2'-ISOPROPENYL-2-(2-METHOXYETHYLAMINO)PROPRIONANILIDE OXALATE see IMX000
(+)-4-ISOPROPENYL-1-METHYLCYCLOHEXENE see LFU000
p-ISOPROPENYL-α-METHYLSTYRENE see IMX100
ISOPROPENYLNITRILE see MGA750
ISOPROPHENAMINE HYDROCHLORIDE see IMX150
ISOPROPHYL METHYLPHOSPHONOFLUORIDATE see IPX000
ISOPROPILAMINA (ITALIAN) see INK000
2-ISOPROPILAMINO-4-METILAMINO-6-METILTIO-1,3,5-TRIAZINA (ITALIAN) see INR000
ISOPROPILBENZENE (ITALIAN) see COE750
1-ISOPROPILBIGUANIDE CLORIDRATO (ITALIAN) see IOF100
(2-ISOPROPILCROTONIL)UREA (ITALIAN) see PMS825
ISOPROPILE (ACETATO di) (ITALIAN) see INE100
ISOPROPIL-N-FENIL-CARBAMMATO (ITALIAN) see CBM000
N-ISOPROPILFTALIMMIDE see IRG000
ISOPROPILIDRAZIDE dell'ac. p-CLORO-FENOSSIACETICO (ITALIAN) see CJN250
(1-ISOPROPIL-3-METIL-1H-PIRAZOL-5-IL)-N,N-DIMETIL-CARBAMMATO (ITALIAN) see DSK200
ISOPROPOXAMINE see IPY500
o-ISOPROPOXYANILINE see INA200
(3)-O-2-ISOPROPOXY-CARBONYL-1-METHYLVINYL-O-METHYL ETHYLPHOSPHORAMIDOTHIOATE see MKA000

11-ISOPROPOXY-15,16-DIHYDRO-17-CYCLOPENTA(a)PHENANTHREN-17-ONE see INA400
4-ISOPROPOXYDIPHENYLAMINE see HKF000
p-ISOPROPOXYDIPHENYLAMINE see HKF000
2-ISOPROPOXYETHANOL see INA500
(2-ISOPROPOXYETHYL)CARBAMATE see IND000
N-(2-ISOPROPOXY-3-HYDROXYMERCURIPROPYL)BARBITAL see INE000
2-ISOPROPOXY-4-HYDROXYPHENYL METHYLCARBAMATE see INE025
3'-ISOPROPOXY-2-METHYLBENZANILIDE see INE050
(ISOPROPOXYMETHYL)OXIRANE see IPD000
ISOPROPOXYMETHYLPHORYL, FLUORIDE see IPX000
2-ISOPROPOXYNITROBENZENE see NIR550
o-ISOPROPOXYNITROBENZENE see NIR550
1-ISOPROPOXYPENTACHLOROBUTADIENE see INE055
N-(4-ISOPROPOXYPHENYL)ANILINE see HKF000
2-ISOPROPOXYPHENYL (METHYL)(T-BUTOXYSULFINYL)CARBAMATE see INE062
o-ISOPROPOXYPHENYL METHYLCARBAMATE see PMY300
2-ISOPROPOXYPHENYL-N-METHYLCARBAMATE see PMY300
o-ISOPROPOXYPHENYL-N-METHYLCARBAMATE see PMY300
2-ISOPROPOXYPHENYL N-METHYLCARBAMATE, nitrosated see PMY310
2-ISOPROPOXYPHENYL (METHYL)(N-HEXOXYSULFINYL)CARBAMATE see INE065
o-ISOPROPOXYPHENYL METHYLNITROSOCARBAMATE see PMY310
o-ISOPROPOXYPHENYL N-METHYL-N-NITROSOCARBAMATE see PMY310
2-ISOPROPOXYPROPANE see IOZ750
1-ISOPROPOXY-2,3-PROPANEDIOL see GGA050
3-ISOPROPOXY-1,2-PROPANEDIOL see GGA050
N,N'-(3-ISOPROPOXY-1,2-PROPANEDIOXYSULFINYL)BIS(S-METHYL-N-METHYLCARBAMOYLOXYTHIOACETAMIDATE) see INE068
N,N'-(3-ISOPROPOXY-1,2-PROPANEDIOXYSULFINYL)BIS(3-METHYLPHENYLMETHYLCARBAMATE) see INE075
N,N'-(3-ISOPROPOXY-1,2-PROPANEDIOXYSULFINYL)BIS(1-NAPHTHYLMETHYLCARBAMATE) see INE085
ISOPROPYDRIN see DMV600
ISOPROPYLACETAAT (DUTCH) see INE100
ISOPROPYLACETAT (GERMAN) see INE100
ISOPROPYL ACETATE see INE100
ISOPROPYL (ACETATE d') (FRENCH) see INE100
ISOPROPYLACETIC ACID see ISU000
ISOPROPYL ACETIC ACID, BENZYL ESTER see ISW000
ISOPROPYLACETONE see HFG500
N-ISOPROPYL-N-(ACETOXYMETHYL)NITROSAMINE see ING300

ISOPROPYL-N-ACETOXY-N-PHENYLCARBAMATE see ING400
ISOPROPYL ACID PHOSPHATE solid see PHE500
2-ISOPROPYL ACRYLALDEHYDE OXIME see ING509
ISOPROPYL ACRYLAMIDE see INH000
N-ISOPROPYLACRYLAMIDE see INH000
ISOPROPYL ADIPATE see DNL800
ISOPROPYLADRENALINE see DMV600
ISOPROPYL ALCOHOL see INJ000
ISOPROPYL ALCOHOL, TITANIUM(4+) SALT see IRN200
ISO-PROPYLALKOHOL (GERMAN) see INJ000
ISOPROPYLALLYLAZETYLKARBAMID (GERMAN) see IQX000
ISOPROPYLALLYLBARBITURIC ACID see AFT000
ISOPROPYLAMID KYSELINY AKRYLOVE see INH000
ISOPROPYLAMINE see INK000
ISOPROPYLAMINE HYDROCHLORIDE see INL000
4-(2-ISOPROPYLAMINE-1-HYDROXYETHYL)METHANESULFOANILIDE HYDROCHLORIDE see CCK250
1-ISOPROPYLAMINE-3-(1-NAPHTHYLOXY)-2-PROPANOL see ICB000
ISOPROPYLAMINOANTIPYRINE see INM000
4-(ISOPROPYLAMINO)ANTIPYRINE see INM000
ISOPROPYLAMINO-4-AZIDO-6-METHYLTHIO-1,3,5-TRIAZIN (GERMAN) see ASG250
ISOPROPYL 3-AMINOCROTONATE see INM100
ISOPROPYL β-AMINOCROTONATE see INM100
1-ISOPROPYLAMINO-3-(2-CYCLOHEXYLPHENOXY)-2-PROPANOL HYDROCHLORIDE see ERE100
4-ISOPROPYLAMINO-2,3-DIMETHYL-1-PHENYL-3-PYRAZOLIN-5-ONE see INM000
4-ISOPROPYLAMINODIPHENYLAMINE see PFL000
ISOPROPYLAMINOETHANOL see INN400
2-ISOPROPYLAMINOETHANOL see INN400
N-ISOPROPYLAMINOETHANOL see INN400
ISOPROPYLAMINO ETHANOL HYDROCHLORIDE see INN500
2-(ISOPROPYLAMINO)ETHANOL HYDROCHLORIDE see INN500
ISOPROPYLAMINO-O-ETHYL-(4-METHYLMERCAPTO-3-METHYLPHENYL)PHOSPHATE see FAK000
4-(2-ISOPROPYLAMINO-1-HYDROXYAETHYL)METHANESULFONALID HYDROCHLORID (GERMAN) see CCK250
1-(ISOPROPYLAMINO)-2-HYDROXY-3-(o-(ALLYLOXY)PHENOXY)PROPANE see CNR500
ISOPROPYLAMINOHYDROXYETHYLMETHANESULFONALIDE HYDROCHLORIDE see CCK250
1-(1-(2-(3-ISOPROPYLAMINO-2-HYDROXYPROPOXY)-3,6-DICHLOROPHENYL)VINYL)-1H-IMIDAZOLE HCl see DFM875
4-((3-ISOPROPYLAMINO)-2-HYDROXYPROPOXY)-3-METHOXY-BENZOIC ACID METHYL ESTER see VFP200
7-(ISOPROPYLAMINOISOPROPYL)THEOPHYLLINE HYDROCHLORIDE see INP100
(±)-1-(ISOPROPYLAMINO)-3-(p-(2-METHOXYETHYL)PHENOXY)-2-

14-ISOPROPYLDIBENZ(a,j)ACRIDINE see IOR000

10-ISOPROPYL-3,4,5,6-DIBENZACRIDINE (FRENCH) see IOR000

ISOPROPYL-4,4'-DIBROMOBENZILATE see IOS000

ISOPROPYL-4,4'-DICHLOROBENZILATE see PNH750

ISOPROPYL 3,4-DIETHOXYCARBANILATE see IOS100

ISOPROPYL 3,4-DIETHOXYPHENYLCARBAMATE see IOS100

ISOPROPYL DIETHYLDITHIOPHOSPHORYLACETAMIDE see IOT000

N-ISOPROPYL-β-DIHYDROXYPHENYL-β-HYDROXYETHYLAMINE see DMV600

4'-ISOPROPYL-4-DIMETHYLAMINOAZOBENZENE see IOT875

α-ISOPROPYL-α-(2-DIMETHYLAMINOETHYL)-1-NAPHTHACETAMIDE see IOU000

α-ISOPROPYL-α-(2-(DIMETHYLAMINO)ETHYL)-1-NAPHTHALENEACETAMIDE see IOU000

α-ISOPROPYL-α-(2-DIMETHYLAMINOETHYL)-1-NAPHTHYLACETAMIDE see IOU000

m-ISOPROPYL-p-DIMETHYL-AMINO-PHENOL-DIMETHYL-URETHANE METHIODIDE see IOU200

α-(ISOPROPYL)-α-(β-DIMETHYLAMINOPROPYL)PHENYLACETONITRIL CITRATE see PCK639

α-(ISOPROPYL)-α-(β-DIMETHYLAMINOPROPYL)PHENYLACETONITRIL CYCLAMATE see PCK669

α-(ISOPROPYL)-α-(β-DIMETHYLAMINOPROPYL)PHENYLACETONITRILE CYCLAMATE see IHO200

(N-(2-ISOPROPYL-4-DIMETHYLCARBAMOYLOXY)PHENYL)TRIMETHYLAMMONIUM IODIDE see HJZ000

ISOPROPYL DIMETHYL CARBINOL see AOK750

ISOPROPYL-O,O-DIMETHYLDITHIOPHOSPHORYL-1-PHENYLACETATE see IOV000

4-ISOPROPYL-2,3-DIMETHYL-1-PHENYL-3-PYRAZOLIN-5-ONE see INY000

N'-ISOPROPYL-N,N'-DIMETHYL-1,3-PROPANE-DIAMINE see IOW000

3-ISOPROPYL-5,7-DIMORPHOLINOMETHYLTROPOLONE DIHYDROCHLORIDE see IOW500

ISOPROPYL-2,4-DINITRO-6-SEC-BUTYLPHENYL CARBONATE see CBW000

4-ISOPROPYL-2,6-DINITROPHENOL see IOX000

ISOPROPYLDIPHENYL see IOF200

p-ISOPROPYLDIPHENYL see IOF250

ISOPROPYLENE BROMIDE see BOA250

ISOPROPYL EPOXYPROPYL ETHER see IPD000

ISOPROPYLESTER KYSELINY BENZOOVE see IOD000

ISOPROPYLESTER KYSELINY DUSITE see IQQ000

ISOPROPYLESTER KYSELINY KARBAMINOVE see IOJ000

ISOPROPYLESTER KYSELINY OCTOVE see INE100

ISOPROPYLESTER KYSELINY SKORICOVE see IOO000

ISOPROPYL ESTER MERCAPTOPHENYLACETIC ACID S-ESTER with O-O-DIMETHYL PHOSPHORODITHIOATE see IOV000

ISOPROPYL ETHANE FLUOROPHOSPHONATE see ELX100

ISOPROPYL ETHANESULFONATE see IOX500

N-ISOPROPYLETHANOLAMINE see INN400

ISOPROPYL ETHER see IOZ750

ISOPROPYL 3-(((ETHYLAMINO)METHOXYPHOSPHINOTHIOYL)OXY)CROTONIC ACID see IOZ800

ISOPROPYL ETHYL URETHAN see IPA000

ISOPROPYL-N-FENYL-CARBAMAAT (DUTCH) see CBM000

N-ISOPROPYL-N'-FENYL-p-FENYLENDIAMIN (CZECH) see PFL000

ISOPROPYL FLUOPHOSPHATE see IRF000

ISOPROPYL γ-FLUOROBUTYRATE see IPA100

ISOPROPYL FORMATE see IPC000

ISOPROPYLFORMIC ACID see IJU000

N-ISOPROPYLFTALIMID see IRG000

α-ISOPROPYL GLYCEROL ETHER see GGA050

ISOPROPYL GLYCIDYL ETHER see IPD000

ISOPROPYL GLYCOL see INA500

ISOPROPYL HEXADECANOATE see IQW000

ISOPROPYL-n-HEXADECANOATE see IQW000

N'-ISOPROPYL HYDRAZINOMETHANESULFONIC ACID, SODIUM SALT see HGW000

ISOPROPYL HYDROPEROXIDE see IPI000

N-ISOPROPYL-2-HYDROXY-2-(o-CHLOROPHENYL)ETHYLAMINE HYDROCHLORIDE see IMX150

3-ISOPROPYL-4-HYDROXYPHENYL METHYLCARBAMATE see IPI100

ISOPROPYL HYPOCHLORITE see IPI350

2-ISOPROPYLIDENEACETAMIDE see SBW965

ISOPROPYLIDENEACETONE see MDJ750

2-ISOPROPYLIDENEAMINO-OXYETHYL (R)-2-(4-(6-CHLOROQUINOXALIN-2-YLOXY)PHENOXY)PROPIONATE see IPI400

4,4'-ISOPROPYLIDENE-BIS(2-tert-BUTYLPHENOL) see IPJ000

4,4'-ISOPROPYLIDENEBIS(2,6-DICHLOROPHENOL) see TBO750

4,4'-ISOPROPYLIDENEBISPHENOL see BLD500

p,p'-ISOPROPYLIDENEBISPHENOL see BLD500

4,4'-ISOPROPYLIDENEDI-o-CRESOL see IPK000

4-ISOPROPYLIDENE-3,3-DIMETHYL-2-OXETANONE see IPL000

2,3-ISOPROPYLIDENEDIOXYPHENYL METHYLCARBAMATE see DQM600

p,p'-ISOPROPYLIDENEDIPHENOL see BLD500

4,4'-ISOPROPYLIDENEDIPHENOL ALKYL(C12-15) PHOSPHITE see IPL500

4,4'-ISOPROPYLIDENEDIPHENOL DIGLYCIDYL ETHER see BLD750

4,4'-ISOPROPYLIDENEDIPHENOL DIMER with 1-CHLORO-2,3-EPOXYPROPANE see IPM000

4,4'-ISOPROPYLIDENEDIPHENOL, MONOMER with 1-CHLORO-2,3-EPOXYPROPANE see IPN000

4,4'-ISOPROPYLIDENE DIPHENOL POLYMER with 1-CHLORO-2,3-EPOXYPROPANE see ECL000

4,4'-ISOPROPYLIDENEDIPHENOL, POLYMER with 1-CHLORO-2,3-EPOXYPROPANE see IPO000

4,4'-ISOPROPYLIDENEDIPHENOL, TETRAMER with 1-CHLORO-2,3-EPOXYPROPANE see IPP000

ISOPROPYLIDENE GLYCEROL see DVR600

1,2-o-ISOPROPYLIDENE GLYCEROL see DVR600

2-(2-ISOPROPYLIDENEHYDRAZONO)-4-(5-NITRO-2-FURYL)-THIAZOLE see NGN000

1-ISOPROPYLIDENE-4-METHYL-2-CYCLOHEXANONE see MCF500

13-ISOPROPYLIDENEPODOCARP-8(14)-EN-15-OIC ACID see NBU800

ISOPROPYLIDINE AZASTREPTONIGRIN see IPR000

ISOPROPYL IODIDE see IPS000

ISOPROPYL ISOBUTYL ARSINIC ACID see IPS100

ISOPROPYL ISOCYANIDE DICHLORIDE see IPS400

N-ISOPROPYL ISONICOTINHYDRAZIDE see ILE000

ISOPROPYL ISOSTEARATE see IPS450

ISOPROPYL KETONE see DTI600

ISOPROPYLKYANID see IJX000

ISOPROPYL LANOLATE see IPS500

ISOPROPYL MANDELATE see IPT000

ISOPROPYL MEPROBAMATE see IPU000

ISOPROPYL MERCAPTAN (DOT) see IMU000

N-ISOPROPYL-2-MERCAPTOACETAMIDE-S-ESTER with O,O-DIETHYL PHOSPHORODITHIOATE see IOT000

ISOPROPYL MERCAPTOPHENYLACETATE-O,O-DIMETHYL PHOSPHORODITHIOATE see IOV000

ISOPROPYLMERCURY HYDROXIDE see IPW000

ISOPROPYL MESYLATE see IPY000

ISOPROPYL METHANEFLUOROPHOSPHONATE see IPX000

N-ISOPROPYL-β-(4-METHANESULFONAMIDOPHENYL)ETHANOLAMINE HYDROCHLORIDE see CCK250

ISOPROPYLMETHANESULFONATE see IPY000

ISOPROPYL METHANE SULPHONATE see IPY000

ISOPROPYLMETHOXAMINE see IPY500

N-ISOPROPYLMETHOXAMINE see IPY500

2'-ISOPROPYL-2-(2-METHOXYETHYLAMINO)-PROPIONANILIDE HYDROCHLORIDE see IPZ000

3-ISOPROPYL-2-(p-METHOXYPHENYL)-3H-NAPHTH(1,2-d)IMIDAZOLE see THG700

ISOPROPYL(2E,4E)-11-METHOXY-3,7,11-TRIMETHYL-2,4-DODECADIENOATE see KAJ000

8-((ISOPROPYLMETHYLAMINO)METHYL)QUINOLINE see IPZ500

N-ISOPROPYL-α-(2-METHYLAZO)-p-TOLUAMIDE see IQB000

4-ISOPROPYL-1-METHYLBENZENE see CQI000

ISOPROPYL 2-METHYL-2-BUTENOATE see IRN100

2-ISOPROPYL-6-(1-METHYLBUTYL)PHENOL see IQD000

ISOPROPYL N-METHYLCARBANILATE see MOS400

ISOPROPYL α-METHYL CROTONATE see IRN100

1-ISOPROPYL-4-METHYLCYCLOHANE HYDROPEROXIDE see IQE000

4-ISOPROPYL-1-METHYL-1,5-CYCLOHEXADIENE see MCC000

5-ISOPROPYL-2-METHYL-1,3-CYCLOHEXADIENE see MCC000

5-ISOPROPYL-2-METHYL-2,5-CYCLOHEXADIENE-1,4-DIONE see IQF000

2-ISOPROPYL-5-METHYLCYCLOHEXANOL see MCF750

4-ISOPROPYL-1-METHYLCYCLOHEXAN-3-OL see MCG000

ISOPROPYL SALICYLATE O-ESTER with O-ETHYLISOPROPYLPHOSPHORAMIDOTHIOATE see IMF300
ISOPROPYL-S HYDROCHLORIDE see IPG000
O-ISOPROPYL SODIUM DITHIOCARBONATE see SIA000
ISOPROPYL STEARATE see IRL100
ISOPROPYL STYRYL KETONE see IRL300
ISOPROPYL SULFATE see DNO900
N-ISOPROPYL TEREPHTHALAMIC ACID see IRN000
ISOPROPYL TETRADECANOATE see IQN000
S-2-ISOPROPYLTHIOETHYL-O,O-DIMETHYL PHOSPHORODITHIOATE see DSK800
ISOPROPYLTHIOL see IMU000
(ISOPROPYLTHIO)-METHANETHIOL-S-ESTER with O,O-DIETHYL PHOSPHORODITHIOATE see DJN600
erythro-p-(ISOPROPYLTHIO)-α-((1-(OCTYLAMINO)ETHYL)BENZYL ALCOHOL see SOU600
erythro-1-(4-ISOPROPYLTHIOPHENYL)-2-n-OCTYLAMINOPROPANOL see SOU600
ISOPROPYL TIGLATE see IRN100
ISOPROPYL TITANATE(IV) see IRN200
2-ISOPROPYLTOLUENE see IRN300
3-ISOPROPYLTOLUENE see IRN400
m-ISOPROPYLTOLUENE see IRN400
o-ISOPROPYLTOLUENE see IRN300
p-ISOPROPYLTOLUENE see CQI000
ISOPROPYL-2-(4-TRIAZOLYL)-5-BENZIMIDAZOLECARBAMATE see CBA100
α-ISOPROPYL-α-(p-(TRIFLUOROMETHOXY)PHENYL)-5-PYRIMIDINEMETHANOL see IRN500
4-ISOPROPYL-2-(α,α,α-TRIFLUORO-m-TOLYL)MORPHOLINE see IRP000
N-ISOPROPYL-4-(3,4,5-TRIMETHOXYCINNAMOYL)-1-PIPERAZINEACETAMIDE MALEATE see IRQ000
4-ISOPROPYL-2,6,7-TRIOXA-1-ARSABICYCLO(2.2.2)OCTANE see IRQ100
β-ISOPROPYLTROPOLON see IRR000
4-ISOPROPYLTROPOLONE see IRR000
ISOPROPYL UNOPROSTONE see IRR050
ISOPROPYLUREA see IRR100
1-ISOPROPYLUREA see IRR100
N-ISOPROPYLUREA see IRR100
ISOPROPYLUREA and SODIUM NITRITE see SIS675
ISOPROPYL VINYL ETHER see IRS000
ISOPROPYLXANTHIC ACID, SODIUM SALT see SIA000
ISOPROPYLXANTHOGENAN SODNY see SIA000
ISOPROPYL XANTHOGEN DISULFIDE see IRS500
(+)-ISOPROPYL, Z-7-((1R,2R,3R,5S)-3,5-DIHYDROXY-2-(3-OXODECYL)CYCLOPENTYL)HEPT-5-ENOATE see IRR050
ISOPROTERENOL see DMV600
l-ISOPROTERENOL see DMV600
racemic ISOPROTERENOL HDYROCHLORIDE see IQS500
ISOPROTERENOL HYDROCHLORIDE see IMR000
(±)-ISOPROTERENOL HYDROCHLORIDE see IQS500
dl-ISOPROTERENOL HYDROCHLORIDE see IQS500
dl(±)-ISOPROTERENOL HYDROCHLORIDE see IQS500
ISOPROTERENOL MONOHYDROCHLORIDE see IMR000
(±)-ISOPROTERENOL SULFATE see IRU000
dl-ISOPROTERENOL SULFATE see IRU000
ISOPROTHIOLANE see IRU500

ISOPROTURON see IRA050
ISOPSORALIN see FQC000
ISOPTIN see IRV000
ISOPTIN see VHA450
ISOPTO CARBACHOL see CBH250
ISOPTO-CARPINE see PIF250
ISOPTO CETAMIDE see SNQ710
ISOPTO FENICOL see CDP250
ISOPTO-HYDROCORTISONE see HHQ800
ISOPTO HYOSCINE see SBG000
ISOPULEGOL (FCC) see MCE750
ISOPURAMIN BC see AHR600
ISOPURAMIN PLUS see AHR600
ISO-PUREN see CCK125
ISOPURINE see POJ250
ISOPYRIN see INM000
ISOPYRINE see INM000
ISOPYRINE see INY000
ISOQUINALDEHYDE THIOSEMICARBAZONE see IRV300
ISOQUINAZEPON see IRW000
ISOQUINOLINE see IRX000
3-ISOQUINOLINECARBOXYLIC ACID, 1,2,3,4-TETRAHYDRO-6,7-DIMETHOXY-2-(2-((1-(ETHOXYCARBONYL)-3-PHENYLPROPYL)AMINO)-1-OXOPROPYL)-, MONOHYDROCHLORIDE, (3S-(2(R*(R*)),3R*))- see MRA260
ISOQUINOLINE, 1-(3,4-DIETHOXYBENZYL)-6,7-DIETHOXY-, HYDROCHLORIDE see PAH260
ISOQUINOLINE, 1-((3,4-DIETHOXYPHENYL)METHYL)-6,7-DIETHOXY-, HYDROCHLORIDE see PAH260
ISOQUINOLINE, 7,8-DIMETHOXY- see DOF900
ISOQUINOLINE, 6,7-DIMETHOXY-1-VERATRYL-, HYDROCHLORIDE see PAH250
2-(1-ISOQUINOLINYMETHYLENE)-HYDRAZINECARBOTHIOAMIDE (9CI) see IRV300
ISOQUINOLIUM, 2-(3-(4-FORMYLPYRIDINIO)PROPYL)-, DIBROMIDE, OXIME see IRX050
1-(2-ISOQUINOLYL)-5-METHYL-4-PYRAZOLYL METHYL KETONE see IRX100
ISORBID see CCK125
ISORDIL see CCK125
ISORDIL TEMBIDS see CCK125
ISOREN see CLY600
ISORENIN see DMV600
ISORETINENE a see VSK975
ISOSAFROEUGENOL see IRY000
ISOSAFROLE see IRZ000
ISOSAFROLE, OCTYL SULFOXIDE see ISA000
ISOSAFROLE-n-OCTYLSULFOXIDE see ISA000
ISOSCOPIL see HOT500
ISOSET CX 11 see PKB100
ISOSORBIDE see HID350
(+)-d-ISOSORBIDE see HID350
ISOSORBIDE DINITRATE see CCK125
ISOSORBIDE 5-MONONITRATE see ISC500
ISOSORBIDE 5-NITRATE see ISC500
ISOSTEARYL NEOPENTANOATE see ISC550
ISOSTENASE see CCK125
ISOSUMITHION see MKC250
ISO SYSTOX SULFOXIDE see ISD000
ISOTACTIC POLYPROPYLENE see PMP500
ISOTAZIN see DIR000
ISOTEBEZID see ILD000
ISOTENIC ACID see AKG250
ISOTENSE see RCA200
ISOTHAN see LBW000
ISOTHAZINE see DIR000
ISOTHEBAINE HYDROCHLORIDE see ISD033
ISOTHIAZINE see DIR000

1,9-ISOTHIAZOLEANTHRONE-2-CARBOXYLIC ACID see ISD043
ISOTHIAZOLINONE CHLORIDE see ISD066
4-ISOTHIAZOLIN-3-ONE, 5-CHLORO-2-METHYL- see CIN100
3(2H)-ISOTHIAZOLONE, 5-CHLORO-2-METHYL- see CIN100
3(2H)-ISOTHIAZOLONE, 5-CHLORO-2-METHYL-, MIXT. WITH 2-METHYL-3(2H)-ISOTHIAZOLONE see ISD066
3(2H)-ISOTHIAZOLONE, 4,5-DICHLORO-2-CYCLOHEXYL- see CPJ600
3(2H)-ISOTHIAZOLONE, 2-METHYL- see MLB700
ISOTHIN see EPQ000
ISOTHIOATE see DSK800
ISOTHIOCYANATE d'ALLYLE (FRENCH) see AGJ250
ISOTHIOCYANATE de METHYLE (FRENCH) see ISE000
1-ISOTHIOCYANATE-NAPHTHALENE see ISN000
ISOTHIOCYANATOBENZENE see ISQ000
1-ISOTHIOCYANATOBUTANE see ISD100
ISOTHIOCYANATOETHANE see ELX525
(2-ISOTHIOCYANATOETHYL)BENZENE see ISP000
(6-ISOTHIOCYANATOHEXYL)BENZENE see ISD200
ISOTHIOCYANATOMETHANE see ISE000
1-ISOTHIOCYANATONAPHTHALENE see ISN000
4-ISOTHIOCYANATO-4'-NITRODIPHENYLAMINE see AOA050
4-ISOTHIOCYANATO-N-(4-NITROPHENYL)-BENZENAMINE (9CI) see AOA050
N-(3-ISOTHIOCYANATOPHENYL)ACETAMIDE see ISG000
3-ISOTHIOCYANATO-1-PROPENE see AGJ250
(ISOTHIOCYANATO)TRIMETHYLSTANNE see ISF000
ISOTHIOCYANATOTRIMETHYLTIN see ISF000
(ISOTHIOCYANATO)TRIPROPYLSTANNANE see TNB750
ISOTHIOCYANIC ACID see ISF100
ISOTHIOCYANIC ACID (DOT) see ISF100
ISOTHIOCYANIC ACID-m-ACETAMIDOPHENYL ESTER see ISG000
ISOTHIOCYANIC ACID BENZYL ESTER see BEU250
ISOTHIOCYANIC ACID, BUTYL ESTER see ISD100
ISOTHIOCYANIC ACID,-p-CHLOROBENZYL ESTER see CEQ750
ISOTHIOCYANIC ACID, p-CHLOROPHENYL ESTER see ISH000
ISOTHIOCYANIC ACID-p-CYANOPHENYL ESTER see ISI000
ISOTHIOCYANIC ACID, CYCLOHEXYL ESTER see ISJ000
ISOTHIOCYANIC ACID, p-DIMETHYLAMINOPHENYL ESTER see ISJ100
ISOTHIOCYANIC ACID, ETHYLENE ESTER see ISK000
ISOTHIOCYANIC ACID, ETHYL ESTER see ELX525
ISOTHIOCYANIC ACID-m-FLUOROPHENYL ESTER see ISL000
ISOTHIOCYANIC ACID-p-FLUOROPHENYL ESTER see ISM000
ISOTHIOCYANIC ACID, METHYL ESTER see ISE000
ISOTHIOCYANIC ACID, 3-(METHYLSULFINYL)PROPYL ESTER see MPN100

ISOTHIOCYANIC ACID-1-NAPHTHYL ESTER see ISN000
ISOTHIOCYANIC ACID, p-(p-NITROPHENOXY)PHENYL ESTER see LIG100
ISOTHIOCYANIC ACID-m-NITROPHENYL ESTER see ISO000
ISOTHIOCYANIC ACID, PHENETHYL ESTER see ISP000
ISOTHIOCYANIC ACID-p-PHENYLENE ESTER see PFA500
ISOTHIOCYANIC ACID, PHENYL ESTER see ISQ000
4-ISOTHIOCYANO-4'-NITRO DIPHENYLAMINE see AOA050
ISOTHIOINDIGO see DNT300
ISOTHIONIC ACID see HKI500
ISOTHIOUREA see ISR000
ISOTHIOURONIUM CHLORIDE, BENZYL see BEU500
ISOTHIPENDYL see OGI075
ISOTHIPENDYL HYDROCHLORIDE see AEG625
ISOTHYMOL see CCM000
ISOTHYMOL see IQJ000
ISOTTIAMIDA see EPQ000
ISOTIOCIANATO di METILE (ITALIAN) see ISE000
ISOTONIL see DRM000
ISOTOX see BBQ500
ISOTRAC see ITD015
ISOTRATE see CCK125
ISOTRETINOIN see VSK955
ISOTRICYCLOQUINAZOLINE NITRATE see TBI750
ISOTRIMETHYLTETRAHYDRO BENZYL ALCOHOL see ISR100
ISOTRON 11 see TIP500
ISOTRON 12 see DFA600
ISOTRON 22 see CFX500
ISOUREA see USS000
ISOURETHANE see PKL500
ISOURON see ISR200
ISOVAL see BNP750
ISOVAL see ITD010
ISOVALERAL see MHX500
ISOVALERALDEHYDE see MHX500
8-ISOVALERATE see FQS000
ISO-VALERIANATE de BORNYLE see HOX100
ISOVALERIANIC AICD see ISU000
ISOVALERIC ACID see ISU000
ISOVALERIC ACID, ALLYL ESTER see ISV000
ISOVALERIC ACID, BENZYL ESTER see ISW000
ISOVALERIC ACID, 2-BORNYL ESTER (7CI,8CI) see HOX100
ISOVALERIC ACID, BUTYL ESTER see ISX000
ISOVALERIC ACID, CINNAMYL ESTER (6CI,7CI,8CI) see PEE200
ISOVALERIC ACID, (4,7-DIMETHYL-1,6-OCTADIEN-3-YL) ESTER see LGC000
(E)-ISOVALERIC ACID-3,7-DIMETHYL-2,6-OCTADIENYL ESTER see GDK000
ISOVALERIC ACID-8-ESTER with 3-FORMAMIDO-N-(7-HEXYL-8-HYDROXY-4,9-DIMETHYL-2,6-DIOXO-1,5-DIOXONAN-3-YL)SALICYLAMIDE see DUO350
ISOVALERIC ACID, ETHYL ESTER see ISY000
(Z)-ISOVALERIC ACID-3-HEXENYL see ISZ000
ISOVALERIC ACID, HEXYL ESTER (8CI) see HFQ575
ISOVALERIC ACID, ISOBUTYL ESTER see ITA000
ISOVALERIC ACID, ISOPENTYL ESTER see ITB000

ISOVALERIC ACID, p-MENTH-3-YL ESTER see MCG900
ISOVALERIC ACID, METHYL ESTER see ITC000
ISOVALERIC ALDEHYDE see MHX500
ISOVALERONE see DNI800
ISOVALERONITRILE see ITD000
ISOVALERYL see ITD015
ISOVALERYL INDANDIONE see ITD010
2-ISOVALERYLINDAN-1,3-DIONE see ITD010
2-ISOVALERYL-1,3-INDANDIONE see ITD010
2-ISOVALERYL-1,3-INDANDIONE CALCIUM SALT see ITD015
2-ISOVALERYL-1,3-INDANEDIONE see ITD010
ISOVALTRATE see ITD018
ISOVANILLIC ACID see HJC000
ISOVANILLIN see FNM000
ISOVANILLINE see FNM000
ISOVITAMIN C see SAA025
ISOWOGONIN see ITD020
ISOXABEN see ENF100
ISOXAL see PCJ325
ISOXAMIN see SNN500
ISOXANTHINE see XCA000
ISOXATHION see DJV600
ISOXAZOL see HOM259
ISOXAZOLE, 3,5-DIMETHYL- see DSK950
ISOXAZOLIDINE, 2-(3,4,5-TRIETHYOXYBENZOYL)- see ITD025
3-ISOXAZOLIDINONE, 2-((2-CHLOROPHENYL)METHYL)-4,4-DIMETHYL- see CEP800
2-ISOXAZOLIDINYL 3,4,5-TRIETHOXYPHENYL KETONE see ITD025
ISOXAZOLO(4,5-c)PYRIDIN-3(2H)-ONE, 4,5,6,7-TETRAHYDRO- see TCU550
ISOXSUPRINE see VGF000
ISOXSUPRINE HYDROCHLORIDE see VGA300
ISOXSUPRIN HYDROCHLORIDE see VGA300
ISOZIDE see ILD000
ISPENORAL see PDT750
ISPHAMYCIN see CMB000
ISRAVIN see DBX400
ISTIN see DMH400
ISTONYL see DRM000
ISUPREL see DMV600
ISUPREL see IMR000
ISUPREL HYDROCHLORIDE see IMR000
ISUPREN see DMV600
ISZILIN see IDF300
IT 40 see SMQ500
IT 066 see PIB650
IT 931 see DLK200
IT 3456 see CDT000
DL-111-IT see EOL600
ITA-104 see PJB500
ITA 312 see LII400
ITACHIGARDEN see HLM000
ITALCHIN see CFU750
ITALIAN ARUM see ITD050
ITAMID see PJY500
ITAMIDONE see DOT000
ITAMO REAL (PUERTO RICO) see SDZ475
ITAMYCIN see HAL000
ITCH WEED see FAB100
ITF 182 see IAM100
ITF 611 see TKF699
ITF 1016 see ITD100
ITINEROL see HGC500
ITIOCIDE see EPQ000
ITOBARBITAL see AGI750
ITOPAZ see EEH600
ITROP see IGG000
ITRUMIL SODIUM see SHX500
ITRUMIL SODIUM SALT see SHX500
ITURAN see NGE000
I.U. 7 see CAL075

IUDR see DAS000
5-IUDR see DAS000
IUGLON see WAT000
IVALON see FMV000
IVALON see PKP750
IVARLAN 3406 see LAU560
IVARLAN 3407 see LAU560
IVAUGAN see CFY000
IVE see IJQ000
IVERMECTIN see ITD875
IVERSAL see AHI875
IVERTOL see AHI875
IVIRON see IHG000
IVORAN see DAD200
IVORIT see EJM500
IVORY see ITE000
IVOSIT see ACE500
IVY see AFK950
IVY ARUM see PLW800
IVY BUSH see MRU359
IXODEX see DAD200
IXOTEN see TNT500
IXPER 25M see MAH750
IYLOMYCIN see ITF000
IYOMYCIN see ITG000
IYOMYCIN B1 see ITH000
IZ 914 see DYC875
IZAA see EHK600
IZADRIN see IMR000
IZOACRIDINA see AHS500
IZOFORON (POLISH) see IMF400
o-IZOPROPOKSYANILINA (POLISH) see INA200
o-IZOPROPOKSYNITROBENZEN (CZECH) see NIR550
IZOPROPYLOWY ETER (POLISH) see IOZ750
IZOPTIN see VHA450
IZOPTIN HYDROCHLORIDE see VHA450
IZOSYSTOX (CZECH) see DAP200
IZSAB see LEK000
J3 see BPP750
J 164 see HKQ100
J 242 see AHG000
J 2Fp see GLU000
J 400 see PMP500
J 455 see EHK600
J 820 see MCB050
J 0852 see ZJS400
JACINTO de PERU (CUBA) see SLH200
JACK BEAN UREASE see UTU550
JACK-IN-THE-PULPIT see JAJ000
JACK WILSON CHLORO 51 (oil) see CKC000
JACOBINE see JAK000
JACOBINE see SBX500
JACODINE HYDROCHLORIDE see SBX525
JACUTIN see BBP750
JACUTIN see BBQ500
JADE GREEN BASE see JAT000
JADINOL PU see SKF600
JADO see CBT500
JA-FA IPM see IQN000
JA-FA IPP see IQW000
JAFFNA TOBACCO see BFW135
JAGUAR see GLU000
JAGUAR 6000 see GLU000
JAGUAR A 20D see GLU000
JAGUAR A 40F see GLU000
JAGUAR A 20 B see GLU000
JAGUAR GUM A-20-D see GLU000
JAGUAR No. 124 see GLU000
JAGUAR PLUS see GLU000
JAIKIN see SFF000
JALAN see EKO500
JAMAICA GINGER EXTRACT see JBA000
JAMAICAN WALNUT see TOA275
JAMBERRY see JBS100
JAMESTOWN WEED see SLV500
JANIMINE see DLH630
JANTARAN SODNY see SJW100
JANUPAP see HIM000
JANUS BLUE see CMM770

JANUS GREEN B see DHM500
JANUS GREEN V see DHM500
JAPAN AGAR see AEX250
JAPAN CAMPHOR see CBA750
JAPANESE AUCUBA see JBS050
JAPANESE BEAD TREE see CDM325
JAPANESE CAMPHOR see CBB250
JAPANESE CAMPHOR OIL see CBB500
JAPANESE LANTERN PLANT see JBS100
JAPANESE LAUREL see JBS050
JAPANESE MEDLAR see LIH200
JAPANESE MINT OIL see CNR990
JAPANESE OIL OF CAMPHOR see CBB500
JAPANESE PLUM see LIH200
JAPANESE POINSETTIA see SDZ475
JAPAN ISINGLASS see AEX250
JAPAN LACQUER see JCA000
JAPAN OIL TREE see TOA275
JAPANOL BLACK BHK see CMN800
JAPANOL BRILLIANT BLUE RWL (6CI) see
CMO600
JAPANOL BROWN M see CMO800
JAPANOL FAST BLACK D see CMN230
JAPANOL FAST RED 1 see CMO870
JAPANOL FAST RED F see CMO870
JAPANOL VIOLET J see CMP000
JAPAN RED 104 see ADG250
JAPAN RED 201 see CMS155
JAPAN RED 203 see CMS150
JAPAN RED 219 see CMS160
JAPAN RED NO. 203 see CMS150
JAPAN RED NO. 505 see CMO885
JAPAN YELLOW 203 see CMM510
JASAD see ZBJ000
JASMINALDEHYDE see AOG500
JASMINE ABSOLUTE see OGM200
JASMIN de NUIT (HAITI) see DAC500
JASMOLIN I or II see POO250
JASMONE see JCA100
(Z)-JASMONE see JCA100
cis-JASMONE see JCA100
JASMONYL see NMX550
JATAMANSONE see VAG250
(−)-JATAMANSONE see VAG250
JATRONEURAL see TKE500
JATRONEURAL see TKK250
JATROPHA CATHARTICA see CNR135
JATROPHA CURCAS see CNR135
JATROPHA GOSSYPIIFOLIA see CNR135
JATROPHA INTEGERRIMA see CNR135
JATROPHA MACRORHIZA see CNR135
JATROPHA MULTIFIDA see CNR135
JATROPHA PODAGRICA see CNR135
JATROPUR see UVJ450
JAUNE AB see FAG130
JAUNE de BEURRE (FRENCH) see DOT300
JAUNE OB see FAG135
JAUNE ORANGE S see FAG150
JAUNE de QUINOLEINE see CMM510
JAUNE SOLEIL see FAG150
JAVA AMARANTH see FAG020
JAVA CHROME BLUE BLACK BN see
CMP880
JAVA CHROME ORANGE EN see CMP882
JAVA CHROME RED PE see CMG750
JAVA CHROME VIOLET B see HLI000
JAVA CHROME YELLOW 3R see NEY000
JAVA CHROME YELLOW GT see SIT850
JAVA METANIL YELLOW G see MDM775
JAVA NAPHTOL RED 6B see CMM400
JAVA NAPHTOL RED G see CMM300
JAVA ORANGE 2G see HGC000
JAVA ORANGE I see FAG010
JAVA ORANGE II see CMM220
JAVA PONCEAU 2R see FMU070
JAVA RUBINE N see HJF500
JAVA SCARLET 3R see FMU080
JAVA UNICHROME ORANGE E see CMP882
JAVA UNICHROME YELLOW 3R see
NEY000
JAVA UNICHROME YELLOW GT see SIT850
JAVEX see SHU500

JAVILLO (DOMINICAN REPUBLIC,
PUERTO RICO) see SAT875
JAYANTI, extract see SCB100
JAYFLEX DINP see PHW585
JAYFLEX DTDP see DXQ200
JAYSOL see EFU000
JAYSOL S see EFU000
J.B. 305 see EOY000
JB-323 see PJA000
JB 329 see DXP800
JB 336 see MON250
JB 340 see CBF000
JB 516 see PDN000
JB 516 see PDN250
JB 8181 see DLS600
177 J.D. see AJD000
JECTOFER see IHK100
JEERA, EXTRACT see COF335
JEFFAMINE AP-20 see MJQ000
JEFFAMINE AP22 see FMW330
JEFFAMINE AP27 see FMW330
JEFFERSOL DB see DJF200
JEFFERSOL EB see BPJ850
JEFFERSOL EE see EES350
JEFFERSOL EM see EJH500
JEFFOX see PJT200
JEFFOX see PJT200
JEFFOX see PKI500
JEFFOX OL 2700 see MKS250
JELLIN see SPD500
JENACAINE see AIL750
JENACILLIN O see PAQ200
JEN-DIRIL see CFY000
JEQUIRITY, EXTRACT see JCA150
JERUSALEM OAK see SAF000
JERVINE see JCS000
JERVINE-3-ACETATE see JDA000
JERVINE, 11-DEOXO-12-β,13-α-DIHYDRO-
11-β-HYDROXY- see DAQ002
JESTRYL see CBH250
JESUIT'S BALSAM see CNH792
JET BEAD see JDA075
JET FUEL HEF-2 see JDA100
JET FUEL HEF-3 see JDA125
JET FUEL JP-4 see JDA135
JET FUELS see JDJ000
JETRIUM see AFJ400
JETRIUM R see AFJ400
JEW BUSH see SDZ475
JEWELER'S ROUGE see IHC450
JF 5705F see RLF350
JH 1 see MJG100
JICAMA de AQUA (CUBA) see YAG000
JICAMILLA (MEXICO, TEXAS) see CNR135
JICAMO (MEXICO) see YAG000
JIFFY GROW see ICP000
JILKON HYDROBROMIDE see GBA000
JIMBAY BEAN (BAHAMAS) see LED500
JIMSON WEED see SLV500
JIRA, EXTRACT see COF335
JISC 3108 see AGX000
JISC 3110 see AGX000
JL 130 see COW700
J-LIBERTY see MDQ250
JM 8 see CCC075
JM-28 see IGG775
JMC 45498 see DAF300
JN-21 see CQH000
JOD (GERMAN, POLISH) see IDM000
JODAIROL see HGB200
JODAMID (GERMAN) see AAI750
JODANTIPYRINE see IEU085
4-JODBENZOESAEURE see IEE025
2-JODBUTAN see IEH000
JODCYAN see COP000
JODDEOXIURIDIN see DAS000
JODETHAN see ELP500
2-JODFENOL see IEV000
o-JODFENOL see IEV000
JODFENPHOS see IEN000
o-JODHIPPURSAEURE NATRIUM
(GERMAN) see HGB200

JODID SODNY see SHW000
JOD-METHAN (GERMAN) see MKW200
1-JOD-3-METHYLBUTAN see IHU200
1-JOD-2-METHYLPROPAN see IIV509
JODOBIL see PDM750
1-JODOKTAN see OFA100
JODOMIRON see AAI750
JODOPAX see AAN000
JODOPYRINE see IEU085
1-JODPENTAN see IET500
3-JODPHENOL see IEV010
JODPHOSPHONIUM (GERMAN) see
PHA000
JOGEN see JDJ100
JOHNKOLOR see TKQ250
JOJOBA LIQUID WAX see JDJ300
JOJOBA OIL see JDJ300
JOKER see EHH600
JOLIPEPTIN see JDS000
JOLT see EIN000
JOMYBEL see JDS200
JONIT see PFA500
JONKMARI, EXTRACT see JDS100
JONNIX see SNQ500
JONQUIL see DAB700
JON-TROL see DXE600
JOOD (DUTCH) see IDM000
JOODMETHAAN (DUTCH) see MKW200
JOPAGNOST see IFY100
JORCHEM 400 ML see PJT000
JORTAINE see GHA050
JOSAMINA see JDS200
JOSAMYCIN see JDS200
JOSCINE see SBG500
JOTA (GERMAN) see IGD000
JOTALAMSAEURE (GERMAN) see IGD000
JOY POWDER see HBT500
JP 5 see JDS300
JP-10 see TLR675
JP 61 see AJU625
JP 428 HYDROCHLORIDE see DAI200
JP 5 JET FUEL see JDS300
JP-5 NAVY FUEL see JDS300
JR see QAT600
JR 1 see QAT600
JR 125 see QAT600
JR 30M see QAT600
JR 400 see QAT600
JR 600A see RSP100
J SOFT C 4 see DTC600
J-SUL see SNN500
JUCA see CCO680
JUDEAN PITCH see ARO500
JUDOLOR see FQJ100
JUGLANE see WAT000
JUGLON see WAT000
JUGLONE see WAT000
JULIN'S CARBON CHLORIDE see HCC500
JULODIN see CKL250
JUMBEE BEADS (VIRGIN ISLANDS) see
RMK250
JUMBLE BEAD see AAD000
JUMEX see DAZ125
JUMP-AND-GO (BAHAMAS) see LED500
JUNGER FUSTIK see FBW000
JUNIPEN see LID100
JUNIPENE see LID100
JUNIPER BERRY OIL see JEA000
JUNIPERIC ACID LACTONE see OKU000
JUNIPER OIL see JEA000
JUNIPER TAR see JEJ000
JUNIPERUS COMMUNIS auct. non. Linn.,
extract excluding roots see JEJ100
JUNIPERUS COMMUNIS Linn. var.
SAXATILIS Pallas, extract excluding roots see
JEJ100
JUNIPERUS PHOENICEA OIL see JEJ200
JUNLON 110 see ADV900
JURIMER AC 10H see ADV900
JURIMER AC 10P see ADV900
JUSONIN see SFC500
JUSQUIAME (CANADA) see HAQ100

JUSTAMIL see AIE750
JUVASON see PLZ000
JUVASTIGMIN see NCL100
JUVELA see TGJ050
JUVENIMICIN A3 see RMF000
JUVOCAINE see AIT250
JZF see BLE500
K 0 see UTU500
K 2 see SAN300
K 2 see SAN300
K-9 see MEC250
06K see AJI250
K 17 see TEH500
K 17 see UTU500
K 28 see TIL270
K 52 see OHU000
6FK see HCZ000
K-82 see CAJ770
K-83 see CAJ772
K 113 see BKL600
K-117 see NCP875
K2$_{20}$ see VTA650
K 250 see CAT775
K 257 see CBT500
K-300 see CQH000
K315 see KAH000
K-315 see THJ750
K 333 see PJQ790
K 351 see NEA100
K-373 see DAP700
K385 see UTU500
K 525 see SMQ500
K 55E see SMR000
K6-30 see ARM268
K 653 see HGW000
K 694 see CDR550
K-708 see AAE625
K 1875 see NCM700
K 1900 see NHH000
K-1902 see PPW000
K 251T see TKP050
K 2680 see AHP125
K-349-3 see LIU500
K3917 see CFY750
K 3926 see BIM775
K 4277 see IDA400
K 4710 see KFK000
K 6451 see CJT750
K 8870 see UTU500
K-10033 see BPP250
K 121-02 see MCB050
K 22023 see DGD800
K-30052 see EGC500
K 411-02 see UTU500
K 421-01 see MCB050
K 421-02 see MCB050
K 421-05 see MCB050
K 423-02 see MCB050
K62-105 see LEN000
K25 (polymer) see PKQ250
Kh15N55M16 see CNA750
Kh15N55M16V see CNA750
K 31 (pharmaceutical) see CME675
KABAT see KAJ000
KABI 1774 see CMS234
KABIKINASE see SLW450
KACHA HALDI see ICC800
KADMIUM (GERMAN) see CAD000
KADMIUMCHLORID (GERMAN) see CAE250
KADMIUMSTEARAT (GERMAN) see OAT000
KADMU TLENEK (POLISH) see CAH500
KADOX-25 see ZKA000
KAEMPFEROL see ICE000
KAEMPHEROL see ICE000
KAERGONA see MMD500
KAFAR COPPER see CNI000
KAFFIR LILLY see KAJ100
KAFIL SUPER see RLF350
KAFOCIN see CCR890
KAINIC ACID see KAJ200

α-KAINIC ACID see KAJ200
l-α-KAINIC ACID see KAJ200
KAISER CHEMICALS 11 see FOO000
KAISER CHEMICALS 12 see DFA600
KAISER NCO 20 see PKB100
KAKEN see CPE750
KAKERBIN see PAK250
KAKO BLUE B SALT see DCJ200
KAKODYLAN DODNY see HKC500
KAKO RED B BASE see NEQ000
KAKO RED RL BASE see MMF780
KAKO RED TR BASE see CLK220
KAKO SCARLET GG SALT see DEO295
KAKO TARTRAZINE see FAG140
KALCIT see CAD800
KALEX see EIV000
KALGAN see PFC750
K'-ALGILINE see SEH000
KALIMAGNESIA see MAI650
KALINITE see AHF200
KALITABS see PLA500
KALIUMARSENIT (GERMAN) see PKV500
KALIUM-BETA see PLG800
KALIUMCARBONAT (GERMAN) see PLA000
KALIUMCHLORAAT (DUTCH) see PLA250
KALIUMCHLORAT (GERMAN) see PLA250
KALIUMCYANAT (GERMAN) see PLC250
KALIUM-CYANID (GERMAN) see PLC500
KALIUMDICHROMAT (GERMAN) see PKX250
KALIUMHYDROXID (GERMAN) see PLJ500
KALIUMHYDROXYDE (DUTCH) see PLJ500
KALIUMNITRAT (GERMAN) see PLL500
KALIUMPERMANGANAAT (DUTCH) see PLP000
KALIUMPERMANGANAT (GERMAN) see CAV250
KALIUMPERMANGANAT (GERMAN) see PLP000
KALKHYDRATE see CAT225
KALLIDIN see BML500
KALLIKREIN-TRYPSIN INACTIVATOR see PAF550
KALLOCRYL K see PKB500
KALLODENT CLEAR see PKB500
KALMETTUMSOMNIFERUM see GFA000
KALMIA ANGUSTIFOLIA see MRU359
KALMIA LATIFOLIA see MRU359
KALMIA MICROPHYLLA see MRU359
KALMOCAPS see LFK000
KALMOCAPS see MDQ250
KALMUS OEL (GERMAN) see OGK000
KALO see ARB750
KALO see EAI600
KALODIL see DNM400
KALOMEL (GERMAN) see MCW000
KALOPANAXSAPONIN A see HAK075
KALPREN see CAS750
KALPUR TE see THR820
KALTOSTAT see CAM200
KALZIUMARSENIAT (GERMAN) see ARB750
KALZIUMZYKLAMATE (GERMAN) see CAR000
KALZON see MCB575
KAM 1000 see SLK000
KAM 2000 see SLK000
KAM 3000 see SLK000
KAMALIN see KAJ500
KAMANI (HAWAII) see MBU780
KAMAVER see CDP250
KAMBAMINE RED TR see CLK220
KAMBAMINE SCARLET GG BASE see DEO295
KAMBOTHOL AS see CMM760
KAMFOCHLOR see CDV100
KAMILLEN OEL see CBA200
KAMILLENOEL see CDH500
KAMILLENOEL (GERMAN) see DRV000
KAMPFER (GERMAN) see CBA750
KAMPHEROL see ICE000
KAMPOSAN see CDS125

KAMPSTOFF "LOST" see BIH250
KAMYCIN see KAM000
KAMYCIN see KAV000
KAMYNEX see KAM000
KAMYNEX see KAV000
KANABRISTOL see KAM000
KANABRISTOL see KAV000
KANACEDIN see KAM000
KANACEDIN see KAV000
KANAMICINA (ITALIAN) see KAL000
KANAMYCIN see KAL000
KANAMYCIN A see KAL000
KANAMYCIN A SULFATE see KAM000
KANAMYCIN A SULFATE see KAV000
KANAMYCIN B see BAU270
KANAMYCIN B SULFATE see KBA100
KANAMYCIN B, SULFATE (1:1) (SALT) see KBA100
KANAMYCIN MONOSULFATE see KAM000
KANAMYCIN MONOSULFATE see KAV000
KANAMYCIN SULFATE see KAM000
KANAMYCIN SULFATE see KAM000
KANAMYCIN SULFATE see KAV000
KANAMYCIN SULFATE (1:1) SALT see KAV000
KANAMYTREX see KAL000
KANAMYTREX see KAM000
KANAMYTREX see KAV000
KANAQUA see KAM000
KANAQUA see KAV000
KANASIG see KAM000
KANASIG see KAV000
KANATROL see KAM000
KANATROL see KAV000
KANDISET see BCE500
KANECHLOR see PJL750
KANECHLOR 300 see PJL750
KANECHLOR 300 see PJO500
KANECHLOR 400 see PJL750
KANECHLOR 400 see PJO750
KANECHLOR 500 see PJP000
KANECHLOR 500 see PJP250
KANECHLOR C see PJP250
KANEKALON see ADY250
KANEKROL 500 see PAV600
KANENDOMYCIN see BAU270
KANENDOMYCIN SULFATE see KBA100
KANEPAR see TIY500
KANESCIN see KAM000
KANESCIN see KAV000
KANICIN see KAM000
KANNASYN see KAM000
KANNASYN see KAV000
KANNIT see XPJ000
KANO see KAM000
KANO see KAV000
KANOCHOL see DYE700
KANONE see MMD500
KANTEC see MAO275
KAN-TO-KA (JAPANESE) see CNH250
KANTREX see KAL000
KANTREX see KAM000
KANTREX see KAV000
KANTREXIL see KAM000
KANTREXIL see KAV000
KANTROX see KAM000
KANTROX see KAV000
KANZO (JAPANESE) see LFN300
KAOCHLOR see PLA500
KAOLIN see KBB600
KAON see PLG800
KAON-Cl see PLA500
KAON ELIXIR see PLG800
KAOPAOUS see KBB600
KAOPHILLS-2 see KBB600
KAPILIN see VTA100
KAPILON see VTA100
β,KAPPA-CAROTEN-6'-ONE, 3,3'-DIHYDROXY-, (3R,3'S,5'R)- see CBF760
KAPPADIONE see NAQ600
KAPPAXAN see MMD500
KAPPAXAN see VTA100

KAPREOMYCIN see CBF635
e-KAPROLAKTAM (CZECH) see CBF700
KAPROLIT see PJY500
KAPROLON see PJY500
KAPROMIN see PJY500
KAPRON see NOH000
KAPRON see PJY500
KAPRONAN DI-N-BUTYLCINICITY (CZECH) see BJX750
KAPRONAN DICYKLOHEXYLAMINU (CZECH) see DGU400
KAPRYLAN DI-N-BUTYLCINICITY (CZECH) see BLB250
KAPSOLAT see MEP750
KAPTAN see CBG000
KAPTAX see BDF000
KARAKHOL see KBB700
KARAMATE see DXI400
KARATAVIC ACID see KBB800
(+−)-KARATAVIC ACID see KBB800
KARATE see LAS200
KARAYA GUM see KBK000
KARBAM BLACK see FAS000
KARBAMOL see UTU500
KARBAMOL B/M see UTU500
KARBAM WHITE see BJK500
KARBANIL see PFK250
KARBARYL (POLISH) see CBM750
KARBASPRAY see CBM750
KARBATION see SIL550
KARBATION see VFU000
KARBATOX see CBM750
KARBATOX 75 see CBM750
KARBATOX ZAWIESINOWY see CBM750
KARBOFOS see MAK700
KARBOKROMEN (RUSSIAN) see CBR500
5-KARBOKSIMETIL-3-p-TOLIL-TIAZOLIDIN-2,4-DION-2-ACETOFENONHIDRAZON (CZECH) see CCH800
KARBORAFIN see CDI000
KARBOSEP see CBM750
1-(4'-KARBOXYLAMIDOFENYL)-3,3-DIMETHYLTRIAZENU (CZECH) see CCC325
1-(4'-KARBOXYLAMIDOPHENYL)-3,3-DIMETHYLTRIAZEN (GERMAN) see CCC325
KARBROMAL see BNK000
KARBUTILATE see DUM800
KARCON see MMD500
KARDIAMID see DJS200
KARDIN see TJL250
KARDONYL see DJS200
KARENZU DK2 see PFU300
KAREON see MMD500
KARIDIUM see SHF500
KARIGEL see SHF500
KARION see SKV200
KARI-RINSE see SHF500
KARLAN see RMA500
KARMESIN see HJF500
KARMEX see DXQ500
KARMEX DIURON HERBICIDE see DXQ500
KARMEX DW see DXQ500
KARMEX MONURON HERBICIDE see CJX750
KARMEX W. MONURON HERBICIDE see CJX750
KARMINOMYCIN see KBU000
KARMINOMYCIN HYDROCHLORIDE see KCA000
KARNOZZN see CCK665
KARO KAROUNDE ABSOLUTE see KCA050
KARPHOS see DJV600
KARSAN see FMV000
KARSULPHAN see SNR000
KARTAN see VBU100
KARTOFIN see NAK525
KARTRYL see BNK000

KARWINSKIA HUMBOLDTIANA see BOM125
KASAL see SEM305
KASEBON see CBW000
KASH, LEAF EXTRACT see KCA100
KASHMIRJA see CNT350
KASIMID see BBW750
KASSIA OEL (GERMAN) see CCO750
KASTAM see CNT350
KASTONE see HIB010
KASUGAMYCIN see AEB500
KASUGAMYCIN see KCA200
KASUGAMYCIN HYDROCHLORIDE see KCK000
KASUGAMYCIN MONOHYDROCHLORIDE see KCK000
KASUGAMYCIN PHOSPHATE see KCU000
KASUGAMYCIN SULFATE see KDA000
KASUMIN see KCK000
KASUMINL see KCA200
KAT 256 see CMW500
KATAGRIPPE see EEM000
KATALYSIN see AKK750
KATAMIN AB see QAT510
KATAMINE AB see AFP250
KATAMINE AB see DTC600
KATAMINE AB see QAT510
KATANOL C12 see BDJ600
KATAPOL O-12 see OBA100
KATAPOL OA-910 see PKE400
KATAPOL VP-532 see KDA012
KATAPYRIN see AFW500
KATCHUNG OIL see PAO000
KATEXOL 300 see KDA025
KATHA see CCP800
KATHON 886 see ISD066
KATHON BIOCIDE see ISD066
KATHON BIOCIDE see KDA035
KATHON CG see ISD066
KATHON LP PRESERVATIVE see OFE000
KATHON LX see ISD066
KATHON MW 886 BIOCIDE see KDA035
KATHON RH 886 see ISD066
KATHON SP 70 see OFE000
KATHON 886 W see ISD066
KATHON 886MW see ISD066
KATHON WT see ISD066
KATHRO see CMD750
KATIGEN BORDEAUX B-CF see CMS257
KATIGEN DEEP BLACK NND-CF see CMS250
KATIGEN DEEP BLACK RND-CF see CMS250
KATINE see NNM510
KATIV-G see MMD500
KATIV N see VTA000
KATIV POWDER see VTA100
KATLEX see CHJ750
KATONIL see CHX250
KATORIN see PLG800
KATOVIT HYDROCHLORIDE see PNS000
KATRON see BCA000
KATRON see PDN000
KATRONIAZID see PDN000
KAURAMIN 542 see MCB050
KAURAMIN 650 see MCB050
KAURAMIN 700 see MCB050
KAURAMIN 782 see MCB050
KAURAN-17-AL, 9,12,13,16-TETRAHYDROXY-11-OXO-, (12α)- see OOS100
KAURAN-17-AL, 9, 12,13,16-TETRAHYDROXY-11-OXO-, (12-α)- see OOS100
KAUR-16-EN-18-OIC ACID, 15-(ACETYLOXY)-, (4-α,15-β)- see XPJ300
KAURESIN K244 see UTU500
KAURIT 420 see UTU500
KAURIT 285 FL see UTU500
KAURITIL see CNK559
KAURIT M 70 see MCB050
KAURIT S see DTG700

KAUTSCHIN see MCC250
KAVAIN see GJI250
(+)-KAVAIN see GJI250
KAVITAMIN see MCB575
KAVITAN see MCB575
KAWAIN see GJI250
KAWITAN see MCB575
KAYA-ACE see DJB500
KAYACRYL GOLDEN YELLOW GL-ED see CMM895
KAYACYL BLUE BR see CMM070
KAYACYL PURE BLUE FGA see ERG100
KAYACYL SKY BLUE R see CMM080
KAYAESTER O see BSD100
KAYAESTER O 50 see BSD100
KAYAFUME see MHR200
KAYAHARD MCD see NAC000
KAYAHOPE see CJD400
KAYAKU ACID BRILLIANT SCARLET 3R see FMU080
KAYAKU ALIZARINE SKY BLUE R see CMM080
KAYAKU AMARANTH see FAG020
KAYAKU BENZOPURPURINE 4B see DXO850
KAYAKU BLUE B BASE see DCJ200
KAYAKU CONGO RED see SGQ500
KAYAKU DIRECT see CMO000
KAYAKU DIRECT BLACK BH see CMN800
KAYAKU DIRECT BRILLIANT BLUE RW see CMO600
KAYAKU DIRECT BROWN M see CMO800
KAYAKU DIRECT DARK GREEN B see CMO830
KAYAKU DIRECT DEEP BLACK EX see AQP000
KAYAKU DIRECT DEEP BLACK GX see AQP000
KAYAKU DIRECT DEEP BLACK S see AQP000
KAYAKU DIRECT FAST BLACK D see CMN230
KAYAKU DIRECT FAST RED F see CMO870
KAYAKU DIRECT FAST SCARLET 4BS see CMO870
KAYAKU DIRECT GREEN B see CMO840
KAYAKU DIRECT LEATHER BLACK EX see AQP000
KAYAKU DIRECT SCARLET 3B see CMO875
KAYAKU DIRECT SKY BLUE 5B see CMO500
KAYAKU DIRECT SKY BLUE 6B see CMN750
KAYAKU DIRECT SPECIAL BLACK AAX see AQP000
KAYAKU FAST RED 3GL BASE see KDA050
KAYAKU FOOD COLOUR YELLOW NO. 4 see FAG140
KAYAKU MORDANT YELLOW GG see SIT850
KAYAKU RED B BASE see NEQ000
KAYAKU RED RL BASE see MMF780
KAYAKU SCARLET G BASE see NMP500
KAYAKU SCARLET GG BASE see DEO295
KAYAKU SULPHUR BLACK BBR 200 see CMS250
KAYAKU SULPHUR BLACK BBX see CMS250
KAYAKU SULPHUR BLACK BDX see CMS250
KAYAKU SULPHUR BLACK BNX see CMS250
KAYAKU SULPHUR BLACK BX see CMS250
KAYAKU SULPHUR BLACK BX 200 see CMS250
KAYAKU SULPHUR BLACK BX 200 FLAKES see CMS250
KAYAKU SULPHUR BLACK G see CMS250
KAYAKU SULPHUR BLACK TB see CMS250
KAYAKU SULPHUR BLACK 3BX see CMS250

KAYAKU SULPHUR BORDEAUX 3B see CMS257
KAYAKU SULPHUR RED BROWN 9R see CMS257
KAYAKU TARTRAZINE see FAG140
KAYALON FAST BLUE BR see TBG700
KAYALON FAST BLUE FN see MGG250
KAYALON FAST RUBINE B see CMP080
KAYALON FAST SCARLET B see ENP100
KAYALON FAST VIOLET BB see AKP250
KAYALON FAST VIOLET BR see DBY700
KAYALON FAST YELLOW 4R see CMP090
KAYALON FAST YELLOW G see AAQ250
KAYALON FAST YELLOW RR see DUW500
KAYALON FAST YELLOW YL see KDA075
KAYALON POLYESTER BLUE EBL-E see CMP070
KAYALON POLYESTER BLUE EBL-E see DBQ220
KAYALON POLYESTER YELLOW 4R-E see CMP090
KAYALON POLYESTER YELLOW RF see CMP090
KAYALON POLYESTER YELLOW YLF see KDA075
KAYALON POLYESTER YELLOW YL-SE see KDA075
KAYANEX see MJN300
KAYANOL MILLING RED PG see CMM320
KAYANOL MILLING RED RS see CMM330
KAYANOL MILLING TURQUOISE BLUE 3G see CMM120
KAYANOL RED PG see CMM320
KAYANOL RED RS see NAO600
KAYAPHOR AB see CMP100
KAYAPHOR B see CMP100
KAYAPHOR LB see CMP100
KAYARAD TC 101 see TCS100
KAYARUS BLACK ARX see CMN300
KAYARUS BLACK G see CMN240
KAYARUS BLACK G CONC. see CMN240
KAYARUS TURQUOISE BLUE GL see COF420
KAYASET YELLOW G see AAQ250
KAYAZINON see DCM750
KAYAZOL see DCM750
KAY CIEL see PLA500
KAYDOL see MQV750
KAYDOL see MQV875
KAYEXALATE see SJK375
KAYKLOT see MMD500
KAYLITE see PKQ059
KAYQUINONE see MMD500
KAYTRATE see PBC250
KAYTWO see VTA650
KAYVISYN see AKX500
KAYVITE see VTA100
KAZOE see SFA000
KB see ISD066
KB-16 see MIH250
KB-53 see CCP875
KB 95 see BCP650
KB 101 see HAQ500
KB 227 see RGP450
KB-227 see TCY260
KB 227 see THJ825
KB-509 see FMR100
KB-944 see DIU500
KB 1585 see LEJ500
KB-2413 see EAL050
KBM 22 see DOE100
KBM 703 see TLC300
KBM 1003 see TLD000
KB (POLYMER) see SMQ500
KBR 3023 see MOT100
K-BRITE see SHR500
KBT-1585 see LEJ500
KC-400 see PJO750
KC-404 see KDA100
KC-500 see PJP000
KC 9147 see MQA000
KCA ACETATE CRIMSON B see CMP080

KCA ACETATE FAST YELLOW G see AAQ250
KCA ACID MILLING YELLOW M see CMM759
KCA CHROME ORANGE R see NEY000
KCA FOODCOL AMARANTH A see FAG020
KCA FOODCOL SUNSET YELLOW FCF see FAG150
KCA FOODCOL TARTRAZINE PF see FAG140
KCA METHYL ORANGE see MND600
KCA SILK ORANGE R see ADG000
KCA SILK RED G see CMM320
KCA TARTRAZINE PF see FAG140
KCO-1 see INE050
KD 83 see DXG625
KD-136 see HAG300
KDM see BAU270
K-DR see HJS900
KE see CNS800
KE 709 see RNU100
KEBILIS see CDL325
KEBUZONE see KGK000
KEDACILLIN see SNV000
KEDAVON see TEH500
KEESTAR see SLJ500
KEFENID see BDU500
KEFGLYCIN see CCR890
KEFLEX see ALV000
KEFLIN see SFQ500
KEFLODIN see TEY000
KEFORAL see ALV000
KEFZOL see CCS250
KEIMSTOP see CBL750
KELACID see AFL000
KELAMERAZINE see ALF250
KELCO GEL LV see SEH000
KELCOLOID see PNJ750
KELCOSOL see SEH000
KELENE see EHH000
KELEVAN see MDT625
KEL-F see KDK000
KEL-F see KDK000
KEL-F 3 see KDK000
KEL-F 81 see KDK000
KEL-F 90 see KDK000
KEL-F 200 see KDK000
KEL-F 6061 see KDK000
KEL-F 95/5 see KDK000
KELFIZIN see MFN500
KELGIN see SEH000
KELGUM see SEH000
KELICORIN see AHK750
KELINCOR see AHK750
KELOFORM see EFX000
KELP see KDK700
KEL-S see KDK000
KELSET see SEH000
KELSIZE see SEH000
KELTANE see BIO750
KELTEX see SEH000
KELTHANE (DOT) see BIO750
p,p'-KELTHANE see BIO750
KELTHANE DUST BASE see BIO750
KELTHANETHANOL see BIO750
KELTONE see SEH000
KEMADRINE see CPQ250
KEMAMIDE S see OAR000
KEMAMINE 9902D see DTC400
KEMAMINE P690 see DXW000
KEMAMINE P 989 see OHM700
KEMAMINE P990 see OBC000
KEMAMINE Q 9702C see QAT550
KEMAMINE QSML2 see QAT550
KEMAMINE S 190 see KDU100
KEMATE see DEV800
KEMDAZIN see MHC750
KEMESTER 105 see OHW000
KEMESTER 115 see OHW000
KEMESTER 205 see OHW000
KEMESTER 213 see OHW000
KEMI see ICC000

KEMICETINE see CDP250
KEMICETINE SUCCINATE see CDP725
KEMIKAL see CAT225
KEMITHAL see TES500
KEMITRACIN 10 see BAC260
KEMODRIN see DBA800
KEMOLATE see PHX250
KEMOVIRAN see MKW250
KEMPFEROL see ICE000
KEMPORE see ASM270
KEMPORE 125 see ASM270
KEMPORE R 125 see ASM270
KENACHROME BLACK 6B see CMP880
KENACHROME BLUE 2R see HJF500
KENACHROME ORANGE see NEY000
KENACHROME RED B see CMG750
KENACORT see AQX250
KENACORT-A see AQX500
KENACORT DIACETATE SYRUP see AQX750
KENALOG see AQX500
KENAPON see DGI400
KENDALL'S COMPOUND B see CNS625
KENDALL'S COMPOUND E see CNS800
KENDALL'S COMPOUND F see CNS750
KENGSHENGMYCIN see AEB500
KENROX 106 see PBP200
KEOBUTANE-JADE see KGK000
KEPHALIS see EES370
KEPHRINE see MGC350
KEPHTON see VTA000
KEPINOL see TKX000
KEPMPLEX 100 see EIV000
KEPONE see KEA000
KEPTAN see KEA300
KER 710 see EAL100
KERACYANIN see KEA325
KERALYT see SAI000
KERAPHEN see TDE750
KERASALICYL see SJO000
KERB see DTT600
KERECID see DAS000
KERLONE see KEA350
KERMAC 218A see RDK075
KERMAC 600W (MINERAL SEAL OIL) see DXG830
KERNECHTROT see AJQ250
KEROCAINE see AIT250
KEROPUR see BAV000
KEROSAL see SJO000
KEROSENE see KEK000
KEROSENE (PETROLEUM), hydrotreated see KEK100
KEROSINE see KEK000
KEROSINE (petroleum) see KEK000
KEROSINE, HYDRODESULFURIZED see KEK110
KESELAN see CLY500
KESSAR see TAD175
KESSCO 40 see OAV000
KESSCO BSC see BSL600
KESSCOCIDE see ISN000
KESSCOFLEX see BHK000
KESSCOFLEX BS see BSL600
KESSCOFLEX MCP see DOF400
KESSCOFLEX TRA see THM500
KESSCO ICS see IKC050
KESSCO ISOPROPYL see IQW000
KESSCO ISOPROPYL MYRISTATE see IQN000
KESSCOMIR see IQN000
KESSOBAMATE see MQU750
KESSODANTEN see DKQ000
KESSODRATE see CDO000
KESTREL (Pesticide) see AHJ750
KESTRIN see ECU750
KESTRONE see EDV000
KETAJECT see CKD750
KETALAR see CKD750
KETALGIN see MDO750
KETALGIN HYDROCHLORIDE see MDP750
KETAMAN see HKR500

KETAMINE see CKD750
KETAMINE see KEK200
KETAMINE HYDROCHLORIDE see CKD750
KETANEST see CKD750
KETANRIFT see ZVJ000
KETASET see CKD750
KETASON see KGK000
KETAVET see CKD750
KETAZONE see KGK000
KETENE see KEU000
KETENE DIMER see KFA000
KETHOXAL-BIS-THIOSEMICARBAZIDE see KFA100
KETJENBLACK EC see CBT750
KETJENFLEX 8 see ENY200
Δ-KETO 153 see KFK200
3-(3-KETO-7-α-ACETYLTHIO-17-β-HYDROXY-4-ANDROSTEN-17-α-YL)PROPIONIC ACID LACTONE see AFJ500
4-KETOAMYLTRIMETHYLAMMONIUM IODIDE see TLY250
KETOBEMIDONE see KFK000
4-KETOBENZOTRIAZINE see BDH000
KETOBUN-A see ZVJ000
β-KETOBUTYRANILIDE see AAY000
KETOCAINE HYDROCHLORIDE see DNN000
KETOCHOL see DAL000
KETOCHOLANIC ACID see DAL000
7-KETOCHOLESTEROL see ONO000
KETOCONAZOL see KFK100
KETOCONAZOLE see KFK100
KETOCYCLOHEPTANE see SMV000
KETOCYCLOPENTANE see CPW500
4-KETOCYCLOPHOSPHAMIDE see KFK125
KETODERM see KFK100
KETODESTRIN see EDV000
1-KETO-2,3-DIHYDROCYCLOPENTINDOLE OXIME see KFK150
Δ-KETOENDRIN see KFK200
2-KETO-3-ETHOXY-BUTYRALDEHYDE-BIS(THIOSEMICARBAZONE) see KFA100
KETO-ETHYLENE see KEU000
KETOGAZE see MGC350
α-KETOGLUTARIC ACID, DISODIUM SALT see KFK300
3-KETO-l-GULOFURANOLACTONE see ARN000
3-KETO-d-GULOFURANOLACTOSE see EDE700
KETOHEPTAMETHYLENE see SMV000
KETOHEXAMETHYLENE see CPC000
2-KETOHEXAMETHYLENIMINE see CBF700
KETOHYDROXY-ESTRATRIENE see EDV000
KETOHYDROXYESTRIN see EDV000
KETOHYDROXYESTRIN BENZOATE see EDV500
KETOHYDROXYOESTRIN see EDV000
2,3-KETOINDOLINE see ICR000
KETOISDIN see KFK100
KETOLAR see CKD750
KETOLE see ICM000
KETOLIN-H see CJU250
γ-KETO-β-METHOXY-Δ-METHYLENE-Δ^α-HvEXENOIC ACID see PAP750
15-KETO-20-METHYLCHOLANTHRENE see MIM250
KETONE, 1-ADAMANTYLDIAZO METHYL see DAB807
KETONE, 2-AMINO-5-BENZIMIDAZOLYL PHENYL see AIH000
KETONE, (4-AMINO-5-(o-CHLOROPHENYL)-2-METHYLPYRROL-3-YL) METHYL see AJI300
KETONE, (4-AMINO-5-(p-CHLOROPHENYL)-2-METHYLPYRROL-3-YL) METHYL see AJI330

KETONE, (4-AMINO-5-(3,4-DICHLOROPHENYL)-2-METHYLPYRROL-3-YL) METHYL see AJM600
KETONE, (4-AMINO-5-(3,4-DIMETHOXYPHENYL)-2-METHYLPYRROL-3-YL) METHYL see AJO800
KETONE, (4-AMINO-1,2-DIMETHYL-5-PHENYLPYRROL-3-YL) METHYL see AJR100
KETONE, 5-AMINO-1,3-DIMETHYLPYRAZOL-4-YL o-FLUOROPHENYL see AJR400
KETONE, (4-AMINO-1-ETHYL-2-METHYL-5-PHENYLPYRROL-3-YL) METHYL see AKA600
KETONE, (4-AMINO-5-(m-FLUOROPHENYL)-2-METHYLPYRROL-3-YL) METHYL see AKC550
KETONE, (4-AMINO-5-(p-FLUOROPHENYL)-2-METHYLPYRROL-3-YL) METHYL see AKC560
KETONE, (4-AMINO-5-(2-METHOXYPHENYL)-2-METHYLPYRROL-3-YL) METHYL see AKO450
KETONE, (4-AMINO-5-(m-METHOXYPHENYL)-2-METHYLPYRROL-3-YL) METHYL see AKO400
KETONE, (4-AMINO-5-(p-METHOXYPHENYL)-2-METHYLPYRROL-3-YL) METHYL see AKO430
KETONE, (4-AMINO-5-(4-METHOXY-m-TOLYL)-2-METHYLPYRROL-3-YL) METHYL see AKO100
KETONE, (4-AMINO-2-METHYL-5-PHENYLPYRROL-3-YL) METHYL, MONOHYDROCHLORIDE see ALA300
KETONE, (4-AMINO-2-METHYL-5-(m-TOLYL)PYRROL-3-YL) METHYL see AKT830
KETONE, (4-AMINO-2-METHYL-5-(o-TOLYL)PYRROL-3-YL) METHYL see AKT800
KETONE, (4-AMINO-2-METHYL-5-(p-TOLYL)PYRROL-3-YL) METHYL see AKT850
KETONE, (4-AMINOPIPERIDINO)METHYL INDOL-3-YL see AMA100
KETONE, p-ANISYL 3-PYRIDYL see MED100
KETONE, p-ANISYL 4-PYRIDYL see MFH930
KETONE, 9-ANTHRACENYL PHENYL see APH600
KETONE, 3-AZABICYCLO(3.2.2)NONYL CHLOROMETHYL see CEC100
KETONE, 3-AZABICYCLO(3.2.2)NONYL IODOMETHYL see IDZ100
KETONE, 1-AZIRIDINYL 3-(BIS(2-CHLOROETHYL)AMINO)-p-TOLYL see BHQ760
KETONE, 2-BENZOFURANYL METHYL see ACC100
KETONE, 2H-1-BENZOPYRAN-3-YL METHYL see BCS500
KETONE, (1,2,4-BENZOTRIAZIN-3-YL)METHYL 1-PYRROLIDINYL see PPS600
KETONE, 1-BENZYL-2-INDOLYL HYDROXYMETHYL- see BES300
KETONE, 5-BENZYLOXY-3-INDOLYL 4-PIPERIDYL see BFC200
KETONE, 4-BENZYLPIPERAZINYL β-(p-CHLOROPHENYL)PHENETHYL see BFE770
KETONE, BENZYL(4-PYRIDYL), THIOSEMICARBAZONE see BFH100
KETONE, BICYCLO(4.2.0)OCTA-1,3,5-TRIEN-7-YL BENZYL see BFZ100
KETONE, BICYCLO(4.2.0)OCTA-1,3,5-TRIEN-7-YL BENZYL, OXIME see BFZ110
KETONE, BICYCLO(4.2.0)OCTA-1,3,5-TRIEN-7-YL BUTYL see BCH800
KETONE, BICYCLO(4.2.0)OCTA-1,3,5-TRIEN-7-YL METHYL see BFZ120

KETONE, BICYCLO(4.2.0)OCTA-1,3,5-TRIEN-7-YL METHYL, O-ACETYLOXIME see BFZ130
KETONE, BICYCLO(4.2.0)OCTA-1,3,5-TRIEN-7-YL METHYL, O-ALLYLOXIME see BFZ140
KETONE, BICYCLO(4.2.0)OCTA-1,3,5-TRIEN-7-YL METHYL, O-BUTYLOXIME see BFZ150
KETONE, BICYCLO(4.2.0)OCTA-1,3,5-TRIEN-7-YL METHYL, O-(2-HYDROXYPROPYL)OXIME see HNT550
KETONE, BICYCLO(4.2.0)OCTA-1,3,5-TRIEN-7-YL METHYL, O-METHYLOXIME see MFX550
KETONE, BICYCLO(4.2.0)OCTA-1,3,5-TRIEN-7-YL METHYL, OXIME see BFZ160
KETONE, BICYCLO(4.2.0)OCTA-1,3,5-TRIEN-7-YL PENTYL, OXIME see BFZ170
KETONE, BICYCLO(4.2.0)OCTA-1,3,5-TRIEN-7-YL PHENYL see BFZ180
KETONE, 4-BIPHENYL ETHYL see BGM100
KETONE, 1-(1,1'-BIPHENYL)-4-YL-2-((4-(DICHLOROACETYL)PHENYL)AMINO)-2-HYDROXY- see BGL400
KETONE, m-(BIS(2-CHLOROETHYL)AMINO)PHENYL PIPERIDINO see BHP150
KETONE, 3-(BIS(2-CHLOROETHYL)AMINO)-p-TOLYL MORPHOLINO- see BHR400
KETONE, 3-(BIS(2-CHLOROETHYL)AMINO)-p-TOLYL PIPERIDYL- see BIA100
KETONE, 3-BROMO-1-ADAMANTYL DIAZOMETHYL see EEE025
KETONE, BROMOMETHYL 4-(DIPHENYLHYDROXYMETHYL)PIPERIDINO see BMS300
KETONE, 7-(3-(sec-BUTYLAMINO)-2-HYDROXYPROPOXY)-2-BENZOFURANYL METHYL see ACN310
KETONE, 4-(3-(tert-BUTYLAMINO)-2-HYDROXYPROPOXY)-2-BENZOFURANYL METHYL see ACN300
KETONE, 7-(3-(tert-BUTYLAMINO)-2-HYDROXYPROPOXY)-2-BENZOFURANYL METHYL see ACN320
KETONE, 4-(p-tert-BUTYLBENZYL)PIPERAZINYL β-(p-CHLOROPHENYL)PHENETHYL see BQK800
KETONE, 4-(p-tert-BUTYLBENZYL)PIPERAZINYL 3,4,5-TRIMETHOXYPHENYL see BQK830
KETONE, t-BUTYL METHYL see DQU000
KETONE, BUTYL 4-PYRIDYL see VBA100
KETONE, 3-CHLORO-1-ADAMANTYL DIAZOMETHYL see CEF800
KETONE, CHLOROMETHYL 4-(DIPHENYLHYDROXYMETHYL)PIPERIDINO see CEC300
KETONE, CHLOROMETHYL 2-FLUORENYL see CEC700
KETONE, 1-(4-CHLORO-3-NITROPHENYL)-2-ETHOXY-2-((4-(METHYLTHIO)PHENYL)AMINO)- see CJD630
KETONE, 1-(3-CHLOROPHENYL)-2-((4-(DICHLOROACETYL)PHENYL)AMINO)-2-HYDROXY- see CJU300
KETONE, m-CHLOROPHENYL 2-FLUORENYL see CKA100
KETONE, p-CHLOROPHENYL 1-(2-MORPHOLINOETHYL)PYRROL-2-YL, HYDROCHLORIDE see CEO100
KETONE, β-(p-CHLOROPHENYL)PHENETHYL 4-(o-CHLOROPHENYL)PIPERAZINYL see CKI020
KETONE, β-(p-CHLOROPHENYL)PHENETHYL 4-(2-

HYDROXYPROPYL)PIPERAZINYL see CKI180

KETONE, β-(p-CHLOROPHENYL)PHENETHYL 4-(o-METHOXYPHENYL)PIPERAZINYL see CKI185

KETONE, β-(p-CHLOROPHENYL)PHENETHYL 4-(m-METHYLBENZYL)PIPERAZINYL see CKI030

KETONE, β-(p-CHLOROPHENYL)PHENETHYL 4-PHENETHYLPIPERAZINYL see CKI040

KETONE, β-(p-CHLOROPHENYL)PHENETHYL 4-(2-PYRIDYL)PIPERAZINYL see CKI050

KETONE, β-(p-CHLOROPHENYL)PHENETHYL 4-(2-PYRIMIDYL)PIPERAZINYL see CKI060

KETONE, β-(p-CHLOROPHENYL)PHENETHYL 4-(2-THIAZOLYL)PIPERAZINYL see CKI070

KETONE, β-(p-CHLOROPHENYL)PHENETHYL 4-(m-TOLYL)PIPERAZINYL see CKI080

KETONE, β-(p-CHLOROPHENYL)PHENETHYL 4-(o-TOLYL)PIPERAZINYL see CKI090

KETONE, β-(p-CHLOROPHENYL)PHENETHYL 4-(p-TOLYL)PIPERAZINYL see CKI190

KETONE, 4-(o-CHLOROPHENYL)-1-PIPERAZINYLMETHYL 3-PYRIDYL- see KGK120

KETONE, 4-(p-CHLOROPHENYL)PIPERAZINYL 3,4,5-TRIMETHOXYPHENYL see CKJ100

KETONE, 5-CHLORO-2-THIENYL METHYL see CDN525

KETONE, CYCLOHEXYL 4-PYRIDYL see CPI375

KETONE, CYCLOPENTYL 3,4-DIHYDROXYPHENYL see CQA100

KETONE, CYCLOPROPYL 4-((2-HYDROXY-3-(1,2-DIHYDRO-2-IMINO-4-METHYLPYRIDINO)PROPOXY)PHENYL) see PMM300

KETONE, DIBENZOFURAN-2-YL PHENYL see DDC100

KETONE, 2-DIBENZOTHIENYL METHYL see ACH090

KETONE, 2-((4-(DIBROMOACETYL)PHENYL)AMINO)-2-ETHOXY-1-(4-NITROPHENYL)- see DDJ850

KETONE, 3,5-DIBROMO-4-HYDROXYPHENYL 2-MESITYL-3-BENZOFURANYL see DDP300

KETONE, 2-((4-(DICHLOROACETYL)PHENYL)AMINO)-2-ETHOXY-1-(4-NITROPHENYL)- see DEN820

KETONE, 2-((4-(DICHLOROACETYL)PHENYL)AMINO)-2-HYDROXY-1-(4-METHOXYPHENYL)- see DEN840

KETONE, 2-((4-(DICHLOROACETYL)PHENYL)AMINO)-2-HYDROXY-1-(4-METHYLPHENYL)- see DEN860

KETONE, 2-((4-(DICHLOROACETYL)PHENYL)AMINO)-2-HYDROXY-1-(4-PHENOXYPHENYL)- see DEN880

KETONE, 2-((4-(DICHLOROACETYL)PHENYL)AMINO)-2-HYDROXY-1-PHENYL- see DEN900

KETONE, 2-((4-(DICHLOROACETYL)PHENYL)AMINO)-2-HYDROXY-1-(4-(PHENYLTHIO)PHENYL)- see DEN910

KETONE, 4,5-DICHLOROPYRROL-2-YL 2,6-DIHYDROXYPHENYL see POK400

KETONE, p-(2-(DIETHYLAMINO)ETHOXY)PHENYL 2-ETHYL-3-BENZOFURANYL see EDV700

KETONE, p-(2-(DIETHYLAMINO)ETHOXY)PHENYL 2-MESITYL-3-BENZOFURANYL see DHR900

KETONE, 1-(2-(DIETHYLAMINO)ETHYL)-2-(p-ETHOXYBENZYL)-5-BENZIMIDAZOLYL METHYL see DHZ050

KETONE, 1-(2-(DIETHYLAMINO)ETHYL)-2-p-PHENETIDINO-5-BENZIMIDAZOLYL METHYL see DIE350

KETONE, DIETHYLAMINO(1,2,3,4-TETRAHYDRO-9-ACRIDINYL) see ADJ550

KETONE, DIETHYLAMINO(7,8,9,10-TETRAHYDRO-11-6H-CYCLOHEPTA(b)QUINOLINYL) see TCO100

KETONE, 2,3-DIHYDRO-1,4-BENZODIOXIN-6-YL 4-FLUOROPHENYL see DKT500

KETONE, 3,5-DIIODO-4-HYDROXYPHENYL 2,5-DIMETHYL-3-FURYL see DNF450

KETONE, 3,5-DIIODO-4-HYDROXYPHENYL 2-ETHYL-3-BENZOFURANYL see EID200

KETONE, 3,5-DIIODO-4-HYDROXYPHENYL 5-ETHYL-2-FURYL see EID250

KETONE, 3,5-DIIODO-4-HYDROXYPHENYL 2-FURYL see DNF500

KETONE, 3,5-DIIODO-4-HYDROXYPHENYL 2-MESITYL-3-BENZOFURANYL see DNF550

KETONE, 4,7-DIMETHOXY-2-BENZOFURANYL METHYL see DOA810

KETONE, 6,7-DIMETHOXY-2-BENZOFURANYL METHYL see DOA815

KETONE, 5,8-DIMETHOXY-2H-1-BENZOPYRAN-3-YL METHYL see DOA820

KETONE, 7,8-DIMETHOXY-2H-1-BENZOPYRAN-3-YL METHYL see DOA830

KETONE, DIMETHYL see ABC750

KETONE, 5-DIMETHYLAMINO-3-INDOLYL PHENYL see DOU700

KETONE, (4-(DIMETHYLAMINO)-2-METHYL-5-PHENYLPYRROL-3-YL) METHYL see DPL950

KETONE, 10-(2-(DIMETHYLAMINO)PROPYL)PHENOTHIAZIN-2-YL METHYL see AAF800

KETONE, 3,3-DIMETHYL-4-(DIMETHYLAMINO)-4-(o-METHOXYPHENYL)BUTYL o-METHOXYPHENYL see DRL450

KETONE, 3,3-DIMETHYL-4-(DIMETHYLAMINO)-4-(p-METHOXYPHENYL)BUTYL p-METHOXYPHENYL see DRL460

KETONE, 3,3-DIMETHYL-4-(DIMETHYLAMINO)-4-(m-TOLYL)BUTYL m-TOLYL see DRM100

KETONE, 3,3-DIMETHYL-4-(DIMETHYLAMINO)-4-(o-TOLYL)BUTYL o-TOLYL see DRM110

KETONE, 3,3-DIMETHYL-4-(DIMETHYLAMINO)-4-(p-TOLYL)BUTYL p-TOLYL see DRM120

KETONE, 2,5-DIMETHYL-3-FURYL p-HYDROXYPHENYL see DSC100

KETONE, 1,3-DIMETHYL-5-(METHYLAMINO)-4-PYRAZOLYL o-FLUOROPHENYL see DSO500

KETONE, 3,5-DIMETHYL-4-NITROSO-1-PIPERAZINYL METHYL see NJI850

KETONE, (2,4-DIMETHYL-5-PHENYLPYRROL-3-YL) METHYL see DTO300

KETONE, 2,5-DIMETHYLPYRROL-3-YL METHYL see ACI550

KETONE, 2,5-DIMETHYL-3-THIENYL METHYL see DUG425

KETONE, 4-(DIPHENYLHYDROXYMETHYL)PIPERIDINO IODOMETHYL see IDZ200

KETONE, DI-1H-2-PYRROLYL see PPY300

KETONE, p-ETHOXYPHENYL 2-PYRIDYL see EFC700

KETONE, p-ETHOXYPHENYL 3-PYRIDYL see EFC800

KETONE, (p-ETHOXYPHENYL) 4-PYRIDYL see EEN600

KETONE, 4-ETHYL-3,5-DIMETHYLPYRROL-2-YL METHYL see EIL200

KETONE, 2-ETHYL-7-(2-HYDROXY-3-(ISOPROPYLAMINO)PROPOXY)-4-BENZOFURANYL METHYL see ELI600

KETONE, ETHYL 4-(m-HYDROXYPHENYL)-1-METHYLPIPERIDYL see KFK000

KETONE, 1-ETHYL-3-INDOLYL HYDROXYMETHYL see ELD100

KETONE, ETHYL 4-PYRIDYL see PMX300

KETONE, 3-(3-ETHYLSULFONYL)PENTYL PIPERIDINO see EPI400

KETONE, ETHYL VINYL see PBR250

KETONE, 2-FLUORENYL p-METHOXYPHENYL see MEC600

KETONE, FLUOREN-2-YL METHYL see FEI200

KETONE, FLUOREN-2-YL PHENYL see FEM100

KETONE, 2-FURYL p-HYDROXYPHENYL see FQL200

KETONE, 2-FURYL METHYL see ACM200

KETONE, 3-FURYL PHENYL see FQO050

KETONE, 4-HEPTYL PIPERIDINO see PNX800

KETONE, HEPTYL 4-PYRIDYL see HBF600

KETONE, α-HYDROXYBENZYL PHENYL see BCP250

KETONE, 7-(2-HYDROXY-3-(ISOPROPYLAMINO)PROPOXY)-2-BENZOFURANYL METHYL see HLK600

KETONE, HYDROXYMETHYL 2-INDOLYL- see HIS100

KETONE, HYDROXYMETHYL 5-INDOLYL see HIS110

KETONE, HYDROXYMETHYL 1-METHYL-2-INDOLYL- see HIS120

KETONE, HYDROXYMETHYL 1-METHYL-3-INDOLYL see HIS140

KETONE, HYDROXYMETHYL 2-METHYL-3-INDOLYL see HIS150

KETONE, HYDROXYMETHYL 3-METHYL-2-INDOLYL- see HIS130

KETONE, (4-HYDROXY-2-METHYL-5-PHENYL-1H-PYRROL-3-YL) METHYL see HMN100

KETONE, 6-HYDROXY-2-NAPHTHYL PHENYL see BDN200

KETONE, p-HYDROXYPHENYL 2-MESITYL-3-BENZOFURANYL see HNK700

KETONE, p-HYDROXYPHENYL 2-MESITYL-4-BENZOFURANYL see HNK800

KETONE, (p-HYDROXYPHENYL) 2-PYRIDYL see HJN700

KETONE, 8-HYDROXY-5-QUINOLYL METHYL see HOE200

KETONE, 3-HYDROXY-2,2,5,5-TETRAMETHYLTETRAHYDRO-3-FURYL METHYL- see HOI400

KETONE, 5-INDOLINYL METHYL see ACO320

KETONE, 2-INDOLYL METHOXYMETHYL- see ICW200

KETONE, INDOL-5-YL METHYL see ACO300

KETONE, 3-INDOLYL MORPHOLINOMETHYL see ICZ150

KETONE, 3-INDOLYL PIPERIDINOMETHYL see ICZ200

KETONE, 3-INDOLYL 4-PIPERIDYL see ILG200
KETONE, 1-(2-ISOQUINOLYL)-5-METHYL-4-PYRAZOLYL METHYL see IRX100
KETONE, 2-MESITYL-3-BENZOFURANYL p-METHOXYPHENYL see AOY300
KETONE, 8-METHOXY-2H-1-BENOZPYRAN-3-YL METHYL see ACQ760
KETONE, 5-METHOXY-2-BENZOFURANYL METHYL see MEC330
KETONE, 6-METHOXY-2-BENZOFURANYL METHYL see MEC300
KETONE, 7-METHOXY-2-BENZOFURANYL METHYL see MEC320
KETONE, 5-METHOXY-2H-1-BENZOPYRAN-3-YL METHYL see MEC340
KETONE, 6-METHOXY-2H-1-BENZOPYRAN-3-YL METHYL see ACQ700
KETONE, 7-METHOXY-2H-1-BENZOPYRAN-3-YL METHYL see ACQ730
KETONE, METHOXYMETHYL 1-METHYL-2-INDOLYL see MDX300
KETONE, METHOXYMETHYL 3-METHYL-2-INDOLYL- see MDX310
KETONE, (4-METHOXY-2-METHYL-5-PHENYLPYRROL-3-YL) METHYL see MEX275
KETONE, 7-METHOXYNAPHTHO(2,1-b)FURAN-2-YL METHYL see ACQ790
KETONE, 4-(o-METHOXYPHENYL)-1-PIPERAZINYLMETHYL 3-PYRIDYL- see KGK130
KETONE, 4-(o-METHOXYPHENYL)PIPERAZINYL 3,4,5-TRIMETHOXYPHENYL see MFH760
KETONE, 4-(p-METHOXYPHENYL)PIPERAZINYL 3,4,5-TRIMETHOXYPHENYL see MFH770
KETONE, p-METHOXYPHENYL 2-PYRIDYL see MFH900
KETONE, (p-METHOXYPHENYL) 3-PYRIDYL see MED100
KETONE, (p-METHOXYPHENYL) 4-PYRIDYL see MFH930
KETONE, 10-(3-(4-METHOXYPIPERIDINO)PROPYL)PHENOTHIAZIN-2-YL METHYL see MFJ200
KETONE, 2-METHYLCYCLOHEXYL 4-METHYLPIPERIDINO see MLM700
KETONE, 2-METHYLCYCLOPROPYL HYDRAZINO see MIV300
KETONE, 2-METHYL-2-IMIDAZOLIN-1-YL 3,4,5-TRIMETHOXYPHENYL- see IAT200
KETONE, METHYL ISOAMYL see MKW450
KETONE, METHYL 2-METHYL-1,3-DITHIOLAN-2-YL see ACR050
KETONE, METHYL 5-METHYL-3-(5-NITRO-2-FURYL)-4-ISOXAZOLYL see MLP300
KETONE, METHYL 10-(3-(4-METHYL-1-PIPERAZINYL)PROPYL)PHENOTHIAZIN-2-YL see ACR500
KETONE, METHYL 5-METHYL-1-(2-QUINOLYL)-4-PYRAZOLYL see MLW600
KETONE, METHYL 5-METHYL-1-(2-QUINOXALINYL)-4-PYRAZOLYL see MLW630
KETONE, METHYL 5-METHYL-2-THIENYL see MPR300
KETONE, METHYL 10-(3-MORPHOLINOPROPYL)PHENOTHIAZIN-2-YL see MMA600
KETONE, METHYL 3-(5-NITRO-2-FURYL)-5-PHENYL-4-ISOXAZOLYL seee MMJ955
KETONE, METHYL 2-NITRONAPHTHO(2,1-b)FURAN-7-YL see NHR100
KETONE, METHYL (4-NITRO-2-PYRROLYL) see ACT300
KETONE, METHYL (5-NITRO-2-PYRROLYL) see ACT330

KETONE, METHYL (5-NITRO-2-THIENYL) see NML100
KETONE-METHYL-5-OXO-5H-(1)BENZOPYRANO(2,3-b)PYRIDYL see ACU125
KETONE METHYL PHENYL see ABH000
KETONE, (2-METHYL-5-PHENYLPYRROL-3-YL) METHYL see MNY800
KETONE, 2-METHYLPIPERIDINO 2-NAPHTHYL see MOH290
KETONE, METHYL 10-(3-PIPERIDINOPROPYL)PHENOTHIAZIN-2-YL see MOL300
KETONE, METHYL 10-(3-PIPERIDINOPROPYL)PHENOXAZIN-2-YL see MOL400
KETONE, 1-METHYL-3-PIPERIDYL PIPERIDINO see MOH310
KETONE, METHYL 2-PYRIDYL see PPH200
KETONE, METHYL 4-PYRIDYL see ADA365
KETONE, METHYL 1,4,5,6-TETRAHYDRO-2-METHYLCYCLOPENTA(b)PYRROL-3-YL see MPO800
KETONE, (α-METHYL-m-TRIFLUOROMETHYLPHENETHYLAMINO METHYL) PIPERIDINO see MQE100
KETONE, METHYL 2,4,5-TRIMETHYLPYRROL-3-YL see ADE050
KETONE, MORPHOLINO(1,2,3,4-TETRAHYDRO-9-ACRIDINYL) see MRR760
KETONE, PENTYL 4-PYRIDYL see HEW050
KETONE, PENTYL 4-PYRIDYL see PBX800
KETONE, 1-PHENETHYL-3-PIPERIDYL PIPERIDINO see PDI550
KETONE, 4-PHENYL-1-PIPERAZINYLMETHYL-3-PYRIDYL- see NDW520
KETONE, 4-PHENYLPIPERAZINYL 3,4,5-TRIMETHOXYPHENYL see PFX600
KETONE, PHENYL (1-PIPERIDINOCYCLOHEXYL) see PFY200
KETONE, PHENYL 2-PYRIDYL see PGE760
KETONE, PHENYL 3-PYRIDYL see PGE765
KETONE, PHENYL 4-PYRIDYL see PGE768
KETONE, PHENYL PYRROL-2-YL see PGF900
KETONE, PIPERIDINO 3-PIPERIDYL see PIR200
KETONE, PIPERIDINO(1,2,3,4-TETRAHYDRO-9-ACRIDINYL) see PIU100
KETONE PROPANE see ABC750
KETONE, PROPYL 4-PYRIDYL see PNV755
KETONE, 4-(2-PYRIDYL)PIPERAZINYL 3,4,5-TRIMETHOXYPHENYL see TKY300
KETONE, 4-(2-PYRIMIDYL)PIPERAZINYL 3,4,5-TRIMETHOXYPHENYL see PPP550
KETONES see KGA000
KETONE, 4-(2-THIAZOLYL)PIPERAZINYL 3,4,5-TRIMETHOXYPHENYL see TEX220
KETONE, 4-(m-TOLYL)PIPERAZINYL 3,4,5-TRIMETHOXYPHENYL see THF300
KETONE, 4-(p-TOLYL)PIPERAZINYL 3,4,5-TRIMETHOXYPHENYL see THF310
4-KETONIRIDAZOLE see KGA100
KETONOX see MKA500
KETOPENTAMETHYLENE see CPW500
4-KETOPENTANOIC ACID BUTYL ESTER see BRR700
4-KETOPENTENAL see KGA200
α-KETOPHENYLACETIC ACID see OOK150
KETOPHENYLBUTAZONE see KGK000
KETOPROFEN see BDU500
KETOPROFEN SODIUM see KGK100
KETOPRON see BDU500
β-KETOPROPANE see ABC750
1-KETOPROPIONALDEHYDE see PQC000
2-KETOPROPIONALDEHYDE see PQC000
α-KETOPROPIONALDEHYDE see PQC000
α-KETOPROPIONIC ACID see PQC100
1-(2-KETO-2-(3'-PYRIDYL)ETHYL)-4-(2'-CHLOROPHENYL)PIPERAZINE see KGK120

1-(2-KETO-2-(3'-PYRIDYL)ETHYL)-4-(2'-METHOXYPHENYL)PIPERAZINE see KGK130
1-(2-KETO-2-(3'-PYRIDYL)ETHYL)-4-(PHENYL)PIPERAZINE see NDW520
2-KETOPYRROLIDINE-1-YLACETAMIDE see NNE400
2-KETO-4-QUINAZOLINONE see QEJ800
KETORUBRATOXIN B see KGK140
4-KETOSTEARIC ACID see KGK150
l-3-KETOTHREOHEXURONIC ACID LACTONE see ARN000
KETOTIFEN FUMARATE see KGK200
8-KETOTRICYCLO(5.2.1.0^{2,6})DECANE see OPC000
2-KETO-1,7,7-TRIMETHYLNORCAMPHANE see CBA750
4-KETOVALERIC ACID see LFH000
γ-KETOVALERIC ACID see LFH000
KET RED 309 see DTV360
KEUTEN see DEE600
KEVADON see TEH500
KEVLAR 29, MONOMER-BASED see BBO750
KEVLAR 49, MONOMER-BASED see BBO750
KEVLAR 149, MONOMER-BASED see BBO750
KEVLAR 49, SRU see BBO750
KEY-SERPINE see RDK000
KEY-TUSSCAPINE see NOA000
KEY-TUSSCAPINE HYDROCHLORIDE see NOA500
KF-32 see TBT175
KF-868 see KGK300
KF 993 see PJR300
KF 994 see OCE100
KF 995 see DAF350
KF-1820 see DOQ400
KF 2 (HERBICIDE) see DOR800
K-FLEBO see PKV600
K-FLEX DP see DWS800
KFP 3 see SLJ550
KF RED 1 see DTV360
KFS see UTU500
KG-2413 see EAL050
K-GRAN see PLA000
KH 360 see TGG760
KHAINI (INDIA) see SED400
KHALADON 22 see CFX500
KHAROPHEN see ABX500
KHAT LEAF EXTRACT see KGK350
KHE 0145 see MIA250
KHIMCOCCID see RLK890
KHIMCOECID see RLK890
KHIMKOKTSID see RLK890
KHIMKOKTSIDE see RLK890
KHINALIZARIN see TDD000
KHINGAMIN see CLD250
KHINOTILIN see KGK400
KHIZLEVUDITE see HAK050
KHLADON 32 see MJQ300
KHLADON 113 see FOO000
KHLADON 125 see PBD400
KHLADON 744 see CBU250
KHLADON 122A see DKI900
KHLADON 114B2 see FOO525
KHLORAKON see BEG000
KHLORIDIN see TGD000
KHLORTRIANIZEN see CLO750
KHLOTAZOL see CMB675
KHOMECIN see ZJS300
KHOMEZIN see ZJS300
KHP 2 see AHE250
KHROMOLAN see CMH300
K-IAO see PLG800
KIATRIUM see DCK759
KICKER FK 101 see KGK500
KID KILL see MRU359
KIDNEY BEAN TREE see WCA450
KIDOLINE see VGP000
KIEFERNADEL OEL (GERMAN) see PIH500

KIESELGUHR see DCJ800
KIESELSAEURE (GERMAN) see SCL000
KIEZELFLUORWATERSTOFZUUR
(DUTCH) see SCO500
K III see DUS700
KIKKER FK 101 see KGK500
KIKUTHRIN see PMN700
KILACAR see PMF550
KILDIP see DGB000
KILEX see CNK559
KILEX 3 see BSQ750
KILL-ALL see SEY500
KILLAX see TCF250
KILL COW see PJJ300
KILLEEN see CCL500
KILL KANTZ see AQN635
KILMAG see ARB750
KILMITE 40 see TCF250
KILMOR see TKU680
KILOSEB see BRE500
KILPROP see CIR500
KILRAT see ZLS000
KILSEM see CIR250
KILVAL see MJG500
KIM-112 see CPB065
KIMAVOXYL see CBK500
KIN-804 see MFB370
KIN-844 see MFB380
KINADION see VTA000
KINAVOSYL see GGS000
KINEKS see AKO500
KINETIN see FPT100
KINETIN (PLANT HORMONE) see FPT100
KINEX see AKO500
KINGCOT see CCU150
KINGCUP see MBU550
KING'S GOLD see ARI000
KING'S GREEN see COF500
KING'S YELLOW see ARI000
KING'S YELLOW see LCR000
KINIDIN DURETTER see QFS100
KINIDIN DURULES see QFS100
KINILENTIN see QFS100
KINNIKINNIK see CCJ825
KINOPRENE see POB000
KINOTOMIN see FOS100
KIPCA see MMD500
KIPCA WATER SOLUBLE see NAQ600
KIRESUTO B see EIX500
KIRESUTO NTB see DXF000
KIRKSTIGMINE BROMIDE see POD000
KIRKSTIGMINE METHYL SULFATE see
DQY909
α-KIRONDRIN see GEW700
β-KIRONDRIN see GEW700
KIR RICHTER see PAF550
KIRTICOPPER see CNJ950
KITASAMYCIN see SLC000
KITASAMYCIN A3 see JDS200
KITASAMYCIN TARTRATE see LEX000
KITAZIN see DIU800
KITAZIN L see BKS750
KITAZIN P see BKS750
KITINE see RDK000
KITON BLUE A see ERG100
KITON CRIMSON 2R see HJF500
KITON FAST BLUE G see APG700
KITON FAST ORANGE G see HGC000
KITON FAST SCARLET 3B see CMG750
KITON FAST YELLOW A see SGP500
KITON GREEN F see FAE950
KITON GREEN FC see FAE950
KITON GREEN S see ADF000
KITON ORANGE II see CMM220
KITON ORANGE MNO see MDM775
KITON PONCEAU R see FMU070
KITON PURE BLUE L see FMU059
KITON PURE BLUE V see ADE500
KITON PURE BLUE V.FQ see ADE500
KITON RED 2G see CMM300
KITON RED 6B see CMM400
KITON RED 7B see CMM400

KITON RED 6B-FQ see CMM400
KITON RED G see CMM300
KITON RUBINE S see FAG020
KITON SCARLET 4R see FMU080
KITON YELLOW EXTRA see CMM758
KITON YELLOW MS see MDM775
KITON YELLOW T see FAG140
KIVATIN see HKR500
KIWAM (INDIA) see SED400
KIWA TURQUOISE BLUE GL see COF420
KIWI LUSTR 277 see BGJ250
K-IXINA see CAC500
KIZUTA SAPONIN K6 see HAK075
KKM 43 see ISD066
KL-001 see BMN750
KL 255 see BQB250
(−)-KL 255 see BQB250
KL 373 see BGD500
KLAVI KORDAL see NGY000
KLEBCIL see KAM000
KLEBCIL see KAV000
KLEER-LOT see AMY050
KLEESALZ (GERMAN) see OLE000
KLEGECELL see PKQ059
KLERAT see TAC800
KLIMANOSID see RDK000
KLIMORAL see EDU500
KLINE see BBK250
KLINGTITE see NAK500
KLINIT see XPJ000
KLINITAMIN see AHR600
KLINOSORB see CMV850
KLION see MMN250
KLOBEN see BRA250
KLOBEN NEBURON see BRA250
KLOFIRAN see ARQ750
KLONDIKE YELLOW X-2261 see PJQ100
K-LOR see PLA500
KLORALFENAZON see SKS700
KLORAMIN see BIE500
KLORAMIN see CDP000
KLORAMINE-T see CDP000
KLOREX see SFS000
KLORINOL see TIX000
KLOROCIN see SHU500
KLOROKIN see CLD000
KLORPROMAN see CKP500
KLORPROMEX see CKP500
KLORT see MQU750
KLOT see AJP250
KLOTOGEN see MCB575
KLOTOGEN F see MCB575
KLOTOGEN F 16 see MCB575
KLOTOGEN F 227 see MCB575
KLOTRIX see PLA500
KLOTTONE see MMD500
KLT 40 see PKF750
KLUCEL see HNV000
KM see KAL000
KM see SMQ500
KM 2 see UTU500
4K-2M see CIR250
KM 200 see THR820
KM-208 see BAC175
KM-1146 see KGU100
KM 2210 see BFV325
KM (the antibiotic) see KAL000
KMC 113 see BKL600
K MCPA see MIH800
KMC-R 113 see BKL600
KMH see DMC600
KM (POLYMER) see SMQ500
KM 2 (POLYMER) see UTU500
KMTS 212 see SFO500
K1-N see PLK500
KN 320 see IHC550
KNEE PINE OIL see PIH400
KNIGHT'S SPUR see LBF000
KNITTEX ASL see DTG700
KNITTEX LE see DTG700
KNITTEX TC see UTU500
KNITTEX TS see UTU500

KNOCKBAL see BSG300
KNOCKMATE see FAS000
KNOLL H75 see FMS875
KNOLLIDE see PLK500
KO 7 see EAL100
KO 08 see SCR400
KO 18 see LBV100
KO 1173 see MQR775
KOA-HAOLE (HAWAII) see LED500
KOAXIN see MMD500
KOBALT (GERMAN, POLISH) see CNA250
KOBALT CHLORID (GERMAN) see CNB599
KOBALT-EDTA (GERMAN) see DGQ400
KOBALT HISTIDIN (GERMAN) see BJY000
KOBAN see EFK000
KOBU see PAX000
KOBUTOL see PAX000
KOCHINEAL RED A FOR FOOD see
FMU080
KOCIDE see CNM500
KOCIDE see SOD500
KO 1366-CL see BON400
KODAFLEX see TIG750
KODAFLEX DBS see DEH600
KODAFLEX DIBP see DNJ400
KODAFLEX DOA see AEO000
KODAFLEX DOP see DVL700
KODAFLEX DOTP see BJS500
KODAFLEX TOTM see TJR600
KODAFLEX TRIACETIN see THM500
KODAK CD-3 see MDQ850
KODAK LR 115 see CCU250
KODAK SILVER HALIDE SOLVENT HS-103
see BGT500
KODOCYTOCHALASIN-1 see PAM775
KOE 1366 CHLORIDE see BON400
KOE 1173 HYDROCHLORIDE see MQR775
KOFFEIN (GERMAN) see CAK500
KOHLENDIOXYD (GERMAN) see CBU250
KOHLENDISULFID
(SCHWEFELKOHLENSTOFF) (GERMAN)
see CBV500
KOHLENMONOXID (GERMAN) see
CBW750
KOHLENOXYD (GERMAN) see CBW750
KOHLENSAEURE (GERMAN) see CBU250
KOJIC ACID see HLH500
KOKAIN see CNE750
KOKAN see CNE750
KOKAYEEN see CNE750
KOKOTINE see BBQ500
KOLALES HALOMTANO (GUAM) see
RMK250
KOLCHAMIN see MIW500
KOLCHICIN see MIW500
KOLI (HAWAII) see CCP000
KOLIMA 10 see AAU300
KOLIMA 35 see AAU300
KOLIMA 75 see AAU300
KOLKAMIN see MIW500
KOLKLOT see MMD500
2,4,6-KOLLIDIN see TME272
KOLLIDON see PKQ250
KOLLIDON VA 64 see AAU300
KOLOFOG see SOD500
KOLOSPRAY see SOD500
KOLPHOS see PAK000
KOLPON see EDV000
KOLTAR see OQU100
KOLTON see PIZ250
KOLTONAL see PIZ250
KOMBE-STROPHANTHIN see SMN000
KOMBETIN see SMN000
KOMEEN see DBU800
KOMPLEXON I see AMT500
KOMPLEXON IV see CPB120
KOMPLXON see EIV000
KONAKION see VTA000
KONDREMUL see MQV750
KONESSIN DIHYDROBROMIDE see
DOX000
KONESTA see TII250

KONLAX see DJL000
KONTRAST-U see SHX000
KOOLMONOXYDE (DUTCH) see CBW750
KOOLSTOFDISULFIDE
(ZWAVELKOOLSTOF) (DUTCH) see CBV500
KOOLSTOFOXYCHLORIDE (DUTCH) see
PGX000
KOOST see CNT350
KOOT see CNT350
KOPFUME see EIY500
KOP KARB see CNJ750
KOPLEN 2 see SMQ500
KOPLEX AQUATIC HERBICIDE see
DBU800
KOP MITE see DER000
KOPOLYMER BUTADIEN STYRENOVY
(CZECH) see SMR000
KOPREZ 87-110 see UTU500
KOPROL see PDO750
KOPROSTERIN (GERMAN) see DKW000
KOPSOL see DAD200
KOP-THIODAN see EAQ750
KOP-THION see MAK700
KORAD see PKB500
KORBUTONE see AFJ625
KORDIAMIN see DJS200
KOREON see CMK415
KOREON see NBW000
KORGLYKON see CNH780
KORIUM see MJM500
KORLAN see RMA500
KORLANE see RMA500
KORMOGRIZEIN see GJU800
KORODIL see CCK125
KORONAROSIDE A see HAK075
KOROSEAL see PKQ059
KOROSTAN RED G see CMM325
KORO-SULF see SNN500
KOROTRIN see CMG675
KORUM see HIM000
KORUND see EAL100
KOSATE see DJL000
KOSMINK see CBT750
KOSMOBIL see CBT750
KOSMOLAK see CBT750
KOSMOS see CBT750
KOSMOTHERM see CBT750
KOSMOVAR see CBT750
KOST see CNT350
KOSTIL see ADY500
KOTAMITE see CAT775
K-OTHRIN see DAF300
KOTION see DSQ000
KOTOL see BBQ750
KOTORAN see KHK000
KOW see HOG000
KP 2 see PAX000
KP 140 see BPK250
KP 201 see DGV700
KPB see KGK000
KPE see KHK100
K PHENETHICILLIN see PDD350
K-PIN see PIB900
K. PNEUMONIAE ENDOTOXIN see EAS260
K 17 (POLYMER) see UTU500
K-PRENDE-DOME see PLA500
K PREPARATION see BJU000
KR 492 see MKK500
KR 2537 see SMQ500
KRAMERIA IXINA see CAC500
KRAMERIA TRIANDRA see RGA000
KRASTEN 1.4 see SMQ500
KRASTEN 052 see SMQ500
KRASTEN SB see SMQ500
KRATEDYN see EAW000
KRATON LIQUID L-207 POLYMER see
KHK200
KREBON see BQL000
KRECALVIN see DGP900
KRECALVIN see PHC750
KREDAFIL 150 EXTRA see CAT775
KREDAFIL RM 5 see CAT775

KREGASAN see TFS350
KRENITE see ANG750
KRENITE BRUSH CONTROL AGENT see
ANG750
KRENITE (OBS.) see DUS700
KRENITE (OBS.) see DUU600
KREOSOL see MEK325
KRESAMONE see DUS700
KRESIDIN see MGO750
m-KRESOL see CNW750
p-KRESOL see CNX250
o-KRESOL (GERMAN) see CNX000
KRESOLE (GERMAN) see CNW500
KRESOLEN (DUTCH) see CNW500
o-KRESOL-GLYCERINAETHER (GERMAN)
see GGS000
KRESONIT E see DUT800
KRESOXIM-METHYL see MLI900
KRESOXIM-METHYL TECHNICAL see
MLI900
KRESOXYPROPANDIOL see GGS000
KREZAMON see DUT800
KREZIDINE see MGO750
KREZOL (POLISH) see CNW500
KREZONE see CIR250
KREZONIT E see DUT800
KREZONITE see DUU600
KREZOTOL 50 see DUS700
KRINO B 15 see DNM400
KRINOCORTS see DAQ800
KRIPLEX see DEO600
KRIPTIN see WAK000
KRISHNA-NEEL EXTRACT see JDS100
KRISOLAMINE see DBP000
KRISTALLOSE see SJN700
KRISTALL-VIOLETT see AOR500
KRMD 58 see MAE000
KRO 1 see SMR000
KROKYDOLITH (GERMAN) see ARM275
KROLOR ORANGE RKO 786D see MRC000
KROMAD see KHU000
KROMFAX SOLVENT see TFI500
KROMON GERANIUM LAKE see CMG750
KROMON GREEN B see CLK235
KROMON HELIO FAST RED see MMP100
KROMON HELIO FAST RED YS see
MMP100
KROMON LAKE ORANGE TONER see
CMM220
KROMON LAKE RED C see CMS150
KROMON ORANGE G see CMS145
KROMON PERMANENT RED 4B see
CMS155
KROMON RED R see CJD500
KROMON SODIUM LITHOL see NAP100
KROMON YELLOW GXR see CMS208
KROMON YELLOW MTB see DEU000
KRONISOL see BHK000
KRONITEX see TNP500
KRONITEX KP-140 see BPK250
KRONITEX TOF see TNI250
KRONOS TITANIUM DIOXIDE see TGG760
KROTENAL see DXH250
KROTILIN see KHU025
KROTILINE see DAA800
KROTILINE see KHU025
KROTONALDEHYD (CZECH) see COB250
KROTYLCHLORID see CEU825
KROVAR II see BMM650
KRUMKIL see ABF500
KRYOGENIN see CBL000
KRYOLITH (GERMAN) see SHF000
KRYPTOAESCIN see COD900
KRYPTOCUR see LIU370
KRYPTOCYANINE IODIDE see KHU050
KRYSID see AQN635
KRZEWOTOKS see BSQ750
KS 11 see UTU500
KS 35 see UTU500
KS 4B see DEJ100
KS 68M see UTU500
KS 1300 see CAT775

KS 1675 see DBA600
KhS 596 see CGW300
KSM see KCA200
KS-M 0.3P see UTU500
K-STROPHANTHIDIN see SMM500
K-STROPHANTHIN-α see CQH750
K-STROPHANTHIN-β see SMN002
KSYLEN (POLISH) see XGS000
KT 35 see CNK559
KT 136 see KHU100
KT-611 see MFG515
KT6149 see GFO072
KTC 1 see RKZ100
K-THROMBYL see MMD500
K-TROMBINA see MCB575
KTS (PHARMACEUTICAL) see KFA100
KU 5-3 see EAL100
KUBACRON see HII500
KUBARSOL see ABX500
KU 13-032-C see DFL200
KUE 13032c see DFL200
KUEMMEL OIL (GERMAN) see CBG500
KUH-833 see CPB065
KUKUI (HAWAII, GUAM) see TOA275
KULU 40 see CAT775
KUMADER see WAT200
KUMIAI see MIB750
KUMORAN see BJZ000
KUMULUS see SOD500
KUPAOA (HAWAII) see DAC500
KUPFERCARBONAT (GERMAN) see CNJ750
KUPFEROXYCHLORID (GERMAN) see
CNK559
KUPFEROXYDUL (GERMAN) see CNO000
KUPFERRON (CZECH) see ANO500
KUPFERSULFAT (GERMAN) see CNP250
KUPFERSULFAT-PENTAHYDRAT
(GERMAN) see CNP500
KUPFERVITRIOL (GERMAN) see CNP500
KUPRABLAU see CNM500
KUPRATSIN see EIR000
KUPRICOL see CNK559
KUPRIKOL see CNK559
KUPROTSIN see ZJS300
KUR see CNT350
KURALON VP see PKP750
KURAN see TIX500
KURARE OM 100 see AAX250
KURARE POVAL 1700 see PKP750
KURARE PVA 205 see PKP750
KURATE POVAL 120 see PKP750
KURCHICINE see KHU136
KURDUMANA, root extract see CNH750
KUREHALON A0 see CGW300
KURKUMIN see ICC800
KUROMATSUEN see LID100
KUROMATSUENE see LID100
KURON see TIX500
KURON see TIX750
KUROSAL see TIX500
KUSAGARD see FBP350
KUSA-TOHRU see SFS000
KUSATOL see SFS000
KUSCIDE see HKI600
KUSHTHA see CNT350
KUSNARIN see EID000
KUSTA see CNT350
KUTH see CNT350
KUTROL see UVJ475
K-VITAN see MMD500
KW-066 see PNX750
KW 110 see AGX125
KW-125 see AES750
06K-50W see AJI250
KW-1062 see MQS579
KW-1070 see FOK000
KW-1100 see BAC325
KW 2149 see GFO072
KW-2307 see VLF400
KW 3049 see BAV400
KW-3149 see FDD080
KW-4354 see OMG000

KW-5338 see DYB875
KWAS BENZYDYNODWUKAROKSYLOWY
(POLISH) see BFX250
KWAS 2,4-
DWUCHLOROFENOKSYOCTOWY see
DAA800
KWAS DWUMETYLO-
DWUTIOFOSFOROWY see PHH500
KWAS METANIOWY (POLISH) see FNA000
KWASU 2,4-
DWUCHLOROFENOKSYOCTOWEGO see
DAA800
KWD 2019 see TAN100
KWD 2019 see TAN250
KWELL see BBQ500
KWELLS see HOT500
KWG 0599 see SCF525
KWIETAL see QCS000
KWIK (DUTCH) see MCW250
KWIK-KIL see SMN500
KWIKSAN see ABU500
KWIT see EEH600
KW-2-LE-T see LFA020
KYAMEPROMAZINE MALEATE see COS899
KYANACETHYDRAZID see COH250
KYANID SODNY (CZECH) see SGA500
KYANID STRIBRNY (CZECH) see SDP000
KYANITE see AHF500
2-KYANMETHYLBENZIMIDAZOL
(CZECH) see BCC000
KYANOSTRIBRNAN DRASELNY (CZECH)
see PLS250
KYANURCHLORID (CZECH) see TJD750
KYBERNIN see AQN550
KYLAR see DQD400
KYNEX see AKO500
KYOCRISTINE see LEZ000
KYONATE see PLV750
KYORIX BT-S see BAP502
KYPCHLOR see CDR750
KYPFOS see MAK700
KYPMAN 80 see MAS500
KYPTHION see PAK000
KYPZIN see EIR000
KYSELINA ADIPOVA (CZECH) see AEN250
KYSELINA AKRYLOVA see ADS750
KYSELINA AMIDOSULFONOVA (CZECH)
see SNK500
KYSELINA-4-AMINOANISOL-3-
SULFONOVA see AIA500
KYSELINA 4-AMINOAZOBENZEN-3,4'-
DISULFONOVA (CZECH) see AJS500
KYSELINA p-AMINOBENZOOVA see
AIH600
KYSELINA 1-AMINO-2-
ETHOXYNAFTALEN-6-SULFONOVA
(CZECH) see AJU500
KYSELINA-3-AMINO-4-
METHOXYBENZOOVA (CZECH) see
AIA250
KYSELINA 1-AMINO-8-NAFTOL-3,6-
DISULFONOVA (CZECH) see AKH000
KYSELINA 1-AMINO-8-NAFTOL-4-
SULFONOVA (CZECH) see AKH750
KYSELINA 2-AMINO-5-NAFTOL-7-
SULFONOVA (CZECH) see AKI000
KYSELINA-p-AMINOSALICYLOVA
(CZECH) see AMM250
KYSELINA ANILIN-2,5-DISULFONOVA
(CZECH) see AIE000
KYSELINA ANILIN-3-SULFONOVA
(CZECH) see SNO000
KYSELINA 4,4'-AZO-BIS-(4-
KYANVALEROVA) see ASL500
KYSELINA BENZIDIN-2,2'-DISULFONOVA
(CZECH) see BBX500
KYSELINA BENZOOVA (CZECH) see
BCL750
KYSELINA BROMOCTOVA see BMR750
KYSELINA C (CZECH) see ALH250
KYSELINA CEROMSALICYLOVA (CZECH)
see BHA000

KYSELINA 2-CHLOR-6-AMINOFENOL-4-
SULFONOVA (CZECH) see AJH500
KYSELINA o-CHLORBENZOOVA (CZECH)
see CEL250
KYSELINA-S-(8-CHLORMETHYL-1-
NAFTYL)THIOGLYKOLOVA (CZECH) see
CIQ000
KYSELINA CHLOROCTOVA see CEA000
KYSELINA-2-CHLORO-4-
NITROBENZOOVA (CZECH) see CJC250
KYSELINA 4-CHLORO-3-
NITROBENZOOVA (CZECH) see CJC500
KYSELINA 2-CHLOR-4-TOLUIDIN-5-
SULFONOVA (CZECH) see AJJ250
KYSELINA CITRAZINOVA see DMV400
KYSELINA CITRONOVA (CZECH) see
CMS750
KYSELINA CLEVE (CZECH) see ALI250
KYSELINA 1,2-
CYKLOHEXYLENDIAMINTETRAOCTOVA
see CPB120
KYSELINA-2,4-
DIAMINOBENZENSULFONOVA (CZECH)
see PFA250
KYSELINA 2,4-DICHLORFENOXYOCTOVA
see DAA800
KYSELINA DICHLORISOKYANUROVA
(CZECH) see DGN200
KYSELINA 2,5-DICHLOR-4-(3'-METHYL-5'-
PYRAZOLON-1'-YL)BENZENSULFONOVA
(CZECH) see DFQ200
KYSELINA DICHLOROCTOVA see DEL000
KYSELINA 3,6-DICHLORPIKOLINOVA see
DGJ100
KYSELINA O,O-
DIETHYLDITHIOFOSFORECNA (CZECH)
see PHG500
KYSELINA DI-(2-
ETHYLHEXYL)FOSFORECNA see BJR750
KYSELINA 2,3-DIHYDROXYBUTANDIOVA
see TAF750
KYSELINA DI-I (CZECH) see IBF000
KYSELINA 3,5-
DIKARBOXYBENZENSULFONOVA
(CZECH) see SNU500
KYSELINA O,O-
DIMETHYLDITHIOFOSFORCNA (CZECH)
see PHH500
KYSELINA-2,4-
DINITROBENZENSULFONOVA (CZECH)
see DUR400
KYSELINA-4,4'-DINITROSTILBEN-2,2'-
DISULFONOVA (CZECH) see DVF600
KYSELINA DUSICNE see NED500
KYSELINA DUSITE see NMR000
KYSELINA 3,6-ENDOMETHYLEN-
3,4,5,6,7,7-HEXACHLOR-Δ⁴-
TETRAHYDROFTALOVA (CZECH) see
CDS000
KYSELINA ETHOXY-CLEVE-1,6 (CZECH)
see AJU500
KYSELINA 3-(2-
ETHYLBUTOXY)PROPIONOVA see EGY000
KYSELINA-1,3-FENYLENDIAMIN-4-
SULFONOVA (CZECH) see PFA250
KYSELINA 2-FENYL-2-
HYDROXYETHANOVA see MAP000
KYSELINA FUMAROVA (CZECH) see
FOU000
KYSELINA GLYOXYLOVA see GIQ000
KYSELINA H (CZECH) see AKH000
KYSELINA HET (CZECH) see CDS000
KYSELINA HYDROXYBUTANDIOVA
(CZECH) see MAN000
KYSELINA 3-HYDROXY-2-NAFTOOVA see
HMX520
KYSELINA 12-HYDROXY-9-
OKTADECENOVA see RJP000
KYSELINA 2-HYDROXYPROPANOVA see
LAG000
KYSELINA ISOFTALOVA (CZECH) see
IMJ000

KYSELINA ISOMASELNA see IJU000
KYSELINA JABLECNA (CZECH) see
MAN000
KYSELINA JANTAROVA see SMY000
KYSELINA KAKODYLOVA see HKC000
KYSELINA KOCHOVA (CZECH) see ALI750
KYSELINA KYANUROVA (CZECH) see
THS000
KYSELINA MANDLOVA see MAP000
KYSELINA MERKAPTOOCTOVA see TFJ100
KYSELINA MESAKONOVA (CZECH) see
MDI250
KYSELINA METANILOVA (CZECH) see
SNO000
KYSELINA METHAKRYLOVA see MDN250
KYSELINA METHANSULFONOVA
(CZECH) see MDR250
KYSELINA 4-METHOXYBENZOOVA see
AOU600
KYSELINA 2-METHYLAMINO-5-NAFTOL-
7-SULFONOVA (CZECH) see HLX000
KYSELINA N-METHYLANTHRANILOVA
(CZECH) see MGQ000
KYSELINA METHYLARSONOVA see
MGQ530
KYSELINA-N-METHYL-I (CZECH) see
HLX000
KYSELINA MLECNA (CZECH) see LAG000
KYSELINA MUKOCHLOROVA see MRU900
KYSELINA-2-NAFTOL-1-SULFONOVA
(CZECH) see HMX000
KYSELINA 2-NAFTYLAMIN-1,5-
DISULFONOVA (CZECH) see ALH000
KYSELINA-2-NAFTYLAMIN-4,8-
DISULFONOVA (CZECH) see ALH250
KYSELINA-1-NAFTYLAMIN-6-
SULFONOVA (CZECH) see ALI250
KYSELINA-2-NAFTYLAMIN-1-
SULFONOVA (CZECH) see ALH750
KYSELINA 2-NAFTYLAMIN-3,6,8-
TRISULFONOVA (CZECH) see ALI750
KYSELINA N-1-NAFTYLFTALAMOVA see
NBL200
KYSELINA NITRILOTRIOCTOVA see
AMT500
KYSELINA-4-NITRO-2-AMINOFENOL-6-
SULFONOVA (CZECH) see HMY500
KYSELINA 6-NITRO-2-AMINOFENOL-4-
SULFONOVA (CZECH) see AKI250
KYSELINA 4-NITRO-4'-AMINOSTILBEN-
2,2'-DISULFONOVA (CZECH) see ALP750
KYSELINA 1-NITROANTHRACHINON-2-
KARBOXYLOVA see NFS502
KYSELINA NITROBENZEN-m-
SULFONOVA (CZECH) see NFB500
KYSELINA-p-NITROBENZOOVA (CZECH)
see CCI250
KYSELINA N-(4-NITROFENYL)OXAMOVA
see NHX100
KYSELINA-4-NITROTOLUEN-2-
SULFONOVA (CZECH) see MMH250
KYSELINA OXALANILOVA see OLW000
KYSELINA PANTOTHENOVA see PAG300
KYSELINA PEROXYOCTOVA see PCL500
KYSELINA PIKROVA see PID000
KYSELINA PROPIONOVA see PMU750
KYSELINA RICINOLOVA see RJP000
KYSELINA STAVELOVA (CZECH) see
OLA000
KYSELINA SULFAMINOVA (CZECH) see
SNK500
KYSELINA SULFANILOVA see SNN600
KYSELINA 1-SULFOMETHYL-2-
NAFTYLAMIN-6-SULFONOVA (CZECH) see
AMP000
KYSELINA SULFO-TOBIAOVA (CZECH) see
ALH000
KYSELINA TERFTALOVA (CZECH) see
TAN750
KYSELINA 1,2,5,6-
TETRAHYDROBENZOOVA see CPC650

KYSELINA-β,β'-THIODIPROPIONOVA (CZECH) see BHM000
KYSELINA THIOGLYKOLOVA see TFJ100
KYSELINA THIOOCTOVA see TFA500
KYSELINA TOBIASOVA (CZECH) see ALH750
KYSELINA p-TOLUENESULFONOVA (CZECH) see TGO000
KYSELINA 2-TOLUIDIN-4-SULFONOVA (CZECH) see AMT000
KYSELINA-3-TOLUIDIN-6-SULFONOVA (CZECH) see AMT250
KYSELINA-4-TOLUIDIN-3-SULFONOVA (CZECH) see AKQ000
KYSELINA o-TOSYL-H (CZECH) see AKH500
KYSELINA TRICHLOISOKYANUROVA (CZECH) see TIQ750
KYSELINA 2,3,6-TRICHLORBENZOOVA DIMETHYLAMONNA SUL see DOR800
KYSELINA TRICHLOROCTOVA see TII250
KYSELINA TRIFLUOROCTOVA see TKA250
KYSELINA VINNA see TAF750
KYSELINA WOLFRAMOVA (CZECH) see TOD000
KYSLICNIK DI-n-AMYLCINICITY (CZECH) see DVV000
KYSLICNIK DI-n-BUTYLCINICITY (CZECH) see DEF400
KYSLICNIK DIISOAMYLCINICITY (CZECH) see DNL400
KYSLICNIK DIISOBUTYLCINICITY (CZECH) see DNJ000
KYSLICNIK DIISOPROPYLCINICITY (CZECH) see DNR200
KYSLICNIK DI-N-PROPYLCINICITY (CZECH) see DWV000
KYSLICNIK TRI-N-BUTYLCINICITY (CZECH) see BLL750
KYURINETT see YCJ200
KZ 3M see SCQ000
KZ 5M see SCQ000
KZ 7M see SCQ000
K-ZINC see ZKA000
L1 see DSI250
L-2 see BIQ250
L16 see AGX000
84L see DIW000
L-99 see CDG250
L-105 see CCS635
L 195 see UTU500
L-310 see LGK000
L343 see IOT000
L-395 see DSP400
L-561 see DRR400
L 1058 see CJR300
L. 1633 see DEE600
L 1718 see DYE700
L 1811 see DJT400
L-2103 see POA250
L2214 see DDP200
L 2329 see EID200
L. 3428 see AJK750
L-5103 see RKP000
L 5300 see THT800
L 6150 see BKB500
L 6504 see MCB050
L 8580 see MNB250
L-01748 see DJT800
L-10492 see MFJ105
L 10495 see CKL200
L 10499 see PEU650
L-10503 see MFF650
L 109-65 see MCB050
L 11204 see EFC259
L 11373 see PGN840
L 11752 see MFG275
L 121-60 see MCB050
L 12236 see CKA630
L 12375 see MKG800
L 12717 see CKL325
L 13891 see BGL450

L 14085 see BGO200
L 14105 see BGO325
L-36352 see DUV600
L 154819 see LII050
L 586.153 see NNX300
L-691,121 see MDQ900
L 706803 see DAK100
L 735524 see ICE100
L741,626 see CKA700
LA see DXY000
LA 1 see DLY000
LA 01 see CCU150
LA96A see PPN100
LA 1221 see PEU000
LA 6023 see DQR600
LAAM see ACQ258
LAAM HYDROCHLORIDE see ACQ260
LA'AU-'AILA (HAWAII) see CCP000
LAB 108 406 see MRQ300
LABA see LGK100
LABAZ see AJK750
2329 LABAZ see EID200
LABAZENE see PNX750
LABDANOL see IIQ000
LABDANUM OIL see LAC000
LABETALOL HYDROCHLORIDE see HMM000
LABICAN see MDQ250
LABILITE see MAP300
LABITON see DHF600
LABOPAL see DKQ000
LABOPRIN see ARP125
LABOR-NR 2683 see DHR800
LABROCOL see HMM500
LABRODA see TKP100
LABRODAX see TKP100
LABRODAX SUPANATE see TKP100
LABURNUM see GIW195
LABURNUM ANAGYROIDES see GIW195
LABYRIN see CMR100
LAC see CMP905
LAC-43 see BOO000
LACCAIC ACID see CMP905
LAC DYE see CMP905
LACHESIN see ELF500
LACHESINE CHLORIDE see ELF500
LAC LAKE see CMP905
LAC LSP-1 see LIJ000
LACO see PPN100
LACOLIN see LAM000
LACQREN 506 see SMQ500
LACQREN 550 see SMQ500
LACQTEN 1020 see PJS750
LACQUER BLACK S see PCJ200
LACQUER BLACK VB see PCJ200
LACQUER DILUENT see ROU000
LACQUER ORANGE V see TGW000
LACQUER ORANGE V 3G see CMP600
LACQUER ORANGE VG see PEJ500
LACQUER ORANGE VR see XRA000
LACQUER RED V see SBC500
LACQUER RED V 2G see CMS238
LACQUER RED V3B see EOJ500
LACQUER RED VS see SBC500
LACQUERS see LAD000
LACQUERS, NITROCELLULOSE see LAE000
LACQUER YELLOW T see CMM758
LACRETIN see FOS100
LACRIMIN see OPI300
LACTASE see GAV100
LACTASE ENZYME PREPARATIONS from KLUYVEROMYCES LACTIS see LAE350
LACTATE DEHYDROGENASE X see LAO300
LACTATED MONO-DIGLYCERIDES see LAE400
LACTATE d'ETHYLE (FRENCH) see LAJ000
LACTIC ACID see LAG000
(+)-LACTIC ACID see LAG010
d-LACTIC ACID see LAG010
(S)-LACTIC ACID see LAG010
dl-LACTIC ACID see LAG000

l-(+)-LACTIC ACID see LAG010
(S)-(+)-LACTIC ACID see LAG010
LACTIC ACID, ACETATE, CYCLOHEXYL ESTER see LAG030
LACTIC ACID, ANTIMONY SALT see AQE250
LACTIC ACID, BERYLLIUM SALT see LAH000
LACTIC ACID, BUTYL ESTER see BRR600
LACTIC ACID, BUTYL ESTER, BUTYRATE see BQP000
LACTIC ACID, CADMIUM SALT see CAG750
LACTIC ACID, COMPD. WITH 1,2,3,4,5,6-HEXAHYDRO-6,11-DIMETHYL-3-(3-METHYL-2-BUTENYL)-2,6-METHANO-3-BENZAZOCIN-8-OL (1:1) see PBP400
LACTIC ACID, COMPD. WITH 9-METHOXY-5,11-DIMETHYL-6H-PYRIDO(4,3-B)CARBAZOLE (1:1) see MEL780
LACTIC ACID, ETHYL ESTER see LAJ000
LACTIC ACID, IRON(2+) SALT (2:1) see LAL000
LACTIC ACID, LEAD(2+) SALT (2:1) see LDL000
LACTIC ACID LITHIUM SALT see LHL000
LACTIC ACID, MAGNESIUM SALT see LAL100
LACTIC ACID, METHYL ESTER see MLC600
LACTIC ACID, 2-METHYL-, ETHYL ESTER see ELH700
LACTIC ACID, MONOSODIUM SALT see LAM000
LACTIC ACID, NEODYMIUM SALT see LAN000
LACTIC ACID SODIUM SALT see LAM000
LACTIC ACID, SODIUM ZIRCONIUM SALT (4:4:1) see LAO000
LACTIC ACID, 3,3,3-TRIFLUORO-, METHYL ESTER, DIBUTYL PHOSPHATE see TKB287
LACTIC ACID, 3,3,3-TRIFLUORO-, METHYL ESTER, DIETHYL PHOSPHATE see TKB289
LACTIC ACID, 3,3,3-TRIFLUORO-, METHYL ESTER, DIHEXYL PHOSPHATE see TKB290
LACTIC ACID, 3,3,3-TRIFLUORO-, METHYL ESTER, DIISOBUTYL PHOSPHATE see TKB292
LACTIC ACID, 3,3,3-TRIFLUORO-, METHYL ESTER, DIMETHYL PHOSPHATE see TKB294
LACTIC ACID, 3,3,3-TRIFLUORO-, METHYL ESTER, DIPENTYL PHOSPHATE see TKB296
LACTIC ACID, 3,3,3-TRIFLUORO-, METHYL ESTER, DIPROPYL PHOSPHATE see TKB298
LACTIC ACID, TRIS(2-HYDROXYETHYL)(PHENYLMERCURI)AMMONIUM derivative see TNI500
LACTIC ACID, ion(1−), TRIS((2-HYDROXYETHYL)PHENYLMERCURIO)AMMONIUM see TNI500
LACTIC ACID, ZIRCONIUM SALT (3:1) see ZRJ000
LACTIC ACID, ZIRCONIUM SALT (4:1) see ZRS000
LACTIC DEHYDROGENASE X see LAO300
LACTIN see LAR000
LACTITOL DIHYDRATE see LAO500
LACTITOL MONOHYDRATE see LAO600
LACTOBACILLUS LACTIS DORNER FACTOR see VSZ000
LACTOBARYT see BAP000
LACTOBIOSE see LAR000
LACTOCAINE see AIT250
LACTOFLAVIN see RIK000
LACTOFLAVINE see RIK000
LACTOGEN see PMH625
LACTOGENIC HORMONE see PMH625
ε-LACTONE HEXANOIC ACID see LAP000
LACTONITRILE see LAQ000
LACTOSCATONE see LAQ100

LACTOSE see LAR000
d-LACTOSE see LAR000
LACTOSOMATOTROPIC HORMONE see PMH625
LACTULOSE see LAR100
LACTYLATED FATTY ACID ESTERS of GLYCEROL and PROPYLENE GLYCOL see LAR400
LACTYLIC ESTERS of FATTY ACIDS see LAR800
LACUMIN see MOQ250
LADAKAMYCIN see ARY000
LADIE'S THIMBLES see FOM100
LADOGAL see DAB830
LADY LAUREL see LAR500
LAE-32 see LJI000
LAETRILE see LAS000
LAEVORAL see LFI000
LAEVOSAN see LFI000
LAEVOXIN see LFG050
LAEVULIC ACID see LFH000
LAEVULINIC ACID see LFH000
LAGISTASE see EAI200
LAGOSIN see FPC000
LAGRIMAS de MARIA see CAL125
LAIDLOMYCIN see DAK000
LAI (HAITI) see WBS850
LAKANA, MIKINOLIA-HIHIU (HAWAII) see LAU600
LAKE BLACK EXTRA see CMS236
LAKE BLUE AFX see ERG100
LAKE BLUE B BASE see DCJ200
LAKE DEVELOPER A see CMM760
LAKE FAST BLUE BS see IBV050
LAKE FAST BLUE GGS see IBV050
LAKE ORANGE A see CMM220
LAKE ORANGE II YS see CMM220
LAKE PONCEAU see FMU070
LAKE RED 4R see MMP100
LAKE RED C see CHP500
LAKE RED C see CMS150
LAKE RED C 18958 see CMS150
LAKE RED CY see CMS150
LAKE RED 4RII see MMP100
LAKE RED KB BASE see CLK225
LAKE RED 2GL see DVB800
LAKE RED R see NAP100
LAKE RED RL see NAP100
LAKE SCARLET 3B see CMG750
LAKE SCARLET G BASE see NMP500
LAKE SCARLET GG BASE see DEO295
LAKE YELLOW see FAG140
LAKE YELLOW GA see CMS210
LALOI (HAITI) see AGV875
LAM see PPT500
LAMAR see FLZ050
LAMBAST see CFW750
LAMBDA-CYHALOTHRIN see LAS200
LAMBDA-CYHALOTHRIN see LAS200
LAMBDA-CYHALOTHRIN TECHNICAL see LAS200
LAMBDAMYCIN see CDK250
LAMBETH see PMP500
LAMB KILL see MRU359
LAMBRATEN see AJT250
LAMBRIL see BBW750
LAMBROL see FDB200
LAMBROL see FIP999
LAMDIOL see EDO000
LAMIDON see IIU000
(L)-3-(2-AMINOETHOXY)ALANINE see OLK200
LAMITEX see SEH000
LAMIUM ALBUM LINN., EXTRACT see LAS500
LAMORYL see GKE000
LAMPIT see NGG000
LAMPRECID see TKD400
LAMPTEROL see LIO600
LAMURAN see AFG750
LANACORT see HHQ800
LANADIGENIN see DKN300

LANADIN see TIO750
LANALENE L see IPS500
LANALENE P see IPS500
LANALENE S see IPS500
LANAPERL BLUE B see CMM070
LANAPERL FAST RED 3G see CMM320
LANAPERL RED G see CMM320
LANAPERL YELLOW BROWN GT see SGP500
LANASYN GREEN BL see CMM200
LANATOSID A (GERMAN) see LAT000
LANATOSID B (GERMAN) see LAT500
LANATOSID C (GERMAN) see LAU000
LANATOSIDE A see LAT000
LANATOSIDE B see LAT500
LANATOSIDE C see LAU000
LANATOSIDES see LAU400
LANATOXIN see DKL800
LANAZINE see DBA800
LANCE see LAU500
LANCER see FDB500
LANCOL see OBA000
LANDALGINE see AFL000
LANDAMYCINE see RIP000
LANDISAN see MEO750
LANDOCAINE see BQA010
LAND PLASTER see CAX750
LANDRAX see CQM325
LANDRIN see TMC750
LANDRIN see TMD000
LANDRIN, NITROSO DERIVATIVE see NLY500
LANDRUMA see NDX500
LANESTA see CDV700
LANESTA L see IPS500
LANESTA P.S. see IPS500
LANETTE WAX-S see SIB600
LANEX see DUK800
LANGFORD see KBB600
LANGORAN see CCK125
LANI-ALI'I (HAWAII) see AFQ625
LANIAZID see ILD000
LANICOR see DKN400
LANIRAPID see MJD300
LANITOP see MJD300
LANNAGOL LF see PJY100
LANNATE see MDU600
LANODOXIN see DNF600
LANOL see CMD750
LANOLIN see LAU520
LANOLIN, anhydrous see LAU550
LANOLIN, ETHOXYLATED see LAU560
LANOPHYLLIN see TEP000
LANOSTABIL see LAU400
LANOXIN see DKN500
LANTANA see LAU600
LANTANA CAMARA see LAU600
LANTHANA see LBA100
LANTHANACETAT (GERMAN) see LAW000
LANTHANIA (La2O3) see LBA100
LANTHANUM see LAV000
LANTHANUM ACETATE see LAW000
LANTHANUM AMMONIUM NITRATE see ANL500
LANTHANUM CHLORIDE see LAX000
LANTHANUM CONCENTRATE see LAX100
LANTHANUM DIHYDRIDE see LAY499
LANTHANUM EDETATE see LAZ000
LANTHANUM HYDRIDE see LBC000
LANTHANUM NITRATE see LBA000
LANTHANUM OXIDE see LBA100
LANTHANUM(3+) OXIDE see LBA100
LANTHANUM(III) OXIDE see LBA100
LANTHANUM SESQUIOXIDE see LBA100
LANTHANUM SULFATE see LBB000
LANTHANUM(III) SULFATE (2:3) see LBB000
LANTHANUM TRIACETATE see LAW000
LANTHANUM TRIHYDRIDE see LBC000
LANTHANUM TRIOXIDE see LBA100
LANTOSIDE see LAU400
LANVIS see AMH250

LAPACHIC ACID see HLY500
LAPACHOL see HLY500
LAPACHOL WOOD see HLY500
β-LAPACHONE see LBC500
LAPAQUIN see CLD000
LAPAV see PAH250
LAPEMIS HARDWICKII VENOM see SBI910
LAPLEN see CGW300
LAPPACONITINE see LBD000
(+)-LAPPACONITINE see LBD000
LAPROL 1601-2-50M see LBD050
LAPYRIUM CHLORIDE see LBD100
LARAHA see LBE000
LARCH GUM see AQR800
LARD FACTOR see VSK600
LARD (UNHYDROGENATED) see LBE300
LAREX see UTU500
LARGACTIL see CKP250
LARGACTIL MONOHYDROCHLORIDE see CKP500
LARGACTILOTHIAZINE see CKP250
LARGACTYL see CKP250
LARGAKTYL see CKP500
LARGON HYDROCHLORIDE see PMT500
LARIXIC ACID see MAO350
LARIXIN see ALV000
LARIXINIC ACID see MAO350
LARKSPUR see LBF000
LAROCAINE see DNY000
LARODON see INY000
LARODOPA see DNA200
LAROMIN C see BGT800
LAROMIN C 260 see BGT800
LAROXIL see EAH500
LAROXYL see EAH500
LARTEN see MQU750
LARVACIDE see CKN500
LARVADEX see CQE400
LARVATROL see BAC040
LARVIN see LBF100
LARVIN THIO DICARB PESTICIDE OVICIDE see LBF100
LAS see AFO500
LAS see CMV325
LASALOCID see LBF500
LASERDIL see CCK125
LASEX see CHJ750
LASIOCARPINE see LBG000
LASIX see CHJ750
LAS-Mg see LGF800
LAS, MAGNESIUM SALT see LGF800
LAS-Na see LGF825
LASODEX see TEP500
LASSO see CFX000
LAS, SODIUM SALT see LGF825
LATAMOXEF SODIUM see LBH200
LATANOPROST see XAA500
LATEX see PJR000
LATEXOL FAST BLUE SD see IBV050
LATEXOL FAST ORANGE J see CMS145
LATEXOL FAST YELLOW JR see CMS208
LATEXOL RED J see CJD500
LATEXOL SCARLET R see CHP500
LATEX SVKh see CGW300
LATHANOL see SIB700
LATHANOL LAL see SIB700
LATHANOL-LAL 70 see SIB700
LATHOSTEROL see CMD000
LATHYRUS ODORATUS, SEEDS see SOZ000
LATIBON see DII200
LATICATOXIN see LBI000
LATICAUDA SEMIFASCIATA VENOM see LBI000
LATKA M-1 see NAK525
LATKA 666 see BBP750
LATKA 7744 see CBM750
LATRODECTUS M. MACTANS VENOM see BLW500
LATSCHENKIEFEROEL see PIH400
LATTULOSIO (ITALIAN) see LAR100
LATUSATE see AFY500
LATYL BLUE BCN see CMP070

LATYL BLUE BCN see DBQ220
LATYL CERISE N see AKI750
LATYL YELLOW YL see KDA075
LAUDICON see DLW600
LAUDOCAINE see BQA010
LAUDRAN DI-n-BUTYLCINICITY (CZECH) see DDV600
LAUGHING GAS see NGU000
LAURAMIDE DEA see BKE500
LAURAMIDOPROPYL BETAINE see LBI500
LAUREL CAMPHOR see CBA750
LAUREL LEAF OIL see BAT500
LAUREL LEAF OIL see LBK000
LAURELWOOD see MBU780
LAURETH see DXY000
LAURETH 10S see PJV100
LAURIC ACID see LBL000
LAURIC ACID, BARIUM CADMIUM SALT see BAI770
LAURIC ACID, CADMIUM SALT (2:1) see CAG775
LAURIC ACID, DIBUTYLSTANNYLENE derivative see DDV600
LAURIC ACID, DIBUTYLSTANNYLENE SALT see DDV600
LAURIC ACID DIETHANOLAMIDE see BKE500
LAURIC ACID-2,3-EPOXYPROPYL ESTER see LBM000
LAURIC ACID ESTER with 2-HYDROXYETHYL THIOCYANATE see LBO000
LAURIC ACID, METHYL ESTER see MLC800
LAURIC ACID, SODIUM SALT see LBN000
LAURIC ACID, 2-THIOCYANATOETHYL ESTER see LBO000
LAURIC ALCOHOL see DXV600
LAURIC DIETHANOLAMIDE see BKE500
LAURIDIT OD see OHU200
LAURIER ROSE (HAITI) see OHM875
LAURINAMINE see DXW000
LAURINE see CMS850
LAURINE DIMETHYL ACETAL see LBO050
LAURINIC ALCOHOL see DXV600
LAURODIN see AJS750
LAUROLINIUM ACETATE see AJS750
LAUROLITSINE see LBO100
LAUROMACROGOL 400 see DXY000
LAURONITRILE see DXT400
LAUROSCHOLTZINE see TKX700
LAUROSTEARIC ACID see LBL000
LAUROTETANIN see LBO200
LAUROTETANINE see LBO200
(+)-LAUROTETANINE see LBO200
LAUROX see LBR000
1-LAUROYLAZIRIDINE see LBQ000
N-(LAUROYLCOLAMENOFORMYLMETHYL)PYRIDINIUM CHLORIDE see LBD100
LAUROYL DIETHANOLAMIDE see BKE500
LAUROYLETHYLENEIMINE see LBQ000
(LAUROYLOXY)TRIBUTYLSTANNANE see TIE750
LAUROYL PEROXIDE see LBR000
LAUROYL PEROXIDE, TECHNICALLY PURE (DOT) see LBR000
LAURTRIMONIUM CHLORIDE see LBX075
LAURUS NOBILIS OIL see OGQ150
LAURYDOL see LBR000
LAURYL 24 see DXV600
N-LAURYLACETAMIDE see LBR100
LAURYL ACETATE see DXV400
LAURYL ALCOHOL CONDENSED with 4 MOLES ETHYLENE OXIDE see LBS000
LAURYL ALCOHOL CONDENSED with 23 MOLES ETHYLENE OXIDE see LBU000
LAURYL ALCOHOL EO (4) see LBS000
LAURYL ALCOHOL EO (7) see LBT000
LAURYL ALCOHOL EO (23) see LBU000
LAURYL ALCOHOL, ETHOXYLATED see DXY000

LAURYL ALCOHOL ETHYLENE OXIDE (23) see LBU050
LAURYL ALCOHOL (FCC) see DXV600
n-LAURYL ALCOHOL, PRIMARY see DXV600
LAURYL ALDEHYDE (FCC) see DXT000
LAURYLAMINE see DXW000
n-LAURYLAMINE see DXW000
LAURYLAMINE HYDROCHLORIDE see DXW060
LAURYLAMMONIUM HYDROCHLORIDE see DXW060
LAURYL AMMONIUM SULFATE see SOM500
LAURYLBENZENESULFONIC ACID see LBU100
LAURYLBETAIN see LBU200
LAURYLBETAINE see LBU200
LAURYL-N-BETAINE see LBU200
LAURYL DIETHANOLAMIDE see BKE500
LAURYLDIETHYLENETRIAMINE see LBV000
LAURYLDIETHYLENETRIAMINE see LBV000
LAURYLDIMETHYLAMINE see DRR800
N-LAURYLDIMETHYLAMINE see DRR800
LAURYLDIMETHYLAMINE OXIDE see DRS200
LAURYLDIMETHYLAMMONIOACETATE see LBU200
LAURYLDIMETHYLBETAINE see LBU200
LAURYLDIMETHYLDICHLOROBENZYLAMMONIUM CHLORIDE see LBV100
LAURYLESTER KYSELINY DUSICNE (CZECH) see NEE000
LAURYLESTER KYSELINYMETHAKRYLOVE (CZECH) see DXY200
LAURYL GALLATE see DXX200
LAURYLGUANIDINE ACETATE see DXX400
LAURYLISOQUINOLINIUM BROMIDE see LBW000
LAURYL MERCAPTAN see LBX000
m-LAURYL MERCAPTAN see LBX000
LAURYL METHACRYLATE see DXY200
LAURYL-N-METHYLSARCOSINE see LBU200
LAURYLNITRAT (CZECH) see NEE000
LAURYL NITRATE see NEE000
LAURYL POLYETHYLENE GLYCOL ETHER see DXY000
LAURYLPYRIDINIUM CHLORIDE see DXY725
LAURYLPYRIDINIUM CHLORIDE see LBX050
1-LAURYLPYRIDINIUM CHLORIDE see DXY725
1-LAURYLPYRIDINIUM CHLORIDE see LBX050
LAURYL RHODANATE see DYA200
LAURYL SODIUM SULFATE see SIB600
LAURYL SULFATE see MRH250
LAURYL SULFATE AMMONIUM SALT see SOM500
LAURYL SULFATE, SODIUM SALT see SIB600
LAURYL SULFURIC ACID see MRH250
LAURYL THIOCYANATE see DYA200
LAURYL 3,3'-THIODIPROPIONATE see TFD500
LAURYLTRIMETHYLAMMONIUM CHLORIDE see LBX075
LAUSIT see IDA000
LAUTHSCHES VIOLETT (GERMAN) see AKK750
LAUXTOL see PAX250
LAUXTOL A see PAX250
LAV see CMY800
LAVAMENTHE see LBX100
LAVANDIN ABSOLUTE see LCA000

LAVANDIN BENZOL ABSOLUTE see LCA000
LAVANDIN OIL see LCA000
LAVANDULYL ACETATE see LCA100
LAVATAR see CMY800
LAVENDEL OEL see LCC000
LAVENDEL OEL (GERMAN) see LCD000
LAVENDER ABSOLUTE see LCC000
LAVENDER OIL see LCD000
LAVENDER OIL, SPIKE see SLB500
LAVOFLAGIN see ABY900
LAVSAN see PKF750
LAWN-KEEP see DAA800
LAWSONE see HMX600
LAWSONE see WAT000
LAWSONITE see PKF750
LAXADIN see PPN100
LAXAGEN see ACD500
LAXAGETTEN see ACD500
LAXAGETTEN see PPN100
LAXANIN N see PPN100
LAXANORM see DMH400
LAXANS see PPN100
LAXANTHREEN see DMH400
LAXESIN see ELF500
LAXIDOGOL see SJJ175
LAXINATE see DJL000
LAXIPUR see DMH400
LAXIPURIN see DMH400
LAXOBERAL see SJJ175
LAXOBERON see SJJ175
LAXOGEN see PDO750
LAXOREX see PPN100
LA XVII see BMN750
LAYOR CARANG see AEX250
LAZETA see CMR100
LAZO see CFX000
LB-46 see VSA000
L 3MB1 see MJM700
LB 502 see CHJ750
(L)-BC-2605 see CQF079
LBF DISULFIDE see PAG150
LBI see LEF200
LB-ROT 1 see FAG040
LC 44 see FMO129
LC-80 see CCK660
"L," CARPSERP see RDK000
LCR see LEY000
LD 400 see PJS750
LD 600 see PJS750
LD-813 see LCE000
L.D. 3055 see IBP200
LDA see BKE500
LDE see BKE500
LDH-X see LAO300
LDPE 4 see PJS750
LD RUBBER RED 16913 see CHP500
Le-100 see EIF000
29060 LE see VLA000
LEA-COV see SHF500
LEAD see LCF000
LEAD ACETATE see LCG000
LEAD ACETATE see LCV000
LEAD(2+) ACETATE see LCV000
LEAD(II) ACETATE see LCV000
LEAD(IV) ACETATE AZIDE see LCG500
LEAD ACETATE, BASIC see LCH000
LEAD ACETATE BROMATE see LCI000
LEAD ACETATE-LEAD BROMITE see LCI600
LEAD ACETATE TRIHYDRATE see LCJ000
LEAD ACETATE(II), TRIHYDRATE see LCJ000
LEAD ACID ARSENATE see LCK000
LEAD ARSENATE see ARC750
LEAD ARSENATE see LCK000
LEAD ARSENATE see LCK100
LEAD ARSENATE, solid (DOT) see LCK000
LEAD ARSENATE (standard) see LCK000
LEAD(II) ARSENITE see LCL000
LEAD ARSENITES (DOT) see LCL000
LEAD(II) AZIDE see LCM000

LEAD(IV) AZIDE see LCN000
LEAD AZIDE (dry) (DOT) see LCM000
LEAD AZIDE, wetted with not <20% water or mixture of alcohol and water, by weight (DOT) see LCM000
LEAD BOTTOMS see LDY000
LEAD BROMATE see LCO000
LEAD BROWN see LCX000
LEAD CARBONATE see LCP000
LEAD(2+) CARBONATE see LCP000
LEAD CHLORIDE see LCQ000
LEAD(2+) CHLORIDE see LCQ000
LEAD(II) CHLORIDE see LCQ000
LEAD(II) CHLORITE see LCQ300
LEAD CHROMATE see LCR000
LEAD CHROMATE(VI) see LCR000
LEAD CHROMATE, BASIC see LCS000
LEAD CHROMATE MOLYBDATE SULFATE RED see MRC000
LEAD CHROMATE OXIDE (MAK) see LCS000
LEAD CHROMATE, RED see LCS000
LEAD CHROMATE, SULPHATE and MOLYBDATE see LDM000
LEAD COMPOUNDS see LCT000
LEAD(II) CYANIDE see LCU000
LEAD CYANIDE (DOT) see LCU000
LEAD DIACETATE see LCV000
LEAD DIACETATE TRIHYDRATE see LCJ000
LEAD DIBASIC ACETATE see LCV000
LEAD DIBASIC PHOSPHITE see LCV100
LEAD DICHLORIDE see LCQ000
LEAD DIFLUORIDE see LDF000
LEAD DIMETHYLDITHIOCARBAMATE see LCW000
LEAD DINITRATE see LDO000
LEAD DIOXIDE see LCX000
LEAD DIPERCHLORATE see LDS499
LEAD DIPHENYL ACID PROPIONATE see LCZ000
LEAD DIPHENYL NITRATE see LDA000
LEAD DIPICRATE see LDA500
LEAD DISODIUM EDTA see LDB000
LEAD DISODIUM ETHYLENEDINITRILOTETRACETATE see LDB000
LEAD DROSS see LDC000
LEAD DROSS (DOT) see LDY000
LEAD(II) EDTA COMPLEX see LDD000
LEAD(2-), ((ETHYLENEDINITRILO)TETRAACETATO)-, DISODIUM see LDB000
LEAD FLAKE see LCF000
LEAD FLUOBORATE see LDE000
LEAD(II) FLUORIDE see LDF000
LEAD FLUORIDE (DOT) see LDF000
LEAD(II) FLUOROSILICATE see LDG000
LEAD GLYCERONITRATE see LDH000
LEAD HYPONITRITE see LDI000
LEAD HYPOPHOSPHITE see LDJ000
LEAD IMIDE see LDK000
LEAD LACTATE see LDL000
LEAD-MOLYBDENUM CHROMATE see LDM000
LEAD MONONITRORESORCINATE (DRY) (DOT) see LDP000
LEAD MONOSUBACETATE see LCH000
LEAD MONOXIDE see LDN000
LEAD NAPHTHENATE see NAS500
LEAD NITRATE see LDO000
LEAD(2+) NITRATE see LDO000
LEAD(II) NITRATE see LDO000
LEAD(II) NITRATE (1:2) see LDO000
LEAD NITRORESORCINATE see LDP000
LEAD(II) OLEATE (1:2) see LDQ000
LEAD ORTHOPHOSPHATE see LDU000
LEAD ORTHOPLUMBATE see LDS000
LEAD OXIDE see LDN000
LEAD(II) OXIDE see LDN000
LEAD(IV) OXIDE see LCX000
LEAD OXIDE BROWN see LCX000

LEAD OXIDE PHOSPHONATE, HEMIHYDRATE see LCV100
LEAD OXIDE RED see LDS000
LEAD OXIDE YELLOW see LDN000
LEAD PERCHLORATE see LDS499
LEAD(2+) PERCHLORATE see LDS499
LEAD(II) PERCHLORATE see LDS499
LEAD PERCHLORATE, solid or solution (DOT) see LDS499
LEAD(II) PERCHLORATE, HEXAHYDRATE (1:2:6) see LDT000
LEAD PEROXIDE (DOT) see LCX000
LEAD PHOSPHATE see LDU000
LEAD(2+) PHOSPHATE see LDU000
LEAD PHOSPHATE (3:2) see LDU000
LEAD(II) PHOSPHATE (3:2) see LDU000
LEAD(II) PHOSPHINATE see LDJ000
LEAD PHOSPHITE, dibasic (DOT) see LCV100
LEAD PICRATE see PID100
LEAD PICRATE (dry) (DOT) see PID100
LEAD POTASSIUM THIOCYANATE see LDV000
LEAD PROTOXIDE see LDN000
LEAD S2 see LCF000
LEAD SCRAP see LDC000
LEAD SILICATE see LDW000
LEAD STEARATE see LDX000
LEAD STYPHNATE see LEE000
LEAD STYPHNATE (dry) (DOT) see LEE000
LEAD STYPHNATE, wetted or lead trinitroresorcinate, wetted with not <20% water or mixt. (DOT) see LEE000
LEAD SUBACETATE see LCH000
LEAD(II) SULFATE (1:1) see LDY000
LEAD SULFATE, solid, containing more than 3% free acid (DOT) see LDY000
LEAD SULFIDE see LDZ000
LEAD SUPEROXIDE see LCX000
LEAD(II) TARTRATE (1:1) see LEA000
LEAD TETRACETATE see LEB000
LEAD TETRACHLORIDE see LEC000
LEAD, TETRAETHYL- see TCF000
LEAD TETRAOXIDE see LDS000
LEAD(II) THIOCYANATE see LEC500
LEAD TITANATE see LED000
LEAD TITANATE ZIRCONATE see LED100
LEAD TITANIUM ZIRCONIUM OXIDE see LED100
LEAD TREE see LED500
LEAD TRINITRORESORCINATE see LEE000
LEAD(II) TRINITROSOBENZENE-1,3,5-TRIOXIDE see LEF000
LEAD(II) TRINITROSOPHLOROGLUCINOLATE see LEF000
LEAD TRIPROPYL see TNA500
LEAD, TRIS(LAUROYLOXY)PHENYL- see PFL000
LEAD ZIRCONATE TITANATE see LED100
LEAD ZIRCONIUM TITANATE see LED100
LEAD ZIRCONIUM TITANIUM OXIDE see LED100
LEAF ALCOHOL see HFE000
LEAF ALDEHYDE see HFA500
LEAF DROP see SFS500
LEAF GREEN see CMJ900
LEALGIN COMPOSITUM see CLY500
LEANDIN see COH250
LEATHER BLUE G see ADE500
LEATHER BUSH see LEF100
LEATHER FAST RED B see CMM330
LEATHER FLOWER see CMV390
LEATHER GREEN B see FAE950
LEATHER GREEN SF see FAF000
LEATHER ORANGE EXTRA see CMM220
LEATHER ORANGE HR see PEK000
LEATHER PURE BLUE HB see BJI250
LEATHER RED G see CMM300
LEATHER RED HT see GJI400
LEATHERWOOD see LEF100

LEAVER WOOD see LEF100
LEBAYCID see FAQ900
LEBON 15 HYDROCHLORIDE see DYA850
LEBON TM 16 see HCQ525
LE CAPTANE (FRENCH) see CBG000
LECASOL see FOS100
LECENINE see HOC500
LECITHIN see LEF180
LECITHIN-BOUND IODINE see LEF200
LECITHIN IODIDE see LEF200
LECTIN see SKW800
LECTOPAM see BMN750
LECTRAPEL see DTS500
LEDAKRIN see LEF300
LEDAKRIN see NFW500
LEDCLAIR see CAR780
LEDERCILLIN VK see PDT750
LEDERFEN see BGL250
LEDERKYN see AKO500
LEDERLE AA223 see AFI625
LEDERMYCIN see MIJ500
LEDERMYCIN HYDROCHLORIDE see DAI485
LE DINITROCRESOL-4,6 (FRENCH) see DUS700
LEDON 11 see TIP500
LEDON 12 see DFA600
LEDON 114 see FOO509
LEDOSTEN see DJT400
LEFEBAR see EOK000
LEGENTIAL see GCU050
LEGUMEX see CLN750
LEGUMEX D see DGA000
LEGUMEX DB see CIR250
LEGUMEX EXTRA see BAV000
LEGURAME see CBL500
LEGURAME TS see UTA300
LEHYDAN see DKQ000
LEINOLEIC ACID see LGG000
LEIOPLEGIL see LEF400
LEIOPYRROLE see LEF400
LEIPZIG YELLOW see LCR000
LEIVASOM see TIQ250
LEKAMIN see TNF500
LEKTAN see ENJ600
LEKUTHERM 2159 see DKM500
LEKUTHERM X 100 see DKM500
LEMAC 1000 see AAX250
LEMBROL see DCK759
LEMISERP see RDK000
LEMMATOXIN see LEF800
LEMOFLUR see SHF500
LEMOL see PKP750
LEMOL 5-88 see PKP750
LEMOL 5-98 see PKP750
LEMOL 12-88 see PKP750
LEMOL 16-98 see PKP750
LEMOL 24-98 see PKP750
LEMOL 30-98 see PKP750
LEMOL 51-98 see PKP750
LEMOL 60-98 see PKP750
LEMOL 75-98 see PKP750
LEMOL GF-60 see PKP750
LEMON AJAX see AFG625
LEMON CHROME see BAK250
LEMONENE see BGE000
LEMONGRAS OEL (GERMAN) see LEG000
LEMONGRASS OIL EAST INDIAN see LEG000
LEMONGRASS OIL WEST INDIAN see LEH000
LEMONILE see LEH100
LEMON OIL see LEI000
LEMON OIL, distilled see LEI030
LEMON OIL, desert type, coldpressed see LEI025
LEMON OIL, COLDPRESSED (FCC) see LEI000
LEMON OIL, EXPRESSED see LEI000
LEMONOL see DTD000
LEMON PETITGRAIN OIL see LEJ000
LEMON YELLOW see BAK250

LEVOMYCIN see EAD500
LEVOMYSOL HYDROCHLORIDE see LFA020
LEVONORADRENALINE see NNO500
LEVONOREPINEPHRINE see NNO500
LEVONORGESTREL see NNQ525
LEVOPHACETOPERANE HYDROCHLORIDE see LFO000
LEVOPHACETOPERAN HYDROCHLORIDE see LFO000
LEVOPHED see NNO500
LEVOPHED HYDROCHLORIDE see NNP000
LEVOPROMAZINE see MCI500
LEVORENIN see VGP000
LEVORIN see LFF000
LEVOROXINE see LFG050
LEVORPHAN see LFG000
LEVORPHANOL see LFG000
LEVORPHANOL TARTRATE see DYF000
LEVORPHAN TARTRATE see DYF000
LEVOSAN see LFG100
LEVOTHROID see LFG050
LEVOTHYL see MDO775
LEVOTHYL see MDP770
LEVOTHYROX see LFG050
LEVOTHYROXINE see TFZ275
LEVOTHYROXINE SODIUM see LFG050
LEVOTHYROXINE SODIUM see LFG050
LEVOTOMIN see MCI500
LEVOXADROL HYDROCHLORIDE see LFG100
LEVUGEN see LFI000
LEVULIC ACID see LFH000
LEVULINIC ACID see LFH000
LEVULINIC ACID, 2-AMINO-5-HYDROXY-, l- see HND250
LEVULINIC ACID, BUTYL ESTER see BRR700
LEVULINIC ACID, ETHYL ESTER see EFS600
LEVULINIC ACID, TRIPHENYLSTANNYL ESTER see TMW250
LEVULOSE see LFI000
LEWISITE see CLV000
LEWISITE (ARSENIC COMPOUND) see CLV000
LEWISITE II see BIQ250
LEWISITE I OXIDE see DEW000
LEWIS-RED DEVIL LYE see SHS000
LEXATOL see SPC500
LEXEIN X 350 see HID100
LEXIBIOTICO see ALV000
LEXOMIL see BMN750
LEXONE see MQR275
LEXOTAN see BMN750
LEXOTANIL see BMN750
LEY-CORNOX see BAV000
LEYMIN see BAV000
LEYSPRAY see CIR250
LEYTOSAN see ABU500
LEYTOSAN see PFP500
LF 62 see POA250
LFA 2043 see GIA000
LF BOWSEI PW 2 see ZJS400
LFP 83 see FJT100
LF-PW 2 see ZJS400
LG 61 see DJR700
LG 50043 see TKX250
L.G. 11,457 HYDROCHLORIDE see DHS200
LGYCOSPERSE S-20 see PKL030
LH see ILE000
LH see LIU300
AS-LH see LIU302
LH 3012 see ZMA000
LH ANTISERUM see LIU302
LHAS see LIU302
LH RELEASING FACTOR see LIU370
LH-RELEASING HORMONE see LIU370
LH-RF see LIU370
LHRH see LIU370
LH-RH see LIU370

LH-RH see LIU380
LHRH DIACETATE see LIU380
LHRH DIACETATE TETRAHYDRATE see LIU400
LH-RH/FSH-RH see LIU370
LH 30/Z see ZMA000
LIALET 123-1 see AFJ155
LIANE BON GARCON (HAITI) see CMV390
LIATRIS see DAJ800
LIATRIX OLEORESIN see DAJ800
LIBAVIUS FUMING SPIRIT see TGC250
LIBERETAS see DCK759
LIBEXIN see LFJ000
LIBIA SCARLET 4BS see CMO870
LIBIOLAN see MQU750
LIBRATAR see CDQ500
LIBRAX see LFK000
LIBRININ see LFK000
LIBRITABS see LFK000
LIBRIUM see LFK000
LIBRIUM see MDQ250
LIBRIUM HYDROCHLORIDE see MDQ250
LICABILE HYDROCHLORIDE see LFK200
LICARAN HYDROCHLORIDE see LFK200
LICARBIN see HAQ570
LICAREOL ACETATE see LFY600
LICHENIC ACID see FOU000
LICHENIFORMIN A see LFL500
LICHENOL see IRL000
LICHTGRUEN (GERMAN) see FAF000
LICIDRIL see DPE000
LICORICE see LFN300
LICORICE COMPONENT FM 100 see LFN000
LICORICE EXTRACT see LFN300
LICORICE ROOT see LFN300
LICORICE ROOT EXTRACT see LFN300
LICORICE VINE see RMK250
LIDA-MANTLE see DHK400
LIDAMYCIN CREME see NCG000
LIDANAR see MON750
LIDANIL see MON750
LIDENAL see BBQ500
LIDEPRAN HYDROCHLORIDE see LFO000
LIDOCAINE see DHK400
LIDOCAINE HYDROCHLORIDE see DHK600
LIDOFENIN see LFO300
LIDOL see DAM600
LIDOL see DAM700
LIDONE see MRB250
LIDOTHESIN HYDROCHLORIDE see DHK600
LIENOMYCIN see LFP000
LIFEAMPIL see AIV500
LIFEAMPIL see AOD125
LIFENE see MNZ000
LIFIBRATE see MON600
LIFRIL see FLZ050
LIGAND 222 see LFQ000
LIGHT ACRYLATE EC-A see EHG030
LIGHT ACRYLATE IB-XA see IHX700
LIGHT ACRYLATE THF-A see TCS100
LIGHT CAMPHOR OIL see CBB500
LIGHT CATALYTICALLY CRACKED DISTILLATE see DXG840
LIGHT CATALYTICALLY CRACKED NAPHTHA see NAQ540
LIGHT CATALYTIC REFORMED NAPHTHA see NAQ550
LIGHT ESTER ECA see EHG030
LIGHT ESTER PO-A see PER250
LIGHT FAST YELLOW ES see SGP500
LIGHT GAS OIL see GBW025
LIGHT GREEN FCF YELLOWISH see FAF000
LIGHT GREEN LAKE see FAF000
LIGHT GREEN N see AFG500
LIGHTHOUSE CHROME BLACK 6B see CMP880

LIGHTHOUSE CHROME BLUE 2R see HJF500
LIGHTHOUSE CHROME ORANGE see NEY000
LIGHTHOUSE CHROME RED B see CMG750
LIGHT NAPHTHENIC DISTILLATE see MQV810
LIGHT NAPHTHENIC DISTILLATE, SOLVENT EXTRACT see MQV860
LIGHT NAPHTHENIC DISTILLATES (PETROLEUM) see MQV810
LIGHT OIL OF CAMPHOR see CBB500
LIGHT ORANGE CHROME see LCS000
LIGHT ORANGE R see DVB800
LIGHT PARAFFINIC DISTILLATE see MQV815
LIGHT PARAFFINIC DISTILLATE, SOLVENT EXTRACT see MQV862
LIGHT POLYMER see VRU300
LIGHT RED see IHC450
LIGHT RED RS see NAP100
LIGHT SF YELLOWISH (BIOLOGICAL STAIN) see FAF000
LIGHT SPAR see CAX750
LIGHT STRAIGHT-RUN NAPHTHA see NAQ560
LIGHT YELLOW JB see DEU000
LIGHT YELLOW JBR see CMS208
LIGHT YELLOW JBV see CMS210
LIGHT YELLOW JBVT see CMS210
LIGNALOE OIL see BMA550
LIGNASAN see EME100
LIGNASAN FUNGICIDE see BJT250
LIGNASAN-X see BJT250
LIGNOCAINE see DHK400
LIGNOCAINE HYDROCHLORIDE see DHK600
LIGNOSULFONIC ACID, SODIUM SALT see LFQ500
LIGNYL ACETATE see DTC000
LIGROIN see PCT750
LIGUSTRUM VULGARE see PMD550
LIHOCIN see CMF400
LIKUDEN see GKE000
LILACILLIN see SNV000
LILAILA see CDM325
LILAS (HAITI, DOMINICAN REPUBLIC) see CDM325
LILAS de NUIT (HAITI) see DAC500
LILIA-O-KE-AWAWA (HAWAII) see LFT700
LILIAL see LFT000
LILIAL-METHYLANTHRANILATE, Schiff's base see LFT100
LILLY 1516 see DQA400
LILLY 01516 see DQA400
LILLY 22113 see AGL875
LILLY 22451 see MDU500
LILLY 22641 see RDF000
LILLY 34,314 see DRP800
LILLY 36,352 see DUV600
LILLY 37231 see LEZ000
LILLY 38253 see SFQ500
LILLY 39435 see CCR890
LILLY 40602 see TEY000
LILLY-68618 see MRX000
LILLY 69323 see FAP100
LILLY 99094 see VGU750
LILLY 99638 see PAG075
LILLY 126714 see BMO300
LILLY COMPOUND LY133314 see TMP175
LILLY 99638 HYDRATE see CCR850
LILLY-OF-THE-VALLEY see LFT700
LILO see PDO750
LILYAL see LFT000
LILYL ALDEHYDE see CMS850
LILY OF THE FIELD see PAM780
LILY-OF-THE-VALLEY BUSH see FBP520
LIMARSOL MALAGRIDE see ABX500
LIMAS see LGZ000
LIMAWOOD EXTRACT see LFT800
LIMBIAL see CFZ000

LIMBUX see CAT225
LIME see CAU500
LIME ACETATE see CAL750
LIME, BURNED see CAU500
LIME CHLORIDE see HOV500
LIME CHLORIDE (CA(CLO)$_{2.2}$CA(OH)$_2$) see CAT235
LIMED ROSIN see CAW500
LIME MILK see CAT225
LIME-NITROGEN (DOT) see CAQ250
LIME OIL see OGM850
LIME OIL, distilled (FCC) see OGM850
LIME PYROLIGNITE see CAL750
LIMESTONE (FCC) see CAO000
LIME-SULFUR see CAX800
LIME SULPHUR see CAX800
LIME, UNSLAKED (DOT) see CAU500
LIME WATER see CAT225
LIMIT see ABY300
LIMIT see DKE800
LIMONENE see MCC250
1-LIMONENE see MCC500
d-LIMONENE see LFU000
(+)-R-LIMONENE see LFU000
d-(+)-LIMONENE see LFU000
dl-LIMONENE see MCC250
LIMONENE DIOXIDE see LFV000
(−)-LIMONENE (FCC) see MCC500
LIMONENE OXIDE see CAL000
LIMONITE see LFW000
LIMONIUM NASHII see MBU750
LIN 1418 see EPD100
LINADRYL HYDROCHLORIDE see LFW300
LINALOL see LFX000
LINALOL ACETATE see LFY600
LINALOOL see LFX000
p-LINALOOL see LFY000
LINALOOL ACETATE see LFY600
LINALOOL, DEHYDRO- see LFY333
LINALOOL ISOBUTYRATE see LGB000
LINALOOL OXIDE see LFY500
LINALOOL TETRAHYDRIDE see LFY510
LINALYL ACETATE see LFY600
LINALYL ALCOHOL see LFX000
LINALYL-o-AMINOBENZOATE see APJ000
LINALYL ANTHRANILATE see APJ000
LINALYL BENZOATE see LFZ000
LINALYL CINNAMATE see LGA000
LINALYL FORMATE see LGA050
LINALYL ISOBUTYRATE see LGB000
LINALYL ISOVALERATE see LGC000
LINALYL PHENYLACETATE see LGC050
LINALYL PROPIONATE see LGC100
LINALYL α-TOLUATE see LGC050
LINAMARIN see GFC100
LINAMIN see FQJ100
LINAMPHETA see AOB250
LINAMPHETA see TEP500
LINARIS see TKX000
LINARODIN see MDW750
LINASEN see EOK000
LINAXAR see PFJ000
LINCOCIN see LGC200
LINCOCIN see LGD000
LINCOCIN see LGE000
LINCOCIN HYDROCHLORIDE HYDRATE see LGC200
LINCOLCINA see LGD000
LINCOLNENSIN see LGD000
LINCOLN RED 1002 see CJD500
LINCOMYCIN see LGD000
LINCOMYCINE (FRENCH) see LGD000
LINCOMYCIN HYDROCHLORIDE see LGE000
LINCOMYCIN, HYDROCHLORIDE, HEMIHYDRATE see LGC200
LINCOMYCIN, HYDROCHLORIDE HYDRATE see LGC200
LINCOMYCIN HYDROCHLORIDE MONOHYDRATE see LGC200
LINCTUSSAL see DEE600
LINDAGRAIN see BBQ500

LINDAN see DGP900
α-LINDANE see BBQ000
β-LINDANE see BBR000
Δ-LINDANE see BFW500
LINDANE (ACGIH, DOT, USDA) see BBQ500
LINDATOX see EEM000
LINDEROL see NCQ820
LIN-DIBENZANTHRACENE see PAV000
LINDOL see TNP500
LINDRIDE see MPP000
LINEAR ALKYLBENZENE A-315 see LGE200
LINEAR ALKYLBENZENE SULFONATE see AFO500
LINEAR ALKYLBENZENE SULFONATE, MAGNESIUM SALT see LGF800
LINEAR ALKYLBENZENE SULFONATE, SODIUM SALT see LGF825
LINEAR ALKYLBENZENE SULPHONATE see AFO500
LINE RIDER see TAA100
LINESTRENOL see NNV000
LINEVOL 7-9 see LGF875
LINEVOL 79 see LGF875
LINEVOL 911 see AFJ150
LINEX 4L see DGD600
LINFADOL see AFJ400
LINFOLIZIN see CDO500
LINFOLYSIN see CDO500
LINGEL see BRF500
LINGRAINE see EDC500
LINGRAN see EDC500
LINGUSORBS see PMH500
LIN-NAPHTHOANTHRACENE see PAV000
LINODIL see HFG550
LINOLEAMIDE see LGF900
LINOLEAMIDE, N-(α-METHYLBENZYL)- see MHO200
LINOLEIC ACID see LGG000
9,12-LINOLEIC ACID see LGG000
LINOLEIC ACID (oxidized) see LGH000
LINOLEIC ACID AMIDE see LGF900
LINOLEIC ACID mixed with OLEIC ACID see LGI000
LINOLEIC DIETHANOLAMIDE see BKF500
LINOLENATE HYDROPEROXIDE see PCN000
LINOLENIC ACID see OAX100
LINOLENIC HYDROPEROXIDE see PCN000
(LINOLEOYLOXY)TRIBUTYLSTANNANE see LGJ000
LINORMONE see CIR250
LINOROX see DGD600
LINSEED OIL see LGK000
LINTON see CLY500
LINTOX see BBQ500
LINUREX see DGD600
LINURON see DGD600
LINURON (herbicide) see DGD600
LIOMYCIN see HGP550
LIONOGEN BLUE R see IBV050
LIONOGEN MAGNETA R see DTV360
LIONOL YELLOW FG 1310 see CMS208
LIONOL YELLOW GGR see CMS210
LION'S BEARD see PAM780
LION'S MOUTH see FOM100
LIORESAL see BAC275
LIOTHYRONIN see LGK050
LIOTHYRONINE see LGK050
l-LIOTHYRONINE see LGK050
LIOXIN see VFK000
LIPAL 30W see PJY100
LIPAL 4LA see DXY000
LIPAL 400-DL see PJY100
LIPAL 20-OA see PJW500
LIPAMID see ARQ750
LIPAMIDE see DXN709
α-LIPAMIDE see DXN709
4-(dl-α-LIPAMIDO)BUTYRIC ACID see LGK100

LIPAMONE see DHB500
LIPAN see DUS700
LIPARITE see CAS000
LIPARON see DPA000
LIPASE see GGA800
LIPASE AP6 see LGK150
LIPAVIL see ARQ750
LIPAVLON see ARQ750
LIPAZIN see GGA800
LIPENAN see LGK200
LIPFEN see FJT100
LIPHADIONE see CJJ000
LIPID CRIMSON see SBC500
LIPIDE 500 see ARQ750
LIPIDSENKER see ARQ750
LIPILISOL see NMV300
LIPIODOL see LGK225
LIPIODOL LAFAY see LGK225
LIPOACIN see DXN709
LIPOAMID see DXN709
LIPOAMIDE see DXN709
α-LIPOAMIDE see DXN709
LIPOCLIN see CMV700
LIPOCOL L-4 see DXY000
LIPOCOL O-n20 see PJW500
LIPOCOL C2 see PJT300
LIPOCTON see DXN709
LIPO-DIAZINE see PPP500
LIPO EGMS see EJM500
LIPOFACTON see ARQ750
LIPO-FLURBIPROFEN AXETIL see FJT100
LIPOGLUTAREN see HMC000
LIPO GMS 410 see OAV000
LIPO GMS 450 see OAV000
LIPO GMS 600 see OAV000
LIPO-HEPIN see HAQ500
LIPOIC ACID see DXN800
α-LIPOIC ACID see DXN800
α-LIPOIC ACID AMIDE see DXN709
LIPOICIN see DXN709
LIPO-LEVAZINE see PPP500
LIPO-LUTIN see PMH500
LIPOMID see ARQ750
LIPONEURINA see DXO300
α-LIPONIC ACID see DXN800
LIPONORM see ARQ750
α-LIPONSAEURE (GERMAN) see DXN800
LIPOPILL see DTJ400
LIPOPOLYSACCHARIDE see PPQ600
LIPOPOLYSACCHARIDE, from B. ABORTUS Bang. see LGK350
LIPOPOLYSACCHARIDE (E. COLI) see LGK375
LIPOPOLYSACCHARIDE (E. COLI SEROTYPE 026:B6) see LGK375
LIPOPOLYSACCHARIDE, -ESC see LGK375
LIPOPOLYSACCHARIDE, ESCHERICHIA COLI see EDK750
LIPOPOLYSACCHARIDE, ESCHERICHIA COLI see LGK400
LIPOREDUCT see ARQ750
LIPORIL see ARQ750
LIPOSAN see DXN800
LIPOSID see ARQ750
LIPOSOMAL FLURBIPROFEN AXETIL see FJT100
LIPOSORB L-20 see PKG000
LIPOSORB O-20 see PKL100
LIPOSORB O-20 see SKV100
LIPOSORB S-20 see PKL030
LIPOSORB S-20 see SKV150
LIPOSORB O see SKV100
LIPOSORB P see MRJ800
LIPOSORB S see SKV150
LIPOSORB SQ0 see SKV170
LIPOSTAT see LGK500
LIPOTHION see DXN800
LIPOTRIL see CMF750
LIPOVAL A see ARW800
LIPOZYME see DXN709
LIPPIA CITRIODORA OIL see VIK500
LIPRIN see ARQ750

LIPRINAL see ARQ750
LIPTAN see IIU000
LIPUR see GCK300
LIQUACILLIN see BDY669
LIQUADIAZINE see PPP500
LIQUAEMIN see HAQ500
LIQUAEMIN SODIUM see HAQ550
LIQUAGESIC see HIM000
LIQUAMYCIN see TBX000
LIQUAMYCIN INJECTABLE see HOI000
LIQUAMYCIN LA 200 see HOH500
LIQUA-TOX see WAT200
LIQUEFIED CARBON DIOXIDE see LGL000
LIQUEFIED PETROLEUM GAS see LGM000
LIQUEFIED PETROLEUM GAS (DOT) see IIC000
LIQUEMIN see HAQ500
LIQUEMIN see HAQ550
LIQUIBARINE see BAP000
LIQUIDAMBAR STYRACIFLUA see SOY500
LIQUID BRIGHT PLATINUM see PJD500
LIQUID CAMPHOR see CBB500
LIQUID CARBONIC GAS see LGL000
LIQUID DERRIS see RNZ000
LIQUID DETERGENTS containing AES see DBB450
LIQUID ETHYLENE see EIO000
LIQUIDOW see CAO750
LIQUID PITCH OIL see CMY825
LIQUID ROSIN see TAC000
LIQUIGEL see AHC000
LIQUIMETH see MDT750
LIQUIPHENE see ABU500
LIQUIPRIN see SAH000
LIQUIPRON see LGM200
LIQUI-SAN see MLH000
LIQUI-STIK see NAK500
LIQUITAL see EOK000
LIQUOPHYLLINE see TEP000
LIQUORICE see GIG000
LIRANOL see DQA600
LIRANOX see CIR500
LIRCAPYL see PFB350
LIRIO see AHI635
LIRIO (SPANISH) see BAR325
LIRIO (SPANISH) see SLB250
LIRIO CALA (SPANISH) see CAY800
LIROBETAREX see CJX750
LIRO CIPC see CKC000
LIROHEX see TCF250
LIROMAT see OPK250
LIROMATIN see ABX250
LIRONION see LGM300
LIRONOX see BQZ000
LIROPON see DGI400
LIROPREM see PAX250
LIROSTANOL see ABX250
LIROTAN see EIR000
LIROTHION see PAK000
LISACORT see PLZ000
LISERGAN see ABH500
LISERGAN HYDROCHLORIDE see DPI000
LISINOPRIL see LGM400
LISKONUM see LGZ000
LISOLIPIN see SKJ300
LISSAMINE AMARANTH AC see FAG020
LISSAMINE BLUE AR see CMM070
LISSAMINE FAST YELLOW AE see SGP500
LISSAMINE FAST YELLOW AES see SGP500
LISSAMINE GREEN B see ADF000
LISSAMINE GREEN BN see ADF000
LISSAMINE GREEN G see FAE950
LISSAMINE GREEN V see CMM100
LISSAMINE GREEN V 200 see CMM100
LISSAMINE LAKE GREEN SF see FAF000
LISSAMINE RED 2G see CMM300
LISSAMINE RED 6B see CMM400
LISSAMINE TURQUOISE VN see ADE500
LISSAMINE ULTRA BLUE AR see CMM070
LISSAMINE YELLOW AE see SGP500
LISSATAN AC see BLX000
LISSENPHAN see GGS000

LISSOLAMINE see HCQ500
LISTENON see HLC500
LISTICA see HNJ000
LISULFEN see AKO500
LISURIDE HYDROGEN MALEATE see LJE500
LITAC see ADY500
LITALER see HOO500
P-LITE 500 see CAT775
LITEX CA see SMR000
LITHALURE see LHQ100
LITHAMIDE see LGT000
LITHANE see LGZ000
LITHARGE see LDN000
LITHARGE YELLOW L-28 see LDN000
LITHIC ACID see UVA400
LITHICARB see LGZ000
LITHIDRONE see AFT500
LITHINATE see LGZ000
LITHIUM see LGO000
LITHIUM ACETATE see LGO100
LITHIUM ACETYLIDE see LGP875
LITHIUM ACETYLIDE COMPLEXED with ETHYLENEDIAMINE see LGQ000
LITHIUM ACETYLIDE-ETHYLENEDIAMINE COMPLEX see LGQ000
LITHIUM ALANATE see LHS000
LITHIUM ALUMINOHYDRIDE see LHS000
LITHIUM ALUMINUM HYDRIDE (DOT) see LHS000
LITHIUM ALUMINUM HYDRIDE, ETHEREAL (DOT) see LHS000
LITHIUM ALUMINUM TETRAHYDRIDE see LHS000
LITHIUM ALUMINUM TRI-tert-BUTOXYHYDRIDE see LGS000
LITHIUM AMIDE see LGT000
LITHIUM AMIDE, POWDERED see LGT000
LITHIUM ANTIMONIOTHIOMALATE see LGU000
LITHIUM ANTIMONY THIOMALATE see LGU000
LITHIUM ANTIMONY THIOMALATE NONAHYDRATE see AQE500
LITHIUM AZIDE see LGV000
LITHIUM BENZENEHEXOXIDE see LGV700
LITHIUM BENZOATE see LGW000
LITHIUM BIS(TRIMETHYLSILYL)AMIDE see LGX000
LITHIUM BOROHYDRIDE (DOT) see LHT000
LITHIUM BROMIDE see LGY000
LITHIUM tert-BUTANOLATE see LGY100
LITHIUM tert-BUTOXIDE see LGY100
LITHIUM tert-BUTYLATE see LGY100
LITHIUM CARBONATE see LGZ000
LITHIUM CARBONATE (2:1) see LGZ000
LITHIUM CARMINE see LHA000
LITHIUM CHLORIDE see LHB000
LITHIUM CHLOROACETYLIDE see LHC000
LITHIUM CHLOROETHYNIDE see LHC000
LITHIUM CHROMATE see LHD000
LITHIUM CHROMATE(VI) see LHD099
LITHIUM CITRATE HYDRATE see LHD150
LITHIUM COMPOUNDS see LHE000
LITHIUM DIAZOMETHANIDE see LHE450
LITHIUM DIETHYL AMIDE see LHE475
LITHIUM-2,2-DIMETHYLTRIMETHYLSILYL HYDRAZIDE see LHE525
LITHIUM FERRO SILICON see LHK000
LITHIUM FLUORIDE see LHF000
LITHIUM FLUORURE (FRENCH) see LHF000
LITHIUM 1,1,2,2,3,3,4,4,5,5,6,6,7,7,8,8,8-HEPTADECAFLUORO-1-OCTANESULFONATE see LHN100
LITHIUM-1-HEPTYNIDE see LHF625
LITHIUM HEXAAZIDOCUPRATE(4-) see LHG000

LITHIUM HYDRIDE see LHH000
LITHIUM HYDRIDE (UN 1414) (DOT) see LHH000
LITHIUM HYDRIDE, fused solid (UN 2805) (DOT) see LHH000
LITHIUM HYDROXIDE see LHI100
LITHIUM HYDROXIDE, solution (DOT) see LHI100
LITHIUM HYDROXIDE, monohydrate or lithium hydroxide, solid (DOT) see LHI100
LITHIUM HYDROXIDE (Li(OH)) (9CI) see LHI100
LITHIUM HYDROXYBUTYRATE see LHM800
LITHIUM γ-HYDROXYBUTYRATE see LHM800
LITHIUM HYPOCHLORITE see LHJ000
LITHIUM HYPOCHLORITE COMPOUND, dry, containing more than 39% available chlorine (DOT) see LHJ000
LITHIUM IRON SILICON see LHK000
LITHIUM LACTATE see LHL000
LITHIUM METAL (DOT) see LGO000
LITHIUM METAL, IN CARTRIDGES (DOT) see LGO000
LITHIUM MONOBROMIDE see LGY000
LITHIUM NITRIDE see LHM000
LITHIUM-4-NITROTHIOPHENOXIDE see LHM750
LITHIUM OCTADECANOATE see LHQ100
LITHIUM OXIDE see LHM770
LITHIUM OXYBUTYRATE see LHM800
LITHIUM PENTAMETHYLTITANATE-BIS(2,2'-BIPYRIDINE) see LHM850
LITHIUM PERCHLORATE see LHM875
LITHIUM PERCHLORATE TRIHYDRATE see LHN000
LITHIUM PERFLUOROOCTANE SULFONATE see LHN100
LITHIUM PEROXIDE see LHO000
LITHIUMRHODANID see TFF500
LITHIUM SILICON see LHP000
LITHIUM SODIUM NITROXYLATE see LHQ000
LITHIUM STEARATE see LHQ100
LITHIUM SULFATE (2:1) see LHR000
LITHIUM SULFOCYANATE see TFF500
LITHIUM SULPHATE see LHR000
LITHIUM TETRAAZIDOALUMINATE see LHR650
LITHIUM TETRAAZIDOBORATE see LHR675
LITHIUM TETRADEUTEROALUMINATE see LHR700
LITHIUM TETRAHYDROALUMINATE see LHS000
LITHIUM TETRAHYDROBORATE see LHT000
LITHIUM TETRAMETHYLBORATE see LHT400
LITHIUM TETRAMETHYL CHROMATE(II) see LHT425
LITHIUM TETRAZIDO ALUMINATE see LHU000
LITHIUM THIOCYANATE see TFF500
LITHIUM-TIN ALLOY see LHV000
LITHIUM TRIETHYLSILYL AMIDE see LHV500
LITHOBID see LGZ000
LITHOCHOLIC ACID see LHW000
LITHOCHOLIC ACID GLYCINE CONJUGATE see HND200
LITHOCHOLIC ACID TAURINE CONJUGATE see LHW100
LITHOCHOLYLGLYCINE see HND200
LITHOCHOLYLTAURINE see LHW100
LITHOFOR BROWN A see NBG500
LITHOGRAPHIC STONE see CAO000
LITHOL see IAD000
LITHOL FAST SCARLET RN see MMP100
LITHOLITE see LHQ100
LITHOL RED see NAP100

LITHOL RED 3580 see NAP100
LITHOL RED 17676 see NAP100
LITHOL RED B see NAP100
LITHOL RED LAKE see NAP100
LITHOL RED R see NAP100
LITHOL RED RB EXTRA see NAP100
LITHOL RED RL 151 see NAP100
LITHOL RED RS see NAP100
LITHOL RED 3GS see NAP100
LITHOL RED SODIUM SALT see NAP100
LITHOL RED TONER see NAP100
LITHOL RUBIN B see CMS155
LITHOL RUBINE see CMS155
LITHOL RUBINE BNA see CMS155
LITHOL SCARLET K 3700 see CMS148
LITHOL TONER see NAP100
LITHOL TONER EXTRA LIGHT 5000 see
NAP100
LITHOL TONER SODIUM SALT RT 314 see
NAP100
LITHOL TONER YA 8003 see NAP100
LITHONATE see LGZ000
LITHOPONE see LHX000
LITHOSOL DEEP BLUE V see BGB275
LITHOSOL ORANGE R BASE see NMP500
LITHOSOL RED C see CMS150
LITHOSOL RED CLM see CMS150
LITHO SOL SCARLET 3BI see CMG750
LITHOSPERMUM RUDERALE, root extract
see LHX300
LITHOTABS see LGZ000
LITICON see DOQ400
LITLURE A see FAS100
LITLURE B see TBX300
LITMOMYCIN see GJS000
LITSEA CUBEBA OIL see LHX325
LITSOEINE see LBO200
LIV 1176 see DVK200
LIVAZONE see FNF000
LIVER STARCH see GHK300
LIVERT see EKS120
LIVIATIN see DYE425
LIVICLINA see CCS250
LIVIDOMYCIN see LHX350
LIVIDOMYCIN B see DAS600
LIVIDOMYCIN SULFATE see LHX360
LIVODYMYCIN see LHX350
LIVONAL see EGQ000
LIVOSTIN see LFA200
LIXIL see BON325
L.J. 48 see MBX800
L.J. 206 see CBR675
LJ 206 see CCH125
LK 36 see HDY500
LK 14304-18 see BMM575
LL 1530 see NAC500
LL 1656 see BOM600
L 2642-LABAZ see EDV700
LLD FACTOR see VSZ000
LLONCEFAL see TEY000
LM-16 see CMS125
LM 91 see CJJ000
LM-209 see MCJ400
LM 280 see IDJ500
LM 433 see GKG300
LM-637 see BMN000
LM-2717 see CIR750
LM 2910 see LHX498
LM 2911 see LHX499
LM 2916 see LHX500
LM 2917 see LHX510
LM 2918 see LHX515
LM 2930 see LHX600
LM 22070 see DWD400
L-1601-2-50M see LBD050
LMB 357 see MCB050
LM SEED PROTECTANT see MLH000
LN 2299 see BOL315
LOBAK see CKF500
LOBAMINE see MDT740
LO-BAX see HOV500
LOBELIA, INDIAN TOBACCO see LHY000

LOBELIA INFLATA see CCJ825
LOBELIA NICOTIANIFOLIA Roth ex R. & S.,
extract see LHX700
LOBELIA SIPHILITICA see CCJ825
LOBELINE see LHY000
LOBELINE HYDROCHLORIDE see LHZ000
(−)-LOBELINE HYDROCHLORIDE see
LHZ000
LOBELIN HYDROCHLORIDE see LHZ000
LOBELLOA CARDINALIS see CCJ825
LOBENZARIT DISODIUM see LHZ600
LOBENZARIT SODIUM see LHZ600
LOBETRIN see ARQ750
LOBNICO see LHY000
LOBULANTINA see SCA750
LOBUTEROL see BQE250
LOCALYN see SPD500
LOCOID see HHQ825
LOCOWEED see LHZ700
LOCRON EXTRA see AHA000
LOCRON FLAKES see AHA000
LOCRON POWDER see AHA000
LOCRON SOLUTION see AHA000
LOCULA see SNQ000
LOCUNDIEZIDE see RFZ000
LOCUNDIOSIDE see RFZ000
LOCUST BEAN GUM see LIA000
LOCUTURINE see MPA050
LODESTONE YELLOW YB-57 see DEU000
LODIBON see DII200
LODITAC see OMY700
LODOPIN see ZUJ000
LODOSIN see CBQ500
LODOSYN see CBQ500
LODRONATE see SFX730
LOFEPRAMINE see DLH630
LOFEPRAMINE HYDROCHLORIDE see
IFZ900
LOFETENSIN see LIA400
LOFETENSIN HYDROCHLORIDE see
LIA400
LOFEXIDINE HYDROCHLORIDE see
LIA400
LOFTYL see BOM600
LOGGERHEAD WEED (BARBADOS) see
PIH800
LOGIC see EOE200
LOGRAN see THQ550
LOHA see IGK800
LOISOL see TIQ250
LOKARIN see DPE200
LOKUNDJOSID (GERMAN) see RFZ000
LOKUNDJOSIDE see RFZ000
LabelOL see HMM500
LOLINE see FBP400
LOLINE, N-METHYL- see MLD050
LOLININE see ACP300
LOLITOL see FON200
LOMADES see BDJ600
LOMAPECT see LFJ000
LOMAR D see BLX000
LOMAR LS see BLX000
LOMAR PW see BLX000
LOMBRICERA (PUERTO RICO) see PIH800
LOMBRICERO (CUBA) see APM875
LOMBRISTOP see TEX000
LO MICRON TALC 1 see TAB750
LO MICRON TALC, BC 1621 see TAB750
LO MICRON TALC USP, BC 2755 see TAB750
LOMIDIN see DBL800
LOMIDINE see DBL800
LOMIDINE ISOETHIONATE see DBL800
LOMOTIL see LIB000
LOMPER see MHL000
LOMUDAL see CNX825
LOMUDAS see CNX825
LOMUPREN see DMV600
LOMUSTINE see CGV250
LOMYCIN see GJY000
S-LON see PKQ059
LON 41 see DTM600
LON 798 see GKU300

LON-954 see PFV000
LONACOL see EIR000
LONAMIN see DTJ400
LONARID see HIM000
LONAVAR see AOO125
LONDAX see BAV600
LONDOMIN see RCA200
LONDOMYCIN see MDO250
LONDON PURPLE see LIC000
LONETHYL see LID000
LONETIL see LID000
LONGACILIAN see BFC750
LONGANOCT see BPF500
LONGATIN see NOA000
LONGATIN HYDROCHLORIDE see
NOA500
LONGICIL see BFC750
LONGIFENE see BOM250
LONGIFENE see HGC500
LONGIFOLEN see LID100
(+)-LONGIFOLENE see LID100
d-LONGIFOLENE see LID100
LONGIFOLENE (6CI) see LID100
β-LONGILOBINE see RFP000
LONGIN see AKO500
LONG PEPPER see PCB275
LONGUM see MFN500
LONIDAMINE see DEL200
LONIN 3 see BHO500
LONITEN see DCB000
LONOCOL M see MAS500
LONOMYCIN see LIF000
LONOMYCIN, MONOSODIUM SALT see
LIG000
LONOMYCIN, SODIUM SALT see LIG000
LONTREL see DGJ100
LONTREL 3 see DGJ100
LONZA G see PKQ059
LONZAINE 16S see CDF450
LOOPLURE see DXU830
LOOPLURE INHIBITOR see DXU800
LO/OVRAL see NNL500
LOPATOL see LIG100
LOPEMID see LIH000
LOPEMIN see LIH000
LOPERAMIDE HYDROCHLORIDE see
LIH000
LOPERYL see LIH000
LOPHINE see TMS750
LOPID see GCK300
LOPIRIN see MCO750
LOPRAMINE HYDROCHLORIDE see
IFZ900
LOPREMONE see TNX400
LOPRESOR see SBV500
LOPRESS see HGP500
LOPRESSOR see SBV500
LOPURIN see ZVJ000
LOQUAT see LIH200
LORAMET see MLD100
LORAMINE AMB 13 see GHA050
L. ORANGE Z2010 see FAG150
LORAX see CFC250
LORAZEPAM see CFC250
LORDS-AND-LADIES see ITD050
LORETIN see IEP200
LOREX see DGD600
LORFAN see AGI000
LORFAN see LIH300
LORFAN TARTRATE see LIH300
LORIDIN see TEY000
LORINAL see CDO000
LORMETAZEPAM see MLD100
LORMIN see CBF250
LORO see DYA200
LOROL see DXV600
LOROL 20 see OEI000
LOROL 22 see DAI600
LOROL 24 see HCP000
LOROL 28 see OAX000
LOROL THIOCYANATE see DYA200
LOROMISIN see CDP250

LOROTHIDOL see TFD250
LOROX see BNM000
LOROX see DGD600
LOROXIDE see BDS000
LOROX LINURON WEED KILLER see DGD600
LORPHEN see TAI500
LORPOX see BAR800
LORSBAN see CMA100
LORSILAN see CFC250
LORVEK see CIC600
LOSANTIN see HOV500
LOSUNGSMITTEL APV see CBR000
LOTRIFEN see CKL325
LOTRIMIN see MRX500
(−)-LOTUCAINE HYDROCHLORIDE see TGJ625
LOTURINE see MPA050
LOTUSATE see AFY500
LOUISIANA LOBELIA see CCJ825
LOVAGE see PMD550
LOVAGE OIL see LII000
LOVASTATIN ACID see LII050
LOVE BEAD see RMK250
LOVENOX see HAQ500
LOVISCOL see CBR675
LOVOSA see SFO500
LOVOZAL see DGA200
LOW BELIA see CCJ825
LOW ERUCIC ACID RAPESEED OIL see RBK200
LOWESERP see RDK000
LOWETRATE see PBC250
LOWPSTRON see CHJ750
LOXACOR see LIA400
LOXACOR HYDROCHLORIDE see LIA400
LOXANOL 95 see OBA000
LOXANOL K see HCP000
LOXANOL M see OBA000
LOXAPINE see DCS200
LOXIGLUMIDE see LII100
LOXIOL G 10 see GGR200
LOXIOL G 21 see HOG000
LOXON see DFH600
LOXOPROFEN SODIUM DIHYDRATE see LII300
LOXURAN see DIW200
LOXYNIL (GERMAN) see HKB500
LOZILUREA see LII400
LOZOL see IBV100
LPA 39 see SEA500
LPC see DXY725
LPC see LBX050
LPE-5 see LII500
LPG see BFC750
LPG see LGM000
LPG ETHYL MERCAPTAN 1010 see EMB100
L.P.G. (OSHA, ACGIH) see LGM000
LPS see EDK700
LPS see PPQ600
LPT see PKB500
LR 115 see CCU250
LR 400 see QAT600
LR-529 see CFO750
LR 300M see QAT600
L RED Z 3000 see CMP620
LRF see LIU370
LRH see LIU370
LS 121 see NAE100
LS 061A see SMQ500
LS 1727 see NNX600
L.S. 3394 see BLL750
LS 4442 see CLU000
LS 1028E see SMQ500
LS 74783 see AHH800
LS 80.1213 see SFV650
LS 83-5002 see PLD050
LS 840606 see FPQ050
LSD see DJO000
d-LSD see DJO000
LSD-25 see DJO000
LS 519 DIHYDROCHLORIDE see GCE500

LSD-25-PYRROLIDATE see LJM000
LSD TARTRATE see LJG000
LSD TARTRATE see LJM600
(+)-LSD TARTRATE see LJM600
d-LSD TARTRATE see LJM600
LSD 25 TARTRATE see LJM600
LSM-775 see LJJ000
LSP 1 see LIJ000
LTAN see DMH400
1,4,5,8-LTOA (RUSSIAN) see TDD250
LT-SOR see CMY240
LU 1631 see MQS100
LUBALIX see CMY525
LUBERGAL see AGQ875
LUBERGAL see EOK000
LUBRICATING OIL see LIK000
LUBRICATING OIL (mainly mineral) see LIK000
LUBRICATING OIL, CYLINDER see LIK000
LUBRICATING OIL, MOTORS see LIK000
LUBRICATING OIL, SPINDLE see LIK000
LUBRICATING OIL, TURBINE see LIK000
LUBRIZOL 2404 see ADS300
LUBRIZOL AMPS see ADS300
LUBROL 12A9 see DXY000
LUBROL W see PJT300
LUCALOX see AHE250
LUCAMIDE see EEM000
LUCANTHON see DHU000
LUCANTHONE see DHU000
LUCANTHONE HYDROCHLORIDE see SBE500
LUCANTHONE METABOLITE see LIM000
LUCANTHONE MONOHYDROCHLORIDE see SBE500
LUCEL see CMA500
LUCEL-3 see BQI400
LUCEL 4 see TAH950
LUCEL 6 see DRU900
LUCEL (polysaccharide) see SFO500
LUCEL ADA see ASM270
LUCENSOMYCIN see LIN000
LUCICULINE see LIN050
LUCIDIN see LIN100
LUCIDIN ETHYL ETHER see DMT200
LUCIDIN (QUINONE) see LIN100
LUCIDOL see BDS000
LUCIDOL DELTAX see MKA500
LUCIDRIL see AAE500
LUCIDRYL see DPE000
LUCIDRYL HYDROCHLORIDE see AAE500
LUCIJET see LIN400
LUCITE see PKB500
LUCKNOMYCIN see LIN600
LUCKY NUT see YAK350
LUCOFEN see ARW750
LUCOFENE see ARW750
LUCOFLEX see PKQ059
LUCORTEUM ORAL see GEK500
LUCORTEUM SOL see PMH500
LUCOSIL see MPQ750
LUCOVYL PE see PKQ059
LUDIGOL see SIU100
LUDIGOL 60 see SIU100
LUDIGOL F see SIU100
LUDIGOL F,60 see NFC500
LUDIOMIL see LIN800
LUDIOMIL see MAW850
LUDOX see SCH002
LUEH 6 see BGS250
LUETOCRIN DEPOT see HNT500
LUFIXAN see VPP100
LUFYLLIN see DNC000
LUGOL'S IODINE see PLW285
LUGOL'S SOLUTION see PLW285
LUH6 see BGS250
LUH⁶ see BGS250
LUH6-CHLORIDE see BGS250
LUH⁶-Cl2 see BGS250
LUH6-DINITRAT (GERMAN) see OPY000
LULAMIN see TEH500
LULAMIN see TEO250

LULIBERIN see LIU370
LULLAMIN see DPJ400
LULLAMIN see TEO250
LUMBANG (GUAM) see TOA275
LUMBRICAL see PIJ000
LUMEN see EOK000
LUMESETTES see EOK000
LUMICREASE BLUE 4GL see CMN750
LUMILAR 100 see PKF750
LUMINAL see EOK000
LUMINAL SODIUM see SID000
LUMINOL see AJO290
LUMIRELAX see GKK000
LUMIRROR see PKF750
LUMNITZERA RACEMOSA Willd., extract excluding roots see LIN850
LUMOFRIDETTEN see EOK000
LUNACOL see ZJS300
LUNAMIN see POA250
LUNAMYCIN see LIO600
LUNAR CAUSTIC see SDS000
LUNARINE HYDROCHLORIDE see LIP000
LUNETORON see BON325
LUNIPAK see DAB800
LUNIS see FDD085
LUPANIN see LIQ000
LUPANINE see LIQ000
(+)-LUPANINE see LIQ000
d-LUPANINE see LIQ000
LUPAREEN see PMP500
LUPERCO see BDS000
LUPERCO see DGR600
LUPERCO CST see BIX750
LUPEROX see DGR600
LUPEROX 500R see DGR600
LUPEROX 500T see DGR600
LUPEROX FL see BDS000
LUPERSOL see MKA500
LUPERSOL 8 see BSC600
LUPERSOL 11 see BSD250
LUPERSOL 70 see BSC250
LUPERSOL 228Z see ACG300
LUPERSOL DDA 30 see MKA500
LUPERSOL DDM see MKA500
LUPERSOL DNF see MKA500
LUPERSOL DSW see MKA500
LUPERSOL Δ-X see MKA500
LUPETAZINE see LIQ500
LUPETIDINE see LIQ550
2,6-LUPETIDINE see LIQ550
LUPIN see LIT000
LUPINIDINE see SKX500
LUPINUS see LIT000
LUPINUS ANGUSTIFOLIUS, seed alkaloid mixture see LIT000
LUPOLEN 1010H see PJS750
LUPOLEN 1800H see PJS750
LUPOLEN 1800S see PJS750
LUPOLEN 1810H see PJS750
LUPOLEN 4261A see PJS750
LUPOLEN 6011H see PJS750
LUPOLEN 6011L see PJS750
LUPOLEN 6042D see PJS750
LUPOLEN KR 1032 see PJS750
LUPOLEN KR 1051 see PJS750
LUPOLEN KR 1257 see PJS750
LUPOLEN L 6041D see PJS750
LUPOLEN N see PJS750
LUPRANATE M 10 see PKB100
LUPRANATE M 70 see PKB100
LUPRANATE M 20S see PKB100
LUPRINATE M 20 see PKB100
LUPROLITE ACETATE see LIU305
LUPRON see LEZ300
LUPRON DEPOT see LEZ300
LUPROSIL see PMU750
β-LUPULIC ACID see LIU000
LUPULON see LIU000
LUPULONE see LIU000
LURAFIX BLUE FFR see MGG250
LURAMIN see EOK000
LURAN see ADY500

LURANTIN LIGHT TURQUOISE BLUE GL see COF420
LURAZOL BLACK BA see AQP000
LURAZOL ORANGE E see CMM220
LURGO see DSP400
LURIDE see SHF500
LURIDINE see CMF800
LUSIL see SNM500
LUSMIT see TFD500
LUSMIT SS see DXG700
LUSTRAN see ADY500
LUSTREX see SMQ500
LUSTREX H 77 see SMQ500
LUSTREX HH 101 see SMQ500
LUSTREX HH 101 see SMQ500
LUSTREX HP 77 see SMQ500
LUSTREX HT 88 see SMQ500
LUTALYSE see POC750
LUTAMIN see LIU100
LUTATE see HNT500
LUTEAL HORMONE see PMH500
LUTEINIZING GONADOTROPIC HORMONE see LIU300
LUTEINIZING HORMONE see LIU300
LUTEINIZING HORMONE ANTISERUM see LIU302
LUTEINIZING HORMONE, GONADOGRAPHON RELEASING HORMONE see GJI075
LUTEINIZING HORMONE-RELEASING FACTOR see LIU370
LUTEINIZING HORMONE-RELEASING FACTOR (PIG), ACETATE (SALT) see LIU305
LUTEINIZING HORMONE-RELEASING FACTOR(PIG), 6-(3-(9-ANTHRACENYL)-d-ALANINE)-,MONOACETATE (SALT) see LIU307
LUTEINIZING HORMONE-RELEASING FACTOR(PIG), 6-(3-(1-BROMO-2-NAPHTHALENYL)-d-ALANINE)-, MONOACETATE (SALT) see LIU309
LUTEINIZING HORMONE-RELEASING FACTOR(PIG), 6-(3-CYCLOHEXYL-d-ALANINE)-, MONOACETATE (SALT) see LIU311
LUTEINIZING HORMONE-RELEASING FACTOR(PIG), 6-(γ-CYCLOHEXYL-l-α-AMINOCYCLOHEXANEBUTANOIC ACID)-, MONOACETIC (SALT) see LIU313
LUTEINIZING HORMONE-RELEASING FACTOR (PIG), 1-DE-(5-OXO-l-PROLINE)-2-(N-(1,1-DIMETHYLETHOXY) CARBONYL)-o-(PHENYLMETHYL)-l-SERINE)-6-d-TRYPTOPHAN-, MONOACETATE (SALT), TRIHYDRATE see LIU315
LUTEINIZING HORMONE-RELEASING FACTOR(PIG), 1-DE(5-OXO-l-PROLINE)-2-(N-((1,1-DIMETHYLETHOXY) CARBONYL)-o-(PHENYLMETHYL)-l-SERINE)-6-d-TRYPTOPHAN-, HYDROCHLORIDE, HYDRATE (2:3:8) see LIU317
LUTEINIZING HORMONE-RELEASING FACTOR(PIG), 1-DE(5-OXO-l-PROLINE)-2-(N-((1,1-DIMETHYLETHOXY) CARBONYL)-o-(PHENYLMETHYL)-l-SERINE)-6-d-TRYPTOPHAN-10-GLYCINE-, METHYL ESTER, HYDROCHLORIDE, HYDRATE (2:3:8) see LIU319
LUTEINIZING HORMONE-RELEASING FACTOR (PIG), 1-DE-(5-OXO-l-PROLINE)-2-(N²-((PHENYLMETHOXY)CARBONYL)-N-(PHENYLMETHYL)-l-GLUTAMINE)-, HYDROCHLORIDE, HYDRATE (2:3:6) see LIU321
LUTEINIZING HORMONE-RELEASING FACTOR(PIG), 1-DE(5-OXO-l-PROLINE)-2-(N-((PHENYLMETHOXY) CARBONYL)-l-LEUCINE)-, HYDROCHLORIDE, HYDRATE (2:3:6) see LIU323
LUTEINIZING HORMONE-RELEASING FACTOR(PIG), 1-DE(5-OXO-l-PROLINE)-2-(N-((PHENYLMETHOXY) CARBONYL)-o-

(PHENYLMETHYL)-l-SERINE)-6-d-PHENYLALANINE-, HYDROCHLORIDE, HYDRATE (2:3:8) see LIU325
LUTEINIZING HORMONE-RELEASING FACTOR(PIG), 6-(3,5-DIMETHOXY-o-METHYL-d-TYROSINE)-,MONOACETATE (SALT) see LIU327
LUTEINIZING HORMONE-RELEASING FACTOR(PIG), 6-(3-(1-NAPHTHALENYL)-d-ALANINE)-, MONOACETATE(SALT) see LIU329
LUTEINIZING HORMONE-RELEASING FACTOR(PIG), 6-(3-(2-NAPHTHALENYL)-d-ALANINE)-7-(N-METHYL-l-LEUCINE)-, MONOACETATE (SALT) see LIU331
LUTEINIZING HORMONE-RELEASING FACTOR(PIG), 6-(3-(2-NAPHTHALENYL)-d-ALANINE)-9-(N-ETHYL-l-PROLINAMIDE)-10-DEGLYCINAMIDE-, MONOACETATE (SALT) see LIU333
LUTEINIZING HORMONE-RELEASING FACTOR(PIG), 6-(3-(2-NAPHTHALENYL)-d-ALANINE)-7-(N-METHYL-l-LEUCINE-9-(N-ETHYL-l-PROLINAMIDE)-10-DEGLYCINAMIDE-, MONOACETATE (SALT) see LIU335
LUTEINIZING HORMONE-RELEASING FACTOR(PIG), 3-(3-(1-NAPHTHALENYL)-l-ALANINE)-6-(3-(2-NAPHTHALENYL)-d-ALANINE)-9-(N-ETHYL-l-PROLINAMIDE)-10-DEGLYCINAMIDE-, MONOACETATE (SALT) see LIU337
LUTEINIZING HORMONE-RELEASING FACTOR(PIG), 6-(2,3,4,5,6-PENTAFLUORO-d-PHENYLALANINE)-, MONOACETATE (SALT) see LIU339
LUTEINIZING HORMONE-RELEASING FACTOR(PIG), 2-d-PHENYLALANINE-3-l-PROLINE-6-d-PHENYLALANINE see LIU341
LUTEINIZING HORMONE-RELEASING FACTOR(PIG), 2-d-PHENYLALANINE-3-l-PROLINE-6-d-TRYPTOPHAN- see LIU342
LUTEINIZING HORMONE-RELEASING FACTOR(PIG), 2-d-PHENYLALANINE-6-d-PHENYLALANINE-,HYDROCHLORIDE, HYDRATE (2:2:7) see LIU343
LUTEINIZING HORMONE-RELEASING FACTOR(PIG), 2-d-PHENYLALANINE-3-l-VALINE-6-d-TRYPTOPHAN- see LIU345
LUTEINIZING HORMONE-RELEASING FACTOR(PIG), 6-(γ-PHENYL-d-α-AMINOBENZENEBUTANOICACID)-9-(N-ETHYL-l-PROLINAMIDE)-10-DEGLYCINAMIDE-, MONOACETATE (SALT) see LIU347
LUTEINIZING HORMONE-RELEASING FACTOR (PIG), 6-(3-(TRIFLUOROMETHYL)-d-PHENYLALANINE), MONOACETATE, (SALT) see TKF100
LUTEINIZING HORMONE-RELEASING FACTOR(PIG), 6-(4-(TRIFLUOROMETHYL)-d-PHENYLALANINE)-, MONOACETATE (SALT) see LIU349
LUTEINIZING HORMONE-RELEASING FACTOR(PIG), 6-(2,4,6-TRIMETHYL-d-PHENYLALANINE)-,MONOACETATE (SALT) see LIU351
LUTEINIZING HORMONE-RELEASING FACTOR (PIG), 6-d-TRYPTOPHAN-9-(N-ETHYL-l-PROLINAMIDE)-10-DEGLYCINAMIDE- see LIU353
LUTEINIZING HORMONE-RELEASING FACTOR (PIG), 6-d-TRYPTOPHAN-7-(N-METHYL-l-LEUCINE)-9-(N-ETHYL-l-PROLINAMIDE)-10-DEGLYCINAMIDE-, MONOACETATE (SALT) see LIU355
LUTEINIZING HORMONE-RELEASING FACTOR (SWINE), ACETATE (SALT) see LIU305
LUTEINIZING HORMONE-RELEASING HORMONE see LIU370

LUTEINIZING HORMONE-RELEASING HORMONE, DIACETATE (SALT) see LIU380
LUTEINIZING HORMONE-RELEASING HORMONE, DIACETATE, TETRAHYDRATE see LIU400
LUTEINIZING HORMONE-RELEASING HORMONE (PIG), 6-(O-(1,1-DIMETHYLETHYL)-d-SERINE)-9-(N-ETHYL-l-PROLINAMIDE)-10-DEGLYCINAMIDE- see LIU420
LUTEINIZING HORMONE-RELEASING HORMONE, (d-SER(TBU)⁶-EA¹⁰)- see LIU420
LUTEOANTINE see FMT100
LUTEOCRIN see HNT500
LUTEOHORMONE see PMH500
LUTEOLIN see TDD550
LUTEOLINE see TDD550
LUTEOLIN-7-o-RUTINOSIDE see LIU450
LUTEOLOL see TDD550
LUTEOMYCIN see LIU500
LUTEOSAN see PMH500
LUTEOSKYRIN see LIV000
(−)-LUTEOSKYRIN see LIV000
LUTEOSTIMULIN see LIU370
LUTEOTROPHIN see PMH625
LUTEOTROPIC HORMONE see PMH625
LUTEOTROPIC HORMONE LTH see PMH625
LUTEOTROPIN see PMH625
LUTEOZIMAN see LIU300
LUTESTRAL (FRENCH) see CNV750
LUTETIA FAST BLUE RS see IBV050
LUTETIA FAST EMERALD J see PJQ100
LUTETIA FAST ORANGE 3R see CJD500
LUTETIA FAST ORANGE R see DVB800
LUTETIA FAST RED 3R see MMP100
LUTETIA FAST SCARLET RF see MMP100
LUTETIA FAST SCARLET RJN see MMP100
LUTETIA ORANGE J see CMS145
LUTETIA ORANGE 3JR see CMM220
LUTETIA RED C see CMS150
LUTETIA RED CLN see CHP500
LUTETIA RED R see NAP100
LUTETIUM CHLORIDE see LIW000
LUTETIUM CITRATE see LIX000
LUTETIUM(III) NITRATE (1:3) see LIY000
LUTEX see PMH500
LUTHONAL A 20 see OJV600
2,4-LUTIDINE see LIY990
2,5-LUTIDINE see LJA000
2,6-LUTIDINE see LJA010
3,4-LUTIDINE see LJB000
α-α'-LUTIDINE see LJA010
LUTIDON ORAL see GEK500
LUTINYL see CBF250
LUTOCYCLIN see PMH500
LUTOCYLOL see GEK500
LUTOFAN see PKQ059
LUTOGAN see DRT200
LUTO-METRODIOL see EQJ500
LUTONAL A 25 see OJV600
LUTONAL A 50 see OJV600
LUTOPRON see HNT500
LUTOSAN see DRT200
LUTOSOL see INJ000
LUTRELEF see LIU370
LUTRELIN ACETATE see LIU355
LUTREPULSE see LIU305
LUTROL see EAE675
LUTROL see EIM000
LUTROL see PJT000
LUTROL-9 see EJC500
LUTROMONE see PMH500
LUTROPIN see LIU300
LUVATRENE see MNN250
LUVICAN M 150 see VNK200
LUVICAN M 170 see VNK200
LUVIPAL 066 see MCB050
LUVISKOL see PKQ250
LUVISKOL VA 64 see AAU300
LUVISKOL VA 28I see AAU300
LUVISKOL VA 37E see AAU300

LUVISKOL VA 37I see AAU300
LUVISKOL VA 55E see AAU300
LUVISKOL VA 55I see AAU300
LUVISKOL VA 73E see AAU300
LUVOX see FMR575
LUWIPAL 012 see MCB050
LUXAN BLACK R see PFU500
LUXISTELM see PHI500
LUXOMNIN see AGQ875
LUXON see DFH600
LV1N 32, BASE OIL see NAT100
LVM see LHX350
O-LVR see OJY200
LW 3170 see DCS200
LX 14-0 see CQH250
LXON see DFH600
LY 61017 see BGE125
LY 127935 see LBH200
LY133314 see TMP175
LY150720 see PIB700
LY 170053 see ZVJ500
LY 188011 see GCK100
LY 188011 see GCK500
LY 188695 see EAL050
LY253351 see EFA200
(−)-LY 253352 see EFA200
LYCACONITINE, METHYL- see MLD140
LYCANOL see GHK200
LYCEDAN see AOA125
LYCINE see GHA050
LYCOBETAINE see LJB800
(+)-LYCOCTONINE see LJB100
LYCOCTONINE (8CI) see LJB100
LYCOCTONINE, ANTHRANILOYL-
(6CI,7CI) see LJB150
LYCOCTONINE, MONOANTHRANILATE
(ESTER) (8CI) see LJB150
LYCOID DR see GLU000
LYCOPERSICIN see THG250
LYCOPERSIN see LJB700
LYCORANIUM, 1,2,3,3a,6,7,12b,12c-
OCTADEHYDRO-2-HYDROXY- see LJB800
LYCORCIDINOL see LJC000
LYCOREMINE HYDROBROMIDE see
GBA000
LYCORICIDIN-A see LJC000
LYCORICIDINOL see LJC000
LYCORIS AFRICANA see REK325
LYCORIS RADIATA see REK325
LYCORIS SQUAMIGERA see REK325
LYCURIM see LJD500
LYDOL see DAM700
LYE see PLJ500
LYE (DOT) see SHS000
LYE, solution see SHS500
LYGODIUM FLEXUOSUM (LINN.)
SWARTZ., EXTRACT see LJD600
LYGOMME CDS see CCL250
LY188011 HYDROCHLORIDE see GCK100
LYKURIM see LJD500
LYMECYCLINE see MRV250
LYMETHOL see HAQ570
LYMPHCHIN see AOQ875
LYMPHOCIN see AOQ875
LYMPHOQUIN see AOQ875
LYMPHOSCAN see AQL500
LYNAMINE see AHK750
LYNDIOL see LJE000
LYNENOL see NNV000
LYNESTRENOL see NNV000
LYNESTRENOL mixed with MESTRANOL
see LJE000
LYNESTROL see EEH550
LYNESTROL mixed with MESTRANOL see
LJE000
LYNOESTRENOL see NNV000
LYNOESTRENOL mixed with
ETHINYLOESTRADIOL see EEH575
LYNOESTRENOL mixed with MESTRANOL
see LJE000
LYNX see TAN050
LYOBEX see NOA000

LYOBEX HYDROCHLORIDE see NOA500
LYOFIX CH see MCB050
LYOFIX CHN see MCB050
LYOFIX CHN-ZA see MCB050
LYOFIX MLF see MCB050
LYOGLYCOGEN see GHK300
LYONIATOXIN see LJE100
LYONIOL A see LJE100
LYONIOL B see DAD700
LYOPECT see CNG250
LYOPHRIN see AES000
LYOPHRIN see VGP000
LYOVAC COSMEGEN see AEB000
LYP 97 see LBR000
LYPOARAN see DXN709
LYRAL see LJE200
LYSALGO see XQS000
LYSANXIA see DAP700
LYSENYL see LJE500
LYSENYL BIMALEATE see LJE500
LYSENYL HYDROGEN MALEATE see
LJE500
LYSERGAMID see DJO000
LYSERGAMIDE, 1-ACETYL-N,N-DIETHYL-
see AFJ450
LYSERGAMIDE, N-(1-HYDROXYETHYL)-
see LJI100
LYSERGAURE DIETHYLAMID see DJO000
d-LYSERGIC ACID DIETHYLAMIDE see
DJO000
d-LYSERGIC ACID DIETHYLAMIDE see
LJF000
LYSERGIC ACID DIETHYLAMIDE-25 see
DJO000
LYSERGIC ACID DIETHYLAMIDE, 1-
ISOMER see LJF000
LYSERGIC ACID DIETHYLAMIDE
TARTRATE see LJG000
d-LYSERGIC ACID DIETHYLAMIDE
TARTRATE see LJG000
d-LYSERGIC ACID DIMETHYLAMIDE see
LJH000
LYSERGIC ACID ETHYLAMIDE see LJI000
LYSERGIC ACID α-
HYDROXYETHYLAMIDE see LJI100
d-LYSERGIC ACID-1-
HYDROXYMETHYLETHYLAMIDE see
LJL000
LYSERGIC ACID METHYL
CARBINOLAMIDE see LJI100
d-LYSERGIC ACID MONOETHYLAMIDE
see LJI000
LYSERGIC ACID MORPHOLIDE see LJJ000
d-LYSERGIC ACID MORPHOLIDE see
LJJ000
d-LYSERGIC ACID MORPHOLIDE,
TARTARIC ACID SALT see LJK000
LYSERGIC ACID PROPANOLAMIDE see
LJL000
d-LYSERGIC ACID-l,2-PROPANOLAMIDE
see LJL000
LYSERGIC ACID PYROLIDATE see LJM000
d-LYSERGIC ACID PYRROLIDIDE see
LJM000
LYSERGIDE see DJO000
LYSERGIDE TARTRATE see LJM600
LYSERGSAEUREDIAETHYLAMID see
DJO000
LYS (HAITI) see SLB250
LYSILAN see HGI100
l-LYSINAMIDE, l-ARGINYL-l-LEUCYL-l-
ISOLEUCYL-l-PROLYL-l-PROLYL-l-
PHENYLALANYL-l-TRYPTOPHYL- see
AQW110
l-LYSINAMIDE, l-LEUCYL-l-ISOLEUCYL-l-
PROLYL-l-PROLYL-l-PHENYLALANYL-l-
TRYPTOPHYL- see NAE900
LYSINE see LJM700
l-(+)-LYSINE see LJM700
l-LYSINE (9CI) see LJM700
l-LYSINE ACETATE see LJM800

dl-LYSINE ACETYLSALICYLIC ACID SALT
see ARP125
LYSINE ACID see LJM700
l-LYSINE, l-GLUTAMINYL-l-ALANYL-l-
THREONYL-l-VALYL-l-
ASPARTYL-l-VALYL-l-ASPARAGINYL-l-
THREONYL-l-α-ASPARTYL-l-ARGINYL-l-
PROLYLGLYCYL-l-LEUCYL-l-LEUCYL-l-α-
ASPARTYL-l-LEUCYL- see OAP400
d-LYSINE HYDROCHLORIDE see LJN000
l-LYSINE HYDROCHLORIDE see LJO000
l-LYSINE, MONOACETATE see LJM800
dl-LYSINE, MONO(2-
(ACETYLOXY)BENZOATE) see ARP125
LYSINE MONOHYDROCHLORIDE see
LJO000
l-LYSINE MONOHYDROCHLORIDE see
LJO000
LYSINE, MONOSALICYLATE ACETATE
(ESTER), DL- see ARP125
l-LYSINE, N⁶)-IMINOETHYL-,
HYDROCHLORIDE see IBL200
N-LYSINOMETHYLTETRACYCLINE see
MRV250
LYSINOPRIL see LGM400
LYSIVANE see DIR000
LYSOCELLIN see LJP000
LYSOCOCCINE see SNM500
LYSOFON see AGW750
LYSOFORM see FMV000
LYSOL see LJP500
2-LYSOPHOSPHATIDYLCHOLINE see
SLM100
LYSOSTAPHIN see LJQ000
LYSSIPOLL see LJR000
LYSTENON see HLC500
LYSTHENONE see HLC500
LYSURON see ZVJ000
12-l-LYS-13-l-VAL-14-l-LEU-NH2 see MBU776
LYTECA SYRUP see HIM000
LYTHIDATHION see DRU400
LYTISPASM see LJS000
LYTRON 810 see SEA500
LYTRON 812 see SEA500
LYTRON 822 see SEA500
LYTRON 5202 see SMR000
LZ-MB 1 see MJM700
M-2 see CAR875
M 2 see UTU500
M 3 see MCB050
01M see AGD250
M-14 see RKZ000
M-30 see DAE695
M-32 see DAE700
M 40 see CIR250
M 60 see UTU500
M 70 see UTU500
M-71 see CNW125
M 73 see DFV600
M-74 see DXH325
M 76 see MCB050
M 81 see PHI500
M-101 see TFY000
M-108 see TAL575
M 140 see CDR750
M 176 see DUD800
M 259 see AAC000
M 410 see CDR750
M 551 see PFB350
M 1028 see IFA000
M 1703 see IOV000
M 2060 see FIP999
M 3180 see CQH650
M-3432 see TGF030
M 4109 see CIC600
M-4209 see CBT250
M-4212 see MAB050
M-4212 see MQX775
M 4888 see CKB500
M 5055 see CDV100
5512-M see PDC000
M5943 see DGD100

M 6/42 see RSU450
M-6029 see BOO630
M-9500 see TND500
M 10797 see MFC100
M-12210 see MAB055
M 13/20 see PKP750
M 13144 see SOU800
M 3/158 see DAP600
M73101 see EAN700
M 4212 (pesticide) (9CI) see MAB050
MA see DAL300
MA see MNQ500
4-MA see DJP700
MA 108 see EQN250
MA-110 see EHP000
MA-117 see EHO700
MA-1214 see DXV600
MA 1291 see QWJ500
MA 1337 see CMX840
MA 144T1 see DAY835
M-4365A2 see RMF000
MAA see AFI850
MAA see MGQ530
MA300A see DEJ100
MA-144A2 see TAH675
1-MA-4-AA see AKP250
1-MA-40EAA (RUSSIAN) see MGG250
MAAC see HFJ000
MAA SODIUM SALT see DXE600
MAB see MNR500
MA144 B2 see TAH650
MABA see AIH500
MABERTIN see CFY750
MABLIN see BOT250
MABUTEROL see MAB300
MA 100 (CARBON) see CBT500
MACASIROOL see CHJ750
MACBAL see DTN200
MACBECIN II see MAB400
MACE (lachrymator) see CEA750
MACE OIL see OGQ100
MACHETE see CFW750
MACHETE (herbicide) see CFW750
MACHETTE see CFW750
MACH-NIC see NDN000
MACID MILLING ORANGE PROPYL see
ADG000
MACKAMIDE O see OHU200
MACKREAZID see COH250
MACLEYINE see FOW000
MACOCYN see HOH500
MACODIN see MDP750
MACQUER'S SALT see ARD250
MACRABIN see VSZ000
MACROBIN see CLG900
MACROCYCLON see TDN750
MACRODANTIN see NGE000
MACRODIOL see EDO000
MACROFUSINE see MAB500
MACROGOL 1000 see PJT250
MACROGOL 4000 see PJT750
MACROGOL 400 BPC see EJC500
MACROGOL OLEATE 600 see PJY100
MACROL see EDO000
MACROLEX BLUE FR see BKP500
MACROMOMYCIN see MAB750
MACRONDRAY see DAA800
MACROPAQUE see BAP000
MACROSE see DBD700
MACROSPHERICAL 95 see AHA000
MACULIN see MEL725
MACULINE see MEL725
MACULOTOXIN see FOQ000
MAD see AOO475
MADAGASCAR LEMONGRASS OIL see
LEH000
MADAME FATE (JAMAICA) see SLJ650
MADAR see CGA500
MADDER COLOR see MAB800
MADDER (DYE) see MAB800
MADECASSOL see ARN500
MADEIRA IVY see AFK950

MADHURIN see SJN700
MADIOL see AOO475
MADIOL see MQU750
MADRESELVA (MEXICO) see YAK100
MADRIBON see SNN300
MADRIGID see SNN300
MADRINE see DBA800
MADRIQID see SNN300
MADROX see MEU300
MADROXIN see SNN300
MADROXINE see SNN300
MADURIT 152 see MCB050
MADURIT 5238N see MCB050
MADURIT MS see MCB050
MADURIT MW 111 see MCB050
MADURIT MW 112 see MCB050
MADURIT MW 150 see MCB050
MADURIT MW 161 see MCB050
MADURIT MW 166 see MCB050
MADURIT MW 392 see MCB050
MADURIT MW 484 see MCB050
MADURIT MW 559 see MCB050
MADURIT MW 630 see MCB050
MADURIT MW 815 see MCB050
MADURIT MW 909 see MCB050
MADURIT OP see MCB050
MADURIT TN see MCB050
MADURIT VMW 3113 see MCB050
MADURIT VMW 3114 see MCB050
MADURIT VMW 3151 see MCB050
MADURIT VMW 3163 see MCB050
MADURIT VMW 3284 see MCB050
MADURIT VMW 3399 see MCB050
MADURIT VMW 3489 see MCB050
MADURIT VMW 3490 see MCB050
MADURIT VMW 3494 see MCB050
MADURIT VMW 3819 see MCB050
MADURIT VMW 3822 see MCB050
MAEKINENITE see MAK400
MAFATATE see MAC000
MAFENIDE ACETATE see MAC000
MAFU see DGP900
MAG see MRF000
MAGACROM VIOLET N see HLI000
MAGADI SODA see SJT750
MAGBOND see BAV750
MAGCAL see MAH500
MAGCHEM 100 see MAH500
MAGECOL see CBT750
MAGENTA see MAC250
MAGENTA BASE see MAC500
MAGENTA DP see MAC250
MAGENTA E see MAC250
MAGENTA G see MAC250
MAGENTA I see MAC250
MAGENTA II see DSB300
MAGENTA PN see MAC250
MAGENTA POWDER N see MAC250
MAGENTA S see MAC250
MAGENTA SUPER FINE see MAC250
MAGIC LILY see REK325
MAGIC METHYL see MKG250
MAGISTER see CEP800
MAGISTERY OF BISMUTH see BKW100
MAGK'S T.B.1 CONTEBEN see FNF000
MAGLITE see MAH500
MAGMASTER see MAC650
MAGNACIDE H see ADR000
MAGNAMYCIN see CBT250
MAGNAMYCIN A see CBT250
MAGNAMYCIN HYDROCHLORIDE see
MAC600
MAGNESIA see MAH500
MAGNESIA ALBA see MAC650
MAGNESIA MAGMA see MAG750
MAGNESIA USTA see MAH500
MAGNESIA WHITE see CAX750
MAGNESIO (ITALIAN) see MAC750
MAGNESITE see MAC650
MAGNESIUM see MAC750
MAGNESIUM ACETATE see MAD000

MAGNESIUM ALLOYS, powder (UN 1418)
(DOT) see MAC750
MAGNESIUM ALLOYS with >50% magnesium
in pellets, turnings or ribbons (UN 1869) (DOT)
see MAC750
MAGNESIUM ALUMINUM PHOSPHIDE
(DOT) see AHD250
MAGNESIUM AMMONIUM ARSENATE see
MAD025
MAGNESIUM AMMONIUM ARSENATE
DIHYDRATE see MAD025
MAGNESIUM ARSENATE see ARD000
MAGNESIUM ARSENATE PHOSPHOR see
ARD000
MAGNESIUM AUREOLATE see MAD050
MAGNESIUM BIS(2,3-
DIBROMOPROPYL)PHOSPHATE see
MAD100
MAGNESIUMBIS(2,3-
DIBROMOPROPYL)PHOSPHATE see
MAD100
MAGNESIUM, BIS(2-
HYDROXYPROPANOATO-O(1),O(2))-, (T-4)-
(9CI) see LAL100
MAGNESIUM, BIS(LACTATO)-(8CI) see
LAL100
MAGNESIUM BORIDE see MAD250
MAGNESIUM CARBONATE see MAC650
MAGNESIUM(II) CARBONATE (1:1) see
MAC650
MAGNESIUM CARBONATE,
PRECIPITATED see MAC650
MAGNESIUM CHLORATE see MAE000
MAGNESIUM CHLORIDE see MAE250
MAGNESIUM CHLORIDE HEXAHYDRATE
see MAE500
MAGNESIUM CLIPPINGS see MAC750
MAGNESIUM COMPOUNDS see MAE750
MAGNESIUM DIACETATE see MAD000
MAGNESIUM DICHLORATE see MAE000
MAGNESIUM DINITRATE
HEXAHYDRATE see NEE100
MAGNESIUM DIPROPYLACETATE see
MAK275
MAGNESIUM DROSS, wet or hot (DOT) see
MAF000
MAGNESIUM DROSS (HOT) see MAF000
MAGNESIUM FLUORIDE see MAF500
MAGNESIUM FLUOROSILICATE (DOT) see
MAF600
MAGNESIUM FLUORURE (FRENCH) see
MAF500
MAGNESIUM FLUOSILICATE see MAF600
MAGNESIUM FLUOSILICATE see MAG250
MAGNESIUMFOSFIDE (DUTCH) see
MAI000
MAGNESIUM FUMARATE see MAF750
MAGNESIUM GLUCONATE see MAG000
MAGNESIUM GLUTAMATE
HYDROBROMIDE see MAG050
MAGNESIUM GLYCEROPHOSPHATE see
MAG100
MAGNESIUM GOLD PURPLE see GIS000
MAGNESIUM GRANULES, coated particle size
not <149 microns (UN 2950) (DOT) see
MAC750
MAGNESIUM HEXAFLUOROSILICATE see
MAG250
MAGNESIUM HYDRATE see MAG750
MAGNESIUM HYDRIDE see MAG500
MAGNESIUM HYDROGEN PHOSPHATE
see MAG550
MAGNESIUM HYDROXIDE see MAG750
MAGNESIUM LACTATE see LAL100
MAGNESIUM LINEAR ALKYLBENZENE
SULFONATE see LGF800
MAGNESIUM, METHOXY-,
MONOMETHYL CARBONATE see MES980
MAGNESIUM NITRATE (DOT) see MAH000
MAGNESIUM(II) NITRATE (1:2) see MAH000
MAGNESIUM NITRATE HEXAHYDRATE
see NEE100

MAGNESIUM(II) NITRATE (1:2), HEXAHYDRATE see MAH250,
MAGNESIUM(II) NITRATE (1:2), HEXAHYDRATE see NEE100
MAGNESIUM OXIDE see MAH500
MAGNESIUM OXIDE FUME (ACGIH) see MAH500
MAGNESIUM-5-OXO-2-PYRROLIDINECARBOXYLATE see PAN775
MAGNESIUM PELLETS see MAC750
MAGNESIUM PEMOLINE see PAP000
MAGNESIUM PERCHLORATE see PCE000
MAGNESIUM PEROXIDE see MAH750
MAGNESIUM PEROXIDE, solid (DOT) see MAH750
MAGNESIUM PHOSPHATE, DIBASIC see MAH775
MAGNESIUM PHOSPHATE, TRIBASIC see MAH780
MAGNESIUM PHOSPHIDE see MAI000
MAGNESIUM PHTHALOCYANINE see MAI250
MAGNESIUM POLYOXYETHYLENE ALKYL ETHER SULFATE see MAI500
MAGNESIUM POTASSIUM ASPARTATE see MAI600
MAGNESIUM POTASSIUM-l-ASPARTATE see MAI600
MAGNESIUM POTASSIUM SULFATE see MAI650
MAGNESIUM POWDERED see MAC750
MAGNESIUM RIBBONS see MAC750
MAGNESIUM SALICYLATE TETRAHYDRATE see MAI700
MAGNESIUM SALT of AUREOLIC ACID see MAD050
MAGNESIUM SILICATE HYDRATE see MAJ000
MAGNESIUM SILICOFLUORIDE see MAG250
MAGNESIUM STEARATE see MAJ030
MAGNESIUM STEARATE see MAJ030
MAGNESIUM STEARATE (ACGIH) see MAJ030
MAGNESIUM SULFATE (1:1) see MAJ250
MAGNESIUM SULFATE and CALOMEL (8:5) see CAZ000
MAGNESIUM SULFATE adduct of 2,2-DITHIO-BIS-PYRIDINE 1-OXIDE see OIU850
MAGNESIUM SULFATE HEPTAHYDRATE see MAJ500
MAGNESIUM SULPHATE see MAJ250
MAGNESIUM TETRAHYDROALUMINATE see MAJ775
MAGNESIUM THIOSULFATE see MAK000
MAGNESIUM THIOSULFATE HEXAHYDRATE see MAK250
MAGNESIUM TURNINGS (DOT) see MAC750
MAGNESIUM (UN 1869) (DOT) see MAC750
MAGNESIUM, powder (UN 1418) (DOT) see MAC750
MAGNESIUM VALPROATE see MAK275
MAGNETIC 70, 90, and 95 see SOD500
MAGNETIC BLACK see IHC550
MAGNETIC OXIDE see IHC550
MAGNETITE see IHC550
MAGNEZU TLENEK (POLISH) see MAH500
MAGNIFLOC 509C see MCB050
MAGNOFENYL see OPK300
MAGNOLIONE see OOO100
MAGNOLOL see BFX520
MAGNOPHENYL see OPK300
MAGNOSULF see MAK000
MAGNUS RED see CLO700
MAGNUS RED ANION SALT see PJD750
MAGOX see MAG750
MAGOX see MAH500
MAGOX 85 see MAH500
MAGOX 90 see MAH500
MAGOX 95 see MAH500

MAGOX 98 see MAH500
MAGOX OP see MAH500
MAGRACROM ORANGE see NEY000
MAGRACROM ORANGE GR see CMP882
MAGRACROM RED A see CMG750
MAGRACROM YELLOW GG see SIT850
MAGRON see MAE000
MAGSALYL see SJO000
MAH see DMC600
MAHOGANY see MAK300
MAHOGANY EMBL see CMO800
MAIKOHIS see FOS100
MAINTAIN 3 see DMC600
MAINTAIN A see CDT000
MAINTAIN CF125 see CDT000
MAIORAD see TGF175
MAIPEDOPA see DNA200
MAIS BOUILLI (HAITI) see GIW200
MAITANSINE see MBU820
MAITENIN see TGD125
MAITOTOXIN see MAK350
MAITOTOXIN 1 see MAK350
MAIZENA see SLJ500
MAJEPTYL see TFM100
MAJIS S 3307D see SOU800
MAJOL PLX see SFO500
MAJSOLIN see DBB200
MAKAHALA (HAWAII) see DAC500
MAKAROL see DKA600
MAKI see BMN000
MAKINENITE see MAK400
MAKONBU M1 see DNB700
MALACHITE see CNJ750
MALACHITE GREEN CARBINOL see MAK500
MALACHITE GREEN G see BAY750
MALACHITE GREEN OXALATE see MAK600
MALACID see TGD000
MALACIDE see MAK700
MALAFOR see MAK700
MALAGRAN see MAK700
MALAKILL see MAK700
MALAMAR see MAK700
MALAMAR 50 see MAK700
MALAMINE, HEXAKIS(METHOXYMETHYL) see HDY500
MALA MUJER (MEXICO) see CNR135
MALANGA CARA de CHIVO (CUBA) see EAI600
MALANGA (CUBA) see XCS800
MALANGA DEUX PALLES (HAITI) see EAI600
MALANGA ISLENA (CUBA) see EAI600
MALANGA de JARDIN (CUBA) see EAI600
MALANGA TREPADORA (CUBA) see PLW800
MALAOXON see OPK250
MALAOXONE see OPK250
MALAPHELE see MAK700
MALAPHOS see MAK700
MALAQUIN see CLD000
MALAREN see CLD000
MALAREX see CLD000
MALARICIDA see CFU750
MALARIDINE see PPQ650
MALASOL see MAK700
MALASPRAY see MAK700
MALATHION see MAK700
MALATHION-O-ANALOG see OPK250
MALATHION-FENITROTHION MIXTURE see FAO300
MALATHION ULV CONCENTRATE see MAK700
MALATHIOZOO see MAK700
MALATHON see MAK700
MALATHYL LV CONCENTRATE & ULV CONCENTRATE see MAM000
MALATION (POLISH) see MAK700
MALATOL see MAK700
MALATOX see MAK700
MALAYAN CAMPHOR see BMD000

MALAZIDE see DMC600
MALBIT see MAO285
MALCOTRAN see MDL000
MALDISON see MAK700
MALDOCIL see TEN500
MALEALDEHYDIC ACID, DICHLORO-4-OXO-, (Z)-(9CI) see MRU900
MALEATE ACIDE de l'ACETYL-3-DIMETHYLAMINO-3-PROPYL-10-PHENOTHIAZINE (FRENCH) see AAF750
MALEATE de N-(p-CHLOROBENZYL)PYROGLUTAMATE de DIMETHYLAMINOETHYLE see PMH910
MALEATE de 2-((p-CHLOROPHENYL)-2-(PYRIDYL)HYDROXY METHYL)IMIDAZOLINE (FRENCH) see SBC800
MALEATE de CINEPAZIDE (FRENCH) see VGK000
MALEATE de CINPROPAZIDE see IRQ000
MALEATE de N-(p-FLUOROBENZYL)PYROGLUTAMATE de DIMETHYLAMINOETHYLE see PMH927
MALEIC ACID see MAK900
MALEIC ACID ANHYDRIDE (MAK) see MAM000
MALEIC ACID, DIALLYL ESTER see DBK200
MALEIC ACID, DIBUTYL ESTER see DED600
MALEIC ACID DI(1,3-DIMETHYLBUTYL) ESTER see DKP400
MALEIC ACID, DIETHYL ESTER see DJO200
MALEIC ACID DIHEXYL ESTER see DKP400
MALEIC ACID, DIMETHYL ESTER see DSL800
MALEIC ACID, DIPENTYL ESTER see MAL000
MALEIC ACID-N-ETHYLIMIDE see MAL250
MALEIC ACID N-ETHYLIMIDE see MAL250
MALEIC ACID HYDRAZIDE see DMC600
MALEIC ACID, METHYL ERGONOVINE see MJV750
MALEIC ACID, MONODODECYL ESTER, DIOCTYLSTANNYLENE DERIV. see DVN100
MALEIC ACID, MONO(HYDROXYETHOXYETHYL) ESTER see MAL500
MALEIC ACID, MONO(2-HYDROXYPROPYL) ESTER see MAL750
MALEIC ANHYDRIDE see MAM000
MALEIC ANHYDRIDE adduct of BUTADIENE see TDB000
MALEIC ANHYDRIDE HOMOPOLYMER see MAM500
MALEIC ANHYDRIDE OLIGOMER see MAM500
MALEIC ANHYDRIDE OZONIDE see MAM250
MALEIC ANHYDRIDE POLYMER see MAM500
MALEIC ANHYDRIDE, POLYMERS see MAM500
MALEIC ANHYDRIDE, POLYMER WITH ETHYLENE see EJM950
MALEIC ANHYDRIDE-STYRENE COPOLYMER see SEA500
MALEIC ANHYDRIDE-STYRENE POLYMER see SEA500
MALEIC HYDRAZIDE see DMC600
MALEIC HYDRAZIDE 30% see DMC600
MALEIC HYDRAZIDE DIETHANOLAMINE SALT see DHF200
MALEIC HYDRAZINE see DMC600
MALEIMIDE see MAM750
MALEIMIDE, N-ETHYL- see MAL250
MALEIMIDE, N,N'-HEXAMETHYLENEDI- see HEG050

MALEIMIDE, N,N'-(METHYLENEDI-p-PHENYLENE)DI- see BKL800
MALEIMIDE, N,N'-(4-METHYL-m-PHENYLENE)DI- see TGY770
MALEIMIDE, N-2,3-XYLYL- see XSS923
MALEIN 30 see DMC600
MALEINAN SODNY (CZECH) see SIE000
MALEINIC ACID see MAK900
MALEINIMIDE see MAM750
MALEINSAEUREHYDRAZID see DMC600
MALENIC ACID see MAK900
N,N-MALEOYLHYDRAZINE see DMC600
MALESTRONE see MPN500
MALESTRONE (AMPS) see TBF500
MALGESIC see BRF500
MALIASIN see CPO500
MALIC ACID see MAN000
MALIC ACID, DIBUTYL ESTER, (+-)- see DED500
MALIC ACID, 3-HYDROXY- see TAF750
MALIC ACID, SODIUM SALT see MAN250
MALIL see AFS500
MALILUM see AFS500
MALIPUR see CBG000
MALIVAN see DNV000
MALIX see EAQ750
MALLERMIN-F see FOS100
MALLOFEEN see PDC250
MALLOPHENE see PDC250
MALLOPHENE see PEK250
MALLOROL see MOO250
MALLOSIDE see MAN400
MALLOTOXIN see KAJ500
MALMED see MAK700
MALOCID see TGD000
MALOCIDE see TGD000
MALOGEN see MPN500
MALOGEN L.A.200 see TBF750
MALON, combustion products see ADX750
MALONAL see BAG000
MALONALDEHYDE see PMK000
MALONALDEHYDE DIETHYL ACETAL see MAN750
MALONALDEHYDE, ION(1-), SODIUM see MAN800
MALONALDEHYDE SODIUM SALT see MAN800
MALONALDEHYDE TETRAETHYL DIACETAL see MAN750
MALONAL, 3,3-DIMETHYL ACETAL see DOM625
MALONAMIDE see MAO000
MALONAMIDE NITRILE see COJ250
MALONAMONITRILE see COJ250
MALONATE see MAO100
MALONATE DIANION see MAO100
MALONATE ION(2⁻) see MAO100
MALONDIALDEHYDE see PMK000
MALONDIAMIDE see MAO000
MALONIC ACID DIAMIDE see MAO000
MALONIC ACID, DICHLORO-, DIMETHYL ESTER see DRK300
MALONIC ACID, DIETHYL ESTER see EMA500
MALONIC ACID DIMETHYL ESTER see DSM200
MALONIC ACID, (ETHOXYMETHYLENE)-, DIETHYL ESTER see EEV200
MALONIC ACID ETHYL ESTER NITRILE see EHP500
MALONIC ACID, ION(2⁻) see MAO100
MALONIC ACID, THALLIUM SALT (1:2) see TEM399
MALONIC ALDEHYDE see PMK000
MALONIC DIALDEHYDE see PMK000
MALONIC DINITRILE see MAO250
MALONIC ESTER see EMA500
MALONIC MONONITRILE see COJ500
MALONITRILE HYDRAZIDE see COH250
MALONOBEN see DED000
MALONODIALDEHYDE see PMK000
MALONONITRILE see MAO250

MALONONITRILE, CARBONYL-, DIETHYL MERCAPTOLE see BJT800
MALONONITRILE, CYCLOHEXYLIDENE- see CPM800
MALONONITRILE, DICHLORO- see DFN850
MALONONITRILE, (1-ETHOXYETHYLIDENE)-(6CI,7CI,8CI) see EET100
MALONONITRILE HYDRAZIDE see COH250
MALONYLDIALDEHYDE see PMK000
MALONYLDIAMIDE see MAO000
MALOPRIM see SOA500
MALOPRIM see TGD000
MALORAN see CES750
MALOTILATE see MAO275
MALOUETIN see MAO280
MALOUETINE see MAO280
MALPHOS see MAK700
MALTA RED X 2284 see NAY000
MALT EXTRACT see MAO525
MALTI MR see MAO285
MALTISORB see MAO285
MALTIT see MAO285
d-MALTITOL see MAO285
MALTITOL (6CI,7CI) see MAO285
MALTOBIOSE see MAO500
MALTODEXTRIN see MAO300
MALTOL see MAO350
MALTOSE see MAO500
d-MALTOSE see MAO500
MALTOX see MAK700
MALTOX MLT see MAK700
MALT SUGAR see MAO500
α-MALT SUGAR see MAO500
MALT SYRUP see MAO525
MALUS (VARIOUS SPECIES) see AQP875
MALVAVISCUS CONZATTI Greenm., flower extract see ADE125
MALVIDIN CHLORIDE see MAO600
MALVIDOL see MAO750
MALYSOL see MKA250
MALZID see DMC600
MAM see HMG000
MAMA see MDQ770
MAM AC see MGS750
MAM ACETATE see MGS750
MAMALLET-A see DOT000
MAMAN LAMAN (HAITI) see JBS100
MAMBNA see MHW350
MA144 MI see MAB250
MAMIVERT see SLC300
MAMMEX see NGE500
MAMMOTROPIN see PMH625
MAMN see AAW000
MAM-TOX see TCY275
MA144 N2 see RHA125
MANADRIN see EAW000
MANAGANESE ZINC FERRITE see IHB677
MANAM see MAS500
MANCENILLIER (HAITI) see MAO875
MANCHESTER YELLOW see DUX800
MANCHINEEL see MAO875
MANCOFOL see DXI400
MANCOKAR see MAO880
MANCOTOP see MAS300
MANCOZEB see DXI400
MANCOZEB-DINOCAP MIXTURE see MAO880
MANCOZEB + KARATHANE MIXTURE see MAO880
MANDARIN G see CMM220
MANDARIN OIL, COLDPRESSED see MAO900
MANDARIN PETITGRAIN OIL see OHJ150
MANDELAMIDE, N,N-DIETHYL- see DJL700
MANDELAMIDE, N-2-PYRIDYL-, HYDROCHLORIDE see PPM725
MANDELIC ACID see MAP000

MANDELIC ACID, ETHYL ESTER see EMB000
MANDELIC ACID, α-ISOPROPYL-, 3-(DIETHYLAMINO)PROPYL ESTER, HYDROCHLORIDE see DIQ200
MANDELIC ACID NITRILE see MAP250
MANDELIC ACID, 3d-TROPANYL ESTER see HGK550
d(−)-MANDELONITRILE-β-d-GENTIOBIOSIDE see AOD500
d-MANDELONITRILE-β-d-GLUCOSIDO-6-β-d-GLUCOSIDE see AOD500
d,l-MANDELONITRILE-β-d-GLUCOSIDO-6-β-GLUCOSIDE see IHO700
1-MANDELONITRILE-β-GLUCURONIC ACID see LAS000
MANDELONITRILE, PENTACHLORO- see PAX775
MANDELONITRILE, 2,3,4,5,6-PENTACHLORO- see PAX775
MANDELSAEUREAETHYLESTER (GERMAN) see EMB000
MANDELYLTROPEINE see HGK550
MANDELYTROPEINE see HGK550
MANDOKEF see CCX300
MANDOL see FOD000
MANDRIN see EAW000
MANEB see MAS500
MANEB 80 see MAS500
MANEBA see MAS500
MANEBE 80 see MAS500
MANEBE (FRENCH) see MAS500
MANEBGAN see MAS500
MANEB-METHYLTHIOPHANATE mixture see MAP300
MANEB PREPARATIONS, stabilized against self-heating (UN 2968) (DOT) see MAS500
MANEB PREPARATIONS with not <60% maneb (UN 2210) (DOT) see MAS500
MANEB (UN 2210) (DOT) see MAS500
MANEB, stabilized (UN 2968) (DOT) see MAS500
MANEB-ZINC see DXI400
MANEB plus ZINC ACETATE (50:1) see EIQ500
MANEB-ZINEB-KOMPLEX (GERMAN) see DXI400
MANEB ZL4 see MAS500
MANESAN see MAS500
MANETOL see MAP600
MANEX see MAS500
MANEXIN see MAW250
MANGAANBIOXYDE (DUTCH) see MAS000
MANGAANDIOXYDE (DUTCH) see MAS000
MANGAAN(II)-(N,N'-ETHYLEEN-BIS(DITHIOCARBAMAAT)) (DUTCH) see MAS500
MANGACAT see MAP750
MANGAN (POLISH) see MAP750
MANGAN(II)-(N,N'-AETHYLEN-BIS(DITHIOCARBAMAT)) (GERMAN) see MAS500
MANGANATE(3-), HEXACYANO-, TRIPOTASSIUM see TMX600
MANGANATE(3-), HEXAKIS(CYANO-C)-, TRIPOTASSIUM, (OC-6-11)- see TMX600
MANGANATE, TETRACHLORO-, BIS(3,4,5-TRIMETHOXYPHENETHYLAMMONIUM) see BKM100
MANGANDIOXID (GERMAN) see MAS000
MANGANESE see MAP750
MANGANESE ACETATE see MAQ000
MANGANESE(2+) ACETATE see MAQ000
MANGANESE(II) ACETATE see MAQ000
MANGANESE ACETATE TETRAHYDRATE see MAQ250
MANGANESE(II) ACETATE TETRAHYDRATE see MAQ250
MANGANESE ACETYLACETONATE see MAQ500

MANGANESE, (4-AMINOBUTANOATO-N,O)(N-(2,4-DIHYDROXY-3,3-DIMETHYL-1-OXOBUTYL)-β-ALANINATO)- see MAQ600

MANGANESE γ-AMINOBUTYRATOPANTOTHENATE see MAQ600

MANGANESE ARSENATE (Mn₃(AsO₄)₂) HEXAHYDRATE (6CI) see MAQ700

MANGANESE BINOXIDE see MAS000

MANGANESE (BIOSSIDO di) (ITALIAN) see MAS000

MANGANESE (BIOXYDE de) (FRENCH) see MAS000

MANGANESE(II) BIS(ACETYLIDE) see MAQ780

MANGANESE, BIS(2-BENZOYLBENZOATO)BIS(3-(1-METHYL-2-PYRROLIDINYL)PYRIDINE)-, TRIHYDRATE see NDN050

MANGANESE, BIS(l-HISTIDINATO-N,O)-, TETRAHYDRATE see BJX800

MANGANESE, BIS(l-HISTIDINATO)-, TETRAHYDRATE see BJX800

MANGANESE, BIS(5-SULFO-8-QUINOLINOLATO)- see BKJ275

MANGANESE BLACK see MAS000

MANGANESE CAPRYLATE see MAQ790

MANGANESE(II) CHLORIDE (1:2) see MAR000

MANGANESE(II) CHLORIDE TETRAHYDRATE see MAR250

MANGANESE CITRATE see MAR260

MANGANESE COMPOUNDS see MAR500

MANGANESE CYCLOPENTADIENYL TRICARBONYL see CPV000

MANGANESE DIACETATE see MAQ000

MANGANESE DIACETATE TETRAHYDRATE see MAQ250

MANGANESE DICHLORIDE see MAR000

MANGANESE DIMETHYL DITHIOCARBAMATE see MAR750

MANGANESE DINITRATE see MAS900

MANGANESE (DIOSSIDO di) (ITALIAN) see MAS000

MANGANESE DIOXIDE see MAS000

MANGANESE (DIOXYDE de) (FRENCH) see MAS000

MANGANESE EDTA COMPLEX see MAS250

MANGANESE, ((1,2-ETHANEDIYLBIS(CARBAMODITHIOATO))(2-))-, mixed with DIMETHYL (1,2-PHENYLENEBIS(IMINOCARBONOTHIOYL))BIS(CARBAMATE) see MAP300

MANGANESE, ((1,2-ETHANEDIYLBIS(CARBAMODITHIOATO))(2-))-, mixed with DIMETHYL (1,2-PHENYLENEBIS(IMINOCARBONOTHIOYL))BIS(CARBAMATE) and ((1,2-ETHANEDIYLBIS(CARBAMODITHIOATO))(2-))ZINC see MAS300

MANGANESE, ((1,2-ETHANEDIYLBIS(CARBAMODITHIOATO))(2-))-, mixed with ((1,2-ETHANEDIYLBIS(CARBAMODITHIOATO))(2-))ZINC and 2(OR 4)-ISOOCTYL-4,6(OR 2,6)-DINITROPHENYL 2-BUTENOATE see MAO880

MANGANESE ETHYLENE-1,2-BISDITHIOCARBAMATE see MAS500

MANGANESE(II) ETHYLENEBIS(DITHIOCARBAMATE) see MAS500

MANGANESE(II) ETHYLENE DI(DITHIOCARBAMATE) see MAS500

MANGANESE complex with N-FLUOREN-2-YL ACETOHYDROXAMIC ACID see HIR000

MANGANESE FLUORIDE see MAS750

MANGANESE(II) FLUORIDE see MAS750

MANGANESE FLUORURE (FRENCH) see MAS750

MANGANESE GLUCONATE see MAS800

MANGANESE GLYCEROPHOSPHATE see MAS810

MANGANESE GREEN see MAT250

MANGANESE HYPOPHOSPHITE see MAS815

MANGANESE LINOLEATE see MAS818

MANGANESE MANGANATE see MAT500

MANGANESE, (METHYLCYCLOPENTADIENYL)TRICARBONYL- see MAV750

MANGANESE, MONOSULFATE, MONOHYDRATE see MAU300

MANGANESE MONOXIDE see MAT250

MANGANESE NAPHTHENATE see MAS820

MANGANESE NITRATE see MAS900

MANGANESE(2+) NITRATE see MAS900

MANGANESE(II) NITRATE see MAS900

MANGANESE NITRATE (DOT) see MAS900

MANGANESE (II) NITRATE, ANHYDROUS see MAS900

MANGANESE OXIDE see MAS000

MANGANESE OXIDE see MAV550

MANGANESE(II) OXIDE see MAT250

MANGANESE(IV) OXIDE see MAS000

MANGANESE(III) OXIDE see MAT500

MANGANESE(VII) OXIDE see MAT750

MANGANESE(II) PERCHLORATE see MAT899

MANGANESE PERCHLORATE HEXAHYDRATE see MAU000

MANGANESE PEROXIDE see MAS000

MANGANESE SESQUIOXIDE see MAT500

MANGANESE(II) SULFATE (1:1) see MAU250

MANGANESE SULFATE MONOHYDRATE see MAU300

MANGANESE(2+) SULFATE MONOHYDRATE see MAU300

MANGANESE(II) SULFATE TETRAHYDRATE (1:1:4) see MAU750

MANGANESE(II) SULFIDE see MAV000

MANGANESE SUPEROXIDE see MAS000

MANGANESE TALLATE see MAV100

MANGANESE(II) TELLURIDE see MAV250

MANGANESE, TETRACARBONYL(TRIFLUOROMETHYLTHIO)-, dimer see TBN200

MANGANESE(II) TETRAHYDROALUMINATE see MAV500

MANGANESE TETROXIDE see MAV550

MANGANESE TRICARBONYL METHYLCYCLOPENTADIENYL see MAV750

MANGANESE TRIFLUORIDE see MAW000

MANGANESE TRIOXIDE see MAT500

MANGANESE(2+), TRIS(OCTAMETHYLDIPHOSPHORAMIDE-Op,Op')-, (OC-6-11)-, DIPERCHLORATE see TNK400

MANGANESE ZINC BERYLLIUM SILICATE see BFS750

MANGANESE ZINC FERRATE see IHB677

MANGANESE ZINC FERRITE see IHB677

MANGANIC OXIDE see MAT500

MANGAN NITRIDOVANY (CZECH) see MAP750

MANGANOMANGANIC OXIDE see MAV550

MANGANOUS ACETATE see MAQ000

MANGANOUS ACETATE TETRAHYDRATE see MAQ250

MANGANOUS ACETYLACETONATE see MAQ500

MANGANOUS CHLORIDE see MAR000

MANGANOUS CHLORIDE TETRAHYDRATE see MAR250

MANGANOUS DIMETHYLDITHIOCARBAMATE see MAR750

MANGANOUS DINITRATE see MAS900

MANGANOUS NITRATE see MAS900

MANGANOUS OXIDE see MAT250

MANGANOUS SULFATE see MAU250

MANGANOUS SULFATE MONOHYDRATE see MAU300

MANGAN-ZINK-AETHYLENDIAMIN-BIS-DITHIO-CARBAMAT (GERMAN) see DXI400

MAN-GRO see MAU250

MANHEXIN see MAW250

MANICOLE see MAW250

MANIHOT ESCULENTA see CCO680

MANIHOT UTILISSIMA see CCO700

MANINIL see CEH700

MANIOC see CCO680

MANIOKA see CCO680

MANIOL see CCR875

MANITE see MAW250

MANNA SUGAR see HER000

MANNEX see MAW250

MANNITE see HER000

MANNIT-LOST (GERMAN) see MAW500

MANNIT-MUSTARD (GERMAN) see MAW500

MANNITOL see HER000

MANNITOL see MAW100

d-MANNITOL see HER000

d-MANNITOL BUSULFAN see BKM500

MANNITOL, 1,6-DIDEHYDROXY-1,6-DIIODO-, d- see DHA375

MANNITOL, 1,6-DIDEOXY-1,6-DIIODO-, (d)- see DHA375

MANNITOL HEXANITRATE see MAW250

d-MANNITOL HEXANITRATE see MAW250

MANNITOL HEXANITRATE (dry) (DOT) see MAW250

MANNITOL HEXANITRATE, wetted with not <40% water, by weight or mixture (NA 0133) (DOT) see MAW250

MANNITOL MUSTARD DIHYDROCHLORIDE see MAW750

MANNITOL MYLERAN see BKM500

MANNITOL NITROGEN MUSTARD see MAW500

MANNITRIN see MAW250

MANNOGRANOL see BKM500

MANNOMUSTINE see MAW500

MANNOMUSTINE DIHYDROCHLORIDE see MAW750

MANNOPYRANOSIDE, STROPHANTHIDIN-3 6-DEOXY-, α-l- see CNH780

MANNOSULFAN see MAW800

MANOLENE 6050 see PJS750

MANOSEB see DXI400

MANOXAL OT see DJL000

MANRO PTSA 65 E see TGO000

MANRO PTSA 65 H see TGO000

MANRO PTSA 65 LS see TGO000

MANSIL see OLT000

MAN'S MOTHERWORT see CCP000

MANTA see KAJ000

MANTADAN see AED250

MANTHELINE see DJM800

MANTHELINE see XCJ000

MANUCOL see SEH000

MANUCOL DM see SEH000

MANUFACTURED IRON OXIDES see IHC450

MANUTEX see SEH000

MANZANA (SPANISH) see AQP875

MANZANILLO (CUBA, PUERTO RICO, DOMINICAN REPUBLIC) see MAO875

MANZATE see MAS500

MANZATE 200 see DXI400

MANZATE 200 see MAS500

MANZATE D see MAS500

MANZATE MANEB FUNGICIDE see MAS500

MANZEB see DXI400

MANZEB see MAS500

MANZIN see MAS500

MANZIN 80 see DXI400

MAOA see QAK000

MAOH see MKW600

MAOLATE see CJQ250

MAO-REM see PFC750
4M2AP see ALC250
MAPHARSAL see ARL000
MAPHARSEN see ARL000
MAPHARSIDE see ARL000
MAPHENIDE ACETATE see MAC000
MAPLE AMARANTH see ARL020
MAPLE BRILLIANT BLUE FCF see FMU059
MAPLE ERYTHROSINE see FAG040
MAPLE INDIGO CARMINE see FAE100
MAPLE LACTONE see HMB500
MAPLE PONCEAU 3R see FAG018
MAPLE SUNSET YELLOW FCF see FAG150
MAPLE TARTRAZOL YELLOW see FAG140
MAPO see TNK250
MAPOLOSE 60SH50 see MIF760
MAPOLOSE M25 see MIF760
MAPOSOL see SIL550
MAPOSOL see VFU000
MAPP (OSHA) see MFX600
MAPRENAL 980 see MCB050
MAPRENAL MF see MCB050
MAPRENAL MF 590 see MCB050
MAPRENAL MF 650 see MCB050
MAPRENAL MF 900 see MCB050
MAPRENAL MF 904 see MCB050
MAPRENAL MF 910 see MCB050
MAPRENAL MF 915 see MCB050
MAPRENAL MF 920 see MCB050
MAPRENAL MF 927 see MCB050
MAPRENAL MF 929 see MCB050
MAPRENAL MF 980 see MCB050
MAPRENAL MP 500 see MCB050
MAPRENAL NPX see MCB050
MAPRENAL RT-MF 650 see MCB050
MAPRENAL TTX see MCB050
MAPRENAL VMF see MCB050
MAPRENAL VMF 3655 see MCB050
MAPRENAL VMF 3925 see MCB050
MAPRENAL VMF 3935 see MCB050
MAPRENAL VMF 52/7 see MCB050
MAPROFIX 563 see SIB600
MAPROFIX TLS 65 see SON000
MAPROFIX WAC-LA see SIB600
MAPROTILINE see LIN800
MAPROTILINE HYDROCHLORIDE see
MAW850
MAPTC see CMP900
MAQBARL see DTN200
MAQBARL, nitrosated (JAPANESE) see
MMW750
MAQUAT MC 1412 see QAT520
MARALATE see MEI450
MARANHIST see WAK000
MARANTA see SLJ500
MARANYL F 114 see PJY500
MARAPLAN see IKC000
MARASMIC ACID see MAW875
MARASMIC ACID METHYL ESTER see
MLE600
MARASPERSE B see LFQ500
MARASPERSE CBS see LFQ500
MARASPERSE N see LFQ500
MARASPERSE N 22 see LFQ500
MARASPERSE N 22 DISPERSANT see
LFQ500
MARATHON see DWS200
MARAVILLA (MEXICO) see CAK325
MARAZINE see EAN600
MAR BATE see MQU750
MARBLE see CAO000
MARBLEWHITE 325 see CAT775
MARBON 9200 see SMR000
MARBORAN see MKW250
MARCAIN see BOO000
MARCELLOMYCIN see MAX000
MARCOPHANE see BGY000
MARETIN see HMV000
MAREVAN (SODIUM SALT) see WAT220
MAREX see HGC500
MAREZINE see EAN600

MAREZINE HYDROCHLORIDE see
MAX275
MARFIL see CAT775
MARFOTOKS see DJI000
MARGARIC ACID see HAS500
MARGONIL see MQU750
MARGONOVINE see LJL020
MARGOSAN-O see NBR900
MARGOSAN O see NBR900
MARGOSA OIL see NBS300
MARIA GRANDE (PUERTO RICO) see
MBU780
MARICAINE see DHK400
MARIHUANA see CBD750
MARIJUANA see CBD750
MARIJUANA, SMOKE RESIDUE see CBD760
MARIMET 45 see CAW850
MARINCO H see MAG750
MARINDININ see DLR000
MARINE DIESEL FUEL see DHE750
MARINEURINA see DXO300
MARISAN see CMR100
MARISILAN see AIV500
MARITUS YELLOW see DUX800
MARJORAM OIL see MAX875
MARJORAM OIL, SPANISH see MBU500
MARK AO 30 see MOS100
MARK AO 60 see DED100
MARKOFANE see BGY000
MARK PEP 24 see COV800
MARK PFK see REA050
MARKS 4-CPA see CJN000
MARKURE UL2 see DEJ100
MARLAMID D 1885 see OHU200
MARLATE see MEI450
MARLEX 9 see PJS750
MARLEX 50 see PJS750
MARLEX 60 see PJS750
MARLEX 960 see PJS750
MARLEX 6003 see PJS750
MARLEX 6009 see PJS750
MARLEX 6015 see PJS750
MARLEX 6050 see PJS750
MARLEX 6060 see PJS750
MARLEX 9400 see PMP500
MARLEX EHM 6001 see PJS750
MARLEX M 309 see PJS750
MARLEX TR 704 see PJS750
MARLEX TR 880 see PJS750
MARLEX TR 885 see PJS750
MARLEX TR 906 see PJS750
MARLIPAL 1217 see DXY000
MARLON AS 3 see LBU100
MARLOPHEN 820 see PKF500
MARMAG see MAH500
MARMELOSIN see IHR300
MARMER see DXQ500
MARMESIN see MBU525
(+)-MARMESIN see MBU525
(S)-MARMESIN see MBU525
S-(+)-MARMESIN see MBU525
MARNITENSION SIMPLE see RDK000
MAROMERA (CUBA) see RBZ400
MAROXOL-50 see DUZ000
MARPLAN see IKC000
MARPLON see IKC000
MARRONNIER (CANADA) see HGL575
MARSALID see ILE000
MARS BROWN see IHC450
MARSHAL see CCC280
MARSHALL see CCC280
MARSH GAS see MDQ750
MARSH MARIGOLD see MBU550
MARSH ROSEMARY EXTRACT see MBU750
MARSILID see ILE000
MARSIN see MNV750
MARS RED see IHC450
MARSTHINE see FOS100
MARTRATE-45 see PBC250
MARUCOTOL see AAE500
MARUNGUEY (PUERTO RICO) see CNH789
MARVEX see DGP900

MARVINAL see PKQ059
MARYGIN-M see DAB875
MARZIN see DXI400
MARZINE see EAN600
MARZULENE S see MBU775
MAS see MGQ750
15-MAS see ECT700
MA144 S2 see APV000
MAS 5382 see CNB825
MASCHITT see CFY000
MASDIOL see AOO475
MASELNAN DI-n-BUTYLCINICITY
(CZECH) see DDX600
MASENATE see TBG000
MASENONE see MPN500
MASEPTOL see HJL500
MASHERI (INDIA) see SED400
MASLETINE see FOS100
MASOTEN see TIQ250
MASSICOT see LDN000
MASSICOTTE see LDN000
MAST CELL DEGRANULATING PEPTIDE
(VESPA BASALIS) see MBU776
MASTESTONA see MPN500
MASTIC ABSOLUTE see MBU777
MASTIC (RESIN) see MBU777
MASTIPHEN see CDP250
MASTOPARAN B see MBU776
MASTWOOD see MBU780
MATACIL see DOR400
MATANOL see AMK250
MATAVEN see ARW000
MATAVEN see FDB500
MATCH see MLJ500
MATENON see MQS225
MATHIESON 466 see CNL310
MATILIT CM 2L see SEA500
MATO AZUL (PUERTO RICO) see CAK325
MATO de PLAYA (PUERTO RICO) see
CAK325
MATOX see HGP525
MATRICARIA CAMPHOR see CBA750
MATRIGON see DGJ100
MATRIX see THT800
MATROMYCIN see OHO200
MATSUTAKE ALCOHOL (JAPANESE) see
ODW000
MATTING ACID (DOT) see SOI500
MATULANE see PME250
MATULANE see PME500
MAUCHERITE see MBU790
MAURYLENE see PMP500
MAVISERPIN see RDK000
MAVRIK see VBU100
MAVRIK AQUAFLOW see VBU100
MAVRIK HR see VBU100
MAXATASE see BAC000
MAXFORCE see HGP525
MAXIBALIN see EJT600
MAXIBOLIN see EJT600
MAXICHEM 1DTM see MCB050
MAXIDEX see SOW000
MAXIFEN see AOD000
MAXILON GOLDEN YELLOW GL see
CMM895
MAXILON YELLOW 3GL see CMM890
MAXIPEN see PDD350
MAXITATE see MAW250
MAXOLON see AJH000
MAXULVET see SNN300
MA 144 Y see ADG425
MAY APPLE see MBU800
MAY & BAKER S-4084 see COQ399
MAY BLOB see MBU550
MAYCOR see CCK125
MAYFLOWER see LFT700
MAYSANINE see MBU820
MAYSERPINE see RDK000
MAYT see MBU820
MAYTANSINE see MBU820
MAYTANSINOL ISOVALERATE see APE529
MAYTENIN see TGD125

2-MAYTHIC ACID see NAV500
MAY THORN see MBU825
MAYVAT BROWN BR see CMU770
MAYVAT GOLDEN YELLOW see DCZ000
MAYVAT OLIVE AR see DUP100
MAZATICOL HYDROCHLORIDE see MBV100
MAZEPTYL see TFM100
MAZIDE 30 see DHF200
MAZINDOL see MBV250
MAZOTEN see TIQ250
MB see MHR200
4HMB see HLY400
M & B 800 see DBL800
M+B 695 see PPO000
M+B 760 see TEX250
MB 2878 see EAK500
MB-3000 see BKL800
MB 6046 see TIL800
MB 9057 see SNQ500
M&B 3046 see CLO000
M & B 4620 see TDS250
M&B 8873 see HKB500
MB 10064 see DDP000
MB 2050A see PBT000
M&B 11,461 see DNG200
M&B 38544 see DKJ210
M & B 39565 see MQY110
M&B 17,803A see AAE125
MBA see BIE250
7-MBA see MGW750
7-MBA-3,4-DIHYDRODIOL see MBV500
MBA HYDROCHLORIDE see BIE500
MB 1 (ANTIOXIDANT) see MJM700
MBAO see CFA500
MBAO HYDROCHLORIDE see CFA750
MBBA see CQK600
MBBA see MEE100
MBBH 766 see CMP200
MBC see BAV575
MBC see MHC750
MBCP see LEN000
M & B 782 DIHYDROCHLORIDE see DBM400
MBDZ see MHL000
MBH see PME500
MBK see HEV000
MBMA see MJQ250
MBNA see MHW500
MBOA see MEC550
6-MBOA see MEC550
MBOCA see MJM200
MBOEA see MJN100
MBOT see MJO250
MBP see MRF525
MBPMC see TAN300
MB PYRETHROID see MBV700
MB PYRETHROID see MNG525
MBR 6168 see DRR000
MBR 8251 see TKF750
MBR 12325 see DUK000
MBR 3092-42 see MBV710
3M MBR 6168 see DRR000
MBT see BDF000
MBTH see MHJ250
MBTH HYDROCHLORIDE see MHJ275
MBTS see BDE750
MBTS RUBBER ACCELERATOR see BDE750
MBX see MHR200
MBY see MHX250
6-MC see MIP750
MeC see MCB050
2M-4C see CIR250
MC 338 see NIW500
MC 474 see DJI000
MC 1053 see CBW000
MC 1108 see DVJ400
MC 1415 see PMB250
MC 1478 see NIW500
MC 1488 see BRU750
MC 1945 see DVI800
MC 2188 see CDY299

MC 2303 see CBK500
MC 2303 see GGS000
MC 2420 see MLJ000
MC 3199 see BCP685
MC-4379 see MJB600
MC6897 see DQM600
MC 7783 see PLD050
MC 10978 see SFV650
MCA see CEA000
3-MCA see MIJ750
MCA-600 see BDG250
MeCsAc see EJJ500
3-MCA-anti-9,10-DIOL-7,8-EPOXIDE see MBV715
MCA-11,12-EPOXIDE see ECA500
MCA-11,12-OXIDE see ECA500
M. CARINICAUDUS DUMERILII VENOM see MQT000
MCB see CEJ125
4-(MCB) see CLN750
MCC see DEV600
Me-CCNU see CHD500
MCDA-11 see MRU756
MC DEFOLIANT see MAE000
(E)-MC 11,12-DIHYDRODIOL see DML800
M.C. DUMERILLI VENOM see MQT000
MCE see EIF000
MCEABN see MEG750
MCEAMN see MEH000
MCF see CIM325
MCF see MIG000
MCH see MFX560
MCH 52 see UTU500
MCI 028 see HGL700
MCI-2016 see MGE200
MCI-C54875 see DXY000
MCIT see CEX275
MCMAMN see MEH750
MCM-JR-4584 see FLK100
MCN-485 see AJF500
MCN 1025 see NNF000
MCN 2559 see TGJ850
MCN-3113 see XGA725
MCN 2783-21-98 see ZUA300
MCN-A 2833-109 see PCJ230
MCNAMEE see KBB600
MCNEIL 481 see DQU200
MCN-JR-2498 see TKK500
MCN-JR-3345 see FHG000
McN-JR 4263 see PDW750
MCN-JR-4584 see FGU000
MCN-JR-4749 see DYF200
McN-JR-6238 see PIH000
MCN-JR-8299 see TDX750
McN-JR-16,341 see PAP250
MCN-JR-4263-49 see PDW750
MCO 8000 see MIF760
M1 (COPPER) see CNI000
M2 (COPPER) see CNI000
MCP see CIR250
MCP 875 see AFJ400
MC 4000 cP see MIF760
MCPA see CIR250
MCPABN see MEI000
MCPA DIETHANOLAMINE SALT see MIH750
MCPA-ETHYL see EMR000
MCPAMN see MEI250
MCPA POTASSIUM SALT see MIH800
MCPA SODIUM SALT see SIL500
MCPB see CLN750
MCPB see CLO000
MCPB-ETHYL see EHL500
MCP-BUTYRIC see CLN750
4-(MCPD) see CLO000
MCPEE see EMR000
MCPP see CIR500
MCPP see RBF500
2-MCPP see CIR500
MCPP-D-4 see CIR500
MCPP 2,4-D see CIR500
MCPP-K-4 see CIR500

MCPP POTASSIUM SALT see CLO200
MCR see MAB750
MCS 198 see TBL600
MC 20000S see MIF760
MC-T see CAT775
MCT see CPV000
MCT see TMO000
MCTR 171-78 see PLD050
M 3 (CURING AGENT) see OMM300
MCZ NITRATE see MQS560
10-MD see MEK700
MD 141 see DIS600
516 MD see CMR100
MD 2028 see HAF400
2028 MD see HAF400
5579 MD see DAB875
MD 67350 see VGK000
68111 M.D. see IRQ000
MD 780515 see MEW800
l-(α-MD) see DNA800
MDA see MJQ000
MDA see MJQ775
MDA 150 see FMW330
MDA 220 see FMW330
MDAB see DUH600
2-MeDAB see TLE750
3'-MDAB see DUH600
MDBA see MEL500
MD BACITRACIN see BAC260
MDBCP see MBV720
MDE see EMS100
(+−)-MDE see EMS100
MDEA see EMS100
MDEA see MKU250
MDI see MJP400
bu-MDI see BRC500
pr-MDI see DPY600
MDI-CR see PKB100
MDI-CR 100 see PKB100
MDI-CR 200 see PKB100
MDI-CR 300 see PKB100
M 141 DIHYDROCHLORIDE PENTAHYDRATE see SKY500
MDI-PC see DGE200
MDL-035 see THG700
MDL 473 see RKZ100
MDL-899 see MBV735
MDP SODIUM see SIL575
MDS see DBD750
MDS see DXL400
2-ME see MCN250
ME 277 see PAM000
ME-1700 see BIM500
ME 3625 see DFD000
MEA see AJT250
MEA see EEC600
MEA see MCN750
MEA 610 see NAC000
MEAD JOHNSON 1999 see CCK250
MEADOW BRIGHT see MBU550
MEADOW GARLIC see WBS850
MEADOW GREEN see COF500
MEADOW ROSE LEEK see WBS850
MEADOW SAFFRON see CNX800
MEA HYDROCHLORIDE see EEC700
MEARLMAID see GLI000
MEASURIN see ADA725
MEAVERIN see CBR250
MEB see MAS500
MEB-6046 see TIL800
MEB 6447 see CJO250
MEBACID see ALF250
MEBALLYMAL see SBM500
MEBALLYMAL SODIUM see SBN000
MEBANAZINE OXALATE see MBV750
MEBANAZINE SULPHATE see MHO000
MEBANE SODIUM SALT see MBV775
MEBARAL see ENB500
MnEBD see MAS500
MEBENDAZOLE (USDA) see MHL000
MEBENVET see MHL000
MEBEREL see ENB500

MEBHYDROLIN NAPADISYLATE see MBW100
MEBICAR see MBW250
MEBICAR-A see MBW500
MEBICHLORAMINE see BIE500
MEBR see MIIR200
ME4 BROMINAL see DDP000
MEBRON see MCH550
MEBROPHENHYDRAMINE see BMN250
MEBROPHENHYDRAMINE HYDROCHLORIDE see BMN250
MEBRYL see BMN250
MEBUBARBITAL see NBT500
MEBUBARBITAL SODIUM see NBU000
MEBUMAL NATRIUM see NBU000
MEBUMAL SODIUM see NBU000
MEBUTAMATE see MBW750
MEBUTINA see MBW750
MEC see MBW775
MECADOX see FOI000
MECAL see RCK730
MECALMIN see CCR875
MECAMILAMINA (ITALIAN) see VIZ400
MECAMINE see VIZ400
MECAMINE HYDROCHLORIDE see MQR500
MECAMYLAMINE see VIZ400
S-(+)-MECAMYLAMINE see MBW778
MECAMYLAMINE HYDROCHLORIDE see MQR500
MECARBAM see DJI000
MECARBENIL see MBW780
MECAROL see EQL000
MECARPHON see MLJ000
MECB see DJG000
ME-CCNU see CHD250
MECHLORETHAMINE see BIE250
MECHLORETHAMINE HYDROCHLORIDE see BIE500
MECHLORETHAMINE OXIDE see CFA500
MECHLORETHAMINE OXIDE HYDROCHLORIDE see CFA750
MECHLORPROP see RBF500
MECHOLIN see MFX560
MECHOLINE see MFX560
MECHOLYL see ACR000
MECHOTHANE see HOA500
MECICLIN see DAI485
MECINARONE see MBX000
MECLASTINE HYDROGEN FUMARATE see FOS100
MECLIZINE see HGC500
MECLIZINE DIHYDROCHLORIDE see MBX250
MECLIZINE HYDROCHLORIDE see MBX500
MECLOFENAMATE SODIUM see SIF425
MECLOFENAMIC ACID see DGM875
MECLOFENOXANE see DPE000
MECLOFENOXATE HYDROCHLORIDE see AAE500
MECLOMEN see SIF425
MECLOPHENAMIC ACID see DGM875
MECLOPHENOXATE see DPE000
MECLOQUALONE see MIH925
MECLOZINE see HGC500
MECLOZINE HYDROCHLORIDE see MBX500
MECOBALAMIN see VSZ050
MECODIN see MDO750
MECODRIN see BBK000
MECOMEC see CIR500
MECOPEOP see CIR500
MECOPER see CIR500
MECOPEX see CIR500
MECOPEX see CLO200
MECOPEX see RBF500
MECOPROP see CIR500
MECOPROP see RBF500
MECOPROP DIETHANOLAMINE SALT see MIH900
MECOPROP POTASSIUM SALT see CLO200

MECOTURF see CIR500
MECPROP see CIR500
MECRAMINE see AJT250
MECRILAT see MIQ075
MECRYL see CFU750
MECRYLATE see MIQ075
MECS see EJH500
MECYSTEINE HYDROCHLORIDE see MBX800
MEDAMYCIN see TBX250
MEDAPAN see COY500
MEDARON see NGG500
MEDARSED see BPF000
MEDAZEPAM see CGA000
MEDAZEPAM HYDROCHLORIDE see MBY000
MEDAZEPOL see CGA000
MEDEMANOL see MAW250
MEDEMYCIN see MBY150
MEDFALAN see SAX200
MEDIAMID see DJS200
MEDIAMYCETINE see CDP250
MEDIBEN see MEL500
MEDICAINE see BQA010
MEDI-CALGON see SHM500
MEDICEL see AKO500
MEDICON see DBE200
MEDIDRYL see BBV500
MEDIFENAC see AGN000
MEDIFLAVIN see DBX400
MEDIFLOR FC 43 see HAS000
MEDIFURAN see FPI000
MEDIGOXIN see MJD300
MEDIHALER-EPI see AES000
MEDIHALER-EPI see VGP000
MEDIHALER-TETRACAINE see BQA010
MEDILLA see MKP500
MEDINAL see BAG250
MEDINOTERB ACETATE see BRU750
MEDIOCONTRIX see IGC100
MEDIQUIL see IPU000
MEDIREX see IJG000
MEDITERRANEAN BAY OIL see OGQ150
MEDITRENE see IEP200
MEDIUM BLUE see IBV050
MEDIUM BLUE EMBL see ADE750
MEDLEXIN see ALV000
MEDOMET see DNA800
MEDOMET see MJE780
MEDOMIN see COY500
MEDOMINE see COY500
α-MEDOPA see MJE780
MEDOPAQUE see HGB200
MEDOPREN see DNA800
MEDOPREN see MJE780
2,2-MEDP see FNW100
2,4-MEDP see MBY500
4,4-MEDP see MBZ000
MEDPHALAN see SAX200
MEDROGESTERONE see MBZ100
MEDROGESTONE see MBZ100
MEDROGLUTARIC ACID see HMC000
MEDROL see MOR500
MEDROL ACETATE see DAZ117
MEDROL DOSEPAK see MOR500
MEDRONE see MOR500
MEDRONIC ACID, SODIUM SALT see SIL575
MEDROXALOL HYDROCHLORIDE see MBZ120
MEDROXYPROGESTERON see MBZ150
MEDROXYPROGESTERONE see MBZ150
MEDROXYPROGESTERONE ACETATE see MCA000
MEDROXYPROGESTERONE ACETATE and ETHINYLESTRADIOL see POF275
MEDROXYPROGESTERONE ACETATE MIXED WITH ESTRADIOL CYPIONATE see EDQ600
MEDTA see PNJ100
MEDULLIN see MCA025
MEE see EDP000

MEERSCHAUM see SBY100
MEERSCHAUM see SBY200
MEETHOBALM see BQA010
MEFEDINA see DAM700
MEFENACET see BDG100
MEFENAL see SNJ350
MEFENAMIC ACID see XQS000
MEFENOXALONA see MFD500
MEFENOXALONE see MFD500
MEFENSINA see GGS000
MEFLUIDIDE see DUK000
MEFOXIN see CCS510
MEFOXITIN see CCS510
MEFRUSID see MCA100
MEFRUSIDE see MCA100
MEFURINA see AHK750
M.E.G. see EJC500
MEGABA see AJC625
MEGABION see VSZ000
MEGABION (JAPANESE) see AOO475
MEGA BT see BAC040
MEGACE see VTF000
MEGACILLIN ORAL see PDT750
MEGACILLIN SUSPENSION see BFC750
MEGACILLIN TABLETS see BFD000
MEGACORT see DAE525
MEGADIURIL see CFY000
MEGAGRISEVIT see CLG900
MEGALOMICIN A see MCA250
MEGALOMYCIN-A see MCA250
MEGALOVEL see VSZ000
MEGAMYCINE see MDO250
MEGAPHEN see CKP250
MEGAPHEN see CKP500
MEGASEDAN see CGA000
MEGATOX see FFF000
MEGEPTIL see TFM100
MEGESTROL ACETATE see VTF000
MEGESTROL ACETATE + ETHINYLOESTRADIOL see MCA500
MEGESTROL ACETATE 4 mg, ETHINYLOESTRADIOL 50 μg see MCA500
MEGESTRYL ACETATE see VTF000
MEGIMIDE see MKA250
4-ME-GLU-P-2 see AKS100
6-ME-GLU-P-2 see AKS250
MEGLUMINE AMIDOTRIZOATE see AOO875
MEGLUMINE CONRAY see IGC000
MEGLUMINE DIATRIZOATE see AOO875
MEGLUMINE IOCARMATE see IDJ500
MEGLUMINE IODAMIDE see IDJ600
MEGLUMINE IODIPAMIDE see BGB315
MEGLUMINE IODOXAMATE see IFP800
MEGLUMINE IOTHALAMATE see IGC000
MEGLUMINE IOTROXATE see IGD075
MEGLUMINE ISOTHALAMATE see IGC000
MEGLUMINE SODIUM IODAMIDE see MCA775
MEGLUTOL see HMC000
MEGUAN see DQR800
MEHENDI see HMX600
MEHP see MRI100
4-ME-I see MKU000
MEICELIN see CCS365
MEIJI HERBIACE see SEO550
MEISEI SCARLET GG SALT see DEO295
MEISEI TERYL DIAZO BLACK CR see DPO200
MEISEI TERYL DIAZO BLUE HR see DCJ200
MEISI FAST RED RL BASE see MMF780
MEITO MY 30 see GGA800
MEK see MKA400
MEKAMINE see VIZ400
MEKAMIN HYDROCHLORIDE see MQR500
MEK-OXIME see EMU500
MEK PEROXIDE see MKA500
MEKP (OSHA) see MKA500
MELABON see ABG750
MELADININ see XDJ000
MELADININE see XDJ000

MELADUR MS 80 see MCB050
MELAFORM see MCB050
MELAFORM 45 see MCB050
MELAFORM 150 see MCB050
MELAFORM E 45 see MCB050
MELAFORM E 50 see MCB050
MELAFORM E 55 see MCB050
MELAFORM M 45S$_1$ see MCB050
MELAFORM WM6 see MCB050
MELAFORM WM 100 see MCB050
MELALEUCA ALTERNIFOLIA OIL see TAI150
MELALIT see MCB050
MELAMINE see MCB000
MELAMINE 20 see MCB050
MELAMINE 366 see MCB050
MELAMINE, BUTYL- see BRR800
MELAMINE, polymer with FORMALDEHYDE see MCB050
MELAMINE-FORMALDEHYDE CONDENSATE see MCB050
MELAMINE-FORMALDEHYDE COPOLYMER see MCB050
MELAMINE-FORMALDEHYDE POLYMER see MCB050
MELAMINE-FORMALDEHYDE RESIN see MCB050
MELAMINE, FORMALDEHYDE, TOLUENESULFONAMIDE POLYMER, BUTYLATED see MCB505
MELAMINE-FORMOL COPOLYMER see MCB050
MELAMINE, HEXACHLORO-(6CI,7CI,8CI) see TNG275
MELAMINE, POLYMER with FORMALDEHYDE (8CI) see MCB050
MELAMINE RESIN see MCB050
MELAMINKYANURAT (CZECH) see THO750
MELAMIN-N,N',N"-TRIMETHYLSULFONSAURES NATRIUM (GERMAN) see THS500
MELAN 11 see UTU500
MELAN 15 see MCB050
MELAN 20 see MCB050
MELAN 22 see MCB050
MELAN 23 see MCB050
MELAN 26 see MCB050
MELAN 27 see MCB050
MELAN 28 see MCB050
MELAN 29 see MCB050
MELAN 125 see MCB050
MELAN 21A see MCB050
MELAN 220 see MCB050
MELAN 243 see MCB050
MELAN 245 see MCB050
MELAN 287 see MCB050
MELAN 28A see MCB050
MELAN 28D see MCB050
MELAN 445 see MCB050
MELAN 523 see MCB050
MELAN 620 see MCB050
MELAN 630 see MCB050
MELAN 2000 see MCB050
MELAN 284A see MCB050
MELAN 8000 see MCB050
MELAN BLACK see BMA000
MELANEX see AKQ250
MELANILINE see DWC600
MELANOL LP20T see SON000
MELANOMYCIN see MCB100
MELANOSPORIN see MCB250
MELANTHERINE BH see CMN800
MELANTHERINE BHX see CMN800
MELAN X 28 see MCB050
MELAN X 65 see MCB050
MELAN X 71 see MCB050
MELAPRET P see MCB050
MELAROM 3 see MCB050
MELARSEN see SIF450
MELARSONYL POTASSIUM SALT see PLK800

MELASIL K 1 see MCB050
MELASIL K 2 see MCB050
MELASIL K3 see MCB050
MELASIL K 1S see MCB050
MELASIL U see MCB050
MELASIL U 1 see MCB050
MELASIL U 2 see MCB050
MELATONIN see MCB350
MELATONINE see MCB350
MELBEX see MRX000
MELBIN see DQR600
MELDIAN see CKK000
MELDONE see CNU750
MELEM see HAQ700
MELENGESTROL ACETATE see MCB375
MELENGESTROL ACETATE see MCB380
MELERIL see MOO250
MELETIN see QCA000
MELEX sce MQR760
MEL-F see MCB050
MELIA AZEDARACH see CDM325
MELIFORM see PKF750
MELILOTAL see MFW250
MELILOTIN see HHR500
MELILOTOL see HHR500
MELIN see RSU000
MELINAMIDE see MHO200
MELINEX see PKF750
MELINITE see PID000
MELIPAN see MCB500
MELIPAX see CDV100
MELIPRAMIN see DLH600
MELIPRAMIN see DLH630
MELIPRAMINE see DLH600
MELIPRAMINE see DLH630
MELIPRAMINE HYDROCHLORIDE see DLH630
MELIPRAMIN HYDROCHLORIDE see DLH630
MEL-IRON A see MCB050
MELITASE see CKK000
MELITOSE see RBA100
MELITOXIN see BJZ000
MELITRACEN see AEG875
MELITRACENE see AEG875
MELITRACENE HYDROCHLORIDE see TDL000
MELITRACEN HYDROCHLORIDE see TDL000
MELITRIOSE see RBA100
MELITTIN see MCB525
MELITTIN-I see MCB525
MELLARIL see MOO250
MELLARIL HYDROCHLORIDE see MOO500
MELLERETTE see MOO250
MELLERETTEN see MOO250
MELLERIL see MOO250
MELLINESE see CKK000
MELLITE 131 see DEH650
MELLITE 825 see DVK200
MELLOSE see MIF760
MELOCHIA TOMENTOSA see BAR500
MELOCOTON (SPANISH) see AQP890
MELOGEL see SLJ500
MELOLAK B see MCB050
MELOLAK B-II see MCB050
MELOLAM see MCB050
MELOLAM 285 see MCB050
MELONEX see AKQ250
MELOPAS AMP 1 see MCB050
MELOPAS 183GF see MCB050
MELOPAS N 37601 see MCB050
MELOPLAST B see MCB050
MELOXINE see XDJ000
MELPHALAN see PED750
MELPHALAN (RUSSIAN) see BHV000
MELPHALAN HYDROCHLORIDE see BHV000
MELPREX see DXX400
MELSEDIN see MDT250
MELSEDIN BASE see QAK000
MELSMON see MCB535

MELSOMIN see QAK000
MELTROL see PDF250
MELUNA see SLJ500
MEL W see PLK800
MELYSIN see MCB550
MEMA see MEO750
MEMA see MLF250
MEMC see MEP250
MEMCOZINE see SNN300
ME-MDA see MJO250
MEMINE see TCY750
MEMMI see MLF500
MEMORY ROOT see JAJ000
MEMPA see HEK000
MENADIOL see MMC250
MENADIOL DIACETATE see VTA100
MENADIOL DIPHOSPHATE TETRASODIUM SALT see NAQ600
MENADIOL SODIUM DIPHOSPHATE see NAQ600
MENADIOL TETRASODIUM DIPHOSPHATE see NAQ600
MENADION see MMD500
MENADIONE see MMD500
MENADIONE DIPHOSPHATE TETRASODIUM SALT see NAQ600
MENADIONE SODIUM BISULFITE see MCB575
MENADIONE SODIUM HYDROGEN SULFITE see MCB575
MENADIONE SODIUM PHOSPHATE see NAQ600
MENADION-NATRIUM-BISULFIT TRIHYDRAT (GERMAN) see MMD750
MENAGEN see EDV000
MENAPHTAM see CBM750
MENAPHTHON see MMD500
MENAPHTHONE SODIUM BISULFITE see MCB575
MENAPHTHONE SODIUM BISULPHITE see MCB575
1-MENAPHTHYL SULFATE SODIUM see SIF475
MENAPHTONE see MMD500
MENAQUINONE-4 see VTA650
MENAQUINONE K$_4$ see VTA650
MENATENSINA see RCA200
MENATETRENONE see VTA650
MENAZON see ASD000
dd-MENCS see MLC000
MENDEL see MQU750
MENDI see HMX600
MENDIAXON see MKP500
MENDON see CDQ250
MENDRIN see EAT500
MENESIA see MAG000
MENEST see ECU750
MENETYL see EJR500
MENFORMON see EDV000
MENHYDRINATE see DYE600
MENICHLOPHOLAN see DFD000
MENIDRABOL see NNX400
MENIPHOS see MQR750
MENISPERMUM CANADENSE see MRN100
MENITE see MQR750
Me$_2$NMOR see DTA000
MENOCIL see ALX250
MENOGAROL see MCB600
MENOGEN see ECU750
MENOMYCIN see MRA250
MENONASAL see TBN000
MENOQUENS see MCA500
MENOSTILBEEN see DKA600
MENOTAB see ECU750
MENOTROL see ECU750
MENOTROPHIN see FMT100
MENOTROPINS see FMT100
MENSISO see APY500
MENTA-BAL see ENB500
MENTHA ARVENSIS, OIL see MCB625
MENTHA ARVENSIS OIL, PARTIALLY DEMENTHOLIZED (FCC) see MCB625

MENTHA CITRATA OIL see BFN990
p-MENTHA-1,8-DIEN-7-AL see DKX100
o-1,4-MENTHADIENE see MCB700
p-MENTHA-1,3-DIENE see MLA250
p-MENTHA-1,4-DIENE see MCB750
p-MENTHA-1,5-DIENE see MCC000
p-MENTHA-1,8-DIENE see LFU000
p-MENTHA-1,8-DIENE see MCE250
1,8(9)-p-MENTHADIENE see MCC250
d-p-MENTHA-1,8-DIENE see LFU000
(S)-(−)-p-MENTHA-1,8-DIENE see MCC500
p-MENTHA-1,8-DIEN-7-OL see PCI550
l-p-MENTHA-6,8-DIEN-2-OL see MKY250
p-MENTHA-6,8-DIEN-2-OL, PROPIONATE
see MCD000
p-MENTHA-6,8-DIEN-2-ONE see MCD250
6,8(9)-p-MENTHADIEN-2-ONE see MCD250
(R)-(−)-p-MENTHA-6,8-DIEN-2-ONE see
CCM120
1-6,8(9)-p-MENTHADIEN-2-ONE see
CCM120
d-p-MENTHA-6,8,(9)-DIEN-2-ONE see
CCM100
1-p-MENTHA-6(8,9)-DIEN-2-YL ACETATE
see CCM750
MENTHADIENYL FORMATE see IHX450
(−)-trans-2-p-MENTHA-1,8-DIEN-3-YL-5-
PENTYLRESORCINOL see CBD599
1-p-MENTHA-6,8(9)-DIEN-2-YL
PROPIONATE see MCD000
MENTHANE DIAMINE see MCD750
p-MENTHANE-1,8-DIAMINE see MCD750
p-MENTHANE, 1,4-EPOXY- see IKC100
p-MENTHANE HYDROPEROXIDE see
MCE000
p-MENTHANE-8-HYDROPEROXIDE see
MCE000
p-MENTHANE HYDROPEROXIDE,
TECHNICALLY PURE see IQE000
p-MENTHAN-3-OL see MCF750
p-MENTHAN-8-OL see MCE100
dl-3-p-MENTHANOL see MCG000
p-MENTHAN-8-OL ACETATE see DME400
l-p-MENTHAN-3-ONE see MCG275
(Z)-p-MENTHAN-3-ONE see IKY000
p-MENTHAN-3-ONE racemic see MCE250
o-1-MENTHENE see MCE275
p-MENTH-3-ENE see MCE500
1-p-MENTHEN-4-OL see TBD825
8-p-MENTHEN-2-OL see DKV150
p-MENTH-1-EN-8-OL see TBD500
p-MENTH-8-EN-1-OL see TBD775
p-MENTH-8-EN-3-OL see MCE750
t-MENTH-1-EN-8-OL see TBD775
8(9)-p-MENTHEN-3-OL see MCE750
p-MENTH-1-EN-8-OL (8CI) see TBD750
p-MENTH-8-EN-2-OL, ACETATE see
DKV160
p-MENTH-1-EN-8-OL, FORMATE (mixed
isomers) see TBE500
8-p-MENTHEN-2-ONE see DKV175
p-MENTH-1-EN-3-ONE see MCF250
p-MENTH-8-EN-2-ONE see DKV175
4(8)-p-MENTHEN-3-ONE see MCF500
p-MENTH-4(8)-EN-3-ONE see MCF500
d-p-MENTH-4(8)-EN-3-ONE see MCF500
p-MENTH-4(8)-EN-3-ONE, (R)-(+)- see
POI615
p-MENTH-8-EN-2-YL ACETATE see DKV160
p-MENTH-1-EN-8-YL ISOBUTYRATE see
MCF515
MENTHENYL KETONE see MCF525
1-(p-MENTHEN-6-YL)-1-PROPANONE see
MCF525
MENTHEN-1-YL-8 PROPIONATE see
TBE600
MENTHOL see MCF750
l-MENTHOL see MCF750
l-MENTHOL see MCG250
3-p-MENTHOL see MCG000
dl-MENTHOL see MCG000
MENTHOL racemic see MCG000

MENTHOL racemique (FRENCH) see MCG000
MENTHOL, ACETATE (8CI) see MCG500
MENTHOL ACETOACETATE see MCG850
MENTHOL, ISOVALERATE see MCG900
MENTHONE see MCG275
p-MENTHONE see MCG275
trans-MENTHONE see MCG275
MENTHONE, racemic see MCE250
l-MENTHONE (FCC) see MCG275
MENTHYL ACETATE see MCG500
(−)-MENTHYL ACETATE see MCG750
dl-MENTHYL ACETATE see MCG500
1-p-MENTH-3-YL ACETATE see MCG750
l-p-MENTH-3-YL ACETATE see MCG750
MENTHYL ACETATE racemic see MCG500
l-MENTHYL ACETATE (FCC) see MCG750
MENTHYL ACETOACETATE see MCG850
(−)-MENTHYL ACETOACETATE see
MCG850
(−)-MENTHYL ALCOHOL see MCG250
p-MENTH-3-YL ESTER-dl-ACETIC ACID see
MCG500
MENTHYL ISOVALERATE see MCG900
MENTH-3-YL ISOVALERATE see MCG900
MENTHYL 3-MENTHYLBUTYRATE see
MCG900
(−)-MENTHYL PHENYLACETATE see
MCG910
MENTOR 28 see MCG950
MEOBAL see XTJ000
MEOBAL, NITROSATED (JAPANESE) see
XTS000
MEONAL see BPF500
MEONINE see MDT740
MEOTHRIN see DAB825
MEP see EOS000
MEPACRINE see ARQ250
MEPACRINE DIHYDROCHLORIDE see
CFU750
MEPACRINE DIMETHANESULFONATE
SALT see QCS900
MEPACRINE HYDROCHLORIDE see
CFU750
MEPACRINE METHANESULFONATE see
QCS900
MEPADIN see DAM700
MEPAMTIN see MQU750
ME-PARATHION see MNH000
MEPARFYNOL CARBAMATE see MNM500
MEPATON see MNH000
MEPAVLON see MQU750
MEPAZIN see MOQ250
MEPAZINE BASE see MOQ250
MEPEDYL see PIZ250
MEPENICYCLINE see MCH525
MEPENTAMATE see MNM500
MEPENTAMATO see EQL000
MEPENTAMATO see MNM500
MEPENTIL see EQL000
MEPENZOLATE see CBF000
MEPENZOLATE BROMIDE see CBF000
MEPERIDIDE see FLV000
MEPERIDINE see DAM600
MEPERIDINE HYDROCHLORIDE see
DAM700
9-ME-PGA see BMM125
MEPHABUTAZONE see BRF500
MEPHACYCLIN see TBX250
MEPHADRYL see BBV500
MEPHANAC see CIR250
MEPHASERPIN see RDK000
MEPHATE see GGS000
MEPHEDAN see GGS000
MEPHEDINE see DAM700
MEPHELOR see GGS000
MEPHENAMINE HYDROCHLORIDE see
OJW000
MEPHENAMIN HYDROCHLORIDE see
OJW000
MEPHENAMINIC ACID see XQS000
MEPHENESIN CARBAMATE see CBK500
MEPHENON see MDP750

MEPHENOXALONE see MFD500
MEPHENSIN see GGS000
MEPHENYTOIN see MKB250
MEPHEXAMIDE see DIB600
MEPHEXAMIDE see MCH250
MEPHOBARBITAL see ENB500
MEPHOBARBITONE see ENB500
MEPHOSAL see GGS000
MEPHOSFOLAN see DHH400
MEPHSON see GGS000
MEPHYTAL see ENB500
MEPHYTON see VTA000
MEPIBEN see LJR000
MEPICYCLINE PENICILLINATE see
MCH525
MEPIOSINE see MQU750
MEPIPRAZOLE DIHYDROCHLORIDE see
MCH535
MEPIPRAZOLE HYDROCHLORIDE see
MCH535
MEPIQUAT CHLORIDE see MCH540
MEPIRESERPATE HYDROCHLORIDE see
MDW100
MEPIRIZOL see MCH550
MEPISERATE HYDROCHLORIDE see
MDW100
MEPITIOSTANE see MCH600
MEPIVACAINE see SBB000
dl-MEPIVACAINE see SBB000
MEPIVACAINE HYDROCHLORIDE see
CBR250
dl-MEPIVACAINE HYDROCHLORIDE see
CBR250
MEPIVASTESIN see CBR250
MEPOSED see MQU750
MEP (Pesticide) see DSQ000
MEPRANIL see MQU750
MEPRIN see MCI375
MEPRIN (detoxicant) see MCI375
MEPRO see CIR500
MEPROBAM see MQU750
MEPROBAMAT (GERMAN) see MQU750
MEPROBAMATE see MQU750
MEPROBAMATO (ITALIAN) see MQU750
MEPROCOMPREN see MQU750
MEPROCON CMC see MQU750
MEPRODIL see MQU750
MEPRODINE (GERMAN) see NOE550
MEPROFEN see BDU500
MEPROLEAF see MQU750
MEPROMAZINE see MCI500
MEPRONIL see INE050
MEPRONIL (PESTICIDE) see INE050
MEPROSAN see MQU750
MEPROSCILLARIN see MCI750
MEPROTABS see MQU750
MEPROZINE see MQU750
MEPRYLCAINE HYDROCHLORIDE see
OJI600
MEPTIN see PME600
MEPTOX see MNH000
MEPTRAN see MQU750
MEPYRAMIN (GERMAN) see WAK000
MEPYRAMINE HYDROCHLORIDE see
MCJ250
MEPYRAMINE MALEATE see DBM800
MEPYRAMINE 7-THEOPHYLLINE
ACETATE see MCJ300
MEPYRAPONE see MCJ370
MEPYREN see WAK000
MEQUIN see QAK000
MEQUINOL see MFC700
MEQUITAZINE see MCJ400
MER 25 see DHS000
MER 29 see TMP500
MER-41 see CMX700
MERABITOL see AFH250
MERACTINOMYCIN see AEB000
MERADAN see AJU625
MERADANE see AJU625
MERAKLON see PMP500
MERALEIN DISODIUM see SIF500

MERALEN see TKH750
MERALLURIDE see MFC000
MERALLURIDE see TEQ000
MERALLURIDE SODIUM see SIG000
MERAMEC M 25 see IHC550
MERANTINE BLUE AF see ERG100
MERANTINE BLUE EG see FAE000
MERANTINE BLUE V see CMM062
MERANTINE BLUE VF see ADE500
MERANTINE GREEN G see FAE950
MERANTINE GREEN SF see FAF000
MERANTINE GREEN V see CMM100
MERATRAN see DWK400
MERBAPHEN see CCG500
MERBENTUL see CLO750
MERBROMIN see MCV000
MERCALEUKIN see POK000
MERCAMINE see AJT250
MERCAMINE DISULFIDE see MCN500
MERCAPTAMINE see AJT250
MERCAPTAN AMYLIQUE (FRENCH) see PBM000
MERCAPTAN METHYLIQUE (FRENCH) see MLE650
MERCAPTAN METHYLIQUE PERCHLORE (FRENCH) see PCF300
MERCAPTANS see MCJ500
MERCAPTAZOLE see MCO500
2-MERCAPTOACETANILIDE see MCK000
α-MERCAPTOACETANILIDE see MCK000
MERCAPTOACETATE see TFJ100
(MERCAPTOACETATO(2-)-O,S)METHYL-MERCURATE(1-), SODIUM see SIM000
MERCAPTOACETIC ACID see TFJ100
2-MERCAPTOACETIC ACID see TFJ100
α-MERCAPTOACETIC ACID see TFJ100
MERCAPTOACETIC ACID ACETATE see ACQ250
MERCAPTOACETIC ACID, DIESTER with DITHIO-p-UREIDOBENZENEARSONOUS ACID see CBI250
MERCAPTOACETIC ACID ETHYL ESTER see EMB200
MERCAPTOACETIC ACID-2-ETHYLHEXYL ESTER see EKW300
MERCAPTOACETIC ACID METHYL ESTER see MLE750
MERCAPTOACETIC ACID, SODIUM-BISMUTH SALT see BKX750
MERCAPTOACETIC ACID SODIUM SALT see SKH500
MERCAPTOACETONITRILE see MCK300
4-(MERCAPTOACETYL)MORPHOLINE O,O-DIMETHYL PHOSPHORODITHIOATE see MRU250
β-MERCAPTOAETHYLAMIN CHLORHYDRAT see MCN750
β-MERCAPTOALANINE see CQK000
6-MERCAPTO-2-AMINOPURINE see AMH250
2-MERCAPTO-5-AMINO-1,3,4-THIADIAZOLE see AKM000
4-MERCAPTOANILINE see AIF750
o-MERCAPTOANILINE see AIF500
p-MERCAPTOANILINE see AIF750
2-MERCAPTOBARBITURIC ACID see MCK500
2-MERCAPTOBARBITURIC ACID see MCK500
7-MERCAPTOBENZ(a)ANTHRACENE see BBH750
2-MERCAPTOBENZIMIDAZOLE see BCC500
MERCAPTOBENZIMIDAZOLE ZINC SALT see ZIS000
2-MERCAPTOBENZIMIDAZOLE ZINC SALT (2:1) see ZIS000
o-MERCAPTOBENZOESAEURE (GERMAN) see MCK750
o-MERCAPTOBENZOIC ACID see MCK750
o-MERCAPTOBENZOIC ACID, DIESTER with DITHIO-p-

UREIDOBENZENEARSONOUS ACID see TFD750
MERCAPTOBENZOIMIDAZOLE see BCC500
2-MERCAPTOBENZOIMIDAZOLE see BCC500
MERCAPTOBENZOTHIAZOLE see BDF000
2-MERCAPTOBENZOTHIAZOLE see BDF000
2-MERCAPTOBENZOTHIAZOLEDISULFIDE see BDE750
2-MERCAPTOBENZOTHIAZOLE MONOETHANOLAMINE SALT see HKO012
2-MERCAPTOBENZOTHIAZOLE SODIUM DERIVATIVE see SIG500
2-MERCAPTOBENZOTHIAZOLE SODIUM SALT see SIG500
2-MERCAPTOBENZOTHIAZOLE ZINC SALT see BHA750
(2-MERCAPTOBENZOTHIAZOLYL)-2-(2-AMINOTHIAZOL-4-YL)-2-METHOXYIMINOACETATE (SYN) see MCK800
2-MERCAPTOBENZOTHIAZYLDISULFIDE see BDE750
2-MERCAPTOBENZOXAZOLE see MCK900
(MERCAPTOBUTANEDIOATO(1-))GOLD DISODIUM SALT see GJC000
(2-MERCAPTOCARBAMOYL)DI-ACETANILIDE see MCL500
MERCAPTOCYCLOPENTANE see CPW300
MERCAPTODIACETIC ACID see MCM750
MERCAPTO DI-ACETIC ACID, ETHYLENE ESTER see MCN000
MERCAPTODIMETHUR (DOT) see DST000
1-MERCAPTODODECANE see LBX000
2-MERCAPTOETHANESULFONIC ACID MONOSODIUM SALT see MDK875
MERCAPTOETHANOL see MCN250
2-MERCAPTOETHANOL see MCN250
β-MERCAPTOETHANOL see MCN250
(2-MERCAPTOETHYL)AMINE see AJT250
β-MERCAPTOETHYLAMINE see AJT250
β-MERCAPTOETHYLAMINE DISULFIDE see MCN500
MERCAPTOETHYLAMINE HYDROCHLORIDE see MCN750
2-MERCAPTOETHYLAMINE HYDROCHLORIDE see MCN750
2-MERCAPTOETHYLAMINE HYDROCHLORIDE see MCN750
β-MERCAPTOETHYLAMINE HYDROCHLORIDE see MCN750
2-MERCAPTOETHYLAMINE (OXIDIZED) see MCN500
N-(2-MERCAPTOETHYLBENZENESULFONAMIDE)-S-(O,O-DIISOPROPYL PHOSPHORODITHIOATE) see DNO800
(2-MERCAPTOETHYL)CARBAMIC ACID, ETHYL ESTER, S-ESTER with O,O-DIMETHYL PHOSPHORODITHIOATE see EMC000
N-(2-MERCAPTOETHYL)DIMETHYLAMINE HYDROCHLORIDE see DOY600
4-MERCAPTOETHYLMORPHOLINE see MCO000
2-MERCAPTOETHYL TRIMETHOXY SILANE see MCO250
(2-MERCAPTOETHYL)TRIMETHYLAMMONIUM S-ESTER with O,O'-DIETHYLPHOSPHOROTHIOATE see PGY600
(2-MERCAPTOETHYL)TRIMETHYLAMMONIUM IODIDE ACETATE see ADC300
(2-MERCAPTOETHYL)TRIMETHYLAMMONI-

UM IODIDE S-ESTER with O,O-DIETHYL PHOSPHOROTHIOATE see TLF500
MERCAPTOFOS (RUSSIAN) see DAO500
1-MERCAPTOGLYCEROL see MRM750
6-MERCAPTOGUANINE see AMH250
6-MERCAPTOGUANOSINE see TFJ500
2-MERCAPTO-4-HYDROXY-6-METHYLPYRIMIDINE see MPW500
2-MERCAPTO-4-HYDROXY-6-N-PROPYLPYRIMIDINE see PNX000
2-MERCAPTO-4-HYDROXYPYRIMIDINE see TFR250
2-MERCAPTOIMIDAZOLE see IAO000
2-MERCAPTOIMIDAZOLINE see IAQ000
MERCAPTOMERIN see TFK260
MERCAPTOMERIN SODIUM see TFK270
(MERCAPTOMETHYL)BENZENE see TGO750
3-(MERCAPTOMETHYL)-1,2,3-BENZOTRIAZIN-4(3H)-ONE-O,O-DIMETHYL PHOSPHORODITHIOATE see ASH500
3-(MERCAPTOMETHYL)-1,2,3-BENZOTRIAZIN-4(3H)-ONE-O,O-DIMETHYL PHOSPHORODITHIOATE-S-ESTER see ASH500
(trans)-2-MERCAPTOMETHYLCYCLOBUTYLAMINE HYDROCHLORIDE see AKL750
2-MERCAPTO-1-METHYLIMIDAZOLE see MCO500
N-(2-MERCAPTO-2-METHYL-1-OXOPROPYL)-l-CYSTEINE see MCO775
N-(2-MERCAPTO-2-METHYL-1-OXOPROPYL)-l-CYSTEINE SODIUM SALT see SAY875
1-(3-MERCAPTO-2-METHYL-1-OXOPROPYL)-l-PROLINE see MCO750
1-(d-3-MERCAPTO-2-METHYL-1-OXOPROPYL)-l-PROLINE (S,S) see MCO750
N-(MERCAPTOMETHYL)PHTHALIMIDE S-(O,O-DIMETHYL PHOSPHORODITHIOATE) see PHX250
3-MERCAPTO-N-METHYLPROPANAMIDE see MLF100
N-(2-MERCAPTO-2-METHYLPROPANOYL)-l-CYSTEINE see MCO775
N-(2-MERCAPTO-2-METHYLPROPANOYL)-l-CYSTEINE SODIUM SALT see SAY875
1-((2S)-3-MERCAPTO-2-METHYLPROPIONYL)-l-PROLINE see MCO750
2-(MERCAPTOMETHYL)PYRIDINE see POR790
2-MERCAPTO-6-METHYLPYRIMID-4-ONE see MPW500
2-MERCAPTO-6-METHYL-4-PYRIMIDONE see MPW500
MERCAPTOMETHYLTRIETHOXYSILANE see MCP000
2-MERCAPTO-N-(2-METHYOXYETHYL)-ACETAMIDE S-ESTER with O,O-DIMETHYL PHOSPHORODITHIOATE see AHO750
2-MERCAPTONAPHTHALENE see NAP500
β-MERCAPTONAPHTHALENE see NAP500
2-MERCAPTO-6-NITROBENZOTHIAZOLE see MCP250
2-MERCAPTO-4(OR 5)-METHYLBENZIMIDAZOLE see MCO400
5-MERCAPTO-1-PHENYLTETRAZOLE see PGJ750
5-MERCAPTO-3-PHENYL-2H-1,3,4-THIADIAZOLE-2-THIONE see MCP500
MERCAPTOPHOS see DAO600
MERCAPTOPHOS see FAQ900
1-MERCAPTOPROPANE see PML500
2-MERCAPTOPROPANE see IMU000
1-MERCAPTO-2,3-PROPANEDIOL see MRM750
3-MERCAPTO-1,2-PROPANEDIOL see MRM750

α-MERCAPTOPROPANOIC ACID see TFK250
β-MERCAPTOPROPANOIC ACID see MCQ000
2-MERCAPTOPROPIONIC ACID see TFK250
3-MERCAPTOPROPIONIC ACID see MCQ000
α-MERCAPTOPROPIONIC ACID see TFK250
MERCAPTOPROPIONIC ACID, DIBUTYLTIN SALT see DEJ200
MERCAPTOPROPIONYLGLYCINE see MCI375
(2-MERCAPTOPROPIONYL)GLYCINE see MCI375
N-(2-MERCAPTOPROPIONYL)GLYCINE see MCI375
α-MERCAPTOPROPIONYLGLYCINE see MCI375
2-MERCAPTO-6-PROPYL-4-PYRIMIDONE see PNX000
2-MERCAPTO-6-PROPYLPYRIMID-4-ONE see PNX000
3-MERCAPTOPROPYLTRIMETHOXYSILANE see TLC000
γ-MERCAPTOPROPYLTRIMETHOXYSILANE see TLC000
6-MERCAPTOPURIN see POK000
6-MERCAPTOPURIN (GERMAN) see POK000
6-MERCAPTOPURINE see POK000
6-MERCAPTOPURINE MONOHYDRATE see MCQ100
MERCAPTOPURINE-3-N-OXIDE see MCQ250
6-MERCAPTOPURINE 3-N-OXIDE see MCQ250
6-MERCAPTOPURINE 3-N-OXIDE MONOHYDRATE see OMY800
MERCAPTOPURINE RIBONUCLEOSIDE see MCQ500
6-MERCAPTOPURINE RIBOSIDE see MCQ500
MERCAPTOPYRIDETHYL BENZIMIDAZOLE see MCQ800
2-MERCAPTOPYRIDINE see TFP300
2-MERCAPTOPYRIDINE MONOXIDE see MCQ700
2-MERCAPTOPYRIDINE-N-OXIDE SODIUM SALT see MCQ750
2-MERCAPTO-1-(β-4-PYRIDYLETHYL) BENZIMIDAZOLE see MCQ800
2-MERCAPTO-4-PYRIMIDINOL see TFR250
2-MERCAPTO-4-PYRIMIDONE see TFR250
2-MERCAPTOPYRIMID-4-ONE see TFR250
2-MERCAPTOQUINOLINE see QOJ100
MERCAPTOSUCCINIC ACID see MCR000
MERCAPTOSUCCINIC ACID ANTIMONATE(III) HEXALITHIUM SALT see LGU000
MERCAPTOSUCCINIC ACID-S-ANTIMONY DERIVATIVE LITHIUM SALT see LGU000
MERCAPTOSUCCINIC ACID DIETHYL ESTER see MAK700
MERCAPTOSUCCINIC ACID DIETHYL ESTER, S-ESTER WITH O,S-DIMETHYLPHOSPHORODITHIOATE see IKX200
MERCAPTOSUCCINIC ACID, GOLD SODIUM SALT see GJC000
MERCAPTOSUCCINIC ACID, THIOANTIMONATE(III), DILITHIUM SALT see LGU000
p-MERCAPTO SULFADIAZINE see MCR250
7-MERCAPTO-1,3,4,6-TETRAZAINDENE see POK000
2-MERCAPTOTHIAZOLINE see TFS250
MERCAPTOTHION see MAK700
MERCAPTOTION (SPANISH) see MAK700
m-MERCAPTOTOLUENE see TGO800
o-MERCAPTOTOLUENE see TGP000

p-MERCAPTOTOLUENE see TGP250
α-MERCAPTOTOLUENE see TGO750
3-MERCAPTO-1H-1,2,4-TRIAZOLE see THT000
d-MERCAPTOVALINE see MCR750
3-MERCAPTO-l-VALINE see PAP300
d,3-MERCAPTOVALINE see MCR750
3-MERCAPTO-dl-VALINE (9CI) see PAP500
dl-α-MERCAPTOVALINE see PAP500
MERCAPTURIC ACID see ACH000
(R)-MERCAPTURIC ACID see ACH000
MERCAPURIN see POK000
MERCARDAN see SIG000
MERCATE 5 see SAA025
MERCATE 20 see EDE700
MERCAZOLYL see MCO500
MERCHLORATE see MEP250
MERCHLORETHANAMINE see BIE500
MERCK 261 see DTS500
MERCKOGEN 6000 see AAX250
MERCLORAN see CHX250
MERCOL 25 see DXW200
MERCORAL see CHX250
MERCUFENOL CHLORIDE see CHW675
MERCUHYDRIN see TEQ000
MERCUMATILIN SODIUM see MCS000
MERCUPURIN see MCV750
MERCURAM see TFS350
MERCURAMIDE see SIH500
MERCURAN see MEO750
MERCURANINE see MCV000
MERCURATE(1-), ACETATOPHENYL-, AMMONIUM SALT see PFO550
MERCURATE(1-), (BENZOATO(2-)-C2,O1)HYDROXY-, HYDROGEN see HLO450
MERCURATE(4-), BIS(N,N-BIS(CARBOXYMETHYL)GLYCINATO(3-)-N,O,O',O'')-, TETRAHYDROGEN see MDF050
MERCURATE(2-), BIS(l-CYSTEINATO(2-)-O,S)-, DIHYDROGEN, (T-4)- see BIS300
MERCURATE(1-), (4-CARBOXYLATOPHENYL)HYDROXY-, HYDROGEN see HLN800
MERCURATE(1-), (l-CYSTEINATO(2-)-S)METHYL-, HYDROGEN see MCS100
MERCURE (FRENCH) see MCW250
MERCURETIN see SIG000
MERCURHYDRIN SODIUM see SIG000
MERCURIACETATE see MCS750
MERCURIALIN see MGC250
2,2'-MERCURIBIS(6-ACETOXYMERCURI-4-NITRO)ANILINE see MCS250
MERCURIBIS(DIETHYL(2,2-DIMETHYL-4-DITHIOCARBOXYAMINO))BUTYLAMMON IUM DICHLORIDE see MCS500
MERCURIBIS-o-NITROPHENOL see MCS600
MERCURIC ACETATE see MCS750
MERCURIC AMMONIUM CHLORIDE, solid see MCW500
MERCURIC ARSENATE see MDF350
MERCURIC BASIC SULFATE see MDG000
MERCURIC BENZOATE see MCX500
MERCURIC BENZOATE, solid (DOT) see MCX500
MERCURIC BROMIDE see MCY000
MERCURIC BROMIDE, solid see MCY000
MERCURIC CHLORIDE (DOT) see MCY475
MERCURIC CHLORIDE, AMMONIATED see MCW500
MERCURIC CHLORIDE-1,4-OXATHIANE see OMA000
MERCURIC CYANIDE, solid (DOT) see MDA250
MERCURIC DIACETATE see MCS750
MERCURIC-8,8-DICAFFEINE see MCT000
MERCURIC DINAPHTHYLMETHANE DISULPHONATE see MCT250
MERCURIC IODIDE see MDD000
MERCURIC IODIDE, solid see MDD000
MERCURIC IODIDE, solution see MDD000

MERCURIC IODIDE, RED see MDD000
MERCURIC LACTATE see MDD500
MERCURIC NITRATE see MDF000
MERCURIC OLEATE, solid (DOT) see MDF250
MERCURIC OXIDE see MCT500
MERCURIC OXIDE, solid (DOT) see MCT500
MERCURIC OXIDE, RED see MCT500
MERCURIC OXIDE, YELLOW see MCT500
MERCURIC OXYCYANIDE see MDA500
MERCURIC OXYCYANIDE, solid (desensitized) (DOT) see MDA500
MERCURIC PEROXYBENZOATE see MCT750
MERCURIC POTASSIUM CYANIDE (DOT) see PLU500
MERCURIC POTASSIUM CYANIDE, solid (DOT) see PLU500
MERCURIC POTASSIUM IODIDE see NCP500
MERCURIC POTASSIUM IODIDE, solid (DOT) see NCP500
MERCURIC SALICYLATE see MCU000
MERCURIC SALICYLATE, solid (DOT) see MCU000
MERCURIC SUBSULFATE, solid see MDG000
MERCURIC SULFATE, solid see MDG500
MERCURIC SULFOCYANATE see MCU250
MERCURIC SULFOCYANATE, solid (DOT) see MCU250
MERCURIC SULFOCYANIDE see MCU250
MERCURIC THIOCYANATE see MCU250
MERCURIC THIOCYANATE, solid (DOT) see MCU250
MERCURIDIACETALDEHYDE see BJW800
N,N'-MERCURIDIANILINE see DCJ000
3,3'-MERCURIDI-2-PROPYN-1-OL see BKJ250
MERCURIDISALICYLIC ACID, DISODIUM SALT see SJR000
MERCURI-HEMATOPORPHYRIN DISODIUM SALT see HAP000
MERCURIO (ITALIAN) see MCW250
MERCURIPHENOLDISULFONATE SODIUM see MCU500
MERCURIPHENYL ACETATE see ABU500
MERCURIPHENYL CHLORIDE see PFM500
MERCURIPHENYL NITRATE see MCU750
MERCURISALICYLIC ACID see MCU000
MERCURITAL see SIH500
MERCUROCHLORID (DUTCH) see MCW000
MERCUROCHROME see MCV000
MERCUROCHROME-220 SOLUBLE see MCV000
MERCUROCOL see MCV000
MERCUROL see MCV250
MERCUROPHAGE see MCV000
MERCUROPHEN see MCV500
MERCUROPHYLLINE see MCV750
MERCUROTHIOLATE see MDI000
MERCUROUS ACETATE see MDE250
MERCUROUS ACETATE, solid (DOT) see MDE250
MERCUROUS AZIDE (DOT) see MCX000
MERCUROUS BROMIDE, solid (DOT) see MCX750
MERCUROUS CHLORIDE see MCW000
MERCUROUS CHLORIDE see MCY300
MERCUROUS GLUCONATE see MDC500
MERCUROUS GLUCONATE, solid (DOT) see MDC500
MERCUROUS IODIDE see MDC750
MERCUROUS NITRATE, solid (DOT) see MDE750
MERCUROUS OXIDE, BLACK, solid (DOT) see MDF750
MERCUROUS SULFATE, solid (DOT) see MDG250
MERCURY see MCW250
MERCURY ACETATE see MCS750
MERCURY ACETATE see MDE250

MERCURY(2+) ACETATE see MCS750
MERCURY(II) ACETATE see MCS750
MERCURY(II) ACETATE, PHENYL- see ABU500
MERCURY, ACETATO(p-(DIETHYLAMINO)PHENYL)- see AAS300
MERCURY, ACETATO(p-(DIMETHYLAMINO)PHENYL)- see AAS310
MERCURY, (ACETATO-O)PHENYL-, AMMONIATE see PFO550
MERCURY, ACETOXY(2-METHOXYETHYL)- see MEO750
MERCURY, ACETOXYPHENYL- see ABU500
MERCURY ACETYLIDE see MCW349
MERCURY(II) ACETYLIDE see MCW350
MERCURY ACETYLIDE (DOT) see MCW349
MERCURY AMIDE CHLORIDE see MCW500
MERCURY AMINE CHLORIDE see MCW500
MERCURY, ((2-AMINO-2-CARBOXYETHYL)THIO)METHYL-, l- see MCS100
MERCURY AMMONIATED see MCW500
MERCURY AZIDE see MCX000
MERCURY(I) AZIDE see MCX000
MERCURY(II) AZIDE see MCX250
MERCURY(II) BENZOATE see MCX500
MERCURY BICHLORIDE see MCY475
MERCURY BINIODIDE see MDD000
MERCURY, BIS(ACETATO)(mu-(3',6'-DIHYDROXY-2',7'-FLUORANDIYL))DI- see FEV100
MERCURY BIS(CHLOROACETYLIDE) see MCX600
MERCURY(II), BIS(l-CYSTEINATO)- see BIS300
MERCURY, BIS(4-HYDROXY-3-NITROPHENYL)- see MCS600
MERCURY, BIS(TRIFLUOROMETHYLTHIO)- see BLQ525
MERCURY BISULFATE see MDG500
MERCURY(I) BROMATE see MCX700
MERCURY(I) BROMIDE (1:1) see MCX750
MERCURY(II) BROMIDE (1:2) see MCY000
MERCURY(II) BROMIDE COMPLEX with TRIS(2-ETHYLHEXYL) PHOSPHITE see MCY250
MERCURY, BROMOHEXYL see HFR100
MERCURY, BROMO(2-HYDROXYETHYL)-, compound with AMMONIA (1:0.8 moles) see BNL275
MERCURY, (3-(α-CARBOXY-o-ANISAMIDO)-2-HYDROXYPROPYL)HYDROXY- see NCM800
MERCURY, (3-(α-CARBOXYMETHOXYPROPYL)HYDROXY) MONOSODIUM SALT, COMPOUNDED with THEOPHYLLINE (1:1) see MDH750
MERCURY, (2-CARBOXYPHENYL)HYDROXY- see HLO450
MERCURY, (o-CARBOXYPHENYL)HYDROXY- see HLO450
MERCURY, (p-CARBOXYPHENYL)HYDROXY- see HLN800
MERCURY, (3-(3-CARBOXY-2,2,3-TRIMETHYLCYCLOPENTANECARBOXAMIDO)-2-METHOXYPROPYL)(HYD ROGEN MERCAPTOACETATO)- see TFK260
MERCURY CHLORIDE see MCY300
MERCURY(I) CHLORIDE see MCW000
MERCURY(II) CHLORIDE see MCY475
MERCURY(II) CHLORIDE COMPLEX with TRIS(2-ETHYLHEXYL) PHOSPHITE see MCY500
MERCURY(I) CHLORITE see MCY750
MERCURY(II) CHLORITE see MCY755
MERCURY (E)-CHLORO(2-(3-BROMOPROPIONAMIDO)CYCLOHEXYL) see CET000

MERCURY, CHLORO(2-HYDROXYPHENYL)- see CHW675
MERCURY COMPOUNDS, INORGANIC see MCZ000
MERCURY COMPOUNDS, ORGANIC see MDA000
MERCURY(I) CYANAMIDE see MDA100
MERCURY(II) CYANATE see MDA150
MERCURY(II) CYANIDE see MDA250
MERCURY CYANIDE OXIDE see MDA500
MERCURY, (3-CYANOGUANIDINO)ETHYL- see EME025
MERCURY DIACETATE see MCS750
MERCURY-O,O-DI-n-BUTYL PHOSPHORODITHIOATE see MDA750
MERCURY DICHLORITE see MCY755
MERCURY, (DIHYDROGEN PHOSPHATO)METHYL- see MLF520
MERCURY, DIISOAMYL- see DNL200
MERCURY, DIMETHYL see DSM450
MERCURY(II) aci-DINITROMETHANIDE see MDA800
MERCURY DITHIOCYANATE see MCU250
MERCURY(II) EDTA COMPLEX see MDB250
MERCURY(II), ETHYL(5-ETHYLMERCURI-3-(1,2,4-THIADIAZOLYL)THIO)- see EJW600
MERCURY, ETHYL(4-MERCAPTOBENZENESULFONATO-S-⁴))-, SODIUM SALT see SKH150
MERCURY, ETHYL((p-SULFOPHENYL)THIO)-, SODIUM SALT see SKH150
MERCURY, ETHYL(TOLUENESULFONATO)- see EME600
MERCURY(II) FLUOROACETATE see MDB500
MERCURY(II) FORMOHYDROXAMATE see MDB775
MERCURY(II) FULMINATE see MDC000
MERCURY FULMINATE, wetted with not <20% water, or mixture (UN 0135) (DOT) see MDC000
MERCURY(I) GLUCONATE see MDC500
MERCURY, HYDROXY(4-HYDROXY-3-NITROPHENYL)- see HLO400
MERCURY, HYDROXY(NITRATO)DIPHENYLDI- see MDH500
MERCURY, HYDROXYPHENYL- see PFN100
MERCURY, (4-HYDROXY-m-PHENYLENE)BIS(ACETATO)- see HNK575
MERCURY, HYDROXYPHENYL-, cmpd. with NITRATOPHENYLMERCURY (1:1) see MDH500
MERCURY(I) IODIDE see MDC750
MERCURY(II) IODIDE see MDD000
MERCURY IODIDE (DOT) see MDC750
MERCURY(II) IODIDE (solution) see MDD250
MERCURY IODIDE, solution (DOT) see MDC750
MERCURYL ACETATE see MCS750
MERCURY(2+) LACTATE see MDD500
MERCURY, METALLIC (DOT) see MCW250
MERCURY, (METHANETHIOLATO)METHYL- see MLX300
MERCURY METHYLCHLORIDE see MDD750
MERCURY METHYLMERCURY SULFIDE see MLX300
MERCURY, METHYL(METHYLTHIO)- see MLX300
MERCURY(II) METHYLNITROLATE see MDE000
MERCURY, (2-METHYL-5-NITROPHENOLATO(2-)-C⁶,O¹)-(9CI) see NHK900
MERCURY, METHYL(PHOSPHATO(1-)-O)-(9CI) see MLF520

MERCURY, METHYL(TOLUENESULFONATO)- see MLH050
MERCURY MONOACETATE see MDE250
MERCURY MONOCHLORIDE see MCW000
MERCURY-2-NAPHTHALENEDIAZONIUM TRICHLORIDE see MDE500
MERCURY NITRATE see MDF000
MERCURY(I) NITRATE (1:1) see MDE750
MERCURY(II) NITRATE (1:2) see MDF000
MERCURY NITRIDE see TKW000
MERCURY, NITRILOTRIACETATE see MDF050
MERCURY(II) 5-NITROTETRAZOLIDE see MDF100
MERCURY NUCLEATE, solid (DOT) see MCV250
MERCURY, (9-OCTADECENOATO-O)PHENYL-, (Z)-(9CI) see PFP100
MERCURY OLEATE see MDF250
MERCURY, (OLEATO)PHENYL- see PFP100
MERCURY(II) ORTHOARSENATE see MDF350
MERCURY(II) OXALATE see MDF500
MERCURY(I) OXIDE see MDF750
MERCURY(II) OXIDE see MCT500
MERCURY OXIDE SULFATE see MDG000
MERCURY OXYCYANIDE see MDA500
MERCURY(II) PERCHLORATE see MDG200
MERCURY PERCHLORIDE see MCY475
MERCURY PERNITRATE see MDF000
MERCURY(II) PEROXYBENZOATE see MCT750
MERCURY PERSULFATE see MDG500
MERCURY(II) POTASSIUM IODIDE see NCP500
MERCURY PROTOCHLORIDE see MCW000
MERCURY PROTOIODIDE see MDC750
MERCURY SALICYLATE see MCU000
MERCURY and SODIUM PHENOLSULFONATE see MCU500
MERCURY SUBCHLORIDE see MCY300
MERCURY SUBSALICYLATE see MCU000
MERCURY(I) SULFATE see MDG250
MERCURY(II) SULFATE (1:1) see MDG500
MERCURY, SULFATOBIS(METHYL- see BKS810
MERCURY(II) SULFIDE see MDG750
MERCURY TETRAVANADATE see MDH000
MERCURY(II) THIOCYANATE see MCU250
MERCURY THIOCYANATE (DOT) see MCU250
MERCURY(I) THIONITROSYLATE see MDH250
MERCURY, (((p-TOLYL)SULFAMOYL)IMINO)BIS(METHYL)- see MLH100
MERCURY ZINC CHROMATE COMPLEX see ZJA000
MERCUSAL see SIH500
MERCUTAL see PFO750
MERCUZANTHIN see MCV750
MEREPRINE see PGE775
MEREX see KEA000
MERFALAN see BHT750
MERFAMIN see MDI000
MERFAZIN see PFM500
MERFEN see PFP250
MERFEN-STYLI see MDH500
MERGAMMA see ABU500
MERGE see MRL750
MERIAN see AIF000
MERIDIL see MNQ000
MERILID see CHX250
MERINAX see POA250
MERIT see CKW400
MERIT see WBJ700
MERITA EARTH see ICC800
MERITAL see NMV725
MERITIN see DDT300
MERIZONE see BRF500
MERKAPOL P see MCB050

MERKAPOL PG see MCB050
MERKAPTOBENZIMIDAZOL (CZECH) see BCC500
2-MERKAPTOBENZOTIAZOL see BDF000
2-MERKAPTOBENZTHIAZOL see BDF000
5-MERKAPTO-3-FENYL-1,3,4-THIADIAZOL-2-THION DRASELNY (CZECH) see MCP500
2-MERKAPTOIMIDAZOLIN (CZECH) see IAQ000
6-MERKAPTOPURIN, MONOHYDRAT see MCQ100
MERKAZIN see BKL250
MERMETH see SNJ000
MERN see POK000
MEROCYANINE 540 see EDD500
MERODICEIN see SIF500
MERONIDAL see MMN250
MEROPENIN see PDT500
MEROPHAN see BHT250
o-MEROPHAN see BHT250
MEROXYL see AOR500
MEROXYLAN see AOR500
MEROXYLAN-WANDER see AOR500
MEROXYL-WANDER see AOR500
MERPACYL BLUE SK see CMM090
MERPAN see CBG000
MERPHALAN see BHT750
o-MERPHALAN see BHT750
MERPHALAN HYDROCHLORIDE see BHV000
MERPHEN see MDH500
MERPHENE see MDH500
MERPHENYL NITRATE see MCU750
MERPHENYL NITRATE see MDH500
MERPHENYL NITRATE see MDH500
MERPHOS see TIG250
MERPHYLLIN see HAP000
MERPHYRIN see HAP000
MERPISAP PD82 see SKF600
MERPOL see EJN500
MERPOL HCS see AFJ165
MERQUAT 100 see DTS500
MERRILLITE see ZBJ000
MERSALIN see SIH500
MERSALYL see SIH500
MERSALYL THEOPHYLLINE see MDH750
MERSALYL with THEOPHYLLINE see SAQ000
MERSOLITE see ABU500
MERSOLITE 1 see PFN100
MERSOLITE 2 see PFM500
MERSOLITE 7 see MCU750
MERSOLITE 8 see ABU500
MERTEC see TEX000
MERTESTATE see TBF500
MERTHIOLATE see MDI000
MERTHIOLATE SALT see MDI000
MERTHIOLATE SODIUM see MDI000
MERTIONIN see MDT740
MERTORGAN see MDI000
MERURAN see MDI200
MERVAMINE see DJL000
MERVAN see AGN000
MERVAN ETHANOLAMINE SALT see MDI225
MERXIN see CCS510
MERZONIN SODIUM see MDI000
MES see MJW250
MES (buffering agent) see MRT150
MESA see TGE300
MESACONATE see MDI250
MESACONIC ACID see MDI250
MESAMATE see MRL750
MESAMATE CONCENTRATE see MRL750
MESANOLON see MJE760
MESANTOIN see MKB250
MESATON see NCL500
MESCAL BEAN see NBR800
MESCALINE see MDI500
MESCALINE ACID SULFATE see MDJ000

MESCALINE HYDROCHLORIDE see MDI750
MESCALINE SULFATE see MDJ000
MESCOPIL see SBH500
MESECLAZONE see CIL500
MESENTOL see ENG500
MESEREIN see MDJ250
MESIDICAINE HYDROCHLORIDE see DHL800
MESIDIN (CZECH) see TLG500
MESIDINE see TLG500
MESIDINE HYDROCHLORIDE see TLH000
MESITALDEHYDE see MDJ745
MESITOIC ACID see IKN300
MESITOL see MDJ740
MESITYL ALCOHOL see MDJ740
MESITYL ALDEHYDE see MDJ745
MESITYLAMINE see TLG500
MESITYLAMINE HYDROCHLORIDE see TLH000
2-MESITYLBENZOFURAN-3-YL (p-METHOXYPHENYL) KETONE see AOY300
MESITYLENE see TLM050
MESITYLENECARBOXALDEHYDE see MDJ745
2-MESITYLENECARBOXALDEHYDE see MDJ745
MESITYLENE, 2,4-DINITRO- see DUW505
α^1,α^3-MESITYLENEDIOL, 2-HYDROXY-(7CI,8CI) see HLX950
2-MESITYLENESULFONIC ACID, 4,4'-(1,4-ANTHRAQUINONYLENEDIIMINO)DI-, DISODIUM SALT see CMM092
MESITYLENE, 2,4,6-TRINITRO- see TMI800
MESITYLENIC ACID see MDJ748
MESITYL ESTER of 1-PIPERIDINEACETIC ACID HYDROCHLORIDE see FAC175
MESITYLOXID (GERMAN) see MDJ750
MESITYL OXIDE see MDJ750
MESITYL OXIDE (1-PHTHALAZINYL)HYDRAZONE see DQU400
MESITYLOXYDE (DUTCH) see MDJ750
2-MESITYLOXYDIISOPROPYLAMINE HYDROCHLORIDE see MDK000
N-(2-MESITYLOXYETHYL)-N-METHYL-2-(2-METHYLPIPERIDINO)ACETAMIDE HYDROCHLORIDE see MDK250
N-(2-MESITYLOXYETHYL)-N-METHYL-2-(MORPHOLINE)ACETAMIDE HYDROCHLORIDE see MRT250
(1-MESITYLOXY-2-PROPYL)-N-METHYLCARBAMIC ACID-2-(DIETHYLAMINO)ETHYL ESTER, HYDROCHLORIDE see MDK500
N-(1-MESITYLOXY-2-PROPYL)-N-METHYL-2-(2-METHYLPIPERIDINO) ACETAMIDE HYDROCHLORIDE see MDK750
MESNA see MDK875
MESNUM see MDK875
MESOCAINE HYDROCHLORIDE see DHL800
MESOFOLIN see MDV000
MESOKAIN HYDROCHLORIDE see DHL800
MESOMILE see MDU600
MESONEX see HFG550
MESOPIN see MDL000
MESORANIL see ASG250
MESORIDAZINE see MON750
MESOTAL see HFG550
MESOXALONITRILE see MDL250
MESOXALONITRILE, (m-CHLOROPHENYL)HYDRAZONE (8CI) see CKA550
MESOXALYLCARBAMIDE see AFT750
MESOXALYLCARBAMIDE MONOHYDRATE see MDL500
MESOXALYLUREA see AFT750
MESOXALYLUREA MONOHYDRATE see MDL500
MESPAFIN see HGP550

MESTALONE see MJE760
MESTANOLONE see MJE760
MESTENEDIOL see AOO475
MESTERONE see MPN500
MESTINON see MDL600
MESTRANOL see MKB750
MESTRANOL mixed with ANAGESTONE ACETATE (1:10) see AOO000
MESTRANOL mixed with CHLORMADINONE ACETATE see CNV750
MESTRANOL mixed with 6-CHLORO-6-DEHYDRO-17-α-ACETOXYPROGESTERONE see CNV750
MESTRANOL mixed with CHLOROETHYNYL NORGESTREL (1:20) see CHI750
MESTRANOL mixed with ETHYNERONE (1:20) see EQJ000
MESTRANOL mixed with ETHYNODIOL see EQK100
MESTRANOL mixed with ETHYNODIOL DIACETATE see EQK010
MESTRANOL mixed with LYNESTRENOL see LJE000
MESTRANOL mixed with LYNESTROL see LJE000
MESTRANOL mixed with NORETHINDRONE see MDL750
MESTRANOL mixed with NORETHISTERONE see MDL750
MESTRANOL mixed with NORETHYNODREL see EAP000
MESTRANOL mixed with NORGESTREL see NNR000
MESTRENOL see MKB750
MESTRENOL mixed with 6-CHLORO-6-DEHYDRO-17-α-ACETOXYPROGESTERONE see CNV750
MESULFA see ALF250
MESUPRINE see MDM000
MESUPRINE HYDROCHLORIDE see MDM000
MESURAL see LFK000
MESUROL see DST000
MESUXIMIDE see MLP800
MESYLCARBAMOYLMETHYLAMINOMET HYLPHOSPHONIC ACID see MID850
MESYL CHLORIDE see MDR300
MESYLITH see CLD000
META see TDW500
METAARSENIC ACID see ARB000
METABARBITAL see DJO800
META BLACK see AQP000
METABOLITE C see SOA500
METABOLITE I see HNI500
METACAINE see EFX500
METACARDIOL see AKT000
METACE see CLO750
METACEN see IDA000
METACETALDEHYDE see TDW500
METACETONE see DJN750
METACETONIC ACID see PMU750
METACHLOR see CFX000
METACHLORPHENPROP see CJQ300
METACHROME ORANGE R see NEY000
METACHROME RED F see CMO870
METACHROME YELLOW see SIT850
METACHROME YELLOW RA see SIT850
METACIDE see MNH000
METACIL see MPW500
METACIN see ORQ000
METACLOPROMIDE see AJH000
METACORTANDRACIN see PLZ000
METACORTANDRALONE see PMA000
METACRATE see MIB750
METADEE see VSZ100
METADELPHENE see DKC800
METADIAZINE see PPO750
METADIAZOL BROWN 450 see CMO825
METADOMUS see MDO250
METAFOS see MNH000
METAFOS see SII500

METAFUME see MHR200
METAHEXAMIDE see AKQ250
METAHYDRIN see HII500
METAHYDROXYPROCAINE see DHO600
METAISOSEPTOX see DAP400
METAISOSYSTOX see DAP400
METAISOSYSTOX-SOLFON 20 315 see DAP600
METAISOSYSTOXSULFOXIDE see DAP000
METAKRYLAN METYLU (POLISH) see MLH750
METALAXIL see MDM100
METALAXYL see MDM100
METALCAPTASE see MCR750
METALCAPTASE see PAP550
METALDEHYD (GERMAN) see TDW500
METALDEHYDE (DOT) see TDW500
METALDEIDE (ITALIAN) see TDW500
METALKAMATE see BTA250
METALLIBURE see MLJ500
METALLIC ARSENIC see ARA750
METALLIC OSMIUM see OKE000
METALLOENDOPEPTIDASE see GIA050
METALLOENDOPROTEASE see GIA050
METALLOPROTEASE see GIA050
METALLOPROTEINASE see GIA050
METALLOPROTEINASE FROM SNAKE VENOM BOTHROPS ASPER see GIA050
METALUTIN see MDM350
METAM see SIL550
METAMFETAMINA see DBA800
METAMID see PJY500
METAMIDOFOS ESTRELLA see DTQ400
METAMIN see TJL250
METAMINE see TJL250
METAMINE ACID FUCHSINE 6B see CMM400
METAMITON see ALA500
METAMITRON (GERMAN) see ALA500
METAMIZOL MONOHYDRATE see MDM500
METAMPHETAMIN see DBA800
METAMPHETAMINE HYDROCHLORIDE see MDT600
METAM-SODIUM (DUTCH, FRENCH, GERMAN, ITALIAN) see VFU000
METANA ALUMINUM PASTE see AGX000
METANABOL see DAL300
METANDIENON see DAL300
METANDIENONE see DAL300
METANDIENONUM see DAL300
METANDIOL see AOO475
METANDREN see MPN500
METANDRIOL see AOO475
METANDROSTENOLON see DAL300
METANDROSTENOLONE see DAL300
METANEPHRIN see VGP000
METANEX see CBT750
METANFETAMINA see DBA800
METANICOTINE see MDM750
METANILAMIDE see MDM760
METANILAN SODNY (CZECH) see AIF250
METANILE YELLOW O see MDM775
METANILIC ACID see SNO000
METANIL YELLOW see MDM775
METANIL YELLOW 1955 see MDM775
METANIL YELLOW C see MDM775
METANIL YELLOW E see MDM775
METANIL YELLOW EXTRA see MDM775
METANIL YELLOW F see MDM775
METANIL YELLOW G see MDM775
METANIL YELLOW GRIESBACH see MDM775
METANIL YELLOW K see MDM775
METANIL YELLOW KRSU see MDM775
METANIL YELLOW M3X see MDM775
METANIL YELLOW O see MDM775
METANIL YELLOW PL see MDM775
METANIL YELLOW S see MDM775
METANIL YELLOW SUPRA P see MDM775
METANIL YELLOW VS see MDM775
METANIL YELLOW WS see MDM775

METANIL YELLOW Y see MDM775
METANIL YELLOW YK see MDM775
METANIN see CPQ250
METANITE see MGR500
METANOLO (ITALIAN) see MGB150
METANTIOLO (ITALIAN) see MLE650
METANTYL see DJM800
METANTYL see XCJ000
METAOKSEDRIN see SPC500
METAOXEDRIN see NCL500
METAOXEDRIN see SPC500
METAOXEDRINUM see SPC500
METAOXON see PHD750
METAPHEN see NHK900
METAPHENYLALANINE MUSTARD see BHU500
METAPHENYLENEDIAMINE see PEY000
METAPHOR see MNH000
METAPHOS see MNH000
METAPHOSPHORIC ACID, CALCIUM SODIUM SALT see CAX260
METAPHOSPHORIC ACID, TETRASODIUM SALT see SKE550
METAPHYLLIN see TEP500
METAPHYLLINE see TEP500
METAPIRAZONE see AIB300
METAPLEXAN see MCJ400
METAPLEX NO see PKB500
METAPREL see MDM800
METAPROTERENOL see DMV800
METAPROTERENOL SULFATE see MDM800
METAQUALON see QAK000
METAQUEST A see EIX000
METAQUEST B see EIX000
METAQUEST C see EIV000
METARADRINE see HNB875
METARAMINOL see HNB875
(−)-METARAMINOL see HNB875
1-METARAMINOL see HNB875
METARAMINOL BITARTRATE see HNC000
METARAMINOL TARTRATE (1:1) see HNC000
METARSENOBILLON see SNR000
METARTRIL see IDA000
METASAP XX see AHH825
METASEOL see IIU100
METASILICIC ACID see SCL000
METASOL see MLH000
METASOL 30 see ABU500
METASOL P-6 see PFO000
METASOL TK-100 see TEX000
METASON see TDW500
METASQUALENE see TMP500
METASTENOL see DAL300
METASYMPATOL see NCL500
METASYNEPHRINE see NCL500
METASYSTEMOX see DAP000
METASYSTOX see MIW100
METASYSTOX FORTE see DAP400
METASYSTOX-R see DAP000
METASYSTOX-S see DSK600
METATENSIN see RDK000
METATHIONE see DSQ000
METATHION, S-METHYL ISOMER see MKC250
METATION see DSQ000
META TOLUYLENE DIAMINE see TGL750
METATOLYLENEDIAMINE DIHYDROCHLORIDE see DCE000
METATSIN see ORQ000
METATYL see MGJ750
d,l-METATYROSINE see TOG275
METAUPON see OHU000
METAUPON PASTE see SIY000
METAXALONE see XVS000
METAXAN see DJM800
METAXAN see XCJ000
METAXANIN see MDM100
METAXITE see ARM268
METAXON see CIR250
METAXONE see SIL500

METAZALONE see XVS000
METAZIN see HDY500
METAZIN see MDM900
METAZIN see SNJ000
METAZIN 6U see MCB050
METAZINE see HDY500
METAZINE see MDM900
METAZINE see MDM900
METAZOL see MLH000
METAZOLO see MCO500
METAZOLONE see XVS000
METEBANYL see ORE000
METELILACHLOR see MQQ450
METENDIOL see AOO475
METENIX see ZAK300
METENOLONE ACETATE see PMC700
METEPA see TNK250
METERAZIN MALEATE see PMF250
METERFER see FBJ100
METERFOLIC see FBJ100
METET see MDN100
METFENOSSIDIOLO see RLU000
METFORMIN see DQR600
METFORMIN HYDROCHLORIDE see DQR800
"METH" see MDQ500
METHAANTHIOL (DUTCH) see MLE650
METHABOL see PAN100
METHACETONE see DJN750
METHACHLOR see CFX000
METHACHLORPHENPROP see CFC750
METHACHOLIN see MFX560
METHACHOLINE see MFX560
METHACHOLINE CHLORIDE see ACR000
METHACHOLINIUM CHLORIDE see ACR000
METHACIDE see TGK750
METHACIN see ORQ000
METHACON see DPJ400
METHACRALDEHYDE (DOT) see MGA250
trans-METHACRIFOS see MDN150
METHACROLEIN see MGA250
METHACROLEIN DIMER see DLI600
METHACRYLALDEHYDE DIMER see DLI600
METHACRYLATE de BUTYLE (FRENCH) see MHU750
METHACRYLATE DE 2,5-DICHLORO-4,6-DINITROPHENYLE see DFE570
METHACRYLATE de METHYLE (FRENCH) see MLH750
METHACRYL CHLORIDE see MDN899
METHACRYLIC ACID see MDN250
METHACRYLIC ACID, inhibited (DOT) see MDN250
METHACRYLIC ACID (ACGIH,OSHA) see MDN250
METHACRYLIC ACID, ALLYL ESTER see AGK500
METHACRYLIC ACID AMIDE see MDN500
METHACRYLIC ACID ANHYDRIDE see MDN699
METHACRYLIC ACID, 1,3-BUTYLENE ESTER (2:1) see BRG100
METHACRYLIC ACID, BUTYL ESTER, polymer with 2-(DIMETHYLAMINO)ETHYL METHACRYLATE and METHYL METHACRYLATE see MDN505
METHACRYLIC ACID, BUTYL ESTER, POLYMERS see PJL500
METHACRYLIC ACID CHLORIDE see MDN899
METHACRYLIC ACID-3,4-DICHLOROANILIDE see DFO800
METHACRYLIC ACID, 3,6-DICHLORO-2,4-DINITROPHENYL ESTER see DFE570
METHACRYLIC ACID, DIESTER with TETRAETHYLENE GLYCOL see TCE400
METHACRYLIC ACID, DIESTER with TRIETHYLENE GLYCOL see MDN510

METHACRYLIC ACID, 2-(DIETHYLAMINO)ETHYL ESTER see DIB300
METHACRYLIC ACID-DIVINYLBENZENE COPOLYMER see MDN525
METHACRYLIC ACID DODECYL ESTER see DXY200
METHACRYLIC ACID, polymer with ETHYL ACRYLATE see MDN600
METHACRYLIC ACID, 2-HYDROXYPROPYL ESTER see HNV500
METHACRYLIC ACID, ISOBUTYLENE ESTER see BRG100
METHACRYLIC ACID, ISOBUTYL ESTER see IIY000
METHACRYLIC ACID, ISODECYL ESTER see IKM000
METHACRYLIC ACID LAURYL ESTER see DXY200
METHACRYLIC ACID, METHYL ESTER (MAK) see MLH750
METHACRYLIC ACID METHYL ESTER POLYMERS see PKB500
METHACRYLIC ACID, 1-METHYL-1,3-PROPYLENE ESTER see BRG100
METHACRYLIC ACID, 3-METHYL-1,3-PROPYLENE ESTER see BRG100
METHACRYLIC ACID, PHENETHYL ESTER (7CI,8CI) see PFP600
METHACRYLIC ACID, POLYMER WITH ACRYLIC ACID, ETHYL ACRYLATE AND METHYL METHACYRLATE see EFT100
METHACRYLIC ACID, 2-(VINYLOXY)ETHYL ESTER see VQA150
METHACRYLIC ALDEHYDE see MGA250
METHACRYLIC AMIDE see MDN500
METHACRYLIC ANHYDRIDE see MDN699
METHACRYLIC CHLORIDE see MDN899
METHACRYLONITRILE, inhibited see MGA750
γ-METHACRYLOXYPROPYLTRIMETHOXYSILANE see TLC250
METHACRYLOYL ANHYDRIDE see MDN699
METHACRYLOYL CHLORIDE see MDN899
α-METHACRYLOYL CHLORIDE see MDN899
METHACRYLOYLOXYETHYL ISOCYANATE see IKG700
METHACRYLSAEUREBUTYLESTER (GERMAN) see MHU750
METHACRYLSAEUREMETHYL ESTER (GERMAN) see MLH750
METHACRYLYL CHLORIDE see MDN899
METHACYCLINE HYDROCHLORIDE see MDO250
METHACYCLINE MONOHYDROCHLORIDE see MDO250
METHADON see MDO760
METHADONE see MDO750
METHADONE see MDO760
(−)-METHADONE see MDO775
(+)-METHADONE see DBE100
d-METHADONE see DBE100
l-METHADONE see MDO775
(±)-METHADONE see MDO760
6s-METHADONE see DBE100
dl-METHADONE see MDO760
l-(+)-METHADONE see DBE100
s-(+)-METHADONE see DBE100
METHADONE HYDROCHLORIDE see MDP000
d-METHADONE HYDROCHLORIDE see MDP240
l-METHADONE HYDROCHLORIDE see MDP770
(±)-METHADONE HYDROCHLORIDE see MDP750
dl-METHADONE HYDROCHLORIDE see MDP750

racemic METHADONE HYDROCHLORIDE see MDP750
METHADRENE see MJV000
METHAFORM see ABD000
METHAFRONE see AHK750
METHAFURILEN FUMARATE see MDP800
METHAFURYLENE FUMARATE see MDP800
METHAGON see FDD075
METHAHEXAMIDE see AKQ250
METHAKRYLALDEHYD see MGA250
METHAKRYLOXYMETHYLSILATRAN (CZECH) see HMS500
METHALLIBURE see MLJ500
METHALLYL ALCOHOL (DOT) see IMW000
2-(4-(METHALLYLAMINO)PHENYL)PROPIONIC ACID see MDP850
METHALLYL CHLORIDE see CIU750
α-METHALLYL CHLORIDE see CEV250
α-METHALLYL CHLORIDE see CIU750
γ-METHALLYL CHLORIDE see CEU825
1-METHALLYL-3-METHYL-6-AMINOTETRAHYDROPYRIMIDINEDIONE see AKL625
METHALLYL-19-NORTESTOSTERONE see MDQ075
17-α-(1-METHALLYL)-19-NORTESTOSTERONE see MDQ100
METHALUTIN see MDM350
METHAMBUCAINE HYDROCHLORIDE see AJA500
METHAMBUTOXYCAINE HYDROCHLORIDE see AJA500
METHAM DIHYDRATE see SIL550
METHAMIDOPHOS see DTQ400
METHAMIN see HEI500
METHAMINOACETOCATECHOL see MGC350
METHAMINODIAZEPINE HYDROCHLORIDE see MDQ250
METHAMINODIAZEPOXIDE see LFK000
METHAMINODIAZEPOXIDE HYDROCHLORIDE see MDQ250
METHAMPHETAMIN see PFP850
METHAMPHETAMINE see DBB000
(+)-METHAMPHETAMINE see PFP850
d-METHAMPHETAMINE see PFP850
l-METHAMPHETAMINE see PFP850
d-(S)-METHAMPHETAMINE see PFP850
S-(+)-METHAMPHETAMINE see PFP850
(+)-METHAMPHETAMINE CHLORIDE see MDT600
METHAMPHETAMINE HYDROCHLORIDE see DBA800
METHAMPHETAMINE HYDROCHLORIDE see MDT600
(+)-METHAMPHETAMINE HYDROCHLORIDE see MDT600
d-METHAMPHETAMINE HYDROCHLORIDE see MDT600
l-METHAMPHETAMINE HYDROCHLORIDE see MDQ500
dl-METHAMPHETAMINE HYDROCHLORIDE see DAR100
METHAMPHETAMINIUM CHLORIDE see MDT600
METHAM SODIUM see VFU000
METHANABOL see AOO475
METHANAL see FMV000
METHANAMIDE see FMY000
METHANAMINE (9CI) see MGC250
METHANAMINE, N-METHYL-, 2,3,6-TRICHLOROBENZOATE see DOR800
METHANAMINE, N-NITRO-(9CI) see NHN500
METHANAMINE, compd. with TRINITROMETHANE (1:1) see MGC300
METHANAMINIUM, 1-CARBOXY-N,N,N-TRIMETHYL-, CHLORIDE see CCH850
METHANAMINIUM NITRATE see MGN150
METHANDIENONE see DAL300

METHANDIOL see AOO475
METHANDRIOL see AOO475
METHANDROLAN see AOO475
METHANDROLONE see DAL300
METHANDROSTENOLONE see DAL300
METHANE see MDQ750
METHANEARSONIC ACID see MGQ530
METHANEARSONIC ACID, AMMONIUM SALT see MDQ770
METHANEARSONIC ACID, COMPD. WITH 2,2',2"-NITRILOTRIETHANOL see TJK900
METHANEARSONIC ACID DIMERCURY SALT see DNW000
METHANEARSONIC ACID, IRON SALT see IHB680
METHANEARSONIC ACID, MONOAMMONIUM SALT see MDQ770
METHANEARSONIC ACID, TRIETHANOLAMINE SALT see TJK900
METHANEARSONOUS ACID, COMPD. WITH OCTYLAMINE (1:1) see OEK500
METHANEARSONOUS ACID, COPPER(II) SALT see CNM660
METHANE BASE see MJN000
METHANE, BIS(4-CHLOROPHENYL)- see BIM800
METHANE, BIS(2,2-DINITROPROPOXY)- see MJO800
METHANE, BIS(2-FLUORO-2,2-DINITROETHOXY)- see FMV050
METHANEBISPHOSPHONIC ACID, SODIUM SALT see SIL575
METHANEBIS(N,N'-(5-UREIDO-2,4-DIKETOTETRAHYDROIMIDAZOLE)-N,N-DIMETHYLOL) see GDO800
METHANE BORONIC ANHYDRIDE-PYRIDINE COMPLEX see MDQ800
METHANE, BROMODIFLUORO- see BND800
METHANE, BROMOTRIPHENYL- see TNP600
METHANECARBONITRILE see ABE500
METHANECARBOTHIOLIC ACID see TFA500
METHANECARBOXAMIDE see AAI000
METHANECARBOXYLIC ACID see AAT250
METHANE, CHLORONITRO- see CJC900
METHANE, CHLOROTRIFLUORO-, mixt. with TRIFLUOROMETHANE (9CI) see FOO562
METHANE, CYANO- see ABE500
METHANEDIAMINE, N,N,N',N'-TETRAMETHYL-, ETHANEDIOATE see TDR800
METHANE, DIAZODIPHENYL- see DVZ100
METHANEDIAZOHYDROXIDE, POTASSIUM SALT, (Z)- see HMB570
METHANEDICARBOXYLIC ACID see CCC750
METHANEDICARBOXYLIC ACID, DIETHYL ESTER see EMA500
METHANE DICHLORIDE see MJP450
METHANE, DICHLORONITRO- see DFT250
METHANE, DIFLUORO- see MJQ300
METHANE, DINITRO- see DUW507
METHANEDIOL, DINITRATE see MJU150
METHANEDIPHOSPHONIC ACID, SODIUM SALT see SIL575
METHANEDISULFONIC ACID, CALCIUM SALT (1:1) see CAT700
METHANEDITHIOL-S,S-DIESTER with O,O-DIETHYL ESTER PHOSPHORODITHIOIC ACID see EEH600
METHANEPEROXOIC ACID see PCM500
METHANE, PHENYL- see TGK750
METHANEPHOSPHONIC ACID, DIPHENYL ESTER see MDQ825
METHANESULFINIC ACID, THIO-, S-METHYL ESTER (6CI,7CI,8CI) see MLH765
METHANESULFINOTHIOIC ACID, S-METHYL ESTER see MLH765

METHANESULFONAMIDE, N-(4-(9-ACRIDINYLAMINO)-3-METHOXYPHENYL)-, MONO(2-HYDROXYPROPANOATE) see AOD425

METHANESULFONAMIDE, N-(2-((4-AMINO-3-METHYLPHENYL)ETHYLAMINO)ETHYL)-, SULFATE (2:3) (9CI) see MDQ850

METHANESULFONAMIDE, N-(1'-2-(5-BENZOFURAZANYL)ETHYL)-3,4-(DIHYDRO-4-OXOSPIRO(2H-1-BENZOPYRAN-2,4'-PIPERIDIN-6-YL)-, MONOHYDROCHLORIDE see MDQ900

METHANESULFONAMIDE, N-(4-(2-(DIETHYLAMINO)-1-OXOPROPYL)PHENYL)-, MONOHYDROCHLORIDE see DHI876

METHANESULFONAMIDE, N-(5-(2-((1,1-DIMETHYL-2-PHENYLETHYL)AMINO)-1-HYDROXYETHYL)-2-HYDROXYPHENYL)-, MONOHYDROCHLORIDE, (+−)- see ZNS300

METHANESULFONAMIDE, N-(4-(4-(ETHYLHEPTYLAMINO)-1-HYDROXYBUTYL)PHENYL)-, (+−)-, (E)-2-BUTENEDIOATE (2:1) (SALT) see IAB200

METHANESULFONAMIDE, N-(((2-HYDROXY-2-(4-HYDROXYPHENYL)ETHYL)AMINO)((2-(((5-((METHYLAMINO) METHYL)-2-FURANYL)METHYL)THIO)ETHYL)AMINO)METHYLENE)-, (E)- see OKK600

METHANESULFONAMIDE, N-METHYL- see MLH760

METHANESULFONAMIDE,N-(2-(4-AMINO-N-ETHYL-m-TOLUIDINO)ETHYL)-, SULFATE(2:3) see MDQ850

METHANESULFONAMIDE, N-PHENYL-N-((1,1,2,2-TETRACHLORO-2-FLUOROETHYL)THIO)- see FLZ025

METHANESULFONAMIDE, N-((1,1,2,2-TETRACHLORO-2-FLUOROETHYL)THIO)- see FLZ025

METHANESULFONANILIDE, N-((TETRACHLORO-2-FLUOROETHYL)THIO)- see FLZ025

METHANESULFON-m-ANISIDIDE, 4'-(9-ACRIDINYLAMINO)-, compounded with LACTIC ACID see AOD425

METHANESULFONIC ACID see MDR250

METHANESULFONIC ACID, (ANTIPYRINYLMETHYLAMINO)-, SODIUM SALT, MONOHYDRATE see MDM500

METHANESULFONIC ACID CHLORIDE see MDR300

METHANESULFONIC ACID CHLOROETHYL ESTER see CHC750

METHANESULFONIC ACID, COMPOUND with 2-(DIETHYLAMINO)-4-METHYLPENTYL PAMINOBENZOATE (1:1) see LEU000

METHANESULFONIC ACID, ((2,3-DIHYDRO-1,5-DIMETHYL-3-OXO-2-PHENYL-1H-PYRAZOL-4-YL)METHYLAMINO)-,SO DIUM SALT, MIX.T WITH 2,2,6,6-TETRAMETHYL-4-PIPERIDINONE 4-METHYLBENZENESULFONATE see TAL335

METHANESULFONIC ACID, METHYLENE ESTER see MJQ500

METHANESULFONIC ACID-1-METHYLETHYL ESTER see IPY000

METHANESULFONIC ACID, NONAMETHYLENE ESTER see MDR275

METHANESULFONIC ACID, 2-PROPENYL ESTER (9CI) see AGK750

METHANESULFONIC ACID, PROPYL ESTER see PNQ000

METHANESULFONIC ACID, SILVER SALT see SDR500

METHANESULFONIC ACID, SILVER(1+) SALT see SDR500

METHANESULFONIC ACID TETRAMETHYLENE ESTER see BOT250

METHANESULFONYL CHLORIDE see MDR300

METHANESULFONYL FLUORIDE see MDR750

1-(3-(2-(METHANESULFONYL)PHENOTHIAZIN-10-YL)PROPYL)-4-PIPERIDINECARBOXAMIDE see MPM750

METHANESULPHONIC ACID ETHYL ESTER see EMF500

METHANESULPHONIC ACID METHYL ESTER see MLH500

METHANESULPHONYL CHLORIDE see MDR300

METHANESULPHONYL FLUORIDE see MDR750

METHANETELLUROL see MDR775

METHANE, TETRABROMIDE see CBX750

METHANE, TETRABROMO- see CBX750

METHANE TETRACHLORIDE see CBY000

METHANE, TETRAFLUORO- see CBY250

METHANE TETRAMETHYLOL see PBB750

N,N'-METHANETETRAYLBIS(2,6-BIS(1-METHYLETHYL)BENZENAMINE) see BJE550

N,N'-METHANETETRAYL BISCYCLOHEXANAMINE see MDR800

N,N'-METHANETETRAYLBIS-2-PROPANAMINE see DNO400

N,N'-METHANE TETRAYLBIS(1,1,1-TRIMETHYLSILANAMINE) see BLQ950

1,1',1'',1'''-METHANETHETRAYLTETRAKISBENZENE see TEA750

METHANETHIOL see MLE650

(METHANETHIOLATO)METHYLMERCURY see MLX300

METHANETHIOL, (ETHYLSULFINYL)-, S-ESTER WITH O,O-DIISOPROPYLPHOSPHORODITHIOATE see EPH500

METHANETHIOL, (ETHYLSULFONYL)-, S-ESTER WITH O,O-DIETHYL PHOSPHORODITHIOATE see TEZ100

METHANETHIOL, PHENYL- see TGO750

METHANE TRICHLORIDE see CHJ500

METHANE, TRIFLUORO-, mixt. with CHLOROTRIFLUOROMETHANE see FOO562

METHANE, TRIFLUOROIODO- see IFM100

METHANE, TRIMETHYLOLNITRO- see HMJ500

METHANE, compressed (UN 1971) (DOT) see MDQ750

METHANE, refrigerated liquid (cryogenic liquid) (UN 1972) (DOT) see MDQ750

METHANIDE see DJM800

METHANIDE see XCJ000

METHANIMIDAMIDE, N'-(2,4-DIMETHYLPHENYL)-N-((2,4-DIMETHYLPHENYL)IMINO)METHYL)-N-METHYL- see MJL250

1,4-METHANOAZULENE, DECAHYDRO-9-METHYLENE-4,8,8-TRIMETHYL-, (1S-(1-α-3a-β,4-α-8a-β))- see LID100

1,4-METHANOAZULENE, DECAHYDRO-4,8,8-TRIMETHYL-9-METHYLENE-, (1S,3aR,4S,8aS)-(+)-(8CI) see LID100

1H-3a,7-METHANOAZULENE, OCTAHYDRO-6-METHOXY-3,6,8,8-TETRAMETHYL-,(3R-(3-α-3a-β, 6-α-7-β,8aα-))- see CCR525

1H-3-α-7-METHANOAZULEN-6-OL, OCTAHYDRO-3,6,8,8-TETRAMETHYL-, FORMATE, (3R-(3-α-3a-β,6-α-7-β,8aα-))- see CCR524

2,6-METHANO-3-BENZAZOCIN-8-OL-3-ALLYL-6-ETHYL-1,2,3,4,5,6-HEXAHYDRO-11-METHYL see AGG250

2,6-METHANO-3-BENZAZOCIN-8-OL,-,6,11-α-DIMETHYL-1,2,3,4,5,6-HEXAHYDRO-, HYDROCHLORIDE see NNS600

2,6-METHANO-3-BENZAZOCIN-8-OL, 1,2,3,4,5,6-HEXAHYDRO-6,11-DIMETHYL-,-trans- see NNS600

2,6-METHANO-3-BENZAZOCIN-8-OL, 1,2,3,4,5,6-HEXAHYDRO-6,11-α-DIMETHYL-, HYDROCHLORIDE see NNS600

2,6-METHANO-3-BENZAZOCIN-8-OL, 1,2,3,4,5,6-HEXAHYDRO-6,11-DIMETHYL-3-PHENETHYL-, HYDROBROMIDE, (+−)- see HKC625

2,6-METHANO-3-BENZAZOCIN-8-OL, 1,2,3,4,5,6-HEXAHYDRO-6,11-DIMETHYL-, (2-α,6-α,11R*)- see NNS600

4,7-METHANOBENZOTRITHIOLE, HEXAHYDRO- see TNO300

2,4-METHANO-4H-FURO(3,2-B)PYRROL-3-AMINE, HEXAHYDRO-N,N-DIMETHYL-, (2R-(2-α,3-α,3A-β,4-α,6A-β))- see MLD050

2,4-METHANO-4H-FURO(3,2-B)PYRROL-3-AMINE, HEXAHYDRO-N-METHYL-, (2R-(2-α,3-α,3A-β,4-α,6A-β))- see FBP400

2,4-METHANO-4H-FURO(3,2-B)PYRROLE, HEXAHYDRO-3-(METHYLAMINO)- see FBP400

3,5A-METHANO-5AH-CYCLOHEPTACYCLODECEN-8(1H)-ONE, DODECAHYDRO-4,7,10,12,14-PENTAHYDROXY-4,9,9-TRIMETHYL-13-METHYLENE-, (3R-(3R*,4R*,5AS*,7S*,10S*,12S*,13AS*,14R*))- see GJU225

3,5A-METHANO-5AH-CYCLOHEPTACYCLODECEN-8(1H)-ONE, DODECAHYDRO-4,7,10,12,14-PENTAHYDROXY-4,9,9-TRIMETHYL-13-METHYLENE-, (3R-(3R*,4R*,5AS*,7R*,10S*,12S*,13AS*,14R*))- see GJU250

5A,8-METHANO-5AH-CYCLOHEPTA(B)NAPHTHALENE-2,4,4A,7,12(5H)-PENTOL, DODECAHYDRO-3,3,7-TRIMETHYL-11-METHYLENE-, (2S-(2-α,4-α,4A-β,5A-β,7-β,8-β,10A-α,11A-α,12S *))- see LEX250

7,9A-METHANO-9AH-CYCLOPENTA(B)HEPTALENE-2,4,8,11,11A,12(1H)-HEXOL, DODECAHYDRO-1,1,4,8-TETRAMETHYL-, (2S,3AS,4R,4AR,7R,8R,9AS,11R,11AR,12R)- see GJU310

7,9A-METHANO-9AH-CYCLOPENTA(B)HEPTALENE-2,4,8,11,11A,12(1H)-HEXOL, DODECAHYDRO-1,1,4,8-TETRAMETHYL-, 6,12-DIACETATE, (2S,3AS,4R,4AR,7R,8R,9AS,11R,11AR,12R)- see GJU315

7,9A-METHANO-9AH-CYCLOPENTA(B)HEPTALENE-2,4,11,11A-β,12(1H)-PENTOL, 2A,3,3A-α,4A-β,5,6,7-β,10,11A-DECAHYDRO-1,1,4-β,8-TETRAMETHYL- see GJU330

7,9A-METHANO-9AH-CYCLOPENTA(B)HEPTALENE-2,8,11,11A,12(1H)-PENTOL, DODECAHYDRO-1,1,8-TRIMETHYL-4-METHYLENE-, (2S,3AS,4AS,7R,8R,9AS,11R,11AR,12R)- see GJU300

7,9A-α-METHANO-9AH-CYCLOPENTA(B)HEPTALENE-4,8,10,11,11A-β,(1H)-PENT OL, 2-β,3-β-EPOXY-2,3,3A-β,4,4A-β,5,6,7-β,8,9,10-β,11-α-DODECAHYDRO-1, 1,4-β,8-β-TETRAMETHYL- see DAD700

METHANOIC ACID see FNA000

4,7-METHANOINDANDIMETHANOL, HEXAHYDRO- see TJG550

4,7-METHANOINDAN, 1,2,4,5,6,7,8,8-OCTACHLORO-2,3-EPOXY-3A,4,7,7A-TETRAHYDRO-, EXO,ENDO- see OPK275

4,7-METHANOINDAN, 2,2,4,5,6,7,8,8-OCTACHLORO-3a,4,7,7a-TETRAHYDRO- see CDR575

4,7-METHANOINDAN, 3A,4,7,7A-TETRAHYDRO-2,3-EPOXY-1,2,4,5,6,7,8,8-OCTACHLORO-, EXO,ENDO- see OPK275

4,7-METHANOINDAN, 3A-β,4,7,7A-β-TETRAHYDRO-1-β,2-α-4-α-5,6,7-α-8,8-OCTACHLORO- see CDR800

4,7-METHANO-1H-INDENE-2-CARBOXALDEHYDE, OCTAHYDRO-5-METHOXY- see OBW200

4,7-METHANOINDENE-6-CARBOXYLIC ACID, 3a,4,5,6,7,7a-HEXAHYDRO-, ETHYL ESTER see TJG600

4,7-METHANO-1H-INDENE-5,?-DIMETHANOL, OCTAHYDRO- see TJG550

4,7-METHANOINDENE, 4,5,6,7,8,8-HEXACHLORO-3a,4,7,7a-TETRAHYDRO- see HCN000

1,6-METHANO-1H-INDENE, 2,3,3A,4,5,8-HEXACHLORO-3A,6,7,7A-TETRAHYDRO-, (1-α-3A-β,6-α-7A-β,8R*)- see CDR860

4,7-METHANO-1H-INDENE, 1,2,3,4,5,6,7,8,8-NONACHLOR-2,3,3A,4,7,7A-HEXAHYDRO-, (1-α-2-β, 3-α-3A-α-4-β,7-β,7Aα)- see NMV755

4,7-METHANO-1H-INDENE, 2,2,4,5,6,7,8,8-OCTACHLORO-2,3,3a,4,7,7a-HEXAHYDRO-(9CI) see CDR575

4,7-METHANOINDEN, 1,2,4,5,6,7,8,9-OCTACHLORO-3A,4,7,7A-TETRAHYDRO- see CDR760

4,7-METHANOISOBENZOFURAN-1,3-DIONE, 3a,4,7,7a-TETRAHYDROMETHYL- see NAC000

4,7-METHANOISOBENZOFURAN, 1,3,3A,4,7,7A-HEXAHYDRO-4,5,6,7,8,8-HEXACHLORO- see HDK200

4,7-METHANOISOBENZOFURAN-1-OL, 1,3,3A,4,7,7A-HEXAHYDRO-4,5,6,7,8,8-HEXACHLORO- see HDK210

4,7-METHANOISOBENZOFURAN-1(3H)-ONE, 3A,4,7,7A-TETRAHYDRO-4,5,6,7,8,8-HEXACHLORO- see EAQ815

METHANOL see MGB150

METHANOLACETONITRILE see HGP000

N-METHANOLACRYLAMIDE see HLU500

METHANOL, BENZYL- see PDD750

METHANOL, CYCLOHEXYL- see HDH200

METHANOL, (METHYL-ONN-AZOXY)-, BENZOATE (ester) (9CI) see MGS925

METHANOL, OXIRANYL- see GGW500

METHANOL, OXYDI-, DINITRATE see OPQ200

METHANOL, PHOSPHINIDYNETRI- see TNI600

METHANOL, PHOSPHINIDYNETRIS- see TNI600

METHANOL, SODIUM SALT see SIK450

2-METHANOL TETRAHYDROPYRAN see MDS500

METHANOL, TRICHLORO-, CHLOROFORMATE see TIR920

METHANOMETHENE see MCE500

4,7-METHANO-2,3,8-METHENOCYCLOPENT(a)INDENE, DODECAHYDRO-, stereoisomer see DLJ500

1,4-METHANONAPHTHALENE-5,8-DIONE, 1,4,4a,8a-TETRAHYDRO- see TCU650

METHANONE, (2-AMINO-1H-BENZIMIDAZOL-5-YL)PHENYL- see AIH000

METHANONE, (2-AMINO-5-CHLOROPHENYL)PHENYL- see AJE400

METHANONE, (5-AMINO-1,3-DIMETHYL-1H-PYRAZOL-4-YL)(2-FLUOROPHENYL)- see AJR400

METHANONE, 9-ANTHRACENYLPHENYL- see APH600

METHANONE, BIS(2-HYDROXY-4-METHOXYPHENYL)-(9CI) see BJY800

METHANONE, BIS(4-HYDROXYPHENYL)-, (2,4-DINITROPHENYL)HYDRAZONE see BKI300

METHANONE, (4-BROMOPHENYL)PHENYL-(9CI) see BMU170

METHANONE, (3-CHLOROPHENYL)-9H-FLUOREN-2-YL- see CKA100

METHANONE, 2-DIBENZOFURANYLPHENYL- see DDC050

METHANONE, 2-DIBENZOTHIENYL(4-NITROPHENYL)- see DDD900

METHANONE, (3,5-DIBROMO-4-HYDROXYPHENYL)(2-(2,4,6-TRIMETHYLPHENYL)-3-BENZOFURANYL)- see DDP300

METHANONE, (4,5-DICHLORO-1H-PYRROL-2-YL)(2,6-DIHYDROXYPHENYL)- see POK400

METHANONE, (4-(2-(DIETHYLAMINO)ETHOXY)PHENYL)(2-ETHYL-3-BENZOFURANYL)-(9CI) see EDV700

METHANONE, (4-(2-(DIETHYLAMINO)ETHOXY)PHENYL)(2-(2,4,6-TRIMETHYLPHENYL)-3-BENZOFURA NYL)-, HCl see DHR900

METHANONE, (2,3-DIHYDRO-1,4-BENZODIOXIN-6-YL)(4-FLUOROPHENYL)- see DKT500

METHANONE, (2,3-DIHYDRO-5-METHYL-3-(4-MORPHOLINOMETHYL)PYRROLO(1,2,3-DE)-1,4-BENZOXAZ IN-6-YL)-1-NAPHTHALENYL-, (3R)- see NAP545

METHANONE, (1,3-DIMETHYL-5-(METHYLAMINO)-1H-PYRAZOL-4-YL)(2-FLUOROPHENYL)- see DSO500

METHANONE, DIPHENYL-, OXIME (9CI) see BCS400

METHANONE, 9H-FLUOREN-2-YL(4-METHOXYPHENYL)- see MEC600

METHANONE, 9H-FLUOREN-2-YLPHENYL- see FEM100

METHANONE, 3-FURFANYLPHENYL-(9CI) see FQO050

METHANONE, (4-HYDROXY-3,5-DIIODOPHENYL)(2-(2,4,6-TRIMETHYLPHENYL)-3-BENZOFURANYL)- see DNF550

METHANONE, (2-HYDROXY-4-METHOXYPHENYL)(2-HYDROXYPHENYL)-(9CI) see DMW250

METHANONE, (2-HYDROXY-4-(OCTYLOXY)PHENYL)PHENYL-(9CI) see HND100

METHANONE, (4-HYDROXYPHENO)PHENYL-/TD TOXICITY DATA: see HJN100

METHANONE, (4-HYDROXYPHENYL)(2-(2,4,6-TRIMETHYLPHENYL)-3-BENZOFURANYL)- see HNK700

METHANONE, (4-HYDROXYPHENYL)(2-(2,4,6-TRIMETHYLPHENYL)-4-BENZOFURANYL)- see HNK800

METHANONE, (4-METHOXYPHENYL)PHENYL- see MEC333

METHANONE, (4-METHOXYPHENYL)(2-(2,4,6-TRIMETHYLPHENYL)-3-BENZOFURANYL)- see AOY300

METHANONE, PHENYL-1H-PYRROL-2-YL-(9CI) see PGF900

6,9-METHANO-8H-PYRIDO(1',2':1,2)AZEPINO(4,5-b)INDOLE-6(6aH)-CARBOXYLIC ACID, 7,8,9,10,12,13-HEXAHYDRO-6-ETHYL-13a-HYDROXY-, METHYL ESTER, HYDROCHLORIDE see CNS200

4,7-METHANOSELENOPHENE, OCTAHYDRO-2,2,3,3-TETRAFLUORO- see TCI100

o-(1,4-METHANO-1,2,3,4-TETRAHYDRO-6-NAPHTHYL)-N-METHYL-N-(m-TOLYL)-THIOCARBAMATE see MQA000

METHANSULFONSAURE-N-METHYLAMID see MLH760

METHANTHELINE BROMIDE see DJM800

METHANTHELINE BROMIDE see XCJ000

METHANTHELINIUM BROMIDE see XCJ000

METHANTHINE BROMIDE see XCJ000

METHANTHIOL (GERMAN) see MLE650

METHAPHENILENE HYDROCHLORIDE see DCJ850

METHAPHOXIDE see TNK250

METHAPYRAPONE see MCJ370

METHAPYRILENE see DPJ200

METHAPYRILENE see TEO250

METHAPYRILENE HYDROCHLORIDE see DPJ400

METHAPYRILENE HYDROCHLORIDE (L.A.) see DPJ400

METHAPYRILENE HYDROCHLORIDE (S.A.) see DPJ400

METHAPYRILENE mixed with SODIUM NITRITE (1:2) see MDT000

METHAQUALONE see QAK000

METHAQUALONE HYDROCHLORIDE see MDT250

METHAQUALONEINONE see QAK000

METHAR see DXE600

METHARBITAL see DJO800

METHARBITONE see DJO800

METHARBUTAL see DJO800

METHARSINAT see DXE600

METHASAN see BJK500

β-METHASONE-17-VALERATE see VCA000

METHASQUIN see DBY500

METHAZATE see BJK500

METHAZINE see IDA000

METHAZINE see MDM900

METHAZOLE see BGD250

METHAZONIC ACID see NEJ600

METHCAINE HYDROCHLORIDE see IJZ000

METHDILAZINE see MPE250

METHDILAZINE HYDROCHLORIDE see MDT500

METHEDRINE see DBA800

METHEDRINE see DBB000

METHEDRINE see MDT600

METHEDRINE HYDROCHLORIDE see DBA800

METHEDRINE HYDROCHLORIDE see MDT600

METHEGRIN see MJV750

METHELINA see DJM800

METHELINA see XCJ000

METHENAMIC ACID see XQS000

METHENAMINE see HEI500

1,3,4-METHENO-1H-CYCLOBUTA(c,d)-PENTALENE-2-LEVULINIC ACID, 1,1A,3,3A,4,5,5A,5B,6-DECACHLOROOCTAHYDRO-2-HYDROXY-, ETHYL ESTER see MDT625

2,5,7-METHENO-3H-CYCLOPENTA(A)PENTALEN-3-ONE, 3B,4,5,6,6,6A-HEXACHLORODECAHYDRO-, (2-α-3A-β,3B-β,4-β,5-β,6A-β,7-α-7A-β,8R*)- see KFK200

METHENOLONE ACETATE see PMC700

N,N'-METHENYL-o-PHENYLENEDIAMINE see BCB750

METHENYL TRIBROMIDE see BNL000

METHENYL TRICHLORIDE see CHJ500

METHERGINE see PAM000

METHETOIN see ENC500

METHEXAMIDE see AKQ250

METHEXENYL see ERD500

METHEXENYL SODIUM see ERE000

METHIACIL see MPW500

METHIAMAZOLE see MCO500
METHIAZIC ACID see MNQ500
METHIAZINIC ACID see MNQ500
METHIDATHION see DSO000
METHILANIN see MDT740
METHIOCARB see DST000
METHIOCIL see MPW500
METHIODAL SODIUM see SHX000
METHIODIDE of N-BENZYLURETHANE of 3-DIMETHYLAMINOPHENOL see HNO000
METHIODIDE of N-METHYLURETHANE of 3-DIETHYLAMINOPHENOL see MID250
METHIODIDE of N-METHYLURETHANE of 3-DIMETHYLAMINOPHENOL see HNO500
METHIODIDE of N-METHYLURETHANE of 4-DIMETHYLAMINOPHENOL see HNP000
METHIOMEPRAZINE HYDROCHLORIDE see MDT650
METHIONAL see MPV400
METHIONAL see TET900
METHIONAMINE see ACQ275
METHIONINE see MDT750
d-METHIONINE see MDT730
l-METHIONINE see MDT750
l-(−)-METHIONINE see MDT750
(±)-METHIONINE see MDT740
dl-METHIONINE see MDT740
METHIONINE, N-ACETYL-, dl- see ACQ270
dl-METHIONINE, N-ACETYL-(9CI) see ACQ270
l-METHIONINE, S-ADENOXYL- see AEM310
l-METHIONINE, N-(N-(3-(p-FLUOROPHENYL)-l-ANANYL)-3-(m-(BIS(2-CHLOROETHYL)AMINO)PHENYL)-l-ALANYL)-, ETHYL ESTER, HYDROCHLORIDE see POI550
METHIONINE HYDANTOIN see MDT800
METHIONINE HYDROXY ANALOG see HMR600
METHIONINE PLATINUM DICHLORIDE see MDT900
METHIONINE SULFOXIDE see ALF600
dl-METHIONINE SULFOXIDE see ALF600
METHIONINE SULFOXIMINE see MDU100
dl-METHIONINE-dl-SULFOXIMINE see MDU100
METHIONINYLADENYLATE see AEM310
METHIOPLEGIUM see TKW500
METHISAZONE see MKW250
METHIUM CHLORIDE see HEA500
METHIXENE HYDROCHLORIDE see THL500
METHOCARBAMOL see GKK000
METHOCEL 10 see MIF760
METHOCEL 15 see MIF760
METHOCEL 181 see MIF760
METHOCEL 400 see MIF760
METHOCEL 4000 see MIF760
METHOCEL A see MIF760
METHOCEL CHG see MIF760
METHOCEL 4000CPS see MIF760
METHOCEL HG see HNX000
METHOCEL MC see MIF760
METHOCEL MC 25 see MIF760
METHOCEL MC4000 see MIF760
METHOCEL MC 8000 see MIF760
METHOCEL 400CPS see MIF760
METHOCEL SM 100 see MIF760
METHOCHLOPRAMIDE see AJH000
METHOCHLORIDE of N-METHYLURETHANE of 3-DIMETHYLAMINOPHENOL see HNN000
METHOCHLORPROMAZINE IODIDE see MIJ300
METHOCILLIN-S see SLJ000
METHOCILLIN S see SPD600
METHOCROTOPHOS see DOL800
METHODICHLOROPHEN see MQR100

METHOFADIN see MDU300
METHOFADIN see MDU300
METHOFAZINE see MDU300
METHOFLURANE see DFA400
METHOGAS see MHR200
METHOHEXITAL SODIUM see MDU500
METHOHEXITONE SODIUM see MDU500
METHOIDAL SODIUM see SHX000
METHOIN see MKB250
METHOLCARB see MIB750
METHOLENE 2095 see MHY650
METHOLENE 2218 see MJW000
METHOLENE 2296 see MLC800
METHOMYL see MDU600
METHOPHENAZATE ACID FUMARATE see MDU750
METHOPHENAZINE DIFUMARATE see MDU750
METHOPHOLINE see MDV000
METHOPHYLLINE see TEP500
METHOPIRAPONE see MCJ370
METHOPLAIN see DNA800
METHOPLAIN see MJE780
METHOPRENE see KAJ000
METHOPROMAZINE MALEATE see MFK750
METHOPROPTRYNE see INQ000
METHOPTERIN see MDV500
METHOPYRAPONE see MCJ370
METHOPYRININE see MCJ370
METHOPYRONE see MCJ370
METHOQUINE see CFU750
METHORATE HYDROBROMIDE see DBE200
d-METHORPHAN see DBE150
Δ-METHORPHAN see DBE150
d-METHORPHAN HYDROBROMIDE see DBE200
METHORPHINAN see MKR250
METHORPHINAN HYDROBROMIDE see MDV250
METHOSCOPYLAMINE BROMIDE see SBH500
METHOSERPEDINE see MEK700
METHOSERPIDINE see MEK700
METHOSTAN see AOO475
METHOTEXTRATE see MDV500
METHOTREXATE see MDV500
METHOTREXATE DISODIUM SALT see MDV600
METHOTREXATE SODIUM see MDV750
METHOTRIMEPRAZINE see MCI500
METHOXA-DOME see XDJ000
METHOXADONE see MFD500
METHOXAMINE HYDROCHLORIDE see MDW000
METHOXANE see DFA400
METHOXCIDE see MEI450
METHOXERPATE HYDROCHLORIDE see MDW100
METHOXIPHENADRIN HYDROCHLORIDE see OJY000
METHOXO see MEI450
METHOXOLONE see XVS000
METHOXONE see CIR250
METHOXONE see CIR500
METHOXONE see SIL500
METHOXONE M see CLO200
METHOXSALEN see XDJ000
METHOXYACETALDEHYDE see MDW250
2-METHOXYACETALDEHYDE see MDW250
α-METHOXYACETALDEHYDE see MDW250
1-METHOXY-2-ACETAMIDOFLUORENE see MER000
2-METHOXYACETANILIDE see AAR000
3-METHOXYACETANILIDE see AAQ750
4-METHOXYACETANILIDE see AAR250
m-METHOXYACETANILIDE see AAQ750
o-METHOXYACETANILIDE see AAR000
p-METHOXYACETANILIDE see AAR250

2'-METHOXYACETANILIDE see AAR000
3'-METHOXYACETANILIDE see AAQ750
4'-METHOXYACETANILIDE see AAR250
METHOXYACETIC ACID see MDW275
2-METHOXYACETIC ACID see MDW275
2-METHOXYACETOACETANILIDE see ABA500
o-METHOXYACETOACETANILIDE see ABA500
2'-METHOXYACETOACETANILIDE see ABA500
4-METHOXYACETOFENON see MDW750
METHOXYACETONE see MDW300
METHOXYACETONE see MFL100
1-METHOXYACETONE see MDW300
1-METHOXYACETONE see MFL100
4-METHOXYACETOPHENONE see MDW750
p-METHOXYACETOPHENONE see MDW750
4'-METHOXYACETOPHENONE see MDW750
1-METHOXY-2-ACETOXYPROPANE see PNL265
METHOXYACETYL CHLORIDE see MDW780
METHOXYACETYLENE see MDX000
2-(METHOXYACETYL)INDOLE see ICW200
(METHOXYACETYL)METHYLCARBAMIC ACID-o-ISOPROPOXYPHENYL ESTER see MDX250
2-(METHOXYACETYL)-1-METHYLINDOLE see MDX300
2-(METHOXYACETYL)-3-METHYLINDOLE see MDX310
5-METHOXY-N-ACETYLTRYPTAMINE see MCB350
4'-(3-METHOXY-9-ACRIDINYLAMINO)METHANESULFONALIDE see MDY000
4'-(1-METHOXY-9-ACRIDINYLAMINO)METHANESULFONANILIDE see MDX350
4'-(4-METHOXY-9-ACRIDINYLAMINO)METHANESULFONANILIDE see MDY250
N-(4-((3-METHOXY-9-ACRIDINYL)AMINO)PHENYL)-METHANESULFONAMIDE see MDY000
p-METHOXYACRYLOPHENONE see ONW100
4'-METHOXYACRYLOPHENONE see ONW100
2-METHOXY-AETHANOL (GERMAN) see EJH500
2-METHOXYAETHYLACETAT (GERMAN) see EJJ500
1-METHOXY-AETHYL-AETHYLNITROSAMIN (GERMAN) see MEO500
1-METHOXY-AETHYL-METHYLNITROSAMIN (GERMAN) see MEP500
METHOXYAETHYLQUECKSILBERCHLORID see MEP250
METHOXYAETHYLQUECKSILBERSILIKAT (GERMAN) see MEP000
p-METHOXYALLYLBENZENE see AFW750
2-METHOXY-4-ALLYLPHENOL see EQR500
2'-METHOXY-4'-ALLYLPHENYL 4-GUANIDINOBENZOATE see MDY300
METHOXYAMINE see MKQ880
METHOXYAMINE HYDROCHLORIDE see MNG500
2-METHOXY-4-AMINOAZOBENZENE see MDY400
2-METHOXY-1-AMINOBENZENE see AOV900
2-METHOXY-1-AMINOBENZENE HYDROCHLORIDE see AOX250
4-METHOXY-2-AMINOBENZOTHIAZOLE see AKM750

6-METHOXY-2-AMINOBENZOTHIAZOLE see MDY750

2-METHOXY-4-AMINO-5-CHLORO-N-β-(DIETHYLAMINOETHYL)BENZAMIDE DIHYDROCHLORIDE MONOHYDRATE see MQQ300

2-METHOXY-3-AMINODIBENZOFURAN see MDZ000

3-METHOXY-4-AMINODIPHENYL see MEA000

3-METHOXY-2-AMINODIPHENYLENE OXIDE see AKN500

2-METHOXY-4-AMINO-5-HYDROXYMETHYLPYRIMIDINE see AKO750

6-METHOXY-8-(4-AMINO-1-METHYLBUTYLAMINO)QUINOLINE see PMC300

1-METHOXY-2-AMINONAPHTHALENE see MFA250

4-METHOXY-4-AMINO-2-PENTANOL see MEA250

3-METHOXY-4-AMINOSTILBENE see MFP500

4-METHOXYAMPHETAMINE HYDROCHLORIDE see MEA500

dl,4-METHOXYAMPHETAMINE HYDROCHLORIDE see MEA500

2-METHOXYANILINE see AOV900

3-METHOXYANILINE see AOV890

4-METHOXYANILINE see AOW000

o-METHOXYANILINE see AOV900

p-METHOXYANILINE see AOW000

2-METHOXYANILINE HYDROCHLORIDE see AOX250

o-METHOXYANILINE HYDROCHLORIDE see AOX250

2-METHOXYANILINIUM NITRATE see MEA600

2-METHOXY-p-ANISIDINE HYDROCHLORIDE see MEA625

4-METHOXY-o-ANISIDINE HYDROCHLORIDE see MEA625

3-METHOXYANISOLE see REF025

1-METHOXY-9,10-ANTHRACENEDIONE see MEA650

1-METHOXYANTHRAQUINONE see MEA650

6-METHOXYARISTOLOCHIC ACID D see MEA700

METHOXYAZOXYMETHANOLACETATE see MEA750

4-METHOXYBENZALACETONE see MLI400

p-METHOXYBENZALACETONE see MLI400

2-METHOXYBENZALDEHYDE see AOT525

3-METHOXYBENZALDEHYDE see AOT300

4-METHOXYBENZALDEHYDE see AOT530

6-METHOXYBENZALDEHYDE see AOT525

m-METHOXYBENZALDEHYDE see AOT300

o-METHOXYBENZALDEHYDE see AOT525

p-METHOXYBENZALDEHYDE (FCC) see AOT530

2-METHOXYBENZAMIDE see AOT750

o-METHOXYBENZAMIDE see AOT750

2-(p-METHOXYBENZAMIDO)ACETOHYDROXAMIC ACID see MEA800

5-METHOXY-BENZ(a)ANTHRACENE see MEB000

7-METHOXY-BENZ(a)ANTHRACENE see MEB500

8-METHOXY-BENZ(a)ANTHRACENE see MEB250

3-METHOXY-1,2-BENZANTHRACENE see MEB000

5-METHOXY-1,2-BENZANTHRACENE see MEB250

10-METHOXY-1,2-BENZANTHRACENE see MEB500

5-METHOXY-1,2-BENZ(a)ANTHRACENE see MEB250

3-METHOXY-7H-BENZ(de)ANTHRACEN-7-ONE see MEB750

3-METHOXYBENZANTHRONE see MEB750

2-METHOXYBENZENAMINE see AOV900

3-METHOXYBENZENAMINE see AOV890

4-METHOXYBENZENAMINE see AOW000

2-METHOXYBENZENAMINE HYDROCHLORIDE see AOX250

METHOXYBENZENE see AOX750

4-METHOXYBENZENEACETIC ACID see MFE250

p-METHOXYBENZENEACETONITRILE see MFF000

4-METHOXYBENZENEAMINE see AOW000

2-METHOXYBENZENEAMINE HYDROCHLORIDE see AOX250

2-METHOXYBENZENECARBOXALDEHYDE see AOT525

2-METHOXY-1,4-BENZENEDIAMINE see MEB800

4-METHOXY-1,3-BENZENEDIAMINE see DBO000

4-METHOXY-1,2-BENZENEDIAMINE (9CI) see MFG000

2-METHOXY-1,4-BENZENEDIAMINE SULFATE see MEB820

4-METHOXY-1,3-BENZENEDIAMINE SULFATE see DBO400

4-METHOXY-1,3-BENZENEDIAMINE SULFATE (1:1) see DBO400

4-METHOXY-1,3-BENZENEDIAMINE SULFATE (1:1) HYDRATE see MEB900

4-METHOXY-1,3-BENZENEDIAMINE SULPHATE see DBO400

3-METHOXY-1,2-BENZENEDIOL see PPQ550

4-METHOXYBENZENEMETHANOL see MED500

2-(p-METHOXYBENZENESULFONAMIDO)-5-ISOBUTYL-1,3,4-THIADIAZOLE see IIY100

2-METHOXYBENZENETHIOL see MFQ300

o-METHOXYBENZENETHIOL see MFQ300

2-METHOXY-4H-1,2,3-BENZODIOXAPHOSPHORINE-2-SULFIDE see MEC250

1-(5-METHOXY-2-BENZOFURANYL)ETHANONE see MEC330

1-(6-METHOXY-2-BENZOFURANYL)ETHANONE see MEC300

1-(7-METHOXY-2-BENZOFURANYL)ETHANONE see MEC320

5-METHOXY-2-BENZOFURANYL METHYL KETONE see MEC330

6-METHOXY-2-BENZOFURANYL METHYL KETONE see MEC300

7-METHOXY-2-BENZOFURANYL METHYL KETONE see MEC320

o-METHOXYBENZOHYDRAZIDE see AOV250

2-METHOXYBENZOIC ACID see MPI000

3-METHOXYBENZOIC ACID see AOU600

4-METHOXYBENZOIC ACID see AOU600

m-METHOXYBENZOIC ACID see AOU500

o-METHOXYBENZOIC ACID see MPI000

p-METHOXYBENZOIC ACID see AOU600

2-METHOXYBENZOIC ACID HYDRAZIDE see AOV250

4-METHOXYBENZOIC ACID HYDRAZIDE see AOV500

o-METHOXYBENZOIC ACID HYDRAZIDE see AOV250

p-METHOXYBENZOIC ACID HYDRAZIDE see AOV500

o-METHOXYBENZOIC ACID METHYL ESTER see MLH800

p-METHOXYBENZOIC ACID 2-PHENYLHYDRAZIDE see MEC332

p-METHOXYBENZOIC HYDRAZIDE see AOV500

4-METHOXYBENZOPHENONE see MEC333

p-METHOXYBENZOPHENONE see MEC333

7-METHOXY-2H-1-BENZOPYRAN-2-ONE see MEK300

1-(5-METHOXY-2H-1-BENZOPYRAN-3-YL)ETHANONE see MEC340

5-METHOXY-2H-1-BENZOPYRAN-3-YL METHYL KETONE see MEC340

6-METHOXYBENZO(a)PYRENE see MEC500

6-METHOXYBENZOXAZOLINONE see MEC550

6-METHOXY-2(3H)-BENZOXAZOLONE see MEC550

METHOXYBENZOYL CHLORIDE see AOY250

2-(4'-METHOXYBENZOYL)FLUORENE see MEC600

2-METHOXYBENZOYL HYDRAZIDE see AOV250

4-METHOXYBENZOYL HYDRAZIDE see AOV500

o-METHOXYBENZOYLHYDRAZIDE see AOV250

2-METHOXYBENZOYLHYDRAZINE see AOV250

4-METHOXYBENZOYLHYDRAZINE see AOV500

(p-METHOXYBENZOYL)HYDRAZINE see AOV500

N-(4-METHOXY)BENZOYLOXYPIPERIDINE see MED000

3-(4-METHOXYBENZOYL)PYRIDINE see MED100

4-(4-METHOXYBENZOYL)PYRIDINE see MFH930

8-METHOXY-3,4-BENZPYRENE see MED250

p-METHOXYBENZYL ACETATE see AOY400

4-METHOXYBENZYLACETONE see MFF580

4-METHOXYBENZYL ALCOHOL see MED500

p-METHOXYBENZYL ALCOHOL see MED500

p-METHOXYBENZYL BUTYRATE see MED750

p-METHOXYBENZYL CHLORIDE see MEE000

p-METHOXYBENZYL CYANIDE see MFF000

N-(p-METHOXYBENZYL)-N',N'-DIMETHYL-N-2-PYRIDYLETHYLENEDIAMINE see WAK000

N-p-METHOXYBENZYL-N',N'-DIMETHYL-N-α-PYRIDYLETHYLENEDIAMINE see WAK000

N-p-METHOXYBENZYL-N'-N'-DIMETHYL-N-α-PYRIDYLETHYLENEDIAMINE MALEATE see DBM800

N-p-METHOXYBENZYL-N',N'-DIMETHYL-N-2-PYRIMIDINYLETHYLENE DIAMINE HYDROCHLORIDE see RDU000

2-METHOXYBENZYL ETHYL ETHER see EMF600

o-METHOXYBENZYL ETHYL ETHER see EMF600

p-METHOXYBENZYL FORMATE see MFE250

4-METHOXYBENZYLIDENEACETONE see MLI400

p-METHOXYBENZYLIDENEACETONE see MLI400

4-METHOXYBENZYLIDENE-4'-N-BUTYLANILINE see MEE100

3-(p-METHOXYBENZYL)-4-(p-METHOXYPHENYL)HEXANE see MEE150

3-(p-METHOXYBENZYL)-4-(p-METHOXYPHENYL)-3-HEXENE see CNH525

4-METHOXYBENZYL METHYL KETONE see AOV875

p-METHOXYBENZYL METHYL KETONE see AOV875

p-METHOXYBENZYL PHENYLACETATE see APE000

p-METHOXYBENZYL PHENYL KETONE see MEC333

2-METHOXY BIPHENYL see PEG000

4-METHOXYBIPHENYL see PEG250

p-METHOXYBIPHENYL see PEG250

3-METHOXYBIPHENYLAMINE see MEA000

1-((4-METHOXY(1,1'-BIPHENYL)-3-YL)METHYL)PYRROLIDINE see MEF400

2-METHOXY-4,6-BIS(ETHYLAMINO)-s-TRIAZINE see BJP250

2-METHOXY-4,6-BIS(ISOPROPYLAMINO)-s-TRIAZINE see MFL250

2-METHOXY-4,6-BIS(ISOPROPYLAMINO)-1,3,5-TRIAZINE see MFL250

4-METHOXY-α-(BIS(4-METHOXYPHENYL)METHYLENE)-N,N-DIMETHYLBENZENEETHANAMINE HCl see TNJ750

exo-2-METHOXYBORNANE see IHX500

2-METHOXYBROMOBENZENE see BMT400

4-METHOXYBROMOBENZENE see AOY450

o-METHOXYBROMOBENZENE see BMT400

p-METHOXYBROMOBENZENE see AOY450

1-METHOXYBUTANE see BRU780

α-METHOXYBUTANE see BRU780

3-METHOXY BUTANOIC ACID see MEF500

4-((1-METHOXYBUTOXY)(5-VINYL-2-QUINUCLIDINYL)METHYL)-6-QUINOLINOL DIHYDROCHLORIDE see EQQ500

3-METHOXYBUTYL ACETATE see MHV750

2-METHOXY-4-sec-BUTYLAMINO-6-AETHYLAMINO-s-TRIAZIN (GERMAN) see BQC250

2-METHOXY-4-tert-BUTYLAMINO-6-AETHYLAMINO-s-TRIAZIN (GERMAN) see BQC500

3-METHOXYBUTYLESTER KYSELINY OCTOVE see MHV750

4-METHOXY-2-tert-BUTYLPHENOL see BRN000

3-METHOXY BUTYRALDEHYDE see MEF750

3-METHOXYBUTYRIC ACID see MEF500

METHOXYCAFFEINE see MEF800

8-METHOXYCAFFEINE see MEF800

4'-METHOXYCARBONYL-N-ACETOXY-N-METHYL-4-AMINOAZOBENZENE see MEG000

2-(METHOXY-CARBONYLAMINO)-BENZIMIDAZOL see MHC750

2-(METHOXYCARBONYLAMINO)-BENZIMIDAZOLE see MHC750

3-METHOXYCARBONYLAMINOPHENYL N-3'-METHYLPHENYLCARBAMATE see MEG250

2-(METHOXYCARBONYL)ANILINE see APJ250

1-(METHOXYCARBONYL)AZIRIDINE see MGR900

4'-METHOXYCARBONYL-N-BENZOYLOXY-N-METHYL-4-AMINOAZOBENZENE see MEG500

3-METHOXYCARBONYL-6-N-BUTYL-7-BENZYLOXY-4-OXOQUINOLINE see NCN600

METHOXYCARBONYL CHLORIDE see MIG000

5-METHOXYCARBONYL-2,6-DIMETHYL-4-(2-NITROPHENYL)-3-PYRIDINECARBOXYLIC ACID see MRJ200

N-(2-METHOXYCARBONYLETHYL)-N-(1-ACETOXYBUTYL)NITROSAMINE see MEG750

N-(2-METHOXYCARBONYLETHYL)-N-(ACETOXYMETHYL)NITROSAMINE see MEH000

METHOXYCARBONYLETHYLENE see MGA500

N-METHOXYCARBONYLETHYL-N-NITROSOUREA see MEH100

4'-METHOXYCARBONYL-N-HYDROXY-N-METHYL-4-AMINOAZOBENZENE see MEH250

N-(METHOXYCARBONYLMETHYL)-N-(1-ACETOXYBUTYL)NITROSAMINE see MEH500

N-(METHOXYCARBONYLMETHYL)-N-(ACETOXYMETHYL)NITROSAMINE see MEH750

N-METHOXYCARBONYLMETHYL-N-NITROSOUREA see MEH760

3-METHOXYCARBONYL-N-(3'-METHYLPHENYL)-CARBAMAT see MEG250

(2-METHOXYCARBONYL-1-METHYL-VINYL)-DIMETHYL-FOSFAAT (DUTCH) see MQR750

(2-METHOXYCARBONYL-1-METHYL-VINYL)-DIMETHYL-PHOSPHAT (GERMAN) see MQR750

2-METHOXYCARBONYL-1-METHYLVINYL DIMETHYLPHOSPHATE see MQR750

1-METHOXYCARBONYL-1-PROPEN-2-YL DIMETHYL PHOSPHATE see MQR750

3-METHOXYCARBONYL PROPEN-2-YL TRIFLUOROMETHANE SULFONATE see MEH775

N-(3-METHOXYCARBONYLPROPYL)-N-(1-ACETOXYBUTYL)NITROSAMINE see MEI000

N-(3-METHOXYCARBONYLPROPYL)-N-(ACETOXYMETHYL)NITROSAMINE see MEI250

N-METHOXYCARBONYLPROPYL-N-NITROSOUREA see MEI300

(1-METHOXYCARBOXYPROPEN-2-YL)PHOSPHORIC ACID, DIMETHYL ESTER see MQR750

3-METHOXYCATECHOL see PPQ550

6-METHOXYCATECHOL see PPQ550

METHOXYCHLOR see MEI450

p,p'-METHOXYCHLOR see MEI450

METHOXYCHLOR mixed with DIAZINON see MEI500

2-METHOXY-6-CHLORO-9-(4-BIS(2-CHLOROETHYL)AMINO-1-METHYLBUTYLAMINO)ACRIDINE DIHYDROCHLORIDE see QDS000

2-METHOXY-6-CHLORO-9-(3-(2-CHLOROETHYL)AMINOPROPYLAMINO) ACRIDINE DIHYDROCHLORIDE see MEJ250

2-METHOXY-6-CHLORO-9-(3-(2-CHLOROETHYL) MERCAPTO PROPYLAMINO) ACRIDINE HYDROCHLORIDE see CFA250

2-METHOXY-6-CHLORO-9-(4-DIETHYLAMINO-1-METHYLBUTYLAMINO)ACRIDINEDIHYDROCHLORIDE see CFU750

2-METHOXY-6-CHLORO-9-DIETHYLAMINOPENTYLAMINOACRIDINE see ARQ250

2-METHOXY-6-CHLORO-9-(3-(ETHYL-2-CHLOROETHYL)AMINOPROPYLAMINO)ACRIDINE DIHYDROCHLORIDE see ADJ875

2-METHOXY-6-CHLORO-9-(3-(ETHYL-2-CHLOROETHYL)AMINOPROPYLAMINO)ACRIDINE DIHYDROCHLORIDE see QDS000

2-METHOXY-5-CHLOROPROCAINAMIDE see AJH000

5-METHOXYCHRYSENE see MEJ500

6'-METHOXYCINCHONAN-9-OL see QFS000

(8-α,9R)-6'-METHOXYCINCHONAN-9-OL see QHJ000

6'-METHOXYCINCHONAN-9-OL DIHYDROCHLORIDE see QIJ000

(9S)-6'-METHOXYCINCHONAN-9-OL SULFATE (1:1) (SALT) see QFS100

(9S)-6'-METHOXYCINCHONAN-9-OL SULFATE (2:1) (SALT) see QHA000

6-METHOXYCINCHONINE see QHJ000

o-METHOXY CINNAMALDEHYDE see MEJ750

cis-2-METHOXYCINNAMIC ACID see MEJ775

cis-o-METHOXYCINNAMIC ACID see MEJ775

o-METHOXYCINNAMIC ALDEHYDE see MEJ750

5-(p-METHOXY-CINNAMOYL)-4,7-DIMETHOXY-6-DIMETHYLAMINOETHOXYBENZOFURAN see MBX000

METHOXYCITRONELLAL METHYL ETHER see DSM800

7-METHOXYCOUMARIN see MEK300

2-METHOXY-p-CRESOL see MEK325

3-METHOXY-17-β-CYANETHOXYESTRA-1,3,5(10)-TRIEN see MBW775

3-METHOXY-17-β-CYANETHOXYOESTRA-1,3,5(10)-TRIEN see MBW775

β-(2-METHOXY-5-CYCLOHEXYLBENZOYL)PROPIONIC ACID SODIUM SALT see MEK350

METHOXYCYCLOPROPANE see CQE750

1-METHOXYCYCLOPROPANE see CQE750

3-METHOXY-DBA see MEK750

9-METHOXY-DBA see MEL000

METHOXY-DDT see MEI450

1-METHOXYDECANE see MIW075

10-METHOXYDESERPIDINE see MEK700

10-METHOXY-11-DESMETHOXYRESERPINE see MEK700

5-METHOXYDIBENZ(a,h)ANTHRACENE see MEK750

7-METHOXYDIBENZ(a,h)ANTHRACENE see MEL000

3-METHOXY-1,2:5,6-DIBENZANTHRACENE see MEK750

9-METHOXY-1,2,5,6-DIBENZANTHRACENE see MEL000

N-METHOXY-3-(3,5-DI-tert-BUTYL-4-HYDROXYBENZYLIDENE)-2-PYRROLIDONE see MEL100

2-METHOXY-3,6-DICHLOROBENZOIC ACID see MEL500

(3-METHOXY-4-((N,N-DIETHYLCARBAMIDO)METHOXY)PHENYL)ACETIC ACID n-PROPYL ESTER see PMM000

2-METHOXY-4-(o-(o,o-DIETHYLPHOSPHOROTHIOYL))BENZAL DOXIMINO-N-METHYLCARBAMATE see MEL525

METHOXYDIGLYCOL see DJG000

11-METHOXY-15,16-DIHDYROCYCLOPENTA(a)PHENANTHREN-17-ONE see DLR200

5-METHOXYDIHYDROSTERIGMATOCYSTIN see MEL550

5-METHOXY-2-(DIMETHOXYPHOSPHINYLTHIOMETHYL)PYRONE-4 see EAS000

3-METHOXY-4-DIMETHYLAMINOAZOBENZENE see DTK800

2'-METHOXY-4-DIMETHYLAMINOAZOBENZENE see DSN000

3'-METHOXY-4-DIMETHYLAMINOAZOBENZENE see DSN200

4'-METHOXY-4-DIMETHYLAMINOAZOBENZENE see DSN400

2-METHOXY-10-(3'-DIMETHYLAMINOPROPYL)PHENOTHIAZINE see MFK500

2-METHOXY-7,12-DIMETHYLBENZ(a)ANTHRACENE see MEL570

3-METHOXY-7,12-DIMETHYLBENZ(a)ANTHRACENE see MEL580

4-METHOXY-7,12-DIMETHYLBENZ(a)ANTHRACENE see MEL600

10-METHOXY-1,6-DIMETHYLERGOLINE-8-METHANOL-5-BROMO-3-PYRIDINECARBOXYLATE (ester) see NDM000

10-METHOXY-1,6-DIMETHYL-ERGOLIN-8-β-METHANOL-(5-BROMNICOTINAT) (GERMAN) see NDM000

1-(8-METHOXY-4,8-DIMETHYLNONYL)-4-(1-METHYLETHYL)BENZENE see MEL700

o-METHOXY-N-α-DIMETHYLPHENETHYLAMINEHYDROCHLORIDE see OJY000

7-METHOXY-α,10-DIMETHYLPHENOTHIAZINE-2-ACETIC ACID see POE100

trans-1-(2-(p-(7-METHOXY-2,2-DIMETHYL-3-PHENYL-4-CHROMANYL)PHENOXY)ETHYL)PYRROLIDINE see CCW725

9-METHOXY-5,11-DIMETHYL-6H-PYRIDO(4,3-b)CARBAZOLE see MEL775

1-METHOXY-2,4-DINITROBENZENE see DUP800

9-METHOXY-1,3-DIOXOLO(4,5-G)FURO(2,3-B)QUINOLINE see MEL725

METHOXYDIURON see DGD600

2-METHOXY-DMBA see MEL570

3-METHOXY-DMBA see MEL580

4-METHOXY-DMBA see MEL600

METHOXYDON(E) see MFD500

9-METHOXYELLIPTICIN see MEL775

METHOXYELLIPTICINE see MEL775

9-METHOXY-ELLIPTICINE see MEL775

9-METHOXYELLIPTICINE see MEL775

9-METHOXYELLIPTICINE LACTATE see MEL780

METHOXY-9-ELLIPTICINE LACTATE see MEL780

METHOXYELLIPTIONE see MEL775

2-METHOXYESTRADIOL see MEL785

2-METHOXYESTRA-1,3,5(10)-TRIENE-3,17-β-DIOL see MEL785

3-METHOXY-ESTRA-1,3,5(10)-TRIENE-17-β-OL see MFB775

(17-β)-3-METHOXY-ESTRA-1,3,5(10)-TRIEN-17-OL (9CI) see MFB775

3-METHOXYESTRA-1,3,5(10)-TRIEN-17-β-OL see MFB775

3-((3-METHOXYESTRA-1,3,5(10)-TRIEN-17-β-YL)OXY)PROPIONITRILE see MBW775

METHOXYETHANAL see MDW250

METHOXYETHANE see EMT000

2-METHOXYETHANOL (ACGIH) see EJH500

2-METHOXYETHANOL, ACETATE see EJJ500

2-METHOXYETHANOL, ACRYLATE see MEM250

2-METHOXYETHANOL, ACRYLATE see MIF750

2-METHOXYETHANOL PHOSPHATE (3:1) see TLA600

2-METHOXYETHANOYL CHLORIDE see MDW780

METHOXYETHENE see MQL750

METHOXY ETHER of PROPYLENE GLYCOL see PNL250

3-METHOXY-17-α-ETHINYLESTRADIOL see MKB750

3-METHOXY-17 α ETHINYLOESTRADIOL see MKB750

2-METHOXYETHOXY ACRYLATE see MEL790

2-(2-METHOXYETHOXY)ETHANOL see DJG000

2-(2-METHOXYETHOXY)ETHANOL ACETATE see MIE750

2-(2-(2-METHOXYETHOXY)ETHOXY)ETHANOL see TJQ750

2-(2-(2-METHOXYETHOXY)ETHOXY)ETHANOL ACETATE see AAV500

2-(2-(2-METHOXYETHOXY)ETHOXY)ETHYL ACETATE see AAV500

2-(2-(2-METHOXYETHOXY)ETHOXY)ETHYLESTER KYSELINY OCTOVE see AAV500

2-(2-METHOXYETHOXY)ETHYL ACETATE see MIE750

2-(2-METHOXYETHOXY)ETHYLESTER KYSELINY OCTOVE see MIE750

α-((2-METHOXYETHOXY)METHYL)-ω-HYDROXYPOLY(OXY-1,2-ETHANEDIYL)COCO ALKYL ETHERS see MEL800

N-(5-(2-METHOXYETHOXY)-2-PYRIMIDINYL)BENZENESULFONAMIDE SODIUM SALT see GHK200

2-METHOXY-ETHYL ACETAAT (DUTCH) see EJJ500

2-METHOXYETHYL ACETATE (ACGIH) see EJJ500

2-METHOXYETHYL-12-ACETOXY-9-OCTADECENOATE see MIF500

2-METHOXYETHYL ACETYL RICINOLEATE see MIF500

METHOXYETHYL ACRYLATE see MEM250

2-METHOXYETHYL ACRYLATE see MEM250

2-METHOXYETHYLAMINE see MEM500

2-METHOXY-4-ETHYLAMINO-6-ISOPROPYLAMINO-s-TRIAZINE see EGD000

2-(2-METHOXYETHYLAMINO)-2'-METHYL-PROPIONANILIDE see MEN000

S-(2-((2-METHOXYETHYL)AMINO-2-OXOETHYL) O,O-DIMETHYL) PHOSPHORODITHIOATE see AHO750

2-(2-METHOXYETHYLAMINO)PROPIONANILIDE see MEM750

2-(2-METHOXYETHYLAMINO)-o-PROPIONOTOLUIDIDE see MEN000

2-(2-METHOXYETHYLAMINO)-2',6'-PROPIONOXYLIDIDE see MEN250

2-(2-METHOXYETHYLAMINO)-2',6'-PROPIONOXYLIDIDE PERCHLORATE see MEN500

3-METHOXY-10-ETHYL-1,2-BENZANTHRACENE see MEN750

2-METHOXYETHYL-BIS(2-CHLOROETHYL)AMINE HYDROCHLORIDE see MEN775

METHOXYETHYL CARBAMATE see MEO000

S-(N-2-METHOXYETHYLCARBAMOYLMETHYL) DIMETHYL PHOPHOROTHIOLOTHIONATE see AHO750

4-METHOXY-9-(3-(ETHYL-2-CHLOROETHYL) see EHJ500

2-METHOXYETHYLE, ACETATE de (FRENCH) see EJJ500

2-METHOXYETHYL ESTER CARBAMIC ACID see MEO000

1-METHOXY ETHYL ETHYLNITROSAMINE see MEO500

METHOXYETHYL MERCURIC ACETATE see MEO750

METHOXYETHYL MERCURIC ACETATE see MEO750

METHOXYETHYL MERCURIC CHLORIDE see MEP250

2-METHOXYETHYLMERCURIC CHLORIDE see MEP250

(β-METHOXYETHYL)MERCURIC CHLORIDE see MEP250

METHOXYETHYL MERCURIC SILICATE see MEP000

METHOXYETHYLMERCURY ACETATE see MEO750

METHOXYETHYLMERCURY CHLORIDE see MEP250

2-METHOXYETHYLMERCURY CHLORIDE see MEP250

β-METHOXYETHYLMERCURY CHLORIDE see MEP250

2-METHOXYETHYLMERKURIACETAT see MEO750

2-METHOXYETHYLMERKURICHLORID see MEP250

1-METHOXY ETHYL METHYLNITROSAMINE see MEP500

2-METHOXYETHYL 2-(m-NITROBENZYLIDENE)ACETOACETATE see MEP600

2-METHOXYETHYL 2-((3-NITROPHENYL)METHYLENE)ACETOACETATE see MEP600

1-(2-METHOXYETHYL)-1-NITROSOUREA see NKO900

N-(2-METHOXYETHYL)-N-NITROSOUREA see NKO900

METHOXYETHYL OLEATE see MEP750

METHOXYETHYL OLEATE see MEP750

(±)-1-(4-(2-METHOXYETHYL)PHENOXY)-3-((1-METHYLTHYL)AMINO)-2-PROPANOL see MQR144

METHOXYETHYL PHTHALATE see MEQ000

2-METHOXYETHYL PHTHALATE see DOF400

2-METHOXYETHYL VINYL ETHER see VPZ000

3-METHOXY-17-α-ETHYNOESTRADIOL see MKB750

3-METHOXYETHYNYLESTRADIOL see MKB750

3-METHOXY-17-α-ETHYNYLESTRADIOL see MKB750

3-METHOXY-17-α-ETHYNYL-1,3,5(10)-ESTRATRIEN-17-β-OL see MKB750

3-METHOXYETHYNYLOESTRADIOL see MKB750

3-METHOXY-17-ETHYNYLOESTRADIOL-17-β see MKB750

3-METHOXY-17-α-ETHYNYL-1,3,5(10)-OESTRATRIEN-17-β-OL see MKB750

1-METHOXY-2-FAA see MER000

7-METHOXY-2-FAA see MER250

METHOXYFENOZIDE see MEQ100

1-p-METHOXYFENYL-3,3-DIMETHYLTRIAZEN (CZECH) see DSN600

METHOXYFLUORAN see DFA400

METHOXYFLUORANE see DFA400

1-METHOXY-2-FLUORENAMINE HYDROCHLORIDE see MEQ750

1-METHOXYFLUOREN-2-AMINE HYDROCHLORIDE see MEQ750

7-METHOXY-N-2-FLUORENYLACETAMIDE see MER250

N-(1-METHOXYFLUOREN-2-YL)ACETAMIDE see MER000

N-(1-METHOXY-2-FLUORENYL)ACETAMIDE see MER000

N-(7-METHOXY-2-FLUORENYL)ACETAMIDE see MER250

N-(7-METHOXYFLUOREN-2-YL)ACETAMIDE see MER250

METHOXYFLURANE see DFA400

8-METHOXY-(FURANO-3'.2':6.7-COUMARIN) see XDJ000

9-METHOXY-7H-FURO(3,2-g)BENZOPYRAN-7-ONE see XDJ000

4-METHOXY-7H-FURO(3,2-g)(1)BENZOPYRAN-7-ONE see MFN275

8-METHOXY-2',3',6,7-FUROCOUMARIN see XDJ000

8-METHOXY-4',5',6,7-FUROCOUMARIN see XDJ000

3-METHOXYGONA-1,3,5(10)-TRIEN-17-ONE see NNP100

6-METHOXYHARMAN see HAI500

METHOXYHYDRASTINE see NOA000

METHOXYHYDRASTINE HYDROCHLORIDE see NOA500

p-METHOXYHYDRATROPALDEHYDE see MLJ050

3-METHOXY-4-HYDROXYBENZALDEHYDE see VFK000

3-METHOXY-4-HYDROXYBENZOIC ACID see VFF000

3-METHOXY-4-HYDROXYBENZOIC ACID DIETHYLAMIDE see DKE200

3-METHOXY-4-HYDROXYBENZOIC ACID, ETHYL ESTER see EQE500

4-METHOXY-2-HYDROXYBENZOPHENONE see MES000

3-METHOXY-4-HYDROXY-BENZYLACETONE see VFP100

β-METHOXY-β'-HYDROXYDIETHYL ETHER see DJG000

METHOXYHYDROXYETHANE see EJH500

N-(2-METHOXY-3-HYDROXYMERCURIPROPYL)BARBITAL see MES250

METHOXYHYDROXYMERCURIPROPYLSUCCINYLUREA see MFC000

l-3-METHOXY-ω-(1-HYDROXY-1-PHENYLISOPROPYLAMINO)PROPIOPHENONE HYDROCHLORIDE see OQU000

2-METHOXY-10-(3-(4-HYDROXYPIPERIDINO)-2-METHYLPROPYL)PHENOTHIAZINE see LEO000

3-METHOXY-10-(3-(4-HYDROXYPIPERIDYL)-2-METHYLPROPYL)PHENOTHIAZINE see LEO000

8-METHOXY-4-HYDROXYQUINOLINE-2-CARBOXYLIC ACID see HLT500

3-METHOXY-4-HYDROXYTOLUENE see MEK325

1-METHOXY IMIDAZOLE-N-OXIDE see MES300

3-METHOXY-5,4'-IMINOBIS(1-BENZAMIDO-ANTHRAQUINONE) see MES500

9-(1-(METHOXYIMINO)ETHYL)-6-OXO-N,N,2,2,5-PENTAMETHYL-7-OXA-3,4-DITHIA-5,8-DIAZADEC-8-EN-10-AMIDE see MES550

(E)-2-METHOXYIMINO-N-METHYL-2-(2-PHENOXYPHENYL)ACETAMIDE see MQQ800

METHOXYINDOLEACETIC ACID see MES850

5-METHOXYINDOLEACETIC ACID see MES850

5-METHOXYINDOLE-3-ACETIC ACID see MES850

5-METHOXYINDOLE-3-ETHANOL see MFT300

5-METHOXY-1H-INDOLE-3-ETHANOL (9CI) see MFT300

N-(2-(5-METHOXY-4-INDOLYL)ETHYL)ACETAMIDE see MES870

5-METHOXY-1H-INDOL-3-YL 4-PIPERIDYLMETHYL KETONE MONOHYDROCHLORIDE see MES900

(5-METHOXY-3-INDOLYL)-(4-PIPERIDYL-METHYL)-KETON HYDROCHLORID see MES900

(5'-METHOXY-3'-INDOLYL)-β-PYRIDYLMETHANOL see MFN700

2-METHOXY-4-ISOPROPYLAMINO-6-ETHYLAMINO-s-TRIAZINE see EGD000

1-METHOXYISOPROPYL BROMOACETATE see MES930

1-METHOXYISOPROPYL CHLOROACETATE see MES940

METHOXYLAMINE HYDROCHLORIDE see MNG500

METHOXYLENE see DPJ400

METHOXYMAGNESIUM-METHYLCARBONATE see MES980

1-(2-METHOXY-4-METHOXYCARBONYL-1-PHENOXY)-3-ISOPROPYLAMINO-2-PROPANOL see VFP200

(METHOXYMETHOXY)CYCLODODECANE see FMV200

2-(METHOXYMETHOXY)ETHANOL see MET000

4-METHOXY-2-(5-METHOXY-3-METHYL-1H-PYRAZOL-1-YL)-6-METHYLPYRIMIDINE see MCH550

METHOXY-2(METHOXY-2-PHENYL)-4 OXADIAZOLINE-1,3,4 ONE-5 see OKU200

2-METHOXY-4-(o-METHOXYPHENYL)-Δ²-1,3,4-OXADIAZOLIN-5-ONE see OKU200

4-METHOXY-3-METHYLACETOPHENONE see MET100

2-METHOXY-3-METHYL-4-ACETYLTETRAHYDROFURAN see MHR100

METHOXYMETHYLACRYLAMIDE see MEX300

N-(METHOXYMETHYL)ACRYLAMIDE see MEX300

METHOXYMETHYL-AETHYLNITROSAMINE (GERMAN) see MEV750

3-METHOXYMETHYLAMINOAZOBENZENE see MNS000

(E)-(3-(METHOXYMETHYLAMINO)-1-METHYL-3-OXO-1-PROPENYL)DIMETHYL PHOSPHATE see DOL800

2-METHOXY-5-METHYLANILINE see MGO750

4-METHOXY-2-METHYLANILINE see MGO500

3-METHOXY-7-METHYLBENZ(c)ACRIDINE see MET875

5-METHOXY-7-METHYL-BENZ(a)ANTHRACENE see MEU000

8-METHOXY-7-METHYL-BENZ(a)ANTHRACENE see MEU250

7-METHOXY-12-METHYLBENZ(a)ANTHRACENE see MEU000

3-METHOXY-10-METHYL-1,2-BENZANTHRACENE see MEU000

5-METHOXY-10-METHYL-1,2-BENZANTHRACENE see MEU250

4-METHOXY-2-METHYLBENZENAMINE see MGO500

2-METHOXY-5-METHYL-BENZENAMINE (9CI) see MGO750

4-METHOXY-α-METHYLBENZENEPROPANAL see MLJ050

o-METHOXY METHYL BENZOATE see MLH800

6-METHOXY-2-METHYL-2H-1-BENZOPYRAN-3-CARBONITRILE see MEU275

2-METHOXY-2-METHYLBUTANE see PBX400

S-(((METHOXYMETHYL-CARBAMOYL)METHYL)) O,O-DIMETHYLPHOSPHORODITHIOATE see FNE000

S-(N-METHOXYMETHYLCARBAMOYLMETHYL) DIMETHYL PHOSPHOROTHIOLOTHIONONATE see FNE000

1-METHOXY-1-METHYLCYCLODODECANE see MEU300

6-(4-METHOXY-3-(1-METHYLCYCLOHEXYL)PHENYL)-2-NAPHTHALENECARBOXYLIC ACID see MEU400

3-METHOXY-17-METHYL-15H-CYCLOPENTAPHENANTHRENE see MEU750

3-METHOXY-17-METHYL 15H-CYCLOPENTA(a)PHENANTHRENE see MEU750

11-METHOXY-17-METHYL-15H-CYCLOPENTA(a)PHENANTHRENE see MEV000

1-METHOXY-1-METHYL-3-(3,4-DICHLOROPHENYL)UREA see DGD600

6-METHOXY-11-METHYL-15,16-DIHYDRO-17H-CYCLOPENTA(a) PHENANTHREN-17-ONE see MEV250

19-METHOXY-1,2-(METHYLENEDIOXY)-6a-α-APORPHIN-11-OL see HLT000

(2-(2-METHOXY METHYL ETHOXY)METHYL ETHOXY)PROPANOL see MEV500

2-(((((METHOXY(1-METHYLETHOXY)PHOSPHINOTHIOYL)THIO)METHYL)THIO)ETHYLETHYLCARBAMATE see MEV600

METHOXYMETHYL ETHYL NITROSAMINE see MEV750

p-METHOXY-α-METHYLHYDROCINNAMALDEHYDE see MLJ050

4-METHOXYMETHYL-5-HYDROXY-6-METHYL-3-PYRIDINEMETHANOL HYDROCHLORIDE see MEY000

2-METHOXY-10-(2-METHYL-3-(4-HYDROXYPIPERIDINO)PROPYL)PHENOTHIAZINE see LEO000

3-METHOXY-6-METHYLINDOLO(3,2-C)QUINOLINE-1,4-DIONE see MEV800

3-METHOXY-6-METHYL-1H-INDOLO(3,2-C)QUINOLINE-1,4(11H)-DIONE see MEV800

7-METHOXYMETHYL-12-METHYLBENZ(a)ANTHRACENE see MEW000

METHOXYMETHYL 1-METHYL-2-INDOLYL KETONE see MDX300

METHOXYMETHYL 3-METHYL-2-INDOLYL KETONE see MDX310

METHOXYMETHYL METHYL KETONE see MDW300

METHOXYMETHYL METHYL KETONE see MFL100

METHOXYMETHYL-METHYLNITROSAMIN (GERMAN) see MEW250

METHOXYMETHYL METHYLNITROSAMINE see MEW250

8-METHOXY-1-METHYL-4-((p-((p-((1-METHYLPYRIDINIUM-4-YL)AMINO)PHENYL)CARBAMOYL)ANILINO)QUINOLINIUM, DIBROMIDE see MEW500

6-METHOXY-1-METHYL-4-((p-((p-((1-METHYLPYRIDINIUM-4-YL)AMINO)PHENYL)CARBAMOYL)ANILINO)QUINOLINIUM, DI-p-TOLUENESULFONATE see MEW750

4-METHOXY-5-METHYL-6-(7-METHYL-8-(TETRAHYDRO-3,4-DIHYDROXY-2,4,5-TRIMETHYL-2-FURANYL)-1,3,5,7-OCTATETRAENYL)-2H-PYRAN-2-ONE see CMS500

3-METHOXY-17-METHYLMORPHINAN HYDROBROMIDE see LFD200

d-3-METHOXY-N-METHYLMORPHINAN HYDROBROMIDE see DBE200

(±)-3-METHOXY-17-METHYLMORPHINAN HYDROBROMIDE see RAF300

3-METHOXY-17-METHYL-9-α,13-α,14-α-MORPHINAN HYDROBROMIDE see DBE200

(+)-6-METHOXY-α-METHYL-2-NAPHTHALENEACETIC ACID see MFA500

(S)-6-METHOXY-α-METHYL-2-NAPHTHALENEACETIC ACID see MFA500

l-(−)-6-METHOXY-α-METHYL-2-NAPHTHALENEACETIC ACID SODIUM SALT see NBO550

4-METHOXY-N-METHYLNAPHTHALIMIDE see MEW760

α-(METHOXYMETHYL)-2-NITROIMIDAZOLE-1-ETHANOL see NHH500

α-(METHOXYMETHYL)-2-NITRO-1H-IMIDAZOLE-1-ETHANOL see NHH500

α-(METHOXYMETHYL)-2-NITRO-1H-IMIDAZOLE-1-ETHANOL (9CI) see NHH500

7-METHOXY-1-METHYL-2-NITRONAPHTHO(2,1-b)FURAN see MEW775

N-METHOXYMETHYL-N-NITROSO-sec-BUTYLAMINE see BRU000

N-METHOXY-N-METHYLNONANAMIDE see EQS500

(METHOXYMETHYL)OXIRANE see GGW600

3-METHOXY-5-METHYL-4-OXO-2,5-HEXADIENOIC ACID see PAP750

3-((4-(5-(METHOXYMETHYL)-2-OXO-3-OXAZOLIDINYL)PHENOXY)METHYL)BENZONITRILE see MEW800

N-(7-METHOXY-3-METHYL-4-OXO-2-PHENYL-4H-CHROMEN-8-YL)METHYL-N,N-DIMETHYLAMINE see DNV000

5-METHOXY-2-METHYL-1-(1-OXO-3-PHENYL-2-PROPENYL)-1H-INDOLE-3-ACETIC ACID (9CI) see CMP950

4-METHOXY-4-METHYL-2-PENTANONE see MEX250

4-METHOXY-4-METHYLPENTAN-2-ONE (DOT) see MEX250

dl-p-METHOXY-α-METHYL-PHENETHYLAMINE HYDROCHLORIDE see MEA500

2-METHOXY-4-METHYLPHENOL see MEK325

N-(2-(2-METHOXY-5-METHYLPHENYL)ETHYL)PROPIONAMIDE see MEX260

N-(2-(2-METHOXY-5-METHYLPHENYL)ETHYL)TRIFLUOROACETAMIDE see MEX265

4-β-METHOXY-1-METHYL-4-α-PHENYL-3-α,5-α-PROPANOPIPERIDINE HYDROGEN CITRATE see ARX150

1-(4-METHOXY-2-METHYL-5-PHENYL-1H-PYRROL-3-YL)ETHANONE see MEX275

9-METHOXY-9-((4-METHYL-1-PIPERAZINYL)METHYL)ACRIDAN see MEX285

2-METHOXY-2-METHYLPROPANE see MHV859

N-(METHOXYMETHYL)-2-PROPENAMIDE see MEX300

2-METHOXY-3(5)-METHYLPYRAZINE see MEX350

2-(3-METHOXY-5-METHYLPYRAZOL-2-YL)-4-METHOXY-6-METHYLPYRIMIDINE see MCH550

7-METHOXY-1-METHYL-9H-PYRIDO(3,4-b)INDOLE see HAI500

3-METHOXY-6-METHYL-11H-PYRIDO(3',4':4,5)PYRROLO(3,2-C)QUINOLINE-1,4-DIONE see MEX400

3-METHOXY-6-METHYL-1H-PYRIDO(3',4':4,5)PYRROLO(3,2-C)QUINOLINE-1,4(11H)-DIONE see MEX400

4-METHOXYMETHYLPYRIDOXINE HYDROCHLORIDE see MEY000

4-METHOXYMETHYLPYRIDOXOL HYDROCHLORIDE see MEY000

1-(4-METHOXY-6-METHYL-2-PYRIMIDINYL)-3-METHYL-5-METHOXYPYRAZOLE see MCH550

p-METHOXY-β-METHYLSTYRENE see PMQ750

6-METHOXY-1-METHYL-1,2,3,4-TETRAHYDRO-β-CARBOLINE see MLJ100

3-METHOXY-6-METHYL-7,8,9,10-TETRAHYDRO-11H-INDOLO(3,2-C)QUINOLINE-1,4-DIONE see MEY100

N-(4-(METHOXYMETHYL)-1-(2-(2-THIENYL)ETHYL)-4-PIPERIDINYL)-N-PHENYLPROPIONAMIDE CITRATE see SNH150

N-(4-(METHOXYMETHYL)-1-(2-(2-THIENYL)ETHYL)-4-PIPERIDYL)PROPIONANILIDE CITRATE see SNH150

(METHOXY(METHYLTHIO)PHOSPHINOTHIOYL)BUTANEDIOIC ACID DIETHYL ESTER see IKX200

2-(METHOXY(METHYLTHIO)PHOSPHINYLIMINO)-3-ETHYL-5-METHYL-1,3-OXAZOLIDINE see MOA750

2-(METHOXY(METHYLTHIO)PHOSPHINYLIMINO)-3-METHYL-1,3-THIAZOLINE see MEY200

4-METHOXY-6-METHYL-1,3,5-TRIAZIN-2-AMINE see MGH800

6-METHOXY-3-METHYL-1,7,8-TRIHYDROXYANTHRAQUINONE see MEY750

3-METHOXY-4-MONOMETHYLAMINOAZOBENZENE see MNS000

METHOXYN see DBA800

METHOXYNAL see AAE500

2-METHOXY-1,4-NAPHTHALENEDIONE see MEY800

4-(6-METHOXY-2-NAPHTHALENYL)-2-BUTANONE see MFA300

3-(6-METHOXY-2-NAPHTHALENYL)-1,2-DIMETHYL (2R,3S)-REL-3-PYRROLIDINOL YDROCHLORIDE see MEY900

2-METHOXYNAPHTHOQUINONE see MEY800

2-METHOXY-1,4-NAPHTHOQUINONE see MEY800

3-(4-METHOXY-1-NAPHTHOYL)PROPIONIC ACID see MEZ300

β-(1-METHOXY-4-NAPHTHOYL)-PROPIONSAEURE (GERMAN) see MEZ300

1-METHOXY-2-NAPHTHYLAMINE see MFA000

1-METHOXY-2-NAPHTHYLAMINE HYDROCHLORIDE see MFA250

4-(6-METHOXY-2-NAPHTHYL)-2-BUTANONE see MFA300

N-(2-(2-METHOXYNAPHTHYL)ETHYL)BUTYRAMIDE see MFA325

N-(2-(2-METHOXY-1-NAPHTHYL)ETHYL)CYCLOPROPYLCARBOXAMIDE see MFA335

N-(2-(2-METHOXY-1-NAPHTHYL)ETHYL)PENTANAMIDE see MFA345

N-(2-(2-METHOXY-1-NAPHTHYL)-1-METHYLETHYL)ACETAMIDE see MFA355

N-(2-(2-METHOXY-1-NAPHTHYL)-1-METHYLETHYL)PROPIONAMIDE see MFA365

N-(2-(2-METHOXY-1-NAPHTHYL)-1-METHYLETHYL)TRIFLUOROACETAMIDE see MFA375

(+)-2-(METHOXY-2-NAPHTHYL)-PROPIONIC ACID see MFA500

(+)-2-(METHOXY-2-NAPHTHYL)-PROPIONIC ACID see MFA500

(+)-2-(METHOXY-2-NAPHTHYL)-PROPIONSAEURE see MFA500

d-2-(6'-METHOXY-2'-NAPHTHYL)-PROPIONSAEURE see MFA500

4-METHOXY-2-NITROANILIN (CZECH) see MFB000

2-METHOXY-4-NITROANILINE see NEQ000

2-METHOXY-5-NITROANILINE see NEQ500

4-METHOXY-2-NITROANILINE see MFB000

3-METHOXY-4-NITROAZOBENZENE see MFB250

2-METHOXY-4-NITROBENZENAMINE see NEQ000

2-METHOXY-5-NITROBENZENAMINE see NEQ500

2-METHOXYNITROBENZENE see NER000

4-METHOXYNITROBENZENE see NER500

p-METHOXYNITROBENZENE see NER500

1-METHOXY-2-NITROBENZENE see NER000

1-METHOXY-4-NITROBENZENE see NER500

2-METHOXY-5-NITRO-1,4-BENZENEDIAMINE see MFB300

5-METHOXY-2-NITROBENZOFURAN see MFB350

5-METHOXY-2-NITROBENZOFURAN see MFB350

1-METHOXY-2-(2-NITROETHENYL)BENZENE see NMR100

N-METHOXY-2-NITRO-1H-IMIDAZOLE-1-ACETAMIDE see MFB370

N-METHOXY-2-NITRO-1H-IMIDAZOLE-1-BUTANAMIDE see MFB380

3-METHOXY-2-(2-NITRO-1-IMIDAZOLYL)-1-PROPANOL see DBA700

7-METHOXY-2-NITRONAPHTHO(2,1-b)FURAN see MFB400

8-METHOXY-2-NITRONAPHTHO(2,1-b)FURAN see MFB410

8-METHOXY-6-NITROPHENANTHOL-(3,4-d)-1,3-DIOXOLE-5-CARBOXYLIC ACID see AQY250

8-METHOXY-6-NITROPHENANTHRO(3,4-d)-1,3-DIOXOLE-5-CARBOXYLIC ACID SODIUM SALT see AQY125

8-METHOXY-6-NITROPHENANTHRO(3,4-D)-1,3-DIOXOLE-5-CARBOXYLIC ACID SODIUM SALT see SEY050

(E)-1-METHOXY-3-(2-(4-NITROPHENYL)ETHENYL)BENZENE see MFB620

(E)-1-METHOXY-4-(2-(4-NITROPHENYL)ETHENYL)BENZENE see MFB600

2-(4-METHOXY-3-NITROPHENYL)-3-NITRO-4H-1-BENZOPYRAN-4-ONE see MFB430

(3-METHOXY-4-NITROPHENYL)PHENYLDIAZENE see MFB250

4-METHOXY-2-NITROPHENYLTHIOCYANATE see MFB500

N-METHOXY-N-NITROSOMETHYLAMINE see DSZ000

1-METHOXY-N-NITROSO-N-PROPYLPROPYLAMINE see DWU800

(E)-METHOXY-4'-NITROSTILBENE see MFB600

(E)-3-METHOXY-4'-NITROSTILBENE see MFB620

3-METHOXY-17-α-19-NORPREGNA-K,3,5(10)-TRIEN-20-YN-17-OL see MKB750

11-β-METHOXY-19-NOR-17-α-PREGNA-1,3,5(10)-TRIEN-20-YNE-3,17-DIOL see MRU600

3-METHOXY-19-NOR-17-α-PREGNA-1,3,5(10)-TRIEN-20-YNE-2,17-DIOL see HLQ050

(17-α)-3-METHOXY-19-NORPREGNA-1,3,5(10)-TRIEN-20-YN-17-OL see MKB750

3-METHOXY-19-NOR-17-α-PREGNA-1,3,5(10)-TRIEN-10-YN-17-OL see MKB750

d-threo-METHOXY-3-(1-OCTENYL-O,N,N-AZOXY)-2-BUTANOL see EAG000

3-METHOXYOESTRADIOL see MFB775

METHOXYOXIMERCURIPROPYLSUCCINYL UREA see MFC000

2-METHOXY-N-(2-OXO-1,3-OXAZOLIDINE-3-YL)-ACET-2,6-XYLIDINE see MFC100

2-METHOXY-2-((1-OXO-2-PROPENYL)AMINO)ACETIC ACID, METHYL ESTER see MGA400

S-5-METHOXY-4-OXOPYRAN-2-YLMETHYL DIMETHYL PHOSPHOROTHIOATE see EAS000

S-((5-METHOXY-2-OXO-1,3,4-THIADIAZOL-3(2H)-YL)METHYL)-O,O-DIMETHYL PHOSPHORODITHIOATE see DSO000

METHOXYPECTIN see PAO150

2-METHOXY-1-(PENTYLOXY)-4-(1-PROPENYL)-BENZENE see AOK000

METHOXYPHENAMINE HYDROCHLORIDE see OJY000

METHOXYPHENAMINIUM CHLORIDE see OJY000

8-METHOXYPHENANTHRO(3,4-d)-1,3-DIOXOLE-5-CARBOXYLIC ACID see AQX825

8-METHOXYPHENANTHRO(3,4-d)-1,3-DIOXOLE-5-CARBOXYLIC ACID METHYL ESTER see MGQ525

6-METHOXY-1-PHENAZINOL 5,10-DIOXIDE see HLT100

p-METHOXYPHENETHYLAMINE see MFC500

1-(p-METHOXYPHENETHYL)HYDRAZINE HYDROGEN SULFATE see MFC600

1-(p-METHOXYPHENETHYL)-HYDRAZINE SULFATE (1:1) see MFC600

4-(β-METHOXYPHENETHYL)-α-PHENYL-1-PIPERAZINEPROPANOL DIHYDROCHLORIDE see ECU600

2-METHOXYPHENOL see GKI000

3-METHOXYPHENOL see REF050

4-METHOXYPHENOL see MFC700

m-METHOXYPHENOL see REF050

o-METHOXYPHENOL see GKI000

p-METHOXYPHENOL see MFC700

METHOXYPHENOTHIAZINE see MCI500

1-(3-(2-METHOXYPHENOTHIAZIN-10-YL)-2-METHYLPROPYL)-4-PIPERIDINOL see LEO000

2-(p-METHOXYPHENOXY)-N-(2-(DIETHYLAMINO)ETHYL)ACETAMIDE see DIB600

2-(p-METHOXYPHENOXY)-N-(2-(DIETHYLAMINO)ETHYL)ACETAMIDE see MCH250

METHOXYPHENOXYDIOL see RLU000

N-(2-o-METHOXYPHENOXYETHYL)-6,7-DIMETHOXY-3,4-DIHYDRO-2-(1H)-ISOQUINOLINE CARBOXAMIDINE HBr see SBA875

(2-(p-METHOXYPHENOXY)ETHYL)HYDRAZINE HYDROCHLORIDE see MFD250

3-(2-METHOXYPHENOXY)-1-GLYCERYL CARBAMATE see GKK000

3-(o-METHOXYPHENOXY)-2-HYDROXYPROPYL CARBAMATE see GKK000

3-(o-METHOXYPHENOXY)-2-HYDROXYPROPYL-1-NICOTINATE see HLT300

(3R-trans)-3-((4-METHOXYPHENOXY)METHYL)-1-METHYL-4-PHENYLPIPERIDINE HYDROCHLORIDE see FAJ200

5-(o-METHOXYPHENOXYMETHYL)-2-OXAZOLIDINONE see MFD500

5-(o-METHOXYPHENOXYMETHYL)-2-OXAZOLIDONE see MFD500

N-(4-(4-METHOXYPHENOXY)PHENYL)-N,N-DIMETHYLUREA see LGM300

3-o-METHOXYPHENOXYPROPANE 1:2-DIOL see RLU000

3-(o-METHOXYPHENOXY)-1,2-PROPANEDIOL see RLU000

3-(o-METHOXYPHENOXY)-1,2-PROPANEDIOL 1-CARBAMATE see GKK000

4-METHOXYPHENYLACETIC ACID see MFE250

p-METHOXYPHENYLACETIC ACID see MFE250

p-METHOXYPHENYLACETONE see AOV875

4-METHOXYPHENYLACETONITRILE see MFF000

p-METHOXYPHENYLACETONITRILE see MFF000

2-METHOXY-2-PHENYLACETOPHENONE see MHE000

β-(o-METHOXYPHENYL)ACROLEIN see MEJ750

o-METHOXYPHENYLAMINE see AOV900

p-METHOXYPHENYLAMINE see AOW000

o-METHOXYPHENYLAMINE HYDROCHLORIDE see AOX250

N-(p-METHOXYPHENYL)-1-AZIRIDINECARBOXAMIDE see MFF250

2-METHOXY-4-PHENYLAZOANILINE see MFF500

p-((p-METHOXYPHENYL)AZO)ANILINE see MFF550

4-((p-METHOXYPHENYL)AZO)-o-ANISIDINE see DNY400

3-METHOXY-4-(PHENYLAZO)BENZENAMINE see MDY400

N-(2-METHOXY-4-(PHENYLAZO)PHENYL)HYDROXYLAMINE see HLR000

2-(4-METHOXYPHENYL)-1H-BENZ(de)ISOQUINOLINE-1,3(2H)-DIONE see MFF560

1-(2-(p-(6-METHOXY-2-PHENYLBENZOFURAN-3-YL)PHENOXY)ETHYL)PYRROLIDINE HYDROCHLORIDE see PGF100

1-(2-(p-(6-METHOXY-2-PHENYL-3-BENZOFURANYL)PHENOXY)ETHYL)PYRROLIDINE HYDROCHLORIDE see PGF100

1-(2-(4-(7-METHOXY-3-PHENYL-2H-1-BENZOPYRAN-4-YL)PHENOXY)ETHYL)PYRROLIDINE see PGF112

1-(2-(4-(6-METHOXY-2-PHENYLBENZO(b)THIEN-3-YL)PHENOXY)ETHYL)PYRROLIDINE HYDROCHLORIDE see MFF575

2-METHOXYPHENYL BROMIDE see BMT400

4-METHOXYPHENYL BROMIDE see AOY450

o-METHOXYPHENYL BROMIDE see BMT400

p-METHOXYPHENYL BROMIDE see AOY450

p-METHOXYPHENYLBUTANONE see MFF580

4-p-METHOXYPHENYL-2-BUTANONE see MFF580

4-(p-METHOXYPHENYL)-2-BUTANONE see MFF580

4-(p-METHOXYPHENYL)-3-BUTEN-2-ONE see MLI400

p-METHOXYPHENYL-N-CARBAMOYLAZIRIDINE see MFF250

(E)-o-METHOXY-α-PHENYL-CINNAMIC ACID see MFG600

α-p-METHOXYPHENYL-α-DI-n-BUTYLAMINOACETAMIDE see DDT300

3-p-METHOXYPHENYL-5-DIETHYLAMINOETHYL-1,2,4-OXADIAZOLE see CPN750

2-(3-METHOXYPHENYL)-5,6-DIHYDROIMIDAZO(2,1-A)ISOQUINOLINE see IAK300

2-(m-METHOXYPHENYL)-5,6-DIHYDROIMIDAZO(2,1-A)ISOQUINOLINE see IAK300

2-(o-METHOXYPHENYL)-5,6-DIHYDROIMIDAZO(2,1-A)ISOQUINOLINE see MFF600

2-(p-METHOXYPHENYL)-5,6-DIHYDROIMIDAZO(2,1-A)ISOQUINOLINE see MFF620

2-(p-(6-METHOXY-2-PHENYL-3,4-DIHYDRO-1-NAPHTHYL)PHENOXY)TRIETHYLAMINE HYDROCHLORIDE see MFF625

1-(4-(p-(6-METHOXY-2-PHENYL-3,4-DIHYDRO-1-NAPHTHYL)PHENYL)BUTYL)PYRROLIDINE HYDROCHLORIDE see MFF635

2-(3-METHOXYPHENYL)-5,6-DIHYDRO-s-TRIAZOLO(5,1-a)ISOQUINOLINE see MFF650

2-(m-METHOXYPHENYL)-5,6-DIHYDRO-s-TRIAZOLO(5,1-a)ISOQUINOLINE see MFF650

1-(m-METHOXYPHENYL)-2-DIMETHYLAMINOMETHYLCYCLOHEXAN-1-OL HYDROCHLORIDE see THJ750

trans-1-(m-METHOXYPHENYL)-2-DIMETHYLAMINOMETHYLCYCLOHEXAN-1-OL HYDROCHLORIDE see THJ750

1-(p-METHOXYPHENYL)-3,3-DIMETHYLTRIAZENE see DSN600

2-(p-METHOXYPHENYL)-3,3-DIPHENYLACRYLONITRILE see MFF750

α-(p-METHOXYPHENYL)-β,β-DIPHENYLACRYLONITRILE see MFF750

5-(p-METHOXYPHENYL)-1,2-DITHIOCYCLOPENTEN-3-THIONE see AOO490

5-(p-METHOXYPHENYL)-3H-1,2-DITHIOLE-3-THIONE see AOO490

5-(4-METHOXYPHENYL)-3H-1,2-DITHIOLE-3-THIONE (9CI) see AOO490

2-METHOXY-p-PHENYLENEDIAMINE see MEB800

4-METHOXY-m-PHENYLENEDIAMINE see DBO000

4-METHOXY-m-PHENYLENEDIAMINE see MFG000

p-METHOXY-m-PHENYLENEDIAMINE see DBO000

4-METHOXY-m-PHENYLENEDIAMINE SULFATE see DBO400

4-METHOXY-m-PHENYLENEDIAMINE SULPHATE see DBO400

p-METHOXY-m-PHENYLENEDIAMINE SULPHATE see DBO400

p-METHOXYPHENYLETHYLAMINE see MFC500

o-METHOXY-β-PHENYLETHYLHYDRAZINE DIHYDROGEN SULFATE see MFG200 p-METHOXY-β-PHENYLETHYLHYDRAZINE DIHYDROGEN SULFATE see MFC600

1-(2-METHOXY-2-PHENYL)ETHYL-4-(2-HYDROXY-3-METHOXY-3-PHENYL)PROPYLPIPERAZINE DIHYDROCHLORIDE see MFG250

1-(2-METHOXY-2-PHENYLETHYL)-4-(3-HYDROXY-3-PHENYLPROPYL)PIPERAZINE DIHYDROCHLORIDE see ECU600

4-(2-METHOXY-2-PHENYLETHYL)-α-PHENYL-1-PIPERAZINEPROPANOL (9CI) see EQY600

o-METHOXYPHENYL GLYCERYL ETHER see RLU000

2-(3-METHOXYPHENYL)IMIDAZO(2,1-A)ISOQUINOLINE see MFG252

2-(4-METHOXYPHENYL)IMIDAZO(2,1-A)ISOQUINOLINE see MFG254

2-(m-METHOXYPHENYL)IMIDAZO(2,1-A)ISOQUINOLINE see MFG252

2-(o-METHOXYPHENYL)IMIDAZO(2,1-A)ISOQUINOLINE see MFG256

2-(p-METHOXYPHENYL)IMIDAZO(2,1-A)ISOQUINOLINE see MFG254

2-(p-(6-METHOXY-2-PHENYL-3-INDENYL)PHENOXY)TRIETHYLAMINE HYDROCHLORIDE see MFG260

2-(p-(6-METHOXY-2-PHENYLINDEN-3-YL)PHENOXY)TRIETHYLAMINE HYDROCHLORIDE see MFG260

β-(o-METHOXYPHENYL)ISO-p-TROPYLMETHYLAMINEHYDROCHLORIDE see OJY000

2-(3-METHOXYPHENYL)-8-METHOXY-5H-s-TRIAZOLO(5,1-a)ISOINDOLE see MFG275

α-(2-METHOXYPHENYL)-β-METHYLAMINOPROPANEHYDROCHLORIDE see OJY000

9-β-METHOXY-9-α-PHENYL-3-METHYL-3-AZABICYCLO(3.3.1)NONANE CITRATE see ARX150

2-(4-METHOXYPHENYL)-3-(1-METHYLETHYL)-3H-NAPHTH(1,2-d)IMIDAZOLE see THG700

2-(((4-METHOXYPHENYL)METHYLHYDRAZONO)METHYL)-1,3,3-TRIMETHYL-3H-INDOLIUM METHYL SULFATE see CMM895

4-METHOXYPHENYL METHYL KETONE see MDW750

p-METHOXYPHENYL METHYL KETONE see MDW750

1-(p-METHOXYPHENYL)-3-METHYL-3-NITROSOUREA see MFG400

1-(4-METHOXYPHENYL)-3-METHYL TRIAZENE see MFG510

4-(2-METHOXYPHENYL)-α-((1-NAPHTHALENYLOXY)METHYL)-1-PIPERAZINEETHANOL see MFG515

1-(2-METHOXYPHENYL)-2-NITROETHENE see NMR100

1-(2-(p-(α-p-METHOXYPHENYL)-β-NITROSTYRYL)PHENOXY)ETHYL)PYRROLIDINE CITRATE (1:1) see NHP500

1-(2-(p-(α-p-METHOXYPHENYL)-β-NITROSTYRYL)PHENOXY)ETHYL)PYRROLIDINE MONOCITRATE see NHP500

4-(4-METHOXYPHENYL)-6H-1,3,5-OXATHIAZINE see MFG520

2-(p-(p-METHOXY-α-PHENYLPHENETHYL)PHENOXY)TRIETHYLAMINE see MFG525

2-(p-(p-METHOXY-α-PHENYLPHENETHYL)PHENOXY)TRIETHYLAMINE HYDROCHLORIDE see MFG530

trans-3-(o-METHOXYPHENYL)-2-PHENYLACRYLIC ACID see MFG600

2-(p-(2-(p-METHOXYPHENYL)-1-PHENYL-1-BUTENYL)PHENOXY)TRIETHYLAMINE CITRATE see HAA310

(4-METHOXYPHENYL)PHENYLMETHANONE see MEC333

2-((3-o-METHOXYPHENYLPIPERAZINO)-PROPYL)-3-METHYL-7-METHOXYCHROMONE see MFH000

N-(2-(4-(2-METHOXYPHENYL)-1-PIPERAZINYL)ETHYL)-N-(2-PYRIDINYL)CYCLOHEXANECARBOXAMIDE see TKN100

4-(o-METHOXYPHENYL)-1-PIPERAZINYL)-p-FLUOROBUTYROPHENONE see HAF400

(+−)-1-(4-(2-METHOXYPHENYL)PIPERAZINYL)-3-(1-NAPHTHYLOXY)PROPAN-2-OL see MFG515

7-(4-(3-METHOXYPHENYL)-1-PIPERAZINYL)-4-NITROBENZOFURAZAN-1-OXIDE see MFH750

6-(3-(4-(o-METHOXYPHENYL)-1-PIPERAZINYL)PROPYLAMINO)-1,3-DIMETHYLURACIL see USJ000

4-(o-METHOXYPHENYL)PIPERAZINYL 3,4,5-TRIMETHOXYPHENYL KETONE see MFH760

4-(p-METHOXYPHENYL)PIPERAZINYL 3,4,5-TRIMETHOXYPHENYL KETONE see MFH770

1-(p-METHOXYPHENYL)-2-PROPANONE see AOV875

3-(2-METHOXYPHENYL)-2-PROPENAL see MEJ750

1-(p-METHOXYPHENYL)PROPENE see PMQ750

(Z)-3-(2-METHOXYPHENYL)-2-PROPENOIC ACID (9CI) see MEJ775

1-(4-METHOXYPHENYL)-2-PROPEN-1-ONE see ONW100

2-(3-METHOXYPHENYL)PYRAZOLO(5,1-a)ISOQUINOLINE see MFH800

2-(m-METHOXYPHENYL)-PYRAZOLO(5,1-a)ISOQUINOLINE see MFH800

p-METHOXYPHENYL 2-PYRIDYL KETONE see MFH900

p-METHOXYPHENYL 3-PYRIDYL KETONE see MED100

p-METHOXYPHENYL 4-PYRIDYL KETONE see MFH930

4-(m-METHOXYPHENYL)SEMICARBAZONE 1-METHYL-1H-PYRROLE-2-CARBOXALDEHYDE see MFH945

4-(p-METHOXYPHENYL)SEMICARBAZONE 1-METHYL-1H-PYRROLE-2-CARBOXALDEHYDE see MFH950

4-(m-METHOXYPHENYL)SEMICARBAZONE-1H-PYRROLE-2-CARBOXALDEHYDE see MFH955

4-(p-METHOXYPHENYL)SEMICARBAZONE-1H-PYRROLE-2-CARBOXALDEHYDE see MFH960

(2-(p-METHOXYPHENYLTHIO)ETHYL)HYDRAZINE MALEATE see MFJ000

4-(m-METHOXYPHENYL)THIOSEMICARBAZONE 1-METHYL-1H-PYRROLE-2-CARBOXALDEHYDE see MFJ010

4-(p-METHOXYPHENYL)THIOSEMICARBAZONE 1-METHYL-1H-PYRROLE-2-CARBOXALDEHYDE see MFJ020

4-(m-METHOXYPHENYL)THIOSEMICARBAZONE-1H-PYRROLE-2-CARBOXALDEHYDE see MFJ030

4-(p-METHOXYPHENYL)THIOSEMICARBAZONE-1H-PYRROLE-2-CARBOXALDEHYDE see MFJ040

5-(m-METHOXYPHENYL-3-(o-TOLYL))-s-TRIAZOLE see MFJ100

2-(3-METHOXYPHENYL)-5H-s-TRIAZOLO(5,1-a)ISOINDOLE see MFJ105

2-(m-METHOXYPHENYL)-s-TRIAZOLO(5,1-a)ISOQUINOLINE see MFJ110

1-(o-METHOXYPHENYL)-4-(3,4,5-TRIMETHOXYBENZOYL)PIPERAZINE see MFH760

1-(p-METHOXYPHENYL)-4-(3,4,5-TRIMETHOXYBENZOYL)PIPERAZINE see MFH770

(4-METHOXYPHENYL)(2-(2,4,6-TRIMETHYLPHENYL)-3-BENZOFURANYL)METHANONE see AOY300

5-(p-METHOXYPHENYL)TRITHIONE see AOO490

4-METHOXY-6-(β-PHENYLVINYL)-5,6-DIHYDRO-α-PYRONE see GJI250

p-METHOXYPHENYL VINYL KETONE see ONW100

5-(m-METHOXYPHENYL)-3-(2,4-XYLYL)-s-TRIAZOLE see MFJ115

10-(3-(4-METHOXYPIPERIDINO)PROPYL)PHENOTHIAZIN-2-YL METHYL KETONE see MFJ200

METHOXY POLYETHYLENE GLYCOL 350 see MFJ750

METHOXY POLYETHYLENE GLYCOL 550 see MFK000

METHOXY POLYETHYLENE GLYCOL 750 see MFK250

α-ω-METHOXYPOLY(ETHYLENE OXIDE) see DOM100

2-METHOXYPROMAZINE see MFK500

METHOXYPROMAZINE MALEATE see MFK750

1-METHOXYPROPANE see MOU830

α-METHOXY PROPANE see MOU830

METHOXYPROPANEDIOL see RLU000

3-METHOXYPROPANENITRILE see MFL750

3-METHOXYPROPANNITRIL see MFL750

2-METHOXYPROPANOL see MFK800

1-METHOXY-2-PROPANOL see PNL250

2-METHOXY-1-PROPANOL see MFK800

2-METHOXYPROPANOL-1 see MFK800

3-METHOXY-1-PROPANOL see MFL000

METHOXY-2-PROPANONE see MDW300

METHOXY-2-PROPANONE see MFL100

1-METHOXY-2-PROPANONE see MDW300

1-METHOXY-2-PROPANONE see MFL100

METHOXYPROPAZINE see MFL250

2-METHOXYPROPENE see MFL300

4-METHOXYPROPENYLBENZENE see PMQ750

1-METHOXY-4-PROPENYLBENZENE see PMQ750

1-METHOXY-4-(2-PROPENYL)BENZENE see AFW750

(E)-1-METHOXY-4-(1-PROPENYL)BENZENE see PMR250

2-METHOXY-4-PROPENYLPHENOL see IKQ000

2-METHOXY-4-PROP-2-ENYLPHENOL see EQR500

2-METHOXY-4-(2-PROPENYL)PHENOL see EQR500

2-METHOXY-4-PROPENYLPHENYL ACETATE see AAX750

3-METHOXYPROPIONIC ACID METHYL ESTER see MFL400

β-METHOXYPROPIONIC ACID, METHYL ESTER see MFL400

3-METHOXYPROPIONITRILE see MFL750

3-METHOXYPROPIONITRILE see MFL750

2'-METHYLACETOACETANILIDE see ABA000
METHYLACETOACETATE see MFX250
METHYL ACETONE (DOT) see MKA400
p-METHYL ACETOPHENONE see MFW250
4'-METHYL ACETOPHENONE see MFW250
METHYL ACETOPHOS see DRB600
METHYLACETOPYRONONE see MFW500
METHYL ACETOXON see DRB600
N-METHYL-N-(α-ACETOXYBENZYL)NITROSAMINE see NKP000
METHYL-β-ACETOXYETHYL-β-CHLOROETHYLAMINE see MFW750
METHYLACETOXYMALONONITRILE see MFX000
METHYL(ACETOXYMETHYL)NITROSAMINE see AAW000
METHYL-12-ACETOXY-9-OCTADECENOATE see MFX750
METHYL-12-ACETOXYOLEATE see MFX750
N-METHYL-N-(1-ACETOXY-2-OXOPROPYL)NITROSAMINE see HMJ600
6-METHYL-17-α-ACETOXYPREGNA-4,6-DIENE-3,20-DIONE see VTF000
6-α-METHYL-17-α-ACETOXYPREGN-4-ENE-3,20-DIONE see MCA000
6-α-METHYL-17-α-ACETOXYPROGESTERONE see MCA000
METHYL ACETYLACETATE see MFX250
METHYL ACETYLACETONATE see MFX250
3-METHYL-4-ACETYLAMINOBIPHENYL see PEB250
N-METHYL-N-ACETYLAMINOMETHYLNITROSAMINE see MFX500
1-METHYL-4-ACETYLBENZENE see MFW250
O-METHYL-1-ACETYLBENZOCYCLOBUTENE OXIME see MFX550
METHYLACETYL CHOLINE see ACR000
β-METHYLACETYLCHOLINE see MFX560
β-METHYLACETYLCHOLINE CHLORIDE see ACR000
METHYL ACETYLENE see MFX590
METHYL ACETYLENEDICARBOXYLATE see DOP400
METHYL ACETYLENE-PROPADIENE MIXTURE see MFX600
METHYL ACETYLENE and PROPADIENE MIXTURES, stabilized (DOT) see MFX600
METHYL 4-((5-ACETYL-2-FURANYL)OXY)-α-ETHYLBENZENEACETATE see MFX650
N-METHYL-N'-(p-ACETYLPHENYL)-N-NITROSOUREA see MFX725
METHYL ACETYL RICINOLEATE see MFX750
N-METHYL-N'-ACETYLUREA see MFY000
4'-(2-METHYL-9-ACRIDINYLAMINO)METHANESULFONANILIDE see MFY750
4'-(3-METHYL-9-ACRIDINYLAMINO)METHANESULFONANILIDE see MFZ000
4'-(4-METHYL-9-ACRIDINYLAMINO)METHANESULFONANILIDE see MGA000
2-METHYLACROLEIN see MGA250
α-METHYLACROLEIN see MGA250
β-METHYLACROLEIN see COB250
β-METHYL ACROLEIN see COB260
METHYLACRYLAAT (DUTCH) see MGA500
METHYLACRYLALDEHYDE see MGA250
METHYLACRYLALDEHYDE see MGA250
2-METHYL ACRYLALDEHYDE OXIME see MGA275
2-METHYLACRYLAMIDE see MDN500
N-METHYLACRYLAMIDE see MGA300

METHYL ACRYLAMIDOGLYCOLATE METHYL ETHER see MGA400
METHYL-ACRYLAT (GERMAN) see MGA500
METHYL ACRYLATE see MGA500
METHYL ACRYLATE, INHIBITED (DOT) see MGA500
3-METHYLACRYLIC ACID see COB500
α-METHYLACRYLIC ACID see MDN250
β-METHYLACRYLIC ACID see COB500
2-METHYLACRYLIC ACID DODECYL ESTER see DXY200
α-METHYL ACRYLIC AMIDE see MDN500
METHYLACRYLONITRILE see MGA750
α-METHYLACRYLONITRILE see MGA750
β-METHYLACRYLONITRILE see COC300
cis-β-METHYLACRYLONITRILE see BOX600
α-METHYL-1-ADAMANTANEMETHYLAMINE HYDROCHLORIDE see AJU625
1,10-(N-METHYL-N-(1'-ADAMANTYL)AMINO)DECANE DIIODOMETHYLATE see DAE500
N-METHYLADENOSINE mixed with SODIUM NITRITE (1:4) see SIS650
N⁶-METHYLADENOSINE mixed with SODIUM NITRITE (1:4) see SIS650
METHYL ADIPATE see DOQ300
N-METHYLADRENALINE see MJV000
4-METHYLAESCULETIN see MJV800
METHYLAETHYLNITROSAMIN (GERMAN) see MKB000
METHYLAL see MGA850
2-METHYLALANINE see MGB000
α-METHYLALANINE see MGB000
METHYL ALCOHOL see MGB150
METHYL ALDEHYDE see FMV000
1-METHYL-2-ALDOXIMINOPYRIDINIUM CHLORIDE see FNZ000
1-METHYL-2-ALDOXIMINOPYRIDINIUM IODIDE see POS750
METHYLALKOHOL (GERMAN) see MGB150
1-METHYL-2-(p-ALLOPHANOYLBENZYL)HYDRAZINE HYDROBROMIDE see MKN750
p-((2-METHYLALLYL)AMINO)HYDRATROPIC ACID see MDP850
(±)-2-(p-((2-METHYLALLYL)AMINO)PHENYL)PROPIONIC ACID see MGC200
2-METHYL-ALLYLCHLORID (GERMAN) see CIU750
1-METHYLALLYL CHLORIDE see CEV250
2-METHYLALLYL CHLORIDE see CIU750
METHYL ALLYL CHLORIDE (DOT) see CIU750
α-METHYLALLYL CHLORIDE see CEV250
β-METHYLALLYL CHLORIDE see CIU750
γ-METHYLALLYL CHLORIDE see CEU825
1-METHYL-5-ALLYL-5-(1-METHYL-2-PENTYNYL)BARBITURIC ACID SODIUM SALT see MDU500
METHYLALLYLNITROSAMIN (GERMAN) see MMT500
METHYLALLYLNITROSAMINE see MMT500
17-α-(2-METHYLALLYL)-19-NORTESTOSTERONE see MDQ075
dl-3-METHYL-3-ALLYL-4-PROPIONOXYPIPERIDINE HYDROCHLORIDE see AGV890
N-((1-METHYLALLYL)THIOCARBAMOYL)-N'-(METHYLTHIOCARBAMOYL)HYDRAZINE see MLJ500
1-α-METHYLALLYLTHIOCARBAMOYL-2-METHYLTHIOCARBAMOYLHYDRAZINE see MLJ500
METHYL ALUMINIUM SESQUIBROMIDE see MGC225

METHYL ALUMINIUM SESQUICHLORIDE see MGC230
METHYL ALUMINUM SESQUIBROMIDE see MGC225
METHYL ALUMINUM SESQUIBROMIDE see MGC225
METHYL ALUMINUM SESQUICHLORIDE see MGC230
METHYLAMBEPYRINE see AIN100
METHYLAMID KYSELINY-2,4,5-TRICHLORBENZEN-SULFONOVE (CZECH) see MQC250
METHYLAMINE see MGC250
METHYLAMINE (ACGIH,OSHA) see MGC250
(2-METHYLAMINE-1-HYDROXYETHYL)METHANESULFONANILIDE METHANESULFONATE see AHL500
METHYLAMINEN (DUTCH) see MGC250
METHYLAMINE NITROFORM (DOT) see MGC300
METHYLAMINE, compd. with TRINITROMETHANE see MGC300
METHYLAMINE, anhydrous (UN 1061) (DOT) see MGC250
METHYLAMINE, aqueous solution (UN 1235) (DOT) see MGC250
4-METHYLAMINOACETOCATECHOL see MGC350
4-METHYLAMINOACETOPYROCATECHOL see MGC350
9-(p-(METHYLAMINO)ANILINO)ACRIDINE HYDROBROMIDE see MGD000
4-METHYL-2-AMINOANISOLE see MGO750
METHYLAMINOANTIPYRINE SODIUM METHANESULFONATE see AMK500
4-(METHYLAMINO)AZOBENZENE see MNR500
N-METHYL-4-AMINOAZOBENZENE see MNR500
N-METHYL-p-AMINOAZOBENZENE see MNR500
3'-METHYL-4-AMINOAZOBENZENE see MNR750
4'-METHYL-4-AMINOAZOBENZENE see TGV750
(METHYLAMINO)BENZENE see MGN750
N-METHYLAMINOBENZENE see MGN750
1-METHYL-2-AMINOBENZENE see TGQ750
2-METHYL-1-AMINOBENZENE see TGQ750
4-METHYLAMINOBENZENE-1,3-BIS(SULFONYL AZIDE) see MGD100
1-METHYL-2-AMINOBENZENE HYDROCHLORIDE see TGS500
2-METHYL-1-AMINOBENZENE HYDROCHLORIDE see TGS500
METHYL-N-(4-AMINOBENZENESULFONYL)CARBAMATE see SNQ500
3-(METHYLAMINO)-2,1-BENZISOTHIAZOLE see MGD200
3-(METHYLAMINO)-2,1-BENZISOTHIAZOLE HYDROCHLORIDE see MGD210
METHYL 2-AMINOBENZOATE see APJ250
METHYL o-AMINOBENZOATE see APJ250
METHYL p-AMINOBENZOATE see AIN150
2-(METHYLAMINO)BENZOIC ACID see MGQ000
o-(METHYLAMINO)BENZOIC ACID see MGQ000
N-METHYL-2-AMINOBENZOIC ACID see MGQ000
N-METHYL-o-AMINOBENZOIC ACID see MGQ000
4-METHYL-2-AMINOBENZOTHIAZOLE see AKQ500
6-METHYL-2-AMINOBENZOTHIAZOLE see MGD500
4-METHYL-2-AMINOBENZOTHIAZOLE HYDROCHLORIDE see MGD750

3-METHYL-4-AMINOBIPHENYL see MGE000

METHYLAMINO-BIS(1-AZIRIDINYL)PHOSPHINE OXIDE see MGE100

1-METHYLAMINO-4-BROMANTHRACHINON see BNN550

1-(METHYLAMINO)-4-BROMOANTHRAQUINONE see BNN550

2-(4-METHYLAMINOBUTOXY)DIPHENYLMETHANE HYDROCHLORIDE see MGE200

N-((METHYLAMINO)CARBONYL)-N-(((METHYLAMINO)CARBONYL)OXY)ACETAMIDE see CBG075

2-(METHYLAMINO)-2-(2-CHLOROPHENYL)CYCLOHEXANONE see KEK200

METHYLAMINOCOLCHICIDE see MGF000

4-(METHYLAMINO)-o-CRESOL, HYDROGEN SULFATE (2:1) see MGF250

4-METHYLAMINO CRESOL-2-SULFATE see MGF250

N-METHYLAMINODIGLYCOL see MKU250

4-METHYLAMINO-1,5-DIMETHYL-2-PHENYL-3-PYRAZOLONE SODIUM METHANESULFONATE see AMK500

2-METHYL-4-AMINODIPHENYL see MGF500

3-METHYL-4-AMINODIPHENYL see MGE000

4'-METHYL-4-AMINODIPHENYL see MGF750

2-METHYLAMINOETHANOL see MGG000

N-METHYLAMINOETHANOL see MGG000

β-(METHYLAMINO)ETHANOL see MGG000

1-METHYLAMINO-4-ETHANOLAMINOANTHRAQUINONE see MGG250

1-METHYLAMINOETHANOLCATHECHOL HYDROCHLORIDE see AES500

m-METHYLAMINOETHANOLPHENOL see NCL500

p-METHYLAMINOETHANOLPHENOL see HLV500

m-METHYLAMINOETHANOLPHENOL HYDROCHLORIDE see SPC500

2-METHYL-4-AMINO-5-ETHOXYMETHYLPYRIMIDINE see MGG300

2-(METHYLAMINO)ETHYLAMINE see MJW100

S-(R,R)-d-(α-(1-METHYLAMINO)ETHYL)BENZENEMETHANOL HYDROCHLORIDE see POH250

(R-(R*,S*))-α-(1-(METHYLAMINO)ETHYL)BENZENEMETHANOL HYDROCHLORIDE see EAX000

(−)-α-(1-METHYLAMINOETHYL)BENZYL ALCOHOL see EAW000

α-(1-(METHYLAMINO)ETHYL)BENZYL ALCOHOL see POH000

1-α-(1-METHYLAMINOETHYL)BENZYL ALCOHOL see EAW000

α-(1-METHYLAMINOETHYL)BENZYL ALCOHOL HYDROCHLORIDE see MGH250

d-(α-(1-METHYLAMINO)ETHYL)BENZYL ALCOHOL HYDROCHLORIDE see POH250

dl-α-(1-(METHYLAMINO)ETHYL) BENZYL ALCOHOL HYDROCHLORIDE see EAX500

1-α-(1-(METHYLAMINO)ETHYL)BENZYL ALCOHOL SULFATE see EAY500

d-α-(1-(METHYLAMINO)ETHYL)BENZYL PHOSPHATE see EAY150

l-α-(1-(METHYLAMINO)ETHYL)BENZYL PHOSPHATE see EAY155

dl-α-(1-(METHYLAMINO)ETHYL)BENZYL PHOSPHATE see EAY175

α-(1-METHYLAMINOETHYL)-p-HYDROXYBENZYL ALCOHOL see HKH500

2-METHYL-3-(β-AMINOETHYL)-5-METHOXYBENZOFURAN see MGH750

N-METHYLAMINOETHYL-2-METHYLBENZHYDRYL ETHER HYDROCHLORIDE see TGJ250

2-(2-(METHYLAMINO)ETHYL)PYRIDINE see HGE820

4-(2-METHYLAMINOETHYL)PYROCATECHOL HYDROCHLORIDE see EAZ000

2-METHYLAMINOFLUORENE see FEI500

2-METHYL-6-AMINOHEPTANE see ILM000

2-METHYLAMINO-HEPTANE, HYDROCHLORIDE see NCL300

2-METHYL-6-AMINO-2-HEPTANOL HYDROCHLORIDE see HBB000

6-METHYL-2-AMINO-6-HEPTANOL HYDROCHLORIDE see HBB000

1-METHYLAMINO-4-(β-HYDROXYETHYLAMINO)ANTHRAQUINONE see MGG250

3'-(2-(METHYLAMINO)-1-HYDROXYETHYL)METHANESULFONANILIDE METHANESULFONATE see AHL500

2-METHYL-4-AMINO-1-HYDROXYNAPHTHALENE see AKX500

β-METHYLAMINO-α-(4-HYDROXYPHENYL)ETHYL ALCOHOL see HLV500

2-METHYLAMINOISOCAMPHANE see VIZ400

3-METHYLAMINOISOCAMPHANE see VIZ400

3-METHYLAMINOISOCAMPHANE HYDROCHLORIDE see MQR500

2-METHYLAMINOISOOCTANE HYDROCHLORIDE see ODY000

2-METHYL-4-AMINO-6-METHOXY-s-TRIAZINE see MGH800

2-METHYLAMINO METHYL BENZOATE see MGQ250

1-METHYLAMINOMETHYLDIBENZO(b,c)BICYCLO(2,2,2)OCTADIENE HYDROCHLORIDE see BCH750

6-METHYLAMINO-2-METHYLHEPTENE see ILK000

METHYLAMINO-METHYLHEPTENE HYDROCHLORIDE see ODY000

METHYLAMINOMETHYL(4-HYDROXYPHENYL)CARBINOL see HLV500

2-(6-(METHYLAMINO)-3-(METHYLIMINO)-3H-XANTHEN-9-YL)BENZOIC ACID MONOPERCHLORATE see RGW100

2-METHYLAMINO-4-METHYLTHIO-6-ISOPROPYLAMINO-1,3,5-TRIAZINE see INR000

2-METHYL-4-AMINO-1-NAPHTHOL see AKX500

o-METHYL-2-AMINO-1-NAPHTHOL HYDROCHLORIDE see MFA250

4-METHYLAMINO-N-NITROSOAZOBENZENE see MMY250

1-METHYLAMINO-4-OXYETHYLAMINOANTHRAQUINONE (RUSSIAN) see MGG250

3-(α-METHYLAMINOPHENETHYL)PHENOL HYDROCHLORIDE see MGJ600

N-(α-METHYLAMINOPHENETHYL)PHENOL HYDROCHLORIDE see MGJ600

3-METHYL-4-AMINOPHENOL see AKZ000

p-METHYLAMINOPHENOLSULFATE see MGJ750

METHYL-p-AMINOPHENOL SULFATE see MGJ750

p-(METHYLAMINO)PHENOL SULFATE (2:1) (SALT) see MGJ750

6-METHYL-2-(p-AMINO PHENYL) see ALX000

METHYLAMINOPHENYLDIMETHYLPYRAZOLONE METHANESULFONATE SODIUM see AMK500

1-2-METHYLAMINO-1-PHENYLPROPANOL see EAW000

d-psi-2-METHYLAMINO-1-PHENYL-1-PROPANOL see POH000

METHYL ((4-AMINOPHENYL)SULFONYL)CARBAMATE see SNQ500

3-METHYL-4-AMINO-6-PHENYL-1,2,4-TRIAZIN(4H)-ON (GERMAN) see ALA500

(+-)-5-(METHYLAMINO)-2-PHENYL-4-(3-(TRIFLUOROMETHYL)PHENYL)-3(2H)-FURANONE see MGJ775

2-METHYL-2-AMINOPROPANE HYDROCHLORIDE see MGJ800

3-(METHYLAMINO)PROPANENITRILE see COM800

β-METHYLAMINOPROPIONITRIL see COM800

1-(3-METHYLAMINOPROPYL)-2-ADAMANTANOL HYDROCHLORIDE see MGK750

1-(3-METHYLAMINOPROPYL)DIBENZO(b,e)BICYCLO(2.2.2)OCTADIENE HYDROCHLORIDE see MAW850

5-(3-METHYLAMINOPROPYL)-5H-DIBENZO(a,d)CYCLOHEPTENE see DDA600

5-(3-METHYLAMINOPROPYL)-5H-DIBENZO(a,d)CYCLOHEPTENE HYDROCHLORIDE see POF250

9-(γ-METHYLAMINOPROPYL)-9,10-DIHYDRO-9,10-ETHANOANTHRACENE HYDROCHLORIDE see MAW850

5-(3-(METHYLAMINO)PROPYLIDENE)DIBENZO(a,e)CYCLOHEPTA(1,5)DIENE see NNY000

4-(3'-METHYLAMINOPROPYLIDENE)-9,10-DIHYDRO-4H-BENZO(4,5)CYCLOHEPTA(1,2-b)THIOPHEN see MGL500

5-(3-METHYLAMINOPROPYLIDENE)-10,11-DIHYDRO-5H-DIBENZO(a,d)CYCLOHEPTENE see NNY000

METHYLAMINOPROPYLIMINODIBENZYL see DSI709

N-(γ-METHYLAMINOPROPYL)IMINODIBENZYL HYDROCHLORIDE see DLS600

p-(2-METHYLAMINOPROPYL)PHENOL see FMS875

METHYLAMINOPTERIN see MDV500

4-METHYL-2-AMINOPYRIDINE see ALC250

METHYL-4-AMINO-2-PYRIDINE see ALC250

1-METHYL-3-AMINO-5H-PYRIDO(4,3-b)INDOLE see ALD500

N-METHYL-4-AMINO-1,2,5-SELENADIAZOLE-3-CARBOXAMIDE see MGL600

2-METHYL-4-AMINOSTILBENE see MPJ250

3-METHYL-4-AMINOSTILBENE see MPJ500

3-β-METHYLAMINO-2,2,3-TRIMETHYLBICYCLO(2.2.1)HEPTANE see VIZ400

2-METHYLAMINO-2,3,3-TRIMETHYLNORBORANE see VIZ400

2-METHYLAMINO-2,3,3-TRIMETHYLNORBORNANE see VIZ400

6-METHYL-4-AMINO-2,3,5-TRIMETHYLQUINOXALINE see TDV400

METHYLAMMONIUM CHLORITE see MGN000

METHYLAMMONIUM NITRATE see MGN150

METHYLAMMONIUM PERCHLORATE see MGN250

METHYLAMPHETAMINE see DBB000

(+)-METHYLAMPHETAMINE see PFP850

d-METHYLAMPHETAMINE see PFP850

N-METHYLAMPHETAMINE see DBB000

6-METHYL-1,2-BENZANTHRACENE see MGW500

7-METHYL-1,2-BENZANTHRACENE see MGW740

8-METHYL-1:2-BENZANTHRACENE see MGX250

9-METHYL-1,2-BENZANTHRACENE see MGX500

1'-METHYL-1,2-BENZANTHRACENE see MGU750

10-METHYL-1,2-BENZANTHRACENE see MGW750

2'-METHYL-1,2-BENZANTHRACENE see MGV000

4'-METHYL-1:2-BENZANTHRACENE see MGV500

10-METHYL-1,2-BENZANTHRACENE-5-CARBONAMIDE see MGX750

7-METHYLBENZ(a)ANTHRACENE-8-CARBONITRILE see MGY000

7-METHYLBENZ(a)ANTHRACENE-10-CARBONITRILE see MGY250

12-METHYLBENZ(a)ANTHRACENE-7-CARBOXALDEHYDE see FNT000

7-METHYLBENZ(a)ANTHRACENE-12-CARBOXALDEHYDE see MGY500

12-METHYL BENZ(A)ANTHRACENE-7-ETHANOL see HKU000

7-METHYLBENZ(a)ANTHRACENE-12-METHANOL see HMF500

12-METHYLBENZ(a)ANTHRACENE-7-METHANOL ACETATE (ESTER) see ABR250

12-METHYLBENZ(a)ANTHRACENE-7-METHANOL BENZOATE (ESTER) see BDQ250

7-METHYLBENZ(a)ANTHRACENE-5,6-OXIDE see MGZ000

7-METHYLBENZ(a)ANTHRACEN-8-YL CARBAMIDE see MGX750

S-(12-METHYL-7-BENZ(a)ANTHRYLMETHYL)HOMOCYSTEINE see MHA000

N-METHYLBENZAZIMIDE, DIMETHYLDITHIOPHOSPHORIC ACID ESTER see ASH500

METHYLBENZEDRIN see DBA800

METHYL 1H-BENZEMEDAZOL-2-YLCARBAMATE see MHC750

N-METHYLBENZENAMIDE see MHA250

2-METHYLBENZENAMINE see TGQ750

3-METHYLBENZENAMINE see TGQ500

4-METHYLBENZENAMINE see TGR000

m-METHYLBENZENAMINE see TGQ500

N-METHYLBENZENAMINE see MGN750

o-METHYLBENZENAMINE see TGQ750

p-METHYLBENZENAMINE see TGR000

2-METHYLBENZENAMINE HYDROCHLORIDE see TGR500

2-METHYLBENZENAMINE HYDROCHLORIDE see TGS500

4-METHYLBENZENAMINE HYDROCHLORIDE see TGS750

o-METHYLBENZENAMINE HYDROCHLORIDE see TGS500

5-METHYL-1,3-BENZENDIOL see MPH500

METHYLBENZENE see TGK750

METHYL BENZENEACETATE see MHA500

2-METHYLBENZENECARBONITRILE see TGT500

METHYL BENZENECARBOXYLATE see MHA750

2-METHYL-1,2-BENZENEDIAMINE see TGM100

2-METHYL-1,4-BENZENEDIAMINE see TGM000

2-METHYL-1,4-BENZENEDIAMINE see TGM400

3-METHYL-1,2-BENZENEDIAMINE see TGY800

4-METHYL-1,2-BENZENEDIAMINE see TGM250

4-METHYL-1,3-BENZENEDIAMINE see TGL750

2-METHYL-1,4-BENZENEDIAMINE DIHYDROCHLORIDE see DCE200

2-METHYL-1,4-BENZENEDIAMINE SULFATE see DCE600

METHYL BENZENEDIAZOATE see MHB000

4-METHYLBENZENEDIAZONIUM SULFATE see TGM500

p-METHYLBENZENEDIAZONIUM SULPHATE see TGM500

4-METHYLBENZENEDIAZONIUM TETRAFLUOROBORATE (1⁻) see TGM450

2-METHYL-1,3-BENZENEDIOL see MPH400

3-METHYL-1,2-BENZENEDIOL see DNE000

4-METHYL-1,2-BENZENEDIOL see DNE200

β-METHYLBENZENEETHANAMINE see PGB760

(S)-α-METHYL-BENZENEETHANAMINE SULFATE (2:1) see BBK500

α-METHYLBENZENEETHANEAMINE see BBK000

α-METHYLBENZENEETHANOL see PGA600

β-METHYLBENZENEETHANOL see HGR600

4-METHYL-BENZENEMETHANOL see MHB250

4-METHYLBENZENEMETHANOL ACETATE see XQJ700

α-METHYLBENZENEMETHANOL ACETATE see SMP600

α-METHYLBENZENEPROPANAMINE see PEE300

(+−)-α-METHYLBENZENEPROPANAMINE see PEE300

2-METHYLBENZENESULFONAMIDE see TGN250

4-METHYLBENZENESULFONAMIDE see TGN500

o-METHYLBENZENESULFONAMIDE see TGN250

p-METHYLBENZENESULFONAMIDE see TGN500

METHYL BENZENESULFONATE see MHB300

4-METHYLBENZENESULFONIC ACID see TGO000

p-METHYLBENZENESULFONIC ACID see TGO000

4-METHYL-BENZENESULFONIC ACID BUTYL ESTER (9CI) see BSP750

4-METHYL-BENZENESULFONIC ACID, SODIUM SALT see SKK500

N-(4-METHYLBENZENESULFONYL)-N'-(3-AZABICYCLO(3.3.0)OCT-3-YL)UREA see DBL700

p-METHYL BENZENE SULFONYL CHLORIDE see TGO250

N-(4-METHYLBENZENESULFONYL)-N'-CYCLOHEXYLUREA see CPR000

4-METHYLBENZENESULFONYL FLUORIDE see TGO500

p-METHYLBENZENESULFONYL FLUORIDE see TGO500

4-METHYLBENZENESULFONYL ISOCYANATE see IKG925

p-METHYLBENZENESULFONYL ISOCYANATE see IKG925

4-METHYL BENZENE SYLFONYL CHLORIDE see TGO250

2-METHYLBENZENETHIOL see TGP000

3-METHYLBENZENETHIOL see TGO800

4-METHYLBENZENETHIOL see TGP250

m-METHYLBENZENETHIOL see TGO800

o-METHYLBENZENETHIOL see TGP000

p-METHYLBENZENETHIOL see TGP250

METHYLBENZETHONIUM CHLORIDE see MHB500

METHYL BENZETHONIUM CHLORIDE MONOHYDRATE see MHB500

10-METHYL-7H-BENZIMIDAZOL(2,1-a)BENZ(de)ISOQUINOLIN-7-ONE see MHC000

METHYL-2-BENZIMIDAZOLE see MHC250

2-METHYLBENZIMIDAZOLE see MHC250

5-METHYLBENZIMIDAZOLE see MHC500

METHYL 2-BENZIMIDAZOLECARBAMATE see MHC750

METHYL-2-BENZIMIDAZOLE CARBAMATE and SODIUM NITRITE see CBN375

METHYL BENZIMIDAZOLE-2-YL CARBAMATE see MHC750

7-METHYLBENZO(c)ACRIDINE see MGT500

METHYL BENZOATE (FCC, DOT) see MHA750

6-METHYL-3,4-BENZOCARBAZOLE see MHD000

9-METHYL-1:2-BENZOCARBAZOLE see MHD250

10-METHYL-7H-BENZO(c)CARBAZOLE see MHD000

N-(2-METHYLBENZODIOXAN)-N'-ETHYL-β-ALANINAMIDE see MHD300

N-(2-METHYL-1,4-BENZODIOXAN)-N'-METHYL-β-ALANINAMIDE see MHD300

4-(2-(2-METHYL-1,3-BENZODIOXOL-2-YL)ETHYL)PIPERAZIN-1-YL-2-ETHANOL DIHYDROCHLORIDE see MHD750

3-METHYLBENZO(b)FLUORANTHENE see MGT400

7-METHYLBENZO(b)FLUORANTHENE see MGT410

8-METHYLBENZO(b)FLUORANTHENE see MGT415

12-METHYLBENZO(b)FLUORANTHENE see MGT420

2-METHYLBENZOIC ACID see TGQ000

3-METHYLBENZOIC ACID see TGP750

m-METHYLBENZOIC ACID see TGP750

o-METHYLBENZOIC ACID see TGQ000

4-METHYLBENZOIC ACID METHYL ESTER see MNR250

METHYL BENZOIN see MHE000

METHYLBENZOL see TGK750

2-METHYLBENZONITRILE see TGT500

3-METHYLBENZONITRILE see TGT250

4-METHYLBENZONITRILE see TGT750

m-METHYLBENZONITRILE see TGT250

o-METHYLBENZONITRILE see TGT500

p-METHYLBENZONITRILE see TGT750

5-METHYLBENZO(rat)PENTAPHENE see MHE250

8-METHYLBENZO(rst)PENTAPHENE-5-CARBOXALDEHYDE see FNU000

METHYL-1,12-BENZOPERYLENE see MHE500

7-METHYLBENZO(a)PHENALENO(1,9-hi)ACRIDINE see MHE750

7-METHYLBENZO(h)PHENALENO(1,9-bc)ACRIDINE see MHF000

4-METHYLBENZO(c)PHENANTHRENE see MHL750

5-METHYLBENZO(c)PHENANTHRENE see MHF250

6-METHYLBENZO(c)PHENANTHRENE see MHF500

4-METHYL BENZOPHENONE see MHF750

6-METHYL-2H-1-BENZOPYRAN-2-ONE see MIP750

7-METHYL-2H-1-BENZOPYRAN-2-ONE (9CI) see MIP775

α-METHYL-5H-(1)-BENZOPYRANO(2,3-b)PYRIDINE-7-ACETIC ACID see PLX400

1-METHYLBENZO(a)PYRENE see MHG250

2-METHYLBENZO(a)PYRENE see MHG500

3-METHYLBENZO(a)PYRENE see MHG250

4-METHYLBENZO(a)PYRENE see MHG750

5-METHYLBENZO(a)PYRENE see MHH200

6-METHYLBENZO(a)PYRENE see MHM000

7-METHYL-2,3:9,10-
BIS(METHYLENEDIOXY)-7,13a-
SECOBERBIN-13a-ONE see FOW000
N-METHYL-N,N-BIS(3-
METHYLSULFONYLOXYPROPYL)AMINE
4,4'-BIPHENYLDISULFONATE see MHQ775
METHYLBISMUTH OXIDE see MHR000
2-METHYL-4,6-
BIS((OCTYLTHIO)METHYL)PHENOL see
MHR025
4-METHYL-2,6-BIS(1-
PHENYLETHYL)PHENOL see MHR050
N-METHYL-BIS(2,4-
XYLYLIMINOMETHYL)AMINE see MJL250
β-METHYLBIVINYL see IMS000
17-α-METHYL-B-NORTESTOSTERONE see
MNC150
METHYL BORATE see TLN000
METHYL BOTRYODIPLODIN see MHR100
1-METHYL-BP see MHG250
5-METHYL-BP see MHH200
METHYLBROMFENVINPHOS see MHR150
METHYLBROMID (GERMAN) see MHR200
METHYL BROMIDE see MHR200
dl-METHYLBROMIDE see MDL000
METHYL BROMIDE and ETHYLENE
DIBROMIDE MIXTURE, liquid (DOT) see
BNM750
METHYL BROMOACETATE see MHR250
METHYL α-BROMOACETATE see MHR250
2-METHYL-4-BROMOANILINE see MHR500
4-METHYL-2-BROMOANILINE see BOG500
2-METHYLBROMOBENZENE see BOG260
3-METHYLBROMOBENZENE see BOG300
m-METHYLBROMOBENZENE see BOG300
METHYL-4-BROMOBENZENEDIAZOATE
see MHR750
METHYL 4-BROMO-2-BUTENOATE see
MHR790
METHYL BROMOCROTONATE see
MHR790
METHYL 4-BROMOCROTONATE see
MHR790
METHYL γ-BROMOCROTONATE see
MHR790
O-METHYL-O-(4-BROMO-2,5-
DICHLOROPHENYL)PHENYL
THIOPHOSPHONATE see LEN000
METHYL-(BROMOMERCURI)FORMATE see
MHS250
METHYL 2-(BROMOMETHYL)ACRYLATE
see MHS300
METHYL α-(BROMOMETHYL)ACRYLATE
see MHS300
1-METHYL-7-
BROMOMETHYLBENZ(a)ANTHRACENE
see BNQ750
METHYL 2-(BROMOMETHYL)-2-
PROPENOATE see MHS300
1-METHYL-3-(p-BROMOPHENYL)-1-
NITROSOUREA see BNX125
1-METHYL-3-(p-BROMOPHENYL)UREA see
MHS375
1-METHYL-3-(p-BROMOPHENYL)UREA
mixed with SODIUM NITRITE see SIQ675
METHYL 1-BROMOVINYL KETONE see
MHS400
METHYLBROMPHENVINPHOS see MHR150
1-METHYL-3-(p-
BROMPHENYL)HARNSTOFF see MHS375
1-METHYL-3-(p-BROMPHENYL)-1-
NITROSOHARNSTOFF (GERMAN) see
BNX125
N-METHYLBUTABARBITAL see BRJ250
1-METHYLBUTADIENE see PBA100
2-METHYLBUTADIENE see IMS000
2-METHYL-1,3-BUTADIENE (DOT) see
IMS000
2-METHYLBUTANAL see MJX500
3-METHYLBUTANAL see MHX500
2-METHYL-1-BUTANAL see MJX500
2-METHYLBUTANAL-4 see MHX500

α-METHYLBUTANAL see MJX500
3-METHYL-2-BUTANAMINE see AOE200
2-METHYLBUTANE see EIK000
2-METHYL-1,4-BUTANEDIOL see MHS500
α-(3-METHYL-BUTANE)-α-(2-
PIPERIDYLETHYL) BENZYL ALCOHOL
HYDROCHLORIDE see HBF000
2-METHYLBUTANE SECONDARY
MONONITRILE see ITD000
2-METHYL-2-BUTANETHIOL see MHS550
METHYL n-BUTANOATE see MHY000
2-METHYLBUTANOIC ACID see MHS600
3-METHYLBUTANOIC ACID see ISU000
3-METHYLBUTANOIC ACID, BUTYL
ESTER see ISX000
3-METHYLBUTANOIC ACID, ETHYL
ESTER see ISY000
2-METHYLBUTANOIC ACID, n-HEXYL
ESTER see HFR200
3-METHYLBUTANOIC ACID, METHYL
ESTER see ITC000
3-METHYLBUTANOIC ACID-2-
METHYLPROPYL ESTER see ITA000
3-METHYLBUTANOIC ACID,
PHENYLETHYL ESTER see ISW000
3-METHYL-BUTANOIC ACID-2-
PHENYLETHYL ESTER see PDF775
3-METHYLBUTANOIC ACID 3-PHENYL-2-
PROPENYL ESTER see PEE200
3-METHYLBUTANOIC ACID, 2-PROPENYL
ESTER see ISV000
endo-3-METHYLBUTANOIC ACID 1,7,7-
TRIMETHYLBICYCLO(2.2.1)HEPT-2-YL
ESTER see HOX100
2-METHYLBUTANOL see MHS750
3-METHYL BUTANOL see IHP000
2-METHYL BUTANOL-1 see MHS750
2-METHYL BUTANOL-2 see PBV000
2-METHYL-2-BUTANOL see PBV000
2-METHYL-4-BUTANOL see IHP000
3-METHYLBUTAN-1-OL see IHP000
3-METHYLBUTAN-3-OL see PBV000
3-METHYL-1-BUTANOL (CZECH) see
IHP000
3-METHYL-1-BUTANOL CARBAMATE see
IHQ000
3-METHYL-1-BUTANOL NITRATE see
ILW100
3-METHYLBUTANOL NITRITE see IMB000
3-METHYL-2-BUTANONE see MLA750
3-METHYL BUTAN-2-ONE (DOT) see
MLA750
METHYLBUTENE see IHR220
2-METHYLBUTENE see IHR220
2-METHYL-1-BUTENE see MHT000
3-METHYL-1-BUTENE see MHT250
2-METHYL-2-BUTENEDIOIC ACID see
CMS320
cis-METHYLBUTENEDIOIC ACID see
CMS320
(Z)-2-METHYL-2-BUTENEDIOIC ACID (9CI)
see CMS320
trans-2-METHYL-2-BUTENEDIOIC ACID see
MDI250
cis-2-METHYL-2-BUTENEDIOIC ACID,
DIMETHYL ESTER see DRF200
METHYL trans-2-BUTENOATE see COB825
3-METHYL-2-BUTENOIC ACID see MHT500
trans-2-METHYL-2-BUTENOIC ACID see
TGA700
2-METHYLBUTENOIC ACID-7-((2,3-
DIHYDROXY-2-(1-METHYLETHYL)-1-
OXOBUTOXY)METHYL)-2,3,5,7a-
TETRAHYDRO-1H-PYRROLIZIN-1-YL
ESTER see SPB500
3-METHYL-2-BUTENOIC ACID ETHYL
ESTER see MIP800
(E)-2-METHYL-2-BUTENOIC ACID ETHYL
ESTER see TGA800
2-METHYL-2-BUTENOIC ACID, HEXYL
ESTER see HFX000
METHYLBUTENOL see MHU100

2-METHYL-3-BUTEN-2-OL see MHU100
3-METHYL-1-BUTEN-3-OL see MHU100
3-METHYL-2-BUTEN-1-OL see MHU110
3-METHYL-BUTEN-(1)-OL-(3) (GERMAN)
see MHU100
3-METHYL-3-BUTEN-2-ON (GERMAN) see
MKY500
2-METHYL-1-BUTEN-3-ONE see MKY500
3-METHYL-2-BUTENYL ACETATE see
DOQ350
3-METHYL-2-BUTENYL BENZOATE see
MHU150
4-(3-METHYL-2-BUTENYL)-1,2-DIPHENYL-
3,5-PYRAZOLIDINEDIONE see PEW000
3-(3-METHYL-2-BUTENYL)-1,2,3,4,5,6-
HEXAHYDRO-6,11-DIMETHYL-2,6-
METHANO-3-BENZAZOCIN-8-OL see
DOQ400
9-((3-METHYL-2-BUTENYL)OXY)-7H-
FURO(3,2-g)(1)BENZOPYRAN-7-ONE see
IHR300
5-(1-METHYL-1-BUTENYL)-5-
PROPYLBARBITURIC ACID see MHU200
3-METHYL-2-BUTENYL SALICYLATE see
PMB600
2-METHYL-1-BUTEN-3-YNE see MHU250
3-METHYL-3-BUTEN-1-
YNYLTRIETHYLLEAD see MHU500
3-METHYL-BUTIN-(1)-OL-(3) (GERMAN) see
MHX250
(2-(3-METHYLBUTOXY)ETHYL)BENZENE
see IHV050
3'-METHYLBUTTERGELB (GERMAN) see
DUH600
1-METHYLBUTYL ACETATE see AOD735
3-METHYLBUTYL ACETATE see IHO850
3-METHYL-1-BUTYL ACETATE see IHO850
2-METHYL-BUTYLACRYLAAT (DUTCH) see
MHU750
2-METHYL-BUTYLACRYLAT (GERMAN) see
MHU750
2-METHYL BUTYLACRYLATE see MHU750
METHYLBUTYLAMINE see MHV000
N-(METHYL) BUTYL AMINE see MHV000
N-METHYL-n-BUTYLAMINE see MHV000
3-METHYLBUTYL α-
AMINOBENZENEACETATE
HYDROCHLORIDE (±)- see PCV750
p-METHYL-tert-BUTYLBENZENE see
BSP500
1-METHYL-4-tert-BUTYLBENZENE see
BSP500
METHYL-5-BUTYL-2-
BENZIMIDAZOLECARBAMATE see
BQK000
1-(3-METHYL)BUTYL BENZOATE see
IHP100
3-METHYLBUTYL BROMIDE see BNP250
3-METHYLBUTYL BUTYRATE see IHP400
METHYL-1-(BUTYLCARBAMOYL)-2-
BENZIMIDAZOLYLCARBAMATE see
BAV575
METHYL-1,3-BUTYLENE GLYCOL
ACETATE see MHV750
1-METHYLBUTYL ESTER CARBAMIC ACID
see MOU500
3-METHYLBUTYL ETHANOATE see
IHO850
METHYL BUTYL ETHER see BRU780
METHYL n-BUTYL ETHER see BRU780
METHYL tert-BUTYL ETHER see MHV859
METHYL tert-BUTYL ETHER (DOT) see
MHV859
3-METHYLBUTYL FORMATE see IHS000
METHYLBUTYL HYDRAZINE see MHW000
1-METHYL-2-BUTYL-HYDRAZINE
DIHYDROCHLORIDE see MHW250
α-METHYL-p-(tert-
BUTYL)HYDROCINNAMALDEHYDE see
LFT000
3-METHYLBUTYL 2-
HYDROXYBENZOATE see IME000

3-METHYLBUTYL IODIDE see IHU200
2-METHYLBUTYL ISOVALERATE see MHW260
METHYL tert-BUTYL KETONE see DQU000
METHYL n-BUTYL KETONE (ACGIH) see HEV000
N-3-METHYLBUTYL-N-1-METHYL ACETONYLNITROSAMINE see MHW350
2-METHYLBUTYL-3-METHYLBUTANOATE see MHW260
METHYL-N-(p-tert-BUTYL-α-METHYLHYDROCINNAMYLIDENE) ANTHRANILATE see LFT100
3-METHYLBUTYL NITRITE see IMB000
METHYL-BUTYL-NITROSAMIN (GERMAN) see MHW500
METHYLBUTYLNITROSAMINE see MHW500
METHYL-N-BUTYLNITROSAMINE see MHW500
METHYL-tert-BUTYLNITROSAMINE see MHW750
m-(1-METHYLBUTYL)PHENYL METHYLCARBAMATE see AON250
α-METHYL-β-(p-tert-BUTYLPHENYL)PROPIONALDEHYDE see LFT000
2-METHYL-2-sec-BUTYL-1,3-PROPANEDIOL DICARBAMATE see MBW750
5-(1-METHYLBUTYL)-5-(2-PROPENYL)-2,4,6(1H,3H,5H)-PYRIMIDINETRIONE MONOSODIUM SALT see SBN000
5-(1-METHYLBUTYL)-5-(2-PROPENYL)-2,4,6(1H,3H,5H)-PYRIMIDINITRIONE see SBM500
5-METHYL-3-BUTYLTETRAHYDROPYRAN-4-YL ACETATE see MHX000
2-((o-(N-METHYL-N-(tert-BUTYLTHIOSULFENYL)CARBAMOYL)OXI MINO))-1,3-DITHIOLANE see MHX100
2-METHYL-3-BUTYN-2-AMINE see MHX200
2-METHYL-3-BUTYN-2-OL see MHX250
2-METHYLBUTYN-3-OL-2 see MHX250
3-METHYL-1-BUTYN-3-OL see MHX250
METHYL-3-BUTYN-2-OL CARBAMATE see CBM875
2-METHYLBUTYRALDEHYDE see MJX500
3-METHYLBUTYRALDEHYDE see MHX500
3-METHYLBUTYRALDEHYDE see MHX500
α-METHYLBUTYRALDEHYDE see MJX500
METHYL BUTYRATE see MHY000
METHYL-n-BUTYRATE see MHY000
2-METHYLBUTYRIC ACID see MHS600
3-METHYLBUTYRIC ACID see ISU000
α-METHYLBUTYRIC ACID see MHS600
β-METHYLBUTYRIC ACID see ISU000
3-METHYLBUTYRIC ACID, ALLYL ESTER see ISV000
(E)-3-METHYLBUTYRIC ACID-3,7-DIMETHYL-2,6-OCTADIENYL ESTER see GDK000
3-METHYLBUTYRIC ACID, ETHYL ESTER see ISY000
3-METHYLBUTYRIC ACID HEXYL ESTER see HFQ575
4-METHYL-γ-BUTYROLACTONE see VAV000
γ-METHYL-γ-BUTYROLACTONE see VAV000
2-METHYLBUTYRONITRILE see ITD000
3-METHYLBUTYRONITRILE see ITD000
8-(3-METHYLBUTYRYLOXY)-DIACETOXYSCIRPENOL see FQS000
METHYL CADMIUM AZIDE see MHY550
METHYL-CALMINAL see ENB500
METHYLCAMPHENOATE see MHY600
METHYL CAPRATE see MHY650
METHYL n-CAPRATE see MHY650
METHYL CAPRINATE see MHY650
METHYL CAPROATE see MHY700

2-METHYLCAPROIC ACID see HEN500
5-METHYLCAPROIC ACID see IKN400
α-METHYLCAPROIC ACID see HEN500
1-METHYLCAPROLACTAM see EGN500
N-METHYLCAPROLACTAM see EGN500
N-METHYL-ε-CAPROLACTAM see MHY750
METHYL CAPRONATE see MHY700
METHYL CAPRYLATE see MHY800
METHYL CARBAMATE see MHZ000
N-METHYLCARBAMATE de 4-DIMETHYLAMINO-3-METHYL PHENYLE (FRENCH) see DOR400
METHYLCARBAMATE-1-NAPHTHALENOL see CBM750
METHYLCARBAMATE-1-NAPHTHOL see CBM750
N-METHYLCARBAMATE de 1-NAPHTYLE see CBM750
N-METHYLCARBAMATE de 1-NAPHTYLE (FRENCH) see CBM750
METHYLCARBAMIC ACID-(4-ALLYLDIMETHYLAMMONIO)-m-TOLYL ESTER, IODIDE see TGH685
METHYLCARBAMIC ACID-3-sec-BUTYL-6-CHLOROPHENYL ESTER see MOU750
METHYLCARBAMIC ACID o-sec-BUTYLPHENYL ESTER see MOV000
METHYLCARBAMIC ACID-2-CHLORO-5-(1-METHYLPROPYL)PHENYL ESTER see MOU750
METHYLCARBAMIC ACID-2-CHLORO-5-tert-PENTYLPHENYL ESTER see MIA000
METHYLCARBAMIC ACID-o-CUMENYL ESTER see MIA250
METHYLCARBAMIC ACID-m-CYM-5-YL ESTER see CQI500
METHYLCARBAMIC ACID-2,4-DICHLORO-5-ETHYL-m-TOLYL ESTER see MIA500
N-METHYL-CARBAMIC ACID-3-DIETHYLAMINOPHENYL ESTER, DIMETHYLSULFATE see HNK550
N-METHYLCARBAMIC ACID-3-DIETHYLAMINOPHENYL ESTER, METHIODIDE see MID250
N-METHYLCARBAMIC ACID-3-DIETHYLAMINOPHENYL ESTER, METHOCHLORIDE see TGH665
N-METHYL-CARBAMIC ACID-3-DIETHYLAMINOPHENYL ESTER, METHOSULFATE see HNK550
METHYLCARBAMIC ACID-m-(DIETHYLMETHYLAMMONIO)PHENYL ESTER, CHLORIDE see TGH665
N-METHYLCARBAMIC ACID-3-(DIETHYLMETHYLAMMONIO)PHENYL ESTER, IODIDE see MID250
METHYL-CARBAMIC ACID-m-(DIETHYLMETHYLAMMONIO)PHENYL ESTER, METHOSULFATE see HNK550
METHYL CARBAMIC ACID 2,3-DIHYDRO-2,2-DIMETHYL-7-BENZOFURANYL ESTER see CBS275
N-METHYLCARBAMIC ACID, m-(α-DIMETHYLAMINOETHYL)PHENYL ESTER, HYDROCHLORIDE see MIC250
N-METHYLCARBAMIC ACID-3-DIMETHYLAMINOPHENYL ESTER, ALLYL BROMIDE see TGH680
N-METHYL-CARBAMIC ACID-3-DIMETHYLAMINOPHENYL ESTER, HYDROCHLORIDE see MID000
N-METHYL CARBAMIC ACID, 4-DIMETHYLAMINOPHENYL ESTER METHIODIDE see HNP000
N-METHYLCARBAMIC ACID-3-DIMETHYLAMINOPHENYL ESTER METHOCHLORIDE see HNN000
N-METHYLCARBAMIC ACID-3-DIMETHYLAMINOPHENYL ESTER, METHOSULFATE see HNR000

N-METHYLCARBAMIC ACID-3-DIMETHYLAMINOPHENYL ESTER, PROPYL BROMIDE see TGH675
METHYLCARBAMIC ACID-5-DIMETHYLAMINO-2,4-XYLYL ESTER, HYDROCHLORIDE see MIA775
N-METHYLCARBAMIC ACID 2,4-DIMETHYL-5-DIMETHYLAMINOPHENYL ESTER, HYDROCHLORIDE see MIA775
N-METHYLCARBAMIC ACID, (3,4-DIMETHYLPHENYL)ESTER see XTJ000
METHYL-CARBAMIC ACID, ESTER with ESEROLINE see PIA500
METHYLCARBAMIC ACID, ETHYL ESTER see EMQ500
METHYLCARBAMIC ACID 4-HYDROXY-m-CUMENYL ESTER see IPI100
METHYLCARBAMIC ACID α-HYDROXY-m-CUMENYL ESTER see HLK900
METHYLCARBAMIC ACID 4-HYDROXY-3,5-DIISOPROPYLPHENYL ESTER see MIA800
METHYLCARBAMIC ACID 4-HYDROXY-2-ISOPROPOXYPHENYL ESTER see INE025
N-METHYLCARBAMIC ACID-3-ISOPROPYL-4-DIMETHYLAMINOPHENYL ESTER, METHOCHLORIDE see HJY500
METHYLCARBAMIC ACID-2,3-(ISOPROPYLIDENEDIOXY)PHENYL ESTER see DQM600
METHYLCARBAMIC ACID-m-((1-METHYL)BUTYL)PHENYL ESTER mixed with CARBAMIC ACID, METHYL-m-(1-ETHYLPROPYL)PHENYL ESTER (3:1) see BTA250
N-METHYLCARBAMIC ACID-3-METHYL-4-DIMETHYLAMINOPHENYL ESTER, ALLYLIODIDE see TGH685
N-METHYLCARBAMIC ACID-2-METHYL-5-DIMETHYLAMINOPHENYL ESTER, METHOSULFATE see TGH650
N-METHYLCARBAMIC ACID 2-METHYL-5-DIMETHYLAMINOPHENYL ESTER, METHOSULFURIC ACID see TGH655
N-METHYLCARBAMIC ACID-4-METHYL-3-DIMETHYLAMINOPHENYL ESTER, METHOSULFATE see TGH660
METHYLCARBAMIC ACID-4-METHYLTHIO-m-CUMENYL ESTER see MIB000
METHYLCARBAMIC ACID-4-METHYLTHIO-m-TOLYL ESTER see MIB250
METHYL CARBAMIC ACID-4-(METHYLTHIO)-3,5-XYLYL ESTER see DST000
METHYLCARBAMIC ACID-1-NAPHTHYL ESTER see CBM750
METHYLCARBAMIC ACID PHENYL ESTER see PFS350
METHYLCARBAMIC ACID-o-(2-PROPYNYLOXY)PHENYL ESTER see MIB500
METHYLCARBAMIC ACID-2,6-PYRIDINEDIYLDIMETHYLENE ESTER see PPH050
N-METHYL CARBAMIC ACID-8-QUINOLINYL ESTER, METHIODIDE see MID500
METHYLCARBAMIC ACID m-TOLYL ESTER see MIB750
METHYLCARBAMIC ACID, (4-TRIMETHYLAMMONIO)-m-CUMENYL ESTER, CHLORIDE see HJY500
METHYLCARBAMIC ACID, (m-(TRIMETHYLAMMONIO)PHENYL) ESTER CHLORIDE see HNN000
METHYLCARBAMIC ACID, (m-(TRIMETHYLAMMONIO)PHENYL)ESTER, IODIDE see HNO500
N-METHYLCARBAMIC ACID-p-(TRIMETHYLAMMONIO)PHENYL ESTER IODIDE see HNP000

N-METHYLCARBAMIC ACID-(3-(TRIMETHYLAMMONIO)PHENYL) ESTER, METHYLSULFATE see HNR000

METHYLCARBAMIC ACID 5-(TRIMETHYLAMMONIO)-o-TOLYL ESTER, BISULFATE see TGH655

METHYLCARBAMIC ACID-5-(TRIMETHYLAMMONIO)-o-TOLYL ESTER, CHLORIDE see HOL000

METHYLCARBAMIC ACID-3-(TRIMETHYLAMMONIO)-p-TOLYL ESTER, METHYLSULFATE see TGH660

METHYLCARBAMIC ACID-5-(TRIMETHYLAMMONIO)-o-TOLYL ESTER, METHYLSULFATE see TGH650

METHYLCARBAMIC ACID, TRIMETHYLPHENYL ESTER see TMC750

METHYLCARBAMIC ACID-3,4-XYLYL ESTER see XTJ000

METHYLCARBAMIC ESTER of α-3-HYDROXYPHENYLETHYLDIMETHYLAMINE HYDROCHLORIDE see MIC250

METHYLCARBAMIC ESTER of 3-OXYPHENYLDIMETHYLAMINE HYDROCHLORIDE see MID000

METHYLCARBAMIC ESTER of OXYPHENYLMETHYLDIETHYLAMMONIUM IODIDE see MID250

METHYLCARBAMIC ESTER of p-OXYPHENYLTRIMETHYLAMMONIUM IODIDE see HNP000

METHYLCARBAMIC ESTER of 3-OXYPHENYLTRIMETHYLAMMONIUM METHYLSULFATE see HNR000

METHYLCARBAMIC ESTER of 8-OXYQUINOLINE METHIODIDE see MID500

METHYLCARBAMOYLETHYL ACRYLATE see MID750

2-METHYL-4-CARBAMOYL-5-HYDROXYIMIDAZOLE see MID800

METHYLCARBAMOYLMETHYLAMINOMETHYLPHOSPHONIC ACID see MID850

S-METHYLCARBAMOYLMETHYL-O,O-DIMETHYL PHOSPHORODITHIOATE see DSP400

o-METHYLCARBAMOYL-2-METHYLPROPENEALDOXIME see MBW780

2-(o-(METHYLCARBAMOYL)OXIMINO)-3,3-DIMETHYLTETRAHYDRO-1,4-THIAZIN-5-ONE see MID860

2-(o-(METHYLCARBAMOYL)OXIMINO)-1,4-DITHIANE see MID870

2-(o-(METHYLCARBAMOYL)OXIMINO)-3-METHYLTETRAHYDRO-1,4-THIAZIN-5-ONE see MPU600

2-(o-(METHYLCARBAMOYL)OXIMINO)-3,3,6-TRIMETHYLTETRAHYDRO-1,4-THIAZIN-5-ONE see TMH300

(4-METHYLCARBAMOYLOXY-o-CUMENYL)TRIMETHYLAMMONIUM CHLORIDE see HJY500

(3-(N-METHYLCARBAMOYLOXY)PHENYL)ALLYLDIMETHYLAMMONIUM BROMIDE see TGH680

(3-(N-METHYLCARBAMOYLOXY)PHENYL)DIETHYLMETHYLAMMONIUM CHLORIDE see TGH665

(3-(N-METHYLCARBAMOYLOXY)PHENYL)DIETHYLMETHYL-AMMONIUM IODIDE see MID250

(3-(N-METHYLCARBAMOYLOXY)PHENYL)DIMETHYLPROPYLAMMONIUM BROMIDE see TGH675

((p-METHYLCARBAMOYLOXY)PHENYL)TRIETHYLAMMONIUM IODIDE see HNP000

(3-(METHYLCARBAMOYLOXY)PHENYL)TRIMETHYLAMMONIUM CHLORIDE see HNN000

(3-(METHYLCARBAMOYLOXY)PHENYL)TRIMETHYLAMMONIUM IODIDE see HNO500

(4-(N-METHYLCARBAMOYLOXY)PHENYL)TRIMETHYLAMMONIUM IODIDE see HNP000

(3-(N-METHYLCARBAMOYLOXY)PHENYL)TRIMETHYLAMMONIUM METHYLSULFATE see HNR000

(3-(N-METHYLCARBAMOYLOXY)PHENYL)TRIMETHYL-ARSONIUM IODIDE see MID900

METHYL N-(CARBAMOYLOXY)THIOACETIMIDATE see MPS250

(4-METHYLCARBAMOYLOXY-o-TOLYL)ALLYLDIMETHYLAMMONIUM IODIDE see TGH685

(3-(METHYLCARBAMOYLOXY)-p-TOLYL)TRIMETHYLAMMONIUM BISULFATE see TGH655

(3-(METHYLCARBAMOYLOXY)-p-TOLYL)TRIMETHYLAMMONIUM METHYLSULFATE see TGH650

o-METHYLCARBANILIC ACID-N-ETHYL-3-PIPERDINYL ESTER see MIE250

9-METHYL-9H-CARBAZOL-3-AMINE see MMI700

9-METHYLCARBAZOLE see MIE300

N-METHYLCARBAZOLE see MIE300

1-METHYL-1-CARBETHOXY-4-PHENYL HEXAMETHYLENIMINE CITRATE see MIE600

1-METHYL-4-CARBETHOXY-4-PHENYLPIPERIDINE HYDROCHLORIDE see DAM700

METHYL CARBETHOXYSYRINGOYL RESERPATE see RCA200

METHYLCARBINOL see EFU000

METHYL CARBITOL see DJG000

METHYL CARBITOL ACETATE see MIE750

3-METHYL-4-CARBOLINE see MPA050

2-METHYL-β-CARBOLINE see MPA050

N-METHYL-β-CARBOLINE-3-CARBOXAMIDE see MPA065

METHYL-4-CARBOMETHOXY BENZOATE see DUE000

METHYL CARBONATE see MIF000

METHYLCARBONYL FLUORIDE see ACM000

METHYLCARBOPHENOTHION see MQH750

17-β-(1-METHYL-3-CARBOXYPROPYL)ETHIOCHOLAN-3-α-OL see LHW000

17-β-(1-METHYL-3-CARBOXYPROPYL)-ETIOCHOLANE-3-α,12-DIOL see DAQ400

17-β-(1-METHYL-3-CARBOXYPROPYL)ETIOCHOLANE-3-α,7-β-DIOL see DMJ200

N-METHYL-N-(3-CARBOXYPROPYL)NITROSAMINE see MIF250

METHYLCATECHOL see GKI000

3-METHYLCATECHOL see DNE000

4-METHYLCATECHOL see DNE200

METHYL-CCNU see CHD250

trans-METHYL-CCNU see CHD250

METHYL CEDRYL ETHER see CCR525

METHYL CELLOSOLVE ACETATE (OSHA, DOT) see EJJ500

METHYL CELLOSOLVE ACETYLRICINOLEATE see MIF500

METHYL CELLOSOLVE ACRYLATE see MEM250

METHYL CELLOSOLVE ACRYLATE see MIF750

METHYL CELLOSOLVEAT OLEATE see MEP750

METHYL CELLOSOLVE (OSHA, DOT) see EJH500

METHYL CELLOSOLYE ACETAAT (DUTCH) see EJJ500

METHYL CELLULOSE see MIF760

METHYL CELLULOSE-A see MIF760

METHYL CELLULOSE ETHER see MIF760

METHYL CENTRALITE see DRB200

4-METHYLCHALCONE see MIF762

p-METHYLCHALCONE see MIF762

β-METHYLCHALCONE see MPL000

METHYL CHAVICOL see AFW750

METHYL CHEMOSEPT see HJL500

6-METHYL-CHINOXALIN-2,3-DITHIOL-CYCLO-CARBONAT (GERMAN) see ORU000

METHYLCHLOORFORMIAT (DUTCH) see MIG000

METHYLCHLOR see MIF763

p,p-METHYLCHLOR see MIF763

METHYL-2-CHLORAETHYLNITROSAMIN (GERMAN) see CIQ500

METHYLCHLORID (GERMAN) see MIF765

METHYL CHLORIDE see MIF765

METHYL CHLORIDE−METHYLENE CHLORIDE MIXTURE see CHX750

p-METHYLCHLOROACETANILIDE see CEB500

4-METHYL-α-CHLOROACETANILIDE see CEB500

METHYL CHLOROACETATE see MIF775

METHYL CHLOROACETATE (DOT) see MIF775

METHYL γ-CHLOROACETOACETATE see MIH600

METHYL-2-CHLOROACRYLATE see MIF800

METHYL-α-CHLOROACRYLATE see MIF800

2-METHYL-4-CHLOROANILINE see CLK220

4-METHYL-2-CHLOROANILINE see CLK210

2-METHYL-4-CHLOROANILINE HYDROCHLORIDE see CLK235

4-METHYL-3-CHLOROANILINE HYDROCHLORIDE see CLK230

2-(2'-METHYL-3'-CHLORO)ANILINONICOTINIC ACID see CMX770

2-METHYLCHLOROBENZENE see CLK100

1-METHYL-2-CHLOROBENZENE see CLK100

N-METHYL-N'-(4-CHLOROBENZHYDRYL)PIPERAZINE DIHYDROCHLORIDE see CDR000

o-METHYL-o-2-CHLORO-4-tert-BUTYLPHENYL-N-METHYLAMIDOPHOSPHATE see COD850

METHYL CHLOROCARBONATE see MIG000

METHYL (CHLOROCARBONYL)(4-(TRIFLUOROMETHOXY)PHENYL)CARBAMATE see MIG100

METHYL-2-CHLORO-3-(4-CHLOROPHENYL)PROPIONATE see CFC750

2-METHYL-5-CHLORO-2-(N,N-DIETHYLAMINOETHOXYETHYL)-1,3-BENZODIOXOLE see CFO750

METHYL-β-CHLOROETHYLAMINE HYDROCHLORIDE see MIG250

N'-METHYL-N'-β-CHLOROETHYLBENZALDEHYDE HYDRAZONE see MIG500

N'-METHYL-N'-β-CHLOROETHYL-(p-DIMETHYLAMINO)-BENZALDEHYDE HYDRAZONE see MIG750

1-METHYL-1-(β-CHLOROETHYL)ETHYLENIMONIUM see MIG800

METHYL-β-CHLOROETHYL-ETHYLENIMONIUM PICRYLSULFONATE see MIG850

1-METHYL-1-(β-CHLOROETHYL)ETHYLENIMONIUM PICRYLSULFONATE see MIG850
METHYL-β-CHLOROETHYL-β-HYDROXYETHYLAMINE HYDROCHLORIDE see MIH000
METHYL(2-CHLOROETHYL)NITROSAMINE see CIQ500
METHYL-N-(β-CHLOROETHYL)-N-NITROSOCARBAMATE see MIH250
4-METHYL-5-(β-CHLOROETHYL)THIAZOLE see CHD750
5-METHYL-3-(2-CHLORO-6-FLUOROPHENYL)-4-ISOXAZOLYLPENICILLIN SODIUM see FDA100
METHYL CHLOROFORM see MIH275
METHYL CHLOROFORMATE (DOT) see MIG000
METHYL-2-CHLORO-9-HYDROXYFLUORENE-9-CARBOXYLATE see CDT000
1-METHYL-5-CHLOROINDOLINE METHYLBROMIDE see MIH300
METHYLCHLOROMETHYL ETHER (DOT) see CIO250
METHYL CHLOROMETHYL ETHER, anhydrous (DOT) see CIO250
4-METHYL-6-(((2-CHLORO-4-NITRO)PHENYL)AZO)-m-ANISIDINE see MIH500
METHYL 4-CHLORO-3-OXOBUTANOATE see MIH600
3-METHYL-4-CHLOROPHENOL see CFD990
2-METHYL-4-CHLOROPHENOXYACETIC ACID see CIR250
(2-METHYL-4-CHLOROPHENOXY)ACETIC ACID, DIETHANOLAMINE SALT see MIH750
2-METHYL-4-CHLOROPHENOXYACETIC ACID POTASSIUM SALT see MIH800
(2-METHYL-4-CHLOROPHENOXY)ACETIC ACID, SODIUM SALT see SIL500
2-METHYL-4-CHLOROPHENOXYBUTYRIC ACID see CLN750
4-(METHYL-4-CHLOROPHENOXY)BUTYRIC ACID see CLN750
γ-2-METHYL-4-CHLOROPHENOXYBUTYRIC ACID see CLN750
4-(2-METHYL-4-CHLOROPHENOXY)BUTYRIC ACID, SODIUM SALT see CLO000
2-(2-METHYL-4-CHLOROPHENOXY)PROPIONIC ACID see CIR500
α-(2-METHYL-4-CHLOROPHENOXY)PROPIONIC ACID see CIR500
2-METHYL-4-CHLOROPHENOXY-α-PROPIONIC ACID see CIR500
2-(2-METHYL-4-CHLOROPHENOXY)PROPIONIC ACID, DIETHANOLAMINE SALT see MIH900
N-(METHYL-3-CHLOROPHENYL)ANTHRANILIC ACID see CLK325
2-METHYL-3-(2-CHLOROPHENYL)CHINAZOLON-4 see MIH925
2-METHYL-3-(4'-CHLOROPHENYL)CHINAZOLON-(4) (GERMAN) see MII250
N'-(2-METHYL-4-CHLOROPHENYL)-N,N-DIMETHYLFORMAMIDINE see CJJ250
3-METHYL-4-((o-CHLOROPHENYL)HYDRAZONE)-4,5-ISOXAZOLEDIONE see MLC250

3-METHYL-4-(o-CHLOROPHENYLHYDRAZONO)-5-ISOXAZOLONE see MLC250
1-((2-METHYL-4-CHLOROPHENYL)METHYL)-INDAZOLE-3-CARBOXYLIC ACID see TGJ875
1-((2-METHYL-4-CHLOROPHENYL)METHYL)-1H-INDAZOLE-3-CARBOXYLIC ACID see TGJ875
N-METHYL-N'-(p-CHLOROPHENYL)-N-NITROSOUREA see MMW775
2-METHYL-4-(p-CHLOROPHENYL)-2,4-PENTANEDIOL see CKG000
METHYL-2-(4-(p-CHLOROPHENYL)PHENOXY)-2-METHYLPROPIONATE see MIO975
2-METHYL-3-(o-CHLOROPHENYL)-4-QUINAZOLINONE see MIH925
2-METHYL-3-(4-CHLOROPHENYL)-4(3H)-QUINAZOLINONE see MII250
METHYL-4-CHLOROPHENYL SULFIDE see CKG500
METHYL-p-CHLOROPHENYL SULFIDE see CKG500
METHYL-4-CHLOROPHENYL SULFONE see CKG750
METHYL-4-CHLOROPHENYL SULFOXIDE see CKH000
METHYL CHLOROPHOS see TIQ250
METHYLCHLOROPINDOL see CMX850
METHYL-2-CHLOROPROPIONATE (DOT) see CKT000
2-METHYL-6-CHLORO-4-QUINAZOLINONE see MII750
METHYL CHLOROSULFONATE see MIJ000
METHYLCHLOROTHIAZIDE see MIV500
METHYLCHLOROTHION see MIJ250
METHYL-3-((5-(2-CHLORO-4-(TRIFLUOROMETHYL)PHENOXY)-2-NITROPHENYL)AMINO)BUTYRATE see MIJ275
4-(2-METHYL-4-CHLORPHENOXY)-BUTTERSAEURE (GERMAN) see CLN750
2-METHYL-4-CHLORPHENOXYESSIGSAEURE (GERMAN) see CIR250
2-(2-METHYL-4-CHLORPHENOXY)-PROPIONSAEURE (GERMAN) see CIR500
N'-(2-METHYL-4-CHLORPHENYL)-FORMAMIDIN-HYDROCHLORID (GERMAN) see CJJ250
METHYLCHLORPINDOL see CMX850
N-METHYLCHLORPROMAZINE IODIDE see MIJ300
METHYL CHLORPYRIFOS see CMA250
METHYLCHLORTETRACYCLINE see MIJ500
METHYLCHOLANTHRENE see MIJ750
3-METHYLCHOLANTHRENE see MIJ750
4-METHYLCHOLANTHRENE see MIK250
5-METHYLCHOLANTHRENE see MIK000
20-METHYLCHOLANTHRENE see MIJ750
22-METHYLCHOLANTHRENE see MIK250
20-METHYLCHOLANTHRENE CHOLEIC ACID see MIK500
3-METHYLCHOLANTHRENE COMPOUND with PICRIC ACID (1:1) see MIL750
3-METHYLCHOLANTHRENE COMPOUND with 1,3,5-TRINITROBENZENE (1:1) see MIM000
(E)-3-METHYLCHOLANTHRENE-11,12-DIHYDRODIOL see DML800
cis-3-METHYLCHOLANTHRENE-1,2-DIOL see MIK750
3-METHYLCHOLANTHRENE-11,12-EPOXIDE see ECA500
3-METHYLCHOLANTHRENE-2-ONE see MIL250
3-METHYLCHOLANTHRENE-11,12-OXIDE see MIL500

20-METHYLCHOLANTHRENE PICRATE see MIL750
20-METHYLCHOLANTHRENE-TRINITROBENZENE see MIM000
3-METHYLCHOLANTHREN-1-OL see HMA000
3-METHYL-1-CHOLANTHRENOL see HMA000
3-METHYLCHOLANTHREN-2-OL see HMA500
3-METHYLCHOLANTHREN-1-ONE see MIM250
3-METHYLCHOLANTHREN-2-ONE see MIL250
20-METHYLCHOLANTHREN-15-ONE see MIM250
3-METHYLCHOLANTHRYLENE see DAL200
20-METHYLCHOLANTHRYLENE see DAL200
β-METHYLCHOLINE CHLORIDE see MIM300
β-METHYLCHOLINE CHLORIDE CARBAMINOYL see HOA500
β-METHYLCHOLINE CHLORIDE URETHAN see HOA500
1-METHYLCHRYSENE see MIM500
2-METHYLCHRYSENE see MIM750
3-METHYLCHRYSENE see MIN000
4-METHYLCHRYSENE see MIN250
5-METHYLCHRYSENE see MIN500
6-METHYLCHRYSENE see MIN750
anti-5-METHYLCHRYSENE-1,2-DIOL-3,4-EPOXIDE see MIN800
α-METHYLCINNAMALDEHYDE see MIO000
METHYL CINNAMATE see MIO500
METHYL CINNAMIC ALCOHOL see MIO750
METHYL CINNAMIC ALDEHYDE see MIO000
α-METHYLCINNAMIC ALDEHYDE see MIO000
α-METHYLCINNAMYL ALCOHOL see MIO750
METHYL CINNAMYLATE see MIO500
α-METHYLCINNIMAL see MIO000
4-METHYL CINNOLINE see MIO770
4-METHYL CINNOLINE see MIO800
METHYLCISTOX see DAO800
METHYL CLOFENAPATE see MIO975
METHYLCLOTHIAZIDE see MIV500
METHYLCOBALAMIN see VSZ050
METHYL COBALAMINE see VSZ050
METHYLCOLCHAMINONE see MGF000
METHYLCOLCHICINE see MIW500
METHYL COPPER see MIP250
4-METHYLCOUMARIN see MIP500
6-METHYLCOUMARIN see MIP750
7-METHYLCOUMARIN see MIP775
6-METHYLCOUMARINIC ANHYDRIDE see MIP750
6-METHYL-m-CRESOL see XKS000
3-METHYLCROTONAMIDE see SBW965
METHYL CROTONATE see COB825
METHYL trans-CROTONATE see COB825
METHYL α-CROTONATE see COB825
(Z)-3-((Z)-2-METHYLCROTONATE)4,9-EPOXYCEVANE-3-β,4-β,12,14,16-β,17,20-HEPTOL see CDG000
3-METHYLCROTONIC ACID see MHT500
(E)-2-METHYLCROTONIC ACID see TGA700
trans-2-METHYLCROTONIC ACID see TGA700
β-METHYLCROTONIC ACID see MHT500
3-METHYLCROTONIC ACID 2-sec-BUTYL-4,6-DINITROPHENYL ESTER see BGB500
3-METHYLCROTONIC ACID, ETHYL ESTER see MIP800
γ-METHYL-α,β-CROTONOLACTONE see MKH500

γ-METHYL-β,γ-CROTONOLACTONE see MKH250

p-METHYL-CUMENE see CQI000

METHYL CYANIDE see ABE500

METHYL CYANOACETATE see MIQ000

METHYL 2-CYANOACETATE see MIQ000

METHYL CYANOACRYLATE see MIQ075

METHYL 2-CYANOACRYLATE see MIQ075

METHYL α-CYANOACRYLATE see MIQ075

9-METHYL-10-CYANO-1,2-BENZANTHRACENE see MIQ250

4-METHYLCYANOBENZENE see TGT750

METHYL 2-CYANO-3-(2-BROMOPHENYL)ACRYLATE see MIQ300

METHYL CYANOCARBAMATE DIMER see MIQ350

METHYL CYANOETHANOATE see MIQ000

METHYL β-CYANOETHYL ETHER see MFL750

d-6-METHYL-8-CYANOMETHYLERGOLINE see MIQ400

METHYL-(E)-2-(2-(6-(2-CYANOPHENOXY)PYRIMIDIN-4-YLOXY)PHENYL)-3-METHOXYACRILATE see ASP525

o-METHYLCYCLIZINE DIHYDROCHLORIDE see MIQ725

1-METHYLCYCLODODECYL METHYL ETHER see MEU300

N-METHYLCYCLOHEXANAMINE see MIT000

METHYLCYCLOHEXANE see MIQ740

1-METHYL-2,4-CYCLOHEXANEDIAMINE see DBY000

4-METHYL-1,3-CYCLOHEXANEDIAMINE see DBY000

(±)-α-METHYLCYCLOHEXANEETHYLAMINE HYDROCHLORIDE see CPG625

α-METHYLCYCLOHEXANEMETHANOL see CPL125

METHYLCYCLOHEXANOL see MIQ745

m-METHYLCYCLOHEXANOL see MIQ750

o-METHYLCYCLOHEXANOL see MIR000

METHYLCYCLOHEXANOL (ACGIH,DOT,OSHA) see MIQ745

METHYL CYCLOHEXANOLS, Fp not >60.5 degrees C (DOT) see MIQ745

2-METHYL-CYCLOHEXANON (GERMAN, DUTCH) see MIR500

METHYLCYCLOHEXANONE see MIR250

2-METHYLCYCLOHEXANONE see MIR500

3-METHYLCYCLOHEXANONE see MIR600

4-METHYLCYCLOHEXANONE see MIR625

o-METHYLCYCLOHEXANONE see MIR500

1-METHYLCYCLOHEXAN-2-ONE see MIR500

METHYL-3-CYCLO-HEXANONE-1 (FRENCH) see MIR600

METHYL-4-CYCLO-HEXANONE-1 (FRENCH) see MIR625

4-METHYLCYCLOHEXENE see MIR750

2-METHYL-4-CYCLOHEXENE-1-CARBOXALDEHYDE see MPO000

6-METHYL-3-CYCLOHEXENE-1-CARBOXALDEHYDE see MPO000

N-METHYL-4-CYCLOHEXENE-1,2-DICARBOXIMIDE see MIS250

1-((6-METHYL-3-CYCLOHEXEN-1-YL)CARBONYL)PIPERIDINE see MIS500

(METHYL-3-CYCLOHEXENYL)METHANOL see MIS750

N-METHYL-5-CYCLOHEXENYL-5-METHYLBARBITURIC ACID see ERD500

METHYLCYCLOHEXYLAMINE see MIT000

N-METHYLCYCLOHEXYLAMINE see MIT000

S-2-((4-(4-METHYLCYCLOHEXYL)BUTYL)AMINO)ETHYL THIOSULFATE see MIT250

METHYLCYCLOHEXYLCARBINOL see CPL125

1-(cis/trans-4-METHYLCYCLOHEXYL)ETHANOL see MIT615

METHYL CYCLOHEXYLFLUOROPHOSPHONATE see MIT600

2-METHYLCYCLOHEXYLMETHANOL see CPL125

N-METHYL-N-(CYCLOHEXYLMETHYL) AMINE see MIT610

4-METHYLCYCLOHEXYLMETHYLCARBINOL see MIT615

METHYLCYCLOHEXYLNITROSAMIN (GERMAN) see NKT500

METHYLCYCLOHEXYLNITROSAMINE see NKT500

N-METHYL-N-(2-CYCLOHEXYL-2-PHENYL-1,3-DIOXOLAN-4-YL-METHYL)-PIPERIDINIUM IODIDE see OLW400

3-METHYLCYCLOPENTADECANONE see MIT625

3-METHYL-1-CYCLOPENTADECANONE see MIT625

2-METHYL-1,3-CYCLOPENTADIENE see MIT700

METHYLCYCLOPENTADIENYL MANGANESE TRICARBONYL see MAV750

2-METHYLCYCLOPENTADIENYL MANGANESETRICARBONYL see MAV750

2-METHYLCYCLOPENTADIENYL MANGANESE TRICARBONYL (ACGIH) see MAV750

METHYLCYCLOPENTADIENYL MANGANESE TRICARBONYL (OSHA) see MAV750

METHYLCYCLOPENTANE see MIU500

METHYL CYCLOPENTANE (DOT) see MIU500

3-METHYLCYCLOPENTANE-1,2-DIONE see HMB500

17-METHYL-15H-CYCLOPENTA(a)PHENANTHRENE see MIU750

METHYL CYCLOPENTENOLONE (FCC) see HMB500

10-METHYL-1,2-CYCLOPENTENOPHENANTHRENE see MIV250

N-METHYL-1-CYCLOPHENE-1,2-DICARBOXIMIDE see MPP250

METHYLCYCLOPROPANECARBONYLHYDRAZINE see MIV300

METHYL CYCLOPROPYL ETHER see CQE750

METHYLCYCLOTHIAZIDE see MIV500

3-METHYLCYKLOHEXANONPEROXID (CZECH) see BKR250

METHYLCYKLOPENTADIENTRIKARBONYLMANGANIUM see MAV750

METHYL CYSTEINE HYDROCHLORIDE see MBX800

2-METHYL-DAB see TLE750

3'-METHYL-DAB see DUH600

3-METHYL-4-DAB see MJF000

METHYL-DBCP see MBV720

METHYL-DDT see MIF763

N-METHYL-N-DEACETYLCOLCHICINE see MIW500

(E)-N-(2-METHYLDECAHYDROISOQUINOL-5-YL)-3,4,5-TRIMETHOXY-BENZAMIDE see DAE700

trans-N-(2-METHYLDECAHYDROISOQUINOL-5-YL)-3,4,5-TRIMETHOXYBENZAMIDE see DAE700

α-METHYL DECALACTONE see MIW050

2-METHYL-1-DECANAL see MIW000

METHYL DECANOATE see MHY650

METHYL n-DECANOATE see MHY650

4-METHYLDECANOLIDE see MIW050

4-METHYL-cis-DECENE γ-LACTONE see MIW060

METHYL DECYL ETHER see MIW075

METHYL n-DECYL ETHER see MIW075

6-METHYL-6-DEHYDRO-17-α-ACETOXYPROGESTERONE see VTF000

6-METHYL-6-DEHYDRO-17-α-ACETYLPROGESTERONE see VTF000

6-METHYL-6-DEHYDRO-17-METHYLPROGESTERONE see MBZ100

N-METHYLDEMECOLCINE see MIW500

METHYL DEMETON see MIW100

O-METHYLDEMETON see DAO800

METHYL DEMETON METHYL see MIW250

METHYL-DEMETON-O see DAO800

METHYL DEMETON-O-SULFOXIDE see DAP000

METHYL DEMETON THIOESTER see DAP400

β-d-METHYL-2-DEOXY-2-(3-METHYL-3-NITROSOUREIDO)GLUCOPYRANOSIDE see MPJ800

5-METHYLDEOXYURIDINE see TFX790

N-METHYL-N-DESACETYLCOLCHICINE see MIW500

N-METHYL-N-DESACETYLCOLCHICINE see MIW500

METHYL 2,4-D ESTER see DAA825

METHYL DIACETOACETATE see MIW750

2-METHYLDIACETYLBENZIDINE see MIX000

2-METHYL-N,N'-DIACETYLBENZIDINE see MIX000

METHYLDIAETHYLCARBINOL (GERMAN) see MNJ100

N-METHYLDIAMINOETHANE see MJW100

2'-METHYL-2,4-DIAMINO-3-METHYLAZOBENZENE see MIX100

2'-METHYL-2,4-DIAMINO-5-METHYLAZOBENZENE see MLP770

METHYLDIAZENE see MIX250

METHYL DIAZEPINONE see DCK759

METHYL DIAZINE see MIX250

3-METHYLDIAZIRINE see MIX500

METHYL DIAZOACETATE see MIX750

N-METHYLDIAZOACETYLGLYCINE AMIDE see DCO200

7-METHYLDIBENZ(c,h)ACRIDINE see MIY000

14-METHYLDIBENZ(a,h)ACRIDINE see MIY200

14-METHYLDIBENZ(a,j)ACRIDINE see MIY250

9-METHYL-3,4,5,6-DIBENZACRIDINE see MIY000

10-METHYL-1,2:5,6-DIBENZACRIDINE see MIY200

10-METHYL-3,4,5,6-DIBENZACRIDINE see MIY250

2-METHYLDIBENZ(a,h)ANTHRACENE see MIY500

3-METHYLDIBENZ(a,h)ANTHRACENE see MIY750

6-METHYL DIBENZ(a,h)ANTHRACENE see MIZ000

10-METHYLDIBENZ(a,c)ANTHRACENE see MJA000

4-METHYL-1,2,5,6-DIBENZANTHRACENE see MIZ000

2'-METHYL-1:2:5:6-DIBENZANTHRACENE see MIY500

3'-METHYL-1:2:5:6-DIBENZANTHRACENE see MIY750

N-METHYL-3:4:5:6-DIBENZCARBAZOLE see MJA250

N-METHYL-7H-DIBENZO(c,g)CARBAZOLE see MJA250

5-METHYL-DIBENZO(b,def)CHRYSENE see MJA500

10-METHYLDIBENZO(def,p)CHRYSENE see MJB000

12-METHYL DIBENZO(def,mno)CHRYSENE see MGP250

14-METHYLDIBENZO(b,def)CHRYSENE-7-CARBOXALDEHYDE see FNV000

6-METHYLDIBENZO(def,mno)CHRYSENE-12-CARBOXALDEHYDE see FNQ000

1-METHYL-4-(5-DIBENZO(a,e)CYCLOHEPTATRIENYLIDENE)PIPERIDINE HYDROCHLORIDE see PCI250

N-METHYL-5H-DIBENZO(a,d)CYCLOHEPTENE-5-PROPYLAMINE HYDROCHLORIDE see POF250

7-METHYLDIBENZO(h,rst)PENTAPHENE see MJA750

5-METHYL-1,2,3,4-DIBENZOPYRENE see MJB000

2'-METHYL-1,2:4,5-DIBENZOPYRENE see MMD000

3'-METHYL-1,2:4,5-DIBENZOPYRENE see MMD250

5-METHYL-3,4:8,9-DIBENZOPYRENE (FRENCH) see MJA500

7-METHYL-1:2:3:4-DIBENZPYRENE see MJB250

5-METHYL-3,4,9,10-DIBENZPYRENE (FRENCH) see MHE250

METHYLDIBORANE see MJB300

2-METHYL-1,2-DIBROMO-3-CHLOROPROPANE see MBV720

METHYLDIBROMOGLUTARONITRILE see DDM500

N-METHYL-DIBROMOMALEINIMIDE see MJB360

4-METHYL-2,6-DI-terc.BUTYLFENOL (CZECH) see BFW750

METHYL-DI-tert-BUTYLPHENOL see BFW750

4-METHYL-2,6-DI-tert-BUTYLPHENOL see BFW750

METHYLDICHLORARSINE see DFP200

METHYL DICHLOROACETATE (DOT) see DEM800

METHYLDICHLOROARSINE (DOT) see DFP200

O-METHYL-O-2,5-DICHLORO-4-BROMOPHENYL PHENYLTHIOPHOSPHONATE see LEN000

METHYL-3,4-DICHLOROCARBANILATE see DEV600

N-METHYL-2,2'-DICHLORODIETHYLAMINE see BIE250

N-METHYL-2,2'-DICHLORODIETHYLAMINE HYDROCHLORIDE see BIE500

N-METHYL-2,2'-DICHLORODIETHYLAMINE-N-OXIDE HYDROCHLORIDE see CFA750

METHYL DICHLOROETHANOATE see DEM800

METHYLDI(2-CHLOROETHYL)AMINE see BIE250

METHYLDI(2-CHLOROETHYL)AMINE HYDROCHLORIDE see BIE500

N-METHYL-DI-2-CHLOROETHYLAMINE HYDROCHLORIDE see BIE500

METHYLDI(β-CHLOROETHYL)AMINE HYDROCHLORIDE see BIE500

N-METHYL-DI-2-CHLOROETHYLAMINE-N-OXIDE see CFA500

METHYLDI(2-CHLOROETHYL)AMINE-N-OXIDE HYDROCHLORIDE see CFA750

p-METHYL-DI-(2-CHLOROETHYL)-BENZYLAMINE HYDROCHLORIDE see EAK100

N-METHYLDICHLOROMALEINIMIDE see DFP800

METHYL 3,4-DICHLORO-5-NITRO-2-FURANCARBOXYLATE see MJB550

METHYL 3,4-DICHLORO-5-NITRO-2-FUROATE see MJB550

METHYL 5-(2,4-DICHLOROPHENOXY)-2-NITROBENZOATE see MJB600

METHYL 2-(4-(2,4-DICHLOROPHENOXY)PHENOXY)PROPIONATE see IAH050

METHYL-N-(3,4-DICHLOROPHENYL)CARBAMATE see DEV600

METHYL α,4-DICHLOROPHENYLPROPANOATE see CFC750

METHYL DICHLOROSILANE (DOT) see DFS000

METHYL-DICHLORSILAN (CZECH) see DFS000

N-METHYLDICYCLOHEXYLAMINE see MJC750

N-METHYL-N-(β-DICYCLOHEXYLAMINOETHYL)PIPERIDINE BROMIDE see MJC775

METHYL(2-α-5-β,6-α-18-β)-3,4-DIDEHYDROIBOGAMINE-18-CARBOXYLATE see CCP925

METHYL DIEPOXYDIALLYLACETATE see MJD000

METHYLDIETHANOLAMINE see MKU250

N-METHYLDIETHANOLAMINE see MKU250

N-METHYLDIETHANOLIMINE see MKU250

METHYL (DIETHOXYPHOSPHINYL)ACETATE see MJD100

2-METHYL-4-DIETHYLAMINOPENTAN-5-OL p-AMINOBENZOATE see LEU000

METHYL 3-(3-(DIETHYLAMINO)PROPYLCARBAMOYLMETHOXY)-p-CYMENE-2-CARBOXYLATE see MJD200

5-METHYL-7-DIETHYLAMINO-s-TRIAZOLO-(1,5-a)PYRIMIDINE see DIO200

2-METHYL-N,N-DIETHYLBENZAMIDE see DKD000

3-METHYL-N,N-DIETHYLBENZAMIDE see DKC800

1-METHYL-4-DIETHYLCARBAMOYLPIPERAZINE CITRATE see DIW200

1-METHYL-4-DIETHYLCARBAMYLPIPERAZINE see DIW000

METHYLDIETHYLCARBINOL see MNJ100

METHYL DIETHYL CARBINOLURETHAN see ENF000

o-METHYL S-(o',o'-DIETHYLPHOSPHORAMIDO)ETHYL PHOSPHOROMORPHOLINOTHIOATE see MJD300

METHYL 2,2-DIFLUOROMALONYL FLUORIDE see MJD400

METHYL DIFLUOROPHOSPHITE see MJD275

METHYLDIGOXIN see MJD300

4'''-METHYLDIGOXIN see MJD300

4'''-o-METHYLDIGOXIN see MJD300

β-METHYLDIGOXIN see MJD300

4'''-β-METHYLDIGOXIN see MJD300

4'''-o-METHYLDIGOXIN ACETONE (2:1) see MJD500

METHYL-DIHYDROARTEMISININE see ARL425

6-METHYL-3,4-DIHYDRO-2H-1-BENZOTHIOPYRAN-7-SULFONAMIDE 1,1-DIOXIDE see MQQ050

trans-3-METHYL-7,8-DIHYDROCHOLANTHRENE-7,8-DIOL see MJD600

trans-3-METHYL-9,10-DIHYDROCHOLANTHRENE-9,10-DIOL see MJD610

trans-3-METHYL-11,12-DIHYDROCHOLANTHRENE-11,12-DIOL see DML800

11-METHYL-15,16-DIHYDRO-17H-CYCLOPENTA(a)PHENANTHRENE see MJD750

11-METHYL-15,16-DIHYDRO-17H-CYCLOPENTA(a)PHENANTHREN-17-ONE see MJE500

3-METHYL-2,3-DIHYDRO-9H-ISOXAZOLO(3,2-b)QUINAZOLIN-9-ONE see MJE250

METHYL DIHYDROJASMONATE see HAK100

11-METHYL-15,16-DIHYDRO-17-OXOCYCLOPENTA(a)PHENANTHRENE see MJE500

2-(8-METHYL-10,11-DIHYDRO-11-OXODIBENZ(b,f)OXEPIN-2-YL)PROPIONIC see DLI650

1-METHYL-1,6-DIHYDROPICOLINALDEHYDE OXIME HYDROCHLORIDE see PMJ550

METHYLDIHYDROPYRAN see MJE750

N-METHYL-1,6-DIHYDROPYRIDINE-2-CARBALDOXIME HYDROCHLORIDE see PMJ550

1-METHYL-1,6-DIHYDRO-2-PYRIDINECARBOXYALDEHYDE OXIME HYDROCHLORIDE see PMJ550

METHYLDIHYDROTESTOSTERONE see MJE760

17-α-METHYLDIHYDROTESTOSTERONE see MJE760

2-METHYL-1,3-DIHYDROXYBENZENE see MPH400

(3-(2R,3S))-4-METHYL-2,3-DIHYDROXY-2-(3-METHYLBUTYL)BUTANEDIOATE (ESTER) CEPHALOTAXINE see IKS500

METHYL (±)-11-α-16-DIHYDROXY-16-METHYL-9-OXOPROST-13-EN-1-OATE see MJE775

α-METHYL-l-3,4-DIHYDROXYPHENYLALANINE see DNA800

α-METHYL-l-3,4-DIHYDROXYPHENYLALANINE see MJE780

α-METHYL-L-3,4-DIHYDROXYPHENYLALANINE see MJE780

l-α-METHYL-3,4-DIHYDROXYPHENYLALANINE see DNA800

α-METHYL-β-(3,4-DIHYDROXYPHENYL)-l-ALANINE see DNA800

l-(−)-α-METHYL-β-(3,4-DIHYDROXYPHENYL)ALANINE see DNA800

3-METHYL-3,4-DIHYDROXY-4-PHENYL-BUTIN-1 (GERMAN) see DMX800

3-METHYL-3,4-DIHYDROXY-4-PHENYL-1-BUTYNE see DMX800

METHYL-(β-(3,4-DIHYDROXY PHENYL ETHYL) AMINE HYDROCHLORIDE see EAZ000

METHYLDIIODOARSINE see MGQ775

o-METHYL S-(o',o'-DIISOPROPYLPHOSPHORAMIDO)ETHYL PHENYLPHOSPHONOTHIOATE see MJE790

4-METHYL-1,2-DIMERCAPTOBENZENE see TGN000

METHYL 6,7-DIMETHOXY-4-ETHYL-β-CARBOLINE-3-CARBOXYLATE see EID600

METHYL 3,3-DIMETHOXY-2-OXA-7,10-DIAZA-3-SILATRIDECAN-13-OATE see MJE793

1-METHYL-2,6-DI-(p-METHOXYPHENETHYL)PIPERIDINE ETHANESULFONATE see MJE800

METHYL-3-(DIMETHOXYPHOSPHINYLOXY)CROTONATE see MQR750

METHYLDIMETHOXYSILANE see MJE900

5-METHYL-10-β-DIMETHYLAMINOAETHYL-10,11-DIHYDRO-11-OXO-5-DIBENZO(b,e)(1,4)DIAZEPIN see DCW800

3'-METHYL-4-DIMETHYLAMINOAZOBENZEN (CZECH) see DUH600

2-METHYL-4-DIMETHYLAMINOAZOBENZENE see TLE750

3-METHYL-4-DIMETHYLAMINOAZOBENZENE see MJF000

2'-METHYL-4-DIMETHYLAMINOAZOBENZENE see DUH800

3'-METHYL-4-DIMETHYLAMINOAZOBENZENE see DUH600

4'-METHYL-4-DIMETHYLAMINOAZOBENZENE see DUH400

o'-METHYL-p-DIMETHYLAMINOAZOBENZENE see DUH800

p'-METHYL-p-DIMETHYLAMINOAZOBENZENE see DUH400

2-METHYL-N,N-DIMETHYL-4-AMINOAZOBENZENE see DUH800

2-METHYL-N,N-DIMETHYL-4-AMINOAZOBENZENE see TLE750

3'-METHYL-N,N-DIMETHYL-4-AMINOAZOBENZENE see DUH600

3'-METHYLDIMETHYLAMINOAZOBENZOL (GERMAN) see DUH600

3-METHYL-4-DIMETHYLAMINO-2,2-DIPHENYLBUTYRAMIDE see DOY400

α-METHYL-β-(N-(2-DIMETHYLAMINOETHYL)-NPHENYL)AMINO-Δ^{α,β}BUTENOLID see MJF250

3-METHYL-4-(N-(2-DIMETHYLAMINOETHYL)-N-PHENYLAMINO)-2(5H) FURANONE see MJF250

METHYL 2-(DIMETHYLAMINO)-N-HYDROXY-2-OXOETHANIMIDOTHIOATE see OLT100

METHYL-2-(DIMETHYLAMINO)-N-(((METHYLAMINO)CARBONYL)OXY)-2-OXOETHANIMIDOTHIOATE see DSP600

2-METHYL-5'-(p-DIMETHYLAMINOPHENYLAZO)QUINOLINE see DQE600

3-METHYL-5'-(p-DIMETHYLAMINOPHENYLAZO)QUINOLINE see DQE400

2'-METHYL-5'-(p-DIMETHYLAMINOPHENYLAZO)QUINOLINE see DPQ600

3'-METHYL-5'-(p-DIMETHYLAMINOPHENYLAZO)QUINOLINE see MJF500

6'-METHYL-5'-(p-DIMETHYLAMINOPHENYLAZO)QUINOLINE see MJF750

7'-METHYL-5'-(p-DIMETHYLAMINOPHENYLAZO)QUINOLINE see DPQ400

8'-METHYL-5'-(p-DIMETHYLAMINOPHENYLAZO)QUINOLINE see MJG000

4-METHYL-3-DIMETHYLAMINOPHENYL ESTER-N-METHYLCARBAMIC ACID see DQE800

(METHYL-2-DIMETHYLAMINO-3-PROPYL)-10-PHENOTHIAZINE HYDROCHLORIDE see TKV200

2'-METHYL-4-DIMETHYLAMINOSTILBENE see TMF750

3'-METHYL-4-DIMETHYLAMINOSTILBENE see TMG000

4'-METHYL-4-DIMETHYLAMINOSTILBENE see TMG250

METHYL-4-DIMETHYLAMINO-3,5-XYLYL CARBAMATE see DOS000

METHYL-4-DIMETHYLAMINO-3,5-XYLYL ESTER of CARBAMIC ACID see DOS000

o-METHYLDIMETHYLANILINE see DUH200

METHYL-2,2-DIMETHYLBICYCLO(2.2.1)HEPTANE-3-CARBOXYLATE see MHY600

METHYL-1,1-DIMETHYLBUTANON(3)-NITROSAMIN (GERMAN) see MMX750

METHYL-1-(DIMETHYLCARBAMOYL)-N-(METHYLCARBAMOYLOXY)THIOFORMIMIDATE see DSP600

S-METHYL-1-(DIMETHYLCARBAMOYL)-N-((METHYLCARBAMOYL)OXY)THIOFORMIMIDATE see DSP600

METHYL 3,11-DIMETHYL-7-ETHYL-10,11-EPOXY-2,6-TRIDECADIENOATE see MJG100

METHYL 1,1-DIMETHYLETHYL ETHER see MHV859

2-METHYL-3-(3,5-DIMETHYL-4-HYDROXYPHENYL)-3,4-DIHYDROQUINAZOLIN-4-ONE see DSH800

METHYL-N',N'-DIMETHYL-N-((METHYLCARBAMOYL)OXY)-1-THIOOXAMIMIDATE see DSP600

(E)2-METHYL-3,7-DIMETHYL-2,6-OCTADIENYL ESTER PROPANOIC ACID see GDI000

METHYL 1,1-DIMETHYLPROPYL ETHER see PBX000

METHYL 2-(((((4,6-DIMETHYL-2-PYRIMIDINYL)AMINO)CARBONYL)AMINO)SULFONYL)BENZOATE see SNW550

N-METHYL-O,O-DIMETHYLTHIOLOPHOSPHORYL-5-THIA-3-METHYL-2-VALERAMIDE see MJG500

4-METHYL-2,6-DINITROANILINE see DVI100

2-METHYL-3,5-DINITROBENZAMIDE see DUP300

2-METHYL-3,5-DINITROBENZENAMINE see AJR750

4-METHYL-2,6-DINITROBENZENAMINE see DVI100

4-METHYL-3,5-DINITROBENZENAMINE see MJG600

METHYLDINITROBENZENE see DVG600

1-METHYL-2,4-DINITROBENZENE see DVH000

1-METHYL-3,5-DINITRO-BENZENE see DVH800

2-METHYL-1,3-DINITROBENZENE see DVH400

2-METHYL-1,4-DINITROBENZENE see DVH200

4-METHYL-1,2-DINITROBENZENE see DVH600

1-METHYL-2,3-DINITRO-BENZENE (9CI) see DVG800

2-METHYL-4,6-DINITROPHENOL see DUS700

3-METHYL-4,6-DINITROPHENOL see DUT700

2-METHYL-4,6-DINITROPHENOL, AMMONIUM SALT see DUT800

2-METHYLDINITROPHENOL SODIUM SALT see SGP550

2-METHYL-4,6-DINITROPHENOL SODIUM SALT see DUU600

N-METHYL-N,4-DINITROSOANILINE see MJG750

N-METHYL-N,p-DINITROSOANILINE see MJG750

N-METHYL-N,4-DINITROSOBENZENAMINE see MJG750

2-METHYLDINITROSOPIPERAZINE see MJH000

N-METHYL-2,4-DINITRO-N-2,4,6-TRIBROMOPHENYL(-6)TRIFLUOROMETHYLBENZENEAMINE see BMO300

METHYLDIOCTYLAMINE see MND125

N-METHYLDIOCTYLAMINE see MND125

N-METHYL-N,N-DIOCTYL-1-OCTANAMINIUM CHLORIDE see MQH000

((2-METHYL-1,3-DIOXALAN-4-YL)METHYL)TRIMETHYLAMMONIUM IODIDE see MJH250

2-METHYL-1,4-DIOXASPIRO(4.5)DECANE see MJH500

METHYL DIOXOLANE see MJH750

METHYLDIOXOLANE see MJH775

2-METHYL-1,3-DIOXOLANE see MJH775

2-METHYL-1,3-DIOXOLANE-2-ACETIC ACID ETHYL ESTER see EFR100

((2-METHYL-1,3-DIOXOLAN-4-YL)METHYL)TRIMETHYLAMMONIUM CHLORIDE see MJH800

6-METHYL-1,11-DIOXY-2-NAPHTHACENECARBOXAMIDE see TBX000

2-METHYLDIPHENHYDRAMINE see MJH900

o-METHYLDIPHENHYDRAMINE see MJH900

4-METHYLDIPHENYL see MJH905

p-METHYLDIPHENYL see MJH905

4'-METHYL-2-DIPHENYLAMINECARBOXYLIC ACID see TGV000

METHYL DIPHENYL ETHER see PEG000

3-METHYLDIPHENYL ETHER see MNV770

METHYL DIPHENYLMETHYL KETONE see MJH910

(±)-γ-METHYL-α,α-DIPHENYL-1-PYRROLIDINEPROPANOL HYDROCHLORIDE see ARN700

2-METHYL-1,2-DI-3-PYRIDINYL-1-PROPANONE (9CI) see MCJ370

N-(6-METHYLDIPYRIDO(1,2-A:3',2'-D)IMIDAZOL-2-YL)HYDROXYLAMINE see HIU550

6-METHYL DIPYRIDO(1,2-a:3',2'-d)IMIDAZOL-2-AMINE see AKS250

2-METHYL-1,2-DIPYRIDYL-(3'-1-OXOPROPANE) DITARTRATE see MJJ000

2-METHYL-1,2-DI-3-PYRIDYL-1-PROPANONE TARTRATE (1:2) see MJJ000

2-METHYL-m-DITHIANE-2-CARBOXALDEHYDE o-(METHYLCARBAMOYL)OXIME see MJJ100

N-METHYL-3,6-DITHIA-3,4,5,6-TETRAHYDROPHTHALIMIDE see MJJ250

1-METHYL-3-(DI-2-THIENYLMETHYLENE)PIPERIDINE CITRATE see ARP875

1-(1-METHYL-3,3-DI-2-THIENYL-2-PROPENYL)PIPERIDINE HYDROCHLORIDE see MJJ500

N-METHYLDITHIOCARBAMATE de SODIUM (FRENCH) see VFU000

METHYLDITHIOCARBAMIC ACID, SODIUM SALT see VFU000

N-METHYLDITHIOCARBAMIC ACID, SODIUM SALT see VFU000

METHYL DITHIOCARBANILATE see MJK500

METHYLDITHIOCYANATOARSINE HOMOPOLYMER see MJK750

METHYLDITHIOKARBAMAN SODNY DIHYDRAT see SIL550

(4-METHYL-1,3-DITHIOLAN-2-YLIDENE)PHOSPHORAMIDIC ACID, DIETHYL ESTER see DHH400

6-METHYL-1,3-DITHIOLO(4,5-b)QUINOXALIN-2-ONE see ORU000

METHYL DIVINYL ACETYLENE see MJL000

2-METHYL-1,3-DI(2,4-XYLYLIMINO)-2-AZAPROPANE see MJL250

2-METHYL-DNPZ see MJH000

METHYL DODECANOATE see MLC800

METHYL DODECYLATE see MLC800

METHYL DODECYL BENZYL AMMONIUM CHLORIDE see DYA600

METHYLDOPA see DNA800

α-METHYL-l-DOPA see DNA800

l-α-METHYLDOPA see DNA800

l-α-METHYLDOPAHYDRAZINE see CBQ500

N-METHYLDOPAMINE HYDROCHLORIDE see EAZ000

METHYL DOPA SESQUIHYDRATE see MJE780

METHYL DURSBAN see CMA250

METHYL-E-600 see PHD500

METHYL-E 605 see MNH000

METHYLE (ACETATE de) (FRENCH) see MFW100

METHYL E-CROTONATE see COB825

O,O-METHYLEEN-BIS-(4-CHLOORFENOL) see MJM500

METHYLEEN-S,S'-BIS(O,O-DIETHYL-DITHIOFOSFAAT) (DUTCH) see EEH600

3,3'-METHYLEEN-BIS-(4-HYDROXY-CUMARINE) (DUTCH) see BJZ500

METHYLE (FORMIATE de) (FRENCH) see MKG750

6-METHYLENANDROSTA-1,4-DIENE-3,17-DIONE see MJL275

METHYL ENANTHATE see MKK100

S,S'-METHYLEN-BIS(O,O-DIAETHYL-DITHIOPHOSPHAT) (GERMAN) see EEH600

3,3'-METHYLEN-BIS-(4-HYDROXY-CUMARIN) (GERMAN) see BJZ000

METHYLENDIC ANHYDRIDE see NAC000

3,4-METHYLENDIOXY-6-PROPYLBENZYL-n-BUTYL-DIAETHYLENGLYKOLAETHER (GERMAN) see PIX250

METHYLENDIRHODANID (CZECH, GERMAN) see MJT500

1,1'-METHYLENDITHIOSEMIKARBAZID see MJN300

METHYLENE ACETONE see BOY500

16-METHYLENE-17-α-ACETOXY-19-NOR-4-PREGNENE-3,20-DIONE see HMB650

METHYLENEAMINE N-OXIDE see FMW400

METHYLENEAMINOACETONITRILE see HHW000

1,12-METHYLENEBENZ(a)ANTHRACENE see MJL300

1',9-METHYLENE-1,2-BENZANTHRACENE see MJL300

endo-α-METHYLENEBENZENEACETIC ACID 8-METHYL-8-AZABICYCLO(3.2.1)OCT-3-YL ESTER see AQO250

METHYLENE BICHLORIDE see MJP450

2,2'-METHYLENEBIPHENYL see FDI100

METHYLENEBISACRYLAMIDE see MJL500

N,N'-METHYLENEBIS(ACRYLAMIDE) see MJL500

METHYLENEBIS(2-AMINO-BENZOIC ACID) DIMETHYL ESTER (9CI) see MJQ250

METHYLENEBIS(4-AMINOCYCLOHEXANE) see MJQ260

N,N'-METHYLENEBIS(2-AMINO-1,3,4-THIADIAZOLE) see MJL750

METHYLENEBIS(ANILINE) see MJQ000

2,4-METHYLENEBIS(ANILINE) see MJP750

4,4'-METHYLENEBISANILINE see MJQ000

4,4'-METHYLENEBIS(BENZENEAMINE) see MJQ000

4,4'-METHYLENEBIS(2,6-BIS(1,1-DIMETHYLETHYL)PHENOL) see MJM700

2,2'-METHYLENEBIS(6-tert-BUTYL-p-CRESOL) see MJO500

2,2'-METHYLENEBIS(6-tert-BUTYL-4-ETHYLPHENOL) see MJN250

4,4'-METHYLENEBIS(2-CARBOMETHOXYANILINE) see MJQ250

4,4'-METHYLENE(BIS)-CHLOROANILINE see MJM200

4,4'-METHYLENE BIS(2-CHLOROANILINE) see MJM200

METHYLENE-4,4'-BIS(o-CHLOROANILINE) see MJM200

4,4'-METHYLENEBIS(o-CHLOROANILINE) see MJM200

p,p'-METHYLENEBIS(o-CHLOROANILINE) see MJM200

p,p'-METHYLENEBIS(α-CHLOROANILINE) see MJM200

4,4'-METHYLENE-BIS(2-CHLOROANILINE) HYDROCHLORIDE see MJM250

4,4'-METHYLENEBIS-2-CHLOROBENZENAMINE see MJM200

2,2'-METHYLENEBIS(4-CHLOROPHENOL) see MJM500

2,2'-METHYLENEBIS(4-CHLOROPHENOL) see MJM500

4,4'-METHYLENEBIS(CYCLOHEXANAMINE) see MJQ260

4,4'-METHYLENEBIS(CYCLOHEXYLAMINE) see MJQ260

METHYLENE BIS(4-CYCLOHEXYLISOCYANATE) see MJM600

METHYLENE BIS(4-CYCLOHEXYLISOCYANATE) (ACGIH,OSHA) see MJM600

METHYLENE-S,S'-BIS(O,O-DIAETHYL-DITHIOPHOSPHAT) (GERMAN) see EEH600

4,4'-METHYLENEBIS(2,6-DI-T-BUTYLPHENOL) see MJM700

METHYLENEBIS(DIETHYLPHOSPHONATE) see MJT100

6,6'-METHYLENEBIS(1,2-DIHYDRO-2,2,4-TRIMETHYLQUINOLINE) see MRU760

4,4'-METHYLENE BIS(N,N'-DIMETHYLANILINE) see MJN000

4,4'-METHYLENEBIS(N,N-DIMETHYL)BENZENAMINE see MJN000

4,4'-METHYLENEBIS(o-ETHYLANILINE) see MJN100

4,4'-METHYLENEBIS(2-ETHYLBENZENAMINE) see MJN100

2,2'-METHYLENEBIS(4-ETHYL-6-tert-BUTYLPHENOL) see MJN250

N,N-METHYLENE-BIS(ETHYL CARBAMATE) see MJT750

2,2'-METHYLENEBISFURAN see FPX028

2,2'-METHYLENEBIS(HYDRAZINECARBOTHIAMIDE) see MJN300

3,3'-METHYLENEBIS(4-HYDROXY-1,2-BENZOPYRONE) see BJZ000

3,3'-METHYLENEBIS(4-HYDROXYCOUMARIN) see BJZ000

3,3'-METHYLENE-BIS-(4-HYDROXYCOUMARINE) (FRENCH) see BJZ000

4,4'-METHYLENEBIS(3-HYDROXY-2-NAPHTHOIC ACID) see PAF100

4,4'-METHYLENEBIS(3-HYDROXY-2-NAPHTHOIC ACID) with TRIS-(p-AMINOPHENYL)CARBONIUM SALT (1:2) see TNC725

4,4'-METHYLENEBIS(3-HYDROXY-2-NAPHTHOIC ACID) with TRIS-(p-AMINOPHENYL)CARBONIUM SALT (1:2) see TNC750

4,4'-METHYLENEBIS(3-HYDROXY-2-NAPHTHOIC ACID) with TRIS-(p-

AMINOPHENYL)METHYLIUM SALT (1:2) see TNC725

4,4'-METHYLENEBIS(3-HYDROXY-2-NAPHTHOIC ACID) with TRIS-(p-AMINOPHENYL)METHYLIUM SALT (1:2) see TNC750

4,4'-METHYLENEBIS(3-HYDROXY-2-NAPTHOIC ACID ESTER) with 2-(2-(4-(p-CHLORO-α-PHENYLBENZYL)-1-PIPERAZINYL)-ETHOXY)ETHANOL see HOR500

METHYLENEBIS(4-ISOCYANATOBENZENE) see MJP400

1,1-METHYLENEBIS(4-ISOCYANATOBENZENE) see MJP400

5,5'-METHYLENEBIS(2-ISOCYANATO)TOLUENE see MJN750

METHYLENE BIS(METHANESULFONATE) see MJQ500

4,4'-METHYLENEBIS(2-METHYLANILINE) see MJO250

4,4'-METHYLENEBIS(N-METHYLANILINE) see MJO000

METHYLENE-BIS(METHYL ANTHRANILATE) see MJQ250

4,4'-METHYLENEBIS(2-METHYLBENZENAMINE) see MJO250

4,4'-METHYLENEBIS(N-METHYLBENZENAMINE) see MJO000

2,2"-METHYLENEBIS(4-METHYL-6-tert-BUTYLPHENOL) see MJO500

2,2'-METHYLENE-BIS(4-METHYL-6-NONYLPHENOL) see MJO750

N,N"-(METHYLENEBIS(N'-(1-(HYDROXYMETHYL)-2,5-DIOXO-4-IMIDAZOLIDINYL)UREA) see GDO800

METHYLENE BIS(NITRAMINE) see MJO775

2,2'-METHYLENEBIS(6-NONYL-p-CRESOL) see MJO750

METHYLENE-BIS-ORTHOCHLOROANILINE see MJM200

2,2'-METHYLENEBIS OXIRANE (9CI) see DHE000

1,1'-(METHYLENEBIS(OXY))BIS(4-CHLORO)BENZENE see NCM700

1,1'-(METHYLENEBIS(OXY)BIS(2-CHLOROETHANE)) see BID750

1,1'-(METHYLENEBIS(OXY))BIS(2,2-DINITROPROPANE) see MJO800

1,1'-(METHYLENEBIS(OXY))BIS(2-FLUORO-2,2-DINITROETHANE) see FMV050

4,4'-METHYLENEBISPHENOL see BKI250

METHYLENEBIS(4-PHENYLENE ISOCYANATE) see MJP400

METHYLENEBIS(p-PHENYLENE ISOCYANATE) see MJP400

METHYLENE BISPHENYL ISOCYANATE see MJP400

METHYLENEBIS(4-PHENYL ISOCYANATE) see MJP400

METHYLENEBIS(p-PHENYL ISOCYANATE) see MJP400

4,4'-METHYLENEBIS(PHENYL ISOCYANATE) see MJP400

p,p'-METHYLENEBIS(PHENYL ISOCYANATE) see MJP400

4,4'-METHYLENEBIS(PHENYLMALEIMIDE) see BKL800

METHYLENEBIS(PHOSPHONIC ACID) SODIUM SALT see SIL575

METHYLENEBISPHOSPHONIC ACID TETRASODIUM SALT see SIL575

3,3'-METHYLENEBIS(1-(PIPERIDINOMETHYL)INDOLE) see MJP420

METHYLENEBIS(1-THIOSEMICARBAZIDE) see MJN300

2,2'-METHYLENEBIS(3,4,6-TRICHLOROPHENOL) see HCL000

METHYLENE BLUE see BJI250

METHYLENE BLUE (medicinal) see BJI250

METHYLENE BLUE A see BJI250
METHYLENE BLUE BB see BJI250
METHYLENE BLUE BB ZINC FREE see BJI250
METHYLENE BLUE CHLORIDE see BJI250
METHYLENE BLUE CHLORIDE (biological stain) see BJI250
METHYLENE BLUE D see BJI250
METHYLENE BLUE I (medicinal) see BJI250
METHYLENE BLUE NF (medicinal) see BJI250
METHYLENE BLUE POLYCHROME see BJI250
METHYLENE BLUE USP (medicinal) see BJI250
METHYLENE BLUE USP XII (medicinal) see BJI250
METHYLENE BROMIDE see DDP800
2-METHYLENEBUTANAL see EFS700
2-METHYLENEBUTYRALDEHYDE see EFS700
(4-(2-METHYLENEBUTYRYL)-2,3-DICHLOROPHENOXY)ACETIC ACID see DFP600
METHYLENEBUTYRYL PHENOXYACETIC ACID see DFP600
METHYLENE CHLORIDE see MJP450
METHYLENE CHLOROBROMIDE see CES650
1,2-α-METHYLENE-6-CHLORO-Δ6-17-α-HYDROXYPROGESTERONE ACETATE see CQJ500
1,2-α-METHYLENE-6-CHLORO-PREGNA-4,6-DIENE-3,20-DIONE 17-α-ACETATE see CQJ500
1,2-α-METHYLENE-6-CHLORO-Δ4,6-PREGNADIENE-17-α-OL-3,20-DIONE 17-α-ACETATE see CQJ500
1,2-α-METHYLENE-6-CHLORO-Δ4,6-PREGNADIENE-17-α-OL-3,20-DIONE ACETATE see CQJ500
4,5-METHYLENECHRYSENE see CPU000
METHYLENE CYANIDE see MAO250
METHYLENECYCLOHEXANE OXIDE see SLD920
2-METHYLENE-CYCLOPENTAENE-3-ONO-4,5-EPOXY-4,5-DIMETHYL-1-CARBOXYLIC ACID see MJU500
2-METHYLENECYCLOPROPANEALANINE see MJP500
2-METHYLENECYCLOPROPANYLALANINE see MJP500
β-(METHYLENECYCLOPROPYL)ALANINE see MJP500
N,N'-METHYLENEDIACRYLAMIDE see MJL500
METHYLENEDIANILINE see MJQ000
2,4'-METHYLENEDIANILINE see MJP750
4,4'-METHYLENEDIANILINE see MJQ000
p,p'-METHYLENEDIANILINE see MJQ000
4,4-METHYLENEDIANILINE (ACGIH) see MJQ000
4,4'-METHYLENEDIANILINE DIHYDROCHLORIDE see MJQ100
p,p'-METHYLENEDIANILINE DIHYDROCHLORIDE see MJQ100
METHYLENEDIANTHRANILIC ACID DIMETHYL ESTER see MJQ250
1',9-METHYLENE-1,2:5,6-DIBENZANTHRACENE see DCR600
METHYLENE DIBROMIDE see DDP800
METHYLENE DICHLORIDE see MJP450
4,4'-METHYLENEDICYCLOHEXANAMINE see MJQ260
4,4'-METHYLENEDICYCLOHEXANEAMINE see MJQ260
4,4'-METHYLENEDICYCLOHEXYLAMINE see MJQ260
METHYLENE DIFLUORIDE see MJQ300

3,4-METHYLENE-DIHYDROXYBENZALDEHYDE see PIW250
METHYLENE DIIODIDE see DNF800
METHYLENE DIISOCYANATE see DNK100
METHYLENEDILITHIUM see MJQ325
METHYLENE DIMETHANESULFONATE see MJQ500
METHYLENE DIMETHYL ETHER see MGA850
4,4'-METHYLENEDIMORPHOLINE see MJQ750
METHYLENEDINAPHTHALENESULFONIC ACID BISPHENYLMERCURI SALT see PFN000
3,3'-METHYLENEDI-2-NAPHTHALENESULFONIC ACID, MERCURY SALT see MCT250
METHYLENE DINITRATE see MJU150
1,5-METHYLENE-3,7-DINITROSO-1,3,5,7-TETRAAZACYCLOOCTAINE see DVF400
1,5-METHYLENE-3,7-DINITROSO-1,3,5,7-TETRAAZACYCLOOCTANE see DVF400
3,4-METHYLENEDIOXY-ALLYLBENZENE see SAD000
1,2-METHYLENEDIOXY-4-ALLYLBENZENE see SAD000
METHYLENEDIOXYAMPHETAMINE see MJQ775
3,4-METHYLENEDIOXY-AMPHETAMINE see MJQ775
3,4-(METHYLENEDIOXY)BENZALACETONE see MJR250
3,4-METHYLENEDIOXYBENZALDEHYDE see PIW250
METHYLENEDIOXYBENZENE see MJR000
1,2-METHYLENEDIOXYBENZENE see MJR000
3,4-METHYLENEDIOXYBENZYL ACETATE see PIX000
3,4-METHYLENEDIOXYBENZYL ACETONE see MJR250
2-(4-(3,4-METHYLENEDIOXYBENZYL)PIPERAZINO)PYRIMIDINE see TNR485
1-(3,4-METHYLENEDIOXYBENZYL)-4-(2-PYRIMIDYL)PIPERAZINE see TNR485
1,2-(METHYLENEDIOXY)-4-(3-BROMO-1-PROPENYL)BENZENE see BOA750
2,3-METHYLENEDIOXY-5,7-DIMETHOXY-10-NITROPHENANTHROIC ACID see MEA700
3,4-METHYLENEDIOXY-N,α-DIMETHYL-β-PHENYLETHYLAMINE HYDROCHLORIDE see MJR500
METHYLENEDIOXYETHAMPHETAMINE see EMS100
METHYLENEDIOXYETHYLAMPHETAMINE see EMS100
3,4-METHYLENEDIOXYETHYLAMPHETAMINE see EMS100
3,4-METHYLENEDIOXY-α-ETHYL-β-PHENYLETHYLAMINE see MJR750
1,2-METHYLENEDIOXY-4-(1-HYDROXYALLYL)BENZENE see BCJ000
7,8-METHYLENEDIOXYISOQUINOLINE see MJR775
3,4-METHYLENEDIOXY-α-METHYL-β-PHENYLETHYLAMINE HYDROCHLORIDE see MJS750
3,4-(METHYLENEDIOXY)-10-NITROPHENANTHRENE-1-CARBOXYLIC ACID see MJR780
3,4-METHYLENEDIOXY-β-NITROSTYRENE see MJR800
1,2-(METHYLENEDIOXY)-4-(2-(OCTYLSULFINYL)PROPYL)BENZENE see ISA000
3,4-METHYLENEDIOXYPHENOL see MJU000

2-(3,4-(METHYLENEDIOXY)PHENOXY)-1-((3,4-(METHYLENEDIOXY)PHENOXY)METHYL)ETHYLAMINE see MJS250
1-(5-(3,4-(METHYLENEDIOXY)PHENOXY)-4-((3,4-(METHYLENEDIOXY)PHENOXY)METHYL)PENTYLPIPERIDINE) CITRATE see MJS500
2-(3,4-METHYLENEDIOXYPHENOXY)-3,6,9-TRIOXOUNDECANE see MJS550
1-(3,4-METHYLENEDIOXYPHENYL)-2-AMINOPROPANE see MJS750
4-(3,4-(METHYLENEDIOXY)PHENYL)-3-BUTEN-2-ONE see MJR250
3,4-METHYLENEDIOXY-β-PHENYLETHYLAMINE HYDROCHLORIDE see MJT000
2-(3,4-(METHYLENEDIOXY)PHENYL)IMIDAZO(2,1-A)ISOQUINOLINE see MJT025
1,2-METHYLENEDIOXY-4-PROPENYLBENZENE see IRZ000
3,4-METHYLENEDIOXY-1-PROPENYL BENZENE see IRZ000
1,2-(METHYLENEDIOXY)-4-PROPYLBENZENE see DMD600
(3,4-METHYLENEDIOXY-6-PROPYLBENZYL)(BUTYL)DIETHYLENE GLYCOL ETHER see PIX250
3,4-METHYLENEDIOXY-6-PROPYLBENZYL-n-BUTYL DIETHYLENEGLYCOL ETHER see PIX250
4,4'-METHYLENE DIPHENOL see BKI250
METHYLENEDIPHENYL-4,4'-DIAMIDINE see MJT050
4,4'-METHYLENEDIPHENYL DIISOCYANATE see MJP400
1,1'-(METHYLENEDI-4,1-PHENYLENE)BIS-1H-PYRROLE-2,5-DIONE see BKL800
METHYLENEDI-p-PHENYLENE DIISOCYANATE see MJP400
METHYLENEDI-p-PHENYLENE ISOCYANATE see MJP400
4,4'-METHYLENEDIPHENYLENE ISOCYANATE see MJP400
METHYLENE DI(PHENYLENE ISOCYANATE) (DOT) see MJP400
4,4'-METHYLENEDIPHENYL ISOCYANATE see MJP400
METHYLENEDIPHOSPHONIC ACID, SODIUM SALT see SIL575
METHYLENEDI(PHOSPHONIC ACID) TETRAETHYL ESTER see MJT100
METHYLENE DITHIOCYANATE see MJT500
1,1'-METHYLENEDI(THIOSEMICARBAZIDE) see MJN300
4,4'-METHYLENE DI-o-TOLUIDINE see MJO250
METHYLENE DIURETHAN see MJT750
METHYLENE ETHER of OXYHYDROQUINONE see MJU000
N-METHYLENE GLYCINONITRILE see HHW000
METHYLENE GLYCOL see FMV000
METHYLENE GLYCOL DINITRATE see MJU150
METHYLENE GREEN see MJU250
METHYLENE IODIDE see DNF800
METHYLENEMAGNESIUM see MJU350
METHYLENEMELAMINE POLYCONDENSATE see MCB050
(1S-cis)-5-METHYLENE-6-(1-METHYLETHENYL)-2-CYCLOHEXEN-1-OL see CCL109
3-METHYLENE-7-METHYL-1,6-OCTADIENE see MRZ150
3-METHYLENE-7-METHYLOCTAN-7-YL ACETATE see DLX100
3-METHYLENE-7-METHYL-1-OCTEN-7-OL see MLO250

3-METHYLENE-7-METHYL-1-OCTEN-7-YL ACETATE see AAW500

4-METHYLENE-2-OXETANONE see KFA000

METHYLENE OXIDE see FMV000

2-METHYLENE-3-OXO-CYCLOPENTANECARBOXYLIC ACID see SAX500

2-METHYLENE-3-OXO-1-CYCLOPENTANECARBOXYLIC ACID compounded with NICOTINIC HYDRAZIDE see SAY000

2-METHYLENE-3-OXOCYCLOPENTANECARBOXYLIC ACID, SODIUM SALT see SJS500

2,5-METHYLENE-6-PROPYL-3-CYCLOHEXENECARBOXALDEHYDE see CML620

METHYLENESUCCINYLOXYBIS(TRIBUTYLSTANNANE) see BLL500

S,S'-METHYLENE O,O,O',O'-TETRAETHYL PHOSPHORODITHIOATE see EEH600

2,5-endo-METHYLENE-Δ³-TETRAHYDROBENZYL ACRYLATE see BFY250

METHYLENETETRAHYDROPYRAN see MJU400

8-METHYLENE-4,11,11-(TRIMETHYL)BICYCLO(7.2.0)UNDEC-4-ENE see CCN000

2-METHYLENE-1,3,3-TRIMETHYLINDOLINE see TLU200

2-METHYLENEVALERALDEHYDE see PNT800

METHYLENE VIOLET BN see AJP300

METHYLENE VIOLET 3RD see AJP300

METHYLENIMINOACETONITRILE see HHW000

METHYLENIUM CERULEUM see BJI250

METHYLENOMYCIN A see MJU500

METHYLEPHEDRIN (GERMAN) see MJU750

METHYLEPHEDRINE see MJU750

METHYLEPHEDRINE HYDROCHLORIDE see DPD200

l-METHYLEPHEDRINE HYDROCHLORIDE see DPD200

N-METHYLEPHEDRINE HYDROCHLORIDE see DPD200

dl-METHYLEPHENDRINE HYDROCHLORIDE see MNN350

N-METHYLEPINEPHRINE see MJV000

METHYL-18-EPIRESERPATE METHYL ETHER HYDROCHLORIDE see MQR200

3-METHYL-11,12-EPOXYCHOLANTHRENE see ECA500

6-METHYL-3,4-EPOXYCYCLOHEXYLMETHYL-6-METHYL-3,4-EPOXYCYCLOHEXANE CARBOXYLATE see ECB000

METHYLERGOBASINE see PAM000

METHYLERGOBASINE MALEATE see MJV750

METHYLERGOBREVIN see PAM000

(8-β)-6-METHYLERGOLINE-8-ACETAMIDE TARTRATE (2:1) see MJV500

d-6-METHYLERGOLINE-8-ACETONITRILE see MIQ400

1-((5R,8S,10R)-6-METHYL-8-ERGOLINYL)-3,3-DIETHYLUREA see DLR100

1-((5R,8S,10R)-6-METHYL-8-ERGOLINYL)-3,3-DIETHYLUREA HYDROGEN MALEATE see DLR150

METHYLERGOMETRIN see PAM000

METHYLERGOMETRINE see PAM000

METHYLERGOMETRINE MALEATE see MJV750

METHYLERGONOVIN see PAM000

METHYLERGONOVINE see PAM000

METHYLERGONOVINE MALEATE see MJV750

6-o-METHYLERYTHROMYCIN see MJV775

6-o-METHYLERYTHROMYCIN A see MJV775

METHYLESCULETIN see MJV800

4-METHYLESCULETIN see MJV800

4-METHYLESCULETOL see MJV800

METHYL ESTER N,N-(DIPHENYL)-4-UREIDO-5,7-DICHLORO-2-CARBOXYQUINOLINE see MJV900

METHYL ESTER FLUOROSULFURIC ACID see MKG250

METHYL ESTER of GABA see AJC625

METHYL ESTER of p-HYDROXYBENZOIC ACID see HJL500

METHYLESTER KISELINY OCTOVE (CZECH) see MFW100

METHYLESTER KYSELINY ANTHRANILOVE see APJ250

METHYLESTER KYSELINY BENZOOVE see MHA750

METHYLESTER KYSELINY BROMOCTOVE see MHR250

METHYLESTER KYSELINY 2-CHLOR-5-NITROBENZOOVE see CJC515

METHYLESTER KYSELINY-2-CHLOR-5-NITROBENZOOVE (CZECH) see CJD625

METHYLESTER KYSELINY CHLOROCTOVE see MIF775

METHYLESTER KYSELINY 4-FLUORMASELNE see MKE000

METHYLESTER KYSELINY FOSFORITE (CZECH) see PGZ950

METHYLESTER KYSELINY METHAKRYLOVE see MLH750

METHYLESTER KYSELINY 3-METHOXYPROPIONOVE see MFL400

METHYLESTER KYSELINY 1-NAFTYLOCTOVE see NAK525

METHYLESTER KYSELINY ORTHOMRAVENCI (CZECH) see TLX600

METHYLESTER KYSELINY p-TOLUENSULFONOVE (CZECH) see MLL250

METHYL ESTER of METHANESULFONIC ACID see MLH500

METHYL ESTER of METHANESULPHONIC ACID see MLH500

METHYL ESTER of SERPENTINIC ACID see SCA475

METHYL ESTER STEARIC ACID see MJW000

METHYL ESTER STREPONIGRIN see SMA500

METHYL ESTER of WOOD ROSIN see MFT500

METHYL ESTER of WOOD ROSIN, partially hydrogenated (FCC) see MFT500

METHYLESTRENOLONE see MDM350

METHYLE (SULFATE de) (FRENCH) see DUD100

N-METHYLETHANEDIAMINE see MJW100

N-METHYL-1,2-ETHANEDIAMINE see MJW100

1,1'-(METHYLETHANEDILIDENEDINITRILO) BIGUANIDINE DIHYDROCHLORIDE DIHYDRATE see MKI000

((((1-METHYL-1,2-ETHANEDIYL)BIS(CARBAMODITHIOATO))(2−))ZINC HOMOPOLYMER see ZMA000

4,4'-(1-METHYL-1,2-ETHANEDIYL)BIS-2,6-PIPERAZINEDIONE see PIK250

METHYL ETHANE SULFONATE see MJW250

METHYL ETHANE SULPHONATE see MJW250

1-METHYLETHANETHIOL see IMU000

N-METHYLETHANOANTHRACENE-9-(10H)-METHYLAMINE HYDROCHLORIDE see BCH750

N-METHYL-9,10-ETHANOANTHRACENE-9(10H)-PROPANAMINE HYDROCHLORIDE see MAW850

METHYL ETHANOATE see MFW100

N-METHYLETHANOLAMINE see MGG000

METHYLETHENE see PMO500

4-(1-METHYLETHENYL)-1-CYCLOHEXENE-1-CARBOXALDEHYDE (9CI) see DKX100

5-(1-METHYLETHENYL)-β,β,2-TRIMETHYL-1-CYCLOPENTENE-1-PROPANOL PROPANOATE see MJW300

METHYL ETHER see MJW500

METHYL ETHER OF 3,5-DI-tert-BUTYL-4-HYDROXYBENZENE see DEE300

METHYL ETHER of PROPYLENE GLYCOL (α) see MJW750

METHYL ETHERTRIMETHYLCOLCHICINIC ACID-l-TARTRATE see DBA175

METHYL ETHER TRIMETHYLCOLCHICINIC ACID-d-TARTRATE, HYDRATE see DBA175

8-METHYL ETHER of XANTHURENIC ACID see HLT500

11-β-METHYL-17-α-ETHINYLESTRADIOL see MJW875

METHYL ETHOXOL see EJH500

2-(1-METHYLETHOXY)-BENZENAMINE see INA200

2-(2-(1-METHYLETHOXY)ETHOXYL)ETHANOL see IOJ500

METHYL 2-((4-ETHOXY-6-METHYLAMINO-1,3,5-TRIAZIN-2-YL)CARBAMOYLSULFAMOYL)BENZOATE (IUPAC) see MJW900

9-METHYL-10-ETHOXYMETHYL-1,2-BENZANTHRACENE see EEX000

9-METHYL-10-ETHOXYMETHYL-1,2-BENZANTHRACENE see MJX000

N,N'-((1-((1-METHYLETHOXY)METHYL)-1,2-ETHANEDIYL)BIS(OXYSULFINYL(METHYLIMINO)CARBONYLOXY))BIS(2-(DIMETHYLAMINO)-2-OXO-ETHANIMIDOTHIOIC ACID, DIMETHYL ESTER see MJX100

((1-METHYLETHOXY)METHYL)OXIRANE see IPD000

2-(1-METHYLETHOXY)PHENOL METHYLCARBAMATE see PMY300

2-(1-METHYLETHOXY)PHENYL (((1,1-DIMETHYLETHOXY)SULFINYL)METHYL) CARBAMATE see INE062

2-(1-METHYLETHOXY)PHENYL ((HEXYLOXY)SULFINYL)METHYLCARBAMATE see INE065

2-(1-METHYLETHOXY)PHENYL (METHOXYACETYL)METHYLCARBAMATE see MDX250

2-(l-(METHYLETHOXY)PHENYL METHYL(METHYLTHIO)ACETYL)CARBAMATE see MLX000ᵃ

N-(3-(1-METHYLETHOXY)PHENYL)-2-(TRIFLUOROMETHYL)BENZAMIDE see TKB286

METHYLETHYLACETALDEHYDE see MJX500

METHYLETHYLACETIC ACID see MHS600

METHYLETHYLACETYLENYLCARBINOL see EQL000

1-METHYLETHYLAMINE see INK000

4-(METHYLETHYL)AMINOAZOBENZENE see ENB000

N-METHYL-N-ETHYL-p-AMINOAZOBENZENE see ENB000

N-METHYL-4'-ETHYL-p-AMINOAZOBENZENE see EOJ000

1-METHYLETHYL 3-AMINO-2-BUTENOATE see INM100

4-(((1-METHYLETHYL)AMINO)CARBONYL)-BENZOIC ACID (9CI) see IRN000

1-METHYLETHYL 3-AMINOCROTONATE see INM100

α-(((1-METHYLETHYL)AMINO)METHYL)-2-NAPHTHALENEMETHANOL, HYDROCHLORIDE see INT000

α-(((1-METHYLETHYL)AMINO)METHYL)-4-NITROBENZENEMETHANOL see INU000

(±)-1-((1-METHYLETHYL)AMINO)-3-(1-NAPHTHALENYLOXY)-2-PROPANOL (9CI) see PNB790

1-((1-METHYLETHYL)AMINO)-3-(1-NAPHTHALENYLOXY)-2-PROPANOL HYDROCHLORIDE see ICC000

2-METHYLETHYLAMINO-1-PHENYL-1-PROPANOL HYDROCHLORIDE see EJR500

1-((1-METHYLETHYL)AMINO)-3-(2-(2-PROPENYLOXY)PHENOXY)-2-PROPANOL see CNR500

1-((1-METHYLETHYL)AMINO)-3-(2-(2-PROPENYL)PHENOXY)-2-PROPANOL see AGW250

9-(3-((1-METHYLETHYL)AMINO)PROPYL)-9H-FLUORENE-9-CARBOXAMIDE HYDROCHLORIDE see IBW400

2-METHYL-6-ETHYL ANILINE see MJY000

2-METHYL-6-ETHYL ANILINE see MJY000

7-METHYL-9-ETHYLBENZ(c)ACRIDINE see MJY250

3-METHYL-10-ETHYLBENZ(c)ACRIDINE see EMM000

10-METHYL-3-ETHYL-7,8-BENZACRIDINE (FRENCH) see MJY250

3-METHYL-10-ETHYL-7,8-BENZACRIDINE (FRENCH) see EMM000

4-(1-METHYLETHYL)-BENZALDEHYDE (9CI) see COE500

2-(1-METHYLETHYL)BENZENAMINE see INW100

4-METHYLETHYLBENZENE see EPT000

o-METHYLETHYLBENZENE see EPS500

p-METHYLETHYLBENZENE see EPT000

4-(1-METHYLETHYL)-1,2-BENZENEDIOL see IOK000

4-(1-METHYLETHYL)BENZENEMETHANOL ACETATE see IOF050

4-(1-METHYLETHYL)BENZONITRILE see IOD050

3-(1-METHYLETHYL)-1H-2,1,3-BENZOTHIAZAIN-4(3H)-ONE-2,2-DIOXIDE see MJY500

2-METHYL-3-ETHYLBENZOXAZOLIUM IODIDE see MJY525

1-METHYLETHYL 4-BROMO-α-(4-BROMOPHENYL)-α-HYDROXYBENZENEACETATE see IOS000

METHYLETHYLBROMOMETHANE see BMX750

(1-METHYLETHYL)CARBAMIC ACID 2-(((AMINOCARBONYL)OXY)METHYL)-2-METHYLPENTYL ESTER see IPU000

METHYLETHYLCARBINOL see BPW750

METHYL ETHYL CELLULOSE see MJY550

1-METHYLETHYL-4-CHLORO-α-(4-CHLOROPHENYL)-α-HYDROXYBENZENEACETATE see PNH750

4-(1-METHYLETHYL)CYCLOHEXADIENE-1-ETHANOL FORMATE see IHX450

4-(1-METHYLETHYL)CYCLOHEXADIENE-1-ETHYL FORMATE see IHX450

6-(1-METHYLETHYL)-2-DECAHYDRO-NAPHTHALENOL see IOO310

1-METHYLETHYL (3,4-DIETHOXYPHENYL)CARBAMATE see IOS100

METHYL 4-ETHYL-6,7-DIMETHOXY-9H-PYRIDO(3,4-B)INDOLE-3-CARBOXYLATE see EID600

4-METHYL-4-ETHYL-2,6-DIOXOPIPERIDINE see MKA250

METHYLETHYLENE see PMO500

1-METHYLETHYLENE CARBONATE see CBW500

N-METHYLETHYLENEDIAMINE see MJW100

METHYLETHYLENE GLYCOL see PML000

METHYL ETHYLENE OXIDE see PNL600

4-METHYLETHYLENETHIOUREA see MJZ000

N-METHYLETHYLENEUREA see EJN400

METHYLETHYLENIMINE see PNL400

2-METHYLETHYLENIMINE see PNL400

METHYL ETHYL ETHER (DOT) see EMT000

1-METHYLETHYL-2-((ETHOXY((1-METHYLETHYL)AMINO)PHOSPHINOTHIOYL)OXY)BENZOATE see IMF300

(E)-1-METHYLETHYL-3-((ETHYLAMINO)METHOXYPHOSPHINOTHIOYL)OXY-2-BUTENOATE see MKA000

1-(METHYLETHYL)-ETHYL 3-METHYL-4-(METHYLTHIO)PHENYL PHOSPHORAMIDATE see FAK000

1-METHYLETHYL-2-(1-ETHYLPROPYL)-4,6-DINITROPHENYL CARBONATE see CBW000

METHYLETHYLETHYNYLCARBINOL see EQL000

3-METHYL-3-ETHYLGLUTARIMIDE see MKA250

β-METHYL-β-ETHYLGLUTARIMIDE see MKA250

2-(1-METHYLETHYL)HYDRAZIDE 4-PYRIDINECARBOXYLIC ACID DIHYDROCHLORIDE (9CI) see IGG600

4,4'-(1-METHYLETHYLIDENE)BIS(2,6-DIBROMOPHENOL) see MKA270

4,4'-(1-METHYLETHYLIDENE)BIS(2,6-DICHLOROPHENOL) see TBO750

4,4'-(1-METHYLETHYLIDENE)BIS(2-METHYLPHENOL) see IPK000

4,4'-(1-METHYLETHYLIDENE)BISPHENOL DISODIUM SALT see BLD800

2,2'-((1-METHYLETHYLIDENE)BIS(4,1-PHENYLENEOXYMETHYLENE))BISOXIRANE see BLD750

N-METHYLETHYLIDENEDIAMINE see MJW100

9-(3'-METHYL-4'-ETHYLIDENE-THIOSEMICARBAZIDO)ACRIDINE see MKA300

METHYLETHYL KETAZINE see EMT600

METHYL ETHYL KETONE see MKA400

METHYL ETHYL KETONE AZINE see EMT600

METHYL ETHYL KETONE HYDROPEROXIDE see MKA500

METHYL ETHYL KETONE KETAZINE see EMT600

METHYL ETHYL KETONE PEROXIDE see MKA500

METHYL ETHYL KETONE PEROXIDE, in solution with >9% by weight active oxygen (DOT) see MKA500

METHYL ETHYL KETONE SEMICARBAZONE see MKA750

METHYLETHYLKETONHYDROPEROXIDE see MKA500

METHYL ETHYL KETOXIME see EMU500

METHYLETHYLMETHANE see BOR500

2-METHYL-3-ETHYL-4-p-METHOXYPHENYL-Δ³-CYCLOHEXENE CARBOXYLIC ACID see CBO625

N-(1-METHYLETHYL)-4-((2-METHYLHYDRAZINO)METHYL)BENZAMIDE MONOHYDROCHLORIDE see PME500

1-METHYLETHYL-2-(1-METHYLPROPYL)-4,5-DINITROPHENYLESTER CARBONIC ACID see CBW200

N-(1-METHYLETHYL)-N'-(1-NITRO-9-ACRIDINYL)-1,3-PROPANEDIAMINE DIHYDROCHLORIDE see INW000

N-(1-METHYLETHYL)-3-(5-NITRO-2-FURANYL)-2-PROPENAMIDE see FPO000

O-METHYL O-ETHYL O-p-NITROPHENYL THIOPHOSPHATE see ENI175

METHYLETHYLNITROSAMINE see MKB000

N,N-METHYLETHYLNITROSAMINE see MKB000

((1-METHYLETHYL)NITROSOAMINO)-METHANOL ACETATE (ESTER) (9CI) see ING300

N-(1-METHYLETHYL)-N-NITROSO-UREA (9CI) see NKO425

1-METHYLETHYL OCTADECANOATE see IRL100

1-METHYL-4-ETHYLOCTYLAMINE see EMY000

METHYLETHYLOLAMINE see MGG000

2-(1-METHYLETHYL)PHENOL see IQX100

4-(1-METHYLETHYL)PHENOL see IQZ000

(1-METHYLETHYL)PHENOL PHOSPHATE (3:1) see TKU300

1-METHYL-5-ETHYL-5-PHENYLBARBITURIC ACID see ENB500

2-METHYL-3-ETHYL-4-PHENYL-4-CYCLOHEXENE CARBOXYLIC ACID see FAO200

2-METHYL-3-ETHYL-4-PHENYL-Δ⁴-CYCLOHEXENECARBOXYLIC ACID see FAO200

2-METHYL-3-ETHYL-4-PHENYL-Δ⁴-CYCLOHEXENE CARBOXYLIC ACID SODIUM SALT see MBV775

METHYLETHYL PHENYLETHYL CARBINYL ACETATE see PFR300

3-(1-METHYLETHYL)PHENYL ((HEXYLOXY)SULFINYL)METHYLCARBAMATE see IRC060

3-METHYL-5-ETHYL-5-PHENYLHYDANTOIN see MKB250

2-(1-METHYLETHYL)PHENYL METHYLCARBAMATE see MIA250

1-METHYLETHYL 3-PHENYLPROPENOATE see IOO000

1-METHYLETHYL 3-PHENYL-2-PROPENOATE see IOO000

α-(5-(1-METHYLETHYL)-1-PHENYL-1H-PYRAZOL-4-YL)-1-PIPERIDINEBUTANOL see MKB300

(1-METHYLETHYL)PHOSPHORAMIDOTHIOIC ACID O-(2,4-DICHLOROPHENYL)-O-METHYL ESTER see DGD800

N-(1-METHYLETHYL)-2-PROPANAMINE see DNM200

N-(1-METHYLETHYL)-2-PROPENAMIDE see INH000

METHYL 3-((ETHYL(PROPYLAMINO)PHOSPHINOTHIOYL)OXY)-2-BUTENOATE see MKB320

2-METHYL-5-ETHYLPYRIDINE see EOS000

6-METHYL-3-ETHYLPYRIDINE see EOS000

METHYL ETHYL PYRIDINE (DOT) see EOS000

2-METHYL-5-ETHYLPYRIDINE (DOT) see EOS000

N-METHYL-N-ETHYL-4-(4'-(PYRIDYL-1'OXIDE)AZO)ANILINE see MKB500

3-METHYL-3-ETHYLPYRROLIDINE-2,5-DIONE see ENG500

γ-METHYL-γ-ETHYL-SUCCINIMIDE see ENG500

METHYLETHYLTHIOFOS see ENI175

(R*,S*)-4-((1-METHYLETHYL)THIO)-α-(1-OCTYLAMINO)ETHYL)BENZENEMETHANOL see SOU600

METHYLETHYLTHIOPHOS see ENI175

O-METHYL-O-ETHYL-O-2,4,5-TRICHLOROPHENYL THIOPHOSPHATE see TIR250

α-(1-METHYLETHYL)-α-(4-(TRIFLUOROMETHOXY)PHENYL)-5-PYRIMIDINEMETHANOL see IRN500

4-(1-METHYLETHYL)-2-(3-(TRIFLUOROMETHYL)PHENYL)MORPHOLINE see IRP000

(1-METHYLETHYL)UREA see IRR100
1-(METHYLETHYL)UREA and SODIUM NITRITE see SIS675
3-METHYLETHYNYLESTRADIOL see MKB750
17-α-METHYLETHYNYL-19-NORTESTOSTERONE see MNC100
18-METHYL-17-α-ETHYNYL-19-NORTESTOSTERONE see NNQ500
18-METHYL-17-α-ETHYNYL-19-NORTESTOSTERONE see NNQ520
3-METHYLETHYNYLOESTRADIOL see MKB750
METHYL EUGENOL (FCC) see AGE250
METHYL EUGENOL GLYCOL see MKC000
S-METHYL FENITROOXON see MKC250
S-METHYL FENITROTHION see MKC250
2-METHYLFLUORANTHENE see MKC500
3-METHYLFLUORANTHENE see MKC750
8-METHYLFLUORANTHENE see MKC775
9-METHYLFLUORENE see MKC800
METHYL FLUORIDE see FJK000
METHYL FLUOROACETATE see MKD000
2-METHYL-4-FLUOROANILINE see FJR900
7-METHYL-9-FLUOROBENZ(c)ACRIDINE see MKD250
7-METHYL-11-FLUOROBENZ(c)ACRIDINE see MKD500
10-METHYL-3-FLUORO-5,6-BENZACRIDINE see MKD750
7-METHYL-2-FLUOROBENZ(a)ANTHRACENE see FJN000
METHYL N-(5-(p-FLUOROBENZOYL)-2-BENZIMIDAZOLYL)CARBAMATE see FDA887
8-METHYL-2-(γ-(p-FLUOROBENZOYL)PROPYL)-2,3,4,4A,5,9B-HEXAHYDRO-1H-PYRIDO(3,4-B)INDOLE 2HCL see MKD800
METHYL-4-FLUOROBUTYRATE see MKE000
METHYL-γ-FLUOROBUTYRATE see MKE000
METHYL-ω-FLUOROBUTYRATE see MKE000
METHYL-γ-FLUOROCROTONATE see MKE250
16-α-METHYL-9-α-FLUORO-1-DEHYDROCORTISOL see SOW000
METHYLFLUOROFORM see TJY900
METHYL 2-(FLUOROFORMYL)DIFLUOROACETATE see MJD400
16-α-METHYL-9-α-FLUORO-Δ¹-HYDROCORTISONE see SOW000
METHYL-γ-FLUORO-β-HYDROXYBUTYRATE see MKE750
17-α-METHYL-9-α-FLUORO-11-β-HYDROXYTESTERONE see AOO275
METHYL-γ-FLUORO-β-HYDROXYTHIOLBUTYRATE see MKF000
2-(2-METHYL-5-FLUOROPHENYLAMINO)-2-IMIDAZOLINE HYDROCHLORIDE see FMR080
METHYLFLUOROPHOSPHORIC ACID, ISOPROPYL ESTER see IPX000
METHYL-FLUORO-PHOSPHORYLCHOLINE see MKF250
16-α-METHYL-9-α-FLUOROPREDNISOLONE see SOW000
16-α-METHYL-6-α-FLUORO-PREDNISOLONE-21-ACETATE see PAL600
16-α-METHYL-9-α-FLUORO-1,4-PREGNADIENE-11-β,17-α,21-TRIOL-3,20-DIONE see SOW000
METHYL FLUOROSULFATE see MKG250
METHYL FLUOROSULFONATE see MKG250
16-α-METHYL-9-α-FLUORO-11-β,17-α,21-TRIHYDROXYPREGNA-1,4-DIENE-3,20-DIONE see SOW000

METHYLFLUORPHOSPHORSAEURECHOLINESTER (GERMAN) see MKF250
METHYLFLUORPHOSPHORSAEUREISOPROPYLESTER (GERMAN) see IPX000
METHYLFLUORPHOSPHORSAEUREPINAKOLYLESTER (GERMAN) see SKS500
METHYL FLUORSULFONATE see MKG250
METHYLFLURETHER see EAT900
9-METHYLFOLIC ACID see HGI525
x-METHYLFOLIC ACID see MKG275
METHYLFORMAMIDE see MKG500
N-METHYLFORMAMIDE see MKG500
METHYL FORMATE see MKG750
METHYLFORMIAAT (DUTCH) see MKG750
METHYLFORMIAT (GERMAN) see MKG750
METHYL FORMYL see DOO600
N-METHYL-N-FORMYLHYDRAZINE see FNW000
N-METHYL-N-FORMYL HYDRAZONE of ACETALDEHYDE see AAH000
1-METHYL-2-FORMYL-5-NITROIMIDAZOLE see MKG800
1-METHYL-2-FORMYLPYRIDINIUM CHLORIDE OXIME see FNZ000
METHYL FOSFERNO see MNH000
METHYLFUMARIC ACID see MDI250
5-METHYL-2-FURALDEHYDE see MKH550
METHYLFURAN see MKH000
2-METHYLFURAN see MKH000
5-METHYL-2-FURANCARBOXALDEHYDE see MKH550
METHYL 2-FURANCARBOXYLATE see MKH600
3-METHYL-2,5-FURANDIONE see CMS322
5-METHYL-2(3H)-FURANONE see MKH250
5-METHYL-2(5H)-FURANONE see MKH500
5-METHYL FURFURAL see MKH550
5-METHYL-2-FURFURAL see MKH550
5-METHYLFURFURALDEHYDE see MKH550
METHYL FUROATE see MKH600
METHYL 2-FUROATE see MKH600
5-METHYL-2H-FURO(2,3-H)-1-BENZOPYRAN-2-ONE see HLV955
2-METHYL-3-FURYLACROLEIN see FPX050
2-METHYL-3-(2-FURYL)ACROLEIN see FPX050
α-METHYLFURYLACROLEIN see FPX050
α-METHYL-β-FURYLACROLEIN see FPX050
METHYL 2-FURYL KETONE see ACM200
5-(5-METHYL-2-FURYL)-1-PHENYL-3-(2(PIPERIDINO)ETHYL)-1H-PYRAZOLINE HYDROCHLORIDE see MKH750
2-METHYL-3-(2-FURYL)PROPENAL see FPX050
METHYL GAG see MKI000
METHYL GALLATE see MKI100
6'-N-METHYLGENTAMICIN C¹ₐ see MQS579
METHYL GLUCAMINE BILIGRAFIN see BGB315
METHYLGLUCAMINE DIATRIXOATE see AOO875
METHYLGLUCAMINE-3,5-DIIODO-4-PYRIDONE-N-ACETATE see DNG400
METHYL GLUCAMINE IODIPAMIDE see BGB315
METHYLGLUCAMINE IOTALAMATE see IGC000
METHYLGLUCAMINE IOTHALAMATE see IGC000
N-METHYLGLUCAMINE and SULFALENE see SNJ400
N-METHYLGLUCAMINE and SULFAMONOMETOXINE see SNL830
α-METHYLGLUCOSIDE TETRANITRATE see MKI125
3-METHYL GLUTARALDEHYDE see MKI250
α-METHYLGLYCEROL TRINITRATE see MKI300
METHYL GLYCIDYL ETHER see GGW600
METHYL GLYCOL see EJH500

METHYL GLYCOL see PML000
METHYL GLYCOL ACETATE see EJJ500
METHYL GLYCOL MONOACETATE see EJJ500
METHYLGLYCYL FP-12 see AFQ820
METHYL GLYKOL (GERMAN) see EJH500
METHYLGLYKOLACETAT (GERMAN) see EJJ500
METHYLGLYOXAL see PQC000
METHYL GLYOXYLATE see MKI550
4-METHYLGUAIACOL see MEK325
p-METHYLGUAIACOL see MEK325
METHYLGUANIDIN (GERMAN) see MKI750
METHYLGUANIDINE see MKI750
METHYLGUANIDINE mixed with SODIUM NITRITE (1:1) see MKJ000
METHYL GUTHION see ASH500
METHYLGUVACINE see AQT625
N-METHYLGUVACINE see AQT625
METHYLHARNSTOFF and NATRIUMNITRIT (GERMAN) see MQJ250
11-METHYL-11H-BENZO(a)CARBAZOLE see MHD250
METHYL HEPTANETHIOL see MKJ250
METHYL HEPTANOATE see MKK100
3-METHYL-5-HEPTANONE see EGI750
5-METHYL-3-HEPTANONE see EGI750
5-METHYL-3-HEPTANONE (OSHA) see ODI000
5-METHYL-3-HEPTANONE OXIME see EMP550
METHYLHEPTENOL see MKJ750
6-METHYL-5-HEPTEN-2-OL see MKJ750
METHYL HEPTENONE see MKK000
6-METHYL-5-HEPTEN-2-ONE see MKK000
METHYL HEPTINE CARBONATE see MND275
METHYL HEPTOATE see MKK100
1-METHYLHEPTYLAMINE see OEK010
2-METHYL-2-HEPTYLAMINE see ILM000
6-METHYL-2-HEPTYLAMINE see ILM000
METHYL n-HEPTYLATE see MKK100
(6-(1-METHYL-HEPTYL)-2,4-DINITRO-FENYL)-CROTONAAT (DUTCH) see AQT500
(6-(1-METHYL-HEPTYL)-2,3-DINITRO-PHENYL)-CROTONAT (GERMAN) see AQT500
2-(1-METHYLHEPTYL)-4,6-DINITROPHENYL CROTONATE see AQT500
6-METHYL-2-HEPTYLHYDRAZINE see MKK250
6-METHYL-2-HEPTYL-ISOPROPYLIDENHYDRAZIN (GERMAN) see MKK500
6-METHYL-2-HEPTYLISOPROPYLIDENHYDRAZINE see MKK500
METHYL HEPTYL KETONE see NMY500
METHYLHEPTYLNITROSAMIN (GERMAN) see HBP000
METHYLHESPERIDIN see MKK600
METHYLHEXABARBITAL see ERD500
METHYLHEXABITAL see ERD500
3-β-14-METHYLHEXADECANOATE-CHOLEST-5-EN-3-OL see CCJ500
2-METHYL-1,5-HEXADIENE-3-YNE see MJL000
METHYL HEXAFLUOROISOBUTYRATE see MKK750
3A-METHYL-4,7-HEXAHYDROEPOXYBENZO(c)THIOPHENE-1,3-DIONE(3a-α,4-β,7-β,7a-α)- see MKK800
N-METHYLHEXAHYDRO-2-PICOLINIC ACID, 2,6-DIMETHYLANILIDE see SBB000
2-METHYLHEXANE see MKL250
5-METHYL-2,3-HEXANEDIONE see MKL300
METHYL HEXANOATE see MHY700

METHYL n-HEXANOATE see MHY700
2-METHYLHEXANOIC ACID see HEN500
5-METHYLHEXANOIC ACID see IKN400
3-METHYL-3-HEXANOL see MKL350
3-METHYL-HEXANOL-(3) see MKL350
2-METHYL-3-HEXANONE see MKL400
2-METHYL-5-HEXANONE see MKW450
5-METHYL-2-HEXANONE see MKW450
5-METHYLHEXAN-2-ONE (DOT) see MKW450
5-METHYL-3-HEXEN-2-ONE see MKM250
2-METHYL-5-HEXEN-3-YN-2-OL see MKM300
METHYL HEXOATE see MHY700
METHYLHEXYLACETALDEHYDE see MNC175
1-METHYLHEXYLAMINE see HBM490
5-METHYL-2-HEXYLAMINE see MKM500
1-METHYLHEXYLAMINE SULFATE see AKD600
METHYL HEXYLATE see MHY700
5-METHYLHEXYLIC ACID see IKN400
METHYL HEXYL KETONE (FCC) see ODG000
1-METHYLHEXYL-β-OXYBUTYRATE see MKM750
METHYL-N-(((((HEXYLOXY)SULFINYL)METHYLAMINO)CARBONYL)OXY)ETHANIMIDOTHIOATE see MKM800
METHYLHOMATROPINE BROMIDE see MDL000
8-METHYLHOMOTROPINIUM BROMIDE see MDL000
α-((N-METHYL-N-HOMOVERATRYL)-γ-AMINOPROPYL)-3,4-DIMETHOXYPHENYLACETONITRILE see IRV000
METHYL HYDANTOIN see MKB250
METHYL HYDRAZINE see MKN000
1-METHYL HYDRAZINE see MKN000
METHYLHYDRAZINE (DOT) see MKN000
METHYLHYDRAZINE ETHANEDIOATE (1:1) see MKN300
METHYLHYDRAZINE HYDROCHLORIDE see MKN250
METHYLHYDRAZINE OXALATE see MKN300
METHYL HYDRAZINE SULFATE see MKN500
METHYLHYDRAZINIUM NITRATE see MRJ250
9-(1-METHYLHYDRAZINO)ACRIDINE MONOHYDROCHLORIDE see MKN600
4-((2-METHYLHYDRAZINO)METHYL)-N-ISOPROPYLBENZAMIDE see PME250
p-(N'-METHYLHYDRAZINOMETHYL)-N-ISOPROPYLBENZAMIDE HYDROCHLORIDE see PME500
(α-(2-METHYLHYDRAZINO)-p-TOLUOYL)UREA, MONOHYDROBROMIDE see MKN750
METHYL HYDRIDE see MDQ750
Δ¹-6-α-METHYLHYDROCORTISONE see MOR500
METHYL HYDROGEN SILOXANE see TDJ100
METHYL HYDROPEROXIDE see MKO000
N-METHYL-N-(HYDROPEROXYMETHYL)NITROSAMINE see HIE700
METHYLHYDROQUINONE see MKO250
2-METHYLHYDROQUINONE see MKO250
METHYL-p-HYDROQUINONE see MKO250
p-METHYL HYDROTROPALDEHYDE see THD750
METHYLHYDROXAMIC ACID see ABB250
METHYL HYDROXIDE see MGB150
METHYL N-HYDROXYACETIMIDOTHIOATE see MPS270

4-(N-METHYLHYDROXYAMINO)-QUINOLINE-1-OXIDE see HLX550
17-α-METHYL-17-β-HYDROXY-1,4-ANDROSTADIEN-3-ONE see DAL300
METHYL-3-HYDROXYANTHRANILATE see HJC500
1-METHYL-4-HYDROXYBENZENE see CNX250
METHYL-o-HYDROXYBENZOATE see MPI000
METHYL p-HYDROXYBENZOATE see HJL500
2-METHYL-3-HYDROXY-4,5-BIS(HYDROXYMETHYL)PYRIDINE see PPK250
2-METHYL-3-HYDROXY-4,5-BIS(HYDROXYMETHYL)PYRIDINE HYDROCHLORIDE see PPK500
N-METHYL-N-(4-HYDROXYBUTYL)NITROSAMINE see MKP000
METHYL-3-HYDROXYBUTYRATE see MKP250
METHYL m-HYDROXYCARBANILATE, m-METHYLCARBANILATE see MEG250
1-METHYL-4-HYDROXY-2-CHINOLON (CZECH) see HMR500
4-METHYL-7-HYDROXYCOUMARIN see MKP500
4-METHYL-7-HYDROXY COUMARIN DIETHOXYTHIOPHOSPHATE see PKT000
2-METHYL-5-HYDROXY-8-(β-DIAETHYLAMINO-AETHOXY)FURANO-6,7:2'3'-CHROMON HYDROCHLORIDE see AHP375
2-METHYL-3-HYDROXY-4,5-DIHYDROXYMETHYL-PYRIDIN (GERMAN) see PPK250
2-METHYL-3-HYDROXY-4,5-DI(HYDROXYMETHYL)PYRIDINE see PPK250
17-α-METHYL-17-β-HYDROXY-4-ESTREN-3-ONE see MDM350
METHYL N-HYDROXYETHANIMIDOTHIOATE see MPS270
(2-METHYL-3-(1-HYDROXYETHOXYETHYL-4-PIPERAZINYL)PROPYL)-10-PHENOTHIAZINE see MKQ000
METHYL(β-HYDROXYETHYL)AMINE see MGG000
N-METHYL-N-HYDROXYETHYLANILINE see MKQ250
N-METHYL-N-β-HYDROXYETHYLANILINE see MKQ250
1-METHYL-1-(β-HYDROXYETHYL)ETHYLENAMMONIUM PICRYLSULFONATE see MKQ500
METHYL-β-HYDROXYETHYL-ETHYLENIMONIUM PICRYLSULFONATE see MKQ500
2-METHYL-1-(2-HYDROXYETHYL)-5-NITROIMIDAZOLE see MMN250
2-METHYL-3-(2-HYDROXYETHYL)-4-NITROIMIDAZOLE see MMN250
METHYL-2-HYDROXYETHYLNITROSAMINE see NKU350
METHYL-2-HYDROXYETHYLNITROSOAMINE see NKU350
2-METHYL-3-HYDROXY-4-FORMYL-5-HYDROXYMETHYLPYRIDINE HYDROCHLORIDE see VSU000
METHYL-3-β-HYDROXY-1-α-H,5-α-H-TROPANE-2-β-CARBOXYLATE BENZOATE (ESTER) see CNE750
1-METHYL-2-HYDROXYIMINOMETHYLPYRIDINIUM IODIDE see POS750

(+−)-(E)-METHYL-2-((E)-HYDROXYIMINO)-5-NITRO-6-METHOXY-3-HEXENEAMIDE see MKQ600
1-METHYL-3-HYDROXY-4-ISOPROPYLBENZENE see TFX810
N-METHYL-4-HYDROXYKARBOSTYRIL (CZECH) see HMR500
METHYLHYDROXYLAMINE see MKQ875
o-METHYLHYDROXYLAMINE see MKQ880
o-METHYLHYDROXYLAMINE see MNG500
o-METHYLHYDROXYLAMINE HYDROCHLORIDE see MNG500
2-METHYL-4-HYDROXYLAMINOQUINOLINE 1-OXIDE see MKR000
5-METHYL-4-HYDROXYLAMINOQUINOLINE-1-OXIDE see HIV000
6-METHYL-4-HYDROXYLAMINOQUINOLINE-1-OXIDE see HIV500
7-METHYL-4-HYDROXYLAMINOQUINOLINE-1-OXIDE see HIW000
8-METHYL-4-HYDROXYLAMINOQUINOLINE-1-OXIDE see HIW500
METHYL 4-HYDROXY-3-METHOXYBENZOATE see VFF100
METHYL-o-(4-HYDROXY-3-METHOXYCINNAMOYL)RESERPATE see MKR100
7-METHYL-12-HYDROXYMETHYLBENZ(a)ANTHRACENE see HMF500
17-α-METHYL-2-HYDROXYMETHYLENE-17-HYDROXY-5-α-ANDROSTAN-3-ONE see PAN100
METHYL 4-(2-HYDROXY-3-((1-METHYLETHYL)AMINO)PROPOXY)BENZENEPROPANOATE see MKR125
1-METHYL-3-(HYDROXYMETHYL)IMIDAZOLIUM CHLORIDE LAURATE see MKR150
1-METHYL-2-HYDROXYMETHYL-5-NITROIMIDAZOLE see MMN500
4-METHYL-5-HYDROXYMETHYLURACIL see HMH300
N-METHYL-3-HYDROXYMORPHINAN see MKR250
2-METHYL-3-HYDROXY-1,4-NAPHTHOQUINONE see HMI000
2-METHYL-5-HYDROXY-1,4-NAPHTHOQUINONE see PJH610
3-METHYL-4-HYDROXY-1-NAPHTHYLAMINE see AKX500
METHYL HYDROXYOCTADECADIENOATE see MKR500
4-METHYL-4-HYDROXY-1-OCTYNE see MKR550
2-METHYL-2-p-HYDROXYPHENYLBUTANE see AON000
METHYL-p-HYDROXYPHENYL KETONE see HIO000
METHYL (4-(m-HYDROXYPHENYL)-1-METHYL)-4-PIPERIDYL KETONE see HNK950
N-METHYL-1-(3-HYDROXYPHENYL)-2-PHENYLETHYLAMINE HYDROCHLORIDE see MGJ600
p-METHYL-m-HYDROXY-PHENYL-PROPANOLAMINE HYDROCHLORIDE see MKS000
α-METHYL-ω-HYDROXYPOLY(OXY(METHYL-1,2-ETHANEDIYL)) see MKS250
6-α-METHYL-17-α-HYDROXYPROGESTERONE see MBZ150
6-α-METHYL-17-α-HYDROXYPROGESTERONE ACETATE see MCA000

6-METHYL-17-α-HYDROXY-Δ⁶-
PROGESTERONE ACETATE see VTF000
6-METHYL-4-HYDROXYPYRON-(2) see
THL800
2-METHYL-3-HYDROXY-4-PYRONE see
MAO350
3-METHYL-4-HYDROXY-3-PYRROLIN-2-
ONE AMMONIUM SALT see MKS750
2-METHYL-8-HYDROXYQUINOLINE see
HMQ600
METHYL-12-HYDROXYSTEARATE see
HOG500
N-METHYLHYOSCINE BROMIDE see
SBH750
N-METHYLHYOSCINE METHYL SULFATE
see DAB750
(−)-α-METHYLHYOSCYAMINE
HYDROCHLORIDE see LFC000
METHYL HYPOCHLORITE see MKT000
N-METHYL 1A see MKW300
N,N'-METHYLIDENEBISACRYLAMIDE see
MJL500
4,4',4"-METHYLIDYNETRIANILINE see
THP000
4,4',4"-
METHYLIDYNETRISBENZENEAMINE see
THP000
1,1',1'-
(METHYLIDYNETRIS(OXY))TRIS(ETHANE
) see ENY500
1-METHYLIMIDAZOLE see MKT500
2-METHYLIMIDAZOLE see MKT750
4-METHYLIMIDAZOLE see MKU000
1-METHYLIMIDAZOLE METHIODIDE see
MKU100
1-METHYLIMIDAZOLE-2-THIOL see
MCO500
4-METHYL-2-IMIDAZOLIDINETHIONE see
MJZ000
2-METHYL-3-(Δ²-
IMIDAZOLINYLMETHYL)BENZO(b)THIOP
HENE HYDROCHLORIDE see MHJ500
1-METHYLIMIDAZO(4,5-
b)(1,5)NAPHTHYRIDIN-2-AMINE see
MKU125
1-METHYLIMIDAZO(4,5-
b)(1,6)NAPHTHYRIDIN-2-AMINE see
MKU130
1-METHYLIMIDAZO(4,5-
B)(1,7)NAPHTHYRIDIN-2-AMINE see
MKU135
1-METHYLIMIDAZO(4,5-
b)(1,8)NAPHTHYRIDIN-2-AMINE see
MKU140
1-METHYL-1H-IMIDAZO(4,5-
b)(1,5)NAPHTHYRIDIN-2-AMINE see
MKU125
1-METHYL-1H-IMIDAZO(4,5-
b)(1,6)NAPHTHYRIDIN-2-AMINE see
MKU130
1-METHYL-1H-IMIDAZO(4,5-
B)(1,7)NAPHTHYRIDIN-2-AMINE see
MKU135
1-METHYL-1H-IMIDAZO(4,5-
b)(1,8)NAPHTHYRIDIN-2-AMINE see
MKU140
1-METHYL-1H-IMIDAZO(4,5-b)QUINOLIN-
2-AMINE see MKU145
1-METHYLIMIDAZO(4,5-b)QUINOLINE-2-
AMINE see MKU145
N-METHYLIMIDODICARBONIMIDIC
DIAMIDE MONOHYDROCHLORIDE see
MHP400
2,2'-(METHYLIMINO)BIS(N-ETHYL-N,N-
DIMETHYLETHANAMINIUM)
DIBROMIDE see MKU750
3,3'-(METHYLIMINO)BIS-1-PROPANOL
DIMETHANESULFONATE (ESTER), (1,1'-
BIPHENYL)-4,4'-DISULFOANTE (1:1) (SALT)
see MHQ775
METHYLIMINODIETHANOL see MKU250

N-METHYLIMINODIETHANOL see
MKU250
2,2'-(METHYLIMINO)DIETHANOL see
MKU250
N-METHYL-2,2'-IMINODIETHANOL see
MKU250
N-METHYL-2,2'-IMINODIETHANOL see
MKU250
((METHYLIMINO)DIETHYLENE)BIS(ETHY
LDIMETHYLAMMONIUM BROMIDE) see
MKU750
N,N'-
((METHYLIMINO)DIMETHYLIDYNE)DI-
2,4-XYLIDINE see MJL250
N-METHYL-3,3'-IMINODIPROPIONITRILE
see MHQ750
2-METHYL-1,3-INDANDIONE see MKV500
α-METHYL-β-INDOLAETHYLAMINE
(GERMAN) see AME500
3-METHYLINDOLE see MKV750
3-METHYL-1H-INDOLE see MKV750
β-METHYLINDOLE see MKV750
N-METHYLINDOLE-2,3-DIONE
THIOSEMICARBAZONE see MKW250
1-METHYLINDOLE-2,3-DIONE-3-
(THIOSEMICARBAZONE) see MKW250
α-METHYL-β-INDOLEETHYLAMINE see
AME500
2-METHYLINDOLINE see MKV800
4-(N-METHYL-3-
INDOLYETHYL)PYRIDINIUM
HYDROCHLORIDE see MKW000
4-(1-METHYL-3-
INDOLYETHYL)PYRIDINE
HYDROCHLORIDE see MKW000
4-(2-(1-METHYL-3-INDOLYL)ETHYL)-1-(3-
(TRIMETHYLAMMONIO)PROPYL)PYRIDI
NIUM DICHLORIDE see MKW100
3-((2-METHYL-1H-INDOL-3-YL)METHYL)-
1-(PHENYLMETHYL)-4-PIPERIDINONE see
MKW150
METHYL IODIDE see MKW200
m-METHYLIODOBENZENE see IFG100
N-METHYLISATIN
THIOSEMICARBAZONE see MKW250
N-METHYLISATIN-3-
(THIOSEMICARBAZONE) see MKW250
N-METHYL-ISATOIC ANHYDRIDE see
MKW300
METHYLISOAMYL ACETATE see HFJ000
METHYL ISOAMYL KETONE see MKW450
METHYL ISOAMYL KETOXIME see
MKW500
N-METHYL-dl-ISOBORNYLAMINE
HYDROCHLORIDE see MQR500
METHYL ISOBUTENYL KETONE see
MDJ750
METHYL ISOBUTYL CARBINOL see
AOK750
METHYL ISOBUTYL CARBINOL see
MKW600
METHYLISOBUTYL CARBINOL see
MKW600
METHYLISOBUTYLCARBINOL ACETATE
see HFJ000
METHYLISOBUTYLCARBINYL ACETATE
see HFJ000
METHYL-ISOBUTYL-CETONE (FRENCH)
see HFG500
METHYLISOBUTYLKETON (DUTCH,
GERMAN) see HFG500
METHYL ISOBUTYL KETONE (ACGIH,
DOT) see HFG500
METHYL ISOBUTYL KETONE PEROXIDE,
in solution with >9% by weight active oxygen
(DOT) see IJB100
METHYL ISOBUTYL KETOXIME see
MKW750
METHYL ISOBUTYRATE see MKX000
N-METHYL-2-ISOCAMPHANAMINE see
VIZ400

METHYLISOCYANAAT (DUTCH) see
MKX250
METHYL ISOCYANAT (GERMAN) see
MKX250
METHYL ISOCYANATE see MKX250
METHYL ISOCYANATE, solutions (DOT) see
MKX250
METHYL ISOCYANIDE see MKX500
METHYL ISOCYANOACETATE see MKX575
METHYL ISOEUGENOL (FCC) see IKR000
METHYLISOMIN see DBA800
METHYLISOMYN see MDT600
METHYL ISONITRILE see MKX500
METHYLISOOCTENYLAMINE see ILK000
METHYL ISOPENTANOATE see ITC000
1-METHYL-4-
ISOPROPENYLCYCLOHEXAN-3-OL see
MCE750
1-METHYL-4-ISOPROPENYL-1-
CYCLOHEXENE see MCC250
1-METHYL-4-ISOPROPENYL-6-
CYCLOHEXEN-2-OL see MKY250
1-1-METHYL-4-ISOPROPENYL-6-
CYCLOHEXEN-2-ONE see CCM120
d-1-METHYL-4-ISOPROPENYL-6-
CYCLOHEXEN-2-ONE see CCM100
Δ-1-METHYL-4-ISOPROPENYL-6-
CYCLOHEXEN-2-ONE see MCD250
METHYL ISOPROPENYL KETONE see
MKY500
METHYL ISOPROPENYL KETONE
INHIBITED (DOT) see MKY500
N-METHYL-2-
ISOPROPOXYPHENYLCARBAMATE see
PMY300
m-METHYLISOPROPYLBENZENE see
IRN400
p-METHYLISOPROPYL BENZENE see
CQI000
1-METHYL-2-ISOPROPYLBENZENE see
IRN300
1-METHYL-3-ISOPROPYLBENZENE see
IRN400
1-METHYL-4-ISOPROPYLBENZENE see
CQI000
1-METHYL-2-ISOPROPYLBENZOL see
IRN300
1-METHYL-2-p-
(ISOPROPYLCARBAMOYL)BENZOHYDRA
ZINE HYDROCHLORIDE see PME500
1-(METHYL-2-(-
ISOPROPYLCARBAMOYL)BENZYL)HYDR
AZINE see PME250
1-METHYL-2-(p-
ISOPROPYLCARBAMOYLBENZYL)HYDRA
ZINE HYDROCHLORIDE see PME500
1-METHYL-4-
ISOPROPYLCYCLOHEXADIENE-1,3 see
MLA250
1-METHYL-4-ISOPROPYL-1,3-
CYCLOHEXADIENE see MLA250
1-METHYL-4-
ISOPROPYLCYCLOHEXADIENE-1,4 see
MCB750
2-METHYL-5-ISOPROPYL-1,3-
CYCLOHEXADIENE see MCC000
1-METHYL-4-ISOPROPYLCYCLOHEXANE-
8-OL see MCE100
6-METHYL-3-
ISOPROPYLCYCLOHEXANOL see DKV150
1-METHYL-4-ISOPROPYL-1-
CYCLOHEXEN-3-ONE see MCF250
5-METHYL-2-ISOPROPYL-2-HEXENAL see
MLA300
α-METHYL-p-
ISOPROPYLHYDROCINNAMALDEHYDE
see COU500
α-METHYL-p-ISOPROPYL
HYDROCINNAMIC ALDEHYDE DIETHYL
ACETAL see COU510
α-METHYL-p-
ISOPROPYLHYDROCINNAMIC

ALDEHYDE DIMETHYL ACETAL see COU525
1-METHYL-4-ISOPROPYLIDENE-3-CYCLOHEXANONE see MCF500
METHYL ISOPROPYL KETONE see MLA750
2-METHYL-5-ISOPROPYLPHENOL see CCM000
5-METHYL-2-ISOPROPYL-1-PHENOL see TFX810
N-METHYL-3-ISOPROPYLPHENYL CARBAMATE see COF250
N-METHYL-m-ISOPROPYLPHENYL CARBAMATE see COF250
(3-METHYL-5-ISOPROPYLPHENYL)-N-METHYLCARBAMAT (GERMAN) see CQI500
3-METHYL-5-ISOPROPYLPHENYL-N-METHYLCARBAMATE see CQI500
2-METHYL-3-(p-ISOPROPYLPHENYL)PROPIONALDEHYDE see COU500
5-METHYL-2-ISOPROPYL-3-PYRAZOLYL DIMETHYLCARBAMATE see DSK200
METHYL(5-ISOPROPYL-N-(p-TOLYL)-o-TOLUENESULFONAMIDO)MERCURY see MLB000
METHYLISOPSEUDOIONONE see TMJ100
METHYL ISOSYSTOX see DAP400
3-METHYL-5-ISOTHIAZOLAMINE HYDROCHLORIDE see MLB600
3-METHYL-5-ISOTHIAZOLAMINE MONOHYDROCHLORIDE see MLB600
2-METHYL-4-ISOTHIAZOLIN-3-ONE see MLB700
2-METHYL-3(2H)-ISOTHIAZOLONE see MLB700
METHYLISOTHIOCYANAAT (DUTCH) see ISE000
METHYL-ISOTHIOCYANAT (GERMAN) see ISE000
d-d-METHYLISOTHIOCYANATE see MLC000
METHYL ISOTHIOCYANATE (DOT) see ISE000
S-METHYLISOTHIOUREA SULFATE (1:1) see MPV750
2-METHYLISOTHIOURONIUM SULFATE see MPV750
S-METHYLISOTHIURONIUM SULFATE see MPV790
METHYL ISOVALERATE see ITC000
METHYLISOVALERATE (DOT) see ITC000
METHYL ISOXATHION see MLC100
3-METHYL-5-ISOXAZOLECARBOXYLIC ACID 2-BENZYLHYDRAZIDE see BEW000
5-METHYL-3-ISOXAZOLECARBOXYLIC ACID-2-BENZYLHYDRAZIDE see IKC000
3-METHYL-4,5-ISOXAZOLEDIONE-4-((2-CHLOROPHENYL)HYDRAZONE) see MLC250
N'-(5-METHYL-3-ISOXAZOLE)SULFANILAMIDE see SNK000
5-METHYL-3-ISOXAZOLOL see HLM000
5-METHYL-3(2H)-ISOXAZOLONE see HLM000
N'-(5-METHYL-3-ISOXAZOLYL)SULFANILAMIDE see SNK000
N'-(5-METHYLISOXAZOL-3-YL)SULPHANILAMIDE see SNK000
N¹-(5-METHYL-3-ISOXAZOLYL)SULPHANILAMIDE see SNK000
METHYLIUM, TRIS(4-AMINOPHENYL)-, 4,4'-METHYLENEBIS(3-HYDROXY-NAPHTHOATE) (2:1) see TNC725
N-METHYLJERVINE see MLC500
METHYLJODID (GERMAN) see MKW200
METHYLJODIDE (DUTCH) see MKW200
7-METHYLJUGLON see RBF200
7-METHYLJUGLONE see RBF200
METHYLKARBITOLACETAT see MIE750

METHYL KETONE see ABC750
METHYLKYANID see ABE500
METHYL LACTATE see MLC600
2-METHYLLACTIC ACID ETHYL ESTER see ELH700
2-METHYLLACTONITRILE see MLC750
METHYL LAURATE see MLC800
METHYL LAURINATE see MLC800
N-METHYLLAUROTETANINE see TKX700
METHYL LEADATE see LCW000
METHYLLITHIUM see MLD000
N-METHYLLOLINE see MLD050
METHYL-LOMUSTINE see CHD250
METHYLLORAZEPAM see MLD100
N-METHYLLORAZEPAM see MLD100
N-METHYL-LOST see BIE250
1-METHYL-LUMILYSERGOL-8-(5-BROMONICOTINATE)-10-METHYL ETHER see NDM000
METHYLLYCACONITINE see MLD140
METHYLLYSERGIC ACID BUTANOLAMIDE see MLD250
1-METHYLLYSERGIC ACID BUTANOLAMIDE see MLD250
1-METHYLLYSERGIC ACID ETHYLAMIDE see MLD500
d-1-METHYL LYSERGIC ACID MONOETHYLAMIDE see MLD500
METHYLMAGNESIUM BROMIDE (ethyl ether solution) see MLE000
METHYL MAGNESIUM BROMIDE in ETHYL ETHER (DOT) see MLE000
METHYLMAGNESIUM IODIDE see MLE250
S-METHYLMALATHION see IKX200
METHYL MALEATE see DSL800
METHYLMALEIC ACID see CMS320
METHYLMALEIC ACID, DIMETHYL ESTER see DRF200
METHYLMALEIC ANHYDRIDE see CMS322
2-METHYLMALEIC ANHYDRIDE see CMS322
3-METHYLMALEIC ANHYDRIDE see CMS322
α-METHYLMALEIC ANHYDRIDE see CMS322
METHYL MALONATE see DSM200
METHYL MARASMATE see MLE600
METHYLMERCAPTAAN (DUTCH) see MLE650
METHYL MERCAPTAN see MLE650
METHYLMERCAPTOACETATE see MLE750
METHYL-2-MERCAPTOACETATE see MLE750
3-METHYLMERCAPTOANILINE see ALX100
METHYLMERCAPTOBENZIMIDAZOLE see MCO400
2-METHYLMERCAPTO-4,6-BIS(ISOPROPYLAMINO)-s-TRIAZINE see BKL250
4-METHYLMERCAPTO-3,5-DIMETHYLPHENYL N-METHYLCARBAMATE see DST000
2-METHYLMERCAPTO-4-ETHYLAMINO-6-ISOPROPYLAMINO-s-TRIAZINE see MPT500
5-(β-METHYLMERCAPTOETHYL)HYDANTOIN see MDT800
dl-5-(β-METHYLMERCAPTOETHYL)HYDANTOIN see MDT800
METHYL-MERCAPTOFOS TEOLOVY see DAP400
1-METHYL-2-MERCAPTOIMIDAZOLE see MCO500
1-METHYL-2-MERCAPTO-5-IMIDAZOLE CARBOXYLIC ACID see MLF000
2-METHYLMERCAPTO-4-ISOPROPYLAMINO-6-ETHYLAMINO-s-TRIAZINE see MPT500

METHYLMERCAPTO-4-ISOPROPYLAMINO-6-METHYLAMINO-s-TRIAZINE see INR000
4-METHYLMERCAPTO-3-METHYLPHENYL DIMETHYL THIOPHOSPHATE see FAQ900
2-METHYLMERCAPTO-10-((2-N-METHYL-2-PIPERIDYL)ETHYL)PHENOTHIAZINE see MOO250
2-METHYLMERCAPTO-10-(2-(N-METHYL-2-PIPERIDYL))ETHYLPHENOTHIAZINE HYDROCHLORIDE see MOO500
4-METHYLMERCAPTOPHENOL see MPV300
METHYLMERCAPTOPHOS see DAO800
METHYL-MERCAPTOPHOS see MIW100
d-2-METHYL-3-MERCAPTOPROPANOYL-l-PROLINE see MCO750
3-(METHYLMERCAPTO)PROPIONALDEHYDE see MPV400
3-(METHYLMERCAPTO)PROPIONALDEHYDE see TET900
β-(METHYLMERCAPTO)PROPIONALDEHYDE see MPV400
β-(METHYLMERCAPTO)PROPIONALDEHYDE see TET900
N-METHYL 3-MERCAPTOPROPIONAMIDE see MLF100
METHYLMERCAPTOPROPIONIC ALDEHYDE see MPV400
METHYLMERCAPTOPROPIONIC ALDEHYDE see TET900
6-METHYLMERCAPTOPURINE RIBONUCLEOSIDE see MPU000
6-METHYLMERCAPTOPURINE RIBOSIDE see MPU000
1-METHYL-5-MERCAPTO-1,2,3,4-TETRAZOLE see MPQ250
4-METHYLMERCAPTO-3,5-XYLYL METHYLCARBAMATE see DST000
N-METHYLMERCURI-BIS-p-TOLUENSULFONAMID (CZECH) see BLK250
METHYLMERCURIC CHLORIDE see MDD750
METHYLMERCURIC CYANOGUANIDINE see MLF250
METHYLMERCURIC CYSTEINE see MCS100
METHYLMERCURIC DICYANDIAMIDE see MLF250
METHYLMERCURIC DICYANDIAMIDE see MLF250
METHYLMERCURICHLORENDIMIDE see MLF500
METHYLMERCURIC HYDROXIDE see MLG000
METHYLMERCURIC PHOSPHATE see MLF520
METHYLMERCURIC SULFATE see BKS810
S-(METHYLMERCURIC)THIOGLYCOLIC ACID SODIUM SALT see SIM000
N-(METHYLMERCURI)-1,4,5,6,7,7-HEXACHLOROBICYCLO(2.2.1)HEPT-5-ENE-2,3-DICARBOXIMIDE see MLF500
8-(METHYLMERCURIOXY)QUINOLINE see MLH000
METHYLMERCURIPENTACHLORFENOLAT (CZECH) see MLG250
N-METHYLMERCURI-1,2,3,6-TETRAHYDRO-3,6-ENDOMETHANO-3,4,5,6,7,7-HEXACHLOROPHTHALIMIDE see MLF500
N-METHYLMERCURI-1,2,3,6-TETRAHYDRO-3,6-METHANO-3,4,5,6,7,7-HEXACHLOROPHTHALIMIDE see MLF500
METHYLMERCURY see MLF550
METHYL-MERCURY(1+) (9CI) see MLF550
METHYLMERCURY(I) CATION see MLF550

METHYLMERCURY CHLORIDE see MDD750

METHYLMERCURY CYSTEINE see MCS100

METHYLMERCURY DICYANDIAMIDE see MLF250

METHYLMERCURY DIMERCAPTOPROPANOL see MLF750

METHYLMERCURY HYDROXIDE see MLG000

METHYLMERCURY β-HYDROXYQUINOLATE see MLH000

METHYLMERCURY 8-HYDROXYQUINOLINATE see MLH000

METHYLMERCURY ION see MLF550

METHYLMERCURY ION(1+) see MLF550

METHYLMERCURY OXINATE see MLH000

METHYLMERCURY OXYQUINOLINATE see MLH000

METHYLMERCURY PENTACHLOROPHENATE see MLG250

METHYLMERCURY PERCHLORATE see MLG500

METHYLMERCURY PHOSPHATE see MLF520

METHYLMERCURY PROPANEDIOLMERCAPTIDE see MLG750

METHYLMERCURY QUINOLINOLATE see MLH000

METHYLMERCURY TOLUENESULFONATE see MLH050

METHYL-MERCURY TOLUENESULPHAMIDE see MLH100

METHYLMERKURIDIKYANDIAMID see MLF250

α-METHYLMESCALINE see MLH250

METHYL MESYLATE see MLH500

METHYLMETHACRYLAAT (DUTCH) see MLH750

METHYL-METHACRYLAT (GERMAN) see MLH750

METHYL METHACRYLATE see MLH750

METHYL METHACRYLATE HOMOPOLYMER see PKB500

METHYL METHACRYLATE MONOMER, INHIBITED (DOT) see MLH750

METHYL METHACRYLATE POLYMER see PKB500

METHYL METHACRYLATE RESIN see PKB500

N-METHYLMETHANAMINE see DOQ800

N-METHYLMETHANAMINE with BORANE (1:1) see DOR200

N-METHYLMETHANAMINE HYDROCHLORIDE see DOR600

METHYLMETHANE see EDZ000

N-METHYLMETHANESULFONAMIDE see MLH760

METHYL METHANESULFONATE see MLH500

METHYL METHANESULPHONATE see MLH500

METHYL METHANETHIOSULFINATE see MLH765

S-METHYL METHANETHIOSULFINATE see MLH765

METHYL METHANOATE see MKG750

METHYL METHANSULFONAT (GERMAN) see MLH500

METHYL METHANSULFONATE see MLH500

METHYL METHANSULPHONATE see MLH500

2-METHYL-4-METHOXYANILINE see MGO500

4-METHYL-1-METHOXYBENZENE see MGP000

METHYL 2-METHOXYBENZOATE see MLH800

METHYL o-METHOXYBENZOATE see MLH800

METHYL-p-METHOXYBENZOATE see AOV750

1-METHYL-7-METHOXY-β-CARBOLINE see HAI500

METHYL-3-METHOXY CARBONYLAZOCROTONATE see MLI350

METHYL-p-METHOXYCINNAMYLKETONE see MLI400

3-METHYL-7-METHOXY-8-(DIMETHYLAMINO-METHYL)-FLAVONE HYDROCHLORIDE see DNV200

2-METHYL-5-METHOXY-N-DIMETHYLTRYPTAMINE see MLI750

METHYL-2-METHOXY-5-N-DIMETHYLTRYPTAMINE see MLI750

(E)-4-METHYL-5-METHOXY-7-HYDROXY-6-(5-CARBOXY-3METHYLPENT-2-EN-1-YL)PHTHALIDE see MRX000

METHYL-3-METHOXY-4-HYDROXY STYRYL KETONE see MLI800

METHYL (E)-2-METHOXYIMINO-(2-(o-TOLYLOXYMETHYL)PHENYL)ACETATE see MLI900

9-METHYL-10-METHOXYMETHYL-1,2-BENZANTHRACENE see MEW000

METHYL(METHOXYMETHYL)NITROSAMINE see MEW250

METHYL(((METHOXYMETHYLPHOSPHINOTHIOYL)THIO)ACETYL)METHYLCARBAMATE see MLJ000

METHYL 2-(3-(4-METHOXY-6-METHYL-1,3,5-TRIAZIN-2-YL)UREIDOSULPHONYL)BENZOATE see MQR400

N-METHYL-4-METHOXYNAPHTHALIMIDE see MEW760

N-METHYL-N'-(p-METHOXYPHENYL)-N-NITROSOUREA see MFG400

2-METHYL-3-(p-METHOXYPHENYL)PROPANAL see MLJ050

N-METHYL-5-METHOXY-3-PIPERIDYLIDENEDITHIENYLMETHANE METHOBROMIDE see TGB160

2-(3-METHYL-5-METHOXY-1-PYRAZOLYL)-4-METHOXY-6-METHYLPYRIMIDINE see MCH550

METHYL p-METHOXYSTYRYL KETONE see MLI400

1-METHYL-6-METHOXY-1,2,3,4-TETRAHYDRO-β-CARBOLINE see MLJ100

3-METHYL-8-METHOXY-3H,1,2,5,6-TETRAHYDROPYRAZINO-(1,2,3-ab)-β-CARBOLINE HYDROCHLORIDE see IBQ300

METHYL-α-METHYLACRYLATE see MLH750

1-METHYL-6-(1-METHYLALLYL)DITHIOBIUREA see MLJ500

1-METHYL-6-(1-METHYLALLYL)-2,5-DITHIOBIUREA see MLJ500

N-METHYL-3-METHYL-p-AMINOAZOBENZENE see MLY250

N-METHYL-2'-METHYL-4-AMINOAZOBENZENE see MPY250

N-METHYL-2'-METHYL-p-AMINOAZOBENZENE see MPY250

N-METHYL-3'-METHYL-4-AMINOAZOBENZENE see MPY000

N-METHYL-3'-METHYL-p-AMINOAZOBENZENE see MPY000

N-METHYL-4'-METHYL-4-AMINOAZOBENZENE see MPY500

N-METHYL-4'-METHYL-p-AMINOAZOBENZENE see MPY500

METHYL METHYLAMINOBENZOATE see MGQ250

2-METHYL-2-(METHYLAMINO)-1,3-BENZODIOXOLE HYDROCHLORIDE see MLJ750

((METHYL N-((METHYLAMINO)CARBONYL)OXY)ETHANIMIDO)THIOATE see MDU600

2-METHYL-4-(2-(METHYLAMINO)ETHOXY)-5-ISOPROPYL-PHENOL see DAD500

2-METHYL-4-(2-(METHYLAMINO)ETHOXY)-5-(1-METHYLETHYL)-PHENOL see DAD500

2-METHYL-6-METHYLAMINO-2-HEPTENE see ILK000

N-METHYL-4'-(p-METHYLAMINOPHENYLAZO)ACETANILIDE see MLK750

N-METHYL-N-(4-((4-(METHYLAMINO)PHENYL)AZO)PHENYLACETAMIDE) see MLK750

5-METHYL-6-METHYLAMINO-3-PHENYLQUINOXALINE see QQS300

5-METHYL-6-METHYLAMINOQUINOXALINE see MLK775

6-METHYL-8-METHYLAMINO-s-TRIAZOLO(4,3-b)PYRIDAZINE see MLK800

METHYL-6-METHYL-AMINO-8-s-TRIAZOLO(4,3b)PYRIDAZINE (FRENCH) see MLK800

METHYL-N-METHYL ANTHRANILATE see MGQ250

METHYL-4-METHYLBENZENESULFONATE see MLL250

METHYL-p-METHYLBENZENESULFONATE see MLL250

METHYL-4-METHYLBENZOATE see MPX850

METHYL-p-METHYLBENZOATE see MPX850

4-METHYL-1-(2-(2-METHYL-1,3-BENZODIOXOL-2-YL)ETHYL)PIPERAZINE HYDROCHLORIDE see MLL500

1-METHYL-5-(4-METHYLBENZOYL)-PYRROLE-2-ACETIC ACID see TGJ850

1-METHYL-5-(4-METHYLBENZOYL)-1H-PYRROLE-2-ACETIC ACID SODIUM SALT see SKJ340

METHYL 1-(α-METHYLBENZYL)IMIDAZOLE-5-CARBOXYLATE see MQQ500

METHYL 2-METHYLBUTANOATE see MLL600

METHYL-3-METHYLBUTANOATE see ITC000

METHYL 2-METHYLBUTYRATE see MLL600

METHYL-3-METHYLBUTYRATE see ITC000

((2-METHYL-5-METHYLCARBAMOYLOXY)PHENYL)TRIMETHYLAMMONIUM METHYLSULFATE see TGH660

METHYL-N-((METHYLCARBAMOYL)OXY)THIOACETIMIDATE see MDU600

S-METHYL N-[METHYLCARBAMOYLOXY]THIOACETIMIDATE see MDU600

cis-1-METHYL-2-METHYL CARBAMOYL VINYL PHOSPHATE see MRH209

METHYL (E)-2-METHYLCROTONATE see MPW700

METHYL trans-2-METHYLCROTONATE see MPW700

METHYL α-METHYLCROTONATE see MPW700

2-METHYL-1-((6-METHYL-3-CYCLOHEXEN-1-YL)CARBONYL)PIPERIDINE see MLL650

3-METHYL-1-((6-METHYL-3-CYCLOHEXEN-1-YL)CARBONYL)PIPERIDINE see MLL655

4-METHYL-1-((6-METHYL-3-CYCLOHEXEN-1-YL)CARBONYL)PIPERIDINE see MLL660

6-METHYL-2-(4-METHYL-3-CYCLOHEXEN-1-YL)-5-HEPTEN-2-OL see BGO775

2-METHYL-1-((2-METHYLCYCLOHEXYL)CARBONYL)PIPERIDINE see MLM500

3-METHYL-1-((2-METHYLCYCLOHEXYL)CARBONYL)PIPERIDINE see MLM600

4-METHYL-1-((2-METHYLCYCLOHEXYL)CARBONYL)PIPERIDINE see MLM700

10-METHYL-1',9-METHYLENE-1,2-BENZANTHRACENE see MLN500

α-METHYL-3,4-(METHYLENEDIOXY)PHENETHYLAMINE see MJQ775

α-METHYL-3,4-METHYLENEDIOXYPHENETHYLAMINE HYDROCHLORIDE see MJS750

1-METHYL-2-(3,4-METHYLENEDIOXYPHENYL)ETHYL OCTYL SULFOXIDE see ISA000

d-2-METHYL-5-(1-METHYLENENYL)-CYCLOHEXANONE see DKV175

7-METHYL-3-METHYLENE-1,6-OCTADIENE see MRZ150

2-METHYL-6-METHYLENE-2-OCTANOL ACETATE (ESTER) see DLX100

2-METHYL-6-METHYLENE-7-OCTEN-2-OL see MLO250

2-METHYL-6-METHYLENE-7-OCTEN-2-OL ACETATE see AAW500

2-METHYL-6-METHYLENE-7-OCTEN-2-YL ACETATE see AAW500

1-METHYL-4-(1-METHYLETHENYL)CYCLOHEXANOL see TBD775

1-METHYL-4-(1-METHYLETHENYL)-(S)-CYCLOHEXENE see MCC500

(R)-1-METHYL-4-(1-METHYLETHENYL)-CYCLOHEXENE see LFU000

2-METHYL-5-(1-METHYLETHENYL)-2-CYCLOHEXEN-1-OL ACETATE see CCM750

2-METHYL-5-(1-METHYLETHENYL)-2-CYCLOHEXEN-1-OL PROPIONATE see MCD000

(S)-2-METHYL-5-(1-METHYLETHENYL)-2-CYCLOHEXEN-1-ONE see CCM100

(R)-2-METHYL-5-(1-METHYLETHENYL)-2-CYCLOHEXEN-1-ONE (9CI) see CCM120

(1R-trans)-2-(3-METHYL-6-(1-METHYLETHENYL)-2-CYCLOHEXEN-1-YL)-5-PENTYL-1,3-BENZENEDIOL see CBD599

2-METHYL-5-(1-METHYLETHENYL)CYCLOHEXYL ACETATE see DKV160

5-METHYL-2-(1-METHYLETHENYL)-4-HEXEN-1-OL ACETATE see LCA100

2-METHYL-N-(3-(1-METHYLETHOXY)PHENYL)BENZAMIDE see INE050

1-METHYL-2-(1-METHYLETHYL)BENZENE see IRN300

1-METHYL-3-(1-METHYLETHYL)BENZENE see IRN400

4-METHYL-α-(1-METHYLETHYL)BENZENEACETIC ACID, (5-(2-FURANYLMETHYL)-2-THIENYL)METHYL ESTER see MLO301

4-METHYL-α-(1-METHYLETHYL)BENZENEACETIC ACID, (4,5,6,7-TETRAHYDROBENZO(B)THIEN-2-YL) METHYL ESTER see MLO320

(1S-1-α,4-α,5-α)-4-METHYL-1-(1-METHYLETHYL)-BICYCLO(3.1.0)HEXAN-3-ONE see TFW000

1-METHYL-2-(1-METHYLETHYL)-1,4-CYCLOHEXADIENE see MCB700

5-METHYL-2-(1-METHYLETHYL)CYCLOHEXANOL see MCF750

(1R-(1-α,2-β,5-α))-5-METHYL-2-(1-METHYLETHYL)CYCLOHEXANOL see MCG250

5-METHYL-2-(1-METHYLETHYL)-CYCLOHEXANOL (1-α,2-β,5-α) see MCG000

(R-(1α,2β,5α))-5-METHYL-2-(1-METHYLETHYL)-CYCLOHEXANOL ACETATE (9CI) see MCG750

5-METHYL-2-(1-METHYLETHYL)CYCLOHEXANONE see IKY000

(Z)-5-METHYL-2-(1-METHYLETHYL)CYCLOHEXANONE see IKY000

trans-5-METHYL-2-(1-METHYLETHYL)-CYCLOHEXANONE see MCG275

1-METHYL-2-(1-METHYLETHYL)-1-CYCLOHEXENE, DIDEHYDRO deriv. see MCE275

4-METHYL-1-(1-METHYLETHYL)-3-CYCLOHEXEN-1-OL (9CI) see TBD825

3-METHYL-6-(1-METHYLETHYL)-2-CYCLOHEXEN-1-ONE see MCF250

1-METHYL-4-(1-METHYLETHYL)-2,3-DIOXABICYCLO(2.2.2)OCT-5-ENE see ARM500

6-METHYL-N-(1-METHYLETHYL)-2-HEPTANAMINE HYDROCHLORIDE see IGG300

5-METHYL-2-(1-METHYLETHYL)-2-HEXEN-1-OL see IQH000

5-METHYL-2-(1-METHYLETHYL)-2-HEXEN-1-YL ACETATE see AAV250

1-METHYL-4-(1-METHYLETHYLIDENE)CYCLOHEXENE see TBE000

1-METHYL-2-(1-METHYLETHYL)-5-NITRO-1H-IMIDAZOLE see IGH000

1-METHYL-4-(1-METHYLETHYL)-7-OXABICYCLO(2.2.1)HEPTANE see IKC100

5-METHYL-2-(1-METHYLETHYL)PHENOL see TFX810

3-METHYL-5-(1-METHYLETHYL)PHENOLMETHYLCARBAMATE see CQI500

METHYL (4-(1-METHYLETHYL)PHENYL)METHYL 3-PYRIDINYLCARBONIMIDODITHIOATE see MLO350

1-METHYL-4-(1-METHYLETHYL)-2-(1-PROPENYL)BENZENE see VIP100

2-METHYL-2-(α-METHYLHEXYLAMINO)PROPYL-p-AMINOBENZOATE HYDROCHLORIDE see MLO750

METHYL 3-METHYLHEXYL KETONE see MLO800

METHYL 3-METHYLHEXYL KETONE see MND075

METHYL METHYLMERCURIC SULFIDE see MLX300

METHYL((METHYL(((5-METHYL-1,3-OXATHIOLAN-4-YLIDENE)AMINO)OXY)CARBONYL)AMINO)THIO) CARBAMIC ACID, ETHYL ESTER see MLO900

METHYL(4-METHYL-N-((4-METHYLPHENYL)SULFONYL)BENZENESULFONAMIDATO-N)-MERCURY see BLK250

d-3-METHYL-N-METHYLMORPHINAN PHOSPHATE see MLP250

METHYL (((METHYL-(4-MORPHOLINOTHIO)AMINO)CARBONYL)OXY)ETHANIMIDOTHIOATE see MLL150

METHYL 5-METHYL-3-(5-NITRO-2-FURYL)-4-ISOXAZOLYL KETONE see MLP300

1-METHYL-3-((1-METHYL-4-NITRO-1H-IMIDAZOL-2-YL)METHYLENE)-2-PYRROLIDINONE see MLP400

2-METHYL-3-(1-METHYL-5-NITRO-1H-IMIDAZOL-2-YL)-2-PROPEN-1-OL see MLP450

O-METHYL O-(4-METHYL-2-NITROPHENYL) (1-METHYLETHYL)PHOSPHORAMIDOTHIOATE see AMX300

5-METHYL-5-(1-METHYL-1-PENTENYL)BARBITURIC ACID see MLP500

1-METHYL-4-(4-METHYLPENTYL)-3-CYCLOHEXENE-1-CARBOXALDEHYDE see VIZ150

2-METHYL-2-(2-(N-METHYL-N-PHENETHYLAMINO)ETHYL)-1,3-BENZODIOXOLE HYDROCHLORIDE see MLP750

2-METHYL-4-((2-METHYLPHENYL)AZO)BENZENAMINE see AIC250

2-METHYL-4-((2-METHYLPHENYL)AZO)-1,3-BENZENEDIAMINE see MIX100

4-METHYL-6-((2-METHYLPHENYL)AZO)-1,3-BENZENEDIAMINE see MLP770

1-((2-METHYL-4-((2-METHYLPHENYL)AZO)PHENYL)AZO)-2-NAPHTHALENOL see SBC500

N-METHYL-2-((o-METHYL-α-PHENYLBENZYL)OXY)ETHYLAMINE HYDROCHLORIDE see TGJ250

1-METHYL-4-(o-METHYL-α-PHENYLBENZYL)PIPERAZINE DIHYDROCHLORIDE see MIQ725

METHYL (3-METHYLPHENYL)CARBAMOTHIOIC ACID, o-2-NAPHTHALENYL ESTER see TGB475

O-METHYL-S-(4-METHYLPHENYL) ETHYLPHOSPHONODITHIOATE see EOO000

3-METHYL-2-((1-METHYL-2-PHENYL-1H-INDOL-3-YL)AZO)THIAZOLIUM CHLORIDE see CMM870

N-METHYL-N'-(p-METHYLPHENYL)-N-NITROSOUREA see MNA650

N-METHYL-2-((2-METHYLPHENYL)PHENYLMETHOXY)ETHANAMINE HYDROCHLORIDE see TGJ250

1-METHYL-4-((2-METHYLPHENYL)PHENYLMETHYL)PIPERAZINE DIHYDROCHLORIDE (9CI) see MIQ725

2-METHYL-3-(2-METHYLPHENYL)-4-QUINAZOLINONE see QAK000

2-METHYL-3-(2-METHYLPHENYL)-4(3H)-QUINAZOLINONE see QAK000

N-METHYL-α-METHYL-α-PHENYLSUCCINIMIDE see MLP800

N-METHYL-α-α-METHYLPHENYLSUCCINIMIDE see MLP800

3-METHYL-1-(4-METHYLPHENYL)TRIAZENE see MQB250

6-METHYL-α-(4-METHYL-1-PIPERAZINYLCARBONYL)ERGOLINE-8-β-PROPIONITRILE see MLR400

2-METHYL-11-(4-METHYL-1-PIPERAZINYL)-DIBENZO(b,f)(1,4)THIAZEPINE see MQQ000

2-METHYL-2-(2-(4-METHYL-1-PIPERAZINYL)ETHYL)-1,3-BENZODIOXOLE HYDROCHLORIDE see MLL500

METHYL 10-(3-(4-METHYL-1-PIPERAZINYL)PROPYL)PHENOTHIAZIN-2-YL KETONE see ACR500

5-METHYL-3-(4-METHYL-1-PIPERAZINYL)-5H-PYRIDAZINO(3,4-b)(1,4)BENZOXAZINE HYDROCHLORIDE see ASC250

2-METHYL-4-(4-METHYL-1-PIPERAZINYL)-10H-THIENO(2,3-B)(1,5)BENZODIAZEPINE see ZVJ500

5-METHYL-4-(4-METHYL-1-PIPERAZINYL)THIENO(2,3-d)PYRIMIDINE HYDROCHLORIDE see MLR500

N-METHYL-2-(2-METHYLPIPERIDINO)-N-(2-PHENOXYETHYL)ACETAMIDE HYDROCHLORIDE see MLS250

N-METHYL-2-(2-METHYLPIPERIDINO)-N-(2-(o-TOLYLOXY)ETHYL)-ACETAMIDE HYDROCHLORIDE see MLS750

METHYL-2-METHYL-2-PROPENOATE see MLH750

α-METHYL-4-((2-METHYL-2-PROPENYL)AMINO)BENZENEACETIC ACID see MDP850

α-METHYL-4-(2-METHYLPROPYL)BENZENEACETIC ACID see IIU000

α-METHYL-4-(2-METHYLPROPYL)BENZENEACETIC ACID 2-PYRIDINYLMETHYL ESTER see IAB000

α-(3-METHYL-1-(2-METHYLPROPYL)BUTYL)-ω-HYDROXYPOLY(OXY-1,2-ETHANEDIYL) see MLS800

2-METHYL-N-(2-METHYLPROPYL)-1-PROPANAMINE see DNH400

1-METHYL-4-((p-(p-((1-METHYLPYRIDINIUM-4-YL)AMINO)BENZAMIDO)ANILINO)QUIN OLINIUM), DI-p-TOLUENESULFONATE see MLT250

1-METHYL-4-(((4-(3-((4-((1-METHYLPYRIDINIUM-4-YL)AMINO)PHENYL)AMINO)3-OXO-1-PROPENYL)PHENYL)AMINO)QUINOLINI UM, DIBROMIDE see MLT500

1-METHYL-4-((p-(p-((1-METHYLPYRIDINIUM-4-YL)AMINO)PHENYL)CARBAMOYL)ANILIN O)QUINOLINIUM), DIBROMIDE see MLU000

1-METHYL-4-((p-(p-((1-METHYLPYRIDINIUM-4-YL)AMINO)PHENYL)CARBAMOYL)ANILIN O)-7-NITROQUINOLINIUM), DI-p-TOLUENE SULFONATE see MLT750

1-METHYL-4-(p-(p-((1-METHYLPYRIDINIUM-4-YL)AMINO)STYRYL)ANILINO)QUINOLIU M DIBROMIDE see MLU250

1-METHYL-4-((p-(p-((p-(1-METHYLPYRIDINIUM-4-YL)PHENYL)CARBAMOYL)ANILINO)QUIN OLINIUM), DI-p-TOLUENE SULFONATE see MLU750

1-METHYL-6-((p-(p-((1-METHYLQUINOLINIUM-6-YL)CARBAMOYL)BENZAMIDO)BENZAMI DO)QUINOLINIUM, DI-p-TOLUENE SULFONATE see MLW250

1-METHYL-6-(((p-(p-((1-METHYLQUINOLINIUM-6-YL)CARBAMOYL)BENZAMIDO)PHENYL)C ARBAMOYL)QUINOLINIUM), DI-p-TOLUENE SULFONATE see MLW500

METHYL 5-METHYL-1-(2-QUINOLYL)-4-PYRAZOLYL KETONE see MLW600

METHYL 5-METHYL-1-(2-QUINOXALINYL)-4-PYRAZOLYL KETONE see MLW630

METHYL α-METHYLSTYRYL KETONE see MNS600

2-METHYL-2-(METHYLSULFINYL)PROPANAL O-((METHYLAMINO)CARBONYL)OXIME see MLW750

2-METHYL-2-(METHYLSULFINYL)PROPIONALDEHYDE O-(METHYLCARBAMOYL)OXIME see MLW750

(((2-METHYL(METHYLSULFONYL)AMINO)-2-

OXOETHYL)AMINO)METHYLPHOSPHONI C ACID see MID850

2-METHYL-2-(METHYLSULFONYL)PROPANAL-o-((METHYLAMINO)CARBONYL)OXIME see AFK000

2-METHYL-2-(METHYLSULFONYL)PROPIONALDEHYD E-o-(METHYLCARBAMOYL)OXIME see AFK000

METHYL((METHYLTHIO)ACETYL)CARBA MIC ACID-o-ISOPROPOXYPHENYL ESTER see MLX000

3-METHYL-2-(METHYLTHIO)-BENZOTHIAZOLIUM-p-TOLUENESULFONATE see MLX250

1-METHYL-2-(METHYLTHIO)-1H-IMIDAZOLE-4,5-DIMETHANOL BIS(METHYLCARBAMATE)HYDROCHLOR IDE see MLX275

METHYL(METHYLTHIO)MERCURY see MLX300

3-METHYL-4-METHYLTHIOPHENOL see MLX750

S-5-METHYL-2-METHYLTHIO-5-PHENYL-3-PHENYLAMINO-3,5-DIHYDRO-4H-IMIDAZOL-4-ONE see FAJ300

2-METHYL-2-(METHYLTHIO)PROPANAL-O-((METHYLAMINO)CARBONYL)OXIME see CBM500

2-METHYL-2-(METHYLTHIO)PROPANAL-o-((METHYLNITROSOAMINO)CARBONYL)O XIME see NJJ500

2-METHYL-2-(METHYLTHIO)PROPANAL OXIME see MLX800

2-METHYL-2-(METHYLTHIO)PROPANOL-o-((N-METHYL-N-MORPHOLINOSULFENYL)CARBAMOYL)O XIME see MLX820

2-METHYL-2-(METHYLTHIO)PROPANOL-o-((METHYL(4-MORPHOLINYLTHIO)AMINO)CARBONYL) OXIME see MLX820

2-METHYL-2-(METHYLTHIO)PROPIONALDEHYDE-O-(METHYLCARBAMOYL)OXIME see CBM500

2-METHYL-2-(METHYLTHIO)PROPIONALDEHYDE-o-((METHYL)(DECOXYSULFINYL)CARBAMO YL)OXIME see MLX830

2-METHYL-2-(METHYLTHIO)PROPIONALDEHYDE-o-(N-METHYL-N-(4-MORPHOLINOSULFENYL)CARBAMOYL)O XIME see MLX820

2-METHYL-2-(METHYLTHIO)PROPIONALDEHYDE-o-((METHYLNITROSO)CARBAMOYL) OXIME see NJJ500

2-METHYL-2-(METHYLTHIO)PROPIONALDEHYDE OXIME see CBM500

2-METHYL-2-METHYLTHIO-PROPIONALDEHYD-O-(N-METHYL-CARBAMOYL)-OXIM (GERMAN) see CBM500

METHYL METIRAM see MLX850

METHYLMITOMYCIN see MLY000

N-METHYLMITOMYCIN C see MLY000

METHYL MONOBROMOACETATE see MHR250

METHYL MONOCHLORACETATE see MIF775

METHYL MONOCHLOROACETATE see MIF775

3-METHYL-4-MONOMETHYLAMINOAZOBENZENE see MLY250

3'-METHYL-4-MONOMETHYLAMINOAZOBENZENE see MPY000

N-METHYLMONOTHIOSUCCINIMIDE see MLY500

(+−)-17-METHYLMORPHINAN-3-OL HYDROBROMIDE see HMH500

(−)-17-METHYLMORPHINAN-3-OL TARTRATE DIHYDRATE see MLZ000

(+)-17-METHYLMORPHINAN-3-OL TARTRATE HYDRATE see MMA000

METHYLMORPHINE see CNF500

N-METHYL MORPHINE CHLORIDE see MRP000

N-METHYLMORPHINIUM CHLORIDE see MRP000

4-METHYLMORPHOLINE see MMA250

N-METHYL MORPHOLINE see MMA250

METHYLMORPHOLINE (DOT) see MMA250

α-METHYL-4-MORPHOLINEACETIC ACID-2,6-XYLYL ESTER HYDROCHLORIDE see MMA525

1-METHYL-MORPHOLINO-3-PHTHALIMIDO-GLUTARIMIDE see MRU080

METHYL 10-(3-MORPHOLINOPROPYL)PHENOTHIAZIN-2-YL KETONE see MMA600

N-METHYL-N-(MORPHOLINOTHIO)CARBAMIC ACID 2,3-DIHYDRO-2,2-DIMETHYL-7-BENZOFURANYL ESTER see MRU050

6-METHYL-MP-RIBOSIDE see MPU000

METHYL MUSTARD OIL see ISE000

METHYLNAFTALEN see MMB500

5-METHYLNAFTOYLBENZIMIDAZOL (CZECH) see MHC000

N-METHYL-1-NAFTYL-CARBAMAAT (DUTCH) see CBM750

METHYL NAMATE see SGM500

2-METHYL-1,4-NAPHTHALENDIONE see MMD500

METHYLNAPHTHALENE see MMB500

1-METHYLNAPHTHALENE see MMB750

2-METHYLNAPHTHALENE see MMC000

α-METHYLNAPHTHALENE see MMB750

β-METHYLNAPHTHALENE see MMC000

METHYL 1-NAPHTHALENEACETATE see NAK525

METHYL 1-NAPHTHALENEACETIC ACID see NAK525

2-METHYL-1,4-NAPHTHALENEDIOL see MMC250

2-METHYL-1,4-NAPHTHALENEDIOL BIS(DIHYDROGEN PHOSPHATE) TETRASODIUM SALT see NAQ600

2-METHYL-1,4-NAPHTHALENEDIONE see MMD500

2-METHYL-1,4-NAPHTHAQUINOL BIS(DISODIUM PHOSPHATE) see NAQ600

2-METHYL-1,4-NAPHTHOCHINON (GERMAN) see MMD500

2-METHYL-1,4-NAPHTHOCHINON-NATRIUM-BISULFIT TRIHYDRAT (GERMAN) see MMD750

5-METHYLNAPHTHO(1,2,3,4-def)CHRYSENE see MMD000

6-METHYLNAPHTHO(1,2,3,4-def)CHRYSENE see MMD250

METHYLNAPHTHOHYDROQUINONE see MMC250

2-METHYL-1,4-NAPHTHOHYDROQUINONE see MMC250

2-METHYL-1,4-NAPHTHOHYDROQUINONE DIACETATE see VTA100

2-METHYL-1,4-NAPHTHOHYDROQUINONE DIPHOSPHORIC ACID ESTER TETRASODIUM SALT see NAQ600

2-METHYL-1,4-NAPHTHOQUINOL see MMC250

2-METHYL-1,4-NAPHTHOQUINOL BIS(DISODIUM PHOSPHATE) see NAQ600

2-METHYL-1,4-NAPHTHOQUINONE see MMD500

3-METHYL-1,4-NAPHTHOQUINONE see MMD500

2-METHYL-1,4-NAPHTHOQUINONE, SODIUM BISULFITE, TRIHYDRATE see MMD750

METHYL 1-NAPHTHYLACETATE see NAK525

METHYL α-NAPHTHYLACETATE see NAK525

3-METHYL-2-NAPHTHYLAMINE see MME500

N-METHYL-1-NAPHTHYLAMINE see MME250

3-METHYL-2-NAPHTHYLAMINE HYDROCHLORIDE see MME750

N-METHYL-1-NAPHTHYL-CARBAMAT (GERMAN) see CBM750

N-METHYL NAPHTHYLCARBAMATE see MME800

N-METHYL-1-NAPHTHYL CARBAMATE see CBM750

N-METHYL-α-NAPHTHYLCARBAMATE see CBM750

N-METHYL-N-(1-NAPHTHYL)FLUOROACETAMIDE see MME809

METHYL 1-NAPHTHYL KETONE see ABC475

METHYL-2-NAPHTHYL KETONE see ABC500

METHYL α-NAPHTHYL KETONE see ABC475

α-METHYL NAPHTHYL KETONE see ABC475

β-METHYL NAPHTHYL KETONE see ABC500

METHYL-β-NAPHTHYL KETONE (FCC) see ABC500

N-METHYL-N-(1-NAPHTHYL)MONOFLUOROACETAMIDE see MME809

N-METHYL-α-NAPHTHYLURETHAN see CBM750

METHYL NICOTINATE see NDV000

N-METHYLNICOTINATE see TKL890

N-METHYLNICOTINIC ACID see TKL890

N'-METHYLNICOTINIC ACID see TKL890

METHYL NIRAN see MNH000

METHYLNITRAMINE see NHN500

METHYL NITRAMINE (dry) (DOT) see NHN500

METHYL NITRATE see MMF500

2-METHYL-1-NITRATODIMERCURIO-2-NITRATOMERCURIO PROPANE see MMF600

1-METHYLNITRAZEPAM see DLV000

METHYL NITRITE see MMF750

1-METHYL-4-NITRO-5-(2'-AMINO-6'-PURINYL)MERCAPTOIMIDAZIDE see AKY250

2-METHYL-4-NITROANILINE see MMF780

2-METHYL-5-NITROANILINE see NMP500

4-METHYL-3-NITROANILINE see NMP000

6-METHYL-3-NITROANILINE see NMP500

N-METHYL-4-NITROANILINE see MMF800

2-METHYL-1-NITRO-9,10-ANTHRACENEDIONE see MMG000

2-METHYL-1-NITROANTHRAQUINONE see MMG000

2-METHYLNITROBENZENE see NMO525

3-METHYLNITROBENZENE see NMO500

4-METHYLNITROBENZENE see NMO550

m-METHYLNITROBENZENE see NMO500

o-METHYLNITROBENZENE see NMO525

p-METHYL NITROBENZENE see NMO550

2-METHYL-5-NITRO-BENZENEAMINE see NMP500

9-METHYL-ω-(p-NITROBENZENEAZO)-3,4-BENZACRIDINE see NIK500

2-METHYL-5-NITRO-1,3-BENZENEDIAMINE see MMG100

2-METHYL-5-NITRO-1,4-BENZENEDIAMINE see MMG200

2-METHYL-6-NITRO-1,4-BENZENEDIAMINE see MMG210

4-METHYL-5-NITRO-1,3-BENZENEDIAMINE see NFV540

METHYL-2-NITROBENZENE DIAZOATE see MMH000

2-METHYL-5-NITROBENZENESULFONIC ACID see MMH250

4-METHYL-3-NITROBENZENE SULFONIC ACID see MMH400

2-METHYL-5-NITROBENZENESULFONYL CHLORIDE see MMH500

1-METHYL-2-NITROBENZIMIDAZOLE see MMH740

METHYL-p-NITROBENZOATE see MMI000

3-METHYL-2-NITROBENZOYL CHLORIDE see MMI250

3-METHYL-4-NITRO-1-BUTEN-3-YL ACETATE see MMI640

3-METHYL-4-NITRO-2-BUTEN-1-YL ACETATE see MMI650

9-METHYL-3-NITROCARBAZOLE see MMI700

9-METHYL-3-NITROCARBAZOLE see MMI710

9-METHYL-3-NITRO-9H-CARBAZOLE see MMI710

1-METHYL-7-NITRO-5-(2-FLUOROPHENYL)-3H-1,4-BENZODIAZEPIN-2(1H)-ONE see FDD100

4-METHYL-1-((5-NITROFURFURYLIDENE)AMINO)-2-IMIDAZOLIDINONE see MMJ000

3-METHYL-4-(5'-NITROFURYLIDENE-AMINO)-TETRAHYDRO-4H-1,4-THIAZINE-1,1-DIOXIDE see NGG000

5-METHYL-3-(5-NITRO-2-FURYL)ISOXAZOLE see MMJ950

N-METHYL-3-(5-NITRO-2-FURYL)-N-NITROSO-1H-1,2,4-TRIAZOL-5-AMINE see MMU000

METHYL 3-(5-NITRO-2-FURYL)-5-PHENYL-4-ISOXAZOLYL KETONE see MMJ955

5-METHYL-3-(5-NITRO-2-FURYL)PYRAZOLE see MMJ960

2-METHYL-4-(5-NITRO-2-FURYL)THIAZOLE see MMJ975

N-(1-METHYL-3-(5-NITRO-2-FURYL)-s-TRIAZOL)-5-YL-ACETAMIDE see MMK000

N-(1-METHYL-3-(5-NITRO-2-FURYL)-1H-1,2,4-TRIAZOL-5-YL)ACETAMIDE see MMK000

N-METHYL-N-NITROGLYCINE see NJH500

1-METHYL-3-NITROGUANIDINE see MML250

N-METHYL-N'-NITROGUANIDINE see MML250

1-METHYL-3-NITROGUANIDINE mixed with SODIUM NITRITE (1:1) see MML500

1-METHYL-3-NITROGUANIDINIUM NITRATE see MML550

1-METHYL-3-NITROGUANIDINIUM PERCHLORATE see MML575

1-METHYL-2-NITROIMIDAZOLE see MML750

1-METHYL-4-NITROIMIDAZOLE see MMM000

2-METHYL-5-NITROIMIDAZOLE see MMM500

4-METHYL-5-NITROIMIDAZOLE see MMM750

1-METHYL-4-NITRO-1H-IMIDAZOLE see MMM000

2-METHYL-5-NITRO-1H-IMIDAZOLE see MMM500

4-METHYL-5-NITRO-1H-IMIDAZOLE see MMM750

1-METHYL-5-NITRO-IMIDAZOLE-2-CARBOXALDEHYDE see MKG800

2-METHYL-5-NITROIMIDAZOLE-1-ETHANOL see MMN250

1-METHYL-5-NITROIMIDAZOLE-2-METHANOL see MMN500

1-METHYL-5-NITROIMIDAZOLE-2-METHANOL CARBAMATE (ESTER) see MMN750

1-METHYL-5-NITRO-1H-IMIDAZOLE-2-METHANOL CARBAMATE ESTER see MMN750

4-((E)-2-(1-METHYL-5-NITRO-1H-IMIDAZOL-2-YL)-AETHENYL)-2-PYRIMIDINAMIN (GERMAN) see TJF000

4-((E)-2-(1-METHYL-5-NITRO-1H-IMIDAZOL-2-YL)-ETHENYL)-2-PYRIMIDINAMINE see TJF000

METHYLNITROIMIDAZOLYLMERCAPTOPURINE see ASB250

6-(1'-METHYL-4'-NITRO-5'-IMIDAZOLYL)-MERCAPTOPURINE see ASB250

5-((1-METHYL-5-NITRO-1H-IMIDAZOL-2-YL)METHYLENE)-2(5H)-FURANONE see FPM200

1-(1-METHYL-5-NITRO-1H-IMIDAZOL-2-YL)-3-(METHYLSULFONYL)-2-IMIDAZOLIDINO NE see SAY950

6-(METHYL-p-NITRO-5-IMIDAZOLYL)-THIOPURINE see ASB250

6-((1-METHYL-4-NITROIMIDAZOL-5-YL)THIO)PURINE see ASB250

6-(1-METHYL-4-NITROIMIDAZOL-5-YLTHIO)PURINE see ASB250

6-(1-METHYL-p-NITRO-5-IMIDAZOLYL)-THIOPURINE see ASB250

6-((1-METHYL-4-NITRO-1H-IMIDAZOL-5-YL)THIO)-1H-PURINE see ASB250

2-METHYL-5-NITRO-1H-ISOINDOLE-1,3(2H)-DIONE see NHN700

METHYLNITROLIC ACID see NHY250

2-METHYL-1-NITRO-NAPHTHALENE see MMN800

3-METHYL-4-NITRO-1-(p-NITROPHENYL)-2-PYRAZOLIN-5-ONE see PIE000

METHYLNITRONITROSOGUANIDINE see MMP000

1-METHYL-3-NITRO-1-NITROSOGUANIDINE see MMP000

N-METHYL-N'-NITRO-N-NITROSOGUANIDINE see MMP000

N-METHYL-N-NITRO-N-NITROSOGUANIDINE see MMP000

5-METHYL-2-NITRO-7-OXA-8-MERCURABICYCLO(4.2.0)OCTA-1,3,5-TRIENE see NHK900

2-METHYL-4-NITROPHENOL see NFV010

4-METHYL-2-NITROPHENOL see NFU500

1-((4-METHYL-2-NITROPHENYL)AZO)-2-NAPHTHALENOL see MMP100

1-METHYL-7-NITRO-5-PHENYL-1,3-DIHYDRO-2H-1,4-BENZODIAZEPIN-2-ONE see DLV000

3-METHYL-4-NITROPHENYL DIMETHYL PHOSPHATE see PHD750

(E)-1-METHYL-4-(2-(4-NITROPHENYL)ETHENYL)BENZENE see MNB100

METHYL 3-NITROPHENYL KETONE see NEL500

METHYL-p-NITROPHENYL KETONE see NEL600

4-METHYL-5-NITRO-1-(PHENYLMETHYL)-1H-IMIDAZOLE see MMP110

5-METHYL-4-NITRO-1-(PHENYLMETHYL)-1H-IMIDAZOLE see MMP113

1-METHYL-5-NITRO-2-(2-PHENYL-1-PROPENYL)-1H-IMIDAZOLE see MMP120

1-METHYL-5-NITRO-2-((PHENYLSULFONYL)METHYL)-1H-IMIDAZOLE see MMP125

METHYLNITROPHOS see DSQ000

N-METHYL-4-NITROPHTHALIMIDE see NHN700

2-METHYL-2-NITROPROPANE see NFQ500

2 METHYL-2-NITROPROPANE-1,3-DIOL see NHO500

2-METHYL-2-NITRO-PROPANOL NITRATE see MMP200

N-(2-METHYL-2-NITROPROPYL)-p-NITROSOANILINE see NHK800

N-(2-METHYL-2-NITROPROPYL)-4-NITROSOBENZAMINE see NHK800

2-METHYL-4-NITROPYRIDINE-1-OXIDE see MMP500

3-METHYL-4-NITROPYRIDINE-1-OXIDE see MMP750

1-METHYL-3-NITRO-5H-PYRIDO(4,3-B)INDOLE see NHP050

2-METHYL-4-NITROQUINOLINE-1-OXIDE see MMQ250

2-METHYL-4-NITROQUINOLINE N-OXIDE see MMQ250

3-METHYL-4-NITROQUINOLINE-1-OXIDE see MMQ500

5-METHYL-4-NITROQUINOLINE-1-OXIDE see MMQ750

6-METHYL-4-NITROQUINOLINE-1-OXIDE see MMR000

7-METHYL-4-NITROQUINOLINE-1-OXIDE see MMR250

8-METHYL-4-NITROQUINOLINE-1-OXIDE see MMR500

1-(4-N-METHYL-N-NITROSAMINOBENZYLIDENE)INDENE see MMR750

1-(METHYLNITROSAMINO)-2-BUTANONE see MMR800

4-(METHYLNITROSAMINO)-2-BUTANONE see MMR810

1-N-METHYL-N-NITROSAMINO-1-DEOXY-d-GLUCITOLE see DAS400

1-(METHYLNITROSAMINO)ETHYL ACETATE see ABR500

METHYLNITROSAMINOMETHYL-d3 ESTER ACETIC ACID see MMS000

2-METHYLNITROSAMINO-2-METHYLPENTANON(4) (GERMAN) see MMX750

3-(METHYLNITROSAMINO)-1,2-PROPANEDIOL see NMV450

3-(METHYLNITROSAMINO)PROPIONALDEHYDE see NKV050

3-METHYLNITROSAMINOPROPIONITRILE see MMS200

6-(METHYLNITROSAMINO)PURINE see MMT250

2-(METHYLNITROSAMINO)PYRIDINE see NKQ000

γ-(METHYLNITROSAMINO)-3-PYRIDINEBUTYRALDEHYDE see MMS250

4-(N-METHYL-N-NITROSAMINO)-4-(3-PYRIDYL)BUTANAL see MMS250

4-(METHYLNITROSAMINO)-1-(3-PYRIDYL)-1-BUTANOL see MMS300

4-(N-METHYL-N-NITROSAMINO)-1-(3-PYRIDYL)-1-BUTANONE see MMS500

4-(4-N-METHYL-N-NITROSAMINOSTYRYL)QUINOLINE see MMS750

METHYLNITROSOACETAMID (GERMAN) see MMT000

METHYLNITROSOACETAMIDE see MMT000

N-METHYL-N-NITROSOACETAMIDE see MMT000

1-METHYL-1-NITROSOACETYLUREA see ACR400

N-METHYL-N-NITROSO-N'-ACETYLUREA see ACR400

N-METHYL-N-NITROSOADENINE see MMT250

N⁶-(METHYLNITROSO)ADENOSINE see MMT300

N-METHYL-N-NITROSO-β-ALANINE see MMT300

N-METHYL-N-NITROSOALLYLAMINE see MMT500

2-(N-METHYL-N-NITROSO)AMINOACETONITRILE see MMT750

α-METHYLNITROSOAMINOBENZYL ALCOHOL ACETATE (ESTER) see NKP000

4-(METHYLNITROSOAMINO)BUTYRIC ACID see MIF250

1-(N-METHYL-N-NITROSOAMINO)-1-DEOXY-d-GLUCITOL see DAS400

1-N-METHYL-N-NITROSOAMINO-1-DESOXY-d-GLUCIT (GERMAN) see DAS400

2-(METHYLNITROSOAMINO)ETHANOL see NKU350

2-(METHYLNITROSOAMINO)ETHANOL 4-METHYLBENZENESULFONATE (ESTER) see NKU420

α-(1-(N-METHYL-N-NITROSOAMINO)ETHYL)BENZYL ALCOHOL see NKC000

5-(N-METHYL-N-NITROSO)AMINO-3-(5-NITRO-2-FURYL)-s-TRIAZOLE see MMU000

2-(N-METHYL-N-NITROSOAMINO)-1-PHENYL-1-PROPANOL see NKC000

1-(METHYLNITROSOAMINO)-2-PROPANOL see NKU500

3-(METHYLNITROSOAMINO)-1-PROPANOL see NKU520

1-(METHYLNITROSOAMINO)2-PROPANONE see NKV000

3-(METHYLNITROSOAMINO)PROPIONIC ACID see MMT300

α-(3-(METHYLNITROSOAMINO)PROPYL)-3-PYRIDINEMETHANOL see MMS300

N-METHYL-N-NITROSO-2-AMINOPYRIDINE see NKQ000

N-METHYL-N-NITROSO-4-AMINOPYRIDINE see NKQ100

γ-(METHYLNITROSOAMINO)-3-PYRIDINEBUTANAL see MMS250

4-(N-METHYL-N-NITROSOAMINO)-4-(3-PYRIDYL)-1-BUTANONE see MMS500

N-METHYL-N-NITROSOANILINE see MMU250

N-METHYL-N-NITROSOBENZAMIDE see MMU500

N-METHYL-N-NITROSOBENZENAMINE see MMU250

o-METHYLNITROSOBENZENE see NLW500

1-METHYL-2-NITROSOBENZENE see NLW500

2-METHYL-N-NITROSO-BENZIMIDAZOLE CARBAMATE and SODIUM NITRITE (1:1) see SIQ700

N-METHYL-N-NITROSO-N'-(2-BENZOTHIAZOLYL)-HARNSTOFF (GERMAN) see NKR000

N-METHYL-N-NITROSO-N'-(2-BENZOTHIAZOLYL)-UREA see NKR000

N-METHYL-N-NITROSOBENZYLAMINE see MHP250

1-METHYL-1-NITROSOBIURET see MMV000

N-METHYL-N-NITROSOBIURET see MMV000

1-METHYL-1-NITROSO-3-(p-BROMOPHENYL)UREA see BNX125

N-METHYL-N-NITROSOBUTYLAMINE see MHW500

METHYLNITROSOCARBAMIC ACID-o-CHLOROPHENYL ESTER see MMV250

METHYLNITROSOCARBAMIC ACID-2,3-DIHYDRO-2,3-DIMETHYL-7-BENZOFURANYL ESTER see NJQ500

METHYLNITROSO-CARBAMIC ACID-2,3-DIHYDRO-2,2-DIMETHYL-3-HYDROXY-7-BENZOFURANYL ESTER see HMZ100

METHYLNITROSOCARBAMIC ACID-3,4-DIMETHYLPHENYL ESTER see DTN400

METHYLNITROSOCARBAMIC ACID o-(1,3-DIOXOLAN-2-YL)PHENYL ESTER see MMV500

N-METHYL-N-NITROSOCARBAMIC ACID, ETHYL ESTER see MMX250

METHYLNITROSOCARBAMIC ACID-α-(ETHYLTHIO)-o-TOLYL ESTER see MMW750

METHYLNITROSOCARBAMIC ACID o-ISOPROPOXYPHENYL ESTER see PMY310

METHYLNITROSOCARBAMIC ACID-o-ISOPROPYLPHENYL ESTER see MMW250

METHYL-NITROSOCARBAMIC ACID-1-NAPHTHYL ESTER see NBJ500

N-METHYL-N-NITROSOCARBAMIC ACID-m-3-PENTYLPHENYL ESTER see PBX750

N-METHYL-N-NITROSOCARBAMIC ACID, PHENYL ESTER see NKV500

N-METHYL-N-NITROSOCARBAMIC ACID, TRIMETHYLPHENYL ESTER see NLY500

METHYLNITROSOCARBAMIC ACID 3,4-XYLYL ESTER see XTS000

METHYLNITROSOCARBAMIC ACID-3,5-XYLYL ESTER see MMW750

N⁵-(METHYLNITROSOCARBAMOYL)-l-ORNITHINE see MQY325

N⁵-(N-METHYL-N-NITROSOCARBAMOYL)-l-ORNITHINE see MQY325

N△-(N-METHYL-N-NITROSOCARBAMOYL)-l-ORNITHINE see MQY325

N-((METHYLNITROSOCARBAMOYL)OXY)-2-METHYLTHIOACETIMIDIC ACID see NKX000

N-METHYL-N-NITROSO-N'-CARBAMOYLUREA see MMV000

1-METHYL-1-NITROSO-3-(p-CHLOROPHENYL)UREA see MMW775

METHYLNITROSOCYANAMIDE see MMX000

N-METHYL-N-NITROSOCYCLOHEXYLAMINE see NKT500

N-METHYL-N-NITROSODECYLAMINE see MMX200

1-METHYL-N-NITROSODIETHYLAMINE see ELX500

N-METHYL-N-NITROSO-ETHAMINE see MKB000

N-METHYL-N-NITROSO-ETHENYLAMINE see NKY000

N-METHYL-N-NITROSOETHYLAMINE see MKB000

N-METHYL-N-NITROSOETHYLCARBAMATE see MMX250

N-METHYL-N-NITROSO-β-d-GLUCOSAMINE see MMX500

N-METHYL-N-NITROSO-β-d-GLUCOSYLAMIN (GERMAN) see MMX500

N-METHYL-N-NITROSO-β-d-GLUCOSYLAMINE see MMX500

N-METHYL-N-NITROSOGLYCINE see NLR500

N-METHYL-N'-NITROSOGUANIDINE see MMX600

METHYLNITROSO-HARNSTOFF (GERMAN) see MNA750

N-METHYL-N-NITROSO-HARNSTOFF (GERMAN) see MNA750

N-METHYL-N-NITROSOHEPTYLAMINE see HBP000

N-METHYL-N-NITROSO-1-HEXANAMINE see NKU400

N-METHYL-N-NITROSOHEXYLAMINE see NKU400

N-METHYL-N-NITROSOLAURYLAMINE see NKU000

N-METHYL-N-NITROSOMETHANAMINE see NKA600

4-METHYL-4-N-(NITROSOMETHYLAMINO)-2-PENTANONE see MMX750

2-METHYL-4-NITROSOMORPHOLINE see NKU550

N-METHYL-N-NITROSONITROGUANIDIN see MMP000

1-METHYL-1-NITROSO-3-NITROGUANIDINE see MMP000

N-METHYL-N-NITROSO-N'-NITROGUANIDINE see MMP000

N-METHYL-N-NITROSOOCTANAMIDE see MMY000

N-METHYL-N-NITROSOOCTYLAMINE see NKU590

2-METHYL-3-NITROSO-1,3-OXAZOLIDINE see NKU600

5-METHYL-3-NITROSO-1,3-OXAZOLIDINE see NKU875

N-METHYL-N-NITROSO-4-OXO-4-(3-PYRIDYL)BUTYL AMINE see MMS500

N-METHYL-N-NITROSOPENTYLAMINE see AOL000

N-METHYL-N-NITROSOPHENETHYLAMINE see MNU250

N-METHYL-N-NITROSO-4-(PHENYLAZO)ANILINE see MMY250

N-METHYL-N-NITROSO-1-PHENYLETHYLAMINE see NKW000

METHYL-(4-NITROSOPHENYL)NITROSAMINE see MJG750

1-METHYL-1-NITROSO-3-PHENYLUREA see MMY500

N-METHYL-N-NITROSO-N'-PHENYLUREA see MMY500

1-METHYL-4-NITROSOPIPERAZINE see NKW500

N'-METHYL-N-NITROSOPIPERAZINE see NKW500

2-METHYLNITROSOPIPERIDINE see NLI000

3-METHYLNITROSOPIPERIDINE see MMY750

4-METHYLNITROSOPIPERIDINE see MMZ000

3-METHYL-1-NITROSO-4-PIPERIDONE see MMZ800

N-METHYL-N-NITROSO-1-PROPANAMINE see MNA000

N-METHYL-N-NITROSO-2-PROPEN-1-AMINE see MMT500

METHYLNITROSO-PROPIONAMIDE see MNA250

N-METHYL-N-NITROSOPROPIONAMIDE see MNA250

METHYL-NITROSOPROPIONSAEUREAMID (GERMAN) see MNA250

METHYLNITROSOPROPIONYLAMIDE see MNA250

N-METHYL-N-NITROSO-1H-PURIN-6-AMINE see MMT250

1-METHYL-3-NITROSO-5H-PYRIDO(4,3-B)INDOLE see MNA300

N-METHYL-N-NITROSO-4-(2-(4-QUINOLINYL)ETHENYL)BENZENAMINE see MMS750

2-METHYL-N-NITROSOTHIAZOLIDINE see MNA500

METHYLNITROSO-p-TOLUENESULFONAMIDE see THE500

1-METHYL-1-NITROSO-3-(p-TOLYL)UREA see MNA650

N-METHYL-N-NITROSOUNDECYLAMINE see NKX500

METHYLNITROSOUREA see MNA750

1-METHYL-1-NITROSOUREA see MNA750

N-METHYL-N-NITROSOUREA see MNA750

METHYLNITROSOUREE (FRENCH) see MNA750

d-1-(3-METHYL-3-NITROSOUREIDO)-1-DEOXYGALACTOPYRANOSE see MNB000

METHYLNITROSOURETHAN (GERMAN) see MMX250

METHYLNITROSOURETHANE see MMX250

N-METHYL-N-NITROSO-URETHANE see MMX250

N-METHYL-N-NITROSOVINYLAMINE see NKY000

(E)-4-METHYL-4'-NITROSTILBENE see MNB100

trans-4'-METHYL-4-NITROSTILBENE see MNB100

4-METHYL-5-(5-NITRO-2-4H-1,2,4-TRIAZOL-3-AMINE) see AKY000

2-METHYL-N-(4-NITRO-3-(TRIFLUOROMETHYL)PHENYL)PROPANAMIDE (9CI) see FMR050

1-METHYL-2-NITRO-5-VINYL-1H-IMIDAZOLE see MNB250

N-METHYL-N-(5-(N'-METHYLANILINO)-2,4-PENTADIENYLIDENE) ANILINIUM CHLORIDE see MLL000

2-(o-(N-METHYL-N-(N'-METHYL-N'-ETHOXYCARBONYLAMINOSULFENYL)CARBAMOYL)OXIMINO)-1,4-DITHIANE see MLL100

S-METHYL N-(N'-METHYL-N'-HEXOXYSULFINYLCARBAMOYLOXY)THIOACETIMIDATE see MKM800

2-METHYL-N-(N'-METHYL-N'-(4-MORPHOLINOSULFENYL)CARBAMOYLOXY)THIOACETIMIDATE see MLL150

N⁶-METHYL-N⁶-NITROSO-1H-PURIN-6-AMINE see MMT250

N⁶-METHYL-N⁶-NITROSO-9b-d-RIBOFURANOSYL-9H-PURIN-6-AMINE see MMT300

o-METHYL NOGALAROL see MNB300

7-o-METHYL NOGALAROL see MNB300

7-o-METHYLNOGAROL see MCB600

3-METHYL-2(3)-NONENENITRILE see MNB500

METHYL-2-NONENOATE see MNB750

METHYLNONYLACETALDEHYDE see MQI550

METHYL n-NONYL ACETALDEHYDE see MQI550

METHYL NONYL ACETALDEHYDE DIMETHYL ACETAL see MNB600

METHYL NONYL ACETIC ALDEHYDE see MQI550

METHYL NONYLENATE see MNB750

METHYL NONYL KETONE see UKS000

METHYL-n-NONYL KETONE see UKS000

1-METHYL-7-(NONYLOXY)-9H-PYRIDO(3,4-b)INDOLE HYDROCHLORIDE see NNC100

METHYL-2-NONYNOATE see MNC000

α-METHYLNORADRENALINE HYDROCHLORIDE see AMB000

N-METHYLNORAPORMORPHINE HYDROCHLORIDE see AQP500

N-METHYL-NORDOCEINE see CNF500

N-METHYLNOREPHEDRINE see EAW000

21-METHYLNORETHISTERONE see MNC110

7(R)-o-METHYLNORGAROL see MCB600

1-METHYLNORHARMAN see MPA050

11-β-METHYL-19-NOR-17-α-PREGNA-1,3,5(10)-TRIEN-20-YNE-3,17-DIOL see MJW875

METHYLNORTESTOSTERONE see MDM350

17-METHYL-19-NORTESTOSTERONE see MDM350

17-α-METHYL-19-NORTESTOSTERONE see MDM350

METHYL OCTADECANOATE see MJW000

METHYL-9-OCTADECENOATE see OHW000

METHYL (Z)-9-OCTADECENOATE see OHW000

METHYL cis-9-OCTADECENOATE see OHW000

2-METHYLOCTANAL see MNC175

α-METHYLOCTANAL see MNC175

2-METHYLOCTANE see MNC250

3-METHYLOCTANE see MNC500

4-METHYLOCTANE see MNC750

METHYL OCTANOATE see MHY800

METHYL OCTANOATE and METHYL DECANOATE see MND000

3-METHYL-3-OCTANOL see MND050

3-METHYLOCTAN-3-OL see MND050

7-METHYL-1-OCTANOL see ILJ000

5-METHYL-2-OCTANONE see MLO800

5-METHYL-2-OCTANONE see MND075

3-METHYL-1-OCTEN-OL see MND100

METHYLOCTENYLAMINE see ILK000

METHYL 2-OCTINATE see MND275

METHYL OCTINE CARBONATE see MNC000

4-METHYLOCTIN-4-OL see MKR550

METHYL OCTYL ACETALDEHYDE see MIW000

METHYLOCTYLDIAZENE 1-OXIDE see MGT000

METHYL OCTYL KETONE see OFE050

METHYL n-OCTYL KETONE see OFE050

N-METHYL-N-OCTYL-1-OCTANAMINE see MND125

METHYL OCTYNE CARBONATE see MNC000

METHYL 2-OCTYNOATE see MND275

4-METHYL-1-OCTYN-4-OL see MKR550

(+−)-4-METHYL-1-OCTYN-4-OL see MKR550

METHYL OENANTHYLATE see MKK100

METHYLOESTRENOLONE see MDM350

35-METHYLOKADAIC ACID see MND300

METHYLOL see MGB150

N-METHYLOLACRYLAMIDE see HLU500

9-METHYLOLANTHRACENE see APG600

N'-METHYLOL-o-CHLORTETRACYCLINE see MND500

N-METHYLOL DIMETHYLPHOSPHONOPROPIONAMIDE see MND550

METHYL OLEATE see OHW000

N-METHYL-N-OLEOYLTAURINE SODIUM SALT see SIY000

METHYLOL FORMALDEHYDE see GHO100

3-METHYLOLPENTANE see EGW000

2-METHYLOLPHENOL see HMK100

o-METHYLOLPHENOL see HMK100

N-METHYLOLPHTHALIMIDE see HMP100

METHYLOLPROPANE see BPW500

METHYLOLUREA RESIN see UTU500

METHYL-4-OMBELLIFERONE SODEE (FRENCH) see HMB000

(METHYL-ONN-AZOXY)METHANOL see HMG000

(METHYL-ONN-AZOXY)METHANOL, ACETATE (ester) see MGS750

(METHYL-ONN-AZOXY)METHYL-β-d-GLUCOPYRANOSIDE see COU000

(METHYL-ONN-AZOXY)METHYL-β-D-GLUCOPYRANOSIDURONIC ACID see MGS700

(METHYL-ONN-AZOXY)METHYL-3-o-β-d-GLUCOPYRANOSYLβ-d-GLUCOPYRANOSIDE see MND575

METHYL ORANGE see MND600

METHYL ORANGE B see MND600

METHYL ORTHOFORMATE see TLX600

METHYL ORTHOSILICATE see MPI750

METHYL ORTHOSILICATE (DOT) see MPI750

9-METHYL-3-OXA-9-AZATRICYCLO(3.3.1.0²,⁴)NONAN-7-OL TROPATE (ester) see SBG000

4-METHYL-7-OXABICYCLO(4.1.0)HEPTANE-3-CARBOXYLIC ACID, ALLYL ESTER see AGF500

2-(3-(5-METHYL-1,3,4-OXADIAZOL-2-YL)-3,3-DIPHENYLPROPYL)-2-AZABICYCLO(2.2.2)OCTANE see DWF700

5-METHYL-1,2-OXATHIOLANE 2,2-DIOXIDE see BOU300

METHYLOXAZEPAM see CFY750

N-METHYLOXAZEPAM see CFY750

3-METHYLOXAZOLIDINE see MND700

N-METHYLOXAZOLIDINE see MND700

3-METHYL-1,3-OXAZOLIDINE see MND700

3-METHYL-2-OXAZOLIDONE see MND750

4-METHYL-2-OXETANONE see BSX000

(+-)-4-METHYL-2-OXETANONE see BSX100

2-METHYLOXINE see HMQ600

(−)-METHYLOXIRANE see ECE700

METHYL OXIRANE see PNL600

(+)-METHYLOXIRANE see PNL650

(R)-METHYLOXIRANE see PNL650

S-METHYLOXIRANE see ECE700

(R)-(+)-METHYLOXIRANE see PNL650

(S)-2-METHYLOXIRANE see ECE700

3-METHYLOXIRANEMETHANOL 4-NITROBENZOATE (R)- see ECC700

2-METHYLOXIRANEMETHANOL 4-NITROBENZOATE (2S)- see ECC702

3-METHYLOXIRANEMETHANOL 4-NITROBENZOATE (2R-trans)- see EBJ600

3-METHYLOXIRANEMETHANOL 4-NITROBENZOATE (2S-trans)- see EBJ700

METHYLOXIRANE, POLYMER with OXIRANE, MONOBUTYL ESTER see MNE250

(2R-cis)-(3-METHYLOXIRANYL)PHOSPHONIC ACID see PHA550

(2R-cis)-(3-METHYLOXIRANYL)PHOSPHONIC ACID DISODIUM SALT see DXF600

METHYL OXITOL see EJH500

6-METHYL-2-OXO-2H-1-BENZOPYRAN-3-CARBOXYLIC ACID see MNE275

1-(2-((4-METHYL-2-OXO-2H-1-BENZOPYRAN-7-YL)OXY)ETHYL)PYRIDINIUM BROMIDE see PPA350

METHYL(1-OXOBUTYL)CARBAMIC ACID 2,2-DIMETHYL-1,3-BENZODIOXOL-4-YL ESTER see MNE300

METHYL-3-OXOBUTYRATE see MFX250

(5-METHYL-2-OXO-1,3-DIOXOLEN-4-YL)METHYL-D-α-AMINOBENZYLPENICILLINATE HYDROCHLORIDE see LEJ500

6-METHYL-2-OXO-1,3-DITHIOLO(4,5-b)QUINOXALINE see ORU000

METHYL OXOETHANOATE see MKI550

1-METHYL-4-OXO-9-(p-NITROPHENACYL)-1H,2,3,4,9-TETRAHYDROPYRROLO(2,3-B)QUINOLINE see MNE400

METHYL-12-OXO-trans-10-OCTADECENOATE see MNF250

N-(2-(2-METHYL-4-OXOPENTYL))ACRYLAMIDE see DTH200

METHYL(1-OXOPENTYL)CARBAMIC ACID 2,2-DIMETHYL-1,3-BENZODIOXOL-4-YL ESTER see MNF300

METHYL 3-OXO-2-PENTYLCYCLOPENTANEACETATE see HAK100

3-METHYL-4-OXO-2-PHENYL-4H-1-BENZOPYRAN-8-CARBOXYLATE 1-PIPERIDINEETHANOL HYDROCHLORIDE see FCB100

3-METHYL-4-OXO-5-PIPERIDINO-Δ²,ᵃ-THIAZOLIDINEACETIC ACID ETHYL ESTER see MNG000

(3-METHYL-4-OXO-5-PIPERIDINO-2-THIAZOLIDINYLIDENE)ACETIC ACID ETHYL ESTER see MNG000

(3-METHYL-4-OXO-5-(1-PIPERIDINYL)-2-THIAZOLIDINYLIDENE)ACETIC ACID ETHYL ESTER see MNG000

11-METHYL-1-OXO-1,2,3,4-TETRAHYDROCHRYSENE see MNG250

METHYL p-OXYBENZOATE see HJL500

METHYLOXYLAMMONIUM CHLORIDE see MNG500

4-(METHYLOXYMETHYL)BENZYL CHRYSANTHEMUMMONOCARBOXYLATE see MBV700

4-(METHYLOXYMETHYL)BENZYL CHRYSANTHEMUM MONOCARBOXYLATE see MNG525

4-METHYL-5-OXYMETHYLURACIL see HMH300

1-(4-METHYLOXYPHENYL)-3,3-DIMETHYLTRIAZINE see DSN600

N-METHYL-β-OXY-β-PHENYLISOPROPYLAMINHYDROCHLORID see EAW500

2-METHYL-3-OXY-γ-PYRONE see MAO350

METHYLPARABEN (FCC) see HJL500

METHYLPARAFYNOL CARBAMATE see MNM500

METHYL PARAHYDROXYBENZOATE see HJL500

METHYL PARAOXON see PHD500

METHYL PARASEPT see HJL500

METHYL PARATHION see MNH000

N-METHYLPAREDRINE see FMS875

METHYL PCT see DTQ600

METHYL PECTIN see PAO150

METHYL PECTINATE see PAO150

METHYL PENTACHLOROPHENATE see MNH250

METHYL(PENTACHLOROPHENOXY)MERCURY see MLG250

METHYL PENTACHLOROPHENYL ESTER see MNH250

METHYLPENTADIENE see MNH500

2-METHYL-1,3-PENTADIENE see MNH750

4-METHYL-1,3-PENTADIENE see MNI000

2-METHYLPENTAMETHYLENEDIAMINE see MNI515

4-METHYLPENTANAL see MQJ500

2-METHYLPENTANE see IKS600

3-METHYLPENTANE see MNI500

3-METHYL PENTANEDIAL see MKI250

2-METHYL-1,5-PENTANEDIAMINE see MNI515

2-METHYL-2,4-PENTANEDIAMINE see MNI525

2-METHYL PENTANE-2,4-DIOL see HFP875

2-METHYL-2,4-PENTANEDIOL see HFP875

3-METHYL-1,5-PENTANEDIOL see MNI750

2-METHYL-2,4-PENTANEDIOL ESTER with BORIC ACID (H3BO3) (1:2) cyclic BIS(1,1,3-TRIMETHYLMETHYLENE) ESTER see TKM250

4-METHYLPENTANENITRILE see MNJ000

METHYL PENTANOATE see VAQ100

2-METHYLPENTANOIC ACID see MQJ750

2-METHYLPENTANOL-1 see AOK750

2-METHYL-4-PENTANOL see MKW600

3-METHYL-3-PENTANOL see MNJ100

4-METHYLPENTANOL-2 see MKW600

3-METHYL-PENTANOL-(3) (GERMAN) see MNJ100

4-METHYL-2-PENTANOL, ACETATE see HFJ000

3-METHYL-3-PENTANOL CARBAMATE see ENF000

4-METHYL-2-PENTANOL (MAK) see MKW600

2-METHYL-2-PENTANOL-4-ONE see DBF750

4-METHYL-2-PENTANON (CZECH) see HFG500

4-METHYL-PENTAN-2-ON (DUTCH, GERMAN) see HFG500

2-METHYL-4-PENTANONE see HFG500

4-METHYL-2-PENTANONE (FCC) see HFG500

3-METHYL-2-n-PENTANYL-2-CYCLOPENTEN-1-ONE see DLQ600

2-METHYL-2-PENTENAL see MNJ750

2-METHYL-2-PENTEN-1-AL see MNJ750

2-METHYLPENTENE see MNK000

2-METHYL-1-PENTENE see MNK000

2-METHYL-PENTENE-1 see MNK000

2-METHYL-2-PENTENE see MNK100

2-METHYL-PENTENE-2 see MNK100

4-METHYL-1-PENTENE see MNK250

4-METHYL-2-PENTENE see MNK500

trans-4-METHYL-2-PENTENE see MNK750

2-METHYL-2-PENTENE-1-AL see MNJ750

4-METHYL-3-PENTENE-2-ONE see MDJ750

3-METHYL-1-PENTEN-3-OL see MNL250

4-METHYL-2-PENTEN-4-OL see MNL500

3-METHYL-PENTEN-(1)-OL-(3) (GERMAN) see MNL250

4-METHYL-3-PENTEN-2-ON (DUTCH, GERMAN) see MDJ750

2-METHYL-1-PENTEN-3-ONE see IMW500

2-METHYL-2-PENTEN-4-ONE see MDJ750

4-METHYL-3-PENTEN-2-ONE see MDJ750

4-METHYL-3-PENTEN-2-ONE (1-PHTHALAZINYL)HYDRAZONE see DQU400

4-(4-METHYL-3-PENTENYL)-3-CYCLOHEXENE-1-CARBOXALDEHYDE see IKT100

3-METHYL-2-(cis-2-PENTEN-1-YL)-2-CYCLOPENTEN-1-ONE see JCA100

1-(4-METHYL-3-PENTENYL)-4-FORMYL-1-CYCLOHEXENE see IKT100

3-METHYL-2-PENTEN-4-YN-1-OL see MNL775

3-METHYLPENTIN-3-OL see EQL000

3-METHYL-PENTIN-(1)-OL-(3) (GERMAN) see MNM500

2-METHYLPENTYL ACETATE see MGN300

4-METHYL-2-PENTYL ACETATE see HFJ000

2-METHYLPENTYL CARBITOL see DJG200

2-METHYLPENTYL CELLOSOLVE see EJK000

METHYL N-(β-PENTYLCINNAMYLIDENE)ANTHRANILATE see AOH100

3-METHYL-2-PENTYL-2-CYCLOPENTEN-1-ONE see DLQ600

4-METHYL-2-PENTYL-DIOXOLANE see MNM450

2-METHYL-2,3-PENTYLENE OXIDE see ECC600

METHYL tert-PENTYL ETHER see PBX400

METHYL PENTYL KETONE see MGN500

METHYL-N-PENTYLNITROSAMINE see AOL000

METHYL (2-PENTYL-3-OXOCYCLOPENTYL)ACETATE see HAK100

2-((2-METHYLPENTYL)OXY)ETHANOL see EJK000

2-(2-((2-METHYLPENTYL)OXY)ETHOXY)ETHANOL see DJG200

3-METHYLPENT-1-YN-3-OL see EQL000

3-METHYL-1-PENTYN-3-OL see EQL000

METHYLPENTYNOL CARBAMATE see MNM500

3-METHYL-1-PENTYN-3-OL CARBAMATE see MNM500

METHYL PERCHLORATE see MNM750

METHYL PERFLUOROMETHACRYLATE see MNN000

METHYLPERIDIDE see FLV000

METHYLPERIDOL see MNN250

METHYLPERIDOL HYDROCHLORIDE see MNN250

METHYLPERONE HYDROCHLORIDE see FKI000

9-METHYL-PGA see BMM125

15(s)-15-METHYL-PGE2 see MOV800

METHYL-PGF2-α see CCC100

15(S)-15-METHYL PGF2-α see CCC100

15-METHYL-PGF2-α-METHYL ESTER see MLO301

15(s)15-METHYL-PGF2-α-METHYL ESTER see MLO301

15(2)15-METHYL PGF2-α TROMETHAMINE SALT see CCC110

dl-N-METHYLPHEDRINE HYDROCHLORIDE see MNN350

1-METHYLPHENANTHRENE see MNN400

2-METHYLPHENANTHRENE see MNN520

2-METHYLPHENANTHRO(2,1-d)THIAZOLE see MNO250

5-METHYLPHENAZINE METHYLSULFATE see MNO500

N-METHYLPHENAZIUM METHOSULFATE see MNO500

5-METHYLPHENAZIUM METHYL SULFATE see MNO500

METHYLPHENAZONIUM METHOSULFATE see MRW000

N-METHYLPHENAZONIUM METHOSULFATE see MNO500

5-N-METHYLPHENAZONIUM METHOSULFATE see MNO500

N-METHYLPHENAZONIUM METHOSULPHATE see MNO500

METHYL PHENCAPTON see MNO750

2-METHYLPHENELZINE see MNO775

4-METHYLPHENELZINE see MNO780

α-METHYL-PHENETHYL ALCOHOL see PGA600

β-METHYLPHENETHYL ALCOHOL see HGR600

α-METHYLPHENETHYLAMINE see AOA250

β-METHYLPHENETHYLAMINE see PGB760

(±)-α-METHYLPHENETHYLAMINE see BBK000

dl-α-METHYLPHENETHYLAMINE see BBK000

α-METHYL-PHENETHYLAMINE compounded with AMBERLITE XE-69 see AOB000

α-METHYLPHENETHYLAMINE, d-FORM see AOA500

dl-α-METHYL-PHENETHYLAMINE HYDROCHLORIDE see AOA500

dl-α-METHYL-PHENETHYLAMINE PHOSPHATE see AOB500

α-METHYLPHENETHYLAMINE PHOSPHATE, dl-MIXTURE see AOB500

d-α-METHYLPHENETHYLAMINE SULFATE see BBK500

(±)-α-METHYLPHENETHYLAMINE SULFATE see AOB250

dl-α-METHYLPHENETHYLAMINE SULFATE see BBK250

2-(α-METHYLPHENETHYL)AMINOETHANOL see PFI750

2-METHYL-2-(2-(PHENETHYLAMINO)ETHYL)-1,3-BENZODIOXOLE HYDROCHLORIDE see MNP250

4-(2-(α-METHYLPHENETHYLAMINO)ETHYL)-3-PHENYL-1,2,4-OXADIAZOLE MONOHYDROCHLORIDE see WBA600

7-(2-(1-METHYL-2-PHENETHYLAMINO)ETHYL)THEOPHYLLINE HYDROCHLORIDE see CBF825

7-(2-((α-METHYLPHENETHYL)AMINO)ETHYL)THEOPHYLLINE HYDROCHLORIDE see CBF825

METHYL PHENETHYL ETHER see PFD325

METHYL 2-PHENETHYL ETHER see PFD325

(o-METHYLPHENETHYL)HYDRAZINE see MNO775

(p-METHYLPHENETHYL)HYDRAZINE see MNO780

(α-METHYLPHENETHYL)HYDRAZINE see PDN000

METHYL PHENETHYL KETONE see PDF800

1-(α-METHYLPHENETHYL)-2-(5-METHYL-3-ISOXAZOLYLCARBONYL)HYDRAZINE see MNP300

N'-4-(4-METHYLPHENETHYLOXY)PHENYL-N-METHOXY-N-METHYLUREA see MNP350

1-(α-METHYLPHENETHYL)-2-PHENETHYLHYDRAZINE see MNP400

4-(1-METHYL-1-PHENETHYL)PHENOL see COF400

N-(α-METHYLPHENETHYL)-2-PHENYLGLYCINONITRILE see AOB300

1-(α-METHYLPHENETHYL)-4-PHENYLPIPERAZINE DIHYDROCHLORIDE see MNP450

3-(α-METHYLPHENETHYL)SYDONE IMINE MONOHYDROCHLORIDE see SPA000

3-(α-METHYLPHENETHYL)-1,4,5,6-TETRAHYDRO-as-TRIAZINE HYDROCHLORIDE see TCV460

2-(p-METHYLPHENETHYL)-3-THIOSEMICARBAZIDE see MNP500

METHYLPHENIDAN see MNQ000

METHYL PHENIDATE see MNQ000

METHYLPHENIDATE HYDROCHLORIDE see RLK000

METHYL PHENIDYL ACETATE see MNQ000

METHYLPHENIDYLACETATE HYDROCHLORIDE see RLK000

2-METHYLPHENISOPROPYLAMINE SULFATE see MNQ250

4-METHYLPHENISOPROPYLAMINE SULFATE see AQQ000

METHYLPHENOBARBITAL see ENB500

1-METHYLPHENOBARBITAL see ENB500

N-METHYLPHENOBARBITAL see ENB500

N-METHYLPHENOBARBITOL see ENB500

METHYLPHENOBARBITONE see ENB500

2-METHYLPHENOL see CNX000

3-METHYLPHENOL see CNW750

4-METHYLPHENOL see CNX250

m-METHYLPHENOL see CNW750

o-METHYLPHENOL see CNX000

p-METHYLPHENOL see CNX250

4-METHYLPHENOL METHYL ETHER see MGP000

METHYLPHENOL SODIUM SALT see SFZ050

3-METHYLPHENOL SODIUM SALT see SJP000

4-METHYLPHENOL, SODIUM SALT see SIM100

10-METHYLPHENOTHIAZINE-2-ACETIC ACID see MNQ500

N-METHYL-3-PHENOTHIAZINYLACETIC ACID see MNQ500

(10-METHYL-2-PHENOTHIAZINYL)ACETIC ACID see MNQ500

2-(2-(4-(2-METHYL-3-PHENOTHIAZIN-10-YLPROPYL)-1-PIPERAZINYL)ETHOXY)ETHANOL DIHYDROCHLORIDE see DXR800

2-(2-(4-(2-METHYL-3-PHENOTHIAZIN-10-YLPROPYL)-1-PIPERAZINYL)ETHOXY)ETHANOL see MKQ000

1-METHYL-3-PHENOXYBENZENE see MNV770

(±)-α-METHYL-3-PHENOXYBENZENEACETIC ACID CALCIUM SALT DIHYDRATE see FAP100

3-(3-METHYLPHENOXY)BENZYL 2-(4-CHLOROPHENYL)-2-METHYLPROPYL ETHER see MNQ600

1-(2-(o-METHYLPHENOXY)ETHYL)HYDRAZINE HYDROGEN SULFATE see MNR100

(1-METHYL-2-PHENOXYETHYL)HYDRAZINE MALEATE see PDV700

(1-METHYL-2-PHENOXYETHYL)HYDRAZINIUM see PDV700

((METHYLPHENOXY)METHYL)OXIRANE see TGZ100

((4-METHYLPHENOXY)METHYL)OXIRANE see GGY175

3-(2-METHYLPHENOXY)-1,2-PROPANEDIOL see GGS000

3-(2-METHYLPHENOXY)-1,2-PROPANEDIOL 1-CARBAMATE see CBK500

α-METHYLPHENSUXIMIDE see MLP800

α-METHYL PHENYLACETALDEHYDE see COF000

4-METHYLPHENYL ACETATE see MNR250

p-METHYLPHENYL ACETATE see MNR250

METHYL PHENYLACETATE (FCC) see MHA500

METHYL(2-PHENYLAETHYL)NITROSAMIN (GERMAN) see MNU250

METHYLPHENYLAMINE see MGN750

N-METHYLPHENYLAMINE see MGN750

2-((4-METHYLPHENYL)AMINO)BENZOIC ACID see TGV000

2-METHYL-4-PHENYLANILINE see MGF500

N-(2-METHYLPHENYL)ANTHRANILIC ACID see MNR300

N-(4-METHYLPHENYL)ANTHRANILIC ACID see TGV000

N-(o-METHYLPHENYL)ANTHRANILIC ACID see MNR300

N-(p-METHYLPHENYL)ANTHRANILIC ACID see TGV000

METHYLPHENYLARSINIC ACID see HMK200

METHYLPHENYLARSONIC ACID see HMK200

N-METHYL-p-(PHENYLAZO)ANILINE see MNR500

p-(3-METHYLPHENYLAZO)ANILINE see MNR750

N-METHYL-4-(PHENYLAZO)-o-ANISIDINE see MNS000

4-((4-METHYLPHENYL)AZO)BENZENAMINE see TGV750

2-METHYL-4-(PHENYLAZO)-1,3-BENZENEDIAMINE see MNS100

4-METHYL-6-(PHENYLAZO)-1,3-BENZENEDIAMINE see CMM800

1-(2-METHYLPHENYL)AZO-2-NAPHTHALENAMINE see FAG135

1-((2-METHYLPHENYL)AZO)-2-NAPHTHALENAMINE see FAG135

1-((2-METHYLPHENYL)AZO)-2-NAPHTHALENOL see TGW000

1-(2-METHYLPHENYL)AZO-2-NAPHTHYLAMINE see FAG135

N-METHYL-N-(p-(PHENYLAZO)PHENYL)HYDROXYLAMINE see HLV000

METHYLPHENYLBARBITURIC ACID see ENB500

7-METHYL-9-PHENYLBENZ(c)ACRIDINE see MNS250

5-METHYL-7-PHENYL-1:2-BENZACRIDINE see MNS250

2-METHYL-N-PHENYLBENZAMIDE see MNS500

4-METHYL-N-PHENYLBENZENESULFONAMIDE see TGN600

2-(4-METHYLPHENYL)-1H-BENZ(de)ISOQUINOLINE-1,3(2H)-DIONE see MNS550

4-METHYLPHENYL BENZOATE see TGX100

o-METHYLPHENYL BROMIDE see BOG260

2-METHYL-4-PHENYL-2-BUTANOL see BEC250

2-METHYL-4-PHENYL-2-BUTANOL ACETATE see MNT000

3-METHYL-4-PHENYL-3-BUTEN-2-ONE see MNS600

3-METHYL-4-PHENYL-3-BUTEN-2-ONE see MNS600

N-METHYL-N-PHENYL-tert-BUTYLACETAMIDO-β-HYDROXYETHYLAMINE HYDROCHLORIDE see EAN650

2-METHYL-4-PHENYL-2-BUTYL ACETATE see MNT000

2-METHYL-1-PHENYL-3-BUTYNE-1,2-DIOL see DMX800

3-(METHYLPHENYL)CARBAMIC ACID 3-((METHOXYCARBONYL)AMINO)PHENYL ESTER see MEG250

METHYLPHENYLCARBAMIC ESTER OF 3-OXYPHENYLTRIMETHYLAMMONIUM METHYLSULFATE see HNR500

N-METHYL-4-PHENYL-4-CARBETHOXYPIPERIDINE see DAM600

N-METHYL-4-PHENYL-4-CARBETHOXYPIPERIDINE HYDROCHLORIDE see DAM700

METHYLPHENYLCARBINOL see PDE000

METHYLPHENYLCARBINOL ACETATE see SMP600

METHYL PHENYLCARBINYL ACETATE see MNT075

METHYL PHENYL CARBINYL ACETATE see SMP600

METHYLPHENYLCARBINYL PROPIONATE see PFR000

1-METHYL-4-PHENYL-4-CARBOETHOXYPIPERIDINE HYDROCHLORIDE see DAM700

1-METHYL-5-PHENYL-7-CHLORO-1,3-DIHYDRO-2H-1,4-BENZODIAZEPIN-2-ONE see DCK759

1-METHYL-3-PHENYL-5-CHLOROIMIDAZO(4,5-b)PYRIDIN-2-ONE see MNT100

p-METHYLPHENYLDIAZONIUM FLUOROBORATE see TGM450

METHYLPHENYLDICHLOROSILANE (DOT) see DFQ800

2-METHYL-9-PHENYL-2,3-DIHYDRO-1-PYRIDINDENE HYDROBROMIDE see NOE525

3-METHYLPHENYL-N-(1,2,5,6-DI-o-ISOPROPYLIDENE-3-o-GLUCOFURANOSYLSULFINYL)-N-METHYLCARBAMATE see MNT200

METHYLPHENYLDIMETHOXYSILANE see DOH400

3-METHYLPHENYL-N-(2,2-DIMETHYL-1,3-DIOXOLANE-5-YLMETHOXYSULFINYL)-N-METHYLCARBAMATE see MNT250

1-(o-METHYLPHENYL)-3,3-DIMETHYL-TRIAZEN (GERMAN) see MNT500

1-(2-METHYLPHENYL)-3,3-DIMETHYLTRIAZENE see MNT500

1-(3-METHYLPHENYL)-3,3-DIMETHYLTRIAZENE see DSR200

1-(m-METHYLPHENYL)-3,3-DIMETHYLTRIAZENE see DSR200

1-(o-METHYLPHENYL)-3,3-DIMETHYL-TRIAZENE see MNT500

2-(METHYLPHENYL)-1,3-DIOXAN-5-OL (mixed isomers) see TGK500

4-METHYL-2-PHENYL-m-DIOXOLANE see MNT600

4-METHYL-2-PHENYL-1,3-DIOXOLANE see MNT600

METHYLPHENYL DIPHENYL PHOSPHATE see TGY750

METHYL PHENYLDITHIOCARBAMATE see MJK500

1-METHYL-2-PHENYL-3-DODECYLBENZIMIDAZOLINIUM FERROCYANIDE see TNH750

1-METHYL-2-PHENYL-3-m-DODECYLBENZIMIDAZOLIUM HEXACYANOFERRATE see TNH750

1,1'-(4-METHYL-1,3-PHENYLENE)BIS(3-(2-CHLOROETHYL)-3-NITROSOUREA) see MNT750

1,1'-(4-METHYL-1,3-PHENYLENE)BIS-1H-PYRROLE-2,5-DIONE see TGY770

METHYLPHENYLENEDIAMINE see TGL500

2-METHYL-p-PHENYLENEDIAMINE see TGM600

4-METHYL-m-PHENYLENEDIAMINE see TGL750

N-METHYL-o-PHENYLENEDIAMINE see MNU000

2-METHYL-p-PHENYLENEDIAMINE SULPHATE see DCE600

4-METHYL-PHENYLENE DIISOCYANATE see TGM750

METHYL-m-PHENYLENE DIISOCYANATE see TGM740

2-METHYL-m-PHENYLENE ESTER, ISOCYANIC ACID see TGM800

METHYLPHENYLENE ISOCYANATE see TGM740

4-METHYL-PHENYLENE ISOCYANATE see TGM750

2-METHYL-m-PHENYLENE ISOCYANATE see TGM800

1-(4-(METHYLPHENYL))ETHANOL see TGZ000

1-(p-METHYLPHENYL)ETHANOL see TGZ000

4-(2-(4-METHYLPHENYL)ETHENYL)BENZENAMINE, (E)- see MNU025

METHYL PHENYL ETHER see AOX750

β-METHYLPHENYLETHYL ACETATE see PGB750

α-METHYL-β-PHENYLETHYL ACETATE see ABU800

α-METHYL PHENYLETHYL ALCOHOL see HGR000

METHYL(2-PHENYLETHYL)ARSINIC ACID see MNU050

N-METHYL-5-PHENYL-5-ETHYLBARBITAL see ENB500

1-METHYL-5-PHENYL-5-ETHYLBARBITURIC ACID see ENB500

4-(1-METHYL-1-PHENYLETHYL)-2,6-BIS-(1-PIPERIDINYLMETHYL)PHENOL DIHYDROBROMIDE see BHR750

as-METHYLPHENYLETHYLENE see MPK250

METHYL PHENYLETHYL ETHER see PFD325

3-METHYL-5,5-PHENYLETHYLHYDANTOIN see MKB250

(2-(2-METHYLPHENYL)ETHYL)-HYDRAZINE see MNO775

o-METHYL-β-PHENYLETHYLHYDRAZINE DIHYDROGEN SULFATE see MNU100

p-METHYL-β-PHENYLETHYLHYDRAZINE DIHYDROGEN SULFATE see MNU150

(1-METHYL-2-PHENYLETHYL)-HYDRAZINEIUM CHLORIDE see PDN250

METHYL PHENYLETHYL KETONE see PDF800

METHYL-2-PHENYLETHYL KETONE see PDF800

METHYL-PHENYLETHYL-NITROSAMINE see MNU250

4-(1-METHYL-1-PHENYLETHYL)PHENOL MIXT. WITH BORAX AND 2,2'-OXYBIS(ETHANOL) see MRW100

N-(1-METHYL-2-PHENYLETHYL)-γ-PHENYLBENZENEPROPANAMINE see PEV750

S-1-METHYL-1-PHENYLETHYL PIPERIDINE 1-CARBOTHIOATE see MNU300

3-METHYL-3-PHENYLGLYCIDIC ACID ETHYL ESTER see ENC000

METHYLPHENYL GLYCIDYL ETHER see GGY175

4-METHYLPHENYLHYDRAZINE HYDROCHLORIDE see MNU500

4-METHYLPHENYL 2-HYDROXYBENZOATE see THD850

4-METHYLPHENYLHYDROXYLAMINE see HJA500

N-(2-METHYLPHENYL)-HYDROXYLAMINE see THA000

N-(4-METHYLPHENYL)HYDROXYLAMINE see HJA500

N-(p-METHYLPHENYL)HYDROXYLAMINE see HJA500

3-METHYL-2-PHENYLIMIDAZO(2,1-A)ISOQUINOLINE see MNU600

N-(2-METHYLPHENYL)IMIDODICARBONIMIDIC DIAMIDE see TGX550

2,2'-((2-METHYLPHENYL)IMINO)BISETHANOL see DMT800

2,2'-((3-METHYLPHENYL)IMINO)BISETHANOL see DHF400

N'-(5-METHYL-3-PHENYL-1-INDOLYL)-N,N,N'-TRIMETHYLETHYLENEDIAMINE HYDROCHLORIDE see MNU750

3-METHYLPHENYL IODIDE see IFG100

2-METHYLPHENYL ISOCYANATE see IKG725

4-METHYLPHENYL ISOCYANATE see IKG735

o-METHYLPHENYL ISOCYANATE see IKG725

p-METHYLPHENYL ISOCYANATE see IKG735

1-METHYL-4-PHENYLISONIPECOTIC ACID, ETHYL ESTER see DAM600

1-METHYL-4-PHENYLISONIPECOTIC ACID ETHYL ESTER HYDROCHLORIDE see DAM700

1-METHYL-4-PHENYLISONIPECOTIC ACID ETHYL ESTER-1-OXIDE HYDROCHLORIDE see DYB250

N-METHYL-β-PHENYLISOPROPYLAMIN (GERMAN) see DBB000

METHYL-β-PHENYLISOPROPYLAMINE see PFP850

N-METHYL-β-PHENYLISOPROPYLAMINE see DBB000

l-N-METHYL-β-PHENYLISOPROPYLAMINE HYDROCHLORIDE see MDQ500

dl-N-METHYL-β-PHENYLISOPROPYLAMINE HYDROCHLORIDE see DAR100

N-METHYL-β-PHENYLISOPROPYLAMINHYDROCHLORID (GERMAN) see DBA800

5-METHYL-3-PHENYLISOXAZOLE-4-CARBOXYLIC ACID see MNV000

5-METHYL-3-PHENYL-4-ISOXAZOLYL-PENICILLIN see DSQ800

5-METHYL-3-PHENYL-4-ISOXAZOLYL PENICILLIN, SODIUM see MNV250

METHYL PHENYL KETONE see ABH000

4-METHYLPHENYLMERCAPTAN see TGP250

p-METHYLPHENYLMERCAPTAN see TGP250

m-METHYLPHENYL METHYLCARBAMATE see MIB750

3-METHYLPHENYL N-METHYLCARBAMATE see MIB750

(4-METHYLPHENYL)METHYL CHLORIDE see MHN300

3-METHYLPHENYL METHYLNITROSOCARBAMATE see MNV500

3-METHYLPHENYL-N-METHYL-N-NITROSOCARBAMATE see MNV500

N-METHYL-4-(2-(PHENYLMETHYL)PHENOXY)-1-BUTANAMINE HYDROCHLORIDE see MGE200

1-(1-METHYL-2-(2-(PHENYLMETHYL)PHENOXY)ETHYL)PIPERIDINE PHOSPHATE (9CI) see PJA130

3-METHYLPHENYLMETHYL(((2-PHENYL-1,3-DIOXAN-5-YL)METHOXY)SULFINYL)CARBAMATE see MNV600

α-METHYL-α-PHENYL N-METHYL SUCCINIMIDE see MLP800

3-METHYL-2-PHENYLMORPHOLINE HYDROCHLORIDE see MNV750

1-METHYL-5-PHENYL-7-NITRO-1,3-DIHYDRO-2H-1,4-BENZODIAZEPIN-2-ONE see DLV000

METHYLPHENYLNITROSAMINE see MMU250

3-METHYLPHENYL N-NITROSO-N-METHYLCARBAMATE see MNV500

METHYLPHENYLNITROSOUREA see MMY500

N-METHYL-N'-PHENYL-N-NITROSOUREA see MMY500

2-METHYL-4-PHENYL-6H-1,3,5-OXATHIAZINE see MNV760

4-(4-METHYLPHENYL)-6H-1,3,5-OXATHIAZINE see MNV765

(4-METHYLPHENYL)OXIRANE see MPK750

2-(4-METHYLPHENYL)OXIRANE see MPK750

N-METHYL-N-PHENYLOXIRANEMETHANAMINE see GGY155

4-METHYL-1-PHENYL-1-PENTEN-3-ONE see IRL300

3-METHYLPHENYL-N-(2-PHENYL-1,3-DIOXANE-5-YLMETHOXYSULFINYL)-N-METHYLCARBAMATE see MNV600

3-METHYLPHENYL PHENYL ETHER see MNV770

m-METHYLPHENYL PHENYL ETHER see MNV770

3-(4-METHYLPHENYL)-1-PHENYL-2-PROPEN-1-ONE see MIF762

3-METHYL-1-PHENYL-2-PHOSPHOLENE 1-OXIDE see MNV800

METHYLPHENYLPHOSPHORAMIDIC ACID DIETHYL ESTER see MNW100

1-METHYL-4-PHENYLPIPERAZINE see MNW150

7-(4-(3-METHYLPHENYL)-1-PIPERAZINYL)-4-NITROBENZOFURAZAN-1-OXIDE see THD275

9-METHYL-2-(3-(4-PHENYL-1-PIPERAZINYLPROPYL))-1,2,3,4-TETRAHYDRO-β-CARBOLIN-1-ONE 2HCL see MNW200

1-METHYL-4-PHENYL-PIPERIDIN-4-CARBON-SAEURE-AETHYLESTER-HYDROCHLORID (GERMAN) see DAM600

METHYL α-PHENYL-2-PIPERIDINEACETATE HYDROCHLORIDE see RLK000

1-METHYL-4-PHENYLPIPERIDINE-4-CARBOXYLIC ACID ETHYL ESTER see DAM600

1-METHYL-4-PHENYL-4-PIPERIDINOL PROPIONATE (ESTER) HYDROCHLORIDE see MNW775

1-METHYL-4-PHENYL-4-PIPERIDINOL PROPIONATE (ester) 1-OXIDE HYDROCHLORIDE see MNW780

METHYL α-PHENYL-α-(2-PIPERIDYL)ACETATE see MNQ000

2-METHYL-1-PHENYLPROPANE see IIN000

2-METHYL-2-PHENYLPROPANE see BQJ250

2-METHYL-3-PHENYL-2-PROPENAL see MIO000

METHYL-3-PHENYLPROPENOATE see MIO500

2-(p-METHYLPHENYL)PROPIONALDEHYDE see THD750

1-METHYL-4-PHENYL-4-PROPIONOXYPIPERIDINE HYDROCHLORIDE see MNW775

1-METHYL-4-PHENYL-4-PROPIONOXYPIPERIDINE N-OXIDE HYDROCHLORIDE see MNW780

α-METHYL-γ-PHENYL-N-PROPYLAMIN see PEE300

α-METHYL-γ-PHENYL-N-PROPYLAMINE see PEE300

1-(4-METHYLPHENYL)-2-PROPYLAMINE SULFATE see AQQ000

N-(2-METHYLPHENYL)-2-(PROPYLAMINO)-PROPANAMIDE MONOHYDROCHLORIDE see CMS260

3-(4-METHYLPHENYL)-N-(4-PROPYLCYCLOHEXYL)-2-PROPENAMIDE see MNW790

2-METHYL-1-PHENYL-2-PROPYL HYDROPEROXIDE see MNX000

N-METHYL-N-(3-PHENYLPROPYL)-2-(PYRROLIDINYL)ACETAMIDE HYDROCHLORIDE see MNX260

N-METHYL-N-(1-PHENYL-2-PROPYL)-2-(PYRROLIDINYL)ACETAMIDE HYDROCHLORIDE see MNX250

1-METHYL-1-PHENYL-2-PROPYNYL CYCLOHEXANECARBAMATE see MNX850

1-METHYL-1-PHENYL-2-PROPYNYL-N-CYCLOHEXYLCARBAMATE see MNX850

N-(2-METHYLPHENYL)-1H-PYRAZOLE-1-ACETAMIDE see MNX300

N-(3-METHYLPHENYL)-1H-PYRAZOLE-1-ACETAMIDE see MNX310

N-(4-METHYLPHENYL)-1H-PYRAZOLE-1-ACETAMIDE see MNX320

3-METHYL-1-PHENYL-1H-PYRAZOLE-5-AMINE see PFQ350

3-METHYL-1-PHENYL-2-PYRAZOLIN-5-ONE see NNT000

3-METHYL-1-PHENYL-5-PYRAZOLONE see NNT000

3-METHYL-1-PHENYL PYRAZOL-5-YL DIMETHYL CARBAMATE see PPQ625

3-METHYL-1-PHENYL-5-PYRAZOLYL DIMETHYL CARBAMATE see PPQ625

α-(5-METHYL-1-PHENYL-1H-PYRAZOL-4-YL)-1-PIPERIDINEBUTANOL see MNX420

1-METHYL-4-PHENYLPYRIDINIUM see CQI550

1-METHYL-4-PHENYLPYRIDINIUM CHLORIDE see MNX900

trans-1-(4'-METHYLPHENYL)-1-(2'-PYRIDYL)-3-PYRROLIDINOPROP-1-N E HYDROCHLORIDE see TMX775

1-METHYL-3-PHENYLPYRROLIDIN-2,5-DIONE see MNZ000

1-METHYL-3-PHENYL-2,5-PYRROLIDINEDIONE see MNZ000

5-METHYL-1-PHENYL-2-(PYRROLIDINYL)IMIDAZOLE see MNY750

METHYL-5 PHENYL-1 (PYRROLIDINYL-1)-2 IMIDAZOLE (FRENCH) see MNY750

(E)-2-(1-(4-METHYLPHENYL)-3-(1-PYRROLIDINYL)-1-PROPENYL)-PYRIDINE MONOHYDROCHLORIDE see TMX775

1-(2-METHYL-5-PHENYL-1H-PYRROL-3-YL)ETHANONE see MNY800

4-METHYLPHENYL SALICYLATE see THD850

METHYLPHENYLSUCCINIMIDE see MNZ000

N-METHYL-2-PHENYL-SUCCINIMIDE see MNZ000

N-METHYL-α-PHENYLSUCCINIMIDE see MNZ000

METHYL 5-(PHENYLSULFINYL)-2-BENZIMIDAZOLECARBAMATE see OMY500

METHYL (5-PHENYLSULFINYL)-1H-BENZIMIDAZOL-2-YL CARBAMATE see OMY500

3-METHYL-1-PHENYL-3-(2-SULFOETHYL)TRIAZENE SODIUM SALT see PEJ250

METHYL PHENYL SULFONE see PFR375

p-METHYLPHENYLSULFONIC ACID see TGO000

p-METHYLPHENYLSULFONYL FLUORIDE see TGO500

4-METHYLPHENYLSULFONYL ISOCYANATE see IKG925

p-METHYLPHENYLSULFONYL ISOCYANATE see IKG925

3-((4-METHYLPHENYL)SULFONYL)-2-PROPENENITRILE see THD875

5-METHYL-1-PHENYL-3,4,5,6-TETRAHYDRO-1H-2,5-BENZOXAZOCINE HYDROCHLORIDE see NBS500

2-METHYL-9-PHENYL-2,3,4,9-TETRAHYDRO-1H-INDENO(2,1-c)PYRIDINE TARTRATE see PDD000

3-METHYL-2-PHENYLTETRAHYDRO-2H-1,4-OXAZINE HYDROCHLORIDE see MNV750

2-METHYL-9-PHENYL-2,3,4,9-TETRAHYDRO-1-PYRIDINDENE HYDROCHLORIDE see TEQ700

2-METHYL-9-PHENYL-2,3,4,9-TETRAHYDRO-1-PYRIDINDENE TARTRATE see PDD000

1-METHYL-4-PHENYL-1,2,3,6-TETRAHYDROPYRIDINE see TCV470

N-METHYL-N'-(4'-PHENYL-THIAZOLYL(2'))-HARNSTOFF (GERMAN) see MOA000

1-METHYL-3-(4-PHENYL-2-THIAZOLYL)UREA see MOA000

N-METHYL-N'-(4-PHENYL-2-THIAZOLYL)UREA see MOA000

N-METHYL-N'-(4'-PHENYL-THIAZOLYL(2'))-UREA see MOA000

1-METHYL-4-(PHENYLTHIO)PYRIDINIUM IODIDE see MOA250

N-METHYL-N'-PHENYL THIOUREA see MOA500

1-(1-METHYL-2-((α-PHENYL-o-TOLYL)OXY)ETHYL)PIPERIDINE see MOA600

3-METHYL-1-PHENYLTRIAZENE see MOA725

4-METHYL-5-PHENYL-2-TRIFLUOROMETHOXAZOLIDINE see MOA750

1-METHYL-3-PHENYL-5-(3-(TRIFLUOROMETHYL)PHENYL)-4(1H)-PYRIDINONE see FMQ200

3-METHYLPHENYLUREA see THF750

4-METHYLPHENYLUREA see THG000

3-METHYL-2-PHENYLVALERIC ACID-2-DIETHYLAMINOETHYL ESTER METHYL BROMIDE see VBK000

3-METHYL-2-PHENYLVALERIC ACID DIETHYL(2-HYDROXYETHYL)METHYLAMMONIUM BROMIDE ESTER see VBK000

2-((3-METHYL-2-PHENYLVALERYL)OXY)-N,N-DIETHYL-N-METHYLETHANAMINIUM BROMIDE see VBK000

METHYL PHOSPHATE see DLO800

METHYL PHOSPHATE see TMD250

o-METHYLPHOSPHATE see DLO800

METHYL PHOSPHINE see MOB000

METHYL PHOSPHITE see TMD500

METHYLPHOSPHODITHIOIC ACID-S-(((p-CHLOROPHENYL)THIO)METHYL)-O-METHYL ESTER see MOB250

METHYL PHOSPHONATE see DSG600

METHYLPHOSPHONIC ACID, (2-(BIS(1-METHYLETHYL)AMINO)ETHYL) ETHYL ESTER see MOB275

METHYLPHOSPHONIC ACID DIMETHYL ESTER see DSR400

METHYL PHOSPHONIC DICHLORIDE see MOB399

METHYLPHOSPHONIC DIFLUORIDE see MJD275

METHYLPHOSPHONODITHIOIC ACID O-METHYL ESTER, S-ESTER with 2-MERCAPTO-N-METHYLACETAMIDE see MOB500

METHYLPHOSPHONOFLUORIDIC ACID, 3,3-DIMETHYL-2-BUTYL ESTER see SKS500

METHYLPHOSPHONOFLUORIDIC ACID ISOPROPYL ESTER see IPX000

METHYLPHOSPHONOFLUORIDIC ACID-1-METHYLETHYL ESTER see IPX000

METHYLPHOSPHONOFLUORIDIC ACID 1,2,2-TRIMETHYLPROPYL ESTER see SKS500

METHYLPHOSPHONOTHIOIC ACID-S-(2-(BIS(METHYLETHYL)AMINO)ETHYL)o-ETHYL ESTER see EIG000

METHYLPHOSPHONOTHIOIC ACID-O,S-DIETHYL ESTER see DJR700

METHYLPHOSPHONOTHIOIC ACID-O-ETHYL O-(p-(METHYLTHIO)PHENYL)ESTER see MOB599

METHYLPHOSPHONOTHIOIC ACID-O-ETHYL O-(4-(METHYLTHIO)PHENYL)ESTER (9CI) see MOB599

METHYLPHOSPHONOTHIOIC ACID-O-(4-NITROPHENYL)-O-PHENYL ESTER see MOB699

METHYLPHOSPHONOTHIOIC ACID-O-(p-NITROPHENYL)-O-PHENYL ESTER see MOB699

METHYLPHOSPHONOTHIOIC ACID-O-PHENYL ESTER, O-ESTER with p-HYDROXYBENZONITRILE see MOB750

METHYL PHOSPHONOTHIOIC DICHLORIDE, pyrophoric liquid (DOT) see MOC000

METHYLPHOSPHONOUS DICHLORIDE see MOC250

METHYLPHOSPHONYLDIFLUORIDE see MJD275

METHYLPHOSPHORAMIDIC-2-CHLORO-4-(1,1-DIMETHYLPROPYL)PHENYL METHYL ESTER see DYE200

METHYL-PHOSPHORAMIDOTHIOIC ACID o-(tert-BUTYL-2-CHLOROPHENYL)ESTER see BQU500

N-METHYL-PHOSPHORAMIDOTHIOIC ACID-O-ISOPROPYL ESTER see DYD800

METHYL PHOSPHORODITHIOATE (6CI,7CI) see PHH500

METHYLPHOSPHOROTHIOATE see DUG500

METHYL PHOXIM see MOC275

METHYL PHTHALATE see DTR200

N-METHYLPHTHALIMIDE see MOC300

METHYL-4-PHTHALIMIDO-dl-GLUTARAMATE see MOC500

N-METHYL-2-PHTHALIMIDOGLUTARIMIDE see MOC750

METHYL PHTHALYL ETHYL GLYCOLATE see MOD000

2-METHYL-3-PHYTHYL-1,4-NAPHTHOCHINON (GERMAN) see VTA000

METHYL PICRATE see TMK300

METHYL PINACOLYLOXY PHOSPHORYLFLUORIDE see SKS500

METHYL PINACOLYL PHOSPHONOFLUORIDATE see SKS500

N-METHYL-2-PIPECOLIC ACID, 2,6-DIMETHYLANILIDE see SBB000

N-METHYL-2-PIPECOLIC ACID, 2,6-XYLIDIDE see SBB000

(±)-1-METHYL-2',6'-PIPECOLOXYLIDIDE see SBB000

1-METHYL-2',6'-PIPECOLOXYLIDIDE HYDROCHLORIDE see CBR250

dl-1-METHYL-2',6'-PIPECOLOXYLIDIDE HYDROCHLORIDE see CBR250

1-METHYLPIPERAZINE see MOD250

2-METHYLPIPERAZINE see MOD100

N-METHYLPIPERAZINE see MOD250

4-METHYL-1-PIPERAZINEACETIC, ACID ((5-NITRO-2-FURANYL)METHYLENE)HYDRAZIDE, ACETATE see NDY360

4-METHYL-1-PIPERAZINEACETIC ACID (5-NITROFURFURYLIDENE)HYDRAZIDE see NDY350

4-METHYL-1-PIPERAZINEACETIC, ACID (5-NITROFURFURYLIDENE)HYDRAZIDE ACETATE see NDY360

4-METHYL-1-PIPERAZINEACETIC, ACID (5-NITROFURFURYLIDENE)HYDRAZIDE DIHYDROCHLORIDE see NDY370

4-METHYL-1-PIPERAZINEACETIC, ACID (5-NITROFURFURYLIDENE)HYDRAZIDE HYDROCHLORIDE see NDY380

4-METHYL-1-PIPERAZINEACETIC ACID, (5-NITROFURFURYLIDENE)HYDRAZIDE MALEATE (1:1) see NDY390

4-METHYL-1-PIPERAZINEACETIC ACID 2,6-XYLYL ESTER DIHYDROCHLORIDE see FAC060

o-(2'-(N"-METHYLPIPERAZINE)-N'-ETHYL)-N-(p-CHLOROBENZOYL)-dl-TYROSYLDI-N-PROPYLAMIDE CITRATE see NMV460

4-METHYL-PIPERAZINO-ACETOHYDRAZONE-5-NITROFURFUROL see NDY350

10-(γ-(N'-METHYLPIPERAZINO)PROPYL)-2-TRIFLUOROMETHYLPHENOTHIOZINE see TKE500

(4-METHYLPIPERAZINYLACETYL)HYDRAZONE-5-NITRO-2-FURALDEHYDE see NDY350

N-METHYL-PIPERAZINYL-N'-AETHYL-PHENOTHIAZIN (GERMAN) see MOE250

p-(5-(5-(4-METHYL-1-PIPERAZINYL)-2-BENZIMIDAZOLYL)-2-BENZIMIDAZOLYL)-PHENOL TRIHYDROCHLORIDE see MOD500

4-(5-(4-METHYL-1-PIPERAZINYL)(2,5'-BI-1H-BENZIMIDAZOL)-2'-YL)-PHENOL TRIHYDROCHLORIDE see MOD500

4-(4-METHYL-1-PIPERAZINYLCARBONYL)-1-PHENYL-2-PYRROLIDINONE HYDROCHLORIDE see MOD750

6-(4-METHYL-1-PIPERAZINYL)-11H-DIBENZ(b,e)AZEPINE see HOU059

10-(2-(4-METHYL-1-PIPERAZINYL)ETHYL)PHENOTHIAZINE see MOE250

3-(4-METHYLPIPERAZINYLIMINOMETHYL)-RIFAMYCIN SV see RKP000

8-(4-METHYLPIPERAZINYLIMINOMETHYL) RIFAMYCIN SV see RKP000

8-(((4-METHYL-1-PIPERAZINYL)IMINO)METHYL)RIFAMYCIN SV see RKP000

2-(4-METHYL-1-PIPERAZINYL)-10-METHYL-3,4-DIAZAPHENOXAZINDIHYDROCHLORID (GERMAN) see ASC250

6-(4-METHYL-1-PIPERAZINYL)MORPHANTHRIDINE see HOU059

N-(γ-(4'-METHYLPIPERAZINYL-1')PROPYL)-3-CHLOROPHENOTHIAZINE see PMF500

N-METHYL-PIPERAZINYL-N'-PROPYL-PHENOTHIAZIN (GERMAN) see PCK500

N-(3-(4-METHYL-1-PIPERAZINYL)PROPYL)PHENOTHIAZINE see PCK500

10-(3-(4-METHYL-1-PIPERAZINYL)PROPYL)-10H-PHENOTHIAZINE (9CI) see PCK500

1-(10-(3-(4-METHYL-1-PIPERAZINYL)PROPYL)PHENOTHIAZIN-2-YL)-1-BUTANONE DIMALEATE see BSZ000

10-(3-(4-METHYL-1-PIPERAZINYL)PROPYL)-2-(TRIFLUOROMETHYL) PHENOTHIAZINE see TKE500

10-(3-(4-METHYL-1-PIPERAZINYL)PROPYL)-2-TRIFLUOROMETHYLPHENOTHIAZINE DIHYDROCHLORIDE see TKK250

4-(4-METHYL-1-PIPERAZINYL)-5,6,7,8-TETRAHYDRO-(1)-BENZOTHIENO(2,3-d) PYRIMIDINE HYDROCHLORIDE see MOF750

4-(4-METHYL-1-PIPERAZINYL)THIENO(2,3-d) PYRIMIDINE HYDROCHLORIDE see MOG000

2-(4-METHYL-1-PIPERAZINYL)-11-(p-TOLYL)-10,11-DIHYDROPYRIDAZINO(3,4-b)(1,4)BENZOXAZEPINE see MOG250

2-METHYLPIPERIDINE see MOG750

3-METHYLPIPERIDINE see MOH000

4-METHYLPIPERIDINE see MOH250

N-METHYLPIPERIDINE see MOG500

1-METHYLPIPERIDINE (DOT) see MOG500

1-METHYL-PIPERIDINE-4-CARBONSAURE-O,O-XYLIDID HYDROCHLORID (GERMAN) see MQO250

(1-METHYL-dl-PIPERIDINE-2-CARBOXYLIC ACID)-2,6-DIMETHYLANILIDE HYDROCHLORIDE see CBR250

γ-(4-METHYLPIPERIDINE)-p-FLUOROBUTYROPHENONE HYDROCHLORIDE see FKI000

2-METHYLPIPERIDINE β-NAPHTHOAMIDE see MOH290

2-METHYL-1-PIPERIDINEPROPANOL BENZOATE HYDROCHLORIDE see IJZ000

1-METHYL-3-(PIPERIDINOCARBONYL)PIPERIDINE see MOH310

β-4-METHYLPIPERIDINOETHYL BENZOATE HYDROCHLORIDE see MOI250

2-METHYL-2-(2-PIPERIDINOETHYL)-1,3-BENZODIOXOLE HYDROCHLORIDE see MOI500

N-(1-METHYL-2-PIPERIDINOETHYL)-N-2-PYRIDYLPROPIONAMIDE FUMARATE see PMX250

(±)-6-((2-METHYLPIPERIDINO)METHYL)-5-INDANOL MALEATE see PJI600

β-METHYL-4-PIPERIDINOPHENETHYLAMINE DIHYDROCHLORIDE see MOJ500

2-METHYL-2-(4-PIPERIDINOPHENYL)ETHYLAMINE DIHYDROCHLORIDE see MOJ500

2-METHYL-1-PIPERIDINOPROPANOL, BENZOATE see PIV750

METHYL-4-(3-PIPERIDINOPROPIONYLAMINO)SALICYLATE, METHIODIDE see MOK000

γ-3-METHYLPIPERIDINOPROPYL-p-AMINOBENZOATE HYDROCHLORIDE see MOK500

(2-METHYLPIPERIDINO)PROPYL BENZOATE see PIV750

3-(2-METHYLPIPERIDINO)PROPYL BENZOATE HYDROCHLORIDE see IJZ000

dl-(2-METHYLPIPERIDINO)PROPYL BENZOATE HYDROCHLORIDE see IJZ000

γ-(2-METHYLPIPERIDINO)PROPYL BENZOATE HYDROCHLORIDE see IJZ000

3-(2-METHYLPIPERIDINO)PROPYL-3,4-DICHLOROBENZOATE see MOL250

γ-(2-METHYLPIPERIDINO)PROPYL-3,4-DICHLOROBENZOATE see MOL250

METHYL 10-(3-PIPERIDINOPROPYL)PHENOTHIAZIN-2-YL KETONE see MOL300

METHYL 10-(3-PIPERIDINOPROPYL)PHENOXAZIN-2-YL KETONE see MOL400

2-METHYL-3-PIPERIDINOPYRAZINE MONOSULFATE see MOM750

2-METHYL-3-PIPERIDINOPYRAZINE SULFATE see MOM750

2-METHYL-3-PIPERIDINO-1-p-TOLYLPROPAN-1-ONE see TGK200

2-METHYL-3-PIPERIDINO-1-p-TOLYLPROPAN-1-ONE HYDROCHLORIDE see MRW125

N-(1-METHYL-2-(1-PIPERIDINYL)ETHYL)-N-2-PYRIDINYLPROPANAMIDE-(E)-2-BUTENEDIOATE (1:1) see PMX250

1-METHYL-4-PIPERIDYL-p-AMINOBENZOATE HYDROCHLORIDE see MON000

N-METHYLPIPERIDYL-(4)-BENZHYDRYLAETHER SALZSAUREN SALZE (GERMAN) see LJR000

N-METHYL-3-PIPERIDYL BENZILATE see MON250

N-METHYL-3-PIPERIDYL BENZILATE METHOBROMIDE see CBF000

1-METHYL-4-PIPERIDYL BENZOATE HYDROCHLORIDE see MON500

1-METHYL-4-PIPERIDYL BIS(p-CHLOROPHENOXY)ACETATE see MON600

N-METHYL-3-PIPERIDYL-α-CYCLOHEXYL MANDELATE see MOQ500

N-METHYL-3-PIPERIDYLDIPHENYLGLYCOLATE METHOBROMIDE see CBF000

9-(4'-(N-METHYLPIPERIDYLENE))THIOXANTHENE MALEATE see MOP000

1-METHYL-3-PIPERIDYL ESTER METHOBROMIDE BENZILIC ACID see CBF000

10-(2-(1-METHYL-2-PIPERIDYL)ETHYL)-2-METHYLSULFINYL PHENOTHIAZINE see MON750

10-(2-(1-METHYL-2-PIPERIDYL)ETHYL)-2-(METHYLTHIO)PHENOTHIAZINE see MOO250

10-(2-(1-METHYL-2-PIPERIDYL)ETHYL)-2-METHYLTHIOPHENOTHIAZINE HYDROCHLORIDE see MOO500

(1-METHYL-4-PIPERIDYLIDENE)-9-ANTHROL-9,10-DIHYDRO-10-HYDROCHLORIDE see WAJ000

1-METHYL-3-PIPERIDYLIDENEDI(2-THIENYL)METHANE see BLV000

9-(1-METHYL-4-PIPERIDYLIDENE)THIOXANTHENE see MOO750

9-(N-METHYL-PIPERIDYLIDEN-4)THIOXANE MALEATE see MOP000

9-(1-METHYL-PIPERIDYL-(2)-METHYL)-CARBAZOL (GERMAN) see MOP500

9-(METHYL-2-PIPERIDYL)METHYLCARBAZOLE see MOP500

(N-METHYL-3-PIPERIDYL)METHYLPHENOTHIAZINE see MOQ250

1-(1-METHYL-2-PIPERIDYL)METHYLPHENOTHIAZINE see MOQ000

10-(1-METHYLPIPERIDYL-3-METHYL)PHENOTHIAZINE see MOQ250

10-(1-METHYL-3-PIPERIDYL)METHYL PHENOTHIAZINE see MOQ250

9-(1-METHYL-3-PIPERIDYLMETHYL)THIAXANTHENE HYDROCHLORIDE see THL500

9-((N-METHYL-3-PIPERIDYL)METHYL)-THIOXANTHENHYDROCHLORID (GERMAN) see THL500

1-METHYL-3-PIPERIDYL-α-PHENYLCYCLOHEXANEGLYCOLATE see MOQ500

d-1-METHYL-3-PIPERIDYL-dl-α-PHENYLCYCLOHEXANEGLYCOLATE HYDROCHLORIDE see PMS800

1-METHYL-3-PIPERIDYL PIPERIDINO KETONE see MOH310

γ-(2-METHYLPIPERIDYL)PROPYL BENZOATE see PIV750

(+−)-γ-(2-METHYLPIPERIDYL)PROPYL BENZOATE HYDROCHLORIDE see IJZ000

METHYL PIRIMIPHOS see DIN800

METHYL POTASSIUM see MOR250

METHYLPREDNISOLONE see MOR500

6-α-METHYLPREDNISOLONE see MOR500

METHYLPREDNISOLONE ACETATE see DAZ117

6-METHYLPREDNISOLONE ACETATE see DAZ117

METHYLPREDNISOLONE 21-ACETATE see DAZ117

6-α-METHYLPREDNISOLONE ACETATE see DAZ117

METHYLPREDNISOLONE SODIUM SUCCINATE see USJ100

6-α-METHYLPREDNISOLONE SODIUM SUCCINATE see USJ100

METHYLPREDNISOLONE SULEPTANATE see MOR600

16-β-METHYL-1,4-PREGNADIENE-9-α-FLUORO-11-β,17-α,21-TRIOL- 3,20-DIONE see BFV750

6-METHYL-$\Delta^{4,6}$-PREGNADIEN-17-α-OL-3,20-DIONE ACETATE see VTF000

6-α-METHYL-4-PREGNENE-3,20-DION-17-α-OL ACETATE see MCA000

METHYLPROMAZINE see AFL500

METHYLPROPAMINE see DBA800

2-METHYLPROPANAL see IJS000

2-METHYL-1-PROPANAL see IJS000

2-METHYL-1-PROPANAL OXIME see IJT000

N-METHYLPROPANAMIDE see MOS900

2-METHYLPROPANE see MOR750

2-METHYLPROPANENITRILE see IJX000

α-METHYLPROPANENITRILE see IJX000

2-METHYLPROPANETHIOL see IIX000

2-METHYL-2-PROPANETHIOL see MOS000

METHYL PROPANOATE see MOT000

(Z)-2-METHYLPROPANOIC ACID 3,7-DIMETHYL-2,6-OCTADIENYL ESTER see NCO200

2-METHYLPROPANOIC ACID 2,4-HEXADIENYL ESTER see HCS700

2-METHYLPROPANOIC ACID 1-METHYL-1-(4-METHYL-3-CYCLOHEXEN-1-YL)ETHYL ESTER see MCF515

2-METHYL-PROPANOIC ACID-3-PHENYL-2-PROPENYL ESTER see CMR750

2-METHYL PROPANOL see IIL000

2-METHYLPROPAN-1-OL see IIL000

2-METHYL-1-PROPANOL see IIL000

2-METHYL-2-PROPANOL see BPX000

2-METHYL-2-PROPANOL LITHIUM SALT see LGY100

2-METHYLPROPANOYL CHLORIDE see IJV100

4,4',4"-(1-METHYL-1-PROPANYL-3-YLIDENE)TRIS(2-(1,1-DIMETHYLETHYL)-5-METHYLPHENOL) see MOS100

N-METHYL-N-PROPARGYLBENZYLAMINE see MOS250

N-METHYL-N-PROPARGYL-3-(2,4-DICHLOROPHENOXY)PROPYLAMINE HYDROCHLORIDE see CMY000

METHYL PROPARGYL ETHER see MFN250

2-METHYLPROPENAL (CZECH) see MGA250

2-METHYLPROPENAMIDE see MDN500

METHYL PROPENATE see MGA500

2-METHYLPROPENE see IIC000

2-METHYL-2-PROPENE-1,1-DIOL DIACETATE see AAW250

2-METHYLPROPENENITRILE see MGA750

2-METHYL-1-PROPENE-1-ONE see DSL289

METHYL PROPENOATE see MGA500

METHYL-2-PROPENOATE see MGA500

2-METHYLPROPENOIC ACID see MDN250

2-METHYL-2-PROPENOIC ACID ANHYDRIDE (9CI) see MDN699

2-METHYLPROPENOIC ACID CHLORIDE see MDN899

2-METHYL-2-PROPENOIC ACID 2-(DIMETHYLAMINO)-1-((DIMETHYLAMINO)METHYL)ETHYL ESTER (9CI) see BJI125

2-METHYL-2-PROPENOIC ACID, ETHYL ESTER see EMF000

2-METHYL-2-PROPENOIC ACID METHYL ESTER see MLH750

2-METHYL-2-PROPENOIC ACID METHYL ESTER HOMOPOLYMER see PKB500

2-METHYL-2-PROPENOIC ACID METHYL ESTER, POLYMER with TRIBUTYL((2-METHYL-1-OXO-2-PROPENYL)OXY)STANNANE see OIY000

2-METHYL-2-PROPENOIC ACID METHYL ESTER, POLYMER with TRIBUTYL((2-METHYL-1-OXO-2-PROPENYL)OXY)STANNANE and ((2-METHYL-1-OXO-2-PROPENYL)OXY)TRIPOPYLSTANNANE see OIW000

2-METHYL-2-PROPENOIC ACID-2-METHYLPROPYL ESTER see IIY000

2-METHYL-2-PROPENOIC ACID-3-(TRIMETHOXYSILYL)PROPYL ESTER see TLC250

1-METHYL PROPENOL see MQL250

2-METHYL-2-PROPEN-1-OL see IMW000

2-METHYL-2-PROPENOYL CHLORIDE see MDN899

2-METHYLPROPENYL CHLORIDE see MDN899

METHYL PROPENYL KETONE see PBR500

4-(2-(2-(2-METHYL-1-PROPENYL)-5-NITRO-1H-IMIDAZOL-1-YL)ETHYL)MORPHOLINE see MOS300

N-METHYLPROPHAM see MOS400

METHYL PROPIOLATE see MOS875

2-METHYLPROPIONALDEHYDE see IJS000

N-METHYLPROPIONAMIDE see MOS900

METHYL PROPIONATE see MOT000

2-METHYLPROPIONIC ACID see IJU000

α-METHYLPROPIONIC ACID see IJU000

N-METHYLPROPIONIC ACID AMIDE see MOS900

2-METHYLPROPIONIC ACID, ETHYL ESTER see ELS000

2-METHYLPROPIONITRILE see IJX000

N-METHYLPROPIONSAEUREAMID see MOS900

2-METHYLPROPIONYL CHLORIDE see IJV100

α-METHYLPROPIONYL CHLORIDE see IJV100

3-(2-METHYLPROPIONYL)-5-INDOLECARBONITRILE see COP525

METHYL 5-n-PROPOXY-2-BENZIMIDAZOLE CARBAMATE see OMY700

4-(2-METHYLPROPOXY)BENZOIC ACID 3-(DIETHYLAMINO)-1,2-DIMETHYLPROPYL ESTER see GBU700

1-(1-METHYLPROPOXYCARBONYL)-2-(2-HYDROXYETHYL)PIPERIDINE see MOT100

N-((2-METHYLPROPOXY)METHYL)-2-PROPENAMIDE see IIE100

1-(2-METHYLPROPOXY)-2-PROPANOL (9CI) see IIG000

2-METHYLPROPYL ACETATE see IIJ000

2-METHYL-1-PROPYL ACETATE see IIJ000

METHYLPROPYLACETIC ACID see MQJ750

2-METHYLPROPYL ALCOHOL see IIL000

1-METHYLPROPYLAMINE see BPY000

(2-METHYLPROPYL)-p-AMINOBENZOATE see MOT750

2-METHYL-2-(PROPYLAMINO)-1-PROPANOL BENZOATE HYDROCHLORIDE see OJI600

2-METHYL-2-(PROPYLAMINO)-1-PROPANOL BENZOATE (ester), HYDROCHLORIDE see OJI600

METHYLPROPYLARSINIC ACID see MOT800

METHYL PROPYLATE see MOT000

12-METHYL-7-PROPYLBENZ(a)ANTHRACENE see MOU250

4-(2-METHYLPROPYL)BENZENEACETIC ACID see IJG000

2-METHYLPROPYL BUTYRATE see BSW500

METHYL PROPYL CARBINOL see PBM750

METHYLPROPYLCARBINOL CARBAMATE see MOU500

METHYL-PROPYL-CETONE (FRENCH) see PBN250

3-(1-METHYLPROPYL)-6-CHLOROPHENYL METHYLCARBAMATE see MOU750

2-(1-METHYLPROPYL)CYCLOHEXANONE see MOU800

4-METHYL-N-(4-PROPYLCYCLOHEXYL)BENZAMIDE see MOU820

METHYL PROPYL DIKETONE see HEQ200

6-(1-METHYL-PROPYL)-2,4-DINITROFENOL (DUTCH) see BRE500

(6-(1-METHYL-PROPYL)-2,4-DINITRO-FENYL)-3,3-DIMETHYL ACRYLAAT (DUTCH) see BGB500

2-(1-METHYLPROPYL)-4,6-DINITROPHENOL see BRE500

2-(1-METHYL-N-PROPYL) 4,6-DINITROPHENOL AMMONIUM SALT see BPG250

2-(1-METHYL-N-PROPYL)-4,6-DINITROPHENOL TRIETHANOLAMINE SALT see BRE750

2-(1-METHYLPROPYL)-4,6-DINITROPHENYL ACETATE see ACE500

2-(1-METHYLPROPYL)-4,6-DINITROPHENYL-β,β-DIMETHACRYLATE see BGB500

(6-(1-METHYL-PROPYL)-2,4-DINITRO-PHENYL)-3,3-DIMETHYL ACRYLAT (GERMAN) see BGB500

2-(1-METHYL-2-PROPYL)-4,6-DINITROPHENYL ISOPROPYLCARBONATE see CBW000

β-METHYLPROPYL ETHANOATE see IIJ000

2-METHYL-2-PROPYLETHANOL see AOK750

METHYL PROPYL ETHER see MOU830

METHYL n-PROPYL ETHER see MOU830

2-METHYLPROPYL HEXANOATE see IIT000

1-METHYLPROPYL 2-(2-HYDROXYETHYL)-1-PIPERIDINECARBOXYLATE see MOT100

2,2'-(2-METHYLPROPYLIDENE)BIS(4,6-DIMETHYLPHENOL) see IIU100

N,N''-(2-METHYLPROPYLIDENE)BISUREA (9CI) see IIV000

2-METHYLPROPYL ISOBUTYRATE see IIW000

2-METHYLPROPYL ISOVALERATE see ITA000

METHYL-n-PROPYL KETONE see PBN250

METHYL PROPYL KETONE (ACGIH, DOT) see PBN250

2-METHYLPROPYL METHACRYLATE see IIY000

2-METHYLPROPYL-3-METHYLBUTYRATE see ITA000

METHYL-N-PROPYLNITROSAMINE see MNA000

METHYLPROPYLNITROSOAMINE see MNA000

N-(2-METHYLPROPYL)-N-NITROSOUREA see IJF000

8-METHYL-3-(2-PROPYLPENTANOYLOXY)TROPINIUM BROMIDE see LJS000

2-(1-METHYLPROPYL)PHENYL METHYLCARBAMATE see MOV000

m-(1-METHYLPROPYL)PHENYLMETHYLCARBAMATE see BSG250

2-METHYL-2-PROPYL-1,3-PROPANEDIOL BUTYLCARBAMATE CARBAMATE see MOV500

2-METHYL-2-PROPYL-1,3-PROPANEDIOL CARBAMATE ISOPROPYLCARBAMATE see IPU000

2-METHYL-2-N-PROPYL-1,3-PROPANEDIOL DICARBAMATE see MQU750

2-METHYLPROPYLPROPANOIC ACID-2-METHYLPROPYL ESTER (9CI) see IIW000

5-(1-METHYLPROPYL)-5-(2-PROPENYL)-2,4,6(1H,3H,5H)-PYRIMIDINETRIONE (9CI) see AFY500

5-(2-METHYLPROPYL)-5-(2-PROPENYL)-2,4,6(1H,3H,5H)-PYRIMIDINETRIONE (9CI) see AGI750

2-METHYLPROPYL PROPIONATE see PMV250

6-METHYL-2-PROPYL-4-PYRIMIDINYL DIMETHYL CARBAMATE see PNQ250

METHYL (5-(PROPYLSULFONYL)-1H-BENZIMIDAZOL-2-YL) CARBAMATE see PNV900

METHYL 5-(PROPYLTHIO)-2-BENZIMIDAZOLECARBAMATE see VAD000

S-(1-METHYLPROPYL) o-(2,2,2-TRIFLUOROETHYL)-FORMYLMETHYLPHOSPHORAMIDODITHIOATE see MOV100

2-METHYL-2-PROPYLTRIMETHYLENE BUTYLCARBAMATE CARBAMATE see MOV500

2-METHYL-2-PROPYLTRIMETHYLENE CARBAMATE see MQU750

3-METHYL-5-PROPYL-1,2,4-TRIOXOLANE see HFC500

METHYL PROPYNOATE see MOS875

N-METHYL-N-2-PROPYNYLBENZYLAMINE see MOS250

N-METHYL-N-(2-PROPYNYL)BENZYLAMINE HYDROCHLORIDE see BEX500

1-METHYL-2-PROPYNYL-m-CHLOROCARBANILATE see CEX250

1-METHYL-2-PROPYNYL-m-CHLOROPHENYLCARBAMATE see CEX250

1-METHYLPROPYNYL 3-CHLOROPHENYLCARMATE see CEX250

1-METHYLPROPYNYL ESTER of 3-CHLOROPHENYLCARBAMIC ACID see CEX250

2-METHYL-5-(2-PROPYNYL)-3-FURYLMETHYL-cis-trans-CHRYSANTHEMATE see PMN700

6-*α*-METHYL-17-(1-PROPYNYL)TESTOSTERONE see DRT200

6-*α*-METHYL-17-*α*-PROPYNYLTESTOSTERONE see DRT200

15(R)-METHYLPROSTAGLANDIN E2 see AQT575

15(R)-15-METHYLPROSTAGLANDIN E2 see AQT575

15(s)-15-METHYL-PROSTAGLANDIN E2 see MOV800

15-METHYLPROSTAGLANDIN F2-α see CCC100

15(S)-METHYLPROSTAGLANDIN F2-α see CCC100

15(S)-15-METHYL-PROSTAGLANDIN F2-α see CCC100

(15S)-15-METHYLPROSTAGLANDIN F2-α see CCC100

15-METHYL-PROSTAGLANDIN-F2-α-METHYL ESTER see MLO301

15(s)15-METHYL-PROSTAGLANDIN-F2-α-METHYL ESTER see MLO301

15(S)15-METHYL PROSTAGLANDIN F2-α TROMETHAMINE see CCC110

N-METHYL-4-PROTOADAMANTANEAMINE HYDROCHLORIDE see MOW000

N-METHYL-4-PROTOADAMANTANEMETHANAMINE MALEATE see MOW250

METHYL PROTOANEMONIN see MOW500

METHYLPROTOCATECHUALDEHYDE see VFK000

S-METHYLPSEUDOTHIOUREA SULFATE see MPV750

9-METHYL PTEROYLGLUTAMIC ACID see BMM125

x-METHYLPTEROYLGLUTAMIC ACID see MKG275

2-METHYL-4-(1H-PURIN-6-YLAMINO)-1-BUTANOL see RAF400

2-METHYLPYRAZINE see MOW750

3-METHYLPYRAZOLE see MOX010

4-METHYLPYRAZOLE see MOX000

5-METHYLPYRAZOLE see MOX010

3(5)-METHYLPYRAZOLE see MOX010

3-METHYL-2-PYRAZOLIN-5-ONE see MOX100

3-METHYL-PYRAZOLON-(5) see MOX100

METHYLPYRAZOLYL DIETHYLPHOSPHATE see MOX250

3-METHYLPYRAZOLYL-5-DIETHYLPHOSPHATE see MOX250

5-METHYL-1H-PYRAZOL-3-YL DIMETHYLCARBAMATE see DQZ000

1-METHYLPYRENE see MOX875

3-METHYLPYRENE see MOX875

α-METHYL-1-PYRENEMETHANOL see HKY700

α-METHYL-2-PYRENEMETHANOL see MOT000

2-METHYLPYRIDINE see MOY000

3-METHYLPYRIDINE see PIB920

4-METHYLPYRIDINE see MOY250

α-METHYLPYRIDINE see MOY000

3-METHYLRODANIN see MPH750
METHYLROSANILINCHLORID see AOR500
METHYLROSANILINE see TJK000
METHYLROSANILINE CHLORIDE see AOR500
METHYLROSANILINUM CHLORATUM see AOR500
METHYL SALICYLATE see MPI000
METHYLSALICYLATE METHYL ETHER see MLH800
4-METHYLSALINOMYCIN see NBO600
3-METHYLSALYCILIC ACID see CNX625
N-METHYLSARCOSINE HYDROCHLORIDE see MPI100
METHYLSCOPOLAMINE BROMIDE see SBH500
METHYLSCOPOLAMINE HYDROBROMIDE see SBH500
N-METHYLSCOPOLAMINE METHOSULFATE see DAB750
METHYLSCOPOLAMINE METHYL SULFATE see DAB750
N-METHYLSCOPOLAMINE METHYL SULFATE see DAB750
METHYL SCOPOLAMINE NITRATE see SBH520
N-METHYLSCOPOLAMMONIUM BROMIDE see SBH500
METHYLSCOPOLAMMONIUM METHYLSULFATE see DAB750
METHYL SELENAC see SBQ000
METHYL SELENIDE see DUB200
METHYL SELENIUM see DUB200
METHYLSELENO-2-BENZOIC ACID see MPI200
o-(METHYLSELENO)BENZOIC ACID see MPI200
o-(METHYLSELENO)BENZOIC ACID SODIUM SALT see MPI205
(METHYLSELENO)TRIS(DIMETHYLAMINO)PHOSPHONIUM IODIDE see MPI225
METHYLSENFOEL (GERMAN) see ISE000
METHYLSERGIDE BIMALEATE see MQP500
1-METHYLSILACYCLOPENTA-2,4-DIENE see MPI600
METHYLSILANE see MPI625
METHYLSILATRAN see MPI650
METHYLSILATRANE see MPI650
1-METHYLSILATRANE see MPI650
METHYL SILICATE see MPI750
METHYL SILICATE see TDJ100
METHYL SILICATE 51 see TDJ100
METHYL SILICONE see PJR000
METHYLSILVER see MPI800
METHYLSILYLIDYNETRIS(2-ETHYLHEXANOATE) see TNI100
O-METHYLSINAPIC ACID see CMQ100
METHYL SODIUM see MPJ000
METHYL STEARATE see MJW000
o-METHYLSTERIGMATOCYSTIN see MPJ050
METHYL STIBINE see MPJ100
2-METHYL-4-STILBENAMINE see MPJ250
3-METHYL-4-STILBENAMINE see MPJ500
METHYL STREPONIGRIN see SMA500
β-METHYLSTREPTOZOTOCIN see MPJ800
α-METHYLSTYREEN (DUTCH) see MPK250
METHYLSTYRENE see VQK650
2-METHYLSTYRENE see VQK660
3-METHYLSTYRENE see VQK670
m-METHYLSTYRENE see VQK670
o-METHYLSTYRENE see VQK660
p-METHYLSTYRENE see VQK700
α-METHYL STYRENE see MPK250
4-METHYLSTYRENE OXIDE see MPK750
p-METHYLSTYRENE OXIDE see MPK750
α-METHYL-STYROL (GERMAN) see MPK250
METHYL trans-STYRYL KETONE see BAY275

METHYL STYRYLPHENYL KETONE see MPL000
6-METHYL-7-SULFAMIDO-THIOCHROMAN-1,1-DIOXIDE see MQQ050
5-METHYL-3-SULFANILAMIDOISOXAZOLE see SNK000
5-METHYL-2-SULFANILAMIDO-1,3,4-THIADIAZOLE see MPQ750
METHYL SULFANILYL CARBAMATE see SNQ500
METHYL SULFATE (DOT) see DUD100
METHYLSULFAZIN see ALF250
METHYL SULFIDE (DOT) see TFP000
METHYL SULFIDE compound with BORANE (1:1) see MPL250
2-((METHYLSULFINYL)ACETYL)PYRIDINE see MPL500
METHYLSULFINYL ETHYLTHIAMINE DISULFIDE see MPL600
METHYLSULFINYLMETHANE see DUD800
METHYL SULFOCHLORIDE see MDR300
METHYL SULFOCYANATE see MPT000
METHYLSULFONAL see BJT750
9-(p-(METHYLSULFONAMIDO)ANILINO)-3-ACRIDINE CARBAMIC ACID METHYL ESTER see MPL750
N-(9-(p-(METHYLSULFONAMIDO)ANILINO)ACRIDIN-3-YL)ACETAMIDE METHANESULFONATE see MPM000
METHYLSULFONIC ACID, ETHYL ESTER see EMF500
N-(9-((4-((METHYLSULFONYL)AMINO)PHENYL)AMINO)-3-ACRIDINYL)ACETAMIDE METHANESULFONATE see MPM000
(METHYLSULFONYL)BENZENE see PFR375
2-METHYLSULFONYL-10-(3-(4-CARBAMOYLPIPERIDINO)PROPYL)PHENOTHIAZINE see MQR000
2-METHYLSULFONYL-10-(3-(4'-CARBAMOYLPIPERIDINO)PROPYL)-PHENOTHIAZINE see MPM750
METHYLSULFONYL CHLORAMPHENICOL see MPN000
METHYLSULFONYL CHLORIDE see MDR300
2-METHYLSULFONYL-o-(N-METHYL-CARBAMOYL)-BUTANON-(3)-OXIM (GERMAN) see SOB500
1-METHYLSULFONYL-3-(1-METHYL-5-NITRO-2-IMIDAZOLYL)-2-IMIDAZOLIDINONE see SAY950
1,4-(METHYLSULFONYLOXYETHYLAMINO)-1,4-DIDEOXY-ERYTHRIOLDIMETHYLSULFONATE see LJD500
((METHYLSULFONYL)OXY)TRIBUTYLSTANNANE see TIF000
((METHYLSULFONYL)OXY)TRIPHENYLSTANNANE see TMW500
1-(3-(2-(METHYLSULFONYL)PHENOTHIAZIN-10-YL)PROPYL)ISONIPECOTAMIDE see MQR000
1-(3-(2-(METHYLSULFONYL)PHENOTHIAZIN-10-YL)PROPYL)-4-PIPERIDINE CARBOXAMIDE see MQR000
1-(3-(2-(METHYLSULFONYL)-10H-PHENOTHIAZIN-10-YL)PROPYL)-4-PIPERIDINE CARBOXAMIDE see MQR000
2-(4-(METHYLSULFONYL)PHENYL)-IMIDAZO(1,2-a)PYRIDINE see ZUA200
METHYL SULFOXIDE see DUD800
5-METHYL-3-SULPHANIL-AMIDOISOXAZOLE see SNK000
METHYL SULPHIDE see TFP000
3-METHYLSULPHINYLPROPYLISOTHIOCYANATE see MPN100

METHYLSULPHONAL see BJT750
METHYLSYSTOX see DAO800
METHYL SYSTOX see MIW100
METHYL TARTRONIC ACID see MPN250
o-(METHYLTELLURO)BENZOIC ACID see ADF600
o-(METHYLTELLURO)BENZOIC ACID see MPN275
METHYLTERT-BUTYL KETONE see DQU000
17-METHYLTESTOSTERON see MPN500
METHYLTESTOSTERONE see MPN500
17-METHYLTESTOSTERONE see MPN500
17-α-METHYLTESTOSTERONE see MPN500
Δ¹-17-METHYLTESTOSTERONE see DAL300
Δ¹-17-α-METHYLTESTOSTERONE see DAL300
2-METHYL-3,4,5,6-TETRABROMOPHENOL see TBJ500
4-o-METHYL-12-o-TETRADECANOYLPHORBOL-13-ACETATE see MPN600
2-METHYL-3,3,4,5-TETRAFLUORO-2-BUTANOL see MPN750
10-METHYL-1,2-TETRAHYDRO-1,2:5,6-BENZACRIDINE see MPO400
10-METHYL-1,2-TETRAHYDRO-1,2:7,8-BENZACRIDINE see MPO390
2-METHYL-1,2,3,6-TETRAHYDROBENZALDEHYDE see MPO000
6-METHYL-1,2,3,4-TETRAHYDROBENZ(a)ANTHRACENE see MPO250
4-METHYL-1',2',3',4'-TETRAHYDRO-1,2-BENZANTHRACENE see MPO250
1-METHYL-1,2,3,4-TETRAHYDRO-β-CARBOLINE-3-CARBOXYLIC ACID see MPO300
7-METHYL-1,2,3,4-TETRAHYDRODIBENZ(c,h)ACRIDINE see MPO390
14-METHYL-8,9,10,11-TETRAHYDRODIBENZ(a,h)ACRIDINE see MPO400
6-METHYL-2,3,5,7-TETRAHYDRO-6H-p-DITHIINO-(2,3-C)PYRROLE-5,7-DIONE see MJJ250
METHYLTETRAHYDROFURAN see MPO500
2-METHYLTETRAHYDROFURAN (CZECH) see MPO500
METHYL TETRAHYDRO-5-HYDROXY-4-METHYL-3-FURYL KETONE see BMK290
N-METHYL-1,2,3,4-TETRAHYDROISOQUINOLINE HYDROCHLORIDE see MPO750
METHYL 1,4,5,6-TETRAHYDRO-2-METHYLCYCLOPENTA(b)PYRROL-3-YL KETONE see MPO800
METHYL-1,2,5,6-TETRAHYDRO-1-METHYLNICOTINATE see AQT750
METHYL-1,2,5,6-TETRAHYDRO-1-METHYLNICOTINATE, HYDROBROMIDE see AQU000
2-METHYL-2-(4-(1,2,3,4-TETRAHYDRO-1-NAPHTHALENYL)PHENOXY)PROPANOIC ACID see MCB500
2-METHYL-2-(4-(1,2,3,4-TETRAHYDRO-1-NAPHTHYL)PHENOXY)PROPANOIC ACID see MCB500
2-METHYL-2-(p-(1,2,3,4-TETRAHYDRO-1-NAPHTHYL)PHENOXY)PROPIONIC ACID see MCB500
α-METHYL-α-(p-1,2,3,4-TETRAHYDRONAPHTH-1-YLPHENOXY)PROPIONIC ACID see MCB500
N-METHYL-Δ-TETRAHYDRONICOTINIC ACID METHYL ESTER see AQT750
METHYLTETRAHYDROPHTHALIC ANHYDRIDE see MPP000

N-METHYL-1,2,3,6-
TETRAHYDROPHTHALIMIDE see MIS250
N-METHYL-3,4,5,6-
TETRAHYDROPHTHALIMIDE see MPP250
N-METHYLTETRAHYDROPYRIDINE-β-
CARBOXYLIC ACID METHYL ESTER see
AQT750
N-METHYLTETRAHYDROPYRROLE see
MPB250
2-METHYL-1,2,3,4-
TETRAHYDROQUINOLINE-5-METHANOL
see MPP300
N-METHYL-
TETRAHYDROTHIAMIDINTHIONE
ACETIC ACID see MPP750
2-METHYL-3-(3,7,11,15-TETRAMETHYL-
2,6,10,14-HEXADECATETRAENYL)-1,4-
NAPHTHOQUINONE see VTA650
2-METHYL-3-(3,7,11,15-TETRAMETHYL-2-
HEXADECENYL)-1,4-
NAPHTHALENEDIONE see VTA000
2-METHYL-3-trans-TETRAMETHYL-1,4-
NAPHTHQUINONE see VTA650
N-METHYL-N,2,4,6-TETRANITROANILINE
see TEG250
N-METHYL-N,2,4,6-
TETRANITROBENZENAMINE see TEG250
1-METHYL-1H-TETRAZOLE-5-THIOL see
MPQ250
α-METHYL TETRONIC ACID see MPQ500
METHYL 5-(2-THENOYL)-2-
BENZIMIDAZOLECARBAMATE see OJD100
METHYLTHEOBROMIDE see CAK500
1-METHYLTHEOBROMINE see CAK500
7-METHYLTHEOPHYLLINE see CAK500
N¹-(5-METHYL-1,3,4-THIADIAZOL-2-YL)-
SULFANILAMIDE see MPQ750
N¹-(5-METHYL-1,3,4-THIADIAZOL-2-YL)-
SULFANILAMIDE see MPQ750
N-METHYL-3-THIA-2-METHYL-
VALERAMID DER O,O-
DIMETHYLTHIOLPHOSPHORSAEURE
(GERMAN) see MJG500
METHYL 3-THIANAPHTHENYL KETONE
see MPQ900
2-METHYLTHIAZOLIDINE see MPR000
METHYL-2-THIAZOLIDINE (FRENCH) see
MPR000
5-METHYL-2-(6-(3-
THIAZOLIDINYL)HEXYLOXY)PYRIDINE
DIHYDROCHLORIDE see MPA250
N-(3-METHYL-2-
THIAZOLIDINYLIDENE)NICOTINAMIDE
see MPR250
METHYL-2 Δ-2 THIAZOLINE see DLV900
α-METHYL-4-(2-
THIENYLCARBONYL)BENZENEACETIC
ACID see TEN750
METHYL (5-(2-THIENYLCARBONYL)-1H-
BENZIMIDAZOLE-2-YL)CARBAMATE see
OJD100
METHYLTHIENYLCETONE see MPR300
2-METHYLTHIIRANE see PNL750
METHYLTHIOACETALDEHYDE-o-
(CARBAMOYL)OXIME see MPS250
1-(METHYLTHIO)ACETALDEHYDE
OXIME see MPS270
2-METHYLTHIO-ACETALDEHYD-O-
(METHYLCARBAMOYL)-OXIM (GERMAN)
see MDU600
1-(METHYLTHIO)ACETALDOXIME see
MPS270
METHYL(THIOACETAMIDO) MERCURY
see MPS300
METHYL THIOACETOHYDROXAMATE see
MPS270
l-γ-METHYLTHIO-α-AMINOBUTYRIC
ACID see MDT750
4-(METHYLTHIO)ANILINE see AMS675
m-(METHYLTHIO)ANILINE see ALX100
3-(METHYLTHIO)BENZENAMINE see
ALX100

4-(METHYLTHIO)BENZENAMINE see
AMS675
(METHYLTHIO)BENZENE see TFC250
2-METHYLTHIOBENZIMIDAZOLE see
MPS500
6-METHYL-2-THIO-2H-1,3-BENZOXAZINE-
2,4(3H)-DIONE see MPS600
2-METHYLTHIO-4,6-
BIS(ISOPROPYLAMINO)-s-TRIAZINE see
BKL250
6-METHYLTHIOCHROMAN-7-
SULFONAMIDE 1,1-DIOXIDE see MQQ050
4-(METHYLTHIO)-m-CRESOL see MLX750
4-(METHYLTHIO)-m-CRESOL-O-ESTER
with O-ETHYL
METHYLPHOSPHORAMIDOTHIOATE see
ENI500
METHYL THIOCYANATE see MPT000
4-METHYLTHIO-3,5-DIMETHYLPHENYL
METHYLCARBAMATE see DST000
17-β-(METHYLTHIO)ESTRA-1,3,5(10)-
TRIEN-3-OL see MPT100
2-(METHYLTHIO)-ETHANETHIOL-O,O-
DIMETHYL PHOSPHOROTHIOATE see
MIW250
2-(METHYLTHIO)-ETHANETHIOL-S-
ESTER with O,O-DIMETHYL
PHOSPHOROTHIOATE see MIW250
2-(METHYLTHIO)ETHANOL ACRYLATE
see MPT250
2-METHYLTHIOETHYL ACRYLATE see
MPT250
2-METHYLTHIO-4-ETHYLAMINO-6-tert-
BUTYLAMINO-s-TRIAZINE see BQC750
2-(2-
METHYLTHIOETHYLAMINO)ETHYLGUA
NIDINE SULFATE see MPT300
2-METHYLTHIO-4-ETHYLAMINO-6-
ISOPROPYLAMINO-s-TRIAZINE see
MPT500
5-(2-(METHYLTHIO)ETHYL)HYDANTOIN
see MDT800
(+−)-5-(2-
(METHYLTHIO)ETHYL)HYDANTOIN see
MDT800
dl-5-(2-
(METHYLTHIO)ETHYL)HYDANTOIN see
MDT800
5-(β-(METHYLTHIO)ETHYL)HYDANTOIN
see MDT800
5-(2-(METHYLTHIO)ETHYL)-2,4-
IMIDAZOLIDINEDIONE see MDT800
7-(2-METHYLTHIO)ETHYL)-
THEOPHYLLINE see MDN100
7-(β-
METHYLTHIOETHYL)THEOPHYLLINE see
MDN100
METHYLTHIOGLYCOLATE see MLE750
2-METHYLTHIO-6-HYDROXY-8-
THIAPURINE see MPT600
METHYLTHIOINOSINE see MPU000
6-METHYLTHIOINOSINE see MPU000
2-METHYLTHIO-4-ISOPROPYLAMINO-6-
METHYLAMINO-s-TRIAZINE see INR000
METHYLTHIOKYANAT see MPT000
METHYLTHIOMETHANE see TFP000
S-METHYL THIOMETHANESULFINATE
see MLH765
3-(METHYLTHIO)-o-
((METHYLAMINO)CARBONYL)OXIME-2-
BUTANONE see MPU250
2-(METHYLTHIO)-4-(METHYLAMINO)-6-
(ISOPROPYLAMINO)-s-TRIAZINE see
INR000
2-METHYLTHIO-o-(N-
METHYLCARBAMOYL)-BUTANONOXIM-3
(GERMAN) see MPU250
8-β-((METHYLTHIO)METHYL)-6-
PROPYLERGOLINE
METHANESULFONATE see MPU500
3-METHYL-2,5-THIOMORPHOLINEDIONE
2-(o-

((METHYLAMINO)CARBONYL)OXIME) see
MPU600
METHYLTHIONINE CHLORIDE see BJI250
METHYLTHIONIUM CHLORIDE see BJI250
(E)-1-(2-(METHYLTHIO)-1-(2-
(PENTYLOXY)PHENYL)ETHENYL)-1H-
IMIDAZOLE MONOHYDROCHLORIDE see
NCP525
METHYL THIOPHANATE see PEX500
METHYLTHIOPHANATE-MANEB mixture
see MAP300
2-METHYLTHIOPHENE see MPV000
3-METHYLTHIOPHENE see MPV250
2-METHYLTHIOPHENOL see TGP000
3-METHYLTHIOPHENOL see TGO800
4-(METHYLTHIO)PHENOL see MPV300
4-METHYLTHIOPHENOL see TGP250
m-METHYLTHIOPHENOL see TGO800
o-METHYLTHIOPHENOL see TGP000
p-(METHYLTHIO)PHENOL see MPV300
p-METHYLTHIOPHENOL see TGP250
4-METHYLTHIOPHENYLDIMETHYL
PHOSPHATE see PHD250
METHYLTHIOPHOS see MNH000
3-(METHYLTHIO)PROPANAL see MPV400
3-(METHYLTHIO)PROPANAL see TET900
3-(METHYLTHIO)PROPIONALDEHYDE see
MPV400
3-(METHYLTHIO)PROPIONALDEHYDE see
TET900
β-(METHYLTHIO)PROPIONALDEHYDE
see MPV400
β-(METHYLTHIO)PROPIONALDEHYDE
see TET900
2-METHYLTHIO-PROPIONALDEHYD-O-
(METHYLCARBAMOYL)-OXIM (GERMAN)
see MDU600
METHYLTHIOPROPIONIC
CYANOHYDRIN see HMR550
METHYL THIOPSEUDOUREA SULFATE
see MPV750
METHYL THIOPSEUDOUREA SULFATE
see MPV750
S-METHYLTHIOPSEUDOUREA SULFATE
see MPV790
2-METHYL-2-THIOPSEUDOUREA
SULFATE see MPV750
2-METHYL-2-THIOPSEUDOUREA
SULFATE see MPV790
2-METHYL-2-THIOPSEUDOUREA
SULFATE (1:1) SODIUM SALT see MPV800
6-(METHYLTHIO)PURINE
RIBONUCLEOSIDE see MPU000
6-METHYLTHIOPURINE RIBOSIDE see
MPU000
6-METHYL-2-THIO-2,4-
(1H3H)PYRIMIDINEDIONE see MPW500
5-METHYLTHIO-(1,2,5)THIADIAZOLO(3,4-
D)PYRIMIDIN-7(6H)-ONE see MPT600
METHYLTHIOURACIL see MPW500
6-METHYLTHIOURACIL see MPW500
4-METHYL-2-THIOURACIL see MPW500
6-METHYL-2-THIOURACIL see MPW500
METHYL THIOUREA see MPW600
1-METHYLTHIOUREA see MPW600
S-METHYLTHIOUREA SULFATE see
MPV750
S-METHYLTHIOURONIUM SULFATE see
MPV790
1-METHYL-3-(THIOXANTHEN-9-
YLMETHYL)-1-PIPERIDINE
HYDROCHLORIDE see THL500
METHYLTHIOXOARSINE see MGQ750
5-METHYL-6-THIOXOTETRAHYDRO-3-
THIADIAZINEACETIC ACID see MPP750
4-(METHYLTHIO)-3,5-XYLENOL-O-ESTER
with O,O-DIETHYL PHOSPHOROTHIOATE
see DJR800
4-(METHYLTHIO)-3,5-XYLENOL
METHYLCARBAMATE see DST000
4-(METHYLTHIO)-3,5-XYLYL
METHYLCARBAMATE see DST000

METHYL THIRAM see TFS350
METHYL THIURAMDISULFIDE see TFS350
S-METHYLTHIURONIUM SULFATE see MPV750
O-METHYLTHYMOL see MPW650
METHYL THYMOL ETHER see MPW650
METHYL THYMYL ETHER see MPW650
METHYL TIGLATE see MPW700
METHYLTIN TRICHLORIDE see MQC750
METHYL-4-TOLUATE see MPX850
METHYL-p-TOLUATE see MPX850
METHYL-α-TOLUATE see MHA500
METHYL TOLUENE see XGS000
o-METHYLTOLUENE see XHJ000
p-METHYLTOLUENE see XHS000
METHYL TOLUENE-4-SULFONATE see MLL250
METHYL-p-TOLUENESULFONATE see MLL250
α-METHYL-α-TOLUIC ALDEHYDE see COF000
2-METHYL-p-TOLUIDINE see XMS000
4-METHYL-o-TOLUIDINE see XMS000
5-METHYL-o-TOLUIDINE see XNA000
6-METHYL-m-TOLUIDINE see XNA000
2-METHYL-p-TOLUIDINE HYDROCHLORIDE see XOJ000
4-METHYL-o-TOLUIDINE HYDROCHLORIDE see XOJ000
5-METHYL-o-TOLUIDINE HYDROCHLORIDE see XOS000
6-METHYL-m-TOLUIDINE HYDROCHLORIDE see XOS000
1-METHYL-5-(p-TOLUOYL)-2-PYRROLEACETIC ACID see SKJ340
1-METHYL-5-p-TOLUOYL-PYRROLE-2-ACETIC ACID see TGJ850
2-METHYL-3-o-TOLYL-6-AMINO-CHINAZOLINON-4 (GERMAN) see AKM125
N-METHYL-p-(m-TOLYLAZO)ANILINE see MPY000
N-METHYL-p-(o-TOLYLAZO)ANILINE see MPY250
N-METHYL-p-(p-TOLYLAZO)ANILINE see MPY500
2-METHYL-4-((o-TOLYL)AZOANILINE) HYDROCHLORIDE see MPY750
2'-METHYL-4'-(o-TOLYLAZO)OXANILIC ACID see OLG000
N-METHYL-2-(α-(2-TOLYLBENZYL)OXY)ETHYLAMINE HYDROCHLORIDE see TGJ250
8-METHYL-3-(α-(o-TOLYL)BENZYLOXY)TROPANIUM IODIDE see MPZ000
1-METHYL-4-(α-o-TOLYLBENZYL)PIPERAZINE DIHYDROCHLORIDE see MIQ725
N-METHYL-N-(m-TOLYL)CARBAMOTHIOIC ACID-(1,2,3,4-TETRAHYDRO-1,4-METHANONAPHTHALEN-6-YL) ESTER see MQA000
METHYL 3-(m-TOLYLCARBAMOYLOXY)PHENYLCARBAMATE see MEG250
METHYL-p-TOLYLCARBINOL see TGZ000
2-METHYL-3-o-TOLYL-4(3H)-CHINAZOLINON (GERMAN) see QAK000
2-METHYL-3-o-TOLYL-4(3H)-CHINAZOLONE see QAK000
2-METHYL-3-TOLYLCHINAZOLON-4 HYDROCHLORIDE (GERMAN) see MDT250
2-METHYL-3-(o-TOLYL)-3,4-DIHYDRO-4-(QUINAZOLINONE) see QAK000
2-METHYL-3-(o-TOLYL)-3,4-DIHYDRO-4-QUINAZOLINONE see QAK000
METHYL-p-TOLYL ETHER see MGP000
METHYL-p-TOLYL KETONE see MFW250
2-METHYL-3-TOLYL-4-OXYBENZDIAZINE see QAK000

3-METHYL-1-p-TOLYL-PYRAZOLIN-5-ONE see THA300
2-METHYL-3-o-TOLYL-4(3H)-QUINAZOLINONE see QAK000
2-METHYL-3-o-TOLYL-4(3H)-QUINAZOLINONE HYDROCHLORIDE see MDT250
2-METHYL-3-(2-TOLYL)QUINAZOL-4-ONE see QAK000
2-METHYL-3-o-TOLYL-4-QUINAZOLONE see QAK000
2-METHYL-3-(o-TOLYL)-4-QUINAZOLONE HYDROCHLORIDE see MDT250
2-METHYL-3-o-TOLYL-6-SULFAMYL-7-CHLORO-1,2,3,4-TETRAHYDRO-4-QUINAZOLINONE see ZAK300
3-METHYL-5-(p-TOLYLSULFONYL)-1,2,4-THIADIAZOLE see MQB100
3-METHYL-1-(p-TOLYL)-TRIAZENE see MQB250
METHYL TOSYLATE see MLL250
METHYL-p-TOSYLATE see MLL250
4-o-METHYL-TPA see MPN600
METHYLTRETAMINE see TNK000
METHYLTRIACETOXYSILANE see MQB500
5-(3-METHYL-1-TRIAZENO)IMIDAZOLE-4-CARBOXAMIDE see MQB750
5-METHYL-1,2,4-TRIAZOLE(3,4-b)BENZOTHIAZOLE see MQC000
5-METHYL-s-TRIAZOLO(3,4-b)BENZOTHIAZOLE see MQC000
2'-METHYL-1,2:4,5:8,9-TRIBENZOPYRENE see MJA750
METHYLTRICAPRYLYLAMMONIUM CHLORIDE see MQH000
METHYL TRICHLORIDE see CHJ500
METHYL TRICHLOROACETATE see MQC150
N-METHYL-2,4,5-TRICHLOROBENZENESULFONAMIDE see MQC250
METHYLTRICHLOROMETHANE see MIH275
2-((N-METHYL-N-TRICHLOROMETHANESULFENYL)CARBAMOYLOXIMINO)-1,4-DITHIANE see MQC300
4-((N-METHYL-N-TRICHLOROMETHANESULFENYL)CARBAMOYLOXIMINO)-1,3-DITHIOLANE see MQC320
METHYLTRICHLOROPLATINATE(2-) see MQC400
METHYLTRICHLOROSILANE see MQC500
METHYLTRICHLOROSTANNANE see MQC750
METHYLTRICHLOROTIN see MQC750
METHYL-TRICHLORSILAN (CZECH) see MQC500
α-METHYLTRICYCLO(3.3.1.1³,⁷)DECANE-1-METHANAMINE HYDROCHLORIDE see AJU625
1-METHYLTRICYCLOQUINAZOLINE see MQD000
3-METHYLTRICYCLOQUINAZOLINE see MQD250
4-METHYLTRICYCLOQUINAZOLINE see MQD500
METHYLTRIETHOXYSILANE see MQD750
METHYL TRIFLUORIDE see CBY750
2-METHYL-3-(2'-TRIFLUORMETHYLPHENYL) CHINAZOLON-4 HYDROCHLORIDE see MQD800
5-METHYL-2-TRIFLUOROMETHYLOXAZOLIDINE see MQE000
(α-METHYL-m-TRIFLUOROMETHYLPHENETHYLAMINOMETHYL) PIPERIDINO KETONE see MQE100

METHYL 2,2,3-TRIFLUORO-3-OXOPROPANOATE see MJD400
8-METHYL-3-(α-(α,α,α-TRIFLUORO-o-TOLYL)BENZYLOXY)TROPANIUM IODIDE see MQF000
METHYL-3,3,3-TRIFLUORO-2-(TRIFLUOROMETHYL)PROPIONATE see MKK750
METHYL TRIFLUOROVINYL ETHER see MQF200
6-METHYL-1,3,8-TRIHYDROXYANTHRAQUINONE see MQF250
METHYL 3,4,5-TRIHYDROXYBENZOATE see MKI100
1-METHYL-3,7,9-TRIHYDROXY-6H-DIBENZO(b,d)PYRAN-6-ONE see AGW476
METHYL 3,4,5-TRIMETHOXYBENZOATE see MQF300
METHYL-18-o-(3,4,5-TRIMETHOXYCINNAMOYL RESERPATE see TLN500
α-METHYL-3,4,5-TRIMETHOXYPHENETHYLAMINE HYDROCHLORIDE see TKX500
METHYLTRIMETHOXYSILANE see MQF500
(2-METHYL-2-(2-(TRIMETHYLAMMONIO)ETHOXY)ETHYL) DIETHYLMETHYL AMMONIUM DIIODIDE see MQF750
METHYLTRIMETHYLENE GLYCOL see BOS500
4-METHYL TRIMETHYLENE SULFITE see MQG500
METHYL TRIMETHYLOL METHANE TRINITRATE see MQG600
METHYL 2,4,5-TRIMETHYLPYRROL-3-YL KETONE see ADE050
N-METHYL-N-TRIMETHYLSILYLTRIFLUOROACETAMIDE see MQG750
1-METHYL-2,4,5-TRINITROBENZENE see TMN400
3-METHYL-2,4,6-TRINITROPHENOL see TML500
METHYLTRIOCTYLAMMONIUM CHLORIDE see MQH000
4-METHYL-2,6,7-TRIOXA-1-ARSABICYCLO(2.2.2)OCTANE see MQH100
1-METHYL-2,8,9-TRIOXO-5-AZA-1-SILABICYCLO(3.3.3)UNDECANE see MPI650
3-METHYL-1,2,4-TRIOXOLANE see PMP250
1-METHYL-4-(3,3,3-TRIS(p-CHLOROPHENYL)PROPIONYLPIPERAZINE MONOHYDROCHLORIDE see MQH250
METHYLTRIS-2-ETHYLHEXENOXYSILANE see TNI100
METHYLTRIS(2-ETHYLHEXYLOXYCARBONYLMETHYLTHIO)STANNANE see MQH500
METHYL TRITHION see MQH750
8-METHYLTROPINIUM BROMIDE 2-PROPYLVALERATE see LJS000
4-METHYL-TROPOLONE see MQI000
α-METHYLTRYPTAMINE see AME500
l-5-METHYLTRYPTOPHAN see MQI250
METHYL TUADS see TFS350
α-METHYL-m-TYRAMINE see AMF500
α-METHYLTYRAMINE HYDROCHLORIDE see HJA600
(±)-α-METHYL-p-TYROSINE METHYL ESTER HYDROCHLORIDE see MQI500
dl-α-METHYL-p-TYROSINE METHYL ESTER HYDROCHLORIDE see MQI500
4-METHYLUMBELLIFERON (CZECH) see MKP500
METHYLUMBELLIFERONE see MEK300
4-METHYLUMBELLIFERONE see MKP500
β-METHYLUMBELLIFERONE see MKP500
4-METHYLUMBELLIFERONE-O,O-DIETHYL THIOPHOSPHATE see PKT000

4-METHYLUMBELLIFERONE 4-GUANIDINOBENZOATE see GLC100
METHYL-4-UMBELLIFERONE SODIUM see HMB000
2-METHYLUNDECANAL see MQI550
METHYL 9-UNDECENOATE see ULS400
METHYL 10-UNDECENOATE see ULS400
METHYL UNDECYLENATE see ULS400
4-METHYLURACIL see MPW500
4-METHYLURACIL see MQI750
5-METHYLURACIL see TFX800
6-METHYLURACIL see MQI750
METHYLUREA see MQJ000
1-METHYLUREA see MQJ000
N-METHYLUREA see MQJ000
METHYL UREA and SODIUM NITRITE see MQJ250
METHYLURETHAN see MHZ000
N-METHYL URETHAN see EMQ500
METHYLURETHANE see EMQ500
METHYLURETHANE see MHZ000
N-METHYLURETHANE see EMQ500
N-METHYLURETHANEBENZENESULFOHYDRAZINE see MQJ300
N-METHYLURETHANE of HYDROCHLORIDE of 2-DIMETHYLAMINO-p-CRESOL see DPL900
N-METHYLURETHANE of HYDROCHLORIDE of 3-DIMETHYLAMINOPHENOL see MID000
N-METHYLURETHANE HYDROCHLORIDE of 6-DIMETHYLAMINO-o-4-XYLENOL see MIA775
4-METHYL VALERALDEHYDE see MQJ500
METHYL VALERATE see VAQ100
METHYL n-VALERATE see VAQ100
METHYL VALERIANATE see VAQ100
2-METHYLVALERIC ACID see MQJ750
α-METHYLVALERIC ACID see MQJ750
2-METHYL-1,5-VALERODINITRILE see MQK000
β-METHYL-Δ-VALEROLACTONE see MQK100
4-METHYLVALERONITRILE see MNJ000
METHYL VANILLATE see VFF100
METHYLVANILLIN see VHK000
4-o-METHYLVANILLIN see VHK000
trans-8-METHYL-N-VANILLYL-6-NONEAMIDE see CBF750
N-METHYL-N-VINYLACETAMIDE see MQK500
METHYLVINYL ACETATE see MQK750
METHYL VINYL ADIPATE see MQL000
1-METHYL-2-VINYLBENZENE see VQK660
1-METHYL-3-VINYLBENZENE see VQK670
α-METHYLVINYL BROMIDE see BOA250
METHYL VINYL CARBINOL see MQL250
METHYL-VINYL-CETONE (FRENCH) see BOY500
METHYL VINYL ETHER see MQL750
METHYLVINYLKETON (GERMAN) see BOY500
METHYL VINYL KETONE see BOY500
METHYL VINYL KETONE see MQM100
METHYLVINYLNITROSAMIN (GERMAN) see NKY000
METHYLVINYLNITROSAMINE see NKY000
METHYL VINYL OXIMINO SILANE see MQM150
METHYLVINYLOXYETHYL SULFIDE see MQM200
2-METHYL-5-VINYLPYRIDINE see MQM500
METHYL VINYL SULFONE see MQM750
2-METHYL-5-VINYL TETRAZOLE see MQM775
4-METHYL-5-VINYL THIAZOLE see MQM800
METHYL VIOLET see MQN025
METHYL VIOLET 2B see MQN025
METHYL VIOLET 6B see MQN000

METHYL VIOLET 10B see AOR500
METHYL VIOLET BB see MQN025
METHYL VIOLET CARBINOL see MQN250
METHYL VIOLET CARBINOL BASE see MQN250
METHYL VIOLET 10BD see AOR500
METHYL VIOLET FN see MQN025
METHYL VIOLET 10BK see AOR500
METHYL VIOLET N see MQN025
METHYL VIOLET 10BN see AOR500
METHYL VIOLET 5BNO see AOR500
METHYL VIOLET 10BNS see AOR500
METHYL VIOLET 5BO see AOR500
METHYL VIOLET 10BO see AOR500
METHYLVIOLETT see AOR500
METHYL-VIOLETT (GERMAN) see MQN025
METHYLVIOLOGEN see PAJ000
METHYL VIOLOGEN (2+) see PAI990
METHYLVIOLOGEN DIBROMIDE see PAI995
METHYL VIOLOGEN DICHLORIDE see PAJ000
METHYL VIOLOGEN (REDUCED) see PAJ000
3-METHYLXANTHINE see MQN500
1-METHYL-N-(2,6-XYLYL)ISONIPECOTAMIDE HYDROCHLORIDE see MQO250
1-METHYL-2-(2,6-XYLYLOXY)-ETHYLAMINE HYDROCHLORIDE see MQR775
N-METHYL-N-((1-(3,5-XYLYLOXY)-2-PROPYL)CARBAMIC ACID-2-DIETHYLAMINO)ETHYL ESTER, HYDROCHLORIDE see MQO500
N-METHYL-N-(1-(3,5-XYLYLOXY)-2-PROPYL)CARBAMIC ACID-2-(2-METHYLPIPERIDINO)ETHYL ESTER HYDROCHLORIDE see MQO750
N-METHYL-N-(1-(3,5-XYLYLOXY)-2-PROPYL)CARBAMIC ACID-2-(PYRROLIDINYL)ETHYL ESTER, HCl see MQP000
N-METHYL-N'-2,4-XYLYL-N-(N-2,4-XYLYLFORMIMIDOYL)FORMAMIDINE see MJL250
METHYL YELLOW see DOT300
METHYL ZIMATE see BJK500
METHYLZINC IODIDE see MQP250
METHYL ZINEB see BJK500
METHYL ZINEB see ZMA000
METHYL ZIRAM see BJK500
METHYPHENYLMETHANOL see PDE000
METHYPROLON see DNW400
METHYPRYLON see DNW400
METHYRIMOL see BRD000
METHYSERGID see MLD250
METHYSERGIDE see MLD250
METHYSERGIDE DIMALEATE see MQP500
METIAPINE see MQQ000
METIAZIC ACID see MNQ500
METIAZINIC ACID see MNQ500
METICORTELANE see PMA100
METICORTELONE see PMA000
METICORTELONE ACETATE see SOV100
METICORTELONE SOLUBLE see PMA100
METICRAN see MQQ050
METICRANE see MQQ050
METI-DERM see PMA000
METIDIONE see AOO475
METIFEX see EDW500
METIFONATE see TIQ250
METIGUANIDE see DQR800
METILACRILATO (ITALIAN) see MGA500
METILAMIL ALCOHOL (ITALIAN) see MKW600
METILAMINE (ITALIAN) see MGC250
2-METIL-6-AMINO-EPTANO (ITALIAN) see ILM000
1-METILBIGUANIDE CLORIDRATO (ITALIAN) see MHP400
3-METIL-BUTANOLO (ITALIAN) see IHP000

METIL CELLOSOLVE (ITALIAN) see EJH500
2-METILCICLOESANONE (ITALIAN) see MIR500
METILCLOR see MIF763
METILCLOROFORMIATO (ITALIAN) see MIG000
METILCLORPINDOL see CMX850
N-METIL-N-(β-DICICLOESILAMINOETIL)PIPERIDINIO BROMURO (ITALIAN) see MJC775
METILDIGOXIN see MJD300
METILDIGOXIN see MJD500
METILDIGOXINA (SPANISH) see MJD300
2'-METIL-3'-DIMETILAMINO-PROPIL-5-IMINODIBENZILE (ITALIAN) see DLH200
METILDIOLO see AOO475
N-METIL-DITIOCARBAMMATO di SODIO (ITALIAN) see VFU000
METILE (ACETATO di) (ITALIAN) see MFW100
METILENBIOTIC see MDO250
O,O-METILEN-BIS(4-CLORO-FENOLO) see MJM500
3,3'-METILEN-BIS(4-IDROSSI-CUMARINA) (ITALIAN) see BJZ000
4,4'-METILENE-BIS-o-CLOROANILINA (ITALIAN) see MJM200
(6-(1-METIL-EPITL)-2,4-DINITRO-FENIL)-CROTONATO (ITALIAN) see AQT500
METILESTER del ACIDO BISDEHIDROISYNOLICO (SPANISH) see BIT000
7-METILETER del ACIDO BISDEHIDRODOISYNOLICO see BIT030
METILETILCHETONE (ITALIAN) see MKA400
METIL (FORMIATO di) (ITALIAN) see MKG750
METILISOBUTILCHETONE (ITALIAN) see HFG500
METIL ISOCIANATO (ITALIAN) see MKX250
METILMERCAPTANO (ITALIAN) see MLE650
METILMERCAPTOFOSOKSID see DAP000
METIL METACRILATO (ITALIAN) see MLH750
α-METIL-β-(2-METILENE-4,5-DIIDROIMIDAZOLIL)BENZOTIOFANE CLORIDRATO (ITALIAN) see MHJ500
N-METIL-1-NAFTIL-CARBAMMATO (ITALIAN) see CBM750
N-(4-(β-(5-METILOSSAZOL-3-CARBOSSAMIDO)-ETIL)-BENZENESOLFONIL)-N¹-CICLOESIL-UREA see DBE885
METILPARATION (HUNGARIAN) see MNH000
4-METILPENTAN-2-OLO (ITALIAN) see MKW600
4-METILPENTAN-2-ONE (ITALIAN) see HFG500
4-METIL-3-PENTEN-2-ONE (ITALIAN) see MDJ750
1-(N-METIL-PIPERIDIL-4')-3-FENIL-4-BENZIL-PIRAZOLONE-5 (ITALIAN) see BCP650
3-γ-(α-METIL-1-PIPERIDINO)PROPIL-5,5-DIFENILTIOIDANTOINA CLORIDRATO (ITALIAN) see DWF875
(6-(1-METIL-PROPIL)-2,4-DINITRO-FENIL)-3,3-DIMETIL-ACRILATO (ITALIAN) see BGB500
6-(1-METIL-PROPIL)-2,4-DINITRO-FENOLO (ITALIAN) see BRE500
α-METIL-STIROLO (ITALIAN) see MPK250
2-METIL-2-TIOMETIL-PROPIONALDEID-O-(N-METIL-CARBAMOIL)-OSSIMA (ITALIAN) see CBM500
6-METIL-TIOURACILE (ITALIAN) see MPW500
METILTRIAZOTION see ASH500

METINDOL see IDA000
METIONE see MDT740
d-METIONIEN (AUSTRALIAN) see MDT730
METIPREGNONE see MCA000
METIPRILONE see DNW400
METIRAM see MQQ250
METIRAM, mixed with SULFUR see BMC800
METISAZONUM see MKW250
METIZOL see MCO500
METIZOLIN see BAV000
METIZOLINE HYDROCHLORIDE see MHJ500
METMERCAPTURON see DST000
METOBROMURON see PAM785
METOCARBAMOL see GKK000
METOCARBAMOLO see GKK000
METOCHLOPRAMIDE see AJH000
METOCLOL see AJH000
METOCLOPRAMIDE DIHYDROCHLORIDE MONOHYDRATE see MQQ300
METOCRYST see AOO475
METOFANE see DFA400
METOFENINA see GKK000
METOFOLINE see MDV000
METOFOS see MQQ350
METOFOS PLYNNY see MQQ350
2-METOKSY-4-ALLILOFENOL (POLISH) see EQR500
METOKSYCHLOR (POLISH) see MEI450
METOKSYETYLOWY ALKOHOL (POLISH) see EJH500
METOLACHLOR see MQQ450
METOLACHLOR-CYANAZINE MIXTURE see MQQ475
METOLAZONE see ZAK300
METOLCARB see MIB750
METOLOSE 60SH see MIF760
METOLOSE 60SH400 see MIF760
METOLOSE MC 8000 see MIF760
METOLOSE SM 15 see MIF760
METOLOSE SM 100 see MIF760
METOLOSE SM 4000 see MIF760
METOLQUIZOLONE see QAK000
METOMEGA CHROME YELLOW GM see SIT850
METOMIDATE see MQQ500
METOMIDATE HYDROCHLORIDE see MQQ750
METOMIL (ITALIAN) see MDU600
METOMINOSTROBIN see MQQ800
METON see HEA500
METOPIMAZINE see MQR000
METOPIRON see MCJ370
METOPIRON see MJJ000
METOPIRONE see MCJ370
METOPIRONE DITARTRATE see MJJ000
METOPRINE see MQR100
METOPROLOL see MQR144
(±)-METOPROLOL see MQR144
METOPROLOL HEMITARTRATE see MQR150
METOPROLOL TARTRATE see MQR150
METOPROLOL TARTRATE see SBV500
METOPRYL see MOU830
METOPYRONE see MCJ370
METOQUINE see CFU750
METOSERPATE HYDROCHLORIDE see MDW100
METOSERPATE HYDROCHLORIDE see MQR200
(2-METOSSICARBONIL-1-METIL-VINIL)-DIMETIL-FOSFATO (ITALIAN) see MQR750
2-METOSSIETANOLO (ITALIAN) see EJH500
2-METOSSIETILACETATO (ITALIAN) see EJJ500
S-((5-METOSSI-4H-PIRON-2-IL)-METIL)-O,O-DIMETIL-MONOTIOFOSFATO (ITALIAN) see EAS000
METOSSIPROPANDIOLO see RLU000
METOSYN see FDD150

METOTHYRINE see MCO500
METOX see CEP000
METOX see MEI450
METOXADONE see MFD500
METOXAL see SNK000
METOXFLURAN see DFA400
METOXIDON see SNN300
METOXIFLURAN see DFA400
METOXON see CKC000
METOXURON see MQR225
METOXYDE see HJL500
4-o-METPA see MQR250
METRACTYL see MQU750
METRAMAC see DJA400
METRAMAK see DJA400
METRAMINOL BITARTRATE see HNC000
(−)-METRAMINOL (+)-BITARTRATE see HNC000
1-METRAMINOL BITARTRATE see HNC000
d-(−)-METRAMINOL BITARTRATE see HNC000
METRANIL see PBC250
METRASPRAY see BQA010
METRIBEN see TIK000
METRIBUZIN see MQR275
METRIFONATE see TIQ250
METRIOL TRINITRATE see MQG600
METRIPHONATE see TIQ250
METRISONE see MOR500
METRIZAMIDE see MQR300
METRIZOATE see MQR350
METRIZOIC ACID see MQR350
METRODIOL see EQJ500
METRODIOL DIACETATE see EQJ500
METROGEN RED FORMER KB SOLN see CLK225
METROGESTONE see MBZ100
METRON see MNH000
METRONE see MPN500
METRONIDAZ see MMN250
METRONIDAZOL see MMN250
METRONIDAZOLO see MMN250
METROPINE see MGR500
METRORAT see DBE200
METRO TALC 4604 see TAB750
METRO TALC 4608 see TAB750
METRO TALC 4609 see TAB750
No. 907 METRO TALC see TAB750
METROXEDRINE see SPC500
METSO 20 see SJU000
METSO BEADS 2048 see SJU000
METSO BEADS, DRYMET see SJU000
METSO PENTABEAD 20 see SJU000
MET-SPAR see CAS000
METSUCCIMIDE see MLP800
METSULFURON-METHYL see MQR400
METURAL see DTG700
METYCAINE see PIV750
METYLAL (POLISH) see MGA850
METYLENO-BIS-FENYLOIZOCYJANIAN see DNJ800
METYLENU CHLOREK (POLISH) see MJP450
METYLESTER KYSELINY SALICYLOVE (CZECH) see MPI000
METYLFENEMAL see ENB500
METYLOAMINA (POLISH) see MGC250
METYLOCYKLOHEKSAN (POLISH) see MIQ740
METYLOCYKLOHEKSANOL (POLISH) see MIQ745
METYLOCYKLOHEKSANON (POLISH) see MIR250
METYLOETYLOKETON (POLISH) see MKA400
METYLOHYDRAZYNA (POLISH) see MKN000
METYLOIZOBUTYLOKETON (POLISH) see HFG500
1-METYLO-2-MERKAPTOIMIDAZOLEM (POLISH) see MCO500

N-METYLO-N'-NITRO-N-NITROZOGUANIDYNY see MMP000
METYLOPARATION (POLISH) see MNH000
METYLOPROPYLOKETON (POLISH) see PBN250
7-(β-METYLOTIOETYLO)-TEOFILINA (POLISH) see MDN100
METYLOWY ALKOHOL (POLISH) see MGB150
METYLPARATION (CZECH) see MNH000
METYLU BROMEK (POLISH) see MHR200
METYLU CHLOREK (POLISH) see MIF765
METYLU JODEK (POLISH) see MKW200
METYNA see ENB500
METYPRAPONE BITARTRATE see MJJ000
METYRAPON see MCJ370
METYRAPONE see MCJ370
METYRAPONE DITARTRATE see MJJ000
MEV see MNB250
MEVALOTIN see LGK500
MEVASINE see VIZ400
MEVASIN HYDROCHLORIDE see MQR500
MEVINFOS (DUTCH) see MQR750
MEVINOLINIC ACID AMMONIUM SALT see LII050
MEVINPHOS see MQR750
MEXACARBATE (DOT) see DOS000
MEXAMINE HYDROCHLORIDE see MFT000
MEXAZOLAM see MQR760
MEXENE see BJK500
MEXEPHENAMIDE see DIB600
MEXEPHENAMIDE see MCH250
MEXICA FLAME LEAF see EQX000
MEXICAN BREADFRUIT see SLE890
MEXICAN FLAME TREE see EQX000
MEXICAN FLOWER PLANT see EQX000
MEXICAN HUSK TOMATO see JBS100
MEXICANIN E see MQR765
MEXICANINE E see MQR765
MEXICO WEED see CCP000
MEXIDE see RNZ000
MEXIDEX see SOW000
MEXILETINE HYDROCHLORIDE see MQR775
MEXITIL see MQR775
MEXOCINE see DAI485
MEXOCINE see MIJ500
MEXOLAMINE see CPN750
MEXPECTIN see PAO150
MEXYL see ABX500
MEYPRALGIN R/LV see SEH000
MEZARONIL see ASG250
MEZATON see NCL500
MEZATON see SPC500
R(−)-MEZATON see NCL500
MEZCALINE see MDI500
MEZCALINE SULFATE see MDJ000
MEZCLINE see MDI500
MEZENE see BJK500
MEZEPAN see CGA000
MEZEREIN see MDJ250
MEZEREUM see LAR500
MEZIDINE see TLG500
MEZINEB see ZMA000
MEZINIUM METHYL SULFATE see MQS100
MEZLOCILLIN see MQS200
MEZOLIN see IDA000
MEZOTOX see DFT800
MEZURON see ASG250
MF see UTU500
MF 1 see UTU500
MF 17 see UTU500
MF 27 see UTU500
MF 004 see MCB050
MF 009 see MCB050
MF-344 see EFK000
MF 565 see MAP300
MF 598 see MAP300
MF 910 see MCB050
MFA see FIC000
MFA see MKD000

MFA see TLX175
MFAS-R 100P see MCB050
M.F. FULVIUS VENOM see CNR150
MFH see FNW000
MFI see IPX000
MFI-PC see FDA100
MFM see TLX175
MFNA see MME809
M 60 (FORMALDEHYDE POLYMER) see UTU500
MFP 8 see MCB050
MFPS 1 see UTU500
MF RESIN see UTU500
M. FULVIUS FULVIUS VENOM see CNR150
MG 02 see CGO550
MG 46 see LGK200
M.G. 8823 see ERE100
MG 8926 see CPP000
MG 18037 see BQX000
MG 18370 see DTJ400
MG 18415 see AJK000
MG 18512 see CPP750
MG 18570 see DTJ400
M.G. 18590 see TEY600
M.G.18755 see IBW100
MGA see MCB380
MGA 100 (STEROID) see MCB380
MGB1 see CGY610
MGB2 see BIE610
M.G. 8948-2HCl see BJW000
M.G. 18001-3HCl see BJV750
MGI 114 see HLU600
M7-GIFTKOERNER see TEM000
MGK 11 see BHJ500
MGK-264 see OES000
MGK 326 see EAU500
MGK DIETHYLTOLUAMIDE see DKC800
MGK DOG AND CAT REPELLENT see UKS000
MGK REPELLENT 11 see BHJ500
MGK REPELLENT-326 see EAU500
MGK REPELLENT 874 see OGA000
MGK REPELLENT 1,207 see CKU750
M 1 (GROWTH REGULATOR) see NAK525
MH see DMC600
M.H. see HAP000
MH see MKK600
M-2H see BPF825
MH 30 see DMC600
MH-40 see DMC600
MH-532 see PGA750
MHA ACID see HMR600
MHA-FA see HMR600
MHA NITRILE see HMR550
MH 36 BAYER see DMC600
MHOROMER see EJH000
MHP see NKU500
MHSK see MLI800
683 M HYDROCHLORIDE see OOE100
3-MI see MKV750
MI 85 see AQN750
217 MI see TLF500
MIA see IDZ000
MIADONE see MDP750
MIAK see MKW450
MIALEX see CKK250
MIANESINA see GGS000
MIANG TEA LEAF EXTRACT see MQS215
MIANSERIN see MQS220
MIANSERINE see MQS220
MIANSERINE HYDROCHLORIDE see BMA625
MIANSERIN HYDROCHLORIDE see BMA625
MIANSERYNA see MQS220
MIARSENOL see NCJ500
MIAZOLE see IAL000
MIBC see MKW600
MIBK see HFG500
MIBOLERON see MQS225
MIBOLERONE see MQS225
MIBP see MRI775

MIBT see MKW250
MIC see ISE000
MIC see MKW600
MIC see MKX250
3-MIC see MKW600
MICA see MQS250
MICALEX see AHD650
MICA SILICATE see MQS250
MICASIN see CDS500
MICASIN see CKL500
MICEVIT see MAS300
MICHLER'S BASE see MJN000
MICHLER'S HYDRIDE see MJN000
p,p'-MICHLER'S HYDROL see TDO750
MICHLER'S KETONE see MQS500
p,p'-MICHLER'S KETONE see MQS500
MICHLER'S METHANE see MJN000
MICHROME No. 226 see AJQ250
MICIDE see EIR000
MICOCHLORINE see CDP250
MICOFENOLICO ACIDO (SPANISH) see MRX000
MICOFUME see DSB200
MICOFUR see NGC000
MICOL see HCQ500
MICONAZOLE see MQS550
MICONAZOLE NITRATE see MQS560
MICREST see DKA600
MICROCAL 160 see CAW850
MICROCAL ET see CAW850
MICROCARB see CAT775
MICRO-CEL see CAW850
MICRO-CEL A see CAW850
MICRO-CEL B see CAW850
MICRO-CEL C see CAW850
MICRO-CEL E see CAW850
MICRO-CEL T see CAW850
MICRO-CEL T26 see CAW850
MICRO-CEL T38 see CAW850
MICRO-CEL T41 see CAW850
MICROCETINA see CDP250
MICRO-CHECK 12 see CBG000
MICRO-CHEK 11 see OFE000
MICRO-CHEK SKANE see OFE000
MICROCID see GFG100
MICROCILLIN see CBO250
MICROCOCCIN see MQS565
MICROCOCCIN P1 see MQS565
MICROCOCCIN P, 13',19'-DIDEHYDRO-19'-DEOXY-28,44-DIHYDRO-44-HYDROXY- see MQS565
MICROCOP see CNK559
MICROCRYSTALLINE QUARTZ see SCK600
MICROCRYSTALLINE WAX see PCT600
MICROCYSTIN-A see AQV990
MICROCYSTIN-LR see AQV990
MICROCYSTIS AERUGINOSA TOXIN see AQV990
MICRO DDT 75 see DAD200
MICRODIOL see EDO000
MICRO DRY see AHA000
MICROEST see DKA600
MICROFLOTOX see SOD500
MICROGRIT WCA see AHE250
MICROGYNON see NNL500
MICROIODIDE see PLW285
MICRO-LEX GREEN 5B see BLK000
MICROLITH GREEN G-FP see PJQ100
MICROLYSIN see CKN500
MICROMIC CR 16 see CAT775
MICROMICIN see MQS579
MICROMYA see CAT775
MICROMYCIN see MQS579
MICRONASE see CEH700
MICRONEX see CBT750
MICRONOMICIN SULFATE see MQS600
MICRONOMYCIN SULFATE see MQS600
MICRO-PEN see PAQ200
MICROPENIN see MNV250
MICROPHOS 90 see ZJS400
MICROSETILE BLUE EB see TBG700
MICROSETILE BLUE FF see MGG250

MICROSETILE DIAZO BLACK G see DPO200
MICROSETILE ORANGE RA see AKP750
MICROSETILE PINK BN see AKE250
MICROSETILE RUBINE 2B see CMP080
MICROSETILE SCARLET B see ENP100
MICROSETILE VIOLET 3R see DBP000
MICROSETILE VIOLET B see AKP250
MICROSETILE YELLOW 2R see DUW500
MICROSETILE YELLOW GR see AAQ250
MICROSUL see MPQ750
MICROTAN PIRAZOLO see AIF000
MICROTEX LAKE RED CR see CHP500
MICROTHENE see PJS750
MICROTHENE 510 see PJS750
MICROTHENE 704 see PJS750
MICROTHENE 710 see PJS750
MICROTHENE F see PJS750
MICROTHENE FN 500 see PJS750
MICROTHENE FN 510 see PJS750
MICROTHENE MN 754-18 see PJS750
MICROTIN see SFQ500
MICROTRIM see TKX000
MICROWHITE 25 see CAT775
MICROZUL see CJJ000
MICRURUS ALLENI YATESI VENOM see MQS750
MICRURUS CARINICAUDUS DUMERILII VENOM see MQT000
MICRURUS FULVIUS FULVIUS VENOM see CNR150
MICRURUS FULVIUS VENOM see MQT100
MICRURUS MIPARTIUS HERTWIGI VENOM see MQT250
MICRURUS NIGROCINCTUS VENOM see MQT500
MICTINE see AFW500
MIDARINE see HLC500
MIDAZOLAM see MQT525
MIDECAMYCIN see LEV025
MIDECAMYCIN see MBY150
MIDECAMYCIN A₁ see MBY150
MIDELID see BGC625
MIDETON FAST RED VIOLET R see DBP000
MI 85 DI see ASA000
MIDICEL see AKO500
MIDIKEL see AKO500
MIDLON ORANGE PROPYL see ADG000
MIDLON RED PG see CMM320
MIDLON RED PRS see CMM330
MIDLON YELLOW PROPYL see CMM759
MIDODRINE see MQT530
(±)-MIDODRINE HYDROCHLORIDE see MQT530
MIDONE see DBB200
MIDOXIN see HGP500
MIDRONAL see CMR100
MIEDZ (POLISH) see CNM000
MIEDZIAN see CNK559
MIEDZIAN 50 see CNK559
MIELUCIN see BOT250
MIERENZUUR (DUTCH) see FNA000
MIFEPRISTONE see HKB700
MIFUROL see CCK630
MI-GEE see DNF800
MIGHTY 150 see NAJ500
MIGLYOL 810 NEUTRAL OIL see CBF710
MIGLYOL 812 NEUTRAL OIL see CBF710
MIGRISTENE see FMU039
MIGUGEN see EFJ600
MIH see PME250
MIH HYDROCHLORIDE see PME500
MIIKE 20 see CBT750
MIK see HFG500
MIKACION BRILLIANT BLUE RS see CMS222
MIKACION BRILLIANT RED 5BS see PMF540
MIKAL see AHH800
MIKAMETAN see IDA000
MIKAMYCIN see VRF000

MIKAMYCIN B see VRA700
MIKAMYCIN IA see VRA700
MIKEDIMIDE see MKA250
MIKELAN see MQT550
MIKEPHOR BP see CMP100
MIKEPHOR BX see CMP100
MIKEPHOR BX CONC see CMP100
MIKEPHOR TB see CMP200
MIKETAZOL DEVELOPER ONS see
HMX520
MIKETHREN BRILLIANT ORANGE GR see
CMU820
MIKETHRENE BLUE 3G see DLO900
MIKETHRENE BLUE BC see DFN300
MIKETHRENE BLUE BCS see DFN300
MIKETHRENE BLUE RSN see IBV050
MIKETHRENE BRILLIANT BLUE R see
IBV050
MIKETHRENE BROWN BR see CMU770
MIKETHRENE BROWN GR see CMU770
MIKETHRENE BROWN R see CMU780
MIKETHRENE GOLD YELLOW see DCZ000
MIKETHRENE GREY K see IBJ000
MIKETHRENE GREY M see CMU475
MIKETHRENE GREY MG see CMU475
MIKETHRENE MARINE BLUE G see
CMU500
MIKETHRENE OLIVE R see DUP100
MIKETHRENE ORANGE GR see CMU820
MIKETHRENE ORANGE R see CMU815
MIKETHRENE RED BROWN 5RF see
CMU800
MIKETHREN NAVY BLUE FRA see VGP100
MIKETON BRILLIANT BLUE B see MGG250
MIKETON FAST BLUE see TBG700
MIKETON FAST PINK FF 3B see DBX000
MIKETON FAST PINK RL see AKO350
MIKETON FAST SCARLET B see ENP100
MIKETON FAST TURQUOISE BLUE G see
DMM400
MIKETON FAST VIOLET B see DBY700
MIKETON FAST YELLOW G see AAQ250
MIKETON FAST YELLOW YL see KDA075
MIKETON POLYESTER BLUE FBL see
CMP070
MIKETON POLYESTER BLUE FBL see
DBQ220
MIKETON POLYESTER PINK RL see
AKO350
MIKETON POLYESTER RED FB see AKI750
MIKETON POLYESTER YELLOW 5R see
CMP090
MIKETON POLYESTER YELLOW YL see
KDA075
MIKETORIN see EAI000
MIKIPALAOA (HAWAII) see CNG825
MIKROKALCIT see CAD800
MIKROLOUR see PJS750
MIKRONAL S 40 see MCB050
MIL see HCP750
MIL see MQU000
MIL-H-19457C see TIA130
MILAXEN see HEG000
MILBAM see BJK500
MILBAN see BJK500
MILBAN F see TFD500
MIL-B-4394-B see CES650
MILBEDOCE see VSZ000
MILBEMYCIN D see MQT600
MILBEX see CKL500
MILBEX mixed with PHOSALON (2:1) see
MQT750
MILBEX mixed with PHOSALONE (2:1) see
MQT750
MILBOL see BIO750
MILBOL 49 see BBQ500
MILCHSAEURE (GERMAN) see LAG000
MIL-COL see MLC250
MILCURB see BRD000
MILCURB see BRI750
MILCURB SUPER see BRI750
MILD ACONITATE see ADH500

MILD ACONITINE see ADH500
MILDIOMYCIN see MQU000
MILDMEN see LFK000
MILDMEN see MDQ250
MILD MERCURY CHLORIDE see MCW000
MILDOFIX see MRQ300
MIL-DU-RID see BGJ750
MILEPSIN see DBB200
MILESTROL see DKA600
MILEZIN see MCI500
MILFARON see CDP750
MILFURAM see CGI550
MILGO see BRI750
MILGO E see BRI750
MILID see BGC625
MILIDE see BGC625
MILK ACID see LAG000
MILK-CLOTTING ENZYME from
BACILLUS CEREUS see MQU075
MILK-CLOTTING ENZYME from
ENDOTHIA PARASITICA see MQU100
MILK-CLOTTING ENZYME from MUCOR
MIEHEI see MQU120
MILK-CLOTTING ENZYME from MUCOR
PUSILLUS see MQU125
MILK OF LIME see CAT225
MILK OF MAGNESIA see MAG750
MILK SUGAR see LAR000
MILK THISTLE EXTRACT see SDX900
MILK WHITE see LDY000
MIL L 7808 see MQU250
MIL L 17535 see MQU500
MILLED CASSAVA POWDER see CCO700
MILLERITE see MQU510
MILLER NU SET see TIX500
MILLER P.C. WEEDKILLER see PLC250
MILLER'S FUMIGRAIN see ADX500
MILLICORTEN see SOW000
MILLING BRILLIANT SCARLET GN see
CMM320
MILLING FAST ORANGE 2R see ADG000
MILLING FAST ORANGE R see ADG000
MILLING FAST RED B see CMM330
MILLING FAST RED G see CMM320
MILLING FAST RED GL see CMM320
MILLING FAST RED PG see CMM320
MILLING FAST RED R see NAO600
MILLING FAST RED RS see NAO600
MILLING ORANGE R see ADG000
MILLING RED A see CMM325
MILLING RED B see CMM330
MILLING RED BB see CMM330
MILLING RED J see CMM320
MILLING RED PRX see NAO600
MILLING RED SWB see CMM330
MILLING RED SWG see CMM320
MILLING SCARLET 2G see CAD800
MILLING SCARLET DH see CMM325
MILLING SCARLET G see CMM320
MILLING SCARLET R see CMM325
MILLING YELLOW 3G see CMM759
MILLING YELLOW 3J see CMM759
MILLING YELLOW RX see CMM759
MILLIONATE 300 see PKB100
MILLIONATE MR see PKB100
MILLIONATE MR 100 see PKB100
MILLIONATE MR 200 see PKB100
MILLIONATE MR 300 see PKB100
MILLIONATE MR 340 see PKB100
MILLIONATE MR 400 see PKB100
MILLIONATE MR 500 see PKB100
MILLIPHYLLINE see CNR125
MILLON'S BASE ANHYDRIDE see DNW200
MILLOPHYLLINE see CNR125
MILMER see BLC250
MILOCEP see MQQ450
MILOGARD see PMN850
MILONTIN see MNZ000
MILORI BLUE see IGY000
MILOXACIN see MQU525
MILPREM see ECU750
MILPREM see MQU750

MILPREX see DXX400
MILRINONE see MQU600
MILSAR see CKL500
MILSTEM see BRI750
MILSTEM SEED DRESSING see BRI750
MILTANN see MQU750
MILTEFOSINE see HCP750
MILTON see SHU500
MILTOWN see MQU750
MILTOX see EIR000
MILTOX see ZJS300
MILTOX SPECIAL see EIR000
MIMEDRAN see PIK625
MIMOSA ABSOLUTE see MQV000
MIMOSA TANNIN see MQV250
MIMOSA Z see CMP050
MIMOSINE see HOC500
MINACIDE see CQI500
MINALFENE see MDP850
MINAPHIL see TEP500
MINCARD see AFW500
MINEPENTATE see DPA800
MINERAL DUSTS see MQV500
MINERAL FIRE RED 5DDS see MRC000
MINERAL FIRE RED 5GS see MRC000
MINERAL GREEN see COF500
MINERAL NAPHTHA see BBL250
MINERAL OIL see MQV750
MINERAL OIL, PETROLEUM
CONDENSATES, VACUUM TOWER see
MQV755
MINERAL OIL, PETROLEUM
DISTILLATES, ACID-TREATED HEAVY
NAPHTHENIC (mild or no solvent-refining or
hydrotreatment) see MQV760
MINERAL OIL, PETROLEUM
DISTILLATES, ACID-TREATED HEAVY
PARAFFINIC (severe solvent-refining and/or
hydrotreatment) see MQV765
MINERAL OIL, PETROLEUM
DISTILLATES, ACID-TREATED LIGHT
NAPHTHENIC (mild or no solvent-refining or
hydrotreatment) see MQV770
MINERAL OIL, PETROLEUM
DISTILLATES, ACID-TREATED LIGHT
PARAFFINIC (mild or no solvent-refining or
hydrotreatment) see MQV775
MINERAL OIL, PETROLEUM DISTILLATES
CATALYTIC DEWAXED HEAVY
NAPHTHENIC (mild or no solvent-refining or
hydrotreatment) see MQV776
MINERAL OIL, PETROLEUM DISTILLATES
CATALYTIC DEWAXED HEAVY
PARAFFINIC (mild or no solvent-refining or
hydrotreatment) see MQV778
MINERAL OIL, PETROLEUM DISTILLATES
CATALYTIC DEWAXED LIGHT
NAPHTHENIC (mild or no solvent-refining or
hydrotreatment) see MQV777
MINERAL OIL, PETROLEUM DISTILLATES
CATALYTIC DEWAXED LIGHT
PARAFFINIC (mild or no solvent-refining or
hydrotreatment) see MQV779
MINERAL OIL, PETROLEUM
DISTILLATES, HEAVY NAPHTHENIC see
MQV780
MINERAL OIL, PETROLEUM
DISTILLATES, HEAVY PARAFFINIC see
MQV785
MINERAL OIL, PETROLEUM
DISTILLATES, HYDROTREATED (mild)
HEAVY NAPHTHENIC see MQV790
MINERAL OIL, PETROLEUM
DISTILLATES, HYDROTREATED (mild)
HEAVY PARAFFINIC see MQV795
MINERAL OIL, PETROLEUM
DISTILLATES, HYDROTREATED (severe)
HEAVY PARAFFINIC see MQV796
MINERAL OIL, PETROLEUM
DISTILLATES, HYDROTREATED (mild)
LIGHT NAPHTHENIC see MQV800

MINERAL OIL, PETROLEUM DISTILLATES, HYDROTREATED (mild) LIGHT PARAFFINIC see MQV805
MINERAL OIL, PETROLEUM DISTILLATES, LIGHT NAPHTHENIC see MQV810
MINERAL OIL, PETROLEUM DISTILLATES, LIGHT PARAFFINIC see MQV815
MINERAL OIL, PETROLEUM DISTILLATES, SOLVENT-DEWAXED HEAVY NAPHTHENIC (mild or no solvent-refining or hydrotreatment) see MQV820
MINERAL OIL, PETROLEUM DISTILLATES, SOLVENT-DEWAXED HEAVY PARAFFINIC (mild or no solvent-refining or hydrotreatment) see MQV825
MINERAL OIL, PETROLEUM DISTILLATES, SOLVENT-DEWAXED LIGHT NAPHTHENIC (mild or no solvent-refining or hydrotreatment) see MQV835
MINERAL OIL, PETROLEUM DISTILLATES, SOLVENT-DEWAXED LIGHT PARAFFINIC (mild or no solvent-refining or hydrotreatment) see MQV840
MINERAL OIL, PETROLEUM DISTILLATES, SOLVENT-REFINED (mild) HEAVY NAPHTHENIC see MQV845
MINERAL OIL, PETROLEUM DISTILLATES, SOLVENT-REFINED (mild) HEAVY PARAFFINIC see MQV850
MINERAL OIL, PETROLEUM DISTILLATES, SOLVENT-REFINED (mild) LIGHT NAPHTHENIC see MQV852
MINERAL OIL, PETROLEUM DISTILLATES, SOLVENT-REFINED (mild) LIGHT PARAFFINIC see MQV855
MINERAL OIL, PETROLEUM EXTRACTS, HEAVY NAPHTHENIC DISTILLATE SOLVENT see MQV857
MINERAL OIL, PETROLEUM EXTRACTS, HEAVY PARAFFINIC DISTILLATE SOLVENT see MQV859
MINERAL OIL, PETROLEUM EXTRACTS, LIGHT NAPHTHENIC DISTILLATE SOLVENT see MQV860
MINERAL OIL, PETROLEUM EXTRACTS, LIGHT PARAFFINIC DISTILLATE SOLVENT see MQV862
MINERAL OIL, PETROLEUM EXTRACTS, RESIDUAL OIL SOLVENT see MQV863
MINERAL OIL, PETROLEUM RESIDUAL OILS, ACID-TREATED see MQV872
MINERAL OIL, WHITE see MQV875
MINERAL OIL, WHITE (FCC) see MQV750
MINERAL ORANGE see LDS000
MINERAL PITCH see ARO500
MINERAL RED see LDS000
MINERAL SPIRITS see MQV900
MINERAL WHITE see CAX750
MINETOIN see DKQ000
MINETOIN see DNU000
MINGIT see DFP600
MINIDRIL see NNL500
MINIHIST see DBM800
MINIHIST see WAK000
MINILYN see EEH575
MINIMYCIN see OMK000
MINIPLANOR see ZVJ000
MINIUM see LDS000
MINIUM NON-SETTING RL-95 see LDS000
MINOALEUIATIN see TLP750
MINOCIN see MQW100
MINOCYCLIN see MQW250
MINOCYCLINE see MQW250
MINOCYCLINE CHLORIDE see MQW100
MINOCYCLINE HYDROCHLORIDE see MQW100
MINOPHAGEN A see AQW000
MINORAN see MEK700
MINORLAR see EEH520
MINOSSIDILE (ITALIAN) see DCB000

MINOVLAR see EEH520
MINOXIDIL see DCB000
MINOZINAN see MCI500
MINPROG see POC350
MINTACO see NIM500
MINTACOL see NIM500
MINTAL see NBU500
MINTEZOL see TEX000
MINT-O-MAG see MAG750
MINTUSSIN see MGR250
MIN-U-GEL 200 see PAE750
MIN-U-GEL 400 see PAE750
MIN-U-GEL FG see PAE750
MINURIC see DDP200
MINUS see DKE800
MINUS see SEH000
MINUSIN see NNW500
MINZIL see CLH750
MINZOLUM see TEX000
MIOARTRINA see IPU000
MIOBLOCK see PAF625
MIOCURIN see RLU000
MIODAR see PGG350
MIOFILIN see TEP500
MIOLAXENE see GKK000
MIOLISODAL see IPU000
MIOLISODOL see IPU000
MIO 40GN see IHG100
MIONAL see EAV700
MIO-PRESSIN see RDK000
MIORATRINA see IPU000
MIORELAX see RLU000
MIORIL see IPU000
MIORILAS see GKK000
MIORILAX see CKF500
MIORIODOL see IPU000
MIO-SED see CKF500
MIOSTAT see CBH250
MIOTICOL see DNR309
MIOTINE see MIC250
MIOTISAL see NIM500
MIOTISAL A see NIM500
MIOTOLON see AOO300
MIOWAS see GKK000
MIPAFOX see PHF750
MIPAX see DTR200
MIPC see MIA250
MIPCIN see MIA250
MIPCINE see MIA250
MI-PILO OPHTH SOL see PIF250
MIPK see MLA750
MIPSIN see MIA250
MIPSIN, nitrosated (JAPANESE) see MMW250
MeIQ see AJQ600
MIRACIL D see DHU000
MIRACIL D see SBE500
MIRACIL D HYDROCHLORIDE see SBE500
MIRACLE see DAA800
MIRACOL see SBE500
MIRADOL see EPD500
MIRAL see PDN000
MIRAL see PHK000
MIRAL 10 G see PHK000
MIRAMEL see EQL000
MIRAMID WM 55 see PJY500
MIRAMINE GT see TAC075
MIRAMINE TOC see TAC050
MIRANOL C2M ANHYDROUS ACID see MQW300
MIRANOL C 2M CONC. see IAT250
MIRANOL C2M CONC. ANHYDROUS ACID see MQW300
MIRANOL C2M CONC. NP PROPYLENE GLYCOL see MQW300
MIRANOL C2M-SF CONC see AOC250
MIRANOL C 2M-SF CONC. see IAT250
MIRANOL MHT see AOC275
MIRAPRONT see DTJ400
MIRASON 9 see PJS750
MIRASON 16 see PJS750
MIRASON M 15 see PJS750
MIRASON M 50 see PJS750

MIRASON M 68 see PJS750
MIRASON NEO 23H see PJS750
MIRATAINE BB see LBI500
MIRATHEN see PJS750
MIRATHEN 1313 see PJS750
MIRATHEN 1350 see PJS750
MIRBANE 850 see MCB050
MIRBANE MR2 see MCB050
MIRBANE OIL see NEX000
MIRBANE SM 607 see MCB050
MIRBANE SM 800 see MCB050
MIRBANE SM 850 see MCB050
MIRBANE SU 118K see UTU500
MIRBANIL see EPD500
MIRCOL see MCJ400
MIRCOSULFON see PPP500
MIREX see MQW500
MIRLON see NOH000
MIROISTONIL see DRM000
MIROMORFALIL see NAG500
MIROPROFEN see IAY000
MIROSERINA see CQH000
MIROTIN see MNZ000
MIRREX MCFD 1025 see PKQ059
MIRTAZAPINE see HDS210
MISASIN see CKL500
MISCLERON see ARQ750
MISHERI (INDIA) see SED400
MISHRI (INDIA) see SED400
MISODINE see DBB200
MISOLYNE see DBB200
MISONIDAZOLE see NHH500
MISOPROSTOL see MJE775
MISPICKEL see ARJ750
MISTABRON see MDK875
MISTABRONCO see MDK875
MISTLETOE (AMERICAN) see MQW525
MISTRAL see MRQ300
MISTRAL T see MRQ300
MISTRON 2SC see TAB750
MISTRON FROST P see TAB750
MISTRON RCS see TAB750
MISTRON STAR see TAB750
MISTRON SUPER FROST see TAB750
MISTRON VAPOR see TAB750
MISTURA C see CBH250
MISULBAN see BOT250
MISULVAN see EPD500
MISURAN see MSC100
MIT see ISE000
MITABAN see MJL250
MITAC see MJL250
MITACIL see DOR400
MITANOLINE see PMC250
MIT-C see AHK500
MITC see ISE000
MITENON see MMD500
MITEXAN see MDK875
MITHRACIN see MQW750
MITHRAMYCIN see MQW750
MITHRAMYCIN A see MQW750
MITHRAMYCIN, MAGNESIUM SALT see MAD000
MITICIDE K-101 see CJT750
MITIGAN see BIO750
MITION see CKM000
MITIS GREEN see COF500
MITOBROMOL see DDP600
MITOBRONITOL see DDP600
MITO-C see AHK500
MITOCHROMIN see MQX000
MITOCIN-C see AHK500
MITOCROMIN see MQX000
MITOLAC see DDJ000
MITOLACTOL see DDJ000
MITOMALCIN see MQX250
MITOMEN see CFA500
MITOMEN see CFA750
MITOMIN see CFA500
MITOMYCIN see AHK500
MITOMYCIN A see MQX500
MITOMYCIN B see MQX750

MITOMYCIN-C see AHK500
MITOMYCINUM see AHK500
MITONAFIDE see MQX775
MITOSTAN see BOT250
MITOTANE see CDN000
MITOX see CEP000
MITOXAN see CQC500
MITOXAN see CQC650
MITOXANA see IMH000
MITOXANTHRONE see MQY090
MITOXANTRONE see MQY090
MITOXANTRONE HYDROCHLORIDE see
MQY100
MITOXINE see BIE500
MITOZOLOMIDE see MQY110
MITRAMYCIN see MQW750
MITRONAL see CMR100
MITSUI ACID PURE BLUE VX see CMM062
MITSUI ALIZARINE B see DMG800
MITSUI ALIZARINE RED S see SEH475
MITSUI ALIZARINE SAPHIROL SE see
APG700
MITSUI ANTHRACENE BROWN see
TKN500
MITSUI AURAMINE O see IBA000
MITSUI BENZOPURPURINE 4BX see
DXO850
MITSUI BLUE B BASE see DCJ200
MITSUI BRILLIANT GREEN G see BAY750
MITSUI CHROME BLUE BLACK BC see
CMP880
MITSUI CHROME ORANGE A see NEY000
MITSUI CHROME ORANGE GR see CMP882
MITSUI CHROME VIOLET BC see HLI000
MITSUI CHROME YELLOW GG see SIT850
MITSUI CONGO RED see SGQ500
MITSUI CROME ORANGE AN see NEY000
MITSUI CRYSTAL VIOLET see AOR500
MITSUI DIRECT BLACK BH see CMN800
MITSUI DIRECT BLACK D see CMN230
MITSUI DIRECT BLACK EX see AQP000
MITSUI DIRECT BLACK GX see AQP000
MITSUI DIRECT BLUE 2BN see CMO000
MITSUI DIRECT BLUE RW see CMO600
MITSUI DIRECT BRILLIANT SCARLET 8B
see CMO880
MITSUI DIRECT BROWN M see CMO800
MITSUI DIRECT DARK GREEN BX see
CMO830
MITSUI DIRECT FAST RED F see CMO870
MITSUI DIRECT FAST SCARLET 4BS see
CMO870
MITSUI DIRECT GREEN BC see CMO840
MITSUI DIRECT SCARLET 3BX see CMO875
MITSUI DIRECT SKY BLUE 5B see CMO500
MITSUI GARNET GBC SALT see MPY750
MITSUI INDIGO PASTE see BGB275
MITSUI INDIGO PURE see BGB275
MITSUI METANIL YELLOW see MDM775
MITSUI METHYLENE BLUE see BJI250
MITSUI MILLING SCARLET G see CMM320
MITSUI NAPHTHOZOL AS see CMM760
MITSUI NYLON FAST SKY BLUE B see
CMM090
MITSUI RED B BASE see NEQ000
MITSUI RED 3GL BASE see KDA050
MITSUI RED RL BASE see MMF780
MITSUI RED TR BASE see CLK220
MITSUI RHODAMINE see RGW000
MITSUI RHODAMINE BX see FAG070
MITSUI RHODAMINE 6GCP see RGW000
MITSUI SAFRANINE see GJI400
MITSUI SCARLET G BASE see NMP500
MITSUI SULPHUR BLACK B see CMS250
MITSUI SULPHUR BLACK BC see CMS250
MITSUI SULPHUR BLACK BF see CMS250
MITSUI SULPHUR BLACK BO see CMS250
MITSUI SULPHUR BLACK BS see CMS250
MITSUI SULPHUR BLACK G see CMS250
MITSUI SULPHUR BLACK GF see CMS250
MITSUI SULPHUR RED BROWN 3B see
CMS257

MITSUI TARTRAZINE see FAG140
MITSUI TSUYA INDIGO 2B see ICU135
MITUSI SCARLET GG BASE see DEO295
MIVIZON see FNF000
MIXTURE of p-METHENOLS see TBD500
MIZODIN see DBB200
MIZOLIN see DBB200
MIZORIBINE see BMM000
MJ 505 see PGG350
MJ 505 see PGG355
MJ 1992 see SKW000
MJ 1992 see SKW500
MJ 1998 see HLX500
MJ 1999 see CCK250
MJ 5022 see MPE250
MJ 5190 see AHL500
MJ 10061 see DDP200
MJ 4309-1 see OPK000
MJF 9325 see IMH000
MJF-12264 see FLZ050
MJ 1999 HYDROCHLORIDE see CCK250
MK₄ see VTA650
MK-33 see GEW700
MK 56 see POL500
MK 75 see NCW100
MK 125 see SOW000
MK 141 see PCI500
MK 142 see MQY125
MK 184 see TAF675
MK-188 see RBF100
MK 196 see IBQ400
(±)-MK 196 see IBQ400
MK-217 see SKL600
MK 231 see SOU550
MK 239 see CGH750
MK 240 see DDA600
MK 244 see EAJ150
MK 351 see DNA800
MK-351 see MJE780
MK 360 see TEX000
MK-366 see BAB625
MK 421 see EAO100
MK 486 see CBQ500
MK 521 see LGM400
MK 522 see LGM400
MK-595 see DFP600
MK639 see ICE100
MK 647 see DKI600
MK 665 see CHP750
MK 819 see LII050
MK-872 see HMS875
MK 905 see CBA100
MK 933 see ITD875
MK 936 see ARW200
MK 950 see TGB185
MK-955 see PHA550
MK 0244 see EAJ150
MK 0521 see LGM400
MK. B51 see DNA800
MK.B51 see MJE780
MKC-442 see EAN525
MK-142 DIMETHANESULFONATE see
MQY125
2M-4KH see CIR250
MKH 52 see UTU500
MK 421 MALEATE see EAO100
ML 21 see MCB050
ML 97 see FAB400
ML 97 see PGX300
ML 045 see MCB050
ML 133 see MCB050
ML 33F see SKV100
ML 55F see SKV100
ML 630 see MCB050
ML 1024 see TEQ500
ML 3120 see MCB050
MLA-74 see MLD500
MLO 81-125 see KDK000
MLO 83-322 see KDK000
MLO-5277 (ORGANO-SILICATE) see
MQY250
MLT see MAK700

MM see BKM500
MM 83 see MCB050
MM 4462 see AOP250
MM 14151 see CMV250
MM 160V30M see MCB050
MMA see MGQ250
MMC see AHK500
MMC see MDD750
MMD see MLF250
MME see MFC700
MME see MLH750
M 3 (MELAMINE POLYMER) see MCB050
M'-METHYL-p-
DIMETHYLAMINOAZOBENZENE see
DUH600
MMH see MKN000
4-MMPD see DBO000
4-MMPD SULPHATE see DBO400
MMS see MLH500
MMT see MAV750
MMTs-BTR see MIF760
MMTP see MLX750
ω-MMYCIN see TBX000
N-2-M see CCK665
MN-1695 see MQY300
MNA see MMU250
MNA see NEN500
6-MNA see MMT250
MNA DIMETHYL ACETAL see MNB600
MNB see MMU500
MNBK see HEV000
MNC see MMX000
MN-CELLULOSE see CCU150
MNCO see MQY325
MNE see NDM000
MNFA see MME809
MNG see MMP000
MNNG see MMP000
MNPA see MFA500
MNPN see MMS200
MNQ see MMD500
MNT see MDQ075
MNT see MNA500
MNT see NMO500
MNU see MMX250
MNU see MNA750
MNU and CYCLOPHOSPHAMIDE (2:1) see
CQC600
MO see NIW500
MO 338 see NIW500
MO-500 see FKM100
MO 709 see DRM000
MOB see MES000
M-2-OB see MMR800
M-3-OB see MMR810
MOBAM see BDG250
MOBAM PHENOL see BDG250
MOBAN see MRB250
MOBAY MRS see PKB100
MOBENOL see BSQ000
MOBILAN see IDA000
MOBILAT see MQY350
MOBILAWN see DFK600
MOBIL DBHP see DEG800
MOBIL MC-A-600 see BDG250
MOBILTHERM LIGHT see AQZ150
MOBIL V-C 9-104 see EIN000
MOBUTAZON see MQY400
MOBUZON see MQY400
MOC-45 see NND100
MOCA see MJM200
MOCAP see EIN000
MOCK AZALEA see DBA450
MOCTYNOL see GGS000
MODALINE SULFATE see MOM750
MODANE see DMH400
MODANE SOFT see DJL000
MODECCIN see MRA000
MODECCIN TOXIN see MRA000
MODENOL see RDK000
MODERAMIN see MFD500
MODERIL see TLN500

MODIFIED HOP EXTRACT. see HGK750
MODIFIED POLYACRYLAMIDE RESINS see MRA075
MODIFIER 113-63 see MRA100
MODIRAX see POA250
MODITEN see FMP000
MODITEN see TJW500
MODITEN ENANTHATE see PMI250
MODITEN-RETARD see PMI250
MODOCOLL 1200 see SFO500
MODOWN see MJB600
MODR ALIZARINOVA CISTA B (CZECH) see AIY750
MODR BRILANTNI ALIZARINOVA BRL (CZECH) see DKR400
MODR BRILANTNI OSTAZINOVA S-R (CZECH) see DGN400
MODR DISPERZNI 56 see CMP070
MODR DISPERZNI 73 see CMP075
MODRENAL see EBY600
MODR FRALOSTANOVA 3G (CZECH) see DNE400
MODR KYPOVA 1 see BGB275
MODR KYPOVA 4 see IBV050
MODR KYPOVA 18 see VGP100
MODR KYSELA 1 see ADE500
MODR KYSELA 41 see CMM070
MODR KYSELA 78 see CMM090
MODR KYSELA 129 see CMM100
MODR METHYLENOVA (CZECH) see BJI250
MODR MIDLONOVA STALA ER (CZECH) see TMA750
MODR NAMORNICKA OSTANTHRENOVA G (CZECH) see CMU500
MODR NAMORNICKA OSTANTHRENOVA RA (CZECH) see CMU750
MODR OSTACETOVA LG (CZECH) see BND250
MODR OSTACETOVA LR see BNC800
MODR OSTACETOVA P3R (CZECH) see MGG250
MODR OSTACETOVA SE-LB (CZECH) see DBT000
MODR PIGMENT 60 see IBV050
MODR POTRAVINARSKA 3 see ADE500
MODR POTRAVINARSKA 4 see IBV050
MODR PRIMA 15 see CMO500
MODR REAKTIVNI 4 see CMS222
MODULEX see CBT750
MOEBIQUIN see DNF600
MOENOMYCIN see MRA250
MOENOMYCIN A see MRA250
MOEXIPRIL HYDROCHLORIDE see MRA260
MOF see DFA400
MOFEBUTAZONE see MQY400
MOGADAN see DLY000
MOGALAROL see MRA275
MOGETON GRANULE see AJI250
MOGUL see CBT750
MOGUL L see CBT750
MOHEPTAN see MDP750
MOHICAN RED A-8008 see CHP500
MOHR'S SALT see IGL200
MOLACCO see CBT750
MOLANTIN P see MRA300
MOLASSES ALCOHOL see EFU000
MOLATOC see DJL000
MOLCER see DJL000
MOLDAMIN see BFC750
MOLDCIDIN B see FPC000
MOLDEX see HJL500
MOLECULAR CHLORINE see CDV750
MOLECULAR SIEVE 13X with 14.6% DI-n-BUTYLAMINE see MRA750
MOLE DEATH see SMN500
MOLEVAC see PQC500
MOLINATE see EKO500
MOLINATE SULFOXIDE see MRA800
MOLINDONE HYDROCHLORIDE see MRB250

MOLINILLO (PUERTO RICO) see SAT875
MOLIPAXIN see CKJ000
MOLIPAXIN see THK880
MOL-IRON see FBO000
MOLIVATE see CMW400
MOLLAN O see DVL700
MOLLINOX see QAK000
MOLLUSCICIDE BAYER 73 see DFV600
MOLMATE see EKO500
MOLOFAC see DJL000
MOLOL see MQV750
MOLSIDOLAT see MRN275
MOLSIDOMINE see MRN275
MOLTEN ADIPIC ACID see AEN250
MOLURAME see BJK500
MOLYBDATE see MRC250
MOLYBDATE, CALCIUM see CAT750
MOLYBDATE, HEXAAMMONIUM (9CI) see ANH600
MOLYBDATE (MOO4(2-)), DIPOTASSIUM, (T-4)- see PLL125
MOLYBDATE ORANGE see MRC000
MOLYBDATE ORANGE Y 786D see MRC000
MOLYBDATE ORANGE YE 421D see MRC000
MOLYBDATE ORANGE YE 698D see MRC000
MOLYBDATE RED see MRC000
MOLYBDATE RED AA3 see MRC000
MOLYBDENITE see MRD750
MOLYBDEN RED see MRC000
MOLYBDENUM see MRC250
MOLYBDENUM AZIDE PENTACHLORIDE see MRC500
MOLYBDENUM AZIDE TRIBROMIDE see MRC600
MOLYBDENUM BORIDE see MRC650
MOLYBDENUM CHLORIDE (MoCl$_3$) see MRD800
MOLYBDENUM-COBALT-CHROMIUM ALLOY see VSK000
MOLYBDENUM COMPOUNDS see MRC750
MOLYBDENUM DIAZIDE TETRACHLORIDE see MRD000
MOLYBDENUM DIOXIDE see MRD250
MOLYBDENUM-LEAD CHROMATE see LDM000
MOLYBDENUM ORANGE see LDM000
MOLYBDENUM OXIDE see MRD250
MOLYBDENUM(IV) OXIDE see MRD250
MOLYBDENUM(VI) OXIDE see MRE000
MOLYBDENUM PENTACHLORIDE see MRD500
MOLYBDENUM RED see MRC000
MOLYBDENUM(IV) SULFIDE see MRD750
MOLYBDENUM, TRICARBONYL(1,3,5-CYCLOHEPTATRIENE)- see COY100
MOLYBDENUM TRICHLORIDE see MRD800
MOLYBDENUM TRIOXIDE see MRE000
MOLYBDIC ACID DIAMMONIUM SALT see ANM750
MOLYBDIC ACID, DIPOTASSIUM SALT see PLL125
MOLYBDIC ACID, DISODIUM SALT see DXE800
MOLYBDIC ACID, HEXAAMMONIUM SALT see ANH600
MOLYBDIC ACID (H$_2$MoO$_4$), CALCIUM SALT (1:1) see CAT750
MOLYBDIC ANHYDRIDE see MRE000
MOLYBDIC SULFIDE see MRD750
MOLYBDIC TRIOXIDE see MRE000
MOLYKOTE see MRD750
MOLYKOTE 522 see TAI250
MOMENTOL see TKX000
MOMENTUM see HIM000
MOMETINE see CJL500
MOMORDICA BALSAMINA see FPD100
MOMORDICA CHARANTIA see FPD100
MOMORDIQUE A FUEILLES de VIGNE see FPD100

MON 39 see GIQ100
MON 097 see CGO550
MON 139 see GIQ100
MON 0573 see PHA500
MON 2139 see PHA500
MON 4620 see ABY300
MON 4621 see ABY300
MON 7200 see POQ600
MON 12000 see CGC150
MON 13200 see TEX550
MON 15100 see POQ600
MONACETYLFERROCENE see ABA750
MONACOLINIC K ACID see LII050
MONACRIN see AHS500
MONACRIN see AHS750
MONACRIN HYDROCHLORIDE see AHS750
MONAGYL see MMN250
MONALIDE see CGL250
MONAM see SIL550
MONAMID 150-LW see BKE500
MONAQUEST see DJG800
MONARCH see CBT750
MONARCH see ZVJ000
MONARCH GREEN WD see PJQ100
MONARGAN see ABX500
MONASIRUP see PMQ750
MONASPOR see CCS550
MONASTRAL FAST GREEN 3Y see CMS140
MONASTRAL FAST GREEN 6Y see CMS140
MONASTRAL FAST GREEN 3YA see CMS140
MONASTRAL FAST GREEN 6YA see CMS140
MONASTRAL FAST GREEN BGNA see PJQ100
MONASTRAL FAST GREEN G see PJQ100
MONASTRAL FAST GREEN GD see PJQ100
MONASTRAL FAST GREEN GF see PJQ100
MONASTRAL FAST GREEN GFNP see PJQ100
MONASTRAL FAST GREEN GN see PJQ100
MONASTRAL FAST GREEN GNA see PJQ100
MONASTRAL FAST GREEN GTP see PJQ100
MONASTRAL FAST GREEN GV see PJQ100
MONASTRAL FAST GREEN GWD see PJQ100
MONASTRAL FAST GREEN GX see PJQ100
MONASTRAL FAST GREEN GXB see PJQ100
MONASTRAL FAST GREEN GYH see PJQ100
MONASTRAL FAST GREEN LGNA see PJQ100
MONASTRAL FAST GREEN 2GWD see PJQ100
MONASTRAL GREEN B see PJQ100
MONASTRAL GREEN B PIGMENT see PJQ100
MONASTRAL GREEN G see PJQ100
MONASTRAL GREEN GFN see PJQ100
MONASTRAL GREEN GH see PJQ100
MONASTRAL GREEN GN see PJQ100
MONASTRAL GREEN Y-GT 805D see CMS140
MONATE see MRL750
MONAWET MD 70E see DJL000
MONAWET MM 80 see DKP800
MONAZAN see MQY400
MONCEREN see UTA400
MONCHROME VIOLET B see HLI000
MONCIDE see HKC000
MONCUT see TKB286
MONDUR E 429 see PKB100
MONDUR E 441 see PKB100
MONDUR E 541 see PKB100
MONDUR MR see PKB100
MONDUR MR 200 see PKB100
MONDUR MRS see PKB100
MONDUR MRS 10 see PKB100
MONDUR P see PFK250

MONDUR-TD see TGM740
MONDUR TD see TGM750
MONDUR-TD-80 see TGM740
MONDUR TD-80 see TGM750
MONDUR TDS see TGM750
MONEL 400 see NCX525
MONELAN see MRE225
MONELGIN see OAV000
MONENSIC ACID see MRE225
MONENSIN A see MRE225
MONENSIN, MONOSODIUM SALT (9CI) see MRE230
MONENSIN SODIUM see MRE230
MONENSIN SODIUM SALT see MRE230
MONENSIN (USDA) see MRE225
MONEX see BJL600
MONGARE see MJK750
MONGUARD see DFQ500
MONHYDRIN see PMJ500
MONILIFORMIN see MRE250
MONISTAT see MQS550
MONITAN see PKL100
MONITE see HJE100
MONITOR see DTQ400
MONKEY PISTOL see SAT875
MONKEY'S DINNER BELL see SAT875
MONKIL WP see MGQ750
MONKSHOOD see MRE275
MONOACETIN see GGO000
MONO-ACETIN see MRE300
1-MONOACETIN see GGO000
α-MONOACETIN see GGO000
MONOACETOXYSCIRPENOL see ECT700
MONOACETYL GLYCERINE see GGO000
MONOACETYLHYDRAZINE see ACM750
MONOACETYL 4-HYDROXYAMINOQUINOLINE see QQA000
MONOACETYLISOSPIRAMYCIN see MRE500
15-MONO-o-ACETYLSCIRPENOL see ECT700
MONO-o-ACETYLSOLANOSIDE see MRE600
MONOACETYLSPIRAMYCIN see MRE750
MONOAETHANOLAMIN (GERMAN) see EEC600
MONOALLYLAMINE see AFW000
MONOALLYL GLYCERIN ETHER see GGA912
MONOALLYLUREA see AGV000
MONOALUMINUM PHOSPHATE see PHB500
MONOAMMONIUM CARBONATE see ANB250
MONOAMMONIUM GLUTAMATE see MRF000
MONOAMMONIUM l-GLUTAMATE see MRF000
MONOAMMONIUM GLUTAMATE MONOHYDRATE see MRF100
MONOAMMONIUM l-GLUTAMATE MONOHYDRATE see MRF100
MONOAMMONIUM GLYCYRRHIZINATE see AMY700
MONOAMMONIUM GLYCYRRHIZINATE see GIE100
MONOAMMONIUM METHANEARSONATE see MDQ770
MONOAMMONIUM SULFAMATE see ANU650
MONOAMMONIUM SULFATE see ANJ500
MONOAMMONIUM SULFIDE see ANJ750
MONOAMMONIUM SULFITE see ANB600
MONOAMYLAMINE see PBV505
MONOANTHRANILOYLLYCOCTONINE see LJB150
MONOAZO see MDM775
MONOBASIC racemic AMPHETAMINE PHOSPHATE see AOB500
MONOBASIC CHROMIUM SULFATE see CMJ565

MONOBASIC CHROMIUM SULFATE see NBW000
MONOBASIC CHROMIUM SULPHATE see NBW000
MONOBASIC LEAD ACETATE see LCH000
MONOBASIC dl-α-METHYLPHENETHYLAMINE PHOSPHATE see AOB500
MONOBENZALPENTAERYTHRITOL see MRF200
MONOBENZONE see AEY000
MONOBENZYLAMINE see BDX750
MONOBENZYL-p-AMINOPHENOL HYDROCHLORIDE see MRF250
MONOBENZYL ETHER HYDROQUINONE see AEY000
MONOBENZYL HYDROQUINONE see AEY000
MONOBENZYL PHTHALATE see MRK100
MONO-N-BENZYL PHTHALATE see MRK100
MONOBOR-CHLORATE see SFS500
MONOBROMESSIGSAEURE (GERMAN) see BMR750
MONOBROMOACETALDEHYDE see BMR000
MONOBROMOACETIC ACID see BMR750
MONOBROMOACETONE see BNZ000
MONOBROMOBENZENE see PEO500
MONOBROMODIFLUOROPHOSPHINE SULFIDE see TFO500
MONOBROMOETHANE see EGV400
MONOBROMOGLYCEROL see MRF275
MONOBROMOISOVALERYLUREA see BNP750
2-MONOBROMOISOVALERYLUREA see BNP750
MONOBROMOMETHANE see MHR200
MONOBUTILAMINA see BPX750
MONOBUTYL see MQY400
MONOBUTYLAMINE see BPX750
MONO-n-BUTYLAMINE see BPX750
MONOBUTYL DIPHENYL SODIUM MONOSULFONATE see AQU750
MONOBUTYL GLYCOL ETHER see BPJ850
MONO-tert-BUTYLHYDROQUINONE see BRM500
MONOBUTYL PHOSPHITE see MRF500
MONOBUTYL PHTHALATE see MRF525
MONO-n-BUTYL PHTHALATE see MRF525
MONOBUTYLTIN TRICHLORIDE see BSO750
MONOBUTYLTIN TRILAURATE see BSO750
MONOCAINE HYDROCHLORIDE see IAC000
MONOCALCIUM ARSENITE see CAM500
MONOCALCIUM CARBONATE see CAT775
MONOCALCIUM DI-l-GLUTAMATE TETRAHYDRATE see MRF550
MONOCALCIUM DISODIUM EDTA see CAR780
MONOCALCIUM PHOSPHATE see CAW110
MONOCALCIUM PHOSPHATE MONOHYDRATE see CAY100
MONOCARBAMIDE DIHYDROGEN SULFATE see UTU600
MONOCARBETHOXYHYDRAZINE see EHG000
MONOCEL SCARLET B see ENP100
MONOCHLOORAZIJNZUUR (DUTCH) see CEA000
MONOCHLOORBENZEEN (DUTCH) see CEJ125
MONOCHLORACETIC ACID see CEA000
MONOCHLORACETONE see CDN200
MONOCHLORALUREA see HOL100
MONOCHLORAMIDE see CDO750
MONOCHLORAMINE see CDO750
MONOCHLORBENZENE see CEJ125
MONOCHLORBENZOL (GERMAN) see CEJ125

MONOCHLORESSIGSAEURE (GERMAN) see CEA000
MONOCHLORETHANE see EHH000
MONOCHLORHYDRIN see CDT750
MONOCHLORHYDRINE du GLYCOL (FRENCH) see EIU800
MONOCHLORIMIPRAMINE see CDU750
MONOCHLOROACETALDEHYDE see CDY500
MONOCHLOROACETIC ACID see CEA000
MONOCHLOROACETIC ACID METHYL ESTER see MIF775
MONOCHLOROACETONE see CDN200
MONOCHLOROACETONE, inhibited (DOT) see CDN200
MONOCHLOROACETONE, stabilized (DOT) see CDN200
MONOCHLOROACETONE, unstabilized (DOT) see CDN200
MONOCHLOROACETONITRILE see CDN500
MONOCHLOROACETYL CHLORIDE see CEC250
MONOCHLOROAMINE see CDO750
MONOCHLOROAMMONIA see CDO750
α-MONOCHLOROANTHRAQUINONE see CEI000
MONOCHLOROBENZENE see CEJ125
MONOCHLOROBIPHENYL see CGM550
3-MONOCHLOROBIPHENYL see CGM770
MONOCHLOROCYCLOHEXANE see CPI400
3-MONOCHLORODIBENZOFURAN see MRF575
MONOCHLORODIBROMOTRIFLUOROETHANE see MRF600
MONOCHLORODIFLUOROMETHANE see CFX500
MONOCHLORODIMETHYL ETHER (MAK) see CIO250
MONOCHLORODIPHENYL OXIDE see MRG000
MONOCHLOROETHANOIC ACID see CEA000
2-MONOCHLOROETHANOL see EIU800
MONOCHLOROETHENE see VNP000
MONOCHLOROETHYLCYCLOPHOSPHAMIDE see MRG025
MONOCHLOROETHYLENE (DOT) see VNP000
MONOCHLOROETHYLENE OXIDE see CGX000
MONOCHLOROHYDRIN see CDT750
α-MONOCHLOROHYDRIN see CDT750
MONOCHLOROISOTHYMOL see CEX275
MONOCHLOROMETHANE see MIF765
MONOCHLOROMETHYL CYANIDE see CDN500
MONO-CHLORO-MONO-BROMO-METHANE see CES650
MONOCHLOROMONOFLUOROETHANE see FOO553
MONOCHLOROMONOFLUOROMETHANE see CHI900
MONOCHLORONAPHTHALENE see MRG050
MONOCHLOROPENTAFLUOROETHANE (DOT) see CJI500
MONOCHLOROPHENYLETHER see MRG000
p-MONOCHLOROPHENYL PHENYL SULFONE see CKI625
MONOCHLOROPHENYLXYLYLETHANE see MRG060
1-MONOCHLOROPINACOLINE see MRG070
MONOCHLOROPOLYOXYETHYLENE see PJU250
β-MONOCHLOROPROPIONIC ACID see CKS500
MONOCHLOROSULFURIC ACID see CLG500

MONOCHLOROTETRAFLUOROETHANE
(DOT) see CLH000
MONOCHLOROTRIFLUOROETHYLENE
see CLQ750
MONOCHLOROTRIFLUOROMETHANE
(DOT) see CLR250
MONOCHORIA VAGINALIS (Burm. f.) Presl.,
extract see MRG100
MONOCHROME BRILLIANT ORANGE GR
see CMP882
MONOCHROME ORANGE GR see CMP882
MONOCHROME ORANGE R see NEY000
MONOCHROME RED 2GL see CMG750
MONOCHROME YELLOW 3R see NEY000
MONOCHROME YELLOW MG see SIT850
MONOCHROMIUM OXIDE see CMK000
MONOCHROMIUM TRIOXIDE see CMK000
MONOCIL see MRH209
MONOCIL 40 see MRH209
"MONOCITE" METHACRYLATE
MONOMER see MLH750
MONOCLAIR see CCK125
MONOCLOROBENZENE (ITALIAN) see
CEJ125
MONOCOBALT OXIDE see CND125
MONOCOPPER MONOSULFIDE see
CNQ000
MONOCORTIN see PAL600
MONOCRATILIN see MRH000
MONOCRESYL DIPHENYL PHOSPHATE
see TGY750
MONOCRON see MRH209
MONOCROTALIC ACID see MRG200
MONOCROTALINE see MRH000
MONOCROTALINE, 3,8-DIDEHYDRO- see
DAL350
MONOCROTOPHOS see ASN000
MONOCROTOPHOS see MRH209
MONOCROTOPHOS see MRH209
MONOCROTOPHOS (ACGIH,OSHA) see
MRH209
MONOCYANOACETIC ACID see COJ500
MONOCYCLOHEXYLTIN ACID see
MRH212
MONODEHYDROSORBITOL
MONOOLEATE see SKV100
MONODEMETHYLIMIPRAMINE see DSI709
MONO(2,3-
DIBROMOPROPYL)AMMONIUM
PHOSPHATE see MRH214
MONO(2,3-
DIBROMOPROPYL)PHOSPHATE see
MRH217
MONO-DIGITOXID (GERMAN) see DKL875
MONO- and DIGLYCERIDES see MRH215
MONO- and DIGLYCERIDES,
MONOSODIUM PHOSPHATE
DERIVATIVES see MRH218
MONO and DIGLYCERIDES, SODIUM
SULFOACETATE DERIVATIVES see SJZ050
MONODION see VTA000
MONO-, DI-, and TRIPOTASSIUM CITRATE
see MRH225
MONO-, DI-, and TRISODIUM CITRATE see
MRH230
MONO-, DI-, and TRISTEARYL CITRATE see
MRH235
MONODODECYLAMINE see DXW000
MONODODECYL ESTER SULFURIC ACID
see MRH250
MONODORM see BPF500
MONODRAL see PBS000
MONODRAL BROMIDE see PBS000
MONODRIN see MRH209
MONOESTER with 4-BUTYL-4-
(HYDROXYMETHYL)-1,2-DIPHENYL-
SUCCINIC ACID 3,5-
PYRAZOLIDINEDIONE see SOX875
MONOETHANEAMINE BENZOATE see
MRH300
MONOETHANOLAMINE see EEC600

MONOETHANOLAMINE
HYDROCHLORIDE see EEC700
MONOETHANOLAMMONIUM 2-
MERCAPTOBENZOTHIAZOLE see HKO012
MONOETHANOLETHYLENEDIAMINE see
AJW000
MONOETHYL ACID PHOSPHATE see
ECI300
MONOETHYL ADIPATE see ELC600
MONOETHYLADIPIC ACID ESTER see
ELC600
N-MONOETHYLAMIDE of O,O-
DIMETHYLDITHIOPHOSPHORYLACETIC
ACID see DNX600
MONOETHYLAMINE (DOT) see EFU400
MONOETHYLAMINE, anhydrous (DOT) see
EFU400
2-N-MONOETHYLAMINOETHANOL see
EGA500
MONOETHYLAMMONIUM CHLORIDE see
EFW000
MONOETHYLARSINE see EGM150
MONOETHYL BUTYLPHOSPHONATE
ANHYDRIDE WITH DIETHYL
PHOSPHATE see MRH400
MONOETHYLDICHLOROTHIOPHOSPHAT
E see MRI000
MONOETHYLENEDIUREA see EJC100
MONOETHYLENE GLYCOL see EJC500
MONOETHYLENE GLYCOL DIMETHYL
ETHER see DOE600
MONOETHYL ETHER of DIETHYLENE
GLYCOL see CBR000
MONOETHYL HEXANEDIOATE see
ELC600
MONOETHYLHEXYL PHTHALATE see
MRI100
MONO(2-ETHYLHEXYL)PHTHALATE see
MRI100
MONO(2-ETHYLHEXYL)SULFATE
SODIUM SALT see TAV500
MONOETHYLISOSELENOURONIUMBRO
MIDE-HYDROBROMIDE see AJY000
MONOETHYLPHENYLTRIAZENE see
MRI250
MONOETHYL PHOSPHATE see ECI300
MONOETHYL PHTHALATE see MRI275
MONOETHYLTIN TRICHLORIDE see
EPS000
MONOFLUORAZIJNZUUR (DUTCH) see
FIC000
MONOFLUORESSIGSAEURE (GERMAN)
see FIC000
MONOFLUORESSIGSAEURE, NATRIUM
(GERMAN) see SHG500
MONOFLUORETHANOL see FID000
MONOFLUOROACETAMIDE see FFF000
MONOFLUOROACETATE see FIC000
MONOFLUOROACETIC ACID see FIC000
MONOFLUOROETHANE see FIB000
MONOFLUOROETHANOL see FID000
MONOFLUOROETHYLENE see VPA000
α-MONOFLUOROMETHYLHISTIDINE see
MRI285
MONOFLUOROPHOSPHORIC ACID,
anhydrous see PHJ250
MONOFLUOROTRICHLOROMETHANE see
TIP500
MONOFURACIN see NGE500
MONOFURFURYLIDENEACETONE see
FPX030
MONOGERMANE see GEI100
MONOGLYCERIDE CITRATE see MRI300
MONOGLYCEROL-p-AMINOBENZOATE
see GGQ000
MONOGLYCERYL OLEATE see GGR200
MONOGLYCIDYL ETHER of N-
PHENYLDIETHANOLAMINE see MRI500
MONO-GLYCOCOARD see DKL800
MONOGLYME see DOE600
MONO-N-HEXYLAMINE see HFK000
8-MONOHYDRO MIREX see MRI750

MONOHYDROTRIETHYLETAIN see TJV050
MONOHYDROXYBENZENE see PDN750
MONO-2-(2-
HYDROXYETHOXY)ETHYLESTER
KYSELINY MALEINOVE (CZECH) see
MAL500
MONO(HYDROXYETHYL) ESTER MALEIC
ACID see EJG500
MONOHYDROXY-MERCURI-DI-
IODORESORCIN-SULPHONPHTHALEIN
DISODIUM see SIF500
MONOHYDROXYMETHANE see MGB150
MONO-(2-HYDROXYPROPYL)ESTER
KYSELINY MALEINOVE (CZECH) see
MAL750
MONOIODOACETAMIDE see IDW000
MONOIODOACETATE see IDZ000
MONOIODOACETIC ACID see IDZ000
MONOIODOMETHANESULFONIC ACID,
SODIUM SALT see SHX000
MONOIODOTRIFLUOROMETHANE see
IFM100
MONOIODURO di METILE (ITALIAN) see
MKW200
MONOISOAMYL MESO-2,3-
DIMERCAPTOSUCCINATE see MRI760
MONOISOBUTYLAMINE see IIM000
MONO-ISO-BUTYL PHTHALATE see
MRI775
MONO-ISO-PROPANOLAMINE see AMA500
N-MONOISOPROPYLAMIDE of O,O-
DIETHYLDITHIOPHOSPHORYLACETIC
ACID see IOT000
MONOISOPROPYLAMINE see INK000
MONOISOPROPYLAMINE
HYDROCHLORIDE see INL000
MONOISOPROPYLAMINOETHANOL see
INN400
4-MONOISOPROPYLAMINO-1-PHENYL-
2,3-DIMETHYL-5-PYRAZOLONE see
INM000
MONO-ISOPROPYLAMMONIOVA SUL see
GIQ100
MONOISOPROPYLBIPHENYL see IOF200
MONOISOPROPYL CITRATE see MRI785
MONOISOPROPYL ETHER of ETHYLENE
GLYCOL see INA500
MONO-KAY see VTA000
1-MONOLAURIN see MRJ000
MONOLAURYL DIMETHYLAMINE see
DRR800
MONOLINURON see CKD500
MONOLITE FAST BLUE 3R see IBV050
MONOLITE FAST BLUE 3RD see IBV050
MONOLITE FAST BLUE RV see IBV050
MONOLITE FAST BLUE SRS see IBV050
MONOLITE FAST BLUE 2RV see DFN300
MONOLITE FAST BLUE 2RVSA see DFN300
MONOLITE FAST GREEN GVSA see PJQ100
MONOLITE FAST NAVY BLUE BV see
BGB275
MONOLITE FAST ORANGE G see CMS145
MONOLITE FAST ORANGE GA see CMS145
MONOLITE FAST ORANGE R see DVB800
MONOLITE FAST RED G see CJD500
MONOLITE FAST SCARLET CA see MMP100
MONOLITE FAST SCARLET GSA see
MMP100
MONOLITE FAST SCARLET RB see MMP100
MONOLITE FAST SCARLET RBA see
MMP100
MONOLITE FAST SCARLET RN see
MMP100
MONOLITE FAST SCARLET RNA see
MMP100
MONOLITE FAST SCARLET RNV see
MMP100
MONOLITE FAST SCARLET RT see MMP100
MONOLITE FAST YELLOW GLV see
CMS208
MONOLITE RED CN see CMS150
MONOLITE RED R see NAP100

MONOLITE RUBINE 3B see DTV360
MONOLITE YELLOW GL see CMS208
MONOLITE YELLOW GLA see CMS208
MONOLITE YELLOW GT see DEU000
MONOLITHIUM ACETYLIDE-AMMONIA see MRJ125
MONOLUPINE see RHZ100
MONOMER FA see FPQ900
MONOMER MG-1 see EJH000
MONOMETHACRYLIC ETHER of ETHYLENE GLYCOL see EJH000
MONOMETHYLACETAMIDE see MFT750
N-MONOMETHYLAMIDE of O,O-DIMETHYLDITHIOPHOSPHORYLACETIC ACID see DSP400
MONOMETHYLAMINE see MGC250
MONOMETHYL-AMINOAETHANOL (GERMAN) see MGG000
4-MONOMETHYLAMINOAZOBENZENE see MNR500
p-MONOMETHYLAMINOAZOBENZENE see MNR500
MONOMETHYLAMINOETHANOL see MGG000
N-MONOMETHYLAMINOETHANOL see MGG000
2-MONOMETHYLAMINOFLUORENE see FEI500
N-MONOMETHYL-2-AMINOFLUORENE see FEI500
MONOMETHYLANILINE see MGN750
N-MONOMETHYLANILINE see MGN750
MONOMETHYL ANILINE (OSHA) see MGN750
MONOMETHYLARSINIC ACID see MGQ530
10-MONOMETHYLBENZO(a)PYRENE see MHH500
MONO(3-METHYLBUTYL) 2,3-DIMERCAPTOBUTANEDIOATE (R*,S*)- see MRI760
MONOMETHYL DIHYDROGEN PHOSPHATE see DLO800
MONOMETHYL 2,6-DIMETHYL-4-(2-NITROPHENYL)-3,5-PYRIDINEDICARBOXYLATE see MRJ200
MONOMETHYL ETHER of ETHYLENE GLYCOL see EJH500
MONOMETHYL ETHER HYDROQUINONE see MFC700
MONOMETHYLFORMAMIDE see MKG500
MONOMETHYLFOSFIT (CZECH) see PGZ950
MONOMETHYL GUANIDIN (GERMAN) see MKI750
MONOMETHYLGUANIDINE see MKI750
MONOMETHYL HYDRAZINE see MKN000
MONOMETHYLHYDRAZINE NITRATE see MRJ250
MONOMETHYLMALEIC ANHYDRIDE see CMS322
MONOMETHYL MERCURY CHLORIDE see MDD750
MONOMETHYLOLACRYLAMIDE see HLU500
MONOMETHYLOLFORMALDEHYDE see GHO100
MONOMETHYLTIN TRICHLORIDE see MQC750
MONOMYCIN see MRJ600
MONOMYCIN A see NCF500
MONONICKEL MONOSULFIDE see NDL100
MONONITROCHLOROBENZENE see CJA950
MONONITRONAPHTHALENE see NHP990
MONONITROSOCIMETIDINE see MRJ700
MONONITROSOPIPERAZINE see MRJ750
N-MONONITROSOPYRIDINOL CARBAMATE see POR250
MONOOCTADECYLAMINE see OBC000
MONOOCTADECYL ETHER of GLYCEROL see GGA915

MONOOCTYL PHOSPHATE see OFQ050
MONOOCTYLPHOSPHORIC ACID see OFQ050
MONO-n-OCTYLTIN TRICHLORIDE see OGG000
MONO-n-OCTYL-TIN-TRIS-(2-ETHYLHEXYLMERCAPTOACETATE) see OGI000
MONO-N-OCTYL-ZINN-TRICHLORID (GERMAN) see OGG000
MONOOLEIN see GGR200
MONOOLEOYLGLYCEROL see GGR200
MONOPALMITATE SORBITAN see MRJ800
MONOPEN see BFD000
MONOPENTEK see PBB750
MONOPERACETIC ACID see PCL500
MONOPEROXY SUCCINIC ACID see MRK000
MONOPHEN see PFC750
MONOPHENOL see PDN750
MONOPHENYLBUTAZONE see MQY400
MONOPHENYLGLYOXIME see HLI400
MONOPHENYLHEPTAMETHYLCYCLOTETRASILOXANE see HBA600
MONO(PHENYLMETHYL) 1,2-BENZENEDICARBOXYLATE see MRK100
MONOPHENYLUREA see PGP250
MONOPHOR see AOB500
MONOPHOS see AOB500
MONOPHYLLINE see HOA000
MONOPLEX DBS see DEH600
MONOPLEX DCP see BLB750
MONOPLEX DOA see AEO000
MONOPLEX DOS see BJS250
MONOPOTASSIUM ARSENATE see ARD250
MONOPOTASSIUM DIHYDROGEN ARSENATE see ARD250
MONOPOTASSIUM aci-1-DINITROETHANE see MRK250
MONOPOTASSIUM d-GLUCARATE see PLJ350
MONOPOTASSIUM GLUTAMATE see MRK500
MONOPOTASSIUM l-GLUTAMATE (FCC) see MRK500
MONOPOTASSIUM GLUTAMATE MONOHYDRATE see MRK525
MONOPOTASSIUM l-GLUTAMATE MONOHYDRATE see MRK525
MONOPOTASSIUM 2-HYDROXYPROPANOATE ACID see PLK650
MONOPOTASSIUM PHOSPHATE see PLQ405
MONOPOTASSIUM SALT of ACETYLENEDICARBOXYLIC ACID see ACJ500
MONOPOTASSIUM SULFATE see PKX750
MONOPOTASSIUM TARTRATE see PKU600
MONOPRIM see TKZ000
MONO-N-PROPYLAMINE see PND250
MONOPROPYLENE GLYCOL see PML000
MONOPROPYL ETHER of ETHYLENE GLYCOL see PNG750
MONOPYRROLE see PPS250
MONORHEIN see RHZ700
MONORHEUMETTEN see MQY400
MONOSAN see DAA800
MONOSILANE see SDH575
MONOSILICIC ACID see SCN600
MONOSIZER W710L see TJR600
MONOSODIOGLUTAMMATO (ITALIAN) see MRL500
MONOSODIUM 5-ACETAMIDO-2,4,6-TRIIODO-N-METHYLISOPHTHALAMATE see IGC100
MONOSODIUM ACETYLIDE see MRK609
MONOSODIUM ACID METHANEARSONATE see MRL750
MONOSODIUM ACID METHARSONATE see MRL750
MONOSODIUM-β-AMINOETHYL THIOPHOSPHATE see AKB500

MONOSODIUM (4-AMINO-1-HYDROXYBUTYLIDENE)BISPHOSPHONATE TRIHYDRATE see SKL600
MONOSODIUM (4-AMINOPHENYL)ARSONATE see ARA500
MONOSODIUM ARSENATE see ARD600
MONOSODIUM ASCORBATE see ARN125
MONOSODIUM BARBITURATE see MRK750
MONOSODIUM CARBONATE see SFC500
MONOSODIUM CEFAZOLIN see CCS250
MONOSODIUM CEROXITIN see CCS510
MONOSODIUM CITRATE see MRL000
MONOSODIUM CYANURATE see SGB550
MONOSODIUM DIHYDROGEN CITRATE see MRL000
MONOSODIUM DIHYDROGEN PHOSPHATE see SJH100
MONOSODIUM 2,5-DIHYDROXYBENZOATE see GCU050
MONOSODIUM 4-((2,4-DIHYDROXYPHENYL)AZO)BENZENESULFONATE see MRL100
MONOSODIUM-5-ETHYL-5-(1-METHYLBUTYL) THIOBARBITURATE see PBT500
MONOSODIUM FERRIC EDTA see EJA379
MONOSODIUM FLUCLOXACILLIN see FDA100
MONOSODIUM GLUCONATE see SHK800
MONOSODIUM GLUTAMATE see MRL500
α-MONOSODIUM GLUTAMATE see MRL500
MONOSODIUM-l-GLUTAMATE (FCC) see MRL500
MONOSODIUM GLYCOLATE see SHT000
MONOSODIUM (1-HYDROXY-2-(3-PYRIDINYL)ETHYLIDENE)BISPHOSPHONATE see RLF400
MONOSODIUM METHANEARSONATE see MRL750
MONOSODIUM METHANEARSONIC ACID see MRL750
MONOSODIUM METHYLARSONATE see MRL750
MONOSODIUM NITRILOTRIACETATE see NEH800
MONOSODIUM NOVOBIOCIN see NOB000
MONOSODIUM OROLATE HYDRATE see SIY100
MONOSODIUM PHOSPHATE see SJH100
MONOSODIUM PHOSPHATE DERIVATIVES of MONO- and DIGLYCERIDES see MRH218
MONOSODIUM SALT of 2,2'-METHYLENE BIS(3,4,6-TRICHLOROPHENOL) see MRM000
MONOSODIUM-2-SULFANILAMIDOPYRIMIDINE see MRM250
MONOSODIUM-2-SULFANILAMIDOTHIAZOLE see TEX500
MONOSODIUM THYROXINE see LFG050
MONOSODIUM URATE see SKO575
MONOSOL SCARLET 3B see CMG750
MONOSORB XP-4 see SJH100
MONOSPAN see SBG500
MONOSTEARIN see OAV000
MONOSTEARYL TRIMETHYL AMMONIUM CHLORIDE see TLW500
MONOSTEOL see SLL000
MONOSULFUR DICHLORIDE see SOG500
MONOSULPH see MRM300
MONOTARD see LEK000
MONOTEN see PFC750
MONOTHIOBENZOIC ACID see TFC550
MONOTHIOETHYLENEGLYCOL see MCN250
MONOTHIOGLYCEROL see MRM750
α-MONOTHIOGLYCEROL see MRM750
MONOTHIOSUCCINIMIDE see MRN000
MONO-THIURAD see BJL600

MONOTHIURAM see BJL600
MONOTRICHLOR-AETHYLIDEN-α-GLUCOSE (GERMAN) see GFA000
MONOTRICHLORO-TETRA(MONOPOTASSIUM DICHLORO)-PENTA-s-TRIAZINETRIONE see MRN050
MONOUREA SULFURIC ACID ADDUCT see UTU600
MONOVAR see NNQ500
MONOVERIN see SDZ000
MONOVINYLHEPTAMETHYLCYCLOTETRASILOXANE see VPF150
MONOVINYL PHOSPHATE see VQA400
MONOXONE see SFU500
β-MONOXYNAPHTHALENE see NAX000
MONSANTO CP-8574 see PGZ900
MONSANTO CP-10502 see PGZ915
MONSANTO CP-13206 see PHG600
MONSANTO CP-16226 see OAL000
MONSANTO CP-19203 see MOB500
MONSANTO CP-19699 see CON300
MONSANTO CP-40294 see MOB699
MONSANTO CP-40507 see MOB750
MONSANTO CP-43858 see TIP750
MONSANTO CP 47114 see DSQ000
MONSANTO CP-48985 see CFC500
MONSANTO CP-49674 see DOP200
MONSANTO CP 51969 see BNL250
MONSOL SCARLET 3BS see CMG750
MONSTERA DELICIOSA see SLE890
MONSUR see CBM750
MONTAN 80 see SKV100
MONTANE 40 see MRJ800
MONTANE 60 see SKV150
MONTANE 83 see SKV170
MONTANOA TOMENTOSA, leaf extract, crude see ZTS600
MONTANOA TOMENTOSA, leaf extract, semi-purified see ZTS625
MONTANOX 80 see PKL100
MONTAR see HKC000
MONTAR see PJL750
MONTEBAN see NBO600
MONTECATINI L-561 see DRR400
MONTHYBASE see EJM500
MONTHYLE see EJM500
MONTMORILLONITE see BAV750
MONTREL see COD850
MONTROSE PROPANIL see DGI000
MONUREX see CJX750
MONURON see CJX750
MONURON-TCA see CJY000
MONUROX see CJX750
MONURUON see CJX750
MONUURON see CJX750
MONZAOMYCIN see TAB300
MONZET see USJ075
MOON see GGA000
MOONSEED see MRN100
MOOSEWOOD see LEF100
MOP see NKV000
8-MOP see XDJ000
MOPA see MFE250
MOPARI see DGP900
MOPAZIN see MFK500
MOPAZINE see MFK500
MOPERONE CHLORHYDRATE see MNN250
MOPERONE HYDROCHLORIDE see MNN250
MOPLEN see PMP500
MOPLENE see MRN125
MOPLEN RO-QG 6015 see PJS750
MOPOL M see MRD750
MOPOL S see MRD750
MOQUIZONE see MRN250
MOQUIZONE HYDROCHLORIDE see PCJ350
MORAMIDE see AFJ400
MORANTEL TARTRATE see MRN260
MORANTREL TARTRATE see MRN260
MORBOCID see FMV000

MORBUSAN see ENB500
MOR-CRAN see NBL200
MORDANT YELLOW 3R see SIU000
MOREPEN see AOD125
MORESTAN see ORU000
MORESTANE see ORU000
MORESTIN see EDV600
MORFAMQUAT see BJK750
MORFAX see BDF750
MORFINA (ITALIAN) see MRO500
MORFLEX 510 see TJR600
MORFOTHION (DUTCH) see MRU250
MORFOXONE see BJK750
MORIAL see MRN275
MORIAMIN S 2 see AHR600
MORIAMIN SN see AHR600
MORIHEPAMINE see AHR600
MORIN see MRN500
MORIN see MRN500
MORINGA OLEIFERA Lamk., extract excluding roots see MRN550
MORIPERAN see AJH000
MORIPRON see AHR600
MORISYLYTE CITRATE see MRN600
MORNIDINE see CJL500
MOROCIDE see BGB500
MORONAL see NOH500
MOROSAN see DCK759
MOROZOL 2 see MRN625
MORPAN CBP see HCP800
MORPAN CHA see HCQ525
MORPAN T see TCB200
MORPHACETIN see HBT500
MORPHACTIN see CDT000
MORPHANQUAT DICHLORIDE see BJK750
MORPHERIDINE see MRN675
MORPHERIDINE DIHYDROCHLORIDE see MRN675
MORPHIA see MRO500
MORPHINA see MRO500
MORPHINAN-3,6-α-DIOL, 7,8-DIDEHYDRO-4,5-α-EPOXY-17-METHYL-, 6-ACETATE see ACR600
MORPHINAN-3,6-DIOL, 7,8-DIDEHYDRO-4,5-EPOXY-17-METHYL-, (5-α-6-α)-, MONOHYDRATE see MRP100
MORPHINAN, 3-METHOXY-17-PHENETHYL-, TARTRATE, (−)- see MRN700
MORPHINAN-6-α-OL, 7,8-DIDEHYDRO-4,5-α-EPOXY-3-METHOXY-17-METHYL-, PHOSPHATE (1:1) see CNG500
MORPHINAN-6-OL, 7,8-DIDEHYDRO-4,5-EPOXY-3-METHOXY-17-METHYL-, (5-α,6-α)-, PHOSPHATE(1:1) (SALT), MIXT. WITH 2-(ACETYLOXY)BENZOIC ACID, 3,7-DIHYDRO-1,3,7-TRIMETHYL-1H-PURINE-2,6-DIONE, N-(4-ETHOXYPHENYL)ACETAMIDE AND 5-ETHYL-5-PHENYL-2,4,6(1H,3H,5H)-PYRIMIDINETRIONE see SBN400
MORPHINAN-6-α-OL, 7,8-DIDEHYDRO-4,5-α-EPOXY-3-METHOXY-17-METHYL-, SULFATE (2:1) (salt) see CNG750
MORPHINAN-6-ONE,17-(CYCLOPROPYLMETHYL)-4,5-α-OXY-3,14-DIHYDOXY-,HYDROCHLORIDE see NAH100
MORPHINAN-6-ONE, 3,14-DIHYDROXY-4,5-α-EPOXY-17-METHYL-, HYDROCHLORIDE see ORG100
MORPHINAN-6-ONE, 4,5-EPOXY-3,14-DIHYDROXY-17-METHYL-, HYDROCHLORIDE, (5-α)- (9CI) see ORG100
MORPHINAN-6-ONE, 4,5-α-EPOXY-3,14-DIHYDROXY-17-(2-PROPENYL)- see NAG550
MORPHINAN-6-ONE, 4,5-α-EPOXY-3-METHOXY-17-METHYL-, TARTRATE (1:1) see DKX050
MORPHINAN, 6,7,8,14-TETRADEHYDRO-4,5-α-EPOXY-3,6-DIMETHOXY-17-METHYL-, HYDROCHLORIDE see TEN100

(−)-MORPHINE see MRO500
MORPHINE see MRO500
MORPHINE 6-ACETATE see ACR600
MORPHINE CHLORHYDRATE see MRO750
MORPHINE CHLORIDE see MRO750
MORPHINE DIACETATE see HBT500
MORPHINE HYDROCHLORIDE see MRO750
MORPHINE METHOCHLORIDE see MRP000
MORPHINE METHYLCHLORIDE see MRP000
MORPHINE-3-METHYL ETHER see CNF500
MORPHINE MONOHYDRATE see MRP100
MORPHINE MONOMETHYL ETHER see CNF500
MORPHINE SULFATE see MRP250
MORPHINE SULPHATE see MRP250
MORPHINISM see MRO500
MORPHINUM see MRO500
MORPHIUM see MRO500
MORPHOCYCLINE see MRP500
MORPHOL see SMP450
MORPHOL (GROWTH REGULATOR) see SMP450
MORPHOLINE see MRP750
MORPHOLINE, N-AMINOPROPYL- see AMF250
MORPHOLINE, AQUEOUS MIXTURE (DOT) see MRP750
MORPHOLINE, 4-(3-(BIS(2-CHLOROETHYL)AMINO)-p-TOLUOYL)- see BHR400
MORPHOLINEBORANE see MRQ250
MORPHOLINE, compounded with BORANE (1:1) see MRQ250
MORPHOLINE, 4-BUTYL- see BRV100
4-MORPHOLINECARBOXALDEHYDE see FOA100
4-MORPHOLINECARBOXAMIDE, N-(2-((2-HYDROXY-3-(4-HYDROXYPHENOXY)PROPYL)AMINO)ETHYL)-, (E)-2-BUTENEDIOATE (2:1) (salt) see XAH000
MORPHOLINE, 4-(1-CYCLOPENTEN-1-YL)- see CPY800
MORPHOLINE, 4-DECANOYL- see CBF725
MORPHOLINE,2,6-DIMETHYL-4-(3-(4-(1,1-DIMETHYLETHYL)PHENYL)-2-METHYLPROPYL)-, cis- see FAQ300
MORPHOLINE, 4-(3-(4-(1,1-DIMETHYLETHYL)PHENYL)-2-METHYLPROPYL)-2,6-DIMETHYL- see MRQ300
MORPHOLINE, 2,6-DIMETHYL-4-NITROSO-, (Z)- see NKA695
MORPHOLINE, 2,6-DIMETHYL-N-NITROSO-, (cis)- see NKA695
MORPHOLINE DISULFIDE see BKU500
MORPHOLINE, 4,4'-(DITHIODICARBONOTHIOYL)BIS-(9CI) see MRR090
4-MORPHOLINEETHANAMINE see AKA750
4-MORPHOLINEETHANESULFONIC ACID see MRT150
MORPHOLINE ETHANOL see MRQ500
4-MORPHOLINEETHANOL, HYDROCHLORIDE see HKW300
MORPHOLINE, 4,4'-(2-ETHYL-2-NITRO-1,3-PROPANEDIYL)BIS- see MRQ525
MORPHOLINE, 4,4'-(2-ETHYL-2-NITROTRIMETHYLENE)DI- see MRQ525
MORPHOLINE HYDROCHLORIDE see MRQ600
MORPHOLINE, N-METHYL- see MMA250
MORPHOLINE, 4-(2-(2-METHYL-1-PROPENYL)-5-NITRO-1H-IMIDAZOL-1-YL)ETHYL)- see MOS300
MORPHOLINE, 4-((MORPHOLINOTHIOCARBONYL)THIO)- see OPQ100

MORPHOLINE, 4-((4-MORPHOLINYLTHIO)THIOXOMETHYL)-(9CI) see OPQ100
MORPHOLINE, 4-(2-NITROBUTYL)- see NFR600
4-MORPHOLINENONYLIC ACID see MRQ750
MORPHOLINE, 4-(1-OXODECYL)-(9CI) see CBF725
MORPHOLINE, 2-(2-(2-PHENOXYPHENYL)ETHYL)-, (Z)-2-BUTENEDIOATE (2:1) see PDV300
MORPHOLINE, 4,4',4''-PHOSPHINOTHIOYLDYNETRIS- see TFN600
MORPHOLINE SALICYLATE see SAI100
MORPHOLINE, componded with SALICYLIC ACID (1:1) see SAI100
MORPHOLINE and SODIUM NITRITE (1:1) see SIN675
4-MORPHOLINE SULFENYL CHLORIDE see MRR075
4-MORPHOLINETHIOCARBONYL DISULFIDE see MRR090
MORPHOLINIUM, 2,2'-(1,1'-BIPHENYL)-4,4'-DIYLBIS(2-HYDROXY-4,4-DIMETHYL- see HAP100
MORPHOLINIUM, 2,2'-(4,4'-BIPHENYLENE)BIS(2-HYDROXY-4,4-DIMETHYL- see HAP100
MORPHOLINIUM, 4,4-DIMETHYL-, CHLORIDE see SMP450
MORPHOLINIUM, 4-ETHYL-4-HEXADECYL-, ETHYL SULFATE see EKN550
MORPHOLINIUM, HEXACHLOROSTANNATE(2-) (2:1) see DUO500
MORPHOLINIUM, (3-INDOLYLMETHYLENE)-, HEXACHLOROSTANNATE(2-) (2:1) see BKJ500
MORPHOLINIUM PERCHLORATE see MRR100
1-MORPHOLINOACETYL-3-PHENYL-2,3-DIHYDRO-4(1H)-QUINAZOLINONE HYDROCHLORIDE see PCJ350
9-(MORPHOLINOAMINO)ACRIDINE see MRR112
9-(MORPHOLINOAMINO)ACRIDINE MONO(METHYL SULFATE) see MRR115
N-MORPHOLINO-β-(2-AMINOMETHYLBENZODIOXAN)-PROPIONAMIDE see MRR125
MORPHOLINOBENZENE see PFS750
MORPHOLINO-2-BENZOTHIAZOLYL DISULFIDE see BDF750
MORPHOLINOCARBONYLACETONITRILE see MRR750
4-MORPHOLINOCARBONYL-2,3-TETRAMETHYLENEQUINOLINE see MRR760
MORPHOLINO-CNU see MRR775
(1-MORPHOLINOCYCLOPENTENE) see CPY800
MORPHOLINODAUNOMYCIN see MRR850
MORPHOLINODAUNOMYCIN see MRR900
14-MORPHOLINODAUNORUBICIN see MRR900
3'-MORPHOLINO-3'-DEAMINODAUNORUBICIN see MRT100
MORPHOLINODISULFIDE see BKU500
2-(MORPHOLINODITHIO)BENZOTHIAZOLE see BDF750
MORPHOLINOETHANESULFONIC ACID see MRT150
2-MORPHOLINOETHANESULFONIC ACID see MRT150
9-((2-MORPHOLINOETHYL)AMINO)ACRIDINE see MRT200

β-MORPHOLINOETHYL BENZHYDRYL ETHER HYDROCHLORIDE see LFW300
3-O-(2-MORPHOLINOETHYL)MORPHINE see TCY750
O³-(2-MORPHOLINOETHYL)MORPHINE see TCY750
β-MORPHOLINOETHYLMORPHINE see TCY750
N-2-MORPHOLINOETHYL-5-NITROIMIDAZOLE see NHH000
MORPHOLINOETHYL NORPETHIDINE DIHYDROCHLORIDE see MRN675
1-(2-MORPHOLINOETHYL)-4-PHENYLISONIPECOTIC ACID ETHYL ESTER DIHYDROCHLORIDE see MRN675
N-(1-(MORPHOLINOMETHYL)-2,6-DIOXO-3-PIPERIDYL)PHTHALIMIDE see MRU080
2-(MORPHOLINO)-N-METHYL-N-(2-MESITYLOXYETHYL)ACETAMIDE HYDROCHLORIDE see MRT250
l-5-(MORPHOLINOMETHYL)-3-((5-NITROFURFURYLIDENE)AMINO)-2-OXAZOLIDINONEHYDROCHLORIDE see FPI150
5-MORPHOLINOMETHYL-3-(5-NITROFURFURYLIDINE)AMINO-2-OXAZOLIDINONE HYDROCHLORIDE see FPI100
5-MORPHOLINOMETHYL-3-(5-NITRO-2-FURFURYLIDINE-AMINO)-2-OXAZOLIDINONE see FPI000
1-MORPHOLINOMETHYLTHALIDOMIDE see MRU080
4-MORPHOLINO-2-(5-NITRO-2-THIENYL)QUINAZOLINE see MRU000
N-MORPHOLINO NONANAMIDE see MRQ750
MORPHOLINOPHOSPHONIC ACID DIMETHYL ESTER see DST800
1-(6-MORPHOLINO-3-PYRIDAZINYL)-2-(1-(tert-BUTOXYCARBONYL)-2-PROPYLIDENE)HYDRAZINE see RFU800
N-(MORPHOLINOSULFENYL)CARBOFURAN see MRU050
3-MORPHOLINOSYDNONE IMINE HYDROCHLORIDE see MRU075
3-MORPHOLINOSYDNONIMINE see MRU076
MORPHOLINO(7,8,9,10-TETRAHYDRO-11-(6H-CYCLOHEPTA(b)QUINOLINYL)) KETONE see MRU077
MORPHOLINO-THALIDOMIDE see MRU080
2-(MORPHOLINOTHIO)BENZOTHIAZOLE see BDG000
4-((MORPHOLINOTHIOCARBONYL)THIO)MORPHOLINE see OPQ100
MORPHOLIN SALICYLAT see SAI100
N-MORPHOLINYL-2-BENZOTHIAZOLYL DISULFIDE see BDF750
4-MORPHOLINYL-2-BENZOTHIAZYL DISULFIDE see BDF750
MORPHOLINYLETHYLMORPHINE see TCY750
3-(2-(4-MORPHOLINYL)ETHYL)MORPHINE see TCY750
1-(2-N-MORPHOLINYLETHYL)-5-NITROIMIDAZOLE see NHH000
1-((2-MORPHOLINYL)ETHYL)-4-PHENYL-4-PIPERIDINECARBOXYLIC ACID ETHYL ESTER DIHYDROCHLORIDE see MRN675
MORPHOLINYLMERCAPTOBENZOTHIAZOLE see BDG000
5-(4-MORPHOLINYL)NAPHTHO(2,3-H)QUINOLINE-7,12-DIONE see MRU090
5-(4-MORPHOLINYL)NAPHTHO(2,3-H)QUINOLINE-7,12-DIONE see MRU090
4-MORPHOLINYLPHOSPHONIC ACID DIMETHYL ESTER see DST800

3-((6-(4-MORPHOLINYL)-3-PYRIDAZINYL)HYDRAZONO)BUTANOIC ACID 1,1- DIMETHYL ETHYL ESTER see RFU800
4-(2-MORPHOLINYL)PYROCATECHOL see MRU100
2-(4-MORPHOLINYLTHIO)BENZOTHIAZOLE see BDG000
4-((4-MORPHOLINYLTHIO)THIOXOMETHYL)MORPHOLINE see OPQ100
3-MORPHOLYLAETHYLMORPHIN (GERMAN) see TCY750
N-MORPHOLYLCYSTEAMIN (GERMAN) see MCO000
MORPHOTHION see MRU250
MORROCID see BGB500
MORSODREN see MLF250
MORSYDOMINE see MRN275
MORTON EP-227 see MLF250
MORTON EP-316 see CQI500
MORTON EP332 see DSO200
MORTON EP 333 see CJJ500
MORTON SOIL DRENCH see MLF250
MORTON SOIL-DRENCH-C see MLF250
MORTON WP-161E see ISE000
MORTOPAL see TCF250
MORYL see CBH250
MOS-708 see BDG250
MOSANON see SEH000
MOSAPRIDE see MRU253
MOSATIL see CAR780
MOSCARDA see MAK700
MOSCHUS KETONE see MIT625
MOSE see MRU255
MOSKENE see MRU300
MOSS GREEN see COF500
MOSTEN see PMP500
MOSYLAN see DNG400
5-MOT see MFS400
MOTAZOMIN see MRN275
MOTH BALLS (DOT) see NAJ500
MOTHER-IN-LAW PLANT see CAL125
MOTHER-IN-LAW'S TONGUE PLANT see DHB309
MOTH FLAKES see NAJ500
MOTIAX see FAB500
MOTILIUM see DYB875
MOTILYN see PAG200
MOTIORANGE R see PEJ500
MOTIROT 2R see OHI200
MOTIROT G see XRA000
MOTOLON see QAK000
MOTOMCO TRACKING POWDER see ITD010
MOTOR BENZOL see BBL250
MOTOR SPIRIT (DOT) see GBY000
MOTOX see CDV100
MOTRIN see IIU000
MOTTENHEXE see HCI000
MOULDRITE A256 see UTU500
MOUNTAIN GREEN see COF500
MOUNTAIN LAUREL see MRU359
MOUNTAIN TOBACCO see AQY500
MOUS-CON see ZLS000
MOUSE-NOTS see SMN500
MOUSE PAK see WAT200
MOUSE-RID see SMN500
MOUSE-TOX see SMN500
MOVINYL 100 see PKQ059
MOVINYL 114 see AAX250
MOWIOL see PKP750
MOWIOL N 30-88 see PKP750
MOWIOL N 50-98 see PKP750
MOWIOL N 70-98 see PKP750
MOXADIL see AOA095
MOXALACTAM DISODIUM see LBH200
MOXAM see LBH200
MOXESTROL see MRU600
MOXIE see MEI450

MOXISYLYTE HYDROCHLORIDE see TFY000
MOXNIDAZOLE see MRU750
MOXONE see DAA800
MOZAMBIN see QAK000
MP see POK000
8-MP see XDJ000
2M-4CP see CIR500
MP 620 see IDJ550
MP 655 see CHE750
MP 1023 see AAN000
MP 12-50 see TAB750
MP-2000 see MRU752
MP 25-38 see TAB750
MP2689 see MRJ200
MP 45-26 see TAB750
MP 1 (refractory) see EAL100
MPA see DAZ117
3-MPA see MFM000
3MPA see MCQ000
3-MPC see MNM500
M.P. CHLORCAPS T.D. see TAI500
MPCM see XTJ000
MPE see MPU500
MPF 2 see UTU500
MPG see MRK500
15-M3-PGF2-α see CCC100
M 4212 (PHARMACEUTICAL) see MQX775
MPI-PC see DSQ800
MPI-PENICILLIN see DSQ800
MPI TC 99M DTPA KIT (CHELATE) see TAI200
MPK see PBN250
MPK 90 see PKQ250
MPN see MNA000
MPN see NKV100
MPNU see MMY500
2'-MePO4' see DSS200
M 2 (POLYMER) see UTU500
M 70 (POLYMER) see UTU500
M 76 (POLYMER) see MCB050
MPP see FAQ900
MPP+ see CQI550
1-MPPN see DWU800
MPS see USJ100
MPT see MPX850
MPTP see TCV470
MPTP HYDROCHLORIDE see TCV500
MR see MPH300
MR 1 see MCB050
MR 56 see BOO630
MR 67 see MCB050
MR 84 see SCK600
MR-100 see WAT211
MR 200 see PKB100
MR 231 see MCB050
MR 2000 see PKB100
MRA-CN see COP765
MRAVENCAN DI-n-BUTYLCINICITY (CZECH) see DDZ000
MRAVENCAN SODNY see SHJ000
MRAVENCAN TRIBENZYLCINICITY (CZECH) see FOE000
MRAVENCAN VAPENATY (CZECH) see CAS250
MRC 910 see GIA000
MRD 108 see DWK400
MRL-37 see MFG525
MRL 41 see CMX700
MRL-37 HYDROCHLORIDE see MFG530
MROWCZAN ETYLU (POLISH) see EKL000
MR 1 (RESIN) see MCB050
MRX III see AHR600
MS 1 see ASM050
3-MS see CNX625
MS 21 see MCB050
MS 33 see SKV150
MS 53 see SNK000
MS 001 see MCB050
MS-222 see EFX500
MS 33F see SKV150
MS. 752 see SBE500

MS 1053 see DJI000
MS-1112 see BFV760
MS 1143 see DJI000
MS-4101 see FMR075
MS-5075 see FBP850
MS-5101 see MRW785
MS-ANTIGEN 40 see MRU755
MSBC see MCB575
MS-BENZANTHRONE see BBI250
MSC-102824 see AQN750
MS 1 (CATALYST) see ASM050
MSDA-11 see MRU756
MSD 803 ACID see LII050
MSD 803 FREE ACID see LII050
MSF see MDR750
MSG see MRL500
MSK-C see CAT775
MSMA see MRL750
MSMED see ECU750
MSP 100F see MCB050
MS-R 100S see MCB050
MSTFA see MQG750
MSZYCOL see BBQ500
MT see HOA000
MTs see MIF760
MT-45 see MRU757
MT 101 see NBA600
MT-141 see CCS365
MT 14-411 see CMY525
MTB see HNY500
MTB 51 see DJM800
MTB 51 see XCJ000
MTBE see MHV859
MTBHQ see BRM500
MTD see DTQ400
MTD see TGL750
MTDQ see MRU760
MTEAI see MRU775
M.T.F. see DUV600
(d-MTF6)-LHRH ACETATE see TKF100
MTI 500 see EQN725
MTIC see MQB750
MTIQ see MPO750
MTMC see MIB750
M.T. MUCORETTES see MPN500
MTN see TGT250
Me-TPA see MPN600
MTQ see QAK000
MTQ HYDROCHLORIDE see MDT250
MTU see MPW500
MTX see MDV500
MTX DISODIUM see MDV600
MTX SODIUM see MDV750
MUCAESTHIN see BQA010
MUCAINE see DTL200
MUCALAN see IHO200
MUCIC ACID see GAR000
MUCICLAR see CBR675
MUCIDRIL see CBR675
MUCIDRINA see VGP000
MUCINOL see AOO490
MUCITUX see ECU550
MUCOCHLORIC ACID see MRU900
MUCOCHLORIC ANHYDRIDE see MRV000
MUCOCIS see CBR675
MUCODYNE see CBR675
MUCOFLUID see MDK875
MUCOLASE see CBR675
MUCOLEX see CBR675
MUCOLYSIN see MCI375
MUCOLYTICUM see ACH000
MUCOLYTICUM LAPPE see ACH000
MUCOMYCIN see MRV250
MUCOMYST see ACH000
(E,E)-MUCONALDEHYDE see HCR600
trans,trans-MUCONALDEHYDE see HCR600
MUCONOMYCIN A see MRV500
MUCOPOLYSACCHARIDE, POLYSULFURIC ACID ESTER see MRV525
MUCOPOLYSACCHARIDIPOLY SCHWEFELSAEUREESTER (GERMAN) see MRV525

MUCOPRONT see CBR675
MUCORAMA see PMJ500
MUCOSOLVAN see AHJ500
MUCOSOLVIN see ACH000
MUCOXIN see DTL200
MUCRONATINE see RFP100
MUCUNA MONOSPERMA DC. ex Wight (extract excluding roots) see MRV600
MUDAR see COD675
MUGAN see CBM750
MUGB see GLC100
MUGIBON see MAS300
MUGIBON 70WP see MAS300
MUGUET (CANADA) see LFT700
MUIRAMID see AAI250
MULDAMINE see MRV750
MUL F 66 see PKL750
MULHOUSE WHITE see LDY000
MULSIFEROL see VSZ100
MULSOPAQUE see ELQ500
MULTACID YELLOW 3R see SGP500
MULTAMAT see DQM600
MULTERGAN see MRW000
MULTERGAN METHYL SULFATE see MRW000
MULTEZIN see MRW000
MULTICHLOR see CDP000
MULTICIDE 2154 see PDR700
MULTICIDE 2167 see MRW080
MULTICUER BROWN MPH see SGP500
MULTIFLEX MM see CAT775
MULTIFUGE CITRATE see PIJ500
MULTIN see HIM000
MULTIPROP see CDT000
MUNDIAL BLACK MO see CMS250
MUNDISAL see CMG000
MU OIL TREE see TOA275
MURACIL see MPW500
MURATOX see DJI000
MURBETOL see OKS500
MURCIL see MDQ250
MUREL see VBK000
MUREX see GFA000
MURFOS see PAK000
MURFOTOX see DJI000
MURFULVIN see GKE000
MURIATE of PLATINUM see PJE000
MURIATIC ACID see HHL000
MURIATIC ETHER see EHH000
MURIOL see CJJ000
MURITAN see DEQ000
MUROTOX see DJI000
MUROTOX see MRW100
MUROX see CHJ300
MURPHOTOX see DJI000
MURUTOX see DJI000
MURVESCO see CJR500
MURVIN see CBM750
MUSARIL see CFG750
MUSCALM see MRW125
MUSCALURE see TJF400
MUSCAMONE see TJF400
MUSCARIN see MRW250
MUSCARINE see MRW250
dl-MUSCARINE see MRW250
MUSCIMOL see AKT750
MUSCLE ADENYLIC ACID see AOA125
MUSCONE see MIT625
MUSCULAMINE see DCC400
MUSCULAMINE TETRAHYDROCHLORIDE see GEK000
MUSCULARON see IHH000
MUSETTAMYCIN see APV000
MUSHROOM AMNITA RUBESCENS TOXIN see AHI500
MUSHROOMS see MRW269
MUSK see MRW250
MUSK 36A see ACL750
MUSK AMBRETTE see BRU500
MUSK AMBRETTE see OKU100
MUSK AMBRETTE (NATURAL) see OKU100
MUSKARIN see MRW250

MUSKEL see CKF500
MUSKEL-TRANCOPAL see CKF500
MUSK KETONE see ACE600
MUSK NATURAL scc OKU100
MUSKONE see MIT625
MUSK R 1 see OKW100
MUSK-T see EJQ500
MUSK TIBETENE see MRW272
MUSK TONALID see EEE100
MUSK XYLENE see TML750
MUSK XYLENE (DOT) see TML750
MUSK XYLOL see TML750
MUSONAL see QCS900
MUSQUASH POISON see WAT325
MUSQUASH ROOT see WAT325
MUSSEL POISON DIHYDROCHLORIDE see
SBA500
N-MUSTARD (GERMAN) see BIE500
MUSTARD CHLOROHYDRIN see CHC000
MUSTARD GAS see BIH250
MUSTARD GAS SULFONE see BIH500
MUSTARD HD see BIH250
MUSTARD OIL see AGJ250
MUSTARD SULFONE see BIH500
MUSTARD VAPOR see BIH250
MUSTARGEN see BIE250
MUSTARGEN see BIE500
MUSTARGEN HYDROCHLORIDE see
BIE500
MUSTINE see BIE250
MUSTINE HYDROCHLOR see BIE500
MUSTINE HYDROCHLORIDE see BIE500
MUSTRON see CFA750
MUSUET SYNTHETIC see CMS850
MUSUETTINE PRINCIPLE see CMS850
MUTABASE see DCQ700
MUTAGEN see BIE250
MUTAMYCIN see AHK500
MUTAMYCIN 6 see EBC000
MUTAMYCIN (MITOMYCIN for
INJECTION) see AHK500
MUTHESA see DTL200
MUTHMANN'S LIQUID see ACK250
MUTOXIN see DAD200
MUW 1193 see MNU300
MUZOLIMINE see EAE675
MV see MAK275
MV-678 see MEL700
MV 770 see CKK530
MV 119A see DLK200
MVEEG (RUSSIAN) see EJL500
MVNA see NKY000
MW 30 see MCB050
M 33W see MCB050
MW 801 see SEO550
MX see CFL100
2M-4X see SIL500
MX 40 see MCB050
MX 705 see MCB050
MX 4500 see SMQ500
MX 5514 see SMQ500
MX 5516 see SMQ500
MX 5517-02 see SMQ500
MXDA see XHS800
2M-4XP see CIR325
MY-1 see MRW785
MY-71 see DGM730
MY 93 see MNU300
MY/68 see FAF000
MY 33-7 see TGJ625
MY 41-6 see AJC000
MY-5116 see IHR200
MYACINE see NCD550
MYACYNE see NCE000
MYALEX see CKK250
MYAMBUTOL see EDW875
MYANIL see GGS000
MYARSENOL see SNR000
MYASUL see AKO500
MYBASAN see ILD000
MY-B-DEN see AOA125
MYBORIN see MRW275

MYCAIFRADIN SULFATE see NCG000
MYCANDEN see IFA000
MYCARDOL see PBC250
MYCELAX see MRX500
MYCELEX scc MRX500
MYCHEL see CDP250
MYCIFRADIN see NCE000
MYCIFRADIN-N see NCG000
MYCIGIENT see NCG000
MYCILAN see IFA000
MYCINAMICIN 1 see MRW500
MYCINAMICINS II see MRW750
MYCINOL see CDP250
MYCIVIN see LGC200
MYCLOBUTANIL see MRW775
MYCOBACIDIN see CMP885
1-MYCOBACIDIN see CCI500
MYCOBACTERIUM BOVIS BCG EXTRACT
see MRW785
MYCOBACTYL see GBB500
MYCOBAN see SJL500
MYCOBUTOL see EDW875
MYCOCURAN see GGS250
MYCODIFOL see TBQ280
MYCOFARM see BFD250
MYCOHEPTIN see MRW800
MYCOHEPTYNE see MRW800
MYCOIN see CMV000
MYCOLUTEIN see DTU200
MYCOPHENOLIC ACID see MRX000
MYCOPHYT see PIF750
MYCO-POLYCID see CDY325
MYCOSHIELD TMQTHC 20 see HOH500
MYCOSPOR see BGA825
MYCOSPORIN see MRX500
MYCOSTATIN see NOH500
MYCOSTATIN 20 see NOH500
MYCOTICIN see MRY000
MYCOTICIN (1:1) see MRY000
MYCOTOXIN F2 see ZAT000
MYCOZOL see TEX000
MYCRONIL see BJK500
MYDECAMYCIN see MBY150
MYDETON see TGK200
MYDFRIN see SPC500
MYDOCALM see MRW125
MYDOCALM see TGK200
MYDRIAL see AOA250
MYDRIASIN see MGR250
MYDRIATIN see NNM000
MYDRIATINE see NNN000
MYDRIATINE see PMJ500
MYEBROL see DDP600
MYELOBROMOL see DDP600
MYELOLEUKON see BOT250
MYELOTRAST see SHX000
MYELOTRAST see TDQ230
MYELOTRAST DI-N-METHYLGLUCAMINE
SALT see IDJ500
MYGAL see BET000
MYKOSTIN see VSZ100
MYLAR see PKF750
MYLAXEN see HEG000
MYLEPSIN see DBB200
MYLEPSINUM see DBB200
MYLERAN see BOT250
MYLIS see DAL040
MYLODORM see AMX750
MYLOFANOL see PGG000
MYLON (CZECH) see DSB200
MYLONE see DSB200
MYLONE 85 see DSB200
MYLOSAR see ARY000
MYLOSUL see AKO500
MYNOSEDIN see IIU000
MYOCAINE see RLU000
MYOCHOLINE see HOA500
MYOCHRYSINE see GJC000
MYOCOL see AEH750
MYOCON see NGY000
MYOCORD see TAL475
MYOCRISIN see GJC000

MYODETENSINE see GGS000
MYODIGIN see DKL800
MYODIL see ELQ250
MYODYL see ELQ500
MYOFER 100 see IGS000
MYOFLEXINE see CDQ750
MYOGLYCERIN see NGY000
MYOHEMATIN see CQM325
MYO-INOSISTOL HEXAKISPHOSPHATE
see PIB250
MYO-INOSITOL, HEXAKIS(HYDROGEN
SULFATE), HEXASODIUM SALT see IDE400
MYO-INOSITOL HEXAPHOSPHATE see
PIB250
MYOLASTAN see CFG750
MYOLAX see GGS000
MYOLAXENE see GKK000
MYOLYSEEN see PIL550
MYOMYCIN B see MRY100
MYOMYCIN SULFATE see MRY250
MYOPAN see GGS000
MYOPLEGINE see HLC500
MYOPONE see WBJ700
MYOPORUM LAETUM see MRY600
MYORDIL see TNJ750
MYORELAX see RLU000
MYOREXON see CCK125
MYOSALVARSAN see SNR000
MYOSCAINE see RLU000
MYOSEROL see GGS000
MYOSPAZ see PFJ000
MYOSTHENINE see VGP000
MYOSTIBIN see AQH800
MYOSTIBIN see AQI250
MYOSTON see AOA125
MYOTOLON see AOO300
MYOTRATE "10" see PBC250
MYOTRIPHOS see ARQ500
MYOXANE see GGS000
MYPROZINE see PIF750
MYRABOLAM TANNIN see MRZ100
MYRAC ALDEHYDE see IKT100
MYRAFORM see PKQ059
MYRCENE see MRZ150
MYRCENOL see MLO250
MYRCENYL ACETATE see AAW500
MYRCIA OIL see BAT500
MYRCIA OIL see LBK000
MYRICA CERIFERA see WBA000
MYRICETIN see HDW150
MYRICETIN 3'-GLUCOSIDE see GFC200
MYRICETIN HEXAACETATE see MRZ200
MYRICETOL see HDW150
MYRICIA OIL see BAT500
MYRICITIN see HDW150
MYRINGACAINE DROPS see CHW675
MYRISIIC ALCOHOL see TBY250
MYRISTALDEHYDE see TBX500
MYRISTAN DI-n-BUTYLCINICITY (CZECH)
see BLH309
MYRISTICA see NOG000
MYRISTIC ACID see MSA250
MYRISTIC ACID, BUTYL ESTER see MSA300
MYRISTIC ACID, ISOPROPYL ESTER see
IQN000
MYRISTIC ACID, SODIUM SALT see SIN900
MYRISTIC ALDEHYDE see TBX500
MYRISTICA OIL see NOG500
MYRISTICIN see MSA500
9-MYRISTOYL-1,7,8-ANTHRACENETRIOL
see MSA750
10-MYRISTOYL-1,8,9-ANTHRACENETRIOL
see MSA750
1-MYRISTOYLAZIRIDINE see MSB000
MYRISTOYLETHYLENEIMINE see MSB000
N-MYRISTOYLOXY-AAF see ACS000
N-MYRISTOYLOXY-AAIF see MSB100
N-MYRISTOYLOXY-N-ACETYL-2-
AMINOFLUORENE see ACS000
N-MYRISTOYLOXY-N-ACETYL-2-AMINO-
7-IODOFLUORENE see MSB100

N-MYRISTOYLOXY-N-MYRISTOYL-2-
AMINOFLUORENE see MSB250
MYRISTYL ALCOHOL (mixed isomers) see
TBY500
MYRISTYL-γ-PICOLINIUM CHLORIDE see
MSB500
MYRISTYL STEARATE see TCB100
MYRISTYL SULFATE, SODIUM SALT see
SIO000
MYRISTYLTRIMETHYLAMMONIUM
BROMIDE see TCB200
MYRITICALORIN see RSU000
MYRITICOALORIN see RSU000
MYRITOL 318 see CBF710
MYROBALANS TANNIN see MSB750
MYRRH OIL see MSB775
MYRTAN TANNIN see MSC000
MYRTENAL see FNK150
MYRTENYL ACETATE see MSC050
(+)-MYRTENYL ACETATE see MSC050
MYRTLE OIL see OGU000
MYSEDON see DBB200
MYSOLINE see DBB200
MYSONE see HHQ800
MYSORITE see ARM262
MYSTECLIN-F see AOC500
MYSTER GRASS see DAE100
MYSTERIA see CNX800
MYSTOX WFA see BGJ750
MYSURAN see MSC100
MYSURAN CHLORIDE see MSC100
MYTAB see TCB200
MYTELASE see MSC100
MYTELASE CHLORIDE see MSC100
MYTOMYCIN see AHK500
MYTRATE see VGP000
MYVAK see VSK900
MYVAX see VSK900
MYVIZONE see FNF000
MYVPACK see VSK600
MYXIN see HLT100
MYXOVIROMYCIN see AHN625
N 5 see EHG100
N-9 see NNB300
N 34 see CAT775
N 50 see UTU500
N 68 see BRS000
N 135 see EHP700
N 252 see OMY825
N-399 see PEM750
N 521 see DSB200
N-553 see MRW125
N 642 see PIR100
N 714 see TAF675
N 715 see THL500
N-746 see CLX250
N 869 see SIL550
N 2038 see ZJS300
N 2404 see CJD650
N-2596 see EHL670
N 2790 see FMU045
N 3051 see BSG000
N 4328 see EOO000
N 4446 see EMR100
N 4548 see MOB250
N 7001 see AEG875
N 7009 see FMO129
N 7009 see FMO150
N 714C see TAF675
Ni 270 see NCW500
N-1544A see PFC750
168N15 see SMQ500
No. 48-80 see PMB800
Ni 4303T see NCW500
NA see EID000
NA see NBE500
NA-22 see IAQ000
NA-53 see BCP000
NA 73 see CJU275
NA 97 see PAF625
NA-101 see TDX000
NA-872 see AHJ250

NA 872 see AHJ500
NA 8318 see ENF100
NAA 800 see NAK500
NAAM see NAK000
NAB see NJK150
NAB 365 see VHA350
NAB-739 see AJM575
NAB 365Cl see VHA350
NABAC see HCL000
NABADIAL see AOO475
NABAM see DXD200
NABAME (FRENCH) see DXD200
NABOLIN see MPN500
NABOR BRILLIANT PINK 28 see CMM850
NABOR ORANGE G see CMM820
NABOR YELLOW 4G see CMM890
NABU see CDK800
NABUMETONE see MFA300
NAC see ACH000
NAC see CBM750
NACARAT A EXPORT see HJF500
NACCANOL NR see DXW200
NACCONATE-100 see TGM740
NACCONATE 100 see TGM750
NACCONATE 300 see MJP400
NACCONATE 400 see BBP000
NACCONATE H 12 see MJM600
NACCONOL 98SA see LBU100
NACCONOL LAL see SIB700
NACELAN BLUE CBG see DMM400
NACELAN BLUE G see TBG700
NACELAN BLUE KLT see MGG250
NACELAN FAST YELLOW CG see AAQ250
NACELAN PINK 3B see DBX000
NACELAN PINK B see AKE250
NACELAN SCARLET CSB see ENP100
NACELAN VIOLET 4B see AKP250
NACELAN VIOLET 4R see DBP000
Na-CEMMIX see BLX000
N¹-ACETYL-4-
AMINOPHENYLSULFONAMIDE see
SNQ710
NACIMYCIN see RKK000
NACLEX see BDE250
NACM-CELLULOSE SALT see SFO500
NACRICLASINE see LJC000
NAC-TB see ACH000
NACYCLYL see EDR000
NAD see CNF390
NAD see NAK000
NAD+ see CNF390
β-NAD see CNF390
NADEINE see DKW800
NA-DESOXYCHOLAT (GERMAN) see
SGE000
NADIC METHYL ANHYDRIDE see NAC000
NADIDE see CNF390
NADISAL see SJO000
NADISAN see BSM000
NADIZAN see BSM000
NADOLOL see CNR675
NADONE see CPC000
NADOXOLOL HYDROCHLORIDE see
NAC500
NADOZONE see BRF500
NADP see CNF400
β-NADP see CNF400
NAD PHOSPHATE see CNF400
NADROTHYRON D see SKJ300
NAEPAINE see PBV750
Na-AESCINAT see EDM000
NAFAMSTAT MESILATE see AHN700
NAFCILLIN SODIUM SALT see SGS500
NAFEEN see SHF500
NAFENOIC ACID see MCB500
NAFENOPIN see MCB500
NAFIDIMIDE see NAC600
NAFIVERINE DIHYDROCHLORIDE see
NAD000
NAFKA see ROH900
NAFKA CRYSTAL GUM see ROH900
NAFKA KRISTALGOM see ROH900

NAFOXIDINE see NAD500
NAFOXIDINE HYDROCHLORIDE see
NAD750
NAFRINE see AEX000
NAFRONYL see NAE000
NAFRONYL OXALATE see NAE100
NAFTALAM see NBL200
NAFTALEN (POLISH) see NAJ500
NAFTALIN-BUTIL-SOLFONATO (ITALIAN)
see NBS700
NAFTALOFOS see HMV000
NAFTIDROFURYL see NAE000
NAFTIDROFURYL OXALATE see NAE100
1-NAFTILAMINA (SPANISH) see NBE700
β-NAFTILAMINA (ITALIAN) see NBE500
1-NAFTIL-TIOUREA (ITALIAN) see AQN635
NAFTIPRAMIDE see IOU000
NAFTIZIN see NAH550
2-NAFTOL (DUTCH) see NAX000
β-NAFTOL (DUTCH) see NAX000
NAFTOLEN ZD see AQZ150
2-NAFTOLO (ITALIAN) see NAX000
β-NAFTOLO (ITALIAN) see NAX000
NAFTOLO MM see CMM760
NAFTOPEN see SGS500
NAFTOPIDIL see MFG515
α-NAFTYLAMIN (CZECH) see NBE700
β-NAFTYLAMIN (CZECH) see NBE500
2-NAFTYLAMIN-5,7-DISULFONAN SODNY
(CZECH) see ALH500
1-NAFTYLAMINE (DUTCH) see NBE700
2-NAFTYLAMINE (DUTCH) see NBE500
4-(2-NAFTYLAMINO)FENOL see NBF500
1-NAFTYLESTER KYSELINY
METHYLKARBAMINOVE see CBM750
α-NAFTYL-N-METHYLKARBAMAT see
CBM750
β-NAFTYLOAMINA (POLISH) see NBE500
1-NAFTYLTHIOUREUM (DUTCH) see
AQN635
NAFTYPRAMIDE see IOU000
NAFUSAKU see NAK500
NAGANOL see BAT000
NAGARMOTHA OIL see NAE505
NAGARMUSTA OIL see NAE505
NAGARSE see BAC000
NAGASE (BACILLUS SUBTILIS) see NAE507
NAGENT see GCU050
NAGENTIS see GCU050
NAGRAVON see VSZ000
NAH see NCQ900
NAH 80 see SHO500
NAHP see NJJ950
NA 65 HYDROCHLORIDE see PIV500
Na III HYDROCHLORIDE see DWF200
NAILAMIDE YELLOW BROWN E-L see
SGP500
NAIRIT see PJQ050
NAIXAN see MFA500
NAJA FLAVA VENOM see NAE510
NAJA HAJE ANNULIFERA VENOM see
NAE512
NAJA HAJE VENOM see NAE515
NAJA MELANOLEUCA VENOM see
NAE875
NAJA MOSSAMBICA MOSSAMBICA
CARDIOTOXIN III see AQW100
NAJA MOSSAMBICA MOSSAMBICA α-
NEUROTOXIN I see NAE900
NAJA MOSSAMBICA MOSSAMBICA
PHOSPHOLIPASE A2 II see AQW110
NAJA NAJA ATRA VENOM see NAF000
NAJA NAJA KAOUTHIA VENOM see
NAF200
NAJA NAJA NAJA VENOM see ICC700
NAJA NAJA PHILIPPINENSIS
NEUROTOXIN see NAF225
NAJA NAJA SIAMENSIS VENOM see
NAF250
NAJA NIGRICOLLIS VENOM see NAG000
NAJA NIVEA VENOM see NAG200
NAK 1654 see FAO220

NAKED LADY LILY see AHI635
NAKO BROWN R see ALT250
NAKO FAST GREY BL see CMU320
NAKO H see PEY500
NAKO TEG scc ALS990
NAKO TGG see REA000
NAKO TMT see TGL750
NAKO TRB see NAW500
NAKO TSA see DBO400
NAKO YELLOW EGA see ALT000
NAKVA see HII500
NALADOR see SOU650
NALCAMINE G-13 see AHP500
NALCAST see SCK600
NALCO 680 see AHG000
NALCOAG see SCH002
NALCON 240 see BMS250
NALCON 243 see DSB200
NALCROM see CNX825
NALDE see AHL500
NALED see NAG400
NALFLOC 636 see ADV900
NALFON see FAP100
NALGESIC see FAP100
NALIDIC ACID see EID000
NALIDICRON see EID000
NALIDIXATE SODIUM see NAG450
NALIDIXIC ACID see EID000
NALIDIXIC ACID SODIUM SALT see
NAG450
NALIDIXIN see EID000
NALINE HYDROCHLORIDE see NAG500
NALITUCSAN see EID000
NALKIL see BMM650
(d-NAL(1)⁶))-LHRH ACETATE see LIU329
NALLINE see AFT500
(NAL(1)³)-d-NAL(2)⁶)-PRO-NHET⁹))-LHRH
ACETATE see LIU337
(NAL(1)³)-d-NAL(2)⁶)-PRO-NHET⁹))-
LUTEINIZING HORMONE-RELEASING
HORMONE ACETATE see LIU337
(N-NAL(2)⁶)-NME-LEU⁷)-LHRH ACETATE
see LIU331
(d-NAL(2)⁶)-NME-LEU⁷)-PRO-NHET⁹))-
LHRH ACETATE see LIU335
NALORFINA see AFT500
NALORPHINE see AFT500
NALORPHINE HYDROCHLORIDE see
NAG500
NALORPHINIUM see AFT500
NALOX see MMN250
NALOXIPHAN see AGI000
NALOXONE see NAG550
l-NALOXONE see NAG550
NALOXONE HYDROCHLORIDE see
NAH000
(d-NAL(2)⁶)-PRO-NHET⁹))-LHRH ACETATE
see LIU333
NALTREXONE see CQF099
NALTREXONE HYDROCHLORIDE see
NAH100
NALTROPINE see AGM250
NALUTORAL see GEK500
NALUTRON see PMH500
NAM see NCR000
NAMATE see DXE600
l-NAME see NDY675
NAMEKIL see TDW500
N4 AMINE see TBI700
NaAMP see AEM750
NAMPHEN see XQS000
NAMURON see TDA500
5-NAN see NEJ500
NANA-HONUA (HAWAII) see AOO825
NANAOMYCIN A see RMK200
NANCHOR see RMA500
NANDERVIT-N see NCR000
NANDROLIN see DYF450
NANDROLON see NNX400
NANDROLONE see NNX400
NANDROLONE DECANOATE see NNE550

NANDROLONE PHENPROPIONATE see
DYF450
NANDROLONE PHENYLPROPIONATE see
DYF450
NANDRON see EAN700
NANI-ALI'I (HAWAII) see AFQ625
NANI-O-HILO (HAWAII) see PCB300
NANKAI ACID ORANGE I see FAG010
NANKAI BRILLIANT FAST BLUE G see
COF420
NANKER see RMA500
NANKOR see RMA500
NANM see AFT500
NANOFIN see LIQ550
NANOPHYN see LIQ550
NANOPLAST FB 101 see MCB050
NANSA 1042P see LBU100
NANSA HS 55 see SKF600
NANSA SSA see LBU100
NANSA YS 94 see BBS275
N,N'''-(2,6-
ANTHRAQUINONYLENE)BIS(N,N-
DIETHYLACETAMIDE) see APL250
NAOP see AGM125
NAOTIN see NCQ900
NAPA see HIM000
NAPACETIN see IIU000
NAPCLOR-G see SJA000
NAPELLINE see LIN050
NAPENTAL see NBU000
NAPHAZOLINE see NAH500
NAPHAZOLINE HYDROCHLORIDE see
NCW000
NAPHAZOLINE NITRATE see NAH550
NAPHCON see NCW000
NAPHCON FORTE see NCW000
NAPHID see NAR000
NAPHTAMINE BLUE 2B see CMO000
NAPHTAMINE BLUE 2B see CMO250
NAPHTAMINE BLUE 10G see CMO500
NAPHTAMINE BLUE RW see CMO600
NAPHTAMINE BORDEAUX B see CMO872
NAPHTAMINE FAST RED F see CMO870
NAPHTAMINE GREEN B see CMO840
NAPHTAMINE SKY BLUE DD see CMN750
NAPHTAMINE VIOLET N see CMP000
NAPHTAZOL A see CMM760
NAPHTHA see NAH600
NAPHTHA see NAH600
NAPHTHA see ROU000
NAPHTHA, hydrotreated see NAH600
NAPHTHACAINE HYDROCHLORIDE see
NAH800
NAPHTH(2,3-a)ACEANTHRYLENE see
NAH830
NAPHTH(2,1-d)ACENAPHTHYLENE see
NAH900
NAPHTHACENE see NAI000
5,12-NAPHTHACENEDIONE, 7,8,9,10-
TETRAHYDRO-8-ACETYL-10-((3-(((S)-2-
AMINO-4-METHYL-1-OXOPENTYL)AMI
NO)-2,3,6-TRIDEOXY-α-l-LYXO-
HEXOPYRANOSYL)OXY)-6,8,11-
TRIHYDROXY-1-MET HOXY-, (8S-CIS)- see
LEX300
5,12-NAPHTHACENEDIONE, 7,8,9,10-
TETRAHYDRO-8-ACETYL-1-METHOXY-10-
((2,3,6-TRIDEOXY-3-(4-MORPHOLINYL)-α-l-
lyxo-HEXOPYRANOSYL)OXY)-6,8,11-
TRIHYDROXY-, (8S-cis)- see MRR850
5,12-NAPHTHACENEDIONE, 7,8,9,10-
TETRAHYDRO-10-((3-AMINO-2,3,6-
TRIDEOXY-α-l-LYXO-
HEXOPYRANOSYL)OXY)-1-METHOXY-8-
(4-MORPHOLINYLACETYL)-6,8,11-
TRIHYDROXY-, (8S-CIS)- see MRR900
5,12-NAPHTHACENEDIONE, 7,8,9,10-
TETRAHYDRO-10-((3-(3-
CYANOMORPHOLINO)-2,3,6-TRIDEOXY-
α-L-lyxo-HEXOPYRANOSYL)OXY)-8-
(HYDROXYACETYL)-1-METHOXY-6,8,11-
TRIHYDR OXY-, (8s-cis)- see COP765

5,12-NAPHTHACENEDIONE, 7,8,9,10-
TETRAHYDRO-8-(HYDROXYACETYL)-1-
METHOXY-10-((2,3,6-TRIDEOXY-
MORPHOLINO-α-L-lyxo-
HEXOPYRANOSYL)OXY)-6,8,11-
TRIHYDROXY- see MRT100
5,12-NAPHTHACENEDIONE, 7,8,9,10-
TETRAHYDRO-8-(HYDROXYACETYL)-1-
METHOXY-10-((2,3,6-TRIDEOXY-3-
(TRIFLUOROACETAMIDO)-α-l-LYXO-
HEXOPYRANOSYL)OXY)-6,8,11-
TRIHYDROXY-, (8S-CIS)- see TJX300
NAPHTH(2,3-e)ACEPHENANTHRYLENE
see NAI100
NAPHTHA COAL TAR (OSHA) see NAH600
α-NAPHTHACRIDINE see BAW750
NAPHTH(2,1,8-mna)ACRIDINE, 1-NITRO-
see NET100
NAPHTH(2,1,8-mna)ACRIDINE, 3-NITRO-
see NET120
NAPHTH(2,1,8-mna)ACRIDINE, 1-NITRO-, 6-
OXIDE see NET130
NAPHTH(2,1,8-mna)ACRIDINE, 3-NITRO-, 6-
OXIDE see NET140
α-NAPHTHAL see NAJ000
2-NAPHTHALAMINE see NBE500
NAPHTHALANE see DAE800
α-NAPHTHALDEHYDE see NAJ000
1-NAPHTHALDEHYDE, 2-HYDROXY- see
HMU200
2-NAPHTHALENAMINE see NBE500
1-NAPHTHALENAMINE, 4-(PHENYLAZO)-
see PEJ600
2-NAPHTHALENAMINE, 1,2,3,4-
TETRAHYDRO-6-((4-
METHOXYPHENYL)SULFONYLMETHYL)-
N,N-DIPROPYL- see DWR350
1-NAPHTHALENAZO-2',4'-
DIAMINOBENZENE see NBG500
NAPHTHALENE see NAJ500
NAPHTHALENE, molten (DOT) see NAJ500
NAPHTHALENE, crude or refined (DOT) see
NAJ500
NAPHTHALENE ACETAMIDE see NAK000
1-NAPHTHALENEACETAMIDE see NAK000
α-NAPHTHALENEACETAMIDE see
NAK000
1-NAPHTHALENEACETIC ACID see
NAK500
NAPHTHALENE-1-ACETIC ACID see
NAK500
α-NAPHTHALENEACETIC ACID see
NAK500
1-NAPHTHALENEACETIC ACID, ETHYL
ESTER see ENL900
2-NAPHTHALENEACETIC ACID, 6-
METHOXY-α-METHYL-, (+)-(8CI) see
MFA500
2-NAPHTHALENEACETIC ACID, 6-
METHOXY-α-METHYL-, (S)-(9CI) see
MFA500
2-NAPHTHALENEACETIC ACID, 6-
METHOXY-α-METHYL-, SODIUM SALT, l-
(−)- see NBO550
2-NAPHTHALENEACETIC ACID, 6-
METHOXY-α-METHYL-, SODIUM SALT,
(R)- (9CI) see NBO550
NAPHTHALENEACETIC ACID METHYL
ESTER see NAK525
1-NAPHTHALENEACETIC ACID, METHYL
ESTER see NAK525
1-NAPHTHALENEACETIC ACID, SODIUM
SALT see NAK550
1,8-NAPHTHALENE-1,2-BENZIMIDAZOLE
see NAK700
NAPHTHALENE, BIS(1-METHYLETHYL)-
see BKL600
NAPHTHALENE, 1-BROMO- see BNS200
1-NAPHTHALENECARBONITRILE see
NAY090
2-NAPHTHALENECARBONITRILE (9CI) see
NAY100

BIPHENYL)-4-YL)AZO)-, DISODIUM SALT see CMO872

2-NAPHTHALENESULFONIC ACID, 6-AMINO-3-((4-((4-AMINOPHENYL)AZO)-2-METHOXY-5-METHYLPHENYL)AZO)-4-HYDROXY-, MONOSODIUM SALT see CMN230

2-NAPHTHALENESULFONIC ACID, 6-AMINO-4-HYDROXY- see AKH800

2-NAPHTHALENESULFONIC ACID, 6-((2-AMINO-4-((2-HYDROXYETHYL)AMINO)PHENYL)AZO)-3-((4-((4-((7-((2-AMINO-4-((2-HYDROXYETHYL)AMINO)PHENYL)AZO)-1-HYDROXY-3-SULFO-2-NAPHTHALENYL)AZO)PHENYL) AMINO)-3-SULFOPHENYL)AZO)-4-HYDROXY-, TRISODIUM SALT see CMN300

2-NAPHTHALENESULFONIC ACID, 7-(BENZOYLAMINO)-4-HYDROXY-3-((4-((4-SULFOPHENYL)AZO)PHENYL)AZO)-,DISODIUM SALT see CMO885

1-NAPHTHALENESULFONIC ACID, 3,3'-((3,3'-DIMETHYL(1,1'-BIPHENYL)-4,4'-DIYL)BIS(AZO))BIS(4-AMINO-), DISODIUM SALT see DXO850

1-NAPHTHALENESULFONIC ACID, 6-((2,4-DIMETHYL-6-SULFOPHENYL)AZO)-5-HYDROXY-, DISODIUM SALT see CMP620

2-NAPHTHALENESULFONIC ACID, 5,7-DINITRO-8-HYDROXY- see FBZ100

2-NAPHTHALENESULFONIC ACID, 8-HYDROXY-5,7-DINITRO-(8CI) see FBZ100

1-NAPHTHALENESULFONIC ACID, 3-HYDROXY-4-((1-HYDROXY-2-NAPHTHALENYL)AZO)-, MONOSODIUM SALT see CMP880

1-NAPHTHALENESULFONIC ACID, 3-HYDROXY-4-((1-HYDROXY-2-NAPHTHALENYL)AZO)-7-NITRO-, MONOSODIUM SALT see EDC625

1-NAPHTHALENESULFONIC ACID, 3-HYDROXY-4-(2-HYDROXY-1-NAPHTHYLAZO)-, SODIUM SALT see HLI100

1-NAPHTHALENESULFONIC ACID, 4-HYDROXY-3-((4'-((1-HYDROXY-5-SULFO-2-NAPHTHALENYL)AZO)- 3,3'-DIMETHYL(1,1'-BIPHENYL)-4-YL)AZO)-, DISODIUM SALT see ASM100

1-NAPHTHALENESULFONIC ACID, 2-((2-HYDROXY-1-NAPHTHALENYL)AZO)-, MONOSODIUM SALT see NAP100

2-NAPHTHALENESULFONIC ACID, 3,3'-METHYLENEDI-, SILVERSALT see SDP200

NAPHTHALENESULFONIC ACID, POLYMER with FORMALDEHYDE, SODIUM SALT (9CI) see BLX000

2-NAPHTHALENESULFONIC ACID, 1,2,3,4-TETRAHYDRO-2-METHYL-1,4-DIOXO-, SODIUM SALT see MCB575

1,4,5,8-NAPHTHALENETETRACARBOXYLIC ACID see NAP250

1,4,5,8-NAPHTHALENETETRACARBOXYLIC ACID-1,8:4,5-DIANHYDRIDE see NAP300

NAPHTHALENE-1,2,3,4-TETRAHYDRIDE see TCX500

2-NAPHTHALENETHIOL see NAP500

2-NAPHTHALENE-2-THIOL see NAP500

β-NAPHTHALENETHIOL see NAP500

NAPHTHALENE, 1,3,5-TRINITRO- see TMM600

1-NAPHTHALENOL see NAW500

2-NAPHTHALENOL see NAX000

1-NAPHTHALENOL, ACETATE (9CI) see NAU600

2-NAPHTHALENOL, DECAHYDRO-, FORMATE see DAF150

2-NAPHTHALENOL, DECAHYDRO-6-(1-METHYLETHYL)- see IOO310

1-NAPHTHALENOL, 7-(DIPROPYLAMINO)-5,6,7,8-TETRAHYDRO- see HKF370

2-NAPHTHALENOL, 1-((2-METHOXYPHENYL)AZO)-(9CI) see CMS238

1-NAPHTHALENOL, METHYLCARBAMATE (9CI) scc CBM750

2-NAPHTHALENOL, 5,6,7,8-TETRAHYDRO-7-DIPROPYLAMINO- see HKF350

2-NAPHTHALENOL, 1-((2,2',3,3'-TETRAMETHYL(1,1'-BIPHENYL)-4-YL)AZO)- see TDM820

1(4H)-NAPHTHALENONE, 2-(ACETYLOXY)-4-((3,4-DIMETHYL-5-ISOXAZOLYL)IMINO)- see ACU200

1(2H)-NAPHTHALENONE, 3,4-DIHYDRO-5-(3-(tert-BUTYLAMINO)-2-HYDROXYPROPOXY)-, HYDROCHLORIDE, (−)- see LFA100

1(2H)-NAPHTHALENONE, OCTAHYDRO-4A,8A-DIMETHYL-7-(1-METHYLETHYL)-, (4A-α,7-β,8A-α)- see VAG200

1(2H)-NAPHTHALENONE, OCTAHYDRO-7-β-ISOPROPYL-4A-α,8A-α-DIMETHYL-, (−)- see VAG250

2-NAPHTHALENOXYACETIC ACID see NBJ700

2-((1-NAPHTHALENYLAMINO)CARBONYL)BENZOIC ACID see NBL200

4-(1-NAPHTHALENYLAZO)-1,3-PHENYLENEDIAMINE see NBG500

N-1-NAPHTHALENYL-1,2-ETHANEDIAMINE DIHYDROCHLORIDE see NBH500

1-(1-NAPHTHALENYL)ETHANONE see ABC475

1-(2-NAPHTHALENYL)ETHANONE see ABC500

4-(2-(1-NAPHTHALENYL)ETHENYL)-PYRIDINE HYDROCHLORIDE (9CI) see NBO525

1-NAPHTHALENYL ((HEXYLOXY)SULFINYL)METHYLCARBAMATE see NAP525

NAPHTHALENYL)METHANONE see NAP545

1-NAPHTHALENYL METHYLCARBAMATE see CBM750

α-NAPHTHALENYL METHYLCARBAMATE see CBM750

1-(2-NAPHTHALENYL)-3-(5-NITRO-2-FURANYL)-2-PROPEN-1-ONE see NAP600

(β-NAPHTHALENYLOXY)ACETIC ACID see NBJ700

2-(2-NAPHTHALENYLOXY)-N-PHENYLPROPANAMIDE see NBA600

4,4'-(3-(1-NAPHTHALENYL)-1,5-PENTANEDIYL)BISMORPHOLINE see DUO600

1-NAPHTHALENYLTHIOUREA see AQN635

1,8-NAPHTHALIC ANHYDRIDE see NAQ000

NAPHTHALIC ANHYDRIDE, 4-CHLORO-(7CI,8CI) see CJA050

NAPHTHALIDINE see NBE700

NAPHTHALIMIDE, 3-AMINO-N-(2-(DIMETHYLAMINO)ETHYL)- see NAC600

NAPHTHALIMIDE, 4-METHOXY-N-METHYL- see MEW760

NAPHTHALIN (DOT) see NAJ500

NAPHTHALINE see NAJ500

NAPHTHALOPHOS see HMV000

NAPHTHALOXIMIDE-O,O-DIETHYL PHOSPHOROTHIOATE see NAQ500

NAPHTHALOXIMIDODIETHYL THIOPHOSPHATE see NAQ500

α-NAPHTHALTHIOHARNSTOFF (GERMAN) see AQN635

NAPHTHAMINE DARK GREEN B see CMO830

NAPHTHANE see DAE800

NAPHTHANIL AS see CMM760

NAPHTHANIL BLUE B BASE see DCJ200

2-NAPHTHANILIDE, 3-HYDROXY- see CMM760

NAPHTHANILIDE OL SUPRA see CMM760

NAPHTHANILIDE RC see CMM760

NAPHTHANILIDE RC SUPRA see CMM760

NAPHTHANIL RED 3G BASE see KDA050

NAPHTHANIL RED B BASE see NEQ000

NAPHTHANIL SCARLET 2G BASE see DEO295

NAPHTHANIL SCARLET G BASE see NMP500

NAPHTHANTHRACENE see BBC250

NAPHTHANTHRONE see BBI250

NAPHTHANTHRONE see BCZ100

Δ⁵,⁷,⁹-NAPHTHANTRIENE see TCX500

NAPHTHA (PETROLEUM), CATALYTIC REFORMED see NAQ510

NAPHTHA (PETROLEUM), HEAVY CATALYTIC CRACKED see NAQ520

NAPHTHA (PETROLEUM), HEAVY CATALYTIC REFORMED see NAQ530

NAPHTHA (PETROLEUM), LIGHT CATALYTIC CRACKED see NAQ540

NAPHTHA (PETROLEUM), LIGHT CATALYTIC REFORMED see NAQ550

NAPHTHA (PETROLEUM), LIGHT STRAIGHT-RUN see NAQ560

NAPHTHA (PETROLEUM), SWEETENED see NAQ570

NAPHTHA (PETROLEUM), THERMAL CRACKED see NAQ580

1:4-NAPHTHAQUINOL-BISDISODIUM PHOSPHATE see NAQ600

1,2 NAPHTHAQUINONE see NBA000

NAPHTHA SAFETY SOLVENT see SLU500

NAPHTHA (UN2553) (DOT) see NAH600

NAPHTHA, solvent (UN1256) (DOT) see NAH600

NAPHTHA, petroleum (UN1255) (DOT) see NAH600

NAPHTHAZINE GREEN S see ADF000

NAPHTHAZINE ROSE 2G see CMM300

NAPHTHAZINE YELLOW RP see MRL100

NAPHTHENATE de COBALT (FRENCH) see NAR500

NAPHTHENE see NAJ500

NAPHTHENIC ACID see NAR000

NAPHTHENIC ACID ALUMINUM SALT see NAR100

NAPHTHENIC ACID, CALCIUM SALT see NAR200

NAPHTHENIC ACID, COBALT SALT see NAR500

NAPHTHENIC ACID, COPPER SALT see NAS000

NAPHTHENIC ACID, LEAD SALT see NAS500

NAPHTHENIC ACID, PHENYLMERCURY SALT see PFP000

NAPHTHENIC ACIDS, ALUMINUM SALT see NAR100

NAPHTHENIC ACIDS, MANGANESE SALTS see MAS820

NAPHTHENIC ACID, ZINC SALT see NAT000

NAPHTHENIC BASE LUBE STOCK see MQV845

NAPHTHENIC OILS see NAT100

NAPHTHENIC OILS (PETROLEUM), CATALYTIC DEWAXED HEAVY (9CI) see MQV776

NAPHTHENIC OILS (PETROLEUM), CATALYTIC DEWAXED LIGHT (9CI) see MQV777

3H-NAPHTH(1,2-D)IMIDAZOL-2-AMINE, 3-METHYL- see AKT900

1H-NAPHTH(2,3-d)IMIDAZOLE, 2-AMINO-see ALJ100

3H-NAPHTH(1,2-D)IMIDAZOLE, 2-AMINO-3-METHYL- see AKT900

NAPHTH(2',3':6,7)INDOLO(2,3-c)DINAPHTHO(2,3-a:2',3'-i)CARBAZOLE-5,10,15, 17,22,24-HEXONE see CMU770
NAPHTHIOMATE T see TGB475
NAPHTHIONIC ACID see ALI000
1,4-NAPHTHIONIC ACID see ALI000
NAPHTHIPRAMIDE see IOU000
NAPHTHISEN see NAH550
NAPHTHIZEN see NAH550
NAPHTHIZINE see NAH500
NAPHTHO(1,2,3-mno)ACEPHENANTHRYLENE see BCI265
NAPHTHO(2,1,8-hij)ACEPHENANTHRYLENE see BCH900
NAPHTHO(2,1-A)PYRENE see BCS460
NAPHTHOCAINE HYDROCHLORIDE see NAH800
NAPHTHO(1,2,3,4-def)CHRYSENE see NAT750
α-β-NAPHTHO-2,3-DIPHENYL-TRIAZOLIUM CHLORID (GERMAN) see DWH875
NAPHTHO(1,8-gh:4,5-g'h')DIQUINOLINE see NAU000
NAPHTHO(1,8-gh:5,4-g'h')DIQUINOLINE see NAU500
NAPHTHOELAN NAVY BLUE see PFU500
NAPHTHOELAN RED B BASE see NEQ000
NAPHTHOELAN RED RL BASE see MMF780
NAPHTHO(2,3-E)PYRENE see NAU510
α-NAPHTHOFLAVONE see NBI100
β-NAPHTHOFLAVONE see NAU525
NAPHTHO(1,2-k)FLUORANTHENE see NAU530
NAPHTHO(2,3-B)FLUORANTHENE see NAU535
NAPHTHO(2',3':2,3)FLUORANTHENE see NAU535
α-NAPHTHOFLUORENE see NAU600
NAPHTHO(1,2-H)QUINOLINE see NAU700
NAPHTHO(2,3-H)QUINOLINE-7,12-DIONE, 5-MORPHOLINO- see MRU090
NAPHTHO(2,3-H)QUINOLINE-7,12-DIONE, 5-PIPERIDINO- see PIT625
NAPHTHO(2,3-H)QUINOLINE-7,12-DIONE, 5-(1-PIPERIDINYL)- see PIT625
2-NAPHTHOHYDROXAMIC ACID see NAV000
2-NAPHTHOHYDROXAMIC ACID-o-ACETATE ESTER see ACS250
2-NAPHTHOHYDROXAMIC ACID-o-PROPIONATE ESTER see NBC500
2-NAPHTHOHYDROXIMIC ACID see NAV000
1-NAPHTHOIC ACID see NAV490
2-NAPHTHOIC ACID see NAV500
α-NAPHTHOIC ACID see NAV490
β-NAPHTHOIC ACID see NAV500
2-NAPHTHOIC ACID, 3-HYDROXY- see HMX520
2-NAPHTHOIC ACID, 4,4'-METHYLENEBIS(3-HYDROXY)-, compounded with (E)-1,4,5,6-TETRAHYDRO-1-METHYL-2-(2-(2-THIENYL)VINYL)PYRIMIDINE (1:1) see POK575
NAPHTHOIDE AS see CMM760
NAPHTHOIDE AS NO. 100 see CMM760
NAPHTHOL see NAW000
1-NAPHTHOL see NAW500
2-NAPHTHOL see NAX000
α-NAPHTHOL see NAW500
β-NAPHTHOL see NAX000
NAPHTHOL AS-A see CMM760
1-NAPHTHOL, ACETATE see NAU600
α-NAPHTHOL ACETATE see NAU600
NAPHTHOL ACNAC see CMM760
NAPHTHOLACTAM see NAX100
1,8-NAPHTHOLACTAM see NAX100
2-NAPHTHOL, 5-AMINO- see ALJ500
NAPHTHOL AS see CMM760
NAPHTHOL AS SUPRA see CMM760

NAPHTHOLATE AS see CMM760
NAPHTHOL B see NAX000
NAPHTHOL B.O.N. see HMX520
2-NAPHTHOL-3-CARBOXYLIC ACID see HMX500
2-NAPHTHOL, DECAHYDRO-, ACETATE see DAF100
2-NAPHTHOL, DECAHYDRO-, FORMATE see DAF150
2-NAPHTHOL-3,6-DISULFONIC ACID SODIUM SALT see ROF300
2-NAPHTHOL ETHYL ETHER see EEY500
β-NAPHTHOL ETHYL ETHER see EEY500
NAPHTHOL GREEN see NAX500
NAPHTHOL GREEN B see NAX500
β-NAPHTHOL ISOBUTYL ETHER see NBJ000
1-NAPHTHOL N-METHYLCARBAMATE see CBM750
NAPHTHOL ORANGE see CMM220
NAPHTHOL ORANGE see FAG010
α-NAPHTHOL ORANGE see FAG010
β-NAPHTHOL ORANGE see CMM220
2-NAPHTHOL ORANGE II see CMM220
NAPHTHOL RED B see FAG020
NAPHTHOL RED B see NAY000
NAPHTHOL RED B 20-7575 see NAY000
NAPHTHOL RED DEEP 10459 see NAY000
NAPHTHOL RED D TONER 35-6001 see NAY000
2-NAPHTHOL-6-SULFONIC ACID see HMU500
β-NAPHTHOL-6-SULFONIC ACID see HMU500
1-NAPHTHOL-3-SULFONIC ACID, 7-AMINO- see AKH800
β-NAPHTHOLSULFONIC ACID S see HMU500
1-NAPHTHOL, 1,2,3,4-TETRAHYDRO- see TDI350
NAPHTHOL YELLOW see DUX800
1-NAPHTHONITRILE see NAY090
2-NAPHTHONITRILE see NAY100
α-NAPHTHONITRILE see NAY090
β-NAPHTHONITRILE see NAY100
1-NAPHTHONITRILE, 5-NITRO- see NHR200
NAPHTHO(2',1':6,7)PHENANTHRO(3,4-B)OXIRENE-2,3-DIOL, 1A,2,3,13C-TETRAHYDRO- see NAY200
3H-NAPHTHO(2,1-b)PYRAN-2-CARBOXYLIC ACID, 3-OXO-, ETHYL ESTER see OOI200
1H,3H-NAPHTHO(1,8-cd)PYRAN-1,3-DIONE, 6-CHLORO- see CJA050
2H-NAPHTHO(1,2-B)PYRAN-5,6-DIONE, 3,4-DIHYDRO-2,2-DIMETHYL- see LBC500
1H-NAPHTHO(2,1-b)PYRAN-1-ONE, 3-PHENYL- see NAU525
4H-NAPHTHO(1,2-b)PYRAN-4-ONE, 2-PHENYL- see NBI100
NAPHTHOPYRIN see NAY500
β-NAPHTHOQUINALDINE see BDB750
NAPHTHO(2,3-f)QUINOLINE see NAZ000
α-NAPHTHOQUINOLINE see BDC000
1,2-NAPHTHOQUINONE see NBA000
1,4-NAPHTHOQUINONE see NBA500
α-NAPHTHOQUINONE see NBA500
β-NAPHTHOQUINONE see NBA000
1,4-NAPHTHOQUINONE, 5,8-DIHYDROXY-6-(1-HYDROXY-4-METHYL-3-PENTENYL)-, (−)- see HLJ650
1,4-NAPHTHOQUINONE, 2,3-DIMETHYL- see DSU310
1,4-NAPHTHOQUINONE, 2-DIMETHYLAMINO- see DPN300
1,4-NAPHTHOQUINONE, 2-HYDROXY- see HMX600
1,4-NAPHTHOQUINONE, 5-HYDROXY- see WAT000
1,4-NAPHTHOQUINONE, 8-HYDROXY- see WAT000

1,4-NAPHTHOQUINONE, 5-HYDROXY-7-METHYL- see RBF200
1,4-NAPHTHOQUINONE, 2-METHOXY- see MEY800
β-NAPHTHOQUINONE-4-SULFONATE SODIUM SALT see DLK000
NAPHTHORESORCINOL see NAN000
NAPHTHOSOL FAST RED KB BASE see CLK225
NAPHTHOSTYRIL see NAX100
NAPHTHO(1,2-e)THIANAPHTHENO(3,2-b)PYRIDINE see BCF500
NAPHTHO(2,1-e)THIANAPHTHENO(3,2-b)PYRIDINE see BCF750
α-NAPHTHOTHIOUREA see AQN635
2H-NAPHTHO(1,2-d)TRIAZOLE, 2-(4-STYRYL-3-SULFOPHENYL)-7-SULFO-, DISODIUM SALT see DXG025
NAPHTHO(1,2-d)TRIAZOLE-7-SULFONIC ACID,2-(4-(2-PHENYLETHENYL)-3-SULFOPHENYL)-, DISODIUM see DXG025
4-(2H-NAPHTHO(1,2-d)TRIAZOL-2-YL)-2-STILBENESULFONIC ACID SODIUM SALT see TGE155
2-NAPHTHOXYACETIC ACID see NBJ700
β-NAPHTHOXYACETIC ACID see NBJ700
α-(2-NAPHTHOXY)PROPIONANILIDE see NBA600
α-(β-NAPHTHOXY)PROPIONANILIDE see NBA600
1-(2-NAPHTHOYL)-AZIRIDINE see NBC000
β-NAPHTHOYLETHYLENEIMINE see NBC000
2-NAPHTHOYLHYDROXAMIC ACID see NAV000
N-(2-NAPHTHOYL)-o-PROPIONYLHYDROXYLAMINE see NBC500
1-NAPHTHYLACETAMIDE see NAK000
α-NAPHTHYLACETAMIDE see NAK000
1-NAPHTHYL ACETATE see NAU600
α-NAPHTHYL ACETATE see NAU600
α-NAPHTHYLACETIC see NAK500
NAPHTHYLACETIC ACID see NAK500
1-NAPHTHYLACETIC ACID see NAK500
α-NAPHTHYLACETIC ACID see NAK500
α-NAPHTHYLACETIC ACID METHYL ESTER see NAK525
N-2-NAPHTHYLACETOHYDROXAMIC ACID see NBD000
α-NAPHTHYL ACETONITRILE see NBD500
α-(1-NAPHTHYL)ACETONITRILE see NBD500
(6-(3-(1-NAPHTHYL)-d-ALANINE))-LHRH ACETATE see LIU329
(6-(3-(2-NAPHTHYL)-d-ALANINE)-7-(N^α)-METHYLEUCINE))-LHRH ACETATE see LIU331
β-NAPHTHYL ALCOHOL see NAX000
1-NAPHTHYLALDEHYDE see NAJ000
α-NAPHTHYLALDEHYDE see NAJ000
1-NAPHTHYLAMIN (GERMAN) see NBE700
2-NAPHTHYLAMIN (GERMAN) see NBE500
β-NAPHTHYLAMIN (GERMAN) see NBE500
1-NAPHTHYLAMINE see NBE700
2-NAPHTHYLAMINE see NBE500
6-NAPHTHYLAMINE see NBE500
α-NAPHTHYLAMINE see NBE700
β-NAPHTHYLAMINE see NBE500
NAPHTHYLAMINE BLUE see CMO250
2-NAPHTHYLAMINE-4,8-DISULFONIC ACID see ALH250
2-NAPHTHYLAMINE-6,8-DISULFONIC ACID see NBE850
β-NAPHTHYLAMINEDISULFONIC ACID see ALH250
β-NAPHTHYLAMINE-4,8-DISULFONIC ACID see ALH250
2-NAPHTHYLAMINE-1-d-GLUCOSIDURONIC ACID see ALL000
1-NAPHTHYLAMINE HYDROCHLORIDE see NBF000

α-NAPHTHYLAMINE HYDROCHLORIDE see NBF000

NAPHTHYLAMINE MUSTARD see BIF250

2-NAPHTHYLAMINE MUSTARD see NBE500

1-NAPHTHYLAMINE-4-SULFONIC ACID see ALI000

1-NAPHTHYLAMINE-6-SULFONIC ACID see ALI250

2-NAPHTHYLAMINE-1-SULFONIC ACID see ALH750

2-NAPHTHYLAMINE-8-SULFONIC ACID see ALI300

5-NAPHTHYLAMINE-2-SULFONIC ACID see ALI250

α-NAPHTHYLAMINE-p-SULFONIC ACID see ALI000

1-NAPHTHYLAMINE, 1,2,3,4-TETRAHYDRO-5-BENZYLOXY-8-CHLORO-N,N-DIMETHYL-, HYDROCHLORIDE see BFA930

2-(2-NAPHTHYLAMINO)BENZOIC ACID see NBF600

4-(2-NAPHTHYLAMINO)PHENOL see NBF500

p-(2-NAPHTHYLAMINO)PHENOL see NBF500

N-(1-NAPHTHYL)ANILINE see PFT250

N-(2-NAPHTHYL)ANILINE see PFT500

N-2 NAPHTHYLANTHRANILIC ACID see NBF600

4-(1-NAPHTHYLAZO)-2-NAPHTHYLAMINE see NBG000

4-(1-NAPHTHYLAZO)-m-PHENYLENEDIAMINE see NBG500

5-(β-NAPHTHYLAZO)-2,4,6-TRIAMINOPYRIMIDINE see NBO500

2-NAPHTHYLBIS(2-CHLOROETHYL)AMINE see BIF250

β-NAPHTHYL-BIS-(β-CHLOROETHYL)AMINE see BIF250

1-NAPHTHYL-N-BUTYL-N-NITROSOCARBAMATE see BRY750

α-NAPHTHYLCARBOXALDEHYDE see NAJ000

α-NAPHTHYLCARBOXYLIC ACID see NAV490

1-NAPHTHYL CHLOROCARBONATE see NBH200

1-NAPHTHYL CHLOROFORMATE see NBH200

α-NAPHTHYL CHLOROFORMATE see NBH200

β-NAPHTHYL-DI-(2-CHLOROETHYL)AMINE see BIF250

1-NAPHTHYL-N-(2,3,4,5-DI-o-ISOPROPYLIDENE-1-α-o-FRUCTOPYRANOSYLSULFINYL)-N-METHYLCARBAMATE see NBH300

α-NAPHTHYLENEACETIC ACID see NAK500

1,2-(1,8-NAPHTHYLENE)BENZENE see FDF000

1,5-NAPHTHYLENEDIAMINE see NAM000

1,5-NAPHTHYLENE DISULFONIC ACID see AQY400

NAPHTHYLENEETHYLENE see AAF275

NAPHTHYLENE YELLOW see DUX800

α-NAPHTHYLESSIGSAEURE (GERMAN) see NAK500

N-(1-NAPHTHYL)ETHYLENEDIAMINE DIHYDROCHLORIDE see NBH500

1-NAPHTHYL-N-ETHYL-N-NITROSOCARBAMATE see NBI000

α-NAPHTHYLFLAVONE see NBI100

O-(2-NAPHTHYL)GLYCOLIC ACID see NBJ700

2-NAPHTHYLHYDROXAMIC ACID see NAV000

β-NAPHTHYL HYDROXIDE see NAX000

1-NAPHTHYLHYDROXYLAMINE see HIX000

2-NAPHTHYLHYDROXYLAMINE see NBI500

N-1-NAPHTHYLHYDROXYLAMINE see HIX000

β-NAPHTHYL ISOBUTYL ETHER see NBJ000

1-NAPHTHYL ISOCYANIDE see IKH760

(2-NAPHTHYL)-1-ISOPROPYLAMINOETHANOL see INS000

NAPHTHYLISOPROTERENOL see INS000

NAPHTHYLISOPROTERENOL HYDROCHLORIDE see INT000

1-NAPHTHYL ISOTHIOCYANATE see ISN000

α-NAPHTHYL ISOTHIOCYANATE see ISN000

2-NAPHTHYL MERCAPTAN see NAP500

β-NAPHTHYL MERCAPTAN see NAP500

1-(1-NAPHTHYLMETHOXY)-2,3-EPOXYPROPANE see NBJ200

1-NAPHTHYL METHYLCARBAMATE see CBM750

1-NAPHTHYL N-METHYLCARBAMATE see CBM750

α-NAPHTHYL METHYLCARBAMATE see CBM750

α-NAPHTHYL N-METHYLCARBAMATE see CBM750

1-NAPHTHYLMETHYL GLYCIDYL ETHER see NBJ200

1-NAPHTHYL (METHYL)(HEXOXYSULFINYL)CARBAMATE see NAP525

2-(1-NAPHTHYLMETHYL)-2-IMIDAZOLINE see NAH500

α-NAPHTHYLMETHYL IMIDAZOLINE see NAH500

2-(α-NAPHTHYLMETHYL)-IMIDAZOLINE see NAH500

2-(1-NAPHTHYLMETHYL)IMIDAZOLINE HYDROCHLORIDE see NCW000

2-(1-NAPHTHYLMETHYL)-2-IMIDAZOLINE HYDROCHLORIDE see NCW000

1-NAPHTHYL-N-METHYL-KARBAMAT see CBM750

1-NAPHTHYL METHYL KETONE see ABC475

2-NAPHTHYL METHYL KETONE see ABC500

α-NAPHTHYL METHYL KETONE see ABC475

β-NAPHTHYL METHYL KETONE see ABC500

1-NAPHTHYL METHYLNITROSOCARBAMATE see NBJ500

1-NAPHTHYL-N-METHYL-N-NITROSOCARBAMATE see NBJ500

2-NAPHTHYL N-METHYL-N-(3-TOLYL)THIONOCARBAMATE see TGB475

o-2-NAPHTHYL m,N-DIMETHYLTHIOCARBANILATE see TGB475

1-NAPHTHYLNITRILE see NAY090

α-NAPHTHYLNITRILE see NAY090

β-NAPHTHYL ORANGE see CMM220

(2-NAPHTHYLOXY)ACETIC ACID see NBJ700

(2-(α-NAPHTHYLOXY)ETHYL)HYDRAZINE MALEATE see NBK000

4-(α-NAPHTHYLOXY)-3-HYDROXYBUTYRAMIDOXIME HYDROCHLORIDE see NAC500

4-α-NAPHTHYLOXY-3-HYDROXY-BUTYRAMIDOXIM-HYDROCHLORID (GERMAN) see NAC500

1-(1-NAPHTHYLOXY)-2-HYDROXY-3-ISOPROPYLAMINOPROPANE HYDROCHLORIDE see ICC000

2-(2-NAPHTHYLOXY)PROPIONANILIDE see NBA600

4,4'-(3-(1-NAPHTHYL)-1,5-PENTAMETHYLENE)DIMORPHOLINE see DUO600

2-NAPHTHYLPHENYLAMINE see PFT500

β-NAPHTHYLPHENYLAMINE see PFT500

2-NAPHTHYL-p-PHENYLENEDIAMINE see NBL000

N-1-NAPHTHYLPHTHALAMATE see NBL200

N-1-NAPHTHYLPHTHALAMIC ACID see NBL200

α-NAPHTHYLPHTHALAMIC ACID see NBL200

N-1-NAPHTHYLPHTHALAMIC ACID SODIUM SALT see SIO500

α-NAPHTHYLPHTHALAMIC ACID SODIUM SALT see SIO500

N-1-NAPHTHYL-PHTHALAMIDSAEURE (GERMAN) see NBL200

1-NAPHTHYL-N-PROPYL-N-NITROSOCARBAMATE see NBM500

α-NAPHTHYL RED see PEJ600

β-NAPHTHYLSULFONIC ACID see NAP000

3-(1-NAPHTHYL)-2-TETRAHYDROFURFURYLPROPIONIC ACID 2-(DIETHYLAMINO)ETHYL ESTER see NAE000

1-NAPHTHYL THIOACETAMIDE see NBN000

α-NAPHTHYLTHIOCARBAMIDE see AQN635

1-NAPHTHYL-THIOHARNSTOFF (GERMAN) see AQN635

2-NAPHTHYL THIOL NAP500

1-(1-NAPHTHYL)-2-THIOUREA see AQN635

N-(1-NAPHTHYL)-2-THIOUREA see AQN635

α-NAPHTHYLTHIOUREA see AQN635

α-NAPHTHYLTHIOUREA (DOT) see AQN635

1-NAPHTHYL THIOUREA (MAK) see AQN635

1-NAPHTHYL-THIOUREE (FRENCH) see AQN635

5-β-NAPHTHYL-2:4:6-TRIAMINOAZOPYRIMIDINE see NBO500

N-1-NAPHTHYL-N,N',N'-TRIETHYLETHYLENEDIAMINE see NBO515

4-(1-NAPHTHYLVINYL)-PYRIDINE HYDROCHLORIDE see NBO525

4-(2-(1-NAPHTHYL)VINYL)PYRIDINE HYDROCHLORIDE see NBO525

NAPHTHYPRAMIDE see IOU000

1,8-NAPHTHYRIDINE-3-CARBOXYLIC ACID, 1-ETHYL-1,4-DIHYDRO-7-METHYL-4-OXO-, SODIUM SALT see NAG450

NAPHTOCARD ORANGE II see CMM220

NAPHTOCARD RED 2G see CMM300

NAPHTOCARD RED 6B see CMM400

NAPHTOCARD YELLOW O see FAG140

NAPHTOELAN A see CMM760

NAPHTOELAN FAST ORANGE GC BASE see CEH690

NAPHTOELAN FAST RED 3GL BASE see KDA050

NAPHTOELAN FAST SCARLET G BASE see NMP500

NAPHTOELAN FAST SCARLET G SALT see NMP500

NAPHTOELAN MITSUI SCARLET GG SALT see DEO295

NAPHTOELAN ORANGE R BASE see NEN500

NAPHTOELAN RED GG BASE see NEO500

2-NAPHTOL (FRENCH) see NAX000

β-NAPHTOL (GERMAN) see NAX000

NAPHTOL AS see CMM760

NAPHTOL AS-KG see TGR000

NAPHTOL AS-KGLL see TGR000

2-NAPHTOL-6-SULFOSAURE (GERMAN) see HMU500

NAPHTO(1,2-c-d-e)NAPHTACENE (FRENCH) see BCP750
NAPHTOX see AQN635
NAPHTYZIN see NAH550
NAPHURIDE see BAT000
NAPHYL-1-ESSIGSAEURE (GERMAN) see NAK500
NAPOLONE see MIF760
NAPOTON see LFK000
NAPOTON see MDQ250
NAPRINOL see HIM000
NAPROANILIDE see NBA600
NAPROPION see SJL500
NAPROSINE see MFA500
NAPROSYN see MFA500
NAPROXEN see MFA500
NAPROXEN SODIUM see NBO550
NAPRUX see MFA500
NAPTALAM see NBL200
NAPTALAME see NBL200
NAPTALAM SODIUM see SIO500
NAPTHAMINE BROWN DC see CMO800
NAPTHOLROT S see FAG020
NAPTRO see NBL200
NAQUA see HII500
NAQUIVAL see RDK000
NARAMYCIN see CPE750
NARASIN see NBO600
NARCAN see NAH000
NARCEOL see MNR250
NARCICLASINE see LJC000
NARCISO (CUBA, MEXICO) see DAB700
NARCISSUS see DAB700
NARCISSUS ABSOLUTE see NBP000
NARCISSUS POETICUS see DAB700
NARCISSUS PSEUDONARCISSUS see DAB700
NARCOGEN see TIO750
NARCOLAN see ARW250
NARCOLAN see THV000
NARCOLO see AFJ400
NARCOMPREN see NOA000
NARCOMPREN HYDROCHLORIDE see NOA500
NARCOSAN see ERD500
NARCOSAN SOLUBLE see ERE000
NARCOSINE see NOA000
NARCOSINE HYDROCHLORIDE see NOA500
NARCOSINE, HYDROCHLORIDE (8CI) see NOA500
NARCOTANE see HAG500
NARCOTANN NE-SPOFA (RUSSIAN) see HAG500
NARCOTILE see EHH000
NARCOTINE see NBP275
NARCOTINE see NOA000
1-α-NARCOTINE see NOA000
1-α-NARCOTINE HYDROCHLORIDE see NOA500
NARCOTINE N-OXIDE HYDROCHLORIDE see NBP300
NARCOTUSSIN see NOA000
NARCOZEP see FDD500
NARCYLEN see ACI750
NARDELZINE see PFC750
NARDIL see PFC500
NARDIL see PFC750
NAREA see HDP500
NARGOLINE see NDM000
NARIGIX see EID000
(−)-NARINGENIN see NBP350
NARINGENIN see NBP350
(S)-NARINGENIN see NBP350
NARINGENINE see NBP350
NARINGETOL see NBP350
NARITHERACIN see VGZ000
NARKOLAN see ARW250
NARKOLAN see THV000
NARKOSOID see TIO750
NARLENE see BQU500
NARPHEN see PMD325

NARSIS see CGA000
NASALIDE see FDD085
NASDOL see TBG000
NASIVINE see ORA100
NASMIL see CNX825
NASOL see EAW000
NASS (IRAN) see SED400
NASTENON see PAN100
NASTYN see EHP000
NASWAR (PAKISTAN and AFGHANISTAN) see SED400
NA TA see TII250
NATA see TII500
NATACYN see PIF750
NATAMYCIN see PIF750
NATASOL FAST RED TR SALT see CLK235
NATERETIN see BEQ625
NATHULANE see PME500
NAT-333 HYDROCHLORIDE see DAI200
NATICARDINA see GAX000
NATIL see DNU100
NATIONAL 120-1207 see AAX250
NATIVE CALCIUM SULFATE see CAX750
NaATP see AEM250
NATRASCORB see ARN125
NATRASCORB INJECTABLE see ARN000
NATREEN see BCE500
NATREEN see SGC600
NATRI-C see ARN125
NATRILIX see IBV100
NATRINAL see BAG250
NATRIN HERBICIDE see SKL000
NATRIONEX see AAI250
NATRI-PAS see SEP000
NATRIPHENE see BGJ750
NATRIUM see SEE500
NATRIUMACETAT (GERMAN) see SEG500
NATRIUMALUMINUMFLUORID (GERMAN) see SHF000
NATRIUMANTIMONYLTARTRAT (GERMAN) see AQI750
NATRIUMARSENIT (GERMAN) see ARJ500
NATRIUMAZID (GERMAN) see SFA000
NATRIUMAZIDE (DUTCH) see SFA000
NATRIUMBARBITALS (GERMAN) see BAG250
NATRIUMBICHROMAAT (DUTCH) see SGI000
NATRIUMCHLORAAT (DUTCH) see SFS000
NATRIUMCHLORAT (GERMAN) see SFS000
NATRIUMCHLORID (GERMAN) see SFT000
NATRIUM CITRICUM (GERMAN) see DXC400
NATRIUMDEHYDROCHOLAT (GERMAN) see SGD500
NATRIUM-2,4-DICHLORPHENOXYATHYLSULFAT (GERMAN) see CNW000
NATRIUMDICHROMAAT (DUTCH) see SGI000
NATRIUMDICHROMAT (GERMAN) see SGI000
NATRIUM-3-α,12-α-DIHYDROXYCHOLANAT (GERMAN) see SGE000
NATRIUMFLUORACETAAT (DUTCH) see SHG500
NATRIUMFLUORACETAT (GERMAN) see SHG500
NATRIUM FLUORIDE see SHF500
NATRIUMGLUTAMINAT (GERMAN) see MRL500
NATRIUMHEXAFLUOROALUMINATE (GERMAN) see SHF000
NATRIUM-HUMAT see HGM100
NATRIUMHYDROXID (GERMAN) see SHS000
NATRIUMHYDROXYDE (DUTCH) see SHS000
NATRIUMHYPOPHOSPHIT (GERMAN) see SHV000

NATRIUM-O-ISOPROPYLDITHIOKARBONAT see SIA000
NATRIUMJODAT (GERMAN) see SHV500
NATRIUMJODID (GERMAN) see SHW000
NATRIUMMALAT (GERMAN) see MAN250
NATRIUM-N-METHYL-DITHIOCARBAMAAT (DUTCH) see VFU000
NATRIUMMETHYLDITHIOCARBAMAT (GERMAN) see SIL550
NATRIUM-N-METHYL-DITHIOCARBAMAT (GERMAN) see VFU000
NATRIUMMOLYBDAT (GERMAN) see DXE800
NATRIUMNICOTINAT (GERMAN) see NDW500
NATRIUMNITRAT UND N-METHYLANILIN (GERMAN) see MGO250
NATRIUM NITRIT (GERMAN) see SIQ500
NATRIUMOXALAT (GERMAN) see SIY500
NATRIUMPERCHLORAAT (DUTCH) see PCE750
NATRIUMPERCHLORAT (GERMAN) see PCE750
NATRIUMPHOSPHAT (GERMAN) see SJH090
NATRIUMPROPIONAT see SJL500
NATRIUMPYROPHOSPHAT see TEE500
NATRIUMRHODANID (GERMAN) see SIA500
NATRIUMSALZ DER 2,2-DICHLORPROPIONSAEURE see DGI600
NATRIUMSELENIAT (GERMAN) see DXG000
NATRIUMSELENIT (GERMAN) see SJT500
NATRIUMSILICOFLUORID (GERMAN) see DXE000
NATRIUMSULFAT (GERMAN) see SJY000
NATRIUMSULFID (GERMAN) see SJZ000
NATRIUMTARTRAT (GERMAN) see SKB000
NATRIUMTRICHLOORACETAAT (DUTCH) see TII500
NATRIUMTRICHLORACETAT (GERMAN) see TII500
NATRIUMTRIPOLYPHOSPHAT (GERMAN) see SKN000
NATRIUMZYKLAMATE (GERMAN) see SGC000
NATRIURAN see CLY600
NATRIURETIC HORMONE (bovine pineal) see AQW125
NATROSOL see HKQ100
NATROSOL 250 see HKQ100
NATROSOL 150L see HKQ100
NATROSOL 180L see HKQ100
NATROSOL 250G see HKQ100
NATROSOL 250H see HKQ100
NATROSOL 250L see HKQ100
NATROSOL 250M see HKQ100
NATROSOL 300H see HKQ100
NATROSOL 250H4R see HKQ100
NATROSOL 250MH see HKQ100
NATROSOL 250HHP see HKQ100
NATROSOL 250HHR see HKQ100
NATROSOL LR see HKQ100
NATROSOL 240JR see HKQ100
NATROSOL 250HR see HKQ100
NATROSOL 250HX see HKQ100
NATULAN see PME250
NATULAN see PME500
NATULANAR see PME500
NATULAN HYDROCHLORIDE see PME500
NATURAL CALCIUM CARBONATE see CAO000
NATURAL GAS CONDENSATES, GASOLINE see GCC200
NATURAL GASOLINE (DOT) see GBY000
NATURAL GAS, compressed (with high methane content) (UN 1971) (DOT) see MDQ750
NATURAL GAS, refrigerated liquid (cryogenic liquid) (with high methane content) (UN 1972) (DOT) see MDQ750

NCI-C03601 see PHW750
NCI-C03612 see BBO500
NCI-C03656 see DDA800
NCI-C03667 see DAC800
NCI-C03678 see OAJ000
NCI-C03689 see DVQ000
NCI-C03703 see HCF500
NCI-C03714 see TAI000
NCI-C03736 see AOQ000
NCI-C03736 see BBL000
NCI-C03747 see AOX250
NCI-C03758 see AOX500
NCI-C03770 see CEC000
NCI-C03781 see TMD250
NCI-C03792 see ENV500
NCI-C03805 see BNK000
NCI-C03816 see DKC400
NCI-C03827 see DQD400
NCI-C03838 see CIV000
NCI-C03849 see SHX000
NCI-C03850 see DVR200
NCI-C03861 see LHW000
NCI-C03907 see PIC000
NCI-C03918 see DQJ200
NCI-C03930 see PEY650
NCI-C03941 see ALL500
NCI-C03952 see NNT000
NCI-C03963 see NEM480
NCI-C03974 see TNL250
NCI-C03974 see TNL500
NCI-C03985 see DGG950
NCI-C04126 see DTH000
NCI-C04137 see DBN200
NCI-C04159 see BKF250
NCI-C04240 see TGG760
NCI-C04251 see TGF250
NCI-C04273 see CMJ582
NCI-C04502 see DGW200
NCI-C04524 see DGV800
NCI-C04535 see DFF809
NCI-C04546 see TIO750
NCI-C04557 see DMF000
NCI-C04568 see IEP000
NCI-C04579 see TIN000
NCI-C04580 see PCF275
NCI-C04591 see CBV500
NCI-C04604 see HCI000
NCI-C04615 see AGB250
NCI-C04626 see MIH275
NCI-C04637 see TIP500
NCI-C04671 see MDV500
NCI-C04682 see AEB000
NCI-C04693 see DAC000
NCI-C04706 see AHK500
NCI-C04717 see DAB600
NCI-C04728 see AQQ750
NCI-C04739 see ADA000
NCI-C04740 see CGV250
NCI-C04762 see DDP600
NCI-C04773 see BIF750
NCI-C04784 see MPU000
NCI-C04795 see DDJ000
NCI-C04819 see DCF200
NCI-C04820 see BIA250
NCI-C04831 see HOO500
NCI-C04842 see VKZ000
NCI-C04853 see PED750
NCI-C04864 see LEY000
NCI-C04875 see DFO000
NCI-C04886 see POK000
NCI-C04897 see PLZ000
NCI-C04900 see CQC650
NCI-C04922 see CQN000
NCI-C04933 see CDN000
NCI-C04944 see BHT750
NCI-C04955 see CHD250
NCI-C05970 see REA000
NCI-C06008 see TAB750
NCI-C06111 see BDX500
NCI-C06155 see BQQ750
NCI-C06224 see EHH000
NCI-C06360 see BEE375

NCI-C06428 see MQW500
NCI-C06462 see SFA000
NCI-C06508 see BDX000
NCI-C07272 see TGK750
NCI-C08628 see OPK250
NCI-C08640 see CBM500
NCI-C08651 see FAQ900
NCI-C08662 see CNU750
NCI-C08673 see DCM750
NCI-C08684 see DEV800
NCI-C08695 see DUK800
NCI-C08899 see GFO080
NCI-C08991 see ARM250
NCI-C08991 see ARM280
NCI-C09007 see ARM275
NCI-C50011 see BCP250
NCI-C50033 see SBT000
NCI-C50044 see BII250
NCI-C50055 see DUE000
NCI-C50077 see PMO500
NCI-C50088 see EJN500
NCI-C50099 see PNL600
NCI-C50102 see MJP450
NCI-C50124 see PDN750
NCI-C50135 see EIU800
NCI-C50146 see OPM000
NCI-C50157 see RDK000
NCI-C50168 see CNU000
NCI-C50191 see SIB600
NCI-C50204 see TAV750
NCI-C50226 see ZAT000
NCI-C50259 see HEJ500
NCI-C50260 see DBR400
NCI-C50282 see BGE325
NCI-C50317 see DCE400
NCI-C50351 see BGJ250
NCI-C50362 see HER000
NCI-C50384 see EFT000
NCI-C50395 see GLU000
NCI-C50419 see LIA000
NCI-C50442 see BJK500
NCI-C50453 see EQR500
NCI-C50464 see AGJ250
NCI-C50475 see AEX250
NCI-C50533 see TGM750
NCI-C50544 see WCB000
NCI-C50602 see BOP500
NCI-C50613 see AMW000
NCI-C50635 see BLD500
NCI-C50646 see CBF700
NCI-C50657 see DBL200
NCI-C50668 see MJP400
NCI-C50680 see MLH750
NCI-C50715 see MCB000
NCI-C50737 see CQK600
NCI-C50748 see AQQ500
NCI-C52459 see TBQ000
NCI-C52733 see DVL700
NCI-C52904 see NAJ500
NCI-C53587 see PAH780
NCI-C53634 see HCA500
NCI-C53781 see AAQ250
NCI-C53792 see CHP500
NCI-C53838 see HGC000
NCI-C53849 see HJF500
NCI-C53894 see PAW500
NCI-C53907 see FAG150
NCI-C53929 see PEJ500
NCI-C54262 see VPK000
NCI-C54364 see GMA000
NCI-C54375 see BEC500
NCI-C54386 see AEO000
NCI-C54546 see SBW000
NCI-C54557 see AQP000
NCI-C54568 see CMO750
NCI-C54579 see CMO000
NCI-C54604 see MJQ100
NCI-C54626 see MRC000
NCI-C54637 see PJQ100
NCI-C54660 see DHF200
NCI-C54706 see FEW000
NCI-C54717 see ISV000

NCI-C54728 see DTD800
NCI-C54739 see RMK020
NCI-C54740 see DST800
NCI-C54751 see TNI250
NCI-C54773 see DSG600
NCI-C54795 see DHE750
NCI-C54808 see ARN000
NCI-C54819 see IKE000
NCI-C54820 see CIU750
NCI-C54831 see TIQ250
NCI-C54842 see PMK000
NCI-C54853 see EES350
NCI-C54886 see CEJ125
NCI-C54897 see HKN875
NCI-C54900 see TBG700
NCI-C54911 see SGP500
NCI-C54922 see ALO750
NCI-C54933 see PAX250
NCI-C54944 see DEP600
NCI-C54955 see DEP800
NCI-C54966 see REF000
NCI-C54977 see EBR000
NCI-C54988 see TCF000
NCI-C54999 see CPD750
NCI-C55005 see CPC000
NCI-C55050 see TDI000
NCI-C55061 see TDH750
NCI-C55072 see CDS000
NCI-C55107 see CEA750
NCI-C55118 see CEQ600
NCI-C55129 see DRS200
NCI-C55130 see BNL000
NCI-C55141 see PNJ400
NCI-C55152 see AQF000
NCI-C55163 see CCP250
NCI-C55174 see DHF000
NCI-C55185 see GEO000
NCI-C55196 see NGE000
NCI-C55209 see OLA000
NCI-C55210 see RNZ000
NCI-C55221 see SHF500
NCI-C55232 see XGS000
NCI-C55243 see BND500
NCI-C55254 see CFK500
NCI-C55265 see TAI500
NCI-C55276 see BBL250
NCI-C55287 see PAU500
NCI-C55298 see QPA000
NCI-C55301 see POP250
NCI-C55312 see CNF330
NCI-C55323 see BKE500
NCI-C55345 see DFX800
NCI-C55367 see BPX000
NCI-C55378 see PAX250
NCI-C55403 see XJJ010
NCI-C55425 see GFQ000
NCI-C55436 see DDS000
NCI-C55447 see MKA500
NCI-C55458 see AJL500
NCI-C55469 see IEL800
NCI-C55470 see WCJ000
NCI-C55481 see EGV400
NCI-C55492 see PEO500
NCI-C55505 see SEM000
NCI-C55516 see DDQ400
NCI-C55527 see BOX750
NCI-C55538 see EBX500
NCI-C55549 see GGW500
NCI-C55550 see TEO250
NCI-C55561 see TBX250
NCI-C55572 see LFU000
NCI-C55583 see NKM000
NCI-C55594 see MHZ000
NCI-C55607 see HCE500
NCI-C55618 see IMF400
NCI-C55630 see IMR000
NCI-C55641 see SPC500
NCI-C55652 see EAY500
NCI-C55663 see AES500
NCI-C55674 see EDJ500
NCI-C55685 see PDE000
NCI-C55696 see SNC000

NCI-C55709 see CDP250
NCI-C55710 see AOB250
NCI-C55721 see DNA800
NCI-C55743 see PBC250
NCI-C55754 see SJH095
NCI-C55765 see DKQ000
NCI-C55776 see PJD000
NCI-C55787 see HFV500
NCI-C55798 see PDO750
NCI-C55801 see HIM000
NCI-C55812 see MIP750
NCI-C55823 see GEM000
NCI-C55834 see HIH000
NCI-C55845 see QQS200
NCI-C55856 see CCP850
NCI-C55867 see MCD250
NCI-C55878 see BOV000
NCI-C55889 see HAP500
NCI-C55890 see HHR500
NCI-C55903 see XDJ000
NCI-C55925 see CFY000
NCI-C55936 see CHJ750
NCI-C55947 see TDY250
NCI-C55958 see NEM500
NCI-C55969 see AOR500
NCI-C55970 see ALO000
NCI-C55981 see ASM270
NCI-C55992 see NIF000
NCI-C56019 see BKC000
NCI-C56031 see DFI210
NCI C56042 see UVJ450
NCI-C56064 see NGE500
NCI-C56075 see BAU750
NCI-C56086 see AOD125
NCI-C56097 see DWW000
NCI-C56100 see BFC750
NCI-C56111 see CMP969
NCI-C56122 see RGW000
NCI-C56133 see BAY500
NCI-C56144 see IDA000
NCI-C56155 see TMN490
NCI-C56166 see BCK250
NCI-C56177 see FPQ875
NCI-C56188 see XNJ000
NCI-C56199 see EID000
NCI-C56202 see FPK000
NCI-C56213 see ABC250
NCI-C56224 see FPU000
NCI-C56235 see IPU000
NCI-C56246 see QFS000
NCI-C56257 see HKR500
NCI-C56279 see COB260
NCI-C56280 see MNQ000
NCI-C56291 see BSU250
NCI-C56304 see PFL000
NCI-C56315 see DQV250
NCI-C56326 see AAG250
NCI-C56337 see BJL000
NCI-C56348 see DTC800
NCI-C56359 see PAH250
NCI-C56360 see DBB200
NCI-C56382 see BIE500
NCI-C56393 see EGP500
NCI-C56406 see VQK650
NCI-C56417 see BMC000
NCI-C56428 see DQF800
NCI-C56439 see IPD000
NCI-C56440 see HCZ000
NCI-C56451 see PKL500
NCI-C56462 see MRH000
NCI-C56473 see HOH500
NCI-C56484 see OGQ100
NCI-C56508 see HMY000
NCI-C56519 see BDF000
NCI-C56520 see MNH250
NCI-C56531 see BRF500
NCI-C56542 see SGB550
NCI-C56553 see BRV500
NCI-C56564 see CBF750
NCI-C56575 see CAL000
NCI-C56586 see CHP250
NCI-C56597 see SNH000

NCI-C56600 see SNJ000
NCI-C56611 see TBO850
NCI-C56633 see TKV000
NCI-C56644 see HOO100
NCI-C56655 see PAX250
NCI-C56666 see AGH150
NCI-C56762 see DSR400
NCI-C60015 see COG000
NCI-C60026 see AHP000
NCI-C60048 see DJX000
NCI-C60060 see MKQ880
NCI-C60066 see MKQ875
NCI-C60071 see MRL750
NCI-C60082 see NEX000
NCI-C60093 see PFJ250
NCI-C60102 see QCJ000
NCI-C60106 see QCA000
NCI-C60117 see TAJ000
NCI-C60128 see CGO500
NCI-C60139 see VOA000
NCI-C60162 see FCA100
NCI-C60173 see MCY475
NCI-C60184 see PAD500
NCI-C60195 see PEC500
NCI-C60208 see VPP000
NCI-C60219 see PGX000
NCI-C60220 see TJB600
NCI-C60231 see CEA000
NCI-C60242 see TIK500
NCI-C60286 see PKL100
NCI-C60297 see CNV000
NCI-C60311 see CNA250
NCI-C60322 see DTG000
NCI-C60333 see HLU500
NCI-C60344 see NDK500
NCI-C60355 see KDA050
NCI-C60366 see MMP100
NCI-C60377 see NAY000
NCI-C60388 see NER000
NCI-C60399 see MCW250
NCI-C60402 see EEA500
NCI-C60413 see DER000
NCI-C60537 see NMO550
NCI-C60559 see CHY250
NCI-C60560 see TCR750
NCI-C60571 see HEN000
NCI-C60582 see PKQ250
NCI-C60582 see PKQ500
NCI-C60582 see PKQ750
NCI-C60582 see PKR000
NCI-C60582 see PKR250
NCI-C60582 see PKR750
NCI-C60582 see PKS000
NCI-C60606 see TAG750
NCI-C60628 see TFD250
NCI-C60639 see DYE600
NCI-C60640 see DPJ200
NCI-C60651 see WAK000
NCI-C60662 see TMP750
NCI-C60673 see DQA400
NCI C60684 see DYE500
NCI-C60695 see TMJ750
NCI-C60708 see NCD500
NCI-C60719 see BEM500
NCI-C60720 see MPE250
NCI-C60742 see DMN400
NCI-C60753 see DUP600
NCI-C60786 see NEO500
NCI-C60797 see PKQ059
NCI-C60800 see CMH300
NCI-C60811 see ZSJ000
NCI-C60822 see ABE500
NCI-C60844 see DUS200
NCI-C60866 see BRR900
NCI C60877 see CMO650
NCI-C60899 see DIX200
NCI-C60902 see TLP500
NCI-C60913 see DSB000
NCI-C60924 see DWC600
NCI-C60935 see LBX000
NCI-C60946 see AFW750
NCI-C60957 see MES000

NCI-C60968 see IJS000
NCI-C60979 see IKQ000
NCI-C60980 see BCC500
NCI-C61018 see NMW500
NCI-C61029 see PMT750
NCI-C61041 see TNP500
NCI-C61052 see IJD000
NCI-C61074 see BAK000
NCI-C61074 see BAK020
NCI C61096 see CMM330
NCI-C61109 see CMN750
NCI-C61110 see CMN800
NCI-C61143 see MAU250
NCI-C61165 see SBX525
NCI-C61176 see ARA500
NCI-C61187 see TIV750
NCI-C61201 see CJU250
NCI-C61289 see CMO250
NCI C61290 see CMO500
NCI-C61325 see ICC800
NCI-C61392 see BGB275
NCI-C61405 see HEO000
NCI-C61494 see BEN000
NCI-C61585 see TBT150
NCI-C60253A see ARM262
NCI-C61223A see ARM268
NCI-CO1752 see CKK000
NCI-CO1763 see BSQ000
NCI-CO2131 see BSS250
NCI-CO4035 see CFK000
NCI-CO5210 see CKP500
NCI-CO03247 see ABB000
(+)-NC-NITROARGININE see OIU900
NCO 20 see PKB100
NCR CE EE DOV7 see DDM200
NCR DR DCN see BGF250
NCS see NBV500
NCS-14210 see BHV000
NCS-27381 see BHU500
NCS 110435 see CDB800
NCS 349174 see XDJ025
NC-319 TECHNICAL see CGC150
ND-4 see DPI710
NDBA see BRY500
NDC 0082-4155 see DAC200
NDC 002-1452-01 see VKZ000
NDD see CGA500
NDEA see NJW500
NDELA see NKM000
NDFDA see PCG725
NDGA see NBR000
NDHU see NJY000
NDMA see DSY600
NDMA see NKA600
NDMI see NJK500
NDPA see DWI000
NDPA see NKB700
NDPEA see NBR100
NDPhA see DWI000
NDR-132 see PML350
NDR A-3682 see FKW000
NDRC-143 see AHJ750
NDR-5998A HYDROCHLORIDE see DAI200
NDS see DKH250
NEA see NKD000
NEAMIN see NCE500
NEANTINE see DJX000
NEARGAL see LJR000
NEASINA see SNJ000
NEATHALIDE see INS000
NEAT OIL OF SWEET ORANGE see OGY000
NEAUFATIN see TEH500
NEAZINE see PPP500
NEBCIN see TGI500
NEBERK see FLZ050
NEBRAMYCIN see NBR500
NEBRAMYCIN FACTOR 5 see BAU270
NEBRAMYCIN FACTOR 6 see TGI250
NEBRAMYCIN V see BAU270
NEBRAMYCIN X see NCE500
NEBRASKA FERN see PJJ300

NEBULARIN(E) see RJF000
NEBULARINE see RJF000
NEBUREA see BRA250
NEBUREX see BRA250
NEBURON see BRA250
NECARBOXYLIC ACID see AFR250
NECATORINA see CBY000
NECATORINE see CBY000
NECCANOL SW see DXW200
NECKLACEPOD SOPHORA see NBR800
NECKLACEWEED see BAF325
NECTADON see NOA000
NECTADON HYDROCHLORIDE see NOA500
NEDCIDOL see DCM750
NEEDLE ANTIMONY see AQL500
NEEM EXTRACT see NBR900
NEEM OIL see NBS300
NEFCO see NGE500
NEFIRACETAM D-2 see DTK400
NEFOPAM see FAL000
NEFOPAM HYDROCHLORIDE see NBS500
NEFRAS 150/200 see GCC200
NEFRAS S 50/170 see GCC200
NEFRECIL see PDC250
NEFRIX see CFY000
NEFROSUL see SNH900
NEFTIN see NGG500
NEFURTHIAZOLE see NDY500
NEFUSAN see DSB200
NEGALIP see ARQ750
NEGAMICIN see NCE500
NEGANOX DLTP see TFD500
NEGRAM see EID000
NEGUVON see TIQ250
NEGUVON A see TIQ250
NEIRINE see VQR300
NEKAL see NBS700
NEKAL WT-27 see DJL000
NEKLACID FAST ORANGE 2G see HGC000
NEKLACID ORANGE 1 see FAG010
NEKLACID ORANGE II see CMM220
NEKLACID RED 3R see FMU080
NEKLACID RED A see FAG020
NEKLACID RED RR see FMU070
NEKLACID RUBINE W see HJF500
NEKLACID YELLOW G see MRL100
NEKLACID YELLOW T see FAG140
NEKLAMIN BLACK BH see CMN800
NEKLAMIN BROWN M see CMO800
NEKTROHAN see ZVJ000
NELBON see DLY000
NELIPRAMIN see DLH600
NELKEN OEL (GERMAN) see CMY475
NELLITE see PEV500
NELPON see TJK100
NEM see ENL000
NEM see MAL250
NEMA see MKB000
NEMA see PCF275
NEMABROM see DDL800
NEMACIDE see DFK600
NEMACTIL see PIW000
NEMACUR see FAK000
NEMAFENE see DGG000
NEMAFOS see EPC500
NEMAFUME see DDL800
NEMAGON see DDL800
NEMAGONE see DDL800
NEMAGON SOIL FUMIGANT see DDL800
NEMALITE see MAG750
NEMALITE see NBT000
NEMALITH (GERMAN) see NBT000
NEMAMORT see BII250
NEMANAX see DDL800
NEMAPAN see TEX000
NEMAPAZ see DDL800
NEMAPHOS see EPC500
NEMASET see DDL800
NEM-A-TAK see DHH200
NEMATOCIDE see DDL800
NEMATOCIDE see EPC500

NEMATOLYT see PAG500
NEMATOX see DDL800
NEMAZENE see PDP250
NEMAZINE see PDP250
NEMAZON see DDL800
NEMBUSEN see GGS000
NEMBU-SERPIN see RDK000
NEMBUTAL see NBT500
NEMBUTAL CALCIUM see CAV000
NEMBUTAL SODIUM see NBU000
NEMEROL see DAM600
NEMICIDE see LFA020
NEMISPOR see DXI400
NENDRIN see EAT500
NENESIN see BNK000
NEO see TEH500
NEOABIETIC ACID see NBU800
NEOANTERGAN see WAK000
NEOANTERGAN HYDROCHLORIDE see MCJ250
NEOANTERGAN MALEATE see DBM800
NEOANTERGAN PHOSPHATE see NBV000
NEOANTICID see CAT775
NEOANTIMOSAN see AQH500
NEO-ANTITENSOL see RDK000
NEO-ANTITERSOL see RDK000
NEOARSOLUIN see NCJ500
NEOARSPHENAMINE see NCJ500
NEOASYCODILE see DXE600
NEO-ATOMID see ARQ750
NEO-AVAGAL see SBH500
NEOBAN see CEW500
NEOBAR see BAP000
NEOBARENE see NBV100
NEOBEE M-5 see CBF710
NEOBEE O see CBF710
NEO-BENODINE see TGJ475
NEOBIOTIC see NCG000
NEOBOR see SFF000
NEOBRETTIN see NCD550
NEOBRIDAL see WAK000
NEOCAINE see AIL750
NEOCAINE see AIT750
NEO-CALMA see CJR909
NEOCARBORANE see NBV100
NEOCARCINOSTATIN see NBV500
NEO-CARDIAMINE see GCE600
NEO-β-CAROTENE see NBV200
NEOCARZINOSTATIN see NBV500
NEOCARZINOSTATIN K see NBV500
NEO-CEBICURE see SAA025
NEO-CEBITATE see EDE700
NEOCHIN see CLD000
NEOCHROMIUM see NBW000
NEOCID see DAD200
NEOCIDOL see DCM750
NEO-CLEANER see SHU500
NEOCLYM see FBP100
NEOCOCCYL see SNM500
NEO-CODEMA see CFY000
NEOCOMPENSAN see PKQ250
NEO-COROVAS see PBC250
NEOCROSEDIN see EHP000
NEOCRYL A-1038 see ADV900
NEO-CULTOL see MQV750
NEOCURDIONE see NBW100
NEOCYCLINE see TBX000
NEOCYCLOHEXIMIDE see CPE750
NEODALIT see DCW800
NEODECADRON see BFW325
NEODECANOIC ACID see NBW500
NEODECANOIC ACID, 2,3-EPOXYPROPYL ESTER see GGY160
NEODECANOIC ACID, HEXADECYL ESTER see HCP600
NEODECANOIC ACID, OXIRANYLMETHYL ESTER (9CI) see GGY160
NEODELPREGNIN see MCA500
NEO-DEMA see CLH750
NEO-DEVOMIT see EAN600
NEODICOUMARIN see BKA000

NEODICOUMAROL see BKA000
NEODICUMARINUM see BKA000
NEODIGOXIN see NBW600
NEO-DIGOXIN see NBW600
NEODOL 91 see AFJ150
NEODOL 23-1 see AFJ155
NEODOL 23-3 see AFJ155
NEODOL 23-5 see AFJ155
NEODOL 23-9 see AFJ155
NEODOL 25-3 see AFJ160
NEODOL 25-7 see AFJ160
NEODOL 25-9 see AFJ160
NEODOL 45-7 see AFJ168
NEODOL 91-6 see AFN900
NEODOL 23-3T see AFJ155
NEODOL 23-6.5 see AFJ155
NEODOL 25-12 see AFJ160
NEODOL 4511 see AFJ168
NEODOL 45-11 see AFJ168
NEODOL 45-13 see AFJ168
NEODOL 45-18 see AFJ168
NEODOL 45-2.25 see AFJ168
NEODORM (NEW) see NBT500
NEO-DOUXAN see NIC200
NEODOX 45-7 see AFJ168
NEODRENAL see DMV600
NEODRINE see DBA800
NEODROL see DME500
NEODURABOLIN see EJT600
NEODYMACETAT (GERMAN) see NBX500
NEODYMIA see NCC000
NEODYMIUM see NBX000
NEODYMIUM ACETATE see NBX500
NEODYMIUM CHLORIDE see NBY000
NEODYMIUM LACTATE see LAN000
NEODYMIUM(III) NITRATE (1:3) see NCB000
NEODYMIUM(III) NITRATE, HEXAHYDRATE (1:3:6) see NCB500
NEODYMIUM OXIDE see NCC000
NEODYMIUM(3+) OXIDE see NCC000
NEODYMIUM(III) OXIDE see NCC000
NEODYMIUM SESQUIOXIDE see NCC000
NEODYMIUM TRIACETATE see NBX500
NEODYMIUM TRIOXIDE see NCC000
NEODYMLACTAT (GERMAN) see LAN000
NEO-EPININE see DMV600
NEO-ERGOTIN see EDC500
NEOESERINE BROMIDE see POD000
NEOESERINE METHYL SULFATE see DQY909
NEO-ESTRONE see ECU750
NEO-FARMADOL see HNI500
NEO-FAT 8 see OCY000
NEO-FAT 10 see DAH400
NEO-FAT 12 see LBL000
NEO-FAT 18-61 see SLK000
NEO-FAT 90-04 see OHU000
NEO-FAT 92-04 see OHU000
NEO-FAT 18-S see SLK000
NEOFEMERGEN see LJL000
NEOFEN see HNI500
NEO-FERRUM see IHG000
NEOFIX FP see CNH125
NEOFLUMEN see CFY000
NEOFOLLIN see EDS100
NEO-FULCIN see GKE000
NEOGEL see CBO500
NEOGEN R see SKF600
NEO GERM-I-TOL see AFP250
NEO-GILURYTMAL see DNB000
NEOGLAUCIT see IRF000
NEOGYNON see NNL500
NEO HELIOPAN 303 see ODY150
NEOHEPTANOIC ACID see NCC100
NEOHETRAMINE see NCD500
NEOHETRAMINE HYDROCHLORIDE see RDU000
NEOHEXANE (DOT) see DQT200
NEO-HIBERNEX see DQA600
NEO-HOMBREOL see TBG000
NEO-HOMBREOL-M see MPN500

NEOHYDRAZID see COH250
NEOHYDRIN see CHX250
NEOISCOTIN see SIJ000
NEO-ISTAFENE see HGC500
NEOL see DTG400
NEOLAMIN see TES800
NEOLEXINA see ALV000
NEOLIN see BFC750
NEOLITE F see CAT775
NEOLOID see CCP250
NEOLUTIN see HNT500
NEO-MANTLE CREME see NCG000
NEOMCIN see NCE000
NEOMETANTYL see HKR500
NEOMETHIODAL see DNG400
NEOMIX see NCD550
NEOMIX see NCG000
NEOMYCIN see NCE000
NEOMYCIN A see NCE500
NEOMYCIN B see NCF500
NEOMYCIN B SULFATE (SALT) see NCD550
NEOMYCIN C see NCF300
NEOMYCIN E see NCF500
NEOMYCINE SULFATE see NCG000
NEOMYCIN SULFATE see NCG000
NEOMYCIN SULPHATE see NCG000
NEOMYSON G HYDROCHLORIDE see
UVA150
NEON see NCG500
NEO-NACLEX see BEQ625
NEONAL see BPF500
NEO-NAVIGAN see DYE600
NEONICOTINE see AON875
NEONIKOTIN see AON875
NEO-NILOREX see DKE800
NEONOL see NCG700
NEONOL 2B1317-12 see NCG842
NEONOL AF-14 see NCG715
NEONOL AF 9-4 see NCG800
NEONOL AF 9-6 see NCG820
NEONOL AF9-25 see NCG830
NEONOL AFM 9-10 (0.5) see NCG840
NEONOL P 1215-3 see AFJ160
NEONOL P 1215-6 see AFJ160
NEONOL P 1215-8 see AFJ160
NEONOL P 1215-9 see AFJ160
NEONOL P 12-16-3 see AFJ165
NEONOL P 1215-12 see AFJ160
NEONOL V1020-3 see NCG850
NEON, compressed (UN 1065) (DOT) see
NCG500
NEON, refrigerated liquid (cryogenic liquid) (UN
1913) (DOT) see NCG500
NEONYL BLUE 4R see CMM100
NEONYL RED 2B see CMO870
NEONYL SCARLET R see CMM320
NEO-OESTRANOL 1 see DKA600
NEO-OESTRANOL II see DKB000
NEO-ORMONAL see AOO275
NEOOXEDRINE see SPC500
NEO-OXYPATE see PQC500
NEOPANTANOYL CHLORIDE see DTS400
NEOPAX see SJJ175
NEOPELLIS see TFD250
NEOPENTANE see NCH000
NEOPENTANETETRAYL NITRATE see
PBC250
1,1',1'',1'''-(NEOPENTANE
TETRAYLTETRAOXY)TETRAKIS(2,2,2-
TRICHLOROETHANOL) see NCH500
NEOPENTANOIC ACID see PJA500
NEOPENTYLENE GLYCOL see DTG400
NEOPENTYL GLYCOL see DTG400
NEOPENTYL GLYCOL DIACRYLATE see
DUL200
NEOPENTYL GLYCOL DIGLYCIDYL
ETHER see NCI300
NEOPEPULSAN see HKR500
NEOPHARMEDRINE see DBA800
NEOPHEDAN see PEC250
NEOPHENAL see PEC250
NEOPHRYN see SPC500

NEOPHYILINE see TEP500
NEOPHYLLIN see DNC000
NEOPHYLLINE see DNC000
NEOPHYLLIN M see DNC000
NEOPINAMINE see NCI400
NEOPLATIN see PJD000
NEOPOL 23-6.5 see AFJ155
NEOPOLEN see PJS750
NEOPOLEN 30N see PJS750
NEOPOL (FUNGICIDE) see BAP260
NEOPRENE see NCI500
NEOPRENE see PJQ050
NEOPREX see CQE400
NEOPROGESTIN see NCI525
NEOPROMA see MFK500
NEOPROSERINE see NCI550
NEOPROTOVERATRIN see NCI600
NEOPROTOVERATRINE see NCI600
NEOPSICAINE HYDROCHLORIDE see
NCJ000
NEORAM BLU see CNK559
NEORESERPAN see RCA200
NEORESTAMIN see TAI500
NEORON see IOS000
NEO-RONTYL see BEQ625
NEOSABENYL see CJU250
NEO-SALICYL see SJO000
NEOSALVARSAN see NCJ500
NEOSAR see CQC650
NEO-SCABICIDOL see BBQ500
NEOSEDYN see TEH500
NEOSEPT see HCL000
NEOSEPTAL CL see SHU500
NEOSERFIN see RDK000
NEOSERINE BROMIDE see POD000
NEO-SERP see RDK000
NEOSETILE BLUE EB see TBG700
NEOSETILE PINK BN see AKE250
NEOSETILE SCARLET B see ENP100
NEOS-HIDANTOINA see DKQ000
NEO SILOX D see SFT500
NEO-SINEFRINA see SPC500
NEO-SKIODAN see DNG400
NEOSLOWTEN see RDK000
NEOSOLANIOL see NCK000
NEOSOLANIOL MONOACETATE see
ACS500
NEOSONE D see MFA250
NEOSORB ND see KDK000
NEOSOREXA see BGO100
NEOSOREXA PP580 see BGO100
NEOSPASMINA see NCK100
NEO-SPECTRA see CBT750
NEO-SPECTRA II see CBT750
NEOSPIRAMYCIN see NCK500
NEOSPIRAN see GCE600
NEOSTAM see NCL000
NEOSTAM STIBAMINE GLUCOSIDE see
NCL000
NEOSTEN see AGN000
NEOSTENE see AOO475
NEOSTENOVASAN see DNC000
NEOSTERON see AOO475
NEOSTIBOSAN see SLP600
NEOSTIGMETH see DQY909
NEOSTIGMINE see NCL100
NEOSTIGMINE BROMIDE see POD000
NEOSTIGMINE METHOSULFATE see
DQY909
NEOSTIGMINE METHYL BROMIDE see
POD000
NEOSTIGMINE METHYL SULFATE see
DQY909
NEOSTIGMINE MONOMETHYLSULFATE
see DQY909
NEOSTON see AGN000
NEOSTREPAL see SNN300
NEOSTREPSAN see TEX250
NEOSULF see NCD550
NEOSUPRANOL HYDROCHLORIDE see
NCL300
NEO-SUPRIMAL see HGC500

NEO-SUPRIMEL see HGC500
NEOSYDYN see TEH500
NEOSYMPATOL see SPC500
NEOSYNEPHRINE see NCL500
NEOSYNEPHRINE see SPC500
NEOSYNEPHRINE HYDROCHLORIDE see
SPC500
NEOSYNESINE see SPC500
NEO-TENEBRYL see DNG400
NEO-TESTIS see TBF500
NEOTEX see CBT750
NEOTHESIN see PIV750
NEOTHESIN HYDROCHLORIDE see IJZ000
NEOTHRAMYCIN see NCM275
NEOTHRAMYCIN A see TCP000
NEOTHYL see MOU830
NEOTHYLLINE see DNC000
NEOTIBIL see FNF000
NEOTILINA see DNC000
NEOTIZIDE see SIJ000
NEO-TIZIDE SODIUM SALT see SIJ000
NEOTONZIL see MCI500
NEOTOPSIN see PEX500
NEO-TRAN see MQU750
NEOTRAN see NCM700
NEO-TRIC see MMN250
NEO-VASOPHYLINE see DNC000
NEOVIRIDOGRISEIN IV see EDW200
NEOVITAMIN A ACID see VSK955
NEOVLETTA see NNL500
NEOXAZOL see SNN500
NEOXIN see ILD000
NEOZEPAM see DLY000
NEOZEX 4010B see PJS750
NEOZEX 45150 see PJS750
NEOZINE see MCI500
NEO-ZINE see MNV750
NEOZONE A see PFT250
NEOZONE D see PFT500
NEPALIN 2 see HAK075
NEPHELOR see GGS000
NEPHENTINE see MQU750
NEPHIS see EIY500
NEPHOCARP see TNP250
NEPHRAMIDE see AAI250
NEPHRAMINE see AHR600
NEPHRIDINE see VGP000
NEPRESOL see OJD300
NEPRESOLIN see OJD300
NEPRESSOL see OJD300
NEPTAL see NCM800
NEPTUNE BLACK X 60 see PCJ200
NEPTUNE BLUE BRA CONCENTRATION
see FMU059
NEPTUNIUM see NCN500
NEQUINATE see NCN600
NERA see ZLS200
NERACID see CBG000
NERA EMULZE (CZECH) see ZLS200
NERAL see DTC800
NERAN BRILLIANT GREEN G see FAE950
NEREB see MAS500
NEREISTOXIN DIBENZENESULFONATE
see NCN650
NERIINE DIHYDRBROMIDE see DOX000
NERIODIN see DEO600
NERIOL see OHQ000
NERIOLIN see OHQ000
NERIOSTENE see OHQ000
NERISONA see DKF130
NERISONE see DKF130
NERIUM OLEANDER see OHM875
NERKOL see DGP900
NEROBIL see DYF450
NEROBIOLIL see DYF450
NEROBOL see DAL300
NEROBOLETTES see DAL300
NEROL ACETATE (6CI) see NCO100
NEROL (FCC) see DTD200
NEROLI BIGARADE OIL, TUNISIAN see
NCO000
NEROLIDOL see NCN700

NEROLIDYL ACETATE see NCN800
NEROLIN see EEY500
NEROLINE see EEY500
NEROLIN FRAGAROL see NBJ000
NEROLIN II see EEY500
NEROLIN NEW see EEY500
NEROLI OIL, ARTIFICIAL see APJ250
NEROLI OIL, TUNISIAN see NCO000
NERONE see MCF525
NERPRUN (CANADA) see MBU825
NERUSIL see NCQ100
NERVACTON see BCA000
NERVANAID B ACID see EIX000
NERVANAID B LIQUID see EIV000
NERVANID B see EIV000
NERVATIL see BCA000
NERVILON see DXO300
NERVOSETON see BAG250
NERVOTON see DOZ000
NERYL ACETATE see NCO100
NERYL FORMATE see FNC000
NERYL ISOBUTYRATE see NCO200
NERYL ISOVALERIANATE see NCO500
NERYL-β-METHYL BUTYRATE see NCO500
NERYL PROPIONATE see NCP000
NESDONAL see PBT250
NESDONAL SODIUM see PBT500
NESOL see MCC250
NESONTIL see CFZ000
NESPOR see MAS500
NESSLER REAGENT see NCP500
NESTON see MDT740
NESTYN see EHP000
NETAGRONE 600 see DAA800
NETAL see BNL250
NETALID see INS000
NETAZOL see CIR250
NETH see INS000
NETHALIDE see INS000
NETHALIDE HYDROCHLORIDE see
INT000
NETHAMINE HYDROCHLORIDE see
EJR500
NETICONAZOLE HYDROCHLORIDE see
NCP525
NETILLIN see NCP550
NETILMICIN see SBD000
NETILMICIN SULFATE see NCP550
NETOBIMIN see NCP600
NETOCYD see DJT800
NETROMYCIN see NCP550
NETROPSIN see NCP875
NETROPSIN HYDROCHLORIDE see
NCQ000
NETSUSARIN see DOT000
NEU see ENV000
NEU see NKE500
NEUCHLONIC see DLY000
NEUCOCCIN see FMU080
NEUFIL see DNC000
NEULACTIL see PIW000
NEULEPTIL see PIW000
NEUMOLISINA see ELC500
NEUPERM GFN see DTG000
NEURACEN see BEG000
NEURACTIL see MCI500
NEURAKTIL see BCA000
NEURALEX see NCQ100
NEURAMINIDASE see NCQ200
α-NEURAMINIDASE see NCQ200
NEURAXIN see GKK000
NEURAZINE see CKP500
NEURIDINE see DCC400
NEURIDINE TETRAHYDROCHLORIDE see
GEK000
NEURIN see VQR300
NEURINE see VQR300
NEURIPLEGE see CLY750
NEUROBARB see EOK000
NEUROBENZIL see BCA000
NEUROCAINE see CNE750
NEUROCIL see MCI500

NEURODYN see TEH500
NEUROLEPTONE see BCA000
NEURONIKA see ADA725
NEUROPROCIN see EHP000
NEUROSEDIN see TEH500
NEUROSTAIN see NCQ500
NEUROTONE see RLU000
NEUROTOXIN A (NAJA NAJA REDUCED)
see NCQ520
NEUROTOXIN, CLOSTRIDIUM
BOTULINUM see BMM292
NEUROTOXIN (NAJA NAJA
PHILIPPINENSIS MAJOR) see NAF225
NEUROTOXIN S4C6 (ASPIDELAPS
SCUTATUS REDUCED) see NCQ530
NEUROTRAST see ELQ500
NEUROZINA see CJR909
NEURVITA see DXO300
NEURYL see SHL500
NEUSTAB see FNF000
NEUT see SFC500
NEUTHION see GFW000
NEUTRAFIL see DNC000
NEUTRAFILLINA see DNC000
NEUTRAL ACRIFLAVINE see DBX400
NEUTRAL ACRIFLAVINE see XAK000
NEUTRAL AMMONIUM CHROMATE see
ANF500
NEUTRAL AMMONIUM CHROMATE see
NCQ550
NEUTRAL AMMONIUM FLUORIDE see
ANH250
NEUTRAL BERBERINE SULFATE see
BFN625
NEUTRAL FAST ORANGE G see CMP882
NEUTRAL POTASSIUM CHROMATE see
PLB250
NEUTRAL PROFLAVINE SULPHATE see
DBN400
NEUTRAL RED see AJQ250
NEUTRAL RED CHLORIDE see AJQ250
NEUTRAL RED PG see CMM320
NEUTRAL RED W see AJQ250
NEUTRAL SAPONINS of PANAX GINSENG
ROOT see GEQ425
NEUTRAL SODIUM CHROMATE see
DXC200
NEUTRAL VERDIGRIS see CNI250
NEUTRAL ZINC PHOSPHATE see ZJS400
NEUTRAPHYLLIN see DNC000
NEUTRAPHYLLINE see DNC000
NEUTRAZYME see SIB600
NEUTROFLAVINE see XAK000
NEUTRONYX 600 see PKF000
NEUTRONYX 605 see PKF500
NEUTRONYX 622 see GHS000
NEUTRORMONE see AOO475
NEUTROSEL NAVY BN see DCJ200
NEUTROSEL RED TRVA see CLK235
NEUTROSTERON see AOO475
NEUTROXANTINA see DNC000
NEUVITAN see GEK200
NEUWIED GREEN see COF500
NEVANAID-B POWDER see TNL250
NEVAX see DJL000
NEVIGRAMON see EID000
NEVIKEN see CAX800
NEVIN see ILD000
NEVIN see SNL850
NEVIN (mixture) see SNL850
NEVIREX see CGO550
NEVRACTEN see CDQ250
NEVRITON see DXO300
NEVRODYN see TEH500
NEW BLITANE see ZJS300
NEW COCCIN see FMU080
NEWCOL 60 see SKV150
NEWCOL 560SF see ANO600
NEWCOL 560SN see SNY100
NEW GREEN see COF500
NEW IMPROVED CERESAN see BJT250
NEW IMPROVED GRANOSAN see BJT250

NEWLASE see NCQ600
NEW MILSTEM see BRI750
NEW PATENT BLUE A-CE EXTRA see
CMM062
NEW PATENT BLUE EXTRA PURE A see
CMM062
NEWPHRINE see SPC500
NEW PINK BLUISH GEIGY see FAG040
NEWPOL LB3000 see BRP250
NEW PONCEAU 4R see FMU070
NEWTOL see XPJ000
NEW VICTORIA GREEN EXTRA I see
AFG500
NEW YORK RED see NAP100
NEXAGAN see EGV500
NEXION see BNL250
NEXION 40 see BNL250
NEXIT see BBQ500
NEXOVAL see CKC000
NF see BLX000
NF see NGE500
NF 6 see TGI250
NF 21 see COM830
NF 44 see PEX500
NF 64 see NGB700
NF 161 see FPI200
NF 246 see NDY000
NF 260 see FPI000
NF-260 see FPI150
NF-269 see FPI100
NF 323 see NDY400
NF 1010 see EAE500
β-NF see NAU525
NF (dispersant) see BLX000
NF 35 (fungicide) see DJV000
NF-A see BLX000
5-NFA see NGI000
NFHAA see FNO000
NF-902 HYDROCHLORIDE see FPI100
NFIP see NGI800
NFN see DLU800
No. 1 FORTHFAST RED R see CJD500
NaFPAK see SHF500
Na FRINSE see SHF500
NFT see FPF000
NG see NGY000
NG-180 see NGG500
NGAI CAMPHOR see NCQ820
(−)-NGAIONE see NCQ900
NGAIONE see NCQ900
cis-NGAIONE see NCQ900
NG-NITROARGININE see OIU900
l-NG-NITROARGININE see OIU900
NH 18 see PKP750
NH 188 see RDU000
NHA see NBI500
NHC see HFN500
NH-LOST see BHN750
NHMI see OBY000
3N4HPA see HMY000
NHSA see HGL930
NHU see HKQ025
NHYD see NKJ000
NIA 249 see AFR250
NIA 2,995 see DEV600
NIA 4556 see DFO800
NIA 5273 see PIX250
NIA 5462 see EAQ750
NIA 5488 see CKM000
NIA-5767 see EAS000
NIA 5996 see DER800
NIA 9044 see BGB500
NIA 9102 see MQQ250
NIA-9241 see BDJ250
NIA 10242 see CBS275
NIA 11092 see DUM800
NIA 16,388 see PNV750
NIA 16824 see PFV760
NIA 17170 see BEP500
NIA-18739 see BEP750
NIA 2995J see DEV600
NIA 33297 see AHJ750

NIACEVIT see NCR000
NIACIDE see FAS000
NIACIN see NCQ900
NIACINAMIDE see NCR000
NIACINAMIDE ASCORBATE see NCR025
NIACINAMIDE ASCORBIC ACID
COMPLEX see NCR025
NIACIN BENZYL ESTER see NCR040
NIAGARA 1,137 see BIT250
NIAGARA 1240 see EEH600
NIAGARA 4512 see SKQ400
NIAGARA 4556 see DFO800
NIAGARA 5006 see DER800
NIAGARA 5,462 see EAQ750
NIAGARA 5767 see EAS000
NIAGARA 5943 see AIX000
NIAGARA 5,996 see DER800
NIAGARA 9044 see BGB500
NIAGARA 9241 see BDJ250
NIAGARA BLUE see CMO250
NIAGARA BLUE 2B see CMO000
NIAGARA BLUE 4B see CMO500
NIAGARA BLUE RW see CMO600
NIAGARAMITE see SOP500
NIAGARA P.A. DUST see NDN000
NIAGARA SKY BLUE see CMO500
NIAGARA-STIK see NAK500
NIAGARATHAL see DXD000
NIAGARATRAN see CJT750
NIAGARIL see BEQ625
NIAGATHAL see TBT150
NIAGRA 10242 see CBS275
NIALAMIDE see BET000
NIALATE see EEH600
NIALK see TIO750
NIAMID see BET000
NIAMIDAL see BET000
NIAMIDE see NCR000
NIAMINE see DJS200
NIAPROOF 4 see SIO000
NIA PROOF 08 see TAV750
NIAQUITIL see BET000
NIAX AFPI see PKB100
NIAX CATALYST AL see BJH750
NIAX CATALYST ESN see NCS000
NIAX E 488, POLYMER WITH 1,3-
DIISOCYANATOMETHYLBENZENE see
DNK250
NIAX FLAME RETARDANT 3 CF see
CGO500
NIAX ISOCYANATE TDI see DNK200
NIAX ISOCYANATE TDI see TGM740
NIAX POLYOL E-558 see DNK250
NIAX POLYOL E-488 TDI COPOLYMER see
DNK250
NIAX POLYOL LG-168 see NCT000
NIAX POLYOL LHT-42 see NCT500
NIAX TDI see TGM750
NIAX TDI see TGM800
NIAX TDI-P see TGM750
NIAX TRIOL 6000 see NCV500
NIAZOL see NCW000
NIBAL see PMC700
NIBREN WAX see TIT500
NIBROL see TEH500
NIBUFIN see NIM000
NICACID see NCQ900
NICAMETATE CITRATE see EQQ100
NICAMETATE DIHYDROGEN CITRATE
see EQQ100
NICAMIDE see DJS200
NICAMIDE see NCR000
NICAMIN see NCQ900
NICAMINA see NCR000
NICAMINDON see NCR000
NICANGIN see NCQ900
NICARB see NCW100
NICARBAZIN see NCW100
NICARBAZINE see NCW100
NICARDIPINE HYDROCHLORIDE see
PCG550
NICASIR see NCR000

NICAZIDE see ILD000
NICCOLITE (8CI) see NDE030
NICEL see MIF760
NICELATE see EID000
N. I. CERESAN see EME100
NICERGOLIN (GERMAN) see NDM000
NICERGOLINE see NDM000
NICERITROL see NCW300
NICETAMIDE see DJS200
NICETHAMIDE see DJS200
NICHEL (ITALIAN) see NCW500
NICHEL TETRACARBONILE (ITALIAN) see
NCZ000
NICHOLIN see CMF350
NICHROME see NCX515
NICIZINA see ILD000
NICKEL see NCW500
NICKEL 270 see NCW500
NICKEL(II) ACETATE (1:2) see NCX000
NICKEL ACETATE TETRAHYDRATE see
NCX500
NICKEL(II) ACETATE TETRAHYDRATE
see NCX500
NICKEL ALLOY, BASE, Ni 63–70, Cu 25–37,
Fe 0–2.5, Mn 0–2, Si 0–0.5, C 0–0.3 see NCX525
NICKEL ALLOY, BASE, Ni 57–62, Fe 22–28,
Cr 14–18, Si 0.8–1.6, Mn 0–1, C 0–0.2 see
NCX515
NICKEL ALLOY, Ni,Be see NCX510
NICKEL ALLOY, Ni 47–59, Co 17–20, Cr
13–17, Mo 4.5–5.7, Al 3.7–4.7, Ti 3–4, Fe 0–1, C
0–0.1 (AISI 687) see NCY135
NICKEL ALUMINIUM ALLOY see SLN300
NICKEL AMMONIUM SULFATE see
NCY050
NICKEL AMMONIUM SULFATE see
NCY051
NICKEL AMMONIUM SULFATE
HEXAHYDRATE see NCY060
NICKEL ANTIMONIDE see NCY100
NICKEL ARSENIDE see NCY110
NICKEL ARSENIDE (As2-Ni5) see NDJ399
NICKEL ARSENIDE (As8-Ni11) see NDJ400
NICKEL ARSENIDE SULFIDE see NCY125
NICKEL-BASED COBALT ALLOY see
NCY135
NICKEL(2+), BIS(N,N'-BIS(2-
AMINOETHYL)-1,2-ETHANEDIAMINE-
N,N',N^N)-, (T-4)-TETRAOXOTUNGSTATE(2-
) (1:1) see BLN100
NICKEL, BIS(CARBONATO(2-
))HEXAHYDROXY PENTA- see NCY525
NICKEL BISCYCLOPENTADIENE see
NDA500
NICKEL,
BIS(TRIETHYLENETETRAMINE)TUNGSTA
TO- see BLN100
NICKEL BISTRIPHENYLPHOSPHINE
DITHIOCYANATE see BLS500
NICKEL BLACK see NDE010
NICKEL BOROFLUORIDE see NDC000
NICKEL(II) CARBONATE (1:1) see NCY500
NICKEL CARBONATE HYDROXIDE see
NCY525
NICKEL CARBONATE HYDROXIDE see
NCY530
NICKEL CARBONATE HYDROXIDE see
NCY600
NICKEL, (CARBONATO(2-
))TETRAHYDROXYTRI- see NCY600
NICKEL, (CARBONATO(2-
))TETRAHYDROXYTRI-, TETRAHYDRATE
see NCY530
NICKEL CARBONYL see NCZ000
NICKEL CARBONYLE (FRENCH) see
NCZ000
NICKEL CHLORIDE see NDH000
NICKEL(II) CHLORIDE (1:2) see NDH000
NICKEL(II) CHLORIDE HEXAHYDRATE
(1:2:6) see NDA000
NICKEL CHROMATE see CMH270

NICKEL CHROMATE see NDA100
NICKEL CHROMATE(VI) see CMH270
NICKEL CHROMITE see NDA100
NICKEL CHROMIUM OXIDE see CMH270
NICKEL CHROMIUM OXIDE see NDA100
NICKEL COMPLEX with N-FLUOREN-2-YL
ACETOHYDROXAMIC ACID see HIR500
NICKEL, COMPOUND with pi-
CYCLOPENTADIENYL (1:2) see NDA500
NICKEL COMPOUNDS see NDB000
NICKEL CYANIDE (DOT) see NDB500
NICKEL CYANIDE (solid) see NDB500
NICKEL DIACETATE TETRAHYDRATE see
NCX500
NICKEL DIAMMONIUM DISULFATE
HEXAHYDRATE see NCY060
NICKEL DIBUTYLDITHIOCARBAMATE see
BIW750
NICKEL DIFLUORIDE see NDC500
NICKEL DIHYDROXIDE see NDE000
NICKEL DINITRITE see NDG550
NICKEL DISULFIDE see NDB875
NICKEL (DUST) see NCW500
NICKEL(II) EDTA COMPLEX see EJA500
NICKEL FERRITE BROWN SPINEL see
CMS137
NICKEL(II) FLUOBORATE see NDC000
NICKEL(II) FLUORIDE (1:2) see NDC500
NICKEL FLUOROBORATE see NDC000
NICKEL(II) FLUOSILICATE (1:1) see
NDD000
NICKEL-GALLIUM ALLOY see NDD500
NICKEL(II) HYDROXIDE see NDE000
NICKEL(III) HYDROXIDE see NDE010
NICKEL HYDROXYCARBONATE see
NCY525
NICKELIC HYDROXIDE see NDE010
NICKELIC OXIDE see NDH500
NICKELINE see NDE030
NICKEL IRON CHROMITE BLACK SPINEL
see CMS135
NICKEL IRON SULFIDE see NDE500
NICKEL-IRON SULFIDE MATTE see
NDE500
NICKEL(II) ISODECYL
ORTHOPHOSPHATE (3:2) see NDF000
NICKEL MONOANTIMONIDE see NCY100
NICKEL MONOARSENIDE see NCY110
NICKEL MONOSELENIDE see NDF400
NICKEL MONOSULFATE HEXAHYDRATE
see NDL000
NICKEL MONOSULFIDE see NDL100
NICKEL MONOXIDE see NDF500
NICKEL NITRATE see NDG000
NICKEL(II) NITRATE (1:2) see NDG000
NICKEL(2+) NITRATE, HEXAHYDRATE see
NDG500
NICKEL(II) NITRATE, HEXAHYDRATE
(1:2:6) see NDG500
NICKEL NITRITE see NDG550
NICKEL NITRITE see NDG550
NICKELOCENE see NDA500
NICKELOUS ACETATE see NCX000
NICKELOUS ACETATE TETRAHYDRATE
see NCX500
NICKELOUS CARBONATE see NCY500
NICKELOUS CHLORIDE see NDH000
NICKELOUS FLUORIDE see NDC500
NICKELOUS HYDROXIDE see NDE000
NICKELOUS OXIDE see NDF500
NICKELOUS SULFATE see NDK500
NICKELOUS SULFATE HEPTAHYDRATE
see NDK990
NICKELOUS SULFIDE see NDL100
NICKELOUS TETRAFLUOROBORATE see
NDC000
NICKEL OXIDE see NDH500
NICKEL(II) OXIDE (1:1) see NDF500
NICKEL OXIDE, IRON OXIDE, and
CHROMIUM OXIDE FUME see IHE000
NICKEL OXIDE (MAK) see NDF500
NICKEL OXIDE PEROXIDE see NDH500

NICKEL PARTICLES see NCW500
NICKEL(2+) PERCHLORATE, HEXAHYDRATE see NDJ000
NICKEL PEROXIDE see NDH500
NICKEL POTASSIUM CYANIDE see NDI000
NICKEL PROTOXIDE see NDF500
NICKEL REFINERY DUST see NDI500
NICKEL(2+) SALT PERCHLORIC ACID HEXAHYDRATE see NDJ000
NICKEL SELENIDE see NDF400
NICKEL SELENIDE see NDJ475
NICKEL SELENIDE (3:2) CRYSTALLINE see NDJ475
NICKEL SESQUIOXIDE see NDH500
NICKEL SPONGE see NCW500
NICKEL SUBARSENIDE see NDJ399
NICKEL SUBARSENIDE see NDJ400
NICKEL SUBSELENIDE see NDJ475
NICKEL SUBSULFIDE see NDJ500
NICKEL SUBSULPHIDE see NDJ500
NICKEL (II) SULFAMATE see NDK000
NICKEL SULFARSENIDE see NCY125
NICKEL SULFATE see NDK500
NICKEL SULFATE (1:1) see NDK500
NICKEL(II) SULFATE see NDK500
NICKEL(2+)SULFATE (1:1) see NDK500
NICKEL(II) SULFATE (1:1) see NDK500
NICKEL2+ SULFATE HEPTAHYDRATE see NDK990
NICKEL(II) SULFATE HEPTAHYDRATE see NDK990
NICKEL SULFATE HEXAHYDRATE see NDL000
NICKEL (II) SULFATE HEXAHYDRATE see NDL000
NICKEL(II) SULFATE HEXAHYDRATE (1:1:6) see NDL000
NICKEL SULFIDE see NDB875
NICKEL SULFIDE see NDJ500
NICKEL SULFIDE see NDL050
NICKEL SULFIDE see NDL100
NICKEL(2+) SULFIDE see NDL100
NICKEL(II) SULFIDE see NDL100
α-NICKEL SULFIDE (1:1) CRYSTALLINE see NDL050
α-NICKEL SULFIDE (3:2) CRYSTALLINE see NDJ500
NICKEL SULPHATE HEXAHYDRATE see NDL000
NICKEL SULPHIDE see NDJ500
NICKEL TELLURIDE see NDL425
NICKEL TETRACARBONYL see NCZ000
NICKEL TETRACARBONYLE (FRENCH) see NCZ000
NICKEL(II) TETRAFLUOROBORATE see NDC000
NICKEL-TITANATE see NDL500
NICKEL TITANIUM OXIDE see NDL500
NICKEL TRIOXIDE see NDH500
NICKEL TRITADISULPHIDE see NDJ500
NICKEL ZINC FERRATE see IHB800
NICKEL ZINC FERRITE see IHB800
NICLOFEN see DFT800
NICLOFOLAN see DFD000
NICLOSAMIDE see DFV400
NICLOSAMIDE see DFV600
NICO see NCQ900
NICO-400 see NCQ900
NICOBID see NCQ900
NICOBION see NCR000
NICOCAP see NCQ900
NICOCIDE see NDN000
NICOCIDIN see NCQ900
NICOCODINE see CNG250
NICOCRISINA see NCQ900
NICODAN see NCQ900
NICODEL see PCG550
NICODELMINE see NCQ900
NICO-DUST see NDN000
NICOFORT see NCR000
NICO-FUME see NDN000
NICOGEN see NCR000

NICOLANE see NOA000
NICOLAR see NCQ900
NICOLEN see NGG500
NICOLIN see CMF350
NICOMETH see NDV000
NICOMIDOL see NCR000
NICOMOL see CME675
NICONACID see NCQ900
NICONAT see NCQ900
NICONAZID see NCQ900
NICOR see DJS200
NICORANDIL see NDL800
NICORDAMIN see DJS200
NICORINE see DJS200
NICOROL see CHJ750
NICOROL see NCQ900
NICORYL see DJS200
NICOSAN 2 see NCR000
NICOSIDE see NCQ900
NICO-SPAN see NCQ900
NICOSYL see NCQ900
NICOTA see NCR000
NICOTAMIDE see NCR000
NICOTAMIN see NCQ900
NICOTENE see NCQ900
NICOTERGOLINE see NDM000
NICOTIANA ATTENUATA see TGI100
NICOTIANA GLAUCA see TGI000
NICOTIANA GLAUCA see TGI100
NICOTIANA LONGIFLORA see TGI100
NICOTIANA RUSTICA see TGI100
NICOTIANA TABACUM see TGH725
NICOTIANA TABACUM see TGI100
NICOTIBINA see ILD000
NICOTIL see NCQ900
NICOTILAMIDE see NCR000
NICOTILILAMIDO see NCR000
NICOTINA (ITALIAN) see NDN000
NICOTINALDEHYDE see NDM100
NICOTINAMIDE-ADENINE DINUCLEOTIDE see CNF390
NICOTINAMIDE, N-(2-INDOL-3-YLETHYL)- see NDW525
2-NICOTINAMIDOETHYL NITRATE see NDL800
NICOTINAMIDOMETHYLAMINOPYRAZOLONE see AQN500
NICOTINE see NDN000
(−)-NICOTINE see NDN000
l-NICOTINE see NDN000
NICOTINE, solid (DOT) see NDN000
NICOTINE, liquid (DOT) see NDN000
NICOTINE ACID see NCQ900
NICOTINE ACID AMIDE see NCR000
NICOTINE ACID TARTRATE see NDS500
NICOTINEALDEHYDE see NDM100
NICOTINE ALKALOID see NDN000
NICOTINEAMIDE ADENINE DINUCLEOTIDE see CNF390
NICOTINE BITARTRATE see NDS500
NICOTINE, COMPD. WITH MANGANESE(II) o-BENZOYL BENZOATE, TRIHYDRATE (2:1) see NDN050
NICOTINE, COMPOUND, with NICKEL(II)-o-BENZOYL BENZOATE TRIHYDRATE (2:1) see BHB000
NICOTINE, 1'-DEMETHYL-1'-NITROSO-, 1-OXIDE see NLD525
NICOTINE DIHYDROCHLORIDE see NDN100
NICOTINE HYDROCHLORIDE see NDP400
(+)-NICOTINE HYDROCHLORIDE see NDQ000
1-NICOTINE HYDROCHLORIDE see NDP500
d-NICOTINE HYDROCHLORIDE see NDQ000
NICOTINE HYDROCHLORIDE (d,l) see NDP400
NICOTINE HYDROCHLORIDE, solution (DOT) see NDP400

NICOTINE HYDROGEN TARTRATE see NDS500
(−)-NICOTINE HYDROGEN TARTRATE see NDS500
NICOTINE MONOSALICYLATE see NDR000
NICOTINE, 1'-NITROSO-1'-DEMETHYL- see NLD500
NICOTINE, 1'-OXIDE see NDR100
NICOTINE-N'-OXIDE see NDR100
NICOTINE 1'-N-OXIDE see NDR100
NICOTINE SALICYLATE (DOT) see NDR000
NICOTINE SULFATE see NDR500
NICOTINE SULFATE see NDR500
NICOTINE SULFATE, solid or solution (DOT) see NDR500
NICOTINE TARTRATE see NDS500
NICOTINE TARTRATE (1:2) see NDS500
NICOTINE TARTRATE (DOT) see NDS500
NICOTINIC ACID see NCQ900
NICOTINIC ACID, ALUMINUM SALT see AHD650
NICOTINIC ACID AMIDE see NCR000
NICOTINIC ACID, BENZYL ESTER see NCR040
NICOTINIC ACID, BUTYL ESTER see NDT500
NICOTINIC ACID-7,8-DIDEHYDRO-4,5-α-EPOXY-3-METHOXY-17-METHYLMORPHINAN-6-α-YL ESTER see CNG250
NICOTINIC ACID DIETHYLAMIDE see DJS200
NICOTINIC ACID, 3-(2,6-DIMETHYLPIPERIDINO)PROPYL ESTER, HYDROCHLORIDE see NDU000
NICOTINIC ACID, ESTER with CODEINE see CNG250
NICOTINIC ACID, HYDRAZIDE see NDU500
NICOTINIC ACID-2-HYDROXY-3-(o-METHOXYPHENOXY)PROPYL ESTER see HLT300
NICOTINIC ACID N-METHYLBETAINE see TKL890
NICOTINIC ACID, METHYL ESTER see NDV000
NICOTINIC ACID, 3-(2-METHYLPIPERIDINO)PROPYL ESTER, HYDROCHLORIDE see NDU000
NICOTINIC ACID NITRILE see NDW515
NICOTINIC ACID, POTASSIUM SALT see PLL200
NICOTINIC ACID, SODIUM SALT see NDW500
NICOTINIC ACID-1,2,5,6-TETRAHYDRO-1-METHYL-, METHYL ESTER, HYDROCHLORIDE see AQU250
NICOTINIC ALCOHOL see NDW510
NICOTINIC ALDEHYDE see NDM100
NICOTINIC AMIDE see NCR000
NICOTINIPCA see NCQ900
NICOTINONITRILE see NDW515
NICOTINONITRILE, 1,2-DIHYDRO-4-METHOXY-1-METHYL-2-OXO- see RJK100
NICOTINOYL HYDRAZINE see NDU500
NICOTINOYL HYDRAZINE see NDU500
1-NICOTINOYLMETHYL-4-PHENYL-PIPERAZINE see NDW520
N-NICOTINOYLTRYPTAMIDE see NDW525
NICOTINSAEURE (GERMAN) see NCQ900
NICOTINSAEUREAMID (GERMAN) see NCR000
NICOTINYL ALCOHOL see NDW510
NICOTINYLHYDRAZIDE see NDU500
NICOTION see EPQ000
NICOTOL see NCR000
NICOTYLAMIDE see NCR000
NICOTYRINE see NDX300
3,2'-NICOTYRINE see NDX300
β-NICOTYRINE see NDX300
NICOULINE see RNZ000

NICOUMALONE see ABF750
NICOVASAN see NCQ900
NICOVASEN see NCQ900
NICOVEL see NCQ900
NICOVEL see NCR000
NICOVIT see NCR000
NICOVITOL see NCR000
NICOXIN see NCW100
NICOZIDE see ILD000
NICOZYMIN see NCR000
NICRAZIN see NCW100
NICRAZINE see NCW100
NICYL see NCQ900
NIDA see MMN250
NIDANTIN see OOG000
NIDORAL see CCY000
NIDRAFUR see NGB700
NIDRAFUR P see NGB700
NIDRAN see ALF500
NIDRANE see BEG000
NIDRAZID see ILD000
NIDROXYZONE see FPE100
NIEA see HKW475
NIERALINE see VGP000
NIESYMETRYCZNA DWU
METYLOHYDRAZYNA (POLISH) see
DSF400
NIFEDIN see AEC750
NIFEDIPINE see AEC750
NIFELAT see AEC750
NIFENALOL see INU000
NIFLAN see PLX400
NIFLEX see SAV000
NIFLUMIC ACID see NDX500
NIFLURIL see NDX500
NIFOS T see TCF250
NIFTHOLIDE see FMR050
NIFTOLIDE see FMR050
NIFULIDONE see NGG500
NIFURADENE see NDY000
NIFURAN see NGG500
NIFURATRONE see HKW450
NIFURDAZIL see EAE500
NIFURHYDRAZONUM see NGB700
NIFUROXAZID see DGQ500
NIFUROXAZIDE see DGQ500
NIFUROXIME see NGC000
NIFURPIPONE see NDY350
NIFURPIPONE ACETATE see NDY360
NIFURPIPONE DIHYDROCHLORIDE see
NDY370
NIFURPIPONE HYDROCHLORIDE see
NDY380
NIFURPIPONE MALEATE (1:1) see NDY390
NIFURPIRINOL see NDY400
NIFURPRAZINE HYDROCHLORIDE see
ALN250
NIFURTHIAZOLE see NDY500
NIFURTIMOX see NGG000
NIFUZON see NGE500
NIGALAX see PPN100
(−)-NIGALDIPINE HYDROCHLORIDE see
NDY550
(R)-NIGALDIPINE HYDROCHLORIDE see
NDY550
NIGERICIN SODIUM SALT see SIO788
NIGHT BLOOMING JESSAMINE see
DAC500
NIGHTCAPS see PAM780
NIGHTSAGE (JAMAICA) see YAK300
NIGHTSHADE see DAD880
NIGLYCON see NGY000
NIGRESCIGENIN see NDY565
NIGRIN see SMA000
NIGROSINE SB see CMS236
NIGROSINE SPIRIT SOLUBLE see CMS236
NIGROSINE SSB see CMS236
NIGROSINE SSBZ 14 see CMS236
NIGROSINE SSBZ 30 see CMS236
NIGROSINE SSJ see CMS236
NIGROSINE SSJJ see CMS236
NIGROSINE WN see CMS236

NIH 204 see DCH300
NIH 204 see TKZ000
NIH 3127 see SBE500
NIH 7519 see HKC625
NIH 7574 see BBU625
NIH 7672 see MDV000
NIH 7958 see DOQ400
NIH 7981 see COV500
NIH 7519 HYDROBROMIDE see PMD325
NIH-4185 HYDROCHLORIDE see DJP500
NIH-5145 HYDROCHLORIDE see EIJ000
NIH 7440 HYDROCHLORIDE see AGV890
NIH-LH-B 9 see LIU300
NIHONTHRENE BLUE BC see DFN300
NIHONTHRENE BLUE RSN see IBV050
NIHONTHRENE BRILLIANT BLUE RCL see
DFN300
NIHONTHRENE BRILLIANT BLUE RP see
IBV050
NIHONTHRENE BRILLIANT VIOLET 4R
see DFN450
NIHONTHRENE BRILLIANT VIOLET RR
see DFN450
NIHONTHRENE BROWN BR see CMU770
NIHONTHRENE BROWN GR see CMU770
NIHONTHRENE BROWN R see CMU780
NIHONTHRENE FAST ORANGE R see
CMU815
NIHONTHRENE GOLDEN YELLOW see
DCZ000
NIHONTHRENE GREY M see CMU475
NIHONTHRENE NAVY BLUE G see
CMU500
NIHONTHRENE OLIVE R see DUP100
NIHONTHRENE RED BB see CMU825
NIHYDRAZON see NGB700
NIHYDRAZONE see NGB700
NIKAFLOC D 1000 see CNH125
NIKALAC 031 see MCB050
NIKALAC MS 11 see MCB050
NIKALAC MS 21 see MCB050
NIKALAC MS 40 see MCB050
NIKALAC MS 001 see MCB050
NIKALAC MW 12 see MCB050
NIKALAC MW 22 see MCB050
NIKALAC MW 30 see MCB050
NIKALAC MW 40 see MCB050
NIKALAC MW 30M see MCB050
NIKALAC MW 10LF see MCB050
NIKALAC MW 12LF see MCB050
NIKALAC MX 40 see MCB050
NIKALAC MX 45 see MCB050
NIKALAC MX 65 see MCB050
NIKALAC MX 032 see MCB050
NIKALAC MX 054 see MCB050
NIKALAC MX 430 see MCB050
NIKALAC MX 485 see MCB050
NIKALAC MX 705 see MCB050
NIKALAC MX 706 see MCB050
NIKALAC MX 750 see MCB050
NIKARDIN see DJS200
NIKARESIN S 176 see MCB050
NIKARESIN S 260 see MCB050
NIKARESIN S 305 see MCB050
NIKARESIN S 306 see MCB050
NIKA-TEMP see PKQ059
NIKAVINYL SG 700 see PKQ059
NIKETAMID see DJS200
NIKETHAROL see DJS200
NIKETHYL see DJS200
NIKETILAMID see DJS200
NIKION see BEQ625
NIKKELTETRACARBONYL (DUTCH) see
NCZ000
NIKKOL BC see PJT300
NIKKOL BC 30 see PJT300
NIKKOL BC 40 see PJT300
NIKKOL BC 15TX see PJT300
NIKKOL BC 20TX see PJT300
NIKKOL BL see DXY000
NIKKOL HCO 60 see CCP310
NIKKOL IPIS see IPS450

NIKKOL MYO 2 see PJY100
NIKKOL MYO 10 see PJY100
NIKKOL OTP 70 see DJL000
NIKKOL S.C.S see HCP900
NIKKOL SNP see ANO600
NIKKOL SO 10 see SKV100
NIKKOL SO-15 see SKV100
NIKKOL SO 15 see SKV170
NIKKOL SO-30 see SKV100
NIKKOL SP10 see MRJ800
NIKKOL SS 30 see SKV150
NIKKOL TO see PKL100
NIKORIN see DJS200
NIKO-TAMIN see NCR000
NIKOTIN (GERMAN) see NDN000
NIKOTINSAEUREAMID (GERMAN) see
NCR000
NIKOTINSULFAT see NDR500
NIKOTYNA (POLISH) see NDN000
NILACID see ABX500
NILE BLUE see AJN250
NILE BLUE A see AJN250
NILE BLUE AX see AJN250
NILE BLUE BASE see AJN250
NILE BLUE CHLORIDE see AJN250
NILE BLUE HYDROCHLORIDE see AJN250
NILERGEX see OGI075
NILERGEX HYDROCHLORIDE see AEG625
NILEVA see ENX600
NILEVAR see ENX600
NILHISTIN see DCJ850
NILODIN see DHU000
NILODIN see SBE500
NILOX see BPF000
NILOX PBNA see PFT500
NILSTAT see NOH500
NILUDIPINE see NDY600
NILVADIPINE see NDY650
NILVERM see TDX750
NILVEROM see TDX750
NIMBECETIN see ICE000
NIMBLE WEED see PAM780
NIMBOSTEROL see SDZ350
NIMCO CHOLESTEROL BASE H see
CMD750
NIMERGOLINE see NDM000
NIMETAZEPAM see DLV000
N(5)-(IMINO(NITROAMINO)METHYL)-l-
ORNITHINE METHYL ESTER see NDY675
NIMITEX see TAL250
NIMITOX see TAL250
NIMODIPINE see NDY700
NIM OIL see NBS300
NIMONIC AP1 see NCY135
NIMORAZOLE see NHH000
NIMORAZOLO see NHH000
NIMOTOP see NDY700
NIMROD see BRJ000
NIMROD T see BRJ000
NIMUSTINE see NDY800
NIMUSTINE HYDROCHLORIDE see
ALF500
NINCALUICOLFLASTINE see VKZ000
NINHYDRIN see DMV200
NINHYDRIN HYDRATE see DMV200
NINOL 4821 see BKE500
NINOL 2012E see CNF330
NINOL AA62 see BKE500
NINOL AA-62 EXTRA see BKE500
NINOL AA-62 EXTRA see LBL000
NINOPTERIN see BMM125
NIOBATE, OCTAPOTASSIUM see PLL250
NIOBE OIL see MHA750
NIOBIA see NEA050
NIOBIUM see NDZ000
NIOBIUM-93 see NDZ000
NIOBIUM CHLORIDE see NEA000
NIOBIUM ELEMENT see NDZ000
NIOBIUM OXIDE see NEA050
NIOBIUM(V) OXIDE see NEA050
NIOBIUM(5+) OXIDE see NEA050
NIOBIUM PENTACHLORIDE see NEA000

NIOBIUM PENTAOXIDE see NEA050
NIOBIUM PENTOXIDE see NEA050
NIOBIUM POTASSIUM FLUORIDE see PLN500
NIOBIUM POTASSIUM OXIDE see PLL250
NIOCINAMIDE see NCR000
NIOFORM see CHR500
NIOGEN ES 160 see PJY100
NIOI-PEPA (HAWAII) see PCB275
NIOMIL see DQM600
NIONATE see FBK000
NIONG see NGY000
NIOPAM see IFY000
NIOZYMIN see NCR000
NIP see DFT800
NIPA 49 see PNM750
NIPABUYL see BSC000
NIPAGALLIN A see EKM100
NIPAGALLIN LA see DXX200
NIPAGALLIN P see PNM750
NIPAGIN see HJL500
NIPAGIN A see HJL000
NIPAHEPTYL see HBO650
NIPAM see INH000
NIPA NO. 48 see EKM100
NIPANTIOX 1-F see BQI000
NIPAR S-20 see NIY000
NIPAR S-20 SOLVENT see NIY000
NIPAR S-30 SOLVENT see NIY000
NIPASOL see HNU500
NIP-A-THIN see NBL200
NIPAXON see NOA000
NIPAZIN A see HJL000
NIPELLEN see NCQ900
NIPEON A 21 see PKQ059
NIPERYT see PBC250
NIPERYTH see PBC250
NIPHANOID see BQA010
NIPLEN see ILD000
NIPODAL see PMF500
NIPOL 407 see SMR000
NIPOL 576 see PKQ059
NIPPAS see SEP000
NIPPON BLUE BB see CMO000
NIPPON BORDEAUX GS see CMO872
NIPPON DARK GREEN B see CMO830
NIPPON DEEP BLACK see AQP000
NIPPON DEEP BLACK GX see AQP000
NIPPON DIRECT SKY BLUE see CMO500
NIPPON FAST SCARLET 4BS see CMO870
NIPPON GREEN B see CMO840
NIPPON KAGAKU SAFRANINE GK see GJI400
NIPPON KAGAKU SAFRANINE T see GJI400
NIPPON ORANGE GG see CMO860
NIPPON ORANGE X-881 see DVB800
NIPPON PURPURINE 8B see CMO880
NIPPON SODA see FBP350
NIPRADILOL see NEA100
NIPRIDE see SIU500
NIPRIDE DIHYDRATE see SIW500
NIPSAN see DCM750
NIQUETAMIDA see DJS200
NIRAN see CDR750
NIRAN see PAK000
NIRATIC HYDROCHLORIDE see LFA020
NIRATIC-PURON HYDROCHLORIDE see LFA020
NIRAZIN see NCW100
NIRIDAZOLE see NML000
NIRIT see DVF800
NIROSAN see TDY120
NIRVANOL see EOL000
(−)-NIRVANOL see EOL050
l-NIRVANOL see EOL050
(R)-NIRVANOL see EOL050
NIRVOTIN see PGE000
NISENTIL see NEA500
NISENTIL see NEB000
NISENTIL HYDROCHLORIDE see NEB000
NISETAMIDE see DJS200

NISIDANA see DCV800
NISIN PREPARATION see NEB050
NISOLONE see SOV100
NISONE see PLZ100
NISOTIN see EPQ000
NISPERO DEL JAPON (CUBA, PUERTO RICO) see LIH200
NISSAN AMINE AB see OBC000
NISSAN AMINE BB see DXW000
NISSAN ANON BL see LBU200
NISSAN CATION AB see TLW500
NISSAN CATION BB see LBX075
NISSAN CATION BB 300 see LBX075
NISSAN CATION FB see LBX075
NISSAN CATION M2-100 see TCA500
NISSAN CATION PB 40 see HCQ525
NISSAN CATION S2-100 see DTC600
NISSAN DIAPION S see SIY000
NISSAN DIAPON T see SIY000
NISSAN NONION OP 83 see SKV170
NISSAN NONION OP 83RAT see SKV170
NISSAN NONION P 208 see PJT300
NISSAN NONION P 210 see PJT300
NISSAN NONION P 213 see PJT300
NISSAN NONION P 220 see PJT300
NISSAN NONION PP40 see MRJ800
NISSAN NONION PP 40R see MRJ800
NISSAN NONION SP 60 see SKV150
NISSAN TRAX N300 see SNY100
NISSEN BLACK BGL see CMS250
NISSEN BLACK BK see CMS250
NISSEN BLACK BX see CMS250
NISSEN BLACK KBX see CMS250
NISSEN BLACK LBX see CMS250
NISSEN BORDEAUX CFB see CMS257
NISSEN BRILLIANT GREEN 4GF see CMS257
NISSEN BRILLIANT GREEN 5GFF see CMS257
NISSEN BRILLIANT GREEN GG see CMS257
NISSOCAINE see AIL750
NISSOL EC see MME809
NISSORUN see CJU275
NISY see DRB400
NITARSONE see NIJ500
NITAZOL see ABY900
NITAZOLE see ABY900
NITEBAN see ILD000
NITER see PLL500
NITHIAMIDE see ABY900
NITHIAZID see ENV500
NITHIAZIDE see ENV500
NITHIOCYAMINE see AOA050
NITICID see CHS500
NITICOLIN see CMF350
NITOBANIL see FLZ050
NITOFEN see DFT800
NITOL see BIE500
NITOL "TAKEDA" see BIE500
NITOMAN see TBJ275
NITORA see NGY000
NITOSORG see GIQ100
NITRADOR see DUS700
NITRADOS see DLY000
NITRAFEN see DFT800
NITRALAMINE HYDROCHLORIDE see NEC000
NITRALDONE see FPI000
NITRAMIDE see NEG000
NITRAMIDE, NITRO-, AMMONIUM SALT see NHR550
NITRAMIN see ALQ000
NITRAMINE see ALQ000
NITRAMINE see TEG250
NITRAMINE, METHYL- see NHN500
NITRAMINOACETIC ACID see NEC500
NITRAMINOACETIC ACID see NGY700
NITRAN see DUV500
4-NITRANILINE see NEO500
m-NITRANILINE see NEN500
p-NITRANILINE see NEO500

NITRANITOL see MAW250
NITRANOL see TJL250
NITRAPHEN see DFT800
NITRAPYRIN (ACGIH) see CLP750
NITRATE d'AMYLE (FRENCH) see AOL250
NITRATE d'ARGENT (FRENCH) see SDS000
NITRATE de BARYUM (FRENCH) see BAN250
NITRATE MERCUREUX (FRENCH) see MDE750
NITRATE MERCURIQUE (FRENCH) see MDF000
NITRATE PHENYL MERCURIQUE see MDH500
NITRATE de PLOMB (FRENCH) see LDO000
NITRATE de PROPYLE NORMAL (FRENCH) see PNQ500
NITRATES see NED000
NITRATE SALT OF ETHYL AURAMINE see EGM200
NITRATE de SODIUM (FRENCH) see SIO900
NITRATE de STRONTIUM (FRENCH) see SMK000
NITRATE de ZINC (FRENCH) see ZJJ000
NITRATINE see SIO900
NITRATION BENZENE see BBL250
NITRAZEPAM see DLY000
NITRAZOL CF EXTRA see NEO500
NITRE see PLL500
NITRE CAKE see SEG800
NITRENDIPINE see EMR600
NITRENE NO see OHU200
NITRENPAX see DLY000
NITRETAMIN see TJL250
NITRETAMIN PHOSPHATE see TJL250
NITRIC ACID see NED500
NITRIC ACID, over 40% (DOT) see NED500
NITRIC ACID other than red fuming with >70% nitric acid (DOT) see NED500
NITRIC ACID other than red fuming with not >70% nitric acid (DOT) see NED500
NITRIC ACID, ALUMINUM SALT see AHD750
NITRIC ACID, ALUMINUM(3+) SALT see AHD750
NITRIC ACID, ALUMINUM SALT, NONAHYDRATE (8CI,9CI) see AHD900
NITRIC ACID, AMMONIUM LANTHANUM SALT see ANL500
NITRIC ACID, AMMONIUM SALT see ANN000
NITRIC ACID, ANHYDRIDE with PEROXYACETIC ACID see PCL750
NITRIC ACID, BARIUM SALT see BAN250
NITRIC ACID, BERYLLIUM SALT see BFT000
NITRIC ACID, BERYLLIUM SALT, TETRAHYDRATE see BFR300
NITRIC ACID, BERYLLIUM SALT, TRIHYDRATE see BFT100
NITRIC ACID, BISMUTH(3+) SALT see BKW250
NITRIC ACID, CADMIUM SALT see CAH000
NITRIC ACID, CADMIUM SALT, TETRAHYDRATE see CAH250
NITRIC ACID, CALCIUM SALT (8CI,9CI) see CAU000
NITRIC ACID, CALCIUM SALT, TETRAHYDRATE see CAU250
NITRIC ACID, CERIUM(3+) SALT, HEXAHYDRATE see CDB250
NITRIC ACID, CERIUM(3+) SALT (8CI, 9CI) see CDB000
NITRIC ACID, CESIUM SALT see CDE250
NITRIC ACID, CHROMIUM (3+) SALT see CMJ600
NITRIC ACID, CHROMIUM(3+) SALT, NONAHYDRATE see CMJ610
NITRIC ACID, COBALT(2+) SALT see CNC500
NITRIC ACID, COBALT(2+) SALT, HEXAHYDRATE see CND010

NITRIC ACID COPPER(2+) SALT TRIHYDRATE see CNN000

NITRIC ACID, DODECYL ESTER see NEE000

NITRIC ACID DYSPROSIUM(3+) SALT HEXAHYDRATE see DYH000

NITRIC ACID, ERBIUM (3+) SALT see ECY500

NITRIC ACID, ERBIUM (3+) SALT, HEXAHYDRATE see ECZ000

NITRIC ACID, ETHYL ESTER see ENM500

NITRIC ACID, 2-ETHYLHEXYL ESTER see EKW600

NITRIC ACID, EUROPIUM(3+) SALT, HEXAHYDRATE see ERC000

NITRIC ACID, FUMING (DOT) see NEE500

NITRIC ACID, GADOLINIUM(3+) SALT see GAL000

NITRIC ACID, GALLIUM(3+) SALT see GBS000

NITRIC ACID, HOLMIUM(3+) SALT, HEXAHYDRATE see HGH000

NITRIC ACID, IRON(3+) SALT see FAY200

NITRIC ACID, IRON(3+) SALT see IHB900

NITRIC ACID, IRON (3+) SALT, NONAHYDRATE see IHC000

NITRIC ACID, ISOPROPYL ESTER see IQP000

NITRIC ACID, LANTHANUM AMMONIUM SALT see ANL500

NITRIC ACID, LEAD(2+) SALT see LDO000

NITRIC ACID, LUTETIUM(3+) SALT see LIY000

NITRIC ACID, MAGNESIUM SALT (2:1) see MAH000

NITRIC ACID, MAGNESIUM SALT, HEXAHYDRATE see NEE100

NITRIC ACID, MANGANESE(2+) SALT see MAS900

NITRIC ACID, MERCURY(I) SALT see MDE750

NITRIC ACID, MERCURY(II) SALT see MDF000

NITRIC ACID METHYL ESTER see MMF500

NITRIC ACID, NEODYMIUM SALT see NCB000

NITRIC ACID, NEODYMIUM (3+) SALT, HEXAHYDRATE see NCB500

NITRIC ACID, NICKEL(II) SALT see NDG000

NITRIC ACID, NICKEL(2+) SALT, HEXAHYDRATE see NDG500

NITRIC ACID, PHENYLMERCURY SALT see MCU750

NITRIC ACID, POTASSIUM SALT see PLL500

NITRIC ACID, PRASEODYMIUM(3+) SALT see PLY250

NITRIC ACID, PROPYL ESTER see PNQ500

NITRIC ACID (RED FUMING) see NEE500

NITRIC ACID, RED FUMING (DOT) see NEE500

NITRIC ACID, SAMARIUM(3+) SALT, HEXAHYDRATE see SAT000

NITRIC ACID, SILVER(1+) SALT see SDS000

NITRIC ACID, SODIUM SALT see SIO900

NITRIC ACID, STRONTIUM SALT see SMK000

NITRIC ACID, THALLIUM(1+) SALT see TEK750

NITRIC ACID, THALLIUM(3+) SALT see TEK800

NITRIC ACID, THORIUM(4+) SALT see TFT500

NITRIC ACID, THULIUM(3+) SALT see TFX100

NITRIC ACID, THULIUM(3+) SALT, HEXAHYDRATE see TFX250

NITRIC ACID TRIESTER OF GLYCEROL see NGY000

NITRIC ACID, URANIUM SALT see UPA100

NITRIC ACID (WHITE FUMING) see NEF000

NITRIC ACID, YTTERBIUM(3+) SALT see YDS800

NITRIC ACID, YTTERBIUM(3+) SALT, HEXAHYDRATE see YEA000

NITRIC ACID, YTTRIUM(3+) SALT see YFJ000

NITRIC ACID, YTTRIUM(3+)SALT, HEXAHYDRATE see YFS000

NITRIC ACID, ZINC SALT see ZJJ000

NITRIC ACID, ZINC SALT, HEXAHYDRATE see NEF500

NITRIC AMIDE see NEG000

NITRIC ETHER see ENM500

NITRIC OXIDE see NEG100

NITRIC OXIDE and NITROGEN TETROXIDE MIXTURES see NGT500

NITRIDAZOLE see NML000

NITRIDES see NEH000

NITRILE ACRILICO (ITALIAN) see ADX500

NITRILE ACRYLIQUE (FRENCH) see ADX500

NITRILE ADIPICO (ITALIAN) see AER250

NITRILES see NEH500

NITRILE TRICHLORACETIQUE (FRENCH) see TII750

NITRIL KISELINY DIETHYLAMINOOCTOVE (CZECH) see DHJ600

NITRIL KYSELINY-o-CHLORBENZOOVE (CZECH) see CEM000

NITRIL KYSELINY p-CHLORBENZOOVE (CZECH) see CEM250

NITRIL KYSELINY ISOFTALOVE (CZECH) see PHX550

NITRIL KYSELINY MALONOVE (CZECH) see MAO250

NITRIL KYSELINY MANDLOVE (CZECH) see MAP250

NITRIL KYSELINY STEAROVE (CZECH) see SLL500

NITRIL KYSELINY TEREFTALOVE (CZECH) see BBP250

NITRIL KYSELINY β,β'-THIODIPROPIONOVE (CZECH) see DGS600

NITRIL KYSELINY m-TOLUYLOVE (CZECH) see TGT250

NITRIL KYSELINY p-TOLUYLOVE (CZECH) see TGT750

NITRIL MASTNE KYSELINY S (CZECH) see AFQ000

NITRILOACETIC ACID TRISODIUM SALT MONOHYDRATE see NEI000

NITRILOACETONITRILE see COO000

1,1'-(Δ,Δ'-NITRILODITETRAMETHYLENE)BIS(γ-BUTYL-γ-(1-NAPHTHYL)PIPERIDINE) see NEH600

1,1'-(Δ,Δ'-NITRILODITETRAMETHYLENE)BIS(γ-BUTYL-γ-(1-NAPHTHYL)PYRROLIDINE) see NEH610

NITRILOMALONAMIDE see COJ250

NITRILOPYRAMINE TOSYLATE see NEH700

NITRILOTRIACETIC ACID see AMT500

NITRILOTRIACETIC ACID, DISODIUM SALT see DXF000

NITRILOTRIACETIC ACID MONOSODIUM SALT see NEH800

NITRILOTRIACETIC ACID SODIUM SALT see SEP500

NITRILOTRIACETIC ACID, TRISODIUM SALT see SIP500

NITRILOTRIACETIC ACID TRISODIUM SALT MONOHYDRATE see NEI000

NITRILOTRIACETONITRILE see NEI600

NITRILO-2,2',2"-TRIETHANOL see TKP500

2,2',2"-NITRILOTRIETHANOL see TKP500

(2,2',2"-NITRILOTRIETHANOL)PHENYLMERCURY(1+) LACTATE (salt) see TNI500

2,2',2"-NITRILOTRIETHANOL TRINITRATE PHOSPHATE see TJL250

NITRILOTRIMETHANEPHOSPHONIC ACID see NEI100

NITRILOTRIMETHYLENEPHOSPHONIC ACID see NEI100

NITRILOTRIMETHYLPHOSPHONIC ACID see NEI100

NITRILOTRIMETHYLPHOSPHONIC ACID ZINC complex TRISODIUM TETRAHYDRATE see ZJJ200

1,1',1"-NITRILOTRI-2-PROPANOL see NEI500

NITRILOTRISACETONITRILE see NEI600

2,2',2"-NITRILOTRISACETONITRILE see NEI600

2,2',2"-NITRILOTRISETHANOL HYDROCHLORIDE see NEI700

NITRILOTRIS(ETHYLAMINE) see NEI800

2,2',2"-NITRILOTRIS(ETHYLAMINE) see NEI800

NITRILOTRISILANE see TNJ250

(NITRILOTRIS(METHYLENE))TRISPHOSPHONIC ACID see NEI100

NITRILOTRIS(METHYLPHOSPHONIC ACID) see NEI100

NITRIMIDAZINE see NHH000

NITRIN see NGY000

NITRINE see NGY000

NITRINE-TDC see NGY000

N-NITRISO-N-(2,3-DIHYDROXYPROPYL)-N-(2-HYDROXYETHYL)AMINE see NBR100

NITRITE see DVF800

NITRITES see NEJ000

NITRITE de SODIUM (FRENCH) see SIQ500

NITRITO see NGR500

NITRITO D'AMILE see ILW100

5-NITROACENAPHTHENE see NEJ500

5-NITROACENAPHTHYLENE see NEJ500

5-NITROACENAPTHENE see NEJ500

2-NITROACETALDEHYDE OXIME see NEJ600

2-NITRO-4-ACETAMINOFENETOL (CZECH) see NEL000

4-NITROACETANILIDE see NEK000

p-NITROACETANILIDE see NEK000

4'-NITROACETANILIDE see NEK000

2-NITRO-p-ACETANISIDIDE see NEK100

5-NITRO-2-ACETILAMINOTIAZOLO see ABY900

3-NITROACETOFENON see NEL500

3-NITRO-p-ACETOPHENETIDE see NEL000

3-NITRO-p-ACETOPHENETIDIDE see NEL000

5-NITRO-p-ACETOPHENETIDIDE see NEL000

3'-NITRO-p-ACETOPHENETIDIN see NEL000

m-NITROACETOPHENONE see NEL500

m-NITROACETOPHENONE see NEL500

o-NITROACETOPHENONE see NEL450

p-NITROACETOPHENONE see NEL600

2'-NITROACETOPHENONE see NEL450

3'-NITROACETOPHENONE see NEL500

4'-NITROACETOPHENONE see NEL600

NITRO ACID 100 percent see HMY000

NITRO ACID SULFITE see NMJ000

2-((1-NITRO-9-ACRIDINYL)AMINO)ETHANOL HYDROCHLORIDE see NHE550

2,2'-((2-(8-NITRO-9-ACRIDINYLAMINO)ETHYL))DIETHANOL HYDROCHLORIDE see NFW350

4'-(3-NITRO-9-ACRIDINYLAMINO)METHANESULFONANILIDE see NEM000

N-(4-((3-NITRO-9-ACRIDINYL)AMINO)PHENYL)METHANESULFONAMIDE see NEM000

N'-(1-NITRO-9-ACRIDINYL)-N,N,1-TRIMETHYL-1,2-ETHANEDIAMINE HYDROCHLORIDE see NFW435

NITROACRONINE see NEM100
NITROACRONYCINE see NEM100
β-NITROALCOHOL see NFY550
NITROALKANES see NEM300
m-NITROAMINOBENZENE see NEN500
2-NITRO-4-AMINODIPHENYLAMINE see NEM350
4-NITRO-2-AMINOFENOL (CZECH) see NEM500
p-NITROAMINOFENOL (POLISH) see NEM500
4'-NITRO-4-AMINO-3-HYDROXYDIPHENYL HYDROCHLORIDE see AKI500
4'-NITRO-4-AMINO-3-HYDROXYDIPHENYL HYDROGEN CHLORIDE see AKI500
2-NITRO-4-AMINOPHENOL see NEM480
3-NITRO-6-AMINOPHENOL see ALO000
5-NITRO-2-AMINOPHENOL see ALO000
o-NITRO-p-AMINOPHENOL see NEM480
p-NITRO-o-AMINOPHENOL see NEM500
2-(N-NITROAMINO)PYRIDINE-N-OXIDE see NEM600
5-N-NITROAMINOTETRAZOLE see NEN000
5-NITRO-2-AMINOTHIAZOLE see ALQ000
2-NITRO-4-AMINOTOLUENE see NMP000
4-NITRO-2-AMINOTOLUENE see NMP500
4-NITROAMINO-1,2,4-TRIAZOLE see NEN300
p-NITROANILINA see NEO500
2-NITROANILINE see NEO000
3-NITROANILINE see NEN500
m-NITROANILINE see NEN500
N-NITROANILINE see NEO510
o-NITROANILINE see NEO000
o-NITROANILINE see NEO000
p-NITROANILINE see NEO500
p-NITROANILINE see NEO500
4-NITROANILINE, 2,6-DICHLORO- see RDP300
p-NITROANILINE MERCURY(II) derivative see NEP000
4-NITROANILINE-2-SULFONIC ACID see NEP500
4-NITROANILINIUM PERCHLORATE see NEP600
(p-(p-NITROANILINO)PHENYL)ISOCYANIC ACID see IKI000
3-NITRO-p-ANISANILIDE see NEP650
4-NITRO-o-ANISIDINE see NEQ000
4-NITRO-o-ANISIDINE see NEQ000
5-NITRO-o-ANISIDINE see NEQ500
p-NITROANISOL see NER500
2-NITROANISOLE see NER000
4-NITROANISOLE see NER500
o-NITROANISOLE see NER000
p-NITROANISOLE see NER500
2-NITROANTHRACENE see NES000
5-NITROANTHRACENE see NES100
9-NITROANTHRACENE see NES100
1-NITRO-9,10-ANTHRACENEDIONE see NET000
1-NITROANTHRACHINON (CZECH) see NET000
4-NITROANTHRANILIC ACID see NES500
1-NITROANTHRAQUINONE see NET000
α-NITROANTHRAQUINONE see NET000
1-NITRO-2-ANTHRAQUINONECARBOXYLIC ACID see NFS502
NITROARGININE see OIU900
NITRO-l-ARGININE see OIU900
N^G)-NITRO-l-ARGININE METHYL ESTER see NDY675
1-NITRO-6-AZABENZO(a)PYRENE see NET100
3-NITRO-6-AZABENZO(a)PYRENE see NET120

1-NITRO-6-AZABENZO(a)PYRENE-N-OXIDE see NET130
3-NITRO-6-AZABENZO(a)PYRENE-N-OXIDE see NET140
3-NITRO-3-AZAPENTANE-1,5-DIISOCYANATE see NHI500
4-NITROAZOBENZENE see NET500
p-NITROAZOBENZENE see NET500
5-NITROBARBITURIC ACID see NET550
o-NITROBENZACETONITRILE see NFN600
2-NITROBENZALDEHYDE see NEU500
3-NITROBENZALDEHYDE see NEV000
4-NITROBENZALDEHYDE see NEV500
m-NITROBENZALDEHYDE see NEV000
o-NITROBENZALDEHYDE see NEU500
p-NITROBENZALDEHYDE see NEV500
2-NITROBENZALDEHYDE DIMETHYLACETAL see NEV510
o-NITROBENZALDEHYDE DIMETHYLACETAL see NEV510
2-NITROBENZAMIDE see NEV523
4-NITROBENZAMIDE see NEV525
o-NITROBENZAMIDE see NEV523
p-NITROBENZAMIDE see NEV525
2-(p-NITROBENZAMIDO)ACETOHYDROXAMIC ACID see NEW550
7-NITRO BENZ(a)ANTHRACENE see NEW600
3-NITROBENZANTHRONE see NEW700
3-NITRO-7H-BENZ(DE)ANTHRACEN-7-ONE see NEW700
NITROBENZEEN (DUTCH) see NEX000
NITROBENZEN (POLISH) see NEX000
3-NITROBENZENAMINE see NEN500
4-NITROBENZENAMINE see NEO500
NITROBENZENE see NEX000
NITROBENZENE, liquid (DOT) see NEX000
2-NITROBENZENEACETIC ACID see NII500
4-NITROBENZENEACETIC ACID see NII510
2-NITROBENZENEACETONITRILE see NFN600
3-NITROBENZENEACETONITRILE see NFN500
4-NITROBENZENEACETONITRILE see NIJ000
p-NITROBENZENEACETONITRILE see NIJ000
4-NITROBENZENEARSONIC ACID see NIJ500
o-NITROBENZENEARSONIC ACID see NEX500
p-NITROBENZENEAZOSALICYLIC ACID see NEY000
m-NITROBENZENEBORONIC ACID see NEY500
m-NITROBENZENECARBOXYLIC ACID see NFG000
2-NITRO-1,4-BENZENEDIAMINE see ALL750
4-NITRO-1,2-BENZENEDIAMINE see ALL500
4-NITROBENZENEDIAZONIUM AZIDE see NEZ000
3-NITROBENZENEDIAZONIUM CHLORIDE see NEZ100
4-NITROBENZENEDIAZONIUM FLUOBORATE see NEZ200
4-NITROBENZENEDIAZONIUM NITRATE see NFA000
3-NITROBENZENEDIAZONIUM PERCHLORATE see NFA500
m-NITROBENZENE DIAZONIUM PERCHLORATE (DOT) see NFA500
2-NITROBENZENEDIAZONIUM SALTS see NFA600
4-NITROBENZENEDIAZONIUM SALTS see NFA625
p-NITROBENZENEDIAZONIUM TETRAFLUOROBORATE see NEZ200

4-NITROBENZENEDIAZONIUM TETRAFLUOROBORATE(1-) see NEZ200
2-NITROBENZENEMETHANOL see NFM550
NITROBENZENE SODIUM SULFONATE see SIU100
p-NITROBENZENESULFONAMIDE see NFB000
3-NITROBENZENESULFONIC ACID see NFB500
m-NITROBENZENESULFONIC ACID see NFB500
m-NITROBENZENESULFONIC ACID, SODIUM SALT see NFC500
NITROBENZEN-m-SULFONAN SODNY (CZECH) see NFC500
2-NITROBENZIMIDAZOLE see NFC700
6-NITRO-BENZIMIDAZOLE see NFD500
5-NITRO-1H-BENZIMIDAZOLE see NFD500
2-NITRO-1H-BENZIMIDAZOLE (9CI) see NFC700
5-NITROBENZIMIDAZOLE NITRATE see NFD600
6-NITROBENZIMIDAZOLE NITRATE see NFD600
2-NITROBENZOFURAN see NFD700
4-NITRO-BENZOFURAZAN see NFD800
2,2'-((7-NITRO-4-BENZOFURAZAZANYL)IMINO)BISETHANOL OXIDE see NFF000
2-NITROBENZOIC ACID see NFG500
3-NITROBENZOIC ACID see NFG000
4-NITROBENZOIC ACID see CCI250
m-NITROBENZOIC ACID see NFG000
o-NITROBENZOIC ACID see NFG500
p-NITROBENZOIC ACID see CCI250
4-NITROBENZOIC ACID CHLORIDE see NFK100
p-NITROBENZOIC ACID CHLORIDE see NFK100
p-NITROBENZOIC ACID HYDRAZIDE see NFH000
p-NITROBENZOIC ACID 2-PHENYLHYDRAZIDE see NFH100
p-NITROBENZOIC ACID, PIPERIDINO ESTER see NFL500
NITROBENZOL (DOT) see NEX000
NITROBENZOL, liquid (DOT) see NEX000
3-NITROBENZONITRILE see NFH500
4-NITROBENZONITRILE see NFI010
m-NITROBENZONITRILE see NFH500
o-NITROBENZONITRILE see NFI000
p-NITROBENZONITRILE see NFI010
1-NITROBENZO(a)PYRENE see NFI100
1-NITROBENZO(e)PYRENE see NFI200
2-NITROBENZO(a)PYRENE see NFI210
3-NITROBENZO(e)PYRENE see NFI220
3-NITROBENZO(a)PYRENE-trans-9,10-DIHYDRODIOL see NFI230
6-NITRO-2-BENZOTHIAZOLAMINE see NFI240
5-NITROBENZOTRIAZOL (DOT) see NFJ000
5-NITROBENZOTRIAZOLE see NFJ000
5-NITRO-1H-BENZOTRIAZOLE see NFJ000
6-NITRO-1H-BENZOTRIAZOLE see NFJ000
3-NITROBENZOTRIFLUORIDE see NFJ500
m-NITROBENZOTRIFLUORIDE (DOT) see NFJ500
7-NITRO-3H-2,1-BENZOXAMERCUROL-3-ONE see NHY500
2-NITROBENZOYL CHLORIDE see NFK990
4-NITROBENZOYL CHLORIDE see NFK100
m-NITROBENZOYL CHLORIDE see NFK500
o-NITROBENZOYL CHLORIDE see NFK990
p-NITROBENZOYL CHLORIDE see NFK100
3-NITROBENZOYL NITRATE see NFL100
8-(4-NITROBENZOYLOXY)-9-HYDROXY-8,9-DIHYDRO-AFLATOXIN B1 see NFL200
N-(4-NITRO)BENZOYLOXYPIPERIDINE see NFL500

1-(p-NITROBENZOYL)-2-THIOBIURET see NFL600
3-NITROBENZ(a)PYRENE see NFM000
6-NITROBENZ(a)PYRENE see NFM500
4-NITROBENZYL ACETATE see NFM100
p-NITROBENZYLACETATE see NFM100
2-NITROBENZYL ALCOHOL see NFM550
4-NITROBENZYL ALCOHOL see NFM560
o-NITROBENZYL ALCOHOL see NFM550
2-NITROBENZYL BROMIDE see NFM700
4-NITROBENZYL BROMIDE see NFN000
p-NITROBENZYL BROMIDE see NFN000
2-NITROBENZYL CHLORIDE see NFN500
3-NITROBENZYL CHLORIDE see NFN010
m-NITROBENZYL CHLORIDE see NFN010
o-NITROBENZYL CHLORIDE see NFN500
p-NITROBENZYL CHLORIDE see NFN400
4-NITRO-BENZYL-CYANID (GERMAN) see NIJ000
3-NITROBENZYL CYANIDE see NFN500
4-NITROBENZYL CYANIDE see NIJ000
m-NITROBENZYL CYANIDE see NFN500
p-NITROBENZYLCYANIDE see NIJ000
p-NITROBENZYLIDENEACETOPHENONE see NFS505
2-((m-NITROBENZYLIDENE)AMINO)ETHANOL N-OXIDE see HKW345
2-((p-NITROBENZYLIDENE)AMINO)ETHANOL N-OXIDE see HKW350
m-NITROBENZYLIDENE-1,2-BENZ-9-METHYLACRIDINE see NMF000
m-NITROBENZYLIDENE-3,4-BENZ-9-METHYLACRIDINE see NMD500
o-NITROBENZYLIDENE-1,2-BENZ-9-METHYLACRIDINE see NMF500
o-NITROBENZYLIDENE-3,4-BENZ-9-METHYLACRIDINE see NME000
p-NITROBENZYLIDENE-1,2-BENZ-9-METHYLACRIDINE see NMG000
p-NITROBENZYLIDENE-3,4-BENZ-9-METHYLACRIDINE see NME500
3-NITROBENZYLIDENE ETHYL ESTER see ENO200
2-NITROBENZYL NITRILE see NFN600
1-(p-NITROBENZYL)-2-NITROIMIDAZOLE see NFO700
1-(p-NITROBENZYL)PYRIDINE see NFO750
o-NITROBIPHENYL see NFP500
p-NITROBIPHENYL see NFQ000
4'-NITRO-4-BIPHENYLAMINE see ALM000
4-NITROBIPHENYL ESTER see NIU000
2-NITRO-1,1-BIS(p-CHLOROPHENYL)PROPANE see BIN500
NITRO BLUE TETRAZOLIUM see NFQ020
NITRO BLUE TETRAZOLIUM see NMK100
p-NITRO BLUE TETRAZOLIUM see NFQ020
p-NITRO BLUE TETRAZOLIUM see NMK100
NITRO BLUE TETRAZOLIUM CHLORIDE see NFQ020
NITRO BLUE TETRAZOLIUM CHLORIDE see NMK100
p-NITRO BLUE TETRAZOLIUM CHLORIDE see NFQ020
p-NITRO BLUE TETRAZOLIUM CHLORIDE see NMK100
NITRO BLUE TETRAZOLIUM SALT see NFQ020
NITRO BLUE TETRAZOLIUM SALT see NMK100
2-NITROBROMOBENZENE see NFQ080
3-NITROBROMOBENZENE see NFQ090
4-NITROBROMOBENZENE see NFQ100
m-NITROBROMOBENZENE see NFQ090
o-NITROBROMOBENZENE see NFQ080
p-NITROBROMOBENZENE see NFQ100
2-NITRO-4-BROMOPHENOL see NFQ200
4-NITRO-2-(5'-BROMOSALICYLIDENAMINO)DIPHENYL AMINE see NFQ300

NITRO BT see NFQ020
NITRO BT see NMK100
tert-NITROBUTANE see NFQ500
2-NITROBUTENE see NFQ550
1-NITRO-3-BUTENE see NFR000
2-NITRO-1-BUTENE see NFQ550
2-NITRO-2-BUTENE see NFR500
4-(2-NITROBUTYL)MORPHOLINE see NFR600
N-(2-NITROBUTYL)MORPHOLINE see NFR600
4-NITROCALONE see NFS505
N-NITROCARBAMIDE see NMQ500
2-NITRO-9H-CARBAZOLE see NFR700
NITROCARBOL see NHM500
NITRO CARBO NITRATE see NFS500
1-NITRO-2-CARBOXYANTHRAQUINONE see NFS502
NITROCARDIOL see TJL250
NITROCELLULOSE E950 see CCU250
NITROCELLULOSE, dry or wetted with <25% water (or alcohol), by weight (UN 0340) (DOT) see CCU250
NITROCELLULOSE, plasticized with not <18% plasticizing substance, by weight (UN 0343) (DOT) see CCU250
NITROCELLULOSE, solution, flammable with not >12.6% nitrogen, by weight (UN 2059) (DOT) see CCU250
NITROCELLULOSE, unmodified or plasticized with <18% plasticizing substance (UN 0341) (DOT) see CCU250
NITROCELLULOSE, wetted with not <25% alcohol, by weight (UN 0342) (DOT) see CCU250
NITROCELLULOSE with alcohol not <25% alcohol by weight, and not >12.6% nitrogen (UN 2556) (DOT) see CCU250
NITROCELLULOSE with plasticizing not <18% plasticizing substance, by weight (UN 2557) (DOT) see CCU250
NITROCELLULOSE with water not <25% water, by weight (UN 2555) (DOT) see CCU250
4-NITROCHALCONE see NFS505
4-NITROCHINOLIN N-OXID (SWEDISH) see NJF000
p-NITROCHLOORBENZEEN (DUTCH) see NFS525
NITROCHLOR see DFT800
2-NITRO-4-CHLOROANILINE see KDA050
4-NITRO-2-CHLOROANILINE see CJA175
1-NITRO-5-CHLOROANTHRAQUINONE see CJA250
NITROCHLOROBENZENE see CJA950
m-NITROCHLOROBENZENE see CJB250
o-NITROCHLOROBENZENE see CJB750
p-NITROCHLOROBENZENE see NFS525
m-NITROCHLOROBENZENE, solid (DOT) see CJB250
p-NITROCHLOROBENZENE solid (DOT) see NFS525
4-NITRO-7-CHLOROBENZOFURAZAN see NFS550
p-NITROCHLOROBENZOL (GERMAN) see NFS525
3-NITRO-4-CHLOROBENZOTRIFLUORIDE see NFS700
4-NITRO-2-(p-CHLOROBENZYLIDENAMINO)DIPHENY LAMINE see NFS800
NITROCHLOROFORM see CKN500
p-NITRO-m-CHLOROPHENYL DIMETHYL THIONOPHOSPHATE see MIJ250
p-NITRO-o-CHLOROPHENYL DIMETHYL THIONOPHOSPHATE see NFT000
4-NITRO-2-(5'-CHLOROSALICYLIDENAMINO)DIPHENY LAMINE see NFT100
3-NITRO-4-CHLORO-α,α,α-TRIFLUOROTOLUENE see NFS700
6-NITROCHRYSENE see NFT400
p-NITROCINNAMALDEHYDE see NFT425

4-NITROCINNAMIC ACID see NFT440
p-NITROCINNAMIC ACID see NFT440
p-NITROCLOROBENZENE (ITALIAN) see NFS525
NITRO COMPOUNDS see NFT459
NITRO COMPOUNDS of AROMATIC HYDROCARBONS see NFT500
NITROCOTTON see CCU250
2-NITRO-p-CRESOL see NFU500
4-NITRO-m-CRESOL see NFV000
4-NITRO-o-CRESOL see NFV010
NITROCRESOLAMINE see DBY700
NITROCYCLOHEXANE see NFV500
NITROCYCLOPENTANE see NFV530
2-NITRO-1,4-DIAMINOBENZENE see ALL750
4-NITRO-1,2-DIAMINOBENZENE see ALL500
2-NITRO-4,6-DIAMINOTOLUENE see NFV540
2-NITRO-6H-DIBENZO(b,d)PYRAN-6-ONE see NFV600
2-NITRODIBENZOTHIOPHENE see NFV700
3-NITRODIBENZOTHIOPHENE see NFV710
NITRO-p-DICHLOROBENZENE see DFT400
4'-NITRO-2,4-DICHLORODIPHENYL ETHER see DFT800
NITRODIETHANOLAMINE DINITRATE see NFW000
N-NITRODIETHANOLAMINE DINITRATE see NFW000
N-NITRODIETHYLAMINE see DJS500
1-NITRO-9-(DIETHYLAMINOETHYLAMINE)-ACRIDINE DIHYDROCHLORIDE see NFW100
1-NITRO-9-(DIETHYLAMINOPROPYLAMINE)-ACRIDINE DIHYDROCHLORIDE see NFW200
4-NITRODIFENYLAMIN (CZECH) see NFY000
4-NITRODIFENYLAMIN-2-SULFONAN SODYN (CZECH) see AOS500
4-NITRODIFENYLETHER (CZECH) see NIU000
8-NITRO-DIHYDRO-1,3-BENZOXAZINE-2-THIONE-4-ONE see NFW210
2-NITRO-4,5-DIHYDROPYRENE see NFW220
1-NITRO-9-(2-DIHYDROXYETHYLAMINO-ETHYLAMINO)-ACRIDINE HYDROCHLORIDE see NFW350
N-NITRODIMETHYLAMINE see DSV200
1-NITRO-9-(DIMETHYLAMINE)-ACRIDINE DIHYDROCHLORIDE see NFW400
1-NITRO-9-(DIMETHYLAMINO)-ACRIDINE HYDROCHLORIDE see NFW425
2'-NITRO-4-DIMETHYLAMINOAZOBENZENE see DSW800
3'-NITRO-4-DIMETHYLAMINOAZOBENZENE see DSW600
1-NITRO-10-(DIMETHYLAMINOETHYL)-9-ACRIDONE HYDROCHLORIDE see NFW430
1-NITRO-9-((2-DIMETHYLAMINO)-1-METHYLETHYLAMINO)-ACRIDINE DIHYDROCHLORIDE see NFW435
1-NITRO-9-(5-DIMETHYLAMINOPENTYLAMINO)-ACRIDINE DIHYDROCHLORIDE see NFW450
1-NITRO-14-(DIMETHYLAMINOPROPYL)-ACRIDONE HYDROCHLORIDE see NFW460
1-NITRO-10-(3-DIMETHYLAMINOPROPYL)-ACRIDONE HYDROCHLORIDE see NFW460

1-NITRO 10-(3-DIMETHYLAMINOPROPYL)-ACRIDON HYDROCHLORIDE see NFW460
1-NITRO-9-(3-DIMETHYLAMINOPROPYLAMINE)ACRIDINE-N¹⁰-OXIDE DIHYDROCHLORIDE see CAB125
1-NITRO-9-(3-DIMETHYLAMINOPROPYLAMINE)-ACRIDINE-N-OXIDE DIHYDROCHLORIDE see NFW470
1-NITRO-9-(3'-DIMETHYLAMINOPROPYLAMINO)-ACRIDINE see NFW500
3-NITRO-9-(3'-DIMETHYLAMINOPROPYLAMINO)ACRIDINE see NFW525
1-NITRO-9-(3-DIMETHYLAMINOPROPYLAMINO)-ACRIDINE DIHYDROCHLORIDE see LEF300
2-NITRO-9-(3'-DIMETHYLAMINOPROPYLAMINO)ACRIDINE DIHYDROCHLORIDE see NFW535
2-NITRO-N,N-DIMETHYLANILINE see NFW600
NITRODIMETHYLBENZENE see NMS000
1-NITRO-3-(2,4-DINITROPHENYL)UREA see NFX500
2-NITRODIPHENYL see NFP500
4-NITRODIPHENYL see NFQ000
o-NITRODIPHENYL see NFP500
4-NITRODIPHENYLAMINE see NFY000
p-NITRODIPHENYLAMINE see NFY000
4-NITRO-2-(p-DIPHENYLAMINOBENZYLIDENAMINO) DIPHENYLAMINE see NFY100
2-NITRODIPHENYL ETHER see NIT500
4-NITRODIPHENYL ETHER see NIU000
p-NITRODIPHENYL ETHER see NIU000
2-(p-(2-NITRO-1,2-DIPHENYLVINYL)PHENOXY)TRIETHYLAMINE CITRATE see EAF100
N-NITRO-DMA see DSV200
4-NITRODRACYLIC ACID see CCI250
NITRO-DUR see NGY000
NITRODURAN see TJL250
dl-1-(2-NITRO-3-EMTHYLPHENOXY)-3-tert-BUTYLAMINO-PROPAN-2-OL see BQF750
NITROETAN (POLISH) see NFY500
NITROETHANE see NFY500
2-NITROETHANOL see NFY550
NITROETHYLENE POLYMER see NFY560
NITROETHYL NITRATE (DOT) see NFY570
4-NITRO-2-ETHYLQUINOLINE-N-OXIDE see NGA500
NITROFAN see DUS700
NITRO FAST GREEN GB see BLK000
NITRO FAST YELLOW SL see CMS240
NITROFEN see DFT800
NITROFENE (FRENCH) see DFT800
m-NITROFENOL see NIE600
o-NITROFENOL see NIF010
4-NITROFENOL (DUTCH) see NIF000
1-p-NITROFENYL-3,3-DIMETHYLTRIAZEN (CZECH) see DSX400
4-NITRO-1,3-FENYLENDIAMIN see NIM550
2-NITROFLUORANTHENE see NGA600
3-NITROFLUORANTHENE see NGA700
4-NITROFLUORANTHENE see NGA700
NITROFLUORENE see FDN100
2-NITROFLUORENE see NGB000
3-NITROFLUORENE see NGB200
NITRO-9H-FLUORENE see FDN100
3-NITRO-9H-FLUORENE see NGB200
2-NITRO-9H-FLUOREN-9-OL see HMY080
3-NITROFLUORENONE see NGB500
2-NITRO-9-FLUORENONE see NGB300
3-NITRO-9-FLUORENONE see NGB500
3-NITRO-9H-FLUOREN-9-ONE see NGB500
NITROFLUORFEN see NGB600
NITROFLUORFENE see NGB600

4-NITROFLUOROBENZENE see FKL000
p-NITROFLUOROBENZENE see FKL000
NITROFORM see TMM500
5-NITRO-2-FURALDEHYDE see NGE775
5-NITRO-2-FURALDEHYDE ACETYLHYDRAZONE see NGB700
5-NITROFURALDEHYDE DIACETATE see NGB800
5-NITRO-2-FURALDEHYDE-2-(2-HYDROXYETHYL)SEMICARBAZONE see FPE100
5-NITRO-2-FURALDEHYDE OXIME see NGC000
5-NITROFURALDEHYDE SEMICARBAZIDE see NGE500
6-NITROFURALDEHYDE SEMICARBAZIDE see NGE500
5-NITRO-2-FURALDEHYDE SEMICARBAZONE see NGE500
5-NITRO-2-FURALDEHYDE THIOSEMICARBAZONE see NGC400
5-NITRO-2-FURALDOXIME see NGC000
5-NITRO-2-FURAMIDE o-ACETOXIME see NGC500
5-NITRO-2-FURAMIDE o-BUTYROXIME see NGC550
5-NITRO-2-FURAMIDE o-PROPIONOXIME see NGC570
5-NITRO-2-FURAMIDOXIME see NGD000
NITROFURAN see NGD400
2-NITROFURAN see NGD400
5-NITROFURAN see NGD400
5-NITRO-2-FURANACRYLAMIDE see NGL500
5-NITRO-2-FURANACRYLIC ACID see NGI000
5-NITROFURAN-2-ALDEHYDE SEMICARBAZONE see NGE500
5-NITRO-5-FURANCARBOXALDEHYDE (9CI) see NGE775
5-NITRO-2-FURANCARBOXALDEHYDE SEMICARBAZONE see NGE500
5-NITROFURANCARBOXYLIC ACID see NGH500
5-NITRO-2-FURANCARBOXYLIC ACID (9CI) see NGH500
5-NITRO-2-FURAN METHANEDIOL DIACETATE see NGB800
5-NITRO-2-FURANMETHANOL see NGD600
NITROFURANTOIN see NGE000
6-(2-(5-NITRO-2-FURANYL)ETHENYL)-2-PYRIDINEMETHANOL (9CI) see NDY400
1-(((5-NITRO-2-FURANYL)METHYLENE)AMINO)-2-IMIDAZOLIDINONE see NDY000
3-(((5-NITRO-2-FURANYL)METHYLENE)AMINO)-2-OXAZOLIDINONE see NGG500
1-((5-NITROFURANYL-2)METHYLENEAMINO)TETRAHYDROPYRIMIDONE-2-ONE see FPO100
N-((5-NITRO-2-FURANYL)METHYLENE)-1H-BENZIMIDAZOL-1-AMINE see NGE100
((5-NITRO-2-FURANYL)METHYLENE)HYDRAZIDEACETIC ACID (9CI) see NGB700
2((5-NITRO-2-FURANYL)METHYLENE)HYDRAZINECARBOXAMIDE see NGE500
N-((5-NITRO-2-FURANYL)METHYLENE)-2H-INDAZOL-2-AMINE see NGE130
N-((5-NITRO-2-FURANYL)METHYLENE)-1H-PYRAZOL-1-AMINE see NGE160
N-((5-NITRO-2-FURANYL)METHYLENE)-4H-1,2,4-TRIAZOL-4-AMINE see NGG550
N-((5-NITRO-2-FURANYL)METHYLENE)TRICYCLO(3.3.1.1³·⁷)DECAN-1-AMINE see NGE777
3-(5-NITRO-2-FURANYL)-2-PROPENOIC ACID see NGI000

5-(5-NITRO-2-FURANYL)-1,3,4-THIADIAZOL-2-AMINE see NGI500
N-(4-(5-NITRO-2-FURANYL)-2-THIAZOLYL)ACETAMIDE see AAL750
2-(4-(5-NITRO-2-FURANYL)-2-THIAZOLYL)-HYDRAZINECARBOXALDEHYDE see NDY500
NITROFURATE see NGH500
NITROFURAZOLIDONE see NGG500
NITROFURAZOLIDONUM see NGG500
NITROFURAZONE see NGE500
NITROFURFURAL see NGE775
5-NITROFURFURAL see NGE775
5-NITROFURFURALACETYLHYDRAZONE see NGB700
3-(5'-NITROFURFURALAMINO)-2-OXAZOLIDONE see NGG500
5-NITROFURFURALDEHYDE see NGE775
5-NITROFURFURAL DIACETATE see NGB800
5-NITROFURFURAL SEMICARBAZONE see NGE500
5-NITROFURFURYL ALCOHOL see NGD600
(1-((5-NITROFURFURYLIDENE)AMINO)ADAMANTANE) see NGE777
2-(5-NITRO-2-FURFURYLIDENE)AMINOETHANOL N-OXIDE see HKW450
1-((5-NITROFURFURYLIDENE)AMINO)HYDANTOIN see NGE000
N-(5-NITROFURFURYLIDENE)-1-AMINOHYDANTOIN see NGE000
N-(5-NITRO-2-FURFURYLIDENE)-1-AMINOHYDANTOIN see NGE000
1-((5-NITROFURFURYLIDENE)AMINO)HYDANTOIN MONOHYDRATE see NGE780
1-((5-NITROFURFURYLIDENE)AMINO)-2-IMIDAZOLIDINONE see NDY000
N-(5-NITRO-2-FURFURYLIDENEAMINO)-2-IMIDAZOLIDINONE see NDY000
N-(5-NITRO-2-FURFURYLIDENE)-1-AMINO-2-IMIDAZOLIDONE see NDY000
1-((5-NITROFURFURYLIDENE)AMINO)-2-METHYLTETRAHYDRO-1,4-THIAZINE-4,4-DIOXIDE see NGG000
4-((5-NITROFURFURYLIDENE)AMINO)-3-METHYLTHIOMORPHOLINE-1,1-DIOXIDE see NGG000
N-(5-NITRO-2-FURFURYLIDENE)-3-AMINOOXAZOLIDINE-2-ONE see NGG500
3-((5-NITROFURFURYLIDENE)AMINO)-2-OXAZOLIDONE see NGG500
N-(5-NITRO-2-FURFURYLIDENE)-3-AMINO-2-OXAZOLIDONE see NGG500
(1-((5-NITROFURFURYLIDENE)AMINO)PYRAZOLE) see NGE160
4-((5-NITROFURFURYLIDENE)AMINO)-4H-1,2,4-TRIAZOLE see NGG550
(5-NITRO-2-FURFURYLIDENEAMINO)UREA see NGE500
5-NITROFURFURYLIDENE DIACETATE see NGB800
(NITRO-5' FURFURYLIDENE-2') HYDROXY-4 BENZHYDRAZIDE (FRENCH) see DGQ500
N-(6-(5-NITROFURFURYLIDENEMETHYL)-1,2,4-TRIAZIN-3-YL)IMINODIMETHANOL see BKH500
NITROFURMETHONE see FPI000
NITROFURMETON see FPI000
5-NITRO-2-FUROHYDRAZIDE IMIDE see NGG600
5-NITROFUROIC ACID see NGH500
5-NITRO-2-FUROIC ACID 2-ACETYLHYDRAZIDE see NGH600
5-NITRO-2-FUROIC ACID 2-BUTYRYLHYDRAZIDE see NGC550

NITROFUROXIME see NGC000
NITROFUROXON see NGG500
NITROFURYLACRYLAMIDE see NGI000
5-NITRO-2-FURYLACRYLAMIDE see NGL500
3-(5-NITRO-2-FURYL)ACRYLAMIDE see NGL500
5-NITRO-2-FURYL ACRYLIC ACID see NGI000
3-(5-NITRO-2-FURYL)ACRYLIC ACID see NGI000
(1-(5-NITRO-2-FURYL)-2-(6-AMINO-3-PYRIDAZYL)-ETHYLENE HYDROCHLORIDE see ALN250
2-(5-NITRO-2-FURYL)-5-AMINO-1,3,4-THIADIAZOLE see NGI500
5-(5-NITRO-2-FURYL)-2-AMINO-1,3,4-THIADIAZOLE see NGI500
N-((5-NITRO-2-FURYL)FORMIMIDOYL)-o-PROPIONYLHYDROXYLAMINE see NGC570
3-((5-NITROFURYLIDENE)AMINO)-2-OXAZOLIDINE see NGG500
3-(5-NITRO-2-FURYL)-IMIDAZO(1,2-a)PYRIDINE see NGI800
(5-NITRO-2-FURYL) METHYL KETONE see ACT250
((3-(5-NITRO-2-FURYL)-1-(2-(5-NITRO-2-FURYL)VINYL)ALLYIDENE)AMINO)GUANIDINE see PAF500
N-(3-(5-NITRO-2-FURYL)-6H-1,2,4-OXADIAZINYL)ACETAMIDE see AAL500
5-(5-NITRO-2-FURYL)-1,3,4-OXADIAZOLE-2-OL see NGK000
N-((3-(5-NITRO-2-FURYL)-1,2,4-OXADIAZOLE-5-YL)METHYL)ACETAMIDE see NGK500
3-(5-NITRO-2-FURYL)-2-PHENYLACRYLAMIDE see NGL000
3-(5-NITRO-2-FURYL)-2-PHENYL-2-PROPENAMIDE see NGL000
3-(5-NITRO-2-FURYL)-2-PROPENAMIDE see NGL500
N-(5-(5-NITRO-2-FURYL)-1,3,4-THIADIAZOL-2-YL)ACETAMIDE see FQJ000
4-(5-NITRO-2-FURYL)THIAZOLE see NGM400
N-(4-(5-NITRO-2-FURYL)-2-THIAZOLYL)ACETAMIDE see AAL750
N-(4-(5-NITRO-2-FURYL)THIAZOL-2-YL)ACETAMIDE see AAL750
N-(4-(5-NITRO-2-FURYL)-2-THIAZOLYL)FORMAMID (GERMAN) see NGM500
N-(4-(5-NITRO-2-FURYL)-2-THIAZOLYL)FORMAMIDE see NGM500
(4-(5-NITRO-2-FURYL)THIAZOL-2-YL)HYDRAZONOACETONE see NGN000
N-(4-(5-NITRO-2-FURYL)-2-THIAZOLYL)-2,2,2-TRIFLUOROACETAMIDE see NGN500
N,N'-(6-(5-NITRO-2-FURYL)-s-TRIAZINE-2,4-DIYL)BISACETAMIDE see DBF400
3-(5-NITRO-2-FURYL)-1H-1,2,4-TRIAZOL-5-AMINE see ALM750
N-(3-(5-NITRO-2-FURYL)-s-TRIAZOL-5-YL)-N-NITROSOETHYLAMINE see ENS000
3-(5-NITRO-2-FURYL)-3',4',5'-TRIMETHOXYACRYLOPHENONE see NGN600
6-(5-NITRO-2-FURYLVINYL)-3-(DIHYDROXYDIMETHYLAMINO)-1,2,4-TRIAZENE see BKH500
6-(2-(5-NITRO-2-FURYL)VINYL-2-PYRIDINE-METHANOL see NDY400
N-(4-(2-(5-NITRO-2-FURYL)VINYL)2-THIAZOLYL)FORMAMIDE see FNG000
((6-(2-(5-NITRO-2-FURYL)VINYL)-as-TRIAZIN-3-YL)IMINO)DIMETHANOL see BKH500

N-(6-(2-(5-NITRO-2-FURYL)VINYL)-1,2,4-TRIAZIN-3-YL)IMINODIMETHANOL see BKH500
NITROGEN see NGP500
NITROGEN BROMIDE see BMP250
NITROGEN CHLORIDE see CDW000
NITROGEN CHLORIDE see NGQ000
NITROGEN CHLORIDE DIFLUORIDE see NGR000
NITROGEN DIOXIDE see NGR500
NITROGEN DIOXIDE, DI- see NGU500
NITROGEN FLUORIDE see NGW000
NITROGEN FLUORIDE OXIDE see NGS500
NITROGEN GAS see NGP500
NITROGEN GLUCOSIDE of SODIUM-p-AMINOPHENYLSTIBONATE see NCL000
NITROGEN HALF MUSTARD see CGW000
NITROGEN IODIDE see IDN000
NITROGEN IODIDE see NGW500
NITROGEN LIME see CAQ250
NITROGEN MONOXIDE see NEG100
NITROGEN MONOXIDE, mixed with NITROGEN TETROXIDE see NGT500
NITROGEN MUSTARD see BIE250
NITROGEN MUSTARD HYDROCHLORIDE see BIE500
NITROGEN MUSTARD OXIDE see CFA500
NITROGEN MUSTARD OXIDE see CFA750
NITROGEN MUSTARD-N-OXIDE see CFA500
NITROGEN MUSTARD-N-OXIDE see CFA750
NITROGEN MUSTARD-N-OXIDE HYDROCHLORIDE see CFA750
NITROGENOL see HCP800
NITROGEN OXIDE see NGU000
NITROGEN OXIDES mixed with OZONE (47%:53%) see ORY000
NITROGEN OXYCHLORIDE see NMH000
NITROGEN OXYFLUORIDE see NMH500
NITROGEN PEROXIDE see NGR500
NITROGEN SELENIDE see SBT100
NITROGEN TETROXIDE see NGU500
NITROGEN TETROXIDE-NITRIC OXIDE MIXTURE see NGT500
NITROGEN TRIBROMIDE HEXAAMMONIATE see NGV500
NITROGEN TRICHLORIDE see NGQ500
NITROGEN TRICHLORIDE (DOT) see NGQ500
NITROGEN TRIFLUORIDE see NGW000
NITROGEN TRIIODIDE see NGW500
NITROGEN TRIIODIDE-AMMONIA see NGX000
NITROGEN TRIIODIDE-SILVER AMIDE see NGX500
NITROGEN, compressed (UN 1066) (DOT) see NGP500
NITROGEN, refrigerated liquid (cryogenic liquid) (UN 1977) (DOT) see NGP500
NITROGLICERINA (ITALIAN) see NGY000
NITROGLICERYNA (POLISH) see NGY000
N-NITROGLICIN see NGY700
NITROGLYCERIN see NGY000
NITROGLYCERIN, liquid, not desensitized (DOT) see NGY000
NITROGLYCERINE see NGY000
NITROGLYCERIN mixed with ETHYLENE GLYCOL DINITRATE (1:1) see NGY500
NITROGLYCERIN, SPIRITS OF see NGY000
NITROGLYCERIN, desensitized, not <40% non-volatile water insoluble phlegmatizer (UN 0143) (DOT) see NGY000
NITROGLYCERIN, solution in alcohol, with >1% but not >5% nitroglycerin (UN 3064) (DOT) see NGY000
NITROGLYCERIN, solution in alcohol, with >1% but not >10% nitroglycerin (UN 0144) (DOT) see NGY000
NITROGLYCERIN, solution in alcohol, with not >1% nitroglycerin (UN 1204) (DOT) see NGY000

NITROGLYCEROL see NGY000
N-NITROGLYCINE see NGY700
NITROGLYCOL see EJG000
NITROGLYKOL (CZECH) see EJG000
NITROGLYN see NGY000
NITROGRANULOGEN see BIE500
NITROGRANULOGEN HYDROCHLORIDE see BIE500
4-NITROGUAIACOL see NHA000
NITROGUANIDINE see NHA500
2-NITROGUANIDINE see NHA500
α-NITROGUANIDINE see NHA500
NITROGUANIDINE, wetted (UN 1336) (DOT) see NHA500
NITROGUANIDINE, dry or wetted with <20% water, by weight (UN 0282) (DOT) see NHA500
NITROGUANIL see NIJ400
2-NITRO-2-HEPTENE see NHB000
3-NITRO-2-HEPTENE see NHB000
2-NITRO-7,8,9,10,11,12-HEXAHYDROCHRYSENE see NHB600
2-NITRO-2-HEXENE see NHD000
3-NITRO-3-HEXENE see NHE000
1-NITRO-9-(3-HEXYLAMINOPROPYLAMINO)ACRIDINE DIHYDROCHLORIDE see HFK500
1-NITROHYDANTOIN see NHE100
NITROHYDRENE see NHE500
NITROHYDROCHLORIC ACID (DOT) see HHM000
NITROHYDROCHLORIC ACID, diluted (DOT) see HHM000
3-NITRO-4-HYDROXYBENZENEARSONIC ACID see HMY000
2-NITRO-1-HYDROXYBENZENE-4-ARSONIC ACID see HMY000
1-NITRO-9-(HYDROXYETHYLAMINO)-ACRIDINE HYDROCHLORIDE see NHE550
2-(3-NITRO-6-(β-HYDROXYETHYLAMINO)PHENOXY)ETHANOL see HKK100
6-NITRO-4-HYDROXYLAMINOQUINOLINE-1-OXIDE see HIX500
7-NITRO-4-HYDROXYLAMINOQUINOLINE-1-OXIDE see HIY000
4-NITRO-5-HYDROXYMERCURIORTHOCRESOL see NHK900
2-NITRO-2-(HYDROXYMETHYL)-1,3-PROPANEDIOL see HMJ500
3-NITRO-4-HYDROXYPHENYLARSENOUS ACID see NHE600
3-NITRO-4-HYDROXYPHENYLARSONIC ACID see HMY000
5-NITRO-8-HYDROXYQUINOLINE see NHF500
6-NITRO IA see NHK600
2-NITROIMIDAZOLE see NHG000
2-NITRO-1H-IMIDAZOLE see NHG000
4-NITRO-1H-IMIDAZOLE (9CI) see NHG100
2-NITROIMIDAZOLE RIBOSIDE see NHG200
1-NITRO-2-IMIDAZOLIDONE see NKL000
N-NITRO-2-IMIDAZOLIDONE see NKL000
4-(2-(5-NITROIMIDAZOL-1-YL)ETHYL)MORPHOLINE see NHH000
1-(2-NITROIMIDAZOL-1-YL)-3-METHOXYPROPAN-2-OL see NHH500
1-(2-NITRO-1-IMIDAZOLYL)-3-METHOXY-2-PROPANOL see NHH500
3-(2-NITROIMIDAZOL-1-YL)-1,2-PROPANEDIOL see NHI000
3-(2-NITRO-1H-IMIDAZOL-1-YL)-1,2-PROPANEDIOL see NHI000
2,2'-(NITROIMINO)BISETHANOL, DINITRATE (ESTER) (9CI) see NFW000
2,2'-NITROIMINOBIS(ETHYLNITRATE) see NFW000
2,2'-NITROIMINODIETHANOL NITRATE see NFW000

NITROIMINODIETHYLENEDIISOCYANIC ACID see NHI500

3,3'-NITROIMINODIPROPIONIC ACID see NHI600

2,2'-(NITROIMINO)ETHANOL DINITRATE see NFW000

2-(5-NITRO-α-IMINOFURFURYL)HYDRAZINE see NGG600

4-NITROINDANE see NHJ000

5-NITROINDANE see NHJ009

7-NITROINDAZOLE see NHK000

7-NITRO-1H-INDAZOLE see NHK000

5-NITROINDOLE see NHK500

5-NITRO-1H-INDOLE see NHK500

6-NITRO-ISATOIC ANHYDRIDE see NHK600

NITRO ISOBUTANE TRIOL TRINITRATE (DOT) see NHK650

NITROISOBUTYL GLYCERYL TRINITRATE see NHK650

NITROISOBUTYL GLYCOL TRINITRATE see NHK650

NITROISOPROPANE see NIY000

1-NITRO-9-(3-ISOPROPYLAMINOPROPYLAMINE)-ACRIDINE DIHYDROCHLORIDE see INW000

1-NITRO-9-(3-ISOPROPYLAMINOPROPYLAMINO)-ACRIDINE DIHYDROCHLORIDE see INW000

NITRO KLEENUP see DUZ000

NITROL see NGY000

NITROL see NHK800

NITROLAN see NGY000

NITRO-LENT see NGY000

NITROLETTEN see NGY000

NITROLIME see CAQ250

NITROLINGUAL see NGY000

NITROLOTRIACETIC ACID DISODIUM SALT MONOHYDRATE see NHK850

NITROLOWE see NGY000

NITROL (PROMOTER) see NHK800

NITRO MANNITE see MAW250

NITROMANNITE (dry) (DOT) see MAW250

NITROMANNITE, wetted with not <40% water, by weight or mixture (NA 0133) (DOT) see MAW250

NITROMANNITOL see MAW250

NITROMEL see NGY000

NITROMERSOL see NHK900

NITROMERSOL SOLUTION see NHK900

NITROMESITYLENE see NHM000

NITROMETAN (POLISH) see NHM500

N-NITROMETHANAMINE see NHN500

NITROMETHANE see NHM500

3-NITRO-6-METHOXYANILINE see NEQ500

5-NITRO-2-METHOXYANILINE see NEQ500

2-NITRO-7-METHOXYNAPHTHO(2,1-b)FURAN see MFB400

2-NITRO-8-METHOXYNAPHTHOL(2,1-b)-FURAN see MFB410

N-NITROMETHYLAMINE see NHN500

3-NITRO-4-METHYLANILINE see NMP000

3-NITRO-6-METHYLANILINE see NMP500

5-NITRO-2-METHYLANILINE see NMP500

1-NITRO-2-METHYLANTHRAQUINONE see MMG000

2-NITRO-3-METHYL-5-CHLOROBENZOFURAN see NHN550

((α-NITROMETHYL)-o-CHLOROBENZYLTHIO)ETHYLAMINE HYDROCHLORIDE see NEC000

2-(NITROMETHYLENE)-1-(PHENYLMETHYL)IMIDAZOLIDINE see NHN575

2-(NITROMETHYLENE)-2,3,4,5-TETRAHYDRO-1H-3-BENZAZEPINE see NHN600

3-(5-NITRO-1-METHYL-2-IMIDAZOLYL)-METHYLENE-AMINO-5-MORPHOLINO-METHYL-2-OXAZOLIDONE HCl see MRU750

1-((2-NITRO-4-METHYLPHENYL)AZO)-2-NAPHTHOL see MMP100

N-(3-NITRO-4-METHYLPHENYL)HYDROXYLAMINE see HMI100

4-NITRO-N-METHYLPHTHALIMIDE see NHN700

2-NITRO-2-METHYL-1,3-PROPANEDIOL see NHO500

2-NITRO-2-METHYL-1-PROPANOL see NHP000

2-NITRO-2-METHYLPROPANOL NITRATE (DOT) see MMP200

3-NITRO-1-METHYL-5H-PYRIDO(4,3-B)INDOLE see NHP050

NITROMIDE and SULFANITRAN see NHP100

NITROMIDINE see TJF000

NITROMIFENE CITRATE see NHP500

NITROMIM see CFA750

NITROMIN see CFA500

NITROMIN HYDROCHLORIDE see CFA750

NITROMIN IDO see ALQ000

NITROMURIATIC ACID (DOT) see HHM000

NITRON see CCU250

NITRONAL see EKW600

1-NITRONAPHTH(2,1,8-mna)ACRIDINE see NET100

3-NITRONAPHTH(2,1,8-mna)ACRIDINE see NET120

1-NITRONAPHTH(2,1,8-mna)ACRIDINE 6-OXIDE see NET130

3-NITRONAPHTH(2,1,8-mna)ACRIDINE 6-OXIDE see NET140

NITRONAPHTHALENE see NHP990

NITRONAPHTHALENE see NHP990

1-NITRONAPHTHALENE see NHQ000

2-NITRONAPHTHALENE see NHQ500

NITRONAPHTHALENE (DOT) see NHP990

α-NITRONAPHTHALENE see NHQ000

β-NITRONAPHTHALENE see NHQ500

5-NITRONAPHTHALENE ETHYLENE see NEJ500

2-NITRONAPHTHO(2,1-b)FURAN see NHQ950

1-(2-NITRONAPHTHO(2,1-b)FURAN-7-YL)ETHANONE see NHR100

5-NITRO-1-NAPHTHONITRILE see NHR200

3-NITRO-2-NAPHTHYLAMINE see NHR500

NITRONET see NGY000

NITRONG see NGY000

NITRONIC 40 see IGL110

NITRONIC 40 STAINLESS STEEL see IGL110

NITRONITRAMIDE AMMONIUM SALT see NHR550

4-NITRO-2-(p-NITROBENZYLIDENAMINO)DIPHENYLAMINE see NHR600

2-NITRO-1-(p-NITROBENZYL)IMIDAZOLE see NFO700

N-NITRO-N-(3-(5-NITRO-2-FURYL)-s-TRIAZOL-5-YL)ETHYLAMINE see ENN500

2-NITRO-1-(4-NITROPHENOXY)-4-(TRIFLUOROMETHYL)BENZENE see NIX000

(E)-1-NITRO-3-(2-(4-NITROPHENYL)ETHENYL)BENZENE see NHR620

4-NITRO-2-(3'-NITROSALICYLIDENAMINO)DIPHENYLAMINE see NHR640

N'-NITRO-N-NITROSO-N-METHYLGUANIDINE see MMP000

3-NITRO-1-NITROSO-1-PROPYLGUANIDINE see NHS000

1-NITRO-6-NITROSOPYRENE see NHS100

1-NITRO-8-NITROSOPYRENE see NHS120

NITRONIUM TETRAFLUOROBORATE(1−) see NHS500

NITRON LAVSAN see PKF750

NITRON (NITROCELLULOSE) see CCU250

4-NITRO-N^2-((4-NITROPHENYL)METHYLENE)-N^1-PHENYL-1,2-BENZENEDIAMINE see NHR600

3-NITRO-N^1,N^1,N^4-TRIS(2-HYDROXYETHYL)-p-PHENYLENEDIAMINE see HKN875

2-NITRO-2-NONENE see NHT000

3-NITRO-3-NONENE see NHU000

5-NITRO-4-NONENE see NHV500

2-NITRO-N1-PHENYL-1,4-BENZENEDIAMINE see NEM350

2-NITRO-N^1)-PHENYL-1,4-BENZENEDIAMINE see NEM350

NITRON (POLYESTER) see PKF750

2-NITRO-2-OCTENE see NHW000

3-NITRO-2-OCTENE see NHW500

3-NITRO-3-OCTENE see NHX000

4'-NITROOXANILIC ACID see NHX100

1-NITRO-1-OXIMINOETHANE see NHY100

NITROOXIMINOMETHANE see NHY250

7-NITRO-3-OXO-3H-2,1-BENZOXAMERCUROLE see NHY500

5-NITRO-N-(2-OXO-3-OXAZOLIDINYL)-2-FURANMETHANIMINE see NGG500

4-NITRO-2-(p-OXYBENZYLIDENAMINO)DIPHENYLAMINE see NHY600

4-NITRO-2-(2'-OXYBENZYLIDENAMINO)DIPHENYLAMINE see NHY700

4-NITRO-2-(2'-OXY-3'-METHOXYBENZYLIDENAMINO)DIPHENYLAMINE see PDN800

NITROPENTA see PBC250

NITROPENTAERYTHRITE see PBC250

NITROPENTAERYTHRITOL see PBC250

NITROPENTAGLYCERIN see MQG600

1-NITROPENTANE see AOL500

2-NITRO-2-PENTENE see NHZ000

3-NITRO-2-PENTENE see NIA000

3-NITROPERCHLORYLBENZENE see NIA700

p-NITROPEROXYBENZOIC ACID see NIB000

3-NITROPERYLENE see NIB500

NITROPHEN see DFT800

1-NITROPHENANTHRENE see NIB600

2-NITROPHENANTHRENE see NIC000

3-NITROPHENANTHRENE see NIC100

NITROPHENE see DFT800

2-NITROPHENETHYL ALCOHOL see NIM560

o-NITROPHENETHYL ALCOHOL see NIM560

5-NITRO-o-PHENETIDINE see NIC200

p-NITROPHENETOL (GERMAN) see NID000

o-NITROPHENETOLE see NIC990

p-NITROPHENETOLE see NID000

2-NITROPHENOL see NIF010

3-NITROPHENOL see NIE600

4-NITROPHENOL see NIF000

o-NITROPHENOL see NIF010

m-NITROPHENOL (DOT) see NIE600

p-NITROPHENOL (DOT) see NIF000

p-NITROPHENOL ACETATE see ABS750

3-(α-(p-NITROPHENOL)-β-ACETYLETHYL)-4-HYDROXYCOUMARIN see ABF750

p-NITROPHENOL, ALUMINUM SALT see NIF100

NITROPHENOLARSONIC ACID see HMY000

P-NITROPHENOL, ESTER with DIETHYL PHOSPHATE see NIM500

3-(4-NITROPHENYL)OXIRANEMETHANOL trans-(−)- see NIS300
3-(4-NITROPHENYL)OXIRANEMETHANOL trans-(+)- see NIS298
3-NITRO-N-PHENYL-4-(PHENYLAMINO)BENZENESULFONAMIDE see KDA075
2-NITROPHENYL PHENYL ETHER see NIT500
4-NITROPHENYL PHENYL ETHER see NIU000
p-NITROPHENYL PHENYLETHER see NIU000
O-(4-NITROPHENYL) O-PHENYLMETHYL PHOSPHONOTHIOATE see MOB699
3-(4-NITROPHENYL)-1-PHENYL-2-PROPEN-1-ONE see NFS505
4-NITRO-5-(4-PHENYL-1-PIPERAZINYL)BENZOFURAZAN OXIDE see NIV000
3-(4-NITROPHENYL)PROPENOIC ACID see NFT440
3-(4-NITROPHENYL)-N-(4-PROPYLCYCLOHEXYL)-2-PROPENAMIDE see NIV050
N-(4-NITROPHENYL)-N'-(3-PYRIDINYLMETHYL)UREA see PPP750
(p-NITROPHENYLSELENYL)ACETIC ACID see NIV100
4-(p-NITROPHENYL)SEMICARBAZONE 1-METHYL-1H-PYRROLE-2-CARBOXALDEHYDE see NIV150
4-(p-NITROPHENYL)SEMICARBAZONE-1H-PYRROLE-2-CARBOXALDEHYDE see NIV200
N-(p-NITROPHENYL)SULFANILAMIDE see SNQ600
2-NITROPHENYL SULFONYL DIAZOMETHANE see NIW300
4-(4-NITROPHENYL)THIAZOLE see NIW400
S-(p-NITROPHENYL) THIOACETATE see NIW450
p-NITROPHENYLTHIOL ACETATE see NIW450
p-NITROPHENYL-2,4,6-TRICHLOROPHENYL ETHER see NIW500
p-NITROPHENYL-α,α,α-TRIFLUORO-2-NITRO-p-TOLYL ETHER see NIX000
4-NITROPHENYL VINYL ETHER see VPZ333
p-NITROPHENYL VINYL ETHER see VPZ333
NITROPHOS see DSQ000
4-NITROPHTHALIMIDE see NIX100
NITROPONE C see BRE500
NITROPORE see ASM270
NITROPORE OBSH see BBS300
NITROPORE OBSH see OPE000
1,1'-(2-NITROPORPYLIDENE)BIS(4-CHLOROBENZENE) see BIN500
NITROPROPANE see NIY000
1-NITROPROPANE see NIX500
2-NITROPROPANE see NIY000
NITROPROPANE (DOT) see NIY000
β-NITROPROPANE see NIY000
2-NITROPROPANE ION see NIY025
2-NITROPROPANE NITRONATE see NIY025
3-NITROPROPANOL see NIY050
1-NITROPROPENE see NIY100
2-NITROPROPENE see NIY200
3-NITROPROPIONIC ACID see NIY500
β-NITROPROPIONIC ACID see NIY500
5-NITRO-2-n-PROPOXYANILINE see NIY525
4-NITRO-N-(4-PROPYLCYCLOHEXYL)BENZAMIDE see NIY600
NITROPRUSSIATE de SODIUM see SIW500

NITROPRUSSIDNATRIUM see SIW500
NITROPRUSSIDNATRIUM (GERMAN) see SIU500
NITROPYRENE see NIY700
1-NITROPYRENE see NJA000
3-NITROPYRENE see NJA000
4-NITROPYRENE see NJA100
1-NITRO-4-PYRENOL see NJA200
1-NITROPYREN-4-OL see NJA200
4-NITROPYRIDINE-1-OXIDE see NJA500
4-NITROPYRIDINE-N-OXIDE see NJA500
5-NITRO-2,4,6(1H,3H,5H)-PYRIMIDINETRIONE see NET550
5-NITROPYROMUCATE see NGH500
4-NITROQUINALDINE-N-OXIDE see MMQ250
2-NITROQUINOLINE see NJB500
5-NITROQUINOLINE see NJC000
6-NITROQUINOLINE see NJC500
8-NITROQUINOLINE see NJD500
4-NITROQUINOLINE-6-CARBOXYLIC ACID-1-OXIDE see CCG000
4-NITRO-6-QUINOLINECARBOXYLIC ACID-1-OXIDE see CCG000
4-NITROQUINOLINE-1-OXIDE see NJF000
4-NITROQUINOLINE-N-OXIDE see NJF000
5-NITRO-8-QUINOLINOL see NHF500
NITRORECTAL see NGY000
2-NITRO-1-β-d-RIBOFURANOSYLIMIDAZOLE see NHG200
5-NITROSALICYLALDEHYDE see HMX700
4-NITROSALICYLIC ACID see NJF100
p-NITROSALICYLIC ACID see NJF100
4-NITRO-SALICYLSAEURE see NJF100
NITROSAMINES see NJH000
N-NITROSAMINO DIACETONITRIL (GERMAN) see NKL500
N-NITROSARCOSINE see NJH500
NITROSATED COAL DUST EXTRACT see NJH750
N-NITROSAZETIDINE see NJL000
NITROSCANATE see LIG100
NITRO-SIL see AMY500
NITROSIMINODIACETONITRILE see NKL500
N-NITROSOACETANILIDE see NJI700
N-NITROSO-N-(1-ACETOXYMETHYL)BUTYLAMINE see BRX500
N-NITROSO-N-(ACETOXYMETHYL)-N-ISOBUTYLAMINE see ABR125
N-NITROSO-N-(ACETOXY)METHYL-N-METHYLAMINE see AAW000
N-NITROSO-N-(1-ACETOXYMETHYL)PROPYL AMINE see PNR250
NITROSO-N-(1-ACETOXYMETHYL)TRIDEUTEROMETHYLAMINE see MMS500
N-NITROSO-2-ACETYLAMINOFLUORENE see FEM050
NITROSO-4-ACETYL-3,5-DIMETHYLPIPERAZINE see NJI850
1-NITROSO-4-ACETYL-3,5-DIMETHYLPIPERAZINE see NJI900
N-NITROSOAETHYLAETHANOLAMIN (GERMAN) see ELG500
NITROSOAETHYLDIMETHYLHARNSTOFF see DRV600
NITROSOALDICARB see NJJ500
N-NITROSOALLYL-2,3-DIHYDROXYPROPYLAMINE see NJY500
N-NITROSOALLYLETHANOLAMINE see NJJ875
NITROSO-ALLYL-2-HYDROXYPROPYLAMINE see NJJ950
N-NITROSOALLYL-2-HYDROXYPROPYLAMINE see NJJ950
N-NITROSO-N-ALLYL-N-(2-HYDROXYPROPYL)AMINE see NJJ950
N-NITROSOALLYLMETHYLAMINE see MMT500

NITROSO-ALLYL-2-OXOPROPYLAMINE see AGM125
N-NITROSOALLYL-2-OXOPROPYLAMINE see AGM125
NITROSOALLYLUREA see NJK000
N-NITROSOAMINODIETHANOL see NKM000
4-(NITROSOAMINO-N-METHYL)-1-(3-PYRIDYL)-1-BUTANONE see MMS500
1-NITROSOANABASINE see NJK150
N-NITROSOANABASINE see NJK150
(+−)-1-NITROSOANABASINE see NJK200
(+−)-N-NITROSOANABASINE see NJK200
N'-NITROSOANABASINE see NJK150
N'-NITROSOANATABINE see NJK300
N-(p-NITROSOANILINOMETHYL)-2-NITROPROPANE see NHK800
N-NITROSOAZACYCLOHEPTANE see NKI000
N-NITROSOAZACYCLONONANE see OCA000
N-NITROSOAZACYCLOOCTANE see OBY000
1-NITROSOAZACYCLOTRIDECANE see NJK500
NITROSO-AZETIDIN (GERMAN) see NJL000
NITROSOAZETIDINE see NJL000
1-NITROSOAZETIDINE see NJL000
N-NITROSOAZETIDINE see NJL000
NITROSO-BAYGON see PMY310
1-NITROSO-4-BENZOYL-3,5-DIMETHYLPIPERAZINE see NJL850
N-NITROSO-4-BENZOYL-3,5-DIMETHYLPIPERAZINE see NJL850
N-NITROSOBENZTHIAZURON see NKR000
N-NITROSOBENZYLMETHYLAMINE see MHP250
NITROSOBENZYLUREA see NJM000
2-NITROSOBIPHENYL see NJM100
3-NITROSOBIPHENYL see NJM395
4-NITROSOBIPHENYL see NJM400
m-NITROSOBIPHENYL see NJM395
o-NITROSOBIPHENYL see NJM100
2-NITROSO-1,1'-BIPHENYL see NJM100
4-NITROSO-1,1'-BIPHENYL see NJM400
N-NITROSOBIS(ACETOXYETHYL)AMINE see NKM500
N-NITROSOBIS(2-ACETOXYPROPYL)AMINE see NJM500
NITROSOBIS(2-CHLOROETHYL)AMINE see BIF500
NITROSOBIS(2-CHLOROPROPYL)AMINE see DFW000
N-NITROSOBIS(2-ETHOXYETHYL)AMINE see BJO250
N-NITROSOBIS(2-HYDROXYETHYL)AMINE see NKM000
N-NITROSOBIS(2-HYDROXYPROPYL)AMINE see DNB200
N-NITROSOBIS(2-METHOXYETHYL)AMINE see BKO000
N-NITROSOBIS(2-OXOBUTYL)AMINE see NKL300
N-NITROSOBIS(2-OXOPROPYL)AMINE see NJN000
N-NITROSO-BIS-(4,4,4-TRIFLUORO-n-BUTYL)AMINE see NJN300
NITROSOBROMOETHYLUREA see NJN500
N-NITROSOBUTANAMINE see NJO000
N-NITROSOBUTYLAMINE see NJO000
N-NITROSO-1-BUTYLAMINO-2-PROPANONE see BRY000
N-NITROSO-N-BUTYLBUTYRAMIDE see NJO150
N-NITROSO-N-(BUTYL-N-BUTYROLACTONE)AMINE see NJO200
N-NITROSO-N-BUTYL-N-(3-CARBOXYPROPYL)AMINE see BQQ250
N-NITROSO-N-BUTYLETHYLAMINE see EHC000

N-NITROSOETHYL-2-
HYDROXYETHYLAMINE see ELG500
N-NITROSO-N-ETHYL-N-(2-
HYDROXYETHYL)AMINE see ELG500
N-NITROSOETHYLISOPROPYLAMINE see
ELX500
N-NITROSOETHYLMETHYLAMINE see
MKB000
1-NITROSO-1-ETHYL-3-METHYLUREA see
NKE100
1-NITROSO-1-ETHYL-3-(2-
OXOPROPYL)UREA see NKE120
N-NITROSOETHYLPHENYLAMINE (MAK)
see NKD000
N-NITROSO-2-ETHYLTHIAZOLIDINE see
ENU500
NITROSOETHYLUREA see ENV000
NITROSOETHYLURETHAN see NKE500
N-NITROSO-N-ETHYLURETHAN see
NKE500
N-NITROSO-N-ETHYLURETHAN see
NKE500
NITROSO-N-ETHYLURETHANE see
NKE500
N-NITROSOETHYLVINYLAMINE see
NKF000
N-NITROSO-N-ETHYLVINYLAMINE see
NKF000
N-NITROSOFENYLHYDROXYLAMIN
AMONNY (CZECH) see ANO500
2-NITROSOFLUORENE see NKF500
N-NITROSO-N-2-
FLUORENYLACETAMIDE see FEM050
NITROSOFLUOROETHYLUREA see
NKG000
NITROSOFOLIC ACID see NKG450
NITROSOFOLIC ACID see NLP000
NITROSOGLYPHOSATE see NKG500
NITROSOGUANIDIN (GERMAN) see
NKH000
NITROSOGUANIDINE see NKH000
N-NITROSOGUANIDINE see NKH000
NITROSOGUANIDINE (DOT) see MMP000
N-NITROSOGUVACINE see NKH500
NITROSOGUVACOLINE see NKH500
N-NITROSOGUVACOLINE see NKH500
NITROSOHEPTAMETHYLENEIMINE see
OBY000
N-NITROSOHEPTAMETHYLENEIMINE see
OBY000
NITROSO-HEPTAMETHYLENIMIN
(GERMAN) see OBY000
N-NITROSO-4,4,4,4',4',4'-
HEXAFLUORODIBUTYLAMINE see NJN300
N-NITROSOHEXAHYDROAZEPINE see
NKI000
N-NITROSOHEXAMETHYLENEIMINE see
NKI000
NITROSOHEXAMETHYLENIMINE see
NKI000
NITROSO-n-HEXYLMETHYLAMINE see
NKU400
NITROSO-N-HEXYLUREA see HFS500
NITROSO HYDANTOIC ACID see NKI500
1-NITROSOHYDANTOIN see NKJ000
1-NITROSO-1-HYDROXYETHYL-3-
CHLOROETHYLUREA see NKJ050
1-NITROSO-1-(2-HYDROXYETHYL)-3-(2-
CHLOROETHYL)UREA see NKJ050
NITROSO-2-HYDROXYETHYLUREA see
HKW500
N-NITROSOHYDROXYETHYLUREA see
HKW500
1-NITROSO-1-(2-HYDROXYETHYL)UREA
see HKW500
N-NITROSOHYDROXYLAMINE see
HOW500
N-NITROSO-3-HYDROXYPIPERIDINE see
NLK000
N-NITROSOHYDROXYPROLINE see
HNB500

1-NITROSO-1-HYDROXYPROPYL-3-
CHLOROETHYLUREA see NKJ100
1-NITROSO-1-(2-HYDROXYPROPYL)-3-(2-
CHLOROETHYL)UREA see NKJ100
N-NITROSO-(2-HYDROXYPROPYL)-(2-
HYDROXYETHYL)AMINE see HKW475
N-NITROSO(2-HYDROXYPROPYL)(2-
OXOPROPYL)AMINE see HNX500
N-NITROSO-2-HYDROXY-n-PROPYL-n-
PROPYLAMINE see NLM500
NITROSO-3-HYDROXYPROPYLUREA see
NKK000
NITROSO-2-HYDROXY-N-PROPYLUREA
see NKO400
NITROSO-3-HYDROXY-N-PROPYLUREA
see NKK000
N-NITROSO-2-HYDROXY-N-PROPYLUREA
see NKO400
N-NITROSO-3-HYDROXYPYRROLIDINE
see NLP700
N-NITROSOIMIDAZOLIDINETHIONE see
NKK500
1-NITROSOIMIDAZOLIDINONE see
NKL000
1-NITROSO-2-IMIDAZOLIDINONE see
NKL000
N-NITROSO-IMIDAZOLIDON (GERMAN)
see NKL000
N-NITROSOIMIDAZOLIDONE see NKL000
2,2'-(NITROSOIMINO)BISACETONITRILE
see NKL500
1,1'-(NITROSOIMINO)BIS-2-BUTANONE see
NKL300
2,2'-(NITROSOIMINO)BISETHANOL see
NKM000
(NITROSOIMINO)DIACETONE see NJN000
2,2'-(N-NITROSOIMINO)DIACETONITRILE
see NKL500
NITROSOIMINO DIETHANOL see NKM000
N-NITROSO-2,2'-
IMINODIETHANOLDIACETATE see
NKM500
1,1'-NITROSOIMINODI-2-PROPANOL see
DNB200
N-NITROSO-1,1'-IMINODI-2-PROPANOL
see DNB200
1,1-(N-NITROSOIMINO)DI-2-PROPANOL,
DIACETATE see NJM500
N-NITROSOINDOLIN (GERMAN) see
NKN000
1-NITROSOINDOLINE see NKN000
N-NITROSOINDOLINE see NKN000
N-NITROSOISOBUTYLTHIAZOLIDINE see
NKN500
N-NITROSO-ISO-BUTYLUREA see IJF000
N-NITROSOISONIPECTOIC ACID see
NKO000
NITROSOISOPROPANOL-
ETHANOLAMINE see HKW475
NITROSOISOPROPANOLUREA see NKO400
NITROSOISOPROPYLUREA see NKO425
NITROSO-LANDRIN see NLY500
NITROSO LINURON see NKO500
NITROSO-METHOMYL see NKX000
N-NITROSO-2-METHOXY-2,6-
DIMETHYLMORPHOLINE see NKO600
1-NITROSOMETHOXYETHYLUREA see
NKO900
NITROSO-2-METHOXYETHYLUREA see
NKO900
N-NITROSOMETHOXYMETHYLAMINE see
DSZ000
N-NITROSO-N-
METHOXYMETHYLMETHYLAMINE see
MEW250
N-NITROSO-N-METHYLACETAMIDE see
MMT000
N-NITROSO-N-METHYL-N-α-
ACETOXYBENZYLAMINE see NKP000
N-NITROSO-N-METHYL-N-
ACETOXYMETHYLAMINE see AAW000

o-(N-NITROSO-N-METHYL-β-ALANYL)-l-
SERINE see NKP500
NITROSOMETHYLALLYLAMINE see
MMT500
N-NITROSOMETHYLALLYLAMINE see
MMT500
N-NITROSOMETHYLAMINACETONITRIL
(GERMAN) see MMT750
N-
NITROSOMETHYLAMINOACETONITRILE
see MMT750
N-NITROSO-4-
METHYLAMINOAZOBENZENE see
MMY250
N-NITROSO-4-
METHYLAMINOAZOBENZOL (GERMAN)
see MMY250
3-(N-
NITROSOMETHYLAMINO)PROPIONALDE
HYDE see NKV050
2-NITROSOMETHYLAMINOPYRIDINE see
NKQ000
4-NITROSOMETHYLAMINOPYRIDINE see
NKQ100
N-NITROSO-N-METHYL-2-
AMINOPYRIDINE see NKQ000
N-NITROSO-N-METHYL-4-
AMINOPYRIDINE see NKQ100
4-(N-NITROSO-N-METHYLAMINO)-4-(3-
PYRIDYL)BUTANAL see MMS250
4-(N-NITROSO-N-METHYLAMINO)-1-(3-
PYRIDYL)-1-BUTANONE see MMS500
3-(N-
NITROSOMETHYLAMINO)SULFOLAN
(GERMAN) see NKQ500
N-NITROSOMETHYLAMINOSULFOLANE
see NKQ500
N-NITROSO-N-METHYL-N-AMYLAMINE
see AOL000
NITROSOMETHYLANILINE see MMU250
N-NITROSO-N-METHYLANILINE see
MMU250
N-NITROSO-N-METHYL-N'-(2-
BENZOTHIAZOLYL)UREA see NKR000
N-NITROSOMETHYLBENZYLAMINE see
MHP250
N-NITROSO-N-(2-METHYLBENZYL)-
METHYLAMIN (GERMAN) see NKR500
N-NITROSO-N-(3-METHYLBENZYL)-
METHYLAMIN (GERMAN) see NKS000
N-NITROSO-N-(4-METHYLBENZYL)-
METHYLAMIN (GERMAN) see NKS500
N-NITROSO-N-(2-
METHYLBENZYL)METHYLAMINE see
NKR500
N-NITROSO-N-(3-
METHYLBENZYL)METHYLAMINE see
NKS000
N-NITROSO-N-(4-
METHYLBENZYL)METHYLAMINE see
NKS500
NITROSOMETHYLBIS(CHLOROETHYL)UR
EA see NKT000
1-NITROSO-1-METHYL-3,3-BIS-(2-
CHLOROETHYL)UREA see NKT000
N-NITROSO-N-METHYLBIURET see
MMV000
N-NITROSOMETHYL-N-BUTYLAMINE see
MHW500
NITROSOMETHYL-d³-n-BUTYLAMINE see
NKT090
N-NITROSO-N-METHYL-N-n-BUTYL-1-d²-
AMINE see NKT100
N-NITROSO-N-METHYLCAPRYLAMIDE
see MMY000
N-NITROSO-N-METHYLCARBAMIDE see
MNA750
N-NITROSOMETHYL-3-
CARBOXYPROPYLAMINE see MIF250
N-NITROSOMETHYL-2-
CHLOROETHYLAMINE see CIQ500

N-NITROSOMETHYLCYCLOHEXYLAMINE see NKT500

N-NITROSO-N-METHYLCYCLOHEXYLAMINE see NKT500

NITROSOMETHYL-n-DECYLAMINE see MMX200

NITROSOMETHYLDIAETHYLHARNSTOFF see DJP600

NITROSOMETHYLDIETHYLUREA see DJP600

1-NITROSO-1-METHYL-3,3-DIETHYLUREA see DJP600

N-NITROSOMETHYL(2,3-DIHYDROXYPROPYL)AMINE see NMV450

N-NITROSO-N-METHYL-N-DODECYLAMIN (GERMAN) see NKU000

NITROSOMETHYL-n-DODECYLAMINE see NKU000

N-NITROSO-N-METHYL-N-DODECYLAMINE see NKU000

N-NITROSOMETHYLETHANOLAMINE see NKU350

N-NITROSOMETHYLETHYLAMINE (MAK) see MKB000

1-NITROSO-1-METHYL-3-ETHYLUREA see NKU370

N-NITROSO-N-METHYL-4-FLUOROANILINE see FKF800

N-NITROSOMETHYLGLYCINE see NLR500

N-NITROSO-N-METHYL-HARNSTOFF (GERMAN) see MNA750

N-NITROSO-N-METHYLHEPTYLAMINE see HBP000

NITROSOMETHYL-n-HEXYLAMINE see NKU400

N-NITROSO-N-METHYL-(4-HYDROXYBUTYL)AMINE see MKP000

N-NITROSOMETHYL-(2-HYDROXYETHYL)AMINE see NKU350

N-NITROSOMETHYL-(2-HYDROXYETHYL)AMINE p-TOLUENESULFONATE ESTER see NKU420

N-NITROSOMETHYL-(2-HYDROXYPROPYL)AMINE see NKU500

N-NITROSOMETHYL-(3-HYDROXYPROPYL)AMINE see NKU520

N-NITROSOMETHYLMETHOXYAMINE see DSZ000

N-NITROSO-N-METHYL-o-METHYLHYDROXYLAMIN (GERMAN) see DSZ000

N-NITROSO-N-METHYL-o-METHYL-HYDROXYLAMINE see DSZ000

NITROSO-2-METHYLMORPHOLINE see NKU550

N-NITROSO-2-METHYLMORPHOLINE see NKU550

NITROSOMETHYLNEOPENTYLAMINE see NKU570

N-NITROSO-N-METHYLNITROGUANIDINE see MMP000

N-NITROSO-N-METHYL-4-NITROSO-ANILINE see MJG750

NITROSOMETHYL-n-NONYLAMINE see NKU580

NITROSO-N-METHYL-n-NONYLAMINE see NKU580

N-NITROSOMETHYL-n-NONYLAMINE see NKU580

NITROSOMETHYL-n-OCTYLAMINE see NKU590

NITROSO-N-METHYL-n-OCTYLAMINE see NKU590

N-NITROSOMETHYL-n-OCTYLAMINE see NKU590

NITROSO-2-METHYL-1,3-OXAZOLIDINE see NKU600

NITROSO-5-METHYL-1,3-OXAZOLIDINE see NKU875

N-NITROSO-2-METHYL-1,3-OXAZOLIDINE see NKU600

N-NITROSO-5-METHYL-1,3-OXAZOLIDINE see NKU875

NITROSO-5-METHYLOXAZOLIDONE see NKU875

N-NITROSOMETHYL(?-OXOBUTYL)AMINE see MMR800

N-NITROSOMETHYL(3-OXOBUTYL)AMINE see MMR810

N-NITROSOMETHYL-2-OXOPROPYLAMINE see NKV000

N-NITROSO-N-METHYL-N-OXOPROPYLAMINE see NKV050

NITROSOMETHYL-N-PENTYLAMINE see AOL000

N-NITROSOMETHYLPENTYLNITROSAMINE see NKV100

N-NITROSOMETHYLPHENYLAMINE (MAK) see MMU250

NITROSOMETHYLPHENYLCARBAMATE see NKV500

N-NITROSO-N-METHYLPHENYLCARBAMATE see NKV500

N-NITROSO-N-p-METHYLPHENYL-1-DEOXY-d-FRUCTOSYLAMINE see NLW600

N-NITROSO-N-METHYL-(1-PHENYL)-ETHYLAMIN (GERMAN) see NKW000

N-NITROSO-N-METHYL-2-PHENYLETHYLAMINE see MNU250

N-NITROSO-N-METHYL-1-(1-PHENYL)ETHYLAMINE see NKW000

NITROSOMETHYLPHENYLUREA see MMY500

N-NITROSO-N'-METHYLPIPERAZIN (GERMAN) see NKW500

1-NITROSO-4-METHYLPIPERAZINE see NKW500

N-NITROSO-N'-METHYLPIPERAZINE see NKW500

R(−)-N-NITROSO-2-METHYL-PIPERIDIN (GERMAN) see NLI500

R(−)-N-NITROSO-2-METHYLPIPERIDINE see NLI500

NITROSO-3-METHYL-4-PIPERIDONE see MMZ800

NITROSOMETHYLPROPYLAMINE see MNA000

NITROSOMETHYL-N-PROPYLAMINE see MNA000

3-NITROSO-1-METHYL-5H-PYRIDO(4,3-B)INDOLE see MNA300

N-NITROSO-N-METHYL-n-TETRADECYLAMINE see NKW800

NITROSO-2-METHYLTHIOPROPIONALDEHYDE-o-METHYL CARBAMOYL-OXIME see NKX000

N-NITROSO-N-METHYL-4-TOLYLSULFONAMIDE see THE500

NITROSOMETHYL-2-TRIFLUOROETHYLAMINE see NKX300

NITROSOMETHYLUNDECYLAMINE see NKX500

NITROSOMETHYLUREA see MNA750

NITROSOMETHYL UREA see MOA500

1-NITROSO-1-METHYLUREA see MNA750

N-NITROSO-N-METHYLUREA see MNA750

NITROSOMETHYLURETHAN (GERMAN) see MMX250

NITROSOMETHYLURETHANE see MMX250

N-NITROSO-N-METHYLURETHANE see MMX250

N-NITROSOMETHYLVINYLAMINE see NKY000

NITROSOMETOXURON see NKY500

N-NITROSOMORPHOLIN (GERMAN) see NKZ000

NITROSOMORPHOLINE see NKZ000

4-NITROSOMORPHOLINE see NKZ000

N-NITROSOMORPHOLINE (MAK) see NKZ000

NITROSO-MPMC see DTN400

NITROSO-MTMC see MNV500

NITROSO-NAC see NBJ500

1-NITROSO-2-NAFTOL see NLB000

α-NITROSO-β-NAFTOL see NLB000

1-NITROSONAPHTHALENE see NLA000

2-NITROSONAPHTHALENE see NLA500

1-NITROSO-2-NAPHTHOL see NLB000

2-NITROSO-1-NAPHTHOL see NLB500

NITROSO-β-NAPHTHOL see NLB000

α-NITROSO-β-NAPHTHOL see NLB000

1-NITROSO-2-NITROAMINO-2-IMIDAZOLINE see NLB700

1-NITROSO-3-NITRO-1-BUTYLGUANIDINE see NLC000

N-NITROSO-N'-NITRO-N-BUTYLGUANIDINE see NLC000

1-NITROSO-3-NITRO-1-PENTYLGUANIDINE see NLC500

N-NITROSO-N-p-NITROPHENYL-d-RIBOSYLAMINE see NLC600

3-NITROSO-1-NITRO-1-PROPYLGUANIDINE see NLD000

N-NITROSO-N'-NITRO-N-PROPYLGUANIDINE see NHS000

1-NITROSO-6-NITROPYRENE see NHS100

1-NITROSO-8-NITROPYRENE see NHS120

N-NITROSO-N-(NITROSOMETHYL)PENTYLAMINE see NKV100

NITROSONIUM BISULFITE see NMJ000

N-NITROSONORNICOTINE see NHR200

N'-NITROSONORNICOTINE see NLD500

(+−)-N'-NITROSONORNICOTINE see NLP800

N'-NITROSONORNICOTINE-1-N-OXIDE see NLD525

N-NITROSOOCTAMETHYLENEIMINE see OCA000

1-NITROSO-1-OCTYLUREA see NLD800

N-NITROSO-O,N-DIAETHYLHYDROXYLAMIN (GERMAN) see NKC500

N-NITROSO-O,N-DIETHYLHYDROXYLAMINE see NKC500

3-NITROSO-7-OXA-3-AZABICYCLO(4.1.0)HEPTANE see NKC300

N-NITROSOOXAZOLIDIN (GERMAN) see NLE000

3-NITROSOOXAZOLIDINE see NLE000

N-NITROSOOXAZOLIDINE see NLE000

N-NITROSO-1,3-OXAZOLIDINE see NLE000

NITROSOOXAZOLIDONE see NLE000

N-NITROSO-N-(2-OXOBUTYL)BUTYLAMINE see BSB500

N-NITROSO-N-(3-OXOBUTYL)BUTYLAMINE see BSB750

N-NITROSO-(2-OXOBUTYL)(2-OXOPROPYL)AMINE see NLE400

N-NITROSO-(2-OXOPROPYL)-N-BUTYLAMINE see BRY000

1-NITROSO-1-OXOPROPYL-3-CHLOROETHYLUREA see NLE440

N-NITROSO-2-OXOPROPYL-2,3-DIHYDROXYPROPYLAMINE see NJY550

N-NITROSO-2-OXO-N-PROPYL-N-PROPYLAMINE see ORS000

1-NITROSO-2-OXOPROPYLUREA see OOO050

N-NITROSO-N-(2-OXOPROPYL)UREA see OOO050

N-NITROSO-N-PENTYLCARBAMIC ACID-ETHYL ESTER see AOL750

N-NITROSO-N-PENTYL-(4-HYDROXYBUTYL)AMINE see NLE500

1-NITROSO-1-PENTYLUREA see PBX500

N-NITROSOPERHYDROAZEPINE see NKI000

N-NITROSOPHENACETIN see NLE550

2-NITROSOPHENANTHRENE see NLF000

1-NITROSO-1-PHENETHYLUREA see NLG000

p-NITROSOPHENETOLE see EEY550
NITROSOPHENOL see NLF200
2-NITROSOPHENOL see NLF300
4-NITROSOPHENOL see NLF200
p-NITROSOPHENOL see NLF200
4-NITROSO-N-PHENYLANILINE see NKB500
N-NITROSO-N-PHENYLANILINE see DWI000
p-NITROSO-N-PHENYLANILINE see NKB500
4-NITROSO-N-PHENYLBENZENAMINE see NKB500
N-NITROSOPHENYLBENZYLAMINE see BFD750
NITROSOPHENYLETHYLUREA see NLG000
N-NITROSO-N-(2-PHENYLETHYL)UREA see NLG000
N-NITROSOPHENYLHYDROXYLAMIN AMMONIUM SALZ (GERMAN) see ANO500
N-NITROSOPHENYLHYDROXYLAMINE AMMONIUM SALT see ANO500
α-(NITROSO(PHENYLMETHYL)AMINO)-BENZENEMETHANOL ACETATE (ester) see ABL875
N-NITROSO-N-(PHENYLMETHYL)UREA see NJM000
NITROSO-4-PHENYLPIPERIDINE see PFT600
1-NITROSO-4-PHENYLPIPERIDINE see PFT600
N-NITROSO-4-PHENYLPIPERIDINE see PFT600
NITROSOPHENYLUREA see NLG500
N-NITROSO-N-PHENYLUREA see NLG500
N-NITROSO-N-(PHOSPHONOMETHYL)GLYCINE see NKG500
N-NITROSO-4-PICOLYLETHYLAMINE see NLH000
1-NITROSOPIPECOLIC ACID see NLH500
N-NITROSOPIPECOLIC ACID see NLH500
1-NITROSO-2-PIPECOLINE see NLI000
1-NITROSO-3-PIPECOLINE see MMY750
1-NITROSO-4-PIPECOLINE see MMZ000
R(−)-N-NITROSO-α-PIPECOLINE see NLI500
1-NITROSOPIPERAZINE see MRJ750
N-NITROSOPIPERAZINE see MRJ750
NITROSOPIPERIDIN see NLJ500
N-NITROSO-PIPERIDIN see NLJ500
1-NITROSOPIPERIDINE see NLJ500
N-NITROSOPIPERIDINE see NLJ500
N-NITROSO-Δ²-PIPERIDINE see NLU480
N-NITROSO-Δ³-PIPERIDINE see NLU500
1-NITROSO-2-PIPERIDINECARBOXYLIC ACID see NLH500
1-NITROSO-4-PIPERIDINECARBOXYLIC ACID see NKO000
NITROSO-3-PIPERIDINOL see NLK000
NITROSO-4-PIPERIDINOL see NLK500
N-NITROSO-3-PIPERIDINOL see NLK000
NITROSO-4-PIPERIDINONE see NLL000
N-NITROSO-4-PIPERIDINONE see NLL000
(+−)-3-(1-NITROSO-2-PIPERIDINYL)PYRIDINE see NJK200
NITROSO-4-PIPERIDONE see NLL000
1-NITROSO-4-PIPERIDONE see NLL000
1-NITROSO-l-PROLINE see NLL500
N-NITROSO-l-PROLINE see NLL500
N-NITROSO-N-2-PROPENYLUREA see NJK000
NITROSOPROPHAM see NJP500
NITROSOPROPOXUR see PMY310
N-NITROSO-N-PROPYLACETAMIDE see NLM000
N-NITROSO-N(N-PROPYL)ACETAMIDE see NLM000
4-(N-NITROSOPROPYLAMINO)BUTYRIC ACID see PNG500

1-(NITROSOPROPYLAMINO)-2-PROPANOL see NLM500
1-(NITROSOPROPYLAMINO)-2-PROPANONE see ORS000
N-NITROSO-N-PROPYL-1-BUTANAMINE see PNF750
N-NITROSO-N-PROPYLBUTYLAMINE see PNF750
N-NITROSO-N-PROPYLCARBAMIC ACID ETHYL ESTER see PNR500
N-NITROSO-N-PROPYLCARBAMIC ACID-1-NAPHTHYL ESTER see NBM500
N-NITROSO-N-PROPYL-(4-HYDROXYBUTYL)AMINE see NLN000
N-NITROSO-N-PROPYLPROPANAMINE see NKB700
N-NITROSO-N-PROPYL-1-PROPANAMINE see NKB700
N-NITROSO-N-PROPYLPROPIONAMIDE see NLN500
N-NITROSO-N-PROPYL-PROPRIONAMID (GERMAN) see NLN500
N-NITROSOPROPYLTHIAZOLIDINE see IQS000
3-NITROSO-2-PROPYLTHIAZOLIDINE see NLO000
N-NITROSO-2-N-PROPYLTHIAZOLIDINE see NLO000
NITROSOPROPYLUREA see NLO500
NITROSO-N-PROPYLUREA see NLO500
N-NITROSO-N-PROPYLUREA see NLO500
N-NITROSO-N-PTEROYL-l-GLUTAMIC ACID see NLP000
1-NITROSOPYRENE see NLP375
N-NITROSO-2-(3'-PYRIDYL)PIPERIDINE see NJK150
1-NITROSO-2-(3-PYRIDYL)PYRROLIDINE see NLD750
N-NITROSOPYRROLIDIN (GERMAN) see NLP500
NITROSOPYRROLIDINE see NLP480
1-NITROSOPYRROLIDINE see NLP500
N-NITROSOPYRROLIDINE see NLP500
N-NITROSO-2-PYRROLIDINE see NLP600
1-NITROSO-3-PYRROLIDINOL see NLP700
1-NITROSO-2-PYRROLIDINONE see NLP600
3-(1-NITROSO-2-PYRROLIDINYL)PYRIDINE see NLD500
(+−)-3-(1-NITROSO-2-PYRROLIDINYL)PYRIDINE see NLP800
NITROSO-3-PYRROLIN (GERMAN) see NLQ000
N-NITROSO-3-PYRROLINE see NLQ000
4-NITROSOQUINOLINE-1-OXIDE see NLQ500
5-NITROSO-8-QUINOLINOL see NLR000
NITROSORBID see CCK125
NITROSORBIDE see CCK125
NITROSORBON see CCK125
N-NITROSOSARCOSINE see NLR500
N-NITROSOSARCOSINE, ETHYL ESTER see NLS000
o-(N-NITROSOSARCOSYL)-l-SERINE see NLS500
NITROSO SARKOSIN see NLR500
NITROSO-SARKOSIN-AETHYLESTER see NLS000
4-NITROSO-trans-STILBENE see NLT000
NITROSOSWEP see DFV800
1-NITROSO-1,2,5,6-TETRAHYDRONICOTINIC ACID METHYL ESTER see NKH500
N-NITROSOTETRAHYDRO-1,2-OXAZIN see NLT500
N-NITROSO-TETRAHYDROOXAZIN-1,2 (GERMAN) see NLT500
N-NITROSOTETRAHYDROOXAZIN-1,3 (GERMAN) see NLU000
3-NITROSO-TETRAHYDRO-1,3-OXAZINE see NLU000

N-NITROSO-TETRAHYDRO-1,2-OXAZINE see NLT500
N-NITROSO-TETRAHYDRO-1,3-OXAZINE see NLU000
2-NITROSOTETRAHYDRO-2H-1,2-OXAZINE see NLT500
1-NITROSO-1,2,3,4-TETRAHYDROPYRIDINE see NLU480
N-NITROSO-1,2,3,4-TETRAHYDROPYRIDINE see NLU480
N-NITROSO-1,2,3,6-TETRAHYDROPYRIDINE see NLU500
N-NITROSO-Δ²-TETRAHYDROPYRIDINE see NLU480
N-NITROSO-Δ³-TETRAHYDROPYRIDINE see NLU500
1-NITROSO-TETRAHYDRO-2(1H)-PYRIMIDINONE see PNL500
1-NITROSO-2,2,6,6-TETRAMETHYLPIPERIDINE see TDS000
N-NITROSOTHIALDINE see NLU600
3-NITROSOTHIAZOLIDINE see NLV500
N-NITROSOTHIAZOLIDINE see NLV500
4-NITROSOTHIOMORPHOLINE see NLW000
N-NITROSOTHIOMORPHOLINE see NLW000
NITROSOTOLUENE see NLW500
2-NITROSOTOLUENE see NLW500
o-NITROSOTOLUENE see NLW500
N-NITROSO-N-p-TOLYL-d-FRUCTOPYRANOSYLAMINE see NLW600
NITROSOTRIAETHYLHARNSTOFF (GERMAN) see NLX500
NITROSOTRIDECYLUREA see NLX000
1-NITROSO-1-TRIDECYLUREA see NLX000
NITROSOTRIETHYLUREA see NLX500
N-NITROSO-2,2,2-TRIFLUORODIETHYLAMINE see NLX700
N-NITROSO-2,2,2-TRIFLUOROETHYL-ETHYLAMINE see NLX700
2-(NITROSO(3-(TRIFLUOROMETHYL)PHENYL)AMINO)-BENZOIC ACID see TKF699
N-NITROSO-N-(α,α,α-TRIFLUORO-m-TOLYL)ANTHRANILIC ACID see TKF699
NITROSOTRIHYDROXY-DIPROPYLAMINE see NOC400
1-NITROSO-2,2,4-TRIMETHYL-1,2,-DIHYDRO-QUINOLINE (POLYMER) see COG250
N-NITROSO-2,2,4-TRIMETHYL-1,2-DIHYDROQUINOLINE,POLYMER see COG250
NITROSOTRIMETHYLENEIMINE see NJL000
N-NITROSOTRIMETHYLENEIMINE see NJL000
N-NITROSO-TRIMETHYLHARNSTOFF (GERMAN) see TLU750
N-NITROSOTRIMETHYLHYDRAZIN (GERMAN) see NLY000
1-NITROSO-1,2,2-TRIMETHYLHYDRAZINE see NLY000
N-NITROSO-N,N',N'-TRIMETHYLHYDRAZINE see NLY000
NITROSOTRIMETHYLPHENYL-N-METHYLCARBAMATE see NLY500
NITROSO-3,4,5-TRIMETHYLPIPERAZINE see NLY750
1-NITROSO-3,4,5-TRIMETHYLPIPERAZINE see NLY750
N-NITROSO-3,4,5-TRIMETHYLPIPERAZINE see NLY750
1-NITROSO-2,2,4-TRIMETHYL-1(2H)QUINOLINE, POLYMER see COG250
N-NITROSO-N-((TRIMETHYLSILYL)METHYL)UREA see TMF600
NITROSOTRIMETHYLUREA see TLU750
N-NITROSOTRIMETHYLUREA see TLU750

NITROSOTRIS(CHLOROETHYL)UREA see NLZ000
1-NITROSO-1,3,3-TRIS-(2-CHLOROETHYL)UREA see NLZ000
NITROSOUNDECYLUREA see NMA000
N-NITROSO-N-UNDECYLUREA see NMA000
NITROSOUREA see NMA500
NITROSTABILIN see NGY000
NITROSTARCH see NMB000
NITROSTARCH, dry or wetted with <20% water, by weight (UN 0146) (DOT) see NMB000
NITROSTARCH, wetted with not <20% water, by weight (UN 1337) (DOT) see NMB000
NITROSTAT see NGY000
NITROSTIGMIN (GERMAN) see PAK000
NITROSTIGMINE see PAK000
4-NITROSTILBEN (GERMAN) see NMC000
4-NITROSTILBENE see NMC000
(E)-4-NITROSTILBENE see NMC050
trans-4-NITROSTILBENE see NMC050
trans-p-NITROSTILBENE see NMC050
β-NITROSTYRENE see NMC100
γ-NITROSTYRENE see NMC100
7-(m-NITROSTYRYL)BENZ(c)ACRIDINE see NMD500
7-(o-NITROSTYRYL)BENZ(c)ACRIDINE see NME000
7-(p-NITROSTYRYL)BENZ(c)ACRIDINE see NME500
12-(m-NITROSTYRYL)BENZ(a)ACRIDINE see NMF000
12-(o-NITROSTYRYL)BENZ(a)ACRIDINE see NMF500
12-(p-NITROSTYRYL)BENZ(a)ACRIDINE see NMG000
p-NITROSTYRYL PHENYL KETONE see NFS505
1-NITRO-5-SULFONATO-ANTHRAQUINONE see DVS200
1-NITRO-6(7)SULFONATOANTHRAQUINONE see DVR400
NITROSYL AZIDE see NMG500
NITROSYL CHLORIDE see NMH000
NITROSYL CYANIDE see NMH275
NITROSYL ETHOXIDE see ENN000
NITROSYLFERRICYTOCHROME C see CQM325
NITROSYL FLUORIDE see NMH500
NITROSYL HYDROGEN SULFATE see NMJ000
NITROSYL HYDROXIDE see NMR000
NITROSYL PERCHLORATE see NMI000
NITROSYLRUTHENIUM TRICHLORIDE see NMI500
NITROSYL SULFATE see NMJ000
NITROSYLSULFURIC ACID see NMJ000
NITROSYL TRIBROMIDE see NMJ400
NITROTEREPHTHALIC ACID see NMJ500
4-NITRO-2,3,5,6-TETRACHLORANISOLE see TBR250
3-NITRO-1,2,4,5-TETRACHLOROBENZENE see TBR750
4-NITRO-1,2,3,5-TETRACHLORO-BENZENE see TBR500
5-NITROTETRAZOL see NMK000
NITROTETRAZOLIUM BLUE see NFQ020
NITROTETRAZOLIUM BLUE see NMK100
p-NITROTETRAZOLIUM BLUE see NFQ020
p-NITROTETRAZOLIUM BLUE see NMK100
NITRO TETRAZOLIUM BT see NFQ020
NITRO TETRAZOLIUM BT see NMK100
NITROTETRAZOLIUM CHLORIDE BLUE see NFQ020
NITROTETRAZOLIUM CHLORIDE BLUE see NMK100
NITROTHIAMIDAZOL see NML000
NITROTHIAMIDAZOLE see NML000
NITROTHIAZOLE see NML000
N-(5-NITRO-2-THIAZOLYL)ACETAMIDE see ABY900

5-NITRO-2-THIAZOLYLAMINE see ALQ000
1-(5-NITRO-2-THIAZOLYL)HYDANTOIN see KGA100
1-(5-NITRO-2-THIAZOLYL)IMIDAZOLIDIN 2-ONE see NML000
1-(5-NITRO-2-THIAZOLYL)-2-IMIDAZOLIDINONE see NML000
1-(5-NITRO-2-THIAZOLYL)-2-IMIDAZOLINONE see NML000
1-(5-NITRO-2-THIAZOLYL)-2-OXOTETRAHYDROIMIDAZOL see NML000
1-(5-NITRO-2-THIAZOLYL)-2-OXOTETRAHYDROIMIDAZOLE see NML000
1-(5-NITRO-2-THIENYL)ETHANONE see NML100
8-NITRO-2-THIO-2H,1,3-BENZOXAZINE-2,4(3H)-DIONE see NFW210
2-NITROTHIOPHENE see NMO000
4-NITROTHIOPHENE-2-SULFONYL CHLORIDE see NMO400
2-NITROTOLUENE see NMO525
3-NITROTOLUENE see NMO500
4-NITROTOLUENE see NMO550
m-NITROTOLUENE see NMO500
o-NITROTOLUENE see NMO525
p-NITROTOLUENE see NMO550
mixo-NITROTOLUENE see NMO510
4-NITRO-o-TOLUENEDIAZONIUM TETRAFLUOROBORATE see NMO700
5-NITRO-o-TOLUENESULFONYL CHLORIDE see MMH500
4-NITROTOLUEN-2-SULFOCHLORID (CZECH) see MMH500
3-NITRO-4-TOLUIDIN (CZECH) see NMP000
3-NITRO-p-TOLUIDINE see NMP000
5-NITRO-2-TOLUIDINE see NMP500
5-NITRO-4-TOLUIDINE see NMP000
5-NITRO-o-TOLUIDINE see NMP500
6-NITRO-m-TOLUIDINE see NMP550
m-NITRO-p-TOLUIDINE see NMP000
5-NITRO-o-TOLUIDINE HYDROCHLORIDE see NMP600
3-NITROTOLUOL see NMO500
4-NITROTOLUOL see NMO550
1-(o-NITRO-p-TOLYLAZO)-2-NAPHTHOL see MMP100
NITROTRIAZOLONE see NMP620
3-NITRO-1,2,4-TRIAZOL-5-ONE see NMP620
NITROTRIBROMOMETHANE see NMQ000
4-NITRO-1,2,3-TRICHLOROBENZENE see NMQ025
NITROTRICHLOROMETHANE see CKN500
2-NITRO-3,4,6-TRICHLOROPHENOL see NMQ050
2-NITRO-α-α-α-TRIFLUORO-p-CRESOL see NMQ100
4'-NITRO-3'-TRIFLUOROMETHYLISOBUTYRANILIDE see FMR050
4-NITRO-3-TRIFLUOROMETHYLPHENOL see TKD400
m-NITROTRIFLUOROTOLUENE see NFJ500
m-NITROTRIFLUORTOLUOL (GERMAN) see NFJ500
NITROUREA see NMQ500
1-NITROUREA see NMQ500
N-NITROUREA see NMQ500
NITROUS ACID see NMR000
NITROUS ACID, AMMONIUM SALT see ANO250
NITROUS ACID-n-BUTYL ESTER see BRV500
NITROUS ACID-sec-BUTYL ESTER see BRV750
NITROUS ACID-1,1-DIMETHYLETHYL ESTER see BRV760
NITROUS ACID, ISOBUTYL ESTER see IJD000
NITROUS ACID-3-METHYL BUTYL ESTER see IMB000

NITROUS ACID, METHYL ESTER see MMF750
NITROUS ACID, 1-METHYLETHYL ESTER (9CI) see IQQ000
NITROUS ACID-1-METHYL PROPYL ESTER see BRV750
NITROUS ACID, 2-METHYLPROPYL ESTER see IJD000
NITROUS ACID, NICKEL(2+) SALT see NDG550
NITROUS ACID, PENTYL ESTER see AOL500
NITROUS ACID, POTASSIUM SALT see PLM500
NITROUS ACID, PROPYL ESTER see PNQ750
NITROUS ACID, REACTION PRODUCTS WITH 4-METHYL-1,3-BENZENEDIAMINE HYDROCHLORIDE see CMM780
NITROUS ACID, SODIUM SALT see SIQ500
NITROUS DIPHENYLAMIDE see DWI000
NITROUS ETHER see ENN000
NITROUS ETHYL ETHER see ENN000
NITROUS FUMES see NEE500
NITROUS OXIDE (DOT) see NGU000
NITROUS OXIDE, compressed (UN 1070) (DOT) see NGU000
NITROUS OXIDE, refrigerated liquid (UN 2201) (DOT) see NGU000
NITROVARFARIAN see ABF750
NITROVIN see DKK100
2-(2-NITROVINYL)ANISOLE see NMR100
β-NITROVINYLFURAN see FQM100
3-NITRO-1-VINYLPYRAZOLECARBOXYLIC ACID see NMR300
NITROWARFARIN see ABF750
NITROX see MNH000
NITROXANTHIC ACID see PID000
NITROXIDE, BIS(1,1-DIMETHYLETHYL) (9CI) see DEF090
NITROXIDE, DI-tert-BUTYL see DEF090
NITROXYL CHLORIDE see NMT000
NITROXYLENE see NMS000
NITROXYLENE see NMS000
NITRO-p-XYLENE see NMS520
2-NITRO-m-XYLENE see NMS500
NITROXYLENES (o-, m-, p-) (DOT) see NMS000
NITROXYNIL see HLJ500
NITROZAN K see MJG750
NITROZELL RETARD see NGY000
NITROZONE see NGE500
NITRUMON see BIF750
NITRYL CHLORIDE see NMT000
NITRYL FLUORIDE see NMT500
NITRYL HYPOCHLORITE see CDX000
NITRYL HYPOFLUORITE see NMU000
NITRYL KWASU NIKOTYNOWEGO see NDW515
NITRYL PERCHLORATE see NMU500
NITTO ACID RED PG see CMM320
NITTO DIRECT SKY BLUE 5B see CMO500
NIVACHINE see CLD000
NIVALENOL see NMV000
NIVALENOL-4-O-ACETATE see FQR000
NIVALIN see GBA000
NIVAQUINE B see CLD000
NIVELONA see DAP800
NIVEMYCIN see NCE000
NIVEUSIN C see NMV200
NIVITIN see SKV200
NIVOCILIN see HGP550
NIVONORM see NMV300
NIVRAL see LBF100
NIX see SIA000
NIXON E/C see EHG100
NIXON N/C see CCU250
NIX-SCALD see SAV000
NIXYN see XLS300
NIXYN HERMES see XLS300
NIZOFENONE FUMARATE see NMV400

NIZORAL see KFK100
NIZOTIN see EPQ000
NK 5 see KHU050
NK 136 see DJT800
NK 171 see EAV500
NK 313 see LFJ500
NK-381 see LEX400
NK 421 see BFV300
NK 483 see BPT300
NK 592 see CJD400
NK 631 see BLY770
NK 631 see PCB000
NK 711 see LEN000
NK-843 see NGY000
NK 0795 see DJB500
NK 1006 see BAU270
NK-1158 see DWU400
NK 2272 see EDD500
NK-15561 see MJN300
NK 104 (ACID) see NMV425
NK 5 (DYE) see KHU050
NK ESTER 3G see MDN510
NK ESTER A-TMPT see TLX175
NK FASTER see MCB050
NKK 100 see IRU500
NKK 105 see MAO275
N-LOST see BIE500
N-LOST (GERMAN) see BIE250
N-LOST-PHOSPHORSAEUREDIAMID
(GERMAN) see PHA750
NLPD see PHA750
NM 11 see PKP750
NM 14 see PKP750
NMA see MMU250
NMA see NAC000
1-N-2-MA (RUSSIAN) see MMG000
NMBA see MHW500
NMBA-d2 see NKT100
NMBA-d3 see NKT090
NMC 50 see CBM750
Na MCPA see SIL500
NMDDA see NKU000
NMDHP see NMV450
NMEA see MKB000
NMH see MNA750
NMHP see NKU500
N-6-MI see NKI000
NMNA see NKU570
NMO see CFA500
NMOP see NKV000
NMOR see NKZ000
NMP see MPF200
NMU see MNA750
NMUM see MMX250
NMUT see MMX250
NMVA see NKY000
NNA see MMS250
N. NAJA NAJA VENOM see ICC700
NNDG see DQR600
N',N''-
DIMETHYLETHYLENEDIHYDRAZINE see
EEB225
NNF-109 see IRU500
NNF-136 see TKB286
1000 NN FERRITE see IHB800
NNI 750 see BOO631
N. NIGRICOLLIS VENOM see NAG000
NNK see MMS500
NNN see NHR200
NNN-1-N-OXIDE see NLD525
NNO see NDR100
NOBACID see SAN000
NOBECUTAN see TFS350
NOBEDON see HIM000
NOBEDORM see QAK000
NOBFELON see IIU000
NOBFEN see IIU000
NOBGEN see IIU000
NOBILEN see TFR250
NOBITOCIN S see ORU500
NOBLEN see PMP500
NOBRIUM see CGA000

NO BUNT see HCC500
NO BUNT 40 see HCC500
NO BUNT 80 see HCC500
NO BUNT LIQUID see HCC500
NOC 18 see DHJ333
NOCA see SLL000
NOCARDICIN A see APT750
NOCARDICIN COMPLEX see NMV480
NOCBIN see DXH250
NOCCELER BG see TGX550
NOCCELER CZ see CPI250
NOCCELER PR see DEK500
NOCILON see IRA050
NOCODAZOLE see OJD100
NO. 56 CONC. PERMANENT ORANGE G
see CMS145
No. 3 CONC. SCARLET see CHP500
No. 55 Conc. PALE YELLOW SF see CMS210
NOCRAC 224 see PJQ750
NOCRAC MMB see MCO400
NOCRAC NS 5 see MJN250
NOCRAC NS 6 see MJO500
NOCTAL see QCS000
NOCTAMID see MLD100
NOCTAN see DNW400
NOCTAZEPAM see CFZ000
NOCTEC see CDO000
NOCTENAL see QCS000
NOCTILENE see QAK000
NOCTIVANE SODIUM see ERE000
NOCTOSEDIV see TEH500
NOCTOSOM see FMQ000
NOCTOVANE see ERD500
NODAPTON see GIC000
NO-DHU see NJY000
NO. 2 DIESEL FUEL see DHE850
NO-DOZ see CAK500
NOE (FRENCH) see CBA100
NO-ETU see NKK500
NOFLAMOL see PJL750
NO. 59 FORTHFAST BENZIDINE YELLOW
see CMS145
NO. 2 FORTHFAST SCARLET see MMP100
NO. 6 FUEL OIL see HAJ750
NOGALAMYCIN see NMV500
NOGAL de LA INDIA (CUBA) see TOA275
NOGALOMYCIN see NMV500
NOGAMYCIN see NMV600
NOGEDAL see DPH600
NOGEST see MCA000
NOGEST see POF275
NOGOS see DGP900
NOGRAM see EID000
NOHDABZ see HKB000
NOHFAA see HIP000
NOHO-MALIE (HAWAII) see YAK350
NOIGEN 160 see DXY000
NOIGEN ET 147 see AFJ155
NOIGEN ET 167 see AFJ155
NOIR BRILLANT BN (FRENCH) see BMA000
NOISETTE des GRANDS-FONDS
(GUADELOUPE) see TOA275
NOIX de BANCOUL (GUADELOUPE) see
TOA275
NOIX des MOLLUQUES (GUADELOUPE)
see TOA275
NOKHEL see AHP375
NOLA see OIU900
NOLEPTAN see FMU000
NOLTAM see TAD175
NOLTRAN see CMA250
NOLUDAR see DNW400
NOLVADEX see TAD175
NOLVASAN see BIM250
NOMATE CHOKEGARD see HCP050
NOMATE PBW see GJK000
NOMERSAN see TFS350
NOMETIC see DWK200
NOMETINE see CJL500
NOMIFENSIN see NMV700
NOMIFENSIN see NMV725
NOMIFENSINE see NMV700

NOMIFENSINE HYDROGEN MALEATE see
NMV725
NOMIFENSINE MALEATE see NMV725
NOMIFENSIN HYDROGEN MALEATE see
NMV725
NONABROMOBIPHENYL see NMV735
NONACARBONYL DIIRON see NMV740
NONACHLAZINE see NMV750
trans-NONACHLOR see NMV755
NONACHLORO-4-PHENOXYPHENOL see
HNB550
NONADECAFLUORODECANOIC ACID see
PCG725
NONADECAFLUORO-n-DECANOIC ACID
see PCG725
2,6-NONADIENAL see NMV760
trans-2,cis-6-NONADIENAL see NMV760
trans,cis-2,6-NONADIENAL see NMV760
trans,cis-2,6-NONADIENAL see NMV760
trans,trans-2,4-NONADIENAL see NMV775
2,6-NONADIENENITRILE, 3,7-DIMETHYL-
see LEH100
2,6-NONADIENOIC ACID, 7-ETHYL-9-(3-
ETHYL-3-METHYLOXIRANYL)-3-METHYL-
, METHYL ESTER see MJG100
NONADIENOL see NMV780
trans,cis-2,6-NONADIENOL see NMV780
2-trans-6-cis-NONADIEN-1-OL see NMV780
1,6-NONADIEN-3-OL, 3,7-DIMETHYL-,
ACETATE see HGI585
NONAFLUOR-terc.BUTANOL see HDF075
NONAFLUOROBUTANESULFONIC ACID
see PCG625
NONAFLUORO-1-BUTANESULFONIC
ACID see PCG625
NONAFLUORO-tert-BUTANOL see HDF075
3,3,4,4,5,5,6,6-NONAFLUORO-1-HEXENE see
NMV800
Δ-NONALACTONE see NMV790
γ-NONALACTONE (FCC) see CNF250
1-NONALDEHYDE see NMW500
NONALOL see NNB500
1,4-NONALOLIDE see CNF250
1,5-NONALOLIDE see NMV790
NONAMETHYLENE
DIMETHANESULFONATE see MDR275
1-NONANAL see NMW500
NONANAMIDE, N-HYDROXY- see NMX600
NONANE see NMX000
1-NONANECARBOXYLIC ACID see
DAH400
4-NONANECARBOXYLIC ACID see
NMX500
NONANE-1,9-DIMETHANESULFONATE
see MDR275
NONANE DIMETHANESULPHONATE see
MDR275
NONANEDIOIC ACID see ASB750
1,3-NONANEDIOL ACETATE see NMX550
NONANEDIOL-1,3 ACETATE see NMX550
NONANEDIOL, DIACETATE see NMX550
NONANEDIOL DIACETATE see NMX550
1,9-NONANEDIOL,
DIMETHANESULFONATE see MDR275
NONANEDIOTIC ACID, DIHEXYL ESTER
see ASC000
NONANENITRILE see OES300
n-NONANENITRILE see OES300
NONANOHYDROXAMIC ACID see
NMX600
NONANOIC ACID see NMY000
NONANOIC ACID, 3,6-DIOXA-2,5-
DI(TRIFLUOROMETHYL)-
UNDECAFLUORO-, ACID FLUORIDE see
HDC435
NONANOIC ACID, ETHYL ESTER see
ENW000
NONANOIC ACID, HEPTADECAFLUORO-,
AMMONIUM SALT see HAS050
NONANOIC ACID OXYDI-3,1-
PROPANEDIYL ESTER (9CI) see DWT000

NONANOIC ACID OXYDIPROPYLENE ESTER see DWT000
NONANOIC ACID, TRIBUTYLSTANNYL ESTER see TIF500
NONAN-1-OL see NNB500
1-NONANOL see NNB500
2-NONANONE see NMY500
NONAN-2-ONE see NMY500
5-NONANONE see NMZ000
NONAN-5-ONE see NMZ000
NONANONITRILE see OES300
n-NONANONITRILE see OES300
NONANOPHENONE, 9-FLUORO- see FKT050
1-NONANOYLAZIRIDINE see NNA000
NONANOYLETHYLENEIMINE see NNA000
4-NONANOYLMORPHOLINE see MRQ750
(NONANOYLOXY)TRIBUTYLSTANNANE see TIF500
NONANOYL PEROXIDE see NNA100
NONASULFAN see MDR275
NONASULPHAN see MDR275
1,3,5,7-NONATETRAENE,3,7-DIMETHYL-9-(2,6,6-TRIMETHYL-2-CYCLOHEXENE-1-YLIDENE)-, (ALL-E)- see AFQ700
2,4,6,8-NONATETRAENOIC ACID, 9-(4-METHOXY-2,3,6-TRIMETHYLPHENYL)-3,7-DIMETHYL-, ETHYL ESTER, (Z,E,E,E)- see EMJ600
2-NONENAL see NNA300
NON-2-ENAL see NNA300
2-NONEN-1-AL see NNA300
6-NONENAL, (Z)- see NNA325
(Z)-6-NONENAL see NNA325
cis-6-NONENAL see NNA325
cis-6-NONEN-1-AL see NNA325
trans-2-NONENAL (FCC) see NNA300
2-NONENENITRILE, 3-METHYL- see MNB500
2-NONENOIC ACID, METHYL ESTER see MNB750
cis-6-NONEN-1-OL see NNA530
trans-2-NONEN-1-OL see NNA532
2-NONEN-2-OL-4-ONE-1,1,1-TRIFLUORODIMETHYLTHALLIUM SALT see NNA600
2-NONEN-2-OL, 4-OXO-1,1,1-TRIFLUORO-, DIMETHYLTHALLIUMSALT see NNA600
2-NONEN-4,6,8-TRIYN-1-AL see NNB000
α-NONENYL ALDEHYDE see NNA300
NONEX 25 see PJY100
NONEX 30 see PJY100
NONEX 52 see PJY100
NONEX 64 see PJY100
NONEX 411 see HKJ000
NON-FER-AL see CAT775
NONFLAMIN see TGE165
NONFLEX RD see PJQ750
NON-FLOCCULATING GREEN G 25 see PJQ100
NONIC 218 see PKE350
NONIDET P40 see GHS000
NONIDET R 7 see AFJ160
NONIDET R 12 see AFJ160
NONION 06 see PJY100
NONION HS 206 see GHS000
NONION O2 see PJY100
NONION O4 see PJY100
NONION OP80R see SKV100
NONION P 208 see PJT300
NONION P 210 see PJT300
NONION PP40 see MRJ800
NONION SP 60 see SKV150
NONION SP 60R see SKV150
NONISOL 200 see PJY100
NONISOLD see CNH125
n-NONOIC ACID see NMY000
NONOX CL see NBL000
NONOX D see PFT500
NONOX DCP see IPK000
NONOX DPPD see BLE500
NONOX TBC see BFW750

NONOXYNOL see NND500
NONOXYNOL-9 see NNB300
NONOX ZA see PFL000
1,6-NONP see NHS100
1,8-NONP see NHS120
NONYL ACETATE see NNB400
NONYL ACETOPHENONE OXIME see NND100
NONYL ALCOHOL see NNB500
n-NONYL ALCOHOL see NNB500
sec-NONYL ALCOHOL see DNH800
1-NONYL ALDEHYDE see NMW500
1-NONYLAMINE, N-METHYL-N-NITROSO- see NKU580
NONYLCARBINOL see DAI600
NONYL (DIETHYLAMINO)OXOACETATE see NNB600
NONYL N,N-DIETHYLOXAMATE see NNB600
o-NONYL-HARMOL HYDROCHLORIDE see NNC100
n-NONYLIC ACID see NMY000
NONYL METHYL KETONE see UKS000
NONYLNAPHTHALENE see NNC600
4-NONYLPHENOL see NNC510
NONYL PHENOL (mixed isomers) see NNC500
p-NONYLPHENOL, BRANCHED see NNC600
NONYLPHENOL ETHOXYLATE see NNC650
NONYLPHENOL POLYETHYLENE GLYCOL ETHER see NNC650
4-NONYLPHENOL POLYMER WITH FORMALDEHYDE AND OXIRANE see FMW343
NONYLPHENOL, POLYOXYETHYLENE ETHER see NND500
23-(4-NONYLPHENOXY)-3,6,9,12,15,18,21-HEPTAOXATRICOSAN-1-OL see NNC700
NONYLPHENOXYPOLY(ETHYLENEOXY)ETHANOL see NNB300
NONYLPHENOXYPOLYETHYLENEOXY ETHANOL-IODINE COMPLEX see NND000
1-(4-NONYLPHENYL)ETHANONE OXIME see NND100
α-(4-NONYLPHENYL)-OMEGA-HYDROXYPOLY(OXY-1,2-ETHANEDIYL)-BRANCHED see NND200
NONYL PHENYL POLYETHYLENE GLYCOL see NND500
NONYL PHENYL POLYETHYLENE GLYCOL ETHER see NND500
NONYLTRICHLOROSILANE see NNE000
2-NONYNAL DIMETHYLACETAL see NNE100
2-NONYNAL, DIMETHYL ACETAL see NNE100
2-NONYN-1-AL DIMETHYLACETAL see NNE100
2-NONYNOIC ACID, METHYL ESTER see MNC000
No. 156 ORANGE CHROME see LCS000
No. 177 ORANGE LAKE see CMM220
NOOTRON see NNE400
NOOTROPIL see NNE400
NOOTROPYL see NNE400
NO-Pip see NLJ500
NO-Pro see NLL500
NOPALCOL 1-0 see PJY100
NOPALCOL 4-0 see PJY250
NOPALCOL 6-0 see PJY100
NOPALCOL 6-L see PJY000
NOPALMATE see PLH500
NOPCAINE see PAQ200
NOPCOCIDE see TBQ750
NOPCOGEN 14-L see DJD500
NOPCO 2272-R see NNE500
NOPCOTE C 104 see CAX350
4-NOPD see ALL500
NO-PEST see DGP900
NO-PEST STRIP see DGP900

NOPIL see TKX000
NOPINEN see POH750
NOPINENE see POH750
NOPOL see DTB800
NOPOL ACETATE see DTC000
NOPOL (POLYMER) see PJS750
NOPOL (TERPENE) see DTB800
NO-PRESS see MBW750
NOPTIL see EOK000
NOPYL ACETATE see DTC000
NO-PYR see NLP500
NOR-1 see MKQ600
NORACYCLINE see LJE000
(−)-NORADREC see NNO500
NORADRENALIN see NNO500
NORADRENALINA (ITALIAN) see NNO500
NORADRENALINE see NNO500
(−)-NORADRENALINE see NNO500
d-(−)-NORADRENALINE see NNO500
l-NORADRENALINE see NNO500
(−)-NORADRENALINE ACID TARTRATE see NNO699
(−)-NORADRENALINE BITARTRATE see NNO699
(+−)-NORADRENALINE BITARTRATE see NNE525
l-NORADRENALINE BITARTRATE see NNO699
dl-NORADRENALINE BITARTRATE see NNE525
(−)-NORADRENALINE BITARTRATE MONOHYDRATE see NNO699
NORADRENALINE HYDROCHLORIDE see NNP000
(±)-NORADRENALINE HYDROCHLORIDE see NNP050
dl-NORADRENALINE HYDROCHLORIDE see NNP050
NORADRENALINE HYDROGEN TARTRATE see NNO699
(−)-NORADRENALINE TARTRATE see NNO699
l-NORADRENALINE TARTRATE see NNO699
NORADRENALINE TARTRATE (1:1) see NNO699
NOR-ADRENALIN HYDROCHLORIDE see NNP000
NORADRENLINE see NNO500
NORAL ALUMINUM see AGX000
NORAL EXTRA FINE LINING GRADE see AGX000
NORAL INK GRADE ALUMINUM see AGX000
NORAL NON-LEAFING GRADE see AGX000
NOR-AM EP 332 see DSO200
NOR-AM EP 333 see CJJ500
NORAMITRIPTYLINE see NNY000
NORAMIUM M 2SH see QAT550
NORAMIUM M 2SH15 see QAT550
NORAM O see OHM700
NORANAT see IBV100
NORANDROLONE PHENYLPROPIONATE see DYF450
NORANDROSTENOLON see NNX400
NORANDROSTENOLONE see NNX400
19-NORANDROSTENOLONE see NNX400
NORANDROSTENOLONE DECANOATE see NNE550
19-NORANDROSTENOLONE DECANOATE see NNE550
NORANDROSTENOLONE PHENYLPROPIONATE see DYF450
NORANTIPYRINE see NNT000
NORARTRINAL see NNO500
NORBILAN see TGZ000
NORBOLDINE see LBO100
(+)-NORBOLDINE see LBO100
NORBOLETHONE see NNE600
NORBORAL see BSM000
NORBORMIDE see NNF000

NORBORNADIENE see NNG000
2,5-NORBORNADIENE see NNG000
2-NORBORNANAMINE, N,2,3,3-TETRAMETHYL-, S-(+)- see MBW778
endo-2-NORBORNANECARBOXYLIC ACID ETHYL ESTER see NNG500
NORBORNANE, 2,2-DIMETHYL-3-METHYLENE-, OCTACHLORO DERIV. see OAH100
2-NORBORNANEMETHANOL see NNH000
2-NORBORNANONE see NNK000
2-NORBORNANONE, 1,3,3-TRIMETHYL-, (1R,4S)-(+)- see FAM300
2-NORBORNENE see NNH500
2-NORBORNENE, 5-(BROMOMETHYL)-1,2,3,4,7,7-HEXACHLORO- see BNQ600
5-NORBORNENE-2-CARBONITRILE see NNH600
5-NORBORNENE-2-CARBOXYLC ACID, 1,4,5,6,7,7-HEXACHLORO-3-(HYDROXYMETHYL)-, γ-LACTONE see EAQ815
trans-5-NORBORNENE-2,3-DICARBONYL CHLORIDE see NNI000
5-NORBORNENE-2,3-DICARBOXIMIDE, 5-(α-HYDROXY-α-2-PYRIDYLBENZYL)-N-(2-METHOXYETHYL)-7-(α-2-PYRIDYLBENZYLIDENE)-, ENDO- see NNI100
5-NORBORNENE-2,3-DICARBOXYLIC ACID, BIS(2-ISOCYANATOETHYL) ESTER see BKJ800
cis-5-NORBORNENE-2,3-DICARBOXYLIC ACID DIMETHYL ESTER see DRB400
5-NORBORNENE-2,3-DICARBOXYLIC ACID, 1,4,5,6,7,7-HEXACHLORO-, DIBUTYL ESTER (8CI) see CDS025
5-NORBORNENE-2,3-DICARBOXYLIC ANHYDRIDE, 1,4,5,6,7,7-HEXACHLORO- see CDS050
5-NORBORNENE-2,3-DIMETHANOL, 1,4,5,6,7,7-HEXACHLORO- see HCC550
5-NORBORNENE-2,3-DIMETHANOL, 1,4,5,6,7,7-HEXACHLORO-, CYCLIC SULFATE see EAQ820
5-NORBORNENE-2,3-DIMETHANOL, 1,4,5,6,7,7-HEXACHLORO-, CYCLIC SULFITE, EXO- see EAQ800
5-NORBORNENE-2,3-DIMETHANOL, 1,4,5,6,7,7-HEXACHLORO-, CYCLIC SULFITE, ENDO- see EAQ810
2-NORBORNENE, 1,2,3,4,7,7-HEXACHLORO-5,6-BIS(CHLOROMETHYL)- see BFY100
5-NORBORNENE-2-METHANOL ACRYLATE see BFY250
5-NORBORNENE-2-METHYLOLACRYLATE see BFY250
α-5-NORBORNEN-2-YL-α-PHENYL-PIPERIDINE PROPANOL HYDROCHLORIDE see BGD750
2-NORBORNYL ACRYLATE see NNI150
NORBORNYLENE see NNH500
NORBUTORPHANOL TARTRATE see NNJ600
NORCAIN see EFX000
NORCAMPHANE see EOM000
2-NORCAMPHANE METHANOL see NNH000
NORCAMPHENE see NNH500
NORCAMPHOR see NNK000
NORCHLORCYCLIZINE see NNK500
NORCHLORCYCLIZINE HYDROCHLORIDE see NNL000
NORCOZINE see CKP500
20-NORCROTALANAN-11,15-DIONE, 3,8-DIDEHYDRO-14,19-DIHYDRO-12,13-DIHYDROXY-, (13-α-14-α)- see DAL350
NORCURON see VGU075
NORDEN see AKT250
NORDETTE see NNL500
NORDHAUSEN ACID (DOT) see SOI500

NORDIALEX see DBL700
NORDIAZEPAM see CGA500
NORDICOL see EDO000
NORDICORT see HHR000
NORDIHYDROGUAIARETIC ACID see NBR000
NORDIHYDROGUAIRARETIC ACID see NBR000
NORDIMAPRIT see NNL400
NORDIOL see NNL500
NORDIOL-28 see NNL500
NORDOPAN see BIA250
NOREA see HDP500
NORENOL see AKT000
NOREPHEDRANE see AOB250
NOREPHEDRANE see BBK000
(−)-NOREPHEDRINE see NNM000
dl-NOREPHEDRINE see NNM500
psi-NOREPHEDRINE see NNM510
NOR-psi-EPHEDRINE see NNM510
d-NOR-psi-EPHEDRINE see NNM510
NOREPHEDRINE HYDROCHLORIDE see NNN000
(−)-NOREPHEDRINE HYDROCHLORIDE see NNN500
(−)-NOREPHEDRINE HYDROCHLORIDE see NNO000
d-NOREPHEDRINE HYDROCHLORIDE see NNO000
l-NOREPHEDRINE HYDROCHLORIDE see NNN500
dl-NOREPHEDRINE HYDROCHLORIDE see PMJ500
NOREPINEPHRINE see NNO500
(−)-NOREPINEPHRINE see NNO500
l-NOREPINEPHRINE see NNO500
(−)-NOREPINEPHRINE BITARTRATE see NNO699
(+−)-NOREPINEPHRINE BITARTRATE see NNE525
l-NOREPINEPHRINE BITARTRATE see NNO699
dl-NOREPINEPHRINE BITARTRATE see NNE525
NOREPINEPHRINE HYDROCHLORIDE see NNP000
(±)-NOREPINEPHRINE HYDROCHLORIDE see NNP050
dl-NOREPINEPHRINE HYDROCHLORIDE see NNP050
(−)-NOREPINEPHRINE TARTRATE see NNO699
NOREPIRENAMINE see NNO500
NORES see HDP500
18-NORESTRONE METHYL ETHER see NNP100
NORETHANDROL see EAP000
NORETHANDROLONE see ENX600
NORETHINDRONE-17-ACETATE see ABU000
NORETHINDRONE ACETATE 3-CYCLOPENTYL ENOL ETHER see QFA275
NORETHINDRONE ACETATE and ETHINYLESTRADIOL see EEH520
NORETHINDRONE mixed with ETHYNYLESTRADIOL see EQM500
NORETHINDRONE mixed with MESTRANOL see MDL750
NORETHINDRONE OXIME see NNQ050
NORETHINDRONE 3-OXIME see NNQ050
NORETHINODREL see EEH550
19-NOR-ETHINYL-4,5-TESTOSTERONE see NNP500
19-NOR-ETHINYL-5,10-TESTOSTERONE see EEH550
NORETHINYNODREL see EEH550
19-NORETHISTERONE see NNP500
19-NORETHISTERONE ACETATE see ABU000
NORETHISTERONE ACETATE mixed with ETHINYL OESTRADIOL see EEH520

NORETHISTERONE ENANTHATE see NNQ000
NORETHISTERONE mixed with ETHINYL OESTRADIOL (60:1) see EQM500
NORETHISTERONE mixed with MESTRANOL see MDL750
NORETHISTERONE OENANTHATE see NNQ000
NORETHISTERONE OXIME see NNQ050
NORETHISTERONE-3-OXIME see NNQ050
19-NORETHYLTESTOSTERONE see ENX600
19-NOR-17-α-ETHYLTESTOSTERONE see ENX600
NORETHYNODRAL see EEH550
NORETHYNODREL see EEH550
19-NORETHYNODREL see EEH550
NORETHYNODREL and ETHINYLESTRADIOL-3-METHYL ETHER (50:1) see EAP000
NORETHYNODREL mixed with MESTRANOL see EAP000
19-NOR-17-α-ETHYNYLANDROSTEN-17-β-OL-3-ONE see NNP500
19-NOR-17-α-ETHYNYL-17-β-HYDROXY-4-ANDROSTEN-3-ONE see NNP500
19-NOR-17-α-ETHYNYLTESTOSTERONE see NNP500
19-NORETHYNYLTESTOSTERONE ACETATE see ABU000
NORETHYSTERONE ACETATE see ABU000
NOREX see CJQ000
NORFENEFRINE HYDROCHLORIDE see NNT100
NORFLEX see DPH000
NORFLOXACIN see BAB625
NORFLURANE see EEC100
NORFLURAZON see NNQ100
NORFORMS see ABU500
NORGAMEM see TEV000
NORGE SALTPETER see CAU000
NORGESTREL see NNQ500
d(−)-NORGESTREL see NNQ500
d-NORGESTREL see NNQ500
d(−)-NORGESTREL see NNQ520
d-NORGESTREL see NNQ520
l-NORGESTREL see NNQ525
(±)-NORGESTREL see NNQ500
dl-NORGESTREL see NNQ500
α-NORGESTREL see NNQ500
NORGESTREL mixed with ETHINYL ESTRADIOL see NNL500
dl-NORGESTREL mixed with ETHINYLESTRADIOL see NNL500
NORGESTREL mixed with MESTRANOL see NNR000
NORGESTRIENONE see NNR125
NORGINE see AFL000
NORGLAUCIN see NNR200
NORGLAUCINE see NNR200
d-N-NORGLAUCINE see NNR200
NORHARMAN see NNR300
NOR-HN2 see BHO250
NOR-HN2 HYDROCHLORIDE see BHO250
NORHOMOEPINEPHRINE HYDROCHLORIDE see AMB000
NORIDIL see UVJ450
NORIDYL see UVJ450
NORIGEST see NNQ000
NORILGAN-S see SNN500
NORIMIPRAMINE see DSI709
NORIMYCIN V see CDP250
NORINYL-1 see MDL750
NORISODRINE see DMV600
NORISODRINE HYDROCHLORIDE see IMR000
NORIT see CBT500
NORIZALPININ see GAZ000
NORKEL see AHK750
NORKOOL see EJC500
NORLANOSTA-5,23-DIENE-3,11,22-TRIONE see COE180

19-NORLANOSTA-5,23-DIENE-3,11,22-TRIONE, 25-(ACETYLOXY)-2,16,20-TRIHYDROXY-9-(HYDROXYMETHYL)-, (2-β,9-β,10-α,16-α,23E)- see COE180
19-NOR-9-β,10-α-LANOSTA-5,23-DIENE-3,11,22-TRIONE, 2-β,16-α,20,25-TETRAHYDROXY-9-METHYL-,25-ACETATE see COE190
NORLESTRIN see EEH520
NORLEUCAMINE see PBV505
NORLEUCINE see AJC950
NORLEUCINE, l- see AJC950
l-(+)-NORLEUCINE see AJC950
l-NORLEUCINE (9CI) see AJC950
ε-NORLEUCINE see AJD000
NORLEUCINE, 6-AMIDINO-, MONOHYDROCHLORIDE, l- see IDA525
NORLEUCINE, 6-AMIDINO-, MONOHYDROCHLORIDE, HYDRATE see IDA525
NOR-LOST HYDROCHLORID (GERMAN) see BHO250
NORLUTATE see ABU000
NORLUTIN see NNP500
NORLUTINE ACETATE see ABU000
NORLUTIN ENANTHATE see NNQ000
NORMABRAIN see NNE400
NORMADATE see HMM500
NORMALIP see ARQ750
NORMAL LEAD ACETATE see LCV000
NORMAL LEAD ORTHOPHOSPHATE see LDU000
NORMAL SUPERPHOSPHATE see SOV500
NORMASTIGMIN see DQY909
NORMASTIGMIN see NCL100
NORMAT see ARQ750
NORMERSAN see TFS350
NORMETANDRONE see MDM350
NORMETHANDROLONE see MDM350
NORMETHANDRONE see MDM350
NORMETHYL EX4442 see POF250
19-NOR-17-α-METHYLTESTOSTERONE see MDM350
NORMETOL see AKT000
NORMI-NOX see QAK000
NORMISON see CFY750
NORMOCYTIN see VSZ000
NORMODYNE see HMM500
NORMO-LEVEL see NNR400
NORMOLIPOL see ARQ750
NORMONSON see CHG000
NORMORESCINA see TLN500
NORMORPHINONE, N-ALLYL-7,8-DIHYDRO-14-HYDROXY-, (−)- see NAG550
NORMOSAN see CHG000
NORMOSON see CHG000
NORMOSTEROL see NNR400
NORMOSTEROLO see NMV300
NORMUSCONE see CPU250
NORNICOTIN see NNR500
NORNICOTINE see NNR500
NORNICOTINE (+)-HYDROCHLORIDE see NNS500
d-NORNICOTINE HYDROCHLORIDE see NNS500
NORNICOTYRINE see PQB500
NOR-NITROGEN MUSTARD see BHN750
NORNITROGEN MUSTARD HYDROCHLORIDE see BHO250
NOROCAINE see AIL750
NORODIN see DBA800
NORODIN see PFP850
NORODIN HYDROCHLORIDE see MDT600
NOROX BZP-250 see BDS000
19-NOR-P see NNT500
NORPACE see RSZ600
NORPENTAZOCINE see NNS600
NORPHEN see AKT250
NORPHENAZONE see NNT000
NORPHENYLEPHRINE see AKT000
(±)-NORPHENYLEPHRINE HYDROCHLORIDE see NNT100

2-NORPINENE-2-CARBOXALDEHYDE, 6,6-DIMETHYL- see FNK150
NORPLANT see NNQ525
NORPRAMIN see DLS600
NORPRAZEPAM see CGA500
(17-α)-19-NORPREGNA-1,3,5(10)-TRIEN-20-YNE-3,17,DIOL see EEH500
19-NOR-17-α-PREGNA-1,3,5(10)-TRIEN-2-YNE-3,17-DIOL see EEH500
19-NORPREGNA-1,3,5(10)-TRIEN-20-YNE-2,17-DIOL, 3-METHOXY-, (17-α)- see HLQ050
19-NOR-17-α-PREGNA-1,3,5(10)-TRIEN-20-YNE-2,17-DIOL, 3-METHOXY- see HLQ050
19-NOR-17-α-PREGNA-4,9,11-TRIEN-20-YN-3-ONE, 17-HYDROXY-, OXIME see OMY815
19-NORPREGNA-4,9,11-TRIEN-20-YN-3-ONE, 17-HYDROXY-, OXIME, (17-α)- see OMY815
19-NORPREGN-4-ENE-3,20-DIONE see NNT500
19-NOR-17-α-PREGN-5(10)-EN-20-YNE-3-α,17-DIOL see NNU000
19-NOR-17-α-PREGN-5(10)-EN-20-YNE-3-β,17-DIOL see NNU500
(3-β,17-α)-19-NORPREGN-4-EN-20-YNE-3,17-DIOL DIACETATE see EQJ500
19-NOR-17-α-PREGN-4-EN-20-YNE-3-β,17-DIOL DIACETATE mixed with 3-METHYOXY-19-NOR-17-α-PREGNA-1,3,5(10)-TRIEN-20-YN-17-OL see EQK010
10-NOR-17-α-PREGN-4-EN-20-YNE-3-β,17-DIOL mixed with 3-METHOXY-17-α-19-NORPREGNA-1-3-5(10)-TRIEN-20-YN-17-OL see EQK100
(17-α)-19-NORPREGN-4-EN-20-YN-17-OL see NNV000
19-NOR-17-α-PREGN-4-EN-20-YN-17-OL see NNV000
19-NOR-17-α-PREGN-4-EN-20-YN-3-ONE, 17-HYDROXY-, 1-ADAMANTANECARBOXYLATE (ESTER), OXIME see HNC500
19-NOR-17-α-PREGN-4-EN-20-YN-3-ONE, 17-HYDROXY-, mixed with 19-NOR-17-α-PREGNA-1,3,5(10)-TRIEN-2-YNE-3,17-DIOL (60:1) see EQM500
19-NOR-17-α-PREGN-4-EN-20-YN-3-ONE, 17-HYDROXY-, OXIME see NNQ050
19-NORPREGN-4-EN-20-YN-3-ONE, 17-HYDROXY-, OXIME, (17-α)- see NNQ050
NOR-PRESS 25 see HGP500
NOR-PROGESTELEA see NCI525
19-NORPROGESTERONE see NNT500
(−)-NORPSEUDOEPHEDRINE see NNV500
NORPSEUDOEPHEDRINE, (+)- see NNM510
d-NORPSEUDOEPHEDRINE see NNM510
(+)-NORPSEUDOEPHEDRINE HYDROCHLORIDE see NNW500
d-NORPSEUDOEPHEDRINE HYDROCHLORIDE see NNW500
1-NOR-PSI-EPHEDRIN (GERMAN) see NNV500
19-NORSPIROXENONE see NNX300
NORSTEARANTHRENE see TCJ500
NORSULFASOL see TEX250
NORSULFAZOLE see TEX250
NORSYMPATHOL see AKT250
NORSYNEPHRINE see AKT000
NORSYNEPHRINE see AKT250
2-NORTALL OIL-1H-IMIDAZOLE-1-ETHANOL, 4,5-DIHYDRO- see TAC050
NORTEC see CDO000
NORTESTONATE see NNX400
NORTESTOSTERONE see NNX400
19-NORTESTOSTERONE see NNX400
(+)-19-NORTESTOSTERONE see NNX400
17-α-NORTESTOSTERONE see ENX600
19-NORTESTOSTERONE-17-N-(2-CHLOROETHYL)-N-NITROSOCARBAMATE see NNX600

NORTESTOSTERONE DECANOATE see NNE550
19-NORTESTOSTERONE DECANOATE see NNE550
19-NORTESTOSTERONE HOMOFARNESATE see NNX650
19-NORTESTOSTERONE PHENYLPROPIONATE see DYF450
NORTHERN COPPERHEAD VENOM see NNX700
NORTIMIL see DLS600
NORTRIPTYLINE see NNY000
NORURON see HDP500
NORVAL see BMA625
NORVAL see DJL000
NORVALAMINE see BPX750
l-NORVALINE, 5-AMINO- see OJS100
NORVALINE, 5-(1,3-DIOXO-2-ISOINDOLINYL)-5-OXO-, dl- see PIA100
l-NORVALINE, 5-HYDROXY-4-OXO- see HND250
NORVALINE, 3-METHYL- see IKX000
NORVALINE, 4-METHYL- see LES000
NORVEDAN see CKI750
NORVINISTERONE see NCI525
NORVINYL see PKQ059
NORVINYL P 6 see AAX175
19-NOR-17-α-VINYLTESTOSTERONE see NCI525
NORWAY SALTPETER see CAU000
NORWEGIAN SALTPETER see CAU000
NORZETAM see NNE400
NO SCALD see DVX800
NOSCAPAL see NOA000
NOSCAPALIN see NOA000
NOSCAPINE see NBP275
NOSCAPINE see NOA000
NOSCAPINE HYDROCHLORIDE see NOA500
NOSCOSED see CLD250
NOSIM see CCK125
NOSOPHENE SODIUM see TDE750
NOSPAN see MOV500
NOSPASM see IPU000
NOSTAL see EHP000
NOSTAL see QCS000
NOSTEL see CHG000
NOSTIN see EHP000
NOSTRAL see QCS000
NOSYDRAST see DNG400
NOTANDRON see AOO475
NOTANDRON-DEPOT see AOO475
NOTARAL see BFD000
NOTARAL see CFZ000
NOTATIN see GFG100
NOTECHIS SCUTATUS VENOM see ARV550
NOTENQUIL see ABH500
NOTENSIL see AAF750
NOTENSIL see ABH500
NOTESIL see ABH500
NOTEZINE see DIW000
NOTICIN see RCK730
NOTOMYCIN A1 see CNV500
NOURALGINE see SEH000
NOURITHION see PAK000
NOURYSET 200 see AGD250
NOVA see MRW775
NOVACRYL RED 2G see BAQ750
NOVACRYSIN see GJG000
NOVADELOX see BDS000
NOVADEX see NOA600
NOVADOX see HGP550
NOVADRAL see AKT000
NOVADRAL see NNT100
NOVAFED see POH000
NOVAFED see POH250
NOVAKOL see FNF000
NOVALICHIN see NOA700
NOVALLYL see AFS500
NOVALON YELLOW 2GN see AAQ250
NOVAMIDON see DOT000
NOVAMIN see APT000

NOVAMIN see DYE600
NOVAMIN see PMF500
NOVAMINE see DYE600
NOVAMONT 2030 see PMP500
NOVANTOINA see DKQ000
NOVANTOINA see DNU000
NOVA-PHENO see EOK000
NOVARSAN see NCJ500
NOVARSENOBENZOL see NCJ500
NOVARSENOBILLON see NCJ500
NOVASMASOL see MDM800
NOVASUROL see CCG500
NOVATEC JUO 80 see PJS750
NOVATEC JVO 80 see PJS750
NOVATHION see DSQ000
NOVATIC BROWN R see CMU780
NOVATONE see MDW750
NOVATRIN see MDL000
NOVATROPINE see MDL000
NOVA W see MRW775
NOVAZOLE see CBA100
NOVECYL see SAH000
NOVEDRIN HYDROCHLORIDE see EJR500
NOVEGE see PBK000
NOVEMBICHIN see NOB700
NOVEMBIKHIN see NOB700
NOVERIL see DCW800
NOVERME see MHL000
NOVERYL see DCW800
NOVESIN see OPI300
NOVESINE see OPI300
NOVIBEN see CBA100
NOVICET see VLF000
NOVICODIN see DKW800
NOVID see ADA725
NOVIDIUM CHLORIDE see HGI000
NOVIDORM see THS800
NOVIGAM see BBQ500
NOVISMUTH see BKW100
NOVIZIR see EIR000
NOVOBIOCIN see SMB000
NOVOBIOCIN MONOSODIUM see NOB000
NOVOBIOCIN, MONOSODIUM SALT see NOB000
NOVOBIOCIN, SODIUM derivative see NOB000
NOVOCAINAMIDE see AJN500
NOVOCAIN-CHLORHYDRAT (GERMAN) see AIT250
NOVOCAINE see AIL750
NOVOCAINE AMIDE see AJN500
NOVOCAINE HYDROCHLORIDE see AIT250
NOVOCAIN HYDROCHLORID (GERMAN) see AIT250
NOVOCAL see CAS750
NOVOCAMID see AJN500
NOVOCAMID HYDROCHLORIDE see PME000
NOVOCEBRIN HYDROCHLORIDE see DXI800
NOVOCHLOROCAP see CDP250
NOVOCILLIN see BFD250
NOVOCOLIN see DAL000
NOVOCONESTRON see ECU750
NOVODIL see DNU100
NOVODIPHENYL see DNU000
NOVODOLAN see TKG000
NOVODRIN see DMV600
NOVOEMBICHIN see NOB700
NOVOHEPARIN see HAQ500
NOVOHETRAMIN see RDU000
NOVOL see OBA000
NOVOLEN see PMP500
NOVOL KETONE see NOB800
NOVOL POE 20 see PJW500
NOVOMAZINA see CKP250
NOVOMYCETIN see CDP250
NOVON see PBK000
NOVON 712 see PKQ059
NOVONAL see DJU200
NOVONIDAZOL see MMN250

NOVOPERM YELLOW FGL see CMS214
NOVOPHENICOL see CDP250
NOVOPHENYL see BRF500
NOVOPHONE see SOA500
NOVO-R see SMB000
NOVOSAXAZOLE see SNN500
NOVOSCABIN see BCM000
NOVOSED see MDQ250
NOVOSERIN see CQH000
NOVOSERPINA see RCA200
NOVOSIR N see EIR000
NOVOSPASMIN see NOC000
NOVOVANILLIN see NOC100
NOVOX see BSC500
NOVYDRINE see BBK000
NOXA see NBJ700
2-NOXA see NBJ700
NOXABEN see DGE200
NOXAL see DXH250
NOXFISH see RNZ000
NOXIPTILINE HYDROCHLORIDE see DPH600
NOXIPTILIN HYDROCHLORID (GERMAN) see DPH600
NOXIPTYLINE HYDROCHLORIDE see DPH600
NOXODYN see TEH500
NOXOKRATIN see EQL000
NOXYLIN see UTU500
NOXYRON see DYC800
NP see NIY700
1-NP see NIX500
2-NP see ALL750
NP 2 see NCW500
2-NP see NIY000
NP-9 see NNB300
NP-48 see FBP350
NP 55 see CDK800
NP 212 see CJR909
NP 246 see CJN750
NP 250 see CJN800
NPA see DNB000
NPA see NBL200
NPA see SIO500
NPA-3 see SIO500
NPAB see PNC925
NPA, SODIUM SALT see SIO500
NPD see TED500
NPG see DTG400
NPGB see NIQ500
NPH 83 see MJG500
NPH-1091 see BDJ250
NPH ILETIN see IDF325
NPH INSULIN see IDF325
NPH 50 INSULIN see IDF325
NPIP see NLJ500
NPLG see PHY400
NP-48 NA see FBP350
NPOH see DVD200
NPP see DYF450
2-NPPD see ALL750
NPRO see NLL500
NaPSt see SMQ500
NPU see NLO500
NPYR see NLP500
4-NQO see NJF000
NR.C 2294 see DNA800
NRD-18 see TBM100
NRDC 104 see BEP500
NRDC 107 see BEP750
NRDC 119 see RDZ875
NRDC 124 see CQD930
NRDC 146 see PCK080
NRDC 147 see PCK075
NRDC 148 see PCK060
NRDC 149 see RLF350
NRDC 157 see CQD900
NRDC 159 see DFG300
NRDC 160 see RLF350
NRDC 161 see DAF300
NRDC 166 see RLF350
NRDC 167 see PCK065

NRL-18-B see TNI000
NS-2 see CEO120
NS-6 see NFL600
NS 11 see DTG000
NS-21 see UTU700
Ni 0901-S see NCW500
NS 200 (filler) see CAT775
NS (carbonate) see CAT775
NS 100 (carbonate) see CAT775
NSAR see NLR500
NSC-185 see CPE750
NSC-339 see DVF200
NSC-423 see DAA800
NSC-729 see PIB925
NSC-739 see AMG750
NSC-740 see MDV500
NSC-741 see HHQ800
NSC-742 see ASA500
NSC-743 see POJ500
NSC-744 see CKV500
NSC 746 see UVA000
NSC-749 see AJO500
NSC-750 see BOT250
NSC-751 see EEI000
NSC-752 see AMH250
NSC-753 see POJ250
NSC-755 see POK000
NSC-756 see THT350
NSC-757 see CNG938
NSC-759 see BCB750
NSC-762 see BIE250
NSC-762 see BIE500
NSC-763 see DUD800
NSC-1026 see AJK250
NSC-1390 see ZVJ000
NSC-1393 see POM600
NSC-1532 see DUZ000
NSC-1895 see DCF200
NSC-2066 see TEP000
NSC-2083 see AMM250
NSC-2100 see NGE500
NSC-2101 see HMY000
NSC-2107 see NGE000
NSC-2666 see DDV800
NSC-2752 see FOU000
NSC-3051 see MKG500
NSC-3053 see AEB000
NSC-3055 see AEI000
NSC-3058 see BDH250
NSC-3060 see PKV500
NSC-3061 see TGD000
NSC-3063 see DAY825
NSC-3069 see CDP250
NSC-3070 see DKA600
NSC-3072 see SFA000
NSC-3073 see FMT000
NSC-3088 see CDO500
NSC-3094 see DTP000
NSC-3095 see IBC000
NSC-3096 see MIW500
NSC-3138 see DOO800
NSC 3233 see TNI300
NSC-3424 see QDJ000
NSC-3425 see THR750
NSC-4170 see MMD500
NSC-4320 see SMN002
NSC-4729 see TES000
NSC-4730 see EGI000
NSC-4911 see MCQ500
NSC-5088 see SAL500
NSC-5159 see CDK250
NSC-5230 see EQA000
NSC-5340 see AHL000
NSC-5354 see PEY250
NSC-5356 see DSB000
NSC-5366 see NBP275
NSC-5366 see NOA000
NSC-6091 see SOA500
NSC 6199 see DDM425
NSC-6396 see TFQ750
NSC-6470 see NDY000
NSC-6738 see DGP900

NSC-7365 see DCQ400
NSC-7571 see AHS750
NSC-7760 see POS750
NSC-7764 see LAQ000
NSC-7778 see MES000
NSC-8028 see IIM000
NSC-8194 see EQL500
NSC-8260 see MGA250
NSC-8491 see THY100
NSC-8652 see DSF300
NSC-8806 see BHV250
NSC-8806 see PED750
NSC-8819 see ADR000
NSC-9166 see TBG000
NSC-9169 see HOI000
NSC-9324 see AFS500
NSC-9369 see MMP000
NSC-9659 see ILD000
NSC-9698 see MAW750
NSC-9701 see MPN500
NSC-9704 see PMH500
NSC 9706 see TND500
NSC 9717 see TND250
NSC-9895 see EDO000
NSC-10023 see PLZ000
NSC-10107 see CFA500
NSC-10107 see CFA750
NSC-10108 see CLO750
NSC-10483 see CNS750
NSC-10815 see AMX750
NSC-10873 see BHN750
NSC-10873 see BHO250
NSC-11595 see PNW800
NSC-11905 see HLY500
NSC-12165 see AOO275
NSC-12169 see EDU500
NSC-13002 see DOS800
NSC-13252 see CMB000
NSC-13875 see HEJ500
NSC-14083 see SLW500
NSC-14210 see BHT750
NSC-14574 see SBE500
NSC-14575 see EMC500
NSC-15193 see DAR600
NSC-15432 see EEH550
NSC-15747 see BFL125
NSC-15780 see AOD500
NSC-16498 see BIA000
NSC-16895 see LGZ000
NSC-17118 see CLD500
NSC-17262 see BDC750
NSC-17591 see TBF750
NSC-17592 see HNT500
NSC-17661 see GHE000
NSC-17663 see BHN500
NSC-17777 see PGG355
NSC-18016 see CHC750
NSC-18268 see AEA750
NSC-18321 see BHB750
NSC-18429 see AOQ875
NSC-18439 see BIC600
NSC-19477 see HEG000
NSC-19488 see BRS500
NSC-19494 see MQR100
NSC-19893 see FMM000
NSC-19962 see CBO625
NSC-19987 see MOR500
NSC-20264 see AOA125
NSC-20526 see THW750
NSC-20527 see BOD600
NSC-21206 see ALL250
NSC-21402 see AKM000
NSC-22314 see TCF000
NSC-22420 see DCP400
NSC-23519 see DCQ600
NSC 23623 see DMV500
NSC-23890 see DSU000
NSC-23891 see BHQ300
NSC-23892 see BCC250
NSC-23909 see MNA750
NSC 23989 see BHD150
NSC-24343 see OLW100

NSC-24559 see MQW750
NSC-24639 see NLC000
NSC-24818 see PJJ225
NSC-24890 see NKE500
NSC-24970 see BMZ000
NSC-25154 see BHJ250
NSC-25855 see TEV000
NSC-25958 see NLB700
NSC-26154 see AJD000
NSC-26198 see PAN100
NSC-26271 see CQC500
NSC-26271 see CQC650
NSC-26805 see EMF500
NSC-26806 see VMA000
NSC-26812 see AQO000
NSC-26980 see AHK500
NSC-27640 see DAR400
NSC-28693 see MRH000
NSC-29215 see TND000
NSC-29421 see BDG325
NSC-29422 see TFJ500
NSC-29630 see DFO000
NSC-29863 see GKS000
NSC-30152 see PMO250
NSC-30211 see TNF500
NSC 30551 see MJM700
NSC-30622 see SBX525
NSC-31083 see AEA109
NSC-31712 see TCQ500
NSC-32065 see HOO500
NSC-32074 see RJA000
NSC-32606 see BOP750
NSC-32743 see ABN000
NSC-32946 see MKI000
NSC-33531 see DMH600
NSC-33659 see LBD100
NSC-33669 see EAL500
NSC-33669 see EAN000
NSC-33674 see NLD000
NSC-34372 see DFH000
NSC-34391 see KHU050
NSC-34462 see BIA250
NSC-34533 see GKE000
NSC-34652 see MKB250
NSC-34699 see NLC500
NSC-35051 see BHU750
NSC-35051 see SAX200
NSC-35770 see CMX700
NSC-36354 see DBA175
NSC-37095 see EHV500
NSC-37448 see PDT250
NSC-37538 see BKM500
NSC-37725 see NNV000
NSC-38191 see ENU000
NSC-38270 see OIU499
NSC-38721 see CDN000
NSC-38887 see AKY250
NSC-39069 see TFU500
NSC-39084 see ASB250
NSC-39147 see SMC500
NSC-39470 see BFV750
NSC-39661 see DAS000
NSC-39690 see DIG400
NSC-40774 see MPU000
NSC-40902 see AOO500
NSC-42076 see BER500
NSC-42722 see DAL300
NSC-44185 see EAM500
NSC-44690 see DIS200
NSC-45383 see SMA000
NSC-45384 see SMA500
NSC-45388 see DAB600
NSC-45403 see ENV000
NSC-45463 see DOY600
NSC-45624 see SKI500
NSC-46015 see TCW500
NSC-47547 see CHE750
NSC 47842 see VKZ000
NSC-48841 see BHP250
NSC-49171 see SAN000
NSC 49842 see VLA000
NSC-50256 see MLH500

NSC-50364 see MMN250
NSC-50413 see AQY250
NSC-50982 see BHW500
NSC-51143 see IAN100
NSC-52695 see CQN000
NSC-52760 see DAK200
NSC-52947 see PAB500
NSC-53396 see AEA250
NSC 56408 see TNY500
NSC-56410 see MLY000
NSC-56654 see ASO510
NSC-57199 see BHT250
NSC-58404 see DCO800
NSC-58514 see CMK650
NSC-58775 see DLY000
NSC-59729 see SKX000
NSC-60195 see MQG500
NSC-60380 see CMA250
NSC-60520 see DGH400
NSC-62209 see BIF250
NSC-62579 see DJB800
NSC-62580 see DRO200
NSC-62709 see IPR000
NSC-62939 see TKX250
NSC-63346 see DQD200
NSC-63701 see VGZ000
NSC-63878 see AQQ750
NSC-63878 see AQR000
NSC-64375 see BDW000
NSC-64393 see SLC000
NSC-65104 see SBZ000
NSC-65346 see SAU000
NSC-65426 see PML250
NSC-66847 see TEH500
NSC-67239 see THM750
NSC-67574 see LEY000
NSC-67574 see LEZ000
NSC-68075 see TEH250
NSC-68626 see ADA000
NSC-69187 see MEL775
NSC-69536 see MLJ500
NSC-69811 see MKW250
NSC-69856 see NBV500
NSC-69945 see PHA750
NSC-70731 see LGD000
NSC-70762 see TGJ500
NSC-70845 see NMV500
NSC-71045 see HKQ025
NSC-71047 see HOM270
NSC-71261 see TFJ250
NSC-71423 see VTF000
NSC-71795 see EAI850
NSC-71936 see CMQ725
NSC-71964 see PCU425
NSC-72005 see TIL500
NSC-73438 see NKL000
NSC-74437 see DHE100
NSC 75520 see TKH325
NSC-76098 see ARW750
NSC-76411 see OIU000
NSC-77213 see PME250
NSC-77213 see PME500
NSC-77471 see MQX000
NSC-77517 see MKN750
NSC-77518 see DCK759
NSC-78409 see BIH325
NSC-78559 see DAB800
NSC-78572 see HIZ000
NSC-79019 see PQC000
NSC-79037 see CGV250
NSC-79389 see ARQ750
NSC-80087 see DOT600
NSC-81430 see CQJ500
NSC-82116 see KFA100
NSC-82151 see DAC000
NSC-82151 see DAC200
NSC-82174 see EID000
NSC 82190 see NJW700
NSC-82196 see IAN000
NSC-82261 see GCS000
NSC-82699 see TKH750
NSC 83265 see TNR475

NTG see NGY000
NTL see SBD000
NTL SULFATE see NCP550
NTM see DTR200
NTN 80 see AMX300
NTN 801 see BDG100
NTN 5006 see AMX400
NTN 6867 see AMX300
NTN-8629 see DGC800
NTN 19701 see UTA400
NTN 33893 see CKW400
NTN 33894 see CKW410
NTO see NMP620
NTO (DOT) see NMP620
NTOI see NML000
NTPA see NOC400
NTPP see DYF450
NTS 40 see PJT300
NU-1196 see NEA500
NU-1196 see NEB000
NU-1932 see NOE550
NU 2017 see ABV250
NU-2121 see NDW510
NU 2206 see MDV250
NU 2206 see MKR250
NU-2221 see DDF200
NU 2222 see TKW500
NUARIMOL see CHJ300
NUARSOL see ARA500
NU-BAIT II see MDU600
NUBARENE see MIH925
NUBIAN BLACK BT see PCJ200
NUBIAN YELLOW TB see PEJ600
NUBILON ORANGE R see CMM220
NUCES NUCISTAE see NOG000
NUCHAR see CBT500
NUCHAR 722 see CDI000
NUCIDOL see DCM750
NUCIFERIN see NOE500
NUCIFERINE see NOE500
(−)-NUCIFERINE see NOE500
1-NUCIFERINE see NOE500
NUCIN see WAT000
NUCLEAR FAST RED see AJQ250
NUCLEASE, DEOXYRIBO- see EDK650
NUCLEASE, RESTRICTION
ENDODEOXYRIBO-, HINCII see REF260
NUCLEASE, RESTRICTION
ENDODEOXYRIBO-, HINDIII see REF262
NUCLEASE, RESTRICTION
ENDODEOXYRIBO-, MSEI see REF265
NUCLEASE, RESTRICTION
ENDODEOXYRIBO-, RSAI see REF270
NUCLEIC ACIDS, RIBO- see RJA600
NUCLEOCARDYL see AEH750
NUCLON ARGENTINIAN see FNF000
NUC SILICONE VS 7158 see DAF350
NUC SILICONE VS 7207 see OCE100
NUCTALON see CKL250
NUDRIN see MDU600
NUEZ de la INDIA (PUERTO RICO) see
TOA275
NUEZ NOGAL (PUERTO RICO) see TOA275
NUFENOXOLE see DWF700
NU-FLOW M see MRW775
NUFLUOR see SHF500
NUGESTORAL see GEK500
NU-1326 HYDROBROMIDE see NOE525
NU 1604 HYDROCHLORIDE see TEQ700
NU/ICRF 500 see DLO970
NUISANCE DUSTS and AEROSOLS see
NOF000
NUITAL see ELQ600
NUJOL see MQV750
NULANS see CFY500
NU-LAWN WEEDER see DDP000
NULLAPON B see EIV000
NULLAPON BF-78 see EIV000
NULLAPON BF ACID see EIX000
NULLAPON BFC CONC see EIV000
NULOGYL see NHH000
NULSA see BGC625

NUMAL see AFT000
NU MAN see MPN500
NU-MANESE see MAT250
NUMBER 2 BURNER FUEL see DHE800
NUMBER 2 FUEL OIL see DHE800
NUMOQUIN see HHR700
NUMOQUIN HYDROCHLORIDE see
ELC500
NUMORPHAN HYDROCHLORIDE see
ORG100
NUODEX V 1525 see DEJ100
NUOPLAZ see DXQ200
NUOPLAZ DOP see DVL700
NUPERCAINAL see DDT200
NUPERCAINE see DDT200
NUPERCAINE HYDROCHLORIDE see
NOF500
NUPOL 1629 see BLD810
NUPOL 46-4005 see BLD810
NUPRIN see AIE750
NURAN see PCI250
NUREDAL see BET000
NURELLE see CIC600
NURELLE see TIV750
NU REXFORM see LCK100
NUSTAR see THS860
NUSYN-NOXFISH see PIX250
NUTINAL see BCA000
NUTMEG see NOG000
NUTMEG OIL see NOG500
NUTMEG OIL, EAST INDIAN see NOG500
NU-TONE see NAK500
NUTRASWEET see ARN825
NUTRIFOS STP see SJH200
NUTROP see SBH500
NUTROSE see SFQ000
NUVA see DGP900
NUVACRON see MRH209
NUVACRON 20 see MRH209
NUVANOL see DSQ000
NUVANOL N see IEN000
NUVAPEN see AIV500
NUVELBI V.C.A. see DXO300
NUX MOSCHATA see NOG000
NUX VOMICA see SMN500
NUX-VOMICA TREE see NOG800
N 4000V see SMQ500
NVC 9025 see PJS750
NVP see NBO525
NYACOL see SCH002
NYACOL 830 see SCH002
NYACOL 1430 see SCH002
NYACOL A 1530 see AQF000
NYAD 10 see WCJ000
NYAD 325 see WCJ000
NYA G see WCJ000
NYANTHRENE BLUE BFP see DFN300
NYANTHRENE BRILLIANT VIOLET 4R see
DFN450
NYANTHRENE BROWN R see CMU780
NYANTHRENE BROWN RB see CMU770
NYANTHRENE GOLDEN YELLOW see
DCZ000
NYANTHRENE OLIVE R see DUP100
NYANTHRENE RED G 2B see CMU825
NYANZA BLUE RW see CMO600
NYANZA FAST BROWN M see CMO800
NYANZA FAST RED FA see CMO870
NYANZA FAST SCARLET 4BSA see CMO870
NYANZA FAST TURQUOISE GL see COF420
NYAZIN see BET000
NYCOLINE see PJT200
NYCO Liquid RED GF see RGW000
NYCOR 200 see WCJ000
NYCOR 300 see WCJ000
NYCOTON see CDO000
NYCTAL see BNK000
NYDRAN see BEG000
NYDRANE see BEG000
NYDRAZID see ILD000
NY IV-34-1 see TEH250
NYLIDRIN HYDROCHLORIDE see DNU200

NYLMERATE see ABU500
NYLOCROM YELLOW 3R see SGP500
NYLOMINE ACID BLUE B-B see CMM070
NYLOMINE ACID RED C-R see CMO870
NYLOMINE ACID RED P4B see HJF500
NYLOMINE ACID SCARLET C-R see
CMM320
NYLOMINE ACID SCARLET P-R see
CMM320
NYLOMINE ACID YELLOW B-RD see
SGP500
NYLOMINE BLUE A 2B see CMM090
NYLOMINE BLUE A 3R see CMM100
NYLOMINE YELLOW A-G see FAL050
NYLON see NOH000
NYLON-6 see PJY500
NYLON SALT see NOH100
NYLON 66 SALT see HEG120
NYLON 610 SALT see HEG130
NYLOQUINONE BLUE 2J see TBG700
NYLOQUINONE BLUE 4J see DMM400
NYLOQUINONE BORDEAUX B see CMP080
NYLOQUINONE LIGHT YELLOW 4JL see
AAQ250
NYLOQUINONE ORANGE JR see AKP750
NYLOQUINONE PINK B see AKO350
NYLOQUINONE PURE BLUE see MGG250
NYLOQUINONE RED N see ENP100
NYLOQUINONE VIOLET R see DBP000
NYLOQUINONE YELLOW 2R see DUW500
NYLOQUINONE YELLOW 4J see AAQ250
NYLOSAN BLUE C-L see CMM092
NYLOSAN BLUE F-L see CMM092
NYLOSAN GREEN F-BL see CMM200
NYLOSAN YELLOW E-3R see SGP500
NYMCEL S see SFO500
NYMCEL WV-L see AGW700
NYSCAPS see WAK000
NYSCONITRINE see NGY000
NYSTAN see NOH500
NYSTATIN see NOH500
NYSTATINE see NOH500
NYSTAVESCENT see NOH500
NYTAL see TAB750
NZ see CAT775
O 250 see SKV100
O-2857 see BDE500
OA-A 1102 see CAT775
OAAT see AIC250
OAP see HED500
OAT EXTRACT see OAD090
OBB see OAH000
OBDZ see OMY700
OBELINE PICRATE see ANS500
OBEPAR see DKE800
OBESIN see PNN300
OBESITABS see AOB500
OBIDOXIME CHLORIDE see BGS250
OBIDOXIME DICHLORIDE see BGS250
OBIDOXIME HYDROCHLORIDE see
BGS250
OBLEVIL see EQL000
OBLITEROL see EMT500
m-OBLIVON see EQL000
OBLIVON C see MNM500
OBLIVON CARBAMATE see MNM500
OBOB see NLE400
OBPA see OMY850
OBRACINE see TGI500
OBRAMYCIN see TGI250
OBSH see OPE000
OBSTON see DJL000
OC 1085 see MQT550
OC-11588 see AKC570
OCDD see OAJ000
OCENOL see OBA000
OCEOL see OBA000
OCHRATOXIN see OAD095
OCHRATOXIN A see CHP250
OCHRATOXIN A SODIUM SALT see
OAD100
OCHRE see IHC450

OCI 56 see SGG500
OCIMENE see DTF200
OCIMENOL see DTC990
OCIMUM BASILICUM OIL see BAR250
OCIMUM SANCTUM LINN., LEAF
EXTRACT see OAD200
OCNA see CJA200
OCOTEA CYMBARUM OIL see OAF000
OCPA see CEH750
OCPNA see CJA175
OCRITEN see DAQ800
OCTA see CPB120
OCTAACETYLSUCROSE see OAF100
OCTABENZONE see HND100
OCTABROMOBIPHENYL see OAH000
ar,ar,ar,ar,ar',ar',ar'-OCTABROMO-1,1'-
BIPHENYL see OAH000
OCTABROMOBIPHENYL ETHER see
OAF200
OCTABROMODIPHENYL see OAH000
OCTABROMODIPHENYL ETHER see
OAF200
OCTABROMODIPHENYL OXIDE see
OAF200
OCTACAINE HYDROCHLORIDE see
MLO750
OCTACARBONYLDICOBALT see CNB500
1,2,4,5,6,7,8,8-OCTACHLOOR-3a,4,7,7a-
TETRAHYDRO-4,7-endo-METHANO-
INDAAN (DUTCH) see CDR750
OCTACHLOR see CDR750
OCTACHLOR EPOXIDE see OPK275
OCTACHLOROCAMPHENE see CDV100
OCTACHLOROCAMPHENE see OAH100
OCTACHLORODIBENZODIOXIN see
OAJ000
OCTACHLORODIBENZO-p-DIOXIN see
OAJ000
OCTACHLORODIBENZO(b,e)(1,4)DIOXIN
see OAJ000
1,2,3,4,6,7,8,9-
OCTACHLORODIBENZODIOXIN see
OAJ000
OCTACHLORODIHYDRODICYCLOPENTA
DIENE see CDR750
OCTACHLORODIPROPYLETHER see
OAL000
1,2,4,5,6,7,8,8-OCTACHLORO-2,3,3a,4,7,7a-
HEXAHYDRO-4,7-METHANOINDENE see
CDR750
1,2,4,5,6,7,8,8-OCTACHLORO-2,3,3a,4,7,7a-
HEXAHYDRO-4,7-METHANO-1H-INDENE
see CDR750
OCTACHLORO-HEXAHYDRO-
METHANOISOBENZOFURAN see OAN000
1,3,4,5,6,8,8-OCTACHLORO-1,3,3a,4,7,7a-
HEXAHYDRO-4,7-
METHANOISOBENZOFURAN see OAN000
1,2,4,5,6,7,8,8-OCTACHLORO-3a,4,7,7a-
HEXAHYDRO-4,7-METHYLENE INDANE
see CDR750
OCTACHLORO-4,7-
METHANOHYDROINDANE see CDR750
OCTACHLORO-4,7-
METHANOTETRAHYDROINDANE see
CDR750
1,2,4,5,6,7,8,8-OCTACHLORO-4,7-
METHANO-3a,4,7,7a-
TETRAHYDROINDANE see CDR750
1,3,4,5,6,7,10,10-OCTACHLORO-4,7-endo-
METHYLENE-4,7,8,9-
TETRAHYDROPHTHALAN see OAN000
OCTACHLORONAPHTHALENE see
OAP000
1,3,4,5,6,7,8,8-OCTACHLORO-2-OXA-
3a,4,7,7a-TETRAHYDRO-4,7-
METHANOINDENE see OAN000
OCTACHLOROSTYRENE see OAP050
1,2,4,5,6,7,8,8-OCTACHLORO-3a,4,7,7a-
TETRAHYDRO-4,7-METHANOINDAN see
CDR750

2,2,4,5,6,7,8,8-OCTACHLORO-3a,4,7,7a-
TETRAHYDRO-4,7-METHANOINDAN see
CDR575
1,2,4,5,6,7,8,8-OCTACHLORO-3a,4,7,7a-
TETRAHYDRO-4,7-METHANOINDANE see
CDR750
1,2,4,5,6,7,10,10-OCTACHLORO-4,7,8,9-
TETRAHYDRO-4,7-METHYLENEINDANE
see CDR750
1,2,4,5,6,7,8,8-OCTACHLOR-3a,4,7,7a-
TETRAHYDRO-4,7-endo-METHANO-
INDAN (GERMAN) see CDR750
OCTACIDE 264 see OES000
9,12-OCTADECADIENAMIDE, N-(1-
PHENYLETHYL)-, (9Z,12Z)- see MHO200
9,12-OCTADECADIENOIC ACID see
LGG000
cis-9,cis-12-OCTADECADIENOIC ACID see
LGG000
cis,cis-9,12-OCTADECADIENOIC ACID see
LGG000
3,13-OCTADECADIEN-1-OL, ACETATE see
OAP070
(Z,Z)-3,13-OCTADECADIENOL ACETATE
see OAP070
(Z,Z)-3,13-OCTADECADIEN-1-OL
ACETATE see OAP070
3,13-OCTADECADIEN-1-OL, ACETATE,
(Z,Z)- see OAP070
OCTADECAFLUORODECAHYDRONAPHT
HALENE see PCG700
OCTADECAFLUOROOCTANE see OAP100
OCTADECANAMINE ACETATE see OAP300
1-OCTADECANAMINIUM, N,N-
DIMETHYL-N-OCTADECYL-, CHLORIDE
(9CI) see DXG625
1-OCTADECANAMINIUM, N,N,N-
TRIETHYL-, IODIDE (9CI) see OBI100
OCTADECANE, 7,8-EPOXY-2-METHYL-,
cis-(8CI) see ECB200
OCTADECANENITRILE see SLL500
OCTADECANEUROPEPTIDE DIAZEPAM-
BINDING see OAP400
OCTADECANNAMIDE see OAR000
OCTADECANOIC ACID see SLK000
OCTADECANOIC ACID, ALUMINUM SALT
see AHH825
OCTADECANOIC ACID, AMMONIUM
SALT see ANU200
OCTADECANOIC ACID, BARIUM
CADMIUM SALT (4:1:1) (9CI) see BAI800
OCTADECANOIC ACID, BARIUM SALT
(9CI) see BAO825
OCTADECANOIC ACID, BUTYL ESTER
(9CI) see BSL600
OCTADECANOIC ACID, CADMIUM SALT
see OAT000
OCTADECANOIC ACID, CALCIUM SALT
see CAX350
OCTADECANOIC ACID, 9,10-EPOXY-, cis-
see OAT100
OCTADECANOIC ACID, 2-ETHYLHEXYL
ESTER (9CI) see OFU300
OCTADECANOIC ACID, ISOHEXADECYL
ESTER (9CI) see IKC050
OCTADECANOIC ACID, LITHIUM SALT see
LHQ100
OCTADECANOIC ACID, MAGNESIUM
SALT see MAJ030
OCTADECANOIC ACID, METHYL ESTER
see MJW000
OCTADECANOIC ACID, 1-METHYLETHYL
ESTER (9CI) see IRL100
OCTADECANOIC ACID, MONOESTER with
1,2-PROPANEDIOL see SLL500
OCTADECANOIC ACID, MONOESTER with
1,2,3-PROPANETRIOL see OAV000
OCTADECANOIC ACID, SODIUM SALT see
SJV500
OCTADECANOIC ACID, TETRADECYL
ESTER (9CI) see TCB100

OCTADECANOIC ACID, ZINC SALT see
ZMS000
OCTADECANOL see OAX000
1-OCTADECANOL see OAX000
n-OCTADECANOL see OAX000
OCTADECANONITRILE see SLL500
5,7,11,13-OCTADECATETRAYNE-1,18-DIOL
see OAX050
9,12,15-OCTADECATRIENEPEROXOIC
ACID (Z,Z,Z) (9CI) see PCN000
9,12,15-OCTADECATRIENOIC ACID see
OAX100
9-OCTADECENAMIDE, N,N-BIS(2-
HYDROXYETHYL)-, (Z)- see OHU200
9-OCTADECENENITRILE, (Z)- see OAX300
(Z)-9-OCTADECENENITRILE see OAX300
9,10-OCTADECENOIC ACID see OHU000
cis-OCTADEC-9-ENOIC ACID see OHU000
cis-9-OCTADECENOIC ACID see OHU000
trans-OCTADEC-9-ENOIC ACID see EAF500
trans-9-OCTADECENOIC ACID see EAF500
cis-Δ9-OCTADECENOIC ACID see OHU000
trans-Δ9-OCTADECENOIC ACID see EAF500
9-OCTADECENOIC ACID (Z)-, compd. with
1-BUTANAMINE (1:1) (9CI) see OHU100
(Z)-9-OCTADECENOIC ACID BUTYL
ESTER see BSB000
9-OCTADECENOIC ACID CALCIUM SALT
see CAU300
9-OCTADECENOIC ACID, 12-HYDROXY-,
(Z)- see RJP000
9-OCTADECENOIC ACID (Z)-, 1-
(HYDROXYMETHYL)-1,2-ETHANEDIYL
ESTER, (S)- see SED525
(Z)-9-OCTADECENOIC ACID METHYL
ESTER see OHW000
9-OCTADECENOIC ACID (Z)-,
MONOESTER with 1,2,3-PROPANETRIOL
(9CI) see GGR200
9-OCTADECENOIC ACID (Z)-, POTASSIUM
SALT see OHY000
(Z)-9-OCTADECENOIC ACID, TIN (2+)
SALT see OAZ000
(Z)-9-OCTADECENOIC ACID mixed with
(Z,Z)-9,12-OCTADECADIENOIC ACID see
LGI000
(Z)-9-OCTADECEN-1-OL see OBA000
cis-9-OCTADECEN-1-OL see OBA000
cis-9-OCTADECENYLAMINE see OHM700
(Z)-α-9-OCTADECENYL-ω-
HYDROXYPOLY(OCY-1,2-ETHANEDIYL)
see OIG000
α-9-OCTADECENYL-ω-
HYDROXYPOLY(OXY-1,2-ETHANEDIYL,
(Z) see PJW500
(Z)-α-9-OCTADECENYL-ω-
HYDROXYPOLY(OXY-1,2-ETHANEDIYL)
see OIG040
α-9-OCTADECENYL-ω-
HYDROXYPOLY(OXY(METHYL-1,2-
ETHANEDIYL)) see UHA000
α,α'-((9-OCTADECENYLIMINO)DI-2,1-
ETHANEDIYL)BIS(ω-HYDROXY-
POLY(OXY-1,2-ETHANEDIYL) see OBA100
OCTA DECYL ALCOHOL see OAX000
n-OCTADECYL ALCOHOL see OAX000
OCTADECYLAMINE see OBC000
1-OCTADECYLAMINE see OBC000
n-OCTADECYLAMINE see OBC000
OCTADECYLAMINE, ACETATE see OAP300
OCTADECYLAMINE HYDROCHLORIDE
see OBE000
N-OCTADECYL-N-BENZYL-N,N-
DIMETHYLAMMONIUMCHLORIDE see
DTC600
OCTADECYLDIMETHYLBENZYLAMMONI
UM CHLORIDE see DTC600
α-OCTADECYLETHER of GLYCEROL see
GGA915
1-O-OCTADECYLGLYCEROL see GGA915
OCTADECYL ISOCYANATE see OBG000

3-(OCTADECYLOXY)-1,2-PROPANEDIOL see GGA915

OCTADECYL SODIUM SULFATE see OBG100

OCTADECYLTRICHLOROSILANE see OBI000

N-OCTADECYLTRIETHYLAMMONIUM IODIDE see OBI100

OCTADECYLTRIMETHYLAMMONIUM CHLORIDE see TLW500

OCTADECYLTRIMETHYLAMMONIUM p-FLUOROBENZENESULFONATE see OBI200

1-OCTADEDCYLAMINE ACETATE see OAP300

1,2,3,3a,6,7,12b,12c-OCTADEHYDRO-2-HYDROXYLYCORANIUM see LJB800

2,6-OCTADIENAL, 3,7-DIMETHYL-, (E)- see GCU100

2,6-OCTADIENAL, 3,7-DIMETHYL-, DIETHYL ACETAL see CMS323

1,7-OCTADIENE see OBK000

2,6-OCTADIENE, 1,1-DIETHOXY-3,7-DIMETHYL-(9CI) see CMS323

1,6-OCTADIENE, 3,7-DIMETHYL- see CMT050

1,6-OCTADIENE, 3,7-DIMETHYL-3-(1-ETHOXYETHOXY)- see ELZ050

2,6-OCTADIENE, 1-ETHOXY-3,7-DIMETHYL- see GDG100

2,6-OCTADIENOIC ACID, 3,7-DIMETHYL-, ISOPENTYL ESTER, (E)- see IHS100

2,6-OCTADIEN-1-OL, 3,7-DIMETHYL-, ACETATE, (Z)- see NCO100

2,6-OCTADIEN-1-OL, 3,7-DIMETHYL-, PROPIONATE, (E)-(8CI) see GDM450

OCTAFLUOROADIPAMIDE see OBK100

OCTAFLUOROADIPONITRILE see PCG600

OCTAFLUOROAMYL ALCOHOL see OBS800

OCTAFLUOROBUTENE-2 see OBO000

OCTAFLUORO-sec-BUTENE see OBM000

OCTAFLUOROBUT-2-ENE (DOT) see OBO000

1,1,1,2,3,4,4,4-OCTAFLUORO-2-BUTENE see OBO000

OCTAFLUOROCYCLOBUTANE (DOT) see CPS000

OCTAFLUOROISOBUTENE see OBM000

OCTAFLUOROISOBUTYLENE see OBM000

OCTAFLUOROPENTANOL see OBS800

OCTAFLUORO-1-PENTANOL see OBS800

2,2,3,3,4,4,5,5-OCTAFLUORO-1-PENTANOL see OBU000

OCTAFLUOROPENTYL ALCOHOL see OBS800

OCTAFLUOROTOLUENE see PCH500

OCTAHEDRITE (MINERAL) see OBU100

OCTAHYDROAZOCINE see SMV500

OCTAHYDROAZOCINE HYDROCHLORIDE see OBU150

OCTAHYDROAZOCINE HYDROCHLORIDE mixed with SODIUM NITRITE (1:1) see SIT000

(2-(OCTAHYDRO-1-AZOCINYL)ETHYL)GUANIDINE SULFATE see GKS000

2-(OCTAHYDRO-1-AZOCINYL)ETHYL GUANIDINE SULPHATE see GKU000

((OCTAHYDRO-2-AZOCINYL)METHYL)GUANIDINE see GKO800

((OCTAHYDRO-2-AZOCINYL)METHYL)GUANIDINE SULFATE HYDRATE see CAY875

OCTAHYDRO-1-BENZOPYRAN-2-ONE see OBV200

OCTAHYDRO-2H-1-BENZOPYRAN-2-ONE (9CI) see OBV200

OCTAHYDRO-2H-BISOXIRENO(a,f)INDENE see BFY750

OCTAHYDROCOUMARIN see OBV200

OCTAHYDRODIBENZ(a,h)ANTHRACENE see OBW000

OCTAHYDRO-1:2:5:6-DIBENZANTHRACENE see OBW000

6,7,9,10,17,18,20,21-OCTAHYDRODIBENZO(b,k)(1,4,7,10,13,16)HEXAOXYCYCLOOCTADECIN see COD575

(1S,cis)1,2,3,4,5,6,7,8-OCTAHYDRO-1,4-DIMETHYL-7-(1-METHYLETHYLIDENE)-AZULENE see GKO000

1,2,3,4,5,6,7,8-OCTAHYDRO-1,4-DIMETHYL-7-(1-METHYLETHYLIDENE)AZULENEMONOEPOXIDE see EBU100

(4A-α,7-β,8A-α)-OCTAHYDRO-4A,8A-DIMETHYL-7-(1-METHYLETHYL)-1(2H)-NAPHTHALENONE see VAG200

(2S-(1(4*(4*)),2-α,3-α-β,7-α-β))-OCTAHYDRO-1-(2-((1-(ETHOXYCARBONYL)-3-PHENYLPROPYL)AMINO)-1-OXYPROPYL)-1H-INDOLE-2-CARBOXYLIC ACID MONOHYDROCHLORIDE see ICL500

1,2,3,4,6,7,12A-OCTAHYDRO-2-(1-(p-FLUOROPHENYL)-1-OXO-4-BUTYL)-PYRAZINO(2,1:6,1)PYRIDO(3,4-B)INDOLE see CCW500

2,5,5a,6,9,10,10a,1a-OCTAHYDRO-4-HYDROXYMETHYL-1,1,7,9-TETRAMETHYL-5,5a-6-TRIHYDROXY-1H-2,8a-METHANOCYCLOPENTA(a)CYCLOPROPA(e)CYCLODECEN-11-ONE-5-HEXADECANOATE see IDB000

1,2,3,3a,4,5,6,8a-OCTAHYDRO-2-ISOPROPYLIDENE-6-AZULENOL-4,8-DIMETHYL ACETATE see AAW750

OCTAHYDRO-4,7-METHANOINDENE-AR,AR'-DIMETHANOL see TJG550

4-(OCTAHYDRO-4,7-METHANO-5H-INDEN-5-YLIDENE)BUTANAL see OBW100

OCTAHYDRO-1,6-METHANONAPHTHALEN-1(2H)-AMINE see AKD875

OCTAHYDRO-5-METHOXY-4,7-METHANO-1H-INDENE-2-CARBOXALDEHYDE see OBW200

1,2,3,4,6,7,7a,11c-OCTAHYDRO-9-METHOXY-2-METHYL-BENZOFURO(4,3,2-efg)(2)BENZAZOCIN-6-OL HBr see GBA000

1,2,6,7,8,9,10,12b-OCTAHYDRO-3-METHYLBENZ(j)ACEANTHRYLENE see HDR500

OCTAHYDRO-1-NITROSOAZOCINE see OBY000

OCTAHYDRO-1-NITROSO-1H-AZONINE see OCA000

OCTAHYDRO-3,6,8,8-TETRAMETHYL-1H-3a,7-METHANOAZULEN-6-OL ACETATE see CCR250

OCTAHYDRO-(1,2,4,5)TETRAZINO(1,2-A)(1,2,4,5)TETRAZINE see OCA500

OCTAHYDRO-3,6,9-TRIMETHYL-3,12-EPOXY-12H-PYRANO(4,3-j)-1,2-BENZODIOXEPIN-10(3H)-ONE see ARL375

OCTAKIS(HYDROXYMETHYL)PHOSPHONIUM SULFATE see TDI000

OCTA-KLOR see CDR750

Δ-OCTALACTONE see ODE300

γ-OCTALACTONE see OCE000

OCTALENE see AFK250

OCTALOX see DHB400

OCTAMETHYLCYCLOTETRASILOXANE see OCE100

OCTAMETHYL-DIFOSFORZUUR-TETRAMIDE (DUTCH) see OCM000

OCTAMETHYLDIPHOSPHORAMIDE see OCM000

OCTAMETHYL-DIPHOSPHORSAEURE-TETRAMID (GERMAN) see OCM000

5,5'-(OCTAMETHYLENEBIS(CARBONYLIMINO)BIS(N-METHYL-2,4,6-

TRIIODOISOPHTHALAMIC ACID) see OCI000

N,N'-OCTAMETHYLENEBIS(2,2-DICHLOROACTAMIDE) see BIX250

OCTAMETHYLENEDIAMINE see OCU200

1,8-OCTAMETHYLENEDIAMINE see OCU200

OCTAMETHYLENE DICYANIDE see SBK500

OCTAMETHYLPYROPHOSPHORAMIDE see OCM000

OCTAMETHYL PYROPHOSPHORTETRAMIDE see OCM000

OCTAMETHYLSILANETETRAMINE see OCM100

OCTAMETHYL TETRAMIDO PYROPHOSPHATE see OCM000

1-OCTANAL see OCO000

OCTANALDEHYDE see OCO000

OCTANAL, 3,7-DIMETHYL- see TCY300

OCTANAL, 7-HYDROXY-3,7-DIMETHYL-, DIMETHYL ACETAL see LBO050

OCTANAL, 2-METHYL- see MNC175

1-OCTANAMINE see OEK015

2-OCTANAMINE see OEK010

1-OCTANAMINE, N-METHYL-N-OCTYL- see MND125

1-OCTANAMINIUM, N,N-DIMETHYL-N-OCTYL-, CHLORIDE see DTF820

1-OCTANAMINIUM, N-METHYL-N,N-DIOCTYL-, CHLORIDE (9CI) see MQH000

OCTAN AMYLU (POLISH) see AOD725

OCTAN ANTIMONITY (CZECH) see AQJ750

OCTAN BARNATY (CZECH) see BAH500

OCTAN n-BUTYLU (POLISH) see BPU750

OCTAN CYKLOHEXYLAMINU (CZECH) see CPG000

OCTAN DRASELNY see PKT750

OCTANE see OCU000

n-OCTANE see OCU000

1-OCTANECARBOXYLIC ACID see NMY000

4-OCTANECARBOXYLIC ACID see OCU100

1,8-OCTANEDIAMINE see OCU200

1,8-OCTANEDIAMINE, DIHYDROCHLORIDE see OCW000

OCTANE-1,8-DIAMINE DIHYDROCHLORIDE see OCW000

1,8-OCTANEDICARBOXYLIC ACID see SBJ500

OCTANE DIMETHANESULPHONATE see OCW025

OCTANEDINITRILE (9CI) see OCW050

1,8-OCTANEDIOL, DIMETHANESULFONATE see OCW025

OCTANE, HEPTADECAFLUORO-1-IODO- see PCH325

OCTANE, 1,1,1,2,2,3,3,4,4,5,5,6,6,7,7,8,8-HEPTADECAFLUORO-8-IODO-(9CI) see PCH325

OCTANE, 1-IODO- see OFA100

OCTANE, 2-IODO- see OFA200

OCTANENITRILE see OCW100

OCTANE-1-NNO-AZOXYMETHANE see MGT000

OCTANE, OCTADECAFLUORO- see OAP100

OCTANEPEROXOIC ACID, 1,1-DIMETHYLETHYL ESTER see BSD100

1-OCTANESULFONAMIDE,1,1,2,2,3,3,4,4,5,5,6,7,7,8,8,8-HEPTADECAFLUORO- see PCH315

1-OCTANESULFONIC ACID,1,1,2,2,3,3,4,4,5,5,6,6,7,7,8,8,8-HEPTADECAFLUORO-, LITHIUM SALT see LHN100

1-OCTANESULFONYL FLUORIDE see OCW200

tert-OCTANETHIOL see MKJ250

tert-OCTANETHIOL see OFE030

OCTAN ETOKSYETYLU (POLISH) see EES400

OCTAN ETYLU (POLISH) see EFR000
OCTAN FENYLRTUTNATY (CZECH) see ABU500
OCTANIL see ILK000
OCTAN KOBALTNATY (CZECH) see CNA500
OCTAN MANGANATY (CZECH) see MAQ000
OCTAN MEDNATY (CZECH) see CNI250
OCTAN METYLU (POLISH) see MFW100
OCTANOIC ACID see OCY000
OCTANOIC ACID, 2-ACETYL-, ETHYL ESTER see EKS150
OCTANOIC ACID ALLYL ESTER see AGM500
OCTANOIC ACID, CADMIUM SALT (2:1) see CAD750
OCTANOIC ACID, 2,2-DIMETHYL-, HEXADECYL ESTER see HCP600
OCTANOIC ACID, ETHYL ESTER see ENY000
OCTANOIC ACID, HEXYL ESTER see HFS600
OCTANOIC ACID, 5-HYDROXY-, Δ-LACTONE see ODE300
OCTANOIC ACID, 5-HYDROXY-, LACTONE (6CI) see ODE300
OCTANOIC ACID, ISOPENTYL ESTER see IHP500
OCTANOIC ACID, METHYL ESTER see MHY800
OCTANOIC ACID, METHYL ESTER mixed with METHYL DECANOATE see MND000
OCTANOIC ACID, PENTADECAFLUORO-see PCH050
OCTANOIC ACID, POTASSIUM SALT see PLN250
OCTANOIC ACID, 1,2,3-PROPANETRIYL ESTER see TMO000
OCTANOIC ACID-2-PROPENYL ESTER see AGM500
OCTANOIC ACID, PROPYL ESTER see PNR600
OCTANOIC ACID, SODIUM SALT see SIX600
OCTANOIC ACID, p-TOLYL ESTER see THB000
OCTANOIC ACID TRIGLYCERIDE see TMO000
OCTANOIC ACID, ZINC SALT (2:1) see ZEJ000
OCTANOIC/DECANOIC ACID TRIGLYCERIDE see CBF710
OCTANOL see OEI000
2-OCTANOL see OCY090
3-OCTANOL see OCY100
OCTANOL-3 see OCY100
n-OCTANOL see OEI000
D-n-OCTANOL see OCY100
OCTANOL, mixed isomers see ODE000
1-OCTANOL ACETATE see OEG000
2-OCTANOL, 8,8-DIMETHOXY-2,6-DIMETHYL-(9CI) see LBO050
3-OCTANOL, 3,7-DIMETHYL- see LFY510
2-OCTANOL, 3,7-DIMETHYL-7-METHOXY-see DLR300
1-OCTANOL (FCC) see OEI000
OCTANOLIC ACID (mixed isomers) see ODE200
5-OCTANOLIDE see ODE300
OCTANOLIDE-1,4 see OCE000
3-OCTANOL, 3-METHYL- see MND050
2-OCTANONE see ODG000
3-OCTANONE see ODI000
1-OCTANONE, 1-(4-(HYDROXYAMINO)PHENYL)- see HIY200
2-OCTANONE, 5-METHYL- see MLO800
2-OCTANONE, 5-METHYL- see MND075
OCTANONITRILE see OCW100
n-OCTANOYL PEROXIDE (DOT) see CBF705

2-OCTANOYL-1,2,3,4-TETRAHYDROISOQUINOLINE see ODO000
OCTAN PROPYLU (POLISH) see PNC250
OCTAN WINYLU (POLISH) see VLU250
n-OCTANYL ACETATE see OEG000
OCTAN ZINECNATY (CZECH) see ZCA000
OCTASULFAN see OCW025
OCTASULPHAN see OCW025
OCTATENSIN see GKQ000
OCTATENZINE see GKQ000
1,3,7-OCTATRIEN-5-YNE see ODQ300
OCTATROPINE METHYLBROMIDE see LJS000
OCTAVE see IAL200
OCTE see ODQ400
(E)-2-OCTENAL see ODQ400
2-OCTENAL, (2E)- see ODQ400
trans-2-OCTEN-1-AL see ODQ800
2-OCTENE, 8-ETHOXY-2,6-DIMETHYL- see EES100
1-OCTEN-3-OL see ODW000
cis-3-OCTEN-1-OL see ODW025
1-OCTEN-3-OL ACETATE see ODW028
6-OCTEN-1-OL, 3,7-DIMETHYL- see DTF410
7-OCTEN-2-OL, 2,6-DIMETHYL-8-(1H-INDOL-1-YL)- see ICS100
1-OCTEN-3-OL, 3-METHYL- see MND100
OCTENYL ACETATE see ODW028
1-OCTEN-3-YL ACETATE see ODW030
1-OCTEN-3-YL BUTYRATE see ODW040
6-OCTEN-1-YN-3-OL, 3,7-DIMETHYL- see LFY333
6-OCTEN-4-YN-2-OL, 1-(DIMETHYLAMINO)-6-ETHYL-, HYDROCHLORIDE see DPH500
OCTHILINONE see OFE000
OCTIC ACID see OCY000
OCTILIN see OEI000
OCTIN see ILK000
OCTIN HYDROCHLORIDE see ODY000
OCTINUM see ILK000
OCTOCLOTHEPIN see ODY100
OCTOCLOTHEPINE see ODY100
OCTOCRILENE see ODY150
OCTOCRYLENE see ODY150
OCTODRINE see ILM000
OCTOGEN see CQH250
OCTOGEN, desensitized (UN 0483) (DOT) see CQH250
OCTOGEN, wetted with not <15% water, by weight (UN 0226) (DOT) see CQH250
n-OCTOIC ACID see OCY000
OCTOIL see DVL700
OCTOIL S see BJS250
OCTON see ILK000
OCTOPAMINE see AKT250
m-OCTOPAMINE see AKT000
dl-m-OCTOPAMINE HYDROCHLORIDE see NNT100
OCTOPIROX see PJA180
OCTOTIAMINE see GEK200
OCTOWY ALDEHYD (POLISH) see AAG250
OCTOWY BEZWODNIK (POLISH) see AAX500
OCTOWY KWAS (POLISH) see AAT250
OCTOXINOL see PKF500
OCTOXYNOL see PKF500
OCTOXYNOL 3 see PKF500
OCTOXYNOL 9 see PKF500
OCTREOTIDE ACETATE see ODY200
(OC-6-22)-TRIS(ACETONITRILE)TRICHLORORHODIUM see TNC600
OCTSETAN see SNQ000
OCTYL ACETATE see OEE000
OCTYL ACETATE see OEG000
1-OCTYL ACETATE see OEG000
3-OCTYL ACETATE see OEI100
n-OCTYL ACETATE see OEG000
β-OCTYL ACROLEIN see DXU280
OCTYL ACRYLATE see ADU250

OCTYL ADIPATE see AEO000
OCTYL ALCOHOL see OEI000
OCTYL ALCOHOL ACETATE see OEG000
OCTYL ALCOHOL, NORMAL-PRIMARY see OEI000
n-OCTYL ALDEHYDE OCO000
2-OCTYLAMINE see OEK010
N-OCTYLAMINE see OEK015
OCTYLAMINE, METHANEARSONATE (1:1) see OEK500
1-OCTYLAMINE, N-METHYL-N-NITROSO-see NKU590
OCTYLAMMONIUM METHYLARSONATE see OEK500
N-n-OCTYLATROPINE BROMIDE see OEM000
N-n-OCTYL-ATROPINIUMBROMID (GERMAN) see OEM000
8-OCTYLATROPINIUM BROMIDE see OEM000
N-OCTYLATROPINIUM BROMIDE see OEM000
N-n-OCTYLATROPINIUM BROMIDE see OEM000
p-OCTYLBENZOIC ACID see OEQ000
4-n-OCTYLBENZOIC ACID see OEQ000
N-OCTYL BICYCLOHEPTENE DICARBOXIMIDE see OES000
N-OCTYLBICYCLO-(2.2.1)-5-HEPTENE-2,3-DICARBOXIMIDE see OES000
n-OCTYL BROMIDE see BNU000
γ-n-OCTYL-γ-n-BUTYROLACTONE see OES100
OCTYL CARBINOL see NNB500
OCTYL CYANIDE see OES300
1-OCTYL CYANIDE see OES300
n-OCTYL CYANIDE see OES300
OCTYL DECYL DIMETHYL AMMONIUM CHLORIDE see OES400
OCTYL DECYL PHTHALATE see OEU000
n-OCTYL n-DECYL PHTHALATE see OEU000
OCTYL 2,4-DICHLOROPHENOXYACETATE see OEU050
OCTYL (DIETHYLAMINO)OXOACETATE see OEU100
OCTYL N,N-DIETHYLOXAMATE see OEU100
OCTYL-DIMETHYL-p-AMINOBENZOIC ACID see AOI500
OCTYL-DIMETHYL-BENZYLAMMONIUM CHLORIDE see OEW000
OCTYLDODECANOL see OEW100
2-OCTYLDODECANOL see OEW100
2-OCTYLDODECYL ALCOHOL see OEW100
1,8-OCTYLENEDIAMINE see OCU200
OCTYLENE EPOXIDE see ECE000
OCTYLENE GLYCOL see EKV000
OCTYL EPOXYTALLATE see FAB920
sec-OCTYL EPOXYTALLATE see FAB900
2,4-D-OCTYL ESTER see OEU050
OCTYL ESTER OF 2,4-D see OEU050
n-OCTYL ESTER of 3,4,5-TRIHYDROXYBENZOIC ACID see OFA000
OCTYL ETHER see OEY000
OCTYL FORMATE see OEY100
n-OCTYL FORMATE see OEY100
OCTYL GALLATE see OFA000
N-OCTYL-o-HYDROXYBENZOATE see SAJ500
γ-OCTYL-β-HYDROXY-Δ^{α,β}-BUTENOLID (GERMAN) see HND000
n-OCTYLIC ACID see OCY000
OCTYL IODIDE see OFA100
1-OCTYL IODIDE see OFA100
2-OCTYL IODIDE see OFA200
n-OCTYL IODIDE see OFA100
1-n-OCTYL IODIDE see OFA100
sec-OCTYL IODIDE see OFA200
n-OCTYLISOSAFROLE SULFOXIDE see ISA000

2-OCTYL-4-ISOTHIAZOLIN-3-ONE see OFE000

2-OCTYL-3(2H)-ISOTHIAZOLONE see OFE000

OCTYL KETONE see OFE020

tert-OCTYLMERCAPTAN see MKJ250

tert-OCTYLMERCAPTAN (DOT) see OFE030

OCTYL METHYL KETONE see OFE050

n-OCTYL NITROSUREA see NLD800

OCTYL-OCTADECYL DIMETHYL ETHYLBENZYL AMMONIUM CHLORIDES see BBA500

cis-3-OCTYL-OXIRANEOCTANOIC ACID see ECD500

cis-3-OCTYLOXIRANE OCTANOIC ACID see OAT100

4-(p-OCTYLOXYPHENYL)SEMICARBAZONE 1-METHYL-1H-PYRROLE-2-CARBOXALDEHYDE, see OFE100

4-(p-OCTYLOXYPHENYL)SEMICARBAZONE-1H-PYRROLE-2-CARBOXALDEHYDE see OFE200

4'-(OCTYLOXY)-3-PIPERIDINOPROPIOPHENONE HYDROCHLORIDE see OFG000

4-OCTYLOXY-β-(1-PIPERIDYL)PROPIOPHENONE HYDROCHLORIDE see OFG000

OCTYL PALMITATE see OFG100

OCTYLPEROXIDE see OFI000

OCTYL PHENOL see OFK000

4-OCTYLPHENOL see OFK100

p-OCTYLPHENOL see OFK100

p-tert-OCTYLPHENOL see TDN500

OCTYL PHENOL CONDENSED with 12–13 MOLES ETHYLENE OXIDE see PKF500

OCTYLPHENOL EO (3) see OFM000

OCTYL PHENOL EO (5) see OFO000

OCTYLPHENOL EO (10) see OFQ000

OCTYL PHENOL EO (16) see OFM900

OCTYL PHENOL EO (20) see OFU000

OCTYL PHENOL condensed with 3 MOLES ETHYLENE OXIDE see OFM000

OCTYL PHENOL condensed with 5 MOLES ETHYLENE OXIDE see OFO000

OCTYL PHENOL condensed with 16 MOLES ETHYLENE OXIDE see OFM900

OCTYL PHENOL condensed with 20 MOLES ETHYLENE OXIDE see OFU000

OCTYL PHENOL condensed with 8-10 MOLES ETHYLENE OXIDE see OFQ000

p-tert-OCTYLPHENOXYETHOXYETHYLDIMET HYLBENZYL AMMONIUM CHLORIDE see BEN000

OCTYLPHENOXYPOLY(ETHOXYETHAN OL) see GHS000

tert-OCTYLPHENOXYPOLY(ETHOXYETHAN OL) see GHS000

p-tert-OCTYLPHENOXYPOLYETHOXYETHANO L see PKF500

OCTYLPHENOXYPOLY(ETHYLENEOXY) ETHANOL see GHS000

tert-OCTYLPHENOXYPOLY(OXYETHYLENE) ETHANOL see GHS000

OCTYLPHOSPHATE see OFQ050

OCTYL PHTHALATE see DVL600

OCTYL PHTHALATE see DVL700

n-OCTYL PHTHALATE see DVL600

4'-OCTYL-3-PIPERIDINOPROPIOPHENONE HYDROCHLORIDE see PIY000

N-OCTYLPYRROLIDINONE see OFU100

1-OCTYL-2-PYRROLIDINONE see OFU100

N-OCTYLPYRROLIDONE see OFU100

OCTYL SALICYLATE see SAJ500

OCTYL SEBACATE see BJS250

OCTYL SODIUM SULFATE see OFU200

OCTYL STEARATE see OFU300

OCTYLSULFONYL FLUORIDE see OCW200

1-OCTYLSULFONYL FLUORIDE see OCW200

2-(OCTYLTHIO)ETHANOL see OGA000

OCTYLTRICHLOROSILANE see OGE000

OCTYLTRICHLOROSTANNANE see OGG000

OCTYLTRIS(2-ETHYLHEXYLOXYCARBONYLMETHYLT HIO)STANNANE see OGI000

4-OCTYN-3,6-DIOL, 3,6-DIMETHYL- see DTF850

1-OCTYNE see OGI025

OCTYNECARBOXYLIC ACID, METHYL ESTER see MNC000

1-OCTYN-3-OL see OGI050

1-OCTYN-4-OL, 4-METHYL- see MKR550

OC-U-MID see SNQ000

OCUSEPTINE see SJL500

OCUSOL see SPC500

OCU-VINC see VLF000

ODA see SJN700

ODANTOL see OGI075

ODB see DEP600

ODB see EDP000

ODCB see DEP600

Z,Z-ODDA see OAP070

O,α-DICHLOROTOLUENE see CEO200

ODISTON see DCK000

ODODIBORANE see OGI100

ODORLESS LIGHT PETROLEUM HYDROCARBONS see OGI200

OE3 see EDU500

OE 7 see EDJ500

OEDROPHANIUM see TAL490

OEKOLP see DKA600

OENANTHAL see HBB500

OENANTHALDEHYDE see HBB500

OENANTHE CROCATA see WAT315

OENANTHIC ACID see HBE000

OENANTHIC ALDEHYDE see HBB500

OENANTHIC ETHER see EKN050

OENANTHOL see HBB500

OENANTHOTOXIN see EAO500

OENANTHYLIC ACID see HBE000

OENETHYL HYDROCHLORIDE see NCL300

OESTERGON see EDO000

OESTRADIOL see EDO000

d-OESTRADIOL see EDO000

cis-OESTRADIOL see EDO000

α-OESTRADIOL see EDO000

β-OESTRADIOL see EDO000

OESTRADIOL-17-α see EDO500

OESTRADIOL-17-β see EDO000

3,17-β-OESTRADIOL see EDO000

d-3,17-β-OESTRADIOL see EDO000

OESTRADIOL BENZOATE see EDP000

OESTRADIOL-3-BENZOATE see EDP000

β-OESTRADIOL BENZOATE see EDP000

β-OESTRADIOL-3-BENZOATE see EDP000

17-β-OESTRADIOL-3-BENZOATE see EDP000

OESTRADIOL DIPROPIONATE see EDR000

OESTRADIOL-3,17-DIPROPIONATE see EDR000

β-OESTRADIOL DIPROPIONATE see EDR000

17-β-OESTRADIOL DIPROPIONATE see EDR000

3,17-β-OESTRADIOL DIPROPIONATE see EDR000

OESTRADIOL 3-METHYL ETHER see MFB775

OESTRADIOL MONOBENZOATE see EDP000

OESTRADIOL MUSTARD see EDR500

OESTRADIOL PHOSPHATE POLYMER see EDS000

OESTRADIOL POLYESTER with PHOSPHORIC ACID see EDS000

OESTRADIOL R see EDO000

OESTRAFORM (BDH) see EDP000

OESTRASID see DAL600

OESTRA-1,3,5(10)-TRIENE-3,17-β-DIOL see EDO000

17-β-OESTRA-1,3,5(10)-TRIENE-3,17-DIOL see EDO000

1,3,5(10)-OESTRATRIENE-3,17-β-DIOL 3-BENZOATE see EDP000

OESTRA-1,3,5(10)-TRIENE-3,16-α,17-β-TRIOL see EDU500

1,3,5-OESTRATRIENE-3-β-3,16-α,17-β-TRIOL see EDU500

(16-α,17-β)-OESTRA-1,3,5(10)-TRIENE-3,16,17-TRIOL see EDU500

1,3,5-OESTRATRIEN-3-OL-17-ONE see EDV000

1,3,5(10)-OESTRATRIEN-3-OL-17-ONE see EDV000

Δ-1,3,5-OESTRATRIEN-3-β-OL-17-ONE see EDV000

OESTRATRIOL see EDU500

OESTRENOLON see NNX400

OESTRILIN see ECU750

OESTRIN see EDV000

OESTRIOL see EDU500

16-α,17-β-OESTRIOL see EDU500

3,16-α,17-β-OESTRIOL see EDU500

OESTRODIENE see DAL600

OESTRODIENOL see DAL600

OESTRO-FEMINAL see ECU750

OESTROFORM see EDV000

OESTROGENINE see DKA600

OESTROGLANDOL see EDO000

OESTROGYNAEDRON see DKB000

OESTROGYNAL see EDO000

OESTROL VETAG see DKA600

OESTROMENIN see DKA600

OESTROMENSIL see DKA600

OESTROMENSYL see DKA600

OESTROMIENIN see DKA600

OESTROMON see DKA600

OESTRONBENZOAT (GERMAN) see EDV500

OESTRONE see EDV000

OESTRONE-3-SULPHATE SODIUM SALT see EDV600

OESTROPAK MORNING see ECU750

OESTROPEROS see EDV000

OESTRORAL see DAL600

OETs see HKQ100

OF see PMI250

OFDZ see OMY500

OFF see DKC800

OFFITRIL see HNI500

OFF-SHOOT-O see MND000

OFHC Cu see CNI000

OFII see OKO600

OFLOCET see OGI300

OFLOCIN see OGI300

OFLOXACIN see OGI300

OFLOXACIN see OGI300

OFLOXACINE see OGI300

OFNACK see POP000

OFNA-PERL SALT RRA see CLK235

OFTALENT see CDP250

OFTALFRINE see SPC500

OFTANOL see IMF300

OFUNACK see POP000

OFURACE see CGI550

OG 1 see SEH000

OGEEN 515 see OAV000

OGEEN GRB see OAV000

OGEEN M see OAV000

OGEN see PIK450

OGINEX see FBP100

OGOSTAL see CBF675

OGYLINE see NNR125

3-OHAA see AKE750

'OHAI-ALI'I (HAWAII) see CAK325

'OHAI (HAWAII) see SBC550
'OHAI-KE'OKE'O (HAWAII) see SBC550
'OHAI-'ULA'ULA (HAWAII) see SBC550
OHB see SAH000
OH-BBN see HJQ350
1-α-OH-CC see HJV000
cis-4-OH-CCNU see CHA500
trans-2-OH-CCNU see CHA750
trans-4-OH CCNU see CHB000
4-OH-CP see HKA200
1-α-OH-D³ see HJV000
3-OHDBTL see BRN600
4-OH-E2 see HKH850
2-OH-ESTRADIOL see HKH600
4-OH-ESTRADIOL see HKH850
17-β-OH-ESTRADIOL see EDO000
OHIO 347 see EAT900
OHIO RED see NAP100
7-OHM-MBA see HMF000
7-OHM-12-MBA see HMF000
17-β-OH-OESTRADIOL see EDO000
1-OHP see OKY100
OHRIC see DGF000
1-α-OH VITAMIN D3 see HJV000
3,3'-(GERMANOIC ANHYDRIDE)
DIPROPANOIC ACID see CCF125
OIL de ACORUS CALAMUS (SPANISH) see
OGL020
OIL of ANISEED see PMQ750
OIL, ARTEMISIA see ARL250
OIL, BITTER ALMOND see BLV500
OIL BLACK BT see PCJ200
OIL BLUE see CNQ000
OIL CAMPHOR SASSAFRASSY see CBB500
OIL CEDAR see CCR000
OIL CITRUS RETICULATA see OHJ150
OIL-DRI see AHF500
OIL-FURNACE BLACK see CBT750
OIL GARLIC see AGS250
OIL GAS see OGI350
OIL GERANIUM REUNION see GDC000
OIL GREEN see CMJ900
OIL, HIBAWOOD see HGA100
OIL MANDARIN see OHJ150
OIL MENTHA ARVENSIS see CNR990
OIL of MIRBANE (DOT) see NEX000
OIL MIST, MINERAL (OSHA, ACGIH) see
MQV750
OIL of MOUNTAIN PINE see PIH400
OIL of MYRBANE see NEX000
OIL of MYRISTICA see NOG500
OIL of NUTMEG see NOG500
OIL OF ABIES ALBA see AAC250
OIL OF ACORUS CALAMUS Linn. see
OGL020
OIL OF ALLSPICE see PIG740
OIL OF AMERICAN WORMSEED see
CDL500
OIL OF ANISE see AOU250
OIL OF ARBOR VITAE see CCQ500
OIL OF ARGEMONE mixed with OIL OF
MUSTARD see OGS000
OIL OF BASIL see BAR250
OIL OF BAY see BAT500
OIL OF BAY see LBK000
OIL OF BERGAMOT, rectified see BFO000
OIL OF BERGAMOT, coldpressed see BFO000
OIL OF BITTER ORANGE see OJK330
OIL OF CALAMUS, GERMAN see OGK000
OIL OF CALAMUS, SPANISH see OGL020
OIL OF CAMPHOR RECTIFIED see CBB500
OIL OF CAMPHOR WHITE see CBB500
OIL OF CARAWAY see CBG500
OIL OF CARDAMON see CCJ625
OIL OF CASSIA see CCO750
OIL OF CEDAR LEAF see CCQ500
OIL OF CELERY see OGL100
OIL OF CHENOPODIUM see CDL500
OIL OF CHINESE CINNAMON see CCO750
OIL OF CINNAMON see CCO750
OIL OF CINNAMON, CEYLON see CCO750
OIL OF CINNAMON, CEYLON see CMQ510

OIL OF CITRONELLA see CMT000
OIL OF CIVET see OGM100
OIL OF CLOVE see CMY475
OIL OF CORIANDER see CNR735
OIL OF CUBEB see COE175
OIL OF EUCALYPTUS see EQQ000
OIL OF FENNEL see FAP000
OIL OF FUR see AAC250
OIL OF GARLIC see GBU825
OIL OF GERANIUM see GDA000
OIL OF GRAPEFRUIT see GJU000
OIL OF HARTSHORN see BMA750
OIL OF HEERABOL-MYRRH see MSB775
OIL OF JASMINE see OGM200
OIL OF JUNIPER BERRY see JEA000
OIL OF LABDANUM see LAC000
OIL OF LAVANDIN see LCA000
OIL OF LAVANDIN, ABRIAL TYPE see
LCA000
OIL OF LAVENDER see LCC000
OIL OF LAVENDER see LCD000
OIL OF LEMON see LEI000
OIL OF LEMON, distilled see LEI030
OIL OF LEMON, desert type, coldpressed see
LEI025
OIL OF LEMONGRASS, EAST INDIAN see
LEG000
OIL OF LEMONGRASS, WEST INDIAN see
LEH000
OIL OF LIME, distilled see OGM850
OIL OF LIME OIL, COLDPRESSED see
OGM800
OIL OF MACE see OGQ100
OIL OF MARJORAM, SPANISH see MBU500
OIL OF MEDITERRANEAN BAY see
OGQ150
OIL OF MUSCATEL see CMU900
OIL OF MUSCATEL see OGQ200
OIL OF MUSTARD, artificial see AGJ250
OIL OF MUSTARD, EXPRESSED mixed with
OIL OF ARGEMONE see OGS000
OIL OF MYRCIA see BAT500
OIL OF MYRCIA see LBK000
OIL OF MYRTLE see OGU000
OIL OF NIOBE see MHA750
OIL OF NUTMEG, expressed see OGQ100
OIL OF ONION see OJD200
OIL OF ORANGE see OGY000
OIL OF ORIGANUM see OJO000
OIL OF PALMA CHRISTI see CCP250
OIL OF PELARGONIUM see GDA000
OIL OF PIMENTA see PIG740
OIL OF PIMENTO see PIG740
OIL OF ROSE see RNA000
OIL OF ROSE BLOSSOM see RNA000
OIL OF ROSE BULGARIAN see RNA000
OIL OF ROSE BULGARIAN see RNF000
OIL OF ROSE GERANIUM see GDA000
OIL OF ROSE MOROCCAN see RNK000
OIL OF RUE see OGY200
OIL OF SANDALWOOD, EAST INDIAN see
OGY220
OIL OF SANTAL see OGY220
OIL OF SASSAFRAS see OHI000
OIL OF SHADDOCK see GJU000
OIL OF SILVER FIR see XRA000
OIL OF SILVER PINE see AAC250
OIL OF SPEARMINT see SKY000
OIL OF SPIKE LAVENDER see SLB500
OIL OF SWEET FLAG see OGK000
OIL OF SWEET ORANGE see OGY000
OIL OF THUJA see CCQ500
OIL OF THYME see TFX500
OIL OF VETIVER see VJU000
OIL OF WHITE CEDAR see CCQ500
OIL OF WINTERGREEN see MPI000
OIL ORANGE see PEJ500
OIL ORANGE 2R see XRA000
OIL ORANGE 4G see CMP600
OIL ORANGE G see CMP600
OIL ORANGE KB see XRA000
OIL ORANGE MO see CMP600

OIL ORANGE MON see CMP600
OIL ORANGE MON EXTRA see CMP600
OIL ORANGE N EXTRA see XRA000
OIL ORANGE O'PEL see TGW000
OIL ORANGE R see XRA000
OIL ORANGE SS see TGW000
OIL ORANGE X see XRA000
OIL ORANGE XO see XRA000
OIL of PALMAROSA see PAE000
OIL of PARSLEY see PAL750
OIL PIMENTA BERRIES see PIG740
OIL of PIMENTA LEAF see PIG730
OIL of PINE see PIH750
OIL PINK see CMS238
OIL RED see CMS238
OIL RED see OHI200
OIL RED 3 see SBC500
OIL RED 7 see SBC500
OIL RED 2B see SBC500
OIL RED 3B see OHI200
OIL RED 3B see SBC500
OIL RED 3G see OHI200
OIL RED 47 see SBC500
OIL RED 4B see SBC500
OIL RED 113 see CMS238
OIL RED 282 see SBC500
OIL RED 6566 see OHI200
OIL RED A see SBC500
OIL RED APT see SBC500
OIL RED AS see OHI200
OIL RED B see OHI200
OIL RED BB see SBC500
OIL RED BS see SBC500
OIL RED BS see WAT000
OIL RED D see SBC500
OIL RED ED see SBC500
OIL RED F see SBC500
OIL RED G see OHI200
OIL RED GO see SBC500
OIL RED GRO see XRA000
OIL RED IV see SBC500
OIL RED O see OHI200
OIL RED O see XRA000
OIL RED OG see CMS238
OIL RED PEL see SBC500
OIL RED RC see SBC500
OIL RED RO see XRA000
OIL RED RR see SBC500
OIL RED S see SBC500
OIL RED TAX see SBC500
OIL RED XO see FAG080
OIL RED XO see XRA000
OIL RED ZD see SBC500
OIL ROSE GERANIUM see GDC000
OIL ROSE GERANIUM ALGERIAN see
GDA000
OIL ROSE TURKISH see OHJ000
OILS, ALLSPICE see PIG740
OILS, AMYRIS see WBJ650
OILS, AMYRIS, ACETYLATED see AON595
OILS, ANGELICA ROOT see AOO760
OIL of SASSAFRAS BRAZILIAN see OAF000
OILS, BASIL see BAR250
OIL SCARLET see OHI200
OIL SCARLET see SBC500
OIL SCARLET see XRA000
OIL SCARLET 48 see SBC500
OIL SCARLET 6G see XRA000
OIL SCARLET 371 see XRA000
OIL SCARLET 389 see CMS238
OIL SCARLET APYO see XRA000
OIL SCARLET AS see OHI200
OIL SCARLET BL see XRA000
OIL SCARLET G see OHI200
OIL SCARLET L see XRA000
OIL SCARLET YS see XRA000
OILS, CARROT see CCL750
OILS, CEDAR LEAF see CCQ500
OILS, CELERY see OGL100
OILS, CHAMOMILE, GERMAN see CDH500
OILS, CINNAMON see CCO750
OILS, CINTONELLA see CMT000

OILS, CLARY SAGE see OGQ200
OILS, CLOVE see CMY475
OILS, CLOVE LEAF see CMY500
OILS, CLOVE STEM see CMY510
OILS, COFFEE see CNG800
OILS, CORIANDER see CNR735
OILS, COSTUS see CNT350
OILS, CUBEB see COE175
OILS, CUMIN see COF325
OILS, DOUGLAS FIR see OJK340
OILS, EUCALYPTUS, E. CITRIODORA,
ACETYLATED see OHJ100
OILS, GARLIC see GBU825
OILS, GLYCERIDIC, POPPYSEED,
IODINATED see LGK225
OILS, GUAIACWOOD, ACETATES see
GKG400
OIL-SHALE PYROLYSE LAC LSP-1 see
LIJ000
OILS, JASMINE see OGM200
OILS, JUNIPER see JEA000
OILS, JUNIPERUS PHOENICEA see JEJ200
OILS, KARO-KAROUNDE see KCA050
OILS, LIME see OGM850
OILS, MINT, MENTHA CITRATA see
BFN990
OILS, NIM see NBS300
OIL SOLUBLE ANILINE YELLOW see
PEI000
OIL SOLUBLE RED S see CMS238
OIL-SOL. YELLOW ZH see CMP600
OILS, ORRIS see OHJ130
OILS, PALM see PAE500
OILS, PETITGRAIN see OHJ150
OILS, PIMENTA see PIG740
OILS, PINE see PIH750
OILS, SANDALWOOD see OGY220
OILS, SPRUCE see SLG650
OILS, SWEET BAY see OGQ150
OILS, SWEET BIRCH see SOY100
OILS, TEA-TREE see TAI150
OILS, VERBENA see VIK500
OILS, WHEAT GERM see WBJ700
OIL THUJA see CCQ500
OIL of TURPENTINE see TOD750
OIL of TURPENTINE, RECTIFIED see
TOD750
OIL VERMILION see CMS238
OIL VERMILION LP see CMS238
OIL VIOLET see EOJ500
OIL VIOLET IRS see HOK000
OIL VIOLET R see DBP000
OIL of VITRIOL (DOT) see SOI500
OIL YELLOW see AIC250
OIL YELLOW see DOT300
OIL YELLOW 21 see AIC250
OIL YELLOW 2R see AIC250
OIL YELLOW 2635 see OHJ875
OIL YELLOW 2681 see AIC250
OIL YELLOW A see AIC250
OIL YELLOW A see FAG130
OIL YELLOW AAB see PEI000
OIL YELLOW AT see AIC250
OIL YELLOW C see AIC250
OIL YELLOW DE see OHJ875
OIL YELLOW DEA see OHJ875
OIL YELLOW E190 see OHJ875
OIL YELLOW ENC see OHJ875
OIL YELLOW GA see OHJ875
OIL YELLOW G EXTRA see CMP600
OIL YELLOW GG see CMP600
OIL YELLOW HA see OHK000
OIL YELLOW I see AIC250
OIL YELLOW NB see OHJ875
OIL YELLOW OB see FAG135
OIL YELLOW OPS see OHK000
OIL YELLOW SIS see CMS240
OIL YELLOW T see AIC250
OJO de CANGREJO (PUERTO RICO) see
RMK250
OK 7 see PJY100
OK 174 see AKC570

OK 622 see PAJ000
OK-1166 see FCC050
OKADAIC ACID see OHK100
OKADAIC ACID, 35-METHYL- see MND300
OKASA-MASCUL see TBG000
OKITEN G 23 see PJS750
OKO see DGP900
OK PRE-GEL see SLJ500
OKSAL see OKY300
OKSAZIL see MSC100
OKSISYKLIN see HOH500
OKTADECYLAMIN (CZECH) see OBC000
ω-H-OKTAFLUORPENTANOL-1
(GERMAN) see OBS800
OKTAMETHYLCYKLOTETRASILOXAN see
OCE100
OKTAMETHYLENDIKYANID see SBK500
OKTAN (POLISH) see OCU000
OKTANEN (DUTCH) see OCU000
terc.OKTANTHIOL see MKJ250
terc.OKTANTHIOL see OFE030
OKTATERR see CDκ750
OKTOGEN see CQH250
OKTYLESTER 2,4-
DICHLORFENOXYOCTOVE see OEU050
OKTYLESTER KYSELINY GALLOVE see
OFA000
p-terc.OKTYLFENOL (CZECH) see TDN500
OKULTIN M see CIR250
OKULTIN MP see RBF500
OKY-046 SODIUM see SHV100
OL 27-400 see CQH130
OLACHINDOX see HKV100
OLAFLUR see BJY800
OLAMINE see EEC600
OLANZAPINE see ZVJ500
OLAQUINDOX see HKV100
OLATE FLAKES see OIA000
OLCADIL see CMY525
OLD 01 see ADV900
OLDHAMITE see CAY000
OLDREN see ZAK300
OLEALKONIUM CHLORIDE see OHK200
OLEAL ORANGE R see PEJ500
OLEAL ORANGE SS see TGW000
OLEAL RED BB see SBC500
OLEAL RED G see CMS238
OLEAL YELLOW 2G see DOT300
OLEAL YELLOW RE see OHK000
OLEAMIDE see OHM600
OLEAMIDE, N,N-BIS(2-HYDROXYETHYL)-
see OHU200
OLEAMIDE DEA see OHU200
OLEAMINE see OHM700
'OLEANA (HAWAII) see OHM875
OLEANDER see OHM875
OLEANDOMYCIN see OHM900
OLEANDOMYCINE see OHM900
OLEANDOMYCIN HYDROCHLORIDE see
OHO000
OLEANDOMYCIN
MONOHYDROCHLORIDE see OHO000
OLEANDOMYCIN PHOSPHATE see
OHO200
OLEANDRIN see OHQ000
OLEANDRINE see OHQ000
OLEAN-12-EN-28-OIC ACID, 3-((2-o-(6-
DEOXY-α-l-manNOPYRANOSYL)-α-l-
ARABINOPYRANOSYL) OXY)-23-
HYDROXY-, (3-β,4-α)- see HAK075
OLEANOGLYCOTOXIN A see OHQ300
OLEANOGLYCOTOXIN B see LEF800
OLEATE of MERCURY see MDF250
OLEFIANT GAS see EIO000
OLEFINS see OHS000
α-OLEFIN SULFONATE see AFN500
α-OLEFIN SULPHONATE see AFN500
OLEIC ACID see OHU000
OLEIC ACID AMIDE see OHM600
OLEIC ACID, compounded with
BUTYLAMINE (1:1) see OHU100
OLEIC ACID, BUTYL ESTER see BSB000

OLEIC ACID CALCIUM SALT see CAU300
OLEIC ACID DIETHANOLAMIDE see
OHU200
OLEIC ACID DIETHANOLAMINE
CONDENSATE see OHU200
OLEIC ACID GLYCEROL MONOESTER see
GGR200
OLEIC ACID GLYCIDYL ESTER see ECJ000
OLEIC ACID, 12-HYDROXY- see RJP000
OLEIC ACID LEAD SALT see LDQ000
OLEIC ACID, LEAD(2+) SALT (2:1) see
LDQ000
OLEIC ACID mixed with LINOLEIC ACID see
LGI000
cis-OLEIC ACID, METHYL ESTER see
OHW000
OLEIC ACID MONOGLYCERIDE see
GGR200
OLEIC ACID NITRILE see OAX300
OLEIC ACID POLY(OXYETHYLENE)
ESTER see PJY100
OLEIC ACID, POTASSIUM SALT see
OHY000
OLEIC ACID, SODIUM SALT see OIA000
OLEIC ACID, ZINC SALT see ZJS000
OLEIC DIETHANOLAMIDE see OHU200
OLEIC NITRILE see OAX300
OLEINAMINE see OHM700
OLEIN, 1,2-DI-, (S)-(−)- see SED525
OLEJ NAPEDOWY III see DHE900
OLEJ NAPEDOWY III see FOP000
OLEOAKARITHION see TNP250
OLEOCUIVRE see CNO000
OLEOFAC see IOT000
OLEOGESAPRIM see ARQ725
OLEOL see OBA000
OLEOMYCETIN see CDP250
OLEONITRILE see OAX300
OLEO NORDOX see CNO000
OLEOPARAPHENE see PAK000
OLEOPARATHION see PAK000
OLEOPHOSPHOTHION see MAK700
OLEORESIN TUMERIC see TOD625
OLEOSUMIFENE see DSQ000
OLEOVITAMIN A see VSK600
OLEOVITAMIN D see VSZ100
OLEOVITAMIN D3 see CMC750
OLEOVOFOTOX see MNH000
OLEOX 5 see PJY100
1-OLEOYLAZIRIDINE see OIC000
OLEOYLETHYLENEIMINE see OIC000
OLEOYLGLYCEROL see GGR200
OLEOYLMETHYLTAURINE SODIUM SALT
see SIY000
OLEOYLNITRILE see OAX300
OLEPAL I see PJY100
OLEPAL III see PJY100
OLEPTAN see FMU000
OLETAC 100 see PMP500
OLETETRIN see TBX000
OLEUM see SOI520
OLEUM ABIETIS see PIH750
OLEUM SINAPIS VOLATILE see AGJ250
OLEUM TIGLII see COC250
OLEYL ALCOHOL see OBA000
OLEYL ALCOHOL EO (2) see OIG000
OLEYL ALCOHOL EO (10) see OIG040
OLEYL ALCOHOL EO (20) see PJW500
OLEYL ALCOHOL condensed with 10 moles
ETHYLENE OXIDE see OIG040
OLEYL ALCOHOL condensed with 2 moles
ETHYLENE OXIDE see OIG000
OLEYL ALCOHOL condensed with 20 MOLES
ETHYLENE OXIDE see PJW500
OLEYLAMIN (GERMAN) see OHM700
OLEYL AMINE see OHM700
OLEYLAMINE-HF see OII200
OLEYLAMINE HYDROFLUORIDE see
OII200
OLEYLAMINHYDROFLUORID (GERMAN)
see OII200

OLEYLMONOGLYCERIDE see GGR200
OLEYL NITRILE see OAX300
OLEYLONITRILE see OAX300
OLEYLPOLYOXETHYLENE GLYCOL
ETHER see OIK000
OLIBANUM GUM see OIM000
OLIBANUM OIL see OIM025
OLICINE see GGR200
OLIGO(DA) see PJQ400
OLIGOMERIC PHOSPHATE ESTER see
REA050
OLIGOMYCIN A, mixed with OLIGOMYCIN
B see OIS000
OLIGOMYCIN B, mixed with OLIGOMYCIN
A see OIS000
OLIGOMYCIN C see OIM100
OLIGO Z see SFO100
OLIMPEN see MCH525
OLIMPLEX see NCK100
OLIN see DXN300
OLIN 53139 see HEF500
'OLINANA (HAWAII) see OHM875
OLIN MATHIESON 2,424 see EFK000
OLIN MO. 2174 see AQO000
OLIO DI CROTON (ITALIAN) see COC250
OLIO DI RICINO IDROGENATO see
HHW502
OLITENSOL see BBW750
OLITREF see DUV600
OLIVE AR, ANTHRIMIDE see IBJ000
OLIVE OIL see OIQ000
OLIVE R BASE see IBJ000
OLIVOMITSIN see OIS000
OLIVOMITSIN see OIU499
OLIVOMYCIN see OIS000
OLIVOMYCIN see OIU499
OLIVOMYCIN A see OIU000
OLIVOMYCIN A see OIU499
OLIVOMYCIN D see OIU499
OLIVOMYCINE see OIU499
OLIVOMYCIN I see OIU000
OLIVOMYCIN, SODIUM SALT see CMK750
OLIV OSTANTHRENOVY R see DUP100
'OLIWA (HAWAII) see OHM875
OLOSED see MNM500
OLOTHORB see PKL100
OLOW (POLISH) see LCF000
OLPISAN see PAX000
OLSAEURENITRIL see OAX300
OLTITOX see CBM750
OLYMP see THS860
OM see EEA550
OM 518 see MFD500
OM-1455 see LIN400
OM-1463 see TIP750
OM 1562 see BLG600
OM-1563 see ZMJ000
OM 2424 see EFK000
OM 2803 see AFA200
OM 53139 see HEF500
OMACIDE 24 see HOC000
OMADINE see HOB500
OMADINE see MCQ700
OMADINE, CUPRIC see BLG600
OMADINE MDS see OIU850
OMADINE ZINC see ZMJ000
OMAFLORA see HHC000
OMAHA see LCF000
OMAHA & GRANT see LCF000
OMAIN see MIW500
OMAINE see MIW500
OMAIT see SOP000
OMAL see TIW000
OMAZENE see CNL310
OMAZINE see CNL310
OMCA see TJW500
OMCHLOR see DFE200
OMEGA see IAL200
OMEGA 127 see MKP500
OMEGA-BENZOYLACETOPHENONE see
PFU300

OMEGA CHROME BLUE BLACK B see
CMP880
OMEGA CHROME BLUE FB see HJF500
OMEGA CHROME DARK VIOLET D see
HLI000
OMEGA CHROME ORANGE G see CMP882
OMEGA CHROME RED SB see CMG750
α,OMEGA-DIAMINO-4,9-
DIOXADODECANE see BGU600
OMEGA-NITROARGININE see OIU900
OMEGA-NITRO-l-ARGININE see OIU900
7-OMEN see MCB600
OMETHOAT see DNX800
OMETHOATE see DNX800
O¹⁴-METHYLDELECTINE see LJB150
OMF 59 see PHA750
OM-HYDANTOINE see DKQ000
OM-HYDANTOINE SODIUM see DNU000
OMI see DEA100
OMI see FAB000
OMIFIN see CMX700
OMITE see SOP000
OMNIBON see SNN300
OMNI-PASSIN see DJT800
OMNIPEN see AIV500
OMNIPEN-N see SEQ000
OMNIZOLE see TEX000
OMNYL see QAK000
O⁶-MONOACETYLMORPHINE see ACR600
OMP-1 see OIW000
OMP-2 see OIY000
OMP-4 see OJA000
OMP-5 see OJC000
OMPA see OCM000
OMPACIDE see OCM000
OMPATOX see OCM000
OMPAX see OCM000
OMPERAN see EPD500
OMS 2 see FAQ900
OMS 14 see DGP900
OMS-15 see COF250
OMS-29 see CBM750
OMS-32 see MIA250
OMS-33 see PMY300
OMS 43 see DSQ000
OMS-47 see DOS000
OMS-93 see DST000
OMS 115 see DGD800
OMS-174 see CGI500
OMS 468 see AFR250
OMS 570 see EAQ750
OMS-597 see TMD000
OMS-658 see BNL250
OMS-659 see EGV500
OMS-708 see BDG250
OMS-771 see CBM500
OMS 834 see MRH209
OMS-978 see IRC050
OMS-0971 see CMA100
OMS 1075 see DRR400
OMS-1092 see IOV000
OMS-1155 see CMA250
OMS-1206 see BEP500
OMS-1211 see IEN000
OMS 1325 see FAB400
OMS 1328 see CDS750
OMS 1804 see CJV250
OMS 1809 see PDR700
OMS 1810 see PDR700
OMS-2005 see MDN150
OMS-2014 see CQE400
OMS 3023 see FAR150
OMSAT see TKX000
OMT see SIY000
OMTAN see OAN000
OMYA see CAT775
OMYA BLH see CAT775
OMYACARB F see CAT775
OMYALENE G 200 see CAT775
OMYALITE 90 see CAT775
ONAONA-IAPANA (HAWAII) see DAC500
ONCB see CJB750

ONCO-CARBIDE see HOO500
ONCO-CARBIDE see TGN250
ONCODAZOLE see OJD100
ONCOL see AKC570
ONCOL 5G see AKC570
No. 49 CONCENTRATED BENZIDINE
YELLOW see DEU000
ONCOSTATIN see AEA500
ONCOSTATIN K see AEB000
ONCOTEPA see TFQ750
ONCOTIOTEPA see TFQ750
ONCOVEDEX see TND000
ONCOVIN see LEY000
ONCOVIN see LEZ000
ONDANSETRON see OJD150
ONDENA see DAC200
O,N-DIETHYL-N-
NITROSOHYDROXYLAMINE see NKC500
ONDOGYNE see FBP100
ONDONID see FBP100
ONECIDE see FDA885
ONECIDE EC see FDA885
ONE-IRON see FBJ100
ONEX see CJT750
ONGROVIL S 165 see PKQ059
ONION see WBS850
ONION OIL see OJD200
ONION TREE see WBS850
ONKOTIN see DBD700
ONL 900 see SLJ550
ONMEX see PAP240
ONO 802 see CDB775
ONOZUKA P 500 see CCU150
ONQUININ see PAF550
ONT see NMO525
ONTOSEIN see OJM400
ONTRACIC 800 see MFL250
ONTRACK see MFL250
ONTRACK 8E see MQQ450
ONTRACK-WE-2 see MFL250
ONYX see SCI500
ONYX see SCJ500
ONYX BTC (ONYX OIL & CHEM CO) see
AFP250
ONYXIDE 200 see THR820
ONYXIDE 3300 see BBA625
ONYXIDE (ONYX OIL & CHEM CO) see
AFN750
ONYXOL 345 see BKE500
O,O'DIANISIDINE see DCJ200
O,O,O'O'-TETRAMETHYL-O,O-THIODI-p-
PHENYLENE PHOSPHOROTHIOATE see
TAL250
OOS see TAB750
OP-7 see OJD230
OP-10 see OJD232
OP 1062 see GHS000
OPACIN see TDE750
OPALINE GREEN G 1 see PJQ100
OPALON see PKQ059
OPALON 400 see AAX175
OPARENOL see DNG400
OPC 1085 see MQT550
OPC 2009 see PME600
OPC-8212 see DLF700
OPC-13013 see CMP825
OPE-3 see OFM000
OPE 30 see PKF500
OPERIDINE see DAM700
OPERTIL see ECW600
OPHIOBOLIN see CNF109
OPHIOBOLIN A see CNF109
OPHIOPHAGUS HANNAH VENOM see
ARV125
OPHTALMOKALIKAN see KAV000
OPHTALMOKALIXAN see KAM000
OPHTHALAMIN see VSK600
OPHTHALMADINE see DAS000
OPHTHAZIN see OJD300
OPHTHOCILLIN see DVU000
OPIAN see NBP275
OPIAN see NOA000

OPIAN HYDROCHLORIDE see NOA500
OPIANINE see NOA000
OPIANINE HYDROCHLORIDE see NOA500
OPILON HYDROCHLORIDE see TFY000
OPINSUL see AKO500
OPIPRAMOL see DCV800
OPIPRAMOL DIHYDROCHLORIDE see IDF000
OPIUM see OJG000
OPIUM, PYROLYZATE see OJG509
OPLOSSINGEN (DUTCH) see FMV000
OPOBALSAM see BAF000
OPOSIM see ICC000
OPP see BGJ250
OPP-Na see BGJ750
OPP-SODIUM see BGJ750
OPRAMIDOL see DCV800
OPREN see OJI750
OPSB see BRP250
OP-SULFA 30 see SNQ710
OPTAL see PND000
OPTALIDON see AGI750
OPTANOL FAST SCARLET GN see CMM320
OPTANOL RED R see NAO600
OPTANOL SCARLET GS see CMM325
OPTANOL YELLOW R see CMM759
OPTEF see CNS750
OPTENYL see PAH250
OP-THAL-ZIN see ZNA000
OPTHEL-S see SNQ710
OPTHOCHLOR see CDP250
OPTICAL BLEACH 13-61 see OOI200
OPTIDASE see CCP525
OPTIMAX see OJG550
OPTIMIL see MDT250
OPTIMINE see DLV800
OPTIMYCIN see MDO250
OPTINOXAN see QAK000
OPTION see FAQ200
OPTIPEN see PDD350
OPTIPHYLLIN see TEP000
OPTISULIN LONG see IDF300
OPTIVAL see PLY275
OPTOCHIN see HHR700
OPTOCHIN HYDROCHLORIDE see ELC500
OPTOCHINIDIN see QFS100
OPTOCIL see SCK600
OPTOCIL (QUARTZ) see SCK600
OPTOQUINE see HHR700
OPTOQUINHYDROCHLORIDE see ELC500
8-OQ see QPA000
OR 1191 see FAB400
OR 1191 see PGX300
OR-1549 see AJP125
OR-1550 see AJL875
OR-1556 see AJN375
OR-1578 see AJO625
OR 66780 see EIP050
ORABET see BSQ000
ORABILEX see EQC000
ORABOLIN see EJT600
ORACAINE HYDROCHLORIDE see OJI600
ORACEF see ALV000
ORACEFAL see DYF700
ORACET RED 3B see AKE250
ORACET SAPPHIRE BLUE G see TBG700
ORACET VIOLET 2R see DBP000
ORACET VIOLET B see AKP250
ORACET VIOLET BN see AKP250
4(OR 6)-(ACETYLOXY)-5(OR 4)-HEXENOIC ACID see ACU600
ORACILLIN see PDT500
α-ORACILLIN see PDD350
ORACIL-VK see PDT750
ORACON see DNX500
ORACONAL see MCA500
ORADEXON see SOW000
ORADEXON PHOSPHATE see BFW325
ORADIAN see CDR550
ORADIAN see CKK000
ORADIL see CLY600
ORAFLEX see OJI750

ORAFURAN see NGE000
ORAGEST see MCA000
ORAGESTON see AGF750
ORAGRAFIN-SODIUM see SKM000
ORAHEXAL see CDT250
ORALCID see ABX500
ORALIN see BSQ000
ORALITH ORANGE PG see CMS145
ORALITH RED see CJD500
ORALITH RED 2GL see DVB800
ORALITH RED P4R see MMP100
ORALITH RED SR WATER SOLUBLE see NAP100
ORALITH SCARLET 3B WATER SOLUBLE see CMG750
ORALOPEN see PDD350
ORALSONE see HHR000
ORALSTERONE see AOO275
ORA-LUTIN see GEK500
ORAMID see SAH000
ORANGE 3 see MND600
L-ORANGE 2 see FAG150
ORANGE #10 see HGC000
1333 ORANGE see FAG010
11550 ORANGE see CMM220
ORANGE ACID G see MRL100
ORANGE A l'HUILE see PEJ500
ORANGE B see OJK325
ORANGE BASE CIBA II see NEO000
ORANGE BASE CIBA IV see CEH690
ORANGE BASE IRGA I see NEN500
ORANGE BASE IRGA II see NEO000
ORANGE BASE IRGA IV see CEH690
ORANGE BASE NGC see CEH690
ORANGE CHROME see LCS000
ORANGE CRYSTALS see ABC500
ORANGE EXTRA N see CMM220
ORANGE EXTRA P see CMM220
ORANGE G see CMS145
ORANGE G (indicator) see HGC000
ORANGE G (biological stain) see HGC000
ORANGE GC BASE see CEH675
ORANGE G DYE see HGC000
ORANGE I see FAG010
ORANGE II see CMM220
ORANGE IIC see CMM220
ORANGE III see MND600
ORANGE II for LAKES see CMM220
ORANGE IIP see CMM220
ORANGE II R see FAG150
ORANGE IIS see CMM220
ORANGE IISM see CMM220
ORANGE II SPECIAL FOR LACQUER see CMM220
ORANGE INSOLUBLE OLG see PEJ500
ORANGE INSOLUBLE OLG see XRA000
ORANGE INSOLUBLE RR see XRA000
ORANGE LEAD see LDS000
ORANGE LEAF OIL, BITTER see OHJ150
ORANGE LEAF WATER, ABSOLUTE see OHJ150
ORANGE No. 203 see DVB800
ORANGE No. 205 see CMM220
ORANGE NITRATE CHROME see LCS000
ORANGE OIL see OGY000
ORANGE OIL see OJK330
ORANGE OIL see PDD750
ORANGE OIL, BITTER, COLDPRESSED see OJK330
ORANGE OIL, DISTILLED see OJK340
ORANGE OIL, coldpressed (FCC) see OGY000
ORANGE OIL KB see XRA000
ORANGE PAL see FAG150
ORANGE PEL see PEJ500
ORANGE PIGMENT X see DVB800
ORANGE RESENOLE No. 3 see PEJ500
ORANGE RGL CONC. SPECIALLY PURE see FAG150
ORANGE ROOT see ICC800
ORANGES see BBK500
ORANGE SALT CIBA II see NEO000
ORANGE SALT IRGA II see NEO000

ORANGE SALT NRD see CEG800
ORANGE 2 SODIUM SALT see CMM220
ORANGE SOLUBLE A l'HUILE see PEJ500
ORANGE 3R SOLUBLE IN GREASE see TGW000
ORANGE TONER GRT see CMM220
ORANGE Y see CMM220
ORANGE Y see CMS145
ORANGE YA see CMM220
ORANGE YELLOW S see FAG150
ORANGE YELLOW S.AF see FAG150
ORANGE YELLOW S.FQ see FAG150
ORANGE YZ see CMM220
ORANGE ZH see CMM820
ORANIL see BSM000
ORANIXON see GGS000
ORANYL see BSM000
ORANZ BRILANTNI OSTAZINOVA S-2R (CZECH) see DGO000
ORANZ G (POLISH) see HGC000
ORANZ KYPOVA 5 see CMU815
ORANZ KYSELA 7 see CMM220
ORANZ POTRAVINARSKA 3 see CMP600
ORANZ ROZPOUSTEDLOVA 1 see CMP600
ORANZ ZASADITA 2 (CZECH) see DBP999
ORAP see PIH000
ORAPEN see PDT750
ORARSAN see ABX500
ORASECRON see NNL500
ORASONE see PLZ000
ORASPOR see CCS530
ORASTHIN see ORU500
ORASULIN see BSM000
ORATESTIN see AOO275
ORA-TESTRYL see AOO275
ORATRAST see BAP000
ORATREN see PDT500
ORATROL see DEQ200
ORAVIRON see MPN500
ORAVUE see IGA000
ORBAMIN BLACK A CONC. see CMN300
ORBAMIN BORDEAUX B see CMO872
ORBAMIN BROWN MS see CMO800
ORBAMIN GREEN B see CMO840
ORBAMIN SCARLET 4BS see CMO870
ORBANTIN TURQUOISE BLUE GL see COF420
ORBENIN SODIUM see SLJ050
ORBENIN SODIUM HYDRATE see SLJ000
ORBICIN see DCQ800
ORBICIN see PAG050
ORBINAMON see NBP500
ORBIT see PMS930
ORBON see OAV000
ORCA 100-2 see MCB050
ORCANON see MPW500
ORCED see DGN200
ORCHARD BRAND ZIRAM see BJK500
ORCHIOL see TBG000
ORCHISTIN see TBG000
ORCIN see MPH500
ORCINOL see MPH500
ORCIPRENALINE see DMV800
ORCIPRENALINE SULFATE see MDM800
2,4(OR 4,6)-DIETHYL-6(or 2)-METHYL-1,3-BENZENEDIAMINE see DJP100
ORDIMEL see ABB000
ORDINARY AZOXYBENZENE see ASO750
ORDINARY LACTIC ACID see LAG000
ORDRAM see EKO500
OREGON BALSAM see OJK340
OREGON HOLLY see HGF100
OREMET see TGF250
ORESOL see RLU000
ORESON see RLU000
ORESTOL see DKB000
2(OR 5)-ETHYL-4-HYDROXY-5(OR 2)-METHYL-3(2H)-FURANONE see HND150
ORETIC see CFY000
ORETON see TBG000
ORETON-F see TBF500
ORETON-M see MPN500

ORETON METHYL see MPN500
ORETON PROPIONATE see TBG000
OREZAN see BSQ000
ORF 2166 see CBO625
ORF 3858 see FAO200
ORF 4563 see MBV775
ORF 5513 see BJG150
ORF 8511 see DWF300
ORF 13811 see OJK300
ORF 18489 see OGI300
ORFENADRINA see DPH000
ORG-483 see EJT600
ORG 2969 see DBA750
ORG 10849 see AQN550
ORG 485-50 see NNV000
ORGA-414 see AMY050
ORGA 3045 see SKE100
ORGABOLIN see EJT600
ORGABORAL see EJT600
ORGAMETIL see NNV000
ORGAMETRIL see NNV000
ORGAMETROL see NNV000
ORGAMIDE see PJY500
ORGANEX see CAK500
ORGANIC GLASS E 2 see PKB500
ORGANIDIN see IEL800
ORGANIL 644 see MAP300
ORGANOFUNCTIONAL SILANE 45-49 see
TLC650
ORGANOL 2R see NBG500
ORGANOL BORDEAUX B see EOJ500
ORGANOL BROWN 2R see NBG500
ORGANOL FAST GREEN J see BLK000
ORGANOL ORANGE see PEJ500
ORGANOL ORANGE 2J see CMP600
ORGANOL ORANGE 2R see TGW000
ORGANOL RED B see SBC500
ORGANOL RED BS see OHI200
ORGANOL SCARLET see OHI200
ORGANOL VERMILION see CMS238
ORGANOL YELLOW see PEI000
ORGANOL YELLOW 25 see AIC250
ORGANOL YELLOW ADM see DOT300
ORGANOMETALS see OJM000
ORGANON see AGF750
ORGANON'S DOCA ACETATE see DAQ800
ORGASEPTINE see SNM500
ORGASTERON see MDM350
ORGASTYPTIN see EDU500
ORGATRAX see HOR470
ORG GB 94 see BMA625
ORG-NA 96 see AOO403
ORG NA 97 see PAF625
ORG NC 45 see VGU075
ORGOTEIN see OJM400
ORGOTEINS see OJM400
2-((2,3,3A,4,7,7A(OR-3A,4,5,6,7,7A)-
HEXAHYDRO-4,7-METHANO-1H-
INDENYL)OXY)ETHANOL see EJJ100
5(OR 4)-HEXENOIC ACID, 4(OR 6)-
(ACETYLOXY)- see ACU600
ORICUR see CHX250
ORIDIN see OJV500
ORIENTAL BERRY see PIE500
ORIENT BASIC MAGENTA see MAC250
ORIENT OIL BLACK HBB see PCJ200
ORIENT OIL ORANGE PS see PEJ500
ORIENT OIL RED OG see CMS238
ORIENT OIL RED RR see SBC500
ORIENT OIL YELLOW GG see DOT300
ORIENTOMYCIN see CQH000
ORIENT SPIRIT BLACK AB see CMS236
ORIENT SPIRIT BLACK SB see CMS236
ORIENT WATER BLACK 100L see CMN240
ORIENT WATER BLACK 200L see CMN240
ORIENT WATER PINK 2 see ADG250
ORIFUNGAL M see KFK100
ORIGANUM OIL see OJO000
ORIGINAL CUBATAO SULPHUR BLACK
see CMS250
ORIMETEN see AKC600
ORIMON see PDP250

ORINASE see BSQ000
ORINAZ see BSQ000
ORION BLUE 3B see CMO250
ORIPAVINE, 6,14-endo-
ETHYLENETETRAHYDRO-7-(1-
HYDROXY-1-METHYLBUTYL)-(7CI) see
EQO450
ORISUL see AIF000
ORISULF see AIF000
γ-ORIZANOL see OJY200
ORIZON see PJS750
ORIZON 805 see PJS750
ORLON, combustion products see ADX750
ORLUTATE see ABU000
ORMETEIN see OJM400
ORMETOPRIM see OJO100
ORNAMENTAL WEED see AJM000
ORNID see BMV750
ORNIDAZOLE see OJS000
ORNITHINE see OJS100
(+)-ORNITHINE see OJS100
ORNITHINE, l- see OJS100
l-(−)-ORNITHINE see OJS100
(S)-ORNITHINE see OJS100
(+)-S-ORNITHINE see OJS100
dl-ORNITHINE, 2-(DIFLUOROMETHYL)-,
MONOHYDROCHLORIDE see EAE775
l-ORNITHINE HYDROCHLORIDE see
OJU000
l-ORNITHINE MONOHYDROCHLORIDE
see OJU000
L-ORNITHINE, N(5)-
(IMINO(NITROAMINO)METHYL)- see
OIU900
l-ORNITHINE, N(5)-
(IMINO(NITROAMINO)METHYL)-,
METHYLESTER see NDY675
ORNITHINE, N(5)-(NITROAMIDINO)-, l- see
OIU900
ORNITROL see ARX800
OROBETINA see DXO300
OROFAR see BDJ600
OROLEVOL see MQU750
ORONOL see ART250
OROPUR see OJV500
OROSOMYCIN see TFC500
OROTIC ACID see OJV500
OROTIC ACID mixed with CHOLESTEROL
and CHOLIC ACID (2:2:1) see OJV525
OROTIC ACID, SODIUM SALT, HYDRATE
(1:1:1) see SIY100
OROTONIN see OJV500
OROTSAURE (GERMAN) see OJV500
OROTURIC see OJV500
OROTYL see OJV500
OROVERMOL see TDX750
OROXIN see ALV000
OROXINE see LFG050
5-(3 OR 6-OXO-1-CYCLOHEXEN-1-YL)-5-
ETHYLBARBITURIC ACID see OJV600
ORPHENADINE see MJH900
ORPHENADRIN see MJH900
ORPHENADRINE see MJH900
ORPHENADRINE CITRATE see DPH000
ORPHENADRINE HYDROCHLORIDE see
OJW000
ORPHENOL see BGJ750
ORPIMENT see ARI000
ORQUISTERONE see TBF500
ORRIS see OHJ130
ORRIS ABSOLUTE see OHJ130
ORRIS CONCRETE see OHJ130
ORRIS CONCRETE OIL see OHJ130
ORRIS EXTRACT see OHJ130
ORRIS Liquid see OHJ130
ORRIS OIL see OHJ130
ORRIS RESIN see OHJ130
ORRIS RESINOID see OHJ130
ORRIS ROOT EXTRACT see OHJ130
ORRIS ROOT OIL see OHJ130
ORRIS ROOT OIL see OJW100
ORSILE see BEQ625

ORSIN see PEY500
ORTAL SODIUM see EKT500
ORTEDRINE see BBK000
ORTHAMINE see PEY250
ORTHANILIC ACID see SNO100
ORTHENE see DOP600
ORTHENE-755 see DOP600
ORTHESIN see EFX000
ORTHO 4355 see NAG400
ORTHO 5353 see BTA250
ORTHO-5655 see MOU750
ORTHO 5865 see CBF800
ORTHO 9006 see DTQ400
ORTHO 12420 see DOP600
ORTHO 20615 see CGI550
l'ORTHO, p-AMINONITROPHENOL, SEL
SODIQUE (FRENCH) see ALO500
ORTHOARSENIC ACID see ARB250
ORTHOARSENIC ACID HEMIHYDRATE
see ARC500
ORTHOBENZOIC ACID, cyclic 7,8,10a-
ESTER with 5,6-EPOXY-
4,5,6,6a,7,8,9,10,10a,10b-DECAHYDRO-
3a,4,7,8,10a-PENTAHYDROXY-5-
(HYDROXYMETHYL)-8-ISOPROPENYL-
2,10-DIMETHYLBEN Z(e)AZULEN-3(3aH)-
ONE see DAB850
ORTHOBORIC ACID see BMC000
ORTHOCAINE see AKF000
ORTHO C-1 DEFOLIANT & WEED KILLER
see SFS000
ORTHOCHLOROPARANISIDINE see
CEH750
ORTHOCHROME ORANGE R see NEY000
ORTHOCHROME YELLOW GGW see
SIT850
ORTHOCIDE see CBG000
ORTHOCRESOL see CNX000
ORTHODERM see AKF000
ORTHODIAZINE see OJW200
ORTHODIBROM see NAG400
ORTHODIBROMO see NAG400
ORTHODICHLOROBENZENE see DEP600
ORTHODICHLOROBENZOL see DEP600
ORTHO-DIQUAT see EJC025
ORTHO N-4 DUST see NDN000
ORTHO N-5 DUST see NDN000
ORTHO EARWIG BAIT see DXE000
ORTHO-FLUORODIPHENYL see FGI050
ORTHOFORM see AKF000
ORTHOFORMIC ACID, ETHYL ESTER see
ENY500
ORTHOFORMIC ACID, TRIETHYL ESTER
see ENY500
ORTHOFORMIC ACID, TRIMETHYL
ESTER see TLX600
ORTHO P-G BAIT see COF500
ORTHO GRASS KILLER see CBM000
ORTHOHYDROXYBENZOIC ACID see
SAI000
ORTHOHYDROXYDIPHENYL see BGJ250
ORTHOIODIN see HGB200
ORTHO-KLOR see CDR750
ORTHO L10 DUST see LCK000
ORTHO L10 DUST see LCK100
ORTHO L40 DUST see LCK000
ORTHO-LM APPLE SPRAY see MLH000
ORTHO LM CONCENTRATE see MLH000
ORTHO LM SEED PROTECTANT see
MLH000
ORTHO MALATHION see MAK700
ORTHO MC see MAE000
ORTHO-MITE see SOP500
ORTHOMRAVENCAN ETHYLNATY
(CZECH) see ENY500
ORTHOMRAVENCAN METHYLNATY
(CZECH) see TLX600
ORTHONAL see QAK000
ORTHO-NOVUM see MDL750
ORTHO PARAQUAT CL see PAJ000
ORTHOPHALTAN see TIT250
ORTHOPHENANTHROLINE see PCY250

ORTHOPHENYLALANINE MUSTARD see BHT250
ORTHOPHENYLPHENOL see BGJ250
ORTHOPHOS see PAK000
ORTHO PHOSPHATE DEFOLIANT see BSH250
ORTHOPHOSPHORIC ACID see PHB250
ORTHOPHOSPHORUS ACID see PGZ899
ORTHORIX see CAX800
ORTHOSAN MB see DTC600
ORTHOSERPINA see RDK000
ORTHOSIL see SJU000
ORTHOSILICIC ACID see SCN600
ORTHOTELLURIC ACID see TAI750
ORTHOTOLUIC ACID see TGQ000
ORTHO-TOLUOL-SULFONAMID (GERMAN) see TGN250
ORTHOTRAN see CJT750
ORTHOVANILLINE see VFP000
ORTHO WEEVIL BAIT see DXE000
ORTHOXENOL see BGJ250
ORTHOXINE HYDROCHLORIDE see OJY000
ORTIN see TJL250
ORTISPORINA see ALV000
ORTIZON see HIB500
ORTODRINEX HYDROCHLORIDE see OJY000
ORTOL see GGS000
ORTONAL see QAK000
ORTRAN see DOP600
ORTRIL see DOP600
3,7,7(OR 4,7,7)-TRIMETHYLBICYCLO(4.1.0)HEPT-3-ENE see CCK510
ORTUDUR see PKQ059
ORUDIS see BDU500
O 4 (RUSSIAN STABILIZER) see MHR025
ORUVAIL see BDU500
ORVAGIL see MMN250
ORVINYLCARBINOL see AFV500
ORVUS WA PASTE see SIB600
ORYZAEMATE see PMD800
ORYZALIN see OJY100
ORYZALIN see OJY100
ORYZANIN see TES750
ORYZANINE see TES750
γ-ORYZANOL see OJY200
ORYZON see PAX775
OS 30 see PJT300
OS 55 see PJT300
OS-0704 see BGC125
OS 1836 see CLV375
OS 1897 see DDL800
OS 2046 see MQR750
OS-3966-A see RMK200
OSACYL see AMM250
OSAGE ORANGE see MRN500
OSAGE ORANGE CRYSTALS see MRN500
OSAGE ORANGE EXTRACT see MRN500
OSALMID see DYE700
OSALMIDE see DYE700
OSARSAL see ABX500
OSARSOLE see ABX500
OSBAC see MOV000
OSBON AC see PCL500
OS-CAL see CAT775
OSCINE see SBG000
OSCOPHEN see ARP250
O. SCUTELLATUS (AUSTRALIA) VENOM see ARV500
OSDMP see DJR700
OSIREN see AFJ500
OSIROL see DLR300
OSMIC ACID see OKK000
OSMITROL see HER000
OSMIUM see OKE000
OSMIUM HEXAFLUORIDE see OKG000
OSMIUM(IV) OXIDE see OKI000
OSMIUM(VIII) OXIDE see OKK000
OSMIUM TETROXIDE see OKK000
OSMOCAINE see OKK100

OSMOFERRIN see IHK000
OSMOSOL EXTRA see PND000
OSNERVAN see CPQ250
OSOCIDE see CBG000
OSPEN see PDT500
OSPENEFF see PDT500
OS-40 (PHOSPHATE ESTER) see OKK500
OSSALIN see SHF500
OSSIAMINA see CFA750
OSSIAN see OOG000
OSSICHLORIN see CFA750
OSSIDO di MESITILE (ITALIAN) see MDJ750
OSSIN see SHF500
OSTACET BRILLIANT RED E-LB see AKI750
OSTACET YELLOW P2G see AAQ250
OSTACET YELLOW SE-LG see KDA075
OSTAMER see CDV625
OSTANTHREN BLUE BCL see DFN300
OSTANTHREN BLUE BCS see DFN300
OSTANTHREN BLUE RS see IBV050
OSTANTHREN BLUE RSN see IBV050
OSTANTHREN BLUE RSZ see IBV050
OSTANTHREN BROWN BR see CMU770
OSTANTHRENE BLUE RS see IBV050
OSTANTHRENE BROWN BR see CMU770
OSTANTHRENE ORANGE GR see CMU820
OSTANTHREN GREY M see CMU475
OSTANTHREN OLIVE R see DUP100
OSTANTHREN ORANGE GR see CMU820
OSTANTHREN REDDISH BROWN 5RF see CMU800
OSTAZIN BLACK H-N see CMS220
OSTAZIN BRILLIANT BLUE S-R see CMS222
OSTAZIN BRILLIANT RED S 5B see PMF540
OSTELIN see VSZ100
OSTENOL see TLG000
OSTENSIN see TLG000
OSTEOBOND SURGICAL BONE CEMENT see PKB500
OSTEO D see HJS900
OSTREOGRYCIN see VRF000
OSTREOGRYCIN B see VRA700
OSUTIDINE see OKK600
OSVARSAN see ABX500
OSWEGO ORANGE X 2065 see CMS145
OSYRITRIN see RSU000
OSYROL see AFJ500
OSYROL see DLR300
OT see HLC000
OTACRIL see DFP600
OTAHEITE WALNUT (VIRGIN ISLANDS) see TOA275
OTAMOL see OKO200
OTAN see OLW100
OTB see ONI000
OTBE (FRENCH) see BLL750
OTC see HOH500
OTERBEN see BSQ000
OTETRYN see HOI000
OTIFURIL see FPI000
OTILONIUM BROMIDE see OKO400
OTOBIOTIC see NCG000
OTODYNE see EQQ500
OTOFURAN see NGE500
OTOKALIXIN see KAM000
OTOKALIXIN see KAV000
OTOPHEN see CDP250
OTOS see OPQ100
OTRACID see CJT750
OTRIVINE HYDROCHLORIDE see OKO500
OTRIVIN HYDROCHLORIDE see OKO500
OTS see TGN250
OTS 11 see DVM800
OTS 15 see DVN400
OTsS 14 see PJT300
OTsS 20 see PJT300
OTTAFACT see CFD990
OTTANE (ITALIAN) see OCU000
OTTASEPT see CLW000
OTTASEPT EXTRA see CLW000

OTTILONIO BROMURO (ITALIAN) see OKO400
1,2,4,5,6,7,8,8-OTTOCHLORO-3A,4,7,7A-TETRAIDRO-4,7-endo-METANO-INDANO (ITALIAN) see CDR750
OTTO FUEL II see OKO600
OTTOMETIL-PIROFOSFORAMMIDE (ITALIAN) see OCM000
OTTO ROSE see RNA000
OTTO of ROSE see RNA000
OTTO of ROSE see RNF000
OTTO ROSE TURKISH see OHJ000
OUABAGENIN-l-RHAMNOSID (GERMAN) see OKS000
OUABAGENIN-l-RHAMNOSIDE see OKS000
OUABAIN see OKS000
OUABAINE see OKS000
OU-B see CBT500
OUBAIN see OKS000
OUDENONE see OKS100
OUDENONE SODIUM SALT see OKS150
OUPLATE see CNT350
OURARI see COF750
OUTFLANK see AHJ750
OUTFLANK-STOCKADE see AHJ750
OUTFOX see CQI750
(OV₁) see HNK600
OVABAN see VTF000
OVADOFOS see DSQ000
OVADZIAK see BBQ500
OVAHORMON see EDO000
OVAHORMON BENZOATE see EDP000
OVALBUMINS see AFI780
OVAMIN 30 see DAP810
OVANON see LJE000
OVARELIN see LIU370
OVARIOSTAT (FRENCH) see LJE000
OVASTEROL see EDO000
OVASTEROL-B see EDP000
OVASTEVOL see EDO000
OVATRAN see CJT750
OVEST see ECU500
OVESTERIN see EDU500
OVESTIN see EDU500
OVESTINON see EDU500
OVESTRION see EDU500
OVETTEN see MNM500
OVEX see CJT750
OVEX see EDP000
OVEX see EDV000
OVIDON see NNL500
OVIFOLLIN see EDV000
OVIN see DNX500
OVINE PITUITARY INTERSTITIAL CELL STIMULATING HORMONE see LIU300
OVISOT see ABO000
OVISOT see CMF250
OVITELMIN see MHL000
OVOCHLOR see CJT750
OVOCICLINA see EDO000
OVOCYCLIN see EDO000
OVOCYCLIN BENZOATE see EDP000
OVOCYCLIN DIPROPIONATE see EDR000
OVOCYCLINE see EDO000
OVOCYCLIN M see EDP000
OVOCYCLIN-MB see EDP000
OVOCYCLIN-P see EDR000
OVOCYLIN see EDO000
OVOSTAT see EEH575
OVOSTAT 1375 see EEH575
OVOSTAT E see EEH575
OVOTOX see CJT750
OVOTRAN see CJT750
OVRAL see NNL500
OVRAL 21 see NNL500
OVRAL 28 see NNL500
OVRAN see NNL500
OVRANETT see NNL500
O-V STATIN see NOH500
OVULEN see EQK100
OVULEN 30 see DAP810

4-OXA-2-THIA-3-PHOSPHA-1-STANNAHEXANE, 3-ETHOXY-1,1,1-TRIPHENYL-, 3-SULFIDE see EFL600
1,3,2-OXATHIASTANNOLANE, 2,2-DIBUTYL- see DEF150
6H-1,3,5-OXATHIAZINE, 4-(4-CHLOROPHENYL)- see CKH120
6H-1,3,5-OXATHIAZINE, 4-(4-METHOXYPHENYL)- see MFG520
6H-1,3,5-OXATHIAZINE, 2-METHYL-4-PHENYL- see MNV760
6H-1,3,5-OXATHIAZINE, 4-(4-METHYLPHENYL)- see MNV765
6H-1,3,5-OXATHIAZINE, 4-PHENYL- see PFT700
1,2-OXATHIETANE-2,2-DIOXIDE see OMC000
1,2-OXATHIOLANE-2,2-DIOXIDE see PML400
1,2-OXATHIOLANE, 5-METHYL-, 2,2-DIOXIDE see BOU300
OXATIMIDE see OMG000
OXATOMIDA see OMG000
OXATOMIDE see OMG000
OXATONE see OLW100
5-OXATRICYCLO(8.2.0.0^{4,6})DODECANE, 4,12,12-TRIMETHYL-9-METHYLENE- , (1R,4R,6R,10S)- see CCN100
3-OXATRICYCLO(3.2.1.0^{2,4})OCTANE-6-CARBOXYLIC ACID, ETHYL ESTER see ENZ300
2H-1,3,2-OXAZAPHOSPHORIN-2-AMINE, N-(2-CHLOROETHYL)TETRAHYDRO-, 2-OXIDE see MRG025
2-H-1,3,2-OXAZAPHOSPHORINANE see CQC650
OXAZEPAM see CFZ000
OXAZIL see MSC100
OXAZIMEDRINE see PMA750
OXAZINOMYCIN see OMK000
OXAZOCILLIN see DSQ800
OXAZOLAM see OMK300
2-OXAZOLAMINE, N-(3,4-DIMETHYLPHENYL)-4,5-DIHYDRO- see XOS500
OXAZOLAZEPAM see OMK300
OXAZOLIDIN see HNI500
OXAZOLIDINE A see DTG750
OXAZOLIDINE, 3-(DICHLOROACETYL)-5-(2-FURANYL)-2,2-DIMETHYL- see FPZ200
2,4-OXAZOLIDINEDIONE, 5,5-DIPHENYL- see DWI300
OXAZOLIDINE, 3-METHYL- see MND700
OXAZOLIDINE T see OMM300
2-OXAZOLIDINETHIONE, 5-ETHENYL-, (R)-(9CI) see VQA100
2-OXAZOLIDINETHIONE, 5-VINYL-, (R)- see VQA100
OXAZOLIDINE, 3-(3,4,5-TRIETHOXYBENZOYL)- see OMM100
OXAZOLIDIN-GEIGY see HNI500
OXAZOLIDINONE see MFD500
2-OXAZOLIDINONE see OMM000
2-OXAZOLIDINONE, 3-(FURFURYLIDENEAMINO)- see FPM300
3-OXAZOLIDINYL 3,4,5-TRIETHOXYPHENYL KETONE see OMM100
OXAZOLIDONE see OMM000
2-OXAZOLIDONE, 3-((2-FURANYLMETHYLENE)AMINO)- see FPM300
2-OXAZOLIDONE, 3-(((1-METHYLPYRROL-2-YL)METHYLENEAMINO)-4-(PIPERIDINOMETHYL)- see MPF300
2-OXAZOLINE, 3,4-XYLIDINO- see XOS500
2-OXAZOLINE, 2-(3,4-XYLIDINO)- see XOS500
1,3,4-OXAZOL-2(3H)-ONE, 3-(2,4-DICHLORO-5-(1-METHYLETHOXY)PHENYL)-5-(1,1-DIMETHYLETHYL)- see OMM200

1H,3H,5H-OXAZOLO(3,4-C)OXAZOLE-7A(7H)-METHANOL see OMM300
4H-1,3,2-OXAZOPHOSPHORIN-4-ONE, 2-(BIS(2-CHLOROETHYL)AMINO)TETRAHYDRO-, 2-OXIDE see KFK125
OXAZYL see MSC100
OXCORD see AEC750
OXEBETRINIL see OKS200
OXEDRINE see HLV500
m-OXEDRINE see NCL500
(−)-m-OXEDRINE see NCL500
m-OXEDRINE see SPC500
p-OXEDRINE see HLV500
OXEDRINE TARTRATE see SPD000
OXELADIN see DHQ200
OXELADIN CITRATE see OMS400
OXELADINE CITRATE see OMS400
OXENDOLONE see ELF100
2-OXEPANONE, 7-BUTYL- see BSB100
2-OXEPANONE, 7-BUTYL-(9CI) see BSB100
2-OXEPANONE (8CI, 9CI) see LAP000
OXEPINAC see OMU000
OXEPINACO (SPANISH) see OMU000
OXEPIN-3,6-ENDOPEROXIDE see EBQ550
OXETACAINE see DTL200
OXETACAINE HYDROCHLORIDE see EAN650
OXETAN see OMW000
OXETANE see OMW000
2-OXETANONE, 4-METHYL-, (+-)- see BSX100
OXETHACAINA (ITALIAN) see DTL200
OXETHAZAINE HYDROCHLORIDE see EAN650
OXETHAZINE see DTL200
OXFENDAZOLE see OMY500
OXIAMIN see IDE000
OXIARSOLAN see ARL000
OXIBENDAZOLE see OMY700
OXIBUTININA HYDROCHLORIDE see OPK000
OXIBUTOL see HNI500
OXICOB see CNK559
OXIDASE GLUCOSE see GFG100
OXIDATE LE see MHA750
OXIDATION BASE see NAN505
OXIDATION BASE 22 see ALL750
OXIDATION BASE 25 see NEM480
OXIDATION BASE 10A see PEY650
OXIDATION BASE 12A see DBO400
OXIDE of CHROMIUM see CMJ900
N^{10}-OXIDE-1-NITRO-9-(3-DIMETHYLAMINOPROPYLAMINO)-DIHYDROCHLORIDE ACRIDINE see CAB125
N-OXIDE de PHENAZINE (FRENCH) see PDB750
3-N-OXIDE PURIN-6-THIOL MONOHYDRATE see OMY800
1-OXIDE-4-QUINOLINAMINE see AML500
3-OXIDIDO 17-α-ETHYNYL 17-β-HYDROXY ESTRA-4,9,11-TRIENE see OMY815
OXIDIMETHIIN see OMY825
10,10'-OXIDIPHENOXARSINE see OMY850
OXIDIZED l-CYSTEINE see CQK325
OXIDIZED LINOLEATE see LGH000
OXIDIZED LINOLEIC ACID see LGH000
OXIDOETHANE see EJN500
α,β-OXIDOETHANE see EJN500
1,8-OXIDO-p-MENTHANE see CAL000
OXIDOPAMINE see HKF875
OXI-FENIBUTOL see HNI500
OXIFENON see ORQ000
OXIFENYLBUTAZON see HNI500
OXIFLAVIL see ELH600
OXIKON see DLX400
OXILAPINE see DCS200
OXILORPHAN see CQF079
α-OXIME BENZOIN see BCP500
OXIME COPPER see BLC250

OXIME,17-HYDROXY-19-NORPREGNA-4,9,11-TRIEN-20-YN-3-ONE see OMY815
OXIMES see OMY899
OXIMETHOLONUM see PAN100
OXIMETOLONA see PAN100
2-OXIMINO-3-BUTANONE see OMY910
OXIMINO OXAMYL see OLT100
1-OXINDENE see BCK250
OXINE see QPA000
OXINE COPPER see BLC250
OXINE CUIVRE see BLC250
OXINE SULFATE see QPS000
OXINOFEN see OPK300
p-OXINOZON see NBF500
2-OXI-PROPYL-PROPYLNITROSAMIN (GERMAN) see ORS000
OXIRAAN (DUTCH) see EJN500
(+−)-OXIRACETAM see HND900
OXIRANE see EJN500
OXIRANE, 2-((1,1'-BIPHENYL-4-YLOXY)METHYL)- see PFU600
OXIRANE, 2,2'-(2,2-BIS((OXIRANYLMETHOXY)METHYL)-1,3-PROPANEDIYLBIS(OXYMETHYLENE))BIS- see PBB850
OXIRANE, 2,2'-(1,4-BUTANEDIYL)BIS-(9CI) see DHD800
OXIRANE, 2,2'-(1,4-BUTANEDIYLBIS(OXYMETHYLENE))BIS-(9CI) see BOS100
OXIRANE, BUTYL- see HFC100
OXIRANE-CARBOXALDEHYDE see GGW000
OXIRANECARBOXALDEHYDE, 3-(1-HYDROXYHEXYL)-, (2-α-3-α(S*))- see HNB700
OXIRANECARBOXALDEHYDE, 3-(1-HYDROXYHEXYL)-, (2-α-3-α(R*))- see EBY550
OXIRANE CARBOXALDEHYDE OXIME see ECG100
OXIRANECARBOXAMIDE (9CI) see OMY912
OXIRANECARBOXAMIDE, 2-METHYL-N-PHENYL- see ECC650
OXIRANECARBOXAMIDE, 2-METHYL-N-(PHENYLMETHYL)- see ECA200
OXIRANECARBOXAMIDE, N-PHENYL- see ECE650
OXIRANECARBOXAMIDE, N-(PHENYLMETHYL)- see EBH890
OXIRANECARBOXYLIC ACID, 3-(((1-(((4-((AMINOIMINOMETHYL)AMINO)BUTYL)AMINO)CARBONYL)-3-METHYLBUTYL)AMINO)CARBONYL)-, (2S-(2-α,3-β(R*)))- see TFK255
OXIRANECARBOXYLIC ACID, 3-(((3-METHYL-1-(((3-METHYLBUTYL)AMINO)CARBONYL)BUTYL)AMINO) CARBONYL)-, ETHYL ESTER, (2S-(2-α-3-β(R*)))- see OMY925
OXIRANE, 2-CHLORO-3-METHYL-, cis-(9CI) see CKS099
OXIRANE, 2-CHLORO-3-METHYL-, trans-(9CI) see CKS100
OXIRANE, (CHLOROMETHYL)-, POLYMER WITH N-(3-AMINOPROPYL)-N-METHYL-1,3-PROPANE DIAMINE see EAZ600
OXIRANE, (CHLOROMETHYL)-, POLYMER WITH α-HYDRO-ω-HYDROXYPOLY(OXY(METHYL-1,2-ETHANEDIYL)) see PKK100
OXIRANE, ((4-CHLOROPHENOXY)METHYL)-(9CI) see CKA200
OXIRANE, 2-DECYL-3-(5-METHYLHEXYL)-, cis- see ECB200
OXIRANE, 2-(3,5-DICHLOROPHENYL)-2-(2,2,2-TRICHLOROETHYL)- see TJK100
OXIRANE, 2,3-DIMETHYL-(9CI) see EBJ100
OXIRANE, 2,3-DIMETHYL-, cis-(9CI) see EBJ200

OXIRANE, ((1,1-DIMETHYLETHOXY)METHYL)- see BRK800

OXIRANE, ((4-(1,1-DIMETHYLETHYL)PHENOXY)METHYL)-(9CI) see BSE600

OXIRANE, 2,2'-(1,2-ETHANEDIYLBIS(OXYMETHYLENE))BIS-(9CI) see EEA600

OXIRANE, (ETHOXYMETHYL)-(9CI) see EBQ700

OXIRANE, (ETHOXYMETHYL)-(9CI) see EKM200

OXIRANE, (((2-ETHYLHEXYL)OXY)METHYL)-(9CI) see GGY100

OXIRANE, ((HEXYLOXY)METHYL)-(9CI) see GGY125

OXIRANEMETHANAMINE, N-METHYL-N-PHENYL- see GGY155

OXIRANEMETHANAMINE, N-(4-(OXIRANYLMETHOXY)PHENYL)-N-(OXIRANYLMETHYL)- see ONE100

OXIRANEMETHANAMINIUM, N,N,N-TRIMETHYL-, CHLORIDE (9CI) see GGY200

OXIRANEMETHANOL, 3-(4-BROMOPHENYL)-, (2R-trans)- see EBH900

OXIRANEMETHANOL, 3-(4-BROMOPHENYL)-, (2S-trans)- see EBH905

OXIRANEMETHANOL, 3-METHYL-, 4-NITROBENZOATE, (R)- see ECC700

OXIRANEMETHANOL, 2-METHYL-, 4-NITROBENZOATE, (2S)- see ECC702

OXIRANEMETHANOL, 3-METHYL-, 4-NITROBENZOATE, (2R-trans)- see EBJ600

OXIRANEMETHANOL, 3-METHYL-, 4-NITROBENZOATE, (2S-trans)- see EBJ700

OXIRANEMETHANOL NITRATE see ECI600

OXIRANEMETHANOL, 3-(4-NITROPHENYL)-, trans-(−)- see NIS300

OXIRANEMETHANOL, 3-(4-NITROPHENYL)-, trans-(+)- see NIS298

OXIRANEMETHANOL, 3-PHENYL-, (2R-trans)- see PFF350

OXIRANEMETHANOL, 3-PHENYL-, (2S-trans)- see ECE600

OXIRANE, (METHOXYMETHYL)-(9CI) see GGW600

OXIRANE, METHYL-, (R)-(9CI) see PNL650

OXIRANE, METHYL-, (S)-(9CI) see ECE700

OXIRANE, ((1-METHYLETHOXY)METHYL)-(9CI) see IPD000

OXIRANE ((METHYLPHENOXY)METHYL) (9CI) see TGZ100

OXIRANE, ((4-METHYLPHENOXY)METHYL)-(9CI) see GGY175

OXIRANE, METHYL-, POLYMER WITH OXIRANE, BLOCK (9CI) see PKG600

OXIRANE, METHYL-, POLYMER WITH OXIRANE, BLOCK (9CI) see TBB800

OXIRANE, ((1-NAPHTHALENYLMETHOXY)METHYL)-(9CI) see NBJ200

OXIRANE, (4-NITROPHENOXY)- see NII100

OXIRANE, ((4-NITROPHENOXY)METHYL)-(9CI) see NIN050

OXIRANEOCTANOIC ACID, 3-OCTYL-, cis- see OAT100

OXIRANE, ((PHENYLMETHOXY)METHYL)-, (R)- see BEP670

OXIRANE, ((PHENYLMETHOXY)METHYL)-, (S)- see BFC225

OXIRANE, 2,2'-(2,5,8,11-TETRAOXADODECANE-1,12-DIYL)BIS-(9CI) see TJQ333

OXIRANE, TRIFLUORO(TRIFLUOROMETHYL)- see HDF050

7-OXIRANYLBENZ(a)ANTHRACENE see ONC000

4-OXIRANYLBENZONITRILE see ONC100

6-OXIRANYLBENZO(a)PYRENE see ONE000

1-OXIRANYL-1,2-ETHANEDIOL see ONE050

N-(4-(OXIRANYLMETHOXY)PHENYL)-N-(OXIRANYLMETHYL)OXIRANEMETHANAMINE see ONE100

OXIRANYLMETHYL ESTER of OCTADECANOIC ACID see SLK500

OXIRANYLMETHYL ESTER of 9-OCTADECENOIC ACID see ECJ000

2-(OXIRANYLMETHYL)-1H-ISOINDOLE-1,3(2H)-DIONE see ECK000

N-(OXIRANYLMETHYL)-N-PHENYL-OXIRANEMETHANAMINE (9CI) see DKM120

3-OXIRANYL-7-OXABICYCLO(4.1.0)HEPTENE see VOA000

1-OXIRANYLPYRENE see ONG000

OXIRENO(H)QUINOLINE, 1A,7B-DIHYDRO-, (+−)- see QNJ200

OXIRENO(f)QUINOLINIUM, 1A,7B-DIHYDRO-4-METHYL-, cis-(+−)- see MPF900

OXISURAN see MPL500

OXITETRACYCLIN see HOH500

OXITOL see EES350

OXITOSONA-50 see PAN100

OXITROPIUM BROMIDE see ONI000

OXIURAN see AOR500

OXIVOR see CNK559

OXLOPAR see HOI000

OXO see TAB750

3'-(3-OXO-7-α-ACETYLTHIO-17-β-HYDROXYANDROST-4-EN-17-β-YL)PROPIONIC ACID LACTONE see AFJ500

6-OXO-6H-ANTHRA(9,1-CD)ISOTHIAZOLE-3-CARBOXYLIC ACID see ISD043

7-OXOBENZ(de)ANTHRACENE see BBI250

α-OXOBENZENEACETIC ACID see OOK150

2-OXOBENZIMIDAZOLE see ONI100

5-OXO-5H-BENZO(E)ISOCHROMENO(4,3-b)INDOLE see DLY800

2-OXO-1,2-BENZOPYRAN see CNV000

γ-OXO(1,1-BIPHENYL)-4-BUTANOIC ACID see BGL250

OXOBIS(SULFATO(2-)-O)-ZIRCONATE(2-), DISODIUM (9CI) see ZTS100

OXOBOI see OOG000

2-OXOBORNANE see CBA750

3-OXO-BUTANOIC ACID BUTYL ESTER see BPV250

3-OXOBUTANOIC ACID ETHYL ESTER see EFS000

3-OXOBUTANOIC ACID METHYL ESTER see MFX250

1-(1-OXOBUTYL)AZIRIDINE see BSY000

4-(3-OXOBUTYL)-1,2-DIPHENYL-3,5-PYRAZOLIDINEDIONE see KGK000

γ-OXO-α-BUTYLENE see BOY500

p-(3-OXOBUTYL)PHENYL ACETATE see AAR500

trans-pi-OXOCAMPHOR see VSF400

19-OXO-CARDOGENEN-(20:22)-TRIOL(3-β,5,14) (GERMAN) see SMM500

4-OXO-2-(β-CHLOROETHYL)-2,3-DIHYDROBENZO-1,3-OXAZINE see CDS250

3-OXO-N-(2-CHLOROPHENYLBUTANAMIDE) see AAY600

OXOCHLORPROMAZINE see ONI300

7-OXOCHOLESTEROL see ONO000

2-OXOCHROMAN see HHR500

8-OXOCOPTISINE see ONO500

18-OXOCORTICOSTERONE see AFJ875

N-(2-OXO-3,5,7-CYCLOHEPTATRIEN-1-YL)AMINOOXOACETIC ACID ETHYL ESTER see ONQ000

((4-OXO-2,5-CYCLOHEXADIEN-1-YLIDENE)AMINO)GUANIDINE THIOSEMICARBAZONE see AHI875

4-OXOCYCLOPHOSPHAMIDE see KFK125

8-OXO-DG see HKA770

OXODIACETIC ACID see ONQ100

N-(6-OXO-6H-DIBENZO(b,d)PYRAN-1-YL)ACETAMIDE see BCH250

5-OXO-5,13-DIHYDROBENZO(E)(2)BENZOPYRANO(4,3-b)INDOLE see DLY800

OXODIOCTYLSTANNANE see DVL400

OXODIPEROXODIPYRIDINECHROMIUM(VI) see ONU000

OXODIPEROXOPYRIDINE CHROMIUM-N-OXIDE see ONU100

α-OXODIPHENYLMETHANE see BCS250

OXODIPHENYLSTANNANE, POLYMER see DWO600

OXODISILANE see ONW000

OXODOLIN see CLY600

1'-OXOESTRAGOLE see ONW100

9-OXO-2-FLUORENYLACETAMIDE see ABY250

N-(9-OXO-2-FLUORENYL)ACETAMIDE see ABY250

8-OXOGUANOSINE see ONW150

3-OXO-l-GULOFURANOLACTONE see ARN000

4-OXOHEPTANEDIOIC ACID, ISONICOTINOYL HYDRAZONE see ILG000

(17-β)17-((1-OXOHEPTYL)OXY)-ANDROST-4-EN-3-ONE see TBF750

6-OXO-trans,trans-2,4-HEXADIENOIC ACID see ONW200

6-OXO-2,4-HEXADIENOIC ACID (E,E)- see ONW200

2-OXOHEXAMETHYLENIMINE see CBF700

1-(5-OXOHEXYL)-3,7-DIMETHYLXANTHINE see PBU100

17-((1-OXOHEXYL)OXY)PREGN-4-ENE-3,20-DIONE see HNT500

1-(5-OXOHEXYL)THEOBROMINE see PBU100

(E)-7-OXO-3-β-HYDROXY-14-α-METHYL-8-β-PODOCARPANE-Δ13-α-ACETIC ACID-2-(DIMETHYLAMINO)ETHYL ESTER HYDROCHLORIDE see CCO675

4-OXO-4H-IMIDAZO(4,5-D)-v-TRIAZINE see ARY500

2-(3-OXO-1-INDANYLIDENE)-1,3-INDANDIONE see ONY000

2-OXO-3-ISOBUTYL-9,10-DIMETHOXY-1,3,4,6,7,11-β-HEXAHYDRO-2H-BENZOQUIN OLIZINE see TBJ275

8-OXO-8H-ISOCHROMENO(4',3':4,5)PYRROLO(2,3-f)QUINOLINE see OOA000

p-(1-OXO-2-ISOINDOLINYL)-HYDRATROPIC ACID see IDA400

2-(p-(1-OXO-2-ISOINDOLINYL)PHENYL)-PROPIONIC ACID see IDA400

α-(4-(1-OXO-2-ISO-INDOLINYL)-PHENYL)-PROPIONIC ACID see IDA400

OXOLAMINA CLORIDRATO (ITALIAN) see OOE100

OXOLAMINE see OOC000

OXOLAMINE CITRATE see OOE000

OXOLAMINE HYDROCHLORIDE see OOE100

OXOLANE see TCR750

OXOLE see FPK000

OXOLINIC ACID see OOG000

OXOMEMAZINE see AFL750

OXOMETHANE see FMV000

6-OXO-3-METHOXY-N-METHYL-4,5-EPOXYMORPHINAN see OOI000

1-OXO-2-(p-((α-METHYL)CARBOXYMETHYL)PHENYL)ISO INDOLINE see IDA400
3-OXO-N-(2,4-METHYLPHENYL)BUTANAMIDE see OOI100
3-OXO-3H-NAPHTHO(2,1-b)PYRAN-2-CARBOXYLIC ACID ETHYL ESTER see OOI200
OXONIUM, TRIETHYL-, TETRAFLUOROBORATE(1-) see TJL600
3-(1-OXO-4,7-NONADIENYL)OXIRANECARBOXAMIDE see ECE500
(2R-(2-α,3-α(4E,7E)))-3-(1-OXO-4,7-NONADIENYL)OXIRANECARBOXAMIDE see ECE500
5-OXONONANE see NMZ000
(E)-12-OXO-10-OCTADECENOIC ACID, METHYL ESTER see MNF250
12-OXO-trans-10-OCTADECENOIC ACID, METHYL ESTER see MNF250
α-1-(OXOOCTADECYL)-ω-HYDROXYPOLY(OXY-1,2-ETHANEDIYL) see PJW750
9-OXO-8-OXATRICYCLO(5.3.1.0²,⁶)UNDECANE see OOK000
8-OXOPENTADECANE see HBO790
4-OXOPENTANOIC ACID see LFH000
OXOPHENARSINE see OOK100
OXOPHENARSINE HYDROCHLORIDE see ARL000
OXOPHENYLACETIC ACID see OOK150
2-OXO-2-(PHENYLAMINO)ETHYL SELENOCYANATE see OOK175
β-OXO-α-PHENYLBENZENEPROPANENITRILE see OOK200
((4-OXO-2-PHENYL-4H-1-BENZOPYRAN-7-YL)OXY)ACETIC ACID ETHYL ESTER see ELH600
3-OXO-N-PHENYLBUTANAMIDE see AAY000
(17-β)-17-(1-OXO-3-PHENYLPROPOXY)-ESTR-4-EN-3-ONE (9CI) see DYF450
3-OXO-Δ¹,⁸-PHTHALANACETIC ACID see CCH150
N-(2-OXO-3-PIPERIDYL)PHTHALIMIDE see OOM300
5-OXO-l-PROLYL-l-HISTIDYL-l-PROLINAMIDE TARTRATE see POE050
2-OXOPROPANAL see PQC000
OXOPROPANEDINITRILE see OOM400
OXOPROPANEDINITRILE (CARBONYL DICYANIDE) see MDL250
2-OXOPROPANOIC ACID see PQC100
2-((1-OXO-2-PROPENYL)OXY)-N,N,N-TRIMETHYLETHANAMINIUM CHLORIDE see OOM500
2-OXOPROPIONIC ACID see PQC100
17-(1-OXOPROPOXY)-(17-β)-ANDROST-4-EN-3-ONE see TBG000
p-(2-OXOPROPOXY)BENZENEARSONIC ACID see OOO000
N-(2-OXOPROPYL)-N-NITROSOUREA see OOO050
3-(2-OXOPROPYL)-2-PENTYLCYCLOPENTANONE see OOO100
1-OXOPROPYLPROPYLNITROSAMIN (GERMAN) see NLN500
1-OXOPROPYLPROPYLNITROSAMINE see NLN500
2-OXO-PROPYL-PROPYLNITROSAMINE see ORS000
(2-OXOPROPYL)PROPYLNITROSOAMINE see ORS000
2-OXOPYRIDINE see OOO200
γ-OXO-3-PYRIDINEBUTANAL see OOO300
4-OXO-4-(3-PYRIDYL)BUTANAL see OOO300
2-OXOPYRROLIDINE see PPT500

2-OXO-PYRROLIDINE ACETAMIDE see NNE400
2-OXO-1-PYRROLIDINEACETAMIDE see NNE400
5-OXO-2-PYRROLIDINECARBOXYLIC ACID MAGNESIUM SALT (2:1) see PAN775
2'-OXOPYRROLIDINO-1-PYRROLIDINO-4-BUTYNE see OOY000
2-OXOPYRROLIDIN-1-YLACETAMIDE see NNE400
2-OXO-2-(1-PYRROLIDINYL)ETHYL-N-(((METHYLAMINO)CARBONYL)OXY)ETH ANIMIDOTHIOATE see OOO500
OXOSILANE see OOS000
2-OXOSPARTEINE see LIQ000
OXOSULFATOVANADIUM PENTAHYDRATE see VEZ100
OXOSUMITHION see PHD750
11-OXO-9,12-α,13,17-TETRAHYDROXYKAURAN-17(S)-AL see OOS100
4-OXO-2,2,6,6-TETRAMETHYLPIPERIDINE see TDT770
1-4-OXO-2-THIAZOLIDINEHEXANOIC ACID see CCI500
4-OXO-2-THIONOTHIAZOLIDINE see RGZ550
OXOTREMORIN see OOY000
OXOTREMORINE see OOY000
OXOTREMORINE FUMARATE (2:3) see OOY100
OXOTREMORINE SESQUIFUMARATE see OOY100
8-OXOTRICYCLO(5.2.1.0²,⁶)DECANE see OPC000
(3-β,5-β,12-α)-11-OXO-3,12,14-TRIHYDROXYBUFA-20,22-DIENOLIDE see POG275
19-OXO-3-β,5,14-TRIHYDROXY-5-β-BUFA-20,22-DIENOLIDE-3-ACETATE see HAN550
(3-β,5-α,15-β)-19-OXO-3,14,15-TRIHYDROXYCARD-20(22)-ENOLIDE see AFS800
5-OXO-1,3,3-TRIMETHYLCYCLOHEXANECARBONITRI LE see IMG500
2,2'-(2-OXO-TRIMETHYLENE)BIS(1,1-DIMETHYLPYRROLIDINIUM) DIBENZENESUFLONATE see COG500
2,2'-(2-OXOTRIMETHYLENE)BIS(1,1-DIMETHYLPYRROLIDINIUM) DIIODIDE see COG750
5-OXO-2,4,8-TRIMETHYL-6-OXA-3,9-DITHIA-2,4,7-TRIAZADEC-7-ENOIC ACID, 2-(2-(2-METHOXYETHOXY) ETHOXY)ETHYL ESTER see OPC025
5-OXO-2,4,8-TRIMETHYL-6-OXA-3,9-DITHIA-2,4,7-TRIAZADEC-7-ENOIC ACID, (1-METHYLETHYLIDENE)DI-4,1-PHENYLENE ESTER see OPC035
1-OXO-2-(2,4,6-TRIMETHYLPHENYL)-1H-INDEN-3-YL DODECANOATE see OPC045
(17-β)-17-((1-OXO-4,8,12-TRIMETHYL-3,7,11-TRIDECATRIENYL)OXY)ESTR-4-EN-3-ONE see NNX650
6-OXOUNDECANE see ULA000
4-OXOVALERIC ACID see LFH000
β-(γ-OXOVALEROYL)FURAN see IGF300
(4-OXOVALERYLOXY)TRIPHENYLSTANNA NE see TMW250
22-OXOVINCALEUKOBLASTINE see LEY000
OXPENTIFYLLINE see PBU100
OXPRENOLOL see AGW000
OXPRENOLOL see CNR500
OXPRENOLOL HYDROCHLORIDE see THK750
OXSORALEN see XDJ000
OXTRIMETHYLLINE see CMG300
OXTRIPHYLLINE see CMG300
OXUCIDE see PIJ500

OXURASIN see HEP000
OXY-5 see BDS000
OXY-10 see BDS000
OXYACANTHAN-7-OL, 6,6',12'-TRIMETHOXY-2,2'-DIMETHYL- see HGI050
p-OXYACETOPHENONE see HIO000
OXY ACID BLACK BASE see PFU500
β-OXYAETHYL-MORPHOLIN (GERMAN) see MRQ500
10-(3-(4-OXYAETHYL-PIPERAZINO)PROPYL-(1))-4-AZAPHENTHIAZIN DIHYDROCHLORID see ORI400
OXYAETHYLTHEOPHYLLIN (GERMAN) see HLC000
OXYALKYLATED NONYL PHENOLIC RESIN see FMW343
OXYAMINE see CFA750
1-OXY-2-AMINO-4-NITROBENZOL AETHYL ETHER see NIC200
3-OXYANTHRANILIC ACID see AKE750
p-OXYBENZALDEHYDE see FOF000
OXYBENZENE see PDN750
p-OXYBENZOESAEUREAETHYLESTER (GERMAN) see HJL000
p-OXYBENZOESAEUREHEPTYLESTER see HBO650
p-OXYBENZOESAEUREMETHYLESTER (GERMAN) see HJL500
p-OXYBENZOESAEUREPROPYLESTER (GERMAN) see HNU500
p-OXYBENZOESAURE (GERMAN) see SAI500
OXYBENZONE see MES000
OXYBENZOPYRIDINE see QPA000
OXYBISACETIC ACID see ONQ100
2,2'-OXYBISACETIC ACID see ONQ100
OXYBIS(4-AMINOBENZENE) see OPM000
OXYBIS((3-AMINOPROPYL)DIMETHYLSILANE) see OPC100
4,4'-OXYBISANILINE see OPM000
p,p'-OXYBIS(ANILINE) see OPM000
4,4'-OXYBISBENZENAMINE see OPM000
p,p'-OXYBISBENZENE DISULFONYLHYDRAZIDE see OPE000
1,1'-OXYBISBENZENE OCTABROMO DERIV. see OAF200
OXYBISBENZENESULFONIC ACID DIHYDRAZIDE see OPE000
OXYBIS(BENZENESULFONYL HYDRAZIDE) see OPE000
p,p'-OXYBIS(BENZENESULFONYL) HYDRAZINE see OPE000
1,1'-OXYBIS(BUTANE) see BRH750
2,2'-OXYBISBUTANE see BRH760
2,2'-OXYBISBUTANE see OPE030
4,4'-OXYBIS(2-CHLOROANILINE) see BGT000
4,4'-OXYBIS(2-CHLORO-BENZENAMINE) see BGT000
1,1'-OXYBIS(4-CHLOROBUTANE) see OPE040
1,1'-OXYBIS(1-CHLOROETHANE) see BID000
1,1'-OXYBIS(2-CHLORO)ETHANE see DFJ050
1,1'-OXYBIS(2-(2-CHLOROETHYL)THIOETHANE see DFK200
OXYBIS(CHLOROMETHANE) see BIK000
2,2'-OXYBIS(1-CHLOROPROPANE) see BII250
OXYBIS(DIBUTYL(2,4,5-TRICHLOROPHENOXY)TIN) see OPE100
5,5'-OXYBIS(3,4-DICHLORO-2(5H))-FURANONE see MRV000
4,4'-OXYBIS(2,3-DICHLORO-4-HYDROXYCROTONIC ACID)-DI-γ-LACTONE see MRV000

OXYBIS(N,N-DIMETHYLACETAMIDETRIPHENYLSTIBONIUM) DIPERCHLORATE see OPG000
1,1'-OXYBISETHANE see EJU000
1,1'-(OXYBIS(2,1-ETHANEDIYLOXY))BISBUTANE see DDW200
(OXYBIS(2,1-ETHANEDIYLOXY))BIS-PROPANAMINE see EBV100
3,3'-(OXYBIS(2,1-ETHANEDIYLOXY))BIS-1-PROPANAMINE see DJD800
OXYBIS(2,1-ETHANEDIYLOXY-2,1-ETHANEDIYL) 3-(DODECYLTHIO)PROPANOATE see OPG050
2,2'-OXYBISETHANOL see DJD600
1,1'-OXYBISETHENE see VOP000
2,2'-(OXYBIS(ETHYLENEOXY))DIETHANOL see TCE250
1,1'-OXYBIS(2-ETHYLHEXANE) see DJK600
1,1'-OXYBISHEPTANE see HBO000
1,1'-OXYBISHEXANE see DKO800
OXYBISMETHANE see MJW500
1,1'-OXYBIS(4-METHYLBENZENE) see THC600
(1,1'-(OXYBIS(METHYLENE))BIS(4-(1,1-DIMETHYLETHYL))-PYRIDINIUM, DICHLORIDE (9CI) see SAB800
2,2'-(OXYBIS(METHYLENE))BISFURAN see FPX025
1,1'-(OXYBIS(METHYLENE))BIS(4-(HYDROXYIMINO)METHYL)PYRIDINIUM DICHLORIDE see BGS250
1,1'-(OXYBIS(METHYLENESULFONYL))BIS(2-CHLOROETHANE) see OPG100
OXYBIS(4-NITROBENZENE) see NIM600
1,1'-OXYBIS(4-NITROBENZENE) see NIM600
2,2'-OXYBIS-6-OXABICYCLO-(3.1.0)HEXANE see BJN250
2,2'-OXYBIS(6-OXABICYCLO(3.1.0)HEXANE) mixed with 2,2-BIS(p-(2,3-EPOXYPROPOXY)PHENYL)PROPANE see OPI200
1,1'-OXYBIS(2,3,4,5,6-PENTABROMOBENZENE) (9CI) see PAU500
1,1'-OXYBISPENTANE see PBX000
10-10' OXYBISPHENOXYARSINE see OMY850
1,1'-OXYBISPROPANE see PNM000
3,3'-OXYBIS(1-PROPENE) see DBK000
OXYBIS(TRIBUTYLTIN) see BLL750
1,1'-OXYBIS(2,4,5-TRICHLOROBENZENE) see HCH485
OXYBIS(TRIMETHYLSILANE) see HEE000
OXYBULAN see PFR325
OXYBUPROCAINE HYDROCHLORIDE see OPI300
OXYBUTANAL see AAH750
β-OXYBUTENE see EBJ100
β-OXYBUTTERSAEURE-p-PHENETIDID see HJS850
OXYBUTYNIN CHLORIDE see OPK000
OXYBUTYRIC ALDEHYDE see AAH750
OXYCAINE see DHO600
OXYCARBON SULFIDE see CCC000
OXYCARBOPHOS see OPK250
OXYCARBOXIN see DLV200
OXYCARBOXINE see DLV200
OXY CHEK 114 see MJO500
OXYCHINOLIN see QPA000
o-OXYCHINOLIN (GERMAN) see QPA000
OXYCHLORDAN see OPK275
OXYCHLORDANE see OPK275
OXYCHLORID FOSFORECNY see PHQ800
OXYCHLORUE de CUIVRE see CNK559
OXYCHLORURE CHROMIQUE (FRENCH) see CML125
OXYCHOLIN see DAL000

OXYCIL see SFS000
OXYCINCHOPHEN see OPK300
OXYCLIPINE see MOQ500
OXYCLOR see CNK559
OXYCLOZANID see DMZ000
OXYCLOZANIDE see DMZ000
OXYCODEINONE see PCG500
OXYCODONE HYDROCHLORIDE see DLX400
OXYCODON HYDROCHLORIDE see DLX400
OXYCOLOR see AOR500
OXYCON see DLX400
7-OXYCOUMARIN see HJY100
OXYCUR see CNK559
OXY DBCP see DDL800
OXYDE d'ALLYLE et de GLYCIDYLE (FRENCH) see AGH150
OXYDE de BARYUM (FRENCH) see BAO000
OXYDE de CALCIUM (FRENCH) see CAU500
OXYDE de CARBONE (FRENCH) see CBW750
OXYDE de CHLORETHYLE (FRENCH) see DFJ050
OXYDE d'ETHYLE (FRENCH) see EJU000
OXYDE de MERCURE (FRENCH) see MCT500
OXYDE de MESITYLE (FRENCH) see MDJ750
OXYDEMETONMETHYL see DAP000
OXYDEMETON-METILE (ITALIAN) see DAP000
OXYDE NITRIQUE (FRENCH) see NEG100
OXYDEPROFOS see DSK600
OXYDE de PROPYLENE (FRENCH) see PNL600
OXYDE de TRIBUTYLETAIN see BLL750
OXYDIACETIC ACID see ONQ100
2,2'-OXYDIACETIC ACID see ONQ100
OXYDIACETIC ACID, DISODIUM SALT see SIZ000
OXYDIAMINE BROWN 3GN see CMO810
OXYDIANILINE see OPM000
4,4'-OXYDIANILINE see OPM000
p,p'-OXYDIANILINE see OPM000
OXYDIAZEPAM see CFY750
OXYDIAZOL see BGD250
1,1'-OXYDI-4-CHLOROBUTANE see OPE040
2,2'-OXYDIETHANOL see DJD600
OXYDIETHANOLIC ACID see ONQ100
N-OXYDIETHYL-2-BENZOTHIAZOLSULFENAMID (CZECH) see BDF750
OXYDIETHYLENE ACRYLATE see ADT250
N-(OXYDIETHYLENE)BENZOTHIAZOLE-2-SULFENAMIDE see BDG000
OXYDIETHYLENE BIS(CHLOROFORMATE) see OPO000
1,1'-(OXYDIETHYLENE)BIS(4-FORMYLPYRIDINIUM BROMIDE), DIOXIME see OPO100
OXYDIETHYLENE CHLOROFORMATE see OPO000
OXYDIETHYLENE DIACRYLATE see ADT250
OXYDIETHYLENEDICARBONIC ACID DIALLYL ESTER see AGD250
3,3'-(2,2'-OXYDIETHYLENEDIOXYBISACETAMIDO)BIS(2,4,6-TRIIODOBENZOIC ACID) see IGD400
(N,N'-OXYDIETHYLENEDIOXYDIETHYLENE)BIS(TRIETHYLAMMONIUM IODIDE) see OPQ000
N-OXYDIETHYLENE THIOCARBAMYL-N-OXYDIETHYLENE SULFENAMIDE see OPQ100
OXYDIFORMIC ACID DIETHYL ESTER see DIZ100
β-(4-OXY-3,5-DIJOD-PHENYL)-α-PHENYL-PROPIONSAEURE (GERMAN) see PDM750

OXYDIMETHANOL DINITRATE see OPQ200
1,1'-OXYDIMETHYLENE BIS(4-tert-BUTYLPYRIDINIUM CHLORIDE) see SAB800
3,3'-(OXYDIMETHYLENEBIS(CARBONYLIMINO))BIS(2,4,6-TRIIODOBENZOIC ACID DISODIUM SALT) see BGB350
1,1'-(OXYDIMETHYLENE)BIS(4-FORMYLPYRIDINIUM)DICHLORIDE DIOXIME see BGS250
1,1'-(OXYDIMETHYLENE)BIS(4-FORMYLPYRIDINIUM) DINITRATE DIOXIME see OPY000
1,1'-(OXYDIMETHYLENE)BIS(4-FORMYLPYRIDINIUM) DIOXIME DICHLORIDE see BGS250
OXYDIMETHYLQUINAZINE see AQN000
4,4'-OXYDIPHENOL see OQI000
p,p'-OXYDIPHENOL see OQI000
p-OXYDIPHENYLAMINE see AOT000
4,4'-OXYDIPHENYLAMINE see OPM000
OXYDI-p-PHENYLENEDIAMINE see OPM000
1,1'-OXYDI-2-PROPANOL see OQM000
3,3'-OXYDI-1-PROPANOL DIBENZOATE see DWS800
OXYDIPROPANOL PHOSPHITE (3:1) see OQO000
3,3'-OXYDIPROPIONITRILE see OQQ000
β,β'-OXYDIPROPIONITRILE see OQQ000
OXYDISULFOTON see OQS000
N-OXYD-LOST see CFA500
N-OXYD-LOST see CFA500
N-OXYD-MUSTARD see CFA500
OXYDOL see HIB050
OXYDRENE see DBA800
OXYEPHEDRINE see HKH500
OXYETHYLATEED TERTIARY OCTYL-PHENOL-FORMALDEHYDE POLYMER see TDN750
OXYETHYLENATED DODECYL ALCOHOL see DXY000
OXYETHYLENATED HEXADECYL ALCOHOL see PJT300
OXYETHYLIDENEDIPHOSPHONIC ACID see HKS780
1-(β-OXYETHYL)-2-METHYL-5-NITROIMIDAZOLE see MMN250
OXYETHYLTHEOPHYLLINE see HLC000
OXYFED see DBA800
1-OXYFEDRIN see OQU000
OXYFENAMATE see HNJ000
OXYFENON see ORQ000
OXYFLAVIL see ELH600
OXYFLUORFEN see OQU100
OXYFLUORFENE see OQU100
OXYFUME see EJN500
OXYFUME 12 see EJN500
OXYFUME 20 see EJO000
OXYFUME 30 see EJO000
OXYFURADENE see NDY000
β-OXY-GABA see AKF375
OXYGEN see OQW000
OXYGEN DIFLUORIDE see ORA000
OXYGEN FLUORIDE see ORA000
OXYGEN, compressed (UN 1072) (DOT) see OQW000
OXYGEN, refrigerated liquid (cryogenic liquid) (UN 1073) (DOT) see OQW000
OXYGERON see VLF000
OXYHYDROCHINON (GERMAN) see BBU250
OXYHYDROQUINONE see BBU250
OXYJECT 100 see HOI000
OXYKODAL see DLX400
OXYKON see DLX400
OXYLAN see DKQ000
OXYLAN see PFR325
OXYLITE see BDS000
OXYMAG see MAH500

OXY MBC see SFS500
OXYMEMAZINE see AFL750
OXYMETAZOLINE see ORA100
OXYMETAZOLINE CHLORIDE see AEX000
OXYMETAZOLINE HYDROCHLORIDE see AEX000
OXYMETEBANOL see ORE000
OXYMETHALONE see PAN100
OXYMETHAZOLINE see ORA100
OXYMETHEBANOL see ORE000
OXYMETHENOLONE see PAN100
OXYMETHOLONE see PAN100
OXY-2-METHOXY-3-BENZALDEHYDE (FRENCH) see VFP000
OXY-3-METHOXY-4 BENZALDEHYDE (FRENCH) see FNM000
OXYMETHUREA see DTG700
OXYMETHYLENE see FMV000
5-OXYMETHYLFURFUROLE see ORG000
OXYMETHYLPHTHALIMIDE see HMP100
OXYMETOZOLINE see ORA100
OXYMORPHINONE HYDROCHLORIDE see ORG100
OXYMORPHONE HYDROCHLORIDE see ORG100
OXYMURIATE OF POTASH see PLA250
OXYMYCIN see CQH000
OXYMYCIN see HOH500
OXYMYKOIN see HOH500
β-OXYNAPHTHOIC ACID see HMX500
OXYNEURINE see GHA050
OXY-NH2 see CFA500
OXYNICOTINE see NDR100
N-OXYNICOTINE see NDR100
OXYOZYL see AOR500
OXYPAAT see HEP000
OXYPANAMINE see ORI300
OXYPANGAM see DNM400
OXYPARATHION see NIM500
OXYPENDYL HYDROCHLORIDE see ORI400
OXYPER see SJB400
OXYPERTIN see ECW600
OXYPERTINE see ECW600
OXYPHENALON see RBU000
OXYPHENAMATE see HNJ000
OXYPHENBUTAZONE see HNI500
OXYPHENIC ACID see CCP850
OXYPHENON see ORQ000
OXYPHENONIUM see ORQ000
OXYPHENONIUM BROMIDE see ORQ000
OXYPHENYLBUTAZONE see HNI500
1-(p-OXYPHENYL)-2-METHYLAMINOPROPAN (GERMAN) see FMS875
α-(p-OXYPHENYL)-β-METHYL AMINOPROPANE see FMS875
3-OXYPHENYL TRIMETHYLAMMONIUM IODIDE see HNN500
OXYPHIONFOS see DSK600
OXYPHOSPHONATE see TIQ300
OXYPHYLLINE see HLC000
OXYPHYLLINE (AMIDO) see HLC000
4-OXYPIPERIDOL see HOI300
OXYPROCAIN see DHO600
OXYPROCAINE see DHO600
p-OXYPROPIOPHENONE see ELL500
2-OXY-1,3-PROPYLENEDIAMINE-N,N,N',N'-TETRAMETHYLENEPHOSPHONIC ACID see DYE550
β-OXYPROPYLPROPYLNITROSAMINE see ORS000
γ-OXYPROPYLTHEOBROMIN (GERMAN) see HNZ000
γ-(γ-OXYPROPYL)-THEOBROMIN (GERMAN) see HNZ000
β-OXYPROPYLTHEOBROMINE see HNY500
β-OXYPROPYLTHEOPHYLLIN see HOA000
OXYPROPYLTHEOPHYLLINE see HOA000
OXYPSORALEN see XDJ000

4-OXYPYRIMIDINE see ORS050
4-OXYPYRIMIDINE see PPP140
OXYPYRRONIUM see IBP200
OXYPYRRONIUM BROMIDE see IBP200
OXYQUINOLINE see QPA000
8-OXYQUINOLINE see QPA000
OXYQUINOLINE BENZOATE see HOE100
OXYQUINOLINE SULFATE see QPS000
OXYQUINOLINOLEATE de CUIVRE (FRENCH) see BLC250
5-OXYRESORCINOL see PGR000
OXYRITIN see RSU000
OXYSONIUM IODIDE see HOJ150
OXYSTEARIN see ORS100
OXYSTIN see ORU500
OXYSULFATOVANADIUM see VEZ000
OXYTERRACIN see HOH500
OXYTERRACINE see HOH500
OXYTERRACYNE see HOH500
OXYTETRACYCLINE see HOH500
OXYTETRACYCLINE AMPHOTERIC see HOH500
OXYTETRACYCLINE HYDROCHLORIDE see HOI000
OXYTHANE see NCM700
OXYTHEONYL see HLC000
OXYTHIAMIN see ORS200
OXYTHIAMINE see ORS200
OXYTHIOQUINOX see ORU000
OXYTOCIC see EDB500
OXYTOCIN see ORU500
OXYTOL ACETATE see EES400
m-OXYTOLUENE see CNW750
o-OXYTOLUENE see CNX000
p-OXYTOLUENE see CNX250
OXYTRIL see HKB500
OXYTRIL M see DDP000
OXYTROPIUM BROMIDE see ONI000
OXYURANUS SCUTELLATUS (AUSTRALIA) VENOM see ARV500
OXYUREA see HOO500
OXYUREA see TGN250
OXY WASH see BDS000
OXYZINE see PIJ500
OXYZIN (TABL.) see HEP000
OZ see OJY200
γ-OZ see OJY200
OZAGREL see SHV100
OZIDE see ZKA000
OZLO see ZKA000
OZOBAN see EDE700
OZOCERITE see ORU900
OZOKERITE see ORU900
OZON (POLISH) see ORW000
OZONE see ORW000
OZONE mixed with NITROGEN OXIDES (53%:47%) see ORY000
OZONIDES see ORY499
OZP 9 see DXB450
P07 see PKO500
P 12 see MNV250
P23 see TIA100
P-25 see SLJ000
P-33 see CBT750
P-40 see DXG000
P 48 see CMV400
P-50 see AIV500
P 55 see VJZ000
P68 see CBT750
134 P see MLK800
P 1-60 see LED100
P-165 see ASA500
P-203 see RCZ200
P 237 see BTA000
P 241 see PEO750
P 253 see LJR000
P-267 see BPR500
P 271 see PAB000
P 284 see DWF000
P 301 see HNJ000
P-314 see DHR000
P-329 see DPU800

P 391 see MOQ250
P-398 see ACX500
501 P see AOO800
P 527 see MOE250
P 652 see PDU250
P 887 see ILE000
P 892 see MOQ000
Pt-05 see DFJ000
Pt-09 see TBO776
Pt-93 see PLW200
P 1011 see DGE200
P 10-59 see BBS275
P-1108 see DVJ400
P 1133 see BET000
P 1134 see COS500
P 1142 see PDN000
P1250 see CBT750
P 1297 see BEQ500
P 1393 see BDE250
P 1488 see BRU750
P1496 see RBF100
P 1531 see PFC750
P-2292 see ADA000
P 2647 see BCL250
P-4000 see NIY525
P-5048 see DRT200
P-5307 see PDY870
6020P see PJS750
P-7138 see NDY400
P1 3419 see FAQ600
P 71-0129 see FAO100
P 286 (contrast medium) see IGD200
PA see PAP750
PA see PEG500
PA 93 see SMB000
PA 94 see CQH000
PA 130 see PJS750
PA 144 see MQW750
PA 190 see PJS750
PA 520 see PJS750
PA 560 see PJS750
PA 775 see OHM900
PA 6 (polymer) see PJY500
PAA see BBL500
PAA see PJK350
PAA-25 see ADV900
PAA-701 DIHYDROCHLORIDE see BFX125
PA'AILA (HAWAII) see CCP000
PAB see SEO500
PA 114B see VRA700
PABA see AIH600
PABANOL see AIH600
PABESTROL see DKA600
PABESTROL see DKB000
PABIALGIN see IGI000
PABRACORT see HHQ800
PABS see SNM500
PABYRIN see HAQ500
PACAMINE HYDROCHLORIDE see NCL300
PACATAL see MOQ250
PACATAL BASE see MOQ250
PACEMO see HIM000
PACETYN see EHP000
PACHAK see CNT350
PACHYCARPINE see PAB250
6-β,7-α,9-α,11-α-PACHYCARPINE see SKX500
PACHYRHIZUS EROSUS see YAG000
PACIENCIA see DAB700
PACIENX see CFZ000
PACIFAN see NBU000
PACINOL see TJW500
PACITANE see BBV000
PACITRAN see DCK759
PACITRAN see MQR200
PACLITAXEL see TAH775
PACLOBUTRAZOL see TMK125
PACM 20 see MJQ260
PACTAMYCIN see PAB500
PAD 522 see PJS750
PADAN see BHL750

PADARYL see BQW000
PADISAL see MRW000
PADOPHENE see PDP250
PADRIN see PAB750
PAE see INN500
PAEONIA MOUTAN see PAC000
PAEONIFLORIN see PAC000
PAEONOL see PAC250
PAEONY ROOT see PAC000
PAFAVIT see VTA100
PAFE see PKE550
PAG see NNR400
PAGANO-COR see DHS200
PAGITANE HYDROCHLORIDE see CQH500
PAGODA TREE see NBR800
PAICER see DES500
PAIDAZOLO see AIF000
PA'INA (HAWAII) see JBS100
PAIN de COULEUVRE (CANADA) see BAF325
PAINTER'S NAPHTHA see PCT250
PAINT WHITE see BKW100
PAISAJE (PUERTO RICO) see PGQ285
PAISLEY POLYMER see PMP500
PAKA (HAWAII) see TGI100
PAKHTARAN see DUK800
PAL see PEC750
PALA see PGY750
PALACET SCARLET B see CMP080
PALACET YELLOW GN see AAQ250
PALACOS see PKB500
PALACRIN see CFU750
PALAFER see FBJ100
PALAFUGE see PFR325
PALANIL BLUE 7G see DMM400
PALANIL BLUE GL see CMP075
PALANIL BLUE R see CMP070
PALANIL BLUE R see DBQ220
PALANILCARRIER A see PAC500
PALANIL PINK RF see AKO350
PALANIL RED BF see AKI750
PALANIL VIOLET 3B see DBY700
PALANIL VIOLET 6R see DBX000
PALANIL YELLOW 5R see CMP090
PALANIL YELLOW G see AAQ250
PALANIL YELLOW GE see KDA075
PALANIL YELLOW 5RX see CMP090
PALANTHRENE BLUE 3G see DLO900
PALANTHRENE BLUE BC see DFN300
PALANTHRENE BLUE BCA see DFN300
PALANTHRENE BLUE GPT see IBV050
PALANTHRENE BLUE GPZ see IBV050
PALANTHRENE BLUE 3GN see DLO900
PALANTHRENE BLUE RPT see IBV050
PALANTHRENE BLUE RPZ see IBV050
PALANTHRENE BLUE RSN see IBV050
PALANTHRENE BRILLIANT BLUE R see IBV050
PALANTHRENE BRILLIANT ORANGE GR see CMU820
PALANTHRENE BROWN BR see CMU770
PALANTHRENE BROWN R see CMU780
PALANTHRENE GOLDEN YELLOW see DCZ000
PALANTHRENE NAVY BLUE G see CMU500
PALANTHRENE NAVY BLUE RB see CMU750
PALANTHRENE OLIVE R see DUP100
PALANTHRENE ORANGE R see CMU815
PALANTHRENE PRINTING BLUE KRS see IBV050
PALANTHRENE RED G 2B see CMU825
PALAPENT see NBU000
PALASH SEED EXTRACT see BOV800
PALASONIN, PIPERAZINE SALT (1:1) see PAC600
PALATINOL A see DJX000
PALATINOL AH see DVL700
PALATINOL BB see BEC500
PALATINOL C see DEH200
PALATINOL DN see PHW585

PALATINOL IC see DNJ400
PALATINOL M see DTR200
PALATINOL N see PHW585
PALATINOL Z see PHW575
PALATONE see MAO350
PALAVALE see EAE000
PALE ORANGE CHROME see LCS000
PALESTROL see DKA600
PALETA de PINTOR (PUERTO RICO) see CAL125
PALFADONNA see AFJ400
PALFIUM see AFJ400
PALIN see PIZ000
PALINUM see TDA500
PALIOFAST BLUE SBL see COF420
PALIOGEN RED 4790 see DTV360
PALIOGEN RED L 4790 see DTV360
PALIUROSIDE see RSU000
PALLADATE(2-), TETRACHLORO-, DIAMMONIUM (8CI) see ANE750
PALLADATE(2-), TETRACHLORO-, DIAMMONIUM, (SP-4-1)-(9CI) see ANE750
PALLADIUM see PAD750
PALLADIUM(2+) ACETATE see PAD300
PALLADIUM(II) ACETATE see PAD300
PALLADIUM ACETATE (Pd(OAc)₂) (7CI) see PAD300
PALLADIUM CHLORIDE see PAD500
PALLADIUM(2+) CHLORIDE see PAD500
PALLADIUM CHLORIDE, DIHYDRIDE see PAD750
PALLADIUM DIACETATE see PAD300
PALLADOUS ACETATE see PAD300
PALLADOUS CHLORIDE see PAD500
PALLETHRINE see AFR250
PALLICID see ABX500
PALMA CHRISTI (HAITI) see CCP000
PALM O-ARA-C see PAE260
PALMAROSA OIL see PAE000
PALMATINE HYDROXIDE see PAE100
PALMATINIUM HYDROXIDE see PAE100
PALM BUTTER see PAE500
PALMITA see PGA750
PALMITA de JARDIN (PUERTO RICO) see CNH789
PALMITAMIDE see PAE240
PALMITIC ACID see PAE250
PALMITIC ACID AMIDE see PAE240
PALMITIC ACID, 5'-ESTER WITH 1-β-d-ARABINOFURANOSYLCYTOSINE see PAE260
PALMITIC ACID, 2-ETHYLHEXYL ESTER see OFG100
PALMITIC ACID, HEXADECYL ESTER see HCP700
PALMITIC ACID, TRIESTER WITH PYRIDOXOL see HMQ575
PALMITOYL, l-ASCORBIC ACID see ARN150
PALMITOYL CYTARABINE see PAE260
PALMITYL ACETATE see HCP100
PALMITYL ALCOHOL see HCP000
PALMITYLAMINE see HCO500
PALMITYLDIMETHYLAMINE see HCP525
PALMITYL PALMITATE see HCP700
PALMITYLTRIMETHYLAMMONIUM CHLORIDE see HCQ525
PALM KERNEL OIL (UNHYDROGENATED) see PAE275
PALMO-ARA-C see AQS875
PALM OIL see PAE500
PALM OIL (UNHYDROGENATED) see PAE300
PALO GUACO (CUBA) see CDH125
PALOHEX see HFG550
PALO de NUEZ (PUERTO RICO) see TOA275
PALOPAUSE see ECU750
PALOSEIN see OJM400
PAL-P see PII100
PALTET see TBX250
PALUDRINE see CKB250
PALUDRINE DIHYDROCHLORIDE see PAE600

PALUDRINE HYDROCHLORIDE see CKB500
PALUSIL HYDROCHLORIDE see CKB500
PALYGORSCITE see PAE750
PALYGORSKIT (GERMAN) see PAE750
PALYTHOATOXIN see PAE875
PALYTOXIN see PAF000
PAM see ALW500
l-PAM see PED750
PA 11M see ADV900
PAM (CZECH) see POS750
PAMA see PGV250
PAMACEL YELLOW G-3 see AAQ250
PAMACYL see AMM250
PAMAQUIN see RHZ000
PAMAZONE see CJR909
2-PAM CHLORIDE see FNZ000
PAMEION see PAH250
PAMINE see SBH500
PAMINE BROMIDE see SBH500
2-PAM IODIDE see POS750
PAMISAN see ABU500
PAMISYL see AMM250
PAMISYL SODIUM see SEP000
2-PAM METHANESULFONATE see PLX250
PAMN see PNR250
PAMOIC ACID see PAF100
PAMOLYN see OHU000
PAMOSOL 2 FORTE see EIR000
PAMOVIN see PQC500
PAM-PERCHLORAT (GERMAN) see FOA000
2-PAM SULFONATE see POP300
PAN see BFW120
PAN see PCL750
PANA see PFT250
PANACAINE see PDU250
PANACEF see CCR850
PANACELAN see POC500
PANACID see PAF250
PANACIDE see MJM500
PANACRYL YELLOW 3GL see CMM890
PANACUR see FAL100
PANADOL see HIM000
PANADON see PAG200
PANALDINE see TGA525
PANAM see CBM750
PANAMIDIN DIHYDROCHLORIDE see DBM400
PANAX see GEQ400
PANAX GINSENG, ROOT EXTRACT see GEQ425
PANAX SAPONIN E see PAF450
PANAZON see DKK100
PANAZONE see PAF500
PANCAL see CAU750
PANCALMA see MQU750
PANCID see SNN500
PANCIL see OFE000
PANCODINE see DLX400
PANCORAL see BQJ500
PANCREATIC BASIC TRYPSIN INHIBITOR see PAF550
PANCREATIC DEOXYRIBONUCLEASE see EDK650
PANCREATIC DORNASE see EDK650
PANCREATIC TRYPSIN INHIBITOR see PAF550
PANCREATIC TRYPSIN INHIBITOR (KUNITZ) see PAF550
PANCREATIN see PAF600
PANCREATOPEPTIDASE E see EAG875
PANCRIDINE see DBN400
PANCURONIUM BROMIDE see PAF625
PANCURONIUM BROMIDE HYDRATE see PAF630
PANCURONIUM DIBROMIDE HYDRATE see PAF630
PANDEX see PKM250
PANDIGAL see LAU400
PANDRINOX see MLF250
PANDUROL see BHD250
PANESTIN see TBG000

PANETS see HIM000
PANEX see HIM000
PANFLAVIN see DBX400
PANFLAVIN see XAK000
PANFORMIN see BQL000
PANFUNGOL see KFK100
PANFURAN see FPF000
PANFURAN-S see BKH500
PANGUL see TEH500
PANHIBIN see SKS600
PANIMYCIN see PAG050
PANINI-'AWA'AWA (HAWAII) see AGV875
PANITHAL see SAH000
PANLANAT see LAU400
PANMYCIN see TBX000
PANMYCIN HYDROCHLORIDE see TBX250
PANOCTINE TRIACETATE see GLQ100
PANO-DRENCH 4 see MLF250
PANODRIN A-13 see MLF250
PANOFEN see HIM000
PANOGEN see MEO750
PANOGEN see MLF250
PANOGEN 15 see MLF250
PANOGEN 43 see MLF250
PANOGEN M see MEO750
PANOGEN METOX see MEO750
PANOGEN PX see MLF250
PANOGEN TURF FUNGICIDE see MLF250
PANOGEN TURF SPRAY see MLF250
PANOLID see DWE800
PANORAL see CCR850
PANORAL see PAG075
PANORAL HYDRATE see CCR850
PANORAM 75 see TFS350
PANORAM D-31 see DHB400
PANOSINE see MMD500
PANOSPRAY 30 see MLF250
PANOXYL see BDS000
PANPARNIT see PET250
PANRONE see FNF000
PANSOIL see EFK000
PANTALGINE see DAM700
PANTAS see HKR500
PANTASOTE R 873 see PKQ059
PANTELMIN see MHL000
PANTETHINE see PAG150
d-PANTETHINE see PAG150
PANTETINA see PAG150
PANTHELINE see HKR500
PANTHENOL see PAG200
d-PANTHENOL see PAG200
d(+)-PANTHENOL (FCC) see PAG200
PANTHER CREEK BENTONITE see BAV750
PANTHESIN see LEU000
PANTHESINE see LEU000
PANTHION see PAK000
PANTHODERM see PAG200
PANTHOJECT see CAU750
PANTHOLIC-L see IDE000
PANTHOLIN see CAU750
PANTOBROMINO see HNY500
PANTOCAINE see BQA010
PANTOCAINE HYDROCHLORIDE see TBN000
PANTOCID see HAF000
PANTOCRIN see PAG225
PANTOGAM see CAT125
PANTOL see PAG200
PANTOLAX see HLC500
PANTOMICINA see EDH500
PANTOMIN see PAG150
PANTONSILETTEN see DBX400
PANTOPAQUE see ELQ500
PANTOPRIM see TKX000
PANTOSEDIV see TEH500
PANTOSIN see PAG150
PANTOTHENATE CALCIUM see CAU750
PANTOTHENATE de ZINC (FRENCH) see ZKS000
PANTOTHENIC ACID see PAG300
(+)-PANTOTHENIC ACID see PAG300
PANTOTHENIC ACID, d- see PAG300

d-PANTOTHENIC ACID see PAG300
(d)-(+)-PANTOTHENIC ACID see PAG300
PANTOTHENIC ACID, CALCIUM SALT see CAU750
(+)-PANTOTHENIC ACID, CALCIUM SALT see CAU750
PANTOTHENIC ACID, ZINC SALT see ZKS000
PANTOTHENOL see PAG200
d-PANTOTHENOL see PAG200
PANTOTHENYL ALCOHOL see PAG200
d-PANTOTHENYL ALCOHOL see PAG200
d(+)-PANTOTHENYL ALCOHOL see PAG200
PANTOVERNIL see CDP250
PANTOZOL 1 see COD000
PAN-TRANQUIL see MQU750
PANURIN see CFY000
PAN-W-29 see MOZ000
PANWARFIN see WAT220
PAP see ALT250
PAP see DRR400
PAP see PDC250
PAP-1 see AGX000
PAPAIN see PAG500
PAPANERINE see PAH000
PAPAO-APAKA (GUAM) see EAI600
PAPAO-ATOLONG (GUAM) see EAI600
PAPAVARINE CHLORHYDRATE see PAH250
PAPAVERINA (ITALIAN) see PAH000
PAPAVERIN CARBOXYLIC ACID, SODIUM SALT see PAG750
PAPAVERINE see PAH000
PAPAVERINE CHLOROHYDRATE see PAH250
PAPAVERINE CHLOROHYDRATE see PAH250
PAPAVERINE HYDROCHLORIDE see PAH250
PAPAVERINE MONOHYDROCHLORIDE see PAH250
PAPAVERIN-HCL (GERMAN) see PAH250
PAPAYOTIN see PAG500
PAPER BLACK BA see AQP000
PAPER BLACK T see AQP000
PAPER BLUE R see AOR500
PAPER DEEP BLACK C see AQP000
PAPER RED 4B see DXO850
PAPER RED HRR see FMU070
PAPER SCARLET 4BS see CMO870
PAPER SCARLET 3BX see CMO875
PAPETHERINE HYDROCHLORIDE see PAH260
PAP H see PAH250
PAPI see PKB100
PAPI 20 see PKB100
PAPI 27 see PKB100
PAPI 135 see PKB100
PAPI 580 see PKB100
PAPI 901 see PKB100
PAPOOSE ROOT see BMA150
PAPP see AMC000
PAPRIKA see PAH275
PAPRIKA OLEORESIN see PAH280
PAPTHION see DRR400
PARA see PEY500
PARAAMINODIPHENYL see AJS100
PARABAR 441 see BFW750
PARABEN see HJL500
PARABEN see HNU500
PARABENCIL see PFR325
PARABENCILFENOL see PFR325
PARABIS see MJM500
PARABROMODYLAMINE MALEATE see BNE750
PARABROMOTOLUENE see BOG255
PARABUTYRALDEHYDE see TNC100
PARACAIN see AIT250
PARACETALDEHYDE see PAI250
PARACETAMOLE see HIM000
PARACETAMOLO (ITALIAN) see HIM000

PARACETANOL see HIM000
PARACETOPHENETIDIN see ABG750
PARACHLORAMINE see HGC500
PARACHLOROCIDUM see DAD200
PARACHLOROPHENOL see CJK750
PARACIDE see DEP800
PARACODIN see DKW800
PARACODINE see DKW800
PARA-COL see PAJ000
PARACORT see PLZ000
PARACORTOL see PMA000
PARACOTOL see PMA000
PARACRESYL ACETATE see MNR250
PARACRESYL ISOBUTYRATE see THA250
PARA CRYSTALS see DEP800
d-PARACURARINE CHLORIDE see TOA000
PARACYMENE see CQI000
PARACYMOL see CQI000
PARADERIL see RNZ000
PARADI see DEP800
PARADIAZINE see POL490
PARADICHLORBENZOL (GERMAN) see DEP800
PARADICHLOROBENZENE see DEP800
PARADICHLOROBENZOL see DEP800
PARA-DIEN see DAL600
PARADIONE see PAH500
PARADISE TREE see CDM325
(0)-PARADOL see VFP100
PARADONE BLUE RC see DFN300
PARADONE BLUE RS see IBV050
PARADONE BRILLIANT BLUE R see IBV050
PARADONE BRILLIANT ORANGE GR see CMU820
PARADONE BRILLIANT ORANGE GR NEW see CMU820
PARADONE DARK BLUE RFW see VGP100
PARADONE GOLDEN YELLOW see DCZ000
PARADONE GREY M see CMU475
PARADONE GREY MG see CMU475
PARADONE NAVY BLUE G see CMU500
PARADONE OLIVE GREEN B see ALT000
PARADONE OLIVE R see DUP100
PARADONE PRINTING BLACK BLSF see CMU320
PARADONE PRINTING BLACK TLSF see CMU320
PARADONE PRINTING BLUE FRS see IBV050
PARADONE RED BROWN 2RD see CMU770
PARADORMALENE see TEO250
PARADOW see DEP800
PARADUST see PAK000
PARAESIN see BQH250
PARAFFIN see PAH750
PARAFFIN HYDROCARBONS see PAH770
PARAFFIN OIL see MQV750
PARAFFIN OILS (PETROLEUM), CATALYTIC DEWAXED HEAVY (9CI) see MQV778
PARAFFIN OILS (PETROLEUM), CATALYTIC DEWAXED LIGHT (9CI) see MQV779
PARAFFIN WAX see PAH750
PARAFFIN WAXES and HYDROCARBON WAXES, CHLORINATED see PAH780
PARAFFIN WAXES and HYDROCARBON WAXES, CHLORINATED (C_{12}, 60% CHLORINE) see PAH800
PARAFFIN WAXES and HYDROCARBON WAXES, CHLORINATED (C_{23}, 43% CHLORINE) see PAH810
PARAFFIN WAX FUME (ACGIH) see PAH750
PARAFLEX see CDQ750
PARAFLU see TKH750
PARAFORM see FMV000
PARAFORMALDEHYDE see PAI000
PARAFORMALDEHYDE-UREA POLYMER see UTU500

PARAFORMALDEHYDE-UREA RESIN see UTU500
PARAFORMALDEHYDE, UREA RESIN, ISOBUTYLATED see IIM300
PARAFORSN see PAI000
PARAFUCHSIN (GERMAN) see RMK020
PARAGLAS see PKB500
PARAGLYINE see MOS250
PARAHEXYL see HGK500
PARAHYDROXYBENZALDEHYDE see FOF000
PARAISO (MEXICO) see CDM325
PARAL see PAI250
PARALACTIC ACID see LAG010
PARALCTIN see PMH625
PARALDEHYD (GERMAN) see PAI250
PARALDEHYDE see PAI250
PARALDEIDE (ITALIAN) see PAI250
PARALERGIN see ARP675
PARALEST see BBV000
PARALKAN see BBA500
PARALYTIC SHELLFISH POISON DIHYDROCHLORIDE see SBA500
PARA M see AKE250
PARA-MAGENTA see RMK020
PARAMAL see DBM800
PARAMANDELIC ACID see MAP000
PARAMAR see PAK000
PARAMEL DC see CNH125
PARAMENTHANE HYDROPEROXIDE see IQE000
PARAMETADIONE see PAH500
PARAMETHADIONE see PAH500
PARAMETHASONE 21-ACETATE see PAL600
PARAMETHAZONE ACETATE see PAL600
PARAMETHYL PHENOL see CNX250
PARAMIBE see DDS600
PARAMICINA see APP500
PARAMID see AKO500
PARAMIDIN see BQW825
PARAMIDINE see BQW825
PARAMID SUPRA see AKO500
PARAMINE BLACK B see AQP000
PARAMINE BLACK BH see CMN800
PARAMINE BLACK E see AQP000
PARAMINE BLUE 2B see CMO000
PARAMINE BLUE 3B see CMO250
PARAMINE FAST BROWN M see CMO800
PARAMINE FAST RED F see CMO870
PARAMINE FAST SCARLET 3B see CMO875
PARAMINE FAST SCARLET 4BS see CMO870
PARAMINE FAST VIOLET N see CMP000
PARAMINE GREEN B see CMO840
PARAMINE GREEN BN see CMO840
PARAMINOL see AIH600
PARAMINOPROPIOPHENONE see AMC000
PARAMINYL see WAK000
PARAMINYL MALEATE see DBM800
PARAMORFAN see DNU310
PARAMORPHAN see DLW600
PARAMORPHINE see TEN000
PARAMOTH see DEP800
PARAMYCIN see AMM250
PARA-MYRAC ALDEHYDE see IKT100
PARANAPHTHALENE see APG500
PARANATE see AIH600
PARANEPHRIN see VGP000
PARANIT ETHANE DISULFONATE see CBG250
PARANITROFENOL (DUTCH) see NIF000
PARANITROFENOLO (ITALIAN) see NIF000
PARANITROPHENOL (FRENCH, GERMAN) see NIF000
PARANITROSODIMETHYLANILIDE see DSY600
PARANOL see ALT250
PARANOL FAST RED 8BL see CMO885
PARA-NONYL PHENOL see NNC510
PARANUGGETS see DEP800
PARA ORANGE see FAG150

PARAOXON see NIM500
PARAOXONE see NIM500
PARAOXON-METHYL see PHD500
PARAPAN see HIM000
PARA-PAS see AMM250
PARAPEST M-50 see MNH000
PARAPHENOLAZO ANILINE see PEI000
PARA PHENYL BENZOIC ACID see PEL600
PARAPHENYLEN-DIAMINE see PEY500
PARAPHOS see PAK000
PARAPLEX P 543 see PKB500
PARAPLEX RG-2 (60%) see PAI750
PARAQUAT see PAI990
PARAQUAT (ACGIH) see PAJ000
PARAQUAT BIS(METHYL SULFATE) see PAJ250
PARAQUAT CHLORIDE see PAJ000
PARAQUAT CL see PAJ000
PARAQUAT DIBROMIDE see PAI995
PARAQUAT DICATION see PAI990
PARAQUAT DICHLORIDE see PAJ000
PARAQUAT, DICHLORIDE see PAJ000
PARAQUAT DIHYDRIDE see PAJ100
PARAQUAT DIMETHYL SULFATE see PAJ250
PARAQUAT DIMETHYL SULPHATE see PAJ250
PARAQUAT I see PAJ250
PARAQUAT ION see PAI990
PARAROSANILINE see RMK020
PARAROSANILINE CHLORIDE see RMK020
PARAROSANILINE HYDROCHLORIDE see RMK020
PARAROSANILINE, N,N,N',N',N'',N''-HEXAMETHYL-, CHLORIDE see AOR500
PARASAL see AMM250
PARASALICIL see AMM250
PARASALINDON see AMM250
PARASAN see BCA000
PARASCORBIC ACID see PAJ500
PARASEPT see BSC000
PARASEPT see HJL500
PARASEPT see HNU500
PARASOL see CNM500
PARASOLV 504 see AQZ150
PARASORBIC ACID see PAJ500
(+)-PARASORBINSAEURE (GERMAN) see PAJ500
PARASPAN see SBH500
PARASPEN see HIM000
PARASTARIN see EJM500
PARASULFONDICHLORAMIDO BENZOIC ACID see HAF000
PARASYMPATOL see HLV500
PARATAF see MNH000
PARATHENE see PAK000
PARATHESIN see EFX000
PARATHIAZINE see PAJ750
PARATHION see PAK000
M-PARATHION see MNH000
PARATHION, liquid (DOT) see PAK000
PARATHION and compressed gas mixture (DOT) see PAK230
PARATHION-ETHYL see PAK000
PARATHION METHYL see MNH000
PARATHION-METILE (ITALIAN) see MNH000
PARATHION S see DJS800
PARATHORMONE see PAK250
PARATHYRIN see PAK250
PARATHYROID HORMONE see PAK250
PARATOX see MNH000
PARAWET see PAK000
PARAXANTHINE see PAK300
PARAXENOL see BGJ500
PARAXIN see CDP250
PARAXIN SUCCINATE see CDP725
PARAZENE see DEP800
PARAZINE see PIJ500
PARAZONE see FNF000
PARBENDAZOLE see BQK000
PARBOCYL-REV see SJO000

PARCIDOL see DIR000
PARCLAY see KBB600
PARCLOID see PKQ059
PARDA see DNA200
PARDIDOL see DIR000
PARDISOL see DIR000
PARDROYD see PAN100
PAREDRINOL see FMS875
PARENTERAL see CJR909
PARENTRACIN see BAC250
PARENZYME see TNW000
PARENZYMOL see TNW000
PAREPHYLLIN see CNR125
PAREST see MDT250
PAREST see QAK000
PAR ESTRO see ECU750
PARETH 25 see AFJ160
PARETH 25-7 see AFJ160
PARETH 45-7 see AFJ168
PARETH 45-13 see AFJ168
PAREZ 607 see MCB050
PAREZ 613 see MCB050
PAREZ 707 see MCB050
PARFENAC see BPP750
PARFENAL see BPP750
PARFEZINE see DIR000
PARFURAN see NGE000
PARGITAN see BBV000
PARGLYAMINE see MOS250
PARGONYL see NCF500
PARGYLINE see MOS250
PARGYLINE HYDROCHLORIDE see BEX500
PARICINA see APP500
PARIDINE RED LCL see CHP500
PARIDOL see HJL500
PARIS FAST BLUE 1605 see COF420
PARIS GREEN see COF500
PARIS RED see LDS000
PARIS VIOLET R see MQN025
PARIS YELLOW see LCR000
PARITOL see SEH450
PARKE DAVIS CI-628 see NHP500
PARKEMED see XQS000
PARKIBLEU see CMO250
PARKIN see DIR000
PARKINSAN see BBV000
PARKIPAN see CMO250
PARKISOL see DIR000
PARKOPAN see BBV000
PARKOPAN see PAL500
PARKOPHYLLIN see TEP000
PARKOSED see BNK000
PARKOTAL see EOK000
PAR KS-12 see PMC250
PARLAY see TMK125
PARLEF see TKH750
PARLIF see TKH750
PARLODEL see BNB325
PARLODION see CCU250
PARMAL see WAK000
PARMATHASONE ACETATE see PAL600
PARMAVERT see NNE100
PARMETOL see CFD990
PARMIDIN see PPH050
PARMIDINE see PPH050
PARMIDINE R see PPH050
PARMINAL see QAK000
PARMOL see HIM000
PARMONE see NAK500
PARNATE see PET500
PARNATE see PET750
PAROIL CHLOREZ see PAH780
PAROL see CFD990
PAROL see MQV750
PAROL see MQV875
PAROLEINE see MQV750
PAROMOMYCIN see NCF500
PAROMOMYCIN SULFATE see APP500
PAROXAN see NIM500
PAROXETINE see PAL650
PAROXON see ELL500

PAROXYL see ABX500
PAROXYPROPIONE see ELL500
PAROZONE see SHU500
PARPANIT see PET250
PARPHEZEIN see DIR000
PARPON see BCA000
PARROT GREEN see COF500
PARRYCOP see CNK559
PARSAL see AJC000
PARSALMIDE see AJC000
PARSIDOL see DIR000
PARSITAN see DIR000
PARSLEY APIOL see AGE500
PARSLEY CAMPHOR see AGE500
PARSLEY HERB OIL (FCC) see PAL750
PARSLEY OIL see PAL750
PARSLEY SEED OIL (FCC) see PAL750
PARTEL see DJT800
PARTERGIN see PAM000
PARTEROL see DME300
PARTHENICIN see PAM175
PARTHENIN see PAM175
PARTOCON see ORU500
PARTREX see TBX250
PARTUSISTEN see FAQ100
PARVOLEX see ACH000
PARZATE see DXD200
PARZATE see EIR000
PARZONE see DKW800
PAS see AMM250
PASA see AMM250
PASADE see SEP000
PASALIN see PAM500
PASALON see AMM250
PASALON-RAKEET see SEP000
PASARA see AMM250
PAS-C see AMM250
PASCO see ZBJ000
PASCO see ZKA000
PASCORBIC see AMM250
PASEM see AMM250
PASEPTOL see HNU500
PASEXON 100T see SHK800
PAS-H 10 see DTS500
PASILLA (PUERTO RICO) see CDM325
PASK see AMM250
PASMED see AMM250
PASNAL see SEP000
PASNODIA see AMM250
PASOLAC see AMM250
PASOTOMIN see PMF250
PASPALIN see PAM775
PASPALIN P I see PAM775
PASQUE FLOWER see PAM780
PASSIFLORAE INCARNATAE
EXTRACTUM see PAM782
PASSIFLORA EXTRACT see PAM782
PASSIFLORA INCARNATA, EXTRACT see
PAM782
PASSIFLORIN see MPA050
PASSIONFLOWER EXTRACT see PAM782
PASSODICO see SEP000
PAT see PKE600
PATAP see BHL750
PATCHOULI OIL see PAM783
PATCHOULY OIL see PAM783
PATCHUK see CNT350
PATENT ALUM see AHG800
PATENTBLAU V see ADE500
PATENT BLUE see ADE500
PATENT BLUE A see ERG100
PATENT BLUE AE see FMU059
PATENT BLUE AF see ERG100
PATENT BLUE V see ADE500
PATENT BLUE V see CMM062
PATENT BLUE V CARMINE BLUE V see
CMM062
PATENT BLUE VF see ADE500
PATENT BLUE VF-CF see ADE500
PATENT BLUE VF SPECIAL see ADE500
PATENT BLUE VS see ADE500
PATENT GREEN see COF500

PATHCLEAR see PAJ000
PATHOCIDIN see AJO500
PATHOCIDINE see AJO500
PATHOCIL see DGE200
PATHOCLON see DXN709
PATHOMYCIN see APY500
PATORAN see PAM785
PATRICIN see VRF000
PATRINOSIDE see PAM789
PATRINOSIDE-AGLYCONE see PAM800
PATROL see FAQ230
PATROVINE see DHX800
PATROVINE see THK000
PATTINA V 82 see PKQ059
PATTONEX see PAM785
PATULIN see CMV000
PAUCIMYCIN see NCF500
PAUSITAL see CKE750
P. AUSTRALIS VENOM see ARV000
PAVABID see PAH250
PAVAGRANT see PAH250
PAVAKEY see PAH250
PAVASED see PAH250
PAVATEST see PAH250
PAVATRIN see CBS250
PAVATRINE see FDK000
PAVATRINEAT see CBS250
PAVISOID see PAN100
PAVULON see PAF625
PAVULON see PAF630
PAXAREL see ACE000
PAXATE see DCK759
PAXILON see BGD250
PAXISTIL see CJR909
PAXISYN see DLY000
PAXITAL see MOQ250
PAXYL see TAF675
PAY-OFF see COQ385
PAY-OFF see DRN200
PAYZE see BLW750
PAYZONE see DKK100
PAYZONE see PAF500
PAZITAL see CGA000
PB see PIX250
PB see POQ250
PB 40 see HCQ525
PB-106 see PIX800
P 2010B see PJS750
PBA see BEA100
PBA, DIMETHYLAMINE SALT see PJQ000
PBA (GROWTH STIMULANT) see BEA100
PBAN-560 see PAN250
PBAN-560 (degassed) see PAN500
PBB see FBU000
PBB see FBU509
PBB see PJL335
PB 89 CHLORIDE see FMU000
PBDP see SAN300
PBDZ see BQK000
PB-89 HYDROCHLORIDE see FMU000
PBK 1 see DTS500
PBNA see PFT500
PBQI see BDD500
PB-1,6-QUINONE see BCU500
PBS see SID000
PBTA-1 see ABX833
PBTA-2 see ABX836
PBX(AF) 108 see CPR800
PBXW 108(E) see CPR800
PBZ see TMP750
PC see PGP250
PC 1 see BQJ500
PC 222 see CKD250
PC 603 see CJN250
PC-1421 see PII500
1PC6115 see MCB050
PCA see MCR750
PCA see PEE750
PCA see TDV300
PC ALCOHOL DF see PAN750
PCB see PJL750
PCB see PJL800

PCB see PJM000
PCB see PJM250
PCB see PJM500
PCB see PJM750
PCB see PJN000
PCB see PJN250
PCB see PJN500
PCB see PJN750
PCB see PJO000
PCB see PJO250
PCB see PJO500
PCB see PJO750
PCB see PJP000
PCB see PME250
PCB2 see HCD500
PCBs see PJL750
PCB 15 see DET850
PCB 18 see TIL258
PCB 26 see TIL262
PCB28 see TIL270
PCB 33 see TIL260
PCB 40 see TBO520
PCB 52 see TBO530
PCB 54 see TBO600
PCB 87 see PAV610
PCB 105 see PAV630
PCB 118 see PAV620
PCB 126 see PAV700
PCB 156 see HCC600
PCB 157 see HCC600
PCBG see GFW050
PCB HYDROCHLORIDE see PME500
PCBS see CJR500
PCC see CDV100
PCC see PIN225
γ-PCCH see PAV780
P.C. 80 CRABGRASS KILLER see PLC250
PCDE 32 see HCH485
PCDE 35 see TIL700
PCDE 37 see DEX270
PCDE 102 see TIL700
PCDF see PJP100
PCEO see TBQ275
PCHO see PAI250
PCI see CJR500
PCL see HCE500
PCM see PCF300
PCMC see CFD990
PCMH see PAN775
PCMN see PAX775
PCMNI see CIQ100
PCMX see CLW000
PCNB see PAX000
PCNU see CGW250
PCON see KDA050
PCONA see KDA050
PCP see PAX250
PCP (anesthetic) see PDC890
PCPA see CJN000
PCPA see CJR125
PCPBS see CJR500
PCPCBS see CJT750
PCP HYDROCHLORIDE see AOO500
PCPI see CKB000
PCT see PJP250
PCT see PJP300
PCTP see PAY500
PD 93 see PAF250
PD 71627 see AJR400
PD 73093 see DSO500
PD 109394 see AAI118
PD-113309 see XDJ025
PD128,907 see TDB250
PD 134298-38A see DMS950
P.D.A.B. see DOT300
p-PDA HCl see PEY650
PDB see DEP800
PDC see DGG800
PDCB see DEP800
PDD see PGT250
PDD 6040I see CJV250
PDH see POK300

p-PD HCl see PEY650
PDMT see DTP000
m-PDN see PHX550
p-PDN see BBP250
PDP see PDC250
PDQ see CLN750
PDT see DTP000
PDTA see PNJ100
PDTA-Sb see AQI500
PDU see DTP400
PE see PBB750
PE-11 see MKV800
PE-043 see PAN800
PE 122 see PKE850
PE 304 see CHF600
PE 512 see PJS750
PE 60A see LED100
PE 60E see LED100
PE 617 see PJS750
PEA see PDD750
β-PEA see PDD750
PEACE PILL see AOO500
PEACH see AQP890
PEACH ALDEHYDE see HBN200
PEACH ALDEHYDE see UJA800
PEACHES see AOB250
PEACH LACTONE see HBN200
PEACHROME see CMJ565
PEACH-THIN see NBL200
PEACOCK BLUE X-1756 see FMU059
PEA FLOWER LOCUST see GJU475
PEANUT OIL see PAO000
PEARL ASH see PLA000
PEARL ASH see PLA250
PEARLPUSS see CCL250
PEARL STEARIC see SLK000
PEARLY GATES see DJO000
PEAR OIL see AOD725
PEAR OIL see IHO850
PEARSALL see AGY750
PEB1 see DAD200
PEBC see PNF500
PEBULATE see PNF500
PECAN SHELL POWDER see PAO000
PECATAL see MOQ250
PECAZINE see MOQ250
PECNON see KGK000
PECTA-DIAZINE, suspension see PPP500
PECTALGINE see SEH000
PECTAMOL see DHQ200
PECTIN see PAO150
PECTINATE see PAO150
PECTINIC ACID see PAO150
PECTINS see PAO150
PECTITE see MBX800
PECTOLIN see TCY750
PECTOX see CBR675
PED see LBT000
PEDALITIN TETRAACETATE see PAO160
PEDIAFLOR see SHF500
PEDIDENT see SHF500
PEDIFLAVONE see ITD020
PEDILANTHUS TITHYMALOIDES see
SDZ475
PEDINEX see CPK550
PEDINEX (FRENCH) see CPK500
PEDIPEN see PDT750
PEDRACZAK see BBQ500
PEDRIC see HIM000
PEERACHROME YELLOW R see NEY000
PEERACID ORANGE II see CMM220
PEERAMINE BLACK E see AQP000
PEERAMINE BLACK GXOO see AQP000
PEERAMINE BRIGHT YELLOW MN see
CMP050
PEERAMINE CONGO RED see SGQ500
PEERLESS see CBT750
PEERLESS see KBB600
PEFURAZOATE see PAO170
PEG 200 see PJT200
PEG 300 see PJT225
PEG 400 see PJT230

PEG 600 see PJT240
PEG 1000 see PJT250
PEG 1500 see PJT500
PEG 4000 see PJT750
PEG 6000 see PJU000
PEGANONE see EOL100
PEG-5 LANOLIN see LAU560
PEG-20 LANOLIN see LAU560
PEG-24 LANOLIN see LAU560
PEG-30 LANOLIN see LAU560
PEG-50 LANOLIN see LAU560
PEG-60 LANOLIN see LAU560
PEG-85 LANOLIN see LAU560
PEG-100 LANOLIN see LAU560
PEG 1000MO see PJY100
PEG-9 NONYL PHENYL ETHER see PKF000
PEG 200MO see PJY100
PEG 600MO see PJY100
PEGOANONE see EOL100
PEG-9 OCTYL PHENYL ETHER see PKF500
PEG-6 OLEATE see PJY100
PEG-20 OLEATE see PJY100
PEG-32 OLEATE see PJY100
PEG-20 OLEYL ETHER see PJW500
PEGOSPERSE 400MO see PJY100
PEGOTERATE see PKF750
PEHA see PBD000
PEHANORM see TEM500
PELADOW see CAO750
PELAGOL 3GA see ALT000
PELAGOL BA see DBO400
PELAGOL CD see PEY650
PELAGOL D see PEY500
PELAGOL DA see DBO000
PELAGOL DR see PEY500
PELAGOL EG see ALS990
PELAGOL GREY see DBO400
PELAGOL GREY C see CCP850
PELAGOL GREY CD see PEY650
PELAGOL GREY D see PEY500
PELAGOL GREY GG see ALT000
PELAGOL GREY J see TGL750
PELAGOL GREY L see DBO000
PELAGOL GREY P BASE see ALT250
PELAGOL GREY RS see REA000
PELAGOL GREY SLA see DBO400
PELAGOL J see TGL750
PELAGOL L see DBO000
PELAGOL P BASE see ALT250
PELAGOL SLA see DBO400
PELA PUERCO (DOMINICAN REPUBLIC)
see DHB309
PELARGIC ACID see NMY000
PELARGIDENOLON see ICE000
PELARGIDENOLON 1497 see ICE000
PELARGIDENON 1449 see CDH250
PELARGOL see DTE600
PELARGON (RUSSIAN) see NMY000
PELARGONE see OFE020
PELARGONHYDROXAMIC ACID see
NMX600
PELARGONIC ACID see NMY000
PELARGONIC ALCOHOL see NNB500
PELARGONIC ALDEHYDE see NMW500
PELARGONIC MORPHOLIDE see MRQ750
PELARGONITRILE see OES300
PELARGONIUM OIL see GDA000
PELARGONONITRILE see OES300
PELARGONOYL PEROXIDE see NNA100
PELARGONYL HYDROXAMIC ACID see
NMX600
PELARGONYL PEROXIDE see NNA100
PELARGONYL PEROXIDE, technically pure
(DOT) see NNA100
PELASPAN 333 see SMQ500
PELASPAN ESP 109s see SMQ500
PELAZID see ILD000
PELENTAN see BKA000
PELIDOL see DWI300
PELIDORM see BNK000
PELIKAN C 11/1431a see CBT500
PELLAGRAMIN see NCQ900

PELLAGRA PREVENTIVE FACTOR see
NCQ900
PELLAGRIN see NCQ900
PELLCAFS see BBK500
PELLCAP see BBK500
PELLCAPS see BBK500
PELLETEX see CBT750
PELLETIERINE see PAO500
(R)-(−)-PELLETIERINE see PAO500
PELLIDOL see ACR300
PELLIDOLE see ACR300
PELLON 2506 see PMP500
PELLUGEL see CCL250
PELMIN see NCR000
PELMINE see NCR000
PELONIN see NCQ900
PELONIN AMIDE see NCR000
PELSON see DLY000
PELT-44 see PEX500
PELTAR see MAP300
PELTOL BR see PFU500
PELTOL BR II see PFU500
PELTOL D see PEY500
PELT SOL see DJV000
PELUCES see CLY500
PELVIRAN see DNG400
PEMAL see ENG500
PEMALIN see ENG500
PEMOLINE MAGNESIUM see PAP000
PEMOLINE MAGNESIUM CHELATE see
PAP000
PEMOLIN and MAGNESIUM HYDROXIDE
see PAP000
PEMPIDINA TARTRATO (ITALIAN) see
PAP110
PEMPIDINE HYDROCHLORIDE see PAP100
PEMPIDINE TARTRATE see PAP110
PEN 100 see PJS750
PEN 200 see PDD350
PEN A see AOD125
PEN-A-BRASIVE see BFD250
PENADUR see BFC750
PENADUR L-A see BFC750
PENAGEN see PDT750
PENALEV see BFD000
d-PENAMINE see MCR750
PEN A/N see SEQ000
PENAR see DRS000
PENATIN see CMV000
PENATIN see GFG100
PENBAR see NBU000
PENBRISTOL see AIV500
PENBRITIN see AIV500
PENBRITIN PAEDIATRIC see AIV500
PENBRITIN-S see SEQ000
PENBRITIN SYRUP see AIV500
PENBROCK see AIV500
PENBUTOLOL see PAP225
(−)-PENBUTOLOL see PAP225
PENBUTOLOL SULFATE see PAP230
PENCAL see ARB750
PENCARD see PBC250
PENCB see PAV630
PENCHLOROL see PAX250
PENCICLOVIR SODIUM see POK100
PENCIL GREEN SF see FAF000
PENCILLIC ACID see PAP750
PENCOGEL see CCL250
PENCOMPREN see PDT750
PENCONAZOLE see PAP240
PENCYCURON see UTA400
PENDEPON see BFC750
PENDEROL see PMB250
PEN-DI-BEN see BFC750
PENDIMETHALIN see DRN200
PENDIOMID see MKU750
PENDIOMIDE BROMIDE see MKU750
PENDIOMIDE DIBROMIDE see MKU750
PENDITAN see BFC750
PENDURAN see BFC750
PENETECK see MQV750
PENETECK see MQV875

PENETRACYNE see MCH525
PENFENATE see TIL800
PENFLURIDOL see PAP250
PENFORD 260 see HLB400
PENFORD 280 see HLB400
PENFORD 290 see HLB400
PENFORD GUM 380 see SLJ500
PENFORD P 208 see HLB400
PENGITOXIN see GEW000
PENGLOBE see BAB250
PENIALMEN see SEQ000
PENICIDIN see CMV000
PENICILLAMIN see MCR750
(S)-PENICILLAMIN see MCR750
PENICILLAMINE see MCR750
d-PENICILLAMINE see MCR750
l-PENICILLAMINE see PAP300
dl-PENICILLAMINE see PAP500
PENICILLAMINE HYDROCHLORIDE see PAP550
d-PENICILLAMINE HYDROCHLORIDE see PAP550
PENICILLANIC ACID DIOXIDE SODIUM SALT see PAP600
PENICILLANIC ACID 1,1-DIOXIDE SODIUM SALT see PAP600
PENICILLANIC ACID SULFONE SODIUM SALT see PAP600
PENICILLIC ACID see PAP750
PENICILLIN see PAQ000
PENICILLIN P-12 see DSQ800
PENICILLIN P-12 see MNV250
PENICILLIN, compounded with 9-AMINOACRIDINE see AHT000
PENICILLIN AT see AGK250
PENICILLINATE de 2,5-DIPHENYL PIPERAZINE see PGQ100
PENICILLIN BT see BRS250
PENICILLIN compounded with CHOLINE CHLORIDE see PAQ060
PENICILLIN-CHOLINESTER CHLORID (GERMAN) see PAQ060
PENICILLIN G see BDY669
PENICILLIN G BENETHAMINE see PAQ100
PENICILLIN G, compounded with N,N'-DIBENZYLETHYLENEDIAMINE (2:1) see BFC750
PENICILLIN G EPHEDRINE SALT see PAQ120
PENICILLIN-G, MONOSODIUM SALT see BFD250
PENICILLIN G POTASSIUM see BFD000
PENICILLIN G POTASSIUM SALT see BFD000
PENICILLIN G PROCAINE see PAQ200
PENICILLIN G SALT of N,N'-DIBENZYLETHYLENEDIAMINE see BFC750
PENICILLIN G, SODIUM see BFD250
PENICILLIN G, SODIUM SALT see BFD250
PENICILLIN O see AGK250
PENICILLIN PHENOXYMETHYL see PDT500
PENICILLIN POTASSIUM PHENOXYMETHYL see PDT750
PENICILLIN V see PDT500
PENICILLIN V POTASSIUM see PDT750
PENICILLIN V POTASSIUM SALT see PDT750
PENICILLIUM ROQUEFORTI TOXIN see PAQ875
PENICILLIUM VIRIDICATUM TOXIN see PAQ900
PENICIN see PCU500
PENICLINE see AIV500
PENIDURAL see BFC750
PENIDURE see BFC750
PENILARYN see BFD250
PENILENTE see BFC750
PENILTETRA see MCH525
PENIMEPICYCLINE see MCH525
PENIN see PCU500

PENIRAZINE see PGQ100
PENISEM see BFD000
PENITE see SEY500
PENITRACIN see BAC250
PENITREM A see PAR250
PENIZILLIN (GERMAN) see PAQ000
PENNAC see SGF500
PENNAC CBS see CPI250
PENNAC CRA see IAQ000
PENNAC MBT POWDER see BDF000
PENNAC MS see BJL600
PENNAC TBBS see BQK750
PENNAC ZT see BHA750
PENNAMINE see DAA800
PENNAMINE D see DAA800
PENNCAP-M see MNH000
PENNFLOAT M see LBX000
PENNFLOAT S see LBX000
PENNSALT TD-72 see DJI000
PENN SALT TD-183 see TBV750
PENNSALT TD 5032 see HEE500
PENNWALT C-4852 see DSQ000
PENNWHITE see SHF500
PENNYROYAL OIL see PAR500
PENNZONE B see DEI000
PENNZONE E see DKC400
PENOCTONIUM BROMIDE see PAR600
PEN-ORAL see PDT500
PENOTRANE see PFN000
PENOXALINE see DRN200
PENPHENE see CDV100
PENPHENE see TBV750
PENRECO see MQV750
PENSIG see PDD350
PEN-SINT see DGE200
PENSTAPHOCID see MNV250
PENSYN see AOD125
PENTA see PAX250
PENTAACETYLGITOXIN see GEW000
PENTA-o-ACETYLGITOXIN see GEW000
PENTAAMMINEAQUACOBALT(III) CHLORATE see PAR750
PENTAAMMINEPYRAZINERUTHENIUM(II) PERCHLORATE see PAR799
PENTAAMMINEPYRIDINERUTHENIUM(II) PERCHLORATE see PAS829
PENTAAMMINE(THEOPHYLLINE)RUTHENIUM(3+) TRICHLORIDE see PAS835
PENTAAMMINETHIOCYANATOCOBALT(III) PERCHLORATE see PAS859
PENTAAMMINETHIOCYANATORUTHENIUM(II) PERCHLORATE see PAS879
PENTAAZAACENAPHTHYLENE-5'-PHOSPHATE ESTER MONOHYDRATE see TJE875
PENTAAZACENTOPTHYLENE see TJE870
1,4,7,10,13-PENTAAZATRIDECANE see TCE500
1,3,3,5,5-PENTAAZIRIDINO-1-THIA-2,4,6-TRIAZA-3,5-DIPHOSPHORINE-1-OXIDE see SED700
PENTABARBITAL SODIUM see NBU000
PENTABARBITONE see NBT500
PENTABORANE(9) see PAT750
PENTABORANE(11) see PAT799
PENTABORANE (ACGIH,DOT,OSHA) see PAT750
PENTABORANE(9)DIAMMONIATE see DCF725
PENTABROMFENOL see PAU250
PENTABROMOCHLOROCYCLOHEXANE see CPB550
1,2,3,4,5-PENTABROMO-6-CHLOROCYCLOHEXANE see CPB550
1,2,3,7,8-PENTABROMODIBENZOFURAN see PAT820
PENTABROMODIPHENYL ETHER see PAT830
PENTABROMODIPHENYL ETHER (DE-71) see PAT830
PENTABROMODIPHENYL OXIDE see PAT830

PENTABROMOETHYLBENZENE see PAT850
2,3,4,5,6-PENTABROMOETHYLBENZENE see PAT850
PENTABROMOPHENOL see PAU250
PENTABROMOPHENYLALLYLETHER see PAU600
PENTABROMOPHENYL ETHER see PAU500
PENTABROMO PHOSPHORANE see PHR250
PENTABROMO PHOSPHORUS see PHR250
PENTABROMO(2-PROPENYLOXY)BENZENE see PAU600
PENTAC see DAE425
PENTACAINE see PPW000
PENTACAINE HYDROCHLORIDE see PPW000
PENTACARBONYLIRON see IHG500
PENTACARBONYL(PIPERIDINE)CHROMIUM see PAU700
PENTACENE see PAV000
PENT-ACETATE see AOD725
PENTACHLOORETHAAN (DUTCH) see PAW500
PENTACHLOORFENOL (DUTCH) see PAX250
PENTACHLORAETHAN (GERMAN) see PAW500
PENTACHLORETHANE (FRENCH) see PAW500
PENTACHLORIN see DAD200
PENTACHLORNITROBENZOL (GERMAN) see PAX000
PENTACHLOROACETONE see PAV225
PENTACHLOROACETOPHENONE see PAV250
2',3',4',5',6'-PENTACHLOROACETOPHENONE see PAV250
PENTACHLOROANISOLE see MNH250
2,3,4,5,6-PENTACHLOROANISOLE see MNH250
PENTACHLOROANTIMONY see AQD000
PENTACHLOROBENZALDEHYDE OXIME see PAV300
PENTACHLOROBENZENE see PAV500
PENTACHLORO-BENZENETHIOL see PAY500
PENTACHLOROBIPHENYL see PAV600
2,2',3,4,5'-PENTACHLOROBIPHENYL see PAV610
2,3,3',4,4',5-PENTACHLOROBIPHENYL see PAV620
2,3,3',4,4'-PENTACHLOROBIPHENYL see PAV630
2,3,4,2',5'-PENTACHLOROBIPHENYL see PAV610
2,3,4,3',4'-PENTACHLOROBIPHENYL see PAV630
3,3',4,4',5-PENTACHLOROBIPHENYL see PAV700
2,2',3,4',5-PENTACHLOROBIPHENYL see PAV610
2,5,2',3',4'-PENTACHLOROBIPHENYL see PAV610
3,4,2',4',5-PENTACHLOROBIPHENYL see PAV630
3,4,2',4',5'-PENTACHLOROBIPHENYL see PAV620
3,4,3',4',5'-PENTACHLOROBIPHENYL see PAV700
2,3,3'4,4'-PENTACHLORO-1,1'-BIPHENYL see PAV630
3,3',4,4',5-PENTACHLORO-1,1'-BIPHENYL see PAV700
S-(1,2,3,4,4-PENTACHLOROBUTADIENYL)GLUTATHIONE see GFW050
PENTACHLOROBUTANE see PAV750
PENTACHLORO-3-BUTENOIC ACID see PAV775

2,2,3,4,4-PENTACHLORO-3-BUTENOIC ACID see PAV775
γ-PENTACHLOROCYCLOHEXENE see PAV780
PENTACHLOROCYCLOPENTADIENE see PAV800
1,2,3,4,5-PENTACHLOROCYCLOPENTADIENE see PAV800
PENTACHLORO-2,4-CYCLOPENTADIEN-1-YL see PAV800
PENTACHLOROCYCLOPROPANE see PAV810
1,2,3,4,7-PENTACHLORODIBENZODIOXIN see PAV820
1,2,3,4,7-PENTACHLORODIBENZO-p-DIOXIN see PAV820
1,2,3,7,8-PENTACHLORODIBENZO-p-DIOXIN see PAV850
1,2,4,7,8-PENTACHLORODIBENZO-p-DIOXIN see PAV840
1,2,3,7,8-PENTACHLORODIBENZOFURAN see PAV860
2,3,4,7,8-PENTACHLORODIBENZOFURAN see PAW100
3,3',5,5',6-PENTACHLORO-2,2'-DIHYDROXYBENZANILIDE see DMZ000
3,5,6,3',5'-PENTACHLORO-2,2'-DIHYDROXYBENZANILIDE see DMZ000
PENTACHLORODIPHENYL see PAV600
2,2',3,4,5'-PENTACHLORODIPHENYL see PAV610
2,2',4,5,6'-PENTACHLORODIPHENYL ETHER see TIL700
PENTACHLORO DIPHENYL OXIDE see PAW250
PENTACHLOROETHANE see PAW500
PENTACHLOROFENOL see PAX250
2',2'',4',4'',5-PENTACHLORO-4-HYDROXY-ISOPHTHALANILIDE see PAW600
3,3',5,5',6-PENTACHLORO-2'-HYDROXYSALICYLANILIDE see DMZ000
PENTACHLOROMANDELONITRILE see PAX775
PENTACHLOROMETHOXYBENZENE see MNH250
1,1,2,3,4-PENTACHLORO-4-(1-METHYLETHOXY)-1,3-BUTADIENE see INE055
PENTACHLORONAPHTHALENE see PAW750
PENTACHLORONITROBENZENE see PAX000
PENTACHLOROPHENATE see PAX250
PENTACHLOROPHENATE SODIUM see SJA000
PENTACHLOROPHENOL see PAX250
2,3,4,5,6-PENTACHLOROPHENOL see PAX250
PENTACHLOROPHENOL (GERMAN) see PAX250
PENTACHLOROPHENOL, DOWICIDE EC-7 see PAX250
PENTACHLOROPHENOL, DP-2 see PAX250
PENTACHLOROPHENOL, SODIUM SALT see SJA000
PENTACHLOROPHENOL, SODIUM derivative mixed with TETRACHLOROPHENOL SODIUM derivative (4:1) see PAX750
PENTACHLOROPHENOL, TECHNICAL see PAX250
PENTACHLOROPHENOXY SODIUM see SJA000
PENTACHLOROPHENYL BUTYL ETHER see BPM690
PENTACHLOROPHENYL CHLORIDE see HCC500
(PENTACHLOROPHENYL)GLYCOLONITRILE see PAX775
PENTACHLOROPHENYL METHYL ETHER see MNH250

PENTACHLOROPHENYL PROPYL ETHER see PAX800
1,1,2,2,3-PENTACHLOROPROPANE see PAY000
1,1,1,3,3-PENTACHLOROPROPANONE see PAV225
1,1,1,3,3-PENTACHLORO-2-PROPANONE see PAV225
1,1,2,3,3-PENTACHLOROPROPENE see PAY200
1,1,2,3,3-PENTACHLORO-1-PROPENE see PAY200
1,1,2,3,3-PENTACHLOROPROPYLENE see PAY200
2,3,4,5,6-PENTACHLOROPYRIDINE see PAY250
PENTACHLORORHODATE(2-) DIPOTASSIUM see PAY300
PENTACHLOROTHIOFENOLAT ZINECNATY (CZECH) see BLC500
PENTACHLOROTHIOPHENOL see PAY500
α-α-α-2,4-PENTACHLOROTOLUENE see PAY550
2,4,α-α-α-PENTACHLOROTOLUENE see PAY550
PENTACHLORTHIOFENOL (CZECH) see PAY500
PENTACHLORURE d'ANTIMOINE (FRENCH) see AQD000
PENTACIN see CAY500
PENTACINE see CAY500
PENTACLOROETANO (ITALIAN) see PAW500
PENTACLOROFENOLO (ITALIAN) see PAX250
PENTACON see PAX250
PENTAC WP see DAE425
PENTACYANONITROSYLFERRATE BARIUM see PAY600
PENTACYANONITROSYLFERRATE COBALT see PAY610
PENTACYANONITROSYLFERRATE ZINC see ZJJ400
PENTADECAFLUOROOCTANOIC ACID see PCH050
PENTADECAFLUORO-n-OCTANOIC ACID see PCH050
1-(2,2,3,3,4,4,5,5,6,6,7,7,8,8,8-PENTADECAFLUOROOCTYL)PYRIDINIUM TRIFLUOROMETHANESULFONATE see MBV710
1-PENTADECANAMINE see PBA000
N-PENTADECANE see PAY750
1-PENTADECANECARBOXYLIC ACID see PAE250
7-PENTADECANECARBOXYLIC ACID see HFP500
PENTADECANOIC ACID see PAZ000
8-PENTADECANONE see HBO790
PENTADECAN-8-ONE see HBO790
PENTADECYCLIC ACID see PAZ000
PENTADECYLAMINE see PBA000
1-PENTADECYLAMINE see PBA000
n-PENTADECYLAMINE see PBA000
1,3-PENTADIENE see PBA100
(E)-1,3-PENTADIENE see PBA250
trans-1,3-PENTADIENE see PBA250
1,3-PENTADIENE-1-CARBOXALDEHYDE see SKT500
1,3-PENTADIENE-1-CARBOXYLIC ACID see SKU000
1:4-PENTADIENE DIOXIDE see DHE000
2,4-PENTADIENOIC ACID, 5-(4a-(ACETYLOXYL)-3,4,4a,5,6,9-HEXAHYDRO-3,7-DIMETHYL-1-OXO-1H-4,9a-ETHANOCYCLOHEPTA(c)PYRAN-3-YL)-2-METHYL-, (3-α(2E,4E),4-α-9a-α)-(−)- (9CI) see POH550
2,4-PENTADIENOL see PBA300
2,4-PENTADIEN-1-OL see PBA300
1,4-PENTADIEN-3-ONE, 1,5-BIS(4-AZIDOPHENYL)-(9CI) see BGW720

1,3-PENTADIYNE see PBB000
1,3-PENTADIYN-1-YL COPPER see PBB229
1,3-PENTADIYN-1-YL SILVER see PBB449
PENTADORM see EQL000
PENTAERYTHRITE see PBB750
PENTAERYTHRITE TETRANITRATE see PBC250
PENTAERYTHRITE TETRANITRATE (DOT) see PBC250
PENTAERYTHRITE TETRANITRATE, dry (DOT) see PBC250
PENTAERYTHRITE TETRANITRATE, desensitized, wet (DOT) see PBC250
PENTAERYTHRITE TETRANITRATE, with not less than 7% wax (DOT) see PBC250
PENTAERYTHRITOL see PBB750
PENTAERYTHRITOL CHLOROL see NCH500
PENTAERYTHRITOL DIBROMIDE see DDQ400
PENTAERYTHRITOL DIBROMOHYDRIN see DDQ400
PENTAERYTHRITOL DICHLOROHYDRIN see ALB000
PENTAERYTHRITOL ESTER of PARTIALLY HYDROGENATED WOOD ROSIN see PBB800
PENTAERYTHRITOL ESTER of WOOD ROSIN see PBB810
PENTAERYTHRITOL GLYCIDYL ETHER see PBB850
PENTAERYTHRITOL, TETRAACETATE see NNR400
PENTAERYTHRITOL TETRABENZOATE see PBC000
PENTAERYTHRITOL TETRANICOTINATE see NCW300
PENTAERYTHRITOL TETRANITRATE see PBC250
PENTAERYTHRITOL TETRANITRATE, diluted see PBC250
PENTAERYTHRITOL TRIACRYLATE see PBC750
PENTAETHYLENEHEXAMINE see PBD000
2,2,4,4,6-PENTAETHYLENIMINO-6-MORPHOLINO-CYCLOTRIPHOSPHAZATRIEN see FOM000
PENTAFIN see PBC250
PENTAFLUOROANILINE see PBD250
2,3,4,5,6-PENTAFLUOROANILINE see PBD250
PENTAFLUOROANTIMONY see AQF250
PENTAFLUOROBENZOIC ACID see PBD275
PENTAFLUOROBENZYL (1R)-trans-(2,2-DICHLOROVINYL)-2,2-DIMETHYLCYCLOPROPANECARBOXYLATE see FAO220
1,2,3,4,5-PENTAFLUOROBICYCLO(2.2.0)HEXA-2,5-DIENE see PBD300
PENTAFLUOROCHLOROBENZENE see PBE100
PENTAFLUOROETHANE see PBD400
1,1,1,2,2-PENTAFLUOROETHANE see PBD400
1,1,2,2,2-PENTAFLUOROETHANE see PBD400
PENTAFLUOROGUANIDINE see PBD500
PENTAFLUOROIODINE see IDT000
PENTAFLUOROPHENOL see PBD750
(6-(3-(PENTAFLUOROPHENYL)-d-ALANINE))-LHRH ACETATE see LIU339
PENTAFLUOROPHENYLALUMINUM DIBROMIDE see PBE000
PENTAFLUOROPHENYLAMINE see PBD250
PENTAFLUOROPHENYL CHLORIDE see PBE100
PENTAFLUOROPHENYLLITHIUM see PBE250

1,5-PENTANEDIAMINE, N-(6-METHOXY-8-QUINOLINYL)-N'-PROPYL-, PHOSPHATE see MFM100

1,5-PENTANEDIAMINE, 2-METHYL- see MNI515

2,4-PENTANEDIAMINE, 2-METHYL- see MNI525

1,4-PENTANEDIAMINE, N4-(6-CHLORO-2-METHOXY-9-ACRIDINYL)-N1,N1-DIETHYL-, DIMETHANESULFONATE see QCS900

1,4-PENTANEDIAMINE, N1-(7-CHLORO-4-QUINOLINYL)-N1,N1-DIETHYL-, DIHYDROCHLORIDE see CLD100

PENTANE, 1,2-DIBROMO-1,1,5-TRICHLORO- see DDT100

1,5-PENTANEDICARBOXYLIC ACID see PIG000

PENTANE, 1,5-DIIODO- see PBH125

PENTANEDINITRILE see TLR500

PENTANEDINITRILE, 2-BROMO-2-(BROMOMETHYL)- see DDM500

PENTANEDINITRILE, 2,4-DINITRO-3-PHENYL-, compounded with 2,2-DICHLORO-N-(2,2-DICHLOROETHYL) ETHANAMINE (1:1) see GFM250

PENTANEDIOIC ACID see GFS000

1,5-PENTANEDIOIC ACID see GFS000

PENTANEDIOIC ACID, 2-(1,3-DIHYDRO-1,3-DIOXO-2H-ISOINDOL-2-YL)-, (S)-(9CI) see PHY400

PENTANEDIOIC ACID, 2-OXO-, DISODIUM SALT see KFK300

1,5-PENTANEDIOL see PBK750

PENTANE-1,5-DIOL see PBK750

2,4-PENTANEDIOL see PBL000

PENTANEDIOL-2,4 see PBL000

1,5-PENTANEDIOL, DIPROPIONATE see PBI300

(2,4-PENTANEDIONATO-O,O')PHENYLMERCURY see PBL250

PENTANEDIONE see ABX750

1,5-PENTANEDIONE see GFQ000

2,3-PENTANEDIONE see PBL350

2,4-PENTANEDIONE, 3-BENZYLIDENE- see DBH900

2,4-PENTANEDIONE (FCC) see ABX750

2,4-PENTANEDIONE, NICKEL(II) DERIVATIVE see PBL500

2,4-PENTANEDIONE PEROXIDE see PBL600

2,4-PENTANEDIONE, PHENYLMERCURIC SALT see PBL250

2,4-PENTANEDIONE, 3-(PHENYLMETHYLENE)-(9CI) see DBH900

2,4-PENTANEDIONE, ZIRCONIUM COMPLEX see PBL750

4,4'-(1,5-PENTANEDIYLBIS(OXY))BIS-BENZENECARBOXIMIDAMIDE, (9CI) see DBM000

PENTANE, 1,2-EPOXY- see ECE525

PENTANE, 1-IODO- see IET500

PENTANEN (DUTCH) see PBK250

PENTANENITRILE (9CI) see VAV300

PENTANE, 1,1'-OXYBIS-(9CI) see PBX000

1,2,3,4,5-PENTANEPENTOL see RIF000

PENTANE, 1,1,1,5-TETRACHLORO- see TBS100

1-PENTANETHIOL see PBM000

2-PENTANETHIOL, 2,4,4-TRIMETHYL- see MKJ250

2-PENTANETHIOL, 2,4,4-TRIMETHYL- see OFE030

PENTANITROANILINE see PBM100

2,3,4,5,6-PENTANITROANILINE see PBM100

PENTANITROANILINE (dry) (DOT) see PBM100

PENTANOCHLOR see SKQ400

PENTANOIC ACID see VAQ000

n-PENTANOIC ACID see VAQ000

tert-PENTANOIC ACID see PJA500

PENTANOIC ACID, 4,4'-AZOBIS(4-CYANO)-(9CI) see ASL500

PENTANOIC ACID, 2,5-DIAMINO-, (S)- see OJS100

PENTANOIC ACID, 4-((3,4-DICHLOROBENZOYL)AMINO)-5-((3-METHOXYPROPYL)PENTYLAMINO)-5-OXO-, (+−)- see LII100

PENTANOIC ACID, 2,2-DIPHENYL, 2-(N,N-DIETHYLAMINO)ETHYL ESTER, HYDROCHLORIDE see PBM500

PENTANOIC ACID, 3-HEXENYL ESTER, (Z)- see HFE800

PENTANOIC ACID, METHYL ESTER (9CI) see VAQ100

PENTANOIC ACID, 4-OXO-, BUTYL ESTER (9CI) see BRR700

PENTANOIC ACID, 4-OXO-, ETHYL ESTER (9CI) see EFS600

PENTANOIC ACID, 2-(2-PROPENYL)- see AGQ100

PENTANOIC ACID, 2-PROPYL-, CALCIUM SALT (9CI) see VCK200

PENTANOIC ACID, 2-PROPYL-, MAGNESIUM SALT see MAK275

PENTANOL-1 see AOE000

PENTAN-1-OL see AOE000

2-PENTANOL see PBM750

PENTANOL-2 see PBM750

3-PENTANOL see IHP010

PENTANOL-3 see IHP010

PENTAN-3-OL see IHP010

N-PENTANOL see AOE000

tert-PENTANOL see PBV000

1-PENTANOL ACETATE see AOD725

2-PENTANOL, ACETATE see AOD735

2-PENTANOL CARBAMATE see MOU500

2-PENTANOL, 3-ETHYL- see EOB300

3-PENTANOL, 3-ETHYL- see TJP550

4-PENTANOLIDE see VAV000

3-PENTANOL, 3-METHYL-1-PHENYL- see PFR200

1-PENTANOL, 2,2,4-TRIMETHYL- see TLZ100

2-PENTANONE see PBN250

PENTANONE-3 see DJN750

3-PENTANONE see DJN750

PENTANONE, 1-BENZOCYCLOBUTYL- see BCH800

PENTANONE, 1-BENZOCYCLOBUTYL-, OXIME see BFZ170

1-PENTANONE, 1-BICYCLO(4.2.0)OCTA-1,3,5-TRIEN-7-YL- see BCH800

1-PENTANONE, 1-BICYCLO(4.2.0)OCTA-1,3,5-TRIEN-7-YL-, OXIME see BFZ170

2-PENTANONE, 5-(DIETHYLAMINO)- see NOB800

2-PENTANONE, 1-(5-(3-FURYL)-2-METHYL-TETRAHYDRO-2-FURYL)-4-METHYL-, (2R,5S)-(−)- see NCQ900

2-PENTANONE, 4-HYDROXY-1-PHENYL-5,5,5-TRIFLUORO-4-(TRIFLUOROMETHYL)- see TKB285

1-PENTANONE, 5-METHOXY-1-(4-(TRIFLUOROMETHYL)PHENYL)-, o-(2-AMINOETHYL)OXIME, (E)- see FMR550

1-PENTANONE, 5-METHOXY-1-(4-(TRIFLUOROMETHYL)PHENYL)-, o-(2-AMINOETHYL)OXIME,(E)-, (Z)-2-BUTENEDIOATE (1:1) see FMR575

2-PENTANONE, 4-METHYL-, PEROXIDE see IJB100

2-PENTANONE, 4-METHYL-1-(2,3,4,5-TETRAHYDRO-5-METHYL-(2,3'-BIFURAN)-5-YL)-, (2S-cis)- (9CI) see NCQ900

1-PENTANONE, 1-(4-PYRIDYL)- see VBA100

N-PENTAN-4-ONE-N,N,N-TRIMETHYLAMMONIUM IODIDE see TLY250

3-PENTANONE, 1-(2,6,6-TRIMETHYL-2-CYCLOHEXEN-1-YL)- see TLO530

PENTANTIN see DAM700

PENTANTIN HYDROCHLORIDE see DAM700

PENTANYL see PDW750

1,4,7,10,13-PENTAOXACYLOPENTADECANE see PBO000

3,6,9,12,15-PENTAOXAHEPTADECANE see PBO250

2,5,8,11,14-PENTAOXAPENTADECANE see PBO500

PENTAPHEN see AON000

PENTAPHENATE see SJA000

PENTAPHENE HYDROCHLORIDE see PET250

PENTAPOTASSIUM TRIPHOSPHATE see PLW400

PENTAPYRROLIDIUM BITARTRATE see PBT000

PENTASILOXANE, DODECAMETHYL- see DXS800

PENTASILVER DIAMIDOPHOSPHATE see PBO800

PENTASILVER DIIMIDOTRIPHOSPHATE see PBP000

PENTASILVER ORTHODIAMIDOPHOSPHATE see PBP100

PENTASODIUM COLISTINMETHANESULFONATE see SFY500

PENTASODIUM NITRILOTRIS(METHYLENEPHOSPHONATE) see PBP200

PENTASODIUM TRIPHOSPHATE see SKN000

PENTASOL see AOE000

PENTASOL see PAX250

PENTASULFURE de PHOSPHORE (FRENCH) see PHS000

PENTAZOCINE see DOQ400

PENTAZOCINE HYDROCHLORIDE see PBP300

PENTAZOCINE LACTATE see PBP400

PENTECH see DAD200

PENTEFENZOL see SOU800

PENTEK see PBB750

2-PENTENAL see PBP500

4-PENTENAL see PBP750

2-(E)-PENTENAL see PBP800

2-PENTENAL, (E)- see PBP800

(E)-2-PENTENAL see PBP800

trans-2-PENTENAL see PBP800

2-PENTENAL, 4,5-EPOXY- see ECE550

2-PENTENAL, 4-OXO- see KGA200

4-PENTENAMIDE, 2-ALLYL-2-PHENYL- see AGR100

1-PENTENE see PBQ000

2-PENTENE see PBQ250

2-PENTENE, 2-METHYL- see MNK100

1-PENTENE, 3-METHYLENE- see EGV600

2-PENTENENITRILE, (Z)- see PBQ275

(Z)-2-PENTENENITRILE see PBQ275

cis-2-PENTENENITRILE see PBQ275

trans-2-PENTENE OZONIDE see PBQ300

4-PENTENOIC ACID see PBQ750

4-PENTENOIC ACID, 2-PROPYL- see AGQ100

4-PENTEN-1-OL see PBR000

2-PENTEN-1-OL, 5-(2,3-DIMETHYLTRICYCLO(2.2.1.0²·⁶)HEPT-3-YL)-2-METHYL-, (R(Z))- see SAU375

1-PENTEN-3-ONE see PBR250

3-PENTEN-2-ONE see PBR500

1-PENTEN-3-ONE, 4-METHYL-1-PHENYL- see IRL300

1-PENTEN-3-ONE, 2-METHYL-` see IMW500

1-PENTEN-3-ONE, 1-(2,6,6-TRIMETHYL-2-CYCLOHEXEN-1-YL)-2-METHYL- see COW780

cis-2-PENTENONITRILE see PBQ275

4-PENTENONITRILE, 3-HYDROXY-, S- see COP400

4-PENTENYL 2-((2-FURANYLMETHYL)(1H-IMIDAZOL-1-YLCARBONYL)AMINO)BUTANOATE see PAO170
PENT-4-ENYL N-FURFURYL-N-IMIDAZOL-1-YLCARBONYL-dl-HOMOALANINATE (IUPAC) see PAO170
2-PENTEN-4-YN-3-OL see PBR750
PENTESTAN-80 see PBC250
PENTETATE TRISODIUM CALCIUM see CAY500
PENTETRATE UNICELLES see PBC250
PENTHAMIL see CAY500
PENTHAMIL see DJG800
PENTHAZINE see PDP250
PENTHIENATE BROMIDE see PBS000
PENTHIOBARBITAL see PBT250
PENTHIOBARBITAL SODIUM see PBT500
PENTHRANE see DFA400
PENTICORT see COW825
PENTID see BFD000
PENTIDS see BFD000
PENTIFORMIC ACID see HEU000
PENTILEN see CFU750
PENTILIUM see PBT000
PENTIN C see MNM500
R-PENTINE see CPU500
PENTINIMID see ENG500
PENTISOMICIN see EBC000
PENTISOMICINA (SPANISH) see EBC000
PENTITOL see RIF000
PENTLANDITE see PBS100
PENTOBARBITAL see NBT500
PENTOBARBITAL see PBS250
PENTOBARBITAL CALCIUM see CAV000
R(+)-PENTOBARBITAL SODIUM see PBS500
PENTOBARBITONE SODIUM see NBU000
PENTOBARBITURATE see NBT500
PENTOBARBITURIC ACID see NBT500
PENTOFRAN see DSI709
PENTOFURYL see DGQ500
PENTOKSIL see HMH300
PENTOLE see CPU500
PENTOLINIUM BITARTRATE see PBT000
PENTOLINIUM DITARTRATE see PBT000
PENTOLINIUM TARTRATE see PBT000
PENTOLITE see PBT050
PENTOLITE, dry or wetted with <15% water, by weight (DOT) see PBT050
PENTOMID see MKU750
PENTONAL see NBU000
PENTOSENUCLEIC ACIDS see RJA600
PENTOSTAM see AQH800
PENTOSTAM see AQI250
PENTOSTATIN see PBT100
PENTOTHAL see PBT250
PENTOTHAL SODIUM see PBT500
PENTOTHIOBARBITAL see PBT250
PENTOXIFYLLIN see PBU100
PENTOXIFYLLINE see PBU100
PENTOXIL see HMH300
PENTOXIPHYLLIUM see PBU100
PENTOXYL see HMH300
1-PENTOXY-2-METHOXY-4-PROPENYLBENZENE see AOK000
PENTOXYPHYLLINE see PBU100
PENTOYL see EAN700
PENTRAN see DFA400
PENTRANE see DFA400
PENTRATE see PBC250
PENTREX see AIV500
PENTREXL see AIV500
PENTRIOL see PBC250
PENTRYATE 80 see PBC250
PENTYDORM see EQL000
PENTYL see NBU000
PENTYL ACETATE see AOD725
1-PENTYL ACETATE see AOD725
2-PENTYL ACETATE see AOD735
n-PENTYL ACETATE see AOD725
PENTYL ALCOHOL see AOE000
sec-PENTYL ALCOHOL see PBM750

tert-PENTYL ALCOHOL see PBV000
1-PENTYLALLYL ACETATE see ODW028
PENTYLAMINE see PBV505
1-PENTYLAMINE see PBV505
n-PENTYLAMINE see PBV505
PENTYLAMINE (mixed isomers) see PBV500
2,N-PENTYLAMINOETHYL-p-AMINOBENZOATE see PBV750
8-PENTYLBENZ(a)ANTHRACENE see AOE750
sec-PENTYLBENZENE see PBV600
tert-PENTYLBENZENE see AOF000
PENTYL BENZOATE see PBV800
n-PENTYL BENZOATE see PBV800
p-PENTYLBENZOIC ACID see PBW000
4-n-PENTYLBENZOIC ACID see PBW000
PENTYLBIPHENYL see AOF500
PENTYL BROMIDE see AOF800
1-PENTYL BROMIDE see AOF800
n-PENTYL BROMIDE see AOF800
PENTYL BUTYRATE see AOG000
PENTYLCANNABICHROMENE see PBW400
PENTYL CAPROATE see PBW450
PENTYLCARBINOL see HFJ500
3-PENTYLCARBINOL see EGW000
sec-PENTYLCARBINOL see EGW000
PENTYL CHLORIDE see PBW500
α-PENTYLCINNAMALDEHYDE see AOG500
α-PENTYL CINNAMYL ACETATE see AOG750
PENTYLCYCLOHEXANOL ACETATE see AOI000
4-tert-PENTYLCYCLOHEXANONE see AOH750
PENTYLCYCLOPENTANONEPROPANONE see OOO100
2-PENTYL-2-CYCLOPENTEN-1-ONE see PBW600
PENTYLDICHLOROARSINE see AOI200
PENTYLENE see AOI800
1,5-PENTYLENE GLYCOL see PBK750
PENTYLENETETRAZOL see PBI750
PENTYL ESTER PHOSPHORIC ACID see PBW750
PENTYL ETHER see PBX000
PENTYL FORMATE see AOJ500
n-PENTYL FORMATE see AOJ500
PENTYLFORMIC ACID see HEU000
2-PENTYLFURAN see AOJ600
2-n-PENTYLFURAN see AOJ600
N-PENTYLHEPTANAMIDE see PBX100
PENTYL HEXANOATE see PBW450
n-PENTYLHYDRAZINE HYDROCHLORIDE see PBX250
tert-PENTYL HYDROPEROXIDE see PBX325
N-PENTYL-N-(4-HYDROXYBUTYL)NITROSAMINE see NLE500
2-PENTYLIDENECYCLOHEXANONE see PBX350
PENTYLIDENE GYROMITRIN see PBJ875
PENTYL IODIDE see IET500
1-PENTYL IODIDE see IET500
n-PENTYL IODIDE see IET500
PENTYL KETONE see ULA000
PENTYL MERCAPTAN see PBM000
tert-PENTYL MERCAPTAN see MHS550
2-PENTYL-3-METHYL-2-CYCLOPENTEN-1-ONE see DLQ600
tert-PENTYL METHYL ETHER see PBX400
PENTYL NITRITE see AOL500
1-PENTYL-3-NITRO-1-NITROSOGUANIDINE see NLC500
4-(PENTYLNITROSAMINO)-1-BUTANOL see NLE500
n-PENTYLNITROSOUREA see PBX500
m-PENTYLOXYCARBANILIC ACID, trans-2-(1-PYRROLIDINYL)CYCLOHEXYL ESTER HYDROCHLORIDE see PPW000

(3-(PENTYLOXY)PHENYL)CARBAMIC ACID, 2-(1-PYRROLIDINYL)CYCLOHEXYL ESTER, HCl, (E)- see PPW000
(E)-(1-(2-PENTYLOXYPHENYL)-1-IMIDAZOLYL-2-METHYLTHIO)ETHYLENE HYDROCHLORIDE see NCP525
PENTYL PENTYLAMINE see DCH200
3-PENTYL-2-((3-PENTYL-2(3H)-BENZOXAZOLYLIDENE)-1-PROPENYL)BENZOXAZOLIUM IODIDE see DVU700
PENTYL PERFLUOROBUTANOATE see PCG625
2-PENTYLPHENOL see AOM325
o-PENTYLPHENOL see AOM325
p-PENTYLPHENOL see AOM250
o-(sec-PENTYL) PHENOL see AOM500
p-(sec-PENTYL) PHENOL see AOM750
p-tert-PENTYLPHENOL see AON000
m-(3-PENTYL)PHENYL-N-METHYL-N-NITROSOCARBAMATE see PBX750
PENTYL PROPANOATE see AON350
n-PENTYL PROPANOATE see AON350
PENTYL PROPIONATE see AON350
PENTYL 4-PYRIDYL KETONE see HEW050
PENTYL 4-PYRIDYL KETONE see PBX800
6-(PENTYLTHIO)PURINE see PBX900
PENTYLTRICHLOROSILANE see PBY750
PENTYLTRIETHOXYSILANE see PBZ000
PENTYLTRIMETHYLAMMONIUM IODIDE see TMA500
1-PENTYL-6,6,9-TRIMETHYL-6a,7,10,10a-TETRAHYDRO-6H-DIBENZO(b,d)PYRAN-3-OL see HNE400
3-PENTYL-6,6,9-TRIMETHYL-6a,7,8,10a-TETRAHYDRO-6H-DIBENZO(b,d)PYRAN-1-OL see TCM250
PENTYMAL see AMX750
PENTYMALUM see AMX750
1-PENTYNE see PCA250
2-PENTYNE see PCA300
m-PENTYNOL see EQL000
4-PENTYN-2-ONE, 5-HYDROXY- see HNE450
1-(4-PENTYNYLOXY)-4-PHENOXYBENZENE see PCA400
PENTYREST see EQL000
PEN V see PDT500
PEN-VEE see PDT500
PEN-VEE-K see PDT750
PEN-VEE-K POWDER see PDT750
PENVIKAL see PDT750
PEN-V-K POWDER see PDT750
PENWAR see PAX250
PENZAL N 300 see PAQ200
PENZYLPENICILLIN SODIUM SALT see BFD250
PEONIA (CUBA) see RMK250
PEP see BLY770
PEP see EDS000
PEP see PCC000
PEP see PJR750
PEP 211 see PJS750
PEPCID see FAB500
PEPCIDINE see FAB500
PEPDUL see FAB500
PEPLEOMYCIN see BLY770
PEPLEOMYCIN SULFATE see PCB000
PEPLOMYCIN see BLY770
PEPPER BUSH see DYA875
PEPPERMINT CAMPHOR see MCF750
PEPPERMINT OIL see PCB250
PEPPER PLANTS see PCB275
PEPPER-ROOT see FAB100
PEPPER TREE see PCB300
PEPPER TURNIP see JAJ000
PEPQ see PCB400
PEPROSAN see CNK559
PEPSTATIN see PCB750
PEPSTATIN A see PCB750
PEPTAZIN BAFD see BDK800

PEPTICHEMIO see PCC000
PEPTISANT 10 see BDK800
PEPTON 22 see BDK800
PER-ABRODIL see DNG400
PER-ABRODIL M see IFA100
PERACETIC ACID see PCL500
PERACON see PCK639
PERAGAL ST see PKQ250
PERAGIT see BBV000
PERANDREN see TBF500
PERANDREN see TBG000
PERATOX see PAX250
PERAWIN see PCF275
PERAZIL see CDR000
PERAZIL see CDR500
PERAZIL DIHYDROCHLORIDE see CDR000
PERAZINE see PCK500
PERAZINE MALEATE see PCC475
PERBENZOATE de BUTYLE TERTIAIRE (FRENCH) see BSC500
PERBENZOIC ACID see PCM000
PERBORIC ACID, SODIUM SALT see SJB100
PERBROMOBIPHENYL see PCC480
PERBROMYL FLUORIDE see PCC750
PERBUNAN C see PJQ050
PERBUTYL H see BRM250
PERCAIN see NOF500
PERCAINE HYDROCHLORIDE see NOF500
PERCA ORANGE GR see CMM220
PERCAPYL see CHX250
PERCARBAMIDE see HIB500
PERCELINE OIL see HCP550
PERCHLOORETHYLEEN, PER (DUTCH) see PCF275
PERCHLOR see PCF275
PERCHLORAETHYLEN, PER (GERMAN) see PCF275
PERCHLORATE ACID, LEAD SALT, HEXAHYDRATE see LDT000
PERCHLORATE de MAGNESIUM (FRENCH) see PCE000
PERCHLORATES see PCD000
PERCHLORATE de SODIUM (FRENCH) see PCE750
PERCHLORETHYLENE see PCF275
PERCHLORETHYLENE, PER (FRENCH) see PCF275
PERCHLORIC ACID see PCD250
PERCHLORIC ACID, >72% acid by weight (DOT) see PCD250
PERCHLORIC ACID, AMMONIUM SALT see PCD500
PERCHLORIC ACID, BARIUM SALT•3H$_2$O see PCD750
PERCHLORIC ACID, CHROMIUM(3+) SALT see CMI300
PERCHLORIC ACID, COBALT(II) SALT, HEXAHYDRATE see CND900
PERCHLORIC ACID, COPPER(II) SALT, DIHYDRATE (8CI, 9CI) see CNO500
PERCHLORIC ACID, ETHYL ESTER see EOD000
PERCHLORIC ACID, LITHIUM SALT, TRIHYDRATE see LHN000
PERCHLORIC ACID, MAGNESIUM SALT see PCE000
PERCHLORIC ACID, MANGANESE(2+) SALT, HEXAHYDRATE see MAU000
PERCHLORIC ACID, MANGANESE(2+) SALT, compounded with 3 mols. of OCTAMETHYLPYROPHOSPHORAMIDE see TNK400
PERCHLORIC ACID, NICKEL(II) SALT compounded with OCTAMETHYL PYROPHOSPHORAMIDE see PCE250
PERCHLORIC ACID, SODIUM SALT see PCE750
PERCHLORIC ACID, SODIUM SALT, MONOHYDRATE see SJB450
PERCHLORIC ACID, TRICHLOROMETHYL ESTER see TIS750

PERCHLORIC ACID, >50% but not >72% acid, by weight (UN 1873) (DOT) see PCD250
PERCHLORIC ACID, not >50% acid, by weight (UN 1802) (DOT) see PCD250
PERCHLORIC ACID, ZINC SALT, HEXAHYDRATE see ZKS100
PERCHLORIDE of MERCURY see MCY475
PERCHLORMETHYLMERKAPTAN (CZECH) see PCF300
PERCHLOROBENZENE see HCC500
PERCHLOROBUTADIENE see HCD250
PERCHLORO-2-CYCLOBUTENE-1-ONE PCF250
PERCHLOROCYCLOPENTADIENE see HCE500
PERCHLORO-2-CYCLOPENTENONE see HCE600
PERCHLOROCYCLOPENTEN-2-ONE see HCE600
PERCHLORODIHOMOCUBANE see MQW500
PERCHLOROETHANE see HCI000
PERCHLOROETHYLENE see PCF275
PERCHLOROMELAMINE see TNG275
PERCHLOROMETHANE see CBY000
PERCHLOROMETHYL MERCAPTAN see PCF300
PERCHLORON see HOV500
PERCHLOROPENTACYCLODECANE see MQW500
PERCHLOROPENTACYCLO(5.2.1.0^{2,6}.0^{3,9}.0^{5,8}) DECANE see MQW500
PERCHLOROPYRIMIDINE see TBU250
PERCHLOROTHIOPHENE see TBV750
PERCHLORURE d'ANTIMOINE (FRENCH) see AQD000
PERCHLORURE de FER see FAU000
PERCHLORYLBENZENE see PCF500
PERCHLORYL FLUORIDE see PCF750
PERCHLORYL HYPOFLUORITE see FFD000
PERCHLORYL PERCHLORATE see PCF775
1-PERCHLORYLPIPERIDINE see PCG000
PERCHROMATES see PCG250
PERCIN see ILD000
PERCLENE see PCF275
PERCLOROETILENE (ITALIAN) see PCF275
PERCLUSON see CMW750
PERCLUSONE see CMW750
PERCOBARB see ABG750
PERCOBARB see PCG500
PERCOCCIDE see ALF250
PERCODAN see ABG750
PERCODAN see PCG500
PERCODAN HYDROCHLORIDE see DLX400
PERCOL 1697 see DTS500
PERCOLATE see PHX250
PERCORAL see DJS200
PERCORTEN see DAQ800
PERCOSOLVE see PCF275
PERCOTOL see DAQ800
PERCUTACRINE see PMH500
PERCUTACRINE ANDROGENIQUE see TBF500
PERCUTATRINE OESTROGENIQUE ISCOVESCO see DKA600
PERCUTINA see SPD500
PERDIPINE see PCG550
PERDOX see SJB400
PEREBRAL see DNU100
PEREGRINA (PUERTO RICO, CUBA, US) see CNR135
PEREMESIN see HGC500
PEREQUIL see MQU750
PERFECTA see MQV750
PERFECTHION see DSP400
PERFENAZINA (ITALIAN) see CJM250
PERFLUIDONE see TKF750
PERFLUORIDE see FEX875
PERFLUOROACETIC ACID see TKA250
PERFLUOROACETIC ANHYDRIDE see TJX000

PERFLUOROACETYL CHLORIDE see TJX500
PERFLUOROADIPIC ACID DINITRILE see PCG600
PERFLUOROADIPINIC ACID DINITRILE see PCG600
PERFLUOROADIPONITRILE see PCG600
PERFLUOROALLYL CHLORIDE see CJI550
PERFLUOROAMMONIUM OCTANOATE see ANP625
PERFLUOROBENZOIC ACID see PBD275
1-PERFLUOROBUTANESULFONIC ACID see PCG625
PERFLUOROBUT-2-ENE see OBO000
PERFLUORO-2-BUTENE (DOT) see OBO000
PERFLUORO(2-BUTOXYPROPYL VINYL ETHER) see PCG635
(PERFLUOROBUTYL)ETHENE see NMV800
PERFLUOROBUTYLETHYLENE (ACGIH) see NMV800
PERFLUORO-tert-BUTYL PEROXYHYPOFLUORITE see PCG650
PERFLUOROCAPRYLIC ACID see PCH050
PERFLUOROCTANOIC ACID see PCH050
PERFLUOROCTYLSULFONAMIDE see PCH315
PERFLUOROCYCLOBUTANE see CPS000
PERFLUORODECAHYDRONAPHTHALENE see PCG700
PERFLUORODECALIN see PCG700
PERFLUORODECANOIC ACID see PCG725
PERFLUORO-N-DECANOIC ACID see PCG725
PERFLUORODECYL IODIDE see IEU075
PERFLUORODIBUTYLETHER see PCG755
PERFLUORO-n-DIBUTYL ETHER see PCG755
PERFLUOROETHANE see HDC100
PERFLUOROETHENE see TCH500
PERFLUORO ETHER see PCG760
PERFLUOROETHYLENE see TCH500
PERFLUOROFORMAMIDINE see PCG775
PERFLUOROGLUTARIC ACID DINITRILE see HDC300
PERFLUOROGLUTARONITRILE see HDC300
PERFLUOROHEPTANE see PCH000
PERFLUORO-n-HEPTANE see PCH000
PERFLUOROHEPTANECARBOXYLIC ACID see PCH050
PERFLUOROHEXYL IODIDE see PCH100
PERFLUORO HYDRAZINE see TCI000
PERFLUORO-1-IODODECANE see IEU075
PERFLUORO-1-IODOOCTANE see PCH325
PERFLUOROISOBUTYLENE (ACGIH) see OBM000
PERFLUOROMETHANE see CBY250
PERFLUOROMETHOXYPROPIONIC ACID METHYL ESTER see PCH275
PERFLUOROMETHYLCYCLOHEXANE see PCH290
PERFLUOROMETHYL IODIDE see IFM100
PERFLUORO(METHYLOXIRANE) see HDF050
PERFLUORO-tert-NITROSOBUTANE see PCH300
PERFLUOROOCTANE see OAP100
n-PERFLUOROOCTANE see OAP100
PERFLUOROOCTANESULFONAMIDE see PCH315
PERFLUOROOCTANESULFONIC ACID see HAS075
PERFLUOROOCTANESULFONIC ACID AMIDE see PCH315
PERFLUOROOCTANESULFONIC ACID, TETRAETHYLAMMONIUM SALT see TCD100
N-PERFLUOROOCTANESULFONYL-N-METHYLCARBAMOYL ((NONADECAETHOXY)BUTOXY)BUTYL ETHER see PCH320
PERFLUOROOCTANOIC ACID see PCH050

PERFLUOROOCTYL IODIDE see PCH325
PERFLUOROPROPENE see HDF000
PERFLUOROPROPIONIC ACID, METHYL
ESTER see PCH350
PERFLUOROPROPYLENE scc HDF000
PERFLUOROPROPYLENE OXIDE see
HDF050
PERFLUOROSUCCINIC ACID see TCJ000
PERFLUOROTOLUENE see PCH500
PERFLUOROTRIBUTYLAMINE see HAS000
PERFMID see BSN000
PERGACID VIOLET 2B see FAG120
PERGANTENE see SHF500
PERGASOL TURQUOISE BLUE GAL
SUPRA see COF420
PERGITRAL see PBC250
PERGLOTTAL see NGY000
PER-GLYCERIN see LAM000
PERGOLIDE MESYLATE see MPU500
PERHEXILINE MALEATE see PCH800
PERHYDRIT see HIB500
PERHYDROAZEPINE see HDG000
2-PERHYDROAZEPINONE see CBF700
PERHYDROGERANIOL see DTE600
PERHYDROL see HIB050
PERHYDROL-UREA see HIB500
PERHYDRONAPHTHALENE see DAE800
PERIACTIN see PCI500
PERIACTIN HYDROCHLORIDE see PCI250
PERIACTINOL see PCI250
PERIACTINOL see PCI500
PERIACTIN SYRUP see PCI250
PERICHLOR see NCH500
PERICHLORAL see NCH500
PERICHTHOL see IAD000
PERICIAZINE see PIW000
PERICLASE see MAH500
PERICLOR see NCH500
PERICYAZINE see PIW000
PERIDAMOL see PCP250
PERIDEX see CDT250
PERIDEX-LA see PBC250
PERI-DINAPHTHALENE see PCQ250
PERIETHYLENENAPHTHALENE see
AAF275
PERIFUNAL see PJB500
PERILAX see GKK000
PERILAX see PPN100
PERILENE see PCQ250
PERILITON BRILLIANT PINK R see
AKO350
PERILLA ALCOHOL see PCI550
PERILLA ALDEHYDE see DKX100
PERILLA FRUTESCENS (Linn.) Britt., extract
see PCI600
PERILLA FRUTESCENS OIL see PCJ100
PERILLA KETONE see PCI750
PERILLAL see DKX100
PERILLALDEHYDE see DKX100
1-PERILLALDEHYDE-α-ANTIOXIME see
PCJ000
PERILLA OCIMOIDES Linn., extract see
PCI600
PERILLA OIL see PCJ100
PERILLARTINE see PCJ000
PERILLA SUGAR see PCJ000
PERILLOL see PCI550
PERILLYL ALCOHOL see PCI550
PERILLYL ALDEHYDE see DKX100
PERIMETAZINE see LEO000
PERIMETHAZINE see LEO000
1H-PERIMIDINE,2,3-DIHYDRO-2,2-
DIMETHYL-6-((4-(PHENYLAZO)-1-
NAPHTHYL)AZO)- see PCJ200
2(1H)-PERINAPHTHAZOLONE see NAX100
PERINAPHTHENONE see PCJ225
7-PERINAPHTHENONE see PCJ225
PERINDOPRIL tert-BUTYLAMINE see
PCJ230
PERINDOPRIL ERBUMINE see PCJ230
trans-PERINONE see CMU820
O-PERIODIC ACID see PCJ250

PERIODIN see PLO500
PERIOGRAF see HJE100
PERIPHERINE see BBW750
PERIPHERMIN see ACR300
PERIPHETOL see BOV825
PERISOLOL see NDL800
PERISOXAL CITRATE see PCJ325
PERISTIL see PCJ350
PERISTON see PKQ250
PERITOL see PCI250
PERITRATE see PBC250
PERITYL see PBC250
PERK see PCF275
PERKADOX AMBN see ASM025
PERKADOX SE 8 see CBF705
PERKE see BBK500
PERKLONE see PCF275
PERLANKROL RN see SNY100
PERLAPINE see HOU059
PERLATAN see EDV000
PERLITE see PCJ400
PERLITON BLUE B see TBG700
PERLITON BLUE FFR see MGG250
PERLITON BLUE GREEN B see DMM400
PERLITON ORANGE 3R see AKP750
PERLITON PINK 3B see AKE250
PERLITON RED VIOLET FFB see DBX000
PERLITON RUBINE 4B see CMP080
PERLITON VIOLET 3R see DBP000
PERLITON VIOLET B see DBY700
PERLITON YELLOW G see AAQ250
PERLITON YELLOW RR see DUW500
PERLON see NOH000
PERLUTEX see MCA000
PERM-A-CHLOR see TIO750
PERMACIDE see PAX250
PERMAFRESH 183 see DTG000
PERMAFRESH 477 see DTG700
PERMAFRESH SW see TCJ900
PERMAGARD see PAX250
PERMAGEL see PAE750
PERMAGEN YELLOW see CMS210
PERMAGEN YELLOW GA see CMS210
PERMA KLEER see DJG800
PERMA KLEER 50 ACID see EIX000
PERMA KLEER 50 CRYSTALS see EIV000
PERMA KLEER 50 CRYSTALS DISODIUM
SALT see EIX500
PERMA KLEER TETRA CP see EIV000
PERMA KLEER 50, TRISODIUM SALT see
TNL250
PERMALUX see DEK500
PERMALUX NEOPRENE ACCELERATOR
see DEK500
PERMANAX 45 see PJQ750
PERMANAX TQ see PJQ750
PERMANENT GREEN TONER GT-376 see
PJQ750
PERMANENT ORANGE see DVB800
PERMANENT ORANGE G see CMS145
PERMANENT ORANGE G EXTRA see
CMS145
PERMANENT PINK E see DTV360
PERMANENT RED 4B see CMS155
PERMANENT RED 4R see MMP100
PERMANENT RED BB see CMS148
PERMANENT RED BBa see CMS148
PERMANENT RED F 6R see CMS155
PERMANENT RED TONER R see CJD500
PERMANENT WHITE see BAP000
PERMANENT WHITE see ZKA000
PERMANENT YELLOW see BAK250
PERMANENT YELLOW 2K see DUW500
PERMANENT YELLOW FGL see CMS214
PERMANENT YELLOW G see CMS210
PERMANENT YELLOW GHG see DEU000
PERMANENT YELLOW GR see CMS208
PERMANENT YELLOW GR01 see CMS208
PERMANENT YELLOW LIGHT see CMS210
PERMANGANATE of POTASH (DOT) see
PLP000

PERMANGANATE de POTASSIUM
(FRENCH) see PLP000
PERMANGANATES see PCJ500
PERMANGANATE de SODIUM (FRENCH)
see SJC000
PERMANGANIC ACID AMMONIUM SALT
see PCJ750
PERMANGANIC ACID, BARIUM SALT see
PCK000
PERMANGANIC ACID(HMnO₄), CALCIUM
SALT (8CI,9CI) see CAV250
PERMANGANIC ACID, SODIUM SALT see
SJC000
PERMANSA RED see CJD500
PERMAPEN see BFC750
PERMASAN see PAX250
PERMASEAL see PJH500
PERMATONE ORANGE see DVB800
PERMATON RED XL 20-7015 see CJD500
PERMATOX DP-2 see PAX250
PERMATOX PENTA see PAX250
PERMEK N see MKA500
cis-PERMETHRIN see PCK050
(−)-cis-PERMETHRIN see PCK055
(+−)-cis-PERMETHRIN see PCK060
(+)-cis-PERMETHRIN see PCK065
(1R)-cis-PERMETHRIN see PCK065
1S-cis-PERMETHRIN see PCK055
trans-PERMETHRIN see PCK070
(−)-trans-PERMETHRIN see PCK085
(+)-trans-PERMETHRIN see PCK075
trans-(+−)-PERMETHRIN see PCK080
(1R)-trans-PERMETHRIN see PCK075
1S-trans-PERMETHRIN see PCK085
PERMETHRIN (USDA) see AHJ750
PERMETRINA (PORTUGUESE) see AHJ750
PERMETRIN (HUNGARIAN) see AHJ750
PERMICORT see CNS750
PERMINAL FC-P see CNH125
PERMITAL see DYE600
PERMITE see PAX250
PERMITIL see TJW500
PERMITIL HYDROCHLORIDE see FMP000
PERMONID see DKX600
PERMONOSULFAMIC ACID see HLN100
PERNAEMON see VSZ000
PERNAEVIT see VSZ000
PERNAMBUCO EXTRACT see LFT800
PERNAZINE see PCK500
PERNETTYA (various species) see PCK600
PERNIPURON see VSZ000
PERNITHRENE BLUE BC see DFN300
PERNITHRENE BLUE RS see IBV050
PERNITHRENE BROWN R see CMU780
PERNITHRENE OLIVE R see DUP100
PERNOCTON see BOR000
PERNOSTON see BOR000
PERNOX see CLY500
PERNSATOR-WIRKSTOFF see PNN300
PEROCAN CITRATE see PCK639
PEROCAN CYCLAMATE see PCK669
PEROLYSEN see PBJ000
PERONE see HIB050
PERONE 30 see HIB010
PERONE 35 see HIB010
PERONE 50 see HIB010
PERONIAS (PUERTO RICO) see RMK250
PEROPAL see THT500
PEROPYRENE see DDC400
PEROSIN see EIR000
PEROSSIDO di BENZOILE (ITALIAN) see
BDS000
PEROSSIDO di BUTILE TERZIARIO
(ITALIAN) see BSC750
PEROSSIDO di IDROGENO (ITALIAN) see
HIB050
PEROXAN see HIB050
PEROXIDE see HIB050
PEROXIDE, ACETYL
CYCLOHEXYLSULFONYL see ACG300
PEROXIDE, BIS(1-OXONONYL) see
NNA100

PEROXIDE, BIS(1-OXOOCTYL) (9CI) see CBF705
PEROXIDE, BIS(1-OXOPROPYL) see DWQ800
PEROXIDE, 1-HYDROPEROXYCYCLOHEXYL 1-HYDROXYCYCLOHEXYL see CPC300
PEROXIDE, OCTANOYL see CBF705
PEROXIDE, (PHENYLENEBIS(1-METHYLETHYLIDENE))BIS(1,1-DIMETHYLETHYL)- see BHL100
PEROXIDE, (PHENYLENEDIISOPROPYLIDENE)BIS(tert-BUTYL)- see BSD100
PEROXIDES, INORGANIC see PCL000
PEROXIDES, ORGANIC see PCL250
PEROXIDE, (1,1,4,4-TETRAMETHYL-1,4-BUTANEDIYL)BIS(1,1-DIMETHYLETHYL) see DRJ800
PEROXIDE, (1,1,4,4-TETRAMETHYLTETRAMETHYLENE)BIS(tert-BUTYL) see DRJ800
PEROXIDE, (VINYLSILYLIDYNE)TRIS(tert-BUTYL- see VQU333
1,4-PEROXIDO-p-MENTHENE-2 see ARM500
PEROXISOMICIN A2 see PCL300
PEROXISOMICINE A2 see PCL300
PEROXOACETIC ACID see PCL500
PEROXOMONOPHOSPHORIC ACID see PCN500
PEROXOMONOSULFURIC ACID see PCN750
PEROXYACETIC ACID see PCL500
PEROXYACETIC ACID, >43% and with >6% hydrogen peroxide (DOT) see PCL500
PEROXYACETYL NITRATE see PCL750
PEROXYACETYL PERCHLORATE see PCL775
PEROXYACRYLIC ACID, 2-METHOXYETHYL ESTER see MEL790
PEROXYBENZOIC ACID see PCM000
4-PEROXY-CPA see HIF000
PEROXYDE de BARYUM (FRENCH) see BAO250
PEROXYDE de BENZOYLE (FRENCH) see BDS000
PEROXYDE de BUTYLE TERTIAIRE (FRENCH) see BSC750
PEROXYDE d'HYDROGENE (FRENCH) see HIB050
PEROXYDE de LAUROYLE (FRENCH) see LBR000
PEROXYDE de PLOMB (FRENCH) see LCX000
PEROXYDICARBONATE d'ISOPROPYLE see DNR400
PEROXYDICARBONIC ACID, BIS(2-ETHYLHEXYL) ESTER see DJK800
PEROXYDICARBONIC ACID, BIS(1-METHYLETHYL) ESTER see DNR400
PEROXYDICARBONIC ACID, DIBENZYL ESTER see DDH200
PEROXYDICARBONIC ACID, DIBUTYL ESTER see BSC800
PEROXYDICARBONIC ACID, DICYCLOHEXYL ESTER see DGV650
PEROXYDICARBONIC ACID, DIETHYL ESTER see DJU600
PEROXYDICARBONIC ACID, DI(2-ETHYLHEXYL) ESTER see DJK800
PEROXYDICARBONIC ACID DIPROPYL ESTER see DWV400
PEROXYDISULFIRIC ANHYDRIDE see DXH600
PEROXYDISULFURIC ACID DIPOTASSIUM SALT see DWQ000
PEROXYDISULFURYL DIFLUORIDE see PCM250
PEROXYFORMIC ACID see PCM500
PEROXYFUROIC ACID see PCM550
PEROXYHEXANOIC ACID see PCM750

PEROXYISOBUTYRIC ACID, tert-BUTYL ESTER see BSC600
PEROXYLINOLEIC ACID, SODIUM SALT see SIC250
PEROXYLINOLENIC ACID see PCN000
PEROXYMONOPHOSPHORIC ACID see PCN500
PEROXYMONOSULFURIC ACID see PCN750
PEROXYNITRIC ACID see PCO000
PEROXYNITRITE see PCO050
PEROXYNITRITE see PCO050
PEROXYOCTANOIC ACID, tert-BUTYL ESTER see BSC600
PEROXYPROPIONIC ACID see PCO100
PEROXYPROPIONYL NITRATE see PCO150
PEROXYPROPIONYL PERCHLORATE see PCO175
PEROXY SODIUM CARBONATE see SJB400
PEROXYSULFURIC ACID, POTASSIUM SALT see PLP750
PEROXYTRIFLUOROACETIC ACID see PCO250
PERPARINE HYDROCHLORIDE see PAH260
PERPARIN HYDROCHLORIDE see PAH260
PERPERINE HYDROCHLORIDE see PAH260
PERPHENAZIN see CJM250
PERPHENAZINE see CJM250
PERPHENAZINE DIHYDROCHLORIDE see PCO500
PERPHENAZINE HYDROCHLORIDE see PCO750
PERPHENAZINE MALEATE see PCO850
PER-RADIOGRAPHOL see DNG400
PERSADOX see BDS000
PERSAMINE see DLH630
PERSANTIN see PCP250
PERSANTINAT see ARQ750
PERSANTINE see PCP250
PERSEC see PCF275
PERSIAN BERRY see MBU825
PERSIAN BERRY LAKE see MQF250
PERSIAN LILAC see CDM325
PERSIAN ORANGE see CMM220
PERSIAN ORANGE LAKE see CMM220
PERSIAN ORANGE X see CMM220
PERSIAN RED see LCS000
PERSIA-PERAZOL see DEP800
PERSICOL see HBN200
PERSIMMON see PCP500
PERSISTEN see DJT400
PERSISTOL see TND500
PERSKLERAN see RDK000
PERSPEX see PKB500
PERSULFATE d'AMMONIUM (FRENCH) see ANR000
PERSULFATE de SODIUM (FRENCH) see SJE000
PERSULFEN see SNN300
PERTESTIS see TBF600
PERTHANE see DJC000
PERTHIOXANTHATE, TRICHLOROMETHYL ALLYL see TIR750
PERTHIOXANTHATE, TRICHLOROMETHYL METHYL see TIS500
PERTOFRAM see DLH630
PERTOFRAN see DLS600
PERTOFRAN see DSI709
PERTOFRANE see DLS600
PERTOFRANE see DSI709
PERTOXIL see CMW500
PERU BALSAM see PCP750
PERU BALSAM OIL see PCQ000
PERUSCABIN see BCM000
PERUVIAN BALSAM see BAE750
PERUVIAN JACINTH see SLH200
PERUVIAN MASTIC TREE see PCB300
PERUVOSID see EAQ050
PERUVOSIDE see EAQ050
PERVAGAL see HKR500

PERVAL see VLF000
PERVERTIN see DBB000
PERVETRAL see ORI400
PERVINCAMINE see VLF000
PERVITIN see DBA800
PERVITIN see MDT600
PERVONE see VLF000
PERYCIT see NCW300
PERYLENE see PCQ250
PERYLENE, 3-NITRO- see NIB500
PES 100 see PJS750
PES 200 see PJS750
PESTAN see DJI000
M-74 (PESTICIDE) see DXH325
PESTMASTER see EIY500
PESTMASTER EDB-85 see EIY500
PESTMASTER (OBS.) see MHR200
PESTON XV see PHF750
PESTOX see OCM000
PESTOX 3 see OCM000
PESTOX 14 see BJE750
PESTOX 15 see PHF750
PESTOX 101 see NIM500
PESTOX III see OCM000
PESTOX IV see BJE750
PESTOX PLUS see PAK000
PESTOX XIV see BJE750
PESTOX XV see PHF750
PETA see PBC750
PETASITENINE see PCQ750
PETASITENINE (neutral) see PCQ750
PETASITES JAPONICUS MAXIM see PCR000
PETEHA see PNW750
PETE-PETE (HAITI) see RBZ400
PETERPHYLLIN see TEP500
PETERSILIENSAMEN OEL (GERMAN) see PAL750
PETHIDINE CHLORIDE see DAM700
PETHIDINETER see DAM600
PETHIDOINE see DAM600
PETHION see PAK000
PETIDIN see DAM700
PETIDION see TLP750
PETIDON see TLP750
PETILEP see TLP750
PETINIMID see ENG500
PETINUTIN see MLP800
PETITGRAIN BIGARADE OIL see OHJ150
PETITGRAIN OIL see OHJ150
PETITGRAIN OIL, PARAGUAY TYPE see PCR100
PETITGRAIN OIL SAPONIFIED see OHJ150
PETIT PRECHEUR (CANADA) see JAJ000
PETNAMYCETIN see CDP250
PETNIDAN see ENG500
PETRICHLORAL see NCH500
PETRIN see PCR150
PETRISUL see AKO500
PETROGALAR see MQV750
PETROHOL see INJ000
PETROL see PCR250
PETROL (DOT) see GBY000
PETROLATUM see PCR200
PETROLATUM, liquid see MQV750
PETROLEUM see PCR250
PETROLEUM ASPHALT see ARO500
PETROLEUM ASPHALT see PCR500
PETROLEUM BENZIN see NAH600
PETROLEUM BITUMEN see ARO500
PETROLEUM CRUDE see PCR250
PETROLEUM CRUDE OIL (DOT) see PCR250
PETROLEUM DERIVED DISTILLATE FUEL, MARINE see DHE750
PETROLEUM-DERIVED NAPHTHA see NAH600
PETROLEUM DISTILLATE see PCS250
PETROLEUM DISTILLATES, CLAY-TREATED HEAVY NAPHTHENIC see PCS260

PETROLEUM DISTILLATES, CLAY-
TREATED LIGHT NAPHTHENIC see
PCS270
PETROLEUM DISTILLATES,
HYDROTREATED (mild) HEAVY
NAPHTHENIC see MQV790
PETROLEUM DISTILLATES (NAPHTHA)
see NAH600
PETROLEUM DISTILLATES, SOLVENT-
DEWAXED HEAVY PARAFFINIC see
MQV825
PETROLEUM ETHER see PCT250
PETROLEUM GASES, liquefied or liquefied
petroleum gas (DOT) see LGM000
PETROLEUM GAS, LIQUEFIED see LGM000
PETROLEUM OIL (UN1270) (DOT) see
NAH600
PETROLEUM PITCH see ARO500
PETROLEUM ROOFING TAR see ARO500
PETROLEUM ROOFING TAR see PCR500
PETROLEUM 60 SOLVENT see PCS750
PETROLEUM 70 SOLVENT see PCT000
PETROLEUM SPIRIT (DOT) see PCT250
PETROLEUM SPIRITS see MQV900
PETROLEUM SPIRITS see PCT250
PETROLEUM 50 THINNER see PCT500
PETROLEUM WAX see PCT600
PETROLEUM WAX, SYNTHETIC (FCC) see
PCT600
PETROL ORANGE Y see PEJ500
PETROL, SYNTHETIC see GCC200
PETROL YELLOW C see CMS240
PETROL YELLOW WT see DOT300
PETROSOL 100 see AQZ150
PETROSULPHO see IAD000
PETROTHENE see PJS750
PETROTHENE LB 861 see PJS750
PETROTHENE LC 731 see PJS750
PETROTHENE LC 941 see PJS750
PETROTHENE NA 219 see PJS750
PETROTHENE NA 227 see PJS750
PETROTHENE XL 6301 see PJS750
PETUNIDOL see PCU000
PETZINOL see TIO750
PEVARYL see EAE000
PEVIKON D 61 see PKQ059
PEVITON see NCQ900
PEXID see PCH800
PEYRONE'S CHLORIDE see PJD000
PF-1 see DSA800
PF-3 see IRF000
PF-26 see DWK700
PF 38 see FQU875
PF-82 see DWK900
PF 1593 see BON325
PFAZ 322 see EML600
PFD see PCG700
PFDA see PCG725
PFEFFERMINZ OEL (GERMAN) see PCB250
PFETTFER'S SUBSTANCE see AMK250
PFH see PNM650
PFIB see OBM000
PFIKLOR see PLA500
PFIZER 1393 see BDE250
PFIZER-E see EDJ500
PFIZERPEN see BFD000
PFIZERPEN A see AIV500
PFIZERPEN VK see PDT750
PFOA see PCH050
PFOS see HAS075
PFPA see FLF000
(d-PFP⁶))-LHRH ACETATE see LIU339
PFT see CML870
PG see PML250
PG 12 see PML000
PG-501 see MBV100
PGA see AHC000
PGA¹ see POC250
PGA2 see MCA025
PGA² see MCA025
(155)-PGA² see MCA025
5,6-cis-PGA² see MCA025

PGABA see PEE500
PhGABA see PEE500
PGB see PJJ250
PGD2 see POC275
PGDN see PNL000
PGE see PFF360
PGE-1 see POC350
PGE2 see DVJ200
PGE2 SODIUM SALT see POC360
PGF 1-α see POC400
PGF2-α see POC500
(±)-PGF2-α see POC525
dl-PGF2-α see POC525
15M-PGF2-α see MLO301
PGF2-α racemic mixture see POC525
PGF2 METHYL ESTER see DVJ100
PGF2-α METHYL ESTER see DVJ100
9-β,11α-PGF(SUB 2-α) see DCI900
PGF2-α THAM see POC750
PGF2-α TRIS SALT see POC750
PGF2-α TROMETHAMINE see POC750
PG 12 (OIL) see AQZ150
P 11H see ADV900
Ph. 458 see SCA525
PH 1882 see BNV500
PH 60-40 see CJV250
PHA see PIB575
PHACETUR see PEC250
PHALDRONE see CDO000
PHALLACIDIN see COV525
PHALLOIDIN see PCU350
PHALLOIDINE see PCU350
PHALLOIDIN, 7-(4-HYDROXY-l-LEUCINE)-
(9CI) see PCU355
PHALLOIN see PCU355
PHALTAN see TIT250
PHANAMIPHOS see FAK000
PHANANTIN see DKQ000
PHANODORM see TDA500
PHANODORN see TDA500
PHANQUINONE see PCY300
PHANQUINONUM see PCY300
PHANQUONE see PCY300
PHARGAN see DQA400
PHARLON see EDS100
PHARMACID GREEN S see ADF000
PHARMACINE YELLOW R see CMM759
PHARMAGEL A see PCU360
PHARMAGEL AdB see PCU360
PHARMAGEL B see PCU360
PHARMAGLO RED G see CMM325
PHARMANIL RED RB see NAO600
PHARMANIL SCARLET Y see CMM320
PHARMANTHRENE GOLDEN YELLOW see
DCZ000
PHARMASORB-COLLOIDAL see PAE750
PHARMATEX YELLOW G see CMM759
PHARMAZOID RED KB see CLK225
PHARMEDRINE see AOB250
PHAROS 100.1 see SMR000
PHARYCIDIN CONCENTRATE see BEL900
PHASEOLUNATIN see GFC100
PHASOLON see BDJ250
PH BC see BQJ500
PHBN see NLN000
PHC see PMY300
M-PHDM see BKL750
PHEASANT'S EYE see PCU375
PHEBUZIN see BRF500
PHELIPAEA CALOTROPIDIS Walp., extract
see CMS245
α-PHELLANDRENE (FCC) see MCC000
PHELLOBERIN A see PCU390
PHEM see PCU400
PHEMERIDE see BEN000
PHE-MER-NITE see MCU750
PHEMERNITE see MDH500
PHEMEROL CHLORIDE see BEN000
PHEMEROL CHLORIDE MONOHYDRATE
see BBU750
PHEMETONE see ENB500
PHEMITHYN see BEN000

PHEMITON see ENB500
PHEMITONE see ENB500
PHENACAINE see BJO500
PHENACALUM see PEC250
PHENACEMIDE see PEC250
PHENACEREUM see PEC250
PHENACETALDEHYDE DIMETHYL
ACETAL see PDX000
p-PHENACETIN see ABG750
PHENACETUR see PEC250
PHENACETUROHYDROXAMIC ACID see
HIY300
PHENACETYLCARBAMIDE see PEC250
PHENACETYLUREA see PEC250
PHENACHLOR see TIW000
PHENACID see PCU425
PHENACIDE see CDV100
PHENACITE see PCV400
PHENACTYL see CKP250
PHENACYLAMINE see AHR250
PHENACYL-6-AMINOPENICILLINATE see
PCU500
PHENACYL CHLORIDE see CEA750
PHENACYLIDENE CHLORIDE see DEN200
PHENACYLPIVALATE see PCV350
PHENADONE see MDO750
PHENADONE HYDROCHLORIDE see
MDP750
PHENADOR-X see BGE000
PHENAEMAL see EOK000
PHENAGLYCODOL see CKE750
PHENAKITE see PCV400
PHENAKITE (Be₂SiO₄) see PCV400
PHENALCO see MCU750
PHENALENONE see PCJ225
PHENALEN-1-ONE see PCJ225
1H-PHENALEN-1-ONE see PCJ225
PHENALENO(1,9-gh)QUINOLINE see
PCV500
PHENALGENE see AAQ500
PHENALGIN see AAQ500
PHENALLYMAL see AGQ875
PHENALLYMALUM see AGQ875
PHENALONE see PCJ225
PHENALZINE see PFC750
PHENALZINE DIHYDROGEN SULFATE
see PFC750
PHENALZINE HYDROGEN SULPHATE see
PFC750
PHENAMACIDE HYDROCHLORIDE see
PCV750
PHENAMIDE see FAJ150
PHENAMIDE see PCV775
PHENAMINE see AOB250
PHENAMINE BLACK BCN-CF see AQP000
PHENAMINE BLACK CL see AQP000
PHENAMINE BLACK E see AQP000
PHENAMINE BLACK E 200 see AQP000
PHENAMINE BLUE BB see CMO000
PHENAMINE BLUE RW see CMO600
PHENAMINE BORDEAUX B see CMO872
PHENAMINE BROWN 3G see CMO825
PHENAMINE BROWN D 3G see CMO810
PHENAMINE BROWN MB see CMO800
PHENAMINE DARK GREEN B see CMO830
PHENAMINE FAST BROWN T see CMO820
PHENAMINE FAST BROWN TWC see
CMO820
PHENAMINE FAST RED F see CMO870
PHENAMINE FAST SCARLET 4BS see
CMO870
PHENAMINE GREEN BG see CMO840
PHENAMINE GREEN C see CMO840
PHENAMINE GREEN G see CMO840
PHENAMINE PURPURINE 4B see DXO850
PHENAMINE SCARLET 3B see CMO875
PHENAMINE SKY BLUE A see CMO500
PHENAMINE VISCOSE BLACK RR see
CMN240
PHENAMIZOLE HYDROCHLORIDE see
DCA600

PHENANTHRA-ACENAPHTHENE see PCW000
9,10-PHENANTHRAQUINONE see PCX250
PHENANTHREN (GERMAN) see PCW250
PHENANTHRENE see PCW250
1-PHENANTHRENECARBOXYLIC ACID, 1,2,3,4,4A,9,10,10A-OCTAHYDRO-1,4A-DIMETHYL-7-(1-METHYLETHYL)-6-SULFO-, (1R-(1-α,4A-β,10A-α))-, MONOSODIUM SALT see EAB560
PHENANTHRENE-1,2-DIHYDRODIOL see DMA000
PHENANTHRENE-3,4-DIHYDRODIOL see PCW500
PHENANTHRENE, 1,6-DINITRO- see DUY225
PHENANTHRENE, 2,6-DINITRO- see DUY230
PHENANTHRENE, 2,7-DINITRO- see DUY232
PHENANTHRENE, 3,5-DINITRO- see DUY235
PHENANTHRENE, 3,6-DINITRO- see DUY240
PHENANTHRENE, 2,10-DINITRO- see DUY245
PHENANTHRENE, 3,10-DINITRO- see DUY250
9,10-PHENANTHRENEDIONE see PCX250
PHENANTHRENE-9,10-EPOXIDE see PCX000
9-PHENANTHRENEMETHANOL, α-((DIPENTYLAMINO)METHYL)-1,2,3,4-TETRAHYDRO- see DCH300
9-PHENANTHRENEMETHANOL, 1,2,3,4-TETRAHYDRO-α-((DIPENTYLAMINO)METHYL)- see DCH300
PHENANTHRENE, 1-METHYL- see MNN400
PHENANTHRENE, 2-METHYL- see MNN520
PHENANTHRENE, 1-NITRO- see NIB600
PHENANTHRENE, 3-NITRO- see NIC100
9,10-PHENANTHRENE OXIDE see PCX000
PHENANTHRENEQUINONE see PCX250
9,10-PHENANTHRENEQUINONE see PCX250
PHENANTHRENETETRAHYDRO-3,4-EPOXIDE see ECR500
PHENANTHRENE, 1,5,9-TRINITRO- see TMM610
PHENANTHRENE, 1,6,9-TRINITRO- see PCX275
PHENANTHRENE, 1,7,9-TRINITRO- see TMM615
PHENANTHRENE, 2,6,9-TRINITRO- see TMM620
PHENANTHRENE, 3,6,9-TRINITRO- see TMM638
PHENANTHRENE, 2,5,10-TRINITRO- see TMM618
PHENANTHRENE, 3,5,10-TRINITRO- see TMM635
PHENANTHRIDINE see PCX300
6-PHENANTHRIDINE see PCX300
PHENANTHRO(3,4-D)-1,3-DIOXOLE-5-CARBOXYLIC ACID, 8,10-DIMETHOXY-6-NITRO- see MEA700
PHENANTHRO(3,4-D)-1,3-DIOXOLE-5-CARBOXYLIC ACID, 8-METHOXY-6-NITRO-, SODIUM SALT see SEY050
PHENANTHRO(3,4-D)-1,3-DIOXOLE-5-CARBOXYLIC ACID, 6-NITRO- see MJR780
PHENANTHRO(3,4-D)-1,3-DIOXOLE-5-CARBOXYLIC ACID, 6-NITRO-, SODIUM SALT see SEY075
4,7-PHENANTHROLENE-5,6-QUINONE see PCY300
o-PHENANTHROLINE see PCY250
1,10-PHENANTHROLINE see PCY250
1,10-o-PHENANTHROLINE see PCY250
β-PHENANTHROLINE see PCY250
4,7-PHENANTHROLINE-5,6-DIONE see PCY300

(1,10-PHENANTHROLINE)ZINC(2+) see ZLJ100
PHENANTHRO(2,1-d)THIAZOLE see PCY400
2-PHENANTHRYLACETAMIDE see AAM250
9-PHENANTHRYLACETAMIDE see PCY750
N-2-PHENANTHRYLACETAMIDE see AAM250
N-(2-PHENANTHRYL)ACETAMIDE see AAM250
N-3-PHENANTHRYLACETAMIDE see PCY500
N-9-PHENANTHRYLACETAMIDE see PCY750
2-PHENANTHRYLACETHYDROXAMIC ACID see PCZ000
N-(2-PHENANTHRYL)ACETOHYDROXAMIC ACETATE see ABK250
N-2-PHENANTHRYLACETOHYDROXAMIC ACID see PCZ000
2-PHENANTHRYLAMINE see PDA250
3-PHENANTHRYLAMINE see PDA500
9-PHENANTHRYLAMINE see PDA750
PHENANTOIN see MKB250
PHENANTRIN see PCW250
PHENAROL see CKF500
PHENARONE see PEC250
PHENARSAZINE CHLORIDE see PDB000
PHENARSAZINE OXIDE see PDB250
10-PHENARSAZINETHIOL, S-ESTER with O,O-DIISOOCTYLPHOSPHORODITHIOATE see PDB300
S-(10-PHENARSAZINYL)-O,O-DIISOOCTYLPHOSPHORODITHIOATE see PDB300
PHENARSEN see OOK100
PHENARSENAMINE see SAP500
PHENASAL see DFV400
PHENATHYL see CKP250
PHENATOINE see DKQ000
PHENATOX see CDV100
PHENAZEPAM see PDB350
PHENAZIN see PDB750
1-PHENAZINAMINE (9CI) see PDB400
PHENAZINE see DKE800
PHENAZINE see PDB500
PHENAZINE, 1-AMINO- see PDB400
2,3-PHENAZINEDIAMINE see PDB600
2,7-PHENAZINEDIAMINE, 3,8-DIMETHYL-(9CI) see DBT550
PHENAZINE, 2,3-DIAMINO- see PDB600
PHENAZINE, 2,7-DIAMINO-3,8-DIMETHYL- see DBT550
PHENAZINE, 1,7-DINITRO- see DUY300
PHENAZINE ETHOSULFATE see EOE000
PHENAZINE METHOSULFATE see MNO500
PHENAZINE-9-OXIDE see PDB750
PHENAZINE-N-OXIDE see PDB750
PHENAZINE-5N-OXIDE see PDB750
PHENAZINIUM, 3-AMINO-7-(DIMETHYLAMINO)-5-PHENYL-, CHLORIDE see AJP300
PHENAZINIUM, 3-AMINO-2,8-DIMETHYL-7-(2-HYDROXY-1-NAPHTHYLAZO)-5-PHENYL-, CHLORIDE see CMM770
PHENAZINIUM, 3,7-DIAMINO-2,8-DIMETHYL-5-PHENYL-, CHLORIDE see GJI400
2-PHENAZINOL, 3-AMINO- see ALS100
2-PHENAZINOL, 8-AMINO-7-METHYL-, REACTION PRODUCTS WITH SODIUM SULFIDE (NA2(A solid.)) see CMS257
1-PHENAZINOL, 6-METHOXY-, 5,10-DIOXIDE see HLT100
PHENAZIN OXIDE see PDB750
PHENAZIN-5-OXIDE see PDB750
PHENAZITE see PCV400
PHENAZO see PDC250

PHENAZO BLACK BH see CMN800
PHENAZO BLACK D see CMN230
PHENAZOCINE HYDROBROMIDE see PMD325
PHENAZODINE see PDC250
PHENAZODINE see PEK250
PHENAZOLINE see PDC000
PHENAZONE (pharmaceutical) see AQN000
PHENAZOPYRIDINE see PEK250
PHENAZOPYRIDINE HYDROCHLORIDE see PDC250
PHENAZOPYRIDINIUM CHLORIDE see PDC250
PHENBENZAMINE see BEM500
PHENBENZAMINE HYDROCHLORIDE see PEN000
PHENBUTAZOL see BRF500
PHENCAPTON see PDC750
PHENCARBAMID see PDC850
PHENCARBAMIDE see PDC850
PHENCARBAMIDE HYDROCHLORIDE see PDC875
PHENCARBAMID HYDROCHLORIDE (GERMAN) see PDC875
PHENCAROL see DWM400
PHENCEN see PMI750
PHENCYCLIDINE see PDC890
PHENCYCLIDINE HYDROCHLORIDE see AOO500
PHENDAL see DRR400
PHENDIMETRAZINE BITARTRATE see DKE800
PHENDIMETRAZINE HYDROCHLORIDE see DTN800
PHENDIMETRAZINE TARTRATE see PDD000
PHENDIPHAM see MEG250
PHENE see BBL250
PHENEDRINE see AOB250
PHENEDRINE see BBK000
PHENEENE GERMICIDAL SOLUTION and TINCTURE see AFP250
PHENEGIC see PDP250
PHENELZIN see PFC750
PHENELZINE see PFC500
PHENELZINE ACID SULFATE see PFC750
PHENELZINE BISULPHATE see PFC750
PHENELZINE SULFATE see PFC750
PHENEMALUM see SID000
(ν-PHENENYLTRIS(OXYETHYLENE))TRIS(TRIETHYLAMMONIUM IODIDE) see PDD300
PHENERGAN see DQA400
PHENERGAN HYDROCHLORIDE see PMI750
PHENESTERINE see CME250
PHENESTRIN see CME250
PHENETAMID see NMV300
PHENETAMIDE see NMV300
PHENETAMINE HYDROCHLORIDE see LFK200
PHENETHAMINE HYDROCHLORIDE see LFK200
PHENETHANOL see PDD750
PHENETHECILLIN POTASSIUM see PDD350
PHENETHECILLIN POTASSIUM SALT see PDD350
PHENETHICILLIN K see PDD350
PHENETHICILLIN K SALT see PDD350
1-PHENETHOXY-1-PROPOXYETHANE see PDD400
β-PHENETHYBIGUANIDE see PDF000
2-PHENETHYL ACETATE see PFB250
sec-PHENETHYL ACETATE see SMP600
β-PHENETHYL ACETATE see PFB250
PHENETHYL ALCOHOL see PDD750
2-PHENETHYL ALCOHOL see PDD750
α-PHENETHYL ALCOHOL see PDE000
β-PHENETHYL ALCOHOL see PDD750

PHENETHYL ALCOHOL, BENZOATE see PFB750
PHENETHYL ALCOHOL, FORMATE see PFC250
PHENETHYL ALCOHOL, α-METHYL- see PGA600
PHENETHYL ALCOHOL, β-METHYL- see HGR600
PHENETHYL ALCOHOL, o-NITRO- see NIM560
β-PHENETHYLAMINE see PDE250
PHENETHYLAMINE, 2,5-DIMETHOXY-α,4-DIMETHYL- see SLU600
PHENETHYLAMINE, 2,5-DIMETHOXY-4-ETHYL-α-METHYL- see SLU600
PHENETHYLAMINE, N,α-DIMETHYL-, (S)-(+)- see PFP850
PHENETHYLAMINE, α-HEPTYL-3,4,5-TRIMETHOXY- see HBP435
β-PHENETHYLAMINE HYDROCHLORIDE see PDE500
PHENETHYLAMINE, β-HYDROXY- see HNF000
PHENETHYLAMINE, N-ISOPROPYL-α-METHYL-, HYDROCHLORIDE see IQH500
PHENETHYLAMINE, β-METHYL- see PGB760
PHENETHYLAMINE, α-METHYL-, SULFATE (2:1) see BBK250
PHENETHYLAMINE, N,N,α-TRIMETHYL- see TMA600
β-PHENETHYL-o-AMINOBENZOATE see APJ500
PHENETHYL ANTHRANILATE see APJ500
PHENETHYL BENZOATE see PFB750
1-PHENETHYLBIGUANIDE see PDF000
β-PHENETHYLBIGUANIDE see PDF000
PHENETHYLBIGUANIDE HYDROCHLORIDE see PDF250
1-PHENETHYLBIGUANIDE HYDROCHLORIDE see PDF250
N'-β-PHENETHYLBIGUANIDE HYDROCHLORIDE see PDF250
PHENETHYL BROMIDE see PFB770
2-PHENETHYL BROMIDE see PFB770
β-PHENETHYL BROMIDE see PFB770
2-PHENETHYL BUTANOATE see PFB800
β-PHENETHYL N-BUTANOATE see PFB800
PHENETHYL BUTYRATE see PFB800
PHENETHYLCARBAMID (GERMAN) see EFE000
PHENETHYL CHLORACETATE see PDF500
PHENETHYL CINNAMATE see BEE250
β-PHENETHYL CINNAMATE see BEE250
PHENETHYL CYANIDE see HHP100
PHENETHYLDIGUANIDE see PDF000
PHENETHYLENE see SMQ000
(4,β-PHENETHYLENEBIS(CARBONYLMETHYL))BIS(DIMETHYL-2-HYDROXYETHYLAMMONIUM) DIBROMIDE see PDF510
PHENETHYLENE OXIDE see EBR000
PHENETHYL ESTER HYDRACRYLIC ACID see PFC100
PHENETHYL ESTER ISOVALERIC ACID see PDF775
N'-β-PHENETHYLFORMAMIDINYLLIMINOUREA see PDF000
PHENETHYL FORMATE see PFC250
PHENETHYLGLUCOSINOLATE see PDF525
2-PHENETHYLGLUCOSINOLATE see PDF525
PHENETHYLHYDRAZINE see PFC500
PHENETHYLHYDRAZINE SULFATE (1:1) see PFC750
1-(1-PHENETHYL)-IMIDAZOLE-5-CARBOXYLIC ACID, METHYL ESTER, HYDROCHLORIDE see MQQ750
PHENETHYL ISOBUTYRATE see PDF750

β-PHENETHYL ISOTHIOCYANATE see ISP000
PHENETHYL ISOVALERATE see PDF775
PHENETHYL METHACRYLATE see PFP600
PHENETHYL 2-METHYLBUTYRATE see PDF780
2-PHENETHYL 2-METHYLBUTYRATE see PDF780
PHENETHYLMETHYLETHYLCARBINOL see PFR200
PHENETHYL METHYL KETONE see PDF800
PHENETHYL-8-OXA-1-DIAZA-3,8-SPIRO(4,5)DECANONE-2-HYDROCHLORIDE see DAI200
PHENETHYL PHENYLACETATE see PDI000
N-(p-PHENETHYL)PHENYLACETOHYDROXAMIC ACID see PDI250
α-(p-PHENETHYLPHENYL)-1-IMIDAZOLEETHANOL MONOHYDROCHLORIDE see DAP880
1-PHENETHYLPIPERIDINE see PDI500
1-PHENETHYL-3-(PIPERIDINOCARBONYL)PIPERIDINE see PDI550
N-(1-PHENETHYL-4-PIPERIDINYL)PROPIONANILIDE DIHYDROGEN CITRATE see PDW750
1-PHENETHYL-4-PIPERIDYL-p-AMINOBENZOATE HYDROCHLORIDE see PDJ000
1-PHENETHYL-4-PIPERIDYL BENZOATE HYDROCHLORIDE see PDJ250
1-PHENETHYL-3-PIPERIDYL PIPERIDINO KETONE see PDI550
N-(1-PHENETHYL-4-PIPERIDYL)PROPIONANILIDE CITRATE see PDW750
N-(1-PHENETHYL-4-PIPERIDYL)PROPIONANILIDE DIHYDROGEN CITRATE see PDW750
PHENETHYL PROPIONATE see PDK000
1-PHENETHYL-4-N-PROPIONYLANILINOPIPERIDINE see PDW500
N-PHENETHYL-4-(N-PROPIONYLANILINO)PIPERIDINE see PDW500
PHENETHYL SALICYLATE see PDK200
1-PHENETHYLSEMICARBAZIDE see PDK300
2-PHENETHYL-3-THIOSEMICARBAZIDE see PDK500
PHENETHYL TIGLATE see PFD250
PHENETHYLUREA see PDK750
PHENETICILLIN POTASSIUM see PDD350
p-PHENETIDIN see PDK790
p-PHENETIDINANTIMONYLTARTRAT (GERMAN) see PDL500
PHENETIDINE see EEL100
PHENETIDINE see PDK790
2-PHENETIDINE see PDK819
m-PHENETIDINE see PDK800
o-PHENETIDINE see PDK819
m-PHENETIDINE ANTIMONYL TARTRATE see PDL000
o-PHENETIDINE ANTIMONYL TARTRATE see PDL100
p-PHENETIDINE ANTIMONYL TARTRATE see PDL500
PHENETIDINE HYDROCHLORIDE see PDL750
p-PHENETIDINE HYDROCHLORIDE see PDL750
o-PHENETIDINE, 5-NITRO-(8CI) see NIC200
PHENETIDINES (DOT) see EEL100
p-PHENETOLCARBAMID (GERMAN) see EFE000
p-PHENETOLCARBAMIDE see EFE000
PHENETOLE see PDM000

p-PHENETOLECARBAMIDE see EFE000
PHENETOLE, p-NITRO- see NIC990
PHENETOLE, p-NITROSO- see EEY550
(−)-PHENETURIDE see PDM100
PHENETURIDE see PFB350
l-PHENETURIDE see PDM100
p-PHENETYLUREA see EFE000
PHENEXAN see NMV300
PHENFLUORAMINE HYDROCHLORIDE see PDM250
PHENFORMINE see PDF000
PHENGLYKODOL see CKE750
PHENHYDREN see PFJ750
PHENIBUT see PEE500
PHENIBUT HYDROCHLORIDE see GAD000
PHENIC ACID see PDN750
PHENICARB see PEC250
PHENICOL see EGQ000
PHENIDONE see PDM500
PHENIDYLATE see MNQ000
PHENIGAM see PEE500
PHENIGAMA see PEE500
PHENIGAMA HYDROCHLORIDE see GAD000
PHENIGAM HYDROCHLORIDE see GAD000
PHENINDAMINE HYDROCHLORIDE see TEQ700
PHENINDAMINE HYDROGEN TARTRATE see PDD000
PHENINDIONE see PFJ750
PHENIODOL see PDM750
PHENIPRAZINE see PDN000
PHENIPRAZINE HYDROCHLORIDE see PDN250
PHENIRAMINE MALEATE see TMK000
PHENISATIN see ACD500
PHENISOBROMOLATE see IOS000
(±)-PHENISOPROPYLAMINE SULFATE see AOB250
PHENISTAN see SPC500
PHENITOL see MCU750
PHENITROTHION see DSQ000
PHENIZIDOLE see PEL250
PHENIZINE see PDN000
PHENLINE see PFC750
PHENMAD see ABU500
PHENMEC see PFS350
PHENMEDIPHAM see MEG250
PHENMEDIPHAME see MEG250
PHENMERZYL NITRATE see MCU750
PHENMERZYL NITRATE see MDH500
PHENMETHYL TRIMETHYLAMMONIUM IODIDE see BFM750
PHENMETRAZIN see PMA750
PHENMETRAZINE see PMA750
PHENMETRAZINE HYDROCHLORIDE see MNV750
PHENOBAL see EOK000
PHENOBAL SODIUM see SID000
PHENOBARBITAL see EOK000
PHENOBARBITAL ELIXIR see SID000
PHENOBARBITAL Na see SID000
PHENOBARBITAL SODIUM see SID000
PHENOBARBITAL SODIUM SALT see SID000
PHENOBARBITOL and DIPHENYLHDANTOIN see DWD000
PHENOBARBITONE see EOK000
PHENOBARBITONE and PHENOBARBITONE see DWD000
PHENOBARBITONE SODIUM see SID000
PHENOBARBITONE SODIUM SALT see SID000
PHENOBARBITURIC ACID see EOK000
PHENO BLACK EP see AQP000
PHENO BLACK SGN see AQP000
PHENO BLUE 2B see CMO000
PHENOBOLIN see DYF450
PHENO BRIGHT GREEN see CMO840
PHENO BROWN MRS see CMO800
PHENOCAINE see BQH250

PHENOCHLOR see PJL750
PHENOCLOR see PJL750
PHENOCLOR DP6 see PJN250
PHENOCYCLIN see BIS750
PHENOCYCLIN see BIS750
PHENODIANISYL see PDN500
PHENODIANISYL HYDROCHLORIDE see PDN500
PHENODIOXIN see DDA800
PHENODODECINIUM BROMIDE see DXX000
PHENODYNE see PFC750
PHENO FAST RED F see CMO870
PHENO FAST SCARLET 4B see CMO875
PHENO FAST SCARLET 9B see CMO885
PHENO FAST SCARLET 4BSY see CMO870
PHENOFORMINE HYDROCHLORIDE see PDF250
PHENOHEP see HCI000
PHENOL see PDN750
PHENOL 25 see AJU900
PHENOL, molten (DOT) see PDN750
PHENOL ACETATE see PDY750
PHENOL ALCOHOL see PDN750
PHENOL, 4-ALLYL-2-METHOXY-, FORMATE (ester) see EQS100
PHENOL, 2-AMINO-4-ARSENOSO- see OOK100
PHENOL, 2-AMINO-4-ARSENOSO-, SODIUM SALT see ARJ900
PHENOL, 5-AMINO-2-CHLORO- see AJI260
PHENOL, 2-AMINO-4,6-DICHLORO- see AJM525
PHENOL, 4-(1-AMINOETHYL)- see AKA800
PHENOL, (((2-((2-AMINOETHYL)AMINO)ETHYL)AMINO)METHYL)- see AJU700
PHENOL, 4-AMINO-3-METHYL- see AKZ000
PHENOL, 5-((AMINOOXY)METHYL)-2-BROMO-(9CI) see BMM600
PHENOL, 4-(2-AMINOPROPYL)-, HYDROCHLORIDE see HJA600
PHENOL, p-(2-AMINOPROPYL)-, HYDROCHLORIDE see HJA600
PHENOL, p-(4-AMINO-m-TOLUIDINO)- see AMT300
PHENOL, 2-(((2-ANILINO-5-NITROPHENYL)IMINO)METHYL)- see NHY700
PHENOL, 4-(((2-ANILINO-5-NITROPHENYL)IMINO)METHYL)- see NHY600
PHENOL, 2-(((2-ANILINO-5-NITROPHENYL)IMINO)METHYL)-4-CHLORO- see NFT100
PHENOL, 2-(((2-ANILINO-5-NITROPHENYL)IMINO)METHYL)-6-METHOXY- see PDN800
PHENOL, 2-(((2-ANILINO-5-NITROPHENYL)IMINO)METHYL)-6-NITRO- see NHR640
PHENOL, p-ARSENOSO- see HNG800
PHENOL, 4-ARSENOSO-2-NITRO- see NHE600
PHENOL-p-ARSONIC ACID see PDO250
PHENOLATED CAMPHOR see PDO275
PHENOL, 4,4'-(3H-2,1-BENZOXATHIOL-3-YLIDENE)BIS(2,6-DIBROMO-3-METHYL-, S,S-DIOXIDE see BNA940
PHENOL, 4,4'-(3H-2,1-BENZOXATHIOL-3-YLIDENE)BIS(2,6-DIBROMO-3-METHYL-, S,S-DIOXIDE see TBJ505
PHENOL, 4,4'-(3H-2,1-BENZOXATHIOL-3-YLIDENE)BIS-, S,S-DIOXIDE (9CI) see PDO800
PHENOL,4,4'-(3H-2,1-BENZOXATHIOL-3-YLIDENE)BIS(5-METHYL-2-(1-METHYLETHYL))-, S,S-DIOXIDE see TFX850
PHENOL, 4,4'-(3H-2,1-BENZOXATHIOL-3-YLIDENE)DI-, S,S-DIOXIDE see PDO800

PHENOL, p-(BENZYLAMINO)- see BDY750
PHENOL, 2,4-BIS(ACETOXYMERCURI)- see HNK575
PHENOL, 2,6-BIS(1,1-DIMETHYLETHYL)-4-(METHOXYMETHYL)- see DEE300
PHENOL, 2,4-BIS(1,1-DIMETHYLETHYL)-6-(1-(4-METHOXYPHENYL)ETHYL)- see BJK580
PHENOL, 2,6-BIS(1,1-DIMETHYLETHYL)-4-METHYL-, METHYLCARBAMATE see TAN300
PHENOL, 2,4-BIS(1,1-DIMETHYLETHYL)-6-(1-PHENYLETHYL)- see BJK650
PHENOL, 2,4-BIS(1,1-DIMETHYLETHYL)-6-(PHENYLMETHYL)- see DEG150
PHENOL, 2,6-BIS(1-METHYLETHYL)-(9CI) see DNR800
PHENOL, 2,4-BIS(1-METHYLETHYL)-6-(((TRIPHENYLSTANNYL)OXY)CARBONYL)- see TMV830
PHENOL, BROMO- see BNU800
PHENOL, 4-BROMO-2-(((2-ANILINO-5-NITROPHENYL)IMINO)METHYL)- see NFQ300
PHENOL, 4-BROMO-2,5-DICHLORO- see LEN050
PHENOL, 4-BROMO-2-NITRO- see NFQ200
PHENOL, 4-BROMO-2-(((5-NITRO-2-(PHENYLAMINO)PHENYL)IMINO)METHYL)- see NFQ300
PHENOL, o-(tert-BUTYL)- see BSE460
PHENOL, 4-(BUTYLAMINO)- see BQG650
PHENOL, 2-sec-BUTYL-4,6-DINITRO-, ACETATE (ESTER) (8CI) see ACE500
PHENOL, 6-t-BUTYL-3-(2-IMIDAZOLIN-2-YLMETHYL)-2,4-DIMETHYL- see ORA100
PHENOL, 3-(1-BUTYL-3-ISOBUTYL-3-PYRROLIDINYL)-, CITRATE see BRQ300
PHENOL, 2-tert-BUTYL-5-METHYL- see BQV600
PHENOL, 2-tert-BUTYL-6-METHYL- see BRU790
PHENOL, 4-tert-BUTYL-2-METHYL- see BRU800
PHENOL, CAMPHORATED see PDO275
PHENOL, 4-(3-CARBAZOLYLAMINO)- see CBN100
PHENOLCARBINOL see BDX500
PHENOL, 4-CHLORO- see CJK750
PHENOL, 2-(((2-CHLOROETHYL)AMINO)METHYL)-4-NITRO- see CGQ280
PHENOL, o-(CHLOROMERCURI)- see CHW675
PHENOL, 2-(2-CHLORO-1-METHOXYETHOXY)-, METHYLCARBAMATE see LAU500
PHENOL, 2-CHLORO-5-METHYL- see CFE500
PHENOL, 4-CHLORO-5-METHYL-2-(1-METHYLETHYL)-(9CI) see CLJ800
PHENOL, 4-CHLORO-2-(((5-NITRO-2-(PHENYLAMINO)PHENYL)IMINO)METHYL)- see NFT100
PHENOL, 4-CHLORO-2-(PHENYLMETHYL)-, SODIUM SALT see SFB300
PHENOL, 3-(2-CHLORO-4-(TRIFLUOROMETHYL)PHENOXY)-, ACETATE see CLR300
PHENOL CONDENSATION PRODUCTS, WITH 1-CHLORO-2,3-EPOXYPROPANE AND FORMALDEHYDE see FMW333
PHENOL CONDENSATION PRODUCTS, WITH ETHYLENEDIAMINE AND FORMALDEHYDE see EEA550
PHENOL, 2-CYCLOHEXYL-4,6-DINITRO-, COMPD WITH DICYLOHEXYLAMINE see CPK550
PHENOL, 4-(CYCLOHEXYLIDENE(4-HYDROXYPHENYL)METHYL)- see CPM770

PHENOL, 2-CYCLOHEXYL-4-METHYL- see CPP050
PHENOL, 2,4-DIBROMO- see DDR150
PHENOL, 2,6-DIBROMO-4-NITRO- see DDQ500
PHENOL, 2,6-DI-tert-BUTYL- see DEG100
PHENOL, 2,6-DI-tert-BUTYL-4-METHOXYMETHYL- see DEE300
PHENOL, 2,5-DICHLORO- see DFX850
PHENOL, 3,4-DICHLORO- see DFY425
PHENOL, 3,5-DICHLORO- see DFY450
PHENOL, 3,6-DICHLORO-2,4-DINITRO-, CROTONATE (ESTER) see DFE560
PHENOL, 4,5-DICHLORO-2-METHOXY- see DFL720
PHENOL, 2,6-DICHLORO-4-OCTYL- see DFW700
PHENOL, 2,6-DIETHYL- see DJU700
PHENOL, m-(DIETHYLAMINO)- see DIO400
PHENOL, 3-(DIETHYLAMINO)-(9CI) see DIO400
PHENOL, 3-(DIETHYLAMINO)-, METHOCHLORIDE see DJN430
PHENOL, 4,4'-(1,2-DIETHYLETHYLENE)BIS(2-AMINO)- see DJI250
PHENOL, 4,4'-(1,2-DIETHYLIDENE-1,2-ETHANEDIYL)BIS-, (E,E)-(9CI) see DHB550
PHENOL, 3-(DIETHYLMETHYLAMMONIO)-, CHLORIDE see DJN430
PHENOL, DIMETHYL- see XKA000
PHENOL, 2,3-DIMETHYL- see XKJ000
PHENOL, m-(DIMETHYLAMINO)- see HNK560
PHENOL,4-(1-(4-(2-(DIMETHYLAMINO)ETHOXY)PHENYL)-2-PHENYL-1-BUTENYL)-, (E)- see HOH100
PHENOL, 4-((DIMETHYLAMINO)METHYL)-2,6-BIS(1,1-DIMETHYLETHYL)-(9CI) see DEA100
PHENOL, 4-((DIMETHYLAMINO)METHYL)-2,6-BIS(1,1-DIMETHYLETHYL)-(9CI) see FAB000
PHENOL, p-(α-α-DIMETHYLBENZYL)- see COF400
PHENOL, 2-(1,1-DIMETHYLETHYL)-4-METHOXY- see BRQ300
PHENOL, 3-(1,1-DIMETHYLETHYL)-, METHYLCARBAMATE (9CI) see BSG300
PHENOL, (1,1-DIMETHYLETHYL)-, PHOSPHATE (3:1) see TIA130
PHENOL, 4-(1,1-DIMETHYLETHYL)-, SODIUM SALT see BSE300
PHENOL, 2,6-DIMETHYL-, PHOSPHATE (3:1) see TNR550
PHENOL, DIMETHYL-, PHOSPHATE (3:1) (9CI) see XLS100
PHENOL, p-(2,4-DINITROANILINO)- see DUW500
PHENOL, 2,6-DINITRO-4-ISOPROPYL- see IOX000
PHENOL, 2,4-DI-tert-PENTYL- see DCI000
PHENOLE (GERMAN) see PDN750
PHENOL, 4-ETHENYL-(9CI) see VQA200
PHENOL, 4-ETHENYL-, ACETATE see ABW550
PHENOL, 4-ETHENYL-2-METHOXY-(9CI) see VPF100
PHENOL, 4-ETHOXY-(9CI) see EFA100
PHENOL, o-ETHYL- see PGR250
PHENOL, p-ETHYL- see EOE100
PHENOL, 3-(ETHYLAMINO)-4-METHYL- see EGF100
PHENOL, 4,4',4"-ETHYLIDYNETRI- see ELO600
PHENOL, 4,4',4"-ETHYLIDYNETRIS- see ELO600
PHENOL, 4-(1-ETHYL-2-(4-METHOXYPHENYL)-1-BUTENYL)-, (E)- see EMI525
PHENOL, o-FLUORO- see FKT100

PHENOL GLUCOSIDE see PDO300
PHENOL-GLYCERINAETHER see GGA950
PHENOL GLYCEROL ETHER see GGA950
PHENOL GLYCERYL ETHER see GGA950
PHENOL-GLYCIDAETHER (GERMAN) see PFF360
PHENOL GLYCIDYL ETHER (MAK) see PFF360
PHENOL, o-HEPTYL- see HBP350
PHENOL, 2-HEPTYL-(9CI) see HBP350
PHENOL, p-(HEPTYLOXY)- see HBP285
PHENOL, o-HEXYL- see HFU600
PHENOL, 2-HEXYL-(9CI) see HFU600
PHENOL, 4,4'-IMINOBIS-(9CI) see IBJ100
PHENOL, 4,4'-IMINODI- see IBJ100
PHENOL, 3-IODO- see IEV010
PHENOL, m-IODO- see IEV010
PHENOL, ISOBUTYLENATED, PHOSPHATE (3:1) see IIQ150
PHENOL, m-ISOPROPYL- see IQX090
PHENOL, o-ISOPROPYL- see IQX100
PHENOL, ISOPROPYLATED, PHOSPHATE (3:1) see DYF500
PHENOL, 4,4'-ISOPROPYLENEDI-, DISODIUM SALT see BLD800
PHENOL, 4,4'-ISOPROPYLIDENEDI-, DISODIUM DERIV. see BLD800
PHENOL, O-ISOPROPYL-, METHYLCARBAMATE see MIA250
PHENOL, 2-METHOXY-4-METHYL- see MEK325
PHENOL, 2-METHOXY-6-(((5-NITRO-2-(PHENYLAMINO)PHENYL)IMINO)METHYL)- see PDN800
PHENOL, 2-METHOXY-4-(OXIRANYLMETHYL)- see EBT500
PHENOL, 2-METHOXY-3,4,5,6-TETRACHLORO- see TBQ290
PHENOL, 2-METHOXY-TRICHLORO- see TIP600
PHENOL, 6-METHOXY-2,3,4-TRICHLORO- see TIP630
PHENOL, 2-METHOXY-4-VINYL- see VPF100
PHENOL, p-(α-METHYLBENZYL)- see PFD400
PHENOL, 2-METHYL-4,6-BIS((OCTYLTHIO)METHYL)- see MHR025
PHENOL, 4-METHYL-2,6-BIS(1-PHENYLETHYL)-(9CI) see MHR050
PHENOL, 2-METHYLDINITRO-, SODIUM SALT (9CI) see SGP550
PHENOL, 4,4'-METHYLENEBIS(2,6-BIS(1,1-DIMETHYLETHYL))- see MJM700
PHENOL, 4,4'-METHYLENEBIS(2,6-DI-tert-BUTYL)- see MJM700
PHENOL, 2,2'-METHYLENEBIS(6-(1,1-DIMETHYLETHYL))-4-ETHYL-(9CI) see MJN250
PHENOL, 2,3-(METHYLENEDIOXY)-, METHYLCARBAMATE see BCJ100
PHENOL, 3-(1-METHYLETHYL)- see IQX090
PHENOL, 2-(1-METHYLETHYL)-(9CI) see IQX100
PHENOL, 4,4'-(1-METHYLETHYLIDENE)BIS(2,6-DIBROMO- see MKA270
PHENOL, 4,4'-(1-METHYLETHYLIDENE)BIS-, DISODIUM SALT see BLD800
PHENOL, 4,4'-(1-METHYLETHYLIDENE)BIS(2-METHYL-(9CI) see IPK000
PHENOL, 2-(1-METHYLETHYL)-, METHYLCARBAMATE (9CI) see MIA250
PHENOL, (1-METHYLETHYL)-, PHOSPHATE (3:1) see TKU300
PHENOL, 4-(1-METHYL-1-PHENETHYL)-(9CI) see COF400
PHENOL, 2-METHYL-4-((4-(PHENYLAZO)PHENYL)AZO)- see CMP090

PHENOL, 4-(1-METHYL-1-PHENYLETHYL)-, MIXT. WITH BORAX AND 2,2'-OXYBIS(ETHANOL) see MRW100
PHENOL, 4,4',4''-(1-METHYL-1-PROPANYL-3-YLIDENE)TRIS(2-(1,1-DIMETHYLETHYL)-5-METHYL- see MOS100
PHENOL, 2-(1-METHYLPROPYL)-4,6-DINITRO-, ACETATE (ESTER) (9CI) see ACE500
PHENOL, 2,2'-(2-METHYLPROPYLIDENE)BIS(4,6-DIMETHYL- see IIU100
PHENOL, 2-(1-METHYLPROPYL)-, METHYLCARBAMATE see MOV000
PHENOL, METHYL-, SODIUM SALT see SFZ050
PHENOL, 3-METHYL-, SODIUM SALT (9CI) see SJP000
PHENOL, 4-METHYL-, SODIUM SALT (9CI) see SIM100
PHENOL, p-(METHYLTHIO)- see MPV300
PHENOL, p-NITRO-, HYDROGEN PHOSPHATE see BLA600
PHENOL, 2-NITRO-6-(((5-NITRO-2-(PHENYLAMINO)PHENYL)IMINO)METHYL)- see NHR640
PHENOL, 2-(((5-NITRO-2-(PHENYLAMINO)PHENYL)IMINO)METHYL)- see NHY700
PHENOL, 4-(((5-NITRO-2-(PHENYLAMINO)PHENYL)IMINO)METHYL)- see NHY600
PHENOL, 2-NITRO-3,4,6-TRICHLORO- see NMQ050
PHENOL, 4-NITRO-3-(TRIFLUOROMETHYL)- see TKD400
PHENOL, p-NONYL- see NNC510
PHENOL, 4-NONYL-, BRANCHED see NNC600
PHENOL, 4-NONYL-, POLYMER WITH FORMALDEHYDE AND OXIRANE see FMW343
PHENOL, 4-OCTYL- see OFK100
PHENOL, o-PENTYL- see AOM325
PHENOL, 2-PENTYL-(9CI) see AOM325
PHENOL, 4-(1-PHENYLETHYL)- see PFD400
PHENOL, 4-(PHENYLMETHYL)-, CARBAMATE see PFR325
PHENOL, o-PHENYL-, SODIUM deriv. see BGJ750
PHENOLPHTHALEIN see PDO750
PHENOLPHTHALEIN, 3',3''-DIMETHYL- see CNX400
PHENOLPHTHALEIN, 4,5,6,7-TETRABROMO-3',3''-DISULFO-, DISODIUM SALT see HAQ600
PHENOLPHTHALOL see PDO775
PHENOL, POLYMER WITH 1,2-ETHANEDIAMINE AND FORMALDEHYDE (9CI) see EEA550
PHENOL, 4,4'-(2-PYRIDINYLMETHYLENE)BIS-, DIACETATE (ESTER) see PPN100
PHENOL, 4,4'-(2-PYRIDYLMETHYLENE)DI-, DIACETATE (ester) see PPN100
PHENOL RED see PDO800
PHENOL SODIUM SALT see SJF000
PHENOLSULFONEPHTHALEIN see PDO800
PHENOLSULFONIC ACID see HJH500
PHENOLSULFONIC ACID, liquid (DOT) see HJH500
1-PHENOL-4-SULFONIC ACID ZINC SALT see ZIJ300
PHENOLSULFONPHTHALEIN see PDO800
PHENOLSULPHONPHTHALEIN see PDO800
PHENOLTETRABROMOPHTHALEINSULFONATE see HAQ600
PHENOL, 2,3,5,6-TETRACHLORO-4-(PENTACHLOROPHENOXY)- see HNB550

PHENOL, THIO- see PFL850
PHENOL, 2,4,6-TRI-tert-BUTYL-(6CI,7CI,8CI) see TIA100
PHENOL, 3,4,5-TRICHLORO- see TIW100
PHENOL, 2,4,6-TRICHLORO-, CHLOROFORMATE see TIY800
PHENOL, m-TRIFLUOROMETHYL- see TKD400
PHENOL, 4,4'-(2,2,2-TRIFLUORO-1-(TRIFLUOROMETHYL)ETHYLIDENE)BIS- see HCZ100
PHENOL, 4,4'-(TRIFLUORO-1-(TRIFLUOROMETHYL)ETHYLIDENE)DI- see HCZ100
PHENOL, 2,4,6-TRIMETHYL-(9CI) see MDJ740
PHENOL, 2,4,6-TRINITRO- see PID000
PHENOL, 2,4,6-TRINITRO-, AMMONIUM SALT (9CI) see ANS500
PHENOL, 2,4,6-TRIS(1,1-DIMETHYLETHYL)- see TIA100
PHENOLURIC see EOK000
PHENOL, p-VINYL- see VQA200
PHENOL, p-VINYL-, ACETATE (6CI,7CI,8CI) see ABW550
PHENOMERCURIC ACETATE see ABU500
PHENOMET see EOK000
PHENO NAVY BLUE see CMN800
PHENONYL see EOK000
PHENOPENICILLIN see PDT500
PHENO-m-PENICILLIN see PDD350
PHENOPERIDINE HYDROCHLORIDE see PDO900
PHENOPLASTE ORGANOL RED B see SBC500
PHENOPROMIN see BBK500
PHENOPROPAZINE see DIR000
PHENOPROZINE see DIR000
PHENOPYRIDINE see QPA000
PHENOPYRINE see BRF500
PHENOQUIN see PGG000
PHENOSAN see PDP250
PHENOSANE see PEE750
PHENOSANE 23 see DED100
PHENOSELENAZIN-5-IUM, 3-AMINO-7-(DIMETHYLAMINO)-2-METHYL-, CHLORIDE see SBU950
PHENOSMOLIN see PDP100
PHENOSUCCIMIDE see MNZ000
PHENOTAN see ACE500
PHENOTAN see BRE500
PHENOTEROL HYDROBROMIDE see FAQ100
PHENOTHIAZINE see PDP250
PHENOTHIAZINE-10-CARBODITHIOIC ACID-2-(DIETHYLAMINO)ETHYL ESTER see PDP500
10-PHENOTHIAZINECARBOXYLIC ACID β-DIETHYLAMINOETHYL ESTER HYDROCHLORIDE see THK600
PHENOTHIAZINE, 2-CHLORO- see CJL100
10H-PHENOTHIAZINE, 2-CHLORO- see CJL100
PHENOTHIAZINE, 2-CHLORO-10-(3-(DIMETHYLAMINO)PROPYL)-, 5,5-DIOXIDE see ONI300
PHENOTHIAZINE, 2-CHLORO-10-(3-(1-METHYL-4-PIPERAZINYL)PROPYL)-, ETHANEDISULFONATE see PME700
PHENOTHIAZINE, 10-(3-(DIMETHYLAMINO)-2-METHYLPROPYL)-2-ETHYL-, MONOHYDROCHLORIDE see ELQ600
PHENOTHIAZINE, 10-(3-DIMETHYLAMINO-2-METHYLPROPYL)-, HYDROCHLORIDE see TKV200
PHENOTHIAZINE, 10-(3-(DIMETHYLAMINO)-2-METHYLPROPYL)-3-METHOXY- see PDP600
PHENOTHIAZINE, 10-(3-(DIMETHYLAMINO)-2-METHYLPROPYL)-

(3-PHENOXYPHENYL)METHYL 4-CHLORO-α-(1-METHYLETHENYL)BENZENEACETATE see PDR670
(3-PHENOXYPHENYL)METHYL 4-CHLORO-α-(1-METHYLETHYL)BENZENEACETATE see PDV330
(3-PHENOXYPHENYL)METHYL-3-(2,2-DICHLORETHENYL)-2,2-DIMETHYLCYCLOPROPANECARBOXYLATE see AHJ750
(3-PHENOXYPHENYL)METHYL α-ETHYLBENZENEACETATE see PDR664
(3-PHENOXYPHENYL)METHYL α-ETHYL-4-METHOXYBENZENEACETATE see PDR660
(3-PHENOXYPHENYL)METHYL α-ETHYL-4-METHOXYBENZENEACETATE see PDV360
(3-PHENOXYPHENYL)METHYL 4-METHOXY-α-(1-METHYLETHENYL)BENZENEACETATE see PDR666
(3-PHENOXYPHENYL)METHYL 4-METHOXY-α-(1-METHYLETHYL)BENZENEACETATE see PDR675
(3-PHENOXYPHENYL)METHYL 4-METHYL-α-(1-METHYLETHENYL)BENZENEACETATE see PDR668
(3-PHENOXYPHENYL)METHYL 4-METHYL-α-(1-METHYLETHYL)BENZENEACETATE see PDR680
dl-2-(3-PHENOXYPHENYL)-PROPIONIC ACID CALCIUM SALT, DIHYDRATE see FAP100
d,l-2-(3-PHENOXYPHENYL)PROPIONIC ACID SODIUM SALT see FAQ000
PHENOXYPHOSPHORUS DICHLORIDE see PHE850
PHENOXYPHOSPHORYL DICHLORIDE see PHE800
1-PHENOXY-2,3-PROPANEDIOL see GGA950
3-PHENOXY-1,2-PROPANEDIOL see GGA950
3-PHENOXY-1,2-PROPANEDIOL DIACETATE see PFF300
2-PHENOXYPROPANOL see PNL300
1-PHENOXY-2-PROPANOL see PDV460
PHENOXYPROPAZINE MALEATE see PDV700
PHENOXYPROPENE OXIDE see PFF360
2-PHENOXYPROPIONIC ACID see PDV725
2-PHENOXYPROPYL ALCOHOL see PNL300
2-(PHENOXY-2-PROPYLAMINO)-1-(p-HYDROXYPHENYL)-1-PROPANOL HYDROCHLORIDE see VGF000
PHENOXYPROPYLENE OXIDE see PFF360
(3-PHENOXYPROPYL)GUANIDINE SULFATE see GLS700
(3-PHENOXYPROPYL)HYDRAZINE MALEATE see PDV750
4-PHENOXY-3-(PYRROLIDINYL)-5-SULFAMOYLBENZOIC ACID see PDW250
16-PHENOXY-17,18,19,20 TETRANOR PROSTAGLANDIN E$_2$ METHYL SULFONYLAMIDE see SOU650
16-PHENOXY-ω-17,18,19,20-TETRANOR PROSTAGLANDIN E2 METHYLSULFONYLAMIDE see SOU650
PHENOXYTHRIN see PDR700
PHENOXYTOL see PER000
3-PHENOXYTOLUENE see MNV770
m-PHENOXYTOLUENE see MNV770
4-(3-(α-PHENOXY-p-TOLYL)PROPYL)-MORPHOLINE see PDU250
PHENOXYTRIETHYLSTANNANE see TJV250

PHENOZIN see ZIJ300
PHENSEDYL see DQA400
PHENTALAMINE see PDW400
PHENTANYL see PDW500
PHENTANYL CITRATE see PDW750
PHENTERMINE see DTJ400
PHENTHIAZINE see PDP250
PHENTHIURAM see FAQ930
PHENTHOATE see DRR400
PHENTHOATE OXON see PDW800
PHENTIN ACETATE see ABX250
PHENTINOACETATE see ABX250
PHENTOLAMINE see PDW400
PHENTOLAMINE MESILATE see PDW950
PHENTOLAMINE MESYLATE see PDW950
PHENTOLAMINE METHANESULFONATE see PDW950
PHENTOLAMINE METHANESULPHONATE see PDW950
PHENTYDRONE see TCR300
PHENTYRIN see BHY500
PHENURIDE see PFB350
PHENURON see PEC250
PHENUTAL see PEC250
PHENVALERATE see FAR100
PHENYBUT HYDROCHLORIDE see GAD000
PHENYCHOLON see EGQ000
PHENYGAM HYDROCHLORIDE see GAD000
PHENYLACETALDEHYDE 2,4-DIHYDROXY-2-METHYLPENTANE ACETAL see PFR400
PHENYLACETALDEHYDE DIMETHYL ACETAL see PDX000
PHENYLACETALDEHYDE ETHYLENEGLYCOL ACETAL see BEN250
PHENYLACETALDEHYDE (FCC) see BBL500
PHENYLACETALDEHYDE GLYCERYL ACETAL see PDX250
2-PHENYLACETAMIDE see PDX750
N-PHENYLACETAMIDE see AAQ500
N-PHENYLACETAMIDE see PDX500
α-PHENYLACETAMIDE see PDX750
2-(2-PHENYLACETAMIDO)ACETOHYDROXAMIC ACID see HIY300
PHENYLACETAMIDOPENICILLANIC ACID see BDY669
p-PHENYLACETANILIDE see PDY500
2'-PHENYLACETANILIDE see PDY000
3'-PHENYLACETANILIDE see PDY250
4'-PHENYLACETANILIDE see PDY500
PHENYL ACETATE see PDY750
PHENYLACETATE SODIUM SALT see SFA200
N'-PHENYLACETHYDRAZIDE see ACX750
PHENYLACETIC ACID see PDY850
ω-PHENYLACETIC ACID see PDY850
PHENYLACETIC ACID ALLYL ESTER see PMS500
PHENYLACETIC ACID AMIDE see PDX750
PHENYLACETIC ACID 2-BENZYLHYDRAZIDE see PDY870
PHENYL ACETIC ACID DIETHYLAMINOETHOXYETHANOL ESTER CITRATE see BOR350
PHENYLACETIC ACID, ETHYL ESTER see EOH000
PHENYLACETIC ACID, p-METHOXYBENZYL ESTER see APE000
PHENYLACETIC ACID, METHYL ESTER see MHA500
PHENYLACETIC ACID, PHENETHYL ESTER see PDI000
PHENYLACETIC ACID SODIUM SALT see SFA200
PHENYLACETIC ALDEHYDE see BBL500
N-PHENYLACETOACETAMIDE see AAY000
PHENYL ACETO-ACETONITRILE see PEA500

2-PHENYLACETOACETONITRILE see PEA500
α-PHENYLACETOACETONITRILE see PEA500
PHENYLACETONE see MHO100
α-PHENYLACETONE see MHO100
PHENYLACETONITRILE see PEA750
2-PHENYLACETONITRILE see PEA750
PHENYLACETONITRILE, liquid (DOT) see PEA750
2-PHENYLACETOPHENONE see PEB000
2-PHENYLACETOPHENONE see PEB000
p-PHENYLACETOPHENONE see MHP500
4'-PHENYLACETOPHENONE see MHP500
4'-PHENYL-o-ACETOTOLUIDE see PEB250
(PHENYLACETOXY)TRIETHYL PLUMBANE see TJT250
PHENYL-β-ACETYLAMINE see PDX750
PHENYLACETYLENE see PEB750
3-(1'-PHENYL-2'-ACETYLETHYL)-4-HYDROXYCOUMARIN see WAT200
3-(α-PHENYL-β-ACETYLETHYL)-4-HYDROXYCOUMARIN see WAT200
(PHENYL-1 ACETYL-2 ETHYL)-3-HYDROXY-4 COUMARINE (FRENCH) see WAT200
PHENYLACETYLGLYCINE DIMETHYLAMIDE see PEB775
N-(PHENYLACETYL)GLYCINOHYDROXAMIC ACID see HIY300
PHENYL ACETYL NITRILE see PEA750
(PHENYLACETYL)UREA see PEC250
α-PHENYLACETYLUREA see PEC250
PHENYLACETYLUREE (FRENCH) see PEC250
PHENYLACROLEIN see CMP969
3-PHENYLACROLEIN see CMP969
3-PHENYLACRYLAMIDE see CMP973
PHENYLACRYLIC ACID see CMP975
3-PHENYLACRYLIC ACID see CMP975
trans-3-PHENYLACRYLIC ACID see CMP980
tert-β-PHENYLACRYLIC ACID see CMP975
3-PHENYLACRYLOPHENONE see CDH000
β-PHENYLACRYLOPHENONE see CDH000
β-PHENYLAETHYLAMIN (GERMAN) see PDE250
1-PHENYL-2-AETHYLAMINO-PROPAN (GERMAN) see EGI500
PHENYLAETHYLBERNSTEINSAEURE-AETHYL-DIAETHYL-AMINOAETHYL-DI-ESTER (GERMAN) see SBC700
1-PHENYLAETHYLBIGUANID HYDROCHLORID (GERMAN) see PDF250
PHENYLAETHYLCARBINOL (GERMAN) see EGQ000
5,5-PHENYL-AETHYL-3-(β-DIAETHYLAMINO-AETHYL)-2,4,6-TRIOXO-HEXAHYDROPYRIMIDIN-HCl (GERMAN) see HEL500
PHENYLAETHYL-HYDRAZIN see PFC750
PHENYLAETHYLMALONSAEURE-AETHYL-DIAETHYLAMINOAETHYL-DI-ESTER (GERMAN) see PCU400
PHENYLAETHYLMALONSAEURE-AETHYLESTER-DIAETHYLAMINOAETHYL-AMID (GERMAN) see FAJ150
PHENYLAETHYLSENFOEL (GERMAN) see ISP000
l-PHENYLALANINAMIDE, N-((PHENYLMETHOXY)CARBONYL)-l-TRYPTOPHYL-l-METHIONYL-l-ASPARTYL- see GCE200
PHENYLALANINE see PEC750
3-PHENYLALANINE see PEC750
d-PHENYLALANINE see PEC500
l-PHENYLALANINE see PEC750
(S)-PHENYLALANINE see PEC750
PHENYL-α-ALANINE see PEC750
β-PHENYLALANINE see PEC750
d-β-PHENYLALANINE see PEC500

l-β-PHENYLALANINE see PEC750
β-PHENYL-α-ALANINE, l- see PEC750
dl-PHENYLALANINE (FCC) see PEC500
dl-PHENYLALANINE, 4-FLUORO-(9CI) see FLD100
PHENYLALANINE MUSTARD see BHU750
d-PHENYLALANINE MUSTARD see BHU750
d-PHENYLALANINE MUSTARD see SAX200
l-PHENYLALANINE MUSTARD see PED750
o-PHENYLALANINE MUSTARD see BHT250
dl-PHENYLALANINE MUSTARD see BHT750
(±)-o-PHENYLALANINE MUSTARD see BHT250
l-PHENYLALANINE MUSTARD HYDROCHLORIDE see BHV250
dl-PHENYLALANINE MUSTARD HYDROCHLORIDE see BHV000
PHENYLALANINE NITROGEN MUSTARD see PED750
l-PHENYLALANINE, N-((PHENYLMETHOXY)CARBONYL)-, ETHENYL ESTER see CBR235
2-d-PHENYLALANINE-3-l-PROLINE-6-d-TRYPTOPHANLUTEINIZING HORMONE-RELEASING FACTOR(PIG) see LIU342
d,l-PHENYLALANINE, PYROLYZATE see ALY000
2-d-PHENYLALANINE-3-l-VALINE-6-d-TRYPTOPHANLUTEINIZING HORMONE-RELEASING-FACTOR(PIG) see LIU345
PHENYLALANIN-LOST (GERMAN) see BHT750
(S)-d-PHENYLALANYL-N-(4-((AMINOIMINOMETHYL)AMINO)-1-FORMYLBUTYL)-l-PROLINAMIDE SULFATE (1:1) see PEE100
d-PHENYLALANYL-l-PHENYLALANYL-l-PHENYLALANYL-d-TRYPTOPHYL-l-LYSYL-l-THREON YL-l-PHENYLALANYL-l-THREONINAMIDE see PED800
d-PHENYLALANYL-l-PHENYLALANYL-l-TYROSYL-d-TRYPTOPHYL-l-LYSYL-l-VALYL-l-PHENYLALANYL-3-(2-NAPHTHALENYL)-d-ALANINAMIDE see PED850
d-PHENYLALANYL-l-PROLYL-l-ARGININE ALDEHYDE SULFATE (1:1) see PEE100
γ-PHENYLALLYL ACETATE see CMQ730
3-PHENYLALLYL ALCOHOL see CMQ740
γ-PHENYLALLYL ALCOHOL see CMQ740
5-PHENYL-5-ALLYLBARBITURIC ACID see AGQ875
PHENYLALLYL CINNAMATE see CMQ850
3-PHENYLALLYL ISOVALERATE see PEE200
PHENYLAMINE see AOQ000
PHENYLAMINE HYDROCHLORIDE see BBL000
PHENYLAMINOACETIC ACID ISOAMYL ESTER HYDROCHLORIDE see PCV750
1-PHENYL-2-AMINO-AETHAN (GERMAN) see PDE250
N-PHENYL-p-AMINOANILINE see PFU500
2-(PHENYLAMINO)BENZOIC ACID see PEG500
1-PHENYL-3-AMINO-BUTAN see PEE300
4-(PHENYLAMINO)BUTANE see BQH850
1-PHENYL-3-AMINOBUTANE see PEE300
PHENYL-α-AMINO-n-BUTYRAMIDE-p-ARSONIC ACID see CBK750
β-PHENYL-γ-AMINOBUTYRATE see PEE500
β-PHENYL-γ-AMINOBUTYRIC ACID see PEE500
N-PHENYLAMINOCARBONYL)AZIRIDINE see PEH250
17-β-PHENYLAMINOCARBONYLOXYOESTRA-1,3,5(10)-TRIENE-3-METHYL ETHER see PEE600

1-PHENYL-4-AMINO-5-CHLOROPYRIDAZON-(6) (GERMAN) see PEE750
1-PHENYL-4-AMINO-5-CHLOROPYRIDAZONE-6 see PEE750
1-PHENYL-4-AMINO-5-CHLORO-6-PYRIDAZONE see PEE750
1-PHENYL-4-AMINO-5-CHLORPYRIDAZ-6-ONE see PEE750
trans-2-PHENYL-1-AMINOCYCLOPROPANE see PET750
2,2'-(PHENYLAMINO)DIETHANOL see BKD500
4-PHENYLAMINODIPHENYLAMINE see BLE500
p-PHENYLAMINODIPHENYLAMINE see BLE500
1-PHENYL-2-AMINOETHANE see PDE250
1-PHENYL-2-AMINOETHANE HYDROCHLORIDE see PDE500
2-(PHENYLAMINO)ETHANOL see AOR750
1-PHENYL-2-AMINO-1-ETHANOL, HYDROCHLORIDE see PFA750
α-PHENYL-β-AMINOETHANOL HYDROCHLORIDE see PFA750
4-(PHENYLAMINO)-PHENOL see AOT000
PHENYL-p-AMINOPHENOL see AOT000
N-PHENYL-p-AMINOPHENOL see AOT000
2-PHENYL-2-(p-AMINOPHENYL)PROPIONAMIDE see ALX500
1-PHENYL-2-AMINO-PROPAN (GERMAN) see AOA250
d-1-PHENYL-2-AMINOPROPAN (GERMAN) see AOA500
1-PHENYL-2-AMINOPROPANE see AOA250
d-1-PHENYL-2-AMINOPROPANE see AOA500
dl-1-PHENYL-2-AMINOPROPANE see BBK000
PHENYL-1-AMINO-1-PROPANE HYDROCHLORIDE see PEF750
1-PHENYL-1-AMINO-PROPANE, HYDROCHLORIDE see PEF750
PHENYL-2-AMINO-1-PROPANE HYDROCHLORIDE see PEF500
1-PHENYL-2-AMINOPROPANE MONOPHOSPHATE see AOB500
1-PHENYL-2-AMINOPROPANE SULFATE see BBK250
d-1-PHENYL-2-AMINOPROPANE SULFATE see BBK500
l-1-PHENYL-2-AMINOPROPANE SULFATE see BBK750
dl-1-PHENYL-2-AMINOPROPANOL-1 see NNM500
dl-1-PHENYL-2-AMINO-1-PROPANOL MONOHYDROCHLORIDE see PMJ500
N-(3-(PHENYLAMINO)-2-PROPENYLIDENE)BENZENAMINE MONOHYDROCHLORIDE see PFJ500
2-PHENYLAMINOPROPIONITRILE see AOT100
3-PHENYL-5-AMINO-1,2,4-TRIAZOLYL-(1)-(N,N'-TETRAMETHYL) DIAMIDOPHOSPHONATE see AIX000
2-PHENYLANILINE see BGE250
N-PHENYLANILINE see DVX800
o-PHENYLANILINE see BGE250
p-PHENYLANILINE see AJS100
2-PHENYLANISOLE see PEG000
o-PHENYLANISOLE see PEG000
p-PHENYLANISOLE see PEG250
PHENYL p-ANISYL KETONE see MEC333
PHENYLANTHRANILIC ACID see PEG500
N-PHENYLANTHRANILIC ACID see PEG500
PHENYL ARSENIC ACID see BBL750
PHENYLARSENOXIDE see PEG750
PHENYLARSINEDICHLORIDE see DGB600
PHENYL ARSINE OXIDE see PEG750
PHENYLARSONIC ACID see BBL750

PHENYLARSONOUS DIBROMIDE see DDR200
PHENYLARSONOUS DICHLORIDE see DGB600
β-PHENYLATHYLAMINHYDROCHLORID (GERMAN) see PDE500
PHENYL AZIDE see PEH000
N-PHENYL-1-AZIRIDINECARBOXAMIDE see PEH250
α-PHENYL-1-AZIRIDINEETHANOL see PEH500
p-PHENYLAZOACETANILIDE see PEH750
4'-PHENYLAZOACETANILIDE see PEH750
4-(PHENYLAZO)ANILINE see PEI000
N-(PHENYLAZO)ANILINE see DWO800
p-(PHENYLAZO)ANILINE see PEI000
p-(PHENYLAZO)ANILINE HYDROCHLORIDE see PEI250
4-(PHENYLAZO)-o-ANISIDINE see MFF500
1-PHENYLAZO-2-ANTHROL see PEI750
4-(PHENYLAZO)-m-ANTISIDINE see MDY400
4-(PHENYLAZO)BENZENAMINE see PEI000
4-(PHENYLAZO)-1,3-BENZENEDIAMINE MONOHYDROCHLORIDE see PEK000
PHENYLAZODIAMINOPYRIDINE HYDROCHLORIDE see PDC250
3-PHENYLAZO-2,6-DIAMINOPYRIDINE HYDROCHLORIDE see PDC250
β-PHENYLAZO-α,α'-DIAMINOPYRIDINE HYDROCHLORIDE see PDC250
PHENYLAZO-α,α'-DIAMINOPYRIDINE MONOHYDROCHLORIDE see PDC250
4-(PHENYLAZO)DIPHENYLAMINE see PEI800
N-PHENYLAZO-N-METHYLTAURINE SODIUM SALT see PEJ250
1-(PHENYLAZO)-2-NAPHTHALENAMINE see FAG130
1-(PHENYLAZO)-2-NAPHTHALENOL see PEJ500
1-(PHENYLAZO)-2-NAPHTHOL see PEJ500
1-PHENYLAZO-β-NAPHTHOL see PEJ500
1-PHENYLAZO-2-NAPHTHOL-6,8-DISULFONIC ACID, DISODIUM SALT see HGC000
1-PHENYLAZO-2-NAPHTHOL-6,8-DISULPHONIC ACID, DISODIUM SALT see HGC000
1-(PHENYLAZO)-2-NAPHTHYLAMINE see FAG130
4-PHENYLAZO-1-NAPHTHYLAMINE see PEJ600
PHENYLAZO α-NAPHTHYLAMINE see PEJ600
4-PHENYLAZOPHENOL see HJF000
p-PHENYLAZOPHENOL see HJF000
p-PHENYLAZOPHENYLAMINE see PEI000
1-(4-PHENYLAZO-PHENYLAZO)-2-ETHYLAMINONAPHTHALENE see EOJ500
(PHENYLAZO-4-PHENYLAZO)-1-ETHYLAMINO-2-NAPHTHALENE see EOJ500
1-((4-(PHENYLAZO)PHENYL)AZO)-2-NAPHTHALENOL see OHI200
1-((p-PHENYLAZO)PHENYL)AZO-2-NAPHTHOL see OHI200
4-(PHENYLAZO)-m-PHENYLENEDIAMINE see DBP999
4-PHENYLAZO-m-PHENYLENEDIAMINE see PEK000
4-(PHENYLAZO)-m-PHENYLENEDIAMINE MONOHYDROCHLORIDE see PEK000
p-(PHENYLAZO)PHENYL ISOCYANIDE see PEK050
4-PHENYLAZOPYRIDINE see PEK100
3-(PHENYLAZO)-2,6-PYRIDINEDIAMINE see PEK250
3-(PHENYLAZO)-2,6-PYRIDINEDIAMINE, HYDROCHLORIDE see PDC250

(DIETHYLAMINO)ETHOXY)ETHYL) ESTER HYDROCHLORIDE see DHQ500

1-PHENYLCYCLOPENTANECARBOXYLIC ACID-2-DIETHYLAMINOETHYL ESTER HYDROCHLORIDE see PET250

1-PHENYLCYCLOPENTANECARBOXYLIC ACID 1-DIETHYLAMINOETHYL ESTER, 1,2-ETHANE DISULFONATE see CBG250

1-PHENYLCYCLOPENTANECARBOXYLIC ACID-1-METHYL-4-PIPERIDINYL ESTER HYDROCHLORIDE see GAC000

1-PHENYLCYCLOPENTANECARBOXYLIC ACID-1-METHYL-4-PIPERIDYL ESTER HYDROCHLORIDE see GAC000

1-PHENYL-1-CYCLOPENTYL-3-PIPERIDINO-1-PROPANOL HYDROCHLORIDE see CQH500

PHENYLCYCLOPROMINE SULFATE. see PET500

trans-2-PHENYLCYCLOPROPYLAMINE see PET750

α-PHENYLDEOXYBENZOIN see TMS100

1-PHENYL-3-(β-DIAETHYLAMINO-AETHYL)-5-FURYL-PYRAZOLIN-HYDROCHLORID (GERMAN) see DIA400

1-PHENYL-3,3-DIAETHYLTRIAZEN (GERMAN) see PEU500

PHENYLDIALLYLACETAMIDE see AGR100

2-PHENYLDIAZENECARBOXAMIDE see CBL000

PHENYLDIAZENECARBOXYLIC ACID 2-PHENYLHYDRAZIDE see DVY950

PHENYL DIAZOMETHYL KETONE see PET800

PHENYLDIAZONIUM FLUOROBORATE (SALT) see BBO325

PHENYLDIAZONIUM TETRAFLUOROBORATE see BBO325

PHENYLDIBROMOARSINE see DDR200

3-PHENYL-5-(DIBUTYLAMINOETHYLAMINO)-1,2,4-OXADIAZOLE HYDROCHLORIDE see PEU000

PHENYL DICHLORARSINE see DGB600

PHENYLDICHLOROARSINE see DGB600

PHENYL DICHLOROPHOSPHATE see PHE800

PHENYLDICHLOROPHOSPHINE see DGE400

PHENYL DICHLOROPHOSPHITE see PHE850

PHENYL-5,6-DICHLORO-2-TRIFLUOROMETHYL-BENZIMIDAZOLE-1-CARBOXYLATE see DGA200

PHENYL DIETHANOLAMINE see BKD500

N-PHENYLDIETHANOLAMINE see BKD500

PHENYL-N,N-DIETHYLACETAMIDE see PEU100

β-PHENYL-o-(DIETHYLAMINOETHOXY)PROPIOPHENONE HYDROCHLORIDE see DHS200

PHENYLDIETHYLAMINO-1'-AMINOACETIC ACID, ISOPENTYL ESTER see NOC000

3-PHENYL-3-DIETHYLAMINOETHYLBENZOFURANONE-2-HYDROCHLORIDE see DIF200

2-PHENYL-9-DIETHYLAMINOETHYL-9H-IMIDAZO(1,2-A)BENZIMIDAZOLE DIHYDROCHLORIDE see DIF300

3-PHENYL-5-(β-(DIETHYLAMINO)ETHYL)-1,2,4-OXADIAZOLE CITRATE see OOE000

3-PHENYL-5-(β-DIETHYLAMINOETHYL)-1,2,4-OXADIAZOLE HYDROCHLORIDE see OOE100

3-PHENYL-5-(β-DIETHYLAMINOETHYL)-1,2,4-OXODIAZOLE see OOC000

1-PHENYL-2-DIETHYLAMINO-1-PROPANONE see DIP400

1-PHENYL-2-DIETHYLAMINOPROPANONE-1-HYDROCHLORIDE see DIP600

1-PHENYL-2-DIETHYLAMINO-1-PROPANONE HYDROCHLORIDE see DIP600

2-PHENYL-3-DIETHYLPYRROLIDINOETHOXY-6-METHOXYBENZOFURAN HYDROCHLORIDE see DJY100

1-PHENYL-3-(O,O-DIETHYL-THIONOPHOSPHORYL)-1,2,4-TRIAZOLE see THT750

1-PHENYL-3,3-DIETHYLTRIAZENE see PEU500

PHENYLDIFLUOROARSINE see DKI400

2-PHENYL-4,5-DIHYDROIMIDAZO(1,2-A)QUINOLINE see PEU525

2-(p-(2-PHENYL-3,4-DIHYDRO-1-NAPHTHYL)PHENOXY)TRIETHYLAMINE see PEU545

2-PHENYL-5,6-DIHYDROPYRAZOLO(5,1-a)ISOQUINOLINE see PEU600

2-PHENYL-5,6-DIHYDRO-s-TRIAZOLO(5,1-a)ISOQUINOLINE see PEU650

2-PHENYL-3,5-DIHYDROXY-4-BUTYLPYRAZOLIDINE see MQY400

2-PHENYL-1,3-DIKETOHYDRINDENE see PFJ750

1-PHENYL-2-DIMETHYLAMINO-PROPAN see TMA600

1-PHENYL-2-DIMETHYLAMINOPROPANE see TMA600

α-PHENYL-β-DIMETHYL AMINO PROPANE see TMA600

1-PHENYL-2-DIMETHYLAMINOPROPANOL see MJU750

2-PHENYL-4-DIMETHYLAMINOPROPYLAMINO-6-ETHOXY-3(2H)-PYRIDAZINONE HYDROCHLORIDE see TGH690

PHENYLDIMETHYLCARBINOL see DTN100

1-PHENYL-2,3-DIMETHYL-4-DIMETHYLAMINOPYRAZOLONE-5 see DOT000

1-PHENYL-2,3-DIMETHYL-4-DIMETHYLAMINOPYRAZOL-5-ONE see DOT000

2-PHENYL-1,1-DIMETHYLETHYLHYDROPEROXIDE see MNX000

1-PHENYL-2,3-DIMETHYL-4-(ISOPROPYLAMINO)-2-PYRAZOLIN-5-ONE see INM000

1-PHENYL-2,3-DIMETHYL-4-ISOPROPYLAMINOPYRAZOLONE see INM000

1-PHENYL-2,3-DIMETHYL-4-ISOPROPYL-3-PYRAZOLIN-5-ONE see INY000

1-PHENYL-2,3-DIMETHYL-4-ISOPROPYLPYRAZOL-5-ONE see INY000

d-2-PHENYL-3,4-DIMETHYLMORPHOLINE HYDROCHLORIDE see DTN800

O-PHENYL-N,N'-DIMETHYL PHOSPHORODIAMIDATE see PEV500

1-PHENYL-2,3-DIMETHYLPYRAZOLE-5-ONE see AQN000

1-PHENYL-2,3-DIMETHYL-5-PYRAZOLONE see AQN000

1-PHENYL-2,3-DIMETHYL-5-PYRAZOLONE-4-METHYLAMINOMETHANESULFONATE SODIUM see AMK500

1-PHENYL-2,3-DIMETHYLPYRAZOLONE-(5)-4-METHYLAMINOMETHANESULFONICACID SODIUM see AMK500

PHENYL DIMETHYL PYRAZOLON METHYL AMINOMETHANE SODIUM SULFONATE see AMK500

2-PHENYL-5-DIMETHYLTETRAHYDRO-1,4-OXAZINE see DTN775

2-PHENYL-5,5-DIMETHYL-TETRAHYDRO-1,4-OXAZINE HYDROCHLORIDE see PEV600

2-PHENYL-5,5-DIMETHYL-TETRAHYDRO-1,4-OXAZIN HYDROCHLORID (GERMAN) see PEV600

1-PHENYL-3,3-DIMETHYLTRIAZENE see DTP000

PHENYLDIMETHYLTRIAZINE see DTP000

1-PHENYL-3,3-DIMETHYLUREA see DTP400

3-PHENYL-1,1-DIMETHYLUREA see DTP400

N-PHENYL-N',N'-DIMETHYLUREA see DTP400

3-PHENYL-1,1-DIMETHYLUREA, TRICHLOROACETATE see FAR050

2-PHENYL-m-DIOXANE-5,5-DIMETHANOL see MRF200

2-PHENYL-m-DIOXAN-5-OL see BBA000

4-PHENYLDIPHENYL see TBC750

1-PHENYL-2-(1',1'-DIPHENYLPROPYL-3'-AMINO)PROPANE see PEV750

4-PHENYL-1,2-DIPHENYL-3,5-PYRAZOLIDINEDIONE see PEW000

PHENYL DISELENIDE see BLF500

PHENYL DISULFIDE see PEW250

PHENYLDITHIOCARBAMIC ACID, AMMONIUM SALT see ANR250

PHENYLDODECAN (GERMAN) see PEW500

1-PHENYLDODECANE see PEW500

PHENYL-DRANE see SPC500

PHENYLEN see PFJ750

3,3'-(1,4-PHENYLENEBIS(CARBONYLIMINO-4,1-PHENYLENECARBONYLIMINO))BIS(1-ETHYLPYRIDINIUM) see PEW550

(p-PHENYLENEBIS(CARBONYLMETHYL))BIS(DIMETHYL-2-HYDROXYETHYLAMMONIUM)DIBROMIDE see PEW600

1,1'-(p-PHENYLENEBIS(CARBONYLMETHYL))DI-3-PICOLINIUM DIBROMIDE see PEW650

PHENYLENE-p,p'-BIS(2-DIMETHYLBENZYLAMMONIUMPROPYL) DICHLORIDE see PEW700

p-PHENYLENEBIS(DIMETHYLSILANE) see PEW725

1,4-PHENYLENEBIS(DIMETHYLSILANE) see PEW725

m-PHENYLENEBIS(DIPHENYL PHOSPHATE) see REA050

(1,2-PHENYLENEBIS(IMINOCARBONOTHIOYL))BISCARBAMIC ACID DIETHYL ESTER see DJV000

(1,2-PHENYLENEBIS(IMINOCARBONOTHIOYL))BIS-CARBAMIC ACID, DIMETHYL ESTER, mixed with 5-ETHOXY-3-(TRICHLOROMETHYL)-1,2,4-THIADIAZOLE see PEW750

N,N'-(m-PHENYLENE)BISMALEIMIDE see BKL750

m-PHENYLENEBIS(METHYLAMINE) see XHS800

p-PHENYLENEBIS(METHYLAMINE) see PEX250

N,N'-1,4-PHENYLENEBIS(4-METHYLBENZENESULFONAMIDE) see PEX275

N,N'-(1,4-PHENYLENEBIS(METHYLENE))BIS(2,2-DICHLORO-N-ETHYL-ACETAMIDE (9CI) see PFA600

m-PHENYLENEBIS(1-METHYLETHYLENE))BIS(BENZYLDIMETHYLAMMONIUM CHLORIDE) see PEX300

(m-PHENYLENEBIS(1-METHYLETHYLENE)BIS(BENZYLDIMETHYLAMMONIUM) DICHLORIDE see PEX300

p-PHENYLENEBIS(1-METHYLETHYLENE)BIS(BENZYLDIMETHYLAMMONIUM CHLORIDE) see PEW700

(p-PHENYLENEBIS(1-METHYLETHYLENE))BIS(BENZYLDIMETHYLAMMONIUM) DICHLORIDE see PEW700

m-PHENYLENEBIS(1-METHYLETHYLENE)BIS(DIMETHYLETHYLAMMONIUM BROMIDE) see PEX325

(m-PHENYLENEBIS(1-METHYLETHYLENE))BIS(DIMETHYLETHYLAMMONIUM) DIBROMIDE see PEX325

p-PHENYLENEBIS(1-METHYLETHYLENE)BIS(DIMETHYLETHYLAMMONIUM) BROMIDE see PEX330

(p-PHENYLENEBIS(1-METHYLETHYLENE))BIS(DIMETHYLETHYLAMMONIUM) DIBROMIDE see PEX330

m-PHENYLENEBIS(1-METHYLETHYLENE)BIS(TRIMETHYLAMMONIUM BROMIDE) see PEX350

p-PHENYLENEBIS(1-METHYLETHYLENE)BIS(TRIMETHYLAMMONIUM BROMIDE) see PEX355

(m-PHENYLENEBIS(1-METHYLETHYLENE))BIS(TRIMETHYLAMMONIUM) DIBROMIDE see PEX350

(p-PHENYLENEBIS(1-METHYLETHYLENE))BIS(TRIMETHYLAMMONIUM) DIBROMIDE see PEX355

4,4'-(1,4-PHENYLENE-BIS(1-METHYLETHYLIDENE))BISANILINE see BGV800

1,1'-(p-PHENYLENEBIS(OXYETHYLENE))DIHYDRAZINE DIHYDROCHLORIDE see HGL600

2,2'-(1,3-PHENYLENEBIS(OXYMETHYLENE))BISOXIRANE see REF000

1,1'-(m-PHENYLENE)BIS-1H-PYROLE-2,5-DIONE (9CI) see BKL750

4,4'-o-PHENYLENEBIS(3-THIOALLOPHANIC ACID)DIMETHYL ESTER see PEX500

PHENYLENE-m,m'-BIS(2-TRIMETHYLAMMONIUMPROPYL)DIBROMIDE see PEX350

PHENYLENE-p,p'-BIS(2-TRIMETHYLAMMONIUMPROPYL)DIBROMIDE see PEX355

1,1'-(p-PHENYLENEBIS(VINYLENE-p-PHENYLENE))BIS(PYRIDINIUM) DI-p-TOLUENESULFONATE see PEX750

1,3-PHENYLENE CYANATE see REA200

m-PHENYLENEDIACETATE see REA100

m-PHENYLENEDIAMINE see PEY000

o-PHENYLENEDIAMINE see PEY250

p-PHENYLENEDIAMINE see PEY500

1,3-PHENYLENEDIAMINE see PEY000

1,4-PHENYLENEDIAMINE see PEY500

m-PHENYLENEDIAMINE (DOT) see PEY000

1,2-PHENYLENEDIAMINE (DOT) see PEY250

p-PHENYLENEDIAMINE, N,N'-DICYCLOHEXYL- see BBN100

p-PHENYLENEDIAMINE, N,N-DIETHYL-, SULFATE (1:1) see DJV250

o-PHENYLENEDIAMINE DIHYDROCHLORIDE see PEY600

p-PHENYLENEDIAMINE DIHYDROCHLORIDE see PEY650

1,3-PHENYLENEDIAMINE DIHYDROCHLORIDE see PEY750

1,4-PHENYLENEDIAMINE DIHYDROCHLORIDE see PEY650

p-PHENYLENEDIAMINE, N-(1,3-DIMETHYLBUTYL)-N'-PHENYL- see DQV250

p-PHENYLENEDIAMINE, N-(1,4-DIMETHYLPENTYL)-N'-PHENYL- see DTI800

m-PHENYLENEDIAMINE, 4-ETHOXY- see EFC300

m-PHENYLENEDIAMINE, 4-(p-ETHOXYPHENYLAZO)- see EEQ600

m-PHENYLENEDIAMINE HYDROCHLORIDE see PEY750

p-PHENYLENEDIAMINE HYDROCHLORIDE see PEY650

PHENYLENEDIAMINE, META, solid (DOT) see PEY000

p-PHENYLENEDIAMINE, 2-METHOXY- see MEB800

m-PHENYLENEDIAMINE, 4-METHOXY-, DIHYDROCHLORIDE see DBO100

p-PHENYLENEDIAMINE, 2-METHOXY-, SULFATE see MEB820

m-PHENYLENEDIAMINE, 1-METHOXY-, SULFATE, HYDRATE see MEB900

m-PHENYLENEDIAMINE, 2-METHYL-4-((2-METHYLPHENYL)AZO)- see MIX100

m-PHENYLENEDIAMINE, 4-METHYL-6-((2-METHYLPHENYL)AZO)- see MLP770

m-PHENYLENEDIAMINE, 2-METHYL-4-(PHENYLAZO)- see MNS100

m-PHENYLENEDIAMINE, 4-METHYL-6-(PHENYLAZO)- see CMM800

m-PHENYLENEDIAMINE, 4-NITRO- see NIM550

o-PHENYLENEDIAMINE, 3-NITRO- see NIM525

p-PHENYLENEDIAMINE, 2-NITRO-N¹)-PHENYL- see NEM350

PHENYLENEDIAMINE, PARA, solid (DOT) see PEY500

m-PHENYLENEDIAMINESULFONIC ACID see PFA250

p-PHENYLENEDIAMINESULFONIC ACID see PEY800

m-PHENYLENEDIAMINE-4-SULFONIC ACID see PFA250

1,3-PHENYLENEDIAMINE-4-SULFONIC ACID see PFA250

1,3-PHENYLENE-DI-4-AMINOPHENYL ETHER see REF070

p-PHENYLENE DIAZIDE see DCL125

1,1'-(1,3-PHENYLENEDICARBONYL)BIS(2-METHYLAZIRIDINE) see BLG400

p-PHENYLENEDICARBONYL DICHLORIDE see TAV250

PHENYLENE-m,m'-DI-(2-DIMETHYLBENZYLAMMONIUMPROPYL)DICHLORIDE see PEX300

PHENYLENE-m,m'-DI(2-DIMETHYLETHYLAMMONIUMPROPYL)DIBROMIDE see PEX325

PHENYLENE-p,p'-DI(2-DIMETHYLETHYLAMMONIUMPROPYL)DIBROMIDE see PEX330

m-PHENYLENE DIISOCYANATE see BBP000

p-PHENYLENE DIISOCYANATE see PFA300

1,4-PHENYLENE DIISOCYANATE see PFA300

(PHENYLENEDIISOPROPYLIDENE)BIS(tert-BUTYLPEROXIDE) see BHL100

PHENYLENE-1,4-DIISOTHIOCYANATE see PFA500

1,4-PHENYLENEDIISOTHIOCYANIC ACID see PFA500

N,N'-(m-PHENYLENEDIMALEIMIDE) see BKL750

N,N'-(p-PHENYLENEDIMETHYLENE)BIS(2,2-DICHLORO-N-ETHYLACETAMIDE) see PFA600

m-PHENYLENEDIMETHYLENE ISOCYANATE see XIJ000

o-PHENYLENEDIOL see CCP850

4,4'-(m-PHENYLENEDIOXY)DIANILINE see REF070

m-PHENYLENE ISOCYANATE see BBP000

p-PHENYLENE ISOCYANATE see PFA300

2,3-PHENYLENEPYRENE see IBZ000

2,3-o-PHENYLENEPYRENE see IBZ000

1,10-(o-PHENYLENE)PYRENE see IBZ000

1,10-(1,2-PHENYLENE)PYRENE see IBZ000

1,3-PHENYLENE TETRAPHENYL PHOSPHATE see REA050

PHENYLENE THIOCYANATE see PFA500

o-PHENYLENETHIOUREA see BCC500

PHENYLEPHRINE see NCL500

(−)-PHENYLEPHRINE see NCL500

R(−)-PHENYLEPHRINE see NCL500

PHENYLEPHRINE HYDROCHLORIDE see SPC500

d-(−)-PHENYLEPHRINE HYDROCHLORIDE see SPC500

1-PHENYL-1,2-EPOXYETHANE see EBR000

(1R,2R)-(+)-1-PHENYL-1,2-EPOXYPROPANE see ECH600

PHENYL-2,3-EPOXYPROPYL ETHER see PFF360

PHENYLESSIGSAEURE NATRIUM-SALZ see SFA200

PHENYLETHANAL see BBL500

PHENYLETHANE see EGP500

PHENYLETHANEDIOL see SMQ100

PHENYLETHANOIC ACID BUTYL ESTER see BQJ350

1-PHENYLETHANOL see PDE000

2-PHENYLETHANOL see PDD750

β-PHENYLETHANOL see PDD750

PHENYL ETHANOLAMINE see AOR750

N-PHENYLETHANOLAMINE see AOR750

PHENYLETHANOL AMINE HYDROCHLORIDE see PFA750

1-PHENYLETHANONE see ABH000

PHENYLETHENE see SMQ000

4-(2-PHENYLETHENYL)BENZENAMINE see SLQ900

4-(2-PHENYLETHENYL)BENZENAMINE,(E) see AMO000

N-(4-(2-PHENYLETHENYL)PHENYL)ACETAMIDE see SMR500

1-((2-PHENYLETHENYL)PHENYL)ETHANONE see MPL000

2-(4-(2-PHENYLETHENYL)-3-SULFOPHENYL)-2H-NAPHTHO(1,2-d)TRIAZOLE SODIUM SALT see TGE155

PHENYL ETHER see PFA850

PHENYL ETHER-BIPHENYL MIXTURE see PFA860

1-PHENYL ETHER-2,3-DIACETATE GLYCEROL see PFF300

PHENYL ETHER, DICHLORO see DFE800

PHENYL ETHER, HEXACHLORO derivative (8CI) see CDV175

PHENYL ETHER MONO-CHLORO see MRG000

PHENYL ETHER, OCTABROMO DERIV. see OAF200

PHENYL ETHER PENTACHLORO see PAW250

PHENYL ETHER TETRACHLORO see TBP250

cis-2-(1-PHENYLETHOXY)CARBONYL-1-METHYLVINYL DIMETHYLPHOSPHATE see COD000

1-(2-PHENYL-2-ETHOXYETHYL)-4-(2-BENZYLOXYPROPYL)PIPERAZINE see PFB000

1-(2-PHENYL-2-ETHOXY)ETHYL-4-(2-BENZYLOXY)PROPYLPIPERAZINE DIHYDROCHLORIDE see ECU550

1-PHENYLETHYL ACETATE see SMP600

2-PHENYLETHYL ACETATE see PFB250

sec-PHENYLETHYL ACETATE see SMP600

PHENYLGLYOXYLIC ACID see OOK150

PHENYLGLYOXYLIC ACID see OOK150

PHENYLGLYOXYLONITRILE OXIME-O,O-DIETHYL PHOSPHOROTHIOATE see BAT750

PHENYLGOLD see PFH275

N-PHENYLHEXANAMIDE see HEV600

2-PHENYLHYDRACRYLIC ACID-3-α-TROPANYL ESTER see ARR000

PHENYL HYDRATE see PDN750

2-PHENYLHYDRAZIDEBENZENECARBOXIMIDIC ACID MONOHYDROCHLORIDE see PEK675

2-PHENYLHYDRAZIDE, CARBAMIC ACID see CBL000

PHENYLHYDRAZIN (GERMAN) see PFI000

PHENYLHYDRAZINE see PFI000

2-PHENYLHYDRAZINECARBOTHIOAMIDE see PGM750

1-PHENYLHYDRAZINE CARBOXAMIDE see CBL000

2-PHENYLHYDRAZINECARBOXAMIDE see CBL000

PHENYLHYDRAZINE HYDROCHLORIDE see PFI250

PHENYLHYDRAZINE MONOHYDROCHLORIDE see PFI250

PHENYLHYDRAZINE-p-SULFONIC ACID see PFI500

PHENYLHYDRAZIN HYDROCHLORID (GERMAN) see PFI250

PHENYLHYDRAZINIUM CHLORIDE see PFI250

1-PHENYL-2-HYDRAZINOPROPANE see PDN000

PHENYL HYDRIDE see BBL250

PHENYLHYDROQUINONE see BGG250

2-PHENYLHYDROQUINONE see BGG250

o-PHENYLHYDROQUINONE see BGG250

PHENYLHYDROXAMIC ACID see BCL500

PHENYL HYDROXIDE see PDN750

PHENYLHYDROXYACETIC ACID see MAP000

threo-1-PHENYL-1-HYDROXY-2-AMINOPROPANE see NNM510

o-(PHENYLHYDROXYARSINO)BENZOIC ACID see PFI600

1-PHENYL-2-(β-HYDROXYETHYL)AMINOPROPANE see PFI750

2-PHENYL-2-HYDROXYETHYL CARBAMATE see PFJ000

2-PHENYL-2-HYDROXYETHYL, m-CHLOROPHENYL ARSINIC ACID see CKA575

N-PHENYLHYDROXYLAMINE see PFJ250

β-PHENYLHYDROXYLAMINE see PFJ250

PHENYLHYDROXYLAMINIUM CHLORIDE see PFJ275

PHENYL HYDROXYMERCURY see PFN100

1-PHENYL-2-HYDROXY-2-METHYL-1-PROPANONE see HMQ100

1-PHENYL-1-HYDROXYPENTANE see BQJ500

1-PHENYL-2-(p-HYDROXYPHENYL)-3,5-DIOXO-4-BUTYLPYRAZOLIDINE see HNI500

PHENYLIC ACID see PDN750

PHENYLIC ALCOHOL see PDN750

PHENYL-IDIUM see PDC250

PHENYL-IDIUM 200 see PDC250

2-PHENYL-2-IMIDAZOLINE see PFJ300

PHENYLIMIDOCARBONYL CHLORIDE see PFJ400

PHENYLIMIDOCARBONYL CHLORIDE see PFJ400

N-PHENYL-IMIDODICARBONIMIDIC DIAMIDE MONOHYDROCHLORIDE see PEO000

N-PHENYLIMIDOPHOSGENE see PFJ400

PHENYLIMINOCARBONYL DICHLORIDE see PFJ400

N-PHENYLIMINOCARBONYL DICHLORIDE see PFJ400

2,2'-(PHENYLIMINO)DIETHANOL see BKD500

5-PHENYL-2-IMINO-4-OXAZOLIDINONE see IBM000

5-PHENYL-2-IMINO-4-OXOOXAZOLIDINE see IBM000

N-(3-PHENYLIMINO-1-PROPENYL)ANILINE HYDROCHLORIDE see PFJ500

2-PHENYLINDAN-1,3-DIONE see PFJ750

2-PHENYL-1,3-INDANDIONE see PFJ750

PHENYLINDIONE see PFJ750

2-PHENYLINDOLE see PFJ760

α-PHENYLINDOLE see PFJ760

PHENYL IODIDE see IEC500

PHENYLIODINE(III) CHROMATE see PFJ775

PHENYLIODINE(III) NITRATE see PFJ780

9-PHENYL-9-IODOFLUORENE see PFK000

PHENYL ISOCYANATE see PFK250

PHENYL ISOCYANIDE, m-NITRO- see NIR200

PHENYL ISOCYANIDE, p-NITRO- see IKH780

PHENYL ISOCYANIDE, p-(PHENYLAZO)- see PEK050

PHENYL ISOHYDANTOIN see IBM000

PHENYLISONITRILE DICHLORIDE see PFJ400

2-PHENYLISOPROPANOL see DTN100

β-PHENYLISOPROPYLAMIN (GERMAN) see AOA250

(PHENYLISOPROPYL)AMINE see AOA250

β-PHENYLISOPROPYLAMINE see AOA250

dl-β-PHENYLISOPROPYLAMINE HYDROCHLORIDE see AOA750

β-PHENYL ISOPROPYLAMINE SULFATE see AOB250

d-β-PHENYLISOPROPYLAMINE SULFATE see BBK500

7-(PHENYL-ISOPROPYL-AMINO-AETHYL)-THEOPHYLLIN-HYDROCHLORID (GERMAN) see CBF825

N-PHENYL ISOPROPYL CARBAMATE see CBM000

PHENYLISOPROPYLHYDRAZINE see PDN000

β-PHENYLISOPROPYLHYDRAZINE see PDN000

PHENYLISOPROPYLHYDRAZINE HYDROCHLORIDE see PDN250

d-PHENYLISOPROPYLMETHYLAMINE see PFP850

(−)-PHENYLISOPROPYLMETHYLPROPYNYLAMINE see DAZ125

(+)-PHENYLISOPROPYLMETHYLPROPYNYLAMINE HYDROCHLORIDE see DAZ120

(±)-PHENYLISOPROPYLMETHYLPROPYNYLAMINE HYDROCHLORIDE see DAZ118

N-PHENYL-N'-ISOPROPYL-p-PHENYLENEDIAMINE see PFL000

3-(β-PHENYLISOPROPYL)-SIDNONIMINE HYDROCHLORIDE see SPA000

PHENYL ISOTHIOCYANATE see ISQ000

3-PHENYL-2-ISOXAZOLINE-5-PHOSPHONIC ACID see PFL100

α-(5-PHENYL-3-ISOXAZOLYL)-1-PIPERIDINEETHANOL CITRATE (2:1) see PCJ325

PHENYLIUM see PDR100

PHENYL KETONE see BCS250

PHENYLLEAD TRILAURATE see PFL200

PHENYLLIN see PFJ750

PHENYLLITHIUM see PFL500

PHENYLMAGNESIUM BROMIDE see PFL600

N-PHENYLMALEIMIDE see PFL750

α-PHENYLMANDELIC ACID-N,N-DIMETHYLPIPERIDINIUM-2-METHYL ESTER METHYLSULFATE see CDG250

PHENYL MERCAPTAN see PFL850

PHENYLMERCAPTOACETONITRILE see COP755

1-PHENYL-5-MERCAPTOTETRAZOLE see PGJ750

1-PHENYL-5-MERCAPTO-1,2,3,4-TETRAZOLE see PGJ750

PHENYLMERCURIACETATE see ABU500

PHENYL MERCURIC ACETATE see ABU500

PHENYLMERCURIC BROMIDE see PFM250

PHENYLMERCURIC CHLORIDE see PFM500

PHENYLMERCURIC DINAPHTHYLMETHANEDISULFONATE see PFN000

PHENYL MERCURIC FIXTAN see PFN000

PHENYLMERCURIC HYDROXIDE see PFN100

PHENYLMERCURIC HYDROXIDE (DOT) see PFN100

PHENYLMERCURIC-8-HYDROXYQUINOLINATE see PFN250

PHENYLMERCURIC 3,3'-METHYLENEBIS(2-NAPHTHALENESULFONATE) see PFN000

PHENYLMERCURIC NITRATE see MCU750

PHENYLMERCURIC NITRATE, BASIC see MDH500

PHENYLMERCURIC OLEATE see PFP100

PHENYLMERCURIC TRIETHANOLAMMONIUM LACTATE see TNI500

PHENYLMERCURIC UREA see PFP500

N-(PHENYLMERCURI)-1,4,5,6,7,7-HEXACHLOROBICYCLO(2.2.1)HEPTENE-5-DICARBOXIMIDE see PFN500

PHENYLMERCURILAURYL THIOETHER see PFN750

8-PHENYLMERCURIOXY-QUIN-OLINE see PFN250

PHENYLMERCURI PROPIONATE see PFO000

PHENYLMERCURIPYROCATECHIN see PFO250

PHENYLMERCURITRIETHANOLAMMONIUM LACTATE see TNI500

PHENYLMERCURIUREA see PFP500

PHENYLMERCURY ACETATE see ABU500

PHENYLMERCURY ACETATE 95% plus ETHYLMERCURY CHLORIDE 5% see PFO500

PHENYLMERCURY AMMONIUM ACETATE see PFO550

PHENYLMERCURY BROMIDE see PFM250

PHENYLMERCURY CATECHOLATE see PFO750

PHENYLMERCURY CHLORIDE see PFM500

PHENYLMERCURY HYDROXIDE see PFN100

PHENYLMERCURY METHYLENEDINAPHTHALENESULFONATE see PFN000

PHENYLMERCURY NAPHTHENATE see PFP000

PHENYLMERCURY NITRATE see MCU750

PHENYLMERCURY OLEATE see PFP100

PHENYLMERCURY OXYQUINOLATE see PFN250

PHENYLMERCURY PROPIONATE see PFO000

PHENYLMERCURY SILVER BORATE see PFP250

PHENYLMERCURY TRIETHANOLAMINE LACTATE see TNI500

PHENYLMERCURY UREA see PFP500

PHENYLMETHACRYLATE see PFP600

PHENYLMETHANAL see BBM500
PHENYLMETHANE see TGK750
PHENYLMETHANESULFONYL FLUORIDE see TGO300
PHENYLMETHANETHIOL see TGO750
PHENYLMETHANOL see BDX500
N^2-((PHENYLMETHOXY)CARBONYL)-l-GLUTAMINYL-l-TRYPTOPHYL-l-SERYL-l-TYROSYL-d-METHIONYL-l-LEUCYL-l-ARGINYL-N-ETHYL-l-PROLINAMIDE, HYDROCHLORIDE HYDRATE (1:1:2) see PFP700
N^2-((PHENYLMETHOXY)CARBONYL)-l-GLUTAMINYL-l-TRYPTOPHYL-l-SERYL-l-TYROSYL-d-PROLYL-l-LEUCYL-l-ARGINYL-N-ETHYL-l-PROLINAMIDE, ACETATE, HYDRATE (2:7:5) see PFP720
N^2-((PHENYLMETHOXY) see PFP730
N^2-((PHENYLMETHOXY)CARBONYL)-l-GLUTAMINYL-l-TRYPTOPHYL-l-SERYL-l-TYROSYL-THREO see PFP740
N-((PHENYLMETHOXY)CARBONYL)-l-PHENYLALANINE ETHENYL ESTER see CBR235
N-((PHENYLMETHOXY)CARBONYL see PFP760
N-((PHENYLMETHOXY)CARBONYL)-o-((PHENYLMETHYL)-l-SERYL-l-TRYPTOPHYL-l-SERYL-l-TYROSYL see PFP780
N-((PHENYLMETHOXY)CARBONYL)-o-(PHENYLMETHYL)-l-SERYL-l-TRYPTOPHYL-l-SERYL-l-TYROSYL see PFP800
N-((PHENYLMETHOXY)CARBONYL)-o-(PHENYLMETHYL)-l-SERYL-l-TRYPTOPHYL-l-SERYL-l-TYROSYL see PFP820
N-((PHENYLMETHOXY)CARBONYL)-o-(PHENYLMETHYL)-l-SERYL-l-TRYPTOPHYL-l-SERYL-l-TYROSYL see PFP840
N-((PHENYLMETHOXY)CARBONYL)-l-TRYPTOPHYL-l-METHIONYL-l-ASPARTYL-l-PHENYLALANINAMIDE see GCE200
1-PHENYL-2-METHOXYDIAZENE see MGR800
1-PHENYL-2-METHOXYDIAZENE see MHB000
1-(2-PHENYL-2-METHOXY)ETHYL-4-(3-PHENYL-3-HYDROXY)PROPYL PIPERAZINE HYDROCHLORIDE see ECU600
4-(PHENYLMETHOXY)PHENOL see AEY000
PHENYLMETHYLACETAMIDE see MFW000
1-PHENYL-2-METHYL-3-(((2-ACETOXYBENZOYL)AMINO)METHYL)PYRAZOLONE see PNR800
PHENYLMETHYL ALCOHOL see BDX500
(PHENYLMETHYL)AMINE see BDX750
N-PHENYLMETHYLAMINE see MGN750
1-PHENYL-2-METHYLAMINE-PROPANOL-1-SULFATE see EAY500
1-PHENYL-1-METHYL-2-AMINO-AETHAN see PGB760
1-PHENYL-1-METHYL-2-AMINOETHANE see PGB760
d-1-PHENYL-2-METHYLAMINOPROPAN see PFP850
1-PHENYL-2-METHYLAMINO-PROPAN (GERMAN) see DBB000
1-PHENYL-2-METHYLAMINOPROPANE see DBB000
d-1-PHENYL-2-METHYLAMINOPROPANE see PFP850
α-PHENYL-β-METHYLAMINOPROPANE see DBB000
1-PHENYL-2-METHYLAMINOPROPANOL see EAW000
1-PHENYL-2-METHYLAMINOPROPANOL-1 see EAX500

1-PHENYL-3-METHYL-5-AMINOPYRAZOLE see PFQ350
3-PHENYL-10-METHYL-7:8 BENZACRIDINE (FRENCH) see MNS250
2-(PHENYLMETHYL)-1H-BENZIMIDAZOLE see BEA825
1-(1-(PHENYLMETHYL)BUTYL)PYRROLIDINE HYDROCHLORIDE see PNS000
PHENYL METHYLCARBAMATE see PFS350
PHENYL N-METHYLCARBAMATE see PFS350
PHENYLMETHYLCARBINOL see PDE000
PHENYLMETHYLCARBINYL PROPIONATE see PFR000
1-PHENYL-5-METHYL-8-CHLORO-1,2,4,5-TETRAHYDRO-2,4-DIOXO-3H-1,5-BENZODIAZEPINE see CIR750
PHENYLMETHYL 2-CHLORO-4-(TRIFLUOROMETHYL)-5-THIAZOLECARBOXYLATE see BEG300
1-PHENYL-2-(METHYL-(β-CYANOETHYL)AMINO)PROPAN-1-OL see CCX600
1-(PHENYLMETHYL)-2(1H)-CYCLOHEPTIMIDAZOLONE see BEH000
PHENYLMETHYLCYCLOSILOXANE, mixed copolymer see PFR100
PHENYLMETHYLDICHLOROSILANE see DFQ800
PHENYLMETHYLDIMETHOXYSILANE see DOH400
N-(PHENYLMETHYL)DIMETHYLAMINE see DQP800
S-(PHENYLMETHYL) (1,2-DIMETHYLPROPYL)ETHYLCARBAMOTHIOATE see PFR120
S-(PHENYLMETHYL) DIPROPYLCARBAMOTHIOATE see PFR130
α-(PHENYLMETHYLENE)BENZENEACETONITRILE see DVX600
2-(PHENYLMETHYLENE)HEXANAL see BQV250
2-(PHENYLMETHYLENE)OCTANOL see HFO500
PHENYLMETHYL ESTER THIOCYANIC ACID (9CI) see BFL000
PHENYL METHYL ETHER see AOX750
1-1-PHENYL-2-METHYLETHYLAMINOPROPAN-1-OL HYDROCHLORIDE see EJR500
1-(2-PHENYL-2-METHYL)ETHYL-4-(3-PHENYL-3-HYDROXY)-PROPYLPIPERAZINE DIHYDROCHLORIDE see ECU600
(5-(PHENYLMETHYL)-3-FURANYL)METHYL 4-CHLORO-α-ETHYLBENZENEACETATE see PFR150
(5-(PHENYLMETHYL)-3-FURANYL)METHYL α-ETHYLBENZENEACETATE see BEP600
(5-(PHENYLMETHYL)-3-FURANYL)METHYL-4-METHOXY-α-(1-METHYLETHYL)BENZENEACETATE see BEP650
1-PHENYL-3-METHYL-3-(2-HYDROXYAETHYL)-TRIAZEN (GERMAN) see HKV000
1-PHENYL-3-METHYL-3-(2-HYDROXYETHYL)TRIAZENE see HKV000
1-PHENYL-3-METHYL-3-HYDROXY-TRIAZEN (GERMAN) see HMP000
1-PHENYL-3-METHYL-3-HYDROXY-TRIAZENE see HMP000
PHENYLMETHYLIMIDAZOLINE see BBW750
((1-(PHENYLMETHYL)-1H-INDAZOL-3-YL)OXY)-ACETIC ACID (9CI) see BAV325
3-PHENYL-5-METHYLISOXAZOL-4-CARBONSAEURE (GERMAN) see MNV000
PHENYL METHYL KETONE see ABH000

PHENYLMETHYL MERCAPTAN see TGO750
PHENYLMETHYL METHYL KETONE see MHO100
PHENYLMETHYLNITROSAMINE see MMU250
PHENYL METHYLNITROSOCARBAMATE see NKV500
3-PHENYL-1-METHYL-1-NITROSOHARNSTOFF (GERMAN) see MMY500
4-PHENYL-2-METHYL-6H-1,3,5-OXATHIAZINE see MNV760
1-PHENYL-3-METHYL-5-OXO-2-PYRAZOLINE see NNT000
(PHENYLMETHYL) PENICILLINIC ACID see BDY669
1-PHENYL-3-METHYL-3-PENTANOL see PFR200
1-PHENYL-3-METHYL-3-PENTANYL ACETATE see PFR300
4-(PHENYLMETHYL)PHENOL CARBAMATE see PFR325
4-PHENYL-α-METHYLPHENYLACETATE-γ-PROPYLSULFONATE SODIUM SALT see PFR350
α-PHENYL-α-(1-METHYL-2-PHENYL)ETHYLAMINOACETONITRILE see AOB300
3-PHENYL-2-METHYL-PROPEN-2-OL-1 see MIO750
N-(PHENYLMETHYL)-1H-PURIN-6-AMINE see BDX090
1-PHENYL-3-METHYLPYRAZOLONE-5 see NNT000
1-PHENYL-3-METHYL-5-PYRAZOLONE see NNT000
1-PHENYL-3-METHYL-5-PYRAZOLYL-N,N-DIMETHYL CARBAMATE see PPQ625
2-(PHENYLMETHYL)PYRIDINE see BFG600
PHENYL p-METHYLSTYRYL KETONE see MIF762
1-PHENYL-3-METHYL-3-(2-SULFOAETHYL) NATRIUM SALZ (GERMAN) see PEJ250
1-PHENYL-3-METHYL-3-(2-SULFOETHYL)TRIAZENE, SODIUM SALT see PEJ250
PHENYL METHYL SULFONE see PFR375
PHENYLMETHYLSULFONYL FLUORIDE see TGO300
dl-2-PHENYL-3-METHYLTETRAHYDRO-1,4-OXAZINE see PMA750
2-PHENYL-3-METHYLTETRAHYDRO-1,4-OXAZINE HYDROCHLORIDE see MNV750
7-(PHENYLMETHYL)-3-THIA-7-AZABICYCLO(3.3.1)NONANE PERCHLORATE see BFK370
6-((PHENYLMETHYL)THIO)-1H-PURIN-2-AMINE (9CI) see BFL125
6-((PHENYLMETHYL)THIO)-1H-PURINE (9CI) see BDG325
1-PHENYL-3-METHYLTRIAZINE see MOA725
2-(PHENYLMETHYL)-4,4,6-TRIMETHYL-1,3-DIOXANE see PFR400
PHENYLMETHYLUREA see BFN125
PHENYLMETHYLVALERIANSAEURE-β-DIAETHYLAMINOAETHYLESTER-BROMMETHYLAT (GERMAN) see VBK000
PHENYLMONOCHLOROTOLYLETHANE see PFR600
1,1-PHENYLMONOCHLOROXYLYLETHANE see MRG060
α-PHENYL MONOGLYCERYL ETHER see GGA950
PHENYLMONOGLYCOL ETHER see PER000
PHENYLMONOMETHYLCARBAMATE see PFS350

1-PHENYL-3-MONOMETHYLTRIAZENE see MOA725
1-PHENYL-3-MONOMETHYLTRIAZENE see PFS500
N-PHENYLMONOTHIOSUCCINIMIDE see PFS600
PHENYL MORPHOLINE see PFS750
N-PHENYLMORPHOLINE see PFS750
PHENYL MUSTARD OIL see ISQ000
PHENYL-β-NAPHTHALAMINE see MFA250
2-PHENYL-1,8-NAPHTHALIMIDE see PEL650
2-PHENYL-4H-NAPHTHO(1,2-b)PYRAN-4-ONE see NBI100
3-PHENYL-1H-NAPHTHO(2,1-b)PYRAN-1-ONE see NAU525
PHENYLNAPHTHYLAMINE see PFT250
PHENYL-2-NAPHTHYLAMINE see PFT500
N-PHENYL-1-NAPHTHYLAMINE see PFT250
N-PHENYL-2-NAPHTHYLAMINE see PFT500
PHENYL-α-NAPHTHYLAMINE see PFT250
α-PHENYLNAPHTHYLAMINE see PFT250
PHENYL-β-NAPHTHYLAMINE see PFT500
N-PHENYL-α-NAPHTHYLAMINE see PFT250
N-PHENYL-β-NAPHTHYLAMINE see PFT500
PHENYLNITRAMINE see NEO510
N-PHENYLNITRAMINE see NEO510
4-PHENYL-NITROBENZENE see NFQ000
p-PHENYL-NITROBENZENE see NFQ000
N-PHENYL-p-NITROSOANILINE see NKB500
PHENYL-NITROSO-HANRSTOFF (GERMAN) see NLG500
4-PHENYLNITROSOPIPERIDINE see PFT600
N-PHENYL-N-NITROSOUREA see NLG500
4-PHENYL-6H-1,3,5-OXATHIAZINE see PFT700
PHENYLOXIRANE see EBR000
1-PHENYLOXIRANE see EBR000
2-PHENYLOXIRANE see EBR000
3-PHENYLOXIRANEMETHANOL (2R-trans)- see PFF350
3-PHENYLOXIRANEMETHANOL (2S-trans)- see ECE600
PHENYLOXOACETIC ACID see OOK150
1-PHENYL-4-OXO-8-(4,4-BIS(4-FLUOROPHENYL)BUTYL)-1,3,8-TRIAZASPIRO(4,5)DECANE see PFU250
1-PHENYL-3-OXOPYRAZOLIDINE see PDM500
1-PHENYL-2-β-OXY-AETHYL-AMINO-PROPAN (GERMAN) see PFI750
m-(PHENYLOXY)BENZALDEHYDE see PDR600
α-PHENYL-β-OXYETHYLAMINOPROPANE see PFI750
β-PHENYL-γ-OXYPROPIONSAEURE-TROPYL-ESTER (GERMAN) see ARR000
S-PHENYL PARATHION see DJS800
PHENYLPENTAMETHYLCYCLOTRISILOXANE see PBI700
PHENYLPENTANOL see BQJ500
1-PHENYLPENTANOL see BQJ500
PHENYL PERCHLORYL see HCC500
PHENYL PHENACYL KETONE see PFU300
2-PHENYLPHENOL see BGJ250
4-PHENYLPHENOL see BGJ500
o-PHENYLPHENOL see BGJ250
p-PHENYLPHENOL see BGJ500
2-PHENYLPHENOL SODIUM SALT see BGJ750
o-PHENYLPHENOL, SODIUM SALT see BGJ750
α-(4-PHENYLPHENOXY)PROPIONIC ACID see BGN100
α-PHENYLPHENYLACETONITRILE see DVX200

o-PHENYLPHENYL BISPHENYL PHOSPHATE see XFS300
o-PHENYLPHENYL DIPHENYL PHOSPHATE see XFS300
N-PHENYL-p-PHENYLENEDIAMINE see PFU500
N-PHENYL-N-(1-(2-PHENYLETHYL)-4-PIPERIDINYL)-PROPANAMIDE (9CI) see PDW500
p-PHENYLPHENYL GLYCIDYL ETHER see PFU600
N-PHENYL-N-(1-(PHENYLIMINO)ETHYL)-N'-2,5-DICHLOROPHENYLUREA see PFV000
α-PHENYL-α(β-PHENYLISOPROPYLAMINO)ACETONITRILE see AOB300
α-PHENYL-α-N-(1-PHENYLISOPROPYL)AMINOACETONITRILE see AOB300
4-PHENYL-1-(2-(PHENYLMETHXY)ETHYL)-4-PIPERIDINECARBOXYLIC ACID ETHYL ESTER see BBU625
(s)-3-PHENYL-1'-(PHENYLMETHYL)-(3,4'-BIPIPERIDINE)-2,6-DIONE HYDROCHLORIDE see DBE000
PHENYL PHOSPHATE ((PhO)₂(HO)PO) see PHE300
PHENYLPHOSPHINE see PFV250
PHENYLPHOSPHINE DICHLORIDE see DGE400
PHENYLPHOSPHINIC ACID see HNL200
PHENYLPHOSPHONIC ACID see PFV500
PHENYLPHOSPHONIC ACID DIOCTYL ESTER see PFV750
PHENYLPHOSPHONIC ACID ISOBUTYL 2-PROPYNYL ESTER see PFV760
PHENYLPHOSPHONIC AZIDE CHLORIDE see PFV775
PHENYLPHOSPHONIC DIAZIDE see PFW000
PHENYLPHOSPHONOCHLORIDOTHIOIC ACID O-ETHYL ESTER see EOM100
PHENYLPHOSPHONOTHIOIC ACID O-(4-BROMO-2,5-BROMO-2,5-DICHLOROPHENYL) O-METHYL ESTER see LEN000
PHENYLPHOSPHONOTHIOIC ACID, O-(2,4-DICHLOROPHENYL), O-ETHYL ESTER see SCC000
PHENYLPHOSPHONOTHIOIC ACID-o-ETHYL ESTER-o-ESTER with p-HYDROXYBENZONITRILE see CON300
PHENYLPHOSPHONOUS ACID see HNL200
PHENYLPHOSPHONOUS ACID DICHLORIDE see DGE400
PHENYLPHOSPHONOUS DICHLORIDE see DGE400
PHENYL PHOSPHONYL DICHLORIDE see PFW100
PHENYLPHOSPHORIC DICHLORIDE see PHE800
PHENYL PHOSPHORODICHLORIDITE see PHE850
PHENYL PHOSPHORUS DICHLORIDE see DGE400
PHENYL PHOSPHORUS DICHLORIDE (DOT) see DGE400
PHENYL PHOSPHORUS THIODICHLORIDE see PFW200
PHENYL PHOSPHORUS THIODICHLORIDE see PFW210
N-PHENYLPHTHALIMIDINE see PFW750
1-PHENYLPIPERAZINE see PFX000
N-PHENYLPIPERAZINE see PFX000
2-(3-(N-PHENYLPIPERAZINO)-PROPYL)-3-METHYLCHROMONE see PFX250
2-(3-(N-PHENYLPIPERAZINO)-PROPYL)-3-METHYL-7-FLUOROCHROMONE see PFX500

4-PHENYLPIPERAZINYL 3,4,5-TRIMETHOXYPHENYL KETONE see PFX600
α-PHENYL-2-PIPERIDINEACETIC ACID METHYL ESTER see MNQ000
4-PHENYL-4-PIPERIDINECARBOXALDEHYDE see PFY100
4-PHENYL-1-PIPERIDINECARBOXAMIDE see PFY105
α-PHENYL-2-PIPERIDINEMETHANOL ACETATE HYDROCHLORIDE see LFO000
1-PHENYL-3-(β-PIPERIDINO-AETHYL)-5-(α'-METHYL-α-)-FURYL-PYRAZOLIN-HCL (GERMAN) see MKH750
5,5-PHENYLPIPERIDINO-1,3-BIS(β-PIPERIDINOAETHYL)-BARBITURSAEURECITRAT see BLG280
PHENYL (1-PIPERIDINOCYCLOHEXYL) KETONE see PFY200
α-PHENYL-α-(2-PIPERIDINOETHYL)-β-ETHYLBUTYRIC ACID NITRILE HYDROCHLORIDE see EQZ000
α-PHENYL-α-(2-PIPERIDINOETHYL)-β-ETHYLBUTYRONITRILE HYDROCHLORIDE see EQZ000
2-PHENYL-6-PIPERIDINOHEXYNOPHENONE see PFY750
1-PHENYL-3-PIPERIDINOPROPAN-1-ONE HYDROCHLORIDE see PIV500
1-PHENYL-1-(2-PIPERIDYL)-1-ACETOXYMETHANE HYDROCHLORIDE see LFO000
PHENYL-(2-PIPERIDYL)METHYL ACETATE HYDROCHLORIDE see LFO000
1-PHENYL-4-((4-PIPERONYL-1-PIPERAZINYLCARBONYL)-2-PYRROLIDINONE HYDROCHLORIDE see PGA250
2-PHENYLPROPANAL see COF000
3-PHENYLPROPANAL see HHP000
3-PHENYL-1-PROPANAL see HHP000
1-PHENYLPROPANE see IKG000
2-PHENYLPROPANE see COE750
2-PHENYL-1,3-PROPANEDIOL DICARBAMATE see FAG200
1-PHENYL-1,2-PROPANEDIONE see PGA500
3-PHENYLPROPANENITRILE see HHP100
1-PHENYLPROPANOL see EGQ000
3-PHENYLPROPANOL see HHP050
1-PHENYL-1-PROPANOL see EGQ000
1-PHENYL-2-PROPANOL see PGA600
2-PHENYLPROPAN-1-OL see HGR600
γ-PHENYLPROPANOL see HHP050
1-PHENYL-2-PROPANOL ACETATE see ABU800
3-PHENYL-1-PROPANOL ACETATE see HHP500
PHENYLPROPANOLAMINE see NNM000
PHENYLPROPANOLAMINE HYDROCHLORIDE see NNN000
PHENYLPROPANOLAMINE HYDROCHLORIDE see PMJ500
3-PHENYL-1-PROPANOL CARBAMATE see PGA750
3-PHENYL-1-PROPANOL (FCC) see HHP050
1-PHENYL-1-PROPANONE see EOL500
1-PHENYL-2-PROPANONE see MHO100
3-PHENYLPROPENAL see CMP969
3-PHENYL-2-PROPENAL see CMP969
(E)-3-PHENYLPROPENAL see CMP971
(E)-3-PHENYL-2-PROPENAL see CMP971
3-PHENYL-2-PROPENAL DIMETHYL ACETAL see PGA775
3-PHENYLPROPENAMIDE see CMP973
3-PHENYL-2-PROPENAMIDE see CMP973
2-PHENYLPROPENE see MPK250
β-PHENYLPROPENE see MPK250
3-PHENYLPROPENOIC ACID see CMP975
3-PHENYL-2-PROPENOIC ACID see CMP975

(E)-3-PHENYL-2-PROPENOIC ACID see CMP980

3-PHENYL-2-PROPENOIC ACID-1,5,DIMETHYL-1-VINYL-4-HEXEN-1-YL ESTER see LGA000

3-PHENYL-2-PROPENOIC ACID-1-ETHENYL-1,5-DIMETHYL-4-HEXENYL ESTER see LGA000

3-PHENYL-2-PROPENOIC ACID METHYL ESTER (9CI) see MIO500

3-PHENYL-2-PROPENOIC ACID, 2-METHYLPROPYL ESTER see IIQ000

3-PHENYL-2-PROPENOIC ACID PHENYLMETHYL ESTER (9CI) see BEG750

3-PHENYL-2-PROPENOIC ACID 3-PHENYLPROPYL ESTER see PGB800

3-PHENYL-2-PROPEN-1-OL see CMQ740

3-PHENYL-2-PROPEN-1-YL ACETATE see CMQ730

3-PHENYL-2-PROPENYLANTHRANILATE see API750

3-PHENYL-2-PROPEN-1-YL ANTHRANILATE see API750

PHENYLPROPENYL n-BUTYRATE see CMQ800

3-PHENYL-2-PROPEN-1-YL CINNAMATE see CMQ850

3-PHENYL-2-PROPEN-1-YL FORMATE see CMR500

3-PHENYL-2-PROPENYL 3-PHENYL-2-PROPENOATE see CMQ850

3-PHENYL-2-PROPENYL PROPIONATE see CMR850

3-PHENYL-2-PROPEN-1-YL PROPIONATE see CMR850

5-PHENYL-5-(2-PROPENYL)-2,4,6(1H,3H,5H)-PYRIMIDINETRIONE (9CI) see AGQ875

PHENYLPROPIOLIQUE NITRILE see PGD100

PHENYLPROPIOLONITRILE see PGD100

α-PHENYLPROPIONALDEHYDE see COF000

β-PHENYLPROPIONALDEHYDE see HHP000

2-PHENYLPROPIONALDEHYDE DIMETHYL ACETAL see HII600

2-PHENYLPROPIONALDEHYDE DIMETHYL ACETAL see PGA800

2-PHENYLPROPIONALDEHYDE (FCC) see COF000

3-PHENYLPROPIONALDEHYDE (FCC) see HHP000

PHENYLPROPIONITRILE see HHP100

3-PHENYLPROPIONITRILE see HHP100

β-PHENYLPROPIONITRILE see HHP100

3-PHENYLPROPIONYL AZIDE see PGB500

17-β-PHENYLPROPIONYLOXY-4-ESTREN-3-ONE see DYF450

PHENYL(PROPIONYLOXY)MERCURY see PFO000

4-PHENYLPROPIOPHENONE see BGM100

PHENYLPROPYL ACETATE see HHP500

2-PHENYLPROPYL ACETATE see PGB750

3-PHENYL-1-PROPYL ACETATE see HHP500

3-PHENYLPROPYL ACETATE (FCC) see HHP500

PHENYLPROPYL ALCOHOL see HHP050

1-PHENYLPROPYL ALCOHOL see EGQ000

2-PHENYLPROPYL ALCOHOL see HGR600

3-PHENYLPROPYL ALCOHOL see HHP050

β-PHENYLPROPYL ALCOHOL see HGR600

γ-PHENYLPROPYL ALCOHOL see HHP050

3-PHENYLPROPYL ALDEHYDE see HHP000

2-PHENYLPROPYLAMINE see PGB760

β-PHENYLPROPYLAMINE see PGB760

α-PHENYL-α-PROPYLBENZENEACETIC ACID-2-(DIETHYLAMINO)ETHYL ESTER HYDROCHLORIDE see PBM500

α-PHENYL-α-PROPYL-BENZENEACETIC ACID-S-(DIETHYLAMINO)ETHYL ESTER see DIG400

γ-PHENYLPROPYLCARBAMAT (GERMAN) see PGA750

γ-PHENYLPROPYL CARBAMATE see PGA750

3-PHENYLPROPYL CHLOROACETATE see PGB770

PHENYLPROPYL CINNAMATE see PGB800

3-PHENYLPROPYL CINNAMATE see PGB800

3-PHENYL-N-(4-PROPYLCYCLOHEXYL)-2-PROPENAMIDE see PGB850

2-PHENYLPROPYLENE see MPK250

β-PHENYLPROPYLENE see MPK250

PHENYLPROPYL FORMATE see HHQ000

3-PHENYL-1-PROPYL FORMATE see HHQ000

α-(β-PHENYL-PROPYL)-β-HYDROXY-Δ^{α,β}-BUTENOLID (GERMAN) see PGC750

3-(3-PHENYLPROPYL)-4-HYDROXY-2(5H)FURANONE see PGC750

3-PHENYLPROPYL ISOBUTYRATE see HHQ500

O-PHENYL-S-PROPYL METHYL PHOSPHONODITHIOATE see PGD000

PHENYLPROPYL PROPIONATE see HHQ550

3-PHENYLPROPYL PROPIONATE see HHQ550

β-PHENYLPROPYL PROPIONATE see HHQ550

α-(1-PHENYL-5-PROPYL-1H-PYRAZOL-4-YL)-1-PIPERIDINEBUTANOL see PGD050

PHENYLPROPYNENITRILE see PGD100

1-PHENYL-2-PROPYNYL CARBAMATE see PGE000

PHENYLPSEUDOHYDANTOIN see IBM000

6-PHENYL-2,4,7-PTERIDINETRIAMINE see UVJ450

1-PHENYL-3-PYRAZOLIDINONE see PDM500

1-PHENYL-3-PYRAZOLIDONE see PDM500

1-PHENYL-4-PYRAZOLIN-3-ONE see PGE250

2-PHENYL-8H-PYRAZOLO(5,1-a)ISOINDOLE see PGE300

2-PHENYL-PYRAZOLO(5,1-a)ISOQUINOLINE see PGE350

p-(3-PHENYL-1-PYRAZOLYL)BENZENESULFONIC ACID see PGE500

α-(1-PHENYL-1H-PYRAZOL-4-YL)-1-PIPERIDINEBUTANOL see PGE550

N'-(1-PHENYLPYRAZOL-5-YL)SULFANILAMIDE see AIF000

N^1-(1-PHENYLPYRAZOL-5-YL)-SULFANILAMIDE (8CI) see AIF000

2-PHENYLPYRIDINE see PGE600

4-PHENYLPYRIDINE see PGE750

o-PHENYLPYRIDINE see PGE600

p-PHENYLPYRIDINE see PGE750

PHENYL(2-PYRIDYL)(β-N,N-DIMETHYLAMINOMETHYL) METHANE MALEATE see TMK000

1-PHENYL-1-(2-PYRIDYL)-3-DIMETHYLAMINOPROPANE see TMJ750

1-PHENYL-1-(2-PYRIDYL)-3-DIMETHYLAMINOPROPANE HYDROCHLORIDE see PMS900

1-PHENYL-1-(2-PYRIDYL)-3-DIMETHYLAMINOPROPANE MALEATE see TMK000

3-PHENYL-3-(2-PYRIDYL)-N,N-DIMETHYLPROPYLAMINE see TMJ750

PHENYL 2-PYRIDYL KETONE see PGE760

PHENYL 3-PYRIDYL KETONE see PGE765

PHENYL 4-PYRIDYL KETONE see PGE768

1-(2-(PHENYL(2-PYRIDYLMETHYL)AMINO)ETHYL)PIPERIDINE HYDROCHLORIDE see CNE375

PHENYL-2-PYRIDYLMETHYL-β-N,N-DIMETHYLAMINOETHYL ETHER see DYE500

PHENYL-2-PYRIDYLMETHYL-β-N,N-DIMETHYLAMINOETHYL ETHER SUCCINATE see PGE775

N-PHENYL-N-(2-PYRIDYLMETHYL)-2-PIPERIDINOETHYLAMINE HYDROCHLORIDE see CNE375

1-PHENYL-3-(p-2-PYRIDYLSULFAMOYLANILINO)-1,3-PROPANEDISULFONIC ACID DISODIUM SALT see DXF400

2-PHENYL-3-p-(β-PYRROLIDINOETHOXY)PHENYL-6-METHOXY-BENZOFURAN HYDROCHLORIDE see PGF100

3-PHENYL-4-(p-(β-PYRROLIDINOETHOXY)PHENYL)-7-METHOXYCHROMENE see PGF112

2-PHENYL-3-p-(β-PYRROLIDINOETHOXY)PHENYL-(2:1-b)NAPHTHOFURAN see PGF125

2-PHENYL-3-p-(β-PYRROLIDINOETHOXY)-PHENYL-NAPHTHO(2:1-b)FURAN see PGF125

4-PHENYL-2-PYRROLIDINONE see PGF800

PHENYL-2-PYRROLIDINOPENTANE HYDROCHLORIDE see PNS000

1-PHENYL-2-N-PYRROLIDINOPENTANE HYDROCHLORIDE see PNS000

2-PHENYL-1-(p-(2-(1-PYRROLIDINYL)ETHOXY)PHENYL)NAPHTHO(2,1-b)FURAN see PGF125

2-PHENYL-N-(2-(1-PYRROLIDINYL)ETHYL)GLYCINE ISOPENTYL ESTER see CBA375

PHENYLPYRROLIDONE see PGF800

4-PHENYLPYRROLIDONE-2 see PGF800

PHENYL PYRROL-2-YL KETONE see PGF900

PHENYL-1H-PYRROL-2-YLMETHANONE see PGF900

PHENYLQUECKSILBERACETAT (GERMAN) see ABU500

PHENYLQUECKSILBERBRENZKATECHIN (GERMAN) see PFO750

PHENYLQUECKSILBERCHLORID (GERMAN) see PFM500

2-PHENYLQUINOLINE-4-CARBOXYLIC ACID see PGG000

2-PHENYL-4-QUINOLINECARBOXYLIC ACID see PGG000

2-PHENYL-4-QUINOLONECARBOXYLIC AICD SODIUM SALT see SJH000

PHENYLQUINONE see PEL750

PHENYLRAMIDOL see PGG350

PHENYLRAMIDOL HYDROCHLORIDE see PGG355

PHENYLRHODANID see TFF600

3-PHENYL-RHODANIN (GERMAN) see PGG500

3-PHENYLRHODANINE see PGG500

N-PHENYLSALICYLAMIDE see SAH500

PHENYL SALICYLATE see PGG750

3-PHENYLSALICYLIC ACID see PGH000

PHENYL SALIGENIN PHOSPHATE see SAN300

PHENYLSALIOXON see SAN300

PHENYL SELENIDE see PGH250

PHENYLSELENONIC ACID see PGH500

PHENYLSEMICARBAZIDE see CBL000

1-PHENYLSEMICARBAZIDE see CBL000

4-PHENYLSEMICARBAZONE 1-METHYL-1H-PYRROLE-2-CARBOXALDEHYDE see PGH600

4-PHENYLSEMICARBAZONE-1H-PYRROLE-2-CARBOXALDEHYDE see PGH650

PHENYLSENFOEL (GERMAN) see ISQ000

PHENYLSILANE see SDX250

PHENYLSILATRANE see PGH750

PHENYLSILICON TRICHLORIDE see TJA750

PHENYLSILVER see PGI000

PHENYL SODIUM see PGI100

PHENYL STYRYL KETONE see CDH000

1-PHENYL-5-SULFANILAMIDOPYRAZOLE see AIF000

PHENYL SULFIDE see PGI500

5-(PHENYLSULFINYL)-2-BENZIMIDAZOLECARBAMIC ACID METHYL ESTER see OMY500

(5-(PHENYLSULFINYL)-1H-BENZIMIDAZOL-2-YL)CARBAMIC ACID METHYL ESTER see OMY500

5-PHENYLSULFINYL-2-CARBOMETHOXYAMINOBENZIMIDAZOLE see OMY500

PHENYLSULFOHYDRAZIDE see BBS300

PHENYL SULFONE see PGI750

PHENYLSULFONIC ACID see BBS250

2-(PHENYLSULFONYLAMINO)-1,3,4-THIADIAZOLE-5-SULFONAMIDE see PGJ000

PHENYLSULFONYL CHLORIDE see BBS750

2-(PHENYLSULFONYL)ETHANOL see BBT000

PHENYLSULFONYL HYDRAZIDE see BBS300

PHENYLSULFONYLHYDRAZINE see BBS300

(PHENYLSULFONYL)METHANE see PFR375

N-(4-PHENYLSULFONYL-o-TOLYL)-1,1,1-TRIFLUOROMETHANESULFONAMIDE see TKF750

4-(PHENYLSULFOXYETHYL)-1,2-DIPHENYL-3,5-PYRAZOLIDINEDIONE see DWM000

PHENYLSUXIMIDE see MNZ000

5'-PHENYL-m-TERPHENYL see TMR000

dl-6-PHENYL-2,3,5,6-TETRAHYDROIMIDAZOLE(2,1-b)THIAZOLE-HYDROCHLORIDE see TDX750

6-PHENYL-2,3,5,6-TETRAHYDROIMIDAZO(2,1-b)THIAZOLE see LFA000

5-PHENYLTETRAZOLE see PGJ500

1-PHENYLTETRAZOLE-5-THIOL see PGJ750

3-PHENYL-1-TETRAZOLYL-1-TETRAZENE see PGK000

PHENYLTHALLIUM DIAZIDE see PGK100

N-PHENYL-N'-1,2,3-THIADIAZOL-5-YL-UREA see TEX600

5-PHENYL-2,4-THIAZOLEDIAMINE MONOHYDROCHLORIDE (9CI) see DCA600

N-PHENYLTHIOACETAMIDE see TFA250

(PHENYLTHIO)ACETONITRILE see COP755

1-(PHENYLTHIO)-9,10-ANTHRACENEDIONE see PGL000

1-(PHENYLTHIO)ANTHRAQUINONE see PGL000

N-PHENYLTHIOBENZAMIDE see PGL250

(5-(PHENYLTHIO))-2-BENZIMIDAZOLECARBAMIC ACID, METHYL ESTER see FAL100

2-(PHENYLTHIO)BENZOIC ACID see PGL270

(4-PHENYLTHIOBUTYL)HYDRAZINE MALEATE see PGL500

PHENYLTHIOCARBAMIDE see PGN250

PHENYL THIOCYANATE see TFF600

4-(PHENYLTHIOETHYL)-1,2-DIPHENYL-3,5-PYRAZOLIDINE-DIONE see DWJ400

((1-PHENYLTHIOMETHYL)ETHYL)HYDRAZINE MALEATE see PGM250

PHENYLTHIONOPHOSPHONIC DICHLORIDE see PFW200

PHENYLTHIONOPHOSPHONIC DICHLORIDE see PFW210

PHENYLTHIOPHOSPHONATE de O-ETHYLE et O-4-NITROPHENYLE (FRENCH) see EBD700

PHENYLTHIOPHOSPHONIC DIAZIDE see PGM500

1-PHENYLTHIOSEMICARBAZIDE see PGM750

4-PHENYLTHIOSEMICARBAZIDE see PGN000

4-PHENYL-3-THIOSEMICARBAZIDE see PGN000

4-PHENYLTHIOSEMICARBAZONE 1-METHYL-1H-PYRROLE-2-CARBOXALDEHYDE see PGN030

4-PHENYLTHIOSEMICARBAZONE-1H-PYRROLE-2-CARBOXALDEHYDE see PGN045

α-(PHENYLTHIO)-p-TOLUIDINE see PGN100

1-PHENYLTHIOUREA see PGN250

N-PHENYLTHIOUREA see PGN250

1-PHENYL-2-THIOUREA see PGN250

α-PHENYLTHIOUREA see PGN250

PHENYLTOLOXAMINE DIHYDROGEN CITRATE see DTO600

PHENYLTOLOXAMINE HYDROCHLORIDE see DTO800

N-PHENYL-p-TOLUENESULFONAMIDE see TGN600

(N-PHENYL-p-TOLUENESULFONAMIDO)ETHYLMERCURY see EME500

PHENYL m-TOLYL ETHER see MNV770

PHENYL p-TOLYL KETONE see MHF750

(2-(PHENYL-o-TOLYLMETHOXY)ETHYL)METHYLAMINE HYDROCHLORIDE see TGJ250

PHENYL-(o-TOLYLMETHYL)METHYLAMINOETHYL ETHER HYDROCHLORIDE see TGJ250

5-PHENYL-3-(o-TOLYL)-s-TRIAZOLE see PGN400

6-PHENYL-2,4,7-TRIAMINOPTERIDINE see UVJ450

2-PHENYL-5H-1,2,4-TRIAZOLO(5,1-a)ISOINDOLE see PGN825

2-PHENYL-s-TRIAZOLO(5,1-a)ISOQUINOLINE see PGN840

2-PHENYL-(1,2,4)TRIAZOLO(5,1-a)ISOQUINOLINE (9CI) see PGN840

1-PHENYL-1,2,4-TRIAZOLYL-3-(O,O-DIETHYLTHIONOPHOSPHATE) see THT750

PHENYLTRICHLOROMETHANE see BFL250

PHENYL(TRICHLOROMETHYL)CARBINOL see TIR800

PHENYL TRICHLOROSILANE (DOT) see TJA750

PHENYLTRIETHOXYSILANE see PGO000

PHENYLTRIFLUOROSILANE see PGO500

1-PHENYL-4-(3,4,5-TRIMETHOXYBENZOYL)PIPERAZINE see PFX600

PHENYLTRIMETHOXYSILANE see PGO750

PHENYL TRIMETHYL AMMONIUM BROMIDE see TMB000

PHENYLTRIMETHYLAMMONIUM CHLORIDE see TMB250

PHENYL TRIMETHYL AMMONIUM HYDROXIDE see TMB500

PHENYLTRIMETHYLAMMONIUM IODIDE see TMB750

4-PHENYL-1,2,5-TRIMETHYL-4-PIPERIDINOL PROPIONATE see IMT000

PHENYL-2,8,9-TRIOXA-5-AZA-1-SILABICYCLO(3.3.3) UNDECANE see PGH750

PHENYLTRIS(DODECANOYLOXY)PLUMBANE see PFL200

PHENYLUREA see PGP250

1-PHENYLUREA see PGP250

N-PHENYLUREA see PGP250

PHENYL UREA-p-DI(CARBOXYMETHYL)THIOARSENITE see CBI250

PHENYLURETHAN see CBL750

PHENYLURETHAN(E) see CBL750

N-PHENYLURETHANE see CBL750

α-PHENYL-VALERATE du DIETHYLAMINO-ETHANOL CHLORHYDRATE (FRENCH) see PGP500

2-PHENYLVALERIC ACID-2-(DIETHYLAMINO)ETHYL ESTER HYDROCHLORIDE see PGP500

PHENYLVANADIUM(V) DICHLORIDE OXIDE see PGP600

N-PHENYL-N-VINYLACETAMIDE see VLU220

PHENYLVINYL KETONE see PMQ250

PHENYL XYLYL KETONE see PGP750

1-PHENYL-1-(3,4-XYLYL)-2-PROPYNYL N-CYCLOHEXYLCARBAMATE see PGQ000

PHENYRACILLIN see PGQ100

PHENYRACILLINE see PGQ100

PHENYRAL see AGQ875

PHENYRAL see EOK000

PHENYRAMIDOL see PGG350

PHENYRAMIDOL HYDROCHLORIDE see PGG355

PHENYRIT see PEC250

PHENYTOIN and PHENOBARBITONE see DWD000

PHENYTOIN SODIUM see DNU000

PHENZIDOL see PEL250

PHENZIDOLE see PEL250

(d-PHE²)-d-PHE⁶))-LHRH HYDROCHLORIDE HYDRATE (2:2:7) see LIU343

d-PHE-PHE-PHE-d-TRP-LYS-THR-PHE-THR-NH2 see PED800

d-PHE-PHE-TYR-d-TRP-LYS-VAL-PHE-d-NAL-NH2 see PED850

(d-PHE²)-PRO³)-d-PHE⁶))-LHRH see LIU341

(d-PHE²)-PRO³)-d-TRP⁶))-LHRH see LIU342

PHERMERNITE see MCU750

PHERMERNITE see MDH500

PHEROCON CL see DXU830

PHEROCON CM see CNG760

PHEROCON MFF see BQT600

PHERODIN SL see FAS100

PHERODIN SL see TBX300

PHETADEX see BBK500

PHETIDINE see DAM600

PHETYLUREUM see PEC250

(d-PHE²)-VAL³)-d-TRP⁶))-LHRH see LIU345

PHGABA HYDROCHLORIDE see GAD000

PHIC see PGQ275

PHILBLACK see CBT750

PHILBLACK N 550 see CBT750

PHILBLACK N 765 see CBT750

PHILBLACK O see CBT750

PHILIPS 2133 see CPW250

PHILIPS 2605 see BIK750

PHILIPS-DUPHAR PH 60-40 see CJV250

PHILIPS-DUPHAR V-101 see CKL750

PHILLIPS 1863 see AMJ250

PHILLIPS 1908 see AMA000

PHILLIPS 2038 see AAM750

R-874 PHILLIPS see OGA000

PHILLIPS R-11 see BHJ500

PHILODENDRON (various species) see PGQ285

PHILODORM see TDA500

PHILOPON see DBA800

PHILOPON see MDT600

PHILOSOPHER'S WOOL see ZKA000

PHILOSTIGMIN BROMIDE see POD000

PHILOSTIGMIN METHYL SULFATE see DQY909

PHIP see AKZ200

PHISODANV see HCL000

PHISOHEX see HCL000

PHIX see ABU500
PHIXIA see CMS850
PHLEBIAKAURANOL ALDEHYDE see OOS100
PHLEOCIDIN see PGQ350
PHLEOMYCIN see PGQ500
PHLEOMYCIN COMPLEX see PGQ750
PHLEOMYCIN COMPLEX see PGQ750
PHLEOMYCIN D2 see BLY760
PHLORHIZIN see PGQ800
PHLORIDZIN see PGQ800
PHLORIDZINE see PGQ800
PHLORIZIN see PGQ800
PHLORIZINE see PGQ800
PHLORIZOSIDE see PGQ800
PHLOROGLUCIN see PGR000
PHLOROGLUCINOL see PGR000
PHLOROGLUCINOL TRIMETHYL ETHER see TKY250
PHLOROL see PGR250
PHLORONE see XQJ000
PHLOROPROPIONONE see TKP100
PHLOROPROPIOPHENONE see TKP100
PHLORRHIZIN see PGQ800
PHLOXIN see CMM000
PHLOXIN B see ADG250
PHLOXINE see CMM000
PHLOXINE 2G see CMM300
PHLOXINE B see ADG250
PHLOXINE BBN SUPRA see SKS400
PHLOXINE G see CMM300
PHLOXINE K see CMM000
PHLOXINE O see SKS400
PHLOXINE P see ADG250
PHLOXINE RED 20-7600 see BNK700
PHLOXINE TONER B see BNH500
PHLOX RED TONER X-1354 see BNH500
PHOB see EOK000
PHOBEX see BCA000
PHOLATE see AQO000
PHOLCODIN see TCY750
PHOLCODINE see TCY750
PHOLEDRINE see FMS875
PHOLETONE see FMS875
PHOMIN see CQM125
PHOMOPSIN see PGR775
PHONURIT see AAI250
PHORADENDRON RUBRUM see MQW525
PHORADENDRON SEROTINUM see MQW525
PHORADENDRON TOMENTOSUM see MQW525
PHORAT (GERMAN) see PGS000
PHORATE see PGS000
PHORATE-10G see PGS000
PHORATE SULFONE see TEZ100
PHORBOL see PGS250
PHORBOL ACETATE, CAPRATE see ACZ000
PHORBOL ACETATE, LAURATE see PGS500
PHORBOL ACETATE LAURATE see PGU250
PHORBOL ACETATE, MYRISTATE see PGV000
PHORBOL-12-o-BUTYROYL-13-DODECANOATE see BSX750
PHORBOL CAPRATE, (+)-(S)-2-METHYLBUTYRATE see PGU500
PHORBOL CAPRATE, TIGLATE see PGU750
PHORBOL-12,13-DIACETATE see PGS750
PHORBOL-12,13-DIBENZOATE see PGT000
PHORBOL-12,13-DIBUTYRATE see PGT100
PHORBOL-12,13-DIDECANOATE see PGT250
PHORBOL-12,13-DIHEXA(Δ-2,4)-DIENOATE see PGT500
PHORBOL-12,13-DIHEXANOATE see PGT750
PHORBOL 12,13-DIMYRISTATE see PGT800
PHORBOL 12,13-DITETRADECANOATE see PGT800
PHORBOL LAURATE, (+)-S-2-METHYLBUTYRATE see PGU000

PHORBOL MONOACETATE MONOLAURATE see PGS500
PHORBOL MONOACETATE MONOLAURATE see PGU250
PHORBOL MONOACETATE MONOMYRISTATE see PGV000
PHORBOL MONODECANOATE (S)-(+)-MONO(2-METHYLBUTYRATE) see PGU500
(E)-PHORBOL MONODECANOATE MONO(2-METHYLCROTONATE) see PGU750
PHORBOL MONOLAURATE MONO(S)-(+)-2-METHYLBUTYRATE see PGU000
PHORBOL MYRISTATE ACETATE see PGV000
PHORBOL-9-MYRISTATE-9a-ACETATE-3-ALDEHYDE see PGV250
PHORBOLOL ACETATE MYRISTATE see PGV500
PHORBOLOL MYRISTATE ACETATE see PGV500
PHORBOL-12-o-TIGLYL-13-BUTYRATE see PGV750
PHORBOL-12-o-TIGLYL-13-DODECANOATE see PGW000
PHORBYOL see CCP250
PHORDENE see DFY800
PHORON (GERMAN) see PGW250
PHORONE see PGW250
PHORTISOLONE see PLY275
PHORTOX see TAA100
PHOS 2 see PCB400
PHOSADEN see AOA125
PHOSALON see BDJ250
PHOSALONE see BDJ250
PHOSALON mixed with MILBEX (1:2) see MQT750
PHOSAN-PLUS see PGW500
PHOSAZETIM see BIM000
PHOSCHLOR R50 see TIQ250
PHOSCLERE T 310 see IKM025
PHOSCOLIC ACID see PGW600
PHOSDRIN (OSHA) see MQR750
PHOSELERE T 26 see IKL200
PHOSETHYL ALUMINUM see AHH800
PHOSFENE see MQR750
PHOSFLEX 179 see XLS100
PHOSFLEX 390 see IKL100
PHOSFLEX 179-C see TMO600
PHOSFLEX T-BEP see BPK250
PHOS-FLUR see SHF500
PHOSFOLAN see DXN600
PHOSFOLAN see PGW750
PHOSFON D see THY500
PHOSFON S see THY200
PHOSGARD C-22-R see CGO525
PHOSGEN (GERMAN) see PGX000
PHOSGENE see PGX000
PHOSGENE, THIO- see TFN500
PHOSHOROTHIOIC ACID-S-(2-(ETHYLAMINO)-2-OXOETHYL)-O,O-DIMETHYL ESTER see DNX600
PHOSKIL see PAK000
PHOSMET see PHX250
PHOSPHACOL see NIM500
PHOSPHADEN see AOA125
PHOSPHALUGEL see PHB500
PHOSPHAM see PGX250
PHOSPHAMID see DSP400
PHOSPHAMIDON see FAB400
(E)-PHOSPHAMIDON see PGX300
(Z)-PHOSPHAMIDON see PGX275
cis-PHOSPHAMIDON see PGX275
trans-PHOSPHAMIDON see PGX300
PHOSPHATE 100 see EAS000
PHOSPHATE de O,O-DIETHYLE et de O-2-CHLORO-1-(2,4-DICHLOROPHENYL) VINYLE (FRENCH) see CDS750
PHOSPHATE de DIETHYLE et de 3-METHYL-5-PYRAZOLYLE (FRENCH) see MOX250

PHOSPHATE de O,O-DIMETHLE et de O-(1,2-DIBROMO-2,2-DICHLORETHYLE) (FRENCH) see NAG400
PHOSPHATE de DIMETHYLE et de (2-CHLORO-2-DIETHYLCARBAMOYL-1-METHYL-VINYLE) (FRENCH) see FAB400
PHOSPHATE de DIMETHYLE et de 2,2-DICHLOROVINYLE (FRENCH) see DGP900
PHOSPHATE de DIMETHYLE et de 2-DIMETHYLCARBAMOYL-1-METHYL VINYLE (FRENCH) see DGQ875
PHOSPHATE de DIMETHYLE et de 2-METHOXYCARBONYL-1 METHYLVINYLE (FRENCH) see MQR750
PHOSPHATE de DIMETHYLE et de 2-METHYLCARBAMOYL 1-METHYL VINYLE see MRH209
PHOSPHATE ROCK see FDH000
PHOSPHATES see PGX500
PHOSPHATE, SODIUM HEXAMETA see SHM500
PHOSPHATE de TRICRESYLE (FRENCH) see TNP500
PHOSPHEMOL see PAK000
PHOSPHENE (FRENCH) see MQR750
PHOSPHENOL see PAK000
PHOSPHENTASIDE see AOA125
PHOSPHESTROL see DKA200
PHOSPHIDES see PGX750
PHOSPHINE see PGY000
PHOSPHINE, DIPHENYLPROPYL- see PNI600
PHOSPHINE OXIDE, BIS(1-AZIRIDINYL)METHYLAMINO- see MGE100
PHOSPHINE OXIDE, (3-METHYLBUTYL)DIOCTYL-(9CI) see DVK709
PHOSPHINE OXIDE, TRIISOPENTYL- see TNJ285
PHOSPHINE OXIDE, TRIOCTYL- see TMO575
PHOSPHINE OXIDE, TRIPROPYL- see TNA800
PHOSPHINE OXIDE, TRIS(p-DIMETHYLAMINOPHENYL)-, compounded with STANNIC CHLORIDE (2:1) see BLT775
PHOSPHINE OXIDE, TRIS(3-METHYLBUTYL)- see TNJ285
PHOSPHINE SELENIDE, TRIMETHYL- see TMD400
PHOSPHINE SELENIDE, TRIPIPERIDINO- see TMX350
PHOSPHINE SULFIDE, MONOCHLORODIFLUORO- see TFO600
PHOSPHINE SULFIDE, TRIBUTYL- see TIA450
PHOSPHINE SULFIDE, TRIMORPHOLINO- see TFN600
PHOSPHINE SULFIDE, TRIPHENYL- see TMU100
PHOSPHINE, TRIBUTYL- see TIA300
PHOSPHINE, TRIBUTYL-, SULFIDE see TIA450
PHOSPHINE, TRIS(DIPROPYLENE GLYCOL)- see OQO000
PHOSPHINIC ACID see PGY250
PHOSPHINIC ACID, BIS(1-AZIRIDINYL)-, p-(BENZYLOXY)PHENYL ESTER see BFC300
PHOSPHINIC ACID, DIETHYL-, ANHYDRIDE WITH DIETHYL PHOSPHOROTHIONATE see DJW100
PHOSPHINIC ACID, DIMORPHOLINO-, PHENYL ESTER see DUO700
PHOSPHINIC ACID, PHENYL- see HNL200
PHOSPHINIC ACID, PHENYL-, ION(1-), TETRABUTYLPHOSPHONIUM see TBM100
PHOSPHINIC AMIDE, P,P-BIS(1-AZIRIDINYL)-N-ETHYL- see BGX775
PHOSPHINIC AMIDE, p,p-BIS(1-AZIRIDINYL)-N-ISOPROPYL- see BGX850
PHOSPHINIC AMIDE, P,P-BIS(1-AZIRIDINYL)-N-METHYL- see MGE100

PHOSPHINIC AMIDE, p,p-BIS(1-AZIRIDINYL)-N-(1-METHYLETHYL)-(9CI) see BGX850

PHOSPHINIC AMIDE, p,p-BIS(1-AZIRIDINYL)-N-PROPYL- see BGY500

2,2'-PHOSPHINICOBIS(2-HYDROXY-PROPANOIC) ACID (9CI) see PGW600

2,2'-PHOSPHINICODILACTIC ACID see PGW600

PHOSPHINIDYNETRISMETHANOL see TNI600

1,1',1"-(PHOSPHINIDYNETRIS((1-METHYLETHYLENE)OXY))TRI-2-PROPANOL see PGY300

PHOSPHINOETHANE see EON000

PHOSPHINOTHIOIC ACID, DIPHENYL-, S-(4-CHLORO-2-BUTYNYL) ESTER see CEV840

1,1',1"-PHOSPHINOTHIOYLIDYNETRISAZIRIDINE see TFQ750

PHOSPHINOTHRICIN MONOAMMONIUM SALT see ANI800

PHOSPHINOX PZ 06 see ZJS400

1,1',1"-PHOSPHINYLIDYNETRISAZIRIDINE see TND250

1,1',1"-PHOSPHINYLIDYNETRIS(2-METHYL)AZRIDINE see TNK250

PHOSPHOCYSTEAMINE see AKA900

o-PHOSPHOETHANOLAMINE see EED000

2-PHOSPHOLENE, 3-METHYL-1-PHENYL-, 1-OXIDE see MNV800

PHOSPHOLINE see PGY600

PHOSPHOLINE (pharmaceutical) see TLF500

PHOSPHOLINE IODIDE see TLF500

PHOSPHOLIPASE C, from PSEUDOMONAS AERUGINOSA see POH640

PHOSPHOMOLYBDENIC ACID, SODIUM SALT see SIM200

PHOSPHOMYCIN see PHA550

N-PHOSPHONACETYL-l-ASPARTATE see PGY750

PHOSPHONACETYL-l-ASPARTIC ACID see PGY750

PHOSPHON D see THY500

PHOSPHONE D see THY500

PHOSPHONIC ACID see PGZ899

PHOSPHONIC ACID, (1-(ACETYLOXY)-2,2,2-TRICHLOROETHYL)-, DIMETHYL ESTER see CDN510

PHOSPHONIC ACID, (1-(ACETYLOXY)-2,2,2-TRICHLOROETHYL)-, DIPHENYL ESTER see DVX400

PHOSPHONIC ACID, ((2-(2-AMINO-1,6-DIHYDRO-6-OXO-9H-PURIN-9-YL)ETHOXY)METHYL)- see PHA725

PHOSPHONIC ACID, (4-AMINO-1-HYDROXYBUTYLIDENE)BIS- see AKF300

PHOSPHONIC ACID, (4-AMINO-1-HYDROXYBUTYLIDENE)BIS-, MONOSODIUM SALT, TRIHYDRATE see SKL600

PHOSPHONIC ACID, ((2-(4-AMINO-2-OXO-1(2H)-PYRIMIDINYL)-1-(HYDROXYMETHYL)ETHOXY)METHYL)-, (S)- see HNS550

PHOSPHONIC ACID, ((2-(6-AMINO-9H-PURIN-9-YL)ETHOXY)METHYL)- see PHA700

PHOSPHONIC ACID, BENZYL-, DIBUTYL ESTER see BFD760

PHOSPHONIC ACID, ((BIS(2-(BIS(PHOSPHONOMETHYL)AMINO)ETHYL)AMINO)METHYL)-, SODIUM SALT (8CI) see DJG700

PHOSPHONIC ACID, ((BIS(2-HYDROXYETHYL)AMINO)METHYL)-,DIETHYL ESTER see DIV300

PHOSPHONIC ACID, BIS(1-METHYLETHYL) ESTER see DNQ600

PHOSPHONIC ACID, BUTYL-, MONOETHYL ESTER, ANHYDRIDE WITH DIETHYL PHOSPHATE see MRH400

PHOSPHONIC ACID, (2-CARBOXYETHYL)- see PHA560

PHOSPHONIC ACID, (2-CHLOROETHYL)-, BIS(2-CHLOROETHYL) ESTER see BIB800

PHOSPHONIC ACID, (DICHLOROMETHYLENE)BIS-, DISODIUM SALT see SFX730

PHOSPHONIC ACID, (DICHLOROMETHYLENE)DI-, DISODIUM SALT see SFX730

PHOSPHONIC ACID, (2-(DIETHOXYMETHYLSILYL)ETHYL)-, DIETHYL ESTER see DJA330

PHOSPHONIC ACID, (4,5-DIHYDRO-3-PHENYL-5-ISOXAZOLYL)- see PFL100

PHOSPHONIC ACID, (DITHIODIMETHYLENE)DI-, TETRAMETHYL ESTER see PGZ900

PHOSPHONIC ACID, DODECYL- see DXT500

PHOSPHONIC ACID, ETHENYL-, BIS(2-((BUTOXYMETHYLPHOSPHINYL)OXY)ETHYL) ESTER see PGZ910

PHOSPHONIC ACID, HEXAMETHYLENEBIS(NITRILODIMETHYLENE)TETRA-(7CI,8CI) see HEQ600

PHOSPHONIC ACID, (1,6-HEXANEDIYLBIS(NITRILOBIS(METHYLENE)))TETRAKIS- see HEQ600

PHOSPHONIC ACID, (1,6-HEXANEDIYLBIS(NITRILOBIS(METHYLENE)))TETRAKIS-, POTASSIUM SALT see HEQ610

PHOSPHONIC ACID, 1-HYDROXY-1,1-ETHANEDIYL ESTER see HKS780

PHOSPHONIC ACID, (1-HYDROXYETHYLIDENE)BIS- see HKS780

PHOSPHONIC ACID, (3-((HYDROXYMETHYL)AMINO)-3-OXOPROPYL)-, DIMETHYL ESTER (9CI) see MND550

PHOSPHONIC ACID, (2-((HYDROXYMETHYL)CARBAMOYL)ETHYL)-, DIMETHYL ESTER see MND550

PHOSPHONIC ACID, (1-HYDROXY-2-(3-PYRIDINYL)ETHYLIDENE)BIS-MONOSODIUM SALT see RLF400

PHOSPHONIC ACID, (1-HYDROXYVINYL)-, DIMETHYL ESTER, DIMETHYL PHOSPHATE see PGZ915

PHOSPHONIC ACID, METHYL-, DIPHENYL ESTER see MDQ825

PHOSPHONIC ACID, METHYLENEBIS-, TETRASODIUM SALT see SIL575

PHOSPHONIC ACID, METHYLENEDI-, TETRAETHYL ESTER see MJT100

PHOSPHONIC ACID, (((2-METHYL(METHYLSULFONYL)AMINO)-2-OXOETHYL)AMINO)METHYL- see MID850

PHOSPHONIC ACID, (3-METHYLOXIRANYL)-, (2R-cis)-(9CI) see PHA550

PHOSPHONIC ACID, MONOETHYL ESTER, ALUMINUM SALT (3:1) see AHH800

PHOSPHONIC ACID, MONOMETHYL see PGZ950

PHOSPHONIC ACID, (NITRILOTRIS(METHYLENE))TRI- see NEI100

PHOSPHONIC ACID, (NITRILOTRIS(METHYLENE))TRI-, PENTASODIUM SALT see PBP200

PHOSPHONIC ACID, PHENYL-, DIPHENYL ESTER see TMU300

PHOSPHONIC ACID, PHENYL-, ISOBUTYL 2-PROPYNYL ESTER see PFV760

PHOSPHONIC ACID, (PHENYLMETHYL)-, DIBUTYL ESTER see BFD760

PHOSPHONIC ACID, PHENYL-, 2-METHYLPROPYL 2-PROPYNYL ESTER (9CI) see PFV760

PHOSPHONIC ACID, (((PHOSPHONOMETHYL)IMINO)BIS(2,1-ETHANEDIYLNITRILOBIS(METHYLENE))) TETRAKIS-, SODIUM SAL see DJG700

PHOSPHONIC ACID, (2,2,2-TRICHLORO-1-HYDROXYETHYL)-, DIMETHYL ESTER, ACETATE (ESTER) see CDN510

PHOSPHONIC ACID, (2,2,2-TRICHLORO-1-HYDROXYETHYL)-, DIPHENYL ESTER see TIQ300

PHOSPHONIC ACID, (2-(TRIETHOXYSILYL)ETHYL)-, DIETHYL ESTER see DKD500

PHOSPHONIC DIAMIDE, P-4-PIPERIDYL-N,N,N',N'-TETRAMETHYL- see TDT850

PHOSPHONIC DIAMIDE, N,N,N',N'-TETRAMETHYL-P-4-PIPERIDINYL- see TDT850

PHOSPHONIC DIAMIDE, N,N,N',N'-TETRAMETHYL-P-4-PIPERIDYL- see TDT850

PHOSPHONIUM, BUTYLTRIPHENYL-, BROMIDE see BSR900

PHOSPHONIUM, ETHYLTRIPHENYL-, IODIDE see EQD150

PHOSPHONIUM IODIDE see PHA000

PHOSPHONIUM, (METHYLSELENO)TRIS(DIMETHYLAMINO)-, IODIDE see MPI225

PHOSPHONIUM PERCHLORATE see PHA250

PHOSPHONIUM S see THY200

PHOSPHONIUM, TETRABUTYL-, PHENYLPHOSPHINATE see TBM100

9-(5-o-PHOSPHONO-β-d-ARABINOFURANOSYL)-9H-PURIN-6-AMINE see AQQ905

2-PHOSPHONOBUTANETRICARBONIC ACID PENTAMETHYL ESTER see PHA300

PHOSPHONOCHLORIDOTHIOIC ACID, PHENYL-, O-ETHYL ESTER see EOM100

PHOSPHONODITHIOIC ACID, ETHYL-, S-(4-CHLORO-3-METHYLPHENYL)o-ETHYL ESTER see EMR100

PHOSPHONODITHIOIC ACID, ETHYL-, S-(p-CHLOROPHENYL) o-ETHYL ESTER see EHL670

PHOSPHONODITHIOIC ACID, ETHYL-, S-(4-CHLORO-m-TOLYL) o-ETHYL ESTER see EMR100

N-(PHOSPHONOMETHYL)GLYCINE see PHA500

PHOSPHONOMYCIN see PHA550

PHOSPHONOMYCIN see PHA550

PHOSPHONOMYCIN DISODIUM SALT see DXF600

PHOSPHONOMYCIN SODIUM see DXF600

21-(PHOSPHONOOXY)PREGNA-1,4-DIENE-3,20-DIONE DISODIUM SALT see CCS675

3-PHOSPHONOPROPANOIC ACID see PHA560

3-PHOSPHONOPROPIONIC ACID see PHA560

β-PHOSPHONOPROPIONIC ACID see PHA560

PHOSPHONOSELENOIC ACID, ETHYL-, Se-(2-(DIETHYLAMINO)ETHYL) O-ETHYL ESTER see SBU900

PHOSPHONOSUCCINIC ACID TETRAMETHYL ESTER see PHA565

PHOSPHONOTHIOIC ACID, PHENYL-, S-(4-CHLORO-2-BUTYNYL) o-ETHYL ESTER see CEV850

PHOSPHONOTHIOIC ACID, PHENYL-, o-METHYL S-(o',o'-DIISOPROPYLPHOSPHORAMIDO)ETHYL ESTER see MJE790

PHOSPHONOTHIOIC DICHLORIDE, ETHYL- see EOP600

PHOSPHONOTHIOIC DICHLORIDE, PHENYL- see PFW200

PHOSPHONOTHIOIC DICHLORIDE, PHENYL- see PFW210

PHOSPHONOUS ACID, (1,1'-BIPHENYL)-4,4'-DIYLBIS-, TETRAKIS(2,4-BIS(1,1-DIMETHYLETHYL)PHENYL) ESTER see PCB400

2-PHOSPHONOXYBENZOESAEURE see PHA575

PHOSPHONOXYBENZOIC ACID see PHA575

2-PHOSPHONOXYBENZOIC ACID see PHA575

PHOSPHON S see THY200

9-(2-PHOSPHONYLMETHOXYETHYL)ADENINE see PHA700

9-((2-PHOSPHONYLMETHOXY)-ETHYL)GUANINE see PHA725

PHOSPHOPYRIDOXAL see PII100

PHOSPHOPYRON see EAS000

PHOSPHOPYRONE see EAS000

PHOSPHORAMIDE see PHB550

PHOSPHORAMIDE MUSTARD CYCLOHEXYLAMINE SALT see PHA750

PHOSPHORAMIDODITHIOIC ACID, FORMYLMETHYL-, S-(1-METHYLPROPYL) o-(2,2,2-TRIFLUOROETHYL) ESTER see MOV100

PHOSPHORAMIDOTHIOIC ACID, ISOPROPYL-, O-ETHYL O-(2-NITRO-P-TOLYL) ESTER see AMX400

PHOSPHORAMIDOTHIOIC ACID, N-ISOPROPYL-, O-METHYL O-(2-NITRO-P-TOLYL) ESTER see AMX300

PHOSPHORAMIDOTHIOIC ACID, (1-METHYLETHYL)-, O-ETHYL-O-(4-METHYL-2-NITROPHENYL) ESTER (9CI) see AMX400

PHOSPHORAMIDOTHIOIC ACID, (3-METHYL-2-THIAZOLIDINYLIDENE)-, O,S-DIMETHYL ESTER see MEY200

PHOSPHORE BLANC (FRENCH) see PHP010

PHOSPHORE(PENTACHLORURE de) (FRENCH) see PHR500

PHOSPHORE(TRICHLORURE de) (FRENCH) see PHT275

PHOSPHORIC ACID see PHB250

PHOSPHORIC ACID, ALUMINUM SALT (1:1), (solution) see PHB500

PHOSPHORIC ACID AMIDE see PHB550

PHOSPHORIC ACID, BERYLLIUM SALT (1:1) see BFS000

PHOSPHORIC ACID, 2-BIPHENYLYL DIPHENYL ESTER see XFS300

PHOSPHORIC ACID, (1,1'-BIPHENYL)-2-YL DIPHENYL ESTER see XFS300

PHOSPHORIC ACID, BIS(4-CHLOROBUTYL)-3-CHLORO-4-METHYL-2-OXO-2H-1-BENZOPYRAN-7-YL ESTER (9CI) see CHO800

PHOSPHORIC ACID, BIS(2-CHLOROETHYL) 4-METHYL-2-OXO-2H-1-BENZOPYRAN-7-YL ESTER (9CI) see HMA600

PHOSPHORIC ACID, BIS(2-CHLOROETHYL) p-NITROPHENYL ESTER see CHF600

PHOSPHORIC ACID, BIS(2-CHLOROPROPYL) p-NITROPHENYL ESTER see PHB600

PHOSPHORIC ACID, BIS(3-CHLOROPROPYL) p-NITROPHENYL ESTER see PHB605

PHOSPHORIC ACID, BIS(2,3-DICHLOROPROPYL)-3-CHLORO-4-METHYL-2-OXO-2H-1-BENZOPYRAN-7-YL ESTER see CHO850

PHOSPHORIC ACID, BIS(2-ETHYLHEXYL)ESTER, SODIUM SALT (8CI,9CI) see TBA750

PHOSPHORIC ACID, BIS(p-NITROPHENYL) ESTER see BLA600

PHOSPHORIC ACID, 2-BROMO-1-(2,4-DICHLOROPHENYL)ETHENYL DIETHYL ESTER (9CI) see BMO310

PHOSPHORIC ACID, 2-BROMO-1-(2,4-DICHLOROPHENYL)ETHENYL DIMETHYL ESTER (9CI) see MHR150

PHOSPHORIC ACID, 2-BROMO-1-(2,4-DICHLOROPHENYL)VINYL DIETHYL ESTER see BMO310

PHOSPHORIC ACID, 2-BROMO-1-(2,4-DICHLOROPHENYL)VINYL DIMETHYL ESTER see MHR150

PHOSPHORIC ACID, (p-tert-BUTYLPHENYL) DIPHENYL ESTER see BSF300

PHOSPHORIC ACID, CALCIUM SALT (2:1), MONOHYDRATE see CAY100

PHOSPHORIC ACID, 2-CHLORO-1-(2,4-DICHLOROPHENYL)ETHENYL DIETHYL ESTER, MIXT. WITH 1,1'-(2,2,2-TRICHLOROETHYLIDENE)BIS(4-METHOXYBENZENE see MQQ350

PHOSPHORIC ACID, 2-CHLORO-3-(DIETHYLAMINO)-1-METHYL-3-OXO-1-PROPENYL-,DIMETHYL ESTER, (Z)- see PGX275

PHOSPHORIC ACID, 2-CHLORO-1-(2,4,5-TRICHLOROPHENYL)ETHENYL DIMETHYL ESTER see TBW100

(Z)-PHOSPHORIC ACID-2-CHLORO-1-(2,4,5-TRICHLOROPHENYL)ETHENYL DIMETHYL ESTER see RAF100

PHOSPHORIC ACID CHROMIUM (III) SALT see CMK300

PHOSPHORIC ACID, CHROMIUM(3+) SALT (1:1) see CMK300

PHOSPHORIC ACID, COBALT(2+) SALT (2:3) see CND920

PHOSPHORIC ACID, CYCLIC METHYLENE-o-PHENYLENE PHENYL ESTER see SAN300

PHOSPHORIC ACID, CYCLOHEXYL DIPHENYL ESTER see CPP800

PHOSPHORIC ACID, DIBUTYL 2-HYDROXYETHYL ESTER see HKQ700

PHOSPHORIC ACID, DIBUTYL PHENYL ESTER see DEG600

PHOSPHORIC ACID-2,2-DICHLOROETHENYL DIMETHYL ESTER see DGP900

PHOSPHORIC ACID-2,2-DICHLOROVINYL METHYL ESTER, CALCIUM SALT mixed with 2,2-DICHLOROVINYL PHOSPHORIC ACID CALCIUM SALT see PHC700

PHOSPHORIC ACID, DIESTER WITH 2-METHYL-1,4-NAPHTHALENEDIOL, TETRASODIUM SALT see NAQ600

PHOSPHORIC ACID, DIETHYL ESTER see DJW500

PHOSPHORIC ACID, DIETHYL ESTER, with 3-CHLORO-7-HYDROXY-4-METHYLCOUMARIN see CIK750

PHOSPHORIC ACID, DIETHYL ESTER, with ETHYL 3-HYDROXYCROTONATE see CBR750

PHOSPHORIC ACID, DIETHYL ESTER-N-NAPHTHALIMIDE derivative see HMV000

PHOSPHORIC ACID, DIETHYL ESTER, NAPHTHALIMIDO derivative see HMV000

PHOSPHORIC ACID-DIETHYL-(3-METHYL-5-PYRAZOLYL) ESTER see MOX250

PHOSPHORIC ACID DIETHYL 4-NITROPHENYL ESTER see NIM500

PHOSPHORIC ACID, 1-(DIMETHOXYPHOSPHINYL)ETHENYL DIMETHYL ESTER (9CI) see PGZ915

PHOSPHORIC ACID, DIMETHYL ESTER see PHC800

PHOSPHORIC ACID, DIMETHYL ESTER with p-CHLOROBENZYL-3-HYDROXYCROTONATE see CEQ500

PHOSPHORIC ACID, DIMETHYL ESTER, ESTER with DIMETHYL 3-HYDROXYGLUTACONATE see SOY000

PHOSPHORIC ACID, DIMETHYL ESTER, ESTER with cis-3-HYDROXY-N-METHYLCROTONAMIDE see MRH209

PHOSPHORIC ACID, DIMETHYL ESTER, ESTER WITH 2-CHLORO-N,N-DIETHYL-3-HYDROXYCROTANAMIDE, (Z)- see PGX275

PHOSPHORIC ACID, DIMETHYL ESTER, ester with N-HYDROXYNAPHTHALIMIDE see DOL400

PHOSPHORIC ACID, DIMETHYL 1-METHYL-3-(METHYLAMINO)-3-OXO-1-PROPENYL ESTER, (E)- see MRH209

PHOSPHORIC ACID DIMETHYL-p-(METHYLTHIO)PHENYL ESTER see PHD250

PHOSPHORIC ACID, DIMETHYL-p-NITROPHENYL ESTER see PHD500

PHOSPHORIC ACID, DIMETHYL-4-NITROPHENYL ESTER (9CI) see PHD500

PHOSPHORIC ACID, DIMETHYL-4-NITRO-m-TOLYL ESTER see PHD750

PHOSPHORIC ACID, DIMETHYL 2,2,2-TRICHLOROETHYL ESTER see TJE885

PHOSPHORIC ACID, DIMETHYL-3,5,6-TRICHLORO-2-PYRIDYL ESTER see PHE250

PHOSPHORIC ACID, DIPHENYL ESTER see PHE300

PHOSPHORIC ACID DIPROPYL-4-METHYLTHIOPHENYL ESTER see DWU400

PHOSPHORIC ACID, DISODIUM SALT see SJH090

PHOSPHORIC ACID, 1,2-ETHANEDIYL TETRAKIS(2-CHLOROETHYL) ESTER see TDG760

PHOSPHORIC ACID, ETHYLENE TETRAKIS(2-CHLOROETHYL) ESTER see TDG760

PHOSPHORIC ACID, 4-HEXYLPHENYL DIPHENYL ESTER see HFU700

PHOSPHORIC ACID, ISODECYL DIPHENYL ESTER see IKL100

PHOSPHORIC ACID, ISODECYL NICKEL(2+) SALT (2:3) see NDF000

PHOSPHORIC ACID, ISOPROPYL ESTER see PHE500

PHOSPHORIC ACID, LEAD(2+) SALT (2:3) see LDU000

PHOSPHORIC ACID, 2-(1-METHYLETHYL)PHENYL DIPHENYL ESTER see IRA100

PHOSPHORIC ACID, 4-(1-METHYLETHYL)PHENYL DIPHENYL ESTER see IRA200

PHOSPHORIC ACID, METHYLPHENYL DIPHENYL ESTER (9CI) see TGY750

PHOSPHORIC ACID, MONO(2,3-DIBROMOPROPYL) ESTER see MRH217

PHOSPHORIC ACID, MONOETHYL ESTER see ECI300

PHOSPHORIC ACID, MONO(1-METHYLETHYL) MONOPHENYL ESTER see IRC100

PHOSPHORIC ACID, MONOOCTYL ESTER see OFQ050

PHOSPHORIC ACID, 1,3-PHENYLENE TETRAPHENYL ESTER see REA050

PHOSPHORIC ACID, PHENYL ESTER, CYCLIC ESTER WITH o-HYDROXYBENZYL ALCOHOL see SAN300

PHOSPHORIC ACID, TRIALLYL ESTER see THN750

PHOSPHORIC ACID TRIAMIDE see PHB550

PHOSPHORIC ACID, 2,2,2-TRICHLOROETHYL ESTER see TIP300

PHOSPHORIC ACID, TRI-o-CRESYL ESTER see TMO600

PHOSPHORIC ACID TRIETHYLENE IMIDE see TND250

PHOSPHORIC ACID TRIETHYLENEIMINE (DOT) see TND250

PHOSPHORIC ACID, TRIISOBUTYL ESTER (8CI) see IJJ200

PHOSPHORIC ACID, TRIMETHYL ESTER see TMD250

PHOSPHORIC ACID, TRIPHENYL ESTER see TMT750

PHOSPHORIC ACID, TRI-2-PROPENYL ESTER see THN750

PHOSPHORIC ACID, TRIS(tert-BUTYLPHENYL) ESTER see TIA130

PHOSPHORIC ACID, TRIS(2-CHLORO-1-METHYLETHYL) ESTER see TNG000

PHOSPHORIC ACID, TRIS(2,3-DIBROMOPROPYL) ESTER see TNC500

PHOSPHORIC ACID, TRIS(2,3-DICHLOROPROPYL) ESTER see TNG750

PHOSPHORIC ACID TRIS(1,3-DICHLORO-2-PROPYL)ESTER see FQU875

PHOSPHORIC ACID, TRIS(2-ETHYLHEXYL) ESTER see TNI250

PHOSPHORIC ACID, TRIS(2-METHYLPHENYL) ESTER see TMO600

PHOSPHORIC ACID, TRIS(2-METHYLPROPYL) ESTER see IJJ200

PHOSPHORIC ACID, TRISODIUM SALT see SJH200

PHOSPHORIC ACID, TRISODIUM SALT, DODEAHYDRATE see SJH300

PHOSPHORIC ACID, TRISODIUM SALT, DODECAHYDRATE see SJH300

PHOSPHORIC ACID, TRIS(TRIMETHYLSILYL) ESTER see PHE750

PHOSPHORIC ACID, TRITOLYL ESTER see TNP500

PHOSPHORIC ACID, ZINC SALT (2:1), DIHYDRATE see ZGJ050

PHOSPHORIC BROMIDE see PHR250

PHOSPHORIC CHLORIDE see PHR500

PHOSPHORIC SULFIDE see PHS000

PHOSPHORIC TRIAMIDE see PHB550

PHOSPHORIC TRIS(DIMETHYLAMIDE) see HEK000

PHOSPHORIDOXAL COENZYME see PII100

PHOSPHOROCHLORIDOTHIOIC ACID-O,O-DIMETHYL ESTER see DTQ600

PHOSPHOROCHLORIDOTHIOIC ACID, cyclic O,O-(2-ETHYL-1-PROPYLTRIMETHYLENE) ESTER see CHH125

PHOSPHORODIAMIDIC FLUORIDE, N,N'-DI-N-BUTYL- see DDU700

PHOSPHORODICHLORIDIC ACID, ETHYL ESTER see EOR000

PHOSPHORODICHLORIDIC ACID, PHENYL ESTER see PHE800

PHOSPHORODICHLORIDOUS ACID, PHENYL ESTER see PHE850

PHOSPHORODIFLUORIDIC ACID see PHF250

PHOSPHORODI(ISOPROPYLAMIDIC) FLUORIDE see PHF750

PHOSPHORODITHIOIC ACID-O,O-BIS(1-METHYLETHYL)-S-((2-((PHENYLSULFONYL)AMINO)ETHYL ESTER see DNO800

PHOSPHORODITHIOIC ACID S-((tert-BUTYLTHIO)METHYL)-O,O-DIETHYL ESTER see BSO000

PHOSPHORODITHIOIC ACID, S-(6-CHLORO-3,4-DIHYDRO-2H-1-BENZOTHIOPYRAN-4-YL) o,o-DIETHYL ESTER see CLH810

PHOSPHORODITHIOIC ACID, S-(6-CHLORO-3,4-DIHYDRO-2H-1-BENZOTHIOPYRAN-4-YL) o,o-DIMETHYL ESTER see CLH820

PHOSPHORODITHIOIC ACID-S-(2-CHLORO-1-(1,3-DIHYDRO-1,3-DIOXO-2H-ISOINDOL-2-YL))ETHYL-O,O-DIETHYL ESTER see DBI099

PHOSPHORODITHIOIC ACID, S-(2-((4-CHLOROPHENYL)(1-METHYLETHYL)AMINO)-2-OXOETHYL)-,o,o-DIMETHYL ESTER see AOT255

PHOSPHORODITHIOIC ACID-S-((2-CHLORO-1-PHTHALIMIDOETHYL)-O,)-DIETHYL ESTER see DBI099

PHOSPHORODITHIOIC ACID, o,o-DIBUTYL ESTER, ZINC SALT see ZGA500

PHOSPHORODITHIOIC ACID, O,O-DIETHYL ESTER see PHG500

(E)-PHOSPHORODITHIOIC ACID-O,O-DIETHYL ESTER-S,S-DIESTER with p-DIOXANE-2,3-DIETHIOL see DVQ600

PHOSPHORODITHIOIC ACID, O,O-DIETHYL ESTER, S,S-DIESTER with THIODIMETHANETHIOL see PHG600

PHOSPHORODITHIOIC ACID, O,O-DIETHYL ESTER, S-ESTER with 6-CHLORO-3-(MERCAPTOMETHYL)-2-BENZOXAZOLINONE mixed with MILBEX (1:2) see MQT750

PHOSPHORODITHIOIC ACID-O,O-DIETHYL ESTER-S-ESTER with ETHYL MERCAPTOACETATE see DIX000

PHOSPHORODITHIOIC ACID-O,O-DIETHYL ESTER-S-ESTER with N-ISOPROPYL-2-MERCAPTOACETAMIDE see IOT000

PHOSPHORODITHIOIC ACID-O,O-DIETHYL ESTER, SODIUM SALT see PHG750

PHOSPHORODITHIOIC ACID, O,O-DIETHYL S-((ETHYLSULFINYL)METHYL) ESTER see TEZ200

PHOSPHORODITHIOIC ACID, O,O-DIETHYL S-((ETHYLSULFONYL)METHYL) ESTER see TEZ100

PHOSPHORODITHIOIC ACID, o,o-DIETHYL S-(2-(ETHYLTHIO)-6-METHYL-4-PYRIMIDINYL) ESTER see DJJ393

PHOSPHORODITHIOIC ACID, S-(3,4-DIHYDRO-2H-1-BENZOTHIOPYRAN-4-YL) o,o-DIMETHYL ESTER see TFE300

PHOSPHORODITHIOIC ACID, O,O-DIISOOCTYL S-(10-PHENARSAZINYL) ESTER see PDB300

PHOSPHORODITHIOIC ACID, S-(1,2-DIMETHOXYCARBONYL)ETHYL o,o-DIMETHYL ESTER see CBS800

PHOSPHORODITHIOIC ACID, O,O-DIMETHYL ESTER see PHH500

PHOSPHORODITHIOIC ACID-O,O-DIMETHYL ESTER-S-ESTER with DIETHYL MERCAPTOSUCCINATE see MAK700

PHOSPHORODITHIOIC ACID, O,O-DIMETHYL ESTER, S-ESTER with N-(2-MERCAPTOETHYL)ACETAMIDE see DOP200

PHOSPHORODITHIOIC ACID-O,O-DIMETHYL ESTER, S-ESTER with 2-MERCAPTO-N-METHYL-N-NITROSO ACETAMIDE see NKA500

PHOSPHORODITHIOIC ACID, o,o-DIMETHYL ESTER,S-ESTER WITH 1,2-BIS(METHOXYCARBONYL)ETHANETHIOL see CBS800

PHOSPHORODITHIOIC ACID, O,O-DIMETHYL ESTER, SODIUM SALT see PHI250

PHOSPHORODITHIOIC ACID, S-(1,1-DIMETHYLETHYL) o-ETHYL S-(1-METHYLPROPYL) ESTER see ENF050

PHOSPHORODITHIOIC ACID, O,O-DIMETHYL-S-(2-ETHYLTHIO)ETHYL ESTER see PHI500

PHOSPHORODITHIOIC ACID S-(((1,1-DIMETHYLETHYL)THIO)METHYL)-O,O-DIETHYL ESTER see BSO000

PHOSPHORODITHIOIC ACID-O,O-DIMETHYL-S-(2-(METHYLAMINO)-2-OXOETHYL) ESTER see DSP400

PHOSPHORODITHIOIC ACID-O,O-DIMETHYL-S-(2-((1-METHYLETHYL)THIO)ETHYL) ESTER see DSK800

PHOSPHORODITHIOIC ACID, o,o-DIMETHYL S-((5-(METHYLTHIO)-1,3,4-THIADIAZOL-2-YL)METHYL) ESTER see DST100

PHOSPHORODITHIOIC ACID, O,O-DIMETHYL S-(MORPHOLINOCARBONYLMETHYL) ESTER see MRU250

PHOSPHORODITHIOIC ACID-S,S'-1,4-DIOXANE-2,3-DIYL-O,O,O',O'-TETRAETHYL ESTER see DVQ709

PHOSPHORODITHIOIC ACID, O,O-DIPROPYL S-(2-PIPECOLINOCARBONYLMETHYL) ESTER see PIX775

PHOSPHORODITHIOIC ACID, O-ETHYL-,S,S-BIS(1-METHYLPROPYL)ESTER see EHY100

PHOSPHORODITHIOIC ACID, S-((ETHYLSULFINYL)METHYL) O,O-DIISOPROPYL ESTER see EPH500

PHOSPHORODITHIOIC ACID, ZINC SALT see ZGS000

PHOSPHORODITHIONIC ACID, S-2-(ETHYLTHIO)ETHYL-O,O-DIETHYL ESTER see DXH325

PHOSPHOROFLUORIDIC ACID see PHJ250

PHOSPHOROFLUORIDIC ACID, DIETHYL ESTER see DJJ400

PHOSPHOROFLUORIDIC ACID, DIISOPROPYL ESTER see IRF000

PHOSPHOROFLUORIDIC ACID, DIMETHYL ESTER see DSA800

PHOSPHOROMORPHOLINOTHIOIC ACID, o-METHYL S-(o',o'-DIETHYLPHOSPHORAMIDO)ETHYL ESTER see MJD300

PHOSPHOROOXYTRIAMIDE see PHB550

PHOSPHOROSELENOIC ACID, Se-(2-(DIETHYLAMINO)ETHYL) O,O-DIETHYL ESTER see DJA325

PHOSPHOROTHIOIC ACID see AMX825

PHOSPHOROTHIOIC ACID, o-(4-(1-((ACETYLOXY)IMINO)ETHYL)-3-METHYLPHENYL) o,o-DIETHYL ESTER see ACU300

PHOSPHOROTHIOIC ACID, S-(2-AMINOETHYL) ESTER see AKA900

PHOSPHOROTHIOIC ACID, O-(4-(AMINOSULFONYL)PHENYL) O,O-DIMETHYL ESTER (9CI) see CQL250

PHOSPHOROTHIOIC ACID-S-BENZYL-O,O-DIETHYL ESTER see DIU800

PHOSPHOROTHIOIC ACID, o,o-BIS(1-METHYLETHYL) ESTER see IRF500

PHOSPHOROTHIOIC ACID, o-(4-(((((BUTYLAMINO)CARBONYL)OXY)IMINO)METHYL)PHENYL) o,o-DIETHYL ESTER see DJW880

PHOSPHOROTHIOIC ACID, S-(4-CHLORO-2-BUTYNYL) O,o-DIETHYL ESTER see CEV830

PHOSPHOROTHIOIC ACID, O-(2-CHLORO-1-ISOPROPYLIMIDAZOL-4-YL) O,O-DIETHYL ESTER see PHK000

PHOSPHOROTHIOIC ACID, S-((6-CHLORO-2-OXOOXAZOLO(4,5-B)PYRIDIN-3(2H)-YL)METHYL) O,O-DIMETHYL ESTER see ARY800

PHOSPHORUS PENTACHLORIDE see PHR500
PHOSPHORUS PENTAFLUORIDE see PHR750
PHOSPHORUS PENTAOXIDE see PHS250
PHOSPHORUS PENTASULFIDE see PHS000
PHOSPHORUS PENTASULFIDE, free from yellow or white phosphorus (DOT) see PHS000
PHOSPHORUS PENTOXIDE see PHS250
PHOSPHORUS PERCHLORIDE see PHR500
PHOSPHORUS PERSULFIDE see PHS000
PHOSPHORUS SESQUISULFIDE see PHS500
PHOSPHORUS SESQUISULFIDE, free from yellow or white phosphorus (DOT) see PHS500
PHOSPHORUS (III) SULFIDE (IV) see PHS500
PHOSPHORUS SULFOBROMIDE see TFN750
PHOSPHORUS THIOCYANATE see PHS750
PHOSPHORUS TRIAMIDE, HEXAMETHYL- see HEK100
PHOSPHORUS TRIAZIDE see PHT000
PHOSPHORUS TRIBROMIDE see PHT250
PHOSPHORUS TRICHLORIDE see PHT275
PHOSPHORUS TRICYANIDE see PHQ250
PHOSPHORUS TRIFLUORIDE see PHQ500
PHOSPHORUS TRIHYDRIDE see PGY000
PHOSPHORUS TRIHYDROXIDE see PGZ899
PHOSPHORUS TRIOXIDE see PHT500
PHOSPHORUS TRISULFIDE see PHT750
PHOSPHORUS TRISULFIDE, free from yellow or white phosphorus (DOT) see PHT750
PHOSPHORUS, white or yellow, dry or under water or in solution (UN 1381) (DOT) see PHP010
PHOSPHORUS WHITE, molten (UN 2447) (DOT) see PHP010
PHOSPHORUS, YELLOW (ACGIH,OSHA) see PHP010
PHOSPHORWASSERSTOFF (GERMAN) see PGY000
PHOSPHORYL AMIDE see PHB550
PHOSPHORYL BROMIDE see PHU000
PHOSPHORYL CHLORIDE see PHQ800
PHOSPHORYLETHANOLAMINE see EED000
PHOSPHORYL FLUORIDE see PHR000
PHOSPHORYL HEXAMETHYLTRIAMIDE see HEK000
O-PHOSPHORYL-4-HYDROXY-N,N-DIMETHYLTRYPTAMINE see PHU500
PHOSPHORYL TRIAMIDE (PO(NH2)3) see PHB550
PHOSPHORYL TRIBROMIDE see PHU000
PHOSPHOSTIGMINE see PAK000
PHOSPHOTEX see TEE500
PHOSPHOTHION see MAK700
PHOSPHOTOX E see EEH600
PHOSPHOTRIAMIDE see PHB550
PHOSPHOTUNGSTIC ACID see PHU750
PHOSPHURE de MAGNESIUM (FRENCH) see MAI000
PHOSPHURES d'ALUMINUM (FRENCH) see AHE750
PHOSPHURE de SODIUM (FRENCH) see SJI500
PHOSPHURE de STRONTIUM (FRENCH) see SML000
PHOSPHURE de ZINC (FRENCH) see ZLS000
PHOSTEN HLP 1 see PKE850
PHOSTEX see BIT250
PHOSTIL see HBU415
PHOSTOXIN see AHE750
PHOSVEL see LEN000
PHOSVEL PHENOL see LEN050
PHOSVIN see ZLS000
PHOSVIT see DGP900
PHOTOBILINE see TDE750
PHOTODIELDRIN see PHV250
PHOTODYN see PHV275
PHOTOL see MGJ750

PHOTOMIREX see MRI750
PHOTOPHOR see CAW250
PHOTRIN see FOM000
PHOTRINE see FOM000
PHOXIME see BAT750
PHOXIM-METHYL see MOC275
PHOXIN see BAT750
PHOZALON see BDJ250
PHP see ELL500
PHPH see BGE000
PHPS see DYE700
Ph-QA 33 see INU200
PHRENOLAN see MDU750
PHRILON see NOH000
PHT see TMB750
PHT 4 see TBJ700
PHTALALDEHYDES (FRENCH) see PHV500
PHTALOPHOS see HMV000
PHTHALALDEHYDE see PHV500
p-PHTHALALDEHYDE see TAN500
PHTHALALDEHYDIC ACID see FNK010
PHTHALAMIC ACID, N-1-NAPHTHYL- see NBL200
PHTHALAMIC ACID, N-1-NAPHTHYL-, SODIUM SALT see SIO500
PHTHALAMIDE see BBO500
p-PHTHALAMIDE see TAN600
PHTHALAMIDE, N,N,N',N'-TETRAETHYL- see GCE600
PHTHALAMODINE see CLY600
PHTHALAMUDINE see CLY600
Δ1,α-PHTHALANACETIC ACID, 3-OXO- see CCH150
1,3-PHTHALANDIONE see PHW750
PHTHALANONE see PHW800
PHTHALAZINE, 1,4-DIHYDRAZINO- see OJD300
1,4-PHTHALAZINEDIONE, 5-AMINO-2,3-DIHYDRO- see AJO290
1,4-PHTHALAZINEDIONE, 2,3-DIHYDRO-, DIHYDRAZONE (9CI) see OJD300
PHTHALAZINOL see PHV725
PHTHALAZINOL (PHOSPHODIESTERASE INHIBITOR) see PHV725
PHTHALAZINONE see PHV750
1(2H)PHTHALAZINONE see PHV750
1(2H)-PHTHALAZINONE, 4-(2-(DIMETHYLAMINO)ETHOXY)-, OXIME, MONOHYDROCHLORIDE see TAC825
1(2H)-PHTHALAZINONE (1,3-DIMETHYL-2-BUTENYLIDENE)HYDRAZONE see DQU400
1(2H)-PHTHALAZINONE HYDRAZONE see HGP495
1(2H)-PHTHALAZINONE HYDRAZONE HYDROCHLORIDE see HGP500
1(2H)-PHTHALAZINONE, HYDRAZONE, MONOHYDROCHLORIDE see HGP500
3-(1-PHTHALAZINYL)CARBAZIC ACID ETHYL ESTER see CBS000
PHTHALAZOL see PHY750
PHTHALAZONE see PHV750
PHTHALETHAMIDE see GCE600
PHTHALIC ACID see PHW250
m-PHTHALIC ACID see IMJ000
PHTHALIC ACID ANHYDRIDE see PHW750
PHTHALIC ACID, BENZYL ESTER see MRK100
PHTHALIC ACID, BIS(2-CHLOROETHYL) ESTER see BIG600
o-PHTHALIC ACID BIS(DIETHYLAMIDE) see GCE600
PHTHALIC ACID BIS(2-METHOXYETHYL) ESTER see DOF400
PHTHALIC ACID, BIS(2-OCTYL) ESTER see BLB750
PHTHALIC ACID, DECYL HEXYL ESTER see PHW500
PHTHALIC ACID, DIALLYL ESTER see DBL200
o-PHTHALIC ACID, DIALLYL ESTER see DBL200

o-PHTHALIC ACID DIAMIDE see BBO500
PHTHALIC ACID, DICAPRYL ESTER see BLB750
PHTHALIC ACID, DICYCLOHEXYL ESTER see DGV700
PHTHALIC ACID, DIDODECYL ESTER see PHW550
PHTHALIC ACID, DIETHYL ESTER see DJX000
PHTHALIC ACID, DIGLYCIDYL ESTER see DKM600
PHTHALIC ACID, DIHEPTYL ESTER see HBP400
PHTHALIC ACID DIHEXYL ESTER see DKP600
PHTHALIC ACID, DIISODECYL ESTER see PHW575
PHTHALIC ACID, DIISONONYL ESTER see PHW585
PHTHALIC ACID, DIISOPROPYL ESTER see PHW600
PHTHALIC ACID DINITRILE see PHY000
PHTHALIC ACID DIOCTYL ESTER see DVL700
PHTHALIC ACID, DI-2-OCTYL ESTER see BLB750
PHTHALIC ACID, DIPENTYL ESTER see AON300
PHTHALIC ACID, DIPROPYL ESTER see DWV500
PHTHALIC ACID, DITRIDECYL ESTER see DXQ200
PHTHALIC ACID, HEXAHYDRO-, BIS(2,3-EPOXYPROPYL) ESTER see DKM500
PHTHALIC ACID, HEXAHYDRO-, DIGLYCIDYL ESTER see DKM500
PHTHALIC ACID METHYL ESTER see DTR200
PHTHALIC ACID, MONOETHYL ESTER see MRI275
PHTHALIC ACID, MONOPOTASSIUM SALT see HIB600
PHTHALIC ACID, TETRACHLORO- see TBT100
PHTHALIC ANHYDRIDE see PHW750
PHTHALIC ANHYDRIDE, TETRABROMO- see TBJ700
PHTHALIC ANHYDRIDE, TETRACHLORO- see TBT150
m-PHTHALIC DICHLORIDE see IMO000
o-PHTHALIC IMIDE see PHX000
PHTHALIDE see PHW800
PHTHALIDE 3,3,-BIS(p-HYDROXYPHENYL)- see PDO750
PHTHALIDE, 3-BUTYLIDENE- see BRQ100
PHTHALIDE, 3-(3-FURYL)-3a,4,5,6-TETRAHYDRO-3a,7-DIMETHYL- see FOM200
PHTHALIDE, 4,5,6,7-TETRACHLORO- see TBT175
PHTHALIMETTEN see PDO750
PHTHALIMIDE see PHX000
PHTHALIMIDE, N-(p-BROMOPHENYL)- see BNX300
PHTHALIMIDE, N-sec-BUTYL- see BSH600
PHTHALIMIDE, N-(CYCLOHEXYLTHIO)- see CPQ700
PHTHALIMIDE, N,N'-ETHYLENEBIS(TETRABROMO- see EIT150
PHTHALIMIDE, N-HYDROXY- see HNS600
PHTHALIMIDE, N-(HYDROXYMETHYL)- see HMP100
PHTHALIMIDE, N-METHYL- see MOC300
PHTHALIMIDE, 4-NITRO- see NIX100
PHTHALIMIDIMIDE see DNE400
PHTHALIMIDO-O,O-DIMETHYL PHOSPHORODITHIOATE see PHX250
2-PHTHALIMIDOGLUTARIC ACID ANHYDRIDE see PHX100
2-PHTHALIMIDOGLUTARIMIDE see TEH500

3-PHTHALIMIDOGLUTARIMIDE see TEH500

α-PHTHALIMIDOGLUTARIMIDE see TEH500

α-(N-PHTHALIMIDO)GLUTARIMIDE see TEH500

PHTHALIMIDOMETHYL ALCOHOL see HMP100

PHTHALIMIDOMETHYL-O,O-DIMETHYL PHOSPHORODITHIOATE see PHX250

PHTHALISOCIZER DINP see PHW585

(PHTHALOCYANINATO(2-))COPPER see CNO800

(PHTHALOCYANINATO(2−))MAGNESIUM see MAI250

PHTHALOCYANINE BLUE 01206 see DNE400

PHTHALOCYANINE BRILLIANT GREEN see PJQ100

(29H,31H-PHTHALOCYANINE-C-SULFONATO(3-)-N29,N30,N31,N32)COBALTATE(1-) HYDROGEN see CND940

PHTHALOCYANINE GREEN see PJQ100

PHTHALOCYANINE GREEN 6G see CMS140

PHTHALOCYANINE GREEN LX see PJQ100

PHTHALOCYANINE GREEN V see PJQ100

PHTHALOCYANINE GREEN VFT 1080 see PJQ100

PHTHALOCYANINE GREEN WDG 47 see PJQ100

PHTHALOCYANINESULFONIC ACID, COBALT COMPLEXES see CND940

PHTHALOCYANINESULFONIC ACID, COBALT DERIV. see CND940

PHTHALODINITRILE see PHY000

m-PHTHALODINITRILE see PHX550

o-PHTHALODINITRILE see PHY000

p-PHTHALODINITRILE see BBP250

PHTHALOGEN see DNE400

PHTHALOL see DJX000

PHTHALOLACTONE see PHW800

PHTHALONITRILE see PHY000

PHTHALOPHOS see PHX250

N-PHTHALOYL-l-ASPARTIC ACID see PHY250

m-PHTHALOYL CHLORIDE see IMO000

p-PHTHALOYL CHLORIDE see TAV250

PHTHALOYL DIAZIDE see PHY275

p-PHTHALOYL DICHLORIDE see TAV250

N-PHTHALOYL-l-GLUTAMIC ACID see PHY400

N-PHTHALOYLGLUTAMIMIDE see TEH500

PHTHALOYL PEROXIDE see PHY500

PHTHALOYLSULFATHIAZOLE see PHY750

PHTHALSAEUREANHYDRID (GERMAN) see PHW750

PHTHALSAEUREDIAETHYLESTER (GERMAN) see DJX000

PHTHALSAEUREDIMETHYLESTER (GERMAN) see DTR200

PHTHALTAN see TIT250

2-(N⁴-PHTHALYAMINOBENZENESULFONAMIDE)THIAZOLE see PHY750

2-(N⁴-PHTHALYAMINOBENZENESULFONAMIDO)THIAZOLE see PHY750

2-(p-PHTHALYLAMINOBENZENESULFAMIDO)THIAZOLE see PHY750

N-PHTHALYL-dl-ASPARTIMIDE see PIA000

o-PHTHALYLBIS(DIETHYLAMIDE) see GCE600

N-PHTHALYLGLUTAMIC ACID see PHY400

N-PHTHALYL-d,l-GLUTAMIC ACID see PHY400

N-PHTHALYLGLUTAMIC ACID IMIDE see TEH500

N-PHTHALYL-dl-GLUTAMIN see PIA100

N-PHTHALYL-d,l-GLUTAMINE see PIA100

N-PHTHALYL-GLUTAMINSAEURE-IMID (GERMAN) see TEH500

N-PHTHALYL-dl-GLUTAMINSAURE see PHY400

4-PHTHALYLGLUTARAMIC ACID see PIA250

α-N-PHTHALYLGLUTARAMIDE see TEH500

N-PHTHALYLISOGLUTAMINE see PIA250

PHTHALYLNORSULFAZOLE see PHY750

2-(N⁴-PHTHALYLSULFANILAMIDO)THIAZOLE see PHY750

2-(p-N-PHTHALYLSULFANILYL)AMINOTHIAZOLE see PHY750

PHTHALYLSULFATHIAZOLE see PHY750

PHTHALYLSULFONAZOLE see PHY750

PHTHALYLSULPHATHIAZOLE see PHY750

PHTHIOCOL see HMI000

1-PHTHLAZINYLHYDRAZONE ACETONE see HHE500

PHTIVAZID see VEZ925

PHTIVAZIDE see VEZ925

PHXA 41 see XAA500

PHYBAN see MRL750

PHYGON see DFT000

PHYGON PASTE see DFT000

PHYGON SEED PROTECTANT see DFT000

PHYGON XL see DFT000

PHYLANTHOSIDE see PIA300

PHYLCARDIN see TEP500

PHYLLEMBLIN see EKM100

PHYLLINDON see TEP500

PHYLLOCHINON (GERMAN) see VTA000

PHYLLOCORMIN N see HLC000

PHYLLOQUINONE see VTA000

trans-PHYLLOQUINONE see VTA000

α-PHYLLOQUINONE see VTA000

PHYMONE see NAK500

PHYOL see PJA250

PHYONE see PJA250

PHYSALIA PHYSALIS TOXIN see PIA375

PHYSALIN-X see PIA400

PHYSALIS (VARIOUS SPECIES) see JBS100

PHYSEPTONE see MDO750

PHYSEPTONE HYDROCHLORIDE see MDP750

PHYSEX see CMG675

PHYSIC NUT see CNR135

PHYSIOMYCINE see MDO250

PHYSOSTIGMINE see PIA500

PHYSOSTIGMINE SALICYLATE (1:1) see PIA750

PHYSOSTIGMINE SO4 see PIB000

PHYSOSTIGMINE SULFATE see PIB000

PHYSOSTIGMINE SULFATE (2:1) see PIB000

PHYSOSTOL see PIA500

PHYSOSTOL SALICYLATE see PIA750

PHYTAR see HKC000

PHYTAR 138 see HKC000

PHYTAR 560 see HKC000

PHYTAR 560 see HKC500

PHYTAR 600 see HKC000

PHYTIC ACID see PIB250

PHYTOGEL BASE see PIB300

PHYTOGERMINE see VSZ450

PHYTOGLYCOGEN see GHK300

PHYTOHAEMAGGLUTININ see PIB575

PHYTOHEMAGGLUTININ see PIB575

PHYTOHEMAGLUTININS see PIB575

PHYTOL see PIB600

trans-PHYTOL see PIB600

PHYTOLACCA AMERICANA see PJJ315

PHYTOLACCA DODECANDRA, extract see EAQ100

PHYTOMELIN see RSU000

PHYTOMENADIONE see VTA000

PHYTOMYCIN see SLY500

PHYTONADIONE see VTA000

PHYTOSOL see EPY000

PIACCAMIDE see DAB875

PIAFOL see MCB050

PIAMID see MCB050

PIANADALIN see BNK000

PIANIZOL see UTU500

PIAPONON see PMH500

PIATHERM see UTU500

PIATHERM D see UTU500

PIAZINE see POL490

PIBECARB see PCV350

PIBUTIDINE HYDROCHLORIDE see PIB650

PIC-CLOR see CKN500

PICCOLASTIC see SMQ500

PICCOLASTIC D-100 see SMQ500

PICCOLASTIC A see SMQ500

PICCOLASTIC A 5 see SMQ500

PICCOLASTIC A 25 see SMQ500

PICCOLASTIC A 50 see SMQ500

PICCOLASTIC A 75 see SMQ500

PICCOLASTIC C 125 see SMQ500

PICCOLASTIC D see SMQ500

PICCOLASTIC D 125 see SMQ500

PICCOLASTIC D 150 see SMQ500

PICCOLASTIC E 75 see SMQ500

PICCOLASTIC E 100 see SMQ500

PICCOLASTIC E 200 see SMQ500

PICEA OIL see SLG650

PICENADOL see PIB700

PICENE see PIB750

PICEOL see HIO000

PICFUME see CKN500

PICHLORAM K see PLQ760

PICHLORAM POTASSIUM SALT see PLQ760

PICHTOSIN see IHX600

PICHTOSINE see IHX600

PICHUCO (MEXICO) see ROU450

PICIS CARBONIS see CMY800

PICLORAM see PIB900

PICLORAM POTASSIUM SALT see PLQ760

PICOLINALDEHYDE, 2-PYRIDYLHYDRAZONE see DWX100

PICOLINAMIDE OXIME see PIB910

PICOLINAMIDOXIME see PIB910

2-PICOLINAMINE see ALB750

2-PICOLINE see MOY000

3-PICOLINE see PIB920

4-PICOLINE see MOY250

m-PICOLINE see PIB920

o-PICOLINE see MOY000

p-PICOLINE see MOY250

α-PICOLINE see MOY000

β-PICOLINE see PIB920

γ-PICOLINE see MOY250

PICOLINE-2-ALDEHYDE THIOSEMICARBAZONE see PIB925

3-PICOLINE, 6-AMINO- see AMA010

4-PICOLINE, 2-CHLORO-6-METHOXY-α-α-α-TRICHLORO- see CIC600

PICOLINIC ACID see PIB930

α-PICOLINIC ACID see ILC000

PICOLINIC ACID AMIDOXIME see PIB910

PICOLINIC ACID, 4-AMINO-3,5,6-TRICHLORO-, MONOPOTASSIUM SALT see PLQ760

PICOLINIC ACID, 5-BUTYL-, CALCIUM SALT, HYDRATE see FQR100

PICOLINIC ACID, 3,6-DICHLORO- see DGJ100

3-PICOLINIUM, 1,1'-(p-PHENYLENEBIS(CARBONYLMETHYL))DI-, DIBROMIDE see PEW650

PICOLINOHYDROXAMIC ACID, N-METHYL- see POQ330

2-PICOLYIDENEBIS(p-PHENYL SODIUM SULFATE) see SJJ175

α-PICOLYL ALCOHOL see POR800

β-PICOLYL ALCOHOL see NDW510

γ-PICOLYL ALCOHOL see POR810

2-PICOLYLAMINE see ALB750

4-PICOLYLAMINE see ALC250

2-PICOLYL CHLORIDE HYDROCHLORIDE see PIC000

4,4'-(2-PICOLYLIDENE)BIS(PHENYLSULFURIC ACID) DISODIUM SALT see SJJ175
N-(2-PICOLYL)-N-PHENYL-N-(2-PIPERIDINOETHYL)AMINE HYDROCHLORIDE see CNE375
N-(2-PICOLYL)-N-PHENYL-N-(2-PIPERIDINOETHYL)AMINE TRIPALMITATE see PIC100
PICOPERIDAMINE HYDROCHLORIDE see CNE375
PICOPERINE HYDROCHLORIDE see CNE375
PICOPERINE TRIPALMITATE see PIC100
PICOSULFATE SODIUM see SJJ175
PICOSULFOL see SJJ175
PICRACONITINE see PIC250
PICRAGOL see PID200
PICRAMIC ACID see DUP400
PICRAMIC ACID, SODIUM SALT see PIC500
PICRAMIC ACID, ZIRCONIUM SALT (WET) see PIC750
PICRAMIDE see PIC800
PICRAMIDE (DOT) see PIC800
PICRATES see PIC899
PICRATOL see ANS500
PICRIC ACID see PID000
PICRIC ACID (ACGIH,OSHA) see PID000
PICRIC ACID, AMMONIUM SALT see ANS500
PICRIC ACID, LEAD SALT see PID100
PICRIC ACID, wet, with not <10% water (NA 1344) (DOT) see PID000
PICRIC ACID, SILVER(1+) SALT see PID200
PICRIC ACID, dry or wetted with <30% water, by weight (UN 0154) (DOT) see PID000
PICRIDE see CKN500
PICRITE (the explosive) see NHA500
PICRITE, dry or wetted with <20% water, by weight (UN 0282) (DOT) see NHA500
PICRITE, wetted with not <20% water, by weight (UN 1336) (DOT) see NHA500
PICROLONIC ACID see PIE500
PICRONITRIC ACID see PID000
PICROPODOPHYLLIN see PIE100
PICROPODOPHYLLOTOXIN see PIE100
PICROTIN, compounded with PICROTOXININ (1:1) see PIE500
PICROTOL see PID200
PICROTOXIN see PIE500
PICROTOXINE see PIE500
PICROTOXININ see PIE510
PICROTOXININ, α-DIHYDRO- see DMB300
PICROTOXININE see PIE510
PICRYL AZIDE see PIE525
PICRYL CHLORIDE see PIE530
PICRYL CHLORIDE (DOT) see TML325
PICRYLMETHYLNITRAMINE see TEG250
PICRYLNITROMETHYLAMINE see TEG250
2-PICRYL-5-NITROTETRAZOLE see PIE550
PICRYL SULFIDE see BLR750
PICTYL see EOE200
PID see DVV600
PID see PFJ750
PIDIFIX 303 see MCB050
PIECIOCHLOREK FOSFORU (POLISH) see PHR500
PIED PIPER MOUSE SEED see SMN500
PIELIK see DAA800
PIELIK E see SGH500
PIE PLANT see RHZ600
PIERIS FLORIBUNDA see FBP520
PIERIS JAPONICA see FBP520
PIETIL see OOG000
PIFARNINE METHANESULFONATE see PIE750
PIFAZINE METHANESULFONATE see PIE750
PIGEON BERRY see GIW200
PIGEON-BERRY see PJJ315
PIGEON BERRY see ROA300

PIGEON PEA PHYTOHEMAGGLUTININ see PIB575
PIGLET PRO-GEN V see ARA500
PIGMENT ANTHRAQUINONE DEEP BLUE see IBV050
PIGMENT BLACK 7 see CBT750
PIGMENT BLUE 60 see IBV050
PIGMENT BLUE ANTHRAQUINONE see IBV050
PIGMENT BLUE ANTHRAQUINONE V see IBV050
PIGMENT DEEP BLUE ANTHRAQUINONE see IBV050
PIGMENT FAST GREEN G see PJQ100
PIGMENT FAST GREEN GN see PJQ100
PIGMENT FAST ORANGE see DVB800
PIGMENT FAST ORANGE G see CMS145
PIGMENT FAST SCARLET 3B see CMG750
PIGMENT FAST YELLOW GP see CMS210
PIGMENT FAST YELLOW 2GP see CMS210
PIGMENT GREEN 7 see PJQ100
PIGMENT GREEN 15 see LCR000
PIGMENT GREEN 38 see CMS140
PIGMENT GREEN PHTHALOCYANINE see PJQ100
PIGMENT GREEN PHTHALOCYANINE V see PJQ100
PIGMENT ORANGE 13 see CMS145
PIGMENT ORANGE ERH see CMS145
PIGMENT ORANGE G see CMS145
PIGMENT ORANGE ZH see CMS145
PIGMENT PONCEAU R see FMU070
PIGMENT RED 3 see MMP100
PIGMENT RED 4 see CJD500
PIGMENT RED 23 see NAY000
PIGMENT RED 53 see CMS150
PIGMENT RED 57 see CMS155
PIGMENT RED 60 see CMG750
PIGMENT RED 122 see DTV360
PIGMENT RED 48:1 see CMS148
PIGMENT RED 64:1 see CMS160
PIGMENT RED BH see NAY000
PIGMENT RED CD see CHP500
PIGMENT RED GG see CMS150
PIGMENT RED RL see MMP100
PIGMENT RUBINE B see CMS155
PIGMENT RUBINE BCL see CMS155
PIGMENT RUBY see MMP100
PIGMENT SCARLET see MMP100
PIGMENT SCARLET 3B see CMG750
PIGMENT SCARLET 25A see CMG750
PIGMENT SCARLET 829 see CMG750
PIGMENT SCARLET B see MMP100
PIGMENT SCARLET CP-1394 see CMG750
PIGMENT SCARLET 25AD see CMG750
PIGMENT SCARLET N see MMP100
PIGMENT SCARLET R see MMP100
PIGMENT SCARLET TONER RB see CMS160
PIGMENT TRANSPARENT YELLOW 2K see PIE800
PIGMENT TRANSPARENT YELLOW O see PIE830
PIGMENT WHITE 7 see ZNJ100
PIGMENT WHITE 18 see CAT775
PIGMENT WHITE 32 see ZJS400
PIGMENT YELLOW 13 see CMS208
PIGMENT YELLOW 14 see CMS210
PIGMENT YELLOW 2G see CMS210
PIGMENT YELLOW 33 see CAP750
PIGMENT YELLOW 97 see CMS214
PIGMENT YELLOW GGP see CMS210
PIGMENT YELLOW GPP see CMS210
PIGMENT YELLOW GT see DEU000
PIGMENT YELLOW MH see CMS208
PIGMENT YELLOW TRANSPARENT 2K see PIE800
PIGMENT YELLOW TRANSPARENT O see PIE830
PIGMEX see AEY000
PIGTAIL PLANT see APM875
PIG-WRACK see CCL250
PIH see PDN000

PIK-OFF see EEA000
PIKRINEZUUR (DUTCH) see PID000
PIKRINSAEURE (GERMAN) see PID000
PIKRYNOWY KWAS (POLISH) see PID000
PILEWORT see FBS100
PILIOPHEN see TDE750
PILLARDRIN see MRH209
PILLARON see DTQ400
PILLARQUAT see PAJ000
PILLARSTIN see MHC750
PILLARTAN see TBQ280
PILLARXONE see PAJ000
PILLARZO see CFX000
PILLS (INDIA) see SED400
PILOCARPINE see PIF000
PILOCARPINE HYDROCHLORIDE see PIF250
PILOCARPINE MONOHYDROCHLORIDE see PIF250
PILOCARPINE MONONITRATE see PIF500
PILOCARPINE MURIATE see PIF250
PILOCARPINE NITRATE see PIF500
PILOCARPOL see PIF000
PILOCEL see PIF250
PILOFRIN see PIF500
PILOMIOTIN see PIF250
PILORAL see FOS100
PILOT see QMA100
PILOT 447 see MKP500
PILOT HD-90 see DXW200
PILOT SF-40 see DXW200
PILOT STS 32 see SIK460
PILOVISC see PIF250
PILPOPHEN see DQA400
PILSICAINIDE HYDROCHLORIDE HEMIHYDRATE see PIF600
PIMACOL-SOL see NAK500
PIMAFUCIN see PIF750
PIMARICIN see PIF750
PIMELIC ACID see PIG000
PIMELIC ACID DINITRILE see HBD000
PIMELIC KETONE see CPC000
PIMELONITRILE see HBD000
4,4'-(PIMELOYLBIS(IMINO-p-PHENYLENEIMINO))BIS(1-ETHYLPYRIDINIUM) DIPERCHLORATE see PIG250
4,4'-(PIMELOYLBIS(IMINO-p-PHENYLENEIMINO))BIS(1-METHYLPYRIDINIUM) DIBROMIDE see PIG500
PIMENTA BERRY OIL see PIG740
PIMENTA LEAF OIL see PIG730
PIMENTA LEAF OIL see PIG740
PIMENTA OIL see PIG740
PIMENTA RACEMOSA OIL see LBK000
PIMENT BOUC (HAITI) see PCB275
PIMENT (HAITI) see PCB275
PIMENTO OIL see PIG740
PIMEPROFEN see IAB000
PIMETON see BJP250
PIMIENTA del BRAZIL (PUERTO RICO) see PCB300
PIMIENTO de AMERICA (CUBA) see PCB300
PIMM see PFN500
(−)-PIMOBENDAN see PIG800
(+)-PIMOBENDAN see PIG804
d-PIMOBENDAN see PIG804
l-PIMOBENDAN see PIG800
PIMOZIDE see PIH000
PIN see EBD700
PINACIDIL see COS500
PINACOL see TDR000
PINACOLIN see DQU000
PINACOLINE see DQU000
PINACOLONE see DQU000
PINACOLOXYMETHYLPHOSPHORYL FLUORIDE see SKS500
PINACOLYL ALCOHOL (6CI) see BRU300
PINACOLYL METHYLFLUOROPHOSPHONATE see SKS500

1-PIPERAZINEETHANOL see HKY500
1-PIPERAZINEETHANOL, 4-(2-METHOXYPHENYL)-α-((1-NAPHTHALENYLOXY)METHYL)- see MFG515
1-PIPERAZINEETHANOL, 4-(3-(10H-PYRIDO(3,2-B)(1,4)BENZOTHIAZIN-10-YL)PROPYL)-,DIHYDROCHLORIDE see ORI400
1-PIPERAZINEETHANOL, 4-(3-(TRIFLUOROMETHYL)PHENYL)-, MONO(2-(ACETYLOXY)BENZOATE) (SALT) see TKF723
PIPERAZINE ETHYLCARBOXYLATE see EHG050
PIPERAZINE HEXAHYDRATE see PIK500
PIPERAZINE HYDROCHLORIDE see PIK000
PIPERAZINE, 1-ISOPROPYL- see IRG025
PIPERAZINE, 1-(o-METHOXYPHENYL)-4-NICOTINOYLMETHYL- see KGK130
PIPERAZINE, 1-(o-METHOXYPHENYL)-4-(3,4,5-TRIMETHOXYBENZOYL)- see MFH760
PIPERAZINE, 1-(p-METHOXYPHENYL)-4-(3,4,5-TRIMETHOXYBENZOYL)- see MFH770
PIPERAZINE, 2-METHYL- see MOD100
PIPERAZINE, 1-(1-METHYLETHYL)-(9CI) see IRG025
PIPERAZINE, 1-(α-METHYLPHENETHYL)-4-PHENYL-, DIHYDROCHLORIDE see MNP450
PIPERAZINE, 1-PHENYL-4-(3,4,5-TRIMETHOXYBENZOYL)- see PFX600
PIPERAZINE, 1-(2-PYRIDYL)-4-(3,4,5-TRIMETHOXYBENZOYL)- see TKY300
PIPERAZINE, 1-(2-PYRIMIDYL)-4-(TRIMETHOXYBENZOYL)- see PPP550
PIPERAZINE and SODIUM NITRITE (4:1) see PIJ630
PIPERAZINE, compounded with 3-(SULFOOXY)ESTRA-1,3,5(10)-TRIEN-17-ONE (1:1) (9CI) see PIK450
PIPERAZINE SULTOSILATE see PIK625
PIPERAZINE, 1-(2-THIAZOLYL)-4-(3,4,5-TRIMETHOXYBENZOYL)- see TEX220
PIPERAZINE, 1-(m-TOLYL)-4-(3,4,5-TRIMETHOXYBENZOYL)- see THF300
PIPERAZINE, 1-(p-TOLYL)-4-(3,4,5-TRIMETHOXYBENZOYL)- see THF310
PIPERAZINIUM, 4-(β-CYCLOHEXYL-β-HYDROXYPHENETHYL)-1,1-DIMETHYL-, METHYL SULFATE see HFG400
2-(1-PIPERAZINYL)ETHANOL see HKY500
2-(1-PIPERAZINYL)-QUINOLINE (Z)-2-BUTENEDIOATE (1:1) (9CI) see QWJ500
2-(1-PIPERAZINYL)-QUINOLINE MALEATE (1:1) see QWJ500
PIPERIDILATE HYDROCHLORIDE see EOY000
PIPERIDIN (GERMAN) see PIL500
PIPERIDINE see PIL500
PIPERIDINE, 4-AMINO-2,2,6,6-TETRAMETHYL- see AMQ800
PIPERIDINE,3-((1,3-BENZODIOXOL-5-YLOXY)METHYL)-4-(4-FLUOROPHENYL)-, trans-(−)- see PAL650
PIPERIDINE, 1-(m-(BIS(2-CHLOROETHYL)AMINO)BENZOYL)- see BHP150
PIPERIDINE, 1-(3-(BIS(2-CHLOROETHYL)AMINO)-p-TOLUOYL)- see BIA100
PIPERIDINE, 1-(3-(3,5-BIS(TRIFLUOROMETHYL)PHENYL)-2-PROPYNYL-4-(1,1-DIMETHYLETHYL)-, HYDROCHLORIDE see PIL510
PIPERIDINE, 4-tert-BUTYL-1-NITROSO see NJO300

PIPERIDINE, 4-CARBETHOXY-1-(3-HYDROXY-3-PHENYLPROPYL)-4-PHENYL-, HYDROCHLORIDE see PDO900
1-PIPERIDINECARBODITHIOIC ACID, compounded with PIPERIDINE see PIY500
PIPERIDINE-N-CARBONIC ACID AMIDE see PIL525
1-PIPERIDINECARBOTHIOIC ACID, S-(1-METHYL-1-PHENYLETHYL) ESTER see MNU300
1-PIPERIDINECARBOXAMIDE see PIL525
1-PIPERIDINECARBOXAMIDE, N-BUTYL-2,6-DIMETHYL- see BRE300
2-PIPERIDINECARBOXAMIDE,1-BUTYL-N-(2,6-DIMETHYLPHENYL)MONOHYDROCHLORIDE MONOHYDRATE see BOO000
2-PIPERIDINECARBOXAMIDE, N-(2,6-DIMETHYLPHENYL)-1-PROPYL-(R)-(9CI) see RMA600
2-PIPERIDINECARBOXYLIC ACID (9CI) see HDS300
4-PIPERIDINECARBOXYLIC ACID (9CI) see ILG100
4-PIPERIDINECARBOXYLIC ACID, 1-(4-CYANO-4-(4-FLUOROPHENYL)CYCLOHEXYL)-3-METHYL-4-PHENYL-, MONOHYDROCHLORIDE, (3S-(1(CIS),3α,4β))- see LFA200
1-PIPERIDINECARBOXYLIC ACID, 2-(2-HYDROXYETHYL)-, 1-METHYLPROPYL ESTER see MOT100
4-PIPERIDINECARBOXYLIC ACID, 1-(3-HYDROXY-3-PHENYLPROPYL)-4-PHENYL-, ETHYL ESTER, HCL see PDO900
(PIPERIDINE)CHROMIUM PENTACARBONYL see PAU700
PIPERIDINE, 1-(3-CYCLOHEXEN-1-YLCARBONYL)-2-METHYL-(9CI) see CPD630
PIPERIDINE, 1-(CYCLOHEXYLCARBONYL)-3-METHYL-see CPI350
PIPERIDINE, 4-(5H-DIBENZO(A,D)CYCLOHEPTEN-5-YLIDENE)- see DBA550
PIPERIDINE, 2,6-DIMETHYL- see LIQ550
PIPERIDINE, 1-(3-(4-(1,1-DIMETHYLETHYL)PHENYL)-2-METHYLPROPYL)- see FAQ230
PIPERIDINE, 2,6-DIMETHYL-1-((2-METHYLCYCLOHEXYL)CARBONYL)- see DSP650
PIPERIDINE, 3,5-DIMETHYL-1-NITROSO-, (E)- see DTA700
PIPERIDINE, 3,5-DIMETHYL-1-NITROSO-, (Z)- see DTA690
2,6-PIPERIDINEDIONE, 3-(4-AMINOPHENYL)-3-ETHYL- see AKC600
3-PIPERIDINE-1,1-DIPHENYL-PROPANOL-(1) METHANESULPHONATE see PIL550
1-PIPERIDINEETHANOL see HKY600
2-PIPERIDINEETHANOL see HKY600
N-PIPERIDINEETHANOL see HKY600
2-PIPERIDINEETHANOL-m-AMINOBENZOATE (ester) HYDROCHLORIDE see AIQ880
2-PIPERIDINEETHANOL-o-AMINOBENZOATE (ester) HYDROCHLORIDE see AIQ885
2-PIPERIDINEETHANOL-p-AMINOBENZOATE (ester) HYDROCHLORIDE see AIQ890
1-PIPERIDINEETHANOL BENZILATE HYDROCHLORIDE see PIM000
2-PIPERIDINEETHANOL CARBANILATE (ester) HYDROCHLORIDE see PIU800
1-PIPERIDINEETHANOL, α-(8-(TRIFLUOROMETHYL)NAPHTHO(2,1-B)THIEN-4-YL)-, HYDROCHLORIDE see TKD390

PIPERIDINE, 2-ETHYL-1-(3-METHYL-1-OXO-2-BUTENYL)- see EMY100
PIPERIDINE HYDROCHLORIDE see HET000
PIPERIDINE, 1-HYDROXY- (9CI) see PIO925
4-PIPERIDINEMETHANOL, 1-BROMOACETYL-α-α-DIPHENYL- see BMS300
4-PIPERIDINEMETHANOL, 1-CHLOROACETYL-α-α-DIPHENYL- see CEC300
4-PIPERIDINEMETHANOL, 1-(IODOACETYL)-α-α-DIPHENYL- see IDZ200
PIPERIDINE, 3-((4-METHOXYPHENOXY)METHYL)-1-METHYL-4-PHENYL-, HYDROCHLORIDE, (3R-trans)- see FAJ200
PIPERIDINE, 2-METHYL-1-((6-METHYL-3-CYCLOHEXEN-1-YL)CARBONYL)- see MLL650
PIPERIDINE, 3-METHYL-1-((6-METHYL-3-CYCLOHEXEN-1-YL)CARBONYL)- see MLL655
PIPERIDINE, 4-METHYL-1-((6-METHYL-3-CYCLOHEXEN-1-YL)CARBONYL)- see MLL660
PIPERIDINE, 4-METHYL-1-((2-METHYLCYCLOHEXYL)CARBONYL)- see MLM700
PIPERIDINE, 1-(1-METHYL-2-(2-(PHENYLMETHYL)PHENOXY)ETHYL)-(9CI) see MOA600
PIPERIDINE, 1-(1-METHYL-2-((α-PHENYL-o-TOLYL)OXY)ETHYL)- see MOA600
PIPERIDINE, 1-((1-METHYL-3-PIPERIDINYL)CARBONYL)- see MOH310
PIPERIDINE, 1,1'-(Δ,Δ'-NITRILODITETRAMETHYLENE)BIS(γ-BUTYL-γ-(1-NAPHTH YL)- see NEH600
PIPERIDINE, 4-(2-OXAZOLIN-2-YLAMINO)-2,2,6,6-TETRAMETHYL- see TDT900
PIPERIDINEPENTACARBONYLCHROMIUM see PAU700
PIPERIDINE, 1-((1-(2-PHENYLETHYL)-3-PIPERIDINYL)CARBONYL)- see PDI550
PIPERIDINE, 1-(3-PIPERIDYL)CARBONYL-see PIR200
1-PIPERIDINEPROPANENITRILE see PIM250
1-PIPERIDINEPROPIONITRILE see PIM250
3-(1-PIPERIDINE)PROPIONITRILE see PIM250
PIPERIDINE, 1-PROPYL- see PNS800
PIPERIDINE, 2-PROPYL-, (S)- see PNT000
PIPERIDINE, 1-(2-PROPYLVALERYL)- see PNX800
PIPERIDINIC ACID see PIM500
PIPERIDINIUM see PIY500
PIPERIDINIUM, 1,1'-(2-β,16-β-(3-α,17-β-DIHYDROXY-5-α-ANDROSTANYLENE))BIS(1-METHYL-, DIBROMIDE see AOO403
PIPERIDINIUM, 1,1-DIMETHYL-, CHLORIDE see MCH540
2-PIPERIDINO-p-ACETOPHENETIDIDE HYDROCHLORIDE see PIM750
2-PIPERIDINO-2',6'-ACETOXYLIDIDE HYDROCHLORIDE see PIN000
4-PIPERIDINOACETYL-3,4-DIHYDRO-2H-1,4-BENZOXAZINE HYDROCHLORIDE see PIN100
β-PIPERIDINOAETHYL-(3-CHLOR-4-n-BUTOXY-5-METHYLPHENYL)KETONHYDROCHLORID (GERMAN) see BPI625
β-PIPERIDINOAETHYL-(3-CHLOR-4-PROPOXY-5-METHYLPHENYL)-KETONHYDROCHLORID (GERMAN) see CIU325

9-(PIPERIDINOAMINO)ACRIDINE see PIN200

3-(PIPERIDINOCARBONYL)PIPERIDINE see PIR200

PIPERIDINOCYCLOHEXANECARBONITRILE see PIN225

1-PIPERIDINOCYCLOHEXANECARBONITRILE see PIN225

α-(1-PIPERIDINO)-CYCLOHEXYL PHENYL KETONE see PFY200

6-PIPERIDINO-2,4-DIAMINOPYRIMIDINE-3-OXIDE see DCB000

β-PIPERIDINOETHANOL see HKY600

2-PIPERIDINOETHANOL HYDROCHLORIDE see PIN275

2-(1-PIPERIDINO)-ETHYL BENZILATE ETHYLBROMIDE see PIX800

2-(1-PIPERIDINO)ETHYL BENZILATE HYDROCHLORIDE see PIM000

N-(2-PIPERIDINOETHYL)CARBAMIC ACID, 6-CHLORO-o-TOLYL ESTER, HYDROCHLORIDE see PIO000

N-(2-PIPERIDINOETHYL)CARBAMIC ACID, MESITYL ESTER, HYDROCHLORIDE see PIO250

N-(2-(PIPERIDINO)ETHYL)CARBAMIC ACID, 2,6-XYLYL ESTER, HYDROCHLORIDE see PIO500

PIPERIDINOETHYL CHLORIDE, HYDROCHLORIDE see CHG500

PIPERIDINOETHYL-2-HEPTOXYPHENYLCARBAMOATE HYDROCHLORIDE see PIO750

PIPERIDINOETHYL-3-METHYLFLAVONE-8-CARBOXYLATE HYDROCHLORIDE see FCB100

2-PIPERIDINOETHYL-3-METHYL-4-OXO-2-PHENYL-4H-1-BENZOPYRAN-8-CARBOXYLATE HYDROCHLORIDE see FCB100

N-(2-PIPERIDINOETHYL)-N-(2-PYRIDYLMETHYL)ANILINE HYDROCHLORIDE see CNE375

N-(2-PIPERIDINOETHYL)-N-(2-PYRIDYLMETHYL)ANILINE TRIPALMITATE see PIC100

PIPERIDINO HEXOSE REDUCTONE see PIO900

1-PIPERIDINOL see PIO925

4-PIPERIDINOL, 1-(3-(2-ACETYLPHENOTHIAZIN-10-YL)PROPYL)- see HNT200

4-PIPERIDINOL, 1-BENZYL-4-ETHYNYL-3-(1-(3-INDOLYL)ETHYL)- see BEO500

4-PIPERIDINOL, 4-(4-CHLOROPHENYL)-1-(1H-INDOL-3-YLMETHYL)- see CKA700

4-PIPERIDINOL, 4-ETHYNYL-3-(1-(1H-INDOL-3-YL)ETHYL)-1-(PHENYLMETHYL)- see BEO500

2-PIPERIDINOMETHYL-1,4-BENZODIOXAN HYDROCHLORIDE see BCI500

PIPERIDINOMETHYLCYCLOHEXANE CAMPHOSULFATE see PIQ750

PIPERIDINOMETHYLCYCLOHEXANE CHLORHYDRATE SALT see PIR000

2-PIPERIDINOMETHYL-4-METHYL-1-TETRALONE HYDROCHLORIDE see PIR100

α-(PIPERIDINOMETHYL)-5-PHENYL-3-ISOXAZOLEMETHANOL CITRATE see HNT075

1-PIPERIDINO-2-METHYL-3-(p-TOLYL)-3-PROPANONE see TGK200

1-PIPERIDINO-2-METHYL-3-(p-TOLYL)-3-PROPANONE HYDROCHLORIDE see MRW125

4-PIPERIDINONE, 3-(1-(1H-INDOL-3-YL)ETHYL)-1-(PHENYLMETHYL)- see ICW150

4-PIPERIDINONE, 3-((2-METHYL-1H-INDOL-3-YL)METHYL)-1-(PHENYLMETHYL)- see MKW150

4-PIPERIDINONE, 2,2,6,6-TETRAMETHYL-(9CI) see TDT770

1-PIPERIDINO-3-(p-OCTYLPHENYL)-3-PROPANONE HYDROCHLORIDE see PIY000

1-PIPERIDINO-3-(4'-OCTYLPHENYL)-PROPAN-3-ON-HYDROCHLORID (GERMAN) see PIY000

PIPERIDINOOXY, 4-HYDROXY-2,2,6,6-TETRAMETHYL- see HOI300

PIPERIDINOOXY, 2,2,6,6-TETRAMETHYL-see TDT800

1-PIPERIDINO-2-PHENYL-AETHAN (GERMAN) see PDI500

3-PIPERIDINO-1-PHENYL-1-BICYCLOHEPTENYL-1-PROPANOL see BGD500

3-PIPERIDINO-1-PHENYL-1-BICYCLO(2.2.1)HEPTEN-(5)-YL-PROPANOL-(1)(GERMAN) see BGD500

1-PIPERIDINO-2-PHENYLETHANE see PDI500

PIPERIDINO 3-PIPERIDYL KETONE see PIR200

3-PIPERIDINO-1,2-PROPANEDIOL DICARBANILATE see DVO819

3-PIPERIDINO-1,2-PROPANEDIOL DICARBANILATE HYDROCHLORIDE see DVV500

3-γ-(1-PIPERIDINO)PROPIL-5,5-DIFENILTIOIDANTOINA CLORIDRATO (ITALIAN) see DWK500

4-(3-PIPERIDINOPROPIONAMIDO)SALICYLIC ACID METHYL ESTER, METHIODIDE see MOK000

3-PIPERIDINOPROPIONITRILE see PIM250

β-PIPERIDINOPROPIONITRILE see PIM250

β-PIPERIDINOPROPIOPHENONE HYDROCHLORIDE see PIV500

3-PIPERIDINO-4'-PROPOXYPROPIOPHENONE HYDROCHLORIDE see PNB250

γ-PIPERIDINOPROPYL-p-AMINOBENZOATE HYDROCHLORIDE see PIS500

α-(3-PIPERIDINOPROPYL)BENZHYDROL see DWK200

2-(1-PIPERIDINO)-2-(2-THENYL)ETHYLAMINE MALEATE see PIT250

1-Δ-3-PIPERIDINO-3-o-TOLOXYPROPAN-2-OL HYDROCHLORIDE see TGK225

PIPERIDINPENTAKARBONYLCHROMIUM see PAU700

N-(PIPERIDIN-1-YL)-5-(4-CHLOROPHENYL)-1-(2,4-DICHLOROPHENYL)-4-METHYL-1H-PYRAZOLE-3- see CCC290

1-(1-PIPERIDINYL)-CYCLOHEXANECARBONITRILE (9CI) see PIN225

2-(1-PIPERIDINYL)ETHANOL see HKY600

2-(1-PIPERIDINYL)ETHANOL HYDROCHLORIDE see PIN275

PIPERIDINYLETHYLMORPHINE see PIT600

N-(2-PIPERIDINYLMETHYL)-2,5-BIS(2,2,2-TRIFLUOROETHOXY)BENZAMIDE see FCC065

5-(1-PIPERIDINYL)NAPHTHO(2,3-H)QUINOLINE-7,12-DIONE see PIT625

1-PIPERIDINYLOXY, 4-HYDROXY-2,2,6,6-TETRAMETHYL-(9CI) see HOI300

1-PIPERIDINYLOXY, 2,2,6,6-TETRAMETHYL-(9CI) see TDT800

3-(1-PIPERIDINYL)-1-(4-PROPOXYPHENYL)-1-PROPANONE HYDROCHLORIDE see PNB250

3-(2-PIPERIDINYL)PYRIDINE see AON875

(S)-3-(2-PIPERIDINYL)PYRIDINE HYDROCHLORIDE see PIT650

(S)-3-(2-PIPERIDINYL)PYRIDINE SULFATE see PIV550

6-(1-PIPERIDINYL)-2,4-PYRIMIDINEDIAMINE-3-OXIDE see DCB000

PIPERIDOLATE HYDROCHLORIDE see EOY000

PIPERIDON (GERMAN) see PIU000

2-PIPERIDONE see PIU000

4-PIPERIDONE, 2,2,6,6-TETRAMETHYL- see TDT770

α-(2-PIPERIDYL)BENZHYDROL see DWK400

α-(2-PIPERIDYL)BENZHYDROL HYDROCHLORIDE see PII750

α-(4-PIPERIDYL)BENZHYDROL HYDROCHLORIDE see PIY750

4-(1-PIPERIDYL)CARBONYL-2,3-TETRAMETHYLENEQUINOLINE see PIU100

3-(1-PIPERIDYL)-1-CYCLOHEXYL-1-PHENYL-1-PROPANOL HYDROCHLORIDE see BBV000

α-PIPERIDYL 6,8-DICHLORO-2-PHENYL-4-QUINOLINE METHANOL HYDROCHLORIDE see PIU200

β-PIPERIDYLETHANOL see HKY600

β-PIPERIDYLETHANOL HYDROCHLORIDE see PIN275

β-2-PIPERIDYLETHYL-m-AMINOBENZOATE HYDROCHLORIDE see AIQ880

β-2-PIPERIDYLETHYL-o-AMINOBENZOATE HYDROCHLORIDE see AIQ885

β-2-PIPERIDYLETHYL-p-AMINOBENZOATE HYDROCHLORIDE see AIQ890

α-(2-PIPERIDYLETHYL)BENZHYDROL HYDROCHLORIDE see PMC250

β-2-PIPERIDYLETHYLPHENYLURETHANE HYDROCHLORIDE see PIU800

2-(1-PIPERIDYLMETHYL)-1,4-BENZODIOXAN HYDROCHLORIDE see BCI500

1-(N-PIPERIDYL)-6-METHYL-3-PHENYL HEPTANOL HYDROCHLORIDE see HBF500

3-(1-PIPERIDYL)-1,2-PROPANE DICARBANILATE see DVO819

3-(1-PIPERIDYL)-1,2-PROPANEDIOL DICARBANILATE HYDROCHLORIDE see DVV500

1-(2-PIPERIDYL)-2-PROPANONE see PAO500

β-(1-PIPERIDYL)PROPIOPHENONE HYDROCHLORIDE see PIV500

3-(2-PIPERIDYL)-PYRIDINE see AON875

1-3-(2'-PIPERIDYL)PYRIDINE see AON875

3-(2-PIPERIDYL)PYRIDYL SULFATE see PIV550

PIPERILATE HYDROCHLORIDE see PIM000

PIPERIN see PIV600

PIPERINE see PIV600

PIPERITONE see MCF250

PIPER LONGUM L., fruit extract see PIV650

PIPEROCAINE see PIV750

PIPEROCAINE HYDROCHLORIDE see IJZ000

PIPEROCAINIUM CHLORIDE see IJZ000

PIPEROCYANOMAZINE see PIW000

PIPEROLINIC ACID see HDS300

PIPERONAL see PIW250

PIPERONALACETONE see MJR250

PIPERONAL BIS(2-(2-BUTOXYETHOXY)ETHYL)ACETAL see PIZ499

PIPERONALDEHYDE see PIW250

PIPERONAL DIETHYL ACETAL see PIW500
PIPERONYL see FHG000
PIPERONYL ACETATE see PIX000
PIPERONYL ACETONE see MJR250
N-PIPERONYLALANINE see PIX100
PIPERONYL ALDEHYDE see PIW250
PIPERONYL BUTOXIDE see PIX300
PIPERONYL CYCLOHEXANONE see
PIX300
PIPERONYLIDENEACETONE see MJR250
2-(4-PIPERONYL-1-PIPERAZINYL)-9H-
PURINE-9-ETHANOL
DIHYDROCHLORIDE see PIX750
2-(4-PIPERONYL-1-
PIPERAZINYL)PYRIMIDINE see TNR485
PIPERONYL SULFOXIDE see ISA000
1-PIPERONYL-4-(3,7,11-TRIMETHYL-2,6,10-
DODECANTRIENYL)-PIPERAZINE
METHANESULFONATE see PIE750
PIPEROPHOS see PIX775
PIPEROXANE HYDROCHLORIDE see
BCI500
1-PIPEROYLPIPERIDINE see PIV600
PIPERSAL see DAM600
PIPERYLENE see PBA100
trans-PIPERYLENE see PBA250
PIPE STEM see CMV390
PIPETHANATE ETHYLBROMIDE see
PIX800
PIPETHANATE HYDROCHLORIDE see
PIM000
PIPIZAN CITRATE SYRUP see PIJ500
PIPOBROMAN see BHJ250
PIPOCTANONE HYDROCHLORIDE see
PIY000
PIPOLPHEN see DQA400
PIP-PIP see PIY500
PIPRACIL see SJJ200
PIPRADOL see DWK400
α-PIPRADOL see DWK400
γ-PIPRADOL see PIY750
PIPRADOL HYDROCHLORIDE see PII750
PIPRADROL HYDROCHLORIDE see PII750
PIPRAM see PIZ000
PIPRIL see SJJ200
PIPRINHYDRINATE see PIZ250
PIPROCTANYLIUMBROMID (GERMAN) see
AGE750
PIPROTAL see PIZ499
PIPROZOLIN see EOA500
PIPROZOLINE see EOA500
PIPTAL see PJA000
PIRABUTINA see HNI500
PIRACAPS see TBX250
PIRACETAM see NNE400
PIRAFLOGIN see HNI500
PIRAMIDON see DOT000
PIRARREUMOL "B" see BRF500
PIRAZETAM see NNE400
PIRAZINON see DJX400
PIRAZOXON (ITALIAN) see MOX250
PIRBUTEROL DIHYDROCHLORIDE see
POM800
PIRECIN see POF500
PIREF see DIZ100
PIRENZEPINE HYDROCHLORIDE see
GCE500
PIRETANIDE see PDW250
PIRETRINA 1 (PORTUGUESE) see POO050
PIREVAN see PJA120
PIREXYL see MOA600
PIREXYL PHOSPHATE see PJA130
PIRFLOXACIN see FLV050
PIRIA'S ACID see ALI000
PIRIBEDIL see TNR485
PIRIBENZIL see TMP750
PIRIBENZIL METHYL SULFATE see
CDG250
PIRID see PDC250
PIRID see PEK250
PIRIDACIL see PDC250
PIRIDANE see CMA100

7-(2-(PIRIDIL)-METILAMMINO-ETIL)-
TEOFILLINO NICOTINATO (ITALIAN) see
PPN000
PIRIDINA (ITALIAN) see POP250
PIRIDINOL CARBAMATO (SPANISH) see
PPH050
PIRIDISIR see PPP500
PIRIDOL see DOT000
PIRIDOLAN see PJA140
PIRIDOLO see AKO500
PIRIDOSAL see DAM600
PIRIDOSAL see DAM700
PIRIDROL see DWK400
PIRIDROL HYDROCHLORIDE see PII750
PIRIEX see TAI500
PIRIMAL-M see ALF250
PIRIMECIDAN see TGD000
PIRIMETAMINA (SPANISH) see TGD000
PIRIMICARB see DOX600
PIRIMIFOSETHYL see DIN600
PIRIMIFOS-METHYL see DIN800
PIRIMIPHOS-ETHYL see DIN600
PIRIMOR see DOX600
PIRINITRAMIDE see PJA140
PIRINIXIL see CLW500
PIRISTIN see POO750
PIRITON see CLX390
PIRITON see TAI500
PIRITRAMIDE see PJA140
PIRLINDOLE see HDS225
PIRLINDOLE HYDROCHLORIDE see
HDS225
PIRLINDON see HDS225
PIRMAZIN see SNJ000
PIRMENOL HYDROCHLORIDE see PJA170
PIROAN see PCP250
PIROCARD BLACK B see CMS250
PIROCARD BLACK IT see CMS250
PIROCARD BLACK PL see CMS250
PIROCARD BLACK PSE see CMS250
PIROCARD BLACK PSR see CMS250
PIROCARD BLACK T see CMS250
PIROCARD BLACK TR see CMS250
PIROCARD CORINTH 4R see CMS257
PIROCARD CORINTH MR see CMS257
PIROCARD CORINTH RB see CMS257
PIROCARD CORINTH SX see CMS257
PIROCARD CORINTH SX EXTRA see
CMS257
PIROCRID see POE100
PIROCTONE OLAMINE see PJA180
PIROD see UNJ800
PIRODAL see PAF250
PIROFOS see SOD100
PIROHEPTINE see PJA190
PIROHEPTINE HYDROCHLORIDE see
TMK150
PIROMEN see PJA200
PIROMIDIC ACID see PAF250
PIROMIDINA see DOT000
PIROSOLVINA see EEM000
PIROXANTRONE see XDJ025
PIROXICAM see FAJ100
PIRPROFEN see PJA220
PIRROLIDINOMETIL-TETRACICLINA
(ITALIAN) see PPY250
PIRROXIL see NNE400
PIRVINIUM PAMOATE see PQC500
PIRYDYNA (POLISH) see POP250
PISCAROL see IAD000
PISCIDEIN see HJO500
PISCIDIC ACID see HJO500
PISCIOL see IAD000
PISS-A-BED (JAMAICA) see CNG825
PISTACIA LENTISCUS ABSOLUTE see
MBU777
PITAYINE see QFS000
PITC see ISQ000
PITCH see CMZ100
PITCH APPLE see BAE325
PITCH, COAL TAR see CMZ100
PITMAL see TLP750

PITOCIN see ORU500
PITON S see ORU500
PITREX see TGB475
PITTCHLOR see HOV500
PITTCIDE see HOV500
PITTCLOR see HOV500
PITTSBURGH PX-138 see DVL700
PITUITARY GLAND ADRENO CORTICO-
TROPIC HORMONE see AES650
PITUITARY GROWTH HORMONE see
PJA250
PITUITARY LACTOGENIC HORMONE see
PMH625
PITUITARY LUTEINIZING HORMONE see
LIU300
PIVACIN see PIH175
PIVADORM see BNP750
PIVADORN see BNP750
PIVAL see PIH175
PIVALDION (ITALIAN) see PIH175
PIVALDIONE (FRENCH) see PIH175
PIVALIC ACID see PJA500
PIVALIC ACID CHLORIDE see DTS400
PIVALIC ACID, HEXYL ESTER see HFP700
PIVALIC ACID LACTONE see DTH000
PIVALIC ACID, SODIUM SALT see SGM600
PIVALIC ACID, TRIMETHYLACETIC ACID
see SGM600
PIVALOLACTONE see DTH000
PIVALOLYL CHLORIDE see DTS400
PIVALONITRILE see PJA750
PIVALOYL AZIDE see PJB000
PIVALOYL CHLORIDE see DTS400
2-PIVALOYL-INDAAN-1,3-DION (DUTCH)
see PIH175
2-PIVALOYL-INDAN-1,3-DION (GERMAN)
see PIH175
2-PIVALOYL-1,3-INDANDIONE see PIH175
2-PIVALOYLINDANE-1,3-DIONE see
PIH175
PIVALOYLMETHYL BROMIDE see PJB100
PIVALOYLOXYMETHYL d-α-
AMINOBENZYLPENICILLINATE
HYDROCHLORIDE see AOD000
PIVALYL CHLORIDE see DTS400
2-PIVALYL-1,3-INDANDIONE see PIH175
PIVALYL VALONE see PIH175
PIVALYN see PIH175
PIVAMPICILLIN HYDROCHLORIDE see
AOD000
PIVATIL see AOD000
PIVMECILLINAM HYDROCHLORIDE see
MCB550
PIX see MCH540
PIXALBOL see CMY800
PIX CARBONIS see CMY800
PIX LITHANTHRACIS see CMY800
PJ185 see LIH000
PK 10169 see HAQ550
PK-MERZ see TJG250
PKhNB see PAX000
PL 218 see PJV100
PL 3468 see DGA850
P 4070L see PJS750
PLACE-PAX see WAT211
PLACIDAL see EQL000
PLACIDAL see MNM500
PLACIDAS see MNM500
PLACIDEX see MFD500
PLACIDIL see CHG000
PLACIDOL see CJR909
PLACIDOL E see DJX000
PLACIDON see MQU750
PLACIDYL see CHG000
PLACODINE see MBU790
PLAC OUT see CDT250
PLAFIBRIDA (SPANISH) see PJB500
PLAFIBRIDE see PJB500
PLANADALIN see BNK000
PLANELON PB 501 see PAT830
PLANIUM see PJS750
PLANOCAINE see AIT250

PLANOCHROME see MCV000
PLANOFIX see NAK500
PLANOMIDE see TEX250
PLANOMYCIN see FBP300
PLANOTOX see DAA800
PLANT DITHIO AEROSOL see SOD100
PLANTDRIN see MRH209
PLANTFUME 103 SMOKE GENERATOR see SOD100
PLANTGARD see DAA800
PLANTIFOG 160M see MAS500
PLANTOMYCIN see SLY500
PLANT PIN see SOB500
PLANT PROTEASE CONCENTRATE see BMO000
PLANT PROTECTION PP511 see DIN800
PLANTULIN see PMN850
PLANTVAX see DLV200
PLANT WAX see DLV200
PLANUM see CFY750
PLAQUENIL see PJB750
PLASDONE see PKQ250
PLASIL see AJH000
PLASKON 201 see PJY500
PLASKON 3369 see MCB050
PLASKON 3381 see MCB050
PLASKON 3382 see MCB050
PLASKON CTFE see KDK000
PLASKON PP 60-002 see PJS750
PLASMA COAGULASE see CMY725
PLASMASTERIL see HLB400
PLASMINOGEN ACTIVATOR see PJB800
PLASMINOGEN ACTIVATOR (HUMAN TISSUE-TYPE PROTEIN MOIETY REDUCED), 84-l-SERINE- see PJB810
PLASMINOGEN ACTIVATOR (HUMAN TISSUE-TYPE PROTEIN MOIETY REDUCED), N-(N2-(N-GLYCYL-l-ALANYL)-l-ARGINYL)-, GLYCOFORM see SDI100
PLASMOCHIN see RHZ000
PLASMOCIDE see RHZ000
PLASMOCOAGULASE see CMY725
PLASMOQUINE see RHZ000
PLASTANOX 2246 see MJO500
PLASTANOX 425 ANTIOXIDANT see MJN250
PLASTANOX LTDP see TFD500
PLASTANOX LTDP ANTIOXIDANT see TFD500
PLASTANOX STDP see DXG700
PLASTANOX STDP ANTIOXIDANT see DXG700
PLASTAZOTE X 1016 see PJS750
PLASTER of PARIS see CAX500
PLASTHALL 503 see BSB000
PLASTIBEST 20 see ARM268
PLASTICIZER BDP see BQX250
PLASTICIZER DIHEXYL ADIPATE see HEO200
PLASTICIZER G-316 see PJC000
PLASTICIZER GPE see PJC500
PLASTICIZER 4GO see PJC250
PLASTICIZER Z-88 see PJC750
PLASTIFIX PC see PJQ050
PLASTOLEIN 9058 see BJQ500
PLASTOLEIN 9214 see FAB920
PLASTOLEIN 9058 DOZ see BJQ500
PLASTOL ORANGE G see CMS145
PLASTOL RUBINE BC see CMS155
PLASTOL YELLOW GG see CMS210
PLASTOL YELLOW GP see CMS210
PLASTOMOLL DOA see AEO000
PLASTOPAL BT see UTU500
PLASTORESIN ORANGE F 3A see CMP600
PLASTORESIN ORANGE F4A see PEJ500
PLASTORESIN RED F see SBC500
PLASTORESIN RED FR see CMS238
PLASTORESIN RED SR see NAP100
PLASTORESIN VIOLET 5BO see AOR500
PLASTRONGA see PJS750
PLASTYLENE MA 2003 see PJS750
PLASTYLENE MA 7007 see PJS750

PLATENOMYCIN B1 see MBY150
PLATENOMYCIN-B3 see LEV025
PLATH-LYSE see MJM500
PLATIBLASTIN see PJD000
cis-PLATIN see PJD000
PLATIN (GERMAN) see PJD500
PLATINATE(1-), DICHLORO(l-METHIONINATO-N,S)-, HYDROGEN, (SP-4-3)- see MDT900
PLATINATE, DINITROTETRACHLORO-, DIPOTASSIUM see PLD720
PLATINATE(2-), HEXACHLORO-, DIHYDROGEN, HEXAHYDRATE see DLO400
PLATINATE(2-), HEXACHLORO-, (OC-6-11)-, DIHYDROGEN, COMPD. WITH 8-(3-(DIMETHYLAMINO)PROPOXY)-3,7-DIHYDRO-1,3,7-TRIMETHYL-1H-PURINE-2,6-DIONE (1:2) see CNG850
PLATINATE (2-), METHYLTRICHLORO- see MQC400
PLATINATE(2-), NITROTRICHLORO-, DIPOTASSIUM see PLN050
PLATINATE(2-), TETRACHLORO-, DIAMMONIUM see ANV800
PLATINATE(2-), TETRAKIS(THIOCYANATO)-, DIPOTASSIUM see PLU590
PLATINATE(1-), TRICHLOROETHYLENE-, DIPOTASSIUM see PLW200
PLATINATE(2-), TRICHLORO(NITRITO-N-), DIPOTASSIUM (SP-4-2)- see PLN050
PLATINEX see PJD000
PLATINIC AMMONIUM CHLORIDE see ANF250
PLATINIC CHLORIDE see CKO750
PLATINIC POTASSIUM CHLORIDE see PLR000
PLATINIC SODIUM CHLORIDE see SJJ500
PLATINOL see PJD000
PLATINOL AH see DVL700
PLATINOL DOP see DVL700
PLATINOUS CHLORIDE see PJE000
cis-PLATINOUS DIAMMINE DICHLORIDE see PJD000
PLATINOUS POTASSIUM CHLORIDE see PJD250
PLATINUM see PJD500
PLATINUM(II) AMMINE TRICHLOROPOTASSIUM see PJD750
PLATINUM, BIS(l-ASCORBATO-O3)(1,2-CYCLOHEXANEDIAMINE-N,N')-, (SP-4-2)- see BGW400
PLATINUM, BIS(3,5-DIMETHYLPYRIDINE)DICHLORO-, cis-, (SP-4-1)- see BJL100
PLATINUM(II), BIS(METHYL SELENIDE)DICHLORO-, cis- see DEU115
PLATINUM (II), BIS(METHYL SELENIDE)SULFATO-, HYDRATE see SNS100
PLATINUM BLACK see PJD500
PLATINUM(IV) CHLORIDE see PJE250
PLATINUM(1+), CHLORO(1,2-ETHANEDIAMINE-N,N')(SULFINYLBIS(METHANE)-O)-, CHLORIDE, (SP-4-3)- see CGN400
PLATINUM COMPOUNDS see PJE500
PLATINUM, (1,2-CYCLOHEXANEDIAMINE-N,N')(ETHANEDIOATO(2-)-O,O')-, (SP-4-2-(1R-trans)- see OKY100
PLATINUM (II), (CYCLOHEXANE-1,2-DIAMINE)DIHYDROXO-, DIHYDRATE see DMG100
PLATINUM(II) (CYCLOHEXANE-1,2-DIAMINE)ISOCITRATO-, (Z)- see PGQ275
PLATINUM (II), DIAMMINE(BENZYLMALONATO)- see DCF720

PLATINUM (II), DIAMMINEDIBROMO-, cis- see DCF760
PLATINUM, DIAMMINEDIBROMO-, cis- (8IC) see DCF760
PLATINUM, DIAMMINEDIBROMO-, (SP-4-2)- see DCF760
cis-PLATINUM(II) DIAMMINEDICHLORIDE see PJD000
trans-PLATINUM(II)DIAMMINEDICHLORIDE see DEX000
cis-PLATINUMDIAMMINE TETRACHLORIDE see TBO776
cis-PLATINUM(IV) DIAMMINOTETRACHLORIDE see TBO776
PLATINUM DIARSENIDE see PJE750
PLATINUM, DICHLOROBIS(α-METHYLBENZENEMETHANAMINE)-, (SP-4-2) see DEU150
PLATINUM, DICHLOROBIS(METHYL SELENIDE)-, cis- see DEU115
PLATINUM, DICHLOROBIS(SELENOBIS(METHANE))-(SP-4-2) see DEU115
PLATINUM, DICHLORO(N,N'-DIETHYL-2,4-PENTANEDIAMINE-N,N')-, (SP-4-2-(2R-(2R*(S*),4R*(S))))- see DEY650
PLATINUM, DICHLORO(N,N'-DIETHYL-2,4-PENTANEDIAMINE-N,N')-, (SP-4-2-(2S-(2R*(S*),4R*(S*))))- see DEY680
PLATINUM, DICHLORO(l-METHIONINE)-(8CI) see MDT900
PLATINUM, DICHLORO(METHIONINE)-(7CI) see MDT900
PLATINUM (II), DICHLORO(2,3-NAPHTHYLENEDIAMMINE)- see DFT033
PLATINUM, DICHLORO(2,3,6,7-TETRAHYDRO-1H,5H-BENZO(i,j)QUINOLIZINE)- see DGL700
PLATINUM(II) DINITRODIAMMINE see DCF800
PLATINUM ETHYLENEDIAMINE DICHLORIDE see DFJ000
PLATINUM FULMINATE see PJF000
PLATINUM (2-), NITROTRICHLORO-, DIPOTASSIUM see PLN050
PLATINUM SPONGE see PJD500
PLATINUM SULFATE see PJF500
PLATINUM(II) SULFATE see PJF500
PLATINUM SULFATE TETRAHYDRATE see PJF750
PLATINUM(II) SULFATE TETRAHYDRATE see PJF750
PLATINUM TETRACHLORIDE see PJE250
cis-PLATINUM-2-THYMINE see PJG000
PLATINUM THYMINE BLUE see PJG000
PLATINUM(1+), TRIAMMINECHLORO-, (SP-4-2)-, (SP-4-2)-AMMINETRICHLOROPLATINATE(1-) (1:1) see CLO700
PLATINUM URACIL BLUE see PJG125
PLATIPHILLIN see PJG135
PLATIPHILLIN HYDROCHLORIDE see PJG150
PLATIPHYLLIN see PJG135
PLATIPHYLLIN HYDROCHLORIDE see PJG150
PLATYPHYLLINE see PJG135
PLAVOLEX see DBD700
PLAXIDOL see CJR909
PLE 1053 see ASO375
PLECYAMIN see VSZ000
PLEGANGIN see VIZ400
PLEGATIL see MQF750
PLEGECYL see ABH500
PLEGICIN see ABH500
PLEGINE see DKE800
PLEGINE see PDD000
PLEGOMAZIN see CKP250
PLEGOMAZIN see CKP500
PLENASTRIL see PAN100
PLENOLIN see DLO875

PLENUR see LGZ000
PLEOCIDE see ABY900
PLESSY'S GREEN (HEMIHEPTAHYDRATE)
see CMK300
PLESTROVIS see QFA250
PLETIL see TGD250
PLEXIGLAS see PKB500
PLEXIGUM M 920 see PKB500
PLH see LIU300
PLICTRAN see CQH650
PLIDAN see DCK759
PLIFENATE see TIL800
PLIMASINE see MNQ000
PLIOFILM see PJH500
PLIOFLEX see SMR000
PLIOGRIP see PKL500
PLIOLITE see BOP100
PLIOLITE S5 see SMR000
PLIOVAC AO see AAX175
PLIOVIC see PKQ059
PLISULFAN see AIF000
PLIVA see BIE500
PLIVAPHEN see ABH500
PLLETIA see PMI750
PLOMB FLUORURE (FRENCH) see LDF000
PLOSTENE see TKH750
PLOYMANNURONIC ACID see AFL000
PLP see PII100
PLTU see MJZ000
PLUCHEA LANCEOLATA (DC.) Cl., extract
excluding roots see PJH550
PLUCKER see NAK500
PLUM see AQP890
PLUMBAGIN see PJH610
PLUMBAGO see CBT500
PLUMBAGO ZEYLANICA Linn., root extract
see PJH615
PLUMBANE, (1-
BENZIMIDAZOLYL)TRIBUTYL- see TIA800
PLUMBANE, BIS(ACETYLOXY)DIHEXYL-
see DKP200
PLUMBANE, (1-IMIDAZOLYL)TRIBUTYL-
see IAT400
PLUMBANE,
PHENYLTRIS(DODECANOYLOXY)- see
PFL200
PLUMBANE, TETRACHLORO-, mixed with
C$_7$-C$_{12}$ ALCOHOLS (9CI) see MRN125
PLUMBANE, TRIBUTYLIMIDAZOL-1-YL-
see IAT400
PLUMBANE,
TRIS(LAUROYLOXY)PHENYL- see PFL200
PLUMBOPLUMBIC OXIDE see LDS000
PLUMBOUS ACETATE see LCJ000
PLUMBOUS ACETATE see LCV000
PLUMBOUS CHLORIDE see LCQ000
PLUMBOUS CHROMATE see LCR000
PLUMBOUS FLUORIDE see LDF000
PLUMBOUS OXIDE see LDN000
PLUMBOUS PHOSPHATE see LDU000
PLUMBOUS SULFIDE see LDZ000
PLUMBYLIUM, TRIETHYL-,
HEXAFLUOROSILICATE (2-) (2:1) see
BLN200
PLURACOL P-410 see PJT000
PLURACOL P-710 see PJT000
PLURACOL P-1010 see PJT000
PLURACOL P-2010 see PJT000
PLURACOL P-3010 see PJT000
PLURACOL P-4010 see PJT000
PLURACOL E see PJT200
PLURACOL E-200 see PJT000
PLURACOL E-300 see PJT000
PLURACOL E-400 see PJT000
PLURACOL E-600 see PJT000
PLURACOL E-1500 see PJT000
PLURACOL E-4000 see PJT000
PLURACOL E-6000 see PJT000
PLURAFAC RA 43 see DXY000
PLURAGARD see MCB000
PLURAGARD C 133 see MCB000
PLUREXID see CDT250

PLURONIC L-81 see PJH630
PLURYLE see BEQ625
PLURYLE see BEQ625
PLUSURIL see BEQ625
PLUTONIUM see PJH750
PLUTONIUM BISMUTHIDE see PJH775
PLUTONIUM COMPOUNDS see PJI000
PLUTONIUM(III) HYDRIDE see PJI250
PLYAMINE HD 1129A see UTU500
PLYAMINE M27 see MCB050
PLYAMINE P 364BL see UTU500
PLYCTRAN see CQH650
PLYMOUTH IPP see IQW000
PLYMOUTM IPM see IQN000
PLYSET TD688 see MCB050
PM245 see BEM500
PM 334 see MNZ000
P.M. 346 see EIT100
P.M. 388 see PBH100
PM 396 see MLP800
PM 671 see ENG500
PM 2763 see DTI600
P.M. 434,526 see DBE885
PMA see ABU500
PMA see PGV000
PMAC see ABU500
PMACETATE see ABU500
PMAL see ABU500
PMAS see ABU500
PM-ATS 380 see NCY135
PMB see ECU750
PMC see PFM500
PMCG see PJI575
PMCG HYDROCHLORIDE see PJI575
PMDT see PBG500
PMEA see PHA700
PMEG see PHA725
P.M.F. see PFN000
PMFP see SKS500
PMH see PAP000
(±)-PMHI MALEATE see PJI600
PMMA see PKB500
PM$_{(1)}$MM$_{(3)}$ see HBA600
PMN 89-234 see REA050
PMO 10 see PFP100
PMP see PHX250
PMP see TGK200
PMP SODIUM GLUCONATE see SHK800
PMS see MNO500
PMS see MRW000
PMS see PNQ000
PMS see SCA750
PMS 1.5 see SCR400
PMS 300 see SCR400
PMS 154A see SCR400
PMS 200A see SCR400
PMSF see TGO300
PMSG see SCA750
PMS No. 1 see PJW750
PMT see MOA725
PMT see PFS500
PM-TC see PMI000
PN see CFC000
PN6 see AQO000
19583RP-Na see KGK100
PNA see NEO500
PNAP see NEL600
PNB see NFQ000
PNBAS see NEY000
PNCB see NFS525
PND see COS500
PNEUMOREL see DAI200
PNNG see NLD000
PNOA see NEQ000
PNOT see NMP500
2-N-p-PDA see ALL750
PNS 25 see SCR400
PNSP see SIV600
PNT see NMO550
PNU see NLO500
PO-20 see PMD800
POA see PER250

POAST see CDK800
POCAN BUSH see PJJ315
PO-DIMETHOATE see DNX800
PODOPHYLLIN see PJJ000
PODOPHYLLINIC ACID LACTONE see
PJJ225
PODOPHYLLOTOXIN see PJJ225
PODOPHYLLOTOXIN-o-BENXYLIDENE-
β-d-GLUCOPYRANOSIDE see BER500
PODOPHYLLOTOXIN-BENZILIDEN-
GLUCOSID (GERMAN) see BER500
PODOPHYLLOTOXIN, 4,6-o-
BENZYLIDENE-β-d-GLUCOPYRANOSIDE
see PJJ250
PODOPHYLLOTOXIN-β-d-BENZYLIDENE
GLUCOSIDE see PJJ250
PODOPHYLLOTOXIN, 4'-o-DEMETHYL-4-
DEOXY- see DAN388
PODOPHYLLOTOXIN GLUCOSIDE see
PJJ250
PODOPHYLLOTOXIN 4-o-GLUCOSIDE see
PJJ250
PODOPHYLLOTOXIN β-d-GLUCOSIDE see
PJJ250
PODOPHYLLUM see PJJ000
PODOPHYLLUM PELTATUM see MBU800
PODOPHYLLUM RESIN see PJJ000
POE 20 SORBITAN MONOLAURATE see
PKG000
POHA (HAWAII) see JBS100
POINSETTIA see EQX000
POINSETTIA PULCHERRIMA GRAH see
EQX000
POINT TWO see SHF500
POIRENATE see EHT600
POIS COCHON (HAITI) see YAG000
POIS MANIOC (HAITI) see YAG000
POISON BULB see SLB250
POISON BUSH see BOO700
POISON de COULEUVRE (CANADA) see
BAF325
POISON HEMLOCK see PJJ300
POISON PARSLEY see PJJ300
POISON ROOT see PJJ300
POISON SEGO see DAE100
POISON TOBACCO see HAQ100
POISON TREE see BOO700
POIS PUANTE (HAITI) see CNG825
POIS VALLIERE (HAITI) see SBC550
POKE see PJJ315
POKEBERRY see PJJ315
POKEWEED see PJJ315
POLAAX see OAX000
POLACARITOX see CKM000
POLACTINE G YELLOW see CMS226
POLAKTYN YELLOW G see CMS226
POLAMIDON see MDP770
POLAMIDONE see MDO750
POLAN NAVY BLUE E 2R see CMM080
POLAN RED FS see CMO870
POLAN YELLOW E-3R see SGP500
POLAPREZINC see ZEJ100
POLARAMIN see PJJ325
POLARAMINE MALEATE see PJJ325
POLAR BRILLIANT BLUE RAW see CMM092
POLAR BRILLIANT BLUE RAWL see
CMM092
POLARIS see BLG250
POLARONIL see CLX300
POLARONIL (GERMAN) see TAI500
POLAR ORANGE R see ADG000
POLAR RED G see CMM320
POLAR RED GBD see NAO600
POLAR RED G SUPRA see CMM320
POLAR RED R see NAO600
POLAR RED RS see CMM330
POLCARB see CAT775
POLCILLIN see PJJ350
POLCOMINAL see EOK000
POLECAT WEED see SDZ450
POLECTRON 430 see VQK595
POLECTRON 450 see VQK595

POLECTRON 845 see AAU300
POLEON see EID000
POLFOS see MHR150
POLFOSCHLOR see TIQ250
POLICAPRAN see PJY500
POLICAR MZ see DXI400
POLICAR S see DXI400
POLIDIM see DOR800
POLIDOCANOL see DXY000
POLIFEN see TAI250
POLIFLOGIL see HNI500
POLIFUNGIN see PJJ500
POLIFURIT see GHY100
POLIGOSTYRENE see SMQ500
POLIK see IFA000
POLIKARBATSIN (RUSSIAN) see MQQ250
POLINALIN see DOT000
POLISEPTIL see TEX250
POLISIN see BKL250
POLITEF see TAI250
POLITEN see PJS750
POLITEN I 020 see PJS750
POLIURON see BEQ625
POLIVAL see TEX000
POLIVINIT see PKQ059
POLKWEED see PJJ315
POLLACID see HCQ500
POLNOKS R see PJQ750
POLOMEL ME 3 see MCB050
POLOMEL MEC 3 see MCB050
POLONIUM see PJJ750
POLONIUM CARBONYL see PJK000
POLOPIRYNA see ADA725
POLOXALENE see PJK150
POLOXALENE FREE-CHOICE LIQUID
TYPE C FEED see PJK151
POLOXAL RED 2B see HJF500
POLOXAMER 331 see PJK200
POLPHOS see MHR150
POLPRETAN K 2 see MCB050
POLSTIGMINE see DQY909
POLY see SKN000
POLYAC see PJR500
POLYACRYLAMIDE see PJK350
POLYACRYLAMIDE see PJK350
POLYACRYLAMIDE RESINS, MODIFIED
see MRA075
POLYACRYLATE see ADV900
POLY(ACRYLIC ACID) see ADV900
POLYACRYLONITRILE see ADX750
POLYACRYLONITRILE, COMBUSTION
PRODUCTS see PJK360
POLYADENYLATE-POLYURIDYLATE see
PJK400
POLYADENYLIC.POLYURIDYLIC ACID see
PJK400
POLYAETHYLEN see PJS750
POLYAETHYLENGLYCOLE 200 (GERMAN)
see PJT200
POLYAETHYLENGLYKOLE 300
(GERMAN) see PJT225
POLYAETHYLENGLYKOLE 400
(GERMAN) see PJT230
POLYAETHYLENGLYKOLE 600
(GERMAN) see PJT240
POLYAETHYLENGLYKOLE 1500
(GERMAN) see PJT500
POLYAETHYLENGLYKOLE 4000
(GERMAN) see PJT750
POLYAETHYLENGLYKOLE 6000
(GERMAN) see PJU000
POLYAETHYLENGLYKOLE #1000
(GERMAN) see PJT250
POLYALUMINIUM CHLORIDE see AGZ998
POLYAMID (GERMAN) see NOH000
POLYAMIDE 6 see PJY500
POLYAMIDE-6 (combustion products) see
PJK750
POLYAMINE D see PJL000
POLYAMINE H SPECIAL see PJL100
POLYAMINE T see FMW330

POLY-p-AMINOBENZALDEHYD (CZECH)
see DXS000
POLY(ε-AMINOCAPROIC ACID) see PJY500
POLY A. POLY U see PJK400
POLYARABINOGALACTAN see AQR800
POLY(A.U) see PJK400
POLYBENZARSOL see BCJ150
POLYBIS(2-ETHYLBUTYL)SILOXANE see
EHC900
POLYBOR see DXF200
POLYBOR see SFF000
POLYBOR 3 see DXF200
POLY[BORANE(1)] see PJL325
POLYBOR-CHLORATE see SFS500
POLYBREME see HCV500
POLYBROMINATED BIPHENYL see
FBU509
POLYBROMINATED BIPHENYL see
HCA500
POLYBROMINATED BIPHENYL (FF-1) see
FBU509
POLYBROMINATED BIPHENYLS see
FBU000
POLYBROMINATED BIPHENYLS see
PJL335
POLYBROMINATED SALICYLANILIDE see
THW750
POLYBROMOETHYLENE see PKQ000
cis-POLY(BUTADIENE) see PJL350
POLY(1,3-BUTADIENE PEROXIDE) see
PJL375
POLYBUTADIENE-POLYSTYRENE
COPOLYMER see SMR000
POLYBUTENE see PJL400
POLYBUTENES see PJL400
POLYBUTYLENE GLYCOL see GHY100
POLY(BUTYLENE OXIDE) see GHY100
POLY(BUTYL METHACRYLATE) see PJL500
POLYCAPROAMIDE see PJY500
POLY(ε-CAPROAMIDE) see PJY500
POLYCAPROLACTAM see PJY500
POLY(ε-CAPROLACTAM) see PJY500
POLYCARBACIN see MQQ250
POLYCARBACINE see MQQ250
POLYCARBAMATE see BJK000
POLYCARBAZIN see MQQ250
POLYCARBAZINE see MQQ250
POLYCARBONATE PC see PKE300
POLY(CARBON MONOFLUORIDE) see
PJL600
POLYCAT 8 see DRF709
POLYCHLORCAMPHENE see CDV100
POLYCHLORINATED BIPHENYL
(AROCLOR 1016) see PJL800
POLYCHLORINATED BIPHENYL
(AROCLOR 1221) see PJM000
POLYCHLORINATED BIPHENYL
(AROCLOR 1232) see PJM250
POLYCHLORINATED BIPHENYL
(AROCLOR 1242) see PJM500
POLYCHLORINATED BIPHENYL
(AROCLOR 1248) see PJM750
POLYCHLORINATED BIPHENYL
(AROCLOR 1254) see PJN000
POLYCHLORINATED BIPHENYL
(AROCLOR 1260) see PJN250
POLYCHLORINATED BIPHENYL
(AROCLOR 1262) see PJN500
POLYCHLORINATED BIPHENYL
(AROCLOR 1268) see PJN750
POLYCHLORINATED BIPHENYL
(AROCLOR 2565) see PJO000
POLYCHLORINATED BIPHENYL
(AROCLOR 4465) see PJO250
POLYCHLORINATED BIPHENYL
(KANECHLOR 300) see PJO500
POLYCHLORINATED BIPHENYL
(KANECHLOR 400) see PJO750
POLYCHLORINATED BIPHENYL
(KANECHLOR 500) see PJP000
POLYCHLORINATED BIPHENYLS see
PJL750

POLYCHLORINATED CAMPHENES see
CDV100
POLYCHLORINATED DIBENZOFURANS
see PJP100
POLYCHLORINATED TERPHENYL see
PJP250
POLYCHLORINATED TERPHENYL see
PJP250
POLYCHLORINATED TRIPHENYL see
PJP300
POLYCHLORINATED TRIPHENYL
(AROCLOR 5442) see PJP750
POLYCHLORINATED TRIPHENYL
(AROCLOR 5460) see PJP800
POLYCHLOROBENZOIC ACID,
DIMETHYLAMINE SALTS see PJQ000
POLYCHLOROBIPHENYL see PJL750
POLY(2-CHLOROBUTADIENE) see PJQ050
POLY(2-CHLORO-1,3-BUTADIENE) see
PJQ050
POLYCHLOROCAMPHENE see CDV100
POLYCHLORO COPPER
PHTHALOCYANINE see PJQ100
POLYCHLORODICYCLOPENTADIENE see
BAF250
POLYCHLORODICYCLOPENTADIENE
ISOMERS see BAF250
POLY(CHLOROETHYLENE) see PKQ059
POLYCHLOROPINENE see PJQ250
POLYCHLOROPRENE see PJQ050
POLY(CHLOROTRIFLUOROETHENE) see
KDK000
POLY(CHLOROTRIFLUOROETHYLENE)
see KDK000
POLYCHLOROTRIPHENYL see PJP300
POLYCHOM see ZJS300
POLYCHOME see ZJS300
POLYCIDAL see MFN500
POLYCILLIN see AIV500
POLYCILLIN see AOD125
POLYCILLIN-N see SEQ000
POLYCIZER 332 see BSL600
POLYCIZER 532 see OEU000
POLYCIZER 562 see OEU000
POLYCIZER 962-BPA see DXQ200
POLYCIZER DBP see DEH200
POLYCIZER DBS see DEH600
POLYCLAR L see PKQ250
POLYCLENE see DGB000
POLYCO see SJK000
POLYCO 2410 see SMR000
POLYCO 2611 see CGW300
POLYCOR DARK GREEN S see CMO830
POLYCOR RED GS see CMM325
POLYCOR RED SE see CMO870
POLYCO 220NS see SMQ500
POLY C POLY I see PJY750
POLYCRON see BNA750
POLYCTYIDYLIC-POLYINOSINIC ACID see
PJY750
POLYCYCLIC MUSK see ACL750
POLYCYCLINE see TBX000
POLYCYCLINE HYDROCHLORIDE see
TBX250
POLYCYCLOPENTADIENYLTITANIUM
DICHLORIDE see PJQ275
POLY(DA) see PJQ400
POLY DAP see PJQ400
POLYDAZOL see PJQ350
POLY(DEOXYADENYLIC ACID) see PJQ400
POLY(2'-DEOXYADENYLIC ACID) see
PJQ400
POLYDEOXYRIBOADENYLIC ACID see
PJQ400
POLYDEOXYRIBOTHYMIDYLIC ACID see
TFX795
POLYDEOXYTHYMIDYLIC ACID see
TFX795
POLYDESIS see PKP750
POLYDEXTROSE see PJQ425
POLYDEXTROSE SOLUTION see PJQ430
POLYDIBROMOSILANE see PJQ500

POLYDIBROMOSILYLENE see PJQ500
POLY(DI.DC) see DAR930
POLY(1,2-DIHYDRO-2,2,4-TRIMETHYLQUINOLINE) see PJQ750
POLY(DIMERCURYIMMONIUM ACETYLIDE) see PJQ775
POLY(DIMERCURYIMMONIUM BROMATE) see PJQ780
POLY(DIMERCURYIMMONIUM HYDOXIDE) see DNW200
POLYDIMETHYLDIPHENYLSILOXANE see PJQ790
POLY((DIMETHYLIMINIO)HEXAMETHYLENE(DIMETHYLIMINO)TRIMETHYLENE DIBROMIDE) see IFT300
POLY((DIMETHYLIMINIO)-1,6-HEXANEDIYL(DIMETHYLIMINIO)METHYLENE(1,1'-BIPHENYL)-4,4'-DIYLMETHYLENE DICHLORIDE) see PJQ800
POLYDIMETHYLSILOXANE see DTR850
POLYDIMETHYL SILOXANE see PJR000
POLYDIMETHYLSILOXANE RUBBER see PJR250
POLYDIMETHYL SILOXY CYCLICS see PJR300
POLY-p-DINITROSOBENZENE see PJR500
POLY(DI).POLY(DC) see DAR930
POLY(DT) see TFX795
POLYDYE DISPERSE YELLOW GSFM see KDA075
POLYDYE YELLOW GSFD see KDA075
POLY-EM 12 see PJS750
POLY-EM 40 see PJS750
POLY-EM 41 see PJS750
POLYESTER DISPERSE BLUE see DBQ220
POLYESTER TGM 3 see MDN510
POLY(ESTRADIOL PHOSPHATE) see EDS000
POLYESTRADIOL PHOSPHATE see PJR750
POLYETHER DIAMINE L-1000 see PJS000
POLYETHYLENE see PJS750
POLYETHYLENE AS see PJS750
POLYETHYLENE 600 DIBENZOATE see PKE750
POLYETHYLENE GLYCOL see PJT000
POLYETHYLENE GLYCOL 200 see PJT200
POLYETHYLENE GLYCOL 300 see PJT225
POLYETHYLENE GLYCOL 400 see PJT230
POLYETHYLENE GLYCOL 600 see PJT240
POLYETHYLENE GLYCOL 1000 see PJT250
POLYETHYLENE GLYCOL 1500 see PJT500
POLYETHYLENE GLYCOL 4000 see PJT750
POLYETHYLENE GLYCOL 6000 see PJU000
POLYETHYLENE GLYCOL CETYL ETHER see PJT300
POLYETHYLENE GLYCOL CHLORIDE 210 see PJU250
POLYETHYLENE GLYCOL DIBENZOATE see PKE750
POLYETHYLENE GLYCOL 220 DIBENZOATE see PKE750
POLYETHYLENE GLYCOL 400, DICHLORIDE see PJU300
POLYETHYLENE GLYCOL 200 DI(2-ETHYLHEXOATE) see FCD500
POLYETHYLENE GLYCOL DIMETHYL ETHER see DOM100
POLYETHYLENE GLYCOL DISTEARATE see PJU500
POLYETHYLENE GLYCOL 300 DISTEARATE see PJU500
POLYETHYLENE GLYCOL 400 (DI) STEARATE see PJU500
POLYETHYLENE GLYCOL 600 (DI) STEARATE see PJU500
POLYETHYLENE GLYCOL DISTEARATE 1000 see PJU750
POLYETHYLENE GLYCOL DODECYL ETHER see DXY000
POLYETHYLENE GLYCOL E 600 see PJV000

POLYETHYLENE GLYCOL (5) LANOLIN see LAU560
POLYETHYLENE GLYCOL (24) LANOLIN see LAU560
POLYETHYLENE GLYCOL-27 LANOLIN see LAU560
POLYETHYLENE GLYCOL (30) LANOLIN see LAU560
POLYETHYLENE GLYCOL-40 LANOLIN see LAU560
POLYETHYLENE GLYCOL (50) LANOLIN see LAU560
POLYETHYLENE GLYCOL (60) LANOLIN see LAU560
POLYETHYLENE GLYCOL-75 LANOLIN see LAU560
POLYETHYLENE GLYCOL (85) LANOLIN see LAU560
POLYETHYLENE GLYCOL (100) LANOLIN see LAU560
POLYETHYLENE GLYCOL 1000 LANOLIN see LAU560
POLYETHYLENE GLYCOL LAURYL THIOETHER see PJV100
POLYETHYLENE GLYCOL MONO(DODECYLTHIO)ETHER see PJV100
POLYETHYLENE GLYCOL MONO(2-(DODECYLTHIO)ETHYL)ETHER see PJV100
POLYETHYLENE GLYCOL MONOETHER with p-tert-OCTYLPHENYL see PKF500
POLYETHYLENE GLYCOL MONOLEYL ETHER see OIG040
POLYETHYLENE GLYCOL MONO(OCTYLPHENYL) ETHER see GHS000
POLYETHYLENE GLYCOL MONO(4-OCTYLPHENYL) ETHER see PKF500
POLYETHYLENE GLYCOL MONO(4-tert-OCTYLPHENYL) ETHER see PKF500
POLYETHYLENE GLYCOL MONO(p-tert-OCTYLPHENYL) ETHER see PKF500
POLYETHYLENE GLYCOL MONOOLEATE see PJY100
POLYETHYLENE GLYCOL MONOSTEARATE see PJV250
POLYETHYLENE GLYCOL MONOSTEARATE 200 see PJV500
POLYETHYLENE GLYCOL MONOSTEARATE 400 see PJV750
POLYETHYLENE GLYCOL MONOSTEARATE 1000 see PJW000
POLYETHYLENE GLYCOL MONOSTEARATE 6000 see PJW250
POLYETHYLENE GLYCOL MONO(p-(1,1,3,3-TETRAMETHYLBUTYL)PHENYL) ETHER see PKF500
POLYETHYLENE GLYCOL 450 NONYL PHENYL ETHER see PKF000
POLYETHYLENE GLYCOL OCTYLPHENOL ETHER see PKF500
POLYETHYLENE GLYCOL OCTYLPHENYL ETHER see GHS000
POLYETHYLENE GLYCOL p-OCTYLPHENYL ETHER see PKF500
POLYETHYLENE GLYCOL p-OCTYLPHENYL ETHER see PKF500
POLYETHYLENE GLYCOL 450 OCTYL PHENYL ETHER see PKF500
POLYETHYLENE GLYCOL p-tert-OCTYLPHENYL ETHER see PKF500
POLYETHYLENE GLYCOL OLEATE see PJY100
POLYETHYLENE GLYCOL 1000 OLEYL ETHER see PJW500
POLYETHYLENE GLYCOL PALMITYL ETHER see PJT300
POLYETHYLENE GLYCOLS MONOHEXADECYL ETHER see PJT300
POLYETHYLENEGLYCOLS MONOSTEARATE see PJW750
POLYETHYLENE GLYCOL p-1,1,3,3,-TETRAMETHYLBUTYLPHENYL ETHER see PKF500

POLYETHYLENEIMIN (CZECH) see PJX000
POLYETHYLENE IMINE see PJX000
POLYETHYLENE LAURYL THIOETHER see PJV100
POLY(ETHYLENE-MALEIC ANHYDRIDE) see EJM950
POLYETHYLENE MALEIC ANHYDRIDE COPOLYMER see EJM950
POLY(ETHYLENE OXIDE) see PJT000
POLYETHYLENE OXIDE CETYL ETHER see PJT300
POLY(ETHYLENE OXIDE) DODECYL ETHER see DXY000
POLYETHYLENE OXIDE HEXADECYL ETHER see PJT300
POLYETHYLENE OXIDE MONOOLEATE see PJY100
POLY(ETHYLENE OXIDE)OCTYLPHENYL ETHER see GHS000
POLY(ETHYLENE OXIDE) OLEATE see PJY100
POLYETHYLENE-POLYPROPYLENE GLYCOLS PLURONIC L-81 see PJH630
POLYETHYLENE RESINS see PJS750
POLYETHYLENE TEREPHTHALATE see PKF750
POLYETHYLENE TEREPHTHALATE FILM see PKF750
POLY(ETHYLENE TETRAFLUORIDE) see TAI250
POLYETHYLENE Y-141-A see PJX750
POLYETHYLENIMINE (10,000) see PJX800
POLYETHYLENIMINE (20,000) see PJX825
POLYETHYLENIMINE (35,000) see PJX835
POLYETHYLENIMINE (40,000) see PJX845
POLY(ETHYLIDENE PEROXIDE) see PJX850
POLY(ETHYL VINYL ETHER) see OJV600
POLYFENE see TAI250
POLYFER see IGS000
POLYFIBRON 120 see SFO500
POLYFIX PM 5 see MCB050
POLYFIX PM 107 see MCB050
POLYFLEX see SMQ500
POLYFLON see TAI250
POLYFOAM PLASTIC SPONGE see PKL500
POLYFOAM SPONGE see PKL500
POLY(FORMALDEHYDE-ISOBUTYL ALCOHOL-UREA) see IIM300
POLYFUNGIN see PJJ500
POLYFURIT see GHY100
POLYFURIT 1000 see GHY100
POLYFURITE see GHY100
POLY-G see PJT200
POLY G 400 see PJT230
POLYGARD see TMX550
POLY-GIRON see TEH500
POLYGLUCIN see DBD700
POLY-l-GLUTAMATE see GFM310
POLYGLUTAMIC ACID see GFM310
POLY(l-GLUTAMIC ACID) see GFM310
POLY(α-l-GLUTAMIC ACID) see GFM310
l-γ-POLYGLUTAMIC ACID see GFM310
POLYGLYCERATE (60) see EEU100
POLYGLYCEROL ESTERS of FATTY ACIDS see PJX875
POLYGLYCOL 1000 see PJT250
POLYGLYCOL 4000 see PJT750
POLYGLYCOL 15-200 see PJX900
POLYGLYCOLAMINE H-163 see AMC250
POLYGLYCOLDIAMINE H 221 see EBV100
POLYGLYCOL DISTEARATE see PJU500
POLYGLYCOL E see PJT200
POLYGLYCOL E1000 see PJT250
POLYGLYCOL E-4000 see PJT750
POLYGLYCOL E-4000 USP see PJT750
POLYGLYCOL LAURATE see PJY000
POLYGLYCOL MONOOLEATE see PJY100
POLYGLYCOL OLEATE see PJY250
POLYGON see SKN000
POLYGONUM HYDROPIPER L., dry powdered whole plant see WAT350

POLYGRIPAN see TEH500
POLY-G SERIES see PJT000
POLY(HEXAMETHYLENE DIISOCYANATE) see HEG300
POLY 1:C see PJY750
POLY(IMINOCARBONYLPENTAMETHYLENE) see PJY500
POLY(IMINO(1-OXO-1,6-HEXANEDIYL)) see PJY500
POLYINOSINATE:POLYCYTIDYLATE see PJY750
POLYINOSINIC:POLYCYTIDYLIC ACID COPOLYMER see PJY750
POLYISOBUTYLENE see PJY800
POLY(GERMANIUM MONOHYDRIDE) see GEG000
POLYMALEIC ACID see PJY850
POLYMALEIC ACID, SODIUM SALT see PJY855
POLY(MALEIC ANHYDRIDE) see MAM500
POLYMARCIN see MQQ250
POLYMARCINE see MQQ250
POLYMARSIN see MQQ250
POLYMARZIN see MQQ250
POLYMARZINE see MQQ250
POLYMAT see MQQ250
POLYMEG see GHY100
POLYMEG 600 see GHY100
POLYMEG 650 see GHY100
POLYMEG 1000 see GHY100
POLYMEG 1020 see GHY100
POLYMEG 2000 see GHY100
POLYMEG 2010 see GHY100
POLYMER 261 see DTS500
POLYMERIC DIALDEHYDE see PKA000
POLYMER JR see QAT600
POLYMER JR 1 see QAT600
POLYMER JR 125 see QAT600
POLYMER JR 30M see QAT600
POLYMER JR 400 see QAT600
POLYMER LR see QAT600
POLYMERS of EPICHLOROHYDRIN and 2,2-BIS(4-HYDROXYPHENYL)PIPERAZINE see ECM500
POLYMERS (PETROLEUM), VISCOUS see VRU300
POLYMERS, WATER-INSOLUBLE see PKA850
POLYMERS, WATER-SOLUBLE see PKA860
POLYMER 261LV see DTS500
POLY(METHIBIS(HYDROXYMETHYL)UREYLENE)AMER see UTU500
POLY(2-METHOXY-5-(2-(METHYLAMINO)ETHYL)-m-PHENYLENEMETHYLENE) see PMB800
POLY-(METHYLBIS(THIOCYANATO)ARSINE) see MJK750
POLY(METHYLENEMAGNESIUM) see PKB000
POLYMETHYLENEPOLYPHENYL ISOCYANATE see PKB100
POLYMETHYLMETHACRYLATE see PKB500
POLYMINE D see BEN000
POLYMIST A12 see PJS750
POLYMO GREEN FBH see PJQ100
POLYMO GREEN FGH see PJQ100
POLYMON BLUE 3R see IBV050
POLYMONE see DGB000
POLYMON GREEN 6G see PJQ100
POLYMON GREEN G see PJQ100
POLYMON GREEN GN see PJQ100
POLY(MONOCHLOROTRIFLUOROETHYLENE) see KDK000
POLYMO ORANGE GR see CMS145
POLYMO RED FGN see MMP500
POLYMO YELLOW GR see CMS208
POLYMUL CS 81 see PJS750
POLYMYCIN see PKB775
POLYMYXIN see PKC000
POLYMYXIN A see PKC250

POLYMYXIN B see PKC500
POLYMYXIN B1 see PKC550
POLYMYXIN B SULFATE see PKC750
POLYMYXIN D1 see PKD050
POLYMYXIN E see PKD250
POLYMYXIN E SULFATE see PKD300
POLYMYXIN E SULFATE (SALT) see PKD300
POLYNAL YELLOW G see KDA075
POLYNOXYLIN see UTU500
POLYOESTRADIOL PHOSPHATE see EDS000
POLYOX see PJT000
POLYOXIETHYLENE (6) ALKYL (13) ETHER see TJJ250
POLYOXIN AL see PKE100
POLYOXIN B see PKE100
POLY(1-(2-OXO-1-PYRROLIDINYL)ETHYLENE) see PKQ250
POLY(1-(2-OXO-1-PYRROLIDINYL)ETHYLENE)IODINE COMPLEX see PKE250
POLY(OXY-1,4-BUTANEDIYL), α-HYDRO-ω-HYDROXY-(9CI) see GHY100
POLY(OXYBUTYLENE) GLYCOL see GHY100
POLY(OXY-1,4-BUTYLENE) GLYCOL see GHY100
POLY(OXYCARBONYLOXY-1,4-PHENYLENE(1-METHYLETHYLIDENE)-1,4-PHENYLENE) see PKE300
POLY(OXY(DIMETHYLSILYLENE)) see PJR000
POLY(OXY-1,2-ETHANEDIYL), α-(2-CHLOROETHYL)-ω-(2-CHLOROETHOXY)- see PJU300
POLY(OXY-1,2-ETHANEDIYL), α-((3-β)-CHOLEST-5-EN-3-YL)-ω-HYDROXY-(9CI) see PKE700
POLY(OXY-1,2-ETHANEDIYL), α-DECYL-ω-HYDROXY- see PKE390
POLY(OXY-1,2-ETHANEDIYL), α-DECYL-ω-HYDROXY-, PHOSPHATE, POTASSIUM SALT see PKE800
POLY(OXY-1,2-ETHANEDIYL), α-DODECYL-ω-HYDROXY-, PHOSPHATE see PKE850
POLY(OXY-1,2-ETHANEDIYL), α-(2-(DODECYLTHIO)ETHYL)-, ω-HYDROXY- see PJV100
POLY(OXY-1,2-ETHANEDIYL), α-(2-(tert-DODECYLTHIO)ETHYL)-ω-HYDROXY- see PKE350
POLY(OXY-1,2-ETHANEDIYL), α-α',α'',α'''-(1,2-ETHANEDIYLBIS(NITRILODI-2,1-ETHANEDIYL))TETRAKIS(ω-HYDROXY)- see EIV750
POLY(OXY-1,2-ETHANEDIYL), α-((((HEPTADECAFLUOROOCTYL)SULFONYL)METHYLAMINO)CARBONYL)-OMEGA-BUTOXY- see PCH320
POLY(OXY-1,2-ETHANEDIYL), α-HEXADECYL-ω-HYDROXY- see PJT300
POLY(OXY-1,2-ETHANEDIYL), α-HYDRO-omega-HYDROXY- see PJT000
POLY(OXY-1,2-ETHANEDIYL), α-ISODECYL-ω-HYDROXY- see PKE370
POLY(OXY-1,2-ETHANEDIYL), α-((2-METHOXYETHOXY)METHYL)-ω-HYDROXY-, COCO ALKYL ETHERS see MEL800
POLY(OXY-1,2-ETHANEDIYL), α-METHYL-ω-METHOXY-(9CI) see DOM100
POLY(OXY-1,2-ETHANEDIYL), α-(3-METHYL-1-(2-METHYLPROPYL)BUTYL)-ω-HYDROXY- see MLS800
POLY(OXY-1,2-ETHANEDIYL)-, α-(4-NONYLPHENYL)-OMEGA-HYDROXY-, BRANCHED see NND200
POLY(OXY-1,2-ETHANEDIYL), α,α'-((9-OCTADECENYLIMINO)DI-2,1-

ETHANEDIYL)BIS(ω-HYDROXY- see PKE400
POLY(OXY-1,2-ETHANEDIYL), α-(1-OXODODECYL)-ω-HYDROXY-(9CI) see PJY000
POLY(OXY-1,2-ETHANEDIYLOXYCARBONYL-1,4-PHENYLENECARBONYL) see PKF750
POLY(OXY-1,2-ETHANEDIYL), α-SULFO-ω-(NONYLPHENOXY)-, AMMONIUM SALT see ANO600
POLYOXYETHYLATED (C9-10) ALKYL THIOETHER see PKE350
POLYOXYETHYLATED (4) ISODECYL ALCOHOL see PKE370
POLYOXYETHYLATED (6) ISODECYL ALCOHOL see PKE390
POLYOXYETHYLATED OLEYL AMINE see PKE400
POLYOXYETHYLATED VEGETABLE OIL see PKE500
POLYOXYETHYLENE (75) see PJT750
POLYOXYETHYLENE 1500 see PJT500
POLYOXYETHYLENE ALKYLAMINE MONO-FATTY ACID ESTER see PKE550
POLYOXYETHYLENE-ALKYL CITRIC DIESTER-TRIETHANOLAMINE see PKE600
POLYOXYETHYLENE (7) ALKYL (14) ETHER see TBY750
POLYOXYETHYLENE-sec-ALKYL ETHER CITRIC DIESTER TRIETHANOLAMINE see PKE600
POLY(OXYETHYLENE)CETYL ETHER see PJT300
POLYOXYETHYLENE CHOLESTERYL ETHER see PKE700
POLYOXYETHYLENE DIBENZOATE see PKE750
POLYOXYETHYLENE DIMETHYL ETHER see DOM100
POLYOXYETHYLENE(4)DOCYL ALCOHOL PHOSPHATE POTASSIUM SALT see PKE800
POLYOXYETHYLENE DODECANOL see LBT000
POLY(OXYETHYLENE)HEXADECYL ETHER see PJT300
POLYOXYETHYLENE (5) LANOLIN see LAU560
POLYOXYETHYLENE (20) LANOLIN see LAU560
POLYOXYETHYLENE (24) LANOLIN see LAU560
POLYOXYETHYLENE (30) LANOLIN see LAU560
POLYOXYETHYLENE (50) LANOLIN see LAU560
POLYOXYETHYLENE (60) LANOLIN see LAU560
POLYOXYETHYLENE (85) LANOLIN see LAU560
POLYOXYETHYLENE (100) LANOLIN see LAU560
POLYOXYETHYLENE LAURIC ALCOHOL see DXY000
POLYOXYETHYLENE LAURYL ETHER see DXY000
POLYOXYETHYLENE LAURYL ETHER PHOSPHATE see PKE850
POLY(OXYETHYLENE)MONOCETYL ETHER see PJT300
POLYOXYETHYLENE (20) MONO- and DIGLYCERIDES of FATTY ACIDS see EEU100
POLYOXYETHYLENE MONOLAURATE see PJY000
POLYOXYETHYLENE-20 MONOLAURATE see PJY000
POLYOXYETHYLENE MONOOCTYLPHENYL ETHER see GHS000

POLYOXYETHYLENE MONO(OCTYLPHENYL) ETHER see PKF500
POLY(OXYETHYLENE) MONOOLEATE see PJY100
POLYOXYETHYLENE MONOSTEARATE see PJW750
POLYOXYETHYLENE-8-MONOSTEARATE see PJV250
POLYOXYETHYLENE NONYLPHENOL see NND500
POLYOXYETHYLENE (9) NONYL PHENYL ETHER see PKF000
POLY(OXYETHYLENE)OCTYLPHENOL ETHER see GHS000
POLYOXYETHYLENE (9) OCTYLPHENYL ETHER see PKF500
POLYOXYETHYLENE (13) OCTYLPHENYL ETHER see PKF500
POLY(OXYETHYLENE)-p-tert-OCTYLPHENYL ETHER see PKF500
POLY(OXYETHYLENE) OLEATE see PJY100
POLY(OXYETHYLENE) OLEIC ACID ESTER see PJY100
POLYOXYETHYLENE (20) OLEYL ETHER see PJW500
POLY(OXYETHYLENEOXYTEREPHTHALOYL) see PKF750
POLY(OXYETHYLENE)PALMITYL ETHER see PJT300
POLY(OXYETHYLENE)-POLY(OXYPROPYLENE)-POLY(OXYETHYLENE) POLYMER see PJK150
POLY(OXYETHYLENE)-POLY(OXYPROPYLENE)-POLY(OXYETHYLENE) POLYMER see PJK151
POLYOXYETHYLENE (20) SORBITAN MONOLAURATE see PKG000
POLYOXYETHYLENE (20) SORBITAN MONOLAURATE see PKG000
POLYOXYETHYLENE (20) SORBITAN MONOLAURATE see PKL000
POLYOXYETHYLENE SORBITAN MONOOLEATE see PKL100
POLYOXYETHYLENE SORBITAN MONOPALMITATE see PKG500
POLYOXYETHYLENE SORBITAN MONOPALMITATE see PKG500
POLYOXYETHYLENE 20 SORBITAN MONOPALMITATE see PKG500
POLYOXYETHYLENE SORBITAN MONOSTEARATE see PKL030
POLYOXYETHYLENE 20 SORBITAN MONOSTEARATE see PKL030
POLYOXYETHYLENE SORBITAN OLEATE see PKL100
POLYOXYETHYLENE (20) SORBITAN TRIOLEATE see TOE250
POLYOXYETHYLENE(8)STEARATE see PJV250
POLYOXYL 10 OLEYL ETHER see OIG040
POLYOXYMETHYLENE see TMP000
POLYOXYMETHYLENE GLYCOLS see FMV000
POLY(OXYPROPYLENE) BUTYL ETHER see BRP250
POLYOXYPROPYLENE MONOBUTYL ETHER see BRP250
POLYOXYPROPYLENE-POLYOXYETHYLENE BLOCK COPOLYMER see PKG600
POLY(OXYTETRAMETHYLENE) see GHY100
POLY(OXYTETRAMETHYLENE)DIOL see GHY100
POLYPEPTIDE, BACILLUS THURINGIENSIS subsp. ISRAELENSIS, crystal preparation see BAC135
POLYPEPTIN see CMS218

POLY(PEROXYISOBUTYROLACTONE) see PKH260
POLYPHENOL FRACTION OF BETEL NUT see BFW010
POLY(p-PHENYLENE TEREPHTHALAMIDE) see BBO750
POLY p-PHENYLENE TEREPTHALAMIDE ARAMID FIBER see PKH850
POLYPHLOGIN see PGG000
POLYPHOS see SHM500
POLYPODINE A see HKG500
POLY (A). POLY (U) COMPLEX see PJK400
POLYPRO 1014 see PMP500
POLYPROPENE see PMP500
POLY(2-PROPYL-m-DIOXANE-4,6-DIYLENE) see PKI000
POLYPROPYLENE see PMP500
POLYPROPYLENE, combustion products see PKI250
POLYPROPYLENE GLYCOL see PKI500
POLYPROPYLENE GLYCOL 425 see PKI550
POLYPROPYLENE GLYCOL 750 see PKI750
POLYPROPYLENE GLYCOL 1200 see PKJ250
POLYPROPYLENE GLYCOL 2025 see PKJ500
POLYPROPYLENE GLYCOL 4025 see PKK000
POLYPROPYLENE GLYCOL, (CHLOROMETHYL)OXIRANE POLYMER see PKK100
POLYPROPYLENE GLYCOL METHYL ETHER see MKS250
POLYPROPYLENE GLYCOL MONOBUTYL ETHER see BRP250
POLYPROPYLENE GLYCOL 400, MONOBUTYL ETHER see PKK500
POLYPROPYLENE GLYCOL 800, MONOBUTYL ETHER see PKK750
POLYPROPYLENE GLYCOL MONOMETHYLETHER see MKS250
POLYPROPYLENGLYKOL (CZECH) see PKI500
POLYQUATERNIUM 6 see DTS500
POLYQUATERNIUM 10 see QAT600
POLYRAM see MQQ250
POLYRAM 80 see MQQ250
POLYRAM COMBI see MQQ250
POLYRAM M see MAS500
POLYRAM 80WP see MQQ250
POLYRAM ULTRA see TFS350
POLYRAM Z see EIR000
POLY(RA). POLY(RU) see PJK400
POLY-RU:RA see PJK400
POLYSILICONE see PJR250
POLYSILYLENE see PKK775
POLYSION N 22 see PJS750
POLYSIZER 173 see PKP750
POLY(SODIUM p-STYRENESULFONATE) see SJK375
POLY-SOLV see CBR000
POLY-SOLV DB see DJF200
POLY-SOLV DM see DJG000
POLY-SOLV EB see BPJ850
POLY-SOLV EE see EES350
POLY-SOLV EE ACETATE see EES400
POLY-SOLV EM see EJH500
POLY-SOLVE MPM see PNL250
POLY-SOLV TB see TKL750
POLY-SOLV TE see EFL000
POLY-SOLV TM see TJQ750
POLYSORBAN 80 see PKL100
POLYSORBATE 20 see PKL000
POLYSORBATE 40 see PKG500
POLYSORBATE 60 see PKL030
POLYSORBATE 65 see PKL050
POLYSORBATE 65 see SKV195
POLYSORBATE 80 see PKL100
POLYSORBATE 80, U.S.P. see PKL100
POLYSTEP A 11 see BBS275
POLYSTICHOCITRIN see FBY000
POLYSTROL D see SMQ500

POLYSTYRENE see SMQ500
POLYSTYRENE-ACRYLONITRILE see ADY500
POLYSTYRENE BW see SMQ500
POLYSTYRENE LATEX see SMQ500
POLY(STYRENESULFONATE) SODIUM SALT see SFO100
POLYSTYROL see SMQ500
POLY(T) see TFX795
POLYTAC see PMP500
POLYTAR BATH see CMY800
POLYTEF see TAI250
POLY-TERGENT S 405LF see AFJ175
POLY-TERGENT SLF 18 see AFJ175
POLYTETRAFLUOROETHENE see TAI250
POLYTETRAFLUOROETHYLENE see TAI250
POLY((6-((1,1,3,3-TETRAMETHYLBUTYL)AMINO)-1,3,5-TRIAZINE-2,4-DIYL)((2,2,6,6-TETRAMETHYL-4-PIPERIDINYL)IMINO)-1,6-HEXANEDIYL((2,2,6,6-TETRAMETHYL-4-PIPERIDINYL)IMINO)) see PKL150
POLY(TETRAMETHYLENE ETHER) see GHY100
POLY(TETRAMETHYLENE ETHER)DIOL see GHY100
POLY(TETRAMETHYLENE ETHER)GLYCOL see GHY100
POLY(TETRAMETHYLENE GLYCOL) see GHY100
POLY(TETRAMETHYLENE OXIDE) see GHY100
POLY(TETRAMETHYLENE OXIDE) GLYCOL see GHY100
POLY(N,N,N',N'-TETRAMETHYL-N-TRIMETHYLENEHEXAMETHYLENEDIAMMONIUM DIBROMIDE) see HCV500
POLYTEX 973 see ADV900
POLYTHENE see PJS750
POLYTHERM see PKQ059
POLYTHIAZIDE see PKL250
POLYTHYMIDINE see TFX795
POLYTHYMIDYLIC ACID see TFX795
POLYTOX see DGB000
POLY(TRIBROMOSTYRENE) see EEE650
POLY(TRIFLUOROCHLOROETHYLENE) see KDK000
POLY(TRIFLUOROETHYLENE CHLORIDE) see KDK000
POLY(TRIFLUOROMONOCHLOROETHYLENE) see KDK000
POLY(TRIFLUOROVINYL CHLORIDE) see KDK000
POLYURETHANE A see PKL500
POLYURETHANE ESTER FOAM see PKL500
POLYURETHANE ETHER FOAM see PKL500
POLYURETHANE FOAM see PKL500
POLYURETHANE SPONGE see PKL500
POLYURETHANE Y-195 see PKL750
POLYURETHANE Y-217 see PKM000
POLYURETHANE Y-218 see PKM250
POLYURETHANE Y-221 see PKM500
POLYURETHANE Y-222 see PKM750
POLYURETHANE Y-223 see PKN000
POLYURETHANE Y-224 see PKN250
POLYURETHANE Y-225 see PKN500
POLYURETHANE Y-226 see PKN750
POLYURETHANE Y-227 see PKO000
POLYURETHANE Y-238 see CDV625
POLYURETHANE Y-290 see PKO500
POLYURETHANE Y-299 see PKO750
POLYURETHANE Y-302 see PKP000
POLYURETHANE Y-304 see PKP250
POLYVIDONE see PKQ250
POLYVINOL see PKP750
POLYVINYL ACETATE CHLORIDE see PKP500
POLYVINYL ACETATE (FCC) see AAX250
POLYVINYL ALCOHOL see PKP750
POLYVINYL ALCOHOL 18/11 see PKP800

POLYVINYL BROMIDE see PKQ000
POLYVINYLBUTYRAL (CZECH) see PKI000
POLYVINYL BUTYRAL RESINS see PKI000
POLY(n-VINYLBUTYROLACTAM) see PKQ250
POLY(VINYLCARBAZOLE) see VNK200
POLY(9-VINYLCARBAZOLE) see VNK200
POLY(N-VINYLCARBAZOLE) see VNK200
POLYVINYLCHLORID (GERMAN) see PKQ059
POLYVINYL CHLORIDE see PKQ059
POLYVINYLCHLORIDE ACETATE see PKP500
POLYVINYL CHLORIDE–POLYVINYL ACETATE see AAX175
POLY(VINYLCYCLOHEXENE DIOXIDE) see PKQ070
POLY(VINYL ETHYL ETHER) see OJV600
POLYVINYLIDENE FLUORIDE (PYROLYSIS) see DKH600
POLY(VINYLIMIDAZOLE) see VPP100
POLY(1-VINYLIMIDAZOLE) see VPP100
POLY(N-VINYLIMIDAZOLE) see VPP100
POLYVINYLPOLYPYRROLIDONE see PKQ150
POLY(VINYLPYRIDINE 1-OXIDE) see PKQ100
POLY(VINYLPYRIDINE N-OXIDE) see PKQ100
POLY(1-VINYL-2-PYRROLIDINONE) Hueper's polymer No. 1 see PKQ500
POLY(1-VINYL-2-PYRROLIDINONE) Hueper's polymer No. 2 see PKQ750
POLY(1-VINYL-2-PYRROLIDINONE) Hueper's polymer No. 3 see PKR000
POLY(1-VINYL-2-PYRROLIDINONE) Hueper's polymer No. 4 see PKR250
POLY(1-VINYL-2-PYRROLIDINONE) Hueper's polymer No. 5 see PKR500
POLY(1-VINYL-2-PYRROLIDINONE) Hueper's polymer No. 6 see PKR750
POLY(1-VINYL-2-PYRROLIDINONE) Hueper's polymer No. 7 see PKS000
POLY(1-VINYL-2-PYRROLIDINONE) HOMOPOLYMER see PKQ250
POLYVINYLPYRROLIDONE see PKQ250
POLYVINYL SULFATE POTASSIUM SALT see PKS250
POLYVIOL see PKP750
POLYVIOL M 13/140 see PKP750
POLYVIOL MO 5/140 see PKP750
POLYVIOL W 25/140 see PKP750
POLYVIOL W 40/140 see PKP750
POLYWAX 1000 see PJS750
POLY-ZOLE AZDN see ASL750
POMALUS ACID see MAN000
POMARSOL see TFS350
POMARSOL Z FORTE see BJK500
POMASOL see TFS350
POMEX see CBM750
POMME EPINEUSE (FRENCH) see SLV500
POMMIER (FRENCH) see AQP875
PONALAR see XQS000
PONALID see DWE800
PONALIDE see DWE800
PONCEAU 3R see FAG018
PONCEAU 4R see FMU080
PONCEAU 4R ALUMINUM LAKE see FMU080
PONCEAU BNA see FMU070
PONCEAU INSOLUBLE OLG see OHI200
PONCEAU INSOLUBLE OLG see XRA000
PONCEAU R (BIOLOGICAL STAIN) see FMU070
PONCEAU SX see FAG050
PONCEAU XYLIDINE (BIOLOGICAL STAIN) see FMU070
PONCYL see GKE000
PONDERAL see PDM250
PONDERAX see PDM250
PONDIMIN see PDM250
PONDINIL see PKS500

PONDOCIL see AOD000
PONDOCILLIN see AOD000
PONECIL see AIV500
PONOLITH FAST VIOLET 4RN see DFN450
PONOLITH ORANGE Y see CMS145
PONOXYLAN see UTU500
PONSOL BLUE BCS see DFN300
PONSOL BLUE BF see DFN300
PONSOL BLUE BFD see DFN300
PONSOL BLUE BFDP see DFN300
PONSOL BLUE BFN see DFN300
PONSOL BLUE BFND see DFN300
PONSOL BLUE BFP see DFN300
PONSOL BLUE GZ see IBV050
PONSOL BLUE RCL see IBV050
PONSOL BLUE RPC see IBV050
PONSOL BRILLIANT BLUE R see IBV050
PONSOL BROWN ARD see CMU780
PONSOL BROWN ARN see CMU780
PONSOL BROWN RBT see CMU770
PONSOL JADE GREEN SUPRA D see VGP100
PONSOL NAVY BLUE RA see VGP100
PONSOL NAVY BLUE RAD see VGP100
PONSOL OLIVE AR see DUP100
PONSOL OLIVE ARD see DUP100
PONSOL RED 2B see CMU825
PONSOL RED 2BD see CMU825
PONSOL RP see IBV050
PONSTAN see XQS000
PONSTEL see XQS000
PONSTIL see XQS000
PONSTYL see XQS000
PONTACHROME BLUE BLACK BB see CMP880
PONTACHROME RED B see CMG750
PONTACHROME VIOLET SW see HLI000
PONTACHROME YELLOW GS see SIT850
PONTACHROME YELLOW 3RN see NEY000
PONTACYL BRILLIANT BLUE see ADE500
PONTACYL BRILLIANT BLUE A see ERG100
PONTACYL BRILLIANT BLUE V see ADE500
PONTACYL CARMINE 2G see CMM300
PONTACYL CARMINE 6B see CMM400
PONTACYL FAST BLUE R see ADE750
PONTACYL GREEN BL see FAE950
PONTACYL RUBINE R see HJF500
PONTACYL SCARLET RR see FMU080
PONTACYL SKY BLUE 4BX see CMO500
PONTAL see XQS000
PONTALITE see PKB500
PONTAMINE BLACK E see AQP000
PONTAMINE BLACK EBN see AQP000
PONTAMINE BLUE BB see CMO000
PONTAMINE BLUE RW see CMO600
PONTAMINE BLUE 3BX see CMO250
PONTAMINE BOND BLUE B see CMO650
PONTAMINE BORDEAUX B see CMO872
PONTAMINE BROWN BCW see CMO820
PONTAMINE BROWN BT see CMO820
PONTAMINE BROWN D 3GN see CMO810
PONTAMINE BROWN N3G see CMO825
PONTAMINE BROWN NCR see CMO810
PONTAMINE BROWN RMR see CMO800
PONTAMINE DEEP BLUE BH see CMN800
PONTAMINE DEVELOPER TN see TGL750
PONTAMINE DIAZO BLACK BHSW see CMN800
PONTAMINE FAST BLUE 7GLN see CMO650
PONTAMINE FAST RED F see CMO870
PONTAMINE FAST RED FCB see CMO870
PONTAMINE FAST RED 8BLX see CMO885
PONTAMINE FAST SCARLET 4BS see CMO870
PONTAMINE GREEN BXN see CMO840
PONTAMINE GREEN GXN see CMO840
PONTAMINE GREEN S see CMO830
PONTAMINE PURE YELLOW see CMP050

PONTAMINE PURE YELLOW MN see CMP050
PONTAMINE SCARLET 3B see CMO875
PONTAMINE SKY BLUE see CMN750
PONTAMINE SKY BLUE 5BX see CMO500
PONTAMINE VIOLET N see CMP000
PONTAMIN FAST TURQUOISE 8GL see COF420
PONTOCAINE see BQA010
PONTOCAINE HYDROCHLORIDE see TBN000
POOA see PJY100
POOGIPHALAM, nut extract see BFW000
POOR MAN'S TREACLE see WBS850
POP see ELL500
POP-DOCK see FOM100
POPO-HAU (HAWAII) see HGP600
POPROLIN see PMP500
POPULAGE see MBU550
POPULNETIN see ICE000
PORAMINE MALEATE see PJJ325
PORAPAK B see VQK595
PORCINE-TRYPSIN see PKS600
PORFIROMYCIN see MLY000
PORFIROMYCINE see MLY000
PORK TRYPSIN see PKS600
POROCEL see BAR900
POROCEL O see BAR900
POROFOR 57 see ASL750
POROFOR 505 see ASM270
POROFOR ADC/R see ASM270
POROFOR BSH see BBS300
POROFOR-BSH-PULVER see BBS300
POROFOR CHKHC-18 see DVF400
POROFOR CHKHZ-5 see MQJ300
POROFOR ChKhZ 9 see BBS300
POROFOR ChKhZ 21 see ASM270
POROFOR ChKhZ 21R see ASM270
POROLEN see PJS750
POROPHOR B see DVF400
POROPHORE 4X3-5 see MQJ300
PORPHYROMYCIN see MLY000
PORTAMYCIN see SLW475
PORTLAND CEMENT see PKS750
PORTLAND CEMENT SILICATE see PKS750
PORTLAND STONE see CAO000
PORTUGUESE MAN-OF-WAR TOXIN see PIA375
POSEDRAN see BEG000
POSEDRINE see BEG000
POSSE see CCC280
POSSUM WOOD see SAT875
POSTAFEN see HGC500
POSTERIOR PITUITARY EXTRACT see ORU500
POSTINOR see NNQ500
POSTINOR see NNQ520
POST LOCUST see GJU475
PO-SYSTOX see DAP200
POTABLAN see CGL250
POTALIUM see PLG800
POTASAN see PKT000
POTASH see PLA000
POTASH CHLORATE (DOT) see PLA250
POTASORAL see PLG800
POTASSA see PLJ500
POTASSE CAUSTIQUE (FRENCH) see PLJ500
POTASSIO (CHLORATO di) (ITALIAN) see PLA250
POTASSIO (IDROSSIDO di) (ITALIAN) see PLJ500
POTASSIO (PERMANGANATO di) (ITALIAN) see PLP000
POTASSIUM see PKT250
POTASSIUM (liquid alloy) see PKT500
POTASSIUM, metal liquid alloy (DOT) see PKT500
POTASSIUM ACESULFAME see AAF900
POTASSIUM ACETATE see PKT750
POTASSIUM ACETYLENE-1,2-DIOXIDE see PKT775

POTASSIUM ACETYLIDE see PKU000
POTASSIUM ACID ARSENATE see ARD250
POTASSIUM ACID FLUORIDE see PKU250
POTASSIUM ACID FLUORIDE (solution) see PKU500
POTASSIUM ACID PHTHALATE see HIB600
POTASSIUM ACID SACCHARATE see PLJ350
POTASSIUM ACID SULFATE see PKX750
POTASSIUM ACID TARTRATE see PKU600
POTASSIUM ALGINATE see PKU700
POTASSIUM ALUM see AHF100
POTASSIUM ALUM see AHF200
POTASSIUM ALUM see PKU725
POTASSIUM ALUM DODECAHYDRATE see AHF200
POTASSIUM ALUMINUM SULFATE see AHF100
POTASSIUM ALUMINUM SULFATE see PKU725
POTASSIUM AMALGAM see PKU750
POTASSIUM AMIDE see PKV000
POTASSIUM, AMMINETRICHLOROPLATINATE (1-) see PJD750
POTASSIUM AMYLXANTHATE see PKV100
POTASSIUM AMYLXANTHOGENATE see PKV100
POTASSIUM n-AMYLXANTHOGENATE see PKV100
POTASSIUM ANTIMONYL TARTRATE see AQG250
POTASSIUM ANTIMONYL-d-TARTRATE see AQG250
POTASSIUM ANTIMONYL-d-TARTRATE see AQG500
POTASSIUM ANTIMONYL-l-TARTRATE see AQH000
POTASSIUM ANTIMONYL-d,l-TARTRATE see AQG750
POTASSIUM ANTIMONYL-meso-TARTRATE see AQH250
POTASSIUM ANTIMONY TARTRATE see AQG250
POTASSIUM ARSENATE see ARD250
POTASSIUM ARSENATE, MONOBASIC see ARD250
POTASSIUM ARSENITE see PKV500
POTASSIUM ASPARTATE see PKV600
POTASSIUM l-ASPARTATE see PKV600
POTASSIUM ASULAM see PKV700
POTASSIUM AZAOROTATE see AFQ750
POTASSIUM AZIDE see PKW000
POTASSIUM AZIDODISULFATE see PKW250
POTASSIUM AZIDOSULFATE see PKW500
POTASSIUM BENZENEHEXOXIDE see PKW550
POTASSIUM BENZENESULFONYLPEROXYSULFATE see PKW750
POTASSIUM BENZOATE see PKW760
POTASSIUM BENZOYLPEROXYSULFATE see PKX000
POTASSIUM BENZYLPENICILLIN see BFD000
POTASSIUM BENZYLPENICILLINATE see BFD000
POTASSIUM BENZYLPENICILLIN G see BFD000
POTASSIUM BICARBONATE see PKX100
POTASSIUM BICHROMATE see PKX250
POTASSIUM BIFLUORIDE see PKU250
POTASSIUM BIFLUORIDE, solution (DOT) see PKU500
POTASSIUM BIFLUORIDE, solid or solution (DOT) see PKU250
POTASSIUM BIPHOSPHATE see PLQ405
POTASSIUM BIPHTHALATE see HIB600
POTASSIUM BISACCHARATE see PLJ350

POTASSIUM BIS(2-HYDROXYETHYL)DITHIOCARBAMATE see PKX500
POTASSIUM BIS(PROPYNYL)PALLADATE see PKX639
POTASSIUM BIS(PROPYNYL)PLATINATE see PKX700
POTASSIUM BISULFATE see PKX750
POTASSIUM BISULPHATE see PKX750
POTASSIUM BITARTRATE see PKU600
POTASSIUM BOROFLUORIDE see PKY000
POTASSIUM BOROHYDRATE see PKY250
POTASSIUM BOROHYDRIDE (DOT) see PKY250
POTASSIUM BROMATE see PKY300
POTASSIUM BROMIDE see PKY500
POTASSIUM (E)-BUTANEDIAZOTATE see HNV100
POTASSIUM (Z)-BUTANEDIAZOTATE see HNV150
POTASSIUM-tert-BUTOXIDE see PKY750
POTASSIUM BUTYLXANTHATE see PKY850
POTASSIUM-o-BUTYL XANTHATE see PKY850
POTASSIUM BUTYLXANTHOGENATE see PKY850
POTASSIUM CANRENOATE see PKZ000
POTASSIUM CAPRYLATE see PLN250
POTASSIUM CARBONATE (2:1) see PLA000
POTASSIUM CARBONYL see PKW550
POTASSIUM CARBONYL (DOT) see CCB500
POTASSIUM CHLORATE see PLA250
POTASSIUM CHLORATE (DOT) see PLA250
POTASSIUM (CHLORATE de) (FRENCH) see PLA250
POTASSIUM CHLORATE (UN 1485) (DOT) see PLA250
POTASSIUM CHLORATE, solution (UN 2427) (DOT) see PLA250
POTASSIUM CHLORIDE see PLA500
POTASSIUM CHLORIDE OXIDE see PLK000
POTASSIUM CHLORITE see PLA525
POTASSIUM 7-CHLORO-2,3-DIHYDRO-2-OXO-5-PHENYL-1H-1,4-BENZODIAZEPINE-3-CARBOXYLATE KOH see CDQ250
POTASSIUM CHLOROPALLADATE see PLA750
POTASSIUM CHLOROPLATINATE see PLR000
POTASSIUM CHLOROPLATINITE see PJD250
POTASSIUM CHROMATE(VI) see PLB250
POTASSIUM CHROMIC SULFATE see PLB500
POTASSIUM CHROMIC SULPHATE see PLB500
POTASSIUM CHROMIUM ALUM see CMG850
POTASSIUM CHROMIUM ALUM see PLB500
POTASSIUM CITRATE see PLB750
POTASSIUM CITRATE TRI(HYDROGEN PEROXIDATE) see PLB759
POTASSIUM CLAVULANATE see PLB775
POTASSIUM CLAVULANATE mixed with AMOXICILLIN (1:2) see ARS125
POTASSIUM COLUMBATE see PLL250
POTASSIUM COMPOUNDS see PLC100
POTASSIUM COPPER(I) CYANIDE see PLC175
POTASSIUM CUPROCYANIDE see PLC175
POTASSIUM CYANATE see PLC250
POTASSIUM CYANIDE see PLC500
POTASSIUM CYANIDE, solution (DOT) see PLC500
POTASSIUM CYANIDE-POTASSIUM NITRITE see PLC775
POTASSIUM CYANONICKELATE HYDRATE see TBW250
POTASSIUM CYCLOHEXANEHEXONE-1,3,5-TRIOXIMATE see PLC780

POTASSIUM CYCLOPENTADIENIDE see PLC800
POTASSIUM DICHLOROISOCYANURATE see PLD000
POTASSIUM 5-(2,4-DICHLOROPHENOXY)-2-NITROBENZOATE see PLD050
POTASSIUM DICHLORO-s-TRIAZINETRIONE see PLD000
POTASSIUM DICHROMATE(VI) see PKX250
POTASSIUM DICHROMATE, ZINC CHROMATE, and ZINC HYDROXIDE (1:3:1) see ZFJ122
POTASSIUM DICYANOCUPRATE(1-) see PLC175
POTASSIUM DIETHYNYLPALLADATE(2-) see PLD100
POTASSIUM DIETHYNYLPLATINATE(2-) see PLD150
POTASSIUM DIFLUOROPHOSPHATE see PLD250
POTASSIUM DIHYDROGEN ARSENATE see ARD250
POTASSIUM DIHYDROGEN PHOSPHATE see PLQ405
POTASSIUM DIMETHYL DITHIOCARBAMATE see PLD500
POTASSIUM-2,5-DINITROCYCLOPENTANONIDE see PLD550
POTASSIUM DINITROMETHANIDE see PLD575
POTASSIUM DINITROOXALATOPLATINATE(2-) see PLD600
POTASSIUM aci-1,1-DINITROPROPANE see PLD700
POTASSIUM-1,1-DINITROPROPANIDE see PLD700
POTASSIUM DINITROSOSULFITE see PLD710
POTASSIUM DINITROTETRACHLOROPLATINATE see PLD720
POTASSIUM-3,5-DINITRO-2-(1-TETRAZENYL)PHENOLATE see PLD730
POTASSIUM DIOXIDE see PLE260
POTASSIUM DIPEROXY ORTHOVANADATE see PLE500
POTASSIUM DISULPHATOCHROMATE(III) see PLB500
POTASSIUM DODECANOATE see PLK750
POTASSIUM (E)-ETHANEDIAZOTATE see HMB560
POTASSIUM ETHOXIDE see PLE575
POTASSIUM O-ETHYL DITHIOCARBONATE see PLF000
POTASSIUM ETHYLMERCURIC THIOGLYCOLLATE see PLE750
POTASSIUM ETHYLXANTHATE see PLF000
POTASSIUM ETHYL XANTHOGENATE see PLF000
POTASSIUM FERRICYANATE see PLF250
POTASSIUM FERRICYANIDE see PLF250
POTASSIUM FERROCYANATE see TEC500
POTASSIUM FERROCYANIDE see PLH100
POTASSIUM FERROCYANIDE see TEC500
POTASSIUM FLUOBORATE see PKY000
POTASSIUM N-FLUOREN-2-YL ACETOHYDROXAMATE see HIS000
POTASSIUM FLUORIDE see PKU250
POTASSIUM FLUORIDE see PLF500
POTASSIUM FLUORIDE, solution (DOT) see PLF500
POTASSIUM FLUORIDE, DIHYDRATE see PLF750
POTASSIUM FLUOROACETATE see PLG000
POTASSIUM FLUOROBORATE see PKY000
POTASSIUM FLUORURE (FRENCH) see PLF500
POTASSIUM FLUOSILICATE see PLH750
POTASSIUM FLUOTANTALATE see PLH000
POTASSIUM FLUOZIRCONATE see PLG500

POTASSIUM PERMANGANATE see PLP000
POTASSIUM (PERMANGANATE de) (FRENCH) see PLP000
POTASSIUM PEROXIDE see PLP250
POTASSIUM PEROXOFERRATE(2−) see PLP500
POTASSIUM PEROXYDISULFATE see DWQ000
POTASSIUM PEROXYDISULPHATE see DWQ000
POTASSIUM PEROXYFERRATE see PLP500
POTASSIUM PEROXYSULFATE see PLP750
POTASSIUM PERRHENATE see PLQ000
POTASSIUM PERRHENATE(1⁻) see PLQ000
POTASSIUM PERSULFATE (DOT) see DWQ000
POTASSIUM PHENETHICILLIN see PDD350
POTASSIUM (1-PHENOXYETHYL)PENICILLIN see PDD350
POTASSIUM-α-PHENOXYETHYL PENICILLIN see PDD350
POTASSIUM PHENOXYMETHYLPENICILLIN see PDT750
POTASSIUM-6-(α-PHENOXYPROPIONAMIDO)PENICILLANATE see PDD350
POTASSIUM PHENYL DINITROMETHANIDE see PLQ275
POTASSIUM PHENYLGLYCINE see PLQ300
POTASSIUM PHOSPHATE see PLQ410
POTASSIUM PHOSPHATE, DIBASIC see PLQ400
POTASSIUM PHOSPHATE, MONOBASIC see PLQ405
POTASSIUM PHOSPHATE, TRIBASIC see PLQ410
POTASSIUM PHOSPHATE, TRIBASIC see PLQ410
POTASSIUM PHOSPHINATE see PLQ750
POTASSIUM PICLORAM see PLQ760
POTASSIUM PICRATE see PLQ775
POTASSIUM PLATINIC CHLORIDE see PLR000
POTASSIUM PLATINOCHLORIDE see PJD250
POTASSIUM POLYMETAPHOSPHATE see PLR125
POTASSIUM POLY(VINYL SULFATE) see PKS250
POTASSIUM (E)-PROPANEDIAZOTATE see ELE700
POTASSIUM-O-PROPIONOHYDROXAMATE see PLR175
POTASSIUM 3-PYRIDINECARBOXYLATE see PLL200
POTASSIUM PYROPHOSPHATE see PLR200
POTASSIUM PYROPHOSPHATE see TEC100
POTASSIUM PYROSULFITE see PLR250
POTASSIUM RHENATE see PLR500
POTASSIUM RHENATE (KReO₄) see PLQ000
POTASSIUM RHENIUM OXIDE (KReO₄) see PLQ000
POTASSIUM RHODANATE see PLV750
POTASSIUM RHODANIDE see PLV750
POTASSIUM SALT of BENZYLPENICILLIN see BFD000
POTASSIUM SALT of SORREL see OLE000
POTASSIUM SELENATE see PLR750
POTASSIUM SELENITE (7CI) see SBO100
POTASSIUM SELENOCYANATE see PLS000
POTASSIUM SELENOCYANOACETATE see PLS100
POTASSIUM SILICOFLUORIDE (DOT) see PLH750
POTASSIUM SILVER CYANIDE see PLS250
POTASSIUM SODIUM ALLOY see PLS500
POTASSIUM SORBATE see PLS750
POTASSIUM STANNATE TRIHYDRATE see PLS760
POTASSIUM STEARATE see PLS775
POTASSIUM SULFATE (2:1) see PLT000

POTASSIUM SULFIDE (2:1) see PLT250
POTASSIUM SULFIDE, anhydrous (UN 1382) (DOT) see PLT250
POTASSIUM SULFIDE, hydrated with not <30% water of crystallization (UN 1847) (DOT) see PLT250
POTASSIUM SULFIDE with <30% water of crystallization (UN 1382) (DOT) see PLT250
POTASSIUM SULFITE see PLT500
POTASSIUM SULFOCYANATE see PLV750
POTASSIUM SULFURDIIMIDATE see PLT750
POTASSIUM SULFUR DIIMIDE see PLT750
POTASSIUM SUPEROXIDE (DOT) see PLE260
POTASSIUM TARTRATE see PKU600
POTASSIUM TELLURITE see PLU000
POTASSIUM TETRABROMOPLATINATE see PLU250
POTASSIUM TETRACHLOROPALLADATE see PLN750
POTASSIUM TETRACHLOROPLATINATE(II) see PJD250
POTASSIUM TETRACYANOMERCURATE(II) see PLU500
POTASSIUM TETRACYANONICKELATE see NDI000
POTASSIUM TETRACYANONICKELATE(II) see NDI000
POTASSIUM TETRACYANOTITANATE(IV) see PLU550
POTASSIUM TETRAETHYNYL NICKELATE(2⁻) see PLU575
POTASSIUM TETRAETHYNYL NICKELATE(4⁻) see PLU580
POTASSIUM TETRAIODOMERCURATE(II) see NCP500
POTASSIUM TETRAKISTHIOCYANATOPLATINATE see PLU590
POTASSIUM-1,1,2,2-TETRANITROETHANDIIDE see PLU600
POTASSIUM TETRAPEROXYCHROMATE see PLU750
POTASSIUM TETRAPEROXYMOLYBDATE see PLV000
POTASSIUM TETRAPEROXYTUNGSTATE see PLV250
POTASSIUM-1-TETRAZOLACETATE see PLV275
POTASSIUM THALLIUM(I)AMIDE AMMONIATE see PLV500
POTASSIUM THIOCYANATE see PLV750
POTASSIUM THIOCYANIDE see PLV750
POTASSIUM THIOSULFATE see PLW000
POTASSIUM TITANIUM OXIDE see PLW150
POTASSIUM-s-TRIAZINE-2,4-DIONE-6-CARBOXYLATE see AFQ750
POTASSIUM, TRICHLOROAMMINEPLATINATE (2) see PJD750
POTASSIUM TRICHLOROETHYLENEPLATINATE see PLW200
POTASSIUM TRICYANODIPEROXOCHROMATE (3⁻) see PLW275
POTASSIUM TRIIODIDE see PLW285
POTASSIUM TRINITROMETHANIDE see PLW300
POTASSIUM-2,4,6-TRINITROPHENOXIDE see PLQ775
POTASSIUM TRIPHOSPHATE see PLW400
POTASSIUM TRIPOLYPHOSPHATE see PLW400
POTASSIUM TROCLOSENE see PLD000
POTASSIUM, metal alloys (UN 1420) (DOT) see PKT250
POTASSIUM VANADIUM TRIOXIDE see PLK810
POTASSIUM WARFARIN see WAT209
POTASSIUM XANTHATE see PLF000

POTASSIUM XANTHOGENATE see PLF000
POTASSIUM XANTHOGENATE BUTYL ETHER see PKY850
POTASSIUM ZINC CHROMATE see CMK400
POTASSIUM ZINC CHROMATE see PLW500
POTASSIUM ZINC CHROMATE HYDROXIDE see PLW500
POTASSIUM Z-PROPANEDIAZOTATE see ELE720
POTASSURIL see PLG800
POTATO ALCOHOL see EFU000
POTATO BLOSSOMS, GLYCOALKALOID EXTRACT see PLW550
POTATO, GREEN PARTS see PLW750
POTAVESCENT see PLA500
POTCRATE see PLA250
POTENTIATED ACID GLUTARALDEHYDE see GFQ000
POTHOS see PLW800
POTIDE see PLK500
POTOMAC RED see CHP500
POUNCE see AHJ750
POVAL 117 see PKP750
POVAL 120 see PKP750
POVAL 203 see PKP750
POVAL 205 see PKP750
POVAL 217 see PKP750
POVAL 1700 see PKP750
POVAL C 17 see PKP750
POVAN see PQC500
POVANYL see PQC500
POVIDONE-IODINE see PKE250
POVIDONE (USP XIX) see PKQ250
POVIMAL ST see SEA500
POWDERED METALS see PLX000
POWDER GREEN see COF500
POWDER and ROOT see RNZ000
POWMYL WP see IOS100
POYAMIN see VSZ000
POZZOLITH 400N see BLX000
PP 5 see PCG700
PP-12 see HMN000
PP 009 see FDA885
PP010 see MHI600
PP021 see CLS050
PP 062 see DOX600
PP 100 see EJC025
PP 145 see CDS800
PP148 see PAJ000
PP149 see BRI750
PP 156 see DXI450
PP175 see ASD000
PP 192 see CEX800
PP-199 see FDB300
PP211 see DIN600
PP 321 see LAS200
PP 333 see TMK125
PP383 see RLF350
PP 450 see FMR200
PP511 see DIN800
PP 523 see HCN050
PP 557 see AHJ750
PP 563 see GJU600
PP 581 see TAC800
PP588 see BRJ000
PP604 see GJU200
PP618 see BOO631
PP 675 see BRD000
PP 745 see BJK750
PP 748 see CHJ400
PP781 see MLC250
PP 910 see PAJ250
PP993 see FMU300
PP-1466 see RSZ375
P 2070P see PJS750
PPA see NNM000
PPBP see BCP650
PPC 3 see MIA250
PPD see PEY500
PPE 2 see PJS750
PPE201 see PKO500
PP FACTOR see NCQ900

PP-FACTOR see NCR000
P.P. FACTOR-PELLAGRA PREVENTIVE
FACTOR see NCQ900
P.P.G. 750 see PKI750
P.P.G. 1200 see PKJ250
PPG-14 BUTYL ETHER see BRP250
PPG-16 BUTYL ETHER see BRP250
PPG-33 BUTYL ETHER see BRP250
P. PHYSALIS TOXIN see PIA375
P. PORPHYRIACUS (AUSTRALIA) VENOM
see ARV250
PPTC see PNI750
PPZEIDAN see DAD200
PQ see PCY300
PQD see DVR200
17190RP see POE100
PR 703-78 see UTU500
PR 934-423A see RCK750
PRACARBAMIN see UVA000
PRACARBAMINE see UVA000
PRACTALOL see ECX100
PRACTOLOL see ECX100
PRADONE KOMBI see UTA300
PRADONE PLUS see UTA300
PRADONE TS see UTA300
PRADUPEN see BDY669
PRAECIRHEUMIN see BRF500
PRAEDYN see CMG675
PRAENITRON see TJL250
PRAENITRONA see TJL250
PRAEQUINE see RHZ000
PRAGMAZONE see CKJ000
PRAGMAZONE see THK880
PRAGMOLINE see CMF260
PRAIRIE CROCUS see PAM780
PRAIRIE HEN FLOWER see PAM780
PRAIRIE RATTLESNAKE VENOM see
PLX100
PRAJMALINE see PNC875
PRAJMALINE BITARTRATE see DNB000
PRAJMALINE HYDROGEN TARTRATE see
DNB000
PRAJMALIUM see PNC875
PRAKTOLOLU (POLISH) see ECX100
PRALIDOXIME CHLORIDE see FNZ000
PRALIDOXIME IODIDE see POS750
PRALIDOXIME MESYLATE see PLX250
PRALIDOXIME METHANESULFONATE see
PLX250
PRALIDOXIME METHIODIDE see POS750
PRALIDOXIME METHOSULFATE see
PLX250
PRALLETHRIN see THI500
PRALUMIN see TDA500
PRAMIL see SBN000
PRAMINDOLE see DPX200
PRAMITOL see MFL250
PRAMIVERINE HYDROCHLORIDE see
SDZ000
PRAMIVERIN HYDROCHLORIDE see
SDZ000
PRAMUSCIMOL see AKG250
PRANDIOL see PCP250
PRANONE see GEK500
PRANONE and DIETHYLSTILBESTROL see
EEI025
PRANONE and STILBESTROL see EEI025
PRANOPROFEN see PLX400
PRANTAL see DAP800
PRANTAL METHYLSULFATE see DAP800
PRAPAGEN WK see QAT550
PRAPAGEN WKT see QAT550
PRAPAR see MKU750
PRAPARAT 5968 see HGP500
PRAPARAT 9295 see MKU750
PRASEODYMIUM see PLX500
PRASEODYMIUM CHLORIDE see PLX750
PRASEODYMIUM(III) NITRATE (1:3) see
PLY250
PRASTERONE see AOO450
PRASTERONE SODIUM SULFATE see
DAL040

PRASTERONE SODIUM SULFATE
DIHYDRATE see DAL030
PRAVACHOL see LGK500
PRAVASTATIN SODIUM see LGK500
PRAXILENE see NAE100
PRAXIS see IDA400
PRAXITEN see CFZ000
PRAYER BEAD see AAD000
PRAYER BEADS see RMK250
PRAZEPAM see DAP700
PRAZEPINE see DLH600
PRAZIL see CKP250
PRAZINIL see CCK780
PRAZIQUANTEL see BGB400
PRAZOSIN see AJP000
PRAZOSIN HYDROCHLORIDE see FPP100
P-RC see PAQ900
PRD EXPERIMENTAL NEMATOCIDE see
DGL200
PREAN see MBW750
PRECEPTIN see PKF500
PRECIPITATED BARIUM SULPHATE see
BAP000
PRECIPITATED CALCIUM PHOSPHATE see
CAW120
PRECIPITATED CALCIUM SULFATE see
CAX750
PRECIPITATED SILICA see SCL000
PRECIPITATED SULFUR see SOD500
PRECIPITE BLANC see MCW000
PRECISION CLEANING AGENT see TJE100
PRECOCENE 1 see DAM830
PRECOCENE 2 see AEX850
PRECOCENE I see DAM830
PRECOCENE II see AEX850
PRECORT see PLZ000
PRECORTANCYL see PMA000
PRECORTISYL see PMA000
PREDALON-S see SCA750
PREDENT see SHF500
PREDIACORTINE see SOV100
PREDICORT see SOV100
PREDNE-DOME see PMA000
PREDNELAN see PMA000
PREDNELAN-N see SOV100
PREDNESOL see PLY275
PREDNICEN-M see PLZ000
PREDNIDOREN see SOV100
PREDNILONGA see PLZ000
PREDNIS see PMA000
PREDNI-SEDIV see TEH500
PREDNISOLONBISUCCINAT see PLY300
PREDNISOLONE see PMA000
PREDNISOLONE ACETATE see SOV100
PREDNISOLONE 21-ACETATE see SOV100
PREDNISOLONE BISUCCINATE see PLY300
PREDNISOLONE DISODIUM PHOSPHATE
see PLY275
PREDNISOLONE HEMISUCCINATE see
PLY300
PREDNISOLONE 21-HEMISUCCINATE see
PLY300
PREDNISOLONE 21-PHOSPHATE
DISODIUM see PLY275
PREDNISOLONE SODIUM
HEMISUCCINATE see PMA100
PREDNISOLONE SODIUM PHOSPHATE
see PLY275
PREDNISOLONE SODIUM SUCCINATE see
PMA100
PREDNISOLONE 21-SODIUM SUCCINATE
see PMA100
PREDNISOLONE SUCCINATE see PLY300
PREDNISOLONE 21-SUCCINATE see
PLY300
PREDNISOLONE 21-SUCCINATE SODIUM
see PMA100
PREDNISOLONE VALERATE ACETATE see
AAF625
PREDNISOLONE-17-VALERATE-21-
ACETATE see AAF625
PREDNISOLUT see PLY300

PREDNISON see PLZ000
PREDNISONE see PLZ000
PREDNISONE ACETATE see PLZ100
PREDNISONE 21-ACETATE see PLZ100
PREDNIZON see PLZ000
PREDONIN see PMA000
PREDONINE see PMA000
PREDONINE INJECTION see SOV100
PREDONINE SOLUBLE see PMA100
PREDSOL see PLY275
PREDSOLAN see PLY275
PREEGLONE see DWX800
PREEGLONE see EJC025
PREFAR see DNO800
PREFAXIL see BBW750
PREFEMIN see CHJ750
PREFERID see BOM520
PREFLAN see BSN000
PREFMID see BSN000
PREFORAN see NIX000
PREFRIN see SPC500
PREGAN-1,4-DIENE-3,20-DIONE, 11,71-
DIHYDROXY-6-METHYL-21-((8-
(METHYL(2-SULFOETHYL) AMINO)-1,8-
DIOXOOCTYL)OXY)-, MONOSODIUM
SALT, (6-α-11-β)- see MOR600
PREGARD see CQG250
PREGICIL see AAF750
PREGLANDIN see CDB775
1,4-PREGNADIEN-17-α-21-DIOL-3,11,20-
TRIONE-21-ACETATE see PLZ100
1,4-PREGNADIENE-17-α,21-DIOL-3,11,20-
TRIONE see PLZ000
9-β,10-α-PREGNA-4,6-DIENE-3,20-DIONE
see DYF759
(9-β,10-α)-PREGNA-4,6-DIENE-3,20-DIONE
(9CI) see DYF759
PREGNA-1,4-DIENE-3,20-DIONE, 21-
(ACETYLOXY)-11,17-DIHYDROXY-, (11-β)-
see SOV100
PREGNA-1,4-DIENE-3,20-DIONE, 21-(3-
CARBOXY-1-OXOPROPOXY)-11,17-
DIHYDROXY-, (11-β)- (9CI) see PLY300
PREGNA-1,4-DIENE-3,20-DIONE, 7-
CHLORO-11-HYDROXY-16-METHYL-17,21-
BIS(1-OXOPROPOXY)-, (7-α-11-β,16-α)- see
AFI980
PREGNA-4,6-DIENE-3,20-DIONE, 6-
CHLORO-17-HYDROXY-1-α,2-α-
METHYLENE-, ACETATE see CQJ500
PREGNA-1,4-DIENE-3,20-DIONE,9-
FLUORO-11,16,17,21-TETRAHYDROXY-,(11-
β,16-α) see AQX250
9-β,10-α-PREGNA-4,6-DIENE-3,20-DIONE
and 17-α-HYDROXYPREGN-4-ENE-3,20-
DIONE (9:10) see PMA250
PREGNA-1,4-DIENE-3,20-DIONE, 11-
β,17,21-TRIHYDROXY-, 21-(DIHYDROGEN
PHOSPHATE), DISODIUM SALT see PLY275
PREGNA-1,4-DIENE-3,20-DIONE, 11-
β,17,21-TRIHYDROXY-, 21-(HYDROGEN
SUCCINATE) see PLY300
PREGNA-1,4-DIENE-3,20-DIONE, 11-
β,17,21-TRIHYDROXY-, 21-(HYDROGEN
SUCCINATE), MONOSODIUM SALT see
PMA100
1,4-PREGNADIENE-3,20-DIONE-11-β,17-
α,21-TRIOL see PMA000
1,4-PREGNADIENE-11-β,17-α,21-TRIOL-
3,20-DIONE see PMA000
PREGNA-1,4-DIENE-3,11,20-TRIONE, 21-
(ACETYLOXY)-17-HYDROXY- (9CI) see
PLZ100
PREGNA-1,4-DIENE-3,11,20-TRIONE, 17,21-
DIHYDROXY-, 21-ACETATE see PLZ100
17-α-2,4-PREGNADIEN-20-YNO(2,3-
d)ISOXAZOL-17-OL see DAB830
17-α-PREGNA-2,4-DIEN-20-YNO(2,3-
d)ISOXAZOL-17-OL see DAB830
5-α-17-α-PREGNA-2-EN-20-YN-17-OL,
ACETATE see PMA450

PREGNANE-3,20-DIAMMINIUM, N,N,N,N',N',N'-HEXAMETHYL-, DIIODIDE, (3-β,5-α,20S)- see MAO280
PREGNANE-3,20-DIONE, 11,17,21-TRIHYDROXY-, (5β,11β)- see PMA500
5-β-PREGNANE-3,20-DIONE, 11-β,17,21-TRIHYDROXY- see PMA500
PREGNAN-20-ONE, 3,11,17,21-TETRAHYDROXY-, (3α,5β,11β)- see UVJ460
5-β-PREGNAN-20-ONE, 3-α,11-β,17,21-TETRAHYDROXY- see UVJ460
PREGNANT MARE SERUM GONADOTROPIN see SCA750
PREGN-4-EN-17α,21-DIOL-3,11,20-TRIONE see CNS800
3,20-PREGNENE-4 see PMH500
17-α-PREGN-4-ENE-21-CARBOXYLIC ACID, 1-HYDROXY-7-α-MERCAPTO-3-OXO-α-LACTONE see AFJ500
PREGN-5-ENE-3-β,20-α-DIAMINE see IGH700
4-PREGNENE-17α,21-DIOL-3,11,20-17,21-DIHYDROXY- see CNS800
4-PREGNENE-11-α,21-DIOL-3,20-DIONE see CNS625
Δ⁴-PREGNENE-17α,21-DIOL-3,11,20-TRIONE see CNS800
4-PREGNENE-17,α21-DIOL-3,11,20-TRIONE 21-ACETATE see CNS825
PREGNENEDIONE see PMH500
PREGNENE-3,20-DIONE see PMH500
PREGN-4-ENE-3,20-DIONE see PMH500
4-PREGNENE-3,20-DIONE see PMH500
Δ⁴-PREGNENE-3,20-DIONE see PMH500
PREGN-4-ENE-3,20-DIONE, 17-(ACETYLOXY)-6-METHYL-, (6-α)-, mixture with (17-β)-3-HYDROXYESTRA-1,3,5(10)-TRIEN-17-YL CYCLOPENTANEPROPANOATE see EDQ600
PREGN-4-ENE-3,20-DIONE, 16-α,17-DIHYDROXY-, CYCLIC ACETAL WITH ACETONE see PMH550
14-α-PREGN-4-ENE-3,20-DIONE, 17-α,21-DIHYDROXY-, 21-IODOACETATE see DNA850
PREGN-4-ENE-3,20-DIONE, 11-β,18-EPOXY-18,21-DIHYDROXY- see ECA100
PREGN-4-ENE-3,20-DIONE, 11β,18-EPOXY-18,21-DIHYDROXY- see ECA100
PREGN-4-ENE-3,20-DIONE, 11,18-EPOXY-10,21-DIHYDROXY-, (11-β)- see ECA100
PREGN-4-ENE-3,20-DIONE, 11,18-EPOXY-18,21-DIHYDROXY-, (11β)- see ECA100
PREGN-4-ENE-3,20-DIONE, 16,17-((1-METHYLETHYLIDENE)BIS(OXY))-, (16-α)- see PMH550
4-PREGNENE-3,20-DIONE-21-OL ACETATE see DAQ800
4-PREGNENE-11-β,17-α,21-TRIOL 3,20-DIONE see CNS750
PREGNENINOLONE see GEK500
4-PREGNEN-21-OL-3,20-DIONE see DAQ600
PREGN-5-EN-20-ONE, 11-(ACETYLOXY)-3,14-DIHYDROXY-12-(3-METHYL-1-OXOBUTOXY)-, (3-β,11-α,12-β,14-β)- see DYE650
PREGN-4-EN-20-ONE, 3-β,17-DIHYDROXY-, 17-ACETATE, and 3-METHOXY-19-NOR-17-α-PREGNA-1,3,5(10)-TRIEN-20-YN-17-OL (20:1) see ABV600
14-β-PREGN-5-EN-20-ONE, 3-β,11-α,12-β,14-TETRAHYDROXY-, 11-ACETATE 12-ISOVALERATE see DYE650
17-α-PREGN-5-EN-20-YNE-3-β,17-DIOL-3-(3-CYCLOHEXYLPROPIONATE) see EDY600
17-α-PREGN-4-EN-20-YNO(2,3-d)ISOXAZOL-17-OL see DAB830
PREGN-2-EN-20-YN-17-OL, ACETATE, (5-α-17α-- see PMA450

17-α-PREGN-4-EN-20-YN-3-ONE, 17-HYDROXY-, and trans-α-α'-DIETHYL-4,4'-STILBENEDIOL see EEI025
PREGNIN see GEK500
PREGNYL see CMG675
PRELIS see SBV500
PRELUDE see IAL200
PRELUDIN see PMA750
PRELUDIN HYDROCHLORIDE see MNV750
PREMALIN see CKD500
PREMALIN see DGD600
PREMALOX see CBM000
PREMARIN see ECU750
PREMARIN see PMB000
PREMAZINE see BJP000
PREMERGE see BRE500
PREMERGE 3 see BRE500
PREMERGE PLUS see BQI000
PREMGARD see BEP500
PREMOBOLAN see PMC700
PREMODRIN see DBA800
PRENAZONE see PEW000
PRE-NCS see PMB300
PRENDEROL see PMB250
PRENDIOL see PMB250
PRENEMA see SOV100
PRENEOCARZINOSTATIN see PMB300
PRENIMON see TND000
PRENITRON see TJL250
PRENOL see MHU110
PRENORMINE see TAL475
PRENOXDIAZINE HYDROCHLORIDE see LFJ000
PRENT see AAE100
PRENTOX see PIX250
PRENTOX see RNZ000
PRENYL ACETATE see DOQ350
PRENYL ALCOHOL see MHU110
PRENYLAMINE see PEV750
PRENYLAMINE LACTATE see DWL200
PRENYL BENZOATE see MHU150
α-PRENYL-α-(2-DIMETHYLAMINOETHYL)-1-NAPHTHYLACETAMIDE see PMB500
4-PRENYL-1,2-DIPHENYL-3,5-PYRAZOLIDINEDIONE see PEW000
PRENYL SALICYLATE see PMB600
PREP see SFV250
PREPALIN see VSK600
PRE-PAR see RLK700
PREPARATION 84 see TCQ260
PREPARATION 125 see DFT800
PREPARATION 48-80 see PMB800
PREPARATION AF see HEI500
PREPARATION C (the Russian Drug) see HEF200
PREPARATION HE 166 see DUS500
PREPARATION S see HEF200
PREPARED CHALK see CAT775
PREP-DEFOLIANT see SFV250
PREQUINE see RHZ000
PREROIDE see MJE760
PRESAL R60 see MCB050
PRESAMINE see DLH630
PRE-SAN see DNO800
PRE-SATE see ARW750
PRESATE HYDROCHLORIDE see ARW750
PRESCRIN see FAQ950
PRESDATE see HMM500
PRESERVAL M see HJL500
PRESERVAL P see HNU500
PRESERV-O-SOTE see CMY825
PRESFERSUL see FBO000
PRESINOL see DNA800
PRESINOL see MJE780
PRESOLISIN see DNA800
PRESOLISIN see MJE780
PRESOMEN see ECU750
PRESOXIN see ORU500
PRESPERSION, 75 UREA see USS000
PRESSAL see MCB050
PRESSALOLO see HMM500

PRESSIMEDIN see RDK000
PRESSITAN see FMS875
PRESSOMIN HYDROCHLORIDE see MDW000
PRESSONEX see HNB875
PRESSONEX see HNC000
PRESSONEX BITARTRATE see HNC000
PRESSOROL see HNC000
PRESTOCICLINA see MCH525
PRESUREN see VJZ000
PRETILACHLOR see PMB850
PRETILACHLORE see PMB850
PRETONINE see HOA575
PREVANGOR see PBC250
PREVENCILINA P see SLJ050
PREVENOL see CKC000
PREVENOL 56 see CKC000
PREVENTAL see MJM500
PREVENTOL see CKC000
PREVENTOL see MJM500
PREVENTOL 1 see SKK500
PREVENTOL 56 see CKC000
PREVENTOL CMK see CFD990
PREVENTOL GD see MJM500
PREVENTOL GDC see MJM500
PREVENTOL I see TIV750
PREVENTOL O EXTRA see BGJ250
PREVENTOL-ON see BGJ750
PREVENTOL ON & ON EXTRA see BGJ750
PREVEX see PNI250
PREVICUR see EIH500
PREVICUR-N see PNI250
PREVOCEL #12 see PMC100
PREWEED see CKC000
PREZA see HBT500
PREZERVIT see DSB200
PR-F 36 Cl see CID825
PRIADEL see LGZ000
PRIAMIDE see DAB875
PRIARIE SMOKE see PAM780
PRIATIN see SCA750
PRIAZIMIDE see DAB875
PRIDE see FMQ200
PRIDE of CHINA see CDM325
PRIDE of INDIA see CDM325
PRIDINOL see PMC250
PRIDINOL HYDROCHLORIDE see PMC250
PRIDINOL MESILATE see PMC275
PRIESTS PENTLE see JAJ000
PRIGLONE see PAI990
PRILEPSIN see DBB200
PRILOCAINE HYDROCHLORIDE see CMS260
PRILTOX see PAX250
PRIM see PMD550
PRIMACAINE see AJA500
PRIMACAINE HYDROCHLORIDE see AJA500
PRIMACHIN (GERMAN) see AKR250
PRIMACIONE see DBB200
PRIMACLONE see DBB200
PRIMACOL see NAK500
PRIMACONE see DBB200
PRIMAGRAM see MQQ450
PRIMAKTON see DBB200
PRIMAL see AHI875
PRIMALAN see MCJ400
PRIMAL ASE 60 see ADV900
PRIMAMYCIN see HAH800
PRIMANTRON see SCA750
PRIMAQUINE see PMC300
PRIMAQUINE DIPHOSPHATE see PMC310
PRIMAQUINE PHOSPHATE see PMC310
PRIMARY AMYL ACETATE see AOD725
PRIMARY AMYL ALCOHOL see AOE000
PRIMARY DECYL ALCOHOL see DAI600
PRIMARY ISOBUTYL IODIDE see IIV509
PRIMARY OCTYL ALCOHOL see OEI000
PRIMARY SODIUM PHOSPHATE see SJH100
PRIMATENE MIST see VGP000
PRIMATOL see ARQ725
PRIMATOL see EGD000

PRIMATOL see MFL250
PRIMATOL 25E see MFL250
PRIMATOL-M80 see BQB000
PRIMATOL P see PMN850
PRIMATOL Q see BKL250
PRIMATOL S see BJP000
PRIMAZE see ARQ725
PRIMAZIN see SNJ000
PRIMBACTAM see ARX875
PRIME+ see FDD078
PRIMENE IMT see AHP752
PRIMENE JM-T see PMC400
PRIMENE 81-R see CAY710
PRIMEXTRA see MQQ450
PRIMICID see DIN600
PRIMIDOLOL HYDROCHLORIDE see
PMC600
PRIMIDON see DBB200
PRIMIDONE see DBB200
PRIMIER DIESEL FUEL see DHE850
PRIMIN see DSK200
PRIMINE see PMI750
PRIMISULFURON see TCI150
PRIMOBOLAN see PMC700
PRIMOBOLONE see PMC700
PRIMOCARCIN see PMC750
PRIMOCORT see DAQ800
PRIMOCORTAN see DAQ800
PRIMODOS see EEH520
PRIMOFOL see EDO000
PRIMOGONYL see CMG675
PRIMOGYN B see EDP000
PRIMOGYN BOLEOSUM see EDP000
PRIMOGYN I see EDP000
PRIMOL 335 see MQV750
PRIMOLUT C see GEK500
PRIMOLUT DEPOT see HNT500
PRIMONABOL see PMC700
PRIMOSTAT see GEK510
PRIMOTEC see DIN600
PRIMOTEST see TBF500
PRIMOVLAR see NNL500
PRIMPERAN see AJH000
PRIMROSE YELLOW see ZFJ100
PRIMUN see FDD100
PRINADOL HYDROBROMIDE see PMD325
PRINALGIN see AGN000
PRINCILLIN see AOD125
PRINCIPAL BILE PIGMENT see HAO900
PRINCIPEN see AIV500
PRINCIPEN/N see SEQ000
PRINICID see DIN600
PRINIVIL see LGM400
PRINODOLOL see VSA000
PRINTEL'S see SMQ500
PRINTEX see CBT750
PRINTEX 60 see CBT750
PRINZONE see SNH900
PRIODAX see PDM750
PRIODERM see MAK700
PRIOSET TD756 see MCB050
PRIOSPEN see PDD350
PRISCOL see BBJ750
PRISCOL see BBW750
PRISCOLINE see BBW750
PRISCOLINE HYDROCHLORIDE see
BBJ750
PRISILIDENE HYDROCHLORIDE see
NEB000
PRISILIDINE see NEA500
PRISMANE see PMD350
PRIST see EJH500
PRISTACIN see CCX000
PRISTANE see PMD500
PRISTIMERIN see PMD525
PRISTINAMYCIN see VRF000
PRISTINAMYCIN IA see VRA700
PRIVADOL see GGQ050
PRIVAPROL see CKL325
PRIVENAL see ERE000
PRIVET see PMD550
PRIVINE see NAH500

PRIVINE see NAH550
PRIVINE HYDROCHLORIDE see NCW000
PRIVINE NITRATE see NAH550
PRIZOLE HYDROCHLORIDE see NCW000
PRJMT see AHP752
PRL-3191 see EHY000
PRM-TC see PPY725
PRN see DTO600
PRO 460 see HEE600
PRO-ACTIDIL see TMX775
PROADIFEN see DIG400
PROADIFEN HYDROCHLORIDE see
PBM500
PROASMA HYDROCHLORIDE see OJY000
(99-126)-hmn PROATRIOPEPTIN see HGL680
PROAZAIMINE see DQA400
PROAZAMINE see DQA400
PROBAN see CQL250
PROBAN 420B see MCB050
PRO-BAN M see TEH500
PRO-BANTHINE see HKR500
PROBARBITAL see ELX000
PROBARBITONE see ELX000
PROBE see BGD250
PROBECID see DWW000
PROBEDRYL see BBV500
PROBEN see DWW000
°PROBENAZOLE
±PROBENAZOLE see PMD800
PROBENECID ACID see DWW000
PROBENECID SODIUM SALT see DWW200
PROBENEMID see DWW000
PROBESE-P see PMA750
PROBESE-P HYDROCHLORIDE see
MNV750
PROBILIN see EOA500
PROBON see PMD825
PROBONAL see PMD825
PROCAINAMIDE see AJN500
PROCAINAMIDE ACRYLOYL MONOMER
see PMD850
PROCAINAMIDE HYDROCHLORIDE see
PME000
PROCAINAMIDE SULFATE see AJN750
PROCAINE see AIL750
PROCAINE ACRYLOYL MONOMER see
PME100
PROCAINE AMIDE see AJN500
PROCAINE AMIDE HYDROCHLORIDE see
PME000
PROCAINE AMIDE SULFATE see AJN750
PROCAINE, BASE see AIL750
PROCAINE BENZYLPENICILLINATE see
PAQ200
PROCAINE FLUOBORATE see TCH000
PROCAINE HYDROCHLORIDE see AIT250
PROCAINE PENICILLIN G see PAQ200
PROCALM see BCA000
PROCALMIDOL see MQU750
PROCAMIDE see AJN500
PROCAMIDE HYDROCHLORIDE see
PME000
PROCAN-SR HYDROCHLORIDE see
PME000
PROCAPAN HYDROCHLORIDE see
PME000
PROCARBAZIN (GERMAN) see PME500
PROCARBAZINE see PME250
PROCARBAZINE HYDROCHLORIDE see
PME500
PROCARDIA see AEC750
PROCARDIN see POB500
PROCARDINE see DJS200
PROCARDYL HYDROCHLORIDE see
PME000
PROCASIL see PNX000
PROCATEROL HYDROCHLORIDE see
PME600
PROCENE UF 1.5 see PJS750
PROCEROSIDE see PME650
PROCEROSIDE (CALOTROPIS) see PME650
PROCETHAZINE DIMALEATE see CCK700

PROCHLORAZ see IAL200
PROCHLOROPERAZINE see PMF500
PROCHLOROPROAZINE HYDROGEN
MALEATE see PMF250
PROCHLORPEMAZINE see PMF500
PROCHLORPERAZINE see PMF500
PROCHLORPERAZINE BIMALEATE see
PMF250
PROCHLORPERAZINE DIMALEATE see
PMF250
PROCHLORPERAZINE EDISYLATE see
PME700
PROCHLORPERAZINE ETHANE
DISULFONATE see PME700
PROCHLORPERAZINE HYDROGEN
MALEATE see PMF250
PROCHLORPERAZINE MALEATE see
PMF250
PROCHLORPERIZINE MALEATE see
PMF250
PROCHLORPROMAZINE see PMF500
PROCHOLON see DAL000
PROCIDLIDINA see CPQ250
PROCILAN see POB500
PROCINOLOL HYDROCHLORIDE see
PMF525
(±)-PROCINOLOL HYDROCHLORIDE see
PMF535
PROCION BLACK H-N see CMS220
PROCION BLUE MX-R see CMS222
PROCION BRILLIANT BLUE MR see
CMS222
PROCION BRILLIANT BLUE MX-R see
CMS222
PROCION BRILLIANT BLUE RS see CMS222
PROCION BRILLIANT RED M 5B see
PMF540
PROCION BRILLIANT RED MX 5B see
PMF540
PROCION BRILLIANT RED 5BS see PMF540
PROCION RED MX 5B see PMF540
PROCION YELLOW MX 4R see RCZ000
PROCIT see DQA400
PROCLONOL see PMF550
PROCOAGULO see NAQ600
PROCOL OA-20 see PJW500
PROCONAZOLE see PMS930
PRO-COR see HNY500
PROCORMAN see DJS200
PROCTIN see SEH000
PROCTODON see DVV500
PROCURAN see DAF800
PROCUTENE see TIL500
PROCYAZINE see PMF600
PROCYCLIDINE see CPQ250
PROCYKLIDIN see CPQ250
PROCYMIDONE see PMF750
PROCYTOX see CQC500
PROCYTOX see CQC500
PRODALUMNOL see SEY500
PRODALUMNOL DOUBLE see SEY500
PRODAN see DXE000
PRODARAM see BJK500
PRODECTINE see PPH050
PRODEL see DYF759
PRODENIALURE B see TBX300
PRODHYBASE ETHYL see EJM500
PRODHYPHORE B see PJY100
PRO-DIABAN see PMG000
PRODIAMINE see DWS200
PRODICTAZIN see DIR000
PRODIERAZINE see DIR000
PRODILIDINE see DTO200
α-PRODINE see NEA500
α-PRODINE HYDROCHLORIDE see
NEB000
PRODIXAMON see HKR500
PRODLURE see FAS100
PRO-DORM see QAK000
PRODOX 131 see IQX100
PRODOX 133 see IQZ000
PRODOX 146 see DEG000

PRODOX 156 see DCI000
PRODOX 340 see BST000
PRODOX 146A-85X see DEG000
PRODOX ACETATE see PMG600
PRODOXAN see GEK500
PRODOXOL see OOG000
PRODROMINE see TCY750
PRO-DRONE see MEL700
PRODROXAN see GEK500
PRODUCER GAS see PMG750
PRODUCT 308 see HCP000
PRODUCT 5022 see MPE250
PRODUCT 5190 see AHL500
PRODUCT DDN see LBU200
PRODUCT HDN see CDF450
PRODUCT No. 161 see SIB600
PRO-DUOSTERONE see NNL500
PRODUXAN see GEK500
PRO 46AE see OPO100
PROEPTATRIENE (ITALIAN) see PMH600
PROFALVINE SULPHATE see DBN400
PROFAM see CBM000
PROFAMINA see BBK000
PROFARMIL see TEH500
PROFAX see PMP500
PROFAX A 60-008 see PJS750
PROFECUNDIN see VSZ450
PROFEMIN see CHJ750
PROFENAMINA (ITALIAN) see DIR000
PROFENAMINUM see DIR000
PROFENID see BDU500
PROFENOFOS see BNA750
PROFENONE see ELL500
PROFERRIN see IHG000
PROFETAMINE see AOB500
PROFETAMINE PHOSPHATE see AOB500
PROFIROMYCIN see MLY000
PROFLAVIN see DBN600
PROFLAVINE see DBN600
PROFLAVINE see PMH100
PROFLAVINE HEMISULPHATE see PMH100
PROFLAVINE HYDROCHLORIDE see PMH250
PROFLAVINE MONOHYDROCHLORIDE see PMH250
PROFLAVINE MONOHYDROCHLORIDE HEMIHYDRATE see DBN200
PROFLAVINE (SULFATE) see DBN400
PROFLAVIN SULFATE see DBN400
PROFLURALIN see CQG250
PROFOLIOL see DBN600
PROFOLIOL see EDO000
PROFORMIPHEN see DBN600
PROFUME A see CKN500
PROFUME (OBS.) see MHR200
PROFUNDOL see AFY500
PROFUNDOL see DBN600
PROFURA see DBN600
PROGABIDE see CKA125
PROGALLIN A see EKM100
PROGALLIN LA see DXX200
PROGALLIN P see PNM750
PROGARMED see DBN600
PRO-GASTRON see HKR500
PROGEKAN see PMH500
PRO-GEN see DBN600
PRO-GEN SODIUM see ARA500
PROGESIC see DBN600
PROGESIC see FAP100
PROGESTAB see GEK500
PROGESTEROL see PMH500
PROGESTERONE see PMH500
β-PROGESTERONE see PMH500
PROGESTERONE 16,17-ACETONIDE see PMH550
PROGESTERONE CAPROATE see HNT500
PROGESTERONE, 16-α,17-DIHYDROXY-, CYCLIC ACETAL WITH ACETONE see PMH550
PROGESTERONE mixed with ESTRADIOL BENZOATE (14:1 moles) see EDQ000

PROGESTERONE mixed with ESTRA-1,3,5(10)-TRIENE-3,17-β-DIOL-3-BENZOATE (14:1 moles) see EDQ000
PROGESTERONE RETARD PHARLON see HNT500
PROGESTERONUM see PMH500
PROGESTIN see GEK500
PROGESTIN P see GEK500
PROGESTOLETS see GEK500
PROGESTONE see PMH500
PROGESTORAL see GEK500
PRO-GIBB see GEM000
PROGLICEM see DCQ700
PROGLUMETACINA (SPANISH) see POF550
PROGLUMETACIN MALEATE see POF550
PROGLUMIDE see BGC625
PROGLUMIDE SODIUM see PMH575
PROGLYDE DMM see DWS900
PROGUANIL see CKB250
PROGUANIL HYDROCHLORIDE see CKB500
PROGYNON see EDO000
PROGYNON see EDS100
PROGYNON B see EDP000
PROGYNON BENZOATE see EDP000
PROGYNON-DEPOT see EDS100
PROGYNON-DH see EDO000
PROGYNON-DP see EDR000
PROGYNOVA see EDS100
PROHEPTADIENE see EAH500
PROHEPTADIEN MONOHYDROCHLORIDE see EAI000
PROHEPTATRIENE see PMH600
PROHEPTATRIENE HYDROCHLORIDE see DPX800
PROHEPTATRIEN MONOHYDROCHLORIDE see DPX800
PROHEXADIONE CALCIUM see CPB065
PROKARBOL see DUS700
PROKAYVIT see MMD500
PROKAYVIT ORAL see VTA100
PROKETAZINE DIMALEATE see CCK700
PROKSIFEIN see POF525
PROLACTIN see PMH625
PROLAN see BIN500
PROLAN B see FMT100
PROLAN (CSC) see BIN500
PROLATE see PHX250
PROLAX see GGS000
PROLIDON see PMH500
PROLIN see WAT211
(−)-PROLINE see PMH905
l-PROLINE see PMH900
PROLINE, l- see PMH905
l-PROLINE see PMH905
l-(−)-PROLINE see PMH905
(−)-(S)-PROLINE see PMH905
(S)-PROLINE see PMH905
l-PROLINE, 1-((4-CHLOROPHENYL)METHYL)-5-OXO- see CKF800
l-PROLINE, 1-((4-CHLOROPHENYL)METHYL)-5-OXO-, AMMONIUM SALT see CKF810
l-PROLINE, 1-((4-CHLOROPHENYL)METHYL)-5-OXO-, 2-(DIMETHYLAMINO)ETHYL ESTER, (Z)-2-BUTENEDIOATE (1:1) see PMH910
L-PROLINE, 1-((4-CHLOROPHENYL)METHYL)-5-OXO-, METHYL ESTER see CKF850
l-PROLINE, 1-((4-CHLOROPHENYL)METHYL)-5-OXO-, compounded with N-(1-METHYLETHYL) BENZENEMETHANAMINE (1:1) see PMH912
l-PROLINE, 1-((4-CHLOROPHENYL)METHYL)-5-OXO-, compounded with N-(1-METHYLETHYL)-2-PROPANAMINE (1:1) see PMH915

l-PROLINE, 1-((2,4-DICHLOROPHENYL)METHYL)-5-OXO- see DGE300
l-PROLINE, 1-((2,6-DICHLOROPHENYL)METHYL)-5-OXO- see DGE305
l-PROLINE, 1-((3,4-DICHLOROPHENYL)METHYL)-5-OXO- see DGE310
l-PROLINE, 1-((2,6-DICHLOROPHENYL)METHYL)-5-OXO-, METHYL ESTER see DGE315
l-PROLINE, 1-((3,4-DICHLOROPHENYL)METHYL)-5-OXO-, METHYL ESTER see DGE320
l-PROLINE, 1-((2,6-DICHLOROPHENYL)METHYL)-5-OXO-, compounded with N-(1-METHYLETHYL)-2-PROPANAMINE (1:1) see PMH917
l-PROLINE, 1-((3,4-DICHLOROPHENYL)METHYL)-5-OXO-, compounded with N-(1-METHYLETHYL)-2-PROPANAMINE (1:1) see PMH920
l-PROLINE, 1-((2,6-DICHLOROPHENYL)METHYL)-5-OXO-, compounded with 2-PROPANAMINE (1:1) see PMH923
l-PROLINE, 1-((3,4-DICHLOROPHENYL)METHYL)-5-OXO-, compounded with 2-PROPANAMINE (1:1) see DGE325
l-PROLINE, 1-((4-FLUOROPHENYL)METHYL)-5-OXO- see FLH200
l-PROLINE, 1-((4-FLUOROPHENYL)METHYL)-5-OXO-, AMMONIUM SALT see FLH300
l-PROLINE, 1-((4-FLUOROPHENYL)METHYL)-5-OXO-, 2-(DIMETHYLAMINO)ETHYL ESTER, (Z)-2-BUTENEDIOATE (1:1) see PMH927
l-PROLINE, 1-((4-FLUOROPHENYL)METHYL)-5-OXO-, (2,2-DIMETHYL-1,3-DIOXOLAN-4-YL)METHYL ESTER see FLH310
l-PROLINE, 1-((4-FLUOROPHENYL)METHYL)-5-OXO-, compounded with N-ETHYLETHANAMINE (1:1) see FLH325
l-PROLINE, 1-((4-FLUOROPHENYL)METHYL)-5-OXO-, METHYL ESTER see FLH315
l-PROLINE, 1-((4-FLUOROPHENYL)METHYL)-5-OXO-, compounded with N-(1-METHYLETHYL) BENZENEMETHANAMINE (1:1) see PMH930
l-PROLINE, 1-((4-FLUOROPHENYL)METHYL)-5-OXO-, compounded with N-(1-METHYLETHYL)-2-PROPANAMINE (1:1) see PMH933
l-PROLINE, 1-((4-FLUOROPHENYL)METHYL)-5-OXO-, (4-OXO-4H-1-BENZOPYRAN-2-YL)METHYL ESTER see FLH320
PROLINE, 4-HYDROXY-, l- see HNT525
l-PROLINE, 4-HYDROXY-, trans- see HNT525
l-PROLINE, 1-(2-METHYL-3-(NITROSOTHIO)-1-OXOPROPYL)-, (S)- see NJP300
l-PROLINE, 1-(N²)-(1-CARBOXY-3-PHENYLPROPYL)-l-LYSYL)-, (S)- see LGM400
l-PROLINE, 1-(N²)-(1-CARBOXY-3-PHENYLPROPYL)-l-LYSYL-, DIHYDRATE, (S)- see CCJ100
PROLINOMETHYLTETRACYCLINE see PMI000
PROLINTANE HYDROCHLORIDE see PNS000
PROLIXAN see AQN750
PROLIXAN see ASA000

PROLIXIN see FMP000
PROLIXINE see TJW500
PROLIXIN ENANTHATE see PMI250
PROLONGAL see IGS000
PROLONGINE see DWW000
PROLOPRIM see TKZ000
PROLOXIN see GGS000
PROLUTOL see GEK500
PROLUTON C see GEK500
PROLUTON DEPOT see HNT500
PROMACID see CKP500
PROMACTIL see CKP250
PROMAMIDE see DTT600
PROMANIDE see AOO800
PROMANTINE see PMI750
PROMAPAR see CKP500
PROMAQUID see FMU039
PROMAR see DVV600
PROMARIT see ECU750
PROMASSOL see AHI875
PROMAXON P60 see CAW850
PROMAZIL see CKP250
PROMAZINAMIDE see DQA400
PROMAZINE see DQA600
PROMAZINE HYDROCHLORIDE see
PMI500
PROMECARB see CQI500
PROMEDOL see IMT000
PROMERAN see CHX250
PROMET see FPO200
PROMETASIN see DQA400
PROMETAZIN see DQA400
PROMET 660SCO see FPO200
PROMETHAZINE N-(2'-
DIMETHYLAMINO-2'-
METHYLETHYL)PHENOTHIAZINE
HYDROCHLORIDE see PMI750
PROMETHAZINE HYDROCHLORIDE see
PMI750
PROMETHIAZIN (GERMAN) see PMI750
PROMETHIAZINE see DQA400
PROMETHIN see FMS875
PROMETHIUM see PMJ000
PROMETIN see FMS875
PROMETON see MFL250
PROMETONE see MFL250
PROMETREX see BKL250
PROMETRIN see BKL250
PROMETRYN see BKL250
PROMETRYNE (USDA) see BKL250
PROMEZATHINE see DQA400
PROMIBEN see DLH600
PROMIBEN see DLH630
PROMIDE (parasympatholytic) see BGC625
PROMIDE HYDROCHLORIDE see PME000
PROMIDIONE see GIA000
PROMIN see AOO800
PROMINAL see ENB500
PROMIN SODIUM see AOO800
PROMIT see DBD700
PROMONTA see MOQ000
PROMOTESTON see TBF500
PROMOTIL see PNS000
PROMOTIN see AOO800
PROMPT INSULIN ZINC SUSPENSION see
LEK000
PROMPTONAL see EOK000
PROMUL 5080 see HKJ000
PROMURIT see DEQ000
PROMURITE see DEQ000
PROMYR see IQN000
PRONABOL see ENX600
PRONAMIDE see DTT600
PRONASE see PMJ100
PRONDOL see DPX200
PRONE see GEK500
PRONESTYL see AJN500
PRONESTYL HYDROCHLORIDE see
PME000
PRONETALOL see INS000
PRONETHALOL see INS000
PRONETHALOL see INT000

PRONETHALOL HYDROCHLORIDE see
INT000
PRONEURIN see DXO300
PRONTALBIN see SNM500
PRONTAMID see SNQ000
PRONTODIN see DUO400
PRONTOSIL I see SNM500
PROPACHLOR see CHS500
PROPACHLORE see CHS500
PROPACIL see PNX000
PROPADERM see AFJ625
PROPADRINE see NNM000
PROPADRINE HYDROCHLORIDE see
NNN000
PROPADRINE HYDROCHLORIDE see
PMJ500
PROPAFENONE HYDROCHLORIDE see
PMJ525
PROPAFENON HYDROCHLORID see
PMJ525
PROPAL see CIR500
PROPAL see IQW000
PROPALDEHYDE see PMT750
PROPALDON see QCS000
PROPALGYL see DPE200
PROPALLYLONAL see QCS000
PRO-PAM see PMJ550
PROPAMIDINE DIHYDROCHLORIDE see
DBM400
PROPAMINE D see TDQ750
PROPAMOCARB HYDROCHLORIDE see
PNI250
PROPANAL see PMT750
PROPANAL, 2-CHLORO-(9CI) see CKP700
PROPANAL, 2-METHYL-2-
(METHYLSULFINYL)-, O-
((METHYLAMINO)CARBONYL)OXIME see
MLW750
PROPANAL, 2-METHYL-2-
(METHYLSULFONYL)-o,o'-
(THIOBIS((METHYLIMINO)CARBONYL))DI
OXIME see BKT300
PROPANAL, 2-METHYL-2-(METHYLTHIO)-,
OXIME see MLX800
PROPANAL, 3-(METHYLNITROSOAMINO)-
see NKV050
PROPANAL, 3-(METHYLTHIO)-(9CI) see
MPV400
PROPANAL, 3-(METHYLTHIO)-(9CI) see
TET900
PROPANALOL see ICB000
PROPANAMIDE, N-(AMINOCARBONYL)-
see AJD800
PROPANAMIDE, N-(5-CHLORO-4-
METHYL-2-THIAZOLYL)-(9CI) see CIU800
PROPANAMIDE, N,N-DIETHYL-(9CI) see
DJX250
PROPANAMIDE, N-(2-(5-ETHYL-2-
METHOXYPHENYL)ETHYL)- see EMI530
PROPANAMIDE, 3-MERCAPTO-N-
METHYL- see MLF100
PROPANAMIDE, N-(2-(2-METHOXY-5-
METHYLPHENYL)ETHYL)- see MEX260
PROPANAMIDE, N-(2-(2-METHOXY-1-
NAPHTHALENYL)-1-METHYLETHYL)- see
MFA365
PROPANAMIDE, N-METHYL-(9CI) see
MOS900
PROPANAMIDE, 2-(2-
NAPHTHALENYLOXY)-N-PHENYL- see
NBA600
PROPANAMINE see PND250
2-PROPANAMINE see INK000
1-PROPANAMINE, 3,3'-(1,4-
BUTANEDIYLBIS(OXY))BIS- see BGU600
1-PROPANAMINE, 2-CHLORO-N,N-
DIMETHYL- see CGJ290
2-PROPANAMINE, 1-CHLORO-N,N-
DIMETHYL-, HYDROCHLORIDE see
CGJ280

2-PROPANAMINE, N-(2-CHLOROETHYL)-
N-(1-METHYLETHYL)-, HYDROCHLORIDE
see CGV600
1-PROPANAMINE, 3-DIBENZ(b,e)OXEPIN-
11(6H)-YLIDENE-N,N-DIMETHYL-,
HYDROCHLORIDE see AEG750
2-PROPANAMINE, N,N-DICHLORO-2-
METHYL- see BQY275
1-PROPANAMINE, 3-(10,11-DIHYDRO-5H-
DIBENZO(A,D)CYCLOHEPTEN-5-
YLIDENE)-N,N-DI-METHYL-N-OXIDE see
AMY000
1-PROPANAMINE, N,N-DIMETHYL-3-
THIENO(2,3-C)(2)BENZOTHIEPIN-4(9H)-
YLIDNEN-,S-OXIDE, (E)-, (Z)-2-
BUTENEDIOATE (1:1) see DXI480
2-PROPANAMINE, HYDROCHLORIDE
(9CI) see INL000
2-PROPANAMINE, N,N'-
METHANETETRAYLBIS-(9CI) see DNO400
1-PROPANAMINE, 2-METHYL- see IIM000
2-PROPANAMINE, N-(1-METHYLETHYL)-
see DNM200
2-PROPANAMINE, 2-METHYL-,
HYDROCHLORIDE see MGJ800
1-PROPANAMINE, 3-(TRIETHYLSILYL)- see
AMG200
1-PROPANAMINE, 3-(TRIMETHOXYSILYL)-
N-(3-(TRIMETHOXYSILYL)PROPYL)- see
TLC650
1-PROPANAMINIUM, 2-(ACETYLOXY)-
N,N,N-TRIMETHYL- see MFX560
1-PROPANAMINIUM, 3-CARBOXY-2-
HYDROXY-N,N,N-TRIMETHYL-,
CHLORIDE, (R)- (9CI) see CCK660
1-PROPANAMINIUM, 3-CARBOXY-2-
HYDROXY-N,N,N-TRIMETHYL-,
CHLORIDE, (±)-(9CI) see CCK655
1-PROPANAMINIUM, 3-CYANO-2-
HYDROXY-N,N,N-TRIMETHYL-,
CHLORIDE, (+−)- see COL050
1-PROPANAMINIUM, 3,3'-(1,2-
ETHANEDIYLBIS(DIMETHYLSILYLENE))B
IS(N,N,N-TRIMETHYL-), DIIODIDE see
EEB200
1-PROPANAMINIUM, 2-HYDROXY-N,N,N-
TRIMETHYL-, CHLORIDE (9CI) see MIM300
1-PROPANAMINIUM, 3,3'-(1,4-
PHENYLENEBIS(OXY))BIS(N-ETHYL-N,N-
DIMETHYL-, DIIODIDE (9CI) see DHA500
1,3-PROPAN-BIS-(4-
HYDROXYIMINOMETHYL-PYRIDINIUM-
(1))-DIBROMIDS (GERMAN) see TLQ500
PROPANE see PMJ750
PROPANE, 1-(BENZYLOXY)-2,3-EPOXY-,
(R)- see BEP670
PROPANE, 1-(BENZYLOXY)-2,3-EPOXY-,
(R)- see BFC225
PROPANE, 1-(BENZYLOXY)-2,3-EPOXY-,
(S)- see BEP670
PROPANE, 1-(BENZYLOXY)-2,3-EPOXY-,
(S)- see BFC225
PROPANE, 1-(4-BIPHENYLOXY)-2,3-
EPOXY- see PFU600
PROPANE, 1,3-BIS(DIAZO)- see DCQ525
PROPANE, 3-BROMO-1,1,1,2,2-
PENTAFLUORO- see PBF600
PROPANE, 1-tert-BUTOXY-2,3-EPOXY- see
BRK800
PROPANE, 1-(p-tert-BUTYLPHENOXY)-2,3-
EPOXY- see BSE600
1-PROPANECARBOXYLIC ACID see
BSW000
PROPANE, 1-CHLORO-1,2-EPOXY-, (E)- see
CKS100
PROPANE, 1-CHLORO-1,2-EPOXY-, (Z)- see
CKS099
PROPANE, 1-CHLORO-1,1,2,2,3,3-
HEXAFLUORO- see CHK800
PROPANE, 3-CHLORO-1,1,1,2,2-
PENTAFLUORO- see CJI525

PROPANE, 1-(p-CHLOROPHENOXY)-2,3-EPOXY- see CKA200
PROPANE, 3-CHLORO-1,1,1-TRIFLUORO- see TJY200
PROPANE, 2-DEUTERO-2-NITRO- see DBC100
PROPANEDIAL see PMK000
1,3-PROPANEDIAL see PMK000
1,3-PROPANEDIALDEHYDE see PMK000
PROPANEDIAL, ION(1-), SODIUM (9CI) see MAN800
1,2-PROPANEDIAMINE see PMK250
1,3-PROPANEDIAMINE see PMK500
PROPANEDIAMINE, N-(3-AMINOPROPYL)- see AIX250
1,3-PROPANEDIAMINE, N-(3-AMINOPROPYL)-N'-(3-AMINOPROPYL)(AMINO)PROPYL)- see TED600
NITROSOHYDRAZINO)BUTYL)- see SLG625
1,3-PROPANEDIAMINE, N-(3-AMINOPROPYL)-N-METHYL- see EAZ600
1,3-PROPANEDIAMINE, N-(3-AMINOPROPYL), POLYMER WITH (CHLOROMETHYL)OXIRANE see AME100
AMINOPROPYL)-2-HYDROXY-2-
NFW525
1,3-PROPANEDIAMINE, N,N-DIMETHYL-N'-(2-NITRO-9-ACRIDINYL)-, DIHYDROCHLORIDE see NFW535
(ETHYLCARBONIMIDOYL)-N,N-DIMETHYL-, MONOHYDROCHLORIDE see EAE100
1,3-PROPANEDIAMINE, N-HEXYL-N'-(1-NITRO-9-ACRIDINYL)-, DIHYDROCHLORIDE see HFK500
1,3-PROPANEDIAMINE, N,N'-1,2-ETHANEDIYLBIS- see TBI700
1,3-PROPANEDIAMINE, N,N'-ETHYLENEBIS- see TBI700
1,3-PROPANEDIAMINE, N,N'-(TRIMETHOXYSILYL)PROPYL)- see TLC600
PROPANE, 1,2-DIBROMO-1,1,2,3,3,3-HEXAFLUORO- see DDO450
PROPANE, 1,2-DIBROMO-2-METHYL- see DDQ150
1,3-PROPANEDICARBOXYLIC ACID see GFS000
PROPANE, DICHLORO- see DGF350
PROPANE, 1,3-DICHLORO-1,1,2,2,3,3-HEXAFLUORO- see DFM110
PROPANE, 1,3-DICHLORO-1,1,2,2,3-PENTAFLUORO- see DFW830
PROPANE, 3,3-DICHLORO-1,1,1,2,2-PENTAFLUORO- see DFW850
PROPANE, DIETHYL SULFONE see ABD500
PROPANE, 2,2-DIFLUORO-1,1,1,3,3,3-HEXACHLORO- see HCH475
PROPANE, 1,3-DIIODO- see TLR050
PROPANE, DIISOTHIOUREA DIHYDROBROMIDE see PMK750
1,3-PROPANEDIMERCAPTAN see PML350
PROPANE-1,3-DIMETHANESULFONATE see TLR250

PROPANEDINITRILE see MAO250
PROPANEDINITRILE, (BIS(ETHYLTHIO)METHYLENE)- see BJT800
(9CI) see TBW750
PROPANEDINITRILE, CYCLOHEXYLIDENE- see CPM800
PROPANEDINITRILE, (1-ETHOXYETHYLIDENE)- see EET100
PROPANEDIOATE see MAO100
PROPANEDIOIC ACID see CCC750
PROPANEDIOIC ACID, ((3,5-BIS(1,1-DIMETHYLETHYL)-4-HYDROXYPHENYL)METHYL)BUTYL-, BIS(1,2,2,6,6-PENTAMETHYL-4-PIPERIDINYL) ESTER see PMK800
PROPANEDIOIC ACID, (CARBOXYMETHOXY)-, TRISODIUM SAL see SFO700
PROPANEDIOIC ACID, DICHLORO-, DIMETHYL ESTER see DRK300
PROPANEDIOIC ACID, DIETHYL ESTER (9CI) see DSM200
PROPANEDIOIC ACID, DITHALLIUM SALT see TEM399
PROPANEDIOIC ACID, 1,3-DITHIOLAN-2-YLIDENE-, BIS(1-METHYLETHYL) ESTER see IRU500
PROPANEDIOIC ACID, ION(2-) see MAO100
PROPANEDIOIC ACID, (1-METHYL-5-NITRO-1H-IMIDAZOL-2-YL)METHYLENE)-, DIMETHYL ESTER see DSP800
1,2-PROPANEDIOL see PML000
1,2-PROPANEDIOL, 2,2'-
PROPANE-1,2-DIOL see PML000
1,3-PROPANEDIOL see PML250
PROPANE-1,3-DIOL see PML250
1,2-PROPANEDIOL-1-ACRYLATE see HNT600
1,2-PROPANEDIOL, ALLYL ETHER see PNK000
PROPANEDIOL, (ALLYLOXY)- see GGA912
1,2-PROPANEDIOL, 3-(6-AMINO-9H-PURIN-9-YL)-, (+)- see AMH850
1,2-PROPANEDIOL, 3-(6-AMINO-9H-PURIN-9-YL)-, (S)- see AMH800
1,2-PROPANEDIOL, 3-(6-AMINO-9H-PURIN-9-YL)-, (R,S)- see AMH850
1,2-PROPANEDIOL, 3-AZIDO- see ASG700
1,2-PROPANEDIOL, 2,2'-BIS((ACETYLOXY)METHYL)-, DIACETATE (9CI) see NNR400
1,3-PROPANEDIOL BIS(α-(p-CHLOROPHENOXY)ISOBUTYRATE) see SDY500
1,3-PROPANEDIOL BIS(2-(4-CHLOROPHENOXY)-2-METHYLPROPIONATE) see SDY500
1,3-PROPANEDIOL, 2,2'-BIS(HYDROXYMETHYL)-, ALLYL ETHER see AGQ050
1,3-PROPANEDIOL, 2,2-BIS((NITROOXY)METHYL)-, DINITRATE (ester; mixt. with 2-METHYL-1,3,5-TRINITROBENZENE see PBT050
1,2-PROPANEDIOL CARBONATE see CBW500
1,2-PROPANEDIOL, 3-CHLORO-, 1-BENZOATE see CKQ500
1,2-PROPANEDIOL, 3-CHLORO-, DIACETATE see CII900
1,2-PROPANEDIOL CYCLIC CARBONATE see CBW500

1,1-PROPANEDIOL, 2,3-DICHLORO-, DIACETATE see DBF875
1,3-PROPANEDIOL, 2,2-DI-p-TOLYL-, MONO(METHYLCARBAMATE) see INE025
1,2-PROPANEDIOL, 3-(1,3,5,7,9-DODECAPENTAENYLOXY)-, (ALL-E)- see AFS100
1,2-PROPANEDIOL, 3-((10-ETHYL-11-(p-HYDROXYPHENYL)DIBENZ(B,F)OXEPIN-3-YL)OXY)-, HYDRATE (4:1) see ELK600
1,3-PROPANEDIOL, FORMAL see DVP000
1,3-PROPANEDIOL, 2-HYDROXY ACETAMIDO-1-p-NITROPHENYL-, d-(-)-THREO- see CDP350
1,3-PROPANEDIOL, 2-METHYL-, TRINITRATE (ESTER) see MQG600
(HYDROXYMETHYL)-2-NITRO-, TRINITRATE (ester) see NHK650
(HYDROXYMETHYL)-2-PROPYL-, CYCLIC ESTER with ANTIMONIC ACID see PNX500
1,2-PROPANEDIOL, 3-ISOPROPOXY- see GGA050
1,3-PROPANEDIOL, 3-(1-METHYLETHOXY)- see GGA050
1,2-PROPANEDIOL, 2-METHYL-2-(NITROOXY)METHYL-, DINITRATE (ESTER; see MQG600
1,2-PROPANEDIOL, 3-(4-(1-METHYL-1-(4-(OXIRANYLMETHOXY)PHENYL)ETHYL)PHENOXY)- see ECE800
1,2-PROPANEDIOL, MONOBUTYL ETHER see PML260
1,2-PROPANEDIOL, MONOSTEARATE see SLL000
1,2-PROPANEDIOL, 3-(NITROSO-2-PROPENYLAMINO)- see NJY500
1,2-PROPANEDIOL, 3-(OCTADECYLOXY)- see GGA915
1,2-PROPANEDIOL, 3,3'-OXYDI-, TETRANITRATE see TDY100
1,2-PROPANEDIOL, 3-PHENOXY- see GGA950
1,3-PROPANEDIOL, 2-PHENYL-, DICARBAMATE see FAG200
PROPANEDIOL, (2-PROPENYLOXY)- see GGA912
PROPANEDIONE see PQC000
1,3-PROPANEDIONE see PMK000
1,3-PROPANEDIONE, 2-BROMO-1,3-DIPHENYL- see BNH100
1,3-PROPANEDIONE, 1,3-DIPHENYL- see PFU300
N,N-(1,3-PROPANEDIOXYSULFINYL)BIS(2,3-DIHYDRO-2,2-DIMETHYLBENZOFURANYL-7-METHYLCARBAMATE) see PML270
1,2-PROPANEDITHIOL see PML300
1,3-PROPANEDITHIOL see PML350
1,1'-(1,3-PROPANEDIYL)BIS(4-(HYDROXYIMINO)METHYLPYRIDINIUM) DIBROMIDE see TLQ500
1,1'-(1,3-PROPANEDIYL)BIS((4-(HYDROXYIMINO)METHYL)PYRIDINIUM) DICHLORIDE see TLQ750
4,4'-(1,3-PROPANEDIYLBIS(OXY))BIS-BENZENECARBOXIMIDAMIDE,DIHYDRO CHLORIDE see DBM400
1,2-PROPANEDIYL CARBONATE see CBW500

PROPANE-2-D, 2-NITRO-(9CI) see DBC100
PROPANE, 1,2-EPOXY-3-ETHOXY- see EBQ700
PROPANE, 1,2-EPOXY-3-ETHOXY- see EKM200
PROPANE, 1,2-EPOXY-3-((2-ETHYLHEXYL)OXY)- see GGY100
PROPANE, 1,2-EPOXY-1,1,2,3,3,3-HEXAFLUORO- see HDF050
PROPANE, 1,2-EPOXY-3-(HEXYLOXY)- see GGY125
PROPANE, 1,2-EPOXY-3-METHOXY- see GGW600
PROPANE, 1,2-EPOXY-2-METHYL- see IIQ600
PROPANE, 1,2-EPOXY-3-(p-NITROPHENOXY)- see NIN050
PROPANE, 1,2-EPOXY-3-(p-TOLYLOXY)- see GGY175
PROPANE, 1-ETHOXY-(9CI) see EPC125
PROPANE, 2-ETHOXY-2-METHYL-(9CI) see EHA550
PROPANE, HEPTAFLUOROIODO- see HAY300
PROPANE, 1,1,1,2,3,3-HEXAFLUORO- see HDE050
PROPANE, 1,1,1,3,3,3-HEXAFLUORO- see HDE100
PROPANE, 1-ISOTHIOCYANATO-3-(METHYLSULFINYL)-(9CI) see MPN100
PROPANE, 1-METHOXY-(9CI) see MOU830
PROPANE, 2-METHOXY-2-METHYL- (9CI) see MHV859
PROPANE, 1,1'-(METHYLENEBIS(OXY))BIS(2,2-DINITRO)- see MJO800
PROPANE, 1-(1-NAPHTHYLMETHOXY)-2,3-EPOXY- see NBJ200
PROPANENITRILE see PMV750
PROPANENITRILE, 3-((2-((2-AMINOETHYL)AMINO)ETHYL)AMINO)-(9CI) see LBV000
PROPANENITRILE, 3-ANILINO- see AOT100
PROPANENITRILE, 3-((4-ETHOXYPHENYL)AMINO)- see COM900
PROPANENITRILE, 2-((4-(ETHYLAMINO)-6-(METHYLTHIO)-1,3,5-TRIAZIN-2-YL)AMINO)-2-METHYL- see COI050
PROPANENITRILE, 3-(ETHYLPHENYLAMINO)-(9CI) see EHQ500
PROPANENITRILE, 3-(PHENYLAMINO)- see AOT100
PROPANENITRILE, TRICHLORO- see TJC800
PROPANENITRILE, 2,2,3-TRICHLORO- see TJC850
PROPANE, 2-NITRO-, ION (1-) see NIY025
PROPANE-2-NITRONATE see NIY025
α-γ-PROPANE OXIDE see OMW000
PROPANE, 1,1'-OXYBIS-(9CI) see PNM000
PROPANE, 2,2'-OXYBIS(1-CHLORO)- see BII250
PROPANE, OXYBIS(METHOXY- see DWS900
PROPANE PEROXOIC ACID see PCO100
PROPANEPEROXOIC ACID, 2-METHYL, 1,1-DIMETHYLETHYL ESTER see BSC600
1-PROPANESULFONANILIDE, 4-(4-CARBAMOYL-9-ACRIDINYLAMINO)-3'-METHOXY, HYDROCHLORIDE see MFN000

1-PROPANESULFONIC ACID, 2-ACRYLAMIDO-2-METHYL- see ADS300
1-PROPANESULFONIC ACID-3-HYDROXY-Y-SULTONE see PML400
1-PROPANESULFONIC ACID, 2-METHYL-2-((1-OXO-2-PROPENYL)AMINO)- see ADS300
PROPANE SULTONE see PML400
1,3-PROPANE SULTONE (MAK) see PML400
PROPANE, 1,2,2,3-TETRACHLORO- see TBT300
1-PROPANETHIOL see PML500
PROPANE-1-THIOL see PML500
2-PROPANETHIOL see IMU000
1,2,3-PROPANETRICARBOXYLIC ACID, 2-(ACETYLOXY)-, TRIBUTYL ESTER see THX100
1,2,3-PROPANETRICARBOXYLIC ACID, 2-(ACETYLOXY), TRIETHYL ESTER (9CI) see ADD750
1,2,3-PROPANETRICARBOXYLIC ACID, 1,1'-(1-(2-AMINO-9,11-DIHYDROXY-2-METHYLTRIDECYL)-2-(1-METHYLPENTYL)-1,2-ETHANEDIYL) ESTER see FPA500
1,2,3-PROPANETRICARBOXYLIC ACID, 1,1'-(1-(12-AMINO-4,9,11-TRIHYDROXY-2-(1-METHYLPENTYL)1,2-ETHANEDIYL)ESTER see MAB500
1,2,3-PROPANETRICARBOXYLIC ACID, 2-HYDROXY-, BISMUTH(3+) POTASSIUM SALT (2:1:3) see BKX800
1,2,3-PROPANETRICARBOXYLIC ACID, 2-HYDROXY-, COMPOUNDS, BISMUTH(3+) SALT, COMPD. WITH N-(2-(((5-(DIMETHYLAMINO)METHYL)-2-FURANYL)METHYL)THIO)ETHYL)-N'-METHYL-2-NIT RO-1,1-ETHENEDIAMINE (1:1:1) see RBF450
1,2,3-PROPANETRICARBOXYLIC ACID, 2-HYDROXY, TIN(2+) SALT (2:3) (9CI) see TGC285
1,2,3-PROPANETRICARBOXYLIC ACID, 2-HYDROXY, TRIBUTYL ESTER see THY100
1,2,3-PROPANETRICARBOXYLIC ACID, 2-HYDROXY, TRILITHIUM SALT, HYDRATE see LHD150
1,2,3-PROPANETRICARBOXYLIC ACID, 2-HYDROXY, ZINC SALT (2:3) (9CI) see ZFJ250
1,2,3-PROPANETRIOL see GGA000
1,2,3-PROPANETRIOL, DIACETATE (9CI) see GGA100
1,2,3-PROPANETRIOL, 1,3-DINITRATE (9CI) see DNK300
1,2,3-PROPANETRIOL, 1,3-DIOXIME see DNK300
N,N',N''-(1,2,3-PROPANETRIOXYSULFINYL)TRIS(2-ISOPROPOXYPHENYLMETHYLCARBAMATE) see CBH770
N,N',N''-(1,2,3-PROPANETRIOXYSULFINYL)TRIS(1-NAPHTHYL METHYLCARBAMATE) see PML650
1,2,3-PROPANETRIYL (BIS(1-AZIRIDINYL)PHOSPHINYL)CARBAMATE see CBH770
1,2,3-PROPANETRIYL NITRATE see NGY000

PROPANIL see DGI000
PROPANIMIDAMIDE, 2,2'-AZOBIS(2-METHYL-), DIHYDROCHLORIDE (9CI) see ASM050
PROPANIMIDOTHIOIC ACID, 2,2-DIMETHYL-N-((METHYLAMINO)CARBONYL)OXY)-, 2-AMINO-2-OXOETHYL ESTER see ALQ640
PROPANOIC ACID see PJA500
PROPANOIC ACID see PMU750
PROPANOIC ACID, 2-(ACETYLAMINO)-3-(ACETYLTHIO)-, ETHYL ESTER see DBG600
PROPANOIC ACID, 2-((1,1'-BIPHENYL)-4-YLOXY)- see BGN100
PROPANOIC ACID, 2-((BIS(HEXYLOXY)PHOSPHINYL)OXY)-3,3,3-TRIFLUORO-, METHYL ESTER see TKB290
PROPANOIC ACID, 2-(BIS(2-METHYLPROPOXY)PHOSPHINYL)OXY)-3,3,3-TRIFLUORO-, METHYL ESTER see TKB292
PROPANOIC ACID, 2-(BIS(PENTYLOXY)PHOSPHINYL)OXY)-3,3,3-TRIFLUORO-, METHYL ESTER see TKB296
PROPANOIC ACID, 3-BROMO-2-OXO-(9CI) see BOD550
PROPANOIC ACID BUTYLESTER (9CI) see BSJ500
PROPANOIC ACID, CALCIUM SALT (9CI) see CAW400
PROPANOIC ACID,2-(4-((6-CHLORO-2-BENZOTHIAZOLYL)OXY)PHENOXY)-, ETHYLESTER see EHH600
PROPANOIC ACID, 2-(4-((6-CHLORO-2-BENZOXAZOLYL)OXY)PHENOXY)-, ETHYL ESTER, (+-)- see FAQ200
PROPANOIC ACID, 2-(4'-CHLORO(1,1'-BIPHENYL)-4-YL)OXY)-2-METHYL-, METHYL ESTER (9CI) see MIO975
PROPANOIC ACID, 2-(4-CHLORO-2-METHYLPHENOXY), (+-)- see RBF500
PROPANOIC ACID, 2-(4-CHLORO-2-METHYLPHENOXY)-, POTASSIUM SALT (9CI) see CLO200
PROPANOIC ACID, 2-(4-CHLOROPHENOXY)-2-METHYL-, ETHYL ESTER, CALCIUM SALT see EHL625
PROPANOIC ACID,2-(4-((6-CHLORO-2-QUINOXALINYL)OXY)PHENOXY)-, ETHYL ESTER see QMA100
PROPANOIC ACID, 2-(4-((6-CHLORO-2-QUINOXALINYL)OXY)PHENOXY)-, METHYL ESTER see HAG860
PROPANOIC ACID, 2-((3-CHLORO-5-(TRIFLUOROMETHYL)-2-PYRIDINYL)OXY)PHENOXY)-, 2-(((1-METHYLETHYLIDENE)AMINO)OXY)ETHYL ESTER, (R)- see IPI400
PROPANOIC ACID, 2-CHLORO-, SODIUM SALT see CKT100
PROPANOIC ACID,2-(4-((3-CHLORO-5-(TRIFLUOROMETHYL)-2-PYRIDINYL)OXY)PHENOXY)-, METHYL ESTER see EHY050
PROPANOIC ACID, 2,3-DIBROMO-, ETHYL ESTER see EHY050
PROPANOIC ACID, 2,3-DIBROMO-, 2-HYDROXYETHYL ESTER see HKQ600
PROPANOIC ACID, 3-((DIBUTOXYPHOSPHINYL)OXY)-3,3,3-TRIFLUORO-, METHYL ESTER see TKB287
PROPANOIC ACID, 2-(4-(2,2-DICHLOROCYCLOPROPYL)PHENOXY)-2-METHYL- see CMS216
PROPANOIC ACID, 2-(2,4-DICHLOROPHENOXY)-, (R)- see DGA880

PROPANEX see DGI000
PROPANID see DGI000
PROPANIDE see PMM000
PROPANIDIDE see PMM000
PROPANID see DGI000
PROPANETRIOL TRINTRATE see NGY000
1,2,3-PROPANETRIOL, TRINITRATE see NGY000
PROPANETRIOL TRIACETATE see THM500
PROPANETRIOL, TRIACETATE see THM500
1,2,3-PROPANETRIOL MONOACETATE see GGO000

PROPANOIC ACID, 2-(2,4-DICHLOROPHENOXY)-, mixture with (2,4-DICHLOROPHENOXY)ACETIC ACID,(2,4,5-TRICHLOROPHENOXY)ACETIC ACID and 2-(2,4,5-TRICHLOROPHENOXY)PROPANOIC ACID see PMM100

PROPANOIC ACID, 2-((DIETHOXYPHOSPHINYL)OXY)-3,3,3-TRIFLUORO-, METHYL ESTER see TKB289

PROPANOIC ACID, 2-((DIMETHOXYPHOSPHINYL)OXY)-3,3,3-TRIFLUORO-, METHYL ESTER see TKB294

PROPANOIC ACID, 2,2-DIMETHYL-, n-HEXYL ESTER see HFP700

PROPANOIC ACID, 2,2-DIMETHYL-, HEXYL ESTER (9CI) see HFP700

PROPANOIC ACID, 2,2-DIMETHYL-, ISOOCTADECYL ESTER see ISC550

PROPANOIC ACID, 2,2-DIMETHYL-, SODIUM SALT see SGM600

PROPANOIC ACID, 2-((DIPROPOXYPHOSPHINYL)OXY)-3,3,3-TRIFLUORO-, METHYL ESTER see TKB298

PROPANOIC ACID, 3-(DODECYLTHIO)-, OXYBIS(2,1-ETHANEDIYLOXY-2,1-ETHANEDIYL) ESTER see OPG050

PROPANOIC ACID, ETHENYL ESTER see VQK000

PROPANOIC ACID, 2-(1,1,2,3,3,3-HEXAFLUORO-2-(HEPTAFLUOROPROPOXY)PROPOXY)-2,3,3,3-TETRAFLUO RO- see HDC435

PROPANOIC ACID, HEXYL ESTER see HFV100

PROPANOIC ACID, 2-HYDROXY- see LAG000

PROPANOIC ACID, 2-HYDROXY-, (S)-(9CI) see LAG010

PROPANOIC ACID, 2-HYDROXY-, CALCIUM SALT see CAT600

PROPANOIC ACID, 2-HYDROXY-, COMPD. WITH 9-METHOXY-5,11-DIMETHYL-6H-PYRIDO(4,3-B)CARBAZOLE see MEL780

PROPANOIC ACID, 2-HYDROXY-2-METHYL-, ETHYL ESTER (9CI) see ELH700

PROPANOIC ACID, 2-HYDROXY-, MONOPOTASSIUM SALT (9CI) see PLK650

PROPANOIC ACID, 2-HYDROXY-, TRIANHYDRIDE with ANTIMONIC ACID (H₃SbO₃) (9CI) see AQE250

PROPANOIC ACID, 2-(IMINO(5-NITRO-2-FURANYL)METHYL)HYDRAZIDE see PMW800

PROPANOIC ACID, 3-MERCAPTO-(9CI) see MCQ000

PROPANOIC ACID, 3-METHOXY-, ETHYL ESTER (9CI) see EMI550

PROPANOIC ACID, 2-METHYL-, 3,7-DIMETHYL-2,6-OCTADIENYL ESTER, (Z)- see NCO200

PROPANOIC ACID, METHYL ESTER see MOT000

PROPANOIC ACID, 2-METHYL-, 2,4-HEXADIENYL ESTER see HCS700

PROPANOIC ACID, 2-METHYL-, 3-HEXENYL ESTER see HFE518

PROPANOIC ACID, 2-METHYL-, 3-HEXENYL ESTER, (Z)- see HFE520

PROPANOIC ACID, 2-METHYL-, HEXYL ESTER (9CI) see HFQ550

PROPANOIC ACID, 2-METHYL-, 1-METHYL-1-(4-METHYL-3-CYCLOHEXEN-1-YL)ETHYL ESTER see MCF515

PROPANOIC ACID, 2-METHYL-, 2-PHENOXYETHYL ESTER see PDS900

PROPANOIC ACID, 2-METHYLPROPYL ESTER see PMV250

PROPANOIC ACID, 2-OXO-(9CI) see PQC100

PROPANOIC ACID, PENTYL ESTER see AON350

PROPANOIC ACID-2-PHENYLETHYL ESTER see PDK000

PROPANOIC ACID, 3-PHOSPHONO- see PHA560

PROPANOIC ACID, 3-(1H-PURIN-6-YLTHIO)- see CCF270

PROPANOIC ACID, SODIUM SALT see SJL500

PROPANOIC ACID, 2,2,3,3-TETRAFLUORO-, SODIUM SALT see SKE100

PROPANOIC ACID, 3,3'-THIOBIS-, BIS(2-ETHYLHEXYL) ESTER see BJS550

PROPANOIC ACID, 3,3'-THIOBIS-, DIDODECYL ESTER see TFD500

PROPANOIC ACID, 3,3'-THIOBIS-, DIOCTADECYL ESTER (9CI) see DXG700

PROPANOIC ACID, 2,2,3-TRIFLUORO-3-OXO-, METHYL ESTER see MJD400

PROPANOIC ANHYDRIDE see PMV500

PROPANOL-1 see PND000

PROPAN-2-OL see INJ000

2-PROPANOL see INJ000

n-PROPANOL see PND000

1-PROPANOL (ACGIH) see PND000

i-PROPANOL (GERMAN) see INJ000

PROPANOLAMINE see PMM250

3-PROPANOLAMINE see PMM250

1,3-PROPANOLAMINE see PMM250

2-PROPANOL, 1-AMINO-3-CHLORO-, (−)- see AJI520

2-PROPANOL, 1-AMINO-3-CHLORO-, (±)- see AJI530

2-PROPANOL, 1-BUTYLAMINO-3-(NAPHTHYLOXY)- see BQG600

2-PROPANOL, 1,1'-(2-BUTYNYLENEDIOXY)BIS(3-CHLORO- see BST900

1-PROPANOL, 3-CHLORO- see CKP600

2-PROPANOL, 1-CHLORO-3-FLUORO-, MIXT. WITH 1,3-DIFLUORO-2-PROPANOL see GEW725

2-PROPANOL, 1-CHLORO-3-ISOPROPOXY- see CHS250

1-PROPANOL, CHLORO-, PHOSPHATE (3:1) see TJC870

1-PROPANOL, 2-CHLORO-, PHOSPHATE (3:1), MIXED WITH 1-CHLORO-2-PROPANOL PHOSPHATE (3:1) see TJC870

2-PROPANOL, 1-(CYCLOHEXYLAMINO)- see CPG700

2-PROPANOL, 1-(CYCLOHEXYLAMINO)-, BENZOATE (ESTER) see OKK100

2-PROPANOL, 1-(2-(1-CYCLOPENTEN-1-YL)PHENOXY)-3-((1,1-DIMETHYLETHYL)AMINO)-, (+−)- see CPY850

2-PROPANOL, 1-(4-CYCLOPROPYLCARBONYLPHENOXY)-3-(1,2-DIHYDRO-2-IMINO-4-METHYLPYRIDINO)- see PMM300

2-PROPANOL, 1-(4-(2-(CYCLOPROPYLMETHOXY)ETHYL)PHENOXY)-3-((1-METHYLETHYL)AMINO)-, HYDROCHLORIDE see KEA350

1-PROPANOL, 2,3-DIBROMO-, DIHYDROGEN PHOSPHATE see MRH217

1-PROPANOL, 2,3-DIBROMO-, HYDROGEN PHOSPHATE, AMMONIUM SALT see MRH214

1-PROPANOL, 2,3-DIBROMO-, HYDROGEN PHOSPHATE, MAGNESIUM SALT see MAD100

1-PROPANOL, 2,3-DIBROMO-, PHOSPHATE (1:1) see MRH217

1-PROPANOL, 3-(DIETHYLAMINO)-2,2-DIMETHYL-, TROPATE, PHOSPHATE see AOD250

1-PROPANOL, 3-(1-(DIFLUORO((TRIFLUOROETHYNYL)OXY)METHYL)-1,2,2,2-TETRAFLUOROETHOXY)-2, 2,3,3-TETRAFLUORO-, CARBAMATE see ERC700

2-PROPANOL, 1-((3,4-DIHYDRO-2H-1-BENZOTHIOPYRAN-8-YL)OXY)-3-((1,1-DIMETHYLETHYL)AMINO)-,(+−)- see TBF400

2-PROPANOL, 1-((3,4-DIMETHOXYPHENETHYL)AMINO)-3-(m-TOLYLOXY)-, MONOHYDROCHLORIDE see BFW400

2-PROPANOL, 1-((2-(3,4-DIMETHOXYPHENYL)ETHYL)AMINO)-3-(3-METHYLPHENOXY)-, HCL see BFW400

2-PROPANOL, 2-((1-DIMETHYLETHYL)AZO)- see BQI400

PROPANOL, 1,1-DIPHENYL-3-PIPERIDINO-, METHANESULFONATE see PIL550

2-PROPANOL, 1-(1,3,5,7,9-DODECAPENTAENYLOXY)-, (ALL-E)- see AFQ820

PROPANOLE (GERMAN) see PND000

PROPANOLEN (DUTCH) see PND000

PROPANOL, ETHOXY-(9CI) see EJV000

PROPANOL, 2-ETHYLBUTOXY-, MIXED ISOMERS see EGX600

2-PROPANOL, 1-(2-ETHYLPHENOXY)-3-(((1S)-1,2,3,4-TETRAHYDRO-1-NAPHTHALENYL)AMINO)-, (2S)-ETHANEDIONATE (1:1) (SALT) see EOG600

2-PROPANOL, 1,1,1,3,3,3-HEXAFLUORO-2-(TRIFLUOROMETHYL)- see HDF075

PROPANOLI (ITALIAN) see PND000

PROPANOLIDE see PMT100

2-PROPANOL, 1-(1H-INDEN-4(OR 7)-YLOXY)-3-((1-METHYLETHYL)AMINO)-, HYDROCHLORIDE, (+−)- see IBY650

2-PROPANOL, 1-(7-INDENYLOXY)-3-(ISOPROPYLAMINO)-, HYDROCHLORIDE AND 1-(4-INDENYLOXY)-3-ISOPROPYLAMINO-2-PROPANOL HYDROCHLORIDE (2:1 TAUTOMERIC MIXTURE) see IBY650

PROPANOL, ISOBUTOXY-, (MIXED ISOMERS) see IIG500

1-PROPANOL, 2-METHOXY- see MFK800

2-PROPANOL, 2-METHYL-, LITHIUM SALT see LGY100

PROPANOL, 2-METHYL-2-(METHYLTHIO)-, o-((METHYL(4-MORPHOLINYLTHIO)AMINO)CARBONYL)OXIME see MLX820

1-PROPANOL, 2-METHYL-2-NITRO-, NITRATE see MMP200

1-PROPANOL, 3-(METHYLNITROSOAMINO)- see NKU520

PROPANOL NITRITE see PNQ750

2-PROPANOL NITRITE see IQQ000

1-PROPANOL, 3-NITRO- see NIY050

PROPANOLONE see PQC000

2-PROPANOL, 1-(4(OR 7)-INDENYLOXY)-3-(ISOPROPYLAMINO)-, HYDROCHLORIDE see IBY650

PROPANOL, OXYBIS- see DWS500

2-PROPANOL, 1,1'-(PENTAMETHYLENEDIOXY)BIS(3-CHLORO- see PBH150

1-PROPANOL, 2-PHENOXY- see PNL300

2-PROPANOL, 1-PHENOXY- see PDV460

1-PROPANOL, 3-PHENYL-, PROPIONATE see HHQ550

2-PROPANOL, 1,1',1"-(PHOSPHINIDYNETRIS((1-METHYLETHYLENE)OXY))TRI- see PGY300

2-PROPANOL, ((2,3,6-TRICHLOROBENZYL)OXY)- see TIL255

2-PROPANOL, 1-((2,3,6-TRICHLOROBENZYL)OXY)- see TJA200

2-PROPANOL, 1-((2,3,6-TRICHLOROPHENYL)METHOXY)- see TJA200

2-PROPANOL, 1-(2-((3,3,5-
TRIMETHYLCYCLOHEXYL)OXY)PROPOX
Y)-(9CI) see TLO600
PROPANONE see ABC750
2-PROPANONE see ABC750
2-PROPANONE, 1-(ACETYLOXY)-1-
(METHYLNITROSOAMINO)-(9CI) see
HMJ600
2-PROPANONE, 1-(4-(CHLOROMETHYL)-
1,3-DIOXOLAN-2-YL)- see CIL850
1-PROPANONE, 1-(4-CHLOROPHENYL)-2-
METHYL- see IOL100
2-PROPANONE, 1,1-DICHLORO- see
DGG500
2-PROPANONE, 1,3-DIHYDROXY- see
OLW100
1-PROPANONE, 1-(1,3-DIHYDROXY-4-((2-
HYDROXYPHENYL)METHYL)-9H-
XANTHEN-2-YL)-3-PHENYL- see PMM400
1-PROPANONE, 1-(3,4-
DIMETHOXYPHENYL)- see PMX600
2-PROPANONE, 1-(3,4-
DIMETHOXYPHENYL)- see VIK300
1-PROPANONE, 3-(DIMETHYLAMINO)-1-
INDOL-3-YL- see DPT300
1-PROPANONE, 3-(DIMETHYLAMINO)-1-
(1H-INDOL-3-YL)- see DPT300
2-PROPANONE, 1,1-DIPHENYL- see MJH910
1-PROPANONE, 1-(2-(β-d-
GLUCOPYRANOSYLOXY)-4,6-
DIHYDROXYPHENYL)-3-(4-
HYDROXYPHENYL)- see PGQ800
2-PROPANONE, HEXACHLORO- see
HCL500
1-PROPANONE,1-(10-(3-(4-(2-
HYDROXYETHYL)-1-
PIPERAZINYL)PROPYL)PHENOTHIAZIN-
2-YL)-, MALEATE (1:2) (SALT) see CCK700
2-PROPANONE, 1-HYDROXY-3-
(METHYLNITROSAMINO)-,
ACETATE(ESTER) see HMJ600
1-PROPANONE, 2-HYDROXY-2-METHYL-
1-PHENYL- see HMQ100
1-PROPANONE, 1-(4-(3-
HYDROXYPHENYL)-1-METHYL-4-
PIPERIDINYL)-(9CI) see KFK000
1-PROPANONE, 1-(4-(m-
HYDROXYPHENYL)-1-METHYL-4-
PIPERIDYL)- see KFK000
1-PROPANONE, 1-(2-(2-HYDROXY-3-
(PROPYLAMINO)PROPOXY)PHENYL)-3-
PHENYL-, HYDROCHLORIDE (9CI) see
PMJ525
1-PROPANONE, 1-p-MENTH-6-EN-2-YL- see
MCF525
2-PROPANONE, 1-METHOXY- see MDW300
2-PROPANONE, 1-METHOXY- see MFL100
2-PROPANONE, 1-(NITROSO-2-
PROPENYLAMINO)-(9CI) see AGM125
2-PROPANONE OXIME see ABF000
2-PROPANONE, 1-PHENYL- see MHO100
2-PROPANONE, o-
((PHENYLAMINO)CARBONYL)OXIME see
PEQ500
1-PROPANONE, 1-(4-PYRIDYL)- see PMX300
1-PROPANONE, 1-PYRROL-2-YL- see
PMX320
2-PROPANONE, 1,1,1-TRICHLORO- see
TJB775
1-PROPANONE, 1-(2,4,6-
TRIHYDROXYPHENYL)- see TKP100
PROPANOSEDYL see RLU000
PROPANOYL CHLORIDE see PMW500
PROPANOYL CHLORIDE, 2-METHYL- see
IJV100
PROPANTAN see PMM000
PROPANTEL see HKR500
PROPANTHELINE BROMIDE see HKR500
PROPAPHEN see CKP500
PROPAPHENIN see CKP250
PROPAPHENIN HYDROCHLORIDE see
CKP500

PROPAQUIZAFOP see IPI400
2-PROPARGILOSSI-5-AMINO-N-(n-BUTIL)-
BENZAMIDE (ITALIAN) see AJC000
PROPARGITE (DOT) see SOP000
3-PROPARGLOXYPHENYL-N-METHYL-
CARBAMATE see PMN250
PROPARGYL ALCOHOL see PMN450
PROPARGYL ALDEHYDE see PMT250
PROPARGYL BROMIDE see PMN500
PROPARGYL CHLORIDE see CKV275
N^{10}-PROPARGYL-5,8-DIDEAZAFOLIC
ACID see PMN550
PROPARGYL ETHER see POA500
5'-PROPARGYL-α'-ETHYNYL-3'-
FURYLMETHYL α-ETHYLPHENYL
ACETATE see PMN600
5-PROPARGYLFURFURYL
CHRYSANTHEMATE see POD875
5-PROPARGYL-2-FURYLMETHYL, dl-
cis,trans-CHRYSANTHEMATE see POD875
PROPARGYLIC ACID see PMT275
PROPARSAMIDE see SJG000
PROPARTHRIN see PMN700
PROPASA see AMM250
PROPASIN see PMN850
PROPASOL B see PML260
PROPASOL P see PNB750
PROPASOL SOLVENT B see BPS250
PROPASOL SOLVENT M see PNL250
PROPASOL SOLVENT P see PML260
PROPASOL SOLVENT P see PNB500
PROPASOL SOLVENT P see PNB750
PROPASTE 6708 see TAV750
PROPASTE T see SON000
PROPATHENE see PMP500
PROPAVAN see IDA500
PROPAX see CFZ000
PROPAXOLIN CITRATE see POF500
PROPAXOLINE CITRATE see POF500
PROPAZINE see PMN850
PROPAZONE see PMO250
PROPEL see LAG000
PROPELLANT 12 see DFA600
PROPELLANT 22 see CFX500
PROPELLANT 114 see FOO509
PROPELLANT C318 see CPS000
2-PROPENAL see ADR000
PROP-2-EN-1-AL see ADR000
PROPENAL (CZECH) see ADR000
2-PROPENAL, 3-(4-AMINO-2-OXO-1(2H)-
PYRIMIDINYL)- see CQM400
2-PROPENAL, 3-(6-AMINO-9H-PURIN-9-
YL)- see AEH300
2-PROPENAL, 2-CHLORO-(9CI) see CED800
PROPENAL DIETHYL ACETAL see DHH800
2-PROPENAL, 3-(2-FURANYL)- see FPK025
2-PROPENAL, 3-(2-FURANYL)-2-METHYL-
see FPX050
2-PROPENAL, 3-(2,4,5,6,7,7A-HEXAHYDRO-
3,7-DIMETHYL-1H-INDEN-4-YL)-2-
METHYL-, (4S-(4-α(E), 7-β,7A-α))- see
VAG300
2-PROPENAL, 2-METHYL- see MGA250
2-PROPENAL, 3-(5-METHYL-2,4-DIOXO-3,4-
DIHYDRO-2H-PYRIMIDIN-1-YL)- see
TFX805
2-PROPENAL, 3-PHENYL-, (E)-(9CI) see
CMP971
PROPENAMIDE see ADS250
2-PROPENAMIDE see ADS250
2-PROPENAMIDE, N-(BUTOXYMETHYL)-
(9CI) see BPM660
2-PROPENAMIDE, N-(1,1-
DIMETHYLETHYL)-(9CI) see BPW050
2-PROPENAMIDE, HOMOPOLYMER see
PJK350
2-PROPENAMIDE, 3-(4-HYDROXY-3,5-
DIMETHOXYPHENYL)-N-(3-(HEXYLOXY)-
1,2-DIHYDRO-4-HYDROXY-1-METHYL-2-
OXO-7-QUINOLINYL)- see HKB550
2-PROPENAMIDE, N-
(METHOXYMETHYL)- see MEX300

2-PROPENAMIDE, N-METHYL- (9CI) see
MGA300
2-PROPENAMIDE, N-(1-METHYLETHYL)-
(9CI) see INH000
2-PROPENAMIDE, N-(1-METHYLETHYL)-3-
(5-NITRO-2-FURANYL)-(9CI) see FPO000
2-PROPENAMIDE, N-((2-
METHYLPROPOXY)METHYL)-(9CI) see
IIE100
2-PROPENAMIDE, 3-PHENYL-(9CI) see
CMP973
2-PROPENAMINE see AFW000
2-PROPEN-1-AMINE see AFW000
2-PROPEN-1-AMINE, N,N,2-TRIMETHYL-
(9CI) see TME255
2-PROPEN-1-AMINIUM, N,N,N-TRIETHYL-,
IODIDE (9CI) see AGU300
PROPENE see PMO500
1-PROPENE see PMO500
PROPENE ACID see ADS750
1-PROPENE, 3-BROMO-1-CHLORO- see
BNA880
PROPENE, 3-CHLOROPENTAFLUORO- see
CJI550
1-PROPENE, 3-CHLORO-1,1,2,3,3-
PENTAFLUORO-(9CI) see CJI550
trans-1-PROPENE-1,2-DICARBOXYLIC
ACID see MDI250
1-PROPENE, DICHLORO- see DGG700
1-PROPENE, DICHLORO- see DGI700
PROPENE, 1,1-DICHLORO- see DGG750
1-PROPENE, 1,1-DICHLORO-(9CI) see
DGG750
PROPENE, 1,3-DICHLORO-2-METHYL- see
DFN330
1-PROPENE, 1,3-DICHLORO-2-METHYL-
see DFN330
1-PROPENE, 3,3-DIETHOXY-(9CI) see
DHH800
PROPENE-1,3-DIOL DIACETATE see
PMO800
1-PROPENE-1,3-DIOL, DIACETATE see
PMO800
2-PROPENE-1,1-DIOL, 2-METHYL-,
DIACETATE see AAW250
1-PROPENE HOMOPOLYMER (9CI) see
PMP500
1-PROPENE, 3-IODO-(9CI) see AGI250
1-PROPENE, 2-METHOXY- see MFL300
PROPENENITRILE see ADX500
2-PROPENENITRILE see ADX500
2-PROPENENITRILE, 3-((4-
CHLOROPHENYL)SULFONYL)- see CKJ850
2-PROPENENITRILE HOMOPOLYMER
(9CI) see ADX750
2-PROPENENITRILE, (2-
HYDROXYETHYL)- see HKY650
2-PROPENENITRILE, 3-(4-
METHYLPHENYL)SULFONYL)- see THD875
2-PROPENENITRILE, POLYMER with 1,3-
BUTADIENE and ETHENYLBENZENE
(9CI) see ADX740
2-PROPENENITRILE, POLYMER with
CHLOROETHENE (9CI) see ADY250
2-PROPENENITRILE POLYMER with
ETHENYLBENZENE see ADY500
2-PROPENE-NITRILE, POLYMER WITH 1,3-
BUTADIENE, 3-CARBOXY-1-CYANO-1-
METHYLPROPYL-TERMINATED,2-
HYDROXY-3-((1-OXO-2-
PROPENYL)OXY)PROPYL ESTER see
PMO900
2-PROPENENITRILE, 2,3,3-TRICHLORO-
(9CI) see TJC100
PROPENE, 1-NITRO- see NIY100
PROPENE OXIDE see PNL600
PROPENE OZONIDE see PMP250
PROPENE POLYMER see PKI250
PROPENE POLYMERS see PMP500
PROPENE TETRAMER see PMP750
2-PROPENE-1-THIOL see AGJ500

5-(1-PROPENYL)-1,3-BENZODIOXOLE see IRZ000
5-(2-PROPENYL)-1,3-BENZODIOXOLE see SAD000
1-PROPENYL BROMIDE see BOA000
4-PROPENYLCATECHOL METHYLENE ETHER see IRZ000
PROPENYL CHLORIDE see PMR750
2-PROPENYL CHLORIDE see AGB250
PROPENYL CINNAMATE see AGC000
1-PROPENYL CYANIDE see COC300
cis-1-PROPENYL CYANIDE see BOX600
2-PROPENYL DISULPHIDE see AGF300
(8-β)-6-(2-PROPENYL)ERGOLINE-8-ACETAMIDE TARTRATE see PMR800
17-(2-PROPENYL)ESTR-4-EN-17-OL see AGF750
PROPENYL ETHER see DBK000
PROPENYLGUAETHOL (FCC) see IRY000
4-PROPENYLGUAIACOL see IKQ000
2-PROPENYL HEPTANOATE see AGH250
2-PROPENYL-N-HEXANOATE see AGA500
2-PROPENYL ISOTHIOCYANATE see AGJ250
2-PROPENYL ISOVALERATE see ISV000
4-(2-PROPENYL)-2-METHOXYPHENYL FORMATE see EQS100
2-PROPENYL 3-METHYLBUTANOATE see ISV000
4-PROPENYL-1,2-METHYLENEDIOXYBENZENE see IRZ000
(2-PROPENYLOXY)BENZENE see AGR000
3-(2-PROPENYLOXY)-1,2-BENZISOTHIAZOLE 1,1-DIOXIDE (9CI) see PMD800
((2-PROPENYLOXY)METHYL)OXIRANE see AGH150
2-PROPENYLPHENOL see PMS250
o-PROPENYLPHENOL see PMS250
2-(1-PROPENYL)PHENOL see PMS250
2-PROPENYL PHENYLACETATE see PMS500
p-PROPENYLPHENYL METHYL ETHER see PMQ750
N-2-PROPENYL-2-PROPEN-1-AMINE see DBI600
(2-PROPENYL)THIOUREA see AGT500
PROP-2-ENYL TRIFLUOROMETHANE SULFONATE see PMS775
(Z)-5-PROPENYL-1,2,4-TRIMETHOXYBENZENE see ARM100
2-PROPENYL 3,5,5-TRIMETHYLHEXANOATE see AGU400
2-PROPENYLUREA see AGV000
N-2-PROPENYLUREA see AGV000
4-PROPENYL VERATROLE see IKR000
PROPENZOLATE HYDROCHLORIDE see PMS800
PROPERICIAZINE see PIW000
PROPERIDOL see DYF200
PROPETAMPHOS see MKA000
PROPHAM see CBM000
PROPHAM see PMS825
PROPHENAL see AGQ875
PROPHENATIN see DEO600
PROPHENPYRIDAMINE HYDROCHLORIDE see PMS900
PROPHENPYRIDAMINE MALEATE see TMK000
PROPHOS see EIN000
PROPICOL see DNR309
PROPICONAZOLE see PMS930
PROPIDIUM DIIODIDE see PMT000
PROPIDIUM IODIDE see PMT000
PROPILDAZINA (ITALIAN) see HHD000
PROPILENTIOUREA see MJZ000
PROPILTHIOURACIL see PNX000
6-PROPIL-TIOURACILE (ITALIAN) see PNX000
PROPINAN DI-n-BUTYLCINICITY (CZECH) see DEB400
PROPINE see MFX590

PROPINEB see ZMA000
PROPINEBE see ZMA000
γ-PROPIOBUTYROLACTONE see HBA550
PROPIOCINE see EDH500
PROPIOLACTONE see PMT100
3-PROPIOLACTONE see PMT100
1,3-PROPIOLACTONE see PMT100
β-PROPIOLACTONE see PMT100
PROPIOLALDEHYDE see PMT250
PROPIOLIC ACID see PMT275
PROPIOLIC ACID, TRIPHENYLSTANNYL ESTER see TMW600
PROPIOLONITRILE, PHENYL- see PGD100
PROPIOLOYL CHLORIDE see PMT300
PROPIOMAZINE HYDROCHLORIDE see PMT500
PROPIOMAZINE MALEATE see IDA500
PROPIONALDEHYDE see PMT750
PROPIONALDEHYDE, 2-CHLORO-3-ETHOXY-, DIETHYL ACETAL see CLP800
PROPIONALDEHYDE, DICROTYL ACETAL see PMU100
PROPIONALDEHYDE, 3-(p-ISOPROPYLBENZYL)-, DIMETHYL ACETAL see COU525
PROPIONALDEHYDE, 2-METHYL-2-(METHYLSULFINYL)-, O-(METHYLCARBAMOYL)OXIME see MLW750
PROPIONALDEHYDE, 2-METHYL-2-(METHYLTHIO)-, OXIME see MLX800
PROPIONALDEHYDE, 3-(METHYLTHIO)- see MPV400
PROPIONALDEHYDE, 3-(METHYLTHIO)- see TET900
PROPIONALDEHYDE, 3,3,3-TRIFLUORO- see TKH020
PROPIONAMIDE see PMU250
PROPIONAMIDE, N-BENZYL-2,3-EPOXY- see EBH890
PROPIONAMIDE, N-BENZYL-2,3-EPOXY-2-METHYL- see ECA200
PROPIONAMIDE, N-((5-CHLORO-4-METHYL)THIAZOLYL)- see CIU800
PROPIONAMIDE, N,N-DIBUTYL- see DEH300
PROPIONAMIDE, N,N-DIETHYL- see DJX250
PROPIONAMIDE, N,N-DIMETHYL-3-(PYRROLIDIN-1-YL)- see DTV330
PROPIONAMIDE, N-METHYL- see MOS900
PROPIONAMIDINE, 2,2'-AZOBIS(2-METHYL-), DIHYDROCHLORIDE see ASM050
PROPIONANILIDE see PMU500
PROPIONANILIDE, 2,3-EPOXY- see ECE650
PROPIONANILIDE, 2,3-EPOXY-2-METHYL- see ECC650
PROPIONAN SODNY see SJL500
PROPIONATE d'ETHYLE (FRENCH) see EPB500
PROPIONATE de METHYLE (FRENCH) see MOT000
PROPIONE see DJN750
PROPIONIC ACID see PMU750
PROPIONIC ACID, 3-(ACETYL-(3-AMINO-2,4,6-TRIIODOPHENYL)AMINO)-2-METHYL- see IDJ550
PROPIONIC ACID (ACGIH,DOT,OSHA) see PMU750
PROPIONIC ACID AMIDE see PMU250
PROPIONIC ACID ANHYDRIDE see PMV500
PROPIONIC ACID, 2-(4-BIPHENYLYLOXY)- see BGN100
PROPIONIC ACID CHLORIDE see PMW500
PROPIONIC ACID, 2-(N-(3-CHLORO-4-FLUOROPHENYL)BENZAMIDO)-, METHYL ESTER see FDB500
PROPIONIC ACID, 2-(m-CHLOROPHENOXY)- see CJQ300

PROPIONIC ACID, 2-((4-CHLORO-O-TOLYL)OXY)-, (+)- see RBF500
PROPIONIC ACID, 2-((4-CHLORO-o-TOLYL)OXY)-, compounded with 2,2'-IMINODIETHANOL (1:1) see MIH900
PROPIONIC ACID, 2-((4-CHLORO-o-TOLYL)OXY)-, POTASSIUM SALT see CLO200
PROPIONIC ACID, CINNAMYL ESTER see CMR850
PROPIONIC ACID, 2,3-DIBROMO- see DDS500
PROPIONIC ACID, 2,3-DIBROMO-, ETHYL ESTER see EHY050
PROPIONIC ACID-3,4-DICHLOROANILIDE see DGI000
PROPIONIC ACID, 2-(2,4-DICHLOROPHENOXY)-, (+)- see DGA880
PROPIONIC ACID, 2-(4-(2,4-DICHLOROPHENOXY)PHENOXY)-, METHYL ESTER see IAH050
PROPIONIC ACID-3,7-DIMETHYL-2,6-OCTADIEN-1-YL ESTER see NCP000
PROPIONIC ACID, α-1,3-DIMETHYL-4-PHENYL-4-PIPERIDYL ESTER see NEA500
PROPIONIC ACID, 3,3'-DISELENODI-, SODIUM SALT see DWZ100
PROPIONIC ACID, ETHYL ESTER see EPB500
PROPIONIC ACID GRAIN PRESERVER see PMU750
PROPIONIC ACID, HEXYL ESTER (6CI,7CI,8CI) see HFV100
PROPIONIC ACID, 2-HYDROXY- see LAG000
PROPIONIC ACID, ISOBUTYL ESTER see PMV250
PROPIONIC ACID, 2-(p-ISOBUTYLPHENYL)-, 2-PYRIDYLMETHYL ESTER see IAB000
PROPIONIC ACID, ISOPENTYL ESTER see ILW000
PROPIONIC ACID, 2-(4-(METHALLYLAMINO)PHENYL)- see MDP850
PROPIONIC ACID, 3-METHOXY-, METHYL ESTER see MFL400
PROPIONIC ACID, 2-METHYL- see IJU000
PROPIONIC ACID, 2-(8-METHYL-10,11-DIHYDRO-11-OXODIBENZ(b,f)OXEPIN-2-YL)- see DLI650
PROPIONIC ACID, 2-METHYLENE- see MDN250
PROPIONIC ACID, 2-(3,4-METHYLENEDIOXYBENZYLAMINO)- see PIX100
PROPIONIC ACID, NERYL ESTER see NCP000
PROPIONIC ACID, 3,3'-NITROIMINODI- see NHI600
PROPIONIC ACID, 2-(5-NITRO-α-IMINOFURFURYL)HYDRAZIDE see PMW800
PROPIONIC ACID, PENTAMETHYLENE ESTER (2:1) see PBI300
PROPIONIC ACID, PENTYL ESTER (6CI,7CI,8CI) see AON350
PROPIONIC ACID, 2-PHENOXY- see PDV725
PROPIONIC ACID-2-PHENOXYETHYL ESTER see EJK500
PROPIONIC ACID, PHENYLMERCURY SALT see PFO000
PROPIONIC ACID, 3-PHOSPHONO- see PHA560
PROPIONIC ACID, PROPYL ESTER see PNU000
PROPIONIC ACID, 3-(PURIN-6-YLTHIO)- see CCF270
PROPIONIC ACID, 3-SELENINO- see SBN510

PROPIONIC ACID, 2,2,3,3-TETRAFLUORO-, SODIUM SALT see SKE100
PROPIONIC ACID, 3,3'-THIODI-, BIS(2-ETHYLHEXYL) ESTER see BJS550
PROPIONIC ACID, 3,3'-THIODI-, DIOCTADECYL ESTER see DXG700
PROPIONIC ACID, 2-(p-((5-(TRIFLUOROMETHYL)-2-PYRIDYL)OXY)PHENOXY)-, BUTYL ESTER see FDA885
PROPIONIC ALDEHYDE see PMT750
PROPIONIC AMIDE see PMU250
PROPIONIC ANHYDRIDE see PMV500
PROPIONIC CHLORIDE see PMW500
PROPIONIC ETHER see EPB500
PROPIONIC NITRILE see PMV750
PROPIONITRILE, 3-(ALKYLAMINO)- see AFO200
PROPIONITRILE, 3-AMINO-N,N-DIBUTYL- see DDU800
PROPIONITRILE, 3-((2-((2-AMINOETHYL)AMINO)ETHYL)AMINO)- (8CI) see LBV000
PROPIONITRILE, 3-ANILINO- see AOT100
PROPIONITRILE, 2-((4-(ETHYLAMINO)-6-(METHYLTHIO)-s-TRIAZIN-2-YL)AMINO)-2-METHYL- see COI050
PROPIONITRILE, 3-(N-ETHYLANILINO)- (6CI,7CI,8CI) see EHQ500
PROPIONITRILE, 2-HYDROXY- see LAQ000
PROPIONITRILE, 2-HYDROXY-2-METHYL-3,3,3-TRIFLUORO-, ACETATE see ABR700
PROPIONITRILE, 3-METHOXY- see MFL750
PROPIONITRILE, 3-p-PHENETIDINO- (6CI,8CI) see COM900
PROPIONITRILE, TRICHLORO- see TJC800
PROPIONITRILE, 2,2,3-TRICHLORO- see TJC850
PROPIONITRILE, α,α,β-TRICHLORO- see TJC850
PROPIONITRILE, 3-(TRIETHOXYSILYL)- see CON250
β-PROPIONOLACTONE see PMT100
PROPIONONITRILE see PMV750
PROPIONYLBENZENE see EOL500
4-(N-PROPIONYL)BENZYLIMINO-1-ETHYL-2,2,6,6-TETRAMETHYLPIPERIDINE see EPM000
PROPIONYL CHLORIDE see PMW500
PROPIONYLCHOLINE IODIDE see PMW750
3-PROPIONYL-10-DIMETHYLAMINO-ISOPROPYLPHENOTHIAZINE MALEATE see IDA500
10-PROPIONYL DITHRANOL see PMW760
N-PROPIONYL-N-2-FLUORENYLHYDROXYLAMINE see FEO000
PROPIONYL HYPOBROMITE see PMW770
N-PROPIONYLINDOLE see ICW100
N²)-PROPIONYL-5-NITRO-2-FUROHYDRAZIDE IMIDE see PMW800
PROPIONYL OXIDE see PMV500
PROPIONYL PEROXIDE (DOT) see DWQ800
p-PROPIONYLPHENOL see ELL500
PROPIONYLPHLOROGLUCINOL see TKP100
N-PROPIONYL-2-(1-PIPERIDINOISOPROPYL)AMINOPYRIDINE FUMARATE see PMX250
PROPIONYLPROMETHAZINE MALEATE see IDA500
4-PROPIONYLPYRIDINE see PMX300
N-PROPIONYL-N-(2-PYRIDYL)-1-PIPERIDINO-2-AMINOPROPANE FUMARATE see PMX250
2-PROPIONYLPYRROLE see PMX320
PROPIOPHENONE see EOL500
PROPIOPHENONE, 2-BROMO- see BOB550
PROPIOPHENONE, 4'-CHLORO-2-METHYL-(6CI,7CI,8CI) see IOL100

PROPIOPHENONE, 3',4'-DIMETHOXY- see PMX600
PROPIOPHENONE, 4'-(DIMETHYLAMINO)-3-(4-PHENYL-1,2,3,6-TETRAHYDRO-1-PYRIDYL)- see DPS700
PROPIOPHENONE, 2-HYDROXY-2-METHYL- see HMQ100
PROPIOPHENONE, 4'-PHENYL- see BGM100
PROPIOPHENONE, 2',4',6'-TRIHYDROXY- see TKP100
PROPIOPHLOROGLUCINE see TKP100
PROPIOPROMAZINE MALEATE see PMX500
PROPIOVERATRONE see PMX600
PROPIRAM FUMARATE see PMX250
PROPISAMINE see BBK000
PROPITAN see FHG000
PROPITAN see PII200
PROPITOCAINE HYDROCHLORIDE see CMS260
PROPIVANE see PGP500
PROP-JOB see DGI000
PROPOFOL see DNR800
PROPOKSURU (POLISH) see PMY300
PROPOLIN see PMP500
PROPON see TIX500
PROPONESIN HYDROCHLORIDE see TGK225
PROPONEX-PLUS see CIR500
PROPONEX-PLUS see RBF500
PROPOPHANE see PMP500
PROPOQUIN DIHYDROCHLORIDE see PMY000
PROPOX see PNA500
PROPOXUR see PMY300
PROPOXUR NITROSO see PMY310
4-PROPOXYBENZALDEHYDE see PMY500
p-PROPOXYBENZALDEHYDE see PMY500
p-(N-PROPOXY)BENZALDEHYDE see PMY500
o-PROPOXYBENZAMIDE see PMY750
2-N-PROPOXYBENZAMIDE see PMY750
N-(PROPOXY-5-BENZIMIDAZOLYL)-2,CARBAMATE de METHYLE (FRENCH) see OMY700
N-(2-(5-PROPOXYBENZIMIDAZOLYL)) METHYL CARBAMATE see OMY700
p-PROPOXYBENZOIC ACID 2-(1-PYRROLIDINYL)ETHYL ESTER HYDROCHLORIDE see PNA200
PROPOXYCHEL see PNA500
2-PROPOXYETHANOL see PNG750
2-PROPOXYETHANOL ACETATE see PNA225
2-PROPOXYETHYL ACETATE see PNA225
PROPOXYPHENE see PNA250
(+)-PROPOXYPHENE see DAB879
d-PROPOXYPHENE see DAB879
α-dl-PROPOXYPHENE CARBINOL see PNA300
PROPOXYPHENE HYDROCHLORIDE see PNA500
(+)-PROPOXYPHENE HYDROCHLORIDE see PNA500
d-PROPOXYPHENE HYDROCHLORIDE see PNA500
α-PROPOXYPHENE HYDROCHLORIDE see PNA500
α-d-PROPOXYPHENE HYDROCHLORIDE see PNA500
d-PROPOXYPHENE MONOHYDROCHLORIDE see PNA500
PROPOXYPHENE N see DYB400
PROPOXYPHENE-2-NAPHTHALENESULFONATE see DYB400
PROPOXYPHENE NAPSYLATE see DYB400
d-PROPOXYPHENE NAPSYLATE HYDRATE see DAB880
PROPOXYPHENYL see PNS750
2-(3-PROPOXYPHENYL)IMIDAZO(2,1-A)ISOQUINOLINE see PNA600

2-(m-PROPOXYPHENYL)IMIDAZO(2,1-A)ISOQUINOLINE see PNA600
4'-PROPOXY-3-PIPERIDINO PROPIOPHENONE HYDROCHLORIDE see PNB250
4-PROPOXY-β-(1-PIPERIDYL)PROPIOPHENONE HYDROCHLORIDE see PNB250
PROPOXYPIPEROCAINE see PNB250
PROPOXYPIPEROCAINE HYDROCHLORIDE see PNB250
1-PROPOXY-2-PROPANOL see PNB500
n-PROPOXYPROPANOL (mixed isomers) see PNB750
PROPRANOLOL see ICB000
PROPRANOLOL see ICC000
(±)-PROPRANOLOL see PNB790
dl-PROPRANOLOL see PNB790
racemic-PROPRANOLOL see PNB790
PROPRANOLOL HYDROCHLORIDE see ICC000
dl-PROPRANOLOL HYDROCHLORIDE see PNB800
β-PROPRIOLACTONE (OSHA) see PMT100
β-PROPROLACTONE see PMT100
PROPROP see DGI400
PROPYCIL see PNX000
n-PROPYL ACETAL see AAG850
PROPYL ACETATE see PNC250
1-PROPYL ACETATE see PNC250
2-PROPYL ACETATE see INE100
n-PROPYL ACETATE see PNC250
PROPYLACETIC ACID see VAQ000
PROPYL ACETOXYMETHYLNITROSAMINE see PNR250
N-PROPYL-N-(ACETOXYMETHYL)NITROSAMINE see PNR250
2-PROPYLACROLEIN see PNT800
α-PROPYLACROLEIN see PNT800
β-PROPYL ACROLEIN see PNC500
PROPYLADIPHENIN see PBM500
S-PROPYL-N-AETHYL-N-BUTYL-THIOCARBAMAT (GERMAN) see PNF500
N-PROPYLAJMALINE see PNC875
N-PROPYLAJMALINE BITARTRATE see DNB000
N-PROPYLAJMALINE BROMIDE see PNC925
N-PROPYLAJMALINE HYDROGEN TARTRATE see DNB000
N-PROPYLAJMALINIUM see PNC875
N⁴-PROPYLAJMALINIUM see PNC875
N-PROPYLAJMALINIUM BITARTRATE see DNB000
1-PROPYLAJMALINIUM BROMIDE see PNC925
N-PROPYLAJMALINIUMHYDROGENTARTRAT (GERMAN) see DNB000
N⁴-PROPYLAJMALINIUM HYDROGEN TARTRATE see DNB000
PROPYL ALCOHOL see PND000
1-PROPYL ALCOHOL see PND000
n-PROPYL ALCOHOL see PND000
sec-PROPYL ALCOHOL (DOT) see INJ000
PROPYL ALDEHYDE see PMT750
i-PROPYLALKOHOL (GERMAN) see INJ000
n-PROPYL ALKOHOL (GERMAN) see PND000
PROPYLAMINE see PND250
2-PROPYLAMINE see INK000
N-PROPYLAMINE see PND250
sec-PROPYLAMINE see INK000
PROPYLAMINE, N,N-BIS(2-CHLOROETHYL)-, HYDROCHLORIDE see PNF050
PROPYLAMINE, N-tert-BUTYL-1-METHYL-3,3-DIPHENYL-, HYDROCHLORIDE see TBC200

PROPYLAMINE, 2-CHLORO-N,N-DIMETHYL- see CGJ290
1-PROPYLAMINE, 3-CHLORO-N,N-DIMETHYL-, HYDROCHLORIDE see CGJ300
PROPYLAMINE, 1,2-DIMETHYL- see AOE200
PROPYLAMINE, 2,3-EPOXY-N,N-DIETHYL- see GGW800
PROPYLAMINE, N-ETHYL-1,2-DIMETHYL- see EIL050
PROPYLAMINE, 3,3'-(ETHYLENEDIOXY)BIS- see DCB200
PROPYLAMINE, 3,3'-IMINOBIS- see AIX250
PROPYLAMINE, 1-METHYL-3-PHENYL- see PEE300
PROPYLAMINE, α-METHYL-γ-PHENYL- see PEE300
α-PROPYLAMINE-2-METHYL-PROPIONANILIDEHYDROCHLORIDE see CMS260
PROPYLAMINE, 3,3'-(TETRAMETHYLENEDIOXY)BIS- see BGU600
PROPYLAMINE, 3-(TRIETHOXYSILYL)- see TJN000
PROPYLAMINE, 3-(TRIETHYLSILYL)- see AMG200
PROPYLAMINO-BIS(1-AZIRIDINYL)PHOSPHINE OXIDE see BGY500
2-(PROPYLAMINO)-o-PROPIONOTOLUIDIDE HYDROCHLORIDE see CMS260
4-PROPYLANILINE see PNE000
p-PROPYLANILINE see PNE000
4-N-PROPYLANILINE see PNE000
p-N-PROPYLANILINE see PNE000
4-PROPYLANISOLE see PNE250
4-n-PROPYLANISOLE see PNE250
p-n-PROPYL ANISOLE see PNE250
4-PROPYLBENZALDEHYDE see PNE500
p-PROPYLBENZALDEHYDE see PNE500
8-PROPYLBENZ(a)ANTHRACENE see PNE750
5-n-PROPYL-1,2-BENZANTHRACENE see PNE750
n-PROPYLBENZENE see IKG000
PROPYL BENZENE (DOT) see IKG000
PROPYL 1,4-BENZODIOXAN-2-CARBOXYLATE see PNE800
5-PROPYL-1,3-BENZODIOXOLE see DMD600
5-PROPYLBENZO(c)PHENANTHRENE see PNF000
2-n-PROPYL-3:4-BENZPHENANTHRENE see PNF000
3-PROPYLBICYCLO(2.2.1)HEPT-5-ENE-2-CARBOXALDEHYDE see CML620
N-PROPYL-BIS(2-CHLOROETHYL)AMINE HYDROCHLORIDE see PNF050
PROPYLBIS(β-CHLOROETHYL)AMINE HYDROCHLORIDE see PNF050
PROPYLBIS(2-HYDROXYETHYL)AMINE see PNH775
PROPYL BROMIDE see BNX750
PROPYL BUTANOATE see PNF100
S-PROPYL BUTYLETHYLTHIOCARBAMATE see PNF500
N-PROPYL-N-BUTYLNITROSAMINE see PNF750
PROPYL BUTYRATE see PNF100
PROPYL CAPRYLATE see PNR600
PROPYL CARBAMATE see PNG250
N-PROPYL CARBAMATE see PNG250
PROPYLCARBINOL see BPW500
N-PROPYLCARBINYL CHLORIDE see BQQ750
N-PROPYL-N-(3-CARBOXYPROPYL)NITROSAMINE see PNG500

PROPYL CELLOSOLVE see PNG750
N-PROPYL CHLORIDE see CKP750
1-PROPYL-3-(p-CHLOROBENZENESULFONYL)UREA see CKK000
N-PROPYL-N'-(p-CHLOROBENZENESULFONYL)UREA see CKK000
PROPYL CHLOROCARBONATE see PNH000
N-PROPYL-N-(2-CHLOROETHYL)-2,6-DINITRO-4-TRIFLUOROMETHYLANILINE see FDA900
N-PROPYL-N-(2-CHLOROETHYL)-α,α,α-TRIFLUORO-2,6-DINITRO-p-TOLUIDINE see FDA900
PROPYL CHLOROFORMATE see PNH000
n-PROPYL CHLOROFORMATE (DOT) see PNH000
S-PROPYL CHLOROTHIOFORMATE see PNH150
N-PROPYL-N'-p-CHLORPHENYLSULFONYLCARBAMIDE see CKK000
n-PROPYL CINNAMATE see PNH250
PROPYLCOPPER(I) see PNH500
PROPYL CYANIDE see BSX250
N-(4-PROPYLCYCLOHEXYL)BENZAMIDE see PNH522
N-(4-PROPYLCYCLOHEXYL)-4-MORPHOLINECARBOXAMIDE see PNH527
N-(4-PROPYLCYCLOHEXYL)-3-(3,4,5-TRIMETHOXYPHENYL)-2-PROPENAMIDE see PNH533
3-PROPYLDIAZIRINE see PNH550
PROPYLDICHLORARSINE see PNH650
PROPYL-p,p'-DICHLOROBENZILATE see PNH750
PROPYLDIETHANOLAMINE see PNH775
N-PROPYLDIETHANOLAMINE see PNH775
PROPYL(4-((DIETHYLCARBAMOYL)METHOXY)-3-METHOXYPHENYL)ACETATE see PMM000
PROPYL-N,N-DIETHYLSUCCINAMATE see PNH800
4'-N-PROPYL-4-DIMETHYLAMINOAZOBENZENE see DTT400
2-PROPYL-3-DIMETHYLAMINO-5,6-METHYLENEDIOXYINDENE HYDROCHLORIDE see DPY600
2-N-PROPYL-3-DIMETHYLAMINO-5,6-METHYLENEDIOXYINDENE HYDROCHLORIDE see DPY600
PROPYL(3-(DIMETHYLAMINO)PROPYL)CARBAMATE MONOHYDROCHLORIDE see PNI250
5-n-PROPYL-9,10-DIMETHYL-1,2-BENZANTHRACENE see PNI500
1-n-PROPYL-3,7-DIMETHYL XANTHINE see PNW250
PROPYLDIPHENYLPHOSPHINE see PNI600
S-PROPYL DIPROPYLTHIOCARBAMATE see PNI750
PROPYL-N,N-DIPROPYLTHIOLCARBAMATE see PNI750
N-PROPYL-DI-N-PROPYLTHIOLCARBAMATE see PNI750
PROPYL DISELENIDE see PNI850
PROPYLENE (DOT, ACGIH) see PMO500
PROPYLENE ALDEHYDE see ADR000
PROPYLENE ALDEHYDE see COB260
1,1'-PROPYLENEBIS(3-(2-CHLOROETHYL)-3-NITROSOUREA) see PNJ000
1,1'-PROPYLENEBIS-CNU see PNJ000
PROPYLENE BISDITHIOCARBAMATE see MLX850
PROPYLENEBIS(DITHIOCARBAMATO)ZINC see ZMA000
1,2-PROPYLENE CARBONATE see CBW500
PROPYLENE CHLORIDE see PNJ400
PROPYLENECHLOROHYDRIN see CKR500
PROPYLENEDIAMINE see PMK250

1,3-PROPYLENEDIAMINE see PMK500
PROPYLENE DIAMINE (DOT) see PMK250
PROPYLENEDIAMINE TETRA-ACETIC ACID see PNJ100
PROPYLENE DIBROMIDE see DDR400
PROPYLENE DIBROMIDE see DDR600
PROPYLENE DICHLORIDE see DGG800
PROPYLENE DICHLORIDE see PNJ400
α,β-PROPYLENE DICHLORIDE see PNJ400
PROPYLENE DIMETHANESULFONATE see TLR250
4,4'-PROPYLENEDI-2,6-PIPERAZINEDIONE see RCA375
PROPYLENE EPOXIDE see PNL600
PROPYLENE FORMAL see PNJ500
1,2-PROPYLENE GLYCOL see PML000
1,3-PROPYLENE GLYCOL see PML250
α-PROPYLENEGLYCOL see PML000
β-PROPYLENE GLYCOL see PML250
PROPYLENE GLYCOL ALGINATE see PNJ750
PROPYLENE GLYCOL, ALLYL ETHER see PNK000
PROPYLENE GLYCOL BUTOXY ETHER see PML260
PROPYLENE GLYCOL-n-BUTYL ETHER see BPS250
PROPYLENE GLYCOL-sec-BUTYL PHENYL ETHER see PNK500
PROPYLENE GLYCOL CYCLIC CARBONATE see CBW500
PROPYLENE GLYCOL DIACETATE see PNK500
α-PROPYLENE GLYCOL DIACETATE see PNK750
PROPYLENE GLYCOL DINITRATE see PNL000
PROPYLENE GLYCOL-1,2-DINITRATE see PNL000
1,2-PROPYLENE GLYCOL DINITRATE see PNL000
PROPYLENE GLYCOL ETHYL ETHER see EFF500
PROPYLENE GLYCOL (FCC) see PML000
PROPYLENE GLYCOL ISOBUTYL ETHER see IIG000
PROPYLENE GLYCOL LACTOSTEARATE see LAR400
PROPYLENE GLYCOL METHYL ETHER see PNL250
PROPYLENE GLYCOL MONOACETATE see PNL100
PROPYLENE GLYCOL MONOACRYLATE see HNT600
PROPYLENE GLYCOL MONO-n-BUTYL ETHER see BPS500
PROPYLENE GLYCOL MONO- and DIESTERS see PNL225
PROPYLENE GLYCOL MONO- and DIESTERS of FATTY ACIDS see PNL225
PROPYLENE GLYCOL MONOETHYL ETHER see EJV000
β-PROPYLENE GLYCOL MONOETHYL ETHER see EFG000
PROPYLENE GLYCOL-β-MONOETHYL ETHER see EFG000
PROPYLENE GLYCOL MONOMETHYL ETHER see PNL250
PROPYLENE GLYCOL MONOMETHYL ETHER see PNL250
α-PROPYLENE GLYCOL MONOMETHYL ETHER see PNL250
β-PROPYLENE GLYCOL MONOMETHYL ETHER see MFL000
PROPYLENE GLYCOL MONOMETHYL ETHER ACETATE see PNL265
PROPYLENE GLYCOL MONOMETHYL ETHER (ACGIH,OSHA) see PNL250
PROPYLENE GLYCOL MONOSTEARATE see PNL225
PROPYLENE GLYCOL MONOSTEARATE see SLL000

PROPYLENE GLYCOL MONOPROPYL ETHER see PNB750
PROPYLENE GLYCOL PHENYL ETHER see PNL300
PROPYLENE GLYCOL n-PROPYL ETHER see PNB500
PROPYLENE GLYCOL USP see PML000
PROPYLENE IMINE see PNL400
1,2-PROPYLENEIMINE see PNL400
PROPYLENE IMINE, INHIBITED (DOT) see PNL400
α-PROPYLENE MONO-N-BUTYL ETHER see PML260
PROPYLENENITROSOUREA see PNL500
(−)-PROPYLENE OXIDE see ECE700
PROPYLENE OXIDE see PNL600
(+)-PROPYLENE OXIDE see PNL650
(R)-PROPYLENE OXIDE see PNL650
(S)-PROPYLENE OXIDE see ECE700
(S)-(−)-PROPYLENE OXIDE see ECE700
PROPYLENE OXIDE, (S)-(−)- see ECE700
1,2-PROPYLENE OXIDE see PNL600
1,3-PROPYLENE OXIDE see OMW000
PROPYLENE OXIDE, (R)-(+)- see PNL650
(R)-(+)-PROPYLENE OXIDE see PNL650
PROPYLENE OXIDE and ETHYLENE OXIDE BLOCK POLYMER see PJK200
PROPYLENE OXIDE HEXAFLUORIDE see HDF050
PROPYLENE OXIDE-METHANOL ADDUCT see MKS250
PROPYLENE PHENOXETOL see PDV460
PROPYLENE POLYMER see PKI250
PROPYLENE POLYMER see PMP500
1,3-PROPYLENE SULFATE see TLR750
PROPYLENE SULFIDE see PNL750
PROPYLENE TETRAMER see PMP750
PROPYLENE THIOUREA see MJZ000
PROPYLENGLYKOL-MONOMETHYLAETHER see PNL250
PROPYLENTHIOHARNSTOFF see MJZ000
(8-β)-6-PROPYLERGOLINE-8-ACETAMIDE TARTRATE (2:1) see PNL800
(8S)-6-PROPYLERGOLINE-8-CARBAMIC ACID ETHYL ESTER see EPC115
1-((5R,8S,10R)-6-PROPYL-8-ERGOLINYL)-3,3-DIETHYLUREA see DJX300
PROPYLESTER KYSELINY DUSICNE see PNQ500
PROPYLESTER KYSELINY MASELNE see PNF100
PROPYLESTER KYSELINY MRAVENCI see PNM500
PROPYLESTER KYSELINY OCTOVE see PNC250
PROPYLESTER KYSELINY SKORICOVE see PNH250
n-PROPYL ESTER of 3,4,5-TRIHYDROXYBENZOIC ACID see PNM750
PROPYL ETHER see PNM000
β-PROPYL-α-ETHYLACROLEIN see BRI000
PROPYL-ETHYLBUTYLTHIOCARBAMATE see PNF500
PROPYLETHYL-N-BUTYLTHIOCARBAMATE see PNF500
PROPYL N-ETHYL-N-BUTYLTHIOCARBAMATE see PNF500
N-PROPYL-N-ETHYL-N-(N-BUTYL)THIOCARBAMATE see PNF500
S-(N-PROPYL)-N-ETHYL-N-N-BUTYLTHIOCARBAMATE see PNF500
PROPYL ETHYLBUTYLTHIOLCARBAMATE see PNF500
N-PROPYL-N-ETHYL-N-(N-BUTYL)THIOCARBAMATE see PNF500
N,N-PROPYL ETHYL CARBAMATE see EPC050
PROPYL ETHYL ETHER see EPC125
n-PROPYL FORMATE see PNM500
PROPYL FORMATE (DOT) see PNM500
PROPYLFORMIC ACID see BSW000

N-n-PROPYL-N-FORMYLHYDRAZINE see PNM650
PROPYL FP-12 see DXT900
PROPYL GALLATE see PNM750
n-PROPYL GALLATE see PNM750
4-PROPYLGUAIACOL see MFM750
p-PROPYLGUAIACOL see MFM750
p-n-PROPYLGUAIACOL see MFM750
2-PROPYLHEPTANIC ACID see NMX500
2-PROPYLHEPTANOL see PNN250
2-PROPYLHEPTANSAEURE (GERMAN) see NMX500
PROPYLHEXADRINE see PNN300
PROPYLHEXADRINE HYDROCHLORIDE see PNN300
2-PROPYLHEXANOIC ACID see OCU100
PROPYLHEXEDRINE see PNN400
N-PROPYLHYDRAZINE HYDROCHLORIDE see PNO000
PROPYL HYDRIDE see PMJ750
N-PROPYL-N-(HYDROPEROXYMETHYL)NITROSAMINE see HIE570
PROPYL p-HYDROXYBENZOATE see HNU500
n-PROPYL p-HYDROXYBENZOATE see HNU500
PROPYL-p-HYDROXYBENZOATE, SODIUM SALT see PNO250
PROPYL(4-HYDROXYBUTYL)NITROSAMINE see NLN000
PROPYLIC ALCOHOL see PND000
PROPYLIC ALDEHYDE see PMT750
n-PROPYLIDENE BUTYRALDEHYDE see HBI800
PROPYLIDENE CHLORIDE see DGF400
3-PROPYLIDENE-1(3H)-ISOBENZOFURANONE see PNO500
PROPYLIDENE PHTHALIDE see PNO500
2,2'-(PROPYLIMINO)BISETHANOL see PNH775
2,2'-(PROPYLIMINO)DIETHANOL see PNH775
i-PROPYL IODIDE see IPS000
n-PROPYL IODIDE see PNO750
PROPYL ISOCYANATE see PNP000
1-PROPYL ISOCYANATE see PNP000
m-PROPYL ISOCYANATE see PNP000
PROPYL ISOMER see PNP250
n-PROPYL ISOMER see PNP250
2-PROPYLISONICOTINYLTHIOAMIDE see PNW750
PROPYL KETONE see DWT600
PROPYL LITHIUM see PNP275
PROPYL MERCAPTAN see PML500
2-PROPYL MERCAPTAN see IMU000
N-PROPYL MERCAPTAN see PML500
6-PROPYLMERCAPTOPURINE see PNW800
PROPYL METHACRYLATE see PNP750
n-PROPYL METHACRYLATE see PNP750
n-PROPYL METHANESULFONATE see PNQ000
PROPYL METHANOATE see PNM500
PROPYLMETHANOL see BPW500
1-PROPYL-3-METHOXY-4-HYDROXYBENZENE see MFM750
PROPYLMETHYLCARBINYLALLYL BARBITURIC ACID SODIUM SALT see SBN000
PROPYLMETHYLCARBINYLETHYL BARBITURIC ACID SODIUM SALT see NBU000
4-PROPYL-1,2-METHYLENEDIOXYBENZENE see DMD600
2-n-PROPYL-4-METHYLPYRIMIDYL-(6)-N,N-DIMETHYL CARBAMATE see PNQ250
6-PROPYL-MP see PNW800
PROPYL NITRATE see PNQ500
n-PROPYL NITRATE see PNQ500
PROPYL NITRITE see PNQ750

N-PROPYL-N'-NITRO-N-NITROSOGUANIDINE see NLD000
4-(PROPYLNITROSAMINO)-1-BUTANOL see NLN000
PROPYLNITROSAMINOMETHYL ACETATE see PNR250
1-(PROPYLNITROSAMINO)PROPYL ACETATE see ABT750
N-PROPYLNITROSOHARNSTOFF (GERMAN) see NLO500
2-PROPYL-N-NITROSOTHIAZOLIDINE see NLO000
N-PROPYLNITROSOUREA see NLO500
1-PROPYL-1-NITROSOUREA see NLO500
N-PROPYL-N-NITROSOURETHANE see PNR500
PROPYL OCTANOATE see PNR600
19-PROPYLORVINOL see EQO450
PROPYLORVINOL HYDROCHLORIDE see EQO500
PROPYLOWY ALKOHOL (POLISH) see PND000
PROPYLOXIRANE see ECE525
PROPYLPARABEN (FCC) see HNU500
PROPYLPARASEPT see HNU500
2-PROPYLPENTAMIDE see PNX600
2-PROPYLPENTANOIC ACID see PNR750
2-PROPYLPENTANOIC ACID CALCIUM SALT see VCK200
2-PROPYLPENTANOYLTROPINIUM METHYLBROMIDE see LJS000
2-PROPYL-4-PENTENOIC ACID see AGQ100
n-PROPYL PERCARBONATE see DWV400
N-(4-PROPYLPHENAZOL-5-YL)-2-ACETOXYBENZAMIDE see PNR800
PROPYL PHENETHYL ACETAL see PDD400
1-(α-PROPYLPHENETHYL)PYRROLIDINE HYDROCHLORIDE see PNS000
o-PROPYLPHENOL see PNS250
p-PROPYLPHENOL see PNS500
PROPYL PHENYL ETHER see PNS750
N-2-PROPYL-N'-PHENYL-p-PHENYLENEDIAMINE see PFL000
PROPYL 3-PHENYL-2-PROPENOATE see PNH250
PROPYLPHYLLIN see DNC000
1-PROPYLPIPERIDINE see PNS800
2-PROPYLPIPERIDINE see PNT000
N-PROPYLPIPERIDINE see PNS800
β-PROPYLPIPERIDINE see PNT000
1-PROPYL-4-PIPERIDYL-p-AMINOBENZOATE HYDROCHLORIDE see PNT500
1-PROPYL-4-PIPERIDYL BENZOATE HYDROCHLORIDE see PNT750
6-(PROPYLPIPERONYL)-BUTYL CARBITYL ETHER see PIX250
6-PROPYLPIPERONYL BUTYL DIETHYLENE GLYCOL ETHER see PIX250
N-PROPYL-1-PROPANAMINE see DWR000
PROPYL PROPANOATE see PNU000
2-PROPYL-2-PROPENAL see PNT800
PROPYL PROPIONATE see PNU000
n-PROPYL PROPIONATE see PNU000
1-PROPYL-6-((p-(p-((1-PROPYLQUINOLINIUM-6-YL)CARBAMOYL)BENZAMIDO)BENZAMIDO)QUINLINIUM), DI-p-TOLUENESULFONATE see PNV250
PROPYL-2-PROPYNYLPHENYLPHOSPHONATE see PNV750
O-n-PROPYL O-(2-PROPYNYL) PHENYLPHOSPHONATE see PNV750
2-PROPYL-4-PYRIDINECARBOTHIOAMIDE see PNW750
PROPYL 4-PYRIDYL KETONE see PNV755
n-PROPYLSELENINIC ACID see PNV760
PROPYL SILANE see PNV775
PROPYL SODIUM see PNV800

5-(PROPYLSULFONYL)-2-BENZIMIDAZOLECARBAMIC ACID METHYL ESTER see PNV900
1-PROPYL THEOBROMINE see PNW250
2-PROPYLTHIAZOLIDINE see PNW300
2-N-PROPYLTHIAZOLIDINE see PNW300
((PROPYLTHIO)-5-1H-BENZIMIDAZOLYL-2) CARBAMATE de METHYLE (FRENCH) see VAD000
(5-(PROPYLTHIO)-1H-BENZIMIDAZOL-2-YL)CARBAMIC ACID METHYL ESTER see VAD000
2-PROPYL-4-THIOCARBAMOYLPYRIDINE see PNW750
5-(PROPYLTHIO)-2-CARBOMETHOXYAMINOBENZIMIDAZOLE see VAD000
PROPYL THIOCYANATOACETATE see TFF150
2-PROPYL-THIOISONICOTINAMIDE see PNW750
PROPYLTHIOL see PML500
N-PROPYLTHIOL see PML500
6-(PROPYLTHIO)PURINE see PNW800
6-PROPYL-2-THIO-2,4(1H,3H)PYRIMIDINEDIONE sce PNX000
PROPYL THIOPYROPHOSPHATE see TED500
PROPYL-THIORIST see PNX000
1-(PROPYLTHIO)-2,3,6-TRIMETHYL-6-HEPTEN-4-YN-3-OL see TME260
PROPYLTHIOURACIL see PNX000
4-PROPYL-2-THIOURACIL see PNX000
6-PROPYL-2-THIOURACIL see PNX000
6-N-PROPYLTHIOURACIL see PNX000
6-N-PROPYL-2-THIOURACIL see PNX000
PROPYLTHIOURACIL and IODINE see PNX100
PROPYL-THYRACIL see PNX000
1-(N-PROPYL-N-(2-(2,4,6-TRICHLOROPHENOXY)ETHYL)CARBAMOYL)IMIDAZOLE see IAL200
N-PROPYL-N-(2-(2,4,6-TRICHLOROPHENOXY)ETHYL)-1H-IMIDAZOLE-1-CARBOXAMIDE see IAL200
n-PROPYLTRICHLOROSILANE see PNX250
PROPYLTRICHLOROSILANE (DOT) see PNX250
n-PROPYL-3,4,5-TRIHYDROXYBENZOATE see PNM750
PROPYLTRIIODOGERMANE see TKR050
5-PROPYL-4-(2,5,8-TRIOXA-DODECYL)-1,3-BENZODIOXOL (GERMAN) see PIX250
4-PROPYL-2,6,7-TRIOXA-1-STIBABICYCLO(2.2.2)OCTANE see PNX500
PROPYLUREA see PNX550
1-PROPYLUREA see PNX550
N-PROPYLUREA see PNX550
1-PROPYLUREA and SODIUM NITRITE see SIT500
n-PROPYLUREA and SODIUM NITRITE see SIT500
PROPYL URETHANE see PNG250
2-PROPYLVALERAMIDE see PNX600
PROPYL-2-VALERAMIDE see PNX600
α-PROPYLVALERAMIDE see PNX600
2-PROPYLVALERIC ACID see PNR750
2-PROPYLVALERIC ACID CALCIUM SALT (2:1) see CAY675
2-PROPYLVALERIC ACID CALCIUM SALT (2:1) see VCK200
2-PROPYLVALERIC ACID SODIUM SALT see PNX750
1-(2-PROPYLVALERYL)PIPERIDINE see PNX800
2-PROPYNAL (9CI) see PMT250
2-PROPYN-1-AMINE see POA000
2-PROPYN-1-AMINE, N-NITROSO-N-2-PROPYNYL- see NKB600
2-PROPYNENITRILE, 3-PHENYL- see PGD100
1-PROPYNE-3-OL see PMN450

PROPYNE (OSHA) see MFX590
3-PROPYNETHIOL see PNY275
PROPYNOIC ACID see PMT275
2-PROPYNOIC ACID see PMT275
3-PROPYNOL see PMN450
2-PROPYN-1-OL see PMN450
2-PROPYN-1-THIOL see PNY750
2-PROPYNYL ALCOHOL see PMN450
2-PROPYNYLAMINE see POA000
2-PROPYNYLAMINE, 1,1-DIMETHYL- see MHX200
1-PROPYNYL COPPER(I) see POA100
PROPYNYLCYCLOHEXANOL CARBAMATE see POA250
1-(2-PROPYNYL)CYCLOHEXANOL CARBAMATE see POA250
1-(2-PROPYNYL)CYCLOHEXYL CARBAMATE see POA250
2-PROPYNYL ETHER see POA500
1-(5-(2-PROPYNYL)-3-FURANYL)-2-PROPYNYL α-ETHYLBENZENEACETATE see PMN600
m-(2-PROPYNYLOXY)PHENYL ESTER METHYLCARBAMIC ACID see PMN250
2-(2-PROPYNYLOXY)PHENYL METHYLCARBAMATE see MIB500
3-(2-PROPYNYLOXY)PHENYL-N-METHYLCARBAMATE see PMN250
PROP-2-YNYL-3,7,11-TRIMETHYL-2,4-DODECADIENOATE see POB000
2-PROPYNYL(2E,4E)-3,7,11-TRIMETHYL-2,4-DODECADIENOATE see POB000
2-PROPYNYL VINYL SULFIDE see POB250
PROPYON see PMY300
PROPYPERONE see FLN000
PROPYPHENAZONE see INY000
N-3'-a-PROPYPHENAZONYL-2-ACETOXYBENZAMIDE see PNR800
PROPYTHIOURACIL see PNX000
PROPYZAMIDE see DTT600
PROQUANIL see MQU750
PROQUAZONE see POB300
P. ROQUEFORTI TOXIN see PAQ875
PRORALONE-MOP see XDJ000
PRORESIDOR see BER500
PROREX see DQA400
PROREX see PMI750
PROSAPOGENIN CP3B see HAK075
PROSCILLAN see POB500
PROSCILLARIDIN see POB500
PROSCOMIDE see SBH500
PROSEPTINE see SNM500
PROSEPTOL see SNM500
PROSERIN see DQY909
PROSERINE see POD000
PROSERINE BROMIDE see POD000
PROSERINE METHYL SULFATE see DQY909
PROSEROUT see AAE500
PROSERYL see DPE000
PROSEVOR 85 see CBM750
PROSIL 248 see TNJ500
PRO-SONIL see TDA500
PROSPASMIN see PGP500
PROSPASMINE see PGP500
PROSPASMINE HYDROCHLORIDE see PGP500
PROSPIDIN see POC000
PROSPIDINE see POC000
PROSPIDIUM CHLORIDE see POC000
PROSTAGLANDIN A1 see POC250
PROSTAGLANDIN A1 see POC250
PROSTAGLANDIN A2 see MCA025
(+)-PROSTAGLANDIN A2 see MCA025
PROSTAGLANDIN D2 see POC275
PROSTAGLANDIN E1 see POC350
PROSTAGLANDIN E2 see DVJ200
(−)-PROSTAGLANDIN E2 see DVJ200
PROSTAGLANDIN E1-217 see POC250
(15S)-PROSTAGLANDIN E2 see DVJ200
PROSTAGLANDIN E2 SODIUM SALT see POC360

PROSTAGLANDIN F1-α see POC400
PROSTAGLANDIN F2-α see POC500
dl-PROSTAGLANDIN F2-α see POC525
PROSTAGLANDIN F2-α, racemic mixture see POC525
PROSTAGLANDIN F2-α METHYL ESTER see DVJ100
PROSTAGLANDIN F2-α-THAM see POC750
PROSTAGLANDIN F2-α THAM SALT see POC750
PROSTAGLANDIN F2a TROMETHAMINE see POC750
PROSTALMON F see POC500
PROSTANDIN see POC350
PROSTAPHILIN see MNV250
PROSTAPHILIN A see SLJ050
PROSTAPHLIN-A see SLJ000
PROSTAPHLYN see DSQ800
PROSTARMON F see POC500
PROSTEARIN see SLL000
PROST-13-EN-1-OIC ACID, 11,16-DIHYDROXY-16-METHYL-9-OXO-, METHYL ESTER, (11-α-13E)-(±)- see MJE775
PROSTETIN see ELF100
PROSTIGMIN see NCL100
PROSTIGMIN BROMIDE see POD000
PROSTIGMINE see NCL100
PROSTIGMINE BROMIDE see POD000
PROSTIGMINE METHYLSULFATE see DQY909
PROSTIN see CCC100
PROSTIN E2 see DVJ200
PROSTIN F2-α see POC500
PROSTIN VR see POC350
PROSTOGLANDIN E2-METHYLHESPERIDIN COMPLEX see KHK100
PROSTOSIN see POB500
PROSTRUMYL see MPW500
PROSULFOCARB see PFR130
PROSULTHIAMINE see DXO300
PROSULTIAMINE see DXO300
PROSYKLIDIN see CPQ250
PROSZIN see POB500
PROTABEN P see HNU500
PROTABOL see TFK300
PROTACELL 8 see SEH000
PROTACHEM 630 see PKF000
PROTACHEM GMS see OAV000
PROTACHEM SMP see MRJ800
PROTACHEM SOC see SKV170
PROTACINE see POF550
PROTACTINIUM see POD500
PROTACTYL see DQA600
PROTAGENT see PKQ250
(−)-PROTAGLANDIN E1 see POC350
PROTAGLANDIN F1 see POC400
PROTALBA see POF000
PROTALBINIC ACID see AFI780
PROTAMINE SULFATE see POD750
PROTAMINE ZINC INSULIN see IDF325
PROTAMINE ZINC INSULIN INJECTION see IDF325
PROTAMINE ZINC INSULIN SUSPENSION see IDF325
PROTANABOL see PAN100
PROTANAL see SEH000
PROTANDREN see AOO475
PROTARS see CAM300
PROTASIN see POB500
PROTASORB L-20 see PKG000
PROTASORB O-20 see PKL100
PROTATEK see SEH000
PROTAZINE see DQA400
PROTAZINE see PMI750
PROTEASE, BACILLUS SUBTILIS NEUTRAL see BAC020
PROTECT see NAQ000
PROTECTON see IBQ100
PROTECTONA see DKA600
PROTEINA see DME500

PROTEINASE, ASPERFILLUS ALKALINE see SBI860
PROTEINASE, ASPERGILLUS TERRICOLA NEUTRAL see TBF350
PROTEINASE INHIBITOR E 64 see TFK255
PROTEINASE, METALLO- see GIA050
PROTEIN HYDROLYSATE FROM COLLAGEN see HID100
PROTEIN HYDROLYZATES, COLLAGEN see HID100
PROTEINS, COLLAGEN, HYDROLYSATE see HID100
PROTEK Q see QAT520
PROTELINE see PMJ100
PROTENAZA 1 see PMJ100
PROTERGURIDE see DJX300
PROTERNOL see DMV600
PROTESINE DMU see DTG700
PROTEX (POLYMER) see AAX250
PROTHAZIN see DQA400
PROTHAZIN METHOSULFATE see MRW000
PROTHEOBROMINE see HNY500
PROTHEOPHYLLINE see DNC000
PROTHIADEN see DYC875
PROTHIADENE HYDROCHLORIDE see DPY200
cis-PROTHIADENE-S-OXIDE HYDROGEN MALEATE see POD800
PROTHIADEN HYDROCHLORIDE see DPY200
PROTHIADEN SPOFA see DYC875
PROTHIDIUM see PPQ000
PROTHIL see MBZ100
PROTHIOCARB see EIH500
PROTHIOFOS-OXON see EEE200
PROTHIONAMIDE see PNW750
PROTHIOPHOS see DGC800
PROTHIPENDYL see DYB600
PROTHIPENDYL HYDROCHLORIDE see DYB800
PROTHIUCIL see PNX000
PROTHIURONE see PNX000
PROTHIZINIC ACID see POE100
PROTHOATE see IOT000
PROTHRIN see POD875
PROTHROMADIN see WAT200
PROTHROMBIN see WAT220
PROTHYCIL see PNX000
PROTHYRAN see PNX000
PROTIOAMPHETAMINE see AOA250
PROTION see PNW750
PROTIONAMID see PNW750
PROTIONAMIDE see PNW750
PROTIONIZINA see PNW750
PROTIRELIN see TNX400
PROTIRELIN TARTRATE see POE050
PROTIURAL see PNX000
PROTIVAR see AOO125
PROTIZINIC ACID see POE100
PROTOAT (HUNGARIAN) see IOT000
PROTOBOLIN see DAL300
PROTOCATECHUALDEHYDE DIMETHYL ETHER see VHK000
PROTOCATECHUIC ACID see POE200
PROTOCATECHUIC ACID, 3-METHYL ESTER see VFF000
PROTOCATECHUIC ALDEHYDE DIMETHYL ETHER see VHK000
PROTOCATECHUIC ALDEHYDE ETHYL ETHER see EQF000
PROTOCATECHUIC ALDEHYDE METHYLENE ETHER see PIW250
PROTOCHLORURE d'IODE (FRENCH) see IDS000
PROTOCOL C see DTG000
PROTOMIN see AOO800
PROTONA see DME500
PROTOPAM CHLORIDE see FNZ000
PROTOPAM IODIDE see POS750
PROTOPET see MQV750
PROTOPHENICOL see CDP500
PROTOPINE see FOW000

PROTOPOLIGONATOZID G see POE300
PROTOPOLYGONATOSIDE G see POE300
PROTOPORPHYRIN DISODIUM see DXF700
PROTOPORPHYRIN SODIUM see DXF700
PROTOPORPHYRIN SODIUM SALT see DXF700
PROTOPYRIN see EEM000
PROTOTYPE III SOFT see PKQ059
PROTOVERATRIN see POF000
PROTOVERATRINE see POF000
PROTOVERATRINE A see POF000
PROTOVERATRINE B see NCI600
PROTOX TYPE 166 see ZKA000
PROTOXYL see ARA500
PROTRIPTYLINE see DDA600
PROTRIPTYLINE HYDROCHLORIDE see POF250
PROTRYPTYLINE see DDA600
PROVADO see CKW400
PROVAMYCIN see SLC000
PROVASAN see EQQ100
PROVENTIL see BQF500
PROVEST see POF275
PROVIGAN see DQA400
PROVITAMIN D see CMD750
PROVITAMIN D₃ see DAK600
PROVITAR see AOO125
PROWL see DRN200
PROXAGESIC see DAB879
PROXAGESIC see PNA500
PROXAN SODIUM see SIA000
PROXAZOLE CITRATE see POF500
PROX DW see DTG000
PROXEL PL see BCE475
PROXEN see MFA500
PROXIFEINE see POF525
PROXIFEN see POF525
PROXIL see POF550
PROXIMPHAM (GERMAN) see PEQ500
PROXIPHYLLINE see HOA000
PROXITANE 4002 see PCL500
PROX M 3R see MCB050
PROXOL see TIQ250
PROX RX see TCJ900
PROXYPHYLLINE see HOA000
PROZIL see CKP250
PROZIME 10 see SBI860
PROZIN see CKP250
PROZINEX see PMN850
PROZOIN see PMU750
PROZORIN see PLY275
PRS 640 see BML500
PRT see POE100
PRT see POF800
PR TOXIN see PAQ875
PR TOXIN see POF800
PR TOXINE see POF800
PRULET see PDO750
PRUMYCIN see AFI500
PRUNACETIN A see POG000
PRUNE see CAR790
PRUNETOL see GCM350
PRUNICYANIN see KEA325
PRUNIT see SOU800
PRUNOLIDE see CNF250
PRUNUS (VARIOUS SPECIES) see AQP890
PRURALGAN see DNX400
PRURALGIN see DNX400
PRUSSIAN BLUE see IGY000
PRUSSIAN BROWN see IHC450
PRUSSIC ACID see HHS000
PRUSSIC ACID, UNSTABILIZED see HHS000
PRUSSITE see COO000
200U/P-RVM see PAE750
PRX 1195 see SMQ500
PRYNACHLOR see CDS275
PRYSKYRICE MH see MCB050
PRYSOLINE see DBB200
PRZEDZIORKOFOS (POLISH) see PDC750
PS see CKN500
PS 1 see AHE250

PS 1 see SMQ500
PS 2 see SMQ500
PS 089 see PJQ790
PS 200 see SMQ500
PS 209 see SMQ500
PS 802 see DXG700
PS 925 see TCB600
PS 2383 see TKX250
PS 454H see SMQ500
PS 100 (carbonate) see CAT775
PSB see SAN300
PS-B see SMQ500
PSB-C see SMQ500
PSB-S-E see SMQ500
PSB-S see SMQ500
PSB-S 40 see SMQ500
PSC-801 see POB500
PSC CO-OP WEEVIL BAIT see DXE000
PSEUDECHIS AUSTRALIS VENOM see ARV000
PSEUDECHIS PORPHYRIACUS (AUSTRALIA) VENOM see ARV250
PSEUDECHIS PORPHYRIACUS VENOM see ARV250
PSEUDOACETIC ACID see PMU750
PSEUDOACONITINE see POG250
PSEUDOBUFARENOGIN see POG275
PSEUDOBUTYLBENZENE see BQJ250
PSEUDOBUTYLENE see BOW255
PSEUDO-BUTYLENE see BOW500
PSEUDOCEF see CCS550
PSEUDOCUMENE see TLL750
PSEUDOCUMIDINE see TLG250
PSEUDOCUMIDINE HYDROCHLORIDE see TLG750
PSEUDOCUMOHYDROQUINONE see POG300
PSEUDOCUMOL see TLL750
p-PSEUDOCUMOQUINONE see POG400
PSEUDOCYANURIC ACID see THS000
PSEUDODIGITOXIN see GEU000
PSEUDOEPHEDRINE see POH000
l-(+)-PSEUDOEPHEDRINE see POH000
(−)-PSEUDOEPHEDRINE HYDROCHLORIDE see POH500
1-PSEUDOEPHEDRINE HYDROCHLORIDE see POH500
d-PSEUDOEPHEDRINE HYDROCHLORIDE see POH250
l(+)-PSEUDOEPHEDRINE HYDROCHLORIDE see POH250
PSEUDOGINSENOSIDE D see PAF450
PSEUDOHEXYL ALCOHOL see EGW000
3H-PSEUDOINDOLIUM, 2-(2-(2,4-DIMETHOXYANILINO)VINYL)-1,3,3-TRIMETHYL-CHLORIDE see CMM890
PSEUDOIONONE see POH525
PSEUDO-α-ISOMETHYL IONONE see TMJ100
PSEUDOLARIC ACID A see POH550
PSEUDOLARIC ACID B see POH600
PSEUDOMETHYLIONONE see DRR700
PSEUDOMONAS AERUGINOSA ENDOTOXIN see POH620
PSEUDOMONAS AERUGINOSA EXOTOXIN see POH630
PSEUDOMONAS AERUGINOSA PHOSPHOLIPASE C see POH640
PSEUDOMONAS AERUGINOSA TOXIN see POH650
PSEUDOMONAS sp. No. 14 POLYSACCHARIDE see POH670
PSEUDOMONAS sp. No. 16 POLYSACCHARIDE see POH680
PSEUDOMONAS POLYSACCHARIDE see PJA200
PSEUDOMONAS PSEUDOMALLEI EXOTOXIN see POH690
PSEUDOMONAS SYRINGAE, ICE-NUCLEATION-ACTIVE, Strain 3/a see IAC100
PSEUDOMONIL see CCS550

PSEUDONAJA TEXTILIS VENOM see TEG650
PSEUDONAJATOXIN B see POH700
PSEUDONOREPHEDRINE see NNM510
PSEUDOPINEN see POH750
PSEUDOPINENE see POH750
PSEUDOTHEOPHYLLINE see TEP000
PSEUDO-THIOHYDANTOIN see IAP000
PSEUDOTHIOUREA see ISR000
PSEUDOTHYMINE see MQI500
PSEUDOTROPINE BENZOATE HYDROCHLORIDE see TNS200
PSEUDOUREA see USS000
PSEUDOUREA, 2-AMINOETHYL-2-THIO-, DIACETATE see AJY300
PSEUDOUREA, 2-((2-AMINO-4-THIAZOLYL)METHYL)-2-THIO-,DIHYDROCHLORIDE see AMS800
PSEUDOUREA, 2-(p-CHLOROBENZYL)-2-THIO-, MONOHYDROCHLORIDE see CEQ800
PSEUDOUREA, 2-DECYL-2-THIO-, MONOHYDROCHLORIDE see DAJ480
PSEUDOUREA, 1,3-DIMETHYL-2-TETRADECYL-2-THIO-, HYDRIODIDE see DUE700
PSEUDOUREA, 2-(1,3-DIOXO-2-ISOINDOLINYL)ETHYLTHIO-,HYDROBROMIDE see DVR500
PSEUDOUREA, 2,2'-(IMINODIETHYLENE)BIS(2-THIO-, DIHYDROBROMIDE see IBJ050
PSEUDOUREA, 2-METHYL-2-THIO-, SULFATE see MPV790
PSEUDOUREA, 2-METHYL-2-THIO-, SULFATE (1:1) see MPV750
PSEUDOUREA, 2-THIO-, 2-AMINOETHYL-, DIACETATE see AJY300
PSEUDOUREA, 2-THIO-, 4-(2-AMINOTHIAZOLYL)METHYL-, DICHLORIDE see AMS800
PSEUDOXANTHINE see XCA000
PSI-BUFARENOGIN see POG275
PSICAINE-NEU HYDROCHLORIDE see NCJ000
PSICAIN-NEW HYDROCHLORIDE see NCJ000
PSICHIAL see MDQ250
PSICODISTEN see BET000
PSICOPAX see CFC250
PSICOPAX see CFZ000
PSICOPERIDOL-R see TKK500
PSICOPLEGIL see MNM500
PSICOSAN see LFK000
PSICOSAN see MDQ250
PSICOSEDINA see MNM500
PSICOSTEN see PDN000
PSICOSTERONE see AOO450
PSICRONIZER see NMV725
PSIDIUM GUAJAVA see GLW000
PSIDIUM GUAJAVA Linn., extract excluding roots see POH800
PSILOCINE see HKE000
PSILOCIN PHOSPHATE ESTER see PHU500
PSILOCIPIN see PHU500
PSILOTSIBIN see PHU500
PSILOTSIN see HKE000
PSIQUIM see CGA000
PSL see LEN000
PSML see PKG000
PSORADERM see MFN275
PSORALEN see FQD000
PSORIACID-STIFT see DMG900
PSP see PDO800
PSP 204 see EPH500
PSP (INDICATOR) see PDO800
PS 2 (POLYMER) see SMQ500
PS 5 (POLYMER) see SMQ500
PSV-L see SMQ500
PSV-L 1 see SMQ500
PSV-L 2 see SMQ500
PSV-L 1S see SMQ500

PSYCHAMINE A 66 see PMA750
PSYCHAMINE A 66 HYDROCHLORIDE see MNV750
PSYCHEDRINE see BBK000
PSYCHEDRINUM see AOB250
PSYCHEDRYNA see AOB250
PSYCHODRINE see BBK500
PSYCHOLIQUID see TEH500
PSYCHOPERIDOL see TKK500
PSYCHO-SOMA see MAG050
PSYCHOTABLETS see TEH500
PSYCHOZINE see CKP500
PSYCOPERIDOL HYDROCHLORIDE see TKK750
PSYMOD see PII500
PSYTOMIN see PCK500
PT-1 see TBX300
PT-2 see FAS100
PT 01 see OGI300
PT 155 see DAD040
PT 515 see PDR700
P 2020T see PJS750
P 2050T see PJS750
P 4007T see PJS750
PTAB see TNI500
PTAP see AON000
PTAQUILOSIDE see POI100
PTB see PJG000
PTB see TMV500
PTC see PCC000
PTC see PGN250
PTEGLU see FMT000
4(1H)-PTERIDINONE, 2-AMINO-6-(1,2-DIHYDROXYPROPYL)-, (S-(R*,S*))-(9CI) see BGD075
4(3H)-PTERIDINONE, 2-AMINO-6-(l-ERYTHRO-1,2-DIHYDROXYPROPYL)- see BGD075
PTERIDIUM AQUILINUM see BML000
PTERIDIUM AQUILINUM TANNIN see BML250
PTERIN H B2 see BGD075
PTERIS AQUALINA see BML000
PTEROFEN see UVJ450
PTEROPHENE see UVJ450
PTEROYLGLUTAMIC ACID see FMT000
PTEROYL-l-GLUTAMIC ACID see FMT000
PTEROYLMONOGLUTAMIC ACID see FMT000
PTEROYL-l-MONOGLUTAMIC ACID see FMT000
PTFE see TAI250
d-(PTF⁶))LHRH ACETATE see LIU349
PTG see EQP000
PTG 100 see GHY100
PTG 200 see GHY100
PTG 300 see GHY100
PTG 400 see GHY100
PTG 500 see GHY100
PTG 500P see GHY100
PTH see PAK250
PTMG see GHY100
PTMG 1000 see GHY100
PTMG 2000 see GHY100
PTMS see AKQ000
PTMSA see AKQ000
PTO see HOB500
PTR-E 3 see EHP700
PTR-E 40 see EHP700
PTS 2 see PJS750
PTT 119 see POI550
PTU see PGN250
PTU (thyreostatic) see PNX000
PT-URACIL see PJG125
PTX see PAE875
PTYCHODISCUS BREVIS TOXIN see BMM150
PU 603 see CJN250
P 4007EU see PJS750
PUA-HOKU (HAWAII) see SLJ650
PUA KALAUNU (HAWAII) see COD675
PUDDING-PIPE TREE see GIW300

PUD (HERBICIDE) see DTP400
PUERARIA PHASEOLOIDES (Roxb.) Benth., extract excluding roots see POI575
PUFFER POISON, HYDROCHLORIDE see POI600
PUKE WEED see CCJ825
PUKIAWE-LEI (HAWAII) see RMK250
PULANOMYCIN see FBP300
PULARIN see HAQ550
PULEGON see POI615
PULEGONE see MCF500
PULEGONE see POI615
(+)-PULEGONE see POI615
d-PULEGONE see MCF500
d-PULEGONE see POI615
(+)-(R)-PULEGONE see POI615
(R)-(+)-PULEGONE see POI615
PULICARIA ANGUSTIFOLIA DC., extract see POI650
PULMICORT see BOM520
PULMOCLASE see CBR675
PULSAN see ICA000
PULSAN (PHARMACEUTICAL) see IBY650
PULSATILLA SAPONIN A see HAK075
PULSOTYL see FMS875
PUMA see FAQ200
PUNCH see THS860
PUNICA GRANATUM Linn., fruit skin see POI800
PUNICINE see PAO500
PURADIN see NGG500
PURAGEL see PCU360
PURALIN see TFS350
PURAPURIDINE see POJ000
PURASAN-SC-10 see ABU500
PURATIZED see TNI500
PURATIZED AGRICULTURAL SPRAY see TNI500
PURATIZEDAT AGRICULTURAL SPRAY see TNI500
PURATIZED B-2 see MDD500
PURATIZED N5E see TNI500
PURATRONIC CHROMIUM CHLORIDE see CMJ250
PURATRONIC CHROMIUM TRIOXIDE see CMK000
PURATURF see TNI500
PURATURF 10 see ABU500
PURECAL see CAT775
PURECALO see CAT775
PURE CHRYSOIDINE YBH see PEK000
PURE CHRYSOIDINE YD see PEK000
PURE EOSINE YY see BNH500
PURE LEMON CHROME L3GS see LCR000
PURE ORANGE CHROME M see LCS000
PURE ORANGE II S see CMM220
PURE QUARTZ see SCI500
PURE QUARTZ see SCJ500
PUREX see SFT000
PURE ZINC CHROME see ZFJ100
PURGA see PDO750
PURGEN see PDO750
PURGING BUCKTHORN see MBU825
PURGING FISTULA see GIW300
PURGOPHEN see PDO750
PURIFILIN see DNC000
PURIMETHOL see POK000
1H-PURIN-2-AMINE see AMH000
1H-PURIN-6-AMINE see AEH000
1H-PURIN-2-AMINE, 6-CHLORO- see CHK300
9H-PURIN-6-AMINE,2-CHLORO-9-(2-DEOXY-2-FLUORO-β-d-ARABINOFURANOSYL)- see CFJ200
1H-PURIN-6-AMINE, N-(PHENYLMETHYL)-(9CI) see BDX090
9H-PURIN-6-AMINE, 9-(PHENYLMETHYL)-(9CI) see BDX100
1H-PURIN-6-AMINE PHOSPHATE see POJ100
1H-PURIN-6-AMINE, SULFATE see AEH500
PURIN B see SHU500

PURINE see POJ250
7H-PURINE see POJ250
9H-PURINE see POJ250
9H-PURINE-9-ACETALDEHYDE, α-(1-FORMYL-2-HYDROXYETHOXY)-1,6-DIHYDRO-6-OXO-, (R,R)- see IDE050
9H-PURINE-9-ACETALDEHYDE, α-(1-FORMYL-2-HYDROXYETHOXY)-1,6-DIHYDRO-6-OXO-, (R,(R*,R*))- see IDE050
1H-PURINE-2-AMINE, 6-((1-METHYL-4-NITRO-1H-IMIDAZOL-5-YL)-THIO)- see AKY250
PURINE, 2-AMINO-6-CHLORO- see CHK300
9H-PURINE, 6-BENZYLAMINO-9-TETRAHYDROPYRAN-2-YL- see BEA100
9H-PURINE-9-BUTANOIC ACID, 6-AMINO-α-β-DIHYDROXY-, (R-(R*,R*))- (9CI) see POJ300
1H-PURINE-2,6-DIAMINE see POJ500
PURINE-2,6-DIOL see XCA000
9H-PURINE-2,6-DIOL see XCA000
2,6(1,3)-PURINEDION see XCA000
PURINE-2,6-(1H,3H)-DIONE see XCA000
1H-PURINE-2,6-DIONE, 3-(ACETYLOXY)-3,7-DIHYDRO- see HOP300
PURINE-6,8(1H,9H)-DIONE, 2-AMINO-9-β-d-RIBOFURANOSYL- see ONW150
1H-PURINE-2,6-DIONE, 3,7-DIHYDRO-1,7-DIMETHYL-(9CI) see PAK300
1H-PURINE-2,6-DIONE, 3,7-DIHYDRO-8-(3-(DIMETHYLAMINO)PROPOXY)-1,3,7-TRIMETHYL-, MONOHYDROCHLORIDE see POF525
1H-PURINE-2,6-DIONE, 3,7-DIHYDRO-1,3-DIMETHYL-7-(3-(METHYLPHENYLAMINO)PROPYL)- see DLI625
1H-PURINE-2,6-DIONE, 3,7-DIHYDRO-1,3-DIMETHYL-, SODIUM SALT (9CI) see SKH000
1H-PURINE-2,6-DIONE, 3,7-DIHYDRO-8-METHOXY-1,3,7-TRIMETHYL-(9CI) see MEF800
1H-PURINE-2,6-DIONE, 8-ETHOXY-3,7-DIHYDRO-1,3,7-TRIMETHYL-(9CI) see EEO000
PURINE-3-OXIDE see POJ750
PURINE, 6-(PENTYLTHIO)- see PBX900
PURINE RIBONUCLEOSIDE see RJF000
9-PURINE RIBONUCLEOSIDE see RJF000
PURINE RIBOSIDE see RJF000
PURINE-6-THIOL see POK000
6-PURINETHIOL see POK000
3H-PURINE-6-THIOL see POK000
9H-PURINE-6(1H)-THIONE, 2-AMINO-9-(2-DEOXY-β-d-ERYTHRO-PENTOFURANOSYL)- see TFJ200
6H-PURINE-6-THIONE, 2-AMINO-1,7-DIHYDRO- (9CI) see AMH250
6H-PURINE-6-THIONE, 1,7-DIHYDRO-, MONOHYDRATE (9CI) see MCQ100
PURINETHOL see POK000
1H-PURINE-2,6,8(3H)-TRIONE, 7,9-DIHYDRO- (9CI) see UVA400
1H-PURINE-2,6,8(3H)-TRIONE, 7,9-DIHYDRO-, MONOSODIUM SALT (9CI) see SKO575
9H-PURIN-6-OL see DMC000
9H-PURIN-2-OL, 6-AMINO-9-β-d-RIBOFURANOSYL- see IKS450
6(1H)-PURINONE see DMC000
PURIN-6(3H)-ONE see DMC000
6H-PURIN-6-ONE, 1,9-DIHYDRO-2-AMINO-9-(3,4-DIHYDROXYBUTYL)- see DMI700
6H-PURIN-6-ONE, 1,9-DIHYDRO-2-AMINO-9-(4-HYDROXY-3-(HYDROXYMETHYL)BUTYL)-, MONOSODIUM SALT see POK100
6H-PURIN-6-ONE, 1,9-DIHYDRO-2-AMINO-9-((2-HYDROXY-1-(HYDROXYMETHYL)ETHOXY)METHYL)-,MONOSOD IUM SALT see GBU200

PURIN-6-THIOL, MONOHYDRAT see MCQ100
6H-PURIN-6-THION, MONOHYDRAT see MCQ100
3-(7H-PURIN-6-YLTHIO)PROPIONIC ACID see CCF270
PURIVEL see MQR225
PUROCYCLINA see TBX000
PURODIGIN see DKL800
PUROMYCIN see AEI000
PUROMYCIN CHLOROHYDRATE see POK300
PUROMYCIN DIHYDROCHLORIDE see POK300
PUROMYCIN HYDROCHLORIDE see POK250
PUROMYCIN HYDROCHLORIDE see POK300
PUROPHYLLIN see HOA000
PUROSIN-TC see POB500
PUROSTROPHAN see OKS000
PUROVERINE see POF000
PURPLE 4R see FAG050
PURPLE ALLAMANDA see ROU450
PURPLE MINT PLANT EXTRACT see PCI750
PURPLE RED see FMU080
PURPUREA B see DAD650
PURPUREA GLYCOSIDE B see DAD650
PURPUREAGLYKOSID B (GERMAN) see DAD650
PURPURID see DKL800
PURPURIN see TKN750
PURPURIN 4B see DXO850
PURPURINE see TKN750
PURPURINE 4B see DXO850
PURPUROCATECHOL see TNV550
PURPUROGALLIN see TDD500
PURSERPINE see RDK000
PURTALC USP see TAB750
PUT 108 see AJM550
PUTCHUK see CNT350
PUTRESCIN see BOS000
PUTRESCINE see BOS000
PUTRESCINE DIHYDROCHLORIDE see POK325
PUTRESCINE HYDROCHLORIDE see POK325
PVA see AAF625
PVA see PKP750
PVA 008 see PKP750
PVBR see PKQ000
P.V. CARBACHOL see CBH250
P. V. CARPINE LIQUIFILM see PIF500
PVC CORDO see AAX175
PVC (MAK) see PKQ059
PV-FAST GREEN G see PJQ100
PV FAST ORANGE GRL see CMU820
PV FAST PINK E see DTV360
P. VIRIDICATUM TOXIN see PAQ900
PVK see PDT750
PVK see VNK200
PV-ORANGE G see CMS145
PVP 1 see PKQ500
PVP 2 see PKQ750
PVP 3 see PKR000
PVP 4 see PKR250
PVP 5 see PKR500
PVP 6 see PKR750
PVP 7 see PKS000
PVP 8T see PJS750
PVP (FCC) see PKQ250
PVP-IODINE see PKE250
PVPO see PKQ100
PVPP see PKQ150
PVP-VA see AAU300
PVP/VA COPOLYMER see AAU300
PVP-VA-E 735 see AAU300
PVP/VA-S 630 see AAU300
PVS 4 see PKP750
PVSK see PKS250
P. VULGARIS see PAM780
PWA 1013 see NCY135

PWE of H. INFLUENZAE see HAB710
PWP 8 see MCB050
PWP 15 see MCB050
PX 104 see DEH200
PX-138 see DVL600
PX-238 see AEO000
PX 404 see DEH600
PX 438 see BJS250
PX-538 see DVK800
PX-806 see FAB920
PXO see OMY850
PY 11 see DGJ175
PY 100 see PJS750
PY2763 see SMQ500
PYBUTHRIN see PIX250
PYCARIL see NCR040
PYCAZIDE see ILD000
PYDOXAL see PII100
PYDRIN see FAR100
PYDT see DJY000
PYELOKON-FR see AAN000
PYELOSIL see DNG400
PYKARYL see NCR040
PYKNOLEPSINUM see ENG500
PYLAPRON see ICC000
PYLOSTROPIN see MGR500
PYLUMBRIN see DNG400
PYMAFED see DBM800
PYMAFED see WAK000
PYMETROZINE see DLV850
PYNACOLYL METHYLFLUOROPHOSPHONATE see SKS500
PYNAMIN see AFR250
PYNAMIN-FORTE see AFR250
PYNDT see DTV200
PYNOSECT see BEP500
PYOCEFAL see CCS550
PYOKTANIN see AOR500
PYOLUTEORIN see POK400
PYOPEN see CBO250
PYOPENE see CBO250
PYOSTACINE see VRF000
PYOVERM see AOR500
PYQUITON see BGB400
N-N-PYR see NLP500
PYRA see WAK000
PYRABENZAMINE see POO750
PYRABITAL see AMK250
PYRACETAM see NNE400
PYRACETON see DNG400
PYRACLOFOS see POK500
PYRACRYL ORANGE Y see PEK000
PYRACRYMYCIN-1 see COV125
PYRADEX see DBI200
PYRADONE see DOT000
PYRAHEXYL see HGK500
PYRALCID see ALF250
PYRALENE see PJL750
PYRALENE 1476 see PAV600
PYRALIN see CCU250
PYRAMAL see WAK000
PYRA MALEATE see DBM800
PYRAMEM see NNE400
PYRAMIDON see DOT000
PYRAMIDONE see DOT000
PYRAMINE see PEE750
PYRAMIN RB see PEE750
PYRAMON see AMK250
PYRAN ALDEHYDE see ADR500
2H-PYRAN, 3,6-DIHYDRO-6-BUTYL-2,4-DIMETHYL- see GMG100
2H-PYRAN, 3,4-DIHYDRO-2-ETHOXY-4-METHYL- see EEV150
2H-PYRAN-2,6(3H)-DIONE, DIHYDRO-2-PHTHALIMIDO- see PHX100
2H-PYRAN, 2-ETHOXY-3,4-DIHYDRO-4-METHYL- see EEV150
2H-PYRAN, 2-ETHOXY-4-METHYL TETRAHYDRO- see EEX100
PYRANICA see CGH750
PYRANILAMINE MALEATE see DBM800

PYRANINYL see DBM800
PYRANISAMINE see WAK000
PYRANISAMINE MALEATE see DBM800
PYRAN, 2-(2-METHYL-1-PROPENYL)-4-
METHYLTETRAHYDRO- see RNU000
7H-PYRANO(2,3-C)ACRIDIN-7-ONE, 3,12-
DIHYDRO-6-METHOXY-3,3,12-
TRIMETHYL-ar-NITRO- see NEM100
PYRANO(3,4-E)BENZIMIDAZOL-7(3H)-
ONE, 2,6-DIAMINO-3,4-DIMETHYL- see
DBT500
PYRANOL see PJL750
PYRANOL 1499 see TIM100
2H-PYRAN-2-ONE, 6-
HEPTYLTETRAHYDRO- see HBP450
2H-PYRAN-2-ONE,6-(6-(HEXAHYDRO-3,4-
DIHYDROXY-3A,4-DIMETHYL-5-
ETHYLFURO(2,3-B)FURAN-2-YL)-1,3,5-
HEXATRIENYL)-4-METHOXY-5-METHYL-,
(2R-(2-α(1E,3E,5E), 3-α,3A-β, 4-β,5-β,6A-β))-
see ARP640
2H-PYRAN-2-ONE, 4-HYDROXY-6-
METHYL- see THL800
4H-PYRAN-4-ONE, TETRAHYDRO-2,6-
DIMETHYL- see TCQ350
2H-PYRAN-2-ONE, TETRAHYDRO-3-
METHYL- see MQK100
2H-PYRAN-2-ONE, TETRAHYDRO-6-
PROPYL- see ODE300
4H-PYRANO(3,2-c)QUINOLINE-2-
CARBOXYLIC ACID, 5,6-DIHYDRO-7,8-
DIMETHYL-4,5-DIOXO-, ISOPENTYL
ESTER see IHR200
1H-PYRANO(3,2-F)QUINOLIN-8(7H)-ONE,
2,3-DIHYDRO- see TDB300
1H-PYRANO(3,2-F)QUINOLIN-8(7H)-ONE,
2,3-DIHYDRO-2-HYDROXY- see HOI222
PYRANTEL PAMOATE see POK575
PYRANTEL TARTRATE see TCW750
2H-PYRAN, TETRAHYDRO-2-ETHOXY-4-
METHYL- see EEX100
2H-PYRAN, TETRAHYDROMETHYLENE-
see MJU400
PYRANTON see DBF750
PYRASEPTIC see EEQ600
PYRATHIAZINE see PAJ750
PYRATHIAZINE HYDROCHLORIDE see
POL000
PYRATHYN see DPJ400
PYRATHYN see TEO250
PYRAZALONE ORANGE NP 215 see
CMS145
PYRAZAPON see POL475
PYRAZIDOL see HDS225
PYRAZIDOLE see HDS225
PYRAZINAMIDE see POL500
PYRAZINE see POL490
PYRAZINEAMIDE see POL500
PYRAZINECARBOXAMIDE see POL500
PYRAZINE CARBOXYLAMIDE see POL500
PYRAZINE-2,3-DICARBOXYLIC ACID
IMIDE see POL750
PYRAZINE HEXAHYDRIDE see PIJ000
PYRAZINETHIOL, 3,6-DIMETHYL- see
DTU825
2(1H)-PYRAZINETHIONE, 3,6-DIMETHYL-
see DTU825
PYRAZINIUM, 3-(DIMETHYLAMINO)-1-
(2,2-DIPHOSPHONOETHYL)-, INNER SALT
see DOY500
PYRAZINO(2,1-A)PYRIDO(2,3-
C)(2)BENZAZEPINE, 1,2,3,4,10,14B-
HEXAHYDRO-2-METHYL- see HDS210
PYRAZINOIC ACID AMIDE see POL500
1H-PYRAZINO(3,2,1-
JK)CARBAZOLE,2,3,3A,4,5,6-HEXAHYDRO-
8-METHYL-, MONOHYDROCHLORIDE see
HDS225
PYRAZINOL-O-ESTER with O,O-DIETHYL
PHOSPHOROTHIOATE see EPC500
2-PYRAZINYLETHANETHIOL see POM000
PYRAZODINE see PDC250

PYRAZOFEN see PDC250
PYRAZOFEN see PEK250
1H-PYRAZOL-1-AMINE, N-((5-NITRO-2-
FURANYL)METHYLENE)- see NGE160
PYRAZOLANTHRONE see APK000
PYRAZOLATE see POM275
PYRAZOL BLUE 3B see CMO250
PYRAZOLE see POM500
PYRAZOLEANTHRONE see APK000
1H-PYRAZOLE-5-CARBOXAMIDE, 4-
CHLORO-N-((4-(1,1-
DIMETHYLETHYL)PHENYL)METHYL)-3-
ETHYL-1-METHYL- see CGH750
1H-PYRAZOLE-3-CARBOXAMIDE, 5-(4-
CHLOROPHENYL)-1-(2,4-
DICHLOROPHENYL)-4-METHYL-N-1-
PIPERIDINYL-, see CCC290
1H-PYRAZOLE-4-CARBOXAMIDE, 5-
CYANO-N-METHYL-1-PHENYL- see
COP752
PYRAZOLECARBOXYLIC ACID, 3-NITRO-
1-VINYL- see NMR300
PYRAZOLE, 3,5-DIMETHYL- see DTU850
PYRAZOLE, 3-METHYL- see MOX010
PYRAZOLE, 5-METHYL- see MOX010
PYRAZOLE, 4-METHYL-1-
((TRICHLOROMETHYL)THIO)- see TIR990
1H-PYRAZOLE-1-PENTANAMIDE, 3-
((IMINO((2,2,2-
TRIFLUOROETHYL)AMINO)METHYL)AMI
NO)- see IBM100
1H-PYRAZOLE-5-SULFONAMIDE, N-(((4,6-
DIMETHOXY-2-
PYRIMIDINYL)AMINO)CARBONYL)-1-
METHYL-4-(2-METHYL-2H-TETRAZOL-5-
YL)- see ASH275
PYRAZOLE, 1-
((TRICHLOROMETHYL)THIO)- see TIS100
PYRAZOL FAST BLACK GS see CMN240
PYRAZOL FAST BRILLIANT BLUE VP see
CMN750
PYRAZOL FAST RED 5BL see CMO885
PYRAZOL FAST RED 8BL see CMO885
PYRAZOLIDIN see BRF500
PYRAZOLINE RED 8BL see CMO885
3-PYRAZOLIN-5-ONE, 4-AMINO-2,3-
DIMETHYL-1-PHENYL- see AIB300
2-PYRAZOLIN-5-ONE, 3-(2,4-
DICHLOROANILINO)-1-(2,4,6-
TRICHLOROPHENYL)- see DEO625
2-PYRAZOLIN-5-ONE, 3-METHYL- see
MOX100
2-PYRAZOLIN-5-ONE, 3-METHYL-1-p-
TOLYL- see THA300
PYRAZOLOADENINE see POM600
1,9-PYRAZOLOANTHRONE see APK000
PYRAZOLO(2,3-a)IMIDAZOLIDINE see
IAN100
3H-PYRAZOL-3-ONE, 4-AMINO-1,2-
DIHYDRO-1,5-DIMETHYL-2-PHENYL- see
AIB300
5-PYRAZOLONE, 3-AMINO-1-PHENYL- see
DKR600
3H-PYRAZOL-3-ONE, 2,4-DIHYDRO-5-
AMINO-2-PHENYL- see DKR600
3H-PYRAZOL-3-ONE, 1,2-DIHYDRO-4-
IODO-1,5-DIMETHYL-2-PHENYL- see
IEU085
5-PYRAZOLONE, 3-METHYL-1-PHENYL-
see NNT000
PYRAZOLONE ORANGE see CMS145
PYRAZOLONE ORANGE YB 3 see CMS145
1H-PYRAZOLO(3,4-d)PYRIMIDIN-4-AMINE
(9CI) see POM600
4H-PYRAZOLO(3,4-d)PYRIMIDIN-4-ONE
see ZVJ000
PYRAZOLO(1,5-a)QUINOLINE, 2-(m-
(BENZYLOXY)PHENYL)- see BFC450
2H-PYRAZOLO(4,3-c)QUINOLINE, 3,3a,4,5-
TETRAHYDRO-2-ACETYL-8-METHOXY-5-
((4-METHYLPHENYL) SULFONYL)-3-
PHENYL-, cis- see POM700

2H-PYRAZOLO(4,3-c)QUINOLINE, 3,3a,4,5-
TETRAHYDRO-2-ACETYL-8-METHOXY-5-
((4-METHYLPHENYL) SULFONYL)-3-(4-
PYRIDINYL)-, trans- see POM710
2H-PYRAZOLO(4,3-c)QUINOLINE, 3,3a,4,5-
TETRAHYDRO-2-ACETYL-5-((4-
METHYLPHENYL)SULFONYL)-3-PHENYL-,
cis- see ACY700
4H-PYRAZOLO(3,4-D)(1,2,4)TRIAZOLO(1,5-
A)PYRIMIDIN-4-ONE, 1,8-DIHYDRO-8-(4-
BROMOPHENYL)- see DKV135
PYRAZON see PEE750
PYRAZONE see PEE750
PYRAZONL see PEE750
PYRAZOXYFEN see DES500
PYRBUTEROL HYDROCHLORIDE see
POM800
PYRCON see PQC500
PYREDAL see PDC250
PYREN (GERMAN) see PON250
1-PYRENAMINE see PON000
PYRENE see PON250
PYRENE, 4,5-DIHYDRO-2,7-DINITRO- see
DLJ100
PYRENE, 4,5-DIHYDRO-2-NITRO- see
NFW220
PYRENE, DINITRO- see DVD300
PYRENE, 2,7-DINITRO- see DVD900
1,6-PYRENEDIONE see PON340
1,8-PYRENEDIONE see PON350
3,8-PYRENEDIONE see PON350
3,10-PYRENEDIONE see PON350
1-PYRENEMETHANOL see PON400
2-PYRENEMETHANOL see HMQ550
1-PYRENEMETHANOL, α-METHYL- see
HKY700
2-PYRENEMETHANOL, α-METHYL- see
MOT000
PYRENE, NITRO- see NIY700
PYRENE, 1-NITRO-6-NITROSO- see NHS100
PYRENE, 1-NITRO-8-NITROSO- see NHS120
1,6-PYRENEQUINONE see PON340
1,8-PYRENEQUINONE see PON350
3,8-PYRENEQUINONE see PON340
3,10-PYRENEQUINONE see PON350
1-PYRENESULFONIC ACID see PON420
PYRENE-1-SULFONIC ACID see PON420
PYRENE-3-SULFONIC ACID see PON420
PYRENE, 4,5,9,10-TETRAHYDRO-2,7-
DINITRO- see TCQ400
PYRENE, 1,3,6,8-TETRANITRO- see TDY650
1-PYRENOL, 6-ACETAMIDO- see HOB050
3-PYRENOL, 1-ACETAMIDO- see HOB025
PYRENOLINE see PCV500
4-PYRENOL, 1-NITRO- see NJA200
PYRENONE 606 see PIX250
N-1-PYRENYLACETAMIDE see AAM650
N-PYREN-2-YLACETAMIDE see PON500
1-(1-PYRENYL)ETHANOL see HKY700
1-(2-PYRENYL)ETHANOL see MOT000
1-PYRENYLOXIRANE see ONG000
4-PYRENYLOXIRANE see PON750
PYREQUAN TARTRATE see TCW750
PYRESIN see AFR250
PYRESYN see AFR250
PYRETHERM see BEP500
PYRETHIA see DQA400
PYRETHIAZINE see DQA400
PYRETHRIN see POO000
PYRETHRIN see POO100
PYRETHRIN I see POO050
PYRETHRIN II see POO100
PYRETHRIN I or II see POO250
PYRETHRINS see POO250
PYRETHROLONE CHRYSANTHEMUM
DICARBOXLIC ACID METHYL ESTER see
POO100
PYRETHROLONE, CHRYSANTHEMUM
MONOCARBOXYLIC ACID ESTER see
POO050

PYRETHROLONE ESTER of CHRYSANTHEMUMDICARBOXYLIC ACID MONOMETHYL ESTER see POO100
PYRETHROL P see MRW080
(+)-PYRETHRONYL (+)-trans-CHRYSANTHEMATE see POO050
(+)-PYRETHRONYL (+)-PYRETHRATE see POO100
PYRETHRUM (ACGIH) see POO250
PYRETHRUM (insecticide) see POO250
PYRETRIN II see POO100
PYRIBENZAMINE see TMP750
PYRIBENZAMINE HYDROCHLORIDE see POO750
PYRIBUTICARB see POO800
PYRICAROYL see DJS200
PYRICIDIN see ILD000
PYRIDACIL see PDC250
PYRIDACIL see PEK250
PYRIDAFENTHION see POP000
PYRIDAMAL-100 see TAI500
PYRIDAPHENTHION see POP000
PYRIDATE see FAQ500
PYRIDAZINE see OJW200
3,3'-(3,6-PYRIDAZINEDIYLBIS(OXY)BIS(N,N-DIETHYL-N-METHYL-1-PROPANAMINIUM)) DIIODIDE (9CI) see BJA825
(3,6-PYRIDAZINEDIYLBIS(OXYTRIMETHYLENE))BIS(DIETHYLMETHYLAMMONIUM IODIDE) see BJA825
6H-PYRIDAZINO(1,2-A)(1,2)DIAZEPINE-1-CARBOXYLIC ACID,OCTAHYDRO-9-((1-(ETHOXYCARBONYL)-3-PHENYLPROPYL)AMINO)-10-OXO-, HYDRATE,(1S-(1-α,9-α(R*)))- see CMP810
3(2H)-PYRIDAZINONE, 2-(3-BROMO-4-CHLOROPHENYL)-4-CHLORO-5-((6-CHLORO-3-PYRIDINYL)METHOXY)- see BNA350
3(2H)-PYRIDAZINONE, 4-BROMO-2-(3,4-DICHLOROPHENYL)-5-((6-IODO-3-PYRIDINYL)METHOXY)- see BND775
3(2H)-PYRIDAZINONE, 2-(4-BROMOPHENYL)-4-CHLORO-5-((4-CHLOROPHENYL)METHOXY)- see BNV800
3(2H)-PYRIDAZINONE, 5-((4-BROMOPHENYL)METHOXY)-4-CHLORO-2-(4-CHLORO-2-FLUOROPHENYL)- see BNX035
3(2H)-PYRIDAZINONE, 5-((4-BROMOPHENYL)METHOXY)-4-CHLORO-2-(4-CHLOROPHENYL)- see BNX040
3(2H)-PYRIDAZINONE, 5-((6-BROMO-3-PYRIDINYL)METHOXY)-4-CHLORO-2-(4-CHLOROPHENYL)- see BOC600
3(2H)-PYRIDAZINONE, 5-((6-BROMO-3-PYRIDINYL)METHOXY)-4-CHLORO-2-(3,4-DICHLOROPHENYL)- see BOC650
3(2H)-PYRIDAZINONE, 4-CHLORO-2-(4-CHLORO-2-FLUOROPHENYL)-5-((4-CHLOROPHENYL)METHOXY)- see CFA800
3(2H)-PYRIDAZINONE, 4-CHLORO-2-(4-CHLOROPHENYL)-5-((4-CHLOROPHENYL)METHOXY)- see CFC050
3(2H)-PYRIDAZINONE, 4-CHLORO-2-(4-CHLOROPHENYL)-5-((6-IODO-3-PYRIDINYL)METHOXY)- see CFC600
3(2H)-PYRIDAZINONE, 4-CHLORO-2-(3,4-DICHLOROPHENYL)-5-((6-IODO-3-PYRIDINYL)METHOXY)- see CFL225
3(2H)-PYRIDAZINONE, 2-(4-CHLOROPHENYL)-5-((4-CHLOROPHENYL)METHOXY)-4-IODO- see CJT800
3(2H)-PYRIDAZINONE, 2-(4-CHLOROPHENYL)-5-((6-CHLORO-3-PYRIDINYL)METHOXY)-4-IODO- see CJU150

3(2H)-PYRIDAZINONE, 5-((4-CHLOROPHENYL)METHOXY)-2-(3,4-DICHLOROPHENYL)-4-IODO- see CKC600
3(2H)-PYRIDAZINONE, 5-((6-CHLORO-3-PYRIDINYL)METHOXY)-2-(3,4-DICHLOROPHENYL)-4-IODO- see CKW330
3(2H)-PYRIDAZINONE, 6-(3,5-DICHLORO-4-METHYLPHENYL)- see DFQ500
3(2H)-PYRIDAZINONE,4,5-DIHYDRO-6-(2-(4-METHOXYPHENYL)-1H-BENZIMIDAZOL-5-YL)-5-METHYL-, (−)- see PIG800
3(2H)-PYRIDAZINONE,4,5-DIHYDRO-6-(2-(4-METHOXYPHENYL)-1H-BENZIMIDAZOL-5-YL)-5-METHYL-, (+)- see PIG804
PYRIDENAL see PDC250
PYRIDENE see PDC250
PYRIDIATE see PDC250
PYRIDIN (GERMAN) see POP250
3-PYRIDINALDEHYDE see NDM100
2-PYRIDINALDOXIM METHOJODID (GERMAN) see POS750
PYRIDIN-2-ALDOXIN (CZECH) see POS750
2-PYRIDINAMIDE, N,N-DIMETHYL- see DQB400
4-PYRIDINAMIDE HYDROCHLORIDE see AMJ000
3-PYRIDINAMINE see AMI250
4-PYRIDINAMINE see AMI500
PYRIDINAMINE (9CI) see POP100
α-PYRIDINAMINE see AMI000
2-PYRIDINAMINE, 3-CHLORO-N-(3-CHLORO-2,6-DINITRO-4-(TRIFLUOROMETHYL)PHENYL)-5-(TRIFLUOROMETHYL)- see CEX800
3-PYRIDINAMINE HYDROCHLORIDE see AMI750
2-PYRIDINAMINE, 5-METHYL- see AMA010
4-PYRIDINAMINE, N-METHYL-N-NITROSO- see NKQ100
PYRIDIN-4-CARBONSAEURE-(DEXAMETHASON-21')-ESTER (GERMAN) see DBC510
PYRIDIN-2,5-DICARBONSAEURE-DI-N-PROPYLESTER (GERMAN) see EAU500
PYRIDINE see POP250
1(4H)-PYRIDINEACETIC ACID, 3,5-DIIODO-4-OXO-, COMPD. WITH 1-DEOXY-1-(METHYLAMINO)GLUCITOL see IFA100
1(4H)-PYRIDINEACETIC ACID, 3,5-DIIODO-4-OXO-, SALT WITH METHYLGLUCAMINE see IFA100
2-PYRIDINEACETONITRILE, α-(2-(BIS(1-METHYLETHYL)AMINO)ETHYL)-α-PHENYL-, 4-METHYLBENZENESULFONATE see NEH700
PYRIDINE, 3-ACETYL- see ABI000
PYRIDINE, 4-ACETYL- see ADA365
PYRIDINE, 5-(ACETYLAMINO)-1-(ACETYLOXY)-2-NAPHTHALENESULFONATE see AAI800
3-PYRIDINEALDEHYDE see NDM100
4-PYRIDINEALDEHYDE see IKZ100
p-PYRIDINEALDEHYDE see IKZ100
2-PYRIDINEALDOXIME CHLORIDE see FNZ000
PYRIDINE-2-ALDOXIME ETHANESULFONATE see POP300
2-PYRIDINE ALDOXIME IODOMETHYLATE see POS750
PYRIDINE-2-ALDOXIME METHIODIDE see POS750
PYRIDINE-2-ALDOXIME METHOCHLORIDE see FNZ000
2-PYRIDINE ALDOXIME METHYL CHLORIDE see FNZ000
PYRIDINE-2-ALDOXIME METHYL IODIDE see POS750
PYRIDINE-2-AMIDOXIME see PIB910

PYRIDINE, AMINO- see POP100
PYRIDINE-3-AZO-p-DIMETHYLANILINE see POP750
PYRIDINE-4-AZO-p-DIMETHYLANILINE see POQ000
PYRIDINE, 2-BENZOYL- see PGE760
PYRIDINE, 3-BENZOYL- see PGE765
PYRIDINE, 4-BENZOYL- see PGE768
PYRIDINE, 2-BENZYL- see BFG600
PYRIDINE BORANE see POQ250
PYRIDINE compounded with BORON HYDRIDE see POQ250
PYRIDINE, 2-BROMO- see BOB600
3-PYRIDINEBUTANAL, γ-OXO- see OOO300
PYRIDINE, 3-BUTYL- see BSJ550
PYRIDINE, 4-BUTYRYL- see PNV755
PYRIDINE-3-CARBALDEHYDE see NDM100
PYRIDINE-4-CARBALDEHYDE see IKZ100
PYRIDINE-2-CARBINOL see POR800
PYRIDINE-3-CARBINOL see NDW510
PYRIDINE-2-CARBO-o-ACETYLHYDROXAMIC ACID see POQ300
PYRIDINE-3-CARBO-o-ACETYLHYDROXAMIC ACID see POQ310
PYRIDINE-4-CARBO-o-ACETYLHYDROXAMIC ACID see POQ320
PYRIDINE-2-CARBO-N-METHYLHYDROXAMIC ACID see POQ330
β-PYRIDINECARBONALDEHYDE see NDM100
PYRIDINE-3-CARBONIC ACID see NCQ900
3-PYRIDINECARBONITRILE (9CI) see NDW515
3-PYRIDINECARBONITRILE, 1,2-DIHYDRO-4-METHOXY-1-METHYL-2-OXO- see RJK100
3-PYRIDINECARBOXALDEHYDE (9CI) see NDM100
4-PYRIDINECARBOXALDEHYDE (9CI) see IKZ100
2-PYRIDINECARBOXALDEHYDE, 2-PYRIDINYLHYDRAZINE see DWX100
2-PYRIDINECARBOXALDEHYDE THIOSEMICARBAZONE see PIB925
PYRIDINE-2-CARBOXALDEHYDE THIOSEMICARBAZONE see PIB925
PYRIDINE-4-CARBOXALDEHYDE THIOSEMICARBAZONE see ILB000
4-PYRIDINECARBOXAMIDE (9CI) see ILC100
γ-PYRIDINECARBOXAMIDE see ILC100
2-PYRIDINECARBOXAMIDE, N-(ACETYLOXY)- see POQ300
3-PYRIDINECARBOXAMIDE, N-(ACETYLOXY)- see POQ310
4-PYRIDINECARBOXAMIDE, N-(ACETYLOXY)- see POQ320
3-PYRIDINECARBOXAMIDE, 6-AMINO-N-HYDROXY- see ALL300
4-PYRIDINECARBOXAMIDE, N-(4-CHLORO-2-(HYDROXYPHENYLMETHYL)PHENYL)- see CHN100
3-PYRIDINECARBOXAMIDE, N-(2,4-DIFLUOROPHENYL)-2-(3-(TRIFLUOROMETHYL)PHENOXY)- see DKJ210
3-PYRIDINECARBOXAMIDE, 1,2-DIHYDRO-2-OXO-N-(2,6-XYLYL)- see XLS300
3-PYRIDINECARBOXAMIDE, N-(2,6-DIMETHYLPHENYL)-1,2-DIHYDRO-2-OXO-(9CI) see XLS300
2-PYRIDINECARBOXAMIDE, N-HYDROXY-N-METHYL- see POQ330
3-PYRIDINECARBOXAMIDE, N-(2-(1H-INDOL-3-YL)ETHYL)-(9CI) see NDW525
2-PYRIDINECARBOXIMIDAMIDE, N-HYDROXY- see PIB910
PYRIDINE-3-CARBOXYDIETHYLAMIDE see DJS200

PYRIDINE-3-CARBOXYLIC ACID see NCQ900

3-PYRIDINECARBOXYLIC ACID see NCQ900

4-PYRIDINECARBOXYLIC ACID see ILC000

o-PYRIDINECARBOXYLIC ACID see PIB930

2-PYRIDINECARBOXYLIC ACID (9CI) see PIB930

α-PYRIDINECARBOXYLIC ACID see PIB930

PYRIDINE-β-CARBOXYLIC ACID see NCQ900

4-PYRIDINECARBOXYLIC ACID-2-ACETYLHYDRAZIDE see ACO750

2-PYRIDINECARBOXYLIC ACID-2-ACETYLHYDRAZIDE (9CI) see ADA000

3-PYRIDINECARBOXYLIC ACID, ALUMINUM SALT see AHD650

PYRIDINE-3-CARBOXYLIC ACID AMIDE see NCR000

3-PYRIDINECARBOXYLIC ACID AMIDE see NCR000

2-PYRIDINECARBOXYLIC ACID, 4-AMINO-3,5,6-TRICHLORO-, MONOPOTASSIUM SALT (9CI) see PLQ760

PYRIDINE-4-CARBOXYLIC ACID-(DEXAMETHASONE-21') ESTER see DBC510

2-PYRIDINECARBOXYLIC ACID, 3,6-DICHLORO-(9CI) see DGJ100

PYRIDINE-3-CARBOXYLIC ACID DIETHYLAMIDE see DJS200

3-PYRIDINECARBOXYLIC ACID, 2-(DIFLUOROMETHYL)-5-(4,5-DIHYDRO-2-THIAZOLYL)-4-(2-METHYLPROPYL)-6-(TRIFLUOROMETHYL)-, METHYL ESTER see TEX550

PYRIDINE-3,5-CARBOXYLIC ACID, 1,4-DIHYDRO-2,6-DIMETHYL-4-(3-NITROPHENYL)-, 2-(4-(4-BENZHYDRYLPIPERAZIN-1-YL)PHENYL)ETHYL METHYL ESTER, (+−)-see WAT230

3,5-PYRIDINECARBOXYLIC ACID, 1,4-DIHYDRO-2,6-DIMETHYL-4-(3-((((3-SPIRO-(1H-INDENE-1,4-PIPERIDIN)-1-YLPROPYL)AMINO)CARBONYL)AMINO)PHENYL)-, DIMETHYL ESTER see DTU852

3-PYRIDINECARBOXYLIC ACID, 3-(2,6-DIMETHYLPIPERIDINO)PROPYL ESTER, HYDROCHLORIDE see NDU000

4-PYRIDINECARBOXYLIC ACID, ETHYL ESTER see ELU000

4-PYRIDINECARBOXYLIC ACID, HYDRAZIDE see ILD000

4-PYRIDINECARBOXYLIC ACID, HYDRAZIDE, HYDRAZONE WITH d-GLUCOSE (9CI) see GFG070

3-PYRIDINECARBOXYLIC ACID, (2-HYDROXY-1,3-CYCLOHEXANEDIYLIDENE)TETRAKIS(METHYLENE) ESTER see CME675

3-PYRIDINECARBOXYLIC ACID-2-HYDROXY-3-(2-METHOXYPHENOXY)PROPYL ESTER see HLT300

4-PYRIDINECARBOXYLIC ACID, ((4-HYDROXY-3-METHOXYPHENYL)METHYLENE)HYDRAZIDE see VEZ925

3-PYRIDINECARBOXYLIC ACID, 3-(2-METHYLPIPERIDINO)PROPYL ESTER, HYDROCHLORIDE see NDW000

4-PYRIDINECARBOXYLIC ACID 2-(3-OXO-3-((PHENYLMETHYL)AMINO)PROPYL)HYDRAZIDE see BET000

3-PYRIDINECARBOXYLIC ACID, PHENYLMETHYL ESTER see NCR040

3-PYRIDINECARBOXYLIC ACID, POTASSIUM SALT see PLL200

3-PYRIDINECARBOXYLIC ACID-1,2,5,6-TETRAHYDRO-1-METHYL ESTER, HYDROCHLORIDE see AQU250

PYRIDINE-CARBOXYLIQUE-2 (FRENCH) see PIB930

PYRIDINE-CARBOXYLIQUE-3 (FRENCH) see NCQ900

PYRIDINE, 2-(p-CHLORO-α-(2-(DIMETHYLAMINO)ETHYL)BENZYL)- see CLX300

PYRIDINE, 4-((4-CHLOROPHENYL)METHYL)- see CEQ900

PYRIDINE, 2-CHLORO-5-(TRIFLUOROMETHYL)- see CLS100

PYRIDINE, 4-(CYCLOHEXYLCARBONYL)-see CPI375

PYRIDINE, 2,6-DIACETAMIDO- see POR300

2,6-PYRIDINEDIAMINE, 3,5-DINITRO-N,N'-BIS(2,4,6-TRINITROPHENYL)- see PQC525

3-PYRIDINEDIAZONIUM TETRAFLUOROBORATE see POQ500

3,5-PYRIDINEDICARBOTHIOIC ACID, 2-(DIFLUOROMETHYL)-4-(2-METHYLPROPYL)-6-(TRIFLUOROMETHYL)-, S,S-DIMETHYL ESTER see POQ600

2,3-PYRIDINEDICARBOXIMIDE see POQ750

3,4-PYRIDINEDICARBOXIMIDE see POR000

2,3-PYRIDINEDICARBOXYLIC ACID see QOJ200

3,5-PYRIDINEDICARBOXYLIC ACID, BIS(1-METHYLETHYL) ESTER see IKC070

3,5-PYRIDINEDICARBOXYLIC ACID, 2-CYANO-1,4-DIHYDRO-6-METHYL-4-(3-NITROPHENYL)-, 3-METHYL-5-(1-METHYLETHYL) ESTER see NDY650

3,5-PYRIDINEDICARBOXYLIC ACID, 1,4-DIHYDRO-2-((2-AMINOETHOXY)METHYL)-4-(2-CHLOROPHENYL)-6-METHYL-, 3-ETHYL 5-METHYL ESTER see AMY100

3,5-PYRIDINEDICARBOXYLIC ACID, 1,4-DIHYDRO-2,6-DIMETHYL-4-(3-NITROPHENYL)-, ETHYL METHYL ESTER see EMR600

3,5-PYRIDINEDICARBOXYLIC ACID, 1,4-DIHYDRO-2,6-DIMETHYL-4-(3-NITROPHENYL)-, METHYL 1-(PHENYLMETHYL)-3-PIPERIDINYL ESTER, MONOHYDROCHLORIDE, (R*,R*)-(+−)- see BAV400

3,5-PYRIDINEDICARBOXYLIC ACID, 1,4-DIHYDRO-2,6-DIMETHYL-4-(3-NITROPHENYL)-,2-(4-(4-(DIPHENYLMETHYL)-1-PIPERAZINYL)PHENYL)ETHYL METHYL ESTER, DIHYDROCHLORIDE see WAT225

3,5-PYRIDINEDICARBOXYLIC ACID, 1,4-DIHYDRO-2,6-DIMETHYL-4-(3-NITROPHENYL)-, 2-(4-(4-(DIPHENYLMETHYL)-1-PIPERAZINYL)PHENYL)ETHYL METHYL ESTER, DIHYDROCHLORIDE, (S)-(+)- see NDY550

3,5-PYRIDINEDICARBOXYLIC ACID, 1,4-DIHYDRO-2,6-DIMETHYL-4-(2-PYRIDYL)-, DIETHYL ESTER see DJB460

2,5-PYRIDINEDICARBOXYLIC ACID, DIISOPROPYL ESTER see IKC070

3,5-PYRIDINEDICARBOXYLIC ACID, 2,6-DIMETHYL-4-(2-NITROPHENYL)-, MONOMETHYL ESTER see MRJ200

2,5-PYRIDINEDICARBOXYLIC ACID, DIPROPYL ESTER see EAU500

2,6-PYRIDINEDIMETHANOL, METHYL CARBAMATE NITROSOCARBAMATE see POR250

PYRIDINE, 2,4-DIMETHYL-(9CI) see LIY990

PYRIDINE, 2,6-DIMETHYL-(9CI) see LJA010

PYRIDINE, 2-DIMETHYLAMINO- see DQB400

PYRIDINE, 2-(α-(2-(DIMETHYLAMINO)ETHOXY)-α-

METHYLBENZYL)-, SUCCINATE (1:1) see PGE775

PYRIDINE, 2-(2-(DIMETHYLAMINO)ETHYL)-, MONOHYDROCHLORIDE see DPI710

PYRIDINE, 2-((2-(DIMETHYLAMINO)ETHYL)(SELENOPHENE-2-YLMETHYL)AMINO)- see DPI750

N,N'-2,6-PYRIDINEDIYLBISACETAMIDE see POR300

2-PYRIDINEETHANAMINE, N,N-DIMETHYL-, MONOHYDROCHLORIDE see DPI710

2-PYRIDINEETHANAMINE, N-METHYL-(9CI) see HGE820

4-PYRIDINEETHANOL see POR500

PYRIDINE, 2-ETHENYL-(9CI) see VQK560

PYRIDINE, 4-(p-ETHOXYBENZOYL)- see EEN600

PYRIDINE, 4-HEPTANOYL- see HBF600

PYRIDINE, HEXAHYDRO-N-NITROSO- see NLJ500

PYRIDINE, 4-HEXANOYL- see HEW050

PYRIDINE, 4-HEXANOYL- see PBX800

PYRIDINE HYDROCHLORIDE see POR750

PYRIDINE, 2-(p-HYDROXYBENZOYL)- see HJN700

PYRIDINE-2-METHANAMIDOXIME see PIB910

3-PYRIDINEMETHANAMINE, N-(2,4-DICHLOROPHENYL)-N-(4,5-DIHYDRO-2-THIAZOLYL)- see DGC050

2-PYRIDINEMETHANETHIOL see POR790

2-PYRIDINEMETHANOL see POR800

3-PYRIDINEMETHANOL see NDW510

4-PYRIDINEMETHANOL see POR810

3-PYRIDINEMETHANOL, 4-(((p-(BIS(2-CHLOROETHYL)AMINO)PHENYL)IMINO)METHYL)-5-HYDROXY-6-METHYL- see BHX275

3-PYRIDINEMETHANOL, α-(3-(METHYLNITROSOAMINO)PROPYL)- see MMS300

PYRIDINE METHIODIDE see MOZ000

PYRIDINE, 3-(4-METHOXYBENZOYL)- see MED100

PYRIDINE, 4-(4-METHOXYBENZOYL)- see MFH930

2-PYRIDINEMETHYLAMINE see ALB750

PYRIDINE, 2-(2-(METHYLAMINO)ETHYL)-see HGE820

PYRIDINE, 4-(METHYLNITROSAMINO)-see NKQ100

PYRIDINE, 3-METHYL-, 1-OXIDE see MOY550

PYRIDINE, 4-METHYL-, 1-OXIDE see MOY790

PYRIDINE, 3-(1-METHYL-2-PYRROLIDINYL)-, DIHYDROCHLORIDE, (S)- see NDN100

PYRIDINE, 3-(1-METHYL-2-PYRROLIDINYL)-, N-OXIDE, (2S)- see NDR100

PYRIDINE, 3-(1-METHYL-2-PYRROLIDINYL)-, (S)-, SULFATE (2:1) see NDR500

3-PYRIDINENITRILE see NDW515

PYRIDINE, 4-(p-NITROBENZYL)- see NFO750

PYRIDINE, 4-NITROSOMETHYLAMINO-see NKQ100

PYRIDINE, 3-(1-NITROSO-2-PIPERIDINYL)-, (+−)- see NJK200

PYRIDINE, 3-(1-NITROSO-2-PIPERIDINYL)-, (S)-(9CI) see NJK150

PYRIDINE, 3-(1-NITROSO-2-PYRROLIDINYL)-, (+−)- see NLP800

PYRIDINE, 3-(1-NITROSO-2-PYRROLIDINYL)-, (S)- see NLD500

PYRIDINE-1-OXIDE see POS000

PYRIDINE N-OXIDE see POS000

PYRIDINE-1-OXIDE-3-AZO-p-
DIMETHYLANILINE see POS250
PYRIDINE-1-OXIDE-4-AZO-p-
DIMETHYLANILINE see POS500
PYRIDINE PERCHROMATE see ONU000
PYRIDINE, 2-PHENYL- see PGE600
PYRIDINE, 4-(PHENYLAZO)- see PEK100
PYRIDINE, 2-(PHENYLMETHYL)-(9CI) see
BFG600
PYRIDINE, 3-(2-PIPERIDINYL)-,
MONOHYDROCHLORIDE, (S)- see PIT650
PYRIDINE, 3-(2-PIPERIDINYL)-, (S)-,
SULFATE (1:1) (9CI) see PIV550
2-PYRIDINEPROPANAMINE, γ-(4-
CHLOROPHENYL)-N,N-DIMETHYL-(9CI)
see CLX300
PYRIDINE, 4-PROPIONYL- see PMX300
2-PYRIDINESULFONAMIDE, N-(((4,6-
DIMETHOXY-2-
PYRIMIDINYL)AMINO)CARBONYL)-3-
(ETHYLSULFONYL)- see DOM700
2-PYRIDINESULFONAMIDE, N-(((4,6-
DIMETHOXY-2-
PYRIMIDINYL)AMINO)CARBONYL)-3-
(TRIFLUOROMETHYL)- see FCC050
PYRIDINE, 1,2,3,6-TETRAHYDRO-1-
(CYCLOHEXYLCARBONYL)- see CPI380
PYRIDINE, 1,2,3,6-TETRAHYDRO-1-((6-
METHYL-3-CYCLOHEXEN-1-
YL)CARBONYL)- see TCV375
PYRIDINE, 1,2,3,6-TETRAHYDRO-1-((2-
METHYLCYCLOHEXYL)CARBONYL)- see
TCV400
PYRIDINE, 1,2,3,6-TETRAHYDRO-1-(2-
METHYL-1-OXOPENTYL)- see AEY406
PYRIDINE, 1,2,3,6-TETRAHYDRO-1-
METHYL-4-PHENYL- see TCV470
PYRIDINE, 1,2,3,6-TETRAHYDRO-1-
METHYL-4-PHENYL-, HYDROCHLORIDE
see TCV500
PYRIDINE, 3-(TETRAHYDRO-1-
METHYLPYRROL-2-YL) see NDN000
PYRIDINE-2-THIOCARBINOL see POR790
2-PYRIDINETHIOL see TFP300
2-PYRIDINETHIOL, 1-OXIDE see MCQ700
2-PYRIDINETHIOL, 1-OXIDE, CU DERIV.
see BLG600
2-PYRIDINETHIOL-1-OXIDE SODIUM
SALT see MCQ750
2-PYRIDINETHIOL-1-OXIDE, ZINC SALT
see ZMJ000
2(1H)-PYRIDINETHIONE see TFP300
PYRIDINE, 2,3,6-TRICHLORO- see TJC900
PYRIDINE, 2,4,6-TRIMETHYL- see TME272
PYRIDINE, 4-VALERYL- see VBA100
PYRIDINE, 2-VINYL- see VQK560
PYRIDINE, 4-VINYL- see VQK590
PYRIDINIUM ALDOXIME
METHOCHLORIDE see FNZ000
PYRIDINIUM-2-ALDOXIME-N-
METHYLIODIDE see POS750
PYRIDINIUM, 2-AMINO-1-METHYL-, p-
TOLUENESULFONATE see ALC600
PYRIDINIUM, 3,3'-(2-
AMINOTEREPHTHALOYLBIS(IMINO-p-
PHENYLENECARBONYLIMINO))BIS(1-
PROPYL)-, DI-p-TOLUENE SULFONATE see
POT000
PYRIDINIUM, 3,3'-
(BICYCLO(2.2.2)OCTANE-1,4-
DIYLBIS(CARBONYLIMINO-4,1-
PHENYLENECARBONYLIMINO))BIS-1-
PROPYL-, salt with 4-
METHYLBENZENESULFONIC ACID (1:2)
see POT500
PYRIDINIUM, (4,4'-BIPHENYLYLENEBIS(2-
OXOETHYLENE))BIS(3-IODO-,
DIBROMIDE see BGM050
PYRIDINIUM, 1-BUTYL-3-(p-(p-((p-((1-
BUTYLPYRIDINIUM-3-
YL)CARBAMOYL)PHENYL)CARBAMOYL)C

INNAMAMIDO)BENZAMIDO)-,DI-p-
TOLUENE SULFONATE see POT750
PYRIDINIUM, 3-CARBAMOYL-1-β-D-
RIBOFURANOSYL-, HYDROXIDE, 5',5'-
ESTER with ADENOSINE 2'-
(DIHYDROGEN PHOSPHATE) 5'-
(TRIHYDROGEN PYROPHOSPHATE), inner
salt see CNF400
PYRIDINIUM, 3-CARBAMOYL-1-β-D-
RIBOFURANOSYL-, HYDROXIDE, 5'-
ESTER with ADENOSINE 5'-5'-
(TRIHYDROGEN PYROPHOSPHATE), inner
salt see CNF390
PYRIDINIUM, 3-CARBOXY-1-METHYL-,
HYDROXIDE, INNER SALT see TKL890
PYRIDINIUM, 3,3'-(2-
CHLOROTEREPHTHALOYLBIS(IMINO-p-
PHENYLENECARBONYLIMINO))BIS(1-
ETHYL)-, DI-p-TOLUENE SULFONATE see
POU000
PYRIDINIUM, 3,3'-(2-
CHLOROTEREPHTHALOYLBIS(IMINO-p-
PHENYLENECARBONYLIMINO))BIS(1-
METHYL-, DI-p-TOLUENE SULFONATE)
see POU250
PYRIDINIUM, 1,1'-(2,3-
DIHYDROXYTETRAMETHYLENE)BIS(4-
FORMYL-, DIIODIDE, DIOXIME see
BKF900
PYRIDINIUM, 1-DODECYL-, CHLORIDE
see DXY725
PYRIDINIUM, 1-DODECYL-, CHLORIDE
see LBX050
PYRIDINIUM, 1,1'-ETHYLENEDI-,
DIBROMIDE see EIT100
PYRIDINIUM, 1-ETHYL-3- p-(-((p-((1-
ETHYLPYRIDINIUM-3-
YL)CARBAMYL)PHENYLOCARBAMOYL)CI
NNAMAMIDO)BENZAMIDO)-, DI-p-
TOLUENE SULFONATE see POV250
PYRIDINIUM, 1-ETHYL-3-(p-(p-((p-(1-
ETHYLPYRIDINIUM-3-
YL)PHENYL)CARBAMOYL)CINNAMAMID
O)PHENYL)-DI-p-TOLUENE SULFONATE
see POV500
PYRIDINIUM, 2-FORMYL-1-METHYL-
,CHLORIDE, OXIME mixed with 1,1'-
TRIMETHYLENEBIS(4-
FORMYLPYRIDINIUM), DICHLORIDE,
DIOXIME (1:1) see POV750
PYRIDINIUM, 2-FORMYL-2'-METHYL-1,1'-
(OXYDIMETHYLENE)DI-, DICHLORIDE,
OXIME see FNW100
PYRIDINIUM, 2-FORMYL-4'-METHYL-1,1'-
(OXYDIMETHYLENE)DI-, DICHLORIDE,
OXIME see MBY500
PYRIDINIUM, 4-FORMYL-1-(3-
(TRIETHYLAMMONIO)PROPYL)-,
DIBROMIDE, OXIME see FOJ050
PYRIDINIUM, 1-HEXADECYL-, BROMIDE
see HCP800
PYRIDINIUM, 1,1'-
HEXAMETHYLENEBIS(2-FORMYL-,
DIBROMIDE, DIOXIME see HEG010
PYRIDINIUM, 1,1'-
HEXAMETHYLENEBIS(3-FORMYL-,
DIBROMIDE, DIOXIME see HEG020
PYRIDINIUM, 1,1'-
HEXAMETHYLENEBIS(4-FORMYL-,
DIBROMIDE, DIOXIME see HEE600
PYRIDINIUM, 2-
((HYDROXYAMINO)CARBONYL)-1-
METHYL-, IODIDE see HJS910
PYRIDINIUM, 3-
((HYDROXYAMINO)CARBONYL)-1-
METHYL-, IODIDE see HJS920
PYRIDINIUM, 4-
((HYDROXYAMINO)CARBONYL)-1-
METHYL-, IODIDE see HJS930
PYRIDINIUM, 2-HYDROXYCARBAMOYL-1-
METHYL-, IODIDE see HJS910

PYRIDINIUM, 3-HYDROXYCARBAMOYL-1-
METHYL-, IODIDE see HJS920
PYRIDINIUM, 4-HYDROXYCARBAMOYL-1-
METHYL-, IODIDE see HJS930
PYRIDINIUM, 1-(2-
HYDROXYETHYLCARBAMOYLMETHYL)-,
CHLORIDE, DODECANOATE see LBD100
PYRIDINIUM,2-
((HYDROXYIMINO)METHYL)-1-(((4-
METHYLPYRIDINIO)METHOXY)METHYL)
-, DICHLORIDE see MBY500
PYRIDINIUM, 3,3'-(2-
METHOXYTEREPHTHALOYLBIS(IMINO-
p-PHENYLENECARBONYLIMINO))BIS(1-
ETHYL)-, DI-p-TOLUENE SULFONATE see
POW000
PYRIDINIUM, 3,3'-(2-
METHOXYTEREPHTHALOYLBIS(IMINO-
p-PHENYLENECARBONYLIMINO))BIS(1-
METHYL)-DI-p-TOLUENE SULFONATE see
POW250
PYRIDINIUM, 3,3'-(2-
METHOXYTEREPHTHALOYLBIS(IMINO-
p-PHENYLENECARBONYLIMINO))BIS(1-
PROPYL)-DI-p-TOLUENE SULFONATE see
POW500
PYRIDINIUM, 1-METHYL-3-(p-(p-((p-((1-
METHYLPYRIDINIUM-3-
YL)AMINO)PHENYL)CARBAMOYL)BENZA
MIDO)BENZAMIDO)-, DIIODIDE see
POX250
PYRIDINIUM, 1-METHYL-3-(p-(p-((p-((1-
METHYLPYRIDINIUM-4-
YL)AMINO)PHENYL)CARBAMOYL)BENZA
MIDO)BENZAMIDO)-, DI-p-TOLUENE
SULFONATE see POX500
PYRIDINIUM, 1-METHYL-4-(p-(p-((p-((1-
METHYLPYRIDINIUM-4-
YL)AMINO)PHENYL)CARBAMOYL)CINNA
MAMIDO)ANILINO)-, DI-p-TOLUENE
SULFONATE see POX750
PYRIDINIUM, 1-METHYL-3-(p-(p-((((1-
METHYLPYRIDINIUM-3-
YL)CARBAMOYL)1,4-
NAPHTHYL)CARBAMOYL)BENZAMIDO)B
ENZAMIDO)-, DI-p-TOLUENE
SULFONATE see POY250
PYRIDINIUM, 1-METHYL-3-(m-(p-((p-((1-
METHYLPYRIDINIUM-3-
YL)CARBAMOYL)PHENYL)CARBAMOYL)B
ENZAMIDO)BENZAMIDO)-, DIIODIDE see
POY750
PYRIDINIUM, 1-METHYL-3-(p-(p-((p-(1-
METHYLPYRIDINIUM-3-
YL)PHENYL)CARBAMOYL)BENZAMIDO)B
ENZAMIDO)-, DI-p-TOLUENE
SULFONATE see PPA000
PYRIDINIUM, 1-METHYL-3-(p-(p-((p-(1-
METHYLPYRIDINIUM-3-
YL)PHENYL)CARBAMOYL)CINNAMAMID
O)PHENYL)-, DI-p-TOLUENE SULFONATE
see PPA250
PYRIDINIUM, N-METHYL-, 2-NITRILE
METHANESULPHONATE see MOZ200
PYRIDINIUM, 1-(2-((4-METHYL-2-OXO-2H-
1-BENZOPYRAN-7-YL)OXY)ETHYL)-,
BROMIDE see PPA350
PYRIDINIUM, 1-METHYL-4-PHENYL- see
CQI550
PYRIDINIUM, 1-METHYL-4-PHENYL-,
CHLORIDE see MNX900
PYRIDINIUM, 3,3'-(2-
METHYLTEREPHTHALOYLBIS(IMINO-p-
PHENYLENECARBONYLIMINO))BIS(1-
ETHYL)-DI-p-TOLUENE SULFONATE see
PPA750
PYRIDINIUM, 3,3'-(2-
METHYLTEREPHTHALOYLBIS(IMINO-p-
PHENYLENECARBONYLIMINO))BIS(1-
METHYL)-DI-p-TOLUENE SULFONATE see
PPB000

PYRIDINIUM, 3,3'-(2-METHYLTEREPHTHALOYLBIS(IMINO-p-PHENYLENECARBONYLIMINO))BIS(1-PROPYL)-, DI-p-TOLUENE SULFONATE see PPB250

PYRIDINIUM, 3,3'-(1,4-NAPHTHYLENEBIS(CARBANYLIMINO-p-PHENYLENECARBONYLIMINO))BIS(1-METHYL)-DI-p-TOLUENE SULFONATE see PPB500

PYRIDINIUM NITRATE see PPB550

PYRIDINIUM, 1-(2-OXO-2-((1-OXODODECYL)OXY)ETHYL)-AMINO)ETHYL)-, CHLORIDE see LBD100

PYRIDINIUM, 1,1'-(OXYBIS(METHYLENE))BIS(2-((HYDROXYIMINO)METHYL)), DICHLORIDE (9CI) see PPC000

PYRIDINIUM, 1,1'-(OXYDI-2,1-ETHANEDIYL)BIS(4-((HYDROXYIMINO)METHYL)-, DIBROMIDE see OPO100

PYRIDINIUM, 1,1'-(OXYDIETHYLENE)BIS(4-FORMYL-, DIBROMIDE, DIOXIME see OPO100

PYRIDINIUM, 1,1'-(OXYDIMETHYLENE)BIS(2-FORMYL)-, DICHLORIDE DIOXIME see PPC000

PYRIDINIUM PERCHLORATE see PPC100

PYRIDINIUM, 4,4'-(p-PHENYLENEBIS(ACRYLOYLIMINO-p-PHENYLENEIMINO))BIS(1-ETHYL, DI-p-TOLUENE SULFONATE) see PPC250

PYRIDINIUM, 3,3'-(1,4-PHENYLENEBIS(CARBONYLIMINO-4,1-PHENYLENECARBONYLIMINO))BIS(1-ETHYL- see PEW550

PYRIDINIUM,-3,3'-(1,4-PHENYLENEBIS(CARBONYLIMINO-4,1-PHENYLENECARBONYLIMINO))BIS(1-ETHYL- see PEW550

PYRIDINIUM, 1-PROPYL-3-(p-(p-((p-((1-PROPYLPYRIDINIUM-3-YL)CARBAMOYL)PHENYL)CARBAMOYL)CINNAMAMIDO)BENZAMIDO)-, DI-p-TOLUENE SULFONATE see PPD000

PYRIDINIUM, 1-PROPYL-3-(p-(p-((p-(1-PROPYLPYRIDINIUM-3-YL)PHENYL)CARBAMOYL)CINNAMAMIDO)PHENYL)-, DI-p-TOLUENE SULFONATE see PPD250

PYRIDINIUM, 4,4'-(SUBEROYLBIS(IMINO-p-PHENYLENEIMINO))BIS(1-METHYL-), DIBROMIDE see PPD500

PYRIDINIUM, 3,3'-(TEREPHTHALOYLBIS(IMINO(3-CHLORO-p-PHENYLENE)CARBONYLIMINO))BIS(1-METHYL-DI-p-TOLUENE SULFONATE) see PPE500

PYRIDINIUM, 3,3'-(TEREPHTHALOYLBIS(IMINO(3-METHOXY-p-PHENYLENE)CARBONYLIMINO))BIS(1-ETHYL-, DI-p-TOLUENE SULFONATE) see PPF250

PYRIDINIUM, 3,3'-(TEREPHTHALOYLBIS(IMINO(3-METHOXY-p-PHENYLENE)CARBONYLIMINO))BIS(1-PROPYL)-, DI-p-TOLUENE SULFONATE see PPG250

PYRIDINIUM, 3,3'-(TEREPHTHALOYLBIS(IMINO-p-PHENYLENECARBONYLIMINO))BIS(1-BUTYL, DI-p-TOLUENESULFONATE) see PPH000

2-PYRIDINOL see OOO200

3-PYRIDINOL see PPH025

4-PYRIDINOL see HOB100

PYRIDINOL CARBAMATE see PPH050

2-PYRIDINOL, 3,5-DICHLORO- see DGJ150

PYRIDINOL NITROSOCARBAMATE see PPH100

2-PYRIDINOL, 3,5,6-TRICHLORO-, O-ESTER with O,O-DIETHYL PHOSPHOROTHIOATE see CMA100

2-PYRIDINONE see OOO200

2(1H)-PYRIDINONE (9CI) see OOO200

2(1H)-PYRIDINONE, 3,5-DICHLORO- see DGJ150

4(1H)-PYRIDINONE, 1,2-DIMETHYL-3-HYDROXY- see DSI250

2(1H)-PYRIDINONE, 4,6-DIPHENYL-1-(((5-NITRO-2-FURANYL)METHYLENE)AMINO)- see DWH900

2(1H)-PYRIDINONE, 1-HYDROXY-4-METHYL-6-(2,4,4-TRIMETHYLPENTYL)-, COMPD. WITH 2-AMINOETHANOL (1:1) see PJA180

2(1H)-PYRIDINONE, 1-METHYL-(9CI) see MPA075

4(1H)-PYRIDINONE, 1-METHYL-3-PHENYL-5-(3-(TRIFLUOROMETHYL)PHENYL)- see FMQ200

2(1H)-PYRIDINONE, 3,5,6-TRICHLORO- see HOL125

α-((2-PYRIDINYLAMINO)METHYL)-BENZENEMETHANOL MONOHYDROCHLORIDE see PGG355

3-PYRIDINYLCARBONIMIDODITHIOIC ACID, (2-CHLOROPHENYL)METHYL (2,4-DICHLOROPHENYL)METHYL ESTER see PPH103

3-PYRIDINYLCARBONIMIDODITHIOIC ACID, 1,1-DIMETHYLETHYL (4-(1,1-DIMETHYLETHYL)PHENYL)METHYL ESTER see PPH110

3-PYRIDINYLCARBONIMIDODITHIOIC ACID, (4-(1,1-DIMETHYLETHYL)PHENYL)METHYL1-METHYLPROPYL ESTER see PPH115

3-PYRIDINYLCARBONIMIDODITHIOIC ACID, (4-(1,1-DIMETHYLETHYL)PHENYL)METHYL 1-METHYLETHYL ESTER see PPH120

3-PYRIDINYLCARBONIMIDODITHIOIC ACID, (4-(1,1-DIMETHYLETHYL)PHENYL)METHYL2-METHYLPROPYL ESTER see PPH125

3-PYRIDINYLCARBONIMIDODITHIOIC ACID, (4-(1,1-DIMETHYLETHYL)PHENYL)METHYL 1,1-DIMETHYLPROPYL ESTER see PPH130

3-PYRIDINYLCARBONIMIDODITHIOIC ACID, (4-(1,1-DIMETHYLETHYL)PHENYL)METHYL 1-ETHYL-1-METHYLPROPYL ESTER see PPH135

3-PYRIDINYLCARBONIMIDOTHIOIC ACID, o-CYCLOHEXYL S-((4-(1,1-DIMETHYLETHYL)PHENYL)METHYL) ESTER see PPH140

3-PYRIDINYLCARBONIMIDOTHIOIC ACID, o-CYCLOPENTYL S-((4-(1,1-DIMETHYLETHYL)PHENYL)METHYL) ESTER see PPH145

3-PYRIDINYLCARBONIMIDOTHIOIC ACID, o-(2,5-DIMETHYLCYCLOHEXYL) S-((4-(1,1-DIMETHYLETHYL)PHENYL)METHYL) ESTER see PPH150

3-PYRIDINYLCARBONIMIDOTHIOIC ACID, o-(3,5-DIMETHYLCYCLOHEXYL) S-((4-(1,1-DIMETHYLETHYL)PHENYL)METHYL) ESTER see PPH155

3-PYRIDINYLCARBONIMIDOTHIOIC ACID, S-((4-(1,1-DIMETHYLETHYL)PHENYL)METHYL) o-OCTYL ESTER see DRV900

3-PYRIDINYLCARBONIMIDOTHIOIC ACID, S-((4-(1,1-DIMETHYLETHYL)PHENYL)METHYL)-o-(1-METHYLBUTYL) ESTER see PPH160

3-PYRIDINYLCARBONIMIDOTHIOIC ACID, S-((4-(1,1-DIMETHYLETHYL)PHENYL)METHYL)o-(1-METHYLPROPYL) ESTER see PPH165

3-PYRIDINYLCARBONIMIDOTHIOIC ACID, S-((4-(1,1-DIMETHYLETHYL)PHENYL)METHYL)o-(2-METHYLCYCLOHEXYL) ESTER see PPH168

3-PYRIDINYLCARBONIMIDOTHIOIC ACID, S-((4-(1,1-DIMETHYLETHYL)PHENYL)METHYL)o-(2-METHYLPROPYL) ESTER see PPH170

3-PYRIDINYLCARBONIMIDOTHIOIC ACID, S-((4-(1,1-DIMETHYLETHYL)PHENYL)METHYL)o-(3-METHYLBUTYL) ESTER see PPH172

3-PYRIDINYLCARBONIMIDOTHIOIC ACID, S-((4-(1,1-DIMETHYLETHYL)PHENYL)METHYL)o-(3-METHYLCYCLOHEXYL) ESTER see PPH174

3-PYRIDINYLCARBONIMIDOTHIOIC ACID, S-((4-(1,1-DIMETHYLETHYL)PHENYL)METHYL)o-(4-METHYLCYCLOHEXYL) ESTER see PPH176

3-PYRIDINYLCARBONIMIDOTHIOIC ACID, S-((4-(1,1-DIMETHYLETHYL)PHENYL)METHYL)o-(1,2-DIMETHYLPROPYL) ESTER see PPH180

3-PYRIDINYLCARBONIMIDOTHIOIC ACID, S-((4-(1,1-DIMETHYLETHYL)PHENYL)METHYL) o-(3,3,5-TRIMETHYLCYCLOHEXYL) ESTER see PPH185

1-PYRIDINYLETHANONE see PPH200

1-(2-PYRIDINYL)ETHANONE see PPH200

4,4'-(2-PYRIDINYLMETHYLENE)BISPHENOL BIS(HYDROGEN SULFATE) (ESTER) DISODIUM SALT see SJJ175

PYRIDIPCA see PPK500

PYRIDIUM see PDC250

PYRIDIUM see PEK500

PYRIDIVITE see PDC250

2-PYRIDIYLAMINE see PPH300

10H-PYRIDO(3,2-B)(1,4)BENZOTHIAZINE, 10-(2-(DIETHYLAMINO)PROPYL)- see DIR100

10H-PYRIDO(3,2-B)(1,4)BENZOTHIAZINE, 10-(2-(DIMETHYLAMINO)PROPYL)- see OGI075

10H-PYRIDO(3,2-B)(1,4)BENZOTHIAZINE-10-ETHANAMINE, N,N,α-TRIMETHYL- see OGI075

(2-(10h-PYRIDO(3,2-b)(1,4)BENZOTHIAZIN-10-YL)PROPYL)TRIMETHYLAMMONIUMBROMIDE see PPH350

6H-PYRIDO(4,3-B)CARBAZOLE, 5,11-DIMETHYL-9-METHOXY-, LACTATE see MEL780

6H-PYRIDO(4,3-B)CARBAZOLE, 5-(HYDROXYMETHYL)-11-METHYL- see HMH050

6H-PYRIDO(4,3-B)CARBAZOLE, 5-(HYDROXYMETHYL)-11-METHYL-, METHYLCARBAMATE see HMH100

6H-PYRIDO(4,3-B)CARBAZOLE-5-METHANOL, 11-METHYL- see HMH050

6H-PYRIDO(4,3-B)CARBAZOLE-5-METHANOL, 11-METHYL-, METHYLCARBAMATE see HMH100

6H-PYRIDO(4,3-B)CARBAZOLE-5-METHANOL, 11-METHYL-, METHYLCARBAMATE (ESTER) see HMH100

PYRIDO(3',2':5,6)CHRYSENE see BCQ000

PYRIDO(3',2':3,4)FLUORANTHENE see FDP000

9H-PYRIDO(2,3-B)INDOL-2-AMINE, 3-ETHYL- see PPH400

PYROCARBONIC ACID, DIETHYL ESTER see DIZ100
PYROCATECHIN see CCP850
PYROCATECHINIC ACID see CCP850
PYROCATECHOL see CCP850
PYROCATECHOL DIMETHYL ETHER see DOA200
PYROCATECHOL-3,5-DISULFONIC ACID see PPQ100
PYROCATECHOL, 3-METHOXY- see PPQ550
PYROCATECHOL, 3,4,6-TRICHLORO- see TIL600
PYROCATECHUIC ACID see CCP850
PYROCELLULOSE see CCU150
PYRO-CHEK 77B see EEB230
PYRO-CHEK 63PB see EEE650
PYRO-CHEK LM see EEE650
PYROCHOL see DAQ400
PYROCLOR 5 see PAV600
PYROD see UNJ800
PYRODIN see ACX750
PYRODINE see ACX750
PYRODONE see OES000
PYRODOXIN see PPK250
PYROGALLIC ACID see PPQ500
PYROGALLOL see PPQ500
PYROGALLOL DIMETHYLETHER see DOJ200
PYROGALLOL-1,3-DIMETHYL ETHER see DOJ200
PYROGALLOL 1-METHYL ETHER see PPQ550
PYROGALLOL 1-MONOMETHYL ETHER see PPQ550
PYROGASTRONE see CBO500
PYROGEN see PPQ600
PYROGENE DEEP BLACK B see CMS250
PYROGENE DEEP BLACK D see CMS250
PYROGENE VIOLET BROWN X see CMS257
PYROGENE VIOLET BROWN X-CF see CMS257
PYROGENS see TGJ850
l-PYROGLUTAMYL-l-HISTIDYL-l-PROLINEAMIDE see TNX400
l-PYROGLUTAMYL-l-HISTIDYL-l-PROLINEAMIDE-l-TARTRATE see POE050
PYROGUAIAC ACID see GKI000
PYROKOHLENSAEURE DIAETHYL ESTER (GERMAN) see DIZ100
2-PYROL see PPT500
M-PYROL see MPF200
PYROLAN see PPQ625
10-(2-(2-PYROLIDYL)ETHYL)PHENOTHIAZINE HYDROCHLORIDE see POL000
PYROLLNITRIN see CFC000
PYROLUSITE BROWN see MAS000
PYROMECAINE see PQB750
PYROMELLITIC ACID see PPQ630
PYROMELLITIC ACID ANHYDRIDE see PPQ635
PYROMELLITIC ACID DIANHYDRIDE see PPQ635
PYROMELLITIC ANHYDRIDE see PPQ635
PYROMELLITIC DIANHYDRIDE see PPQ635
PYROMEN see PJA200
PYROMET 538 see IGL110
PYROMET 800 see IGL100
PYROMIJIN see PII100
PYROMUCIC ACID see FQF000
PYROMUCIC ACID METHYL ESTER see MKH600
PYROMUCIC ALDEHYDE see FPQ875
PYRON see FAQ600
PYRONALORANGE see PEJ500
PYRONALROT B see OHI200
PYRONALROT R see XRA000
PYRONARIDINE see PPQ650
PYRONIN B see DIS200
PYRONINE see PPQ750

PYRONIN G see PPQ750
PYRONIN YELLOW see PPQ750
PYROPENTYLENE see CPU500
PYROPHOSPHATE see TEE500
PYROPHOSPHATE de TETRAETHYLE (FRENCH) see TCF250
PYROPHOSPHORIC ACID, DIMETHYL DIISOPROPYL ESTER see DRL300
PYROPHOSPHORIC ACID OCTAMETHYLTETRAAMIDE see OCM000
PYROPHOSPHORIC ACID, TETRAMETHYL ESTER see TDV000
PYROPHOSPHORODITHIOIC ACID, TETRAETHYL ESTER see SOD100
PYROPHOSPHORODITHIOIC ACID, O,O,O,O-TETRAETHYL ESTER see SOD100
PYROPHOSPHORYLTETRAKISDIMETHYL AMIDE see OCM000
PYROPLASMIN see PJA120
PYRORACEMIC ACID see PQC100
PYRORACEMIC ALDEHYDE see PQC000
PYROSET FLAME RETARDANT TKP see TDH500
PYROSET TKO see TDI000
PYROSET TKP see TDH500
PYROSTIB see AQH500
PYROSULFUROUS ACID, DIPOTASSIUM SALT see PLR250
PYROSULFURYL CHLORIDE see PPR500
PYROSULPHURIC ACID see SOI520
PYROTARTARIC ACID NITRILE see TLR500
PYROTONE RED TONER RA-5520 see CJD500
PYROTROPBLAU see CMO250
PYROVATEX 3805 see MND550
PYROVATEX CP see MND550
PYROXYCHLOR see CIC600
PYROXYLIC SPIRIT see MGB150
PYROXYLIN see CCU250
PYRROCYCLINE N see SPE000
PYRROLAMIDOL see AFJ400
PYRROLAMIDOLUM see AFJ400
PYRROLAZOATE see PAJ750
PYRROLAZOTE ABERGIC see POL000
PYRROLAZOTE HYDROCHLORIDE see POL000
PYRROLE see PPS250
PYRROLE, 2-ACETYL-4-NITRO- see ACT300
PYRROLE, 2-ACETYL-5-NITRO- see ACT330
1H-PYRROLE-3-CARBONITRILE, 4-BROMO-2-(4-CHLOROPHENYL)-1-(ETHOXYMETHYL)-5-(TRIFLUOROMETHYL)- see BNA600
1H-PYRROLE-3-CARBONITRILE, 4-(2,3-DICHLOROPHENYL)- see FAQ220
1H-PYRROLE-2-CARBOXALDEHYDE, THIOSEMICARBAZONE see PPS300
1H-PYRROLE-2-CARBOXYLIC ACID, 2,5-DIHYDRO-, (S)- see DAL375
(2S-(2-α,7-β,7A-β,14-β,14a-α))-1H-PYRROLE-2-CARBOXYLIC ACID-DODECAHYDRO-11-OXO-7,14-METHANO-2H,6H-DIPYRIDO(1,2-α:1',2'-e)(1,5)DIAZOCIN-2-YL) ESTER see CAZ125
PYRROLE-2,5-DIONE see MAM750
1H-PYRROLE-2,5-DIONE, 3-(BROMOMETHYL)-4-CHLORO- see BNQ050
1H-PYRROLE-2,5-DIONE, 3-CHLORO-4-(CHLOROMETHYL)- see CFB730
1H-PYRROLE-2,5-DIONE, 3-CHLORO-4-(DICHLOROMETHYL)- see CFL120
1H-PYRROLE-2,5-DIONE, 1-(2,3-DIMETHYLPHENYL)- see XSS923
1H-PYRROLE-2,5-DIONE, 1,1'-(METHYLENEDI-4,1-PHENYLENE)BIS-(9CI) see BKL800
1H-PYRROLE-2,5-DIONE, 1,1'-(4-METHYL-1,3-PHENYLENE)BIS- see TGY770
PYRROLE, 1-FURFURYL- see FPX100
PYRROLIDINE see PPS500
1-PYRROLIDINEACETAMIDE, N-(2,6-DIMETHYLPHENYL)-N-(2-((2,6-

DIMETHYLPHENYL)AMINO)-2-OXOETHYL)-2-OXO- see PPS550
1-PYRROLIDINEACETAMIDE, N-(2,6-DIMETHYLPHENYL)-2-HYDROXY-5-OXO- see DTM900
1-PYRROLIDINEACETAMIDE, N-(2,6-DIMETHYLPHENYL)-3-METHYL-2-OXO- see DTN500
1-PYRROLIDINEACETAMIDE, N-(2,6-DIMETHYLPHENYL)-4-METHYL-2-OXO- see DTN520
1-PYRROLIDINEACETAMIDE, 4-HYDROXY-2-OXO-, (+−)- see HND900
3-PYRROLIDINEACETIC ACID,2-CARBOXY-4-(5-CARBOXY-1-METHYL-1,3-HEXADIENYL)-,(2S-(2-α,3-β,4-β(1Z,3E,5S*)))- see DYB825
3-PYRROLIDINEACETIC ACID, 2-CARBOXY-4-ISOPROPENYL- see KAJ200
3-PYRROLIDINEACETIC ACID, 2-CARBOXY-4-(1-METHYLETHENYL)-, (2S-(2-α,3-β,4-β))- see KAJ200
PYRROLIDINE, 1-((1,2,4-BENZOTRIAZIN-3-YL)ACETYL)- see PPS600
PYRROLIDINE, 1-BUTYL-3-(m-HYDROXYPHENYL)-3-ISOBUTYL-, CITRATE see BRQ300
PYRROLIDINE, 1,1'-(2-BUTYNE-1,4-DIYL)BIS-, DIHYDROCHLORIDE see THL575
PYRROLIDINE, 1,1'-(2-BUTYNYLENE)DI-, DIHYDROCHLORIDE see THL575
2-PYRROLIDINECARBOXAMIDE, 1-((4-FLUOROPHENYL)METHYL)-5-OXO- see FGI025
2-PYRROLIDINECARBOXAMIDE, 1-((4-FLUOROPHENYL)METHYL)-5-OXO-N-(4-OXO-4H-1-BENZOPYRAN-2-YL)- see FLH150
2-PYRROLIDINECARBOXAMIDE, 1-((4-FLUOROPHENYL)METHYL)-5-OXO-N-(3-(TRIFLUOROMETHYL)PHENYL)- see FLH400
2-PYRROLIDINECARBOXYLIC ACID see PMH905
(−)-2-PYRROLIDINECARBOXYLIC ACID see PMH905
2-PYRROLIDINECARBOXYLIC ACID, (S)- see PMH905
l-2-PYRROLIDINECARBOXYLIC ACID see PMH900
l-PYRROLIDINE-2-CARBOXYLIC ACID see PMH905
(S)-2-PYRROLIDINECARBOXYLIC ACID see PMH905
l-α-PYRROLIDINECARBOXYLIC ACID see PMH905
1,2-PYRROLIDINECARBOXYLIC ACID, 1-BENZYL-2-VINYL ESTER see CBR247
3-PYRROLIDINECARBOXYLIC ACID, 3-(m-HYDROXYPHENYL)-1-METHYL-, ETHYL ESTER see ELL600
1,2-PYRROLIDINEDICARBOXAMIDE, N,N'-BIS(4-OXO-4H-1-BENZOPYRAN-2-YL)-5-OXO- see CBT600
1,2-PYRROLIDINEDICARBOXYLIC ACID-1-BENZYL 2-(1,2-DIBROMOETHYL) ESTER see CBR245
2,5-PYRROLIDINEDIONE see SND000
1-PYRROLIDINEETHANAMINE, β-((2-METHYLPROPOXY)METHYL)-N-PHENYL-N-(PHENYLMETHYL)-, MONOHYDROCHLORIDE, MONOHYDRATE see IIG600
1-PYRROLIDINEETHANOL see PPS700
PYRROLIDINEETHANOL, 3,4-DIMETHYL- see HKR550
PYRROLIDINE, 1-(4-ETHOXY-2-BUTYNYL)- see EEN700
PYRROLIDINE, 1-(p-((p-ETHYLPHENYL)AZO)PHENYL)- see EPC950

PYRROLIDINE, 1-((4-METHOXY(1,1'-BIPHENYL)-3-YL)METHYL)- see MEF400

PYRROLIDINE, 1-(2-(4-(7-METHOXY-3-PHENYL-2H-1-BENZOPYRAN-4-YL)PHENOXY)ETHYL)- see PGF112

PYRROLIDINE, 1-(2-(p-(7-METHOXY-3-PHENYL-2H-1-BENZOPYRAN-4-YL)PHENOXY)ETHYL)- see PGF112

PYRROLIDINE, 1-(4-(p-(6-METHOXY-2-PHENYL-3,4-DIHYDRO-1-NAPHTHYL)PHENYL)BUTYL)-, HYDROCHLORIDE see MFF635

PYRROLIDINE, 1-METHYL-2-(3-PYRIDYL)-, SULFATE see NDR500

PYRROLIDINE, 1,1'-(Δ,Δ'-NITRILODITETRAMETHYLENE)BIS(γ-BUTYL-γ-(1-NAPHTH YL)- see NEH610

PYRROLIDINE, NITROSO- see NLP480

PYRROLIDINIUM, 1-CARBOXYMETHYL-1-METHYL-, IODIDE, METHYL ESTER see CCH300

PYRROLIDINO-AETHYLPHENTHIAZIN (GERMAN) see PAJ750

2-PYRROLIDINOETHANOL see PPS700

2-(1-PYRROLIDINO)ETHANOL see PPS700

4-(2,1-PYRROLIDINOETHOXY)QUINOLINE HYDROCHLORIDE see QUJ300

N-(2-PYRROLIDINOETHYL)PHENOTHIAZINE HYDROCHLORIDE see POL000

2-(2-PYRROLIDINOETHYL)-3a,4,7,7a-TETRAHYDRO-4,7-ETHANEISOINDOLINE DIMETHIODIDE see PPT325

3-PYRROLIDINOMETHYL-4-HYDROXYBIPHENYL see PPT400

3-PYRROLIDINO-N-METHYL-4-METHOXYBIPHENYL see MEF400

PYRROLIDINO-METHYL-TETRACYCLINE see PPY250

N-(PYRROLIDINOMETHYL)TETRACYCLINE see PPY250

2-PYRROLIDINONE see PPT500

α-PYRROLIDINONE see PPT500

2-PYRROLIDINONEACETAMIDE see NNE400

2-PYRROLIDINONE, 4-(AMINOMETHYL)-1-(PHENYLMETHYL)-, (E)-2-BUTENEDIOATE (2:1) see AKR100

2-PYRROLIDINONE, 3-((3,5-BIS(1,1-DIMETHYLETHYL)-4-HYDROXYPHENYL)METHYLENE)-1-METHOXY- see MEL100

2-PYRROLIDINONE, 1-COCO ALKYL derivs. see CNF340

2-PYRROLIDINONE, 1-CYCLOHEXYL- see CPQ275

2-PYRROLIDINONE, 1-(3-(DIMETHYLAMINO)PROPYL)- see DQA800

2-PYRROLIDINONE, 1-(3-(DIMETHYLAMINO)PROPYL)-3-((1-METHYL-5-NITRO-1H-IMIDAZOL-2-YL)METHYLENE)- see PPT600

2-PYRROLIDINONE, 1-(4-(DIPROPYLAMINO)-2-BUTYNYL)- see DWR300

2-PYRROLIDINONE, 1-DODECYL- see DXY750

2-PYRROLIDINONE, 1-ETHENYL-, POLYMER with ETHENYLBENZENE see VQK595

2-PYRROLIDINONE, 3-((1-(2-ETHOXYETHYL)-5-NITRO-1H-IMIDAZOL-2-YL)METHYLENE)-1-METHYL- see EET200

2-PYRROLIDINONE, 1-ETHYL- see EPC700

2-PYRROLIDINONE, 1-ETHYL-4-(2-MORPHOLINOETHYL)-3,3-DIPHENYL- see ENL100

2-PYRROLIDINONE, 1-ETHYL-4-(2-(4-MORPHOLINYL)ETHYL)-3,3-DIPHENYL-(9CI) see ENL100

2-PYRROLIDINONE, 1-(2-HYDROXYETHYL)- see HLB380

2-PYRROLIDINONE, 3-((1-(2-HYDROXYETHYL)-5-NITRO-1H-IMIDAZOL-2-YL)METHYLENE)-1-METHYL- see HKW460

2-PYRROLIDINONE, 3-HYDROXY-1-METHYL-5-(3-PYRIDINYL)- see HJX700

2-PYRROLIDINONE, 1-ISOPROPYL- see IRG100

2-PYRROLIDINONE, METHYL- see MPB600

2-PYRROLIDINONE, 1-(1-METHYLETHYL)-(9CI) see IRG100

2-PYRROLIDINONE, 1-METHYL-3-((1-METHYL-4-NITRO-1H-IMIDAZOL-2-YL)METHYLENE)- see MLP400

2-PYRROLIDINONE, 1-METHYL-5-(3-PYRIDINYL)-, N-OXIDE, (S)- see CNT645

2-PYRROLIDINONE, 1-OCTYL- see OFU100

2-PYRROLIDINONE, 1-(4-(1-PYRROLIDINYL)-2-BUTENYL)-, (Z)- see PPU800

2-PYRROLIDINONE, 1-(4-(1-PYRROLIDINYL)BUTYL)-, ETHANEDIOATE (1:1) see PPU850

2-PYRROLIDINONE, 1-(4-(1-PYRROLIDINYL)-2-BUTYNYL)-, (E)-2-BUTENEDIOATE (2:3) see OOY100

2-PYRROLIDINONE, 1-(4-(1-PYRROLIDINYL)-2-BUTYNYL)-, FUMARATE (2:3) see OOY100

2-PYRROLIDINONE, 1-(4-(1-PYRROLIDINYL)-2-BUTYNYL)-, SESQUIFUMARATE see OOY100

2-PYRROLIDINONE, 1-TALLOW ALKYL derivs. see TAC200

2-PYRROLIDINONE, 5,5'-((1,1,3,3-TETRABUTYL-1,3-DISTANNOXANEDIYL)BIS(OXYCARBONYL))BIS-, (S-(R*,R*))- see PPT630

2-PYRROLIDINONE, 1-VINYL-, POLYMER with STYRENE (8CI) see VQK595

2-PYRROLIDINYL-p-ACETOPHENETIDIDE HYDROCHLORIDE see PPU750

3-(1-PYRROLIDINYL)ANDROSTA-3,5-DIEN-17-ONE see PPU780

1-(4-(1-PYRROLIDINYL)-2-BUTENYL)-2-PYRROLIDINONE (Z)- see PPU800

cis-1-(4-PYRROLIDIN-1'-YLBUT-2-ENYL)PYRROLID-2-ONE see PPU800

1-(4-(1-PYRROLIDINYL)BUTYL)-2-PYRROLIDINONE ETHANEDIOATE (1:1) see PPU850

1-(4-PYRROLIDIN-1-YLBUTYL)-PYRROLID-2-ONE HYDROGEN OXALATE see PPU850

1-(4-(1-PYRROLIDINYL)-2-BUTYNYL)-2-PYRROLIDINONE see OOY000

1-(N-(1-PYRROLIDINYL)-2-BUTYNYL)-2-PYRROLIDINONE SESQUIFUMARATE see OOY100

4-PYRROLIDINYLCARBONYLMETHYLPIPERAZINYL-3,4,5-TRIMETHOXYCINNAMYL KETONE MALEATE see PPV750

1-(1-PYRROLIDINYLCARBONYL)METHYL)-4-(3,4,5-TRIMETHOXYCINNAMOYL)PIPERAZINE MALEATE see VGK000

trans-2-(1-PYRROLIDINYL)CYCLOHEXYL-3-PENTYLOXYCARBANILATE HYDROCHLORIDE see PPW000

2-(1-PYRROLIDINYL)ETHANOL see PPS700

N-(2-PYRROLIDINYLETHYL)CARBAMIC ACID, 6-CHLORO-o-TOLYL ESTER, HYDROCHLORIDE see PPW750

N-(2-PYRROLIDINYLETHYL)CARBAMIC ACID, MESITYL ESTER, HYDROCHLORIDE see PPX000

N-(2-(PYRROLIDINYL)ETHYL)CARBAMIC ACID, 2,6-XYLYL ESTER HYDROCHLORIDE see PPX250

2-(PYRROLIDINYL)ETHYL-2-CHLORO-6-METHYLCARBANILATE HYDROCHLORIDE see CIK500

2-(2-(PYRROLIDINYL)ETHYL)-5-NITRO-1H-BENZ(de)ISOQUINOLINE-1,3(2H)-DIONE see MAB055

10-(2-(1-PYRROLIDINYL)ETHYL)PHENOTHIAZINE see PAJ750

3-(1-PYRROLIDINYLMETHYL)(1,1'-BIPHENYL)-4-OL see PPT400

N-(1-PYRROLIDINYLMETHYL)-TETRACYCLINE see PPY250

1-PYRROLIDINYLOXY, 3-CARBOXY-2,2,5,5-TETRAMETHYL- see TDV300

3-(2-PYRROLIDINYL)PYRIDINE see NNR500

3-(2-PYRROLIDINYL)-(s)-PYRIDINE see NNR500

trans-2-(3-(1-PYRROLIDINYL)-1-p-TOLYLPROPENYL)PYRIDINE MONOHYDROCHLORIDE see TMX775

4-(1-PYRROLIDINYL)-1-(2,4,6-TRIMETHOXYPHENYL)-1-BUTANONE HYDROCHLORIDE see BOM600

PYRROLIDON (GERMAN) see PPT500

PYRROLIDONE see PPT500

2-PYRROLIDONE see PPT500

α-PYRROLIDONE see PPT500

2-PYRROLIDONEACETAMIDE see NNE400

PYRROLIDONE CARBOXYLATE de MAGNESIUM HYDRATE (FRENCH) see PAN775

10-(2-(1-PYRROLIDYL)ETHYL)PHENOTHIAZINE see PAJ750

10-(2-(1-PYRROLIDYL)ETHYL)PHENOTHIAZINE HYDROCHLORIDE see POL000

1-PYRROLINE-2-ACRYLAMIDE see COV125

3-PYRROLINE-2-CARBOXYLIC ACID, l-(8CI) see DAL375

3-PYRROLINE-2,5-DIONE see MAM750

1H-PYRROLIZINE-8-ACETAMIDE, HEXAHYDRO-N-(2,6-DIMETHYLPHENYL)-HYDROCHLORIDE, HEMIHYDRATE see PIF600

1H-PYRROLIZINE-7A(5H)-ACETAMIDE, TETRAHYDRO-N-(2,6-DIMETHYLPHENYL)-, HYDROCHLORIDE,HYDRATE (2:2:1) see PIF600

PYRROLNITRIN see CFC000

(E)-1H-PYRROLO(2,1-C)(1,4)BENZODIAZEPINE-2-ACRYLAMIDE, 5,10,11,11a-TETRAHYDRO-9,11-DIHYDROXY-8-METHYL-5-OXO see API000

5H-PYRROLO(2,2-c)(1,4)BENZODIAZEPIN-5-ONE, 2-ETHYLIDENE-1,2,3,10,11,11A-HEXAHYDRO-8-HYDROXY-7,11-DIMETHOXY- see THG500

7H-PYRROLO(2,3-D)PYRIMIDINE, 6,7-DIMETHYL-2,4-DI-1-PYRROLIDINYL-, SULFATE (1:1) see DRQ300

1H-PYRROLO(2,3-B)QUINOLIN-4-ONE, 2,3,4,9-TETRAHYDRO-1-METHYL-9-(p-NITROPHENACYL)- see MNE400

PYRROLYLENE see BOP500

PYRROL-2-YL KETONE see PPY300

7-(((2-PYRROLYL)METHYL)AMINO)-ACTINOMYCIN D see PQB275

7-((1H-PYRROL-2-YLMETHYL)AMINO)ACTINOMYCIN D see PQB275

PYRROL-2-YLPHENYL KETONE see PGF900

3-PYRROL-2-YLPYRIDINE see PQB500

PYRROMECAINE HYDROCHLORIDE see PQB750

PYRROMELITIC ACID DIANHYDRIDE see PQB800
PYRRO(b)MONAZOLE see IAL000
2-PYRRYL ETHYL KETONE see PMX320
α-PYRRYL ETHYL KETONE see PMX320
PYRUVALDEHYDE see PQC000
PYRUVIC ACID see PQC100
PYRUVIC ACID, BROMO- see BOD550
PYRUVIC ALDEHYDE see PQC000
PYRUVOHYDROXIMOYL CHLORIDE, OXIME see PQC200
PYRVINIUM EMBONATE see PQC500
PYRVINIUM-4,4'-METHYLENEBIS(3-HYDROXYNAPHTHALENE-2-CARBOXYLATE) see PQC500
PYRVINIUM PAMOATE see PQC500
PYSOCOCCINE see SNM500
PY-TETRAHYDROSERPENTINE see AFG750
PYX see PQC525
PYX EXPLOSIVE see PQC525
PZ see CAT775
PZ-13 see PDB750
PZ 177 see CJW500
PZ 222 see CKD250
PZ 1511 see CCK780
PZh2M see IGK800
PZhO see IGK800
PZO see PDB250
PZT see LED100
PZT 5 see LED100
PZT 8 see LED100
PZT 574 see LED100
PZT-C see LED100
6Q8 see NBL200
Q-137 see DJC000
Q 174 see TNJ500
Q 2-1401 see PJR300
QA 71 see PNW800
QB-1 see CMX920
QCB see PAV500
Q-D 86P see DXG625
QDO see DVR200
QG 100 see SCK600
QGH see BDD000
QIDAMP see AIV500
QIDMYCIN see EDJ500
QIDPEN G see BFD000
QIDPEN VK see PDT750
QIDTET see TBX250
QIKRON see BIN000
QINGHAOSU (CHINESE) see ARL375
QING HAU SAU (CHINESE) see ARL375
QM 589 see IHX700
QM 657 see DGW450
QM-1143 see MOF750
QM-1148 see MLR500
QM-1149 see MOG000
QM-6008 see BAV625
QNB see QVA000
QO POLYMEG see GHY100
QO POLYMEG 1000 see GHY100
QO THFA see TCT000
QPB see PJA000
QR 483 see MCB050
QR 819 see BLX000
QSAH 7 see PKQ059
QUAALUDE see QAK000
QUADRACYCLINE see TBX250
QUADRATIC ACID see DMJ600
QUADROL see QAT000
QUADROSILAN see CMS234
QUAMAQUIL see PGA750
QUAMONIUM see HCQ500
QUANTALAN see CME400
QUANTRIL see BCL250
QUANTROVANIL see EQF000
QUANTRYL see BCL250
QUARTAMIN 24P see LBX075
QUARTAMIN 24W see LBX075
QUARTZ see SCJ500
QUARTZ GLASS see SCK600

QUARTZ SAND see SCK600
QUATERNARIO CPC see CCX000
QUATERNARIO LPC see DXY725
QUATERNARIO LPC see LBX050
QUATERNARY AMMONIUM COMPOUNDS, ALKYLBENZYLDIMETHYL, CHLORIDES see AFP500
QUATERNARY AMMONIUM COMPOUNDS, BENZYL-C$_{10-18}$-ALKYLDIMETHYL, CHLORIDES see QAT510
QUATERNARY AMMONIUM COMPOUNDS, BENZYL-C$_{12}$-C$_{16}$-ALKYLDIMETHYL, CHLORIDES see QAT520
QUATERNARY AMMONIUM COMPOUNDS, BIS(HYDROGENATED TALLOW ALKYL)DIMETHYL CHLORIDES see QAT550
QUATERNARY AMMONIUM COMPOUNDS, (CARBOXYMETHYL)(3-COCOAMIDOPROPYL)DIMETHYL-, HYDROXIDES, INNER SALTS see CNF185
QUATERNARY AMMONIUM COMPOUNDS, TRI-(C$_{8-10}$)-ALKYLMETHYL-, CHLORIDES see QAT565
QUATERNARY AMMONIUM COMPOUNDS, TRIMETHYLTALLOW ALKYL-, CHLORIDE see TAC300
QUATERNIUM 5 see DXG625
QUATERNIUM-10 see TLW500
QUATERNIUM-12 see DGX200
QUATERNIUM 13 see TCB200
QUATERNIUM 15 see CEG550
QUATERNIUM-18 see QAT550
QUATERNIUM 19 see QAT600
QUATERNIUM 40 see DTS500
QUATERNOL 1 see DTC600
QUATRACHLOR see BEN000
QUATRESIN see MSB500
QUATREX see TBX250
QUAZO PURO (ITALIAN) see SCJ500
QUEBRACHIN see YBJ000
QUEBRACHINE see YBJ000
QUEBRACHOL see SDZ350
QUEBRACHO TANNIN see QBJ000
QUECKSILBER (GERMAN) see MCW250
QUECKSILBER CHLORID (GERMAN) see MCY475
QUECKSILBER(I)-CHLORID (GERMAN) see MCW000
QUECKSILBER CHLORUER (GERMAN) see MCW000
QUECKSILBEROXID (GERMAN) see MCT500
QUECKSILBEROXID (GERMAN) see MDF750
QUEENSGATE WHITING see CAT775
QUELAMYCIN see QBS000
QUELETOX see FAQ900
QUELICIN see CMG250
QUELICIN see HLC500
QUELICIN CHLORIDE see HLC500
QUELLADA see BBQ500
QUEMICETINA see CDP250
QUEN see CFZ000
QUERCETIN see QCA000
QUERCETIN, 3-(6-DEOXY-α-l-MANNOPYRANOSIDE) see QCJ000
QUERCETIN-3-(6-o-(6-DEOXY-α-l-MANNOPYRANOSYL)-β-d-GLUCOPYRANOSIDE) see RSU000
QUERCETIN DIHYDRATE see QCA175
QUERCETINE see QCA000
QUERCETIN RHAMNOGLUCOSIDE see RSU000
QUERCETIN-3-RHAMNOGLUCOSINE see RSU000
QUERCETIN-3-(6-o-α-l-RHAMNOPYRANOSYL-β-d-GLUCOPYRANOSIDE) see RSU000
QUERCETIN-3-l-RHAMNOSIDE see QCJ000

QUERCETIN-3-RUTINOSIDE see RSU000
QUERCETOL see QCA000
QUERCITIN see QCA000
QUERCITRIN see QCJ000
QUERCUS AEGILOPS L. TANNIN see VCK000
QUERCUS FALCATA PAGODAEFOLIA see CDL750
QUERTINE see QCA000
QUERTON 28CL see DTF820
QUERYL see CNR125
QUESTEX 4 see EIV000
QUESTEX 4H see EIX000
QUESTIOMYCIN A see QCJ275
QUESTRAN see CME400
QUESYL see DEL300
QUETIAPINE FUMARATE see QCJ300
QUETIMID see TEH500
QUETINIL see DCK759
QUIACTIN see OLU000
QUIADON see MCH535
QUIATRIL see DCK759
QUICK see CJJ000
QUICKLIME (DOT) see CAU500
QUICKPHOS see AHE750
QUICKSAN see ABU500
QUICKSET EXTRA see MKA500
QUICKSET SUPER see MKA500
QUICK SILVER see MCW250
QUIDE see PII500
QUIESCIN see RDK000
QUIETAL see QCS000
QUIETALUM see QCS000
QUIETIDON see MQU750
QUIETOPLEX see TEH500
QUI-LEA see QFA250
QUILIBREX see CFZ000
QUILONE see LGO100
QUILON S see CMH300
QUILONUM RETARD see LGZ000
QUIMAR see CML880
QUIMOTRASE see CML880
QUINACETOPHENONE see DMG600
QUINACHLOR see CLD000
QUINACRIDONE MAGNETA see DTV360
QUINACRINE see ARQ250
QUINACRINE DIHYDROCHLORIDE see CFU750
QUINACRINE, DIMETHANESULFONATE see QCS900
QUINACRINE ETHYL M/2 see QCS875
QUINACRINE ETHYL MUSTARD see DFH000
QUINACRINE HYDROCHLORIDE see CFU750
QUINACRINE METHANESULFONATE see QCS900
QUINACRINE MUSTARD see QDJ000
QUINACRINE MUSTARD see QDS000
QUINACRINE MUSTARD DIHYDROCHLORIDE see QDS000
QUINADOME see DNF600
QUINAGAMINE see CLD000
QUINAGLUTE see QDS225
QUINALBARBITAL see SBM500
QUINALBARBITONE see SBM500
QUINALBARBITONE SODIUM see SBN000
QUINALDIC ACID see QEA000
QUINALDINE see QEJ000
QUINALDINE, 4,4'-(DECAMETHYLENEDIIMINO)DI-, DIACETATE see SAP600
QUINALDINIC ACID see QEA000
QUINALDINIUM, 1-ETHYL-, IODIDE see EPD600
QUINALIZARIN see TDD000
QUINALIZARINE see TDD000
QUINALPHOS see DJY200
QUINALSPAN see SBN000
QUINAMBICIDE see CHR500
QUINAPYRAMINE see AQN625

QUINAPYRAMINE METHYL SUFLATE see AQN625
QUINAZINE see QRJ000
QUINAZOLINE, 4-(2-(1,1'-BIPHENYL)-4-YLETHOXY)- see BGM070
QUINAZOLINE, 4-((4-(1,1-DIMETHYLETHYL)PHENYL)ETHOXY)- see FAK200
QUINAZOLINEDIONE see QEJ800
QUINAZOLINE-2,4-DIONE see QEJ800
2,4-QUINAZOLINEDIONE see QEJ800
(1H,3H)QUINAZOLINE DIONE-2,4 see QEJ800
2,4(1H,3H)-QUINAZOLINEDIONE see QEJ800
4-QUINAZOLINOL, 2-METHYL-(8CI) see MPF500
4-QUINAZOLINONE see QFA000
4(3H)-QUINAZOLINONE see QFA000
4(3H)-QUINAZOLINONE, 3-(2-CHLOROPHENYL)-2-METHYL- see MIH925
4(3H)-QUINAZOLINONE, 3-(o-CHLOROPHENYL)-2-METHYL- see MIH925
4(3H)-QUINAZOLINONE, 2-METHYL- see MPF500
4(1H)-QUINAZOLINONE, 2-METHYL-(7CI,9CI) see MPF500
4(3H)-QUINAZOLINONE, 2-METHYL-3-(o-TRIFLUOROMETHYLPHENYL)-, HYDROCHLORIDE see MQD800
QUINAZOLINO(3',2':1,6)PYRIDO(2,3-B)(1,4)BENZODIAZEPINE-9,16-DIONE, 6,7,7A,8-TETRAHYDRO-, (−)- see ARS200
QUINCLORAC see QMS100
QUINCLORAC TECH see QMS100
10-(3-QUINCULIDINYLMETHYL)PHENOTHIAZINE see MCJ400
QUINDO MAGNETA RV 6803 see DTV360
QUINDO MAGNETA RV 6831 see DTV360
QUINDOXIN see QSA000
QUINERCYL see CLD000
QUINESTROL see QFA250
QUINEX see PFN250
QUINEXIN see SAN600
QUINGESTANOL ACETATE see QFA275
QUINGFENGMYCIN see ARP000
QUINHYDRONE see QFJ000
QUINICARDINE see QFS000
QUINICARDINE see QHA000
QUINICINE OXALATE see QFJ300
QUINIDATE see QHA000
QUINIDEX see QFS000
QUINIDEX see QHA000
QUINIDINE see QFS000
(+)-QUINIDINE see QFS000
α-QUINIDINE see CMP910
QUINIDINE BISULFATE see QFS100
QUINIDINE GLUCONATE see QDS225
QUINIDINE-d-GLUCONATE (salt) see QDS225
QUINIDINE MONO-d-GLUCONATE (salt) see QDS225
QUINIDINE MONOSULFATE see QHA000
QUINIDINE POLYGALACTURONATE see GAX000
QUINIDINE SULFATE see QHA000
QUINIDINE SULFATE (1:1) (salt) see QFS100
QUINIDINE SULFATE (2:1) (salt) see QHA000
QUINIDINE SULPHATE see QHA000
QUINIDINE YELLOW KT see CMM510
QUINILON see CLD000
QUININE see QHJ000
(−)-QUININE see QHJ000
β-QUININE see QFS000
QUININE BIMURIATE see QIJ000
QUININE BISULFATE see QMA000
QUININE CHLORIDE see QJS000
QUININE DIHYDROCHLORIDE see QIJ000
(−)-QUININE DIHYDROCHLORIDE see QIJ000

QUININE ETHIODIDE see QIS000
QUININE FORMATE see QIS300
QUININE, FORMATE (SALT) see QIS300
QUININE HYDROBROMIDE see QJJ100
QUININE HYDROCHLORIDE see QJS000
QUININE HYDROGEN SULFATE see QMA000
QUININE MONOHYDROCHLORIDE see QJS000
QUININE MURIATE see QJS000
QUININE SULFATE see QMA000
QUINIPHEN see IEP200
QUINITEX see QHA000
QUINIZARIN see DMH000
QUINIZARINE GREEN BASE see BLK000
QUINO(2,3-B)ACRIDIN-7,14-DIONE, 5,12-DIHYDRO-2,9-DIMETHYL- see DTV360
QUINODIMETHAN, TETRACYANO- see TBW750
QUINOFEN see PGG000
QUINOFOP-ETHYL see QMA100
QUINOFORM see QIS300
QUINOL see HIH000
8-QUINOL see QPA000
β-QUINOL see HIH000
QUINOL DIMETHYL ETHER see DOA400
2-QUINOLINAMINE (9CI) see AMK700
4-QUINOLINAMINE, 8-FLUORO-N-(2-(4-PHENYL-2-THIAZOLYL)ETHYL)- see FLL200
4-QUINOLINAMINE, 8-FLUORO-N-(2-(2-THIENYL)ETHYL)- see FLZ090
4-QUINOLINAMINE, N-HYDROXY-N-METHYL-, 1-OXIDE see HLX550
4-QUINOLINAMINE, N,N'-1,10-DECANEDIYLBIS(2-METHYL)-, DIACETATE see SAP600
7-QUINOLINAMINE, 2,4,6-TRIMETHYL-(9CI) see AMV780
QUINOLINE see QMJ000
QUINOLINE, 6-ACETAMIDO- see QPS200
QUINOLINE-2-ALDEHYDECHLOROMETHYLATE-(p-(BIS-β-CHLOROETHYL)-AMINO)ANIL) see IAK100
3-QUINOLINEAMINE see AMK725
QUINOLINE, 2-AMINO- see AMK700
QUINOLINE, 3-AMINO- see AMK725
QUINOLINE, 4-((4-AMINO-1-METHYLBUTYL)AMINO)-7-CHLORO-, DIHYDROCHLORIDE see CLD100
QUINOLINE, 7-AMINO-2,4,6-TRIMETHYL- see AMV780
QUINOLINE-6-AZO-p-DIMETHYLANILINE see DPR000
QUINOLINE, 5-BROMO-6-METHOXY-8-NITRO- see BNN200
QUINOLINE, 3-BROMO-8-NITRO- see BNT300
6-QUINOLINE CARBONYL AZIDE see QMS000
QUINOLINE-2-CARBOXYLIC ACID see QEA000
2-QUINOLINECARBOXYLIC ACID see QEA000
8-QUINOLINECARBOXYLIC ACID, 3,7-DICHLORO- see QMS100
2-QUINOLINECARBOXYLIC ACID, 5,7-DICHLORO-4-((DIPHENYLAMINO)CARBONYLAMINO)-, METHYL ESTER see MJV900
3-QUINOLINECARBOXYLIC ACID, 1,4-DIHYDRO-1-(2,4-DIFLUOROPHENYL)-6-FLUORO-7- (3-METHYL-1-PIPERAZINYL)-4-OXO- see TAK850
3-QUINOLINECARBOXYLIC ACID, 1,4-DIHYDRO-1-ETHYL-6-FLUORO-4-OXO-7-(1H-PYRROL-1-YL)- see FLV050
3-QUINOLINECARBOXYLIC ACID, 2-(4,5-DIHYDRO-4-METHYL-4-(1-METHYLETHYL)-5-OXO-1H-IMIDAZOL-2-YL)- see IAH100

3-QUINOLINECARBOXYLIC ACID, 1-ETHYL-6,8-DIFLUORO-1,4-DIHYDRO-4-OXO-7-(4-PYRIDINYL)- (9CI) see DLB500
4-QUINOLINECARBOXYLIC ACID, 3-ETHYL-, ETHYL ESTER see EHM200
3-QUINOLINECARBOXYLIC ACID, 6-FLUORO-1-(5-FLUORO-2-PYRIDINYL)-1,4-DIHYDRO-7-(4-METHYL-1-PIPERAZINYL)-4-OXO-, MONOHYDROCHLORIDE see FLT200
QUINOLINE, 2-(2-CHLOROETHYLAMINO)-, HYDROCHLORIDE see QMS200
QUINOLINE, 8-CHLORO-4-(2-METHYLPHENOXY)- see CIR600
QUINOLINE, 4,7-DICHLORO- see DGJ250
QUINOLINE, 5,7-DIFLUORO- see DKI850
QUINOLINE, 1,2-DIHYDRO-2,2,4-TRIMETHYL-, HOMOPOLYMER see PJQ750
2,4-QUINOLINEDIOL see QNA000
trans-QUINOLINE-5,6,7,8-DIOXIDE see QNA100
QUINOLINE, 6-DODECYL-1,2-DIHYDRO-2,2,4-TRIMETHYL- see SAU480
2-QUINOLINEETHANOL see QNJ000
QUINOLINE, 2-ETHYL-4-NITRO-, 1-OXIDE see NGA500
QUINOLINE, 5-FLUORO- see FLV060
QUINOLINE, 7-FLUORO- see FLV070
4-QUINOLINEMETHANOL, 6,8-DICHLORO-2-PHENYL-α-2-PIPERIDINYL- see PIU200
4-QUINOLINEMETHANOL, 6,8-DICHLORO-2-PHENYL-α-2-PIPERIDYL- see PIU200
4-QUINOLINEMETHANOL, 6,8-DICHLORO-2-PHENYL-α-(2-PIPERIDYL)-, MONOHYDROCHLORIDE see PIU200
5-QUINOLINEMETHANOL, 1,2,3,4-TETRAHYDRO-2-METHYL- see MPP300
QUINOLINE, 6-METHOXY-8-((5-(PROPYLAMINO)PENTYL)AMINO)-, PHOSPHATE see MFM100
QUINOLINE, 2-METHYL-4-NITRO, 1-OXIDE see MMQ250
QUINOLINE-7,8-OXIDE see QNJ200
5-QUINOLINESULFONIC ACID, 8,8'-((HYDROXYSTIBYLENE)BIS(OXY))BIS(7-FORMYL)-, DISODIUM SALT see AQE320
2-QUINOLINE THIOACETAMIDE HYDROCHLORIDE see QOJ000
2-QUINOLINETHIOL see QOJ100
2(1H)-QUINOLINETHIONE see QOJ100
QUINOLINE YELLOW see CMM510
QUINOLINE YELLOW A SPIRIT SOLUBLE see CMS240
QUINOLINE YELLOW BASE see CMS240
QUINOLINE YELLOW EXTRA see CMM510
QUINOLINE YELLOW SPIRIT SOLUBLE see CMS240
QUINOLINE YELLOW SS see CMS240
QUINOLINIC ACID see QOJ200
QUINOLINIC ACID IMIDE see POQ750
QUINOLINIUM, 6-AMINO-1-METHYL-4-((4-(((((1-METHYLPYRIDINIUM-4-YL)AMINO)PHENYL)CARBAMOYL)PHENYL)AMINO)-, DIBROMIDE see QOJ250
QUINOLINIUM DIBROMIDE see QOJ250
QUINOLINIUM, 1-ETHYL-4-(3-(1-ETHYL-4(1H)-QUINOLINYLIDENE)-1-PROPENYL)-, IODIDE (9CI) see KHU050
2-QUINOLINOL see CCC250
8-QUINOLINOL see QPA000
8-(QUINOLINOLATO)METHYL MERCURY see MLH000
(8-QUINOLINOLATO)TRIBUTYLSTANNANE see TIB000
8-QUINOLINOL, BENZOATE (ester) see HOE100
8-QUINOLINOL, 5-CHLORO- see CLD600

R 2063 see EHT500
R 2113 see DBA875
R 2167 see HAF400
R 2170 see DAP000
R 225A see DFW830
R 225B see DFW850
R 2267 see AGW675
R-2323 see ENX575
R-2498 see TKK500
R 2498 see TKK750
R 2858 see MRU600
R 3345 see FHG000
R 3365 see PJA140
R-3588 see CNR125
R 3612 see SMQ500
R 40B1 see MHR200
R 4263 see PDW500
R 4263 see PDW750
R 4310 see DLY900
R 4318 see TKG000
R-4461 see DNO800
R-4572 see EKO500
R 4584 see FGU000
R 4584 see FLK100
R 4749 see DYF200
R-4845 see BCE825
R 4929 see BBU800
R-5,158 see DJA400
R 5240 see PDW750
R 5255 see MFB350
R-5461 see BJD000
R 6238 see PIH000
R6597 see NHQ950
R 6700 see OAN000
R 6790 see MEV600
R6998 see MFB410
R7000 see MFB400
R 7158 see FLJ000
R 7237 see ACQ790
R 7372 see MEW775
R 7542 see NHR100
R 8284 see PMF550
R 8299 see TDX750
R 9298 see CLS250
R 9985 see MDV500
Rh-110 see RHF150
Rh 116 see RGW100
R-10044 see FLZ025
R 11333 see BNU725
R 114B2 see FOO525
R-12,564 see LFA020
R12 (DOT) see DFA600
R13 (DOT) see CLR250
R 14487 see DJW890
R 14493 see DJW880
R 14789 see MEL525
R 14805 see DJO420
R 14,827 see EAE000
R 14889 see MQS560
R 14950 see FDD080
R14 (DOT) see CBY250
R 15018 see EEQ200
R 15,175 see ASD000
R 15201 see ACU300
R 15574 see PFR130
R 15996 see DJW900
R 16341 see PAP250
R 16470 see DBE000
R 16659 see HOU100
R 17635 see MHL000
R 17899 see FDA887
R 17934 see OJD100
R 18134 see MQS550
R 18986 see AIH000
R21 (DOT) see DFL000
R22 (DOT) see CFX500
R 23979 see FPB875
R 25061 see TEN750
R-25788 see DBI300
R-28627 see DNT200
R 31520 see CFC100
R-33865 see DJV700

R 35443 see OMG000
R 38486 see HKB700
R 41400 see KFK100
R 51619 see CMS232
R114 (DOT) see DGL600
R 151885 see BJW600
R 152450 see FMR200
R500 (DOT) see DFB400
R502 (DOT) see FOO560
R503 (DOT) see FOO562
Ro 1-7977 see DNV800
Ro 2-7113 see AGV890
Ro 7-0582 see NHH500
R12B1 (DOT) see BNA250
R12B2 (DOT) see DKG850
R13B1 (DOT) see TJY100
R142B (DOT) see CFX250
Ro 04-7683 see INY100
Ro 05-9129 see NHG000
Ro 10-9359 see EMJ500
Ro 215535 see DMJ400
R1132a (DOT) see VPP000
Ro-5-6901/3 see FMQ000
R 8 (fungicide) see MLF250
R 20 (refrigerant) see CHJ500
R 31 (refrigerant) see CHI900
RA 7 see PIB910
RA 8 see PCP250
13-RA see VSK955
RA 1010 see DED100
RA 1226 see ALC250
RA-2258 see EKS200
β-RA see VSK950
RABANO (PUERTO RICO) see DHB309
RABBIT ANTISERUM TO OVINE
LUTEINIZING HORMONE see LIU302
RABBIT-BELLS see RBZ400
RABBIT FLOWER see FOM100
RABCIDE see TBT175
RABON see RAF100
RABOND see RAF100
RACCOON BERRY see MBU800
RACEMETHORPHAN HYDROBROMIDE
see RAF300
RACEMIC AMLODIPINE see AMY100
RACEMIC DIHYDROZEATIN see RAF400
RACEMIC LACTIC ACID see LAG000
RACEMIC MANDELIC ACID see MAP000
RACEMIC α-METHYLAZIDOALANINE see
ASG300
RACEMOMYCIN A see RAG300
RACEMOMYCIN E see RAG400
RACEMORPHAN see MKR250
RACEMORPHAN HYDROBROMIDE see
MDV250
RACEPHEDRINE see EAW100
RACEPHEDRINE HYDROCHLORIDE see
EAX500
RACEPHEN see AOB250
RACEPHEN see AOB500
(4"R)-4"-(ACETYLAMINO)-26-
(BENZOYLOXY)-5-o-DEMETHYL-4"-
DEOXYAVERMECTIN A1A see ABX830
(4"R)-4"-(ACETYLAMINO)-5-o-DEMETHYL -
4"-DEOXY-26-METHOXYAVERMECTIN
A1A see ABX840
RACRYL see ADV900
RACUMIN see EAT600
RACUSAN see DSP400
RADAPON see DGI400
RADAPON see DGI600
RADAR see PMS930
RADAZIN see ARQ725
RADDLE see IHC450
RAD-E-CATE see HKC500
RAD-E-CATE 16 see HKC500
RAD-E-CATE 25 see HKC000
RAD-E-CATE 25 see HKC500
RAD-E-CATE 35 see HKC500
RADEDORM see DLY000
RADEPUR see LFK000
RADIANT YELLOW see CMS210

RADIASURF 7125 see SKV000
RADIASURF 7145 see PKG500
RADIASURF 7155 see SKV100
RADIATION see RAQ000
RADIATION, IONIZING see RAQ010
RADICALISIN see RRA000
RADIOCIN see SPD500
RADIOGRAPHOL see SHX000
RADIOL GERMICIDAL SOLUTION see
EKN500
RADIOSTOL see VSZ100
RADIOTETRANE see TDE750
RADIUM see RAV000
RADIUM F see PJJ750
RADIZINE see ARQ725
RADOCON see BJP000
RADOKOR see BJP000
RADON see RBA000
RADONIL see SNK000
RADONIN see SNN300
RADONNA see CHJ750
RADOSAN see MEO750
RADOX see CFK000
RADOXONE TL see AMY050
RADSTERIN see VSZ100
RAFEX see DUS700
RAFFINOSE see RBA100
d-RAFFINOSE see RBA100
d-(+)-RAFFINOSE see RBA100
RAFLUOR see SHF500
RAGADAN see HBK700
RAGUAR (GUAM) see TOA275
RAGWORT see RBA400
RAIN LILY see RBA500
RAISIN de COULEUVRE (CANADA) see
MRN100
RAK 1 see GJU050
RAKUTO AMARANTH see FAG020
RALABOL see RBF100
RALGRO see RBF100
RALLY see MRW775
(R+)-5-ALLYL-5-(1-
METHYLBUTYL)BARBITURIC ACID
SODIUM SALT see SBL500
RALONE see RBF100
RAM-326 see ORE000
RAMAPO see PJQ100
RAMENTACEONE see RBF200
RAMETIN see HMV000
RAMIK see DVV600
RAMIZOL see AMY050
RAMOR see TEI000
R/AMP see RKP000
RAMPART see PGS000
RAMROD see CHS500
RAM'S CLAWS see FBS100
RAMUCIDE see CJJ000
RAMYCIN see FQU000
RANAC see CJJ000
RANBAXY see PBP400
RANCOSIL see SCK600
RANDOLECTIL see BSZ000
RANDONOS see CMG675
RANDOX see CFK000
RANEY ALLOY see NCW500
RANEY COPPER see CNI000
RANEY NICKEL see NCW500
RANGOON YELLOW see DEU000
RANIDIL see RBF400
RANITIDINE BISMUTH CITRATE see
RBF450
RANITIDINE HYDROCHLORIDE see
RBF400
RANITOL see AFG750
RANKOTEX see CIR500
RANKOTEX see RBF500
RANTOX T see CFK000
RANTUDIL see AAE625
RANUNCULUS ACRIS see FBS100
RANUNCULUS BULBOSUS see FBS100
RANUNCULUS SCELERATUS see FBS100
RAPACODIN see DKW800

RAPAMYCIN see RBK000
RAPESEED OIL see RBK200
RAPE SEED OIL see RBK200
RAPHATOX see DUS700
RAPHETAMINE see BBK000
RAPHETAMINE PHOSPHATE see AOB500
RAPHONE see CIR250
RAPI-CURE CHVE see BLU600
RAPI-CURE DVE-3 see TDY800
RAPID see DOX600
RAPIDOSEPT see DET000
RAPISOL see DJL000
RAPYNOGEN see ICC000
RARE EARTH OXIDES see LAX100
RARE EARTHS see RBP000
RAS-26 see NIJ500
RASCHIT see CFD990
RASEN-ANICON see CFD990
RASIKAL see SFS000
RASORITE 65 see DXG035
RASPBERIN see SAH000
RASPBERRY KETONE see RBU000
RASPBERRY KETONE METHYL ETHER see MFF580
RASPBERRY RED for JELLIES see FAG020
RASTINON see BSQ000
RAT ACTH-RELEASING HORMONE see HGL700
RATAFIN see ABF500
RATAK see BGO100
RATAK see TAC800
RATAK PLUS see TAC800
RAT-A-WAY see ABF500
RAT-A-WAY see WAT200
RATBANE 1080 see SHG500
RAT-B-GON see WAT200
RAT-O-CIDE RAT BAIT see RCF000
RAT CORTICOTROPIN-RELEASING FACTOR see HGL700
RAT CORTICOTROPIN-RELEASING FACTOR-41 see HGL700
RAT CRF see HGL700
RAT CRF(1-41) see HGL700
RAT-GARD see WAT200
RAT HYPOTHALAMIC CRF see HGL700
RATICATE see NNF000
RATIMUS see BMN000
RATINDAN 1 see DVV600
RAT & MICE BAIT see WAT200
RAT-NIP see PHP010
RATO see TMO000
RATOMET see CJJ000
RATON see RBZ000
RATOX see TEL750
RAT'S END see RCF000
RATS-NO-MORE see WAT200
RATSUL SOLUBLE see WAT220
RATTENGIFTKONSERVE see TEM000
RATTEX see CNV000
RATTLE BOX see RBZ400
RATTLER see GIQ100
RATTLESNAKE WEED see FAB100
RATTLEWEED (JAMAICA) see RBZ400
RATTRACK see AQN635
RAUBASINE see AFG750
RAUBASNE HYDROCHLORIDE see AFH000
RAUCAP see RDK000
RAUCUMIN 57 see EAT600
RAUDIFORD see RDK000
RAUDIXIN see RDK000
RAUDIXOID see RDK000
RAUGAL see RDK000
RAUGALLINE see AFH250
RAULEN see RDK000
RAULOYCIN see RDK000
RAULOYDIN see RDK000
RAUMALINA see AFG750
RAUMANON see ILD000
RAUMORINE see RDK000
RAUNERVIL see RDK000
RAUNORINE see RDF000
RAUNORINE see RDK000

RAUNORMIN "ORZAN" see RDK000
RAUNOVA see RCA200
RAUNOVA see RDK000
RAUPASIL see RDK000
RAUPOID see RDK000
RAUPYROL see TLN500
RAURESCINE see TLN500
RAURINE see RDK000
RAUSAN see RDK000
RAU-SED see RDK000
RAUSEDAN see RDK000
RAUSEDIL see RDK000
RAUSEDYL see RDK000
RAUSERPEN-ALK see RDK000
RAUSERPIN see RDK000
RAUSERPIN-ALK see RDK000
RAUSERPINE see RDK000
RAUSERPOL see RDK000
RAUSINGLE see RDK000
RAUTRAX see RCA275
RAUTRIN see RDK000
RAUVILID see RDK000
RAUVLID see RDK000
RAUWASEDIN see RDK000
RAUWILID see RDK000
RAUWILOID see RDK000
RAUWILOID+ see RDK000
RAUWIPUR see RDK000
RAUWOLEAF see RDK000
RAUWOLFIN see AFH250
RAUWOLFINE see AFH250
RAUWOLSCINE HYDROCHLORIDE see YCA000
RAUWOPUR "BYK" see RDK000
RAV 7 see AGD250
RAVAGE see RCA300
RAVEN see CBT750
RAVEN 30 see CBT750
RAVEN 420 see CBT750
RAVEN 500 see CBT750
RAVEN 8000 see CBT750
RAVIAC see CJJ000
RAVINYL see PKQ059
RAVONA see CAV000
RAVONAL see PBT500
RAVYON see CBM750
RAWETIN see HMV000
RAWILID see RDK000
RAW SHALE OIL see COD750
RAYBAR see BAP000
RAY-GLUCIRON see FBK000
RAYON BLACK G see CMN240
RAYON BLACK GSN see CMN240
RAYON BLACK M see CMN240
RAYON FAST BLACK B see CMN240
RAYOPHANE see CCU150
RAYOX see TGG760
RAYWEB Q see CCU150
RAZIOSULFA see AIF000
RAZOL DOCK KILLER see CIR250
RAZOXANE see RCA375
RAZOXIN see PIK250
RAZOXIN see RCA375
RB see RJF400
R-242-B see CKI625
R-451-B see ERD000
1489 RB see PIZ000
RB 1489 see PIZ000
RB 1509 see CGV250
RBA 777 see CLW000
R-BASE see CBN100
R 74 BASE see LJD500
RB-BL see IHC550
1-RBO-12-A see PNA200
RC 17 see CAY300
RC 20 see EHG030
RC 146 see CNG250
R-C 318 see CPS000
RC 5626 see DTF820
RC 5629 see DRR800
RC 61-91 see IAG600
RC 61-96 see BEU800

RC 72-01 see RCA435
RC 72-02 see RCA450
R.C. 27-109 see DGQ500
RC 30-109 see DGQ500
R 225CA see DFW850
RCA WASTE NUMBER P105 see SFA000
RCA WASTE NUMBER U203 see SAD000
RCA WASTE NUMBER U205 see SBR000
R 225CB see DFW830
RC COMONOMER DBM see DED600
RC COMONOMER DOF see DVK600
RC COMONOMER DOM see BJR000
RC 172DBM see AHE250
RCH 1000 see PJS750
R-CHLORODIPHENYL SULPHONE see CKI625
R 30 730 CITRATE SALT see SNH150
RC PLASTICIZER B-17 see BSL600
RC PLASTICIZER DOP see DVL700
RCRA WASTE NUMBER P001 see WAT200
RCRA WASTE NUMBER P002 see ADD250
RCRA WASTE NUMBER P003 see ADR000
RCRA WASTE NUMBER P004 see AFK250
RCRA WASTE NUMBER P005 see AFV500
RCRA WASTE NUMBER P006 see AHE750
RCRA WASTE NUMBER P007 see AKT750
RCRA WASTE NUMBER P008 see AMI500
RCRA WASTE NUMBER P008 see POO050
RCRA WASTE NUMBER P009 see ANS500
RCRA WASTE NUMBER P010 see ARB250
RCRA WASTE NUMBER P011 see ARH500
RCRA WASTE NUMBER P012 see ARI750
RCRA WASTE NUMBER P013 see BAK750
RCRA WASTE NUMBER P014 see PFL850
RCRA WASTE NUMBER P015 see BFO750
RCRA WASTE NUMBER P016 see BIK000
RCRA WASTE NUMBER P017 see BNZ000
RCRA WASTE NUMBER P018 see BOL750
RCRA WASTE NUMBER P020 see BRE500
RCRA WASTE NUMBER P021 see CAQ500
RCRA WASTE NUMBER P022 see CBV500
RCRA WASTE NUMBER P023 see CDY500
RCRA WASTE NUMBER P024 see CEH680
RCRA WASTE NUMBER P026 see CKL000
RCRA WASTE NUMBER P027 see CKT250
RCRA WASTE NUMBER P028 see BEE375
RCRA WASTE NUMBER P029 see CNL000
RCRA WASTE NUMBER P030 see COI500
RCRA WASTE NUMBER P031 see COO000
RCRA WASTE NUMBER P033 see COO750
RCRA WASTE NUMBER P034 see CPK500
RCRA WASTE NUMBER P036 see DGB600
RCRA WASTE NUMBER P037 see DHB400
RCRA WASTE NUMBER P039 see DXH325
RCRA WASTE NUMBER P040 see DJX400
RCRA WASTE NUMBER P042 see VGP000
RCRA WASTE NUMBER P043 see IRF000
RCRA WASTE NUMBER P044 see DSP400
RCRA WASTE NUMBER P045 see DAB400
RCRA WASTE NUMBER P046 see DTJ400
RCRA WASTE NUMBER P047 see DUS700
RCRA WASTE NUMBER P048 see DUZ000
RCRA WASTE NUMBER P049 see DXL800
RCRA WASTE NUMBER P050 see EAQ750
RCRA WASTE NUMBER P051 see EAT500
RCRA WASTE NUMBER P054 see EJM900
RCRA WASTE NUMBER P056 see FEZ000
RCRA WASTE NUMBER P057 see FFF000
RCRA WASTE NUMBER P058 see SHG500
RCRA WASTE NUMBER P059 see HAR000
RCRA WASTE NUMBER P060 see IKO000
RCRA WASTE NUMBER P062 see HCY000
RCRA WASTE NUMBER P063 see HHS000
RCRA WASTE NUMBER P064 see MKX250
RCRA WASTE NUMBER P065 see MDC000
RCRA WASTE NUMBER P066 see MDU600
RCRA WASTE NUMBER P067 see PNL400
RCRA WASTE NUMBER P068 see MKN000
RCRA WASTE NUMBER P069 see MLC750
RCRA WASTE NUMBER P070 see CBM500
RCRA WASTE NUMBER P071 see MNH000
RCRA WASTE NUMBER P072 see AQN635

RCRA WASTE NUMBER U210 see PCF275
RCRA WASTE NUMBER U211 see CBY000
RCRA WASTE NUMBER U212 see TBT000
RCRA WASTE NUMBER U213 see TCR750
RCRA WASTE NUMBER U214 see TEI250
RCRA WASTE NUMBER U215 see TEJ000
RCRA WASTE NUMBER U216 see TEJ250
RCRA WASTE NUMBER U217 see TEK750
RCRA WASTE NUMBER U218 see TFA000
RCRA WASTE NUMBER U219 see ISR000
RCRA WASTE NUMBER U220 see TGK750
RCRA WASTE NUMBER U221 see TGL500
RCRA WASTE NUMBER U221 see TGL750
RCRA WASTE NUMBER U222 see TGS500
RCRA WASTE NUMBER U223 see TGM740
RCRA WASTE NUMBER U223 see TGM750
RCRA WASTE NUMBER U225 see BNL000
RCRA WASTE NUMBER U226 see MIH275
RCRA WASTE NUMBER U227 see TIN000
RCRA WASTE NUMBER U228 see TIO750
RCRA WASTE NUMBER U230 see TIV750
RCRA WASTE NUMBER U231 see TIW000
RCRA WASTE NUMBER U232 see TAA100
RCRA WASTE NUMBER U233 see TIX500
RCRA WASTE NUMBER U234 see TMK500
RCRA WASTE NUMBER U235 see TNC500
RCRA WASTE NUMBER U236 see CMO250
RCRA WASTE NUMBER U237 see BIA250
RCRA WASTE NUMBER U238 see UVA000
RCRA WASTE NUMBER U239 see XGS000
RCRA WASTE NUMBER U240 see DAA800
RCRA WASTE NUMBER U242 see PAX250
RCRA WASTE NUMBER U244 see TFS350
RCRA WASTE NUMBER U246 see COO500
RCRA WASTE NUMBER U247 see MEI450
RCTA see RJA500
RD 8 see SCK600
RD 120 see SCK600
RD 174 see TEC600
RD 292 see CKG000
RD 406 see DGB000
R.D. 1403 see EPD500
RD 1572 see DBV400
RD 2195 see CEP000
R.D. 2786 see WAK000
RD 4593 see CIR500
RD 4593 see RBF500
RD-6584 see RDP300
RD7693 see BAV000
RD 11654 see IJG000
R.D. 13621 see IIU000
R.D. 27419 see MJL250
RDGE see REF000
RDX see CPR800
RDX 76679 see REA200
RDX and HMX MIXTURES, desensitized with
not <10% phlegmatizer by weight (UN 0391)
(DOT) see CPR800
RDX and HMX MIXTURES, wetted with not
<15% water by weight (UN 0391) (DOT) see
CPR800
RDX, desensitized (UN 0483) (DOT) see
CPR800
RDX, wetted with not <15% water by weight
(UN 0072) (DOT) see CPR800
RE-4355 see NAG400
RE 5030 see BSG300
RE 5454 see MIA000
RE 5655 see MOU750
RE 1-0185 see ELH600
RE 12420 see DOP600
RE 20615 see CGI550
RE 40885 see MGJ775
RE-45550 see EPR100
RE 45601 see CMV485
R236EA see HDE050
REACID see COH250
REACTHIN see AES650
REACTIVE BLACK 8 see CMS220
REACTIVE BLUE 4 see CMS222
REACTIVE BLUE 19 see BMM500
REACTIVE BLUE 21 see CMS224

REACTIVE BRILLIANT RED 5SKH see
PMF540
REACTIVE TURQUOISE K see RCA500
REACTOMER RC 20 see EHG030
REACTROL see CMV400
READPRET KPN see DTG000
REALGAR see ARF000
REANIMIL see DNV000
REAZID see COH250
REAZIDE see COH250
REBELATE see DSP400
REBEMID see BCM250
REBONEX see CBT750
REBRAMIN see VSZ000
REBUGEN see IIU000
REC 1-0060 see AJE350
REC 1-0185 see ELH600
REC 7-0040 see FCB100
REC 7-0518 see DNN000
REC 7-0591 see AJO750
REC 15-1533 see DAP880
REC 7/0267 see DNV000
REC 7/0267 see DNV200
REC 7/0268 see DRL600
RE 5305 (CALIFORNIA CHEMICAL) see
BSG250
RECANESCIN see RDF000
RECHETON see KGK000
RECININE see RJK100
RECINNAMINE see TLN500
RECIPIN see RDK000
RECITENSINA see TLN500
RECOFNAN see CMF350
RECOGNAN see CMF350
RECOIL see MFC100
RECOLIP see ARQ750
RECOLITE FAST RED RBL see MMP100
RECOLITE FAST RED RL see MMP100
RECOLITE FAST RED RYL see MMP100
RECOLITE FAST YELLOW B 2T see CMS210
RECOLITE FAST YELLOW BLF see CMS208
RECOLITE FAST YELLOW BLT see CMS208
RECOLITE ORANGE G see CMS145
RECOLITE RED LAKE C see CHP500
RECOLITE RED LYS see NAP100
RECOLITE YELLOW GB see DEU000
RECOMBINANT HUMAN TUMOR
NECROSIS FACTOR α see HGL920
RECOMBINANT HUMAN TUMOR
NECROSIS FACTOR-α see HGL920
RECOMBINANT TISSUE-TYPE
PLASMINOGEN ACTIVATOR see PJB800
RECONIN see FOS100
RECONOX see PDP250
RECOP see CNK559
RECORDIL see ELH600
RECREIN see DPA000
RECTALAD-AMINOPHYLLINE see TEP500
RECTHORMONE OESTRADIOL see EDP000
RECTHORMONE TESTOSTERONE see
TBG000
RECTODELT see PLZ000
RECTULES see CDO000
RED 2G see CMM300
RED 6B see CMM400
RED #14 see HJF500
RED 102 see CMS155
RED 104 see ADG250
RED 10B see CMS228
RED 203 see CMS150
RED 219 see CMS160
1306 RED see FAG050
1379 RED see CMM300
1424 RED see CMS228
1425 RED see CMM400
1427 RED see FAG040
1671 RED see FAG040
1695 RED see FMU070
1860 RED see CHP500
11067 RED see CMS160
11070 RED see CMS155
11411 RED see FAG070

11427 RED see CMS228
11445 RED see BNH500
11554 RED see IHC450
11935 RED see NAP100
11938 RED see CMS150
11959 RED see HJF500
11969 RED see ADG250
12094 RED see CJD500
12101 RED see FAG050
12418 RED see MAC250
111440 RED see OHI200
RED 2B ACID see AJJ250
RED 3B ACID see CLO600
RED 4B ACID see AKQ000
RED ALGAE see AFK930
REDAMINA see VSZ000
REDAX see DWI000
RED B see XRA000
RED BALL see CAT775
RED 2G BASE see NEO500
RED 3G BASE see KDA050
RED BASE CIBA IX see CLK220
RED BASE CIBA IX see CLK235
RED BASE CIBA V see NEQ000
RED BASE CIBA VI see KDA050
RED BASE CIBA X see MMF780
RED BASE 3 GL see KDA050
RED BASE IRGA IX see CLK220
RED BASE IRGA IX see CLK235
RED BASE IRGA V see NEQ000
RED BASE IRGA VI see KDA050
RED BASE IRGA X see MMF780
RED BASE NB see NEQ000
RED BASE NRL see MMF780
RED BASE NTR see CLK220
RED B BASE see NEQ000
RED BEAD VINE see RMK250
RED BEAN see NBR800
REDBIRD FLOWER OR CACTUS see SDZ475
RED CEDARWOOD OIL see CCR000
RED COPPER OXIDE see CNO000
REDDON see TAA100
REDDOX see TAA100
RED DYE No. 2 see FAG020
RED EMB see CMO870
RED EMBL see CMO870
REDERGIN see DLL400
RED FOR LAKE C TONER see CMS150
RED FUMING NITRIC ACID see NEE500
RED GTL see BAQ750
RED HOTS see NBR800
REDICOTE E 5 see LBX075
REDIFAL see SNN300
REDI-FLOW see BAP000
RED INK PLANT see PJJ315
RED IRON ORE see HAO875
RED IRON OXIDE see IHC450
REDISOL see VSZ000
RED KB BASE see CLK225
RED LEAD see LDS000
RED LEAD CHROMATE see LCS000
RED LEAD OXIDE see LDS000
RED LOBELIA see CCJ825
RED MERCURIC IODIDE see MDD000
RED MOROCCO see PCU375
RED No. 1 see FAG050
RED No. 2 see FAG020
RED No. 4 see FAG050
RED No. 5 see XRA000
RED No. 104 see ADG250
RED No. 205 see NAP100
RED No. 227 see CMS228
RED No. 228 see CJD500
RED NO. 40 see FAG100
RED NO. 213 see FAG070
RED NO. 219 see CMS160
RED OCHRE see IHC450
RED OIL see OHU000
RED OIL see TOD500
RED OXIDE of MERCURY see MCT500
RED PEPPER see PCB275
RED PRECIPITATE see MCT500

RED R see FMU070
RED RL BASE see MMF780
RED ROOT see AHJ875
RED SALT CIBA IX see CLK235
RED SALT IRGA IX see CLK235
RED SCARLET see CHP500
RED-SEAL-9 see ZKA000
REDSKIN see AGJ250
RED 3R SOLUBLE IN GREASE see SBC500
RED SPIDER LILY see REK325
RED SQUILL see RCF000
RED TETRAZOLIUM see TMV500
RED TR BASE see CLK220
RED TRS SALT see CLK235
REDUCED LACTOSE WHEY see WBL160
REDUCED MINERALS WHEY see WBL165
REDUCED-d-PENICILLAMINE see MCR750
REDUCTO see DKE800
REDUCTONE see SHR500
REDUFORM see NNW500
REDUL see GHK200
REDUSTEROL see NMV300
REDUTON see RLU000
RED VIOLET WASHABLE see HGE925
RED WEED see PJJ315
RED ZH see OHI200
REE see TFF100
REED AMINE 400 see DFY800
REED 10-51 BRUSH KILLER see TKU680
REED LV 2,4-D see ILO000
REED LV 400 2,4-D see ILO000
REED LV 600 2,4-D see ILO000
REELON see PAF250
REFINED BLEACHED SHELLAC see
SCC705
REFINED PETROLEUM WAX see PCT600
REFINED SOLVENT NAPHTHA see PCT250
REFLEX see CLS050
REFLEXYN see GKK000
REFORMIN see DJS200
REFOSPOREN see RCK000
REFRACTORY CERAMIC FIBERS see
RCK725
REFRIGERANT 12 see DFA600
REFRIGERANT 14 see CBY250
REFRIGERANT 22 see CFX500
REFRIGERANT 112 see TBP050
REFRIGERANT 113 see FOO000
REFRIGERANT 502 see FOO560
REFRIGERANT 112a see TBP000
REFRIGERANT R 14 see CBY250
REFRIGERANT R134A see EEC100
REFUGAL see CMW700
REFUNGINE see RCK730
REFUSAL see DXH250
REGAL see CBT750
REGAL 99 see CBT750
REGAL 300 see CBT750
REGAL 330 see CBT750
REGAL 600 see CBT750
REGAL 400R see CBT750
REGAL SRF see CBT750
REGARDIN see ARQ750
REGELAN see ARQ750
REGENERATED CELLULOSE see HHK000
REGENON see DIP600
REGENT see CBT750
REGIANIN see WAT000
REGIM 8 see TKQ250
REGIN 8 see TKQ250
REGION see SNQ710
REGITIN see PDW400
REGITINE see PDW400
REGITINE MESYLATE see PDW950
REGITINE METHANESULFONATE see
PDW950
REGITIN METHANESULPHONATE see
PDW950
REGITIPE see PDW400
REGLISSE (GUADELOUPE and HAITI) see
RMK250
REGLON see DWX800

REGLON see EJC025
REGLONE see DWX800
REGLONE see EJC025
REGONAL see MDL600
REGONOL see GLU000
REGONON HYDROCHLORIDE see DIP600
REGROTON see RDK000
REGULAR BLEACHED SHELLAC see
SCC700
REGULIN see BFW250
REGULIN see TBJ275
REGULOX see DMC600
REGULOX W see DMC600
REGULOX 50 W see DMC600
REGULTON see MQS100
REGUTOL see DJL000
REHIBIN see FBP100
REHORMIN see DJS200
REICHSTEIN'S F see MIW500
REICHSTEIN'S SUBSTANCE FA see CNS800
REICHSTEIN'S SUBSTANCE H see CNS625
REICHSTEIN'S SUBSTANCE M see CNS750
REICHSTEIN X see AFJ875
REIN GUARIN see GLU000
REISE-ENGLETTEN see DYE600
REKAWAN see PLA500
RELA see IPU000
RELACT see DLY000
RELAMINAL see DCK759
RELAN BETA see BEQ625
RELANE see DVP400
RELANIUM see DCK759
RELASOM see IPU000
RELAX see GKK000
RELAX see IPU000
RELAXAN see PDD300
RELAXANT see GGS000
RELAXAR see GGS000
RELAXIN see RCK740
RELAXYL-G see RLU000
RELBAPIRIDINA see PPO000
RELDAN see CMA250
RELEASIN see RCK740
RELEFACT LH-RH see LIU370
RELESTRID see GKK000
RELIBERAN see MDQ250
RELICOR see DHS200
RELICOR HYDROCHLORIDE see DHS200
RELITON SCARLET BA see ENP100
RELITON YELLOW C see AAQ250
RELITON YELLOW R see DUW500
RELIVERAN see AJH000
RELON P see PJY500
RELUTIN see HNT500
REMACEMIDE HYDROCHLORIDE see
RCK750
REMADERM YELLOW HPR see MDM775
REMALAN BRILLIANT BLUE R see BMM500
REMANTADIN see AJU625
REMASAN CHLOROBLE M see MAS500
REMASOL TURQUOISE BLUE B see CMS224
REMASOL TURQUOISE BLUE G see
CMS224
REMAZIN see HNI500
REMAZOL BLACK B see RCU000
REMAZOL BRILLIANT BLUE R see BMM500
REMAZOL YELLOW G see RCZ000
REMEFLIN see DNV000
REMEFLIN see DNV200
REMESTAN see CFY750
REMESTYP see MGC350
REMICYCLIN see TBX250
REMID see ZVJ000
REMIN see CKE750
REMKO see IGK800
REMOL TRF see BGJ250
REMONOL see RDZ900
REMSED see PMI750
REMTAL see TJL500
REMYLINE Ac see SLJ500
REMZYME PL 600 see GGA800
RENACIT 1 see NAP500

RENAFUR see NDY000
RENAGLADIN see VGP000
RENAL AC see ALT250
RENAL EG see ALS990
RENALEPTINE see VGP000
RENALINA see VGP000
RENAL MD see TGL750
RENAL PF see PEY500
RENAL SLA see DBO400
RENAL SO see CEG625
RENAMYCIN see ACJ250
RENARCOL see ARW250
RENARCOL see GGS000
RENARDIN see DMX200
RENARDINE see DMX200
RENBORIN see DCK759
RENE 77 see NCY135
RENEGADE see RCZ050
RENESE R see RDK000
RENGASIL see PJA220
RENNET see RCZ100
RENO-M-DIP see AOO875
RENOFORM see VGP000
RENOGRAFFIN M-76 see AOO875
RENOGRAFIN see AOO875
RENOLBLAU 3B see CMO250
RENOL MOLYBDATE RED RGS see
MRC000
RENO M see AOO875
RENO M 60 see AOO875
RENON see CLY600
RENONCULE (CANADA) see FBS100
RENOSTYPRICIN see VGP000
RENOSTYPTIN see VGP000
RENOSULFAN see SNN500
REN O-SAL see HMY000
RENSTAMIN see DBM800
RENTIAPRIL see FAQ950
RENTOVET see TAF675
RENTYLIN see PBU100
RENUMBRAL see HGB200
RENURIX see AOO875
REOFOS 95 see TKU300
REOMAX see DFP600
REOMOL D 79P see DVL700
REOMOL DOA see AEO000
REOMOL DOP see DVL700
REOMUCIL see CBR675
REORGANIN see RLU000
REOXYL see HFF500
REP see BCM250
REPAIRSIN see PKB500
REPARIL see EDK875
β-REPARIL see EDL500
REPARIL SODIUM SALT see EDM000
REPEL see DKC800
REPELLENT 612 see EKV000
β-REPELLIN see RCZ200
REPELTIN see AFL500
REPHOXITIN see CCS500
REPICIN see BEQ625
REPOC see PJS750
REPOCAL see CAV000
REPOISE see MFD500
REPOISE MALEATE see BSZ000
REPOSO-TMD see TBF750
REPPER 333 see EAU500
REPPER-DET see DKC800
REPRISCAL see SHL500
REPRODAL see AQH500
REPROMIX see MCA000
REPROTEROL HYDROCHLORIDE see
DNA600
REPTILASE see RDA350
REPTILASE R see RDA350
REPTILASE S see CMY725
REPUDIN-SPECIAL see DKC800
REPULSON see PAF550
REQUTOL see DJL000
RERANIL see TBO500
R-E-S see RDK000
76 RES see ADV900

RESACETOPHENONE see DMG400
β-RESACETOPHENONE see DMG400
RESACTIN A see SMC500
RESALTEX see RDK000
RESAMIN 155F see UTU500
RESAMINE FAST ORANGE G see CMS145
RESAMINE FAST YELLOW GGP see CMS210
RESAMINE PINK 3B see CMG750
RESAMINE RED GB see CMS150
RESAMINE RED RB see NAP100
RESAMINE RED RC see NAP100
RESAMINE RUBINE BC see CMS155
RESAMINE YELLOW GP see CMS210
RESAMIN HW 505 see UTU500
RESAMIN MW 811 see MCB050
RESANTIN see RDA375
RESARIT 4000 see PKB500
RESAZOIN see HNG500
RESAZURIN see HNG500
RESAZURINE see HNG500
RESCALOID see TLN500
RESCAMIN see TLN500
RESCIDAN see TLN500
RESCIN see TLN500
RESCINNAMINE see TLN500
RESCINPAL see TLN500
RESCINSAN see TLN500
RESCITEN see TLN500
RESCUE SQUAD see SHF500
RESCULA see IRR050
RESEDA BODY see PFR400
RESEDIN see RDK000
RESEDREX see RDK000
RESEDRIL see RDK000
RESE-LAR see RDK000
RESER-AR see RDK000
RESERBAL see RDK000
RESERCAPS see RDK000
RESERCEN see RDK000
RESERCRINE see RDK000
RESERFIA see RDK000
RESERJEN see RDK000
RESERLOR see RDK000
RESERP see RDK000
RESERPAL see RDK000
RESERPAMED see RDK000
RESERPANCA see RDK000
RESERPATE de METHYLE (FRENCH) see MPH300
RESERPENE see RDK000
RESERPEX see RDK000
RESERPIC ACID-METHYL ESTER, ESTER with 4-HYDROXY-3,5-DIMETHOXYBENZOIC ACID ETHYL CARBONATE see RCA200
RESERPIDEFE see RDK000
RESERPIDINE see RDF000
RESERPIL see RDK000
RESERPIN see RDK000
RESERPINA see RDK000
RESERPINE see RDK000
RESERPINENE see TLN500
RESERPINE PHOSPHATE see RDK050
RESERPINUM see RDK000
RESERPKA see RDK000
RESERPOID see RDK000
RESERPUR see RDK000
RESERP "WANDER" see RDK000
RESERSANA see RDK000
RESERUTIN see RDK000
RESIATRIC see RDK000
RESIBUFOGENIN see BOM650
RESIDINE see RDK000
RESIDUAL(HEAVY) FUEL OIL see FOP200
RESIDUAL OIL SOLVENT EXTRACT see MQV863
RESIDUAL OILS (PETROLEUM), ACID-TREATED (9CI) see MQV872
RESIDUES (PETROLEUM), CATALYTIC REFORMER FRACTIONATOR see RDK075
RESIDUES (PETROLEUM), THERMAL CRACKED see RDK100

RESIDUES (PETROLEUM), VACUUM see RDK200
RESIL see RLU000
RESIMENE 714 see MCB050
RESIMENE 717 see MCB050
RESIMENE 730 see MCB050
RESIMENE 731 see MCB050
RESIMENE 740 see MCB050
RESIMENE 745 see MCB050
RESIMENE 746 see MCB050
RESIMENE 747 see MCB050
RESIMENE 750 see MCB050
RESIMENE 753 see MCB050
RESIMENE 755 see MCB050
RESIMENE 817 see MCB050
RESIMENE 841 see MCB050
RESIMENE 842 see MCB050
RESIMENE R-881 see MCB075
RESIMENE RF 4518 see MCB050
RESIMENE RF 5306 see MCB050
RESIMENE RS 466 see MCB050
RESIMENE X 712 see MCB050
RESIMENE X 714 see MCB050
RESIMENE X 720 see MCB050
RESIMENE X 730 see MCB050
RESIMENE X 735 see MCB050
RESIMENE X 740 see MCB050
RESIMENE X 745 see MCB050
RESIMENE X 764 see MCB050
RESIMENE X 970 see UTU500
RESIMENE X 975 see UTU500
RESIMENE X 980 see UTU500
RESIMINE 975 see UTU500
RESIN 516 see MCB050
RESIN (solution) see RDP000
RESIN ACIDS and ROSIN ACIDS, CALCIUM SALTS see CAW500
S-RESIN AER 20 see UTU500
RESINATED INDO BLUE B 85 see DFN300
RESINA X see UTU500
RESINE see RDK000
RESINIFERATOXIN see RDP100
RESINOID IRIS see OHJ130
RESINOID ORRIS see OHJ130
RESINOL BROWN RRN see NBG500
RESINOL ORANGE G see CMP600
RESINOL ORANGE R see PEJ500
RESINOL RED 2B see SBC500
RESINOL RED G see CMS238
RESINOL RRN see NBG500
RESINOL YELLOW GR see DOT300
RESINO RED K see CMS148
RESIN SCARLET 2R see XRA000
RESIN, SMA 1440-H see SEA500
RESIN SOLUTION, flammable (DOT) see RDP000
RESIN T see CNH125
RESIN TOLU see BAF000
RESIPAL see TLN500
RESIREN RED TB see AKI750
RESIREN VIOLET TR see DBP000
RESIREN YELLOW TG see AAQ250
RESISAN see RDP300
RESISTAB see RDU000
RESISTAMINE see POO750
RESISTAMINE see TMP750
RESISTOFLEX see PKP750
RESISTOMYCIN see HAL000
RESISTOMYCIN (BAYER) see KAM000
RESISTOMYCIN (BAYER) see KAV000
RESISTOPHEN see MNV250
RESITAN see VBK000
RESKINNAMIN see TLN500
RESLOOM HP see MCB050
RESLOOM HP 50 see MCB050
RESLOOM M 75 see HDY000
RESMETHRIN see BEP500
(+)-cis-RESMETHRIN see RDZ875
(−)-trans-RESMETHRIN see RDZ885
(+)-trans-RESMETHRIN see BEP750
d-trans-RESMETHRIN see BEP750
RESMETRINA (PORTUGUESE) see BEP500

RESMIT see CGA000
RESOACETOPHENONE see DMG400
RESOBANTIN see DJM800
RESOBANTIN see XCJ000
RESOCALM see RDK000
RESOCHIN see CLD000
RESOCHIN see CLD250
RESOCHIN DIPHOSPHATE see CLD250
RESOFORM ORANGE G see PEJ500
RESOFORM ORANGE R see XRA000
RESOFORM RED G see SBC500
RESOFORM YELLOW GGA see DOT300
RESOIDAN see CCK125
RESOLIN BLUE BSL see CMP075
RESOLIN BLUE FBL see CMP070
RESOLIN BLUE FBL see DBQ220
RESOLIN BLUE I-FBL see DBQ220
RESOLIN RED FB see AKI750
RESOLIN RED FBE see AKI750
RESOLIN YELLOW 5R see CMP090
RESOMINE see RDK000
RESOQUINA see CLD000
RESOQUINE see CLD000
RESOQUINE see CLD250
β-RESORCALDEHYDE see REF100
RESORCIN see REA000
RESORCIN ACETATE see RDZ900
β-RESORCINALDEHYDE see REF100
RESORCINE see REA000
RESORCINE BROWN J see XMA000
RESORCINE BROWN R see XMA000
RESORCINE YELLOW see MRL100
RESORCINE YELLOW O EXTRA see MRL100
RESORCIN MONOACETATE see RDZ900
RESORCINOL see REA000
RESORCINOL, 2-AMINO-, HYDROCHLORIDE see AML600
RESORCINOL, 4-BENZYL- see BFI400
RESORCINOL BIS(DIPHENYL PHOSPHATE) see REA050
RESORCINOL BIS(2,3-EPOXYPROPYL)ETHER see REF000
RESORCINOL, DIACETATE see REA100
RESORCINOL, DIBENZOATE see BHB100
RESORCINOL DICYANATE see REA200
RESORCINOL DIGLYCIDYL ETHER see REF000
RESORCINOL DIMETHYL ETHER see REF025
RESORCINOL, 2,4-DINITROSO- see DVF300
β-RESORCINOLIC ACID see HOE600
RESORCINOL, 2-METHYL- see MPH400
RESORCINOL METHYL ETHER see REF050
RESORCINOL, MONOACETATE see RDZ900
RESORCINOL, MONOBENZOATE see HNH500
RESORCINOL MONOMETHYL ETHER see REF050
RESORCINOL OXYDIANILINE see REF070
RESORCINOLPHTHALEIN see FEV000
RESORCINOL PHTHALEIN SODIUM see FEW000
RESORCINOL, 2,4,6-TRINITRO-, BARIUM SALT, HYDRATE (2:1:1) see BAO900
RESORCINOL YELLOW see MRL100
RESORCINOL YELLOW A see MRL100
RESORCIN YELLOW see MRL100
RESORCINYL DIGLYCIDYL ETHER see REF000
RESORCITATE see RDZ900
RESORCY see REA200
β-RESORCYLALDEHYDE see REF100
α-RESORCYLIC ACID see REF200
β-RESORCYLIC ACID see HOE600
β-RESORCYLIC ALDEHYDE see REF100
RESORIN RED FBE see AKI750
RESOTROPIN see HEI500
RESOXOL see SNN500
RESPAIRE see ACH000
RESPENYL see RLU000

RESPENYL see RLU000
RESPERIN see RDK000
RESPERINE see RDK000
RESPIFRAL see DMV600
RESPILENE see MFG250
RESPIRIDE see DAI200
RESPITAL see RDK000
RESPONSAR see REF250
RESPRAMIN see AJD000
RESTAMIN see BBV500
RESTAMINE see BBV500
RESTAS see FMR100
RESTENIL see MQU750
REST-ON see TEO250
RESTORIL see CFY750
RESTOVAR see LJE000
RESTRAN see RDK000
RESTRICTION
ENDODEOXYRIBONUCLEASE HINCII see
REF260
RESTRICTION
ENDODEOXYRIBONUCLEASE HINDIII
see REF262
RESTRICTION
ENDODEOXYRIBONUCLEASE MSEI see
REF265
RESTRICTION ENDONUCLEASE ECOVIII
see REF262
RESTRICTION ENDONUCLEASE HINCII
see REF260
RESTRICTION ENDONUCLEASE HINDIII
see REF262
RESTRICTION ENDONUCLEASE MSEI see
REF265
RESTRICTION ENDONUCLEASE RSAI see
REF270
RESTRICTION ENDONUCLEASE TRU9I see
REF265
RESTROL see DAL600
RESTROPIN see SBH500
RESTRYL see TEO250
RESULFON see AHO250
RESURRECTION LILY see REK325
RESVERATROL see TKP200
RESYDROL WM 501 see MCB050
RESYDROL WM 461E see MCB050
RETABOLIL see NNE550
RETACEL see CMF400
RETALON see DAL600
RETALON-ORAL see DHB500
RETAMA see YAK350
RETAMID see AKO500
RETAR-B$_1$ see FQJ100
RETARCYL see SAI100
RETARD see DMC600
RETARDER AK see PHW750
RETARDER BA see BCL750
RETARDER ESEN see PHW750
RETARDER J see DWI000
RETARDER PD see PHW750
RETARDER W see SAI000
RETARDEX see BCL750
RETARDILLIN see PAQ200
RETARPEN see BFC750
RETASULFIN see AKO500
RETENS see HGP550
RETENSIN see PDD300
R-(−)-5-(2-((2-(2-
ETHOXYPHENOXY)ETHYL)AMINO)PROP
YL)-2-
METHOXYBENZENESULFONAMIDE
HYDROCHLORIDE see EFA200
R-(+)-5-ETHYL-5-(1-
METHYLBUTYL)BARBITURIC ACID
SODIUM SALT see PBS500
R-(+)-ETHYL-1-(1-PHENYLETHYL)-1H-
IMIDAZOLE-5-CARBOXYLATE SULFATE
see HOU100
RETICULIN see SLX500
RETIN-A see VSK950
RETINAL (9CI) see VSK985
9-cis-RETINAL see VSK975

trans-RETINAL see VSK985
RETINAL, all-trans- see VSK985
all-trans-RETINAL see VSK985
RETINALDEHYDE see VSK985
9-cis-RETINALDEHYDE see VSK975
RETINAMIDE, N-ETHYL- see REK330
RETINAMIDE, N-ETHYL-, ALL-trans- see
REK330
RETINAMIDE, N-(2-HYDROXYETHYL)- see
REK350
RETINENE see VSK985
RETINENE 1 see VSK985
α-RETINENE see VSK985
RETINOIC ACID see VSK950
13-cis-RETINOIC ACID see VSK955
β-RETINOIC ACID see VSK950
all-trans-RETINOIC ACID see VSK950
RETINOIC ACID, 7,8-DIDEHYDRO- see
DHA200
RETINOIC ACID, 7,8-DIHYDRO- see
DMD100
RETINOIC ACID ETHYL AMIDE see
REK330
RETINOIC ACID, 2-
HYDROXYETHYLAMIDE see REK350
RETINOIC ACID, SODIUM SALT see SJN000
RETINOID ETRETIN see REP400
RETINOL see VSK600
all-trans RETINOL see VSK600
RETINOL ACETATE see VSK900
RETINOL, ANHYDRO- see AFQ700
RETINOL PALMITATE see VSP000
RETINYL ACETATE see VSK900
all-trans-RETINYL ACETATE see VSK900
all-trans-RETINYLIDENE METHYL
NITRONE see REZ200
RETINYL PALMITATE see VSP000
RETOZIDE see ILD000
RETRANGOR see EID200
RETRO-6-DEHYDROPROGESTERONE see
DYF759
RETRONE see DYF759
cis-RETRONECIC ACID ESTER of
RETRONECINE see RFP000
cis-RETRONECIC ACID ESTER of
RETRONECINE-N-OXIDE see RFU000
RETRONECINE HYDROCHLORIDE see
RFK000
Δ^6-RETROPROGESTERONE see DYF759
(14R)-4,14-RETRO-RETINOL-14-HYDROXY-
see RFK100
4,14-RETRO-RETINOL-14-HYDROXY-,
(14R)- see RFK100
RETRORSINE see RFP000
(15E)-RETRORSINE see RFP100
trans-RETRORSINE see RFP100
RETRORSINE-N-OXIDE see RFU000
RETROVIR see ASE900
RETROVITAMIN A see VSK600
RETZOLATE 1075 see SNY100
REUBLONIL see BCP650
REUDO see BRF500
REUFENAC see AGN000
REULATT S.S. see GFM000
REUMACHLOR see CLD000
REUMACIDE see IDA000
REUMALON see OPK300
REUMAQUIN see CLD000
REUMARTRIL see OPK300
REUMASYL see BRF500
REUMATOX see MQY400
REUMAZOL see BRF500
REUMOFENE see IDA400
REUMOFIL see SOU550
REUMOX see HNI500
REUMYCIN see DAN450
REUPOLAR see BRF500
REVAC see DJL000
REVACRYL A 191 see ADV900
REVELOX-WIRKSTOFF see RCA200
REVENGE see DGI400
REVERIN see PPY250

REVERTINA see VIZ400
REVIDEX see BBK500
REVIENTA CABALLOS (CUBA) see SLJ650
REVONAL see QAK000
REWOMID DLMS see BKE500
REWOPOL HV-9 see PKF000
REWOPOL NLS 30 see SIB600
REWOPOL TLS 40 see SON000
REWOQUAT B 18 see LBX075
REXALL 413S see PMP500
REXAN see CKF500
REXCEL see CCU150
REXENE see EJA379
REXENE see PMP500
REXENE 106 see ADY500
REXOCAINE see BQA010
REXOLITE 1422 see SMQ500
REX REGULANS see GGS000
REYCHLER'S ACID see RFU100
REZERPIN see RDK000
REZIFILM see TFS350
REZIPAS see AMM250
RF 10 see CCU250
RF 46-790 see THH350
R 236FA see HDE100
RFCNU see RFU600
RFNA see NEE500
RG 235 see AMV375
RG 600 see ARM268
R-GENE see AQW000
RGH-1106 see PII250
RGH 2957 see YGA700
RGH-2958 see PEE100
RGH-4405 see EGM100
RGH-5526 see RFU800
RH see RHF000
RH 123 see RGP600
RH-124 see BPU000
RH 315 see DTT600
RH-787 see PPP750
RH 893 see OFE000
RH 948 see CPJ600
RH-0265 see FIW100
RH 0994 see CKK530
RH-2512 see NGB600
RH-2915 see OQU100
RH 3421 see DKV710
RH 3866 see MRW775
RH 6201 see SFV650
RH-8218 see EPC175
RH-24,299 see DES450
RH-40,994 see CKK530
RH-46197 see ASA800
RH-46920 see BRE300
RH-52593 see MOV100
RH-53385 see MIJ275
RH-53,866 see MRW775
RH 63421 see DKV710
RH-75.992 2F see TAI175
RHABARBERONE see DMU600
RHABDOPHIS TIGRINUS TIGRINUS
VENOM see RFU875
RHAMNOL see SDZ350
RHAMNOLUTEIN see ICE000
RHAMNOLUTIN see ICE000
3-β(α-l-RHAMNOPYRANOSIDE)-5,11-α,14-β-
TRIHYDROXY-5-β-CARD(20,22)ENOLIDE
see RFZ000
RHAMNOSIDE, STROPHANTHIDIN-3, α-l-
see CNH780
3-β-RHAMNOSIDO-14-β-HYDROXY-Δ^{4,20,22}-
BUFATRIENOLIDE see POB500
RHAMNUS CALIFORNICA see MBU825
RHAMNUS CATHARTICA see MBU825
RHAMNUS FRANGULA see MBU825
RHATHANI see RGA000
RHEIC ACID see RHZ700
RHEIN see RHZ700
RHEMATAN see PGG000
RHENATE (ReO$_{41}$-)), POTASSIUM, (T-4)- see
PLQ000

RHENIC ACID, POTASSIUM SALT see PLR500
RHENIUM see RGF000
RHENIUM(VII) SODIUM OXIDE see SJD500
RHENIUM(VII) SULFIDE see RGK000
RHENIUM TRICHLORIDE see RGP000
RHENOCURE CA see DWN800
RHENOCURE TP see ZGA500
RHENOCURE TP/S see ZGA500
RHENOSORB C see CAU500
RHENOSORB F see CAU500
RHEODOL SP-P 10 see MRJ800
RHEONINE B see FAG070
RHEOPYRINE see IGI000
RHEOSMIN see RBU500
RHEUMATOL see CCD750
RHEUM EMODIN see MQF250
RHEUMIN TABLETTEN see ADA725
RHEUMON see HKK000
RHEUMON GEL see HKK000
RHEUMOX see AQN750
RHEUM RHABARBARUM see RHZ600
RHINALAR see FDD085
RHINALL see SPC500
RHINANTIN see NCW000
RHINASPRAY see RGP450
RHINATHIOL see CBR675
RHINAZINE see NAH500
RHINE BERRY see MBU825
RHINOCORT see BOM520
RHINOGUTT see RGP450
RHINOPERD see NCW000
RHINOSPRAY see RGP450
RHIZOCTOL see MGQ750
RHIZOPIN see ICN000
RHODACAL 330 see BBS275
RHODACRYST see VSZ000
RHODALLIN see AGT500
RHODALLINE see AGT500
RHODAMIN 6G see RGW000
RHODAMINE see FAG070
RHODAMINE 116 see RGW100
RHODAMINE 123 see RGP600
RHODAMINE 6GB see RGW000
RHODAMINE 6G (BIOLOGICAL STAIN) see RGW000
RHODAMINE 6GBN see RGW000
RHODAMINE 6G CHLORIDE see RGW000
RHODAMINE 590 CHLORIDE see RGW000
RHODAMINE 6GCP see RGW000
RHODAMINE 4GD see RGW000
RHODAMINE 6GD see RGW000
RHODAMINE 5GDN see RGW000
RHODAMINE 6Zh-DN see RGW000
RHODAMINE 6GEX ETHYL ESTER see RGW000
RHODAMINE 6G EXTRA see RGW000
RHODAMINE 6G EXTRA BASE see RGW000
RHODAMINE 6G EXTRA BASE see RGW000
RHODAMINE F4G see RGW000
RHODAMINE F5G see RGW000
RHODAMINE F5G CHLORIDE see RGW000
RHODAMINE F 5GL see RGW000
RHODAMINE GDN see RGW000
RHODAMINE 6 GDN see RGW000
RHODAMINE 6 GDN EXTRA see RGW000
RHODAMINE 4GH see RGW000
RHODAMINE 6GH see RGW000
RHODAMINE 6JH see RGW000
RHODAMINE 6ZH see RGW000
RHODAMINE 7JH see RGW000
RHODAMINE J see RGW000
RHODAMINE 5GL see RGW000
RHODAMINE 6G LAKE see RGW000
RHODAMINE LAKE RED 6G see RGW000
RHODAMINE 69DN EXTRA see RGW000
RHODAMINE 6GO see RGW000
RHODAMINE 116 PERCHLORATE see RGW100
RHODAMINE S (RUSSIAN) see FAG070
RHODAMINE WT see RGZ100
RHODAMINE 6GX see RGW000

RHODAMINE Y 20-7425 see RGW000
RHODAMINE ZH see RGW000
RHODANDINITROBENZOL see DVF800
RHODANIC ACID see RGZ550
RHODANID see ANW750
RHODANIDE see ANW750
RHODANIDE see PLV750
RHODANIN (CZECH) see RGZ550
RHODANINE see RGZ550
RHODANINE, N-METHYL- see MPH750
RHODANINIC ACID see RGZ550
RHODAQUAT M 242C29 see LBX075
RHODAQUAT 7LUF see OHK200
RHODASURF 25-3 see AFJ160
RHODASURF DA 630 see PKE370
RHODATE(3-), HEXACHLORO-, TRIAMMONIUM see HCM050
RHODATE (3-), HEXACHLORO-, TRIAMMONIUM, (OC-6-11)- see HCM050
RHODATE(2-), PENTACHLORO-, DIPOTASSIUM see PAY300
RHODIA see DAA800
RHODIA-6200 see DJA400
RHODIACHLOR see HAR000
RHODIACID see BJK500
RHODIACIDE see EEH600
RHODIACUIVRE see CNK559
RHODIANEBE see MAS500
RHODIA RP 11974 see BDJ250
RHODIASOL see PAK000
RHODIASTAB 83 see PFU300
RHODIATOX see PAK000
RHODIATROX see PAK000
RHODIFAX 16 see CPI250
RHODINE see ADA725
RHODINOL see CMT250
RHODINOL see DTF410
RHODINOL ACETATE see RHA000
RHODINOL (FCC) see DTF400
RHODINYL ACETATE see RHA000
RHODINYL BUTYRATE see DTF800
RHODINYL FORMATE see RHA100
RHODIRUBIN A see RHA125
RHODIRUBIN B see RHA150
RHODIRUBIN C see TAH675
RHODIRUBIN E see MAX000
RHODIUM see RHF000
RHODIUM(II) ACETATE see RHF150
RHODIUM, (BENZENAMINE) CHLORO((1,2,5,6-ETA)-1,5-CYCLOOCTADIENE)- see BBK800
RHODIUM(1+), (N,N'-BIS(2-AMINOETHYL)-1,2-ETHANEDIAMINE-N,N',N'',N''')DICHLORO-, CHLORIDE see DGO250
RHODIUM(1+), BIS(2,2'-BIPYRIDINE)DICHLORO-, CHLORIDE, (Z)- see BHB400
RHODIUM(II) BUTYRATE see RHK250
RHODIUM CHLORIDE see RHK000
RHODIUM CHLORIDE see RHP000
RHODIUM(III) CHLORIDE (1:3) see RHK000
RHODIUM CHLORIDE, TRIHYDRATE see RHP000
RHODIUM, (1,5-CYCLOOCTADIENE)(2,4-PENTANEDIONATO)- see CPR840
RHODIUM, ((1,2,5,6-eta)-1,5-CYCLOOCTADIENE)(2,4-PENTANEDIONATO-O,O')- see CPR840
RHODIUM DIACETATE see RHF150
RHODIUM(1+), DIBROMOTETRAKIS(PYRIDINE)-, BROMIDE, (E)- see DDT050
RHODIUM DIBUTYRATE see RHK250
RHODIUM(1+), DICHLOROBIS(1,10-PHENANTHROLINE-N1),N10))-, CHLORIDE, (OC-6-22)- see DEU160
RHODIUM(1+), DICHLOROTETRAKIS(3-PICOLINE)-, CHLORIDE, (E)- see DGL825
RHODIUM(1+), DICHLOROTETRAKIS(PYRIDINE)-, CHLORIDE, (E)- see DGL835

RHODIUM(1+), DICHLORO(TRIETHYLENETETRAMINE)-, CHLORIDE, (Z)- see DGO250
RHODIUM METAL (OSHA) see RHF000
RHODIUM(II) PROPIONATE see RHK850
RHODIUM(1+), TETRAAMMINEDICHLORO-, CHLORIDE see TBH300
RHODIUM (III), TRIAMMINETRINITRATATO- see THQ100
RHODIUM TRICHLORIDE see RHK000
RHODIUM TRICHLORIDE TRIHYDRATE see RHP000
RHODIUM, TRICHLOROTRIS(ACETONITRILE)- see TNC600
RHODIUM, TRICHLOROTRIS(PYRIDINE)-, SESQUIHYDRATE see TJE300
RHODIUM, TRIS(ACETONITRILE)TRICHLORO- see TNC600
RHODOCIDE see EEH600
RHODODENDRON see RHU500
RHODOJAPONIN IV see GJU315
RHODOL see MGJ750
RHODOLNE see SMQ500
RHODOMYCIN see FPD050
RHODOPAS 6000 see AAX175
RHODOPAS M see AAX250
RHODOQUINE see RHZ000
RHODORA (CANADA) see RHU500
RHODOTOXIN see AOO375
RHODOTYPOS SCANDENS see JDA075
RHODOVIOL see PKP750
RHODOVIOL 4/125 see PKP750
RHODOVIOL 4-125P see PKP750
RHODOVIOL 16/200 see PKP750
RHODOVIOL R 16/20 see PKP750
RHODULINE ORANGE see BJF000
RHODULINE ORANGE NO see BAQ250
RHOMBIC see ELC500
RHOMBININE see RHZ100
RHOMELLOSE see MIF760
RHOMENE see CIR250
RHOMEX see PIJ500
RHONOX see CIR250
RHOPLEX AC-33 (Rohm and Haas) see EMF000
RHOPLEX B 85 see PKB500
RHOTEX GS see SJK000
RHOTHANE see BIM500
RHOTHANE D-3 see BIM500
RH-75.992 TECHNICAL see TAI175
RHUBARB see RHZ600
RHUBARBE (CANADA) see RHZ600
RHUBARB YELLOW see RHZ700
RHUS COPALLINA see SCF000
R 18553 HYDROCHLORIDE see LIH000
RHYTHMONORM see PMJ525
RHYTMATON see AFH250
RHYUNO OIL see SAD000
RI-331 see HND250
RIANIL see CLO750
RIBALL see ZVJ000
RIBAVIRIN see RJA500
RIBBON CACTUS see SDZ475
RIBIPCA see RIK000
RIBITOL see RIF000
RIBO-AZAURACIL see RJA000
RIBO-AZURACIL see RJA000
RIBODERM see RIK000
RIBOFLAVIN see RIK000
RIBOFLAVIN-ADENINE DINUCLEOTIDE see RIF100
RIBOFLAVINE see RIK000
RIBOFLAVINE-ADENINE DINUCLEOTIDE see RIF100
RIBOFLAVINEQUINONE see RIK000
RIBOFLAVINE SULFATE see RIP000
RIBOFLAVIN 5'-PHOSPHATE ESTER MONOSODIUM SALT see RIFP00

RIBOFLAVIN 5'-PHOSPHATE SODIUM see RIFP00
RIBOFLAVIN 5'-(TRIHYDROGEN DIPHOSPHATE), 5'-5'-ESTER with ADENOSINE (9CI) see RIF100
β-D-RIBOFURANOSIDE, 2,3-DIHYDROXYPROPYL-5-DEOXY-5-(DIMETHYLARSINYL)-, (R)- see DNB700
RIBOFURANOSIDE, GUANINE-9, β-D- see GLS000
RIBOFURANOSIDE, 9H-PURINE-6-THIOL-9 see MCQ500
9-β-d-RIBOFURANOSIDOADENINE see AEH750
1-β-RIBOFURANOSYLCYTOSINE see CQM500
9-β-D-RIBOFURANOSYLGUANINE see GLS000
2-β-d-RIBOFURANOSYLMALEIMIDE see RIU000
5-β-d-RIBOFURANOSYL-1,3-OXAZINE-2,4-DIONE see OMK000
5-β-d-RIBOFURANOSYL-2H-1,3-OXAZINE-2,4(3H)-DIONE see OMK000
9-(β-d-RIBOFURANOSYL)PURINE see RJF000
9-(β-d-RIBOFURANOSYL)-9H-PURINE see RJF000
9-β-d-RIBOFURANOSYL-9H-PURINE-6-THIOL see TFJ825
3-β-d-RIBOFURANOSYL-1H-PYRROLE-2,5-DIONE see RIU000
7-β-d-RIBOFURANOSYL-7H-PYRROLO(2,3-D)PYRIMIDINE-4-AMINE see TNY500
7-β-d-RIBOFURANOSYL-7H-PYRROLO(2,3-d)PYRIMIDIN-4-OL see DAE200
2-β-d-RIBOFURANOSYLTHIAZOLE-4-CARBOXAMIDE see RJF500
2-β-d-RIBOFURANOSYL-4-THIAZOLECARBOXAMIDE see RJF500
2-β-d-RIBOFURANOSYL-1,2,4-TRIAZINE-3,5(2H,4H)-DIONE see RJA000
2-β-d-RIBOFURANOSYL-as-TRIAZINE-3,5(2H,4H)-DIONE see RJA000
2-β-d-RIBOFURANOSYL-as-TRIAZINE-3,5(2H,4H)-DIONE 2',3',5'-TRIACETATE see THM750
1-β-d-RIBOFURANOSYL-1,2,4-TRIAZOLE-3-CARBOXAMIDE see RJA500
1-β-d-RIBOFURANOSYLURACIL see UVJ000
RIBOMYCINE see RIP000
RIBONOSINE see IDE000
RIBONUCLEIC ACIDS see RJA600
d-RIBOPYRANOSYLAMINE, N-(p-NITROPHENYL)-N-NITROSO- see NLC600
RIBOSE see RJA700
d-RIBOSE see RJA700
RIBOSE, d- see RJA700
RIBOSTAMIN see RIP000
RIBOSTAMYCIN see XQJ650
RIBOSTAMYCIN SULFATE see RIP000
β-d-RIBOSYL-6-METHYLTHIOPURINE see MPU000
RIBOSYLPURINE see RJF000
RIBOSYLTHIOGUANINE see TFJ500
RIBOSYL-6-THIOPURINE see MCQ500
RIBOTIDE see RJF400
RIBOXAMIDE see RJF500
RICE BRAN WAX see RJF800
RICE STARCH see SLJ500
RICE SYN WAX see HHW502
RICHAMIDE 6310 see BKE500
RICHONATE 1850 see DXW200
RICHONIC ACID B see LBU100
RICHONOL C see SIB600
RICHONOL T see SON000
RICID see DIU800
RICID II see BKS750
RICID P see BKS750
RICIFON see TIQ250
RICIN see RJK000
RICIN D see RJK050

RICIN (HAITI) see CCP000
RICINIC ACID see RJP000
RICININ see RJK100
RICININE see RJK100
RICINOLEIC ACID see RJP000
RICINOLEIC ACID, BARIUM SALT see RJU000
RICINOLEIC ACID, SODIUM SALT see SJN500
RICINOLEIC ACID TRIESTER with GLYCEROL 12-ETHER with TRIDECAETHYLENE GLYCOL see EAN800
RICINOLIC ACID see RJP000
RICINO (PUERTO RICO) see CCP000
RICINS, D see RJK050
RICIN-TOXIN CON A see CNH625
RICINUS COMMUNIS see CCP000
RICINUS OIL see CCP250
RICIRUS OIL see CCP250
RICKAMICIN see SDY750
RICKAMICIN SULFATE see APY500
RICKETON see CMC750
RICO see AOT255
RICON 100 see SMR000
RICORTEX see CNS825
RICYCLINE see TBX250
RIDAURA see ARS150
RIDAZOLE see MMN750
RIDDELLIIN see RJZ000
RIDDELLINE see RJZ000
RIDLITE MMT see MCB050
RIDOMIL MDM100
RIDOMIL 2E see MDM100
RIDZOL P see MMN750
RIFA see RKP000
RIFA ACID ORANGE II see CMM220
RIFADINE see RKP000
RIFAGEN see RKP000
RIFALDAZINE see RKP000
RIFALDIN see RKP000
RIFAMATE see RKP000
RIFAMETANE see RJZ100
RIFAMICINE SV see RKZ000
RIFAMIDE see RKA000
RIFAMPICIN see RKP000
RIFAMPICINE (FRENCH) see RKP000
RIFAMPICIN M/14 see RKA000
RIFAMPICINUM see RKP000
RIFAMPIN see RKP000
RIFAMYCIN see RKK000
RIFAMYCIN see RKZ000
RIFAMYCIN AF/ACPP see RKZ100
RIFAMYCIN AMP see RKP000
RIFAMYCIN B see RKK000
RIFAMYCIN B N,N-DIETHYLAMIDE see RKA000
RIFAMYCIN, 3-(((4-CYCLOPENTYL-1-PIPERAZINYL)IMINO)METHYL)- see RKZ100
RIFAMYCIN DIETHYLAMIDE see RKA000
RIFAMYCIN, 3-(((1-(DIETHYLAMINO)ETHYLIDENE)HYDRAZONO)METHYL)- see RJZ100
RIFAMYCIN M14 see RKA000
RIFAMYCIN O see RKP400
RIFAMYCIN S see RKU000
RIFAMYCIN SV see RKZ000
RIFAPENTINE see RKZ100
RIFAPRODIN see RKP000
RIFATHYROIN see TNX400
RIFINAH see RKP000
RIFIT see PMB850
RIFLE POWDER see ERF500
RIFLE POWDER see PLL750
RIFLOC RETARD see CCK125
RIFOBAC see RKP000
RIFOCIN see RKZ000
RIFOCINA M see RKA000
RIFOCYN see RKZ000
RIFOLDIN see RKP000
RIFOMIDE see RKA000
RIFOMYCIN B see RKK000

RIFOMYCIN B DIETHYLAMIDE see RKA000
RIFOMYCIN O see RKP400
RIFOMYCIN S see RKU000
RIFOMYCIN SV see RKZ000
RIFORAL see RKP000
RIGEDAL see CCK125
RIGENICID see EPQ000
RIGETAMIN see EDC500
RIGEVIDON see NNL500
RIGIDEX see PJS750
RIGIDEX 35 see PJS750
RIGIDEX 50 see PJS750
RIGIDEX TYPE 2 see PJS750
RIGIDIL see BBV500
RIGIDYL see BBV500
RIKAVARIN see AJV500
RIKELATE CALCIUM see CAR780
RIKEMAL O 71D see GGR200
RIKEMAL OL 100 see GGR200
RIKEMAL S 250 see SKV150
RIKEN RESIN MA 31 see MCB050
RIKER 52G see PAF550
RIKER 548 see TIH800
RIKER 595 see BSZ000
RIKER 601 see TND000
RILANSYL see CKF500
RILAQUIL see CKF500
RILASSOL see CKF500
RILAX see CKF500
RILLASOL see CKF500
RILOF see PIX775
RIMACTAN see RKP000
RIMACTAZID see RKP000
RIMADYL see CCK800
RIMANTADINE HYDROCHLORIDE see AJU625
RIMAON see EDW500
RIMAZOLIUM METHYL SULFATE see PMD825
RIMICID see ILD000
RIMIDIN see FAK100
RIMITSID see ILD000
RIMSO-50 see DUD800
RINATIOL see CBR675
RINAZIN see NAH550
RINDEX see MDO250
RINEPTIL see HBM490
RINGER'S GLUCOSE SOLUTION see GFG200
RINO-CLENIL see AFJ625
RIOL see FLZ050
RIOMITSIN see HOH500
RIOMYCIN see FMR500
RIP-15830 see DNG200
RIPAZEPAM see POL475
RIPCORD see RLF350
RIPENTHOL see DXD000
RIPERCOL see TDX750
RIPERCOL-L see LFA020
RIPEREOL see TDX750
RIPOSON see EQL000
RIPOST see MFC100
RIRILIM see BBW500
RIRIPEN see BBW500
RISE see CKA000
RISEDRONATE SODIUM see RLF400
RISELECT see DGI000
RISELF see MFD500
RISEPTIN see LBV100
RISERPA see RDK000
RISPASULF see PDC850
RISTAT see HMY000
RISTOGEN see TKH750
RITALIN see MNQ000
RITALIN see RLK000
RITALINE see MNQ000
RITALIN HYDROCHLORIDE see RLK000
RITANSERIN see RLK100
RITCHER WORKS see MNQ000
RITMENAL see DKQ000
RITMODAN see DNN600

RITMOS see AFH250
RITODRINE HYDROCHLORIDE see RLK700
RITOSEPT see HCL000
RITROSULFAN see LJD500
RITSIFON see TIQ250
RITUSSIN see RLU000
RIVADORM see NBT500
RIVADORN see NBU000
RIVAL see IAL200
RIVANOL see EDW500
RIVASED see RDK000
RIVASIN see RDK000
RIVASTATIN see RLK750
RIVINA HUMILIS see ROA300
RIVINOL see EDW500
RIVIVOL see ILE000
RIVOMYCIN see CDP250
RIVOTRIL see CMW000
RIZABEN see RLK800
RJ 5 see DLJ500
R JUTAN see CAT775
RL-50 see MRL500
RMC see MAC750
RMI 9918 see TAI450
RMI 11002 see FDN000
RMI 12,936 see HLX600
RMI 71782 see DBE835
RMI 71782 see DKH875
RMI9,384A see DLS600
RMI 10,482A see MHJ500
RMI-14042A see LIA400
RMI 10024DA see BJA500
RMI 10874DA see BJG500
RMI 11002 DA see FDN000
RMI 11513 DA see XBA500
RMI 11567 DA see DDB400
RMI 11877 DA see DDE000
R830 (MINERAL) see RSP100
RNA see RJA600
RO 1-5130 see MDL600
RO-1-5155 see NDW510
RO 1-5431 see MDV250
RO 1-5431 see MKR250
RO 1-5470 see RAF300
RO 1-6463 see DNW400
RO 1-6794 see DBE825
RO-1-7700 see AGI000
RO 1-7788 see LFD200
RO-1-9213 see CQH000
RO 1-9569 see TBJ275
RO 2-1160 see CCH250
RO 2-2222 see TKW500
RO 2-2979 see TMB000
RO 2-2980 see HNK000
RO 2-3208 see FDK000
RO 2-3245 see ARX750
RO 2-3248 see AGD500
RO 2-3308 see QVA000
RO 2-3308 see QWJ000
RO 2-3599 see DLP000
RO 2-3742 see AGE000
RO 2-3951 see ARX770
RO 2-4572 see ILE000
RO 2-5803 see BHR750
RO 2-9757 see FMM000
RO 2-9915 see FHI000
RO 2-9945 see TCQ500
RO 4-0403 see TAF675
RO 4-1284 see HKS900
RO 4-1385 see GFO200
RO 4-1778 see MDV000
RO 4-2130 see SNK000
RO 4-3476 see SNL800
RO-4-3780 see VSK955
RO 4-3816 see DBK400
RO 4-4393 see AIE500
RO 4-4602 see SCA400
RO 4-5360 see DLY000
RO 4-6316 see DNA200
RO 4-6467 see PME250
RO 4-6467 see PME500

RO 4-6861 see CIF250
RO 4-8180 see CMW000
RO 4-9253 see BMN750
RO 5-0360 see DAR400
RO 5-0690 see MDQ250
RO 5-0831 see IKC000
RO 5-1226 see MNP300
RO 5-2180 see CGA500
RO 5-3059 see DLY000
RO 5-3072 see AJO280
RO 5-3307 see DAI475
RO 5-3350 see BMN750
RO 5-3438 see FDB100
RO 5-4023 see CMW000
RO 5-4200 see FDD100
RO 5-4864 see CFK150
RO 5-5345 see CFY750
RO 5-5516 see MLD100
RO 5-6789 see CFZ000
RO 5-6901 see DAB800
RO 5-9963 see NHI000
RO 6-4563 see GFY100
RO 6-9550 see MJG100
RO 7-0207 see OJS000
RO 7-1554 see IGH000
RO 7-2340 see BMA650
RO 7-5050 see MIA250
RO 8-4969 see REK350
RO 10-1670 see REP400
RO 10-6338 see BON325
RO 11-1781 see TGA275
RO 11-8958 see EBB700
RO 12-0068 see TAL485
RO 12-3049 see FAQ230
RO 13-5223 see EOE200
RO 13-6298 see AQZ400
RO 13-7410 see AQZ200
RO 13-7837 see EMJ600
RO 13-8320 see AQZ300
RO 13-9904 see CCS588
RO 15-1297 see DGJ160
RO 17-3664 see IPI400
RO 21-5816 see HJS900
RO 21-9738 see DYE415
RO 23-4194 see HDB100
RO 4-12884 see HKS900
RO 47-0203 see BMH800
RO 1-5431/7 see DYF000
RO 1-5470/5 see DBE200
RO 1-5470/6 see LFD200
RO-5-0810/1 see TJE880
RO 5-3307/1 see IKB000
RO-6-4563/8 see GFY100
RO 8-6270/9 see TNX400
RO 1-9569/12 see TBJ275
RO 13-9904/001 see CCS588
RO 14-3169/000 see MRQ300
RO 31-8959/003 see FOL025
RO 47-0203/039 see BMH800
ROACH SALT see SHF500
ROAD ASPHALT see PCR500
ROAD ASPHALT (DOT) see ARO500
ROAD ASPHALT (DOT) see ARO750
ROAD TAR (DOT) see ARO500
ROAD TAR, liquid (DOT) see ARO750
RO-AMPEN see AIV500
RO-AMPEN see AOD125
ROASTED COFFEE RESINOID see CNG800
ROBAMATE see MQU750
ROBANUL see GIC000
ROBARB see AMX750
ROBAVERON see RLK875
ROBAXAN see GKK000
ROBAXIN see GKK000
ROBAXINE see GKK000
ROBAXON see GKK000
ROBENIDINE see RLK890
ROBICILLIN VK see PDT750
ROBIGENIN see ICE000
ROBIGRAM see ARQ750
ROBIMYCIN see EDH500
ROBINAX see GKK000

ROBINETIN see RLP000
ROBINIA PSEUDOACACIA see GJU475
ROBINUL see GIC000
ROBIOCINA see SMB000
ROBISELLIN see ILD000
ROBISON ESTER see GFG150
ROBITET see TBX000
ROBITUSSIN see RLU000
ROBORAL see PAN100
ROC-101 see RLU550
ROCALTROL see DMJ400
ROCCAL see AFP100
ROCCAL see BBA500
ROCEPHIN see CCS588
ROCHE 5-9000 see API125
ROCHELLE SALT see SJK385
ROCHIPEL see DIR000
RO-CILLIN see PDD350
ROCILLIN-VK see PDT750
ROCIPEL see DIR000
ROCK CANDY see SNH000
ROCK OIL see PCR250
ROCK SALT see SFT000
ROCORNAL see DIO200
RO-CYCLINE see TBX250
RODALON see AFP250
RODALON see BBA500
RODALON see BEL900
RODANCA see PLV750
RODANIN S-62 (CZECH) see IAQ000
RODATOX 60 see DVF800
RODENTIN see EAT600
RO-DETH see WAT200
RODEX see FFF000
RO-DEX see SMN500
RODILONE see SNY500
RODINAL see ALT250
RODINE see RCF000
RODINOL see CMT250
RODINOL see DTF410
RODINOLONE see AQX250
RODIPAL see DIR000
RODOCID see EEH600
RODOL 42 see NEM500
RODOL YBA see ALO000
RODY see DAB825
ROE 101 see DBB200
ROELLER COMPOUND see MJG100
ROERIDORM see CHG000
ROFOB 3 see AHH825
ROGERSINE see TKX700
ROGITINE see PDW400
ROGODIAL see DRR400
ROGODIAL see DSP400
ROGOR see DSP400
ROGUE see DGI000
ROHAGIT SD 15 see ADV900
ROHAGIT SD 15 see MDN600
ROHM & HAAS RH-218 see EPC175
ROHYDRA see BAU750
RO-HYDRAZIDE see CFY000
ROHYPNOL see FDD100
ROIDENIN see IIU000
ROIPNOL see FDD100
ROKACET see PJY100
ROKACET O 7 see PJY100
ROKANOL L see DXY000
RO-KO see RNZ000
ROKON see BDF000
ROL see RAF100
ROLAMID CD see BKE500
ROLAZINE see HGP500
ROLAZOTE see PAJ750
ROLAZOTE see POL000
ROLICTON see AKL625
ROLITETRACYCLINE see PPY250
ROLITETRACYCLINE see SPE000
ROLL-FRUCT see CDS125
ROLODIN see RLZ000
ROLODINE see RLZ000
ROLQUAT CDM/BC see QAT520
ROLSERP see RDK000

ROM 203 see IAG700
ROMACRYL see PKB500
ROMANTHRENE BLUE FRS see IBV050
ROMANTRENE BLUE FBC see DFN300
ROMANTRENE BLUE FRS see IBV050
ROMANTRENE BLUE GGSL see IBV050
ROMANTRENE BLUE RSZ see IBV050
ROMANTRENE BRILLIANT BLUE FR see IBV050
ROMANTRENE BRILLIANT BLUE R see IBV050
ROMANTRENE BROWN FBR see CMU770
ROMANTRENE BROWN FGR see CMU770
ROMANTRENE BROWN FR see CMU780
ROMANTRENE GOLDEN YELLOW see DCZ000
ROMANTRENE GREY K see IBJ000
ROMANTRENE NAVY BLUE FG see CMU500
ROMANTRENE NAVY BLUE FRA see VGP100
ROMANTRENE OLIVE FR see DUP100
ROMAN VITRIOL see CNP250
ROMAN VITRIOL see CNP500
ROMERGAN see DQA400
ROMETIN see CHR500
ROMEZIN see ALF250
ROMHIDROL M 501 see MCB050
ROMICIL see OHM900
ROMILAR see DBE200
ROMILAR HYDROBROMIDE see DBE200
RO 31-2848 MONOHYDRATE see CMP810
ROMOPAL O see PJT300
ROMOSOL see ART250
ROMOTAL see TCJ075
ROMPARKIN see BBV000
ROMPHENIL see CDP250
ROMPUN see DMW000
ROMULGIN O see PKL100
RONDAR see CFZ000
RONDIS R see RNU100
RONDOMYCIN see MDO250
RO-NEET see EHT500
RONGALIT see FMW000
RONGALITE C see FMW000
RONIACOL see NDW510
RONIDAZOLE see MMN750
RONIDAZOL-PHARMACHIM (Bulgarian) see MMN750
RONILAN see RMA000
RONIN see PPO000
RONIT see EHT500
RONNEL see RMA500
RONONE see RNZ000
RONSTAR see OMM200
RONTON see ENG500
RONTYL see TKG750
ROOTONE see ICP000
ROOTONE see NAK000
ROOTONE see NAK500
ROPE BARK see LEF100
ROP 500 F see GIA000
ROPION see FJT100
ROPIVACAINE see RMA600
ROPOL see PJS750
ROPOTHENE OB.03-110 see PJS750
ROPTAZOL see NGG500
ROQUESSINE DIHYDROBROMIDE see DOX000
ROQUINE see CLD000
RORASUL see ASB250
RORASUL see PPN750
RORER 148 see QAK000
ROSACETOL see TIT000
ROSA FRANCESA (CUBA) see OHM875
ROSA LAUREL (MEXICO) see RHU500
ROSAMICIN see RMF000
ROSANIL see DGI000
ROSANILINE see MAC250
p-ROSANILINE see MAC500
p-ROSANILINE see RMK000
ROSANILINE BASE see MAC500

ROSANILINE CHLORIDE see MAC250
p-ROSANILINE HCL see RMK020
ROSANILINE HYDROCHLORIDE see MAC250
p-ROSANILINE HYDROCHLORIDE see RMK020
ROSANILINIUM CHLORIDE see MAC250
ROSANOMYCIN A see RMK200
ROSARAMICIN see RMF000
ROSARY PEA see RMK250
ROSCOPENIN see PDT750
ROSCOSULF see SNN300
ROSE ABSOLUTE FRENCH see RMP000
ROSE BAY see OHM875
ROSEBAY see RHU500
ROSE BENGAL SODIUM see RMP175
ROSE CRYSTALS see TIT000
ROSE ETHER see PER000
ROSE-FLOWERED JATROPHA see CNR135
ROSE GERANIUM OIL ALGERIAN see GDA000
ROSE de GRASSE see RNA000
ROSE LAUREL (MEXICO) see OHM875
ROSE de MAI see RNA000
ROSE de MAI ABSOLUTE see RMP000
ROSEMARIE OIL see RMU000
ROSEMARY OIL see RMU000
ROSEMIDE see CHJ750
ROSEN OEL (GERMAN) see RNA000
ROSENOL see RNA000
ROSENOXIDE see RNU000
ROSENSTHIEL see MAT250
ROSEOFUNGIN see RMU100
ROSE OIL see PDD750
ROSE OIL see RNA000
ROSE OIL BULGARIAN see RNA000
ROSE OIL BULGARIAN see RNF000
ROSE OIL MOROCCAN see RNK000
ROSE OTTO see RNA000
ROSE OXIDE see RNU000
ROSE OXIDE LEVO see RNU000
ROSE OXIDE LEVO see RNU000
ROSE QUARTZ see SCI500
ROSE QUARTZ see SCJ500
ROSES see AOB250
ROSETONE see NAK000
ROSIN see RNU100
ROSIN CORE SOLDER PYROLYSIS PRODUCTS see RNU100
ROSIN WW see RNU100
ROSMARIN OIL (GERMAN) see RMU000
ROSOXIDE see RNU000
ROSPAN see PNH750
ROSPIN see PNH750
ROSTONE 2150 see MCB050
RO-SULFIRAM see DXH250
ROTATE see DQM600
ROTAX see BDF000
ROT B see XRA000
ROT C see OHI200
ROTEFIVE see RNZ000
ROTEFOUR see RNZ000
ROTENONA (SPANISH) see RNZ000
ROTENONE see RNZ000
ROTERSEPT see BIM250
ROTERSEPT see CDT250
ROTESAR see BOV825
ROTESSENOL see RNZ000
RO 40-60554-((5,6,7,8-TETRAHYDRO-5,5,8,8-TETRAMETHYL-2-NAPHTHALENYL)CARBONYL)AMINOBENZOIC ACID see TDB765
ROT G see OHI200
ROT GG FETTLOESLICH see XRA000
ROTHANE see BIM500
ROTOCIDE see RNZ000
ROTOX see MHR200
ROTTLERIN see KAJ500
ROUGE see IHC450
ROUGE CERASINE see OHI200
ROUGE de COCHENILLE A see FMU080
ROUGE PLANT see ROA300

ROUGH & READY MOUSE MIX see WAT200
ROUGH & READY RAT BAIT & RAT PASTE see RCF000
ROUGOXIN see DKN400
ROUNDUP see GIQ100
ROUQUALONE see QAK000
ROUSSIN RED METHYL ESTER see BKM530
ROUSSIN'S RED METHYL ESTER see BKM530
ROUTRAX see TKG750
ROVAMICINA see SLC000
ROVRAL see GIA000
ROWACHOL see ROA400
ROWALIND see NDM100
ROWATIN see ROA425
ROWMATE see DET400
ROWMATE see DET600
ROXARSONE (USDA) see HMY000
ROXBURY WAXWORK see AHJ875
ROXEL see RDK000
ROXICAM see FAJ100
R 2010 OXIME see OMY815
ROXIMYCIN see HGP550
ROXINOID see RDK000
ROXION U.A. see DSP400
ROXOSUL TABLETS see SNN500
ROXYNOID see RDK000
ROYAL BLUE see DJO000
ROYAL CBTS see CPI250
ROYAL MBTS see BDE750
ROYAL MH-30 see DMC600
ROYAL SLO-GRO see DMC600
ROYAL SPECTRA see CBT750
ROYALTAC see DAI600
ROYALTAC see ODE000
ROYAL TMTD see TFS350
ROYAL WHITE LIGHT see CAT775
ROYLIN see LJB100
ROYLINE see LJB100
ROZEOFUNGIN see RMU100
ROZOL see CJJ000
ROZTOZOL see CKM000
RP-093 see CMY725
866 R.P. see CFU750
RP 2145 see MPQ750
2259 R.P. see GEW750
RP 2259 see GEW750
RP 2275 see AHO250
2325 RP see DRV850
2339 RP see BEM500
2512 R.P. see DBL800
R.P. 2512 see DBL800
R.P. 2591 see DFX400
RP 2616 see PPP500
RP 2632 see ALF250
2643-RP see ALF250
RP 2786 see WAK000
RP 2929 see TFH500
2987 R.P. see DII200
RP 2990 see TEX250
RP 3203 see DNG400
3277 RP see DQA400
3277 R.P. see PMI750
RP 3356 see DIR000
3359 RP see CKB500
RP 3377 see CLD000
3389 R.P. see DQA400
RP 3554 see MRW000
RP 3602 see GGS000
RP 3697 see PDD300
3718 RP see CGX625
3735 R.P. see SBE500
RP 3799 see DIW000
4182 R.P. see DQA400
RP 4207 see FNF000
RP 4632 see MFK500
4753 R.P. see TGD000
RP 4909 see CLY750
RP5171 see DIG400
5171 RP see PBM500
5278 R.P. see SMA000
5337 R.P. see SLC000

6140 RP see PMF500
6549 RP see TKV200
6710 RP see CIO500
6847 R.P. see AFL750
6909 RP see PIW000
7162 RP see DLH200
7204 RP see COS899
RP7293 see VRF000
RP 7522 see AKO500
7843 R.P. see TFM100
RP 7843 see TFM100
RP 7891 see GFM200
RP 8167 see EEH600
RP 8228 see LFO000
8595 R.P. see DSV800
RP 8599 see FMU039
RP 8823 see MMN250
RP 8908 see PIW000
RP 9159 see LEO000
9159 RP see LEO000
9260 RP see DSS600
9778 R.P. see PNW750
RP-9921 see PAF550
9965 RP see MPM750
RP 9965 see MQR000
RP 10192 see MIJ500
10257 R.P. see TND000
10633 RP see TMK100
11,561 RP see CBL500
RP 11650 see TIT100
11,670 RP see TIT050
RP 13057 see DAC000
13,057 R.P. see DAC000
13245 R. P. see PMM000
13907 R.P. see TKP100
RP 13907 see TKP100
RP 16,091 see MNQ500
RP 17623 see OMM200
RP 18,429 see AJV500
19583 RP see BDU500
22050 R.P. see ROZ000
RP 22410 see PMG000
RP 23465 see UTA300
RP 26019 see GIA000
27267 R.P. see ZUA450
32545 RP see AHH800
32861 R.P. see OKU200
35689 R.P. see IRA050
RP 54780 see OKY100
RP 56976 see TAH800
RP 020630 see BRA300
RPA 2 see NAP500
RPA NO. 2 see NAP500
RPC 1022 see EJM950
RPCNU see ROF200
2339 R.P. HYDROCHLORIDE see PEN000
4560 RP HYDROCHLORIDE see CKP500
RP 13057 HYDROCHLORIDE see DAC200
RP 22,050 HYDROCHLORIDE see ROZ000
2786 R.P. MALEATE see DBM800
RP 8599 MESYLATE see FMU039
RPR-5 see EIB600
RPR-V see EIB600
19583 RP SODIUM see KGK100
RR 15-12-120 see MCB050
R 14 (REFRIGERANT) see CBY250
R 32 (REFRIGERANT) see MJQ300
RS see CCU250
RS 141 see CJJ250
RS 1280 see CBF250
RS-2177 see FDD075
RS-2196 see MBW775
RS-2290 see EDU100
RS 2874 see BIS750
RS 2874 see BIS750
R-3422-S see EMC000
RS-3540 see MFA500
RS-8858 see OMY500
RS 44872 see SNH480
RS-79216 see PCG550
RS 10085-197 see MRA260
RS 84135-004 see EAU200

RSAI see REF270
(R,S)-ALANINE see AFH600
R SALT see ROF300
RS-1401 AT see SPD500
R.S. NITROCELLULOSE see CCU250
R105 SODIUM see RMP175
RS 21592 SODIUM see GBU200
RTE see DIQ125
RTEC (POLISH) see MCW250
RTI-113 see CKM800
R-010-TK see BIX250
RT-PA see PJB800
RU 13 see DIF300
RU 486 see HKB700
RU 1697 see HKI100
RU 2010 see NNR125
RU 2323 see ENX575
RU 2858 see MRU600
RU-4723 see CIR750
RU 486-6 see HKB700
RU-11484 see BEP750
RU 15060 see SOX400
RU 15750 see TKG000
RU 22090 see PCK075
RU 22974 see DAF300
RU 23603 see CQD900
RU 24756 see CCR950
RU 25472 see THJ300
RU 25474 see THJ300
RU 26979 see DAM430
RU 38486 see HKB700
RU 38702 see DTH450
RU 43715 see POB300
RUBAN see NCN650
RUBATONE see BRF500
RUBBER see ROH900
"522" RUBBER ACCELERATOR see PIY500
RUBBER CEMENT see CCW250
RUBBER FAST YELLOW GA see CMS210
RUBBER FAST YELLOW GRA see CMS208
RUBBER HYDROCHLORIDE see PJH500
RUBBER HYDROCHLORIDE POLYMER see PJH500
RUBBER, NATURAL see ROH900
RUBBER RED R EXTRA see CJD500
RUBBER SOLVENT see ROU000
RUBBER VINE see ROU450
RUBEANE see DXO200
RUBEANIC ACID see DXO200
RUBENS BROWN see MAT500
RUBERON see EME100
RUBESCENCE RED MT-21 see NAY000
RUBESCENSLYSIN see AHI500
RUBESOL see VSZ000
RUBIADINPRIMEVEROSIDE see ROU500
RUBIADIN-PRIMVEROSIDE see ROU500
RUBIAZOL A see SNM500
RUBIDAZON see ROU800
RUBIDAZONE see ROU800
RUBIDAZONE MONOHYDROCHLORIDE see ROZ000
RUBIDIUM see RPA000
RUBIDIUM ACETYLIDE see RPB100
RUBIDIUM AZIDE see RPB150
RUBIDIUM CARBONATE see RPB200
RUBIDIUM CHLORIDE see RPF000
RUBIDIUM DICHROMATE see RPK000
RUBIDIUM FLUORIDE see RPP000
RUBIDIUM HYDRIDE see RPU000
RUBIDIUM HYDROXIDE see RPZ000
RUBIDIUM HYDROXIDE SOLUTION (UN 2677) (DOT) see RPZ000
RUBIDIUM HYDROXIDE (UN 2678) (DOT) see RPZ000
RUBIDIUM IODIDE see RQA000
RUBIDIUM NITRIDE see RQF000
RUBIDIUM PENTAIODOTETRAARGENTATE see RQF100
RUBIDIUM SILVER IODIDE (Ag₄RbI₅) (8CI) see RQF100

RUBIDIUM TETRASILVER PENTAIODIDE see RQF100
RUBIDOMYCIN see DAC000
RUBIDOMYCIN see DAC200
RUBIDOMYCINE see DAC000
RUBIDOMYCIN HYDROCHLORIDE see DAC200
RUBIFLAVIN see RQF350
RUBIGAN see FAK100
RUBIGEN see TBJ275
RUBIGINE see HHU500
RUBIGO see IHC450
RUBIJERVINE see SKR600
RUBINATE 44 see MJP400
RUBINATE M see PKB100
RUBINATE MF 178 see PKB100
RUBINATE MF 182 see PKB100
RUBINATE TDI see TGM740
RUBINATE TDI 80/20 see TGM740
RUBINATE TDI 80/20 see TGM750
RUBINE RED RR 1253 see CMS155
RUBINE TONER B see CMS148
RUBINE TONER BA see CMS148
RUBINE TONER BT see CMS148
RUBITOX see BDJ250
RUBOMYCIN see RQK000
RUBOMYCIN C see DAC000
RUBOMYCIN C see DAC200
RUBOMYCIN C 1 see DAC000
RUBOUT see ANI800
RUBRAMIN see VSZ000
RUBRATOXIN B see RQP000
RUBRATOXIN B, DIHYDRO- see RQP050
RUBRATOXIN B, KETO- see KGK140
RUBRATOXIN B, MONOSODIUM see RQP100
RUBRIMENT see NCR040
RUBRINE C see GHA050
RUBRIPCA see VSZ000
RUBROCITOL see VSZ000
RUBRUM SCARLATINUM see SBC500
RUBUS ELLIPTICUS Smith, extract excluding roots see RQU300
RUCAINA see DHK400
RUCAINA HYDROCHLORIDE see DHK600
RUCOFLEX PLASTICIZER DOA see AEO000
RUCON B 20 see PKQ059
RUDBECKIA BICOLOR nutt., extract see RQU650
RUDOTEL see CGA000
RUELENE see COD850
RUELENE 25E see COD850
RUELENE DRENCH see COD850
RUE OIL see OGY200
RUE OIL and HERB see RQU750
RUFAST see DTH450
RUFOCHROMOMYCIN see SMA000
RUFOCROMOMYCINE see SMA000
RUFOL see MPQ750
RUGBY see EHY100
RUGULOSIN see RRA000
(+)-RUGULOSIN see RRA000
RUKSEAM see DAD200
RULENE see COD850
RUMAPAX see HNI500
RUMENSIN see MRE230
RUMESTROL 1 see DKA600
RUMESTROL 2 see DKA600
RUMETAN see ZLS000
RUM ETHER see EOB050
RUN see PDN000
RUNA RH20 see TGG760
RUNCATEX see CIR500
RUOCID see AHO250
RU 13 (PHARMACEUTICAL) see DIF300
RU SILICATE see SCN700
RUSPOL see CCU075
RUSSELL'S VIPER VENOM see VQZ635
RUSSIAN COMFREY LEAVES see RRK000
RUSSIAN COMFREY ROOTS see RRP000
RUTACRIDON see RRP100
(−)-RUTACRIDONE see RRP100

RUTACRIDONE see RRP100
RUTA GRAVEOLENS, extract see RRP675
RUTAPOX SL see BGT800
RUTERGAN HYDROCHLORIDE see DPI000
RUTGERS 612 see EKV000
RUTHENIUM see RRU000
RUTHENIUM(3+),
(ADENOSINE)PENTAAMMINE-,
TRIBROMIDE see AEL800
RUTHENIUM CHLORIDE see RRZ000
RUTHENIUM CHLORIDE HYDROXIDE see
RSA000
RUTHENIUM COMPOUNDS see RSF000
RUTHENIUM(3+),
(DEOXYGUANOSINE)PENTAAMMINE-,
TRICHLORIDE see DAR900
RUTHENIUM(3+),
(GUANINE)PENTAAMMINE-,
TRICHLORIDE see GLO100
RUTHENIUM(3+), HEXAAMMINE-,
TRICHLORIDE see HBY200
RUTHENIUM(3+), HEXAAMMINE-,
TRICHLORIDE, HYDRATE see HBY200
RUTHENIUM(3+), HEXAAMMINE-,
TRICHLORIDE, (OC-6-11)- see HBY200
RUTHENIUMHYDROXIDE TRICHLORIDE
see RSA000
RUTHENIUM HYDROXYCHLORIDE see
RSA000
RUTHENIUM(3+),
(HYPOXANTHINE)PENTAAMMINE-,
TRICHLORIDE see HOW600
RUTHENIUM(3+),
(INOSINE)PENTAAMINE-, TRICHLORIDE
see IDE100
RUTHENIUM OXIDE see RSF875
RUTHENIUM(VIII) OXIDE see RSK000
RUTHENIUM(3+),
PENTAAMMINE(THEOPHYLLINE)-,
TRICHLORIDE see PAS835
RUTHENIUM RED see RSK100
RUTHENIUM SALT of
TETRAMETHYLPHENANTHRENE see
RSP000
RUTHENIUM TETRAOXIDE see RSK000
RUTHENIUM TRICHLORIDE see RRZ000
RUTHENIUM TRICHLORIDE HYDROXIDE
see RSA000
RUTILE see RSP100
RUTILE see TGG760
β-RUTILE see RSP100
RUTIN see RSU000
RUTINIC ACID see RSU000
RUTINOSYL-3-CYANIDINE see KEA325
RUTOSIDE see RSU000
RUTRALIN see BQN600
RUVAZONE see RSU450
RV 1 see SIT850
R-5-VINYL-2-OXAZOLIDINETHIONE see
VQA100
RVK see DXO200
RVM-FG see PAE750
RW3 see SMP450
RX 2557 see CAT775
RX 6029M see TAL325
RYANEXEL see RSZ000
RYANIA see RSZ000
RYANIA POWDER see RSZ000
RYANIA SPECIOSA see RSZ000
RYANICIDE see RSZ000
RYANODINE see RSZ000
RYCELAN see OJY100
RYCELON see OJY100
RYEMYCIN-B2 see TAH675
RYLAM see CBM750
RYLUX BSP (CZECH) see BHN250
RYLUX PRS (CZECH) see BKO750
RYNACROM see CNX825
RYODIPINE see RSZ375
RYOMYCIN see HOH500
RYSER see RDK000
RYTHMODAN see RSZ600

RYTHMOL see BHR750
RYTHRITOL see PBC250
RYTMONORM see PMJ525
RYUTAN (JAPANESE) see RSZ675
RYZELAN see OJY100
RZ 142 see EAN725
S-6 see MIH300
S 07 see SOU800
S 13 see BHD250
S 22 see SAK500
S 30 see PJT300
S-33 see CIF250
S 46 see PFB350
S51 see BBV500
S-62 see ARW750
S 77 see DRC600
S 84 see BNF750
S 049 see SDY625
S115 see NCQ900
S 140 see DAM700
S 142 see PNB250
S 151 see EJM500
S 151 see SAA000
S 154 see BPR500
S-165 see IOS100
S 173 see SMQ500
S 187 see DPI700
S 190 see KDU100
S 202 see DHK600
S 222 see BKB250
S 260 see MCB050
336S see AHP765
S410 see DSK600
S 421 see OAL000
S-578 see DYF700
S 596 see BQE000
S 62-2 see ARW750
S 630 see AAU300
S 650 see BQH250
S 75M see SFO500
S 767 see FAQ800
S-805 see DCS200
S 816 see CNA750
S-847 see CEW500
S 852 see DBL700
S 940 see HMV000
Sb-57 see AQI500
Sb-71 see AQC000
S-1000 see DXS375
S 1006 see PJY100
S-1046 see XTJ000
S 1065 see MIB750
S 1096 see GGR200
S 1097 see GGR200
S 112A see DSQ000
S 1132 see PJY100
S 1520 see IBV100
S 1530 see DLV000
S 1544 see PFC750
S-1605 see IOS100
S 1688 see AMJ500
S 1702 see DBL700
S 1707 see MCB050
S 1708 see MCB050
S 1710 see MCB050
S 1711 see MCB050
S 1752 see FAQ900
S 1844 see FAR150
S 1942 see BNL250
S 2225 see EGV500
S 2539 see PDR700
S 2852 see EQN250
S 2940 see DRR400
S-3151 see AHJ750
S 32-06 see DAB825
S 327D see SOU800
S 3307D see SOU800
S 3308 see DGC100
S-3440 see BFV765
S-3552 see MNP350
S 4004 see YCJ200
S 4068 see THI500

S 4084 see COQ399
S 4087 see CON300
S 5057 see MCB050
S 5439 see PDV330
S 5602 see FAR100
S 5660 see DSQ000
6059S see LBH200
S-6115 see CQI750
6315-S see FCN100
S 640P see APT250
S 6437 see ALV000
S 6900 see DRR200
S-6,999 see NNF000
S 7131 see PMF750
S 8527 see CMV700
S-9115 see CQI750
S-9700 see DAB880
Sa 267 see DGW600
L-S 519 see GCE500
S 10165 see DGI000
10275-S see EBD500
10364S see MCH600
S 1096R see GGR200
S-15126 see DED000
S18327 see FGA200
S18616 see CGB100
S 21634 see MNX900
S 2852F see EQN250
S30066 see FPO000
31252-S see HNT075
31252-S see PCJ325
S 32165 see IOS100
S 3307D see SOU800
S 53482 see FLZ075
S 9490-3 see PCJ230
(R)-S 3308 see THS920
450191-S see AHP875
S 50154-9 see CIQ100
710674-S see CMW550
711389-S see DFM875
S 73 4118 see PDW250
S 65 (polymer) see PKQ059
S6 (pharmaceutical) see MIH300
SA see SAI000
SA 79 see PMJ525
SA96 see MCO775
SA 97 see HGI100
S 99 A see SLD800
SA 111 see SNJ000
SA 217 see CBI000
SA 446 see FAQ950
SA 504 see TGB160
SA 546 see OMY850
SA 1500 see AHH825
SA 20.16 see MCB050
SA 42-548 see MBV250
SAA see SAN000
S 5602 A ALPHA see FAR150
SAATBEIZFUNGIZID (GERMAN) see
HCC500
SABACIDE see VHZ000
SABADILLA see VHZ000
SABADININE see EBL000
SABANE DUST see VHZ000
SABARI see HGC500
SABILA (CUBA, PUERTO RICO) see AGV875
SABLIER (HAITI) see SAT875
SACABUCHE (PUERTO RICO) see JBS100
SACANDAGA YELLOW X 2476 see CMS210
SACARINA see BCE500
SACCAHARIMIDE see BCE500
SACCHARATED FERRIC OXIDE see IHG000
SACCHARATED IRON see IHG000
SACCHARIN see SJN700
SACCHARINA see BCE500
SACCHARIN ACID see BCE500
SACCHARIN AMMONIUM see ANT500
SACCHARINATE AMMONIUM see ANT500
SACCHARIN CALCIUM see CAM750
SACCHARINE see BCE500
SACCHARINE SOLUBLE see SJN700
SACCHARINNATRIUM see SJN700

SACCHARINOL see BCE500
SACCHARINOSE see BCE500
SACCHARIN, SODIUM see SJN700
SACCHARIN, SODIUM SALT see SJN700
SACCHARIN SOLUBLE see SJN700
SACCHAROIDUM NATRICUM see SJN700
SACCHAROL see BCE500
SACCHAROLACTIC ACID see GAR000
SACCHAROSE see SNH000
SACCHAROSE ACETATE ISOBUTYRATE see SNH050
SACCHAROSONIC ACID see SAA025
SACCHARUM see SNH000
SACCHARUM LACTIN see LAR000
SACERIL see DKQ000
SACERIL scc DNU000
SACERNO see MKB250
SACHSISCHBLAU see FAE100
SACHTOLITH see ZNJ100
SACHTOLITH HD-S see ZNJ100
SACROMYCIN see AHL000
SAD see EDA500
SAD-128 see SAB800
SADAMIN see XCS000
SADH see DQD400
SA 97 DIHYDROCHLORIDE see CJR809
SADOFOS see MAK700
SADOPHOS see MAK700
SADOPLON see TFS350
SADOREUM see IDA000
SAEURE DES PHYTINS (GERMAN) see PIB250
SAEURE FLUORIDE (GERMAN) see FEZ000
SAFARITONE SCARLET B see ENP100
SAFARITONE YELLOW G see AAQ250
SAFE-N-DRI see AHF500
SAFFAN see AFK500
SAFFLOWER OIL see SAC000
SAFFLOWER OIL (UNHYDROGENATED) (FCC) see SAC000
SAFFRON see SAC100
SAFFRON YELLOW see DUX800
SAFRANIN see GJI400
SAFRANIN BLUISH see AJP300
SAFRANINE see GJI400
SAFRANINE 6B see AJP300
SAFRANINE 8B see AJP300
SAFRANINE A see GJI400
SAFRANINE B see GJI400
SAFRANINE G see GJI400
SAFRANINE GF see GJI400
SAFRANINE J see GJI400
SAFRANINE O see GJI400
SAFRANINE OK see GJI400
SAFRANINE SUPERFINE G see GJI400
SAFRANINE T see GJI400
SAFRANINE TH see GJI400
SAFRANINE TN see GJI400
SAFRANINE Y see GJI400
SAFRANINE YN see GJI400
SAFRANINE ZH see GJI400
SAFRANIN T see GJI400
SAFRASIN see SAC875
SAFROL see SAD000
SAFROLE see SAD000
SAFROLE MF see SAD000
SAFROTIN see MKA000
SAFROTIN see SAD100
SAFSAN see DXE000
SAGAMICIN see MQS579
SAGAMICIN (OBS.) see MQS579
SAGATAL see NBU000
SAGENITE see RSP100
SAGE OIL see SAE500
SAGE OIL see SAE550
SAGE OIL CLARY see OGQ200
SAGE OIL, DALMATIAN TYPE see SAE500
SAGE OIL, SPANISH TYPE see SAE550
SAGO CYCAS see CNH789
SAGRADO see SAF000
SAH see SAG000
SAH 22 see SEM500

SAH-2348 see MON600
SAH-42-348 see MON600
SAIB see SNH050
SAICLATE see DNU100
SAIKOSIDE, CRUDE see SAF300
SAIPHOS see ASD000
SAISAN see MLC250
SAKOLYSIN (GERMAN) see BHT750
SAKURAI No. 864 see YCJ000
SAL see HMK100
SALACETIN see ADA725
SALACHLOR see SHJ000
SALAMID see SAH000
SALAMIDE see SAH000
SAL AMMONIA see ANE500
SAL AMMONIAC see ANE500
SALAMMONITE see ANE500
SALAZOLON see AQN250
SALAZOPYRIN see PPN750
SALAZOSULFAPYRIDINE see PPN750
SALBEI OEL (GERMAN) see SAE500
SALBEI OEL (GERMAN) see SAE550
SALBUTAMOL see BQF500
SALBUTAMOL HEMISULFATE see SAF400
SALBUTAMOL SULFATE see SAF400
SALCETOGEN see ADA725
SALCOMIN see BLH250
SALCOMINE POWDER see BLH250
SAL ENIXUM see PKX750
SALETIN see ADA725
SALI see SAF500
SALICAL see SAN000
SALICILAMIDE (ITALIAN) see SAH000
SALICIM see SAH000
SALICINE BLUE BLACK AEF see CMP880
SALICINE CHROME RED B see CMG750
SALICOL see CMG000
SALICOYL HYDRAZIDE see HJL100
SALICRESIN FLUID see CHW675
SALICYCLOHYDRAZINE see HJL100
SALICYLADEHYDE see SAG000
SALICYLAL see SAG000
SALICYL ALCOHOL see HMK100
SALICYLALDEHYDE see EOB050
SALICYLALDEHYDE see SAG000
SALICYLALDEHYDE, 5-CHLORO- see CLD800
SALICYLALDEHYDE, 4-(DIETHYLAMINO)-(6CI,8CI) see DIJ230
SALICYLALDEHYDE DIMETHYL ACETAL CARBAMATE see SAG100
SALICYLALDEHYDE ETHYLENEDIIMINE COBALT see BLH250
SALICYLALDEHYDE METHYL ETHER see AOT525
SALICYLALDEHYDE, 5-NITRO-(6CI,7CI,8CI) see HMX700
SALICYLALDEHYDE OXIME see SAG500
SALICYLALDOXIME see SAG500
SALICYLAMID see SAH000
SALICYLAMIDE see SAH000
SALICYLAMIDE, 5-(1-HYDROXY-2-((1-METHYL-3-PHENYLPROPYL)AMINO)ETHYL)-, HYDROCHLORIDE see HMM500
SALICYLANILID see SAH500
SALICYLANILIDE see SAH500
SALICYLANILIDE, 4',5-DIBROMO- see BOD600
SALICYLANILIDE, TRIBROMO- see THV800
SALICYLAZOSULFAPYRIDINE see PPN750
SALICYL-DIAETHYL (GERMAN) see BQH250
SALICYLDIETHYLAMIDE see DJY400
SALICYL-DIMETHYL (GERMAN) see BQH500
SALICYLDIMETHYLAMIDE see DUA800
SALICYL HYDRAZIDE see HJL100
SALICYLHYDROXAMIC ACID see SAL500
SALICYLIC ACID see SAI000
m-SALICYLIC ACID see HJI100
p-SALICYLIC ACID see SAI500

SALICYLIC ACID, 4-AMINO-, 2-(DIETHYLAMINO)ETHYL ESTER, HYDROCHLORIDE see AMM750
SALICYLIC ACID ANILIDE see SAH500
SALICYLIC ACID, BIMOLECULAR ESTER see SAN000
SALICYLIC ACID, (BIS(8-HYDROXYQUINOLYL)AMINO)- see BKJ300
SALICYLIC ACID, BISMUTH BASIC SALT see SAI600
SALICYLIC ACID, 5-CHLORO- see CLD825
SALICYLIC ACID CHOLINE SALT see CMG000
SALICYLIC ACID, 3,6-DICHLORO- see DGK250
SALICYLIC ACID, DIHYDROGEN PHOSPHATE see PHA575
SALICYLIC ACID, 3,5-DINITRO- see HKE600
SALICYLIC ACID-2-ETHYLHEXYL ESTER see ELB000
SALICYLIC ACID, FLUOROACETATE see FFT000
SALICYLIC ACID-3-HEXEN-1-YL ESTER see SAJ000
SALICYLIC ACID, HYDRAZIDE see HJL100
SALICYLIC ACID, ISOBUTYL ESTER see IJN000
SALICYLIC ACID, ISOPENTYL ESTER see IME000
SALICYLIC ACID ISOPROPYL ESTER O-ESTER with O-ETHYL ISOPROPYLPHOSPHORAMIDOTHIOATE see IMF300
SALICYLIC ACID, MAGNESIUM SALT (2:1), TETRAHYDRATE see MAI700
SALICYLIC ACID, METHYL ESTER see MPI000
SALICYLIC ACID, METHYL ESTER, ESTER with p-GUANIDINOBENZOIC ACID see CBT175
SALICYLIC ACID, MONOAMMONIUM SALT see ANT600
SALICYLIC ACID, compounded with MORPHOLINE (1:1) see SAI100
SALICYLIC ACID, 4-NITRO- see NJF100
SALICYLIC ACID OCTYL ESTER see SAJ500
SALICYLIC ACID PENTYL ESTER see SAK000
SALICYLIC ACID-3-PHENYLPROPYL ESTER see SAK500
SALICYLIC ACID with PHYSOSTIGMINE (1:1) see PIA750
SALICYLIC ACID, SODIUM SALT see SJO000
SALICYLIC ACID, 5-(p-(p-SULFOPHENYLAZO)PHENYLAZO)-, DISODIUM SALT see CMP882
SALICYLIC ACID, p-TOLYL ESTER see THD850
SALICYLIC ALDEHYDE see SAG000
SALICYLIC ETHER see SAL000
SALICYLIC ETHYL ESTER see SAL000
SALICYLIC HYDRAZIDE see HJL100
SALICYLOHYDROXAMIC ACID see SAL500
SALICYLOHYDROXIMIC ACID see SAL500
SALICYLOYLGLYCINE see SAN200
SALICYLOYLOXYTRIBUTYLSTANNANE see SAM000
SALICYLOYLSALICYLIC ACID see SAN000
SALICYL PHOSPHATE see PHA575
SALICYLSALICYLIC ACID see SAN000
o-SALICYLSALICYLIC ACID see SAN000
SALICYLSULFONIC ACID see SOC500
SALICYLURIC ACID see SAN200
SALICYL-VASOGEN see ANT600
SALICYL YELLOW see SIT850
SALIFEBRIN see SAH500
SALIGENIN see HMK100
SALIGENIN CYCLIC PHENYL PHOSPHATE see SAN300
SALIGENOL see HMK100
SALIMED see CPR750
SALIMID see CPR750

SALINA see SAN000
SALINE see SFT000
SALINIDE see SAH500
SALINIDOL see SAH500
SALINOMYCIN see SAN500
SALIOXON-PHENYL see SAN300
SALIPHENAZON see AQN250
SALIPRAN see SAN600
SALIPUR see SAH000
SALIPUROL see NBP350
SALIPURPOL see NBP350
SALIPYRAZOLAN see AQN250
SALIPYRINE see AQN250
SALISAN see CLH750
SALISOD see SJO000
SALITHION see MEC250
SALITHION-SUMITOMO see MEC250
SALIX see CHJ750
SALIX TETRA-SPRA ROXB., EXTRACT see
BAD300
SALIZELL see SAH000
SAL de MERCK see CNF000
SALMETEROL see HNI600
SALMIAC see ANE500
SALMIDOCHOL see DYE700
SALMINE SULFATE see SAO200
SALMINE SULFATE (1:1) see SAO200
SALMONELLA ENTERITIDIS
ENDOTOXIN see SAO250
SALMONELLA TYPHI ENDOTOXIN see
EAS100
SALMONELLA TYPHOSA ENDOTOXIN see
EAS100
SALNIDE see SAH500
SALOL see PGG750
SALOXIUM see SAN000
SALP see SEM300
SALPETERSAEURE (GERMAN) see NED500
SALPETERZUUROPLOSSINGEN (DUTCH)
see NED500
SALPIX see AAN000
SALRIN see SAH000
SALSALATE see SAN000
SALSOCAIN see SAO475
SALSONIN see SJO000
SALSTRIP-N-1 see SIU100
SALT see SFT000
SALT BATHS (NITRATE OR NITRITE) see
SAO500
SALT CAKE see SJY000
SALT OF TARTAR see PLA250
SALTPETER see PLL500
SALT of SATURN see LCV000
SALTS of FATTY ACIDS see SAO550
SALUFER see DXE000
SALUNIL see CLH750
SALUPRES see RDK000
SALURAL see BEQ625
SALURES see BEQ625
SALURETIL see CLH750
SALURETIN see CLY600
SALURIC see CLH750
SALURIN see HII500
SALURIN see SIH500
SALUTENSIN see RDK000
SALVACARD see DJS200
SALVACORIN see DJS200
SALVADERA (CUBA) see SAT875
SALVARSAN see SAP500
SALVICYKLIN see SAP600
SALVIS see SEP000
SALVO see DAA800
SALVO see HKC000
SALVO LIQUID see BCL750
SALVO POWDER see BCL750
SALVURON see EEQ600
SALYMID see SAH000
SALYRGAN see SIH500
SALYRGAN THEOPHYLLINE see MDH750
SALYRGAN THEOPHYLLINE see SAQ000
SALYSAL see SAN000
SALYZORON see BBW500

SALZBURG VITRIOL see CNP500
SAM see DUA800
SAM see SAH000
SAMARIUM see SAQ500
SAMARIUMACETAT (GERMAN) see SAR000
SAMARIUM ACETATE see SAR000
SAMARIUM(III) CHLORIDE see SAR500
SAMARIUM CITRATE see SAS000
SAMARIUM NITRAT (GERMAN) see SAT000
SAMARIUM NITRATE see SAT200
SAMARIUM(III) NITRATE, HEXAHYDRATE
(1:3:6) see SAT000
SAMARIUM TRINITRATE see SAT200
SAMARON BLUE 5G see DMM400
SAMARON BLUE FBL see CMP070
SAMARON BLUE FBL see DBQ220
SAMARON BRILLIANT VIOLET B see
DBY700
SAMARON PINK FBL see AKI750
SAMARON PINK RFL see AKO350
SAMARON RED VIOLET F3B see DBX000
SAMARON YELLOW 5RL see CMP090
SAMARON YELLOW PA3 see AAQ250
SAMBUCIN see KEA325
SAMBUCUS CAERULEA see EAI100
SAMBUCUS CANADENSIS see EAI100
SAMBUCUS MELANOCARPA see EAI100
SAMBUCUS MEXICANA see EAI100
SAMBUCUS RACEMOSA see EAI100
SAME see AEM310
SAMECIN see HGP550
SAMID see SAH000
S,2-(3-AMINOPROPYLAMINO)ETHYL-
PHOSPHOROTHIOIC ACID see AMD000
SAMURON see INR000
SAN see ALQ625
SAN 155 see TFH750
SAN 230 see PHI500
SAN 239 see BAC040
SAN 371 see MFC100
SAN 1551 see TFH750
SAN 371F see MFC100
SAN 619F see CQJ150
SANAI, seed extract see CNX827
SANASEED see SMN500
SANASPASMINA see CBA375
SANASTHYMYL see AFJ625
SANATRICHOM see MMN250
SANCAP see EPN500
SANCELER CM-PO see CPI250
SANCHINOSIDE E1 see PAF450
SANCLOMYCINE see TBX000
SANCOAT PW701 see MCB050
SAN-CYAN see COI250
SANCYCLAN see DNU100
SAND see SCI500
SAND see SCJ500
SAND ACID see SCO500
SANDAL see SAU375
SANDALWOOD OIL, WEST INDIAN OIL
see AON600
SANDAMYCIN see AEA750
SANDBOX TREE see SAT875
SAND CORN see DAE100
SANDESIN see AEH750
SANDIMMUN see CQH100
SANDIMMUNE see CQH100
SANDIX see LDS000
SANDOCRYL BLUE BRL see BJI250
SANDOCRYL GOLDEN YELLOW B-GRL
see CMM895
SANDOCRYL ORANGE B-G see CMM820
SANDOCRYL YELLOW BGL see CMM890
SANDOFAN see MFC100
SANDOGLOBULIN see IBP250
SANDOLAN BLUE P-ARL see CMM090
SANDOLAN RED N-RS see CMM330
SANDOLAN TURQUOISE E-AS see ERG100
SANDOLIN see DUS500
SANDOPARIN see HAQ500
SANDOPEL BLACK EX see AQP000
SANDOPEL DARK GREEN B see CMO830

SANDOPTAL see AGI750
SANDORIN GREEN 8GLS see CMS140
SANDORMIN see TEH500
SANDOSCILL see POB500
SANDOSTAB P-EPQ see PCB400
SANDOSTAB PEPQ see PCB400
SANDOTHRENE BLUE NG see DFN300
SANDOTHRENE BLUE NGR see DFN300
SANDOTHRENE BLUE NGW see DFN300
SANDOTHRENE BLUE NRSC see IBV050
SANDOTHRENE BLUE NRSN see IBV050
SANDOTHRENE BROWN NBR see CMU770
SANDOTHRENE BROWN NR see CMU780
SANDOTHRENE BROWN NRF see CMU780
SANDOTHRENE DARK BLUE NR see
VGP100
SANDOTHRENE OLIVE N2R see DUP100
SANDOTHRENE ORANGE R see CMU815
SANDOTHRENE PRINTING YELLOW see
DCZ000
SANDOTHRENE RED N 6B see CMU825
SANDOTHRENE VIOLET 4R see DFN450
SANDOTHRENE VIOLET N 4R see DFN450
SANDOTHRENE VIOLET N 2RB see
DFN450
SANDOXYLAT A 25-7 see AFJ160
SANDOZ 6538 see DJY200
SANDOZ 43-715 see POB300
SANDOZ 52139 see MKA000
SANDRIL see RDK000
SANDRIX see DAB750
SANDRON see RDK000
SANDUVOR 3039 see ODY150
SANDUVOR P-EPQ see PCB400
SANEDRINE see EAW000
SANEGYT see CAY875
SAN-EI BRILLIANT SCARLET 3R see
FMU080
SAN-EI TARTRAZINE see FAG140
SANEPIL see DKQ000
SANFURAN see NGE500
SANG gamma see BBQ500
SANGIVAMYCIN see SAU000
SANGUICILLIN see AOD000
SANGUINARINE, CHLORIDE see SAU100
SANGUINARINE HYDROCHLORIDE see
SAU100
SANGUINARIUM CHLORIDE see SAU100
SANGUIRITRIN see SAU350
SANGUIRITRINE see SAU350
SANGUIRYTHRINE see SAU350
SAN 155 I see TFH750
SAN 244 I see DRR200
SAN 6538 I see DJY200
SAN 6913 I see DRR200
SAN 7107 I see DRR200
SAN 52 139 I see MKA000
SANICLOR 30 see PAX000
SANICRO 31 see IGL100
SANIPIROL see SEP000
SANIPRIOL-4 see AMM250
SANITIZED SPG see ABU500
SANKUMIN see HGI100
SANLOSE SN 20A see SFO500
SANMARTON see FAR100
SANMORIN OT 70 see DJL000
SANOCHOLEN see DAL000
SANOCHRYSINE see GJG000
SANOCID see HCC500
SANOCIDE see HCC500
SANOCRISIN see FBP100
SANODIAZINE see PPP500
SANODIN see CBO500
SANOFLAVIN see DBN400
SANOHIDRAZINA see ILD000
SANOMA see IPU000
SANOQUIN see CLD000
SANOQUIN see CLD250
SANORIN see NAH500
SANORIN-SPOFA see NCW000
SANPRENE LQX 31 see PKP000
SANQUINON see DFT000

SANREX see ADY500
SANSDOLOR see GGS000
SANSEL ORANGE G see PEJ500
SANSERT see MQP500
SANSOCIZER DINP see PHW585
SANSPOR see CBF800
SANTAL OIL see OGY220
(+)-α-SANTALOL see SAU375
d-α-SANTALOL see SAU375
(Z)-α-SANTALOL see SAU375
cis-α-SANTALOL see SAU375
SANTALOL A see SAU375
α-SANTALOL (FCC) see OGY220
SANTALOZONE see DAF150
SANTALYL ACETATE see SAU400
SANTAR see MCT500
SANTAVY'S SUBSTANCE F see MIW500
SANTHEOSE see TEO500
SANTICIZER 8 see ENY200
SANTICIZER 1H see CPQ800
SANTICIZER 160 see BEC500
SANTICIZER 711 see DXQ400
SANTICIZER M-17 see MOD000
SANTICIZER 141 (MONSANTO) see DWB800
SANTICIZIER B-16 see BQP750
SANTOBANE see DAD200
SANTOBRITE see PAX250
SANTOBRITE see SJA000
SANTOCEL see SCH002
SANTOCHLOR see DEP800
SANTOCURE see CPI250
SANTOCURE MOR see BDG000
SANTOCURE VULCANIZATION
ACCELERATOR see CPI250
SANTOFLEX 13 see DQV250
SANTOFLEX 14 see DTI800
SANTOFLEX 17 see BJT500
SANTOFLEX 36 see PFL000
SANTOFLEX 77 see BJL000
SANTOFLEX 134 see SAU475
SANTOFLEX A see SAV000
SANTOFLEX 14 ANTIOZONANT see
DTI800
SANTOFLEX AW see SAV000
SANTOFLEX DD see SAU480
SANTOFLEX IC see PEY500
SANTOGARD PVI see CPQ700
SANTOMERSE 3 see DXW200
SANTONIN see SAU500
α-SANTONIN see SAU500
l-α-SANTONIN see SAU500
SANTONINIC ANHYDRIDE see SAU500
SANTONOX see TFC600
SANTOPHEN see CJU250
SANTOPHEN see PAX250
SANTOPHEN 20 see PAX250
SANTOPHEN I GERMICIDE see CJU250
SANTOQUIN see SAV000
SANTOQUINE see SAV000
SANTOQUIN EMULSION see TMJ000
SANTOQUIN MIXTURE 6 see TMJ000
SANTOSOL 150 see DQL820
SANTOTHERM see PJL750
SANTOTHERM 66 see TEQ800
SANTOTHERM FR see PJL750
SANTOUAR A see DCH400
SANTOVAR A see DCH400
SANTOWAX see TBC750
SANTOWAX M see TBC620
SANTOWHITE CRYSTALS see TFC600
SANTOWHITE POWDER see BRP750
SANTOX see EBD700
SANULCIN see DAB875
SANVEX see BHL750
SANWAPHYLLIN see HOA000
SANWAX 161P see PJS750
SANYO BENZIDINE ORANGE see CMS145
SANYO BENZIDINE YELLOW see CMS210
SANYO BENZIDINE YELLOW-B see
DEU000
SANYO BRILLIANT CARMINE see CMG750
SANYO CARMINE L2B see DMG800

SANYO CYANINE GREEN see PJQ100
SANYO FAST BLUE SALT B see DCJ200
SANYO FAST ORANGE GC BASE see
CEH690
SANYO FAST ORANGE SALT RD see
CEG800
SANYO FAST RED 2B see CMS148
SANYO FAST RED 10B see NAY000
SANYO FAST RED B BASE see NEQ000
SANYO FAST RED 2BE see CMS148
SANYO FAST RED RL BASE see MMF780
SANYO FAST RED SALT RL see MMF780
SANYO FAST RED SALT TR see CLK235
SANYO FAST RED TR BASE see CLK220
SANYO FAST SCARLET GG BASE see
DEO295
SANYO GUM ORANGE A see CMM220
SANYO LAKE RED C see CHP500
SANYO PERMANENT ORANGE D 213 see
CMU820
SANYO PERMANENT ORANGE D 616 see
CMU820
SANYO PHTHALOCYANINE GREEN F6G
see PJQ100
SANYO PHTHALOCYANINE GREEN FB
PURE see PJQ100
SANYO SCARLET PURE see MMP100
SANYO SCARLET PURE NO. 1000 see
MMP100
SANYO THRENE BLUE IRN see IBV050
SAOLAN see DSK200
SAPECRON see CDS750
SAPEP see AMD000
SAPHICOL see ASD000
SAPHIZON see ASD000
SAPHIZON-DP see ASD000
SAPHOS see ASD000
SAPHRAZINE see SAC875
SAPILENT see DLH200
SAPINDOSIDE A see HAK075
SAPONA RED LAKE RL-6280 see NAY000
SAPONIN see SAV500
SAPONIN-GYPSOPHILA see GMG000
SAPPILAN see CJT750
SAPPIRAN see CJT750
SAPRECON C see DRP600
SAPROL see TKL100
SAPWOOD of LINDEN see SAW300
SAQUINAVIR MESYLATE see FOL025
SARAN see SAX000
SARAN 683 see CGW300
SARAN 746 see CGW300
SARANAC YELLOW X 2838 see CMS208
SARAN RESIN 683 see CGW300
SARCELL TEL see SFO500
SARCLEX see DGD600
SARCOCLORIN see BHT750
SARCOLACTIC ACID see LAG010
l-SARCOLYSIN see PED750
dl-SARCOLYSIN see BHT750
p-l-SARCOLYSIN see PED750
o-dl-SARCOLYSIN see BHT250
d-SARCOLYSINE see SAX200
dl-SARCOLYSINE see BHT750
m-l-SARCOLYSINE see SAX210
l-SARCOLYSINE HYDROCHLORIDE see
BHV350
dl-SARCOLYSINE HYDROCHLORIDE see
BHV000
SARCOLYSIN HYDROCHLORIDE see
BHV000
SARCOMYCIN see SAX500
SARCOSET GM see DTG000
SARGOTT see YAG000
SARIFA, seed extract see SAX275
SARIN see IPX000
SARIN II see IPX000
SARKOKLORIN see BHV000
SARKOMYCIN see SAX500
SARKOMYCIN B*, SODIUM SALT see SJS500
SARKOMYCIN compounded with NICOTINIC
HYDRAZIDE see SAY000

SARKOMYCIN, SODIUM SALT see SJS500
SARLACH GN see CMP620
SARLACH R see SBC500
SARMENTOSIGENIN A see NDY565
SARODORMIN see DYC800
SAROLEX see DCM750
SAROMET see DCK759
SAROTEN see EAI000
SAROTENE see EAI000
SARPAGAN see RDK000
SARPAGEN see RDK000
SARPAN see AFG750
SARPIFAN HP 1 see AAX175
SARTOMER 256 see EHG030
SARTOMER 285 see TCS100
SARTOMER 302 see TCS100
SARTOMER 506 see IHX700
SARTOMER SR 206 see BKM250
SARTOMER SR 220 see PMP800
SARTOMER SR 256 see EHG030
SARTOMER SR 351 see TLX175
SARTOSONA see ALV000
SARYUURIN see DYE700
S.A.S.-500 see PPN750
SAS 538 see CFY500
SAS 546 see BCI300
SAS 560 see VAQ200
SAS 561 see CCM050
SAS 643 see DKV400
SAS 2185 see BHL800
SASAPIRIN see SAN000
SASAPYRIN see SAN000
SASAPYRINE see SAN000
SASAPYRINUM see SAN000
SA 96 SODIUM SALT see SAY875
SASP see PPN750
SASSAFRAS see SAY900
SASSAFRAS ALBIDUM see SAY900
SASSAFRAS OIL see OHI000
SASTRIDEX see TKH750
SATANOLON see CCR875
SATECID see CHS500
SATIAGEL GS 350 see CCL250
SATIAGEL GS 350 see CCL350
SATIAGUM 3 see CCL250
SATIAGUM STANDARD see CCL250
SATILAN PLZ see CNH125
SATILAN RL 2 see CNH125
SATINITE see CAX750
SATIN SPAR see CAX750
SATOL see OBA000
SATOX 20WSC see TIQ250
SATRANIDAZOLE see SAY950
SATURN see SAZ000
SATURN BROWN LBR see CMO750
SATURNO see SAZ000
SATURN RED see LDS000
SATURN RED B see CMO885
SATURN TURQUOISE BLUE GL see COF420
SAUCO BLANCO (CUBA) see EAI100
SAUCO (MEXICO, PUERTO RICO) see
EAI100
SAUROL see IAD000
SAUTERALGYL see DAM700
SAUTERZID see ILD000
SAVAC see DNG400
SAVEMIX C 100 see MCB050
SAVENTRINE see DMV600
SAVENTRINE see IMR000
SAVEY see CJU275
SAVIT see CBM750
SAVORY OIL (summer variety) see SBA000
SAX see SAI000
SAXIN see BCE500
SAXIN see SJN700
SAXITOXIN (8CI) see SBA600
SAXITOXIN DIHYDROCHLORIDE see
SBA500
SAXITOXIN HYDRATE see SBA600
SAXITOXIN HYDROCHLORIDE see SBA500
SAXOL see MQV750
SAXOSOZINE see SNN500

SA 50 Y see TGB160
SAYFOR see ASD000
SAYFOS see ASD000
SAYPHOS see ASD000
SAYTEX 102 see PAU500
SAYTEX 111 see MKA270
SAYTEX 125 see PAT830
SAYTEX 102E see PAU500
SAYTEX BCL 462 see DDM300
SAYTEX BT 93 see EIT150
SAYTEX BT 93W see EIT150
SAYTEX RB 49 see TBJ700
SAYTEX RB 100PC see MKA270
SAZZIO see AFL000
SB-8 see HJZ000
SB-23 see DQY909
SB 502 see BOO650
SB 475K see SMQ500
SB 5833 see CFY250
SB 202235 see DMA600
SB 210661 see DLB600
SBa 0108E see BAK000
S.B.A. see BPW750
SBO see DJL000
SBP-1382 see BEP500
SBP-1390 see BEP750
SBP-1513 see AHJ750
S.B. PENICK 1382 see BEP500
SBS see SMR000
SBTPC see SOU675
SC 9 see SCQ000
SC10 see DCD000
SC-110 see ABU500
SC. 121 see PAW600
SC 201 see SCQ000
SC 567 see CIM100
SC 0574 see PFR130
SC-1950 see DRK600
SC-2292 see MEK350
SC 2538 see DIR000
SC-2644 see DOB300
SC-2657 see DOA875
SC-2798 see DOB325
SC 2910 see DJM800
SC 2910 see XCJ000
SC 2957 see PFR120
SC-3171 see HKR500
SC-3296 see BKP300
SC-3402 see BKI750
S.C. 3497 see AFW500
SC-3950 see TDQ075
SC 4641 see NCI525
SC-4642 see EEH550
SC 4722 see ONO000
SC 5914 see ENX600
SC 7031 see DNN600
SC 7801 see ELH800
SC 8016 see CJL500
SC-8117 see MDQ100
SC 9022 see MDQ075
SC-9369 see PDO900
SC 9387 see CJL500
SC 9420 see AFJ500
SC 9794 see PII500
SC/3123 see SBA875
SC 10295 see MMN250
SC10363 see VTF000
SC 11927 see CCP750
SC 11977 see MGN600
SC 12937 see ARX800
SC 13957 see RSZ600
SC-14266 see PKZ000
SC 15090 see TEO500
SC 15983 see AFJ500
SC 27166 see DWF700
SC 28762 see VRP200
SC-29333 see MJE775
SC-11800M see SBA625
SCABANCA see BCM000
SCALDIP see DVX800
SCANBUTAZONE see BRF500
SCANDICAIN see CBR250

SCANDICAIN see SBB000
SCANDICAINE see SBB000
SCANDICANE see SBB000
SCANDISIL see SNN300
SCANDIUM see SBB500
SCANDIUM CHLORIDE see SBC000
SCANDIUM(3+) CHLORIDE see SBC000
SCAPUREN see SBE500
SCARCLEX see DGD600
SCARLET BASE CIBA I see DEO295
SCARLET BASE CIBA II see NMP500
SCARLET BASE IRGA II see NMP500
SCARLET BASE NSP see NMP500
SCARLET B FAT SOLUBLE see OHI200
SCARLET G BASE see NMP500
SCARLET GN see CMP620
SCARLET GN SPECIALLY PURE see CMP620
SCARLET LOBELIA see CCJ825
SCARLET PIGMENT RN see MMP100
SCARLET R see FMU070
SCARLET RED see SBC500
SCARLET RED, BIEBRICH see SBC500
SCARLET RELITON BA see ENP100
SCARLET R (MICHAELIS) see SBC500
SCARLET 4BS see CMO870
SCARLET TONER H3B see CMG750
SCARLET TR BASE see CLK200
SCARLET WISTERIA TREE see SBC550
SCATOLE see MKV750
SC 9 (CARBIDE) see SCQ000
SCE 129 see CCS550
SC 11800 EE see DAP810
SCE 1365 HYDROCHLORIDE see CCS300
SCENTENAL see OBW200
SCEPTER see IAH100
SCF see TEE225
SCH see FMR050
SCH 286 see SOU650
SCH 412 see DJO800
Sch 1000 see IGG000
Sch 4831 see BFV750
SCH 5472 see BBW000
Sch 5705 see PCU400
Sch 5706 see FAJ150
Sch 5712 see BRJ125
SCH 6673 see ABG000
SCH 7307 see IPU000
SCH 9384 see AEX000
SCH 9724 see GCS000
SCH 10015 see HIE525
SCH 10159 see TIP300
SCH 10304 see CMX770
SCH 10649 see DLV800
SCH 11527 see TEY000
SCH 12650 see SBC300
SCH 13475 see SDY750
SCH 13521 see FMR050
Sch 14947 see RMF000
SCH-17726 see GAA100
SCH 20569 see NCP550
SCH 20569 see SBD000
SCH 22219 see AFI980
SCH 15719W see HMM500
SCH 18020W see AFJ625
SCHAARDINGER α-DEXTRIN see COW925
SCHAEFFER'S-β-NAPHTHOLSULFONIC
ACID see HMU500
SCHARLACH B see PEJ500
SCHARLACHROT see SBC500
SCH 5802 B see SBC700
SCH CHLORIDE see HLC500
SCHEELES GREEN see CNN500
SCHEELE'S MINERAL see CNN500
SCHERCEMOL 85 see ISC550
SCHERCOMEL M see MCB050
SCHERCOMID ODA see OHU200
SCHERING 34615 see CQI500
SCHERING-35830 see CGL250
SCHERING 36056 see DSO200
SCHERING 36103 see FNE500
SCHERING 36268 see CJJ250
SCHERING 36268 see CJJ500

SCHERING 38107 see EEO500
SCHERING-38584 see MEG250
SCHERISOLON see PMA000
SCHEROSON see CNS800
SCHEROSON see CNS825
SCHEROSON F see CNS750
SCH 31846 HYDROCHLORIDE see ICL500
(−)-SCH 39166 HYDROCHLORIDE see
BCP690
SCHINOPSIS LORENTZII TANNIN see
QBJ000
SCHINUS MOLLE see PCB300
SCHINUS MOLLE OIL see SBE000
SCHINUS TEREBINTHIFOLIUS see PCB300
SCHISANDRIN see SBE400
SCHISANDROL B see SBE450
SCHISTOSOMICIDE see SBE500
SCHIWANOX see AMX750
SCHIZANDRIN see SBE400
SCHIZANDROL B see SBE400
SCHIZANDROL B see SBE450
SCHLAFEN see CAV000
SCHLEIMSAURE see GAR000
SCHOENOCAULON DRUMMONDII see
GJU460
SCHOENOCAULON OFFICIANLIS see
GJU460
SCHOENOCAULON TEXANUM see GJU460
SCHOENOPLECTUS ARTICULATUS (Linn.)
Palla, extract see SBF550
SCHRADAN see OCM000
SCHRADANE (FRENCH) see OCM000
SCH 21420 SULFATE see SBE800
SCHULTENITE see LCK000
SCHULTZ No. 39 see HGC000
SCHULTZ No. 95 see FMU070
SCHULTZ No. 770 see FMU059
SCHULTZ No. 826 see CMM062
SCHULTZ No. 853 see PPQ750
SCHULTZ No. 918 see CMM510
SCHULTZ Nr. 826 see ADE500
SCHULTZ Nr. 836 see ADF000
SCHULTZ No. 1038 see BJI250
SCHULTZ No. 1041 see AJP250
SCHULTZ No. 1228 see IBV050
SCHULTZ Nr. 208 (GERMAN) see HJF500
SCHULTZ Nr. 1309 (GERMAN) see FAE100
SCHULTZ NO. 31 see OHI200
SCHULTZ NO. 541 see SBC500
SCHULTZ NO. 737 see FAG140
SCHULTZ-TAB No. 779 (GERMAN) see
RMK020
SCHUTTGELB see MQF250
L-SCHWARZ 1 see BMA000
SCHWEFELDIOXYD (GERMAN) see
SOH500
SCHWEFELKOHLENSTOFF (GERMAN) see
CBV500
SCHWEFEL-LOST see BIH250
SCHWEFELSAEURELOESUNGEN
(GERMAN) see SOI500
SCHWEFELWASSERSTOFF (GERMAN) see
HIC500
SCHWEFLIGESAEURE (GERMAN) see
SOO500
SCHWEINFURTERGRUEN see COF500
SCHWEINFURT GREEN see COF500
SCIFLUORFEN see SFV650
SCILLACRIST see POB500
SCILLA "DIDIER" see POB500
SCILLAGLYKOSID A (GERMAN) see
GFC000
SCILLAREN A see GFC000
SCILLARENIN-3,6-DEOXY-4-o-β-d-
GLUCOPYRANOSYL-α-l-
MANNOPYRANOSIDE see GFC000
α-l-SCILLARENIN-3-
RHAMNOPYRANOSIDE see POB500
SCILLARENIN-RHAMNOSE (GERMAN) see
POB500
SCILLAREN & RHAMNOSE & GLUCOSE
(GERMAN) see GFC000

SCILLA (VARIOUS SPECIES) see SLH200
SCILLIROSID see SBF500
SCILLIROSIDE see SBF500
(3-β,6-β)SCILLIROSIDE see SBF500
SCILLIROSIDE GLYCOSIDE see RCF000
SCILLIROSIDIN + GLUCOSE (GERMAN) see SBF500
SCILLIRUBROSIDE see SBF510
3-β-SCILLOBIOSIDO-14-β-HYDROXY-Δ-4,20,22-BUFATRIENOLID (GERMAN) see GFC000
SCINNAMINA see TLN500
SCINTILLAR see XHS000
SCIRPENETRIOL TRIACETATE see SBF525
SCIRP-9-ENE-3-α,4-β,15-TRIOL, TRIACETATE see SBF525
SCIRPUS ARTICULATUS Linn., extract see SBF550
SCLAIR 59 see PJS750
SCLAIR 11K see PJS750
SCLAIR 19A see PJS750
SCLAIR 59C see PJS750
SCLAIR 79D see PJS750
SCLAIR 96A see PJS750
SCLAIR 19X6 see PJS750
SCLAIR 2911 see PJS750
SCLAVENTEROL see NGG500
S-CMC see CCH125
SCOBRO see SBG500
SCOBRON see SBG500
SCOBUTIL see SBG500
SCOBUTYL see BDX500
SC 100 (OIL) see AQZ150
SCOKE see PJJ315
S-COL see SCK600
SCOLINE see HLC500
SCOLINE CHLORIDE see HLC500
SCON 5300 see PKQ059
SCONATEX see AAX175
SCONATEX see VPK000
SCOPAMIN see HOT500
SCOPARON see DRS800
SCOPARONE see DRS800
SCOPINE TROPATE see SBG000
SCOPOLAMINE see SBG000
(−)-SCOPOLAMINE see SBG000
SCOPOLAMINE BROMIDE see HOT500
(−)-SCOPOLAMINE BROMIDE see HOT500
SCOPOLAMINE BROMOBUTYLATE see SBG500
SCOPOLAMINE BUTOBROMIDE see SBG500
SCOPOLAMINE BUTYL BROMIDE see SBG500
SCOPOLAMINE-N-BUTYL BROMIDE see SBG500
SCOPOLAMINE HYDROBROMIDE see HOT500
(−)-SCOPOLAMINE HYDROBROMIDE see HOT500
SCOPOLAMINE HYDROBROMIDE TRIHYDRATE see SBG600
SCOPOLAMINE METHOBROMIDE see SBH500
SCOPOLAMINE METHYLBROMIDE see SBH500
(−)-SCOPOLAMINE METHYL BROMIDE see SBH500
SCOPOLAMINE METHYL NITRATE see SBH520
SCOPOLAMINE compounded with METHYL NITRATE (1:1) see SBH520
SCOPOLAMINIUM BROMIDE see HOT500
SCOPOLAMMONIUM BROMIDE see HOT500
SCOPOLAN see SBG500
SCOPOS see HOT500
SCORBAMIC ACID see AJL125
l-SCORBAMIC ACID see AJL125
SCORBAMINIC ACID see AJL125
SCOT see HBT500
SCOTCH ATTORNEY see BAE325

SCOTCH PAR see PKF750
SCOTCH PINE NEEDLE OIL see PIH500
SCOTCIL see BFD000
SCOUT X-TRA see THJ300
SC 7031 PHOSPHATE see RSZ600
SCREEN see BEG300
SCROBIN see ARQ750
SCTZ see CHD750
SCURENALINE see VGP000
SCUROCAINE see AIL750
SCUROCAINE see AIT250
SCUTL see ABU500
SCW 1 see SCQ000
SCYAN see SIA500
SD 25 see DHY200
40 SD see IMF300
SD 188 see SMQ500
SD-345 see ADR250
SD 354 see SMR000
SD 709 see DRM000
SD-1750 see DGP900
SD 1836 see CLV375
SD 1897 see DDL800
SD 2580 see EHY700
SD 2614 see KFK200
SD 2774 see BNQ600
SD 3450 see HCI475
SD 3562 see DGQ875
SD 4,239 see CEQ500
SD 4294 see COD000
SD 4402 see OAN000
SD 4901 see DBX090
SD 5532 see CDR750
SD 7438 see BES250
SD 7727 see DGD400
SD 7961 see DGM600
SD 8339 see BEA100
SD 8447 see RAF100
SD 8530 see TMC750
SD 8530 see TMD000
SD 8988 see MHR150
SD 8989 see BMO310
SD 9098 see DIX600
SD 9129 see MRH209
SD-10576 see SCD500
SD 14114 see BLU000
SD 15418 see BLW750
SD 16389 see COI000
SD 171-02 see MQQ050
SD 210-32 see PIR000
SD 210-37 see PIQ750
SD 270-31 see OKW000
SD 30,053 see EGS000
S-3307D (+−) see SOU800
SD 417-06 see DAB825
SD-43775 see COQ387
SD 43775 see FAR100
SD 91779 see CQE350
SD95481 see CMP955
SD 208304 see FOL050
SDA 13-060-00 see AFJ100
SD ALCOHOL 23-HYDROGEN see EFU000
S DC 200 see SCR400
SDD see DXH300
SDDC see SGM500
SDDS see AOT125
SDEH see DJL400
SD-GP 6000 see SCQ000
SD-GP 8000 see SCQ000
SD 2124-01 HYDROCHLORIDE see PMF525
(±)-SD 2124-01 HYDROCHLORIDE see PMF535
SDIC see SGG500
(3aS-cis)-3a,12a-DIHYDRO-4,6-DIHYDROXY-ANTHRA(2,3-b)FURO(3,2-d)FURAN-5,10-DIONE (9CI) see DAZ100
SDM see SNN300
SDMH see DSF600
SDM No. 5 see MAW250
SDM No. 23 see PBC250
SDMO see AIE750
SDMO see SNN300

SDmP see SNI500
SDP 640 see PJS750
SDPH see DNU000
SD 8988 (SHELL) see MHR150
SDT 1041 see FNF000
S. DYSENTERIAE TOXIN see SCD750
S. DYSENTERIAE TOXIN see SCD800
SE-1520 see IBV100
SE 1702 see DBL700
SEA ANEMONE TOXIN II see ARS000
SEA ANEMONE VENOM see SBI800
SEA COAL see CMY760
SEACYL VIOLET R see DBP000
SEA DAFFODIL see BAR325
SEAKEM CARRAGEENIN see CCL250
SEA-LEGS see HGC500
SEA ONION see SLH200
SEAPROSE S see SBI860
SEARLE 703 see DNN600
SEARLE 27166 see DWF700
SEARLEQUIN see DNF600
SEARS HEAVY DUTY LAUNDRY DETERGENT see SBI875
SEA SALT see SFT000
SEA SNAKE VENOM, AIPYSURUS LAEVIS see SBI880
SEA SNAKE VENOM, ASTROTIA STOKESII see SBI890
SEA SNAKE VENOM, HYDROPHIS CYANOCINCTUS see SBI900
SEA SNAKE VENOM, LAPEMIS HARDWICKII see SBI910
SEA SNAKE VENOM, MICROCEPHALOPHIS GRACILIS see SBI929
SEATREM see CCL250
SEAWATER MAGNESIA see MAH500
SEAZINA see SNJ000
SEBACIC ACID see SBJ500
SEBACIC ACID, DIBUTYL ESTER see DEH600
SEBACIC ACID, DIETHYL ESTER see DJY600
SEBACIC ACID, DIHYDRAZIDE see SBK000
SEBACIC ACID, compd. with 1,6-HEXANEDIAMINE (1:1) see HEG130
SEBACIL see BAT750
SEBACONITRILE see SBK500
SEBACOYLDIOXYBIS(TRIBUTYLSTANNANE) see BLL000
SEBAKAN HEXAMETHYLENDIAMINU see HEG130
SEBAQUIN see DNF600
SEBAR see SBN000
SEBATROL see FMR050
SEBIZON see SNQ710
SEBUTHYLAZINE see THR600
SECACORNIN see LJL000
SECAGYN see EDC500
SECALCIFEROL see HJS900
SECALONIC ACID D see EDA500
SECBUBARBITAL see BPF000
SECBUBARBITAL SODIUM see BPF250
SECBUTABARBITAL see BPF000
SECBUTOBARBITONE see BPF000
SECHVITAN see PII100
SECLAR see BEG000
SECLIN see NCK100
SECO 8 see SBN000
SECOBARBITAL see SBM500
SECOBARBITAL SODIUM see SBN000
R(+)-SECOBARBITAL SODIUM see SBL500
SECOBARBITONE see SBM500
SECOBARBITONE see SBN000
SECOBARBITONE SODIUM see SBN000
21,22-SECOCAMPTOTHECIN-21-OIC ACID LACTONE see CBB870
(5Z,7E)-9,10-SECOCHESTA-5.7.10(19)-TRIENE-1-α,3-β,25-TRIOL see DMJ400
9,10-SECOCHOLESTA-5,7,10(19)-TRIENE-1-α,3-β-DIOL see HJV000
(1-α,3-β,5Z,7E)-9,10-SECOCHOLESTA-5,7,10(19)-TRIENE-1,3,25-TRIOL see DMJ400

(1-α,3-β,5Z,7E,24R)-9,10-SECOCHOLESTA-5,7,10(19)-TRIENE-1,3,24-TRIOL see SBL600
(3-β,5Z,7E,24R)-9,10-SECOCHOLESTA-5,7,10(19)-TRIENE-3,24,25-TRIOL see HJS900
9,10-SECOCHOLESTA-5,7,10(19)-TRIENE-1,3,24-TRIOL, (1-α,3-β,5Z,7E,24R)- see SBL600
9,10-SECOCHOLESTA-5,7,10(19)-TRIENE-3,24,25-TRIOL, (3-β,5Z,7E,24R)- see HJS900
9,10-SECOCHOLESTA-5,7,10(19)-TRIENE-1,3,25-TRIOL, 26,26,26,27,27,27-HEXAFLUORO-,(1-α,3-β,5Z,7E)- see HDB100
9,10-SECOCHOLESTA-5,7,10(19)-TRIEN-3-β-OL see CMC750
9,10,SECOERGOSTA-5,7,10(19),22-TETRAEN-3-β-OL see VSZ100
16,17-SECOESTRA-1,3,5(10),6,8-PENTAEN-17-OIC ACID, 3-METHOXY- see BIT030
16,17-SECO-13-α-ESTRA-1,3,5,6,7,9-PENTAEN-17-OIC ACID, METHYL ESTER see BIT000
SECOMETRIN see LJL000
SECONAL see SBM500
SECONAL SODIUM see SBN000
SECONDARY AMMONIUM ARSENATE see DCG800
SECONDARY AMMONIUM PHOSPHATE see ANR500
SECONDARY ZINC PHOSPHITE see ZLS200
SECONESINZ see GGS000
SECOPAL OP 20 see GHS000
SECOVERINE see SBN300
SECOVERINE HYDROCHLORIDE see SBN300
SECROSTERON see DRT200
SECROVIN see DNX500
SECT HYDROCHLORIDE see EEQ500
SECTRAL see AAE125
SECUPAN see EDC500
SECURINAN-11-ONE (9CI) see SBN350
SECURININ see SBN350
(−)-SECURININE see SBN350
SECURININE see SBN350
SECURITY see ARB750
SECURITY see LCK000
SEDABAMATE see MQU750
SEDAFORM see ABD000
SEDALANDE see HAF400
SEDALGIN see SBN400
SEDALIS SEDI-LAB see TEH500
SEDALON see AMK250
SEDAMYL see ACE000
SEDANTOINAL see MKB250
SEDAPRAN see DAP700
SEDAPSIN see CKE750
SEDARAUPIN see RDK000
SEDARAUPINA see RDK000
SEDA-RECIPIN see RDK000
SEDAREX see EHP000
SEDA-SALUREPIN see RDK000
SEDA-TABLINEN see EOK000
SEDAVIC see HAF400
SEDERAUPIN see RDK000
SEDESTRAN see DKA600
SEDETINE see OAV000
SEDETOL see EJM500
SEDEVAL see BAG000
SEDICAT see EOK000
SEDIMIDE see TEH500
SEDIN see TEH500
SEDIPAM see DCK759
SEDIRANIDO see DBH200
SEDISPERIL see TEH500
SEDIZORIN see EOK000
SEDMYNOL see ACE000
SEDNOTIC see AMX750
SEDOFEN see EOK000
SEDOMETIL see DNA800
SEDOMETIL see MJE780
SEDONAL see EOK000
SEDONEURAL see SFG500
SEDOPHEN see EOK000
SEDOR see SKS700

SEDORMID see IQX000
SED OSTANTHRENOVA M see CMU475
SEDOVAL see TEH500
SEDRENA see BBV000
SEDRESAN see MEP250
SEDSERP see RDK000
SED-TEMS see MDL000
SEDTRAN see ACE000
SEDURAL see PDC250
SEDURAL see PEK250
SEDUTAIN see SBN000
SEDUXEN see DCK759
SEED of FIDDLENECK see TAG250
SEEDRIN see AFK250
SEEDTOX see ABU500
SEEDVAX see AKR500
SEEKAY WAX see TIT500
SEENOX DS see DXG700
SEFACIN see TEY000
SEFFEIN see CBM750
SEFONA see IEP200
SEFRIL see SBN440
SEFRIL HYDRATE see SBN450
SEGNALE BRONZE RED P see CMS150
SEGNALE LIGHT GREEN G see PJQ100
SEGNALE LIGHT ORANGE G see CMS145
SEGNALE LIGHT ORANGE PG see CMS145
SEGNALE LIGHT ORANGE RNG see DVB800
SEGNALE LIGHT RED 2B see MMP100
SEGNALE LIGHT RED B see MMP100
SEGNALE LIGHT RED BR see MMP100
SEGNALE LIGHT RED C4R see MMP100
SEGNALE LIGHT RED PRG see CJD500
SEGNALE LIGHT RED RL see MMP100
SEGNALE LIGHT RUBINE RG see NAY000
SEGNALE LIGHT YELLOW GRX see CMS208
SEGNALE LIGHT YELLOW 2GR see DEU000
SEGNALE RED 3R see CMS155
SEGNALE RED C see CMS150
SEGNALE RED GS see CMS148
SEGNALE RED LC see CHP500
SEGNALE RED R see NAP100
SEGNALE YELLOW 2GR see CMS210
SEGONTIN see PEV750
S. EGRELTRI ATUNUN (TURKISH) see BML000
SEGUREX see BON325
SEGURIL see CHJ750
SEIKAFAST RED 8040 see CMS148
SEIKAFAST YELLOW 2200 see CMS210
SEIYO AKANE see MAB800
SELACRYN see TGA600
SELECRON see BNA750
SELECTIVE see BPG250
SELECTOMYCIN see SLC000
SELECTROL see SBN475
SELEKTIN see BKL250
SELEKTON B 2 see EIX500
SELEN (POLISH) see SBO500
1,2,5-SELENADIAZOLE-3-CARBOXAMIDE, 4-AMINO- see AMN300
1,2,5-SELENADIAZOLE-3-CARBOXAMIDE, 4-AMINO-N-METHYL- see MGL600
SELENIC ACID see SBN500
SELENIC ACID, liquid (DOT) see SBN500
SELENIC ACID, DIPOTASSIUM SALT see PLR750
SELENIC ACID, DISODIUM SALT, DECAHYDRATE see SBN505
SELENIC ACID, compd. with HYDRAZINE see HGW100
SELENIDE, DIALLYL- see DBL300
SELENIDE, DIETHYL- see DJY700
SELENINIC ACID, PROPYL- see PNV760
β-SELENINOPROPIONIC ACID see SBN510
SELENINYL BIS(DIMETHYLAMIDE) see SBN525
SELENINYL BROMIDE see SBN550
SELENINYL CHLORIDE see SBT500

SELENIOUS ACID see SBO000
SELENIOUS ACID, DIPOTASSIUM SALT (9CI) see SBO100
SELENIOUS ACID, DISODIUM SALT see SJT500
SELENIOUS ACID DISODIUM SALT PENTAHYDRATE see SJT600
SELENIOUS ACID, MONOSODIUM SALT see SBO200
SELENIOUS ANHYDRIDE see SBQ500
SELENIUM see SBO500
SELENIUM (colloidal) see SBP000
SELENIUM ALLOY see SBO500
SELENIUM BASE see SBO500
SELENIUM CHLORIDE see SBS500
SELENIUM(IV) CHLORIDE (1:4) see SBU000
SELENIUM CHLORIDE OXIDE see SBT500
SELENIUM COMPOUNDS see SBP500
SELENIUM CYSTINE see DWY800
SELENIUM CYSTINE see SBP600
SELENIUM-CYSTINE see SBV100
SELENIUM, DICHLOROBIS(2-CHLOROCYCLOHEXYL)- see DEU100
SELENIUM, DICHLOROBIS(2-ETHOXYCYCLOHEXYL)- see DEU125
SELENIUM DIETHYLDITHIOCARBAMATE see DJD400
SELENIUM DIETHYLDITHIOCARBAMATE see SBP900
SELENIUM DIMETHYLDITHIOCARBAMATE see SBQ000
SELENIUM DIOXIDE see SBO000
SELENIUM DIOXIDE see SBQ500
SELENIUM(IV) DIOXIDE (1:2) see SBQ500
SELENIUM DIOXIDE mixed with ARSENIC TRIOXIDE (1:1) see ARJ000
SELENIUM(IV) DISULFIDE (1:2) see SBR000
SELENIUM DISULFIDE (2.5%) SHAMPOO see SBR500
SELENIUM(IV) DISULFIDE SHAMPOO (2.5%) see SBR500
SELENIUM DISULPHIDE (DOT) see SBR000
SELENIUM DUST see SBO500
SELENIUM ELEMENTAL see SBO500
SELENIUM FLUORIDE see SBS000
SELENIUM HEXAFLUORIDE see SBS000
SELENIUM HOMOPOLYMER see SBO500
SELENIUM HYDRIDE see HIC000
SELENIUM METAL POWDER, NON-PYROPHORIC (DOT) see SBO500
SELENIUM METHIONINE see SBU800
SELENIUM MONOCHLORIDE see SBS500
SELENIUM MONONITRIDE see SBT100
SELENIUM MONOSULFIDE see SBT000
SELENIUM NITRIDE see SBT100
SELENIUM NITRIDE (DOT) see SBT100
SELENIUM OXIDE see SBQ500
SELENIUM OXIDE see SBT200
SELENIUM OXIDE (DOT) see SBT200
SELENIUM OXYCHLORIDE see SBT500
SELENIUM SULFIDE see SBR000
SELENIUM SULFIDE see SBT000
SELENIUM SULPHIDE see SBT000
SELENIUM TETRACHLORIDE see SBU000
SELENOASPIRINE see SBU100
2,2'-SELENOBIS(BENZOIC ACID) see SBU150
o,o'-SELENOBIS(BENZOIC ACID) see SBU150
SELENOCYANIC ACID, ESTER WITH α-HYDROXYACETANILIDE see OOK175
SELENOCYANIC ACID, 2-OXO-2-(PHENYLAMINO)ETHYL ESTER see OOK175
SELENOCYANIC ACID, POTASSIUM SALT see PLS000
d,l-SELENOCYSTEINE see SBV100
SELENO-dl-CYSTEINE see SBV100
SELENOCYSTINE see DWY800
SELENOCYSTINE see SBP600
SELENOCYSTINE see SBU200

d-SELENOCYSTINE see SBU300
l-SELENOCYSTINE see SBU303
d,l-SELENOCYSTINE see SBU200
SELENO-dl-CYSTINE see SBU200
4,4'-(SELENODI-2,1-
ETHANEDIYL)BISMORPHOLINE
DIHYDROCHLORIDE see MRU255
SELENODIGLUTATHIONE see SBU500
2-SELENOETHYLGUANIDINE see SBU600
SELENOFORMALDEHYDE see SBU650
6-SELENOGUANOSINE see SBU700
6-SELENO-GUANOSINE (9CI) see SBU700
SELENOHOMOCYSTINE see SBU710
SELENOMERCAPTOAETHYLGUANIDIN
(GERMAN) see SBU600
SELENOMETHIONINE see SBU725
(+−)-SELENOMETHIONINE see SBU800
dl-SELENOMETHIONINE see SBU800
SELENO-dl-METHIONINE see SBU800
SELENONIUM, TRIMETHYL- see TME500
SELENONIUM, TRIMETHYL-, CHLORIDE
see TME600
SELENONIUM, TRIPHENYL-, CHLORIDE
see TMU775
SELENOPHOS see SBU900
SELENOSULFURIC ACID, 2-AMINOETHYL
ESTER see AJS950
SELENO-TOLUIDINE BLUE see SBU950
SELENOUREA see SBV000
SELENSULFID (GERMAN) see SBT000
3-SELENYL-dl-ALANINE see SBV100
SELEPHOS see PAK000
SELES BETA see TAL475
SELEXID see MCB550
SELEXOL see DOM100
SELFER see IDE000
SELF ROCK MOSS see CCL250
SELINON see DUS700
SELLACID YELLOW AEN see SGP500
SELLA FAST RED RS see CMM330
SELLAITE see MAF500
SELOKEN see SBV500
SELSUN see SBW000
SELSUN BLUE see SBR000
SEL-TOX SSO2 and SS-20 see DXG000
SEM (cytostatic) see TND500
SEMAP see PAP250
SEMBRINA see DNA800
SEMBRINA see MJE780
SEMDOXAN see CQC500
SEMDOXAN see CQC650
SE-MEG see SBU600
SEMERON see INR000
SEMESAN see CHO750
SEMEVIN see LBF100
SEMICARBAZIDE see HGU000
SEMICARBAZIDE, 4-(9-ACRIDINYL)-2-
METHYL-3-THIO- see ADQ560
SEMICARBAZIDE, 4-(9-ACRIDINYL)-3-
THIO- see ADQ610
SEMICARBAZIDE, 4-(p-BROMOPHENYL)-1-
((1-METHYL-2-PYRROLYL)METHYLENE)-
see BNX400
SEMICARBAZIDE, 4-(p-BROMOPHENYL)-1-
(2-PYRROLYLMETHYLENE)- see BNX420
SEMICARBAZIDE, 4-(p-BUTOXYPHENYL)-
1-((1-METHYL-2-
PYRROLYL)METHYLENE)- see BPQ300
SEMICARBAZIDE, 4-(p-BUTOXYPHENYL)-
1-(2-PYRROLYLMETHYLENE)- see BPQ330
SEMICARBAZIDE, 2-(o-
CHLOROPHENETHYL)-3-THIO- see CJK100
SEMICARBAZIDE, 4-(m-CHLOROPHENYL)-
1-((1-METHYL-2-
PYRROLYL)METHYLENE)- see CKJ760
SEMICARBAZIDE, 4-(p-CHLOROPHENYL)-
1-((1-METHYL-2-
PYRROLYL)METHYLENE)- see CKJ765
SEMICARBAZIDE, 4-(p-CHLOROPHENYL)-
1-((1-METHYL-2-
PYRROLYL)METHYLENE)-3-THIO- see
CKK600

SEMICARBAZIDE, 4-(m-CHLOROPHENYL)-
1-(2-PYRROLYLMETHYLENE)- see CKJ770
SEMICARBAZIDE, 4-(p-CHLOROPHENYL)-
1-(2-PYRROLYLMETHYLENE)- see CKJ775
SEMICARBAZIDE, 4-(p-CHLOROPHENYL)-
1-(2-PYRROLYLMETHYLENE)-3-THIO- see
CKK700
SEMICARBAZIDE HYDROCHLORIDE see
SBW500
SEMICARBAZIDE, 4-(m-
METHOXYPHENYL)-1-((1-METHYL-2-
PYRROLYL)METHYLENE)- see MFH945
SEMICARBAZIDE, 4-(p-
METHOXYPHENYL)-1-((1-METHYL-2-
PYRROLYL)METHYLENE)- see MFH950
SEMICARBAZIDE, 4-(m-
METHOXYPHENYL) 1 ((1-METHYL-2-
PYRROLYL)METHYLENE)-3-THIO- see
MFJ010
SEMICARBAZIDE, 4-(p-
METHOXYPHENYL)-1-((1-METHYL-2-
PYRROLYL)METHYLENE)-3-THIO- see
MFJ020
SEMICARBAZIDE, 4-(m-
METHOXYPHENYL)-1-(2-
PYRROLYLMETHYLENE)- see MFH955
SEMICARBAZIDE, 4-(p-
METHOXYPHENYL)-1-(2-
PYRROLYLMETHYLENE)- see MFH960
SEMICARBAZIDE, 4-(m-
METHOXYPHENYL)-1-(2-
PYRROLYLMETHYLENE)-3-THIO- see
MFJ030
SEMICARBAZIDE, 4-(p-
METHOXYPHENYL)-1-(2-
PYRROLYLMETHYLENE)-3-THIO- see
MFJ040
SEMICARBAZIDE, METHYLENEBIS(THIO-
see MJN300
SEMICARBAZIDE, 2-(p-
METHYLPHENETHYL)-3-THIO- see
MNP500
SEMICARBAZIDE, 1-((1-METHYL-2-
PYRROLYL)METHYLENE)-4-(p-
NITROPHENYL)- see NIV150
SEMICARBAZIDE, 1-((1-METHYL-2-
PYRROLYL)METHYLENE)-4-PHENYL-3-
THIO- see PGN030
SEMICARBAZIDE, 1-((1-METHYL-2-
PYRROLYL)METHYLENE)-4-(p-
SULFAMOYLPHENYL)- see SNL900
SEMICARBAZIDE, 4-(p-NITROPHENYL)-1-
(2-PYRROLYLMETHYLENE)- see NIV200
SEMICARBAZIDE, 4-(p-
OCTYLOXYPHENYL)-1-((1-METHYL-2-
PYRROLYL)METHYLENE)- see OFE100
SEMICARBAZIDE, 4-(p-
OCTYLOXYPHENYL)-1-(2-
PYRROLYLMETHYLENE)- see OFE200
SEMICARBAZIDE, 1-PHENETHYL- see
PDK300
SEMICARBAZIDE, 2-PHENETHYL-3-THIO-
see PDK500
SEMICARBAZIDE, 4-PHENYL-1-((1-
METHYL-2-PYRROLYL)METHYLENE)- see
PGH600
SEMICARBAZIDE, 1-PHENYL-4-
(PHENYLIMINO)-3-THIO- see DWN200
SEMICARBAZIDE, 4-PHENYL-1-(2-
PYRROLYLMETHYLENE)- see PGH650
SEMICARBAZIDE, 4-PHENYL-1-(2-
PYRROLYLMETHYLENE)-3-THIO- see
PGN045
SEMICARBAZIDE, 1-(2-
PYRROLYLMETHYLENE)-4-(p-
SULFAMOYLPHENYL)- see SNL920
SEMICARBAZIDE, 4-(2-
PYRROLYLMETHYLENE)-2-THIO- see
PPS300
SEMICILLIN see AIV500
p-SEMIDINE see PFU500
SEMIKON see DPJ400

SEMIKON see TEO250
SEMIKON HYDROCHLORIDE see DPJ400
SEMILLA de CULEBRA (MEXICO) see
RMK250
SEMINOLE BEAD see RMK250
SEMINOLE BREAD see CNH789
SEMIOXAMAZID see OLM310
SEMIOXAMAZIDE see OLM310
SEMOPEN see PDD350
SEMOXYDRINE see DBA800
SEMPERVIVUM (JAMAICA) see AGV875
SEMUSTINE see CHD250
SENAQUIN see CLD000
SENARMONTITE see AQF000
SENARMONTITE see SBW600
SENCEPHALIN see ALV000
SENCOR see MQR275
SENCORAL see MQR275
SENCORER see MQR275
SENCOREX see MQR275
SENDOXAN see CQC500
SENDRAN see PMY300
SENDUXAN see CQC500
SENDUXAN see CQC650
SENECA OIL see PCR250
SENECIA JACOBAEA see RBA400
SENECIA LONGILOBUS see RBA400
SENECIO CANNABIFOLIUS, leaves and stalks
see SBW950
SENECIOIC ACID see MHT500
SENECIOIC ACID AMIDE see SBW965
SENECIO LONGILOBUS see SBX000
SENECIONAN-11,16-DIONE, 13,19-
DIDEHYDRO-12,18-DIHYDROXY-(9CI) see
RJZ000
SENECIONAN-11,16-DIONE, 1,2-
DIHYDRO-12-HYDROXY-, (1-α)- see PJG135
SENECIONAN-11,16-DIONE, 12,18-
DIHYDROXY-, (15E)- see RFP100
SENECIONANIUM, 12-(ACETYLOXY)-
14,15,20,21-TETRADEHYDRO-15,20-
DIHYDRO-8-HYDROXY-4-METHYL-11,16-
DIOXO-, (8-xi,12-β,14-Z)- see CMV950
SENECIO NEMORENSIS FUCHSII, alkaloidal
extract see SBX200
SENECIONINE see ARS500
SENECIO VULGARIS see RBA400
SENECIPHYLLIN see SBX500
SENECIPHYLLINE see SBX500
SENECIPHYLLIN HYDROCHLORIDE see
SBX525
SENEGAL GUM see AQQ500
SENEGENIN see SBY000
SENEGIN see SBY000
SENF OEL (GERMAN) see AGJ250
SENFOEL (GERMAN) see ISK000
SENIRAMIN see RCA200
SENKIRKIN see DMX200
SENKIRKINE see DMX200
SENNOSIDE B see SBY050
SENTIPHENE see CJU250
SENTONIL see PDW500
SENTRY see HOV500
SENTRY GRAIN PRESERVER see PMU750
SEOMINAL see RDK000
SEOTAL see SBN000
SEPAN see CMR100
SEPAZON see CMY525
SEPERIDOL see CLS250
SEPEROL see CLS250
SEPIOLITE see SBY100
SEPIOLITE (MG$_2$H$_2$(SIO$_3$)$_3$•H$_2$O) see SBY200
SEPPIC MMD see CIR250
SEPPUDY see CNT350
SEPRISAN see HCP800
SEPSINOL see FPI000
SEPTACIDIN see SBZ000
SEPTACIL see ALF250
SEPTAMIDE ALBUM see SNM500
SEPTAMYCIN see SCA000
SEPTEAL see CDT250
SEPTENE see CBM750

SEPTICOL see CDP250
SEPTINAL see SNM500
SEPTIPULMON see PPO000
SEPTISOL see HCL000
SEPTOCHOL see DAQ400
SEPTOFEN see HCL000
SEPTOPLEX see SNM500
SEPTOS see HJL500
SEPTOTAN see PFN000
SEPTRA see SNK000
SEPTRA see TKX000
SEPTRAN see SNK000
SEPTRAN see TKX000
SEPTRIM see TKX000
SEPTRIN see TKX000
SEPTURAL see PAF250
SEPUDDY see CNT350
SEPYRON see DNU100
SEQ 100 see EIX000
SEQUAMYCIN see SLC000
SEQUESTRENE 30A see EIV000
SEQUESTRENE AA see EIX000
SEQUESTRENE Na3 see TNL250
SEQUESTRENE Na 4 see EIV000
SEQUESTRENE NaFe IRON CHELATE see EJA379
SEQUESTRENE SODIUM 2 see EIX500
SEQUESTRENE ST see EIV000
SEQUESTRENE TRISODIUM see TNL250
SEQUESTRENE TRISODIUM SALT see TNL250
SEQUESTRIC ACID see EIX000
SEQUESTROL see EIX000
SEQUILAR see NNL500
SEQUION 40Na32 see DJG700
SEQUOSTAT see NNL500
SERAGON see SCA750
SERAGONIN see SCA750
SERAX see CFZ000
(d-SER(BU¹⁶))-LH-RH-(1-9)NONAPEPTIDE-ETHYLAMIDE see LIU420
SERDOX NES see AFJ160
SERDOX NES 8 see AFJ160
SEREEN see HOT500
SERENACE see CLY500
SERENACK see DCK759
SERENAL see CFZ000
SERENAL see OMK300
SERENID see CFZ000
SERENID-D see CFZ000
SERENIL see CHG000
SERENIUM see CGA000
SERENIUM see EEQ600
SERENSIL see CHG000
SERENTIL see MON750
SEREN VITA see MDQ250
SERENZIN see DCK759
SEREPAX see CFZ000
SERESTA see CFZ000
SERFIN see RDK000
SERFOLIA see RDK000
SERGETYL see ELQ600
SERGOSIN see SHX000
SERIAL see MCA500
SERIBAK see MRM000
SERICOSOL-N see DVR909
SERIEL see GJS200
SERIL see MQU750
SERILENE BRILLIANT RED 2BL see AKI750
SERILENE RED 2BL see AKI750
SERINE see SCA355
l-SERINE see SCA355
(l)-SERINE see SCA355
(S)-SERINE see SCA355
dl-SERINE see SCA350
l-SERINE DIAZOACETATE see ASA500
l-SERINE DIAZOACETATE (ester) see ASA500
l-SERINE, N-(9,10-DIHYDRO-4-HYDROXY-9,10-DIOXO-1-ANTHRACENYL)-, HYDRAZIDE see DLO970

l-SERINE, ESTER with N-METHYL-N-NITROSOGLYCINE see NLS500
84-l-SERINEPLASMINOGEN ACTIVATOR (HUMAN TISSUE-TYPE PROTEIN MOIETY REDUCED) see PJB810
dl-SERINE 2-(2,3,4-TRIHYDROXYBENZYL)HYDRAZINE HYDROCHLORIDE see SCA400
dl-SERINE 2-((2,3,4-TRIHYDROXYPHENYL)METHYL)HYDRAZIDE MONOHYDROCHLORIDE (9CI) see SCA400
SERINGINA see RCA200
SERINYL BLUE 2G see TBG700
SERINYL HOISERY SCARLET BD see ENP100
SERINYL HOISERY BLUE see MGG250
SERINYL HOISERY YELLOW GD see AAQ250
SERIPLAS BLUE GREEN BW see DMM400
SERIPLAS RED X3B see DBX000
SERIPLAS YELLOW GD see AAQ250
SERISOL BLILLIANT RED X 3B see DBX000
SERISOL BRILLIANT BLUE BG see MGG250
SERISOL BRILLIANT VIOLET 2R see DBP000
SERISOL FAST BLUE BGLW see CMP060
SERISOL FAST BLUE GREEN B see DMM400
SERISOL FAST BLUE GREEN BW see DMM400
SERISOL FAST CRIMSON BD see CMP080
SERISOL FAST RED 2B see AKE250
SERISOL FAST SCARLET BD see ENP100
SERISOL FAST VIOLET 6B see AKP250
SERISOL FAST VIOLET B see DBY700
SERISOL FAST YELLOW A see DUW500
SERISOL FAST YELLOW GD see AAQ250
SERISOL FAST YELLOW N 5RD see CMP090
SERISOL ORANGE YL see AKP750
SERISTAN BLACK B see AQP000
SERITARD see CHN100
SERITOX 50 see DGB000
SERMION see NDM000
SERNAS see CLY500
SERNEL see CLY500
SERNEVIN see EPD500
SERNYL see AOO500
SERNYLAN see AOO500
SERNYL HYDROCHLORIDE see AOO500
SERODEN see FNF000
SEROFINEX see ARQ750
SEROGAN see SCA750
SEROLFIA see RDK000
SEROMYCIN see CQH000
SEROQUEL see QCJ300
SEROTIN CREATININE SULFATE see AJX750
SEROTINEX see ARQ750
SEROTONIN see AJX500
SEROTONIN BENZYL ANALOG see BEM750
SEROTONIN CREATININE SULFATE MONOHYDRATE see AJX750
SEROTROPIN see SCA750
SERP see RDK000
SERP-AFD see RDK000
SERPAGON see RCA200
SERPALAN see RDK000
SERPALOID see RDK000
SERPANEURONA see RDK000
SERPANRAY see RDK000
SERPASIL see RDK000
SERPASIL APRESOLINE see RDK000
SERPASIL APRESOLINE No. 2 see HGP500
SERPASIL-ESIDREX see RDK000
SERPASIL-ESIDREX K see RDK000
SERPASIL-ESIDREX NO. 1 see RDK000
SERPASIL-ESIDREX NO. 2 see RDK000
SERPASIL PREMIX see RDK000
SERPASOL see RDK000
SERPATE see RDK000

SERPATONE see RDK000
SERPAX see CFZ000
SERPAX see RDK000
SERPAZIL see RDK000
SERPAZOL see RDK000
SERPEDIN see RDK000
SERPEN see RDK000
SERPENA see RDK000
SERPENT see YAK350
SERPENTIL see RDK000
SERPENTIN see RDK000
SERPENTINA see RDK000
SERPENTINE see ARM250
SERPENTINE see ARM268
SERPENTINE see SCA475
SERPENTINE CHRYSOTILE see ARM268
SERPENTINE HYDROXIDE, inner salt see SCA475
SERPENTINE (alkaloid) HYDROXIDE, inner salt see SCA475
SERPENTINE "PHARBIL" see RDK000
SERPENTINIC ACID ISOBUTYL ESTER see SCA525
SERPENTINTARTRAT (GERMAN) see SCA550
SERPICON see RDK000
SERPIL see RDK000
SERPILOID see RDK000
SERPILUM see RDK000
SERPINE see RDK000
SERPINE (Pharmaceutical) see RDK000
SERPIPUR see RDK000
SERPIVITE see RDK000
SERPLEX K see RDK000
SERPOGEN see RDK000
SERPOID see RDK000
SERPONE see RDK000
SERPRESAN see RDK000
SERPYRIT see RDK000
SERRAL see DKA600
SERRATIA EXTRACELLULAR PROTEINASE see SCA625
SERRATIA MARCESCENS ENDOTOXIN see SCA600
SERRATIA PISCVATORUM POLYSACCHARIDE see SCA610
SERRATIO PEPTIDASE see SCA625
SERTABS see RDK000
SERTAN see DBB200
(d-SER(TBU)⁶-EA¹⁰)-LHRH see LIU420
(d-SER(TBU)⁶-EA¹⁰)-LUTEINIZING HORMONE-RELEASING HORMONE see LIU420
d-SER(TBU¹⁶)-LH-RH-(1-9)-NONAPEPTIDE ETHYLAMIDE see LIU420
SERTENS see RDK000
SERTENSIN see RDK000
SERTINA see RDK000
SERTINON see EPQ000
SERTOFRAN see DSI709
SERUM GONADOTROPHIN see SCA750
SERUM GONADOTROPIC HORMONE see SCA750
SERUM GONADOTROPIN see SCA750
N-SERVE NITROGEN STABILIZER see CLP750
SERVISONE see PLZ000
SES see CNW000
SESAGARD see BKL250
SESAME OIL see SCB000
SESAMEX see MJS550
SESAMOL see MJU000
SESBANIA SESBAN (L.) Merr. var. BICOLOR W. & A., extract excluding roots see SCB100
SESBANIA (VARIOUS SPECIES) see SBC550
SESDEN see TGB160
SESIN see SCB200
SESONE (ACGIH) see CNW000
SESO VEGETAL (CUBA, PUERTO RICO) see ADG400
SESOXANE see MJS550

SESQUIETHYLALUMINUM CHLORIDE see TJP775
SESQUIMUSTARD see SCB500
SESQUIMUSTARD Q see SCB500
SESQUISODIUM 2-(4'-ARSONOANILINO)-4,6-DIAMINO-s-TRIAZINE see SIF450
SESQUISULFURE de PHOSPHORE (FRENCH) see PHS500
SET see MDI000
SETACYL BLUE BN see MGG250
SETACYL BLUE GREEN P-BS see DMM400
SETACYL BLUE 6GN see DMM400
SETACYL DIAZO NAVY R see DCJ200
SETACYL PINK 3B see AKE250
SETACYL RED 2B see CMP080
SETACYL RED P-3B see DBX000
SETACYL SCARLET B see ENP100
SETACYL SCARLET 2BD see ENP100
SETACYL SCARLET RNA see ENP100
SETACYL TURQUOISE BLUE 2G see DMM400
SETACYL TURQUOISE BLUE 4G see DMM400
SETACYL TURQUOISE BLUE G see DMM400
SETACYL TURQUOISE BLUE GD see DMM400
SETACYL VIOLET R see DBP000
SETACYL YELLOW P-BS see DUW500
SETACYL YELLOW FL see KDA075
SETACYL YELLOW P-FL see KDA075
SETACYL YELLOW G see AAQ250
SETACYL YELLOW P-2GL see AAQ250
SETACYL YELLOW 2GN see AAQ250
SETAMINE US 132 see MCB050
SETAMINE US 141 see MCB050
SETAMINE US 138BB70 see MCB050
SETAMINE US 139BB70 see MCB050
SETHOXYDIM see CDK800
SETHYL see MDL000
S(−)-5-ETHYL-5-(1-METHYLBUTYL)BARBITURIC ACID SODIUM SALT see PBS750
SETILE VIOLET 3R see DBP000
SETONIL see DCK759
SETRETE see PFO550
SETTIMA see DAP700
SEVACARB see CBT750
SEVAL see CBT750
SEVAR 2000 see SKN100
S-SEVEN see SCC000
SEVENAL see EOK000
SEVEN BARK see HGP600
SEVICAINE see AIT250
SEVIMOL see CBM750
SEVIN see CBM750
SEVIN 4 see CBM750
SEVINOL see TJW500
SEVIN (OSHA) see CBM750
SEVRON ORANGE G see CMM820
SEVRON RED GL see BAQ750
SEVRON YELLOW 1 see CMM890
SEVRON YELLOW R see CMM890
SEWIN see CBM750
SEXADIEN see DAL600
SEXADIENO see FBP100
SEXOCRETIN see DKA600
SEXOVAR see FBP100
SEXOVID see FBP100
SEXTONE see CPC000
SEXTONE B see MIQ740
SEXTRA see SCB000
SF 60 see MAK700
SF-337 see AJO500
SF 733 see XQJ650
SF 837 see MBY150
SF 1153 see PJQ790
SF 1154 see PJQ790
SF 1173 see OCE100
SF 1179 see PJQ790
SF 1202 see DAF350
SF 1293 see SEO550

SF-6505 see HLM000
SF6539 see HDC000
SF-7531 see DFQ500
S 7481F1 see CQH100
S-4068 SF see THI500
SF 733 ANTIBIOTIC SULFATE see RIP000
SFERICASE see SCC550
SFK 70 see UTU500
S 2852 FORTE see EQN250
SF-2052 SULFATE see DAB630
SG-67 see SCH002
SGA see SCK600
SGD-SCHA 1059 see BGC500
S. GRISEUS PROTEASE see PMJ100
S. GRISEUS PROTEINASE see PMJ100
SH 100 see OLW400
SH 261 see EGQ000
SH 393 see NNQ000
SH 514 see SKM000
SH 567 see PMC700
SH 582 see GEK510
SH 714 see CQJ500
SH 717 see GHK200
SH 742 see FDA925
SH 771 see AMV375
SH 850 see NNQ500
SH 926 see AAI750
SH 567a see PMC700
SH 6030 see TNJ500
SH 6076 see TLC300
SH 30858 see FAJ150
SH 66752 see PNI250
SH 70850 see NNQ500
SH 71121 see NNL500
SH 80582 see GEK510
SaH 43-715 see POB300
SaH 46-790 see THH350
SHA see SAL500
SHA CHONG DAN see DXC900
SHA CHONG SHUANG see DXC900
SHADOCOL see TDE750
SHAKE-SHAKE see RBZ400
SHALE OIL (DOT) see COD750
SHAMMAH (SAUDI ARABIA) see SED400
SHAMROX see CIR250
SHARSTOP 204 see SGM500
SHAWINIGAN ACETYLENE BLACK see CBT750
SHB 286 see SOU650
SH 213AB see IGD100
SHB 261AB see NNL500
SHB 264AB see NNL500
SHED-A-LEAF see SFS000
SHED-A-LEAF "L" see SFS000
SHEEP DIP see ARE250
SHEEP LAUREL see MRU359
SHELL 40 see BQZ000
SHELL 300 see SMQ500
SHELL 345 see ADR250
SHELL 4072 see CDS750
SHELL 4402 see OAN000
SHELL 5520 see PMP500
SHELLAC, BLEACHED see SCC700
SHELLAC, BLEACHED, WAX-FREE see SCC705
SHELL ATRAZINE HERBICIDE see ARQ725
SHELL CARBON see CBT750
SHELL GOLD see GIS000
SHELL MIBK see HFG500
SHELL OS 1836 see CLV375
SHELLOYNE H see DLJ500
SHELL SD 345 see ADR250
SHELL SD-2580 see EHY700
SHELL SD-3450 see HCI475
SHELL SD-3562 see DGQ875
SHELL SD 4,239 see CEQ500
SHELL SD 4294 see COD000
SHELL SD-5532 see CDR750
SHELL SD 7,438 see BES250
SHELL SD 7,727 see DGD400
SHELL SD-8530 see TMD000
SHELL SD-8988 see MHR150

SHELL SD-9098 see DIX600
SHELL SD 9129 see MRH209
SHELL SD-10576 see SCD500
SHELL SD-14114 see BLU000
SHELL SILVER see SDI500
SHELLSOL 140 see NMX000
SHELL UNDRAUTTED A see AFV500
SHELL WL 1650 see OAN000
SHENG BAI XIN (CHINESE) see CMV375
SHER, EXTRACT see EQY100
SHERSTAT SLN see SAH500
SHERWOOD GREEN A 4436 see PJQ100
(22R,25RS)-2-β,3-β,14,20,22,26-HEXAHYDROXY-5-β-CHOLEST-7-EN-6-ONE see IDD000
SHIBAGEN see FCC050
SHIGATOX see AHO250
SHIGA TOXIN see SCD750
SHIGA TOXIN see SCD800
SHIGELLA DYSENTERIAE TOXIN see SCD750
SHIGELLA DYSENTERIAE TOXIN see SCD800
SHIGRODIN see BRF500
SHIKIMATE see SCE000
SHIKIMIC ACID see SCE000
SHIKIMOLE see SAD000
SHIKISO ACID ANTHRACENE RED G see CMM325
SHIKISO ACID BRILLIANT BLUE 6B see EQG550
SHIKISO ACID FAST YELLOW MR see CMM759
SHIKISO ACID RED PG see CMM320
SHIKISO ACID RED RS see NAO600
SHIKISO AMARANTH see FAG020
SHIKISO DIRECT BRILLIANT BLUE RW see CMO600
SHIKISO DIRECT DARK GREEN B see CMO830
SHIKISO DIRECT SCARLET 3B see CMO875
SHIKISO DIRECT SKY BLUE 5B see CMO500
SHIKISO DIRECT SKY BLUE 6B see CMN750
SHIKISO METANIL YELLOW see MDM775
SHIKO see CMP905
SHIKOMOL see SAD000
SHIMAZAKI PATENT BLUE AFX see ERG100
SHIMMEREX see ABU500
SHINGLE PLANT see SLE890
SHINING SUMAC see SCF000
SHINKOLITE see PKB500
SHIN-NAITO S see TEH500
SHINNIBROL see TEH500
SHINNIPPON FAST RED B BASE see NEQ000
SHINNIPPON FAST RED GG BASE see NEO500
SHINNIPPON FAST RED 3GL BASE see KDA050
SHIONOGI 6059S see LBH200
SHIPRON A see CAT775
SHIRAGIKU ROSIN see RNU100
SHIRLAN see SAH500
SHIRLAN AG see SAH500
SHIRLAN EXTRA see SAH500
SHIRLAN FLOW see CEX800
SHIRLAN (ZENECA) see CEX800
S6F HISTYRENE RESIN see SMR000
SHM see SLD900
SHMP see SHM500
SHOALLOMER see PMP500
SHOCK-FEROL see VSZ100
SHOGUN see IPI400
SHOLEX 5003 see PJS750
SHOLEX 5100 see PJS750
SHOLEX 6002 see PJS750
SHOLEX ET 182 see EJM950
SHOLEX F 171 see PJS750

SHOLEX F 6050C see PJS750
SHOLEX F 6080C see PJS750
SHOLEX L 131 see PJS750
SHOLEX 4250HM PJS750
SHOLEX S 6008 see PJS750
SHOLEX SUPER see PJS750
SHOLEX XMO 314 see PJS750
SHOWA CHROME YELLOW GG see SIT850
SHOWA FAST RED B BASE see NEQ000
SHOWDOMYCIN see RIU000
SHOXIN see NNF000
cis-SHP see SCF025
trans(−)-SHP see SCF050
trans(+)-SHP see SCF075
SHRUBBY BITTERSWEET see AHJ875
SHRUB VERBENA see LAU600
SHS see HCP900
S(−)-α-HYDRAZINO-3,4-DIHYDROXY-α-
METHYLHYDROCINNAMIC ACID
MONOHYDRATE see CBQ529
SI see LCF000
SI-6711 see DJV600
SIACARB see SAZ000
SIALIDASE see NCQ200
SIARKI CHLOREK (POLISH) see SON510
SIARKI DWUTLENEK (POLISH) see SOH500
SIARKOWODOR (POLISH) see HIC500
SIBEPHYLLIN see DNC000
SIBEPHYLLINE see DNC000
SIBIROMYCIN see SCF500
SIBOL see DKA600
SIBUTOL see SCF525
SICILIAN CERISE TONER A-7127 see
FAG070
SICOCLOR see FAM000
SICO FAST YELLOW D 1355 see CMS208
SICO FAT RED BG NEW see CMS238
SICOL 150 see DVL700
SICOL 160 see BEC500
SICOL 184 see PHW575
SICOL 250 see AEO000
SICO LAKE RED 2L see CHP500
SICO LAKE RED CU see CMS150
SICOMET 8400 see EHP700
SICOR ZNP/M see ZJS400
SICOR ZNP/S see ZJS400
SICRON see PKQ059
SIDDIQUI see AFH250
SI (DIOL) see SLD570
SIDNOFEN see SPA000
SIDVAX see AKR500
501 SIEGFRIED see AOO800
SIEGLE FAST GREEN G see PJQ100
SIEGLE ORANGE S see CMS145
SIEGLE RED 1 see MMP100
SIEGLE RED B see MMP100
SIEGLE RED BB see MMP100
SIENNA see IHC450
SIERRA C-400 see TAB750
SIFERRIT see IHG100
SIGACALM see CFZ000
SIGAPRIN see TKX000
SIGETIN see SPA650
SIGMACELL see CCU150
SIGMAFON see MBW750
SIGMAMYCIN see TBX000
SIGMART see NDL800
SIGNAL ORANGE ORANGE Y-17 see
DVB800
SIGNAL RED see NAP100
SIGNOPAM see CFY750
SIGOPHYL see HOA000
SIGURAN see HGC500
SIHS SODIUM see SDX800
SIKHAT see CDU100
SILA-ACE S 620 see TLC300
SILACYCLOBUTANE see SCF550
SILAK M 10 see SCR400
SILANAMINE, 1-ETHENYL-N-
(ETHENYLDIMETHYLSILYL)-1,1-
DIMETHYL- see TDQ050

SILANAMINE, N,N'-METHANE
TETRAYLBIS(1,1,1-TRIMETHYL)- see
BLQ950
SILANE see SDH575
SILANE 32-61 see VQU333
SILANE 40-43 see DJA330
SILANE 40-47 see TLC600
SILANE A-163 see MQF500
SILANE, (3-AMINOPROPYL)TRIETHOXY-
see TJN000
SILANE, γ-AMINOPROPYLTRIETHOXY-
see TJN000
SILANE, (3-AMINOPROPYL)TRIETHYL- see
AMG200
SILANE, BENZYLCHLORODIMETHYL- see
BEE800
SILANE, CHLORO-tert-BUTYLDIMETHYL-
see BQS300
SILANE, CHLORO(1,1-
DIMETHYLETHYL)DIMETHYL- see BQS300
SILANE, 2-CHLOROETHYLTRIS(2-
METHOXYETHOXY)- see CHI200
SILANE, (3-
CHLOROPROPYL)TRIMETHOXY- see
TLC300
SILANE, CHLOROTRIPHENYL- see TMU800
SILANE, (3-
CYANOPROPYL)DIETHOXY(METHYL)- see
COR500
SILANE, (3-CYANOPROPYL)TRICHLORO-
see COR750
SILANE, (3-CYANOPROPYL)TRIETHOXY-
see COR800
SILANE, DICHLOROMETHYL(3,3,3-
TRIFLUOROPROPYL)- see DFS700
SILANE, DIHYDROXYDIPHENYL- see
DMN450
SILANE, DIMETHOXYDIMETHYL- see
DOE100
SILANE, DIMETHOXYMETHYL- see
MJE900
SILANE, ETHENYLTRIMETHOXY-(9CI) see
TLD000
SILANE, ETHOXYDIMETHYL- see EES200
SILANE, ETHYL TRIETHOXY A-15 see
EQA000
SILANE, HYDROXYTRIPHENYL- see
HOM300
SILANE, OXYBIS((3-
AMINOPROPYL)DIMETHYL)- see OPC100
SILANE, PHENYL- see SDX250
SILANE, p-PHENYLENEBIS(DIMETHYL)-
see PEW725
SILANE, 1,4-PHENYLENEBIS(DIMETHYL)-
(9CI) see PEW725
SILANE, PHENYLTRIFLUORO- see PGO500
SILANE, T-BUTYLCHLORODIMETHYL- see
BQS300
SILANE 48-12 TETRAKIS see OCM100
SILANETETRAMINE, OCTAMETHYL- see
OCM100
2,2',2'',2'''-
SILANETETRAYLTETRAKISETHANOL see
TDH100
SILANETRIAMINE, N,N,N',N',N'',N''-
HEXAMETHYL- see TNH300
SILANE, TRI-tert-BUTOXYVINYL- see
SDX300
SILANE, TRICHLOROALLYL- see AGU250
SILANE, TRICHLORO(3-CYANOPROPYL)-
see COR750
SILANE, TRICHLOROCYCLOHEXYL- see
CPR250
SILANE, TRICHLOROETHYL- see EPY500
SILANE, TRICHLOROMETHYL- see
MQC500
SILANE, TRIETHOXY(3-CYANOPROPYL)-
see COR800
SILANE, TRIETHOXYETHYL- see EQA000
SILANE, TRIETHOXY(3-
ISOCYANATOPROPYL)- see IKG900

SILANE, TRIETHOXYMETHYL- see
MQD750
SILANE, TRIETHOXYPHENYL- see PGO000
SILANE, TRIMETHOXY- see TLB750
SILANE, TRIMETHOXYMETHYL- see
MQF500
SILANE, TRIMETHYLCHLORO- see TLN250
SILANE, TRIS(tert-BUTYLDIOXY)VINYL-
see VQU333
SILANE, TRIS(1,1-
DIMETHYLETHOXY)ETHENYL-(9CI) see
SDX300
SILANE, TRIS((1,1-
DIMETHYLETHYL)DIOXY)ETHENYL- see
VQU333
SILANE, VINYL TRICHLORO 1-150 see
TIN750
SILANE, VINYL TRIETHOXY 1-151 see
TJN250
SILANE Y-4086 see EBO000
SILANE-Y-4087 see ECH000
SILANOL, DIPHENYLMETHYL- see
DWH550
SILANTIN see DKQ000
SILASTIC see PJR250
SILATRANE see TMO750
SILBER (GERMAN) see SDI500
SILBERNITRAT see SDS000
SILBESAN see CLD000
SILDENAFIL CITRATE see SCF600
SILENE EF see CAW850
SILIBININ-2',3-DIHYDROGENSUCCINATE
SODIUM SALT see SDX800
SILICA AEROGEL see SCI000
SILICA, AMORPHOS-FUME (ACGIH) see
SCH001
SILICA, AMORPHOUS see SCH002
SILICA, AMORPHOUS-DIATOMACEOUS
EARTH (UNCALCINED) (ACGIH) see
DCJ800
SILICA, AMORPHOUS FUME see SCH001
SILICA, AMORPHOUS FUMED see SCH002
SILICA, AMORPHOUS-FUSED (ACGIH) see
SCK600
SILICA, AMORPHOUS HYDRATED see
SCI000
SILICA, COLLOIDAL see SCN600
SILICA, CRYSTALLINE see SCI500
SILICA, CRYSTALLINE−CRISTOBALITE see
SCJ000
SILICA, CRYSTALLINE−QUARTZ see SCJ500
SILICA, CRYSTALLINE−TRIDYMITE see
SCK000
SILICA, CRYSTALLINE-TRIDYMITE
(ACGIH, OSHA) see SCK000
SILICA FLOUR see SCI500
SILICA FLOUR see SCK500
SILICA FLOUR (powdered crystalline silica) see
SCJ500
SILICA, FUSED see SCK600
SILICA, FUSED see SCK600
SILICA, FUSED (OSHA) see SCK600
SILICA GEL see SCI000
SILICA GEL see SCL000
SILICA, GEL and
AMORPHOUS−PRECIPITATED see SCL000
SILICANE see SDH575
SILICANE, CHLOROTRIMETHYL- see
TLN250
SILICANE, TRICHLOROETHYL- see EPY500
SILICASOL see SCN600
SILICATE D'ETHYLE (FRENCH) see EPF550
SILICATE(2-), HEXAFLUORO-, ALUMINUM
(3:2) see AHB400
SILICATE(2-), HEXAFLUORO-, BARIUM see
BAO750
SILICATE(2-), HEXAFLUORO-, BARIUM
(1:1) (9CI) see BAO750
SILICATE(2-), HEXAFLUORO-, CADMIUM
(8CI,9CI) see CAG500
SILICATE(2-), HEXAFLUORO-, CALCIUM
(1:1) (9CI) see CAX250

SILICATE(2-), HEXAFLUORO-, MAGNESIUM (1:1) see MAF600
SILICATES see SCM500
SILICATE SOAPSTONE see SCN000
SILICA, VITREOUS (9CI) see SCK600
SILICA XEROGEL see SCI000
SILICETANE see SCF550
SILICIC ACID see SCI000
SILICIC ACID see SCL000
SILICIC ACID ALUMINUM SALT see AHF500
SILICIC ACID, BERYLLIUM SALT see SCN500
SILICIC ACID, CYCLIC NITRILOTRIETHYLENE ETHYL ESTER see EFJ600
SILICIC ACID, 2-ETHYLBUTYL ESTER see EHC900
SILICIC ACID, METHYL ESTER see TDJ100
SILICIC ACID, METHYL ESTER of ortho- see MPI750
SILICIC ACID (ORTHO) see SCN600
SILICIC ACID (SI(OH)4) see SCN600
SILICIC ACID, SODIUM SALT see SCN700
SILICIC ACID TETRAETHYL ESTER see EPF550
SILICIC ACID, TETRAKIS(1,1-DIMETHYLPENTYL) ESTER see TBL600
SILICIC ACID, ZIRCONIUM(4+) SALT (1:1) see ZSS000
SILICIC ANHYDRIDE see SCH002
SILICIC ANHYDRIDE see SCJ500
SILICI-CHLOROFORME (FRENCH) see TJD500
SILICIO(TETRACLORURO di) see SCQ500
SILICIUMCHLOROFORM (GERMAN) see TJD500
SILICIUMTETRACHLORID (GERMAN) see SCQ500
SILICIUMTETRACHLORIDE (DUTCH) see SCQ500
SILICIUM(TETRACHLORURE de) (FRENCH) see SCQ500
SILICOBROMOFORM see THX000
SILICOCHLOROFORM see TJD500
SILICOETHANE see DXA000
SILICOFLUORIC ACID see SCO500
SILICON see SCP000
SILICON BROMIDE see SCP500
SILICON CARBIDE see SCQ000
SILICON CARBIDE (ACGIH, OSHA) see SCQ000
SILICON CHLORIDE see SCQ500
SILICON CHLORIDE HYDRIDE see DGK300
SILICON DIBROMIDE SULFIDE see SCR100
SILICON DIOXIDE see SCI500
SILICON DIOXIDE see SCK600
SILICON DIOXIDE (FCC) see SCH002
SILICONE 360 see SCR400
SILICONE L-75 see SCR450
SILICONE L-76 see SCR455
SILICONE L-77 see SCR460
SILICONE L-79 see SCR465
SILICONE 1-174 see TLC250
SILICONE L-522 see SCR470
L-522 SILICONE see SCR470
SILICONE L-7001 see SCR475
SILICONE A-143 see TLC300
SILICONE A-172 see TNJ500
SILICONE A-186 see EBO000
SILICONE A-187 see ECH000
SILICONE A-189 see TLC000
SILICONE A-1100 see TJN000
SILICONE A-1120 see TLC500
SILICONE DC 200 see SCR400
SILICONE DC 360 see SCR400
SILICONE DC 360 FLUID see SCR400
SILICONE DIOXIDE see SCK600
SILICONE RELEASE L 45 see SCR400
SILICONE RUBBER see PJR250
SILICONES see SDC000

SILICONES see SDF000
SILICONE SF 1173 see OCE100
SILICONE Y-6607 see SDF000
SILICON FLUORIDE see SDF650
SILICON FLUORIDE BARIUM SALT see BAO750
SILICON METHYLATE see TDJ100
SILICON MONOCARBIDE see SCQ000
SILICON ORGANIC LACQUER MODIFICATOR 113-63 see MRA100
SILICON OXIDE see SDH000
SILICON PHENYL TRICHLORIDE see TJA750
SILICON POWDER, amorphous (DOT) see SCP000
SILICON SF 1202 see DAF350
SILICON SODIUM FLUORIDE see DXE000
SILICON TETRAAZIDE see SDH500
SILICON TETRACHLORIDE (DOT) see SCQ500
SILICON TETRAFLUORIDE (DOT) see SDF650
SILICON TETRAHYDRIDE see SDH575
SILICON TETRAHYDROXIDE see SCN600
SILICON TRIETHANOLAMIN see SDH670
SILICON ZINC FLUORIDE see ZIA000
SILIKILL see SCH002
SILIKON ANTIFOAM FD 62 see SCR400
SILK see SDI000
SILK FAST GREEN B see CMM200
SILMOS T see CAW850
SILMURIN see RCF000
SILMURIN see SBF500
SILOGOMMA FAST YELLOW 2G see CMS210
SILOGOMMA ORANGE G see CMS145
SILOGOMMA RED RLL see MMP100
SILOMAT see CMW500
SILON see NOH000
SILONIST see CMW500
SILOPOL ORANGE R see DVB800
SILOPOL RED G see CJD500
SILOSAN see DIN800
SILOSOL RED RBN see MMP100
SILOSOL RED RN see MMP100
SILOSUPER PINK B see RGW000
SILOTERMO CARMINE G see CMS155
SILOTERMO ORANGE G see CMS145
SILOTERMO SCARLET B see CMG750
SILOTERMO YELLOW G see CMS210
SILOTON ORANGE GT see CMS145
SILOTON RED 3B see CMG750
SILOTON RED BRLL see MMP100
SILOTON RED R see NAP100
SILOTON RED RLL see MMP100
SILOTON RUBINE 2B see CMS155
SILOTON RUBINE B see CMS155
SILOTON YELLOW GTX see DEU000
SILOTRAS BROWN TRN see NBG500
SILOTRAS ORANGE TR see PEJ500
SILOTRAS RED T3B see SBC500
SILOTRAS RED TG see CMS238
SILOTRAS RUBINE TSB see CMP080
SILOTRAS SCARLET TB see OHI200
SILOTRAS SCARLET TSR see ENP100
SILOTRAS YELLOW T2G see DOT300
SILOTRAS YELLOW TSG see AAQ250
SILOXANES see SDC000
SILOXANES see SDF000
SILOXANES AND SILICONES DI-ME, DI-PH see PJQ790
SILOXANES and SILICONES, DI Me see SCR400
SILTEPLASE see SDI100
SILTEX see SCK600
SILUBIN see BQL000
SILUNDUM see SCQ000
SILVADENE see PPP500
SILVADENE see SNI425
SILVAN (CZECH) see MKH000
SILVER see SDI500
SILVER (colloidal) see SDI750

SILVER ACETATE see SDI800
SILVER(I) ACETATE see SDI800
SILVER(1+) ACETATE see SDI800
SILVER ACETYLIDE see SDJ000
SILVER ACETYLIDE (dry) (DOT) see SDJ000
SILVER ACETYLIDE-SILVER NITRATE see SDJ025
SILVER AMIDE see SDJ500
SILVER 5-AMINOTETRAZOLIDE see SDK000
SILVER AMMONIUM COMPOUNDS see SDK500
SILVER AMMONIUM LACTATE see SDL000
SILVER AMMONIUM NITRATE see SDL500
SILVER AMMONIUM SULFATE see SDM000
SILVER ARSENITE see SDM100
SILVER ATOM see SDI500
SILVER AZIDE see SDM500
SILVER AZIDE (dry) (DOT) see SDM500
SILVER 2-AZIDO-4,6-DINITROPHENOXIDE see SDM525
SILVER AZIDODITHIOFORMATE see SDM550
SILVER BENZO-1,2,3-TRIAZOLE-1-OXIDE see SDM575
SILVER BETA-STANNATE see DXA800
SILVER BOROFLUORIDE see SDN000
SILVER BUSH see NBR800
SILVER BUTEN-3-YNIDE see SDN100
SILVER CARBONATE see SDN200
SILVER(1) CARBONATE see SDN200
SILVER CHAIN see GJU475
SILVER CHLORATE see SDN399
SILVER CHLORITE, dry see SDN500
SILVER CHLOROACETYLIDE see SDN525
SILVER COMPOUNDS see SDO500
SILVER CUP see CDH125
SILVER CYANATE see SDO525
SILVER CYANIDE see SDP000
SILVER 3-CYANO-1-PHENYLTRIAZEN-3-IDE see SDP025
SILVER CYCLOPROPYLACETYLIDE see SDP100
SILVER DIFLUORIDE see SDQ500
SILVER DINAPHTHYLMETHANEDISULPHONATE see SDP200
SILVER DINITRITODIOXYSULFATE see SDP500
SILVER DINITROACETAMIDE see SDP550
SILVER 3,5-DINITROANTHRANILATE see SDP600
SILVER FIR NEEDLE OIL see AAC250
SILVER FIR OIL see AAC250
SILVER(I) FLUORIDE see SDQ000
SILVER(II) FLUORIDE see SDQ500
SILVER FLUOROBORATE see SDN000
SILVER FULMINATE, dry see SDR000
SILVER HALIDE SOLVENT (HS103) see BGT750
SILVER HALIDE SOLV HS103 see DTE100
SILVER 1,3,5-HEXATRIENIDE see SDR150
SILVER 3-HYDROXYPROPYNIDE see SDR175
SILVER MALONATE see SDR350
SILVER MATT POWDER see TGB250
SILVER METHANESULFONATE see SDR500
SILVER 3-METHYLISOXAZOLIN-4,5-DIONE-4-OXIMATE see SDR400
SILVER METHYLSULFONATE see SDR500
SILVER MONOACETATE see SDI800
SILVER MONOACETYLIDE see SDR759
SILVER(1+) NITRATE see SDS000
SILVER NITRATE (DOT) see SDS000
SILVER(I) NITRATE (1:1) see SDS000
SILVER NITRIDE see SDS500
SILVER NITRIDOOSMITE see SDT000
SILVER 4-NITROPHENOXIDE see SDT300
SILVER NITROPRUSSIDE see SDT500
SILVER OSMATE see SDT750
SILVER OXALATE, dry see SDU000
SILVER(1+) OXIDE see SDU500

SILVER PERCHLORATE see SDV000
SILVER PERCHLORYL AMIDE see SDV500
SILVER N-PERCHLORYL BENZYLAMIDE
see SDV600
SILVER PEROXIDE see SDV700
SILVER PEROXYCHROMATE see SDW000
SILVER PHENOXIDE see SDW100
SILVER PICRATE (dry) (DOT) see PID200
SILVER PICRATE, wetted with not <30%
water, by weight (UN1347) (DOT) see PID200
SILVER PINE OIL see AAC250
SILVER POTASSIUM CYANIDE see PLS250
SILVER RUBIDIUM IODIDE (AG4RBI5) see
RQF100
SILVER SULFADIAZINE see SNI425
SILVER SULPHADIAZINE see SNI425
SILVER TETRAFLUOROBORATE see
SDN000
SILVER TETRAZOLIDE see SDW500
SILVER
TRICHLOROMETHANEPHOSPHONATE
see SDW600
SILVER TRIFLUORO METHYL ACETYLIDE
see SDX000
SILVER TRIFLUOROPROPYNIDE see
SDX100
SILVER TRINITROMETHANIDE see
SDX200
SILVER W see CAT775
SILVEX (USDA) see TIX500
SILVIC ACID see AAC500
SILVI-RHAP see TIX500
SILVISAR see HKC500
SILVISAR 510 see HKC000
SILVISAR 550 see MRL750
SILYBIN DIHEMISUCCINATE DISODIUM
see SDX800
SILYBIN DISODIUM HEMISUCCINATE see
SDX800
SILYBIN HEMISUCCINATE SODIUM SALT
see SDX800
SILYBIN SODIUM DIHEMISUCCINATE see
SDX800
SILYBIN SODIUM HEMISUCCINATE see
SDX800
SILYBUM MARIANUM (LINN.) GAERTN.,
EXTRACT see SDX900
SILYLBENZENE see SDX250
SILYL BROMIDE see BOE750
SILYL PEROXIDE Y-5712 see SDX300
SILYMARIN see SDX625
SILYMARIN HYDROGEN
BUTANEDIOATE SODIUM SALT see
SDX630
SILYMARIN SODIUM HEMISUCCINATE see
SDX630
SIM see SNK000
SIMAN see CGA000
SIMANEX see BJP000
SIMARUBACEAE see GEW700
SIMATIN(E) see ENG500
SIMAZIN see BJP000
SIMAZINE 80W see BJP000
SIMAZINE (USDA) see BJP000
SIMAZOL see AMY050
SIMEON see POB500
SIMESKELLINA see AHK750
SIMETON see BJP250
SIMETONE see BJP250
SIMFIBRATE see SDY500
SIMPADREN see SPD000
SIMPALON see HLV500
SIMPAMINA-D see BBK500
SIMPATEDRIN see BBK000
SIMPATOBLOCK see HEA000
SIMPATOL see HLV500
SIMPLA see SGG500
SIMPLOTAN see TGD250
SIMULOL 330 M see DXY000
SIN-1 see MRU076
SIN-10 see MRN275
SINAFID M-48 see MNH000

SINALAR see SPD500
SINALGICO see NBS500
SINALOST see TNF500
SINAN see GGS000
SINANOMYCIN see NCP875
SINAXAR see PFJ000
SINBAR see BQT750
SINCALINE see CMF800
SINCICLAN see DKB000
SINCODEEN see BOR350
SINCODEX see BOR350
SINCODIN see BOR350
SINCODIX see BOR350
SINCORTEX see DAQ800
SINCOUMAR see ABF750
SINDERESIN see CJN250
SINDESVEL see QAK000
SINDIATIL see BQL000
SINDRENINA see VGP000
SINECOD see BOR350
SINEFLUTTER see GAX000
SINEFRINA TARTRATO (ITALIAN) see
SPD000
SINEQUAN see AEG750
SINESALIN see BEQ625
SINESALIN COMPOSITION see RDK000
SINESTROL see DLB400
SINFIBRATE see SDY500
SINFORIL see CKE750
SINGLE SUPERPHOSPHATE see SOV500
SINGOSERP see RCA200
SINIFIBRATE see SDY500
SINIGRIN see AGH125
SINITUHO see PAX250
SINKALIN see CMF800
SINKALINE see CMF800
SINKLE BIBLE (JAMAICA) see AGV875
SINKUMAR see ABF750
SINNAMIN see AQN750
SINNOESTER OGC see GGR200
SINNOQUAT BL 80 see BEO000
SINNOQUAT BL 95 see BEO000
SINNOZON NCX 70 see CAR790
SINOACTINOMYCIN see AEA625
SINOFLUROL see FLZ050
SINOGAN see MCI500
SINOMENI CAULIS et RHIZOMA (LATIN)
see SDY600
SINOMENINE A BIS(METHYL IODIDE) see
TDX835
SINOMENINE A DIMETHIODIDE see
TDX835
SINOMENIUM ACUTUM, crude extract see
SDY600
SINOMIN see SNK000
SINORATOX see DSP400
SINORPHAN see SDY625
SINOX see DUS700
SINOX see DUU600
SINOX GENERAL see BRE500
SINOX W see BPG250
SINTECORT see PAL600
SINTESPASMIL see NOC000
SINTESTROL see DKA600
SINTHROM see ABF750
SINTHROME see ABF750
SINTOFENE see OPK300
SINTOMICETINA see CDP250
SINTOSIAN see DYF200
SINTOTIAMINA see DXO300
SINTROM see ABF750
SINTROMA see ABF750
SINUFED see POH250
SINURON see DGD600
SINUTAB see ABG750
SIOCARBAZONE see FNF000
SIONIT see SKV200
SIONON see SKV200
SIPCAVIT see PEX500
SIPERIN see RLF350
SIPEX BOS see TAV750
SIPEX OLS see OFU200

SIPEX OP see SIB600
SIPLARIL see FMO129
SIPLAROL see FMO129
SIPOL L8 see OEI000
SIPOL L10 see DAI600
SIPOL L12 see DXV600
SIPOL O see OBA000
SIPOL S see OAX000
SIPOMER B-CEA see HGO700
SIPOMER DAM see DBK200
SIPOMER DMM see DSL800
SIPONATE 330 see BBS275
SIPONIC L see DXY000
SIPONIC 218 (OBS.) see PJV100
SIPONIC Y-501 see PJW500
SIPON LT see SON000
SIPONOL S see OAX000
SIPON WD see SIB600
SIPPR-113 see SDY675
SIPTOX I see MAK700
SIRAGAN see CLD000
SIRAN N,N-DIETHYL-p-
FENYLENDIAMINU see DJV250
SIRAN HYDRAZINU (CZECH) see HGW500
SIRBIOCINA see SMB000
SIRINGAL see RCA200
SIRINGINA see RCA200
SIRINGONE see RCA200
SIRISERPIN see RCA200
SIRIUS RED 4B see CMO885
SIRIUS RED 4BA see CMO885
SIRIUS SUPRA TURQUOISE BLUE GL see
COF420
SIRLEDI see NHH000
SIRLENE see PML000
SIRMATE see DET400
SIRMATE see DET600
SIRNIK AMONNY see ANJ750
SIRNIK FOSFORECNY (CZECH) see PHS000
SIRNIK TRIBENZYLCINICTY (CZECH) see
BLK750
SIROKAL see PLG800
SIROMYCIN see VGZ000
SIROTOL see RLU000
SIRUP see GFG000
SISEPTIN see SDY750
SISOLLINE see SDY750
SISOMICIN see SDY750
SISOMICIN HYDROCHLORIDE see SDY755
SISOMICIN SULFATE see APY500
SISOMIN see APY500
SISTALGIN see SDZ000
SISTAN see VFU000
SISTOMETRENOL see LJE000
SISTRURUS MILARIUS BARBOURI VENOM
see SDZ300
SITFAST see FBS100
β-SITOSTERIN see SDZ350
β-SITOSTEROL see SDZ350
SITOSTEROL SULFATE see SDZ370
β-SITOSTEROL SULFATE see SDZ370
β-SITOSTERYL SULFATE see SDZ370
β-SITOSTERYL SULPHATE see SDZ370
SIX HUNDRED SIX see SAP500
SIXTY-THREE SPECIAL E.C. INSECTICIDE
see MNH000
SIXTY-THREE SPECIAL E.C. INSECTICIDE
see PAK000
SK 1 see MCB050
SK 1 see NCW300
SK 65 see DAB879
SK 74 see CMW700
SK 75 see UTU500
SK-100 see TNF500
SK 555 see BHO250
SK-598 see CFA750
SK 75V see UTU500
SK-106N see NGY000
SK1133 see TND500
SK 1150 see AJO500
SK-3818 see TND250
SK 4119 see MGR900

SK 6048 see CKV500
SK 6882 see TFQ750
SK-15673 see PED750
SK 18615 see TFJ500
SK-19849 see BIA250
SK 20501 see CQC650
SK 22591 see HOO500
SK 27702 see BIF750
SK 29836 see DBY500
SK 331 A see XCS000
SK-AMITRIPTYLINE see EAI000
SK-AMPICILLIN see AIV500
SKANE M8 see OFE000
SKATOL see MKV750
SKATOLE see MKV750
ω-SKATOLE CARBOXYLIC ACID see ICN000
SK-BISACODYL see PPN100
SK-CHLORAL HYDRATE see CDO000
SK-CHLOROTHIAZIDE see CLH750
SK-DEXAMETHASONE see SOW000
SK-DIGOXIN see DKN400
SK-DIPHENHYDRAMINE see BAU750
SKDN see WBS675
SKEDULE see CBF250
SKEKhG see EAZ500
SKELAXIN see XVS000
SKELLYSOLVE F see PCT250
SKELLYSOLVE G see PCT250
SKELLY-SOLVE-L see ROU000
SKEROLIP see ARQ750
SK-ERYTHROMYCIN see EDJ500
SK-ESTROGENS see ECU750
SKF 51 see ILM000
SKF 250 see DBA175
SKF 385 see PET750
SKF 478 see DWK200
SKF 1045 see MGD500
SKF 1498 see DQA400
SKF 1717 see AOO490
SKF-183A see ALW000
SKF 2170 see AOO425
SKF 2538 see DIR000
SKF-2601 see CKP250
SKF 2847 see HJB250
SKF 4740 see DAB875
SKF 5019 see TKK250
SKF 5116 see MCI500
SKF 5137 see AFJ400
SKF 525A see PBM500
SKF 5654 see CNS650
SKF 5883 see TFM100
SKF 6539 see HDC000
SKF 6612 see CIQ600
SKF 688A see DDG800
SKF 7261 see FKW000
SKF 7690 see MNC150
SKF 7988 see VRA700
SKF 7988 see VRF000
SKF 8542 see UVJ450
SK+F 1340 see DRX400
SKF 12141 see BGK250
SKF 14463 see SDZ350
SKF 20,716 see PIW000
SKF 2208K see CDF400
SKF 24260 see FOO875
SKF 2680J see MIJ300
SKF 28175 see DMF800
SKF 29044 see BQK000
SKF 30310 see OMY700
SKF 41588 see CCS250
SKF 60771 see APT250
SKF 62698 see TGA600
SKF 62979 see VAD000
SKF-69,634 see CMX860
SKF 83088 see CCS350
SKF-88373 see EBE100
SKF 91487 see NNL400
SKF 92334 see TAB250
SK&F 14287 see DAS000
SK&F 36914 see CLQ500
SK&F 39162 see ARS150

SK&F 92676 see IBQ075
SKF 16805A see BQB250
SKF 38095J2 see BJY800
SKF-525-A see DIG400
SKF 70643-A see DPM200
SKF 16214-A2 see ADI750
SKF 83959 HYDROBROMIDE see CLH050
(+/-)-SKF 77434 HYDROBROMIDE see CLH160
(+/-)-SKF 81297 HYDROBROMIDE see CLH150
SKF-501 HYDROCHLORIDE see FEE100
SK&F No. 478-A see DWK200
SKF No. 769-J² see CBG250
SKF 83088 SODIUM see CCS360
SKG see CBT500
SKI 21739 see BHV000
SKI 24464 see MNA750
SKI 27013 see SAU000
SKI 28404 see CHE750
SKIMMETIN see HJY100
SKIMMETINE see HJY100
SKINO #1 see BSU500
SKINO #2 see EMU500
SKIODAN see SHX000
SKLEROMEX see ARQ750
SKLEROMEXE see ARQ750
SKLERO-TABLINEN see ARQ750
SKLERO-TABULS see ARQ750
SK-LYGEN see MDQ250
SKhN6 see ADY250
SK-NIACIN see NCQ900
SKOLIN see HLC500
SK-PENICILLIN G see BFD000
SK-PENICILLIN VK see PDT750
SK-PHENOBARBITAL see EOK000
SK 1 (PLASTICIZER) see MCB050
SK-PRAMINE see DLH630
SK-PRAMINE HYDROCHLORIDE see DLH630
SK-PREDNISONE see PLZ000
SK-RESERPINE see RDK000
SKS 85 see SMR000
SKT see CBT500
SKT (ADSORBENT) see CBT500
SK-TETRACYCLINE see TBX000
SK-TETRACYCLINE see TBX250
SK-TOLBUTAMIDE see BSQ000
SK-TRIAMCINOLONE see AQX250
SKUNK CABBAGE see FAB100
SKUNK CABBAGE see SDZ450
SKY BLUE 4B see CMO500
SKY BLUE 5B see CMO500
SKYBOND 700 see BBO625
SKYBOND 1028 see CBY800
SKYBOND 2595 see BBO625
SKYBOND 700 POLYIMIDE RESIN see BBO625
SKY FLOWER see GIW200
SL 31 see MNO775
SL 49 see DES500
SL-90 see RLU000
SL-160 see FCC050
SL-236 see FDA885
SL 501 see CMW700
SL-512 see CQF125
SL 700 see CAT775
1000SL see HKS780
SL-6057 see CMB125
SL-75212 see KEA350
SL 76-002 see CKA125
SLAB OIL (OBS.) see MQV875
SLAKED LIME see CAT225
SLAM see CJR300
SLB 261 see TKF723
SL-D.212 see KEA350
SLEEPAN see TEH500
SLEEPING NIGHTSHADE see DAD880
SLEEPWELL see TEO250
S-426-S (LEPETIT) see EAD500
SLIMACIDE V 10 see BHD150
SLIMICIDE see ADR000

SLIMICIDE 78P see BNA300
SLIMICIDE C 77P see BNA300
SLIPPER FLOWER see SDZ475
SLIPPER PLANT see SDZ475
SLIPRO see TEH500
SLJ 0312 see FDA890
SLOE see AQP890
SLO-GRO see DHF200
SLO-GRO see DMC600
SLOMELAM 2 see MCB050
SLO-PHYLLIN see TEP000
S-LOST see BIH250
SLOSUL see AKO500
SLOVASOL 239 see AFJ155
SLOVASOL 255 see AFJ155
SLOVASOL 458 see AFJ168
SLOVASOL 459 see AFJ168
SLOVASOL 2530 see AFJ160
SLOVASOL 3520R see AFJ160
SLOVASOL A see PJY100
SLOW-FE see FBN100
SLOW-K see PLA500
SLS see SIB600
SLUG-TOX see TDW500
SM 67 see MCB050
SM-307 see DLG000
SM 700 see MCB050
SMA see SFU500
SMA 1000 see SEA500
SMA 1440 see SEA500
SMA 2000 see SEA500
SMA 3000 see SEA500
SMA 1440A see SEA500
SMA 1440H see SEA500
SMA 1725A see SEA500
SMA 17352 see SEA500
SMA 2625A see SEA500
SMA 2000A see SEA500
SMA 3000A see SEA500
SMA 4000A see SEA500
SMAC-A see SEA500
SMA 1440-H RESIN see SEA500
SMA 1440-H RESIN see SEA500
SMA 1420AL see SEA500
SMALL JACK-IN-THE-PULPIT see JAJ000
S. MARCESCENS LIPOPOLYSACCHARIDE see SEA400
SMA (STYRENE POLYMER) see SEA500
SMCA see SFU500
SMD 3500 see SMQ500
SMDC see SIL550
SMDC see VFU000
SMEDOLIN see AHP125
SMEESANA see AQN635
SMFPD see MCB050
SMIDAN see PHX250
S. MILARIUS BARBOURI VENOM see SDZ300
SMITHKO KALKARB WHITING see CAT775
SMOG see SEB000
SMOKE BROWN G see TKN750
SMOKE CONDENSATE, cigarette see SEC000
SMOKELESS POWDER see SED000
SMOKELESS TOBACCO see SED400
SMOP see AKO500
SMP see AKO500
SMS 201-995 see ODY200
S MUSTARD see BIH250
SMUT-GO see HCC500
SMY 1500 see ENJ600
SN see SMA000
SN 20 see ADY500
SN 46 see CPK500
S.N. 112 see PPP500
S. N. 166 see AOO800
SN 186 see ADI750
SN 203 see CMW400
SN 390 see CFU750
SN 407 see CCS625
SN 513 see GLQ100
SN 612 see EEQ600
SN 1796 see DCH300

SN 3,363 see POR300
SN 4075 see MEG250
SN-4395 see PQC500
SN 5754 see PEW550
SN 6718 see CLD000
SN 7618 see CLD000
SN 10275 see PIU200
SN 12,837 see CKB500
SN 13,272 see PMC300
SN 35830 see CGL250
SN 36056 see DSO200
SN 36268 see CJJ250
SN 38107 see EEO500
SN-38584 see MEG250
SN-41703 see EIH500
SN 49537 see TEX600
SN 81742 see CDK800
SN 108266 see CQJ150
SN-1-3778-95 see BQI400
SNAKEBERRY see BAF325
SNAKE FLOWER see VQZ675
SNAKEROOT OIL, CANADIAN see SED500
SNAKE VENOM BITIS ARIETANS see BLV075
SNAKE VENOM BITIS GABONICA see BLV080
SNAKE WEED see PJJ300
SNAPPING HAZEL see WCB000
S,N-DIACETYLCYSTEINE MONOETHYL ESTER see DBG600
SN 6771 DIHYDROCHLORIDE see BFX125
SN-1,2-DIOLEIN see SED525
SN-1,2-DIOLEOYLGLYCEROL see SED525
SNEEZING GAS see CGN000
S.N.G see NGY000
SNIECIOTOX see HCC500
SNIP see DQZ000
SNIP FLY see DQZ000
SNIP FLY BANDS see DQZ000
S(+)-N-NITROSO-2-METHYL-PIPERIDIN (GERMAN) see NLJ000
S(+)-N-NITROSO-2-METHYLPIPERIDINE see NLJ000
S(+)-N-NITROSO-α-PIPECOLINE see NLJ000
SNOMELT see CAO750
SNOW ALGIN H see SEH000
SNOWBERRY see SED550
SNOWCAL see CAT775
SNOWCAL 5SW see BKW100
SNOWDROP see SED575
SNOWFLAKE CRYSTALS see SJT750
SNOWFLAKE WHITE see CAT775
SNOWGOOSE see TAB750
SNOW TEX see AHF500
SNOW TEX see KBB600
SNOW TOP see CAT775
SNOW WHITE see ZKA000
SNP see PAK000
SN 401 PENTAHYDRATE see CCQ200
SNUFF see SED400
SO see LCF000
SO 15 see SKV170
SO-33 see NNQ050
SO 57 see OMY815
SOAz see SED700
SOAMIN see ARA500
SOAP see CNF175
SOAP PLANT see DAE100
SOAPSTONE see SCN000
SOAP YELLOW F see FEV000
SOBATUM see SDZ350
SOBELIN see CMV675
SOBELIN see CMV690
SOBENATE see SFB000
SOBIC ALDEHYDE see SKT500
SOBIODOPA see DNA200
SOBITAL see DJL000
SOBRIL see CFZ000
SOC see CIL775
SOCAL see CAT775
SOCAL E 2 see CAT775

SOCAREX see PJS750
SOCLIDAN see EQQ100
SODA ALUM see AHG500
SODA ASH see SFO000
SODA CHLORATE (DOT) see SFS000
SODA LIME see SEE000
SODA LIME with >4% sodium hydroxide (DOT) see SEE000
SODA LYE see SHS000
SODA LYE see SHS500
SODAMIDE see SEN000
SODA MINT see SFC500
SODANIT see SEY500
SODA NITER see SIO900
SODANTON see DKQ000
SODANTON see DNU000
SODA PHOSPHATE see SJH090
SODAR see DXE600
SODASCORBATE see ARN125
SODESTRIN-H see ECU750
SODIO (CLORATO di) (ITALIAN) see SFS000
SODIO (DICROMATO di) (ITALIAN) see SGI000
SODIO, FLUORACETATO di (ITALIAN) see SHG500
SODIO(IDROSSIDO di) (ITALIAN) see SHS000
3-SODIO-5-(5'-NITRO-2'-FURFURYLIDENAMINO)IMIDAZOLIDIN-2,4-DIONE see SEE250
SODIOPAS see SEP000
SODIO (PERCLORATO DI) (ITALIAN) see PCE750
SODIO(TRICLOROACETATO di) (ITALIAN) see TII500
SODITAL see NBU000
SODIUM see SEE500
SODIUM P-50 see SEQ000
1719 SODIUM see ASO510
SODIUM (dispersions) see SEF500
SODIUM (liquid alloy) see SEF600
SODIUM, metal liquid alloy (DOT) see SEF600
SODIUM-4-(2-ACETAMIDOETHYLDITHIO)BUTANESULFINATE see AAK250
SODIUM-3-ACETAMIDO-2,4,6-TRIIODOBENZOATE see AAN000
SODIUM ACETARSONE see SEG000
SODIUM ACETATE see SEG500
SODIUM ACETATE, anhydrous (FCC) see SEG500
SODIUM ACETATE MONOHYDRATE see SEG650
SODIUM ACETAZOLAMIDE see AAS750
SODIUM ACETRIZOATE see AAN000
SODIUM p-ACETYLAMINOPHENYLANTIMONATE see SLP500
SODIUM-3-ACETYLAMINO-2,4,6-TRIIODOBENZOATE see AAN000
SODIUM ACETYLARSANILATE see AQZ900
SODIUM ACETYL ARSANILATE see ARA000
SODIUM ACETYLIDE see SEG700
SODIUM ACID ARSENATE see ARC000
SODIUM ACID ARSENATE, HEPTAHYDRATE see ARC250
SODIUM ACID CARBONATE see SFC500
SODIUM ACID HEPARIN see HAQ550
SODIUM ACID METHANEARSONATE see MRL750
SODIUM ACID PHOSPHATE see SJH100
SODIUM ACID PYROPHOSPHATE (FCC) see DXF800
SODIUM ACID SULFATE see SEG800
SODIUM ACID SULFATE (solid) see SEG800
SODIUM ACID SULFITE see SFE000
SODIUM ACYCLOVIR see AEC725
SODIUM ADENOSINE-5'-MONOPHOSPHATE see AEM750
SODIUM ADENOSINE TRIPHOSPHATE see AEM250

SODIUM ADENOSINE-5'-TRIPHOSPHATE see AEM250
SODIUM AESCINATE see EDM000
SODIUM ALBAMYCIN see NOB000
SODIUM ALBUCID see SNQ000
SODIUM ALGINATE see SEH000
SODIUM ALGINATE SULFATE see SEH450
SODIUM ALIZARINESULFONATE see SEH475
SODIUM ALIZARIN-3-SULFONATE see SEH475
SODIUM n-ALKYLBENZENE SULFONATE see SEH500
SODIUM ALLYLBENZYL THIOBARBITURATE see AFX500
SODIUM-5-ALLYL-5-(1-(BUTYLTHIO)ETHYL) BARBITURATE see AGA250
SODIUM 5-ALLYL-5-(Δ²)-CYCLOHEXENYL)-2-THIOBARBITURATE see SEH600
SODIUM-5-ALLYL-5-ISOPROPYLBARBITURATE see BOQ750
SODIUM-5-ALLYL-5-(1-METHYLBUTYL)BARBITURATE see SBN000
SODIUM-5-ALLYL-5-(1-METHYLBUTYL)-2-THIOBARBITURATE see SOX500
SODIUM-dl-5-ALLYL-1-METHYL-5-(1-METHYL-2-PENTYNYL)BARBITURATE see MDU500
SODIUM 5,5-ALLYL-(2'-METHYLPROPYL)THIOBARBITURATE see SFG700
SODIUM ALUMINATE, solid (UN 2812) (DOT) see AHG000
SODIUM ALUMINATE, solution (UN 1819) (DOT) see AHG000
SODIUM ALUMINOFLUORIDE see SHF000
SODIUM ALUMINOSILICATE see SEM000
SODIUM ALUMINUM FLUORIDE see SHF000
SODIUM ALUMINUM HEXAFLUORIDE see TNK450
SODIUM ALUMINUM HYDRIDE (DOT) see SEM500
SODIUM ALUMINUM OXIDE see AHG000
SODIUM ALUMINUM PHOSPHATE, ACIDIC see SEM300
SODIUM ALUMINUM PHOSPHATE, BASIC see SEM305
SODIUM ALUMINUM SULFATE see AHG500
SODIUM ALUMINUM TETRAHYDRIDE see SEM500
SODIUM AMAZOLENE see ADE750
SODIUM AMIDE see SEN000
SODIUM AMIDOTRIZOATE see SEN500
SODIUM AMINARSONATE see ARA500
SODIUM p-AMINOBENZENEARSONATE see ARA500
SODIUM-p-AMINOBENZENESTIBONATE GLUCOSIDE see NCL000
SODIUM-4-AMINOBENZOATE see SEO500
SODIUM p-AMINOBENZOATE see SEO500
SODIUM (2-AMINO-3-BENZOYLPHENYL)ACETATE MONOHYDRATE see AHK625
SODIUM d-(−)-α-AMINOBENZYLPENICILLIN see SEQ000
SODIUM l-2-AMINO-4-((HYDROXY)(METHYL)PHOSPHINOYL)BUTYRYL-l-ALANYL-l-ALANINE see SEO550
SODIUM AMINOPHENOL ARSONATE see ARA500
SODIUM p-AMINOPHENYLARSONATE see ARA500
SODIUM AMINOPTERIN see SEO600
SODIUM AMINOSALICYLATE see SEP000
SODIUM p-AMINOSALICYLATE see SEP000
SODIUM p-AMINOSALICYLIC ACID see SEP000

SODIUM-7-(2-(2-AMINO-4-THIAZOLYL)-2-METHOXYIMINOACETAMIDO) CEPHALOSPORANATE see CCR950
SODIUM AMINOTRIACETATE see SEP500
SODIUM AMP see AEM750
SODIUM AMPICILLIN see SEQ000
SODIUM AMYLOBARBITONE see AON750
SODIUM-ANALINE ARSONATE see ARA500
SODIUM ANAZOLENE see ADE750
SODIUM ANDROST-5-EN-17-ONE-3-β-YL SULFATE DIHYDRATE see DAL030
SODIUM ANILARSONATE see ARA500
SODIUM-2-ANTHRACHINONESULPHONATE see SER000
SODIUM ANTHRAQUINONE-1,5-DISULFONATE see DLJ700
SODIUM ANTHRAQUINONE-1-SULFONATE see DLJ800
SODIUM-2-ANTHRAQUINONESULFONATE see SER000
SODIUM-9,10-ANTHRAQUINONE-2-SULFONATE see SER000
SODIUM-β-ANTHRAQUINONESULFONATE see SER000
SODIUM ANTIMONATE see AQB250
SODIUM ANTIMONY see AQB250
SODIUM ANTIMONY BIS(PYROCATECHOL-2,4-DISULFONATE) see AQH500
SODIUM ANTIMONY (III) BIS-PYROCATECHOL-3,5-DISULFONATE HEPTAHYDRATE see AQH500
SODIUM ANTIMONY(III)-3-CATECHOL THIOSALICYLATE see SEU000
SODIUM ANTIMONY-2,3-meso-DIMERCAPTOSUCCINATE see AQD750
SODIUM ANTIMONY ERYTHRITOL see SER500
SODIUM ANTIMONY GLUCONATE see AQI000
SODIUM ANTIMONY(V) GLUCONATE see AQI250
SODIUM ANTIMONY(III) GLUCONATE see AQI000
SODIUM ANTIMONYL ADONITOL see SES000
SODIUM ANTIMONYL-d-ARABITOL see SES500
SODIUM ANTIMONYL BISCATECHOL see SET000
SODIUM ANTIMONYL tert-BUTYL CATECHOL see SET500
SODIUM ANTIMONYL CATECHOL THIOSALICYLATE see SEU000
SODIUM ANTIMONYL CITRATE see SEU500
SODIUM ANTIMONYL DIMETHYLCYSTEINE TARTRATE see AQH750
SODIUM ANTIMONYL-d-FUNCITOL see SEV000
SODIUM ANTIMONYL GLUCO-GULOHEPTITOL see SEV500
SODIUM ANTIMONYL GLYCEROL see SEW000
SODIUM ANTIMONYL-d-MANNITOL see SEW500
SODIUM ANTIMONYL-2,5-METHYLENE-d-MANNITOL see SEX000
SODIUM ANTIMONYL-2,4-METHYLENE-d-SORBITOL see SEX500
SODIUM ANTIMONYL TARTRATE see AQI750
SODIUM ANTIMONYL XYLITOL see SEY000
SODIUM ANTIMONY TARTRATE see AQI750
SODIUM ANTIMOSAN see AQH500
SODIUM ARISTOLOCHATE I see SEY050

SODIUM ARISTOLOCHATE II see SEY075
SODIUM ARSANILATE see ARA500
SODIUM p-ARSANILATE see ARA500
SODIUM ARSANILATE (DOT) see ARA500
SODIUM ARSENATE see ARC000
SODIUM ARSENATE see ARD500
SODIUM ARSENATE see ARD600
SODIUM ARSENATE see SEY100
SODIUM ARSENATE see SEY150
SODIUM ARSENATE (DOT) see ARD750
SODIUM ARSENATE (DOT) see SEY100
SODIUM ARSENATE DIBASIC, anhydrous see ARC000
SODIUM ARSENATE, DIBASIC, HEPTAHYDRATE see ARC250
SODIUM ARSENATE HEPTAHYDRATE see ARC250
SODIUM ARSENITE see SEY200
SODIUM ARSENITE see SEY500
SODIUM ARSENITE, solid (DOT) see SEY500
SODIUM ARSENITE, liquid (solution) (DOT) see SEY500
SODIUM ARSENITE, solid (UN 2027) (DOT) see SEY200
SODIUM ARSENITE, aqueous solutions (UN 1686) (DOT) see SEY200
SODIUM ARSONILATE see ARA500
SODIUM-l-ASCORBATE see ARN125
SODIUM ASCORBATE (FCC) see ARN125
SODIUM ASPARTATE see SEZ350
SODIUM l-ASPARTATE see SEZ350
SODIUM l-ASPARTATE see SEZ355
SODIUM ASPIRIN see ADA750
SODIUM ATP see AEM100
SODIUM ATP see AEM250
SODIUM AUROTHIOMALATE see GJC000
SODIUM AUROTHIOPROPANOLSULFONATE see SEZ400
SODIUM AUROTHIOSULPHATE DIHYDRATE see GJG000
SODIUM AZAPROPAZONE see ASA250
SODIUM AZIDE see SFA000
SODIUM-5-AZIDOTETRAZOLIDE see SFA100
SODIUM AZO-α-NAPHTHOLSULFANILATE see FAG010
SODIUM AZO-α-NAPHTHOLSULPHANILATE see FAG010
SODIUM AZORESORCINOLSULFANILATE see MRL100
SODIUM AZOTOMYCIN see ASO510
SODIUM, AZOTURE de (FRENCH) see SFA000
SODIUM, AZOTURO di (ITALIAN) see SFA000
SODIUM BARBITAL see BAG250
SODIUM BARBITONE see BAG250
SODIUM BARBITURATE see MRK750
SODIUM BENZENEACETATE see SFA200
SODIUM BENZENE HEXOIDE see SFA600
SODIUM BENZENESULFONATE see SJH050
SODIUM-1,2 BENZISOTHIAZOLIN-3-ONE-1,1-DIOXIDE see SJN700
SODIUM BENZOATE see SFB000
SODIUM BENZOATE and CAFFEINE see CAK800
SODIUM BENZOIC ACID see SFB000
SODIUM o-BENZOSULFIMIDE see SJN700
SODIUM BENZOSULFONATE see SJH050
SODIUM BENZOSULPHIMIDE see SJN700
SODIUM-2-BENZOSULPHIMIDE see SJN700
SODIUM-o-BENZOSULPHIMIDE see SJN700
SODIUM-2-BENZOTHIAZOLYLSULFIDE see SFB100
SODIUM BENZYL ALCOHOL SULFATE see SFB150
SODIUM o-BENZYL-p-CHLOROPHENATE see SFB200
SODIUM o-BENZYL-p-CHLOROPHENOLATE see SFB200

SODIUM BENZYLPENICILLIN see BFD250
SODIUM BENZYLPENICILLINATE see BFD250
SODIUM BENZYLPENICILLIN G see BFD250
SODIUM BERYLLIUM MALEATE see SFB500
SODIUM BERYLLIUM TARTRATE see SFC000
SODIUM BIBORATE see DXG035
SODIUM BIBORATE see SFF000
SODIUM BIBORATE DECAHYDRATE see SFF000
SODIUM BICARBONATE see SFC500
SODIUM BICHROMATE see SGI000
SODIUM BINOTAL see SEQ000
SODIUM 2-BIPHENYLOLATE see BGJ750
SODIUM (1,1'-BIPHENYL)-2-OLATE see BGJ750
SODIUM p-BIPHENYLSULFONATE see SFC600
SODIUM, (2-BIPHENYLYLOXY)- see BGJ750
SODIUM BIPHOSPHATE see SJH100
SODIUM BIPHOSPHATE anhydrous see SJH100
SODIUM BIS(2-ETHYLHEXYL)PHOSPHATE see TBA750
SODIUM BIS(2-ETHYLHEXYL) SULFOSUCCINATE see DJL000
SODIUM BISMUTHATE see SFD000
SODIUM BISMUTH THIOGLYCOLATE see BKX750
SODIUM BISMUTH THIOGLYCOLLATE see BKX750
SODIUM-3,5-BIS(aci-NITRO)CYCLOHEXENE-4,6-DIIMINIDE see SFD300
SODIUM BISPROPYLACETATE see PNX750
SODIUM BISULFATE, FUSED see SEG800
SODIUM BISULFIDE see SHR000
SODIUM BISULFITE see SFE000
SODIUM BISULFITE see SFE000
SODIUM BISULFITE (1:1) see SFE000
SODIUM BISULFITE, solid (DOT) see SFE000
SODIUM BISULFITE, solution (DOT) see SFE000
SODIUM BLUE VRS see ADE500
SODIUM BORATE see SFE500
SODIUM BORATE see SJB100
SODIUM BORATE anhydrous see SFE500
SODIUM BORATE DECAHYDRATE see SFF000
SODIUM BORATE, PENTAHYDRATE see SKC550
SODIUM BOROHYDRIDE see SFF500
SODIUM BROMATE see SFG000
SODIUM BROMEBRATE see SIK000
SODIUM BROMIDE see SFG500
SODIUM BROMOACETYLIDE see SFG600
SODIUM-5-(2-BROMOALLYL)-5-sec-BUTYLBARBITURATE see BOR250
SODIUM BROMOSULFALEIN see HAQ600
SODIUM BROMOSULFOPHTHALEIN see HAQ600
SODIUM BROMSULPHALEIN see HAQ600
SODIUM BROMSULPHTHALEIN see HAQ600
SODIUM BUTABARBITAL see BPF250
SODIUM BUTANOATE see SFN600
SODIUM BUTAZOLIDINE see BOV750
SODIUM BUTHALITAL see SFG700
SODIUM BUTHALITON see SFG700
SODIUM BUTHALITONE see SFG700
SODIUM tert-BUTYLCARBOXYLATE see SGM600
SODIUM-5-sec-BUTYL-5-ETHYLBARBITURATE see BPF250
SODIUM BUTYLMERCURIC THIOGLYCOLLATE see SFJ500
SODIUM-5-(1-(BUTYLTHIO)ETHYL)-5-ETHYLBARBITURATE see SFJ875

SODIUM-3-BUTYRAMIDO-α-ETHYL-2,4,6-TRIIODOCINNAMATE see EQC000
SODIUM-3-BUTYRAMIDO-α-ETHYL-2,4,6-TRIIODOHYDROCINNAMATE see SKO500
SODIUM BUTYRATE see SFN600
SODIUM n-BUTYRATE see SFN600
SODIUM CACODYLATE (DOT) see HKC500
SODIUM CALCIUM ALUMINOSILICATE, HYDRATED see SFN700
SODIUM CALCIUM EDETATE see CAR780
SODIUM CAPRATE see SGC100
SODIUM CAPRINATE see SGC100
SODIUM CAPRYLATE see SIX600
SODIUM CAPRYL SULFATE see OFU200
SODIUM CARBENICILLIN see CBO250
SODIUM CARBOLATE see SJF000
SODIUM CARBONATE (2:1) see SFO000
SODIUM CARBONATE PEROXIDE see SJB400
SODIUM CARBONATE STABILIZED SULFONATED POLYSTYRENE SODIUM SALT see SFO100
SODIUM CARBOXYMETHYL CELLULOSE see SFO500
SODIUM CARBOXYMETHYL GUAR see SFO600
SODIUM CARBOXYMETHYL GUAR GUM see SFO600
SODIUM o-CARBOXYMETHYLTARTRONATE see SFO700
SODIUM CARRAGEENAN see SFP000
SODIUM CARRAGEENATE see SFP000
SODIUM CARRAGHEENATE see SFP000
SODIUM CARRIOMYCIN see SFP500
SODIUM CASEINATE see SFQ000
SODIUM CEFACETRIL see SGB500
SODIUM CEFAMANDOLE see CCR925
SODIUM CEFAPIRIN see HMK000
SODIUM CEFAZOLIN see CCS250
SODIUM CEFUROXIME see SFQ300
SODIUM CELLULOSE GLYCOLATE see SFO500
SODIUM CEPHACETRILE see SGB500
SODIUM CEPHALOTHIN see SFQ500
SODIUM CEPHALOTIN see SFQ500
SODIUM CEPHAPIRIN see HMK000
SODIUM CEPHAZOLIN see CCS250
SODIUM CETYL SULFATE see HCP900
SODIUM CEZ see CCS250
SODIUM CHAULMOOGRATE see SFR000
SODIUM CHENODEOXYCHOLATE see CDL375
SODIUM CHENODESOXYCHOLATE see CDL375
SODIUM CHLORAMBUCIL see CDO625
SODIUM CHLORAMINE T see CDP000
SODIUM CHLORAMPHENICOL SUCCINATE see CDP500
SODIUM CHLORATE see SFS000
SODIUM (CHLORATE de) (FRENCH) see SFS000
SODIUM CHLORATE, aqueous solution (DOT) see SFS000
SODIUM CHLORATE BORATE see SFS500
SODIUM CHLORIDE see SFT000
SODIUM CHLORIDE OXIDE see SHU500
SODIUM CHLORITE see SFT500
SODIUM CHLORITE (UN 1496) (DOT) see SFT500
SODIUM CHLORITE, solution with >5% available chlorine (UN 1908) (DOT) see SFT500
SODIUM CHLOROACETATE see SFU500
SODIUM-4-CHLOROACETOPHENONE OXIMATE see SFU600
SODIUM CHLOROACETYLIDE see SFV000
SODIUM cis-3-CHLOROACRYLATE see SFV250
SODIUM cis-β-CHLOROACRYLATE see SFV250
SODIUM CHLOROAURATE see GIZ100

SODIUM N-CHLOROBENZENESULFONAMIDE see SFV275
SODIUM-5-(4-CHLOROBENZOYL)-1,4-DIMETHYL-1H-PYRROLE-2-ACETATE DIHYDRATE see ZUA300
SODIUM 1-(p-CHLOROBENZOYL)-5-METHOXY-2-METHYLINDOLE-3-ACETATE TRIHYDRATE see IDA100
SODIUM CHLOROETHYNIDE see SFV000
SODIUM-4-CHLORO-2-METHYL PHENOXIDE see SFV300
SODIUM (4-CHLORO-2-METHYLPHENOXY)ACETATE see SIL500
SODIUM-2-CHLORO-6-PHENYL PHENATE see SFV500
SODIUM CHLOROPLATINATE see SJJ500
SODIUM (Z)-3-CHLORO-2-PROPENOATE see SFV250
SODIUM-2-CHLOROPROPIONATE see CKT100
SODIUM N-CHLORO-4-TOLUENE SULFONAMIDE see SFV550
SODIUM 5-(2-CHLORO-4-(TRIFLUOROMETHYL)PHENOXY)-2-NITROBENZOATE see SFV650
SODIUM CHOLATE see SFW000
SODIUM CHOLIC ACID see SFW000
SODIUM CHONDROITIN POLYSULFATE see SFW300
SODIUM CHONDROITIN SULFATE see SFW300
SODIUM CHROMATE see SGI000
SODIUM CHROMATE (VI) see DXC200
SODIUM CHROMATE (DOT) see DXC200
SODIUM CHROMATE DECAHYDRATE see SFW500
SODIUM CINCHOPHEN see SJH000
SODIUM CINNAMATE see SFX000
SODIUM CITRATE see MRL000
SODIUM CITRATE, anhydrous see TNL000
SODIUM CITRATE (FCC) see DXC400
SODIUM CITRATE, POTASSIUM CITRATE, CITRIC ACID (2:2:1) see SFX725
SODIUM CLODRONATE see SFX730
SODIUM CLOXACILLIN see SLJ050
SODIUM CLOXACILLIN MONOHYDRATE see SLJ000
SODIUM CMC see SFO500
SODIUM CM-CELLULOSE see SFO500
SODIUM COBALTINITRITE see SFX750
SODIUM COCOMETHYLAMINOETHYL-2-SULFONATE see SFY000
SODIUM COCO METHYL TAURIDE see SFY000
SODIUM COLISTIMETHATE see SFY500
SODIUM COLISTINEMETHANESULFONATE see SFY500
SODIUM COLISTIN METHANESULFONATE see SFY500
SODIUM COMPOUNDS see SFZ000
SODIUM COUMADIN see WAT220
SODIUM CRESOLATE see SFZ050
SODIUM m-CRESOLATE see SJP000
SODIUM p-CRESOLATE see SIM100
SODIUM CRESOXIDE see SFZ050
SODIUM m-CRESOXIDE see SJP000
SODIUM p-CRESOXIDE see SIM100
SODIUM CRESYLATE see SFZ050
SODIUM CROMOGLYCATE see CNX825
SODIUM CROMOLYN see CNX825
SODIUM CUMENEAZO-β-NAPHTHOL DISULPHONATE see FAG018
SODIUM CUPROCYANIDE (DOT) see SFZ100
SODIUM CUPROCYANIDE, solid (UN 2316) (DOT) see SFZ100
SODIUM CUPROCYANIDE, solution (UN 2317) (DOT) see SFZ100
SODIUM CYANIDE see SGA500

SODIUM-7-(2-CYANOACETAMIDO)CEPHALOSPORANIC ACID see SGB500
SODIUM CYANURATE see SGB550
SODIUM CYCLAMATE see SGC000
SODIUM CYCLOHEXANESULFAMATE see SGC000
SODIUM CYCLOHEXANESULPHAMATE see SGC000
SODIUM-5-(1-CYCLOHEXEN-1-YL)-1,5-DIMETHYLBARBITURATE see ERE000
SODIUM CYCLOHEXYL AMIDOSULPHATE see SGC000
SODIUM CYCLOHEXYL SULFAMATE see SGC000
SODIUM CYCLOHEXYL SULFAMIDATE see SGC000
SODIUM CYCLOHEXYL SULPHAMATE see SGC000
SODIUM-2,4-D see SGH500
SODIUM DALAPON see DGI600
SODIUM DBDT see SGF500
SODIUM DECANOATE see SGC100
SODIUM-n-DECANOATE see SGC100
SODIUM DECANOIC ACID see SGC100
SODIUM DECYLBENZENESULFONAMIDE see DAJ000
SODIUM DECYLBENZENESULFONATE see DAJ000
SODIUM DECYL SULFATE see SOK000
SODIUM DEDT see SGJ000
SODIUM DEHYDROACETATE (FCC) see SGD000
SODIUM DEHYDROACETIC ACID see SGD000
SODIUM DEHYDROCHOLATE see SGD500
SODIUM DEHYDROEPIANDROSTERONE SULFATE see DAL040
SODIUM DELVINAL see VKP000
SODIUM DEOXYCHOLATE see SGE000
SODIUM DEOXYCHOLIC ACID see SGE000
SODIUM DESOXYCHOLATE see SGE000
SODIUM DEXAMETHASONE PHOSPHATE see DAE525
SODIUM DEXTROTHYROXINE see SKJ300
SODIUM-3,5-DIACETAMIDO-2,4,6-TRIIODOBENZOATE see SEN500
SODIUM DIACETATE see SGE400
SODIUM DIACETYLDIAMINETRIIODOBENZOATE see SEN500
SODIUM N-(p-((2,4-DIAMINO-6-PTERIDYLMETHYL)AMINO)BENZOYL)ASPARTATE see SGE500
SODIUM DIARSENATE see SEY100
SODIUM DIATRIZOATE see SEN500
SODIUM-1,2:5,6-DIBENZANTHRACENE-9,10-endo-α,β-SUCCINATE see SGF000
SODIUM DIBUTYLDITHIOCARBAMATE see SGF500
SODIUM DIBUTYLNAPHTHALENE SULFATE see NBS700
SODIUM DIBUTYLNAPHTHALENESULFONATE see NBS700
SODIUM-2,6-DI-tert-BUTYLNAPHTHALENESULFONATE see DEE600
SODIUM DIBUTYLNAPHTHYLSULFONATE see NBS700
SODIUM DICHLORISOCYANURATE see SGG500
SODIUM DICHLOROACETATE see SGG000
SODIUM (o-(2,6-DICHLOROANILINO)PHENYL)ACETATE see DEO600
SODIUM DICHLOROCYANURATE see SGG500
SODIUM 2,6-DICHLOROINDOPHENOL see SGG650

SODIUM 2,6-
DICHLOROINDOPHENOLATE see SGG650
SODIUM DICHLOROISOCYANURATE see
SGG500
SODIUM DICHLOROISOCYANURATE
DIHYDRATE see THS050
SODIUM-2-((2,6-DICHLORO-3-
METHYLPHENYL)AMINO)BENZOATE see
SIF425
SODIUM-2,4-
DICHLOROPHENOXYACETATE see
SGH500
SODIUM-2-(2,4-
DICHLOROPHENOXY)ETHYL SULFATE
see CNW000
SODIUM-2,4-DICHLOROPHENOXYETHYL
SULPHATE see CNW000
SODIUM (o-((2,6-
DICHLOROPHENYL)AMINO)PHENYL)AC
ETATE see DEO600
SODIUM-2,4-DICHLOROPHENYL
CELLOSOLVE SULFATE see CNW000
SODIUM-2,2-DICHLOROPROPIONATE see
DGI600
SODIUM-α,α-DICHLOROPROPIONATE see
DGI600
SODIUM-1,3-DICHLORO-1,3,5-TRIAZINE-
2,4-DIONE-6-OXIDE see SGG500
1-SODIUM-3,5-DICHLORO-s-TRIAZINE-
2,4,6-TRIONE see SGG500
1-SODIUM-3,5-DICHLORO-1,3,5-TRIAZINE-
2,4,6-TRIONE see SGG500
SODIUM DICHLORO-s-TRIAZINETRIONE,
dry, containing more than 39% available chlorine
(DOT) see SGG500
SODIUM DICHROMATE see SGI000
SODIUM DICHROMATE(VI) see SGI000
SODIUM DICHROMATE de (FRENCH) see
SGI000
SODIUM DICHROMATE DIHYDRATE see
SGI500
SODIUM DICLOXACILLIN see DGE200
SODIUM DICLOXACILLIN
MONOHYDRATE see DGE200
SODIUM DIETHYLBARBITURATE see
BAG250
SODIUM-5,5-DIETHYLBARBITURATE see
BAG250
SODIUM DIETHYLDITHIOCARBAMATE
see SGJ000
SODIUM N,N-
DIETHYLDITHIOCARBAMATE see SGJ000
SODIUM DIETHYLDITHIOCARBAMATE
TRIHYDRATE see SGJ500
SODIUM DI(2-ETHYLHEXYL)PHOSPHATE
see TBA750
SODIUM DI-(2-ETHYLHEXYL)
SULFOSUCCINATE see DJL000
SODIUM DIETHYL OXALOACETATE see
SGJ550
SODIUM DIFORMYLNITROMETHANIDE
HYDRATE see SGK600
SODIUM DIHEXYL SULFOSUCCINATE see
DKP800
SODIUM DIHYDROBIS(2-
METHOXYETHOXY)ALUMINATE see
SGK800
SODIUM DIHYDROGEN ARSENATE see
ARD600
SODIUM DIHYDROGEN CITRATE see
MRL000
SODIUM DIHYDROGEN
NITRILOTRIACETATE see NEH800
SODIUM DIHYDROGEN
ORTHOARSENATE see ARD600
SODIUM DIHYDROGEN PHOSPHATE
(1:2:1) see SJH100
SODIUM DIHYDROGENPHOSPHIDE see
SGM000
SODIUM 2,5-DIHYDROXYBENZOATE see
GCU050

SODIUM-m-DIISOPROPYLBENZOL (Na-m)
DIHYDROPEROXIDE (RUSSIAN) see
DNN840
SODIUM-p-DIISOPROPYLBENZOL (Na-p)
DIHYDROPEROXIDE (RUSSIAN) see
DNN850
SODIUM-2,3-DIMERCAPTOPROPANE-1-
SULFONATE see DNU860
SODIUM-4,4-DIMETHOXY-1-aci-NITRO-3,5-
DINITRO-2,5-CYCLOHEXADIENE see
SGM100
SODIUM-4-
(DIMETHYLAMINO)BENZENEDIAZOSUL
FONATE see DOU600
SODIUM-p-
(DIMETHYLAMINO)BENZENEDIAZOSUL
FONATE see DOU600
SODIUM-4-
(DIMETHYLAMINO)BENZENEDIAZOSUL
PHONATE see DOU600
SODIUM-p-
(DIMETHYLAMINO)BENZENEDIAZOSUL
PHONATE see DOU600
SODIUM-3-
(DIMETHYLAMINOMETHYLENEAMINO)-
2,4,6-TRIIODOHYDROCINNAMATE see
SKM000
SODIUM-(4-
(DIMETHYLAMINO)PHENYL)DIAZENESU
LFONATE see DOU600
SODIUM DIMETHYLARSINATE see
HKC500
SODIUM DIMETHYLARSINIC ACID
TRIHYDRATE see HKC550
SODIUM DIMETHYLARSONATE see
HKC500
SODIUM
DIMETHYLBENZENESULFONATE see
XJJ010
SODIUM N,N-
DIMETHYLDITHIOCARBAMATE see
SGM500
SODIUM 2,2-DIMETHYLPROPANOATE see
SGM600
SODIUM 2,2-DIMETHYLPROPIONATE see
SGM600
SODIUM-4-(2,4-
DINITROANILINO)DIPHENYLAMINE-2-
SULFONATE see SGP500
SODIUM DINITRO-o-CRESOLATE, dry or
wetted with <15% water, by weight (UN 0234)
(DOT) see SGP550
SODIUM DINITRO-o-CRESOLATE, wetted
with not <15% water, by weight (UN 1348)
(DOT) see SGP550
SODIUM-4,6-DINITRO-o-CRESOXIDE see
DUU600
SODIUM DINITRO-o-CRESYLATE see
SGP550
SODIUM DINITROMETHANIDE see
SGP600
SODIUM-5-
DINITROMETHYLTETRAZOLIDE see
SGQ000
SODIUM-2,4-DINITROPHENOL see DVA800
SODIUM-2,4-DINITROPHENOLATE see
DVA800
SODIUM-2,4-DINITROPHENOXIDE see
SGQ100
SODIUM DIOCTYL SULFOSUCCINATE see
DJL000
SODIUM DIOCTYL SULPHOSUCCINATE
see DJL000
SODIUM DIOXIDE see SJC500
SODIUM-1,1-DIOXOPENICILLANATE see
PAP600
SODIUM DIPHENYL-4,4'-BIS-AZO-2"-8"-
AMINO-1"-NAPHTHOL-3",6 "
DISULPHONATE see CMO000
SODIUM DIPHENYLDIAZO-BIS(α-
NAPHTHYLAMINESULFONATE) see
SGQ500

SODIUM DIPHENYLHYDANTOIN see
DNU000
SODIUM DIPHENYL HYDANTOINATE see
DNU000
SODIUM-5,5-DIPHENYLHYDANTOINATE
see DNU000
SODIUM-5,5-DIPHENYL-2,4-
IMIDAZOLIDINEDIONE see DNU000
SODIUM DIPROPYLACETATE see PNX750
SODIUM-α,α-DIPROPYLACETATE see
PNX750
SODIUM DISULFIDE see SGR500
SODIUM DISULFITE see SII000
SODIUM DITHIOCARBAMATE see SGR600
SODIUM-2,3-
DITHIOLPROPANESULFONATE see
DNU860
SODIUM DITHIONITE (DOT) see SHR500
SODIUM DITOLYLDIAZOBIS-8-AMINO-1-
NAPHTHOL-3,6-DISULFONATE see
CMO250
SODIUM DITOLYLDIAZOBIS-8-AMINO-1-
NAPHTHOL-3,6-DISULPHONATE see
CMO250
SODIUM DNP see DVA800
SODIUM DODECANOATE see LBN000
SODIUM
DODECYLBENZENESULFONATE (DOT)
see DXW200
SODIUM
DODECYLBENZENESULFONATE, dry see
DXW200
SODIUM DODECYL SULFATE see SIB600
SODIUM EDETATE see EIV000
SODIUM EDTA see EIV000
SODIUM ENOLATE MONOHYDRATE see
EDE700
SODIUM EOSINATE see BNH500
SODIUM EQUILIN 3-MONOSULFATE see
ECW520
SODIUM EQUILIN SULFATE see ECW520
SODIUM ERYTHORBATE see EDE700
SODIUM ERYTHORBATE see SGR700
SODIUM d-ERYTHRO-3-OXOHEXONATE
LACTONE see EDE700
SODIUM ESTRONE SULFATE see EDV600
SODIUM ESTRONE-3-SULFATE see EDV600
SODIUM ETASULFATE see TAV750
SODIUM ETHAMINAL see NBU000
SODIUM ETHANEPEROXOATE see SJB000
SODIUM ETHASULFATE see TAV750
SODIUM ETHIDRONATE see DXD400
SODIUM ETHOXIDE see SGR800
SODIUM ETHOXYACETYLIDE see SGS000
SODIUM-6-(2-ETHOXY-1-
NAPHTHAMIDO)PENICILLANATE see
SGS500
SODIUM ETHYDRONATE see DXD400
SODIUM ETHYLBARBITAL see BAG250
SODIUM ETHYL-N-BUTYL BARBITURATE
see BPF750
SODIUM-5-ETHYL-5-sec-
BUTYLBARBITURATE see BPF250
SODIUM-2-ETHYLCAPROATE see SGS600
SODIUM
ETHYLENEDIAMINETETRAACETATE see
EIV000
SODIUM
ETHYLENEDIAMINETETRAACETIC ACID
see EIV000
SODIUM-4-ETHYL-1-(3-ETHYLPENTYL)-1-
OCTYL SULFATE see DKD400
SODIUM 2-ETHYLHEXANOATE see SGS600
SODIUM (+−)-2-ETHYLHEXANOATE see
SGS600
SODIUM(2-ETHYLHEXYL)ALCOHOL
SULFATE see TAV750
SODIUM-5-ETHYL-5-
HEXYLBARBITURATE see EKT500
SODIUM-2-ETHYLHEXYL SULFATE see
TAV750

SODIUM-2-ETHYLHEXYLSULFOSUCCINATE see DJL000
SODIUM ETHYLISOAMYLBARBITURATE see AON750
SODIUM ETHYLMERCURIC THIOSALICYLATE see MDI000
SODIUM p-((ETHYLMERCURI)THIO)BENZENESULFONATE see SKH150
SODIUM-o-(ETHYLMERCURITHIO)BENZOATE see MDI000
SODIUM ETHYLMERCURITHIOSALICYLATE see MDI000
SODIUM-5-ETHYL-5-(1-METHYL-1-BUTENYL) BARBITURATE see VKP000
SODIUM-5-ETHYL-5-(1-METHYLBUTYL)BARBITURATE see NBU000
SODIUM-5-ETHYL-5-(1-METHYLBUTYL)-2-THIOBARBITURATE see PBT500
SODIUM-5-ETHYL-5-(1-METHYLPROPYL)BARBITURATE see BPF250
SODIUM-7-ETHYL-2-METHYL-4-UNDECANOL SULFATE see EMT500
SODIUM-7-ETHYL-2-METHYLUNDECYL-4-SULFATE see EMT500
SODIUM-5-ETHYL-5-PHENYLBARBITURATE see SID000
SODIUM ETHYLXANTHATE see SHE500
SODIUM ETHYLXANTHOGENATE see SHE500
SODIUM ETHYNIDE see SEG700
SODIUM ETIDRONATE see DXD400
SODIUM EVIPAL see ERE000
SODIUM EVIPAN see ERE000
SODIUM FEREDETATE see EJA379
SODIUM FERRIC EDTA see EJA379
SODIUM FERRIC PYROPHOSPHATE see SHE700
SODIUM FERROCYANIDE see SHE350
SODIUM FLUCLOXACILLIN see FDA100
SODIUM FLUOACETATE see SHG500
SODIUM FLUOACETIC ACID see SHG500
SODIUM FLUOALUMINATE see SHF000
SODIUM FLUOARALUMINATE(3-) see TNK450
SODIUM FLUORACETATE de (FRENCH) see SHG500
SODIUM FLUORESCEIN see FEW000
SODIUM FLUORESCEINATE see FEW000
SODIUM FLUORIDE see SHF500
SODIUM FLUORIDE, solid and solution (DOT) see SHF500
SODIUM FLUOROACETATE see SHG500
SODIUM FLUOROALUMINATE see TNK450
SODIUM o-FLUOROBENZOATE see SHG600
SODIUM γ-FLUORO-β-HYDROXYBUTYRATE see SHI000
SODIUM FLUOROPHOSPHATE (Na₂PO₃F) see DXD600
SODIUM FLUOROSILICATE see DXE000
SODIUM FLUORURE (FRENCH) see SHF500
SODIUM FLUOSILICATE see DXE000
SODIUM FORMALDEHYDE BISULFITE see SHI500
SODIUM FORMALDEHYDE SULFOXYLATE see FMW000
SODIUM FORMALDEHYDE SULFOXYLATE of 3-AMINO-4-HYDROXYPHENYLARSONIC ACID see SHI625
SODIUM FORMATE see SHJ000
SODIUM FOSFOMYCIN see DXF600
SODIUM FOSFOMYCIN HYDRATE see FOL200
SODIUM FULMINATE see SHJ500
SODIUM FUMARATE see DXD800

SODIUM FUSIDATE see SHK000
SODIUM FUSIDIN see SHK000
SODIUM-GENT see GCU050
SODIUM GENTISATE see GCU050
SODIUM GERMANIDE see SHK500
SODIUM GLUCONATE see SHK800
SODIUM d-GLUCONATE see SHK800
SODIUM GLUCOSULFONE see AOO800
SODIUM GLUTAMATE see MRL500
SODIUM l-GLUTAMATE see MRL500
l(+) SODIUM GLUTAMATE see MRL500
SODIUM GMP see GLS800
SODIUM GOLD CHLORIDE see GIZ100
SODIUM GUANOSINE-5'-MONOPHOSPHATE see GLS800
SODIUM GUANYLATE see GLS800
SODIUM-5'-GUANYLATE see GLS800
SODIUM HEPARIN see HAQ550
SODIUM HEPARINATE see HAQ550
SODIUM HEXABARBITAL see ERE000
SODIUM HEXACHLOROPLATINATE HEXAHYDRATE see HCL300
SODIUM HEXACYCLONATE see SHL500
SODIUM HEXADECANOATE see SIZ025
SODIUM HEXADECYL SULFATE see HCP900
SODIUM HEXAFLUOROALUMINATE see SHF000
SODIUM HEXAFLUOROALUMINATE see TNK450
SODIUM HEXAFLUOROARSENATE see SHM000
SODIUM HEXAFLUOROSILICATE see DXE000
SODIUM HEXAFLUOSILICATE see DXE000
SODIUM HEXAMETAPHOSPHATE see SHM500
SODIUM HEXAMETAPHOSPHATE see SII500
SODIUM HEXANITROCOBALTATE see SFX750
SODIUM HEXAVANADATE see SHN000
SODIUM HEXESTROL DIPHOSPHATE see SHN150
SODIUM HEXETHAL see EKT500
SODIUM HEXOBARBITONE see ERE000
SODIUM-N-HEXYLETHYL BARBITURATE see EKT500
SODIUM HEXYLETHYL THIOBARBITURATE see SHN275
SODIUM HIPPURATE see SHN500
SODIUM HUMATE see HGM100
SODIUM HYALURONATE see HGN600
SODIUM HYDRATE (DOT) see SHS000
SODIUM HYDRATE, solution see SHS500
SODIUM HYDRAZIDE see SHO000
SODIUM HYDRIDE see SHO500
SODIUM HYDROCORTISONE SUCCINATE see HHR000
SODIUM HYDROCORTISONE-21-SUCCINATE see HHR000
SODIUM HYDROFLUORIDE see SHF500
SODIUM HYDROGEN-S-(2-AMINOETHYL)PHOSPHOROTHIOATE see AKB500
SODIUM HYDROGEN-S-(2-AMINOETHYL)PHOSPHOROTHIOIC ACID see AKB500
SODIUM HYDROGEN CARBONATE see SFC500
SODIUM HYDROGEN S-((N-CYCLOOCTYLMETHYLAMIDINO)METHYL) PHOSPHOROTHIOATE HYDRATE (4:4:5) see SHQ000
SODIUM HYDROGEN DIACETATE see SGE400
SODIUM HYDROGEN DIFLUORIDE see SHQ500
SODIUM HYDROGEN FLUORIDE see SHQ500
SODIUM HYDROGEN FLUORIDE see SHQ500

SODIUM HYDROGEN FLUORIDE, solution (DOT) see SHQ500
SODIUM HYDROGEN PHOSPHATE see SJH090
SODIUM HYDROGEN SULFATE, solid (UN 1821) (DOT) see SEG800
SODIUM HYDROGEN SULFATE, solution (UN 2837) (DOT) see SEG800
SODIUM HYDROGEN SULFIDE see SHR000
SODIUM HYDROGEN SULFITE see SFE000
SODIUM HYDROGEN SULFITE, solid (DOT) see SFE000
SODIUM HYDROGEN SULFITE, solution (DOT) see SFE000
SODIUM HYDROGEN TRILACTATOZIRCONYLATE see ZTA000
SODIUM HYDROGEN TRIOXOSELENITE see SBO200
SODIUM HYDROQUINONE MONOSULFONATE see SHQ600
SODIUM HYDROQUINONE SULFONATE see SHQ600
SODIUM HYDROSELENITE see SBO200
SODIUM HYDROSULFIDE see SHR000
SODIUM HYDROSULFIDE, solution (NA 2922) (DOT) see SHR000
SODIUM HYDROSULFIDE, with <25% water of crystallization (UN 2318) (DOT) see SHR000
SODIUM HYDROSULFIDE, with not <25% water of crystallization (UN 2949) (DOT) see SHR000
SODIUM HYDROSULFITE (DOT) see SHR500
SODIUM HYDROSULPHITE see SHR500
SODIUM HYDROXIDE see SHS000
SODIUM HYDROXIDE, dry (DOT) see SHS000
SODIUM HYDROXIDE (liquid) see SHS500
SODIUM HYDROXIDE, bead (DOT) see SHS000
SODIUM HYDROXIDE, flake (DOT) see SHS000
SODIUM HYDROXIDE, solid (DOT) see SHS000
SODIUM HYDROXIDE, granular (DOT) see SHS000
SODIUM HYDROXIDE, solution (FCC) see SHS500
SODIUM HYDROXYACETATE see SHT000
SODIUM α-HYDROXYACETATE see SHT000
SODIUM-o-HYDROXYBENZOATE see SJO000
SODIUM-4-HYDROXYBUTYRATE see HJS500
SODIUM-γ-HYDROXYBUTYRATE see HJS500
SODIUM (HYDROXYDE de) (FRENCH) see SHS000
SODIUM 2-HYDROXYDIPHENYL see BGJ750
SODIUM-1-HYDROXYETHANESULFONATE see AAH500
SODIUM-2-HYDROXYETHOXIDE see SHT100
SODIUM o-HYDROXYMERCURIBENZOATE see SHT500
SODIUM p-HYDROXYMERCURIBENZOATE see SHU000
SODIUM o-((3-(HYDROXYMERCURI)-2-METHOXYPROPYL)CARBAMOYL)PHENOXY ACETATE see SIH500
SODIUM-o-((3-HYDROXYMERCURI-2-METHOXYPROPYL)CARBAMYL)PHENOXYACETATE and THEOPHYLLINE see MDH750
SODIUM-3-HYDROXYMERCURIO-2,6-DINITRO-4-aci-NITRO-2,5-CYCLOHEXADIENONIDE see SHU175

SODIUM-2-HYDROXYMERCURIO-4-aci-
NITRO-2,5-CYCLOHEXADIENONIDE see
SHU250
SODIUM-2-HYDROXYMERCURIO-6-
NITRO-4-aci-NITRO-2,5-
CYCLOHEXADIENONIDE see SHU275
SODIUM
HYDROXYMETHANESULFINATE see
FMW000
SODIUM-1-
(HYDROXYMETHYL)CYCLOHEXANEACE
TATE see SHL500
SODIUM-3-β-HYDROXY-11-OXO-12-
OLEANEN-30-OATE SODIUM SUCCINATE
see CBO500
SODIUM-5(5'-HYDROXYTETRAZOL-3'-
YLAZO)TETRAZOLIDE see HOF500
SODIUM-5-(5'-HYDROXYTETRAZOL-3'-
YLAZO)TETRAZOLIDE see SHU300
SODIUM HYPOBORATE see SHU475
SODIUM HYPOCHLORITE see SHU500
SODIUM HYPOCHLORITE
HEPTAHYDRATE see SHU510
SODIUM HYPOCHLORITE HYDRATE (2:5)
see SHU515
SODIUM HYPOCHLORITE
PENTAHYDRATE see SHU525
SODIUM HYPOCHLORITE PHOSPHATE
(Na13(CLO)(PO4)4) (9CI) see SJH095
SODIUM HYPOPHOSPHITE see SHV000
SODIUM HYPOSULFITE see SKI000
SODIUM HYPOSULFITE see SKI500
SODIUM IBUPROFEN see IAB100
SODIUM (E)-3-(p-(1H-IMIDAZOL-1-
YLMETHYL)PHENYL)-2-PROPENOATE see
SHV100
SODIUM-5,5'-INDIGOTIDISULFONATE see
FAE100
SODIUM INOSINATE see DXE500
SODIUM-5'-INOSINATE see DXE500
SODIUM IODATE see SHV500
SODIUM IODIDE see SHW000
SODIUM IODINE see SHW000
SODIUM-o-IODOBENZOATE see IEE050
SODIUM-p-IODOBENZOATE see IEE100
SODIUM IODOHIPPURATE see HGB200
SODIUM-2-IODOHIPPURATE see IEE050
SODIUM o-IODOHIPPURATE see HGB200
SODIUM IODOMETHANESULFONATE see
SHX000
SODIUM-5-IODO-2-THIOURACIL see
SHX500
SODIUM IOPODATE see SKM000
SODIUM IOTHALAMATE see IGC100
SODIUM IPODATE see SKM000
SODIUM IRON EDTA see EJA379
SODIUM IRON PYROPHOSPHATE see
SHE700
SODIUM ISOAMYLETHYL BARBITURATE
see AON750
SODIUM d-ISOASCORBATE see EDE700
SODIUM-5-ISOBUTYL-5-
(METHYLTHIOMETHYL)BARBITURATE
see SHY000
SODIUM ISOCYANATE see COI250
SODIUM ISOCYANURATE see SGB550
SODIUM ISONICOTINYL HYDRAZINE
METHANSULFONATE see SIJ000
SODIUM ISOPROPOXIDE see SHY300
SODIUM O-ISOPROPYL
DITHIOCARBONATE see SIA000
SODIUM ISOPROPYLETHYL
THIOBARBITURATE see SHY500
SODIUM, (ISOPROPYLIDENEBIS(p-
PHENYLENEOXY))DI- see BLD800
SODIUM ISOPROPYLXANTHATE see
SIA000
SODIUM O-ISOPROPYL XANTHATE see
SIA000
SODIUM ISOPROPYLXANTHOGENATE
see SIA000
SODIUM ISOTHIOCYANATE see SIA500

SODIUM-α-KETOBUTYRATE see SIY600
SODIUM KETOPROFEN see KGK100
SODIUM LACTATE see LAM000
SODIUM LAURATE see LBN000
SODIUM LAURYLBENZENESULFONATE
see DXW200
SODIUM LAURYL ETHER SULFATE see
SIB500
SODIUM LAURYL ETHOXYSULPHATE see
DYA000
SODIUM-N-LAURYL SARCOSINE see
DXZ000
SODIUM LAURYL SULFATE see SIB600
SODIUM LAURYL SULFOACETATE see
SIB700
SODIUM LAURYL TRIOXYETHYLENE
SULFATE see SIC000
SODIUM LEVOTHYROXINE see LFG050
SODIUM LIGNINSULFONATE see LFQ500
SODIUM LINOLEATE HYDROPEROXIDE
see SIC250
SODIUM LINOLEIC ACID
HYDROPEROXIDE see SIC250
SODIUM LITHOCHOLATE see SIC500
SODIUM LITHOL see NAP100
SODIUM LITHOL RED see NAP100
SODIUM LITHOL RED 20-4018 see NAP100
SODIUM LUMINAL see SID000
SODIUM MALEATE see SIE000
SODIUM MALONATE see SIE500
SODIUM MALONDIALDEHYDE see
MAN800
SODIUM MALONYLUREA see BAG250
SODIUM MANNITOL ANTIMONATE see
SIF000
SODIUM MCPA see SIL500
SODIUM MECLOFENAMATE see SIF425
SODIUM MECLOPHENATE see SIF425
SODIUM MELARSEN see SIF450
SODIUM MENADIONE BISULFITE see
MCB575
SODIUM 1-MENAPHTHYL SULFATE see
SIF475
SODIUM MERALEIN see SIF500
SODIUM MERALLURIDE see SIG000
SODIUM MERCAPTAN see SHR000
SODIUM MERCAPTIDE see SHR000
SODIUM MERCAPTOACETATE see SKH500
SODIUM-2-MERCAPTOBENZOTHIAZOLE
see SIG500
SODIUM-2-
MERCAPTOETHANESULFONATE see
MDK875
SODIUM MERCAPTOMERIN see TFK270
SODIUM MERSALYL see SIH500
SODIUM MERTHIOLATE see MDI000
SODIUM METAARSENATE see ARD500
SODIUM METAARSENATE see ARD750
SODIUM METAARSENITE see SEY500
SODIUM METABISULFITE see SII000
SODIUM METABISULFITE see SII000
SODIUM METABISULPHITE see SII000
SODIUM METABORATE see SII100
SODIUM METAL (DOT) see SEE500
SODIUM, METAL DISPERSION IN
ORGANIC SOLVENT see SEF500
SODIUM METAPERIODATE see SJB500
SODIUM METAPHOSPHATE see SII500
SODIUM METAPHOSPHATE see TKP750
SODIUM METASILICATE see SJU000
SODIUM METASILICATE, anhydrous see
SJU000
SODIUM METAVANADATE see SKP000
SODIUM METHANALSULFOXYLATE see
FMW000
SODIUM METHANEARSONATE see
DXE600
SODIUM METHANEARSONATE see
MRL750
SODIUM METHANESULFONATE see SIJ000
4-SODIUM METHANESULFONATE
METHYLAMINE-ANTIPYRINE see AMK500

SODIUM METHARSONATE see DXE600
SODIUM METHIODAL see SHX000
SODIUM METHOHEXITAL see MDU500
SODIUM METHOHEXITONE see MDU500
SODIUM METHOTREXATE see MDV600
SODIUM METHOXIDE see SIK450
SODIUM METHOXYACETYLIDE see SIJ600
SODIUM β-4-METHOXYBENZOYL-β-
BROMOACRYLATE see SIK000
SODIUM dl-1-METHYL-5-ALLYL-5-(1-
METHYL-2-PENTYNYL)BARBITURATE see
MDU500
SODIUM METHYLAMINOANTIPYRINE
METHANESULFONATE see AMK500
SODIUM-4-METHYLAMINO-1,5-
DIMETHYL-2-PHENYL-3-PYRAZOLONE 4-
METHANESULFONATE see AMK500
SODIUM METHYLARSONATE see DXE600
SODIUM METHYLATE see SIK450
SODIUM METHYLATE SOLUTIONS in
alcohol (UN 1289) (DOT) see SIK450
SODIUM METHYLATE (UN 1431) (DOT) see
SIK450
SODIUM-4-METHYLBENZENESULFINATE
see SKJ500
SODIUM METHYLBENZENESULFONATE
see SIK460
SODIUM-p-
METHYLBENZENESULFONATE see
SKK000
SODIUM (2-METHYL-4-
CHLOROPHENOXY)ACETATE see SIL500
SODIUM-N-METHYL
CYCLOHEXENYLMETHYBARBITURATE
see ERE000
SODIUM METHYLDITHIOCARBAMATE see
VFU000
SODIUM N-METHYLDITHIOCARBAMATE
see VFU000
SODIUM N-METHYLDITHIOCARBAMATE
DIHYDRATE see SIL550
SODIUM METHYLENEDIPHOSPHONATE
see SIL575
SODIUM-2-METHYL-7-
ETHYLUNDECANOL-4-SULFATE see
EMT500
SODIUM-2-METHYL-7-ETHYLUNDECYL
SULFATE-4 see EMT500
SODIUM METHYLHEXABITAL see ERE000
SODIUM METHYLHEXABITOL see ERE000
SODIUM-3-METHYLISOXAZOLIN-4,5-
DIONE-4-OXIMATE see SIL600
SODIUM METHYLMERCURIC
THIOGLYCOLLATE see SIM000
SODIUM-2-(N-
METHYLOLEAMIDO)ETHANE-1-
SULFONATE see SIY000
SODIUM METHYL OLEOYL TAURATE see
SIY000
SODIUM N-METHYL-N-OLEOYLTAURATE
see SIY000
SODIUM 4-METHYLPHENOLATE see
SIM100
SODIUM m-METHYLPHENOLATE see
SJP000
SODIUM p-METHYLPHENOLATE see
SIM100
SODIUM 3-METHYLPHENOXIDE see
SJP000
SODIUM 4-METHYLPHENOXIDE see
SIM100
SODIUM 1-METHYL-5-SEMICARBAZONO-
6-OXO-2,3,5,6-TETRAHYDROINDOLE-3-
SULFONATE TRIHYDRATE see AER666
SODIUM-1-METHYL-5-p-
TOLUOYLPYRROLE-2-ACETATE
DIHYDRATE see SKJ350
SODIUM MOLYBDATE see DXE800
SODIUM MOLYBDATE(VI) see DXE800
SODIUM MOLYBDATE DIHYDRATE see
DXE875

SODIUM MOLYBDOPHOSPHATE see SIM200

SODIUM MONENSIN see MRE230

SODIUM MONOCHLORACETATE see SFU500

SODIUM MONO- and DIMETHYL NAPHTHALENE SULFONATE see SIM400

SODIUM MONODODECYL SULFATE see SIB600

SODIUM MONOFLUORIDE see SHF500

SODIUM MONOFLUOROACETATE see SHG500

SODIUM MONOHEXADECYL SULFATE see HCP900

SODIUM MONOHYDROGEN ARSENATE see ARD500

SODIUM MONOHYDROGEN PHOSPHATE (2:1:1) see SJH090

SODIUM MONOIODIDE see SHW000

SODIUM MONOIODOACETATE see SIN000

SODIUM MONOIODOMETHANESULFONATE see SHX000

SODIUM MONOOCTADECYL SULFATE see OBG100

SODIUM MONOSTEARYL SULFATE see OBG100

SODIUM MONOSULFIDE see SJY500

SODIUM MONOXIDE see SIN500

SODIUM MONOXIDE, solid (DOT) see SIN500

SODIUM MORPHOLINECARBODITHIOATE see SIN650

SODIUM MORPHOLINEDITHIOCARBAMATE see SIN650

SODIUM MORPHOLINE and NITRITE (1:1) see SIN675

SODIUM MORPHOLINODITHIOCARBAMATE see SIN650

SODIUM MYCOPHENOLATE see SIN850

SODIUM MYRISTATE see SIN900

SODIUM MYRISTYL SULFATE see SIO000

SODIUM,(N¹)-ACETYLSULFANILAMIDO- see SNQ000

SODIUM NAFCILLIN see SGS500

SODIUM NALIDIXATE see NAG450

SODIUM 1-NAPHTHALENEMETHANOL SULFATE see SIF475

SODIUM NAPHTHIONATE see ALI500

SODIUM-1,2-NAPHTHOQUINONE-4-SULFONATE see DLK000

SODIUM-β-NAPHTHOQUINONE-4-SULFONATE see DLK000

SODIUM-α-NAPHTHYLAMINE-6-SULPHONATE see ALI500

SODIUM N-1-NAPHTHYLPHTHALAMATE see SIO500

SODIUM N-1-NAPHTHYLPHTHALAMIC ACID see SIO500

SODIUM-N⁶,2'-o-DIBYTYRYLADENOSINE 3',5'-CYCLIC PHOSPHATE see DEJ300

SODIUM NEMBUTAL see NBU000

SODIUM NEOPENTANOATE see SGM600

SODIUM-22 NEOPRENE ACCELERATOR see IAQ000

SODIUM NICOTINATE see NDW500

SODIUM NIGERICIN see SIO788

SODIUM NILIXALATE see NAG450

SODIUM NITRATE (1:1) see SIO900

SODIUM NITRATE (DOT) see SIO900

SODIUM NITRIDE see SIP000

SODIUM NITRILOACETATE see SEP500

SODIUM NITRILOTRIACETATE see SEP500

SODIUM NITRILOTRIACETATE see SIP500

SODIUM NITRITE see SIQ500

SODIUM NITRITE mixed with AMINOPYRINE (1:1) see DOT200

SODIUM NITRITE and BENLATE see BAV500

SODIUM NITRITE mixed with BIS(DIETHYLTHIOCARBAMOYL)DISULFIDE see SIS200

SODIUM NITRITE mixed with 1-(p-BROMOPHENYL)-3-METHYLUREA see SIQ675

SODIUM NITRITE and CARBENDAZIM (1:5) see CBN375

SODIUM NITRITE and CARBENDAZIME (1:1) see SIQ700

SODIUM NITRITE mixed with CHLORDIAZEPOXIDE (1:1) see SIS000

SODIUM NITRITE mixed with CIMETIDINE (1:4) see CMP875

SODIUM NITRITE and l-CITRULLINE (1:2) see SIS100

SODIUM NITRITE mixed with 4-(DIMETHYLAMINO)ANTIPYRINE (1:1) see DOT200

SODIUM NITRITE mixed with DIMETHYLDODECYLAMINE (8:7) see SIS150

SODIUM NITRITE and DIMETHYLUREA see DUM400

SODIUM NITRITE mixed with DISULFIRAM see SIS200

SODIUM NITRITE with DODECYL GUANIDINE ACETATE (5:3) see DXX600

SODIUM NITRITE mixed with ETHAMBUTOL (1:1) see SIS500

SODIUM NITRITE mixed with ETHYLENETHIOUREA see IAR000

SODIUM NITRITE and ETHYL UREA (1:1) see EQD900

SODIUM NITRITE and ETHYLUREA (1:2) see EQE000

SODIUM NITRITE mixed with HEPTAMETHYLENEIMINE see SIT000

SODIUM NITRITE and ISOPROPYLUREA see SIS675

SODIUM NITRITE mixed with METHAPYRILENE (2:1) see MDT000

SODIUM NITRITE mixed with N-METHYLADENOSINE (4:1) see SIS650

SODIUM NITRITE and METHYLANILINE (1:1.2) see MGO000

SODIUM NITRITE mixed with N-METHYLANILINE (35:1) see MGO250

SODIUM NITRITE and METHYL-2-BENZIMIDAZOLE CARBAMATE see CBN375

SODIUM NITRITE mixed with METHYLBENZYLAMINE (3:2) see MHN250

SODIUM NITRITE mixed with N-METHYLBENZYLAMINE (1:1) see MHN000

SODIUM NITRITE mixed with 1-METHYL-3-(p-BROMOPHENYL)UREA see SIQ675

SODIUM NITRITE and 1-(METHYLETHYL)UREA see SIS675

SODIUM NITRITE mixed with METHYLGUANIDINE (1:1) see MKJ000

SODIUM NITRITE mixed with 1-METHYL-3-NITROGUANIDINE (1:1) see MML500

SODIUM NITRITE and 2-METHYL-N-NITROSO-BENZIMIDAZOLE CARBAMATE (1:1) see SIQ700

SODIUM NITRITE and 1-METHYL-1-NITROSO-3-PHENYLUREA see SIS700

SODIUM NITRITE and METHYL UREA see MQJ250

SODIUM NITRITE mixed with OCTAHYDROAZOCINE HYDROCHLORIDE (1:1) see SIT000

SODIUM NITRITE and PIPERAZINE (1:4) see PIJ630

SODIUM NITRITE and 1-PROPYLUREA see SIT500

SODIUM NITRITE and n-PROPYLUREA see SIT500

SODIUM NITRITE, mixed with SODIUM NITRATE and POTASSIUM NITRATE see SIT750

SODIUM NITRITE and TRIFORINE see SIT800

SODIUM m-NITROBENZENEAZOSALICYLATE see SIT850

SODIUM p-NITROBENZENEAZOSALICYLATE see SIU000

SODIUM NITROBENZENESULFONATE see SIU100

SODIUM NITROCOBALTATE (III) see SFX750

SODIUM NITROFERRICYANIDE see SIU500

SODIUM NITROMALONALDEHYDE see SIV000

SODIUM aci-NITROMETHANE see SIV500

SODIUM aci-NITROMETHANIDE see SIV500

SODIUM-4-NITROPHENOXIDE see SIV600

SODIUM NITROPRUSSATE see SIU500

SODIUM NITROPRUSSIATE see SIW000

SODIUM NITROPRUSSIDE see SIU500

SODIUM NITROPRUSSIDE DIHYDRATE see SIW500

SODIUM NITROSOPENTACYANOFERRATE (3) see SIW000

SODIUM-4-NITROSOPHENOXIDE see SIW550

SODIUM NITROSYLPENTACYANOFERRATE see SIU500

SODIUM NITROSYLPENTACYANOFERRATE(III) see SIU500

SODIUM NITROSYLPENTACYANOFERRATE(III) DIHYDRATE see SIW500

SODIUM-5-NITROTETRAZOLIDE see SIW600

SODIUM-2-NITROTHIOPHENOXIDE see SIW625

SODIUM NITROXYLATE see SIX000

SODIUM NONYL SULFATE see SIX500

SODIUM NORAMIDOPYRINE METHANESULFONATE see AMK500

SODIUM NORSULFAZOLE see TEX500

SODIUM NOVOBIOCIN see NOB000

SODIUM NPA see SIO500

SODIUM NTA see SEP500

SODIUM OCTADECANOATE see SJV500

SODIUM OCTADECYL SULFATE see OBG100

SODIUM OCTAHYDROTRIBORATE see SIX550

SODIUM OCTANOATE see SIX600

SODIUM n-OCTANOATE see SIX600

SODIUM OCTYL SULFATE see OFU200

SODIUM OCTYL SULPHATE see OFU200

SODIUM OLEATE see OIA000

SODIUM N-OLEOYL-N-METHYLATAURINE see SIY000

SODIUM N-OLEOYL-N-METHYLTAURATE see SIY000

SODIUM OLEYLMETHYLTAURIDE see SIY000

SODIUM OMADINE see HOC000

SODIUM ORBENIN see SLJ050

SODIUM OROTATE HYDRATE see SIY100

SODIU-4-MORPHOLINECARBODITHIOATE see SIN650

SODIUM ORTHOARSENATE see ARD750

SODIUM ORTHOARSENATE see SEY150

SODIUM ORTHOARSENITE see ARJ500

SODIUM ORTHO PHENYLPHENATE see BGJ750

SODIUM ORTHOVANADATE see SIY250

SODIUM OXACILLIN see MNV250

SODIUM OXALATE see SIY500

SODIUM OXALATE see SIY500

SODIUM OXIDE see SIN500

SODIUM OXIDE (Na₂O₂) see SJC500

SODIUM 2-OXOBUTANOATE see SIY600

SODIUM 2-OXOBUTYRATE see SIY600

SODIUM OXYBATE see HJS500

SODIUM OXYCHLORIDE see SHU500

SODIUM OXYDIACETATE see SIZ000

SODIUM PALMITATE see SIZ025

SODIUM PALMITYL SULFATE see HCP900

SODIUM PANTOTHENATE see SIZ050

SODIUM PARATOLUENE SULPHONATE see SKK000

SODIUM PATENT BLUE V see ADE500

SODIUM PCP see SJA000

SODIUM PENICILLANATE 1,1-DIOXIDE see PAP600

SODIUM PENICILLIN see BFD250

SODIUM PENICILLIN G see BFD250

SODIUM PENICILLIN II see BFD250

SODIUM-PENT see NBU000

SODIUM PENTABARBITAL see NBU000

SODIUM PENTABARBITONE see NBU000

SODIUM PENTACARBONYL RHENATE see SIZ100

SODIUM PENTACHLOROPHENATE see SJA000

SODIUM PENTACHLOROPHENATE (DOT) see SJA000

SODIUM PENTACHLOROPHENOL see SJA000

SODIUM PENTACHLOROPHENOLATE see SJA000

SODIUM PENTACHLOROPHENOXIDE see SJA000

SODIUM PENTADECANECARBOXYLATE see SIZ025

SODIUM PENTAFLUOROSTANNITE see SJA500

SODIUM-β,β-PENTAMETHYLENE-γ-HYDROXYBUTYRATE see SHL500

SODIUM PENTHIOBARBITAL see PBT500

SODIUM PENTOBARBITAL see NBU000

SODIUM (R)(+)-PENTOBARBITAL see PBS500

SODIUM PENTOBARBITONE see NBU000

SODIUM PENTOBARBITURATE see NBU000

SODIUM PENTOTHAL see PBT500

SODIUM PENTOTHIOBARBITAL see PBT500

SODIUM PERACETATE see SJB000

SODIUM PERBORATE see SJB100

SODIUM PERBORATE see SJD000

SODIUM PERBORATE TETRAHYDRATE see SJB350

SODIUM PERCARBONATE see SJB400

SODIUM PERCHLORATE see PCE750

SODIUM PERCHLORATE (DOT) see PCE750

SODIUM PERCHLORATE MONOHYDRATE see SJB450

SODIUM PERFLUOROACETATE see SKL100

SODIUM PERIODATE see SJB500

SODIUM PERMANGANATE see SJC000

SODIUM PEROXIDE see SJC500

SODIUM PEROXOBORATE see SJB100

SODIUM PEROXOBORATE see SJD000

SODIUM PEROXYACETATE see SJB000

SODIUM PEROXYBORATE see SJD000

SODIUM PEROXYDISULFATE see SJE000

SODIUM PERRHENATE see SJD500

SODIUM PERSULFATE see SJE000

SODIUM PHENATE see SJF000

SODIUM PHENOBARBITAL see SID000

SODIUM PHENOBARBITONE see SID000

SODIUM PHENOLATE, solid (DOT) see SJF000

SODIUM PHENOL TETRABROMOPHTHALEIN see HAQ600

SODIUM PHENOXIDE see SJF000

SODIUM-α-PHENOXYCARBONYLBENZYLPENICILLIN see CBO000

SODIUM PHENYLACETATE see SFA200

SODIUM PHENYLACETYLIDE see SJF500

SODIUM PHENYL-β-AMINOPROPIONAMIDE-p-ARSONATE see SJG000

SODIUM PHENYLBUTAZONE see BOV750

SODIUM-2-PHENYLCINCHONINATE see SJH000

SODIUM-1-PHENYL-2,3-DIMETHYL-4-METHYLAMINOPYRAZOLON-N-METHANESULFONATE see AMK500

SODIUM-1-PHENYL-2,3-DIMETHYL-5-PYRAZOLONE-4-METHYLAMINO METHANESULFONATE see AMK500

SODIUM PHENYLDIMETHYLPYRAZOLONMETHYL AMINOMETHANE SULFONATE see AMK500

SODIUM PHENYLETHYLBARBITURATE see SID000

SODIUM PHENYLETHYLMALONYLUREA see SID000

SODIUM-N-PHENYLGLYCINAMIDE-p-ARSONATE see CBJ750

SODIUM 2-PHENYLPHENATE see BGJ750

SODIUM o-PHENYLPHENATE see BGJ750

SODIUM o-PHENYLPHENOL see BGJ750

SODIUM o-PHENYLPHENOLATE see BGJ750

SODIUM o-PHENYLPHENOXIDE see BGJ750

SODIUM 3-PHENYL-2-PROPENOATE see SFX000

SODIUM PHENYL SULFINATE DIHYDRATE see SJH025

SODIUM PHENYLSULFONATE see SJH050

SODIUM PHOSPHATE see HEY500

SODIUM PHOSPHATE see SJH200

SODIUM PHOSPHATE see TKP750

SODIUM PHOSPHATE, anhydrous see SJH200

SODIUM PHOSPHATE, DIBASIC see SJH090

SODIUM PHOSPHATE DODECAHYDRATE see SJH300

SODIUM PHOSPHATE HYPOCHLORITE see SJH095

SODIUM PHOSPHATE, MONOBASIC see SJH100

SODIUM PHOSPHATE, TRIBASIC see SJH200

SODIUM PHOSPHATE TRIBASIC DODECAHYDRATE see SJH300

SODIUM PHOSPHIDE see SJI500

SODIUM PHOSPHINATE see SHV000

SODIUM PHOSPHOMOLYBDATE see SIM200

SODIUM PHOSPHOROFLUORIDATE see DXD600

SODIUM PHOSPHOROFLURIDATE see DXD600

SODIUM PHOSPHOROTHIOATE see TNM750

SODIUM PHOSPHOTUNGSTATE see SJJ000

SODIUM PICOSULFATE see SJJ175

SODIUM PICRAMATE, dry or wetted with <20% water, by weight (UN 0235) (DOT) see PIC500

SODIUM PICRAMATE, wetted with not <20% water, by weight (UN 1349) (DOT) see PIC500

SODIUM PICRATE see SJJ190

SODIUM PIPERACILLIN see SJJ200

SODIUM PIVALATE see SGM600

SODIUM PLATINIC CHLORIDE see SJJ500

SODIUM POLYACRYLATE see SJK000

SODIUM POLYALUMINATE see AHG000

SODIUM POLYANHYDROMANNURONIC ACID SULFATE see SEH450

SODIUM POLYANTIMONATE see AQB250

SODIUM POLYMANNURONATE see SEH000

SODIUM POLYOXYETHYLENE ALKYL ETHER SULFATE see SJK200

SODIUM POLYPHOSPHATES, GLASSY see SII500

SODIUM POLYSTYRENESULFONATE see SFO100

SODIUM POLYSTYRENE SULFONATE see SJK375

SODIUM POTASSIUM ALLOY, liquid and solid (DOT) see PLS500

SODIUM POTASSIUM BISMUTH TARTRATE (SOLUBLE) see BKX250

SODIUM POTASSIUM TARTRATE see SJK385

SODIUM PRASTERONE SULFATE see SJK400

SODIUM PRASTERONE SULFATE DIHYDRATE see SJK410

SODIUM PREDNISOLONE PHOSPHATE see PLY275

SODIUM p-2-PROPANOL-OXY-PHENYLARSONATE see SJK475

SODIUM PROPIONATE see SJL500

SODIUM-2-PROPYLPENTANOATE see PNX750

SODIUM-2-PROPYLVALERATE see PNX750

SODIUM PYRIDINETHIONE see HOC000

SODIUM PYRITHIONE see MCQ750

SODIUM PYRITHIONE SQ 3277 see HOC000

SODIUM PYROARSENATE see SEY100

SODIUM PYROBORATE see SFF000

SODIUM PYROBORATE DECAHYDRATE see SFF000

SODIUM PYROPHOSPHATE see DXF800

SODIUM PYROPHOSPHATE (FCC) see TEE500

SODIUM PYROSULFATE see SEG800

SODIUM PYROSULFITE see SII000

SODIUM PYROVANADATE see SJM500

SODIUM QUINALBARBITONE see SBN000

SODIUM RETINOATE see SJN000

SODIUM RHODANATE see SIA500

SODIUM RHODANIDE see SIA500

SODIUM RICINOLEATE see SJN500

SODIUM RIFOMYCIN SV see SJN650

SODIUM (R,R)-5-(2-((2-(3-CHLOROPHENYL)-2-HYDROXYETHYL)AMINO)PROPYL)-1,3-BENZODIOXOLE-2,2- see SJN675

SODIUM SACCHARIDE see SJN700

SODIUM SACCHARIN see SJN700

SODIUM SACCHARINATE see SJN700

SODIUM SACCHARINE see SJN700

SODIUM SALICYLATE see SJO000

SODIUM SALICYL-(γ-HYDROXYMERCURI-β-METHOXYPROPYL)AMIDE-o-ACETATE see SIH500

SODIUM SALICYLIC ACID see SJO000

SODIUM SALT of 1-AMINO-2-NAPHTHOL-3,6-DISULPHONIC ACID see AKH250

SODIUM SALT of CACODYLIC ACID see HKC500

SODIUM SALT of CARBOXYMETHYLCELLULOSE see SFO500

SODIUM SALT-m-CRESOL see SJP000

SODIUM SALT of DICHLORO-s-TRIAZINETRIONE see SGG500

SODIUM SALT of N,N-DIETHYLDITHIOCARBAMIC ACID see SGJ000

SODIUM SALT of 3-(3-DIMETHYLAMINOMETHYLENEAMINO-2,4,6-TRIIODOPHENYL) PROPIONIC ACID see SKM000

SODIUM SALT of 4,6-DINITRO-o-CRESOL see DUU600

SODIUM SALT of ETHYLENEDIAMINETETRAACETIC ACID see EIV000

SODIUM SALT of HEXADECANOIC ACID see SIZ025

SODIUM SALT of HYDROXY-o-CARBOXY-PHENYL-FLUORONE see FEW000

SODIUM SALT of HYDROXYMERCURIACETYLAMINOBENZOIC ACID see SJP500

SODIUM SALT of HYDROXYMERCURIPROPANOL PHENYLACETIC ACID see SJQ000
SODIUM SALT of HYDROXYMERCURIPROPANOL PHENYL MALONIC ACID see SJQ500
SODIUM SALT of ISONICOTINIC ACID see ILF000
SODIUM SALT of MERCURI-BIS SALICYLIC ACID see SJR000
SODIUM SALT of MERCURY DITHIOCARBAMATE PIPERAZINE ACETIC ACID see SJR500
SODIUM SALT of β-4-METHOXYBENZOYL-β-BROMOACRYLIC ACID see SIK000
SODIUM SALT of 4-MORPHOLINECARBODITHIOIC ACID see SIN650
SODIUM SALT OF ACIFLUORFEN see SFV650
SODIUM SALT of PHENYLBUTAZONE see BOV750
SODIUM SALT of SULFONATED NAPHTHALENEFORMALDEHYDE CONDENSATE see BLX000
SODIUM SALT of 1,1,3,3-TETRACYANOPROPENE see TAI050
SODIUM SALT of 2,4,5-TRICHLOROPHENOL see SKK500
SODIUM SARKOMYCIN see SJS500
SODIUM SECOBARBITAL see SBN000
SODIUM SECONAL see SBN000
SODIUM SELENATE see DXG000
SODIUM SELENIDE see SJT000
SODIUM SELENITE see SBO200
SODIUM SELENITE see SJT500
SODIUM SELENITE PENTAHYDRATE see SJT600
SODIUM SESQUICARBONATE see SJT750
SODIUM SILICATE see SCN700
SODIUM SILICATE see SJU000
SODIUM SILICIDE see SJU500
SODIUM SILICOALUMINATE see SEM000
SODIUM SILICOFLUORIDE (DOT) see DXE000
SODIUM SORBATE see SJV000
SODIUM SOTRADECOL see EMT500
SODIUM SOTRADECOL see SIO000
SODIUM STANNOUS TARTRATE see TGF000
SODIUM STEARATE see SJV500
SODIUM STEAROYL LACTYLATE see SJV700
SODIUM STEARYL FUMARATE see SJV710
SODIUM STEARYL SULFATE see OBG100
SODIUM STIBANILATE GLUCOSIDE see NCL000
SODIUM STIBINIVANADATE see SJW000
SODIUM STIBOGLUCONATE see AQH800
SODIUM STIBOGLUCONATE see AQI250
SODIUM-2-(4-STYRYL-3-SULFOPHENYL)-2H-NAPHTHO-(1,2-d)-TRIAZOLE see TGE155
SODIUM SUCARYL see SGC000
SODIUM SUCCINATE see SJW100
SODIUM SULBACTAM see PAP600
SODIUM SULFACETAMIDE see SNQ000
SODIUM SULFACHLOROPYRAZINE MONOHYDRATE see SJW200
SODIUM SULFACYL see SNQ000
SODIUM SULFADIAZINE see MRM250
SODIUM SULFAETHYLTHIADIAZOLE see SJW300
SODIUM SULFAMERAZINE see SJW475
SODIUM SULFAMETAZINE see SJW500
SODIUM SULFAMETHAZINE see SJW500
SODIUM SULFAMETHIAZINE see SJW500
SODIUM SULFAMEZATHINE see SJW500
SODIUM-2-SULFANILAMIDOTHIAZOLE see TEX500

SODIUM SULFANILYLACETAMIDE see SNQ000
SODIUM N-SULFANILYLACETAMIDE see SNQ000
SODIUM SULFAPYMONOHYDRATE see PPO250
SODIUM SULFAPYRIDINE see PPO250
SODIUM SULFAPYRIMIDINE see MRM250
SODIUM SULFATE (2:1) see SJY000
SODIUM SULFATEn anhydrous see SJY000
SODIUM SULFATHIAZOLE see TEX500
SODIUM SULFHYDRATE see SFE000
SODIUM SULFHYDRATE see SHR000
SODIUM SULFIDE see SJY500
SODIUM SULFIDE, anhydrous (DOT) see SJY500
SODIUM SULFIDE with <30% water of crystallization (DOT) see SJY500
SODIUM SULFITE (2:1) see SJZ000
SODIUM SULFITE, anhydrous see SJZ000
SODIUM SULFOACETATE derivatives of MONO and DIGLYCERIDES see SJZ050
SODIUM SULFOBROMOPHTHALEIN see HAQ600
SODIUM SULFOCYANATE see SIA500
SODIUM SULFOCYANIDE see SIA500
SODIUM SULFODI-(2-ETHYLHEXYL)SULFOSUCCINATE see DJL000
SODIUM SULFOXYLATE see SHR500
SODIUM SULFOXYLATE FORMALDEHYDE see FMW000
SODIUM SULPHAMERAZINE see SJW475
SODIUM SULPHAPYRIDINE see PPO250
SODIUM SULPHATE see SJY000
SODIUM SULPHIDE see SJY500
SODIUM SULPHITE see SJY500
SODIUM SULPHOBROMOPHTHALEIN see HAQ600
SODIUM SUPEROXIDE see SJZ100
SODIUM SUPEROXIDE (DOT) see SJZ100
SODIUM SYNTARPEN see SLJ050
SODIUM d-T4 see SKJ300
SODIUM TARTRATE see SKB000
SODIUM l-(+)-TARTRATE see BLC000
SODIUM TARTRATE (FCC) see BLC000
SODIUM TAUROCHOLATE see SKB500
SODIUM TCA INHIBITED see TII500
SODIUM TCA SOLUTION see TII250
SODIUM TELLURATE see SKC000
SODIUM TELLURATE(IV) see SKC500
SODIUM TELLURATE, DIHYDRATE see SKC100
SODIUM TELLURATE VI see SKC000
SODIUM TELLURITE see SKC500
SODIUM TETRABORATE see DXG035
SODIUM TETRABORATE see SFF000
SODIUM TETRABORATE DECAHYDRATE see SFF000
SODIUM TETRABORATE (Na₂B₄O₇) see DXG035
SODIUM TETRABORATE PENTAHYDRATE see SKC550
SODIUM TETRACHLOROAURATE see GIZ100
SODIUM TETRACHLOROAURATE(1-) see GIZ100
SODIUM TETRACHLOROAURATE(3+) see GIZ100
SODIUM TETRACYANATOPALLADATE(II) see SKD600
SODIUM TETRADECANOATE see SIN900
SODIUM TETRADECYL SULFATE see SIO000
SODIUM TETRAFLUOROBORATE see SKE000
SODIUM TETRAFLUOROPROPIONATE see SKE100
SODIUM 2,2,3,3-TETRAFLUOROPROPIONATE see SKE100
SODIUM TETRAHYDROALUMINATE(1−) see SEM500

SODIUM TETRAHYDROBORATE(1-) see SFF500
SODIUM TETRAHYDROGALLATE see SKE500
SODIUM TETRAMETAPHOSPHATE see SKE550
SODIUM TETRAPEROXYCHROMATE see SKF000
SODIUM TETRAPEROXYMOLYBDATE see SKF500
SODIUM TETRAPHOSPHATE see HEY500
SODIUM TETRAPOLYPHOSPHATE see HEY500
SODIUM TETRAPOLYPHOSPHATE see SII500
SODIUM TETRAPROPYLBENZENE SULFONATE see SKF600
SODIUM TETRAVANADATE see SKG500
SODIUM THEOPHYLLINE see SKH000
SODIUM THIACETARSAMIDE see SKH100
SODIUM THIAMYLAL see SOX500
SODIUM THIMERFONATE see SKH150
SODIUM THIOCYANATE see SIA500
SODIUM THIOCYANIDE see SIA500
SODIUM THIOGLYCOLATE see SKH500
SODIUM THIOGLYCOLLATE see SKH500
SODIUM THIOPENTAL see PBT500
SODIUM THIOPENTOBARBITAL see PBT500
SODIUM THIOPENTONE see PBT500
SODIUM THIOPHOSPHATE see TNM750
SODIUM THIOSULFATE see SKI000
SODIUM THIOSULFATE, anhydrous see SKI000
SODIUM THIOSULFATE, PENTAHYDRATE see SKI500
SODIUM THIOSULFATOAURATE(I) DIHYDRATE see TNK500
SODIUM THYROXIN see LFG050
SODIUM THYROXINATE see LFG050
SODIUM THYROXINE see LFG050
SODIUM d-THYROXINE see SKJ300
SODIUM l-THYROXINE see LFG050
SODIUM l-THYROXINE PENTAHYDRATE see SKJ325
SODIUM TOLMETIN see SKJ340
SODIUM TOLMETIN DIHYDRATE see SKJ350
SODIUM-4-TOLUENESULFINATE see SKJ500
SODIUM-p-TOLUENESULFINATE see SKJ500
SODIUM-p-TOLUENE SULFINIC ACID see SKJ500
SODIUM TOLUENESULFONATE see SIK460
SODIUM-p-TOLUENESULFONATE see SKK000
SODIUM p-TOLUENESULFONYLCHLORAMIDE see CDP000
SODIUM-p-TOLYL SULFONATE see SKK000
SODIUM TOSYLATE see SKK000
SODIUM TOSYLCHLORAMIDE see CDP000
SODIUM TRIAZIDOAURATE see SKK100
SODIUM (TRICHLORACETATE de) (FRENCH) see TII500
SODIUM TRICHLOROACETATE see TII500
SODIUM-2,4,5-TRICHLOROPHENATE see SKK500
SODIUM-2,4,5-TRICHLOROPHENOXYETHYL SULFATE see SKL000
SODIUM TRIFLUOROACETATE see SKL100
SODIUM TRIFLUOROSTANNITE see SKL500
SODIUM TRIHYDROGEN (4-AMINO-1-HYDROXYBUTYLIDENE)DIPHOSPHATE, TRIHYDRATE see SKL600
SODIUM 2,3,5-TRIIODOBENZOATE see SKL700

SODIUM TRIIODOHYDROCINNAMATE see SKM000
SODIUM-3,7,12-TRIKETOCHOLANATE see SGD500
SODIUM TRIMETAPHOSPHATE see SKM500
SODIUM TRIMETAPHOSPHATE see TKP750
SODIUM TRIMETHYLACETATE see SGM600
SODIUM-2,4,6-TRINITROPHENOXIDE see SJJ190
SODIUM-3,7,12-TRIOXO-5-β-CHOLANATE see SGD500
SODIUM TRIPHOSPHATE see SKN000
SODIUM TRIPOLYPHOSPHATE see SKN000
SODIUM TRITHIOCYANURATE see SKN100
SODIUM TUNGSTATE see SKN500
SODIUM TUNGSTATE, DIHYDRATE see SKO000
SODIUM TUNGSTOPHOSPHATE see SJJ000
SODIUM TYROPANOATE see SKO500
SODIUM URATE see SKO575
SODIUM VALPROATE see PNX750
SODIUM VANADATE see SIY250
SODIUM VANADATE see SKP000
SODIUM VANADITE see SKQ000
SODIUM VANADIUM OXIDE see SIY250
SODIUM VERONAL see BAG250
SODIUM VERSENATE see EIX500
SODIUM VINBARBITAL see VKP000
SODIUM WARFARIN see WAT220
SODIUM WOLFRAMATE see SKN500
SODIUM XANTHATE see SHE500
SODIUM XANTHOGENATE see SHE500
SODIUM XYLENESULFONATE see XJJ010
SODIUM ZIRCONIUM LACTATE see LAO000
SODIUM ZIRCONIUM OXIDE SULFATE see ZTS100
SODIUM ZIRCONYL SULPHATE see ZTS100
SODIUM ZOMEPIRAC see ZUA300
SODIURETIC see BEQ625
SODIZOLE see SNN500
SODNA SUL KYSELINY cis-β-4-METHOXYBENZOYL-β-BROMAKRYLOVE see CQK600
SODOTHIOL see SKI500
SODUXIN see SJW100
SODYESUL BLACK 2B see CMS250
SODYESUL BLACK PACF see CMS250
SODYESUL BLACK R see CMS250
SODYESUL RED BROWN BCF see CMS257
SOFALCONE see IMS300
SO-FLO see SHF500
SOFRAL BLACK FP see CMS250
SOFRAL PRUNE S EXTRA see CMS257
SOFRAMYCIN see NCF000
SOFRIL see SOD500
SOFTENIL see TEH500
SOFTENON see TEH500
SOFTIL see DJL000
SOFTON 1000 see CAT775
SOFTRAN see BOM250
SOG-4 see CJN250
SOHNHOFEN STONE see CAO000
SOILBROM-40 see EIY500
SOILBROM-85 see EIY500
SOILFUME see EIY500
SOIL FUNGICIDE 1823 see CJA100
SOILSIN see EME100
SOIL STABILIZER 661 see SMR000
SOK see CBM750
SOL see PKB500
SOLABAR see BAO300
SOLACEN see MOV500
SOLACIL see POF500
SOLACIN see MOV500
SOLACTHYL see AES650
SOLACTOL see LAJ000
SOLADREN see VGP000
SOLAESTHIN see MJP450
SOLAMINE see BEN000

SOLAMINE LIGHT TURQUOISE BLUE GL see COF420
SOLAN see SKQ400
SOLANCARPIDINE see POJ000
SOLANDRA (VARIOUS SPECIES) see CDH125
(22r,25s)-5-α-SOLANIDANINE-3-β-OL see SKQ510
(22r,25s)-5-α-SOLANIDAN-3-β-OL see SKQ510
(22s,25r)-5-α-SOLANIDAN-3-β-OL see SKQ515
SOLANID-5-ENE-3-β, 12-α-DIOL see SKR600
(22s,25r)-SOLANID-5-EN-3-β-OL see SKR500
SOLANIDINE-S see POJ000
SOLANILE YELLOW E see SGP500
SOLANIN see SKR875
α-SOLANIN see SKS000
SOLANINE see SKS000
α-SOLANINE see SKS000
SOLANINE HYDROCHLORIDE see SKS100
SOLANIN HYDROCHLORIDE see SKS100
SOLANOSIDE see SKS150
SOLANOSIDE, MONO-o-ACETYL- see MRE600
SOLANTAL see CJH750
SOLANTHRENE BLUE B see DFN300
SOLANTHRENE BLUE F-SBA see DFN300
SOLANTHRENE BLUE RS see IBV050
SOLANTHRENE BLUE RSN see IBV050
SOLANTHRENE BLUE SB see DFN300
SOLANTHRENE BRILLIANT ORANGE JR see CMU820
SOLANTHRENE BRILLIANT VIOLET F 2R see DFN450
SOLANTHRENE BRILLIANT YELLOW see DCZ000
SOLANTHRENE BROWN BR see CMU770
SOLANTHRENE BROWN FR see CMU780
SOLANTHRENE BROWN JR see CMU770
SOLANTHRENE BROWN R see CMU780
SOLANTHRENE GREY BL see CMU320
SOLANTHRENE OLIVE R see DUP100
SOLANTHRENE PRINTING BLACK BL see CMU320
SOLANTHRENE PRINTING BLACK BLN see CMU320
SOLANTHRENE PRINTING BLACK TL see CMU320
SOLANTHRENE PRINTING BLACK TLN see CMU320
SOLANTHRENE R FOR SUGAR see IBV050
SOLANTIN see DKQ000
SOLANTINE BLUE 10GL see CMO650
SOLANTINE RED 8BL see CMO885
SOLANTINE TURQUOISE G see COF420
SOLANTOIN see DNU000
SOLANTYL see DNU000
SOLANUM SODOMEUM, extract see AQP800
SOLANUM TUBEROSUM L see PLW750
SOLAPRET see MCB050
SOLAPRET MH see MCB050
SOLAR 40 see DXW200
SOLAR BLACK G see CMN240
SOLAR BROWN PL see CMO750
SOLAR FAST RED 3G see CMM300
SOLAR FAST RED 6B see CMM400
SOLAR LIGHT ORANGE GX see HGC000
SOLAR ORANGE see CMM220
SOLAR PURE BLUE VX see CMM062
SOLAR PURE YELLOW 8G see CMM750
SOLAR RED B see CMO885
SOLAR RUBINE see HJF500
SOLARSON see CEI250
SOLAR TURQUOISE BLUE GLL see COF420
SOLAR VIOLET 5BN see FAG120
SOLAR WINTER BAN see PML000
SOLASKIL see LFA020
SOLASOD-5-EN-3-β-OL see POJ000
SOLASODINE see POJ000
SOLASON see CEI250
SOLAXIN see CDQ750

SOLBAR see BAO300
SOLBAR see BAP000
SOLBASE see PJT200
SOLBROL A see HJL000
SOLBROL B see BSC000
SOLBROL M see HJL500
SOLBUTAMOL see BQF500
SOLCOSERYL see ADZ125
SOLCOSERYL see SKS299
SOLDACTONE see PKZ000
SOLDEP see TIQ250
SOLDESAM see DAE525
SOLDIER'S BUTTONS see MBU550
SOLDIER'S CAP see MRE275
SOLEAL see OLW100
SOLESTRIL see POB500
SOLESTRO see EDP000
SOLEVAR see ENX600
SOLEX see CAW850
SOLFAC see REF250
2-SOLFAMONYL-4,4'-DIAMINOPHENYLSULFONE see AOT125
SOLFARIN see WAT200
SOLFAST GREEN see PJQ100
SOLFAST GREEN 63102 see PJQ100
SOLFO BLACK B see DUZ000
SOLFO BLACK BB see DUZ000
SOLFO BLACK G see DUZ000
SOLFO BLACK SB see DUZ000
SOLFO BLACK 2B SUPRA see DUZ000
SOLFO BROWN BNR SUPRA see CMS257
SOLFO BROWN 6R SUPRA see CMS257
SOLFOCRISOL see GJG000
SOLFO SERPINE see RDK000
SOLFURO di CARBONIO (ITALIAN) see CBV500
SOLGANAL see ART250
SOLGANAL B see ART250
SOLGOL see CNR675
SOLID GREEN CRYSTALS O see AFG500
SOLID GREEN FCF see FAG000
SOLIMIDIN see ZUA200
SOLINDENE BLUE 2BD see ICU135
SOLINDENE ORANGE R see CMU815
SOLIPHYLLINE see CMG300
SOLIUS RED 4B see CMO885
SOLIWAX see DJL000
SOLKA-FIL see CCU150
SOLKA-FLOC see CCU150
SOLKA-FLOC BW see CCU150
SOLKA-FLOC BW 20 see CCU150
SOLKA-FLOC BW 100 see CCU150
SOLKA-FLOC BW 200 see CCU150
SOLKA-FLOC BW 2030 see CCU150
SOLKETAL see DVR600
SOLLICULIN see EDV000
SOLMETHINE see MJP450
SOLNAPYRIN-A see AIB300
SOLNET see PMB850
SOLO see NBL200
SOLOCALM see FAJ100
SOLOCHROMATE BLACK 6BN see CMP880
SOLOCHROME BLACK see EDC625
SOLOCHROME BLACK 6BN see CMP880
SOLOCHROME BLACK T see EDC625
SOLOCHROME BLACK WDFA see EDC625
SOLOCHROME BLUE FB see HJF500
SOLOCHROME LEATHER ORANGE GR see CMP882
SOLOCHROME LEATHER PINK B see CMG750
SOLOCHROME ORANGE M see NEY000
SOLOCHROME ORANGE RS see CMP882
SOLOCHROME RED B see CMG750
SOLOCHROME RED BS see CMG750
SOLOCHROME VIOLET R see HLI000
SOLOCHROME VIOLET RS see HLI000
SOLOCHROME YELLOW WN see SIT850
SOLOCROM VIOLET RS see HLI000
SOLOMON ISLAND IVY see PLW800

SOLOMON'S LILY see ITD050
SOLON see IMS300
SOLOPHENYL TURQUOISE BLUE GL
SUPRA see COF420
SOLOSIN see TEP000
SOLOZONE see SJC500
SOL PHENOBARBITAL see SID000
SOL PHENOBARBITONE see SID000
SOLPIRIN see ARP125
SOLPRENE 300 see SMR000
SOLPRINA see CLD000
SOLPYRON see ADA725
SOL SODOWA KWASU
LAURYLOBENZENOSULFONOWEGO
(POLISH) see DXW200
SOLSOL NEEDLES see SIB600
SOL. SULFUR BLUE 10 see CMM080
SOLTRIOL see DMJ400
SOLTROL see MQV900
SOLTROL 50 see MQV900
SOLTROL 100 see MQV900
SOLTROL 180 see MQV900
SOLUBACTER see TIL500
SOLUBLE BARBITAL see BAG250
SOLUBLE FLUORESCEIN see FEW000
SOLUBLE GLUSIDE see SJN700
SOLUBLE GUN COTTON see CCU250
SOLUBLE INDIGO see FAE100
SOLUBLE IODOPHTHALEIN see TDE750
SOLUBLE PENTOBARBITAL see NBU000
SOLUBLE PHENOBARBITAL see SID000
SOLUBLE PHENOBARBITONE see SID000
SOLUBLE PHENYTOIN see DNU000
SOLUBLE SACCHARIN see SJN700
SOLUBLE SULFACETAMIDE see SNQ000
SOLUBLE SULFACYL see SNQ000
SOLUBLE SULFADIAZINE see MRM250
SOLUBLE SULFAMERAZINE see SJW475
SOLUBLE SULFAPYRIDINE see PPO250
SOLUBLE SULFATHIAZOLE see TEX500
SOLUBLE TARTRO-BISMUTHATE see
BKX250
SOLUBLE THIOPENTONE see PBT500
SOLUBLE VAN DYKE BROWN see MAT500
SOLUBOND 0-869 see RDP000
SOLUBOND 3520 see RDP000
SOLUCORT see PLY275
SOLU-CORTEF see HHR000
SOLUDAGENAN see PPO250
SOLU-DECADRON see DAE525
SOLU-DECORTIN see PMA100
SOLUFILIN see DNC000
SOLUFILINA see CNR125
SOLUFILINA see DIH600
SOLUFYLLIN see DNC000
SOLUGLACIT see NIM500
SOLU-GLYC see HHR000
SOLULAN 16 see LAU560
SOLULAN 25 see LAU560
SOLULAN C-24 see PKE700
SOLULAN L 575 see LAU560
SOLUMEDINE see SJW475
SOLU-MEDROL see USJ100
SOLU-MEDRONE see USJ100
SOLUMIN FP 85SD see SNY100
SOLUNAPTOL A see CMM760
SOLUPHYLINE see CNR125
SOLUPHYLLINE see HLC000
SOLUPYRIDINE see DXF400
SOLUROL see SOX875
SOLUSEDIV 2% see HAF400
SOLUSOL-75% see DJL000
SOLUSOL-100% see DJL000
SOLUSTIBOSAN see AQH800
SOLUSTIBOSAN see AQI250
SOLUSTIN see AQH800
SOLUSTIN see AQI250
SOLUSURMIN see AQH800
SOLUSURMIN see AQI250
SOLUTION CONCENTREE T271 see
AMY050

SOLUTION GLYCERYL TRINITRATE see
NGY000
SOLUTRAST see IFY000
SOLVACTANT 7 SOLVENT see MLS800
SOLVANOL see DJX000
SOLVANOM see DTR200
SOLVANT YELLOW 33 see CMS240
SOLVAR see PKP750
SOLVARONE see DTR200
SOLVAT 14 see BFL000
SOLVENT 111 see MIH275
SOLVENT BLACK 3 see PCJ200
SOLVENT BLACK 5 see CMS236
SOLVENT BLUE 78 see BKP500
SOLVENT BLUE 93 see BKP500
SOLVENT-DEWAXED HEAVY
NAPHTHENIC DISTILLATE see MQV820
SOLVENT-DEWAXED HEAVY
PARAFFINIC DISTILLATE see MQV825
SOLVENT-DEWAXED LIGHT
NAPHTHENIC DISTILLATE see MQV835
SOLVENT-DEWAXED LIGHT PARAFFINIC
DISTILLATE see MQV840
SOLVENT ETHER see EJU000
SOLVENT NAPHTHA (PETROLEUM),
LIGHT AROMATIC see SKS350
SOLVENT ORANGE 1 see CMP600
SOLVENT ORANGE 15 see BJF000
SOLVENT RED 1 see CMS238
SOLVENT RED 19 see EOJ500
SOLVENT RED 48 see SKS400
SOLVENT RED 72 see DDO200
SOLVENT REFINED COAL-II, HEAVY
DISTILLATE see SLH300
SOLVENT REFINED COAL MATERIALS,
HEAVY DISTILLATE see SLH300
SOLVENT-REFINED (mild) HEAVY
NAPHTHENIC DISTILLATE see MQV845
SOLVENT-REFINED (mild) HEAVY
PARAFFINIC DISTILLATE see MQV850
SOLVENT-REFINED (mild) LIGHT
NAPHTHENIC DISTILLATE see MQV852
SOLVENT-REFINED (mild) LIGHT
PARAFFINIC DISTILLATE see MQV855
SOLVENT VIOLET 26 see DBX000
SOLVENT YELLOW 1 see PEI000
SOLVENT YELLOW 14 see PEJ500
SOLVESSO 100 see SKS410
SOLVIC see PKQ059
SOLVIC 523KC see AAX175
SOLVIREX see DXH325
SOLVOSOL see CBR000
SOLWAY BLUE RB see CMM100
SOLWAY SKY BLUE B see CMM090
SOLWAY SKY BLUE BA see CMM090
SOLYACORD see DJS200
SOLYUSURMIN see AQH800
SOLYUSURMIN see AQI250
SOMA see IPU000
SOMACTON see PJA250
SOMADRIL see IPU000
SOMAGARD see LIU353
SOMALGIT see IPU000
SOMALIA ORANGE 2R see XRA000
SOMALIA ORANGE A2R see XRA000
SOMALIA ORANGE I see PEJ500
SOMALIA RED III see OHI200
SOMALIA RED IV see SBC500
SOMALIA RED PG see CMS238
SOMALIA YELLOW A see DOT300
SOMALIA YELLOW R see AIC250
SOMAN see SKS500
SOMANIL see IPU000
SOMAR see DXE600
SOMATOSTATIN see SKS600
SOMATOSTATIN 14 see SKS600
SOMATOSTATIN-28 (SHEEP) see SKS650
SOMATOTROPIC HORMONE see PJA250
SOMATOTROPIN see PJA250
SOMAZINA see CMF350
SOMBEROL see QAK000
SOMBREVIN see PMM000

SOMBUCAPS see ERD500
SOMBULEX see ERD500
SOMBUTOL see EOK000
SOMELIN see HAG800
SOMILAN see ENE500
SOMINAT see SKS700
SOMIO see GFA000
SOMIPRONT see DUD800
SOMLAN see DAB800
SOMNAFAC see MDT250
SOMNAFAC see QAK000
SOMNAL see AMX750
SOMNALERT see ERD500
SOMNASED see DLY000
SOMNESIN see EQL000
SOMNEVRIN see CHD750
SOMNIBEL see DLY000
SOMNICAPS see DPJ400
SOMNI SED see CDO000
SOMNITE see DLY000
SOMNOLETTEN see EOK000
SOMNOMED see QAK000
SOMNOPENTYL see NBU000
SOMNOS see CDO000
SOMNOSAN see EOK000
SOMNUROL see BNP750
SOMONAL see EOK000
SOMONIL see DSO000
SOMOPHYLLIN see TEP500
SOMOPHYLLIN O see TEP500
SOMSANIT see HJS500
SONACIDE see GFQ000
SONACON see DCK759
SONAFORM see TDA500
SONAL see QAK000
SONALAN see ENE500
SONALEN see ENE500
SONAPAX see MOO250
SONAR see FMQ200
SONATE see ARA500
SONAX see EDW100
SONAX see VFA100
SONAZINE see CKP500
SONBUTAL see BOR000
αSONE ACETONIDE see PMH550
SONEBON see DLY000
SONERILE see BPF500
SONERYL see BPF500
SONG-SAM, ROOT EXTRACT see GEQ425
SONILYN see SNH900
SONISTAN see NBU000
SONNOLIN see DLY000
SONTEC see CDO000
SONTOBARBITAL NABITONE see NBU000
SOOT see SKS750
SOPANOX see TGX550
SOPAQUIN see CLD000
SOPENTAL see NBU000
SOPHIA see MDL750
SOPHIAMIN see MDQ250
SOPHOCARPINE HYDROBROMIDE see
SKS800
SOPHORA SECUNDIFLORA see NBR800
SOPHORA TOMENTOSA see NBR800
SOPHORETIN see QCA000
SOPHORICOL see GCM350
SOPHORIN see RSU000
SOPHORINE see CQL500
SOPP see BGJ750
SOPRABEL see LCK000
SOPRABEL see LCK100
SOPRACOL see MLC250
SOPRACOL 781 see MLC250
SOPRANEBE see MAS500
SOPRATHION see EEH600
SOPRATHION see PAK000
SOPRINAL see BAG250
SOPRINTIN see ABH500
SOPROCIDE see BBQ750
SOPROFOR S 70 see CAR790
SOPRONTIN see AAF750
SOPRONTIN see ABH500

SOPROTIN see ABH500
SORBALDEHYDE see SKT500
SORBANGIL see CCK125
SORBA-SPRAY Mn see MAU250
SORBESTER P 17 see SKV100
SORBIC ACID see SKU000
SORBIC ACID, COPPER SALT see CNP000
SORBIC ACID, POTASSIUM SALT see PLS750
SORBIC ACID, SODIUM SALT see SJV000
SORBIC ALCOHOL see HCS500
SORBIC OIL see PAJ500
SORBICOLAN see SKV200
SORBID see CCK125
SORBIDE NITRATE see CCK125
SORBIDILAT see CCK125
SORBIDINITRATE see CCK125
SORBIMACROGOL LAURATE 300 see PKG000
SORBIMACROGOL OLEATE see PKL100
SORBIMACROGOL TRIOLEATE 300 see TOE250
SORBIMACROGOL TRISTEARATE 300 see SKV195
SORBINIC ALCOHOL see HCS500
SORBISLO see CCK125
SORBISTAT see SKU000
SORBISTAT-K see PLS750
SORBISTAT-POTASSIUM see PLS750
SORBITAL O 20 see PKL100
SORBITAN C see SKV150
SORBITAN MONODODECANOATE see SKV000
SORBITAN, MONODODECANOTE, POLY(OXY-1,2-ETHANEDIYL) DERIVATIVES see PKG000
SORBITAN, MONOHEXADECANOATE, POLY(OXY-1,2-ETHANEDIYL) DERIVS. see PKG500
SORBITAN MONOLAURATE see SKV000
SORBITAN MONOOCTADECANOATE see SKV150
SORBITAN, MONOOCTADECANOATE, POLY(OXY-1,2-ETHANEDIYL) DERIVATIVES see PKL030
SORBITAN MONOOLEATE see SKV100
SORBITAN MONOOLEIC ACID ESTER see SKV100
SORBITAN MONOSTEARATE see SKV150
SORBITAN O see SKV100
SORBITAN, (Z)-9-OCTADECENOATE (2:3) see SKV170
SORBITAN OLEATE see SKV100
SORBITAN PALMITATE see MRJ800
SORBITAN, SESQUIOLEATE see SKV170
SORBITAN STEARATE see SKV150
SORBITAN, TRIOCTADECANOATE, POLY(OXY-1,2-ETHANEDIYL) derivs. (9CI) see SKV195
SORBITAN, TRISTEARATE, POLYOXYETHYLENE derivatives see SKV195
SORBITE see SKV200
SORBITOL see SKV200
d-SORBITOL see SKV200
SORBITRATE see CCK125
SORBO see SKV200
SORBO-CALCIAN see CAL750
SORBO-CALCION see CAL750
SORBOL see SKV200
SORBONIT see CCK125
SORBON S 60 see SKV150
SORBOSE see SKV400
SORBOSTYL see SKV200
SORBYL ACETATE see AAU750
SORBYL ALCOHOL see HCS500
SORCI see FPD100
SORDENAC see CLX250
SORDINOL see CLX250
SOREFLON 604 see TAI250
SORETHYTAN (20) MONOOLEATE see PKL100
SORGEN 30 see SKV170

SORGEN 40 see SKV100
SORGEN 50 see SKV150
SORGEN 70 see MRJ800
SORGHUM GUM see SLJ500
SORGOA see TGB475
SORGOPRIM see BQB000
SORIDERMAL see MNQ500
SORIPAL see MNQ500
SORLATE see PKL100
SORMETAL see CAR780
SORQUAD see CCK125
SORQUAT see CCK125
SORQUETAN see TGD250
SORREL SALT see OLE000
SORROSIE (HAITI) see FPD100
SORSAKA see SKV500
l-SORSAKA, LEAF and STEM EXTRACT see SKV500
SORVILANDE see SKV200
SOS see OFU200
SOSIGON see DOQ400
SOSPITAN see PPH050
SOTACOR see CCK250
SOTALEX see CCK250
SOTALOL see CCK250
SOTALOL HYDROCHLORIDE see CCK250
SOT-BLACK 6 see PCJ200
SOTERENOL see SKW000
SOTERENOL HYDROCHLORIDE see SKW500
SOTIPOX see TIQ250
SOTRADECOL see EMT500
SOTYL see NBU000
SOUCHET see ICC800
SOUCI d'EAU (CANADA) see MBU550
SOUDAN I see PEJ500
SOUDAN II see XRA000
SOUDAN III see OHI200
SOUFRAMINE see PDP250
SOUP see NGY000
SOUTHERN BAYBERRY see WBA000
SOUTHERN BENTONITE see BAV750
SOUTHERN COPPERHEAD VENOM see SKW775
SOVCAIN see NOF500
SOVCAINE see DDT200
SOVCAINE HYDROCHLORIDE see NOF500
SOVERIN see QAK000
SOVIET TECHNICAL HERBICIDE 2M-4C see CIR250
SOVIOL see AAX250
SOVKAIN see NOF500
SOVOCAINE HYDROCHLORIDE see NOF500
SOVOL see PJL750
SOVPRENE see PJQ050
SOWBUG & CUTWORM BAIT see COF500
SOXINAL PZ see BJK750
SOXINOL CZ see CPI250
SOXINOL PZ see BJK500
SOXISOL see SNN500
SOXYSYMPAMINE see MDT600
SOYBEAN LECTIN see SKW800
SOYBEAN OIL (UNHYDROGENATED) see SKW825
SOYBEAN PHYTOHEMAGGLUTININ see SKW800
SOYEX see NIX000
SOY PROTEIN, ISOLATED see SKW840
SP see AIF000
75 SP see DOP600
SP 104 see TCM250
S.P. 147 see LEU000
SP 489 see CGW300
SP 725 see PEF500
SP-52223 see IIE200
SPA see DWA600
SPA-S-753 see DOS400
SPACTIN see BTA325
SPAMOL see CPQ250
SPAN 20 see SKV000
SPAN 40 see MRJ800

SPAN 55 see SKV150
SPAN 60 see SKV150
SPAN 80 see SKV100
SPAN 83 see SKV170
SPANBOLET see SNJ000
SPANISH BROOM see GCM000
SPANISH CARNATION see CAK325
SPANISH FLY see CBE250
SPANISH MARJORAM OIL see MBU500
SPANISH THYME OIL see TFX750
SPANISH WHITE see BKW100
SPANON see CJJ250
SPANONE see CJJ250
SPANTOL see PGA750
SPANTRAN see MQU750
SPARIC see BRE500
SPARICON see SBG500
SPARINE see DQA600
SPARINE HYDROCHLORIDE see PMI500
SPARSOMYCIN see SKX000
SPARTAKON see TDX750
SPARTASE see MAI600
SPARTEINE see SKX500
(−)-SPARTEINE see SKX500
1-SPARTEINE see SKX500
d-SPARTEINE see PAB250
SPARTEINE, d-ISOMER see PAB250
SPARTEINE SULFATE see SKX750
SPARTOCIN see SKX750
SPARTOLOXYN see GGS000
SPARTOSE OM-22 see CCU150
SPA-S 565 see RJZ100
SPASMAMIDE see FAJ150
SPASMEDAL see DAM700
SPASMENTRAL see BBU800
SPASMEX see KEA300
SPASMEXAN HYDROCHLORIDE see LFK200
SPASMIONE see DNU100
SPASMO 3 see KEA300
SPASMOCAN see NOC000
SPASMOCYCLON see DNU100
SPASMOCYCLONE see DNU100
SPASMODOLIN see DAM700
SPASMOLYSIN see HOA000
SPASMOLYTIN see DHX800
SPASMOLYTON see THK000
SPASMO-NIT see PAH250
SPASMOPHEN see ORQ000
SPASURET see FCB100
SPATHE FLOWER see SKX775
SPATHIPHYLLUM (VARIOUS SPECIES) see SKX775
SPATONIN see DIW000
SPAVIT see VSZ450
SPB 50 (SILOXANE) see PJQ790
SP 60 (CHLOROCARBON) see PKQ059
(SP-4-2)-trans(+)-DICHLORO(1,2-CYCLOHEXANEDIAMINE-N,N')- (9CI) see DAD050
SPD PHOSPHATE see AMD500
SPEARMINT OIL see SKY000
SPEARWORT see FBS100
SPECIAL BLACK 1V & V see CBT750
SPECIAL BLUE X 2137 see EOJ500
SPECIAL M see AHA275
SPECIAL ORANGE GR see CMM220
SPECIAL ORANGE H see CMM220
SPECIAL SCHWARZ see CBT750
SPECIAL TERMITE FLUID see DEP600
SPECIFEN see EID000
SPECILLINE G see BDY669
SPECTAZOLE see EAE000
SPECTINOMYCIN DIHYDROCHLORIDE see SLI325
SPECTINOMYCIN, DIHYDROCHLORIDE, PENTAHYDRATE see SKY500
SPECTINOMYCIN HYDROCHLORIDE see SLI325
SPECTOGARD see SKY500
SPECTRACIDE see DCM750
SPECTRAR see INJ000

SPECTRA-SORB UV 9 see MES000
SPECTRA-SORB UV 24 see DMW250
SPECTRA-SORB UV 531 see HND100
SPECTROBID see BAB250
SPECTROLENE BLUE B see DCJ200
SPECTROLENE GARNET G see MPY750
SPECTROLENE RED KB see CLK225
SPECTROLENE RED RL see MMF780
SPECTROLENE SCARLET 2G see DEO295
SPECTROSIL see SCK600
SPECULAR IRON see IHC450
SPEED see DBA800
"SPEED" see MDQ500
SPENCER 401 see PJY500
SPENCER S-6538 see DJY200
SPENCER S-6900 see DRR200
SPENKEL see PKL500
SPENLITE see PKL500
S(−)-PENTOBARBITAL SODIUM see PBS750
SPENT OXIDE see SKZ000
SPENT SULFURIC ACID (DOT) see SOI500
SPERGON I see TBO500
SPERGON TECHNICAL see TBO500
SPERLOX-S see SOD500
SPERLOX-Z see EIR000
1RS,cis-PERMETHRIN see PCK060
1RS-trans-PERMETHRIN see PCK080
SPERM, HUMAN see HGM000
SPERMIDINE see SLA000
SPERMIDINE PHOSPHATE see AMD500
SPERMINE see DCC400
SPERMINE, PURISS see DCC400
SPERMINE TETRAHYDROCHLORIDE see GEK000
SPERM SPECIFIC ISOZYME of LACTATE DEHYDROGENASE see LAO300
SPER/NO see SLG625
SPERSADOX see DAE525
SPERSUL see SOD500
SPERSUL THIOVIT see SOD500
SP 60 ESTER see AAX250
SP G see BER500
SPG 827 see BER500
SP-G IIA see PJJ250
SPH see HGN600
SPHEROIDINE see FOQ000
SPHEROMYCIN see SMB000
SPHERON see CBT750
SPHERON 6 see CBT750
SPHYGMOGENIN see VGP000
SPICEBUSH see CCK675
SPICLOMAZINE HYDROCHLORIDE see SLB000
SPICY JATROPHA see CNR135
SPIDER LILY see BAR325
SPIDER LILY see REK325
SPIDER LILY see SLB250
SPIGELIA ANTHELMIA see PIH800
SPIGELIA MARILANDICA see PIH800
SPIKE see BSN000
SPIKE LAVENDER OIL see SLB500
SPILAN see BEL900
SPINACEN see SLG800
SPINACENE see SLG800
SPINNAKER see CJN300
SPINOCAINE see AIL750
SPINOSAD see SLB600
SPINOSAD TECHNICAL see SLB600
SPINOSYN A see LEK500
SPIPERONE see SLE500
SPIRAMIN see AJV500
SPIRAMYCIN see SLC000
SPIRAMYCIN ADIPATE see SLC300
SPIRAMYCIN, HEXANEDIOATE see SLC300
SPIRAMYCINS see SLC000
SPIRESIS see AFJ500
SPIRIDON see AFJ500
SPIRIT see EFU000
SPIRIT BLACK SB see CMS236
SPIRIT of HARTSHORN see AMY500
SPIRIT NIGROSINE SSB see CMS236
SPIRIT of NITER (sweet) see SLD500

SPIRIT OF GLONOIN see NGY000
SPIRIT OF GLYCERYL TRINITRATE see NGY000
SPIRIT OF TRINITROGLYCERIN see NGY000
SPIRIT ORANGE see PEJ500
SPIRITS of SALT see HHL000
SPIRITS of TURPENTINE see TOD750
SPIRITS of WINE see EFU000
SPIRIT of TURPENTINE see TOD750
SPIRIT YELLOW I see PEJ500
SPIRO-32 see SLD800
SPIRO(1,3-BENZODIOXOLE-2,1'-CYCLOHEXAN)-4-OL, METHYLCARBAMATE see SLD550
2,2'-SPIROBI(1,3-BENZODIOXABOROLE), COMPD. WITH 1,3-DI-o-TOLYLGUANIDINE see DEK500
1,1'-SPIROBI(INDAN)-6,6'-DIOL, 3,3,3',3'-TETRAMETHYL- see SLD570
SPIROBIINDANE see SLD570
1,1'-SPIROBI(1H-INDENE)-6,6'-DIOL, 2,2',3,3'-TETRAHYDRO-3,3,3',3'-TETRAMETHYL- see SLD570
SPIROBROMIN see SLD600
SPIROCID see ABX500
SPIROCTANIE see AFJ500
SPIRO(CYCLOHEXANE-1,5'-HYDANTOIN) see DVO600
SPIRO(17H-CYCLOPENTA(a)PHENANTHRENE-17,2'-(3'H)-FURAN) see AFJ500
SPIRO(17H-CYCLOPENTA(a)PHENANTHRENE-17,2'(5'H)FURAN), PREGN-4-ENE-21-CARBOXYLIC ACID DERIV. see AFJ500
SPIRO(CYCLOPROPANE-1,5'(5H)-INDEN)-7'(6'H)-ONE, 6'-HYDROXY-3'-(HYDROXYMETHYL)-2',4',6'-TRIMETHYL-, (R)- see HLU600
SPIRO(6,16-(EPITHIOPROPANOXYMETHANO)-7,13-IMINO-12H-1,3-DIOXOLO(7,8)ISOQUINO(3,2,-B)(3) BENZAZOCINE-20,1'(2'H)-ISOQUINOLIN)-19-ONE, 3',4',6,6A,7,13,14,16-OCTAHYDRO-5-(ACETYLOXY)-6', 8,14-TRIHYDROXY-7',9-DIMETHOXY-4,10,23-TRIMETHYL-, (6R-(6-α,6A-β,7-β,13-β,14-β,16-α,20R*))- see EAE050
SPIROFULVIN see GKE000
SPIROGERMANIUM DIHYDROCHLORIDE see SLD800
SPIROGERMANIUM HYDROCHLORIDE see SLD800
SPIROHYDANTOIN MUSTARD see SLD900
SPIRO(ISOBENZOFURAN)-1(3H),9'-(9H)XANTHENE-3-ONE, 3',6'-DIHYDROXY-DISODIUM SALT see FEW000
SPIRO(ISOBENZOFURAN-1(3H),9'-(9H)XANTHEN)-3-ONE, 4',5'-DIBROMO-3',6'-DIHYDROXY- (9CI) see DDO200
SPIROLACTONE see AFJ500
SPIROLAKTON see AFJ500
SPIROLANG see AFJ500
SPIROMUSTINE see SLD900
SPIRO(NAPHTHALENE-2(1H),2'-OXIRANE)-3'-CARBOXALDEHYDE, 3,5,6,7,8,8a-HEXAHYDRO-7-ACETOXY-5,6-EPOXY-3',8,8a-TRIMETHYL-3-OXO- see POF800
SPIRONE see AFJ500
SPIRONOLACTONE see AFJ500
SPIRONOLACTONE A see AFJ500
SPIROOXIRANECYCLOHEXANE see SLD920
SPIROPENT see VHA350
SPIROPERIDOL see SLE500
SPIROPITAN see SLE500
SPIROPLATIN see SLE875
SPIROZID see ABX500
SPIZOFURONE see SLE880
SPLENDOR see GJU200

SPLIT LEAF PHILODENDRON see SLE890
SPLOTIN see EPD500
SPM 925 see MRA260
SPMF 4 see MCB050
SPMF 6 see MCB050
SPMF 7 see MCB050
SPOFADAZINE see AKO500
SPOFADRIZINE see PPP500
SPOLAPRET OS see MND550
SPONGIOFORT see PCU360
SPONGOADENOSINE see AEH100
SPONGOCYTIDINE see AQQ750
SPONGOCYTIDINE HYDROCHLORIDE see AQR000
SPONTOX see TAA100
SPOONWOOD IVY see MRU359
SPORAMIN see SBG500
SPORARICIN A see SLF500
SPORGON see IAL200
SPORILINE see TGB475
SPOR-KIL see ABU500
SPOROSTATIN see GKE000
SPORTAK see IAL200
SPORTAK α see IAL200
SPORTAK Δ see IAL200
SPORTAK PF see IAL200
SPOTRETE see TFS350
SPOTTED ALDER see WCB000
SPOTTED COWBANE see WAT325
SPOTTED HEMLOCK see PJJ300
SPOTTED PARSLEY see PJJ300
SPOTTON see FAQ900
SPP see AIF000
SPRACAL see ARB750
SPRAY-DERMIS see NGE500
SPRAY-FORAL see NGE500
SPRAY-HORMITE see SGH500
SPRAYSET MEKP see MKA500
SPRAY-TROL BRANCH RODEN-TROL see WAT200
SPRENGEL EXPLOSIVES see SLG500
SPRIDINE HYDROCHLORIDE see SLG600
SPRIDINE TRIHYDROCHLORIDE see SLG600
SPRINE/NO see SLG625
SPRING-BAK see DXD200
SPRINT see IAL200
SPRITZ-HORMIN/2,4-D see DAA800
SPRITZ-HORMIT see SGH500
SPROUT NIP see CKC000
SPROUT-NIP EC see CKC000
SPROUT-OFF see DAI800
SPROUT-OFF see ODE000
SPROUT/OFF see DMC600
SPROUT-STOP see DMC600
SPRUCE OIL see SLG650
SPS 600 see SMQ500
SPT 50 CPS see EHG100
SPUD-NIC see CKC000
SPUD-NIE see CKC000
SPULMAKO-LAX see PDO750
SPUR see VBU100
SPURGE see BRE500
SPURGE LAUREL see LAR500
SPURGE OLIVE see LAR500
SPUTOLYSIN see DXO800
SQ 1089 see HOO500
SQ 1156 see GGS000
SQ 1489 see TFS350
SQ 2113 see HOB500
SQ 2303 see CBK500
SQ 2321 see FNF000
SQ 3277 see HOC000
SQ 4609 see BDX090
SQ 5439 see PDV330
SQ 8388 see AQO000
SQ 9453 see DUD800
SQ 10,269 see CBQ625
SQ 10,643 see CMP900
SQ 11725 see CNR675
SQ 13396 see IFY000
SQ 14055 see TET800

SQ 14,225 see MCO750
SQ 15,659 see PPY250
SQ-15761 see SKM000
SQ 16,114 see PMI250
SQ 16360 see SHK000
SQ 16423 see MNV250
SQ 16496 see PMC700
SQ 16603 see FQU000
SQ 16819 see HKQ025
SQ-18,566 see HAF300
SQ 21065 see POJ500
SQ 21611 see BDX100
SQ 21977 see MPU000
SQ-21983 see IGA000
SQ 22451 see AMH000
SQ 22947 see DAR000
SQ 22947 see TET800
SQ 26,776 see ARX875
SQ 31000 see LGK500
SQUALANE see SLG700
SQUALEN see SLG800
SQUALENE see SLG800
(E,E,E,E)-SQUALENE see SLG800
trans-SQUALENE see SLG800
SQUALIDIN see IDG000
SQUALIDINE see IDG000
SQUARIC ACID see DMJ600
SQUAW ROOT see BMA150
SQUIBB see HNT500
SQUIBB 4918 see FMP000
SQUILL see RCF000
SQUILL see SBF500
SQUILL see SLH200
SQUIRREL FOOD see DAE100
SR 3 see HIT610
SR07 see AAJ600
SR 41 see HIT620
SR 73 see DFV600
SR-202 see CMU875
SR 206 see BKM250
SR 209 see TCE400
SR 213 see TDQ100
SR 220 see PMP800
SR 247 see DUL200
SR 256 see EHG030
SR 285 see TCS100
SR 339 see PER250
SR 351 see TLX175
SR406 see CBG000
SR 506 see IHX700
SR 1354 see NHH500
SR 2508 see HKW455
SR-52223 see IIE200
SR 720-22 see ZAK300
SR141716A see CCC290
SRA 5172 see DTQ400
SRA 7312 see DJY200
SRA 7847 see EIM000
SRA 12869 see IMF300
SR 506 (ACRYLATE) see IHX700
SRA GOLDEN YELLOW VIII see DUW500
SRC-7 see DWQ850
SRC-15 see BHJ625
SRC-16 see BLD325
SRC-226 see DSH800
SRC-II HEAVY DISTILLATE see CMY750
SRC-II, HEAVY DISTILLATE see SLH300
SR E2 see SMF000
SRG 95213 see DCQ700
SRI 702 see BFL125
SRI 753 see BRS500
SRI 759 see TFJ500
SRI 859 see MNA750
SRI 1354 see NHH500
SRI 1666 see DRN400
SRI 1720 see BIF750
SRI 1869 see NKL000
SRI 2200 see CGV250
SRI 2489 see IAN000
SRI 2619 see CPL750
SRI 2619 see FIJ000
SRI 2638 see CHE500

SRI 2656 see CFH750
SRI 5631-96 see FLZ080
SRI 10163-71 see HFT550
SRIF see SKS600
SRM 705 see SMQ500
SRM 706 see SMQ500
SRM 1475 see PJS750
SRM 1476 see PJS750
SS see CCU250
SS-094 see SKS299
SS 578 see DNF600
SS717 see NCP525
SS 2074 see ORU000
SS320A see HNV050
SS 11946 see IRU500
SS 30 (carbonate) see CAT775
SS 50 (carbonate) see CAT775
SSB₁ see TES800
SSB 100 see CAT775
SSH 43 see ISR200
SSH 0860 see AJR200
S.T. 37 see HFV500
ST 90 see SMQ500
ST 155 see CBF250
ST-155 see CMX760
ST 630 see HDB100
ST 1085 see MQT530
ST-1191 see EQO000
ST-1435 see HMB650
1-ST-2121 see SLI200
ST 720 (FRENCH) see CGB000
ST A 307 see TFK300
STABAXOL P see SLI250
STABICILLIN see PDT500
STABILAN see CMF400
STABILENE see BKA000
STABILENE see BRP250
STABILENE FLY REPELLENT see BRP250
STABILISATOR C see DWN800
STABILISATOR SCD see OAT000
STABILISATOR VH see PGP250
STABILIZATOR AR see PFT500
STABILIZER D-22 see DDV600
STABILIZER DLT see TFD500
STABILIZER I see PFJ760
STABILIZER SCD see OAT000
STABILIZER VH see PGP250
STABILLIN VK SYRUP 125 see PDT750
STABILLIN VK SYRUP 62.5 see PDT750
STABINEX NW 7PS see CAW850
STABINOL see CKK000
STABINOL see IIY100
STABISOL see SAI600
STABLE RED KB BASE see CLK225
STABOXOL 1 see BJE550
STAC see TLW500
STADADORM see AMX750
STADALAX see PPN100
STADERM see IAB000
STADOL see BPG325
STAFAC see VRA700
STAFAC see VRF000
STAFAST see NAK500
STA-FAST see TIX500
STAFF TREE see AHJ875
STAFLEN E 650 see PJS750
STAFLEX 500 see OEU000
STAFLEX DBM see DED600
STAFLEX DBP see DEH200
STAFLEX DBS see DEH600
STAFLEX DOP see DVL700
STAFLEX DOX see BJQ500
STAFLEX TOTM see TJR600
STAFOAM DO see OHU200
STA-FRESH 615 see SGM500
STAGGER WEED see LBF000
STAGNO (TETRACLORURO di) (ITALIAN)
see TGC250
STAINLESS STEEL 21-6-9 see IGL110
STALFEX DTDP see DXQ200
STALFLEX DOS see BJS250
STALLIMYCIN see SLI300

STALLIMYCIN HYDROCHLORIDE see
DXG600
STAM see DGI000
STAM M-4 see DGI000
STAM F 34 see DGI000
STAMINE see WAK000
STAM LV 10 see DGI000
STAMPEDE see DGI000
STAMPEDE 3E see DGI000
STAMPEN see DGE200
STAM SUPERNOX see DGI000
STAMYLAN 900 see PJS750
STAMYLAN 1000 see PJS750
STAMYLAN 1700 see PJS750
STAMYLAN 8200 see PJS750
STAMYLAN 8400 see PJS750
STANAPROL see DME500
STANCLERET 157 see DEJ100
STANDACOL CARMOISINE see HJF500
STANDACOL ORANGE G see HGC000
STANDACOL SUNSET YELLOW FCF see
FAG150
STANDAK see AFK000
STANDAMIDD LD see BKE500
STANDAMUL 1616 see HCP700
STANDAMUL 7061 see IKC050
STANDAMUL DIPA see DNL800
STANDAMUL O20 see PJW500
STANDAPOL 112 CONC see SIB600
STANDAPOL TLS 40 see SON000
STANDARD LEAD ARSENATE see LCK000
STANDARD LECTIN see SKW800
STANDOPAL see MCB050
STANEPHRIN see SPC500
STANGEN see WAK000
STANGEN MALEATE see DBM800
STAN-GUARD 156 see BKO250
STANILO see SKY500
STANILO see SLI325
STAN-MAG MAGNESIUM CARBONATE see
MAC650
STANNACYCLOPENT-3-ENE-2,5-DIONE
see SLI330
STANNANE, ((p-
ACETAMIDOBENZOYL)OXY)TRIPHENYL-
see TMV800
STANNANE, ACETOXYDIPHENYLETHYL-
see EIM100
STANNANE, ACETOXYTRICYCLOHEXYL-
see ABW600
STANNANE, ACETOXYTRIOCTYL- see
ABX150
STANNANE, ACETOXYTRIPENTYL- see
ABX175
STANNANE, ACETOXYTRIS(β,β-
DIMETHYLPHENETHYL)- see TMK200
STANNANE,
(ACETYLENECARBONYLOXY)TRIPHENY
L- see TMW600
STANNANE,
(ACETYLOXY)ETHYLDIPHENYL- see
EIM100
STANNANE, (ACETYLOXY)TRIPROPYL-
(9CI) see TNB000
STANNANE, (ACETYLOXY)TRIS(2-
METHYL-2-PHENYLPROPYL)- see TMK200
STANNANE, BIS((3-
CARBOXYACRYLOYL)OXY)DIOCTYL-,
DIDODECYL ESTER, (Z,Z)- see DVN100
STANNANE, BROMOTRIETHYL-,
compounded with 2-PIPECOLINE (1:1) see
TJU850
STANNANE, BUTYLDIETHYLIODO- see
BRA550
STANNANE, (BUTYLTHIO)TRIOCTYL- see
BSO200
STANNANE, (BUTYLTHIO)TRIPROPYL- see
TMY850
STANNANE, CHLOROTRIISOPROPYL- see
TKT800
STANNANE, CHLOROTRIOCTYL- see
TMO585

STANNANE, CHLOROTRIS(1-METHYLETHYL)- see TKT800
STANNANE, CYANATOTRIMETHYL- see TMI100
STANNANE, (CYANOACETOXY)TRIPHENYL- see TMV825
STANNANE, CYCLOHEXYLHYDROXYOXO- see MRH212
STANNANE, DIBUTYL- see DEI700
STANNANE, DIBUTYLBIS((1-OXOOCTADECYL)OXY)-(9CI) see DEJ250
STANNANE, DIBUTYLBIS(STEAROYLOXY)- see DEJ250
STANNANE, DICHLORODIDODECYL- see DHA600
STANNANE, DICHLORODIETHENYL- see DXR450
STANNANE, DICHLORODIMETHYL- see DUG825
STANNANE, DICHLORODIOCTYL- see DVN300
STANNANE, DICHLORODIVINYL- see DXR450
STANNANE, DICYCLOHEXYLHYDROXYPHENYL- see DGV670
STANNANE, DICYCLOHEXYLOXO- see DGV900
STANNANE, DIMETHYL(2,3-QUINOXALINYLDITHIO)- see QSJ800
STANNANE, ((DIMETHYLTHIOCARBAMOYL)THIO)TRIBUTYL- see TID150
STANNANE, DIOCTYLDICHLORO- see DVN300
STANNANE, DIOCTYLDIPHENYL- see DVK500
STANNANE, (ETHYLTHIO)TRIOCTYL- see EPR200
STANNANE, ETHYNYLENEBIS(CARBONYLOXY)BIS(TRIPHENYL- see BLS900
STANNANE, FLUOROTRIPHENYL- see TMV850
STANNANE, FLUOROTRIS(p-CHLOROPHENYL)- see TNG050
STANNANE, (ISOCYANATO)TRIBUTYL- see THY750
STANNANE, (ISOPROPYLSUCCINYLOXY)TRIBUTYL- see TIE600
STANNANE, MERCAPTOTRIPHENYL-, O,O-DIETHYL PHOSPHORODITHIOATE (8CI) see EFL600
STANNANE, (p-NITROPHENOXY)TRIBUTYL- see NII200
STANNANE, OXYBIS(DIBUTYL(2,4,5-TRICHLOROPHENOXY))- see OPE100
STANNANE, TETRAOCTYL- see TDY790
STANNANE, TETRAOCTYL- see TDY795
STANNANE, TRIBUTYL((4-CHLOROBUTYRYL)OXO)- see TID000
STANNANE, TRIBUTYLISOCYANATO- see THY750
STANNANE, TRIBUTYLSALICYLOYLOXY see SAM000
STANNANE, TRICHLOROOCTYL- see OGG000
STANNANE, TRIETHYL- see TJV050
STANNANE, TRIISOPROPYL(UNDECANOYLOXY)- see TKT850
STANNIC BROMIDE see TGB750
STANNIC CHLORIDE, anhydrous (DOT) see TGC250
STANNIC CHLORIDE, pentahydrate (DOT) see TGC282
STANNIC CHLORIDE PENTAHYDRATE see TGC282
STANNIC CITRIC ACID see SLI335

STANNIC DICHLORIDE DIIODIDE see TGC280
STANNIC IODIDE see TGD750
STANNIC PHOSPHIDE (DOT) see TGE500
STANNOCHLOR see TGC275
1H-STANNOLE-2,5-DIONE, 1,1-DIBUTYL- see SLI330
STANN OMF see DVK200
STANNOPLUS see DAE600
STANNORAM see DAE600
STANNOUS CHLORIDE DIHYDRATE see TGC275
STANNOUS CHLORIDE (FCC) see TGC000
STANNOUS CITRIC ACID see TGC285
STANNOUS DIBUTYLDITRIFLUOROACETATE see BLN750
STANNOUS FLUORIDE see TGD100
STANNOUS IODIDE see TGD500
STANNOUS OLEATE see OAZ000
STANNOUS OXALATE see TGE250
STANNOUS POTASSIUM TARTRATE see TGE750
STANNOUS STEARATE see SLI350
STANNOUS SULFATE see TGF010
STANNOXYL see TGE300
STANNPLOUS see DAE600
STANN RC 40F see DEJ100
STANOLONE see DME500
STANOMYCETIN see CDP250
STANOZOLOL see AOO401
STANSIN see SNN500
ST. ANTHONY'S TURNIP see FBS100
STANWHITE 500 see CAT775
STANZAMINE see POO750
STAPENOR see DSQ800
STAPENOR see MNV250
STAPHCILLIN V see MNV250
STAPHOBRISTOL-250 see SLJ000
STAPHYBIOTIC see SLJ000
STAPHYBIOTIC I see SLJ050
STAPHYLEX see CHJ000
STAPHYLOCOAGULASE see CMY725
STAPHYLOCOCCAL ENTEROTOXIN B see EAV025
STAPHYLOCOCCAL PHAGE LYSATE see SLJ100
STAPHYLOMYCIN see VRA700
STAPHYLOMYCIN see VRF000
STAPHYLOMYCINE see VRA700
STAPYOCINE see VRF000
STAR see GGA000
STARAGEL 90 see SLJ550
STARAGEL 90V see SLJ550
STARAMIC 747 see SLJ500
STAR ANISE OIL see AOU250
STAR CACTUS (HAWAII) see AGV875
STARCH see SLJ500
α-STARCH see SLJ500
STARCH, CORN see SLJ500
STARCH DIPHOSPHATE see SLJ550
STARCH DUST see SLJ500
STARCH GUM see DBD800
STARCH HYDROGEN PHOSPHATE see SLJ550
STARCH HYDROXYETHYL ETHER see HLB400
STARCH (OSHA) see SLJ500
STARCHWORT see JAJ000
STAR DUST see CNE750
STARFOL BS-100 see BSL600
STARFOL GMS 450 see OAV000
STARFOL GMS 600 see OAV000
STARFOL GMS 900 see OAV000
STARFOL IPM see IQN000
STARFOL IPP see IQW000
STAR HYACINTH see SLH200
STARIFEN see EOK000
STARLEX L see CAW850
STARLICIDE see CLK230
STAR-OF-BETHLEHEM see SLJ650
STARSOL No. 1 see AQQ500

STARTER DISTILLATE see SLJ700
STARVE-ACRE see FBS100
STA-RX 1500 see SLJ500
ST52-ASTA see DKA200
STATEX see CBT750
STATEX N 550 see CBT750
STATHION see PAK000
STATOBEX see DKE800
STATOCIN see CCK550
STATOMIN see WAK000
STATOMIN MALEATE see DBM800
STATYL see NCN600
STAUFFER N-2404 see CJD650
STAUFFER N-2596 see EHL670
STAUFFER N-3049 see EPY000
STAUFFER N-3051 see BSG000
STAUFFER N-4328 see EOO000
STAUFFER N-4446 see EMR100
STAUFFER N-4548 see MOB250
STAUFFER ASP-51 see TED500
STAUFFER B-10094 see CON300
STAUFFER B-11163 see DTQ800
STAUFFER CAPTAN see CBG000
STAUFFER MV242 see MLG250
STAUFFER MV-119A see DLK200
STAUFFER N 521 see DSB200
STAUFFER N 2790 see FMU045
STAUFFER R-1,303 see TNP250
STAUFFER R-1492 see MQH750
STAUFFER R 1504 see PHX250
STAUFFER R-1571 see PHL750
STAUFFER R 1608 see EIN500
STAUFFER R-1910 see EID500
STAUFFER R-2061 see PNF500
STAUFFER R-3413 see DTV400
STAUFFER R-4,572 see EKO500
STAUFFER R-6790 see MEV600
STAUFFER R 10044 see FLZ025
STAUFFER R 14487 see DJW890
STAUFFER R 14493 see DJW880
STAUFFER R 14789 see MEL525
STAUFFER R 15018 see EEQ200
STAUFFER R 15996 see DJW900
STAUFFER R-25788 see DBI300
STAUFFER R-3442-S see EMC000
STAURODERM see FMQ000
STAVELAN CINATY (CZECH) see TGE250
STAVELAN SODNY see SIY500
STAVINOR see LHQ100
STAVINOR 30 see CAX350
STAVINOR 40 see BAO825
STAVINOR 1300SN see DEJ100
STAVINOR SN 1300 see DEJ100
STAVUDINE see SLJ800
STAVUDINUM (INN-LATIN) see SLJ800
STAY-FLO see SHF500
STAYPRO WS 7 see HBO650
STAZEPIN see DCV200
STB see SKE000
ST. BENNET'S HERB see PJJ300
STCA see TII500
ST 600-CL see FMR080
ST 1396 CLORHIDRATO (SPANISH) see SBN475
ST-1512 DIHYDROCHLORIDE see HFG600
STEADFAST see CCP000
STEAPSIN see GGA800
STEARALKONIUM CHLORIDE see DTC600
STEARAMIDE see OAR000
STEARAMINE see OBC000
STEARANABOL see CLG900
STEARATE CHROMIC CHLORIDE see CMH300
STEARATO CHROMIC CHLORIDE see CMH300
STEARATO-CHROMIC CHLORIDE COMPLEX see CMH300
STEARATOCHROMIUM CHLORIDE see CMH300
STEAREX BEADS see SLK000
STEARIC ACID see SLK000

STEARIC ACID, ALUMINIUM SALT see AHA250
STEARIC ACID, ALUMINUM SALT see AHH825
STEARIC ACID, AMMONIUM SALT see ANU200
STEARIC ACID, BARIUM CADMIUM SALT (4:1:1) see BAI800
STEARIC ACID, BARIUM SALT see BAO825
STEARIC ACID, CADMIUM SALT see OAT000
STEARIC ACID with CYCLOHEXYLAMINE (1:1) see CPH500
STEARIC ACID-2,3-EPOXYPROPYL ESTER see SLK500
STEARIC ACID, ETHYL ESTER see EPF700
STEARIC ACID, 2-ETHYLHEXYL ESTER see OFU300
STEARIC ACID, ISOBUTYL ESTER see IJN100
STEARIC ACID, ISOHEXADECYL ESTER see IKC050
STEARIC ACID, ISOPROPYL ESTER see IRL100
STEARIC ACID, LEAD SALT see LDX000
STEARIC ACID, LITHIUM SALT see LHQ100
STEARIC ACID, MONOESTER with ETHYLENE GLYCOL see EJM500
STEARIC ACID, MONOESTER with GLYCEROL see OAV000
STEARIC ACID, MONOESTER with 1,2-PROPANEDIOL see SLL000
STEARIC ACID POTASSIUM SALT. see PLS775
STEARIC ACID, SODIUM SALT see SJV500
STEARIC ACID, TETRADECYL ESTER see TCB100
STEARIC ACID, ZINC SALT see ZMS000
STEARIC MONOGLYCERIDE see OAV000
STEARIX ORANGE see PEJ500
STEARIX RED 4B see SBC500
STEARIX RED 4S see SBC500
STEARIX SCARLET see OHI200
STEAROL see OAX000
γ-STEAROLACTONE see SLL400
STEARONITRILE see SLL500
STEAROPHANIC ACID see SLK000
1-STEAROYLAZIRIDINE see SLM000
STEAROYL ETHYLENEIMINE see SLM000
STEAROYL LYSOLECITHIN see SLM100
STEAROYL LYSOPHOSPHATIDYLCHOLINE see SLM100
1-STEAROYLLYSOPHOSPHATIDYLCHOLINE see SLM100
STEAROYL PROPLENE GLYCOL HYDROGEN SUCCINATE see SNG600
STEAR YELLOW JB see DOT300
STEARYL ALCOHOL see OAX000
STEARYL ALCOHOL EO (10) see SLM500
STEARYL ALCOHOL EO (20) see SLN000
STEARYL ALCOHOL condensed with 10 moles ETHYLENE OXIDE see SLM500
STEARYL ALCOHOL condensed with 20 moles ETHYLENE OXIDE see SLN000
STEARYLAMINE see OBC000
n-STEARYLAMINE see OBC000
STEARYL CITRATE see SLN100
STEARYLDIMETHYLAMINE see DTC400
STEARYLDIMETHYLBENZYLAMMONIUM CHLORIDE see DTC600
STEARYL-2-LACTYLIC ACID see SLN175
STEARYL MONOGLYCERIDYL CITRATE see SLN200
STEARYLTRIMETHYLAMMONIUM CHLORIDE see TLW500
STEAWHITE see TAB750
STEBAC see DTC600
STECKAPFUL (GERMAN) see SLV500
STECKER ASC-4 see THW750
STECLIN see TBX000

STECLIN HYDROCHLORIDE see TBX250
STEDIRIL see NNL500
STEDIRIL D see NNL500
STEEL 21-6-9 see IGL110
STEEL, ALUMINUM NICKEL see SLN300
STEINAMID DL 203 S see BKE500
STEINAMID DO 280SE see OHU200
STEINAPHAT EAK 8190 see PKE850
STEINAPOL NOS 25 see SNY100
STEINAPOL TLS 40 see SON000
STEINBUHL YELLOW see BAK250
STEINBUHL YELLOW see CAP750
STELADONE see CDS750
STELAZINE see TKE500
STELAZINE see TKK250
STELAZINE DIHYDROCHLORIDE see TKK250
STELLAMINE see DJS200
STELLARID see POB500
STELLAZINE see TKE500
STELLITE see VSK000
STELLITE 8 see CNA750
STELLITE 21 see CNA750
STELLITE 23 see CNA750
STELLITE 25 see CNA750
STELLITE 27 see CNA750
STELLITE 30 see CNA750
STELLITE 31 see CNA750
STELLITE 31 see CNB825
STELLITE 36 see CNA750
STELLITE 8A see CNA750
STELLITE C see CNA750
STELLITE X40 see CNB825
STELLITE 31 X 40 see CNB825
STELLON PINK see PKB500
STEMAX see PAL600
STEMETIL see PMF500
STEMETIL DIMALEATE see PMF250
STEMONE see EMP550
STEMPOR see MHC750
STENANDIOL see AOO410
STENDOMYCIN A see SLN509
STENDOMYCIN SALICLATE see SLO000
STENEDIOL see AOO475
STENIBELL see AOO475
STENOLANA BRILLIANT BLUE BL see CMM092
STENOLON see DAL300
STENOLON see MPN500
STENOLONE see DAL300
STENOSINE see DXE600
STENOSTERONE see AOO475
STENOVASAN see TEP500
STENTAL EXTENTABS see EOK000
STEPAN D-50 see IQN000
STEPAN D-70 see IQW000
STEPANATE X see XJJ010
STEPAN C40 see MLC800
STEPANOL WAQ see SIB600
STEPANOL WAT see SON000
STEPHANIA HERNANDIFOLIA Walp., extract see SLO100
STERABOL see CLG900
STERAFFINE see OAX000
STERAL see HCL000
STERAMIDE see SNQ710
STERANABOL see CLG900
STERANDRYL see TBG000
STERANE see PMA000
STERAQ see DAQ800
STERASKIN see HCL000
STERAZINE see PPP500
STERCORIN see DKW000
STERCULIA FOETIDA OIL see SLO450
STERCULIA GUM see KBK000
STERCULIA GUM see SLO500
STERICOL see XKA000
STERIDO see BIM250
STERIGMATOCYSTIN see SLP000
STERIGMATOCYSTIN, o-METHYL- see MPJ050

STERILE TETRACAINE HYDROCHLORIDE see TBN000
STERILIZING GAS ETHYLENE OXIDE 100% see EJN500
STERINOL see BEO000
STERINOLU (POLISH) see BEO000
STERISEAL LIQUID #40 see SGM500
STERLING see CBT750
STERLING see SFT000
STERLING N 765 see CBT750
STERLING NS see CBT750
STERLING SO 1 see CBT750
STERLING WAQ-COSMETIC see SIB600
STERLING WAT see SON000
STERNITE 30 see SMQ500
STERNITE ST 30VL see SMQ500
STEROGENOL see HCP800
STEROGYL see VSZ100
STEROIDS see SLP199
STEROLAMIDE see TKP500
STEROLONE see PMA000
STERONYL see MPN500
STEROX DJ SULFACTANT see TAX500
STESIL see CKE750
STESOLID see DCK759
STEVIA REBAUDIANA Bertoni, extract see SLP350
ST 1396 HYDROCHLORID (GERMAN) see SBN475
ST 1396 HYDROCHLORIDE see SBN475
STIBACETIN see SLP500
STIBAMINE GLUCOSIDE see NCL000
STIBANATE see AQH800
STIBANATE see AQI250
STIBANILIC ACID see SLP600
STIBANOSE see AQH800
STIBANOSE see AQI250
STIBATIN see AQH800
STIBATIN see AQI250
STIBENYL see SLP500
STIBIC ANHYDRIDE see AQF750
2,2′,2″-(STIBILIDYNETRIS(THIO))TRIS-BUTANEDIOIC ACID HEXALITHIUM SALT see LGU000
STIBILIUM see DKA600
STIBINE see SLQ000
STIBINE OXIDE, TRIPHENYL- see TMQ550
STIBINE SULFIDE, TRIPHENYL- see TMQ600
STIBINE, TRICHLORO- see AQC500
STIBINE, TRIFLUORO- (9CI) see AQE000
STIBINE, TRIS(DODECYLTHIO)- see TNH850
STIBINOL see AQH800
STIBINOL see AQI250
STIBIUM see AQB750
STIBNAL see AQI750
STIBNITE see SLQ100
STIBOCAPTATE see AQD750
((2-STIBONOPHENYL)THIO)ACETIC ACID see CCH250
((2-STIBONOPHENYL)THIO)ACETIC ACID DIETHANOLAMINE SALT see SLQ500
STIBOPHEN see AQH500
STIBOSAMIN see SLP600
STIBSOL see SEU000
STIBUNAL see AQI750
STICKMONOXYD (GERMAN) see NEG100
STICKSTOFFDIOXID (GERMAN) see NGR200
STICKSTOFFLOST see BIE500
STICKSTOFFWASSERSTOFFSAEURE (GERMAN) see HHG500
STIEDEX see DBA875
STIGMANOL BROMIDE see POD000
STIGMANOL METHYL SULFATE see DQY909
5-α-STIGMASTANE-3-β,5,6-β-TRIOL 3-MONOBENZOATE see SLQ625
STIGMAST-5-EN-3-β-OL see SDZ350
STIGMAST-5-EN-3-OL, (3-β)- see SDZ350
Δ5-STIGMASTEN-3-β-OL see SDZ350

STIGMAST-5-EN-3-β-OL, HYDROGEN SULFATE see SDZ370
STIGMAST-5-EN-3-OL, HYDROGEN SULFATE, (3-β)- see SDZ370
STIGMATELLIN see SLQ650
STIGMOSAN BROMIDE see POD000
STIGMOSAN METHYL SULFATE see DQY909
STIK see NAK500
STIKSTOFDIOXYDE (DUTCH) see NGR500
STIL see DKA600
STILALGIN see GGS000
STILBAMIDINE DIHYDROCHLORIDE see SLS500
STILBEN (GERMAN) see SLR000
4-STILBENAMINE see SLQ900
4-N-STILBENAMINE see SLQ900
trans-4-N-STILBENAMINE see AMO000
STILBENE see SLR000
STILBENE 3 see TGE150
STILBENE, (E)- see SLR100
(E)-STILBENE see SLR100
trans-STILBENE see SLR100
trans-4-STILBENE see AMO000
STILBENE, 4-(BENZYLOXY)-2'-FLUORO-α-PHENYL- see BFC400
STILBENE, α'-BROMO-α-PHENYL- see BOK500
α-STILBENECARBONITRILE see DVX600
4,4'-STILBENEDIAMINE see SLR500
4,4'-STILBENEDICARBOXAMIDINE see SLS000
4,4'-STILBENEDICARBOXAMIDINE, DIHYDROCHLORIDE see SLS500
STILBENE, α-α'-DIETHYL-4,4'-DIMETHYL-, (E)- see DRK500
STILBENE, 4,4'-DINITRO-, (E)- see EEE300
4,4'-STILBENEDIOL, α-ETHYL- see EPF800
2,2'-STILBENEDISULFONIC ACID,-4,4'-BIS((4-ANILINO-6-ETHYLAMINO-s-TRIAZIN-2-YL)AMINO)-, 2NA see DXB500
2,2'-STILBENEDISULFONIC ACID, 4,4'-BIS((4-ANILINO-6-((2-HYDROXYETHYL)AMINO)-s-TRIAZIN-2-YL) AMINO)-, DISODIUM SALT see CMP100
2,2'-STILBENEDISULFONIC ACID, 4,4'-BIS((4-ANILINO-6-((2-HYDROXYETHYL)METHYLAMINO)-s-TRIAZIN- 2-YL)AMINO)-, DISODIUM SALT see TGE100
2,2'-STILBENEDISULFONIC ACID, 4,4'-BIS((4-ANILINO-6-MORPHOLINO-s-TRIAZIN-2-YL)AMINO)-, DISODIUM SALT see CMP200
2,2'-STILBENEDISULFONIC ACID, 4,4'-BIS((4,6-DIANILINO-s-TRIAZIN-2-YL)AMINO)-, DISODIUM SALT see DXB450
2,2'-STILBENEDISULFONIC ACID, 4,4'-BIS((4-(2,5-DISULFOANILINO)-6-(DIETHYLAMINO)-s-TRIAZIN-2-YL)AMINO)-, HEXASODIUM SALT see TGE152
2,2'-STILBENEDISULFONIC ACID, 4,4'-BIS(4-PHENYL-1,2,3-TRIAZOL-2-YL), DIPOTASSIUM SALT see BLG100
2,2'-STILBENEDISULFONIC ACID, 4,4'-BIS(4-PHENYL-2H-1,2,3-TRIAZOL-2-YL)-, DIPOTASSIUM SALT see BLG100
2,2'-STILBENEDISULFONIC ACID, 4,4'-DIAMINO- see FCA100
2,2'-STILBENEDISULFONIC ACID, 4,4'-DIAMINO-, DISODIUM SALT see FCA200
STILBENE, 4-NITRO-, (E)- see NMC050
2-STILBENESULFONIC ACID, 4-(7-SULFO-2H-NAPHTHO(1,2-d)TRIAZOL-2-YL)-, DISODIUM SALT see DXG025
4-STILBENOL, α-α'-DIETHYL-4'-METHOXY-, (E)-(8CI) see EMI525
4-STILBENYL-N,N-DIETHYLAMINE see DKA000
STILBENYL-N,N-DIMETHYLAMINE see DUB800

STILBESTROL see DKA600
STILBESTROL DIETHYL DIPROPIONATE see DKB000
STILBESTROL DIMETHYL ETHER see DJB200
STILBESTROL DIPHOSPHATE see DKA200
STILBESTROL DIPROPIONATE see DKB000
STILBESTROL and PRANONE see EEI025
STILBESTROL PROPIONATE see DKB000
STILBESTRONATE see DKB000
STILBESTRONE see DKA600
STILBETIN see DKA600
STILBIOCINA see SMB000
STILBOEFRAL see DKA600
STILBOESTROFORM see DKA600
STILBOESTROL see DKA600
STILBOESTROL DIPROPIONATE see DKB000
STILBOFAX see DKB000
STILBOFOLLIN see DKA600
STILBOL see DKA600
STILBRON see SBG500
STILKAP see DKA600
STILNY see CGA500
STILON see PJY500
STILPHOSTROL see DKA200
STIL-ROL see DKA600
STILRONATE see DKB000
STIMAMIZOL HYDROCHLORIDE see LFA020
STIMATONE see FMS875
STIMINOL see DJS200
STIMULAN see AOB250
STIMULEX see DBA800
STIMULEX see DBB000
STIMULEXIN see SLU000
STIMULIN see DJS200
STIMULINA see GFO050
STIMULIN RW 3 see SMP450
STINERVAL see PFC500
STINERVAL see PFC750
STINK DAMP see HIC500
STINKING NIGHTSHADE see HAQ100
STINKING WEED see CNG825
STINKING WILLIE see RBA400
STIPEND see CMA100
STIPINE see SEH000
STIPOLAC see TDE750
STIPTANON see EDU500
STIRAMATO see PFJ000
STIROFOS see RAF100
STIROLO (ITALIAN) see SMQ000
STIROPHOS see RAF100
ST. JOHN'S BREAD see LIA000
ST 30UL see SMQ500
STOCKADE see RLF350
STODDARD SOLVENT see SLU500
STOIKON see BCA500
STOMACAIN see DTL200
STOMP see DRN200
STONE RED see IHC450
STOPAETHYL see DXH250
STOP-DROP see NAK500
STOPETHYL see DXH250
STOPETYL see DXH250
STOPGERME-S see CKC000
STOP-MOLD see SFV500
STOPMOLD B see BGJ750
STOP-SCALD see SAV000
STOPSPOT see PFM500
STOPTON ALBUM see SNM500
STOVAINE see AOM000
STOVARSAL see ABX500
STOVARSOL see ABX500
STOVARSOLAN see ABX500
STOXIL see DAS000
STP see SLU600
STP (HALLUCINOGEN) see SLU600
STPHCILLIN A BANYU see DGE200
STPP see SKN000
STR see SMD000
STRABOLENE see DYF450

STRAIGHT-CHAIN ALKYL BENZENE SULFONATE see LGF825
STRAIGHT-RUN KEROSENE see KEK000
STRAIGHT RUN MIDDLE DISTILLATE see GBW000
STRAMONA (ITALIAN) see SLV500
STRAMONIUM see SLV500
STRATHION see PAK000
STRAWBERRY ALDEHYDE see ENC000
STRAWBERRY BUSH see CCK675
STRAWBERRY RED A GEIGY see FMU080
STRAWBERRY TOMATO see JBS100
STRAW OIL see LIK000
STRAZINE see ARQ725
STREL see DGI000
STREPAMIDE see SNM500
STREPCEN see SLW500
STREPCIN see SLY500
STREP-GRAN see SLY500
STREPSULFAT see SLY500
STREPTAGOL see SNM500
d-STREPTAMINE, o-6-AMINO-6-DEOXY-α-d-GLUCOPYRANOSYL-(1-4)-o-(3-DEOXY-4-C-METHYL-3-(METHYLAMINO)-β-l-ARABINOPYRANOSYL-(1-6))-N¹)-(3-AMINO-2-HYDROXY-1-OXOPROPYL)-2-DEOXY-, (S)-, SULFATE (1:2) (SALT) see IHN300
d-STREPTAMINE, O-3-AMINO-3-DEOXY-α-d-GLUCOPYRANOSYL-(1-6)-O-(2,6-DIAMINO-2,3,4,6-TETRADEOXY-α-d-erythro-HEXOPYRANOSYL-(1-4))-N'-(4-AMINO-2-HYDROXY-1-OXOBUTYL)-2-DEOXY-, (S)-, SULFATE (salt) see HAI600
STREPTASE see SLW450
STREPTOCLASE see SNM500
STREPTOCOCCAL FIBRINOLYSIN see SLW450
STREPTOGRAMIN see VRF000
STREPTOGRAMIN B see VRA700
STREPTOKINASE see SLW450
STREPTOKINASE (ENZYME-ACTIVATING) see SLW450
STREPTOL see SNM500
STREPTOLYDIGIN see SLW475
STREPTOMICINA (ITALIAN) see SLW500
STREPTOMYCES GRISEUS PROTEASE see PMJ100
STREPTOMYCES GRISEUS PROTEINASE see PMJ100
STREPTOMYCES PEUCETIUS see DAC000
STREPTOMYCIN see SLW500
STREPTOMYCIN A see SLW500
STREPTOMYCIN C see SLX500
STREPTOMYCIN CALCIUM CHLORIDE see SLY000
STREPTOMYCIN compounded with CALCIUM CHLORIDE (1:1) see SLY000
STREPTOMYCIN and DIHYDROSTREPTOMYCIN see SLY200
STREPTOMYCINE see SLW500
STREPTOMYCIN SESQUISULFATE see SLY500
STREPTOMYCIN SULFATE see SLY500
STREPTOMYCIN, SULFATE (1:3) SALT see SLZ000
STREPTOMYCIN SULPHATE see SLZ000
STREPTOMYCIN SULPHATE B.P. see SLY500
STREPTOMYCINUM see SLW500
STREPTOMYZIN (GERMAN) see SLW500
STREPTONIGRAN see SMA000
STREPTONIGRIN see SMA000
STREPTONIGRIN METHYL ESTER see SMA500
STREPTONIVICIN see SMB000
STREPTOREX see SLY500
STREPTOSIL see SNM500
STREPTOSILTHIAZOLE see TEX250
STREPTOTHRICIN B see RAG400
STREPTOTHRICIN F see RAG300
STREPTOTHRICIN VI see RAG300

STREPTOVARICIN C see SMB850
STREPTOVIRUDIN see SMC375
STREPTOVITACIN A see SMC500
STREPTOVITACIN E 73 see ABN000
STREPTOZOCIN see SMD000
STREPTOZONE see SNM500
STREPTOZOTICIN see SMD000
STREPTROCIDE see SNM500
STREPVET see SLY500
STRESNIL see FLU000
STRESSON see BON400
STREUNEX see BBQ500
STRIADYNE see ARQ500
STRICNINA (ITALIAN) see SMN500
STRICYLON see NCW000
STRIPED ALDER see WCB000
STROBANE see MIH275
STROBANE see PJQ250
STROBANE see TBC500
STROBANE-T-90 see CDV100
STRONCYLATE see SML500
STRONTIUM see SMD500
STRONTIUM ACETATE see SME000
STRONTIUM ACETYLIDE see SME100
STRONTIUM ARSENITE see SME500
STRONTIUM ARSENITE, solid (DOT) see
SME500
STRONTIUM BROMIDE see SMF000
STRONTIUM CHLORATE see SMF500
STRONTIUM CHLORATE, solid or solution
(DOT) see SMF500
STRONTIUM CHLORIDE see SMG500
STRONTIUM CHLORIDE, HEXAHYDRATE
see SMH525
STRONTIUM CHROMATE (1:1) see SMH000
STRONTIUM CHROMATE (VI) see SMH000
STRONTIUM CHROMATE 12170 see SMH000
STRONTIUM COMPOUNDS see SMH500
STRONTIUM DICHLORIDE
HEXAHYDRATE see SMH525
STRONTIUM FLUOBORATE see SMI000
STRONTIUM FLUORIDE see SMI500
STRONTIUM FLUOSILICATE see SMJ000
STRONTIUM IODIDE see SMJ500
STRONTIUM MONOSULFIDE see SMM000
STRONTIUM NITRATE (DOT) see SMK000
STRONTIUM(II) NITRATE (1:2) see SMK000
STRONTIUM PEROXIDE see SMK500
STRONTIUM PHOSPHIDE see SML000
STRONTIUM SALICYLATE see SML500
STRONTIUM SULFIDE see SMM000
STRONTIUM SULPHIDE see SMM000
STRONTIUM YELLOW see SMH000
STROPHANDIOL-d-CYMAROSID see
SMM100
STROPHANTHIDIN see SMM500
STROPHANTHIDIN-d-CYMAROSID
(GERMAN) see CQH750
STROPHANTHIDIN-3-(6-DEOXY-α-l-
MANNOPYRANOSIDE) see CNH780
STROPHANTHIDIN-β-d-DIGITOXOSID
(GERMAN) see HAN800
STROPHANTHIDINE see SMM500
k-STROPHANTHIDINE see SMM500
STROPHANTHIDIN-GLUCOCYMAROSID
(GERMAN) see SMN002
STROPHANTHIDIN-d-GLUCOSE see
SMN200
STROPHANTHIDIN, 3-β-d-GLUCOSIDE see
SMN200
STROPHANTHIDIN-α-l-RHAMNOSIDE see
CNH780
α-l-STROPHANTHIDOL-3,6-DEOXY-
MANNOPYRANOSIDE see CNH785
STROPHANTHIN see SMN000
k-STROPHANTHIN see SMN000
β-k-STROPHANTHIN see SMN002
STROPHANTHIN G see OKS000
STROPHANTHIN K see SMN002
STROPHANTHIN K (crystalline) see SMN002
β-STROPHANTHOBIOSIDE,
STROPHANTHIDIN-3 see SMN002

STROPHANTHUS GRATUS Franch., leaf and
stem bark extract see SMN100
STROPHANTHYL β-d-GLUCOSIDE see
SMN200
STROPHOPERM see OKS000
STROSPESID (GERMAN) see SMN275
STROSPESIDE see SMN275
STROSPESIDE-16-FORMATE see VIZ200
STROSPESIDE TRIFORMATE see TKL175
STRUMACIL see MPW500
STRUMAZOLE see MCO500
STRYCHININE SULFATE see SMP000
STRYCHNIDIN-10-ONE see SMN500
STRYCHNIDIN-10-ONE, 2,3-DIMETHOXY-
(9CI) see BOL750
STRYCHNIDIN-10-ONE, SULFATE (2:1) see
SMP000
STRYCHNIN (GERMAN) see SMN500
STRYCHNINE see NOG800
STRYCHNINE see SMN500
STRYCHNINE, solid and liquid (DOT) see
SMN500
STRYCHNINE, 2,3-DIMETHOXY- see
BOL750
STRYCHNINE MONONITRATE see SMO000
STRYCHNINE NITRATE see SMO000
STRYCHNINE SALT (solid) see SMO500
STRYCHNINE SALTS (DOT) see SMO500
STRYCHNINE SULFATE (2:1) see SMP000
STRYCHNOS see SMN500
STRYCHNOS NUX-VOMICA see NOG800
STRYCIN see SLZ000
STRYON 686 see SMQ500
STRYPHNON see MGC350
STRYPHNONE see MGC350
STRYPTIRENAL see VGP000
STRZ see SMD000
STS see EMT500
STS see SIO000
STS 153 see PEE600
STS 557 see SMP400
ST 5066/S (GERMAN) see AMU750
ST-1512 SULFATE see HFG650
STUDAFLUOR see SHF500
STUGERON see CMR100
STUNTMAN see DMC500
STUPENONE see DLX400
STURCAL D see CAT775
STUTGERON see CMR100
STUTGIN see CMR100
STX DIHYDROCHLORIDE see SBA500
STYLOMYCIN see AEI000
STYLOMYCIN see POK250
STYMULEN see SMP450
STYPHNIC ACID see SMP500
STYPHNONE see MGC350
STYPNON see MGC350
STYPTIC WEED see CNG825
STYRACIN see CMQ850
STYRAFOIL see SMQ500
STYRAGEL see SMQ500
STYRALLYL ACETATE see SMP600
STYRALLYL ALCOHOL see PDE000
STYRALLYL PROPIONATE see PFR000
STYRALYL ALCOHOL see PDE000
STYRAMATE see PFJ000
STYREEN (DUTCH) see SMQ000
STYREN (CZECH) see SMQ000
STYREN-ACRYLONITRILEPOLYMER see
ADY500
STYRENATED DIPHENYLAMINE see
SMP700
STYRENE see SMQ000
STYRENE-ACRYLONITRILE COPOLYMER
see ADY500
STYRENE, β-BROMO-β-NITROSO- see
BNT600
STYRENE-BUTADIENE COPOLYMER see
SMR000
STYRENE-1,3-BUTADIENE COPOLYMER
see SMR000

STYRENE-BUTADIENE POLYMER see
SMR000
STYRENE, CHLORO- see CLE600
STYRENE EPOXIDE see EBR000
STYRENE GLYCOL see SMQ100
STYRENE and MALEIC ANHYDRIDE,
alternate co-polymer see SEA500
STYRENE-MALEIC ANHYDRIDE
COPOLYMER see SEA500
STYRENE-MALEIC ANHYDRIDE
POLYMER see SEA500
STYRENE-MALEIC ANHYDRIDE RESIN see
SEA500
STYRENE, m-METHYL- see VQK670
STYRENE, o-METHYL- see VQK660
STYRENE, p-METHYL- see VQK700
STYRENE, 3,4-METHYLENEDIOXY-β-
NITRO- see MJR800
STYRENE MONOMER (ACGIH) see SMQ000
STYRENE MONOMER, inhibited (DOT) see
SMQ000
STYRENE, β-NITRO- see NMC100
STYRENE, OCTACHLORO- see OAP050
STYRENE OXIDE see EBR000
STYRENE-7,8-OXIDE see EBR000
STYRENE POLYMER see SMQ500
STYRENE POLYMER with 1,3-BUTADIENE
see SMR000
STYRENE POLYMERS see SMQ500
STYRENE, α-β,β-TRIFLUORO- see TKH310
STYRENE-VINYLPYRROLIDINONE
POLYMER see VQK595
STYRENE-VINYLPYRROLIDONE
COPOLYMER see VQK595
STYRENE-N-VINYLPYRROLIDONE
POLYMER see VQK595
STYREX C see SMQ500
STYROCELL PM see SMQ500
STYROFAN 2D see SMQ500
STYROFLEX see SMQ500
STYROFOAM see SMQ500
STYROL (GERMAN) see SMQ000
STYROLE see SMQ000
STYROLENE see SMQ000
STYROLFENOL see PFD400
STYROLUX see SMQ500
STYROLYL ACETATE see SMP600
STYROLYL ALCOHOL see SMQ100
STYROLYL PROPIONATE see PFR000
STYROMAL see SEA500
STYROMAL 5 see SEA500
STYROMAL 30 see SEA500
STYRON see SMQ000
STYRON see SMQ500
STYRON 475 see SMQ500
STYRON 492 see SMQ500
STYRON 666 see SMQ500
STYRON 678 see SMQ500
STYRON 679 see SMQ500
STYRON 683 see SMQ500
STYRON 685 see SMQ500
STYRON 690 see SMQ500
STYRON 440A see SMQ500
STYRON 470A see SMQ500
STYRON 475D see SMQ500
STYRON 69021 see SMQ500
STYRONE see CMQ740
STYRON GP see SMQ500
STYRON 666K27 see SMQ500
STYRON PS 3 see SMQ500
STYRON T 679 see SMQ500
STYRON 666U see SMQ500
STYRON 666V see SMQ500
STYROPIAN see SMQ500
STYROPIAN FH 105 see SMQ500
STYROPOL HT 500 see SMQ500
STYROPOL IBE see SMQ500
STYROPOL JQ 300 see SMQ500
STYROPOL KA see SMQ500
STYROPOR see SMQ000
STYROPOR see SMQ500
STYRYL 430 see AMO250

trans-4'-STYRYLACETANILIDE see SMR500
ar-STYRYLACETOPHENONE see MPL000
p-STYRYLANILINE see SLQ900
6-STYRYL-BENZO(a)PYRENE see SMS000
5-STYRYL-3,4-BENZOPYRENE see SMS000
STYRYL CARBINOL see CMQ740
STYRYL METHYL KETONE see SMS500
STYRYL OXIDE see EBR000
N-(p-STYRYLPHENYL)ACETOHYDROXAMIC
ACETATE see ABW500
N-(p-STYRYLPHENYL)ACETOHYDROXAMIC
ACID see SMT000
trans-N-(p-STYRYLPHENYL)ACETOHYDROXAMIC
ACID see SMT500
N-(p-STYRYLPHENYL)ACETOHYDROXAMIC
ACID ACETATE see ABW500
trans-N-(p-STYRYLPHENYL)ACETOHYDROXAMIC
ACID, COPPER(2+) COMPLEX see SMU000
(E)-N-(p-STYRYLPHENYL)HYDROXYLAMINE see
HJA000
trans-N-(4-STYRYLPHENYL)HYDROXYLAMINE see
HJA000
trans-N-(p-STYRYLPHENYL)HYDROXYLAMINE see
HJA000
STYRYLPYRIDINIUM CHLORIDE,
DIETHYLCARBAMAZINE see SMU100
STZ see SMD000
SU-88 see IMS300
SU 2000 see CBT500
SU 3088 see CDY000
SU 3118 see RCA200
SU-4885 see MCJ370
SU 5864 see GKQ000
SU-5864 see GKS000
SU 5879 see CFY000
SU 8341 see CPR750
SU 9064 see MDW100
SU-9064 see MQR200
SU-10568 see DRC600
SU 11927 see CCP750
SU-13320 see TCN250
SU-13437 see MCB500
SU 21524 see PJA220
SUANOVIL see SLC300
SUAVITIL see BCA000
SUBACETATE LEAD see LCH000
SUBAMYCIN see TBX250
SUBARI see HGC500
SUBCHLORIDE of MERCURY see MCW000
SUBDUE see MDM100
SUBDUE 2E see MDM100
SUBDUE 5SP see MDM100
SUBERANE see COX500
SUBERON see SMV000
SUBERONE see SMV000
SUBERONE OXIME see SMV500
SUBERONISOXIM (GERMAN) see SMV500
SUBERONITRILE see OCW050
4,4'-(SUBEROYLBIS(IMINO-p-
PHENYLENEIMINO))BIS(1-
ETHYLPYRIDINIUM) DIBROMIDE see
SMW000
4,4'-(SUBEROYLBIS(IMINO-p-
PHENYLENEIMINO))BIS(1-
METHYLPYRIDINIUM)DIBROMIDE see
PPD500
SUBICARD see PBC250
SUBITEX see BRE500
SUBITOL see IAD000
SUBLIMAT (CZECH) see MCY475
SUBLIMAZE see PDW750
SUBLIMAZE CITRATE see PDW750
SUBLIMED SULFUR see SOD500
SUBLINGULA see HAQ500

SU-BRONTINE see EAG100
SUBSTANCE DS see AJX500
SUBSTANCE F see MIW500
SUBSTANCE H 20 see MFG600
SUBSTANCE H 36 see MEJ775
SUBSTANCE II see AFG750
SUBSTANZ DS see AJX500
SUBSTANZ NR. 2135 see CKE250
SUBSTANZ NR. 1602 (GERMAN) see CJR959
SUBSTANZ NR. 1766 (GERMAN) see CIS000
SUBSTANZ NR. 1925 (GERMAN) see CKE000
SUBSTANZ NR. 1934 (GERMAN) see DSQ600
SUBSTERINA see NMV300
SUBTIGEN see SMW400
SUBTILIN see SMW500
SUBTILISIN CARLSBURG see BAC000
SUBTILISIN (9CI, ACGIH) see BAC000
SUBTILISIN NOVO see BAC000
SUBTILISINS (ACGIH) see BAB750
SUBTILISINS BPN see BAB750
SUBTILOPEPTIDASE A see BAC000
SUBTILOPEPTIDASE B see BAC000
SUBTILOPEPTIDASE BPN' see BAC000
SUBTILOPEPTIDASE C see BAC000
SUBTOSAN see PKQ250
SUCARYL see CPQ625
SUCARYL ACID see CPQ625
SUCARYL CALCIUM see CAR000
SUCARYL SODIUM see SGC000
SUCCARIL see SGC000
SUCCARIL see SJN700
SUCCICURAN see HLC500
SUCCIMAL see ENG500
SUCCIMER see DNV800
SUCCIMITIN see ENG500
SUCCINALDEHYDE DISODIUM
BISULFITE see SMX000
SUCCINAMIC ACID, N,N-DIETHYL-,
PROPYL ESTER see PNH800
SUCCINAMIC ACID, N,N-DIPROPYL-,
ETHYL ESTER see DWV600
SUCCINATE OF 8-OXYQUINOLINE see
HOE150
SUCCINATE SODIQUE de 21-
HYDROXYPREGNANDIONE (FRENCH) see
VJZ000
SUCCINATO ACIDO DI 1-p-
CLOROFENILPENTILE (ITALIAN) see
CKI000
SUCCINATO de CLORANFENICOL
(SPANISH) see CDP725
SUCCINBROMIMIDE see BOF500
SUCCINCHLORIMIDE see SND500
SUCCINIBROMIMIDE see BOF500
SUCCINIC ACID see SMY000
SUCCINIC ACID ANHYDRIDE see SNC000
SUCCINIC ACID BIS(β-
DIMETHYLAMINOETHYL) ESTER
BISMETHIODIDE see BJI000
SUCCINIC ACID BIS(β-
DIMETHYLAMINOETHYL) ESTER,
DIHYDROCHLORIDE see HLC500
SUCCINIC ACID BIS(β-
DIMETHYLAMINOETHYL)ESTER
DIMETHOCHLORIDE see HLC500
SUCCINIC ACID, BIS(2-
(HEXYLOXY)ETHYL) ESTER see SMZ000
SUCCINIC ACID-α-BUTYL-p-
CHLOROBENZYL ESTER see CKI000
SUCCINIC ACID, CADMIUM SALT (1:1) see
CAI750
SUCCINIC ACID, DIBUTYL ESTER see
SNA500
SUCCINIC ACID DI-n-BUTYL ESTER see
SNA500
SUCCINIC ACID DICHLORIDE see SNG000
SUCCINIC ACID DIESTER with CHOLINE
see CMG250
SUCCINIC ACID DIESTER with CHOLINE
CHLORIDE see HLC500
SUCCINIC ACID, DIESTER with CHOLINE
IODIDE see BJI000

SUCCINIC ACID, DIETHYL ESTER see
SNB000
SUCCINIC ACID, DI-2-HEXYLOXYETHYL
ESTER see SMZ000
SUCCINIC ACID, 2,3-DIHYDROXY- see
TAF750
SUCCINIC ACID, 2,3-DIMERCAPTO- see
DNV610
SUCCINIC ACID, 2,3-DIMERCAPTO-,
DIETHYL ESTER see DJB100
SUCCINIC ACID, DIMETHYL ESTER see
SNB100
SUCCINIC ACID-2,2-
DIMETHYLHYDRAZIDE see DQD400
SUCCINIC ACID DINITRILE see SNE000
SUCCINIC ACID DIPROPYL ESTER see
DWV800
SUCCINIC ACID, DISODIUM SALT see
SJW100
SUCCINIC ACID, (4-ETHOXY-1-
NAPTHYLCARBONYLMETHYL) ESTER see
SNB500
SUCCINIC ACID, HYDROXY- see MAN000
SUCCINIC ACID, O-ISOPROPYL-O'-
TRIBUTYLSTANNYL ESTER see TIE600
SUCCINIC ACID, MERCAPTO-, DIETHYL
ESTER, S-ESTER with O,O-
DIMETHYLPHOSPHOROTHIOATE see
OPK250
SUCCINIC ACID, MERCAPTO-, DIETHYL
ESTER, S-ESTER WITH O,S-DIMETHYL
PHOSPHORODITHIOATE see IKX200
SUCCINIC ACID, MERCAPTO-, DIMETHYL
ESTER, o,o-DIMETHYL
PHOSPHORODITHIOATE see CBS800
SUCCINIC ACID MONOESTER with 7-
CHLORO-1,3-DIHYDRO-3-HYDROXY-5-
PHENYL-2H-1,4-BENZODIAZEPIN-2-ONE
see CFY500
SUCCINIC ACID-α-MONOESTER with d-
threo-(−)-2,2-DICHLORO-N-(β-HYDROXY-
α-(HYDROXYMETHYL)-p-
NITROPHENETHYL)ACETAMIDE see
CDP725
SUCCINIC ACID MONOESTER with N-(2-
ETHYLHEXYL)-3-
HYDROXYBUTYRAMIDE see BPF825
SUCCINIC ACID PEROXIDE (DOT) see
SNC500
SUCCINIC ACID, PHOSPHONO-,
TETRAMETHYL ESTER see PHA565
SUCCINIC ACID, SULFO-, 1,4-DIHEXYL
ESTER, SODIUM SALT (8CI) see DKP800
SUCCINIC ACID, SULFO-, 1,4-DIOCTYL
ESTER, SODIUM SALT see SOD050
SUCCINIC ANHYDRIDE see SNC000
SUCCINIC ANHYDRIDE, DECYL- see
DAJ475
SUCCINIC ANHYDRIDE,
(TETRAPROPENYL)- see TEC600
SUCCINIC CHLORIDE see SNG000
SUCCINIC-1,1-DIMETHYL HYDRAZIDE see
DQD400
SUCCINIC DINITRILE see SNE000
SUCCINIC IMIDE see SND000
SUCCINIC PEROXIDE see SNC500
SUCCINIMIDE see SND000
SUCCINIMIDE, N,2-DIMETHYL-2-PHENYL-
see MLP800
SUCCINIMIDE, N-PHENYL-2-THIO- see
PFS600
SUCCINOCHLORIMIDE see SND500
SUCCINOCHOLINE see CMG250
SUCCINODINITRILE see SNE000
SUCCINONITRILE see SNE000
4'-SUCCINOYLAMINO-2,3'-
DIMETHYLAZOBENZENE see SNF500
SUCCINOYL CHLORIDE see SNG000
SUCCINOYLCHOLINE see CMG250
SUCCINOYLCHOLINE CHLORIDE see
HLC500
SUCCINOYL DIAZIDE see SNF000

4'-SUCCINYLAMINO-2,3'-DIMETHYLAZOBENZOL see SNF500
SUCCINYL-ASTA see HLC500
SUCCINYLATED MONOGLYCERIDES see SNF700
SUCCINYLBISCHOLINE see CMG250
SUCCINYL BISCHOLINE CHLORIDE see HLC500
SUCCINYLBISCHOLINE DICHLORIDE see HLC500
SUCCINYL CHLORIDE see SNG000
SUCCINYLCHOLINE CHLORIDE see HLC500
SUCCINYLCHOLINE DICHLORIDE see HLC500
SUCCINYLCHOLINE HYDROCHLORIDE see HLC500
SUCCINYL DICHLORIDE see SNG000
SUCCINYLDICHOLINE see CMG250
SUCCINYLDICHOLINE CHLORIDE see HLC500
SUCCINYLDICHOLINE IODIDE see BJI000
o,o-SUCCINYLDICHOLINE IODIDE see BJI000
SUCCINYLFORTE see HLC500
SUCCINYLNITRILE see SNG500
SUCCINYL OXIDE see SNC000
SUCCINYL PEROXIDE see SNC500
SUCCISTEARIN see SNG600
SUCCITIMAL see MNZ000
SUCHAR 681 see CBT500
SUCKER PLUCKER see ODE000
SUCKER-STUFF see DMC600
SUCOSTRIN see HLC500
SUCOSTRIN CHLORIDE see HLC500
SUCRA see SJN700
SUCRAPHEN see SPC500
SUCRE EDULCOR see BCE500
SUCRETS see HFV500
SUCRETTE see BCE500
SUCROFER see IHG000
SUCROL see EFE000
SUCROSA see SGC000
SUCROSE see SNH000
SUCROSE ACETATE ISOBUTYRATE see SNH050
SUCROSE ACETOISOBUTYRATE see SNH050
SUCROSE ALLYL ETHER see SNH075
SUCROSE, DIACETATE HEXAISOBUTYRATE see SNH050
SUCROSE FATTY ACID ESTERS see SNH100
SUCROSE, OCTAACETATE see OAF100
SUCROSE, OCTANITRATE see SNH125
SUCROSE OCTANITRATE (dry) (DOT) see SNH125
SUCROSE, POLYALLYL ETHER, POLYMER with ACRYLIC ACID see ADV950
SUDAFED see POH000
SUDAFED see POH250
SUDAN AX see XRA000
SUDAN BLACK B see PCJ200
SUDAN BLACK X 60 see PCJ200
SUDAN BROWN RR see NBG500
SUDAN BROWN YR see NBG500
SUDAN DEEP BLACK BB see PCJ200
SUDAN DEEP BLACK BN see PCJ200
SUDAN G see CMP600
SUDAN G see OHI200
SUDAN G III see OHI200
SUDAN GREEN 4B see BLK000
SUDAN III see OHI200
SUDAN III (G) see OHI200
SUDAN IV see SBC500
SUDAN ORANGE see XRA000
SUDAN ORANGE G see CMP600
SUDAN ORANGE R see PEJ500
SUDAN ORANGE RPA see XRA000
SUDAN ORANGE RRA see XRA000
SUDAN P see SBC500
SUDAN P III see OHI200
SUDAN R see CMS238

SUDAN RED see XRA000
SUDAN RED 7B see EOJ500
SUDAN RED 290 see CMS238
SUDAN RED 4BA see SBC500
SUDAN RED BB see SBC500
SUDAN RED BBA see SBC500
SUDAN RED G (6CI) see CMS238
SUDAN RED III see OHI200
SUDAN RED IV see SBC500
SUDANROT 7B see EOJ500
SUDAN SCARLET 6G see XRA000
SUDAN X see XRA000
SUDAN YELLOW see DOT300
SUDAN YELLOW AR see CMP600
SUDAN YELLOW GGN see OHJ875
SUDAN YELLOW R see PEI000
SUDAN YELLOW RRA see AIC250
SUDINE see SNN300
SU 4885 DITARTRATE see MJJ000
SUESSETTE see SGC000
SUESSTOFF see EFE000
SUESTAMIN see SGC000
SUFENTA see SNH150
SUFENTANIL CITRATE see SNH150
SUFENTANYL see SNH150
SUFENTANYL CITRATE see SNH150
SUFFIX see EGS000
SUFFIX 25 see EGS000
SUFFIX PLUS see FDB500
SUFRALEM see AOO490
SUGAI CHRYSOIDINE see PEK000
SUGAI CONGO RED see SGQ500
SUGAI DIRECT BROWN M see CMO800
SUGAI FAST SCARLET G BASE see NMP500
SUGAI TARTRAZINE see FAG140
SUGANYL see AHO250
SUGAR see SNH000
SUGAR-BOWLS see CMV390
SUGARIN see SGC000
SUGAR of LEAD see LCV000
SUGARON see SGC000
SUGRACILLIN see BFD000
SU 8629 HYDROCHLORIDE see AKL000
SU 8842 HYDROCHLORIDE see MQR200
SUICALM see FLU000
SUISYNCHRON see MLJ500
SUKHTEH see SNH450
SULADYNE see PDC250
SULAMYD see SNQ710
SULAMYD SODIUM see SNQ000
SUL ANILINOVA (CZECH) see BBL000
SULBENICILLIN see SNV000
SULBENICILLIN DISODIUM see SNV000
SULBENTIN see RCK730
SULBENTINE see RCK730
SULCEPHALOSPORIN see CCS550
SULCOLON see PPN750
SULCONAZOLE NITRATE see SNH480
SULDIXINE see SNN300
SULDURA BLACK PG see CMS250
SULEMA (RUSSIAN) see MCY475
SULESTREX see PIK450
SULF-10 see SNQ710
SULFAAFENAZOLO (ITALIAN) see AIF000
SULFABENZAMIDE see SNH800
SULFABENZID see SNH800
SULFABENZIDE see SNH800
SULFABENZOYLAMIDE see SNH800
SULFABENZPYRAZINE see QTS000
SULFABID see SNH800
SULFABROMOMETHAZINE SODIUM see SNH875
SULFABUTIN see AIE750
SULFACANT 9D-297 see AFJ175
SULFACARBAMIDE see SNQ550
4-SULFACARBAMIDE see SNQ550
SULFACET see SNQ710
SULFACETAMIDE see SNQ710
SULFACETAMIDE, SODIUM see SNQ000
SULFACETAMIDE, SODIUM SALT see SNQ000
SULFACETIMIDE see SNQ710

SULFACHLORPYRIDAZINE see SNH900
SULFACID BRILLIANT BLUE 6J see ADE500
SULFACID BRILLIANT GREEN 1B see FAE950
SULFACID LIGHT ORANGE J see HGC000
SULFACID LIGHT YELLOW 5RL see SGP500
SULFACOMBIN see SNI000
SULFACTIN see BAD750
SULFACYL see SNQ710
SULFACYL SODIUM see SNQ000
SULFACYL SODIUM SALT see SNQ000
SULFACYL SOL see SNQ000
SULFACYL SOLUBLE see SNQ000
SULFADENE see BDF000
SULFADIAMINE see SNY500
SULFADIAZINE see PPP500
SULFADIAZINE SILVER see SNI425
SULFADIAZINE SILVER SALT see SNI425
SULFADIMERAZINE see SNJ000
SULFADIMETHOXIN see SNN300
SULFADIMETHOXINE see SNN300
SULFADIMETHOXYDIAZINE see SNN300
SULFADIMETHOXYPYRIMIDINE see SNI500
SULFADIMETHYLDIAZINE see SNJ000
SULFADIMETHYLISOXAZOLE see SNN500
SULFADIMETHYLOXAZOLE see AIE750
SULFADIMETHYLPYRIMIDINE see SNJ000
4-SULFA-2,6-DIMETHYLPYRIMIDINE see SNJ350
SULFADIMETINE see SNJ000
SULFADIMETINE see SNJ350
SULFADIMETOSSINA (ITALIAN) see SNN300
SULFADIMETOXIN see SNN300
SULFADIMEZINE see SNJ000
SULFADIMIDINE see SNJ000
SULFADIMIDINE SODIUM see SJW500
SULFADINE see SNJ000
SULFADOXINE see AIE500
SULFADSIMESINE see SNJ000
SULFAETHOXYPYRIDAZINE see SNJ100
SULFAETHYLTHIADIAZOLE SODIUM see SJW300
SULFAFURAZOL see SNN500
SULFAGAN see SNN500
SULFAGUANIDINE see AHO250
SULFAGUINE see AHO250
SULFA-ISODIMERAZINE see SNJ000
SULFAISODIMERAZINE see SNJ350
SULFAISODIMIDINE see SNJ000
SULFAISODIMIDINE see SNJ350
SULFALENE see MFN500
SULFALENE N-METHYLGLUCAMINE SALT see SNJ400
SULFALEX see AKO500
SULFALLATE see CDO250
SULFALONE see SNW500
SULFAMATE see ANU650
SULFAMERADINE see ALF250
SULFAMERAZIN see ALF250
SULFAMERAZINE SODIUM see SJW475
SULFAMETER SODIUM see MFO000
SULFAMETHALAZOLE see SNK000
SULFAMETHAZINE SODIUM see SJW500
SULFAMETHIAZINE see SNJ000
SULFAMETHIN see SNJ000
SULFAMETHIN see SNJ350
SULFAMETHIZOL see MPQ750
SULFAMETHIZOLE see MPQ750
SULFAMETHOMIDINE see MDU300
SULFAMETHOPYRAZINE see MFN500
SULFAMETHOXAZOL see SNK000
SULFAMETHOXAZOLE see SNK000
SULFAMETHOXAZOL-TRIMETHOPRIM see TKX000
SULFAMETHOXYDIAZINE SODIUM see MFO000
SULFAMETHOXYPYRAZINE see MFN500
3-SULFA-6-METHOXYPYRIDAZINE see AKO500

SULFA-5-METHOXYPYRIMIDINE SODIUM SALT see MFO000
SULFAMETHYLDIAZINE see ALF250
SULFAMETHYLISOXAZOLE see SNK000
SULFAMETHYLIZOLE see MPQ750
SULFAMETHYLTHIADIAZOLE see MPQ750
SULFAMETOMIDINE see MDU300
SULFAMETOPYRAZINE see MFN500
SULFAMETORINE SODIUM see MFO000
SULFAMETOSSIPIRIDAZINA (ITALIAN) see MFN500
SULFAMETOXIPIRIDAZINE see AKO500
SULFAMETOXYPYRIDAZIN (GERMAN) see MFN500
SULFAMEZATHINE see SNJ000
SULFAMIC ACID see SNK500
SULFAMIC ACID, CYCLOHEXYL-, CALCIUM SALT (2:1), DIHYDRATE see CAQ600
SULFAMIC ACID, DIMETHYL-, 5-BUTYL-2-(ETHYLAMINO)-6-METHYL-4-PYRIMIDINYL ESTER see BRJ000
SULFAMIC ACID, MONOAMMONIUM SALT see ANU650
SULFAMIC ACID, TETRAMETHYLENE ESTER see BOU100
SULFAMIDE, N,N'-DIMETHYL- see DUD900
SULFAMIDIC ACID see SNK500
p-SULFAMIDOANILINE see SNM500
SULFAMIDYL see SNM500
SULFAMINSAEURE (GERMAN) see ANU650
SULFAMONOMETHOXIN see SNL800
SULFAMONOMETHOXINE see SNL800
SULFAMONOMETOXINE N-METHYLGLUCAMINE SALT see SNL830
SULFAMOXOLE see AIE750
SULFAMOXOLE and TRIMETHOPRIM see SNL850
SULFAMOXOLE-TRIMETHOPRIM mixture see SNL850
SULFAMOXOLUM see AIE750
4-SULFAMOYLBENZOIC ACID see SNL890
p-SULFAMOYLBENZOIC ACID see SNL890
5'-SULFAMOYL-2-CHLOROADENOSINE see CEF125
3-SULFAMOYLMETHYL-1,2-BENZISOXAZOLE see BCE750
4-(p-SULFAMOYLPHENYL)SEMICARBAZONE 1-METHYL-1H-PYRROLE-2-CARBOXALDEHYDE see SNL900
4-(p-SULFAMOYLPHENYL)SEMICARBAZONE-1H-PYRROLE-2-CARBOXALDEHYDE see SNL920
N-(5-SULFAMOYL-1,3,4-THIADIAZOL-2-YL)ACETAMIDE see AAI250
SULFAMUL see TEX250
N-SULFAMYLBENZAMIDE see SNH800
p-SULFAMYLBENZOIC ACID see SNL890
SULFAMYLON ACETATE see MAC000
SULFAN see SOR500
SULFANA see SNM500
SULFANALONE see SNM500
SULFAN BLUE see ADE500
SULFANIL see SNM500
SULFANILACETAMIDE see SNQ710
SULFANILAMIDE see SNM500
3-SULFANILAMIDE-6-METHOXYPYRIDAZINE see AKO500
SULFANILAMIDE, N^1-(5-tert-BUTYL-1,3,4-THIADIAZOL-2-YL)- see GEW750
SULFANILAMIDE, N^1-(5-ETHYL-1,3,4-THIADIAZOL-2-YL)-, MONOSODIUM SALT see SJW300
SULFANILAMIDE, N$_4$,N$_4$-BIS(2-IODOETHYL)- see BKJ650
SULFANILAMIDE, N^1-2-PYRIDYL-, MONOSODIUM SALT see PPO250
SULFANILAMIDE, N^1-2-PYRIDYL-, N^1-SODIUM deriv. see PPO250

4-SULFANILAMIDO-3,6-DIMETHOXYPYRIDAZINE see DON700
4-SULFANILAMIDO-5,6-DIMETHOXYPYRIMIDINE see AIE500
6-SULFANILAMIDO-2,4-DIMETHOXYPYRIMIDINE see SNN300
5-SULFANILAMIDO-3,4-DIMETHYL-ISOXAZOLE see SNN500
2-SULFANILAMIDO-4,6-DIMETHYLPYRIMIDINE see SNJ000
4-SULFANILAMIDO-2,6-DIMETHYLPYRIMIDINE see SNJ350
6-SULFANILAMIDO-2,4-DIMETHYLPYRIMIDINE see SNJ350
2-SULFANILAMIDO-3-METHOXYPYRAZINE see MFN500
3-SULFANILAMIDO-6-METHOXYPYRIDAZINE see AKO500
6-SULFANILAMIDO-3-METHOXYPYRIDAZINE see AKO500
2-SULFANILAMIDO-5-METHOXYPYRIMIDINE SODIUM SALT see MFO000
3-SULFANILAMIDO-5-METHYLISOXAZOLE see SNK000
2-SULFANILAMIDO-5-METHYL-1,3,4-THIADIAZOLE see MPQ750
5-SULFANILAMIDO-1-PHENYLPYRAZOLE see AIF000
SULFANILAMIDOPYRIMIDINE see PPP500
2-SULFANILAMIDOPYRIMIDINE SODIUM SALT see MRM250
2-SULFANILAMIDOQUINOXALINE see QTS000
2-SULFANILAMIDOTHIAZOLE see TEX250
2-SULFANILAMIDOTHIAZOLE SODIUM SALT see TEX500
SULFANILANILIDE, 3-NITRO-N^4-PHENYL-(7CI,8CI) see KDA075
SULFANILCARBAMID see SNQ550
N-SULFANILCARBAMIDE see SNQ550
SULFANILGUANIDINE see AHO250
SULFANILIC ACID see SNN600
m-SULFANILIC ACID see SNO000
o-SULFANILIC ACID see SNO100
SULFANILSAEURE see SNN600
N-SULFANILYLACETAMIDE SODIUM see SNQ000
N-SULFANILYLACETAMIDE, SODIUM SALT see SNQ000
N-SULFANILYLACETAMIDE, SODIUM SALT see SNQ000
2-SULFANILYL AMINOPYRIDINE see PPO000
2-SULFANILYLAMINOPYRIMIDINE see PPP500
2-(SULFANILYLAMINO)THIAZOLE see TEX250
N-SULFANILYLBENZAMIDE see SNH800
SULFANILYLCARBAMIC ACID, METHYL ESTER see SNQ500
SULFANILYL CHLORIDE, N-ACETYL-(6CI,7CI,8CI) see AAM600
SULFANILYLGUANIDINE see AHO250
N^1-SULFANILYL-N^2-BUTYLCARBAMIDE see BSM000
N^1-SULFANILYL-N^2-BUTYLUREA see BSM000
N-SULFANILYL-N'BUTYLUREE (FRENCH) see BSM000
SULFANILYLUREA see SNQ550
SULFANITRAN see SNQ600
SULFANO see AIE750
SULFANOL BLACK B see CMS250
SULFANOL BLACK MBS see CMS250
SULFANOL BLACK MRS see CMS250
SULFANOL NP 1 see SNQ700
SULFANTHRENE BLUE 2B see ICU135
SULFANTHRENE ORANGE R see CMU815
SULFANTHRENE ORANGE RS see CMU815
N-SULFANYLACETAMIDE see SNQ710
SULFANYLHARNSTOFF see SNQ550

SULFANYLUREE see SNQ550
SULFAPHENAZOLE see AIF000
SULFAPHENAZON see AIF000
SULFAPHENYLPIPAZOL see AIF000
SULFAPHENYLPYRAZOLE see AIF000
SULFAPOL see DXW200
SULFAPOLU (POLISH) see DXW200
SULFAPYRAZINEMETHOXINE see MFN500
SULFAPYRAZINEMETHOXYNE see MFN500
SULFAPYRIDAZINE see AKO500
SULFAPYRIDINE see PPO000
2-SULFAPYRIDINE see PPO000
SULFAPYRIDINE NEUTRAL SOLUBLE see DXF400
SULFAPYRIDINE SODIUM see PPO250
SULFAPYRIDINE, SODIUM SALT see PPO250
SULFAPYRIMIDIN (GERMAN) see PPP500
2-SULFAPYRIMIDINE see PPP500
SULFAQUINOXALINE see QTS000
SULFARLEM see AOO490
SULFARSENOBENZENE see SNR000
SULFARSENOL see SNR000
SULFARSPHENAMINE see SNR000
SULFARSPHENAMINE BISMUTH see BKV250
SULFASALAZINE see PPN750
SULFASAN see BKU500
SULFASAN R POWDER see BKU500
SULFASOL see SNN300
SULFASOMIDINE see SNJ350
SULFASOXAZOLE see SNN500
SULFASTOP see SNN300
SULFATE d'ATROPINE (FRENCH) see ARR500
SULFATE de CUIVRE (FRENCH) see CNP250
SULFATED AMYLOPECTIN see AOM150
SULFATED CASTOR OIL see TOD500
SULFATE DIMETHYLIQUE (FRENCH) see DUD100
SULFATE ESTER of N-HYDROXY-N-2-FLUORENYL ACETAMIDE see FDV000
SULFATE MERCURIQUE (FRENCH) see MDG500
SULFATE de METHYLE (FRENCH) see DUD100
SULFATE de NICOTINE (FRENCH) see NDR500
SULFATEP see SOD100
SULFATE de PLOMB (FRENCH) see LDY000
SULFATES see SNS000
SULFATE de ZINC (FRENCH) see ZNA000
SULFATHALIDINE see PHY750
SULFATHIAZOL see TEX250
2-SULFATHIAZOLE SODIUM see TEX500
SULFATHIAZOLE (USDA) see TEX250
SULFATOBIS(DIMETHYLSELENIDE)PLATINUM(II) HYDRATE see SNS100
cis-SULFATO-1,2-DIAMINOCYCLOHEXANEPLATINUM(II) see SCF025
trans(−)-SULFATO-1,2-DIAMINOCYCLOHEXANEPLATINUM(II) see SCF050
trans(+)-SULFATO-1,2-DIAMINOCYCLOHEXANEPLATINUM(II) see SCF075
SULFAUREA see SNQ550
SULFAVIGOR see AIE750
SULFAZOLE see SNN500
SULFDURAZIN see AKO500
SULFENAMIDE M see BDG000
SULFENAMIDE TS see CPI250
SULFENAX see CPI250
SULFENAX CB see CPI250
SULFENAX CB 30 see CPI250
SULFENAX CB/K see CPI250
SULFENAX MOB (CZECH) see BDF750
SULFENAZIN see TFM100
SULFENONE see CKI625
SULFENTANIL see SNH150

SULFER BLACK BB see CMS250
SULFER BLACK G see CMS250
SULFER BLACK GB see CMS250
SULFER BLACK KBR see CMS250
SULFER RED BROWN RN see CMS257
SULFERROUS see FBN100
SULFERROUS see FBO000
SULFIDAL see SOD500
SULFIDE, ALLYL 2-CHLOROETHYL see AGB600
SULFIDE, BENZYL 2-CHLOROETHYL see BFL100
SULFIDE, m-CHLOROPHENYL PHENYL see CKI612
SULFIDE, ETHYL o-NITROPHENYL see ENQ600
SULFIDE, METHYLVINYLOXYETHYL- see MQM200
SULFIDES see SNT000
SULFIDINE see PPO000
SULFIDOTRICHLORID FOSFORECNY see TFO000
SULFIMEL DOS see DJL000
SULFINPYRAZINE see DWM000
1,1'-SULFINYLBIS(1,2-DICHLOROETHANE) see SNT100
SULFINYLBIS(METHANE) see DUD800
SULFINYL BROMIDE see SNT200
SULFINYL CHLORIDE see TFL000
SULFINYL CYANAMIDE see SNT300
SULFISIN see SNN500
SULFISOMEZOLE see SNK000
SULFISOMIDIN see SNJ000
SULFISOMIDIN see SNJ350
SULFISOMIDINE see SNJ000
SULFISOMIDINE see SNJ350
SULFISOXAZOLE see SNN500
SULFITE CELLULOSE see CCU150
SULFITE LYE see HIC600
SULFITES see SNT500
SULFIZOLE see SNN500
SULFMETHMETON-METHYL see THT800
SULFMETHOXIPIRIDAZINE see AKO500
SULFMIDIL see AIE750
SULFOACETIC ACID see SNU000
SULFOACETIC ACID 1-DODECYL ESTER, SODIUM SALT see SIB700
SULFOACETIC ACID DODECYL ESTER S-SODIUM SALT see SIB700
2-SULFOANTHRAQUINONE SODIUM SALT see SER000
5-SULFOBENZEN-1,3-DICARBOXYLIC ACID see SNU500
SULFOBENZIDE see PGI750
3-SULFOBENZIDINE see BBY250
o-SULFOBENZIMIDE see BCE500
o-SULFOBENZOIC ACID IMIDE see BCE500
SULFOBENZYLPENICILLIN see SNV000
α-SULFOBENZYLPENICILLIN DISODIUM see SNV000
SULFOBROMOPHTHALEIN see HAQ600
SULFOBROMOPHTHALEIN SODIUM see HAQ600
SULFOBROMPHTHALEIN see HAQ600
p-SULFOCARBANILIC ACID, N-METHYL ESTER, S-HYDRAZIDE see MQJ300
SULFOCARBANILIDE see DWN800
SULFOCARBATHION see SNV050
SULFOCARBATHION K see SNV050
SULFOCARBOLIC ACID see HJH500
SULFOCIDINE see SNM500
SULFOCILLIN see SNV000
SULFODEHYDROABIETIC ACID MONOSODIUM SALT see EAB560
12-SULFODEHYDROABIETIC ACID MONOSODIUM SALT see EAB560
3-o-SULFODEHYDROEPIANDROSTERONE see SNV100
SULFODIAMINE see SNY500
SULFODIMESIN see SNJ000
SULFODIMEZINE see SNJ000

SULFODOR (CZECH) see EPH000
SULFOETHANOIC ACID see SNU000
2-SULFOETHYLAMINE see TAG750
SULFOGAL see AOO490
SULFOGEN BLACK BWLE see CMS250
SULFOGEN BORDEAUX 5CFN see CMS257
SULFOGENE BORDEAUX CF see CMS257
SULFOGENE BORDEAUX 5CF see CMS257
SULFOGENE CARBON HCF GRAINS see CMS250
SULFOGENOL see IAD000
SULFOGEN RED BROWN BK see CMS257
SULFO GREEN J see FAF000
SULFOGUANIDINE see AHO250
SULFOGUENIL see AHO250
SULFOLAN see SNW500
SULFOLANE see SNW500
SULFOL-3-ENE see DMF000
3-SULFOLENE see DMF000
β-SULFOLENE see DMF000
SULFO-MERTHIOLATE see SKH150
4-SULFOMETANILIC ACID see AIE000
N-SULFOMETHYL-POLYMYXIN B SODIUM SALT see SNW600
SULFOMETURON METHYL see SNW550
SULFOMYXIN see SNW600
SULFONA see SOA500
SULFONAL see ABD500
SULFONAL NP 1 see SNQ700
SULFONAMIDE see SNM500
4-SULFONAMIDE-4'-DIMETHYLAMINOAZOBENZENE see SNW800
SULFONAMIDE P see SNM500
p-SULFONAMIDOBENZOIC ACID see SNL890
4-SULFONAMIDO-3'-METHYL-4'-AMINOAZOBENZENE see SNX000
2-SULFONAMIDOTHIAZOLE see TEX250
SULFONA P see AOO800
SULFO-3-NAPHTHALENEFURANE see SNX250
2-(4-SULFO-1-NAPHTHYLAZO)-1-NAPHTHOL-4-SULFONIC ACID, DISODIUM SALT see HJF500
SULFONATED CASTOR OIL see TOD500
SULFONATES see SNY000
o-SULFONBENZOIC ACID IMIDE SODIUM SALT see SJN700
p-SULFONDICHLORAMIDOBENZOIC ACID see HAF000
SULFONE, BIS(2-ACETOXYETHYL) see BGQ100
SULFONE, BIS(4-FLUORO-3-NITROPHENYL) see DKH250
SULFONE, p-BROMOPHENYL CHLOROMETHYL see BNV850
SULFONE, METHYL PHENYL see PFR375
SULFONE-2,4,4',5-TETRACHLORODIPHENYL see CKM000
SULFONETHYLMETHANE see BJT750
SULFONIC ACID, MONOCHLORIDE see CLG500
SULFONIMIDE see CBF800
SULFONINE ACID BLUE R see ADE750
SULFONINE ORANGE R see ADG000
SULFONINE RED G see CMM320
SULFONINE RED GN see CMM320
SULFONINE RED GS see CMM320
SULFONINE RED RS see NAO600
SULFONINE RED SG see CMM320
SULFONINE YELLOW CSR see CMM759
SULFONIUM, (2-((CYCLOHEXYLHYDROXYPHENYLACETYL)OXY)ETHYL)DIMETHYL-, IODIDE see HOJ150
SULFONIUM, (2-HYDROXYETHYL)DIMETHYL-, IODIDE, α-PHENYLCYCLOHEXANEGLYCOLATE see HOJ150
SULFONIUM, TRIMETHYL-, IODIDE, OXIDE see TMG800

SULFONMETHANE see ABD500
α-SULFO-ω-(NONYLPHENOXY) POLY(OXY-1,2-ETHANEDIYL) SODIUM SALT see SNY100
N-SULFONOXY-AAF see FDV000
N-SULFONOXY-N-ACETYL-2-AMINOFLUORENE see FDV000
SULFONPHTHAL see PDO800
4,4'-SULFONYLBISACETANILIDE see SNY500
p,p'-SULFONYLBISACETANILIDE see SNY500
4',4"-SULFONYLBIS(ACETANILIDE) see SNY500
1,1'-SULFONYLBIS(4-AMINOBENZENE) see SOA500
3,3'-SULFONYLBIS(ANILINE) see SOA000
4,4'-SULFONYLBISANILINE see SOA500
p,p-SULFONYLBISBENZAMINE see SOA500
4,4'-SULFONYLBISBENZAMINE see SOA500
p,p-SULFONYLBISBENZENAMINE see SOA500
1,1'-SULFONYLBISBENZENE see PGI750
1,1'-(SULFONYLBIS(4,1-PHENYLENEIMINO))BIS(1-DEOXY-1-SULFO-d-GLUCITOL) DISODIUM SALT see AOO800
4,4'-(SULFONYLBIS(4,1-PHENYLENEOXY))BISBENZENAMINE see SNZ000
4,4'-SULFONYLBIS(4-PHENYLENEOXY)DIANILINE see SNZ000
SULFONYL CHLORIDE see SOT000
SULFONYL CHLORIDE FLUORIDE see SOT500
3,3'-SULFONYLDIANILINE see SOA000
4,4'-SULFONYLDIANILINE see SOA500
p,p'-SULFONYLDIANILINE see SOA500
p,p'-SULFONYLDIANILINE-N,N'-DI-d-GLUCOSE SODIUM BISULFITE see AOO800
p,p'-SULFONYLDIANILINE N,N'-DIGLUCOSIDE DISODIUM DISULFONALTE see AOO800
4,4'-SULFONYLDIPHENOL see SOB000
O,O'-(SULFONYLDI-1,4-PHENYLENE) O,O,O',O'-TETRAMETHYL DIPHOSPHOROTHIOATE see PHN250
3-(SULFONYL)-o-((METHYLAMINO)CARBONYL)OXIME-2-BUTANONE see SOB500
N-(SULFONYL-p-METHYLBENZENE)-N'-N-BUTYLUREA see BSQ000
(3-β)-3-(SULFOOXY)ANDROST-5-EN-17-ONE see SNV100
(3-β)-3-(SULFOOXY)-ANDROST-5-EN-17-ONE SODIUM SALT (9CI) see SJK400
6-SULFOOXYMETHYLBENZO(A)PYRENE see SOB550
SULFOPARABLUE see SOB600
SULFOPERAMIDIC ACID see HLN100
1-p-SULFOPHENYLAZO-2-HYDROXYNAPHTHALENE-6-SULFONATE, DISODIUM SALT see FAG150
4-p-SULFOPHENYLAZO-1-NAPHTHOL MONOSODIUM SALT see FAG010
1-p-SULFOPHENYLAZO-2-NAPHTHOL-6-SULFONIC ACID, DISODIUM SALT see FAG150
1-(4-SULFOPHENYL)-3-ETHYLCARBOXY-4-(4-SULFONAPHTHYLAZO)-5-HYDROXYPYRAZOLE see OJK325
1-(p-SULFOPHENYL)-3-PHENYL-PYRAZOL (GERMAN) see PGE500
SULFOPLAN see SNN300
SULFOPON WA 1 see SIB600
SULFOQUANIDINE see AHO250
SULFORON see SOD500
SULFORTHOMIDINE see AIE500
SULFOSALICYLIC ACID see SOC500
5-SULFOSALICYLIC ACID see SOC500
SULFOSFAMIDE see SOD000

SULFOSUCCINIC ACID 1,4-DIOCTYL
ESTER SODIUM SALT see SOD050
SULFOTEP see SOD100
SULFOTEPP see SOD100
SULFOTEX WALA see SIB600
SULFOTHIORINE see SKI500
5-SULFO-TOBIAS ACID see ALH000
SULFOTRIM see TKX000
SULFOTRIMIN see TKX000
SULFOTRINAPHTHYLENOFURAN,
SODIUM SALT see SOD200
SULFOX-CIDE see ISA000
SULFOXIDE see ISA000
SULFOXIDE, BIS(p-CHLOROPHENYL) see
BIN900
SULFOXIDE, BIS(1,2-DICHLOROVINYL) see
SNT100
SULFOXONIUM, TRIMETHYL-, IODIDE see
TMG800
SULFOXYL see BDS000
SULFOXYL see ISA000
2-(6-SULFO-2,4-XYLYLAZO)-1-NAPHTHOL-
4-SULFONIC ACID, DISODIUM SALT see
FAG050
SULFOXYPHENYLPYRAZOLIDINE see
DWM000
SULFOZONA see AKO500
SULFRALEM see AOO490
SULFRAMIN 85 see DXW200
SULFRAMIN ACID 1298 see LBU100
SULFRAMIN 40 FLAKES see DXW200
SULFRAMIN 40 GRANULAR see DXW200
SULFRAMIN 1238 SLURRY see DXW200
SULFSTAT see MPQ750
SULFTECH see SJZ000
SULFUNE see AIE750
SULFUNO see AIE750
SULFUR see SOD500
SULFUR, mixture with BARIUM SULFIDE see
BAP260
SULFUR BLACK see CMS250
SULFUR BLACK 1 see CMS250
SULFUR BLACK 2B see CMS250
SULFUR BLACK B see CMS250
SULFUR BLACK BGL see CMS250
SULFUR BLACK BGND see CMS250
SULFUR BLACK BRW see CMS250
SULFUR BLACK T see CMS250
SULFUR BORDEAUX B-CF see CMS257
SULFUR BROMIDE see SOE000
SULFUR CHLORIDE see SOG500
SULFUR CHLORIDE see SON510
SULFUR CHLORIDE (DI) (DOT) see SON510
SULFUR CHLORIDE FLUORINE see SOF000
SULFUR CHLORIDE (MONO) see SOG500
SULFUR CHLORIDE OXIDE see TFL000
SULFUR CHLORIDE PENTAFLUORIDE see
SOF000
SULFUR CHLOROPENTAFLUORIDE see
SOF000
SULFUR COMPOUNDS see SOF500
SULFUR DECAFLUORIDE see SOQ450
SULFUR DICHLORIDE see SOG500
SULFUR DIFLUORIDE MONOXIDE see
TFL250
SULFUR DIFLUORIDE OXIDE see TFL250
SULFUR DINITRIDE see SOH000
SULFUR DIOXIDE see SOH500
SULFUR DIOXIDE, solution see SOO500
SULFURE de 4-CHLOROBENZYLE et de 4-
CHLOROPHENYLE (FRENCH) see CEP000
SULFURE de METHYLE (FRENCH) see
TFP000
SULFURETED HYDROGEN see HIC500
SULFUR FLOWER (DOT) see SOD500
SULFUR FLUORIDE see SOI000
SULFUR FLUORIDE OXIDE see BLD000
SULFUR HALF-MUSTARD see CHC000
SULFUR HEXAFLUORIDE see SOI000
SULFUR HYDRIDE see HIC500
SULFURIC ACID see SOI500
SULFURIC ACID (mist) see SOI530

SULFURIC ACID, fuming see SOI520
SULFURIC ACID, aromatic see SOI510
SULFURIC ACID, fuming <30% free sulfur
trioxide (DOT) see SOI520
SULFURIC ACID, fuming > or =30% free
sulfur trioxide (DOT) see SOI520
SULFURIC ACID, sec-ALKYL ESTER,
SODIUM SALT see SOJ000
SULFURIC ACID, ALUMINUM POTASSIUM
SALT (2:1:1) see PKU725
SULFURIC ACID, ALUMINUM POTASSIUM
SALT (2:1:1), DODECAHYDRATE see
AHF200
SULFURIC ACID, ALUMINUM SALT (3:2) see
AHG750
SULFURIC ACID, ALUMINUM SALT (3:2),
OCTADECAHYDRATE see AHG800
SULFURIC ACID, AMMONIUM IRON(2+)
SALT, HEXAHYDRATE see IGL200
SULFURIC ACID, AMMONIUM NICKEL(2+)
SALT (2:2:1) see NCY050
SULFURIC ACID, AMMONIUM NICKEL²⁺
SALT (3:2:2) see NCY051
URIC ACID, AMMONIUM NICKEL²⁺ SALT
(2:2:1), HEXAHYDRATE see NCY060
SULFURIC ACID, BARIUM SALT (1:1) see
BAP000
SULFURIC ACID, BERYLLIUM SALT (1:1) see
BFU250
SULFURIC ACID, BERYLLIUM SALT (1:1),
TETRAHYDRATE see BFU500
SULFURIC ACID, BIS(METHYLMERCURY)
SALT see BKS810
SULFURIC ACID, BISMUTH(3+) SALT (3:2)
see BKY600
SULFURIC ACID, CADMIUM(2+) SALT see
CAJ000
SULFURIC ACID, CADMIUM SALT,
HYDRATE see CAJ250
SULFURIC ACID, CADMIUM SALT,
TETRAHYDRATE see CAJ500
SULFURIC ACID, CALCIUM(2+) SALT,
DIHYDRATE see CAX750
SULFURIC ACID, CERIUM SALT (2:1) see
CDB400
SULFURIC ACID, CHROMIUM (3+)
POTASSIUM SALT (2:1:1) see PLB500
SULFURIC ACID,
CHROMIUM(3+)POTASSIUM SALT(2:1:1),
DODECAHYDRATE see CMG850
SULFURIC ACID, CHROMIUM SALT see
CMK405
SULFURIC ACID, CHROMIUM(3+) SALT
(3:2) see CMK415
SULFURIC ACID, CHROMIUM SALT, BASIC
see NBW000
SULFURIC ACID, CHROMIUM(3+) SALT
(3:2), PENTADECAHYDRATE see CMK425
SULFURIC ACID, COBALT(2+) SALT (1:1)
see CNE125
SULFURIC ACID, COBALT(2+) SALT (1:1),
HEPTAHYDRATE (8CI,9CI) see CNE150
SULFURIC ACID, COPPER(2+) SALT (1:1) see
CNP250
SULFURIC ACID, COPPER(2+) SALT,
PENTAHYDRATE see CNP500
SULFURIC ACID, CYCLIC ETHYLENE
ESTER see EJP000
SULFURIC ACID, DECYL ESTER, SODIUM
SALT see SOK000
SULFURIC ACID, DIAMMONIUM SALT see
ANU750
SULFURIC ACID, DICESIUM SALT see
CDE500
SULFURIC ACID, DILITHIUM SALT see
LHR000
SULFURIC ACID, DIMETHYL ESTER see
DUD100
SULFURIC ACID, DIPOTASSIUM SALT see
PLT000
SULFURIC ACID, DISODIUM SALT see
SJY000

SULFURIC ACID, DITHALLIUM(1+) SALT
(8CI, 9CI) see TEM000
SULFURIC ACID, DODECYL ESTER,
TRIETHANOLAMINE SALT see SON000
SULFURIC ACID, GALLIUM SALT (3:2) see
GBS100
SULFURIC ACID, INDIUM SALT see ICJ000
LFURIC ACID, IRON(2+) SALT (1:1) see
FBN100
LFURIC ACID, IRON (3+) SALT (3:2) see
FBA000
SULFURIC ACID LANTHANUM(3+) SALT
(3:2) see LBB000
SULFURIC ACID, LAURYL ESTER,
AMMONIUM SALT see SOM500
SULFURIC ACID, LEAD(2+) SALT (1:1) see
LDY000
SULFURIC ACID, LITHIUM SALT (1:2) see
LHR000
SULFURIC ACID, MAGNESIUM
POTASSIUM SALT (3:2:2) see MAI650
SULFURIC ACID, MAGNESIUM SALT (1:1),
compounded with 2,2'-DITHIOBIS(PYRIDINE)
1,1'-OXIDE see OIU850
SULFURIC ACID, MAGNESIUM SALT (1:1)
HEPTAHYDRATE see MAJ500
SULFURIC ACID, MANGANESE(2+) SALT
see MAU250
SULFURIC ACID, MANGANESE(2+) SALT
(1:1), MONOHYDRATE see MAU300
SULFURIC ACID, MANGANESE (2+) SALT
(1:1) TETRAHYDRATE see MAU750
SULFURIC ACID, MERCURY(2+) SALT (1:1)
see MDG500
SULFURIC ACID MIXTURE with SULFUR
TRIOXIDE see SOI520
SULFURIC ACID, MONOAMMONIUM SALT
see ANJ500
SULFURIC ACID, MONOANHYDRIDE with
NITROUS ACID see NMJ000
SULFURIC ACID, MONOBENZYL ESTER,
SODIUM SALT see SFB150
SULFURIC ACID, MONODECYL ESTER,
SODIUM SALT see SOK000
SULFURIC ACID, MONODODECYL ESTER,
AMMONIUM SALT see SOM500
SULFURIC ACID, MONODODECYL ESTER,
compounded with 2,2',2"-
NITRILOTRIETHANOL (1:1) see SON000
SULFURIC ACID, MONODODECYL ESTER,
compounded with 2,2',2"-
NITRILOTRIS(ETHANOL) see SON000
SULFURIC ACID, MONODODECYL ESTER,
SODIUM SALT see SIB600
SULFURIC ACID, MONO(2-(2-(2-
(DODECYLOXY)ETHOXY)ETHOXY)ETHY
L) ESTER, SODIUM SALT see SIC000
SULFURIC ACID, MONOETHENYL ESTER,
HOMOPOLYMER, POTASSIUM SALT see
PKS250
SULFURIC ACID, MONOETHYL ESTER,
ION(1-), 4-ETHYL-4-
HEXADECYLMORPHOLINIUM see EKN550
SULFURIC ACID, MONO(2-
ETHYLHEXYL)ESTER see ELB400
SULFURIC ACID, MONO(2-
ETHYLHEXYL)ESTER, SODIUM SALT (8CI)
see TAV750
SULFURIC ACID, MONONONYL ESTER,
SODIUM SALT see SIX500
SULFURIC ACID, MONOOCTADECYL
ESTER, SODIUM SALT see OBG100
SULFURIC ACID, MONOOCTYL ESTER,
SODIUM SALT see OFU200
SULFURIC ACID,
MONO(PHENYLMETHYL) ESTER,
SODIUM SALT see SFB150
SULFURIC ACID, MONOPOTASSIUM SALT
see PKX750
SULFURIC ACID, MONOSODIUM SALT see
SEG800

SULFURIC ACID, MONOTETRADECYL ESTER, SODIUM SALT see SIO000
SULFURIC ACID, MYRISTYL ESTER, SODIUM SALT see SIO000
SULFURIC ACID, NICKEL(2+)SALT see NDK500
SULFURIC ACID, NICKEL(2+) SALT (1:1) see NDK500
SULFURIC ACID, NICKEL²⁺ SALT (1:1), HEPTAHYDRATE see NDK990
SULFURIC ACID, NICKEL(2+) SALT, HEXAHYDRATE see NDL000
SULFURIC ACID, THALLIUM SALT see TEL750
SULFURIC ACID, THALLIUM(2+) SALT see TEM050
SULFURIC ACID, THALLIUM(3+) SALT see TEM100
SULFURIC ACID, THALLIUM(1+) SALT (1:2) see TEM000
SULFURIC ACID, TIN(2+) SALT (1:1) see TGF010
SULFURIC ACID, TITANIUM(4+) SALT see TGF220
SULFURIC ACID, TITANIUM(4+) SALT (2:1) see TGH250
SULFURIC ACID, TITANIUM(4+) SALT (2:1) see TGH252
SULFURIC ACID, VANADIUM SALT see VEA100
SULFURIC ACID, ZINC SALT (1:1) see ZNA000
SULFURIC ACID, ZINC SALT (1:1), HEPTAHYDRATE see ZNJ000
SULFURIC ACID, ZIRCONIUM(4+) SALT (2:1) see ZTJ000
SULFURIC AND HYDROFLUORIC ACIDS, MIXTURE (DOT) see HHV000
SULFURIC ANHYDRIDE see SOR500
SULFURIC CHLOROHYDRIN see CLG500
SULFURIC OXIDE see SOR500
SULFURIC OXYCHLORIDE see SOT000
SULFURIC OXYFLUORIDE see SOU500
SULFURINE see MPQ750
SULFUR MONOBROMIDE see SOE000
SULFUR MONOCHLORIDE see SON510
SULFUR MUSTARD see BIH250
SULFUR MUSTARD GAS see BIH250
SULFURMYCIN A see SON520
SULFURMYCIN B see SON525
SULFUR NITRIDE see SOO000
SULFUROUS ACID see SOO500
SULFUROUS ACID ANHYDRIDE see SOH500
SULFUROUS ACID, 2-(p-tert-BUTYLPHENOXY)CYCLOHEXYL-2-PROPYNYL ESTER see SOP000
SULFUROUS ACID, 2-(p-tert-BUTYLPHENOXY)-1-METHYLETHYL-2-CHLOROETHYL ESTER see SOP500
SULFUROUS ACID, CALCIUM SALT (2:1) (8CI,9CI) see CAN000
SULFUROUS ACID, DIBENZYL ESTER see BFK000
SULFUROUS ACID, DIPOTASSIUM SALT see PLT500
SULFUROUS ACID, cyclic ester with 1,4,5,6,7,7-HEXACHLORO-5-NORBORNENE-2,3-DIMETHANOL see EAQ750
SULFUROUS ACID, MONOAMMONIUM SALT see ANB600
SULFUROUS ACID, MONOSODIUM SALT see SFE000
SULFUROUS ACID, MONOSODIUM SALT, REACTION PRODUCTS WITH BISPHENOL A DISODIUM SALT AND FORMALDEHYDE see SOP600
SULFUROUS ACID, SODIUM SALT (1:2) see SJZ000
SULFUROUS ANHYDRIDE see SOH500
SULFUROUS DICHLORIDE see TFL000
SULFUROUS OXIDE see SOH500

SULFUROUS OXYCHLORIDE see TFL000
SULFUROUS OXYFLUORIDE see TFL250
SULFUR OXIDE see SOH500
SULFUR OXIDE (N-FLUOROSULPHONYL)IMIDE see SOQ000
SULFUR PENTAFLUORIDE see SOQ450
SULFUR PHOSPHIDE see PHS000
SULFUR SELENIDE see SBT000
SULFUR SUBCHLORIDE see SON510
SULFUR TETRACHLORIDE see SOQ500
SULFUR TETRAFLUORIDE see SOR000
SULFUR THIOCYANATE see SOR200
SULFUR TRIOXIDE see SOR500
SULFUR TRIOXIDE, uninhibited (NA 1829) (DOT) see SOR500
SULFUR TRIOXIDE, inhibited (UN 1829) (DOT) see SOR500
SULFURYL AZIDE CHLORIDE see SOS500
SULFURYL CHLORIDE see SOT000
SULFURYL CHLORIDE FLUORIDE see SOT500
SULFURYL CHLOROFLUORIDE see SOT500
SULFURYL DIAZIDE see SOU000
SULFURYL FLUORIDE see SOU500
SULFURYL FLUOROCHLORIDE see SOT500
SULGIN see AHO250
SULINDAC see SOU550
SULINOL see SOU550
SULISOBENZONE see HLR700
SULKA see CAX800
SULKOL see SOD500
SULMET see SJW500
SULMET see SNJ000
SULOCTIDIL see SOU600
SULOCTIDYL see SOU600
SULOCTON see SOU600
SULODYNE see PDC250
SULOUREA see ISR000
SULPELIN see SNV000
SULPHABUTIN see BOT250
SULPHACETAMIDE see SNQ710
SULPHACETAMIDE SODIUM see SNQ000
SULPHADIAZINE see PPP500
SULPHADIMETHOXINE see SNN300
SULPHADIMETHYLISOXAZOLE see SNN500
SULPHADIMETHYLPYRIMIDINE see SNJ000
SULPHADIMIDINE see SNJ000
SULPHADIONE see SOA500
SULPHAFURAZ see SNN500
SULPHAGUANIDINE see AHO250
SULPHAMERAZINE see ALF250
SULPHAMETHALAZOLE see SNK000
SULPHAMETHIZOLE see MPQ750
SULPHAMETHOXAZOL see SNK000
SULPHAMETHOXAZOLE see SNK000
SULPHAMETHOXYPYRIDAZINE see AKO500
SULPHAMETHYLISOXAZOLE see SNK000
SULPHAMIC ACID (DOT) see SNK500
SULPHAMOPRINE see SNI500
SULPHAN BLUE see ADE500
SULPHANILAMIDE see SNM500
5-SULPHANILAMIDO-3,4-DIMETHYL-ISOXAZOLE see SNN500
3-SULPHANILAMIDO-5-METHYLISOXAZOLE see SNK000
SULPHANILIC ACID see SNN600
SULPHASALAZINE see PPN750
SULPHASIL see SNQ710
SULPHASOMIDINE see SNJ350
SULPHATHIAZOLE see TEX250
SULPHAUREA see SNQ550
SULPHEIMIDE see CBF800
SULPHENAZOLE see AIF000
SULPHENONE see CKI625
SULPHENTAL see PDO800
SULPHISOMEZOLE see SNK000
SULPHISOXAZOL see SNN500
2-SULPHOBENZOIC IMIDE see BCE500

SULPHOBENZOIC IMIDE CALCIUM SALT see CAM750
SULPHOBENZOIC IMIDE, SODIUM SALT see SJN700
SULPHO BLACK see CMS250
SULPHOBROMOPHTHALEIN see HAQ600
SULPHOBROMOPHTHALEIN SODIUM see HAQ600
SULPHOCARBONIC ANHYDRIDE see CBV500
SULPHOFURAZOLE see SNN500
SULPHO GREEN 2B see FAE950
SULPHOLANE see SNW500
SULPHOL BLACK B 150 see CMS250
SULPHOL BLACK BS see CMS250
SULPHOL BLACK G CONC see CMS250
SULPHOL BLACK GSP PASTE see CMS250
SULPHOL BLACK XG see CMS250
SULPHOL DARK RED R see CMS257
SULPHOL Liquid BLACK QG see CMS250
SULPHOL PRUNE B see CMS257
SULPHOL RED BROWN 3B see CMS257
SULPHOL RED BROWN 5B see CMS257
SULPHOL RED BROWN 3RB see CMS257
SULPHON ACID BLUE R see ADE750
SULPHON ACID BLUE RA see ADE750
SULPHONAL NP 1 see SNQ700
1-(4-SULPHO-1-NAPHTHYLAZO)-2-NAPHTHOL-3,6-DISULPHONIC ACID, TRISODIUM SALT see FAG020
SULPHON-MERE see SOA500
SULPHONOL FAST RED R see CMM330
SULPHONOL ORANGE R see ADG000
SULPHONOL RED PG see CMM320
SULPHONOL RED R see CMM330
SULPHONTHAL see PDO800
SULPHON YELLOW RS-CF see CMM759
1,1'-SULPHONYLBIS(4-AMINOBENZENE) see SOA500
p,p-SULPHONYLBISBENZAMINE see SOA500
4,4'-SULPHONYLBISBENZAMINE see SOA500
p,p-SULPHONYLBISBENZENAMINE see SOA500
4,4'-SULPHONYLBISBENZENAMINE see SOA500
SULPHONYLDIANILINE see SOA500
p,p-SULPHONYLDIANILINE see SOA500
1-p-SULPHOPHENYLAZO-2-NAPHTHOL-6-SULPHONIC ACID, DISODIUM SALT see FAG150
SULPHORMETHOXINE see AIE500
SULPHOS see PAK000
SULPHOXALINE see SNW500
SULPHOXIDE see ISA000
SULPHUR (DOT) see SOD500
SULPHUR, molten (DOT) see SOD500
SULPHUR, lump or powder (DOT) see SOD500
SULPHUR BLACK see CMS250
SULPHUR BLACK 1 see CMS250
SULPHUR BLACK 911 see CMS250
SULPHUR BLACK B see CMS250
SULPHUR BLACK BRW see CMS250
SULPHUR BLACK BT see CMS250
SULPHUR BLACK 911 EXTRA see CMS250
SULPHUR BLACK G see CMS250
SULPHUR BLACK HD GRAINS see CMS250
SULPHUR BLACK HD GRAINS 115 see CMS250
SULPHUR BLACK HD GRAINS 125 see CMS250
SULPHUR BLACK LA see CMS250
SULPHUR BLACK M see CMS250
SULPHUR BLACK P see CMS250
SULPHUR BLACK PA see CMS250
SULPHUR BLACK PAN see CMS250
SULPHUR BLACK T see CMS250
SULPHUR BLACK TC see CMS250
SULPHUR BLACK WLX see CMS250
SULPHUR BROWN 5RP see CMS257

SULPHUR DIOXIDE, LIQUEFIED (DOT) see SOH500

SULPHUR FAST BLACK BG see CMS250

SULPHURIC ACID see SOI500

SULPHURIC ACID, CADMIUM SALT (1:1) see CAJ000

SULPHUR MUSTARD GAS see BIH250

SULPHUR PRUNE MX see CMS257

SULPHUR RED BROWN 3B HIGHLY CONC see CMS257

SULPIRID see EPD500

SULPIRIDE see EPD500

SUL-PO-MAG see MAI650

SULPRIM see TKX000

SULPROFOS see SOU625

SULPROSTONE see SOU650

SULSOL see SOD500

SULTAMICILLIN TOSILATE see SOU675

SULTANOL see BQF500

SULTIRENE see AKO500

SULTOPRIDE see EPD100

SULTOPRIDE HYDROCHLORIDE see SOU725

SULTOSILATO de PIPERACINA (SPANISH) see PIK625

SULTOSILIC ACID, PIPERAZINE SALT see PIK625

SULTROPRIDE CHLORHYDRATE see SOU725

SULXIN see SNN300

SULZOL see TEX250

SUM 3170 see DCS200

SUMAC TANNIN see SOU750

SUMADIL see DVW700

SUMAGIC see SOU800

SUMAPEN VK see PDT750

SUMATRA CAMPHOR see BMD000

SUMATRA YELLOW X 1940 see CMS210

SUMATRIPTAN see DPH300

SUMEDINE see ALF250

SUMETROLIM see TKX000

SUMIACRYL ORANGE G see CMM820

SUMIACRYL RED G see BAQ750

SUMIACRYL RED GT see BAQ750

SUMIACRYL YELLOW 3G see CMM890

SUMI-ALFA see FAR150

SUMI-ALPHA see FAR150

SUMICIDIN see FAR100

SUMICIDIN A ALPHA see FAR150

SUMICURE M see MJQ000

SUMIDUR 44V10 see PKB100

SUMIDUR 44V20 see PKB100

SUMIDUR 44VM see PKB100

SUMIFIX TURQUOISE BLUE G see CMS224

SUMIFLOC CL8 see MCB050

SUMIFLY see FAR100

SUMIKANOL 508 see MCB050

SUMIKAPRINT YELLOW GFN see CMS210

SUMIKARON BLUE S-BG see CMP075

SUMIKARON BLUE E-BL see CMP070

SUMIKARON BLUE E-BL see DBQ220

SUMIKARON BLUE E-FBL see CMP070

SUMIKARON BLUE E-FBL see DBQ220

SUMIKARON BLUE R see CMP070

SUMIKARON BLUE R see DBQ220

SUMIKARON RED E-FBL see AKI750

SUMIKATHENE see PJS750

SUMIKATHENE F 702 see PJS750

SUMIKATHENE F 101-1 see PJS750

SUMIKATHENE F 210-3 see PJS750

SUMIKATHENE G 201 see PJS750

SUMIKATHENE G 202 see PJS750

SUMIKATHENE G 701 see PJS750

SUMIKATHENE G 801 see PJS750

SUMIKATHENE G 806 see PJS750

SUMIKATHENE HARD 2052 see PJS750

SUMILEX see PMF750

SUMILIGHT BLACK AB see CMN300

SUMILIGHT BLACK G see CMN240

SUMILIGHT RED 4B see CMO885

SUMILIGHT SUPRA TURQUOISE BLUE G see COF420

SUMILIT BBM see LRP750

SUMILIT EXA 13 see PKQ059

SUMILIT PCX see AAX175

SUMILIZER TPS see DXG700

SUMIMAL 100 see MCB050

SUMIMAL 40S see MCB050

SUMIMAL 100C see MCB050

SUMIMAL M see MCB050

SUMIMAL M 22 see MCB050

SUMIMAL M 55 see MCB050

SUMIMAL M 70 see MCB050

SUMIMAL M 30W see MCB050

SUMIMAL M 40S see MCB050

SUMIMAL M 40W see MCB050

SUMIMAL M 50W see MCB050

SUMIMAL M 62W see MCB050

SUMIMAL M 65B see MCB050

SUMIMAL M 668 see MCB050

SUMIMAL M 100C see MCB050

SUMIMAL M 100D see MCB050

SUMIMAL M 504C see MCB050

SUMINE 2005 see BDX750

SUMINE 2006 see BDX750

SUMINE 2015 see DQP800

SUMINOL BRILLIANT SCARLET DH see CMM325

SUMINOL FAST BLUE PR see CMM100

SUMINOL FAST SKY BLUE B see CMM090

SUMINOL LEVELLING SKY BLUE R see CMM080

SUMINOL MILLING RED GRS see NAO600

SUMINOL MILLING RED RS see CMM330

SUMINOL MILLING YELLOW MR see CMM759

SUMINOL RED PG see CMM320

SUMINOL RED RS see NAO600

SUMIOXON see PHD750

SUMIOXONE see PHD750

SUMIPLEX LG see PKB500

SUMIPOWER see FAR100

SUMIREZ 607 see MCB050

SUMIREZ 613 see MCB050

SUMIREZ 614 see UTU500

SUMIREZ 615 see MCB050

SUMIREZ M613 see MCB050

SUMIREZ RESIN 613 see MCB050

SUMISCLEX see PMF750

SUMISET D see CNH125

SUMISEVEN see SOU800

SUMISORB 130 see HND100

SUMISOYA see FLZ075

SUMITAL see AMX750

SUMITARD XL see PHB550

SUMITEKKUSU REJIN 810 see UTU500

SUMITEX m³ see MCB050

SUMITEX 260 see UTU500

SUMITEX 810 see UTU500

SUMITEX FSK see DTG000

SUMITEX H 10 see PKP750

SUMITEX M6 see MCB050

SUMITEX M10 see MCB050

SUMITEX MC see MCB050

SUMITEX MK see MCB050

SUMITEX MW see MCB050

SUMITEX NF 113 see UTU500

SUMITEX NS see DTG000

SUMITEX RESIN 810 see UTU500

SUMITEX RESIN MC see MCB050

SUMITHIAN see DSQ000

SUMITHION S-ISOMER see MKC250

SUMITHRIN see PDR700

SUMITOL see BQC250

SUMITOL 80W see BQC250

SUMITOMO FAST SCARLET G see CMM325

SUMITOMO LIGHT GREEN SF YELLOWISH see FAF000

SUMITOMO PATENT PURE BLUE VX see ADE500

SUMITOMO PX 11 see PKQ059

SUMITOMO S 4084 see COQ399

SUMITOMO WOOL GREEN S see ADF000

SUMITOX see MAK700

SUMMETRIN see PAG500

SUMMIT see CJN300

SUMPOCAINE see SPB800

SUN 1165 see PIF600

SUN-4936 see HGL680

SUNAPTIC ACID B see NAR000

SUNAPTIC ACID C see NAR000

SUNBRELLA see AIH600

SUNBURST RED see NAP100

SUNCHOLIN see CMF350

SUNCHOMINE ORANGE GR see CMP882

SUNCHROMINE BLACK ET see EDC625

SUNCHROMINE BLUE BLACK B see CMP880

SUNCHROMINE VIOLET B see HLI000

SUNCHROMINE YELLOW GG see SIT850

SUNCIDE see PMY300

SUNCIDE, nitrosated see PMY310

SUNETTE see AAF900

SUNFAST MAGENTA see DTV360

SUNFLOWER OIL (UNHYDROGENATED) see SOU875

SUNFRAL see FLZ050

SUNITOMO S 4084 see COQ399

SUNLIGHT see SOV000

SUNLIGHT 700 see CAT775

SUN ORANGE A GEIGY see FAG150

SUNRABIN see EAU075

SUNSET YELLOW see FAG150

SUNSET YELLOW BSS see FAG150

SUNSET YELLOW FCF see FAG150

SUNSET YELLOW FCF SUPRA see FAG150

SUNSET YELLOW FU see FAG150

SUNSET YELLOW FU SUPRA see FAG150

SUNSET YELLOW LAKE see FAG150

SUNSOFT O 30B see GGR200

SUNSOLT WA see SNY100

SUNTOP M 300 see MCB050

SUNTOP M 420 see MCB050

SUNTOP M700 see MCB050

SUNTOP M701 see MCB050

SUNWAX 151 see PJS750

SUN YELLOW see FAG150

SUN YELLOW A-CE see FAG150

SUN YELLOW A-FDC see FAG150

SUN YELLOW EXTRA CONC. A EXPORT see FAG150

SUN YELLOW EXTRA PURE A see FAG150

SUN YELLOW FCF see FAG150

SUPACAL see CNG980

SUPARI, nut extract see BFW000

SUPARI (INDIA) see AQT650

SUPEOL see GGR200

SUPER 3S see CAT775

SUPER 1500 see CAT775

SUPER 3-1000 see EHP700

SUPERACRYL AE see PKB500

SUPER AMIDE L-9A see BKE500

SUPERANABOLON see DYF450

SUPERAN BLUE AR see CMM070

SUPERBA see CBT750

SUPERBARNON see FBW135

SUPER-BECKAMINE see MCB050

SUPER-BECKAMINE G 821 see MCB050

SUPER-BECKAMINE J 820 see MCB050

SUPER-BECKAMINE J 840 see MCB050

SUPER-BECKAMINE J 1600 see MCB050

SUPER-BECKAMINE L 101 see MCB050

SUPER-BECKAMINE L 105 see MCB050

SUPER-BECKAMINE L 117 see MCB050

SUPER-BECKAMINE L 121 see MCB050

SUPER BECKOSOL ODL 131-60 see MCB050

SUPER BLAZER see FIW100

SUPERBRESILINE see LFT800

SUPER-CAID see BMN000

SUPER-CARBOVAR see CBT750

SUPERCEL 3000 see USS000

SUPERCHROME BLACK TS see EDC625

SUPERCHROME BLUE BC see CMP880

SUPERCHROME BLUE BG see CMP880

SUPERCHROME RED B see CMG750
SUPERCHROME VIOLET B see HLI000
SUPERCICLIN see PPY250
SUPERCOAT see CAT775
SUPER COBALT see CNA250
SUPERCOL see LIA000
SUPERCOL G.F. see GLU000
SUPERCOL U POWDER see GLU000
SUPERCORTIL see PLZ000
SUPERCORTYL see SOV100
SUPER COSAN see SOD500
SUPER CRAB-E-RAD-CALAR see CAM000
SUPER DAL-E-RAD see CAM000
SUPER DAL-E-RAD-CALAR see CAM000
SUPER-DENT see SHF500
SUPER-DE-SPROUT see DMC600
SUPER D WEEDONE see DAA800
SUPER D WEEDONE see TAA100
SUPER DYLAN see PJS750
SUPERELGETOL see DUT800
SUPERFLAKE ANHYDROUS see CAO750
SUPERFLOC see PKF750
SUPERFUSTEL see FBW000
SUPERFUSTEL K see FBW000
SUPERGAN see RDK000
SUPER GLUE see EHP700
SUPER GLUE see MIQ075
SUPERGLYCERINATED FULLY
HYDROGENATED RAPESEED OIL see
RBK200
SUPER HARTOLAN see CMD750
SUPERIAN YELLOW R see SGP500
SUPERINONE see TDN750
SUPERIOR OIL see MQV796
SUPERIOR OIL see MQV855
SUPERIUONE (FRENCH) see TDN750
SUPERLYSOFORM see FMV000
SUPERMAN MANEB F see MAS500
SUPERMIKROKALCIT see CAD800
SUPERMITE see CAT775
SUPER MOSSTOX see MJM500
SUPER MULTIFEX see CAT775
SUPERNOX see DGI000
SUPERNYLITE BLUE BR see CMM090
SUPEROL see GGA000
SUPEROL RED C RT-265 see CHP500
SUPERORMONE CONCENTRE see DAA800
SUPEROX see BDS000
SUPEROXOL see HIB050
SUPER-PFLEX see CAT775
SUPERPHOSPHATE see SOV500
SUPERPREDNOL see SOW000
SUPER PRODAN see DXE000
SUPER RODIATOX see PAK000
SUPER-ROZOL see BMN000
SUPERSEPTIL see SNJ000
SUPERSORBON IV see CBT500
SUPERSORBON S 1 see CBT500
SUPER-SPECTRA see CBT750
SUPER SPROUT STOP see DMC600
SUPER SSS see CAT775
SUPER SUCKER-STUFF see DMC600
SUPER SUCKER-STUFF HC see DMC600
SUPERTAH see CMY800
SUPERTITE 10 see HJE100
SUPER-TREFLAN see DUV600
SUPER VMP see NAH600
SUPICAINE AMIDE SULFATE see AJN750
SUPICANE AMIDE HYDROCHLORIDE see
PME000
SUPININ see SOW500
SUPININE see SOW500
SUPLEXEDIL see BPM750
SUPONA see CDS750
SUPONE see CDS750
SUP'OPERATS see BMN000
SUPOTRAN see CKF500
SUPRA see IHC450
SUPRACAPSULIN see VGP000
SUPRACET BLUE GREEN B see DMM400
SUPRACET BRILLIANT BLUE BG see
MGG250

SUPRACET BRILLIANT BLUE 2GN see
TBG700
SUPRACET BRILLIANT RED 2B see AKE250
SUPRACET BRILLIANT VIOLET 3R see
DBP000
SUPRACET DIAZO BLACK A see DPO200
SUPRACET FAST CRIMSON B see CMP080
SUPRACET FAST GREEN BLUE B see
DMM400
SUPRACET FAST PINK 2R see AKO350
SUPRACET FAST PINK 3B see DBX000
SUPRACET FAST SCARLET B see ENP100
SUPRACET FAST VIOLET B see DBY700
SUPRACET FAST YELLOW 2R see DUW500
SUPRACET FAST YELLOW 4R see CMP090
SUPRACET FAST YELLOW G see AAQ250
SUPRACET ORANGE R see AKP750
SUPRACET VIOLET 2B see AKP250
SUPRACET YELLOW RR see DUW500
SUPRACHOL see SGD500
SUPRACID see SNQ000
SUPRADIN see VGP000
SUPRAMIKE see BAP000
SUPRAMYCIN see TBX250
SUPRANEPHRANE see VGP000
SUPRANEPHRINE see VGP000
SUPRANEPHRIN SOLUTION see AES500
SUPRANOL see VGP000
SUPRANOL FAST RED 3G see CMM330
SUPRANOL FAST RED GG see CMM330
SUPRANOL FAST RED RX see NAO600
SUPRANOL FAST SCARLET GN see
CMM320
SUPRANOL ORANGE RA see ADG000
SUPRANOL RED PBX-CF see CMM330
SUPRANOL RED PG-CF see CMM320
SUPRANOL RED R see CMM330
SUPRANOL SCARLET BN see CMM320
SUPRANOL SCARLET GS see CMM325
SUPRANOL YELLOW R see CMM759
SUPRARENIN see AES000
SUPRARENIN see VGP000
SUPRARENIN HYDROCHLORIDE see
AES500
SUPRASEC see LIH000
SUPRASEC 1042 see PKB100
SUPRASEC DC see PKB100
SUPRASIL see SCK600
SUPRASIL W see SCK600
SUPRASTIN see CKV625
SUPRATHEN see PJS750
SUPRATHEN C 100 see PJS750
SUPRATONIN see MQS100
SUPRAZO RED 4B see CMO885
SUPREFACT see LIU420
SUPREL see VGP000
SUPREMAL see DYE600
SUPREME DENSE see TAB750
SUPREXCEL RED 8BL see CMO885
SUP'R FLO see DXQ500
SUP'R FLO see MAS500
SUP'R FLO FERBAM FLOWABLE see FAS000
SUPRIFEN see HKH500
SUPRIFENE see HKH500
SUPRIFEN PSB HYDROCHLORIDE see
DNU200
SUPRILENT see VGA300
SUPRIMAL see HGC500
SUPRIN see TKX000
SUPRISTOL see SNL850
SUPROFEN see TEN750
SUPROTAN see CKF500
SURAMETHINIUM see HLC500
SURAMIN see BAT000
SURAMINE see BAT000
SURAUTO see IDW000
SURCHLOR see SHU500
SURCOPUR see DGI000
SURCO SXS see XJJ010
SUREAU (CANADA, HAITI) see EAI100
SURECIDE see CON300
SUREM see DLY000

SUREM see PEU000
SURENINE see VGP000
SURE-SET see CJN000
SURESTRINE see BIS750
SURESTRINE see BIS750
SURESTRINE see BIT000
SURESTRYL see BIS750
SURESTRYL see BIS750
SURESTRYL see BIT000
SURESTRYL see MRU600
SURFACTANT NF see BLX000
SURFACTANT RW 20 see TNP512
SURFACTANT WK see DXY000
SURFEX MM see CAT775
SURFIL S see CAT775
SURFLAN see OJY100
SURFLON S-III-S see HAS050
SURFONIC L 24-4 see AFJ155
SURFROYAL CTAC see HCQ525
SURGAM see SOX400
SURGEX see BET000
SURGICAL SIMPLEX see PKB500
SURGI-CEN see HCL000
SURHEME see PEU000
SURIKA see TKH750
SURINAM GREENHEART WOOD see
HLY500
SURIRENE see AKO500
SURITAL see AGL375
SURITAL see SOX500
SURITAL SODIUM see SOX500
SURITAL SODIUM (derivative) see SOX500
SURITAL SODIUM SALT see SOX500
SURMONTIL see DLH200
SURMONTIL MALEATE see SOX550
SUROFENE see HCL000
SURPLIX see DLH600
SURPLIX see DLH630
SURPRACIDE see DSO000
SURPUR see DGI000
SURRECTAN see AJY250
SURSUM see CJN250
SURSUMID see EPD500
SU SEGURO CARPIDOR see DUV600
SUSPEN see PDT750
SUSPENDOL see ZVJ000
SUSPENSO see CAT775
SUSPHRINE see VGP000
SUSTANE see BFW750
SUSTANE see BQI000
SUSTANE see BRM500
SUSTANE 1-F see BQI000
SUSTANONE see TBF500
SUSVIN see MRH209
SUTAN see EID500
SUTICIDE see HCQ500
SUTOPROFEN see TEN750
SUVREN see BRS000
SUXAMETHIONIUM CHLORIDE see
HLC500
SUXAMETHONIUM see CMG250
SUXAMETHONIUM CHLORIDE see HLC500
SUXAMETHONIUM DICHLORIDE see
HLC500
SUXAMETHONIUM IODIDE see BJI000
SUXCERT see HLC500
SUXEMETHONIUM see CMG250
SUXETHONIUM CHLORIDE see HLC500
SUXIBUZONE see SOX875
SUXIL see SNE000
SUXILEP see ENG500
SUXIMAL see ENG500
SUXIMER see DNV610
SUXIN see ENG500
SUXINUTIN see ENG500
SUXINYL see HLC500
SUY-B 2 see IGK800
SUZORITE MICA see MQS250
SUZU see ABX250
SUZU H see HON000
SV-052 see SMW400
SV 1017 see DKP800

SV-1522 see ABH500
SVC see ABX500
SVKh 1 see CGW300
SVITPREN see PJQ050
SVKh 40 see CGW300
SVO 9 see PKL100
SW 400 see CAW850
SW-751 see POM275
SWALLOW WORT see CCS650
SWAMP HELLEBORE see FAB100
SWAMP WOOD see LEF100
SWANOL AM 301 see LBU200
SWANOL CA 2150 see LBX075
SWANOL CA 2350 see HCQ525
SWARTZIOL see ICE000
SWAT see SOY000
SWEBATE see TAL250
SWEDISH GREEN see CNN500
SWEDISH GREEN see COF500
SWEENEY'S ANT-GO see ARD750
SWEEP see PAJ000
SWEEP see TBQ750
SWEETA see SJN700
SWEET BELLS see DYA875
SWEET BETTIE see CCK675
SWEET BIRCH OIL see MPI000
SWEET BIRCH OIL see SOY100
SWEET DIPEPTIDE see ARN825
SWEETENED NAPHTHA (PETROLEUM) see NAQ570
SWEET GUM see SOY500
SWEET MYRTLE see WBA000
SWEET ORANGE OIL see OGY000
SWEET PEA SEEDS see SOZ000
SWEET POTATO PLANT see CCO680
SWEET SHRUB see CCK675
SWEETWOOD BARK OIL see CCO500
SWEP see DEV600
SWERTIANOLIN see SOZ100
SWIETENIA MAHAGONI see MAK300
SWINE PROSTATE EXTRACT see RLK875
SWISS BLUE see BJI250
SWISS CHEESE PLANT see SLE890
SWITCH IVY see DYA875
SWP (ANTIOXIDANT) see BRP750
SWS 03314 see PJR300
SWS-F 222 see PJR300
SY-83 see LAG000
SYBIROMYCIN see SCF500
SYBR GREEN I see SOZ300
SYCOTROL see PIM000
SYD 230 see CGB250
SYDNONE, 3-(o-CHLOROPHENYL)- see CKK025
SYDNONE IMINE, 3-(1-METHYL-2-PHENYLETHYL)-, MONOHYDROCHLORIDE see SPA000
SYDNOPHENE see SPA000
SYDNOPHEN HYDROCHLORIDE see SPA000
SYEP see SPA500
SYGETHIN see SPA650
SYGETIN see SPA650
SYKOSE see BCE500
SYKOSE see SJN700
SYLACAUGA 88B see CAT775
SYLANTOIC see DKQ000
SYLANTOIC see DNU000
SYLGARD 184 CURING AGENT see SPB000
SYLLIT see DXX400
SYLODEX see ARM268
SYMBIO see SNN300
SYMMETREL see AED250
SYMMETREL see TJG250
SYMMETRIC DIMETHYLUREA see DUM200
SYMPAMINA-D see BBK500
SYMPAMINE see BBK000
SYMPATEDRINE see BBK000
SYMPATEKTOMAN see TCC000
SYMPATHIN E see NNO500
SYMPATHIN I see VGP000
SYMPATHOL see HLV500

SYMPATHOL see SPD000
m-SYMPATHOL see NCL500
m-SYMPATHOL see SPC500
SYMPATHOLYTIN see DCR200
SYMPATHOLYTIN see DCT050
SYMPATOL see SPD000
m-SYMPATOL see NCL500
m-SYMPATOL see SPC500
SYMPATOL TARTRATE see SPD000
SYMPHORICARPOS (various species) see SED550
SYMPHYTINE see SPB500
SYMPHYTUM OFFICINALE L see RRK000
SYMPHYTUM OFFICINALE L see RRP000
SYMPLOCARPUS FOETIDUS see SDZ450
SYMPOCAINE HYDROCHLORIDE see SPB800
SYMPROPAMIN see FMS875
SYMPTOM 2 see POH250
SYMULER BRILLIANT SCARLET G see CMS160
SYMULER EOSIN TONER see BNH500
SYMULER FAST BLUE 6011 see IBV050
SYMULER FAST ORANGE GRD see CMU820
SYMULER FAST SCARLET 4R see MMP100
SYMULER FAST VIOLET R see DFN450
SYMULER FAST YELLOW 4090G see CMS210
SYMULER FAST YELLOW 5GF see CMS210
SYMULER FAST YELLOW GF see DEU000
SYMULER FAST YELLOW GRF see CMS208
SYMULER FAST YELLOW GRTF see CMS208
SYMULER LAKE RED C see CHP500
SYMULER ORANGE LAKE 43 see CMM220
SYMULER RED 3023 see CMS148
SYMULER RED NRY see CMS148
SYMULEX MAGENTA F see FAG070
SYMULEX PINK F see FAG070
SYMULON ACID BRILLIANT SCARLET 3R see FMU080
SYMULON ACID FAST YELLOW MR see CMM759
SYMULON ACID ORANGE II see CMM220
SYMULON ACID RED PG see CMM320
SYMULON CHROME VIOLET B see HLI000
SYMULON DIRECT BLACK BH see CMN800
SYMULON DIRECT BORDEAUX NS see CMO872
SYMULON METANIL YELLOW see MDM775
SYMULON ORANGE GC BASE see CEH690
SYMULON RED B BASE see NEQ000
SYMULON RED RL BASE see MMF780
SYMULON SCARLET G BASE see NMP500
SYMULON SCARLET 2G SALT see DEO295
SYNACRYL FAST RED 2G see BAQ750
SYNACRYL RED 2G see BAQ750
SYNADENYLIC ACID see SPB100
SYNADRIN see PEV750
SYNALAR see SPD500
SYNALGOS see DQA400
SYNAMOL see SPD500
SYNANCEJA HORRIDA Linn. VENOM see SPB875
SYNANDONE see SPD500
SYNANDRETS see MPN500
SYNANDROL see TBG000
SYNANDROL F see TBF500
SYNANDRONE see SPD500
SYNANDROTABS see MPN500
SYNANTHIC see OMY500
SYNAPAUSE see EDU500
SYNAPEN see PDD350
SYNAPHORIN see CMG675
SYNASAL see SPC500
SYNASTERON see PAN100
SYNATE see SBN000
SYNBETAN P see MEG250
SYNCAINE see AIT250
SYNCAL see BCE500
SYNCELOSE see MIF760
SYNCHOMATE ORANGE G see CMP882

SYNCHROCEPT B see FAQ500
SYNCHROMATE ORANGE AOR see NEY000
SYNCILLIN see PDD350
SYNCL see ALV000
SYNCORDAN see FMS875
SYNCORT see DAQ800
SYNCORTA see DAQ800
SYNCORTYL see DAQ800
SYNCOUMAR see ABF750
SYNCUMA see HAQ570
SYNCUMAR see ABF750
SYNCURARINE see PDD300
SYNCURINE see DAF600
SYNDIOL see EDO000
SYNDIOTACTIC POLYPROPYLENE see PMP500
SYNDROX see MDQ500
SYNDROX see MDT600
SYNEKTAN NPP see SOP600
SYNELAUDINE see DAM700
SYNEPHRIN see HLV500
m-SYNEPHRINE see NCL500
m-SYNEPHRINE see SPC500
p-SYNEPHRINE see HLV500
dl-SYNEPHRINE see SPD000
dl-p-SYNEPHRINE see SPD000
m-SYNEPHRINE HYDROCHLORIDE see SPC500
SYNEPHRINE TARTRATE see SPD000
(+−)-SYNEPHRINE TARTRATE see SPD000
SYNERGID R see DWW000
SYNERGIST 264 see OES000
SYNERONE see TBG000
SYNERPENIN see PDD350
SYNESTRIN see DKA600
SYNESTRIN see DKB000
SYNESTROL see DAL600
SYNESTROL see DLB400
SYNETHENATE see SPC500
SYNFAT 1006 see CAU000
SYNFEROL AH EXTRA see SPD100
SYNFLORAN see DUV600
SYNFUELS see GCC200
SYNGACILLIN see AJJ875
SYNGESTERONE see PMH500
SYNGESTROTABS see GEK500
SYNGUM D 46D see GLU000
SYNGYNON see HNT500
SYNHEXYL see HGK500
SYNISTAMIN see TAI500
SYNKAMIN see AKX500
SYNKAMIN BASE see AKX500
SYNKAVITE see NAQ600
SYNKAY see MMD500
SYNKAYVITE see NAQ600
SYNKLOR see CDR750
SYNMIOL see DAS000
SYNOESTRON see DKB000
SYNOPEN see CKV625
SYNOPEN R see CKV625
SYNOTODECIN see PPY250
SYNOTOL L-60 see BKE500
SYNOVEX S see PMH500
SYNOX 5LT see MJO500
SYNOX TBC see BSK000
SYNPEN see CKV625
SYNPENIN see AIV500
SYNPERONIC OP see GHS000
SYNPITAN see ORU500
SYNPOL 1500 see SMR000
SYNPOR see CCU250
SYNPREN-FISH see PIX250
SYNPRO STEARATE see CAX350
SYNSAC see SPD500
SYNSTIGMIN BROMIDE see POD000
SYNSTIGMINE see DER600
SYNTAR see CMY800
SYNTARIS see FDD085
SYNTARPEN see DGE200
SYNTARPEN see SPD600
SYNTASE 62 see MES000

SYNTASE 100 see DMI600
SYNTEDRIL see BBV500
SYNTEFIX see CNH125
SYNTEN BLUE P-BGL see CMP075
SYNTEN YELLOW 2G see AAQ250
SYNTEN YELLOW P 2R see DUW500
SYNTES 12A see EIV000
SYNTESTRIN see DKB000
SYNTESTRINE see DKB000
SYNTETREX see PPY250
SYNTETRIN see PPY250
SYNTETRIN NITRATE see SPE000
SYNTEXAN see DUD800
SYNTHALEN K see PIB300
SYNTHALINE GREEN see PJQ100
SYNTHAMIDE 5 see SPE500
SYNTHARSOL see ACN250
SYNTHECILLIN see PDD350
SYNTHECILLINE see PDD350
SYNTHEMUL 90-588 see ADV900
SYNTHENATE see HLV500
SYNTHETIC 3956 see CDV100
SYNTHETIC BRADYKININ see BML500
SYNTHETIC β-CAROTENE CRYSTAL see NBV200
SYNTHETIC CRUDE OIL see COD725
SYNTHETIC EUGENOL see EQR500
SYNTHETIC GASOLINE see GCC200
SYNTHETIC GLYCERIN see GGA000
SYNTHETIC INDIGO see BGB275
SYNTHETIC INDIGO TS see BGB275
SYNTHETIC IRON OXIDE see IHC450
SYNTHETIC ISOPARAFFINIC PETROLEUM HYDROCARBONS see ILR150
SYNTHETIC LH-RH see LIU370
SYNTHETIC MUSTARD OIL see AGJ250
SYNTHETIC OXYTOCIN see ORU500
SYNTHETIC PARAFFIN and SUCCINIC DERIVATIVES see SPE600
SYNTHETIC PYRETHRINS see AFR250
SYNTHETIC TRF see TNX400
SYNTHETIC TRH see TNX400
SYNTHETIC TSH-RELEASING FACTOR see TNX400
SYNTHETIC TSH-RELEASING HORMONE see TNX400
SYNTHETIC WINTERGREEN OIL see MPI000
SYNTHILA see DJB200
SYNTHOBILIN see HAQ570
SYNTHOESTRIN see DKA600
SYNTHOFOLIN see DKA600
SYNTHOMYCINE see CDP250
SYNTHOPHYLLINE see DNC000
SYNTHOSTIGMINE BROMIDE see POD000
SYNTHOSTIGMINE METHYL SULFATE see DQY909
SYNTHOVO see DLB400
SYNTHRIN see BEP500
SYNTHROID see LFG050
SYNTHROID SODIUM see LFG050
SYNTOCIN see ORU500
SYNTOCINON see ORU500
SYNTOCINONE see ORU500
SYNTODRIL see BBV500
SYNTOFOLIN see DKA600
SYNTOLUTAN see PMH500
SYNTOMETRINE see LJL000
SYNTON YELLOW 2G see AAQ250
SYNTOPHEROL see VSZ450
SYNTOPHEROL ACETATE see TGJ055
SYNTOSTIGMIN see DER600
SYNTOSTIGMIN (tablet) see POD000
SYNTOSTIGMIN BROMIDE see POD000
SYNTOSTIGMINE BROMIDE see POD000
SYNTROGENE see DLB400
SYNTROM see ABF750
SYNTRON B see EIV000
SYNTROPAN see AOD250
SYN-U-TEX 4113E see MCB050
SYPHOS see ASD000
SYRAP see CMG000

SYRAPRIM see TKZ000
SYRIAN BEAD TREE see CDM325
SYRINGALDEHYDE see DOF600
SYRINGEALDEHYDE see DOF600
SYRINGIC ACID see SPE700
SYRINGIC ACID ETHYL CARBONATE ESTER with METHYL RESERPATE see RCA200
SYRINGIC ALDEHYDE see DOF600
SYRINGOL see DOJ200
SYRINGOPINE see RCA200
SYRINGYLALDEHYDE see DOF600
SYROSINGOPIN see RCA200
SYROSINGOPINE see RCA200
SYRUP of IPECAC, U.S.P. see IGF000
SYS 67ME see SIL500
SYS 67MPROP see CLO200
SYSTAM see OCM000
SYSTAMEX see OMY500
SYSTANATE MR see PKB100
SYSTANAT MR see PKB100
SYSTEMOX see DAO600
SYSTHANE see MRW775
SYSTHANE 6 FLO see MRW775
SYSTODIN see QHA000
SYSTOGENE see TOG250
SYSTOPHOS see OCM000
SYSTOX see DAO600
SYSTOX SULFONE see SPF000
SYSTRAL see CIS000
SYTAM see OCM000
SYTASOL see CBW000
SYTLOMYCIN AMINONUCLEOSIDE see ALQ625
SYTOBEX see VSZ000
SYTON FAST GERANIUM 3B see CMG750
SYTON FAST ORANGE G see CMS145
SYTON FAST RED 2G see DVB800
SYTON FAST RED R see CJD500
SYTON FAST SCARLET RB see MMP100
SYTON FAST SCARLET RD see MMP100
SYTON FAST SCARLET RN see MMP100
SYTRON see EJA379
SYZYGIUM JAMBOS (Linn.) Alston, extract excluding roots see SPF200
SZ 6300 see TLD000
SZAZ see MNB000
SZESCIOMETYLENODWUIZOCYJANIAN see DNJ800
SZKLARNIAK see DGP900
T-2 see TIK500
T3 see LGK050
T₃ see LGK050
T4 see CPR800
T10 see DSF300
T 40 see TGF250
T 72 see PNX000
L-T3 see LGK050
l-T4 see TFZ275
T 100 see EHG100
T 100 see TGM740
T 101 see UTU500
T-113 see BPG000
T-125 see TBX000
T-144 see IPX000
2,4,5-T see TAA100
T 593 see OKK600
864T see IBQ100
Ts 14 see PJT300
Ts 20 see PJT300
Ts 30 see PJT300
Ts 35 see PJT300
Ts 40 see PJT300
Ts 55 see PJT300
Ts 62 see PJT300
T-1035 see DSA800
T-1036 see DJJ400
T-1088 see HNP000
T-1123 see TGH665
T-1125 see HNO000
T-1152 see HNO500
T 1220 see SJJ200

T 1258 see TAK850
T-1384 see NCP875
T-1551 see CCS369
T-1690 see HNN000
T-1703 see IRF000
T-1768 see DPL900
T-1770 see MIA775
T 1824 see BGT250
T-1835 see DEC200
T-1843 see MIC250
T-1982 see TAA400
T-2002 see BJE750
T-2104 see EIF000
T-2106 see IPX000
T-2588 see TAA420
T-3689 see PCG635
T 514' see PCL300
79T61 see SBE500
T30177 see ZNS400
T-42082 see SFP500
Th 1165a see FAQ100
Ts36Khr see ZFJ130
α-T see MIH275
β-T see TIN000
T 130-2500 see CAT775
T4 (hormone) see TFZ275
Ts 35 (polymer) see PJT300
TA see THS855
TA 1 see MRN675
TA 12 see TAN750
T-4CA see TEV000
TA 064 see DAP850
Ts A 16 see PJT300
TA-2711 see EAB560
TAA see TFA000
TA-AZUR see THM750
TABAC (FRENCH) see TGI100
TABAC du DIABLE (CANADA) see SDZ450
TABACHIN (MEXICO) see CAK325
TABACO (SPANISH) see TGI100
TABALGIN see HIM000
TABASCO HOT PEPPER SAUCE see TAA875
TABASCO PEPPER see PCB275
TABATREX see SNA500
TABILIN see BFD000
TABLE SALT see SFT000
TABLOID see AMH250
TABOON A see EIF000
TABUN see EIF000
TABUTREX see SNA500
TAC-28 see VIK150
TAC 121 see TGG250
TAC 131 see TGG250
TACALCITOL see SBL600
TACARYL see MDT500
TACARYL see MPE250
TACAZYL see MPE250
TACE see CLO750
TACE-FN see CLO750
TACHIGAREN see HLM000
TACHIONIN see HII500
TACHMALIN see AFH250
TACITIN see BCH750
TACKLE see CLS075
TACOSAL see DKQ000
TACOSAL see DNU000
TACP see TNC725
TACRINE see TCJ075
TACROLIMUS HYDRATE see TAA900
TACRYL see MPE250
TACTARAN see TAF675
TACUMIL see TKX000
TAD see TEP500
TAENIATOL see MJM500
TAFASAN see MEP250
TAFASAN 6W see MEP250
TAFAZINE see BJP000
TAFIL see XAJ000
TAG see ABU500
TAG-39 see ECU750
TAG 331 see ABU500
TAGAMET see TAB250

TAGAT see SEH000
TAGATHEN see CHY250
TAGETES MEAL and EXTRACT see TAB260
TAGETES OIL see TAB275
TAG FUNGICIDE see ABU500
TAG HL 331 see ABU500
TAGN see THN800
TAHMABON see DTQ400
TAI-284 see CFH825
TAI 284 see CMV500
(±)-TAI 284 see CFH825
dl-TAI 284 see CFH825
TAIC see THS100
TAIFEN see ZMA000
TAIFUN see EHH600
TAIGUIC ACID see HLY500
TAIGU WOOD see HLY500
TAIL FLOWER see APM875
TAJMALIN see AFH250
TAK see MAK700
TAKACIDIN see TAB300
TAKACILLIN see LEJ500
TAKAMINA see VGP000
TAKAMINE HT (BACILLUS SUBTILIS) see TAB400
TAKANARUMIN see ZVJ000
TAKAOKA ACID RED RS see NAO600
TAKAOKA AMARANTH see FAG020
TAKAOKA BRILLIANT SCARLET 3R see FMU080
TAKAOKA METANIL YELLOW see MDM775
TAKAOKA RHODAMINE B see FAG070
TAKATHENE see PJS750
TAKATHENE P 3 see PJS750
TAKATHENE P 12 see PJS750
TAKE 20 see CBS800
TAKEDO 1969-4-9 see GGA800
TAKENATE see XIJ000
TAKENATE 500 see XIJ000
TAKENATE 300C see PKB100
TAKESULIN see CCS550
TAKILON see PKQ059
TAKTIC see MJL250
TAKYCOR see AFH250
TALADREN see DFP600
TALAMO see AMX750
TALAN see CBW000
TALARGAN see TEH500
TALATROL see TEM500
TALBOT see LCK000
TALBOT see LCK100
TALBUTAL see AFY500
TALC see TAB750
TALC, containing asbestos fibers see TAB775
TALCORD see AHJ750
TALCUM see TAB750
TALIBLASTIN see TEH250
TALIBLASTINE see TEH250
TALIMOL see TEH500
TALISOMYCIN see TAC500
TALISOMYCIN B see TAC750
TALISOMYCIN S[10b] see TAB785
TALL OIL see TAC000
TALL OIL HYDROXYETHYL IMIDAZOLINE see TAC050
TALL OIL, 1-(2-HYDROXYETHYL)-2-IMIDAZOLINE-2-YLNOR- see TAC075
TALL OIL IMIDAZOLINE see TAC075
TALLOL see TAC000
TALLOW see TAC100
TALLOW BENZYL DIMETHYLAMMONIUM CHLORIDE see DTC600
N-TALLOWPYRROLIDINONE see TAC200
(TALLOW)TRIMETHYLAMMONIUM CHLORIDE see TAC300
N-TALLOW-TRIMETHYLAMMONIUM SURFACTANT see TAC300
TALLYSOMYCIN A see TAC500
TALLYSOMYCIN B see TAC750
TALLYSOMYCIN S[10b] see TAB785

TALMON see MAO350
TALODEX see FAQ900
TALOFLOC see DXG625
TALON see TAC800
TALON RODENTICIDE see TAC800
α-l-TALO-OCT-4-ENOFURANURONIC ACID,1-(6-AMINO-9H-PURIN-9-YL)-3,6-ANHYDRO-6-C-CARBOXY-1,5-DIDEOXY- see GJU900
TALOXIMINE HYDROCHLORIDE see TAC825
TALPHENO see EOK000
TALSTAR see TAC850
TALUCARD see POB500
TALUSIN see POB500
TALWAN see DOQ400
TALWIN see DOQ400
TAMA PEARL TP 121 see CAT775
TAMARIZ see SAZ000
TAMARON see DTQ400
TAMAS see BBW500
TAMBALISA (CUBA) see NBR800
TAMCHA see AJV500
TAME see PBX400
TAME (ETHER) see PBX400
TAMETIN see TAB250
TAMEX see BQN600
TAMILAN see PAP000
TAMOFEN see TAD175
TAMOL L see BLX000
TAMOL SN see BLX000
TAMOXASTA see TAD175
TAMOXIFEN see NOA600
TAMOXIFEN (E) see TAC880
CIS-TAMOXIFEN see TAC880
TAMOXIFEN CITRATE see TAD175
TAMOXIFEN CITRATE see TAD175
TAMPOVAGAN STILBOESTROL see DKA600
TAMPULES see CDP000
TAMRAGHOL see CNK559
TAMUSLOSIN HYDROCHLORIDE see EFA200
TANAFOL see CKF500
TANAGER RED X-761 see CJD500
TANAK m³ see MCB050
TANAKAN see CLD000
TANAKAN see CLD250
TANAK MRX see MCB050
TANAMICIN see HGP550
TANAN see TDT800
TANANE see TDT800
TANASUL see MDU300
TANCAL 100 see CAT775
TANDACOTE see HNI500
TANDALGESIC see HNI500
TANDEARIL see HNI500
TANDEM see TJK100
TANDERAL see HNI500
TANDEX see DUM800
TANDIX see IBV100
TANGANTANGAN OIL see CCP250
TANGARINE LAKE X-917 see CMM220
TANGELO OIL see TAD250
TANGERINE OIL see TAD500
TANGERINE OIL, COLDPRESSED (FCC) see TAD500
TANGERINE OIL, EXPRESSED (FCC) see TAD500
TANICAINE see BJO500
TANIDIL see DQA400
TANNEX see IDA000
TANNIC ACID see TAD750
TANNIN see TAD750
TANNIN from ACORN see ADI625
TANNIN from BETEL NUT see BFW050
TANNIN from BRACKEN FERN see BML250
TANNIN from CHERRY BARK OAK see CDL750
TANNIN from CHESTNUT see CDM250
TANNIN-FREE FRACTION of BRACKEN FERN see TAE250

TANNIN from LIMONIUM NASHII see MBU750
TANNIN from MARSH ROSEMARY see MBU750
TANNIN from MIMOSA see MQV250
TANNIN from MYRABOLAM see MRZ100
TANNIN from MYROBALANS see MSB750
TANNIN from MYRTAN see MSC000
TANNIN from PERSIMMON see PCP500
TANNIN from QUEBRACHO see QBJ000
TANNIN from SUMAC see SOU750
TANNIN from SWEET GUM see SOY500
TANNIN from VALONEA see VCK000
TANNIN from WAX MYRTLE see WBA000
TANOL see HOI300
TANONE see DRR400
TANRUTIN see RSU000
TANSTON see XQS000
TANSY OIL see TAE500
TANSY RAGWORT see RBA400
TANTALIC ACID ANHYDRIDE see TAF500
TANTALUM see TAE750
TANTALUM-181 see TAE750
TANTALUM CHLORIDE see TAF000
TANTALUM FLUORIDE see TAF250
TANTALUM OXIDE see TAF500
TANTALUM(V) OXIDE see TAF500
TANTALUM PENTACHLORIDE see TAF000
TANTALUM PENTAFLUORIDE see TAF250
TANTALUM PENTAOXIDE see TAF500
TANTALUM PENTOXIDE see TAF500
TANTALUM POTASSIUM FLUORIDE see PLH000
TANTAN (PUERTO RICO) see LED500
TANTARONE see MJE760
TANTUM see BBW500
TAOMYCIN see HOH500
TAOMYXIN see HOH500
TAORYL see CBG250
TAP see MPN000
TAP 85 see BBQ500
TAP-144 see LEZ300
TA-33MP see TAN750
TAPAR see HIM000
TAPAZOLE see MCO500
T.A.P.E. see NNR400
TAPHAZINE see BJP000
TAPIOCA see CCO680
TAPIOCA see DBD800
TAPIOCA STARCH see SLJ500
TAPIOCA STARCH HYDROXYETHYL ETHER see HLB400
TAPON see SLJ500
TAP 9VP see DGP900
TAP-144-SR see LEZ300
TAR see CMY800
TAR, from tobacco see CMP800
TARACTAN see TAF675
TARA GUM see GMA000
TARAPACAITE see PLB250
TARAPON K 12 see SIB600
TARARACO see AHI635
TARARACO BLANCO (CUBA) see BAR325
TARARACO DOBLE (CUBA) see AHI635
TARASAN see TAF675
TAR CAMPHOR see NAJ500
TAR, COAL see CMY800
TARDAMID see AIE750
TARDAMIDE see AIE750
TARDEX 80 see OAF200
TARDEX 100 see PAU500
TARDIGAL see DKL800
TARDOCILLIN see BFC750
TAREDAN see EHY100
TARFLEN see TAI250
TARGA see QMA100
TARGET MSMA see MRL750
TARICHATOXIN see FOQ000
TARIMYL see SOA500
TARIVID see OGI300
TARLON XB see PJY500
TARNAMID T see PJY500

TARO see EAI600
TAROCTYL see CKP500
TARODYL see GIC000
TARODYN see GIC000
TAR OIL see CMY825
TARO VINE see PLW800
TARPAN see CMW250
TAR, PINE see PIH775
TARRAGON see AFW750
TARRAGON OIL see TAF700
TARTAGO (PUERTO RICO) see CNR135
TARTAN see PHK250
TARTAR see PKU600
TARTAR CREAM see PKU600
TARTAR EMETIC see AQG250
TARTARIC ACID see TAF750
l-(+)-TARTARIC ACID see TAF750
l-TARTARIC ACID, AMMONIUM SALT see DCH000
l-TARTARIC ACID, ANTIMONY POTASSIUM SALT see AQH000
dl-TARTARIC ACID, ANTIMONY POTASSIUM SALT see AQG750
TARTARIC ACID, DIAMMONIUM SALT see DCH000
TARTARIC ACID, DIBENZOATE see DDE300
TARTARIC ACID, DIISOPROPYL ESTER see TAF760
meso-TARTARIC ACID, ION(2−) see TAF775
TARTARIC ACID, MONOSODIUM SALT see SKB000
TARTARIZED ANTIMONY see AQG250
TARTAR YELLOW FS see FAG140
TARTAR YELLOW N see FAG140
TARTAR YELLOW PF see FAG140
TARTAR YELLOW S see FAG140
TARTRAN YELLOW see FAG140
TARTRAPHENINE see FAG140
meso-TARTRATE see TAF775
TARTRATE ANTIMONIO-POTASSIQUE (FRENCH) see AQG250
TARTRATED ANTIMONY see AQG250
TARTRATE de NICOTINE (FRENCH) see NDS500
TARTRAZINE see FAG140
TARTRAZINE A EXPO T see FAG140
TARTRAZINE B see FAG140
TARTRAZINE B.P.C. see FAG140
TARTRAZINE EXTRA PURE A see FAG140
TARTRAZINE FD & C YELLOW #5 see FAG140
TARTRAZINE FQ see FAG140
TARTRAZINE G see FAG140
TARTRAZINE LAKE see FAG140
TARTRAZINE LAKE YELLOW N see FAG140
TARTRAZINE M see FAG140
TARTRAZINE MCGL see FAG140
TARTRAZINE N see FAG140
TARTRAZINE NS see FAG140
TARTRAZINE O see FAG140
TARTRAZINE T see FAG140
TARTRAZINE XX see FAG140
TARTRAZINE XXX see FAG140
TARTRAZINE YELLOW see FAG140
TARTRAZOL BPC see FAG140
TARTRAZOL YELLOW see FAG140
TARTRINE YELLOW O see FAG140
TARWEED see TAG250
TARZOL see DGA200
TASK see DGP900
TASK TABS see DGP900
TASMIN see AOO800
TAT-1 see DIK000
TATB see TMK900
TATBA see AQY375
TAT CHLOR 4 see CDR750
TATD see DXH250
TATD see GEK200
TATERPEX see CKC000
TATHIONE see GFW000

TAT-3 HYDROCHLORIDE see CNE375
TATTOO see DQM600
TAT-3 TRIPALMITATE see PIC100
TATURIL see UVJ450
TAUPHON see TAG750
TAURE(o)DON see GJC000
TAURIN see GJM300
TAURINE see TAG750
TAURINE, N-(3-α-HYDROXY-5-β-CHOLAN-24-OYL)-(8CI) see LHW100
TAURINOPHENETIDINE HYDROCHLORIDE see TAG875
TAUROCHOLATE see TAH250
TAUROCHOLIC ACID see TAH250
TAUROLITHOCHOLIC ACID (6CI,7CI) see LHW100
TAUROMYCETIN see TAH500
TAUROMYCETIN-III see TAH650
TAUROMYCETIN-IV see TAH675
TAUROSIDE E see HAK075
TAUTUBA (PUERTO RICO) see CNR135
TAVEGIL see FOS100
TAVEGYL see FOS100
TA-VERM see PIJ500
TAVOR see CFC250
TAVROMYCETIN III see TAH650
TAVROMYCETIN-IV see TAH675
TAXIFOLIN see DMD000
TAXIFOLIOL see DMD000
TAXILAN see PCK500
TAXIN (GERMAN) see TAH750
TAXINE see TAH750
TAXOL see TAH775
TAXOTERE see TAH800
TAXUS BACCATA LINN., LEAF EXTRACT see KCA100
TAXUS (VARIOUS SPECIES) see YAK500
TAYO BAMBOU (HAITI) see EAI600
TAYSSATO see MEP250
TAZEPAM see CFZ000
TAZICEF see CCQ200
TAZIDIME see CCQ200
TAZONE see BRF500
TB see CMO250
TB 2 see TFD000
TB 220 see CCI525
2,3,6-TBA (herbicide) see TIK500
TBAS-Q see ADS300
TBB see THX500
TBBA see BQK500
TB 1 (BAYER) see FNF000
TBDMS CHLORIDE see BQS300
TBDZ see TEX000
TBE see ACK250
TBEP see BPK250
TBF-43 see TKP850
TBGF see EPW600
TBHP-70 see BRM250
TBHQ (FCC) see BRM500
TBOT see BLL750
TBP see TFD250
TBP see TIA250
TBPMC see BSG300
TBS see THW750
TBS 95 see THW750
TBSM see BPI400
TBT see BSP500
TBTO see BLL750
2,4,5-T BUTOXYETHANOL ESTER see TAH900
2,4,5-T BUTOXYETHYL ESTER see TAH900
T-BUTYLAZO-2-HYDROXYBUTANE see TAH950
T-BUTYLETHANOLAMINE see BRH300
2,4,5-TC see TIX500
TC 3-30 see SMQ500
T 1 (Catalyst) see DBF800
TCA see TII250
TCA see TII500
TCAB see TBN500
TCAOB see TBN550
TCA-PE see AMU000

TCA-PR see AMU500
T-250 CAPSULES see TBX250
TCA SODIUM see TII500
TCAT see TIJ500
TCB see TBO700
TCB see TIL275
2,3,6-TCB see TIK500
2,2',5,5'-TCB see TBO530
2,3,6-TCBA see TIK500
TCBC see TIL250
TCBN see TBS000
TCC see TIL500
TCC see TIQ000
TCC mixed with TFC (2:1) see TIL526
TCD-ALCOHOL DM see TJG550
TCDBD see TAI000
TCDD see TAI000
1,3,6,8-TCDD see TBO800
2,3,7,8-TCDD see TAI000
TCE see TBQ100
1,1,1-TCE see MIH275
TCEO see ECT600
TCH see TFE250
TC HYDROCHLORIDE see TBX250
TCIN see TBQ750
TCM see CHJ500
TCM see TNE775
TCMTB see BOO635
TCNA see TBR250
TCNB see TBR750
TCNP SODIUM SALT see TAI050
TCNQ see TBW750
TCP see HOL125
TCP see TBT000
TCPA see TIY500
TCPE see TIX000
m-TCPN see TBQ750
o-TCPN see TBT200
TCPO see TJB750
TCPP see FQU875
2,4,5-TCPPA see TIX500
TCSA see TBV000
TCT see TFZ000
TCTH see CQH650
TCTNB see TJE200
TCTP see TBV750
TCV-3B see EGM100
TD-183 see TBV750
TD-758 see TBC200
TD 1771 see PEX500
TD-5032 see HEE500
TDA see TGL750
TDBP (CZECH) see TNC500
TDCPP see FQU875
TDCPP see TNG750
TDE see TJQ333
o,p-TDE see CDN000
TDE (DOT) see BIM500
o,p'-TDE see CDN000
p,p'-TDE see BIM500
T-DET see DXY600
TDI see TGM740
2,4-TDI see TGM750
2,6-TDI see TGM800
TDI-80 see TGM740
TDI-80 see TGM750
TDI 80-20 see TGM740
T-82 DIFUMARATE see MDU750
TDI (OSHA) see TGM750
TDOT see TCQ275
TDPA see BHM000
TDS see CHK825
TDS see TES800
2,5-TDS see TGM400
TDS (NEUROTROPE) see TES800
TE see TBF750
TE-031 see MJV775
TE 114 see HNM000
TEA see TCB725
TEA see TCC000
TEA see TJN750
TEAB see TCC000

TEABERRY OIL see MPI000
TEAC see TCC250
TEA CATECHIN see EBB200
TEA CHLORIDE see TCC250
TEAEI see TAI100
TEA LAURYL SULFATE see SON000
T.E.A.S. see SDH670
TEA TREE OIL see TAI150
TEB see TDG500
TEBAC see BFL300
TEBALON see FNF000
TEBE see HNY500
TEBECID see ILD000
TEBECURE see FNF000
TEBEFORM see PNW750
TEBEMAR see FNF000
TEBERUS see EPQ000
TEBESONE I see FNF000
TEBETHIONE see FNF000
TEBEXIN see ILD000
TEBEZON see FNF000
TEBLOC see LIH000
TEBRAZID see POL500
TEBUFENOZIDE see TAI175
TEBUFENOZIDE TECHNICAL see TAI175
TEBUFENPYRAD see CGH750
TEBULAN see BSN000
TEBUTHIURON see BSN000
TEC see TJP750
TECACIN see HGP550
TECELEUKIN see IDG100
TRANS-TECH D 38 see BAP500
TRANS-TECH D 8512 see BAP500
TECH DDT see DAD200
TECHNEPLEX see TAI200
TECHNETATE(1-)-99TC, (N,N-BIS(2-
(BIS(CARBOXYMETHYL)AMINO)ETHYL)G
LYCINATO(5-))-, SODIUM see TAI200
TECHNETIUM TC 99M SULFUR COLLOID
see SOD500
TECHNICAL BHC see BBQ750
90 TECHNICAL GLYCERINE see GGA000
TECHNICAL HCH see BBQ750
TECHNOPOR see PKQ059
TECH PET F see MQV750
TECNAZEN (GERMAN) see TBR750
TECNAZENE see TBR750
TECODIN see DLX400
TECODINE see DLX400
TECOFLEX HR see PKN000
TECOMIN see HLY500
TECPOL see ADV900
TECQUINOL see HIH000
TECRAMINE see TKH750
TECSOL see EFU000
TECTILON ORANGE 3GT see SGP500
TECTO see TEX000
TECTORIGENIN see TKP050
TECTORIGENINE see TKP050
TECZA see TJR000
TEDIMON 31 see PKB100
TEDION see CKM000
TEDION V-18 see CKM000
TEDMA see MDN510
TEDP see SOD100
TEDP (OSHA) see SOD100
TEDTP see SOD100
TEEBACONIN see ILD000
TEF see TND250
TEFAMIN see TEP000
TEFAMIN see TEP500
TEFASERPINA see RDK000
TEFESTROL see TAI220
TEFILAN see DNC000
TEFILIN see TBX250
TEFLEX see KDK000
TEFLON see TAI250
TEFLON (various) see TAI250
TEFLUTHRIN see FMU300
TEFLUTHRINE see FMU300
TEFSIEL C see FLZ050
TEG see TJQ000

TEGAFUR see FLZ050
TEGAFUR mixture with URACIL (1:4) see
UNJ810
TEGDH see TCE375
TEGDN see TJQ500
TEGENCIA HYDROCHLORIDE see DAI200
TEGESTER see IQN000
TEGESTER BUTYL STEARATE see BSL600
TEGESTER ISOPALM see IQW000
TEGIN see OAV000
TEGIN 503 see OAV000
TEGIN 515 see OAV000
TEGIN P see SLL000
TEGO 51 see DYA850
TEGOLAN see CMD750
TEGON see TJQ500
TEGO-OLEIC 130 see OHU000
TEGOPEN see AOC500
TEGOPEN see SLJ000
TEGOPEN see SLJ050
TEGOSEPT B see BSC000
TEGOSEPT E see HJL000
TEGOSEPT M see HJL500
TEGOSEPT P see HNU500
TEGO-STEARATE see EJM500
TEGOSTEARIC 254 see SLK000
TEGRETAL see DCV200
TEGRETOL see DCV200
TEIB see TND000
TEICHOMYCIN A2 see TAI400
TEKKAM see NAK500
TEKODIN see DLX400
TEKRESOL see CNW500
TEKWAISA see MNH000
TEL see TCF000
TELAGAN see TEH500
TELAR see CMA700
TELARGAN see TEH500
TELARGEAN see TEH500
TELCOTENE see PJS750
TELDANE see TAI450
TELDRIN see CLD250
TELDRIN see TAI500
TELEBRIX 38 see TAI600
TELEBRIX 300 see MQR300
TELECOTHENE see PJS750
TELEFOS see IOT000
TELEMID see MDU300
TELEMIN see PPN100
TELEMIN-SOFT see TAI725
TELEPAQUE see IFY100
TELEPATHINE see HAI500
TELEPRIN see TKX000
TELESMIN see DCV200
TELETRAST see IFY100
TELGIN-G see FOS100
TELIDAL see HNI500
TELINE see TBX250
TELIPEX see TBG000
TELL see TCI150
TELLOY see TAJ000
TELLUR (POLISH) see TAJ000
TELLURANE-1,1-DIOXIDE see TAI735
TELLURATE see TAI750
TELLURIC ACID see TAI750
TELLURIC(VI) ACID see TAI750
TELLURIC ACID, AMMONIUM SALT see
ANV750
TELLURIC ACID, DISODIUM SALT see
SKC500
TELLURIC ACID, DISODIUM SALT,
PENTAHYDRATE see TAI800
TELLURIC CHLORIDE see TAJ250
TELLURIUM see TAJ000
TELLURIUM (dust or fume) see TAJ600
TELLURIUM CHLORIDE see TAJ250
TELLURIUM COMPOUNDS see TAJ500
TELLURIUM
DIETHYLDITHIOCARBAMATE see EPJ000
TELLURIUM DIOXIDE see TAJ750
TELLURIUM HEXAFLUORIDE see TAK250
TELLURIUM HYDROXIDE see TAI750

TELLURIUM NITRIDE see TAK500
TELLURIUM OXIDE see TAJ750
TELLURIUM TETRABROMIDE see TAK600
TELLURIUM TETRACHLORIDE see TAJ250
1,1'-TELLUROBISETHANE see DKB150
TELLUROUS ACID, DISODIUM SALT see
SKC500
TELMICID see DJT800
TELMID see DJT800
TELMIDE see DJT800
TELMIN see MHL000
TELOCIDIN B see TAK750
TELODRIN see OAN000
TELODRON see CLD250
TELOMYCIN see TAK800
TELON see BCP650
TELON BLUE BL see CMM090
TELON BLUE RRL see CMM080
TELON CHROME ORANGE G see CMP882
TELONE see DGG000
TELONE see DGG950
TELONE II SOIL FUMIGANT see DGG950
TELON FAST BLACK E see AQP000
TELON FAST BLACK PE see CMN230
TELON FAST RED GG see CMM330
TELON FAST SCARLET N see CMM320
TELON YELLOW NLL see FAL050
TELOTREX see TBX250
TELTOZAN see CAL750
TELVAR see CJX750
TELVAR see DXQ500
TELVAR DIURON WEED KILLER see
DXQ500
TELVAR MONURON WEEDKILLER see
CJX750
TEMAFLOXACIN see TAK850
TEMARIL see TAL000
TEMASEPT see BOD600
TEMASEPT see THW750
TEMASEPT II see THW750
TEMAZEPAM see CFY750
TEMECHINE see TAL275
TEMED see TDQ750
TEMEFOS see TAL250
TEMENTIL see PMF500
TEMEPHOS see TAL250
TEMEQUINE see TAL275
TEMESTA see CFC250
TEMETEX see DKF130
TEMGESIC see TAL325
TEM-HISTINE see DPJ400
TEMIC see CBM500
TEMIK see CBM500
TEMIK G10 see CBM500
TEMIK OXIME see MLX800
TEMIK SULFOXIDE see MLW750
TEMISKAMITE see MBU790
TEMLO see HIM000
TEMOPHOS see TAL250
TEMPALGIN see TAL335
TEMPANAL see HIM000
TEMPARIN see BJZ000
TEMPLIN OIL see AAC250
TEMPO see REF250
TEMPO see TDT800
TEMPODEX see BBK500
TEMPOL see HOI300
TEMPONITRIN see NGY000
TEMPO-RESERPINA see RDK000
TEMPOSERPINE see RDK000
TEMPRA see HIM000
TEMUR see TDX250
TEMUS see BMN000
TEN see TJO000
TENAC see DGP900
TENALIN see TEO250
TENAMENE see BJL000
TENAMENE 1 see BQG650
TENAMENE 2 see DEG200
TENAMENE 31 see BJT500
TENAMINE see TLN500
TENAPLAS see PJS750

TENATHAN see BFW250
TENCILAN see CDQ250
TENDEARIL see HNI500
TENDIMETHALIN see DRN200
TENDOR see IKB000
TENDOSCEN-COMPR. see RDK000
TENDUST see NDN000
TENEBRIMYCIN see NBR500
TENEBRIMYCIN see TAL350
TENEMYCIN see NBR500
TENEMYCIN see TAL350
TENESDOL see CCR875
TENFIDIL see DPJ200
TENIATHANE see MJM500
TENIATOL see MJM500
TENICID see PDM750
TENIPOSIDE see EQP000
TENITE 423 see PMP500
TENITE 800 see PJS750
TENITE 1811 see PJS750
TENITE 2910 see PJS750
TENITE 2918 see PJS750
TENITE 3300 see PJS750
TENITE 3340 see PJS750
TENNECETIN see PIF750
TENNECO 1742 see PKQ059
TENN-PLAS see BCL750
TENNUS 0565 see AAX175
TENORAN see CJQ000
TENORMIN see TAL475
TENOSIN-WIRKSTOFF see TOG250
TENOX BHA see BQI000
TENOX BHT see BFW750
TENOX HQ see HIH000
TENOXICAM see TAL485
TENOX PG see PNM750
TENOX P GRAIN PRESERVATIVE see
PMU750
TENOX TBHQ see BRM500
TENSANYL see RDK000
TENSERLIX see RDK000
TENSERPINE "ASSIA" see RDK000
TENSERPINIE see RDK000
TENSIBAR see DIH000
TENSICOR see DNM400
TENSIFEN see HNJ000
TENSILON see EAE600
TENSILON see TAL490
TENSILON BROMIDE see TAL490
TENSILON CHLORIDE see EAE600
TENSINASE D see DWF200
TENSINYL see MDQ250
TENSIONAL see RDK000
TENSIONORME see RDK000
TENSIVAL see TEH500
TENSOL 7 see PKB500
TENSOPAM see DCK759
TENSYL see AFG750
TENTAROME see FBW110
TENTON see MFK500
TENTONE see MFK500
TENTONE MALEATE see MFK750
TENTRATE-20 see PBC250
TENUATE see DIP600
TENUATE HYDROCHLORIDE see DIP600
TENUAZONIC ACID see VTA750
l-TENUAZONIC ACID see VTA750
TENULIN see TAL550
TENURID see DXH250
TENUTEX see DXH250
TEOBROMIN see TEO500
TEODRAMIN see DYE600
TEOF see TJL600
TEOFILCOLINA see CMG300
2-(7'-TEOFILLINMETIL)-1,3-DIOSSOLANO
(ITALIAN) see TEQ175
TEOFYLLAMIN see TEP000
TEOHARN see MCB000
TEOKOLIN see CMG300
TEONANACATL see PHU500
TEONICON see PPN000
TEOPROLOL see TAL560

TEOS see EPF550
TEP see TJT750
TEPA see TND250
TEPANIL see DIP600
TEPERINE see DLH630
TEPIDONE see SGF500
TEPIDONE RUBBER ACCELERATOR see
SGF500
TEPILTA see DTL200
TEPOGEN see SLJ000
TEPP (ACGIH) see TCF250
TEPSERPINE see RDK000
TEQU see TBQ800
TEQUINOL see HIH000
TERABOL see MHR200
TERACOL see GHY100
TERACOL 30 see GHY100
TERACOL 650 see GHY100
TERACOL 1000 see GHY100
TERACOL 2000 see GHY100
TERALEN see AFL500
TERALIN see DPJ400
TERALLETHRIN see TAL575
TERALUTIL see HNT500
TERAMYCIN HYDROCHLORIDE see
HOI000
TERANOL see ECX100
TERAPINYL see FMS875
TERAPRINT see AKI750
TERASIL BLUE GREEN CB see DMM400
TERASIL BLUE R see CMP075
TERASIL BRILLIANT BLUE 3RL see CMP070
TERASIL BRILLIANT BLUE 3RL see DBQ220
TERASIL BRILLIANT PINK 4BN see DBX000
TERASIL BRILLIANT VIOLET 3B see
DBY700
TERASIL TURQUOISE BLUE G see DMM400
TERASIL YELLOW 2GC see AAQ250
TERASIL YELLOW GBA EXTRA see AAQ250
TERASIL YELLOW GWL see KDA075
TERATHANE see GHY100
TERATHANE 1000 see GHY100
TERATHANE 2900 see GHY100
TERAZOSIN see FPP100
TERBACIL see BQT750
TERBAM see BSG300
TERBENOL see NOA000
TERBENOL HYDROCHLORIDE see
NOA500
TERBENZENE see TBD000
TERBIUM see TAL750
TERBIUM CHLORIDE see TAM000
TERBIUM CHLORIDE, HEXAHYDRATE see
TAM100
TERBIUM CITRATE see TAM500
TERBIUM OXIDE see TAN000
TERBIUM TRICHLORIDE HEXAHYDRATE
see TAM100
TERBOLAN see RDK000
TERBUCARB see TAN300
(−)-TERBUCLOMINE see PAP230
TERBUCONAZOLE see TAN050
TERBUFOS see BSO000
TERBUMETON see BQC500
TERBUTALIN see TAN100
TERBUTALINE see TAN100
TERBUTALINE SULFATE see TAN250
TERBUTALINE SULPHATE see TAN250
TERBUTHYLAZINE see BQB000
TERBUTOL see TAN300
TERCININ see CMV000
TERCYL see CBM750
TEREBENTHINE see TOD750
TEREFTALODINITRIL (CZECH) see BBP250
TEREPHTAHLIC ACID-ETHYLENE
GLYCOL POLYESTER see PKF750
TEREPHTALDEHYDE see TAN500
TEREPHTALDEHYDES (FRENCH) see
TAN500
TEREPHTHALALDEHYDE see TAN500
TEREPHTHALALDEHYDONITRILE see
COK250

TEREPHTHALAMIDE see TAN600
TEREPHTHALAMIDIC ACID see TAN600
TEREPHTHALDIAMIDE see TAN600
TEREPHTHALIC ACID see TAN750
TEREPHTHALIC ACID, BIS(2-
ETHYLHEXYL)ESTER see BJS500
TEREPHTHALIC ACID CHLORIDE see
TAV250
TEREPHTHALIC ACID DIAMIDE see
TAN600
TEREPHTHALIC ACID, DIBUTYL ESTER
see DEH700
TEREPHTHALIC ACID DICHLORIDE see
TAV250
TEREPHTHALIC ACID, DIETHYL ESTER
see DKB160
TEREPHTHALIC ACID, DIOCTYL ESTER
(6CI,7CI,8CI) see BJS300
TEREPHTHALIC ACID, DISODIUM SALT
see TAN755
TEREPHTHALIC ACID ISOPROPYLAMIDE
see IRN000
TEREPHTHALIC ACID METHYL ESTER see
DUE000
TEREPHTHALIC ACID, POLYAMIDE with
p-PHENYLENEDIAMINE (8CI) see BBO750
TEREPHTHALIC ALDEHYDE see TAN500
TEREPHTHALIC DIAMIDE see TAN600
TEREPHTHALIC DICHLORIDE see TAV250
TEREPHTHALONITRILE see BBP250
5,5'-(TEREPHTHALOYLBIS(IMINO-p-
PHENYLENE))BIS(2,4-DIAMINO-1-
ETHYLPYRIMIDINIUM)-DI-p-
TOLUENESULFONATE see TAO750
5,5'-(TEREPHTHALOYLBIS(IMINO-p-
PHENYLENE))BIS(2,4-DIAMINO-1-
METHYLPYRIMIDINIUM)-DI-p-
TOLUENESULFONATE see TAP000
5,5'-(TEREPHTHALOYLBIS(IMINO-p-
PHENYLENE))BIS(2,4-DIAMINO-1-
PROPYLPYRIMIDINIUM)-DI-p-
TOLUENESULFONATE see TAP250
3,3'-(TEREPHTHALOYLBIS(IMINO-p-
PHENYLENE))BIS(1-
PROPYLPYRIDINIUM)-DI-p-
TOLUENESULFONATE see TAR000
4,4'-(TEREPHTHALOYLBIS(IMINO-p-
PHENYLENE))BIS(1-
PROPYLPYRIDINIUM)-DI-p-
TOLUENESULFONATE see TAR250
TEREPHTHALOYL DICHLORIDE see
TAV250
3,3'-(TEREPHTHALOYLDIIMINOBIS(p-
PHENYLENE))BIS(1-
PROPYLPYRIDINIUM)-DI-p-
TOLUENESULFONATE see TAR000
TERETON see MFW100
TERFAN see PKF750
TERFENADINE see TAI450
TERFLUZINE see TKE500
TERFLUZINE see TKK250
TERFLUZINE DIHYDROCHLORIDE see
TKK250
TERGAL see PKF750
TERGEMIST see TAV750
TERGIMIST see TAV750
TERGITOL see EMT500
TERGITOL 4 see SIO000
TERGITOL 7 see DKD400
TERGITOL 08 see TAV750
TERGITOL 25L see AFJ160
TERGITOL 12-P-9 see TAX500
TERGITOL 15-S-5 see TAY250
TERGITOL 25L3 see AFJ160
TERGITOL 25L5 see AFJ160
TERGITOL 25L7 see AFJ160
TERGITOL 25L9 see AFJ160
TERGITOL 12M10 see PJV100
TERGITOL 15-S-20 see TBA500
TERGITOL 25L12 see AFJ160
TERGITOL 15-S-9 (nonionic) see TAZ000
TERGITOL 15-S-12 (nonionic) see TAZ100

TERGITOL ANIONIC 4 see EMT500
TERGITOL ANIONIC 7 see HCP900
TERGITOL ANIONIC 08 see TAV750
TERGITOL ANIONIC P-28 see TBA750
TERGITOL MIN-FOAM 1X see TBA800
TERGITOL NONIONIC XD see GHY000
TERGITOL NONIONIC XH see PKG600
TERGITOL NONIONIC XH see TBB800
TERGITOL NP-14 see TAW250
TERGITOL NP-27 see TAW500
TERGITOL NP-35 (nonionic) see TAX000
TERGITOL NP-40 (nonionic) see TAX250
TERGITOL NP-4 SURFACTANT see NNC650
TERGITOL NP-4 SURFACTANT see NND200
TERGITOL NPX see NND500
TERGITOL PENETRANT 4 see EMT500
TERGITOL PENETRANT 7 see TBB750
TERGITOL TMN-6 see GHU000
TERGITOL TMN-10 see TBB775
TERGITOL TP-9 (NONIONIC) see PKF000
TERGITOL XD (nonionic) see GHY000
TERGITOL XH see PKG600
TERGITOL XH see TBB800
TERGURID see DLR100
TERGURIDE see DLR100
TERGURIDE HYDROGEN MALEATE see
DLR150
TERIAM see UVJ450
TERIC 12A see AFJ160
TERIC 12A8 see AFJ160
TERIC 12A23 see AFJ160
TERIC 15A11 see AFJ168
TERIC 12A23B see AFJ160
TERIC G 12A see AFJ160
TERIC G 12A6 see AFJ160
TERIC G 12A8 see AFJ160
TERIC LA 4 see AFJ160
TERIC LA 8 see AFJ160
TERIDAX see IFZ800
TERIDIN see UVJ450
TERIDOX see DSM500
TERIMON see TAD175
TERININ see CMV000
TERIT see BAR800
TERMIL see TBQ750
TERMINALIA CHEBULA RETZ TANNING
see MSB750
TERMITKIL see DEP600
TERM-I-TROL see PAX250
TERMOFLEKS A see BKO600
TERMOSOLIDO GREEN FG SUPRA see
PJQ100
TERMOSOLIDO RED LCG see CHP500
TERODILINE CHLORIDE see TBC200
TERODILINE HYDROCHLORIDE see
TBC200
TEROFENAMATE see DGN000
TEROLUT see DYF759
TEROM see PKF750
TEROXIRONE see TBC450
Δ6,8-(9)-TERPADIENONE-2 see MCD250
TERPENE HYDROCHLORIDE see BMD300
TERPENE POLYCHLORINATES see TBC500
TERPENE RESIN see TBC550
TERPENE RESIN, NATURAL see TBC575
TERPENE RESIN, SYNTHETIC see TBC580
TERPENTINOEL (GERMAN) see PIH750
TERPENTIN OEL (GERMAN) see TOD750
TERPHAN see PKF750
m-TERPHENYL see TBC620
o-TERPHENYL see TBC640
p-TERPHENYL see TBC750
1,3-TERPHENYL see TBC620
TERPHENYLS see TBD000
p-TERPHENYL-4-YLACETAMIDE see
TBD250
α-TERPINENE (FCC) see MLA250
γ-TERPINENE (FCC) see MCB750
TERPINENOL-4 see TBD825
TERPINEOL see TBD500
4-TERPINEOL see TBD825
α-TERPINEOL see TBD750

β-TERPINEOL see TBD775
α-TERPINEOL ACETATE see TBE250
α-TERPINEOL (FCC) see TBD500
TERPINEOLS see TBD500
TERPINEOL SCHLECHTHIN see TBD750
TERPINOLENE see TBE000
TERPINYL ACETATE see TBE250
TERPINYL FORMATE see TBE500
TERPINYL PROPIONATE see TBE600
TERPINYL THIOCYANOACETATE see
IHZ000
Δ-1,8-TERPODIENE see MCC250
TERRA ALBA see CAX750
TERRACHLOR see PAX000
TERRACHLOR-SUPER X see EFK000
TERRACOAT see EFK000
TERRACOTTA 2RN see NEY000
TERRA COTTA RRN see NEY000
TERRACUR P see FAQ800
TERRAFLO see EFK000
TERRAFUN see PAX000
TERRAFUNGINE see HOH500
TERRAKLENE see PAJ000
TERRAMITSIN see HOH500
TERRAMYCIN see HOH500
TERRAMYCIN SODIUM see TBF000
TERRA-SYSTAM see BJE750
TERRA-SYTAM see BJE750
TERRASYTUM see BJE750
TERRAZOLE see EFK000
TERREIC ACID see TBF325
TERR-O-GAS 100 see MHR200
TERRILITIN see TBF350
TERRILYTIN see TBF350
TERSAN see TFS350
TERSAN 1991 see BAV575
TERSAN-LSR see MAS500
TERSAN-SP see CJA100
TERSASEPTIC see HCL000
TERSAVIN see PAQ120
TERSETILE BLUE 2BL see CMP075
TERSETILE BLUE RBL see CMP070
TERSETILE BLUE RBL see DBQ220
TERSETILE RUBINE FL see AKI750
TERSETILE YELLOW 5R see CMP090
TERSETILE YELLOW GL see KDA075
TERSETILE YELLOW 5RL see CMP090
dl-TERTATALOL see TBF400
TERTATOLOL see TBF400
(+−)-TERTATOLOL see TBF400
TERTRACID BRILLIANT LIGHT BLUE R see
CMM080
TERTRACID CARMINE BLUE A see ERG100
TERTRACID FAST BLUE SR see ADE750
TERTRACID FAST BLUE TR see CMM100
TERTRACID LIGHT BLUE SE see APG700
TERTRACID LIGHT ORANGE G see
HGC000
TERTRACID LIGHT YELLOW 2R see
SGP500
TERTRACID MILLING RED AGE see
CMM325
TERTRACID MILLING RED G see CMM320
TERTRACID MILLING YELLOW R see
CMM759
TERTRACID ORANGE I see FAG010
TERTRACID ORANGE II see CMM220
TERTRACID PONCEAU 2R see FMU070
TERTRACID RED A see FAG020
TERTRACID RED CA see HJF500
TERTRACID YELLOW M see MDM775
TERTRACID YELLOW TRO see MRL100
TERTRAL D see PEY500
TERTRAL EG see ALS990
TERTRAL ERN see NAW500
TERTRAL G see TGL750
TERTRAL P BASE see ALT250
TERTRANESE SCARLET N-B see ENP100
TERTRANESE YELLOW P-GL see KDA075
TERTRANESE YELLOW N-2GL see AAQ250
TERTROCHROME BLUE B see CMP880
TERTROCHROME BLUE FB see HJF500

TERTROCHROME RED AB see CMG750
TERTROCHROME VIOLET N see HLI000
TERTROCHROME YELLOW 3R see NEY000
TERTRODIRECT BLACK BH see CMN800
TERTRODIRECT BLACK BHS see CMN800
TERTRODIRECT BLACK E see AQP000
TERTRODIRECT BLACK ZD see CMN230
TERTRODIRECT BLUE 2B see CMO000
TERTRODIRECT BLUE F see CMO500
TERTRODIRECT BLUE FF see CMN750
TERTRODIRECT BLUE RW see CMO600
TERTRODIRECT BROWN TB see CMO820
TERTRODIRECT FAST RED 5B see CMO885
TERTRODIRECT GREEN B see CMO840
TERTRODIRECT GREEN BG see CMO830
TERTRODIRECT RED 4B see DXO850
TERTRODIRECT RED C see SGQ500
TERTRODIRECT RED F see CMO870
TERTRODIRECT VIOLET N see CMP000
TERTROGRAS ORANGE SG see CMP600
TERTROGRAS ORANGE SV see PEJ500
TERTROGRAS RED N see SBC500
TERTROPHENE BRILLIANT GREEN G see
BAY750
TERTROPHENE BROWN CG see PEK000
TERTROPIGMENT FAST YELLOW VGR see
CMS208
TERTROPIGMENT ORANGE LRN see
DVB800
TERTROPIGMENT ORANGE PG see
CMS145
TERTROPIGMENT PGR see CMS208
TERTROPIGMENT RED HAB see MMP100
TERTROPIGMENT RED LC see CMS150
TERTROPIGMENT RED LR see NAP100
TERTROPIGMENT SCARLET LRN see
MMP100
TERTROSULPHUR BLACK PB see DUZ000
TERTROSULPHUR PBR see DUZ000
TERTROSULPHUR PRUNE B see CMS257
TERTROSULPHUR PRUNE BG see CMS257
TERTROSULPHUR RED BROWN 3BT
EXTRA see CMS257
TERULAN KP 2540 see ADY500
2,4,5-TES see SKL000
TESAZIN 3105-60 see MCB050
TESCOL see EJC500
TESERENE see DAL600
TESLEN see TBF500
TESMIN 210 see MCB050
TESMIN 201-80 see MCB050
TESMIN 250-60 see MCB050
TESMIN 251-60 see MCB050
TESMIN ME 50L see MCB050
TESNOL see ICC000
TESOPREL see BNU725
TESPAMINE see TFQ750
TESTAFORM see TBG000
TEST-ANABOL see CLG900
TESTANDRONE see TBF500
TESTATE see TBF750
TESTAVOL see VSK600
TESTEX see TBG000
TESTHORMONE see MPN500
TESTICULOSTERONE see TBF500
TESTOBASE see TBF500
TESTODET see TBG000
TESTODIOL see AOO475
TESTODRIN see TBG000
TESTODRIN PROLONGATUM see TBF600
TESTOGEN see TBG000
TESTOMED see CLG900
TESTONIQUE see TBG000
TESTOPROPON see TBF500
TESTORA see MPN500
TESTORAL see AOO275
TESTORMOL see TBG000
TESTOSTEROID see TBF500
TESTOSTERONE see TBF500
trans-TESTOSTERONE see TBF500
TESTOSTERONE, 4-CHLORO-, ACETATE
see CLG900

TESTOSTERONE,
CYCLOPENTANEPROPIONATE see TBF600
TESTOSTERONE
CYCLOPENTYLPROPIONATE see TBF600
TESTOSTERONE CYPIONATE see TBF600
TESTOSTERONE 17-β-CYPIONATE see
TBF600
TESTOSTERONE ENANTHATE see TBF750
TESTOSTERONE ENANTHATE and DEPO-
MEDROXYPROGESTERONE ACETATE see
DAM325
TESTOSTERONE ENANTHATE and DEPO-
PROVERA see DAM325
TESTOSTERONE-17-β-ESTRADIOL MIXT.
see TBF650
TESTOSTERONE ETHANATE see TBF750
TESTOSTERONE HEPTANOATE see
TBF750
TESTOSTERONE HEPTOATE see TBF750
TESTOSTERONE HEPTYLATE see TBF750
TESTOSTERONE HYDRATE see TBF500
TESTOSTERONE OENANTHATE see
TBF750
TESTOSTERONE PROPIONATE see
TBG000
TESTOSTERONE PROPIONATE see
TBG000
TESTOSTERONE-17-PROPIONATE see
TBG000
TESTOSTERONE-17-β-PROPIONATE see
TBG000
TESTOSTERONE UNDECANOATE see
TBG050
TESTOSTERONE, UNDECANOATE
(ESTER) see TBG050
TESTOSTERONE UNDECYLATE see
TBG050
TESTOSTOSTERONE see TBF500
TESTOSTROVAL see TBF750
TESTOVIRON see MPN500
TESTOVIRON see TBG000
TESTOVIRON SCHERING see TBF500
TESTOVIRON T see TBF500
TESTOXYL see TBG000
TESTRED see MPN500
TESTREX see TBG000
TESTRONE see TBF500
TESTRYL see TBF500
TESULOID see SOD500
TETA see TJR000
TETACIN see CAR780
TETACIN-CALCIUM see CAR780
TETAINE see BAC175
TETAMON IODIDE see TCC750
TETANUS TOXIN see TBG100
TETAZINE see CAR780
TETD see DXH250
TETIDIS see DXH250
TETIOTHALEIN SODIUM see TDE750
TETLEN see PCF275
TETRAACETIL PENTOETRIOL see NNR400
TETRAACETYL PENTESTRIOL see NNR400
2,3,4,6-TETRA-o-ACETYL-1-THIO-β-d-
GLUCOPYRANOSATO-S-
(TRIETHYLPHOSPHINE)GOLD see ARS150
TETRAACRYLONITRILECOPPER(I)
PERCHLORATE see TBG250
TETRAACRYLONITRILECOPPER(II)
PERCHLORATE see TBG500
O,O,O',O'-TETRAAETHYL-
BIS(DITHIOPHOSPHAT) (GERMAN) see
EEH600
O,O,O,O-TETRAAETHYL-DIPHOSPHAT,
BIS(O,O-DIAETHYLPHOSPHORSAEURE)-
ANHYDRID (GERMAN) see TCF250
O,O,O,O-TETRAAETHYL-
DITHIONOPYROPHOSPHAT (GERMAN)
see SOD100
TETRAALKOFEN BPE see DED100
TETRAAMINE-2,3-BUTANEDIIMINE
RUTHENIUM(III) PERCHLORATE see
TBG600

TETRAAMINEDITHIOCYANATO
COBALT(III) PERCHLORATE see TBG625
1,4,5,8-TETRAAMINO-9,10-
ANTHRACENEDIONE see TBG700
1,4,5,8-TETRAAMINOANTHRAQUINONE
see TBG700
3,3',4,4'-TETRAAMINOBIPHENYL see
BGK500
3,3',4,4'-TETRAAMINOBIPHENYL
TETRAHYDROCHLORIDE see BGK750
TETRA-(4-AMINOPHENYL)ARSONIUM
CHLORIDE see TBG710
TETRAAMMINECOPPER(II) AZIDE see
TBG750
TETRAAMMINECOPPER(II) NITRATE see
TBH000
TETRAAMMINECOPPER(II) NITRITE see
TBH250
TETRAAMMINEDICHLOROPLATINUM(II)
see TBI100
TETRAAMMINEDICHLORORHODIUM
CHLORIDE see TBH300
TETRAAMMINEHYDROXYNITRATOPLAT
INUM(IV) NITRATE see TBH500
TETRAAMMINELITHIUM
DIHYDROGENPHOSPHIDE see TBH750
TETRAAMMINEPALLADIUM(II) NITRATE
see TBI000
TETRAAMMINEPLATINNIUM(II)
DICHLORIDE (SP-4-1)- (9CI) see TBI100
TETRAAMMINEPLATINUM DICHLORIDE
see TBI100
TETRAAMMINEPLATINUM(2+)
DICHLORIDE see TBI100
TETRAAMMINEPLATINUM(II)
DICHLORIDE see TBI100
TETRAAMMINEPLATINUM(2+)
DICHLORIDE (SP-4-1)- see TBI100
TETRA-N-AMYLAMMONIUM BROMIDE
see TEA250
TETRAAMYLAMMONIUM CHLORIDE see
TBI200
TETRA-N-AMYLAMMONIUM CHLORIDE
see TBI200
TETRAAMYLSTANNANE see TBI600
1,3,5,7-TETRAAZAADAMANTANE see
HEI500
1,4,7,10-TETRAAZADECANE see TJR000
1,5,8,12-TETRAAZADODECANE see TBI700
1H-2,5,10,10B-
TETRAAZAFLUORANTHENE, 4-AMINO-6-
METHYL- see ALF650
9,10,14c-15-TETRAAZANAPHTHO(1,2,-3-
fg)NAPHTHACENE NITRATE see TBI750
3,6,9,12-TETRAAZATETRADECANE-1,14-
DIAMINE see PBD000
4a,8a,9a,10a-TETRAAZA-2,3,6,7-
TETRAOXAPERHYDROANTHRACENE see
TBI775
1,4,6,9-TETRAAZA-TRICYCLO-(4.4.1.1⁹,⁴)
DODECAN (GERMAN) see TBJ000
1,3,6,8-
TETRAAZATRICYCLO(4.4.1.1³,⁸)DODECAN
E see TBJ000
TETRAAZIDO BENZENE QUINONE
(DOT) see TBJ250
TETRAAZIDO-p-BENZOQUINONE see
TBJ250
TETRABAKAT see TBX250
TETRA-BASE see MJN000
TETRABENAZINE see TBJ275
TETRABENZAINE see TBJ275
TETRABENZINE see TBJ275
TETRABLET see TBX250
TETRA BLUE 2B see ICU135
TETRABON see TBX000
TETRABORANE(10) see TBJ300
TETRA(BORON
NITRIDE)FLUOROSULFATE see TBJ400
TETRABORON TETRACHLORIDE see
TBJ475

1,1,2,2-TETRABROMAETHAN (GERMAN)
see ACK250
TETRABROMIDE METHANE see CBX750
TETRABROMOACETYLENE see ACK250
3,3',5,5'-TETRABROMO-2,2'-BIPHENYLDIOL
see DAZ110
TETRABROMOBIPHENYL ETHER see
TBJ550
TETRABROMOBISPHENOL A see MKA270
2,2',6,6'-TETRABROMOBISPHENOL A see
MKA270
3,3',5,5'-TETRABROMOBISPHENOL A see
MKA270
3,5,3',5'-TETRABROMOBISPHENOL A see
MKA270
2,4,5,7-TETRABROMO-9-o-
CARBOXYPHENYL-6-HYDROXY-3-
ISOXANTHONE, DISODIUM SALT see
BNH500
3,4,5,6-TETRABROMO-o-CRESOL see TBJ500
TETRABROMO-m-CRESOLPHTHALEIN
SULFONE see BNA940
TETRABROMO-m-CRESOLPHTHALEIN
SULFONE see TBJ505
3',3",5',5"-TETRABROMO-m-
CRESOLSULFONEPHTHALEIN see BNA940
3',3",5',5"-TETRABROMO-m-
CRESOLSULFONEPHTHALEIN see TBJ505
TETRABROMODIAN see MKA270
2,3,7,8-TETRABROMODIBENZODIOXIN
see TBJ510
2,3,7,8-TETRABROMODIBENZO-p-DIOXIN
see TBJ510
2,3,7,8-TETRABROMODIBENZOFURAN see
TBJ525
2,4,5,7-TETRABROMO-12,15-
DICHLOROFLUORESCEIN, DIPOTASSIUM
SALT see CMM000
2',4',5',7'-TETRABROMO-4,7-DICHLORO-
FLUORESCEIN DIPOTASSIUM SALT see
CMM000
TETRABROMODIPHENYL ETHER see
TBJ550
TETRABROMODIPHENYLOPROPANE see
MKA270
1,1,2,2-TETRABROMOETANO (ITALIAN)
see ACK250
S-TETRABROMOETHANE see ACK250
1,1,1,2-TETRABROMOETHANE see TBJ560
1,1,2,2-TETRABROMOETHANE see ACK250
(1R,3S)3[(1'RS)(1',2',2',2'-
TETRABROMOETHYL)]-2,2-
DIMETHYLCYCLOPROPANECARBOXYLIC
ACID (S)-α-CYANO-3-PHENOXYBENZYL
ESTER see TBJ600
2,4,5,7-TETRABROMO-3,6-FLUORANDIOL
see BMO250
2,4,5,7-TETRABROMO-3,6-FLUORANDIOL
see BNH500
TETRABROMOFLUORESCEIN see BMO250
TETRABROMOFLUORESCEIN see BNH500
2',4',5',7'-TETRABROMOFLUORESCEIN see
BMO250
2',4',5',7'-TETRABROMOFLUORESCEIN
DISODIUM SALT see BNH500
TETRABROMOFLUORESCEIN S see
BNH500
TETRABROMOFLUORESCEIN SOLUBLE
see BNH500
2-(2,4,5,7-TETRABROMO-6-HYDROXY-3-
OXO-3H-XANTHENE-9-YL)BENZOIC
ACID, DISODIUM SALT see BNH500
TETRABROMOINDIGO see ICU135
5,5',7,7'-TETRABROMOINDIGO see ICU135
TETRABROMOMETHANE see CBX750
TETRABROMOPHENOLSULFOPHTHALEI
N see HAQ600
TETRABROMOPHTHALIC ACID
ANHYDRIDE see TBJ700
TETRABROMOPHTHALIC ANHYDRIDE
see TBJ700

3,4,5,6-TETRABROMOPHTHALIC ANHYDRIDE see TBJ700
TETRABROMO-PLATINUM(2-), DIPOTASSIUM see PLU250
TETRABROMOSILANE see SCP500
TETRABROMOSILICANE see SCP500
TETRABROMOSULFOPHTHALEIN see HAQ600
TETRABROMO-p-XYLEN (CZECH) see TBK250
TETRABROMO-p-XYLENE see TBK250
α,α,α',α'-TETRABROMO-m-XYLENE see TBK000
TETRABROMSULFTHALEIN see HAQ600
1,1,2,2-TETRABROOMETHAAN (DUTCH) see ACK250
TETRABUTYLAMMONIUM BROMIDE see TBK500
TETRA-N-BUTYLAMMONIUM BROMIDE see TBK500
TETRABUTYLAMMONIUM HYDROXIDE see TBK750
TETRA-n-BUTYLAMMONIUM HYDROXIDE see TBK750
TETRABUTYLAMMONIUM IODIDE see TBL000
TETRA-N-BUTYLAMMONIUMJODID (CZECH) see TBL000
TETRABUTYLAMMONIUM NITRATE see TBL250
TETRA-n-BUTYLAMMONIUM NITRATE see TBL250
TETRA-n-BUTYLCIN (CZECH) see TBM250
TETRABUTYL DICHLOROSTANNOXANE see TBL500
TETRA(2-BUTYLISOPROPYL) ORTHOSILICATE see TBL600
2,2',6,6'-TETRA-tert-BUTYL-4,4'-METHYLENEDIPHENOL see MJM700
TETRA-N-BUTYLPHOSPHONIUM BROMIDE see TBL750
TETRA-N-BUTYLPHOSPHONIUM CHLORIDE see TBM000
TETRABUTYLPHOSPHONIUM PHENYLPHOSPHINATE see TBM100
TETRA-N-BUTYLPHOSPHONIUM PHENYLPHOSPHINATE see TBM100
TETRABUTYLSTANNANE see TBM250
TETRABUTYLTHIURAM DISULPHIDE see TBM750
TETRABUTYLTIN see TBM250
TETRABUTYLTITANATE (CZECH) see BSP250
TETRABUTYLUREA see TBM850
1,1,3,3-TETRABUTYLUREA see TBM850
TETRACAINE see BQA010
TETRACAINE HYDROCHLORIDE see TBN000
TETRACAP see PCF275
TETRACAPS see TBX250
TETRACARBON MONOFLUORIDE see TBN100
TETRACARBONYLMOLYBDENUM DICHLORIDE see TBN150
TETRACARBONYL NICKEL see NCZ000
TETRACARBONYL(TRIFLUOROMETHYLTHIO)MANGANESE dimer see TBN200
1,2,4,5-TETRACARBOXYBENZENE see PPQ630
TETRACEMATE DISODIUM see EIX500
TETRACEMIN see EIV000
TETRACENE see NAI000
TETRACENE see TEF500
TETRACENE EXPLOSIVE see TEF500
2,4,4',5-TETRACHLOOR-DIFENYL-SULFON (DUTCH) see CKM000
1,1,2,2-TETRACHLOORETHAAN (DUTCH) see TBQ100
TETRACHLOORETHEEN (DUTCH) see PCF275
TETRACHLOORKOOLSTOF (DUTCH) see CBY000

TETRACHLOORMETAAN see CBY000
1,1,2,2-TETRACHLORAETHAN (GERMAN) see TBQ100
TETRACHLORAETHEN (GERMAN) see PCF275
N-(1,1,2,2-TETRACHLORAETHYLTHIO)CYCLOHEX-4-EN-1,4-DIACARBOXIMID (GERMAN) see CBF800
N-(1,1,2,2-TETRACHLORAETHYLTHIO)TETRAHYDROPHTHALAMID (GERMAN) see CBF800
1,2,4,5-TETRACHLORBENZOL see TBN220
TETRACHLORDIAN (CZECH) see TBO750
3,4,6,R'-TETRACHLOR-DIPHENYLSULFID (GERMAN) see CKL750
2,4,4',5-TETRACHLOR-DIPHENYL-SULFON (GERMAN) see CKM000
TETRACHLORETHANE see TBQ100
1,1,2,2-TETRACHLORETHANE (FRENCH) see TBQ100
TETRACHLORKOHLENSTOFF, (GERMAN) see CBY000
TETRACHLORMETHAN (GERMAN) see CBY000
2,3,5,6-TETRACHLOR-3-NITROBENZOL (GERMAN) see TBR750
TETRACHLOROACETONE see TBN250
1,1,3,3-TETRACHLOROACETONE see TBN300
N,N,N',N'-TETRACHLOROADIPAMIDE see TBN400
TETRACHLOROAURIC(3+) ACID, SODIUM SALT see GIZ100
3,3',4,4'-TETRACHLOROAZOBENZENE see TBN500
3,4,3',4'-TETRACHLOROAZOBENZENE see TBN500
3,3',4,4'-TETRACHLOROAZOXYBENZENE see TBN550
3,4,3',4'-TETRACHLOROAZOXYBENZENE see TBN550
S-TETRACHLOROBENZENE see TBN220
1,2,3,4-TETRACHLOROBENZENE see TBN740
1,2,3,5-TETRACHLOROBENZENE see TBN745
1,2,4,5-TETRACHLOROBENZENE see TBN220
1,2,4,5-TETRACHLOROBENZENE see TBN750
3,4,5,6-TETRACHLORO-1,2-BENZENEDICARBOXYLIC ACID see TBT100
2,3,5,6-TETRACHLORO-1,4-BENZENEDICARBOXYLIC ACID, DIMETHYL ESTER see TBV250
3,4,5,6-TETRACHLORO-1,2-BENZENEDIOL see TBU500
TETRACHLOROBENZIDINE see TBO000
2,2',5,5'-TETRACHLOROBENZIDINE see TBO000
3,3',6,6'-TETRACHLOROBENZIDINE see TBO000
TETRACHLOROBENZOQUINONE see TBO500
TETRACHLORO-p-BENZOQUINONE see TBO500
TETRACHLORO-1,4-BENZOQUINONE see TBO500
2,3,5,6-TETRACHLORO-p-BENZOQUINONE see TBO500
2,3,5,6-TETRACHLORO-1,4-BENZOQUINONE see TBO500
2,2',3,3'-TETRACHLOROBIPHENYL see TBO520
2,2',5,5'-TETRACHLOROBIPHENYL see TBO530
2,2',6,6'-TETRACHLOROBIPHENYL see TBO600
2,5,2',5'-TETRACHLOROBIPHENYL see TBO530

2,6,2',6'-TETRACHLOROBIPHENYL see TBO600
3,3',4,4'-TETRACHLOROBIPHENYL see TBO700
3,4,3',4'-TETRACHLOROBIPHENYL see TBO700
2,2',5,5'-TETRACHLORO-(1,1'-BIPHENYL)-4,4'-DIAMINE, (9CI) see TBO000
2,2',6,6'-TETRACHLOROBISPHENOL A see TBO750
1,1,2,3-TETRACHLORO-1,3-BUTADIENE see TBO760
1,2,3,4-TETRACHLOROBUTANE see TBO762
1,1,4,4-TETRACHLOROBUTATRIENE see TBO765
TETRACHLOROCARBON see CBY000
9-(3',4',5',6'-TETRACHLORO-o-CARBOXYPHENYL)-6-HYDROXY-2,4,5,7-TETRAIODO-3-ISOXANTHONE•2Na see RMP175
TETRACHLOROCATECHOL see TBU500
TETRACHLOROCOBALTATE(2-) BIS(TETRAETHYLAMMONIUM) see BLH315
2,4,5,6-TETRACHLORO-3-CYANOBENZONITRILE see TBQ750
2,3,4,5-TETRACHLORO-2-CYCLOBUTEN-1-ONE see PCF250
2,3,5,6-TETRACHLORO-2,5-CYCLOHEXADIENE-1,4-DIONE see TBO500
TETRACHLOROCYCLOPENTADIENE see TBO768
1,2,3,4-TETRACHLOROCYCLOPENTADIENE see TBO768
TETRACHLOROCYCLOPENTANE see TBO770
2,2',5,5'-TETRACHLORO-4,4'-DIAMINODIPHENYL see TBO000
cis-TETRACHLORODIAMMINE PLATINUM(IV) see TBO776
trans-TETRACHLORODIAMMINE PLATINUM(IV) see TBO777
TETRACHLORODIAZOCYCLOPENTADIENE see TBO778
2,3,7,8-TETRACHLORODIBENZO(b,e)(1,4)DIOXAN see TAI000
1,2,3,4-TETRACHLORODIBENZODIOXIN see TBO780
1,2,3,8-TETRACHLORODIBENZODIOXIN see TBO790
1,3,6,8-TETRACHLORODIBENZODIOXIN see TBO800
1,3,7,8-TETRACHLORODIBENZODIOXIN see TBO810
2,3,6,7-TETRACHLORODIBENZODIOXIN see TBO795
1,2,3,4-TETRACHLORODIBENZO-p-DIOXIN see TBO780
1,2,3,8-TETRACHLORODIBENZO-p-DIOXIN see CDV125
1,2,3,8-TETRACHLORODIBENZO-p-DIOXIN see TBO790
1,2,7,8-TETRACHLORODIBENZO-p-DIOXIN see TBO795
1,3,6,8-TETRACHLORODIBENZO-p-DIOXIN see TBO800
1,3,7,8-TETRACHLORODIBENZO-p-DIOXIN see TBO810
2,3,6,7-TETRACHLORODIBENZO-p-DIOXIN see TAI000
2,3,7,8-TETRACHLORODIBENZO-p-DIOXIN see TAI000
2,3,7,8-TETRACHLORODIBENZO-1,4-DIOXIN see TAI000
2,3,7,8-TETRACHLORODIBENZOFURAN see TBO850
1,1,1,2-TETRACHLORO-2,2-DIFLUOROETHANE see TBP000
1,1,1,2-TETRACHLORO-2,2-DIFLUOROETHANE see TBP000

1,1,2,2-TETRACHLORO-1,2-DIFLUOROETHANE see TBP050

3,3,4,5-TETRACHLORO-3,6-DIHYDRO-1,2-DIOXIN see TBP075

4,5,6,7-TETRACHLORO-2-(2-DIMETHYLAMINOETHYL)-ISOINDOLINE DIMETHOCHLORIDE see CDY000

TETRACHLORODIPHENYLETHANE see BIM500

2',3,4,6'-TETRACHLORODIPHENYL ETHER see DEX270

TETRACHLORODIPHENYL OXIDE see TBP250

2,4,4',5-TETRACHLORODIPHENYL SULFIDE see CKL750

2,4,5,4'-TETRACHLORODIPHENYL SULFIDE see CKL750

2,4,4',5-TETRACHLORODIPHENYL SULFONE see CKM000

2,4,5,4'-TETRACHLORODIPHENYLSULPHONE see CKM000

TETRACHLORODIPHOSPHANE see TBP500

TETRACHLOROEPOXYETHANE see TBQ275

TETRACHLOROETHANE see TBP750

sym-TETRACHLOROETHANE see TBQ100

1,1,1,2-TETRACHLOROETHANE see TBQ000

1,1,2,2-TETRACHLOROETHANE see TBQ100

TETRACHLOROETHENE see PCF275

1,1,2,2-TETRACHLOROETHYLENE see PCF275

TETRACHLOROETHYLENE (DOT, ACGIH) see PCF275

TETRACHLOROETHYLENE CARBONATE see TBQ255

TETRACHLOROETHYLENE OXIDE see TBQ275

N-1,1,2,2-TETRACHLOROETHYLMERCAPTO-4-CYCLOHEXENE-1,2-CARBOXIMIDE see CBF800

N-((1,1,2,2-TETRACHLOROETHYL)SULFENYL)-cis-4-CYCLOHEXENE-1,2-DICARBOXIMIDE see CBF800

N-(1,1,2,2-TETRACHLOROETHYLTHIO)-4-CYCLOHEXENE-1,2-DICARBOXIMIDE see CBF800

cis-N-((1,1,2,2-TETRACHLOROETHYL)THIO)-4-CYCLOHEXENE-1,2-DICARBOXYMIDECIS-3A,4,7,7A-TETRAHYDRO-2-((1,1,2,2-TETRACHLOROETHYL)THIO)-1H-ISOINDOLE-1,3(2H)-DIONE see TBQ280

N-((TETRACHLORO-2-FLUOROETHYL)THIO)METHANESULFOANILIDE see FLZ025

TETRACHLOROGUAIACOL see TBQ290

2,3,4,5-TETRACHLOROHEXATRIENE see TBQ300

TETRACHLOROHYDROQUINONE see TBQ500

4,5,6,7-TETRACHLORO-1,3-ISOBENZOFURANDIONE see TBT150

TETRACHLOROISOPHTHALONITRILE see TBQ750

TETRACHLORO(ISOQUINOLINE)((TETRAHYDROTHIOPHENE-KAPPAS) 1-OXIDE)-RUTHENATE(1-), SODIUM, (OC-6-11)- see TBQ800

TETRACHLOROMETHANE see CBY000

1,2,4,5-TETRACHLORO-3-METHOXY-6-NITROBENZENE (9CI) see TBR250

TETRACHLORONAPHTHALENE see TBR000

TETRACHLORONITROANISOLE see TBR250

2,3,5,6-TETRACHLORO-4-NITROANISOLE see TBR250

2,3,4,5-TETRACHLORONITROBENZENE see TBS000

2,3,4,6-TETRACHLORONITROBENZENE see TBR500

2,3,5,6-TETRACHLORONITROBENZENE see TBR750

1,2,3,4-TETRACHLORO-5-NITROBENZENE see TBS000

1,2,3,5-TETRACHLORO-4-NITROBENZENE see TBR500

1,2,4,5-TETRACHLORO-3-NITROBENZENE see TBR750

TETRACHLOROPALLADATE(2-) DIPOTASSIUM see PLN750

2,3,5,6-TETRACHLORO-4-(PENTACHLOROPHENOXY)PHENOL see HNB550

1,1,1,5-TETRACHLOROPENTANE see TBS100

TETRACHLOROPHENOL see TBS250

2,3,4,5-TETRACHLOROPHENOL see TBS500

2,3,4,6-TETRACHLOROPHENOL see TBT000

2,3,5,6-TETRACHLOROPHENOL see TBS750

2,4,5,6-TETRACHLOROPHENOL see TBT000

TETRACHLOROPHENYL ETHER see TBP250

TETRACHLOROPHTHALIC ACID see TBT100

TETRACHLOROPHTHALIC ANHYDRIDE see TBT150

4,5,6,7-TETRACHLOROPHTHALIDE see TBT175

o-TETRACHLOROPHTHALODINITRILE see TBT200

TETRACHLOROPHTHALONITRILE see TBT200

m-TETRACHLOROPHTHALONITRILE see TBQ750

3,4,5,6-TETRACHLOROPHTHALONITRILE see TBT200

1,1,1,3-TETRACHLOROPROPANE see TBT250

1,2,2,3-TETRACHLOROPROPANE see TBT300

1,1,3,3-TETRACHLOROPROPANONE see TBN300

1,1,3,3-TETRACHLORO-2-PROPANONE see TBN300

1,1,2,3-TETRACHLOROPROPENE see TBT500

2,3,4,5-TETRACHLOROPYRIDINE see TBT750

2,3,5,6-TETRACHLOROPYRIDINE see TBU000

TETRACHLOROPYRIMIDINE see TBU250

2,4,5,6-TETRACHLOROPYRIMIDINE see TBU250

TETRACHLOROPYROCATECHOL see TBU500

TETRACHLOROPYROCATECHOL HYDRATE see TBU750

TETRACHLOROQUINONE see TBO500

TETRACHLORO-p-QUINONE see TBO500

5,6,7,8-TETRACHLOROQUINOXALINE see CMA500

3,3',4',5-TETRACHLOROSALICYLANILIDE see TBV000

3,5,3',4'-TETRACHLOROSALICYLANILIDE see TBV000

TETRACHLOROSILANE see SCQ500

TETRACHLOROSTANNANE PENTAHYDRATE see TGC282

TETRACHLOROTELLURIUM see TAJ250

TETRACHLOROTEREPHTHALIC ACID DIMETHYL ESTER see TBV250

TETRACHLOROTETRABROMOFLUORESCEIN see SKS400

3,3,4,4-TETRACHLOROTETRAHYDROTHIOPHENE-1,1-DIOXIDE see TBV300

4,5,6,7-TETRACHLORO-2',4',5',7'-TETRAIODOFLUORESCEIN DISODIUM SALT see RMP175

TETRACHLOROTHIOFENE see TBV750

TETRACHLOROTHIOPHENE see TBV750

2,3,4,5-TETRACHLOROTHIOPHENE see TBV750

TETRACHLOROTHORIUM see TFT000

α-α-α-4-TETRACHLOROTOLUENE see TIR900

p,α-α-α-TETRACHLOROTOLUENE see TIR900

4,5,6,7-TETRACHLORO-2-(TRIFLUOROMETHYL)BENZIMIDAZOLE see TBW000

TETRACHLOROTRIFLUOROMETHYLPHOSPHORANE see TBW025

5-endo,6-exo-2,2,5,6-TETRACHLORO-1,7,7-TRIS(CHLOROMETHYL)-BICYCLO(2.2.1)HEPTANE see THH575

2,3,5,6-TETRACHLORPHTHALSAURE-DIMETHYLESTER (GERMAN) see TBV250

TETRACHLORPYROKATECHIN (CZECH) see TBU750

TETRACHLORURE d'ACETYLENE (FRENCH) see TBQ100

TETRACHLORURE de CARBONE (FRENCH) see CBY000

TETRACHLORURE de SILICIUM (FRENCH) see SCQ500

TETRACHLORURE de TITANE (FRENCH) see TGH350

TETRACHLORVINPHOS see TBW100

TETRACICLINA CLORIDRATO (ITALIAN) see TBX250

TETRACICLINA-l-METILENLISINA (ITALIAN) see MRV250

TETRACID see MPQ750

TETRACID CARMINE BLUE V see ADE500

TETRACID MILLING RED B see CMM330

TETRACID MILLING RED G see CMM330

2,4,4',5-TETRACLORO-DIFENIL-SOLFONE (ITALIAN) see CKM000

1,1,2,2-TETRACLOROETANO (ITALIAN) see TBQ100

TETRACLOROETENE (ITALIAN) see PCF275

TETRACLOROMETANO (ITALIAN) see CBY000

TETRACLORURO di CARBONIO (ITALIAN) see CBY000

TETRACOBALT DODECACARBONYL see CNB510

TETRACOMPREN see TBX250

2,6,10,14,18,22-TETRACOSAHEXENE, 2,6,10,15,19,23-HEXAMETHYL-, (all-E)- see SLG800

TETRACOSAHYDRO DIBENZ(b,n)(1,4,7,10,13,16,19,22)OCTAOXACYCLOTETRACOSIN see DGT300

TETRACOSANE, 2,6,10,15,19,23-HEXAMETHYL- see SLG700

(1,1,3,3-TETRACYANOALLYL) SODIUM see TAI050

TETRACYANOETHENE see EEE500

1,1,2,2-TETRACYANOETHENE see EEE500

TETRACYANOETHYLENE see EEE500

1,1,2,2-TETRACYANOETHYLENE see EEE500

N,N,N',N'-TETRACYANOMETHYLAETHYLENEDIAMIN (GERMAN) see EJB000

TETRACYANONICKELATE(2−) DIPOTASSIUM, HYDRATE see TBW250

TETRACYANOOCTAETHYLTETRAGOLD see TBW500

TETRACYANOQUINODIMETHAN see TBW750

TETRACYANOQUINODIMETHAN(e) see TBW750

TETRACYANO-p-QUINODIMETHANE see TBW750

7,7,8,8-TETRACYANOQUINODIMETHANE see TBW750

7,7',8,8'-TETRACYANOQUINODIMETHANE see TBW750

TETRACYCLINE see TBX000

TETRACYCLINE CHLORIDE see TBX250

TETRACYCLINE, 7-CHLORO-6-DEMETHYL- see MIJ500

TETRACYCLINE HYDROCHLORIDE see TBX250

TETRACYCLINE, 5-HYDROXY- see HOH500

TETRACYCLINE I see TBX000

TETRACYCLINE-l-METHYLENE LYSINE see MRV250

TETRACYDIN see ABG750

TETRACYN see TBX000

TETRA-D see TBX250

9Z,12E-TETRADECADIENYL ACETATE see TBX300

9,11-TETRADECADIEN-1-YL, ACETATE, (E,Z)- see FAS100

9,12-TETRADECADIEN-1-YL, ACETATE, (E,Z)- see TBX300

N-TETRADECAFLUOROHEXANE see FJA100

TETRADECANAL see TBX500

1-TETRADECANAL see TBX500

1-TETRADECANAMINIUM, N,N,N-TRIMETHYL-, BROMIDE (9CI) see TCB200

TETRADECANE see TBX750

TETRADECANOATE de ((METHYLENEDIOXY-3,4 PHENYL)-1 DIMETHYL-4,4 PENTENE-1)YLE-3 see BCJ090

TETRADECANOIC ACID see MSA250

TETRADECANOIC ACID, BUTYL ESTER see MSA300

TETRADECANOIC ACID, ISOPROPYL see IQN000

TETRADECANOIC ACID, 1-METHYLETHYL ESTER see IQN000

TETRADECANOIC ACID, SODIUM SALT (9CI) see SIN900

1-TETRADECANOL see TBY250

N-TETRADECANOL-1 see TBY250

TETRADECANOL, mixed isomers see TBY500

TETRADECANOL condensed with 7 moles ETHYLENE OXIDE see TBY750

2-TETRADECANOL, 1-(2-HYDROXYETHYLAMINO)-, and 1-(2-HYDROXYETHYL)-2-DODECANOL (1:1) see HKS400

TETRADECANOYLETHYLENEIMINE see MSB000

12-TETRADECANOYLPHORBOL-13-ACETATE see PGV000

N-TETRADECANOYL-N-TETRADECANOYLOXY-2-AMINOFLUORENE see MSB250

12-o-TETRADECA-2-cis-4-trans,6,8-TETRAENOYLPHORBOL-13-ACETATE see TCA250

4,8,12-TETRADECATRIENOIC ACID, 5,9,13-TRIMETHYL-, 3,7-DIMETHYL-2,6-OCTADIENYLESTER, (E,E,E)- see GDG200

9-TETRADECENAL, (Z)- see TCA300

(Z)-9-TETRADECENAL see TCA300

11-TETRADECENAL, (E)- see TCA330

(E)-11-TETRADECENAL see TCA330

(Z)-9-TETRADECEN-1-AL see TCA300

(Z)-9-TETRADECENOL ACETATE see TCA400

9-TETRADECEN-1-OL, ACETATE, (Z)- see TCA400

TETRADECIN see TBX000

n-TETRADECOIC ACID see MSA250

TETRADECYL ALCOHOL see TBY250

TETRADECYL ALCOHOL see TBY500

N-TETRADECYL ALCOHOL see TBY250

1-TETRADECYL ALDEHYDE see TBX500

TETRADECYL DIMETHYL BENZYLAMMONIUM CHLORIDE see TCA500

TETRADECYLHEPTAETHOYLATE see TCA750

TETRADECYL OCTADECANOATE see TCB100

TETRADECYL PHOSPHONIC ACID see TCB000

TETRADECYL SODIUM SULFATE see SIO000

TETRADECYL STEARATE see TCB100

TETRADECYL SULFATE, SODIUM SALT see SIO000

TETRADECYLTRIMETHYLAMMONIUM BROMIDE see TCB200

1,2,6,7-TETRADEHYDRO-14,17-DIHYDRO-3-METHOXY-16(15H)-OXAERYTHRINAN-15-ONE, HYDROCHLORIDE see EDH000

7,8,13,13A-TETRADEHYDRO-9,10-DIMETHOXY-2,3-(METHYLENEDIOXY)BERBINIUM SULFATE TRIHYDRATE see BFN750

TETRADEHYDRODOISYNOLIC ACID METHYL ETHER see BIT000

6,7,8,14-TETRADEHYDRO-4,5-α-EPOXY-3,6-DIMETHOXY-17-METHYLMORPHINAN see TEN000

6,7,8,14-TETRADEHYDRO-4,5-α-EPOXY-3,6-DIMETHOXY-17-METHYLMORPHINAN HYDROCHLORIDE see TEN100

4,5,6,6a-TETRADEHYDRO-9-HYDROXY-1,2,10-TRIMETHOXYNORAPORPHIN-7-ONE see ARQ325

(3-β)-1,2,6,7-TETRADEHYDRO-3-METHOXYERYTHRINAN-16-OL HYDROCHLORIDE see EDE500

(3-β)-1,2,6,7-TETRADEHYDRO-3-METHOXY-15,16-(METHYLENEBIS(OXY))ERYTHRINAN, HYDROBROMIDE see EDF000

7,8,13,13a-TETRADEHYDRO-2,3,9,10-TETRAMETHOXYBERBINIUM HYDROXIDE see PAE100

12-o-TETRADEKANOYLPHORBOL-13-ACETAT (GERMAN) see PGV000

2,3,4,6-TETRADEOXY-4-((5,6,11,12,13,14,15,16,16A,17,17A,18,21,22-TETRADECAHYDRO-2,12,14,16-TETRAHYDROXY-10-METHOXY-3,6,11,13,15,20-HEXAMETHYL-5,18,21,26-TETRAOXO-4,6-EPOXY-1,23-METHANOBANZO(d)CYCLOPROP(n)(1,9)O XAAZACYCLOTETRACOSIN-25(10H)-YLIDENE)AMINO)-l-erythro-HEXOPYRANOSE, 12-ACETATE see THA600

TETRADICHLONE see CKM000

TETRADIFON see CKM000

TETRADIN see DXH250

TETRADINE see DXH250

TETRADIOXIN see TAI000

TETRADIPHON see CKM000

TETRADONIUM BROMIDE see TCB200

TETRADOX see HGP550

TETRAETHANOL AMMONIUM HYDROXIDE see TCB500

2,4,6,8-TETRAETHENYL-2,4,6,8-TETRAMETHYLCYCLOTETRASILOXANE see TCB600

TETRAETHOXY PROPANE see MAN750

1,1,3,3-TETRAETHOXYPROPANE see MAN750

TETRAETHOXYSILANE see EPF550

TETRAETHYLAMMONIUM see TCB725

TETRAETHYLAMMONIUM BOROHYDRIDE see TCB750

TETRAETHYL AMMONIUM BROMIDE see TCC000

TETRAETHYLAMMONIUM CHLORIDE see TCC250

TETRAETHYLAMMONIUM HEPTADECAFLUOROOCTANESULFONATE see TCD100

TETRAETHYLAMMONIUM HYDROXIDE see TCC500

TETRAETHYLAMMONIUM IODIDE see TCC750

TETRAETHYLAMMONIUM PERCHLORATE see TCD000

TETRAETHYLAMMONIUM PERCHLORATE (dry) (DOT) see TCD002

TETRAETHYLAMMONIUM PERFLUOROOCTANESULFONATE see TCD100

TETRAETHYLAMMONIUM PERFLUORO-1-OCTANESULFONATE see TCD100

TETRAETHYLAMMONIUM PERFLUOROOCTYLSULFONATE see TCD100

N,N,N',N'-TETRAETHYL-1,2-BENZENEDICARBOXAMIDE see GCE600

TETRA(2-ETHYLBUTOXY) SILANE see TCD250

TETRA(2-ETHYLBUTYL) ORTHOSILICATE see TCD250

N,N,N',N'-TETRAETHYL-2-BUTYNYLENEDIAMINE see BJA250

TETRAETHYLCYSTAMINE see TCD300

N,N,N',N'-TETRAETHYLCYSTAMINE see TCD300

TETRAETHYLDIAMINO-o-CARBOXY-PHENYL-XANTHENYL CHLORIDE see FAG070

TETRAETHYLDIARSANE see TCD500

O,O,O,O-TETRAETHYL-DIFOSFAAT (DUTCH) see TCF250

N,N,N,N'-TETRAETHYL-N',N'-DIMETHYL-3-OXAPENTANE-1,5-DIAMMONIUMDI-MONOHYDROGEN TARTRATE see TCD600

TETRAETHYLDITHIODIFOSFAT see SOD100

TETRAETHYL DITHIONOPYROPHOSPHATE see SOD100

TETRAETHYL DITHIOPYROPHOSPHATE see SOD100

O,O,O,O-TETRAETHYL DITHIOPYROPHOSPHATE see SOD100

TETRAETHYL DITHIO PYROPHOSPHATE, liquid (DOT) see SOD100

TETRAETHYL DITHIOPYROPHOSPHATE and GAS MIXTURES, LC50 < or =2000 ppm (DOT) see SOD100

TETRAETHYLENE GLYCOL see TCE250

TETRAETHYLENE GLYCOL DIACRYLATE see ADT050

TETRAETHYLENE GLYCOL, DIBUTYL ETHER see TCE350

TETRAETHYLENE GLYCOL DIETHYL ETHER see PBO250

TETRAETHYLENE GLYCOL DI(2-ETHYLHEXOATE) see FCD500

TETRAETHYLENE GLYCOL-DI-n-HEPTANOATE see TCE375

TETRAETHYLENE GLYCOL DIMETHACRYLATE see TCE400

TETRAETHYLENE GLYCOL DIMETHYL ETHER see PBO500

TETRAETHYLENE GLYCOL MONOPHENYL ETHER see TCE450

TETRAETHYLENEIMIDEPIPERAZINE-N,N'-DIPHOSPHORIC ACID see BJC250

2,3,5,6-TETRA-ETHYLENEIMINO-1,4-BENZOQUINONE see TDG500

TETRAETHYLENEPENTAMINE see TCE500

TETRAETHYL GERMANE see TCE750

TETRAETHYL GERMANIUM see TCE750

TETRA(2-ETHYLHEXOXY)SILANE see EKQ500

TETRA(2-ETHYLHEXYL)ORTHOSILICATE see EKQ500

(−)-Δ¹-3,4-trans-
TETRAHYDROCANNABINOL see TCM250
(−)-Δ⁶-3,4-trans-
TETRAHYDROCANNABINOL see TCM000
5,6,7,8-TETRAHYDRO-3-
CARBAZOLECARBOXYLIC ACID 2-
(DIETHYLAMINO)ETHYL ESTER
HYDROCHLORIDE see TCM400
6,7,8,9-TETRAHYDRO-3-CARBOXY-1,6-
DIMETHYL-4-OXO-4H-PYRIDO(1,2-
a)PYRIMIDINIUM ETHYL ESTER, METHYL
SULFATE see PMD825
5,9,10,14B-TETRAHYDRO-2-CHLORO-5-
METHYLISOQUINO(2,1-
D)(1,4)BENZODIAZEPIN-6(7H)-ONE see
IRW000
TETRAHYDRO-2-(p-CHLOROPHENYL)-3-
METHYL-4H-1,3-THIAZIN-4-ONE-1,1-
DIOXIDE see CKF500
2-(p-(1,2,3,4-TETRAHYDRO-2-(p-
CHLOROPHENYL)NAPHTHYL)PHENOXY)
TRIETHYLAMINE see TCN250
trans-1,2,3,4-TETRAHYDROCHRYSENE-1,2,-
DIOL see DNC600
1A,2,3,11B-TETRAHYDROCHRYSENO(1,2-
B)OXIRENE-2,3-DIOL (1a-α-2-β,3-α-11bα)-
see TCN300
1A,2,3,11B-TETRAHYDROCHRYSENO(1,2-
B)OXIRENE-2,3-DIOL(1a-α,2-α,3-β,11B-α)-
see CML820
TETRAHYDROCOMPOUND F see UVJ460
TETRAHYDROCORTISOL see UVJ460
5-β-TETRAHYDROCORTISOL see UVJ460
2,7,8,9-TETRAHYDRO-1H-
CYCLOPENT(j)ACEANTHRYLENE see
CPY500
1,2,3,4-
TETRAHYDROCYCLOPENT(B)INDOL-3-
ONE OXIME see KFK150
1,2,3,4-TETRAHYDRO-1-(2-DEOXY-β-d-
RIBOFURANOSYL)-2,4-DIOXO-5-
PYRIMIDINECARBOXALDEHYDE see
FNK033
1,2,3,4-
TETRAHYDRODIBENZ(a,h)ANTHRACENE
see TCN750
1,2,3,4-
TETRAHYDRODIBENZ(a,j)ANTHRACENE
see TCO000
TETRAHYDRO-3,5-DIBENZYL-2H-1,3,5-
THIADIAZINE-2-THIONE see RCK730
exo-
TETRAHYDRODI(CYCLOPENTADIENE)
see TLR675
1-
(TETRAHYDRODICYCLOPENTADIENYL)-
3,3-DIMETHYLUREA see HDP500
7,8,9,10-TETRAHYDRO-N,N-DIETHYL-6H-
CYCLOHEPTA(b)QUINOLINE-11-
CARBOXAMIDE, see TCO100
(+)-Z-7,8,9,10-TETRAHYDRO-7-α,8-β-
DIHDYROXY-9-α,10-α-
EPOXYBENZO(a)PYRENE see DMP600
(−)-Z-7,8,9,10-TETRAHYDRO-7-α,8-β-
DIHDROXY-9-α,10-α-
EPOXYBENZO(a)PYRENE see DMP800
1,4,5,6-TETRAHYDRO-3-(10(11-
DIHYDRODIBENZO(a,d)CYCLOHEPTEN-5-
YL)-1-METHYL-as-TRIAZINE
HYDROBROMIDE see TCO200
(E)-8,9,10,11-TETRAHYDRO-8-β,9-α-
DIHYDROXY-10-α,11-α-
BENZ(a)ANTHRACENE see DMO600
8,9,10,11-TETRAHYDRO-10,11-
DIHYDROXY-8-α,9-α-
EPOXYBENZ(a)ANTHRACENE see BAD000
(E)-1,2,3,4-TETRAHYDRO-3-α,4-β-
DIHYDROXY-1-α,2-α-
EPOXYBENZ(a)ANTHRACENE see DLE000
(E)-(±)-8,9,10,11-TETRAHYDRO-8-β,9-α-
DIHYDROXY-10-β,11-β-
EPOXYBENZ(a)ANTHRACENE see DMP200

(+)-cis-7,8,9,10-TETRAHYDRO-7-β,8-α-
DIHYDROXY-9-β,10-β-
EPOXYBENZO(a)PYRENE see DMP600
(−)-7,8,9,10-TETRAHYDRO-7-β,8-α-
DIHYDROXY-9-α,10-α-EPOXY-
BENZO(a)PYRENE see DVO175
(−)-Z-7,8,9,10-TETRAHYDRO-7-β,8-α-
DIHYDROXY-9-β,10-β-
EPOXYBENZO(a)PYRENE see DMP800
(±)-7,8,9,10-TETRAHYDRO-7-α,8-β-
DIHYDROXY-9-β,10-β-
EPOXYBENZO(a)PYRENE see DMQ600
(±)-7,8,9,10-TETRAHYDRO-7-α,8-β-
DIHYDROXY-9-α,10-α-
EPOXYBENZO(a)PYRENE see DMR000
1,2,3,11a-TETRAHYDRO-3,8-DIHYDROXY-
7-METHOXY-5H-PYRROLO(2,1-
c)(1,4)BENZODIAZEPIN-5-ONE-(3S-cis)- see
TCP000
TETRAHYDRO-4,6-DIIMINO-s-TRIAZINE-
2(1H)-THIONE see DCF000
(s)-5,8,13,13a-TETRAHYDRO-9,10-
DIMETHOXY-6H-BENZO(g)-1,3-
BENZODIOXOLO(5,6-a)QUINOLIZINE see
TCJ800
(R)-5,6,6a,7-TETRAHYDRO-1,2-
DIMETHOXY-6-METHYL-4H-
DIBENZO(de,g)QUINOLINE see NOE500
(S)-5,6,6a,7-TETRAHYDRO-1,10-
DIMETHOXY-6-METHYL-4H-
DIBENZO(de,g)QUINOLINE-2,9-DIOL see
DNZ100
1,2,3,4-TETRAHYDRO-7,12-
DIMETHYLBENZ(a)ANTHRACENE see
TCP600
8,9,10,11-TETRAHYDRO-7,12-
DIMETHYLBENZ(a) ANTHRACENE see
DUF000
1,2,3,6-TETRAHYDRO-1,3-DIMETHYL-2,6-
DIOXO-7H-PURINE-7-ACETIC ACID
compounded with N-((4-
METHOXYPHENYL)METHYL)-N',N'-
DIMETHYL-N-2-PYRIDINYL-1,2-
ETHANEDIAMINE (1:1) see MCJ300
TETRAHYDRODIMETHYLFURAN see
TCQ250
TETRAHYDRODIMETHYL FURANE see
TCQ250
2,3,4,5-TETRAHYDRO-2,8-DIMETHYL-5-(2-
(6-METHYL-3-PYRIDYL)ETHYL)-1H-PYRID
O(4,3-b)INDOLE see TCQ260
TETRAHYDRO-3,5-DIMETHYL-4H,1,3,5-
OXADIAZINE-4-THIONE see TCQ275
TETRAHYDRO-2,6-DIMETHYL-4H-PYRAN-
4-ONE see TCQ350
TETRAHYDRO-2H-3,5-DIMETHYL-1,3,5-
THIADIAZINE-2-THIONE see DSB200
TETRAHYDRO-3,5-DIMETHYL-2H-1,3,5-
THIADIAZINE-2-THIONE see DSB200
TETRAHYDRO-2,4-
DIMETHYLTHIOPHENE-1,1-DIOXIDE see
DUD400
4,5,9,10-TETRAHYDRO-2,7-
DINITROPYRENE see TCQ400
TETRAHYDRO-p-DIOXIN see DVQ000
TETRAHYDRO-1,4-DIOXIN see DVQ000
1,2,3,4-TETRAHYDRO-2,4-DIOXO-5-
FLUORO-N-HEXYL-1-
PYRIMIDINECARBOXAMIDE see CCK630
1,2,3,6-TETRAHYDRO-2,6-DIOXO-5-
FLUORO-4-PYRIMIDINECARBOXYLIC
ACID see TCQ500
TETRAHYDRO-2,5-DIOXOFURAN see
SNC000
1,2,3,6-TETRAHYDRO-3,6-
DIOXOPYRIDAZINE see DMC600
α-(TETRAHYDRO-4,6-DIOXO-2-THIOXO-
5(2H)PYRIMIDINYLIDENE)-o-TOLUIC
ACID see TCQ550
1,4,5,6-TETRAHYDRO-4,6-DIOXO-s-
TRIAZINE-2-CARBOXYLIC ACID,
POTASSIUM SALT see AFQ750

5',6',7',8'-
TETRAHYDRODISPIRO(CYCLOHEXANE-
1,2'(3'H)-QUINAZOLINE-4',1''(4 a'H)-
CYCLOHEXANE) see TCQ600
TETRAHYDRO-2,5-DIVINYLPYRAN see
DXR400
TETRAHYDRO-2,5-DIVINYL-2H-PYRAN see
DXR400
1,2,3,4-TETRAHYDRO-DMBA see TCP600
1,2,3,4-TETRAHYDRO-1,2-EPOXY
BENZ(a)ANTHRACENE see ECP500
7,8,9,10-TETRAHYDRO-9,10-EPOXY-
BENZO(a)PYRENE see ECQ100
9,10,11,12-TETRAHYDRO-9,10-EPOXY-
BENZO(e)PYRENE see ECQ150
(±)-1,2,3,4-TETRAHYDRO-3,α,4,α-EPOXY-
1,β,2,α-CHRYSENEDIOL see DMS200
(±)-1,2,3,4-TETRAHYDRO-3,β,4,β-EPOXY-1-
β,2-α-CHRYSENEDIOL see DMS400
(±)-7,8,9,10-TETRAHYDRO-9-β,10-β-EPOXY-
7-α,8-β-DIHYDROXYBENZO(a)PYRENE see
DMQ600
3A,4,7,7A-TETRAHYDRO-1,2-EPOXY-
4,5,6,7,8,8-HEXACHLORO-4,7-
METHANOINDAN see OPK275
TETRAHYDRO-3,4-EPOXYTHIOPHENE-
1,1-DIOXIDE see ECP000
1,2,3,4-TETRAHYDRO-1,2-
EPOXYTRIPHENYLENE see ECT000
3,4,5,6-TETRAHYDRO-2-(α-
ETHOXYBENZYL)-5-ETHYL-5-
METHYLPYRIMIDINE see AJY300
3,4,5,6-TETRAHYDRO-2-(α-
ETHOXYBENZYL)-5-ETHYL-5-
METHYLPYRIMIDINE see TCQ700
1,2,3,4-TETRAHYDRO-6-ETHYL-1,1,4,4-
TETRAMETHYLNAPHTHALENE see
TCR250
5,6,7,8-TETRAHYDRO-3-ETHYL-5,5,8,8-
TETRAMETHYL-2-
NAPHTHALENECARBOXALDEHYDE see
FNK200
TETRAHYDRO F see UVJ460
1,2,3,4-TETRAHYDRO-9-FLUORENONE see
TCR300
TETRAHYDROFOLIC ACID see TCR400
5,6,7,8-TETRAHYDROFOLIC ACID see
TCR400
TETRAHYDROFTALANHYDRID (CZECH)
see TDB000
TETRAHYDROFURAAN (DUTCH) see
TCR750
TETRAHYDROFURAN see TCR750
TETRAHYDRO-2-FURANCARBINOL see
TCT000
TETRAHYDRO-3-
FURANCARBOXALDEHYDE see FOI100
TETRAHYDROFURAN HOMOPOLYMER
SRU see GHY100
TETRAHYDRO-2-FURANMETHANOL see
TCT000
TETRAHYDROFURANNE (FRENCH) see
TCR750
TETRAHYDRO-3-FURANOL see HOI200
TETRAHYDRO-2-FURANONE see BOV000
TETRAHYDRO-2-FURANPROPANOL see
TCS000
2',3'-α-TETRAHYDROFURAN-2'-SPIRO-17-
(ESTER-4-EN-3-ONE) see NNX300
1-(TETRAHYDROFURAN-2-YL)-5-
FLUOROURACIL see FLZ050
(TETRAHYDRO-2-FURANYL)METHYL 2-
PROPENOATE see TCS100
TETRAHYDROFURFURYL ACRYLATE see
TCS100
TETRAHYDROFURFURYL ALCOHOL see
TCT000
TETRAHYDROFURFURYLAMINE see
TCS500
TETRAHYDROFURFURYL BROMIDE see
TCS550

α-TETRAHYDROFURFURYL-1-NAPHTHALENEPROPIONIC ACID 2-(DIETHYLAMINO)ETHYL ESTER see NAE000

TETRAHYDROFURYLALKOHOL (CZECH) see TCT000

3-(2-TETRAHYDROFURYL)-5-FLUOROURACIL see FLZ065

N¹-(2-TETRAHYDROFURYL)-5-FLUOROURACIL see FLZ050

1-(2-TETRAHYDROFURYL)-5-FLUOROURACIL mixture with URACIL (1:4) see UNJ810

2-TETRAHYDROFURYL HYDROPEROXIDE see TCS750

TETRAHYDRO-2-FURYLMETHANOL see TCT000

TETRAHYDROGERANIOL see DTE600

TETRAHYDROGERANYL ACETATE see DTE800

TETRAHYDROGERANYL BUTYRATE see DTF000

1',3'-α,4',7'-α-TETRAHYDRO-7'-α-β-3-(GLUCOPYRANOSYLOXY)-4'-HYDROXY-2',4',6'-TRIMETHYL-SPIRO(CYCLOPROPANE-1,5'-(5H)INDEN)-3',(2'H)-ONE see POI100

TETRAHYDROHARMINE see TCT250

3A,4,7,7A-TETRAHYDRO-4,5,6,7,8,8-HEXACHLORO-4,7-METHANOISOBENZOFURAN-1(3H)-ONE see EAQ815

1-(5,6,7,8-TETRAHYDRO-3,5,5,5,6,8,8-HEXAMETHYL-2-NAPHTHALENYL)ETHANONE see FBW110

TETRAHYDROHYDROCORTISONE see UVJ460

2,3-cis-1,2,3,4-TETRAHYDRO-5-((2-HYDROXY-3-tert-BUTYLAMINO)PROPOXY)-2,3-NAPHTHALENEDIOL see CNR675

(E)-5,10,11,11a-TETRAHYDRO-9-HYDROXY-11-METHOXY-8-METHYL-5-OXO-1H-PYRROLO(2,1-c)(1,4)BENZODIAZEPINE-2-ACRYLAMIDE MONOHYDRATE see API125

TETRAHYDRO-7-α-(1-HYDROXY-1-METHYLBUTYL)-6,14-endo-ETHENOORIPAVINE see EQO450

1-(TETRAHYDRO-5-HYDROXY-4-METHYL-3-FURANYL)-ETHANONE (9CI) see BMK290

1,2,3,4-TETRAHYDRO-5-(HYDROXYMETHYL)-2-(ISOPROPYLAMINO)-1,6-NAPHTHALENEDIOL SESQUIHYDRATE (6-E), HYDROCHLORIDE see TCU000

4-(1,2,3,4-TETRAHYDRO-4-(4-HYDROXY-2-OXO-2H-1-BENZOPYRAN-3-YL)-2-NAPHTHALENYL)BENZONITRILE see TCU100

4'-(1,2,3,4-TETRAHYDRO-4-(4-HYDROXY-2-OXO-2H-1-BENZOPYRAN-3-YL)-2-NAPHTHALENYL)(1,1'-BIPHENYL)-4-CARBONITRILE, cis- see TCU110

4'-(1,2,3,4-TETRAHYDRO-4-(4-HYDROXY-2-OXO-2H-1-BENZOPYRAN-3-YL)-2-NAPHTHALENYL)(1,1'-BIPHENYL)-4-CARBONITRILE, trans- see TCU120

4-((4-(1,2,3,4-TETRAHYDRO-4-(4-HYDROXY-2-OXO-2H-1-BENZOPYRAN-3-YL)-2-NAPHTHALENYL)PHENOXY)BENZONITRILE see TCU130

TETRAHYDRO-7-α-(2-HYDROXY-2-PENTYL)-6,14-endo-ETHENOORIPAVINE see EQO450

7,8,9-TETRAHYDROINDENE see TCU250

3a,4,7,7a-TETRAHYDROINDENE see TCU250

TETRAHYDROINDENE (RUSSIAN) see TCU250

3a,4,7,7a-TETRAHYDRO-1H-INDENE see TCU250

TETRAHYDRO-3-IODOTHIOPHENE-1,1-DIOXIDE see IFG000

3a,4,7,7a-TETRAHYDRO-1,3-ISOBENZOFURANDIONE see TDB000

1,2,3,4-TETRAHYDRO-2-((ISOPROPYLAMINO)METHYL)-7-NITRO-6-QUINOLINEMETHANOL see OLT000

TETRAHYDROISOQUINOLINE see TCU500

1,2,3,4-TETRAHYDROISOQUINOLINE see TCU500

TETRAHYDRO-p-ISOXAZINE see MRP750

TETRAHYDRO-1,4-ISOXAZINE see MRP750

4,5,6,7-TETRAHYDROISOXAZOLO(4,5-c)PYRIDIN-3-OL see TCU550

TETRAHYDROLINALOOL see LFY510

TETRAHYDROLINALOOL see TCU600

TETRAHYDRO-4,7-METHANOINDAN-5(4H)-ONE see OPC000

3a,4,7,7a-TETRAHYDRO-4,7-METHANOINDENE see DGW000

TETRAHYDRO-4,7-METHANOINDENOL see HKB200

1,4,4a,8a-TETRAHYDRO-1,4-METHANONAPHTHALENE-5,8-DIONE see TCU650

1,2,3,4-TETRAHYDRO-1,4-METHANONAPHTHALEN-6-YL N-METHYL-N-(m-TOLYL)CARBAMOTHIOATE see MQA000

o-(1,2,3,4-TETRAHYDRO-1,4-METHANONAPHTHALEN-6-YL)-m,N-DIMETHYL-THIO-CARBANILATE see MQA000

o-(5,6,7,8-TETRAHYDRO-5,8-METHANO-2-NAPHTHYL)-N-METHYL-N-(m-METHYLPHENYL)THIOCARBAMATE see MQA000

8,9,10,11-TETRAHYDRO-3-METHOXY-1H-INDOLE(3,2-C)QUINOLINE-1,4(7H)-DIONE see TCU660

5,6,7,8-TETRAHYDRO-4-METHOXY-6-METHYL-1,2-DIOXOLO-(4,5-g)ISOQUINOLIN-5-OL (9CI) see CNT625

8,9,10,11-TETRAHYDRO-3-METHOXY-6-METHYL-1H-INDOLO(3,2-C)QUINOLINE-1,4(7H)-DIONE see MEY100

1,2,3,4-TETRAHYDRO-7-METHOXY-1-METHYL-9H-PYRIDO(3,4-b)INDOLE HYDROCHLORIDE see TCT250

1-((1,2,3,4-TETRAHYDRO-1-(4-(4-(2-METHOXYPHENYL)-1-PIPERAZINYL)BUTYL)-7-METHYL-2,4-DIOXO-3-PHENYLPYRIDO(2,3-D)PYRIMIDIN-5-YL)CARBONYL)PYRROLIDINE see TCU670

1,2,3,4-TETRAHYDRO-6-METHOXYQUINOLINE see TCU700

4,6,7,14-TETRAHYDRO-5-METHYL-BIS(1,3)BENZODIOXOLO(4,5-c:5',6'-g)AZECIN-13(5H)-ONE see FOW000

1,2,3,4-TETRAHYDRO-11-METHYLCHRYSEN-1-ONE see MNG250

(1-α,2-β,2A-β,3A-β)-1,2,2A,3A-TETRAHYDRO-4-METHYLCHRYSENO(3,4-B)OXIRENE-1,2-DIOL see CML832

(1-α,2-β,2A-β,3A-β)-1,2,2A,3A-TETRAHYDRO-4-METHYLCHRYSENO(3,4-B)OXIRENE-1,2-DIOL see TCU750

1,2,3,6-TETRAHYDRO-1-((6-METHYL-3-CYCLOHEXEN-1-YL)CARBONYL)PYRIDINE see TCV375

1,2,3,6-TETRAHYDRO-1-((2-METHYLCYCLOHEXYL)CARBONYL)PYRIDINE see TCV400

5,6,7,8-TETRAHYDRO-6-METHYL-1,3-DIOXOLO-(4,5-g)ISOQUINOLIN-5-OL see HGR000

1,2,3,4-TETRAHYDRO-2-METHYL-1,4-DIOXO-2-NAPHTHALENESULFONIC ACID SODIUM SALT see MCB575

1-((1,2,3,4-TETRAHYDRO-7-METHYL-2,4-DIOXO-3-PHENYL-1-(4-(4-PHENYL-1-PIPERA ZINYL) BUTYL)PYRIDO(2,3-D)PYRIMIDIN-5-YL)CARBONYL)YRROLIDINE see TCV420

1,2,3,4-TETRAHYDRO-2-(((1-METHYL-ETHYL)AMINO)METHYL)-7-NITRO-6-QUINOLINEMETHANOL see OLT000

4a,5,7a,8-TETRAHYDRO-12-METHYL-9H-9,9c-IMINOETHANOPHENANTHRO(4,5-bcd)FURAN-3,5-DIOL see MRO500

3a,4,7,7a-TETRAHYDRO-3a-METHYL-1,3-ISOBENZOFURANDIONE see EBW000

1,2,3,9-TETRAHYDRO-9-METHYL-3-((2-METHYL-1H-IMIDAZOL-1-YL)METHYL)-4H-CAR BAZOL-4-ONE see OJD150

1-((1,2,3,4-TETRAHYDRO-7-METHYL-1-(4-(4-METHYL-1-PIPERAZINYL)BUTYL)-2,4-DI OXO-3-PHENYLPYRIDO(2,3-D)PYRIMIDIN-5-YL)CARBONYL)PYRROLIDINE see TCV425

TETRAHYDRO-4-METHYL-2-(2-METHYLPROPEN-1-YL)PYRAN see RNU000

1,2,5,6-TETRAHYDRO-1-METHYLNICOTINIC ACID, METHYL ESTER see AQT750

1,2,5,6-TETRAHYDRO-1-METHYLNICOTINIC ACID, METHYL ESTER, HYDROBROMIDE see AQU000

TETRAHYDRO-3-METHYL-4-((5-NITROFURFURYLIDENE)AMINO)-2H-1,4-THIAZINE-1,1-DIOXIDE see NGG000

TETRAHYDRO-N-METHYL-N-NITROSO-3-THIOPHENAMINE 1,1-DIOXIDE see NKQ500

1,2,3,6-TETRAHYDRO-1-(2-METHYL-1-OXOPENTYL)PYRIDINE see AEY406

1,4,5,6-TETRAHYDRO-3-(α-METHYLPHENETHYL)-as-TRIAZINE HYDROCHLORIDE see TCV460

3,4,5,6,7-TETRAHYDRO-5-METHYL-1-PHENYL-1H-2,5-BENZOXAZOCINE see FAL000

2,3,4,9-TETRAHYDRO-2-METHYL-9-PHENYL-1H-INDENO(2,1-c)PYRIDINE HYDROCHLORIDE see TEQ700

1,2,3,4-TETRAHYDRO-2-METHYL-4-PHENYL-8-ISOQUINOLINAMINE (9CI) see NMV700

2,3,4,5-TETRAHYDRO-2-METHYL-5-(PHENYLMETHYL)-1H-PYRIDO(4,3-b)INDOLE see MBW100

1,2,3,6-TETRAHYDRO-1-METHYL-4-PHENYLPYRIDINE see TCV470

1,2,3,6-TETRAHYDRO-1-METHYL-4-PHENYLPYRIDINE HYDROCHLORIDE see TCV500

3,4,5,6-TETRAHYDRO-N-METHYLPHTHALIMIDE see MPP250

1,2,3,4-TETRAHYDRO-1-METHYL-3-PYRIDINECARBOXYLIC ACID see AQT625

2,3,4,9-TETRAHYDRO-1-METHYL-1H-PYRIDO(3,4-B)INDOLE-3-CARBOXYLIC ACID see TCV600

TETRAHYDRO-2-METHYL-6-(TETRAHYDRO-2,5-DIOXO-3-FURYL)PYRAN-3,4-DICARBOXYLIC ANHYDRIDE POLYMER see TCW500

(E)-4,5,6-TETRAHYDRO-1-METHYL-2-(2-(2-THIENYL)ETHENYL)PYRIMIDINE see TCW750

(E)-1,4,5,6-TETRAHYDRO-1-METHYL-2-(2-(2-THIENYL)VINYL)PYRIMIDINE TARTRATE (1:1) see TCW750

TETRAHYDRO-5-METHYL-1,3,5-TRIAZINE-2(1H)-THIONE see TCW800

5,6,7,8-TETRAHYDRO-1-NAFTYL-N-METHYLKARBAMAT (CZECH) see TCY275

TETRAHYDRO-α-(1-NAPHTALENYLMETHYL)-2-FURANPROPANOIC ACID 2-(DIETHYLAMINO)ETHYL ESTER see NAE000

TETRAHYDRONAPHTHALENE see TCX500

6,7,3',4'-TETRAMETHOXY-1-
BENZYLISOQUINOLINE
HYDROCHLORIDE see PAH250
2,3,9,10-TETRAMETHOXY-13a,α-BERBINE
HYDROCHLORIDE see GEO600
dl-2,3,9,10-TETRAMETHOXYBERBIN-1-OL
see TDI750
3,6,11,14-
TETRAMETHOXYDIBENZO(g,p)CHRYSEN
E see TDJ000
(1-β)-6,6',7,12-TETRAMETHOXY-2,2'-
DIMETHYL-BERBAMAN (9CI) see TDX830
3,3,14,14-TETRAMETHOXY-2,15-DIOXA-
3,14-DISILAHEXADECANE see BLQ550
3,3,6,6-TETRAMETHOXY-2,7-DIOXA-3,6-
DISILAOCTANE see EIT200
6',7',10,11-TETRAMETHOXYEMETAN see
EAL500
6',7',10,11-TETRAMETHOXYEMETAN
DIHYDROCHLORIDE TETRAHYDRATE
see EAN500
2,3,10,11-TETRAMETHOXY-8-METHYL-
DIBENZO(a,g)QUINOLIZINIUM SALT with
SULFOACETIC ACID (1:1) see CNR250
1,2,9,10-TETRAMETHOXY-6a-α-
NORAPORPHINE see NNR200
TETRAMETHOXYSILANE see MPI750
TETRAMETHOXYSILANE POLYMER see
TDJ100
3,3',4,4'-TETRAMETHOXYSTILBENE see
TDJ200
(1-β)-6,6',7,12-TETRAMETHOXY-2,2,2',2'-
TETRAMETHYL-BERBAMANIUM
DIIODIDE (9CI) see TDX835
6,6',7',12'-TETRAMETHOXY-2,2,2',2'-
TETRAMETHYLTUBOCURARANIUM
DIIODIDE see DUM000
(1-β)-6,6',7',12'-TETRAMETHOXY-2,2,2',2'-
TETRAMETHYLTUBOCURARANIUM
DIIODIDE see TDO100
N,N,N'-TETRAMETHYL-3,6-
ACRIDINEDIAMINE see BJF000
N,N,N',N'-TETRAMETHYL-3,6-
ACRIDINEDIAMINE
MONOHYDROCHLORIDE (9CI) see BAQ250
((R-R*,R*))-N,N,9,9-TETRAMETHYL-10(9H)-
ACRIDINEPROPANAMINE-2,3-
DIHYDROXYBUTANEDIOATE (1:1) see
DRM000
3,2',4',6'-TETRAMETHYLAMINODIPHENYL
see TDJ250
TETRAMETHYLAMMONIUM AMIDE see
TDJ300
TETRAMETHYLAMMONIUM
AZIDOCYANATOIODATE(I) see TDJ325
TETRAMETHYLAMMONIUM
AZIDOCYANOIODATE(I) see TDJ350
TETRAMETHYLAMMONIUM
AZIDOSELENOCYANATOIODATE(I) see
TDJ375
TETRAMETHYLAMMONIUM
BOROHYDRATE see TDJ500
TETRAMETHYLAMMONIUM BROMIDE see
TDJ750
TETRAMETHYLAMMONIUM CHLORIDE
see TDK000
TETRAMETHYLAMMONIUM CHLORITE
see TDK250
TETRAMETHYLAMMONIUM
DIAZIDOIODATE(I) see TDK300
TETRAMETHYLAMMONIUM HYDROXIDE
see TDK500
TETRAMETHYLAMMONIUM
HYDROXIDE, liquid (DOT) see TDK500
TETRAMETHYLAMMONIUM IODIDE see
TDK750
2,3,9,10-TETRAMETHYLANTHRACENE see
TDK885
N,N,10,10-TETRAMETHYL-Δ$^{9(10),Y}$-
ANTHRACENEPROPYLAMINE see AEG875

N,N,10,10-TETRAMETHYL-Δ$^{9(10H),Y}$-
ANTHRACENEPROPYLAMINE
HYDROCHLORIDE see TDL000
TETRAMETHYLARSONIUM see TDL100
TETRAMETHYLARSONIUM IODIDE see
TDL250
7,8,9,11-TETRAMETHYLBENZ(c)ACRIDINE
see TDL500
1,3,4,10-TETRAMETHYL-7,8-
BENZACRIDINE (FRENCH) see TDL500
7,8,9,12-
TETRAMETHYLBENZ(a)ANTHRACENE see
TDL750
5,6,9,10-TETRAMETHYL-1,2-
BENZANTHRACENE see TDL750
6,7,9,10-TETRAMETHYL-1,2-
BENZANTHRACENE see TDM000
7,9,10,12-
TETRAMETHYLBENZ(a)ANTHRACENE see
TDM000
1,2,3,4-TETRAMETHYLBENZENE see
TDM250
1,2,3,5-TETRAMETHYLBENZENE see
TDM500
1,2,4,5-TETRAMETHYLBENZENE see
TDM750
3,3',5,5'-TETRAMETHYLBENZIDINE see
TDM800
3,5,3',5'-TETRAMETHYLBENZIDINE see
TDM800
N,N,N',N'-TETRAMETHYLBENZIDINE see
BJF600
N,N,N',N'-TETRAMETHYL-p,p'-BENZIDINE
see BJF600
TETRAMETHYL-p-BENZOQUINONE see
TDM810
TETRAMETHYL-1,4-BENZOQUINONE see
TDM810
2,3,5,6-TETRAMETHYLBENZOQUINONE
see TDM810
3,2',4',6'-TETRAMETHYLBIPHENYLAMINE
see TDJ250
1-((2,2',3,3'-TETRAMETHYL(1,1'-BIPHENYL)-
4-YL)AZO)-2-NAPHTHALENOL see TDM820
α,α,α',α'-TETRAMETHYL-β,β'-
BISCHLOROETHYL SULFIDE see BIL000
1,3,5,8-TETRAMETHYL-2,4-BIS(α-
HYDROXYETHYL)PROPHINE-6,7-
DIPROPIONIC ACID see PHV275
TETRAMETHYL BUTANEDIAMINE see
TDN000
N,N,N',N'-TETRAMETHYL-1,3-
BUTANEDIAMINE see TDN000
1,1,3,3-TETRAMETHYLBUTYLAMINE see
TDN250
p-(1',1',3',3'-TETRAMETHYLBUTYL)FENOL
see TDN500
p-(1,1,3,3-TETRAMETHYLBUTYL)PHENOL
see TDN500
p-(1,1,3,3-TETRAMETHYLBUTYL)PHENOL,
polymer with ETHYLENE OXIDE and
FORMALDEHYDE see TDN750
4-(1,1,3,3-TETRAMETHYLBUTYL)PHENOL,
polymer with FORMALDEHYDE and
OXIRANE see TDN750
α-((1,1,3,3-
TETRAMETHYLBUTYL)PHENYL)-ω-
HYDROXY-POLY(OXY-1,2-ETHANEDIYL)
see GHS000
(2-(2,2,3,3-
TETRAMETHYLBUTYLTHIO)ACETOXY)T
RIBUTYLSTANNANE see TDO000
N,N',o,o'-TETRAMETHYLCURINIUM
DIIODIDE see TDO100
TETRAMETHYLCYCLOBUTA-1,3-DIONE
see TDO250
TETRAMETHYL-1,3-
CYCLOBUTANEDIONE see TDO250
2,2,4,4-TETRAMETHYL-1,3-
CYCLOBUTANEDIONE see TDO250

α-2,2,6-
TETRAMETHYLCYCLOHEXENEBUTANAL
see TDO255
4-(2,5,6,6-TETRAMETHYL-2-CYCLO-
HEXEN-1-YL)-3-BUTEN-2-ONE see IGW500
2,6,6,9-TETRAMETHYL-1,4,8-
CYCLOUNDECATRIENE (E,E,E)- see
HGM550
2,2,9,9-TETRAMETHYL-1,10-DECANEDIOL
see TDO260
N,N,N',N'-
TETRAMETHYLDEUTEROFORMAMIDINI
UM PERCHLORATE see TDO275
TETRAMETHYLDIALUMINUM
DIHYDRIDE see TDO500
N,N,N',N'-TETRAMETHYL-p,p'-
DIAMIDODIPHOSPHORIC ACID DIETHYL
ESTER see DIV200
N,N,N',N'-TETRAMETHYL-DIAMIDO-
FOSFORZUUR-FLUORIDE (DUTCH) see
BJE750
TETRAMETHYLDIAMIDOPHOSPHORIC
FLUORIDE see BJE750
N,N,N',N'-TETRAMETHYL-DIAMIDO-
PHOSPHORSAEURE-FLUORID (GERMAN)
see BJE750
TETRAMETHYLDIAMINOBENZHYDROL
see TDO750
TETRAMETHYLDIAMINOBENZOPHENO
NE see MQS500
TETRAMETHYLDIAMINODIPHENYLACE
TIMINE see IBB000
TETRAMETHYLDIAMINODIPHENYLMET
HANE see MJN000
p,p-
TETRAMETHYLDIAMINODIPHENYLMET
HANE see MJN000
4,4'-
TETRAMETHYLDIAMINODIPHENYLMET
HANE see MJN000
N,N'-(N,N-TETRAMETHYL)-1-
DIAMINODIPHENYLNAPHTHYLAMINOM
ETHANE HYDROCHLORIDE see VKA600
N,N,N',N'-TETRAMETHYL-1,2-
DIAMINOETHANE see TDQ750
N,N,N',N'-
TETRAMETHYLDIAMINOMETHAN
(GERMAN) see TDR750
TETRAMETHYL DIAPARA-AMIDO-
TRIPHENYL CARBINOL see AFG500
TETRAMETHYLDIARSANE see TDP250
TETRAMETHYLDIBORANE see TDP275
TETRAMETHYL 1,2-
DICARBOXYETHYLPHOSPHONATE see
PHA565
TETRAMETHYLDIGALLANE see TDP300
TETRAMETHYLDIGOLD DIAZIDE see
TDP500
TETRAMETHYL-1,2-DIOXETANE see
TDP325
N,N,N,2-TETRAMETHYL-1,3-DIOXOLANE-
4-METHANAMINIUM IODIDE (9CI) see
MJH250
TETRAMETHYLDIPHOSPHANE see
TDP525
TETRAMETHYLDIPHOSPHINE see TDP525
N,N,N',N'-TETRAMETHYL-
DIPROPYLENETRIAMINE see TDP750
N,N'-(4,4,7,7-TETRAMETHYL-4,7-
DISILADECAMETHYLEN)BIS(TRIMETHYL
AMMONIUMIODID) see EEB200
sym-TETRAMETHYLDISILOXANE see
TDP775
1,1,3,3-TETRAMETHYLDISILOXANE see
TDP775
4,4'-(1,1,3,3-TETRAMETHYL-1,3-
DISILOXANEDIYL)BIS-1-BUTANAMINE see
OKU300
TETRAMETHYLDISTIBINE see TDQ000
TETRAMETHYLDIURANE SULPHITE see
TFS350

1,1,3,3-TETRAMETHYL-1,3-DIVINYLDISILAZANE see TDQ050
TETRAMETHYL-1,3-DIVINYLDISILOXANE see TDQ075
sym-TETRAMETHYLDIVINYLDISILOXANE see TDQ075
1,4-TETRAMETHYLEN-BIS-(N-AETHYLPIPERIDINIUM)-DIBROMID (GERMAN) see TDQ255
1,4-TETRAMETHYLEN-BIS-(N-AETHYLPYRROLIDINIUM)-DIBROMID (GERMAN) see TDQ260
1,4-TETRAMETHYLEN-BIS-(N-METHYLPIPERIDINIUM)-DIBROMID (GERMAN) see TDQ263
1,4-TETRAMETHYLEN-BIS-(N-METHYLPYRROLIDINIUM)-DIBROMID (GERMAN) see TDQ265
1,4-TETRAMETHYLEN-BIS-(PYRIDINIUM)-DIBROMID (GERMAN) see TDQ325
1,1'-TETRAMETHYLENBIS-PYRIDINIUM DIBROMIDE see TDQ325
TETRAMETHYLENE see COW000
TETRAMETHYLENE ACRYLATE see TDQ100
N,N'-TETRAMETHYLENEBIS(1-AZIRIDINECARBOXAMIDE) see TDQ225
5,5'-(TETRAMETHYLENEBIS(CARBONYLIMINO))BIS(N-METHYL-2,4,6-TRIIODOISOPHTHALAMIC ACID) see TDQ230
1,1'-TETRAMETHYLENEBIS(3-(2-CHLOROETHYL)-3-NITROSOUREA) see TDQ250
1,1'-TETRAMETHYLENEBIS(1-ETHYLPIPERIDINIUM)DIBROMIDE see TDQ255
1,1'-TETRAMETHYLENEBIS(1-ETHYL-PYRROLIDINIUM) DIBROMIDE see TDQ260
TETRAMETHYLENE BIS(METHANESULFONATE) see BOT250
1,1'-TETRAMETHYLENEBIS(1-METHYLPIPERIDINIUM)DIBROMIDE see TDQ263
1,1'-TETRAMETHYLENEBIS(1-METHYL-PYRROLIDINIUM) DIBROMIDE see TDQ265
1,1'-TETRAMETHYLENEBIS(PYRIDINIUM BROMIDE) see TDQ325
TETRAMETHYLENE CHLOROHYDRIN see CEU500
TETRAMETHYLENE CYANIDE see AER250
TETRAMETHYLENE DIACRYLATE see TDQ100
TETRAMETHYLENEDIAMINE see BOS000
1,4-TETRAMETHYLENEDIAMINE see BOS000
TETRAMETHYLENEDIAMINE, N,N-DIETHYL-4-METHYL- see DJQ850
TETRAMETHYLENE DIIODIDE see TDQ400
TETRAMETHYLENE DIMETHANE SULFONATE see BOT250
3,3'-(TETRAMETHYLENEDIOXY)BIS(PROPYLAMINE) see BGU600
3,3'-(TETRAMETHYLENEDIOXY)DI(PROPANAMINE) see BGU600
TETRAMETHYLENEDISULFOTETRAMINE see TDX500
1,4-TETRAMETHYLENE GLYCOL see BOS750
TETRAMETHYLENE GLYCOL DIACRYLATE see TDQ100
TETRAMETHYLENE IODIDE see TDQ400
TETRAMETHYLENE OXIDE see TCR750
TETRAMETHYLENEOXIRANE see CPD000

6,7-TETRAMETHYLENE-5-PHENYL-1,2-DIHYDRO-3H-THIENO(2,3-e)(1,4)DIAZEPIN-2-ONE see BAV625
TETRAMETHYLENESULFIDE see TDC730
TETRAMETHYLENE SULFONE see SNW500
TETRAMETHYLENE SULPHOXIDE see TDQ500
TETRAMETHYLENETETRANITRAMINE see CQH250
TETRAMETHYLENETHIURAM DISULPHIDE see TFS350
TETRAMETHYLENIMINE see PPS500
3,3,5,5-TETRAMETHYL-4-ETHOXYVINYLCYCLOHEXANONE see EES370
TETRAMETHYL ETHYLENE DIAMINE see TDQ750
N,N,N',N'-TETRAMETHYLETHYLENEDIAMINE see TDQ750
TETRAMETHYLETHYLENE GLYCOL see TDR000
TETRAMETHYLETHYLFORMYLTETRALIN see FNK200
3',6',10,11b-TETRAMETHYL-2,3'a,4',5,5',6,6',6a,6b,7,7',7'a,8,11,11a,11b-HEXADECAHYDRO-SPIRO(9H-BENZO(a)FLUORENE-9,2'(3'H)-FURO(3,2-b)PYRIDIN)-3(1H)-ONE see DAQ125
3,7,11,15-TETRAMETHYL-2-HEXADECEN-1-OL see PIB600
N,N,N',N'-TETRAMETHYLHEXAMETHYLENE DIAMINE see TDR100
N,N,N',N'-TETRAMETHYLHEXAMETHYLENEDIAMINE-1,3-DIBROMOPROPANE copolymer see HCV500
N,N,N',N'-TETRAMETHYL-1,6-HEXANEDIAMINE see TDR100
TETRAMETHYLHYDRAZINE HYDROCHLORIDE see TDR250
3,4,5,8-TETRAMETHYL-3H-IMIDAZO(4,5-F)QUINOXALIN-2-AMINE see AMQ600
3,4,7,8-TETRAMETHYL-3H-IMIDAZO(4,5-F)QUINOXALIN-2-AMINE see AMQ620
TETRAMETHYL LEAD see TDR500
N,N,N'-TETRAMETHYLMETHANEDIAMINE see TDR750
N,N,N',N'-TETRAMETHYLMETHANEDIAMINE ETHANEDIOATE see TDR800
TETRAMETHYL METHYLENE DIAMINE (DOT) see TDR750
2,2,6,6-TETRAMETHYLNITROSOPIPERIDINE see TDS000
N,2,3,3-TETRAMETHYL-2-NORBORNAMINE see VIZ400
N,N,2,3-TETRAMETHYL-2-NORBORNANAMINE see TDS250
N,2,3,3-TETRAMETHYL-2-NORBORNANAMINE HYDROCHLORIDE see MQR500
N,2,3,3-TETRAMETHYL-2-NORCAMPHANAMINE see VIZ400
1,(5,5,7,7-TETRAMETHYL-2-OCTANYL)-2-METHYL-5-ETHYLPYRIDINIUM CHLORIDE see TDS500
p-1',1',4',4'-TETRAMETHYLOKTYLBENZENSULFONAN SODNY (CZECH) see DXW200
TETRAMETHYLOLMETHANE see PBB750
2,6,10,14-TETRAMETHYLPENTADECANE see PMD500
1:2:3:4-TETRAMETHYLPHENANTHRENE see TDS750
N,N,N-α-TETRAMETHYL-10H-PHENOTHIAZINE-10-ETHANAMINIUM METHYL SULFATE see MRW000

2-(2,3,5,6-TETRAMETHYLPHENOXY)PROPIONIC ACID see BHM000
TETRAMETHYL-p-PHENYLENEDIAMINE see BJF500
N,N,N',N'-TETRAMETHYL-o-PHENYLENEDIAMINE see TDT000
N,N,N',N'-TETRAMETHYL-p-PHENYLENEDIAMINE see BJF500
N,N,N',N'-TETRAMETHYL-p-PHENYLENEDIAMINE DIHYDROCHLORIDE see TDT250
(2,3,5,6-TETRAMETHYLPHENYL)MERCURY ACETATE see AAS500
TETRAMETHYLPHOSPHONIUM IODIDE see TDT500
TETRAMETHYL PHOSPHONOSUCCINATE see PHA565
TETRAMETHYLPHOSPHORODIAMIDIC FLUORIDE see BJE750
N,N,N,N-TETRAMETHYLPHOSPHORODIAMIDIC FLUORIDE see BJE750
2,2,6,6-TETRAMETHYLPIPERIDINE see TDT750
TETRAMETHYLPIPERIDINE NITROXIDE see TDT800
2,2,6,6-TETRAMETHYLPIPERIDINE-1-OXYL see TDT800
TETRAMETHYLPIPERIDINOL N-OXYL see HOI300
2,2,6,6-TETRAMETHYLPIPERIDINONE see TDT770
2,2,6,6-TETRAMETHYL-4-PIPERIDINONE see TDT770
TETRAMETHYLPIPERIDINOOXY see TDT800
2,2,6,6-TETRAMETHYLPIPERIDINOOXY see TDT800
2,2,6,6-TETRAMETHYLPIPERIDINOOXYL see TDT800
N,N,N',N'-TETRAMETHYL-P-PIPERIDINOPHOSPHONIC DIAMIDE see TDT850
2-(2,2,6,6-TETRAMETHYL-4-PIPERIDINYL)-2-OXAZOLINE see TDT900
2,2,6,6-TETRAMETHYLPIPERIDINYLOXY see TDT800
2,2,6,6-TETRAMETHYL-1-PIPERIDINYLOXY see TDT800
2,2,6,6-TETRAMETHYLPIPERIDONE see TDT770
2,2,6,6-TETRAMETHYL-4-PIPERIDONE see TDT770
2,2,6,6-TETRAMETHYL-4-PIPERIDONE OXIME see TDU000
2,2,6,6-TETRAMETHYLPIPERIDOXYL see TDT800
TETRAMETHYLPLATINUM see TDU250
TETRAMETHYLPLUMBANE see TDR500
N,N,N',N'-TETRAMETHYL-1,3-PROPANEDIAMINE see TDU500
N,N,N',N'-TETRAMETHYL PROPENE-1,3-DIAMINE see TDU750
N,N,N',N'-TETRAMETHYL-1,3-PROPENYLDIAMINE see TDU750
TETRAMETHYLPYRAZINE see TDU800
2,3,5,6-TETRAMETHYL PYRAZINE (FCC) see TDU800
N,N,2,6-TETRAMETHYL-4-(4'-(PYRIDYL-1'-OXIDE)AZO)ANILINE see DQF600
N,N,2,3-TETRAMETHYL-4-(4'-(PYRIDYL-1'-OXIDE)AZO)ANILINE see DQF200
N,N,2,5-TETRAMETHYL-4-(4'-(PYRIDYL-1'-OXIDE)AZO)ANILINE see DQF400
TETRAMETHYL PYRONIN see PPQ750
TETRAMETHYLPYROPHOSPHATE see TDV000
2,2,5,5-TETRAMETHYLPYRROLIDINE see TDV250

1,1,2,2-TETRAPHENYLETHYLENE GLYCOL see TEA600

TETRAPHENYL METHANE see TEA750

TETRAPHENYLRESORCINOL DIPHOSPHATE see REA050

TETRAPHOSPHATE HEXAETHYLIQUE (FRENCH) see HCY000

TETRAPHOSPHOR (GERMAN) see PHP010

TETRAPHOSPHORUS IODIDE see TEB750

TETRAPHOSPHORUS TRISELENIDE see TEC000

TETRAPHOSPHORUS TRISULFIDE see PHS500

TETRA PINK B see DNT300

TETRAPION see SKE100

TETRAPOM see TFS350

TETRAPOTASSIUM DIPHOSPHORATE see TEC100

TETRAPOTASSIUM ETIDRONATE see TEC250

TETRAPOTASSIUM FERROCYANIDE see TEC500

TETRAPOTASSIUM HEXACYANOFERRATE see TEC500

TETRAPOTASSIUM HEXACYANOFERRATE(4-) see TEC500

TETRAPOTASSIUM HEXACYANOFERRATE(II) see TEC500

TETRAPOTASSIUM PYROPHOSPHATE see PLR200

TETRAPROPENYLSUCCINIC ANHYDRIDE see TEC600

TETRAPROPYLAMMONIUM BROMIDE see TEC750

TETRAPROPYLAMMONIUM HYDROXIDE see TED000

TETRA-N-PROPYLAMMONIUM HYDROXIDE see TED000

TETRAPROPYLAMMONIUM IODIDE see TED250

TETRA-N-PROPYLAMMONIUM IODIDE see TED250

TETRAPROPYLAMMONIUM OXIDE see TED000

TETRA-n-PROPYL DITHIONOPYROPHOSPHATE see TED500

TETRA-n-PROPYL DITHIOPYROPHOSPHATE see TED500

O,O,O,O-TETRAPROPYL DITHIOPYROPHOSPHATE see TED500

TETRAPROPYLENE see PMP750

TETRAPROPYLENEBENENESULPHONATE, SODIUM SALT see SKF600

TETRAPROPYLENE BENZENESULFONATE see TED550

TETRAPROPYLENEPENTAMINE see TED600

TETRAPROPYLGERMANE see TED650

TETRAPROPYLGERMANIUM see TED650

TETRAPROPYL LEAD see TED750

N,4':N',4":N",4"'-TETRA(PYRROLE-2-CARBOXAMIDE), N"'-(2-AMIDINOETHYL)-4-FORMAMIDO-1,1',1",1"'-TETRAMETHYL-, HYDROCHLORIDE see DXG450

TETRASELENIUM TETRANITRIDE see TEE000

TETRASEPTAN see BEL900

TETRASILVER DIIMIDODIOXOSULFATE see SDP500

TETRASILVER DIIMIDOTRIPHOSPHATE see TEE100

TETRASILVER ORTHODIAMIDOPHOSPHATE see TEE125

TETRASILVER RUBIDIUM PENTAIODIDE see RQF100

TETRASIPTON see TFS350

TETRASODIUM BIS(CITRATE(3-)FERRATE(4-)) see TEE225

TETRASODIUM BISCITRATO FERRATE see TEE225

TETRASODIUM CYCLOTETRAPHOSPHATE see SKE550

TETRASODIUM DIARSENATE see SEY100

TETRASODIUM DIETHYLSTILBESTROL PHOSPHATE see TEE300

TETRASODIUM DIPHOSPHATE see TEE500

TETRASODIUM EDTA see EIV000

TETRASODIUM ETHYLENEDIAMINETETRAACETATE see EIV000

TETRASODIUM ETHYLENEDIAMINETETRACETATE see EIV000

TETRASODIUM (ETHYLENEDINITRILO)TETRAACETATE see EIV000

TETRASODIUM ETIDRONATE see TEE250

TETRASODIUM FOSFESTROL see TEE300

TETRASODIUM HYDROGEN 2-PHOSPHONATOBUTANE-1,2,4-TRICARBOXYLATE see TEE400

TETRASODIUM 2-METHYL-1,4-NAPHTHAHYDROQUINONE DIPHOSPHORIC ACID ESTER see NAQ600

TETRASODIUM 2-METHYL-1,4-NAPHTHALENEDIOL BIS(DIHYDROGEN PHOSPHATE) see NAQ600

TETRASODIUM PYROPHOSPHATE see TEE500

TETRASODIUM PYROPHOSPHATE, ANHYDROUS see TEE500

TETRASODIUM SALT of EDTA see EIV000

TETRASODIUM SALT of ETHYLENEDIAMINETETRACETIC ACID see EIV000

TETRASODIUM TETRAMETAPHOSPHATE see SKE550

TETRASOL see CBY000

TETRASTIGMINE see TCF250

TETRASUL see CKL750

TETRASULE see PBC250

TETRASULFIDE, BIS(ETHOXYTHIOCARBONYL) see BJU250

TETRASULFUR DINITRIDE see TEE750

TETRASULFUR TETRANITRIDE see SOO000

TETRA SYTAM see BJE750

TETRATELLURIUM TETRANITRIDE see TNO250

3,4,7,8-TETRATHIA-2,9-DIAZADECANEDIOIC ACID, 2,9-DIMETHYL-, BIS(2,3-DIHYDRO-2,2-DIMETHYL-7-BENZOFURANYL) ESTER, 3,8-DIOXIDE see EEB025

TETRATHIIN see OMY825

TETRATHIIN (DESICCANT) see OMY825

TETRATHIURAM DISULFIDE see TFS350

TETRATHIURAM DISULPHIDE see TFS350

TETRATRIMETHYLENEPENTAMINE see TED600

TETRAVEC see PCF275

TETRAVERINE see TBX000

TETRAVINYLLEAD see TEF250

TETRAVINYL-TETRAMETHYLCYCLO-TETRASILOXANE see TCB600

TETRAVOS see DGP900

TETRA-WEDEL see TBX250

7,8,9,10-TETRAZABICYCLO(5.3.0)-8,10-DECADIENE see PBI500

1,2,3,3A-TETRAZACYCLOHEPTA-8A,2-CYCLOPENTADIENE see PBI500

TETRAZENE see TEF500

1-TETRAZENE, 4-AMIDINO-1-(NITRSOAMINOAMIDINO)-(8CI) see TEF500

TETRAZEPAM see CFG750

TETRAZIDO-1,4-BENZOQUINONE see TBJ250

1,2,4,5-TETRAZINE see TEF600

TETRAZINE (dry) (DOT) see TEF600

S-TETRAZINO(1,2-A)-S-TETRAZINE, OCTAHYDRO- see OCA500

(1,2,4,5)TETRAZINO(1,2-A)(1,2,4,5)TETRAZINE, OCTAHYDRO- see OCA500

TETRAZOBENZENE-β-NAPHTHOL see OHI200

TETRAZO DEEP BLACK G see AQP000

TETRAZO DEEP BLACK GC EXTRA see CMN240

TETRAZOL see TEF650

1H-TETRAZOL-5-AMINE see AMR000

TETRAZOLE-5-DIAZONIUM CHLORIDE see TEF675

2H-TETRAZOLIUM, 3,3'-(3,3'-DIMETHOXY(1,1'-BIPHENYL)-4,4'-DIYL)BIS(2-(4-NITROPHENYL)-5-PHEN YL-,DICHLORIDE see NFQ020

2H-TETRAZOLIUM, 3,3'-(3,3'-DIMETHOXY(1,1'-BIPHENYL)-4,4'-DIYL)BIS(2-(4-NITROPHENYL)-5-PHEN YL-,DICHLORIDE see NMK100

TETRAZOLIUM NITRO BLUE see NFQ020

TETRAZOLIUM NITRO BLUE see NMK100

TETRAZOLIUM NITRO BT see NFQ020

TETRAZOLIUM NITRO BT see NMK100

TETRAZOSIN HYDROCHLORIDE DIHYDRATE see TEF700

TETRIL see TEG250

TETRINE see EIV000

TETRINE ACID see EIX000

TETROCARCIN A see TEF725

TETROCHROME ORANGE GR see CMP882

TETRODIRECT BLACK EFD see AQP000

TETRODIRECT BLACK V see CMN240

TETRODIRECT BORDEAUX B see CMO872

TETRODIRECT BROWN MJ see CMO800

TETRODIRECT FAST TURQUOISE GL see COF420

TETRODIRECT SCARLET 4BS see CMO870

TETRODONTOXIN see FOQ000

TETRODOTOXIN see FOQ000

TETRODOXIN see FOQ000

TETROFAN see TEF775

TETROGUER see PCF275

TETROLE see FPK000

TETRON see TCF250

TETRON-100 see TCF250

TETRONE A see TEF750

TETROPHAN see TEF775

TETROPHENE GREEN M see AFG500

TETROPHINE see TEF775

TETROPIGMENT FAST YELLOW BG see CMS210

TETROPIGMENT FAST YELLOW VG see CMS210

TETROPIL see PCF275

TETROSAN see AFP750

TETROSIN OE see BGJ250

TETROSOL see TBX250

TETRYL see TEG250

2,4,6-TETRYL see TEG250

TETRYLAMMONIUM see TCB725

TETRYLAMMONIUM BROMIDE see TCC000

TETRYL FORMATE see IIR000

TETRZOLIUM CHLORIDE see TMV500

TETTERWORT see CCS650

TETURAM see DXH250

TETURAMIN see DXH250

TEVCOCIN see CDP250

TEVCODYNE see BRF500

TEXACO LEAD APPRECIATOR see BPV100

TEXANOL see TEG500

TEXAN RED TONER D see CHP500

TEXAPON T-42 see SON000

TEXAPON ZHC see SIB600

TEXAS see CBT750

TEXAS MOUNTAIN LAUREL see NBR800

TEXAS SARSAPARILLA see MRN100

TEXAS UMBRELLA TREE see CDM325

TEXCRYL see ADV900

TEXIN 192A see PKO500

TEXIN 445D see PKM250

TEXTILE see SFT500

TEXTILE RED WD-263 see NAY000
TEXTILON (OBS.) see TEG650
TEXTILOTOXIN see TEG650
TEXTONE see SFT500
TEX WET 1001 see DJL000
TF see GJS200
TF see TJE100
TF 1169 see FDA885
TFA see TJX900
2,3,4-TFA see TJX900
TFC mixed with TCC (1:2) see TIL526
TFDU see TKH325
TFE see TKA350
TFF see TNT500
T. FLAVOVIRIDIS VENOM see TKW100
T-FLUORIDE see SHF500
TFP see SKE100
TFPHM see TCI500
TFT THILO see TKH325
TG see AMH250
ThG see AMH250
TGA 2 see ADT250
TGA-EXTRACT see PLW550
T-GAS see EJN500
TGD 5161 see SMQ500
TGDR see TFJ200
β-TGDR see TFJ250
T-GELB BZW, GRUN 1 see QCA000
TGM 3 see MDN510
TGM 4 see TCE400
TGM 3S see MDN510
TGM 3PC see MDN510
TGR see TFJ500
TGS see TFJ500
TGT see GGY150
TH-152 see MDM800
TH-218 see EPC175
TH 564 see DTU850
TH/100 see TEP500
1064 TH see DUO400
TH 1314 see EPQ000
1321 TH see PNW750
TH-1395 see GFM200
2180 TH see PMM000
TH 346-1 see DRR400
TH 4128 see NMV300
TH 6040 see CJV250
THA see TCJ075
THACAPZOL see MCO500
THAIMYCIN B see TEG750
THAIMYCIN C see TEH000
THALAMONAL see DYF200
THALIBLASTINE see TEH250
THALICARPIN see TEH250
THALICARPINE see TEH250
(+)-THALICARPINE see TEH250
THALIDOMIDE see TEH500
(−)-THALIDOMIDE see TEH520
(+,−)-THALIDOMIDE see DVT200
(+)-THALIDOMIDE see TEH510
s-(−)-THALIDOMIDE see TEH520
(±)-THALIDOMIDE see TEH520
R-(+)-THALIDOMIDE see TEH510
THALIN see TEH500
THALINETTE see TEH500
THALLIC NITRATE see TEK800
THALLIC OXIDE see TEL050
THALLIC SULFATE see TEM100
THALLINE see TCU700
THALLIUM see TEI000
THALLIUM ACETATE see TEI250
THALLIUM(I) ACETATE see TEI250
THALLIUM(1+) ACETATE see TEI250
THALLIUM(I) AZIDE see TEI500
THALLIUM(I) AZIDODITHIOCARBONATE
see TEI600
THALLIUM BROMATE see TEI625
THALLIUM BROMIDE see TEI750
THALLIUM(I) CARBONATE (2:1) see TEJ000
THALLIUM CHLORATE see TEJ100
THALLIUM CHLORIDE see TEJ250
THALLIUM(1+) CHLORIDE see TEJ250

THALLIUM COMPOUNDS see TEJ500
THALLIUM, DIMETHYL(1,1,1-TRIFLUORO-
2,4-NONANEDIONATO-O,O')-, (T-4)- (9CI)
see NNA600
THALLIUM(I) DITHIOCARBONAZIDATE
see TEI600
THALLIUM(I) FLUOBORATE see TEJ750
THALLIUM(I) FLUORIDE see TEK000
THALLIUM (I) FLUOROACATATE see
TEK100
THALLIUM(I) FLUOSILICATE see TEK250
THALLIUM FULMINATE see TEK300
THALLIUM IODIDE see TEK500
THALLIUM(I) IODIDE see TEK500
THALLIUM(1+) IODIDE see TEK500
THALLIUM(I) IODOACETYLIDE see
TEK525
THALLIUM MALONATE see TEM399
THALLIUM MONOACETATE see TEI250
THALLIUM MONOCHLORIDE see TEJ250
THALLIUM MONOFLUORIDE see TEK000
THALLIUM MONOIODIDE see TEK500
THALLIUM MONONITRATE see TEK750
THALLIUM MONOSELENIDE see TEL500
THALLIUM NITRATE see TEK750
THALLIUM NITRATE see TEK800
THALLIUM(3+) NITRATE see TEK800
THALLIUM(III) NITRATE see TEK800
THALLIUM NITRATE (TL(NO3)3) see
TEK800
THALLIUM(I) NITRIDE see TEL000
THALLIUM OXIDE see TEL040
THALLIUM OXIDE see TEL050
THALLIUM(3+) OXIDE see TEL050
THALLIUM(III) OXIDE see TEL050
THALLIUM PEROXIDE see TEL050
THALLIUM(I) PEROXODIBORATE see
TEL100
THALLIUM aci-
PHENYLNITROMETHANIDE see TEL150
THALLIUM SELENIDE see TEL500
THALLIUM SESQUIOXIDE see TEL050
THALLIUM SULFATE see TEL750
THALLIUM(I) SULFATE (2:1) see TEM000
THALLIUM(III) SULFATE see TEM100
THALLIUM(II) SULFATE (1:1) see TEM050
THALLIUM SULFATE, solid (DOT) see
TEL750
THALLIUM TRINITRATE see TEK800
THALLOUS ACETATE see TEI250
THALLOUS CARBONATE see TEJ000
THALLOUS CHLORIDE see TEJ250
THALLOUS FLUORIDE see TEK000
THALLOUS IODIDE see TEK500
THALLOUS MALONATE see TEM399
THALLOUS NITRATE see TEK750
THALLOUS OXIDE see TEL040
THALLOUS SULFATE see TEM000
THALO GREEN No. 1 see PJQ100
THALRUGOSAMINE see HGI050
(+)-THALRUGOSAMINE see HGI050
THAM see POC750
THAM see TEM500
THAM-E see TEM500
THAM SET see TEM500
THANATE P 210 see PKB100
THANATE P 220 see PKB100
THANATE P 270 see PKB100
THANCAT AN 10 see MGR600
THANISOL see IHZ000
THANITE see IHZ000
THANOL C 150 see HKS550
THANOL R-350X POLYOL see HID050
THBP see BCS325
THBP see TKO250
THC see TCM250
Δ¹-THC see TCM250
Δ⁶-THC see TCM000
Δ⁸-THC see TCM000
Δ⁹-THC see TCM250
TH-DMBA see TCP600
THE-7 see HKG500

THEACITIN see TEQ250
THEAL see DNC000
THEAL AMPULES see DNC000
THEAL TABL. see TEP000
THEAN see HOA000
THEBAINE see TEN000
THEBAINE HYDROCHLORIDE see TEN100
(−)-THEBAINE HYDROCHLORIDE see
TEN100
THEBAIN HYDROCHLORID see TEN100
THECODIN see DLX400
THECODINE see DLX400
THEELIN see EDV000
THEELOL see EDU500
THEFANIL see DPJ200
THEFYLAN see DNC000
THEIN see CAK500
THEINE see CAK500
THEKODIN see DLX400
THELESTRIN see EDV000
THELYKININ see EDV000
THEMALON HYDROCHLORIDE see
DJP500
THENALTON see PAG200
THENARDITE see SJY000
THENARDOL see HIB500
THENFADIL see DPJ200
THENFADIL HYDROCHLORIDE see
TEO000
THENOBARBITAL see EOK000
2-THENOIC ACID see TFM600
N-(2-THENOYL)GLYCINOHYDROXAMIC
ACID see TEN725
p-(2-THENOYL)HYDRATROPIC ACID see
TEN750
(±)-2-(p-(2-THENOYL)PHENYL)PROPIONIC
ACID see TEN750
2-THENYLAMINE, 5-CHLORO-N-(2-
(DIMETHYLAMINO)ETHYL)-N-2-
PYRIDYL- see CHY250
THENYLDIAMINE see DPJ200
THENYLDIAMINE CHLORIDE see TEO000
THENYLDIAMINE HYDROCHLORIDE see
TEO000
THENYLENE see DPJ400
THENYLENE see TEO250
THENYLENE HYDROCHLORIDE see
DPJ400
THENYLIDENEHYDRAZON BENZOIC
ACIDE see TEO100
THENYLPYRAMINE see TEO250
THENYLPYRAMINE HYDROCHLORIDE
see DPJ400
THEOBROMINE see TEO500
THEOBROMINE SODIUM SALICYLATE see
TEO750
THEOCIN see TEP000
THEOCIN SOLUBLE see TEQ250
THEOCOR see HNY500
THEODEN see HOA000
THEODROX see TEP500
THEOESBERIVEN see TEO800
THEOFIBRATE see TEQ500
THEOFOL see TEP000
THEOGEN see ECU750
THEOGRAD see TEP000
THEOHARN see MCB000
THEOLAIR see TEP000
THEOLAMINE see TEP500
THEOLIX see TEP000
THEOMIN see TEP500
THEOMINAL see EOK000
THEON see HOA000
THEOPHILCHOLINE see TEH500
THEOPHORIN TARTRATE see PDD000
THEOPHYL-225 see TEP000
THEOPHYLDINE see TEP500
THEOPHYLLAMINE see TEP500
THEOPHYLLAMINIUM see TEP500
THEOPHYLLIN see TEP000
THEOPHYLLIN AETHYLENDIAMIN
(GERMAN) see TEP500

THEOPHYLLINE see TEP000
THEOPHYLLINE, anhydrous see TEP000
THEOPHYLLINE CHOLINATE see CMG300
7-THEOPHYLLINEETHANOL see HLC000
THEOPHYLLINE ETHYLENEDIAMINE see TEP500
THEOPHYLLINE, compound with ETHYLENEDIAMINE (2:1) see TEP500
8-THEOPHYLLINE MERCURIC ACETATE see TEP750
THEOPHYLLINE METHOXYOXIMERCURIPROPYL SUCCINYLUREA see TEQ000
2-(7'-THEOPHYLLINEMETHYL)-1,3-DIOXOLANE see TEQ175
THEOPHYLLINE SALT of CHOLINE see CMG300
THEOPHYLLINE SODIUM see SKH000
THEOPHYLLINE SODIUM ACETATE see TEQ250
THEOPHYLLINE, compound with SODIUM ACETATE (1:1) see TEQ250
THEOPHYLLIN ETHYLENEDIAMINE see TEP500
THEOPHYLLINUM NATRIUM ACETICUM (GERMAN) see TEQ250
1-(THEOPHYLLIN-7-YL)AETHYL-2-(2-(p-CHLORPHENOXY))-2-METHYLPROPIONAT (GERMAN) see TEQ500
1-(THEOPHYLLIN-7-YL)ETHYL-2-(2-(p-CHLOROPHENOXY))-2-METHYLPROPIONATE see TEQ500
THEOPHYLLINYL-7 PROPANE-3 SULFONATE DE N-BUTYL-HYOSCINE see OKS170
THEOSALVOSE see TEO500
THEOSTENE see TEO500
THEOXYLLINE see CMG300
THEPHORIN HYDROCHLORIDE see TEQ700
THEPHORIN TARTRATE see PDD000
THEPHYLDINE see TEP500
THERABLOAT see PJK150
THERABLOAT see PJK151
THERACANZAN see SNN300
THERADERM see BDS000
THERADIAZINE see PPP500
THERA-FLUR-N see SHF500
THERALAX see PPN100
THERALEPTIQUE see DUO400
THERAMINE see HGD000
THERAPAV see PAH250
THERAPOL see SNM500
THERAPTIQUE see DUO400
THERAZONE see BRF500
THERMA-ATOMIC BLACK see CBT750
THERMACURE see MKA500
THERMAL ACETYLENE BLACK see CBT750
THERMALLY CRACKED RESIDUE see RDK100
THERMALOX see BFT250
THERMATOMIC see CBT750
THERMAX see CBT750
THERMAX 4876 see IGL100
THERMBLACK see CBT750
THERM CHEK 820 see DDV600
THERMINOL 55 see BBM100
THERMINOL 66 see TEQ750
THERMINOL FR-1 see PJL750
THERMINOL 66 HEAT TRANSFER FLUID see TEQ800
"THERMITE" see TER000
THERMOASE PC-10 see BAC000
THERMOGUARD B see AQF000
THERMOGUARD S see AQF000
THERMOLIN 101 see TDG760
THERMOLITE 813 see DVK200
THERMOLITE 831 see BKK750
THERMOLYSIN (BACILLUS THERMOPROTEOLYTICUS) see TER100
THERMOPLASTIC 125 see SMR000

THERUHISTIN HYDROCHLORIDE see AEG625
THESAL see TEO500
THESODATE see TEO500
THETAMID see ENG500
THEVETIA PERUVIANA see YAK350
THEVETIGENIN see DMJ000
THEVETIN see TER250
THEVETIN B see CCX625
THF see TCR750
THFA see TCR400
THF-A see TCS100
THFA see TCT000
THF POLYMER, SRU see GHY100
THFU see FLZ050
TH 1165a HYDROBROMIDE see FAQ100
4-THIA-1-AZABICYCLO(3.2.0)HEPTANE-2-CARBOXYLIC ACID, 6-((AMINOPHENYLACETYL)AMINO)-3,3-DIMETHYL-7-OXO-, (((3,3-DIMETHYL-7-OXO-4-THIA-1-AZABICYCLO(3.2.0)HEPT-2-YL)CARBOXYL)OXY) METHYL ESTER, S,S-DIOXIDE, p-TOLUENESULFONATE, (2S-(2-α(2R*,5S*),5-α-6-β(S*)))- see SOU675
4-THIA-1-AZABICYCLO(3.2.0)HEPTANE-2-CARBOXYLIC ACID, 6-(3-(o-CHLOROPHENYL)-S-METHYL-4-ISOXAZOLECARBOXAMIDO)-3,3-DIMETHYL-7-OXO- see SPD600
4-THIA-1-AZABICYCLO(3.2.0)HEPTANE-2-CARBOXYLIC ACID, 3,3-DIMETHYL-7-OXO-, 4,4-DIOXIDE, SODIUM SALT, (2S-cis)- see PAP600
4-THIA-1-AZABICYCLO(3.2.0)HEPTANE-2-CARBOXYLIC ACID, 3,3-DIMETHYL-7-OXO-6-(2-PHENYLACETAMIDO)-, compd. with 2-(DIETHYLAMINO)ETHYL p-AMINOBENZOATE (1:1) see PAQ200
3-THIA-7-AZABICYCLO(3.3.1)NONANE, 7-(PHENYLMETHYL)-, PERCHLORATE see BFK370
5-THIA-1-AZABICYCLO(4.2.0)OCT-2-ENE-2-CARBOXYLIC ACID, 3-((ACETYLOXY)METHYL)-7-(((2-AMINO-4-THIAZOLYL)METHOXYIMINO)ACETYL)AMINO)-8-OXO-, (6R-(6α,7β(Z)))- see CCS372
5-THIA-1-AZABICYCLO(4.2.0)OCT-2-ENE-2-CARBOXYLIC ACID, 3-(((AMINOCARBONYL)OXY)METHYL)-7-((2-FURANYL(METHOXYIMINO)ACETYL)AMINO)-8-OXO-, 1-(ACETYLOXY)ETHYL ESTER, (6R-(6-α-7-β (Z)))- see CCS625
5-THIA-1-AZABICYCLO(4.2.0)OCT-2-ENE-2-CARBOXYLIC ACID, 7-(((4-(2-AMINO-1-CARBOXY- 2-OXOETHYLIDENE)-1,3-DITHIETAN-2-YL)CARBONYL)AMINO)-7-METHOXY-3-(((1-METHYL-1H-TETRAZOL-5-YL)THIO)METHYL)-8-OXO-, (6R-(6-α-7-α))- see CCS373
5-THIA-1-AZABICYCLO(4.2.0)OCT-2-ENE-2-CARBOXYLIC ACID, 7-((AMINO(4-HYDROXYPHENYL)ACETYL)AMINO)-8-OXO-3-(1-PROPENYL)-, (6R-(6-α,7β(R*)))- see CCS527
5-THIA-1-AZABICYCLO(4.2.0)OCT-2-ENE-2-CARBOXYLIC ACID, 7-(((2-AMINO-4-THIAZOLYL)(METHOXYIMINO)ACETYL)AMINO)-3-((5-METHYL-2H-TETRAZOL-2-YL)METHYL)-8-OXO-, (2,2-DIMETHYL-1-OXOPROPOXY)METHYL ESTER, (6R-(6-α-7-β(Z)))- see TAA420
5-THIA-1-AZABICYCLO(4.2.0)OCT-2-ENE-2-CARBOXYLIC ACID, 7-(((2-AMINO-4-THIAZOLYL)(METHOXYIMINO)ACETYL)AMINO)-3-(((5-(CARBOXYMETHYL)-4-METHYL-2-THIAZOLYL)THIO)METHYL)-8-OXO-, DISODIUM SALT, (6R-(6-α,7-β(Z)))- see CCS367
5-THIA-1-AZABICYCLO(4.2.0)OCT-2-ENE-2-CARBOXYLIC ACID, 7-(((2-AMINO-4-THIAZOLYL)

(METHOXYIMINO)ACETYL)AMINO)-8-OXO-3-((1,2,3-THIADIAZOL-5-YLTHIO)METHYL)-, SODIUM SALT, (6R-(6-α-7-β(Z))- see CCS635
5-THIA-1-AZABICYCLO(4.2.0)OCT-2-ENE-2-CARBOXYLIC ACID, 7-((2-AMINO-4-THIAZOLYL)(METHOXYIMINO)ACETYL)A MINO)-8-OXO-3-(((1,2,5,6-TE TRAHYDRO-2-METHYL-5,6-DIOXO-1,2,4-TRIAZIN-3-YL)THIO)METHYL)-, SODIUM SALT, HYDRATE (2:4:7) (6R-(6-α,7-β(Z)))- see CCS588
5-THIA-1-AZABICYCLO(4.2.0)OCT-2-ENE-2-CARBOXYLIC ACID, 3-(HYDROXYMETHYL)-8-OXO-7-(2-(2-IHIENYL)ACETAMIDO)-, ACETATE, MONOSODIUM SALT see SFQ500
5-THIA-1-AZABICYCLO(4.2.0)OCT-2-ENE-2-CARBOXYLIC ACID, 7-((HYDROXYPHENYLACETYL)AMINO)-3-(((1-METHYL-1H-TETRAZOL-5-YL)THIO)METHYL)-8-OXO-, (6R-(6-α-7-β(R*)))- see CCX300
1-THIA-3-AZAINDENE see BDE500
THIABEN see TEX000
THIABENDAZOLE HYDROCHLORIDE see TER500
THIABENDAZOLE (USDA) see TEX000
THIABENZOLE see TEX000
3-THIABUTAN-2-ONE, O-(METHYLCARBAMOYL)OXIME see MDU600
THIACETAMIDE see TFA000
THIACETARSAMIDE see TFA350
THIACETARSAMIDE SODIUM see SKH100
THIACETAZONE see FNF000
THIACETIC ACID see TFA500
THIACLOPRID see CKW410
THIACOCCINE see TEX250
THIACTIN see TFQ275
THIACYCLOPENTADIENE see TFM250
THIACYCLOPENTANE see TDC730
THIACYCLOPENTANE DIOXIDE see SNW500
THIACYCLOPENTANONE-2 see TDC800
THIACYCLOPENTAN-2-ONE see TDC800
THIACYCLOPROPANE see EJP500
2H-1,3,5-THIADIAZINE-2-THIONE, TETRAHYDRO-3,5-BIS(PHENYLMETHYL)- see RCK730
2H-1,3,5-THIADIAZINE-2-THIONE, TETRAHYDRO-3,5-DIBENZYL- see RCK730
4H-1,3,5-THIADIAZIN-4-ONE, 2-((1,1-DIMETHYLETHYL)IMINO)TETRAHYDRO-3-(1-METHYLETHYL)-5-PHENYL- see BOO631
1,3,4-THIADIAZOL-2-AMINE see AMR250
1,3,4-THIADIAZOLE-2-ACETAMIDO see TES000
1,3,4-THIADIAZOLE, 2,5-BIS(3,4-DIMETHOXYPHENYL)- see BJE600
1,3,4-THIADIAZOLE-2,5-DITHIOL see TES250
1,3,4-THIADIAZOLE, 5-ISOBUTYL-2-(p-METHOXYBENZENESULFONAMIDO)- see IIY100
(1,2,5)THIADIAZOLO(3,4-D)PYRIMIDIN-7(6H)-ONE, 5-METHYLTHIO- see MPT600
(N-1,2,3-THIADIAZOLYL-5)-N'-PHENYLUREA see TEX600
THIADIPONE see BAV625
9-THIAFLUORENE see TES300
4-THIAHEPTANEDIOIC ACID see BHM000
THIALBARBITAL see TES500
THIALBARBITAL SODIUM see SEH600
THIALBARBITONE see TES500
THIALBARBITONE SODIUM see SEH600
THIALBUTONE SODIUM see SFG700
THIALISOBUMAL see SFG700
THIALPENTON see TES500
THIAMAZOLE see MCO500

THIAMBUIENE HYDROCHLORIDE see DJP500

THIAMETON see PHI500

THIAMIDIN F see TES800

THIAMIN see TES750

THIAMIN DISULFIDE see TES800

THIAMINDISULFID-MONOOROTAT (GERMAN) see TET250

THIAMINE CHLORIDE see TES750

THIAMINE CHLORIDE HYDROCHLORIDE see TET300

THIAMINE DICHLORIDE see TET300

THIAMINE, DIHYDROGEN PHOSPHATE (ESTER), MONOHYDROCHLORIDE, DIHYDRATE see TET600

THIAMINE DISULFIDE see TES800

THIAMINE DISULFIDE, OROTATE see TET250

THIAMINE HYDROCHLORIDE see TET300

THIAMINE MONOCHLORIDE see TES750

THIAMINE MONONITRATE see TET500

THIAMINE MONOPHOSPHATE MONOHYDROCHLORIDE DIHYDRATE see TET600

THIAMINE MONOPHOSPHORIC MONOHYDROCHLORIDE BIS-HYDRATED ESTER see TET600

THIAMINE NITRATE see TET500

THIAMINE PHOSPHATE HYDROCHLORIDE DIHYDRATE see TET600

THIAMINE PROPYL DISULFIDE see DXO300

THIAMINE PROPYL DISULFIDE HYDROCHLORIDE see DXO400

THIAMINEPYROPHOSPHATE see TET750

THIAMINEPYROPHOSPHATECHLORIDE see TET750

THIAMINEPYROPHOSPHORICESTER see TET750

THIAMINE TETRAHYDROFURFURYL DISULFIDE see FQJ100

THIAMINE, TRIHYDROGEN PYROPHOSPHATE (ESTER) see TET750

THIAMIN HYDROCHLORIDE see TET300

THIAMINIUM CHLORIDE HYDROCHLORIDE see TET300

THIAMIN PROPYL DISULFIDE see DXO300

THIAMINPYROPHOSPHATE see TET750

THIAMIPRINE see AKY250

THIAMIZIDE see CIP500

THIAMPHENICOL see MPN000

THIAMPHENICOL AMINOACETATE HYDROCHLORIDE see TET780

THIAMPHENICOL GLYCINATE HYDROCHLORIDE see UVA150

THIAMUTILIN see TET800

THIAMYLAL see AGL375

THIAMYLAL SODIUM see SOX500

THIANIDE see EPQ000

5-THIANONANE see BSM125

THIANONANE-5 see BSM125

THIANTAN see DHF600

THIANTAN see DII200

THIA-4-PENTANAL (DOT) see MPV400

THIA-4-PENTANAL (DOT) see TET900

3-THIAPENTANE see EPH000

THIAPHENE see TFM250

2-THIAPROPANE see TFP000

THIARETIC see CFY000

THIASIN see SNN500

THIATE H see DKC400

THIATE U see DEI000

THIATON see TGF075

3,7-THIAXANTHENEDIAMINE-5,5-DIOXIDE see TEU000

THIAZAMIDE see TEX250

THIAZESIM HYDROCHLORIDE see TEU250

THIAZESIUM HYDROCHLORIDE see TEU250

THIAZIDE see CLH750

THIAZINAMIUM METHYL SULFATE see MRW000

THIAZIPIDICO see BEQ625

2-THIAZOLAMINE see AMS250

2-THIAZOLAMINE, 4,5-DIHYDRO-(9CI) see TEV600

THIAZOLE see TEU300

THIAZOLE, 2-ACETAMIDO- see TEW100

5-THIAZOLECARBOXYLIC ACID, 2-AMINO-4-(TRIFLUOROMETHYL)-, ETHYL ESTER see AMU550

5-THIAZOLECARBOXYLIC ACID, 2-CHLORO-4-(TRIFLUOROMETHYL)-, PHENYLMETHYL ESTER see BEG300

THIAZOLE, 5-CHLORO-4-METHYL-2-PROPIONAMIDO- see CIU800

THIAZOLE, 4,5-DIHYDRO-2-METHYL- see DLV900

2,4(3H,5H)-THIAZOLEDIONE see TEV500

4-THIAZOLEETHANETHIOIC ACID, 2-AMINO-α-(METHOXYIMINO)-, S-2-BENZOTHIAZOLYL ESTER, (Z)- see MCK800

THIAZOLE, 4-METHYL-5-VINYL- see MQM800

THIAZOLE YELLOW see CMP050

THIAZOLE YELLOW G see CMP050

3-THIAZOLIDINECARBOXAMIDE, 5-(4-CHLOROPHENYL)-N-CYCLOHEXYL-4-METHYL-2-OXO-, trans- see CJU275

THIAZOLIDINECARBOXYLIC ACID see TEV000

4-THIAZOLIDINECARBOXYLIC ACID see TEV000

THIAZOLIDINE-4-CARBOXYLIC ACID see TEV000

4-THIAZOLIDINECARBOXYLIC ACID, 2-(2-HYDROXYPHENYL)-3-(3-MERCAPTO-1-OXOPROPYL)-, (2R-cis)- see FAQ950

2,4-THIAZOLIDINEDIONE see TEV500

THIAZOLIDINEDIONE-2,4 see TEV500

1,3-THIAZOLIDINE, 2-(METHOXY(METHYLTHIO)PHOSPHINYLIMINO)-3-METHYL- see MEY200

THIAZOLIDINE, 2-PROPYL- see PNW300

2-THIAZOLIDINIMINE see TEV600

1-(2-THIAZOLIDINYL)1,2,3,4,5-PENTANEPENTOL see TEW000

4-THIAZOLIDONE-2-CAPROIC ACID see CCI500

ε-(2-(4-THIAZOLIDONE))HEXANOIC ACID see CCI500

2-THIAZOLINE, 2-AMINO- see TEV600

2-THIAZOLINE, 2-METHYL- see DLV900

THIAZOLIUM, 3-((4-AMINO-2-METHYL-5-PYRIMIDINYL)METHYL)-4-METHYL-5-(2-(PHOSPHONOOXY)ETHYL)-, CHLORIDE, MONOHYDROCHLORIDE, DIHYDRATE see TET600

THIAZOLIUM, 5-(2-HYDROXYETHYL)-3-((4-HYDROXY-2-METHYL-5-PYRIMIDINYL)METHYL)-4-METHYL- see ORS200

THIAZOLIUM, 3-METHYL-2-((1-METHYL-2-PHENYL-1H-INDOL-3-YL)AZO)-, CHLORIDE see CMM870

5H-THIAZOLO(3,2-A)PYRIMIDIN-5-ONE, 6-(2-(4-(BIS(4-FLUOROPHENYL)METHYLENE)-1-PIPERIDINYL)ETHYL)-7-METHYL- see RLK100

THIAZOL YELLOW see CMP050

THIAZOL YELLOW G see CMP050

THIAZOL YELLOW GGM see CMP050

THIAZOL YELLOW R see CMP050

THIAZOL YELLOW Z see CMP050

N-2-THIAZOLYLACETAMIDE see TEW100

2-THIAZOLYLAMINE see AMS250

2-(THIAZOL-4-YL)BENZIMIDAZOLE see TEX000

2-(4-THIAZOLYL)BENZIMIDAZOLE see TEX000

2-(4'-THIAZOLYL)BENZIMIDAZOLE see TEX000

2-(4-THIAZOLYL)-1H-BENZIMIDAZOLE see TEX000

2-(4-THIAZOLYL)-5-BENZIMIDAZOLECARBAMIC ACID METHYL ESTER see TEX200

2-(4-THIAZOLYL)-BENZIMIDAZOLE, HYDROCHLORIDE see TER500

(THIAZOLYL-4)-2 BENZIMIDAZOLYL CARBAMATE-5 D'ISOPROPYLE (FRENCH) see CBA100

N-((THIAZOLYL-4)-2-BENZIMIDAZOLYL)-5-CARBAMATE D'ISOPROPYLE (FRENCH) see CBA100

(2-(4-THIAZOLYL)-1H-BENZIMIDAZOL-5-YL)-CARBAMIC ACID 1-METHYLETHYL ESTER (9CI) see CBA100

3-(N-2-THIAZOLYLFORMIMIDOYL)INDOLE see TEX210

4-(2-THIAZOLYL)PIPERAZINYL 3,4,5-TRIMETHOXYPHENYL KETONE see TEX220

4'-(2-THIAZOLYLSULFAMOYL)PHTHALANILIC ACID see PHY750

4'-(2-THIAZOLYLSULFAMYL)PHTHALANILIC ACID see PHY750

N^1-2-THIAZOLYLSULFANIDAMIDE SODIUM SALT see TEX500

(N^1-2-THIAZOLYLSULFANIDAMIDO)SODIUM see TEX500

N^1-2-THIAZOLYLSULFANILAMIDE see TEX250

1-(2-THIAZOLYL)-4-(3,4,5-TRIMETHOXYBENZOYL)PIPERAZINE see TEX220

N^1-2-THIAZOLYSULFANILAMIDE SODIUM SALT see TEX500

THIAZON see DSB200

THIAZONE see DSB200

THIAZOPHYR see TEX550

THIAZOPYR see TEX550

2-THIAZYLAMINE see AMS250

THIBENZOLE see TEX000

THIBETINE see TJL250

THIBONE see FNF000

2,3,3-THICHLORO-2-PROPENE-1-THIOL, DIISOPROPYLCARBAMATE see DNS600

THIDIAZURON see TEX600

THIDICUR see MPQ750

10H-THIENO(2,3-B)(1,5)BENZODIAZEPINE, 2-METHYL-4-(4-METHYL-1-PIPERAZINYL)- see ZVJ500

1H-THIENO(3,4-d)IMIDAZOLE-4-PENTANOIC ACID, HEXAHYDRO-2-OXO-, (3aS-(3a-α-4-β, 6a-α))- see VSU100

4H-THIENO(2,3-B)THIOPYRAN-2-SULFONAMIDE, 5,6-DIHYDRO-4-AMINO-6-METHYL-, 7,7-DIOXIDE, MONOHYDROCHLORIDE, (4S-trans)- see DAK100

5-(2-THIENOYL)-2-BENZIMIDAZOLECARBAMIC ACID METHYL ESTER see OJD100

N-(5-(2-THIENOYL)-2-BENZIMIDAZOLYL)CARBAMIC ACID METHYL ESTER see OJD100

7-(2-THIENYLACETAMIDO)CEPHALOSPORANIC ACID see CCX250

7-((2-THIENYL)ACETAMIDO)-3-(1-PYRIDYLMETHYL)CEPHALOSPORANIC ACID see TEY000

7-(α-(2-THIENYL)ACETAMIDO)-3-(1-PYRIDYLMETHYL)-3-CEPHEM-4-CARBOXYLIC ACID BETAINE see TEY000

2-THIENYLALANINE see TEY250

β-2-THIENYLALANINE see TEY250

dl-β-2-THIENYLALANINE see TEY250
2-THIENYLALDEHYDE see TFM500
5-(2-THIENYLCARBONYL)-2-
BENZIMIDAZOLECARBAMIC ACID
METHYL ESTER see OJD100
(5-(2-THIENYLCARBONYL)-1H-
BENZIMIDAZOL-2-YL)-CARBAMIC ACID
METHYL ESTER see OJD100
2-THIENYLCARBOXALDEHYDE see
TFM500
(THIENYL)HYDRAZINE see TEY300
THIENYLIC ACID see TGA600
4-(2-THIENYLKETO)-2,3-
DICHLOROPHENOXYACETIC ACID see
TGA600
2-(2-THIENYL)MORPHOLINE
HYDROCHLORIDE see TEY600
1-α-THIENYL-1-PHENYL-3-N-
METHYLMORPHOLINIUM-1-PROPANOL
IODIDE see HNM000
2-THIEPANONE see TEY750
THIEPAN-2-ONE see TEY750
THIERGAN see DQA400
THIETHYLPERAZINE DIMALEATE see
TEZ000
THIETHYLPERAZINE MALEATE see
TEZ000
THIFOR see EAQ750
THIIRANE see EJP500
THILANE see TDC730
THILAVEN see IAD000
THILLATE see TFS350
THILOPEMAL see ENG500
THILOPHENYL see DKQ000
THILOPHENYT see DNU000
THIMBLES see FOM100
THIMBLEWEED see PAM780
THIMECIL see MPW500
THIMER see TFS350
THIMERFONATE SODIUM see SKH150
THIMEROSALATE see MDI000
THIMEROSOL see MDI000
THIMET see PGS000
THIMET SULFONE see TEZ100
THIMET SULFOXIDE see TEZ200
THIMUL see EAQ750
50 THINNER see PCT500
THIOACETAMIDE see TFA000
THIOACETANILIDE see TFA250
THIOACETARSAMIDE see TFA350
THIOACETAZONE see FNF000
THIOACETIC ACID see TFA500
THIOALKOFEN BM see TFD000
THIOALKOFEN BM 4 see TFC600
THIOALLATE see CDO250
THIOALLYL ETHER see AGS250
THIOAMI see DXN709
THIOAMIDE see EPQ000
THIOANILINE see TFI000
4,4'-THIOANILINE see TFI000
p-THIOANISIDINE see AMS675
THIOANISOLE see TFC250
THIOANTIMONIC(III) ACID, TRIESTER
with MERCAPTO SUCCINIC ACID
DILITHIUM SALT, NONAHYDRATE see
AQE500
THIOARSENITE see TFA350
(THIOARSENOSO)METHANE see MGQ750
THIOARSMINE see SNR000
THIOAURIN see TFC500
THIOBARBITURIC ACID see MCK500
2-THIOBARBITURIC ACID see MCK500
THIOBEL see BHL750
THIOBENCARB see SAZ000
THIOBENCARB SULFOXIDE see FMY050
THIOBENZAMIDE see BBM250
THIOBENZANILIDE see PGL250
THIOBENZENESULFONIC ACID S,S'-(2-
(DIMETHYLAMINO)TRIMETHYLENE)
ESTER see NCN650
THIOBENZOIC ACID see TFC550

2-THIO-2H-1,3-BENZOXAZINE-2,4(3H)-
DIONE see TFC570
THIOBENZYL ALCOHOL see TGO750
2-THIO-2-BENZYL-PSEUDOUREA
HYDROCHLORIDE see BEU500
4,4'-THIOBIS(ANILINE) see TFI000
4,4'-THIOBISBENZENAMINE see TFI000
1,1'-THIOBIS(BENZENE) see PGI500
4,4'-THIOBIS(6-tert-BUTYL-m-CRESOL) see
TFC600
4,4'-THIOBIS(6-tert-BUTYL-o-CRESOL) see
TFD000
4,4'-THIOBIS(2-tert-BUTYL-5-
METHYLPHENOL) see TFC600
4,4'-THIOBIS(6-tert-BUTYL-3-
METHYLPHENOL) see TFC600
1,1'-THIOBIS(2-CHLOROETHANE) see
BIH250
2,2'-THIOBIS(4-CHLORO-6-
METHYLPHENOL) see DMN000
2,2'-THIOBIS(4,6-DICHLOROPHENOL) see
TFD250
4,4'-THIOBIS(2-(1,1-DIMETHYLETHYL))-6-
METHYLPHENOL see TFD000
1,1'-THIOBIS(N,N-
DIMETHYLTHIO)FORMAMIDE see BJL600
THIOBIS(DODECYL PROPIONATE) see
TFD500
1,1'-THIOBISETHANE see EPH000
4,4'-THIOBIS(3-METHYL-6-tert-
BUTYLPHENOL) see TFC600
1,1'-THIOBIS(2-METHYL-4-HYDROXY-5-
tert-BUTYLBENZENE) see TFC600
N,N'-
(THIOBIS((METHYLIMINO)CARBONYLOX
Y))BISETHANIMIDOTHIOIC ACID
DIMETHYL ESTER see LBF100
THIOBISMOL see BKX750
THIOBISMOL see BKY500
3,3-THIOBIS(1-PROPENE) see AGS250
THIOBIS(TRIBENZYL-TIN) (8CI) see
BLK750
THIOBORIC ACID, ESTER with 2,2'-
((DIBUTYLSTANNYLENE)DIOXY)DIETHA
NETHIOL (2:3) see TNG250
4-THIOBUTYROLACTONE see TDC800
γ-THIOBUTYROLACTONE see TDC800
ε-THIOCAPROLACTONE see TEY750
THIOCARB see SGJ000
THIOCARBAMATE see ISR000
THIOCARBAMIC ACID-S,S-(2-
(DIMETHYLAMINO)TRIMETHYLENE)EST
ER HYDROCHLORIDE see BHL750
THIOCARBAMIC ACID, N-(6-METHOXY-2-
PYRIDYL)-N-METHYL-, o-3-TERT-
BUTYLPHENYL ESTER see POO800
THIOCARBAMIDE see ISR000
THIOCARBAMISIN see TFD750
THIOCARBAMIZINE see TFD750
THIOCARBAMYLHYDRAZINE see TFQ000
THIOCARBANIL see ISQ000
THIOCARBANILIDE see DWN800
THIOCARBARSONE see CBI250
THIOCARBAZIDE see TFE250
THIOCARBAZIL see FNF000
THIOCARBOHYDRAZIDE see TFE250
THIOCARBONIC DICHLORIDE see TFN500
THIOCARBONIC DIHYDRAZIDE see
TFE250
THIOCARBONOHYDRAZIDE see TFE250
THIOCARBONYL AZIDE THIOCYANATE
see TFE275
THIOCARBONYL CHLORIDE see TFN500
THIOCARBONYL DICHLORIDE see TFN500
THIOCARBOSTYRIL see QOJ100
THIOCHLORID FOSFORECNY see TFO000
THIOCHROMAN-4-ONE, OXIME see TEJ000
4-THIOCHROMANYL-o,o-DIMETHYL
DITHIOPHOSPHATE see TFE300
THIOCHRYSINE see GJG000
THIOCOLCHICOSIDE see TFE325
10-THIOCOLCHICOSIDE see TFE325

THIO-COLCIRAN see DBA200
3-THIOCRESOL see TGO800
4-THIOCRESOL see TGP250
m-THIOCRESOL see TGO800
o-THIOCRESOL see TGP000
p-THIOCRESOL see TGP250
THIOCRON see AHO750
THIOCTACID see DXN800
THIOCTAMID see DXN709
THIOCTAMIDE see DXN709
THIOCTIC ACID see DXN800
6-THIOCTIC ACID see DXN800
6,8-THIOCTIC ACID see DXN800
THIOCTIC ACID AMIDE see DXN709
THIOCTIDASE see DXN800
THIOCTSAN see DXN800
THIOCYAN see CAY250
THIOCYANATES see TFE500
THIOCYANATE SODIUM see SIA500
THIOCYANATOACETIC ACID
CYCLOHEXYL ESTER see TFF000
THIOCYANATOACETIC ACID ETHYL
ESTER see TFF100
THIOCYANATOACETIC ACID
ISOBORNYL ESTER see IHZ000
THIOCYANATOACETIC ACID PROPYL
ESTER see TFF150
THIOCYANATOBENZENE see TFF600
1-THIOCYANATODODECANE see DYA200
THIOCYANATOETHANE see EPP000
α-THIOCYANATOTOLUENE see BFL000
THIOCYANIC ACID, ALLYL ESTER see
AGS750
THIOCYANIC ACID, AMYL ESTER see
AON500
THIOCYANIC ACID, 2-
(BENZOTHIAZOLYLTHIO)METHYL
ESTER see BOO635
THIOCYANIC ACID, 1-CHLORO-1,2-
ETHANEDIYL ESTER see CGW275
THIOCYANIC ACID, CHLOROETHYLENE
ESTER (8CI) see CGW275
THIOCYANIC ACID, DIESTER WITH
DIETHYLENE GLYCOL see TFF200
THIOCYANIC ACID, DIESTER WITH p,p'-
IMINODIPHENOL see DXM200
THIOCYANIC ACID-p-
DIMETHYLAMINOPHENYL ESTER see
TFH500
THIOCYANIC ACID, DODECYL ESTER see
DYA200
THIOCYANIC ACID, 1,2-ETHANEDIYL
ESTER see EJC035
THIOCYANIC ACID, 1,2-ETHANEDIYL
ESTER see VOF300
THIOCYANIC ACID, ETHYLENE ESTER
see EJC035
THIOCYANIC ACID, ETHYL ESTER see
EPP000
THIOCYANIC ACID compounded with
GUANIDINE (1:1) see TFF250
THIOCYANIC ACID, 2-HYDROXYETHYL
ESTER, LAURATE see LBO000
THIOCYANIC ACID, IMINODI-4,1-
PHENYLENE ESTER see DXM200
THIOCYANIC ACID, LITHIUM SALT (1:1)
see TFF500
THIOCYANIC ACID, MERCURY(2+) SALT
see MCU250
THIOCYANIC ACID, 4-METHOXY-2-
NITROPHENYL ESTER see MFB500
THIOCYANIC ACID, OXYDI-2,1-
ETHANEDIYL ESTER see TFF200
THIOCYANIC ACID, OXYDIETHYLENE
ESTER see TFF200
THIOCYANIC ACID, PHENYL ESTER see
TFF600
THIOCYANIC ACID, TRICHLOROMETHYL
ESTER see TFF700
THIOCYANIC ACID,
TRIMETHYLSTANNYL ESTER see TMI750

THIOCYANIC ACID, TRIPHENYLSTANNYL ESTER see TMV750

THIOCYANIC ACID, VINYLENE ESTER (6CI,7CI,8CI) see VOF300

THIOCYANINE BLACK FR GRAINS see CMS250

1-THIOCYANOBUTANE see BSN500

p-THIOCYANODIMETHYLANILINE see TFH500

4-THIOCYANO-N,N-DIMETHYLANILINE see TFH500

THIOCYANO-ESSIGSAEURE-AETHYL-ESTER (GERMAN) see TFF100

THIOCYANO-ESSIGSAEURE-N-PROPYL-ESTER see TFF150

2-THIOCYANOETHYL COCONATE see LBO000

2-THIOCYANOETHYL DODECANOATE see LBO000

2-THIOCYANOETHYL LAURATE see LBO000

β-THIOCYANOETHYL LAURATE see LBO000

THIOCYANOGEN see TFH600

2-(THIOCYANOMETHYLTHIO)BENZOTHIAZOLE, 60% see BOO635

THIOCYCLAM (ETHANEDIOATE 1:1) see TFH750

THIOCYCLAM HYDROGEN OXALATE see TFH750

THIOCYCLOPENTANE-1,1-DIOXIDE see SNW500

THIOCYMETIN see MPN000

THIOCYNAMINE see AGT500

THIOCYTOSINE see TFH800

2-THIOCYTOSINE see TFH800

THIODAN see EAQ750

α-THIODAN see EAQ810

β-THIODAN see EAQ800

THIODANDIOL see HCC550

THIODELONE see MCH600

THIODEMETON see DAO500

THIODEMETON see DXH325

6-THIODEOXYGUANOSINE see TFJ500

THIODERON see MCH600

p,p-THIODIANILINE see TFI000

4,4'-THIODIANILINE see TFI000

THIODICARB see LBF100

2,2'-THIODIETHANOL see TFI500

2,2'-THIODIETHANOL DIACETATE see TFI100

THIODIETHYLENE GLYCOL see TFI500

THIODIFENYLAMINE (DUTCH) see PDP250

THIODIGLYCOL see TFI500

β-THIODIGLYCOL see TFI500

THIODIGLYCOLIC ACID see MCM750

2,2'-THIODIGLYCOLIC ACID see MCM750

β,β'-THIODIGLYCOLIC ACID see MCM750

THIODIGLYCOLLIC ACID see MCM750

2-THIO-3,5-DIMETHYLTETRAHYDRO-1,3,5-THIADIAZINE see DSB200

4,4'-THIODIPHENOL see TFJ000

THIODIPHENYLAMIN (GERMAN) see PDP250

THIODIPHENYLAMINE see PDP250

O,O'-(THIODI-4,1-PHENYLENE)BIS(O,O-DIMETHYL PHOSPHOROTHIOATE) see TAL250

THIODI-p-PHENYLENEDIAMINE see TFI000

O,O'-(THIODI-p-PHENYLENE)-O,O,O',O'-TETRAMETHYL BIS(PHOSPHOROTHIOATE) see TAL250

THIODIPROPIONIC ACID see BHM000

3,3'-THIODIPROPIONIC ACID see BHM000

β,β'-THIODIPROPIONIC ACID see BHM000

β,β'-THIODIPROPIONITRILE see DGS600

THIODOW see EIR000

THIODRIL see CBR675

THIODROL see EBD500

THIOETHANOL see EMB100

2-THIOETHANOL see MCN250

THIOETHANOLAMINE see AJT250

THIOETHYL ALCOHOL see EMB100

(THIOETHYLENE)BIS(BIS(2-HYDROXYETHYL)SULFONIUM) DICHLORIDE see BHD000

THIOETHYL ETHER see EPH000

THIOFACO M-50 see EEC600

THIOFACO T-35 see TKP500

THIOFAN see TDC730

THIOFANATE see DJV000

THIOFANOX see DAB400

THIOFENOL see PFL850

THIOFIDE see BDE750

THIOFOR see EAQ750

THIOFORM (CZECH) see TLS500

THIOFORMIC ACID, PHENYLAZO-, PHENYLHYDRAZIDE see DWN200

THIOFOSGEN (CZECH) see TFN500

THIOFOZIL see TFQ750

THIOFURAM see TFM250

THIOFURAN see TFM250

THIOFURFURAN see TFM250

(1-THIO-d-GLUCOPYRANOSATO)GOLD see ART250

5-THIO-α-d-GLUCOPYRANOSE see GFK000

1-THIOGLUCOPYRANOSE 1-HYDROCINNAMOHYDROXIMATE NO-(HYDROGEN SULFATE) β-d- see PDF525

1-THIO-GLUCOPYRANOSE, MONOGOLD(1+) SALT see ART250

1-THIO-β-d-GLUCOPYRANOSE 1-(N-(SULFOOXY-3-BUTENIMIDATE)), MONOPOTASSIUM see AGH125

5-THIO-d-GLUCOSE see GFK000

THIOGLUCOSE d'OR (FRENCH) see ART250

THIOGLYCERIN see MRM750

1-THIOGLYCEROL see MRM750

α-THIOGLYCEROL see MRM750

THIOGLYCOL (DOT) see MCN250

THIOGLYCOLANILIDE see MCK000

THIOGLYCOLATESODIUM see SKH500

THIOGLYCOLIC ACID see TFJ100

2-THIOGLYCOLIC ACID see TFJ100

THIOGLYCOLIC ACID ANILIDE see MCK000

THIOGLYCOLIC ACID ETHYL ESTER see EMB200

THIOGLYCOLIC ACID-2-ETHYLHEXYL ESTER see EKW300

THIOGLYCOLIC ACID METHYL ESTER see MLE750

THIOGLYCOLIC ACID, SODIUM SALT see SKH500

THIOGLYCOLLIC ACID see TFJ100

THIOGLYCOLLIC ACID, AMMONIUM SALT see ANM500

THIOGLYKOLSAEURE-AETHYLESTER (GERMAN) see EMB200

THIOGLYKOLSAEURE-2-AETHYLHEXYL ESTER (GERMAN) see EKW300

THIOGLYKOLSAEURE-METHYLESTER (GERMAN) see MLE750

THIOGUAIACOL see MFQ300

THIOGUANINE see AMH250

6-THIOGUANINE see AMH250

THIOGUANINE DEOXYRIBOSIDE see TFJ200

β-THIOGUANINE DEOXYRIBOSIDE see TFJ200

β-THIOGUANINE DEOXYRIBOSIDE see TFJ250

6-THIOGUANINE RIBONUCLEOSIDE see TFJ500

THIOGUANINE RIBOSIDE see TFJ500

THIOGUANOSINE see TFJ500

6-THIOGUANOSINE see TFJ500

THIOHEXAM see CPI250

THIOHEXITAL see AGL875

THIO-HMPA see HEK050

2-THIOHYDANTOIN see TFJ750

2-THIO-4-HYDRAZINOURACIL see HHF500

2-THIO-6-HYDROXY-8-AZAPURINE see HOJ100

THIOHYPOXANTHINE see POK000

THIOINDIGO see DNT300

THIOINDIGO BLACK see CMU320

THIOINDIGO ORANGE KKh see CMU815

THIOINDIGO RED B see DNT300

THIOINDIGO RED S see DNT300

THIOINOSINE see TFJ825

THIOISONICOTINAMIDE see TFK000

THIOKARBONYLCHLORID (CZECH) see TFN500

2-THIO-4-KETOIHIAZOLIDINE see RGZ550

THIOKOL NVT see ROH900

THIOKTSAFURE (GERMAN) see DXN800

THIOLA see MCI375

THIOLACETIC ACID see TFA500

2-THIOLACTIC ACID see TFK250

THIOLANE see TDC730

THIOLANE-1,1-DIOXIDE see SNW500

THIOLAN-2-ONE see TDC800

THIOLDEMETON see DAP200

2-THIOL-DIHYDROGLYOXALINE see IAQ000

THIOLE see TFM250

THIOLIN see IAD000

THIOLITE see CAX500

THIOLMECAPTOPHOS see DAO500

THIOL PROTEASE INHIBITOR see TFK255

THIOL SYSTOX see DAP200

THIOLUTIN see ABI250

THIOLUX see SOD500

THIOMALIC ACID see MCR000

2-THIO-MALIC ACID see MCR000

THIOMEBUMAL see PBT250

THIOMEBUMAL SODIUM see PBT500

THIOMECIL see MPW500

THIOMERIN see TFK260

THIOMERIN SODIUM see TFK270

THIOMERSALATE see MDI000

THIOMESTERONE see TFK300

THIOMESTRONE see TFK300

THIOMETAN see DSK600

THIOMETHANOL see MLE650

p-THIOMETHOXYANILINE see AMS675

2-THIO-6-METHYL-1,3-PYRIMIDIN-4-ONE see MPW500

6-THIO-4-METHYLURACIL see MPW500

THIOMETON see PHI500

THIOMICID see FNF000

THIOMIDIL see MPW500

THIOMONOGLYCOL see MCN250

2,5-THIOMORPHOLINEDIONE, 3,3-DIMETHYL-, 2-(o-((METHYLAMINO)CARBONYL)OXIME) see MID860

2,5-THIOMORPHOLINEDIONE, 3-METHYL-, 2-(o-((METHYLAMINO)CARBONYL)OXIME) see MPU600

2,5-THIOMORPHOLINEDIONE, 3,3,6-TRIMETHYL-, 2-(o-((METHYLAMINO)CARBONYL)OXIME) see TMH300

THIOMUL see EAQ750

THIOMYLAL SODIUM see SOX500

THIONAL BLACK 5G-CF see CMS250

THIONAL BLACK 7CF see CMS250

THIONAL DEEP BLACK B see CMS250

THIONAL DEEP BLACK D see CMS250

THIONAL MAROON VR see CMS257

THIONAL PURPLE B-CF see CMS257

THIONAPHTHOL see NAP500

2-THIONAPHTHOL see NAP500

THIO-β-NAPHTHOL see NAP500

β-THIONAPHTHOL see NAP500

THIONAZIN see EPC500

THIONEMBUTAL see PBT500

THIONEX see BJL600

THIONEX see EAQ750

α-THIONEX see EAQ800

β-THIONEX see EAQ810
THIONEX RUBBER ACCELERATOR see BJL600
THIONICID see FNF000
THIONIDEN see EPQ000
THIONIN see AKK750
THIONINE see AKK750
THIONOACETIC ACID see TFA500
THIONOBENZENEPHOSPHONIC ACID ETHYL-p-NITROPHENYL ESTER see EBD700
THIONODEMETON SULFONE see SPF000
THIONOL BLACK B see CMS250
THIONOL BLACK JN see CMS250
THIONONE BLACK 36254 see CMS250
THIONONE BLACK BN EXTRA CONC see CMS250
THIONONE BLACK FA EXTRA CONC see CMS250
THIONONE BLACK PASTE SPECIAL see CMS250
THIONOSINE see MCQ500
5-THIONO-v-TRIAZOLO(4,5-d)PYRIMIDIN-7(4H,6H)-ONE see HOJ100
THIONTAN see DHF600
THIONYLAN see TEO250
THIONYL BROMIDE see SNT200
THIONYL CHLORIDE see TFL000
THIONYL DICHLORIDE see TFL000
THIONYL DIFLUORIDE see TFL250
THIONYL FLUORIDE see TFL250
THIOOCTANOIC ACID see DXN800
THIOORATIN see TET250
2-THIO-4-OXO-6-METHYL-1,3-PYRIMIDINE see MPW500
2-THIO-4-OXO-6-PROPYL-1,3-PYRIMIDINE see PNX000
2-THIO-6-OXYPURINE see TFL500
2-THIO-6-OXYPYRIMIDINE see TFR250
THIOPARAMIZONE see FNF000
THIOPENTAL see PBT250
THIOPENTAL SODIUM see PBT500
THIOPENTAL SODIUM SALT see PBT500
THIOPENTEX see CPK000
THIOPENTOBARBITAL see PBT250
THIOPENTONE see PBT250
THIOPENTONE SODIUM see PBT500
THIOPERAZINE see TFM100
THIOPEROXYDICARBONIC ACID, BIS(1-METHYLETHYL) ESTER- see IRS500
THIOPEROXYDICARBONIC ACID, DIBUTYL ESTER see BSS550
THIOPEROXYDICARBONIC ACID DIETHYL ESTER see BJU000
THIOPEROXYDICARBONIC ACID DIMETHYL ESTER see DUN600
THIOPEROXYDICARBONIC DIAMIDE, N,N'-DIETHYL-N,N'-DIPHENYL- see DJC200
THIOPEROXYDICARBONIC DIAMIDE, TETRAMETHYL-, mixture with (1-α-2-α-3-β,4-α- 5-α-6-β)-1,2,3,4,5,6-HEXACHLOROCYCLOHEXANE and TRICHLOROPHENOL COPPER(2+) SALT see FAQ930
THIOPHAL see TIT250
THIOPHANAT (GERMAN) see DJV000
THIOPHANATE ETHYL see DJV000
THIOPHANATE METHYL-MANEB mixture see MAP300
THIOPHANE see TDC730
THIOPHANE DIOXIDE see SNW500
THIOPHAN SULFONE see SNW500
THIOPHEN see TFM250
THIOPHENE see TFM250
7-(THIOPHENE-2-ACETAMIDO)CEPHALOSPORANIC ACID see CCX250
7-(THIOPHENE-2-ACETAMIDO)CEPHALOSPORANIC ACID SODIUM SALT see SFQ500

7-(THIOPHENE-2-ACETAMIDO)-3-(1-PYRIDYLMETHYL)-3-CEPHEM-4-CARBOXYLIC ACID BETAINE see TEY000
THIOPHENE, 2-ACETYL-5-NITRO- see NML100
2-THIOPHENEALANINE see TEY250
2-THIOPHENEALDEHYDE see TFM500
2-THIOPHENECARBOXALDEHYDE see TFM500
α-THIOPHENECARBOXALDEHYDE see TFM500
THIOPHENECARBOXYLATE de ((METHYLENEDIOXY-3,4 PHENYL)-1 DIMETHYL-4,4-PENTENE-1)YLE-3 see TFM625
2-THIOPHENECARBOXYLIC ACID see TFM600
α-THIOPHENECARBOXYLIC ACID see TFM600
2-THIOPHENECARBOXYLIC ACID, 3-(1,3-BENZODIOXOL-5-YL)-1-(1,1-DIMETHYLETHYL)-2-PROPENYL ESTER see TFM625
(THIOPHENE-2,5-DIYL)BIS(HYDROXYMERCURY)MONOACETATE see HLQ000
THIOPHENE, TETRAHYDRO-3,3,4,4-TETRACHLORO-, 1,1-DIOXIDE see TBV300
2-THIOPHENIC ACID see TFM600
THIOPHENICOL see MPN000
THIOPHENICOL GLYCINATE HYDROCHLORIDE see UVA150
THIOPHENIT see MNH000
THIOPHENITE see DJV000
THIOPHENOL (DOT) see PFL850
2(3H)-THIOPHENONE, DIHYDRO-(8CI,9CI) see TDC800
2-THIOPHENONE, TETRAHYDRO- see TDC800
1-THIOPHENYLANTHRAQUINONE see PGL000
THIOPHOS see PAK000
THIOPHOSGENE see TFN500
THIOPHOSPHAMIDE see TFQ750
THIOPHOSPHATE de S-N-(1-CYANO-1-METHYLETHYL)CARBAMOYLMETHYLE et de O,O-DIETHYLE (FRENCH) see PHK250
THIOPHOSPHATE de O-2,4-DICHLOROPHENYLE et de O,O-DIETHYLE (FRENCH) see DFK600
THIOPHOSPHATE de O,O-DIETHYLE et de O-(3-CHLORO-4-METHYL-7-COUMARINYLE) (FRENCH) see CNU750
THIOPHOSPHATE de O,O-DIETHYLE et de O-(2,5-DICHLORO-4-BROMO) PHENYLE (FRENCH) see EGV500
THIOPHOSPHATE de O,O-DIETHYLE et de S-2-ETHYLTHIO-ETHYLE) (FRENCH) see DAP200
THIOPHOSPHATE de O,O-DIETHYLE et de o-2-ISOPROPYL-4-METHYL-6-PYRIMIDYLE (FRENCH) see DCM750
THIOPHOSPHATE de O,O-DIETHYLE et de S-(N-METHYLCARBAMOYL) METHYLE (FRENCH) see DNX800
THIOPHOSPHATE de O,O-DIETHYLE et de O-(4-METHYL-7-COUMARINYLE) (FRENCH) see PKT000
THIOPHOSPHATE de O,O-DIETHYLE et de O-(4-NITROPHENYLE) (FRENCH) see PAK000
THIOPHOSPHATE de O,O-DIMETHYLE et de O-4-BROMO-2,5-DICHLOROPHENYLE (FRENCH) see BNL250
THIOPHOSPHATE de O,O-DIMETHYLE et de O-3-CHLORO-4-NITROPHENYLE (FRENCH) see MIJ250
THIOPHOSPHATE de O,O-DIMETHYLE et de O-4-CHLORO-3-NITROPHENYLE (FRENCH) see NFT000

THIOPHOSPHATE de O,O-DIMETHYLE et de S-2-ETHYLSULFINYLETHYLE (FRENCH) see DAP000
THIOPHOSPHATE de O,O-DIMETHYLE et de O-2-ETHYLTHIO-ETHYLE (FRENCH) see DAO800
THIOPHOSPHATE de O,O-DIMETHYLE et de S-2-ETHYLTHIOETHYLE (FRENCH) see DAP400
THIOPHOSPHATE de O,O-DIMETHYLE et de S-2-(ISOPROPYLSULFINYL)-ETHYLE see DSK600
THIOPHOSPHATE de O,O-DIMETHYLE et de S-((5-METHOXY-4-PYRONYL)-METHYLE) (FRENCH) see EAS000
THIOPHOSPHATE de O,O-DIMETHYLE et de O-(3-METHYL-4-METHYLTHIOPHENYLE) (FRENCH) see FAQ900
THIOPHOSPHATE de O,O-DIMETHYLE et de O-(3-METHYL-4-NITROPHENYLE) (FRENCH) see DSQ000
THIOPHOSPHATE de O,O-DIMETHYLE et de O-(4-NITROPHENYLE) (FRENCH) see MNH000
THIOPHOSPHATE de O,O-DIMETHYLE et de O-(2,4,5-TRICHLOROPHENYLE) (FRENCH) see RMA500
THIOPHOSPHATE de O,S-DIMETHYL et de O-(3-METHYL-4-NITROPHENYLE) (FRENCH) see MKC250
THIOPHOSPHORIC ACID TRIMORPHOLIDE see TFN600
THIOPHOSPHORIC ACID, TRISODIUM SALT see TNM750
THIOPHOSPHORIC ANHYDRIDE see PHS000
THIOPHOSPHORIC TRICHLORIDE see TFO000
THIOPHOSPHOROUS TRIBROMIDE see TFN750
THIOPHOSPHORSAEURE-O,O-DIMETHYL-S-(6-CHLOR-OXAZOLO(4,5-B)PYRIDIN-2(3H)-O-N-3-YL)METHYL-ESTER see ARY800
THIOPHOSPHORSAEURE-O,S-DIMETHYLESTERAMID (GERMAN) see DTQ400
THIOPHOSPHORYL BROMIDE see TFN750
THIOPHOSPHORYL BROMIDE FLUORIDE (8CI,9CI) see TFO250
THIOPHOSPHORYL CHLORIDE see TFO000
THIOPHOSPHORYL CHLORODIFLUORIDE see TFO600
THIOPHOSPHORYL DIBROMIDEMONOFLUORIDE see TFO250
THIOPHOSPHORYL DIFLUORIDEMONOBROMIDE see TFO500
THIOPHOSPHORYL DIFLUORIDE MONOCHLORIDE see TFO600
THIOPHOSPHORYL DIFLUOROCHLORIDE see TFO600
THIOPHOSPHORYL FLUORIDE see TFO750
THIOPHOSPHORYL FLUORODIBROMIDE see TFO250
THIOPHOSPHORYL MONOBROMODIFLUORIDE see TFO500
THIOPHOSPHORYL MONOCHLORODIFLUORIDE see TFO600
THIOPHOSPHORYL TRICHLORIDE see TFO000
THIOPHYLLINE CHOLINATE see CMG300
THIOPHYLLINE with CHOLLINE see CMG300
THIOPOL see HEK050
THIOPROLINE see TEV000
THIOPRONIN see MCI375
THIOPRONINE see MCI375
2-THIOPROPANE see TFP000
3-THIOPROPANOIC ACID see MCQ000

THIOPROPAZATE DIHYDROCHLORIDE see TFP250
THIOPROPAZATE HYDROCHLORIDE see TFP250
THIOPROPERAZIN see TFM100
THIOPROPERAZINE see TFM100
(THIOPROPERAZINE)-2-DIMETHYLSULFAMIDO-(10-3-1-METHYLPIPERAZINYL-4)PROPYL)-PHENOTHIAZINE see TFM100
3-THIOPROPIONIC ACID see MCQ000
β-THIOPROPIONIC ACID see MCQ000
2-THIO-6-PROPYL-1,3-PYRIMIDIN-4-ONE see PNX000
6-THIO-4-PROPYLURACIL see PNX000
β-THIOPSEUDOUREA see ISR000
6-THIOPURINE RIBONUCLEOSIDE see MCQ500
6-THIOPURINE RIBOSIDE see MCQ500
2-THIOPYRIDINE see TFP300
THIOPYRIDONE-2 see TFP300
2-THIO-1,3-PYRIMIDIN-4-ONE see TFR250
THIORIDAZIEN THIOMETHYL SULFOXIDE see MON750
THIORIDAZIN see MOO250
THIORIDAZINE see MOO250
THIORIDAZINE HYDROCHLORIDE see MOO500
THIORYL see MPW500
THIOSALICYLIC ACID see MCK750
THIOSAN see DXH250
THIOSAN see TFS350
THIOSCABIN see DXH250
THIOSECONAL see AGL375
THIOSEMICARBARZONE see FNF000
THIOSEMICARBAZIDE see TFQ000
3-THIOSEMICARBAZIDE see TFQ000
THIOSEMICARBAZONE ACETONE see TFQ250
THIOSEMICARBAZONE (PHARMACEUTICAL) see FNF000
THIOSEPTAL see PPO000
THIOSERINE see CQK000
THIOSINAMIN see AGT500
THIOSINAMINE see AGT500
THIOSOL see MCI375
THIOSTOP N see SGM500
THIOSTREPTON see TFQ275
THIOSUCCINIMIDE see MRN000
THIOSULFAN see EAQ750
THIOSULFAN TIONEL see EAQ750
THIOSULFATES see TFQ500
THIOSULFIL see MPQ750
THIOSULFIL-A FORTE see PDC250
THIOSULFURIC ACID, DIAMMONIUM SALT see ANK600
THIOSULFURIC ACID, S,S'-(2-(DIMETHYLAMINO)-1,3-PROPANEDIYL) ESTER, DISODIUM SALT see DXC900
THIOSULFURIC ACID, S-(2-(((3-(4,5-DIMETHYL-2-HYDROXYPHENYL)TRICYCLO(3.3.1.1³,⁷)DEC-1-YL)METHYL)AMINO)-2-IMINOETHYL) ESTER see HKC700
THIOSULFURIC ACID, DIPOTASSIUM SALT see PLW000
THIOSULFURIC ACID, DISODIUM SALT, PENTAHYDRATE see SKI500
THIOSULFURIC ACID, GOLD(1+) SODIUM SALT (2:1:3) see GJE000
THIOSULFURIC ACID, GOLD(1+) SODIUM SALT(2:1:3), DIHYDRATE see GJG000
THIOSULFURIC ACID, S-(2-((4-(p-TOLYLOXY)BUTYL)AMINO)ETHYL) ESTER see THB250
THIOSULFUROUS DICHLORIDE see SON510
THIOTEBESIN see FNF000
THIO-TEP see TFQ750
THIOTEPP see SOD100
THIOTETROLE see TFM250
THIOTEX see TFS350

THIOTHAL see PBT250
THIOTHAL SODIUM see PBT500
2-THIOTHIAZOLIDONE see TFS250
2-THIO-1-(THIOCARBAMOYL)UREA see DXL800
THIOTHIXENE see NBP500
THIOTHIXENE DIHYDROCHLORIDE see TFQ600
THIOTHIXINE see NBP500
THIOTHYMIN see MPW500
THIOTHYMINE see TFQ650
2-THIOTHYMINE see TFQ650
THIOTHYRON see MPW500
6-THIOTIC ACID see DXN800
6,8-THIOTIC ACID see DXN800
3-THIOTOLENE see MPV250
THIOTOMIN see DXN709
THIOTOX see TFS350
5-THIO-1H-υ-TRIAZOLO(4,5-d)PYRIMIDINE-5,7(4H,6H)-DIONE see MLY000
THIOTRIETHYLENEPHOSPHORAMIDE see TFQ750
THIOTRITHIAZYL NITRATE see TFR000
THIOURACIL see TFR250
2-THIOURACIL see TFR250
6-THIOURACIL see TFR250
2-THIOUREA see ISR000
THIOUREA (DOT) see ISR000
THIOUREA, N,N'-BIS(1-METHYLETHYL)-(9CI) see DNS800
THIOUREA, N-(2,6-BIS(1-METHYLETHYL)-4-PHENOXYPHENYL)-N'-(1,1-DIMETHYLETHYL)- see BRB100
THIOVANIC ACID see TFJ100
THIOVANOL see MRM750
THIOVAT BRILLIANT INDIGO 4BR see ICU135
THIOVIT see SOD500
THIOXAMYL see DSP600
p-THIOXANE see OLY000
THIOXANTHEN-9-ONE, 1-((2-(ETHYLAMINO)ETHYL)AMINO)-4-(HYDROXYMETHYL)- see EGA600
9H-THIOXANTHEN-9-ONE, 1-((2-(ETHYLAMINO)ETHYL)AMINO)-4-(HYDROXYMETHYL)- see EGA600
THIOXANTHEN-9-ONE, 1-((2-(ETHYLAMINO)ETHYL)AMINO)-4-METHYL- see EGA650
9H-THIOXANTHEN-9-ONE, 1-((2-(ETHYLAMINO)ETHYL)AMINO)-4-METHYL- see EGA650
3-(THIOXANTHEN-9-YL)METHYL-1-PIPECOLINE, HYDROCHLORIDE see THL500
THIOXIDIL see GGS000
THIOXIDRENE see TCZ000
6-THIOXOPURINE see POK000
2-THIOXO-4-THIAZOLIDINONE see RGZ550
4-(THIOXO(3,4,5-TRIMETHOXYPHENYL)METHYL)MORPHOLINE see TNP275
THIOZAMIDE see TEX250
THIOZIN see IAD000
THIOZIN see ZJS300
2-THIOZOLIDINETHIONE see TFS250
THIPENTAL SODIUM see PBT500
THIPHEN see DHY400
THIPHENAMIL HYDROCHLORIDE see DHY400
THIRAM see TFS350
THIRAMAD see TFS350
THIRAME (FRENCH) see TFS350
THIRASAN see TFS350
THIRERANIDE see DXH250
THISTROL see CLN000
THISTROL see CLO000
THIULIX see TFS350
THIURAD see TFS350
THIURAGYL see PNX000

THIURAM see TFS350
THIURAM DISULFIDE see TFS500
THIURAM E see DXH250
THIURAM EF see DJC200
THIURAMIN see TFS350
THIURAMYL see TFS350
THIURANIDE see DXH250
THIURETIC see CFY000
THIURYL see MPW500
THIXOKON see AAN000
THIZONE see FNF000
THLARETIC see CFY000
THN see TCY250
THOMAPYRIN see ARP250
THOMAS BALSAM see BAF000
THOMBRAN see CKJ000
THOMBRAN see THK880
THOMIZINE see THG600
THOMPSON-HAYWARD TH6040 see CJV250
THOMPSON'S WOOD FIX see PAX250
THONZYLAMINE HYDROCHLORIDE see RDU000
THONZYLAMINIUM CHLORIDE see RDU000
THORAZINE see CKP250
THORAZINE see CKP500
THORAZINE HYDROCHLORIDE see CKP500
THORIA see TFT750
THORIDAZINE HYDROCHLORIDE see MOO500
THORIUM see TFS750
THORIUM-232 see TFS750
THORIUM CHLORIDE see TFT000
THORIUM DICARBIDE see TFT100
THORIUM DIOXIDE see TFT750
THORIUM HYDRIDE see TFT250
THORIUM METAL, pyrophoric (DOT) see TFS750
THORIUM (4+) NITRATE see TFT500
THORIUM(IV) NITRATE see TFT500
THORIUM OXIDE see TFT750
THORIUM OXIDE SULFIDE see TFU000
THORIUM TETRACHLORIDE see TFT000
THORIUM TETRANITRATE see TFT500
THORN APPLE see SLV500
THOROTRAST see TFT750
THORTRAST see TFT750
THPA see TDB000
THPC see TDH750
THPS see TDI000
THQ see MKO250
THR-221 see CCS367
THREADLEAF GROUNDSEL see RBA400
THREAMINE see AMA500
THREARIC ACID see TAF750
THREE ELEPHANT see BMC000
THREE-LEAVED INDIAN TURNIP see JAJ000
l-THREITOL-1,4-BISMETHANESULFONATE see TFU500
THRENE BRILLIANT ORANGE GR see CMU820
d-THREO-α-BENZYL-N-ETHYLTETRAHYDROFURFURYLAMINE see ZVA000
D(−)THREO-2-HYDROXYACETAMIDO-1-p-NITROPHENYL-1,3-PROPANEDIOL see CDP350
THREONINE see TFU750
l-THREONINE see TFU750
THREOSE see TKO100
THRETHYLENE see TIO750
THROATWORT see FOM100
THROMBASE see TFU800
THROMBIN see TFU800
THROMBIN-C see TFU800
THROMBIN COAGULASE see CMY725
THROMBOCID see SEH450
THROMBOCYTIN see AJX500
THROMBOFORT see TFU800
THROMBOLIQUINE see HAQ500

THROMBOTONIN see AJX500
THS-101 see DBC500
THS-201 see HAG325
THS-839 see DAP812
THU see ISR000
(1S,4R,5R)-(−)-3-THUJANONE see TFW000
THUJA OIL see CCQ500
α-THUJAPLICIN see TFU900
β-THUJAPLICIN see IRR000
γ-THUJAPLICIN see TFV750
β-THUJAPLICINE see IRR000
γ-THUJAPLICINE see TFV750
THUJON see TFW000
THUJONE see TFW000
(−)-THUJONE see TFW000
l-THUJONE see TFW000
α-THUJONE see TFW000
THULIUM see TFW250
THULIUM CHLORIDE see TFW500
THULIUM EDETATE see TFX000
THULIUM NITRATE see TFX100
THULIUM(3+) NITRATE see TFX100
THULIUM(III) NITRATE, HEXAHYDRATE
(1:3:6) see TFX250
THULIUM TRINITRATE see TFX100
THULOL see EDU500
THURICIDE see BAC040
THURINGIENSIN see BAC125
THURINGIENSIN A see BAC125
THURINGIN see BAC040
THURINTOX see BAC125
THX see TFZ275
THYALONE see BAG250
THYCAPSOL see MCO500
THYLAKENTRIN see FMT100
THYLATE see TFS350
THYLFAR M-50 see MNH000
THYLOGEN see WAK000
THYLOGEN MALEATE see DBM800
THYLOKAY see NAQ600
THYLOQUINONE see MMD500
THYME CAMPHOR see TFX810
THYME OIL see TFX500
THYME OIL RED see TFX750
THYMEOL see AEG875
THYMIAN OEL (GERMAN) see TFX500
THYMIC ACID see TFX810
THYMIDIN see TFX790
THYMIDINE see TFX790
THYMIDINE, 3'-AZIDO-3'-DEOXY- see
ASE900
THYMIDINE, 2',3'-DIDEHYDRO-3'-DEOXY-
see SLJ800
5'-THYMIDYLIC ACID, HOMOPOLYMER
see TFX795
THYMIDYLIC ACID POLYMER see TFX795
5'-THYMIDYLIC ACID, POLYMERS see
TFX795
THYMINE see TFX800
THYMINEDEOXYRIBOSIDE see TFX790
THYMINE-2-DEOXYRIBOSIDE see TFX790
THYMINE, 1-(2,3-DIDEOXY-β-d-GLYCERO-
PENT-2-ENOFURANOSYL)- see SLJ800
THYMINE PROPENAL see TFX805
THYMINE, 2-THIO- see TFQ650
THYMIN (PURINE BASE) see TFX800
THYM OIL see TFX500
THYMOL see TFX810
m-THYMOL see TFX810
o-THYMOL see CCM000
THYMOL, 6,6'-(3H-2,1-BENZOXATHIOL-3-
YLIDENE)DI-, S,S-DIOXIDE see TFX850
THYMOL BLUE see TFX850
THYMOL, 6-CHLORO- see CLJ800
THYMOL METHYL ETHER see MPW650
THYMOLSULFONEPHALEIN see TFX850
THYMOLSULFONEPHTHALEIN see
TFX850
THYMOLSULFOPHTHALEIN see TFX850
THYMOLSULPHONPHTHALEIN see
TFX850
THYMOQUINONE see IQF000

THYMOSULFONPHTHALEIN see TFX850
THYMOXAMINE HYDROCHLORIDE see
TFY000
THYMUS KOTSCHYANUS, OIL EXTRACT
see TFY100
THYMYL METHYL ETHER see MPW650
THYMYL 2-PROPYLVALERATE see VAQ200
THYNESTRON see EDV000
THYNON see DLK200
THYRADIN see TFZ100
THYREOIDEUM see TFZ275
THYREONORM see MPW500
THYREOSTAT see MPW500
THYREOSTAT II see PNX000
THYRIL see MPW500
THYROCALCITONIN see TFZ000
THYROID see TFZ100
THYROID RELEASING HORMONE see
TNX400
THYROLIBERIN see TNX400
THYROTROPIC-RELEASING FACTOR see
TNX400
THYROTROPIC RELEASING HORMONE
see TNX400
THYROTROPIN-RELEASING FACTOR see
TNX400
THYROTROPIN-RELEASING HORMONE
see TNX400
THYROXEVAN see LFG050
THYROXIN see TFZ275
l-THYROXIN see TFZ275
THYROXINE see TFZ275
(−)-THYROXINE see TFZ275
THYROXINE see TFZ300
l-THYROXINE see TFZ275
l-THYROXINE MONOSODIUM SALT see
LFG050
d-THYROXINE SODIUM see SKJ300
THYROXINE SODIUM SALT see LFG050
d-THYROXINE SODIUM SALT see SKJ300
l-THYROXINE SODIUM SALT see LFG050
TI-8 see TCA250
TI-78 see NCN650
TI-1258 see BHL750
TI-1671 see NCN650
TIA 230 see POK500
TIACETAZON see FNF000
TIADIPONE see BAV625
TIAMIPRINE see AKY250
TIAMIZID see CIP500
TIAMIZIDE see CIP500
TIAMULIN see TET800
TIAMULINA (ITALIAN) see TET800
TIAMUTIN see DAR000
TIAPAMIL see TGA275
TIAPRIDE HYDROCHLORIDE see TGA375
TIAPROFENIC ACID see SOX400
TIARAMIDE HYDROCHLORIDE see CJH750
TIAZOFURIN see RJF500
TIAZON see DSB200
TIB see TKQ250
TIBA see TKQ250
2,3,5-TIBA see TKQ250
TIBAMATO see MOV500
TIBA SODIUM SALT see SKL700
TIBAZIDE see ILD000
TIBENZATE see BFK750
TIBERAL see OJS000
TIBEXIN see LFJ000
TIBEY (PUERTO RICO) see SLJ650
TIBICUR see FNF000
TIBINIDE see ILD000
TIBIONE see FNF000
TIBIVIS see ILD000
TIBIZAN see FNF000
TIBRIC ACID see CGJ250
TIBUTOL see TGA500
TIC see TKJ250
TICARCILLIN SODIUM see TGA520
TICKLE WEED see FAB100
TICLID see TGA525
TICLOBRAN see ARQ750

TICLODIX see TGA525
TICLODONE see TGA525
TICLOPIDINE HYDROCHLORIDE see
TGA525
TIC MUSTARD see IAN000
TICOLIN see DXN709
TICREX see TGA600
TICRYNAFEN see TGA600
TIDEMOL see BQL000
TIEMONIUM IODIDE see HNM000
TIEMOZYL see HNM000
TIEMPE see TKZ000
TIENILIC ACID see TGA600
2-(2-TIENYL)MORFOLINA CLORIDRATE
(ITALIAN) see TEY600
TIEZENE see EIR000
TIFEMOXONE see PDU750
TIFEN see DHY400
TIFERRON see DXH300
TIFOMYCINE see CDP250
TIGAN see TKW750
TIGAN HYDROCHLORIDE see TKW750
TIGASON see EMJ500
TIGER ORANGE see CJD500
TIGLIC ACID see TGA700
TIGLIC ACID, ETHYL ESTER (6CI,7CI) see
TGA800
TIGLIC ACID, GERANIOL ESTER see
GDO000
TIGLIC ACID, METHYL ESTER (6CI,7CI) see
MPW700
TIGLINIC ACID see TGA700
12-o-TIGLYL-PHORBOL-13-BUTYRATE see
PGV750
12-o-TIGLYL-PHORBOL-13-
DODECANOATE see PGW000
7-TIGLYLRETRONECINE VIRIDIFLORATE
see SPB500
7-TIGLYL-9-
VIRIDIFLORYLRETRONECINE see SPB500
TIG N800 see IGL100
TIGUVON see FAQ900
TIKLID see TGA525
TIKOFURAN see NGG500
TIL4K8 see TOC250
TILCAREX see PAX000
TILCOTIL see TAL485
TILDIEM see DNU600
TILDIN see BNK000
TILIDATE see EIH000
TILIDINE HYDROCHLORIDE see EIH000
TILLAM (RUSSIAN) see PNF500
TILLAM-6-E see PNF500
TILLANTOX see BDD000
TILLRAM see DXH250
TILMAPOR see CCS550
TILORONE HYDROCHLORIDE see TGB000
TILT see PMS930
TIMACOR see TGB185
TIMAZIN see FMM000
TIMBER RATTLESNAKE VENOM see
TGB150
TIMELIT see IFZ900
TIMEPIDIUM BROMIDE see TGB160
TIMET see PGS000
TIMIPERONE see TGB175
TIMOLET see DLH600
TIMOLET see DLH630
TIMOLOL MALEATE see TGB185
l-TIMOLOL MALEATE see TGB185
TIMONIL see DCV200
TIMONOX see AQF000
TIMOPED see TGB475
TIMOPTOL see TGB185
TIMOSIN see MDQ250
TIN see TGB250
TIN (α) see TGB250
TINA BLUE 2B see ICU135
TIN, ACETOXYTRIPROPYL-(7CI) see
TNB000
TIN, ACETOXYTRIS(β,β-
DIMETHYLPHENETHYL)- see TMK200

TINACTIN see TGB475
TINADERM see TGB475
TINA GREY BL see CMU320
TINAJA (PUERTO RICO) see CNR135
TINA ORANGE R see CMU815
TINA PINK B see DNT300
TINA PRINTING BLACK BL see CMU320
TIN AZIDE TRICHLORIDE see TGB500
TIN BIFLUORIDE see TGD100
TIN, BIS(DIETHYLCARBAMODITHIOATO)- see TGB600
TIN, BIS(DIETHYLCARBAMODITHIOATO-S,S')-, (T-4)- see TGB600
TIN BIS(DIETHYLDITHIOCARBAMATE) see TGB600
TIN(IV) BROMIDE (1:4) see TGB750
TIN(II) CHLORIDE (1:2) see TGC000
TIN(IV) CHLORIDE (1:4) see TGC250
TIN CHLORIDE, fuming (DOT) see TGC250
TIN(II) CHLORIDE DIHYDRATE (1:2:2) see TGC275
TIN CHLORIDE IODIDE see TGC280
TIN(IV) CHLORIDE, PENTAHYDRATE (1:4:5) see TGC282
TIN, CHLOROTRIOCTYL- see TMO585
TIN(2+) CITRATE see TGC285
TIN CITRATE (7CI) see TGC285
TIN, COMPLEX with PYROCATECHOL see TGC300
TIN COMPOUNDS see TGC500
TINDAL see ABG000
TIN DIBUTYL DILAURATE see DDV600
TIN DIBUTYLDITRIFLUOROACETATE see BLN750
TIN DIBUTYL MERCAPTIDE see DEI200
TIN DICHLORIDE see TGC000
TIN, DICHLORODIDODECYL- see DHA600
TIN, DICHLORODIVINYL- see DXR450
TIN, DIETHYL-, DIIODIDE see DJB000
TIN DIFLUORIDE see TGD100
TIN, (DIHYDROGEN PHOSPHORODITHIOATO-S)TRIPHENYL-, O,O-DIETHYL ESTER see EFL600
TIN, DIMETHYL-, DICHLORIDE see DUG825
TIN, DIOCTYL-, DICHLORIDE see DVN300
TIN, DIOCTYLDIPHENYL- see DVK500
TINDURIN see TGD000
TINESTAN see ABX250
TINESTAN 60 WP see ABX250
TIN FLAKE see TGB250
TIN FLUORIDE see TGD100
TIN, FLUOROTRIPHENYL- see TMV850
TINGENIN A see TGD125
TINGENON see TGD125
TINGENONE see TGD125
TINIC see NCQ900
TINIDAZOL see TGD250
TINIDAZOLE see TGD250
TIN(II) IODIDE see TGD500
TIN(IV) IODIDE (1:4) see TGD750
TINMATE see CLU000
TIN(II) NITRATE OXIDE see TGE000
TIN, OCTYL-, TRICHLORIDE see OGG000
TIN, OCTYL-, TRIS(ISOOCTYLTHIO GLYCOLLATE) see OGI000
TINOFEDRINE HYDROCHLORIDE see DXI800
TINOLITE see CBT750
TINON BLUE GF see DFN300
TINON BLUE GL see DFN300
TINON BLUE RS see IBV050
TINON BLUE RSN see IBV050
TINON BRILLIANT ORANGE GR see CMU820
TINON BROWN BR see CMU770
TINON BROWN GR see CMU780
TINON BROWN GRF see CMU780
TINON GOLDEN YELLOW see DCZ000
TINON NAVY BLUE RA see VGP100
TINON OLIVE 2R see DUP100

TINON RED 6B see CMU825
TINON VIOLET 4R see DFN450
TINON VIOLET 2RB see DFN450
TINON VIOLET B 4RP see DFN450
TINOPAL AMS see CMP200
TINOPAL BHS see FCA100
TINOPAL CBS see TGE150
TINOPAL CBS-X see TGE150
TINOPAL CH 3669 see TGE152
TINOPAL EMS see CMP200
TINOPAL 5BM see TGE100
TINOPAL RBS see TGE155
TINOPAL RBS 200 see TGE155
TINORIDINE HYDROCHLORIDE see TGE165
TINOSTAT see DDV600
TINOX see MIW250
TIN OXALATE see TGE250
TIN(2+) OXALATE see TGE250
TIN(II) OXALATE see TGE250
TIN OXIDE see TGE300
TIN P see HML500
TIN PERBROMIDE see TGB750
TIN PERCHLORIDE (DOT) see TGC250
TIN(IV) PHOSPHIDE see TGE500
TIN POTASSIUM TARTRATE see TGE750
TIN POWDER see TGB250
TIN PROTOCHLORIDE see TGC000
TIN-SAN see TIC500
TINSET see OMG000
TIN SODIUM TARTRATE see TGF000
TIN STEARATE see SLI350
TIN(2+) SULFATE see TGF010
TIN(II) SULFATE see TGF010
TIN TETRABROMIDE see TGB750
TINTETRACHLORIDE (DUTCH) see TGC250
TIN TETRACHLORIDE, anhydrous (DOT) see TGC250
TIN TETRAIODIDE see TGD750
TIN, TETRAOCTYL- see TDY790
TIN, TETRAOCTYL- see TDY795
TINTORANE see WAT220
TIN, TRIBUTYL-, ISOTHIOCYANATE see THY750
TIN TRIPHENYL ACETATE see ABX250
TINUVIN 144 see PMK800
TINUVIN 1130 see TGF025
TINUVIN P see HML500
TIOACETAZON see FNF000
TIOALKOFEN BM see TFD000
TIOBENZAMIDE (ITALIAN) see BBM250
TIOCARBAZIL see TGF030
TIOCARONE see FNF000
TIOCONAZOLE see TGF050
TIOCTACID see DXN800
TIOCTAN see DXN709
TIOCTIDASI see DXN800
TIOCTIDASI ACETATE REPLACING FACTOR see DXN800
TIODIFENILAMINA (ITALIAN) see PDP250
TIOFANATE ETILE (ITALIAN) see DJV000
TIOFINE see TGG760
TIOFOS see PAK000
TIOFOSFAMID see TFQ750
TIOFOZIL see TFQ750
TIOGUANIN see AMH250
TIOGUANINE see AMH250
TIOINOSINE see MCQ500
TIOMAN V see MAS300
TIOMERACIL see MPW500
TIOMESTERONE see TFK300
TIONAL see BJT750
TIOPENTALE (ITALIAN) see PBT250
TIOPENTAL SODIUM see PBT500
TIOPRONIN see MCI375
TIOPROPEN see AOO490
TIORALE M see MPW500
TIORIDAZIN see MOO500
TIOTIAMINE see DXO300
TIOTIRON see MPW500
TIOTIXENE see NBP500

TIOTRIFAR see AOO490
TIOURACYL (POLISH) see TFR250
TIO-URASIN SODIUM see AMR750
TIOVEL see EAQ750
TIOVIT-B$_1$ see DXO300
TIOXANONA see TLP750
TIOXIDE see TGG760
TIOXIDE A-HR see OBU100
TIOXIN see ZJS300
TIPEDINE see BLV000
TIPEPIDINE see BLV000
TIPHEN see DHY400
TIPIDI see DXO300
TIP-OFF see NAK500
TIPOL see TGF060
TIPPON see TAA100
TIPPS see TDE750
TIQUIZIUM BROMIDE see TGF075
TIRAMPA see TFS350
TIRANTIL see SBG500
TIRIAN HYDROCHLORIDE see CKB500
TIROIDINA see TFZ100
TIRON see DXH300
TIRON FREE ACID see PPQ100
TIROPRAMIDA (SPANISH) see TGF175
TIROPRAMIDE see TGF175
TIRPATE see DRR000
TISERCIN see MCI500
TISIN see ILD000
TISKAN (CZECH) see NFC500
TISOKINASE see PJB800
TISOMYCIN see CQH000
2,4,5-T ISOOCTYL ESTER see TGF200
2,4,5-T, ISOPROPYL ESTER see TGF210
TISPERSE MB-2X see NBL000
TISPERSE MB-58 see BHA750
TISSONOX 206 see PKQ070
TISULAC see LAG010
TITAANTETRACHLORID (DUTCH) see TGH350
TITANATE see TGF250
TITANATE, BARIUM (1:1) see BAP500
TITANATE BARIUM (1:1) see BAP502
TITANATE CALCIUM (1:1) see CAY300
TITANATE (TI8O17(2-)), DIPOTASSIUM see FQU200
TITANDIOXID (SWEDEN) see TGG760
TITANE (TETRACHLORURE de) (FRENCH) see TGH350
TITANIC ACID, LEAD SALT see LED000
TITANIC SULFATE see TGF220
TITANIC SULPHATE see TGF220
TITANIO TETRACHLORURO di (ITALIAN) see TGH350
TITANIUM see TGF250
TITANIUM 50A see TGF250
TITANIUM (wet powder) see TGF500
TITANIUM ACETONYL ACETONATE see BGQ750
TITANIUM ALLOY see TGF250
TITANIUM AZIDE TRICHLORIDE see TGF750
TITANIUM compounded with BERYLLIUM (1:12) see BFR000
TITANIUM CARBIDE see TGG000
TITANIUM CHLORIDE see TGG250
TITANIUM CHLORIDE see TGH350
TITANIUM(III) CHLORIDE see TGG250
TITANIUM COMPOUNDS see TGG500
TITANIUM DIAZIDE DIBROMIDE see TGG600
TITANIUM DIBROMIDE see TGG625
TITANIUM DICHLORIDE see TGG750
TITANIUM DIOXIDE see TGG760
TITANIUM DISULFATE see TGH250
TITANIUM DISULFATE see TGH252
TITANIUM FERROCENE see DGW200
TITANIUM(4+) ISOPROPOXIDE see IRN200
TITANIUM ISOPROPYLATE see IRN200
TITANIUM METAL POWDER, WET (DOT) see TGF500
TITANIUM(III) METHOXIDE see TGG775

(d-TMO⁶))-LHRH ACETATE see LIU327
TMP see HDF300
TMP see TMD250
TMP (ALCOHOL) see HDF300
TMPD see BJF500
TMPD see TLY750
TMPH see HNN500
(d-TMP⁶))-LHRH ACETATE see LIU351
TMPN see HOI300
TMPTA see TLX175
TMS 480 see TKX000
TMSM see MAW800
TMSN see TDW250
TMTD see TFS350
TMTDS see TFS350
TMTM see BJL600
TMTMS see BJL600
TMTU see TDX000
TMU see TDX250
TMV-4 see TLQ500
m-TMXDI see TDX300
p-TMXDI see TDX400
m-TMXDI/TMP ADDUCT see TDX320
T-2 MYCOTOXIN see FQS000
TN 3J see DEJ100
TN 762 see TEN750
TNA see TDY000
TNB see TMK500
TNBS see TMK800
TNCB see PIE530
TNCS 53 see CNP250
TNF-α see HGL920
TNG see NGY000
TNM see TDY250
TNO 6 see SLE875
TN 80 (PESTICIDE) see TBQ280
α-TNT see TMN490
TNT (OSHA) see TMN490
TNT-TOLITE (FRENCH) see TMN490
TNT, dry or wetted with <30% water, by weight
(UN 0209) (DOT) see TMN490
TO 125 see SAN600
TOA BLA-S see BLX500
TOABOND 40H see AAX250
TOBACCO see TGH725
TOBACCO LEAF ABSOLUTE see TGH750
TOBACCO LEAF, AQUEOUS EXTRACT see
TGH800
TOBACCO LEAF, NICOTIANA GLAUCA see
TGI000
TOBACCO LEAF, NICOTIANA TABACUM
see TGH725
TOBACCO PLANT see TGI100
TOBACCO REFINED TAR see CMP800
TOBACCO SMOKE CONDENSATE see
SEC000
TOBACCO SUCKER CONTROL AGENT 148
see DAI800
TOBACCO SUCKER CONTROL AGENT 148
see ODE000
TOBACCO SUCKER CONTROL AGENT 504
see DAI800
TOBACCO SUCKER CONTROL AGENT 504
see ODE000
TOBACCO TAR see CMP800
TOBACCO TAR see SEC000
TOBACCO WOOD see WCB000
TOBIAS ACID see ALH750
TOBRA see TGI500
TOBRADISTIN see TGI250
TOBRAMYCIN see TGI250
TOBRAMYCIN SULPHATE see TGI500
TOBREX see TGI250
TOCAINIDE see TGI699
TOCAMPHYL see HAQ570
TOCE see DJT400
TOCEN see DJT400
TOCHERGAMINE see TGI725
TOCHLORINE see CDP000
TOCOFEROL-2-(p-CHLOROPHENOXY)-2-
METHYLPROPIONATE see TGJ000
TOCOFIBRATE see TGJ000

TOCOFIBRATO (SPANISH) see TGJ000
TOCOPHEREX see TGJ050
α-TOCOPHEROL see VSZ450
(R,R,R)-α-TOCOPHEROL see VSZ450
(2R,4'R,8'R)-α-TOCOPHEROL see VSZ450
α-TOCOPHEROL ACETATE see TGJ050
(+)-TOCOPHEROL ACETATE see TGJ050
(+-)-α-TOCOPHEROL ACETATE see TGJ055
d-α-TOCOPHEROL ACETATE see TGJ050
d,l-α-TOCOPHEROL ACETATE see TGJ055
d-α-TOCOPHEROL (FCC) see VSZ450
dl-α-TOCOPHEROL (FCC) see VSZ450
α-TOCOPHERYL ACETATE see TGJ050
(+)-α-TOCOPHERYL ACETATE see TGJ050
d-α-TOCOPHERYL ACETATE see TGJ050
d-α-TOCOPHERYL ACETATE see TGJ050
dl-α-TOCOPHERYL ACETATE see TGJ055
(R,R,R)-α-TOCOPHERYL ACETATE see
TGJ050
(2R,4'R,8'R)-α-TOCOPHERYL ACETATE see
TGJ050
dl-α-TOCOPHERYL ACID SUCCINATE see
TGJ060
TOCOPHRIN see TGJ050
TOCOSINE see TOG250
TOCP see TMO600
T-OCTYL MERCAPTAN see MKJ250
T-OCTYL MERCAPTAN see OFE030
TODALGIL see BRF500
TODRALAZINA (ITALIAN) see CBS000
TODRALAZINE see CBS000
TODRALAZINE HYDROCHLORIDE see
TGJ150
TODRALAZINE HYDROCHLORIDE
HYDRATE see TGJ150
TOE see TIH800
TOF see TNI250
TOFAXIN see TGJ050
TOFENACINE HYDROCHLORIDE see
TGJ250
TOFENACIN HYDROCHLORIDE see
TGJ250
TOFISOPAM see GJS200
TOFK see TMO600
TOFRANIL see DLH600
TOFRANIL see DLH630
TOFRANILE see DLH630
TOFURON see FQN000
TOGAL see TGJ350
TOGAMYCIN see SKY500
N-TOIN see NGE000
TOIN UNICELLES see DKQ000
TOK see DFT800
TOK-2 see DFT800
TOKAMINA see VGP000
TOKAWHISKER see SCQ000
TOK E see DFT800
TOK E-25 see DFT800
TOK E 40 see DFT800
TOKIOCILLIN see AIV500
TOKKORN see DFT800
TOKOFEROL ACETATE see TGJ050
TOKOKIN see EDV000
TOKOPHARM see VSZ450
TOKSOBIDIN see BGS250
TOKUNOL M see AMX300
TOKUTHION see DGC800
TOKUTHION OXON see EEE200
TOK WP-50 see DFT800
TOKYO ANILINE ASTRAPHLOXINE FF see
CMM840
TOKYO ANILINE BRILLIANT GREEN see
BAY750
TOLADRYL see TGJ475
TOLAMINE see CDP000
2,4-TOLAMINE see TGL750
TOLANSIN see GGS000
TOLAPIN see PQC500
TOLAVAD see BBJ750
TOLAZAMIDE see TGJ500
TOLAZOLINE see BBW750
TOLAZOLINE CHLORIDE see BBJ750

TOLAZOLINE HYDROCHLORIDE see
BBJ750
TOLAZUL see AJP250
TOLBAN see CQG250
TOLBUSAL see BSQ000
TOLBUTAMID see BSQ000
TOLBUTAMIDE see BSQ000
TOLBUTAMIDE SODIUM SALT see BSQ250
TOLCAINE HYCROCHLORIDE see TGJ625
TOLCASONE see HII500
TOLCICLATE see MQA000
TOLCIL see GGS000
TOLCYCLAMIDE see CPR000
TOLECTIN see SKJ340
TOLERON see FBJ100
TOLESTAN see CMY525
TOLFENAMIC ACID see CLK325
TOLFERAIN see FBJ100
TOLHEXAMIDE see CPR000
o-TOLIDIN see TGJ750
2-TOLIDIN (GERMAN) see TGJ750
2-TOLIDINA (ITALIAN) see TGJ750
TOLIDINE see TGJ750
2-TOLIDINE see TGJ750
o-TOLIDINE see TGJ750
3,3'-TOLIDINE see TGJ750
o,o'-TOLIDINE see TGJ750
o-TOLIDINE DIHYDROCHLORIDE see
DQM000
TOLIFER see FBJ100
TOLIMAN see CMR100
TOLINASE see TGJ500
TOLIT see TMN490
TOLITE see TMN490
TOLKAN see IRA050
TOLL see MNH000
TOLMETIN see TGJ850
TOLMETINE see TGJ850
TOLMETIN SODIUM see SKJ340
TOLMETIN SODIUM DIHYDRATE see
SKJ350
TOLNAFTATE see TGB475
TOLNAPHTHATE see TGB475
TOLNIDAMIDE see TGJ875
TOLOFREN see GGS000
TOLOMOL see AIV500
TOLONIDINE NITRATE see TGJ885
TOLONIUM CHLORIDE see AJP250
TOLOPELON see TGB175
(−)-1-(o-TOLOSSI-3-(2,2,5,5-TETRAMETIL-
PIRROLIDIN-1-IL))-PROPAN-2-OLO
CLORIDRATO (ITALIAN) see TGJ625
3-o-TOLOXY-2-HYDROXYPROPYL-1-
CARBAMATE see CBK500
3-o-TOLOXY-1,2-PROPANEDIOL see
GGS000
3-o-TOLOXY-1,2-PROPANEDIOL-1-
CARBAMIC ACID ESTER see CBK500
TOLPAL see BBJ750
TOLPERISONE see TGK200
TOLPERISONE HYDROCHLORIDE see
MRW125
TOLPRONINE HYDROCHLORIDE see
TGK225
TOLSANIL see TGB475
TOLSERAM see CBK500
TOLSEROL see GGS000
TOLSERON see RLU000
TOLU see BAF000
TOLUALDEHYDE see TGK250
4-TOLUALDEHYDE see FOJ025
p-TOLUALDEHYDE (8CI) see FOJ025
α-TOLUALDEHYDE see BBL500
TOLUALDEHYDE GLYCERYL ACETAL see
TGK500
α-TOLUAMIDE see PDX750
TOLU BALSAM GUM see BAF000
TOLU BALSAM TINCTURE see BAF000
p-TOLUBENZYL ACETATE see XQJ700
TOLUBUTEROL HYDROCHLORIDE see
BQE250
TOLUEEN (DUTCH) see TGK750

TOLUEEN-DIISOCYANAAT see TGM750
TOLUEN (CZECH) see TGK750
TOLUEN-DISOCIANATO see TGM750
TOLUENE see TGK750
TOLUENE, p-ARSENOSO- see TGV100
TOLUENE-2-AZONAPHTHOL-2 see TGW000
o-TOLUENE-1-AZO-2-NAPHTHYLAMINE see FAG135
o-TOLUENEAZO-o-TOLUENEAZO-β-NAPHTHOL see SBC500
o-TOLUENEAZO-o-TOLUENE-β-NAPHTHOL see SBC500
o-TOLUENEAZO-o-TOLUIDINE see AIC250
TOLUENE-2,4-BISMALEIMIDE see TGY770
N,N'-2,4-TOLUENEBISMALEIMIDE see TGY770
TOLUENE, m-BROMO- see BOG300
TOLUENE, o-BROMO- see BOG260
TOLUENE, p-BROMO- see BOG255
TOLUENECARBOXALDEHYDE see TGK250
TOLUENE, ar-CHLORO- see CLK130
TOLUENE, 4-CHLORO-3,5-DINITRO-α-α-α-TRIFLUORO- see CGM225
p-TOLUENE CHLOROMETHYLAMIDE see CEB500
TOLUENE, CHLORO(α-METHYLBENZYL)- see PFR600
TOLUENE, 2-CHLORO-4-NITRO- see CJG800
TOLUENE, 2-CHLORO-6-NITRO- see CJG825
TOLUENE, α-CHLORO-m-NITRO- see NFN010
TOLUENE, m-CHLORO-α,α,α-TRIFLUORO- see CDQ330
TOLUENEDIAMINE see TGL500
m-TOLUENEDIAMINE see TGL750
p-TOLUENEDIAMINE see TGM000
TOLUENE-2,3-DIAMINE see TGY800
TOLUENE-2,4-DIAMINE see TGL750
2,4-TOLUENEDIAMINE see TGL750
TOLUENE-2,5-DIAMINE see TGM000
TOLUENE-2,6-DIAMINE see TGM100
TOLUENE-3,4-DIAMINE see TGM250
2,4-TOLUENEDIAMINE (DOT) see TGL750
p-TOLUENEDIAMINE DIHYDROCHLORIDE see DCE200
TOLUENE-2,4-DIAMINE, 5-(PHENYLAZO)-(6CI) see CMM800
p-TOLUENEDIAMINE SULFATE see DCE600
p-TOLUENEDIAMINE SULFATE see TGM400
2,5-TOLUENEDIAMINE SULFATE see DCE600
TOLUENE-2,5-DIAMINE, SULFATE (1:1) (8CI) see DCE600
p-TOLUENEDIAMINE SULPHATE see DCE600
TOLUENE-2,5-DIAMINE SULPHATE see DCE600
TOLUENE-2,4-DIAMINE, 5-(o-TOLYLAZO)-(8CI) see MLP770
TOLUENE-3,5-DIAMINE, α-α-α-TRIFLUORO-(8CI) see TKB775
2-TOLUENEDIAZONIUM BROMIDE see TGM425
p-TOLUENEDIAZONIUM FLUOROBORATE see TGM450
p-TOLUENEDIAZONIUM HYDROGEN SULFATE see TGM500
o-TOLUENEDIAZONIUM, 4-NITRO-, TETRAFLUOROBORATE(1⁻) see NMO700
2-TOLUENEDIAZONIUM PERCHLORATE see TGM525
o-TOLUENEDIAZONIUM PERCHLORATE see TGM525
2-TOLUENEDIAZONIUM SALTS see TGM550

3-TOLUENEDIAZONIUM SALTS see TGM575
4-TOLUENEDIAZONIUM SALTS see TGM580
p-TOLUENEDIAZONIUM, SULFATE (1:1) see TGM500
p-TOLUENEDIAZONIUM, TETRAFLUOROBORATE (1⁻) see TGM450
p-TOLUENEDIAZONIUM TETRAFLUOROBORATE (7CI) see TGM450
4-TOLUENEDIAZONIUM TRIIODIDE see TGM590
TOLUENE, 2,4-DICHLORO- see DGM700
TOLUENE, p,α-DICHLORO- see CFB600
TOLUENE, α-o-DICHLORO-(8CI) see CEO200
TOLUENE, α-α-DICHLORO- see BAY300
TOLUENE, α-α-DICHLORO-α-FLUORO- see DFL100
TOLUENE, 3,4-DICHLORO-α,α,α-TRIFLUORO- (6CI,8CI) see DGO300
TOLUENE DIISOCYANATE see TGM740
TOLUENE DIISOCYANATE see TGM750
TOLUENE-1,3-DIISOCYANATE see TGM740
TOLUENE-2,4-DIISOCYANATE see TGM750
2,4-TOLUENEDIISOCYANATE see TGM750
TOLUENE-2,6-DIISOCYANATE see TGM800
2,6-TOLUENE DIISOCYANATE see TGM800
TOLUENE, ar,ar-DINITRO- see DVG600
2,3-TOLUENEDIOL see DNE000
2,5-TOLUENEDIOL see MKO250
TOLUENE-3,4-DIOL see DNE200
TOLUENE-3,4-DITHIOL see TGN000
TOLUENE-α,α-DITHIOL BIS(O,O-DIMETHYL PHOSPHORODITHIOATE) see BES250
TOLUENE, o-FLUORO- see FLZ100
TOLUENE HEXAHYDRIDE see MIQ740
TOLUENE, m-IODO-(7CI,8CI) see IFG100
o-TOLUENE ISOCYANATE see IKG725
p-TOLUENE ISOCYANATE see IKG735
m-TOLUENENITRILE see TGT250
p-TOLUENENITRILE see TGT750
TOLUENE, α-α-α-2,4-PENTACHLORO- see PAY550
4-TOLUENESULFANAMIDE see TGN500
p-TOLUENESULFANAMIDE see TGN500
p-TOLUENESULFANILIDE see TGN600
4-TOLUENESULFINIC ACID SODIUM SALT see SKK000
4-TOLUENESULFINYL AZIDE see TGN100
TOLUENE-2-SULFONAMIDE see TGN250
TOLUENE-4-SULFONAMIDE see TGN500
o-TOLUENESULFONAMIDE see TGN250
p-TOLUENESULFONAMIDE see TGN500
p-TOLUENESULFONAMIDE, N-(α-(CHLOROACETYL)PHENETHYL)-, (−)- see THH450
p-TOLUENESULFONAMIDE, N-(3-CHLORO-1-sec-BUTYLACETONYL)- see THH470
p-TOLUENESULFONAMIDE, N-(3-CHLORO-1-ISOPROPYLACETONYL)- see THH555
p-TOLUENESULFONAMIDE, N-(3-CHLORO-1-METHYLACETONYL)-, l- see THH360
p-TOLUENESULFONAMIDE, N-(4-CHLORO-3-OXOBUTYL)- see THH355
p-TOLUENESULFONAMIDE, N-CYCLOHEXYL- see CPQ800
p-TOLUENESULFONAMIDE, N-(1-(DIAZOACETYL)-2-METHYLBUTYL)- see THH480
p-TOLUENESULFONAMIDE, N-(3-DIAZO-1-ISOBUTYLACETONYL)- see THH490
p-TOLUENESULFONAMIDE, N-(3-DIAZO-1-METHYLACETONYL)- see THH380
p-TOLUENESULFONAMIDE, N-(4-DIAZO-3-OXOBUTYL)- see THH375
p-TOLUENESULFONAMIDE, N-(7-DIAZO-6-OXOHEPTYL)- see THH460

p-TOLUENESULFONAMIDE, N,N'-p-PHENYLENEBIS- see PEX275
p-TOLUENESULFONANILIDE see TGN600
TOLUENESULFONIC ACID see TGO000
4-TOLUENESULFONIC ACID see TGO000
p-TOLUENESULFONIC ACID see TGO000
m-TOLUENESULFONIC ACID, 6-((4-AMINO-3-BROMO-1-ANTHRAQUINONYL)AMINO)-, MONOSODIUM SALT see CMM090
p-TOLUENESULFONIC ACID, ANHYDRIDE WITH ISOCYANIC ACID see IKG925
p-TOLUENE SULFONIC ACID CHLORIDE see TGO250
p-TOLUENESULFONIC ACID HYDRAZIDE see THE250
TOLUENESULFONIC ACID SODIUM SALT see SIK460
ar-TOLUENESULFONIC ACID, SODIUM SALT (8CI) see SIK460
o-TOLUENESULFONYLAMIDE see TGN000
p-TOLUENESULFONYLAMIDE see TGN500
p-TOLUENESULFONYLANILIDE see TGN600
1-p-TOLUENESULFONYL-3-BUTYLUREA see BSQ000
p-TOLUENESULFONYL CHLORIDE see TGO250
TOLUENE-4-SULFONYL FLUORIDE see TGO500
p-TOLUENESULFONYL FLUORIDE see TGO500
α-TOLUENESULFONYL FLUORIDE see TGO300
N-(p-TOLUENESULFONYL)-N'-HEXAMETHYLENIMINOUREA see TGJ500
4-TOLUENESULFONYL HYDRAZIDE see TGO550
o-4-TOLUENE SULFONYL HYDROXYLAMINE see TGO575
4-TOLUENESULFONYL ISOCYANATE see IKG925
p-TOLUENESULFONYL ISOCYANATE see IKG925
TOLUENE-p-SULFONYLMETHYLNITROSAMIDE see THE500
TOLUENE-p-SULPHONAMIDE see TGN500
p-TOLUENESULPHONIC ACID see TGO000
TOLUENE-p-SULPHONYL FLUORIDE see TGO500
p-TOLUENESULPHONYL ISOCYANATE see IKG925
TOLUENE, α-α-α-p-TETRACHLORO- see TIR900
2-TOLUENETHIOL see TGP000
4-TOLUENETHIOL see TGP250
m-TOLUENETHIOL see TGO800
o-TOLUENETHIOL see TGP000
p-TOLUENETHIOL see TGP250
α-TOLUENETHIOL see TGO750
TOLUENE-2,4,6-TRIAMINE see TGO800
TOLUENE TRICHLORIDE see BFL250
TOLUENE, 2,3,6-TRICHLORO- see TJD600
TOLUENE, α-α',p-TRICHLORO- see TJD650
TOLUENE, 3,4,5-TRINITRO- see TMN550
TOLUENE, VINYL (mixed isomers) see VQK650
3-TOLUENKARBONITRIL see TGT250
o-TOLUENO-AZO-β-NAPHTHOL see TGW000
ar-TOLUENOL see CNW500
α-TOLUENOL see BDX500
o-TOLUENSULFAMID (CZECH) see TGN000
p-TOLUHYDROQUINOL see MKO250
m-TOLUIC ACID see TGP750
o-TOLUIC ACID see TGQ000
p-TOLUIC ACID see TGQ250
α-TOLUIC ACID see PDY850
m-TOLUIC ACID DIETHYLAMIDE see DKC800

α-TOLUIC ACID, ETHYL ESTER see EOH000
α-TOLUIC ACID, α-HYDROXY- see MAP000
α-TOLUIC ALDEHYDE see BBL500
o-TOLUIC NITRILE see TGT500
p-TOLUIC NITRILE see TGT750
TOLUIDENE BLUE O CHLORIDE see AJP250
m-TOLUIDIN (CZECH) see TGQ500
o-TOLUIDIN (CZECH) see TGQ750
p-TOLUIDIN (CZECH) see TGR000
TOLUIDINE see CEX800
2-TOLUIDINE see TGQ750
3-TOLUIDINE see TGQ500
4-TOLUIDINE see TGR000
m-TOLUIDINE see TGQ500
o-TOLUIDINE see TGQ750
p-TOLUIDINE see TGR000
o-TOLUIDINE, 4-((p-AMINOPHENYL)(4-IMINO-2,5-CYCLOHEXADIEN-1-YLIDENE)METHYL)- see MAC500
m-TOLUIDINE ANTIMONYL TARTRATE see TGR250
o-TOLUIDINE ANTIMONYL TARTRATE see TGR500
p-TOLUIDINE ANTIMONYL TARTRATE see TGR750
TOLUIDINE BLUE see AJP250
TOLUIDINE BLUE see TGS000
TOLUIDINE BLUE O see AJP250
p-TOLUIDINE, 2-BROMO- see BOG500
o-TOLUIDINE, 6-CHLORO- see CLK227
m-TOLUIDINE, 6-CHLORO-α-α-α-TRIFLUORO- see CEG800
m-TOLUIDINE, N,N-DIMETHYL- see TLG700
p-TOLUIDINE, N,N-DIMETHYL- see TLG150
p-TOLUIDINE, 3,5-DINITRO- see MJG600
o-TOLUIDINE, 4,6-DINITRO-N-METHYL-N-(2,4,6-TRIBROMOPHENYL)-α-α-α-TRIFLUOR O- see BMO300
o-TOLUIDINE, 6-ETHYL- see MJY000
2-TOLUIDINE HYDROCHLORIDE see TGS500
m-TOLUIDINE HYDROCHLORIDE see TGS250
o-TOLUIDINE HYDROCHLORIDE see TGS500
p-TOLUIDINE HYDROCHLORIDE see TGS750
m-TOLUIDINE, 6-NITRO- see NMP550
o-TOLUIDINE, 4-NITRO- see MMF780
o-TOLUIDINE, 5-NITRO-, HYDROCHLORIDE see NMP600
p-TOLUIDINE, α-(PHENYLTHIO)- see PGN100
TOLUIDINE RED see MMP100
TOLUIDINE RED 3B see MMP100
TOLUIDINE RED 4R see MMP100
TOLUIDINE RED 10451 see MMP100
TOLUIDINE RED BFB see MMP100
TOLUIDINE RED BFGG see MMP100
TOLUIDINE RED D 28-3930 see MMP100
TOLUIDINE RED LIGHT see MMP100
TOLUIDINE RED M 20-3785 see MMP100
TOLUIDINE RED R see MMP100
TOLUIDINE RED RT-115 see MMP100
TOLUIDINE RED TONER see MMP100
TOLUIDINE RED XL 20-3050 see MMP100
p-TOLUIDINE-m-SULFONIC ACID see AKQ000
TOLUIDINE TONER see MMP100
TOLUIDINE TONER DARK 5040 see MMP100
TOLUIDINE TONER HR X-2741 see MMP100
TOLUIDINE TONER KEEP HR X-2742 see MMP100
TOLUIDINE TONER L 20-3300 see MMP100
TOLUIDINE TONER RT-252 see MMP100
TOLUIDINE TONER 4R X-2700 see MMP100
p-TOLUIDINIUM CHLORIDE see TGS750

2-o-TOLUIDINOETHANOL see TGT000
o-TOLUIDYNA (POLISH) see TGQ750
TOLUILENODWUIZOCYJANIAN see TGM750
α-TOLUIMIDIC ACID see PDX750
TOLUINA see BSQ000
TOLULOX see GGS000
TOLUMETIN SODIUM DIHYDRATE see SKJ350
TOLUMID see BSQ000
p-TOLUNITRIL (CZECH) see TGT750
3-TOLUNITRILE see TGT250
4-TOLUNITRILE see TGT750
m-TOLUNITRILE see TGT250
o-TOLUNITRILE see TGT500
p-TOLUNITRILE see TGT750
α-TOLUNITRILE see PEA750
m-TOLUOL see CNW750
o-TOLUOL see CNX000
p-TOLUOL see CNX250
TOLUOL (DOT) see TGK750
o-TOLUOL-AZO-o-TOLUIDIN (GERMAN) see AIC250
TOLUOLO (ITALIAN) see TGK750
p-TOLUOLSULFONSAEUREAETHYL ESTER (GERMAN) see EPW500
p-TOLUOLSULFONSAEURE METHYL ESTER (GERMAN) see MLL250
N-TOLUOLSULPHONYL HYDRAZINE see THE250
TOLUON see BMA625
m-TOLUONITRILE see TGT250
o-TOLUONITRILE see TGT500
p-TOLUONITRILE see TGT750
TOLUOYL CHLORIDE see TGU010
3-TOLUOYL CHLORIDE see TGU010
m-TOLUOYL CHLORIDE see TGU010
p-TOLUOYL CHLORIDE see TGU020
TOLUQUINOL see MKO250
p-TOLUQUINOLINE see MPF800
p-TOLUQUINONE see MHI250
1,4-TOLUQUINONE see MHI250
TOLU RESIN see BAF000
TOLUSAFRANINE see GJI400
TOLU-SOL see TGK750
TOLUVAN see BSQ000
m-TOLUYL CHLORIDE see TGU010
m-TOLUYLENDIAMIN (CZECH) see TGL750
p-TOLUYLENDIAMINE see TGM000
TOLUYLENE BLUE (BASIC DYE) see DCE800
TOLUYLENE BLUE MONOHYDRATE see TGU500
m-TOLUYLENEDIAMINE see TGL750
2,3-TOLUYLENEDIAMINE see TGY800
2,4-TOLUYLENEDIAMINE see TGL750
TOLUYLENE-2,5-DIAMINE see TGM000
2,6-TOLUYLENEDIAMINE see TGM100
3,4-TOLUYLENEDIAMINE see TGM250
2,4-TOLUYLENEDIAMINE (DOT) see TGL750
p-TOLUYLENEDIAMINE SULPHATE see DCE600
TOLUYLENE-2,5-DIAMINE SULPHATE see DCE600
TOLUYLENE-2,4-DIISOCYANATE see TGM750
TOLUYLENE ORANGE G see CMO860
TOLUYLENE RED see AJQ250
m-TOLUYLIC ACID see TGP750
o-TOLUYLIC ACID see TGQ000
p-TOLUYLIC ACID see TGQ250
TOLVIN see BMA625
TOLVON see BMA625
TOLYCAINE HYDROCHLORIDE see BMA125
p-TOLYCARBAMIDE see THG000
TOLYDRIN see GGS000
m-TOLYENEDIAMINE see TGL750
p-TOLYHYDROXYLAMINE see HJA500
N-p-TOLYHYDROXYLAMINE see HJA500
m-TOLYLACETAMIDE see ABI750

N-m-TOLYLACETAMIDE see ABI750
TOLYL ACETATE see MHM100
p-TOLYL ACETATE see MNR250
p-TOLYL ALCOHOL see CNX250
TOLYL ALDEHYDE see TGK250
α-TOLYL ALDEHYDE DIMETHYL ACETAL see PDX000
TOLYLALDEHYDE GLYCERYL ACETAL see TGK500
TOLYLAMINE see TGR000
m-TOLYLAMINE see TGQ500
o-TOLYLAMINE see TGQ750
p-TOLYLAMINE see TGR000
o-TOLYLAMINE HYDROCHLORIDE see TGS500
N-(2-TOLYL)ANTHRANILIC ACID see MNR300
N-o-TOLYLANTHRANILIC ACID see MNR300
N-(o-TOLYL)ANTHRANILIC ACID see MNR300
N-(p-TOLYL)ANTHRANILIC ACID see TGV000
4-TOLYLARSENOUS ACID see TGV100
N-(p-TOLYL)-1-AZIRIDINECARBOXAMIDE see TGV250
m-TOLYLAZOACETANILIDE see TGV500
5-(o-TOLYLAZO)-2-AMINOTOLUENE see AIC250
p-(p-TOLYLAZO)-ANILINE see TGV750
p-(o-TOLYLAZO)-o-CRESOL see HJG000
4-o-TOLYLAZO-o-DIACETOTOLUIDE see ACR300
4'-(o-TOLYLAZO)-o-DIACETOTOLUIDIDE see ACR300
1-(o-TOLYLAZO)-2-NAPHTHOL see TGW000
1-(o-TOLYLAZO)-β-NAPHTHOL see TGW000
1-(o-TOLYLAZO)-2-NAPHTHYLAMINE see FAG135
2-(o-TOLYLAZO)-p-TOLUIDINE see TGW500
4-(o-TOLYLAZO)-o-TOLUIDINE see AIC250
4-(p-TOLYLAZO)-m-TOLUIDINE see AIC500
4-(p-TOLYLAZO)-o-TOLUIDINE see TGW750
1-((4-TOLYLAZO)TOLYLAZO)-2-NAPHTHOL see TGX000
o-TOLYLAZO-o-TOLYLAZO-2-NAPHTHOL see SBC500
1-(4-o-TOLYLAZO-o-TOLYLAZO)-2-NAPHTHOL see SBC500
o-TOLYLAZO-o-TOLYLAZO-β-NAPHTHOL see SBC500
p-TOLYL BENZOATE see TGX100
3-(α(o-TOLYL)BENZYLOXY)TROPANE HYDROBROMIDE see TGX500
o-TOLYLBIGUANIDE see TGX550
1-o-TOLYLBIGUANIDE see TGX550
2-TOLYL BROMIDE see BOG260
m-TOLYL BROMIDE see BOG300
o-TOLYL BROMIDE see BOG260
m-TOLYLCARBAMIDE see THF750
p-TOLYL-N-CARBAMOYLAZIRIDINE see TGV250
4-TOLYLCARBINOL see MHB250
p-TOLYLCARBINOL see MHB250
TOLYL CHLORIDE see BEE375
o-TOLYL CHLORIDE see CLK100
p-TOLYL CHLORIDE see TGY075
2-TOLYLCOPPER see TGY250
m-TOLYLDIETHANOLAMINE see DHF400
o-TOLYLDIETHANOLAMINE see DMT800
o-TOLYLDIGUANIDE see TGX550
2-p-TOLYL-5,6-DIHYDROIMIDAZO(2,1-A)ISOQUINOLINE see TGY300
TOLYL DIPHENYL PHOSPHATE see TGY750
2,4-TOLYLENEBIS(MALEIMIDE) see TGY770
m-TOLYLENEDIAMINE see TGL750

TOPICLOR 20 see CDR750
TOPICON see HAG325
TOPICORTE see DBA875
TOPICYCLINE see TBX250
TOPIDERM see DBA875
TOPISOLON see DBA875
TOPITOX see CJJ000
TOPITRACIN see BAC250
TOPO see TMO575
TOPOKAIN see AIT250
TOPOREX 500 see SMQ500
TOPOREX 830 see SMQ500
TOPOREX 550-02 see SMQ500
TOPOREX 850-51 see SMQ500
TOPOREX 855-51 see SMQ500
TOPSIN see DJV000
TOPSIN m-DITHANE M45 MIXTURE see
MAS300
TOPSIN WP METHYL see PEX500
TOPSYM see DUD800
TOPUSYN see INR000
TOPZOL see RCF000
TORAK see DBI099
TORATE see BPG325
TORAZINA see CKP250
TORAZINA see CKP500
TORBIN see EIN500
TORBUTROL see BPG325
TORCH see XPJ000
TORCH BRAND see CBT750
TORDON see PIB900
TORDON 10K see PIB900
TORDON 10K see PLQ760
TORDON 22K see PIB900
TORDON 22K see PLQ760
TORDON K see PLQ760
TORDON 101 MIXTURE see PIB900
TORECAN see TEZ000
TORECAN BIMALEATE see TEZ000
TORECAN DIMALEATE see TEZ000
TORECAN MALEATE see TEZ000
TORELLE see PHE250
TORENTAL see PBU100
TORESTEN see TEZ000
TORINAL see QAK000
TORMONA see BSQ750
TORMONA see TAA100
TORMOSYL see THH350
TOROMYCIN see GEO200
TORQUE see BLU000
TORSITE see BGJ250
TORULOX see GGS000
TORYN see CBG250
TOSIC ACID see TGO000
5-TOSILOXI 2-HYDROXIBENCENO
SULFONATO de PIPERACINA (SPANISH)
see PIK625
TOSMICIL see DAM625
TOSMILEN see DAM625
TOSMILENE see DAM625
TOSTRIN see TBG000
TOSYL see CDO000
N-TOSYL-l-ALANINE CHLOROMETHYL
KETONE see THH360
N-TOSYL-β-ALANINE CHLOROMETHYL
KETONE see THH355
N-TOSYL-l-ALANINE DIAZOMETHYL
KETONE see THH380
N-TOSYL-β-ALANINE DIAZOMETHYL
KETONE see THH375
TOSYLAMIDE see TGN500
p-TOSYLAMIDE see TGN500
l-1-TOSYLAMIDO-2-PHENYLETHYL
CHLOROMETHYL KETONE see THH450
6-(N-TOSYL)AMINOCAPROIC ACID
DIAZOMETHYL KETONE see THH460
N-TOSYLANILINE see TGN600
TOSYLCHLORAMIDE SODIUM see CDP000
TOSYL CHLORIDE see TGO250
TOSYL FLUORIDE see TGO500
TOSYL ISOCYANATE see IKG925
p-TOSYL ISOCYANATE see IKG925

N-TOSYL-d,l-ISOLEUCINE
CHLOROMETHYL KETONE see THH470
N-TOSYL-d,l-ISOLEUCINE DIAZOMETHYL
KETONE see THH480
N-TOSYL-l-LEUCINE DIAZOMETHYL
KETONE see THH490
TOSYLLYSINE CHLOROMETHYL
KETONE see THH500
TOSYL-L-LYSINE CHLOROMETHYL
KETONE see THH500
TOSYLLYSYL CHLOROMETHYL KETONE
see THH500
α-TOSYL-L-LYSYLCHLOROMETHYL
KETONE see THH500
N-α-TOSYL-l-LYSYL-
CHLOROMETHYLKETONE see THH500
N-TOSYL-l-PHENYLALANINE
CHLOROMETHYL KETONE see THH550
TOSYL-l-
PHENYLALANYLCHLOROMETHYL
KETONE see THH550
N-TOSYL-l-VALINE CHLOROMETHYL
KETONE see THH555
TOTACEF see CCS250
TOTACILLIN see AIV500
TOTACOL see PAJ000
TOTAL see ANI800
TOTAL see PAJ000
TOTALCICLINA see AIV500
TOTAPEN see AIV500
TOTAZINA see PKD250
TOTM see TJR600
TOTOCAINE HYDROCHLORIDE see
AIT750
TOTOMYCIN see TBX250
TOTP see TMO600
TOTRIL see DNG200
TOTRIL see HKB500
TOUGH see FAQ600
TOUKALIDE see HDY600
2,5-TOULENEDIAMINE SULFATE see
TGM400
TOWK see RGP450
TOX 47 see PAK000
TOXADUST see CDV100
TOXAFEEN (DUTCH) see CDV100
TOXAKIL see CDV100
TOXALBUMIN see AAD000
TOXAN see CBM750
TOXAPHEN (GERMAN) see CDV100
TOXAPHENE see CDV100
TOXAPHENE TOXICANT A see THH560
TOXAPHENE TOXICANT B see THH575
TOXB-DIF see CMY090
TOXER TOTAL see PAJ000
TOXICHLOR see CDR750
TOXIFERINE DICHLORIDE see THI000
TOXIFREN see FBY000
TOXILIC ACID see MAK900
TOXILIC ANHYDRIDE see MAM000
TOXIN, ADENIA DIGITATA (MODECCA
DIGITATA), MODECCIN see MRA000
TOXIN, ANABAENA FLOS-AQUAE NRC-
44-1 see AON825
TOXIN, BACTERIUM CLOSTRIDIUM
TETANI, TETANUS see TBG100
TOXIN, BACTERIUM CORYNE-
BACTERIUM DIPHTHERIAE, DIPTHERIA
see DWP300
TOXIN B, CLOSTRIDIUM DIFFICILE see
CMY090
TOXIN B, ENTERO-, (STAPHYLOCOCCUS
REDUCED) see EAV025
TOXIN, BLUE GREEN ALGA,
MICROCYSTIS AERUGINOSA see AQV990
TOXIN C (DENDROASPIS POLYLEPIS
POLYLEPIS REDUCED) see THI100
TOXIN, CLOSTRIDIUM BOTULINUM see
CMY030
TOXIN, CLOSTRIDIUM DIFFICILE see
CMY070

α-TOXIN, CLOSTRIDIUM NOVYI see
CMY130
TOXIN, CLOSTRIDIUM OEDEMATIENS,
TYPE A see CMY150
β-TOXIN, CLOSTRIDIUM PERFRINGENS
see CMY190
TOXIN, CLOSTRIDIUM SORDELLII see
CMY240
TOXINE TETANIQUE see TBG100
TOXIN FS 2 (DENDROASPIS POLYLEPIS
POLYLEPIS REDUCED) see THI130
TOXIN HT 2 see THI250
TOXIN I (DENDROASPIS POLYLEPIS
POLYLEPIS REDUCED) see THI252
TOXIN I (MICROCYSTIS AERUGINOSA) see
AQV990
TOXIN, JELLYFISH, CHIRONEX
FLECKERI see CDM700
TOXIN, JELLYFISH, SCYPHOZOAN,
CYANEA CAPILLATA see COI125
TOXIN-LR see AQV990
TOXIN, NEMATOCYST, PHYSALIA
PHYSALIS see PIA375
TOXIN, PENICILLIUM ROQUEFORTI see
PAQ875
TOXIN (PENICILLIUM ROQUEFORTII) see
POF800
TOXIN, PENICILLIUM VIRIDICATUM see
PAQ900
TOXIN PR see POF800
TOXIN, SHIGELLA DYSENTERIAE see
SCD750
TOXIN, SHIGELLA DYSENTERIAE see
SCD800
TOXIN T2 see FQS000
TOXIN T-17 see THI255
TOXIN TA2 see THI260
TOXIN TA2 (DENDROASPIS
ANGUSTICEPS REDUCED) see THI260
TOXIN T 17 (MICROCYSTIS AERUGINOSA)
see AQV990
TOXIN T-2 TRIOL see DAD600
TOXIN VI (BURGARUS FASCIATUS
REDUCED) see THI300
TOX-α-NOV see CMY130
TOXOBIDIN see BGS250
TOXOFACTOR see THI425
TOXOGONIN see BGS250
TOXOGONIN DICHLORIDE see BGS250
TOXOGONINE see BGS250
TOXON 63 see CDV100
TOXYLON POMIFERUM see MRN500
TOXYNON see SJP500
TOXYPHEN see CDV100
2,2'-(o-TOYLYIMINO)DIETHANOL see
DMT800
TOYO ACID PHLOXINE see CMM000
TOYO AMARANTH see FAG020
TOYOCAMYCIN NUCLEOSIDE see VGZ000
TOYO CUPRO CYANINE BLUE GL see
COF420
TOYO EOSINE G see BNH500
TOYOFINE A see CAW850
TOYOFINE TF-X see CAT775
TOYOMYCIN see CMK650
TOYO OIL ORANGE see PEJ500
TOYO OIL RED BB see SBC500
TOYO OIL YELLOW G see DOT300
TOYO ORIENTAL OIL BLUE G see BLK000
TP see TBG000
TP-21 see MOO250
TP-21 see MOO500
TP-95 see DDT500
TP 222 see CAT775
2,4,5-TP see TIX500
TP540 see DTQ800
TP 90B see BHK750
TP 121 (filler) see CAT775
TPA see PGV000
TPA-3-β-OL see PGV500
TPB see TEA400
TsPB see HCP800

TPBO see BAY275
TPBS see SKF600
TPD see DXO300
TP DIOL-EPOXIDE-1 see DMT000
TP DIOL EPOXIDE-2 see DMS800
2,4,5-T PGBEE see THJ100
TPG HYDROCHLORIDE see UVA150
TP H4-1,2-DIOL see DND000
TP H4-1,2-EPOXIDE see ECT000
TPIA see MCB500
TPN see CNF400
β-TPN see CNF400
TPN (pesticide) see TBQ750
TPNC see MOS100
TPN (NUCLEOTIDE) see CNF400
T 45 (POLYGLYCOL) see GHS000
T 100 (POLYSACCHARIDE) see EHG100
TPP see TMT750
19NTPP see DYF450
d,d-T80-PRALLETHRIN see THI500
2,4,5-T PROPYLENE GLYCOL BUTYL
ETHER ESTER see THJ100
TPS23 see MON750
TPTA see ABX250
TPTC see CLU000
TPTH see HON000
TPTZ see TMV500
TPU 2T see PKO500
TPU 10M see PKM250
TPZA see ABX250
TR 201 see SMR000
TR-700 see RSP100
TR5379M see COW675
TRABEST see FOS100
TRACHOSEPT see DBX400
TRADENAL see POB500
TRAFARBIOT see AOD125
TRAGACANTH see THJ250
TRAGACANTH GUM see THJ250
TRAGAYA see SEH000
TRA-KILL TRACHEAL MITE KILLER see
MCF750
TRAKIPEAL see MDQ250
TRAL see HFG000
TRALGON see HIM000
TRALKOXYDIM see GJU200
TRALKOXYDIME see GJU200
TRALOMETHRIN see THJ300
TRALOMETHRINE see THJ300
TRAMACIN see AQX500
TRAMADOL see THJ500
TRAMADOL (2) see THJ600
TRAMADOL HYDROCHLORIDE see THJ750
TRAMAL see THJ500
TRAMAL see THJ750
TRAMAZOLINE HYDROCHLORIDE see
THJ825
TRAMAZOLINE HYDROCHLORIDE
MONOHYDRATE see RGP450
TRAMETAN see TFS350
TRAMISOL see LFA020
TRAMISOLE see LFA020
TRANCALGYL see EEM000
TRANCOLON see CBF000
TRANCOPAL see CKF500
TRANCYLPROMINE SULFATE see PET500
TRANDATE see HMM500
TRANEX see AJV500
TRANEXAMIC ACID see AJV500
TRANHEXAMIC ACID see AJV500
TRANID see CFF250
TRANILAST see RLK800
TRANILCYPROMINE see PET750
TRANIMUL see DCK759
TRANITE D-LAY see PBC250
TRANK see AOO500
TRAN-Q see CJR909
TRAN-Q DIHYDROCHLORIDE see HOR470
TRANQUIL see PGA750
TRANQUILAN see MQU750
TRANQUILAN see TAF675
TRANQUILAX see CGA000

TRANQUILLIN see BCA000
TRANQUINIL see RDF000
TRANQUIRIT see DCK759
TRANQUIZINE DIHYDROCHLORIDE see
HOR470
TRANQUO-BUSCOPAN-WIRKSTOFF see
CFZ000
TRANSALLYL CR 39 see AGD250
TRANSAMINE see DAA800
TRANSAMINE see PET750
TRANSAMINE see TAA100
TRANSAMINE SULFATE see PET500
TRANSAMLON see AJV500
TRANSANNON see ECU750
TRANSBRONCHIN see CBR675
TRANSCYCLINE see PPY250
TRANSENE see CDQ250
TRANSENTINE see DHX800
TRANSENTINE see THK000
TRANSERGAN see THK600
TRANSERPIN see RDK000
TRANSETILE BLUE P-FER see MGG250
TRANSETILE RUBINE P-FL see AKI750
TRANSETILE VIOLET P 3R see DBP000
TRANSILIUM see CDQ250
TRANSIT see CHJ750
TRANSITHAL see SFG700
TRANSMIX see DFG300
TRANSPARENT BRONZE SCARLET see
CHP500
TRANSPARENT PIGMENT YELLOW O see
PIE830
TRANSPARENT YELLOW 2K see PIE800
TRANSPARENT YELLOW O see PIE830
TRANSPLANTONE see NAK500
TRANSPOISE see MFD500
TRANSVAALIN see GFC000
TRANXENE see CDQ250
TRANXILEN see CDQ250
TRANXILENE see CDQ250
TRANXILIUM see CDQ250
TRANYLCYPRAMINE see PET750
TRANYLCYPRAMINE SULFATE see PET500
TRANYLCYPROMINE see PET750
TRANYLCYPROMINE SULFATE see PET500
TRANYLCYPROMINE SULPHATE see
PET500
TRANZER see EHP000
TRANZETIL see DHX800
TRANZINE see CKP500
TRAPANAL see PBT500
TRAPANAL SODIUM see PBT500
TRAPEX see ISE000
TRAPEX see VFU000
TRAPEX-40 see ISE000
TRAPEXIDE see ISE000
TRAPIDIL see DIO200
TRAPYMIN see DIO200
TRAQUILAN see TAF675
TRAQUIZINE see CJR909
TRASAN see CIR250
TRASENTIN see DHX800
TRASENTINE see DHX800
TRASICOR see THK750
TRASULPHANE see IAD000
TRASYLOL see PAF550
TRATUL see TAB250
TRAUMANASE see BMO000
TRAUMASEPT see PKE250
TRAUSABUM see AEG875
TRAUSABUN see AEG875
TRAUSABUN see TDL000
TRAVAD see BAP000
TRAVELERS JOY see CMV390
TRAVELIN see DYE600
TRAVELMIN see DYE600
TRAVELON see HGC500
TRAVENON see SKJ300
TRAVEX see SFS000
TRAWOTOX see CDO000
TRAXANOX SODIUM PENTAHYDRATE
see THK850

TRAZENTYNA see DHX800
TRAZENTYNA (POLISH) see THK000
TRAZININ see PAF550
TRAZODON see THK875
TRAZODONE see THK875
TRAZODONE HYDROCHLORIDE see
CKJ000
TRAZODONE HYDROCHLORIDE see
THK880
TRAZOLINONE see EQO000
TREBON see EQN725
TRECALMO see CKA000
TRECATOR see EPQ000
TREDIONE see TLP750
TREDUM see CKG000
TREEMOSS CONCRETE see THK900
TREESAIL see GJU475
TREFANOCIDE see DUV600
TREFICON see DUV600
TREFLAM see DUV600
TREFLAN see DUV600
TREFLANOCIDE ELANCOLAN see DUV600
TRE-HOLD see NAK500
TRELMAR see MQU750
TREMARIL HYDROCHLORIDE see THL500
TREMARIL WONDER see THL500
TREMBLEX see DBE000
TREMERAD see CGB250
TREMIN see BBV000
TREMOLITE see THL550
TREMOLITE ASBESTOS see ARM280
TREMOLITENA see THL550
TREMOLITE (NON-ASBESTIFORM) see
THL550
TREMORINE see DWX600
TREMORINE DICHLOROHYDRATE see
THL575
TREMORINE DIHYDROCHLORIDE see
THL575
TREMORTIN A see PAR250
TREN see NEI800
TRENAMINE D-200 see OHY000
TRENAMINE D-201 see OHY000
TRENBOLONE see THL600
TRENBOLONE ACETATE see HKI100
TREN HP see NEI800
TRENIMON see TND000
TRENTADIL see THL750
TRENTADIL HYDROCHLORIDE see
THL750
TRENTAL see PBU100
TREOMICETINA see CDP250
TREOSULFAN see TFU500
TREPENOL WA see SIB600
TREPIBUTONE see CNG980
TREPIDAN see DAP700
TREPIDONE see MFD500
TREPOL (FRENCH) see BKX250
TRESANIL see TNP275
TRESCATYL see EPQ000
TRESCAZIDE see EPQ000
TRESITOPE see LGK050
TRESOCHIN see CLD000
TRESORTIL see GKK000
TRESPAPHAN see PMP500
TREST see THL500
TRESTEN see TEZ000
TRESULFAN see TFU500
TRETAMINE see TND500
TRETINOIN see VSK950
TRET-O-LITE WF 88 see QAT520
TRET-O-LITE WF 828 see QAT520
TRET-O-LITE XC 503 see OHK200
TRET-O-LITE XC 511 see AFP100
TREVI see CJU275
TREVINTIX see PNW750
TREXAN see NAH100
TREX-SAN see TKU680
TRF see TNX400
TRH-T see POE050
TRI-4 see DUV600
TRIABARB see EOK000

TRI-ABRODIL see AAN000
TRIACETALDEHYDE (FRENCH) see PAI250
TRIACETATE d'HYDROXYHYDROQUINONE see HLF600
TRIACETIC ACID LACTONE see THL800
TRIACETINASE see GGA800
TRIACETIN (FCC) see THM500
TRIACETONAMIN see TDT770
TRIACETONAMINE see TDT770
TRIACETONE AMINE see TDT770
TRIACETONITRILE TUNGSTEN TRICARBONYL see THM250
1,8,9-TRIACETOXYANTHRACENE see APH500
1,2,4-TRIACETOXYBENZENE see HLF600
4',5,7-TRIACETOXY FLAVONE see THM300
4-β,8-α-15-TRIACETOXY-3-α-HYDROXY-12,13-EPOXYTRICHOTHEC-9-ENE see ACS500
3-α,4-β,15-TRIACETOXYSCIRPENE see SBF525
3,4,15-TRIACETOXYSCIRPENOL see SBF525
TRIACETYL APIGENIN see THM300
TRIACETYL-6-AZAURIDINE see THM750
2',3',5'-TRIACETYL-6-AZAURIDINE see THM750
2',3',5'-TRI-o-ACETYL-6-AZAURIDINE see THM750
TRIACETYLDIPHENOLISATIN see ACD500
TRIACETYL GLYCERIN see THM500
2-(2',3',5'-TRIACETYL-β-d-RIBOFURANOSYL)-as-TRIAZINE-3,5-(2H,4H)-DIONE see THM750
3,4,15-TRI-o-ACETYLSCIRPENOL see SBF525
TRIACHOTHEC-9-ENE-3-α,4-β,15-TRIOL, 12,13-EPOXY-, TRIACETATE see SBF525
TRIACRYLFORMAL see THM900
TRIACRYLOYLHEXAHYDROTRIAZINE see THM900
TRI(N-ACRYLOYL)HEXAHYDROTRIAZINE see THM900
TRIACRYLOYLHEXAHYDRO-s-TRIAZINE see THM900
1,3,5-TRIACRYLOYLHEXAHYDROTRIAZINE see THM900
TRIACRYLOYLPERHYDROTRIAZINE see THM900
TRIACYLGLYCEROL HYDROLASE see GGA800
TRIACYLGLYCEROL LIPASE see GGA800
TRIAD see TIO750
TRIADENYL see ARQ500
TRIADIMEFON see CJO250
TRIADIMENOL see CJN300
TRIADIMENOL see MEP250
TRIAETHANOLAMIN-NG see TKP500
TRIAETHYLAMIN (GERMAN) see TJO000
TRIAETHYLENMELAMIN (GERMAN) see TND500
TRIAETHYLENPHOSPHORSAEUREAMID (GERMAN) see TND250
TRIAETHYLZINNACETAT (GERMAN) see ABW750
TRIAETHYLZINNSULFAT (GERMAN) see BLN500
TRIAFOL see CJN300
TRIALLAT (GERMAN) see DNS600
TRIALLATE see DNS600
TRIALLYL ACONITATE see THM910
TRIALLYLAMINE see THN000
TRIALLYL BORATE see THN250
TRIALLYL CYANURATE see THN500
TRIALLYL ISOCYANURATE see THS100
1,3,5-TRIALLYLISOCYANURATE see THS100
1,3,5-TRIALLYLISOCYANURIC ACID see THS100
TRIALLYLOXYPHOSPHINE see PHN300
TRIALLYL PHOSPHATE see THN750
TRIALLYL PHOSPHITE see PHN300

TRIAMCINCOLONE ACETONIDE see AQX500
TRIAMCINOLONE see AQX250
TRIAMCINOLONE ACETONIDE see AQX500
TRIAMCINOLONE-16,17-ACETONIDE see AQX500
TRIAMCINOLONE DIACETATE see AQX750
TRIAMCINOLONE HEXACETONIDE see AQY375
TRIAMELIN see TND500
TRIAMIDOPHOSPHATE see PHB550
TRIAMIFOS (GERMAN, DUTCH, ITALIAN) see AIX000
TRIAMINE BLACK D see CMN230
s-TRIAMINOBENZENE see THN775
1,3,5-TRIAMINOBENZENE see THN775
sym-TRIAMINOBENZENE see THN775
TRI(2-AMINOETHYL)AMINE see NEI800
2,4,7-TRIAMINO-6-FENILPTERIDINA (ITALIAN) see UVJ450
TRIAMINOGUANIDINE NITRATE see THN800
TRIAMINOGUANIDINIUM PERCHLORATE see THO250
TRIAMINO-s-HEPTAZINE see HAQ700
2,4,6-TRIAMINOPHENOL TRIHYDROCHLORIDE see THO500
TRIAMINOPHENYL PHOSPHATE see THO550
2,4,7-TRIAMINO-6-PHENYLPTERIDINE see UVJ450
TRIAMINOPHOSPHINE OXIDE see PHB550
TRIAMINOTOLUENE see TGO800
2,4,6-TRIAMINO-s-TRIAZINE see MCB000
2,4,6-TRIAMINO-1,3,5-TRIAZINE see MCB000
2,4,6-TRIAMINO-s-TRIAZINE compounded with s-TRIAZINE-TRIOL see THO750
2,2',2"-TRIAMINOTRIETHYLAMINE see NEI800
β,β',β"-TRIAMINOTRIETHYLAMINE see NEI800
S-TRIAMINOTRINITROBENZENE see TMK900
1,3,5-TRIAMINOTRINITROBENZENE see THO775
1,3,5-TRIAMINO-2,4,6-TRINITROBENZENE see TMK900
TRIAMINOTRIPHENYLMETHANE see THP000
4,4',4"-TRIAMINOTRIPHENYLMETHANE see THP000
p,p',p"-TRIAMINOTRIPHENYLMETHANE see THP000
4,4'4"-TRIAMINOTRIPHENYLMETHAN-HYDROCHLORID (GERMAN) see RMK020
2,2',2"-TRIAMINOTRIS(ETHYLAMINE) see NEI800
TRIAMIPHOS see AIX000
TRIAMMINEDIPEROXOCHROMIUM(IV) see THP250
TRIAMMINE GOLDTRIHYDROXIDE see THP500
TRIAMMINENITRATOPLANTINUM(II) NITRATE see THP750
cis-TRIAMMINETRICHLOROORUTHENIUM(III) see THQ000
TRIAMMINETRINITRATATORHODIUM (III) see THQ100
TRIAMMINETRINITROCOBALT(III) see THQ250
TRIAMMONIUM ALUMINUM HEXAFLUORIDE see THQ500
TRIAMMONIUM AURINTRICARBOXYLATE see AGW750
TRIAMMONIUM HEXACHLORORHODATE see HCM050
TRIAMMONIUM HEXACHLORORHODATE(3-) see HCM050

TRIAMMONIUM HEXAFLUOROALUMINATE see THQ500
TRIAMMONIUM TRIS-(ETHANEDIOATO(2-)-O,O')FERRATE(3-1) see ANG925
TRIAMPHOS see AIX000
TRIAMPUR see UVJ450
TRIAMTEREN see UVJ450
TRIAMTERENE see UVJ450
TRIAMTERIL see UVJ450
TRIAMTERIL COMPLEX see UVJ450
TRI-n-AMYL BORATE see TMQ000
TRIANATE see TJL250
TRIANGLE see CBT750
TRIANGLE see CNP250
TRIANGLE see CNP500
TRI-p-ANISYLCHLOROETHYLENE see CLO750
TRIANOL DIRECT BLUE 3B see CMO250
TRIANON see PPO000
1,4-TRIANTHRIMID (CZECH) see APL750
1,5-TRIANTHRIMID (CZECH) see APM000
TRIANTINE FAST RED 4BN see CMO885
TRIANTINE FAST TURQUOISE GL see COF420
TRIANTINE LIGHT BROWN 3RN see FMU059
TRIANTINE LIGHT RED 4BN see CMO885
TRIANTOIN see MKB250
TRIAPHOL see CJN300
TRIAPINE see AMI600
TRIARIMOL see THQ525
TRIASOL see TIO750
TRIASULFURON see THQ550
TRIASYN see DEV800
TRIATIX see MJL250
TRIATOMIC OXYGEN see ORW000
TRIATOX see MJL250
3,5,7-TRIAZA-1-AZONIAADAMANTANE, 1-(3-CHLOROALLYL)-, CHLORIDE see CEG550
1,2,3-TRIAZAINDENE see BDH250
3,5,7-TRIAZAINDOLE see POJ250
3,6,9-TRIAZATETRACYCLO[6.1.0.0^{2,4}.O^{5,7}]NONANE see THQ600
TRIAZENE see THQ750
1-TRIAZENE see THQ750
1-TRIAZENE, 1,3-DIETHYL-(9CI) see DKD200
p,p'-TRIAZENYLENEDIBENZENESULFONAMIDE see THQ900
TRIAZICHON (GERMAN) see TND000
TRIAZIDOBORANE see BMG325
TRIAZIDOMETHYLIUM HEXACHLOROANTIMONATE see THR100
2,4,6-TRIAZIDO-1,3,5-TRIAZINE see THR250
TRIAZIN see DEV800
1,2,4-TRIAZIN-3-AMINE, 6-(2-(5-NITRO-2-FURANYL)ETHENYL)-(9CI) see FPF000
TRIAZINATE see THR500
1,3,5-TRIAZINE see THR525
sym-TRIAZINE see THR525
1,3,5-TRIAZINE (9CI) see THR525
TRIAZINE (pesticide) see DEV800
TRIAZINE A 1294 see ARQ725
1,3,5-TRIAZINE-2-AMINE, 4-METHOXY-6-METHYL-(9CI) see MGH800
s-TRIAZINE, 2-AMINO-4-METHOXY-6-METHYL- see MGH800
s-TRIAZINE, 2-(sec-BUTYLAMINO)-4-CHLORO-6-(ETHYLAMINO)- see THR600
s-TRIAZINE, 2-CHLORO-4-(DIETHYLAMINO)-6-(ISOPROPYLAMINO)- see IKM050
1,3,5-TRIAZINE-2,4-DIAMINE, 6-CHLORO-N,N-DIETHYL-N'-(1-METHYLETHYL)- see IKM050
1,3,5-TRIAZINE-2,4-DIAMINE, 6-CHLORO-N-ETHYL-N'-(1-METHYLETHYL)-, MIXT. WITHN-ETHYL-N'-(1-METHYLETHYL)-6-

(METHYLTHIO)-1,3,5-TRIAZINE-2,4-
DIAMINE see HBT100
1,3,5-TRIAZINE-2,4-DIAMINE, 6-CHLORO-
N-ETHYL-N'-(1-METHYLPROPYL)-(9CI) see
THR600
1,3,5-TRIAZINE-2,4-DIAMINE, 6-CHLORO-
N,N,N',N'-TETRAETHYL- (9CI) see AKQ250
1,3,5-TRIAZINE-2,4-DIAMINE, 6-CHLORO-
N,N,N'-TRIETHYL- see TJL500
1,3,5-TRIAZINE-2,4-DIAMINE, 1,6-
DIHYDRO-1-(4-BUTYLPHENYL)-6,6-
DIMETHYL- see BSF275
1,3,5-TRIAZINE-2,4-DIAMINE, 6-(2-FURYL)-
see FQQ500
1,3,5-TRIAZINE-2,4-DIAMINE, 6-(2-
PYRIDINYL)- see DCD050
s-TRIAZINE, 2,6-DIAMINO-4-
BUTYLAMINO- see BRR800
s-TRIAZINE, 2,4-DIAMINO-6-(2-FURYL)- see
FQQ500
s-TRIAZINE, 2,6-DIAMINO-4-(2-PYRIDYL)-
see DCD050
s-TRIAZINE,-4,6-DIAZIRIDINYL-2-(2-
FURYL-5-HYDROXYMETHYL-m-DIOXAN-
5-YLAMINO)- see FQB100
s-TRIAZINE, 1,2-DIHYDRO-1-(p-
BUTYLPHENYL)-4,6-DIAMINO-2,2- see
BSF275
s-TRIAZINE, 1,2-DIHYDRO-1-(p-
CHLOROPHENYL)-4,6-DIAMINO-2,2-
DIMETHYL- see COX400
S-TRIAZINE, 2-((1,2-
DIMETHYLPROPYL)AMINO)-4-
ETHYLAMINO-6-METHYLTHIO- see
ARW775
S-TRIAZINE, 2-((1,2-
DIMETHYLPROPYL)AMINO)-4-
ETHYLAMINO-6-METHYLTHIO- see
DTS700
s-TRIAZINE-3,5(2H,4H)-DIONE see THR750
1,3,5-TRIAZINE-2,4(1H,3H)-DIONE, 1-
AMINO-3-(2,2-DIMETHYLPROPYL)-6-
(ETHYLTHIO)- see AJR200
s-TRIAZINE, HEXAHYDRO-1,3,5-
TRIACRYLOYL- see THM900
s-TRIAZINE, HEXAHYDRO-1,3,5-
TRIMETHYL- see HDW100
1,3,5-TRIAZINE, HEXAHYDRO-1,3,5-
TRIMETHYL- see HDW100
1,3,5-TRIAZINE, HEXAHYDRO-1,3,5-
TRINITRO-(9CI) see CPR800
1,3,5-TRIAZINE, HEXAHYDRO-1,3,5-TRIS(1-
OXO-2-PROPENYL)-(9CI) see THM900
as-TRIAZINE, 3-(α-METHYLPHENETHYL)-
1,4,5,6-TETRAHYDRO-, HYDROCHLORIDE
see TCV460
as-TRIAZINE, 1,4,5,6-TETRAHYDRO-3-(α-
METHYLPHENETHYL)-,
HYDROCHLORIDE see TCV460
S-TRIAZINE-2-THIOL, 1,4,5,6-
TETRAHYDRO-5-METHYL- see TCW800
TRIAZINETHION see TCW800
S-TRIAZINE-2(1H)-THIONE,
TETRAHYDRO-5-METHYL- see TCW800
1,3,5-TRIAZINE-2(1H)-THIONE,
TETRAHYDRO-5-METHYL- see TCW800
1,3,5-TRIAZINE-2,4,6-TRIAMINE see MCB000
1,3,5-TRIAZINE-2,4,6-TRIAMINE, N-BUTYL-
see BRR800
1,3,5-TRIAZINE-2,4,6-TRIAMINE, N-
CYCLOPROPYL- see CQE400
1,3,5-TRIAZINE-2,4,6-TRIAMINE, polymer
with FORMALDEHYDE (9CI) see MCB050
1,3,5-TRIAZINE-2,4,6-TRIAMINE, POLYMER
WITH FORMALDEHYDE, METHYLATED
see THR790
s-TRIAZINE, 2,4,6-TRIAMINO- see MCB000
s-TRIAZINE TRICHLORIDE see TJD750
s-TRIAZINE-1,3,5(2H,4H,6H)-TRIETHANOL
see THR820
1,3,5-TRIAZINE-1,3,5(2H,4H,6H)-
TRIETHANOL (9CI) see THR820

1,3,5-TRIAZINE-2,4,6-TRIMERCAPTAN see
THS250
sym-TRIAZINETRIOL see THS000
s-TRIAZINE-2,4,6-TRIOL see THS000
s-2,4,6-TRIAZINETRIOL see THS000
s-TRIAZINE-2,4,6-TRIOL, TRI(2,3-
EPOXYPROPYL) ESTER see TKL250
s-TRIAZINE-2,4,6(1H,3H,5H)-TRIONE see
THS000
s-TRIAZINE-2,4,6(1H,3H,5H)-TRIONE,
DICHLORO-, POTASSIUM DERIV see
PLD000
1,3,5-TRIAZINE-2,4,6(1H,3H,5H)-TRIONE,
1,3-DICHLORO-, POTASSIUM SALT see
PLD000
1,3,5-TRIAZINE-2,4,6(1H,3H,5H)-TRIONE,
1,3-DICHLORO-, SODIUM SALT,
DIHYDRATE see THS050
s-TRIAZINE-2,4,6(1H,3H,5H)-TRIONE,
MONOSODIUM SALT see SGB550
1,3,5-TRIAZINE-2,4,6(1H,3H,5H)-TRIONE,
MONOSODIUM SALT (9CI) see SGB550
s-TRIAZINE-2,4,6(1H,3H,5H)-TRIONE,
TRIALLYL- see THS100
s-TRIAZINE-2,4,6(1H,3H,5H)-TRIONE, 1,3,5-
TRIALLYL- see THS100
1,3,5-TRIAZINE-2,4,6(1H,3H,5H)-TRIONE,
1,3,5-TRICHLORO-, COMPD. WITH 1,3-
DICHLORO-1,3,5-TRIAZINE-
2,4,6(1H,3H,5H)-TRIONE POTASSIUM SALT
(1:4) see MRN050
1,3,5-TRIAZINE-2,4,6(1H,3H,5H)-TRIONE,
1,3,5-TRI-2-PROPENYL-(9CI) see THS100
s-TRIAZINE-2,4,6(1H,3H,5H)-TRIONE,
TRIS(2,3-EPOXYPROPYL) see GGY150
s-TRIAZINE-2,4,6(1H,3H,5H)-TRIONE, 1,3,5-
TRIS(2,3-EPOXYPROPYL)- see GGY150
s-TRIAZINE-2,4,6(1H,3H,5H)-TRIONE, 1,3,5-
TRIS(OXIRANYLMETHYL)-(9CI) see
GGY150
2,4,6-TRIAZINETRITHIOL see THS250
s-TRIAZINE-2,4,6-TRITHIOL see THS250
S-TRIAZINE-2,4,6-TRITHIOL, TRISODIUM
SALT see SKN100
1,3,5-TRIAZINE-2,4,6(1H,3H,5H)-
TRITHIONE see THS250
1,3,5-TRIAZINE-2,4,6(1H,3H,5H)-
TRITHIONE, TRISODIUM SALT see SKN100
(1,3,5-TRIAZINE-2,4,6-
TRIYLTRINITRILO)HEXAKIS METHANOL
see HDY000
(s-TRIAZINE-2,4,6-
TRIYLTRINITRILO)HEXAMETHANOL see
HDY000
1,1',1''-s-TRIAZINE-2,4,6-
TRIYLTRISAZIRIDINE see TND500
1,2,4-TRIAZIN-5(4H)ONE, 4-AMINO-6-(1,1-
DIMETHYLETHYL)-3-(ETHYLTHIO)- see
ENJ600
1,3,5-TRIAZIN-2(1H)-ONE, TETRAHYDRO-
1,3-BIS(HYDROXYMETHYL)-5-ETHYL- see
TCJ900
s-TRIAZIN-2,4,6-
TRIYLTRIAMINOMETHANESULFONIC
ACID TRISODIUM SALT see THS500
(s-TRIAZIN-2,4,6-
TRIYLTRIAMINO)TRISMETHANESULFONI
C ACID TRISODIUM SALT see THS500
TRIAZIQUINONE see TND000
TRIAZIQUONE see TND000
TRIAZIRIDINOPHOSPHINE OXIDE see
TND250
2,3,5-TRI-(1-AZIRIDINYL)-p-
BENZOQUINONE see TND000
TRI(AZIRIDINYL)PHOSPHINE OXIDE see
TND250
TRI(-1-AZIRIDINYL)PHOSPHINE OXIDE
see TND250
TRIAZIRIDINYLPHOSPHINE SULFIDE see
TFQ750
TRIAZIRIDINYL TRIAZINE see TND500
TRIAZOFOSZ (HUNGARIAN) see THT750

TRIAZOIC ACID see HHG500
TRIAZOLAM see THS800
TRIAZOLAMINE see AMY050
1H-1,2,4-TRIAZOL-3-AMINE see AMY050
4H-1,2,4-TRIAZOL-4-AMINE, 3,5-
BIS(METHYLTHIO)-N-((5-NITRO-2-
FURANYL)METHYLENE)- see BKU150
4H-1,2,4-TRIAZOL-4-AMINE, N-((5-NITRO-
2-FURANYL)METHYLENE)-` see NGG550
TRIAZOL BROWN B see CMO820
TRIAZOLE see THS822
s-TRIAZOLE see THS855
1,2,3-TRIAZOLE see THS850
1H-1,2,4-TRIAZOLE (9CI) see THS855
s-TRIAZOLE, 5-(m-(ALLYLOXY)PHENYL)-3-
(o-TOLYL)- see AGP400
1H-1,2,4-TRIAZOLE, 3,5-BIS(3,4-
DIMETHOXYPHENYL)- see BJE700
1H-1,2,4-TRIAZOLE, 1-((BIS(4-
FLUOROPHENYL)METHYLSILYL)METHYL
)- see THS860
s-TRIAZOLE, 3,5-BIS(o-TOLYL)- see BLK500
1H-1,2,4-TRIAZOLE, 1-((4-BROMO-2-(2,4-
DICHLOROPHENYL)TETRAHYDRO-2-
FURANYL)METHYL)- see BNA920
s-TRIAZOLE, 3-(o-BUTYLPHENYL)-5-(m-
METHOXYPHENYL)- see BSG100
s-TRIAZOLE, 3-(o-BUTYLPHENYL)-5-
PHENYL- see BSH075
1H-1,2,4-TRIAZOLE-1-CARBOXAMIDE, 4,5-
DIHYDRO-3-METHOXY-4-METHYL-5-
OXO-N-(2-
(TRIFLUOROMETHOXY)PHENYL)SULFON
YL)-, SODIUM SALT see FDA895
1H-1,2,4-TRIAZOLE, 1-((2-CHLORO-4-(4-
CHLOROPHENOXY)PHENYL)-4-METHYL-
1,3-DIOXOLAN-2-YL)METHYL)- see THS900
s-TRIAZOLE, 5-(o-CHLOROPHENYL)-3-(o-
TOLYL)- see CKL100
1H-1,2,4-TRIAZOLE, 1-((2-(2,4-
DICHLOROPHENYL)-4-ETHYL-1,3-
DIOXOLAN-2-YL) METHYL)- see EDW100
1H-1,2,4-TRIAZOLE, 1-(2-(2,4-
DICHLOROPHENYL)-4-ETHYL-1,3-
DIOXOLAN-2-YL)METHYL)- see VFA100
1H-1,2,4-TRIAZOLE, 1-(2-(2,4-
DICHLOROPHENYL)PENTYL)- see PAP240
1H-1,2,4-TRIAZOLE, 1-((2-(2,4-
DICHLOROPHENYL)-4-PROPYL-1,3-
DIOXOLAN-2-YL)METHYL)- see PMS930
1H-1,2,4-TRIAZOLE, 1-((2-(2,4-
DICHLOROPHENYL)TETRAHYDRO-5-
(2,2,2-TRIFLUOROETHOXY)-2-FURANYL)
METHYL)-, CIS- see FPQ050
s-TRIAZOLE, 3,5-DIPHENYL- see DWO950
1H-1,2,4-TRIAZOLE, 3,5-DIPHENYL- see
DWO950
1H-1,2,4-TRIAZOLE-1-ETHANOL, β-((1,1'-
BIPHENYL)-4-YLOXY)-α-(1,1-
DIMETHYLETHYL)- see SCF525
1H-1,2,4-TRIAZOLE-1-ETHANOL, α-α-BIS(4-
FLUOROPHENYL)- see BJW600
1H-1,2,4-TRIAZOLE-1-ETHANOL, α-
BUTYL-α-(2,4-DICHLOROPHENYL)-, (+-)-
see HCN050
1H-1,2,4-TRIAZOLE-1-ETHANOL, α-(4-
CHLOROPHENYL)-α-(1-
CYCLOPROPYLETHYL)- see CQJ150
1H-1,2,4-TRIAZOLE-1-ETHANOL, α-(2-(4-
CHLOROPHENYL)ETHYL)-α-(1,1-
DIMETHYLETHYL)-, (+-)- see TAN050
1H-1,2,4-TRIAZOLE-1-ETHANOL, β-((4-
CHLOROPHENYL)METHYL)-α-(1,1-
DIMETHYLETHYL)-,(R*,R*)-(+−)- see
TMK125
1H-1,2,4-TRIAZOLE-1-ETHANOL, β-((4-
CHLOROPHENYL)METHYLENE)-α-(1,1-
DIMETHYLETHYL)-,(E)- see SOU800
1H-1,2,4-TRIAZOLE-1-ETHANOL, β-((2,4-
DICHLOROPHENYL)METHYLENE)-α-(1,1-
DIMETHYLETHYL)-, (E)- see DGC100

1H-1,2,4-TRIAZOLE-1-ETHANOL, β-((2,4-DICHLOROPHENYL)METHYLENE)-α-(1,1-DIMETHYLETHYL)-, (R-(E))- see THS920

1H-1,2,4-TRIAZOLE-1-ETHANOL, α-(2-FLUOROPHENYL)-α-(4-FLUOROPHENYL)- see FMR200

4H-1,2,4-TRIAZOLE, 4-((5-NITROFURFURYLIDENE)AMINO)- see NGG550

1H-1,2,4-TRIAZOLE-1-PROPANENITRILE, α-(2-(4-CHLOROPHENYL)ETHYL)-α-PHENYL- see CKI790

1H-1,2,4-TRIAZOLE-3-THIOL see THT000

TRIAZOL FAST RED 8B see CMO880

TRIAZOL FAST SCARLET 3B see CMO875

5H-s-TRIAZOLO(5,1-A)ISOINDOLE, 2-(4-BIPHENYLYL)- see BGO200

5H-(1,2,4)TRIAZOLO(5,1-A)ISOINDOLE, 2-(1,1'-BIPHENYL)-4-YL- see BGO200

5H-s-TRIAZOLO(5,1-A)ISOINDOLE, 2-(p-CHLOROPHENYL)- see CKL200

5H-(1,2,4)TRIAZOLO(5,1-A)ISOINDOLE, 2-(4-CHLOROPHENYL)- see CKL200

(1,2,4)TRIAZOLO(5,1-A)ISOQUINOLINE, 2-(1,1'-BIPHENYL)-4-YL-5,6-DIHYDRO- see BGL450

s-TRIAZOLO(5,1-A)ISOQUINOLINE, 5,6-DIHYDRO-2-(4-BIPHENYLYL)- see BGL450

TRIAZOLOGUANINE see AJO500

1,2,4-TRIAZOL-5-ONE, 3-NITRO- see NMP620

1,2,4-TRIAZOLO[4,3-a]PYRIDINE-SILVER NITRATE see THT250

s-TRIAZOLO(4,3-a)PYRIDIN-3(2H)-ONE, 2-(3-(4-(m-CHLOROPHENYL)-1-PIPERAZINYL)PROPYL)-, MONOHYDROCHLORIDE see THK880

1H-v-TRIAZOLO(4,5-d)PYRIMIDIN-7-AMINE see AIB340

(1,2,4)TRIAZOLO(1,5-a)PYRIMIDINE see THT275

(1,2,4)TRIAZOLO(4,3-a)PYRIMIDINE see THT280

v-TRIAZOLO(4,5-d)PYRIMIDINE-5,7-DIOL see THT350

1H-v-TRIAZOLO(4,5-d)PYRIMIDINE-5,7(4H,6H)-DIONE see THT350

1H-1,2,3-TRIAZOLO(4,5-D)PYRIMIDINE-5,7-(4H,6H)-DIONE, 4-(ACETYLOXY)- see ABL650

7H-1,2,3-TRIAZOLO(4,5-D)PYRIMIDIN-7-ONE, 1,4-DIHYDRO-5-(2-PROPOXYPHENYL)- see ZAK200

7H-1,2,3-TRIAZOLO(4,5-d)PYRIMIDIN-7-ONE, 1,4,5,6-TETRAHYDRO-5-THIOXO-(9CI) see HOJ100

3H-v-TRIAZOLO(4,5-d)PYRIMIDIN-7(4H,6H)-ONE, 5-THIONO- see HOJ100

(1,2,4)TRIAZOLO(4,3-a)QUINOLINE see THT400

(1H-1,2,4-TRIAZOLYL)TRICYCLOHEXYLSTANNANE see THT500

1H-1,2,4-(TRIAZOL-1-YL)TRIPHENYLSTANNANE see TMX000

(1H-1,2,4-TRIAZOLYL-1-YL)TRICYCLOHEXYLSTANNANE see THT500

TRIAZOPHOS see THT750

TRIAZOTION (RUSSIAN) see EKN000

TRIAZURE see THM750

TRIB see SNK000

TRIBAC see TIK500

TRI-BAN see PIH175

TRIBASENALDEHYDE see FBV050

TRIBASIC ALUMINUM STEARATE see AHH825

TRIBASIC COPPER SULFATE see CNM600

TRIBASIC SODIUM PHOSPHATE see SJH200

TRIBASIC ZINC PHOSPHATE see ZJS400

TRIBENOSIDE see EPW600

TRIBENURON METHYL see THT800

TRIBENZOIN see GGU000

TRIBENZO(a,e,i)PYRENE see DDC200

(1,2,4,5,7,8)TRIBENZOPYRENE see DDC200

(1,2,4,5,8,9)TRIBENZOPYRENE see DDC200

1,2:4,5:8,9-TRIBENZOPYRENE see DDC200

TRIBENZOSIDE see EPW600

TRIBENZYLARSINE see THU000

TRIBENZYLCHLOROSTANNANE see CLP000

TRIBENZYLSULFONIUM IODIDE MERCURIC IODIDE see THU250

TRIBENZYLSULFONIUM IODIDE, compounded with MERCURY IODIDE (1:1) see THU250

TRIBENZYLTIN CHLORIDE see CLP000

TRIBENZYLTIN FORMATE see FOE000

TRIBONATE see DVC200

TRIBORON PENTAFLUORIDE see THU275

TRIBROMAMINE HEXAAMMONIATE see NGV500

TRIBROMETHANOL see ARW250

TRIBROMETHANOL see THV000

TRIBROMINATED POLYSTYRENE see EEE650

TRIBROMMETHAAN (DUTCH) see BNL000

TRIBROMMETHAN (GERMAN) see BNL000

TRIBROMOALDEHYDE HYDRATE see THU500

TRIBROMOALUMINUM see AGX750

2,4,6-TRIBROMOANILINE see THU750

sym-TRIBROMOANILINE see THU750

TRIBROMOARSINE see ARF250

TRIBROMOETHANOL see THV000

2,2,2-TRIBROMOETHANOL see ARW250

TRIBROMOETHENE see THV100

TRIBROMOETHYL ALCOHOL see ARW250

TRIBROMOETHYL ALCOHOL see THV000

2,2,2-TRIBROMOETHYL ALCOHOL see ARW250

TRIBROMOETHYLENE see THV100

1,1,2-TRIBROMOETHYLENE see THV100

1,3,7-TRIBROMO-2-FLUORENAMINE see THV250

1,3,7-TRIBROMOFLUOREN-2-AMINE see THV250

2,4,5-TRIBROMOIMIDAZOLE see THV450

2,4,5-TRIBROMOIMIDAZOLE CADMIUM SALT (2:1) see THV500

TRIBROMOMETAN (ITALIAN) see BNL000

TRIBROMOMETHANE see BNL000

TRIBROMONITROMETHANE see NMQ000

TRIBROMOPHENOL see THV750

2,4,6-TRIBROMOPHENOL see THV750

1-(2,4,6-TRIBROMOPHENYL)-3,3-DIMETHYLTRIAZENE see DUI800

TRIBROMOPHOSPHINE see PHT250

1,2,3-TRIBROMOPROPANE see GGG000

sym-TRIBROMOPROPANE see GGG000

TRIBROMOSALICYLANILIDE see THV800

3,4',5-TRIBROMOSALICYLANILIDE see THW750

TRIBROMOSILANE see THX000

TRIBROMOSTIBINE see AQK000

TRIBROMOTRIMETHYLDIALUMINUM see MGC225

TRIBROMSALAN see THW750

TRIBUFON see BPG000

TRIBURON see TJE880

TRIBURON CHLORIDE see TJE880

TRIBUTILFOSFATO (ITALIAN) see TIA250

TRIBUTON see DAA800

TRIBUTON see TAA100

TRIBUTOXYBORANE see THX750

TRI-n-BUTOXYBORANE see THX750

TRI(2-BUTOXYETHANOL PHOSPHATE) see BPK250

TRIBUTOXYETHYL PHOSPHATE see BPK250

TRI(2-BUTOXYETHYL) PHOSPHATE see BPK250

TRIBUTYL ACETYLCITRATE see THX100

TRIBUTYL O-ACETYLCITRATE see THX100

TRIBUTYL 2-(ACETYLOXY)-1,2,3-PROPANETRICARBOXYLATE see THX100

TRIBUTYLAMINE see THX250

TRI-n-BUTYLAMINE see THX250

TRI-n-BUTYL BORANE see THX500

TRIBUTYL BORATE see THX750

TRI-n-BUTYL BORATE see THX750

TRI-sec-BUTYL BORATE see THY000

TRIBUTYLBORINE see THX500

N,N,N-TRIBUTYL-1-BUTANAMINIUM HYDROXIDE see TBK750

N-N-N-TRIBUTYL-1-BUTANAMINIUM NITRATE see TBL250

TRIBUTYL CELLOSOLVE PHOSPHATE see BPK250

TRIBUTYLCHLOROGERMANE see CLP250

TRIBUTYLCHLOROSTANNANE see CLP500

TRIBUTYL CITRATE see THY100

TRI-n-BUTYL CITRATE see THY100

TRIBUTYL CITRATE ACETATE see THX100

N,N,N-TRIBUTYL-2,4-DICHLOROBENZENEMETHANAMINIUM CHLORIDE see THY200

TRIBUTYL(2,4-DICHLOROBENZYL)AMMONIUM CHLORIDE see THY200

TRIBUTYL(2,4-DICHLOROBENZYL)PHOSPHONIUM CHLORIDE see THY500

TRI-n-BUTYL-2,4-DICHLOROPHENOXYTIN see DGB400

TRIBUTYLE (PHOSPHATE de) (FRENCH) see TIA250

TRIBUTYL((2-ETHYLHEXANOYL)OXY)STANNANE see TID250

TRIBUTYL((2-ETHYL-1-OXOHEXYL)OXY)STANNANE see TID250

TRIBUTYLFOSFAAT (DUTCH) see TIA250

TRIBUTYLFOSFIN see TIA300

TRIBUTYLFOSFINSULFID see TIA450

TRIBUTYL(GLYCOLOYLOXY)STANNANE see GHQ000

TRIBUTYL(GLYCOLOYLOXY)TIN see GHQ000

TRIBUTYL 2-HYDROXY-1,2,3-PROPANETRICARBOXYLATE see THY100

TRIBUTYLHYDROXYSTANNANE see TID500

TRIBUTYLIMIDAZOLE LEAD see IAT400

TRIBUTYLISOCYANATOSTANNANE see THY750

TRIBUTYLLEAD ACETATE see THY850

TRIBUTYLLEAD IMIDAZOLE see IAT400

TRIBUTYL(METHACRYLOXY)STANNANE see THZ000

TRIBUTYL(METHACRYLOYLOXY)STANNANE see THZ000

TRIBUTYL(METHACRYLOYLOXY)-STANNANE POLYMER with METHYL METHACRYLATE (8CI) see OIY000

TRIBUTYL((2-METHYL-1-OXO-2-PROPENYL)OXY)STANNANE see THZ000

TRIBUTYL(NEODECANOYLOXY)STANNANE see TIF250

TRIBUTYL(OLEOYLOXY)STANNANE see TIA000

TRIBUTYL(OLEOYLOXY)TIN see TIA000

TRIBUTYL((1-OXODODECYL)OXY)STANNANE (9CI) see TIE750

(TRIBUTYL)PEROXIDE see BSC750

2,4,6-TRI-tert-BUTYLPHENOL see TIA100

TRIBUTYL-o-PHENYLPHENOXYTIN see BGK000

TRI(tert-BUTYLPHENYL) PHOSPHATE see TIA130

TRIBUTYLPHOSPHAT (GERMAN) see TIA250

TRIBUTYL PHOSPHATE see TIA250

TRI-n-BUTYL PHOSPHATE see TIA250

TRIBUTYLPHOSPHINE see TIA300

TRI-n-BUTYLPHOSPHINE see TIA300
TRIBUTYL-PHOSPHINE compounded with NICKELCHLORIDE (2:1) see BLS250
TRIBUTYLPHOSPHINE SULFIDE see TIA450
TRIBUTYL PHOSPHITE see TIA750
S,S,S-TRIBUTYL PHOSPHOROTRITHIOATE see BSH250
TRIBUTYL PHOSPHOROTRITHIOITE see TIG250
S,S,S-TRIBUTYL PHOSPHOROTRITHIOITE see TIG250
TRI-n-BUTYLPLUMBYL ACETATE see THY850
N-(TRIBUTYLPLUMBYL)BENZIMIDAZOLE see TIA800
N-(TRIBUTYLPLUMBYL)IMIDAZOLE see IAT400
1-(TRI-N-BUTYLPLUMBYL)-IMIDAZOLE see IAT400
TRIBUTYL(8-QUINOLINOLATO)TIN see TIB000
TRIBUTYLSTANNANECARBONITRILE see TIB250
TRIBUTYLSTANNANE FLUORIDE see FME000
TRI-n-BUTYLSTANNANE HYDRIDE see TIB500
TRI-n-BUTYL-STANNANE OXIDE see BLL750
TRIBUTYLSTANNIC HYDRIDE see TIB500
TRIBUTYLSTANNYL ISOCYANATE see THY750
TRIBUTYLSTANNYL METHACRYLATE see THZ000
TRIBUTYLTIN-p-ACETAMIDOBENZOATE see TIB750
TRIBUTYLTIN ACETATE see TIC000
TRIBUTYLTIN BENZOATE see BDR750
TRI-n-BUTYLTIN BROMIDE see TIC250
TRI-n-BUTYLTIN CHLORIDE see CLP500
TRIBUTYLTIN CHLORIDE COMPLEX see TIC500
TRIBUTYLTIN CHLOROACETATE see TIC750
TRIBUTYLTIN-γ-CHLOROBUTYRATE see TID000
TRIBUTYLTIN CYANATE see COI000
TRI-n-BUTYLTIN CYANIDE see TIB250
TRIBUTYLTIN CYCLOHEXANECARBOXYLATE see TID100
TRIBUTYLTIN-S,S'-DIBUTYLDITHIOCARBAMATE see DEB600
TRIBUTYLTIN DIMETHYLDITHIOCARBAMATE see TID150
TRIBUTYLTIN DODECANOATE see TIE750
TRIBUTYLTIN-2-ETHYLHEXANOATE see TID250
TRIBUTYLTIN FLUORIDE see FME000
TRIBUTYLTIN HYDRIDE see TIB500
TRI-n-BUTYLTIN HYDRIDE see TIB500
TRIBUTYLTIN HYDROXIDE see TID500
TRI-N-BUTYL TIN IODIDE see IFM000
TRIBUTYLTIN IODOACETATE see TID750
TRIBUTYLTIN-o-IODOBENZOATE see TIE000
TRIBUTYLTIN-p-IODOBENZOATE see TIE250
TRIBUTYLTIN-β-IODOPROPIONATE see TIE500
TRIBUTYLTIN ISOCYANATE see THY750
TRI-n-BUTYLTIN ISOCYANATE see THY750
TRIBUTYLTIN ISOOCTYLTHIOACETATE see TDO000
TRIBUTYLTIN ISOPROPYLSUCCINATE see TIE600
TRIBUTYLTIN ISOTHIOCYANATE see THY750
TRIBUTYLTIN LAURATE see TIE750
TRIBUTYL TIN LINOLEATE see LGJ000

TRIBUTYLTIN METHACRYLATE see THZ000
TRI-n-BUTYLTIN METHANESULFONATE see TIF000
TRIBUTYLTIN MONOLAURATE see TIE750
TRIBUTYLTIN NEODECANOATE see TIF250
TRIBUTYLTIN NONANOATE see TIF500
TRI-n-BUTYLTIN OLEATE see TIA000
TRIBUTYLTIN OXIDE see BLL750
TRIBUTYLTIN-o-PHENYLPHENOXIDE see BGK000
TRIBUTYLTIN SALICYLATE see SAM000
TRI-N-BUTYLTIN SALICYLATE see SAM000
TRIBUTYLTIN SULFATE see TIF600
TRIBUTYLTIN SULFIDE see HCA700
TRIBUTYLTIN-α-(2,4,5-TRICHLOROPHENOXY)PROPIONATE see TIF750
TRI-n-BUTYLTIN UNDECYLATE see TIG500
TRIBUTYLTIN UNDECYLENATE see TIG500
TRIBUTYL(2,4,5-TRICHLOROPHENOXY)STANNANE see TIG000
TRIBUTYL(2,4,5-TRICHLOROPHENOXY)TIN see TIG000
S,S,S-TRIBUTYL TRITHIOPHOSPHATE see BSH250
S,S,S-TRIBUTYL TRITHIOPHOSPHITE see TIG250
TRIBUTYL(UNDECANOYLOXY)STANNANE see TIG500
TRI-n-BUTYL-ZINN-ACETAT (GERMAN) see TIC000
TRI-N-BUTYL-ZINN BENZOATE (GERMAN) see BDR750
TRI-n-BUTYLZINN-CHLORID (GERMAN) see CLP500
TRI-n-BUTYLZINN-LAURAT (GERMAN) see TIE750
TRI-n-BUTYL-ZINN OLEAT (GERMAN) see TIA000
TRI-N-BUTYL-ZINN SALICYLAT (GERMAN) see SAM000
TRI-n-BUTYL-ZINN UNDECYLAT (GERMAN) see TIG500
TRIBUTYRASE see GGA800
TRIBUTYRIN see TIG750
TRIBUTYRINASE see GGA800
TRIBUTYRIN ESTERASE see GGA800
TRIBUTYROIN see TIG750
TRICADMIUM DINITRIDE see TIH000
TRICAINE see EFX500
TRICAINE METHANE SULFONATE see EFX500
TRICALCIUMARSENAT (GERMAN) see ARB750
TRICALCIUM ARSENATE see ARB750
TRICALCIUM DINITRIDE see TIH250
TRICALCIUM PHOSPHATE see CAW120
TRICALCIUM SILICATE see TIH600
TRICAMBA see TIK000
TRICAPROIN see GGK000
TRICAPRONIN see GGK000
TRICAPROYLGLYCEROL see GGK000
TRICAPRYLIC GLYCERIDE see TMO000
TRICAPRYLIN see TMO000
TRICAPRYLMETHYLAMMONIUM CHLORIDE see MQH000
TRICAPRYLYLAMINE see DVL000
TRICAPRYLYLMETHYLAMMONIUM CHLORIDE see MQH000
TRICARBALLYLIC ACID, β-ACETOXYTRIBUTYL ESTER see ADD750
TRICARBAMIX Z see BJK500
TRICARBONYL(METHYLCYCLOPENTADIENYL)MANGANESE see MAV750
1,2,4-TRICARBOXYBENZENE see TKU700
TRICARNAM see CBM750
TRICESIUM NITRIDE see TIH750

TRICESIUM TRICHLORIDE see CDD000
TRICESIUM TRIFLUORIDE see CDD500
TRICESIUM TRIIODIDE see CDE000
TRICETAMIDE see TIH800
TRICETO 3-7-12 CHOLANATE de Na (FRENCH) see SGD500
TRICHAZOL see MMN250
TRICHLAMIDE see BPT300
TRICHLOORAZIJNZUUR (DUTCH) see TII250
1,1,1-TRICHLOOR-2,2-BIS(4-CHLOOR FENYL)-ETHAAN (DUTCH) see DAD200
2,2,2-TRICHLOOR-1,1-BIS(4-CHLOOR FENYL)-ETHANOL (DUTCH) see BIO750
1,1,1-TRICHLOORETHAAN (DUTCH) see MIH275
TRICHLOORETHEEN (DUTCH) see TIO750
TRICHLOORETHYLEEN (DUTCH) see TIO750
(2,4,5-TRICHLOOR-FENOXY)-AZIJNZUUR (DUTCH) see TAA100
2-(2,4,5-TRICHLOOR-FENOXY)-PROPIONZUUR (DUTCH) see TIX500
O-(2,4,5-TRICHLOOR-FENYL)-O,O-DIMETHYL-MONOTHIOFOSFAAT (DUTCH) see RMA500
TRICHLOORFON (DUTCH) see TIQ250
TRICHLOORMETHAAN (DUTCH) see CHJ500
TRICHLOORMETHYLBENZEEN (DUTCH) see BFL250
TRICHLOORNITROMETHAAN (DUTCH) see CKN500
TRICHLOORSILAAN (DUTCH) see TJD500
TRICHLORACETALDEHYD-HYDRAT (GERMAN) see CDO000
TRICHLOR-ACETONITRIL (GERMAN) see TII750
TRICHLORACRYLYL CHLORIDE see TIJ500
TRICHLORAD see ABY900
1,1,1-TRICHLORAETHAN (GERMAN) see MIH275
TRICHLORAETHEN (GERMAN) see TIO750
TRICHLORAETHYLEN (GERMAN) see TIO750
2,3,3-TRICHLORALLYL-N,N-(DIISOPROPYL)-THIOCARBAMAT (GERMAN) see DNS600
TRICHLORAMINE see NGQ500
TRICHLORAN see TIO750
2,3,6-TRICHLORBENZOESAEURE (GERMAN) see TIK500
1,1,1-TRICHLOR-2,2-BIS(4-CHLOR-PHENYL)-AETHAN (GERMAN) see DAD200
1,1,1-TRICHLOR-2,2-BIS(4-CHLORPHENYL)-AETHANOL (GERMAN) see BIO750
2,2,2-TRICHLOR-1,1-BIS(4-CHLOR-PHENYL)-AETHANOL (GERMAN) see BIO750
1,1,1-TRICHLOR-2,2-BIS(4-METHOXY-PHENYL)-AETHAN (GERMAN) see MEI450
TRICHLORESSIGSAEURE (GERMAN) see TII250
TRICHLORESSIGSAURES NATRIUM (GERMAN) see TII500
1,1,2-TRICHLORETHANE see TIN000
TRICHLORETHANOL see TIN500
TRICHLORETHENE (FRENCH) see TIO750
TRICHLORETHYLENE (FRENCH) see TIO750
TRICHLORETHYL PHOSPHATE see CGO500
2,4,6-TRICHLORFENOL (CZECH) see TIW000
2,3,5-TRICHLORFENOXYOCTAN 1-FENYL-2,2-DIMETHYL-4,6-DIAMINO-1,2-DIHYDRO-1,3,5-TRIAZINU (CZECH) see DMF400
TRICHLORFENSON (OBS.) see CJT750
2,4,6-TRICHLORFENYLESTER KYSELINY OCTOVE (CZECH) see TIY250
TRICHLORFON (USDA, ACGIH) see TIQ250

TRICHLORINATED ISOCYANURIC ACID
see TIQ750
TRICHLORINE NITRIDE see NGQ500
TRICHLOR-3-KYANPROPYLSILAN see
COR750
TRICHLORMETAFOS-3 see TIR250
TRICHLORMETAZID see HII500
TRICHLORMETHAN (CZECH) see CHJ500
TRICHLORMETHIAZIDE see HII500
TRICHLORMETHINE see TNF250
TRICHLORMETHINE see TNF500
TRICHLORMETHINIUM CHLORIDE see
TNF500
TRICHLORMETHYLBENZOL (GERMAN)
see BFL250
TRICHLORMETHYLESTER KYSELINY
CHLORMRAVENCI see TIR920
TRICHLORMETHYLESTER KYSELINY
DICHLORMETHANTHIOSULFONOVE
(CZECH) see DFS600
TRICHLOR-METHYLSILAN see MQC500
N-(TRICHLOR-METHYLTHIO)-
PHTHALAMID (GERMAN) see TIT250
N-(TRICHLOR-METHYLTHIO)-
PHTHALIMID (GERMAN) see CBG000
TRICHLORNITROMETHAN (GERMAN) see
CKN500
TRICHLOROACETALDEHYDE see CDN550
2,2,2-TRICHLOROACETALDEHYDE see
CDN550
TRICHLOROACETALDEHYDE HYDRATE
see CDO000
TRICHLOROACETALDEHYDE
MONOETHYLACETAL see TIO000
TRICHLOROACETALDEHYDE
MONOHYDRATE see CDO000
TRICHLOROACETALDEHYDE OXIME see
TIH825
TRICHLOROACETAMIDE see TII000
2,2,2-TRICHLOROACETAMIDE see TII000
α-α-α-TRICHLOROACETAMIDE see TII000
TRICHLOROACETIC ACID see TII250
TRICHLOROACETIC ACID CHLORIDE see
TIJ150
TRICHLOROACETIC ACID compounded with
N'-(4-CHLOROPHENYL)-N,N-
DIMETHYLUREA (1:1) see CJY000
TRICHLOROACETIC ACID SODIUM SALT
see TII500
TRICHLOROACETIC ACID-2-(2,4,5-
TRICHLOROPHENOXY)ETHYL ESTER see
TIX250
TRICHLOROACETIC ACID
TRIPROPYLSTANNYL ESTER see TNC000
TRICHLOROACETIC ACID (UN 1839)
(DOT) see TII250
TRICHLOROACETIC ACID, solution (UN
2564) (DOT) see TII250
TRICHLOROACETOCHLORIDE see TIJ150
1,1,1-TRICHLOROACETONE see TJB775
1,1,3-TRICHLOROACETONE see TII550
α-α-α-TRICHLOROACETONE see TJB775
α,α',α'-TRICHLOROACETONE see TII550
TRICHLOROACETONITRILE see TII750
(TRICHLOROACETOXY)TRIPROPYLSTAN
NANE see TNC000
10-TRICHLOROACETYL-1,2-
BENZANTHRACENE see TIJ000
TRICHLOROACETYL CHLORIDE see TIJ150
TRICHLOROACETYL FLUORIDE see TIJ175
2,3,3-TRICHLOROACROLEIN see TIJ250
TRICHLOROACRYLONITRILE see TJC100
α-β,β-TRICHLOROACRYLONITRILE see
TJC100
TRICHLOROACRYLOYL CHLORIDE see
TIJ500
2,3,3-TRICHLOROACRYLOYL CHLORIDE
see TIJ500
TRICHLOROACRYLYL CHLORIDE see
TIJ500

2,3,3-TRICHLOROALLYL
DIISOPROPYLTHIOCARBAMATE see
DNS600
S-2,3,3-TRICHLOROALLYL-N,N-
DIISOPROPYLTHIOCARBAMATE see
DNS600
TRICHLOROALLYLSILANE see AGU250
TRICHLOROALUMINUM see AGY750
3,5,6-TRICHLORO-4-AMINOPICOLINIC
ACID see PIB900
2,4,6-TRICHLOROANILINE see TIJ750
sym-TRICHLOROANILINE see TIJ750
3,5,6-TRICHLORO-o-ANISIC ACID see
TIK000
TRICHLOROARSINE see ARF500
2,4,6-TRICHLOROBENZENAMINE see
TIJ750
s-TRICHLOROBENZENE see TIK300
1,2,3-TRICHLOROBENZENE see TIK100
1,2,4-TRICHLOROBENZENE see TIK250
1,2,5-TRICHLOROBENZENE see TIK250
1,2,6-TRICHLOROBENZENE see TIK100
1,3,4-TRICHLOROBENZENE see TIK250
1,3,5-TRICHLOROBENZENE see TIK300
sym-TRICHLOROBENZENE see TIK300
vic-TRICHLOROBENZENE see TIK100
unsym-TRICHLOROBENZENE see TIK250
2,3,6-TRICHLOROBENZENEACETIC ACID
see TIY500
1,2,4-TRICHLOROBENZENE
(ACGIH,OSHA) see TIK250
2,4,5-TRICHLOROBENZENEDIAZO p-
CHLOROPHENYL SULFIDE see CDS500
3,4,6-TRICHLORO-1,2-BENZENEDIOL see
TIL600
TRICHLOROBENZOIC ACID see TIK500
2,3,6-TRICHLOROBENZOIC ACID see
TIK500
2,3,6-TRICHLOROBENZOIC ACID,
DIMETHYLAMINE SALT see DOR800
1,2,4-TRICHLOROBENZOL see TIK250
N,2,6-TRICHLORO-p-BENZOQUINONE
IMINE see CHR000
TRICHLOROBENZYL CHLORIDE see
TIL250
((2,3,6-TRICHLOROBENZYL)OXY)-2-
PROPANOL see TIL255
TRICHLOROBIPHENYL see TIM100
TRICHLORO-1,1'-BIPHENYL see TIM100
2,3,5-TRICHLOROBIPHENYL see TIL262
2',3,4-TRICHLOROBIPHENYL see TIL260
2,2',5-TRICHLOROBIPHENYL see TIL258
2,4',5-TRICHLOROBIPHENYL see TIL275
2,4,4'-TRICHLOROBIPHENYL see TIL270
2,5,2'-TRICHLOROBIPHENYL see TIL258
2,5,3'-TRICHLOROBIPHENYL see TIL262
3,4,2'-TRICHLOROBIPHENYL see TIL260
2',4,4'-TRICHLOROBIPHENYL see TIL270
2,2',5'-TRICHLOROBIPHENYL see TIL258
2,3',4'-TRICHLOROBIPHENYL see TIL260
3,2',5'-TRICHLOROBIPHENYL see TIL262
4,2',4'-TRICHLOROBIPHENYL see TIL270
4,2',5'-TRICHLOROBIPHENYL see TIL275
2,4',5-TRICHLORO-1,1'-BIPHENYL see
TIL275
1,1,1-TRICHLORO-2,2-BIS(p-
ANISYL)ETHANE see MEI450
TRICHLOROBIS (4-CHLOROPHENYL)
ETHANE see DAD200
2,2,2-TRICHLORO-1,1-BIS(4-
CHLOROPHENYL)-ETHANOL (FRENCH)
see BIO750
2,2,2-TRICHLORO-1,1-BIS(4-CLORO-
FENIL)-ETANOLO (ITALIAN) see BIO750
1,1,1-TRICHLORO-2,2-BIS(p-
FLUOROPHENYL) ETHANE see FHJ000
1,1,1-TRICHLORO-2,2-BIS(p-
HYDROXYPHENYL)ETHANE see BKI500
1,1,1-TRICHLORO-2,2-BIS(p-
METHOXYPHENOL)ETHANOL see MEI450
1,1,1-TRICHLORO-2,2-BIS(p-
METHOXYPHENYL)ETHANE see MEI450

1,1,2-TRICHLOROBUTADIENE see TIL350
1,1,2-TRICHLORO-1,3-BUTADIENE see
TIL350
2,3,4-TRICHLOROBUTENE-1 see TIL360
TRICHLORO-tert-BUTYL ALCOHOL see
ABD000
tert-TRICHLOROBUTYL ALCOHOL see
ABD000
β,β,β-TRICHLORO-tert-BUTYL ALCOHOL
see ABD000
TRICHLOROBUTYLENE OXIDE see
ECT500
TRICHLOROBUTYLSILANE see BSR000
3,4,4'-TRICHLOROCARBANILIDE see TIL500
3,4,4'-TRICHLOROCARBANILIDE mixed with
3-TRIFLUOROMETHYL-4,4'-
DICHLOROCARBANILIDE (2:1) see TIL526
3,4,6-TRICHLOROCATECHOL see TIL600
2,4,4-TRICHLORO-3-
CHLOROMETHYLBUTENOIC ACID see
TIL610
2,4,5-TRICHLORO-α-
(CHLOROMETHYLENE)BENZYL
PHOSPHATE see TBW100
2,4,5-TRICHLORO-α-
(CHLOROMETHYLENE)BENZYL
PHOSPHATE ESTER see RAF100
TRICHLORO(CHLOROMETHYL)SILANE
(9CI) see CIY325
1,1,1-TRICHLORO-2-(o-CHLOROPHENYL)-
2-(p-CHLOROPHENYL)ETHANE see BIO625
1,2,4-TRICHLORO-5-((4-
CHLOROPHENYL)SULFONYL)-BENZENE
see CKM000
1,2,4-TRICHLORO-5-((4-
CHLOROPHENYL)THIO)-BENZENE see
CKL750
TRICHLOROCHROMIUM see CMJ250
TRICHLOROCTAN SODNY (CZECH) see
TII500
TRICHLOROCYANIDINE see TJD750
TRICHLOROCYANURIC ACID see TIQ750
TRICHLORO-3-CYCLOHEXENYLSILANE
see CPE500
1,2,4-TRICHLORODIBENZODIOXIN see
TIL620
1,2,4-TRICHLORO DIBENZO-p-DIOXIN see
CDV125
1,2,4-TRICHLORODIBENZO-p-DIOXIN see
TIL620
2,3,7-TRICHLORODIBENZO-p-DIOXIN see
TIL630
N-(2,2,2-TRICHLORO-1-(3,4-
DICHLOROANILINO))ETHYLFORMAMID
E see CDP750
2,3,5-TRICHLORO-N-(3,5-DICHLORO-2-
HYDROXYPHENYL)-6-
HYDROXYBENZAMIDE see DMZ000
1,2,4-TRICHLORO-5-(2,6-
DICHLOROPHENOXY)BENZENE see
TIL700
4,5,6-TRICHLORO-2-(2,4-
DICHLOROPHENOXY)PHENOL see TIL750
1,1,1-TRICHLORO-2,2-DI(4-
CHLOROPHENYL)-ETHANE see DAD200
2,2,2-TRICHLORO-1,1-DI-(4-
CHLOROPHENYL)ETHANOL see BIO750
2,2,2-TRICHLORO-1-(3,4-
DICHLOROPHENYL)ETHANOL ACETATE
see TIL800
TRICHLORO(DICHLOROPHENYL)SILANE
see DGF200
1,1,2-TRICHLORO-2,2-DIFLUOROETHANE
see TIM000
1,1,1-TRICHLORO-2,2-DI(4-
METHOXYPHENYL)ETHANE see MEI450
1,2,3-TRICHLORO-4,6-DINITROBENZENE
see DVI600
TRICHLORODIPHENYL see TIM100
TRICHLORO DIPHENYL ETHER see
CDV175

TRICHLORO DIPHENYL OXIDE see CDV175

3,4,4'-TRICHLORODIPHENYLUREA see TIL500

TRICHLORODODECYLSILANE see DYA800

4,4,4-TRICHLORO-1,2-EPOXYBUTANE see ECT500

1,1,2-TRICHLOROEPOXYETHANE see ECT600

1,1,1-TRICHLORO-2-3-EPOXYPROPANE see TJC250

TRICHLORO ESTERTIN see TIM500

TRICHLOROETHANAL see CDN550

1,1,1-TRICHLOROETHANE see MIH275

1,1,2-TRICHLOROETHANE see TIN000

1,2,2-TRICHLOROETHANE see TIN000

α-TRICHLOROETHANE see MIH275

β-TRICHLOROETHANE see TIN000

TRICHLORO-1,1,1-ETHANE (FRENCH) see MIH275

2,2,2-TRICHLORO-1,1-ETHANEDIOL see CDO000

TRICHLOROETHANOIC ACID see TII250

TRICHLOROETHANOL see TIN500

2,2,2-TRICHLOROETHANOL see TIN500

2,2,2-TRICHLOROETHANOL CARBAMATE see TIO500

TRICHLOROETHENE see TIO750

TRICHLOROETHENYLSILANE see TIN750

2,2,2-TRICHLORO-1-ETHOXYETHANOL see TIO000

TRICHLOROETHYL ALCOHOL see TIN500

2,2,2-TRICHLOROETHYL ALCOHOL see TIN500

TRI-(2-CHLOROETHYL)AMINE see TNF250

TRI-(2-CHLOROETHYL)AMINE HYDROCHLORIDE see TNF500

TRI(β-CHLOROETHYL)AMINE HYDROCHLORIDE see TNF500

TRICHLOROETHYL CARBAMATE see TIO500

TRICHLOROETHYLENE see TIO750

1,2,2-TRICHLOROETHYLENE see TIO750

TRICHLOROETHYLENE EPOXIDE see ECT600

TRICHLOROETHYLENE OXIDE see ECT600

1,1'-(2,2,2-TRICHLOROETHYLIDENE)BIS(4-CHLOROBENZENE) see DAD200

1,1'-(2,2,2-TRICHLOROETHYLIDENE)BIS(4-METHOXYBENZENE) see MEI450

1,2-o-(2,2,2-TRICHLOROETHYLIDENE)-α-d-GLUCOFURANOSE see GFA000

α-TRICHLOROETHYLIDENE GLYCEROL see TIP000

β-TRICHLOROETHYLIDENE GLYCEROL see TIP250

TRICHLOROETHYL PHOSPHATE see TIP300

TRI(2-CHLOROETHYL)PHOSPHATE see CGO500

2,2,2-TRICHLOROETHYL PHOSPHATE see TIP300

TRI-β-CHLOROETHYL PHOSPHATE see CGO500

TRICHLOROETHYLSILANE see EPY500

TRICHLOROETHYLSILICANE see EPY500

TRICHLOROETHYLSTANNANE see EPS000

TRICHLOROETHYLTIN see EPS000

2,3,5-TRICHLOROFENOLAT ZINECNATY (CZECH) see BLN000

TRICHLOROFLUOROETHYLENE see TIP400

TRICHLOROFLUOROMETHANE see TIP500

TRICHLOROFORM see CHJ500

TRICHLOROGUAIACOL see TIP600

4,5,6-TRICHLOROGUAIACOL see TIP630

TRICHLOROHEXADECYLSILANE see HCQ000

TRICHLOROHYDRIN see TJB600

4',4'',5-TRICHLORO-2-HYDROXY-3-BIPHENYLCARBOXANILIDE see TIP750

2,4,4'-TRICHLORO-2'-HYDROXYDIPHENYL ETHER see TIQ000

(2,2,2-TRICHLORO-1-HYDROXYETHYL) DIMETHYLPHOSPHONATE see TIQ250

2,2,2-TRICHLORO-1-HYDROXYETHYL-PHOSPHONATE, DIMETHYL ESTER see TIQ250

(2,2,2-TRICHLORO-1-HYDROXYETHYL)PHOSPHONIC ACID DIMETHYL ESTER see TIQ250

(2,2,2-TRICHLORO-1-HYDROXYETHYL)PHOSPHORIC ACID, DIPHENYL ESTER see TIQ300

3,5,6-TRICHLORO-2-HYDROXYPYRIDINE see HOL125

TRICHLOROISOCYANIC ACID see TIQ750

TRICHLOROISOCYANURIC ACID see TIQ750

1,3,5-TRICHLOROISOCYANURIC ACID see TIQ750

N,N',N''-TRICHLOROISOCYANURIC ACID see TIQ750

TRICHLOROISOCYANURIC ACID-POTASSIUM DICHLOROISOCYANURATE (1:4) see MRN050

TRICHLOROISOPROPANOL see IMQ000

1,1,1-TRICHLOROISOPROPYL ALCOHOL see IMQ000

TRICHLOROMELAMINE see TNE775

TRICHLOROMETAFOS see RMA500

TRICHLOROMETAPHOS-3 see TIR250

TRICHLOROMETHANE see CHJ500

TRICHLOROMETHANE SULFENYL CHLORIDE see PCF300

TRICHLORO-3-METHAPHOS see TIR250

TRICHLOROMETHIADIAZIDE see HII500

TRICHLOROMETHIAZIDE see HII500

2,3,5-TRICHLORO-6-METHOXYBENZOIC ACID see TIK000

3,5,6-TRICHLORO-2-METHOXYBENZOIC ACID see TIK000

TRICHLOROMETHYL ALLYL PERTHIOXANTHATE see TIR750

TRICHLOROMETHYLBENZENE see BFL250

1-(TRICHLOROMETHYL)BENZENE see BFL250

α-(TRICHLOROMETHYL)BENZENEMETHANOL see TIR800

α-(TRICHLOROMETHYL)BENZENEMETHANOL, ACETATE (9CI) see TIT000

α-(TRICHLOROMETHYL)BENZYL ALCOHOL see TIR800

α-(TRICHLOROMETHYL)BENZYL ALCOHOL, ACETATE (9CI) see TIT000

p-TRICHLOROMETHYLCHLOROBENZENE see TIR900

TRICHLOROMETHYL CHLOROFORMATE see TIR920

TRICHLOROMETHYL CYANIDE see TII750

α-2-(TRICHLOROMETHYL)-1,3-DIOXOLANE-4-METHANOL see TIP000

β-2-(TRICHLOROMETHYL)-1,3-DIOXOLANE-4-METHANOL see TIP250

3-(TRICHLOROMETHYL)-5-ETHOXY-1,2,4-THIADIAZOLE see EFK000

N-TRICHLOROMETHYLMERCAPTO-4-CYCLOHEXENE-1,2-DICARBOXIMIDE see CBG000

1-(TRICHLOROMETHYLMERCAPTO)-4-METHYLPYRAZOLE see TIR990

N-(TRICHLOROMETHYLMERCAPTO)PHTHALIMIDE see TIT250

1-(TRICHLOROMETHYLMERCAPTO)PYRAZOLE see TIS100

N-(TRICHLOROMETHYLMERCAPTO)-Δ⁴-TETRAHYDROPHTHALIMIDE see CBG000

TRICHLOROMETHYL METHYL PERTHIOXANTHATE see TIS500

TRICHLOROMETHYLNITRILE see TII750

(TRICHLOROMETHYL)OXIRANE see TJC250

TRICHLOROMETHYL PERCHLORATE see TIS750

TRICHLOROMETHYLPHENYL CARBINOL see TIR800

TRICHLOROMETHYLPHENYLCARBINYL ACETATE see TIT000

1,1,1-TRICHLORO-2-METHYL-2-PROPANOL see ABD000

TRICHLOROMETHYLSTANNANE see MQC750

TRICHLOROMETHYLSULFENYL CHLORIDE see PCF300

TRICHLOROMETHYLSULPHENYL CHLORIDE see PCF300

3-TRICHLOROMETHYLTHIOBENZOTHIAZOLONE see TIT050

TRICHLOROMETHYLTHIO-3-BENZOTHIAZOLONE see TIT050

N-TRICHLOROMETHYLTHIOBENZOTHIAZOLONE see TIT050

3-((TRICHLOROMETHYL)THIO)-2-BENZOXAZOLINONE see TIT100

N-TRICHLOROMETHYLTHIOCYCLOHEX-4-ENE-1,2-DICARBOXIMIDE see CBG000

N-((TRICHLOROMETHYL)THIO)-4-CYCLOHEXENE-1,2-DICARBOXIMIDE see CBG000

N-TRICHLOROMETHYLTHIO-cis-Δ⁴-CYCLOHEXENE-1,2-DICARBOXIMIDE see CBG000

2-((TRICHLOROMETHYL)THIO)-1H-ISOINDOLE-1,3(2H)-DIONE see TIT250

N-(TRICHLOROMETHYLTHIO)PHTHALIMIDE see TIT250

TRICHLOROMETHYLTHIO-1,2,5,6-TETRAHYDROPHTHALAMIDE see CBG000

N-((TRICHLOROMETHYL)THIO)TETRAHYDROPHTHALIMIDE see CBG000

N-TRICHLOROMETHYLTHIO-3A,4,7,7A-TETRAHYDROPHTHALIMIDE see CBG000

TRICHLOROMETHYLTIN see MQC750

5-TRICHLOROMETHYL-1-TRIMETHYLSILYLTETRAZOLE see TIT275

TRICHLOROMOLYBDENUM see MRD800

TRICHLOROMONOFLUOROMETHANE see TIP500

TRICHLOROMONOSILANE see TJD500

TRICHLORONAPHTHALENE see TIT500

TRICHLORONAT see EPY000

2,3,4-TRICHLORONITROBENZENE see NMQ025

2,4,5-TRICHLORONITROBENZENE see TIT750

1,2,3-TRICHLORO-4-NITROBENZENE see NMQ025

1,2,4-TRICHLORO-5-NITROBENZENE see TIT750

2',4',6'-TRICHLORO-4-NITROBIPHENYL ETHER see NIW500

2,4,6-TRICHLORO-4'-NITRODIPHENYL ETHER see NIW500

TRICHLORONITROMETHANE see CKN500

3,4,6-TRICHLORO-2-NITROPHENOL see NMQ050

1,3,5-TRICHLORO-2-(4-NITROPHENOXY)BENZENE see NIW500

TRICHLOROOCTADECYLSILANE see OBI000

TRICHLORO-OXIRANE see ECT600

TRICHLOROOXOVANADIUM see VDP000

TRICHLOROPEROXYACETIC ACID see TIV275

TRICHLOROPHENE see HCL000

2,3,6-TRICHLOROPHENOL see TIV500
2,4,5-TRICHLOROPHENOL see TIV750
2,4,6-TRICHLOROPHENOL see TIW000
3,4,5-TRICHLOROPHENOL see TIW100
2,4,6-TRICHLOROPHENOL ACETATE see TIY250
2,4,5-TRICHLOROPHENOL, O-ESTER with O,O-DIMETHYL PHOSPHOROTHIOATE see RMA500
2,4,5-TRICHLOROPHENOL-O-ESTER with O-ETHYL ETHYLPHOSPHONOTHIOATE see EPY000
2,4,5-TRICHLOROPHENOL, SODIUM SALT see SKK500
2,4,5-TRICHLOROPHENOXYACETIC ACID see TAA100
(2,4,5-TRICHLOROPHENOXY)ACETIC ACID-2-BUTOXYETHYL ESTER see TAH900
2,4,5-TRICHLOROPHENOXYACETIC ACID, BUTYL ESTER see BSQ750
2,4,5-TRICHLOROPHENOXY, ACETIC ACID, ISOOCTYL ESTER see TGF200
(2,4,5,-TRICHLOROPHENOXY)ACETIC ACID, ISOPROPYL ESTER see TGF210
(2,4,5,-TRICHLOROPHENOXY)ACETIC ACID-1-METHYL ESTER (9CI) see TGF210
2,4,5-TRICHLOROPHENOXYACETIC ACID, PROPYLENE GLYCOL BUTYL ETHER ESTERS see THJ100
4-(2,4,5-TRICHLOROPHENOXY)BUTYRIC ACID see TIW750
2-(2,4,5-TRICHLOROPHENOXY)ETHANOL see TIX000
2-(2,4,5-TRICHLOROPHENOXY)ETHYL-2,2-DICHLOROPROPIONATE see PBK000
2,4,5-TRICHLOROPHENOXYETHYL-α,α-DICHLOROPROPIONATE see PBK000
2,4,5-TRICHLOROPHENOXYETHYL-α,α,α-TRICHLOROACETATE see TIX250
2-(2,4,5-TRICHLOROPHENOXY)PROPIONIC ACID see TIX500
α-(2,4,5-TRICHLOROPHENOXY)PROPIONIC ACID see TIX500
2,4,5-TRICHLOROPHENOXY-α-PROPIONIC ACID see TIX500
2-(2,4,5-TRICHLOROPHENOXY)PROPIONIC ACID PROPYLENE GLYCOL BUTYL ETHER ESTER see TIX750
2-(2,4,5-TRICHLOROPHENOXY)PROPIONIC ACID TRIBUTYLSTANNYL ESTER see TIF750
(2,4,5-TRICHLOROPHENOXY)SODIUM see SKK500
2,4,6-TRICHLOROPHENYL ACETATE see TIY250
2,3,6-TRICHLOROPHENYLACETIC ACID see TIY500
2,4,5-TRICHLOROPHENYLAZO-4'-CHLOROPHENYL-SULFIDE see CDS500
TRI-o-CHLOROPHENYL BORATE see TIY750
2,4,6-TRICHLOROPHENYL CHLOROFORMATE see TIY800
2,4,6-TRICHLORO-PHENYLDIMETHYLTRIAZENE see TJA000
1-(2,4,6-TRICHLOROPHENYL)-3,3-DIMETHYLTRIAZENE see TJA000
α-(2,4,6-TRICHLOROPHENYL)HYDRAZONO BENZOYL CHLORIDE see TJA100
2,4,5-TRICHLOROPHENYL IODOPROPARGYL ETHER see IFA000
2,4,5-TRICHLOROPHENYL-γ-IODOPROPARGYL ETHER see IFA000
TRICHLOROPHENYLMETHANE see BFL250
1-((2,3,6-TRICHLOROPHENYL)METHOXY)-2-PROPANOL see TJA200

1-(2,4,6-TRICHLOROPHENYL)-3-p-NITROANILINO-2-PYRAZOLIN-5-ONE see TJA500
2,4,6-TRICHLOROPHENYL-4-NITROPHENYL ETHER see NIW500
TRICHLOROPHENYLSILANE see TJA750
TRICHLOROPHON see TIQ250
TRICHLOROPHOSPHINE SULFIDE see TFO000
2,4,6-TRICHLORO-PMDT see TJA000
1,1,1-TRICHLOROPROPANE see TJB000
1,1,2-TRICHLOROPROPANE see TJB250
1,2,2-TRICHLOROPROPANE see TJB500
1,2,3-TRICHLOROPROPANE see TJB600
TRICHLOROPROPANE OXIDE see TJC250
1,2,3-TRICHLOROPROPANE-2,3-OXIDE see TJB750
1,1,1-TRICHLORO-2-PROPANOL see IMQ000
1,1,1-TRICHLOROPROPANONE see TJB775
1,1,3-TRICHLORO-2-PROPANONE see TII550
2,3,3-TRICHLOROPROPENAL see TIJ250
2,3,3-TRICHLORO-2-PROPENAL see TIJ250
1,1,3-TRICHLOROPROPENE see TJB800
1,2,3-TRICHLOROPROPENE see TJC000
3,3,3-TRICHLOROPROPENE see TJC050
3,3,3-TRICHLORO-1-PROPENE see TJC050
2,3,3-TRICHLORO-2-PROPENENITRILE see TJC100
TRICHLOROPROPENE OXIDE see TJC250
1,1,1-TRICHLOROPROPENE OXIDE see TJC250
3,3,3-TRICHLOROPROPENE OXIDE see TJC250
1,1,1-TRICHLOROPROPENE-2,3-OXIDE see TJC250
2,3,3-TRICHLORO-2-PROPEN-1-OL see TJC500
2,3,4-TRICHLORO-2-PROPENOYL CHLORIDE see TIJ500
2,2,3-TRICHLOROPROPIONALDEHYDE see TJC750
TRICHLOROPROPIONITRILE see TJC800
2,2,3-TRICHLOROPROPIONITRILE see TJC850
TRICHLOROPROPYLENE see TJB750
3,3,3-TRICHLOROPROPYLENE see TJC050
1,1,1-TRICHLOROPROPYLENE OXIDE see TJC250
3,3,3-TRICHLOROPROPYLENE OXIDE see TJC250
TRI(CHLOROPROPYL) PHOSPHATE see TJC870
TRICHLOROPROPYLSILANE see PNX250
2,3,6-TRICHLOROPYRIDINE see TJC900
3,5,6-TRICHLOROPYRIDINE-2-OL see HOL125
3,5,6-TRICHLORO-2-PYRIDINOL see HOL125
3,5,6-TRICHLORO-2(1H)-PYRIDINONE see HOL125
3,5,6-TRICHLORO-2-PYRIDYLOXYACETIC ACID see TJE890
2,3,6-TRICHLOROQUINOXALINE see QSJ050
TRICHLOROSILANE see TJD500
TRICHLOROSTIBINE see AQC500
2,2,2-TRICHLORO-1-(2-THIAZOLYLAMINO)ETHANOL see CMB675
TRICHLOROTITANIUM see TGG250
2,3,6-TRICHLOROTOLUENE see TJD600
α,α,α-TRICHLOROTOLUENE see BFL250
α-α-p-TRICHLOROTOLUENE see TJD650
ω,ω,ω-TRICHLOROTOLUENE see BFL250
N,N',N"-TRICHLORO-2,4,6-TRIAMINE-1,3,5-TRIAZINE see TNE775
TRICHLORO-s-TRIAZINE see TJD750
1,3,5-TRICHLOROTRIAZINE see TJD750
2,4,6-TRICHLOROTRIAZINE see TJD750
sym-TRICHLOROTRIAZINE see TJD750
2,4,6-TRICHLORO-s-TRIAZINE see TJD750

2,4,6-TRICHLORO-1,3,5-TRIAZINE see TJD750
TRICHLORO-s-TRIAZINETRIONE see TIQ750
1,3,5-TRICHLORO-1,3,5-TRIAZINETRIONE see TIQ750
TRICHLORO-s-TRIAZINE-2,4,6(1H,3H,5H)-TRIONE see TIQ750
2,2',2"-TRICHLOROTRIETHYLAMINE see TNF250
2,2',2"-TRICHLOROTRIETHYLAMINE HYDROCHLORIDE see TNF500
TRICHLOROTRIETHYLDIALUMINIUM see TJP775
TRICHLOROTRIETHYLDIALUMINUM see TJP775
1,3,5-TRICHLORO-2,4,6-TRIFLUOROBORAZINE see TJE050
TRICHLOROTRIFLUOROETHANE see FOO000
TRICHLOROTRIFLUOROETHANE see TJE100
1,1,1-TRICHLORO-2,2,2-TRIFLUOROETHANE see TJE100
1,1,2-TRICHLORO-1,2,2-TRIFLUOROETHANE (OSHA, ACGIH, MAK) see FOO000
TRICHLOROTRIMETHYLDIALUMINUM see MGC230
TRICHLORO-1,3,5-TRINITROBENZENE see TJE200
sym-TRICHLOROTRINITROBENZENE see TJE200
1,3,5-TRICHLORO-2,4,6-TRINITROBENZENE see TJE200
1,3,5-TRICHLORO-2,4,6-TRIOXOHEXAHYDRO-s-TRIAZINE see TIQ750
TRICHLOROTRIS(PYRIDINE)RHODIUM SESQUIHYDRATE see TJE300
TRICHLOROVINYLPENTACHLOROBENZENE see OAP050
TRICHLORO(VINYL)SILANE see TIN750
TRICHLOROVINYL SILICANE see TIN750
TRICHLORPHENE see TIQ250
(2,4,5-TRICHLOR-PHENOXY)-ESSIGSAEURE (GERMAN) see TAA100
2-(2,4,5-TRICHLOR-PHENOXY)-PROPIONSAEURE (GERMAN) see TIX500
O-(2,4,5-TRICHLOR-PHENYL)-O,O-DIMETHYL-MONOTHIOPHOSPHAT (GERMAN) see RMA500
2,3,6-TRICHLORPHENYLESSIGSAEURE (GERMAN) see TIY500
TRICHLORPHON see TIQ250
TRICHLORPHON FN see TIQ250
2,2,3-TRICHLORPROPANNITRIL see TJC850
TRICHLORSILAN (GERMAN) see TJD500
TRICHLOR-TRIAETHYLAMIN-HYDROCHLORID (GERMAN) see TNF500
TRICHLORURE d'ANTIMOINE see AQC500
TRICHLORURE d'ARSENIC (FRENCH) see ARF500
sym-TRICHLOTRIAZIN (CZECH) see TJD750
TRICHLOURACETONITRIL (DUTCH) see TII750
TRICHOCHROMOGENIC FACTOR see AIH600
TRICHOCID see ABY900
TRICHOCIDE see MMN250
TRICHOFURON see NGG500
TRICHOMAN see ABY900
TRICHOMOL see MMN250
TRICHOMONACID "PHARMACHIM" see MMN250
TRICHOPOL see MMN250
TRICHORAD see ABY900
TRICHORAL see ABY900
TRICHOSANTHIN see TJF350
TRICHOTHECAN-8-ONE, 12,13-EPOXY-3-α,4-β,7-α,15-TETRAHYDROXY- see DLY300

TRICHOTHECAN-8-ONE, 12,13-EPOXY-3,4,7,15-TETRAHYDROXY-, (3-α,4-β,7-α)- see DLY300

TRICHOTHEC-9-ENE, 12,13-EPOXY-4-β,8-α-15-TRIACETOXY-3-α-HYDROXY- see ACS500

TRICHOTHEC-9-ENE-3-α,4-β,15-TRIOL, 12,13-EPOXY-, 15-ACETATE see ECT700

TRICHOTHEC-9-ENE-3,4,15-TRIOL, 12,13-EPOXY-, 15-ACETATE,- (3-α,4-β)- see ECT700

TRICHOTHEC-9-ENE-3,4,15-TRIOL, 12,13-EPOXY-, TRIACETATE, (3-α,4-β)- see SBF525

TRICHOTHECIN see TJE750

TRICICLIDINA see CPQ250

TRICILOID see CPQ250

TRICIONE see TLP750

TRICIRIBINE see TJE870

TRICIRIBINE PHOSPHATE HYDRATE see TJE875

TRICLABENDAZOLE see CFL200

TRI-CLENE see TIO750

TRICLOBISONIUM CHLORIDE see TJE880

TRICLOCARBAN see TIL500

TRICLOFOS see TIP300

TRICLOFOS-METHYL see TJE885

TRICLOPYR see TJE890

TRI-CLOR see CKN500

TRICLORDIURIDE see HII500

TRICLORETENE (ITALIAN) see TIO750

TRICLORMETIAZIDE (ITALIAN) see HII500

1,1,1-TRICLORO-2,2-BIS(4-CLORO-FENIL)-ETANO (ITALIAN) see DAD200

1,1,1-TRICLOROETANO (ITALIAN) see MIH275

TRICLOROETILENE (ITALIAN) see TIO750

O-(2,4,5-TRICLORO-FENIL)-O,O-DIMETIL-MONOTIOFOSFATO (ITALIAN) see RMA500

TRICLOROMETANO (ITALIAN) see CHJ500

TRICLOROMETILBENZENE (ITALIAN) see BFL250

TRICLORO-NITRO-METANO (ITALIAN) see CKN500

TRICLOROSILANO (ITALIAN) see TJD500

TRICLOROTOLUENE (ITALIAN) see BFL250

TRICLOS see TIP300

TRICLOSAN see TIQ000

TRICLOSE see TJF000

TRICOBALT TETRAOXIDE see CND020

TRICOBALT TETROXIDE see CND020

TRICOFURON see NGG500

TRICOGEN see ABY900

TRICOLAM see TGD250

TRICOLAVAL see ABY900

TRICOLOID see CPQ250

TRICOLOID CHLORIDE see EAI875

TRICOM see MMN250

TRICON BW see EIX000

TRICOP 50 see CNK559

TRICORAL see ABY900

TRI-CORNOX SPECIAL see BAV000

TRICORYL see TJL250

12-TRICOSANONE see TJF250

TRICOSANTHIN see TJF350

9-TRICOSENE see TJF400

9-TRICOSENE, (Z)- see TJF400

(Z)-9-TRICOSENE see TJF400

cis-9-TRICOSENE see TJF400

TRICOSTERIL see ABY900

TRICOWAS B see MMN250

2,4,6-TRICl-PDMT see TJA000

TRICRESILFOSFATI (ITALIAN) see TNP500

TRICRESOL see CNW500

TRI-o-CRESYL BORATE see TJF500

TRICRESYLFOSFATEN (DUTCH) see TNP500

TRICRESYL PHOSPHATE see TMO600

TRICRESYL PHOSPHATE see TNP500

TRI-o-CRESYL PHOSPHATE see TMO600

TRICRESYLPHOSPHATE, with more than 3% ortho isomer (DOT) see TNP500

TRI-o-CRESYL PHOSPHITE see TJF750

TRI-p-CRESYL PHOSPHITE see TJG000

TRICTAL see TAF675

TRICUFIX TURQUOISE BLUE 2GL see COF420

TRICURAN see PDD300

TRICYANIC ACID see THS000

TRICYANOGEN CHLORIDE see TJD750

TRICYCLAMOL see CPQ250

TRICYCLAMOL CHLORIDE see EAI875

TRICYCLAMOL METHOCHLORIDE see EAI875

TRICYCLAMOL SULFATE see TJG225

TRICYCLAZOLE see MQC000

TRICYCLAZONE see MQC000

TRICYCLIC ANTIDEPRESSANTS see TJG239

TRICYCLO(3.3.1.1³,⁷)DECAN-1-AMINE see TJG250

TRICYCLO(3.3.1.1.⁽³,⁷⁾)DECAN-1-AMINE, HYDROCHLORIDE (9CI) see AED250

TRICYCLO(3.3.1.1³,⁷))DECAN-1-AMINE, N-((5-NITRO-2-FURANYL)METHYLENE)- see NGE777

exo-TRICYCLO(5.2.1.0²,⁶)DECANE see TLR675

TRICYCLODECANE(5.2.1.0²,⁶)-3,10-DIISOCYANATE see TJG500

TRICYCLODECANEDIMETHANOL see TJG550

TRICYCLODECANEDIMETHYLOL see TJG550

TRICYCLODECEN-4-YL-8-ACETATE see DLY400

TRICYCLODECENYL PROPIONATE see TJG600

4-TRICYCLODECYLIDENE BUTANAL see OBW100

1,1'-(TRICYCLO(3.3.2.2³,⁷)DEC-1-YLPHOSPHINYLIDENE)BIS-AZIRIDINE see BJP325

TRICYCLO(6.3.1.02,5))DODECAN-1-OL, 4,4,8-TRIMETHYL-, ACETATE, (1R-(1-α-2-α-5-β,8-β))- see CCN050

TRI(2-CYCLOHEXYLCYCLOHEXYL)BORATE see TJG750

TRICYCLOHEXYLHYDROXYSTANNANE see CQH650

TRICYCLOHEXYLHYDROXYTIN see CQH650

1-(TRICYCLOHEXYLSTANNYL)-1H-1,2,4-TRIAZOLE see THT500

TRICYCLOHEXYLTIN HYDROXIDE see CQH650

TRICYCLOHEXYLZINNHYDROXID (GERMAN) see CQH650

TRICYCLOQUINAZOLINE see TJH250

2,6-TRIDECADIENOIC ACID, 3,11-DIMETHYL-10,11-EPOXY-7-ETHYL-, METHYL ESTER see MJG100

1-TRIDECANAMINE, N,N-DITRIDECYL- see TNN760

TRIDECANE see TJH500

n-TRIDECANE see TJH500

TRIDECANE, 1-BROMO- see BOI500

1-TRIDECANECARBOXYLIC ACID see MSA250

1-TRIDECANECARBOXYLIC ACID, ISOPROPYL ESTER see IQN000

TRIDECANEDIOIC ACID, CYCLIC ETHYLENE ESTER see EJQ500

TRIDECANENITRILE see TJH750

TRIDECANITRILE (mixed isomers) see TJI000

TRIDECANOIC ACID see TJI250

TRIDECANOIC ACID-2,3-EPOXYPROPYL ESTER see TJI500

TRIDECANOL see TJI750

1-TRIDECANOL see TJI750

n-TRIDECANOL see TJI750

TRIDECANOL condensed with 6 moles ETHYLENE OXIDE see TJJ250

1-TRIDECANOL PHTHALATE see DXQ200

7-TRIDECANONE see TJJ300

2-TRIDECENAL see TJJ400

n-TRIDECOIC ACID see TJI250

TRIDECYL ACRYLATE see ADX000

TRIDECYL ALCOHOL see TJI750

n-TRIDECYL ALCOHOL see TJI750

N-TRIDECYL-2,6-DIMETHYLMORPHOLIN (GERMAN) see DUJ400

4-TRIDECYL-2,6-DIMETHYLMORPHOLINE see DUJ400

N-TRIDECYL-2,6-DIMETHYLMORPHOLINE see DUJ400

TRIDECYLIC ACID see TJI250

TRIDEMORF see TJJ500

TRIDEMORPH see TJJ500

TRIDESTRIN see EDU500

TRIDEUTEROMETHYL ACETOXYMETHYLNITROSAMINE see MMS000

TRIDEX see DUN600

TRIDEZIBARBITUR see EOK000

1,2,3-TRI(β-DIETHYLAMINOETHOXY)BENZENE TRIETHIODIDE see PDD300

TRI(β-DIETHYLAMINOETHOXY)-1,2,3-BENZENE TRI-IODOETHYLATE see PDD300

TRI-DIGITOXOSIDE (GERMAN) see DKL800

TRI(DIISOBUTYLCARBINYL) BORATE see TJJ750

TRIDILONA see TLP750

2,4,6-TRI(DIMETHYLAMINOMETHYL)PHENOL see TNH000

TRI(p-DIMETHYLAMINOPHENYL)METHANOL see TJK000

TRI(DIMETHYLAMINO)PHOSPHINE OXIDE see HEK000

TRIDIMITE (FRENCH) see SCK000

TRIDIONE see TLP750

TRIDIPAM see TFS350

TRIDIPHANE see TJK100

TRI-n-DODECYL BORATE see TJK250

TRI-(3-DODECYL-1-METHYL-2-PHENYLBENZIMIDAZOLIUM) FERRICYANIDE see TNH750

TRIDONE see TLP750

TRIDYMITE see SCI500

TRIDYMITE see SCK000

TRIDYMITE 118 see SCK000

α-TRIDYMITE see SCK000

TRIELINA (ITALIAN) see TIO750

TRIEN see TJR000

TRIENBOLONE see THL600

TRIENBOLONE ACETATE see HKI100

TRI-ENDOTHAL see DXD000

TRI-ENDOTHAL see EAR000

TRIENOLONE see THL600

TRIENTINE see TJR000

1,2,4,5,9,10-TRIEPOXYDECANE see TJK500

TRI-ERVONUM see MCA500

TRIESIFENIDILE see BBV000

TRIESTE FLOWERS see POO250

TRIETAZINE see TJL500

TRI-ETHANE see MIH275

N,N',N"-TRI-1,2-ETHANEDIYL PHOSPHORIC TRIMIDE see TND250

N,N',N"-TRI-1,2-ETHANEDIYLPHOSPHOROTHIOIC TRIAMIDE see TFQ750

N,N',N"-TRI-1,2-ETHANEDIYLTHIOPHOSPHORAMIDE see TFQ750

TRIETHANOLAMIN see TKP500

TRIETHANOLAMINE (ACGIH) see TKP500

TRIETHANOLAMINE BORATE see TJK750

TRIETHANOLAMINE DODECYLBENZENE SULFONATE see TJK800

TRIETHANOLAMINE DODECYL SULFATE see SON000

TRIETHANOLAMINE HYDROCHLORIDE see NEI700

TRIETHANOLAMINE LAURYL SULFATE see SON000

TRIETHANOLAMINE METHANEARSONATE see TJK900

TRIETHANOLAMINE SILICIEE (FRENCH) see SDH670

TRIETHANOLAMINE TRINITRATE BIPHOSPHATE see TJL250

TRIETHANOLAMINE TRINITRATE DIPHOSPHATE see TJL250

TRIETHANOLAMMONIUM CHLORIDE see NEI700

TRIETHANOMELAMINE see TND500

TRIETHAZINE see TJL500

TRIETHOXONIUM FLUOROBORATE see TJL600

TRIETHOXY(3-AMINOPROPYL)SILANE see TJN000

3-(2,4,5-TRIETHOXYBENZOYL)PROPIONIC ACID see CNG980

1,1,3-TRIETHOXYBUTANE see TJL700

TRIETHOXYDIALUMINUM TRIBROMIDE see TJL775

TRIETHOXY-ETHYLSILANE see EQA000

TRIETHOXYFENYLSILAN see PGO000

1,1,3-TRIETHOXYHEXANE see TJM000

TRIETHOXY(3-ISOCYANATOPROPYL)SILANE see IKG900

TRIETHOXY-2-KYANETHYLSILAN see CON250

TRIETHOXY-3-KYANPROPYLSILAN see COR800

TRIETHOXYMETHANE see ENY500

TRIETHOXYMETHYLSILANE see MQD750

2,4,5-TRIETHOXY-γ-OXOBENZENEBUTANOIC ACID see CNG980

TRIETHOXYPHENYLSILANE see PGO000

1,3,3-TRIETHOXYPROPANE see TJM250

1,3,3-TRIETHOXYPROPENE see TJM500

1,3,3-TRIETHOXY-1-PROPENE see TJM500

TRIETHOXYSILANE see TJM750

4-(TRIETHOXYSILYL)BUTYLAMINE see AJC250

TRIETHOXYSILYLMETHANE see MQD750

3-(TRIETHOXYSILYL)-1-PROPANAMINE see TJN000

3-(TRIETHOXYSILYL)PROPYLAMINE see TJN000

TRIETHOXYVINYLSILANE see TJN250

TRIETHOXYVINYLSILICANE see TJN250

TRIETHYL ACETYLCITRATE see ADD750

TRIETHYLALUMINIUM see TJN750

TRIETHYLALUMINUM see TJN750

TRIETHYLALUMINUM SESQUICHLORIDE see TJP775

TRIETHYLAMINE see TJO000

TRIETHYLAMINE, 2″-CHLORO-1,1′-DIMETHYL-, HYDROCHLORIDE see CGV600

TRIETHYLAMINE, 2,2‴-DITHIOBIS-(8CI) see TCD300

TRIETHYLAMINE, HYDROCHLORIDE see TJO050

TRIETHYLAMINE, 2-(p-(2-PHENYL-3,4-DIHYDRO-1-NAPHTHYL)PHENOXY)- see PEU545

s-(2-TRIETHYLAMINOETHYL)ISOTHIURONIUM BROMIDE HYDROBROMIDE see TAI100

S-(2-TRIETHYLAMINOETHYL)-1′-METHYLISOTHIURONIUM BROMIDE HYDROBROMIDE see MRU775

2-(2-TRIETHYLAMINOETHYLTHIO)-Δ²-IMIDAZOLINE BROMIDE HYDROBROMIDE see EDW300

TRIETHYL(3-AMINOPROPYL)SILANE see AMG200

TRIETHYL AMMONIUM NITRATE see TJO100

TRIETHYLANTIMONY see TJO250

TRIETHYLARSINE see TJO500

TRIETHYLBENZENE see TJO750

TRIETHYL-BENZENE (mixed isomers) see TJO750

N,N,N-TRIETHYLBENZENEMETHANAMINIUM CHLORIDE see BFL300

TRIETHYLBENZYLAMMONIUM CHLORIDE see BFL300

TRIETHYLBISMUTH see TJP000

TRIETHYLBORANE see TJP250

TRIETHYL BORATE see TJP500

TRIETHYLBORINE see TJP250

TRIETHYLCARBINOL see TJP550

TRIETHYLCHLOROPLUMBANE see TJS250

TRIETHYLCHLOROSTANNANE see TJV000

TRIETHYLCHLOROTIN see TJV000

TRIETHYL CITRATE see TJP750

TRIETHYL CITRATE, ACETATE see ADD750

TRIETHYLDIALUMINUM TRICHLORIDE see TJP775

TRIETHYLDIBORANE see TJP780

TRIETHYLENEDIAMINE see DCK400

TRIETHYLENE GLYCOL see TJQ000

TRIETHYLENE GLYCOL-n-BUTYL ETHER see TKL750

TRIETHYLENE GLYCOL, DIACETATE see EJB500

TRIETHYLENE GLYCOL DIACRYLATE see TJQ100

TRIETHYLENE GLYCOL DICHLORIDE see TKL500

TRIETHYLENE GLYCOL DIETHYL BUTYRATE see TJQ250

TRIETHYLENE GLYCOL DI(2-ETHYL-BUTYRATE) see TJQ250

TRIETHYLENE GLYCOL DI(2-ETHYLHEXOATE) see FCD525

TRIETHYLENE GLYCOL DIGLYCIDYL ETHER see TJQ333

TRIETHYLENE GLYCOL DIMETHACRYLATE see MDN510

TRIETHYLENE GLYCOL DIMETHYL ETHER see TKL875

TRIETHYLENE GLYCOL, DINITRATE see TJQ500

TRIETHYLENE GLYCOL DIVINYL ETHER see TDY800

TRIETHYLENE GLYCOL ETHYL ETHER see EFL000

TRIETHYLENE GLYCOL MONOBUTYL ETHER see TKL750

TRIETHYLENE GLYCOL MONOETHYL ETHER see EFL000

TRIETHYLENE GLYCOL MONOHEXYL ETHER see HFT550

TRIETHYLENE GLYCOLMONOMETHYL ETHER see TJQ750

2,3,5-TRIETHYLENEIMINO-1,4-BENZOQUINONE see TND000

TRI(ETHYLENEIMINO)THIOPHOSPHORAMIDE see TFQ750

2,4,6-TRIETHYLENEIMINO-s-TRIAZINE see TND500

2,4,6-TRI(ETHYLENEIMINO)-1,3,5-TRIAZINE see TND500

TRIETHYLENEMELAMINE see TND500

N,N′,N″-TRIETHYLENEPHOSPHOROTHIOIC TRIAMIDE see TFQ750

TRIETHYLENEPHOSPHOROTRIAMIDE see TND250

TRIETHYLENETETRAMINE see TJR000

N,N′,N″-TRIETHYLENETHIOPHOSPHAMIDE see TFQ750

N,N′,N″-TRIETHYLENETHIOPHOSPHORAMIDE see TFQ750

TRIETHYLENETHIOPHOSPHOROTRIAMIDE see TFQ750

TRIETHYLENIMINOBENZOQUINONE see TND000

2,4,6-TRIETHYLENIMINO-s-TRIAZINE see TND500

2,4,6-TRIETHYLENIMINO-1,3,5-TRIAZINE see TND500

TRIETHYLESTER KYSELINY ACETYLCITRONOVE see ADD750

TRIETHYLESTER KYSELINY BORITE see TJP500

O,O,O-TRIETHYLESTER KYSELINY THIOFOSFORECNE (CZECH) see TJU000

N,N,N-TRIETHYL-ETHANAMINIUM (9CI) see TCB725

N,N,N-TRIETHYLETHANAMINIUM CHLORIDE see TCC250

N,N,N-TRIETHYLETHANAMINIUM IODIDE see TCC750

TRIETHYLGALLIUM see TJR250

N,N,N-TRIETHYL-1-HEXADECANAMINIUM BROMIDE see CDF500

TRIETHYLHEXADECYLAMMONIUM BROMIDE see CDF500

TRI(2-ETHYLHEXYL) BORATE see TJR500

TRIETHYLHEXYL PHOSPHATE see TNI250

TRI(2-ETHYLHEXYL)PHOSPHATE see TNI250

TRI-(2-ETHYLHEXYL)PHOSPHITE see TNI300

TRI-2-ETHYLHEXYL TRIMELLITATE see TJR600

TRIETHYL(2-HYDROXYETHYL)-AMMONIUM BROMIDE DICYCLOPENTYLACETATE see DGW600

TRIETHYL(2-HYDROXYETHYL)AMMONIUM-p-TOLUENESULFONATE 3,4,5-TRIMETHOXYBENZOATE see TNV625

TRIETHYLHYDROXY-STANNANE SULFATE (2:1) (8CI) see BLN500

TRIETHYLHYDROXYTIN SULFATE see BLN500

N,N,N-TRIETHYL-2-((IMINO(METHYLAMINO)METHYL)THIO)-ETHANAMINIUM BROMIDE, MONOHYDROBROMIDE see MRU775

TRIETHYL INDIUM see TJR750

TRIETHYL LEAD see TJS000

TRIETHYL LEAD CHLORIDE see TJS250

TRIETHYL LEAD FLUOROACETATE see TJS500

TRIETHYL LEAD FUROATE see TJS750

TRIETHYL LEAD OLEATE see TJT000

TRIETHYL LEAD PHENYL ACETATE see TJT250

TRIETHYL LEAD PHOSPHATE see TJT500

TRIETHYLMETHANOL see TJP550

1,1,3-TRIETHYL-3-NITROSOUREA see NLX500

TRIETHYLOCTADECYLAMMONIUM IODIDE see OBI100

TRIETHYLOLAMINE see TKP500

TRIETHYL ORTHOFORMATE see ENY500

TRIETHYLOXONIUM BOROFLUORIDE see TJL600

TRIETHYLOXONIUM FLUOBORATE see TJL600

TRIETHYLOXONIUM FLUOROBORATE see TJL600

TRIETHYLOXONIUM TETRAFLUOROBORATE see TJL600

TRIETHYL PHOSPHATE see TJT750

TRIETHYL PHOSPHINE see TJT775

TRIETHYLPHOSPHINEAUROUS CHLORIDE see CLQ500
TRIETHYLPHOSPHINE GOLD see ARS150
TRIETHYL PHOSPHINE GOLD NITRATE see TJT780
TRIETHYL PHOSPHITE see TJT800
TRIETHYL PHOSPHONOACETATE see EIC000
O,S,S-TRIETHYL PHOSPHORODITHIOATE see TJT900
TRIETHYL PHOSPHOROTHIOATE see TJU000
O,O,O-TRIETHYL PHOSPHOROTHIOATE see TJU000
O,O,S-TRIETHYL PHOSPHOROTHIOATE see TJU800
TRIETHYLPLUMBYL ACETATE see TJU150
TRIETHYLPROPYL GERMANE see TJU250
TRIETHYLSILYL PERCHLORATE see TJU500
TRIETHYLSTANNANE see TJV050
TRIETHYLSTANNIUM BROMIDE see BOI750
TRIETHYLSTANNYL CHLORIDE see TJV000
TRIETHYLSULFONIUM IODIDE BIS(MERCURIC IODIDE) addition compound see TJU600
TRIETHYLSULFONIUM IODIDE MERCURIC IODIDE addition compound see TJU750
TRIETHYL-SULFONIUM, IODIDE, compound with MERCURY IODIDE (1:2) see TJU600
TRIETHYLSULFONIUM, IODIDE compounded with MERCURY IODIDE (1:1) see TJU750
TRIETHYLTHIOFOSFAT (CZECH) see TJU000
O,O,S-TRIETHYL THIOPHOSPHATE see TJU800
TRIETHYLTIN see TJV050
TRIETHYLTIN ACETATE see ABW750
TRIETHYL TIN BROMIDE see BOI750
TRIETHYLTIN BROMIDE-2-PIPECOLINE see TJU850
TRIETHYLTIN CHLORIDE see TJV000
TRIETHYLTIN HYDRIDE see TJV050
TRIETHYLTIN HYDROPEROXIDE see TJV100
TRIETHYLTIN PHENOXIDE see TJV250
TRIETHYLTIN SULPHATE see BLN500
3,6,9-TRIETHYL-3,6,9-TRIAZAUNDECANE see TJV500
TRIETHYLTRICHLORODIALUMINUM see TJP775
* TRIETHYL(TRIFLUOROACETOXY) STANNANE see TJX250
2,4,6-TRIETHYL-1,3,5-TRIISOPROPYLBORAZINE see TKU000
TRIETHYL-2-(3,4,5-TRIMETHOXYBENZOYLOXY)ETHYLAMM ONIUM-p-TOLUENESULFATE see TNV625
TRIETHYL-2-(3,4,5-TRIMETHOXYBENZOYLOXY)ETHYLAMM ONIUM TOSYLATE see TNV625
TRIETHYNYL ALUMINUM see TJV775
TRIETHYNYL ANTIMONY see TJV785
TRIETHYNYLARSINE see TJW000
TRIETHYNYLPHOSPHINE see TJW250
TRIETILAMINA (ITALIAN) see TJO000
TRIEXIFENIDILA see BBV000
TRIFARON see CJM250
TRI-FEN see TTY500
TRIFENOFOS see EPC175
TRIFENOXYFOSFIN (CZECH) see TMU250
TRIFENSON see CJR500
TRIFENYLFOSFIT (CZECH) see TMU250
TRIFENYLTINACETAAT (DUTCH) see ABX250
TRIFENYL-TINHYDROXYDE (DUTCH) see HON000

TRIFERRIC ADRIAMYCIN see QBS000
TRIFERRIC DOXORUBICIN see QBS000
TRIFLIC ACID see TKB310
TRIFLORAN see DUV600
TRIFLORPERAZINE DIHYDROCHLORIDE see TKK250
TRIFLUMEN see HII500
TRIFLUMETHAZINE see TJW500
TRIFLUOMETHYLTHIAZIDE see TKG750
TRIFLUOPERAZINA (ITALIAN) see TKE500
TRIFLUOPERAZINE see TKE500
TRIFLUOPERAZINE DIMALEATE see TJW600
TRIFLUOPERAZINE HYDROCHLORIDE see TKK250
TRIFLUORACETIC ACID see TKA250
TRIFLUORALIN (USDA) see DUV600
4-TRIFLUORMETHOXY-N-CHLOROCARBOXYPHENYLURETHAN see MIG100
3-(5-TRIFLUORMETHYLPHENYL)-,1-DIMETHYLHARNSTOFF (GERMAN) see DUK800
TRIFLUOROACETALDEHYDE HYDRATE see TJZ000
2-(2,2,2-TRIFLUOROACETAMIDO)-4-(5-NITRO-2-FURYL)THIAZOLE see NGN500
TRIFLUOROACETIC ACID (DOT) see TKA250
TRIFLUOROACETIC ACID ANHYDRIDE see TJX000
TRIFLUOROACETIC ACID SODIUM SALT see SKL100
TRIFLUOROACETIC ACID TRIETHYLSTANNYL ESTER see TJX250
TRIFLUOROACETIC ANHYDRIDE see TJX000
N-TRIFLUOROACETYLADRIAMYCIN see TJX300
TRIFLUOROACETYLADRIAMYCIN-14-VALERATE see TJX350
N-TRIFLUOROACETYLADRIAMYCIN-14-VALERATE see TJX350
2-TRIFLUOROACETYLAMINOFLUORENE see FER000
2-TRIFLUOROACETYLAMINOFLUOREN-9-ONE see TKH000
TRIFLUOROACETYL ANHYDRIDE see TJX000
TRIFLUOROACETYL AZIDE see TJX375
TRIFLUOROACETYL CHLORIDE see TJX500
2-TRIFLUOROACETYL-1,3,4-DIOXAZALONE see TJX600
N-TRIFLUOROACETYLDOXORUBICIN see TJX300
O-TRIFLUOROACETYL-S-FLUOROFORMYL THIOPEROXIDE see TJX625
TRIFLUOROACETYL HYPOCHLORITE see TJX650
TRIFLUOROACETYL HYPOFLUORITE see TJX750
TRIFLUOROACETYLIMINOIODOBENZEN E see TJX775
TRIFLUOROACETYL NITRITE see TJX780
TRIFLUOROACETYL TRIFLUOROMETHANE SULFONATE see TJX800
TRIFLUOROACRYLOYL FLUORIDE see TJX825
TRIFLUOROAMINE OXIDE see NGS500
2,3,4-TRIFLUOROANILINE see TJX900
TRIFLUOROANTIMONY see AQE000
TRIFLUOROARSINE see ARI250
1,1,1-TRIFLUORO-2-BROMO-2-CHLOROETHANE see HAG500
1,1,2-TRIFLUORO-1-BROMO-2-CHLOROETHANE see TJY000
TRIFLUOROBROMOETHYLENE see BOJ000
TRIFLUOROBROMOMETHANE see TJY100

1,1,1-TRIFLUORO-2-CHLORO-2-BROMOETHANE see HAG500
2,2,2-TRIFLUORO-1-CHLORO-1-BROMOETHANE see HAG500
2,2,2-TRIFLUOROCHLOROETHANE see TJY175
1,1,1-TRIFLUORO-2-CHLOROETHANE see TJY175
TRIFLUOROCHLOROETHYLENE (DOT) see CLQ750
1,1,2-TRIFLUORO-2-CHLOROETHYLENE see CLQ750
TRIFLUOROCHLOROETHYLENE POLYMER see KDK000
TRIFLUOROCHLOROMETHANE (DOT) see CLR250
1,1,1-TRIFLUORO-3-CHLOROPROPANE see TJY200
α,α,α-TRIFLUORO-4-CHLOROTOLUENE see CEM825
α,α,α-TRIFLUORO-m-CRESOL see TKE750
5-TRIFLUORO-2'-DEOXYTHYMIDINE see TKH325
2,2,2-TRIFLUORODIAZOETHANE see TJY275
1,1,1-TRIFLUORO-2,2-DICHLOROETHANE see TJY500
1,1,2-TRIFLUORO-1,2-DICHLOROETHANE see TJY750
2',4',6'-TRIFLUORO-4-DIMETHYLAMINOAZOBENZENE see DUK200
α,α,α-TRIFLUORO-2,6-DINITRO-N,N-DIPROPYL-p-TOLUIDINE see DUV600
1,1,1-TRIFLUOROETHANE see TJY900
1,1,2-TRIFLUOROETHANE see TJY950
2,2,2-TRIFLUORO-1,1-ETHANEDIOL see TJZ000
TRIFLUOROETHANOIC ACID see TKA250
2,2,2-TRIFLUOROETHANOL see TKA350
TRIFLUOROETHENE see TKA400
(TRIFLUOROETHENYL)BENZENE see TKH310
(2,2,2-TRIFLUOROETHOXY)ETHENE see TKB250
2,2,2-TRIFLUOROETHYLAMINE see TKA500
TRIFLUOROETHYLAMINE HYDROCHLORIDE see TKA750
N-(2,2,2-TRIFLUOROETHYL)ANILINE see TKA950
1,1,1-TRIFLUOROETHYL CHLORIDE see TJY175
o-(2,2,2-TRIFLUOROETHYL) S-(1-METHYLPROPYL)-N-HYDROGENCARBONYL-N-METHYLPHOSPHOROAMIDODITHIOATE see MOV100
2,2,2-TRIFLUOROETHYL VINYL ETHER see TKB250
2,2,2-TRIFLUORO-N-(FLUOREN-2-YL)ACETAMIDE see FER000
1,1,1-TRIFLUOROFORM see TJY900
N,N,N'-TRIFLUOROHEXANAMIDINE see TKB275
(3,3,3-TRIFLUORO-2-HYDROXY-2-(TRIFLUOROMETHYL))PROPYL BENZYL KETONE see TKB285
TRIFLUOROIODOMETHANE see IFM100
α-α-α-TRIFLUORO-3'-ISOPROPOXY-o-TOLUANILIDE see TKB286
3,3,3-TRIFLUOROLACTIC ACID METHYL ESTER DIBUTYL PHOSPHATE see TKB287
3,3,3-TRIFLUOROLACTIC ACID METHYL ESTER DIETHYL PHOSPHATE TKB289
3,3,3-TRIFLUOROLACTIC ACID METHYL ESTER DIHEXYL PHOSPHATE see TKB290
3,3,3-TRIFLUOROLACTIC ACID METHYL ESTER DIISOBUTYL PHOSPHATE see TKB292

4-(3-(2-TRIFLUOROMETHYLTHIOXANTH-9-YLIDENE)PROPYL)-1-PIPERAZINEETHANOL DIHYDROCHLORIDE see FMO150

(TRIFLUOROMETHYL)TRIFLUOROOXIRANE see HDF050

TRIFLUOROMETHYL TRIFLUOROVINYL ETHER see TKG800

1-TRIFLUOROMETHYL-1,2,2,3,3,4,4,5,5,6,6-UNDECAFLUORO CYCLOHEXANE see PCH290

TRIFLUOROMONOBROMOMETHANE see TJY100

TRIFLUOROMONOCHLOROCARBON see CLR750

TRIFLUOROMONOCHLOROETHYLENE see CLQ750

2,2,2-TRIFLUORO-N-(4-(5-NITRO-2-FURYL)-2-THIAZOLYL)ACETAMIDE see NGN500

α,α,α-TRIFLUORO-m-NITROTOLUENE see NFJ500

2,2,2-TRIFLUORO-N-(9-OXOFLUOREN-2-YL)ACETAMIDE see TKH000

TRIFLUOROPERAZINE DIHYDROCHLORIDE see TKK250

TRIFLUOROPHOSPHINE see PHQ500

1,1,1-TRIFLUOROPROPENE see TKH015

3,3,3-TRIFLUOROPROPENE see TKH015

3,3,3-TRIFLUORO-1-PROPENE see TKH015

TRIFLUOROPROPIONALDEHYDE see TKH020

3,3,3-TRIFLUOROPROPIONALDEHYDE see TKH020

N,N,N'-TRIFLUOROPROPIONAMIDINE see TKH025

3,3,3-TRIFLUOROPROPYLENE see TKH015

3,3,3-TRIFLUOROPROPYNE see TKH030

17-(3,3,3-TRIFLUORO-1-PROPYNYL)ESTRA-1,3,5(10)-TRIEN-3,17-β-DIOL see TKH050

TRIFLUOROPYRAZIN DIHYDROCHLORIDE see TKK250

TRIFLUORO SELENIUM HEXAFLUORO ARSENATE see TKH250

TRIFLUOROSTANNITE HEXADECYLAMINE see TKH300

TRIFLUOROSTANNITE OF HEXADECYLAMINE see TKH300

TRIFLUOROSTIBINE see AQE000

α-β,β-TRIFLUOROSTYRENE see TKH310

TRIFLUOROTHYMIDINE see TKH325

α,α,α-TRIFLUOROTHYMIDINE see TKH325

ω-TRIFLUOROTOLUENE see BDH500

α,α,α-TRIFLUOROTOLUENE see BDH500

α-α-α-TRIFLUOROTOLUENE-3,5-DIAMINE see TKH775

α,α,α-TRIFLUORO-m-TOLUIC ACID THALLIUM(I) SALT see TKH500

α,α,α-TRIFLUORO-p-TOLUIDINE see TKB750

N-(α,α,α-TRIFLUORO-m-TOLYL)ANTHRANILIC ACID see TKH750

N-(α,α,α-TRIFLUORO-m-TOLYL)ANTHRANILIC ACID BUTYL ESTER see BRJ325

4-(2-(α-(α,α,α-TRIFLUORO-m-TOLYL)BENZYLOXY)ETHYL)MORPHOLINE FUMARATE see TKI000

3-(α-(α,α,α-TRIFLUORO-p-TOLYL)BENZYLOXY)TROPANE FUMARATE see TKI750

(α,α,α-TRIFLUORO-m-TOLYL) ISOCYANATE see TKJ250

2-(α,α,α-TRIFLUORO-m-TOLYL)MORPHOLINE see TKJ500

5-(((α,α,α-TRIFLUORO-m-TOLYL)OXY)METHYL)-2-OXAZOLIDINETHIONE see TKJ750

5-(α,α,α-TRIFLUORO-m-TOLYOXYMETHYL)-2-OXAZOLIDINETHIONE see TKJ750

2,4,6-TRIFLUORO-s-TRIAZINE see TKK000

4,4'-((2,2,2-TRIFLUORO-1-(TRIFLUOROMETHYL)ETHYLIDENE)BIS(4,1-PHENYLENEOXY))BISBENZENAMINE see TKK025

4,4'-(TRIFLUORO-1-(TRIFLUOROMETHYL)ETHYLIDENE)DIPHENOL see HCZ100

TRIFLUORO(TRIFLUOROMETHYL)OXIRANE see HDF050

3,3,3-TRIFLUORO-2-(TRIFLUOROMETHYL)PROPENE see HDC450

1,3,5-TRIFLUOROTRINITROBENZENE see TKK050

TRIFLUOROVINYLBROMIDE see BOJ000

TRIFLUOROVINYL CHLORIDE see CLQ750

TRIFLUORURE de CHLORE (FRENCH) see CDX750

TRIFLUPERAZINE see TKE500

TRIFLUPERAZINE DIHYDROCHLORIDE see TKK250

TRIFLUPERIDOL see TKK500

TRIFLUPERIDOL HYDROCHLORIDE see TKK750

TRIFLUPERIDOLO (ITALIAN) see TKK500

TRIFLUPROMAZINE see TKL000

TRIFLURALIN see DUV600

TRIFLURALINA 600 see DUV600

TRIFLURALINE see DUV600

TRIFLURIDINE see TKH325

TRIFOCIDE see DUS700

TRIFOLEX see CLN750

TRIFOLITIN see ICE000

TRIFORINE see TKL100

TRIFORINE and SODIUM NITRITE see SIT800

TRIFORMOL see PAI000

TRIFORMYL-STROSPESIDE see TKL175

TRIFOSFAMIDE see TNT500

TRIFRINA see DUS700

TRIFTAZIN see TKK250

TRIFUNGOL see FAS000

TRIFUREX see DUV600

TRIGARD see CQE400

TRIGEN see TJQ000

TRIGENOLLINE see TKL890

TRIGLYCERIDE HYDROLASE see GGA800

TRIGLYCERIDE LIPASE see GGA800

TRIGLYCIDYL CYANURATE see TKL250

TRIGLYCIDYL ISOCYANURATE see GGY150

1,3,5-TRIGLYCIDYL ISOCYANURATE see GGY150

N,N',N''-TRIGLYCIDYL ISOCYANURATE see GGY150

α-TRIGLYCIDYL ISOCYANURATE see TBC450

1,3,5-TRIGLYCIDYLISOCYANURIC ACID see GGY150

1,3,5-TRIGLYCIDYL-s-TRIAZINETRIONE see GGY150

TRIGLYCINE see AMT500

TRIGLYCOL see TJQ000

TRIGLYCOL, DIACETATE see EJB500

TRIGLYCOL DICAPROATE see TJQ250

TRIGLYCOL DICHLORIDE see TKL500

TRIGLYCOL DIHEXOATE see TJQ250

TRIGLYCOLLAMIC ACID see AMT500

TRIGLYCOL MONOBUTYL ETHER see TKL750

TRIGLYCOL MONOETHYL ETHER see EFL000

TRIGLYCOL MONOMETHYL ETHER see TJQ750

TRIGLYME see TKL875

TRIGONELLIN see TKL890

TRIGONELLINE see TKL890

TRIGONOX 40 see PBL600

TRIGONOX 25/75 see BSD250

TRIGONOX 101-101/45 see DRJ800

TRIGONOX A-75 (CZECH) see BRM250

TRIGONOX C see BSC500

TRIGONOX 25-C75 see BSD250

TRIGONOX F-C50 see BSC250

TRIGONOX HM 80 see IJB100

TRIGONYL see TKX000

TRIGOSAN see ABU500

TRIHERBICIDE CIPC see CKC000

TRIHERBIDE see CBM000

TRIHERBIDE-IPC see CBM000

TRIHEXANOIN see GGK000

TRIHEXANOYLGLYCEROL see GGK000

TRIHEXYL BORATE see TKM000

TRI-n-HEXYL BORATE see TKM000

TRIHEXYLENE GLYCOL BIBORATE see TKM250

TRIHEXYLPHENIDYL see PAL500

TRIHEXYLPHENIDYL HYDROCHLORIDE see BBV000

TRI-n-HEXYLPHOSPHINE OXIDE see TKM500

TRIHEXYLTIN ACETATE see ABX000

TRI-N-HEXYLZINNACETAT (GERMAN) see ABX000

TRIHISTAN see CDR000

TRIHYDRATED ALUMINA see AHC000

TRIHYDRAZINECOBALT(II) NITRATE see TKM750

TRIHYDRAZINENICKEL(II) NITRATE see TKN000

TRIHYDROCHLORIDE see TKN100

TRIHYDROGENPYROPHOSPHATE(ESTER)THIAMINE see TET750

2,3,4-TRIHYDROXYACETOPHENONE see TKN250

3,7,15-TRIHYDROXY-4-ACETOXY-8-OXO-12,13-EPOXY-Δ9-TRICHOTHECENE see FQR000

1,8,9-TRIHYDROXYANTHRACENE see APH250

1,2,3-TRIHYDROXY-9,10-ANTHRACENEDIONE see TKN500

1,2,4-TRIHYDROXY-9,10-ANTHRACENEDIONE see TKN750

1,2,4-TRIHYDROXYANTHRACHINON (CZECH) see TKN750

1,2,3-TRIHYDROXYANTHRAQUINONE see TKN500

1,2,3-TRIHYDROXYANTHRAQUINONE see TKN500

1,2,4-TRIHYDROXYANTHRAQUINONE see TKN750

2,3,4-TRIHYDROXYBENZALDEHYDE see TKN800

1,2,3-TRIHYDROXYBENZEN (CZECH) see PPQ500

s-TRIHYDROXYBENZENE see PGR000

1,2,3-TRIHYDROXYBENZENE see PPQ500

1,2,4-TRIHYDROXYBENZENE see BBU250

1,3,5-TRIHYDROXYBENZENE see PGR000

sym-TRIHYDROXYBENZENE see PGR000

3,4,5-TRIHYDROXYBENZENE-1-PROPYLCARBOXYLATE see PNM750

2,4,6-TRIHYDROXYBENZOIC ACID see TKO000

3,4,5-TRIHYDROXYBENZOIC ACID see GBE000

3,4,5-TRIHYDROXYBENZOIC ACID, n-PROPYL ESTER see PNM750

2-(2,3,4-TRIHYDROXYBENZYL)HYDRAZIDE SERINE MONOHYDROCHLORIDE, dl- see SCA400

3-β,14,16-β-TRIHYDROXY-5-β-BUFA-20,22-DIENOLIDE see BOM655

3-β,14,16-β-TRIHYDROXY-5-β-BUFA-20,22-DIENOLIDE-16-ACETATE see BON000

(R*,R*)-2,3,4-TRIHYDROXYBUTANAL see TKO100

2,4,5-TRIHYDROXYBUTYROPHENONE see TKO250

2',4',5'-TRIHYDROXYBUTYROPHENONE see TKO250

TRIISOBUTYL BORATE see TKR750
TRIISOBUTYLENE OXIDE see TKS000
TRIISOBUTYL ORTHOVANADATE see VDK000
TRIISOBUTYL PHOSPHATE see IJJ200
TRIISOBUTYLTIN CHLORIDE see CLS500
TRIISOBUTYL VANADATE see VDK000
TRIISOCYANATOISOCYANURATE, solution, 70%, by weight (DOT) see IMG000
TRIISODECYL PHOSPHATE see IKM025
TRIISODECYL PHOSPHITE see IKM025
TRIISOOCTYLAMINE see TKS500
TRIISOOCTYL PHOSPHITE see TKT000
TRIISOPENTYLPHOSPHINE see TKT100
TRIISOPENTYLPHOSPHINE OXIDE see TNJ285
TRIISOPROPANOLAMINE see NEI500
TRIISOPROPANOLAMINE BORATE see TKT200
TRIISOPROPOXYALUMINUM see AHC600
TRIISOPROPOXYVANADIUM OXIDE see VDK000
TRIISOPROPYL BORATE (DOT) see IOI000
N-TRI-ISOPROPYL-B-TRIETHYL BORAZOLE see TKU000
TRIISOPROPYLCHLOROSTANNANE see TKT800
TRI(ISOPROPYLPHENYL) PHOSPHATE see TKU300
TRIISOPROPYL PHOSPHITE see TKT500
TRIISOPROPYLTIN ACETATE see TKT750
TRIISOPROPYLTIN CHLORIDE see TKT800
TRIISOPROPYLTIN UNDECYLENATE see TKT850
TRIISOVALERIN see GGM000
2,4,6-TRIJOD-3-ACETAMINOBENZOSAEURE NATRIUM (GERMAN) see AAN000
β-(2,4,6-TRIJOD-3-DIMETHYLAMINOMETHYLENAMINO-PHENYL)-PROPIONSAEURE (GERMAN) see SKM000
TRIJOD-PROPYLGERMAN see TKR050
TRIKEPIN see DUV600
TRIKETOCHOLANIC ACID see DAL000
3,7,12-TRIKETOCHOLANIC ACID see DAL000
TRIKETOHYDRINDENE HYDRATE see DMV200
TRIKHLORFENOLYAT MEDI (RUSSIAN) see CNQ375
TRIKOJOL see MMN250
TRIKOLAVAL see ABY900
TRIKRESYLPHOSPHATE (GERMAN) see TNP500
o-TRIKRESYLPHOSPHATE (GERMAN) see TMO600
TRIKSAZIN see TKX250
TRILAFON see CJM250
TRILAN see EPD500
TRILAX see PDO750
TRILEAD DINITRIDE see TKU250
TRILEAD PHOSPHATE see LDU000
TRILEAD TETROXIDE see LDS000
TRILENE see TIO750
TRILIDONA see TLP750
N,N,4-TRILITHIOANILINE see TKU275
TRILITHIUM CITRATE see TKU500
TRILLEKAMIN see TNF500
TRILOFOSFAMIDA see TNT500
TRILON 83 see EIF000
TRILON A see AMT500
TRILON AO see TNL250
TRILON B see EIV000
TRILON BD see EIX500
TRILON BW see EIX000
TRILONE 46 see IPX000
TRILOPHOSPHAMIDE see TNT500
TRILOSTANE see EBY600
TRILUDAN see TAI450
TRIM see DUV600
TRIMAGNESIUM PHOSPHATE see MAH780

TRIMANGANESE TETRAOXIDE see MAV550
TRIMANGANESE TETROXIDE see MAV550
TRIMANGOL see MAS500
TRIMANGOL 80 see MAS500
TRIMANYL see TKZ000
TRIMAR see TIO750
TRIMATON see SIL550
TRIMATON see VFU000
TRI-ME see MQH750
TRIMEBUTINE see TKU650
TRIMEBUTINE MALEATE see TKU675
TRIMEC see TKU680
TRIMECAINE see DHL800
TRIMECAINE HYDROCHLORIDE see DHL800
TRIMEDAL see TLP750
TRIMEDLURE see BQT600
TRIMEDONE see TLP750
TRIMEDOXIME DICHLORIDE see TLQ750
TRIMEGLAMIDE see TIH800
TRIMEKAIN HYDROCHLORIDE see DHL800
TRIMEKS see MMN250
TRIMELARSEN see PLK800
TRIMELLIC ACID ANHYDRIDE see TKV000
TRIMELLIC ACID-1,2-ANHYDRIDE see TKV000
TRIMELLITIC ACID see TKU700
TRIMELLITIC ACID CYCLIC-1,2-ANHYDRIDE see TKV000
TRIMELLITIC ANHYDRIDE see TKV000
TRIMEPERDINE see IMT000
TRIMEPRAMINE HYDROCHLORIDE see TAL000
TRIMEPRAZINE see AFL500
TRIMEPRAZINE HYDROCHLORIDE see TKV200
TRIMEPRIMINA (ITALIAN) see DLH200
TRIMEPRIMINE MALEATE see SOX550
TRIMEPRIMINE MONOMALEATE see SOX550
TRIMEPROPIMINE see DLH200
TRIMEPROPIMINE HYDROCHLORIDE see TAL000
TRIMEPROPRIMINE HYDROCHLORIDE see TAL000
1,3,5-TRIMERCAPTOTRIAZINE see THS250
2,4,6-TRIMERCAPTO-S-TRIAZINE see THS250
TRIMERCAPTO-s-TRIAZINE TRISODIUM SALT see SKN100
2,4,6-TRIMERCAPTO-s-TRIAZINE TRISODIUM SALT see SKN100
TRIMERCURY DINITRIDE see TKW000
TRIMERESURUS FLAVOVIRIDIS VENOM see TKW100
TRIMESULF see TKX000
TRIMETADIONE see TLP750
TRIMETAPHAN CAMPHORSULFONATE see TKW500
TRIMETAPHAN CAMSILATE see TKW500
TRIMETAPHOSPHATE SODIUM see SKM500
TRIMETAZIDINE DIHYDROCHLORIDE see YCJ200
TRIMETAZIDINE HYDROCHLORIDE see YCJ200
TRIMETHACARB see TMC750
TRIMETHADIONE see TLP750
TRIMETHAPHAN CAMPHORSULFONATE see TKW500
TRIMETHAPHAN-10-CAMPHORSULFONATE see TKW500
TRIMETHAPHAN CAMSYLATE see TKW500
TRIMETHIDINIUM BIMETHOSULFATE see TLG000
TRIMETHIDINIUM METHOSULFATE see TLG000
TRIMETHIN see TLP750
TRIMETHIOPHANE see TKW500
TRIMETHOATE see IOT000

TRIMETHOBENZAMIDE HYDROCHLORIDE see TKW750
TRIMETHOPRIM see TKZ000
TRIMETHOPRIMSULFA see TKX000
TRIMETHOPRIM and SULFAMOXOLE see SNL850
TRIMETHOPRIM-SULFAMOXOLE mixture see SNL850
TRIMETHOPRIM and SULPHAMETHOXAZOLE see TKX000
TRIMETHOPRIOM see TKZ000
TRIMETHOQUINOL see IDD100
(−)-TRIMETHOQUINOL see IDD100
(±)-TRIMETHOQUINOL see TKX125
TRIMETHOXAZINE see TKX250
3,4,5-TRIMETHOXPHENETHYLAMINE SULFATE see MDJ000
3,4,5-TRIMETHOXYAMPHETAMINE see MLH250
3,4,5-TRIMETHOXYAMPHETAMINE HYDROCHLORIDE see TKX500
1,2,10-TRIMETHOXY-6a-α-APORPHIN-9-OL see TKX700
3,4,5-TRIMETHOXYBENZAMIDE see TKY000
ε-(3,4,5-TRIMETHOXYBENZAMIDO)CAPROIC ACID SODIUM SALT see CBF625
ε-(3,4,5-TRIMETHOXYBENZAMIDO)CAPRONSAE URE NATRIUM (GERMAN) see CBF625
5-(3,4,5-TRIMETHOXYBENZAMIDO)-2-METHYL-trans-DECAHYDROISOQUINOLINE see DAE700
cis-5,8,10-H-5-(3,4,5-TRIMETHOXYBENZAMIDO)-2-METHYL DECAHYDROISOQUINOLINE see DAE695
trans-9,10-t-5-H-5-(3,4,5-TRIMETHOXYBENZAMIDO)-2-METHYL DECAHYDROISOQUINOLINE see DAE700
1,3,5-TRIMETHOXYBENZENE see TKY250
3,4,5-TRIMETHOXYBENZENEETHANAMINE see MDI500
3,4,5-TRIMETHOXY-BENZOIC ACID 2-(4-(3-(2-CHLOROPHENOTHIAZIN-10-YL)PROPYL)-1-PIPERAZINYL)ETHYL ESTER, DIFUMARATE see MDU750
3,4,5-TRIMETHOXYBENZOIC ACID 3-((3,3-DIPHENYLPROPYL)AMINO)PROPYL ESTER HYDROCHLORIDE see DWK700
3,4,5-TRIMETHOXYBENZOIC ACID, METHYL ESTER see MQF300
6-((3,4,5-TRIMETHOXYBENZOYL)AMINO)HEXAN OIC ACID SODIUM SALT see CBF625
N-(3,4,5-TRIMETHOXYBENZOYL)GLYCINE DIETHYLAMIDE see TIH800
3,4,5-TRIMETHOXYBENZOYL METHYL RESERPATE see RDK000
4-(3,4,5-TRIMETHOXYBENZOYL)MORPHOLINE see TKX250
N-(3,4,5-TRIMETHOXYBENZOYL)MORPHOLINE see TKX250
1-(3,4,5-TRIMETHOXYBENZOYL)-4-(2-PYRIDYL)PIPERAZINE see TKY300
N-(3,4,5-TRIMETHOXYBENZOYL)TETRAHYDRO-1,4-OXAZINE see TKX250
3,4,5-TRIMETHOXYBENZOYL-N-TETRAHYDROXAZINE see TKX250
3,4,5-TRIMETHOXY-N-BENZOYLTETRAHYDROXAZINE see TKX250
5-(3,4,5-TRIMETHOXYBENZYL)-2,4-DIAMINOPYRIMIDINE see TKZ000
1-1-(3,4,5-TRIMETHOXYBENZYL)-6,7-DIHYDROXY-1,2,3,4-

TETRAHYDROISOQUINOLINE
HYDROCHLORIDE see IDD100
dl-1-(3,4,5-TRIMETHOXYBENZYL)-6,7-
DIHYDROXY-1,2,3,4-
TETRAHYDROISOQUINOLINE
HYDROCHLORIDE see TKX125
1-(2,3,4-
TRIMETHOXYBENZYL)PIPERAZINE
DIHYDROCHLORIDE see YCJ200
TRIMETHOXYBORINE see TLN000
TRIMETHOXYBOROXIN see TKZ100
TRIMETHOXYBOROXINE see TKZ100
1,1,3-TRIMETHOXYBUTANE see TLA000
TRIMETHOXY(3-CHLOROPROPYL)SILANE
see TLC300
3,4,5-TRIMETHOXYCINNAMALDEHYDE
see TLA250
3,4,5-TRIMETHOXYCINNAMIC ACID see
CMQ100
4-(3',4',5'-TRIMETHOXYCINNAMOYL)-1-
(ETHOXYCARBONYLMETHYL)PIPERAZI
NE HYDROCHLORIDE see TLA500
(3',4',5'-TRIMETHOXYCINNAMOYL)-1-(N-
ISOPROPYL AMINO CARBONYL
METHYL)-4 PIPERAZINE MALEATE see
IRQ000
TRIMETHOXYCINNAMOYL METHYL
RESERPATE see TLN500
4-(3,4,5-TRIMETHOXYCINNAMOYL)-1-
PIPERAZINEACETIC ACID ETHYL ESTER
HYDROCHLORIDE see TLA500
4-(3,4,5-TRIMETHOXYCINNAMOYL)-1-
PIPERAZINEACETIC ACID ETHYL ESTER
MALEATE see TLA525
((TRIMETHOXY-3',4',5' CINNAMOYL)-4
PIPERAZINYL)-2 ACETATE D'ETHYLE
(MALEATE) (FRENCH) see TLA525
1-(4-((3',4',5'-TRIMETHOXYCINNAMOYL)-1-
PIPERAZINYL)ACETYL)PYRROLIDINE
MALEATE see VGK000
4-(3,4,5-TRIMETHOXYCINNAMOYL)-1-(1-
PYRROLIDINYL)CARBONYLMETHYLPIP
ERAZINE MALEATE see VGK000
6,6',12'-TRIMETHOXY-2,2'-
DIMETHYLOXYACANTHAN-7-OL see
HGI050
TRI-(2-METHOXYETHANOL)PHOSPHATE
see TLA600
TRIMETHOXYFOSFIN see TMD500
TRIMETHOXYMETHANE see TLX600
3,4,5-TRIMETHOXY-α-METHYL-β-
PHENYLETHYLAMINE
HYDROCHLORIDE see TKX500
TRIMETHOXYMETHYLSILANE see
MQF500
1,2,10-TRIMETHOXY-6a-α-NORAPORPHIN-
6-OL see LBO200
3,4,5-TRIMETHOXYPHENETHYLAMINE
see MDI500
3,4,5-TRIMETHOXYPHENETHYLAMINE
HYDROCHLORIDE see MDI750
3,4,5-TRIMETHOXYPHENYLACRYLIC
ACID see CMQ100
(6-(3-(3,4,5-TRIMETHOXYPHENYL)-d-
ALANINE))-LHRH ACETATE see LIU327
1-(3,4,5-TRIMETHOXYPHENYL)-2-
AMINOPROPANE see TKX500
TRIMETHOXYPHENYL-β-
AMINOPROPANE see MLH250
3,4,5-TRIMETHOXYPHENYL-β-
AMINOPROPANE see MLH250
3,4,5-TRIMETHOXY-β-
PHENYLETHYLAMINE
HYDROCHLORIDE see MDI750
1-((2,3,4-
TRIMETHOXYPHENYL)METHYL)PIPERAZ
INE DIHYDROCHLORIDE (9CI) see YCJ200
5-(3,4,5-TRIMETHOXYPHENYL)-METHYL)-
2,4-PYRIMIDINEDIAMINE see TKZ000
3-(3,4,5-TRIMETHOXYPHENYL)-2-
PROPENAL see TLA250

3-(3,4,5-TRIMETHOXYPHENYL)-2-
PROPENOIC ACID see CMQ100
TRIMETHOXYPHOSPHINE see TMD500
(Z)-1,2,4-TRIMETHOXY-5-(1-
PROPENYL)BENZENE see ARM100
3,4,5-TRIMETHOXY-N-(4-
PROPYLCYCLOHEXYL)BENZAMIDE see
TLA650
TRIMETHOXY SILANE see TLB750
TRIMETHOXYSILANE (DOT) see TLB750
2-(TRIMETHOXYSILYL)ETHANETHIOL see
MCO250
(TRIMETHOXYSILYL)ETHENE see TLD000
TRIMETHOXYSILYLPROPANETHIOL see
TLC000
3-(TRIMETHOXYSILYL)-1-PROPANOL
METHACRYLATE see TLC250
3-(TRIMETHOXYSILYL)PROPYL
CHLORIDE see TLC300
TRIMETHOXYSILYL-3-PROPYLESTER
KYSELINY METHAKRYLOVE (CZECH) see
TLC250
3-(TRIMETHOXYSILYL)PROPYL ESTER
METHACRYLIC ACID see TLC250
N-(3-TRIMETHOXYSILYLPROPYL)-
ETHYLENEDIAMINE see TLC500
N-(3-(TRIMETHOXYSILYL)PROPYL)-1,3-
PROPANEDIAMINE see TLC600
3-(TRIMETHOXYSILYL)-N-(3-
(TRIMETHOXYSILYL)PROPYL)-1-
PROPANAMINE see TLC650
4-(3,4,5-
TRIMETHOXYTHIOBENZOYL)MORPHOLI
NE see TNP275
4-(3,4,5-
TRIMETHOXYTHIOBENZOYL)TETRAHY
DRO-1,4-OXAZINE see TNP275
3,4,5-TRIMETHOXY-α-VINYLBENZYL
ALCOHOL ACETATE see TLC850
TRIMETHOXYVINYLSILANE see TLD000
2,4,6-TRIMETHYLACETANILIDE see
TLD250
2',4',6'-TRIMETHYLACETANILIDE see
TLD250
TRIMETHYLACETHYDRAZIDE
AMMONIUM CHLORIDE see GEQ500
TRIMETHYLACETIC ACID see PJA500
TRIMETHYLACETIC ACID SODIUM SALT
see SGM600
TRIMETHYLACETONITRILE see PJA750
TRIMETHYL-β-
ACETOXYPROPYLAMMONIUM
CHLORIDE see ACR000
TRIMETHYLACETYL AZIDE see PJB000
TRIMETHYL ACETYL CHLORIDE (DOT)
see DTS400
TRIMETHYLADIPIC ACID see TLT250
TRIMETHYLALANE see TLD272
TRIMETHYLALUMINIUM see TLD272
TRIMETHYLALUMINUM see TLD272
TRIMETHYLAMINE see TLD500
TRIMETHYLAMINE BORANE see BMB250
TRIMETHYLAMINE, COMPOUND with
BORANE (1:1) see BMB250
TRIMETHYLAMINE HYDROCHLORIDE see
TLF750
TRIMETHYLAMINE, ISOPROPYLIDENE-
see TME255
TRIMETHYLAMINE OXIDE see TLE100
TRIMETHYLAMINE OXIDE, DIHYDRATE
see TLE250
TRIMETHYLAMINE OXIDE
PERCHLORATE see TLE500
TRIMETHYLAMINE-N-OXIDE
PERCHLORATE see TLE500
α-α',α''-
TRIMETHYLAMINETRICARBOXYLIC
ACID see AMT500
TRIMETHYLAMINE, anhydrous (UN 1083)
(DOT) see TLD500

TRIMETHYLAMINE, aqueous solutions not
>50% trimethylamine, by weight (UN 1297)
(DOT) see TLD500
TRIMETHYLAMINOACETOHYDRAZIDE
CHLORIDE see GEQ500
2,N,N-TRIMETHYL-4-
AMINOAZOBENZENE see TLE750
1,2,4-TRIMETHYL-5-AMINOBENZENE see
TLG250
1,2,4-TRIMETHYL-5-AMINOBENZENE
HYDROCHLORIDE see TLG750
3,2',5'-TRIMETHYL-4-AMINODIPHENYL see
TLF000
S-(2-TRIMETHYLAMINOETHYL)-1'-
ETHYLISOTHIURONIUM BROMIDE
HYDROBROMIDE see EQN700
TRIMETHYLAMINOMETHANE see BPY250
TRIMETHYL-3-AMINOPHENYLARSONIUM
CHLORIDE see TLF250
S-(2-(N,N,N-
TRIMETHYLAMMONIO)ETHYL)-O,O-
DIETHYLPHOSPHOROTHIOLATE
IODIDE see TLF500
((4-TRIMETHYLAMMONIO)-3-
ISOPROPYL)PHENYL ESTER,
DIMETHYLCARBAMIC ACID IODIDE see
HJZ000
3-(TRIMETHYLAMMONIO)PHENOL
IODIDE see HNN500
TRIMETHYLAMMONIUM ACETYL
HYDRAZIDE CHLORIDE see GEQ500
TRIMETHYLAMMONIUM CHLORIDE see
TLF750
TRIMETHYLAMMONIUMOXID HYDRAT
(CZECH) see TLE250
TRIMETHYLAMMONIUM PERCHLORATE
see TLF800
N-(3-TRIMETHYLAMMONIUMPROPYL)-N-
METHYLCAMPHIDINIUM
BIS(METHYLSULFATE) see TLG000
N-(γ-TRIMETHYLAMMONIUMPROPYL)-N-
METHYLCAMPHIDINIUM DIMETHYL
SULFATE see TLG000
N-(γ-TRIMETHYLAMMONIUMPROPYL)-N-
METHYL-CAMPHIDINIUM-
DIMETHYLSULPHATE (GERMAN) see
TLG000
N-(γ-TRIMETHYLAMMONIUMPROPYL)-N-
METHYLCAMPHIDINIUM METHYL
SULFATE (GERMAN) see TLG000
4,4,17-α-TRIMETHYLANDROST-5-ENO(2,3-
d)ISOXAZOL-17-OL see HOM259
4,4',6-TRIMETHYLANGELICIN plus
ULTRAVIOLET A RADIATION see FQD140
2,4,5-TRIMETHYLANILIN (CZECH) see
TLG250
2,4,5-TRIMETHYLANILINE see TLG250
2,4,6-TRIMETHYLANILINE see TLG500
m,N,N-TRIMETHYLANILINE see TLG700
N,N,2-TRIMETHYLANILINE see DUH200
N,N,3-TRIMETHYLANILINE see TLG700
N,N,4-TRIMETHYLANILINE see TLG150
p,N,N-TRIMETHYLANILINE see TLG150
2,4,5-TRIMETHYLANILINE
HYDROCHLORIDE see TLG750
2,4,6-TRIMETHYLANILINE
HYDROCHLORIDE see TLH000
TRIMETHYLANILINIUM CHLORIDE see
TMB250
N,N,N-TRIMETHYLANILINIUM
CHLORIDE see TMB250
TRIMETHYLANILINIUM HYDROXIDE see
TMB500
TRIMETHYLANILINIUM IODIDE see
TMB750
N,N,N-TRIMETHYLANILINIUM IODIDE
see TMB750
2,9,10-TRIMETHYLANTHRACENE see
TLH050
TRIMETHYL ANTIMONY see TLH100
TRIMETHYL ARSINE see TLH150
TRIMETHYLARSINE OXIDE see TLH170

TRIMETHYLARSINE SELENIDE see TLH250

3,5,9-TRIMETHYL-1:2-BENZACRIDINE see TLH350

3,8,12-TRIMETHYLBENZ(a)ACRIDINE see TLH500

5,6,9-TRIMETHYL-1:2-BENZACRIDINE see TLI000

5,7,11-TRIMETHYLBENZ(c)ACRIDINE see TLH350

5,7,8-TRIMETHYL-3:4-BENZACRIDINE see TLH750

7,8,11-TRIMETHYLBENZ(c)ACRIDINE see TLI000

7,9,10-TRIMETHYLBENZ(c)ACRIDINE see TLI250

7,9,11-TRIMETHYLBENZ(c)ACRIDINE see TLI500

8,10,12-TRIMETHYLBENZ(a)ACRIDINE see TLI750

1,3,10-TRIMETHYL-5,6-BENZACRIDINE (FRENCH) see TLI750

1,3,10-TRIMETHYL-7,8-BENZACRIDINE (FRENCH) see TLI500

1,4,10 TRIMETHYL,7:8 BENZACRIDINE (FRENCH) see TLI000

1,6,10-TRIMETHYL,7:8 BENZACRIDINE (FRENCH) see TLH350

2,3,10 TRIMETHYL,5:6 BENZACRIDINE (FRENCH) see TLH750

2,3,10-TRIMETHYL-7:8-BENZACRIDINE (FRENCH) see TLI250

1,10,3'-TRIMETHYL-5,6-BENZACRIDINE (FRENCH) see TLH500

2,4,6-TRIMETHYLBENZALDEHYDE see MDJ745

6,7,8-TRIMETHYLBENZ(a)ANTHRACENE see TLK000

4,5,10-TRIMETHYLBENZ(a)ANTHRACENE see TLJ250

4,7,12-TRIMETHYLBENZ(a)ANTHRACENE see TLJ500

6,7,12-TRIMETHYLBENZ(a)ANTHRACENE see TLJ750

6,8,12-TRIMETHYLBENZ(a)ANTHRACENE see TLK500

7,8,12-TRIMETHYLBENZ(a)ANTHRACENE see TLK750

7,9,12-TRIMETHYLBENZ(a)ANTHRACENE see TLK600

4,9,10-TRIMETHYL-1,2-BENZANTHRACENE see TLJ750

5:9:10-TRIMETHYL-1:2-BENZANTHRACENE see TLK750

6,9,12-TRIMETHYL-1,2-BENZANTHRACENE see TLK600

7,10,12-TRIMETHYLBENZ(a)ANTHRACENE see TLL000

2,4,5-TRIMETHYLBENZENAMINE see TLG250

2,4,6-TRIMETHYLBENZENAMINE see TLG500

N,N,3-TRIMETHYLBENZENAMINE see TLG700

2,4,5-TRIMETHYLBENZENAMINE HYDROCHLORIDE see TLG750

2,4,6-TRIMETHYLBENZENAMINE HYDROCHLORIDE see TLH000

N,N,N-TRIMETHYLBENZENAMINIUM IODIDE see TMB750

TRIMETHYL BENZENE see TLL250

TRIMETHYL BENZENE see TLL500

as-TRIMETHYL BENZENE see TLL750

1,2,3-TRIMETHYL BENZENE see TLL500

1,2,4-TRIMETHYL BENZENE see TLL750

1,2,5-TRIMETHYL BENZENE see TLL750

1,3,5-TRIMETHYL BENZENE see TLM050

sym-TRIMETHYLBENZENE see TLM050

TRIMETHYL BENZENE (ACGIH) see TLM050

TRIMETHYL BENZENE (mixed isomers) see TLL250

1,2,4-TRIMETHYL BENZENE mixed with MESITYLENE (5:3) see TLM000

N,N,N-TRIMETHYLBENZENEMETHANAMINIUM CHLORIDE see BFM250

N,N,N-TRIMETHYLBENZENEMETHANAMINIUM METHOXIDE see TLM100

TRIMETHYL-N-(p-BENZENESULPHONAMIDO)PHOSPHORIMIDATE see TLM250

2,4,6-TRIMETHYLBENZOIC ACID see IKN300

TRIMETHYL BENZOL see TLM050

TRIMETHYLBENZOLOXYPIPERIDINE see EQP500

1,3,6-TRIMETHYLBENZO(a)PYRENE see TLM500

TRIMETHYLBENZOQUINONE see POG400

TRIMETHYL-p-BENZOQUINONE see POG400

TRIMETHYL-1,4-BENZOQUINONE see POG400

2,3,5-TRIMETHYLBENZOQUINONE see POG400

2,3,6-TRIMETHYLBENZOQUINONE see POG400

2,3,5-TRIMETHYL-p-BENZOQUINONE see POG400

N,N,α-TRIMETHYLBENZYLAMINE see DSO800

TRIMETHYLBENZYLAMMONIUM CHLORIDE see BFM250

TRIMETHYLBENZYLAMMONIUM HYDROXIDE see BFM500

endo-1,7,7-TRIMETHYL-BICYCLO(2.2.1)HEPTAN-2-OL see BMD000

(1R-endo)-1,3,3-TRIMETHYLBICYCLO(2.2.1)HEPTAN-2-OL see TLW000

(1-α-2-α-5α--2,6,6-TRIMETHYLBICYCLO(3.1.1)HEPTAN)-2-OL see PIH050

1,3,3-TRIMETHYLBICYCLO(2.2.1)HEPTAN-2-ONE see FAM300

1,7,7-TRIMETHYLBICYCLO(2.2.1)-2-HEPTANONE see CBA750

(1R)-1,3,3-TRIMETHYLBICYCLO(2.2.1)HEPTAN-2-ONE see FAM300

(1R)-1,7,7-TRIMETHYL-BICYCLO(2.2.1)HEPTAN-2-ONE see CBB250

3,7,7-TRIMETHYLBICYCLO(4.1.0)-3-HEPTENE see CCK500

2,6,6-TRIMETHYLBICYCLO(3.1.1)-2-HEPT-2-ENE see PIH250

1-(3,7,7-TRIMETHYLBICYCLO(4.1.0)HEPTENYL)ETHANONE see ACF150

endo-1,7,7-TRIMETHYLBICYCLO(2.2.1)HEPT-2-YL 3-METHYLBUTANOATE see HOX100

4-((1,7,7-TRIMETHYLBICYCLO(2.2.1)HEPT-2-YL)OXY)BENZENAMINE see IHY500

4,6,6-TRIMETHYLBICYKLO(3,1,1)HEPT-3-EN see PIH250

3,2',5'-TRIMETHYLBIPHENYLAMINE see TLF000

TRIMETHYL BISMUTH see TLM750

TRIMETHYLBISMUTHINE see TLM750

TRIMETHYLBORANE see TLM775

TRIMETHYL BORATE see TLN000

2,4,6-TRIMETHYLBORAZINE see TLN100

TRIMETHYLBROMOMETHANE see BQM250

β,γ,γ-TRIMETHYLCAPROALDEHYDE THIOSEMICARBAZONE see TLN125

1,4,9-TRIMETHYLCARBAZOLE see TLN133

TRIMETHYLCARBINOL see BPX000

TRIMETHYLCETYLAMMONIUM BROMIDE see HCQ500

TRIMETHYLCETYLAMMONIUM PENTACHLOROPHENATE see TLN150

TRIMETHYLCHINON see POG400

TRIMETHYL-β-CHLORETHYLAMMONIUMCHLORID see CMF400

TRIMETHYLCHLOROMETHANE see BQR000

TRIMETHYL CHLOROSILANE see TLN250

TRIMETHYLCHLOROSILANE (DOT) see TLN250

TRIMETHYLCHLOROSTANNANE see CLT000

TRIMETHYLCHLOROTIN see CLT000

3,4,5-TRIMETHYLCINNAMIC ACID, ESTER with METHYL RESERPATE see TLN500

3,4,5-TRIMETHYLCINNAMOYL METHYL RESERPATE see TLN500

TRIMETHYLCOLCHICINIC ACID see TLN750

TRIMETHYLCOLCHICINIC ACID METHYL ETHER see TLO000

TRIMETHYLCOLCHICINSAEUREMETHYL ESTER (GERMAN) see TLO000

3,3,5-TRIMETHYL-5-CYANOCYCLOHEXANONE see IMG500

2,3,5-TRIMETHYL-2,5-CYCLOHEXADIEN-1,4-DIONE see POG400

3,3,5-TRIMETHYLCYCLOHEXANECARBOXALDEHYDE see TLO250

α-α-4-TRIMETHYLCYCLOHEXANEMETHANOL see MCE100

3,3,5-TRIMETHYLCYCLOHEXANOL see TLO500

3,5,5-TRIMETHYLCYCLOHEXANOL see TLO500

3,3,5-TRIMETHYL-1-CYCLOHEXANOL see TLO500

cis-3,3,5-TRIMETHYLCYCLOHEXANOL see TLO510

3,3,5-TRIMETHYLCYCLOHEXANOL-α-PHENYL-α-HYDROXYACETATE see DNU100

α,α,4-TRIMETHYL-3-CYCLOHEXENE-1-METHANOL see TBD750

1,2,4 al 1,3,5-TRIMETHYL-3-CYCLOHEXENE-1-METHANOL MIXTURE see ISR100

4-(2,6,6-TRIMETHYL-2(and/or 3)-CYCLOHEXENE)-2-METHYLBUTANAL see TDO255

1,1,3-TRIMETHYL-3-CYCLOHEXENE-5-ONE see IMF400

3,5,5-TRIMETHYL-2-CYCLOHEXENE-1-ONE see IMF400

3,5,5-TRIMETHYL-2-CYCLOHEXEN-1-ON (GERMAN, DUTCH) see IMF400

4-(2,6,6-TRIMETHYL-2-CYCLOHEXEN-1-YL)-2-BUTANONE see DLP800

4-(2,6,6-TRIMETHYL-1-CYCLOHEXEN-1-YL)-3-BUTEN-2-OL see IFT500

4-(2,6,6-TRIMETHYL-2-CYCLOHEXEN-1-YL)-3-BUTEN-2-OL see IFT400

4-(2,6,6-TRIMETHYL-2-CYCLOHEXEN-1-YL)-3-BUTEN-2-OL ACETATE see IFX300

4-(2,4,6-TRIMETHYL-3-CYCLOHEXEN-1-YL)-3-BUTEN-2-ONE see IGJ600

4-(2,6,6-TRIMETHYL-1-CYCLOHEXEN-1-YL)-3-BUTEN-2-ONE see IFX000

4-(2,6,6-TRIMETHYL-2-CYCLOHEXEN-1-YL)-3-BUTEN-2-ONE see IFW000

trans-4-(2,6,6-TRIMETHYL-1-CYCLOHEXEN-1-YL)-3-BUTEN-2-ONE see IFX100

1-(2,6,6-TRIMETHYL-2-CYCLOHEXEN-1-YL)-1,6-HEPTADIEN-3-ONE see AGI500

1-(2,6,6-TRIMETHYL-2-CYCLOHEXEN-1-YL)-3-PENTANONE see TLO530

3,5,5-TRIMETHYLCYCLOHEXYL AMYGDALATE see DNU100

TRIMETHYL LEAD CHLORIDE see TLU175

N²,N⁴,N⁶-TRIMETHYLMELAMINE see TNJ825

TRIMETHYL((2-METHYL-1,3-DIOXOLAN-4-YL)METHYL)AMMONIUM IODIDE (8CI) see MJH250

1,3,3-TRIMETHYL-2-METHYLENEINDOLINE see TLU200

4,11,11-TRIMETHYL-8-METHYLENE-5-OXATRICYCLO(8.2.0.0(4,6))DODECANE see CCN100

TRIMETHYL (1-METHYL-2-PHENOTHIAZIN-10-YLETHYL)AMMONIUM METHYL SULFATE see MRW000

TRIMETHYL(1-METHYL-2-(10-PHENOTHIAZINYL)ETHYL)AMMONIUM METHYL SULFATE see MRW000

N,N,N'-TRIMETHYL-N'-(5-METHYL-3-PHENYL-1H-INDOL-1-YL)-1,2-ETHANEDIAMINE HYDROCHLORIDE see MNU750

N,N,2-TRIMETHYL-4-(4'-(3'-METHYLPYRIDYL-1'-OXIDE)AZO)ANILINE see DQE000

N,N,3-TRIMETHYL-4-(4'-(3'-METHYLPYRIDYL-1'-OXIDE)AZO)ANILINE see DQE200

TRIMETHYLMYRISTYLAMMONIUM BROMIDE see TCB200

1,3,3-TRIMETHYL-6'-NITROINDOLINE-2-SPIRO-2'-BENZOPYRAN see TLU500

TRIMETHYLNITROSOHARNSTOFF (GERMAN) see TLU750

N,2,2-TRIMETHYL-N-NITROSO-1-PROPYLAMINE see NKU570

N-TRIMETHYL-N-NITROSOUREA see TLU750

1,1,3-TRIMETHYL-3-NITROSOUREA see TLU750

2,6,8-TRIMETHYLNONANOL-4 see TLV000

2,6,8-TRIMETHYL-4-NONANOL see TLV000

TRIMETHYL NONANONE see TLV250

2,6,8-TRIMETHYLNONYL VINYL ETHER see VQU000

(−)-endo-1,3,3-TRIMETHYL-2-NORBORNANOL see TLW000

1,3,3-TRIMETHYL-2-NORBORNANOL ACETATE see FAO000

1,3,3-TRIMETHYL-2-NORBORNANONE see TLW250

d-1,3,3-TRIMETHYL-2-NORBORNANONE see FAM300

1,3,3-TRIMETHYL-2-NORBORNANYL ACETATE see FAO000

1,3,3-TRIMETHYL-2-NORCAMPHANONE see TLW250

d-1,3,3-TRIMETHYL-2-NORCAMPHANONE see FAM300

1,7,7-TRIMETHYLNORCAMPHOR see CBA750

3,7,7-TRIMETHYL-3-NORCARENE see CCK500

4,7,7-TRIMETHYL-3-NORCARENE see CCK500

TRIMETHYLOCTADECYLAMMONIUM CHLORIDE see TLW500

TRIMETHYLOLAMINOMETHANE see TEM500

TRIMETHYLOLETHANE TRINITRATE see MQG600

1,1,1-TRIMETHYLOLETHANE TRINITRATE see MQG600

TRIMETHYLOLMELAMINE see TLW750

TRIMETHYLOLNITROMETHANE see HMJ500

TRIMETHYLOL NITROMETHANE TRINITRATE (DOT) see NHK650

TRIMETHYLOLPHOSPHINE see TNI600

TRIMETHYLOLPROPANE see HDF300

1,1,1-TRIMETHYLOLPROPANE see HDF300

TRIMETHYLOLPROPANE DIALLYL ETHER see TLX000

TRIMETHYLOLPROPANE ETHOXYTRIACRYLATE see TLX100

TRIMETHYLOLPROPANE TRIACETOACETATE see ELJ600

TRIMETHYLOLPROPANE TRIACRYLATE see TLX175

TRIMETHYLOLPROPANE TRIMETHACRYLATE see TLX250

TRIMETHYLOLPROPANE TRIMETHANCRYLATE see TLX250

TRIMETHYLOPROPANE MONOALLYL ETHER see TLX110

TRIMETHYL ORTHOFORMATE see TLX600

1,3,3-TRIMETHYL-2-OXABICYCLO(2.2.2)OCTANE see CAL000

TRIMETHYLOXACYCLOPROPANE see TLY175

3,5,5-TRIMETHYL-2,4-OXAZOLIDINEDIONE see TLP750

2,4,5-TRIMETHYL Δ-3-OXAZOLINE see TLX800

TRIMETHYL-2-OXEPANONE (mixed isomers) see TLY000

TRIMETHYLOXIRANE see TLY175

2,2,3-TRIMETHYLOXIRANE see TLY175

N,N,N-TRIMETHYLOXIRANEMETHANAMINIUM CHLORIDE see GGY200

(3,5,5-TRIMETHYL-1-OXOHEXYL)FERROCENE see TLT300

2,4,8-TRIMETHYL-5-OXO-6-OXA-3,9-DITHIA-2,4,7-TRIAZADEC-7-ENOIC ACID ETHYL ESTER see TLY200

N,N,N-TRIMETHYL-4-OXO-1-PENTANAMINIUM IODIDE see TLY250

N,N,N-TRIMETHYL-2-(1-OXOPROPOXY)ETHANAMINIUM IODIDE see PMW750

TRIMETHYL(4-OXYPENTHYL)AMMONIUM IODIDE see TLY250

N,N,N-TRIMETHYL-1-PENTANAMINIUM IODIDE see TMA500

2,2,4-TRIMETHYLPENTANE see TLY500

2,2,4-TRIMETHYL-1,3-PENTANEDIOL see TLY750

TRI(2-METHYL-2,4-PENTANEDIOL)BIBORATE see TKM250

2,2,4-TRIMETHYL-1,3-PENTANEDIOL DIISOBUTYRATE see TLZ000

2,2,4-TRIMETHYLPENTANEDIOL-1,3-DIISOBUTYRATE see TLZ000

2,2,4-TRIMETHYL-1,3-PENTANEDIOL MONOISOBUTYRATE see TEG500

2,4,4-TRIMETHYL-2-PENTANETHIOL see MKJ250

2,4,4-TRIMETHYL-2-PENTANETHIOL see OFE030

2,2,4-TRIMETHYLPENTANOL see TLZ100

2,4,4-TRIMETHYL PENTENE see TMA250

TRIMETHYLPENTYLAMMONIUM IODIDE see TMA500

6,6,9-TRIMETHYL-3-PENTYL-6H-DIBENZO(b,d)PYRAN-1-OL see CBD625

6,6,9-TRIMETHYL-3-PENTYL-7,8,9,10-TETRAHYDRO-6H-DIBENZO(B,D)PYRAN-1-OL see TCM250

N⁽ᴮ⁾,N⁽ᴮ⁾,3-TRIMETHYL-2,8-PHENAZINEDIAMINE MONOHYDROCHLORIDE see AJQ250

N,N,α-TRIMETHYLPHENETHYLAMINE see TMA600

2,4,6-TRIMETHYLPHENOL see MDJ740

N,N,α-TRIMETHYL-10H-PHENOTHIAZINE-10-ETHANAMINE MONOHYDROCHLORIDE see PMI750

N,N,β-TRIMETHYL-10H-PHENOTHIAZINE-10-PROPANAMINE see AFL500

N-(2,4,6-TRIMETHYLPHENYL)ACETAMIDE see TLD250

(6-(3-(2,4,6-TRIMETHYLPHENYL)-d-ALANINE))-LHRH ACETATE see LIU351

1-(2,4,6-TRIMETHYLPHENYLAMINO)ANTHRAQUINONE see TMA750

TRIMETHYLPHENYLAMMONIUM BROMIDE see TMB000

TRIMETHYLPHENYL AMMONIUM CHLORIDE see TMB250

TRIMETHYLPHENYLAMMONIUM HYDROXIDE see TMB500

TRIMETHYLPHENYLAMMONIUM IODIDE see TMB750

N-sym-TRIMETHYLPHENYLDIETHYLAMINOACETAMIDE HYDROCHLORIDE see DHL800

N,N,N'-TRIMETHYL-N'-(3-PHENYL-1H-INDOL-1-YL)-1,2-ETHANEDIAMINE MONOHYDROCHLORIDE see BGC500

TRIMETHYLPHENYLMETHANE see BQJ250

TRIMETHYLPHENYL METHYLCARBAMATE see TMC750

3,4,5-TRIMETHYLPHENYL METHYLCARBAMATE see TMD000

TRI-2-METHYLPHENYL PHOSPHATE see TMO600

1,2,5-TRIMETHYL-4-PHENYL-4-PIPERIDINOL, PROPIONATE (ESTER) see IMT000

1,2,5-TRIMETHYL-4-PHENYL-4-PROPIONOXYPIPERIDINE see IMT000

TRIMETHYL PHOSPHATE see TMD250

O,O,O-TRIMETHYL PHOSPHATE see TMD250

TRIMETHYLPHOSPHINE see TMD275

TRIMETHYLPHOSPHINE SELENIDE see TMD400

TRIMETHYL PHOSPHITE see TMD500

TRIMETHYL PHOSPHITE see TMD500

O,S,S-TRIMETHYL PHOSPHORODITHIOATE see TMD625

TRIMETHYL PHOSPHOROTHIOATE see TMH525

O,O,S-TRIMETHYL PHOSPHOROTHIOATE see DUG500

TRIMETHYL PHOSPHOROTRITHIOATE see TMD650

S,S,S-TRIMETHYL PHOSPHOROTRITHIOATE see TMD699

2,2,6-TRIMETHYL-4-PIPERIDINOL BENZOATE (ESTER) see EQP500

TRIMETHYLPLATINUM HYDROXIDE see TME250

N,N,2-TRIMETHYLPROPENYLAMINE see TME255

1,2,2-TRIMETHYLPROPYL METHYLPHOSPHONOFLUORIDATE see SKS500

2,5,6-TRIMETHYL-7-PROPYLTHIOHEPT-1-EN-3-YN-5-OL see TME260

2',4,8-TRIMETHYLPSORALEN see HOM270

4,5',8-TRIMETHYLPSORALEN see HOM270

TRIMETHYLPYRAZINE see TME270

2,3,5-TRIMETHYLPYRAZINE see TME270

2,4,6-TRIMETHYLPYRIDINE see TME272

2,4,6-TRIMETHYLPYRILIUM PERCHLORATE see TME275

TRIMETHYLQUINONE see POG400

TRIMETHYL-p-QUINONE see POG400

2,3,5-TRIMETHYLQUINONE see POG400

2,3,5-TRIMETHYL-6-QUINOXALINAMINE see QQS340

N,2,5-TRIMETHYL-6-QUINOXALINAMINE see TME300

N,3,5-TRIMETHYL-6-QUINOXALINAMINE see TME305

TRIMETHYLSELENONIUM see TME500

TRIMETHYLSELENONIUM CHLORIDE see TME600

TRIMETHYLSELENONIUM ION see TME500

N-TRIMETHYLSILYLACETAMIDE see TME750

N-(TRIMETHYLSILYL)ACETIMIDIC ACID, TRIMETHYLSILYL ESTER see TMF000

TRIMETHYLSILYL AZIDE see TMF100

TRIMETHYL SILYL-1-CHLORO-7-DIHYDRO-1,3-PHENYL-5,2H-BENZODIAZEPINE-1,4-ONE-2 (FRENCH) see CGB000

N-(TRIMETHYLSILYL)ETHANIMIDIC ACID, TRIMETHYLSILYL ESTER see TMF000

TRIMETHYL SILYL HYDROPEROXIDE see TMF125

N-(TRIMETHYLSILYL)-IMIDAZOL see TMF250

(TRIMETHYLSILYL)IMIDAZOLE see TMF250

1-(TRIMETHYLSILYL)IMIDAZOLE see TMF250

N-(TRIMETHYLSILYL)IMIDAZOLE see TMF250

1-(TRIMETHYLSILYL)-1H-IMIDAZOLE see TMF250

TRIMETHYLSILYLMETHANOL see TMF500

N-TRIMETHYLSILYLMETHYL-N-NITROSOUREA see TMF600

TRIMETHYLSILYL PERCHLORATE see TMF625

TRIMETHYLSTANNANE SULPHATE see TMI500

TRIMETHYLSTANNYL CHLORIDE see CLT000

TRIMETHYLSTANNYL IODINE see IFN000

TRIMETHYLSTEARYLAMMONIUM CHLORIDE see TLW500

TRIMETHYL STIBINE see AQL000

N,N,2'-TRIMETHYL-4-STILBENAMINE see TMF750

N,N,3'-TRIMETHYL-4-STILBENAMINE see TMG000

N,N,4'-TRIMETHYL-4-STILBENAMINE see TMG250

TRIMETHYLSTILBINE see TLH100

TRIMETHYLSULFONIUM IODIDE see TMG500

TRIMETHYLSULFONIUM, IODIDE, COMPOUND with MERCURY IODIDE (1:1) see TMG750

TRIMETHYLSULFONIUM IODIDE MERCURIC IODIDE addition compound see TMG750

TRIMETHYLSULFOXONIUM BROMIDE see TMG775

TRIMETHYL SULPHOXONIUM IODIDE see TMG800

N,N,N-TRIMETHYL-1-TETRADECANAMINIUM BROMIDE see TCB200

(E,E,E)-5,9,13-TRIMETHYL-4,8,12-TETRADECATRIENOIC ACID 3,7-DIMETHYL-2,6-OCTADIENYL ESTER see GDG200

TRIMETHYLTETRADECYLAMMONIUM BROMIDE see TCB200

TRIMETHYL(TETRAHYDRO-4-HYDROXY-5-METHYLFURFURYL)AMMONIUM see MRW250

TRIMETHYLTHALLIUM see TMH250

3,3,6-TRIMETHYL-2,5-THIOMORPHOLINEDIONE-2-(o-((METHYLAMINO)CARBONYL)OXIME) see TMH300

TRIMETHYL THIOPHOSPHATE see TMH525

TRIMETHYLTHIOUREA see TMH750

1,1,3-TRIMETHYL-2-THIOUREA see TMH750

N,N,N'-TRIMETHYLTHIOUREA see TMH750

TRIMETHYLTIN ACETATE see TMI000

TRIMETHYLTIN CHLORIDE see CLT000

TRIMETHYLTIN CYANATE see TMI100

TRIMETHYL TIN HYDROXIDE see TMI250

TRIMETHYLTIN IODIDE see IFN000

TRIMETHYLTIN ISOTHIOCYANATE see ISF000

TRIMETHYLTIN SULPHATE see TMI500

TRIMETHYLTIN THIOCYANATE see TMI750

5,7,8-TRIMETHYLTOCOL see VSZ450

6,6',6"-TRIMETHYLTRIHEPTYLAMINE see TKS500

2,2'-((1,1,3-TRIMETHYLTRIMETHYLENE)DIOXY)BIS(4,4,6-TRIMETHYL)1,3,2-DIOXABORINANE see TKM250

α,α,α'-TRIMETHYLTRIMETHYLENE GLYCOL see HFP875

1,1,1-TRIMETHYL-N-(TRIMETHYLSILYL)SILANAMINE see HED500

1,3,5-TRIMETHYL-2,4,6-TRINITROBENZENE (DOT) see TMI800

2,4,6-TRIMETHYL-1,3,5-TRIOXAAN (DUTCH) see PAI250

2,4,6-TRIMETHYL-s-TRIOXANE see PAI250

2,4,6-TRIMETHYL-1,3,5-TRIOXANE see PAI250

s-TRIMETHYLTRIOXYMETHYLENE see PAI250

1,3,5-TRIMETHYL-2,4,6-TRIS(3,5-DI-tert-BUTYL-4-HYDROXYBENZYL) BENZENE see TMJ000

2,4,6-TRIMETHYL-2,4,6-TRIS(3,3,3-TRIFLUOROPROPYL)CYCLOTRISILOXANE see FLV100

2,6,9-TRIMETHYLUNDECA-2,6,8-TRIEN-10-ONE see TMJ100

3,6,10-TRIMETHYL-3,5,9-UNDECATRIEN-2-ONE see TMJ100

2,6,10-TRIMETHYL-9-UNDECENAL see TMJ150

TRIMETHYL UNDECYLENIC ALDEHYDE see TMJ150

TRIMETHYLUREA see TMJ250

1,1,3-TRIMETHYLUREA see TMJ250

TRIMETHYL(p-UREIDOPHENYL)AMMONIUM IODIDE see TMJ300

TRIMETHYL VINYL AMMONIUM HYDROXIDE see VQR300

1,3,7-TRIMETHYLXANTHINE see CAK500

TRIMETHYL(2-(2,6-XYLYLOXY)ETHYL)AMMONIUM BROMIDE see XSS900

2-(TRIMETIL-ACETIL)-INDAN-1,3-DIONE (ITALIAN) see PIH175

3,5,5-TRIMETIL-2-CICLOESEN-1-ONE (ITALIAN) see IMF400

2,4,6-TRIMETIL-1,3,5-TRIOSSANO (ITALIAN) see PAI250

TRIMETIN see TLP750

TRIMETION see DSP400

3,7,11-TRIMETNYL-1,6,10-DODECATRIEN-3-OL see NCN700

TRIMETOGINE see TKX250

TRIMETON see PMS900

TRIMETON see TMJ750

TRIMETON MALEATE see TMK000

TRIMETOPRIM-SULFA see SNK000

TRIMETOQUINOL see IDD100

1-TRIMETOQUINOL see IDD100

TRIMETOQUINOL HYDROCHLORIDE see IDD100

(+)-TRIMETOQUINOL HYDROCHLORIDE see TMJ800

(±)-TRIMETOQUINOL HYDROCHLORIDE see TKX125

dl-TRIMETOQUINOL HYDROCHLORIDE see TKX125

R-(+)-TRIMETOQUINOL HYDROCHLORIDE see TMJ800

TRIMETOSE see TMK000

TRIMETOZIN see TKX250

TRIMETOZINA see TKX250

TRIMETOZINE see TKX250

2,4,6-TRIMETYLOFENOL see MDJ740

TRIMEX T 08 see TJR600

TRIMFORTE see TKX000

TRIMIDAL see CHJ300

TRIMIFRUIT SC see CHJ300

TRI-MINO see MQW100

TRI-MINOCYCLINE see MQW100

TRIMINOL see CHJ300

TRIMIPRAMINE see DLH200

(+)-TRIMIPRAMINE see TMK100

TRIMIPRAMINE ACID MALEATE see SOX550

TRIMIPRAMINE MALEATE see SOX550

TRIMIPROMINE HYDROCHLORIDE see TAL000

TRIMITAN see TNF500

TRI(MIXED MONO- and DINONYLPHENYL)PHOSPHITE see TMX550

TRIMMIT see TMK125

TRIMOL see TMK150

TRIMON see AQH500

TRIMONASE see TGD250

TRIMOPAN see TKZ000

TRIMORFAMID see FMY100

TRIMORFAMIDE see FMY100

TRIMORPHAMID see FMY100

TRIMORPHAMIDE see FMY100

TRIMOSULFA see TKX000

TRIMPEX see TKZ000

TRIMSTAT see DKE800

TRIMTABS see DKE800

TRI-MU-IODODIIODOTETRAARGENTATE(1-) RUBIDIUM see RQF100

TRIMUSTINE see TNF500

TRIMUSTINE HYDROCHLORIDE see TNF500

TRIMYSTEN see MRX500

TRINALGON see NGY000

TRINATRIUMPHOSPHAT (GERMAN) see SJH200

TRINEOPHYLTIN ACETATE see TMK200

TRINEX see TIQ250

TRINITRIN see NGY000

TRINITROACETONITRILE see TMK250

1,3,5-TRINITRO-2-ACETYLPYRROLE see TMN100

TRINITROANILINE (DOT) see PIC800

2,4,6-TRINITROANISOLE see TMK300

TRINITROBENZEEN see TMK500

TRINITROBENZENE see TMK500

1,3,5-TRINITROBENZENE see TMK500

TRINITROBENZENE-ANILINE COMPLEX see BBL100

2,3,5-TRINITROBENZENEDIAZONIUM-4-OXIDE see TMK775

2,4,6-TRINITROBENZENE-1,3-DIOL see SMP500

2,4,6-TRINITRO-1,3-BENZENEDIOL see SMP500

2,4,6-TRINITROBENZENESULFONIC ACID see TMK800

2,4,6-TRINITRO-1,3,5-BENZENETRIAMINE see TMK900

2,4,6-TRINITROBENZENE-1,3,5-TRIOL see TMM775

TRINITROBENZENE, dry or wetted with <30% water, by weight (UN 0214) (DOT) see TMK500

TRINITROBENZENE, wetted with not <30% water, by weight (UN 1354) (DOT) see TMK500

TRINITROBENZOIC ACID (dry) see TML000

TRINITROBENZOIC ACID, dry or wetted with <30% water, by weight (UN 0215) (DOT) see TML000

TRINITROBENZOIC ACID, wetted with not <30% water, by weight (UN 1355) (DOT) see TML000

TRINITROBENZOL (GERMAN) see TMK500
4,4,4-TRINITROBUTYRIC ACID see TML100
γ,γ,γ-TRINITROBUTYRIC ACID see TML100
TRINITROCHLOROBENZENE see TML325
2,4,6-TRINITROCHLOROBENZENE see PIE530
TRINITRO-m-CRESOL see TML500
2,4,6-TRINITRO-m-CRESOL see TML500
TRINITRO-m-CRESOLIC ACID see TML500
TRINITROCYCLOTRIMETHYLENE TRIAMINE see CPR800
2,4,6-TRINITRO-3,5-DIMETHYL-tert-BUTYLBENZENE see TML750
2,4,6-TRINITRO-1,3-DIMETHYL-5-tert-BUTYLBENZENE see TML750
2,2,2-TRINITROETHANOL see TMM000
TRINITROETHANOL (DOT) see TMM000
2,4,6-TRINITROFENOL (DUTCH) see PID000
2,4,6-TRINITROFENOLO (ITALIAN) see PID000
2,4,7-TRINITROFLUOREN-9-ONE see TMM250
2,4,7-TRINITRO-9-FLUORENONE see TMM250
2,4,7-TRINITROFLUORENONE (MAK) see TMM250
TRINITROGLYCERIN see NGY000
TRINITROGLYCEROL see NGY000
TRINITROL see NGY000
TRINITROMETACRESOL see TML500
TRINITROMETHANE see TMM500
1,3,5-TRINITRONAPHTHALENE see TMM600
1,5,9-TRINITROPHENANTHRENE see TMM610
1,6,9-TRINITROPHENANTHRENE see PCX275
1,7,9-TRINITROPHENANTHRENE see TMM615
2,6,9-TRINITROPHENANTHRENE see TMM620
3,6,9-TRINITROPHENANTHRENE see TMM638
2,5,10-TRINITROPHENANTHRENE see TMM618
3,5,10-TRINITROPHENANTHRENE see TMM635
1,3,5-TRINITROPHENOL see PID000
2,4,6-TRINITROPHENOL AMMONIUM SALT see ANS500
2,4,6-TRINITROPHENOL COMPOUND with 1,2-DIHYDRO-3-METHYLBENZ(j)ACEANTHRYLENE see MIL750
TRINITROPHENOL (UN 0154) (DOT) see PID000
TRINITROPHENOL, wetted with not <30% water, by weight (UN 1344) (DOT) see PID000
2,4,6-TRINITROPHENYL AZIDE see PIE525
TRINITROPHENYLMETHYLNITRAMINE see TEG250
2,4,6-TRINITROPHENYLMETHYLNITRAMINE see TEG250
2,4,6-TRINITROPHENYL-N-METHYLNITRAMINE see TEG250
2,4,6-TRINITROPHENYL NITRAMINE (DOT) see TDY075
2,4,6-TRINITROPHENYL (OSHA) see PID000
TRINITROPHLOROGLUCINOL see TMM775
TRINITROPYRENE see TMN000
1,3,6-TRINITROPYRENE see TMN000
1-(1,3,5-TRINITRO-1H-PYRROL-2-YL)ETHANONE see TMN100
2,4,6-TRINITRORESORCINOL see SMP500
TRINITRORESORCINOL (DOT) see SMP500
TRINITRORESORCINOL, wetted with less than 20% water (DOT) see SMP500
TRINITRORESORCINOL, DRY (DOT) see SMP500
1,3,5-TRINITROSO-1,3,5-TRIAZACYCLOHEXANE see HDV500

TRINITROSOTRIMETHYLENETRIAMINE see HDV500
TRINITROSOTRIMETHYLENTRIAMIN (GERMAN) see HDV500
2,4,6-TRINITROTOLUEEN (DUTCH) see TMN490
TRINITROTOLUENE see TMN490
s-TRINITROTOLUENE see TMN490
2,4,5-TRINITROTOLUENE see TMN400
2,4,6-TRINITROTOLUENE see TMN490
2,4,6-TRINITROTOLUENE see TMN490
3,4,5-TRINITROTOLUENE see TMN550
sym-TRINITROTOLUENE see TMN490
2,4,6-TRINITROTOLUENE (ACGIH,OSHA) see TMN490
TRINITROTOLUENE (UN 0209) (DOT) see TMN490
TRINITROTOLUENE, wetted with not <30% water, by weight (UN 1356) (DOT) see TMN490
s-TRINITROTOLUOL see TMN490
sym-TRINITROTOLUOL see TMN490
2,4,6-TRINITROTOLUOL (GERMAN) see TMN490
1,3,5-TRINITRO-1,3,5-TRIAZACYCLOHEXANE see CPR800
TRINITROTRIETHANOLAMINE see TJL250
TRINOXOL see DAA800
TRINOXOL see TAA100
TRINOXOL see TAH900
TRIOCTADECYL BORATE see TMN750
TRIOCTANOIN see TMO000
TRIOCTANOYLGLYCEROL see TMO000
TRI-n-OCTYLAMINE see DVL000
TRI(2-OCTYL)BORATE see TMO500
TRI-n-OCTYL BORATE see TMO550
TRIOCTYL(BUTYLTHIO)STANNANE see BSO200
TRIOCTYL(ETHYLTHIO)STANNANE see EPR200
TRIOCTYLMETHYLAMMONIUM CHLORIDE see MQH000
TRIOCTYL PHOSPHATE see TNI250
TRIOCTYLPHOSPHINE OXIDE see TMO575
TRI-N-OCTYLPHOSPHINE OXIDE see TMO575
TRIOCTYL PHOSPHITE see TNI300
TRIOCTYLTIN CHLORIDE see TMO585
TRI-N-OCTYLTIN CHLORIDE see TMO585
TRIOCTYLTIN CHLORIDE (6 CI) see TMO585
TRIODURIN see EDU500
TRIOLEIN HYDROLASE see GGA800
TRIOLEYL BORATE see BMC500
TRIOMBRIN see SEN500
TRIONAL see BJT750
TRIOPAC 200 see AAN000
TRIOPAS see AAN000
TRIOPEPTIN (HUMAN α-COMPONENT) see HGL680
TRIORTHOCRESYL PHOSPHATE see TMO600
TRIOSSIMETELENE (ITALIAN) see TMP000
TRIOTHYRONE see LGK050
TRIOVEX see EDU500
2,6,7-TRIOXA-1-ARSABICYCLO(2.2.2)OCTANE, 4-ETHYL-see EQD100
2,6,7-TRIOXA-1-ARSABICYCLO(2.2.2)OCTANE, 4-ISOPROPYL- see IRQ100
2,6,7-TRIOXA-1-ARSABICYCLO(2.2.2)OCTANE, 4-METHYL-see MQH100
2,8,9-TRIOXA-5-AZA-1-SILABICYCLO(3.3.3)UNDECANE see TMO750
2,8,9-TRIOXA-5-AZA-1-SILABICYCLO(3.3.3)UNDECANE, 1-ETHENYL- see VQK600
2,8,9-TRIOXA-5-AZA-1-SILABICYCLO(3.3.3)UNDECANE, 1-ETHOXY- see EFJ600

2,8,9-TRIOXA-5-AZA-1-SILABICYCLO(3.3.3)UNDECANE, 1-METHYL- see MPI650
2,8,9-TRIOXA-5-AZA-1-SILABICYCLO(3.3.3)UNDECANE, 1-VINYL-see VQK600
TRIOXACARCIN C see TMO775
3,6,9-TRIOXADECYLESTER KYSELINY OCTOVE see AAV500
TRIOXALEN see HOM270
TRIOXANE see TMP000
s-TRIOXANE see TMP000
1,3,5-TRIOXANE see TMP000
sym-TRIOXANE see TMP000
s-TRIOXANE, 2,4,6-TRIPROPYL- see TNC100
1,3,5-TRIOXANE, 2,4,6-TRIPROPYL- see TNC100
TRIOXANONA see PEC250
5,8,11-TRIOXAPENTADECANE see DDW200
3,5,9-TRIOXA-4-PHOSPHAHEPTACOSAN-1-AMINIUM, 4,7-DIHYDROXY-N,N,N-TRIMETHYL-10-OXO-, HYDROXIDE, INNER SALT, 4-OXIDE see SLM100
5,7,12-TRIOXA-6-STANNATETRACOSA-2,9-DIENOIC ACID, 6,6-DIOCTYL-4,8,11-TRIOXO-, DODECYL ESTER see DVN100
3,6,9-TRIOXAUNDECANE see DIW800
3,3'-(3,6,9-TRIOXAUNDECANEDIOYLDIAMINO)BIS(2,4,6-TRIIODOBENZOIC ACID) see IGD100
3,6,9-TRIOXAUNDECANE, 2-(3,4-(METHYLENEDIOXY)PHENOXY)- see MJS550
TRIOXAZIN see TKX250
TRIOXIDE(S) see TGG760
TRIOXIFENE MESYLATE see TMP175
3,7,12-TRIOXOCHOLANIC ACID see DAL000
3,7,12-TRIOXO-5-β-CHOLAN-24-OIC ACID see DAL000
3,7,12-TRIOXO-5-β-CHOLAN-24-OIC ACID MONOSODIUM SALT see SGD500
1,2,4-TRIOXOLANE see EJO500
1,2,4-TRIOXOLANES see ORY499
TRIOXON see TAA100
TRIOXONE see BSQ750
TRIOXONE see TAA100
2,6,8-TRIOXOPURINE see UVA400
(1,5-(2,4,6-TRIOXO-(1H,3H,5H)-PYRIMIDYLENE))BIS((2-METHOXYPROPYL)HYDROXYMERCURY SODIUM SALT see BKG500
(2,4,6-TRIOXO)-s-TRIAZINETRIYLTRIS(TRIBUTYLSTANNAN E) see TMP250
TRIOXSALEN see HOM270
3-β,12,14-TRIOXY-CARDEN-(20:22)-OLID (GERMAN) see DKN300
3-β,12,14-TRIOXY-DIGEN-(20:22)-OLID (GERMAN) see DKN300
TRIOXYMETHYLEEN (DUTCH) see TMP000
TRIOXYMETHYLEN (GERMAN) see TMP000
TRIOXYMETHYLENE see PAI000
TRIOXYMETHYLENE see TMP000
3-β,5,14-TRIOXY-19-OXO-CARDEN-(20:22)-OLID (GERMAN) see SMM500
3-β,5,14-TRIOXY-19-OXO-DIGEN-(20:22)-OLID (GERMAN) see SMM500
2,6,8-TRIOXYPURINE see UVA400
TRIOXYSALEN see HOM270
TRIOZANONA see TLP750
TRIPAN BLUE see CMO250
TRIPARANOL see TMP500
TRIPCHLOROLIDE see TNC200
TRI-PCNB see PAX000
TRI-PE see DUN600
TRIPELENNAMINE see TMP750
TRIPELENNAMINA (ITALIAN) see TMP750
TRIPELENNAMINE see TMP750
TRIPELENNAMINE HYDROCHLORIDE see POO750

TRIPELENNAMINE MONOHYDROCHLORIDE see POO750
TRI-PENAR see DRS000
TRI-n-PENTYL BORATE see TMQ000
N,N,N-TRIPENTYL-1-PENTANAMINIUM CHLORIDE see TBI200
TRI-N-PENTYLTIN BROMIDE see BOK250
TRI(PERFLUOROBUTYL)AMINE see HAS000
TRIPERIDOL see TKK500
TRIPERIDOL see TKK750
TRIPERIDOL HYDROCHLORIDE see TKK750
TRIPERVAN see VLF000
TRIPHACYCLIN see TBX250
TRIPHEDINON see BBV000
TRIPHENIDYL see BBV000
TRIPHENIDYL see PAL500
TRIPHENYL see TBD000
m-TRIPHENYL see TBC620
p-TRIPHENYL see TBC750
TRIPHENYLACETO STANNANE see ABX250
TRIPHENYLACRYLONITRILE see TMQ250
2,3,3-TRIPHENYLACRYLONITRILE see TMQ250
α,β,β-TRIPHENYLACRYLONITRILE see TMQ250
TRIPHENYLAMINE see TMQ500
TRIPHENYLANTIMONY see TMV250
TRIPHENYLANTIMONY DICHLORIDE see DGO800
TRIPHENYLANTIMONY OXIDE see TMQ550
TRIPHENYL ANTIMONY SULFIDE see TMQ600
TRIPHENYLBENZENE see TMR000
1,3,5-TRIPHENYLBENZENE see TMR000
sym-TRIPHENYLBENZENE see TMR000
TRIPHENYLBISMUTH see TMR250
TRIPHENYLBISMUTHINE see TMR250
TRIPHENYLBORANE see TMR300
TRIPHENYL BORATE see TMR500
TRIPHENYLBORINE see TMR300
TRIPHENYLBORON see TMR300
TRIPHENYLCHLOROSTANNANE see CLU000
TRIPHENYLCHLOROTIN see CLU000
TRIPHENYLCYANOETHYLENE see TMQ250
TRIPHENYLCYCLOHEXYL BORATE see TMR750
TRIPHENYLENE see TMS000
1,2,2-TRIPHENYLETHANONE see TMS100
TRIPHENYLETHYLENE see TMS250
1,1,2-TRIPHENYLETHYLENE see TMS250
TRIPHENYLGUANIDINE see TMS500
N,N',N"-TRIPHENYLGUANIDINE see TMS500
2,4,5-TRIPHENYLIMIDAZOLE see TMS750
TRIPHENYLLEAD ACETATE see TMT000
TRIPHENYL LEAD(1+) HEXAFLUOROSILICATE see TMT250
TRIPHENYLMETHANETHIOL see TMT500
S-TRIPHENYLMETHYL-l-CYSTEINE see TNR475
TRIPHENYL PHOSPHATE see TMT750
TRIPHENYLPHOSPHINE see TMU000
TRIPHENYLPHOSPHINE MONOSULFIDE see TMU100
TRIPHENYLPHOSPHINE SULFIDE see TMU100
TRIPHENYL PHOSPHITE see TMU250
TRIPHENYL PHOSPHONATE see TMU300
TRIPHENYL-2-PROPENYL-STANNANE (9CI) see AGU500
1,3,4-TRIPHENYLPYRAZOLE-5-ACETIC ACID SODIUM SALT see TMU750
TRIPHENYLSELENONIUM CHLORIDE see TMU775
TRIPHENYLSILYL CHLORIDE see TMU800
TRIPHENYLSTANNANE see TMV775

TRIPHENYLSTANNANE SULFATE (2:1) see BLT000
TRIPHENYLSTANNYL BENZOATE see TMV000
TRIPHENYLSTANNYL HYDRIDE see TMV775
TRIPHENYL STIBINE see TMV250
2,3,5-TRIPHENYL-2H-TETRAZOLIUM CHLORIDE see TMV500
TRIPHENYLTHIOCYANATOSTANNANE see TMV750
TRIPHENYLTIN see TMV775
TRIPHENYLTIN p-ACETAMIDOBENZOATE see TMV800
TRIPHENYLTIN ACETATE see ABX250
TRIPHENYLTIN BENZOATE see TMV000
TRIPHENYLTIN-BIS(DIETHYL)DITHIOPHOSPHATE see EFL600
TRIPHENYLTIN CHLORIDE see CLU000
TRIPHENYLTIN CYANOACETATE see TMV825
TRIPHENYLTIN 3,5-DI-ISOPROPYLSALICYLATE see TMV830
TRIPHENYLTIN FLUORIDE see TMV850
TRIPHENYLTIN HYDRIDE see TMV775
TRIPHENYLTIN HYDROPEROXIDE see TMW000
TRIPHENYLTIN HYDROXIDE (USDA) see HON000
TRIPHENYLTIN IODIDE see IFO000
TRIPHENYLTIN LEVULINATE see TMW250
TRIPHENYLTIN METHANESULFONATE see TMW500
TRIPHENYLTIN OXIDE see HON000
TRIPHENYLTIN PROPIOLATE see TMW600
TRIPHENYL TIN THIOCYANATE see TMV750
TRIPHENYL-1H-1,2,4-TRIAZOL-1-YL TIN see TMX000
TRIPHENYL-ZINNACETAT (GERMAN) see ABX250
TRIPHENYL-ZINNHYDROXID (GERMAN) see HON000
TRIPHOSADEN see ARQ500
TRIPHOSPHADEN see ARQ500
TRIPHOSPHOPYRIDINE NUCLEOTIDE see CNF400
TRIPHOSPHORIC ACID ADENOSINE ESTER see ARQ500
TRIPHOSPHORIC ACID, SODIUM SALT see SKN000
TRIPHTHAZINE see TKE500
TRIPHTHAZINE see TKK250
TRIPHTHAZINE DIHYDROCHLORIDE see TKK250
TRIPIPERAZINE DICITRATE see PIJ500
TRIPIPERIDINOPHOSPHINE OXIDE see TMX250
TRIPIPERIDINOPHOSPHINE SELENIDE see TMX350
TRIPLA-ETILO see DBX400
TRIPLEX III see EIX500
TRI-PLUS see TIO750
TRIPOLI see SCI500
TRIPOLI see TMX500
TRIPOLY see SKN000
TRI(POLYNONYLPHENYL)PHOSPHITE see TMX550
TRIPOLYPHOSPHATE see SKN000
TRIPOMOL see TFS350
TRIPOTASSIUM CITRATE MONOHYDRATE see PLB750
TRIPOTASSIUM DICITRATOBISMUTHATE see BKX800
TRIPOTASSIUM HEXACYANOFERRATE see PLF250
TRIPOTASSIUM HEXACYANOMANGANATE(3-) see TMX600
TRIPOTASSIUM NITRILOTRIACETATE see TMX750
TRIPOTASSIUM PHOSPHATE see PLQ410

TRIPOTASSIUM TRICHLORIDE see PLA500
TRIPROLIDINE HYDROCHLORIDE see TMX775
TRIPROPARGYL CYANURATE see THN500
TRIPROPIONIN see TMY000
TRIPROPIONINE see TMY000
TRIPROPYL ALUMINUM see AHH750
TRIPROPYLALUMINUM see TMY100
TRIPROPYLALUMINUM (DOT) see TMY100
TRI-N-PROPYLAMINE see TMY250
TRIPROPYLAMINE (DOT) see TMY250
TRI-n-PROPYL BORATE see TMY750
TRIPROPYL(BUTYLTHIO)STANNANE see TMY850
TRIPROPYLENE GLYCOL see TMZ000
TRIPROPYLENEGLYCOL DIACRYLATE see TMZ100
TRIPROPYLENE GLYCOL, METHYL ETHER see TNA000
TRIPROPYLENE GLYCOL MONOMETHYL ETHER see TNA000
TRIPROPYLEN GLYKOL MONOMETHYL ETHER see TNA000
TRIPROPYL INDIUM see TNA250
TRIPROPYL LEAD see TNA500
TRIPROPYL LEAD CHLORIDE see TNA750
TRI-n-PROPYL LEAD CHLORIDE see TNA750
TRIPROPYLPHOSPHINE OXIDE see TNA800
TRIPROPYL PLUMBANE see TNA500
N,N,N-TRIPROPYL-1-PROPANAMINIUM HYDROXIDE see TED000
TRIPROPYLSTANNIUM ACETATE see TNB000
TRIPROPYLTIN ACETATE see TNB000
TRI-N-PROPYLTIN BROMIDE see BOK750
TRIPROPYLTIN CHLORIDE see CLU250
TRI-n-PROPYLTIN CHLORIDE see CLU250
TRIPROPYLTIN IODIDE see TNB250
TRIPROPYLTIN IODOACETATE see TNB500
TRIPROPYLTIN-o-IODOBENZOATE see IEF000
TRIPROPYLTIN ISOTHIOCYANATE see TNB750
TRIPROPYLTIN TRICHLOROACETATE see TNC000
2,4,6-TRIPROPYL-s-TRIOXANE see TNC100
TRI-n-PROPYLZINNACETAT see TNB000
2,4,6-TRIPROP-2-YNYLOXY-s-TRIAZINE see THN500
TRIPTIDE see GFW000
TRIPTIL see DDA600
TRIPTIL HYDROCHLORIDE see POF250
TRIPTOLID see TNC175
TRIPTOLIDE see TNC175
TRIPTOLIDE CHLORHYDRIN see TNC200
TRIPTOLIDE 12,13-CHLORHYDRIN see TNC200
TRIPTONE see HOT500
(TRI-2-PYRIDYL)STIBINE see TNC250
TRIQUILAR see NNL500
TRIQUINOL see IDD100
TRIRODAZEEN see DVF800
TRI-RODAZENE see DVF800
TRIS see TEM500
TRIS see TNC500
TRIS (flame retardant) see TNC500
TRIS(ACETONITRILE)TRICARBONYLTUNGSTEN see THM250
TRIS(ACETONITRILE)TRICHLORORHODIUM see TNC600
TRIS(ACETYLACETONATO)ALUMINUM see TNN000
TRIS(ACETYLACETONATO)CHROMIUM see TNN250
TRIS(ACETYLACETONATO)CHROMIUM(III) see TNN250
TRIS(ACETYLACETONE)ALUMINUM see TNN000

TRIS(ACETYLACETONYL)ALUMINUM see TNN000

TRIS(N-ACRYLOYL)HEXAHYDROTRIAZINE see THM900

TRIS(ACRYLOYL)HEXAHYDRO-s-TRIAZINE see THM900

TRISAETHYLENIMINOBENZOCHINON (GERMAN) see TND000

TRISALKOFEN BMB see MOS100

2,4,6-TRIS(ALLYLOXY)TRIAZINE see THN500

TRISAMINE see TEM500

TRIS-AMINO see TEM500

TRIS(AMINOETHYL)AMINE see NEI800

TRIS(β-AMINOETHYL)AMINE see NEI800

TRIS-4-AMINOFENYLMETHAN (CZECH) see THP000

TRISAMINOL see TEM500

TRIS(p-AMINOPHENYL)CARBONIUM PAMOATE see TNC725

TRIS(p-AMINOPHENYL)CARBONIUM PAMOATE see TNC750

TRIS(p-AMINOPHENYL)CARBONIUM SALT with 4,4-METHYLENEBIS(3-HYDROXY-2-NAPHTHOIC ACID) (2:1) see TNC725

TRIS-(p-AMINOPHENYL)CARBONIUM SALT with 4,4-METHYLENEBIS(3-HYDROXY-2-NAPHTHOIC ACID) (2:1) see TNC750

TRIS(p-AMINOPHENYL)METHYLIUM SALT with 4,4'-METHYLENEBIS(3-HYDROXY-2-NAPHTHOIC ACID) (2:1) see TNC725

TRIS(p-AMINOPHENYL)METHYLIUM SALT with 4,4'-METHYLENEBIS(3-HYDROXY-2-NAPHTHOIC ACID) (2:1) see TNC750

TRIS(4-AMINOPHENYL) PHOSPHATE see THO550

TRIS-AMINOPHOSPHINE OXIDE see PHB550

TRISATIN see ACD500

TRIS(2-AZIDOETHYL)AMINE see TNC800

1,1,1-TRIS(AZIDOMETHYL)ETHANE see TNC825

TRIS(1-AZIRIDINE)PHOSPHINE OXIDE see TND250

2,3,5-TRIS(AZIRIDINO)-1,4-BENZOQUINONE see TND000

2,3,5-TRIS(1-AZIRIDINO)-p-BENZOQUINONE see TND000

TRIS(AZIRIDINYL)-p-BENZOQUINONE see TND000

TRIS(1-AZIRIDINYL)-p-BENZOQUINONE see TND000

2,3,5-TRIS(AZIRIDINYL)-1,4-BENZOQUINONE see TND000

2,3,5-TRIS(1-AZIRIDINYL)-p-BENZOQUINONE see TND000

2,3,5-TRIS(1-AZIRIDINYL)-2,5-CYLOHEXADIENE-1,4-DIONE see TND000

TRIS-(1-AZIRIDINYL)PHOSPHINE OXIDE see TND250

TRIS(1-AZIRIDINYL)PHOSPHINE SULFIDE see TFQ750

TRISAZIRIDINYLTRIAZINE see TND500

2,4,6-TRIS(1-AZIRIDINYL)-s-TRIAZINE see TND500

2,4,6-TRIS(1'-AZIRIDINYL)-1,3,5-TRIAZINE see TND500

TRIS(BENZOYLACETONATO)CHROMIUM see TNN500

TRIS-2,2'-BIPYRIDINE CHROMIUM see TND750

TRIS-2,2'-BIPYRIDINE CHROMIUM(II) PERCHLORATE see TNE000

TRIS-2,2'-BIPYRIDINESILVER(II) PERCHLORATE see TNE250

TRIS BIS-BIFLUOROAMINO DIETHOXY PROPANE (DOT) see TOE200

2,4,6-TRIS(BIS(HYDROXYMETHYL)AMINO)-s-TRIAZINE see HDY000

2,4,6-TRIS(BROMOAMINO)-1,3,5-TRIAZINE see TNE275

TRIS(1-BROMO-3-CHLOROISOPROPYL)PHOSPHATE see TNE500

TRIS(2-BROMO-1-CHLORO-2-PROPYL) PHOSPHATE see TNE500

TRIS(2-BROMOETHYL)PHOSPHATE see TNE600

TRIS BUFFER see TEM500

TRIS(2-BUTOXYETHYL) ESTER PHOSPHORIC ACID see BPK250

TRIS(2-BUTOXYETHYL) PHOSPHATE see BPK250

TRIS(tert-BUTOXY)VINYLSILANE see SDX300

TRIS-N-BUTYLAMINE see THX250

TRIS(tert-BUTYLPEROXY)VINYLSILANE see VQU333

TRISBUTYLPHOSPHINE OXIDE see TNE750

1,3,5-TRIS(CARBONYL-2-ETHYL-1-AZIDINE)BENZENE see HGN000

3,6,9-TRIS(CARBOXYMETHYL)-3,6,9-TRIAZAUNDECANEDIOIC ACID see DJG800

N,N,N'-TRIS(2-CHLORAETHYL)-N',O-PROPYLEN-PHOSPHORSAEUREESTER-DIAMID (GERMAN) see TNT500

2,4,6-TRIS(CHLOROAMINO)-1,3,5-TRIAZINE see TNE775

TRIS(2-CHLOROETHOXY)SILANE see TNF000

TRIS(2-CHLOROETHYL)AMINE see TNF250

TRIS(β-CHLOROETHYL)AMINE see TNF250

TRIS(2-CHLOROETHYL)AMINE HYDROCHLORIDE see TNF500

TRIS(β-CHLOROETHYL)AMINE HYDROCHLORIDE see TNF500

TRIS(2-CHLOROETHYL)AMINE MONOHYDROCHLORIDE see TNF500

TRIS(2-CHLOROETHYL)AMMONIUM CHLORIDE see TNF500

TRIS(2-CHLOROETHYL)ESTER PHOSPHORIC ACID see CGO500

TRIS(2-CHLOROETHYL)ESTER of PHOSPHORUS ACID see PHO000

N,N,N'-TRIS(2-CHLOROETHYL)-N',O-PROPYLENE PHOSPHORIC ACID ESTER DIAMIDE see TNT500

TRIS(2-CHLOROETHYL) PHOSPHATE see CGO500

TRIS(β-CHLOROETHYL) PHOSPHATE see CGO500

TRIS(2-CHLOROETHYL)PHOSPHITE see PHO000

N,N,3-TRIS(2-CHLOROETHYL)TETRAHYDRO-2H-1,3,2-OXAPHOSPHORIN-2-AMINE-2-OXIDE see TNT500

TRIS(2-CHLOROISOPROPYL)PHOSPHATE see TNG000

TRIS-1,2,3-(CHLOROMETHOXY)PROPANE see GGI000

TRIS(1-CHLOROMETHYL-2-CHLOROETHYL)PHOSPHATE see FQU875

TRIS(p-CHLOROPHENYL)PHOSPHINE COMPLEX with MERCURIC CHLORIDE (2:1) see BLT500

1-(β,β,β-TRIS(p-CHLOROPHENYL)PROPIONYL)-4-METHYLPIPERAZINE HYDROCHLORIDE see MQH250

TRIS(p-CHLOROPHENYL)TIN FLUORIDE see TNG050

TRIS(CHLOROPROPYL)PHOSPHATE see TJC870

TRIS(o-CRESYL)-PHOSPHATE see TMO600

TRIS(DECYL)AMINE see TJI000

TRIS(1,2-DIAMINOETHANE)CHROMIUM(III) PERCHLORATE see TNG100

TRIS(1,2-DIAMINOETHANE) COBALT(III) NITRATE see TNG150

TRIS(DIBROMOPROPYL)PHOSPHATE see TNC500

TRIS(2,3-DIBROMOPROPYL) PHOSPHATE see TNC500

TRIS(2,3-DIBROMOPROPYL) PHOSPHORIC ACID ESTER see TNC500

TRIS-2,3-DIBROMPROPYL ESTER KYSELINY FOSFORECNE (CZECH) see TNC500

TRIS(DIBUTYLBIS(HYDROXYETHYLTHIO) TIN) BIS(BORIC ACID ESTER) see TNG250

2,4,6-TRIS(DICHLOROAMINO)-1,3,5-TRIAZINE see TNG275

TRIS(1,3-DICHLOROISOPROPYL)PHOSPHATE see FQU875

TRIS-DICHLOROPROPYLPHOSPHATE see TNG750

TRIS(2,3-DICHLOROPROPYL)PHOSPHATE see TNG750

TRIS-(1,3-DICHLORO-2-PROPYL)-PHOSPHATE see FQU875

1,2,3-TRIS(2-DIETHYLAMINOETHOXY)BENZENE TRIETHIODIDE see PDD300

1,2,3-TRIS(2-DIETHYLAMINOETHOXY)BENZENE TRIS(ETHYLIODIDE) see PDD300

TRIS(DIETHYLDITHIOCARBAMATO)IRON see IGT100

TRIS(DIFLUOROAMINE)FLUOROMETHANE see TNG775

2,4,6-TRIS(DI(HYDROXYMETHYL)AMINO)-1,3,5-TRIAZINE see HDY000

TRIS(DIMETHYLAMINO)ANTIMONY see TNG780

2,4,6-TRIS-N,N-DIMETHYLAMINOMETHYLFENOL (CZECH) see TNH000

2,4,6-TRIS(DIMETHYLAMINOMETHYL)PHENOL see TNH000

TRIS(p-DIMETHYLAMINOPHENYL) PHOSPHINE COMPLEX with MERCURIC CHLORIDE (2:1) see BLT750

TRIS(DIMETHYLAMINO)PHOSPHINE OXIDE see HEK000

TRIS(DIMETHYLAMINO)PHOSPHORUS OXIDE see HEK000

N,N',N"-TRIS(DIMETHYLAMINOPROPYL)-s-HEXAHYDROTRIAZINE see TNH250

TRIS(DIMETHYLAMINO)SILANE see TNH300

1,2-(TRISDIMETHYLAMINOSILYL)ETHANE see TNH500

2,4,6-TRIS(DIMETHYLAMINO)-s-TRIAZINE see HEJ500

2,4,6-TRIS(DIMETHYLAMINO)-1,3,5-TRIAZINE see HEJ500

TRIS(DIMETHYLCARBAMODITHIOATO-S,S')IRON see FAS000

TRIS(DIMETHYLDITHIOCARBAMATO)BISMUTH see BKW000

TRIS(DIMETHYLDITHIOCARBAMATO)IRON see FAS000

TRIS(N,N-DIMETHYLDITHIOCARBAMATO) IRON(III) see FAS000

TRIS((1,1-DIMETHYLETHYL)DIOXY)ETHENYLSILANE see VQU333

TRIS(3,5-DIMETHYL-4-HEPTYL) BORATE see TJJ750

TRIS(2,6-DIMETHYLPHENYL)PHOSPHATE see TNR550

TRIS(DIPROPYLENE GLYCOL)PHOSPHINE see OQO000
TRIS(DIPROPYLENE GLYCOL) PHOSPHINE see PGY300
TRIS(DIPROPYLENE GLYCOL)PHOSPHONATE see OQO000
TRIS(1-DODECIL-3-METIL-2-FENIL-1,3-BENZIMIDAZOLIO)-FERRICIANURO (ITALIAN) see TNH750
TRIS(1-DODECYL-3-METHYL-2-FENYL-1,3-BENZIMIDAZOLIUM)-HEXACYANOFERRAAT(III) (DUTCH) see TNH750
TRIS(1-DODECYL-3-METHYL-2-PHENYLBENZIMIDAZOLIUM)FERRICYANIDE see TNH750
TRIS(1-DODECYL-3-METHYL-2-PHENYL-1,3-BENZIMIDAZOLIUM)-HEXACYANOFERRAT(III) (GERMAN) see TNH750
TRIS(DODECYLTHIO)ANTIMONY see TNH850
TRIS(EPOXYPROPYL)ISOCYANURATE see GGY150
TRIS(2,3-EPOXYPROPYL)ISOCYANURATE see GGY150
1,3,5-TRIS(2,3-EPOXYPROPYL)TRIAZINE-2,4,6-TRIONE see TKL250
α-1,3,5-TRIS(2,3-EPOXYPROPYL)-s-TRIAZINE-2,4,6-(1H,3H,5H)-TRIONE see TBC450
TRISERPIN see RDK000
1,3,5-TRIS((2-ETHYL-AZIRIDINYL)-CARBONYL)BENZENE see HGN000
2,4,6-TRIS-(1-(2-ETHYLAZIRIDINYL))-1,3,5-TRIAZINE see TNI000
2,3,5-TRISETHYLENEIMINOBENZOQUINONE see TND000
TRISETHYLENEIMINOQUINONE see TND000
TRIS(ETHYLENEIMINO)TRIAZINE see TND500
TRISETHYLENEIMINO-1,3,5-TRIAZINE see TND500
2,4,6-TRIS(ETHYLENEIMINO)-s-TRIAZINE see TND500
TRIS(N-ETHYLENE)PHOSPHOROTRIAMIDATE see TND250
2,3,5-TRIS(ETHYLENIMINO)BENZOQUINONE see TND000
2,3,5-TRIS(ETHYLENIMINO)-p-BENZOQUINONE see TND000
2,3,5-TRIS(ETHYLENIMINO)-1,4-BENZOQUINONE see TND000
TRIS(ETHYLENIMINO)THIOPHOSPHATE see TFQ750
2,4,6-TRIS(ETHYLENIMINO)-s-TRIAZINE see TND500
TRIS((2-ETHYLHEXANOYL)OXY)METHYLSILANE see TNI100
TRIS(2-ETHYLHEXYL)PHOSPHATE see TNI250
TRIS(2-ETHYLHEXYL)PHOSPHITE see TNI300
TRIS(2-ETHYLHEXYL) TRIMELLITATE see TJR600
TRISFOSFAMIDE see TNT500
TRIS(2-HYDROXYETHYL)AMINE see TKP500
TRIS(2-HYDROXYETHYL)PHENYLMERCURIAMMONIUM LACTATE see TNI500
TRIS-HYDROXYMETHYL-AMINOMETHAN (GERMAN) see TEM500
TRIS-HYDROXYMETHYLAMINOMETHANE see TEM500

TRIS(HYDROXYMETHYL)METHYLAMINE see TEM500
1,1,1-TRIS(HYDROXYMETHYL)METHYLAMINE see TEM500
TRIS(HYDROXYMETHYL)NITROMETHANE see HMJ500
TRIS(HYDROXYMETHYL)PHOSPHINE see TNI600
TRIS(HYDROXYMETHYL)PROPANE see HDF300
1,1,1-TRIS(HYDROXYMETHYL)PROPANE see HDF300
1,1,1-TRISHYDROXYMETHYLPROPANE BICYCLIC PHOSPHITE see TNI750
N,N',N"-TRIS(HYDROXYMETHYL)-1,3,5-TRIAZINE-2,4,6-TRIAMINE see TLW750
TRIS(4-HYDROXYPHENYL)ETHANE see ELO600
TRIS(2-HYDROXYPROPYL)AMINE see NEI500
TRIS(2-HYDROXY-1-PROPYL)AMINE see NEI500
TRISILYLAMINE see TNJ250
TRISILYLARSINE see TNJ275
TRIS(ISOAMYL)PHOSPHINE OXIDE see TNJ285
TRIS(ISOCYANATOHEXYL)BIURET see TNJ300
1,1,1-TRIS(4-ISOCYANATOPHENOXYMETHYL)PROPANE see TNJ400
TRIS(ISOPROPYLPHENYL) PHOSPHATE see TKU300
TRIS-N-LOST see TNF500
TRIS(MERCAPTOACETATO(2-1))BISMUTHATE(3-) TRISODIUM see BKY500
TRIS(METHANOL)PHOSPHINE see TNI600
TRIS(METHOXYETHOXY)VINYLSILANE see TNJ500
TRIS(METHOXYETHOXY)VINYLSILANE see TNJ500
(TRIS(β-METHOXYETHOXY))VINYLSILANE see TNJ500
TRIS-(2-METHOXYETHYL)FOSFAT see TLA600
TRIS(p-METHOXYPHENYL)CHLOROETHYLENE see CLO750
2,3,3-TRIS(p-METHOXYPHENYL)-N,N-DIMETHYALLYLAMINE HYDROCHLORIDE see TNJ750
TRIS(p-METHOXYPHENYL) PHOSPHINE COMPLEX with MERCURIC CHLORIDE (2:1) see BLU250
2,4,6-TRIS(METHYLAMINO)-s-TRIAZINE see TNJ825
2,4,6-TRIS-METHYLAMINO-1,3,5-TRIAZINE see TNJ825
TRIS(2-METHYL-1-AZIRIDINYL)PHOSPHINE OXIDE see TNK250
TRIS(2-METHYLAZIRIDIN-1-YL)PHOSPHINE OXIDE see TNK250
2,4,6-TRIS(2-METHYL-1-AZIRIDINYL)-s-TRIAZINE see TNK000
2,4,6-TRIS((1-(2-METHYLAZIRIDINYL))-1,3,5-TRIAZINE) see TNK000
TRIS(3-METHYLBUTYL)PHOSPHINE see TKT100
N,N',N"-TRIS(1-METHYLETHYLENE)PHOSPHORAMIDE see TNK250
TRIS(1-METHYLETHYLENE)PHOSPHORIC TRIAMIDE see TNK250
TRIS(o-METHYLPHENYL)PHOSPHATE see TMO600
TRIS(2-METHYLPROPYL)ALUMINUM see TKR500

TRIS(p-METHYLTHIOPHENYL) PHOSPHINE COMPLEX with MERCURIC CHLORIDE (2:1) see BLU500
TRIS(MONOCHLOROPROPYL) PHOSPHATE see TJC870
TRIS(NICTINATO)ALUMINUM see AHD650
TRIS NITRO see HMJ500
TRIS(NONAFLUOROBUTYL)AMINE see HAS000
TRIS(OCTAMETHYLPYROPHOSPHORAMIDE)MANGANESE(2+), DIPERCHLORATE see TNK400
TRIS(OCTAMETHYLPYROPHOSPHORAMIDE)NICKEL(2+), DIPERCHLORATE see PCE250
TRISODIUM ALUMINUM HEXAFLUORIDE see TNK450
TRISODIUM-4'-ANILINO-8-HYDROXY-1,1'-AZONAPHTHALENE-3,6,5'-TRISULFONATE see ADE750
TRISODIUM ARSENATE see SEY150
TRISODIUM ARSENATE, HEPTAHYDRATE see ARE000
TRISODIUM ARSENITE see SEY200
TRISODIUM BIS(MONOTHIOSULFATO)AURATE(3⁻) DIHYDRATE see TNK500
TRISODIUM 3-CARBOXY-5-HYDROXY-1-p-SULFOPHENYL-4-p-SULFOPHENYLAZOPYRAZOLE see FAG140
TRISODIUM (CARBOXYMETHOXY)PROPANEDIOATE see SFO700
TRISODIUM (α-CARBOXY)OXYDIACETATE see SFO700
TRISODIUM CITRATE see TNL000
TRISODIUM EDETATE see TNL250
TRISODIUM EDTA see TNL250
TRISODIUM ETHYLENEDIAMINETETRAACETATE see TNL250
TRISODIUM ETHYLENEDIAMINETETRAACETATE TRIHYDRATE see TNL500
TRISODIUM ETIDRONATE see TNL750
TRISODIUM HEXAFLUOROALUMINATE see TNK450
TRISODIUM HEXANITRITOCOBALTATE see SFX750
TRISODIUM HEXANITRITOCOBALTATE(3-) see SFX750
TRISODIUM HEXANITROCOBALTATE see SFX750
TRISODIUM HEXANITROCOBALTATE(3-) see SFX750
TRISODIUM HYDROGEN ETHYLENEDIAMINETETRAACETATE see TNL250
TRISODIUM HYDROGEN (ETHYLENEDINITRILO)TETRAACETATE see TNL250
TRISODIUM-1-HYDROXY-3,6,8-PYRENETRISULFONATE see TNM000
TRISODIUM MONOTHIOPHOSPHATE see TNM750
TRISODIUM NITRILOTRIACETATE see SIP500
TRISODIUM NITRILOTRIACETATE MONOHYDRATE see NEI000
TRISODIUM NITRILOTRIACETIC ACID see SIP500
TRISODIUM ORTHOPHOSPHATE see SJH200
TRISODIUM ORTHOVANADATE see SIY250
TRISODIUM PHOSPHATE see SJH200
TRISODIUM PHOSPHOROTHIOATE see TNM750
TRISODIUM SALT of 3-CARBOXY-5-HYDROXY-1-

SULFOPHENYLAZOPYRAZOLE see
FAG140
TRISODIUM SALT of 1-(4-SULFO-1-
NAPHTHYLAZO)-2-NAPHTHOL-3,6-
DISULFONIC ACID see FAG020
TRISODIUM THIOPHOSPHATE see
TNM750
TRISODIUM 1,3,5-TRIAZINE-2,4,6-
TRITHIOLATE see SKN100
TRISODIUM TRIFLUORIDE see SHF500
TRISODIUM TRIMETAPHOSPHATE see
TKP750
TRISODIUM TRITHIOCYANURATE see
SKN100
TRISODIUM TRITHIOCYANURIC ACID see
SKN100
TRISODIUM VERSENATE see TNL250
TRISODIUM ZINC DTPA see TNM850
TRISOMNIN see SBM500
TRISOMNIN see SBN000
TRISORALEN see HOM270
(α)-1,3,5-TRIS(OXIRANYLMETHYL)-1,3,5-
TRIAZINE-2,4,6-(1H,3H,5H)-TRIONE see
TBC450
TRI-SPAN see UVJ450
TRIS(2,4-PENTANEDIONATO)ALUMINUM
see TNN000
TRIS(2,4-PENTANEDIONATO)CHROMIUM
see TNN250
TRIS(2,4-
PENTANEDIONATO)CHROMIUM(3+) see
TNN250
TRIS(2,4-PENTANEDIONATO)INDIUM see
ICG000
TRIS(2,4-PENTANEDIONATO)IRON see
IGL000
TRIS(2,4-PENTANEDIONE)ALUMINUM see
TNN000
TRIS(1-PHENYL-1,3-
BUTANEDIONATO)CHROMIUM see
TNN500
TRIS(1-PHENYL-1,3-
BUTANEDIONATO)CHROMIUM(3+) see
TNN500
TRIS(1-PHENYL-1,3-BUTANEDIONATO-
O,O')CHROMIUM see TNN500
TRIS(1-PHENYL-1,3-
BUTANEDIONO)CHROMIUM(III) see
TNN500
TRISPHOSPHAMIDE see TNT500
TRIS(PHOSPHONOMETHYL)AMINE see
NEI100
TRISPUFFER see TEM500
TRIS-STERIL see TEM500
TRISTAR see DUV600
TRISTEARYL BORATE see TMN750
TRIS(TOLYLOXY)PHOSPHINE OXIDE see
TNP500
TRIS(o-TOLYL)-PHOSPHATE see TMO600
TRIS-(p-TOLYL)PHOSPHINE see TNN750
1,3,5-TRIS(TRIBUTYLTIN)-s-TRIAZINE-
2,4,6-TRIONE see TMP250
TRIS(TRIDECYL)AMINE see TNN760
1,2,3-TRIS(2-TRIETHYLAMMONIUM
ETHOXY)BENZENE TRIIODIDE see
PDD300
TRIS(TRIFLUOROMETHYL)NITROSOMET
HANE see PCH300
TRIS(TRIFLUOROMETHYL)PHOSPHINE
see TNN775
1,3,5-
TRIS(TRIFLUOROPROPYL)TRIMETHYLCY
CLOTRISILOXANE see FLV100
TRIS(N-(α,α,α-TRIFLUORO-m-
TOLYL)ANTHRANILATO)ALUMINUM see
AHA875
TRIS(3,3,5-
TRIMETHYLCYCLOHEXYL)ARSINE see
TNO000
TRIS-(TRIMETHYLSILYL)FOSFAT (CZECH)
see PHE750

TRISULFIDE,
BIS(ETHOXYTHIOCARBONYL)- see
DKE400
TRISULFON BLUE RW see CMO600
TRISULFON CONGO RED see SGQ500
TRISULFON VIOLET N see CMP000
TRISULFURATED PHOSPHORUS see
PHS500
TRISULPHONE BR see CMO820
TRISUSTAN see TJL250
TRITAC see TJA200
TRITELLURIUM TETRANITRIDE see
TNO250
TRITEREN see UVJ450
TRITERPENE SAPONINS mixture from
AESCULUS HIPPOCASTONUM see TNO275
TRITEX-EXTRA see CDK800
TRITHENE see CLQ750
TRITHEON see ABY900
1,3,5-TRITHIACYCLOHEXANE see TLS500
sym-TRITHIAN (CZECH) see TLS500
1,3,5-TRITHIANE see TLS500
3,4,5-TRITHIATRICYCLO(5.2.1.0²,⁶)DECANE
see TNO300
TRITHIO see AOO490
TRITHIOACETONE see HEL000
TRITHIOANETHOLE see AOO490
TRITHIOBIS(TRICHLOROMETHANE) see
BLM750
TRITHIOCARBONIC ACID, CYCLIC
ETHYLENE ESTER see EJQ100
TRITHIOCYANURIC ACID see THS250
TRITHIOCYANURIC ACID TRISODIUM
SALT see SKN100
TRITHIOFORMALDEHYDE see TLS500
TRITHIO-(p-METHOXYPHENYL)PROPENE
see AOO490
TRITHION see TNP250
TRITHION MITICIDE see TNP250
TRITHIOZINE see TNP275
TRITICOL see MHC750
TRITICONAZOLE see TNP260
TRITIOZINA (ITALIAN) see TNP275
TRITIOZINE see TNP275
TRITISAN see PAX000
TRITOFTOROL see EIR000
TRITOL see TMN490
TRITOLYL PHOSPHATE see TNP500
TRI-2-TOLYL PHOSPHATE see TMO600
TRI-o-TOLYL PHOSPHATE see TMO600
TRITON N-100 see NND500
TRITON A-20 see TDN750
TRITON GR-5 see DJL000
TRITON K-60 see AFP250
TRITON RW 20 see TNP512
TRITON RW 50 see TNP515
TRITON RW 75 see TNP517
TRITON RW 100 see TNP519
TRITON RW 125 see TNP522
TRITON RW 150 see TNP535
TRITON WR 1339 see TDN750
TRITON X 15 see GHS000
TRITON X 35 see PKF500
TRITON X-40 see DTC600
TRITON X 45 see PKF500
TRITON X100 see OFQ000
TRITON X 100 see PKF500
TRITON X 102 see PKF500
TRITON X 165 see PKF500
TRITON X-200 see TNP550
TRITON X 305 see PKF500
TRITON X 405 see PKF500
TRITON X 705 see PKF500
TRITON XL 80N see AFJ172
TRITOX see TII750
TRITRIDECYL AMINE see TNN760
TRI-N-TRIDECYLAMINE see TNN760
TRITROL see CLN750
TRITTICO see CKJ000
TRITTICO see THK880
TRITYL BROMIDE see TNP600
TRITYL CHLORIDE see CLT500

S-TRITYLCYSTEINE see TNR475
S-TRITYL-l-CYSTEINE see TNR475
3-TRITYLTHIO-l-ALANINE see TNR475
3-(TRITYLTHIO)-l-ALANINE see TNR475
TRITYLTHIOL see TMT500
TRIULOSE see OLW100
TRIUMBREN see AAN000
TRIUMPH see PHK000
TRIUROL see AAN000
TRIUROPAN see AAN000
TRIVALENT SODIUM ANTIMONYL
GLUCONATE see AQI000
TRIVASTAL see TNR485
TRIVASTAN see TNR485
TRIVAZOL see MMN250
TRI-VC 13 see DFK600
TRIVINYLANTIMONY see TNR490
TRIVINYLBISMUTH see TNR500
TRIVINYLTIN CHLORIDE see CLU500
TRIVITAN see CMC750
TRIXYLENYL PHOSPHATE see XLS100
TRIXYLYL PHOSPHATE see XLS100
TRI-2,6-XYLYL PHOSPHATE see TNR550
TRIZILIN see DFT800
TRIZIMAN see DXI400
TRIZIMAN D see DXI400
TRIZINC DIPHOSPHATE see ZJS400
TROBICIN see SKY500
TROCHIN see CLD000
TROCINAT see DHY400
TROCINATE see DHY400
TROCINATE HYDROCHLORIDE see
DHY400
TROCLOSENE see DGN200
TROCLOSENE POTASSIUM see PLD000
TROCOSONE see ECU750
TRODAX see HLJ500
TROFORMONE see AOO475
TROFOSFAMID see TNT500
TROFURIT see CHJ750
TROGAMID T see NOH000
TROGUM see SLJ500
TROJACETONOAMINY see TDT770
TROJCHLOREK FOSFORU (POLISH) see
PHT275
TROJCHLOROBENZEN see TIK250
TROJCHLOROBENZEN (POLISH) see
TIK250
TROJCHLOROETAN(1,1,2) (POLISH) see
TIN000
TROJCHLOROWODOREK 4-ACETYLO-4-
(3-CHLOROFENYLO)-1-(3-(4-
METYLOPIPERAZYNO)-PROPYLO)-
PIPERYDYNY see ACG125
TROJKREZYLU FOSFORAN (POLISH) see
TMO600
TROJNITROTOLUEN (POLISH) see TMN490
TROLAMINE see TKP500
TROLEN see RMA500
TROLENE see RMA500
TROLITUL see SMQ500
TROLMINE see TJL250
TROLNITRATE PHOSPHATE see TJL250
TROLOVOL see MCR750
TROLOX see TNR625
TROLOX C see TNR625
TROMASEDAN see BEA825
TROMASIN see PAG500
TROMBARIN see BKA000
TROMBAVAR see SIO000
TROMBIL see BKA000
TROMBOLYSAN see BKA000
TROMBOSAN see BJZ000
TROMBOVAR see EMT500
TROMBOVAR see SIO000
TROMEDONE see TLP750
TROMETAMOL see TEM500
TROMETE see SJH200
TROMETHAMINE see TEM500
TROMETHAMINE PROSTAGLANDIN F2-α
see POC750
TROMETHANE see TEM500

TROMETHANMIN see TEM500
TROMEXAN see BKA000
TROMEXAN ETHYL ACETATE see BKA000
TRONA see BMG400
TRONA see SFO000
TRONA see SJT750
TRONA see SJY000
TRONAMANG see MAP750
TRONOX see TGG760
TROPACAINE HYDROCHLORIDE see TNS200
TROPACOCAINE HYDROCHLORIDE see TNS200
TROPAEOLIN see MND600
TROPAEOLIN 1 see FAG010
TROPAEOLIN D see DOU600
TROPAEOLINE see MRL100
TROPAEOLIN G see MDM775
TROPAEOLIN O see MRL100
TROPAEOLIN OOO see CMM220
TROPAEOLIN OOO 2 see CMM220
TROPAEOLIN R see MRL100
TROPAKOKAIN HYDROCHLORID (GERMAN) see TNS200
2-β-TROPANECARBOXYLIC ACID, 3-β-HYDROXY-, METHYL ESTER, BENZOATE (ESTER) see CNE750
TROPANE, 3-α-((10,11-DIHYDRO-5H-DIBENZO(A,D)CYCLOHEPTEN-5-YL)OXY)-, CITRATE (1:1) see EAG100
3-TROPANYLBENZOATE-2-CARBOXYLIC ACID METHYL ESTER see CNE750
dl-TROPANYL-2-HYDROXY-1-PHENYLPROPIONATE see ARR000
dl-TROPANYL-2-HYDROXY-1-PHENYLPROPIONATE SULFATE see ARR500
3-α-TROPANYL (−)-2-METHYL-2-PHENYLHYDRACRYLATE HYDROCHLORIDE see LFC000
dl-TROPASAEUREESTER DES 3-DIAETHYLAMINO-2,2-DIMETHYL-1-PROPANOL PHOSPHAT (GERMAN) see AOD250
TROPASAEUREESTER DES 3-TRIAETHYLAMMONIUM-2,2-DIMETHYL-1-PROPANOLBROMID (GERMAN) see DSI200
(− +)-TROPATE-3-α-HYDROXY-8-OCTYL-1-α-H,5-α-H-TROPANIUM see OEM000
TROPAX see OPK000
TROPEOLIN see BFL000
TROPEOLIN O see MRL100
TROPHICARD see MAI600
TROPHICARDYL see IDE000
TROPHOSPHAMID see TNT500
TROPHOSPHAMIDE see TNT500
TROPIC ACID, ESTER with SCOPINE see SBG000
TROPIC ACID, ESTER with TROPINE see ARR000
(−)-TROPIC ACID ESTER with TROPINE see HOU000
TROPIC ACID, 9-METHYL-3-OXA-9-AZATRICYCLO(3.3.1.0²,⁴)NON-7-YL ESTER see SBG000
TROPIC ACID, 3-α-TROPANYL ESTER see AQO250
TROPIC ACID-3-α-TROPANYL ESTER see ARR000
TROPIDECHIS CARINATUS VENOM see ARV375
TROPILIDENE see COY000
TROPILIDIN see COY000
TROPIN see MGR250
TROPINE, ATROPATE (ESTER) see AQO250
TROPINE BENZOHYDRYL ETHER METHANESULFONATE see TNU000
TROPINE-4-CHLOROBENZHYDRYL ETHER HYDROCHLORIDE see CMB125
TROPINE, MANDELATE (ESTER) see HGK550

TROPINE (−)-α-METHYLTROPATE HYDROCHLORIDE see LFC000
TROPINE TROPATE see ARR000
TROPINIUM METHOBROMIDE MANDELATE see MDL000
TROPINTRAN see ARR500
TROPISETRON see TNU100
TROPISTON see PGP500
TROPITAL see PIZ499
TROPOCOCAIN HYDROCHLORIDE see TNS200
TROPOLONE see TNV550
TROPOTASIN see TFU800
TROPOTOX see CLN750
TROPOTOX see CLO000
TROPYLIUM PERCHLORATE see TNV575
(±)-TROPYL TROPATE see ARR000
dl-TROPYLTROPATE see ARR000
TROSINONE see GEK500
TROSPIUM CHLORIDE see KEA300
TROTOX see CLN750
TROTYL see TMN490
TROTYL OIL see TMN490
TROVIDUR see PKQ059
TROVIDUR see VNP000
TROVIDUR PE see PJS750
TROVITHERN HTL see PKQ059
TROXIDONE see TLP750
TROXILAN see AFJ400
TROXONIUM TOSILATE see TNV625
TROXONIUM TOSYLATE see TNV625
TROYKYD ANTI-SKIN B see EMU500
TROYKYD ANTI-SKIN BTO see BSU500
TROYSAN 142 see DSB200
TROYSAN COPPER 8% see NAS000
TROYSAN ANTI-MILDEW O see TIT250
TRP-P-1 see TNX275
TRP-P-2 see ALD500
TRP-P-1 (ACETATE) see AJR500
TRP-P-2(ACETATE) see ALE750
D-TRP-LH-RH see GJI100
d-TRP LHRH-PEA see LIU353
d-TRP⁶)-PRO⁹)-N-ETHYLAMIDE-LHRH see LIU353
(d-TRP⁶)-PRO⁹))-LHRH ETHYLAMIDE see LIU353
(d-TRP⁶)-PRO⁹))-LUTEINIZING HORMONE-RELEASING HORMONE ETHYLAMIDE see LIU353
(d-TRP(SUB₆)-PRO⁹))-NET)-GNRH see LIU353
(d-TRP(SUB₆)-PRO⁹))-NET)-GONADOTROPIN RELEASING HORMONE see LIU353
TRU see PQC500
TRUBAN see EFK000
TRUCIDOR see MJG500
TRUE AMMONIUM SULFIDE see ANJ750
TRUFLEX DOA see AEO000
TRUFLEX DOP see DVL700
TRUFLEX DOX see BJQ500
TRUFLEX DTDP see DXQ200
TRUMPET PLANT see CDH125
TRUSONO see EQL000
TRUSONO see MNM500
TRUXAL see TAF675
TRUXALETTEN see TAF675
TRUXIL see TAF675
TRYBEN see TIK500
TRYCITE 1000 see SMQ500
TRYCOL HCS see PJW500
TRYCOL LAL SERIES see DXY000
TRYCOL LF 1 see PKE370
TRYCOL NP-1 see NND500
TRYDET OS SERIES see PJY100
TRYDET SA SERIES see PJV250
TRYFAC 325A see PKE850
TRYFAC 525A see PKE850
TRYOPANOATE SODIUM see SKO500
TRYPAFLAVIN see XAK000
TRYPAFLAVINE see DBX400
TRYPAFLAVINE HYDROCHLORIDE see TNV700

TRYPANBLAU (GERMAN) see CMO250
TRYPAN BLUE see CMO250
TRYPAN BLUE SODIUM SALT see CMO250
TRYPARSAMIDE see CBJ750
TRYPOXYL see ARA500
TRYPSIN see TNW000
TRYPSIN INHIBITOR see PAF550
TRYPSIN INHIBITOR, PANCREATIC BASIC see PAF550
TRYPSIN-KALLIKREIN INHIBITOR (KUNITZ) see PAF550
TRYPTAMIDE see NDW525
TRYPTAMINE see AJX000
TRYPTAMINE HYDROCHLORIDE see AJX250
TRYPTAR see TNW000
TRYPTAZINE DIHYDROCHLORIDE see TKK250
TRYPTIZOL see EAH500
TRYPTIZOL see EAI000
TRYPTIZOL HYDROCHLORIDE see EAI000
(−)-TRYPTOPHAN see TNX000
d-TRYPTOPHAN see TNW250
l-TRYPTOPHAN see TNX000
dl-TRYPTOPHAN see TNW500
l-TRYPTOPHAN, pyrolyzate see TNX100
dl-TRYPTOPHAN, pyrolyzate 1 see TNX275
TRYPTOPHANE see TNX000
d-TRYPTOPHANE see TNW250
l-TRYPTOPHAN (FCC) see TNX000
l-TRYPTOPHAN, 5-HYDROXY-, (9CI) see HOA600
l-TRYPTOPHAN, MIXT. WITH l-ASCORBIC ACID AND 5-HYDROXY-6-METHYL-3,4-PYRIDINEDIMETHANOLHYDROCHLORIDE see OJG550
TRYPTOPHAN P1 see TNX275
TRYPTOPHAN P2 see ALD500
9-l-TRYPTOPHAN-TRYCODINE A HYDROCHLORIDE (9CI) see TOF825
TRYPURE see TNW000
TRYSBEN 200 see DOR800
TRYSBEN 200 see TIK500
TS 16 see AGD250
TS 160 see TNF250
TS-160 see TNF500
TS 219 see NIM500
TS 1801 see AHA875
TS-7236 see FDA885
TSAA 291 see ELF100
TSAA-328 see ELF110
TSA-HP see TGO000
TSA-MH see TGO000
TSAPOLAK 964 see CCU250
TSC see TFQ000
TSD see MCR250
TSELATOX see TNX375
TSEMERIN see HBT100
TSERENOX see BDD000
T-SERP see RDK000
TSH-888 see POO800
TSH-RELEASING FACTOR see TNX400
TSH-RELEASING HORMONE see TNX400
TSH-RF see TNX400
TSIAZID see COH250
TSIDIAL see DRR400
TSIKLAMID see ABB000
TSIKLODOL see BBV000
TSIKLOMETIAZID see CPR750
TSIKLOMITSIN see TBX000
TSIM see TMF250
TSIMAT see BJK500
TSINEB (RUSSIAN) see EIR000
TSIPROMAT (RUSSIAN) see ZMA000
TSIRAM (RUSSIAN) see BJK500
TSITREX see DXX400
TSIZP 34 see ISR000
TSL 8112 see DOE100
TSL 8117 see DOE100
TSL 8123 see MQD750
TSL 8370 see TNJ500
TSLT see TNX375

d-T4 SODIUM see SKJ300
TSP see SJH200
TSPA see TFQ750
TSPP see TEE500
TST see EIV000
TSTS 19 see LED100
TSTS 21 see LED100
TSTS 22 see LED100
TSTS 23 see LED100
T-STUFF see HIB050
TSTX see TGH520
TSUDOHMIN see DEO600
TSUGA OIL see SLG650
TSUKUBAENOLIDE HYDRATE see TAA900
TSUMACIDE see MIB750
TSUMACIDE, nitrosated see MNV500
TSUMAUNKA see MIB750
TSUSHIMYCIN see TNX650
TSUYA INDIGO 2B see ICU135
TT see TMV500
TTC see TMV500
TTD see DXH250
TTD see TFS350
TTFB see TBW000
TTFD see FQJ100
T-17 TOXIN see THI255
T²-TRICHOTHECENE see FQS000
TTS see DXH250
TTT see HDV500
TTX see FOQ000
TU see TFR250
2-TU see TFR250
TUADS see TFS350
TUAMINE see HBM490
TUAMINE SULFATE see AKD600
TUAMINOHEPTANE see HBM490
TUAMINOHEPTANE SULFATE see AKD600
TUAREG see TLN500
TUASOL 100 see THW750
TUATUA (PUERTO RICO) see CNR135
TUAZOLE see MDT250
TUAZOLE see QAK000
TUAZOLONE see QAK000
TUBADIL see TOA000
TUBARINE see TOA000
TUBA ROOT see DBA000
TUBATOXIN see RNZ000
TUBAZIDE see ILD000
TUBERACTINOMYCIN B see VQZ000
TUBERACTINOMYCIN-N SULFATE see VQZ100
TUBERACTIN SULFATE see VQZ100
TUBERCID see ILD000
TUBERCIDIN see TNY500
TUBEREX see PNW750
TUBERGAL see AGQ875
TUBERIT see CBM000
TUBERITE see CBM000
TUBERMIN see EPQ000
TUBEROID see EPQ000
TUBEROSON see EPQ000
TUBERSAN see SEP000
TUBEX see AES650
TUBICON see ILD000
TUBIGAL see FNF000
TUBIN see FNF000
TUBOCIN see RKP000
TUBOCURARANIUM, 6,6',7',12'-
TETRAMETHOXY-2,2,2',2'-TETRAMETHYL-
, DIIODIDE, (1-β)- see TDO100
TUBOCURARIN see TNY750
TUBOCURARINE see TNY750
(+)-TUBOCURARINE see TNY750
d-TUBOCURARINE see TNY750
TUBOCURARINE CHLORIDE see TOA000
(+)-TUBOCURARINE CHLORIDE see
TOA000
d-TUBOCURARINE CHLORIDE see TOA000
TUBOCURARINE, CHLORIDE,
HYDROCHLORIDE, (+)- (8CI) see TOA000
d-TUBOCURARINE CHLORIDE
PENTAHYDRATE see TNZ000

d-TUBOCURARINE DICHLORIDE see
TOA000
(+)-TUBOCURARINE DICHLORIDE
PENTAHYDRATE see TNZ000
TUBOCURARINE DIMETHYL ETHER
IODIDE see DUM000
TUBOCURARINE HYDROCHLORIDE see
TOA000
(+)-TUBOCURARINE HYDROCHLORIDE
see TOA000
d-TUBOCURARINE HYDROCHLORIDE see
TOA000
d-TUBOCURARINE IODIDE DIMETHYL
ETHER see DUM000
TUBOPHAN see DWW000
TUBOTHANE see MAS500
TUBOTIN see ABX250
TUBOTIN see HON000
TU CILLIN see BFD000
TUEX see TFS350
TUFF-LITE see PMP500
TUFT ROOT see DHB309
TUGON see TIQ250
TUGON FLIEGENKUGEL see PMY300
TUGON FLY BAIT see TIQ250
TUGON STABLE SPRAY see TIQ250
TULABASE FAST GARNET GB see AIC250
TULABASE FAST GARNET GBC see AIC250
TULABASE FAST RED RL see MMF780
TULABASE FAST RED TR see CLK220
TULADISPERSE FAST YELLOW 2G see
AAQ250
TULA (PUERTO RICO) see COD675
TULASTERON FAST YELLOW 5R-B see
CMP090
TULASTERON FAST YELLOW G-C see
KDA075
TULATHOL AS see CMM760
TULISAN see TFS350
TULLIDORA see BOM125
TULUYLENDIISOCYANAT see TGM750
TULYL see RLU000
TUMBLEAF see SFS000
TUMENOL see IAD000
TUMESCAL OPE see BGJ250
TUMEX see QPA000
TUMOR NECROSIS FACTOR-α see HGL920
TUNG NUT see TOA275
TUNG NUT MEALS see TOA500
TUNG NUT OIL see TOA510
TUNG OIL TREE see TOA275
TUNGSTEN see TOA750
TUNGSTEN ANTIMONATE see TOA800
TUNGSTEN AZIDE PENTABROMIDE see
TOB000
TUNGSTEN AZIDE PENTACHLORIDE see
TOB250
TUNGSTEN BLUE see TOC750
TUNGSTEN CARBIDE see TOB500
TUNGSTEN CARBIDE, mixed with COBALT
(92%:8%) see TOC000
TUNGSTEN CARBIDE, mixed with COBALT
(85%:15%) see TOB750
TUNGSTEN CARBIDE, mixed with COBALT
and TITANIUM (78%:14%:8%) see TOC250
TUNGSTEN COMPOUNDS see TOC500
TUNGSTEN FLUORIDE see TOC550
TUNGSTEN HEXAFLUORIDE see TOC550
TUNGSTEN-NICKEL CATALYST DUST see
TOC600
TUNGSTEN OXIDE see TOC750
TUNGSTEN STILBONATE see TOA800
TUNGSTEN TRIOXIDE see TOC750
TUNGSTIC ACID see TOD000
TUNGSTIC ACID, SODIUM SALT,
DIHYDRATE see SKO000
TUNGSTIC ANHYDRIDE see TOC750
TUNGSTIC OXIDE see TOC750
TUNGSTOPHOSPHORIC ACID (8CI) see
PHU750
TUNGSTOPHOSPHORIC ACID, SODIUM
SALT see SJJ000

TUNIC see BGD250
TUNICIN see CCU150
TUPHETAMINE see BBK500
TUR see CMF400
TURBINAIRE see DAE525
TURBSVIL see BQT750
TURBULETHYLAZIN (GERMAN) see
BQB000
TURCAM see DQM600
TURF-CAL see ARB750
3336 TURF FUNGICIDE see DJV000
TURGEX see HCL000
TURIMYCIN A3 see LEV025
TURIMYCIN A5 see JDS200
TURIMYCIN P₃ see MBY150
TURINABOL see CLG900
TURINAL see AGF750
TURINGIN-1 see BAC125
TURISYNCHRON see MLJ500
TURKEY RED see DMG800
TURKEY RED see MAB800
TURKEY-RED OIL see TOD500
TURKISH ROSE OIL see OHJ000
TURMERIC see TOD625
TURMERIC rhizome extract see COF850
TURMERIC OIL see COG000
TURMERIC OLEORESIN see COG000
TURMERIC YELLOW see ICC800
TURPENTINE see TOD750
TURPENTINE CAMPHOR see BMD300
TURPENTINE OIL see TOD750
TURPENTINE OIL, RECTIFIER see TOD750
TURPENTINE STEAM DISTILLED see
TOD750
TURPENTINE SUBSTITUTE (UN 1300)
(DOT) see TOD750
TURPENTINE (UN 1299) (DOT) see TOD750
TURPETH MINERAL see MDG000
TURPINAL SL see HKS780
TURQUOISE GLL see COF420
TURQUOISE 8GL see COF420
TUS-1 see XAJ000
TUSILAN see DBE200
TUSSADE see DBE200
TUSSAL see MDP750
TUSSAPAP see HIM000
TUSSAPHED see POH250
TUSSCAPINE see NOA000
TUSSCAPINE HYDROCHLORIDE see
NOA500
TUSSILAGO FARFARA L see CNH250
TUTANE see BPY000
TUTIN see TOE175
TUTINE see TOE175
TUTOCAINE see TOE150
TUTOCAINE HYDROCHLORIDE see
AIT750
TUTOFUSIN TRIS see TEM500
TUTTOMYCIN see NCD550
TUTU see TOE175
TUVICAL 210 see VNK200
TUZET see USJ075
TV 02 see SBL600
TV 485 see HKK000
TV 1322 see AAE625
TVOPA see TOE200
TVOPA (DOT) see TOE200
825TV-PS see SMQ500
TVS 8105 see DVK200
TVS-MA 300 see DEJ100
TVS-N 2000E see DEJ100
TVX 485 see HKK000
TW 30 see LAU560
T-WD602 see TJE100
TWEEN 20 see PKG000
TWEEN 40 see PKG500
TWEEN 60 see PKL030
TWEEN 65 see SKV195
TWEEN 80 see PKL100
TWEEN 85 see TOE250
TWEENASE see GGA800
TWEEN ESTERASE see GGA800

1-UNDECANECARBOXYLIC ACID see LBL000
1,1'-UNDECANEDICARBOXYLIC ACID ESTER with ETHYLENE GLYCOL see EJQ500
UNDECANE, 1,1-DIMETHOXY-2-METHYL- see MNB600
UNDECANOIC ACID see UKA000
UNDECANOIC ACID, 4-HYDROXY-, γ-LACTONE see HBN200
UNDECANOIC ACID, TRIBUTYLSTANNYL ESTER see TIG500
UNDECANOIC ACID, TRIISOPROPYLSTANNYL ESTER see TKT850
n-UNDECANOL see UNA000
γ-UNDECANOLACTONE see HBN200
4-UNDECANOLIDE see HBN200
1,4-UNDECANOLIDE see HBN200
UNDECANOLIDE-1,5 see UKJ000
γ-UNDECANOLIDE see HBN200
2-UNDECANONE see UKS000
6-UNDECANONE see ULA000
UNDECAN-6-ONE see ULA000
1,3,5-UNDECATRIENE see ULA100
3,5,9-UNDECATRIEN-2-ONE, 6,10-DIMETHYL- see POH525
3,5,9-UNDECATRIEN-2-ONE, 3,6,10-TRIMETHYL- see TMJ100
10-UNDECENAL see ULJ000
1-UNDECEN-10-AL see ULJ000
10-UNDECENAL DIGERANYL ACETAL see UNA100
9-UNDECENAL, 2,6,10-TRIMETHYL- see TMJ150
1-UNDECENE, 11,11-BIS((3,7-DIMETHYL-2,6-OCTADIENYL)OXY)- see UNA100
10-UNDECENOIC ACID see ULS000
10-UNDECENOIC ACID, BUTYL ESTER see BSS100
10-UNDECENOIC ACID, ETHYL ESTER see EQD200
9-UNDECENOIC ACID, METHYL ESTER see ULS400
2-UNDECENOL see ULS875
10-UNDECENOL see UMA000
1-UNDECEN-11-OL see UMA000
10-UNDECENOYL CHLORIDE see UMJ000
UNDECENYL ACETATE see UMS000
10-UNDECENYL ACETATE see UMS000
ω-UNDECENYL ALCOHOL see UMA000
n-UNDECOIC ACID see UKA000
UNDECYL ALCOHOL see UNA000
UNDECYL ALDEHYDE see UJJ000
n-UNDECYL ALDEHYDE see UJJ000
UNDECYLENALDEHYDE see ULJ000
10-UNDECYLENEALDEHYDE see ULJ000
UNDECYLENIC ACID see ULS000
9-UNDECYLENIC ACID see ULS000
UNDECYL-10-ENIC ACID see ULS000
10-UNDECYLENIC ACID see ULS000
ω-UNDECYLENIC ACID CHLORIDE see UMJ000
UNDECYLENIC ALCOHOL see UMA000
UNDECYLENIC ALDEHYDE see ULJ000
UNDECYLENIC ALDEHYDE DIGERANYL ACETAL see UNA100
10-UNDECYLENOYL CHLORIDE see UMJ000
UNDECYLIC ACID see UKA000
UNDECYLIC ALDEHYDE see UJJ000
γ-UNDEKALAKTON see HBN200
UNDEN see EDV000
UNDEN see PMY300
UNDEVELOPED LITHOL TONER see NAP100
UNFINISHED LUBRICATING OIL see COD750
UNGARISCHES GELBHOLZ see FBW000
UNGEREMINE see LJB800
UNIBARYT see BAP000
UNIBOLDINA see DNZ100

UNIBUR 70 see CAT775
UNICA 380K see MCB050
UNICA F 730 see MCB050
UNICA RESIN 380K see MCB050
UNICELLES see BPF000
UNICEL-ND see DVF400
UNICEL NDX see DVF400
UNICHEM see PKQ059
UNICHLOR see PAH780
UNICHLOR 50 see PAH780
UNICIN see TBX250
UNICOCYDE see ILD000
UNICONAZOLE see SOU800
UNICONAZOLE-P see SOU800
UNICROP CIPC see CKC000
UNICROP DNBP see BRE500
UNICROP MANEB see MAS500
UNIDERM WGO see WBJ700
UNIDIGIN see DKL800
UNIDOCAN see DAQ800
UNIDRON see DXQ500
UNIFLEX BYO see BSB000
UNIFLEX BYS see BSL600
UNIFLEX DOS see BJS250
UNIFOAM AZ see ASM270
UNIFORM AZ see ASM270
UNIFOS DYOB S see PJS750
UNIFOS EFD 0118 see PJS750
UNIFUME see EIY500
UNIFUR see FPI000
UNI-GUAR see GLU000
UNILAX see ACD500
UNILORD see RDK000
UNIMATE GMS see OAV000
UNIMATE IPM see IQN000
UNIMATE IPP see IQW000
UNIMOLL 66 see DGV700
UNIMOLL BB see BEC500
UNIMYCETIN see CDP250
UNIMYCIN see TBX250
UNION BLACK EM see AQP000
UNION BORDEAUX B see CMO872
UNION BROWN DR see CMO800
UNION CARBIDE 1-174 see TLC250
UNION CARBIDE 1-189 see TLC000
UNION CARBIDE 7207 see OCE100
UNION CARBIDE 7,744 see CBM750
UNION CARBIDE 19786 see CBW000
UNION CARBIDE 20299 see SFV250
UNION CARBIDE A-15 see EQA000
UNION CARBIDE A-150 see TIN750
UNION CARBIDE A-151 see TJN250
UNION CARBIDE A-162 see MQD750
UNION CARBIDE A-163 see MQF500
UNION CARBIDE A-186 see EBO000
UNION CARBIDE A-187 see ECH000
UNION CARBIDE LIQUID G see SCR400
UNION CARBIDE 7158 SILICONE FLUID see DAF350
UNION CARBIDE UC-8305 see CGM400
UNION CARBIDE UC-8454 see TCY275
UNION CARBIDE UC-9880 see CQI500
UNION CARBIDE UC-10,854 see COF250
UNION CARBIDE UC 20047 see CFF250
UNION DARK GREEN B see CMO830
UNION FAST NAVY BLUE DS see CMN800
UNION FAST RED 3B see CMO870
UNION FAST SCARLET 3B see CMO875
UNIPAQUE see AOO875
UNIPEN see SGS500
UNIPHAT A20 see MHY800
UNIPHAT A30 see MHY650
UNIPHAT A40 see MLC800
UNIPINE see PIH750
UNIPON see DGI400
UNIPON see DGI600
UNIPROFEN see OJI750
UNIROYAL see DFT000
UNIROYAL D014 see SOP000
UNIROYAL F849 see AKR500
UNISEDIL see DCK759
UNISEPT see CDT250

UNISOL 4-O see PJY100
UNISOL RH see SFO500
UNISOM see PGE775
UNISOMNIA see DLY000
UNISPIRAN see GCE600
UNISTRADIOL see EDP000
UNISULF see SNN500
UNITANE O-110 see TGG760
UNITANE OR see RSP100
UNITED see CBT750
UNITED CHEMICAL DEFOLIANT No. 1 see SFS000
UNITENE see MCC250
UNITENSEN see CKP500
UNITENSEN see RDK000
UNITERTRACID GREEN BS see ADF000
UNITERTRACID LIGHT BLUE AB see CMM070
UNITERTRACID LIGHT ORANGE G see HGC000
UNITERTRACID LIGHT YELLOW RR see SGP500
UNITERTRACID RED 2G see CMM300
UNITERTRACID RED 6B see CMM400
UNITERTRACID YELLOW TE see FAG140
UNITESTON see TBG000
UNITHIOL see DNU860
UNITINA HR see HHW502
UNITIOL see DNU860
UNITOL see DNU860
UNITOX see CDS750
UNIVASC see MRA260
UNIVERM see CBY000
UNIVOL U 316S see MSA250
UNJECOL 50 see OBA000
UNJECOL 70 see OBA000
UNJECOL 90 see OBA000
UNJECOL 110 see OBA000
UNLEADED GASOLINE see GCE100
UNLEADED MOTOR GASOLINE see GCE100
UNON P see TAI250
UNOPROSTONE ISOPROPYL ESTER see IRR050
UNOSPASTON see DGW600
UNOX 201 see ECB000
UNOX 206 see PKQ070
UNOX 207 see BGA250
UNOXAT 207X see BGA250
UNOXAT EPOXIDE 269 see LFV000
UNOX EP 206 see PKQ070
UNOX EPOXIDE 201 see ECB000
UNOX EPOXIDE 206 see PKQ070
UNOX EPOXIDE 206 see VOA000
UNOX EPOXIDE 207 see BGA250
UNYSH A see FAQ930
UOP 26 see BBN100
UOP 88 see BJT500
UP 1 see SMQ500
UP 2 see SMQ500
UP 27 see SMQ500
UP 83 see NDX500
UP 583 see AJU700
UP 925 see CGW300
UPALET see CNT350
U 46DP see DAA800
UP-583D see AJU700
UP 1E see SMR000
UPIOL see BNP750
UPJOHN U-12,927 see CGI500
UPJOHN U-18120 see MDX250
UPJOHN U-22023 see MLX000
UPJOHN U-32714 see CON300
UPJOHN U-36059 see MJL250
UPM see SMQ500
UPM703 see SMQ500
UPM508L see SMQ500
UR 003 see PAO170
UR 606 see CEH700
UR 1522 see PHA575
URAB see FAR050
URACIL see UNJ800

URACIL 634 see CPR600
URACIL, 5-AMINO- see AMW100
URACIL, 6-AMINO-2-THIO- see AMS750
6-URACILCARBOXYLIC ACID see OJV500
URACIL DESOXYURIDINE see DAZ050
URACIL, 5-FLUORO-3-(TETRAHYDRO-2-
FURYL)- see FLZ065
URACIL mixture with FT (4:1) see UNJ810
URACIL, 5-(HYDROXYMETHYL)-6-
METHYL- see HMH300
URACILLOST see BIA250
URACILMOSTAZA see BIA250
URACIL MUSTARD see BIA250
URACIL, 6-PROPYL-2-THIO-, and IODINE
see PNX100
URACIL RIBOSIDE see UVJ000
URACIL mixture with TEGAFUR (4:1) see
UNJ810
URACIL mixture with 1-(2-
TETRAHYDROFURYL)-5-FLUOROURACIL
(4:1) see UNJ810
URACTONE see AFJ500
URACTYL see SNQ550
URADAL see BNK000
URAGAN see BMM650
URAGAN see SIH500
URAGON see BMM650
URALENIC ACID see GIE000
URALGIN see EID000
URALITE see UTU500
URALITE (POLYMER) see UTU500
URALYT-U see SFX725
URAMID see SNQ550
URAMINE T 80 see HLU500
URAMINE T101 see UTU500
URAMINE T105 see UTU500
URAMINE TSL 58 see UTU500
U-RAMIN P 6100 see MCB050
U-RAMIN P 6300 see MCB050
U-RAMIN T 33 see MCB050
U-RAMIN T 34 see MCB050
URAMITE see UTU500
URAMUSTIN see BIA250
URAMUSTINE see BIA250
URAMYCIN B see VGZ000
URANIN see FEW000
URANINE A EXTRA see FEW000
URANINE USP XII see FEW000
URANINE YELLOW see FEW000
URANIUM see UNS000
URANIUM ACETATE see UPS000
URANIUM AZIDE PENTACHLORIDE see
UOA000
URANIUM, BIS(ACETATO)DIOXO-,
DIHYDRATE see UQT700
URANIUM, BIS(ACETO-O)DIOXO-,
DIHYDRATE (9CI) see UQT700
URANIUM, BIS(NITRATO-O,O')DIOXO-,
(OC-6-11)- see URA100
URANIUM CARBIDE see UOB100
URANIUM(IV) CHLORIDE see UQJ000
URANIUM DICARBIDE see UOC200
URANIUM FLUORIDE see UOC300
URANIUM FLUORIDE (fissile) see UOJ000
URANIUM FLUORIDE OXIDE see UQA000
URANIUM FLUORIDE (U₂F₈) see UOC300
URANIUM HEXAFLUORIDE, fissile excepted
or non-fissile (UN 2978) (DOT) see UOJ000
URANIUM HEXAFLUORIDE, fissile
(containing >1% U-235) (UN 2977) (DOT) see
UOJ000
URANIUM(III) HYDRIDE see UPA000
URANIUM METAL, pyrophoric (DOT) see
UNS000
URANIUM NITRATE see UPA100
URANIUM(IV) OXIDE see UPJ000
URANIUM OXYACETATE see UPS000
URANIUM OXYFLUORIDE see UQA000
URANIUM TETRACHLORIDE see UQJ000
URANIUM TETRAFLUORIDE see UOC300
URANIUM(IV) TETRAHYDROBORATE see
UQT300

URANIUM(III) TETRAHYDROBORATE see
UQS000
URANYL ACETATE see UPS000
URANYL ACETATE DIHYDRATE see
UQT700
URANYL CHLORIDE see URA000
URANYL FLUORIDE see UQA000
URANYL NITRATE see URA100
URANYL NITRATE (solid) see URA200
URANYL NITRATE HEXAHYDRATE see
URS000
URANYL NITRATE HEXAHYDRATE,
solution (DOT) see URS000
URAPIDIL see USJ000
URAPRINT 62-126 see CAW500
URARI see COF750
URAZIUM see PDC250
URBACID see USJ075
URBACIDE see USJ075
URBANYL see CIR750
URBASON see MOR500
URBASON CRYSTAL SUSPENSION see
DAZ117
URBASONE see MOR500
URBASON SOLUBLE see USJ100
URBASULF see MGQ750
URBAZID see USJ075
URBIL see MQU750
URBOL see ZVJ000
dURD see DAZ050
UREA see USS000
UREA, 1-(1-ADAMANTYL)-3-(2-
FLUOROETHYL)-1-NITROSO- see AEF600
UREA, AMIDINO- see GLS900
UREA, 1-AMIDINO-3-(p-NITROPHENYL)-,
MONOHYDROCHLORIDE see NIJ400
UREA ANTIMONYL TARTRATE see UTA000
UREA, N-(1,3-BIS(HYDROXYMETHYL)-2,5-
DIOXO-4-IMIDAZOLIDINYL)-N,N'-
BIS(HYDROXYMETH YL)- see IAS100
UREA, N,N'-BIS(4-NITROPHENYL)-, compd.
with 4,6-DIMETHYL-2(1H)-PYRIMIDINONE
(1:1) (9CI) see NCW100
UREA, N-(4-BROMOPHENYL)-N'-METHYL-
(9CI) see MHS375
UREA, sec-BUTYL- see BSS300
UREA, tert-BUTYL- see BSS310
UREA, 3-(4-(2-tert-BUTYL-5-OXO-Δ²)-1,3,4-
(OXADIAZOLIN-4-YL)-3-
CHLOROPHENYL)-1,1-DIMETHYL- see
UTA300
UREA, N'-(2-CHLOROETHYL)-N-NITROSO-
N-(2-OXOPROPYL)- see NLE440
UREA, N-(4-CHLOROPHENYL)-N'-(3,4-
DICHLOROPHENYL)-(9CI) see TIL500
UREA, N-((4-CHLOROPHENYL)METHYL)-
N-CYCLOPENTYL-N'-PHENYL- see UTA400
UREA, 1-((p-
CHLOROPHENYL)SULFONYL)-3-
CYCLOHEXYL- see CDR550
UREA, 1-((p-
CHLOROPHENYL)SULFONYL)-3-
ISOPROPYL- see CDY100
UREA, 1-((O-
CHLOROPHENYL)SULFONYL)-3-(4-
METHOXY-6-METHYL-S-TRIAZIN-2-YL)-
see CMA700
UREA, 1-(p-CHLOROPHENYL)-2-THIO- see
CKL050
UREA, 1-(p-CHLOROPHENYL)-2-THIO- see
UTA450
UREA, 3-p-CUMENYL-1,1-DIMETHYL- see
IRA050
UREA, N-CYCLOPROPYL-N'-(2,5-
DIFLUOROPHENYL)- see CQE350
UREA, N-(3,4-DICHLOROPHENYL)-N'-
HYDROXY- see DGD085
UREA, 1,1-DIETHYL- see DKD650
UREA, N,N-DIETHYL-(9CI) see DKD650
UREA, N,N-DIETHYL-N'-((8-α)-6-
METHYLERGOLIN-8-YL)- see DLR100

UREA, N,N-DIETHYL-N'-((8-α)-6-
METHYLERGOLIN-8-YL)-, (Z)-2-
BUTENEDIOATE (1:1) see DLR150
UREA, N,N'-DIETHYL-N-NITROSO- see
NJW700
UREA, N,N-DIETHYL-N'-((8-α)-6-
PROPYLERGOLIN-8-YL)- see DJX300
UREA, N,N-DIETHYL-N'-((8-α)-6-
PROPYLERGOLIN-8-YL)-, (Z)-2-
BUTENEDIOATE (1:1) see DJX350
UREA, N-(6-((2,6-
DIFLUOROPHENYL)METHOXY)-2,3-
DIHYDRO-3-BENZOFURANYL)-N-
HYDROXY-, (S)- see DLB600
UREA, 1,1-DIMETHYL- see DUM150
UREA, N,N-DIMETHYL-N'-(5-(1,1-
DIMETHYLETHYL)-3-ISOXAZOLYL)- see
ISR200
UREA, N,N'-DIMETHYL-N,N'-DIPHENYL-
(9CI) see DRB200
UREA, (1,1-DIMETHYLETHYL)-(9CI) see
BSS310
UREA, 1,1-DIMETHYL-3-(p-
ISOPROPYLPHENYL)- see IRA050
UREA, 1,1-DIMETHYL-3-(p-(p-
METHOXYPHENOXY)PHENYL)- see
LGM300
UREA, 1,1-DIMETHYL-3-PHENYL-,
TRICHLOROACETATE see FAR050
UREA DIOXIDE see HIB500
UREA, ETHYL- see EQD875
UREA, 1-ETHYL- see EQD875
UREA, 1,1'-ETHYLENEDI- see EJC100
UREA, N'-ETHYL-N-METHYL-N-NITROSO-
see NKU370
UREA, N-ETHYL-N'-METHYL-N-NITROSO-
see NKE100
UREA, N-ETHYL-N-NITROSO-N'-(2-
OXOPROPYL)- see NKE120
UREA, N-(2-FLUOROETHYL)-N-NITROSO-
N'-TRICYCLO(3.3.1.1³·⁷))DEC-1-YL- see
AEF600
UREA-FORMALDEHYDE ADDUCT see
UTU500
UREA-FORMALDEHYDE CONDENSATE
see UTU500
UREA-FORMALDEHYDE COPOLYMER see
UTU500
UREA-FORMALDEHYDE OLIGOMER see
UTU500
UREA-FORMALDEHYDE POLYMER see
UTU500
UREA, FORMALDEHYDE POLYMER,
ISOBUTYLATED see IIM300
UREA-FORMALDEHYDE
PRECONDENSATE see UTU500
UREA-FORMALDEHYDE PREPOLYMER
see UTU500
UREA-FORMALDEHYDE RESIN see
UTU500
UREA, GUANYL- see GLS900
UREA HYDROGEN PEROXIDE (DOT) see
HIB500
UREA HYDROGEN PEROXIDE SALT see
HIB500
UREA HYDROPEROXIDE see HIB500
UREA, (1-HYDROXY-2,2,2-
TRICHLOROETHYL)- see HOL100
UREA, ISOBUTYL ALCOHOL,
FORMALDEHYDE POLYMER see IIM300
UREA, ISOPROPYL- see IRR100
UREA, N-(2-METHOXYETHYL)-N-
NITROSO- see NKO900
UREA, N-METHOXY-N-METHYL-N'-(4-(2-
(4-METHYLPHENYL)ETHOXY)PHENYL)-
see MNP350
UREA, (1-METHYLETHYL)-(9CI) see IRR100
UREA, 1-METHYL-2-THIO- see MPW600
UREA, MONONITRATE (8CI,9CI) see UTJ000
UREA, N,N''-1,2-ETHANEDIYLBIS-(9CI) see
EJC100
UREA NITRATE see UTJ000

UREA NITRATE (wet) see UTJ000
UREA NITRATE, dry or wetted with <20% water, by weight (UN 0220) (DOT) see UTJ000
UREA NITRATE, wetted with not <20% water, by weight (UN 1357) (DOT) see UTJ000
UREA, N-NITROSO-N-(2-OXOPROPYL)- see OOO050
UREAPAP W see UTU500
UREA, PARAFORMALDEHYDE, ISOBUTYL ALCOHOL, FORMALDEHYDE POLYMER see IIM300
UREA PERCHLORATE see UTU400
UREA PEROXIDE (DOT) see HIB500
UREA, (2-PHENYLBUTYRYL)-, (–)- see PDM100
UREAPHIL see USS000
UREA, POLYMER with FORMALDEHYDE see UTU500
UREA, POLYMER WITH FORMALDEHYDE, ISOBUTYLATED see IIM300
UREA, 2-PROPENYL-(9CI) see AGV000
UREA, PROPIONYL- see AJD800
UREA, PROPYL- see PNX550
UREASE see UTU550
UREA, SULFANILYL- see SNQ550
UREA, SULFATE (1:1) see UTU600
UREA, TETRABUTYL- see TBM850
UREA, 1,1,3,3-TETRABUTYL- see TBM850
URECHITES LUTEA see YAK300
URECHOLINE see HOA500
URECHOLINE CHLORIDE see HOA500
URECOLI S see UTU500
URECOLL K see UTU500
URECOLL KL see UTU500
UREGIT see DFP600
p-UREIDOBENZENEARSONIC ACID see CBJ000
(p-UREIDOBENZENEARSYLENEDITHIO)DI-o-BENZOIC ACID see TFD750
4-UREIDO-1-PHENYLARSONIC ACID see CBJ000
(p-UREIDOPHENYLARSYLENEDITHIO)DIACETIC ACID see CBI250
(p-UREIDOPHENYLARSYLENEDITHIO)DI-o-BENZOIC ACID see TFD750
UREIDOUREAARBOXAMIDE see HGU050
URELIT C see UTU500
URELIT HM see UTU500
URELIT R see UTU500
URENIL see SNQ550
UREOL P see DTG700
UREOPHIL see USS000
UREPRET see UTU500
URESE see BDE250
URESPAN see UTU700
URETAN ETYLOWY (POLISH) see UVA000
URETHAN see UVA000
URETHANE see UVA000
URETHANE POLYMERS see PKL500
URETHYLANE see MHZ000
UREVERT see USS000
UREX see CHJ750
6,6'-UREYLENEBIS(1,1'-DIMETHYLQUINOLINIUM) SULFATE see PJA120
6,6'-UREYLENEBIS(1-METHYLQUINOLINIUM)BIS(METHOSULFATE) see PJA120
URFAMICIN HYDROCHLORIDE see UVA150
URGENEA MARITIMA see RCF000
URGILAN see POB500
URI see SPC500
URIBEN see EID000
URIBEST see NBA600
URIC ACID see UVA400
URIC ACID, MONOSODIUM SALT see SKO575

URICEMIL see ZVJ000
URICOSID see DWW000
URICOVAC see DDP200
URIDINAL see PDC250
URIDINAL see PEK250
URIDINE see UVJ000
β-URIDINE see UVJ000
URIDINE, 2'-DEOXY- see DAZ050
URIDINE, 2'-DEOXY-5-ETHYL- see EHV200
URIDINE, 2'-DEOXY-5-FORMYL- see FNK033
URIDINE 5'-(DIHYDROGEN PHOSPHATE) see UVJ410
URIDINE MONOPHOSPHATE see UVJ410
URIDINE 5'-MONOPHOSPHATE see UVJ410
URIDINE, 4-OXIME see HKA270
URIDINE PHOSPHATE see UVJ410
URIDINE 5'-PHOSPHATE see UVJ410
URIDINE 5'-PHOSPHORIC ACID see UVJ410
URIDION see UVJ400
URIDYLIC ACID see UVJ410
5'-URIDYLIC ACID see UVJ410
URINARY HEBIN see FMT100
URINARY INDICAN see ICD000
URINEX see CLH750
URIODONE see DNG400
URIPLEX see PDC250
URISOXIN see SNN500
URISPAS see FCB100
URITAS see ZVJ000
URITONE see HEI500
URITRATE see OOG000
URITRISIN see SNN500
URIZEPT see NGE000
URLEA see BEQ625
URNER'S LIQUID see DEL000
URO-ALVAR see OOG000
UROANTHELONE see UVJ475
UROBENYL see ZVJ000
UROBIOTIC-250 see PDC250
UROCALUM see UVJ425
UROCANIC ACID see UVJ440
UROCANINIC ACID see UVJ440
UROCARMIN see EEQ600
UROCAUDAL see UVJ450
UROCHECK see TMV500
UROCONTRAST see HGB200
UROCORTISOL see UVJ460
UROCYDAL see MPQ750
URODIATON see MPQ750
URODIAZIN see CFY000
URODINE see PDC250
URODINE see PEK250
URODIXIN see EID000
UROENTERONE see UVJ475
UROFEEN see PDC250
UROFIX see UTU500
UROGAN see SNN500
UROGASTRON see UVJ475
UROGASTRONE see UVJ475
UROGRAFIN ACID see DCK000
UROGRANOIC ACID see DCK000
UROKINASE see UVS500
UROKINASE (ENZYME-ACTIVATING) see UVS500
UROKON SODIUM see AAN000
UROLUCOSIL see MPQ750
UROMALINE see MAP000
UROMAN see EID000
UROMIDE see PDC250
UROMIDIN see PIZ000
UROMIRO see AAI750
UROMIRO 380 see MCA775
UROMIRON see AAI750
UROMITEXAN see MDK875
UROMUCAESTHIN see BQA010
UROMYCINE see GCO000
URONAL see BAG000
URONEG see EID000
URONIUM PERCHLORATE see UTU400
UROPHENYL see PDC250
UROPYRIDIN see PDC250

UROPYRINE see PDC250
UROROST see UVS600
UROSCREEN see TMV500
UROSEMIDE see CHJ750
URO-SEPTRA see TKX000
UROSIN see ZVJ000
UROSULFAN see SNQ550
UROSULFANE see SNQ550
UROSULFON see SNQ710
UROSULFONE see SNQ710
UROTRAST see DCK000
UROTRATE see OOG000
UROTROPIN see HEI500
UROTROPINE see HEI500
UROVISON see SEN500
UROVIST see AOO875
UROX 379 see CJY000
UROX B WATER SOLUBLE CONCENTRATE WEED KILLER see BMM650
UROX HX GRANULAR WEED KILLER see BMM650
UROXIN see AFU250
UROXOL see OOG000
URSACOL see DMJ200
(+)-URSAMINE see RFP100
URSIN see HIH100
URSO see DMJ200
URSOCHOL see DMJ200
URSODEOXYCHOL see DMJ200
URSODEOXYCHOLIC ACID see DMJ200
URSODESOXYCHOLIC ACID see DMJ200
URSOFALK see DMJ200
URSOFERRAN see IGS000
URSOL BROWN O see CEG600
URSOL BROWN RR see ALL750
URSOL D see PEY500
URSOL EG see ALS990
URSOL ERN see NAW500
URSOL OLIVE 6G see CFK125
URSOL P see ALT250
URSOL P BASE see ALT250
URSOL SLA see DBO400
URSOLVAN see DMJ200
URSOL YELLOW BROWN A see ALO000
URTOSAL see SAH000
US 2 see CNH125
USACERT BLUE No. 1 see FAE000
USACERT BLUE No. 2 see FAE100
USACERT FD & C RED No. 4 see FAG050
USACERT FD & C YELLOW NO. 6 see FAG150
USACERT RED No. 1 see FAG018
USACERT RED No. 3 see FAG040
USACERT YELLOW NO. 5 see FAG140
USACERT YELLOW NO. 6 see FAG150
USAF D-1 see IDW000
USAF D-3 see CHU500
USAF D-4 see POD750
USAF D-5 see BSO500
USAF D-9 see CBM000
USAF H-1 see QSA000
USAF M-2 see MCR000
USAF M-4 see ISQ000
USAF M-5 see ALI000
USAF M-6 see FOF000
USAF M-7 see CEI500
USAF P-2 see BJK500
USAF P-5 see TFS350
USAF P-7 see DXQ500
USAF P-8 see CJX750
USAF S-1 see GLK000
USAF P-220 see QQS200
USAF A-233 see AIR250
USAF A-1149 see AMS800
USAF A-3701 see TEV000
USAF A-3803 see PIK500
USAF A-4600 see MAO250
USAF A-6598 see DXP200
USAF A-8354 see TES250
USAF A-8564 see BIQ500
USAF A-8565 see HIM500

USAF A-8798 see CKT250
USAF A-9230 see EEW000
USAF A-9442 see SNE000
USAF A-9789 see DVX600
USAF A-11074 see DXN400
USAF A-14980 see MIV300
USAF A -15972 see DGX000
USAF A-19120 see BEX500
USAF AB-315 see DXJ800
USAF AM-1 see DJI400
USAF AM-2 see DNP700
USAF AM-3 see EMU500
USAF AM-4 see MKW750
USAF AM-5 see AAH250
USAF AM-6 see BSU500
USAF AM-7 see MKW500
USAF AM-8 see IJT000
USAF AN-7 see MFC700
USAF AN-8 see TDK000
USAF AN-9 see DOA400
USAF AN-11 see CAM000
USAF AN-16 see MDQ770
USAF B-7 see ADC750
USAF B-12 see HGI300
USAF B-13 see AJY300
USAF B-15 see TFC600
USAF B-17 see BKU500
USAF B-19 see DWC600
USAF B-21 see DCH400
USAF B-22 see TFD250
USAF B-24 see SAV000
USAF B-30 see TFS350
USAF B-31 see TEF750
USAF B-32 see BJL600
USAF B-33 see BDE750
USAF B-33 see DXH250
USAF B-35 see SGF500
USAF B-40 see MRM750
USAF B-44 see BLJ250
USAF B-44 see DXL800
USAF B-45 see DCF000
USAF B-51 see PAY500
USAF B-58 see FPM000
USAF B-59 see TGN000
USAF B-74 see DTG600
USAF B-100 see DJY800
USAF B-121 see MAL250
USAF BE-25 see TFJ750
USAF BE-4-5 see IAP000
USAF BE-0405 see ADC750
USAF BO-1 see TKM250
USAF BO-2 see BBM000
USAF BV-8 see CJF500
USAF C-1 see TEV000
USAF CB-2 see ILD000
USAF CB-7 see BAC250
USAF CB-10 see AEH750
USAF CB-11 see GLS000
USAF CB-13 see FMT000
USAF CB-17 see XCA000
USAF CB-18 see AEH000
USAF CB-19 see NCG000
USAF CB-20 see TET300
USAF CB-21 see TFA000
USAF CB-22 see NBE500
USAF CB-25 see HOJ100
USAF CB-26 see THT350
USAF CB-27 see RDK000
USAF CB-29 see ICW000
USAF CB-30 see THR750
USAF CB-34 see CQJ750
USAF CB-35 see TFJ100
USAF CB-36 see MCM750
USAF CB-37 see MRM750
USAF CB-96 see HOO100
USAF CF-2 see QCJ000
USAF CF-3 see HBU000
USAF CF-5 see RSU000
USAF CS-1 see DPM400
USAF CS-4 see AKS750
USAF CS-6 see NNM000
USAF CY-2 see CAQ250

USAF CY-4 see NAP500
USAF CY-5 see BDE750
USAF CY-6 see MJN250
USAF CY-7 see BDG000
USAF CY-9 see MES000
USAF CY-10 see NBA500
USAF CY-14 see DCF000
USAF CZ-1 see LFH000
USAF DM-1 see IAP000
USAF DO-1 see CEB250
USAF DO-3 see BMU170
USAF DO-4 see DES000
USAF DO-5 see HHW000
USAF DO-6 see BMY500
USAF DO-10 see GHA100
USAF DO-11 see ELB000
USAF DO-12 see HHR500
USAF DO-14 see HKI500
USAF DO-17 see PDQ750
USAF DO-19 see CPG700
USAF DO-20 see BQW000
USAF DO-21 see EKR500
USAF DO-22 see HKY500
USAF DO-23 see AGR000
USAF DO-28 see DMI600
USAF DO-29 see CDY850
USAF DO-30 see BGG500
USAF DO-32 see TCC000
USAF DO-36 see DVF200
USAF DO-37 see HKM500
USAF DO-38 see DVR000
USAF DO-40 see BMR100
USAF DO-41 see TNC500
USAF DO-42 see DDS000
USAF DO-43 see THU750
USAF DO-44 see ABD250
USAF DO-44 see EMU500
USAF DO-45 see AAG000
USAF DO-46 see AKB000
USAF DO-47 see AGO000
USAF DO-49 see HDH200
USAF DO-50 see MGG000
USAF DO-51 see BOB500
USAF DO-52 see MKU250
USAF DO-53 see BHB300
USAF DO-54 see MHF750
USAF DO-55 see IBH000
USAF DO-59 see PGH000
USAF DO-61 see BHJ000
USAF DO-62 see TBQ500
USAF DO-63 see DWY200
USAF DO-65 see HCE000
USAF DO-68 see DGK200
USAF E-1 see PEW250
USAF E-2 see MCM750
USAF E-4 see DHO400
USAF EA-1 see NGG500
USAF EA-2 see NGE000
USAF EA-3 see FPE100
USAF EA-4 see NGE500
USAF EA-5 see NGC000
USAF EA-14 see NGC400
USAF EE-3 see MCN750
USAF EK see TKO250
USAF EK-3 see AAQ500
USAF EK-206 see PEY750
USAF EK-218 see QMJ000
USAF EK-245 see DWN800
USAF EK-338 see DOT300
USAF EK-356 see HIH000
USAF EK-394 see PEY500
USAF EK-442 see BBL000
USAF EK-488 see ABE500
USAF EK-496 see ABH000
USAF EK-497 see ISR000
USAF EK-510 see TGP250
USAF EK-534 see CBM250
USAF EK-572 see IEE000
USAF EK-600 see CBN000
USAF EK-631 see AHR240
USAF EK-660 see MCK500
USAF EK-678 see PEY600

USAF EK-695 see CCC750
USAF EK-704 see ASL250
USAF EK-705 see TFF250
USAF EK-743 see DVY000
USAF EK-749 see GKY000
USAF EK-794 see QPA000
USAF EK-906 see EMU500
USAF EK-982 see MHK500
USAF EK-1047 see DJC400
USAF EK-1231 see MPV750
USAF EK-1235 see EOL500
USAF EK-1239 see AAY000
USAF EK-1270 see DWC600
USAF EK-1275 see TFQ000
USAF EK-1375 see PEI000
USAF EK-1509 see TGO750
USAF EK-1569 see PGN250
USAF EK-1597 see EEC600
USAF EK-1651 see DXP600
USAF EK-1719 see TFA000
USAF EK-1803 see DKC400
USAF EK-1853 see MHI750
USAF EK-1860 see TFM250
USAF EK-1902 see TFA250
USAF EK-1995 see COH500
USAF EK-2010 see PGL250
USAF EK-2070 see EMB200
USAF EK-2089 see TFS350
USAF EK-2122 see HBD500
USAF EK-2124 see BEU500
USAF EK-2138 see DEI000
USAF EK-2219 see BGJ250
USAF EK-2596 see SGJ000
USAF EK-2635 see DJD000
USAF EK-2676 see TGP000
USAF EK-2680 see TGO800
USAF EK-2784 see CEM500
USAF EK-3092 see DWN200
USAF EK-3110 see DWN400
USAF EK-3302 see ELL500
USAF EK-3941 see AIS500
USAF EK-3967 see HHB500
USAF EK-3991 see MCP250
USAF EK-4037 see AIG000
USAF EK-4196 see MCN250
USAF EK-4376 see AIF500
USAF EK-4394 see DXO200
USAF EK-4628 see HES000
USAF EK-4733 see IAL000
USAF EK-4812 see BDE500
USAF EK-4890 see ADD250
USAF EK-5017 see BDI500
USAF EK-5185 see MMD500
USAF EK-5199 see SKH500
USAF EK-5296 see CKT500
USAF EK-5426 see PGN000
USAF EK-5429 see BDJ000
USAF EK-5432 see BDE750
USAF EK-5496 see TEV500
USAF EK-6232 see QRS000
USAF EK-6279 see SIN000
USAF EK-6326 see DIY100
USAF EK-6326 see DIY150
USAF EK-6454 see MPW500
USAF EK-6540 see BCC500
USAF EK-6561 see ALQ000
USAF EK-6583 see MCK000
USAF EK-6754 see HNI000
USAF EK-6775 see DWI200
USAF EK-7087 see DWO000
USAF EK-7094 see QRJ000
USAF EK-7119 see MLE750
USAF EK-7162 see ENM000
USAF EK-7317 see QSJ000
USAF EK-7372 see TFE250
USAF EK-7423 see DTM000
USAF EK-8413 see AJT500
USAF EK-P-433 see ANW750
USAF EK-P-583 see FOU000
USAF EK-P-737 see TFA500
USAF EK-T-434 see SIA500
USAF EK-P-4382 see CIH100

USAF EK-P-5430 see ACQ250
USAF EK-P-5501 see AMS250
USAF EK-P-5976 see AQN635
USAF EK-P-6255 see BJL600
USAF EK-P-6281 see BLJ250
USAF EK-P-6281 see DXL800
USAF EK-P-6297 see MCR000
USAF EK-T-2805 see MCK750
USAF EK-T-6645 see BKU500
USAF EL-23 see PAP550
USAF EL-30 see MCO500
USAF EL-42 see EEH000
USAF EL-44 see CKE750
USAF EL-45 see PGN000
USAF EL-50 see IMX150
USAF EL-52 see MFC500
USAF EL-54 see BEQ500
USAF EL-57 see IAO000
USAF EL-62 see IAQ000
USAF EL-76 see PDE500
USAF EL-78 see DQQ000
USAF EL-82 see BDY000
USAF EL-97 see MLF000
USAF EL-101 see BQP250
USAF EL-108 see MNM500
USAF FA-4 see TES250
USAF FA-5 see BCP500
USAF FO-1 see BQP250
USAF GE-1 see MNV750
USAF GE-12 see PCP250
USAF GE-13 see DWM000
USAF GE-14 see HNI500
USAF GY-1 see DXP400
USAF GY-2 see BLE500
USAF GY-3 see BDF000
USAF GY-5 see BIX000
USAF GY-7 see BHA750
USAF HA-2 see RGZ550
USAF HA-5 see DGS600
USAF HC-1 see SBJ500
USAF IN-399 see MKW000
USAF KE-3 see MEQ000
USAF KE-5 see IQW000
USAF KE-7 see OAV000
USAF KE-8 see HKJ000
USAF KE-9 see PJW000
USAF KE-11 see EJM500
USAF KE-12 see PJV500
USAF KE-13 see SLL000
USAF KE-14 see PJW250
USAF KE-20 see HIN500
USAF KF-1 see COK250
USAF KF-2 see MCK750
USAF KF-3 see PES750
USAF KF-4 see BCK750
USAF KF-5 see CDN500
USAF KF-10 see EEW000
USAF KF-11 see CEQ600
USAF KF-13 see DVX200
USAF KF-14 see COJ250
USAF KF-15 see HIO000
USAF KF-17 see COJ500
USAF KF-18 see COH250
USAF KF-21 see PEA750
USAF KF-22 see MIQ000
USAF KF-25 see EHP500
USAF KF-26 see MAN750
USAF LO-3 see MCR250
USAF MA-1 see NEL500
USAF MA-2 see BDH000
USAF MA-3 see CEK000
USAF MA-4 see AID500
USAF MA-5 see NFJ500
USAF MA-6 see TKD400
USAF MA-8 see CJA140
USAF MA-9 see BLA800
USAF MA-10 see CDQ750
USAF MA-12 see AJF500
USAF MA-13 see CEG800
USAF MA-16 see BDH500
USAF MA-17 see MLX750
USAF ME-1 see BAD750

USAF MK-1 see DDF700
USAF MK-3 see BHM250
USAF MK-5 see BKD800
USAF MK-6 see DXO200
USAF MK-43 see BJJ000
USAF MO-2 see ANM500
USAF NB-1 see CEG750
USAF ND-09 see PHY000
USAF ND-54 see DMI600
USAF ND-57 see DDE100
USAF ND-59 see DMH400
USAF ND-60 see SOB600
USAF PD-18 see BLF600
USAF PD-20 see DOT800
USAF PD-25 see AKM000
USAF PD-57 see TEV600
USAF PD-58 see EEU500
USAF PE-1 see PEA500
USAF Q-1 see FPW000
USAF Q-2 see TCS500
USAF RH-1 see MDN500
USAF RH-3 see DPG600
USAF RH-4 see MCD750
USAF RH-6 see OFK000
USAF RH-7 see HGP000
USAF RH-8 see MLC750
USAF SC-2 see GKE500
USAF SE-3 see TML100
USAF SN-9 see TEX250
USAF SN-29 see DSU310
USAF SN-31 see DVB850
USAF SO-1 see EGL500
USAF SO-2 see EPK000
USAF ST-40 see MGA750
USAF SZ-1 see BNM250
USAF SZ-3 see MOO500
USAF SZ-B see MOO500
USAF T-1 see CKJ750
USAF T-2 see MPH750
USAF TH-3 see THS250
USAF TH-9 see DXM600
USAF UCTL-7 see AES639
USAF UCTL-8 see MJV750
USAF UCTL-958 see BHY750
USAF UCTL-974 see BII500
USAF UCTL-1766 see MCL500
USAF UCTL-1791 see DOA400
USAF UCTL-1856 see AMC000
USAF VI-6 see PDF250
USAF WI-1 see MRN000
USAF WI-2 see PFS600
USAF WI-3 see MLY500
USAF XF-21 see BCC500
USAF XR-10 see DBL800
USAF XR-17 see AIF800
USAF XR-19 see PFL850
USAF XR-20 see PPQ630
USAF XR-22 see AMY050
USAF XR-27 see AIS500
USAF XR-29 see BDF000
USAF XR-30 see AIS550
USAF XR-31 see AJY250
USAF XR-32 see AJY500
USAF XR-35 see MCK750
USAF XR-41 see CJX750
USAF XR-42 see DXQ500
USALAKE FD & C YELLOW NO. 6 LAKE see
FAG150
USAN-DIAZIQUONE see ASK875
USARAMIN see RFP100
USARAMINE see RFP100
USB-3153 see DWS200
USB-3584 see CNE500
U.S. BLENDED LIGHT TOBACCO
CIGARETTE REFINED TAR see CMP800
USCHARIN see UVS700
USCHARINE see UVS700
USNEIN see UWJ000
USNIACIN see UWJ000
USNIC ACID see UWJ000
(+)-USNIC ACID see UWJ100
d-USNIC ACID see UWJ100

USNIC ACID, (R)-(8CI) see UWJ100
USNINIC ACID see UWJ000
(+)-USNINIC ACID see UWJ100
d-USNINIC ACID see UWJ100
USNINSAEURE (GERMAN) see UWJ000
U.S.P. MENTHOL see MCG250
USP METHYLCELLULOSE see MIF760
USP SODIUM CHLORIDE see SFT000
USPULUM see CHW675
USP XIII STEARYL ALCOHOL see OAX000
USR 604 see DFT000
U.S. RUBBER 604 see DFT000
U.S. RUBBER D-014 see SOP000
UST see UTU500
USTINEX see CIR250
SYMULER FAST PYRAZOLONE ORANGE
G see CMS145
UTAL see GIQ100
UTEDRIN see ORU500
UTERACON see ORU500
UTERAMINE see TOG250
UTIBID see OOG000
UTICILLIN see CBO000
UTICILLIN VK see PDT750
UTIL see EQL000
UTOPAR see RLK700
UTOSTAN see PDC250
UTRASUL see MPQ750
UV 1 see HND100
UV 24 see DMW250
UV 531 see HND100
UV 1173 see HMQ100
UVA 1 see HND100
UV ABSORBER-1 see HML500
UV ABSORBER-2 see DGX000
UVALERAL see BNP750
U-VAN 28 see MCB050
U-VAN 62 see MCB050
U-VAN 102 see MCB050
U-VAN 120 see MCB050
U-VAN 122 see MCB050
U-VAN 128 see MCB050
U-VAN 20S see MCB050
U-VAN 21R see MCB050
U-VAN 220 see MCB050
U-VAN 221 see MCB050
U-VAN 225 see MCB050
U-VAN 22R see MCB050
U-VAN 28N see MCB050
U-VAN 60R see MCB050
U-VAN 2020 see MCB050
U-VAN 20N60 see MCB050
U-VAN 20SA see MCB050
U-VAN 20SB see MCB050
U-VAN 20SE see MCB050
U-VAN 28SE see MCB050
U-VAN 20SE50 see MCB050
U-VAN 20SE60 see MCB050
U-VAN 20HS see MCB050
U-VAN 21HV see MCB050
UVASOL see HIH100
UV CHEK AM 104 see BIW750
UVINOL D-50 see BCS325
UVINUL 400 see DMI600
UVINUL 408 see HND100
UVINUL D-50 see BCS325
UVINUL 3039 see ODY150
UVINUL D 49 see BJY800
UVINUL M 40 see MES000
UVINUL N 539 see ODY150
UVITEX OB see BHK600
UV 20SR see MCB050
U46KW see BSQ750
17-α-UZARIGENIN see UWJ200
UZONE see BRF500
V 7 see TKX250
V-18 see CKM000
V 50 see ASM050
V 59 see ASM025
V 252 see CGD000
V 255 see CKN750
V 315 see CIT750

V 316 see CFU500
V 317 see DIK800
V331 see DIM400
V 340 see DIS000
V 343 see DHJ400
V 346 see DIK600
V 374 see DTR800
V 375 see CGI750
V 377 see DNM600
261LV see DTS500
V 4917 see TLD000
825TV see SMQ500
V 17004 see XTJ000
V 53482 see FLZ075
VA see VFF000
VA 0112 see AAX250
VABEN see CFZ000
VABROCID see NGE500
VAC see VLU250
VACATE see CIR250
VACOR see PPP750
VACUUM RESIDUUM see MQV755
VACUUM RESIDUUM (PETROLEUM) see RDK200
VADEBEX see NOA000
VADEBEX HYDROCHLORIDE see NOA500
VADILEX see IAG625
VADOSILAN see VGA300
VADROCID see NGE500
VAESITE see VAC000
VAFLOL see VSK600
VAGAMIN see DJM800
VAGAMIN see XCJ000
VAGANTIN see DJM800
VAGANTIN see XCJ000
VAGD see AAX175
VAGESTROL see DKA600
VAGILEN see MMN250
VAGILIA see SNN500
VAGIMID see MMN250
VAGISEPT see ABX500
VAGOFLOR see ABX500
VAGOPHEMANIL see DAP800
VAGOPHEMANIL METHYL SULFATE see DAP800
VAGOSIN see CPQ250
VAGOSTIGMIN see DER600
VAGOSTIGMINE see NCL100
VAGOSTIGMINE BROMIDE see POD000
VAGOSTIGMINE METHYL SULFATE see DQY909
VAL 13081 see VCK100
VALACIDIN see SMA000
VALADOL see HIM000
VALAMINA see EEH000
VALAMINE see IIM000
VALAMINETTEN see EEH000
VALAN see MRW000
VALANAS see MFD500
VALBAZEN see VAD000
VALBIL see BPP250
VALCOR see CPP000
VALDRENE see BAU750
VAL-DROP see SFS000
VALECOR see PEV750
VALENTINITE see AQF000
VALENTINITE see VAD100
VALEO see DCK759
VALEPOTRIATE see VAD200
VALERAL see VAG000
n-VALERALDEHYDE see VAG000
VALERALDEHYDE, 2-METHYLENE- see PNT800
VALERAMIDE, 2-(2-ACETAMIDO-4-METHYLVALERAMIDO)-N-(1-FORMYL-4-GUANIDINOBUTYL)-4-METHYL-(S)- see AAL300
VALERAMIDE, N-HEXYL- see VAG100
VALERAMIDE-OM see DOY400
VALERAN DI-n-BUTYLCINICITY (CZECH) see DEA600
VALERANONE see VAG200

(−)-VALERANONE see VAG250
VALERENAL see VAG300
VALERENIC ACID see VAG400
VALERIANIC ACID see VAQ000
VALERIANIC ALDEHYDE see VAG000
VALERIC ACID see VAQ000
n-VALERIC ACID see VAQ000
VALERIC ACID ALDEHYDE see VAG000
VALERIC ACID, 2-AMINO-3-METHYL- see IKX000
VALERIC ACID, 2-AMINO-4-METHYL- see LES000
VALERIC ACID, 4,4'-AZOBIS(4-CYANO)- see ASL500
VALERIC ACID, 3-HEXENYL ESTER, (Z)- see HFE800
VALERIC ACID, 5-HYDROXY-3-METHYL-, Δ-LACTONE see MQK100
VALERIC ACID, METHYL ESTER see VAQ100
VALERIC ACID, 2-PROPYL-, CALCIUM SALT (2:1) see VCK200
VALERIC ACID, 2-PROPYL-, CARVACRYL ESTER see CCM050
VALERIC ACID, 2-PROPYL-, THYMYL ESTER see VAQ200
VALERIC ALDEHYDE see VAG000
4-VALEROLACTONE see VAV000
γ-VALEROLACTONE (FCC) see VAV000
VALERON see PJS750
VALERONE see DNI800
VALERONITRILE see VAV300
n-VALERONITRILE see VAV300
Δ-VALEROSULTONE see BOU250
VALERYLALDEHYDE see VAG000
VALERYL CHLORIDE see VBA000
4-VALERYLPYRIDINE see VBA100
VALETAN see DEO600
VALETHAMATE see VBK000
VALETHAMATE BROMIDE see VBK000
VALEXONE see BAT750
VALFLON see TAI250
VALFOR see AHF500
VALGESIC see HIM000
VALGIS see TEH500
VALGRAINE see TEH500
VALI FAST RED 1308 see RGW000
VALINE see VBP000
d-VALINE see VBU000
VALINE ALDEHDYE see IJS000
dl-VALINE, N-(2-CHLORO-4-(TRIFLUOROMETHYL)PHENYL)-, CYANO(3-PHENOXYPHENYL)METHYL ESTER see VBU100
l-VALINE (FCC) see VBP000
VALINE, 3-MERCAPTO-, l- see PAP300
l-VALINE, 3-MERCAPTO- see PAP300
VALINOMYCIN see VBZ000
VALIOIL see HNI500
VALISONE see VCA000
VALITRAN see DCK759
VALIUM see DCK759
VALKACIT CA see DWN800
VALLADAN see BCA000
VALLAROSOLANOSIDE see VCA100
VALLENE see MBW750
VALLERGINE see DQA400
VALMAGEN see TAB250
VALMETHAMIDE see ENJ500
VALMID see EEH000
VALMIDATE see EEH000
VALOID see EAN600
VALONE see ITD010
VALONEA TANNIN see VCK000
VALONTIN see CAY675
VALONTIN see VCK200
VALOPRIDE see VCK100
VALORON HYDROCHLORIDE see EIH000
VALPIN see LJS000
VALPROATE see PNR750
VALPROATE SODIUM see PNX750
VALPROIC ACID see PNR750

VALPROIC ACID CALCIUM SALT see CAY675
VALPROIC ACID CALCIUM SALT see VCK200
VALPROIC ACID HEMI-CALCIUM SALT see CAY675
VALPROIC ACID HEMI-CALCIUM SALT see VCK200
VALPROIC ACID SODIUM SALT see PNX750
VALPROMIDE see PNX600
VALSPEX 155-53 see PJS750
VALSYN see FPI000
VALTRATE see VAD200
VALTRATS see VAD200
VALTRATUM see VAD200
VALYL GRAMICIDIN A see GJO025
VALZIN see EFE000
VAMIDOATE see MJG500
VAMIDOTHION see MJG500
V. AMMODYTES VENOM see VQZ425
VANADIC ACID, AMMONIUM SALT see ANY250
VANADIC ACID, MONOSODIUM SALT see SKP000
VANADIC ACID, POTASSIUM SALT see PLK810
VANADIC(II) ACID, TRISODIUM SALT see SIY250
VANADIC ANHYDRIDE see VDU000
VANADIC OXIDE see VEA000
VANADIO, PENTOSSIDO di (ITALIAN) see VDU000
VANADIOUS(4+) ACID, DISODIUM SALT see SKQ000
VANADIUM see VCP000
VANADIUM AZIDE TETRACHLORIDE see VCU000
VANADIUM compounded with BERYLLIUM (1:12) see BFR250
VANADIUM BROMIDE see VEK000
VANADIUM CHLORIDE see VEF000
VANADIUM(III) CHLORIDE see VEP000
VANADIUM CHLORIDE OXIDE see DFW800
VANADIUM COMPOUNDS see VCZ000
VANADIUM DICHLORIDE see VDA000
VANADIUM DICHLORIDE OXIDE see DFW800
VANADIUM DUST and FUME (ACGIH) see VDU000
VANADIUM DUST and FUME (ACGIH) see VDZ000
VANADIUM ORE see VDF000
VANADIUM (OSHA) see VDZ000
VANADIUM OXIDE see VEA000
VANADIUM(V) OXIDE see VDU000
VANADIUM OXIDE TRIISOBUTOXIDE see VDK000
VANADIUM OXYCHLORIDE see DFW800
VANADIUM OXYDICHLORIDE see DFW800
VANADIUM OXYTRICHLORIDE see VDP000
VANADIUM PENTAOXIDE see VDU000
VANADIUMPENTOXID (GERMAN) see VDU000
VANADIUM PENTOXIDE (dust) see VDU000
VANADIUM PENTOXIDE (fume) see VDZ000
VANADIUM PENTOXIDE, non-fused form (DOT) see VDU000
VANADIUM PENTOXIDE, nonfused form (DOT) see VDZ000
VANADIUMPENTOXYDE (DUTCH) see VDU000
VANADIUM, PENTOXYDE de (FRENCH) see VDU000
VANADIUM SESQUIOXIDE see VEA000
VANADIUM SULFATE see VEA100
VANADIUM SULPHATE see VEA100
VANADIUM TETRACHLORIDE see VEF000

VANADIUM TRIBROMIDE see VEK000
VANADIUM TRIBROMIDE OXIDE see VEK100
VANADIUM TRICHLORIDE see VEP000
VANADIUM TRICHLORIDE OXIDE see VDP000
VANADIUM TRIOXIDE see VEA000
VANADYL AZIDE DICHLORIDE see VEU000
VANADYL SULFATE see VEZ000
VANADYL SULFATE PENTAHYDRATE see VEZ100
VANADYL TRICHLORIDE see VDP000
VANANOTE see MDW750
VANAY see THM500
VANCENASE see AFJ625
VANCERIL see AFJ625
VANCIDA TM-95 see TFS350
VANCIDE see VEZ925
VANCIDE 40 see NFR600
VANCIDE 89 see CBG000
VANCIDE 20S see HKO012
VANCIDE BL see TFD250
VANCIDE F 5386 see NFR600
VANCIDE FE95 see FAS000
VANCIDE KS see HON000
VANCIDE MANEB 80 see MAS500
VANCIDE MZ-96 see BJK500
VANCIDE P see ZMJ000
VANCIDE PA see BLG500
VANCIDE PA DISPERSION see BLG500
VANCIDE PB see DVI600
VANCIDE TH see HDW000
VANCIDE TM see TFS350
VANCOCIN see VFA000
VANCOCINE HYDROCHLORIDE see VFA050
VANCOCIN HYDROCHLORIDE see VFA050
VANCOMYCIN see VFA000
VANDEX see SBO500
VAN DYK 264 see OES000
VAN DYKE BROWN see MAT500
VANGARD see EDW100
VANGARD see MEG250
VANGARD see VFA100
VANGARD K see CBG000
VANGUARD GF see FAZ000
VANGUARD N see BIW750
VANICID see VEZ925
VANICIDE see CBG000
VANILLA see VFK000
VANILLABERON see VEZ925
VANILLAL see EQF000
VANILLALDEHYDE see VFK000
VANILLA PLANT see DAJ800
VANILLA TINCTURE see VFA200
VANILLIC ACID see VFF000
o-VANILLIC ACID see HJB500
p-VANILLIC ACID see VFF000
VANILLIC ACID DIETHYLAMIDE see DKE200
VANILLIC ACID-N,N-DIETHYLAMIDE see DKE200
VANILLIC ACID, METHYL ESTER see VFF100
VANILLIC ALDEHYDE see VFK000
VANILLIN see VFK000
o-VANILLIN see VFP000
p-VANILLIN see VFK000
VANILLIN METHYL ETHER see VHK000
VANILLINSAEURE-DIAETHYLAMID (GERMAN) see DKE200
VANILLYL ACETONE see VFP100
VANILOL see VFP200
VANIROM see EQF000
VANITROPE see IRY000
VANIZIDE see VEZ925
VANLUBE PCX see BFW750
VANNAX NS see BQK750
VANOBID see LFF000
VANOXIDE see BDS000
VANOX MTI see MCO400

VANQUIN see PQC500
VANSIL see OLT000
VANSIL W 10 see WCJ000
VANSIL W 20 see WCJ000
VANSIL W 30 see WCJ000
VANTYL see PGG000
VANZOATE see BCM000
VAPAM see SIL550
VAPAM see VFU000
VAPIN see LJS000
VAPO-ISO see IMR000
VAPO-N-ISO see DMV600
VAPONA see DGP900
VAPONEFRIN see VGP000
VAPONITE see DGP900
VAPOPHOS see PAK000
VAPOROLE see IMB000
VAPOROOTER see SIL550
VAPORPAC see ILM000
VAPORTHRIN see EQN250
VAPOTONE see TCF250
VARACILLIN see LEJ500
VARAMID ML 1 see BKE500
VARDHAK see NAK500
VARFINE see WAT220
VARIAMINE BLUE SALT RT see PFU500
VARICOL see EMT500
VARIEGATED PHILODENDRON see PLW800
VARIKILL see EOE200
VARIOFORM I see ANN000
VARIOFORM II see USS000
VARIQUART E 228 see HCQ525
VARIQUAT 60LC see QAT510
VARISOFT 100 see DXG625
VARISOFT SDC see DTC600
VARITOL see FMS875
VARITON see DAP800
VARITOX see TII250
VARITOX see TII500
VARNISH MARKER'S NAPHTHA see PCT250
VARNOLINE see SLU500
VAROX see DRJ800
VAROX DCP-R see DGR600
VAROX DCP-T see DGR600
VASAL see PAH250
VASALGIN see ELX000
VASAPRIL see PPH050
VASAZOL see DJS200
VASCARDIN see CCK125
VASCOPIL see HNY500
VASCUALS see VSZ450
VASCULAT see BOV825
VASCULIT see BOV825
VASCULOPATINA see DNM400
VASCUNICOL see BOV825
VASE FLOWER see CMV390
VASE VINE see CMV390
VASICIN (GERMAN) see VGA000
VASICINE see VGA000
VASICINONE HYDROCHLORIDE see VGA025
VASIMID see BBW750
VASIODONE see DNG400
VASITOL see PBC250
VASKULAT see BOV825
VASOBRIX 32 see VGA100
VASOCIL see PPH050
VASOCON see NCW000
VASOCONSTRICTINE see VGP000
VASOCONSTRICTOR see VGP000
VASODIATOL see PBC250
VASODIL see BBW750
VASODILAN see VGA300
VASODILAN see VGA300
VASODILATAN see BBW750
VASODILATATEUR 2249F see FMX000
VASODILIAN see VGF000
VASODISTAL see VGK000
VASODRINE see VGP000
VASOFILINA see TEP500
VASOGLYN see NGY000

VASOKELLINA see AHK750
VASOLAN see IRV000
VASOLAN see VHA450
VASOMED see TJL250
VASOPERIF see CBH250
VASOPLEX see VGA300
VASOPRESSIN, 8-l-ARGININE- see AQW050
VASORBATE see CCK125
VASORELAX see POD750
VASOROME see AOO125
VASOSPAN see PAH250
VASOTON see VGP000
VASOTONIN see VGP000
VASOTRAN see VGA300
VASOTRATE see CCK125
VASO-80 UNICELLES see PBC250
VASOVERIN see PPH050
VASOXINE see MDW000
VASOXINE HYDROCHLORIDE see MDW000
VASOXYL HYDROCHLORIDE see MDW000
VASTAREL see YCJ200
VASTCILLIN see AJJ875
VAT BLACK 1 see CMU320
VAT BLACK K see DUP830
VAT BLACK S see DUP830
VAT BLUE 1 see BGB275
VAT BLUE 4 see IBV050
VAT BLUE 5 see ICU135
VAT BLUE 6 see DFN300
VAT BLUE 18 see VGP100
VAT BLUE 4B see ICU135
VAT BLUE KD see DFN300
VAT BLUE O see IBV050
VAT BLUE OD see IBV050
VAT BRIGHT VIOLET K see DFN450
VAT BRILLIANT BLUE Z see DLO900
VAT BRILLIANT ORANGE see CMU820
VAT BRILLIANT SKY BLUE Z see DLO900
VAT BRILLIANT VIOLET K see DFN450
VAT BRILLIANT VIOLET KD see DFN450
VAT BRILLIANT VIOLET KP see DFN450
VAT BROWN 3 see CMU780
VAT BROWN K see CMU780
VATERITE see CAO000
VAT FAST BLUE BCS see DFN300
VAT FAST BLUE R see IBV050
VAT GOLDEN YELLOW see DCZ000
VAT GRAY S see CMU475
VAT GREEN 3 see CMU810
VAT GREEN B see DFN300
VAT GREY S see CMU475
VAT ORANGE R see CMU815
VAT ORANGE RF see CMU815
VAT PRINTING ORANGE R see CMU815
VATRACIN see AJJ875
VATRAN see DCK759
VAT RED 13 see CMU825
VAT RED 5B see DNT300
VATROLITE see SHR500
VAT SCARLET 2Zh see CMU820
VAT SKY BLUE K see DFN300
VAT SKY BLUE KD see DFN300
VAT SKY BLUE KP 2F see DFN300
VATSOL OT see DJL000
VAT YELLOW 2 see VGP200
VAZADRINE see ILD000
VAZO 64 see ASL750
VAZO 67 see ASM025
VAZO 64-A see ASM025
VAZOFIRIN see PBU100
V-BRITE see SHR500
VC see VNP000
V-C 3-670 see DWU200
V-C 3-764 see PGD000
V-C CHEMICAL V-C 9-104 see EIN000
V-CIL see PDT500
V-CIL-K see PDT750
V-CILLIN see PDT500
V-CILLIN K see MNV250
V-CILLIN K see PDT750
VCM see VNP000

VCN see ADX500
VC13 NEMACIDE see DFK600
VCR see LEY000
VCR SULFATE see LEZ000
VCS 438 see BGD250
VDC see VPK000
VDF see VPP000
VDM see VFU000
VD 1827 p-TOLUENESULFONATE see SOU675
VEATCHINE HYDROCHLORIDE see VGU000
VEBECILLIN see PDT500
VECTAL see ARQ725
VECTAL SC see ARQ725
VECTREN see CIP500
VECURONIUM BROMIDE see VGU075
VEDERON see ILD000
VEDITA 250 see THR525
VEDRIL see PKB500
VEE GEE GELATIN see PCU360
VEETIDS see PDT750
VEGABEN see AJM000
VEGADEX see CDO250
VEGADEX SUPER see CDO250
VEGANTINE see DHX800
VEGANTINE see THK000
VEGASERPINA see RCA200
VEGENTINE FAST BROWN B see CMO820
VEGETABLE GUM see DBD800
VEGETABLE-HUMMING-BIRD see SBC550
VEGETABLE OIL see VGU200
VEGETABLE OIL 1400 see CBF710
VEGETABLE OIL MIST (OSHA) see VGU200
VEGETABLE PEPSIN see PAG500
VEGETABLE (SOYBEAN) OIL, brominated see BMO825
VEGETOX see BHL750
VEGFRU see PGS000
VEGFRU FOSMITE see EEH600
VEGFRU MALATOX see MAK700
VEGOLYSEN see HEA000
VEGOLYSIN see HEA000
VEHAM-SANDOZ see EQP000
VEHEM see EQP000
VEJIGA de PERRO (CUBA) see JBS100
VEL 3973 see DUK000
VEL 4283 see MKA000
VEL 4284 see DRR200
VEL-5026 see RCA300
VELACICLINE see PPY250
VELACYCLINE see PPY250
VELARDON see PAG500
VELBAN see VLA000
VELBE see VLA000
VELDOPA see DNA200
VELFLON see TAI250
VELLORITA (CUBA) see CNX800
VELMOL see DJL000
VELON see CGW300
VELO de NOVIA (MEXICO) see GIW200
VELOSEF see SBN440
VELOSEF HYDRATE see SBN450
VELPAR see HFA300
VELPAR WEED KILLER see HFA300
VELSICOL see TIK000
VELSICOL 104 see HAR000
VELSICOL 506 see LEN000
VELSICOL 1068 see CDR750
VELSICOL COMPOUND C see TIK000
VELSICOL COMPOUND "R" see MEL500
VELSICOL 53-CS-17 see EBW500
VELSICOL 58-CS-11 see MEL500
VELSICOL FCS-303 see BND750
VELSICOL VCS 506 see LEN000
VELUSTRAL KPA see PJS750
VELVETEX see CBT750
VENA see BBV500
VENACIL see VGU700
VENCIPON see EAW000
VENDACID LIGHT ORANGE 2G see HGC000

VENDESINE SULFATE see VGU750
VENDEX see BLU000
VENETIAN RED see IHC450
VENETLIN see BQF500
VENGEANCE see BMO300
VENGICIDE see VGZ000
VENOM, AUSTRALIAN ELAPIDAE SNAKE, ACANTHOPHIS ANTARCTICUS see ARU875
VENOM, AUSTRALIAN ELAPIDAE SNAKE, AUSTRELAPS SUPERBA see ARU750
VENOM, AUSTRALIAN ELAPIDAE SNAKE, OXYURANUS SCUTELLATUS see ARV500
VENOM, AUSTRALIAN ELAPIDAE SNAKE, PSEUDECHIS PORPHYRIACUS see ARV250
VENOM, AUSTRALIAN SNAKE, AUSTRELAPS SUPERBA see ARV625
VENOM, AUSTRALIAN SNAKE, NOTECHIS SCUTATUS see ARV550
VENOM, AUSTRALIAN SNAKE, OPHIOPHAGUS HANNAH see ARV125
VENOM, AUSTRALIAN SNAKE, PSEUDECHIS see ARV000
VENOM, AUSTRALIAN SNAKE, PSEUDONAJA TEXTILIS see TEG650
VENOM, AUSTRALIAN SNAKE, TROPIDECHIS CARINATUS see ARV375
VENOM, COSTA RICAN SNAKE, BOTHROPS ASPER see BMI000
VENOM, COSTA RICAN SNAKE, BOTHROPS ATROX see BMI125
VENOM, COSTA RICAN SNAKE, BOTHROPS GODMANI see BMI500
VENOM, COSTA RICAN SNAKE, BOTHROPS LATERALIS see BMI750
VENOM, COSTA RICAN SNAKE, BOTHROPS NASUTUS see BMJ000
VENOM, COSTA RICAN SNAKE, BOTHROPS NIGROVIRIDIS NIGROVIRIDIS see BMJ250
VENOM, COSTA RICAN SNAKE, BOTHROPS NUMMIFER MEXICANUS see BMJ500
VENOM, COSTA RICAN SNAKE, BOTHROPS OPHRYOMEGA see BMJ750
VENOM, COSTA RICAN SNAKE, BOTHROPS PICADOI see BMK000
VENOM, COSTA RICAN SNAKE, BOTHROPS SCHLEGLII see BMK250
VENOM, COSTA RICAN SNAKE, CROTALUS (CROTALUS) DURISSUS DURISSUS see CNX835
VENOM, COSTA RICAN SNAKE, MICRURUS ALLENI YATESI see MQS750
VENOM, COSTA RICAN SNAKE, MICRURUS MIPARTITUS HERTWIGI see MQT250
VENOM, COSTA RICAN SNAKE, MICRURUS NIGROCINCTUS see MQT500
VENOM, EGYPTIAN COBRA, NAJA HAJE ANNULIFERA see NAE512
VENOM, FORMOSAN BANDED KRAIT, BUNGARUS MULTICINCTUS see BON370
VENOM, INDIAN COBRA, NAJA NAJA NAJA see ICC700
VENOM, LIZARD, HELODERMA SUSPECTUM see HAN625
VENOM, MIDGET FADED RATTLESNAKE, CROTALUS VIRIDIS CONCOLOR see CNY750
VENOM, MIDGET FADED RATTLESNAKE, CROTALUS VIRIDIS HELLERI see COA000
VENOM, MIDGET FADED RATTLESNAKE, CROTALUS VIRIDIS LUTOSUS see COA250
VENOM, MIDGET FADED RATTLESNAKE, CROTALUS VIRIDIS OREGANUS see COA500
VENOM, MIDGET FADED RATTLESNAKE, CROTALUS VIRIDIS VIRIDIS see COA750

VENOM, MIDGET FADED RATTLESNAKE, CROTALUS VIRIDUS CERBERUS see CNY390
VENOM, ORIENTAL HORNET, VESPA ORIENTALIS see VJP900
VENOM, SCORPION, ANDROCTONUS AMOREUXI see AOO250
VENOM, SCORPION, ANDROCTONUS AUSTRALIS HECTOR see AOO265
VENOM, SCORPION, CENTRUROIDES SUFFUSUS SUFFUSUS see CCW925
VENOM, SEA ANEMONE, AIPTASIA PALLIDA see SBI800
VENOM, SEA SNAKE, AIPYSURUS LAEVIS see SBI880
VENOM, SEA SNAKE, AIPYSURUS LAEVIS (AUSTRALIA) see AFG000
VENOM, SEA SNAKE, ASTROTIA STOKESII see SBI890
VENOM, SEA SNAKE, ENHYDRINA SCHISTOSA see EAU000
VENOM, SEA SNAKE, HYDROPHIS CYANOCINCTUS see SBI900
VENOM, SEA SNAKE, HYDROPHIS ELEGANS see HIG000
VENOM, SEA SNAKE, LAPEMIS HARDWICKII see SBI910
VENOM, SEA SNAKE, LATICAUDA SEMIFASCIATA see LBI000
VENOM, SEA SNAKE, MICROCEPHALOPHIS GRACILIS see SBI929
VENOM, SEA SNAKE, NAJA NAJA ATRA see NAF000
VENOM, SNAKE, AGKISTRODON ACUTUS see HGM600
VENOM, SNAKE, AGKISTRODON CONTORTRIX see AEY125
VENOM, SNAKE, AGKISTRODON CONTORTRIX CONTORTRIX see SKW775
VENOM, SNAKE, AGKISTRODON CONTORTRIX MOKASEN see NNX700
VENOM, SNAKE, AGKISTRODON PISCIVORUS see AEY130
VENOM, SNAKE, AGKISTRODON PISCIVORUS PISCIVORUS see EAB200
VENOM, SNAKE, AGKISTRODON RHODOSTOMA see AEY135
VENOM, SNAKE, ANCISTRODON PISCIVORUS see AOO135
VENOM, SNAKE, BOTHROPS COLOMBIENSIS see BMI250
VENOM, SNAKE, BUNGARUS CAERULEUS see BON365
VENOM, SNAKE, BUNGARUS FASCIATUS see BON367
VENOM, SNAKE, CERASTES CERASTES see CCX620
VENOM, SNAKE, CROTALUS ADAMANTEUS see EAB225
VENOM, SNAKE, CROTALUS ATROX see WBJ600
VENOM, SNAKE, CROTALUS CERASTES see CNX830
VENOM, SNAKE, CROTALUS DURISSUS TERRIFICUS see CNY000
VENOM, SNAKE, CROTALUS HORRIDUS see CNY300
VENOM, SNAKE, CROTALUS HORRIDUS HORRIDUS see TGB150
VENOM, SNAKE, CROTALUS RUBER RUBER see CNY325
VENOM, SNAKE, CROTALUS SCUTULATUS SCUTULATUS see CNY350
VENOM, SNAKE, CROTALUS SCUTULATUS SCUTULATUS see CNY375
VENOM, SNAKE, CROTALUS SCUTULATUS SCUTULATUS (CALIFORNIA) see CNY339
VENOM, SNAKE, CROTALUS VIRIDIS VIRIDIS see PLX100
VENOM, SNAKE, DENDROASPIS ANGUSTICEPS see DAP815

VENOM, SNAKE, DENDROASPIS JAMESONI see DAP820
VENOM, SNAKE, DENDROASPIS VIRIDIS see GJU500
VENOM, SNAKE, DISPHOLIDUS TYPHUS see DXG150
VENOM, SNAKE, ECHIS CARINATUS see EAD600
VENOM, SNAKE, ECHIS COLORATUS see EAD650
VENOM, SNAKE, HEMACHATUS HAEMCHATES see HAO500
VENOM, SNAKE, NAJA FLAVA see NAE510
VENOM, SNAKE, NAJA HAJE see NAE515
VENOM, SNAKE, NAJA MELANOLEUCA see NAE875
VENOM, SNAKE, NAJA NAJA KAOUTHIA see NAF200
VENOM, SNAKE, NAJA NIGRICOLLIS see NAG000
VENOM, SNAKE, NICRURUS FULVIUS see MQT100
VENOM, SNAKE, RHABDOPHIS TIGRINUS TIGRINUS see RFU875
VENOM, SNAKE, SISTRURUS MILARIUS BARBOURI see SDZ300
VENOM, SNAKE, TRIMERESURUS FLAVOVIRIDIS see TKW100
VENOM, SNAKE, VIPERA AMMODYTES see VQZ425
VENOM, SNAKE, VIPERA ASPIS see VQZ475
VENOM, SNAKE, VIPERA BORNMULLERI see VQZ500
VENOM, SNAKE, VIPERA LATIFII see VQZ525
VENOM, SNAKE, VIPERA LEBETINA see VQZ550
VENOM, SNAKE, VIPERA PALESTINAE see VQZ575
VENOM, SNAKE, VIPERA RUSSELLII see VQZ635
VENOM, SNAKE, VIPERA RUSSELLII FORMOSENSIS see VQZ625
VENOM, SNAKE, VIPERA XANTHINA PALAESTINAE see VQZ650
VENOM, SNAKE, WALTERINNESIA AEGYPTIA see WAT100
VENOM, SOUTH AFRICAN CAPE COBRA, NAJA NIVEA see NAG250
VENOM, STONEFISH, SYNANCEJA HORRIDA Linn. see SPB875
VENOM, THAILAND COBRA, NAJA NAJA SIAMENSIS see NAF250
VENOPEX see TAB250
VENOPIRIN see ARP125
VENOPIRIN see VHA275
VENTILAT see ONI000
VENTIN SUMACH see FBW000
VENTIPULMIN see VHA350
VENTOLIN see BQF500
VENTOX see ADX500
VENTRAMINE see DJS200
VENTUROL see DXX400
VENZAR see CPR600
VENZONATE see BCM000
VEON 245 see TAA100
VEPEN see PDT750
VEPESID see EAV500
VER A see MRV500
VERACILLIN see DGE200
VERACTIL see MCI500
VERAMID see AMK250
VERAMIN ED 4 see EIV750
VERAMIN ED 40 see EIV750
VERAMON see AMK250
VERANTEROL see PPH050
VERANTIN see TKN750
VERAPAMIL see IRV000
VERAPAMIL HYDROCHLORIDE see VHA450
VERAPRET DH see DTG000

VERATENSINE see VHF000
VERATRALDEHYDE see VHK000
VERATRAMINE see VHP500
VERATRIC ACID see VHP600
VERATRIC ALDEHYDE see VHK000
VERATRIDINE see VHU000
VERATRIDINE see VHZ000
VERATRIN see NCI600
VERATRIN (GERMAN) see VHZ000
VERATRINE see CDG000
VERATRINE see NCI600
VERATRINE see VHZ000
VERATRINE (amorphous) see VHU000
VERATRINE (crystallized) see CDG000
VERATROL see DOA200
VERATROLE see DOA200
VERATROLE METHYL ETHER see AGE250
N-(o-VERATROYL)GLYCINOHYDROXAMIC ACID see VIA875
3-VERATROYLVERACEVINE see VHU000
VERATRUM ALBUM see FAB100
VERATRUM CALIFORNICUM see FAB100
VERATRUM VIRIDE see FAB100
VERATRUM VIRIDE see VIZ000
VERATRUM VIRIDE ALKALOIDS EXTRACT see VIZ000
VERATRYL ALCOHOL, α-VINYL-(6CI,8CI) see HMB700
VERATRYL ALDEHYDE see VHK000
VERATRYLAMINE see VIK050
VERATRYL CHLORID (GERMAN) see BKM750
VERATRYL CHLORIDE see BKM750
VERATRYL CYANIDE see VIK100
VERATRYLHYDRAZINE see VIK150
1-VERATRYLPIPERAZINE DIHYDROCHLORIDE see VIK200
VERATRYL-2-PROPANONE see VIK300
VERAX see BBW500
VERAZINC see ZNA000
VERBENA ABSOLUTE see VIK500
VERBENA HYBRIDA Cornol. & Rpl., extract see VIK400
VERBENA OIL see VIK500
d-VERBENONE see VIP000
VERBINDUNG S 557 HCl see AMM750
VERCIDON see DJT800
VERCYTE see BHJ250
VERDANTIOL see LFT100
VERDICAN see DGP900
VERDICT see HAG860
VERDIPOR see DGP900
VERDONE see CIR250
VERDORACINE see VIP100
VERDOX see BQW490
VERDYL ACETATE see DLY400
VEREDEN see MNM500
VERESENE DISODIUM SALT see EIX500
VERGEMASTER see DAA800
VERGFRU FORATOX see PGS000
VERILOID see RDK000
VERILOID see VIZ000
VERINA see DNU200
3427 VERI PUR PINK see ADG250
VERITANIN see FMS875
VERITOL see FMS875
VERMAGO see PIJ500
VERMICID see AOR500
VERMICIDE BAYER 2349 see TIQ250
VERMICIDIN see MHL000
VERMICOMPREN (TABL.) see HEP000
VERMICULIN see VIZ100
VERMICULINE see VIZ100
VERMILASS see HEP000
VERMINUM see BQK000
VERMIRAX see MHL000
VERMISOL see PIK750
VERMITIN see DFV400
VERMITIN see PDP250
VERMIZYM see PAG500
VERMOESTRICID see CBY000

VERMOX see MHL000
VERNALDEHYDE see VIZ150
VERNAM see PNI750
VERNAMYCIN see VRF000
VERNAMYCIN BA see VRA700
VERNINE see GLS000
VERNOLATE see PNI750
VERODIGEN see GES000
VERODOXIN see VIZ200
VEROL see MGJ750
VEROLETTIN see BAG000
VERONAL see BAG000
VERONAL SODIUM see BAG250
VERONA PAPER RED see CMM840
VERONA YELLOW X-1791 see DEU000
VERON P 130/1 see PKQ059
VEROPHEN see DQA600
VEROS 030 see VIZ250
VEROSPIRON see AFJ500
VEROSPIRONE see AFJ500
VERPANYL see MHL000
VERROL see VSZ450
VERRUCARIN A see MRV500
VERSAL BLUE GGSL see IBV050
VERSAL FAST RED R see CJD500
VERSAL FAST YELLOW PG see CMS210
VERSAL GREEN G see PJQ100
VERSALIDE see ACL750
VERSAL ORANGE RNL see DVB800
VERSAL SCARLET 3BBA see CMG750
VERSAL SCARLET PRNL see MMP100
VERSAL SCARLET RNL see MMP100
VERSAMINE see VIZ400
VERSAR DSMA LQ see DXE600
VERSATIC 10 see DAH425
VERSATIC 9-11 see VIZ500
VERSATIC 10 ACID see DAH425
VERSATIC 9-11 ACID see VIZ500
VERSATIC ACID 911 see VIZ500
VERSA TL 71 see SFO100
VERSA TL 400 see SFO100
VERSA TL 500 see SFO100
VERSENE 9 see TNL250
VERSENE 100 see EIV000
VERSENE ACID see EIX000
VERSENE CA see CAR780
VERSENE CA see CAR800
VERSENE NTA ACID see AMT500
VERSENE POWDER see EIV000
VERSENE SODIUM 2 see EIX500
VERSENOL see HKS000
VERSENOL 120 see HKS000
VERSICOL E 7 see ADV900
VERSICOL E9 see ADV900
VERSICOL E15 see ADV900
VERSICOLORIN A see TKO500
VERSICOLORIN B see VJP800
VERSICOL S 25 see ADV900
VERSIDYNE see MDV000
VERSNELLER NL 63/10 see DQF800
VERSOMNAL see EOK000
VERSOTRANE see PFN000
VERSTRAN see DAP700
VERSULIN see CDH250
VERSUS see BAV325
TRANS-VERT see MRL750
VERTAC see BQI000
VERTAC see DGI000
VERTAC 90% see CDV100
VERTAC DINITRO WEED KILLER see BRE500
VERTAC GENERAL WEED KILLER see BRE500
VERT ACIDE BRILLIANT BS see ADF000
VERTAC METHYL PARATHION TECHNISCH 80% see MNH000
VERTAC SELECTIVE WEED KILLER see BRE500
VERTAC TOXAPHENE 90 see CDV100
VERTAVIS see VIZ000
VERTENEX see BQW500
VERTHION see DSQ000

VERTIGON see PMF250
VERTIMEC see ARW200
VERTISAL see MMN250
VERTOFIX COEUR see ACF250
VERTOLAN see SNJ000
VERTON 2D see DAA800
VERTON 2T see TAA100
VERTON D see DAA800
VERTREL XF see DLY900
VERTRON 2D see DAA800
VESADIN see SJW500
VESAKONTUHO MCPA see CIR250
VESALIUM see CLY500
VESAMIN see AAN000
VESPA ORIENTALIS VENOM see VJP900
VESPARAX-WIRKSTOFF see HHK050
VESPARAZ-WIRKSTOFF see CJR909
VESPERAL see BAG000
VESPRIN see TKL000
VESTIN see PDC250
VESTINOL AH see DVL700
VESTINOL DZ see PHW575
VESTINOL NN see PHW585
VESTINOL OA see AEO000
VESTOLEN see PJS750
VESTOLEN A 616 see PJS750
VESTOLEN A 6016 see PJS750
VESTOLEN P 5232G see SMQ500
VESTOLIT B 7021 see PKQ059
VESTROL see TIO750
VESTYRON see SMQ500
VESTYRON 512 see SMQ500
VESTYRON 114-12 see SMQ500
VESTYRON HI see SMR000
VESTYRON MB see SMQ500
VESTYRON N see SMQ500
VETACALM see TAF675
VETAFLAVIN see DBX400
VETALAR see CKD750
VETALGINE see ARP125
VETALOG see AQX500
VETA-MERAZINE see ALF250
VETAMOX see AAI250
VETARCILLIN see BFC750
VETARSENOBILLON see NCJ500
VETBUTAL see NBU000
VETERINARY NITROFURAZONE see NGE500
VETICILLIN see BFD250
VETICOL see CDP250
VETIDREX see CFY000
VETIOL see MAK700
VETIVER ACETATE see AAW750
VETIVEROL ACETATE see AAW750
VETIVERT ACETATE see AAW750
VETIVERT OIL see VJU000
VETIVERYL ACETATE see AAW750
VETKALM see DYF200
VETOL see MAO350
VETOX see CBM750
VETQUAMYCIN-324 see TBX250
VETRANQUIL see ABH500
VETRAZIN see CQE400
VETRAZIN see VIK150
VETRAZINE see CQE400
VETRAZINE see VIK150
VETRAZIN (PESTICIDE) see CQE400
VETREN see HAQ500
VETSIN see MRL500
VETSPEN see PAQ200
VETSTREP see SLY500
VETSULID see SNH900
VEVETONE see CAT775
VFR 3801 see PKF750
VH see PGP250
VIADENT see SAU100
VIADRIL see VJZ000
VIAGRA see SCF600
VIALIDON see XQS000
VI-ALPHA see VSK600
VIAMIN MF 514 see MCB050
VIAMIN MF 754 see MCB050

VIANIN see AOR500
VIANSIN see MDQ250
VIARESPAN HYDROCHLORIDE see DAI200
VIAROX see AFJ625
VIBALT see VSZ000
VIBATEX S see PKP750
VIBAZINE see BOM250
VIBAZINE see HGC500
VIBAZINE see VJZ050
VIBISONE see VSZ000
VIBRADOX see HGP550
VIBRAMYCIN see DYE425
VIBRAMYCIN HYCLATE see HGP550
VIBRA-TABS see HGP550
VIBRIO CHOLERAE ENDOTOXIN see VJZ100
VIBRIO CHOLERAE EXOTOXIN see VJZ200
VIBRIO CHOLERAE EXOTOXIN see VJZ200
VI-CAD see CAE250
VICALIN see BKW100
VICCILLIN see AIV500
VICCILLIN S see AIV500
VICILAN see VKF000
VICILLIN see AIV500
VICIN see BFC750
VICKNITE see PLL500
VICRON see CAT775
VICRON 31-6 see CAT775
VICTAN see EKF600
VICTENON see NCN650
VICTORIA BLUE R see VKA600
VICTORIA BLUE RS see VKA600
VICTORIA GREEN see AFG500
VICTORIA LAKE BLUE R see VKA600
VICTORIA ORANGE see DUT600
VICTORIA RUBINE O see FAG020
VICTORIA SCARLET 3R see FMU080
VICTORIA YELLOW see DUT600
VICTOR TSPP see TEE500
VIDANGA DRIED BERRY EXTRACT see VKA650
VIDARABIN see AQQ900
VIDARABINE see AEH100
VIDARABINE see AQQ900
VIDDEN D see DGG000
VIDDEN D see DGG950
VIDEOBIL see IGA000
VIDEOPHEL see TDE750
VIDINE see CMF800
VIDLON see PJY500
VIDON 638 see DAA800
VIDOPEN see AOD125
VIDR-2GD see VKA675
VIENNA GREEN see COF500
VIENNA WHITE see CAT775
VIGANTOL see VSZ100
VIGORSAN see CMC750
VIGOT 15 see CAT775
VIKh 65 see CGW300
VIKANE see SOU500
VIKANE FUMIGANT see SOU500
VIKASOL see MCB575
VIKROL RQ see AFP250
VILESCON see PNS000
VILLESCON see PNS000
VILLIAUMITE see SHF000
VILLIAUMITE see SHF500
VILOXAZIN see VKA875
VILOXAZINE see VKA875
VILOXAZINE HYDROCHLORIDE see VKF000
VINAC B 7 see AAX250
VINACETIN A see VQZ000
VINACOL MH see PKP750
VINADINE see OMY850
VINALAK see PKP750
VINAMAR see EQF500
VINAMUL N 710 see SMQ500
VINAMUL N 7700 see SMQ500
VINAROL see PKP750
VINAROL DT see PKP750
VINAROLE see PKP750

VINAROL ST see PKP750
VINAVILOL 2-98 see PKP750
VINBARBITAL SODIUM see VKP000
VINBLASTIN see VKZ000
VINBLASTINE see VKZ000
VINBLASTINE SULFATE see VLA000
VINCADAR see VLF000
VINCAFOLINA see VLF000
VINCAFOR see VLF000
VINCAGIL see VLF000
VINCAIN see AFG750
VINCALEUCOBLASTIN see VKZ000
VINCALEUKOBLASTINE see VKZ000
VINCALEUKOBLASTINE, 3-(AMINOCARBONYL)-O⁴-DEACETYL-3-DE(METHOXYCARBONYL)- see VLF350
VINCALEUKOBLASTINE, O⁴-DEACETYL-3-DE(METHOXYCARBONYL)-4'-DEOXY-3-(((2-ETHOXY-1-(1H- INDOL-3-YLMETHYL)-2-OXOETHYL)AMINO)CARBONYL)-, (3(S))- see VKZ100
VINCALEUKOBLASTINE SULFATE see VLA000
VINCALEUKOBLASTINE SULFATE (1:1) (SALT) see VLA000
VINCAMIDOL see VLF000
VINCAMIN COMPOSITUM see ARQ750
VINCAMINE see VLF000
(+)-VINCAMINE see VLF000
VINCA MINOR L., TOTAL ALKALOIDS see VLF300
VINCAPAN see VLF000
VINCAPRONT see VLF000
VINCASAUNIER see VLF000
VINCEINE see AFG750
VINCES see AKO500
VINCIMAX see VLF000
VINCLOZOLIN (GERMAN) see RMA000
VINCOBLASTINE see VKZ000
VINCRISTINE see LEY000
VINCRISTINE SULFATE ONCORIN see LEZ000
VINCRISTINSULFAT (GERMAN) see LEZ000
VINCRISUL see LEZ000
VINCRYSTINE see LEY000
VINCUBINA see TDT770
VINCUBINE sec TDT770
VINDESINA SULFATO (SPANISH) see VGU750
VINDESINE see VLF350
VINEGAR ACID see AAT250
VINEGAR NAPHTHA see EFR000
VINEGAR SALTS see CAL750
VINESTHENE see VOP000
VINESTHESIN see VOP000
VINETHEN see VOP000
VINETHENE see VOP000
VINETHER see VOP000
VINFOS see RAF100
VINICIZER 80 see DVL700
VINICIZER 85 see DVL600
VI-NICOTYL see NCR000
VI-NICTYL see NCR000
VINIDEN 60 see CGW300
VINIDYL see VOP000
VINIKA KR 600 see PKQ059
VINIKULON see PKQ059
VINILE (ACETATO di) (ITALIAN) see VLU250
VINILE (BROMURO di) (ITALIAN) see VMP000
VINILE (CLORURO di) (ITALIAN) see VNP000
VINIPLAST see PKQ059
VINIPLEN P 73 see PKQ059
VINISIL see PKQ250
VINKEIL 100 see EIX000
VINKRISTIN see LEY000
VINNAROL see PKP750
VINNOL E 75 see PKQ059
VINNOL H 10/60 see AAX175
VINODREL RETARD see VLF000

VINOFLEX see PKQ059
VINOFLEX MO 400* see IJQ000
VINOL see PKP750
VINOL 125 see PKP750
VINOL 205 see PKP750
VINOL 351 see PKP750
VINOL 523 see PKP750
VINOL UNISIZE see PKP750
VINORELBINE DITARTRATE see VLF400
VINOTHIAM see TES750
VINPOCETINE see EGM100
VINSTOP see SGM500
VINTHIONINE see VLU200
l-VINTHIONINE see VLU210
VINYDAN see VOP000
VINYLACETAAT (DUTCH) see VLU250
N-VINYLACETANILIDE see VLU220
VINYLACETAT (GERMAN) see VLU250
VINYL ACETATE see VLU250
VINYL ACETATE, inhibited (DOT) see
VLU250
VINYL ACETATE HOMOPOLYMER see
AAX250
VINYL ACETATE H.Q. see VLU250
VINYL ACETATE OZONIDE see VLU310
VINYL ACETATE POLYMER see AAX250
VINYL ACETATE RESIN see AAX250
VINYL ACETATE–VINYL CHLORIDE
COPOLYMER see AAX175
VINYL ACETATE–VINYL CHLORIDE
POLYMER see AAX175
VINYLACETONITRILE see BOX500
VINYL ACETYLENE see BPE109
VINYL ACRYLATE see EEE800
α-VINYL AE see VMA000
VINYL ALCOHOL POLYMER see PKP750
VINYL AMIDE see ADS250
VINYL A MONOMER see VLU250
VINYL AZIDE see VLY300
1-VINYL AZIRIDINE see VLZ000
α-VINYL-1-AZIRIDINEETHANOL see
VMA000
7-VINYLBENZ(a)ANTHRACENE see
VMF000
VINYLBENZEN (CZECH) see SMQ000
VINYLBENZENE see SMQ000
VINYLBENZENE POLYMER see SMQ500
VINYL BENZOATE see VMK000
VINYLBENZOL see SMQ000
VINYL(β-BIS(β-
CHLOROETHYL)AMINO)ETHYL
SULFONE see BHP500
VINYLBROMID (GERMAN) see VMP000
VINYL BROMIDE see VMP000
VINYL BROMIDE, inhibited (DOT) see
VMP000
VINYL 2-BUTENOATE see VNU000
VINYL-2-(BUTOXYETHYL) ETHER see
VMU000
VINYL BUTYL ETHER see VMZ000
VINYL-n-BUTYL ETHER see VMZ000
VINYL 2-(BUTYLMERCAPTOETHYL)
ETHER see VNA000
VINYL BUTYRATE see VNF000
VINYL BUTYRATE, INHIBITED (DOT) see
VNF000
VINYLBUTYROLACTAM see EEG000
N-VINYLBUTYROLACTAM POLYMER see
PKQ250
VINYL CARBAMATE see VNK000
VINYLCARBAZOLE see VNK100
9-VINYLCARBAZOLE see VNK100
N-VINYLCARBAZOLE see VNK100
9-VINYLCARBAZOLE HOMOPOLYMER see
VNK200
N-VINYLCARBAZOLE HOMOPOLYMER
see VNK200
VINYLCARBAZOLE POLYMER see VNK200
VINYLCARBINOL see AFV500
VINYL CARBINYL BUTYRATE see AFY250
VINYL CARBINYL CINNAMATE see
AGC000

VINYLCHLON 4000LL see PKQ059
VINYLCHLORID (GERMAN) see VNP000
VINYL CHLORIDE see VNP000
VINYL CHLORIDE ACETATE
COPOLYMER see PKP500
VINYL CHLORIDE COPOLYMER with
VINYLIDENE CHLORIDE see CGW300
VINYL CHLORIDE-1,1-
DICHLOROETHYLENE COPOLYMER see
CGW300
VINYL CHLORIDE HOMOPOLYMER see
PKQ059
VINYL CHLORIDE MONOMER see VNP000
VINYL CHLORIDE POLYMER see PKQ059
VINYL CHLORIDE VINYL ACETATE
COPOLYMER see PKP500
VINYL CHLORIDE–VINYL ACETATE
POLYMER see AAX175
VINYL CHLORIDE-VINYLIDENE
CHLORIDE COPOLYMER see CGW300
VINYL CHLORIDE-VINYLIDENE
CHLORIDE POLYMER see CGW300
VINYL-2-CHLOROETHYL ETHER see
CHI250
VINYL-β-CHLOROETHYL ETHER see
CHI250
VINYLCIZER 90 see PHW585
VINYL C MONOMER see VNP000
VINYL CROTONATE see VNU000
VINYL CYANIDE see ADX500
VINYLCYCLOHEXANE MONOXIDE see
VNZ000
1-VINYLCYCLOHEXENE see VNZ990
4-VINYLCYCLOHEXENE see CPD750
1-VINYLCYCLOHEXENE-3 see CPD750
1-VINYLCYCLOHEX-3-ENE see CPD750
4-VINYLCYCLOHEXENE-1 see CPD750
4-VINYL-1-CYCLOHEXENE see CPD750
VINYL CYCLOHEXENE DIEPOXIDE see
VOA000
4-VINYLCYCLOHEXENE DIEPOXIDE see
VOA000
4-VINYL-1-CYCLOHEXENE DIEPOXIDE
see VOA000
4-VINYL-1,2-CYCLOHEXENE DIEPOXIDE
see VOA000
VINYL CYCLOHEXENE DIOXIDE see
VOA000
4-VINYLCYCLOHEXENE DIOXIDE see
VOA000
1-VINYL-3-CYCLOHEXENE DIOXIDE see
VOA000
4-VINYL-1-CYCLOHEXENE DIOXIDE
(MAK) see VOA000
4-VINYLCYCLOHEXENE-1,2-EPOXIDE see
VNZ000
VINYLCYCLOHEXENE MONOXIDE see
VNZ000
4-VINYLCYCLOHEXENE MONOXIDE see
VNZ000
5-VINYL-DEOXYURIDINE see VOA550
VINYLDIAZOMETHANE see DCQ550
VINYL DIHYDROGEN PHOSPHATE see
VQA400
VINYL-2-(N,N-DIMETHYLAMINO)ETHYL
ETHER see VOF000
VINYLE (ACETATE de) (FRENCH) see
VLU250
VINYLE (BROMURE de) (FRENCH) see
VMP000
VINYLE (CHLORURE de) (FRENCH) see
VNP000
VINYLENE BISTHIOCYANATE see VOF300
VINYLENE CARBONATE see VOK000
4,4'-VINYLENEDIANILINE see SLR500
4,4'-VINYLENEDIBENZAMIDINE,
DIHYDROCHLORIDE see SLS500
N,N'-VINYLENEFORMAMIDINE see IAL000
1-VINYL-3,4-EPOXYCYCLOHEXANE see
VNZ000
VINYLESTER KYSELINY MASELNE see
VNF000

VINYLESTER KYSELINY OCTOVE see
VLU250
VINYL ETHANOATE see VLU250
VINYL ETHER see VOP000
VINYLETHYLENE see BOP500
N-VINYLETHYLENEIMINE see VLZ000
VINYL ETHYL ETHER see EQF500
VINYL ETHYL ETHER, inhibited (DOT) see
EQF500
VINYL ETHYL ETHER HOMOPOLYMER
see OJV600
VINYL ETHYL ETHER POLYMER see
OJV600
VINYL-2-ETHYLHEXOATE see VOU000
VINYL-2-ETHYLHEXYL ETHER see ELB500
VINYLETHYLNITROSAMIN (GERMAN) see
NKF000
VINYLETHYLNITROSAMINE see NKF000
VINYL FLUORIDE see VPA000
VINYL FLUORIDE, inhibited (DOT) see
VPA000
VINYL FORMATE see VPF000
VINYLFORMIC ACID see ADS750
VINYLGLYCOLONITRILE see HJQ000
4-VINYLGUAIACOL see VPF100
p-VINYLGUAIACOL see VPF100
VINYLHEPTACYCLOTETRASILOXANE see
VPF150
1-VINYL-1,3,3,5,5,7,7-
HEPTAMETHYLCYCLOTETRASILOXANE
see VPF150
S-VINYL-l-HOMOCYSTEINE see VLU210
S-VINYL-dl-HOMOCYSTEINE see VLU200
VINYLIDENE BROMIDE see VPF200
VINYLIDENE CHLORIDE see VPK000
VINYLIDENE CHLORIDE (II) see VPK000
VINYLIDENE CHLORIDE-BUTYL
ACRYLATE COPOLYMER see VPK333
VINYLIDENE CHLORIDE-VINYL
CHLORIDE POLYMER see CGW300
VINYLIDENE DICHLORIDE see VPK000
VINYLIDENE DIFLUORIDE see VPP000
VINYLIDENE FLUORIDE see VPP000
VINYLIDINE CHLORIDE see VPK000
1-VINYLIMIDAZOLE HOMOPOLYMER see
VPP100
N-VINYLIMIDAZOLE HOMOPOLYMER see
VPP100
N-VINYLIMIDAZOLE POLYMER see
VPP100
VINYL ISOBUTYL ETHER (DOT) see IJQ000
VINYL ISOBUTYL ETHER, inhibited (DOT)
see IJQ000
VINYLITE VYDR 21 see AAX175
N-VINYLKARBAZOL see VNK100
VINYLKYANID see ADX500
VINYLLITHIUM see VPU000
VINYL-2-METHOXYETHYL ETHER see
VPZ000
N-VINYLMETHYLACETAMIDE see
MQK500
N-VINYL-N-METHYLACETAMIDE see
MQK500
VINYL METHYLADIPATE see MQL000
VINYL METHYL ETHER (DOT) see MQL750
VINYL METHYL KETONE see BOY500
VINYL p-NITROPHENYL ETHER see
VPZ333
VINYLNORBORNENE see VQA000
2-VINYLNORBORNENE see VQA000
5-VINYLNORBORNENE see VQA000
5-VINYL-2-NORBORNENE see VQA000
17-VINYL-19-NORTESTOSTERONE see
NCI525
17-α-VINYL-19-NORTESTOSTERONE see
NCI525
VINYLOFOS see DGP900
VINYLON FILM 2000 see PKP750
VINYLOPHOS see DGP900
3-VINYL-7-OXABICYCLO(4.1.0)HEPTANE
see VNZ000
VINYLOXIRANE see EBJ500

VINYLOXYETHANOL see EJL500
2-(VINYLOXY)ETHANOL see EJL500
2-(VINYLOXY)ETHYL METHACRYLATE
see VQA150
2-(VINYLOXY)ETHYL NITRATE see EJK100
VINYLPHATE see CDS750
4-VINYLPHENOL see VQA200
p-VINYLPHENOL see VQA200
p-VINYLPHENOL ACETATE see ABW550
4-VINYLPHENYL ACETATE see ABW550
VINYLPHOSPHATE see RAF100
VINYL PHOSPHATE see VQA400
VINYLPHOSPHONIC ACID BIS(2-
CHLOROETHYL) ESTER see VQF000
5-VINYL-2-PICOLINE see MQM500
α-VINYLPIPERONYL ALCOHOL see BCJ000
VINYL PRODUCTS R 3612 see SMQ500
VINYL PRODUCTS R 10688 see AAX250
VINYL PROPIONATE see VQK000
1-VINYLPYRENE see EEF000
3-VINYLPYRENE see EEF000
4-VINYLPYRENE see EEF500
VINYL PYRIDINE see VQK550
2-VINYLPYRIDINE see VQK560
4-VINYLPYRIDINE see VQK590
α-VINYLPYRIDINE see VQK560
N-VINYLPYRROLIDINONE see EEG000
1-VINYL-2-PYRROLIDINONE see EEG000
N-VINYL-2-PYRROLIDINONE see EEG000
1-VINYL-2-PYRROLIDINONE POLYMER,
compounded with IODINE see PKE250
1-VINYL-2-PYRROLIDINONE POLYMER
with STYRENE see VQK595
VINYLPYRROLIDINONE-STYRENE
POLYMER see VQK595
VINYLPYRROLIDONE see EEG000
N-VINYLPYRROLIDONE see EEG000
1-VINYL-2-PYRROLIDONE see EEG000
N-VINYL-2-PYRROLIDONE (ACGIH) see
EEG000
1-VINYL-2-PYRROLIDONE CROSSLINKED
INSOLUBLE POLYMER see PKQ150
N-VINYLPYRROLIDONE POLYMER see
PKQ250
α-(5-VINYL-2-QUINOLYL)-2-
QUINUCLIDINEMETHANOL see CMP925
VINYLSILATRAN see VQK600
VINYLSILATRANE see VQK600
1-VINYLSILATRANE see VQK600
VINYLSILICON TRICHLORIDE see TIN750
VINYLSTYRENE see DXQ740
m-VINYLSTYRENE see DXQ745
VINYL SULFONE see DXR200
VINYL TOLUENE see VQK650
2-VINYLTOLUENE see VQK660
3-VINYLTOLUENE see VQK670
4-VINYLTOLUENE see VQK700
m-VINYLTOLUENE see VQK670
o-VINYLTOLUENE see VQK660
p-VINYLTOLUENE see VQK700
3- and 4-VINYL TOLUENE (mixed isomers) see
VQK650
VINYL TOLUENE, inhibited mixed isomers
(DOT) see VQK650
VINYL TRICHLORIDE see TIN000
VINYL TRICHLOROSILANE (DOT) see
TIN750
VINYL TRICHLOROSILANE, INHIBITED
(DOT) see TIN750
VINYLTRIETHOXYSILANE see TJN250
VINYL TRIMETHOXY SILANE see TLD000
VINYLTRIMETHYLAMMONIUM
HYDROXIDE see VQR300
VINYL-2,6,8-TRIMETHYLNONYL ETHER
see VQU000
1-VINYL-2,8,9-TRIOXA-5-AZA-1-
SILABICYCLO(3.3.3)UNDECANE see
VQK600
VINYLTRIS(tert-BUTYLDIOXY)SILANE see
VQU333
VINYLTRIS(tert-BUTYLPEROXY)SILANE
see VQU333

VINYLTRIS(METHOXYETHOXY)SILANE
see TNJ500
VINYLTRIS(2-METHOXYETHOXY)SILANE
see TNJ500
VINYLTRIS(β-METHOXYETHOXY)SILANE
see TNJ500
VINYON N see ADY250
VINYZENE see OMY850
VINYZENE bp 5 see OMY850
VINYZENE bp 5-2 see OMY850
VINYZENE (pesticide) see OMY850
VINYZENE SB 1 see OMY850
VIOACTANE see VQZ000
VIOCID see AOR500
VIOCIN see VQZ000
VIOFORM see CHR500
VIOFORM N.N.R. see CHR500
VIOFURAGYN see NGG500
VIOLACETIN see VQU450
VIOLANTHRONE see DCU800
VIOLAQUERCITRIN see RSU000
VIOLET 3 see XGS000
VIOLET 2S see DBY700
11092 VIOLET see HOK000
12416 VIOLET see AOR500
VIOLET BNP see VQU500
VIOLET CP see AOR500
VIOLET DISPERZNI 4 see AKP250
VIOLET GENCIANOVA see AOR500
VIOLET K see HGE925
VIOLET KRYSTALOVA see AOR500
VIOLET KYPOVA 1 see DFN450
VIOLET LEAF ALCOHOL see NMV780
VIOLET LEAF ALDEHYDE see NMV760
VIOLET 6BN see AOR500
VIOLET 5BO see AOR500
VIOLET PIGMENT 31 see DFN450
VIOLET POWDER H 2503 see MQN025
VIOLET R see HGE925
VIOLET ROZPOUSTEDLOVA 12 see
AKP250
VIOLET ROZPOUSTEDLOVA 26 see
DBX000
VIOLET XXIII see AOR500
VIOLET ZASADITA 3 see AOR500
VIOLET ZASADITA 14 see MAC250
VIOLOGEN, METHYL- see PAJ000
VIOLURIC ACID see AFU000
VIOMICINAE (ITALIAN) see VQZ000
VIOMYCIN see VQZ000
VIOMYCIN, 1-(threo-4-HYDROXY-L-3,6-
DIAMINOHEXANOIC ACID)-6-(L-2-(2-
AMINO-1,4,5,6-TETRAHYDRO- 4-
PYRIMIDINYL)GLYCINE)-, (R)-,
SESQUISULFATE see VQZ100
VIOMYCIN SULFATE SALT (2:3) see VQZ100
VIOPSICOL see LFK000
VIOPSICOL see MDQ250
VIO-SERPINE see RDK000
VIOSORB 930 see ODY150
VIOSTEROL see VSZ100
VIOXAN see CBM750
VIOZENE see RMA500
VI-PAR see CIR500
VIPERA AMMODYTES VENOM see VQZ425
VIPERA BERUS VENOM see VQZ475
VIPERA BORNMULLERI VENOM see
VQZ500
VIPERA LATIFII VENOM see VQZ525
VIPERA LEBETINA VENOM see VQZ550
VIPERA PALESTINAE VENOM see VQZ575
VIPERA RUSSELLII FORMOSENSIS
VENOM see VQZ625
VIPERA RUSSELLII VENOM see VQZ635
VIPERA XANTHINA PALAESTINAE
VENOM see VQZ650
VIPERINE (CANADA) see VQZ675
VIPER'S BUGLOSS see VQZ675
VI-PEX see CIR500
VI-PEX see CLO200
VIPICIL see AJJ875
VIPRYNIUM EMBONATE see PQC500

VIRA-A see AEH100
VIRACTIN see VRA000
VIRAMID see RJA500
VIRAZOLE see RJA500
VIRCHEM see SHR500
VIRCHEM 931 see ZJS400
VIRGIMYCIN see VRA700
VIRGIMYCIN see VRF000
VIRGINIA-CAROLINA 3-670 see DWU200
VIRGINIA CAROLINA VC 9-104 see EIN000
VIRGINIAMYCIN see VRF000
VIRGIN'S BOWER see CMV390
VIRICUIVRE see CNK559
VIRIDICATUMTOXIN see VRP000
VIRIDINE see PDX000
VIRIDITOXIN see VRP200
VIRIDOGRISEIN see EDW200
VIRIDOGRISEIN I see EDW200
VIROFRAL see AED250
VIROPHTA see TKH325
VIROPTIC see TKH325
VIRORMONE see TBF500
VIROSIN see AQM250
VIROSTERONE see TBF500
VIRSET 656-4 see MCB000
VIRTEX CC see SHR500
VIRTEX D see SHR500
VIRTEX L see SHR500
VIRTEX RD see SHR500
VIRUBRA see VSZ000
VIRUSTOMYCIN A see VRP775
VIRUZONA see MKW250
VISADRON see NCL500
VISADRON see SPC500
VISCALEX HV 30 see ADV900
VISCARIN see CCL250
VISCARIN see SFP000
VISCARIN 402 see CAO250
VISCERALGIN see HNM000
VISCERALGINE see HNM000
VISCLAIR see MBX800
VISCO 31 see VRP780
VISCO 34 see VRP790
VISCOAT 150 see TCS100
VISCOAT 190 see EHG030
VISCO BLACK N see CMN240
VISCOL see MIF760
VISCOL 350P see PMP500
VISCOLEO OIL see VGU200
VISCON 103 see ADV900
VISCONTRAN L52 see MIF760
VISCOSE BLACK G see CMN240
VISCOSE BLACK GNA see CMN240
VISCOSE BLACK J see CMN240
VISCOSE BLACK N see CMN240
VISCOSE BLACK NG see CMN240
VISCOSOL see MIF760
VISCOTOXIN see VRU000
VISCOUS POLYMER (PETROLEUM) see
VRU300
VISCUM ALBUM see ERA100
VISET see FBW000
VISINE see VRZ000
VISINE HYDROCHLORIDE see VRZ000
VISIREN see OGI300
VISKEN see VSA000
VISKING CELLOPHANE see CCT250
VISKO 31 see VRP780
VISKO-RHAP see DAA800
VISKO-RHAP see DGB000
VISKO-RHAP DRIFT HERBICIDES see
DAA800
VISKO-RHAP LOW DRIFT HERBICIDES see
PMM100
VISKO-RHAP LOW VOLATILE 4L see
DAA800
VISKO RHAP LOW VOLATILE ESTER see
TAA100
VISNAGALIN see AHK750
VISOTRAST see AAN000
VISTABAMATE see MQU750
VISTAGAN see LFA100

VISTAMYCIN see RIP000
VISTAMYCIN see XQJ650
VISTAR see DUK000
VISTAR HERBICIDE see DUK000
VISTARIL HYDROCHLORIDE see VSF000
VISUBUTINA see HNI500
VISUMATIC see DAM625
VISUMIOTIC see DAM625
VITABLEND see PAQ200
VITACAMPHER see VSF400
VITACIN see ARN000
VITAHEXIN P see PII100
VITALLIUM see CNA750
VITALLIUM see VSK000
VITALOID see VQR300
VITAMIN A see VSK600
VITAMIN A1 see VSK600
VITAMIN A ACETATE see VSK900
trans-VITAMIN A ACETATE see VSK900
VITAMIN A ACID see VSK950
13-cis-VITAMIN A ACID see VSK955
VITAMIN A1 ALCOHOL see VSK600
all-trans-VITAMIN A ALCOHOL see VSK600
VITAMIN A ALCOHOL ACETATE see
VSK900
VITAMIN A ALDEHYDE see VSK985
VITAMIN A1 ALDEHYDE see VSK985
9-cis-VITAMIN A ALDEHYDE see VSK975
trans-VITAMIN A ALDEHYDE see VSK985
VITAMIN A1, ANHYDRO- see AFQ700
VITAMIN AB see HAQ500
VITAMIN A PALMITATE see VSP000
VITAMIN B1 see TES750
VITAMIN B^1 see TET300
VITAMIN B2 see RIK000
VITAMIN B3 see NCR000
VITAMIN B3 see PAG300
VITAMIN B4 see AEH000
VITAMIN B-5 see CAU750
VITAMIN B5 see PAG300
VITAMIN B6 see PPK250
VITAMIN B7 see BGD100
VITAMIN B_7 see VSU100
VITAMIN Bc see FMT000
VITAMIN B_{12} COMPLEX see VSZ000
VITAMIN B_1 DISULFIDE see TES800
VITAMIN B12 (FCC) see VSZ000
VITAMIN B HYDROCHLORIDE see TET300
VITAMIN B6-HYDROCHLORIDE see
PPK500
VITAMIN B_6 HYDROCHLORIDE see VSU000
VITAMIN B_{12} METHYL see VSZ050
VITAMIN B1 MONONITRATE see TET500
VITAMIN B1 NITRATE see TET500
VITAMIN B_1 PROPYL DISULFIDE see
DXO300
VITAMIN B6 TRIPALMITATE see HMQ575
VITAMIN BX see AIH600
VITAMIN C see ARN000
VITAMIN C see ARN125
VITAMIN C SODIUM see ARN125
VITAMIN D see VSZ095
VITAMIN D2 see VSZ100
VITAMIN D3 see CMC750
VITAMIN D^3 see HJV000
VITAMIN E see VSZ450
VITAMIN E ACETATE see TGJ050
VITAMIN G see RIK000
VITAMIN H see AIH600
VITAMIN H see BGD100
VITAMIN H see VSU100
VITAMIN K see VSZ500
VITAMIN K1 see VTA000
VITAMIN K3 see MMD500
VITAMIN K4 see VTA100
VITAMIN K5 see AKX500
VITAMIN K_{220} see VTA650
VITAMIN K DIACETATE see VTA100
VITAMIN K INJECTION see MCB575
VITAMIN K3-NATRIUM-BISULFIT
TRIHYDRAT (GERMAN) see MMD750
VITAMIN K2(O) see MMD500

VITAMIN K3 SODIUM BISULFITE see
MCB575
VITAMIN L see API500
VITAMIN M see FMT000
VITAMIN MK 4 see VTA650
VITAMIN P see RSU000
VITAMIN PP see NCR000
VITAMISIN see ARN000
VITANEURON see TES750
VITAPLEX E see VSZ450
VITAPLEX N see NCQ900
VITARUBIN see VSZ000
VITA-RUBRA see VSZ000
VITASCORBOL see ARN000
VITASTAIN see TMV500
VITAVAX see CCC500
VITAVEL-A see VSK600
VITAVEL-D see VSZ100
VITAVEL K see VTA100
VITAVEX see DLV200
VITAYONON see VSZ450
VITAZECHS see PII100
VITEOLIN see VSZ450
VITESTROL see DLB400
VITEXIN see GFC050
VITICOLOR see OLW100
VITICOSTERONE see HKG500
VITIGRAN see CNK559
VITIGRAN BLUE see CNK559
VITINC DAN-DEE-3 see CMC750
VITON see BBQ500
VITPEX see VSK600
VITRAL see VSZ000
VITRAN see TIO750
VITREOSIL IR see SCK600
VITREOUS QUARTZ see SCK600
VITREOUS SILICA see SCK600
VITREX see PAK000
VITRIFIED SILICA see SCK600
VITRIOL BROWN OIL see SOI500
VITRIOL, OIL OF (DOT) see SOI500
VITRIOL RED see IHC450
VITRUM AB see HAQ500
VITUF see PKF750
VI-TWEL see VSZ000
VIVACTIL see DDA600
VIVACTIL see POF250
VIVAL see DCK759
VIVALAN see VKF000
VIVICIL see FMR500
VIVIFUL see CPB065
VIVOTOXIN see VTA750
VK 4 see BAP502
VK-738 see DWE800
VKhVD 40 see CGW300
VK 4 (OXIDE) see BAP502
VLB see VKZ000
VLB MONOSULFATE see VLA000
VLVF see AAX175
VM-26 see EQP000
VMCC see AAX175
VMI 10-3 see AMI500
VML 2 see MCB050
VM & P NAPHTHA see PCT250
VM&P NAPHTHA see PCT250
VM and P NAPHTHA see PCT250
VM & P NAPHTHA (ACGIH,OSHA) see
PCT250
VN 1 see NDZ000
VOCOL see ZGA500
VOCOL 5 see ZGA500
VOFATOX see MNH000
VOGALENE see MQR000
VOGAN see VSK600
VOGAN-NEU see VSK600
VOGEL'S IRON RED see IHC450
VOGLIBOSE see DNC100
VOLAMIN see EEH000
VOLATILE OIL OF MUSTARD see AGJ250
VOLATON see BAT750
VOLCLAY see BAV750
VOLCLAY BENTONITE BC see BAV750

VOLDYS see MCA500
VOLFARTOL see TIQ250
VOLFAZOL see COD000
VOLID see TAC800
VOLIDAN see MCA500
VOLIDAN see VTF000
VOLPO 20 see PJW500
VOLTAGE see POK500
VOLTALEF see KDK000
VOLTALEF 202 see KDK000
VOLTALEF 300LD see KDK000
VOLTALEF 300UF see KDK000
VOLTALEF IS see KDK000
VOLTAREN see DEO600
VOLTAROL see DEO600
VOLUNTAL see TIO500
VOMEX A see DYE600
VOMISSELS see HGC500
VOMITOXIN see VTF500
VONAMYCIN POWDER V see NCE000
VONDACEL BLACK D see CMN230
VONDACEL BLACK N see AQP000
VONDACEL BLACK VG see CMN240
VONDACEL BLACK VN see CMN240
VONDACEL BLUE 2B see CMO000
VONDACEL BLUE FF see CMN750
VONDACEL BLUE HH see CMO500
VONDACEL BORDEAUX B see CMO872
VONDACEL BROWN M see CMO800
VONDACEL BROWN S see CMO820
VONDACEL BROWN SP see CMO820
VONDACEL DARK BLUE BH see CMN800
VONDACEL GREEN B see CMO840
VONDACEL GREEN DB see CMO830
VONDACEL RED CL see SGQ500
VONDACEL RED FN see CMO870
VONDACEL SCARLET 4BS see CMO870
VONDACID BLUE SE see APG700
VONDACID FAST BLUE BR see CMM090
VONDACID FAST YELLOW AE see SGP500
VONDACID GREEN L see FAE950
VONDACID GREEN S see ADF000
VONDACID LIGHT RED NG see CMM300
VONDACID LIGHT YELLOW AE see
SGP500
VONDACID METANIL YELLOW G see
MDM775
VONDACID NAPHTHOL RED 6B see
CMM400
VONDACID ORANGE II see CMM220
VONDACID TARTRAZINE see FAG140
VONDALHYDE see DMC600
VONDAMOL BRILLIANT RED G see
CMM325
VONDAMOL FAST BLUE R see ADE750
VONDAMOL FAST RED G see CMM320
VONDAMOL FAST RED RS see CMM330
VONDCAPTAN see CBG000
VONDODINE see DXX400
VONDRAX see DMC600
VONDURON see DXQ500
VONEDRINE see DBA800
VONTERYL VIOLET 2B see DBY700
VONTERYL YELLOW 3R see CMP090
VONTERYL YELLOW G see AAQ250
VONTERYL YELLOW R see AAQ250
VONTIL see TFM100
VONTROL see DWK200
VOPCOLENE 27 see OHU000
VORANIL see DRC600
VOREN see DBC510
V. ORIENTALIS VENOM see VJP900
VORLEX see ISE000
VORLEX see MLC000
VORONITE see FQK000
VOROX see AMY050
VOROX AA see AMY050
VOROX AS see AMY050
VORTEX see ISE000
VOTEXIT see TIQ250
VP16 see EBB600
VP 1940 see ABX250

VP 16213 see EAV500
VPK 402 see DTS500
VPM see SIL550
VPM see VFU000
V-PYROL see EEG000
VROKON see AAN000
VS-6B see DOY500
VS 7158 see DAF350
VS 7207 see OCE100
V-SERP see RDK000
VT 1 see TGF250
VT 2809 see UTA300
VTI 1 see AOT000
VTS 1 see TEA550
VTS-M see TLD000
VU 51-3N see MCB050
VU 59-3N see MCB050
VU 5711N see MCB050
VUAGT 1-4 see DXH325
VUAGT 866 see FMY100
VUAGT 1964 see DXH325
VUAGT 866/72 see FMY100
VUAGT-I-4 see TFS350
VUCINE see EDW500
VUFB 3511 see PCI250
6605 VUFB see MIQ400
VUFB 6638 see DLR150
VULCACEL B-40 see DVF400
VULCACEL BN see DVF400
VULCACID D see DWC600
VULCACURE see BIX000
VULCACURE see BJC000
VULCACURE see SGF500
VULCAFIX BLUE R see BGB275
VULCAFIX FAST BLUE SD see IBV050
VULCAFIX ORANGE J see CMS145
VULCAFIX ORANGE JV see CMS145
VULCAFIX RED R see NAP100
VULCAFIX SCARLET R see CHP500
VULCAFOR BLUE A see BGB275
VULCAFOR BSM see BDG000
VULCAFOR CBS see CPI250
VULCAFOR FAST BLUE 3R see IBV050
VULCAFOR FAST ORANGE G see CMS145
VULCAFOR FAST ORANGE GA see CMS145
VULCAFOR FAST YELLOW 2G see CMS210
VULCAFOR FAST YELLOW GTA see
DEU000
VULCAFOR HBS see CPI250
VULCAFOR ORANGE R see CJD500
VULCAFOR SCARLET A see MMP100
VULCAFOR TMTD see TFS350
VULCALENT A see DWI000
VULCAL FAST GREEN F2G see PJQ100
VULCAMEL TBN see NAP500
VULCAN see CBT750
VULCAN FAST ORANGE G see CMS145
VULCAN FAST ORANGE GA see CMS145
VULCAN FAST ORANGE GN see CMS145
VULCAN FAST YELLOW G see CMS210
VULCAN FAST YELLOW GR see CMS208
VULCAN FAST YELLOW GRA see CMS208
VULCAN FAST YELLOW GRN see CMS208
VULCANOSINE DARK BLUE L see BGB275
VULCANOSINE FAST BLUE GG see IBV050
VULCANOSINE FAST GREEN G see PJQ100
VULCAN RED LC see CHP500
VULCAN RED LCN see CMS150
VULCAN RED R see CJD500
VULCATARD see DWI000
VULCOL BLUE BZ see ERG100
VULCOL FAST BLUE S see IBV050
VULCOL FAST GREEN F2G see PJQ100
VULCOL FAST ORANGE G see CMS145
VULCOL FAST RED C see CMS150
VULCOL FAST RED L see CHP500
VUL-CUP see BHL100
VUL-CUP 40KE see BHL100
VUL-CUP R see BHL100
VULCUREN 2 see BDF750
VULKACIT 1000 see TGX550
VULKACIT 4010 (CZECH) see PET000

VULKACIT C see CPI250
VULKACIT CZ see CPI250
VULKACIT CZ/C see CPI250
VULKACIT CZ/K see CPI250
VULKACIT D/C see DWC600
VULKACIT DM see BDE750
VULKACIT DM/MGC see BDE750
VULKACIT DOTG/C see DXP200
VULKACIT LDA see BJC000
VULKACIT LDB/C see BIX000
VULKACIT MERCAPTO see BDF000
VULKACIT MTIC see TFS350
VULKACIT NPV/C2 see IAQ000
VULKACIT P EXTRA N see ZHA000
VULKACIT THIURAM see TFS350
VULKACIT THIURAM/C see TFS350
VULKACIT THIURAM MS/C see BJL600
VULKACIT ZM see BHA750
VULKALENT A (CZECH) see DWI000
VULKANOX 4020 see DQV250
VULKANOX 4020 see PEY500
VULKANOX BKF see MJO500
VULKANOX HS/LG see TLP500
VULKANOX HS/POWDER see TLP500
VULKANOX MB 2 see MCO400
VULKANOX NKF see IIU100
VULKANOX PAN see PFT250
VULKASIL see SCH002
VULKAZIT see DWC600
VULKLOR see TBO500
VULNOC AB see ANB100
VULNOPOL NM see SGM500
VULTROL see DWI000
VULVAN see TBG000
VUMON see EQP000
VX see EIG000
VYAC see VLU250
VYDATE see DSP600
VYDATE L INSECTICIDE/NEMATICIDE
see DSP600
VYDATE L OXAMYL
INSECTICIDE/NEMATOCIDE see DSP600
VYDYNE see NOH000
VYGEN 85 see PKQ059
VYNAMON BLUE 3R see IBV050
VYNAMON BLUE A see BGB275
VYNAMON CLARET Y see CMS155
VYNAMON GREEN 6Y see CMS140
VYNAMON GREEN BE see PJQ100
VYNAMON GREEN BES see PJQ100
VYNAMON GREEN GNA see PJQ100
VYNAMON ORANGE CR see LCS000
VYNAMON ORANGE G see CMS145
VYNAMON SCARLET BY see MRC000
VYNAMON YELLOW 2G see CMS210
VYNAMON YELLOW 2GE see CMS210
VYNAMON YELLOW GRE see CMS208
VYNAMON YELLOW GRES see CMS208
VYNW see AAX175
W 37 see BRA625
W 45 see ALC250
W 70 see UTU500
W 101 see PMP500
W 483 see DIR000
W 491 see CCP500
W 524 see TKL100
W 554 see FAG200
W 713 see MOV500
W-801 see CMY650
W 1544 see PFC500
W 1655 see PDC250
W 1655 see PEK250
W 1803 see PGJ000
W 1929 see SFY500
2317-W see DVF800
W-2395 see CIL500
W 2426 see ARW750
W-2429 see DLQ400
W 2451 see MJE250
W 3395 see PMH550
W 3566 see QFA250
W 3699 see EOA500

W 4565 see OOG000
W 4744 see MIH925
W 5219 see BGC625
W5769 see CFC750
W 6309 see DKJ300
W 7320 see AGN000
W 7618 see CLD000
W 1164-3 see RCK740
W-2946M see DNA600
W3207B see MOM750
W 5759A see EIH000
W 769034 see DHU700
W 2900 A see MNG000
W 7000A see LFA100
WACHOLDERBEER OEL (GERMAN) see
JEA000
WACKER S 14/10 see BJE750
W. AEGYPTIA VENOM see WAT100
WA 335 HYDROCHLORIDE see WAJ000
WAIT'S GREEN MOUNTAIN
ANTIHISTAMINE see WAK000
WAKE ROBIN see JAJ000
WAKIL see MFC100
WALKO-NESIN see GGS000
WALNUT EXTRACT see WAT000
WALNUT STAIN see MAT500
WALSRODER MC 20000S see MIF760
WALTERINNESIA AEGYPTIA VENOM see
WAT100
WAMPOCAP see NCQ900
WANADU PIECIOTLENEK (POLISH) see
VDU000
WANDAMIN HM see MJQ260
WANSAR see CCR875
WAPNIOWY TLENEK (POLISH) see CAU500
WARAN see WAT220
WARBEX see FAB600
WARBEXOL see FAB600
WARCO F 71 see CNH125
WARCOUMIN see WAT220
WARDAMATE see MQU750
WARDUZIDE see CLH750
WARECURE C see IAQ000
WAREFLEX see DDT500
WARF see DDW500
WARF ANTIRESISTANT see DDW500
WARFARIN see WAT200
WARFARINE (FRENCH) see WAT200
WARFARIN K see WAT209
WARFARIN POTASSIUM see WAT209
WARFARIN Q CONCENTRATE see WAT211
WARFARIN-S see WAT211
WARFARIN SODIUM see WAT220
WARFILONE see WAT220
WARKEELATE ACID see EIX000
WARKEELATE PS-43 see EIV000
WASH OIL see CMY825
WASSERINA see CQH000
WASSERSTOFFPEROXID (GERMAN) see
HIB050
WATANIDIPINE DIHYDROCHLORIDE see
WAT225
WATANIDIPINE HYDROCHLORIDE see
WAT230
(S)-(+)-WATANIDIPINE HYDROCHLORIDE
see NDY550
WATANIDIPINE HYDROCHLORIDE (S)-
ENANTIOMER see NDY550
WATAPANA SHIMARON see AAD750
WATCHUNG RED Y see CMS148
WATER see WAT259
WATER-d2 (9CI) see HAK000
WATER ARUM see WAT300
WATER BLACK 100 see CMN240
WATER BLACK 200L see CMN240
WATER BLACK P 200 see CMN240
WATERCARB see CBT500
WATER CROWFOOT see FBS100
WATER DRAGON see WAT300
WATER DROPWART see WAT315
WATER GLASS see SJU000
WATER GOGGLES see MBU550

WATER GREEN SX see ADF000
WATER²-H2 see HAK000
WATER HEMLOCK see WAT325
WATER LILY see DAE100
WATER-PEPPER HERB see WAT350
WATER QUENCH PYROLYSIS FUEL OIL see FOP100
WATERSOL S 683 see MCB050
WATERSOL S 685 see MCB050
WATERSOL S 695 see MCB050
WATERSTOFPEROXYDE (DUTCH) see HIB050
WATERWEED see PIH800
WATTLE GUM see AQQ500
WAXAKOL ORANGE GL see PEJ500
WAXAKOL RED BL see SBC500
WAXAKOL VERMILION L see XRA000
WAXAKOL YELLOW NL see AIC250
WAXBERRY see SED550
WAX LE see PJS750
WAX MYRTLE see WBA000
WAXOLINE GREEN see BLK000
WAXOLINE ORANGE A see BJF000
WAXOLINE PURPLE A see HOK000
WAXOLINE RED A see MAC500
WAXOLINE RED O see SBC500
WAXOLINE RED OM see SBC500
WAXOLINE RED OS see SBC500
WAXOLINE YELLOW AD see DOT300
WAXOLINE YELLOW ED see OHJ875
WAXOLINE YELLOW I see PEJ500
WAXOLINE YELLOW O see IBB000
WAXOLINE YELLOW T see CMS240
WAXSOL see DJL000
WAY100635 see TKN100
WAYNECOMYCIN see LGC200
WAYNE RED X-2486 see CHP500
WAYPLEX 55S see DJG700
WBA 8107 see BGO100
WBA 8119 see TAC800
WCZ 628 see EJM950
WD 67/2 see WBA600
WE 941 see LEJ600
WEAK ACID BRILLIANT BLUE RAW see CMM092
WEATHERBEE MUSTARD see BHB750
WEATHER COAT 1000 see ZJS400
WEATHER PLANT see RMK250
WEB 1881 see AKR100
WE 941-BS see LEJ600
WEB 1881FU see AKR100
WEC 50 see TIQ250
WECKAMINE see BBK000
WECOLINE 1295 see LBL000
WECOLINE OO see OHU000
WEDDING BELLS see DJO000
WEDELIATOXIN see WBA700
WEDELOSIDE see WBA700
WEECON see COI250
WEED 108 see MRL750
WEED-AG-BAR see DAA800
WEEDANOL CYANOL see PLC250
WEEDAR see TAA100
WEEDAR-64 see DAA800
WEEDAR ADS see AMY050
WEEDAR AT see AMY050
WEEDAR MCPA CONCENTRATE see CIR250
WEEDAZIN see AMY050
WEEDAZIN ARGINIT see AMY050
WEEDAZOL see AMY050
WEEDAZOL GP2 see AMY050
WEEDAZOL SUPER see AMY050
WEEDAZOL T see AMY050
WEEDAZOL TL see AMY050
WEEDAZOL TL see ANW750
WEEDBEADS see SJA000
WEED-B-GON see DAA800
WEED-B-GON see TIX500
WEED BROOM see DXE600
WEED DRENCH see AFV500
WEED-E-RAD see DXE600

WEED-E-RAD see MRL750
WEEDEX A see ARQ725
WEEDEX GRANULAT see AMY050
WEEDEZ WONDER BAR see DAA800
WEED-HOE see DXE600
WEED-HOE see MRL750
WEEDMASTER GRASS KILLER see TII500
WEEDOCLOR see AMY050
WEEDOL see PAI995
WEEDOL see PAJ000
WEEDONE see PAX250
WEEDONE see TAA100
WEEDONE 40 see EHY600
WEEDONE 128 see IOY000
WEEDONE 170 see DGB000
WEEDONE CONCENTRATE 48 see EHY600
WEEDONE CRAB GRASS KILLER see PLC250
WEEDONE DP see DGB000
WEEDONE LV4 see DAA800
WEEDONE MCPA ESTER see CIR250
WEED-RHAP see CIR250
WEED TOX see DAA800
WEEDTRINE-D see DWX800
WEEDTRINE-D see EJC025
WEEDTRINE-II see ILO000
WEEDTROL see DAA800
WEEVILTOX see CBV500
WEG 147 see CJP500
WEG 148 see MNR100
WEGANTYNA see DHX800
WEGANTYNA (POLISH) see THK000
WEGLA DWUSIARCZEK (POLISH) see CBV500
WEGLA TLENEK (POLISH) see CBW750
WEHYDRYL see BAU750
WEIFACODINE see TCY750
WEISS PHOSPHOR (GERMAN) see PHP010
WEISSPIESSGLANZ see AQF000
WELDING FUMES see WBJ000
WELD LAKE see TDD550
WELFURIN see NGE000
WELLBATRIN see WBJ500
WELLCIDE see PDR700
WELLCOME-248U see AEC700
WELLCOME PREPARATION 47-83 see EAN600
WELLCOME U3B see AMH250
WELLCOPRIM see TKZ000
WELLDORM see SKS700
WELVIC G 2/5 see PKQ059
WEMCOL see IOF200
WEPSIN see AIX000
WEPSYN see AIX000
WEPSYN 155 see AIX000
WESCOZONE see BRF500
WESPURIL see MJM500
WESTERN DIAMONDBACK RATTLESNAKE VENOM see WBJ600
WEST INDIAN BAY OIL see LBK000
WEST INDIAN LEMONGRASS OIL see LEH000
WEST INDIAN LILAC see CDM325
WEST INDIAN PINKROOT see PIH800
WEST INDIAN SANDALWOOD OIL see WBJ650
WESTOCAINE see AIT250
WESTON 439 see IPL500
WESTON 626 see COV800
WESTON DPDP see IKL200
WESTON MDW 626 see COV800
WESTRON see TBQ100
WESTROSOL see TIO750
WETAID SR see DJL000
WET-TONE B see MSB500
WEX 1242 see PMP500
WF-104 see OKK500
WFNA see NEF000
W-GUM see SLJ500
WH 5668 see PMM000
WH 7286 see DMW000
WH 7508 see PIW000

WHATMAN CC-31 see CCU150
WHEAT GERM OIL see WBJ700
WHEAT GLUTEN see WBL100
WHEAT HUSK OIL see WBJ700
WHEY, DRY see WBL150
WHEY FACTOR see OJV500
WHEY, PROTEIN CONCENTRATE see WBL155
WHEY, REDUCED LACTOSE see WBL160
WHEY, REDUCED MINERALS see WBL165
WHICA BA see CAT775
WHIP see FAQ200
WHISKEY see WBS000
WHITCARB W see CAT775
1700 WHITE see TGG760
WHITE ACID (DOT) see ANG250
WHITE ANTHURIUM see SKX775
WHITE ARSENIC see ARI750
WHITE ASBESTOS see ARM268
WHITE ASBESTOS (chrysotile, actinolite, anthophyllite, tremolite) (DOT) see ARM268
WHITE CAMPHOR OIL see CBB500
WHITE CAUSTIC see SHS000
WHITE CAUSTIC, solution see SHS500
WHITE CEDAR see CDM325
WHITE CEDAR OIL see CCQ500
WHITE COPPERAS see ZNA000
WHITE CRYSTAL see SFT000
WHITE FUMING NITRIC ACID see NEF000
WHITE HELLEBORE see FAB100
WHITE HONEY FLOWER see GJU475
WHITE KERRIA see JDA075
WHITE LEAD see LCP000
WHITE LOCUST see GJU475
WHITE MERCURY PRECIPITATED see MCW500
WHITE MINERAL OIL see MQV750
WHITE MINERAL OIL see MQV875
WHITE OIL OF CAMPHOR see CBB500
WHITE OSTER see DYA875
WHITE PETROLATUM see PCR200
WHITE PHOSPHORUS see PHP010
WHITE POPINAC see LED500
WHITE-POWDER see CAT775
WHITE PRECIPITATE see MCW500
WHITE SEAL-7 see ZKA000
WHITESET see MCB050
WHITE SHELLAC see SCC700
WHITE SPIRIT see WBS675
WHITE SPIRIT, DILUTINE 5 see WBS700
WHITE SPIRITS see SLU500
WHITE STAR see AQF000
WHITE STREPTOCIDE see SNM500
WHITE STUFF see HBT500
WHITE TAR see NAJ500
WHITE VITRIOL see ZNA000
WHITE VITRIOL see ZNJ000
WHITEX BH see CMP100
WHITEX BK see CMP100
WHITEX LBH see CMP100
WHITING see CAT775
WHITON 450 see CAT775
WHORTLEBERRY RED see FAG020
W-53 HYDROCHLORIDE see DPJ400
WHYO TREE see GJU475
WICKENOL 101 see IQN000
WICKENOL 111 see IQW000
WICKENOL 116 see DNL800
WICKENOL 122 see BSL600
WICKENOL 127 see IRL100
WICKENOL 131 see IPS450
WICKENOL 141 see MSA300
WICKENOL 155 see OFG100
WICKENOL 156 see OFU300
WICKENOL 158 see AEO000
WICKENOL 303 see AHA000
WICKENOL 321 see AHA000
WICKENOL 323 see AHA000
WICKENOL 324 see AHA000
WICKENOL CPS 325 see AGZ998
WICKERBY BUSH see LEF100
WICKUP see LEF100

WICKY see MRU359
WICOPY see LEF100
WICRON see HGI100
WIDLON see PJY500
WIE OBEN see DJY200
WIJS' CHLORIDE see IDS000
WIKSTROEMIA INDICA (LINN.) C. A. MEY,
EXTRACT EXCLUDING ROOTS see WBS730
WIKSTROEMIA VIRIDIFLORA, EXTRACT
EXCLUDING ROOTS see WBS730
WILD ALLAMANDA (FLORIDA) see YAK300
WILD BALSAM APPLE see FPD100
WILD CALLA see WAT300
WILD COFFEE see CNG825
WILD CROCUS see PAM780
WILD GARLIC see WBS850
WILD GINGER OIL see SED500
WILD JALAP see MBU800
WILD LEMON see MBU800
WILD LICORICE see RMK250
WILD MAMEE see BAE325
WILD NIGHTSHADE see YAK300
WILD ONION see DAE100
WILD ONION see WBS850
WILD TAMARIND (HAWAII, PUERTO
RICO) see LED500
WILD TOBACCO see CCJ825
WILD UNCTION (BAHAMAS) see YAK300
WILD YAM BEAN see YAG000
WILELAIKI (HAWAII) see PCB300
WILKINITE see BAV750
WILLBUTAMIDE see BSQ000
WILLESTROL see DKB000
WILLNESTROL see DAL600
WILLOSETTEN see SJN700
WILPO see DTJ400
WILT PRUF see PKQ059
WILTZ-65 see NAS000
WILTZ-65 see NBW500
WIN 244 see CLD000
WIN 357 see BEL900
WIN 833 see HLK000
WIN 1258 see PJB750
WIN 1539 see KFK000
WIN 1593 see CPM750
WIN 1701 see DOS800
WIN 2022 see AMM750
WIN 2661 see WBS855
WIN 2663 see WBS860
WIN-2848 see DPJ200
WIN-2848 see TEO000
WIN 3074 see PPM550
WIN 3706 see SPB800
WIN 4369 see PBS000
WIN 4981 see BJA825
WIN 5037 see MFM100
WIN 5095 see AFY500
WIN 5162 see DMV600
WIN 5501 see AKT500
WIN 5512 see AKT250
WIN 5606 see BCA000
WIN 6703 see MGJ600
WIN 8077 see MSC100
WIN 11450 see SAN600
WIN-1161-3 see DWC100
WIN 12267 see DEM000
WIN 13,099 see PFA600
WIN 14833 see AOO401
WIN 16568 see DAP812
WIN 17,416 see HEF300
WIN 17625 see HOM259
WIN 17757 see DAB830
WIN 18,320 see EID000
WIN 18,441 see BIX250
WIN 18,446 see BIX250
WIN 20228 see DOQ400
WIN 20740 see COV500
WIN 22005 see UVS500
WIN 24450 see EBY600
WIN 24933 see LIM000
WIN 27262 see EGA600
WIN 27,262 see EGA600

WIN 32729 see EBH400
WIN 32729 see EBY100
WIN 32784 see BLV125
WIN 35833 see CMS216
WIN 39103 see MQR300
WIN 40350 see HJE100
WIN 40680 see AOD375
WIN 47203 see MQU600
WIN-5063-2 see MPN000
WIN 55212 see NAP545
WIN 5522-2 see CPB500
WIN 8308-3 see SEN500
WIN 8851-2 see SKO500
WIN 18320-3 see NAG450
WIN 18501-2 see ECW600
WINACET D see AAX250
WINCORAM see AOD375
WINDFLOWER see PAM780
WINDOWLEAF see SLE890
WINDOW PLANT see SLE890
WINE see WCA000
WINE ETHER see ENW000
WINE PALM see FBW100
WINE PLANT see RHZ600
WINE YEAST OIL see CNG920
WINGSTAY SN 1 see OPG050
WING STOP B see SGM500
WIN-2299 HYDROCHLORIDE see DHW400
WIN 2848 HYDROCHLORIDE SALT see
DPJ400
WINIDEN 60 see CGW300
WINIDUR see PKQ059
WIN-KINASE see UVS500
WINNOFIL S see CAT775
WINOBANIN see DAB830
WINSTROL see AOO401
WINSTROL V see AOO401
WINTER BLOOM see WCB000
WINTER CHERRY see JBS100
WINTER FERN see PJJ300
WINTERGREEN OIL (FCC) see MPI000
WINTERGREEN OIL, SYNTHETIC see
MPI000
WINTERMIN see CKP250
WINTERSTEINER'S COMPOUND F see
CNS800
WINTERSWEET see BOO700
WINTERWASH see DUS700
WINTOMYLON see EID000
WINYLU CHLOREK (POLISH) see VNP000
WINZER SOLUTION see BEL900
WIPEOUT see HGP525
WIRKSTOFF 37289 see EPY000
WIRNESIN see POB500
WISTARIA see WCA450
WISTERIA see WCA450
WISTERIA FLORIBUNDA see WCA450
WISTERIA SINENSIS see WCA450
WITAMINA PP see NCR000
WITAMOL 150 see PHW585
WITAMOL 320 see AEO000
WITCAMIDE 511C see OHU200
WITCARB see CAT775
WITCARB 940 see CBT500
WITCARB P see CAT775
WITCARB REGULAR see CAT775
WITCHES' THIMBLES see FOM100
WITCH HAZEL see WCB000
WITCIZER 200 see BSL600
WITCIZER 201 see BSL600
WITCIZER 300 see DEH200
WITCIZER 312 see DVL700
WITCO see CBT750
WITCO 31 see PJY100
WITCOBLAK NO. 100 see CBT750
WITCO G 339S see CAX350
WITCONATE see AFO250
WITCONATE 60L see TJK800
WITCONATE 60T see TJK800
WITCONATE 79S see TJK800
WITCONATE 5725 see TJK800
WITCONATE S-1280 see TJK800

WITCONATE P 10-59 see BBS275
WITCONATE TAB see TJK800
WITCONOL MS see OAV000
WITCONOL MST see OAV000
WITCO 1298 SULFONIC ACID see LBU100
WITEPSOL E-75 see WCB100
WITHAFERIN A see EDB300
WITHAFERINE A see EDB300
WITTOX C see NAS000
WJG 11 see PJS750
WL 19 see PDK500
WL 20 see NNL500
WL 20 see PDK300
WL 23 see MNP400
WL 28 see CJK000
WL 29 see MFG200
WL 31 see MNU100
WL 32 see MNU150
WL 33 see NNL500
WL 34 see CJK100
WL 39 see MNP500
WL 1650 see OAN000
WL-5792 see DGM600
WL 17731 see KBB700
WL 18236 see MDU600
WL 19805 see BLW750
WL 29761 see FDB500
WL 417-06 see DAB825
WL 43 425 see FBW135
WL 43467 see RLF350
WL 43479 see AHJ750
WL 43775 see FAR100
WL 63611 see COI050
WL 85871 see RCZ050
WL 105305 see BPT300
WL 115110 see FDB400
WM 100 see MCB050
WMO see ELB000
WN 12 see ISE000
WNF 15 see PJS750
WNM see DTG000
WNYESTRON see EDV000
WO 4 see ENP100
WOCHEM No. 320 see OHU000
WODE WHISTLE see PJJ300
WOFACAIN A see BQH250
WOFATOS see MNH000
WOFATOX see MNH000
WOFAVERDIN see CCK000
WOFOTOX see MNH000
WOGENAL SCARLET CB see CMO870
WOJTAB see PLZ000
WOLFEN see CNL500
WOLFRAM see TOA750
WOLFRAMITE see TOC750
WOLFSBANE see AQY500
WOLFSBANE see MRE275
WOLLASTOKUP see WCJ000
WOLLASTONITE see WCJ000
WONDER BULB see CNX800
WONDER TREE see CCP000
WONUK see ARQ725
WOOD ALCOHOL (DOT) see MGB150
WOOD DUST see WCJ100
WOOD ETHER see MJW500
WOOD LAUREL see MRU359
WOOD NAPHTHA see MGB150
WOOD OIL see GME000
WOOD SPIRIT see MGB150
WOOD VINE see YAK100
WOOL BLUE RL see ADE750
WOOL BORDEAUX 6RK see FAG020
WOOL BRILLIANT GREEN SF see FAF000
WOOL FAST BLUE R see ADE750
WOOL FAST ORANGE GA-CF see CMP882
WOOL FAT see LAU550
WOOL GREEN 5 see ADF000
WOOL GREEN B see ADF000
WOOL GREEN BS see ADF000
WOOL GREEN BSNA see ADF000
WOOL GREEN MS see ADF000
WOOL GREEN S see ADF000

XENON(II) PERCHLORATE see XES000
XENON TETRAFLUORIDE see XFA000
XENON(II) TRIFLUOROACETATE see XFJ000
XENON TRIOXIDE see XFS000
XENON (UN 0236) (DOT) see XDS000
XENON, refrigerated liquid (cryogenic liquids) (UN 2591) (DOT) see XDS000
XENYLAMIN (CZECH) see AJS100
XENYLAMINE see AJS100
1-XENYL-3,3-DIMETHYLTRIAZIN (CZECH) see BGL500
o-XENYL DIPHENYL PHOSPHATE see XFS300
XENYTROPIUM BROMIDE see PEM750
XERAC see BDS000
XERAL see GGS000
XERENE see MFD500
XEROSIN see XFS600
XF-408 see XGA000
XF-1015 see PJQ790
XF-13-563 see SCR400
XF 4175L see CBT500
m-XHQ see DSG700
XI-2287 see TCB600
XIBENOL HYDROCHLORIDE see XGA500
XILENOLI (ITALIAN) see XKA000
XILIDINE (ITALIAN) see XMA000
XILOBAM see XGA725
XILOCAINA (ITALIAN) see DHK400
XILOLI (ITALIAN) see XGS000
XIPAMID see CLF325
XIPAMIDE see CLF325
XITIX see ARN000
XK-62-2 see MQS579
XK-70-1 see FOK000
XL 7 see TFD250
XL-50 see PDP250
XL-90 see RLU000
XL 1246 see PJS750
XL 335-1 see PJS750
XL ALL INSECTICIDE see NDN000
XM 1116 see MCB050
XM 1130 see MCB050
3,5-XMC see DTN200
XMNA see NAC000
XN see XCS000
XNM 68 see PJS750
XO 440 see PJS750
XOE 2982 (RUSSIAN) see HBK700
XOLAMIN see FOS100
XORU-OX see CJY000
XP-470 see CJI000
XPA see ADV900
X-539-R see PJY100
XRD 473 see HCY600
XRM 3972 see DGJ100
XS-89 see SMM500
X465A SODIUM SALT see CDK500
XTI 5 see PJQ790
XU 292A see BKL800
XUMBRADIL see DNG400
XX 212 see DAD050
XYCAINE HYDROCHLORIDE see DHK600
XYCIANE see DHK400
XYDE see BGC625
XYDURIL see ARQ750
XYLAMIDE see BGC625
XYLAMIDE (gastroprotective agent) see BGC625
XYLAZINE (USDA) see DMW000
XYLENE see XGS000
m-XYLENE see XHA000
o-XYLENE see XHJ000
p-XYLENE see XHS000
1,2-XYLENE see XHJ000
1,3-XYLENE see XHA000
1,4-XYLENE see XHS000
XYLENE ACID MILLING GREEN BL see CMM200
XYLENE BLACK F see BMA000
XYLENE BLUE AS see ERG100

XYLENE BLUE VS see ADE500
XYLENE BLUE VSG see FMU059
m-XYLENE, 5-tert-BUTYL-2,4,6-TRINITRO- see TML750
m-XYLENE, 5-CHLORO- see CLV500
XYLENE, CHLORO(α-METHYLBENZYL)- see MRG060
m-XYLENE-α,α'-DIAMINE see XHS800
m-XYLENE DIISOCYANATE see XIJ000
XYLENE FAST ORANGE G see HGC000
XYLENE FAST YELLOW ES see SGP500
XYLENE FAST YELLOW GT see FAG140
XYLENE MILLING RED G see CMM325
XYLENE MILLING YELLOW SH see CMM759
XYLENE MUSK see TML750
XYLENEN (DUTCH) see XGS000
p-XYLENE, 2-NITRO- see NMS520
XYLENE RED see ADG125
XYLENESULFONIC ACID see XJA000
m-XYLENESULFONIC ACID see XJJ000
2,4-XYLENESULFONIC ACID see XJJ000
m-XYLENE-4-SULFONIC ACID see XJJ000
3,5-XYLENESULFONIC ACID, 4-HYDROXY- see XJJ005
XYLENESULFONIC ACID, SODIUM SALT see XJJ010
XYLENOL see XKA000
m-XYLENOL see XKJ500
o-XYLENOL see XKJ000
p-XYLENOL see XKS000
2,3-XYLENOL see XKJ000
2,4-XYLENOL see XKJ500
2,5-XYLENOL see XKS000
2,6-XYLENOL see XLA000
3,4-XYLENOL see XLJ000
3,5-XYLENOL see XLS000
1,2,5-XYLENOL see XKS000
1,3,4-XYLENOL see XLJ000
1,3,5-XYLENOL see XLS000
2,6-XYLENOL, 4-BROMO- see BOL303
XYLENOLEN (DUTCH) see XKA000
2,4-XYLENOL, 6,6'-ISOBUTYLIDENEDI- see IIU100
3,5-XYLENOL METHYLCARBAMATE see DTN200
XYLENOL, PHOSPHATE (3:1) see XLS100
2,6-XYLENOL, PHOSPHATE (3:1) (8CI) see TNR550
XYLENOLS (DOT) see XKA000
3,5-XYLENYL-N-METHYLCARBAMATE see DTN200
XYLESTESIN see DHK400
XYLESTESIN HYDROCHLORIDE see DHK600
m-XYLIDENE DIISOCYANATE see XIJ000
2,4-XYLIDENE (MAK) see XMS000
2,6-XYLIDIDE of 2-PYRIDONE-3-CARBOXYLIC ACID see XLS300
2,6-XYLIDID der 2-PYRIDON-3-CARBOXYLSAURE see XLS300
XYLIDINE see XMA000
m-XYLIDINE see XMS000
o-XYLIDINE see XMJ000
o-XYLIDINE see XNJ000
2,3-XYLIDINE see XMJ000
2,4-XYLIDINE see XMS000
2,5-XYLIDINE see XNA000
2,6-XYLIDINE see XNJ000
3,4-XYLIDINE see XNS000
3,5-XYLIDINE see XOA000
m-4-XYLIDINE see XMS000
p-XYLIDINE (DOT) see XNA000
m-XYLIDINE HYDROCHLORIDE see XOJ000
p-XYLIDINE HYDROCHLORIDE see XOS000
2,4-XYLIDINE HYDROCHLORIDE see XOJ000
2,5-XYLIDINE HYDROCHLORIDE see XOS000
XYLIDINEN (DUTCH) see XMA000

3,4-XYLIDINE, N-(2-OXAZOLIN-2-YL)- see XOS500
XYLIDINE PONCEAU see FMU070
3,4-XYLIDINO-2-OXAZOLINE see XOS500
XYLITE see XPJ000
XYLITE (SUGAR) see XPJ000
XYLITOL see XPJ000
XYLITON see XPJ000
l-XYLOASCORBIC ACID see ARN000
XYLOCAIN see DHK400
XYLOCAINE HYDROCHLORIDE see DHK600
XYLOCARD see DHK600
XYLOCHOLINE see XSS900
XYLOCHOLINE BROMIDE see XSS900
XYLOCITIN see DHK400
XYLOCITIN HYDROCHLORIDE see DHK600
XYLOFOP-ETHYL see QMA100
m-XYLOHYDROQUINONE see DSG700
2,6-XYLOHYDROQUINONE see DSG700
XYLOIDIN see CCU250
o-XYLOL see XHJ000
XYLOL (DOT) see XGS000
m-XYLOL (DOT) see XHA000
p-XYLOL (DOT) see XHS000
XYLOLE (GERMAN) see XGS000
5-(3,5-XYLOLOXYMETHYL)OXAZOLIDIN-2-ONE see XVS000
XYLOLTENIN see XPJ100
XYLOMETAZOLINE HYDROCHLORIDE see OKO500
XYLONEURAL see DHK600
XYLOPIC ACID see XPJ300
2,6-XYLOQUINOL see DSG700
p-XYLOQUINONE see XQJ000
(D)-XYLOSE see XQJ300
XYLOSE, D- see XQJ300
XYLOSTATIN see XQJ650
XYLOTOX see DHK400
XYLOTOX HYDROCHLORIDE see DHK600
p-XYLYL ACETATE see XQJ700
2,3-XYLYLAMINE see XMJ000
2,6-XYLYLAMINE see XNJ000
3,4-XYLYLAMINE see XNS000
3,5-XYLYLAMINE see XOA000
N-(2,3-XYLYL)-2-AMINOBENZOIC ACID see XQS000
N-(2,3-XYLYL)ANTHRANILIC ACID see XQS000
1-XYLYLAZO-2-NAPHTHOL see FAG080
1-XYLYLAZO-2-NAPHTHOL see XRA000
1-(o-XYLYLAZO)-2-NAPHTHOL see XRA000
1-(2,4-XYLYLAZO)-2-NAPHTHOL see XRA000
1-(2,5-XYLYLAZO)-2-NAPHTHOL see FAG080
1-XYLYLAZO-2-NAPHTHOL-3,6-DISULFONIC ACID, DISODIUM SALT see FMU070
1-XYLYLAZO-2-NAPHTHOL-3,6-DISULPHONIC ACID, DISODIUM SALT see FMU070
1-(2,4-XYLYLAZO)-2-NAPHTHOL-3,6-DISULPHONIC ACID, DISODIUM SALT see FMU070
XYLYL BROMIDE see XRS000
(((2,6-XYLYLCARBAMOYL)METHYL)IMINO)DI-ACETIC ACID see LFO300
p-XYLYL CHLORIDE see MHN300
m-XYLYLENDIAMIN (CZECH) see XHS800
p-XYLYLENDIAMINE (CZECH) see PEX250
m-XYLYLENDIISOKYANAT see XIJ000
XYLYLENDIISOKYANAT (CZECH) see XIJ000
1,1'-(p-XYLYLENE)BIS(3-(1-AZIRIDINYL)UREA) see XSJ000
p-XYLYLENEBIS(TRIPHENYLPHOSPHONIUM CHLORIDE) see XSS000
XYLYLENE CHLORIDE see XSS250

XYLYLENE DICHLORIDE see XSS250
m-XYLYLENE DICHLORIDE see DGP200
o-XYLYLENE DICHLORIDE see DGP400
p-XYLYLENE DICHLORIDE see DGP600
m-XYLYLENE DIISOCYANATE see XIJ000
p-XYLYLENE ISOCYANATE see XSS260
s,6-XYLYL ESTER of 1-PIPERIDINEACETIC
ACID HYDROCHLORIDE see XSS375
2,6-XYLYL ETHER BROMIDE see XSS900
N-2,3-XYLYLMALEIMIDE see XSS923
3,4-XYLYL METHYLCARBAMATE see
XTJ000
3,5-XYLYL-N-METHYLCARBAMATE see
DTN200
3,4-XYLYL-N-METHYLCARBAMATE,
nitrosated see XTS000
3,5-XYLYL-N-METHYLCARBAMATE,
NITROSATED see MMW750
p-XYLYLMETHYLCARBINAMINE
SULFATE see AQQ000
(2-(3,5-XYLYLOXY)ETHYL)HYDRAZINE
HYDROCHLORIDE see XVA000
(2-(3,4-XYLYLOXY)ETHYL)HYDRAZINE
MALEATE see XVJ000
5-((3,5-XYLYLOXY)METHYL)-2-
OXAZOLIDINONE see XVS000
2-(2,6-XYLYLOXY)TRIETHYLAMINE
HYDROCHLORIDE see XWS000
XYLYL PHOSPHATE see XLS100
2,6-XYLYL PHOSPHATE (7CI) see TNR550
((2,6-
XYLYL)SULFINYL)METHYLCARBAMIC
ACID-2,3-DIHYDRO-2,2-DIMETHYL-7-
BENZOFURANYL ESTER see XWS100
XYLZIN see DMW000
XYMOSTANOL see DKW000
Y 2 see CBM000
Y 3 see CKC000
Y-4 see FAG140
Y 40 see SCK600
Y 195 see PKL750
Y 218 see PKM250
Y 221 see PKM500
Y-223 see PKN000
Y-238 see CDV625
Y 299 see PKO750
Y 302 see PKP000
Y-4153 see DCV400
Y 4302 see TLD000
Y 5350 see CCK550
Y 5712 see SDX300
Y 6047 see CKA000
Y-7131 see EQN600
Y-8004 see PLX400
Y-9000 see ACU125
Y 9030 see IKG900
Y-9213 see IAY000
YADALAN see CLH750
YAGEINE see HAI500
YAJEINE see HAI500
YAJIAOTONG see GEQ100
YALAN see EKO500
YALTOX see CBS275
YAMADA FAST RED RL BASE see MMF780
YAMAFUL see CCK630
YAMA KARIYASU see TDD550
YAMAMOTO METHYLENE BLUE B see
BJI250
YAM BEAN see YAG000
YANOCK see FFF000
YARMOR see PIH750
YARMOR PINE OIL see PIH750
YASOKNOCK see SHG500
YATANSIN see YAG500
YATREN see IEP200
YATROCIN see NGE500
YATROZIDE see ILE000
YAUPON see HGF100
YAUTIA MALANGA (PUERTO RICO) see
EAI600
YAUTIA (PUERTO RICO) see XCS800
YAZUMYCIN A see RAG300

dl-YB2 see ICA000
YuB-25 see NBO525
YC 93 see PCG550
YE 5626 see SCQ000
YEAST (active) see YAK000
YEAST, EXT. see YAK050
YEAST EXTRACT see YAK050
YEDRA (CUBA) see AFK950
YEH-YAN-KU see ARW000
YELLON see IEP200
YELLOW see DCZ000
YELLOW 6Z see CCP950
YELLOW 204 see CMS240
1310 YELLOW see FAG140
1351 YELLOW see FAG150
1409 YELLOW see FAG140
1504 YELLOW see CMP600
1899 YELLOW see FAG150
11363 YELLOW see MDM775
11712 YELLOW see FEV000
11824 YELLOW see FEW000
12417 YELLOW see FEW000
YELLOW AAMX see CMS208
YELLOW AB see FAG130
YELLOW ACID see CMM758
YELLOW ALLAMANDA see AFQ625
YELLOW CROSS LIQUID see BIH250
YELLOW CUPROCIDE see CNO000
YELLOW FALSE JESSAMINE see YAK100
YELLOW FAST DYE 4K see CMP090
YELLOW FERRIC OXIDE see IHC450
YELLOW GINGER see ICC800
YELLOW GINSENG see BMA150
YELLOW GOWAN see FBS100
YELLOW G SOLUBLE in GREASE see
DOT300
YELLOW HENBANE see JBS100
YELLOW JESSAMINE see YAK100
YELLOW LAKE 69 see FAG140
YELLOW LEAD OCHER see LDN000
YELLOW LOCUST see GJU475
YELLOW MERCURIC OXIDE see MCT500
YELLOW MERCURY IODIDE see MDC750
YELLOW M SOLUBLE IN GREASE see
CMP600
YELLOW No. 2 see FAG130
YELLOW NIGHTSHADE see YAK300
YELLOW NO. 5 see FAG140
YELLOW NO. 6 see FAG150
YELLOW NO. 5 FDC see FAG140
YELLOW OB see FAG135
YELLOW OLEANDER see YAK350
YELLOW ORANGE S see FAG150
YELLOW ORANGE SPECIALLY PURE 85
see FAG150
YELLOW ORANGE S SPECIALLY PURE see
FAG150
YELLOW OXIDE of IRON see IHC450
YELLOW OXIDE of MERCURY see MCT500
YELLOW PARILLA (CANADA) see MRN100
YELLOW PETROLATUM see PCR200
YELLOW PHOSPHORUS see PHP010
1903 YELLOW PINK see BNH500
YELLOW POLAKTIN G see CMS226
YELLOW PRECIPITATE see MCT500
YELLOW PRUSSIATE of SODA see SHE350
YELLOW PUCCOON see ICC800
YELLOW PYOCTANINE see IBB000
YELLOW RELITON G see AAQ250
YELLOW RESIN see RNU100
YELLOW ROOT see ICC800
YELLOW SAGE see LAU600
YELLOW SARSAPARILLA see MRN100
YELLOW SF FOR FOOD see FAG150
YELLOW STABLE 4K see CMP090
YELLOW SUN see FAG150
YELLOW SY FOR FOOD see FAG150
YELLOW T see MRL100
YELLOW TONER YB5 see CMS208
YELLOW TRANSPARENT 2K see PIE800
YELLOW TRANSPARENT O see PIE830

YELLOW TRANSPARENT PIGMENT K see
PIE800
YELLOW TRANSPARENT PIGMENT O see
PIE830
YELLOW ULTRAMARINE see CAP500
YELLOW Z see AAQ250
YERBA HEDIONDA (CUBA) see CNG825
YERBA de PORDIOSEROS (CUBA) see
CMV390
YERMONIL see EEH575
YESDOL see CCR875
YESPAZINE see TJW500
YESQUIN (HAITI) see FPD100
YETRIUM see AFJ400
YEW see YAK500
Y-GUTTIFERIN see GME300
YH 9030 see IKG900
Y-3642-HCl see AJU000
Y-3642-HCl see TGE165
Y-3642 HYDROCHLORIDE see TGE165
YL-704 A3 see JDS200
YLANG YLANG OIL see YAT000
YL 704 B$_1$ see MBY150
YL 704 B3 see LEV025
YM 018 see MQU600
YM617 see EFA200
YM-08054 see IBW500
YM-08310 see AMD000
YM-09229 see FAP100
YM 09330 see CCS371
YM 09330 see CCS373
YM-09538 see AOA075
YM-11170 see FAB500
(−)-YM12617 see EFA200
YM-31636 see IAT275
YM-08054-1 see IBW500
YOCLO see ARQ750
YODOCHROME FAST ORANGE GR see
CMP882
YODOCHROME METANIL YELLOW see
MDM775
YODOCHROME ORANGE A see NEY000
YODOCHROME VIOLET B see HLI000
YODOMIN see TEH500
YODOXIN see DNF600
YOHIMBAN-16-CARBOXYLIC ACID
derivative of BENZ(g)INDOLO(2,3-
a)QUINOLIZINE see RDK000
3-β,20-α-YOHIMBAN-16-β-CARBOXYLIC
ACID, 18-β-HYDROXY-10,17-α-
DIMETHOXY-, METHYL ESTER, 3,4,5-
TRIMETHOXYBENZOATE (ester) see
MEK700
3-β,20-α-YOHIMBAN-16-β-CARBOXYLIC
ACID, 18-β-HYDROXY-11,17-α-
DIMETHOXY-METHYL ESTER, 3,4,5-
TRIMETHOXYBENZOATE (ESTER),
TARTRATE see SCA550
YOHIMBAN-16-CARBOXYLIC ACID, 17-
HYDROXY-, METHYL ESTER,
MONOHYDROCHLORIDE, (16α,17β)- see
YBS500
YOHIMBIC ACID METHYL ESTER see
YBJ000
YOHIMBINE see YBJ000
Δ-YOHIMBINE see AFG750
YOHIMBINE HYDROCHLORIDE see
YBS000
β-YOHIMBINE HYDROCHLORIDE see
YBS500
γ-YOHIMBINE HYDROCHLORIDE see
AFH000
YOHIMBINE MONOHYDROCHLORIDE see
YBS000
α-YOHIMBIN HYDROCHLORIDE see
YCA000
β-YOHIMBIN HYDROCHLORIDE see
YBS500
YO-KIN see ICC800
YOMESAN see DFV400
YORACRYL YELLOW 4G see CMM890
YORACRYL YELLOW RL see CMM895

YORK WHITE see CAT775
YOSHI 864 see YCJ000
YOSHINOX 425 see MJN250
YOSHINOX DSTDP see DXG700
YOSHINOX S see TFC600
YOSIMILON see YCJ200
YOUNG FUSTIC see FBW000
YOUNG FUSTIC CRYSTALS see FBW000
YPERITE see BIH250
YPERITE SULFONE see BIH500
YRC 2894 see CKW410
Y-5712 SILYL PEROXIDE see SDX300
YTTERBIUM see YDA000
YTTERBIUM CHLORIDE see YDJ000
YTTERBIUM NITRATE see YDS800
YTTERBIUM(III) NITRATE,
HEXAHYDRATE (1:3:6) see YEA000
YTTERBIUM TRICHLORIDE see YDJ000
YTTRIA see YGA000
YTTRIUM see YEJ000
YTTRIUM-89 see YEJ000
YTTRIUM CHLORIDE see YES000
YTTRIUM CITRATE see YFA000
YTTRIUM EDETATE complex see YFA100
YTTRIUMNITRAT (GERMAN) see YFS000
YTTRIUM(III) NITRATE (1:3) see YFJ000
YTTRIUM(III) NITRATE HEXAHYDRATE
(1:3:6) see YFS000
YTTRIUM OXIDE see YGA000
YTTRIUM TRICHLORIDE see YES000
YU 7802 see BDN125
YUBAN 10S see UTU500
YUBAN 10HV see UTU500
YUCA see CCO680
YUCA BRAVA see CCO680
YUDOCHROME YELLOW GGN see SIT850
YUGILLA (CUBA) see CNH789
YUGOVINYL see PKQ059
YUKALON EH 30 see PJS750
YUKALON HE 60 see PJS750
YUKALON K 3212 see PJS750
YUKALON LK 30 see PJS750
YUKALON MS 30 see PJS750
YUKALON PS 30 see PJS750
YUKALON YK 30 see PJS750
YUKAMATE see MNU300
YUKAWIDE see DGM730
YULAN see EKO500
YUNIHOMU AZ see ASM270
YUROTIN A see ERC600
YUTAC see YGA700
YUTOPAR see RLK700
YV 100AN see BAP502
Z-4 see DOX100
Z 6 see PKV100
Z 11 see SIA000
Z-78 see EIR000
Z 88 see PJC750
Z 102 see BEQ500
Z 103 see ZEJ100
Z-134 see BRE255
Z-905 see PIH100
Z 4828 see TNT500
Z 4942 see IMH000
Z 6030 see TNJ500
Z 6076 see TLC300
Z 7557 see ARP625
ZABILA (MEXICO, DOMINICAN
REPUBLIC) see AGV875
ZACLONDISCOIDS see HHS000
ZACTANE CITRATE see MIE600
ZACTIRIN COMPOUND see ABG750
ZACTRAN see DOS000
ZADINE see DLV800
ZADITEN see KGK200
ZADONAL see EOK000
Z-(3-AETHYL-4-OXO-5-PIPERIDINO-
THIAZOLIDIN-2-YLIDEN)-ESSIGSAURE-
AETHYLESTER (GERMAN) see EOA500
ZAFFRE see CND125
ZAGREB see BIE500
ZAHARINA see BCE500

ZAHLREICHE BEZEICHNUNGEN
(GERMAN) see DUS700
ZAMBESI BLACK D see CMN230
ZAMBESI BLACK DA-CF see CMN230
ZAMBESI DARK BLUE BH see CMN800
ZAMBESIL see CLY600
ZAMI 905 see PIH000
ZAMI 1305 see BQF750
dl-ZAMI 1305 see BQF750
ZAMIA DEBILIS see ZAK000
ZAMIA PUMILA see CNH789
ZAMINE see BBK500
ZANIL see DMZ000
ZANILOX see DMZ000
ZANOSAR see SMD000
ZANTAC see RBF400
ZANTEDESCHIA AETHIOPICA see CAY800
ZANTE FUSTIC see FBW000
ZAPRAWA NASIENNA PLYNNA see
MLF250
ZAPRINAST see ZAK200
ZARAONDAN see ENG500
ZARCILLA (PUERTO RICO) see LED500
ZARDA (INDIA) see SED400
ZARDANCHU see RQU300
ZARDEX see HCP500
ZARODAN see ENG500
ZARONDAN-SAFT see ENG500
ZARONTIN see ENG500
ZAROXOLYN see ZAK300
ZARTALIN see ENG500
ZASSOL see COI250
ZAVILA (PUERTO RICO) see AGV875
Z-9-DDA see GJU050
Z-(±)-2,6-DIMETHYL-α-PHENYL-α-(2-
PYRIDYL)-1-PIPERIDINEBUTANOL
HYDROCHLORIDE see PJA170
ZEAPUR see BJP000
ZEARALANOL see RBF100
ZEARALENONE see ZAT000
(−)-ZEARALENONE see ZAT000
(s)-ZEARALENONE see ZAT000
(10s)-ZEARALENONE see ZAT000
trans-ZEARALENONE see ZAT000
ZEARANOL see RBF100
ZEAZIN see ARQ725
ZEAZINE see ARQ725
ZEBENIDE see EIR000
ZEBTOX see EIR000
ZECTANE see DOS000
ZECTRAN see DOS000
ZEDOARONDIOL see ZAT050
ZEFRAN, combustion products see ADX750
ZEIDANE see DAD200
ZEIN see ZAT100
ZELAN see CIR250
ZELAZA TLENKI (POLISH) see IHF000
ZELDOX see CJU275
ZELEN ALIZARINOVA BRILANTNI G-
EXTRA (CZECH) see APL500
ZELEN KYPOVA 3 see CMU810
ZELEN KYSELA 3 see FAE950
ZELEN KYSELA 40 see CMM200
ZELEN KYSELA 50 see ADF000
ZELEN KYSELA BS see ADF000
ZELEN MIDLONOVA BLS (CZECH) see
APM250
ZELEN MORIDLOVA 4 see NLB000
ZELEN OLIVOVA OSTANTHRENOVA B
see CMU810
ZELEN OSTANTHRENOVA BRILANTNI
FFB (CZECH) see JAT000
ZELEN POTRAVINARSKA 1 see FAE950
ZELEN POTRAVINARSKA 4 see ADF000
ZELIO see TEL750
ZELLEK see HAG850
ZEMERIN see HBT100
ZENADRID (VETERINARY) see PLZ000
ZENALOSYN see PAN100
ZENITE see BHA750
ZENITE SPECIAL see BHA750
ZENTAL see VAD000

ZENTINIC see PAG200
ZENTRONAL see DKQ000
ZENTROPIL see DKQ000
ZENTROPIL see DNU000
ZEOGEL see PAE750
ZEOLITE PARTICLES see ZAT300
ZEOLITES see ZAT300
ZEOTIN see MBX800
ZEPC see EOK550
ZEPELIN see PEW000
ZEPHIRAN CHLORIDE see AFP250
ZEPHIRAN CHLORIDE see BBA500
ZEPHIROL see BBA500
ZEPHROL see EAW000
ZEPHYR see ARW200
ZEPHYRANTHES ATAMASCO see RBA500
ZEPHYR LILY see RBA500
ZERANOL (USDA) see RBF100
ZERDANE see DAD200
ZERIT see SLJ800
ZERLATE see BJK500
ZERTELL see CMA250
ZESET T see VLU250
ZEST see MRL500
ZESTE see ECU750
ZESTRIL see LGM400
ZET see ZBA000
ZETAR see CMY800
ZETAR EMULSION see ZBA000
ZETAX see BHA750
ZETIFEX ZN see TNC500
ZETTYN CHLORIDE see BEL900
ZEXTRAN see DOS000
ZF 36 see PJS750
Z 6 (Flotation agent) see PKV100
ZG 301 see CAT775
Z-(GLN^1)-DES-HIS^2)-d-MET^6)-DES-GLY^{10}))-
LHRH ETHYLAMIDE HCL H2-O (1:1:2) see
PFP700
Z-(GLN^1)-DES-HIS^2)-d-PRO^6)-DES-GLY^{10}))-
LHRH ETHYLAMIDE ACETATE HYDRATE
see PFP720
Z-(GLN^1)-DES-HIS^2)-d-PSE^6)-DES-GLY^{10}))-
LHRH ETHYLAMIDE ACETATE 4H2-O see
PFP740
Z-(GLN^1)-DES-HIS^2)-d-PSE^6)-DES-GLY^{10}))-
LHRH ETHYLAMIDE HCL H2-O see PFP730
Z-(GLN(NH-BZL)1-DES-HIS2)-LHRH
HYDROCHLORIDE HYDRATE (2:3:6) see
LIU321
Z-(GLN(NH-BZL)1-DES-HIS2)-
LUTEINIZING HORMONE-RELEASING
HORMONE HYDROCHLORIDE H2-O
(2:3:6) see LIU321
Z-GLY see CBR125
ZHENGGUANGMYCIN A2 (CHINESE) see
BLY250
ZHENGGUANGMYCIN A5 (CHINESE) see
BLY500
Z-3-HEXENYL VALERATE see HFE800
3ZhP see IGK800
ZIARNIK see ABU500
ZIAVETINE see BQL000
ZIDAN see EIR000
ZIDE see CFY000
ZIDOVUDINE see ASE900
ZIELEN KWASOWA CZYSTA V see ANG135
ZIGADENUS (VARIOUS SPECIES) see
DAE100
ZIKHOM see ZJS300
ZILDASAC see BAV325
ZIMALLOY see CNA750
ZIMANAT see DXI400
ZIMANEB see DXI400
ZIMATE see BJK500
ZIMATE see EIR000
ZIMATE METHYL see BJK500
ZIMCO see VFK000
ZIMELIDINE see ZBA500
(Z)-ZIMELIDINE see ZBA500
cis-ZIMELIDINE see ZBA500

ZIMELIDINE DIHYDROCHLORIDE see ZBA525
ZIMELIDINE HYDROCHLORIDE see ZBA525
ZIMET 3106 see IAK100
ZIMMAN-DITHANE see DXI400
ZIMTALDEHYDE see CMP969
ZIMTSAEURE (GERMAN) see CMP975
ZINACEF see CCS600
ZINADON see ILD000
ZINC see ZBJ000
ZINC ACETATE see ZBS000
ZINC ACETATE, DIHYDRATE see ZCA000
ZINC ACETATE plus MANEB (1:50) see EIQ500
ZINC ACETOACETONATE see ZCJ000
ZINC ACID PHOSPHATE see ZJS400
ZINC, (N-β-ALANYL-l-HISTIDINATO(2-)-(N,N N),O⁰)- see ZEJ100
ZINC ALLYL DITHIO CARBAMATE see ZCS000
ZINC AMMONIUM NITRITE see ZDA000
ZINC ARSENATE see ZDJ000
ZINC ARSENATE, BASIC see ZDJ000
ZINC ARSENATE FLUORIDE see ZDJ100
ZINC-m-ARSENITE see ZDS000
ZINC ARSENITE, solid (DOT) see ZDS000
ZINC ASHES (UN 1435) (DOT) see ZBJ000
ZINCATE(2-), ((ETHYLENEDINITRILO)TETRAACETATO)- see ZGW200
ZINCATE(2-), ((N,N'-1,2-ETHANEDIYLBIS(N-(CARBOXYMETHYL)GLYCINATO)) (4-)-N,N'O,O',ON,ON')-, (OC-6-21)- see ZGW200
ZINC BENZENESULFINATE see ZDS500
ZINC BENZIMIDAZOLE-2-THIOLATE see ZIS000
ZINC-2-BENZOTHIAZOLETHIOLATE see BHA750
ZINC BENZOTHIAZOLYL MERCAPTIDE see BHA750
ZINC BENZOTHIAZOL-2-YLTHIOLATE see BHA750
ZINC BENZOTHIAZYL-2-MERCAPTIDE see BHA750
ZINC BERYLLIUM SILICATE see BFV250
ZINC-BIBUTYLDITHIOCARBAMATE see BIX000
ZINC BICHROMATE see ZFA102
ZINC BIS(BENZENESULFINATE) see ZDS500
ZINC BIS(1H-BENZIMIDAZOLE-2-THIOLATE) see ZIS000
ZINC, BIS(DIBUTYL DITHIOPHOSPHATE) see ZGA500
ZINC, BIS(o,o-DIBUTYL PHOSPHORODITHIOATO-S,S')-, (T-4)- see ZGA500
ZINC BIS(DIMETHYLDITHIOCARBAMATE) see BJK500
ZINC BIS(DIMETHYLDITHIOCARBAMATE)CYCLOHEXYLAMINE COMPLEX see ZEA000
ZINC BIS(DIMETHYLDITHIOCARBAMOYL)DISULPHIDE see BJK500
ZINC,((BIS(d-GLUCONATO-O¹),O²))- (9CI) see ZIA750
ZINC BUTTER see ZFA000
ZINC CAPRYLATE see ZEJ000
ZINC CARBONATE (1:1) see ZEJ050
ZINC l-CARNOSINE see ZEJ100
ZINC CHLORATE see ZES000
ZINC CHLORIDE see ZFA000
ZINC CHLORIDE (ACGIH,OSHA) see ZFA000
ZINC CHLORIDE, anhydrous (UN 2331) (DOT) see ZFA000
ZINC CHLORIDE, solution (UN 1840) (DOT) see ZFA000

ZINC (CHLORURE de) (FRENCH) see ZFA000
ZINC CHROMATE see ZFA100
ZINC CHROMATE see ZFA102
ZINC CHROMATE see ZFJ100
ZINC CHROMATE HYDROXIDE see CMK500
ZINC CHROMATE(VI) HYDROXIDE see CMK500
ZINC CHROMATE(VI) HYDROXIDE see ZFJ100
ZINC CHROMATE, POTASSIUM DICHROMATE, and ZINC HYDROXIDE (3:1:1) see ZFJ122
ZINC CHROMATE with ZINC HYDROXIDE and CHROMIUM OXIDE (9:1) see ZFJ125
ZINC CHROME see PLW500
ZINC CHROME YELLOW see ZFJ100
ZINC CHROMITE see ZFA100
ZINC CHROMITE see ZFJ130
ZINC CHROMIUM OXIDE see ZFA100
ZINC CHROMIUM OXIDE see ZFA102
ZINC CHROMIUM OXIDE see ZFJ100
ZINC CITRATE see ZFJ250
ZINC, COMPLEX WITH 1,10-PHENANTHROLINYLENE see ZLJ100
ZINC COMPOUNDS see ZFS000
ZINC-COPPER CHROMATE COMPLEX see CNQ750
ZINC CYANIDE see ZGA000
ZINC DIACETATE see ZBS000
ZINC DIACETATE, DIHYDRATE see ZCA000
ZINC-DIBUTYLDITHIOCARBAMATE see BIX000
ZINC-N,N-DIBUTYLDITHIOCARBAMATE see BIX000
ZINC, DIBUTYLDITHIOPHOSPHATE see ZGA500
ZINC o,o-DIBUTYL DITHIOPHOSPHATE see ZGA500
ZINC DIBUTYL PHOSPHORODITHIOATE see ZGA500
ZINC o,o-DIBUTYL PHOSPHORODITHIOATE see ZGA500
ZINC DICHLORIDE see ZFA000
ZINC DICHROMATE see ZFA102
ZINC DICHROMATE (VI) see ZFA102
ZINC DICYANIDE see ZGA000
ZINC DIETHYLDITHIOCARBAMATE see BJC000
ZINC-N,N-DIETHYLDITHIOCARBAMATE see BJC000
ZINC DIHYDRAZIDE see ZGJ000
ZINC DIHYDROGEN PHOSPHATE DIHYDRATE see ZGJ050
ZINC DIMETHYLDITHIOCARBAMATE see BJK500
ZINC N,N-DIMETHYLDITHIOCARBAMATE see BJK500
ZINC, DIMETHYLDITHIOCARBAMATE CYCLOHEXYLAMINE COMPLEX see ZEA000
ZINC DISTEARATE see ZMS000
ZINC DITHIONITE see ZGJ100
ZINC DITHIONITE see ZIJ100
ZINC DITHIONITE (DOT) see ZIJ100
ZINC DITHIOPHOSPHATE see ZGS000
ZINC DUST see ZBJ000
ZINC DUST (DOT) see ZBJ000
ZINC-EDTA see ZGW200
ZINC-EDTA COMPLEX see ZGW200
ZINC(II) EDTA COMPLEX see ZGS100
ZINC ETHIDE see DKE600
ZINC ETHOXIDE see ZGW100
ZINC ETHYL (DOT) see DKE600
ZINC ETHYLENEBISDITHIOCARBAMATE see EIR000
ZINC ETHYLENE-1,2-BISDITHIOCARBAMATE see EIR000

ZINC ETHYLENEDIAMINETETRAACETATE see ZGW200
ZINC ETHYLPHENYLDITHIOCARBAMATE see ZHA000
ZINC ETHYLPHENYLTHIOCARBAMATE see ZHA000
ZINC-N-FLUOREN-2-YLACETOHYDROXAMATE see ZHJ000
ZINC FLUORIDE see ZHS000
ZINC FLUOROSILICATE see ZIA000
ZINC FLUOROSILICATE (DOT) see ZIA000
ZINC FLUORURE (FRENCH) see ZHS000
ZINC FLUOSILICATE see ZIA000
ZINC GLUCONATE see ZIA750
ZINC HEXAFLUOROSILICATE see ZIA000
ZINC HYDRIDE see ZIJ000
ZINC HYDROSULFITE see ZGJ100
ZINC HYDROSULFITE see ZIJ100
ZINC HYDROSULFITE see ZIJ100
ZINC HYDROSULFITE (DOT) see ZGJ100
ZINC p-HYDROXYBENZENESULFONATE see ZIJ300
ZINC HYDROXYCHROMATE see CMK500
ZINC HYDROXYCHROMATE see ZFJ100
ZINC INSULIN see LEK000
ZINCITE see ZKA000
ZINC MANGANESE BERYLLIUM SILICATE see BFS750
ZINCMATE see BJK500
ZINC MERCAPTOBENZIMIDAZOLATE see ZIS000
ZINC MERCAPTOBENZIMIDAZOLE see ZIS000
ZINC MERCAPTOBENZOTHIAZOLATE see BHA750
ZINC-2-MERCAPTOBENZOTHIAZOLE see BHA750
ZINC MERCAPTOBENZOTHIAZOLE SALT see BHA750
ZINC MERCURY CHROMATE COMPLEX see ZJA000
ZINC METAARSENITE see ZDS000
ZINC METHARSENITE see ZDS000
ZINC METHIONINE SULFATE see ZJA100
ZINC METIRAM see MQQ250
ZINC MONOSULFIDE see ZNJ100
ZINC MURIATE, solution (DOT) see ZFA000
ZINC NAPHTHENATE see NAT000
ZINC NITRATE see ZJJ000
ZINC(II) NITRATE, HEXAHYDRATE (1:2:6) see NEF500
ZINC NITRILOTRIMETHYLPHOSPHONIC ACID TRISODIUM TETRAHYDRATE see ZJJ200
ZINC NITROSYLPENTACYANOFERRATE see ZJJ400
ZINCO (CLORURO di) (ITALIAN) see ZFA000
ZINC OCTADECANOATE see ZMS000
ZINCO (FOSFURO di) (ITALIAN) see ZLS000
ZINCOID see ZKA000
ZINC OLEATE (1:2) see ZJS000
ZINC OMADINE see ZMJ000
ZINCOP see ZJS300
ZINC ORTHOPHOSPHATE see ZJS400
ZINC OXIDE see ZKA000
ZINC OXIDE (ointment) see ZKJ000
ZINC OXIDE FUME (MAK) see ZKA000
ZINC PANTOTHENATE see ZKS000
ZINC PENTACYANONITROSYLFERRATE.(2−) see ZJJ400
ZINC 2,4-PENTANEDIONATE see ZCJ000
ZINC PERCHLORATE HEXAHYDRATE see ZKS100
ZINC PERMANGANATE see ZLA000
ZINC PEROXIDE see ZLJ000
ZINC-o-PHENANTHROLINE COMPLEX see ZLJ100

ZINC-1,10-PHENANTHROLINE COMPLEX see ZLJ100
ZINC(2+), (1,10-PHENANTHROLINE)-, ION see ZLJ100
ZINC(2+), (1,10-PHENANTHROLINE-N1,N10)- see ZLJ100
ZINC PHENOLSULFONATE see ZIJ300
ZINC p-PHENOL SULFONATE see ZIJ300
ZINC PHOSPHATE (3:2) see ZJS400
ZINC PHOSPHIDE see ZLS000
ZINC PHOSPHITE see ZLS200
ZINC (PHOSPHURE de) (FRENCH) see ZLS000
ZINC POLYACRYLATE see ADW100
ZINCPOLYANEMINE see ZMJ000
ZINC POLYCARBOXYLATE see ADW100
ZINC POTASSIUM CHROMATE see CMK400
ZINC POWDER see ZBJ000
ZINC POWDER (DOT) see ZBJ000
ZINC (N,N'-PROPYLENE-1,2-BIS(DITHIOCARBAMATE)) see ZMA000
ZINC PROTAMINE INSULIN see IDF325
ZINC PT see ZMJ000
ZINC PYRIDINE-2-THIOL-1-OXIDE see ZMJ000
ZINC PYRIDINETHIONE see ZMJ000
ZINC PYRION see ZMJ000
ZINC PYRITHIONE see ZMJ000
ZINC RESINATE see ZMJ100
ZINC STEARATE see ZMS000
ZINC SULFATE see ZNA000
ZINC SULFATE see ZNJ000
ZINC SULFATE (1:1) HEPTAHYDRATE see ZNJ000
ZINC SULFATE HEPTAHYDRATE (1:1:7) see ZNJ000
ZINC SULFIDE see ZNJ100
ZINC SULFOCARBOLATE see ZIJ300
ZINC SULFOPHENATE see ZIJ300
ZINC SULPHATE see ZNA000
ZINC SULPHIDE see ZNJ100
ZINC SUPEROXIDE see ZLJ000
ZINC TETRAOXYCHROMATE 76A see ZFJ100
ZINC-TOX see ZLS000
ZINC TRIFLUOROSTANNITE, HEPTAHYDRATE see ZNS000
ZINC TRISODIUM DIETHYLENETRIAMINEPENTAACETATE see TNM850
ZINC TRISODIUM DTPA see TNM850
ZINC UVERSOL see NAT000
ZINC VITRIOL see ZNA000
ZINC VITRIOL see ZNJ000
ZINC WHITE see ZKA000
ZINC YELLOW see CMK500
ZINC YELLOW see PLW500
ZINC YELLOW see ZFJ100
ZINC YELLOW see ZFJ125
ZINEB see EIR000
ZINEB-COPPER OXYCHLORIDE mixture see ZJS300
ZINEB-ETHYLENE THIURAM DISULFIDE ADDUCT see MQQ250
ZINGERONE see VFP100
ZINGIBERONE see VFP100
ZINGIBER ROSEUM (Roxb.) Rosc., extract see ZNS200
ZINK-(N,N'-AETHYLEN-BIS(DITHIOCARBAMAT)) (GERMAN) see EIR000
ZINK-BIS(N,N-DIMETHYL-DITHIOCARBAMAAT) (DUTCH) see BJK500
ZINK-BIS(N,N-DIMETHYL-DITHIOCARBAMAT) (GERMAN) see BJK500
ZINKCARBAMATE see BJK500
ZINKCHLORID (GERMAN) see ZFA000
ZINKCHLORIDE (DUTCH) see ZFA000
ZINK-(N,N-DIMETHYL-DITHIOCARBAMAT) (GERMAN) see BJK500
ZINKFOSFIDE (DUTCH) see ZLS000
ZINKOSITE see ZNA000

ZINKPHOSPHID (GERMAN) see ZLS000
ZINK-(N,N'-PROPYLEN-1,2-BIS(DITHIOCARBAMAT)) (GERMAN) see ZMA000
ZINN (GERMAN) see TGB250
ZINNTETRACHLORID (GERMAN) see TGC250
ZINOCHLOR see DEV800
ZINOPHOS see EPC500
ZINOSAN see EIR000
ZINOSTATIN see NBV500
ZINPOL see ADV900
ZINPOL see PJS750
(+−)-ZINTEROL HYDROCHLORIDE see ZNS300
ZINTEVIR see ZNS400
ZIPAN see DCK759
ZIPEPROL see MFG250
ZIPEPROL DIHYDROCHLORIDE see MFG250
ZIPROMAT see ZMA000
ZIRAM see BJK500
ZIRAM CYCLOHEXYLAMINE COMPLEX see ZEA000
ZIRAMVIS see BJK500
ZIRASAN see BJK500
ZIRBERK see BJK500
ZIRCAT see ZOA000
ZIRCON see ZSS000
ZIRCONATE, BARIUM (1:1) see BAP750
ZIRCONATE(2-), OXODISULFATO-, DISODIUM (8CI) see ZTS100
ZIRCONIUM see ZOA000
ZIRCONIUM (ACGIH,OSHA) see ZOA000
ZIRCONIUM CHLORIDE see ZPA000
ZIRCONIUM(II) CHLORIDE see ZQJ000
ZIRCONIUM(IV) CHLORIDE (1:4) see ZPA000
ZIRCONIUM CHLORIDE HYDROXIDE see ZPJ000
ZIRCONIUM CHLORIDE OXIDE OCTAHYDRATE see ZPS000
ZIRCONIUM CHLOROHYDRATE see ZPJ000
ZIRCONIUM COMPOUNDS see ZQA000
ZIRCONIUM DIBROMIDE see ZQB100
ZIRCONIUM DICARBIDE see ZQC200
ZIRCONIUM DICHLORIDE see ZQJ000
ZIRCONIUM, DICHLORO-DI-pi-CYCLOPENTADIENYL- see ZTK400
ZIRCONIUM FLUORIDE see ZQS000
ZIRCONIUM GLUCONATE see ZQS100
ZIRCONIUM HYDRIDE see ZRA000
ZIRCONIUM HYDROXYCHLORIDE see ZPJ000
ZIRCONIUM(IV) LACTATE see ZRS000
ZIRCONIUM(III) LACTATE (1:3) see ZRJ000
ZIRCONIUM METAL, dry, chemically produced, finer than 20 mesh particle size (UN 2008) see ZOA000
ZIRCONIUM NITRATE see ZSA000
ZIRCONIUM OXYCHLORIDE see ZSJ000
ZIRCONIUM PICRAMATE, dry or wetted with <20% water, by weight (UN 0236) (DOT) see PIC750
ZIRCONIUM PICRAMATE, wetted with not <20% water, by weight (UN 1517) (DOT) see PIC750
ZIRCONIUM POTASSIUM FLUORIDE see PLG500
ZIRCONIUM POWDER, dry (UN 2008) (DOT) see ZOA000
ZIRCONIUM POWDER, wetted with not <25% water (UN 1358) (DOT) see ZOA000
ZIRCONIUM SCRAP (UN 1932) (DOT) see ZOA000
ZIRCONIUM(IV) SILICATE (1:1) see ZSS000
ZIRCONIUM SODIUM LACTATE see LAO000
ZIRCONIUM SODIUM LACTATE see ZTA000
ZIRCONIUM(IV) SULFATE (1:2) see ZTJ000

ZIRCONIUM TETRACHLORIDE (DOT) see ZPA000
ZIRCONIUM TETRACHLORIDE, solid (DOT) see ZPA000
ZIRCONIUM TETRAFLUORIDE see ZQS000
ZIRCONIUM(IV) TETRAHYDROBORATE see ZTK300
ZIRCONIUM, dry, coiled wire, finished metal sheets, strip or coiled wire (UN 2858) (DOT) see ZOA000
ZIRCONIUM, dry, finished sheets, strip or coiled wire (UN 2009) (DOT) see ZOA000
ZIRCONOCENE, DICHLORIDE see ZTK400
ZIRCONYL ACETATE see ZTS000
ZIRCONYL CHLORIDE see ZSJ000
ZIRCONYL CHLORIDE OCTAHYDRATE see ZPS000
ZIRCONYL HYDROXYCHLORIDE see ZPJ000
ZIRCONYL NITRATE see BLA000
ZIRCONYL SODIUM SULPHATE see ZTS100
ZIRCONYL SULFATE see ZTJ000
ZIREX 90 see BJK500
ZIRIDE see BJK500
ZIRPON see MQU750
ZIRTHANE see BJK500
ZITEX H 662-124 see TAI250
ZITHIOL see MAK700
ZITOSTOP see MAW800
ZITOX see BJK500
ZITOXIL see MFG250
ZITRONELL OEL (GERMAN) see CMT000
ZITRONEN OEL (GERMAN) see LEI000
ZIZALON see ZTS200
ZK 26173 see DPB200
ZK 28943 see MKG800
ZK 31224 see DLR100
ZK 57671 see SOU650
ZK 112119 see IOF075
ZK 136295 see CQH275
ZK-Nr.26173 see DPB200
Z-(LEU¹)-DES-HIS²))-LHRH HYDROCHLORIDE HYDRATE (2:3:6) see LIU323
Z-(LEU¹)-DES-HIS²))-LUTEINIZING HORMONE-RELEASING HORMONE HCL H₂O (2:3:6) see LIU323
ZLUT CHINOLONOVA see CMM510
ZLUT DISPERZNI 3 see AAQ250
ZLUT KYPOVA 2 see VGP200
ZLUT KYSELA 3 see CMM510
ZLUT KYSELA 9 see CMM758
ZLUT KYSELA 23 see FAG140
ZLUT MARCIOVA see DUX800
ZLUT MASELNA (CZECH) see DOT300
ZLUT MASELNA OB see FAG135
ZLUT NAFTOLOVA see DUX800
ZLUT OSTANTHRENOVA GC see VGP200
ZLUT PIGMENT 100 see FAG140
ZLUT POTRAVINARSKA 2 see CMM758
ZLUT POTRAVINARSKA 3 see FAG150
ZLUT POTRAVINARSKA 4 see FAG140
ZLUT POTRAVINARSKA 13 see CMM510
ZLUT PRIRODNI 3 see ICC800
ZLUT PRIRODNI 8 see MRN500
ZLUT PRIRODNI 11 see MRN500
ZLUT ROZPOUSTEDLOVA 6 see FAG135
ZLUT ROZPOUSTEDLOVA 77 see AAQ250
ZMA see ZDS000
ZMBT see BHA750
Z-2-(METHYL(1-OXO-9-OCTADECENYL)AMINO)-ETHANESULFONIC ACID, SODIUM SALT see SIY000
Z-(3-METHYL-4-OXO-5-PIPERIDINO-THIAZOLIDIN-2-YLIDEN)-ESSIGSAEUREAETHYLESTER (GERMAN) see MNG000
Z-METHYLPROPYL ACRYLATE see IIK000
ZN 6 see SHK000
ZnMB see BHA750
ZN 6-NA see SHK000
ZN-0312 T 1/4" see ZFA100

ZOALENE see DUP300

ZOAMIX see DUP300

ZOAPATANOLIDE A see ZTS300

ZOAPATLE, crude leaf extract see ZTS600

ZOAPATLE, semi-purified leaf extract see ZTS625

ZOAQUIN see DNF600

ZOBA 4R see DUP400

ZOBA 3GA see ALT000

ZOBA BLACK D see PEY500

ZOBA BROWN P BASE see ALT250

ZOBA BROWN RR see ALL750

ZOBA EG see ALS990

ZOBA ERN see NAW500

ZOBA GKE see TGL750

ZOBAR see PJQ000

ZOBAR see TIK500

ZOBA SLE see DBO400

ZOGEN DEVELOPER H see TGL750

ZOLAMIDE SOLUTION see AMR750

ZOLAMINE HYDROCHLORIDE see ZUA000

ZOLAPHEN see BRF500

ZOLCIENI POLACTYNOWEJ G (POLISH) see CMS226

ZOLICEF see CCS250

ZOLIDINUM see BRF500

ZOLIMIDIN see ZUA200

ZOLIMIDINE see ZUA200

ZOLIRIDINE see ZUA200

ZOLON see BDJ250

ZOLONE see BDJ250

ZOLONE PM see BDJ250

ZOMAX see ZUA300

ZOMEPIRAC see ZUA300

ZOMEPIRAC SODIUM see ZUA300

ZOMEPIRAC SODIUM SALT see ZUA300

ZONAZIDE see ILD000

ZONDEL see NNT100

ZONGNON (HAITI) see WBS850

ZONIFUR see NGB700

ZOOCOUMARIN (RUSSIAN) see WAT200

ZOOFURIN see NGE000

ZOOLON see BDJ250

ZOPAQUE see TGG760

ZOPICLONE see ZUA450

ZORANE see BRF500

ZORANE see XVS000

ZORIFLAVIN see DBX400

ZORUBICIN see ROU800

ZORUBICIN HYDROCHLORIDE see ROZ000

ZOTEPINE see ZUJ000

ZOTHELONE see PJA120

ZOTIL see ABF750

ZOTOX see ARB250

ZOTOX see ARH500

ZOTOX CRAB GRASS KILLER see ARB250

ZOVIRAX see AEC700

ZOVIRAX SODIUM see AEC725

ZOXAMIDE TECHNICAL see ZUJ500

ZOXAMIN see AJF500

ZOXAZOLAMINE see AJF500

ZOXINE see AJF500

ZP see ZLS000

Z 3 (PESTICIDE) see PLF000

ZPF see ZJS400

ZP-SB see ZJS400

ZR 512 see EQD000

ZR 515 see KAJ000

ZR-777 see POB000

ZR-856 see HCP500

ZR 3210 see VBU100

Z-(SER(BZ1)^1)-DES-HIS^2)-d-PHE^6))-LHRH HYDROCHLORIDE HYDRATE (2:3:8) see LIU325

Z-(SER(BZL)1-DES-HIS2-d-PHE2-DES-GLY10)-l-HRH ETHYLAMIDE

HYDROCHLORIDE HYDRATE (2:3:4) see PFP800

Z10-TR see CFZ000

ZUMARIL see AGN000

ZUTRACIN see BAC250

ZWAVELWATERSTOF (DUTCH) see HIC500

ZWAVELZUUROPLOSSINGEN (DUTCH) see SOI500

ZWITSALAX see DMH400

ZY 15021 see BDJ600

ZYBAN see MAS300

ZYBAN see PEX500

ZYGADENUS (VARIOUS SPECIES) see DAE100

ZYGOMYCIN A1 see NCF500

ZYGOSPORIN A see ZUS000

ZYKLOHEXANDIAMINETETRAESSIGSAEURE KADMIUMKOMPLEXE see CAD900

N-(ZYKLOHEXYL)-SYDNONIMIN HYDROCHLORID (GERMAN) see CPQ650

ZYKLOPHOSPHAMID (GERMAN) see CQC650

ZYLOFURAMINE see ZVA000

ZYLOPRIM see ZVJ000

ZYLORIC see ZVJ000

ZYMOFREN see PAF550

ZYNOPLEX see TAD175

ZYPREXA see ZVJ500

ZYTEL 211 see PJY500

ZYTOX see MHR200

ZYTRON see DGD800

ZZL-0810 see ZVS000

ZZL-0814 see EBU500

ZZL-0816 see EBV000

ZZL-0822 see EBV100

ZZL-0854 see EBV500

ZZLA-0334 see EBW000

Section 4
References

This section lists the general toxicological bibliography of 2250 keyed references to an alpha numeric system.

AAAHAN Australian Journal of Experimental Agriculture and Animal Husbandry. (Commonwealth Scientific and Industrial Research Organization, POB 89, E. Melbourne, Victoria 3002, Australia) V.1- 1961-

AAATAP Annali dell'Accademia di Agricoltura di Torino (Academia de Agricoltura, Via Andrea Doria 10, Turin, Italy) V.119- 1976/77-

AABIAV Annals of Applied Biology. (Biochemical Society, P.O. Box 32, Commerce Way, Whitehall Industrial Estate, Colchester CO2 8HP, Essex, England) V.1- 1914-

AACHAX Antimicrobial Agents and Chemotherapy. (Ann Arbor, MI) 1961-70. For publisher information, see AMACCQ

AACRAT Anesthesia and Analgesia; Current Research. (International Anesthesia Research Society, 3645 Warrensville Center Rd., Cleveland, OH 44122) V.36- 1957-

AAGAAW Antimicrobial Agents Annual. (New York, NY) 1960-60. For publisher information, see AMACCQ

AAGEAA Arkhiv Anatomii, Gistologii i Embriologii. Archives of Anatomy, Histology and Embryology. (v/o Mezhdunarodnaya Kniga, Kuznetskii Most 18, Moscow G-200, USSR) V.1- 1916-

AAHEAF Archives d'Anatomie, d'Histologie et d'Embryologie. (Editions Alsatia, 10 rue Bartholdi, 68001 Colmar, France) V.1- 1922-

AAJRDX AJR, American Journal of Roentgenology. (Charles C. Thomas Publisher, 301-327 E. Lawrence Ave., Springfield, IL 62717) V.126- 1976-

AAMLAR Atti della Accademia Medica Lombarda. (The Academy, c/o Prof. W. Montorsi, Festa del Perdono 3, Milan, Italy) V.20-29, 1931-40; V.15- 1960-

AAMMAU Archives d'Anatomie Microscopique et de Morphologie Experimentale. (Masson Publishing USA, Inc. 14 E. 60th St., New York, NY 10022) V.36- 1947-

AANEAB Acta Anaesthesiologica Scandinavica. (Munksgaard, 35 Noerre Soegade, DK-1370, Copenhagen K, Denmark) V.1- 1957-

AANFAE Annales de l'Anesthesiologie Francaise. (Doin Editeurs, 8 Place de l'Odeon, F-75006 Paris, France) V.1- 1960-

AANLAW Atti della Accademia Nazionale dei Lincei, Rendiconti della Classe di Scienze Fisiche, Matematiche e Naturali. (Academia Nazionale dei Lincei, Ufficio Pubblicazioni, Via della Lungara, 10, I-00165 Rome, Italy) V.1- 1946-

AAOPAF AMA Archives of Ophthalmology. (Chicago, IL) V.44, No. 4-63, 1950-60. For publisher information, see AROPAW

AAREAV Anesthesie, Analgesie, Reanimation. (Masson et Cie, Editeurs, 120 Blvd. Saint-Germain, P-75280, Paris 06, France) V.14- 1957-

ABABAC Annales de Biologie Animale, Biochimie, Biophysique. (Versailles, France) V.1-19, 1961-79.

ABAHAU Acta Biologica Academiae Scientiarum Hungaricae. (Kultura, P.O. Box 149, H-1389 Budapest, Hungary) V.1- 1950-

ABANAE Antibiotics Annual. (New York, NY) 1953-60. For publisher information, see AMACCQ

ABBIA4 Archives of Biochemistry and Biophysics. (Academic Press, 111 5th Ave., New York, NY 10003) V.31- 1951-

ABCHA6 Agricultural and Biological Chemistry. (Maruzen Co. Ltd., P.O.Box 5050 Tokyo International, Tokyo 100-31, Japan) V.25- 1961-

ABEMAV Annals of Biochemistry and Experimental Medicine. (Calcutta, India) V.1-23, 1941-63. For publisher information, see IJEBA6

ABHUE6 Acta Biologica Hungarica. (Kultura, POB 149, H-1389 Budapest, Hungary) V.34- 1983-

ABHYAE Abstracts on Hygiene. (Bureau of Hygiene and Tropical Diseases, Keppel St., London WC1E 7HT, England) V.1- 1926-

ABILAE Archives de Biologie. (Vaillant-Carmanne, SA rue Fond Saint-Servais, 14 B.P.22, B-4000 Liege, Belgium) V.1- 1880-

ABMGAJ Acta Biologica et Medica Germanica. (Berlin, Germany) V.1-41, 1958-82. For publisher information, see BBIADT

ABMHAM Archives Belges de Medicine Social, Hygiene, Medicine du Trevail et Medicine Legale. (Publications Acta Medica Belgica, rue des Champs-Elysees 43, 1050 Brussels 5, Belgium) V.4- 1946-

ABMPAC Advances in Biological and Medical Physics. (Academic Press, 111 5th Ave., New York, NY 10003) V.1- 1948-

ABPAAG Acta Biologiae Experimentalis (Warsaw). (Warsaw, Poland) V.1-13, 1928-39; V.14-29, 1947-69

ACATA5 Acta Anatomica. (S. Karger AG, Arnold-Boecklin-St 25, CH-4000 Basel 11, Switzerland) V.1- 1945-

ACCBAT Acta Clinica Belgica. (Association des Societes Scientifiques Medicales Belges, rue des Champs-Elysees, 43, 1050 Brussels 5, Belgium) V.1- 1946-

ACEDAB Acta Endocrinologica, Supplementum. (Periodica, Skolegade 12E, 2500 Copenhagen-Valby, Denmark) No. 1- 1948-

ACENA7 Acta Endocrinologica. (Periodica, Skolegade 12E, 2500 Copenhagen-Valby, Denmark) V.1- 1948-

ACGHD2 Annals of the American Conference of Governmental Industrial Hygienists. (American Conference of Governmental Industrial Hygienists, Inc., 6500 Glenway Ave., Bldg. D-5, Cincinnati, OH, 54211) V.1- 1981-

ACHAAH Acta Haematologica. (S. Karger AG, Arnold-Boecklin-St 25, CH-4000 Basel 11, Switzerland) V.1- 1948-

ACHCBO Acta Histochemica et Cytochemica. (Japan Society of Histochemistry and Cytochemistry, Dept. Anatomy, Faculty of Medicine, Kyoto University, Konoe-cho, Yoshida, Sakyo-ku, Kyoto, 606, Japan) V.1- 1968-

ACHTA6 Antibiotica et Chemotherapia. (Basel, Switzerland) V.1-16, 1954-70. For publisher information, see ANBCB3

ACIEAY Angewandte Chemie, International Edition in English. (Verlag Chemie GmbH, Postfach 1260/1280, D6940, Weinheim, Germany) V.1- 1962-

ACLRBL Annals of Clinical Research. (Finnish Medical Society, Duodecim Runeberginkatu 47A, SF-00260 Helsinki 26, Finland) V.1- 1969-

ACLSCP Annals of Clinical Laboratory Science. (1833 Delancey Pl., Philadelphia, PA 19103) V.1- 1971-

ACMDBI Acta Medica. (Instituto Politecnico Nacional, Escvele Superior de Medicina, Prolongacion de Diaz Miron y Plan de San Luis, Mexico) V.1- 1965-

ACNSAX Activitas Nervosa Superior. (Avicenum, Malostranske nam. 28, Prague 1, Czechoslovakia) V.1- 1959-

ACPAAN Acta Paediatrica. (Stockholm, Sweden) V.1-53, 1921-64. For publisher information, see APSVAM

ACPADQ Acta Pathologica, Microbiologica et Immunologica Scandinavica, Section A: Pathology. (Munksgaard, 35 Noerre Soegade, DK-1370, Copenhagen K, Denmark) V.90- 1982-

ACPMAP Actualites Pharmacologiques. (Paris, France) No. 1-33, 1949-81. For publisher information, see JNPHAG

ACRAAX Acta Radiologica. (Stockholm, Sweden) V.1-58, 1921-62. For publisher information, see ACRDA8

ACRDA8 Acta Radiologica: Diagnosis. (Box 7449, S-10391 Stockholm, Sweden) NS.V.1- 1963-

ACRSAJ Advances in Cancer Research. (Academic Press, 111 5th Ave., New York, NY 10003) V.1- 1953-

ACRSDM Alcoholism: Clinical and Experimental Research. (Grune and Stratton, Inc., 111 5th Ave., New York, NY 10003) V.1- 1977-

ACTRAQ Acta Tropica. (Schwabe and Co., Steintorstr., 13, CH-4010 Basel, Switzerland) V.1 1944-

ACTTDZ Acta Therapeutica. (Pl. de Bastogne 8, B13, 1080 Brussels, Belgium) V.1- 1975-

ACVIA9 Acta Vitaminologica. (Milan, Italy) V.1-20, 1947-66. For publisher information, see AVEZA6

ACVTA8 Acta Veterinaria. (Jugoslovenska Knjiga, Terazije 27/II, Belgrade, Yugoslavia) V.1- 1951-

ACYTAN Acta Cytologica. (Science Printers and Pub., Inc., 2 Jacklynn Ct., St. Louis, MO 63132) V.1- 1957-

ADBBBW Advances in Behavioral Biology. (Plenum Publishing Corp., 233 Spring Spring St., New York, NY 10013) V.1- 1971-

ADCHAK Archives of Disease in Childhood. (British Medical Journal, 1172 Commonwealth Avenue, Boston, MA 02134) V.1- 1926-

ADCSAJ Advances in Chemistry Series. (American Chemical Society, 1155 16th St., N.W., Washington, DC 20036) No. 1- 1950-

ADEVE* Acta Dermatology and Venerology (Division of Dermatopathology, University of Uppsala, Sweden. Rudbecklaboratoriet, 751 85 UPPSALA)

ADIRDF Advances in Inflammation Research. (Raven Press, 1140 Avenue of the Americas, New York, NY 10036) V.1- 1979-

ADMFAU Archiv fuer Dermatologische Forschung. (Secaucus, NJ) V.240-252, No. 4, 1971-75. For publisher information, see ADREDL

ADRCAC Annali Italiani di Dermatologia Clinica e Sperimentale. (Clinica Dermosifilopatica Universita degli Studi, Policlinica Monteluce, 06100 Perugia, Italy) V.16- 1962-

ADREDL Archives of Dermatological Research. (Springer-Verlag New York, Inc., Service Center, 44 Hartz Way, Secaucus, NJ 07094) V.253- 1975-

ADRPBI Advances in Reproductive Physiology. (Academic Press, 111 Fifth Ave., New York, NY 10003) V.1-6, 1966-73, Discontinued

ADSYAF Archives of Dermatology and Syphilology. (Chicago, IL) V.1-62, 1920-60. For publisher information, see ARDEAC

ADTEAS Advances in Teratology. (Academic Press, 111 5th Ave., New York, NY 10003) V.1-5, 1966-72, Discontinued

ADVEA4 Acta Dermato-Venereologica. (Almqvist and Wiksell International, Box 62, S-101 20 Stockholm, Sweden) V.1- 1920-

ADVED7 Annales de Dermatologie et de Venereologie. (Masson Publishing USA Inc., 133 E. 58th St., New York, NY 10022) V.104- 1977-

ADVPA3 Advances in Pharmacology. (New York, NY) V.1-6, 1962-68. For publisher information, see AVPCAQ

ADVPB4 Advances in Planned Parenthood. (Excerpta Medica, Inc., POB 3085 Princeton, NJ 08540) V.1- 1967-

ADWMAX ADWMAX Abhandlungen der Deutschen Akademie der Wissenschaften zu Berlin, Klasse fuer Medizin. (Berlin, Ger. Dem. Rep.) 1950-68. Discontinued.

AECTCV Archives of Environmental Contamination and Toxicology. (Springer-Verlag New York, Inc., Service Center, 44 Hartz Way, Secaucus, NJ 07094) V.1- 1973-

AEEDDS Adverse Effects of Environmental Chemicals and Psychotropic Drugs. (Elsevier North Holland, Inc., 52 Vanderbilt Ave., New York, NY 10017) V.1- 1973-

AEEXAH Acta Embryologiae Experimentalis. (Via Archirafi 18, 90123 Palermo, Italy) No. 1- 1969-

AEFTAA Acta Europaea Fertilitatis. (Piccin Medical Books, Via Brunacci, 12, 35100 Padua, Italy) V.1- 1969-

AEHA** U.S. Army, Environmental Hygiene Agency Reports. (Edgewood Arsenal, MD 21010)

AEHLAU Archives of Environmental Health. (Heldreff Publications, 4000 Albemarle St., N.W., Washington, DC 20016) V.1- 1960-

AEMBAP Advances in Experimental Medicine and Biology. (Plenum Publishing Corp., 233 Spring St., New York, NY 10013) V.1- 1967-

AEMED3 Annals of Emergency Medicine. (American College of Emergency Physicians, 1125 Executive Circle, Irving, TX 75038)

AEMIDF Applied and Environmental Microbiology. (American Society for Microbiology, 1913 I St., N.W., Washington, DC 20006) V.31- 1976-

AEPPAE Naunyn-Schmiedeberg's Archiv fuer Experimentelle Pathologie und Pharmakologie. (Berlin, Germany) V.110-253, 1925-66. For publisher information, see NSAPCC

AESAAI Annals of the Entomological Society of America. (Entomological Society of America, 4603 Calvert Road, College Park, MD 20740) V.1- 1908-

AETODY Advances in Modern Environmental Toxicology. (Senate Press, Inc., P.O. Box 252, Princeton Junction, NJ 08550) V.1- 1980-

AEXPBL Archiv fuer Experimentelle Pathologie und Pharmakologie. (Leipzig, Germany) V.1-109, 1873-1925. For publisher information, see NSAPCC

AEZRA2 Advances in Enzyme Regulation. (Pergamon Press, Headington Hill Hall, Oxford OX3 OBW, England) V.1- 1963-

AFCPDR Annual Symposium on Fundamental Cancer Research, Proceedings. (University of Texas System Cancer Center, M.D. Anderson Hospital and Tumor Institute, Houston, TX 77030) V.30- 1978-

AFDOAQ Association of Food and Drug Officials of the United States, Quarterly Bulletin. (Editorial Committee of the Association, P.O. Box 20306, Denver, CO 80220) V.1- 1937-

AFECAT Annales des Falsifications et de l'Expertise Chimique. (Paris, France) V.53, No. 613-No. 779, 1960-79

AFPCAG Acta Facultatis Pharmaceuticae, Universitatis Comenianae. (Ustredna Kniznica Farmaceuticka Fakulta Univerzity Komenskeho, Kolinciakova 8, 886 34 Bratislava, Czechoslovakia) V.14- 1967-

AFPEAM Archives Francaises de Pediatrie. (Doin Editeurs, 8 Place de l'Odeon, F-75006 Paris, France) V.1- 1942-

AFPYAE American Family Physician. (American Academy of Family Physicians, 1740 W. 92nd Street, Kansas City, MO 64114) V.2-17, 1962-1969: New series: V.2- 1970-

AFREAW Advances in Food Research. (Academic Press, 111 5th Ave., New York, NY 10003) V.1- 1948-

AFSPA2 Archivo di Farmacologia Sperimentale e Scienze Affini. (Rome, Italy) V.1-82, 1902-1954. Discontinued

AFTOD7 Archivos de Farmacologia y Toxicologia. (Universidad Complutense, Facultad de Medicina, Departamento Coordinado de Farmacologia, Ciudad Universitaria, Madrid 3, Spain) V.1- 1975-

AGACBH Agents and Actions, A Swiss Journal of Pharmacology. (Birkhaeuser Verlag, P.O. Box 34, Elisabethenst 19, CH-4010, Basel, Switzerland) V.1- 1969/70-

AGDJAI Journal of the Academy of General Dentistry. (Academy of General Dentistry, 211 E. Chicago Ave., Suite 1200, Chicago, IL 60611) V.1- 1952(?)-

AGGHAR Archiv fuer Gewerbepathologie und Gewerbehygiene. (Berlin, Germany) V.1-18, 1930-61. For publisher information, see IAEHDW

AGJAEP Agrochemcicals Japan. (Japan Plant Protection Association, 1-43-11, Komagome, Toshima-ku, Tokyo 170, Japan) No.62- 1993-

AGMGAK Acta Geneticae Medicae et Gemellologiae. (Cappelli Editore, Via Marsili 9, I-40124 Bologna, Italy) V.1- 1952-

AGPSA3 AMA Archives of General Psychiatry. (Chicago, IL) V.1-2, 1959-60. For publisher information, see ARGPAQ

AGRIJA Agrochemcicals Japan. (Japan Plant Protection Association, 1-43-11, Komagome, Toshima-ku, Tokyo 170, Japan) No.62-1993-

AGSOA6 Agressologie. Revue Internationale de Physio-Biologie et de Pharmacologie Appliquees aux Effets de l'Agression. (Masson et Cie, Editeurs, 120 Blvd. Saint-Germain, P-75280, Paris 06, France) V.1- 1960-

AGTQAH Annales de Genetique. (Expansion Scientifique Francaise, 5 rue Saint- Benoit, F-75278, Paris 06, France) V.1- 1958-

AHBAAM Archiv fuer Hygiene und Bakteriologie. (Munich, Germany) V.101-154, 1929-71. For publisher information, see ZHPMAT

AHEMA5 Anatomia, Histologia, Embryologia. (Verlag Paul Parey, Linderstr. 44-47, D-1000 Berlin 61, Germany) V.2- 1973-

AHJOA2 American Heart Journal. (C.V. Mosby Co., 11830 Westline Industrial Dr., St. Louis, MO 63141) V.1- 1925-

AHRAAY Arhiv za Higijenu Rada. Archives of Industrial Hygiene. (Zagreb, Yugoslavia) V.1-6, 1950-55. For publisher information, see AHRTAN

AHRTAN Arhiv za Higijenu Rada i Toksikologiju. Archives of Industrial Hygiene and Toxicology. (Jugoslovenska Knjiga, P.O. Box 36, Terazije 27, 11001, Belgrade, Yugoslavia) V.7- 1956-

AHYGAJ Archiv fuer Hygiene. (Munich, Germany) V.1-100, 1883-1928. For publisher information, see ZHPMAT

AICCA6 Acta Unio Internationalis Contra Cancrum. (Louvain, Belgium) V.1-20, 1936-64. For publisher information, see IJCNAW

AIDZAC Aichi Ika Daigaku Igakkai Zasshi. Journal of the Aichi Medical Univ. Assoc. (Aichi Ika Daigaku, Yazako, Nagakute-machi, Aichi-gun, Aichi- Ken 480-11, Japan) V.1- 1973-

AIHAAP American Industrial Hygiene Association Journal. (AIHA, 475 Wolf Ledges Pkwy., Akron, OH 44311) V.19- 1958-

AIHAM* Annual Meeting of American Industrial Hygiene Association. For publisher information, see AIHAAP

AIHOAX Archives of Industrial Hygiene and Occupational Medicine. (Chicago, IL) V.1-2, No. 3, 1950. For publisher information, see AEHLAU

AIHQA5 American Industrial Hygiene Association Quarterly. (Baltimore, MD) V.7-18, 1946-57. For publisher information, see AIHAAP

AIMDAP Archives of Internal Medicine. (American Medical Association, 535 N. Dearborn St., Chicago, IL 60610) V.1- 1908-

AIMEAS Annals of Internal Medicine. (American College of Physicians, 4200 Pine St., Philadelphia, PA 19104) V.1- 1927-

AIMJA9 Ain Shams Medical Journal (Ain Shams Clinical and Scientific Society, c/o Ain Shams University Hospital, Abbassia, Cairo, Egypt) V.1- 1950-

AINCB* Anaesthesia and Intensive Care (Edgecliff Nsw Australian Society of Anaesthesists, P.O. Box 600, Edgecliff, N.S.W. 2027, Australia) V.1- 1972-.

AIPAAV Annales de l'Institut Pasteur. (Paris, France) V.1-123, 1887-1972. For publisher information, see ANMBCM

AIPSAH Archives de l'Institut Pasteur d'Algerie. (Bibliotheque, Institut Pasteur d'Algerie, rue du Docteur-Laveran, Algiers, Algeria) V.1- 1923-

AIPTAK Archives Internationales de Pharmacodynamie et de Therapie. (Editeurs, Institut Heymans de Pharmacologie, De Pintelaan 135, B-9000 Ghent, Belgium) V.4- 1898-

AIPUAN Archivio Italiano di Patologia e Clinica dei Tumori. (Istituto di Farmacologia della Universita, Via Vanvitelli 32, 20129 Milan, Italy) V.1-15, 1957-72, Discontinued

AISFAR Archivio Italiano di Scienze Farmacologiche. (Modena, Italy) 1932-65. Discontinued

AISMAE Archivio Italiano di Scienze Mediche Tropical e di Parassitologia. Italian Archives of the Science of Tropical Medicine and Parasitology. (Rome, Italy) V.1-54, 1920-73. Suspended

AISSAW Annali dell'Istituto Superiore di Sanita. (Istituto Poligrafico dello Stato, Libreria dello Stato, Piazza Verdi, 10 Rome, Italy) V.1- 1965-

AITDAQ Archiwum Immunologii i Terapii Doswiadczalnej. (Warsaw, Poland) V.1-9, 1953-61. For publisher information, see AITEAT

AITEAT Archivum Immunologiae et Therapiae Experimentalis. (Ars Polona-RUCH, P.O. Box 1001, P-00 068 Warsaw, 1, Poland) V.10- 1962-

AIVMBU Archivos de Investigacion Medica. (POB 12,631, Col. Narvarte, Deleg. Benito Juarez, 03020 Mexico City, DF, Mexico, V.1- 1970-

AJANA2 American Journal of Anatomy. (Alan R. Liss Inc., 150 5th Ave., New York, NY 10011) V.1- 1901-

AJBSAM Australian Journal of Biological Sciences. (Commonwealth Scientific and Industrial Research Organization, POB 89, E. Melbourne, Victoria, Australia) V.6- 1953-

AJCAA7 American Journal of Cancer. (New York, NY) V.15-40, 1931-40. For publisher information, see CNREA8

AJCDAG American Journal of Cardiology. (Technical Publishing Co., 666 5th Ave., New York, NY 10103) V.1- 1958-

AJCHAS Australian Journal of Chemistry. (Commonwealth Scientific and Industrial Research Organization, 314 Albert St., P.O. Box 89, E. Melbourne, Victoria, Australia) V.6- 1953-

AJCNAC American Journal of Clinical Nutrition. (American Society for Clinical Nutrition, Inc., 9650 Rockville Pike, Bethesda, MD 20814) V.2- 1954-

AJCPAI American Journal of Clinical Pathology. (J.B. Lippincott Co., Keystone Industrial Park, Scanton, PA 18512) V.1- 1931-

AJDCAI American Journal of Diseases of Children. (American Medical Association, 535 N. Dearborn St., Chicago, IL 60610) V.1-80(3), 1911-50; V.100- 1960-

AJDDAL American Journal of Digestive Diseases. (Plenum Publishing Corp., 233 Spring St., New York, NY 10013) V.5-22, 1938-55, V.1-23, 1956-78

AJDEBP Australasian Journal of Dermatology. (Australasian College of Dermatologists, 271 Bridge Rd., Glebe, NSW 2037, Australia) V.9- 1967-

AJEBAK Australian Journal of Experimental Biology and Medical Science. (Univ. of Adelaide Registrar, Adelaide, S.A. 5000, Australia) V.1- 1924-

AJEMEN American Journal of Emergency Medicine. (WB Saunders, Philadelphia, PA) V.1- 1983-

AJEPAS American Journal of Epidemiology. (Johns Hopkins Univ., 550 N. Broadway, Suite 201, Baltimore, MD 21205) V.81- 1965-

AJGAAR American Journal of Gastroenterology. (American College of Gastroenterology, Inc., 299 Broadway, New York, NY 10007) V.21- 1954-

AJHEAA American Journal of Public Health. (American Public Health Association, Inc., 1015 15th St., N.W., Washington, DC 20005) V.2-17, 1912-27; V.61- 1971-

AJHPA9 American Journal of Hospital Pharmacy. (American Society of Hospital Pharmacists, 4630 Montgomery Ave., Washington, DC 20014) V.15- 1958-

AJHYA2 American Journal of Hygiene. (Baltimore, MD) V.1-80, 1921-64. For publisher information, see AJEPAS

AJIMD8 American Journal of Industrial Medicine. (Alan R. Liss, Inc., 150 Fifth Ave., New York, NY 10011) V.1- 1980-

AJINO* Ajinomoto Co., Inc. (9 W. 57th St., Suite 4625, New York, NY 10019)

AJKDDP American Journal of Kidney Diseases. (Grune & Stratton, Inc., Journal Subscription Department, POB 6280, Duluth, MN 55806) V.1- 1981-

AJMEAZ American Journal of Medicine. (Yorke Medical Group, 666 5th Ave., New York, NY 10103) V.1- 1946-

AJMSA9 American Journal of the Medical Sciences. (Charles B. Slack, Inc., 6900 Grove Rd., Thorofare, NJ 08086) V.1- 1841-

AJNED9 American Journal of Nephrology. (S. Karger Pub., Inc., 79 Fifth Ave., New York, NY 10003) V.1- 1981-

AJOGAH American Journal of Obstetrics and Gynecology. (C.V. Mosby Co., 11830 Westline Industrial Dr., St. Louis, MO 63141) V.1- 1920-

AJOMAZ Alabama Journal of Medical Sciences. (University of Alabama Medical Center, Univ. Station, Birmingham, AL 35294) V.1- 1964-

AJOPAA American Journal of Ophthalmology. (Ophthalmic Publishing Co., 435 N. Michigan Ave., Chicago, IL 60611) V.1- 1918-

AJPAA4 American Journal of Pathology. (Lippincott/Harper, Journal Fulfillment Dept., 2350 Virginia Ave., Hagerstown, MD 21740) V.1- 1925-

AJPEAG American Journal of Public Health and the Nation's Health. (New York, NY) V.18-60, 1928-60. For publisher information, see AJHEAA

AJPHAP American Journal of Physiology. (American Physiological Society, 9650 Rockville Pike, Bethesda, MD 20814) V.1- 1898-

AJPRAL American Journal of Pharmacy and the Sciences Supporting Public Health. (Philadelphia College of Pharmacy and Science, 43rd St., Woodland Ave. and Kingsessing Mall, Philadelphia, PA 19104) V.109-150, 1937-78

AJPSAO American Journal of Psychiatry. (American Psychiatric Association, Publications Services Div., 1700 18th St., N.W., Washington, DC 20009) V.78- 1921-

AJRRAV American Journal of Roentgenology, Radium Therapy and Nuclear Medicine. (Springfield, IL) V.67-125, 1952-75. For publisher information, see AAJRDX

AJSGA3 American Journal of Syphilis, Gonorrhea and Venereal Diseases. (St. Louis, MO) V.20-38, 1936-54. Discontinued

AJSNAO American Journal of Syphilis and Neurology. (St. Louis, MO) V.1-19, 1917-35. For publisher information, see AJSGA3

AJTHAB American Journal of Tropical Medicine and Hygiene. (Allen Press, 1041 New Hampshire St., Lawrence, KS 66044) V.1- 1952-

AJTMAQ American Journal of Tropical Medicine. (Baltimore, MD) V.1-31, 1921-51. For publisher information, see AJTHAB

AJVRAH American Journal of Veterinary Research. (American Veterinary Medical Association, 930 N. Meacham Road, Schaumburg, IL 60196) V.1- 1940-

AKBNAE Arkhiv Biologicheskikh Nauk. Archives of Biological Sciences. (Moscow, USSR) V.1-64, 1892-1941. Discontinued

AKEDAX Archiv fuer Klinische und Experimentelle Dermatologie. (Berlin, Germany) V.201-239, 1955-71. For publisher information, see ADMFAU

AKGIAO Akushcherstvo i Ginekologiya. (v/o Mezhdunarodnaya Kniga, Kuznetskii Most 18, Moscow G-200, USSR) No. 1- 1936-

ALACBI Aldrichimica Acta. (Aldrich Chemical Co., Inc., 940 W. St. Paul Ave., Milwaukee, WI 53233) V.1- 1968-

ALBRW* Albright and Wilson Inc., P.O. Box 26229, Richmond, VA 23260-6229

ALEPA8 Acta Leprol. (Publisher unknown) V.1- 1920(?)-

ALLVAR Alimentation et la Vie. (Societe Scientifique d'Hygiene Alimentaire, 16, rue de l'Estrapade, 75005 Paris, France) V.39- 1951-

ALXXAP Patent Specification (Australia). (Commissioner of Patents, POB 200, Woden, ACT 2606, Australia)

AMACCQ Antimicrobial Agents and Chemotherapy. (American Society for Microbiology, 1913 I St., N.W., Washington, DC 20006) V.1- 1972-

AMASA4 Acta Medica Academiae Scientiarum Hungaricae. (Kultura, POB 149 H-1389, Budapest, Hungary) V.1- 1950-

AMBED4 AMI-Berichte. (Berlin, Fed. Rep. Ger.) No. 1-2, 1978-82

AMBNAS Acta Medica et Biologica. (Niigata University School of Medicine, 1 Asahi-machi-dori, Niigata 951, Japan) V.1- 1953-

AMBOCX Ambio. A Journal of the Human Environment, Research and Management. (Pergamon Press, Inc., Maxwell House, Fairview Park, Elmsford, NY 10523) V.1- 1972-

AMBPBZ Acta Pathologica et Microbiologica Scandinavica, Section A: Pathology. (Copenhagen K, Denmark) V.78-89, 1970-81. For publisher information, see ACPADQ

AMBUCH Acta Medica Bulgarica. (Durzhavno Izdatelstvo Meditsina i Fizkultura, Pl. Slaveikov 11, Sofia, Bulgaria) V.1- 1973-

AMCTAH Antibiotic Medicine and Clinical Therapy. (New York, NY) V.3-8, 1956-61. For publisher information, see CLMEA3

AMCYC* Toxicological Information on Cyanamid Insecticides. (American Cyanamid Co., Agricultural Division, Princeton, NJ 08540)

AMDCA5 AMA Journal of Diseases of Children. (Chicago, IL) V.91-99, 1956-60. For publisher information, see AJDCAI

AMEBA7 Annales Medicinae Experimentalis et Biologiae Fenniae. (Mikonkatu 8, Helsinki, Finland) V.25- 1947-

AMICCW Archives of Microbiology. (Springer-Verlag New York, Inc., Service Center, 44 Hartz Way, Secaucus, NJ 07094) V.95- 1974-

AMIGB9 Acta Microbiologica Polonica, Series A: Microbiologica Generalis. (Warsaw, Poland) V.1-7, No.4, 1969-75, For publisher information, see AMPOAX

AMIHAB AMA Archives of Industrial Health. (Chicago, IL) V.11-21, 1955-60. For publisher information, see AEHLAU

AMIHBC AMA Archives of Industrial Hygiene and Occupational Medicine. (Chicago, IL) V.2-10, 1950-54. For publisher information, see AEHLAU

AMILAN Annales de Medecine Legale. Criminologie, Police Scientifique et Toxicologie. (Paris, France) V.1, 1920-51; V.39-47, 1958-67. For publisher information, see MLDCAS

AMIUAG Acta Medica Iugoslavica. (Akademija Zbora Lijecnika Hrvatske, Subiceva 9/1, Zagreb, Yugoslavia) V.1- 1947-

AMJPA6 American Journal of Pharmacy (1835-1936). (Philadelphia, PA) V.1-108, 1835-1936. For publisher information, see AJPRAL

AMLTAS Annales de Medecine Legale et de Criminologie. (Paris, France) V. 31-8, 1951-8. For publisher information, see MLDCAS

AMNIB6 Acta Manilana, Series A: Natural and Applied Sciences. (University of Santo Tomas Research Center, Manila, D-403, Philippines) V.4- 1968-

AMNTA4 American Naturalist. (University of Chicago Press, 5801 S. Ellis Ave., Chicago, IL 60637) V.1- 1867-

AMOKAG Acta Medicia Okayama. (Okayama University Medical School, 2-5-1 Shikata-cho, Okayama 700, Japan) V.8- 1952-

AMONDS Applied Methods in Oncology. (Elsevier North Holland, Inc., 52 Vanderbilt Ave., New York, NY 10017) V.1- 1978-

AMPLAO AMA Archives of Pathology. (American Medical Association, 535 N. Dearborn St., Chicago, IL 60610) V.50,No. 4-V.69, 1950-60

AMPMAR Archives des Maladies Professionnelles de Medecine du Travail et de Securite Sociale. (Masson et Cie, Editeurs, 120 Blvd. Saint-Germain, P-75280, Paris 06, France) V.7- 1946-

AMPYAT Annales Medico-Psychologiques. (Masson et Cie, Editeurs, 120 Blvd. Saint-Germain, P-75280, Paris 06, France) V.80- 1922-

AMRL** Aerospace Medical Research Laboratory Report. (Aerospace Technical Div., Air Force Systems Command, Wright-Patterson Air Force Base, OH 45433)

AMSHAR Acta Morphologica Academiae Scientiarum Hungaricae. (Akademiai Kiado, P.O. Box 24, H-1389 Budapest 502, Hungary) V.1- 1951-

AMSSAQ Acta Medica Scandinavica, Supplement. (Almqvist and Wiksell, P.O. Box 62, 26 Gamla Brogatan, S-101, 20 Stockholm, Sweden) No. 1- 1921-

AMSVAZ Acta Medica Scandinavica. (Almqvist and Wiksell, P.O. Box 62, 26 Gamla Brogatan, S-101, 20 Stockholm, Sweden) V.52- 1919-

AMTUA3 Acta Medica Turcica. (Dr. Ayhan Okcuoglu, Cocuk Hastalikari Klinig i, c/o Ankara University Tip Facultesi, PK 48, Cebeci, Ankara, Turkey) V.1- 1964-

AMUK** Acta Medica University Kyoto. (Kyoto, Japan)

AMZOAF American Zoologist. (American Society of Zoologists, Box 2739, California Lutheran College, Thousand Oaks, CA 91360) V.1- 1961-

ANAEA3 Annals of Allergy. (American College of Allergists, Box 20671, Bloomington, MN 55420) V.1- 1943-

ANANAU Anatomischer Anzeiger. (VEB Gustav Fischer Verlag, Postfach 176, DDR-69, Jena, Germany) V.1 1886-

ANASAB Anaesthesia. (Blackwell Scientific, Osney Mead, Oxford OX2 OEL, England) V.1- 1946-

ANATAE Anaesthesist. (Springer-Verlag, Heidelberger Pl. 3, D-1000 Berlin 33, Germany) V.1- 1952-

ANBCA2 Analytical Biochemistry. (Academic Press, 111 5th Ave., New York, NY 10003) V.1- 1960-

ANBCB3 Antibiotics and Chemotherapy. (S. Karger, AG, Arnold-Boecklin-St 25, Postfach CH-4009 Basel, Switzerland) V.17- 1971-

ANCHAM Analytical Chemistry. (American Chemical Society, 1155 16th St., N.W., Washington, DC 20036) V.19-1947-

ANDRDQ Andrologia. (Grosse Verlag, Kurfuerstendamm 152, D-1000 Berlin 31, Germany) V.6- 1974-

ANENAG Annales d'Endocrinologie. (Masson et Cie, Editeurs, 120 Blvd. Saint-Germain, P-75280, Paris 06, France) V.1- 1939-

ANESAV Anesthesiology. (J.B. Lippincott Co., Keystone Industrial Park, Scranton, PA 18512) V.1- 1940-

ANGIAB Angiology. (Williams & Wilkins Co., 428 E. Preston St., Baltimore, MD 21202) V.1- 1950-

ANIFAC Annales de Chirurgie Infantile. (Paris, France) V.1- 1960(?)-

ANMBCM Annales de Microbiologie (Paris). (Masson et Cie, Editeurs, 120 Blvd. Saint-Germain, P-75280, Paris 06, France) V.124- 1973-

ANNPHR Annales Pharmaceutic. (Poznan, Poland) V.3-13, 1965-78. Discontinued.

ANOBAU Bulletin de l'Association des Anatomistes. (U.E.R. Sciences Medicales A., Boite Postale 1080, 54019 Nancy Cedex, France)

ANONE2 Annals of Oncology. (European Society for Medical Oncology, Distributed by Kluwer Academic Publishers, Postbus 17, 3300 AA Dordrecht, Netherlands) V.1-1990-

ANOPB5 Annals of Ophthalmology. (American Society of Contemporary Ophthalmology, 211 E. Chicago Ave., Suite 1044, Chicago, IL 60611) V.1- 1969-

ANPBAZ Acta Neurologia et Psychiatrica Belgica. (Brussels, Belgium) V.48-69, 1948-69. For publisher information, see ANUBBR

ANPSAI Archives of Neurology and Psychiatry. (Chicago, IL) V.1-64, 1919-50. For publisher information, see ARNEAS

ANPTAL Acta Neuropathologica. (Springer-Verlag New York, Inc., Service Center, 44 Hartz Way, Secaucus, NJ 07094) V.1- 1961-

ANREAK Anatomical Record. (Alan R. Liss, Inc., 150 5th Ave., New York, NY 10011) V.1- 1906/08-

ANSUA5 Annals of Surgery. (J.B. Lippincott Co., Keystone Industrial Park, Scranton, PA 18512) V.1- 1885-

ANTBAL Antibiotiki. (Moscow, USSR) V.1-29, 1956-84. For publisher information, see AMBIEH

ANTCAO Antibiotics and Chemotherapy. (Washington, DC) V.1-12, 1951-62. For publisher information, see CLMEA3

ANTDEV Anti-Cancer Drugs. (Rapid Communications of Oxford Ltd., The Old Malthous, Paradise St., Oxford OX1 1LD, UK) V.1 1990-

ANTRD4 Anticancer Research. (Anticancer Research, 5 Argyropoulou St., Kato Patissia, Athens 907, Greece) V.1-1981-

ANUBBR Acta Neurologica Belgica. (Association des Societes Scientifiques, Medicales Belges, rue des Champs-Elysees 43, B-1050 Brussels, Belgium) V.1- 1900-

ANYAA9 Annals of the New York Academy of Sciences. (The Academy, Exec. Director, 2 E. 63rd St., New York, NY 10021) V.1- 1877-

ANZJA7 Australian and New Zealand Journal of Surgery. (Blackwell Scientific Publications, College of Surgeons' Gardens, 99 Barry St., Carlton 3053, Australia) V.1- 1931-

ANZJB8 Australian and New Zealand Journal of Medicine. (Modern Medicine of Australia Pty., Ltd., 100 Pacific Highway, North Sydney, 2060, Australia) V.1- 1971-

AOBIAR Archives of Oral Biology. (Pergamon Press, Headington Hill Hall, Oxford OX3 OBW, England) V.1- 1959-

AOGLAR Acta Obstetrica et Gynaecologica Japonica, English Edition. (Tokyo, Japan) V.16-23, 1969-76. For publisher information, see NISFAY

AOGMAU Annali di Ostetricia, Ginecologia, Medicina Perinatale. (Via Commenda, 12, 20122 Milan, Italy) V.93- 1972-

AOGNAX Archivio di Ostetricia e Ginecologia. (C.C. Postale N 6/19773, Naples, Italy) V.1- 1937-

AOGSAE Acta Obstetricia et Gynecologica Scandinavica. (Kvinnokliniken, Lasarettet, Lund, Sweden) V.1- 1921-

AOHYA3 Annals of Occupational Hygiene. (Pergamon Press, Headington Hill Hall, Oxford OX3 OBW, England) V.1- 1958-

AOISDR Annual Report of Osaka City Institute of Public Health and Environmental Sciences. (Osaka-shiritsu Kankyo Kagaku Kenkyusho, 8-34 Tojo-cho, Tennoji-ku, Osaka 543, Japan) No. 43- 1980-

AORLCG Archives of Oto-Rhino-Laryngology. (Springer-Verlag New York, Inc., Service Center, 44 Hartz Way, Secaucus, NJ 07094) V.206- 1973-

AOUNAZ Archiv fuer Orthopaedische und Unfall-Chirurgie. (Munich, Germany) V.1-90 1903-77. see AOTSDE

APACAB Acta Physiologica Academiae Scientiarum Hungaricae. (Akademiai Kiado, P.O. Box 24, H-1389 Budapest 502, Hungary) V.1- 1950-

APAVAY Virchows Archiv fuer Pathologische, Anatomie und Physiologie, und fuer Klinische Medizin. (Berlin, Germany) V.1-343, 1847-1967. For publisher information, see VAAPB7

APBDAJ Archiv der Pharmazie und Berichte der Deutschen Pharmazeutischen Gesellschaft. (Weinheim, Germany) V.262-304, 1924-71. For publisher information, see ARPMAS

APBOAI Acta Facultatis Pharmaceuticae Bohemoslovenicae. (Bratislava, Czechoslovakia) V.4-13, 1961-67. For publisher information, see AFPCAG

APCRAW Advances in Pest Control Research. (New York, NY) V.1-8, 1957-68. Discontinued

APDCDT Advances in Tumour Prevention, Detection and Characterization. (Elsevier North Holland, Inc., 52 Vanderbilt Ave., New York, NY 10017) V.1- 1974-

APDJBE Acta Paediatrica Japonica (Overseas Edition). (Societas Pediatrica Japonica, 1-1-5-Koraku, Bunkyo-ku, Tokyo, 112, Japan) V.1- 1958-

APEPA2 Naunyn-Schmiedebergs Archiv fuer Pharmakologie und Experimentelle Pathologie. (Berlin, Germany) V.254-263, 1966-69. For publisher information, see NSAPCC

APFRAD Annales Pharmaceutiques Francaises. (Masson et Cie, Editeurs, 120 Blvd. Saint-Germain, P-75280, Paris 06, France) V.1- 1943-

APHGAO Acta Pharmaceutica Hungarica. (Kultura, POB 149, H-1389 Budapest, Hungary) 1953- Adopts V.24 in 1955

APHGBP Acta Pathologica (Belgrade). (Belgrade Univerzitet, Institute de Pathologie, Belgrade, Yugoslavia) V.2/3- 1938-

APHRER Annals of Pharmacotherpy. (Harvey Whitney Books Co., POB 42696, Cincinnati, OH 45242) V. 26- 1992-

APJAAG Acta Pathologica Japonica. (Nippon Byori Gakkai, 7-3-1, Hongo, Bunkyo-Ku, Tokyo 113, Japan) V.1- 1951-

APJUA8 Acta Pharmaceutica Jugoslavica. (Jugoslovenska Knjiga, P.O. Box 36, Terazije 27, YU-11001 Belgrade, Yugoslavia) V.1- 1951-

APLAAQ Acta Paediatrica Latina. (Editrice Arti Grafiche Emiliane, postale 10110427, 42100 Reggio Emilia, Italy) V.1- 1948-

APLMAS Archives of Pathology and Laboratory Medicine. (American Medical Association, 535 N. Dearbon St., Chicago, IL. 60610) V.1-5, No. 2, 1926-28, V.100- 1976-

APMBAY Applied Microbiology. (Washington, DC) V.1-30, 1953-75. For publisher information, see AEMIDF

APMIAL Acta Pathologica et Microbiologica Scandinavica. (Copenhagen, Denmark) V.1-77, 1924-69. For publisher information, see AMBPBZ

APMIBM Acta Pathologica et Microbiologica Scandinavica, Section B: Microbiology and Immunology. (Copenhagen, Denmark) V.78B-82B, 1970-74. For publisher information, see APSCD2

APMUAN Acta Pathologica et Microbiologica Scandinavica, Supplementum. (Munksgaard, 35 Noerre Soegade, DK-1370 Copenhagen K, Denmark) No. 1- 1926-

APPBDI Acta Physiologica et Pharmacologica Bulgarica. (Izdatelstvo na Bulgarskata Akademiya na Naukite, St. Geo Milev ul 36, Sofia 13, Bulgaria) V.1- 1974-

APPHAX Acta Poloniae Pharmaceutica. (Ars Polona-RUCH, P.O. Box 1001, P-00 068 Warsaw, 1, Poland) V.1- 1937-

APPNAH Acta Physiologica et Pharmacologica Neerlandica. (Amsterdam, Netherlands) V.1-15, 1950-69. For publisher information, see EJPHAZ

APPYAG Annual Review of Phytopathology. (Annual Reviews, Inc., 4139 El Camino Way, Palo Alto, CA 94306) V.1- 1963-

APRCAS American Perfumer and Cosmetics. (Oak Park, IL) V.77-86, 1962-71. For publisher information, see CSPEAX

APSCAX Acta Physiologica Scandinavica. (Karolinska Institutet, S-10401 Stockholm, Sweden) V.1- 1940-

APSCD2 Acta Pathologica et Microbiologica Scandinavica, Section C: Immunology. (Munksgaard, 35 Noerre Soegade, DK-1370, Copenhagen CK, Denmark) V.83C- 1975-

APSVAM Acta Paediatrica Scandinavica. (Almqvist and Wiksell, P.O. Box 62, 26 Gamla Brogatan, S-101, 20 Stockholm, Sweden) V.54- 1965-

APSXAS Acta Pharmaceutica Suecdca. (Apotekarsocieteten, Wallingatan 26, Box 1136, S-111, 81 Stockholm, Sweden) V.1- 1964-

APTOA6 Acta Pharmacologica et Toxicologica. (Munksgaard, 35 Noerre Soegade, DK-1370, Copenhagen K, Denmark) V.1- 1945-

APTOD9 Abstracts of Papers, Society of Toxicology. Annual Meetings. (Academic Press, 111 5th Ave., New York, NY 10003)

APTRDI Advances in Prostaglandin and Thromboxane Research. (Raven Press 1140 Ave. of the America, New York, NY 10036) V.1- 1976-

APTSAI Acta Pharmacologica et Toxicologica, Supplementum. (Munksgaard, 35 Noerre Soegade, DK-1370, Copenhagen K, Denmark) No. 1- 1947-

APYPAY Acta Physiologica Polonica. (Panstwowy Zaklad Wydawnictw Lekarskich, ul. Dluga 38-40, P-00 238 Warsaw, Poland) V.1- 1950-

AQMOAC Air Quality Monographs. (American Petroleum Institute, 2101 L St., N.W., Washington, DC 20037) No. 69-1- 1969-

ARANDR Archives of Andrology. (Elsevier North Holland, Inc., 52 Vanderbilt Ave., New York, NY 10017) V.1- 1978-

ARCGDG Archives of Gynecology. (Springer-Verlag New York, Inc., Service Center, 44 Hartz Way, Secaucus, NJ 07094) V.226- 1978-

ARCVBP Annales des Recherches Veterinaires. (Institut National de la Recherche Agronomique, Service des Publ., route de Saint-Cyr, 78000 Versailles, France) V.1- 1970-

ARDEAC Archives of Dermatology. (American Medical Association, 535 N. Dearborn St., Chicago, IL 60610) V.82-1960-

ARDIAO Annals of the Rheumatic Diseases. (BMA or British Medical Journal, 1172 Commonwealth Ave., Boston, MA. 02134) V.1- 1939-

ARDSBL American Review of Respiratory Disease. (American Lung Association, 1740 Broadway, New York, NY 10019) V.80- 1959-

AREAD8 Anesteziologiya i Reanimatologiya. (v/o Mezhdunarodnaya Kniga, Kuznetskii Most 18, Moscow G-200, USSR) No. 1- 1977-

ARGEAR Archiv fuer Geschwulstforschung. (VEB Verlag Volk und Gesundheit Neue Gruenstr. 18, DDR-102 Berlin, Germany) V.1- 1949-

ARGPAQ Archives of General Psychiatry. (American Medical Association, 535 N. Dearborn St., Chicago, IL 60610) V.3- 1960-

ARGYAJ Archiv fuer Gynaekologie. (Munich, Germany) V.1-225, 1870-1978. For publisher information, see ARCGDG

ARHEAW Arthritis and Rheumatism. (Arthritis Foundation, 3400 Peachtree Road, N.E., Atlanta, GA 30326) V.1- 1958-

ARINAU Annual Report of the Research Institute of Environmental Medicine, Nagoya University. (Nagoya, Japan) V.1-25, 1951-80

ARKIAP Archiv fuer Kinderheilkunde. (Stuttgart, Germany.) V.1-183, 1880-1971

ARMCAH Annual Review of Medicine. (Annual Reviews, Inc., 4139 El Camino Way, Palo Alto, CA 94306) V.1-1950-

ARMIAZ Annual Review of Microbiology. (Annual Reviews, Inc., 4139 El Camino Way, Palo Alto, CA 94306) V.1- 1947-

ARMKA7 Archiv fuer Mikrobiologie. (Springer (Berlin)) V.1-13, 1930-43; V.14-94, 1948-73. For publisher information, see AMICCW

ARNEAS Archives of Neurology. (American Medical Association, 535 N. Dearborn St., Chicago, IL 60610) V.3-1960-

AROPAW Archives of Ophthalmology. (American Medical Association., 535 N. Dearborn St., Chicago, IL 60610) V.1-44, No. 3, 1929-50; V.64- 1960-

AROTAA Archives of Otolaryngology. (American Medical Association, 535 N. Dearborn St., Chicago, IL 60610) V.1-52, 1925-50; V.72- 1960-

ARPAAQ Archives of Pathology. (American Medical Association., 535 N. Dearborn St., Chicago, IL 60610) V.5, No. 3-V.50, No. 3, 1928-50; V.70-99, 1960-75

ARPMAS Archiv der Pharmazie. (Verlag Chemie GmbH, Postfach 1260/1280 D-6940 Weinheim, Germany) V.51-261, 1835-1923; V.305- 1972-

ARPTAF Arkhiv Patologii. Archives of Pathology. (v/o Mezhdunarodnaya Kniga, Kuznetskii Most 18, Moscow G-200 USSR.) V.1- 1959-

ARPTDI Annual Review of Pharmacology and Toxicology. (Annual Reviews, Inc., 4139 El Camino Way, Palo Alto, CA 94306) V.16- 1976-

ARSIM* Agricultural Research Service, USDA Information Memorandum. (Beltsville, MD 20705)

ARSRDR Antiviral Research. (An Official Publication of the International Society for Antiviral Research, Publisher Elsevier Science, Molenwerf 1 1014 AG Amsterdam The Netherlands) V.1- 1981-

ARSUAX Archives of Surgery. (American Medical Association, 535 N. Dearborn St., Chicago, IL 60610) V.1-61, 1920-50; V.81- 1960-

ARTODN Archives of Toxicology. (Springer-Verlag, Heidelberger Pl. 3, D-1 Berlin 33, Germany) V.32- 1974-

ARTUA4 American Review of Tuberculosis. (New York, NY) V.1-70, 1917-54. For publisher information, see ARDSBL

ARVPAX Annual Review of Pharmacology. (Palo Alto, CA) V.1-15, 1961-75. For publisher information, see ARPTDI

ARZFAN Aerztliche Forschung. (Munich, Germany) V.1-26, 1947-72. Discontinued

ARZNAD Arzneimittel-Forschung. Drug Research. (Editio Cantor Verlag, Postfach 1255, W-7960 Aulendorf, Germany) V.1- 1951-

ASBDD9 Advances in the Study of Birth Defects. (University Park Press, 233 E. Redwood St., Baltimore, MD 21202) V.1- 1979-

ASBIAL Archivio di Science Biologiche. (Cappelli Editore, Via Marsili 9, I-40124 Bologna, Italy) V.1- 1919-

ASBUAN Archives des Sciences Biologiques. (Leningrad, USSR.) V.1-22, 1892-1922. Discontinued

ASMUAA Acta Scholae Medicinalis Universitatis in Kyoto. (Kyoto, Japan) V.27-40, 1947-70. Discontinued

ASPHAK Archives des Sciences Physiologique. (Paris, France) V.1-28, 1947-74. Discontinued

ASSUA6 Acta Psychiatrica Scandinavica, Supplementum. (Munksgaard International Publishers, POB 2148, DK-1016 Copenhagen K, Denmark) No.160- 1961-

ASTTA8 ASTM Special Technical Publication. (American Society for Testing Materials, 1916 Race St., Philadelphia, PA 19103) No. 1- 1911-

ASUPAZ Acta Societatis Medicorum Upsaliensis. (Uppsala, Sweden) V.55-76, 1950-71. For publisher information, see UJMSAP

ATAREK AAMI Technology Assessment Report. (Association for the Advancement of Medical Instrumentation, 1901 N. Ft. Myer Dr., Suite 602, Arlington, VA 22209) No. 1-81- 1981-

ATDAEI Acute Toxicity Data. Journal of the American College of Toxicology, Part B. (Mary Ann Liebert, Inc., 1651 Third Ave., New York, NY 10128) V.1- 1990-

ATENBP Atmospheric Environment. Air Pollution, Industrial Aerodynamics, Micrometerology, Aerosols. (Pergamon Press, Headington Hill Hall, Oxford OX3 OBW, England) V.1- 1967-

ATHBA3 Acta Radiologica, Therapy, Physics, Biology. (P.O. Box 7449, S-103 91 Stockholm, Sweden) V.1-16, No. 6, 1963-77

ATHSBL Atherosclerosis (Shannon, Ireland). (Elsevier/North-Holland Scientific Publishers Ltd., POB 85, Limerick, Ireland) V.11- 1970-

ATMPA2 Annals of Tropical Medicine and Parasitology. (Academic Press, 24-28 Oval Rd., London NWl 7DX, England) V.1- 1907-

ATPNAB Ateneo Parmense, Acta Naturalia. (Ospedale Maggiore, Via Gramsci 14, 43100 Parma, Italy) V.1- 1965-

ATSUDG Archives of Toxicology, Supplement. (Springer-Verlag, Heidelberger Pl. 3, D-1000 Berlin 33, Germany) No. 1- 1978-

ATTHEH Acta Toxicologica et Therapeutica. (Casa Editrice Maccari, Casella Postale 120, I-43100 Parma, Italy) V.1- 1980-

ATXKA8 Archiv fuer Toxikologie. (Berlin, Germany) V.15-31, 1954-74. For publisher information, see ARTODN

AUAAB7 Acta Universitatis Agriculturae, Facultas Agronomica (Brno) (Ustredni Knihovna Vysoke Skoly Zemedelske, Zemedelska 1, 662-65 Brno, Czechoslovakia) V.15- 1967-

AUCMBJ Acta Universitatis Carolinae, Medica, Monographia. (Univerzita Karlova, Ovocny Trh.3, CS-116-36 Prague 1, Czechoslovakia) No. 1- 1954-

AUODDK Acta Universitatis Ouluensis, Series D: Medica. (Oulu University Library, Box 186, SF-90101 Oulu 10, Finland) No. 1- 1972-

AUPJB7 Australian Paediatric Journal. (Royal Childrens Hospital, Parkville, Victoria 3052, Australia) V.1- 1965-

AUPMAF Acta Universitatis Palackianae Olomucensis, Facultatis Medicae. (Statni Pedagogicke Nakladatelstvi, Ostrovni 30, 1 Nove Mesto, 113 01 Prague 1, Czechoslovakia) V.24-

AUVJA2 Australian Veterinary Journal. (Australian Veterinary Association, Executive Director, 134-136 Hampden Road, Artarmon, NSW 2064, Australia) V.2- 1927-

AVBIB9 Advances in the Biosciences. (Pergamon Press Ltd., Headington Hill Hall, Oxford OX3 0BW, England) V.1- 1969-

AVBNAN Arhiv Bioloskih Nauka. (Jugoslovenska Knjigu, P.O. Box 36, Terazije 27, 11001, Belgrade, Yugoslavia)

AVERAG American Veterinary Review. (Chicago, IL) V.1-47, 1877-1915. For publisher information, see JAVMA4

AVEZA6 Acta Vitaminologica et Enzymologica. (Gruppo Lepetit SpA, Via R. Lepetit N. 8, 20124 Milan, Italy) V.21- 1967-

AVPCAQ Advances in Pharmacology and Chemotherapy. (Academic Press, 111 5th Ave., New York, NY 10003) V.7- 1969-

AVSUAR Acta Dermato-Venereologica, Supplementum. (Almqvist and Wiksell Periodical Co., P.O. Box 62, 26 Gamla Brogatan, S-101 20 Stockholm, Sweden) No. 1- 1929-

AWLRAO Australian Wildlife Research. (Commonwealth Scientific and Industrial Research Organization, POB 89, E. Melbourne, Victoria 3002, Australia) V.1- 1974-

AXVMAW Archiv fuer Experimentelle Veterinaermedizin. (S. Hirzel Verlag, Postfach 506, DDR-701 Leipzig, Germany) V.6- 1952-

AZMZA6 Azerbaidzhanskii Meditsinskii Zhurnal. Azerbaidzhan Medical Journal. (v/o Mezhdunarodnaya Kniga, Kuznetskii Most 18, Moscow G-200, USSR.) 1928-41; 1955-

BACCAT Bulletin of the Academy of Sciences of the USSR, Division of Chemical Science (English Translation). Translation of IASKA6. (Plenum Pub. Corp., 233 Spring St., New York, NY 10013) 1952-

BAFEAG Bulletin de l'Association Francaise pour l'Etude du Cancer. (Paris, France) V.1-52, 1908-65. For publisher information, see BUCABS

BANMAC Bulletin de l'Academie Nationale de Medicine. (Masson et Cie, Editeurs, 120 Blvd. Saint-Germain, P-75280, Paris 06, France) V.1- 1836-

BANRDU Banbury Report. (Cold Spring Harbor Laboratory, POB 100, Cold Spring Harbor, NY 11724) V.1- 1979-

BAPBAN Bulletin de l'Academie Polonaise des Sciences, Series des Sciences Biologiques. (Ars Polona-RUCH, P.O. Box 1001 P-00 068 Warsaw 1, Poland) V.5- 1957-

BATTL* Reports produced for the National Institute for Occupational Safety and Health by Battelle Pacific Northwest Laboratories, Richland, WA 99352

BAXXDU British UK Patent Application. (U.S. Patent Office, Science Library, 2021 Jefferson Davis Highway, Arlington, VA 22202)

BBACAQ Biochimica et Biophysica Acta. (Elsevier Publishing Co., POB 211, Amsterdam C, Netherlands) V.1- 1947-

BBGED3 Revista Brasileira de Genetica. Brazilian Journal of Genetics. (Dr. F.A. Moura Duarte, Dep. de Genetica,

Faculdade de Medicina de Ribeirao Preto, 14.100 Ribeirao Preto, Sao Paulo, Brazil) V.1- 1978-

BBIADT Biomedica Biochimica Acta. (Akademie-Verlag GmbH, Postfach 1233, DDR-1086 Berlin, Germany) V.42- 1983-

BBMS** "Medicaments du Systeme Nerveux Vegetalif" Bovet, D., and F. Bovet-Nitti, New York, NY, S. Karger, 1948

BBRCA9 Biochemical and Biophysical Research Communications. (Academic Press Inc., 111 5th Ave., New York, NY 10003) V.1- 1959-

BCFAAI Bollettino Chimico Farmaceutico. (Societa Editoriale Farmaceutica, Via Ausonio 12, 20123 Milan, Italy) V.33- 1894-

BCPCA6 Biochemical Pharmacology. (Pergamon Press Inc., Maxwell House, Fairview Park, Elmsford, NY 10523) V.1- 1974-

BCPHBM British Journal of Clinical Pharmacology. (Macmillan Journals, Houndmills Estate, Basingstoke, Hants RG21 2XS, England) V.1- 1974-

BCSTB5 Biochemical Society Transactions. (Biochemical Society, P.O. Box 32, Commerce Way, Whitehall Rd., Industrial Estate, Colchester CO2 8HP, Essex, England) V.1- 1973-

BCSYDM Bristol-Myers Cancer Symposia. (Academic Press, 111 Fifth Ave., New York, NY 10003) V.1- 1979-

BCTKAG Bromatologia i Chemia Toksykologiczna. (Ars Polona-RUCH, P.O. Box 1001, P-00 068 Warsaw, 1, Poland) V.4- 1971-

BCTRD6 Breast Cancer Research and Treatment. (Kluwer Academic Publishers Group, Distribution Center, POB 322, 3300 AH Dordrecht, Netherlands) V.1- 1981-

BDCGAS Berichte der Deutschen Chemischen Gesellschaft. (Leipzig/Berlin) V.1-61, 1868-1928. For publisher information, see CHBEAM

BDHU** Nachprufung der Toxicitat von Novocain und Tutocain bei Chlorali-sierten Tieren, August Barke Dissertation. (Pharmakologischen Institut der Tierarztlichen Hochschule zu Hannover, Germany, 1936)

BDKS** Studien uber die Pharmakologie des Pinakolins und einiger seiner Derivate mit besonderer Berucksichtigung der Kreislaufwirkung, Lennart Bang Dissertation. (Pharmakologischen Abteilung des Karolinischen Instituts, Stockholm, Sweden, 1934)

BDVU** Uber die Pharmakologische Wirkung des 1,3-Dioxy-2-Nicotinsaureamid-Tetrazols. Eine Experimentelle Studie zur Kenntnis der Wirkung neuer Tetrazolderivate, Rudolf Barwanietz Dissertation. (Institut fuer Veterinar-Pharmakolgie der Universitat Berlin, Germany, 1937)

BEBMAE Byulleten' Eksperimental'noi Biologii i Meditsiny. Bulletin of Experimental Biology and Medicine. (v/o Mezhdunarodnaya Kniga, Kuznetskii Most 18, Moscow G-200, USSR.) V.1- 1936-

BECCAN British Empire Cancer Campaign Annual Report. (Cancer Research Campaign, 2 Carlton House Terrace, London SW1Y 5AR, England) V.1- 1924-

BECTA6 Bulletin of Environmental Contamination and Toxicology. (Springer-Verlag New York, Inc., Service Center, 44 Hartz Way, Secaucus, NJ 07094) V.1- 1966-

BEGMA5 Beitraege zur Gerichtlichen Medizin. (Verlag Franz Deuticke, Helferstorferstr 4, A-1010 Vienna, Austria) V.1- 1911-

BEMTAM Berliner und Muenchener Tieraerztliche Wochenschrift. (Verlag Paul Parey, Lindenstr. 44-47, D-1000 Berlin 61, Germany) 1938-43; No. 1- 1946-

BENPBG Behavioral Neuropsychiatry. (Behavioral Neuropsychiatry Medical Publ., Inc., 61 E. 86th St., New York, NY 10028) V.1- 1969-

BESAAT Bulletin of the Entomological Society of America. (The Society, 4603 Calvert Rd., College Park, MD 20740) V.1- 1955-

BEXBAN Bulletin of Experimental Biology and Medicine. Translation of BEBMAE. (Plenum Publishing Corp., 233 Spring St., New York, NY 10013) V.41- 1956-

BEXBBO Biochemistry and Experimental Biology. (Piccin Medical Books, Via Brunacci, 12, 35100 Padua, Italy) V.10- 1971/72-

BHJUAV British Heart Journal. (British Medical Journal, 1172 Commonwealth Ave., Boston, MA 02134) V.1- 1939-

BIALAY Biochimica Applicata. (Parma, Italy) V.1-19, 1954-72(?). Discontinued

BIANA6 Bibliotheca Anatomica. (S. Karger AG, Postfach CH-4009, Basel, Switzerland) No. 1- 1961-

BIATDR Bulletin of the International Association of Forensic Toxicologists

BIBIAU Biotechnology and Bioengineering. (John Wiley & Sons, 605 3rd Ave., New York, NY 10016) V.4- 1962-

BIBUBX Biological Bulletin (Woods Hole, MA). (Biological Bulletin, Marine Biological Laboratories, Woods Hole, MA 02543) V.1- 1898-

BIBUDZ Biology Bulletin of the Academy of Sciences of the USSR. English translation of IANBAM. (Plenum Publishing Corp., 227 W. 17th St., New York, NY 10011) V.1- 1974-

BICHAW Biochemistry. (American Chemical Society Publications, 1155 16th St., N.W., Washington, DC 20036) V.1- 1962-

BICHBX Bioinorganic Chemistry. (Elsevier North Holland, Inc., 52 Vanderbilt Ave., New York, NY 10017) V.1-9, 1971-78

BICMBE Biochimie. (Masson et Cie, Editeurs, 120 Blvd. Saint-Germain, P-75280, Paris 06, France) V.53- 1971-

BIHAA2 Bibliotheca Haematologica. (S. Karger AG, Arnold-Boecklin-St 25, CH-4000, Basel 11, Switzerland) No. 1- 1955-

BIJOAK Biochemical Journal. (Biochemical Society, P.O. Box 32, Commerce Way, Whitehall Rd., Industrial Estate, Colchester CO2 8HP, Essex, England) V.1- 1906-

BIMADU Biomaterials. (Quadrant Subscription Services Ltd., Oakfield House, Perrymount Rd., Haywards Heath, W. Sussex, RH16 3DH, UK) V.1- 1980-

BIMDA2 Biochemical Medicine. (Academic Press, 111 5th Ave., New York, NY 10003) V.1- 1967-

BIMDB3 Biomedicine. (Masson et Cie, Editeurs, 120 Blvd. Saint-Germain, P-75280, Paris 06, France) V.18- 1973-

BIMEA5 Biologie Medicale. (Paris) V.1-60, 1903-71; new series 1972-V.4,1975

BINEAA Biologia Neonatorum. (Basel, Switzerland) V.1-14, No. 5/6, 1956-1969, For publisher information, see BNEOBV

BINKBT Biologicheskie Nauki. (v/o Mezhdunarodnaya Kniga, Kuznetskii Most 18, Moscow G-200, USSR) No. 3-1965-

BIOFX* BIOFAX Industrial Bio-Test Laboratories, Inc., Data Sheets. (1810 Frontage Rd., Northbrook, IL 60062)

BIOGAL Biologico. (Instituto Biologica, Av. Rodriques Alves, 1252, C.P. 4185, Sao Paulo, Brazil) V.1- 1935-

BIOJAU Biophysical Journal. (Rockefeller Univ. Press, 1230 York Ave., New York, NY 10021) V.1- 1960-

BIOKHI Bioorganicheskaya Khimiya. Bioorganic Chemistry. For English translation, see SJBCD5. (V/O Mezhdunarodnaya Kniga, 113095 Moscow, USSR) V.1-1975

BIORAK Biochemistry. Translation of BIOHAO. (Plenum Publishing Corp., 233 Spring St., New York, NY 10013) V.21- 1956-

BIPCBF Biological Psychiatry. (Plenum Publishing Corp., 233 Spring St., New York, NY 10013) V.1- 1969-

BIPMAA Biopolymers. (John Wiley & Sons, 605 3rd Ave., New York, NY 10158) V.1- 1963-

BIPBU* Biological & pharmaceutical bulletin. (Pharmaceutical Society of Japan, 2-12-15, Shibuya, Shibuya-ku, Tokyo 150-0002, Japan) V.1- 1993-

BIREBV Biology of Reproduction. (Society for the Study of Reproduction, 309 West Clark Street, Champaign, IL 61820) V.1- 1969-

BIRSB5 Biology of Reproduction, Supplement. (Champaign, IL) For publisher information, see BIREBV

BIRUAA Biologische Rundschau. (VEB Gustav Fischer Verlag, Postfach 176, Villengang 2, DDR-69 Jena, Germany) V.1- 1963-

BIZEA2 Biochemische Zeitschrift. (Berlin, Germany) V.1-346, 1906-67. For publisher information, see EJBCAI

BIZNAT Biologisches Zentralblatt. (VEB Georg Thieme, Hainst 17/19, Postfach 946, 701 Leipzig, Germany) V.1-1881-

BJANAD British Journal of Anesthesia. (Macmillan Press Ltd., Houndmills, Basingstoke, Hants. RG21 2XS, UK) V.1-1923-

BJCAAI British Journal of Cancer. (H.K. Lewis and Co., 136 Gower St., London WC1E 6BS, England) V.1- 1947-

BJCPAT British Journal of Clinical Practice. (Medical News Group, 1 Bedford St., London WC2E 9HD, UK) V.10(10)- 1956-

BJDCAT British Journal of Diseases of the Chest. (Bailliere Tindall, 33 The Avenue, Eastbourne BN21 3UN, UK) V.53- 1959-

BJDEAZ British Journal of Dermatology. (Blackwell Scientific Publications, Osney Mead, Oxford OX2 OEL, England) V.63- 1951-

BJEPA5 British Journal of Experimental Pathology. (H.K. Lewis and Co., 136 Gower St., London WC1E 6BS, England) V.1- 1920-

BJHEAL British Journal Haematology. (Blackwell Scientific Pub. Ltd., POB 88, Oxford, UK) V.1- 1955-

BJIMAG British Journal of Industrial Medicine. (British Medical Journal, 1172 Commonwealth Ave., Boston, MA 02134) V.1- 1944-

BJLSAF Botanical Journal of the Linnean Society. (Academic Press, 111 5th Ave., New York, NY 10003) V.62- 1969-

BJOGAS British Journal of Obstetrics and Gynaecology. (British Journal of Obstetrics and Gynaecology, 27 Sussex Place, Regent's Park, London NW1 4RG, England) V.82-1975-

BJOPAL British Journal of Ophthalmology. (British Medical Journal, 1172 Commonwealth Ave., Boston, MA 02134) V.1- 1917-

BJPCAL British Journal of Pharmacology and Chemotherapy. (London, England) V.1-33, 1946-68. For publisher information, see BJPCBM

BJPCBM British Journal of Pharmacology. (Macmillan Journals Ltd., Houndmills Estate, Basingstroke, Hampshire RG21 2XS, England) V.34- 1968-

BJPYAJ British Journal of Psychiatry. (Headley Brothers, Ashford TN24 8HH, Kent, England) V.109- 1963-

BJRAAP British Journal of Radiology. (British Institute of Radiology, 36 Portland Place, London W1N 3DG, England) V.1- 1928-

BJRHDF British Journal of Rheumatology. (Bailliere Tinball, 33 The Ave., Eastborne BN21 3UN, UK) V.1-1928-

BJSUAM British Journal of Surgery. (John Wright and Sons Ltd., 42-44 Triangle West, Bristol BS8 1EX, England) V.1- 1913-

BJURAN Journal of Urology. (Williams & Wilkins Co., 428 E. Preston St., Baltimore, MD 21202) V.1- 1917-

BKNJA5 Biken Journal. (Research Institute for Microbial Diseases, Osaka Univ., Yamada-Kami, Suita, Osaka, Japan) V.1- 1958-

BLFSBY Basic Life Sciences. (Plenum Publishing Corp., 227 W. 17th St., New York, NY 10011) V.1- 1973-

BLLIAX Bratislavske Lekarske Listy. (PNS-Ustredna Expedicia Tlace, Gottwaldovo Namestie 48/7, CS-884 19 Bratislava, Czechoslovakia) V.1- 1921-

BLOOAW Blood. (Grune and Stratton, 111 5th Ave., New York, NY 10003) V.1- 1946-

BLUTA9 Blut. (Springer-Verlag New York, Inc., Service Center, 44 Hartz Way, Secaucus, NJ 07094) V.1- 1955-

BMAOA3 Biologicheskii Zhurnal. Biological Journal. (Moscow, USSR.) V.1-7, No. 6, 1932-38. For publisher information, see ZOBIAU

BMBIES Biochemistry and Molecular Biology International. (Academic Press, Locked BAG 16, Marrickville NSW 2204, Australia) V.29- 1993-

BMBUAQ British Medical Bulletin. (Churchill Livingstone, Robert Stevenson House, 1-3 Baxter's Place, Leith Walk, Edinburgh, EH1 3AF, UK) V.1- 1943-

BMJOAE British Medical Journal. (British Medical Association, BMA House, Travistock Square, London WC1H 9JR, England) V.1- 1857-

BMRII* "U.S. Bureau of Mines Report of Investigation No. 2979" Patty, F.A., and W.P. Yant, 1929

BNEOBV Biology of the Neonate. (S. Karger AG, Postfach, CH-4009 Basel, Switzerland) V.15- 1970-

BOCKAE Bochu Kagaku. Scientific Pest Control. (Kyoto, Japan) V.1-42, 1937-77. Discontinued

BOCKAF Bochu Kagaku. Scientific Pest Control. (Kyoto, Japan) V.1-42, 1937-77. Discontinued.

BPAAAG Beitraege zur Pathologischen Anatomie und Allgemeinen Pathologie. (Stuttgart, Germany) V.3-109, No. 2, 1888-1944; V.109, No. 3-V.140, No. 4, 1947-70, For publisher information, see BTPGAZ

BPBLEO Biological and Pharmaceutical Bulletin. (Pharmaceutical Society of Japan, 2-12-15-201 Shibuya Shibuya-ku, Tokyo 150, Japan) V.16- 1993-

BPJLAQ Bangladesh Pharmaceutical Journal. (Bangladesh Pharmaceutical Society, Department of Pharmacy, University of Dacca, Dacca 2, Bangladesh) V.1- 1972-

BPNSBY Bulletin of the Psychonomic Society. (Psychonomic Society, 1108 W. 34th St., Austin, TX 78705) V.1- 1973-

BPOSA4 British Poultry Science. (Longman Group Ltd., Journals Division, 43-45 Annandale St., Edinburgh EH47 4AT, Scotland) V.1- 1960-

BPYKAU Biophysik. (Berlin, Germany) V.1-10, 1963-73

BRAIAK Brain; Journal of Neurology. (Oxford Univ. Press, Walton St., Oxford OX2 6DP, England) V.1- 1878-

BRCAB7 Basic Research in Cardiology. (Dr. Dietrich Steinkopff Verlag, Postfach 11-10-08, D-6000 Darmstadt 11, Germany) V.68- 1973-

BRGOAY Bulletin de la Societa Royale Belge de Gynecologie et d'Obstetrique. (Brussels, Belgium) V.1-39, 1925-69.

BRREAP Brain Research. (Elsevier Scientific Publishing Co., P.O. Box 211, Amsterdam, Netherlands) V.1- 1966-

BRXXAA British Patent Document. (U.S. Patent Office, Science Library, 2021 Jefferson Davis Highway, Arlington, VA 22202)

BSAMA5 Bulletin der Schweizerische Akademie der Medizinischen Wissenschaften. (Schwabe und Co., Steintorstr. 13, 4000 Basel 10, Switzerland) V.1- 1944-

BSBGAQ Berichte der Schweizerischen Botanischen Gesellschaft. (KRYPTO F. Flueck-Wirth, CH-9053 Teufen, Switzerland) V.1- 1891-

BSBSAS Boletin de la Sociedad de Biologia de Santiago de Chile. (Santiago, Chile) V.1-12, 1943-55. Discontinued

BSCFAS Bulletin de la Societe Chimique de France. (Masson et Cie, Editeurs, 120 Blvd. Saint-Germain, P-75280, Paris 06, France) V.1- 1864-

BSCIA3 Bulletin de la Societe de Chimie Biologique. (Paris, France) V.1-52, 1914-70. For publisher information, see BICMBE

BSECBU Biochemical Systematics and Ecology. (Pergamon Press Inc., Maxwell House, Fairview Park, Elmsford, NY 10523) V.2- 1974-

BSFDA3 Bulletin de la Societe Francaise de Dermatologie et de Syphiligraphie. (Paris, France) V.1-83, 1890-1976. Discontinued

BSIBAC Bolletino della Societe Italiana di Biologia Sperimentale. (Casa Editrice Libraria V. Idelson, Via Alcide De Gasperi, 55, 80133 Naples, Italy) V.2- 1927-

BSPBAD Bulletin de la Societe de Pharmacie de Bordeaux. (Societe de Pharmacie de Bordeaux, Faculte de Medecine et de Pharmacie, 91, rue Leyteire, 33000 Bordeaux, France) V.89- 1951-

BSPHAV Bulletin des Sciences Pharmacologiques. (Paris, France) V.1-49, 1899-1942. For publisher information, see APFRAD

BSPII* SPI Bulletin. (Society of the Plastics Industry, 250 Park Ave., New York, NY 10017)

BSRSA6 Bulletin de la Society Royale des Sciences de Liege. (Societe Royale des Sciences de Liege, Universite de Liege, 15, Ave des Tilleuls, B-4000 Liege, Belgium) V.1- 1932-

BSVMA8 Bulletin de la Societe des Sciences Veterinaires et de Medecine Compare de Lyon. (Societe des Sciences Veterinaires et de Medecine Comparee de Lyon, Ecole National Veterinaire de Lyon, Route de Sain-Bel, Marcy-l'Etoile, 69260 Charbonnieres-les-Bains, France) V.1- 1898-

BTDCAV Bulletin of Tokyo Dental College. (Tokyo Dental College, Misaki-cho, Chiyoda-Ku, Tokyo 101, Japan) V.1- 1960-

BTERDG Biological Trace Element Research. (The Humana Press, Inc., Crescent Manor, P.O. Box 2148, Clifton, NJ 07015) V.1- 1979-

BTMNA7 Bitamin. (Nippon Bitamin Gakkai, 4 Ushinomiya-cho, Yoshida, Sakyo-Ku, Kyoto 606, Japan) V.1- 1948-

BTPGAZ Beitraege zur Pathologie. (Gustav Fischer Verlag, Postfach 72-0143 D-7000 Stuttgart 70, Germany) V.141- 1970-

BTSRAF Biochimica e Terapia Sperimentale. (Milan, Italy) V.1-30, 1909-43. For publisher information, see NIVAAY

BUCABS Bulletin du Cancer. (Masson et Cie, Editeurs, 120 Blvd. Saint-Germain, P-75280, Paris 06, France) V.53- 1966-

BUCKL* Buckman Laboratories, Inc. (1256 N. McLean, Memphis, TN 38108)

BUMMAB Bulletin of the University of Miami School of Medicine and Jackson Memorial Hospital. (Coral Gables, FL) V.1-20, 1939-66. Discontinued

BurLW# Personal Communication from Mr. L.W. Burnette, Material Safety Dept., GAF Corp., 1361 Alps Rd., Wayne, NJ 07470, to Dr. A. Friedman, Tracor Jitco, Inc., November 2, 1978

BUYRAI Bulletin of Parenteral Drug Association. (The Association, Western Saving Fund Bldg., Broad and Chestnut Sts., Philadelphia, PA 19107) V.1- 1946-

BVIPA7 Bulletin of the Veterinary Institute in Pulawy. (Instytut Weterynaryjny, Aleja Partyzantow 55, Pulawy, Poland) V.1- 1957-

BVJOA9 British Veterinary Journal. (Bailliere Tindall, 35 Red Lion Sq., London WClR 4SG, England) V.105- 1949-

BWHOA6 Bulletin of the World Health Organization. (WHO, 1211 Geneva 27, Switzerland) V.1- 1947-

BYYADW Byoin Yakugaku. Hospital Pharmacology. (Yakuji Nippon Sha, 1-11 Izumi-cho, Kanda, Chiyoda-ku, Tokyo 101, Japan) V.1- 1975-

BZARAZ Biologicheskii Zhurnal Armenii. Biological Journal of Armenia. (v/o Mezhdunarodnaya Kniga, Kuznetskii Most 18, Moscow G-200, USSR) V.19- 1966-

CAANAT Comptes Rendus de l'Association des Anatomistes. (Paris, France) V.1- 1916(?)-

CAES** Reports supported by the Pennwalt Corp. and the Center for Air Environment Studies at the Pennsylvania State University

CAJPBD Proceedings of the Congenital Anomalies Research Association of Japan. (Kyoto, Japan) No. 1- 1961-

CALEDQ Cancer Letters (Shannon, Ireland). (Elsevier Scientific Pub. Ireland Ltd., POB 85, Limerick, Ireland) V.1- 1975-

CAMEAS California Medicine. (San Francisco, CA) V.65-119, 1946-73

CANCAR Cancer. (J.B. Lippincott Co., E. Washington Sq., Philadelphia, PA 19105) V.1- 1948-

CANJAE Canadian Anaesthetist's Society Journal. (178 St. George St., Toronto 5, Ontario M5R 2M7, Canada) V.1- 1954-

CAREBK Caries Research. (S. Karger AG, Postfach, CH-4009, Basel, Switzerland) V.1- 1967-

CARYAB Caryologia. (Caryologia, Via Lamarmora 4, 50121 Florence, Italy) V.1- 1948-

CAXXA4 Canadian Patents. (U.S. Patent Office, Science Library, 2021 Jefferson Davis Highway, Arlington, VA 22202)

CBCCT* "Summary Tables of Biological Tests" National Research Council Chemical-Biological Coordination Center. (National Academy of Science Library, 2101 Constitution Ave., N.W., Washington, DC 20418)

CBINA8 Chemico-Biological Interactions. (Elsevier Publishing, P.O. Box 211, Amsterdam C, Netherlands) V.1- 1969-

CBPBB8 Comparative Biochemistry and Physiology, B: Comparative Biochemistry. (Pergamon Press, Headington Hill Hall, Oxford 0X3 0BW, England) V.38- 1971-

CBPCBB Comparative Biochemistry and Physiology, C: Comparative Pharmacology. (Oxford, England) V.50-73, 1975-82

CBTIAE Contributions from Boyce Thompson Institute. (Yonkers, NY) V.1-24, 1925-71. Discontinued

CBTOE2 Cell Biology and Toxicology. (Princeton Scientific Publishers, Inc., 301 N. Harrison St., CN 5279, Princeton, NJ 08540) V.1- 1984-

CCCCAK Collection of Czechoslovak Chemical Communications. (Academic Press, 24-28 Oval Rd., London NW1 7DX, England) V.1- 1929-

CCECAU Criticial Reviews in Environmental Control. (Chemical Rubber Company, Cleveland, OH) V.1-1970-

CCHCDE Zhonghua Jiehe He Huxixi Jibing Zazhi. Chinese Journal of Tuberculosis and Respiratory Diseases. (China International Book Trading Corp., POB 2820, Beijing, People's Republic of China) V.1- 1978-

CCLCDY Zhonghua Zhongliu Zazhi. Chinese Journal of Oncology. (Guozi Sudian, Beijing, People's Republic of China) V.1- 1978-

CCPHDZ Cancer Chemotherapy and Pharmacology. (Springer-Verlag, Heidelberger Pl. 3, D-1 Berlin 33, Germany) V.1- 1978-

CCPTAY Contraception. (Geron-X, Publishers, P.O. Box 1108, Los Altos, CA 94022) V.1- 1970-

CCROBU Cancer Chemotherapy Reports, Part 1. (Washington, DC) V.52, No. 6-V.59, 1968-75. For publisher information, see CTRRDO

CCSUBJ Cancer Chemotherapy Reports, Part 2. (Washington, DC) V.1-5, 1964-75. For publisher information, see CTRRDO

CCSUDL Carcinogenesis-A Comprehensive Survey (Raven Press, 1140 Ave. of the Americas, New York, NY 10036) V.1- 1976-

CCYPBY Cancer Chemotherapy Reports, Part 3. (Washington, DC) V.1-6, 1968-75. For publisher information, see CTRRDO

CDESDK Contraceptive Delivery Systems. (Kluwer Academic Publishers Group Distribution Centre, POB 322, 3300 AH Dordrecht, Netherlands) V.1- 1980-

CDGU** Uber die Pharmakologische Wirkung eines dem Pentamethylentetrazol (Cardiazol) nahestehenden stickstoffhaltigen Kampherab-Kommlings, Hans Hermann Czygan Dissertation. (Hessischen Ludwigs-Universitat zu Giessen, Germany, 1934)

CDPRD4 Cancer Detection and Prevention. (Marcel Dekker, Inc., POB 11305, Church St. Station, New York, NY 10249) V.1- 1979-

CDREEA Cardiovascular Drug Reviews. (Raven Press, 1185 Avenue of the Americas, New York, NY 10036) V.6-1988-

CECED9 Commission of the European Communities, Report EUR. (Office for Official Publications, POB 1003, Luxembourg 1, Luxembourg) 1967-

CEDEDE Clinical and Experimental Dermatology. (Blackwell Scientific Publications Ltd., Osney Mead, Oxford OX2 0EL, England) V.1- 1976-

CEFYAD Ceskoslovenska Fysiologie. (Academia, Vodickova 40, Prague 1, Czechoslovakia) V.1- 1952-

CEHYAN Ceskoslovenska Hygiena. Czechoslovak Hygiene. (Ve Smeckach 30, Prague 1, Czechoslovakia). V.1-1956-

CEKNA5 Chiba-ken Eisei Kenkyusho Nenpo. (666-2 Nitona-cho, Chiva 280, Japan)

CELLB5 Cell (Cambridge, MA.). (Massachusetts Institute of Technology Press, 28 Carleton St., Cambridge, MA 02142) V.1- 1974-

CENEAR Chemical and Engineering News. (American Chemical Society, 1155 16th St., N.W., Washington, DC 20036) V.20- 1942-

CESTAT Ceskoslovenska Stomatologie. (PNS-Ustredni Expedice Tisku, Jindriska 14, Prague 1, Czechoslovakia) V.36- 1936-

CEXIAL Clinical and Experimental Immunology. (Blackwell Scientific Publications Ltd., Osney Mead, Oxford OX2 0EL, England) V.1- 1966-

CFRGBR Code of Federal Regulations. (U.S. Government Printing Office, Supterintendent of Documents, Washington, DC 20402)

CGCGBR Cytogenetics and Cell Genetics. (S. Karger AG, Arnold-Boecklin Str. 25, CH-4011 Basel, Switzerland) V.12-1973-

CGCYDF Cancer Genetics and Cytogenetics. (Elsevier North Holland, Inc., 52 Vanderbilt Ave., New York, NY 10017) V.1- 1979-

CHABA8 Chemical Abstracts. (Chemical Abstracts Service, Box 3012, Columbus, OH 43210) V.1- 1907-

ChaEB# Personal Communication from Dr. E.B. Chappel, Abbott Labs., N. Chicago, IL, to I. Pigman, Tracor Jitco, Inc., January 19, 1978

CHBEAM Chemische Berichte. (Verlag Chemie GmbH, Postfach 129/149, Pallelallee 3, D-6940, Weinheim/Bergst, Germany) V.80- 1947-

CHDDAT Comptes Rendus Hebdomadaires des Seances de l'Academie des Sciences, Serie D. (Centrale des Revues Dunod-Gauthier-Villars, 24-26 Blvd. de l'Hopital, 75005 Paris, France) V.262- 1966-

CHETBF Chest: The Journal of Circulation, Respiration and Related Systems. (American College of Chest Physicians, 911 Busse Hwy., Park Ridge, IL 60068) V.57-1970-

CHHTAT Zhonghua Yixue Zazhi. Chinese Medical Journal. (Guozi Shudian, Beijing, People's Republic of China) V.1- 1915-

CHIMAD Chimia. (Sauerlaender AG, 5001 Aarau, Switzerland) V.1- 1947-

CHINAG Chemistry and Industry. (Society of Chemical Industry, 14 Belgrave Sq., London SW1X 8PS, England) V.1-21, 1923-43; No. 1- 1944-

CHIP** Chemical Hazard Information Profile. Draft Report. (U.S. Environmental Protection Agency, Office of Toxic Substances, 401 M St., S.W., Washington, DC 20460)

CHMBAY Chemistry in Britain. (Chemical Society, Publications, Sales Office, Blackhorse Rd., Letchworth SG6 1HN, Herts, England) V.1- 1965-

CHMTBL Chemical Technology. (American Chemical Society, 1155 16th St., N.W., Washington, DC 20036) V.1-1917-

CHPUA4 Chemicky Prumysl. Chemical Industry. (ARTIA, Ve Smeckach 30, 111-27 Prague 1, Czechoslovakia) V.1-1951-

CHREAY Chemical Reviews. (American Chemical Society, 1155 16th St., N.W., Washington, DC 20036) V.1-1924-

CHROAU Chromosoma. (Springer-Verlag, Heidelberger Pl. 3, D-1000 Berlin 33, Germany) V.1- 1939-

CHRTBC Chromosomes Today. (Elsevier Scientific Publishing Co., P.O. Box 211, Amsterdam, Netherlands) V.1- 1966-

CHTHBK Chemotherapy. (S. Karger AG, Arnold-Boecklin-St 25, CH-4000, Basel 11, Switzerland) V.13-1968-

CHTPBA Chimica Therapeutica. (Editions DIMEO, Arcueil, France) V.1-8, 1965-73

CHWKA9 Chemical Week. (McGraw-Hill, Inc., Distribution Center, Princeton Rd., Hightstown, NJ 08520) V.68- 1951-

CHYCDW Zhonghua Yufangyixue Zazhi. Chinese Journal of Preventive Medicine. (42 Tung Szu Hsi Ta Chieh, Beijing, People's Republic of China) Beginning history not known

CIGET* Ciba-Geigy Toxicology Data/Indexes, 1977. (Ciba-Geigy Corp., Ardsley, NY 10502)

CIGZAF Chiba Igakkai Zasshi. Journal of the Chiba Medical Society. (Chiba, Japan) V.1-49, 1923-73. For publisher information, see CIZAAZ

CIHPDR Zhongguo Yixue Kexueyuan Xuebao. Journal of the Chinese Academy of Medicine. (China Book Trading Corp., POB 2820, Beijing, People's Republic of China) V.1-1979-

CIIT** Chemical Industry Institute of Toxicology, Docket Reports. (POB 12137, Research Triangle Park, NC 27709)

CIRUAL Circulation Research. (American Heart Association, Publishing Director, 7320 Greenville Ave., Dallas, TX 75231) V.1- 1953-

CISCB7 CIS, Chromosome Information Service. (Maruzen Co. Ltd., POB 5050, Tokyo International, Tokyo 100-31, Japan) No. 1- 1961-

CIWYAO Carnegie Institute of Washington, Year Book. (Carnegie Institution of Washington, 1530 P St., N.W., Washington, DC 20005) V.1- 1902-

CIYPDA Sichuan Yixueyuan Xuebao. Acta Akademiae Medicinae Sichuan. (Guozi Sudian, Beijing, People's Republic of China) V.1- 1970-

CIZAAZ Chiba Igaku Zasshi. Chiba Medical Journal. (Chiba Igakkai, Inohana 1-8-8, Chiba 280, Japan) V.50-1974-

CJBBDU Canadian Journal of Biochemistry and Cell Biology. (National Research Council of Canada, Ottawa, Ontario, KIA OR6 Canada) V.61- 1983-

CJBIAE Canadian Journal of Biochemistry. (Ontario, Canada) V.42-60, 1964-82. For publisher information, see CJBBDU

CJBPAZ Canadian Journal of Biochemistry and Physiology. (Ottawa, Canada) V.32-41, 1954-63. For publisher information, see CJBIAE

CJCHAG Canadian Journal of Chemistry. (National Research Council of Canada, Ottawa, Ontario, K1A OR6 Canada) V.29- 1951-

CJCMAV Canadian Journal of Comparative Medicine. (360 Bronson Ave., Ottawa, Ontario, K1R 6J3 Canada) V.1-3, 1937-39; V.32- 1968-

CJMIAZ Canadian Journal of Microbiology. (National Research Council of Canada, Administration, Ottawa, Ontario K1A OR6 Canada) V.1- 1954-

CJPEA4 Canadian Journal of Public Health. (Canadian Public Health Association, 1335 Carling Avenue, Suite 210, Ottawa, Ontario K1Z 8N8, Canada) V.20- 1929-

CJPPA3 Canadian Journal of Physiology and Pharmacology. (National Research Council of Canada, Ottawa, Ontario, K1A OR6 Canada) V.42- 1964-

CJPSDF Canadian Journal of Psychiatry. (Keith Health Care Communications, 289 Rutherford Road S., Suite 11, Brampton, Ontario L6W 3R9, Canada)

CJVRE9 Canadian Journal of Veterinary Research. (339 Booth St. Ottawa, Ontario K1R 7K1, Canada) V.50- 1986-

CKFRAY Ceskoslovenska Farmacie. (PNS-Ustredni Expedice Tisku, Jindriska 14, Prague 1, Czechoslovakia) V.1- 1952-

CKSCDN Zhonghua Minguo Shouyi Xuehui Zashi. (Chinese Society of Veterinary Science, c/o Kuo Li Tai-wan Ta Hsueh, 142 Chou Shan Lu, Taipei, 106, Taiwan) V.1- 1975-

CLBIAS Clinical Biochemistry. (Canadian Society of Clinical Chemists, 151 Slater St., Ottawa, Ontario, K1P 5H3 Canada) V.1- 1967-

CLCEAL Casopis Lekaru Ceskych. Journal of Czech Physicians. (ARTIA, Ve Smeckach 30, Prague 1, Czechoslovakia) V.1- 1862-

CLCHAU Clinical Chemistry. (American Association of Clinical Chemists, 1725 K St., N.W., Washington, DC 20006) V.1- 1955-

CLDFAT Cell Differentiation. (Elsevier/North-Holland Scientific Publishers Ltd., POB 85, Limerick, Ireland) V.1- 1972-

CLDND* Compilation of LD50 Values of New Drugs. (J.R. MacDougal, Dept. of National Health and Welfare, Food and Drug Divisions, 35 John St., Ottawa, Ontario, Canada)

CLECAP Clinical Endocrinology (Oxford). (Blackwell Scientific Pub. Ltd., POB 88, Oxford, UK) V.1- 1972-

CLMEA3 Clinical Medicine. (Clinical Medicine Publications, 444 Frontage Rd., Northfield, IL 60093) V.69- 1962-

CLNHBI Clinical Nephrology. (Dustri-Verlag Dr. Karl Feistle, Postfach 49, D-8024 Munich-Deisenhofen, Germany) V.1- 1973-

CLONEA Clinics in Oncology. (W.B. Saunders Co., W. Washington Sq., Philadelphia, PA 19105) V.1- 1982-

CLPTAT Clinical Pharmacology and Therapeutics. (C.V. Mosby Co., 11830 Westline Industrial Dr., St. Louis, MO 63141) V.1- 1960-

CLREAS Clinical Research. (American Federation for Clinical Research, 6900 Grove Rd., Thorotare, NJ 08086) V.6- 1958-

CMAJAX Canadian Medical Association Journal. (CMA House, Box 8650, Ottawa, Ontario, K1G OG8 Canada) V.1- 1911-

CMBID4 Cellular and Molecular Biology. (Pergamon Press Ltd., Headington Hill Hall, Oxford OX3 0BW, England) V.22- 1977-

CMDT** Christensen, Herbert E., Thesis, National Institute for Occupational Safety and Health, Rockville, MD 20852

CMEP** "Clinical Memoranda on Economic Poisons" US Dept. HEW, Public Health Service, Communicable Disease Center, Atlanta, GA, 1956

CMJOAP Chinese Medical Journal. (Peking, China) V.46-85, 1932-66. For publisher information, see CHMEBA

CMJODS Chinese Medical Journal (Beijing, English Edition). New Series. (Guozi Shudian, Beijing, People's Republic of China) V.1- 1975- (Adopted vol. No. 92 in 1979)

CMJRAY Calcutta Medical Journal. (Calcutta Medical Club, CMC House, 91-B Chittaranjan Ave., Calcutta, India) V.1- 1906-

CMLMDW Collection de Medecine Legale et de Toxicologie Medicale. (Masson et Cie, Editeurs, 120 Blvd. Saint-Germain, F-75280 Paris 06, France) No. 53- 1970-

CMMUAO Chemical Mutagens. Principles and Methods for Their Detection (Plenum Publishing Corp., 233 Spring St., New York, NY 10013) V.1- 1971-

CMROCX Current Medical Research and Opinion. (Clayton-Wray Publications, 27 Sloane Sq., London SW1W 8AB, England) V.1- 1973-

CMSHAF Chemosphere. (Pergamon Press Inc., Maxwell House, Fairview Park, Elmsford, NY 10523) V.1- 1971-

CMTRAG Chemotherapia. (Basel, Switzerland) V.1-12, 1960-67. For publisher information, see CHTHBK

CNCRA6 Cancer Chemotherapy Reports. (Bethesda, MD) V.1-52, 1959-68. For publisher information, see CCROBU

CNJGA8 Canadian Journal of Genetics and Cytology. (Genetics Society of Canada, 151 Slater St., Suite 907, Ottawa, Ontario, K1P 5H4 Canada) V.1- 1959-

CNJMAQ Canadian Journal of Comparative Medicine and Veterinary Science. (Gardenvale, Quebec, Canada) V.4-32, 1940-68. For publisher information, see CJCMAV

CNREA8 Cancer Research. (Public Ledger Building, Suit 816, 6th & Chestnut Sts., Philadelphia, PA 19106) V.1- 1941-

CNRMAW Canadian Journal of Research, Section E, Medical Sciences. (Ottawa, Canada) V.22-28, 1944-50. For publisher information, see CJBIAE

CNVJA9 Canadian Veterinary Journal. (Canadian Veterinary Journal, 360 Bronson Ave., Ottawa, Ontario K1R 6J3, Canada) V.1- 1960-

CODEDG Contact Dermatitis. Environmental and Occupational Dermatitis. (Munksgaard, 35 Norre Sogade, DK 1370 Copenhagen K, Denmark) V.1- 1975-

COINAV Colloques Internationaux du Centre National de la Recherche Scientifique. (Centre National de la Recherche Scientifique, 15, Quai Anatole-France, F-75700 Paris, France) V.1- 1946-

CONANO Congenital Anomalies. (Nippon Senten Ijo Gakkai, 377-2 Ono-higashi, Osakasayama, Osaka-Fu 589 Japan) V.27- 1987-

CONEAT Confinia Neurologica. (S. Karger AG, Arnold-Boecklin-St 25, CH-4000, Basel 11, Switzerland) V.1- 1938-

COREAF Comptes Rendus Hebdomadaires des Seances de l'Academie des Sciences. (Paris, France) V.1-261, 1835-1965. For publisher information, see CHDDAT

CORTBR Clinical Orthopaedics and Related Research. (J.B. Lippincott Co., E. Washington Sq., Philadelphia, PA 19105) No. 26- 1963-

COTODO Journal of Fire and Flammability/Combustion Toxicology Supplement. (Technomic Publishing Co., 265 Post Rd. W., Westport, CT 06880) V.1-2, 1974-75. For Publisher information, see JCTODH

COVEAZ Cornell Veterinarian. (Cornell University, New York State College of Veterinary Medicine, Ithaca, NY 14853) V.1- 1911-

CPAJAK Canadian Psychiatric Association Journal. (Suite 103, 225 Lisgar St., Ottawa, Ontario, K2P OC6, Canada) V.1- 1956-

CPBTAL Chemical and Pharmaceutical Bulletin. (Pharmaceutical Society of Japan, 12-15-501, Shibuya 2-chome, Shibuya-ku, Tokyo, 150, Japan) V.6- 1958-

CPCHAO Clinical Proceedings of the Children's Hospital of the District of Columbia. (Washington, DC) V.1-27, 1971. For publisher information, see CPNMAQ

CPEDAM Clinical Pediatrics. (J.B. Lippincott Co., E. Washington Sq., Philadelphia, PA 19105) V.1- 1962-

CPGPAY Comparative and General Pharmacology. (New York, NY) V.1-5, 1970-74. For publisher information, see GEPHDP

CPHADV Clinical Pharmacy. (American Society of Hospital Pharmacists, 4630 Montgomery Ave., Bethesda, MD 20814) V.1- 1982-

CPHPA5 Jiepou Xuebao. Journal of Anatomy. (Guozi Shudian, Beijing, People's Republic of China) V.1-9, 1953-66; V.10- 1979-

CPNMAQ Clinical Proceedings, Children's Hospital National Medical Center. (2125 13th St., N.W., Washington, DC 20009) V.28- 1972-

CRAAA7 Current Researches in Anesthesia and Analgesia. (Cleveland, OH) V.1-35, 1922-56. For publisher information, see AACRAT

CRBCAI CRC Critical Reviews in Biochemistry. (CRC Press, Inc., 2000 Corporate Blvd., NW, Boca Raton, FL 33431) V.1- 1972-

CRDLP* U.S. Army Chemical Research and Development Laboratory, Special Publication. (Edgewood Arsenal, MD 21010)

CRDLR* U.S. Army Chemical Research and Development Laboratory, Technical Report. (Edgewood Arsenal, MD 21010)

CRNGDP Carcinogenesis. (Information Retrieval, 1911 Jefferson Davis Highway, Arlington, VA 22202) V.1- 1980-

CroHP# Personal Communication from H.P. Crocker, Ultimate Holding Co.-Reckitt and Colman Ltd., London, England, to R.L. Tatken, NIOSH, Cincinnati, OH, September 2, 1980

CRSBAW Comptes Rendus des Seances de la Societe de Biologie et de Ses Filiales. (Masson et Cie, Editeurs, 120 Blvd. Saint-Germain, P-75280, Paris 06, France) V.1- 1849-

CRSHAG Circulatory Shock. (Alan R. Liss Inc., 150 5th Ave., New York, NY 10011) V.1- 1974-

CRSUBM Cancer Research Supplement (Williams & Wilkins Company, 428 E. Preston St., Baltimore, MD 21202) No. 1-4, 1953-56

CRTBB2 CRC Critical Reviews in Biochemistry. (CRC Press, Inc., 2000 Corporate Blvd., NW, Boca Raton, FL 33431) V.1- 1972-

CRTOEC Chemical Research in Toxicology. (American Chemical Soc., Distribution Office Dept. 223, POB 57/36, West End Station, Washington, DC 20037) V.1- 1988-

CRTXB2 CRC Critical Reviews in Toxicology. (CRC Press, Inc., 2000 N.W. 24th St., Boca Raton, FL 33431) V.1-1971-

CSHCAL Cold Spring Harbor Conferences on Cell Proliferation. (Cold Spring Harbor Laboratory, POB 100, Cold Spring Harbor, NY 11724) V.1- 1974-

CSHSAZ Cold Spring Harbor Symposia on Quantitative Biology. (Cold Spring Harbor Laboratory of Quantitative Biology, Cold Spring Harbor, NY 11724) V.1- 1933-

CSLNX* U.S. Army Armament Research & Development Command, Chemical Systems Laboratory, NIOSH Exchange Chemicals. (Aberdeen Proving Ground, MD 21010)

CSPEAX Cosmetics and Perfumery. (Allured Publishing Corp., Box 318, Wheaton, IL 60187) V.88- 1973-

CTCEA9 Current Therapeutic Research, Clinical and Experimental. (Therapeutic Research Press, P.O. Box 514, Tenafly, NJ 07670) V.1- 1959-

CTKIAR Cell and Tissue Kinetics. (Blackwell Scientific Publications Ltd., Osney Mead, Oxford OX2 0EL, England) V.1- 1968-

CTOIDG Cosmetics and Toiletries. (Allured Publishing Corp., P.O. Box 318, Wheaton, IL 60187) V.91- 1976-

CTRRDO Cancer Treatment Reports. (U.S. Government Printing Office, Superintendent of Documents, Washington, DC 20402) V.60- 1976-

CTOXAO Clinical Toxicology. (New York, NY) V.1-18, 1968-81. For publisher information, see JTCTDW

CTSRCS Cell and Tissue Research. (Springer-Verlag New York, Inc., Service Center, 44 Hartz Way, Secaucus, NJ 07094) V.148- 1974-

CTYAD8 Zhongcaoyao. Chinese Herbal Medicine. (Tsa Chih Pien Chi Pu, Hu-nan Yao Kung Yeh Yen Chiu So, Shao-Yang, Hu-nan, People's Republic of China) V.11-1980-

CUMIDD Current Microbiology. (Springer-Verlag New York, Inc., Service Center, 44 Hartz Way, Secaucus, NJ 07094) V.1- 1978-

CVREAU Cardiovascular Research. (British Medical Journal, Box 560B, Kennebunkport, ME 04046) V.1- 1967-

CURTOX Current Toxicology. (Nova Science Publishers, Inc., 6080 Jericho Turnpike, Suite 207, Commack, NY 11725) V.1 1993-

CUSCAM Current Science. (Current Science Association, Mgr., Raman Research Institute, Bangalore 6, India) V.1-1932-

CUTIBC CUTIS; Cutaneous Medicine for the Practitioner. (Technical Pub., 875 Third Ave., New York, NY 10022) 1965-

CUTOEX Current Toxicology. (Nova Science Publishers, Inc., 6080 Jericho Turnpike, Suite 207, Commack, NY 11725) V.1 1993-

CWLTM* Chemical Warfare Laboratories Technical Memorandum. (U.S. Army Chemical Center, Edgewood Arsenal, MD 21010)

CYGEDX Cytology and Genetics. English Translation of Tsitologiya i Genetika. (Allerton Press, Inc., 150 Fifth Ave., New York, NY 10011) V.8- 1974-

CYLPDN Zhongguo Yaoli Xuebao. Acta Pharmacologica Sinica. (Shanghai K'o Hsueh Chi Shu Ch'u Pan She, 450 Shui Chin Erh Lu, Shanghai 200020, People's Republic of China) V.1- 1980-

CYTBAI Cytobios. (The Faculty Press, 88 Regent St., Cambridge, England) V.1- 1969-

CYTOAN Cytologia. (Maruzen Co. Ltd., P.O. Box 5050, Tokyo International, Tokyo 100-31, Japan) V.1- 1929-

CYTZAM Cytobiologie. (Stuttgart 1, Germany) V.1-18, 1969-79. For publisher information, see EJCBDN

DABBBA Dissertation Abstracts International, B: The Sciences and Engineering. (University Microfilms, A Xerox Co., 300 N. Zeeb Rd., Ann Arbor, MI 48106) V.30- 1969-

DABSAQ Dissertation Abstracts, B: The Sciences and Engineering. (University Microfilms, A Xerox Co., 300 N. Zeeb Rd., Ann Arbor, MI 48106) V.27-29, 1966-69

DADEDV Drug and Alcohol Dependence. (Elsevier Sequoia SA, POB 851, CH-1001 Lausanne, Switzerland) V.1- 1975-

DAKMAJ Deutsches Archiv fuer Klinische Medizin. (Munich, Germany) V.1-211, 1865-1965. For publisher information, see EJCIB8

DANKAS Doklady Akademii Nauk S.S.S.R. (v/o Mezhdunarodnaya Kniga, Kuznetskii Most 18, Moscow G-200, USSR.) V.1- 1933-

DANND6 Dopovidi Akademii Nauk Ukrains'koi RSR, Seriya B: Geologichini, Khimichni ta Biologichni Nauki. (v/o Mezhdunarodnaya Kniga, 121200 Moscow, USSR) 1976-

DARUEE Dokl Akad Nauk resp. Uzb.

DAZEA2 Deutsche Apotheker-Zeitung. (Deutscher Apotheker-Verlag, Postfach 40, D-7000 Stuttgart 1, Germany) V.49- 1934-

DAZRA7 Doklady Akademii Nauk Azerbaidzhnskoi SSR. Proceedings of the Academy of Sciences of the Azerbaidzhan SSR. (v/o Mezhdunarodnaya Kniga, Kuznetskii Most 18, Moscow G-200, USSR) V.1- 1945-

DBABEF Doga Bilim Dergisi, Seri A2: Biyoloji. Natural Science Journal, Series A2. (Turkiye Bilimsel ve Teknik Arastirma Kurumu, Ataturk Bul. No. 221, Kavaklidere, Ankara, Turkey) V.8- 1984-

DBANAD Doklady Bolgarskoi Akademii Nauk. (Hemus, Blvd. Russki 6, Sofia, Bulgaria) V.1- 1948-

DBLRAC Doklady Akademii Nauk BSSR. Proceedings of the Academy of Sciences of the Belorussian SSR. (v/o Mezhdunarodnaya Kniga, Kuznetskii Most 18, Moscow G-200, USSR) V.1- 1957-

DBTEAD Diabete. (Le Raincy, France) V.1-22, 1953-1974

DBTGAJ Diabetologia. (Springer-Verlag New York, Inc., Service Center, 44 Hartz Way, Secaucus, NJ 07094) V.1-1965-

DCTODJ Drug and Chemical Toxicology. (Marcel Dekker, POB 11305, Church St. Station, New York, NY 10249) V.1- 1977/78-

DDEVD6 Drug Development and Evaluation. (Gustav Fischer Verlag, Postfach 720143, D-7000 Stuttgart 70, Germany) V.1- 1977-

DDIPD8 Drug Development and Industrial Pharmacy. (Marcel Dekker, POB 11305, Church St. Station, New York, NY 10249) V.3- 1977-

DDREDK Drug Development Research. (Alan R. Liss, Inc., 150 5th Ave., New York, NY 10011) V.1- 1981-

DDSCDJ Digestive Diseases and Sciences. (Plenum Publishing Corp., 233 Spring St., New York, NY 10013) V.24- 1979-

DEBIAO Developmental Biology. (Academic Press, Inc., 111 5th Ave., New York, NY 10003) V.1- 1959-

DEBIDR Developments in Biochemistry. (Elsevier North Holland, Inc., 52 Vanderbilt Ave., New York, NY 10017) V.1- 1978-

DECRDP Drugs under Experimental and Clinical Research. (J.R. Prous Pub., Apartado de Correos 1179, Barcelona, Spain) V.1- 1977-

DEGEA3 Deutsche Gesundheitswesen. (VEB Verlag Volk und Gesundheit, Neue Gruenstr 18, 102 Berlin, Germany) V.1- 1946-

DEMAEP Dental Materials. (Munksgaard International Pub., POB 2148, DK-1016 Copenhagen K, Denmark) V.1-1985-

DEPBA5 Developmental Psychobiology. (John Wiley & Sons Ltd., Baffins Lane, Chichester, Sussex P01 1UD, England) V.1- 1968-

DEVEAA Defense des Vegetaux. (Federation Nationale des Groupements de Protection des Cultures, 149, rue de Bercy, 75595 Paris Cedex, 12, France) V.1- 1947-

DERAAC Dermatologica. (Albert J. Phiebig, Inc., P.O. Box 352, White Plains, NY 10602) V.79- 1939-

DFSCDX Developments in Food Science. (Elsevier Science Pub. Co., Inc., 52 Vanderbilt Ave., New York, NY 10017) V.1- 1978-

DGDFA5 Development, Growth and Differentiation. (Maruzen Co. Ltd., P.O. Box 5050, Tokyo International, Tokyo 100-31, Japan) V.11- 1969-

DHEFDK HEW Publication (FDA. United States). (Washington, DC) 19??-1979(?). For publisher information, see HPFSDS

DIAEAZ Diabetes. (American Diabetes Association, 600 5th Ave., New York, NY 10020) V.1- 1952-

DICHAK Diseases of the Chest. (Chicago, IL) V.1-56, 1935-69. For publisher information, see CHETBF

DICPBB Drug Intelligence and Clinical Pharmacy. (Drug Intelligence and Clinical Pharmacy, Inc., University of Cincinnati, Cincinnati, OH 45267) V.3- 1969-

DICRAG Diseases of the Colon and Rectum. (Harper and Row, Publishers, 10 E. 53rd St., New York, NY 10022) V.1-1958-

DIGEBW Digestion. (S. Karger AG, Arnold-Boecklin Street 25, CH-4011 Basel, Switzerland) V.1- 1968-

DIMCAL Developments in Industrial Microbiology. (Society for Industrial Microbiology, 1401 Wilson Blvd., Arlington, VA 22209) V.1- 1960-

DIPHAH Dissertationes Pharmaceuticae. (Warsaw, Poland) V.1-17, 1949-65. For publisher information, see PJPPAA

DKBSAS Doklady Biological Sciences (English Translation). (Plenum Publishing Corp., 233 Spring St., New York, NY 10013) V.112- 1957-

DLRUAJ Deutsche Lebensmittel - Rundschau. (Wissenschaftliche Verlagsgesellschaft Postfach 101061, D-70009 Stutgart, Germany) V. 34 - 42 n.10, 1936-1944, (1944 - 1947 suspended) V 43 n.1- July 1947-

DMBUAE Danish Medical Bulletin. (Ugeskrift for Laeger, Domus Medica, 2100 Copenhagen, Denmark) V.1- 1954-

DMDSAI Drug Metabolism and Disposition. (Williams and Wilkins Co., 428 E. Preston St., Baltimore, MD 21202) V.1- 1973-

DMWOAX Deutsche Medizinische Wochenschrift. (Georg Thieme Verlag, Herdweg 63, Postfach 732, 7000 Stuttgart 1, Germany) V.1- 1875-

DNEUD5 Developments in Neuroscience (Amsterdam). (Elsevier North Holland, Inc., 52 Vanderbilt Ave., New York, NY 10017) V.1- 1977-

DNSSAW Diseases of the Nervous System, Supplement. (Irvington, NJ) For publisher information, see DNSYAG

DNSYAG Diseases of the Nervous System. (Physicians Postgraduate Press, Box 38293, Memphis, TN 38138) V.1- 38, 1940-77

DOEAAH Down to Earth. A Review of Agricultural Chemical Progress. (Dow Chemical U.S.A., 1703 S. Saginaw Rd., Midland, MI 48640) V.1- 1945-

DOESD6 DOE Symposium Series. (NTIS, 5285 Port Royal Rd., Springfield, VA 22161) No. 45- 1978-

DOVEAA Defense des Vegetaux. (Federation Nationale des Groupements de Protection des Cultures, 149, rue de Bercy, 75595 Paris Cedex, 12, France)V.1-1947-

DOWCC* Dow Chemical Company Reports. (Dow Chemical U.S.A., Health and Environment Research, Toxicology Research Lab., Midland, MI 48640)

DPHFAK Dissertationes Pharmaceuticae et Pharmacologicae. (Warsaw, Poland) V.18-24, 1966-72. For publisher information, see PJPPAA

DPIRDU Dangerous Properties of Industrial Materials Report. (Van Nostrand Reinhold Co., Inc., 115 Fifth Avenue, New York, NY 10003) V.1- 1981-

DPTHDL Developmental Pharmacology and Therapeutics. (S. Karger AG, Postfach CH-4009 Basel, Switzerland) V.1- 1980-

DRFUD4 Drugs of the Future. (J.R. Prous, S.A. International Publishers, Apartado de Correos 1641, Barcelona, Spain) V.1- 1975/76-

DRISAA Drosophila Information Service. (Cold Spring Harbor Laboratory, POB 100, Cold Spring Harbor, NY 11724) No. 1- 1934-

DRSAEA Drug Safety. (Adis International Ltd., Private Bag 65901, Mairangi Bay, Auckland 10, New Zealand) V.5- 1990-

DRSTAT Drug Standards. (Washington, DC) V.19-28, 1951-60. For publisher information, see JPMSAE

DRUGAY Drugs. International Journal of Current Therapeutics and Applied Pharmacology Reviews. (ADIS Press Ltd., 18/F., Tung Sun Commercial Centre, 194-200 Lockhart Road, Wanchai, Hong Kong)

DTESD7 Developments in Toxicology and Environmental Science. (Elsevier, Scientific Publishing Co., POB 211, 1000 AE Amsterdam, Netherlands) V.1- 1977-

DTLVS* "Documentation of Threshold Limit Values for Substances in Workroom Air." For publisher information, see 85INA8

DTLWS* "Documentation of the Threshold Limit Values for Substances in Workroom Air" Supplements. For publisher information, see 85INA8

DTTIAF Deutsche Tieraerztliche Wochenschrift. (Verlag M. und H. Schaper, Postfach 260669, 3 Hanover 26, Germany) V.1- 1893-

DUPON* E.I. Dupont de Nemours and Company, Technical Sheet. (1007 Market St., Wilmington, DE 19898)

DZZEA7 Deutsche Zahnaerztliche Zeitschrift. (Carl Hanser Verlag, Postfach 860420, D-8000 Munich 86, Germany) V.1- 1946-

EAGRDS Experimental Aging Research. (Beech Hill Publishing Co., P.O. Box 136, Southwest Harbor, ME 04679) V.1- 1975-

EAMJAV East African Medical Journal (P.O. Box 41632, Nairobi, Kenya) V.9- 1932-

EAPHA6 Eastern Pharmacist. (Eastern Pharmacist, 507, Ashok Bhawan, 93, Neru Place, New Delhi 110019, India) V.1- 1958-

EATR** U.S. Army, Edgewood Arsenal Technical Report. (Aberdeen Proving Ground, MD 21010)

EbeAG# Personal Communication to NIOSH from A.G. Ebert, International Glutamate Technical Committee, 85 Walnut St., Watertown, MA 02172

ECBUDQ Ecological Bulletins. (Swedish National Science Research Council, Stockholm, Sweden) Number 19- 1975-

ECEBDI Experimental Cell Biology. (Phiebig Inc., POB 352, White Plains, NY 10602) V.44- 1976-

ECJPAE Endocrinologia Japonica. (Japan Publications Trading Co., 1255 Howard St., San Francisco, CA 94103) V.1- 1954-

ECNEAZ Electroencephalography and Clinical Neurophysiology. (Elsevier Scientific Publishing Co., POB 211, 1000 AE Amsterdam, Netherlands) V.1- 1949-

ECREAL Experimental Cell Research. (Academic Press, 111 5th. Ave., New York, NY 10003) V.1- 1950-

EDRCAM Endocrine Research Communications. (Marcel Dekker, POB 11305, Church St. Station, New York, NY 10249) V.1- 1974-

EDWU** Beitrag zur Toxikologie Technischer Weichmachungsmittel, Heinrich Eller Dissertation. (Pharmakologischen Institut der Universitat Wurzburg, Germany, 1937)

EESADV Ecotoxicology and Environmental Safety. (Academic Press, 111 5th Ave., New York, NY 10003) V.1- 1977-

EGESAQ Egeszsegtudomany. (Kultura, POB 149, H-1389 Budapest, Hungary) V.1- 1957-

EGJBAY Egyptian Journal of Botany. (National Information and Documentation Centre, Al-Tahrir St., Awgaf P.O. Dokki, Cairo, Egypt) V.1- 1958-

EISOAU Eiyo to Shokuryo. Food and Nutrition. (Nippon Eiyo, Shokuryo Gakkai, C/o Nippon Gakkai Jimu Senta 2-4-16 Yayoi, Bunkyo-ku, Tokyo 113, Japan) V.10- 1957-

EJBCAI European Journal of Biochemistry. (Springer-Verlag, Heidelberger Pl. 3, D-1 Berlin 33, Germany) V.1- 1967-

EJBLAB Egyptian Journal of Bilharziasis. (National Information Documentation Center, Tahrir St., Dokki, Cairo, Egypt) V.1- 1974-

EJCAAH European Journal of Cancer. (Pergamon Press, Headington Hill Hall, Oxford OX3 OEW, England) V.1- 1965-

EJCBDN European Journal of Cell Biology. (Wissenschaftliche Verlagsgesellschaft mbH, Postfach 40, D-7000, Stuttgart l, Germany) V.19- 1979-

EJCIB8 European Journal of Clinical Investigation. (Springer-Verlag, Heidelberger Pl. 3, D-1 Berlin 33, Germany) V.1- 1970-

EJCODS European Journal of Cancer and Clinical Oncology. (Pergamon Press Ltd., Headington Hill Hall, Oxford OX3 0BW, England) V.17, No. 7- 1981-

EJCPAS European Journal of Clinical Pharmacology. (Springer-Verlag, Heidelberger Pl. 3, D-1 Berlin 33, Germany) V.3- 1970-

EJGCA9 Egyptian Journal of Genetics and Cytology. (Egyptian Soc. of Genetics, c/o Alexandria Univ., Faculty of Agriculture, Dept. of Genetics, Alexandria, Egypt) V.1- 1972-

EJMBA2 Egyptian Journal of Microbiology. (National Information and Documentation Centre, A1-Tahrir St., Awqaf P.O. Dokki, Cairo, Egypt) V.7- 1972-

EJMCA5 European Journal of Medicinal Chemistry. Chimie Therapeutique. (Center National de la Recherche Scientifique, 3 rue J.B. Clement, F-92290 Chatenay-Malabry, France) V.9- 1974-

EJNMD9 European Journal of Nuclear Medicine. (Springer-Verlag New York, Inc., Service Center, 44 Hartz Way, Secaucus, NJ 07094) V.1- 1975-

EJPEDT European Journal of Pediatrics. (Springer-Verlag New York, Inc., Service Center, 44 Hart Way, Secaucus, NJ 07094) V.121- 1975-

EJPHAZ European Journal of Pharmacology. (North-Holland Publishing, POB 211, 1000AE Amsterdam, Netherlands) V.1- 1967-

EJRDD2 European Journal of Respiratory Diseases. (Munksgaard, 35 Norre Sogade, DK 1370 Copenhagen K, Denmark) V.61- 1980-

EJTXAZ European Journal of Toxicology and Environmental Hygiene. (Paris, France) V.7-9, 1974-76. For publisher information, see TOERD9

EKFAE9 Eksperimental'naya i Klinicheskaya Farmakologiya. Experimental and Clinical Pharmacology. (Mezhdunarodnaya Kniga, ul. B. Yakimanka, 39, 117049 Moscow, Russia) V.55- 1992-

EKFMA7 Eksperimental'naya i Klinicheskaya Farmakoterapiya. (Akademiya Nauk Latviiskoi SSR, Inst., Organicheskogo Sinteza, ul. Aizkraukles 21, Riga, USSR.) No. 1- 1970-

EKMMA8 Eksperimentalna Meditsina i Morfologiya. (Hemus, Blvd. Russki 6, Sofia, Bulgaria) V.1- 1962-

EKSODD Eksperimental'naya Onkologiya. Experimental Oncology. (v/o Mezhdunarodnaya Kniga, 121200 Moscow, USSR) V.1- 1979-

EMERY* Emery Industries, Inc., Data Sheets. (4900 Este Ave., Cincinnati, OH 45232)

EMMUEG Environmental and Molecular Mutagenesis. (Alan R. Liss, Inc., 4 E. 11th St., New York, NY 10003) V.10- 1987-

EMPSAL Experimental and Molecular Pathology, Supplement. (Academic Press, 111 5th Ave., New York, NY 10003) No. 1- 1963-

EMSUA8 Experimental Medicine and Surgery. (Brooklyn Medical Press, 600 Lafayette Ave., Brooklyn, NY 11216) V.1- 1943-

ENDKAC Endokrinologie. (Leipzig, Germany) V.1-80, 1928-82. For publisher information, see EXCEDS

ENDOAO Endocrinology (Baltimore). (Williams & Wilkins Co., 428 E. Preston St., Baltimore, MD 21203) V.1- 1917-

ENMUDM Environmental Mutagenesis. (New York, NY) V.1-9, 1979-87. For publisher information, see EMMUEG. V.1- 1979-

ENPBBC Environmental Physiology and Biochemistry. (Copenhagen, Denmark) V.2-5, No. 6, 1972-5, Discontinued

ENVIDV Environmental Internationl. (Pergamon Press, Ltd., Headington Hill Hall, Oxford OX3 0BW, England) V.1- 1978-

ENVRAL Environmental Research. (Academic Press, 111 5th Ave., New York, NY 10003) V.1- 1967-

ENZYAS Enzymologica. (Dr. W. Junk bv Publishers, POB 13713, 2501 ES The Hague, Netherlands) V.1-43, 1936-72

EOGRAL European Journal of Obstetrics, Gynecology and Reproductive Biology. (Elsevier Science Publishing BV, POB 211,1000 AE Amsterdam, Netherlands) V3- 1973-

EPASR* United States Environmental Protection Agency, Office of Pesticides and Toxic Substances. (U.S. Environmental Protection Agency, 401 M St., S.W., Washington, DC 20460) History Unknown

EPILAK Epilepsia. (Elsevier Publishing Co., P.O. Box 211, Amsterdam C, Netherlands) V.1- 1959-

EPXXDW European Patent Application. (U.S. Patent Office, Science Library, 2021 Jefferson Davis Highway, Arlington, VA 22202)

EQSFAP Environmental Quality and Safety. (Academic Press, 111 5th Ave., New York, NY 10003) V.1- 1972-

EQSSDX Environmental Quality and Safety, Supplement. (Academic Press, 111 5th Ave., New York, NY 10003) V.1- 1975-

ERNFA7 Ernaehrungsforschung. (Akademie-Verlag GmbH, Leipziger St. 3-4, 108 Berlin, Germany) V.1- 1956-

ESDBAK Eisei Dobutsu. Sanitary Zoology. (Japan Society of Sanitary Zoology, c/o Institute of Medical Science, Unversity of Tokyo, 4-6- 1 Shiroganedai, Minato-ku, Tokyo 108, Japan) V.1- 1950-

ESKGA2 Eisei Kagaku. (Nippon Yakugakkai, 2-12-15 Shibuya, Shibuya-Ku, Tokyo 150, Japan) V.1- 1953-

ESKHA5 Eisei Shikenjo Hokoku. Bulletin of the National Hygiene Sciences. (Kokuritsu Eisei Shikenjo, 18-1 Kamiyoga 1 chome, Setagaya-ku, Tokyo, Japan) V.1- 1886-

ETATAW Eesti N.S.V. Teaduste Akadeemia Toimetised, Biologia. (v/o Mezhdunarodnaya Kniga, Kuznetskii Most 18, Moscow G-200, USSR.) V.5- 1956-

ETOPFR Environmental Toxicology and Pharmacology (Elsevier Science, P.O.Box 7247-7682,Philadelphia,PA 19170 -7682,USA OR Elsevier Science B.V.,P.O.Box 1270,1000 BG Amsterdam,The Netherlands) V.1- Feb.1996-

ETPAEK Experimental and Toxicologic Pathology. (Gustav Fischer Verlag Jena, Postfach 100537, D-07705 Jena, Germany) V.44 1992-

EURNE* European Neuropsychopharmacology (Elsevier Science, P.O.Box 7247-7682,Philadelphia,PA 19170 - 7682,USA OR Elsevier Science B.V.,P.O.Box 1270,1000 BG Amsterdam,The Netherlands) V.1- Nov.1990-

EUURAV European Urology. (S. Karger Publishers, Inc., 150 Fifth Ave., Suite 1105, New York, NY 10011) V.1- 1975-

EVETBX Environmental Entomology. (Entomological Society of America, 4603 Calvert Rd., College Park, MD 20740) V.1- 1972-

EVHPAZ EHP, Environmental Health Perspectives. Subseries of DHEW Publications. (U.S. Government Printing Office, Superintendent of Documents, Washington, DC 20402) No. 1- 1972-

EVMJB9 Journal of the Egyptian Veterinary Medical Association. (Egyptian Veterinary Medical Association, 8, 26 July St., Cairo, Egypt) V.27- 1967-

EVPHBI Environmental Physiology. (Copenhagen, Denmark) V.1 (No. 1-4), 1971, For publisher information, see ENPBBC

EVSRBT Environmental Science Research. (Plenum Publishing Corp., 233 Spring St., New York, NY 10013) V.1- 1972-

EVSSAV Environmental Space Science. English Translation of Kosmicheskaya Biologiya Meditsina. 1967-70

EXCEDS Experimental and Clinical Endocrinology. (Johann Ambrosius Barth Verlag, Postfach 109, DDR-7010 Leipzig, Germany) V.81- 1983-

EXERA6 Experimental Eye Research. (Academic Press, Inc., 1 E. 1st., Duluth, MN 55802) V.1- 1961-

EXMDA4 International Congress Series - Excerpta Medica. (Elsevier North Holland, Inc., 52 Vanderbilt Ave., New York, NY 10017) No. 1- 1952-

EXMPA6 Experimental and Molecular Pathology. (Academic Press, 111 5th Ave., New York, NY 10003) V.1- 1962-

EXNEAC Experimental Neurology. (Academic Press, 111 5th Ave., New York, NY 10003) V.1- 1959-

EXPAAA Experimental Parasitology. (Academic Press, 111 5th Ave., New York, NY 10003) V.1- 1951-

EXPADD Experimental Pathology. (Elsevier Scientific Publishers Ireland Ltd., POB 85, Limerick, Ireland) V.19- 1981-

EXPEAM Experientia. (Birkhaeuser Verlag, P.O. Box 34, Elisabethenst 19, CH-4010, Basel, Switzerland) V.1- 1945-

EXPTAX Experimentelle Pathologie. (Jena, Germany) V.1-18, 1967-80. For publisher information, see EXPADD

EXXON* Exxon Chemical Company. (P.O. Box 3272, Houston, Texas 77001)

FAATDF Fundamental and Applied Toxicology. (Academic Press, Inc., 1 E. First St., Duluth, MN 55802) V.1- 1981-

FACOEB Food Additives and Contaminants. (Taylor and Francis Inc., 242 Cherry St., Philadelphia, PA 19106) V.1- 1984-

FAONAU Food and Agriculture Organization of United Nations, Report Series. (FAO-United Nations, Room 101, 1776 F Street, NW, Washington, DC 20437)

FATOAO Farmakologiya i Toksikologiya (Moscow). (v/o Mezhdunarodnaya Kniga, Kuznetskii Most 18, Moscow G-200, USSR.) V.2- 1939- For English translation, see PHTXA6 and RPTOAN

FATOBP Farmakologiya i Toksikologiya (Kiev). Pharmacology and Toxicology. (Kievskii Nauchno-Issledovatel'skii Institut Farmakologii i Toksikologii, Kiev, USSR.) No. 1- 1964-

FAVUAI Fiziologicheski Aktivnye Veshchestva. Physiologically Active Substances. (Akademiya Nauk Ukrainskoi S.S.R., Kiev, USSR.) No. 1- 1966-

FAZMAE Fortschritte der Arzneimittelforschung. (Birkhauser Verlag, P.O. Box 34, Elisabethenst 19, CH-4010, Basel, Switzerland) V.1- 1959-

FCHNA9 Food Chemical News. (Food Chemical News, Inc., 1101 Pennsylvania Ave., S.E., Washington, DC 20003) V.1- 1959-

FCTOD7 Food and Chemical Toxicology. (Pergamon Press, Headington Hill Hall, Oxford OX3 0BW, England) V.20- 1982-

FCTXAV Food and Cosmetics Toxicology. (London, UK) V.1-19, 1963-81. For publisher information, see FCTOD7

FDHU** Uber die lokalanasthetischen Eigenschaften des salzsauren Salzes des Benzoyl-d-ekgonin-n-propylesters, Gerda Felgner Dissertation. (Pharmakologischen Institut der Martin-Luther-Universitat Halle-Wittenburg, Germany, 1933)

FDMU** Zur Pharmakologie des Octins, Adolf Fiegenbaum Dissertation. (Pharmakologischen Institut der Westfalischen Wilhelms-Universitat, Munster, Germany, 1935)

FDRLI* Food and Drug Research Labs., Papers. (Waverly, NY 14892)

FDWU** Uber die Wirkung Verschiedener Gifte Auf Vogel, Ludwig Forchheimer Dissertation. (Pharmakologischen Institut der Universitat Wurzburg, Germany, 1931)

FEBLAL FEBS Letters. (Elsevier Scientific Publishing Co., POB 211, 1000 AE Amsterdam, Netherlands) V.1- 1968-

FEPRA7 Federation Proceedings, Federation of American Societies for Experimental Biology. (9650 Rockville Pike, Bethesda, MD 20014) V.1- 1942-

FEREAC Federal Register. (U.S. Government Printing Office, Supt. of Documents, Washington, DC 20402) V.1- 1936-

FESTAS Fertility and Sterility. (American Fertility Society, 1608 13th Ave. S., Birmingham, AL 35205) V.1- 1950-

FGIGDO Fujita Gakuen Igakkaishi. Journal of the Fujita Gakuen Medical Society. (Nagoya Hoken Eisei Daigaku

Fujita Gakuen Igakkai, 1-98 Dengakuga, Kubo, Kutsukake-Cho, Toyoake, Aichi-Ken 470-11, Japan) V. 1- 1977-

FHCYAI Folia Histochemica et Cytochemica. (Ars Polona-RUCH, P.O. Box 1001, P-00 068 Warsaw, 1, Poland) V.1- 1963-

FKGCAR Fortschritte der Kiefer und Gesichts-Chirurgie. (Georg Thieme Verlag, Stuttgart, Germany) V.1- 1955-

FKIZA4 Fukuoka Igaku Zasshi. (Fukuoka Igakkai, c/o Kyushu Daigaku Igakubu, Tatekasu, Fukuoka-shi, Fukuoka, Japan) V.33- 1940-

FLABAZ Fluoride Abstracts. (Kettering Lab., University of Cincinnati, College of Medicine, Dept. of Environmental Health, Cincinnati, OH 45219) No. 1- 1955/58-

FLCRAP Fluorine Chemistry Reviews. (Marcel Dekker Inc., 305 E. 45th St., New York, NY 10017) V.1- 1967-

FLUOA4 Fluoride. (International Society for Fluoride Research, P.O. Box 692, Warren, MI 48090) V.3- 1970-

FMCHA2 Farm Chemicals Handbook. (Meister Publishing, 37841 Euclid Ave., Willoughy, OH 44094)

FMDZAR Fortschritte der Medizin. (Dr. Schwappach und Co., Wessobrunner Str 4, 8035 Gauting vor Muenchen, Germany) V.1- 1883-

FMLED7 FEMS Microbiology Letters. (Elsevier Scientific Publishing Co., POB 211, Amsterdam, Netherlands) V.1- 1977-

FMORAO Folia Morphologica (Prague). (Plenum Publishing Corp. 233 Spring St., New York, NY 10013) V.13- 1965-

FMTYA2 Farmatsiya (Sofia). (Hemus, Blvd. Russki 6, Sofia, Bulgaria) V.1- 1951-

FNSCA6 Forensic Science. (Elsevier Sequoia SA, P.O. Box 851, CH 1001, Lausanne 1, Switzerland) V.1- 1972-

FMXXAJ French Medicament Patent Document. (U.S. Patent and Trademark Office, Foreign Patents, Washington, DC 20231) Discontinued.

FOBGA8 Folia Biologica (Krakow). (Ars Polona-RUCH, POB 154, Warsaw 1, Poland) V.1- 1953-

FOBLAN Folia Biologica (Prague). (Academic Press, 111 Fifth Ave., New York, NY 10003) V.1- 1955-

FOHEAW Folia Haematologica (Leipzig). (Buchexport, Postfach 160, Leninstr. 16, DDR-701, Leipzig, Germany) V.69- 1949-

FOMDAK Folia Medica. (Via Raffaele de Caesare, 31 Naples, Italy) V.1- 1915-

FOMIAZ Folia Microbiologica (Prague). (Academia, Vodickova 40, CS-112 29 Prague 1, (Czechoslavakia) V.4- 1959-

FOMOAJ Folia Morphologica (Warsaw). (Panstwowy Zaklad Wydawnictw Lekarskich, ul. Druga 38-40, P-00 238 Warsaw, Poland) V.1- 1929-

FOREAE Food Research. (Champaign, IL) V.1-25, 1936-60. For publisher information, see JFDSAZ

FOTEAO Food Technology. (Institute of Food Technologists, 221 N. La Salbe St., Chicago, IL 60601) V.1- 1947-

FPNJAG Folia-Psychiatrica et Neurologica Japonica. (Folia Publishing Society, Todai YMCA Bldg., 1-20-6 Mukogaoka, Bunkyo-Ku, Tokyo 113, Japan) 1947-

FRBGAT Frontiers of Biology. (Elsevier North Holland Inc., 52 Vanderbilt Ave., New York, NY 10017) V.1- 1966-

FRMBAZ Farmacia. (Rompresfilatelia, P.O. Box 2001, Calea Grivitei 64-66, Bucharest, Romania) V.1- 1953-

FRMCE8 Farmaco. (Societa Chimica Italiana, Corso Strada Nova, 86, Casella Postale 227, 27100 Pavia, Italy) V.44- 1989-

FRMTAL Farmatsiya (Moscow). Pharmacy. (v/o Mezhdunarodnaya Kniga, 121200 Moscow, USSR) V.16- 1967-

FRPPAO Farmaco, Edizione Pratica. (Casella Postale 114, 27100 Pavia, Italy) V.8- 1953-

FRPSAX Farmaco, Edizione Scientifica. (Casella Postale 114, 27100 Pavia, Italy) V.8- 1953-

FRXXBL French Demande Patent Document. (Commissioner of Patents and Trademarks, Washington, DC 20231)

FRZKAP Farmatsevtichnii Zhurnal (Kiev). (v/o Mezhdunarodnaya Kniga, Kuznetskii Most 18, Moscow G-200, USSR) V.3- 1930-

FSASAX Fette, Seifen, Anstrichmittel. (Industrieverlag von Hernhaussen KG, Roedingsmarkt 24, 2 Hamburg 11, Germany) V.55- 1953-

FSDZD4 Fukuoka Shika Daigaku Gakkai Zasshi. Journal of Fukuoka Dental College. (700 Ta, O-aza, Sawara-ku, Fukuoka, 814-01, Japan) V.1- 1974-

FSINDR Forensic Science International. (Elsevier Scientific Pub. Ireland Ltd., POB 85, Limerick, Ireland) V.12- 1978-

FSIZAQ Fukushima Igaku Zasshi. (c/o Fukushima-Kenritsu Ika Daigaku Toshokan, 5175 Sugizuma-cho, Fukushima, Japan) V.1- 1951-

FSTEAI Farmaco (Pavia); Scienza e Tecnica. (Pavia, Italy) V.1-7, 1946-52. For publisher information, see FRPSAX

FZPAAZ Frankfurter Zeitschrift fuer Pathologie. (Munich, Germany) V.1-77, 1907-67. For publisher information, see VAAZA2

GAFCC* GAF Material Safety Data Sheet. (GAF Chemicals Corporation, 1361 Alps Road, Wayne, NJ 07470)

GANMAX Gann Monograph. (Tokyo, Japan) No. 1-10, 1966-71. For publisher information, see GMCRDC

GANNA2 Gann. Japanese Journal of Cancer Research. (Tokyo, Japan) V.1-75, 1907-84. For publisher information, see JJCREP

GANRAE Gan No Rinsho. Cancer Clinics. (Ishiyaku Publishers, Inc., 7-10 Honkomagome 1-chome, Bunkyo-ku, Tokyo, Japan) V.1- 1954-

GASTAB Gastroenterology. (Elsevier North Holland, Inc., 52 Vanderbilt Avenue, New York, NY 10017) V.1- 1943-

GCENA5 General and Comparative Endocrinology. (Academic Press, 111 Fifth Ave., New York, NY 10003) V.1- 1961-

GCTB** Givaudan Corporation, Corporate Communications. (100 Delawanna Ave., Clifton, NJ 07014)

GDIKAN Gifu Daigaku Igakubu Kiyo. Papers of the Gifu University School of Medicine. (Gifu Daigaku Igakubu, 40 Tsukasa-cho, Gifu, Japan) V.15- 1967-

GEFRA2 Geburtshilfe und Frauenheilkunde. (Intercontinental Medical Books Corp., 381 Park Ave., S., New York, NY 10016) V.1- 1939-

GEIRDK Gendai Iryo. Modern Medical Care. (Gendai Iryosha, Kandabashi Bldg., 1-21 Nishiki-cho, Kanada Chiyoda-ku, Tokyo 101, Japan) V.1- 1969-

GENEA3 Genetica (The Hague). (Dr. W. Junk bv Publishers, POB 13713, 2501 ES The Hague, Netherlands) V.1- 1919-

GENRA8 Genetical Research. (Cambridge University Press, P.O. Box 92, Bentley House, 200 Euston Rd., London NW1 2DB, England) V.1- 1960-

GENTAE Genetics. (P.O. Drawer U, University Station, Austin, TX 78712) V.1- 1916-

GEPHDP General Pharmacology. (Pergamon Press Inc., Maxwell House Fairview Park, Elmsford, NY 10523) V.1- 1970-

GERNDJ Gerontology. (S. Karger AG, Postfach CH-4009, Basel, Switzerland) V.22- 1976-

GESKAC Genetika i Selektsiya. Genetics and Plant Breeding. (Izdatelstvo na Naukite, ul. Akad. G. Bonchev, 1113 Sofia, Bulgaria) V.1- 1968-

GETRE8 Gematologiya i Transfuziologiya. Hematology and Transfusion Science. (v/o Mezhdunarodnaya Kniga, Kuznetskii Most 18, Moscow G-200, USSR) V.28- 1983-

GEXXA8 German (East) Patent Document. (U.S. Patent Office, Science Library, 2021 Jefferson Davis Highway, Arlington, VA 22202)

GHREAA Guy's Hospital Reports. (London) V.1-123, 1836-1974. Discontinued

GICTAL Giornale Italiano di Chemioterapia. Italian Journal of Chemotherapy. (Edizioni Minerva Medica, Casella Postale 491, I-10126 Turin, Italy) V.1-5, 1954-58; V.5- 1962-

GISAAA Gigiena i Sanitariya. For English translation, see HYSAAV. (V/O Mezhdunarodnaya Kniga, 113095 Moscow, USSR) V.1- 1936-

GJSCDQ Georgia Journal of Science. (Georgia Academy of Science, Augusta College, Augusta, GA 30910) V.35-1977-

GMCRDC Gann Monograph on Cancer Research. (Japan Scientific Societies Press, Hongo 6-2-10, Bunkyo-ku, Tokyo 113, Japan) No. 11- 1971-

GMITAB Gazzetta Medica Italiana. (Minerva Medica, Casella Postale 491, Turin, Italy) V.95- 1936-

GMJOAZ Glasgow Medical Journal. (Glasgow, Scotland) 1828-1955. For publisher information, see SMDJAK

GNAMAP Gigiena Naselennykh Mest. Hygiene in Populated Places. (Kievskii Nauchno-Issledovatel'skii Institut Obshchei i Kommunol'noi Gigieny, Kiev, USSR.) V.7- 1967-

GNKAA5 Genetika. (see SOGEBZ for English Translation) (v/o "Mexhdunarodnaya Kniga" Kuznetskii Most 18, Moscow G-200, USSR) No. 1- 1965-

GNRIDX Gendai no Rinsho. (Tokyo, Japan) V.1-10, 1967-76(?)

GOBIDS Gynecologic and Obstetric Investigation. (S. Karger AG, Postfach CH-4009, Basel, Switzerland) V.9-1978-

GRCSB* Goodrich Chemical Co., Service Bulletin (B.F. Goodrich Chemical Group, 6100 Oak Tree Blvd., Cleveland, OH 44131)

GROWAH Growth. (Southern Bio-Research Institute, Florida Southern College, Lakeland, FL 33802) V.1- 1937-

GSLNAG Gaslini(Genoa). Rivista de Pediatria e di Specialita Pediatriche. (Istituto "Giannina Gaslini", Via Cinque Maggio, 39, 16148 Genoa, Italy) V.1- 1969-?

GTKRDX Gan to Kagaku Ryoho. Cancer and Chemotherapy. (Gan to Kagaku Ryohosha, Yaesu Bldg., 1-8-9 Yaesu, Chuo-Ku, Tokyo 103, Japan) V.1- 1974-

GTPPAF Gigiena Truda i Professional'naya Patologiya v Estonskoi SSR. Labor Hygiene and Occupational Pathology in the Estonian SSR. (Institut Eksperimental'noi i Klinicheskoi Meditsiny Ministerstva Zdravookhraneniya Estonskoi SSR, Tallinn, USSR) V.8- 1972-

GTPZAB Gigiena Truda i Professional'nye Zabolevaniia. Labor Hygiene and Occupational Diseases. (v/o Mezhdunarodnaya Kniga, Kuznetskii Most 18, Moscow G-200, USSR.) V.1- 1957-

GUCHAZ "Guide to the Chemicals Used in Crop Protection" Information Canada, 171 Slater St., Ottawa, Ontario, Canada

GUTTAK Gut. (British Medical Journal, 1172 Commonwealth Ave., Boston, MA 02134) V.1- 1960-

GWXXAW German Patent Document. (U.S. Patent Office, Science Library, 2021 Jefferson Davis Highway, Arlington, VA 22202)

GWXXBX German Offenlegungsschrift Patent Document. (U.S. Patent Office, Science Library, 2021 Jefferson Davis Highway, Arlington, VA 22202)

GWZHEW Gongye Weisheng Yu Zhiyebing. Industrial Health and Occupational Diseases. (China International Book Trading Corp., POB 2820, Beijing, People's Republic of China) V.1- 1973-

GYNOA3 Gynecologic Oncology. (Academic Press, 111 Fifth Ave., New York, NY 10003) V.1- 1972-

HAEMAX Haematologica. (Il Pensiero Scientifico, Via Panama 48, I-00198, Rome, Italy) V.1- 1920-

HAONDL Hematological Oncology. (John Wiley & Sons Ltd., Baffins Lane, Chichester, Sussex, PO19 1UD, UK) V.1- 1983-

HAREA6 Harefuah Medicine. (Israel Medical Association, 39 Shaul Hamlech Blvd., Tel Aviv-Jaffa, Israel) V.1- 1924

HarPN# Personal Communication to Henry Lau, Tracor Jitco, from Paul N. Harris, M.D., Eli Lily Co., Greenfield, IN 46140

HAZL** Hazelton Laboratories, Reports. (Falls Church, VA)

HBAMAK "Abdernalden's Handbuch der Biologischen Arbeitsmethoden." (Leipzig, Germany)

HBPTO* Handbook of pesticide toxicology. Robert Krieger ed, Academic press, 2001

HBTXAC "Handbook of Toxicology, Volumes I-V." W.B. Saunders, Philadelphia, PA, 1956-59

HCACAV Helvetica Chimica Acta. (Verlag Helvetica Chimica Acta, Postfach, CH-4002 Basel, Switzerland) V.1-1918-

HCMYAL Histochemistry. (Springer-Verlag New York, Inc., Service Center, 44 Hartz Way, Secaucus, NJ 07094) V.38- 1974-

HDIZAB Hiroshima Daigaku Zasshi, Medical Journal of the Hiroshima University (Hiroshima Daigaku Igakubu Saikingaku Kyoshitsu, Kasumi-cho, Hiroshima, Japan) V.10- 1962-

HDKU** Die Krampfmischung der Lokalanaesthetica, ihre Beeinflussung durch Mineralsalze und Adrenalin, Gerhard Hoppe Dissertation. (Pharmakologischen Institut der Albertus-Universitat zu Konigsberg Pr., Germany, 1933)

HDSKEK Hiroshima Daigaku Sogo Kagakubu Kiyo, 4: Kiso, Kankyo Kagaku Kenkyu. Bulletin of the Faculty of Integrated Arts and Sciences, Hiroshima University, Series 4: Fundamental and Environmental Sciences. (1-1-89 Higashisenda-machi, Naka-ku, Hiroshima, 730, Japan) V.6- 1980-

HDTU** Pharmakologische Prufung von Analgetika, Gunter Herrlen Dissertation. (Pharmakologischen Institut der Universitat Tubingen, Germany, 1933)

HDWU** Beitrage zur Toxikologie des Athylenoxyds und der Glykole, Arnold Hofbauer Dissertation. (Pharmakologischen Institut der Universitat Wurzburg, Germany, 1933)

HEADAE Headache. (American Association for the Study of Headache, Rm 537, 621 S. New Ballas Rd., St. Louis, MO 63141) V.1- 1961-

HEGAD4 Hepatogastroenterology. (Georg Thieme Verlag, Postfach 732, Herdweg 63, D-7000 Stuttgart 1, Germany) V.1- 1980-

HEPHD2 Handbook of Experimental Pharmacology. (Springer-Verlag, Heidelberger Pl. 3, D-1000 Berlin 33, Germany) V.50- 1978-

HERBU* Hercules Incorporated, Hercules Bulletin. (Hercules Incorporated, Pine and Paper Chemicals Department, Wilmington, DE)

HEREAY Hereditas. (J.L. Toernqvist Book Dealers, S-26122 Landskrona, Sweden) V.1- 1947-

HETOEA Human & Experimental Toxicology. (Macmillan Press Ltd., Brunel Road, Houndmills, Basingstoke, Hampshire, RG21 2XS, UK) V.9- 1990-

HGANAO Haigan. Lung Cancer. (Nippon Haigan Gakkai, c/o Chiba Daigaku Igakubu Haigan Kenkyusho, 1-8-1 Inohona, Chiba 280, Japan)

HIFUAG Hifu. Skin. (Nihon Hifuka Gakkai Osaka Chihokai, c/o Osaka Daigaku Hifuka Kyoshitsu, 3-1, Dojima Hamadori, Fukushima-ku, Osaka 553, Japan) V.1- 1959-

HIKYAJ Hinyokika Kiyo. (Acta Urologica Japonica). (Kyoto University Hospital, Department of Urology) V.1- 1955-

HINAAU Hindustan Antibiotics Bulletin. (Hindustan Antibiotics, Pimpri, India) V.1- 1958-

HINEL* Hine Laboratories Reports. (San Francisco, CA)

HIRIA6 Hirosaki Igaku. (Hirosaki Daigaku Igakubu, Hirosaki, Japan) V.1- 1950-

HIUN** Department of Cancer Research, Research Institute for Nuclear Medicine and Biology, Hiroshima University, Kasumi 1-2-3, Hiroshima 734, Japan

HKXUDL Huanjing Kexue Xuebao. Environmental Sciences Journal. (Guoji Shudian, POB 399, Beijing, People's Republic of China) V.1- 1981-

HLSCAE Health Laboratory Science. (American Public Health Assoc., 1015 18th St., N.W., Washington, DC 20036) V.1- 1964-

HLTPAO Health Physics. (Pergamon Press, Maxwell House, Fairview Park, Elmsford, NY 10523) V.1- 1958-

HMACAX Helvetica Medica Acta. (Basel, Switzerland) V.1-37, 1934-1974. For publisher information, see SMWOAS.

HMMRA2 Hormone and Metabolic Research. (Georg Thieme Verlag, Postfach 732, Herdweg 63, D-7000 Stuttgart 1, Germany) V.1- 1969-

HOBEAO Hormones and Behavior. (Academic Press, 111 5th Ave., New York, NY 10003) V.1- 1969-

HOEKAN Hokkaidoritsu Eisei Kenkyushoho. (Hokkaidoritsu Eisei Kenkyusho, Nishi-12-chome, Kita-19-jo, Kita-ku, Sapporo, Japan) No. 1- 1951-

HOIZAK Hokkaido Igaku Zasshi. Hokkaido Journal of Medical Science. (Hokkaido Daigaku Igakubu, Nishi-5-chome, Kita-12-jo, Sapporo, Japan) V.1- 1923-

HOKBAQ Hoken Butsuri. Journal of the Japan Health Physics Society. (Nihon Hoken Butsuri Kyogikai, Sakamoto Bldg., Kitaaoyama 2-12-4, Minato-ku, Tokyo, Japan) V.1- 1966-

HPCQA4 Human Pathology. (W. B. Saunders Co., W. Washington Sq., Philadelphia, PA 19105) V.1- 1970-

HPFSDS HHS Publication (FDA. United States). (U.S. Government Printing Office, Superintendent of Documents, Washington, DC 20402) 1980-

HPPAAL Helvetica Physiologia et Pharmacologica Acta. (Basel, Switzerland) V.1-26, 1943-68. Discontinued

HRMRA3 Hormone Research. (S. Karger AG, Postfach CH-4009, Basel, Switzerland) V.4- 1978-

HSZPAZ Hoppe-Seyler's Zeitschrift fuer Physiologische Chemie. (Walter de Gruyter and Co., Genthiner Street 13, D-1000, Berlin 30, Germany) V.21- 1895/96-

HUGEDQ Human Genetics. (Springer-Verlag, Neuenheimer Landst 28-30, D-6900 Heidelberg 1, Germany) V.31- 1976-

HUMAA7 Humangenetik. (Heidelberg, Germany) V.1-30, 1964-1975. For publisher information, see HUGEDQ

HunNJ# Personal Communication from Nancy J. Hunt, E.I. DuPont de Nemours and Co., 1007 Market St., Wilmington, DE 19898, to Henry Lau, Tracor Jitco, Inc., May 10, 1977

HUPSEC Human Psychopharmacology. (John Wiley & Sons Ltd., Baffins Lane, Chichester, W. Sussex PO19 1UD, UK) V.1-1986-

HURC** Huntingdon Research Center Reports. (Box 527, Brooklandville, MD 21022)

HUREEE Human Reproduction. (IRL Press Inc., POB Q, McLean, VA 22101) V.1- 1986-

HUTODJ Human Toxicology. (Macmillan Press Ltd., Houndmills, Bassingstoke, Hants., RG21 2XS, UK) V.1- 1981-

HXPHAU Handbuch der Experimentellen Pharmakologie. (Berlin, Germany) V.1-49, 1935-78. For publisher information, see HEPHD2

HYDKAK Hoshi Yakka Daigaku Kiyo. Proceedings of the Hoshi College of Pharmacy. (2-chome, Ebara, Shinagawa-ku, Tokyo, Japan) No. 1- 1951-

HYDRDA Hydrometallurgy. (Elsevier Science Publishers B.V., POB 211, 1000 AE Amsterdam, Netherlands) V.1- 1975-

HYDXET Huaxi Yike Daxue Xuebao. Journal of West China University of Medical Sciences. (China International Book Trading Corp., POB 2820, Beijing, People's Republic of China) V.17- 1986-

HYSAAV Hygiene and Sanitation: English Translation of Gigiena Sanitariya. (Springfield, VA) V.29-36, 1964-71. Discontinued

IAAAAM International Archives of Allergy and Applied Immunology. (S. Karger, Postfach CH-4009, Basel, Switzerland) V.1- 1950-

IAANBS Internationales Archiv fuer Arbeitsmedizin. (Berlin, Germany) V.26-34, 1970-74. For publisher information, see IAEHDW

IAEC** Interagency Collaborative Group on Environmental Carcinogenesis, National Cancer Institute, Memorandum, June 17, 1974

IAEHDW International Archives of Occupational and Environmental Health. Springer-Verlag, Heidelberger Pl 3, D-1000 Berlin 33, Germany) V.35- 1975-

IAPUDO IARC Publications. (World Health Organization, CH-1211 Geneva 27, Switzerland) No. 27- 1979-

IAPWAR International Journal of Air and Water Pollution. (London, England) V.4, No. 1-4, 1961. For publisher information, see ATENBP

IARC** IARC Monographs on the Evaluation of Carcinogenic Risk of Chemicals to Man. (World Health Organization, Internation Agency for Research on Cancer, Lyon, France) (Single copies can be ordered from WHO Publications Centre U.S.A., 49 Sheridan Avenue, Albany, NY 12210)

IARCCD IARC Scientific Publications. (Geneva Switzerland) V.1-No. 26, 1971-78, For publisher information, see IAPUDO

ICHAA3 Inorganica Chimica Acta. (Elsevier Sequoia SA, POB 851, CH-1001 Lausanne 1, Switzerland) V.1- 1967-

ICMED9 Intensive Care Medicine. (Springer-Verlag New York, Inc., Service Center, 44 Hart Way, Secaucus, NJ 07094) V.3- 1977-

IDZAAW Idengaku Zasshi. Japanese Journal of Genetics. (Genetics Society of Japan, Nippon Iden Gakkai, Tanida 111, Mishima-shi, Shizuoka 411, Japan) V.1- 1921-

IECHAD Industrial and Engineering Chemistry. (Washington, DC) V.15-62, 1923-70. For publisher information, see CHMTBL

IEDIEP Internal Medicine. (The Japanese Society of Internal Medicine, 34-3, 3-chome, Hongo, Bunkyo-ku, Tokyo 113, Japan) V.31-

IGAYAY Igaku No Ayumi. Progress in Medicine. (Ishiyaku Shuppan K.K., 7-10, Honkomagone 1 chome, Bunkyo-ku, Tokyo, Japan) V.1- 1946-

IGIBA5 Igiena. (Editura Medicala, St 13 Decembrie 14, Bucharest, Romania) V.5- 1956-

IGKEAO Igaku Kenkyu. (Daido Gakkan Shuppanbu, Kyushu Daigaku Igakubu, Hoigaku Kyoshitsu, Fukuoka 812, Japan) V.1- 1927-

IGSBAL Igaku to Seibutsugaku. Medicine and Biology. (1-11-4 Higashi-Kanda, Chiyoda, Tokyo 101, Japan) V.1- 1942-

IGSBDO Izvestiya Akademii Nauk Gruzinskoi SSR, Seriya Biologicheskaya. Proceedings of the Academy of Sciences of the Georgian SSR, Biological Series. (v/o Mezhdunarodnaya Kniga, Kuznetskii Most 18, Moscow G-200, USSR) V.1- 1975-

IHFCAY Industrial Hygiene Foundation of America, Chemical and Toxicological Series, Bulletin. (5231 Centre Ave., Pittsburgh, PA 15232) 1947-69

IIFBA4 Izvestiya na Instituta po Fiziologiya, Bulgarska Akademiya na Naukite. (Izdatelstvo na Bulgarskata Akademiya na Naukite, St. Geo. Milev ul 36, Sofia 13, Bulgaria) V.4- 1960-

IIZAAX Iwate Igaku Zasshi. Journal of the Iwate Medical Association. (Iwate Igakkai, c/o Iwate Ika Daigaku, Uchimaru, Morioka, Japan) V.1- 1947-

IJANDP International Journal of Andrology. (Scriptor Publisher ApS, 15 Gasvaerksvej, DK-1656 Copenhagen V, Denmark) V.1- 1978-

IJBBBQ Indian Journal of Biochemistry and Biophysics. (Council of Scientific and Industrial Research, Publication and Information Director, Hillside Rd., New Delhi 110012, India) V.8- 1971-

IJBOBV International Journal of Biochemistry. (Pergamon Press Inc., Maxwell House, Fairview Park, Elmsford, NY 10523) V.1- 1970-

IJCAAR Indian Journal of Cancer. (Dr. D.J. Jussawalla, Hospital Ave, Parel, Bombay 12, India) V.1- 1963-

IJCBDX International Journal of Clinical Pharmacology and Biopharmacy. (Urban and Schwarzenberg, Pettenkoferst 18, D-8000 Munich 15, Germany) V.11- 1975-

IJCNAW International Journal of Cancer. (International Union Against Cancer, 3 rue du Conseil-General, 1205 Geneva, Switzerland) V.1- 1966-

IJCPB5 International Journal of Clinical Pharmacology, Therapy and Toxicology. (Munich, Germany) V.6-10, 1972-74. For publisher information, see IJCBDX

IJCREE International Journal of Crude Drug Research. (Swets and Zeitlinger B.V., POB 825, 2160 SZ Lisse, Netherlands) V.20- 1982-

IJEBA6 Indian Journal of Experimental Biology. V.1- 1963-. For publisher information, see IJBBBQ

IJEVAW International Journal of Environmental Studies. (Gordon and Breach Science Publishers Ltd., 41-42 William IV St., London WC2N 4DE, England) V.1- 1970-

IJHYEQ International Journal of Hyperthermia. (Taylor & Francis, 1900 Frost Rd., Suite 101, Bristol, PA 19007) V.1- 1985-

IJIMDS International Journal of Immunopharmacology. (Pergamon Press Ltd., Headington Hill Hall, Oxford OX3 0BW, England) V.1- 1979-

IJLEAG International Journal of Leprosy. (Box 39088, Washington, DC 20016) V.1- 1933-

IJMDAI Israel Journal of Medical Sciences. (P.O. Box 2296, Jerusalem, Israel) V.1- 1965-

IJMRAQ Indian Journal of Medical Research. (Indian Council of Medical Research, P.O. Box 4508, New Delhi 110016, India) V.1- 1913-

IJMSAT Irish Journal of Medical Science. (Royal Academy of Medicine, 6 Kildare St., Dublin, Ireland) V.1- 1968-

IJNEAQ International Journal of Neuropharmacology. (Oxford, England/New York, NY) V.1-8, 1962-69. For publisher information, see NEPHBW

IJNMCI International Journal of Nuclear Medicine and Biology. (Pergamon Press Inc., Maxwell House, Fairview Park, Elmsford, NY 10523) V.1- 1973-

IJNUB7 International Journal of Neuroscience. (Gordon and Breach Science Publication S.A., 50 W. 23rd St., New York, NY 10010) V.1- 1970-

IJOCAP Indian Journal of Chemistry. (Council of Scientific and Industrial Research, Publication and Information Director, Hillside Rd., New Delhi 110012, India) V.1-13, 1963-75

IJOTO* International Journal of Toxicology (Continous: Journal of the American College of Toxicology,Taylor & Francis Health Sciences) V.16- 1997-

IJPAAO Indian Journal of Pharmacy. (Bombay, India) V.1- 40, No. 1, 1939-78. For publisher information, see IJSIDW

IJPBAR Indian Journal of Pathology and Bacteriology. (Dr. S.G. Deodhare, Medical College, Miraj Maharashtra, India) V.1- 1958-

IJPPAZ Indian Journal of Physiology and Pharmacology. (Indian Journal of Physiology and Pharmacology, All India Institute of Medical Sciences, Department of Physiology, Ansari Nagar, New Delhi 110016, India) V.1- 1957-

IJPPC3 International Journal of Peptide & Protein Research. (Munksgaard International Publishers, 238 Main St., Cambridge, MA 02142) V.4n.2- 1972-

IJRBA3 International Journal of Radiation Biology and Related Studies in Physics, Chemistry and Medicine. (Taylor and Francis Ltd., 4 John Street, London WC1N 2ET, UK) V.1- 1959-

IJSBDB Indian Journal of Chemistry. Section B: Organic Chemistry, Including Medicinal Chemistry. (Council of Scientific and Industrial Research, Publication and Information Director, Hillside Rd., New Delhi 110012, India) V.14- 1976-

IJSIDN Journal of Traditional Chinese Medicine (English Edition). (All-China Association of Traditional Chinese Medicine, Academy of Traditional Chinese Medicine) V.1- 1981-

IJSIDW Indian Journal of Pharmaceutical Sciences. (Indian Journal of Pharmaceutical Sciences, Kalina, Santa Cruz (East), Bombay 400 029, India) V.40, No. 2- 1978-

IJTEDP International Journal on Tissue Reactions. Cell and Tissue Injuries, Experimental and Clinical Aspects. (Biosciences Ediprint, Rue Winkelried 8, 1211 Geneva 1, Switzerland) V.1- 1979-

IJTOFN International Journal of Toxicology. (Taylor & Francis, 47 Runway Rd., Suite g, Levittown, PA 19057) V.16- 1997-

IJVNAP International Journal for Vitamin and Nutrition Research. (Verlag Hans Huber, Laenggasstr. 76, CH-3000 Bern 9, Switzerland) V.41- 1971-

IMEMDT IARC Monographs on the Evaluation of Carcinogenic Risk of Chemicals to Man. (World Health Organization, Internation Agency for Research on Cancer, Lyon, France) (Single copies can be ordered from WHO Publications Centre U.S.A., 49 Sheridan Avenue, Albany, NY 12210)

IMGAAY Indian Medical Gazette. (Indian Medical Gazette, Block F, 105-C, New Alipore, Calcutta 700053, India) V.1-90, 1866-1955; new series V.1- 1961-

IMLCAV Immunological Communications. (Marcel Dekker, POB 11305, Church St. Station, New York, NY 10249) V.1- 1972-

IMMUAM Immunology. (Blackwell Scientific Publications, Osney Mead, Oxford OX2 OEL, England) V.1- 1958-

IMNGBK Immunogenetics (New York). (Springer-Verlag New York, Inc., Service Center, 44 Hartz Way, Secaucus, NJ 07094) V.1- 1974-

IMSCE2 IRCS Medical Science. (Lancaster, UK) V.12-14 1984-86

IMSUAI Industrial Medicine and Surgery. (Chicago, IL/Miami, FL) V.18-42, 1949-73. For publisher information, see IOHSA5

INDRBA Indian Drugs. (Indian Drugs Manufacturers' Associations c/o Dr. A. Patani, 102B, Poonam Chambers, Dr. A.B. Rd., Worli Bobmay 400 018, India) V.1- 1963-

INFIBR Infection and Immunity. (American Society for Microbiology, 1913 I St., N.W., Washington, DC 20006) V.1- 1970-

INHEAO Industrial Health. (2051 Kizukisumiyoshi-cho, Nakahara-ku, Kawasaki, Japan) V.1- 1963-

INHTE5 Inhalation Toxicology. (Hemisphere Publishing Corp., c/o Taylor & Francis Inc., 1900 Frost Rd., Suite 101, Bristol, PA 19007) V.1- 1989-

INJFA3 International Journal of Fertility. (Allen Press, 1041 New Hampshire St., Lawrence, KS 66044) V.1- 1955-

INJHA9 Indian Journal of Heredity. (Genetic Association of India, Indian Veterinary Research Institute, Izatnagar, India) V.1- 1969-

INJPD2 Indian Journal of Pharmacology. (Indian Pharmacological Society, Jawaharlal Institute of Postgraduate Medical Education and Research, Pharmacology Dept., Pondicherry 605006, India) V.1- 1968(?)-

INMEAF Industrial Medicine. (Chicago, IL) V.1-18, 1932- 49. For publisher information, see IOHSA5

INNDDK Investigational New Drugs. The Journal of New Anticancer Agents. (Kluwer Boston Inc., 190 Old Derby St., Hingham, MA 02043) V.1- 1983-

INOPAO Investigative Ophthalmology. (C.V. Mosby Co., 11830 Westline Industrial Dr., St. Louis, MO 63141) V.1-1962-

INPDAR Indian Pediatrics. (46-B Garcha Rd., Calcutta, India) V.1- 1964-

INPHB6 International Pharmacopsychiatry. (S. Karger AG, Postfach CH-4009, Basel, Switzerland) V.1- 1968-

INSSDM INSERM Symposium. (Elsevier North Holland, Inc., 52 Vanderbilt Ave., New York, NY 10017) No. 1-1975-

INTSAO International Surgery. (International College of Surgeons, 1516 N. Lake Shore Dr., Chicago, IL 60610) V.45- 1966-

INURAQ Investigative Urology. (Williams & Wilkins Co., 428 E. Preston St., Baltimore, MD 21202) V.1- 1963-

INVRAV Investigative Radiology. (J.B. Lippincott Co., E. Washington Sq., Philadelphia, PA 19105) V.1- 1966-

IOBPD3 International Journal of Radiation Oncology, Biology and Physics. Pergamon Press Ltd., Headington Hill Hall, Oxford OX3 0BW, England) V.1- 1975-

IOHSA5 International Journal of Occupational Health and Safety. (Medical Publications, Inc., 3625 Woodhead Dr., Northbrook, IL 60093) V.43- 1974-

IOVSDA Investigative Ophthalmology and Visual Science. (C.V. Mosby Co., 11830 Westline Industrial Dr., St. Louis, MO 63146) V.16- 1977-

IPCLBZ International Pest Control. (McDonald Publications, 268 High St., Uxbridge UB8 1UA, Middlesex., England) V.5- 1962-

IPPABX Iugoslavica Physiologica et Pharmacologica Acta. (Unija Bioloskih Naucnik Drustava Jugoslavije, Postanski fah 127, Belgrade-Zemun, Yugoslavia) V.1- 1965-

IPRAA8 Indian Practitioner. (231 Dr. Dadabhoy Naoroji Rd., Bombay 1, India) V.1- 1947/48-

IPSTB3 International Polymer Science and Technology. (Rapra Technology Ltd., Shawbury, Shrewsbury, Shropshire SY4 4NR, UK)

IRGGAJ Internationales Archiv fuer Gewerbepathologie und Gewerbehygiene. (Heidelberg, Germany) V.19-25, 1962-69. For publisher information, see IAEHDW

IRLCDZ IRCS Medical Science: Library Compendium. (MTP Press Ltd., St. Leonards House, St. Leonards Gate, Lancaster LA1 3BR, England) V.3-11, 1975-83

IRMEA9 International Record of Medicine. (New York, NY) V.170-174, 1957-61. For publisher information, see CLMEA3

IRNEAE International Review of Neurobiology. (Academic Press, 111 5th Ave., New York, NY 10003) V.1-1959-

ISBNBN Izvestiya Sibirskogo Otdeleniya Akademii Nauk S.S.S.R., Seriya Biologicheskikh Nauk. (v/o Mezhdunarodnaya Kniga, Kuznetskii Most 18, Moscow G-200, USSR.) No. 1- 1969-

ISMJAV Israel Medical Journal. (Jerusalem, Israel) V.17-23, 1958-64. For publisher information, see IJMDAI

ISYAM* "Amphetamines and Related Compounds" E. Costa, ed., Proceedings of the Mario Negri Institute for Pharmacological Research, Milan, Italy, New York, NY, Raven Press, 1970

ITCSAF In Vitro. (Tissue Culture Association, 12111 Parklawn Dr., Rockville, MD 20852) V.1- 1965-

ITMZBJ Intensivmedizin. (Dr. Dietrich Steinkopff Verlag, Postfach 11-10-08, D-6100 Darmstadt 11, Germany) V.1-1964-

ITOBAO Izvestiya Akademii Nauk Tadzhikskoi SSR, Otdelenie Biologicheskikh Nauk. Proceedings of the Academy of Sciences of the Tadzhik SSR, Department of Biological Sciences. (v/o Mezhdunarodnaya Kniga, Kuznetskii Most 18, Moscow G-200, USSR) No. 1- 1962-

IUSMDJ ICN-UCLA Symposia on Molecular and Cellular Biology. (Academic Press, 111 5th Ave., New York, NY 10003) V.1- 1974-

IVEBBN Beiheft zur Internationalen Zeitschrift fuer Vitamin-und Ernaehrungsforschung. (Verlag Hans Huber Laenggastr. 76 CH-3000 Bern 9, Switzerland) No. 12- 1972-

IVEJAC Indian Veterinary Journal. (Indian Veterinary Journal, Dr. R. Krishnamurti, G.M.V.C., 10 Avenue Rd., Madras 600 034, India) V.1- 1924-

IVIVE4 In Vivo. (POB 51359, Kiffisia, GR-145 10, Greece) V.1- 1987-

IYKEDH Iyakuhin Kenkyu. Study of Medical Supplies. (Nippon Koteisho Kyokai, 12-15, 2-chome, Shibuya, Shibuya-ku, Tokyo 150, Japan) V.1- 1970-

IZKPAK Internationale Zeitschrift fuer Klinische Pharmakologie, Therapie und Toxikologie. (Munich, Germany) V.1-5, 1967-72. For publisher information, see IJCBDX

IZSBAI Izvestiya Sibirskogo Otdeleniya Akademii Nauk S.S.S.R., Seriya Biologomeditsinskikh Nauk. (Novosibirsk, USSR) 1963-68. For publisher information, see ISBNBN

IZVIAK Internationale Zeitschrift fuer Vitaminforschung. International Review of Vitamin Research. (Bern, Switzerland) V.19-40, 1947-70. For publisher information, see IJVNAP

JAADDB Journal of the American Academy of Dermatology. (C.V. Mosby Co., 11830 Westline Industrial Dr., St. Louis, MO 63141) 1979-

JAAPEE Journal of the American Academy of Child and Adolescent Psychiatry. (Williams & Wilkins, 351 W. Camden St., Baltimore, MD 21201) V.26- 1987-

JACHDX Journal of Antimicrobial Chemotherapy. (Academic Press, 24-28 Oval Rd., London NW1 7DX, England) V.1- 1975-

JACIBY Journal of Allergy and Clinical Immunology. (C.V. Mosby Co., 11830 Westline Industrial Dr., St. Louis, MO 63141) V.48- 1971-

JACSAT Journal of the American Chemical Society. (American Chemical Society Publications, 1155 16th St., N.W., Washington, DC 20036) V.1- 1879-

JACTDZ Journal of the American College of Toxicology. (Mary Ann Liebert, Inc., 500 East 85th St., New York, NY 10028) V.1- 1982-

JADSAY Journal of the American Dental Association. (American Dental Association, 211 E. Chicago Ave., Chicago, IL 60611) V.9- 1922-

JAFCAU Journal of Agricultural and Food Chemistry. (American Chemical Society Publications, 1155 16th St., N.W., Washington, DC 20036) V.1- 1953-

JAGRAC Journal of Agricultural Research. (Washington, DC) V.1-78, 1913-49. Discontinued

JAGSAF Journal of the American Geriatrics Society. (American Geriatrics Society, Inc., 10 Columbus Circle, New York, NY 10019) V.1- 1953-

JAINAA Journal of the Anatomical Society of India. (Dept. of Anatomy, Medical College, Aurangabad MS, India) V.1- 1952-

JAJAAA Journal of Antibiotics, Series A. (Tokyo, Japan) V.6-20, 1953-67. For publisher information, see JANTAJ

JAMAAP JAMA, Journal of the American Medical Association. (American Medical Association, 535 N. Dearborn St., Chicago, IL 60610) V.1- 1883-

JAMPA2 Journal of Animal Morphology and Physiology. (Society of Animal Morphologists and Physiologists, Dept. of Zoology, Faculty of Science, M.S. Univ. of Baroda, Baroda 2, India) V.1- 1954-

JANCA2 Journal of the Association of Official Analytical Chemists. (Association of Official Analytical Chemists, Box 540, Benjamin Franklin Station, Washington, DC 20044) V.49- 1966-

JANSAG Journal of Animal Science. (American Society of Animal Science, 309 West Clark Street, Champaign, IL 61820) V.1- 1942-

JANTAJ Journal of Antibiotics. (Japan Antibiotics Research Association, 2-20-8 Kamiosaki, Shinagawa-ku, Tokyo, Japan) V.2-5, 1948-52; V.21- 1968-

JAOCA7 Journal of the American Oil Chemists' Society. (American Oil Chemists' Society, 508 South 6th St., Champaign, IL 61820) V.24-56, 1947-79

JAPHAR Journal of Anatomy and Physiology. (London, England) V.1-50, 1916. For publisher information, see JOANAY

JAPMA8 Journal of the American Pharmaceutical Association, Scientific Edition. (Washington, DC) V.29-49, 1940-60. For publisher information, see JPMSAE

JAPRAN Journal of Small Animal Practice. (Blackwell Scientific Publications Ltd., POB 88, Oxford, UK) V.1- 1960-

JAPTO* Journal of Applied Toxicology (John Wiley & Sons, Ltd., Oldlands Way Bognor Regis West Sussex, PO22 9SA England) V.1- 1981-

JAPYAA Journal of Applied Physiology. (American Physiological Society, 9650 Rockville Pike, Bethesda, MD 20014) V.1- 1948-

JASIAB Journal of Agricultural Science. (Cambridge University Press, London or Cambridge University Press, New York) V.1- 1905-

JATOD3 Journal of Analytical Toxicology. (Preston Publications, P.O. Box 48312, Niles, IL 60648) V.1- 1977-

JAVMA4 Journal of the American Veterinary Medical Association. (American Veterinary Medical Association, 600 S. Michigan Ave., Chicago, IL 60605) V.48- 1915-

JBCHA3 Journal of Biological Chemistry. (American Society of Biological Chemists, Inc., 428 E. Preston St., Baltimore, MD 21202) V.1- 1905-

JBJSA3 Journal of Bone and Joint Surgery. American Volume. (10 Shattuck St., Boston, MA 02115) V.30- 1948-

JBJSB4 Journal of Bone and Joint Surgery. (Boston, MA) V.20-45, 1922-47. For publisher information, see JBJSA3

JBMRBG Journal of Biomedical Materials Research. (John Wiley & Sons, 605 3rd Ave., New York, NY 10016) V.1- 1967-

JCCCA5 Journal of the Society of Cosmetic Chemists. (Soc. of Cosmetic Chemists, 1995 Broadway, Suite 1701, New York, NY 10023) V.1- 1947-

JCCPAY Journal of Cellular and Comparative Physiology. (Philadelphia, PA) V.1-66, 1932-65. For publisher information, see JCLLAX

JCEMAZ Journal of Clinical Endocrinology and Metabolism. (Williams & Wilkins Co., 428 East Preston St., Baltimore, MD 21202) V.12- 1952-

JCENA4 Journal of Clinical Endocrinology. (Springfield, IL) V.1-11, 1941-51. For publisher information, see JCEMAZ

JCGADC Journal of Clinical Gastroenterology. (Raven Press, 1185 Avenue of the Americas, New York, NY 10036)

JCGBDF Journal of Craniofacial Genetics and Developmental Biology. (Alan R. Liss, Inc., 150 Fifth Ave., New York, NY 10011) V.1- 1980-

JCGEDO Journal of Cytology and Genetics. (Society of Cytologists and Geneticists, Treasurer, Professor M.S. Chennaveeraiah, Karnatak University, Department of Botany, Dharwar 580003, India) V.1- 1966-

JCHAAE Journal of Chemotherapy and Advanced Therapeutics. (Philadelphia, PA) V.3,1926; V.11(2-15), 1934-39

JCHEEU Journal of Chemotherapy. (Edizioni Riviste Scientifiche, via R. Giuliani, 419, 50141 Florence, Italy) V.1- 1989-

JCHOAM Journal of Chemotherapy. (Philadelphia, PA) V.4-11(1), 1927-34.

JCHODP Journal of Clinical Hematology and Oncology. (Wadley Institutes of Molecular Medicine, 9000 Harry Hines Blvd., Dallas, TX 75235) V.5, No. 3- 1975-

JCINAO Journal of Clinical Investigation. (Rockefeller University Press, 1230 York Avenue, New York, NY 10021) V.1- 1924-

JCLBA3 Journal of Cell Biology. (Rockefeller University Press, 1230 York Avenue, New York, NY 10021) V.12- 1962-

JCLLAX Journal of Cellular Physiology. (Alan R. Liss, Inc., 150 Fifth Avenue, New York, NY 10011) V.67- 1966-

JCLPDE Journal of Clinical Psychiatry. (Physician Postgraduate Press, Inc., POB 24008, Memphis TN 38124) V.39- 1978-

JCNDBK Journal of Clinical Pharmacology and New Drugs. (Stamford, CT) V.11-13, No. 4, 1971-3, For publisher information, see JCPCBR

JCPAAK Journal of Clinical Pathology. (British Medical Association, Tavistock Sq., London WC1H 9JR, England) V.1- 1947-

JCPCBR Journal of Clinical Pharmacology. (Hall Associates, P.O. Box 482, Stamford, CN 06904) V.13- 1973-

JCPCDT Journal of Cardiovascular Pharmacology. (Raven Press, 1140 Ave. of the Americas, New York, NY 10036) V.1- 1979-

JCPHB8 Journal of Clinical Pharmacology and Journal of New Drugs. (Albany, NY) V.7-10, 1967-70. For publisher information, see JCPCBR

JCPPAV Journal of Comparative and Physiological Psychology. (American Psychological Association, 1200 17th St., N.W., Washington, DC 20036) V.40- 1947-

JCPRB4 Journal of the Chemical Society, Perkin Transactions 1. (Chemical Society, Publications Sales Office, Burlington House, London W1V 0BN, England) No. 1- 1972-

JCPTA9 Journal of Comparative Pathology and Therapeutics. (Liverpool, England) V.1-74, 1883-1964. For publisher information, see JCVPAR

JCPYDR Journal of Clinical Pyschopharmacology. (Williams & Wilkins Co., 428 E. Preston St., Baltimore, MD 21202) V.1- 1981-

JCREA8 Journal of Cancer Research. (Baltimore, MD) V.1-14, 1916-30. For publisher information, see CNREA8

JCROD7 Journal of Cancer Research and Clinical Oncology. (Springer-Verlag, Heidelberger Pl. 3, D-1 Berlin 33, Germany) V.93- 1979-

JCSOA9 Journal of the Chemical Society. (London, England) 1926-65. For publisher information, see JCPRB4

JCTODH Journal of Combustion Toxicology. (Technomic Publishing Co., 265 Post Rd. W., Westport, CT 06880) V.3, No. 1- 1976-

JCUPBN Journal of Cutaneous Pathology. (Munksgaard, 35 Noerre Soegade, DK-1370 Copenhagen K, Denmark) V.1- 1974-

JCVPAR Journal of Comparative Pathology. (Academic Press Inc., Ltd., 24-28 Oval Rd., London NW1 7DX, England) V.75- 1965-

JDGRAX Journal of Drug Research. (Drug Research and Control Center, 6, Abou Hazem St., Pyramids Ave., POB 29, Cairo, Egypt) V.2- 1969-

JDREAF Journal of Dental Research. (American Association for Dental Research, 734 15th St., NW, Suite 809, Washington, DC 20005) V.1- 1919-

JDSCAE Journal of Dairy Science. (American Dairy Science Association, 309 W. Clark St., Champaign, IL 61820). V.1- 1917-

JEDIDP Journal of Environmental Biology. (Academy of Environmental Biology, India, 657/5, Civil Lines (South), Muzaffarnagar, 251001, India) V.1- 1980-

JEEMAF Journal of Embryology and Experimental Morphology. (P.O. Box 32, Commerce Way, Colchester CO2 8HP, Essex. UK) V.1- 1953-

JEENAI Journal of Economic Entomology. (Entomological Society of America, 4603 Calvert Rd., College Park, MD 21201) V.1- 1908-

JEMAAJ Journal of the Egyptian Medical Association. (The Egyptian Medical Association, 42 Sharia Kasr-el-aini, Cairo, Egypt) V.11- 1928-

JEMEAV Journal of Experimental Medicine. (Rockefeller University Press, 1230 York Avenue, New York, NY 10021) V.1- 1896-

JEPOEC Journal of Environmental Pathology, Toxicology and Oncology. (Chem-Orbital, POB 134, Park Forest, IL 60466) V.5(4)- 1984-

JEPTDQ Journal of Environmental Pathology and Toxicology. (Park Forest South, IL) V.1-5, 1977-81

JESEDU Journal of Environmental Science and Health, Part A: Environmental Science and Engineering. (Marcel Dekker, POB 11305, Church St. Station, New York, NY 10249) V.A11- 1976-

JETHDA Journal of Ethnopharmacology. (Elsevier Sequoia SA, POB 851, CH-1001 Lausanne 1, Switzerland) V.1- 1979-

JETOAS Journal Europeen de Toxicologie. (Paris, France) V.1-6, 1968-72. For publisher information, see TOERD9

JETPEZ Journal of Engineering for Gas Turbines and Power. (American Society of Mechanical Engineers. Order Dept., POB 2300, Fairfield, NJ 07007) V.106- 1984-

JEZOAO Journal of Experimental Zoology (Alan. R. Liss, Inc., 150 Fifth Avenue, New York, NY 10011) V.1- 1904-

JFALAX Journal of the Faculty of Agriculture, Tottori University. (Maruzen Co., Ltd., Export Dep., P.O. Box 5050, Tokyo Int., 100-31 Tokyo, Japan) V.1- 1951-

JFDSAZ Journal of Food Science. (Institute of Food Technologists, Subscrip. Dept., Suite 2120, 221 N. La Salle St., Chicago, IL 60601) V.26- 1961-

JFIBA9 Journal of Fish Biology. (Academic Press, 24-28 Oval Rd., London NW1 7DX, England) V.1- 1969-

JFLCAR Journal of Fluorine Chemistry. (Elsevier Sequoia SA, P.O. Box 851, CH 1001, Lausanne 1, Switzerland) V.1- 1971-

JFMAAQ Journal of the Florida Medical Association. (Florida Medical Association, POB 2411, Jacksonville, FL 32203) V.1- 1914-

JFSCAS Journal of Forensic Sciences. (American Society for Testing and Materials, 1916 Race St., Philadelphia, PA 19103) V.1- 1956-

JGAMA9 Journal of General and Applied Microbiology. (Microbiology Research Foundation, c/o Japan Academic Societies Center Bldg., 4-16, Tokyo 113, Japan) V.1- 1955-

JGHUAY Journal de Genetique Humaine. (Editeurs Medecine et Hygiene, 78, ave de la Rosaraie, 1211 Geneva 4, Switzerland) V.1- 1952-

JGMIAN Journal of General Microbiology (P.O. Box 32, Commerce Way, Colchester CO2 8HP, UK) V.1- 1947-

JGOBAC Journal de Gynecologie Obstetrique et Biologie de la Reproduction. (Masson et Cie, Editeurs, 120 Blvd. Saint-Germain, P-75280, Paris 06, France) V.1- 1972-

JHCYAS Journal of Histochemistry and Cytochemistry. (Elsevier Science Pub. Co., Inc., 52 Vanderbilt Ave., New York, NY 10017) V.1- 1953-

JHEMA2 Journal of Hygiene, Epidemiology, Microbiology and Immunology. (Avicenum, Zdravotnicke Nakladatelstvi, Malostranske namesti 28, Prague 1, Czechoslovakia) V.1- 1957-

JHHBAI Bulletin of the Johns Hopkins Hospital. (Baltimore, MD) V.1-119, 1889-1966. For publisher information, see JHMJAX

JHMJAX Johns Hopkins Medical Journal. (Journals Dept., Johns Hopkins Univ. Press, Baltimore, MD 21218) V.120-151, 1967-82. Discontinued

JHTCAO Journal of Heterocyclic Chemistry. (Hetero Corporation, POB 16000 MH, Tampa, FL 33687) V.1-1964-

JICSAH Journal of the Indian Chemical Society. (Indian Chemical Society, 92, Acharya Prafulla Chandra Rd., Calcutta 700009, India) V.5- 1928-

JIDEAE Journal of Investigative Dermatology. (Williams & Wilkins Co., 428 E. Preston St., Baltimore, MD 21202) V.1- 1938-

JIDHAN Journal of Industrial Hygiene. (Baltimore, MD/New York, NY) V.1-17, 1919-35. For publisher information, see AEHLAU

JIDIAQ Journal of Infectious Diseases. (University of Chicago Press, 5801 S. Ellis Ave., Chicago, IL 60637) V.1-1904-

JIDOAA Jikken Dobutsu. Experimental Animals. (Nippon Jikken Dobutsu Kenkyukai, c/o Tokyo Daigaku Ikngaku Kenkyusho, 6-1, Shiroganedai, Minato-ku, Tokyo 108, Japan) V.1- 1952-

JIDXA3 Journal of the Indiana State Medical Association. (3935 N. Meridian, Indianapolis, IN 46208) V.1- 1908-

JIHTAB Journal of Industrial Hygiene and Toxicology. (Baltimore, MD/New York, NY) V.18-31, 1936-49. For publisher information, see AEHLAU

JIMAAD Journal of the Indian Medical Association. (Indian Medical Association, 53, Creek Row, Calcutta 700014, India) V.1- 1931-

JIMMBG Journal of Immunological Methods. (Elsevier North Holland Inc., 52 Vanderbilt Ave., New York, NY 10017) V.1- 1971-

JIMRBV Journal of International Medical Research. (Secy., Cambridge Medical Publications, Ltd., 435/437 Wellingborough Rd., Northampton NN1 4EZ, England) V.1- 1972-

JIMSAX Journal of the Irish Medical Association. (Irish Medical Association., I.M.A. House, 10 Fitzwilliam Pl., Dublin, Ireland) V.28- 1951-

JINCAO Journal of Inorganic Nuclear Chemistry. (Pergamon Press Ltd., Headington Hill Hall, Oxford OX3 0BW, England) V.1- 1955-

JISMAB Journal of the Iowa State Medical Society. (Iowa Medical Society, 1001 Grand Ave., West Des Moines, IA 50265) V.1- 1911-

JJANAX Japanese Journal of Antibiotics. (Japan Antibiotics Research Association, 2-20-8 Kamiosaki, Shinagawa-ku, Tokyo, Japan) V.21- 1968-

JJATDK JAT, Journal of Applied Toxicology. (Heyden and Son, Inc., 247 S. 41st St., Philadelphia, PA 19104) V.1-1981-

JJCREP Japanese Journal of Cancer Research (Gann). (Elsevier Science Publishers B.V., POB 211, 1000 AE Amsterdam, Netherlands) V.76- 1985-

JJEMAG Japanese Journal of Experimental Medicine. (Editorial Office, The Institute of Medical Science, University of Tokyo, Shirokanedai, Minatoku, Tokyo, Japan) V.7- 1928-

JJIND8 JNCI, Journal of the National Cancer Institute. (U.S. Government Printing Office, Superintendent of Documents, Washington, DC 20402) V.61- 1978-

JJMCAQ Japanese Journal of Medical Science and Biology. (National Institute of Health, 2-chome, Kamiosaki, Shinagawa-ku, Tokyo 141, Japan) V.5- 1952-

JJMDAT Japanese Journal of Medicine. (Nankodo Co., Ltd., POB 5272, Tokyo International 100-31, Japan) V.1-1962-

JJPAAZ Japanese Journal of Pharmacology. (Nippon Yakuri Gakkai, c/o Kyoto Daigaku Igakubu Yakurigaku Kyoshitu Sakyo-ku, Kyoto 606, Japan) V.1- 1951-

JJPHAM Japanese Journal of Physiology. (Business Center for Academic Societies Japan, 4-16, Yayoc 2-chome, Bunkyo-ku, Tokyo 113, Japan) V.1- 1950-

JJTHEC Japanese Journal of Toxicology and Environmental Health. (Nippon Yakugakkai, 2-12-15 Shibuya, Shibuya-ku, Tokyo 150, Japan) V.38- 1992-

JJTOEX Japanese Journal of Toxicology. (Yakugyo Jihosha, Hokushin Bldg., 2-36 Jinbo-cho, Kanda, Chiyoda-ku, Tokyo, 101, Japan) V.1- 1988-

JKXXAF Japanese Kokai Tokkyo Koho Patents. (U.S. Patent Office, Science Library, 2021 Jefferson Davis Highway, Arlington, VA 22202)

JLCMAK Journal of Laboratory and Clinical Medicine. (C.V. Mosby Co., 11830 Westline Industrial Dr., St. Louis, MO 63141) V.1- 1915-

JLSMAW Journal of the Louisiana State Medical Society. (New Orleans, LA) V.105-134, 1953-82

JMCMAR Journal of Medicinal Chemistry. (American Chemical Society Pub., 1155 16th St., N.W., Washington, DC 20036) V.6- 1963-

JMEJAS Jikeikai Medical Journal. (Pharmacological Institute, The Jikei University School of Medicine, Minato-Ku, Tokyo, Japan) V.1- 1954-

JMGZAI Japan Medical Gazette. (Tokyo, Japan) V.1-18(?), 1964-81. Discontinued

JMIABM Journal of Medicine and International Medical Abstracts and Reviews. (Hemangini Publications, P.O. Box 16705, Calcutta, India) V.19-31, 1956-66(?)

JMICAR Journal of Microscopy. (Blackwell Scientific Publications, Ltd., Osney Mead, Oxford OX2 OEL, England) V.89- 1969-

JMMIAV Journal of Medical Microbiology. (Longman Group Journal Division, 4th Ave., Harlow Essex CM20 2JE, England) V.1- 1968-

JMOBAK Journal of Molecular Biology. (Academic Press, 111 Fifth Ave., New York, NY 10003) V.1- 1959-

JMPCAS Journal of Medicinal and Pharmaceutical Chemistry. (Washington, DC) V.1-5, 1959-62. For publisher information, see JMCMAR

JMSCA9 Journal of Mental Science. (London, England) V.4-108, 1857-1962. For publisher information, see BJPYAJ

JMSHAO Journal of the Mount Sinai Hospital. (New York, NY) V.1-36, 1934-69. For publisher information, see MSJMAZ

JMSUAT Journal of the Madras Agricultural Students' Union. (Madras Agricultural Journal, Tamil Nadu

Agricultural University Campus, Coimbatore 641003, India) V.1-16, 1912-28

JMTHBU Journal of the Medical Association of Thailand. (Medical Association of Thailand, New Pejburi Rd., Bangkok 10, Thailand) V.1- 1918-

JMXSAE Journal of Maxillofacial Surgery. (Georg Thieme Verlag, Postfach 732, Herdweg 63, D-7000 Stuttgart 1, Germany)

JNCIAM Journal of the National Cancer Institute. (Washington, DC) V.1-60, No. 6, 1940-78. For publisher information, see JJIND8

JNCSAI Journal of Cell Science. (P.O. Box 32, Commerce Way, Colchester CO2 8HP. Essex. UK) V.1- 1966-

JNDRAK Journal of New Drugs. (Albany, NY) V.1-6, 1961-66. For publisher information, see JCPCBR

JNENAD Journal of Neuropathology and Experimental Neurology. (American Association of Neuropathologists, Inc., 1041 New Hapshire St., Lawrence, KS 66044) V.1-1942-

JNEUAY Journal of Neuropsychiatry. (Chicago, IL) V.1-5, 1959-64. For publisher information, see BENPBG

JNMAAE Journal of the National Medical Association. (Appelton and Lange, 25 Van Zant St., E. Norwalk, CT 06855) V.1- 1909-

JNMDAN Journal of Nervous and Mental Disease. (Williams & Wilkins Co., 428 E. Preston St., Baltimore, MD 21202) V.3- 1876-

JNMDBO Journal of Medicine. (PJD Publications, POB 966, Westbury, NY 11590) V.1- 1970-

JNNPAU Journal of Neurology, Neurosurgery and Psychiatry. (British Medical Association, Tavistock Square, London WC1H 9JR, UK) V.7- 1944-

JNPHAG Journal de Pharmacologie. (Masson Publishing USA, Inc. (Journals,) 211 E. 43rd St., Suite 1306, New York, NY 10017) V.1- 1970-

JNRYA9 Journal of Neurology. (Springer-Verlag New York, Inc., Service Center, 44 Hartz Way, Secaucus, NJ 07094) V.207- 1974-

JNSVA5 Journal of Nutritional Science and Vitaminology. (Business Center Acad. Soc., Japan) V.19- 1973-

JOALAS Journal of Allergy, Including Allergy Abstracts. (St. Louis, MO) V.1-47, 1929-71. For publisher information, see JACIBY

JOANAY Journal of Anatomy. (Cambridge University Press, The Pitt Building, Trumpington Street, Cambridge CB2 1RP, UK) V.51- 1916-

JOAND3 Journal of Andrology. (Lippincott/Harper, Journal Fulfillment Dept., 2350 Virginia Ave., Hagerstown, MD 21740) V.1- 1980-

JOBAAY Journal of Bacteriology. (American Society for Microbiology, 1913 I St., N.W., Washington, DC 20006) V.1- 1916-

JOBIAO Journal of Biochemistry (Tokyo). (Japanese Biochemical Society, Gakkai Senta Bldg., 2-4-16 Yayoi, Bunkyo-Ku, Tokyo 113, Japan or Japan Publication USA or Maruzen) V.1- 1922-

JOBSDN Journal of Biological Sciences. (Indian Academy of Sciences, POB 8005, Bangalore 560080, India) V.1-1979-

JOCEAH Journal of Organic Chemistry. (American Chemical Society Pub., 1155 16th St., N.W., Washington, DC 20036) V.1- 1936-

JOCHFV Journal of Occupational Health. (Japan Society for Occupational Health, 1-29-8 Shinjuku, Shinjuku-ku, Tokyo, 160, Japan) V.38- 1996-

JOCMA7 Journal of Occupational Medicine. (American Occupational Medical Association, 150 N. Wacker Dr., Chicago, IL 60606) V.1- 1959-

JODUA2 Journal of Osaka Dental University. (Osaka Dental Univ., 1-47, Kyobashi Higashi-Ku Osaka 540, Japan) V.1- 1967-

JOENAK Journal of Endocrinology. (Biochemical Society Publications, P.O. Box 32, Commerce Way, Whitehall Industrial Estate, Colchester CO2 8HP, Essex, England) V.1- 1939-

JOETD7 Journal of Ethnopharmacology. (Elsevier Sequoia SA, POB 1304, CH-1001, Lausanne, Switzerland) V.1- 1979-

JOGBAS Journal of Obstetrics and Gynaecology of the British Commonwealth. (London, England) V.68-81, 1961-74. For publisher information, see BJOGAS

JOGNAU Journal of Genetics. (Asit Kr. Bhattacharyya, M.A., 18/1, Barrackpore Trunk Rd, Belghoria, India) V.1-1910-

JOHEA8 Journal of Heredity. (American Genetic Association, 818 18th St. N.W., Washington, DC 20006) V.5- 1914-

JOHYAY Journal of Hygiene. (Cambridge University Press, P.O. Box 92, Bentley House, 200 Euston Rd., London NW1 2DB, England) V.1- 1901-

JOIMA3 Journal of Immunology. (Williams & Wilkins Co., 428 E. Preston St., Baltimore, MD 21202) V.1- 1916-

JOIMD6 Journal of Immunopharmacology. (Marcel Dekker, POB 11305, Church St. Station, New York, NY 10249) V.1- 1978/79-

JOMAAL Journal of Mammalogy. (The American Museum of Natural History, Central Park W at 79th St., New York, NY 10024) V.1- 1919-

JONRA9 Journal of Neurochemistry. (Pergamon Press Ltd., Headington Hill Hall, Oxford OX3 0BW, England) V.1- 1956-

JONUAI Journal of Nutrition. (Journal of Nutrition, Subscription Dept., 9650 Rockville Pike, Bethesda, MD 20014) V.1- 1928-

JOPDAB Journal of Pediatrics. (C.V. Mosby Co., 11830 Westline Industrial Dr., St. Louis, MO 63141) V.1- 1932-

JOPHAN Journal de Physiologie. (Masson et Cie, Editeurs, 120 Blvd. Saint-Germain, P-75280, Paris 06, France) V.39- 1946-

JOPHBO Journal of Oral Pathology. (Munksgaard, 35 Norre Sogade, DK 1370 Copenhagen K, Denmark) V.1-1972-

JOPHDQ Journal of Pharmacobio-Dynamics. (Pharmaceutical Society of Japan, 12-15-501, Shibuya 2-chome, Shibuya-Ku, Tokyo 150, Japan) V.1- 1978-

JOPRAJ Journal of Periodontology. (American Academy of Periodontology, 211 E. Chicago Ave., Chicago, IL 60611) V.1- 1930-

JOPSAM Journal of Psychology. (Journal Press, 2 Commercial St., Provincetown, MA 02657) V.1- 1935/36-

JORCAI Journal of Organometallic Chemistry. (Elsevier Sequoia SA, POB 851, CH-1001 Lausanne l, Switzerland) V.1- 1963-

JOSCDQ Journal of SCCJ (Society of Cosmetic Chemists of Japan). (Nippon Keshohin Gijutsushakai, c/o Shiseido, 7-5-5 Ginza, Chuoku, Tokyo, 104, Japan) V.10- 1976-

JOSUA9 Journal of Oral Surgery. (American Dental Association, 211 E. Chicago Ave., Chicago, IL 60611) V.1-16, 1943-58; V.23- 1965-

JOUOD4 Journal of UOEH (University of Occupational Environmental Health). (University of Occupational and Environmental Health, 1-1 Iseigaoka, Yahata-nishi-ku, Kitakyushu, 807, Japan) V.1- 1979-

JOURAA Journal of Urology. (Williams & Wilkins Co., 428 E. Preston St., Baltimore, MD 21202) V.1- 1917-

JOVIAM Journal of Virology. (American Society for Microbiology, 1913 I St., N.W., Washington, DC 20006) V.1- 1967-

JPBAA7 Journal of Pathology and Bacteriology. (London, England) V.1-96, 1892-1968. For publisher information, see JPTLAS

JPBEAJ Journal de Pharmacie de Belgique. (Masson et Cie, Editeurs, 120 Blvd. Saint-Germain, P-75280, Paris 06, France) V.1- 1945/46-

JPCAAC Journal of the Air Pollution Control Association. (Air Pollution Control Association, 4400 5th Ave., Pittsburgh, PA 15213) V.5- 1955-

JPCEAO Journal fuer Praktische Chemie. (Johann Ambrosius Barth Verlag, Postfach 109, DDR-701, Leipzig, Germany) V.1- 1834- (Several new series, but continuous vol. nos. also used)

JPEMAO Journal of Perinatal Medicine. (Walter de Gruyter, Inc., 200 Sawmill River Rd., Hawthorne, MA 10532) V.1- 1973

JPETAB Journal of Pharmacology and Experimental Therapeutics. (Williams & Wilkins Co., 428 E. Preston St., Baltimore, MD 21202) V.1- 1909/10-

JPFCD2 Journal of Environmental Science and Health, Part B: Pesticides, Food Contaminants, and Agricultural Wastes. (Marcel Dekker, POB 11305, Church St. Station, New York, NY 10249) V.B11- 1976-

JPHAA3 Journal of the American Pharmaceutical Association. (American Pharmaceutical Association, 2215 Constitution Ave., N.W., Washington, DC 20037) V.1-28, 1912-39; New Series V.1-17, 1961-77

JPHYA7 Journal of Physiology. (Cambridge University Press, P.O. Box 92, Bentley House, 200 Euston Rd., London NW1 2DB, England) V.1- 1878-

JPIFAN Japan Pesticide Information. (Japan Plant Protection Association, 1-43-11, Komagome, Toshima-ku, Tokyo 170, Japan) No. 1-1969-

JPMRAB Japanese Journal of Medical Sciences, Part IV: Pharmacology. (Tokyo, Japan) V.1-16, 1926-43. For publisher information, see JJPAAZ

JPMSAE Journal of Pharmaceutical Sciences. (American Pharmaceutical Association, 2215 Constitution Ave., N.W., Washington, DC 20037) V.50- 1961-

JPPGAR Journal de Physiologie et de Pathologie Generale. (Paris, France) V.1-38, 1899-1945. For publisher information, see JOPHAN

JPPMAB Journal of Pharmacy and Pharmacology. (Pharmaceutical Society of Great Britain, 1 Lambeth High Street, London SEI 5JN, England) V.1- 1949-

JPROAR Journal of Protozoology. (Society of Protozoologists, P.O. Box 368, Lawrence, KS 66044) V.1-1954-

JPTLAS Journal of Pathology. (Longman Group Ltd., Subscriptions Journals Department, Fourth Avenue, Harlow, Essex, CM19 5AA, UK) V.97- 1969-

JRARAX Journal of Radiation Research. (c/o Tokai Daigaku Igakubu Bunshi Seibutsugaku Kyoshitsu, Tenbodai, Isehara, 259-11, Japan) V.1- 1960-

JRBED2 Journal of Reproductive Biology and Comparative Endocrinology. (P.G. Institute of Basic Medical Sciences, Dept. of Endocrinology, Taramani, 600 113, India) V.1- 1981-

JRFSAR Journal of Reproduction and Fertility, Supplement. (Journal of Reproduction and Fertility Ltd., 22 New Market Rd., Cambridge CB5 8D7, England) No. 1-1966-

JRHUA9 Journal of Rheumatology. (Business Office, Suite 420, 920 Yonge St., Toronto, Ontario M4W 3J7 Canada) V.1- 1974-

JRIHDC Journal of Research in Indian Medicine Yoga and Homoeopathy. (Banaras Hindu University, Varanasi-5, India) V.11- 1976-

JRIMAO Journal of Research in Indian Medicine. (Varanasi, India) V.1-10, 1966-75. For publisher information, see JRIHDC

JRMSAS Journal of the Royal Microscopical Society. (London, England) V.47-88, 1927-68. For publisher information, see JMICAR

JRPFA4 Journal of Reproduction and Fertility. (Journal of Reproduction and Fertility Ltd., 22 New Market Rd., Cambridge CB5 8D7, England) V.1- 1960-

JRPMAP Journal of Reproductive Medicine. (Box 56, 5841 Maryland Ave., Chicago, IL 60637) V.3- 1969-

JRSMD9 Journal of the Royal Society of Medicine. (Academic Press Inc., Ltd., 24-28 Oval Road, London NW1 7DX, England) V.71- 1978-

JSCCA5 Journal of the Society of Cosmetic Chemists. (Society of Cosmetic Chemists, 50 E. 41st St., New York, NY 10017) V.1- 1947-

JSFAAE Journal of the Science of Food and Agriculture. (Society of Chemical Industry, 14 Belgrave Sq., London SW1X 8PS, England) V.1- 1950-

JSGIED Journal of the Soceity for Gynecologic Investigation. (Elsevier Science, POB 945, New York, NY 10159) V.1- 1994-

JSGRA2 Journal of Surgical Research. (Academic Press, 111 Fifth Ave., New York, NY 10003) V.1- 1961-

JSICAZ Journal of Scientific and Industrial Research, Section C: Biological Sciences. (New Delhi, India) V.14-21, 1955-62. For publisher information, see IJEBA6

JSOMBS Journal of the Society of Occupational Medicine. (John Wright and Sons, 42-44 Triangle W., Bristol BS8 1EX, England)

JSONAU Journal of Surgical Oncology. (Alan R. Liss, Inc., 150 5th Ave., New York, NY 10011) V.1- 1969-

JSONDX Journal of Soviet Oncology. (New York) V. 1-5, 1980-84

JSOOAX Journal of the Chemical Society, Section C: Organic. (London, England) No.1-24, 1966-71. For publisher information, see JCPRB4

JSTBBK Journal of Steroid Biochemistry. (Pergamon Press Ltd., Headington Hill Hall, Oxford OX3 0BW, England) V.1- 1969-

JSXRAJ Journal of Sex Research. (Society for the Scientific Study of Sex, Inc., Mt. Royal and Guilford Aves., Baltimore, MD 21202) V.1- 1965-

JTASAG Journal of the Tennessee Academy of Science. (Tennessee Academy of Science, Box 153, George Peabody College, Nashville, TN 37203) V.1- 1926-

JTBIDS Journal of Thermal Biology. (Pergamon Press Ltd., Headington Hill Hall, Oxford OX3 0BW, England) V.1- 1975-

JTCEEM Journal de Toxicologie Clinique et Experimentale. (SPPIF, B. P. 22, F-41353 Vineuil, France) V.5- 1985-

JTCSAQ Journal of Thoracic and Cardiovascular Surgery. (C.V. Mosby Co., 11830 Westline Industrial Dr., St. Louis, MO 63141) V.38- 1959-

JTCTDW Journal of Toxicology, Clinical Toxicology. (Marcel Dekker, POB 11305, Church St. Station, New York, NY 10249) V.19- 1982-

JTEHD6 Journal of Toxicology and Environmental Health. (Hemisphere Publ., 1025 Vermont Ave., N.W., Washington, DC 20005) V.1- 1975/76-

JTEHF8 Journal of Toxicology and Environmental Health, Part A. (Taylor & Francis, 47 Runway Rd., Suite G, Levittown, PA 19057) V.53- 1998-

JTMHA9 Journal of Tropical Medicine and Hygiene. (Staples and Staples Ltd., 94-96 Wigmore St., London, England) V.10- 1907-

JTOTDO Journal of Toxicology, Cutaneous and Ocular Toxicology. (Marcel Dekker, POB 11305, Church St. Station, New York, NY 10249) V.1- 1982-

JTSCDR Journal of Toxicological Sciences. (Editorial Office, Higashi Nippon Gakuen Univ., 7F Fuji Bldg., Kita 3, Nishi 3, Sapporo 060, Japan) V.1- 1976-

JUIZAG Juntendo Igaku. Juntendo Medicine. (Juntendo Igakkai, 2-1-1, Hongo, Bunkyo-ku, Tokyo, 113, Japan) V.1- 1955-

JULRA7 Journal of Ultrastructure Research. (Academic Press, 111 Fifth Ave., New York, NY 10003) V.1- 1975-

JVPTD9 Journal of Veterinary Pharmacology and Therapeutics. (Blackwell Scientific Publications Ltd., POB 88, Oxford, UK) V.1- 1978-

JWMAA9 Journal of Wildlife Management. (Wildlife Society, Suite 611, 7101 Wisconsin Ave., Washington, DC 20014) V.1- 1937-

JYKYEL Junshi Yixue Kexueyuan Yuankan. (Junshi Yixue Kexueyuan Yuankan Bianjibu, 27, Taipinglu Haidianqu, Beijing, 100850, Peoples Rep. China) V.1- 1983-

JZKEDZ Jitchuken, Zenrinsho Kenkyuho. Central Institute for Experimental Animals, Preclinical Reports. (The Institute, 1433 Nogawa, Takatsu-Ku, Kawasaki 211, Japan) V.1- 1974/75-

KAEMAW Kitasato Archives of Experimental Medicine. (Kitasato Institute, 5-9-1 Shirokane, Minato-ku, Tokyo 108, Japan) V.1- 1917-

KAIZAN Kaibogaku Zasshi. Journal of Anatomy. (Nihon Kaibo Gakkai, c/o Tokyo Daigaku Igakubu Kaibogaku Kyoshitsu, 7-3-1, Hongo, Bunkyo-ku, Tokyo, Japan) V.1- 1928-

KAMJDW Kawasaki Medical Journal. (Kawasaki Medical School, Kurashiki 701-01, Japan) V.1- 1975-

KBAMAJ Kosmicheskaya Biologiya I Aviakosmicheskaya Meditsina Space Biology and Aerospace Medicine. (Mezhdunarodnaya Kniga, Kuznetskii Most 18, Moscow G-200, USSR) V.8- 1974-

KBIJEK Korean Biochemical Journal. (Biochemical Society of the Republic of Korea, POB 226, Mapo, Seoul, 121-600, South Korea) V. 26(4)- 1993-

KBMEAL Kosmicheskaya Biologiya i Meditsina. Space Biology and Medicine. (Moscow, USSR) V.1-7, 1967-73. For publisher information, see KBAMAJ

KCRZAE Kauchuk i Rezina. (v/o Mezhdunarodnaya Kniga, Kuznetskii Most 18, Moscow G-200, USSR) V.11- 1937-

KDIZAA Kagoshima Daigaku Igaku Zasshi. Medical Journal of Kagoshima University. (Kagoshima Daigaku Igakkai, 1208-1, Usuki-cho, Kagoshima 890, Japan) V.1- 1945-

KDPU** Beitrag zur Toxikologischen Wirkung Technischer Losungsmittel, Otto Klimmer Dissertation. (Pharmakologischen Institut der Universitat Wurzburg, Germany, 1937)

KEKHA7 Koshu Eiseiin Kenkyu Hokoku. Bulletin of the Institute of Public Health. (Kokuritsu Koshu Eiseiin, 4-6-1 Shirokanedai Minato-ku, Tokyo, 108, Japan) V.1- 1951-

KEKHB8 Kanagawa-ken Eisei Kenkyusho Kenkyu Hokoku. Bulletin of Kanagawa Prefectural Public Health Laboratories. (Kanagawa Prefectural Public Health Laboratories, 52-2, Nakao-cho, Asahi-ku, Yokohama 221, Japan) No. 1- 1971-

KFIZAO Kyoto-furitsu Ika Daigaku Zasshi. Journal of the Kyoto Prefectural School of Medicine. (Kyoto-furitsu Ika Daigaku Igakkai, Hirokoji, Kawaramachidori, Kamiyoku, Kyoto, Japan) V.1- 1927-

KGGZAL Koku Geka Gakkai Zasshi. Oral Surgery. (Tokyo, Japan)

KHFKDF K'o Hsueh Fa Chan Yueh K'an. Progress in Sciences. (Kuo Chia K'o Hsueh Wei Yuan Hui, 2 Canton St., Taipei 107, Taiwan) V.1- 1973-

KHFZAN Khimiko-Farmatsevticheskii Zhurnal. Chemical Pharmaceutical Journal. (v/o Mezhdunarodnaya Kniga, Kuznetskii Most 18, Moscow G-200, USSR.) V.1- 1967-

KHKKA3 Kachiku Hanshoku Kenkyu Kaishi. Journal of the Society of Animal Reproduction. (Tokyo, Japan) V.1-22, 1955-77. For publisher information, see KHZADH

KHZDAN Khigiena i Zdraveopazvane. (Hemus, blvd Russki 6, Sofia, Bulgaria) V.9- 1966-

KIDZAK Kansai Ika Daigaku Zasshi. Journal of the Kansai Medical School. (Kansai Ika Daigaku Igakkai, 1, Fumizono-cho, Moriguchi 570, Osaka, Japan) V.8- 1956-

KIHSDM Guangxi Yixue. Kuanghsi Medicine. (Kuang-hsi I Hsueh K'o Hsueh Ch'ing Pao Yen Chin So, 19 T'ien T'ao Lu, Nan-ning, Kuang-hsi, People's Republic of China) V.?- 1979-

KIKNAJ Kokuritsu Idengaku Kenkyusho Nempo. Annual Report of the National Institute of Genetics. (Kokuritsu Idengaku Kenkyusho, 1111 Yata, Mishima, Shizuoka-ken 411, Japan) No. 1- 1949-

KIZAAL Kurume Igakkai Zasshi. Journal of the Kurume Medical Association. (Kurume Igakkai, c/o Kurume Daigaku Igakubu, 67, Asahi-machi, Kurume 830, Japan) V.9- 1946-

KIZSB8 Kyorin Igakukai Zasshi. Journal of the Kyorin Medical Society. (6-20-2 Shinkawa, Mitaka City, Tokyo, Japan) V.1- 1970-

KJMDA6 Kobe Journal of the Medical Sciences. (Kobe Daigaku Igakubu, Editorial Board, Kobe, Japan) V.1- 1951-

KJMEA9 Keio Journal of Medicine. (Keio University, School of Medicine, 35 Shinano-machi, Shinjuku-Ku, Tokyo 160, Japan) V.1- 1952-

KJMSAH Kyushu Journal of Medical Science. (Fukuoka, Japan) V.6-15, 1955-64. Discontinued

KLWOAZ Klinische Wochenscrift. (Springer-Verlag, Heidelberger Pl. 3, D-1 Berlin 33, Germany) V.1- 1922-

KNZOAU Kanzo. Liver. (Nippon Kanzo Gakkai, c/o Toyo Bunko, 28-21, 2-chome, Hon Komagome, Bunkyo-ku, Tokyo 113, Japan) V.1- 1960-

KOBUA3 Kobunshi. High Polymers. (Kobunshi Gakkai, 5-12-8, Ginzai, Chuo-Ku, Tokyo 104, Japan) V.1- 1952-

KODAK* Kodak Company Reports. (343 State St., Rochester, NY 14650)

KOKABN Kobunshi Kako. Polymer Applications. (Kobunshi Kankokai, Chiekoin Maruta-machi Kundaru, Kamigyo-ku, Kyoto, 602, Japan) V.13- 1964-

KONODE Kongetsu No Noyaku. Agricultural Chemicals Monthly. (Kagoku Kogyo Nipposha, 3-19-16 Shibaura, Minato-ku, Tokyo 108, Japan) V.1- 1953(?)-

KorCJ# Personal Communication to NIOSH, from C.J. Korpics, Sherwin-Williams Chemicals, 1310 Expressway Drive, Toledo, OH 43608, August 22, 1974

KPJBAR Kalikasan. The Philippine Journal of Biology. (National Publishing Cooperative Inc., 2nd Fl., Santander Bldg., 20 M Hemady St. Cor Aurora Blvd., Quezon City, Philippines) V.1- 1972-

KRANAW Krankheitsforschung. (Leipzig, Germany) V.1-9, 1925-32. Discontinued

KRKRDT Kriobiologiya i Kriomeditsina. (Izdatel'stvo Naukova Dumka, ul Repina 3, Kiev, USSR) No. 1- 1975-

KRMJAC Kurume Medical Journal. (Kurume Igakkai, c/o Kurume Daigaku Igakubu, 67, Asahi-machi, Kurume, Japan) V.1- 1954-

KSGZA3 Kyushu Shika Gakkai Zasshi. Journal of the Kyushu Dental Society. (c/o Kyushu Shika Daigaku, 2-6-1 Manazuru, Kokurakita-ku, Kitakyushu, Japan) V.1- 1933-35; 1939-40; 1951-

KSKZAN Khimiya v Sel'skom Khozyaistve. Chemistry in Agriculture. (v/o Mezhdunarodnaya Kniga, Kuznetskii Most 18, Moscow G-200, USSR.) No. 1- 1963-

KSRNAM Kiso to Rinsho. Clinical Report. (Yubunsha Co., Ltd., 1-5, Kanda Suda-Cho, Chiyoda-ku, KS Bldg., Tokyo 101, Japan) V.1- 1960-

KTHYAC Kuo Li Tai-Wan Ta Hsueh I Hsueh Yuan Yen Chiu Pao Kao. The memoirs of the College of Medicine of the National Taiwan University. (Kuo Li Tai-Wan Ta Hsueh I Hsueh Yuan, Taipei, Taiwan) 1947-

KTUNAA K'at'ollik Taehak Uihakpu Nonmunjip. Journal of Catholic Medical College. (Catholic Medical College, Seoul, South Korea) V.1- 1957-

KTUWDD Koryo Taehakkyo Uikwa Taehak Nonmunjip. Collection of Papers of the Medical School, Korea University

KUIZAR Kumamoto Igakkai Zasshi. Journal of the Kumamoto Medical Society. (Kumamoto Igakkai, C/o Kumamoto Daigaku Igakubu, 2-1, 2-chome Honjo-machi, Kumamoto, Japan) V.1- 1925-

KUMJAX Kumamoto Medical Journal. (Kumamoto Daigaku Igakubu, Library, Kumamoto, Japan) V.1- 1938-

KYDKAJ Kyoritsu Yakka Daigaku Kenkyu Nempo. Annual Report of the Kyoritsu College of Pharmacy. (Kyoritsu Yakka Daigaku, 1-5-30 Shibakoen Minato-ku, Tokyo, Japan) V.1- 1955-

LacHB# Personal Communication from Mr. H.B. Lackey, Chemical Products Div., Crown Zellerbach, Camas, WA 98607, to Dr. H.E. Christensen, NIOSH, Rockville, MD 20852, June 9, 1978

LACHDL Liebigs Annalen der Chemie. (Verlag Chemie International, Inc. 175 Fifth Ave., New York, NY 10010) No. 1- 1979-

LAINAW Laboratory Investigation. (Williams & Wilkins Co., 428 E. Preston St., Baltimore, MD 21202) V.1- 1952-

LAKAA3 Laekartidningen. Medical News. (Swedish Medical Association, POB 5610, 11486 Stockholm 5, Sweden) V.62- 1965-

LAMEDS LARC Medical. (19 Bis, Rue d' Inkermann, 59000 Lille, France) V.1- 1981-

LANCAO Lancet. (7 Adam St., London WC2N 6AD, England) V.1- 1823-

LAPPA5 Lavori dell'Istituto di Anatomia e Istologia Patologica della Universita Degli Studi Perugia. (Istituto di Anatomia e Isologia Patologica, Caselle Postale 327, 06100 Perugia, Italy) V.1- 1939-

LARYA8 Laryngoscope. (222 Pine Lake Rd., Collinsville, IL 62234) V.1- 1896-

LBANAX Laboratory Animals. (Biochemical Society Book Depot, POB 32, Commerce Way, Colchester, Essex CO2 8HP, England) V.1- 1967-

LBASAE Laboratory Animal Science. (American Association for Laboratory Animal Science, 210 N. Hammes Ave., Suite 205, Joliet, IL 60435) V.21- 1971-

LDBU** Langer Dissertation. (Breslow, 1932)

LDTU** Narkoseversuche mit Hoheren Alkoholen und Stickstoffderivaten, Fritz Leube Dissertation. (Pharmakologischen Institut der Universitat Tubingen, Germany, 1931)

LIFSAK Life Sciences. (Pergamon Press, Maxwell House, Fairview Park, Elmsford, NY 10523) V.1-8, 1962-69; V.14-1974-

LilPW# Personal Communication from P.W. Lilley, Sun Company, Inc., P.O. Box 1135, Marcus Hook, PA 19061, to Richard Lewis, NIOSH, November 1, 1983

LitL## Personal Communication from Larry Little, Occidental Chemical Corp., Hooker Chemical Center, 360 Rainbow Blvd. South, Box 728, Niagara Falls, NY 14302, to Doris Sweet, NIOSH, Cincinnati, OH 45226, May 6, 1985

LITRC* Literature Research Co. Translation. (Annandale, VA)

LLOYA2 Lloydia. (Cincinnati, OH) V.1-41, 1938-78. For publisher information, see JNPRDF.

LMSED6 L.E.R.S. Monograph Series. Laboratories d'Etudes et de Recherches Synthelabo Monograph Series. (Raven Press, 1185 Ave. of Americas, New York, NY 10036) V.1- 1983-

LONZA# Personal Communication from LONZA Ltd., CH-4002, Basel, Switzerland, to NIOSH, Cincinnati, OH 45226

LPDSAP Lipids. (American Oil Chemists' Society, 508 South Sixth St., Champaign, IL 61820) V.1- 1966-

LPPTAK Labo-Pharma-Problemes et Techniques (19, rue Louis-le-Grand, Paris, France) V.12- 1965-

LROTDW Laryngo-Rhinologie und Otolaryngologie. (Georg Thieme Verlag, Postfach 732, Herdweg 63, 7000 Stuttgart, Germany) V.54- 1975-

LSBGAY Life Sciences, Part 2: Biochemistry, General and Molecular Biology. (New York, NY) V.9-13, 1970-73. For publisher information, see LIFSAK

LSPPAT Life Sciences, Part 1: Physiology and Pharmacology. (New York, NY) V.9-13, 1970-73. For publisher information, see LIFSAK

LYPHAD Lyon Pharmaceutique. (Edition Paul Chatelain, 63 rue de la Republique, 69 Lyons 2, France) V.1- 1950-

MACPAJ Mayo Clinic Proceedings. (Mayo Foundation, Plummer Bldg., Mayo Clinic, Rochester, MN 55901) V.39-1964-

MADCAJ Medical Annals of the District of Columbia. (Washington, DC) V.1-43, 1932-74. Discontinued

MAGDA3 Mechanisms of Ageing and Development. (Elsevier Sequoia SA, POB 851, Ch 1001, Lausanne 1, Switzerland) V.1- 1972-

MAGJAL Malaysian Agricultural Journal. (Ministry of Agriculture and Fisheries, Business Mgr., Kuala Lumpur, Malaysia) V.45- 1965-

MahWM# Personal Communication from W.M. Mahlburg, Hopkins Agricultural Chemical Co., P.O. Box 7532, Madison, WI 53707, to NIOSH, Cincinnati, OH 45226, November 16, 1982

MAIKD3 Maikotokishin (Tokyo). Mycotoxin. (Maikotokishin Kenkyukai, c/o Tokyo Rika Daigaku Yakugakubu, 12 Funagawara-machi, Ichigaya, Shinjuku-ku, Tokyo 162, Japan) No. 1- 1975-

MAIZAB Manshu Igaku Zasshi. Manthuria Medical Journal. (Dairen/Shimmeicho, South Manchuria) V.1-33, 1923-40. Discontinued

MarJV# Personal Communication from Josef V. Marhold, VUOS, 539-18, Pardubice, Czechoslovakia, to the Editor of RTECS, Cincinnati, OH, March 29, 1977

MASODV Masui to Sosei. Anesthesia and Resuscitation. (Hiroshima Masui Igakkai Hiroshima Daigaku Igakubu Masuigaku Kyoshitsu, 1-2-3 Kasumi, Hiroshima 734, Hiroshima, Japan) V.6- 1970-

MBADEI Metal-Based Drugs. (Freund Publishing House, Suite 500, Chesham House, 150 Regent St., London W1R 5FA, UK) V.1- 1994-

MCBIA7 Microbios. (Faculty Press, 88 Regent St., Cambridge, England) V.1- 1969-

MccSB# Personal Communication from Susan B. McCollister, Dow Chemical U.S.A., Midland, MI 48640, to NIOSH, Cincinnati, OH 45226, June 15, 1984

MCEBD4 Molecular and Cellular Biology. (American Society for Microbiology, 1913 I St., NW, Washington, DC 20006) V.1- 1981-

MDACAP Medicamentos de Actualidad. (Medicamentos de Actualidad, Apartado de Correos 540, Barcelona, Spain) V.1- 1965-

MDCHAG Medicinal Chemistry: A Series of Monographs. (Academic Press, 111 5th Ave., New York, NY 10003) V.1-1963-

MDMIAZ Medycyna Doswiadczalna i Mikrobiologia. (Ars Polona-RUCH, POB 1001, 1, P-00068 Warsaw, 1, Poland) V.1- 1949- For English Translation, see EXMMAV

MDREP* U.S. Army, Chemical Corps Medical Division Reports. (Army Chemical Center, MD)

MDSR** U.S. Army, Chemical Corps Medical Division Special Report. (Army Chemical Center, MD)

MDZEAK Medizin und Ernaehrung. (Stuttgart, Germany) V.1-13, 1959-72. Discontinued

MECHAN Medicinal Chemistry, A Series of Reviews. (New York, NY) V.1-6, 1951-63. Discontinued

MEDIAV Medicine. (Williams & Wilkins, 428 E. Preston St., Baltimore, MD 21202) V.1- 1922-

MEHYDY Medical Hypotheses. (Churchill Livingstone Inc., 19 W. 44 St., New York, NY 10036) V.1- 1975-

MEIEDD Merck Index. (Merck and Co., Inc., Rahway, NJ 07065) 10th ed. 1983-

MEKLA7 Medizinische Klinik. (Urban and Schwarzenberg, Pettenkoferst 18, D-8000 Munich 15, Germany) V.1- 1904-

MELAAD Medicina del Lavoro. Industrial Medicine. (Via S. Barnaba, 8 Milan, Italy) V.16- 1925-

MEMOAQ Medizinische Monatsschrift. (Stuttgart, Germany) V.1-31, 1947-77

MEPAAX Medycyna Pracy. Industrial Medicine. (ARs-Polona-RUSH, POB 1001, 00-068 Warsaw 1, Poland) V.1-1950-

MEPHDN Medical Pharmacy. (Daiichi Seiyaku K.K., 3-14-10 Nihonbashi, Chuo-ku, Tokyo 103, Japan) V.1- 1966-

METAAJ Metabolism, Clinical and Experimental. (Grune and Stratton, Inc., 111 Fifth Ave., New York, NY 10003) V.1- 1952-

METOEV Medical Toxicology. (Auckland, New Zealand) V.1-3, 1986-1988.

METRA2 Medecine Tropicale (Marseille). (Parc du Pharo, 13007 Marseille, France) V.1- 1941-

MEWEAC Medizinische Welt. (F.K. Schattauer Verlag, Postfach 2945, D-7000 Stuttgart 1, Germany) V.1-18, 1927-1944; V.1-1950-

MEXPAG Medicina Experimentalis. (Basel, Switzerland) V.1-11, 1959-64; V.18-19, 1968-69. For publisher information, see JNMDBO

MFEPDX Methods and Findings in Experimental and Clinical Pharmacology. (Methods and Findings, Sub. Dept., Apartado Correos 1179, Barcelona, Spain) V.1- 1979-

MFLRA3 Mededelingen van de Faculteit Landbouwwetenschappen, Rijksuniversiteit Gent. Communications of the Faculty of Agricultural Sciences, State University of Ghent. (Bibliotheek, Faculteit Labdbouwwettenschappen, Rijksuniversiteit Gent, Coupure 533, B-9000, Ghent, Belgium) V.35- 1970-

MGBUA3 Microbial Genetics Bulletin. (Ohio State University, College of Biological Science, Dept. of Microbiology, 484 W. 12th Ave., Columbus, OH 43210) No. 1- 1950-

MGGEAE Molecular and General Genetics. (Springer-Verlag, Heidelberger Pl. 3, D-1 Berlin 33, Germany) V.99-1967-

MGLHAE Mitteilungen Aus Dem Gebiete Der Lebensmitteluntersuchung und Hygiene. (Eidgenoessiche Drucksachen und Materialzentrale, 3000 Bern, Switzerland) V.1- 1910-

MGONAD Magyar Onkologia. Hungarian Onkology. (Kultura, P.O. Box 149, H-1389 Budapest, Hungary) V.1-1957-

MIBLAO Microbiology (Moscow). (Plenum Publishing Corp., 233 Spring St., New York, NY 10013) V.26- 1957-

MIDAD4 NIDA Research Monograph. (National Institute on Drug Abuse, Division of Research, 5600 Fishers Lane, Rockville, MD 20857) No. 1- 1975-

MIFAAB Minerva Farmaceutica. (Turin, Italy) V.1-13, 1952-64. Discontinued

MIIMDV Microbiology and Immunology. (Center for Academic Publications Japan, 4-16, Yayoi 2-chome, Bunkyo-ku, Tokyo 113, Japan) V.21- 1977-

MIKBA5 Mikrobiologiya. (v/o Mezhdunarodnaya Kniga, Kuznetskii Most 18, Moscow G-200, USSR) V.1- 1932-

MILEDM Microbios Letters. (Faculty Press, 88 Regent St., Cambridge, England) V.1- 1976-

MIMEAO Minerva Medica. (Edizioni Minerva Medica, Casella Postale 491, Turin, Italy) V.1- 1909-

MIPEA5 Minerva Pediatrica. (Edizioni Minerva Medica, Cassella Postale 491, I-10126 Turin, Italy) V.1- 1949-

MIVRA6 Microvascular Research. (Academic Press, 111 Fifth Ave., New York, NY 10003) V.1- 1968-

MJAUAJ Medical Journal of Australia. (P.O. Box 116, Glebe 2037, NSW 2037, Australia) V.1- 1914-

MJDHDW Mukogawa Joshi Daigaku Kiyo, Yakugaku Hen. Bulletin of Mukogawa Women's College, Food Science Series. (Mukogawa Joshi Daigaku, 6-46, Ikebiraki-cho, Nishinomiya 663, Japan) No. 19- 1972-

MJOUAL Medical Journal of Osaka University. (The University, 33, Joan-cho, Kita-Ku, Osaka, Japan) V.1- 1949-

MLDCAS Medecine Legale et Dommage Corporel. (Paris, France) V.1-7, 1968-74. Discontinued

MLSR** U.S. Army, Chemical Corps Medical Laboratories Special Reports. (Army Chemical Center, MD)

MMAPAP Mycopathologia et Mycologia Applicata. (The Hague, Netherlands) V.5-54, No. 4, 1950-74, For publisher information, see MYCPAH

MMDPA6 Materia Medica Polona (English Edition). (Ars Polona-RUCH, P.O. Box 1001, P-00 068 Warsaw, 1, Poland) V.1- 1969-

MMEDA9 Military Medicine. (Association of Military Surgeons of the United States, Box 104, Kensington, MD 20795) V.116- 1955-

MMIYAO Medical Microbiology and Immunology. (Springer-Verlag, Heidelberger Pl. 3, D-1 Berlin 33, Germany) V.157- 1971-

MMJJAI Mie Medical Journal. (Mie Medical Society, Mie Prefectual Univ., School of Medicine, Tsu, Japan) V.3-1952-

MMWOAU Muenchener Medizinische Wochenscrift. (Munich, Germany) V.33-115, 1886-1973

MOLAAF Monatsschrift fuer Ohrenheilkunde und Laryngo-rhinologie. (Vienna, Austria) V.1-108, 1867-1974. For publisher information, see LROTDW

MONS** Monsanto Co. Toxicity Information. (Monsanto Industrial Chemicals Co., Bancroft Bldg., Suite 204, 3411 Silverside Rd., Wilmington, DE 19810)

MOPMA3 Molecular Pharmacology. (The American Society for Pharmacology and Experimental Therapeutics, 9650 Rockville Pike, Bethesda, MD 20014) V.1- 1965-

MosJN# Personal Communication from J.N. Moss, Toxicology Department, Rohm and Haas Co., Spring House, PA 19477, to R. J. Lewis, Sr., NIOSH, Cincinnati, OH 45226, August 15, 1979

MPHEAE Medicina et Pharmacologia Experimentalis. (Basel, Switzerland) V.12-17, 1965-67. For publisher information, see PHMGBN

MPPBAB Meditsinskaya Parazitologiya i Parazitarnye Bolezni. Medical Parasitology and Parasitic Diseases. (v/o Mezhdunarodnaya Kniga, Kuznetskii Most 18, Moscow G-200, USSR.) V.1- 1932-

MPPPBK Modern Problems of Pharmacopsychiatry. (S. Karger AG, Postfach, CH-4009, Basel, Switzerland) V.1-1968-

MRBUDF MARDI Research Bulletin. (MARDI, Secy., Publication Committee, POB 208, Serdang, Selangor, Malaysia) V.1- 1973-

MRCSAB Medical Research Council, Special Report Series. (Her Majesty's Stationery Office, P.O. Box 569, London SE1 9NH, England) SRS1- 1915-

MRLAB3 Mededelingen Rijksfaculteit Landbouwwetenschappen, Gent. Communications of the State University of Agricultural Sciences, Ghent. (Ghent, Belgium) V.31-34, 1966-69. For publisher information, see MFLRA3

MRLEDH Mutation Research Letters. (Elsevier/North-Holland Biomedical Press, POB 211, 1000 AE Amsterdam, Netherlands)

MRLR** U.S. Army, Chemical Corps Medical Laboratories Research Reports. (Army Chemical Center, Edgewood Arsenal, MD)

MSCREJ Medical Science Research. (Elsevier Applied Science Publishing Ltd., Crown House, Linton Rd., Barking, Essex IG11 8JU, UK) V.15 1987-

MSERDS Microbiology Series. (Marcel Dekker, Inc., POB 11305, Church St. Station, New York, NY 10249) V.1- 1973-

MTPEEI Meditsina Truda i Promyshlennaya Ekologiya. Industrial Medicine and Ecology. (Mezhdunarodnaya Kniga, ul.B. Yakimanka, 39 117049 Moscow, Russia) No.1- 1993-

MSJMAZ Mount Sinai Journal of Medicine, New York. (The Annenberg Building, Room 10-35, 5th Ave. and 100th St., New York, NY 10029) V.37- 1970-

MUREAV Mutation Research. (Elsevier Science Publications B.V., POB 211, 1000 AE Amsterdam, Netherlands) V.1- 1964-

MTPEKO Meditsina Truda i Promyshlennaya Ekologiya. Industrial Medicine and Ecology. (Mezhdunarodnaya Kniga, ul.B. Yakimanka, 39 117049 Moscow, Russia) No.1- 1993-

MUTAEX Mutagenesis. (IRL Press Ltd. 1911 Jefferson Davis Highway, Suite 907, Arlington, VA 22202) V.1- 1986-

MVCRB3 "Fluorescent Whitening Agents, Proceedings of A Symposium." MVC-Report, Miljoevardscentrum, Stockholm Center for Environmental Sciences, No. 2, 1973

MVMZA8 Monatshefte fuer Veterinaermedizin. (VEB Gustav Fischer Verlag, Postfach 176, Villengang 2, 69 Jena, Germany) V.1- 1946-

MYCPAH Mycopathologia. (Dr. W. Junk bv Publishers, POB 13713, 2501 ES The Hague, Netherlands) V.1- 1938-

MZHUDX Mikrobiologicheskii Zhurnal (Kiev). Journal of Microbiology. (v/o Mezhdunarodnaya Kniga, Kuznetskii Most 18, Moscow G-200, USSR.) V.40- 1978-

MZUZA8 Meditsinskii Zhurnal Uzbekistana (v/o Mezhdunarodnaya Kniga, Kuznetskii Most 18, Moscow G-200, USSR.) No. 1- 1957-

NAGZAC Nagasaki Igakkai Zasshi. Journal of Nagasaki Medical Association. (Nagasaki Igakkai, c/o Nagasaki Daigaku Igakubu, 12-4 Sakamoto-machi, Nagasaki 852, Japan) V.1- 1923-

NAHRAR Nahrung. Chemistry, Biochemistry, Microbiology, Technology, Nutrition. (Akademie-Verlag GmBH, Leipziger St. 3-4, 108 Berlin, Germany) V.1- 1957-

NAIZAM Nara Igaku Zasshi. Journal of the Nara Medical Association. (Nara Kenritsu Ika Daigaku, Kashihara, Nara, Japan) V.1- 1950-

NALSDJ NATO Advanced Study Institute Series, Series A: Life Sciences. (Plenum Publishing Corp., 233 Spring St., New York, NY 10013) V.53- 1983-

NAREA4 Nutrition Abstracts and Reviews. (Central Sales Branch, Commonwealth Agricultural Bureaux, Farnham Royal, Slough SL2 3BN, England) V.1- 1931-

NARHAD Nucleic Acids Research. (Information Retrieval Inc., 1911 Jefferson Davis Highway, Arlington, VA 22202) V.1- 1974-

NASDA6 Nagoya Shiritsu Daigaku Igakkai Zasshi. Journal of the Nagoya City University Medical Association. (The University, Nagoya, Japan) V.1- 1950-

NASGEJ NATO ASI Series, Series G: Ecological Sciences. (Springer-Verlag New York, Inc., Service Center, 44 Hartz Way, Secaucus, NJ 07094) No.1- 1983-

NATUAS Nature. (Macmillan Journals Ltd., Brunel Rd., Basingstoke RG21 2XS, UK) V.1- 1869-

NATWAY Naturwissenschaften. (Springer-Verlag, Heidelberger Platz 3, D-1000 Berlin 33, Germany) V.1- 1913-

NCDREP New Cardiovascular Drugs. (Ravencrest, 1185 Avenue of the Americas, New York, NY 10036) 1985-

NCIAL* Progress Report Submitted to the National Cancer Institute by Arthur D. Little, Inc. (15 Acorn Park, Cambridge, MA 02140)

NCIBR* Progress Report for Contract No. NIH-NCI-E-68-1311, Submitted to the National Cancer Institute by Bio-Research Consultants, Inc. (9 Commercial Ave., Cambridge, MA 02141)

NCICP* Progress Report Submitted to the National Cancer Institute by Charles Pfizer and Company

NCIHL* Progress Report Submitted to the National Cancer Institute by Hazelton Laboratories, Inc

NCIIR* Progress Report for Contract No. N01-CP-12338, Submitted to the National Cancer Institute by IIT Research Institute. (Chicago, IL)

NCILB* Progress Report for Contract No. NIH-NCI-E-C-72-3252, Submitted to the National Cancer Institute by Litton Bionetics, Inc. (Bethesda, MD)

NCIMAV National Cancer Institute, Monograph. (U.S. Government Printing Office, Superintendent of Documents, Washington, DC 20402) No. 1- 1959-

NCIMR* Progress Report Submitted to the National Cancer Institute by Mason Research Institute. (Worcester, MA)

NCINS* National Cancer Institute Report. (Bethesda, MD 20014)

NCIRI* Progress Report Submitted to the National Cancer Institute by Piason Research Institute

NCISA* Progress Report for Contract No. PH-43-63-1132, Submitted to the National Cancer Institute by Scientific Associates, Inc. (6200 S. Lindberg Blvd., St. Louis, MO 63123)

NCISP* National Cancer Institute Screening Program Data Summary, Developmental Therapeutics Program, Bethesda, MD 20205

NCISS* Progress Report Submitted to the National Cancer Institute by South Shore Analytical and Research Laboratory

NCITR* National Cancer Institute Carcinogenesis Technical Report Series. (Bethesda, MD 20014) No. 0-205. For publisher information, see NTPTR*

NCIUS* Progress Report for Contract No. PH-43-64-886, Submitted to the National Cancer Institute by the Institute of Chemical Biology, University of San Francisco. (San Francisco, CA 94117)

NCNSA6 National Academy of Sciences, National Research Council, Chemical-Biological Coordination Center, Review. (Washington, DC)

NCPBBY National Clearinghouse for Poison Control Centers, Bulletin. U.S. Department of Health, Education, and Welfare (Washington, DC)

NDADD8 New Drugs Annual: Cardiovascular Drugs. (New York, NY) V.1-2 1983-84. For publisher information, see NCDREP

NDKIA2 Kankyo Igaku Kenkyusho Nenpo (Nagoya Daigaku). Annual Report of the Research Institute of Environmental Medicine, Nagoya University. (The University, Furo-cho, Chikosa-Ku, Nagoya, Japan) V.1- 1947/48(Pub. 1949)-

NDRC** National Defense Research Committee, Office of Scientific Research and Development, Progress Report

NEACA9 News Edition, American Chemical Society (Easton, PA) V.18-19, 1940-41. For publisher information, see CENEAR

NEAGDO Neurobiology of Aging. (ANKHO International Inc., POB 426, Fayetteville, NY 13066) V.1-1980-

NEJMAG New England Journal of Medicine. (Massachusetts Medical Society, 10 Shattuck St., Boston, MA 02115) V.198- 1928-

NEOLA4 Neoplasma. (Karger-Libri AG, Scientific Booksellers, Arnold-Boecklin-Strasse 25, CH-4000 Basel 11, Switzerland) V.4- 1957-

NEPHBW Neuropharmacology. (Pergamon Press, Headington Hill Hall, Oxford OX3 OBW, England) V.9-1970-

NEPSBV Neuropsychopharmacology, Proceedings of the Meeting of the Collogium Internationale, Neuropsychopharmacologicum. (Excerpta Medica Foundation, P.O. Box 1126, Amsterdam, Netherlands) V.1-1959-

NEREDZ Neurochemical Research. (Plenum Publishing Corp., 233 Spring St., New York, NY 10013) V.1- 1976-

NETEEC Neurotoxicology and Teratology. (Pergamon Press Inc., Maxwell House, Fairview Park, Elmsford, NY 10523) V.9- 1987-

NETOD7 Neurobehavioral Toxicology. (ANKHO International, Inc., P.O. Box 426, Fayetteville, NY 13066) V.1-2, 1979-80, For publisher information, see NTOTDY

NEURAI Neurology. (Modern Medicine Publications, Inc., 757 Third Avenue, New York, NY 10017) V.1- 1951-

NEZAAQ Nippon Eiseigaku Zasshi. Japanese Journal of Hygiene. (Nippon Eisei Gakkai, c/o Kyoto Daigaku Igakubu Koshu Eiseigaku Kyoshita, Yoshida Konoe-cho, Sakyo-ku, Kyoto, Japan) V.1- 1946-

NEZTAF New Zealand Veterinary Journal. (New Zealand Veterinary Association, Massey Univ., Palmerston North, New Zealand) V.1- 1952-

NFGZAD Nippon Funin Gakkai Zasshi. Japanese Journal of Fertility and Sterility. (Nippon Funin Gakkai, 1-1 Sadohara-cho, Ichigaya, Shinjuku-ku, Toyko 162, Japan) V.1- 1956-

NGCJAK Nippon Gan Chiryo Gakkai-shi. Journal of Japan Society for Cancer Therapy. (Nihon Gan Chiryo Gakkai, Kyoto, Japan) V.1- 1966-

NGGKED Nippon Gan Gakkai Sokai Kiji. Proceedings of the Annual Meeting of the Japanese Cancer Association. (Japanese Cancer Association, Tokyo, Japan) V.1- 1956-

NGGZAK Nippon Geka Gakkai Zasshi. Journal of the Japanese Surgical Society. (Nippon Geka Gakkai, 2-3-10 Koraku, Bunkyo-ku, Tokyo 112, Japan) V.8- 1908-

NHOZAX Nippon Hoigaku Zasshi. Japanese Journal of Legal Medicine. (Nippon Hoi Gakkai, c/o Tokyo Daigaku Igakubu Hoigaku Kyoshitsu, 7-3-1, Hongo, Bunkyo-ku, Tokyo, Japan) V.1- 1944-

NHTIA7 Nordisk Hygienisk Tidskrift. Scandinavian Journal of Hygiene. (Stockholm, Sweden) V.1-55, 1920-74. For publisher information, see SWEHDO.

NIAND5 NIPH Annals. (National Institute of Public Health, Postuttak, Oslo, 1, Norway) V.1- 1978-

NIBKAW Nippon Byori Gakkai Kaishi. Journal of the Japanese Pathological Society. (c/o Tokyo Daigaku Igakubu Byorigaku Kyoshitsu, 7-3-1 Hongo, Bunkyo-ku, Tokyo 113, Japan) V.1- 1911-

NICHAS Nichidai Igaku Zasshi. (Nihon Daigaku Igakkai, 30, Oyaguchi kami-machi, Itabashi-ku, Tokyo 173, Japan) V.1- 1937-

NIGHAE Archiv Fuer Japanische Chirurgie. (Nippon Geka Hokan Henshushitsu, c/o Kyoto Daigaku Igakubu Geka Seikei Geka Kyoshitsu, 54 Kawara- machi, Shogoin, Sakyo-ku, Kyoto 606, Japan) V.1- 1924-

NIGZAY Niigata Igakkai Zasshi. Niigata Medical Journal. (Niigata Daigaku, 1 Asahi-Machi, Niigata, Japan) V.60-1946-

NIHBAZ National Institutes of Health, Bulletin. (Bethesda, MD)

NIIHAO Nichidoku Iho. Japanese-German Medical Reports. (Nihon Schering Co., Ltd., 6-64, Nishimiyahara, 2-chome, Yodogawa-ku, Osaka 532, Japan) V.1- 1956-

NIIRDN "Drugs in Japan. Ethical Drugs, 6th Edition 1982" Edited by Japan Pharmaceutical Information Center. (Yakugyo Jiho Co., Ltd., Tokyo, Japan)

NIOSH* National Institute for Occupational Safety and Health, U.S. Dept. of Health, Education, and Welfare, Reports and Memoranda

NIPAA4 Nippon Shokakibyo Gakkai Zasshi. Journal of the Japanese Society of Gastroenterology. (Nippon Shokakibyo Gakkai, 4-12 7-Chome, Ginza, Chuo-Ku, Tokyo 104, Japan) V.1- 1899-

NIPDAD Nihon Daigaku No-Juigakubu Gakujutsu Kenkyu Hokoku. Research Reports of the College of Agriculture and Veterinary Medicine, Nihon Univ. (The Univ., 34-1 Shimouma, 3-chome, Setagaya-ku, Tokyo 154, Japan) No. 1- 1953-

NIPOAC Nippon Shonika Gakkai Zasshi. (Nippon Shonika Gakkai, 1-29-8 Shinjuku, Shinjuku-Ku, Tokyo 160, Japan) V.55- 1951-

NISFAY Nippon Sanka Fujinka Gakkai Zasshi. Journal of Japanese Obstetrics and Gynecology. (Nippon Sanka Fujinka Gakkai, c/o Hoken Kaikan Building., 1-1 Sadohara-cho, Ichigaya, Shinjuku-ku, Tokyo 162, Japan) V.1- 1949-

NIVAAY Notiziario dell'Istituto Vaccinogeno Antitubercolare. Bulletin of the Institute for Antitubercular Vaccinogens. (l'Instituto, Via Clericetti, 45 Milan, Italy) V.1-11, 1951-61

NJGKBV Snake. (Japan Snake Institute, Hon-machi, Yaduzuka, Nitta-gun, Gunma-ken, Japan) V.1- 1969-

NJMSAG Nagoya Journal of Medical Science. (Nagoya University School of Medicine, 65 Tsuruma-cho, Showa-ku, Nagoya 466, Japan) V.2- 1927-

NJUZA9 Japanese Journal of Veterinary Science. (Nippon Jui Gakkai, 1-37-20, Yoyogi, Shibuya-ku, Tokyo 151, Japan) V.1- 1939-

NKEZA4 Nippon Koshu Eisei Zasshi. Japanese Journal of Public Health. (Nippon Koshu Eisei Gakkai, 1-29-8 Shinjuku, Shinjuku-ku, Tokyo 160, Japan) V.1- 1954-

NKGZAE Nippon Ketsueki Gakkai Zasshi. Journal of Japan Haematological Society. (Kyoto Univ. Hospital, Faculty of Medicine, Kyoto, Japan) V.1- 1937-

NKOGAV Nippon Kokuka Gakkai Zasshi. Journal of the Japanese Stomatological Society. (7-3-1, Hongo, Bunkyo-Ku, Tokyo, Japan) V.1- 1952-

NKRZAZ Chemotherapy (Tokyo). (Nippon Kagaku Ryoho Gakkai, 2-20-8 Kamiosaki, Shinagawa-Ku, Tokyo, 141, Japan) V.1- 1953-

NLETDU Neuroendocrinology Letters. (Brain Research Promotion, D-7400, Tubingen, Germany) V.1- 1979-

NLJMAV Netherlands Journal of Medicine. (Elsevier Science, POB 211, 1000 AE Amsterdam. Netherlands) V.16- 1973-

NMJOAA Nagoya Medical Journal. (Nagoya City University Medical School, 2-38 Nagarekawa-machi, Naka-ku, Nagoya 460, Japan) V.1- 1953-

NNAPBA Naunyn-Schmiedebergs Archiv fuer Pharmakologie. (Berlin, Germany) V.264-271, 1969-71. For publisher information, see NSAPCC

NNBYA7 Nature: New Biology. (Macmillan Journals Ltd., Houndmills Estate, Basingstoke, Hants RG21 2XS, England) V.229-246, 1971-73

NNGAAS Nippon Naika Gakkai Zasshi. Journal of the Japanese Society of Internal Medicine. (Nippon Naika Gakkai, 33-5, 3-Chome, Hongo, Bunkyo-ku, Tokyo 13) V.1- 1913-

NNGADV Nippon Noyaku Gakkaishi. (Pesticide Science Society of Japan, 43-11, 1-Chome, Komagome, Toshima-ku, Tokyo 170, Japan) V.1- 1976-

NNGZAZ Nippon Naibumpi Gakkai Zasshi. Journal of the Japan Endocrine Society. (Nippon Naibumpi Gakkai, c/o Kyoto-Furitsu Ika Daigaku, Kojinbashi Nishizume-Sagaru, Kamigyo-ku, Kyoto 602, Japan) V.1- 1925-

NOMDA6 Northwest Medicine. (Seattle, WA) V.1-73, 1903-73. Discontinued

NONSYP Nongsa Sihom Yongu Pogo. Research Reports on Agricultural Experimentation. (Suwon, S. Korea) V.1-26, 1958-84.

NPIRI* Raw Material Data Handbook, Vol.1: Organic Solvents, 1974. (National Association of Printing Ink Research Institute, Francis McDonald Sinclair Memorial Laboratory, Lehigh Univ., Bethlehem, PA 18015)

NPMDAD Nouvelle Presse Medicale. (Masson et Cie, Editeurs, 120 Blvd. Saint-Germain, P-75280, Paris 06, France) V.1- 1972-

NPRNAY Nephron. (S. Karger Publishers, Inc., 150 Fifth Ave., Suite 1105, New York, NY 10011) V.1- 1964-

NRINA3 Nippon Rinsho. Japanese Clinical Medicine. (Nippon Rinsho Sha, 3-1 Dosho-machi, Higashi-ku, Osaka 541, Japan) V.1- 1943-

NRSCDN Neuroscience. (Pergamon Press Ltd., Headington Hill Hall, Oxford OX3 OBW, England) V.1- 1976-

NRTTA8 NRC Technical Translation. (The National Research Council of Canada, Ottawa, Ontario K1A 0R6, Canada) No. 1- 1949-

NRTXDN Neurotoxicology. (Pathotox Publishers, Inc., 2405 Bond St., Park Forest South, IL 60464) V.1- 1979-

NSAPCC Naunyn-Schmiedeberg's Archives of Pharmacology. (Springer-Verlag, Heidelberger Pl. 3, D-1 Berlin 33, Germany) V.272- 1972-

NSMZDZ Nippon Shika Masui Gakkai Zasshi. (Tokyo Ikashika Daigaku Shika Masuigaku Kyoshitsu, 1-5-45 Yushima, Bunkyo-Ku, Tokyo 113, Japan) V.1- 1973-

NTIMBF Nauchnye Trudy, Irkutskii Gosudarstvennyi Meditsinskii Institut. Scientific Works, Irkutsk State Medical Institute. (Irkutskii Gosudarstvennyi Meditsinskii Institut, Irkutsk, USSR) No. 80- 1967-

NTIS** National Technical Information Service. (Springfield, VA 22161) (Formerly U.S. Clearinghouse for Scientific and Technical Information)

NTOTDY Neurobehavioral Toxicology and Teratology. (ANKHO International Inc., P.O. Box 426, Fayetteville, NY 13066) V.3- 1981-

NTPTB* NTP Technical Bulletin. (National Toxicology Program, Landow Bldg. 3A-06, 7910 Woodmont Ave., Bethesda, MD 20205)

NTPTR* National Toxicology Program Technical Report Series. (Research Triangle Park, NC 27709) No. 206-

NUCADQ Nutrition and Cancer. (Franklin Institute Press, POB 2266, Philadelphia, PA 19103) V.1- 1978-

NULSAK Nucleus (Calcutta). (Dr. A.K. Sharma, c/o Cytogenetics Laboratory, Department of Botany, University of Calcutta, 35 Ballygunge Circular Rd., Calcutta 700 019, India) V.1- 1958-

NUNDAJ Neuroendocrinology. (S. Karger AG, Postfach CH-4009 Basel, Switzerland) V.1- 1965/66-

NUPOBT Neuropatologia Polska. (Ars-Polona-RUCH, POB 1001, 00-068 Warsaw 1, Poland) V.1- 1963-

NURIBL Nutrition Reports International. (Geron-X, Inc., POB 1108, Los Altos, CA 94022) V.1- 1970-

NWSCAL New Scientist. (IPC Magazines Ltd., Tower House, Southampton St., London WC2E 9QX, England) V.1- 1956-

NYKGA7 Noyaku Kagaku. Pesticide Science. (Pesticide Science Society of Japan, 43-11, 1-Chome, Komagome, Toshima-ku, Tokyo 170, Japan) V.1-3, 1973-76

NYKZAU Nippon Yakurigaku Zasshi. Japanese Journal of Pharmacology. (Nippon Yakuri Gakkai, 2-4-16, Yayoi, Bunkyo-Ku, Tokyo 113, Japan) V.40- 1944-

NYSJAM New York State Journal of Medicine. (Medical Society of the State of New York, 420 Lakeville Rd., Lake Success, NY 11040) V.1- 1901-

NZMJAX New Zealand Medical Journal. (Otago Daily Times and Witness Newspapers, P.O. Box 181, Dunedin C1, New Zealand) V.1- 1900-

OBGNAS Obstetrics and Gynecology. (Elsevier/North Holland, Inc., 52 Vanderbilt Avenue, New York, NY 10017) V.1- 1953-

OCHRAI Occupational Health Review. (Ottawa, Canada) V.4-22, 1953-71. Discontinued

OCMJAJ Osaka City Medical Journal. (Osaka City Medical Center, 1-4-54, Asahimachi, Abenoku, Osaka 545, Japan) V.1- 1954-

OCRAAH Organometallic Chemistry Reviews, Section A: Subject Reviews. (Lausanne, Switzerland) V.3-8, 1968-72. For publisher information, see JORCAI

ODFU** Vergleichende pharmakologische Prufung Zweier neuer Lokalanas-thetika: Perkain und Pantokain, Wilhelm Ost Dissertation. (Pharmakologischen Institut der Universitat zu Frankfurt am Main, Germany, 1931)

OEKSDJ Osaka-furitsu Koshu Eisei Kenkyusho Kenkyu Hokoku, Shokuhin Eisei Hen. (Osaka-furitsu Koshu Eisei Kenkyusho, 1-3-69 Nakamichi, Higashinari-ku, Osaka 537, Japan) No. 1- 1970-

OEMEEM Occupational and Environmental Medicine. (BMJ Publishing Group, POB 299, London WC1H 9TD, United Kingdom) V.51- 1994-

OFAJAE Okajimas Folia Anatomica Japonica. (Japan Publications Trading Co., 175 5th Ave., New York, NY 10010) V.14- 1936-

OGSUA8 Obstetrical and Gynecological Survey. (Williams & Wilkins, 428 E. Preston St., Baltimore, MD 21202) V.1- 1946-

OHSLAM Occupational Health and Safety Letter. (Environews, Inc., 1097 National Press Bldg., Washington, DC 20045) V.1- 1970-

OIGZDE Osaka-shi Igakkai Zasshi. Journal of Osaka City Medical Association. (Osaka-shi Igakkai, c/o Osaka-shiritsu Daigaku Igakubu, 1-4-54 Asahi-cho, Abeno-ku, Osaka, 545, Japan) V.24- 1975-

OIGZSE Osaka-shi Igakkai Zasshi. Journal of Osaka City Medical Association. (Osaka-shi Igakkai, c/o Osaka-shiritru Daigaku Igakubu, 1-4-54 Asahi-cho, Abeno-ku, Osaka, 545, Japan) V.24- 1975-

OIZAAV Osaka Igaku Zasshi. (Osaka, Japan)

OJVRAZ Onderstepoort Journal of Veterinary Research. (Div. of Agricultural Inform., Dept. of Agricultural Tech. Ser., Private Bag X-144, Pretoria, S. Africa) V.25- 1951-

OJVSA4 Onderstepoort Journal of Veterinary Science and Animal Industry. (Pretoria, S. Africa) V.1-24, 1933-50. For publisher information, see OJVRAZ

OKEHDW Osaka-furitsu Koshu Eisei Kenkyusho Kenkyu Hokoku, Koshu Eisei Hen. Research Reports of the Osaka Prefectural Institute of Public Health, Public Health Section. (1-3-69 Nakamichi, Higashinari-ku, Osaka, 537, Japan) No. 3- 1966-

OMCDS* Olin Chemicals Data Sheet. (Industrial Development, 745 5th Ave., New York, NY 10022)

ONCOAR Oncologia. (Basel, Switzerland) V.1-20, 1948-66. For publisher information, see ONCOBS

ONCOBS Oncology. (S. Karger AG, Postfach CH-4009 Basel, Switzerland) V.21- 1967-

ONCODU Revista de Chirurgui, Oncologie, Radiologie, ORL, Oftalmologie, Stomatologie, Seria: Oncologia.(Rompresfilatelia, ILEXIM, POB 136-137, Bucharest, Romania) V.13, No. 4- 1974-

ONGZAC Ontogenez (Moscow). For English translation, see SJDBA9. (v/o Mezhdunarodnaya Kniga, 113095, Moscow, USSR) V.1- 1970-

OPHTAD Ophtalmologica. (S. Karger AG, Postfach CH-4009 Basel, Switzerland) V.96- 1978-

OSDIAF Osaka Shiritsu Daigaku Igaku Zasshi. Journal of the Osaka City Medical Center. (Osaka-Shiritsu Daigaku Igakubu, 1-4-54, Asahi-cho, Abeno-Ku, Osaka, Japan) V.4-23, 1955-74

OSMJAT Ohio State Medical Journal. (Ohio State Medical Association, 600 S. High St., Suite 500, Columbus, OH 43215) V.1- 1905-

OSOMAE Oral Surgery, Oral Medicine and Oral Pathology. (C.V. Mosby Co., 11830 Westline Industrial Dr., St Louis, MO. 63141) V.1- 1948-

OYYAA2 Oyo Yakuri. Pharmacometrics. (Oyo Yakuri Kenkyukai, Tohoku Daigaku, Kitayobancho, Sendai 980, Japan) V.1- 1967-

OZSEDS Ozone: Science & Engineering. (Pergamon Press Inc., Maxwell House, Fairview Park, Elmsford, NY 10523) V.1- 1979-

PAACA3 Proceedings of the American Association for Cancer Research. (Waverly Press, 428 E. Preston St., Baltimore, MD 21202) V.1- 1954-

PAASAH Publication, American Association for the Advancement of Science. (AAAS, 1515 Massachusetts Ave., N.W., Washington, DC 20005) No. 1-94, 1934-73. Discontinued

PABIAQ Pathologie et Biologie. (Paris, France) V.1-16, 1953-68. For publisher information, see PTBIAN

PACHAS Pure and Applied Chemistry. (Butterworth and Co., Ltd., Borough Green, Sevenoaks, Kent TN15 8PH, England) V.1- 1960-

PAFEAY Patologicheskaya Fiziologiya i Eksperimental'naya Terapiya. (v/o Mezhdunarodnaya Kniga, Kuznetskii Most 18, Moscow G-200, USSR) V.1- 1957-

PAHEAA Pharmaceutica Acta Helvetiae. (Case Postale 210, CH 1211 Geneva 1, Switzerland) V.1- 1926-

PAMIAD Pathologia et Microbiologia. V.23-43, 1960-75. For publisher information, see ECEBDI

PAPOAC Patologia Polska. (Ars-Polona-RUSH, POB 1001, 00-068 Warsaw 1, Poland) V.1- 1950-

PAREAQ Pharmacological Reviews. (Williams & Wilkins, 428 E. Preston St., Baltimore, MD 21202) V.1- 1949-

PARPDS Pathology, Research and Practice. (Gustav Fisher Verlag, Postfach 72 01 43, D-7000 Stuttgart 70, Germany) V.162- 1978-

PARWAC Polskie Archiwum Weterynaryjne. Polish Archives of Veterinary Medicine. (Panstwowe

Wydawnictwo Naukowe, POB 391, P-00251 Warsaw, Poland) V.1- 1951-

PATHAB Pathologica. (Via Alessandro Volta, 8 Casella Postale 894, 16128 Genoa, Italy) V.1- 1908-

PAVEAC Pathologia Veterinaria. (Basel, Switzerland) V.1-7, 1964-70. For publisher information, see VTPHAK

PBBHAU Pharmacology, Biochemistry and Behavior. (ANKHO International Inc., P.O. Box 426, Fayetteville, NY 13066) V.1- 1973-

PBCDDQ Proceedings-British Crop Protection Conference-Pests and Diseases. (British Crop., Protection Council, 2A Kidderminster Rd., Croydon, CRO 2UE, UK) 1977-

PBCWDF Proceedings of the British Crop Protection Conference - Weeds. (British Crop Protection Council, 2A Kidderminster Rd., Croydon, CRO 2UE, UK) 1976-

PBFMAV Problemi na Farmatsiyata. Problems in Pharmacy. (Durzhavno Izdatel'stvo Meditsina i Fizkultura, Pl. Slaveikov II, Sofia, Bulgaria) V.1- 1973-

PBCPCP Proceedings-British Crop Protection Conference-Pests and Diseases. (British Crop., Protection Council, 2A Kidderminster Rd., Croydon, CRO 2UE, UK) 1977-

PBPHAW Progress in Biochemical Pharmacology. (S. Karger AG, Postfach CH-4009 Basel, Switzerland) V.1-1965-

PBPSDY "Pharmacological and Biochemical Properties of Drug Substances" Morton E. Goldberg, ed., Washington, DC, American Pharmaceutical Association, V.1- 1977-

PCBPBS Pesticide Biochemistry and Physiology. (Academic Press, 111 5th Ave., New York, NY 10003) V.1-1971-

PCBRD2 Progress in Clinical and Biological Research. (Allan R. Liss, Inc., 150 5th Ave., New York, NY 10011) V.1- 1975-

PCCRA4 Proceedings of the Canadian Cancer Research Conference. (Univ. of Toronto Press, Toronto, Ontario M55 1A6, Canada) V.1- 1954-

PCIPDV Beijing Yixueyuan Xuebao. Journal of Peking Medical College. (Beijing Yixueyuan, Beijiaohaidianqu, Beijing, People's Republic of China) Beginning history not known

PCJOAU Pharmaceutical Chemistry Journal (English Translation). Translation of KHFZAN. (Plenum Publishing Corp., 233 Spring St., New York, NY 10013) No. 1- 1967-

PCOC** Pesticide Chemicals Official Compendium, Association of the American Pesticide Control Officials, Inc. (Topeka, KS, 1966)

PECAE5 Pediatric Emergency Care. (Williams & Wilkins, 428 E. Preston St., Baltimore, MD 21202) V.1- 1985-

PEDIAU Pediatrics. (P.O. Box 1034, Evanston, IL 60204) V.1- 1948-

PEMNDP Pesticide Manual. (The British Crop Protection Council, 20 Bridport Rd., Thornton Heath CR4 7QG, UK) V.1- 1968-

PENDAV Polish Endocrinology. English translation of Endokrynologia Polska. (Springfield, VA) V.13-23, 1962-72. Discontinued

PENNS* Pennsalt Chemicals Corporation, Technical Div., New Products. (Philadelphia, PA)

PEREBL Pediatric Research. (Williams & Wilkins Co., 428 E. Preston St., Baltimore, MD 21202) V.1- 1967-

PESTC* Pesticide and Toxic Chemical News. (Food Chemical News, Inc., 400 Wyatt Bldg., 777 14th St. N.W., Washington, DC 20005) V.1- 1972-

PESTD5 Proceedings of the European Society of Toxicology. (Amsterdam, Netherlands) V.16-18, 1975-77. Discontinued

PetKP# Personal Communication from Dr. K.P. Petersen, DAK Labs., 59 Lergravsvej, DK-2300, Copenhagen, Demark, to B. Jones, Tracor Jitco, Inc., December 22, 1977

PEXTAR Progress in Experimental Tumor Research. (S. Karger AG, Postfach CH-4009 Basel, Switzerland) V.1-1960-

PFLABK Pfluegers Archiv. European Journal of Physiology. (Springer-Verlag, Heidelberger Pl. 3, 1 Berlin 33, Germany) V.302- 1968-

PGMJAO Postgraduate Medical Journal. (Blackwell Scientific Publications, Osney Mead, Oxford OX2 OEL, England) V.1- 1925-

PGPKA8 Problemy Gematologii i Perelivaniia Krovi. Problems of Hematology and Blood Transfusion. (v/o Mezhdunarodnaya Kniga, Kuznetskii Most 18, Moscow G-200, USSR.) V.1- 1956-

PGTCA4 Pigment Cell. (S. Karger AG, Arnold-Boecklin St., 25 CH-4011 Basel, Switzerland) V.1- 1973-

PHABDI Proceedings of the Hungarian Annual Meeting for Biochemistry. (Magyar Kemikusok Egyesulete, Anker Koz 1, 1061 Budapest, Hungary) V.1- 1961-

PHARAT Pharmazie. (VEB Verlag Volk und Gesundheit, Neue Gruenstr 18, 102 Berlin, Germany) V.1- 1946-

PHARES Pharmacological Research. (Academic Press, 24-28 Oval Rd. London NW1 7DX, UK) V.21-1989-

PHBHA4 Physiology and Behavior. (Pergamon Press Inc., Maxwell House, Fairview Park, Elmsford, NY 10523) V.1-1966-

PHBTH* "Pharmacology: Basis of Therapy" Goodman, L.S., ed., 3rd ed., New York, NY, Macmillan, 1967

PHBUA9 Pharmaceutical Bulletin. (Tokyo, Japan) V.1-5, 1953-57. For publisher information, see CPBTAL

PHINDQ Pharmacy International. (Elsevier Science Publications Co., Inc., 52 Vanderbilt Ave., New York, NY 10017) V.1- 1990-

PHJOAV Pharmaceutical Journal. (Pharmaceutical Press, 17 Bloomsbury Sq., London WC1A 2NN, England) V.131-1933-

PHLIDQ Pharmacochemistry Library. (Elsevier Science Pub. Co., Inc., 52 Vanderbilt Ave., New York, NY 10017) V.1- 1977-

PHMCAA Pharmacologist. (American Society for Pharmacology and Experimental Therapeutics, 9650 Rockville Pike, Bethesda, MD 20014) V.1- 1959-

PHMGBN Pharmacology: International Journal of Experimental and Clinical Pharmacology. (S. Karger AG, Postfach CH-4009 Basel, Switzerland) V.1- 1968-

PHPHA6 Phytiatrie-Phytopharmacie. (Societe Francaise de Phytiatrie et de France) V.1- 1952-. Phytopharmacie, Etoile de Choisy, Route de St-Cyr, 78000 Versailles.

PHPYDQ Pharmacotherapy (Carlisle, MA). (Pharmacotherapy Pub., Inc., 112 School St., Carlisle, MA 01741) V.1- 1981-

PHREA7 Physiological Reviews. (9650 Rockville Pike, Bethesda, MD 20014) V.1- 1921-

PHREEB Pharmaceutical Research. (Thieme Inc., 381 Park Ave. S, New York, NY 10016) No. 1- 1984-

PHRPA6 Public Health Reports. (U.S. Government Printing Office, Superintendent of Documents, Washington, DC 20402) V.1- 1878-

PHTHDT Pharmacology and Therapeutics. (Pergamon Press Ltd., Headington Hill Hall, Oxford OX3 0BW, England) V.4- 1979-

PHTOEH Pharmacology and Toxicology (Copenhagen). (Munksgaard International Pub., POB 2148, DK-1016 Copenhagen K, Denmark) V.60- 1987-

PHTXA6 Pharmacology and Toxicology. Translation of FATOAO. (New York, NY) V.20-22, 1957-59. Discontinued

PHUZBI Pharmazie in Unserer Zeit. (VCH Pub., Inc., 303 N.W. 12th Ave., Deerfield Beach, FL 33441) V.1- 1972-

PHYTAJ Phytopathology. (Phytopathological Society, 3340 Pilot Knob Rd., St. Paul, MN 55121) V.1- 1911-

PIAIA9 Proceedings of the Iowa Academy of Science. (Iowa Academy of Science, Univ. of Northern Iowa, Cedar Falls, IA 50613) V.1- 1887/93-

PIATA8 Proceedings of the Imperial Academy of Tokyo. (Tokyo, Japan) V.1-21, 1912-45. For publisher information, see PJACAW

PISCAD Proceedings of the Indian Science Congress. (Indian Science Congress Association, 14, Dr. Biresh Guha St., Calcutta, 700017, India) 1st- 1914-

PIXXD2 PCT (Patent Cooperation Treaty) International Application. (U.S. Patent and Trademark Office, Foreign Patents, Washington, DC 20231)

PJABDW Proceedings of the Japan Academy, Series B: Physical and Biological Sciences. (Maruzen Co., Ltd., P.O. Box 5050, Tokyo International 100-31, Japan) V.53- 1977-

PJACAW Proceedings of the Japan Academy. (Tokyo, Japan) V.21-53, 1945-77. For publisher information, see PJABDW

PJPAE3 Polish Journal of Pharmacology. (Polish Academy of Sciences, Institute of Pharmacology, Smetna 12, 31-343 Krakow, Poland) V.45- 1993-

PJPHEO Pakistan Journal of Pharmacology. (c/o Univ. of Karachi, Faculty of Pharmacy, Karachi, 32, Pakistan) V.1- 1984-

PJPPAA Polish Journal of Pharmacology and Pharmacy. (ARS-Polona-Rush, POB 1001, 00-068 Warsaw 1, Poland) V.25- 1973-

PLCHB4 Physiological Chemistry and Physics. (U.S. Government Printing Office, Superintendent of Documents, Washington, DC 20402) V.1- 1969-

PLENBW Pollution Engineering. (1301 S. Grove Ave., Barrington, IL 60010) V.1- 1969-

PLIRDW Progress in Lipid Research. (Pergamon Press Ltd., Headington Hill Hall, Oxford OX3 OBW, England) V.17- 1978-

PLMEAA Planta Medica. (Hippokrates-Verlag GmbH, Neckarstr 121, 7 Stuttgart, Germany) V.1- 1953-

PLMEDD Prostaglandins, Leukotrienes and Medicine. (Longman Inc., 19 W. 44th St., New York, NY 10036) V.8- 1982-

PLPSAX Physiological Psychology. (Psychonomic Society, Inc., 1108 W. 34th Ave., Austin, TX 78705) V.1- 1973-

PLRCAT Pharmacological Research Communications. (Academic Press, 111 5th Ave., New York, NY 10003) V.1- 1969-

PMARAU Prensa Medica Argentina. (Junin 845, Buenos Aires, Argentina) V.1- 1914-

PMDCAY Progress in Medical Chemistry. (American Elsevier Publishing Co., 52 Vanderbilt Ave., New York, NY 10017) V.1- 1961-

PMDFA9 Pakistan Medical Forum. (Karachi, Pakistan) V.1-8, 1966-73. Discontinued

PMJMAQ Proceedings of the Annual Meeting of the New Jersey Mosquito Extermination Association. (New Brunswick, NJ) V.1-61, 1914-74

PMRSDJ Progress in Mutation Research. (Elsevier North Holland, Inc., 52 Vanderbilt Ave., New York, NY 10017) V.1- 1981-

PMSBA4 Progress in Molecular and Subcellular Biology. (Springer-Verlag, Heidelberger Pl. 3, D-1000 Berlin 33, Germany) V.1- 1969-

PNASA6 Proceedings of the National Academy of Sciences of the United States of America. (The Academy, Printing and Publishing Office, 2101 Constitution Ave., Washington, DC 20418) V.1- 1915-

PNCCA2 Proceedings, National Cancer Conference. (Philadelphia, PA) V.1-7, 1949-72, For publisher information, see CANCAR

POASAD Proceedings of the Oklahoma Academy of Science. (Oklahoma Academy of Science, c/o James F. Lowell, Executive Secretary-Treasurer, Southwestern Oklahoma State University, Weatherford, OK 73096) V.1- 1910/1920-

POKLA8 Problemi na Onkologiyata. Problems of Oncology. (Durzhavno Izdatelstvo Meditsina i Fizkultura, Pl. Slaveikov 11, Sofia, Bulgaria) V.1- 1973-

POLMAG Polymer. The Chemistry, Physics and Technology of High Polymers. (IPC Science and Technology Press, POB 63, Guildford, Surrey GU2 5BH, England) V.1- 1960-

POMDAS Postgraduate Medicine. (McGraw-Hill, Inc., Distribution Center, Princeton Rd., Hightstown, NJ 08520) V.1- 1947-

POMJAC Polish Medical Journal. (Warsaw, Poland) V.1-11, 1962-72. Discontinued

POSCAL Poultry Science. (Poultry Science Association, Texas A and M University, College Station, TX 77843) V.1- 1921-

PPASAK Proceedings of the Pennsylvania Academy of Science. (Pennsylvania Academy of Science, c/o Stanley

Zagorski, Tr., Gannon College, Perry Sq., Erie, PA 16501)
V.1- 1924/1926-

PPHAD4 Pediatric Pharmacology. (Alan R. Liss, Inc., 150 5th Ave., New York, NY 10011) V.1- 1980-

PPRPAS Produits and Problemes Pharmaceutiques. (Paris, France) V.17-28, 1962-73. Suspended

PPTCBY Proceedings of the International Symposium of the Princess Takamatsu Cancer Research Fund. (Japan Scientific Societies Press, 2-10, Hongo 6-chome, Bunkyo-ku, Tokyo 113, Japan) (1st)- 1970(Pub. 1971)-

PRACAK Practitioner. (5 Bentinck St., London, England) V.1- 1868-

PRBIDC Progress in Reproductive Biology. (S. Karger AG, Arnold-Boecklin Str. 25, CH-4011, Basel, Switzerland) V.1- 1976-

PREBA3 Proceedings of the Royal Society of Edinburgh, Section B. (Royal Society of Edinburgh, 22 George St., Edinburgh, Scotland) V.61- 1943-

PREPAB Przeglad Epidemiologiczny. (Panstwowy Zaklad Wydawnictw Lekarskich, ul. Dluga 38-40, 00-238 Warsaw, Poland) V.1-1947-

PRGLBA Prostaglandins. (Geron-X, Inc., P.O. Box 1108, Los Altos, CA 94022) V.1- 1972-

PRKHDK Problemi na Khigienata. Problems in Hygiene. (Durzhavno Izdatel'stvo Meditsina i Zizkultura, Pl. Slaveikov 11, Sofia, Bulgaria) V.1- 1975-

PRLBA4 Proceedings of the Royal Society of London, Series B, Biological Sciences. (The Society, 6 Carlton House Terrace, London SW1Y 5AG, England) V.76- 1905-

PROEAS Problemy Endokrinologii. (v/o Mezhdunarodnaya Kniga, Kuznetskii Most 18, Moscow G-200, USSR.) V.1-6, 1936-41; V.13- 1967-

PROMDL Prostaglandins and Medicine. (The Longman Group Ltd., Journals Div., 43-45 Annandale St., Edinburgh EH47 4AT, Scotland) V.1- 1978(?)-

PROTA* "Problemes de Toxicologie Alimentaire," Truhaut, R., Paris, France, L'evolution Pharmaceutique, (1955?)

PRPHA8 Produits Pharmaceutiques. (Paris, France) V.1-16, 1946-61. For publisher information, see PPRPAS

PRSMA4 Proceedings of the Royal Society of Medicine. (Grune and Stratton Inc., 111 5th Ave., New York, NY 10003) V.1- 1907-

PSCBAY Psychopharmacology Service Center, Bulletin. (Bethesda, MD) V.1-3, 1961-65. For publisher information, see PSYBB9

PSCHDL Psychopharmacology (Berlin). (Springer-Verlag New York, Inc., Service Center, 44 Hartz Way, Secaucus, NJ 07094) V.47- 1976-

PSDAA2 Proceedings of the South Dakota Academy of Science. (Univ. of South Dakota, Vermillion, SD 57069) V.1- 1916-

PSDTAP Proceedings of the European Society for the Study of Drug Toxicity. (Princeton, NJ 08540) V.1-15, 1963-74. For publisher information, see PESTD5

PSEBAA Proceedings of the Society for Experimental Biology and Medicine. (Academic Press, 111 5th Ave., New York, NY 10003) V.1- 1903/04-

PSSCBG Pesticide Science. (Blackwell Scientific Publications Ltd., Osney Mead, Oxford OX2 OEL, England) V.1- 1970-

PSSID2 Pergamon Series on Environmental Science. (Pergamon Press Ltd., Headington Hill Hall, Oxford OX3 0BW, England) V.1- 1978-

PSSYDG Proceedings of the Serono Symposia. (Academic Press Inc. Ltd., 24-28 Oval Rd., London NW1 7DX, England) V.1- 1973-

PSTDAN Pesticides. (Colour Publications Pvt. Ltd., 126-A Dhuruwadi, Off Dr. Nariman Rd., Bombay 400-025, India) V.1- 1967-

PSYCDE Psychoneuroendocrinology. (Elsevier Science, 660 White Plains Road Tarrytown, NY 10591) V.1- 1975-

PSTGAW Proceedings of the Scientific Section of the Toilet Goods Association. (The Toilet Goods Association, Inc., 1625 I St., N.W., Washington, DC 20006) No. 1-48, 1944-67. Discontinued

PSYBB9 Psychopharmacology Bulletin. (U.S. Government Printing Office, Superintendent of Documents, Washington, DC 20402) V.3- 1966-

PSYPAG Psychopharmacologia (Berlin). (Berlin, Ger.) V.1-46, 1959-76. For publisher information, see PSCHDL

PTBIAN Pathologie-Biologie. (Expansion Scientifique Francaise, 15 rue St. Benoit, Paris 6, France) V.17- 1969-

PTEUA6 Pathologia Europaea. (Presses Academiques Europeennes, 98, Chaussee de Charleroi, Brussels, Belgium) V.1- 1966-

PTLGAX Pathology. (Royal College of Pathologists of Australia, 82 Windmill St., Sydney, NSW 2000, Australia) V.1- 1969-

PTRMAD Philosophical Transactions of the Royal Society of London, Series A: Mathematical and Physical Sciences. (Royal Society of London, 6 Carlton House Terrace, London SW1Y 5AG, England) V.178- 1887-

PUMTAG Trace Substances in Environmental Health. Proceedings of University of Missouri's Annual Conference on Trace Substances in Environmental Health. (Environmental Trace Substances Research Center, Univ. of Missouri, Columbia, MO 65201) V.1- 1967-

PUOMA5 Proceedings of the University of Otago Medical School. (Otago Medical School Research Society, Box 913, Dunedin, New Zealand) V.1- 1922-

PUPHE* Pulmonary pharmacology & therapeutics (London Academic Press, 32 Jamestown Road, London NW1 7BY United Kingdom) V.10- Feb. 1997-

PWPSA8 Proceedings of the Western Pharmacology Society. (Univ. of California, Dept. of Pharmacology, Los Angeles, CA 94122) V.1- 1958-

PYRTAZ Psychological Reports. (Southern Universities Press, Baton Rouge, LA 70813) V.1- 1952-

PYTCAS Phytochemistry. An International Journal of Plant Biochemistry. (Pergamon Press Ltd., Headington Hill Hall, Oxford OX3 OEW, England) V.1- 1961-

QJDRAZ Quarterly Journal of Crude Drug Research. (Lisse, Netherlands) V.1-19, 1961-81 For publisher information, see IJCREE

QJMEA7 Quarterly Journal of Medicine. (Oxford University Press, Press Road, Neasden, London NW 10 0DD, England) V.1- 1932-

QJPPAL Quarterly Journal of Pharmacy and Pharmacology. (London, England) V.2-21, 1929-48. For publisher information, see JPPMAB

QJSAAP Quarterly Journal of Studies on Alcohol. (New Brunswick, NJ) V.1-28, 1940-67. For publisher information, see QJSOAX

QJSOAX Quarterly Journal of Studies on Alcohol, Part A: Originals. (Rutgers State University, New Brunswick, NJ 08903) V.29- 1968-

QUNUAZ Quaderni della Nutrizione. (Moore-Cottrell Subscription Agencies Inc., North Cohocton, NY 14868) V.1- 1934-

RADLAX Radiology. (Radiological Society of North America, 20th and Northampton Sts., Easton, PA 18042) V.1- 1923-

RADOA8 Radiobiologiya. (v/o Mezhdunarodnaya Kniga, Kuznetskii Most 18, Moscow G-200, USSR) V.1- 1961-

RalRL# Personal Communication to NIOSH from Robert L. Raleigh, M.D., Assistant Director of Health and Safety Lab., Eastman Kodak Company, Rochester, NY

RAMAAB Revista de la Asociacion Medica Argentina. (Santa Fe 1171, Buenos Aires, Argentina) V.1- 1915-

RAREAE Radiation Research. (Academic Press, 111 Fifth Ave., New York, NY 10003) V.1- 1954-

RARIAQ Record of Agricultural Research. (Dept. of Agricultural for Northern Ireland, Dundonald House, Upper Newtonards Rd., Belfast BT4 3SB, N. Ireland, UK) V.13- 1963-

RARSAM Radiation Research, Supplement. (Academic Press, 111 5th Ave., New York, NY 10003) No. 1- 1959-

RBBIAL Revista Brasileira de Biologia. (Caixa Postal 1587, ZC-00 Rio de Janeiro, Brazil) V.1- 1941-

RBPMAZ Revue Belge de Pathologie et de Medecine Experimentale. (Brussels, Belgium) V.18-31, 1947-65. For publisher information, see PTEUA6

RCBIAS Revue Canadienne de Biologie. (Les Presses de l'Universite de Montreal, P.O. Box 6128, 101 Montreal 3, Quebec, Canada) V.1- 1942-

RCCRC* Research and Consulting Company, Technical Reports. (RCC NOTOX B.V., Hambakenwetering 7, 5231 DD 's-Hertogenbosch, Netherlands)

RCOCB8 Research Communications in Chemical Pathology and Pharmacology. (PJD Publications, P.O. Box 966, Westbury, NY 11590) V.1- 1970-

RCPBDC Research Communications in Psychology, Psychiatry and Behavior. (PJD Publications, P.O. Box 966, Westbury, NY 11590) V.1- 1976-

RCPRAN Record of Chemical Progress. (Detroit, MI) V.1-32, 1939-71. Discontinued

RCRVAB Russian Chemical Reviews. (Chemical Society, Publications Sales Office, Burlington House, London W1V 0BN, England) V.29- 1960-

RCSADO Research Communications in Substances Abuse. (PJD Publications Ltd., Box 966, Westbury, NY 11590) V.1- 1980-

RCTEA4 Rubber Chemistry and Technology. (Div. of Rubber Chemistry, American Chemical Society, University of Akron, Akron, OH 44325) V.1- 1928-

RDBGAT Radiobiologia, Radiotherapia. (VEB Verlag Volk und Gesundheit, Neue Gruenstr. 18, DDR-102 Berlin, Germany) V.1- 1960-

RDCNBM Reproduction. (Calle Puerto de Bermeo, 11 Madrid 34, Spain) V.1- 1974-

RDMIDP Quarterly Reviews on Drug Metabolism and Drug Interactions. (Freund Pub. House, Suite 500, Chestham House, 150 Regent St., London W1R 5FA, UK) V.3- 1980-

RDWU** Beitrage zur Pharmakologie des Berylliums, Ursula Richter Dissertation. (Pharmakologischen Institut der Universitat Wurzburg, Germany, 1930)

REANBJ Revista Espanola de Anestesiologia y Reanimacion. (Sociedad Espanola de Anestesiologia y Reanimacion, Mallorca, 189 Barcellona 36, Spain) V.1- 1954-

RECYAR Revue Roumaine d'Embryologie et de Cytologie, Serie d'Embryologie. (Bucharest) V.1-8, 1964-71

REDH** J.D. Riedel-E. de Haen A.-G., Laboratory. (Berlin, Germany)

REEBB3 Revue Europeenne d'Etudes Cliniques et Biologiques. European Journal of Clinical and Biological Research. (Paris, France) V.15-17, 1970-72. For publisher information, see BIMDB3

REMBA8 Revista Ecuatoriana de Medicina y Ciencias Biologicas. (Facultad de Ciencias Medicas, Quito, Ecuador) V.1- 1963-

REMVAY Revue d'Elevage et de Medicine Veterinaire des Pays Tropicaux. (Editions Vigot Freres, 23 rue de l'Ecole-de-Medecine, 75006 Paris 6, France) V.1- 1947-

REONBL Revista Espanola de Oncologia. (Instituto Nacional de Oncologia del Cancer, Ciudad Universitaria, Madrid 3, Spain) V.1- 1952-

REPMBN Revue d'Epidemiologie, Medecine Sociale et Sante Publique. (Masson et Cie, Editeurs, 120 Blvd. Saint-Germain, P-75280, Paris 06, France) V.1- 1953-

REPTED Reproductive Toxicology. (Pergamon Press Inc., Maxwell House, Fairview Park, Elmsford, NY 10523) V.1- 1987

RESJAS Journal of the Reticuloendothelial Society. (1964) (Res: Journal of the Reticuloendothelial Society, New York, NY) V.1(1)-14(6), 1964-73

RETOAE Research Today. (Eli Lilly Co., Indianapolis, IN) V.1-16, 1944-60. Discontinued

REXMAS Research in Experimental Medicine. (Springer-Verlag, Heidelberger Pl. 3, D-1 Berlin 33, Germany) V.157- 1972-

RFCTAJ Rassegna di Fisiopatologia Clinica e Terapeutica. Review of Clinical and Therapeutic Physiopathology. (Rome, Italy) V.9-42, 1937-70. Discontinued

RFECAC Revue Francaise d'Etudes Cliniques et Biologiques. (Paris, France) V.1-14, 1956-69. For publisher information, see BIMDB3

RFGOAO Revue Francaise de Gynecologie et d'Obstetrique. (Masson Publishing USA, Inc., 14 E. 60th St., New York, NY 10022) V.20- 1920-

RHPC** Rohm and Haas Company Petroleum Chemicals. (Philadelphia, PA 19105)

RIHYAC Rinsho Hinyokika. Clinical Urology. (Igakushoin Medical Publishers, Inc., 1140 Avenue of the Americas, New York, NY 10036) V.21- 1967-

RIMAAX Rivista di Malariologia. (Rome, Italy) V.5-46, 1926-67. Discontinued

RISSAF Rendiconti Istituto Superiore di Sanita (Italian Edition). (Rome, Italy) V.4-27, 1941-64

RJARAV Rhodesian Journal of Agricultural Research. (Rhodesian Ministry of Agriculture, P.O. Box 8108, Causeway, Salisbury, Rhodesia) V.1- 1963-

RKGEDW Rinsho Kyobu Geka. Japanese Annals of Thoracic Surgery

RMCHAW Revista Medica de Chile. (Sociedad Medica de Santiago, Esmeralda 678, Casilla 23-d, Santiago, Chile) V.1- 1872-

RMEMDQ Revue Roumaine de Morphologie, d'Embryologie et de Physiologie, Serie Morphologie et Embryologie. (ILEXIM, POB 136-137, Bucharest, Romania) V.21- 1975-

RMISDU International Congress and Symposium Series-Royal Society of Medicine Services Limited. (Royal Society of Medicine Services Ltd., 7 E. 60th St., New York, NY 10022) No. 1- 1978-

RMLIAC Revue Medicale de Liege. (Hopital de Baviere, Universite de Liege, Pl. du XX Aout 7, Liege, Belgium) V.1- 1946-

RMMJAK Rocky Mountain Medical Journal. (Denver, CO) V.35-76, 1938-79

RMNIBN Revista Medico-Chirurgicala. (Societatea de Medici si Naturalisti, Bulevardul Independentei, Iasi, Romania) V.35- 1924-

RMSRA6 Revue Medicale de la Suisse Romande. (Societe Medicale de La Suisse Romande, 2 Bellefontaine, 1000 Lausanne, Switzerland) V.1- 1881-

RMVEAG Recueil de Medecine Veterinaire. (Editions Vigot Freres, 23 rue de l'Ecole-de-Medecine, Paris 6, France) V.1- 1824-

ROHM** Rohm and Haas Company Data Sheets (Philadelphia, PA 19105)

RPHRA6 Recent Progress in Hormone Research. Proceedings of the Laurentian Hormone Conference. (Academic Press, 111 5th Ave., New York, NY 10003) V.1- 1947-

RPOBAR Research Progress in Organic Biological and Medicinal Chemistry. (New York, NY) V.1-3, 1964-72. Discontinued

RPTOAN Russian Pharmacology and Toxicology. Translation of FATOAO. (Euromed Publications, 97 Moore Park Rd., London SW6 2DA, England) V.30- 1967-

RPZHAW Roczniki Panstwowego Zakladu Higieny. (Ars Polona-RUSH, POB 1001, 00-068 Warsaw 1, Poland) V.1- 1950-

RRBCAD Revue Roumaine de Biochimie. (Romprestilatelia, POB 2001, Calea Grivitei 64-66, Bucharest, Romania) V.1- 1964-

RRCRBU Recent Results in Cancer Research. (Springer-Verlag New York, Inc., Service Center, 44 Hartz Way, Secaucus, NJ 07094) V.1- 1965-

RRENAR Revue Roumaine d'Endocrinologie. (Bucharest, Romania) V.1-11, 1964-74. For publisher information, see RRENDU

RRESA8 Rastitel'nye Resursy. Plant Resources. (V/O Mezhdunarodnaya Kniga, 113095 Moscow, USSR) V.1- 1965-

RRENDU Revue Roumaine de Medecine, Serei Endocrinologie. (ILEXIM, POB 136-137, R-70116 Bucharest, Romania) V.13- 1975-

RREVAH Residue Reviews. (Springer-Verlag New York, Inc., Service Center, 44 Hartz Way, Secaucus, NJ 07094) V.1- 1962-

RSABAC Revista de la Sociedad Argentina de Biologia. (Associacion Medica Argentina, Santa Fe 1171, Buenos Aires, Argentina) V.1- 1925-

RSPSA2 Rassegna di Studi Psichiatrici. Review of Psychiatric Studies. (Ospedale Psichiatrico, Calata CapodichiNo. 232, 80141 Naples, Italy) V.1- 1911-

RSTUDV Rivista di Scienza e Technologia degli Alimenti e di Nutrizione Umana. Review of Science and Technology of Food and Human Nutrition. (Bologna, Italy) V.5-6, 1975-76. Discontinued

RTOPDW Regulatory Toxicology and Pharmacology. (Academic Press Inc., 111 Fifth Ave., New York, NY 10003) V.1- 1981-

RTPCAT Rassegna di Terapia e Patologia Clinica. (Rome, Italy) V.1-8, 1929-36. For publisher information, see RFCTAJ

RVFTBB Rivista di Farmacologia e Terapia. (Prof. William Ferrari, Via G. Campi 287, 41100 Modena, Italy) V.1- 1970-

RVTSA9 Research in Veterinary Science. (Blackwell Scientific Publications, Ltd., Osney Mead, Oxford OX2 OEL, England) V.1- 1960-

RZOVBM Rivista di Zootecnia e Veterinaria. Zootechnical and Veterinary Review. (Rivista di Zootecnia e Veterinaria, Via Viotti 3/5, 20133 Milan, Italy) No. 1- 1973-

SAAMDZ Substance and Alcohol Actions/Misuse. (Pergamon Press Inc., Maxwell House, Fairview Park, Elmsford, NY 10523) V.1- 1980-

SACAB7 South African Cancer Bulletin. (National Cancer Association of South Africa, 9 Jubillee Rd., Parktown, Johannesburg, S. Africa) V.1- 1957-

SAIGAK Saishin Igaku. Modern Medicine. (Saishin Igakusha, Senchuri Bldg., 3-6-1 Hirano-machi, Higashi-ku, Osaka 541, Japan) V.19- 1945-

SAIGBL Sangyo Igaku. Japanese Journal of Industrial Health. (Japan Association of Industrial Health, c/o Public Health Building, 78 Shinjuku 1-29-8, Shinjuku-Ku, Tokyo, Japan) V.1- 1959-

SAKNAH Soobshcheniia Akademii Nauk Gruzinskoi S.S.R. (v/o Mezhdunarodnaya Kniga, Kuznetskii Most 18, Moscow G-200, USSR.) V.2- 1941-

SAMJAF South African Medical Journal. (Medical Association of South Africa, Secy., P.O. Box 643, Cape Town, S. Africa) V.6- 1932-

SAPHAO Skandinavisches Archiv fuer Physiologic. (Karolinska Institutet, Editorial Office, Stockholm, Sweden) V.1-83, 1899-1940. Superseded by APSCAX

SAVEAB Sammlung von Vergiftungsfaellen. (Berlin, Germany) V.1-13, 1930-1944; V.14, 1952-54. For publisher information, see ARTODN

SBLEA2 Sbornik Lekarsky. (PNS-Ustredni Expedice Tisku, Jindriska 14, Prague 1, Czechoslovakia) V.1- 1887-

SCALA9 Scalpel. (Brussels, Belgium) 1920-V.124, 1971. Discontinued

SCAMAC Scientific American. (Scientific American, Inc., 415 Madison Ave., New York, NY 10017) V.1- 1859-

SCCSC8 Soap, Cosmetics, Chemical Specialties. (MacNair-Dorland, 101 W. 31st St., New York, NY 10001) V.47- 1971-

SCCUR* Shell Chemical Company. Unpublished Report. (2401 Crow Canyon Rd., San Romon, CA 94583)

SchF## Personal Communication from F.W. Schaller, Inco Ltd., Park 80 West Plaza Two, Saddle Brook, NJ 07662, May 16, 1986

SchP## Personal Communication from Dr. P. Schmitz, Bayer AG, 5090 Leverkusen, Bayerwerk, Germany, to NIOSH, April 4, 1986

SCHSAV Soap and Chemical Specialties. (New York, NY) V.30-47, 1954-71. For publisher information, see SCCSC8

SCIEAS Science. (American Association for the Advancement of Science, 1515 Massachusetts Ave., NW, Washington, DC 20005) V.1- 1895-

SCJUAD Science Journal. (London, England) V.1-7, 1965-71. For publisher information, see NWSCAL

SCMBE9 Spokane County Medical Society Bulletin. (Spokane, WA) 1929-81

SCMGDN Somatic Cell and Molecular Genetics. (Plenum Publishing Corp., 233 Spring St., New York, NY 10013) V.10- 1984-

SCNEBK Science News. (Science Service, Inc., 1719 N. St., NW, Washington, DC 20036) V.1- 1921-

SCPHA4 Scientia Pharmaceutica. (Oesterreichische Apotheker-Verlagsgesellschaft MBH, Spitalgasse 31, 1094 Vienna 9, Austria) V.1- 1930-

SCYYDZ Shengzhi Yu Biyun. Reproduction and Contraception. (China International Book Trading Corp., POB 2820, Beijing, People's Republic of China) 1980-

SDMEAL South Dakota Journal of Medicine. (South Dakota Medical Association, 608 West Ave. N., Sioux Falls, SD 57104) V.18- 1965-

SDMU* Zur Pharmakologie einiger Triazoliumverbindungen I, Erhart Schulze Dissertation. (Pharmakologischen Institut der Martin Luther- Universitat Halle-Wittenberg, Germany, 1936)

SDSTBT Scientific and Technical Report-Soap and Detergent Association. (Soap and Detergent Association, 485 Madison Ave., New York, NY 10022) No. 1- 1965-

SDUU* Ueber die Beeinflussung der Salvarsantoxizitat im Tierversuch, Jurgen Steudemann Dissertation. (University of Hamburg, Germany, 1934)

SEIJBO Senten Ijo. Congenital Anomalies. (Nihon Senten Ijo Gakkai, Kyoto 606, Japan) V.1- 1960-

SEMEAS Semana Medica. (Sociedad Argentina de Gastroenterologica, Santa Fe 1171, Buenos Aires, Argentina) V.1- 1894-

SFCRAO Symposium on Fundamental Cancer Research (Williams & Wilkins Co., 428 E. Preston St., Baltimore, MD 21202) V.1- 1947-

SFTIAE Svensk Farmaceutisk Tidskrift. Swedish Journal of Pharmacy. (Apotekarsocieteten, Upplandsgatan 6A, Stockholm C, Sweden) V.1- 1897-

SGOBA9 Surgery, Gynecology and Obstetrics. (Franklin H. Martin Memorial Foundation, 55 E. Erie St., Chicago, IL 60611) V.1- 1905-

SHBOAO Shokubutsu Boeki. (Nippon Shokubutsu Boeki Kyokai, 43-11, Komagome 1-chome, Toshima-Ku, Tokyo, Japan) V.1- 1947-

SheCW# Personal Communication from Dr. C.W. Sheu, Genetic Toxicology Branch, FDA, to H. Lau, Tracor Jitco, March 25, 1977

SHELL* Shell Chemical Company, Technical Data Bulletin. (Agricultural Div., Shell Chemical Co., 2401 Crow Canyon Rd., San Ramon, CA 94583)

SHGKA3 Shika Gakuho. Journal of Dentistry. (Tokyo Shika Daigaku Gakkai, 9-18, 2-chome, Sanzaki-cho, Chiyuoda-ku, Tokyo, Japan) V.1- 1895-

SHHUE8 Shiyou Huagong. Petrochemical Technology. (Beijing Huagon Yanjiuyan, Beikou, Hepingjie, Beijing, People's Republic of China)

SHIGAZ Shigaku. Ondotology. (Nippon Shika Daigaku Shigakkai, 1-9-20 Fujimi, Chiyodaku, Tokyo 102, Japan) V.38- 1949-

SHKKAN Shika Kiso Igakkai Zasshi. Journal of the Japanese Association for Basic Dentistry. (Shika Kiso Igakkai, Nihon Daigaku Shigakubu, 1-8 Surugadai, Kanda, Chiyoda-ku, Tokyo 101, Japan) V.1- 1959-

SHNSAS Shinryo to Shinyaku. Medical Consultation and New Remedies. (Iji Shuppan Sha, 1-2-8 Shinkawa, Chuo-Ku, Tokyo 104, Japan) V.1- 1964-

SHYCD4 Shengwu Huaxue Yu Shengwu Wuli Jinzhan. Progress in Biochemistry and Biophysics. (Kexue Chubanshe, 137 Zhaoyangmennei Dajie, Beijing, People's Republic of China) No. 1- 1974(?)-

SHZAAY Shoyakugaku Zasshi. (Kyoto Daigaku Yakugakubu, Yoshida Shimoadachi-machi, Sakyo-Ku, Kyoto, Japan) V.6- 1952-

SIGAAE Shika Igaku. Odontology. (Osaka Shika Gakkai, 1 Kyobashi, Higashi-Ku, Osaka, Japan) V.1- 1930-

SIGZAL Showa Igakkai Zashi. Journal of the Showa Medical Association. (Showa Daigaku, Showa Igakki, 1-5-8, Hatanodai, Shinagawa-ku, Tokyo 142, Japan) V.1- 1939-

SIHSD8 Shang-hai I Hsueh. (Shanghi, China) V.1- 1978(?)-

SIIPD4 Shanghai Diyi Yixueyuan Xuebao. Journal of the Shanghai First Medical College. (Beijing, Pep. Rep China) 1956- V.12 1985

SinJF# Personal Communication from J.F. Sina, Merck Institute for Therapeutic Research, West Point, PA 19486, to the Editor of RTECS, Cincinnati, OH, on October 26, 1982

SIZSAR Sapporo Igaku Zasshi. Sapporo Medical Journal. (Sapporo Igaku Daigaku, Nishi-17-Chome, Minami-1-jo, Chuo-ku Sapporo 060, Japan) V.3- 1952-

SJDBA9 Soviet Journal Developmental Biology (English Translation). (Plenum Publishing Corp., 233 Spring St., New York, NY 10013) V.1- 1970-

SJHAAQ Scandinavian Journal of Haematology. (Munksgaard, 35 Norre Sogade, DK 1370 Copenhagen K, Denmark) V.1- 1964-

SJHSBD Scandinavian Journal of Haematology. Supplementum. (Munksgaard, Noerre Soegade, DK-1370 Copenhagen K, Denmark) No. 1- 1964-

SJRDAH Scandinavian Journal of Respiratory Diseases. (Munksgaard, 6 Norregade, DK-1165 Copenhagen K, Denmark) V.47- 1966-

SJRHAT Scandinavian Journal of Rheumatology. (Almqvist and Wiksell, POB 45150, S-10430 Stockholm, Sweden) V.1- 1972-

SJUNAS Scandinavian Journal of Urology and Nephrology. (Almqvist and Wiksell Periodical Co., POB 62, 26 Gamla Brogatan, S-101-20 Stockholm, Sweden) V.1- 1967-

SKEZAP Shokuhin Eiseigaku Zasshi. Journal of the Food Hygiene Society of Japan. (Nippon Shokuhin Eisei Gakkai, c/o Kokuritsu Eisei Shikenjo, 18-1, Kamiyoga 1-chome, Setagaya-Ku, Tokyo, Japan) V.1- 1960-

SKIZAB Shikoku Igaku Zasshi. Shikoku Medical Journal. (Tokushima Daigaku Igakubu, Kuramoto-cho, Tokushima, Japan) V.1- 1950-

SKKNAJ Sankyo Kenkyusho Nempo. Annual Report of Sankyo Research Laboratories (Sankyo K.K. Chuo Kenkyusho, 1-2-58 Hiro-machi, Shinagawa-ku, Tokyo 140, Japan) No. 15- 1963-

SKNEA7 Shionogi Kenkyusho Nempo. Annual Report of Shionogi Research Laboratory. (Shionogi Seiyaku K.K. Kenkyusho, 2-47 Sagisukami, Fukushima, Osaka, Japan) No. 1- 1951-

SMBUA9 Stanford Medical Bulletin. (Palo Alto, CA) V.1-20, 1942-62. Discontinued

SMDJAK Scottish Medical Journal. (Longman Group Journal Division, 4th Ave., Harlow, Essex CM20 2JE, England) V.1- 1956-

SMEZA5 Sudebno-Meditsinskaya Ekspertiza. Forensic Medical Examination. (v/o Mezhdunarodnaya Kniga, Kuznetskii Most 18, Moscow G-200, USSR.) V.1- 1958-

SMJOAV Southern Medical Journal. (Southern Medical Association, 2601 Highland Ave., Birmingham, AL 35205) V.1- 1908-

SMSJAR Scottish Medical and Surgical Journal. (Edinburgh, Scotland). For publisher information, see SMDJAK

SMWOAS Schweizerische Medizinische Wochenschrift. (Schwabe and Co., Steintorst 13, 4000 Basel 10, Switzerland) V.50- 1920-

SNSHBT Senshokutai. Chromosome. (Senshokutai Gakkai, Kaokusai Kirisutokyo Daigaku, Mitaka, Tokyo-to, Japan) No. 1- 1946- Adopted new numbering in 1976

SOGEBZ Soviet Genetics. Translation of GNKAA5. (Plenum Publishing Corp., 233 Spring St., New York, NY 10013). V.2- 1966-

SOLGAV Seminars in Oncology. (Grune and Stratton Inc., 111 Fifth Ave., New York, NY 10003) V.1- 1974-

SOMEAU Sovetskaia Meditsina. (v/o Mezhdunarodnaya Kniga, Kuznetskii Most 18, Moscow G-200, USSR.) V.1-1937-

SOVEA7 Southwestern Veterinarian. (College of Veterinary Medicine, Texas A and M University, College Station, TX 77843) V.1- 1948-

SPCOAH Soap, Perfumery and Cosmetics. (United Trade Press, Ltd., 33 Bowling Green Lane, London EC1R0DA, England) V.8, No. 7- 1935-

SPEADM Special Publication of the Entomological Society of America. (4603 Calvert Rd., College Park, MD 20740)

SPERA2 Sperimentale. (Universita degli Studi Piazza San Marco, 4, 50121 Florence, Italy) V.50- 1896-

SpiEW# Personal Communication to Roger Tatken, NIOSH, from Dr. E.W. Spitz, E.P.A. on February 13, 1980

SRTCDF Scientific Report. (Dr. C.R. Krishna Murti, Director, Industrial Toxicology Research Centre, Luchnow-226001, India). History unknown

SSBSEF Scientia Sinica, Series B: Chemical, Biological, Agricultural, Medical, and Earth Sciences (English Edition). (Scientific and Technical Books Service Ltd., POB 197, London WC2N 4DE, England) V.25- 1982-

SSCHAH Soap and Sanitary Chemicals. (New York, NY) V.14-30, 1938-54. For publisher information, see SCCSC8

SSCMBX Scripta Scientifica Medica. (Durzhavno Izdatelstvo Meditsina i Fizkultura, Pl. Slaveikov II, Sofia, Bulgaria) V.4(2)- 1965-

SSEIBV Sumitomo Sangyo Eisie. Sumitomo Industrial Health. (Sumitomo Byoin Sangyo Eisei Kenkyushitsu, 5-2-2, Nakanoshima, Kita-ku, Osaka, 530, Japan) No. 1- 1965-

SSHZAS Seishin Shinkeigaku Zasshi. (Nippon Seishin Shinkei Gakkai, c/o Toyo Bunko, 2-28-21, Honkomagome, Bunkyo-Ku, Tokyo 113, Japan) V.39- 1935-

SSINAV Scientia Sinica. (Peking, People's Republic of China) V.3-24, 1954-81. For publisher information, see SSBSEF

SSSEAK Studi Sassaresi, Sezione 2. (Societa Sassarese di Scienze Mediche e Naturali, Viale S, Pietro 12, I-07100 Sassari, Italy) V.42- 1964-

SSZBAC Sbornik Vysoke Skoly Zemedelske v Brne, Rada B: Spisy Veterinarni. (Brno, Czechoslovakia) 1953-66. For publisher information, see AUAAB7

STBIBN Studia Biophysica. (Akademie-Verlag GmbH, Liepziger Str. 3-4, DDR-108 Berlin, Germany) V.1- 1966-

STCC** Stauffer Chemical Company, Product Safety Information Sheets. (Stauffer Chemical Company, Agricultural Chemical Division, Westport, CT 06880)

STEDAM Steroids. (Holden-Day Inc., 500 Sansome St., San Francisco, CA 94111) V.1- 1963-

STEVA8 Science of the Total Environment. (Elsevier Scientific Publishing Co., P.O. Box 211, Amsterdam C, Netherlands) V.1- 1972-

STGNBT "Spravochnik po Toksikologii i Gigienicheskim Normativam (PDK) Potentsial'no Opasnykh Khimicheskikh Veshchestv" Kushneva, V.S., and R.B. Gorshkova, eds. 46, Zhivopisnaya St., 123182, Moscow, Russia, Izdat 1999

STOAAT Stomatologiia (Moscow). (v/o Mezhdunarodnaya Kniga, Kuznetskii Most 18, Moscow G-200, USSR) V.1- 1923-

StoGD# Personal Communication to NIOSH from Fr. Gary D. Stoner, Dept. of Community Medicine, School of Medicine, University of California, La Jolla, CA 92037, May 25, 1975

STRAAA Strahlentherapie. (Urban and Schwarzenberg, Pattenkoferst 18, D-8000 Munich 15, Germany) V.1- 1912-

STRHAV Staub-Reinhaltung der Luft. (VDI-Verlag GmbH, Postfach 1139, D-4000 Duesseldorf 1, Germany) V.26- 1966-

SUFOAX Surgical Forum. (American College of Surgeons, 55 E. Erie St., Chicago, IL 60611) V.1- 1950-

SUNCO* Sun Company Material Safety Data Sheets. (1608 Walnut St., Philadelphia, PA 19103)

SURGAZ Surgery. (C.V. Mosby Co., 11830 Westline Industrial Dr., St. Louis, MO 63141) V.1- 1937-

SURR** Szechuan University Research Report, Natural Science. (Szechuan University, Szechuan, China) No. 2/3- 1978-

SVLKAO Sbornik Vedeckych Praci Lekarske Fakulty Univerzity Karlovy v Hradci Kralove. Collection of Scientific Works of the Charles University Faculty of Medicine in Hradec Kralove. (Univerzita Karlova, Ovocny Trh. 3, CS-116-36 Prague 1, Czechoslovakia) V.1- 1958-

SWEHDO Scandinavian Journal of Work, Environment and Health. (Haartmaninkatu 1, FIN-00290 Helsinki 29, Finland) V.1 1975-

SWSPBE Proceedings, Southern Weed Science Society. (Southern Weed Science Society, 309 W. Clark St., Champaign, IL 61820) V.22-1969-

SYHJAM Saengyak Hakhoechi. Journal of the Society of Pharmacognosy. (c/o Natural Products Research Institute, Seoul National University, 28 Yunkeon-Dong, Chong-ro-ku, Seoul 110, S. Korea) V.9- 1978-

SYSWAE Shiyan Shengwu Xuebao. Journal of Experimental Biology. (Guozi Shudian, China Publications Center, P.O. Box 399, Peking, People's Republic of China) V.1- 1953- (Suspended 1966-77)

SYXUD2 Shandong Yixueyuan Xuebao. Journal of Shandong Medical College (China International Book Trading Corp., POB 2820, Beijing, People's Republic of China)

SYXUE3 Shenyang Yaoxueyuan Xuebao. Journal of Shenyang College of Pharmacy. (Shenyang Yaoxueyuan, 7, 2-duan Wenhualu, Shenhequ, Shenyang, People's Republic of China) V.1- 1984-

SZAPAC Schweizerische Zietschrift fuer Allgemeine Pathologie und Bakteriologie. (White Plains, NY) V.1-3, 1938-40; V.13-22, 1950-59. For publisher information, see ECEBDI

SZPMAA Sozial-und Praeventivmedizin. (Art. Institut Orell Fuessli AG, Dietzingerstrasse 3, 8003 Zurich, Switzerland) V.19- 1974-

SZTPA5 Schweizerische Zeitschrift fuer Tuberkulose und Pneumonologie. (Basel, Switzerland) V.1- 18, 1944-1961

TABIA2 Tabulae Biologicae. (The Hague, Netherlands) V.1-22, 1925-63. Discontinued

TAFSAI Transactions of the American Fisheries Society. (American Fisheries Society, 5410 Grosvenor Lane, Bethesda, MD 20014) V.1- 1870-

TAGTBR Trudy Azerbaidzhanskogo Nauchno-Issledovatel'skogo Instituta Gigieny Truda i Professional'nykh Zabolevanii. Transactions of the Azerbaidzhan Scientific Research Institute of Labor Hygiene and Occupational Diseases. (The Institute, Baku, USSR.) No. 1- 1966-

TAKHAA Takeda Kenkyusho Ho. Journal of the Takeda Research Laboratories. (Takeda Yakuhin Kogyo K. K., 4-54 Juso-nishino-cho, Higashi Yodogawa-Ku, Osaka 532, Japan) V.29- 1970-

TCMUD8 Teratogenesis, Carcinogenesis, and Mutagenesis. (Alan R. Liss, Inc., 150 Fifth Ave., New York, NY 10011) V.1- 1980-

TCMUE9 Topics in Chemical Mutagenesis. (Plenum Publishing Corp., 233 Spring Street, New York, NY 10013) V.1- 1984-

TCPHP* Toxicology and Clinical Pharmacology of Herbal Products Melanie Johns Cupp ed., Humana press, 2000.

TDBU** Ueber die Wirkung des Braunsteins, Paul Thissen Dissertation. (Pharmakologischen Institut der Tierarztlichen Hochschule zu Berlin, Germany, 1933)

TDKNAF Takeda Kenkyusho Nempo. Annual Report of the Takeda Research Laboratories. (Osaka, Japan) V.1-28, 1936-69. For publisher information, see TAKHAA

TEARAI Terapevticheskii Archiv. Archives of Therapeutics. v/o Mezhdunarodnaya Kniga, Kuznetskii Most 18, Moscow G-200, USSR) V.1-19, 1923-41; V.20- 1948-

TECSDY Toxicological and Environmental Chemistry. (Gordon and Breach Science Pub. Inc., 1 Park Ave., New York, NY 10016) V.3(3/4)- 1981-

TeiD## Personal Communication to NIOSH from Dian Teigler, Director of Biological Research, Lee Pharmaceuticals, 1444 Santa Anita Ave., South El Monte, CA 91733, June 16, 1975

TELEAY Tetrahedron Letters. (Pergamon Press Ltd., Headington Hill Hall, Oxford OX3 OBW, England) 1959-

TFAKA4 Trudy Instituta Fiziologii, Akademiya Nauk Kazakhshoi SSR. Transactions of the Institute of Physiology, Academy of Sciences of the Kazakh. SSR. (The Academy, Alma-Ata, USSR) V.1- 1955-

TGANAK Tsitologiya i Genetika. (v/o Mezhdunarodnaya Kniga, Kuznetskii Most 18, Moscow G-200, USSR) V.1- 1967-

TGMEAJ Tropical and Geographical Medicine. (DeErven F. Bohn NV, Box 66, Harlem, Netherlands) V.10- 1958-

TGNCDL "Handbook of Organic Industrial Solvents" 2nd ed., Chicago, IL, National Association of Mutual Casualty Companies, 1961

THAGA6 Theoretical and Applied Genetics. (Springer-Verlag New York, Inc., Service Center, 44 Hartz Way, Secaucus, NJ 07094) V.38- 1968-

THBRAA Thrombosis Research. (Pergamon Press, Maxwell House, Fairview Park, Elmsford, NY 10523) V.1- 1972-

THERAP Therapie. (Doin, Editeurs, 8 Place de l'Odeon, Paris 6, France) V.1- 1946-

THEWA6 Therapiewoche. (G. Braun GmbH, Postfach 1709, 7500 Karlsruhe 1, Germany) V.1- 1950-

THGNBO Theriogenology. (Geron-X, Inc., P.O. Box 1108, Los Altos, CA 94022) V.1- 1974-

TIDZAH Tokyo Ika Daigaku Zasshi. Journal of Tokyo Medical College. (Tokyo Ika Daigaku Igakkai, 1-412, Higashi Okubo, Shinjuku-ku, Tokyo, Japan) 1918-

TIEUA7 Tieraerztliche Umschau. (Terra-Verlag, Neuhauser. Str. 21, 7750 Constance, Germany) V.1- 1946-

TIHHAH Tai-wan I Hsueh Hui Tsa Chih. Journal of the Formosan Medical Association. (Tai-wan I Hsueh Hui, 1, Sect.1 Jen-ai Road, Taipei, Taiwan) V.45- 1946-

TIUSAD Tin and Its Uses. (Tin Research Institute, 483 W. 6th Ave., Columbus, OH 43201) No. 1- 1939-

TIVIEQ Toxicology In Vitro. (Pergamon Press Inc., Maxwell House, Fairview Park, Elmsford, NY 10523) V.1- 1987-

TIVSAI Trudy, Vsesoyuznyi Nauchno-Issledovatel'skii Institut Veterinarnoi Sanitarii. Transactions, All-Union. Scientific Research Institute of Veterinary Sanitation. (The Institute, Moscow, USSR.) V.11- 1957-

TIYADG Tianjin Yiyao. Tianjin Medical Journal. (Guoji Shudian, China Publications Center, POB 399, Peking, China) V.1- 1973-

TJADAB Teratology, A Journal of Abnormal Development. (Wistar Institute Press, 3631 Spruce St., Philadelphia, PA 19104) V.1- 1968-

TJEMAO Tohoku Journal of Experimental Medicine. (Maruzen Co., Export Dept., P.O. Box 5050, Tokyo Int., 100-31 Tokyo, Japan) V.1- 1920-

TJIDAH Tokyo Jikeikai Ika Daigaku Zasshi. Tokyo Jikeikai Medical Journal. (3-25-8, Nishi Shinbashi, Minato-ku, Tokyo, Japan) V.66- 1951-

TJIZAF Tokyo Joshi Ika Daigaku Zasshi. Journal of Tokyo Women's Medical College. (Society of Tokyo Women's Medical College, c/o Tokyo Joshi Ika Daigaku Toshokan, 10, Kawada-cho, Shinjuku-ku, Tokyo 162, Japan) V.1- 1931-

TJSGA8 Tijdschrift Voor Sociale Geneeskunde. (B.V. Uitgeversmaatschappij. Reflex, Mathenesserlaan 310, Rotterdam 3003, Netherlands) V.1- 1923-

TJXMAH Tokushima Journal of Experimental Medicine. (Tokushima Diagaku Igakubu, 3, Kuramoto-cho, Tokushima, Japan) V.1- 1954-

TKIZAM Tokyo Igaku Zasshi. Tokyo Journal of Medical Sciences. (Tokyo, Japan) V.59-76, 1951-68

TKORAS Trudy Kazakhskogo Nauchno-Issledovatel'skogo Instituta Onkologii i Radiologii. (The Institute, Alma-Ata, USSR.) V.1- 1965-

TMPBAX Trudy Moskovskogo Obshchestva Ispytatelei Prirody. Transactions of the Moscow Society of Naturalists.

(Moskovskoe Obshchestvo Ispytaltelei, Prirody, Moscow, USSR) V.1- 1951-

TMPRAD Tropenmedizin und Parasitologie. (Georg Thieme Verlag, Postfach 732, Herdweg 63, 7000 Stuttgart, Germany) V.25- 1974-

TNEOAO Transactions of the New England Obstetrical and Gynecological Society. (Publisher unknown)

TNICS* Toxicology of New Industrial Chemical Substances. (USSR. Academy of Medical Sciences, Moscow-Meditsina)

TNONF* TNO-report. (TNO Nutrition and Food Research, P.O. Box 360, 3700 AJ Zeist, Utrechtseweg 48, 3704 HE Zeist, Netherlands)

TOANDB Toxicology Annual. (Marcel Dekker, POB 11305, Church St. Station, New York, NY 10249) 1974/75-

TOBHB* TA:Toxikologische Bewertung. Heidelberg, Berufsgenossenschaft der chemischen Industrie

TobJS# Personal Communication to the Editor, Toxic Substances List from Dr. J.S. Tobin, Director, Health and Safety, FMC Corporation, New York, NY 14103, November 9, 1973

TOERD9 Toxicological European Research. (Editions Ouranos, 12 bis, rue Jean-Jaures, 92807 Puteaux, France) V.1- 1978-

TOFOD5 Tokishikoroji Foramu. Toxicology Forum. (Saiensu Foramu, c/o Kida Bldg., 1-2-13 Yushima, Bunkyo-ku, Tokyo, 113, Japan) V.6- 1983-

TOIZAG Toho Igakkai Zasshi. Journal of Medical Society of Toho University. (Toho Daigaku Igakubu Igakkai, 5-21-16, Omori, Otasku, Tokyo, Japan) V.1- 1954-

TOKSVE Toksikologicheskii Vestnik. (18-20 Vadkovskii per. Moscow, 101479, Russia) History Unknown

TOLED5 Toxicology Letters. (Elsevier Scientific Publishing Co., P.O. Box 211, Amsterdam, Netherlands) V.1- 1977-

TOPADD Toxicologic Pathology. (E.I. du Pont de Nemours Co., Inc., Elkton Rd., Newark, DE 19711) V.6(3/4)- 1978-

TORAAK El Torax. (Montevideo, Uruguay) V.1- 1952-

TOSCF2 Toxicological Sciences (Oxford University Press, 6277 Sea Harbor Drive, Orlando, FL 32887) V. 41, Jan. 1998-

TOSUAH Transactions of the Ophthalmological Societies of the UK. (Churchill Livingstone Inc., 19 W. 44th St., New York, NY 10036) V.1- 1880-

TOVEFN Toksikologicheskii Vestnik. (18-20 Vadkovskii per. Moscow, 101479, Russia) History Unknown

TOXIA6 Toxicon. (Pergamon Press, Headington Hill Hall, Oxford OX3 0BW, England) V.1- 1962-

TOXID9 Toxicologist. (Soc. of Toxicology, Inc., 475 Wolf Ledge Parkway, Akron, OH 44311) V.1- 1981-

TPHSDY Trends in Pharmacological Sciences. (Elsevier North Holland, Inc., 52 Vanderbilt Ave., New York, NY 10017) V.1- 1979-

TPKVAL Toksikologiya Novykh Promyshlennykh Khimicheskikh Veshchestv. Toxicology of New Industrial Chemical Sciences. (Akademiya Meditsinskikh Nauk S.S.R., Moscow, USSR.) No. 1- 1961-

TRBMAV Texas Reports on Biology and Medicine. (Texas Reports, University of Texas Medical Branch, Galveston, TX 77550) V.1- 1943-

TRENAF Kenkyu Nenpo-Tokyo-toritsu Eisei Kenkyusho. Annual Report of Tokyo Metropolitan Research Laboratory of Public Health. (24-1, 3 Chome, Hyakunin-cho, Shin-Juku-Ku, Tokyo, Japan) V.1- 1949/50-

TRIPA7 Tin Research Institute, Publication. (The Institute, Fraser Rd., Greenford UB6 7AQ, Middlesex, England) No. 1- 1934-

TRPLAU Transplantation. (Williams & Wilkins, Co., 428 E. Preston St., Baltimore, MD 21202) V.1- 1963-

TRSTAZ Transactions of the Royal Society of Tropical Medicine and Hygiene. (The Society, Manson House, 26 Portland Pl., London W1N 4EY, England) V.14- 1920-

TSCAT* Office of Toxic Substances Report. (U.S. Environmental Protection Agency, Office of Toxics Substances, 401 M Street SW, Washington, DC 20460)

TSITAQ Tsitologiya. Cytology. (v/o Mezhdunarodnaya Kniga, Kuznetskii Most 18, Moscow G-200, USSR) V.1 1959-

TSPMA6 Travaux de la Societe de Pharmacie de Montpellier. (Soc. de Pharmacie de Montpellier, Faculte de Pharmacie de Montpellier, Ave. Ch.-Flahault, 34060 Montpellier, France) V.1- 1942-

TUBEAS Tubercle. (Williams & Wilkins, 428 E. Preston Street, Baltimore, MD 21202) V.1- 1919-

TUMOAB Tumori. (Casa Editrice Ambrosiana, Via G. Frua 6, 20146 Milan, Italy) V.1- 1911-

TVLAA5 Trudy Vsesoyuznogo Nauchno-Issledovatel'skogo Instituta Lekarstvennykh Rastenii. Transactions of the All-Union Scientific Research Institute of Medicinal Plants. (Vsesoyuznyi Nauchno-Issledovatel'skii Institut Lekarstvennykh Rastenii, Bittsa, USSR.) No. 5- 1936-

TWHPA3 Tung Wu Hsueh Pao. Journal of Zoology. (Guozi Shudian, China Publications Center, POB 399, Peking, People's Republic of China) V.1- 1949-

TXAPA9 Toxicology and Applied Pharmacology. (Academic Press, 111 5th Ave., New York, NY 10003) V.1- 1959-

TXCYAC Toxicology. (Elsevier Scientific Pub. Ireland, Ltd., POB 85, Limerick, Ireland) V.1- 1973-

TXMDAX Texas Medicine. (Texas Medical Association, 1905 N. Lamar Blvd., Austin, TX 78705) V.60- 1964-

TYKNAQ Tohoku Yakka Daigaku Kenkyu Nempo. Annual Report of the Tohoku College of Pharmacy. (77, Odawara Minamikotaku, Hara-machi, Sendai, Japan) No. 10- 1963-

UBZHAZ Ukrains'kii Biokhimichnii Zhurnal (1946-77). Ukranian Biochemical Journal. (Kiev, USSR) V.18-49, 1946-77. For publisher information, see UBZHD4

UBZHD4 Ukrainskii Biokhimicheskii Zhurnal. Ukranian Biochemical Journal. v/o Mezhdunarodnaya Kniga, Kuznetskii Most 18, (Moscow G-200, USSR) V.50- 1978-

UCDS** Union Carbide Data Sheet. (Industrial Medicine and Toxicology Dept., Union Carbide Corp., 270 Park Ave., New York, NY 10017)

UCMH** Union Carbide Manual Hazard Sheet. (Industrial Medicine and Toxicology Dept., Union Carbide Corp., 270 Park Ave., New York, NY 10017)

UCPHAQ University of California, Publications in Pharmacology. (Berkeley, CA) V.1-3, 1938-57. Discontinued

UCREAR Urologic and Cutaneous Review. (St. Louis, MO) V.17-56, 1913-52. Discontinued

UCRR** Union Carbide Research Report. (Union Carbide Corp., Old Ridgebury Rd., Danbury, CT 06817)

UDHU** Zur Frage der Entgiftung des Thalliums, Gerhart Urban Dissertation. (Pharmakologischen Institut der Martin Luther-Universitat Halle-Wittenberg, Germany, 1936)

UGLAAD Ugeskrift for Laeger. (Almindelige Danske Laegeforening, Kristianiagade 12A, 2100 Copenhagen, Denmark) V.1- 1839-

UICMAI UICC Monograph Series. (Springer-Verlag New York, Inc., Service Center, 44 Hartz Way, Secaucus, NJ 07094)

UJMSAP Upsala Journal of Medical Sciences. (Almqvist and Wiksell, Periodical Co., POB 62, 26 Gamla Brogatan, S-101, 20 Stockholm, Sweden) V.77- 1972-

UMJOAJ Ulster Medical Journal. (Ulster Medical Society, Belfast City Hospital, Belfast BT9 7AD, N. Ireland) V.1- 1932-

UNMI** University Microfilms International. (300 N. Zeeb Rd., Ann Arbor, MI 48106)

UPJOH* "Compounds Available for Fundamental Research, Volume II-6, Antibiotics, A Program of Upjohn Company Research Laboratory." (Kalamazoo, MI 49001)

URGABW Urologe Ausgabe A. (Springer-Verlag, Heidelberger Pl. 3, D-1000 Berlin 33, Germany) V.9- 1970-

URLRA5 Urological Research. (Springer-Verlag New York, Inc., Service Center, 44 Hartz Way, Secaucus, NJ 07094) V.1- 1973-

UROTAQ Urologia. (Libreria Editrice Canova, Via Panciera, 3b, 31100 Treviso, Italy) V.1- 1934-

URXXAF U.S.S.R. Patent Document. (Gosudarstvennyi Komitet SSSR po Delam Izobretenii i Otkrytii, Malyi Tcherkasski pereulok 2/6, Moscow, Russia).

USBCC* U.S. Borax and Chemical Company Reports. (New York, NY)

USXXAM United States Patent Document. (Commissioner of Patents and Trademarks, Washington, DC 20231)

UZBZAZ Uzbekskii Biologicheskii Zhurnal. (v/o Mezhdunarodnaya Kniga, Kuznetskii Most 18, Moscow G-200, USSR) No. 1- 1958-

VAAPB7 Virchows Archiv, Abteilung A: Pathologische Anatomie. (Berlin, Germany) V.344-361, 1968-73. For publisher information, see VAPHDQ

VAAZA2 Virchows Archiv, Abteilung B: Zellpathologie, Cell Pathology. (Springer-Verlag, Heidelberger Pl. 3, D-1 Berlin 33, Germany) V.1- 1968-

VAMNAQ Vestnik Akademii Meditsinskikh Nauk S.S.S.R. Journal of the Academy of Medical Sciences of the USSR. (v/o Mezhdunarodnaya Kniga, Kuznetskii Most 18, Moscow G-200, USSR.) V.1- 1946-

VAPHDQ Virchows Archiv, Abteilung A: Pathological Anatomy and Histology. (Springer-Verlag, Heidelberger pl.3, D-1 Berlin. 33, Germany) V.362- 1974-

VCTDC* Vanderbilt, R.T., Co., Technical Data Sheet. (230 Park Ave., New York, NY 10017)

VCVGH* "Vrednie chemichescie veshestva, galogenproisvodnie uglevodorodov". (Hazardous substances: Galogenated hydrocarbons) Bandman A.L. et al., Chimia, 1994.

VCVGK* "Vrednie chemichescie veshestva, galogen I kislorod sodergashie organicheskie soedinenia". (Hazardous substances. Galogen and oxygen containing substances), Bandman A.L. et al., Chimia, 1994.

VCVN1* "Vrednie chemichescie veshestva. Neorganicheskie soedinenia elementov I-IV groopp" (Hazardous substances. Inornanic substances containing I-IV group elements), Filov V.A., Chimia, 1988.

VCVPS* "Vrednie chemichescie veshestva. Prirodnie organicheskie soedinenia" (Hazardous substances. Nature products.) Volkova N.V. et al., Sankt-Peterburg, 1998.

VDGIA2 Verhandlungen der Deutschen Gesellschaft fuer Innere Medizin. (J.F. Bergmann Verlag, Trogerstrasse 56, 8000 Munich 80, Germany) V.33-52, 1921-40; V.54- 1948-

VDGPAN Verhandlungen der Deutschen Gesellschaft fuer Pathologie. (Gustav Fischer Verlag, Postfach 53, Wollgrasweg 49, 7000 Stuttgart-Hohenheim, Germany) V.1-1898-

VEARA6 Veterinarski Arhiv. (Jugoslovenska Knjiga, P.O. Box 36, Terazije 27, 11001, Belgrade, Yugoslavia) V.1-1931-

VELPB* Velsicol Chemical Corporation, Product Bulletin. (341 E. Ohio St., Chicago, IL 60611)

VEMJA8 Veterinary Medical Journal. (Cairo Univ., Faculty of Veterinary Medicine, Giza, Egypt) V.1- 1953-

VETNAL Veterinariya. (v/o Mezhdunarodnaya Kniga, Kuznetskii Most 18, Moscow G-200, USSR.) V.1- 1924-

VETRAX Veterinary Record. (7 Mansfield St., London, W1M 0AT, England) V.1- 1888-

VHAGAS Verhandlungen der Anatomischen Gesellschaft. (VEB Gustav Fischer, Postfach 176, DDR-69 Jena, Germany) V.1- 1887-

VHTODE Veterinary and Human Toxicology. (American College of Veterinary Toxicologists, Office of the Secretary-Treasurer, Comparative Toxicology Laboratory, KS State University, Manhatten, Kansas 66506) V.19- 1977-

VIHOAQ Vitamins and Hormones. (Academic Press, 111 5th Ave., New York, NY 10003) V.1- 1943-

VINIT* Vsesoyuznyi Institut Nauchnoi i Tekhnicheskoi Informatsii (VINITI). All-Union Institute of Scientific and Technical Information. (Moscow, USSR)

VIRLAX Virology. (Academic Press, 111 Fifth Ave., New York, NY 10003) V.1- 1955-

VKMGA7 Voprosy Kommunal'noi Gigieny. Problems of Communal Hygiene. (Kiev, USSR.) V.1-6, 1963-66. For publisher information, see GNAMAP

VMDNAV Veterinarno-Meditsinski Nauki. (Hemus, Blvd. Russki 6, Sofia, Bulgaria) V.1- 1964-

VOONAW Voprosy Onkologii. Problems of Onkology. (v/o Mezhdunarodnaya Kniga, Kuznetskii Most 18, Moscow G-200, USSR.) V.1- 1955-

VPITAR Voprosy Pitaniya. Problems of Nutrition. (v/o Mezhdunarodnaya Kniga, Kuznetskii Most 18, Moscow G-200, USSR.) V.1- 1932-

VRDEA5 Vrachebnoe Delo. Medical Profession. (v/o Mezhdunarodnaya Kniga, Kuznetskii Most 18, Moscow G-200, USSR.) No. 1- 1918-

VRRAAT Vestnik Rentgenologii i Radiologii. Bulletin of Roentgenology and Radiology. (v/o Mezhdunarodnaya Kniga, Kuznetskii Most 18, Moscow G-200, USSR.) V.1-1920-

VTPHAK Veterinary Pathology. (S. Karger AG, Postfach CH-4009 Basel, Switzerland) V.8- 1971-

VTYMAC VM/SAC, Veterinary Medicine and Small Animal Clinician. (Veterinary Medicine Publishing Co., 690 S. 4th St., POB 13265, Edwardsville, KS 66113) V.59-1964-

VVIRAT Voprosy Virusologii. (v/o Mezhdunarodnaya Kniga, Kuznetskii Most 18, Moscow G-200, USSR) V.1-1956-

WADTAA Wright Air Development Center, Technical Report, U.S. Air Force. (Wright-Patterson Air Force Base, OH 45433)

WATRAG Water Research. (Pergamon Press Ltd., Headington Hill Hall, Oxford OX3 0BW, England) V.1-1967-

WDMBAM Wadley Medical Bulletin. (Dallas, TX) V.1-5, 1970-75. For publisher information, see JCHODP

WDMU** Vergleichend-pharmakologische Untersuchungen uber die haemolytische Wirkung und die Toxizitat der Lokalanaesthetika, Martin Walther Dissertation. (Pharmakologischen Institut der Martin Luther-Universitat Halle-Wittenburg, Germany, 1936)

WDZAEK Weisheng Dulixue Zazhi. Journal of Health Toxicology. (Weisheng Dulixue Zazhi Bianjibu, Dongdaqiao, Chaoyang Menwai, Beijing, Peop. Rep. China) V.1- 1987

WEHRBJ Work, Environment, Health. (Helsinki, Finland) V.1-11, 1962-74. For publisher information, see SWEHDO

WHOTAC World Health Organization, Technical Report Series. (1211 Geneva, 27, Switzerland) No. 1- 1950-

WHYHAQ Wuhan Yixueyuan Xuebao. (China International Book Trading Corp., POB 2820, Beijing, People's Republic of China) No. 1-1957-Suspended 1959-76(?)

WILEAR Wiadomosci Lekarskie. (Ars Polona-RUSH, POB 1001, P-00068 Warsaw 1, Poland) V.1- 1948-

WIMJAD West Indian Medical Journal. (West Indian Medical Journal, University of the West Indies, Faculty of Medicine, Kingston 7, Jamaica) V.1- 1951-

WJMDA2 Western Journal of Medicine. (California Medical Association, 731 Market St., San Francisco, CA 94103) V.120- 1974-

WKMRAH Wakayama Medical Report. (Wakayama Medical College Library, 9,9-bancho, Wakayama, 640, Japan) V.1- 1953-

WKWOAO Wiener Klinische Wochenschrift. (Springer-Verlag, Postfach 367, A-1011 Vienna, Austria) V.1- 1888-

WMWOA4 Wiener Medizinische Wochenschrift. (Brueder Hollinek, Gallgasse 40A, A-1130 Vienna, Austria) V.1- 1851-

WolMA# Personal Communication to Henry Lau, Tracor Jitco, Inc., from Mark A. Wolf, Dow Chemical Co., Midland, MI 48640

WQCHM* "Water Quality Characteristics of Hazardous Materials" W. Hann, and P.A. Jensen, Environmental Engineering Division, Civil Engineering Department, Texas A and M University, vol. 1-4, 1974

WRABDT Wilhelm Roux's Archives of Developmental Biology. (Springer-Verlag New York, Inc., Service Center, 44 Hartz Way, Secaucus, NJ 07094) V.177- 1975-

WRPCA2 World Review of Pest Control. (London, England) V.1-10, 1962-71. Discontinued

WTMOA3 Wiener Tieraerztliche Monatsschrift. (Ferdinand Berger and Soehne OHG, Wiene Str. 21-23, A-3580 Horn, Austria) V.1- 1914-

WZERDH Wissenschaftliche Zeitschrift der Ernst-Moritz-Arndt-Universitaet Greifswald, Medizinische Reihe. (Buchexport, Postfach 160, Leipzig DDR-7010, Germany) V.21- 1972-

XAWPA2 U.S. Army Chemical Warfare Laboratories. (Army Chemical Center, MD)

XENOBH Xenobiotica. (Taylor and Francis Ltd., 4 John St., London WC1N 2ET, England) V.1- 1971-

XEURAQ U.S. Atomic Energy Commission, University of Rochester, Research and Development Reports. (Rochester, NY)

XPHBAO U.S. Public Health Service, Public Health Bulletin. (Washington, DC)

XPHCI* U.S. Public Health Service, Current Intelligence Bulletin. (National Institute for Occupational Safety and Health, 5600 Fishers Lane, Rockville, MD 20857)

XPHPAW U.S. Public Health Service Publication. (U.S. Government Printing Office, Superintendent of Documents, Washington, DC 20402) No. 1- 1950-

YACHDS Yakuri to Chiryo. Pharmacology and Therapeutics. (Raifu Saiensu Pub. Co., 5-3-3 Yaesu, Chuo-ku, Tokyo 104, Japan) V.1- 1973-

YAHOA3 Yakhak Hoeji. Journal of the Pharmaceutical Society of Korea. (Seoul Taehakkyo Yakhak Taehak, 28 YonGondong, Chongnogu, Seoul, S. Korea) V.1- 1956(?)-

YAKUD5 Gekkan Yakuji. Pharmaceuticals Monthly. (Yakugyo Jihosha, Inaoka Bldg., 2-36 Jinbo-cho, Kandu, Chiyoda-ku, Tokyo 101, Japan) V.1- 1959-

YFZADL Yaowu Fenxi Zazhi. Journal of Pharmaceutical Analysis. (Guoji Shudian, POB 2820, Beijing, People's Republic of China) V.1- 1981-

YHHPAL Yaoxue Xuebao. Acta Pharmaceutica Sinica. Pharmaceutical Journal. (China International Book Trading Corp., POB 2820, Beijing, People's Republic of China) V.1- 1953- (Suspended 1966-78)

YHTPAD Yaoxue Tongbao. Bulletin of Pharmacology. (China International Book Trading Corp., POB 2820, Beijing, People's Republic of China) V.1-12, 1953-66; V.13- 1978-

YIGODN Yiyao Gongye. Pharmaceutical Industry. (c/o Shanghai Yiyao Gongye Yanjiuyuan, 1320 Beijing-Xilu,

Shanghai 200040, People's Republic of China). Beginning history not known

YIKUAO Yamaguchi Igaku. (Yamaguchi Daigaku Igakkai, Kogushi Ube, Yamaguchi-Ken 755, Japan) V.1- 1953-

YJBMAU Yale Journal of Biology and Medicine. (The Yale Journal of Biology and Medicine, Inc., 333 Cedar Street, New Haven, CT 06510) V.1- 1928-

YKGKAM Yukagaku. Journal of Japan Oil Chemists' Society. (Nippon Yukagaku Kyokai, c/o Yushi Kogyo Kaigan, 3-13-11 Nihonbashi Chuo-Ku, Tokyo, Japan) V.5- 1956-

YKIGAK Yokohama Igaku. (Yokohama-Shiritsu Daigaku Igakkai, 2-33 Urafune-cho, Minami-Ku, Yokohama 232, Japan) V.1- 1948-

YKKZAJ Yakugaku Zasshi. Journal of Pharmacy. (Nippon Yakugakkai, 12-15-501, Shibuya 2-chome, Shibuya-ku, Tokyo 150, Japan) No. 1- 1881-

YKRYAH Yakubutsu Ryoho. Medicinal Treatment. (Iji Nipposha, Kamiya Bldg., 4-11-7 Hatchobori, Chuo-Ku, Tokyo 104, Japan) V.1- 1968-

YKYUA6 Yakkyoku. Pharmacy. (Nanzando, 4-1-11, Yushima, Bunkyo-ku, Tokyo, Japan) V.1- 1950-

YMBUA7 Yokohama Medical Bulletin. (Npg. Yokohama Ika Daigaku, Urafune-cho, Minato-ku, Yokohama, Japan) V.1- 1950-

YOIZA3 Yonago Igaku Zasshi. Journal of the Yonago Medical Association. (c/o Tottori Daigaku Igakubu, Nishi-machi, Yonago, Japan) V.1- 1948-

YOMJA9 Yonsei Medical Journal. (Yonsei University College of Medicine, P.O. Box 71, Seoul, S. Korea) V.1- 1960-

YRTMA6 Yonsei Reports on Tropical Medicine. (Yonsei University, Institute of Tropical Medicine, Box 17, Seoul, S. Korea) V.1- 1970-

ZAARAM Zentralblatt fuer Arbeitsmedizin und Arbeitsschutz. (Dr. Dietrich Steinkopff Verlag, Saalbaustr 12,6100 Darmstadt, Germany) V.1- 1951-

ZACCAL Zacchia. Archivio di Medicina Legale, Sociale e Criminologica. Zacchia. Archives of Forensic, Social and Criminological Medicine. (Zacchia, Viale Regina Elena 336, Rome 00161, Italy) V.1- 1921-

ZANZA9 Zeitschrift Angewandte Zoologie. (Duncker und Humblot, Postfach 410329, D-1000 Berlin 41, Germany) V.41- 1954-

ZAPOAK Zeitschrift fuer Allgemeine Mikrobiologie. Morphologie, Physiologie, Genetik und Oekologie der Mikroorganismen. (Akademie-Verlag GmbH, Liepziger Str. 3-4, DDR-108 Berlin, Germany) V.1- 1960-

ZAPPAN Zentralblatt fuer Allgemeine Pathologie und Pathologische Anatomie. (VEB Gustav Fischer Verlag, Postfach 176, Villengang 2, 69 Jena, Germany) V.1- 1890-

ZBPHA6 Zentralblatt fuer Bakteriologie, Parasitenkunde, Infektionskrankheiten und Hygiene, Abteilung I: Medizinisch-Hygienische Bakteriologie, Virusforschung und Parasitologie, Originale. (Stuttgart, Germany) V.31-151, 1902-45; V.152-216, 1947-71. For publisher information, see ZMMPAO

ZBPIA9 Zentralblatt fuer Bakteriologie, Parasitenkunde, Infektionskrankheiten und Hygiene, Abteilung II.

Naturwissenschattliche: Allgemeine, Landwirtschattliche und Technische Miknobiologie. (VEB Gustav Fischer Verlag, Postfach 176, DDR-69 Jena, Germany) V.107-132, 1952-77

ZBPMAL Zentralblatt fuer Pharmazie. (Nurenburg, Germany) V.1-29, 1904-33. Discontinued

ZDBEA9 Zdravookhranenie Belorussii. Public Health Service of White Russia. (v/o Mezhdunarodnaya Kniga, Kuznetskii Most 18, Moscow G-200, USSR) V.1- 1955-

ZDKAA8 Zdravookhranenie Kazakhstana. Public Health of Kazakhstan. (v/o Mezhdunarodnaya Kniga, Kuznetskii Most 18, Moscow G-200, USSR.) V.1- 1941-

ZDTUAB Zdravookhranenie Turkmenistana. Public Health of Turkmenistan. (v/o Mezhdunarodnaya Kniga, 113095 Moscow, USSR) V.1- 1957-

ZDVEA7 Zdravstveni Vestnik. Medical Journal. (Jugoslovenska Knjiga, POB 36, Terazije 27, YU-11001 Belgrade, Yugoslavia) V.15- 1946-

ZDVKAP Zdravookhranenie. Public Health. (v/o Mezhdunarodnaya Kniga, Kuznetskii Most 18, Moscow G-200, USSR) V.1- 1958-

ZEGYAX Zentralblatt fuer Gynaekologie. (Johann Ambrosius Barth Verlag, Postfach 109, DDR-7010, Leipzig, Germany) V.1- 1877-

ZEKBAI Zeitschrift fuer Krebsforschung. (Berlin, Germany) V.1-75, 1903-71. For publisher information, see JCROD7

ZEKIA5 Zeitschrift fuer Kinderheilkunde. (Springer-Verlag, Heidelberger Pl. 3, D-1 Berlin 33, Germany) V.1- 1910-

ZENBAX Zeitschrift fuer Naturforschung, Ausgabe B, Chemie, Biochemie, Biophysik, Biologie und Verwandten Gebieten. (Verlag der Zeitschrift fuer Naturforschung, Postfach 2645, D-7400, Tuebingen, Germany) V.2- 1947-

ZEPHAR Zentralblatt fuer Physiologie. (Leipzig, Germany) V.1-34, 1887-1921. For publisher information, see PFLABK

ZEPRAN Zeitschrift fuer Praeventivmedizin. (Zurich, Switzerland) V.1-13, 1956-68. For publisher information, see SZPMAA

ZEPTAT Zeitschrift fuer Experimentelle Pathologie und Therapie. (Berlin, Germany) V.1-22, 1905-21. For publisher information, see REXMAS

ZERNAL Zeitschrift fuer Ernaehrungswissenschaft. (Steinkopff Verlag, Postfach 1008, 6100 Darmstadt, Germany) V.1- 1960-

ZEVBA5 Zeitschrift fuer Verebungslehre. (Springer-Verlag, Heidelberger Pl. 3, D-1 Berlin 33, Germany) V.89-98, 1958-66

ZGEMAZ Zeitschrift fuer die Gesamte Experimentelle Medizin. (Berlin, Germany) V.1-139, 1913-65. For publisher information, see REXMAS

ZGIMAL Zeitschrift fuer die Gesamte Innere Medizin und Ihre Grenzgebiete. (VEB Georg Thieme, Hainst 17/19, Postfach 946, 701 Leipzig, Germany) V.1- 1946-

ZGSHAM Zeitschrift fuer Gesundheitstechnik und Staedtehygiene. (Berlin, Germany) V.21-27, 1929-35. For publisher information, see ZANZA9

ZHINAV Zeitschrift fuer Hygiene und Infektionskrankheiten. (Berlin, Germany) V.11-151, 1892-1965. For publisher information, see MMIYAO

ZHOYAN "Zhongliu Yanjiu" Cancer Review, Yu, R., et al., eds., Shanghai Science/Technology Publisher, Peop. Rep. China, 1994

ZHPMAT Zentralblatt fuer Bakteriologie, Parasitenkunde, Infektionskrankran- heiten und Hygiene, Abteilung 1: Originale, Reihe B: Hygiene, Praeventive Medizin. (Gustav Fischer Verlag, Postfach 72-01-43, D-7000 Stuttgart 70, Germany) V.155- 1971-

ZHYGAM Zeitschrift fuer die Gesamte Hygiene und Ihre Grenzgebiete. (VEB Georg Thieme, Hainst 17/19, Postfach 946, 701 Leipzig, Germany) V.1- 1955-

ZIEKBA Zeitschrift fuer Immunitaetsforschung, Experimentelle und Klinische Immunologie. (Fischer, Gustav, Verlag, Postfach 53, Wollgrasweg 49, 7000 Stuttgart-Hohenheim, Germany) V.141- 1970-

ZIETA2 Zeitschrift fuer Immunitaetsforschung und Experimentelle Therapie. (Stuttgart, Germany) 1924-V.124, 1962. For publisher information, see ZIEKBA

ZKKOBW Zeitschrift fuer Krebsforschung und Klinische Onkologie. (Berlin, Germany) V.76-92, 1971-78. For publisher information, see JCROD7

ZKMAAX Zhurnal Eksperimental'noi i Klinicheskoi Meditsiny. (v/o Mezhdunarodnaya Kniga, 121200 Moscow, USSR) V.2- 1962-

ZKMEAB Zeitschrift fuer Klin Medicine. (Berlin, Germany) V.1-158, 1879-1965. For publisher information, see EJCIB8

ZLUFAR Zeitschrift fuer Lebensmittel-Untersuchung und-Forschung. (Springer-Verlag, Heidelberger Pl. 3, D-1 Berlin 33, Germany) V.86- 1943-

ZMEIAV Zhurnal Mikrobiologii, Epidemiologii i Immunobiologii. (v/o Mezhdunarodnaya Kniga, Kuznetskii Most 18, Moscow G-200, USSR.) V.14- 1935-

ZMMPAO Zentralblatt fuer Bakteriologie, Parasitenkunde, Infektionskrankh eiten und Hygiene, Abteilung 1: Originale, Reihe A: Medizinische Mikrobiologie und Parasitologie. (Gustav Fischer Verlag, Postfach 72-01-43, D-7000 Stuttgart 70, Germany) V.217- 1971-

ZMWIAJ Zentralblatt fuer die Medizinischen Wissenschaften. (Berlin, Germany)

ZNCBDA Zeitschrift fuer Naturforschung, Section C: Biosciences. (Verlag der Zeitschrift fuer Naturforschung, Postfach 2645, D-7400 Tuebingen, Germany) V.29C- 1974-

ZNTFA2 Zeitschrift fuer Naturforschung. (Wiesbaden, Germany) V.1, No. 1-12 1946. For publisher information, see ZENBAX

ZOBIAU Zhurnal Obshchei Biologii. Journal of General Biology. (v/o Mezhdunarodnaya Kniga, Kuznetskii Most 18, Moscow G-200, USSR.) V.1- 1940-

ZolH## Personal Communication from H. Zollner, Institut fur Biochemie, Der Universitat, Gras, to H. Lau, Tracor Jitco, Inc., October 23, 1975

ZPPLBF Zentralblatt fuer Pharmazie, Pharmakotherapie und Laboratoriums-diagnostik. (VEB Verlag Volk und

Gesundheit, Neue Gruenstr 18, 102 Berlin, Germany)
V.109- 1970-

ZSNUAI Zeitschrift fuer Neurologie. (Berlin) V.198-206,
1970-74

ZSRSBX Zashchita Rastenii (Moscow). Plant Protection.
(v/o Mezhdunarodnaya Kniga, Kuznetskii Most 18,
Moscow G-200, U.S.S.R.) V.11- 1966-

ZTMPA5 Zeitschrift fuer Tropenmedizin und
Parasitologie. (Stuttgart, Germany) V.1-24, 1949-73. For
publisher information, see TMPRAD

ZUBEAQ Zeitschrift fuer Unfallmedizin und
Berufskrankheiten. (Revue de Medecine des Accidents et
des Maladies Professionelles. (Verlag Berichthaus,
Zwingliplatz 3, Postfach, CH-8022, Zurich, Switzerland)
V.1- 1907-

ZVKOA6 Zhurnal Vsesoyuznogo Khimicheskogo
Obshchestva Imeni D.I. Mendeleeva. Journal of the D.I.
Mendeleeva All-Union Chemical Society. (v/o
Mezhdunarodnaya Kiga, Kuznetskii Most 18, Moscow G-
200, USSR.) V.5- 1960-

ZVRAAX Zentralblatt fuer Veterinaermedizin, Reihe A.
(Paul Parey, Linderst 44-47, 1000 Berlin 61, Germany)
V.10- 1963-

ZYDXDM Zhejiang Yike Daxue Xuebao. Journal of
Zhejiang Medical University. (Zhejiang Yike Daxue, Yan'an
Lu, Hanzhou, Zhejiang, People's Republic of China) V.1-
1972-

ZYZAEU Zhongguo Yaoxue Zazhi. Chinese
Pharmacuetical Journal. (China International Book Trading
Corp., POB 2820, Beijing, Peop. China) V.24- 1989-

11FYAN "Fluorine Chemistry, Volume III. Biological
Effects of Organic Fluorides" J.H. Simons, ed., New York,
NY, Academic Press, 1963

13BYAH "Morphological Precursors of Cancer" Lucio
Severi ed., Proceedings of an International Conference,
Universita Degli Studi, Perugia, Italy, June 26-30, 1961,
Perugia, Italy, Univ. of Perugia, 1962

13OPAL "Control of Ovulation" Proceedings of the
Conference, Massachusetts Institute of Technology, Edicott
House, Pergamon Press, 1961

13ZGAF "Gigiena i Toksikologiya Novykh Pestitsidov i
Klinika Otravlenii, Doklady Vsesoyuznoi Konferentsii"
Komiteta Po Izucheniiu i Reglamentatsii Indokhimikatov
Glavnoi Gosudarstvennoi Sanitarnoi Inspektsii USSR., 1962

14CYAT "Industrial Hygiene and Toxicology, 2nd rev.
ed.," Patty, F.A., ed., New York, NY, Interscience
Publishers, 1958-63

14FHAR "Venomous and Poisonous Animals and
Noxious Plants of the Pacific Region" Keegan, H.L., and
W.V. Macfarlane, eds., New York, NY, Pergamon Press
Inc., 1963

14JTAF "Mycotoxins in Foodstuffs" Proceedings of the
Symposium held at the Massachusetts Institute of
Technology, Mar. 18-19, 1964, Wogan, G.N., ed.,
Cambridge, MA, MIT Press, 1965

14KTAK "Boron, Metallo-Boron Compounds and
Boranes" R.M. Adams, New York, NY, Wiley, 1964

14XBAV "International Congress of Chemotherapy"
Proceedings of the 3rd Congress, July 22-27, 1963, Stuttgart,
Thieme, 1964

14ZYA8 "Induction of Mutations and the Mutation
Process" Veleminsky, Jiri, ed., Proceedings of a Symposium
held in Prague, Czechoslavok, 26-28, 1963, Prague,
Czechoslavokia , Publishing House of the Czechoslavok
Academy of Science, 1965

15MBAH "Proceedings of the International Symposium
on Chemotherapy of Cancer" Lugano, Switzerland, April
28-May 1, 1964, Plattner, P.A., ed., New York, NY,
American Elsevier Publishing Co., 1964

15QWAW "Agents Affecting Fertility, a Symposium" 1964,
Austin, C.R., and J.S. Perry, eds., Boston, MA, Little, Brown
and Co., 1965

15TRAW "Preimplantation Stages of Pregnancy, Ciba
Foundation Symposium, London, 1965" Wolstenholme,
G.E.W., and M. O'Connor, eds., Boston, Little, Brown and
Co., 1965

17QLAD "Ratsionl'noe Pitanie" (Rational Nutrition)
Leshchenko, P.D., ed., USSR, 1967

17TVAO "Radiation-Induced Cancer, Proceedings of a
Symposium, Athens, 1969," Lanham, MD, Bernan-
UNIPUB, 1969

17XKAB "International Symposium on Amphetamines and
Related Compounds, Proceedings, Mario Negri Institute for
Pharmacological Research, Milan, 1969," Costa, E., and S.
Garattini, eds., New York, NY, Raven Press, 1970

19DDA6 "Animal Toxins, a Collection of Papers Presented
at the International Symposium on Animal Toxins, 1st,
1966" Russell, F.E., and P.R. Saunders, eds., New York,
NY, Pergamon Press Inc., 1967

19FKA3 "Biochemistry of Some Foodborne Microbial
Toxins" Symposium on Microbial Toxins, New York, NY,
Cambridge, MA, MIT Press, 1967

19UQAS "Khimicheskie Paktory Vneshnei Sredy I Ikh
Gigienicheskoe Znachenie, Materialy Nauchnoi
Konferentsii" 2nd, Pervyi Moskovskii Meditsinskii Institut,
Moscow, USSR, June 22, 1965

20PKA3 "Novye Dannye Po Toksikologii Redkikh Metalov
Ikh Soedinenii" 1967

20ZJAG "Toksikologiia Novykh Khimicheskikh
Veshchesty, Vnedriaemykh V" Rezinovuiu I Shinnaiu
Promyshlennost, 1968

21ACAB "Imifos" Giller, S.A., ed., Izd, Riga, USSR., 1968

21NDAB "Symposium on Carbenoxolone Sodium"
Robson, J.M., and F.M. Sullivan, eds., London, UK,
Butterworth, 1968

22XWAN "Chemical Mutagenesis in Mammals and Man"
Vogel, F. and G. Roehrborn, eds., Berlin, Germany,
Springer-Verlag, 1970

23EIAT "Venomous Animals and Their Venoms"
Buecherl, W., and E.E. Buckley, eds., New York, NY,
Academic Press, Inc., 3 vols., 1968-71

23HZAR "Carcinoma of the Colon and Antecedent
Epithelium" W.J. Burdette, Springfield, IL, C.C. Thomas,
1970

25NJAN "International Cancer Congress, Abstracts, 10th"
Houston, TX, May 22-29, 1970, Cumley, R.W. and J.E.

McCay, eds., Chicago, IL, Year Book Medical Publishers, Inc., 1970

26QZAP "Advances in Antimicrobial and Antineoplastic Chemotherapy, Progress in Research and Clinical Application" 7th Proceedings of the International Congress of Chemotherapy, Hejzlar, M. et al., eds., Prague, Czechoslovakia, Aug. 23-28, 1971, 3 vols., Prague, Czechoslovakia, University Press, 1972

26RAAN "Benzodinzenines" S. Garattini, ed., New York, NY, Raven Press, 1973

26UYA8 "Psoriasis" Proceedings of the International Symposium, Stanford, CA, July 7-10, 1971, Farber, E.M., and A.J. Cox, eds., Stanford, CA, Stanford University Press, 1971

26UZAB "Pesticides Symposia; Collection of Papers Presented at the Sixth & Seventh Inter-American Conferences on Toxicology and Occupational Medicine, Miami, FL," Miami, FL, J.L. Halos, 1968-70

27CWAL "Medical Primatology" Selected Papers from the 3rd Conference on Experimental Medicine and Surgery in Primates, Lyons, France, June 21-23, 1972, Goldsmith, E.I., and J. Moor-Jankowski, eds., White Plains, NY, Phiebig, 1972

27ZIAQ "Drug Dosages in Laboratory Animals-A Handbook" C.D. Barnes and L.G. Eltherington, Berkeley, CA, Univ. of California Press, 1965, 1973

27ZMA4 "Psychopharmacological Agents, Volume 2" Gorden, M., ed., New York, NY, Academic Press, 1967

27ZQAG "Psychotropic Drugs and Related Compounds" E. Usdin and D.H. Efron, 2nd ed., Washington, DC, 1972

27ZTAP "Clinical Toxicology of Commercial Products-Acute Poisoning" Gleason, et al., 3rd Ed., Baltimore, MD, Williams, and Wilkins, 1969

27ZWAY "Heffter's Handbuch der Experimentelle Pharmakologie."

27ZXA3 "Handbook of Poisoning: Diagnosis and Treatment" Dreisbach, R.H., Los Altos, CA, Lange Medical Publications

27ZZA9 "Manual of Pharmacology and Its Applications to Therapeutics and Toxicology" Sollman, T., 8th ed., PA, W.B. Saunders, 1957

28QFAD "Laboratory Animal in Drug Testing" Symposium of the 5th International Committee on Laboratory Animals, Hanover, Sept. 19-21, 1972, Speigel, A., ed., Stuttgart, Germany, Fischer, 1973

28ZEAL "Pesticide Index," Frear, E.H., ed., State College, PA, College Science Pub., 1969

28ZLA8 "Toxicity of Industrial Metals" Browning, E., London, UK, Butterworths, 1961

28ZNAE "Public Health Reports, Supplement." (Baltimore, MD/Washington, DC)

28ZOAH "Chemicals in War" Prentiss, A.M., 1937

28ZPAK "Sbornik Vysledku Toxixologickeho Vysetreni Latek A Pripravku" Marhold, J.V., Institut Pro Vychovu Vedoucicn Pracovniku Chemickeho Prumyclu Praha, Czechoslovakia, 1972

28ZRAQ "Toxicology and Biochemistry of Aromatic Hydrocarbons" Gerarde, H., New York, NY, Elsevier, 1960

28ZSAT "Drugs in Research" DeHaen, P., New York, NY, Paul DeHaen, 1964

29QHAQ "Principles of Medicinal Chemistry" Foye, W.O., eds., Philadelphia, PA, Lea and Febiger, 1974

29QKAZ "Toxins of Animal and Plant Origin" de Vries, A., and E. Kochva, eds., New York, NY, Gordon and Breach Science Publishers, 1971

29ZSA2 "Trifluoperazine: Further Clinical and Laboratory Studies" Moyer, J.E., ed., Philadelphia, PA, Lea and Febiger, 1959

29ZRAX "Pharmacology of Oriental Plants," Chen, K.K., and B. Mukerji, Oxford, UK, Pergamon Press Ltd., 1965

29ZUA8 "Toxicity and Metabolism of Industrial Solvents" Browning, E., New York, NY, Elsevier, 1965

29ZVAB "Handbook of Analytical Toxicology," Sunshine, I., ed., Cleveland, OH, Chemical Rubber Co., 1969

29ZWAE "Practical Toxicology of Plastics" Lefaux, R., Cleveland, OH, Chemical Rubber Company, 1968

30FDAB "Neuropoisons: Their Pathophysiological Actions, 1971-1974" Simpson, L.L., and D.R. Curtis, eds., New York, NY, Plenum Pub. Corp., 1974

30ZDA9 "Chemistry of Pesticides." Melnikov, N.N., New York, NY, Springer-Verlag, 1971

30ZFAF "Toxic Aliphatic Fluorine Compounds" Pattison, F.L.M., London, UK, Elsevier Publishing, 1959

30ZIAO "Vanadium Toxicology and Biological Significance" Faulkner, T.G., Hudson, NY, Elsevier Publishing Company, 1964

31BYAP "Experimental Lung Cancer: Carcinogenesis Bioassays, International Symposium, 1974" New York, NY, Springer, 1974

31TFAO "Progress in Chemotherapy" Proceedings of the 8th International Congress on Chemotherapy, Athens, Greece, 1973, Daikos, G.K. ed., Athens, Greece. Hill. Society of Chemotherapy, 1974

31ZOAD "Pesticide Manual" Worcesteshire, England, British Crop Protection Council, 1968

31ZPAG "Synthetic Analgesics" Oxford, UK, Pergamon Press, 1966

31ZTAS "Alcohols: Their Chemistry, Properties and Manufacture" J.A. Monick, New York, Reinhold Book, 1968

31ZUAV "Investigations in the Field of Organogermanium Chemistry" Rijkens, F., and G.J.M. Van Der Kerk, Utrecht, Netherlands, Schotanus & Jens Utrecht N.V., 1964

32OAAP "Pesticide and the Environment: a Continuing Controversy" Papers Presented and Papers Reviewed at the 8th Inter-American Conference on Toxicology and Occupational Medicine, Miami, FL, July 1973, Deichmann, W.B., ed., New York, NY, Stratton, 1973

32XPAD "Teratology" C.L. Berry, ed., New York, NY, Springer-Verlag, 1975

32YWA5 "Radiation Research, Biomedical, Chemical, and Physical Perspectives, Proceedings of the International Congress of Radiation Research, 5th, Seattle, WA, July 14-20, 1974 (Pub 1975)" Nygaard, O.F., et al., ed., New York, NY, Academic, 1975

32ZCAI "Ganglion-Blocking and Ganglion-Stimulation Agents" Kharkevich, D.A, Moscow, USSR, First Moscow Medical Institute, 1967

32ZDAL "Aldrin Dieldrin Endrin and Telodrin: An Epidemiological and Toxicological Study of Long-Term Occupational Exposure" Jager, K.W., New York, NY, Elsevier, 1970

32ZWAA "Handbook of Poisoning: Diagnosis and Treatment" Dreishach, R.H, 8th ed., Los Altos, CA Lange Medical Publications, 1974

32ZXAD "Transactions of the 36th Annual Meeting of the American Conference of Governmental Industrial Hygienists" Miami Beach, FL., May 12-17, 1974

33IUAS "Mycotoxins" Purchase, I.F.H., ed., Amsterdam, Netherlands, Elsevier, 1974

33NFA8 "Animal Models in Dermatology" Maibach, H., ed., New York, NY, Churchill Livingston, 1975

34LXAP "Insecticide Biochemistry and Physiology" Wilkinson, C.F., ed., New York, NY, Plenum, 1976

34TRAD "Food-Drugs from the Sea" Proceedings of the Conference, 4th, University of Puerto Rico, Mayaguez, PR, November 17-21, 1974, Webber, H.H., and G.D. Ruggieri, eds., Washington, DC, Marine Technology Society, 1976

34ZHAD "Symposium on Mycotoxins in Human Health" Purchase, I.F.H., London, UK, Macmillan, 1971

34ZIAG "Toxicology of Drugs and Chemicals," Deichmann, W.B., New York, NY, Academic Press, Inc., 1969

34ZRA9 "Proceedings Quadrennial Conference on Cancer" Severi, L., ed., Perugia, Italy, University of Perugia, 1966

35FUAR "Animal, Plant, and Microbial Toxins" Ohsaka, Akira, et al., eds., New York, NY, Plenum Press, 1976

35WYAM "In Vitro Metabolic Activation in Mutagenic Testing" Proceedings of the Symposium on the Role of Metabolic Activation in Producing Mutagenic and Carcinogenic Environmental Chemicals, Research Triangle Park, NC, Feb. 9-11, 1976, De Serres, F.J. et al., eds., New York, NY, Elsevier North Holland, 1976

35ZRAG "Medicinal Chemistry, Proceedings of the International Symposium on Medicinal Chemistry, Main Lectures, 4th, Noordwijkerhout, Netherlands, 1974," Maas, J., ed., New York, Elsevier Science Pub. Co., 1974

36PYAS "Pharmacology of Steroid Contraceptive Drugs" Garattini, S. and H.W. Berendes, eds., New York, NY, Raven Press, 1977. (Monograph Mario Negri Inst. Pharm. Res.)

36SBA8 "Antifungal Compounds" Siegel, M.R. and H.D. Sisler, eds., 2 vols., New York, NY, Marcel Dekker, 1977

36THAV "Management of the Poisoned Patient" Rumack, B.H., and A.R. Temple, eds., Princeton, NJ, Science Press, 1977

36YFAG "Biological Reactive Intermediates, Formation, Toxicity and Inactivation" Proceedings of the International Conference on Active Intermediates, Formation, Toxicity and Inactivation, University of Turku, Turku, Finland, July 26-27, 1975. New York, NY, Plenum, 1977

37ASAA "Kirk-Othmer Encyclopedia of Chemical Technology, 3rd Edition" Grayson, M. and D. Eckroth, eds., New York, NY, Wiley, 1978

37KXA7 "Water Chlorination: Environmental Impact and Health Effects" Proceedings of the Conferences on the Environment, Ann Arbor, MI, Ann Arbor Science Pub. Inc., 1975-

37QLAZ "Current Approaches in Toxicology" Ballantyne, B., ed., Bristol, UK, John Wright & Sons, Ltd., 1977

37XLA2 "Current Chemotherapy" Proceedings of the 10th International Congress of Chemotherapy, Zurich, Switzerland, Sept. 18-23, 1977, Siegenthaler,W. and R. Luethy, eds., 2 vols., Washington, DC, American Society for Microbiology, 1978

38KLAC "Structure-Activity Relationship among the Semisynthetic Antibiotics." Perlman, D., ed., New York, NY, Academic Press, 1977

38MKAJ "Patty's Industrial Hygiene and Toxicology" 3rd rev. ed., Clayton, G.D., and F.E. Clayton, eds., New York, NY, John Wiley & Sons, Inc., 1978-82. Vol. 3 originally pub. in 1979; pub. as 2nd rev. ed. in 1985

40QBA3 "Proceedings of the International Congress on Toxicology, Toxicology as a Predictive Science, 1st, Toronto, 1977" Plaa, G.L., and W.A. Duncan, eds., New York, NY, Academic Press, Inc., 1978

40RMA7 "Health and Sugar Substitutes" Proceedings of the ERGOB Conference on Sugar Substitutes, Geneva, Switzerland, 1978, Guggenheim, B., ed., Basel, Switzerland, S. Karger AG, 1979

40WDA5 "Bleomycin: Current Status and New Development" Papers Presented in a Symposium, Oakland, CA, 1977, Carter, S.K., et al., eds., New York, NY, Academic Press, 1978

40YJAX "Evaluation of Embryotoxicity, Mutagenicity and Carcinogenicity Risks In New Drugs" Proceedings of the Symposium on Toxicological Testing for Safety of New Drugs, 3rd, Prague, Czechoslovakia, Apr. 6-8, 1976, Benesova, O., et al., eds., Prague, Czechoslovakia, Univerzita Karlova, 1979

41CIAR "Effects of Poisonous Plants on Livestock" Proceedings of a Joint U.S.-Australian Symposium on Poisonous Plants, Logan, UT, 1978, Keeler, R.F., et al. eds., New York, NY, Academic Press, 1978

41HTAH "Aktual'nye Problemy Gigieny Truda. Current Problems of Labor Hygiene" Tarasenko, N.Y., ed., Moscow, USSR, Pervyi Moskovskii Meditsinskii Inst., 1978

41KEAL "Toxicology, Biochemistry and Pathology of Mycotoxins", Uraguchi, K. and M. Yamazaki, eds., New York, NY, Wiley, 1978

43FLAV "Burger's Medicinal Chemistry" Wolff, M.E., ed., 4th ed., New York, NY, John Wiley & Sons, 1981

43GRAK "Dusts and Disease" Proceedings of the Conference on Occupational Exposures to Fibrous and Particulate Dust and Their Extension into the Environment, 1977, Lemen, R., and J.M. Dement, eds., Park Forest South, IL, Pathotox Publishers, 1979

43MKAT "Current Chemotherapy and Infectious Diseases" Proceedings of the 11th International Congress of Chemotherapy and the 19th Interscience Conference on

Antimicrobial Agents and Chemotherapy, Boston, MA, Oct. 1-5, 1979, American Society for Microbiology, Washington, DC, 1980

43NDAD "Pharmacology of Antihypertensive Drugs," Scriabine, A., ed., New York, Raven Press, 1980

43XWAI "Brain Tumors and Chemical Injuries to the Central Nervous System" Proceedings of the International Neuropathological Symposium, Warsaw, Poland, Sep. 23-26, 1976, Mossakowski, M.J., ed., Warsaw, Poland, Panstwowe Wydawnictwo Naukowe, 1978

43ZRAD "Hemoperfusion: Kidney and Liver Support and Detoxification" Proceedings of the International Symposium, Haifa, Israel, 1979, Sideman, S., and T.M.S. Chang, eds., Washington, DC, Hemisphere Pub. Corp., 1980

44LQAF "International Sulprostone Symposium, (Papers)" Vienna, Austria, 1978, Friebel, K., et al., eds., Berlin, Germany, Schering AG, 1979

45ICAX "Nickel Toxicology" 2nd International Conference on Nickel Toxicology. Swansea, UK, Sept. 3-5, 1980. London, UK, Academic Press, 1980

45JVAR "Research Frontiers in Fertility Regulation" Proceedings of an International Workshop, 1980, Zatuchni, G.I., et al., eds., Hagerstown, MD, Harper and Row Medical Dep., 1980

45OHAA "Short-Term Test Systems for Detecting Carcinogens" Proceedings of the Symposium, 1978, Norpoth, K.H., and R.G. Garner, eds., Berlin, Germany, Springer-Verlag, 1980

45OJAG "Radiation Sensitizers: Their Use in the Clinical Management of Cancer" Proceedings of the Conference, Key Biscayne, FL., Oct. 3-6, 1979, Brady, L.W., ed., New York, NY, Masson Publishing USA, Inc., 1980

45REAG "Teratology of the Limbs" Symposium on Prenatal Development, 4th, 1980, Merker, H.J., et al., eds., Berlin, Germany. Walter de Gruyter and Co., 1980

46GFA5 "Safety Problems Related to Chloramphenicol and Thiamphenicol Therapy" Najean, Y., et al., eds., New York,, NY, Raven Press, 1981

46IZA7 "Retinoids, Advances in Basic Research and Therapy" Proceedings of the International Dermatology Symposium, Berlin, Germany, 1980, Orfanos, C.E., et al., eds., New York, NY, Springer-Verlag, 1981

46OJAN "Chemical Analysis and Biological Fate: Polynuclear Aromatic Hydrocarbons" 5th International Symposium, Battelle's Labs., Oct. 1980. Columbus, OH, Battelle Press, 1981

47JMAE "Cytogenetic Assays of Environmental Mutagens" Hsu, T.C., ed., Totowa, NJ, Allanheld, Osmun and Co., 1982

47YKAF "Sporulation and Germination" Proceedings of the Eighth International Spore Conference, 1980, Washington, DC, American Society for Microbiology, 1981

48ECAY "Spurenelement-Symposium: Nickel, 3rd" Jena, Germany, Anke, M., et al., eds., Jena, Germany, Friedrich-Schiller-Universitaet, 1980

48HGAR "Current Chemotherapy and Immunotherapy" Proceedings of the International Congress of Chemotherapy, 12th, 1981, Periti, P., and G. Gialdroni-

Grassi, eds., Washington, DC, American Society for Microbiology, 1982

48NTAS "Advances in Genetics, Development, and Evolution of Drosophila" Proceedings of the European Drosophila Research Conference, 7th, 1981, Lakovaara, S., ed., New York, NY, Plenum Publishing Corp., 1982

48RKAL "Industrial and Environmental Xenobiotics: Metabolism and Pharmacokinetics of Organic Chemicals and Metals, Proceedings of an International Conference, Prague," 1980, Gut, I., et al., Berlin, Fed. Rep, Ger., Springer-Verlag, 1981

48THAM "Proceedings of the International Symposium on Cyclodextrins" 1st, 1981, Szejtli, J., ed., Hingham, MA, Kluwer Boston Inc., 1982

49RQAC "Chemistry and Biology of Hydroxamic Acids" Proceedings of the International Symposium, 1st, Dayton, OH, 1981, Kehl, H., ed., Basel, Switzerland, S. Karger AG, 1982

49VWAG "Female Transcervical Sterilization" Proceedings of an International Workshop on Non-Surgical Methods for Female Tubal Occlusion, Chicago, IL, June 22-24, 1982, Zatuchni, G.I., et al., eds. New York, NY, Harper and Row, 1983

49WEAZ "Pesticide Chemistry: Human Welfare and the Environment" Proceedings of the International Congress of Pesticide Chemistry, 5th, Kyoto, Japan, Aug. 29-Sept 4, 1982, Miyamoto, J. and P.C. Kearney, eds., Oxford, UK, Pergamon Press Ltd., 1983

49ZAA4 "Ceftriaxon (Rocephin) ein Neues Parenterales Cephalosporin, Hahnenklee - Symposion, 9th, 1981," Dettli, L., et al. eds., Harz, Fed. Rep. Ger.

50EXAK "Formaldehyde Toxicity" Conference, 1980, Gibson, J.E., ed., Washington, DC, Hemisphere Publishing Corp., 1983

50EYAN "Structure-Activity Correlation as a Predictive Tool in Toxicology" Papers from a Symposium, 1981, Golberg, L., ed., Washington, DC, Hemisphere Publishing Corp., 1983

50NNAZ "Polynuclear Aromatic Hydrocarbons: Mechanisms, Methods and Metabolism, Papers of the 8th International Symposium, Columbus, OH, 1983," Cooke, M., and A.J. Dennis, eds., Columbus, OH, Battelle Press, 1985

50WKA3 "Reproduction and Developmental Toxicity of Metals" Proceedings of a Joint Meeting, Rochester NY, 1982, Clarkson, T.W., et al., eds., New York, NY, Plenum Publishing Corp., 1983

51UDAB "Heavy Metals in the Environment, International Conference, 4th, 1983" Heidelberg, S., 2 vols., Edinburgh, UK, CEP Consultants Ltd., 1983

51ZKAW "The Retinoids, Vol. 2" Sporn, M.B., et al., eds., New York, Academic Press, Inc., 1984

52UAA9 "International Congress of Plant Protection, Proceedings of a Conference: Plant Protection for Human Welfare, 10th, Brighton, UK, Nov. 20-25, 1983" (British Crop Protection Council, Bear Farm, Binfield, Bracknell. Berkshire RG12 5QE, UK), 3 vols, 1983.

52EDA6 "Nitrendipine Present Int Workshop Nitrendipine 1st", New York, NY, 1982, Scriabine, A., et al., eds., Baltimore, MD, Urban & Schwarzenberg, 1984

52GUAX "Anorganische Stoffe in der Toxikologie und Kriminalistik, Symposium, Mosbach, Germany, 1983" Heppenheim, Germany, Verlag Dr. Dieter Helm, 1983

52MLA2 "Toxicology of Petroleum Hydrocarbons, Proceedings of the Symposium, 1st, 1982" MacFarland, H.N., et al., eds., Washington, DC, American Petroleum Institute, 1983

52OLAC "Veterinarnaya Entomologiya i Akarologiya" Veterinary Entomology and Acarology, Nepoklonov, A.A., ed., Moscow, USSR, Izdatel'stovo Kolos, 1983

53HJAC "Mouse Liver Neoplasia: Current Perspectives" Popp, J.A., ed., New York, NY, Hemisphere Pub. Corp., 1984

54DEAI "Occupational Health in the Chemical Industry, Proceedings of the International Congress, 11th, Alberta, Canada, 1983" Oxford, R.R., et al., eds., Calgary, Alberta, Canada, Univ. of Calgary, 1984

55CXA9 "Trichothecenes and Other Mycotoxins, Proceedings of the International Mycotoxin Symposium, 1984" Lacey, J., ed., Chichester, UK, John Wiley & Sons Ltd., 1985

55DXAE "Progress in Nickel Toxicology, Proceedings of the International Conference on Nickel Metabolism and Toxicology, 3rd, 1984" Brown, S.S., and F.W. Sunderman, Jr., eds., Oxford, UK, Blackwell Scientific Pub. Ltd., 1985

85AEA9 "Environmental Hazards of Metals" I.T. Brakhnova, trans. by Slep, J.H., New York, NY, Consultants Bureau, 1975

85AGAF "Effects and Dose-response Relationships of Toxic Metals" Nordberg, G.F., ed., Proceedings from an International Meeting Organized by the Subcommittee on the Toxicology of Metals of the Permanent Commission and International Association on Occupational Health, Tokyo, Japan, November 18-23, 1974, New York, NY, Elsevier Scientific, 1976

85AIAL "Toxicity of Pure Foods" Boyd, E.M., Cleveland, OH, CRC Press, 1973

85ALAU "Carbamate Insecticides: Chemistry, Biochemistry, and Toxicology" Kuhr, R.J. and H.W. Dorough, Cleveland, OH, CRC Press, 1976

85ARAE "Agricultural Chemicals, Books I, II, III, and IV" W.T. Thomson, Fresno, CA, Thomson Publications, 1976/77 revision

85AREA "Agricultural Chemicals," Thomson, W.T., 4 vols., Fresno, CA, Thomson Publications, 1976/77 revision

85CVA2 "Oncology 1970" Proceedings of the Tenth International Cancer Congress, Chicago, IL, Year Book Medical Publishers, 1971

85CYAB "Chemistry of Industrial Toxicology" H.B. Elkins, 2nd Ed., New York, John Wiley & Sons, 1959

85DAAC "Bladder Cancer, A Symposium" Lampe K.F. et al., eds., Fifth Inter-American Conference on Toxicology and Occupational Medicine, University of Miami, School of Medicine, Coral Gables, FL, Aesculapius Pub., 1966

85DCAI "Poisoning; Toxicology, Symptoms, Treatments" Arena, J.M. 2nd ed., Springfield, IL, C. C. Thomas, 1970

85DGAU "Catalog of Teratogenic Agents" Shepard, T.H., 2nd Ed., Baltimore, MD, Johns Hopkins University Press, 1976

85DHAX "Medical and Biologic Effects of Environmental Pollutants Series" Washington, DC, National Academy of Sciences, 1972-77

85DJA5 "Malformations Congenitales des Mammiferes" Tuchmann-Duplessis, H., Paris, France, Masson et Cie, 1971

85DKA8 "Cutaneous Toxicity" Drill, V.A. and P. Lazar, eds., New York, NY, Academic Press, 1977

85DLAB "Studies on Chemical Carcinogenesis by Diels-Alder Adducts of Carcinogenic Aromatic Hydrocarbons" Earhart, Jr., R.H., Ann Arbor, MI, Xerox Univ. Microfilms, 1975

85DPAN "Wirksubstanzen der Pflanzenschutz und Schadlingsbekampfungsmittel" Werner Perkow, Berlin, Germany, Verlag Paul Parey, 1971-1976

85DUA4 "Chemical Tumour Problems" Nakahara, W., ed., Tokyo, Japan, Japanese Society for the Promotion of Science, 1970

85DZAJ "Principles of Insect Chemosterilization" Labrecque, G.C. and C.N. Smith, eds., New York, NY, Appleton-Century-Crofts, 1968

85ECAN "Metal Toxicity in Mammals, Vol. 2: Chemical Toxicity of Metals and Metalloids" Venugopal, B. and T.D. Luckey, New York, NY, Plenum Press, 1978

85EGD4 "Toxins: Animal, Plant and Microbial. Proceedings of the International Symposium" Rosenberg, P., ed., New York, NY, Pergamon Press, 1978

85ELDJ "Die Herzwirksamen Glykoside" Baumgarten, G. and W. Forster, Leipzig, Germany, VEB Georg Thieme, 1963

85ERAY "Antibiotics: Origin, Nature, and Properties" Korzyoski, T., Z. Kowszyk-Gindifer, and W. Kurylowicz, eds., Washington DC, American Society for Microbiology, 1978

85ESA3 "Merck Index; an Encyclopedia of Chemicals, Drugs, and Biologicals," 11th ed., Rahway, NJ 07065, Merck & Co., Inc. 1989

85FZAT "Index of Antibiotics from Actinomycetes" Umezawa, H., et al., eds., Tokyo, Japan, University of Tokyo Press, 1967

85GDA2 "CRC Handbook of Antibiotic Compounds, Volumes 1-9" Berdy, Janos. Boca Raton, FL, CRC Press, 1980

85GLAQ "A Survey of Antimalarial Drugs, 1941-1945" Wiselogle, F.Y., 2 vols., Ann Arbor, MI, J.W. Edwards, 1946

85GMAT "Toxicometric Parameters of Industrial Toxic Chemicals Under Single Exposure" Izmerov, N.F., et al., Moscow, USSR, Centre of International Projects, GKNT, 1982

85GNAW "Pharmacodynamie de l'Acide Dipropylacetique (ou Propyl-2-Pentanoique) et de ses Amides" Carraz, G., Imprimerie Eymond, 1968

85GRAA "The Control of Fertility" Pincus, G., New York, NY, Academic Press, 1965

85GUAJ "Antifertility Compounds in the Male and Female" Jackson, H., Springfield, IL, Charles C. Thomas, 1966

85GYAZ "Pflanzenschutz-und Schaedlingsbekaempfungsmittel: Abriss einer Toxikologie und Therapie von Vergiftungen" 2nd ed., Klimmer, O.R., Hattingen, Germany, Hundt-Verlag, 1971

85HBAZ "Environmental Variables in Oral Disease;" A Symposium Presented at the Montreal Meeting of the American Association for the Advancement of Science, 1964, Kreshover, S.J., et al., eds., Washington, DC, American Association for the Advancement of Science, 1966

85INA8 "Documentation of the Threshold Limit Values and Biological Exposure Indices" 5th ed., Cincinnati, OH, American Conference of Governmental Industrial Hygienists, Inc., 1986

85IPAE "Modern Pharmaceuticals of Japan, V" Tokyo, Japan, Pharmaceutical, Medical and Dental Supply Exporters' Association, 1975

85IVAW "Possible Long-Term Health Effects of Short-Term Exposure to Chemical Agents" National Research Council, 3 vols., Washington, DC, National Academy Press, 1982-85

85IXA4 "Structure et Activite Pharmacodyanmique des Medicaments du Systeme Nerveux Vegetatif" Bovet, D., and F. Bovet-Nitti, New York, NY, S. Karger, 1948

85JCAE "Prehled Prumyslove Toxikologie; Organicke Latky" Marhold, J., Prague, Czechoslovakia, Avicenum, 1986

85JDAH "Organophosphorus Pesticides: Organic and Biological Chemistry" Eto, M., Cleveland, OH, CRC Press, Inc., 1974

85JFAN "Agrochemicals Handbook," with updates, Hartley, D., and H. Kidd, eds., Nottingham, UK, Royal Soc. of Chemistry, 1983-86

85KYAH "Merck Index; an Encyclopedia of Chemicals, Drugs, and Biologicals", 11th ed., Rahway, NJ 07065, Merck & Co., Inc. 1988